CHURCHILL LIVINGSTONE
Medical Division of Longman Group UK Limited

Distributed in the United States of America by Churchill Livingstone
Inc., 650 Avenue of the Americas, New York, N.Y. 10011, and by
associated companies, branches and representatives throughout the world.

© Longman Group UK Limited 1989

Thirty-seventh edition 1989
  Reprinted 1989
  Reprinted 1992
  Reprinted 1993

ISBN 0 443 02588 6

British Library Cataloguing in Publication Data

Gray, Henry, *1827–1861*
  Gray's anatomy. — 37th ed.
  1. Man. Anatomy
  I. Title II. Williams, Peter L. (Peter
  Llewellyn)
  611

Library of Congress Cataloging in Publication Data

Gray, Henry, 1827–1861
  Gray's anatomy/edited by Peter L. Williams . . . [et al.]. — 37th
  ed.
    p.   cm.
    Rev. ed. of: Anatomy, descriptive and surgical. 35th ed. 1973.
    Includes index.
    1. Anatomy, Human. I. Williams, Peter L. (Peter Llewellyn)
  II. Gray, Henry, 1827–1861. Anatomy, descriptive and surgical.
  III. Title. IV. Title: Anatomy.
  QM23.2.G73 1989
  611—dc19   88-25625

Printed in Great britain at The Bath Press, Avon

GRAY'S ANATOMY

EDITED BY

PETER L. WILLIAMS
DSc MA MB BChir

ROGER WARWICK
BSc PhD MD ChB

PROFESSORS EMERITI, UNIVERSITY OF LONDON
FORMERLY PROFESSORS OF ANATOMY,
GUY'S HOSPITAL MEDICAL SCHOOL, UNIVERSITY OF LONDON

MARY DYSON
BSc PhD

LAWRENCE H. BANNISTER
BSc PhD

READERS IN ANATOMY, GUY'S HOSPITAL MEDICAL SCHOOL,
UNITED MEDICAL AND DENTAL SCHOOLS OF
GUY'S AND ST THOMAS'S HOSPITALS,
UNIVERSITY OF LONDON

THIRTY-SEVENTH EDITION

CHURCHILL LIVINGSTONE
EDINBURGH LONDON MELBOURNE AND NEW YORK 1989

PREVIOUS
EDITIONS AND EDITORS

First Edition 1858
Second Edition 1860

By HENRY GRAY FRS FRCS
St George's Hospital

Third Edition 1863
Fourth Edition 1865
Fifth Edition 1869
Sixth Edition 1872
Seventh Edition 1875
Eighth Edition 1877
Ninth Edition 1880

By TIMOTHY HOLMES MA FRCS
St George's Hospital

Tenth Edition 1883
Eleventh Edition 1887
Twelfth Edition 1890
Thirteenth Edition 1893
Fourteenth Edition 1897

By T PICKERING PICK FRCS
St George's Hospital

Fifteenth Edition 1901
Sixteenth Edition 1905

By T PICKERING PICK FRCS
St George's Hospital
and R HOWDEN MA MB CM DSc LLD
University of Durham

Seventeenth Edition 1909
Eighteenth Edition 1913
Nineteenth Edition 1916
Twentieth Edition 1918
Twenty-first Edition 1920
Twenty-second Edition 1923
Twenty-third Edition 1926

By R HOWDEN MA MB CM DSc LLD
University of Durham

Twenty-fourth Edition 1930
Twenty-fifth Edition 1932
Twenty-sixth Edition 1935
Twenty-seventh Edition 1938

By T B JOHNSTON CBE MD
Guy's Hospital Medical School
University of London

Twenty-eighth Edition 1942
New Impression 1944
New Impression 1945
Twenty-ninth Edition 1946
New Impression 1947
Thirtieth Edition 1949 (Completely Reset)
New Impression 1950
Thirty-first Edition 1954
New Impression 1956

By T B JOHNSTON CBE MD
and J WHILLIS MD MS
Guy's Hospital Medical School
University of London

Thirty-second (Centenary) Edition 1958
New Impression 1960

By T B JOHNSTON CBE MD
University of London
D V DAVIES MA(Cantab) MB BS
University of London
and F DAVIES MD DSc(Lond) FRCS
University of Sheffield

Thirty-third Edition 1962
New Impression 1964

By D V DAVIES
DSc(Lond) MA(Cantab) MB BS FRCS
University of London
and F DAVIES MD DSc(Lond) FRCS
University of Sheffield

Thirty-fourth Edition 1967

By D V DAVIES
DSc(Lond) MA(Cantab) MB BS FRCS
St Thomas's Hospital Medical School
University of London

Thirty-fifth Edition 1973 (Completely Reset)
New Impressions 1975, 1978

By ROGER WARWICK BSc PhD MD
Professor of Anatomy
Guy's Hospital Medical School
University of London
and PETER L WILLIAMS DSc MA MB BChir
Professor of Anatomy
Guy's Hospital Medical School
University of London

Thirty-sixth Edition 1980 (Completely Reset)

By PETER L WILLIAMS DSc MA MB BChir
Professor of Anatomy
Guy's Hospital Medical School
University of London
and ROGER WARWICK BSc PhD MD
Professor of Anatomy
Guy's Hospital Medical School
University of London

Timothy Holmes 1863–1880

T Pickering Pick 1883–1905

R Howden 1901–1926

T B Johnston 1930–1958

Henry Gray, 1858–1860, seen here in the Dissecting Room
of St George's Hospital 1860

J Whillis 1942–1954

D V Davies 1958–1967

Francis Davies 1958–1962

PREFACE

TO THE THIRTY-SEVENTH EDITION

Human Anatomy is a part, not merely of medical science, but also, beyond that, of biological knowledge, and beyond that also a part of the totality of mankind's understanding of his universe. But it has been customarily isolated, and usually still is in many schools and texts. Yet it is impossible to pursue anatomy intelligently in intellectual isolation. Even in our two previous editions we did not merely press for the amalgamation of structure with function; we also recognized that mature structures must be viewed in the light of their development, evolution and kinetics, *and* at all levels down to cytological, even molecular, detail, if one is to *understand* them. In this, our third joint edition, this outlook has further expanded, partly to emphasize, wherever possible, the intimate physico-chemical processes which collectively constitute an individual life—the arena of much current research. We have in particular essayed to review—in the full sense of the word—much that is and goes on in the human frame as a *living* complex.

It has never been the aim of scientific thought to explain the 'Whys' of existence, though many philosophers and divines over the ages have attempted to do so. But even if we confine our thinking to the 'Hows' of existence, it is stultifying to confine enquiry or description to the artificially bounded 'subjects' of 'education'. Anatomy has been a particular sufferer in this compartmentalization; the structure of things, including the human body, must include physics, chemistry, biochemistry, kinetics, mathematics and much more. There are, in fact, no real boundaries between 'anatomy' and *all* other fields of human enquiry, including not only the supposedly separate Sciences but also, to some extent, the Arts; the human individual has not only applied himself to the creation of innumerable skills and crafts for mere survival, but has, in the unceasing development of mind, evolved the arts—painting, literature, poetry, music, sculpture and drama —to enrich our culture. Even this is not the end of the expansion of a structured being into its environment, because the human being has never been content to exploit its ever-burgeoning talents in accomplishment.

Human beings are social animals—like their forebears—and mere aesthetics were not long sufficient; individuals are parts of communities, themselves parts of races, continents, a world, an earthly biological system, beyond which extend planets, stars and nebulae. For the sake of order in a community, morals, ethics and religions inevitably evolved, side by side with attempted explanations of the Universe, the totality of what *is*. All these developments stimulate a duality of impulse: to seek the factual truth, i.e. scientific research, and to look for explanation, i.e. philosophy, a term of special significance to us. Philosophy—the love of knowledge—has engaged many of the best human minds over millenia of time. It has one peculiarity, which takes it a little beyond what we commonly regard as Science: it is not content with the accumulation of provable observations alone, though this approach has led to greatest change, applied or theoretical, in the human situation. But philosophy, even more than science, which at times uses the same mental gymnastics, is dependent upon ideas and ideals. In fact, in both fields of endeavour, it is human imagination —the refusal blindly to accept traditional knowledge—which has been the most fruitful attitude to *expansion* of knowledge.

So much of our knowledge, our 'science', is open to further investigative expansion, given the ability to form new ideas and tests for them—to blend even the best established knowledge with scepticism—a freedom to think again. This is the philosophy of our third joint edition.

London, 1989

PLW, RW, MD and LHB

ACKNOWLEDGEMENTS

During the preparation of this edition, we have had the great advantage of the meticulous professional assistance of:

DOUGLAS LUKE BSc BDS PhD FDSRCS—*Dental Anatomy*
ANDREW G S LUMSDEN MA PhD—*Dental Anatomy*
JENNIFER HALSTEAD MMAA—*Illustrations*
KEVIN MARKS MMAA—*Illustrations*
KEVIN J FITZPATRICK FIScT FIMBI MIBiol ARPS—*Photography*
MICHAEL C E HUTCHINSON MBBS BDS—*Special Preparations*
E LOWELL REES MBBS—*Index*
JEAN HALL—*Bibliography*

In addition, we owe special thanks to the following: Professor J Szentágothai (University of Budapest), Dr T P S Powell (University of Oxford), Professor J Osborne (Alberta, Canada), Professor R Nieuwenhuys (University of Nijmegen), Professor J Gosling (University of Manchester), Professor Ruth E M Bowden (London), Professor G Ruskell (City University, London), Dr J A Bultner-Ennever (University of Zurich), Dr B Darazs (Medical University of South Africa, Pretoria), Professor H Duvernoy (University of Franche-Compte, France), Dr P D Edmunds (SRI International, Menlo Park, California), Mr J Erian (Farnborough and Chelsfield Park Hospitals, Kent), Professor M A Gaygum (University of Riyadh), Dr Gail ter Haar (Institute of Cancer Research, University of London), Miss Marion Haigh (University of Manchester), Dr T D Hawkins (Addenbrooke's Hospital, Cambridge), Dr Margaret Hutchings (University of Southampton), Miss Valerie Martin (Royal Free Medical School, London), Mr D Negus (Lewisham Hospital, London), Philips Medical Systems, Dr N N Rizk (University of Cairo), Dr med H-D Roth (University of Erlangen-Nurnberg), Professor J P Royle (Victoria, Australia), Professor J F Scott (University of Manchester), Siemens (Erlangen, West Germany), Professor P V Tobias (University of Witwatersrand), Dr G I Verney (Addenbrooke's Hospital, Cambridge), Professor R O Weller (University of Southampton) and Mr J C Bayliss.

We are also indebted to many colleagues in the United Medical and Dental Schools of Guy's and St Thomas's Hospitals (University of London).

On the Guy's Campus to: Professor M Berry, Professor M Brookes, Dr M Daker, Dr Hilary Dodson, Dr S Gallagher, Miss Sheila Gates, Mrs Rachel Hickman, Dr A Hayward, Mr A Kent, Professor S Leibowitz, Mr D Lovell, Professor M Maisey, Dr Moya Meredith-Smith, Miss Samantha Negus, Dr R Poston, Dr L Rayfield, Mr D Ristow, Dr M Sadler, Dr P Shepherd, Miss Sarah Smith, Mr A Thomas, Dr Jane Ward, Dr Caroline Wigley and Dr S R Young.
On the St Thomas's Campus to: Professor M Day, Dr Marilyn Jones and Professor Marion Kendall.

Our relations with our publishers (Churchill Livingstone, Edinburgh) have remained most practical and cordial, due very much to the understanding and exhortation of their staff.

Finally, we record with gratitude the patient help and support of Mrs Irene Williams, Mrs Mary Bannister and Dr Walter R Dyson.

London, 1989   PLW, RW, MD and LHB

HENRY GRAY

FRS FRCS

Since readers of *Gray's Anatomy* will be interested to learn something of the original author, Henry Gray, the following information as to his career has been extracted from an article which appeared in the *St George's Hospital Gazette* of 21 May 1908.

Gray, whose father was private messenger to George IV, and also to William IV, was born in 1827, but of his childhood and early education nothing is known.

On 6 May 1845, he entered as a perpetual student at St George's Hospital, London, and he is described by those who who knew him as 'a most painstaking and methodical worker, and one who learnt his anatomy by the slow but invaluable method of making dissections for himself'.

While still a student he secured, in 1848, the triennial prize of the Royal College of Surgeons for an essay entitled 'The origin, connexions and distribution of the nerves to the human eye and its appendages, illustrated by comparative dissections of the eye in other vertebrate animals'.

At the early age of twenty-five he was, in 1852, elected a Fellow of the Royal Society, and in the following year he obtained the Astley Cooper prize of three hundred guineas for a dissertation 'On the structure and use of the spleen'.

He held successively the posts of demonstrator of anatomy, curator of the museum, and Lecturer on anatomy at St George's Hospital, and was in 1861 a candidate for the post of assistant surgeon. Unfortunately he was struck down by an attack of confluent smallpox, which he contracted while looking after a nephew who was suffering from that disease, and died at the early age of thirty-four. A career of great promise was thus untimely cut short. Writing on 15 June 1861, Sir Benjamin Brodie said, 'His death, just as he was on the point of obtaining the reward of his labours . . . is a great loss to the Hospital and School'.

In 1858 Gray published the first edition of his *Anatomy*, which covered 750 pages and contained 363 figures. He had the good fortune to secure the help of his friend, Dr H Vandyke Carter, a skilled draughtsman and formerly a demonstrator of anatomy at St George's Hospital. Carter made the drawings from which the engravings were executed, and the success of the book was, in the first instance, undoubtedly due in no small measure to the excellence of its illustrations. This edition was dedicated to Sir Benjamin Collins Brodie, Bart, FRS, DCL. A second edition was prepared by Gray and published in 1860.

The portrait of Gray published in the present section is a reproduction of one which appeared in the *St George's Hospital Gazette* of 21 May 1908, where the original is described as being 'a very faded photograph taken by Mr. Henry Pollock, second son of the late Lord Chief Baron Sir Frederick Pollock, and one of the earliest members of the photographic society of London'.

CONTENTS

GRAY'S ANATOMY

INTRODUCTION

1

AN ANALYSIS OF HUMAN ANATOMY

ANATOMY: ITS STATUS AS A DISCIPLINE

As the terminal decade of the 20th century approaches, the revision, sculpture and modulation of the 37th British Edition of *Gray's Anatomy* (published *in continuo* since 1858) has proceeded apace. Limited and leisurely in earlier editions, such activities perceptibly increased in tempo and included occasional bibliographic footnotes over the intermediate years; overall objectives were, however, largely unchanged. Close scrutiny led the present editors to attempt a complete reappraisal of content, approach and format; their views, initiated and advanced in the 35th and subsequent editions, involved a great broadening of scope to reflect the richness of modern anatomy in its various aspects. The origin and current suitability of the name *Anatomy* merits consideration; as we shall appreciate, a commonly held view is that its content is 'axiomatic', 'self-evident' and 'complete and wholly understood for centuries': all false propositions. Similarly, disciplined caution should obtain when contemplating the use of many terms, e.g. 'form', 'structure', 'morphology', 'permanence', 'change', 'dynamism', 'function', 'mechanism', 'vitalism', 'evolution'. Attempts at *definition* of any terms should be made, together with a clear statement of any known, but ever present, constraints on that definition.

Over the millenia there occurred an exponential arborescent growth of recorded observation, contemplation, dialogue, action, prediction and experimentation. The total record, however imperfect and incomplete,—human *Philosophy*—constitutes the broadest encompassing study constructed by mankind. The trunk, main branches, small growing terminal twigs etc. are analogous to the principal and lesser (although sometimes extensive) partial divisions and subdivisions. After the initial dichotomy into natural philosophy and moral philosophy, the next order includes, amongst others, ethics, metaphysics, mathematics, science, philosophies of religion, logic and the contemporary schools of philosophy. The divisions between these segments, zones (or 'whirlpools') of human thought constructs are incomplete in that they are artificially imposed on a continuum; each overlaps, interlaces with but most prominently *interacts with* its neighbours. Attempts to study a discipline in isolation contribute little; greater significance accrues when the state of advancement of its surrounding interacting disciplines is included.

On this basis, recent decades have seen Human Anatomy firmly placed as an essential facet of the core of the *Philosophy of Natural Science*; within *Biology*, the study of the biosphere, it is that area of Zoology directed towards the phylum *Chordata*, and, within the eutherian mammals, the *Primates*. Human anatomy is, therefore, a crucial feature of mankind's unending study of man, which embraces the arts, sciences, mathematics, politics, religions, ergonomics and environmental manipulation—all percepts of human cerebration. Assessing the structure (form or morphology) of the parts, or whole, of the human body is the province of anatomy. However, since all material systems known to man show dynamic changes, some apparently random, others cyclical or directional (ontogeny, learning, evolution), the structural elements selected for comparison should be assessed (in theory) simultaneously. Thus, pedantically, the *structure* of a system is 'the array of parametric variables measured at one instant in time'. Anatomy is no longer merely descriptive and anecdotal, but a central *science* that is numerate (involving measurement, data storage, statistical analysis, mathematical modelling) and, whenever possible, *experimental*. Rote repetition of 'acceptable' answers to specific questions, although engrained in many vocational examinations, is an actual impediment to progress. Uncertainty, in the face of insufficient soundly based data, should be stated and will only lessen as these accumulate.

THE DEVELOPMENT OF GRAY'S ANATOMY

Since its publication in 1858 Gray's Anatomy has evolved, under successive editors, reflecting to varying degrees many recorded facets of international anatomy available to them at any period.

Thus, the following pages initially give a concise historical survey of this evolving volume. This is necessarily viewed against the principal landmarks of anatomy as a whole and those of neighbouring and interacting philosophical disciplines.

Henry Gray's *Anatomy—Descriptive and Surgical* was published in the same year, 1858, as Virchow's *Cellular Pathologie*. Microscopes were improving and Schwann (1839) and Schleiden (1838) had established a cell theory. Broca, Bell, Magendie and others were founding neurology. In 1858, 'The Year' as we might say, Darwin and Russell Wallace presented papers at the same meeting of the Linnean Society, though Wallace was still in the Malay Archipelago, having left his partner, the collector of fossils par excellence Henry Bates, on the Amazon for another ten years.

Gray's book was indeed sandwiched among great events, between Karl Marx' *Communist Manifesto* (1857) and Charles Darwin's *Origin of Species* in 1859, a year which also saw Florence Nightingale's epochal *Management of Hospitals*. The Crimean War and Indian Mutiny were just over and Britain had become 'the Workshop of the World'. The Railway Age was still in full swing; the Liverpool/Manchester link was made in 1850. Victoria Regina was near to becoming Imperatrix and about a third through her reign. Prince Albert, aided by Paxton and Brunel, had promoted the 'Great Exhibition' in Hyde Park in 1851 and Brunel's *Great Eastern* was launched in the same year as Gray's *Anatomy*.

The Industrial Revolution, arising in Britain, was reaching a peak (but could not reach countries such as Japan for half a century). All the Western World, led and supplied technologically by British factories, was in a state of increasing dynamism. The loss of American colonies by European powers was turning Britain, at least, to a second, larger, empire in all other continents and many islands elsewhere. The Napoleonic wars and Trafalgar were memories; Wordsworth, Shelley, Byron and Keats were becoming part of the literary heritage.

Into this stirring era Henry Gray was born (1827). Matriculating at 18, he entered St George's Hospital (1845), qualified and became a demonstrator in the Hospital's department of anatomy (1850). But in his third student year he had already taken the Royal College of Surgeons' triennial prize for his work on the development of the optic and auditory nerves; when still only 25 he was elected to the Royal Society, probably due to his study of the developing endocrine glands. In the next year (1853) he won Guy's Hospital's Astley Cooper Prize with his dissertation *On the Structure and Uses of the Spleen*.

If this were all, Henry Gray had already accomplished far more than most professional men at 27 years but, four years later, as a result of his careful dissections and a clear appreciation of the needs of students and surgeons, came his greatest triumph, the book which became an instant success, despite bad reviews in the *Lancet* and *British Medical Journal*, a book which was to carry his name throughout the international medical world and beyond.

His publisher was John Parker, who also undertook a second edition (1860); but Gray, while still a mere candidate for an assistant surgeonship at St George's, succumbed to smallpox in 1861, while nursing a nephew with the same infection.

Gray had procured the help of a talented fellow demonstrator, Dr H Vandyke Carter, as his illustrator; his work, a pleasing compromise between diagram and reality, was a large contribution to the book's success.

Longmans bought out Parker's business in 1862 and the 3rd edition came out under their imprint; the copyright has remained in their hands. Like their earlier successes, Macaulay's *History of England* and Roget's *Thesaurus*, Gray's *Anatomy* has become a major factor in Longman's empire, although Gray's Centenary volume (1958) was scarcely mentioned in Longman's 250th anniversary volume in 1974. Gray's book, even more significantly, has become globally perhaps the best known textbook in the world of medicine, not only in its original British version but also in an American format since 1859 and in many translations, e.g.

Chinese, Portuguese, Italian, Spanish, and most recently Japanese.

It is worthy of note that Henry Gray was writing about three centuries after the *De Fabrica Humani* of Vesalius had established a tradition of accurate *description*, with little or no reference to function—a tradition still with us, though in Gray's time and even more so today the advent of microscopy and other techniques had added vastly to descriptive detail. It is curious that although much of Gray's research, especially the 'uses' of the spleen, recorded much more than structure, this side of his mind did not enter his book.

Subsequent editors were at first largely surgeons. The 3rd to 9th editions (1863–1880) and the 10th to 14th (1883–1897) were prepared respectively by Timothy Holmes and T Pickering Pick, both surgeons of St Georges. But in the 15th and 16th editions (1901–1905) Pick was assisted by a professional anatomist, R Howden of the University of Durham, who carried on as sole editor of the 17th to 23rd editions (1909–1926). The book was then about halfway through its 128 years of continuous publication. T B Johnston (Guy's Hospital Medical School) was responsible for the 24th to 27th editions (1930–1938) and was assisted in the 28th to 31st editions (1942–1954) by his colleague James Whillis. In 1954 Peter L Williams began his close association with Gray's Anatomy, initially as Indexer and, since 1968, as joint editor with Roger Warwick. In the 32nd Centenary edition (1958), Johnston was joined by D V Davies (St Thomas' Medical School) and F Davies (University of Sheffield), the latter two preparing the 33rd edition. D V Davies revised the 34th edition (1967). The current editors (P L Williams and R Warwick) have produced the 35th (1973) and 36th (1980) editions.

Editing by professional anatomists has certainly changed Gray's text, but initially at a very slow pace, the main burden of change being accumulations of more and more structural detail, especially at the histological and cytological levels. There has also been an increase in quotation of the actual origins of new descriptions, especially under D V Davies' editorship. The book's subtitle in more recent editions, 'Descriptive and Applied', still approximated its content and priority. But in the 35th edition the present editors deliberately expunged this subtitle, since they regard human anatomy, as do some others, not as a mere structural commentary but as a science in its own right, worthy of independent pursuit but at all times inseparable from other studies, particularly physiology, embryology, biochemistry, and so forth. The Vesalian tradition has served its turn; in fact its protraction has led to the division, amongst other ill effects, of function and structure. Physiologists are more fortunate; all functions occur in structures and the basic medical discipline, anatomy, necessarily provides a basis for all functional studies. But the anatomist may, and often does, stop the machine, describing it, even ad nauseum, without appeal to its function. This policy, though fortunately not often nowadays pursued to the ultimate, deprives the subject of its content of rationality. It is difficult to imagine investigation of structure without a concomitant desire to elucidate function, development or any other aspects of structure. Long ago Descartes described Man as a complex of machinery —biochemical, micro- and macro-anatomical—and machines work; all *living* structures are constantly changing at one level or another.

Of course, detached anatomical detail may often, but not always, serve the needs of surgeons and other clinical workers; and this may lead to such spurious concepts as 'surgical anatomy', 'radiological anatomy', and the like. But this obscures the fact that anatomy is concerned with a continuum of knowledge. Moreover the entire subject is merely a subdivision of the larger continuum of science, embracing much knowledge. This is our conviction and hence our policy.

Thus, as intimated (p. 2), human anatomy is approached, in this volume, as a numerate, reference-based, where possible experimental, natural science, notionally directed at structural parameters (previously defined). Structural correlations with many other disciplines may prove of theoretical or practical help; however, the *structure/function* relationship is *indivisible* and central to our thesis. The term function, although commonplace, is infrequently defined, other than 'actions proper to a part';

lexicographers are unhelpful, prominent sources avoid the semantics. For our present purposes, and including our structural definition, *function* may be considered as 'the sequential changes in the array of parametric variables with the passage of time'. This formalized definition offers few constraints; spatial coordinates may span vast distances across the universe, or the minute interparticles, clefts of subatomic units; temporal coordinates extend from the inception of the universe to the time of observation, and range from thousands of millions of years to microseconds. Diverse phenomena, all with their distinctive evolutionary histories, are relevant to our present theme. Some major ones are outlined briefly in the following pages and include: theories of the origin of the universe, stars, star clusters, galaxies, supernovae and planetary systems; the changing nature of matter with the emergence of wide varieties of atomic and molecular species and chemical compounds; alternative (perhaps dualistic and parallel) views of the origin of terrestrial life, its diversification, increasing complexity, dispersion; organic evolution and sexual reproduction; a survey of the animal kingdom with particular reference to the primates and man's place within them; and evidence concerning man's origins. These opening subsections are intended to provide a foundation, before a more detailed specific study of mankind's fabric—cells, intercellular material, tissues—and *functional topography* is undertaken.

ORIGIN OF LIFE ON EARTH

Our knowledge of human origins remains largely conjectural. It relies on fragmentary evidence from many different disciplines and is liable to major reassessment as new clues arrive, which they will no doubt continue to do and with increasing frequency. Relevant data arise from two major sources: the study of the fossilised remains of extinct forms, and comparative investigations of the anatomy, biochemistry and genetics of living species.

Of these, the first is the only reasonably direct means of establishing human ancestry; but because the fossil record is incomplete and it is difficult to interpret skeletal remains in their widely varying states of preservation, this approach has limitations. The more indirect comparative methods have also provided many useful clues about man's relationships, and the younger disciplines of molecular biology, particularly those dealing with molecular genetics, hold out great promise of unravelling the mechanisms by which the form and function of the human body are and were shaped. Of course the processes which gave rise to mankind have operated over immense periods of time, with widely varying influences of environment, both geological and biological. To understand the significance of these events we must also think of the physical forces which created the possibility of human emergence, of the preceding aeons and cosmic background from which life arose at first. Although these considerations may seem out of place in a textbook of anatomy, they are very relevant to our appreciation of the living substrate of the body, of its principles of construction and operation at all of its levels of organization including the cellular, tissue, developmental, systemic and integrated whole body grades, which are the themes of this book.

Present astronomical evidence points to emergence of the material universe from a single event 10–20 thousand million years ago, when the components of matter and perhaps with them the dimensions of space and time first appeared (1.1). In mathematical models of this event, it is envisaged that it occurred with unimaginable explosive force, generating temperatures of many millions of degrees and at first permitting the free existence of the fundamental, subatomic particles of matter. Within microseconds, as they began to cool, these assembled themselves into larger aggregates and finally atoms and then molecules of hydrogen. This matter expanded rapidly outwards and continues to do so even at the present time. From this singularity, the 'big bang', hydrogen molecules were first scattered fairly evenly in space but with some locational variations in density. Then under gravitational influences, they drifted together in denser clusters which eventually became massive and compact enough to convert the gravitational energy into temperatures sufficiently high to start thermonuclear fusion reactions and mass-energy conversion, so creating stars. (Currently, some theoreticians are even

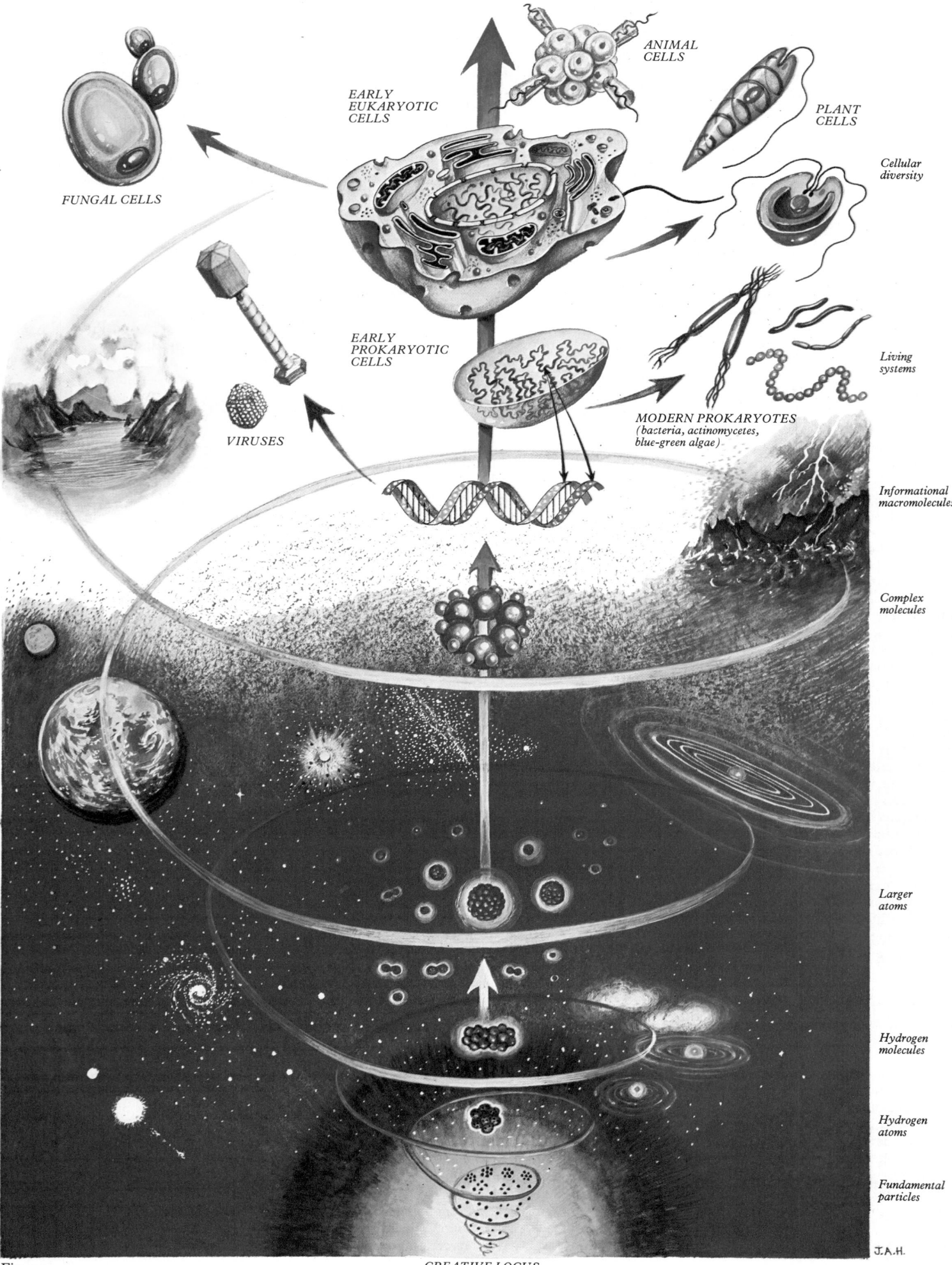

ANIMAL CELLS

EARLY EUKARYOTIC CELLS

PLANT CELLS

Cellular diversity

FUNGAL CELLS

EARLY PROKARYOTIC CELLS

VIRUSES

Living systems

MODERN PROKARYOTES
(bacteria, actinomycetes, blue-green algae)

Informational macromolecules

Complex molecules

Larger atoms

Hydrogen molecules

Hydrogen atoms

Fundamental particles

Fig. 1.1

CREATIVE LOCUS

J.A.H.

exchanging views of innovative, but speculative, mathematical models which may assist the approach to the problems of the prevailing conditions preceding, and at the inception of, the 'big bang'.)

Clusters of stars were drawn together into galaxies, and as they aged they sequentially, but asynchronously, passed through various changes in physical state, depending on their original dispersion and mass. At the enormous temperatures they generated, hydrogen atoms fused into progressively heavier atoms, forming the varieties of elements we know today. Some particularly massive stars became unstable and eventually exploded as supernovae, strewing the products of their fusion reactions into interstellar space where they cooled enough to form complex chemical compounds. It may have been that such material was captured by the gravitational fields of other stars, coalescing into larger masses which circled around them as planetary systems. Our earth appears to have been formed in the solar system at about five thousand million years ago as a hot mass of rocks rich in elements such as carbon, silicon, oxygen, sulphur and phosphorus as well as many metals, e.g. sodium, calcium, potassium and iron. This world appears to have been at first highly volcanic but, as it began to cool, water vapour could condense, forming the early seas, rivers and lakes, and generating the processes of erosion and sedimentary deposition which dominate the geology of our planet today. This early period was now cool enough to allow hitherto impossible chemical reactions in aqueous solution and the early water masses may have teemed with a wide variety of molecules in a dilute solution of inorganic ions. In particular, many carbon-based compounds were stable at these temperatures, yet could react with each other rapidly enough to create a rich carbon chemistry. It is likely that life arose in such an environment.

All matter, living or non-living, consists of the same basic subatomic units, and, varying quantitatively, many common atomic species. The most obvious distinction between the living and non-living is the dynamic level of molecular organization of this matter, that exhibited by the living being greater. Furthermore, development, growth and maintenance of living organisms involves a temporary increase in order, the ability to effect this increasing with the organism's level of complexity. Living organisms are usually ascribed a common series of attributes, e.g. the ability to sense and respond to environmental changes, to move, to grow and to reproduce. These vital attributes are considered further on p. 6. To describe is not to define, however, and what distinguishes the living from the non-living eludes precise definition.

How living matter originated is still speculative. It is generally assumed, but rarely challenged among scientists, that life evolved exclusively *on Earth* from non-living matter. The commonest hypothesis is that when our planet had cooled sufficiently for the water vapour in its atmosphere to condense and begin to erode its rocks, the elements which dissolved in it combined to form simple inorganic molecules which, in response to solar energy, reacted chemically to produce the amino acid units of the enzymes essential to the life process. These interacted further, some evolving into 'living' self-replicating units, resembling viruses, in which the information necessary to ensure the accurate duplication of enzymes and other proteins was encoded in nucleic acids. Enclosure of some such units within proteolipid membranes might have produced the first unicellular organisms, perhaps like bacteria, each able to exert a measure of control over its internal environment, leading to increases in local order, longevity and reproductive capacity.

The results of experiments by Miller & Urey in 1953 suggested that amino acids could have been formed in the terrestrial atmosphere presumed to have existed some 4500 million years ago; this strengthens the hypothesis that life could have originated from non-living systems on earth, but the next and literally vital evolutionary step, the chance interaction of these amino acids to form the proteinaceous enzymes necessary for self replication, has

not been demonstrated. Lack of such a demonstration does not, of course, invalidate the hypothesis but equally the demonstration that amino acids could have formed in the then prevalent conditions does not prove that they actually did.

During the last two decades a novel, and controversial, alternative to the traditional hypothesis that life arose on Earth from non-living matter has emerged. According to Crick & Orgel (1973), Hoyle & Wickramasinghe (1981) and others, the concept of the Earth as the biological centre of the Universe is as erroneous as the pre-Copernican notion that it was its physical, geometrical centre, and the possibility of an extraterrestrial origin of living matter has been raised. This proposal is claimed to be supported by the identification in meteorites of what is presumed to be once-living matter preserved as microfossils possibly resembling viruses and bacteria. A modern bacterium, *Pedomicrobium*, has features suggesting what such organisms may have been like. This species obtains energy by oxygen transfer from salts to either ferrous iron or manganese, a process which could explain how deposits of ferric iron came to be laid down in rock formations as long ago as 3800 million years, well before sufficient free oxygen had been produced by photosynthesis for atmospheric oxidation to have given rise to them. If organisms resembling *Pedomicrobium* were responsible for such processes, would there have been sufficient time, and were conditions suitable, for such organisms to have evolved de novo on Earth? Until 1979 the earliest geological evidence for microbial, prokaryotic life was dated at about 3200 million years ago; since this predated the earliest eukaryotic fossils, it accorded with the theory that eukaryotes had evolved from prokaryotes. However, in 1979 Pflug et al found what they interpreted to be yeast-like and therefore eukaryotic fossils in rock deposited about 3800 million years ago, throwing some doubt on both the theory of their evolution and on the possibility that such complexity could have been reached by chance in the time available. An alternative explanation (Hoyle 1983, and others) is that both prokaryotes and primitive eukaryotes could have had an extraterrestrial origin, existing as components of the interstellar dust and arriving on Earth and on other planets in an already living, though dormant, form after being carried in, and protected by, the frozen water which forms the bulk of comets. According to Bruch (1967) the hazards to the survival of dormant microbial spores under such conditions have been exaggerated, for they could be shielded from the chief dangers, solar ultraviolet irradiation and the solar proton wind, by a thin layer of overlying material, and could also have survived descent through the Earth's atmosphere. If this hypothesis of an extraterrestrial origin of life and its mode of transport to Earth is correct, then the continual bombardment of the Earth with such material may have influenced, and still be affecting, the progress of evolution. The fundamental, fascinating problem of how life, whether originally terrestrial, extraterrestrial *or both*, began remains unresolved. Clearly the temporospatial coordinates of these events also await resolution.

It may also be noted that the idea of an extraterrestrial origin does not solve many of the central problems of life's origins but displaces it to some other cosmic site. However, the theory is notable for its stimulating, controversial nature rather than, as yet, the strength of evidence in its support.

EVOLUTION OF LIFE ON EARTH

However life arose, many remain persuaded that it appears to have done so only once, since the highly complex macromolecules of different living organisms and their metabolic interactions share too many common features to have arisen independently. (More strictly, whatever view of life's origin is envisaged, the 'only once' of the preceding paragraph should, perhaps, imply *one common mechanism*, whatever its possible multiplicity of sites and times.) The simplest living cells are the bacteria, blue-green algae and similar forms (collectively termed Prokaryotes), which have most of the essentials of other cells, except that they do not carry their genetic material in a nuclear compartment and, apart from the outer cell membrane, they lack membranous organelles. More complex cells (Eukaryotes) are likely to have arisen from these (however, vide supra) but differ from them in having nuclei

1.1 The origins of biological organization as envisaged from present cosmological, biochemical and palaeontological data.

containing larger quantities of genetic material in multiple chromosomes sequestered away from metabolic damage. Such cells needed a special (mitotic) apparatus to handle their more elaborate chromosomes at cell division and also perfected ways of exchanging genes in meiosis which increased their variability and therefore the potential to exploit a variable environment. These cells also had complex membranous organelles within their cytoplasm so that different areas of the cell could be allocated to specialized functions. These developments appear to have been crucial because they allowed the formation of progressively more elaborate genetic material, structure and metabolism and consequently more complex form and function. It is likely that there were many different types of cell during this early experimental period, only some of them ancestral to the forms of life we know today, e.g. protozoa, unicellular algae, fungi, multicellular plants and animals (1.1,2) (see Vidal 1984). Some of these may have arisen as combinations of cells giving greater metabolic potential; thus both mitochondria and the chloroplasts of green plants have their own genetic apparatus distinct from the rest of the cell and may have arisen as independent organisms (see Margulis 1981). It is also likely that many of the earliest eukaryotes were able to use sunlight for their energy needs and had photosynthetic pigments. Certainly once the early nutrients in the environment had been depleted this would have been the only source of energy for the biological world, herbivorous and carnivorous animals depending for their supply of chemical energy on green plants. From this point onwards there must have been a great proliferation of different forms and the development of multicellular organisms in which different cells could carry out specialist tasks (see Romer 1968, Attenborough 1979).

THE ANIMAL KINGDOM: SELECTED PHYLA

The fossil record of these events begins with any clarity only when many of the major groups of invertebrates had already appeared, by about 500 million years ago, and the line of descent (1.2) can only be surmised. It is thought that the first truly multicellular animals (*Metazoa*) possessed two layers of cells (i.e. they were diploblastic), but that later a third layer was added giving an outer, ectodermal layer, a middle mesoderm and an inner endodermal lining to the gut (enteron). The outer layer provided an interface with the environment, the inner was concerned mainly with nutrition while the middle layer provided mechanical support, muscular systems and a route for nervous and circulatory systems. In the simplest *triploblasts*, such as flatworms (Platyhelminths), the mesoderm is usually a solid mass but in more complex forms it splits into an outer (somatic) and an inner (splanchnic or visceral) layer, the latter surrounding the gut which is now free to move within the cavity (coelom) so formed within the mesoderm. In these animals (coelomates), an efficient vascular system is essential to convey nutrients absorbed in the gut across the coelom to the outer parts of the body and we find that they all possess propulsile heart-like structures as well as branching vascular channels. The freeing of the outer mesoderm from the gut also made further locomotor advances possible and many such animals are organized on a segmental pattern of repeated muscle blocks and attendant nerves and blood vessels which facilitated many different types of movement and specialized local body functions. This segmental pattern is fundamental to many groups of animals including ourselves and it is significant that at least some of the genetic mechanisms for generating segmentation and local specialization of segments (homiotic genes) appear to be much the same in both insects (e.g. *Drosophila*) and mammals.

From these segmented coelomates, which include annelid worms, arthropods, molluscs and echinoderms (i.e. starfish and their allies), are thought to have arisen the *chordate phylum*, possibly from an echinoderm-like stock. The simplest chordate-like animals we know today are aquatic filter-feeders such as seasquirts (Urochordates) and the fish-like Cephalochordates (the lancelets *Branchiostoma* or *Amphioxus*). These filter their food from water through a series of slits flanked by supporting bars in their pharyngeal walls, which appear to be the forerunners of the embryonic pharyngeal clefts and arches of all vertebrates. Many of these primitive creatures also have hollow dorsal central

nervous systems, derived perhaps by the invagination of superficial nerve nets such as those we see in starfish. They have bilateral symmetry and, at some stage in their lives, segmental muscle blocks grouped around a longitudinal stiff but flexible rod (notochord) permitting sinusoidal swimming movements. At least some of them also develop an additional embryonic group of cells, the neural crest (see p. 132), which is of great significance in the development of many typical chordate features.

The origin of the true vertebrates from such groups appears to have been relatively late, the first well preserved fossil forms, armour-plated fishes called Ostracoderms, occurring in Silurian rocks about 410 million years old, although fragments of fish-like scales also occur in much older strata. Ostracoderms lacked a true jaw mechanism (the agnathan condition) and were apparently rather poor swimmers, judging from the rudimentary state of their fins and the presence of only two semicircular canals in their inner ears. However, they had all of the typical vertebrate features including a brain and surrounding skull, a ventral heart and effective high pressure blood system. Their pharyngeal arches supported gills and the clefts between them allowed the flow of water in a respiratory current. The heart pumped blood through a series of vessels lying in the pharyngeal arches, the basic number being six pairs. Although these were used for perfusion of the gills in respiratory gas exchange, they created a pattern which persists even in human embryonic development, where it is modified to form the great vessels leaving the heart (see p. 213). The modern cyclostomes (e.g. lampreys) are the descendants of these fishes.

The next advance was the modification of the most anterior pharyngeal arch to support the mouth, its ventral part becoming the mandibular arch (see p. 165); this allowed the development of more efficient and varied feeding (the gnathostome grade of organization) and teeth could be anchored to the jaw to increase its effectiveness. (For the upper jaw and teeth see p. 167 et seq., and p. 1299.) Often the next (hyoidean) arch was used to support the first, as well as to contribute to the tongue apparatus. At the same time, swimming ability improved greatly with the formation of complex fins, a third semicircular canal and no doubt advances in brain organization to control movement more efficiently. These steps proved very successful and most later vertebrates arose from this stock (1.2). These early fishes included many different types, most of which became extinct although some, like the cartilaginous sharks and rays (Chondrichthyes), persisted and thrived. Another group of fishes with bony skeletons (Osteichthyes) divided into a line leading to the majority of modern ray-finned fish and another, the Sarcopterygians, in which the fins were highly muscular and had an axial skeleton, a situation we see in modern lungfish and the Coelacanth *Latimeria*, anticipating the limbs of early terrestrial vertebrates. Some of these fishes may have been able to breathe air as well as to use their gills for respiration. Their upper jaws (p. 340) were fused with their brain case (neurocranium) to increase the skull's mechanical stability and various plate-like bones derived from scales in the dermis reinforced the neurocranium, the skeleton of the jaws and other pharyngeal arches. These 'scale bones' correspond to the membrane bones of the human skull, while the more primitive neurocranial and viscerocranial elements are represented by its endochondral bones (see p. 304).

From such fishes emerged the earliest land-living vertebrates, during the Upper Devonian era about 270 million years ago. These were amphibians with many fish-like features but instead of pectoral and pelvic fins they had muscular limbs, each with five digits and a characteristic pattern of axial bones (the pentadactyl limb) as in all terrestrial vertebrates, or tetrapods. They were therefore able to move around on the swampy ground of the coal forests, where their remains became fossilized. Lacking gills except during larval life, they respired through diverticula of the oesophagus, i.e. simple lungs, a common method of accessory breathing even in fishes which live in tropical anoxic waters. Judging from the condition of modern amphibians such as salamanders, frogs and toads, these animals had hearts with partially divided ventricles, allowing them partial separation of blood for systemic and pulmonary circulations and therefore a more efficient respiratory transport permitting increased active movement. With the loss of gills, the arterial arcades of the pharyngeal

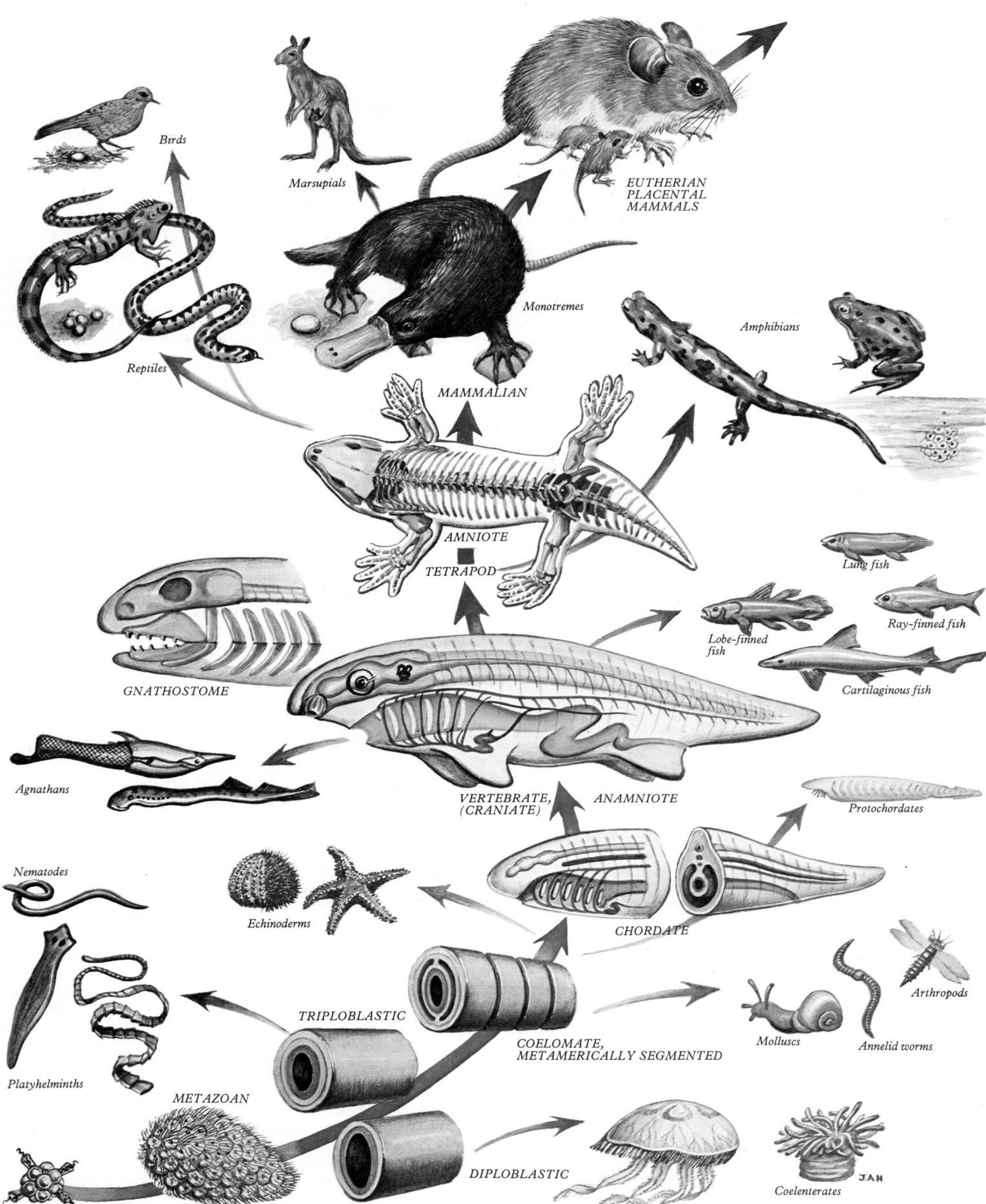

Birds

Marsupials

EUTHERIAN
PLACENTAL
MAMMALS

Monotremes

Amphibians

Reptiles

MAMMALIAN

AMNIOTE

Lung fish

TETRAPOD

Ray-finned fish

Lobe-finned
fish

GNATHOSTOME

Cartilaginous fish

Agnathans

VERTEBRATE,   ANAMNIOTE
(CRANIATE)

Protochordates

Nematodes

Echinoderms

CHORDATE

Arthropods

TRIPLOBLASTIC

COELOMATE,
METAMERICALLY SEGMENTED

Molluscs

Annelid worms

Platyhelminths

METAZOAN

DIPLOBLASTIC

Coelenterates

JAH

1.2 The relationships and different levels of organization of the major
groups of animals leading to the mammalian class, based on currently
available evidence.

arches were reduced in complexity and number and now were simple, but large, arteries leading to the head, body and lungs. Amphibians were, however, still dependent on an aqueous environment since they laid their eggs in water and probably dehydrated easily through a relatively permeable skin.

Early reptiles probably emerged soon after the amphibia. Two major advances freed the new group from slavish dependence on water: the skin became keratinized and water resistant; reproduction could take place entirely on land as the young developed in an egg enclosed in a shell which prevented water loss but allowed gaseous exchange. Within the shell the embryo floated in a fluid-filled space; the amniotic cavity and the membranes immediately surrounding it (amnion and chorion) acted as a gill. Except for the egg-shell, these modifications were retained in all descendants of the reptiles including mammals.

Reptiles were now free to exploit the dry places of the earth as well as the humid ones and they became the dominant vertebrate group for many millions of years. Various internal changes took place including: complete separation of systemic and pulmonary circulations; the formation of a vena cava making the return of blood more efficient; development of a high pressure kidney with more effective filtration; various modifications of the brain which increased its ability to analyse sensory data, organize motor activity and to establish more complex behaviour patterns in general.

Although many of the most successful reptilian groups such as the dinosaurs eventually became extinct, some persisted as the turtles, lizards, snakes and crocodiles of the present or became highly modified for flight, the birds. Another group of great significance comprised relatively unspecialized animals emerging quite early in reptilian history. These had a unique skull construction with a fossa in the middle of its lateral surface allowing the expansion of masticatory muscles, the Synapsid condition. Along this fossil line, several remarkable changes occur. These include the expansion of the brain and its braincase, an alteration in the jaw mechanism so that several of the bones which hitherto had formed part of its articular mechanism became reduced in size, adapted for picking up auditory signals and progressively incorporated into the middle ear. Complex teeth of different types and with more than one cusp were also inserted into the jaws, a sign of more effective mastication, and a new, more robust joint formed between the mandibular condyle and the temporal bone. Locomotion was also modified by a rotation of the limbs bringing their intermediate and distal joints *under* the trunk, thus introducing an ability to lift the body off the ground, a change suggesting a cursorial habit commensurate with a high metabolic rate and therefore warm-bloodedness. (The manner of limb rotation and attendant modifications contrasts sharply in the fore- i.e. upper, and hind- i.e. lower limbs: see pp. 174, 399.) Hair probably covered their bodies in an insulating layer.

Such creatures co-existed with the ruling reptiles from the Permian era from about 200 million years ago and, when the latter became extinct, they replaced them as the dominant land vertebrates. By this time the full mammalian condition had been achieved, with a greatly expanded brain, fully functional temporo-mandibular joint, teeth specialized for different functions (incisors, molars, etc.) and showing the primitive pattern of mammalian cusps (see p. 1299). To judge from the nearest descendants of some of these early mammals, the *Prototheria* (Monotremes) of Australasia such as the Platypus, *Ornithorhynchus*, these creatures were certainly homeothermic, covered in hair, suckled their young, although they still laid eggs, and had many other non-reptilian features such as a division of the trunk into a thorax and an abdomen, with a diaphragm allowing thoracic breathing. Their pharyngeal arterial arches were more streamlined, the third being developed only on the left to form the aorta. Such animals were obviously well equipped to deal with the wide variety of climatic conditions and of food sources available in a drastically changing environment. Their more advanced brains could no doubt enable them to exploit their environment more fully and better maternal care, including suckling, favoured the survival of offspring and permitted prolonged development after birth. Many different groups of such animals rapidly spread over the land masses; one of them, which dispensed with laying eggs by retaining the fetus in utero for the early part of its development,

became dominant. These were the *Metatheria* (Marsupials), including the pouched forms such as the kangaroos and non-pouched species like the opossums. They had relatively larger and less 'reptilian' brains and a more effective feeding apparatus including highly specialized teeth. Again, from the many different types of marsupials a third mammalian group the *Eutheria* (Placentals) arose and, in turn, supplanted them. In these new mammals, the developing young were retained for much longer inside their mothers, attached, as were marsupial embryos, by a placenta which provided nutrition, respiratory exchange, removal of excretions and other metabolic needs. The placenta of the Eutherians was much more elaborate and effective than that of the marsupials, allowing birth at a very advanced stage of development and in particular permitting the brain to achieve considerable complexity before its use for survival in the outside world. These mammals rapidly replaced most of the other groups. This basic eutherian design which first emerged during the Cretaceous period about 120 million years ago, or possibly earlier, was extremely successful at adapting itself to many different evolutionary niches, radiating into the various mammalian forms we know today.

The closest living relatives of these early *Eutheria* are Insectivores such as the Shrews, small highly active animals with primitive tooth patterns and relatively unspecialized body form. It is thought that from species similar to these arose a group of animals which were ancestral to the Primates.

EVOLUTION OF PRIMATES

From this relatively unspecialized mammalian line arose a group of animals with enlarged brains, well-developed vision progressing to a binocular, stereoscopic condition, strongly developed hearing but reduced olfactory sense and an unusual facility in grasping with their fore- and hind-feet (i.e. manus and pes). We see these features, developed and improved upon to varying degrees, in a range of living and fossil species which constitute the *Primate Order* (1.3).

PROSIMIAN PRIMATES

The earliest true primates are thought to have emerged during the Palaeocene era as long ago as 60 million years, or even earlier. Since recognizable fossils of these animals have not yet been found, ideas of what they were like rely heavily on the most primitive living primates, classified as the Suborder Prosimii. To judge from these, the first primates had a cranium expanded to carry a brain with particularly well-developed cerebral and cerebellar cortices. Their eyes were large and a protective bar of bone separated the orbit from the temporal fossa, while the nasal cavity was reduced. Their limbs had very mobile joints and their first digits (pollex and hallux) were somewhat rotated and separate from the others and could be opposed to them for grasping. In the more advanced species the claws were flattened into spatulate nails which assisted manipulation. Dentition was of a primitive mammalian type with unspecialized tooth-cusps, allowing their possessors to eat a wide range of food.

The earliest of these species were probably arboreal, as are the most primitive living primates, the Lemurs, long-tailed tree-dwellers of Madagascan forests, with large brains, highly mobile limbs and well-developed prehensile hands and feet. In their close relatives, the Lorises, which include the bush-baby (*Galago*), the large eyes are turned forward, giving good stereoscopic, binocular vision allowing precise judgement of distance; the brain case is high and the snout much reduced in length, like the more advanced apes. These different groups, collectively known as the Lemuroid primates, and their more specialized relatives the Tarsiers, agile creatures with huge, forward-looking eyes, are all adapted to a nocturnal, arboreal life-style and a varied diet such as fruit, nuts, leafy shoots and insects which they can hold with their grasping digits to bite and chew with their relatively unspecialized teeth; this type of condition may have been typical of all early primate species. Like all primates, they show a high level of parental care, the offspring being few and suckled by means of two thoracically-sited mammary glands.

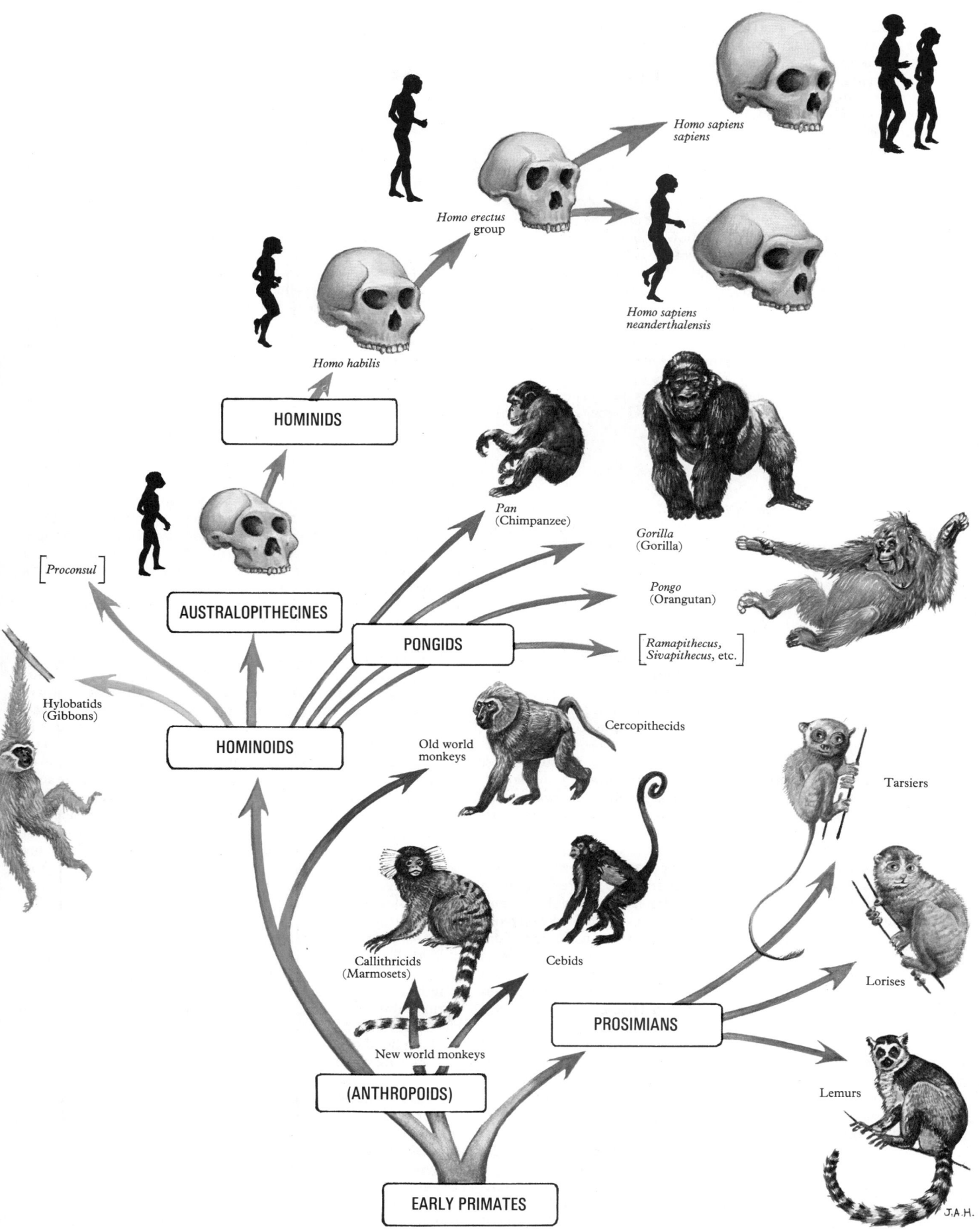

Homo sapiens
sapiens

Homo erectus
group

Homo sapiens
neanderthalensis

Homo habilis

HOMINIDS

Pan
(Chimpanzee)

Gorilla
(Gorilla)

[Proconsul]

AUSTRALOPITHECINES

Pongo
(Orangutan)

PONGIDS

[Ramapithecus,
Sivapithecus, etc.]

Hylobatids
(Gibbons)

HOMINOIDS

Old world
monkeys

Cercopithecids

Tarsiers

Callithricids
(Marmosets)

Cebids

Lorises

PROSIMIANS

New world monkeys

(ANTHROPOIDS)

Lemurs

EARLY PRIMATES

J.A.H.

1.3 A scheme outlining possible evolutionary relationships between *Homo sapiens* and other living and fossil primates. For simplicity and because of present uncertainties about classification, details of species of fossil primates have not been included (see text).

The considerable variety of remaining primates (previously grouped as the *Suborder Anthropoidea*) appeared and approached the human pattern more closely. Their brains are larger relative to body size and their vision fully binocular, with forward-facing eyes and appropriate orbital geometry permitting conjunct eye movements; instead of a thin bar of bone between the eye and the temporal fossa, there is a solid post-orbital plate which forms a protective posterior part of the orbit. Olfaction is further reduced, the snout being even shorter and the face dominated by the brain case. All such species can sit upright, so that the hand is free for more complex manipulative tasks, and there is a correspondingly more precise neuromuscular control of the fore-limb. With this change of body posture goes a change to a less obtuse angle between the axis of the cranial base and that of the vertebral column so that the foramen magnum is now near the base of the skull rather than at its rear and faces downwards (with the trunk erect). Behaviourally, extant forms are highly explorative, active creatures, usually living in groups and possessing a varied repertoire of communication skills, amongst which are vocalization and a corresponding auditory analysis in the brain. They also use various muscles of the face (those supplied by the facial nerve) to convey mood and, having highly developed colour vision (unlike most other mammals), they often employ brightly coloured areas of skin to signal social and sexual status. These primates include two distinct modern groups (Ceboidea and Higher Primates) with their many subgroups.

THE CEBOIDEA

The *New World Monkeys* (*Platyrrhini* or *Ceboidea*) of South America are a prosimian offshoot which probably separated quite late from the main stock (1.3) and is not considered further here.

THE HIGHER PRIMATES

The *Higher Primates* (*Catarrhini*), originally of Africa and Asia, differ from the prosimians in having fewer teeth (a total of 32 rather than the primitive 36), reducing the number of premolars to two in each quadrant and developing an unusual pattern of four blunt cusps on their molars; their nostrils are also close together and face downwards. The functional significance of these changes is not clear but they point to a common origin of this group of animals.

The higher primates are further divisible into two subgroups: first, the *Old World Monkeys* or *Cercopithecoidea* (e.g. macaques, baboons, mandrills and langurs); and second, the *Hominoidea* which include the great apes (gibbons, orang-utans, gorillas, chimpanzees) and humans. Of these different species, the great apes and related fossil forms are on many counts our closest relatives and are clearly of great importance in our search for human ancestry. Like Man, they differ in several respects from the other anthropoids, their brains being larger, their skulls more dome-like and provided with frontal sinuses and their body size generally greater. All have long arms, powerfully mobile shoulder girdles braced by strong clavicles and long fingers which enable the more arboreal types such as gibbons to swing with great ease and speed through treetops; legs tend to be short, pelvic girdles reduced, with ilia flattened in the coronal plane, and tails absent, thus minimizing the weight of the body which in the more arboreal species is transmitted mainly to the shoulder girdle and arms through a rather straight vertebral column. It is likely that all apes emerged from such tree-dwelling, brachiating forms, although it is improbable that they were as well-adapted to this mode of life as are the modern gibbons.

Structurally, the orang-utans, gorillas and chimpanzees, grouped as the family *Pongidae*, are much more similar than the gibbons (the *Hylobatidae*) are to Man. The orang-utans appear to have remained arboreal, while the line leading to gorillas, chimpanzees and humans became adapted to moving on the ground and the pelvic girdle and legs became more robust as the weight was now carried increasingly on the hind limbs. In the gorillas and chimpanzees walking is mainly quadripedal, the long fore-limbs carrying the anterior body weight on the knuckles; but the hands are also used effectively for manipulation and, in the case of chimpanzees, even to employ sticks as simple tools.

Various types of evidence from comparative anatomy, biochemistry (including comparisons of DNA sequences) and immunology suggest that the line leading to the gibbons was the first to diverge, followed by the orang-utans then the gorillas and finally the chimpanzees, which are generally regarded as closest to Man among living creatures. It is interesting that the gorilla's and chimpanzee's chromosomal patterns and the base sequences in their nucleic acids are almost identical to those of humans, although the great apes have 48 chromosomes rather than the human number, 46, probably because of chromosomal fusion in the human karyotype. However, these relationships are still under review, and some authorities place the orang-utan closer to the human line than the chimpanzee.

The fossil evidence of the emergence and separation of these different anthropoid lines is scant; best known are the skull of *Parapithecus*, a monkey-like fossil with some ape-like features, from Egyptian Oligocene deposits about 30 million years old and, from the same deposits, the more characteristically anthropoid jaw of *Propliopithecus*, which may be regarded as the earliest true ape yet found.

FINAL STEPS TO MODERN MAN

Between the Pongidae and modern humans are important differences in structure. A gradual progression from apes to Man is also fragmentarily detectable in the few fossil forms which span the period between about 20 million to 35 000 years ago, during which the human form probably evolved from apes. During this period, the essential hominid features emerged, the most important being a remarkable increase in brain size (the largest brain in the apes is that of the gorilla at 650 ml, whereas the human cranial capacity has a mean of about 1300 ml) and in the folding of the cerebral cortex, the development of bipedal walking and use of the hands for complex manipulation, especially the making of tools (see Pilbeam 1984).

With these major developments came a variety of anatomical alterations. Bipedal walking was made possible by changes in the shape, strength and orientation of the pelvic girdle, hip joint, femur and lower limb skeleton which allowed the placing of one foot in front of the other during striding locomotion. With upright posture, the vertebral column became strengthened against compressive forces and secondarily curved to absorb the mechanical shock of walking, while in the skull the foramen magnum came to lie almost centrally beneath the cranium as the angle between the base of the skull and the cervical vertebrae approached 90°. The occipital ridges, which in the apes are massive, providing reinforced attachments for the large cervical muscles holding up the head, became small as the skull now balanced on an upright vertebral column. The foot changed from a prehensile structure with an opposable first toe into a coherent though flexible plate with parallel digits and a highly arched general structure suited to prolonged standing, walking and running, sometimes on uneven surfaces.

With this bipedal habit, the upper limb was now completely freed for manipulative tasks; changes in articular angle and lengthening of the thumb made it fully opposable to the other digits; subtle alterations in the length and planes of movements of the other fingers and their muscles resulted in a complex, multi-purpose hand capable of delicate tasks such as those involved in tool-making while preserving the powerful grasping action of the apes.

There were also transformations in teeth and jaws, and simplification of the nasal region continued as the olfactory sense further lessened in importance. The maxillae and mandible as a whole protruded less and the tooth rows they bore changed from a rectangular to a more rounded arcade; the canines decreased in size, becoming more spatulate, and the mandibular rami became taller and thinner. However, externally the chin increased in prominence, while on its internal face a ridge for the attachment of muscles, the simian shelf in apes, was reduced to four small genial tubercles (mental spines).

The increase in cerebral size gave the forehead a more domed appearance and fetal cranial enlargement was reflected in a widening of the female pelvic girdle to allow the passage of the larger head of the young at birth.

Although skeletal remains of possible transitionary species are tantalisingly few, painstaking excavations during the last 50 years, particularly in the East African Rift Valley where the strata are exposed in an orderly sequence of dateable volcanic deposits, have provided increasing numbers of specimens, some of them excellently preserved. Inevitably, much controversy has surrounded the interpretation of these bones but there is general agreement over the major trends, which point to the origin of a distinctively human form by about 100 000 years ago.

Amongst the earliest skulls in this progression is that of a small ape, *Proconsul*, dating from about 14 million years ago in the Miocene of Kenya. Its features suggest that it was already some way along a divergent line leading to the gorilla and chimpanzee rather then a direct human antecendent. Another fossil group, found in deposits 10 to 20 million years old in sites widely scattered in Asia, Africa and Asia Minor, includes three rather similar genera, *Dryopithecus*, *Ramapithecus* and *Sivapithecus*, known mainly from fragments of tooth rows and jaw bones. At first they were considered good candidates for a central role in human evolution but the recent find of larger parts of a *Sivapithecus* skull suggests that this group may be more closely related to orang utans.

THE HOMINIDS

The first undoubtedly bipedal fossil anthropoids yet discovered are the *Australopithecines*, a group of ape-like animals found in the rocks of Eastern and South Africa, dating from 1–3.6 million years ago. They are a heterogeneous collection of different species, some, termed robust Australopithecines, having quite massive skulls, and others with more delicate bones (gracile types). However, some of these variations may reflect differences in gender or maturity. The Australopithecines have many features in common with the great apes, e.g. a relatively small braincase and a large dental arcade, but there are also some human affinities; e.g. an upright, bipedal gait is indicated by the shape of the pelvis and femur and by the form of fossil footprints preserved in volcanic mud at Laetoli in Tanzania, dated to 3.6 million years ago and attributed to *Australopithecus afarensis*.

The number of species within the Australopithecines and their relationships to each other and to mankind are still uncertain but it is very likely that some at least of these forms were close to the ancestral human lineage.

Next came the *Habilines*, found in East African deposits dating from 1.35–1.9 million years ago. This group also contains various forms, some with larger brains but ape-like jaws, others with more humanoid teeth but small brains. Their maximum cranial capacity is estimated at 680 ml, which has been considered sufficient to place them with Man in the genus *Homo*. Stone tools have also appeared in the same strata as their fossils and the hand is sufficiently human to suggest that these creatures were their creators, hence the specific name *Homo habilis* ('Handy Man'). However, the Australopithecines are also suspected of making simple stone tools and the distinction between these and the Habilines has not been made without controversy.

Another more modern series of fossils with many undoubtedly human features is grouped under the name *Homo erectus* (Upright Man). This category includes quite a wide range of skeletal forms of various ages between 500 000 and 1.5 million years, differing widely in geographical location, including Indonesia, China, Africa and Europe. However, they all have large crania, with capacities from 850 ml in Java Man to 1225 ml in Peking Man; there is also evidence of advanced behaviour, e.g. the use of fire at camp sites and the making of advanced stone tools (of the Chellean culture). They also had various non-human skeletal specializations such as massive bones and skulls with immense brow ridges and a sagittal crest, suggesting that at least the most typical of them were not ancestral to mankind but some off-shoot from the main humanoid stock.

Neanderthal Man, found in much more recent cave deposits in several parts of Europe dating from about 100 000–35 000 years ago, shows many features closely resembling those of modern humans, including large cranial capacities, equal to or even exceeding ours. They showed complex social behaviour, including, it appears, ceremonial burial of their dead and the making of quite advanced stone tools. But there were differences; like *Homo erectus*, their brow ridges were large and their skeletons comparatively robust. For these reasons they are accorded the sub-specific status of *Homo sapiens neanderthalensis*.

Remains of modern man *Homo sapiens sapiens* (often termed Cro-Magnon Man after the French location where some of the earliest bones were originally found) have been discovered in strata dating from as early as 35 000 years ago in eastern Europe. This date strongly suggests that the Neanderthals co-existed for some time with modern humans, although there is no evidence of interbreeding. These latest fossils are indistinguishable from modern skeletons, hence the human body which is the subject of this book must have at least this degree of antiquity.

The precise relationships between modern humans and the fossil forms briefly described in this account are far from settled; there is no general agreement on the line which human evolution has followed during most of the last four million years. But it appears likely that the human stock arose in Africa from a Habiline-like group of Australopithecines, although the gap between these and ourselves is so large and devoid of relevant fossils that, until further evidence is at hand, we can only speculate as to our lineage during this period.

The human races living today vary in skeletal, muscular and other details but the relationships between the different ethnic groups and their special physical characteristics are beyond the scope of this account. It is noteworthy, though, that anthropological and biochemical studies point to the emergence of all present-day humans from a small group of individuals, perhaps somewhere in the Middle East. Their descendants soon migrated into all habitable continents and adapted to their differing environments and climatic conditions. However, these developments were relatively recent and, in spite of minor differences, e.g. proportions of blood groups in different human populations, the genetic 'distances' between the races are in general insignificant compared with the genetic variation within them.

Analysis of mitochondrial DNA variation (which is transmitted only through the female line) in different races indicates that modern people descended from a female who lived perhaps 200 000 years ago in Africa, and that several waves of emigration from Africa subsequently colonized other land masses (Cann et al 1986).

The inheritance of form and function from the different stages of evolution, culminating in the supreme development of mental self-awareness and abstract thought, is one of the greatest stories that science has to tell, although understood in only the barest outline. Here we study the functional anatomy of *Homo sapiens*, a form which is not only superbly adapted to a multitude of present activities, but also bears within it a memory of its past; an awareness of this helps us to understand many of the otherwise inexplicable patterns expressed in the human frame, and perhaps to appreciate both its frailty and the most wonderfully effective complexity which it has attained.

Anatomical Nomenclature

A subject such as anatomy, with its accent on description, necessarily requires a very large number of names for structure and processes. For effective communication such words should be as simple as possible and used with unfailing precision. Unfortunately both these aims have only been partially accomplished.

In the first place, as in other sciences, common words do not exist in any language for thousands of structures which require to be named, and the need for new names never ceases. For this reason the manufacture of new names has been based, like the Linnean classification, upon stems of Latin and Greek words. When Latin and often also Greek were a usual part of general education, the actual meanings of these stems were familiar. Even in these days, when classical education is the privilege of only a small minority, many of the words used are familiar because they have to some extent crept into ordinary language, sometimes with a little distortion: musculus, tendo, arteria, vena, cranium, etc. are so like their vernacular equivalents that use of the Latin form is easy.

Words derived from Latin and Greek have also the advantage of being suitable for international usage and, to ensure uniform usage, efforts have been made since 1895, when the Basle Nomina Anatomica was introduced. In 1950 at the Oxford (5th) International Congress of Anatomists, a new body was instituted, the International Anatomical Nomenclature Committee, whose first Honorary Secretary was T B Johnston (one of our editorial predecessors), the current Secretary being R Warwick.

It must be said at once that the officially and internationally agreed terms (see Nomina Anatomica, Nomina Histologica, Nomina Embryologica, 5th Edn 1983) are not always adhered to in textbooks or atlases. They are, of course, subject to revision at each quinquennial World Congress of Anatomists and doubtless many are still inept and many are far too complex. Who will stringently adhere to 'aponeurosis musculi bicipitis brachii' for what we in English call the 'bicipital aponeurosis'? Of course, each national language is permitted, and even encouraged, to vernacularize Latin terms, as long as the relation to the official Latin remains clear. No one expects a sudden reference to 'biceps femoris' when dealing with the arm, and hence 'biceps' for 'musculus biceps brachii' is safe to use, as long as no misunderstanding occurs.

This is the policy which we have adopted, not, unfortunately, with complete uniformity. We recognize the difficulty of pure Latinity for a large number of our readers and we prefer to anglicize Latin terms but to give the official form, at least once, in parentheses.

Coming also under the topic of nomenclature are the names of lines, planes and directions, used in describing structures. Conventionally the body is regarded as being in the so-called 'anatomical position' (1.4) whenever such descriptions are applied. This position might be called one of supplication—the body upright, facing forwards (anteriorly), with the palms also anterior. There may appear only two real objections to this. In the natural standing position with pendant arms, the forearms are usually well pronated. It is hence sometimes more apposite to speak of radial and ulnar aspects of the forearm, rather than medial and lateral. (However, some Anatomists regret the rigid adherence to the full anatomical position both morphologically and, from some aspects, functionally. Close inspection reveals it as an energy consuming position, seldom actively adopted and involving some scapular rotation and adduction, full lateral rotation of the humerus, direct mediolateral disposition of the elbow joint's axis, full supination of forearm and hand and with the pollex laterally placed! From this stem the somewhat unexpected courses of the radial and ulnar nerves and the disposition of the carpal and digital flexors and extensors. It proves most instructive to compare these and other features with their arrangement in natural standing or in many habitual postures or complex 'precision' activities. Prominent exceptions are placing objects in the mouth or, for examination, before the eyes.) Secondly, due to rotation of the leg at the hip in all but very primitive quadripeds, the originally dorsal extensor aspect has become anterior or ventral with respect to the knee and ankle joints. Furthermore the extensor and flexor aspects of the hip joint are the converse of those at the knee joint.

However, once a little comprehension of the broader facts of mammalian and hence human morphology is established, these difficulties are easily overcome. We have freely used the synonyms: anterior, ventral, flexor, palmar; also posterior, dorsal, extensor are apposite for the trunk and upper limb but they are not always satisfactory synonyms, e.g. the extensor aspect of the leg is *anterior* with respect to the knee and ankle joints and *superior* in the foot and digits; the plantar (flexor) aspect of the foot is, of course, *inferior*, and so on. (Further details and discussion of these problems may be found at appropriate locations throughout Arthrology and Myology.)

There is also a group of terms defining positions and directions along the axis of the body which is, of course, upright in humans. Structures which are nearer the head end, i.e. *cranial*, are officially *superior* and those nearer the tail end, i.e. *caudal*, are *inferior*.

Medial and *lateral*, meaning nearer or further from the body's midline axis, are complemented by *median*, meaning in the vertical, anteroposterior (sagittal) midline plane. Other recognized planes are transverse (at right angles to the median axis, or the putative axes of limbs) and vertical (coronal), orthogonal to the sagittal plane.

The whole system can be equated with a cubical reference grid, anterior, posterior, superior, inferior, right and left aspects corresponding to the six faces of the cube, whose orthogonal sections are transverse, coronal and sagittal (1.4). Various degrees of obliquity must, of course, add slight complications, compound terms e.g. posterolateral, being employed; but it is a well tried system and most useful, as long as the somewhat artificial 'anatomical position' is kept clearly in mind.

A variety of other positional terms exist and are used occasionally in this text, e.g. *distal* and *proximal* are useful in describing some structures in limbs, the datum point being the limb's attachment to the trunk. They are also sometimes used in connection with nerves, arteries, veins, lymphatics, bronchi and other branching structures. Other terms are often found appropriate e.g. superficial, deep, radial and tangential.

We recognize that our readers may at first find some of these conventions unfamiliar and even irritating but familiarity with this positional system and reasonable adherence to it and its terms are as essential to clear, unambiguous communication (in clinical, academic and *all* such circumstances) as is the use of internationally recognized unequivocal terms for structures and processes.

The Study of Cells and Tissues

HISTORICAL BACKGROUND

As already stated, it is the aim of Anatomy as a scientific discipline not only to describe the form of the body but to gain a profound understanding of the biological principles and processes which underlie that form. The body is essentially a cellular structure: it begins its existence as a single cell, the fertilized ovum, it develops by multiplication and differentiation of cells, it matures as the cells and the substances they generate achieve their mature state; senescence is the decay and death the final cessation of cellular activities. It is therefore highly appropriate to consider the body's general construction in the context of its microscopic cellular anatomy.

The study of cells—cytology—and of their aggregations to form tissues and organs—histology—embraces many complimentary approaches, including the study of cell and tissue structure, physiology, biochemistry, biophysics and biometrics, all of which disciplines have greatly contributed to and continue to enrich our comprehension of cellular life.

Our present view of the body's cellular organization has a history spanning at least three centuries (for accounts of this period see the works by Singer 1931, Singer & Underwood 1962, Taton 1966) and, like most scientific advances, it has closely followed developments in technology, in this case, chiefly the design and construction of optical equipment. Simple systems of multiple lenses were first made in the Netherlands during the early seventeenth century and these primitive microscopes gave access to a hitherto totally unknown world of minute objects. There followed the early period of graphic description of the microscopic features of many animals and plants, begun by the members of the Accademia dei Lincei (1609–1630) in Italy, whose number included, amongst others, Galileo, Cesi, Stelluti and Faber of Bambourg (the inventor of the term 'microscope').

During the second half of the seventeenth century, a number of scientists in different countries (the 'Classical Microscopists') carried out more comprehensive investigations of the world of microscopic biology, which were published as a series of beautifully illustrated monographs. In Bologna, Marcello Malpighi investigated a great range of microscopic objects and was the first to examine the microstructure of the skin, kidney and spleen and to establish that capillary networks intervened between arteries and veins. He also laid the earliest foundations of microscopic embryology. In London, Robert Hooke in his great treatise

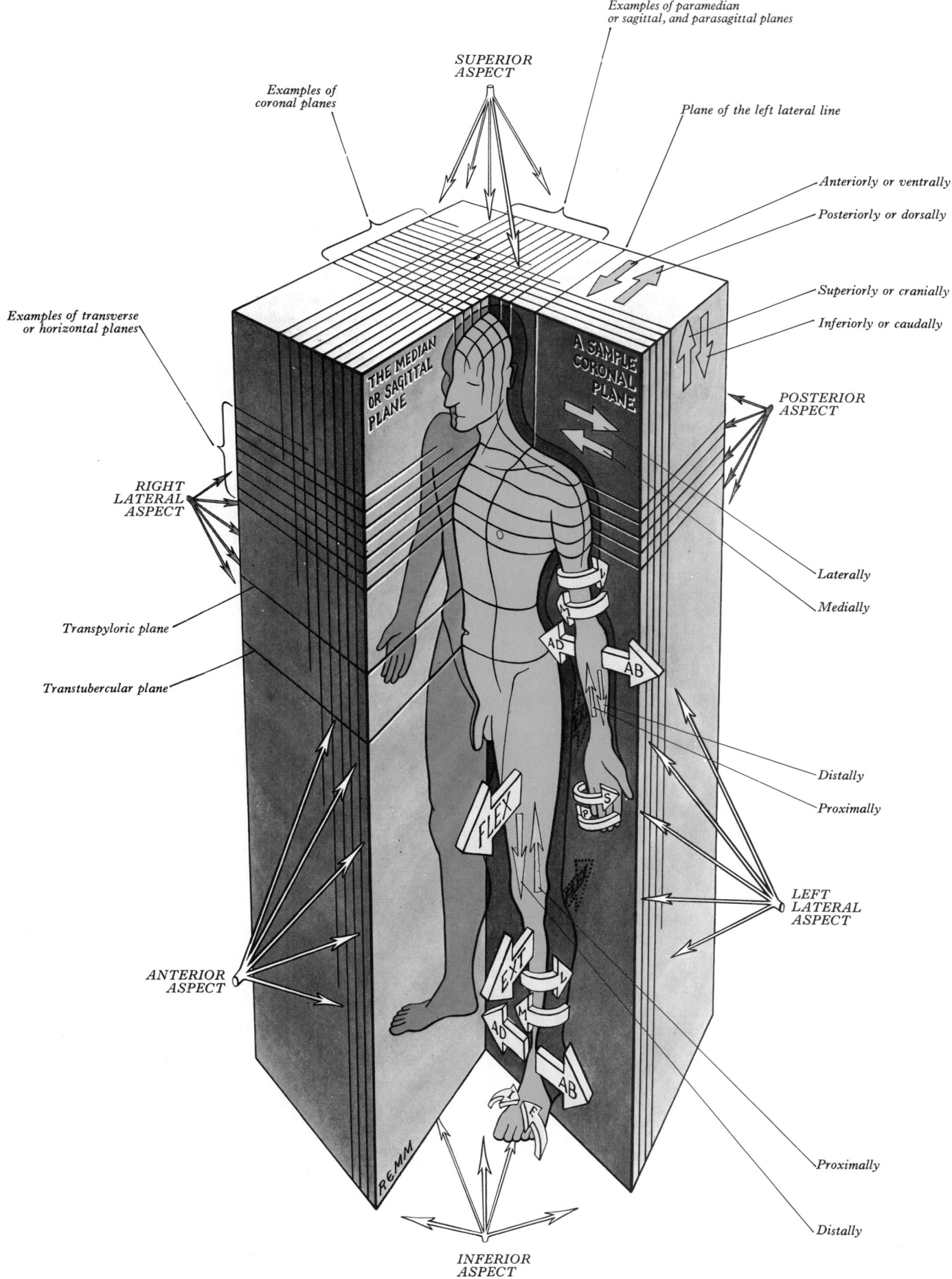

Examples of paramedian
or sagittal, and parasagittal planes

SUPERIOR
ASPECT

Plane of the left lateral line

Examples of
coronal planes

Anteriorly or ventrally

Posteriorly or dorsally

Superiorly or cranially

Inferiorly or caudally

Examples of transverse
or horizontal planes

THE MEDIAN
OR SAGITTAL
PLANE

A SAMPLE
CORONAL
PLANE

POSTERIOR
ASPECT

RIGHT
LATERAL
ASPECT

Laterally

Medially

Transpyloric plane

Transtubercular plane

AD

AB

Distally

Proximally

LEFT
LATERAL
ASPECT

ANTERIOR
ASPECT

FLEX

EXT

AD

AB

Proximally

Distally

INFERIOR
ASPECT

1.4 The terminology widely used in descriptive anatomy The abbreviations on the solid arrows: AD—adduction. AB—abduction, FLEX—flexion (of the thigh at the hip joint), EXT—extension (of the leg at the knee joint), M and L—medial and lateral rotation. P and S—pronation and supination. I and E—inversion and eversion.

13

Micrographia (1665) delineated with meticulous care a great richness of microscopic life and first used the term 'cells', although these were actually the network of dead cell walls in cork wood. Anton van Leeuwenhoek in Delft, Nehemia Grew in London and Jan Swammerdam in the Netherlands and France also made great contributions to animal and plant microscopy during this period.

Thereafter, except for embryologists such as de Graaf, Stensen and Wolff, there were few major advances in microscopic investigation until the nineteenth century, when better microscopes began to be made, achromatic lenses appearing in about 1830 and, somewhat later, immersion objectives. At this time there was a great increase in scientific activity in many disciplines and from this period come the first of the modern histologists such as Henle, Schwann, Deiters and Purkinje and later Remak, Ranvier, Leydig, Virchow, Metchinikov, Golgi and Hertwig, among many others. The work of Kölliker in Würzburg, in particular, was a major contribution to histology and this subject as a separate discipline dates from his publication of the first histology textbook in 1852.

However, these developments depended on new insights into the nature of biological organization. A major conceptual advance was the formulation of the cell theory, independently, by the botanist Schleiden in 1838 and the zoologist Schwann in 1839; by this time it was realized that the whole body is composed of aggregations of microscopic living units, cells, each with a nucleus and possessing a measure of independence as well as being subservient to the body as a whole. The physiology, mechanics and pathological reactions of the body reflect the properties and activities of the cells of which it is composed.

Some years before, the French anatomist Bichat (1771–1802) had coined the term 'tissues' (fabrics) to categorize the textures of the different parts of the body, distinguishing 21 different classes. Schwann reduced these to five (in modern times we have four; see however p. 15). In 1844, Owen used the term 'histology' for their study (from the Greek: *histos* = something woven). Different methods of preparation were now being used to demonstrate the microscopic structure of tissues; separation of tissues and cells by maceration and microdissection were largely supplanted by the more refined methods of wax sectioning and staining with dyes appropriated from the burgeoning dye-stuffs industry.

Developments in the theory and technology of lenses culminated in the design of a high resolution compound microscope by Abbe and its construction by Messrs. Zeiss at Jena in the 1870s, setting the field for a rapid expansion of cytology and histology as serious disciplines. Histology was now established as a major branch of Anatomy and, during the following few decades, most of what we know today as classic light microscope histology was established, including the elucidation of chromosome behaviour in mitosis, meiosis and fertilization and the exploration of the cellular structure of the nervous system. The first successes in the culture of cells in artificial media were also achieved during this period, initially in 1907 by Harrison in Baltimore and subsequently developed by Carrel in New York. In a later period, from about 1930–1950, other optical refinements were introduced to allow viewing of living cells, including dark field, phase, interference contrast and ultraviolet microscopy. However, by this time, histology had lost some of its initial impetus and had long since reached its theoretical limit of point-to-point resolution, i.e. twice the wavelength of visible light (about 0.2μm), limiting effective magnification to about a thousand times.

The renaissance of cell biology was heralded by the development in the 1950s of effective electron microscopy, increasing resolution a thousandfold and permitting useful magnifications of up to a million to be reached. An era of great activity followed, in which appropriate methods of tissue preparation were explored and applied to a wide variety of cells, tissues and organs. The result was the discovery of a great richness of structural detail which has revolutionized our understanding of cells and their functional roles. At the same time, cell fractionation methods for studying the biochemistry of organelles were being developed and used in combination with electron microscopy. The powerful techniques of X-ray diffraction were solving many problems of macromolecular structure, the most celebrated success being the double helical organization of deoxyribonucleic acid (DNA)

demonstrated by Crick, Watson, Wilkins and Franklin in 1952, leading to the unravelling of the genetic code and its role in protein synthesis.

Various other methods have also proved of great value: e.g. autoradiographic analysis, in which the positions of isotopically-labelled molecules assimilated and transported by cells can be detected in sectioned material by their effects on photographic emulsion; various immunological methods for detecting different types of complex molecules by means of specific antibody binding, greatly assisted by the production of monoclonal antibodies (see p. 675); other major contributions with special structural methods including scanning electron microscopy to study the surfaces of cells, freeze-fracture or freeze-etching methods to expose the interiors of their membranes and related structures, microanalysis of the elemental composition of tissues and cells, computer-dependent analysis of light microscope images to visualize organelle behaviour in living cells, quantitative (stereological) analysis of cellular components and many other applications. Most recently, the growth of recombinant DNA technology to detect and determine the precise composition of particular genes and to follow their expression in the cell (see p. 39) has provided a particularly powerful set of tools not only for the analysis of cell function but also to synthesize cellular products for clinical and commercial use.

Inevitably, during the early decades of electron microscopy, a strong emphasis was placed on the description of microscopic structure, so that by the mid-1970s the forms and intracellular contents of most of the major cell types, tissues and organs of the body had been systematically investigated at the ultrastructural level. Since that time, a more experimental approach has prevailed and we are steadily gaining fresh insights into the principles governing the dynamic organization of cellular systems and their interactions. Many major problems remain to be solved and, judging from past experience, more have yet to be recognized. But at least many of the fundamental questions of biological existence are being addressed seriously and we can begin to expect a rich return from the investment of so much intellectual labour over the past century.

THE GENERAL ORGANIZATION OF CELLS

Before proceeding to describe detailed cellular structure, it is desirable to make a brief review of their general composition. The living matter of which cells are made, protoplasm, is composed mainly of water (70% or more by volume), with dissolved inorganic cations (ions of hydrogen, potassium, sodium, calcium, magnesium, iron, etc.) and anions (chloride, bicarbonate, hydroxyl, phosphate, sulphate, etc.), but is also permeated with assemblies of large organic molecules which compose the cellular structure and the system of enzymes and energy carriers providing the basis of living processes within cells. Of the large molecules, the most abundant are those of lipids, carbohydrates, proteins and nucleic acids. The first three of these provide structural materials and enzymes; nucleic acids are important in directing the activities of the cell and in passing on this ability to new cells and to subsequent generations. Numerous smaller organic molecules also abound in the protoplasm, engaged in the teeming biochemical traffic which comprises cell function.

The cell interior is relatively unstable in composition, and must be held ionically, osmotically and electrically within a narrow range of parameters for it to function effectively. However, each cell is also in a constant dynamic interchange with its external environment, including other cells, and continuous expenditure of energy is needed to maintain a steady internal state. It this is lost, the cell undergoes degeneration and eventually dies.

In the living state, most individual cells are greyish in appearance in transmitted light and each is bounded by a deformable, selectively permeable membrane which separates it from the environment, and confers on cells many properties essential to the maintenance of life. Within, the physical properties of the cytoplasm vary from those of a stiff gel to a highly fluid state, depending on the presence of complex filamentous macromolecules and their interactive states. However, individual cells are composed of many distinctive assemblies of macromolecules which form centres of functional cooperation; these *organelles* will be described in some detail in this section, as an appreciation of their roles in cell activities is fundamental to an understanding of the organization of cells and their aggregates.

Cell size

Most mammalian cells lie within the size range 5–50 μm in diameter; e.g. resting lymphocytes are amongst the smallest, at about 6 μm across, red blood cells are about 7.5 μm and columnar epithelial cells about 20 μm tall and 10 μm wide. Some cells are much larger than this: mature ova may be 80 μm across and megakaryocytes of the bone marrow over 200 μm in diameter. Large neurons and striated muscle cells have relatively enormous volumes because of their highly attenuated forms—some may be over a metre in length. Generally speaking, limitations of cell size are determined by problems with rates of diffusion, either of substances into and out of cells or within them. The major advantage of cellularity is that diffusion of materials over short distances of up to 50 μm is relatively rapid so that metabolic needs of active cells can be sustained easily and cellular aggregates react rapidly to the control systems of the body. As a cell increases in size, its mass increase outstrips its surface area unless its shape changes, since the mass varies by the cube of the diameter, whereas the area only increases by the square (see the discussion of this problem by D'Arcy Thompson 1942). Processes depending on the surface area (e.g. diffusion of gases, transport of nutrients, etc.) therefore become increasingly hard to maintain at adequate levels. The distance of the cell periphery from the nucleus also becomes greater, so that exertion of nuclear control on the cytoplasm becomes more difficult. In large cells these problems are to some extent overcome by increasing the relative surface area, either by folding or flattening, and nuclear control can be facilitated by creating more nuclei in each cell either by fusion of mononuclear cells (a *syncytium*), as in skeletal muscle, or more rarely by the multiplication of nuclei without corresponding cytoplasmic division (a *plasmodium*, in the human an unusual and irregular finding in some epithelial cells, e.g. hepatocytes, p. 1391). Intracellular diffusion can also be much accelerated by processes of active transport across membranes and motile mechanisms within the fluid regions of the cell.

Motility is a characteristic of most cells, taking the form of movements of cytoplasm or specific organelles from one part of the cell to another (e.g. cytoplasmic streaming). They also include the extension of parts of cells such as pseudopodia, ruffled borders and microvilli from the surface, locomotion of whole cells by complex streaming interactions with their environment (amoeboid locomotion, etc.) or the beating of flagella (p. 32), cell division, muscle contraction and ciliary beating which moves fluid over internal body surfaces. Cell movements are also involved in the uptake of materials from their environment (endocytosis, phagocytosis) and the reciprocal passage of large molecular complexes out of cells (exocytosis, secretion).

Cell and tissues

Cells rarely operate independently of each other and tend to form aggregates by reciprocal adhesion, often assisted by the formation of special structural attachments. They may also communicate with each other either by releasing and detecting chemical messages diffusing through intercellular spaces or more rapidly by membrane contact, in many cases involving small, temporary transmembrane channels (p. 24). Cohesive or spatially aggregated groups of cells constitute tissues and more complex assemblies of tissues form functional systems such as the visceral organs, whose development and maintenance depend on as yet poorly understood cellular interactions.

In the following section we will first consider the general structure of cells, describing briefly our present understanding of the form and functions of their component organelles, their reproductive activities and motility. Then we will consider two types of cellular aggregations: epithelia and general connective tissue, both of which are distributed widely in many parts of the body. Following this, the structure of the skin (a combination of epithelium and connective tissue) will be described. Other tissues, and microscopic aspects of organs and major systems of the body are described where appropriate in later sections: reproductive tissues in Embryology; skeletal tissues (cartilage and bone) in Osteology; muscle tissue in Myology; blood, lymphoid and vascular tissue in Angiology; nervous tissue in Neurology; and various histological aspects of the viscera and endocrine systems in Splanchnology.

CELL STRUCTURE

Although great modifications of form and activities may take place in different cells during their development and maturation, a common pattern of organization can be seen in their structure and functions. In describing this pattern, as seen with the methods of electron microscopy coupled with many other techniques, we will consider first the outer parts of the cell, the *cytoplasm*, and then the central *nucleus*. Within the cytoplasm are present several distinct systems of organelles (1.5,8,9). These include a series of membrane-bound structures which form separate compartments within the cytoplasm, such as the endoplasmic reticula, Golgi complexes, lysosomes, peroxisomes, mitochondria and transport, secretory and storage vesicles. They also comprise various structures lying outside these membranous organelles in the portion of the cytoplasm known as the *cytosol*, including ribosomes, several types of filamentous protein assemblies (collectively, the *cytoskeleton*), some assisting to determine general cell shape or supporting special extensions of the cell surface (microvilli, cilia, flagella); others are involved in the assembly of new filamentous organelles (e.g. centrioles) or internal movements of the cytoplasm. Also in the cytosol lie many soluble proteins and metabolites of various kinds. The whole cell is bounded externally by a specialized membrane, the plasma membrane. Within the cytoplasm is the nucleus, a special membrane-lined compartment containing the genetic instructions of the cell, the chromosomes; other nuclear organelles, lying in the nuclear sap, e.g. the nucleolus, are involved in their expression. These different structures and an outline of their known activities will now be briefly described. For further details the reader is referred to

other lengthier texts, such as those by Kristić (1979), Fawcett (1981), Alberts et al (1983), P Weiss (1983), Brachet (1985) and Junqueira et al (1986).

Cytoplasm

THE MEMBRANE SYSTEMS OF THE CELL

With the advent of electron microscopy, it was confirmed that cells are bounded by a distinct membrane and internally are permeated by membrane-lined vacuoles and channels. Both external and internal membranes (*cytomembranes*) have many common features. All are composed of phospholipids and proteins, usually in an approximately 3:2 ratio, with a small amount of carbohydrate. The amount of lipid in an external membrane permits a layer two molecules thick to cover the cell surface. As long ago as 1925 Gorter & Grendel proposed that all cell membranes are bilayers of lipid, with the hydrophobic ends of each lipid molecule pointing towards the interior of the membrane and the hydrophilic ends pointing outwards. Later, Danielli & Davson (1935) suggested that the protein might line both sides of the lipid to form a protein/lipid sandwich but more recently the *fluid mosaic* model has been proposed (and widely confirmed) in which the proteins are envisaged as embedded or floating in the lipid bilayer (Singer & Nicolson 1972, see also Bretscher 1985). Some proteins, because of the extensive hydrophobic portions of their polypeptide chains, span the entire width of the membrane

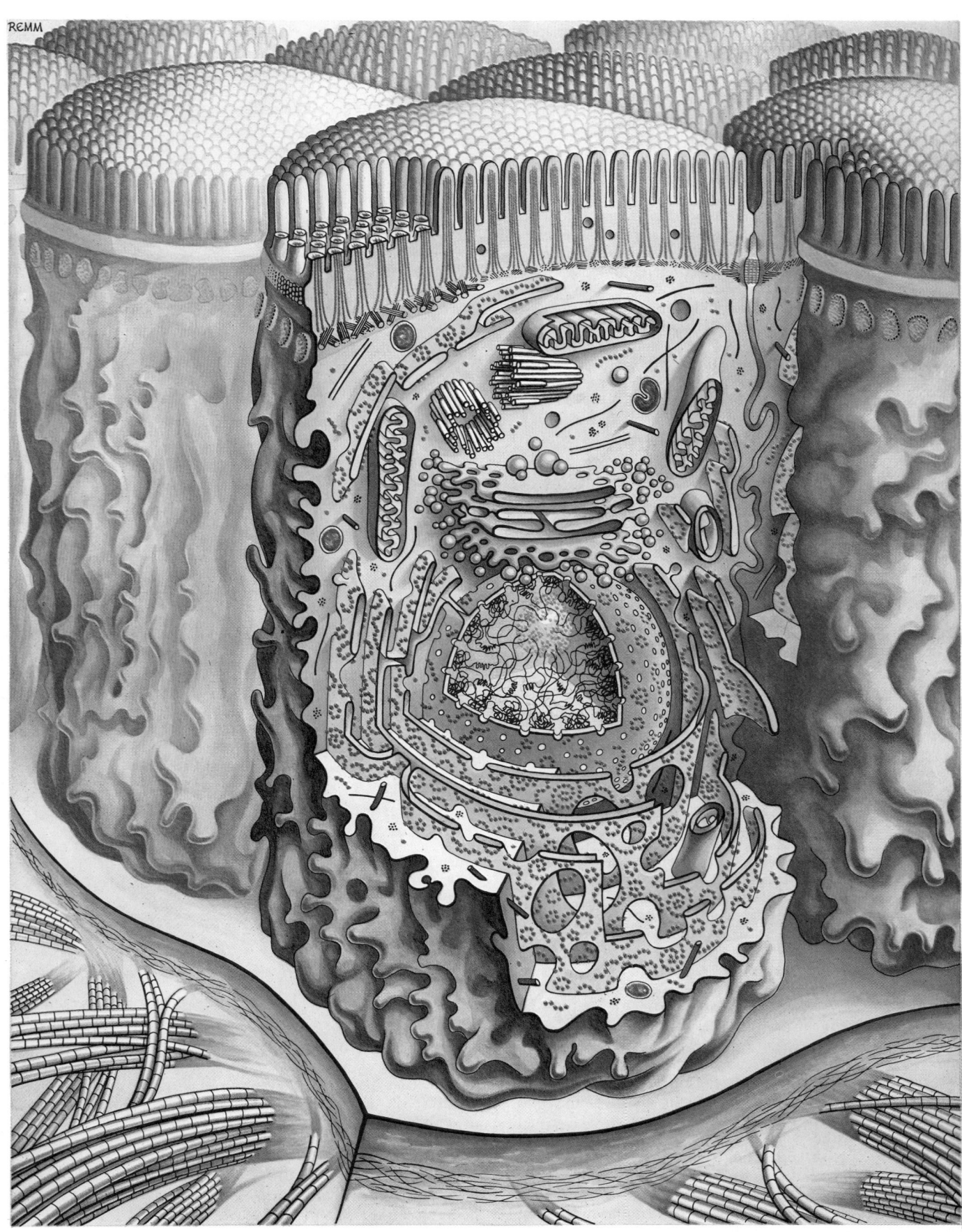

1.5 A three-dimensional reconstruction of some of the principal architectural features of an absorptive cell lying in the simple columnar epithelium of the small intestine. Part of the cell is cut away to expose the nuclear envelope (green); nuclear contents are the nucleolus (red) and chromosomes (black). Outside in the cytoplasm lie the endoplasmic reticulum (yellow) with ribosomes (red) in clusters; the Golgi apparatus (pink) is shown in a supranuclear position, various cytoplasmic vesicles including lysosomes (purple), microtubules (blue), microfilaments, and a centriole pair (grey) are also shown. The apical surface of the cell is covered with microvilli supported by microfilaments which are inserted into a filamentous terminal web. Junctional complexes are seen at the lateral borders of cells apically. A lamina basalis (purple) forms the boundary of the epithelium basally, and lies in close relation to the underlying reticulum and collagen.

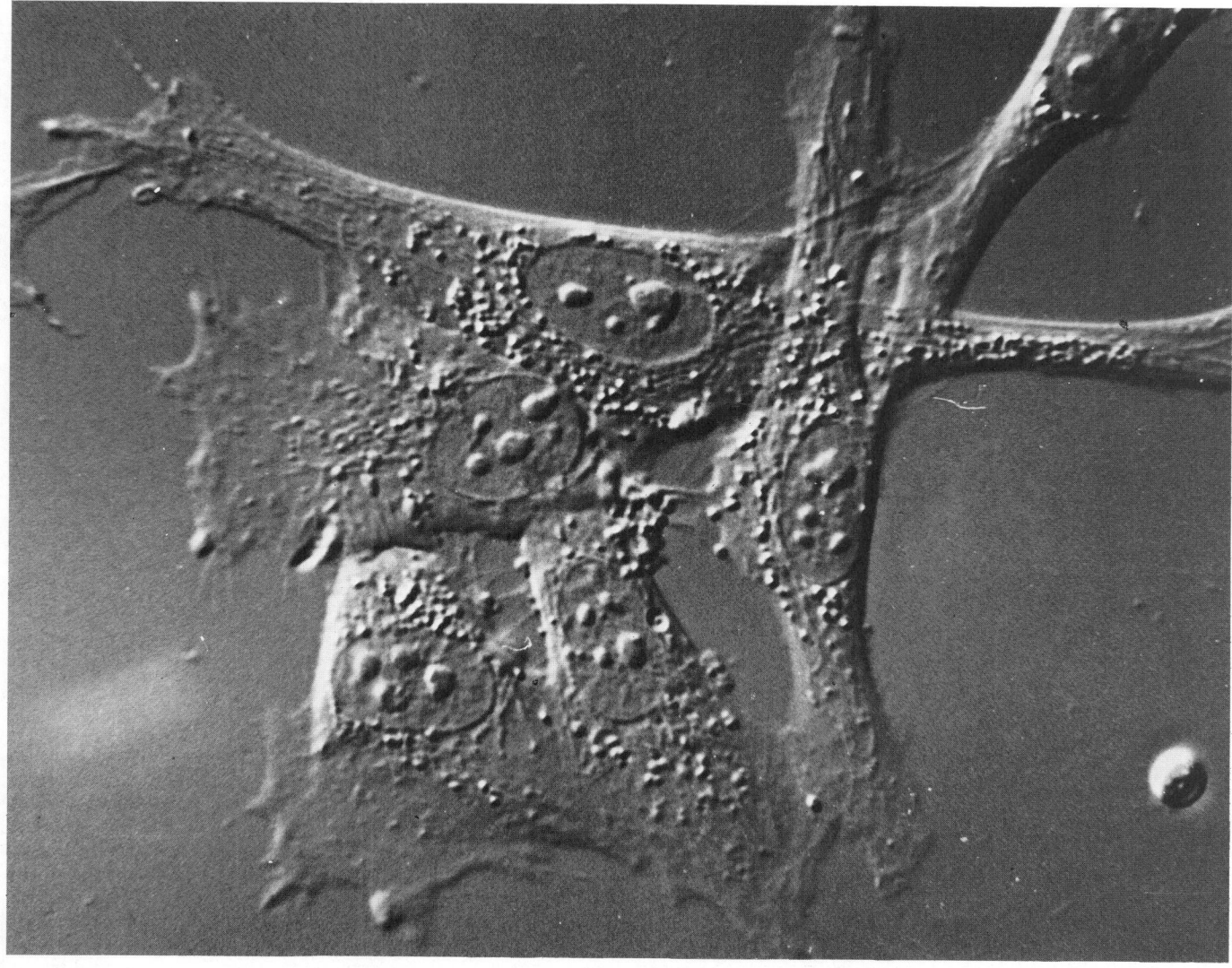

1.6 A high-power micrograph of living cells (fibroblasts in tissue culture), viewed by Normarski interference microscopy. Provided by the Paediatric Research Unit, Guy's Hospital Medical School, London. Magnification × 2000.

(*transmembrane proteins*) whilst others are only superficially embedded (1.10). Carbohydrates in the form of oligosaccharides and polysaccharides are attached either to proteins (glycoproteins) or to lipids (glycolipids), projecting outwards from the surface of the membrane (Nicolson & Poste 1976).

These features, deduced partly from biochemical and biophysical data, can be correlated with electron microscopic appearances. Membranes, suitably fixed and stained by heavy metals, show in section two densely stained layers separated by an electron-translucent zone (1.11), the total thickness being about 7.5 nm (the classic '*unit membrane*' of Robertson 1959), probably reflecting binding of stain by the 'heads' of the phospholipid molecules (see also Robertson 1981). Freeze-fractured and/or etched specimens (Branton 1971, Stolinski & Breathnach 1975), in which the deeply frozen sample is cleaved to expose membrane interiors and of which a metal-shadowed carbon replica is then made and viewed by electron microscopy (1.10,12,A,B), have also demonstrated a bilaminar structure in membranes. Cleavage planes usually pass along the midline of each membrane where the hydrophobic 'tails' of phospholipids meet. This method has also demonstrated 'particles' embedded in the lipid layers; these particles are in the 5–15 nm range and in most cases represent large transmembrane protein molecules or assemblies of such molecules. Particles are distributed asymmetrically between the two half-membranes, usually adhering more to one face than the other. In plasma membranes (which form the cell surface), the *inner* or protoplasmic half-membrane carries most particles, exposed at its externally-facing surface (the P face). The corresponding inwardly-directed (E)

surface of the *external* half-membrane usually shows pits into which the particles fit (1.10,12,14). Not all the proteins of membranes are *visible* as particles, however; some are either too small or not compact enough to appear in this form and have, as yet, only been demonstrated biochemically. Where they have been identified, particles represent channels for the transmembrane passage of ions or molecules.

Biophysical measurements show the phospholipid bilayer to be highly fluid, allowing diffusion *along the plane* of the membrane at rates as high as 2 µm/sec (Bretscher 1975). Thus proteins are able to move freely along such planes unless anchored from within the cell. Some internal membranes possess much more protein than the external cell membrane, e.g. the inner mitochondrial membrane which is rich in enzyme activity; the fluidity of such membranes is correspondingly much reduced.

The functions of cell membranes are many: they form boundaries selectively limiting diffusion and forming physiologically distinct compartments inside the cell, dividing those regions within the channel system (*vacuoplasm*) from those outside it (*cytosol* or *hyaloplasm*). Membranes actively control the passage of electrolytes and small organic molecules, generate bioelectric potentials and provide surfaces for the attachment of enzymes often associated with the movement of reaction products across membranes (*vectorial metabolism*). Membranes also serve as sites for the reception of external stimuli, including hormones and other chemical agents, and for the recognition and attachment of other cells. Conversely, they may alter the activities of other cells by transmitting to them chemical or physical messages of various kinds. Lastly, they can act as points of attachment of intracellular

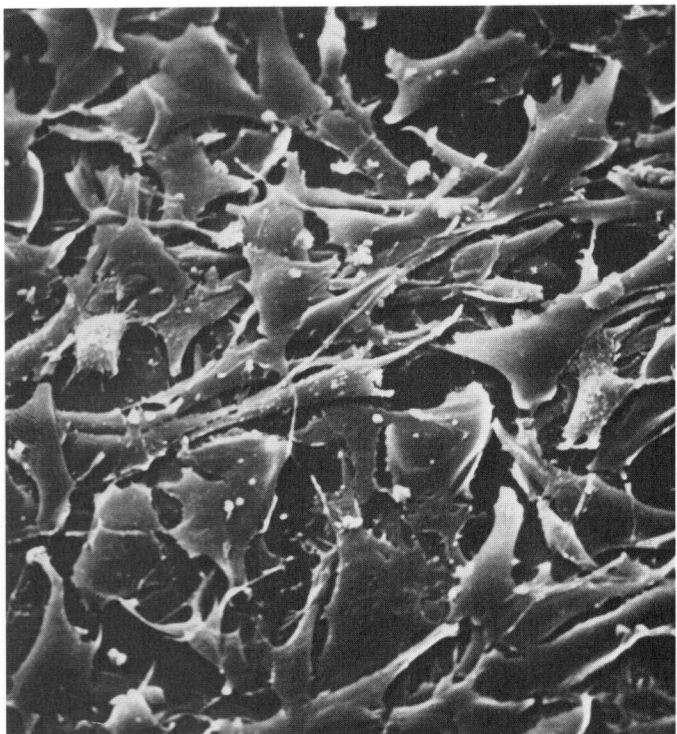

1.7 Flattened fibroblasts growing in tissue culture viewed with the scanning electron microscope. Material provided by L Mallucci. Magnification × 500.

structures—the basis for locomotor activity and for cytoskeletal stability.

Membranes within cells can, on occasion, fuse with each other and so form a potentially continuous system. However, there are barriers to the indiscriminate mixing of membrane components so that each maintains its unique chemical and functional features. So, although the membrane systems of a cell can be viewed as a single entity, they are highly distinctive and localized in their activities.

Cell membranes are synthesized by the granular endoplasmic reticulum (p. 25), usually in collaboration with the Golgi apparatus (Poste & Nicolson 1977).

PLASMA MEMBRANE (PLASMALEMMA OR CELL MEMBRANE)

This differs from other membranes in that it forms the external boundary of the cell and, as such, possesses many distinctive structural features, e.g. it bears a diffuse carbohydrate-rich coat, the *cell coat* or *glycocalyx*, externally. In certain sites it also forms *intercellular junctions* of various types with other cells or *adhesive plaques* with extracellular structures (see p. 24). Within the cell, filaments of the cell skeleton (p. 30) are also anchored to the plasma membrane proteins, immobilizing them or transmitting motile forces from within the cell to the cell surface and thus causing changes of cell shape or locomotion. The plasma membrane is *selectively permeable*, allowing the free passage of some gases and water but restricting the movements of larger ions such as those of sodium, potassium, calcium, chloride and bicarbonate to special proteinaceous channels, which can be opened or closed by the cell to regulate transmembrane traffic. The passage of many other substances such as glucose, amino acids and nucleic acid precursors is also limited to such routes. In all cases, each channel will only admit one species of ion or molecule; substances either move passively through these apertures along diffusion gradients, or by energy-consuming active transport. Movements of such materials into and out of the cell vary greatly, depending on their local concentrations, cell requirements, the availability of chemical energy, action of external hormones and neurotransmitters and many other factors. However, lipid-soluble substances may be able to pass through the lipidic portion of membranes directly so that, e.g. steroid hormones can enter the cytoplasm without passing through protein channels. The uptake of larger molecules involves the invagination and rounding up of the plasma membrane to form small vacuoles termed *endocytic vesicles* (1.8), which are transported to other regions within the cell. The reverse process—extrusion of organic molecules—is achieved by *exocytic vesicles* which fuse with the plasma membrane and release their contents to the exterior.

The plasma membrane, like other cytomembranes, is in constant flux (see Pearse & Bretscher 1981), the whole surface being regularly changed by subtraction through endocytosis or the loss of components externally and by addition of new membrane from exocytic vesicles. Endocytosed membrane may either be degraded in the lysosomal system of the cell or components of it re-cycled back to the cell surface in vesicular form; there is thus a continual traffic between the cell interior and surface, allowing the cell to modify the properties of its plasma membrane in response to metabolic needs or to external stimuli, add new membrane during growth, repair surface damage, take in large molecular complexes from outside and engage in many other interactions with the cell environment.

The plasma membrane also has special roles in co-ordinating many cellular activities by signalling changes in the cell's environment to the cell interior and in maintaining the cell's shape and coherence. It therefore acts as a sensory surface, with a wide variety of special receptor molecules, some responding only to a narrow range of stimuli (e.g. the receptors for insulin, acetyl choline and low density lipoprotein), others being activated by more general factors such as the contact with other cells or inorganic surfaces. Stimulation of the cell surface may result in changes in the bioelectric transmembrane potential causing fluxes of inorganic ions; this is most striking in the *excitable* plasma membranes of nerve and muscle cells in which the 'resting' voltage can change transiently from as much as 100 mV (negative inside) to 50 mV (positive inside) when suitably stimulated, a result of the opening and subsequent closure of channels selectively permeable to sodium and potassium (see p. 878).

Stimulation of surface receptors also often activates a 'second messenger' which may profoundly change the metabolism or motility of the whole cell. Adenylate cyclase, an enzyme associated with the plasma membrane of probably all nucleated cells, is prominent in this process; its activation results in changes in concentrations of cyclic AMP (cyclic adenosine monophosphate) within the cell, leading to alterations in DNA synthesis, gene expression, protein synthesis, actin and myosin interactions and many other intracellular events. Cyclic GMP (cyclic guanidine monophosphate) is controlled by similar enzyme systems and may have effects antagonistic to those of cyclic AMP. Some hormones and neurotransmitters act in this manner and it is probable that many types of intercellular communication use this or similar systems (p. 886). Another mechanism of much current interest involves the phospholipid phosphoinositol and its derivatives in calcium-regulating processes within the cell, which lead to the activation of phosphokinases and phosphorylation of various cellular components, a step with far-reaching metabolic and structural consequences (see Berridge 1985).

Although these considerations apply to the plasma membranes of all cells, those of specialized cells are often highly developed in some particular respect. Thus, sensory membranes derived from the plasma membrane, e.g. in retinal photoreceptors, have numerous photoreceptive proteins embedded in their surfaces; the acetylcholine-receptive sole-plate membranes of skeletal myocytes likewise are studded with protein particles able to bind the transmitter. Other specialized cell membranes, such as that of erythrocytes, have shape-determining components. Research in this field is indeed of great potential since it may reveal much about cellular functions in general and also give a deeper understanding of the actions of drugs on cells; the cell membrane is undoubtedly the primary target of a wide variety of chemotherapeutic agents, anaesthetics and other substances of medical importance.

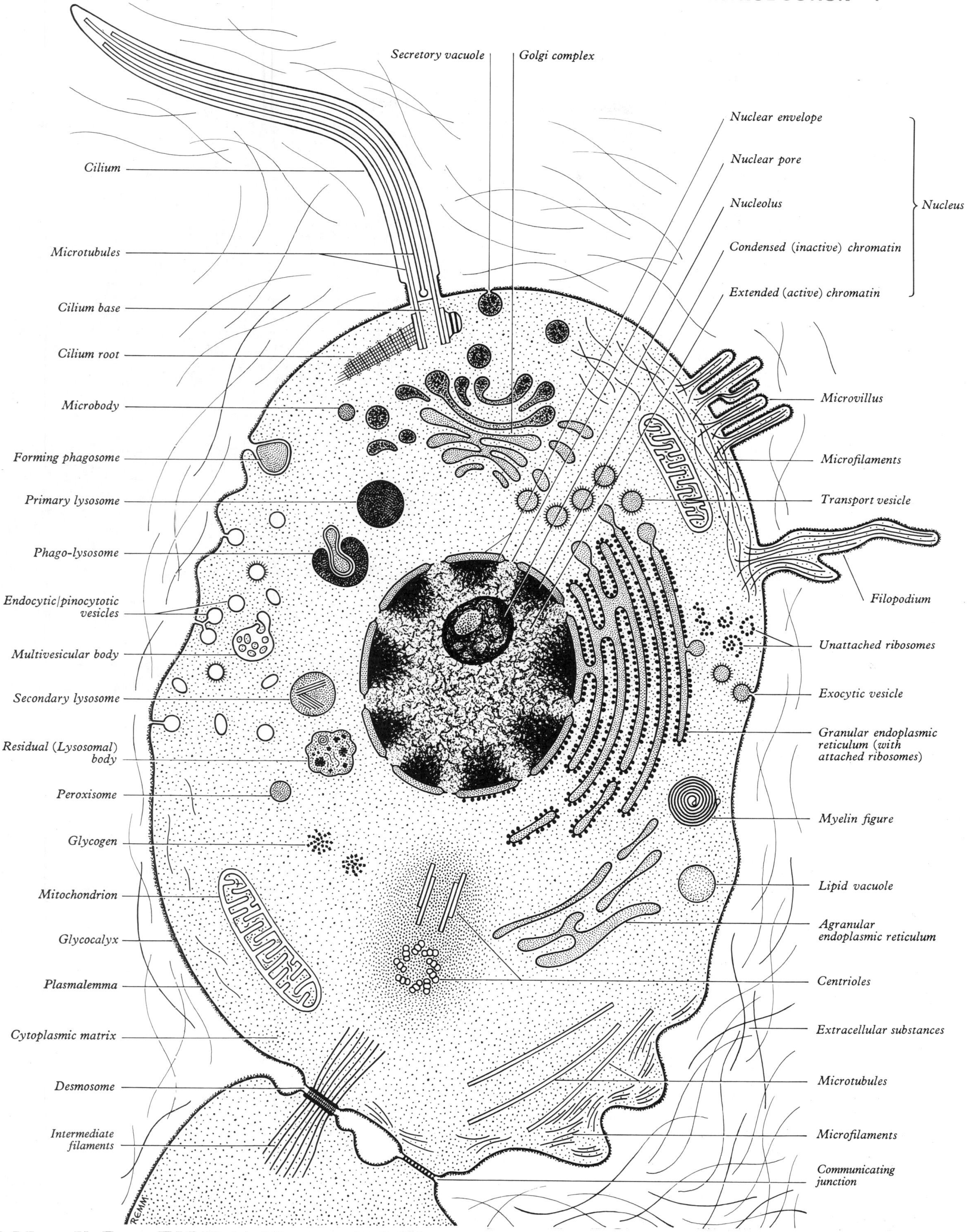

Secretory vacuole

Golgi complex

Nuclear envelope

Nuclear pore

Nucleolus

Condensed (inactive) chromatin

Extended (active) chromatin

} Nucleus

Cilium

Microtubules

Cilium base

Cilium root

Microbody

Forming phagosome

Primary lysosome

Phago-lysosome

Endocytic/pinocytotic vesicles

Multivesicular body

Secondary lysosome

Residual (Lysosomal) body

Peroxisome

Glycogen

Mitochondrion

Glycocalyx

Plasmalemma

Cytoplasmic matrix

Desmosome

Intermediate filaments

Microvillus

Microfilaments

Transport vesicle

Filopodium

Unattached ribosomes

Exocytic vesicle

Granular endoplasmic reticulum (with attached ribosomes)

Myelin figure

Lipid vacuole

Agranular endoplasmic reticulum

Centrioles

Extracellular substances

Microtubules

Microfilaments

Communicating junction

1.8 A composite diagram showing the principal structures found within
tissue cells. Only a proportion of the features illustrated will be present in
any specific cell type.

19

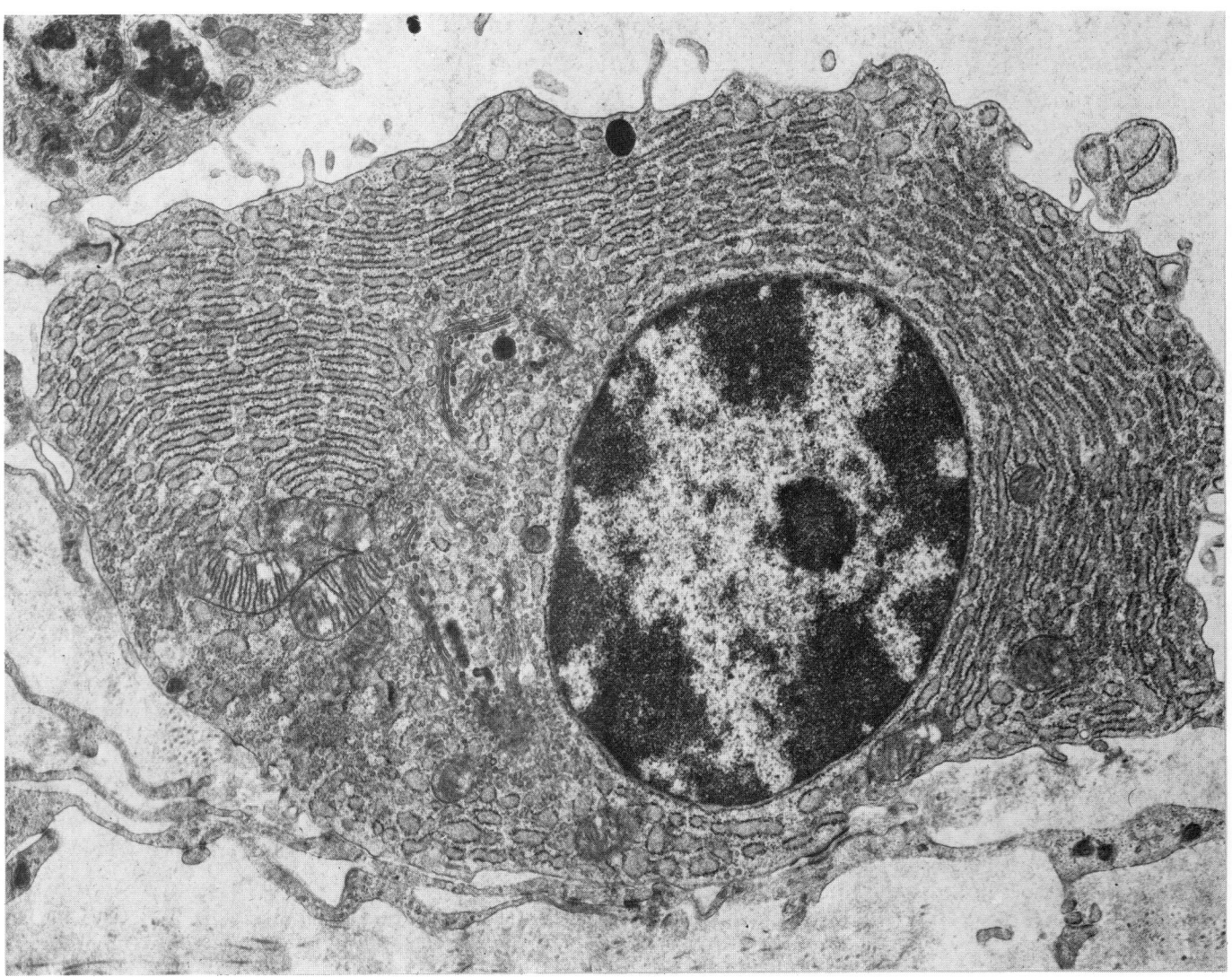

1.9 An electron micrograph of a protein-synthesizing cell (plasmacyte: mature B lymphocyte) in section showing abundant rough endoplasmic reticulum, Golgi apparatus, mitochondria, lysosomes and a cell membrane with surface protrusions (filopodia). Magnification × 12 000.

The Cell Coat (glycocalyx)

As mentioned above, this forms an integral part of the plasma membrane, projecting as a diffuse filamentous layer 2–20 nm or more from the lipoprotein surface. It is composed of the carbohydrate portions of glycoproteins and glycolipids embedded in the plasma membrane (1.10,11), consisting of much branched oligosaccharides and polysaccharides, the terminal residues of which are usually negatively charged sialic acids, such as n-acetyl neuraminic acid, but are also often rich in galactose residues. These and other carbohydrates can be readily demonstrated by electron microscopy, using dyes such as ruthenium red, or more specifically with plant-derived chemical probes termed *lectins* (e.g. concanavalin A, wheat germ agglutinin, phytohaemagglutinin) which bind to particular carbohydrate groups. By conjugating lectins with fluorescent molecules, or with an electron microscopic tracer such as ferritin, horseradish peroxidase or colloidal gold, the surface carbohydrates can readily be visualized and even quantitated (Nicolson 1974, 1976).

Its precise composition varies with cell type; many tissue antigens are located in the coat, including the major histocompatibility antigen (MHC) systems and, in the case of erythrocytes, blood group antigens (p. 681). Special adhesive molecules enabling cells to adhere selectively to other cells or extracellular material are also present and are of utmost importance in maintaining the integrity of cellular assemblies of all kinds. They are also vital to a wide range of developmental movements and interactions between cells, e.g. the formation of intercommunicating neural networks in the nervous system (see p. 865).

Because of the predominance of negatively-charged carbohydrates at cell surfaces, cells tend to repel each other if they approach too closely. Thus, except at special junctions, there is a distance of at least 20 nm between the plasma membranes of adjacent cells. But some positively-charged molecules also exist at cell surfaces and can form intercellular links with negative charges across the intercellular gap.

Since the glycoproteins and glycolipids are usually free to move in the plane of the membrane, addition of lectins can cause the aggregation of these carbohydrate-rich molecules which become cross-linked to form raft-like groups or 'patches'. If these are further aggregated by motile activities of the cell (an energy-dependent process), they merge to form a 'cap' at one pole of the cell. In other cells (e.g. erythrocytes) the carbohydrate-rich molecules are prevented from wandering or forming patches by internal anchoring proteins.

INTERCELLULAR CONTACTS

The plasma membrane is, of course, the surface which establishes contact with other cells and promotes various kinds of cellular interaction. Structurally there are two main classes of contact: (1) those without obvious cytological specializations in the areas of contact, and (2) those with relatively long-lasting junctional features which establish and maintain intercellular contacts. In the first category (Moores & Partridge 1974), by far the most common, cells adhere to one another at any part of their surfaces, enabling them to form loose aggregations whilst permitting some

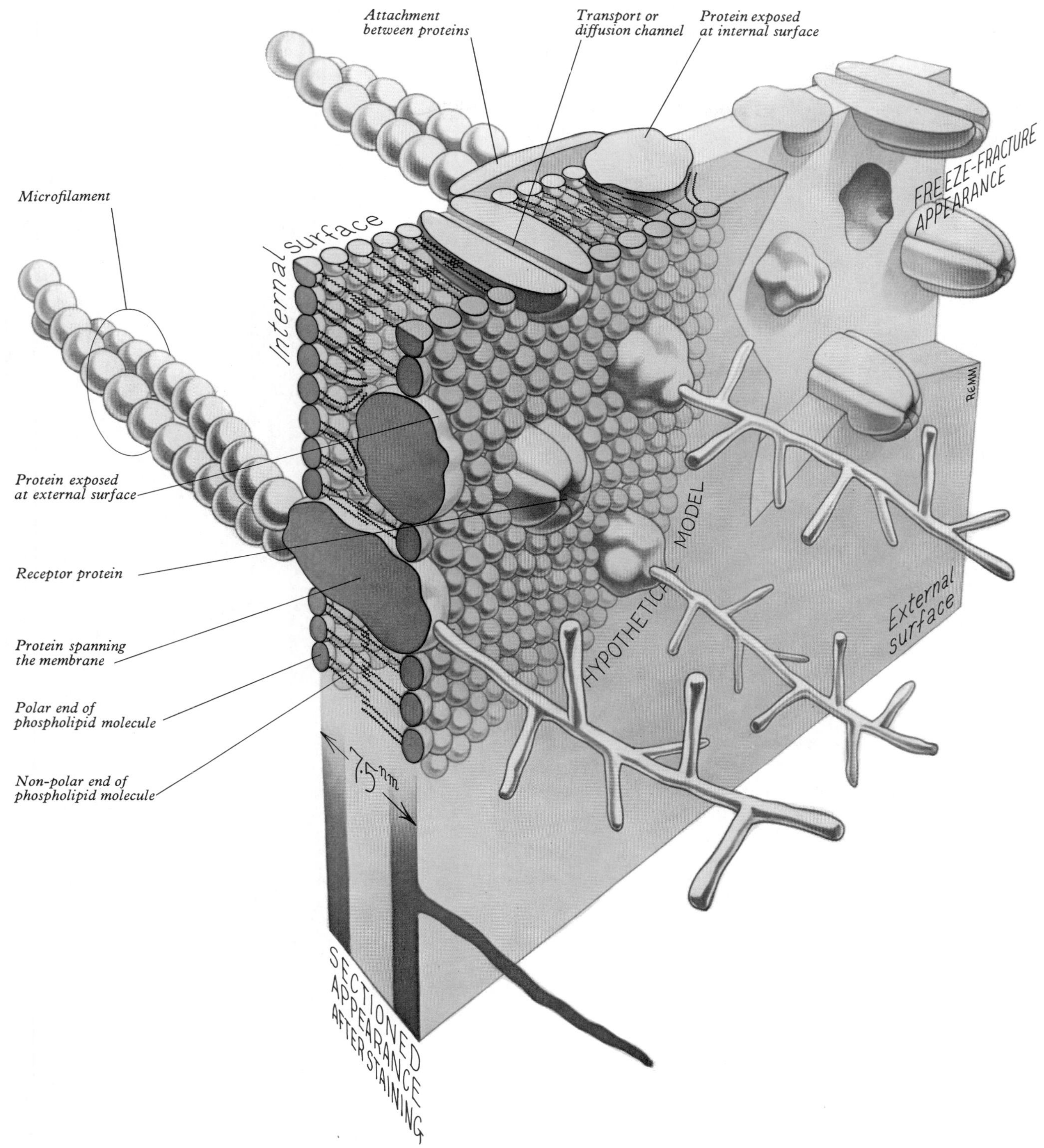

Attachment
between proteins

Transport or
diffusion channel

Protein exposed
at internal surface

Microfilament

FREEZE-FRACTURE
APPEARANCE

Internal surface

Protein exposed
at external surface

Receptor protein

Protein spanning
the membrane

Polar end of
phospholipid molecule

Non-polar end of
phospholipid molecule

HYPOTHETICAL MODEL

External surface

7.5 nm

SECTIONED
APPEARANCE
AFTER STAINING

1.10 The various 'appearances' of the plasma membrane as studied with different electron microscope techniques, including sectioning and freeze-fracturing together with current interpretations of the results of these and other biophysical methods. In this diagram, membrane proteins (green) are either confined to a single leaflet of the lipid bilayer, or they span both layers. Branched carbohydrate chains (grey) are shown attached to some transmembrane proteins on the external surface of the membrane, and a double helical actin filament (pink) is attached to the inside (left). The hydrophobic tails of the lipid molecules are shown as thin black lines and their hydrophilic heads as spheres: blue for the outer surface of the membrane and yellow for the inner surface. These various shapes are, of course, only schematic, as the various protein, lipid and carbohydrate molecules of which cell membranes are constituted have differing detailed shapes.

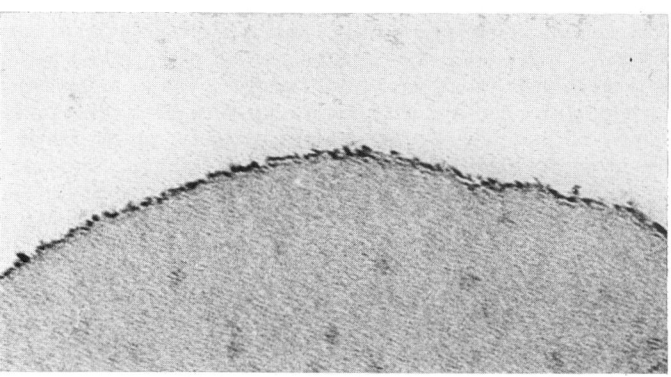

1.11 Electron micrograph of a section through the surface of a fibroblast stained with ruthenium red to show the thick glycocalyx. Magnification × 40 000.

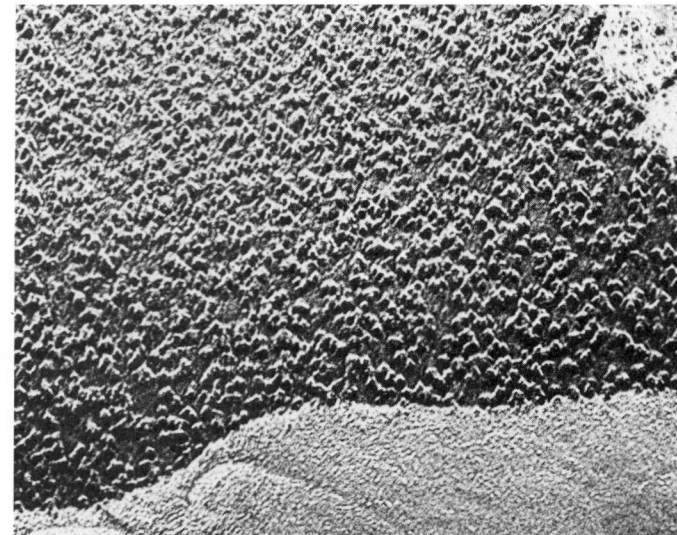

A

B

1.12 Electron micrographs of the plasma membrane of an erythrocyte prepared by freeze-fracturing. A shows the external (E) fracture face, bearing few particles and some pit-like depressions; in B the internal (P) fracture face is visible, showing numerous particles. Magnification × 80 000.

relative movement. Membrane surfaces are separated by a gap of about 20 nm, a distance probably determined by the interplay of *electrostatic repulsion* and *adhesive forces* and perhaps by the thickness of cell coats on the adjacent membrane surfaces. It has been suggested that attraction between oppositely polarized surface groups may be important in adhesion, that divalent cations (e.g. Ca^{2+}) may form bridges between adjacent surfaces or that other *non-specific attractive forces* may operate (Curtis & Pitts 1980). There is increasing evidence that in many cases more *specific interactions* may occur between cell surfaces, with complex sites for mutual recognition and adhesion. Such specific adhesion is shown by cells from the same tissues or embryonic germ layers reaggregating after they have been experimentally separated and in migration of cells and growth of cell extensions (e.g. in the developing nervous system) to contact specific types of 'target' cell (see e.g. Edelman & Rutishauser 1981). But the degree of adhesion varies considerably in different cell varieties; epithelial cells normally adhere to one another strongly, but macrophages —often very motile—may move freely over other cells. Reduction of the normal adhesive properties of cells in malignant neoplasms favour their rapid local spread and their formation of secondary colonies (metastases) elsewhere.

Specialized junctional structures are regions of reciprocally adherent cells where distinctive structures are visible by

electron microscopy (**1.13–15**). A number of different types have been demonstrated (Staehelin & Hull 1978), including those at which the cell membranes are separated by a 20 nm gap or more and those where membranes approach closely or are actually in contact. The former include the *macula adherens* and *zonula*

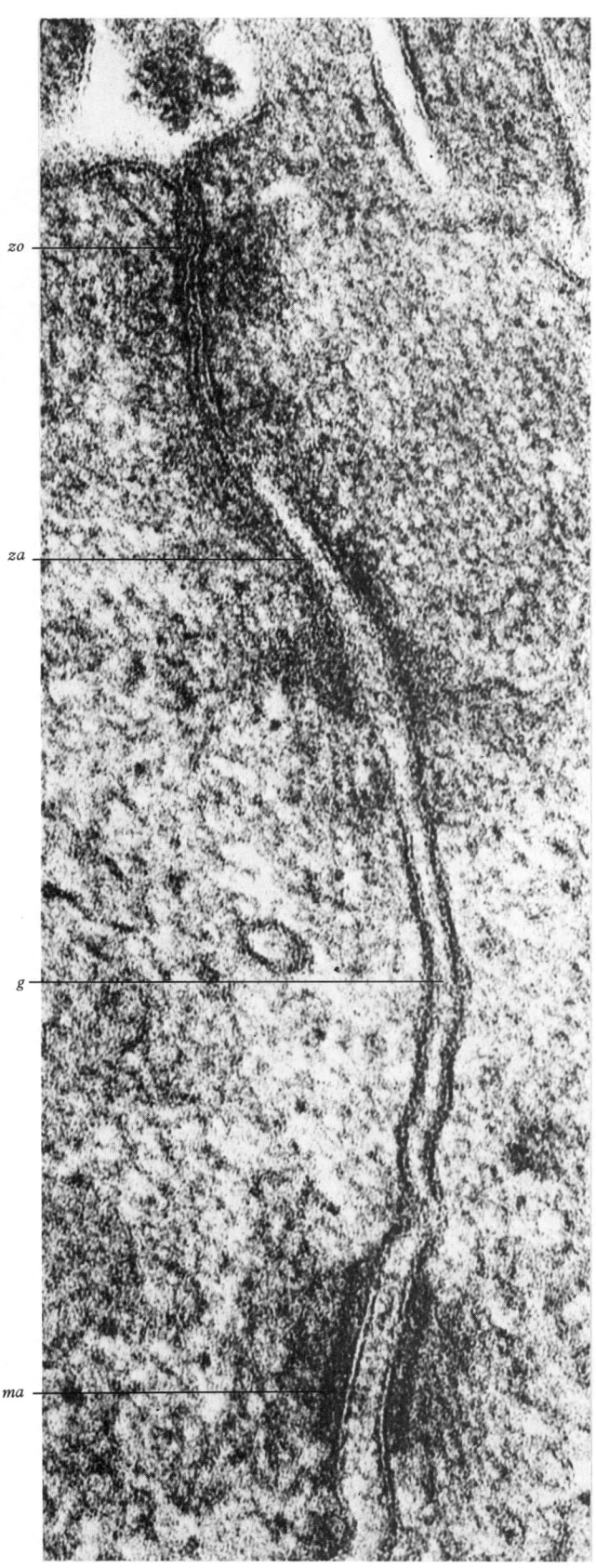

1.13 A high-power electron micrograph of a junctional complex between two epithelial cells showing a zonula occludens (tight junction) (*zo*), a zonula adherens (*za*), a macula adherens (desmosome) (*ma*), and a normal intercellular gap (*g*). Magnification × 130 000.

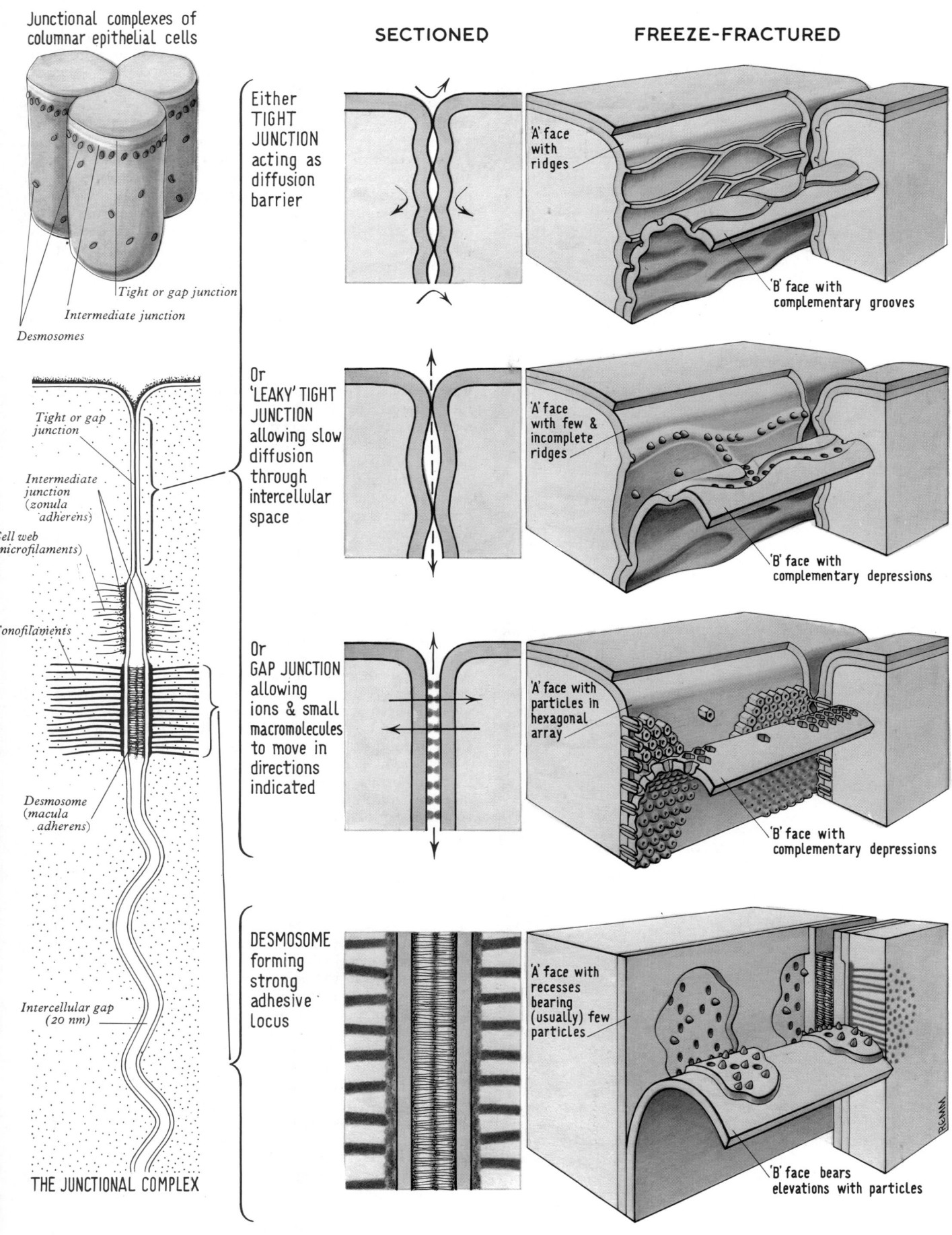

Junctional complexes of columnar epithelial cells

Tight or gap junction

Intermediate junction

Desmosomes

Tight or gap junction

Intermediate junction (zonula adherens)

Cell web (microfilaments)

Tonofilaments

Desmosome (macula adherens)

Intercellular gap (20 nm)

THE JUNCTIONAL COMPLEX

SECTIONED

FREEZE-FRACTURED

Either TIGHT JUNCTION acting as diffusion barrier

'A' face with ridges

'B' face with complementary grooves

Or 'LEAKY' TIGHT JUNCTION allowing slow diffusion through intercellular space

'A' face with few & incomplete ridges

'B' face with complementary depressions

Or GAP JUNCTION allowing ions & small macromolecules to move in directions indicated

'A' face with particles in hexagonal array

'B' face with complementary depressions

DESMOSOME forming strong adhesive locus

'A' face with recesses bearing (usually) few particles

'B' face bears elevations with particles

1.14 Schemata of various junctions commonly occurring between cells, showing current interpretations of the electron microscopic appearances seen after sectioning and with freeze-fracturing techniques. For clarity, the freeze-fracturing data are presented as though it were possible to separate and fold back the two leaflets of the plasma membrane to expose the intramembranous protein particles inserted on their two faces. The *A* face = the *P* fracture face and the *B* = the *E* fracture face in current terminology.

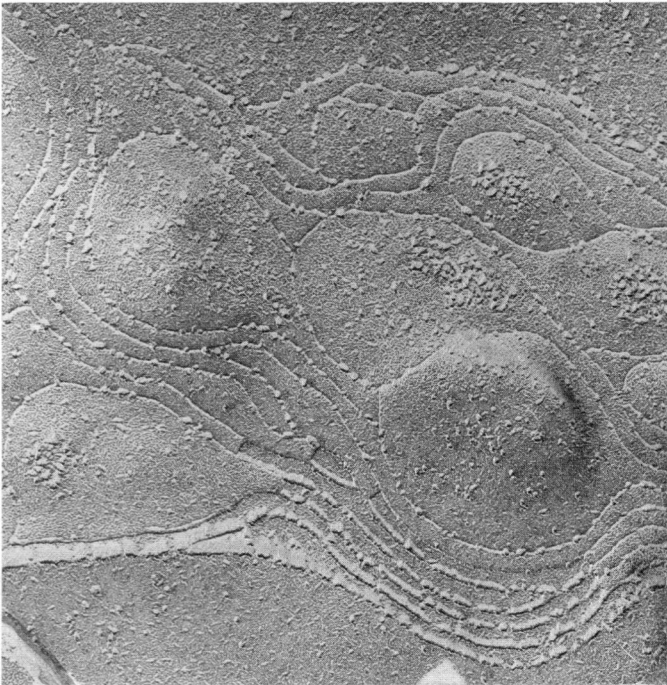

1.15A Freeze-fractured zonula occludens (tight junction) between two epidermal cells of fetal skin showing the anastomotic network of lines representing contacts between membranes. Compare with 1.14. Provided by Andrew Kent, Department of Anatomy, UMDS, Guy's Campus, London. Magnification × 25 000.

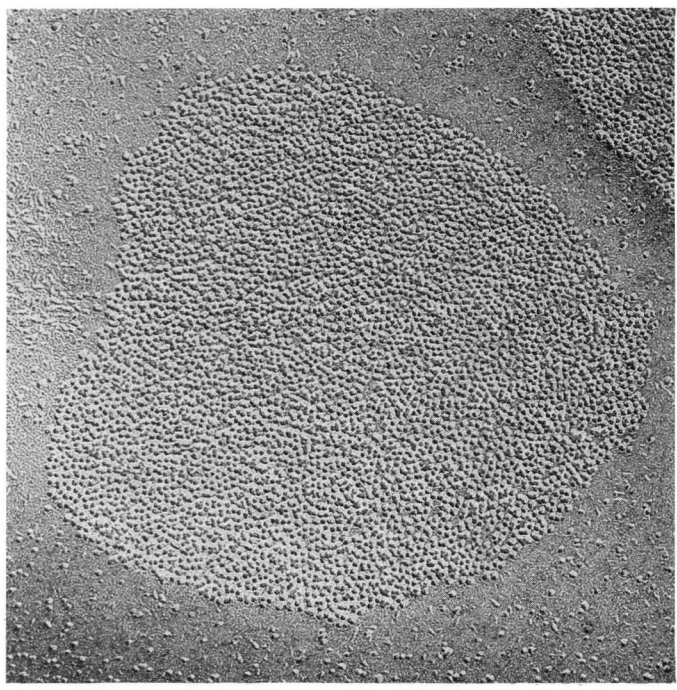

1.15B Freeze-fractured communicating (gap) junctions between two developing epidermal cells. The tightly packed intramembranous particles represent channels for electrotonic coupling and diffusion of small molecules between cells. Provided by Andrew Kent, Department of Anatomy, UMDS, Guy's Campus, London. Magnification × 25 000.

adherens, the latter the *zonula occludens* and the 'communicating junction' or *junctio communicans*, also called the '*gap junction*'.

The *macula adherens (desmosome)* is a plaque-like structure (1.13,14) sited anywhere on the cell surface, where the plasma membrane is coated internally with a layer of dense protein into which intracytoplasmic tonofilaments are inserted. The

intercellular gap, a region rich in a glycoprotein, *desmoglea* (Gorbsky & Steinberg 1981), is bridged by thin interlocked filaments, traversed by densely staining bands. This structure forms anchorage points between cells, particularly where strong cohesion is needed, as in the stratum spinosum of the epidermis; here they are responsible for the post-fixation prickly appearances of the cells. Single-sided or *hemi-desmosomes* are also found between epidermal (and other) cells and their underlying basal laminae.

The *zonula adherens* (1.13,14) is a continuous belt formed around the apices of epithelial cells, or between non-striated myocytes, and as a 'single-sided' junction between muscle cells and their collagenous sheaths. Filaments (usually microfilaments) are often inserted into a dense coat on the cytoplasmic side of the membrane. Unlike the macula adherens, no filaments span the gap between the cell surfaces but an adhesive, non-stainable material may intervene at this point. They probably serve to anchor cells to adjacent structures. Extensive areas of similar structure also occur between the myocytes of cardiac muscle, at intercalated discs (p. 557), where they are termed *fasciae adherentes*.

The *zonula occludens* (tight or occluding junction) is a continuous zone (1.13,14) present around the apices of epithelial cells, between endothelial cells and at other sites where there is a barrier to diffusion through the intercellular space (Pinto da Silva & Kachar 1982). Colloidal tracers such as ferritin and horseradish peroxidase fail to move from the lumen of the intestine into the spaces between the epithelial cells or deeper into the mucosa. At a zonula occludens the membranes of the adjacent cells come into contact, obliterating the gap between them. Freeze-etching shows that contacts between the membranes lie along branching and anastomosing ridges formed by the incorporation of chains of particles within the membranes (1.14,15A), distorting and stiffening them at points of contact. Occluding junctions prevent the leakage of toxic substances from the lumina of viscera into the surrounding tissues, retain colloids within the bloodstream and form important diffusion barriers in many other sites. They also form barriers within the plasma membrane, preventing the diffusion of membrane components, e.g. from the apical surface of a cell to its more basal region. In this way, specialized regions of membrane can be maintained in an appropriate position, e.g. the secretory surface of a gland cell or the ion-transporting surface of a kidney tubule cell (see Simons & Fuller 1985).

The junctio communicans (communicating 'gap' or electrical junction, nexus or *macula communicans*) resembles occluding junctions in sections but apposed membranes are separated by an apparent gap of 3 nm, which is traversed by numerous dense beads arranged in hexagonal arrays on the two surfaces (1.14,15B). Such junctions form limited attachment plaques (Goodenough & Revel 1970) rather than continuous zones, thereby allowing free passage of substances along the cleft between cells. They occur in numerous tissues including the liver, epidermis, connective tissues, cardiac muscle and non-striated muscle (see the review by Hertzberg et al 1981); they are also common in embryonic tissues. In the central nervous system (Staehelin 1974) they are found in the ependyma, in neuroglia, and they form electrical synapses between some types of neurons (see p. 888).

Communicating junctions form channels for the diffusion of ions and larger particles (up to a relative molecular mass (Mr) of about 1000) *from one cell to the next*. Thus, in some excitable tissues (e.g. cardiac and non-striated muscle), one cell can affect another by electrotonic current flow without the intervention of a chemical transmitter, in contrast to the usual mode of synaptic transmission. Elsewhere their functions are not certain; experimentally they have been shown to be permeable to various dyes and to form pathways with low resistance to the flow of ionic current. Communicating junctions probably permit metabolic cooperation between adjacent cells or groups of cells. Thus, in embryonic life, such junctions may aid the establishment of pattern and the coordinated differentiation of the whole blastula, within and between germ layers, or of more localized tissues (Fraser 1985). This may involve the movement of regulatory substances involved in gene blocking or gene repression and derepression diffusing freely or creating morphogenetic gradients (see p. 109). Communicating junctions may also assist in the control of cell division since in damaged tissues they disappear, while cells undergo repair by mitosis, to reappear when regeneration ceases (Bennett 1973, Fusijawa et al 1976).

Freeze-fracturing and -etching methods (Staehelin 1974) and recently, computer-aided high resolution electron microscopy (Unwin & Ennis 1984) show hexagonal arrays of membrane particles on both sides of communicating junctions, each particle (or 'connection') composed of six protein subunits surrounding a central channel (cf. **1**.10,14) which, when apposed to a similar unit in an adjacent cell, forms a small communicating passage between the two. Such particles may also exist in smaller numbers, or singly, elsewhere on the cell, so that intercellular communication may not be restricted to special junctional *areas*. Changes in pH and calcium ion concentrations, etc. can cause narrowing or closure of such channels, so intercellular communication can change with the metabolic alterations in the participating cells (Unwin & Zampighi 1980).

Junctional complexes are present at the apex of epithelial cells (Farquhar & Palade 1963), where a combination of a zonula occludens, zonula adherens and macula adherens (desmosome) is a regular feature (**1**.13,14). This array corresponds partly to the *terminal bar* of light microscopy.

Synapses and *neuromuscular junctions* are specialized areas providing interneuronal and nerve–muscle transmission, and will be described under these headings in a later chapter.

With these two exceptions, all intercellular junctions are labile, being resorbed or reformed in suitable circumstances and so permit slow migration and division of cells and the passage of other cells along intercellular clefts (e.g. leucocytes between endothelial cells).

Other specializations of the plasma membrane include various extensions—microvilli, filopodia, microspikes, cilia and flagella; and less regular, often rounded or leaf-like extensions—'blebs', pseudopodia and ruffled membranes. Indentations include endocytic and pinocytotic vesicles associated with the uptake of materials by the cell and exocytic vesicles involved in the outward transport of materials. These various structures will be discussed below.

ENDOPLASMIC RETICULUM

This is the system of interconnecting membrane-lined channels (Palade 1975) within the cytoplasm (**1**.5,8,16). These channels take various forms, including cisternae (flattened sacs), tubules and vesicles. The membranes divide the cytoplasm into two major compartments: that inside the channel system, the *vacuoloplasm*; and that outside, the *hyaloplasm* or *cytosol*. The former constitutes the space in which secretory products are stored or transported to the Golgi complex and cell exterior; the latter is made up of the colloidal proteins such as enzymes, carbohydrates and small molecules, together with ribosomes and ribonucleic acid.

Structurally, the channel system can be divided into *granular (rough) endoplasmic reticulum* (**1**.16A,B), to the exterior of which ribosomes are attached, and *agranular (smooth) endoplasmic reticulum* (**1**.16C), lacking ribosomes. When cells are disrupted and centrifuged, both endoplasmic reticula break up into vesicles respectively termed *granular* and *agranular microsomes*.

The granular endoplasmic reticulum can synthesize proteins, because of its attached ribosomes. Such proteins are passed *through* the membranes to which the ribosomes are bound and accumulate within the cisternae of this system; thereafter they remain in membrane-bound bodies such as lysosomes or else are secreted to the exterior of the cell. Some carbohydrates are also synthesized by enzymes within the cavities of the granular endoplasmic reticulum and may be attached to newly-formed protein (glycosylation). *Agranular endoplasmic reticulum* is associated with carbohydrate metabolism and many other metabolic processes, including detoxification and synthesis of lipids and of cholesterol and other steroids. The membranes of the agranular endoplasmic reticulum serve as convenient surfaces for the attachment of many enzyme systems (e.g. important detoxification mechanisms involving the enzyme cytochrome

P_{450}) which are thus accessible to the substrates in solution within the cell. They also cooperate with the granular endoplasmic reticulum and Golgi apparatus to elaborate new membranes, the protein, carbohydrate and lipid components being added in different regions.

Highly specialized types of endoplasmic reticulum are present in some cells. In striated muscle cells the agranular endoplasmic reticulum (*sarcoplasmic reticulum*) stores calcium ions, which are liberated to initiate contraction on appropriate stimulation (p. 551).

In embryonic cells, endoplasmic reticulum is scant and ribosome clusters lie mostly unattached in the hyaloplasm. During differentiation, membranes usually increase greatly and ribosomes may become attached to form granular endoplasmic reticulum.

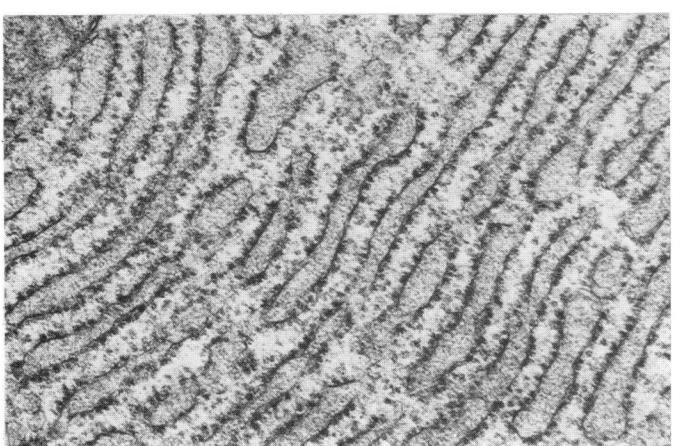

A

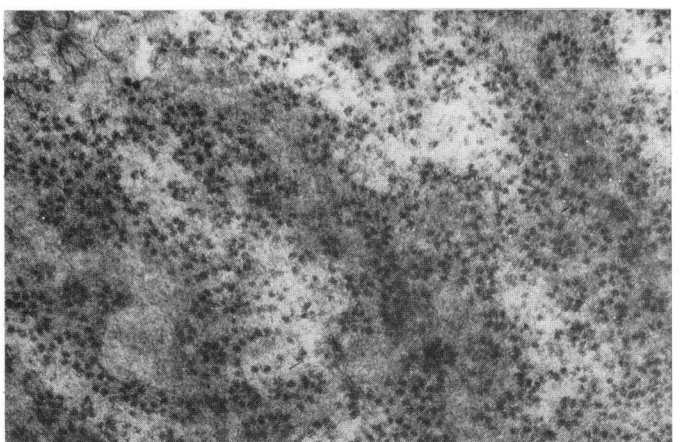

B

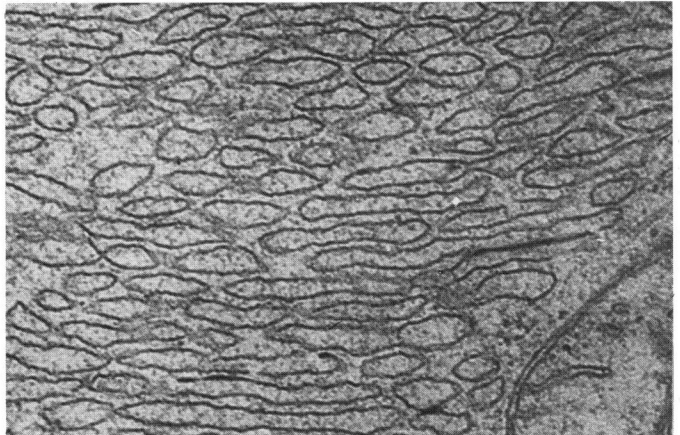

C

1.16 Ultrastructural details of the endoplasmic reticulum: A. rough endoplasmic reticulum; B. polysome groups attached to obliquely sectioned cisternae showing 'rosette' configuration; C. smooth endoplasmic reticulum. B provided by D R Turner, Guy's Hospital Medical School. Magnification × 30 000.

RIBOSOMES

Ribosomes (ribonucleoprotein particles) are granules about 15 nm across (**1.16A,B**), composed of equal parts, by weight, of protein and (ribosomal) RNA; they are responsible for the synthesis of proteins from amino acids (Palade 1955, Siekevitz & Palade 1960). Each ribosome is made of two subunits (Lake 1981), one slightly larger than the other, sedimenting in the centrifuge at different rates (in nucleated cells mainly at 60S and 40S, where S = the Svedberg unit of sedimentation rate, a function of density, shape, etc.; the whole ribosome sediments at 80S). The subunits can be further dissociated into about 73 different proteins (40 in the large subunit and 33 in the small), with structural and enzymatic functions. Three small RNA strands (28, 5.8 and 5S), highly convoluted, lie in the large subunit, and one in the small subunit (18S). Most of these RNA molecules are derived from the nucleolus (p. 40). The large and small subunits are separate from each other when not engaged in protein synthesis.

Ribosomes may be solitary, relatively inactive *monosomes*, or form groups (*polyribosomes* or *polysomes*) attached to messenger RNA which they translate during protein synthesis (**1.33**). Polysomes can be attached to membranes, constituting granular endoplasmic reticulum (vide supra) or may lie free in the hyaloplasm where they synthesize proteins for use outside the channel system, including enzymes of the hyaloplasm, structural proteins of the cell (e.g. actin, tubulin) and haemoglobin in erythroblasts.

In a mature polysome all the attachment sites of the messenger RNA are occupied as ribosomes move along it, synthesizing protein according to its instructions, so the number of ribosomes in a polysome indicates the length of the messenger RNA molecule and hence the size of the protein being made.

The two major subunits have separate roles in protein synthesis. The smaller is the site of attachment and translation of messenger RNA; the larger is responsible for the release of the new protein and, where appropriate, attachment to the endoplasmic reticulum, directing the protein through its membrane into the cisternal cavity.

Subunit proteins of the ribosomes themselves are synthesized in the cytoplasm by other ribosomes. They then apparently enter the nucleus where they bind to ribosomal RNA from the nucleolus to form the two major subunits; these then pass separately back into the cytoplasm and only associate to form a complete ribosome when they attach themselves to a messenger RNA molecule (Nomura 1984). When protein synthesis is over, the two subunits dissociate but may serve more than once.

Mitochondrial ribosomes differ from those of the general cytoplasm, being somewhat smaller (55S). They will be discussed later (see mitochondria, vide infra). The ribosomes of fungi and prokaryote organisms such as bacteria are also smaller than those of the nucleated cells of animals (and plants), which may perhaps reflect the less complex regulatory mechanisms of these relatively primitive species.

MITOCHONDRIA

Mitochondria are membrane-bound organelles of great metabolic significance (Ernster & Schatz 1981, Fawcett 1981, Tzagoloff 1982) producing energy from the oxidative breakdown of large molecules and carrying out various other chemical processes. They were first observed with the light microscope as thread-like, spherical or ellipsoidal bodies in the cytoplasm of most cells, particularly those with a high metabolic rate such as secretory cells in exocrine glands. In living cells, when viewed by bright field, phase contrast or interference microscopy, they have been seen to move, change size and shape and to divide. In size most mitochondria range from 0.5–2.0 μm long; those of cardiac muscle are particularly large.

Each mitochondrion is lined by an outer and an inner membrane, separated by a variable gap (**1.17**). The outer is smooth and sometimes attached to other organelles (e.g. microtubules); the inner membrane is deeply folded to form incomplete transverse or longitudinal septa or tubular invaginations, *cristae mitochondriales*, which thus create a relatively large surface area of

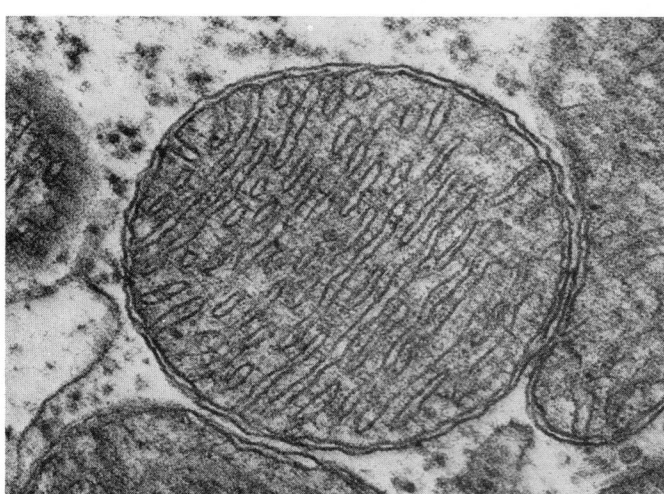

1.17 Electron micrograph showing details of mitochondrial structure: the outer membrane, the inner membrane folded to form cristae and their related intra-mitochondrial spaces are visible. Magnification × 30 000.

membrane. Cristae are more numerous and complex in cells with a high metabolic rate than in relatively inactive ones; in heart muscle, for instance, they are numerous and show complex pleats. In mitochondria of the lipocytes of brown fat, cristae are particularly conspicuous and their chemical activities are diverted to the direct production of heat rather than ATP (see p. 62). Cristae of cells in the adrenal cortex are typically tubular. The significance of these different arrangements is not clear.

In the inner lumen of the mitochondrion lies the finely granular *matrix,* of variable density. Within the matrix are many soluble enzymes and a variety of inclusions have also been described, including calcium salts, organic crystals, glycogen, some ribosomes and the nucleic acids RNA and DNA. The ribosomes are smaller and quite distinct from those of the rest of the cell. The DNA thread forms a complete ring and has a ratio between its nitrogen bases different from that of chromosomal DNA. There are also chemical differences in the RNA types and the precise genetic code used in protein synthesis (Grivell 1983). Mitochondria are able to multiply by division during interphase, some of the inner mitochondrial membrane components being specified by the mitochondrial DNA. To some extent, therefore, this is an example of *cytoplasmic inheritance.* It follows from this that all mitochondria in the body are in effect descended from those in the cytoplasm of the fertilized ovum. Further, it has recently been shown that these are entirely of maternal origin, so that mitochondria (and any of their genetic variants) are passed only through the female line; the mitochondrion of the sperm is not incorporated into the ovum at fertilization. The ribosomes and nucleic acids are, interestingly, similar to those of bacteria, suggesting the origin of mitochondria in the far distant past as symbiotic bacteria which lost their ability to lead an independent existence (see Margulis 1981).

Mitochondria are the principal site of a number of enzyme systems, particularly those of oxidative phosphorylation associated with the tricarboxylic acid (Kreb's) cycle and the cytochrome electron transport sequences of respiration. They are the chief sites where chemical energy is derived from breakdown of organic compounds in respiration to form high-energy organic phosphate compounds (particularly adenosine triphosphate, ATP, and guanosine triphosphate, GTP) by an unusual chemical mechanism, the chemi-osmotic process, which involves the pumping of hydrogen ions out of the mitochondria to drive the synthesis of ATP (Mitchell 1961). These energy-rich compounds pass to other parts of the cell where they fuel energy-consuming reactions. The various enzymes of the Kreb's cycle occur in the mitochondrial matrix, while those of the cytochrome system and oxidative phosphorylation are localized chiefly in the inner mitochondrial membrane. Some of these ATPases form an enzyme assembly (ATP-synthetase) which, when mitochondria are hypotonically disrupted and negatively stained, become visible as minute spheres about 9 nm across, supported by stalks projecting from the inner membrane (*elementary, submitochondrial* or *stalked particles*). Fatty acid metabolism, calcium concentration and various other chemical activities are also carried out in mitochondria.

It is interesting that these organelles are distributed within the cell according to regional energy requirements, e.g. near the bases of cilia in certain epithelia and at the base of the cells of proximal convoluted renal tubules, where considerable active transport occurs. Estimates of mitochondrial numbers in cells also show variability; in mammalian hepatocytes they number about 1000–2000 per cell; in lymphocytes only a few may be present.

Various syndromes are known in which patients have defective mitochondrial enzyme systems, affecting, e.g. the efficiency of muscles, kidney and neural function, etc. (Morgan-Hughes 1986).

LYSOSOMES

Lysosomes are membrane-bound, spheroidal or ellipsoidal bodies, 0.08–0.8 μm in diameter (**1.8,18**), which contain hydrolases capable of degrading a wide variety of substances (de Duve 1965, Holtzman 1976, Bainton 1981, Brachet 1985). They occur in all cells except mature erythrocytes and are essential to cellular catabolism (degradative activities), being particularly numerous in cells with a rapid metabolism and in phagocytes. So far, more than 40 lysosomal enzymes are known including many varieties of proteases, lipases, carbohydrases, esterases and nucleases. They are maintained at a low pH, mainly conjugated to carbohydrates. Cytochemically, the enzyme acid phosphatase (β-glycerylphosphatase) has been widely used as a marker of lysosomes for light and electron microscopy, though it is not invariably present.

Lysosomes have a complex life history, each stage being structurally distinct, and several forms may appear in the same cell. Their proteins are first formed in the granular endoplasmic reticulum where carbohydrate is also added to the enzymes; then they are passed to the Golgi complex for packaging in vesicular form and final processing. Enzyme retention within lysosomes rather than immediate secretion depends on the presence of a specific carbohydrate group added to the enzyme in the granular endoplasmic reticulum. When first released as *primary lysosomes* they are relatively small but may fuse to form larger primary lysosomes up to 0.8 μm across, typically dense and granular within. Primary lysosomes can be used either to degrade materials taken into the cell by endocytosis (*heterophagy*) or to degrade worn out, damaged or 'unwanted' organelles in the cell (*autophagy*). In heterophagy, phagocytic vacuoles (*phagosomes*), containing objects such as bacteria, fuse with one or many primary lysosomes to form *secondary lysosomes* which show signs of digestion—membranous whorls, fibrillar debris, etc. The term *phagolysosome* is given to a phagosome in the process of fusing with a lysosome. Smaller pinocytotic and endocytic vesicles, containing fluids or small particles, fuse with primary lysosomes to form *multivesicular bodies* (**1.18B**), another form of secondary lysosome. In both heterophagy and autophagy, lysosomes become smaller and denser as digestion proceeds; the final *residual bodies*, containing insoluble remains, may be ejected from the cell by exocytosis or may remain within the cell ('*storage excretion*'). In non-dividing cells such as neurons, residual bodies (**1.18C**) may fuse to form much larger aggregates, retained permanently, becoming conspicuous in old age in the form of the lipid-rich pigment lipofuscin ('*senility pigment*') within the nervous system.

In some cells, lysosomes are a dominant feature, as in neutrophil leucocytes, where the primary granules are lysosomes which assist in killing phagocytosed bacteria. Lysosomal enzymes may also be expelled to damage neighbouring foreign organisms or alter extracellular materials; the erosion of bone is in part caused by such lysosomal action from osteoclasts (see p. 295). Abnormal release of enzymes can cause tissue damage, as in certain types of arthritis. Enzymes released to the exterior can also be taken up by other cells and incorporated into their own lysosomes.

Lysosomal membranes, normally impermeable to their enclosed enzymes, may allow the enzymes to leak. When exposed to ionizing radiations, some carcinogens, silica or asbestos

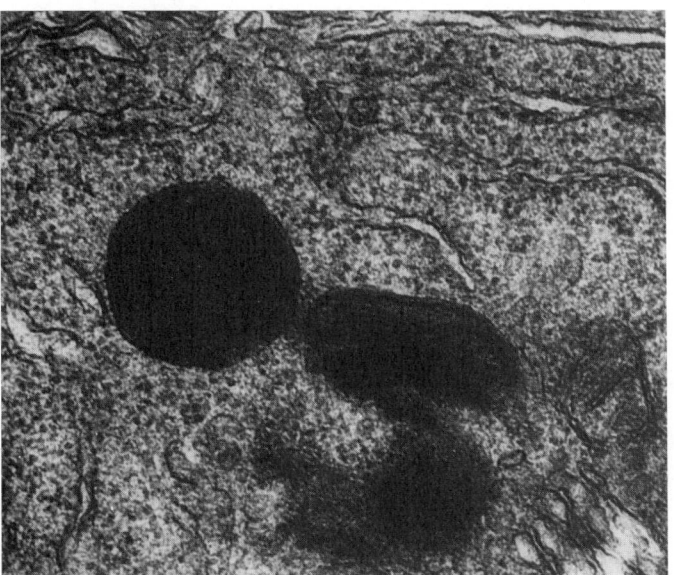

A

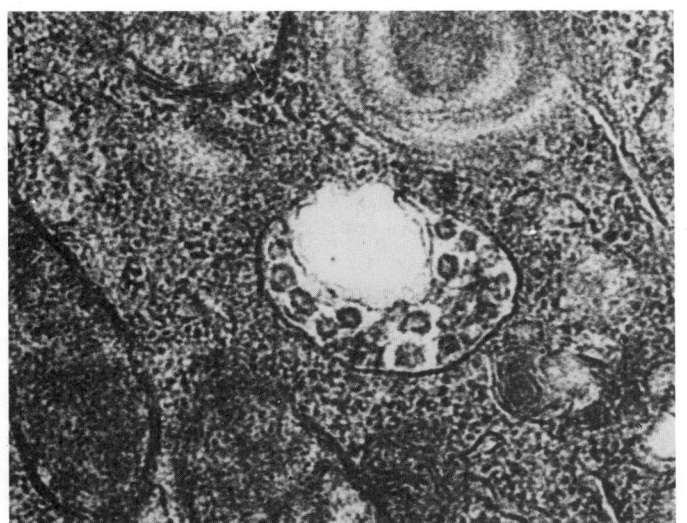

B

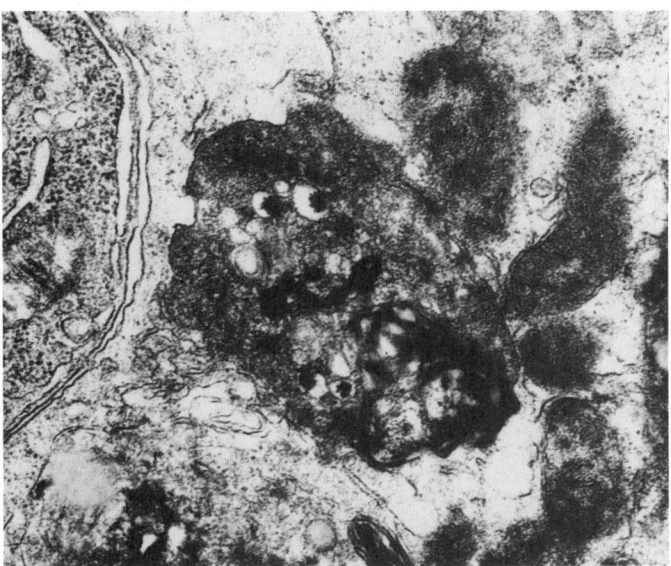

C

1.18 Electron micrographs of lysosomes in various stages: A. a group of lysosomes in an olfactory receptor cell showing small primary lysosomes (left), and larger secondary lysosomes containing lamellar débris; B. a multivesicular body with which endocytic vesicles are in the process of fusing; C. a residual body, the end stage of lysosomal hydrolysis of engulfed cellular organelles. All magnifications × 30 000.

particles, anoxia, heat and many drugs cause such effects, with consequent cellular damage or death or, if the genes are affected, perhaps neoplastic transformation. Conversely, some drugs, e.g. cortisone, can stabilize lysosomal membranes and may therefore inhibit their fusion with phagocytic vesicles. Lysosomal leakage may also be involved in the *programmed death* of redundant cells during embryonic growth or in facilitating their removal once dead (see also p. 103).

By their involvement in breakdown processes in metabolism, defects in lysosomal enzymes can cause profound morphogenetic changes; a wide range of genetic disorders (*lysosomal storage diseases*) has also been described, e.g. Tay-Sachs' disease in which a faulty carbohydrase leads to the accumulation of glycolipid in the central nervous system, with consequent early death, and Hurler's syndrome, where failure to metabolize certain glycosaminoglycans (mucosubstances) causes abnormalities in the formation of connective tissues.

PEROXISOMES

Peroxisomes (microbodies) are membrane-bound vacuoles (1.8), about 0.5–0.15 μm across, often with dense cores or crystalline interiors (de Duve 1973, 1983, Tolbert & Essner 1981). Large (0.5 μm) peroxisomes are found, e.g. in hepatocytes and kidney tubule cells, where they appear to detoxify various substances including ethanol by oxidizing them with molecular oxygen. Oxidation of a wide variety of compounds also occurs in smaller peroxisomes, which probably occur in all nucleated cells and form a vital system for detoxification of materials produced by or taken into cells. The oxidative enzyme *catalase* is particularly important in these processes, breaking down hydrogen peroxide produced by other peroxisomal enzymes, e.g. D-amino acid oxidase and urate oxidase which are also characteristically present. Some fatty acid breakdown also occurs in peroxisomes.

GOLGI COMPLEX (GOLGI APPARATUS OR DICTYOSOME)

Beginning with the Italian Camillo Golgi, optical microscopists of the nineteenth and early twentieth centuries recognized a distinct cytoplasmic region near the nucleus, particularly prominent in secretory cells and staining well with metallic salts. This was long considered an artefact by many but with the advent of electron microscopy it was authenticated as a genuine cellular organelle of great importance (see Farquhar & Palade 1981, Brachet 1985, Farquhar 1985).

The Golgi complex consists of one or more zones of smooth endoplasmic reticulum (1.5,19,20) arranged as stacks of flattened interconnected cisternae; small vesicles are found clustered at their borders. In glandular cells with an apical secretory zone, the Golgi complex is positioned between the secretory surface and the nucleus; in fibroblasts, where secretory activity is more general,

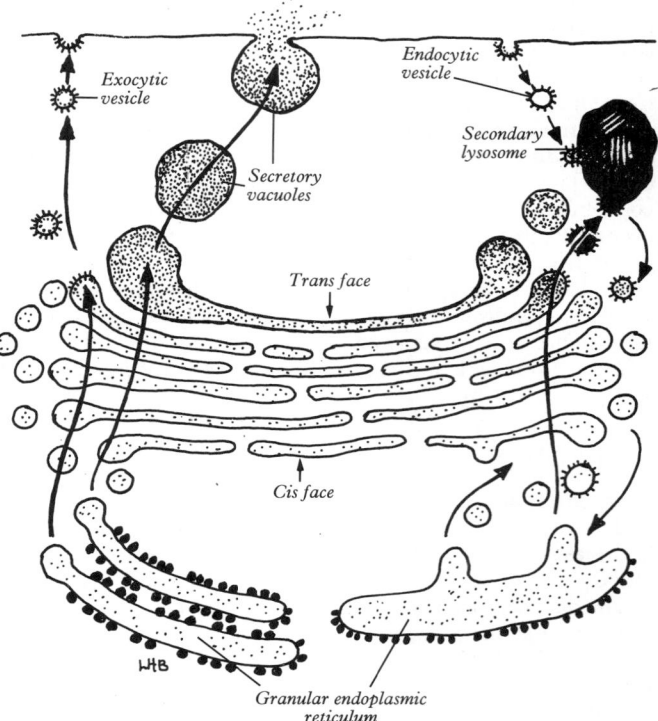

1.20 Some of the main routes of intracellular transport associated with the Golgi complex. The arrows indicate (lower picture) the origin of proteins from the granular endoplasmic reticulum and (left) their passage through the Golgi cisternae to the cell surface, either via exocytic vesicles as surface molecules (e.g. receptors for cholesterol) or as larger vacuoles for secretion. On the right, pathways involving lysosomes are shown: primary lysosomes in the form of small or larger vesicles are budded off from the *trans* face of the Golgi complex and fuse with material to be degraded as secondary lysosomes. Vesicles derived from lysosomes also pass back to the Golgi cisternae which in turn can recycle membrane between its cisternae and also back to the endoplasmic reticulum.

there are two or more groups of flattened vesicles and in liver cells up to 50.

The Golgi complex is often cup-shaped, usually with the convex side nearest the nucleus. Biochemical and autoradiographic studies have shown that the organelle is a metabolically active region where secretory products and lysosomal enzymes are accumulated by fusion of *transport vesicles*. These coalesce with one (usually convex) surface of the complex, termed the '*forming surface*,' or *cis* aspect and, after their contents have moved through the pile of cisternae, other vesicles are budded off at the other (usually the concave) surface, the *condensing* or *trans* surface, as secretory or storage vacuoles (*condensing vacuoles*). The forming surface is usually on the side closer to the nucleus (Jamieson & Palade 1967). In traversing the complex, substances can be altered by various processes including addition of carbohydrates and fatty acids, sulphation and by selective removal of amino acids. In mucus-secreting cells, for instance, the proteins synthesized in the rough endoplasmic reticulum are concentrated in the condensing vacuoles by subtraction of water, and carbohydrates rich in sulphur are added. In other cells, too, much of the carbohydrate in their secretions is added at the Golgi complex, whose membranes bear enzymes associated with carbohydrate synthesis, such as glycosyl-transferase, particularly within the cisternae of the *trans* aspect. The enzyme thiamine pyrophosphatase is also a valuable histochemical marker of Golgi complex activity. It has also been proposed that, by selecting and passing on only purified proteins from the cis to the *trans* surfaces, successive cisternae act collectively rather like a molecular distillation apparatus to concentrate specific proteins free of contaminants (Rothman 1981).

The Golgi complex is also active in primary lysosome production (vide supra) which in neurons and other cells may occur in a special region, the Golgi-endoplasmic reticulum-lysosome

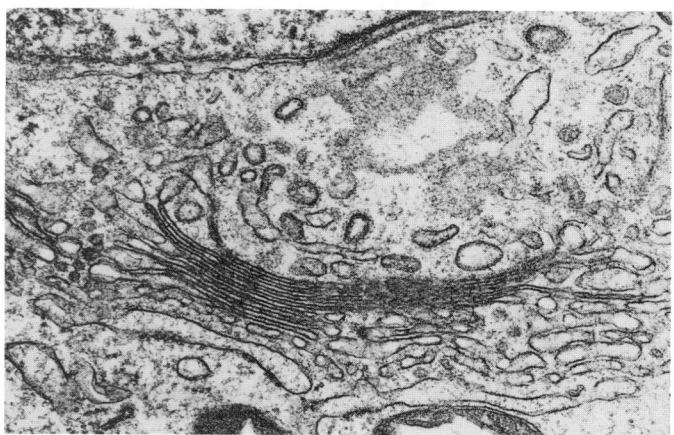

1.19 Electron micrograph of a Golgi apparatus in a spermatocyte showing the central multilamellar cisternae and clusters of peripheral vesicles. Magnification × 15 000.

complex *(G.E.R.L.)* (see Novikoff 1973). The synthesis of cell membranes and glycocalyx also involves this organelle. Various types of modification and sorting of different substances for appropriate destinations may also occur simultaneously in different regions of the same complex.

ANNULATED LAMELLAE

Annulated lamellae are arrays of parallel cisternae which sometimes appear in the cytoplasm and occasionally nuclei of cells undergoing rapid protein synthesis (Kessel 1968), e.g. oöcytes, some embryonic and neoplastic cells. Adjacent membranes fuse at intermittent sites where they are punctuated by structures closely resembling nuclear pores (p. 34), reflecting their origin from the outer nuclear membrane. They may serve as attachments for stored RNA but are otherwise obscure.

INTRACELLULAR VESICLES

These include membrane-bound vacuoles, usually less than 0.1 μm in diameter, involved in the movement of fluids and particles into, out of and within the cell. The process of uptake is *endocytosis* and the passage of materials in vesicles to the cell surface and their release, *exocytosis*. Endocytosis involves different types of vesicle; *phagocytic vesicles* take in relatively large particles, e.g. bacteria, by an energy-using process. Smaller particles, including macromolecules, can be endocytosed in *coated vesicles* without the use of direct energy. Macromolecules binding to surface receptors trigger attachment of tetrahedral assemblies of the protein *clathrin* to the inner face of the plasma membrane, forming a *coated pit*, then a *coated vesicle* (Pearse and Bretscher 1981) which passes into the cytoplasm, to fuse with other organelles (e.g. lysosomes, golgi complexes) after removal of the clathrin coat. Membrane can then be returned to the cell surface in vesicular form. Small vesicles are also used extensively by cells to shuttle substances internally between organelles (*transport vesicles*). *Pinocytosis* involves the bulk uptake of fluid into large, complex vesicles.

Exocytic vesicles, carrying materials out of cells, vary greatly in size and may only be distinguished from endocytic vesicles experimentally by using extracellular tracers, which will enter the latter but not the former (see p. 50). Vesicles of these various types are present in all cells, but are particularly numerous where rapid exchange of materials between the cell and its environment is occurring, e.g. endothelial cells, synaptic terminals of neurons and in secretory cells.

THE CYTOSKELETON

This term is commonly used to denote a heterogeneous collection of filamentous structures which play important roles in maintaining or altering cell shape, generating motion within cells and promoting movement relative to their external environment, i.e. they form the cell's skeleton which, like that of the whole body, performs supportive and motile functions. The components of the cytoskeleton include several types of proteinaceous assemblies, all of them deployed within the cytosolic regions of the cytoplasm lying outside the system of membrane-lined compartments, although they are often attached to them. The way in which movements are created by these structures and their importance both in the daily lives of cells and in more complex processes of development will be considered later (p. 49); here we will give a brief account of their form and composition (see also the reviews by Alberts et al 1983, Brachet 1985, Schliwa 1986, Lackie 1986).

The roll-call of cytoskeletal structures is lengthening rapidly as electron microscopy combined with immunocytochemistry and classic protein chemistry continues to define new forms. Those which we know best are the *microtubules, microfilaments* and *intermediate filaments*; other important structures are *myosin filaments* and there is also a galaxy of minor components involved in linking the foregoing structures together or regulating their activities in some way (see e.g. Weber and Osborn 1985).

Microtubules

These are elongated cylinders about 24 nm in diameter (**1.5,22B**),

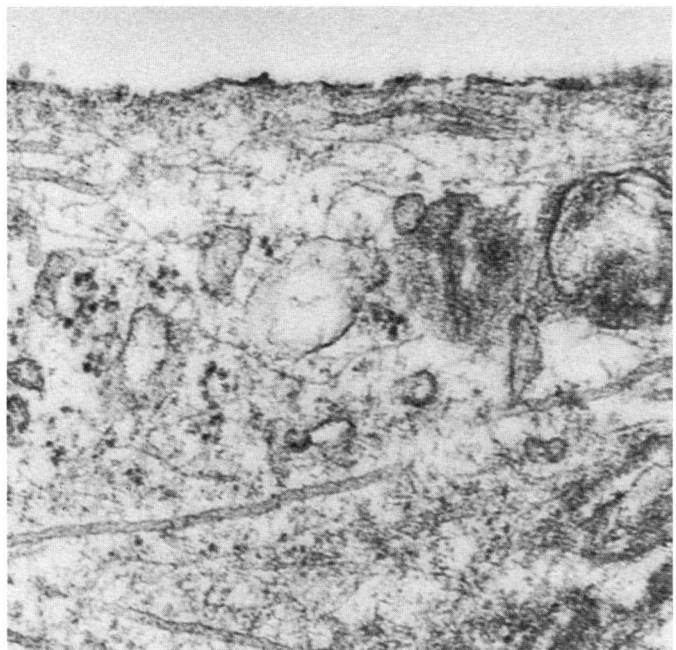

1.21A Electron micrograph of microtubules and microfilaments in a tangential section through a fibroblast. Magnification × 50 000.

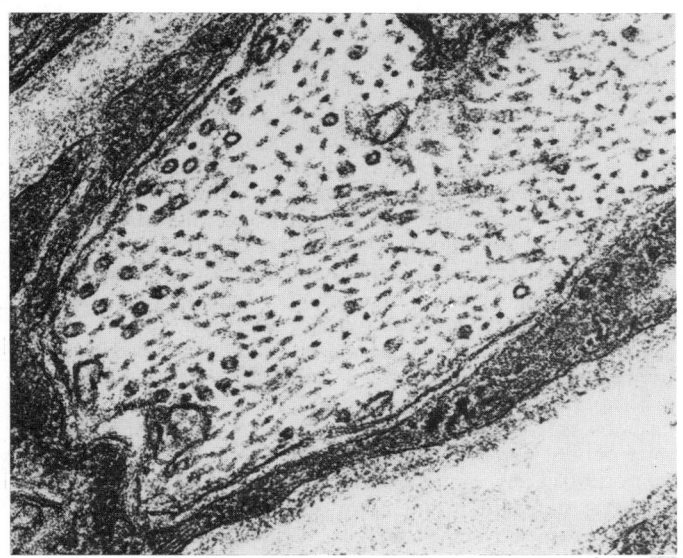

1.21B Microtubules and neurofilaments in a transverse/oblique section of part of a small nerve fibre. The microtubules have a circular cross-section whereas the smaller neurofilaments appear solid. Magnification × 50 000.

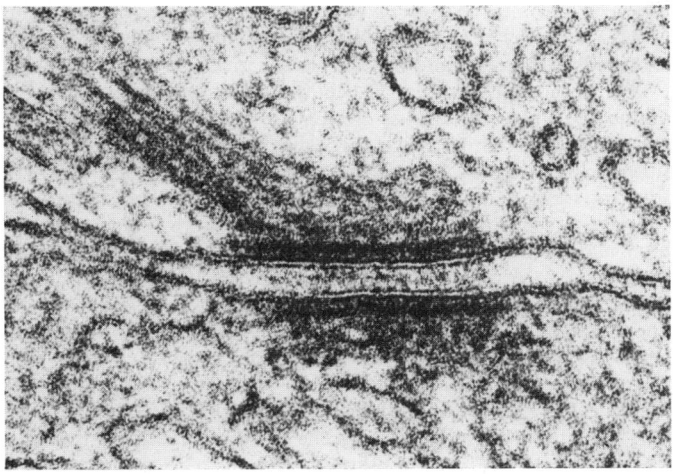

1.21C Intermediate filaments associated with a desmosome (macula adherens) between two epithelial cells. Magnification × 100 000.

of varying length (up to some 70 μm in spermatozoan flagella) and made of protein. They are frequent components of most cell types but are abundant in neurons, leucocytes, blood platelets and in the mitotic spindles of dividing cells. They also form part of the structure of cilia, flagella and centrioles (vide infra). Helical microtubules occur in Schwann cells and oligodendrocytes (pp. 893, 901). Massed parallel bundles of inter-connected microtubules are found in the pillar cells supporting the sensory cells of the cochlea. Microtubules rarely show acute bends, indicating a degree of stiffness.

These structures are composed of two proteins, *tubulins* α and β, arranged as dimers (see e.g. Soifer 1986) with a combined Mr of 100 000 (50 000 each). Each protein subunit is about 5 nm across and microtubules in transverse section are made up of a ring of 12 or 13 such globular subunits which are arranged in long columns ('protofilaments') of alternating tubulins α and β, forming the wall of a cylinder. Each column is slightly shifted on its neighbour, so that a spiral pattern of alternating α and β subunits is also generated (Amos & Klug 1974). Microtubules are formed by the simple polymerization of the tubulin dimers and there is in some cells an equilibrium between polymerized and unpolymerized tubulin, so that some types of microtubule may break down and reform rapidly, although stability varies. Consequently, chemicals which bind to tubulin cause the equilibrium to shift, bringing about the depolymerization of microtubules; the drugs colchicine, colcemide, vinblastin and griseofulvin are well known as anti-microtubule agents acting in this manner.

Formation of microtubules can take place by addition to pre-existing microtubules, or formation de novo in, as yet, poorly understood *microtubular organizing centres* such as those surrounding (and including) centrioles, around which spindle microtubules polymerize during cell division (Pickett-Heaps 1975) and from which cilia can grow (see p. 31). In other instances such centres have no obvious structural basis, thus centrioles are not always essential to microtubule formation.

Because tubulin dimers are asymmetrical, microtubules themselves are also asymmetrical, i.e. they have an integral *direction*, one end being chemically different from the other. It is therefore possible to add tubulin at one end while removing it at the other (a phenomenon called 'treadmilling', see also p. 51).

Attached to the walls of the microtubules are other minor components known as *microtubule-associated proteins (MAPs)*, which are important in linking microtubules to each other or to membranous organelles (e.g. mitochondria) or otherwise modifying microtubule functions. In the case of cilia, some of their microtubules have projecting arms whose movements cause sliding between neighbouring microtubules and thus ciliary beating; a protein (kinesin) associated with the walls of other microtubules appears to create the flow of cytoplasm along their surfaces, e.g. in the movements of substances along neuronal axons (for further details see p. 881).

Microfilaments

These are thinner (6–8 nm) than microtubules and appear solid (1.21,22B); they are prominent components of all cells except mature erythrocytes and are intimately involved in cell motility (Wessels et al 1971). They are composed of the protein *actin*, of which the 6 nm subunits of globular *g-actin* (Mr:43 000) are arranged in a staggered helical form to form filamentous *f-actin*. A number of different varieties of actin are known, some of them forming long relatively stable microfilaments, others assembling into shorter and more transient ones. Some types are primarily associated with cell plasma membranes; others lie deeper within the cytosol. Muscle cells contain particularly robust forms of microfilament (p. 550). Actin binds to the protein *myosin* when an energy source (ATP) is available and when appropriately organized. This leads to various types of shearing movement which can produce cell motility of different kinds, including muscular contraction (p. 550). Actin microfilaments can be detected electron microscopically by incubating cells, made permeable to proteins, with heavy meromyosin which binds to actin and decorates the filaments with arrowhead formations. Actin can also be readily detected by labelling with anti-actin antibodies

conjugated to a fluorescent dye for light microscopy; (see 1.22A) or to an electron-dense marker (e.g. colloidal gold) for electron microscopy. It also binds the fungal derivative phalloidin which can be fluorescently labelled, too. The cytochalasins (fungal products) cause depolymerization of some actin microfilaments (see Spudich & Lin 1972), and therefore provide a valuable experimental tool in research into their functions. Often associated with actin microfilaments are a number of other proteins (e.g. troponin, tropomyosin B, actinin) which assist in various motile activities of these organelles (Lazarides 1976). In cells they are often linked together by shorter filaments, e.g. of the *actin bundling protein, fodrin*, or to cell membranes by such molecules as *vinculin* (in fibroblasts) or *ankyrin* (in erythrocytes, see p. 664).

Intermediate Filaments

These include a family of protein filaments about 10 nm thick, found in different cell types and often present in large numbers where structural strength is needed (1.8,22). Recently it has been shown by immunochemical and other means that several types of intermediate filament occur, all with similar though not identical chemical structure. These include *keratin* filaments of epithelial cells, *vimentin* filaments of connective tissue cells, *desmin* filaments of muscle cells, *neurofilaments* of nerve cells and *glial fibrillary acidic protein* filaments of neuroglial cells.

All have diameters of about 10 nm and are composed of filamentous subunit proteins in the 50 000–100 000 daltons molecular weight range. They tend to form bundles linked together by lateral extensions which appear to be an integral part of each filament.

Keratin filaments include many subtypes with different roles, according to developmental and functional status of the cells containing them. Best known structurally are those of epidermal keratinocytes. In these cells, keratin filaments are at first arranged in bundles forming a network within the cytoplasm and anchored to desmosomes at the cell surface. As the cells mature, the filaments aggregate in flattened layers to give the now scale-like cells considerable strength (see p. 74). However, studies with specific monoclonal antibodies have distinguished many varieties of such filaments, even within keratinocytes; it is possible to discern a changing pattern of keratin chemistry as cells mature. Likewise, in other epithelial tissues, keratin chemistry is equally complex but this diversity is proving of great advantage in detecting pathological changes and in diagnosing the origins of epithelial carcinomas when these have spread from their parent tissue (for a review, see Clausen et al 1986).

Intermediate filaments also occur in the nucleus of the cell, attached to the inner nuclear membrane and providing mechanical strength to the nuclear wall. They form a complex interlacing network or *nuclear membrane skeleton*, which supports the pore complexes (p. 34) and help to determine nuclear shape.

MICROVILLI

These are finger-like extensions of cell surfaces (1.5,8,23) usually about 0.1 μm in diameter and up to about 5 μm long. When arranged in a regular parallel series, they constitute a *striated border*, as at the absorptive surfaces of the epithelial cells of the small intestine; when less regular, as in the gall-bladder epithelium, the term *brush border* is applied.

Internally bundles of actin microfilaments support them (see Mooseker 1985); these are implanted within the apical cytoplasm of the cell amongst other transversely running filaments to create a meshwork, the *terminal web*, which forms a sheet of filaments across the apical cytoplasm, anchored peripherally to the zonula adhaerens. The protein myosin also occurs in the terminal web and is thought to bind to the actin to stiffen this part of the cell. In vitro this meshwork can be stimulated to contract, experimentally. The microfilaments are attached to the membrane of each microvillus by finer lateral filaments along the margins and are inserted apically into a dense mass of the protein α-actinin (Mooseker & Tilney 1975).

Microvilli greatly multiply the area of surface (up to 40 times, in striated borders) and are found at sites of active absorption, such as in the proximal and distal convoluted tubules of the

30

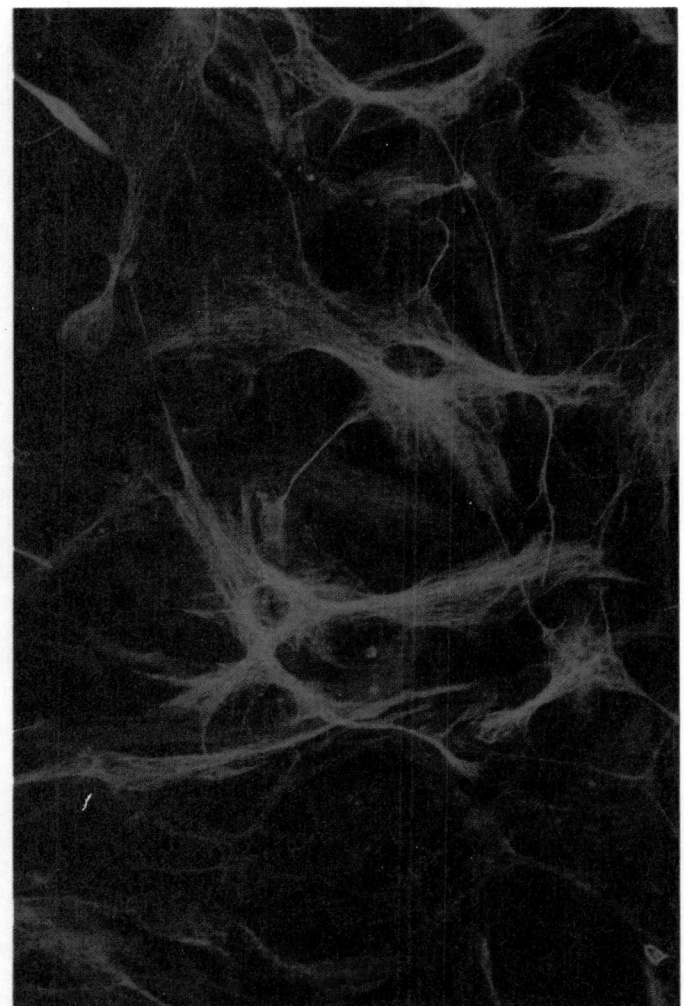

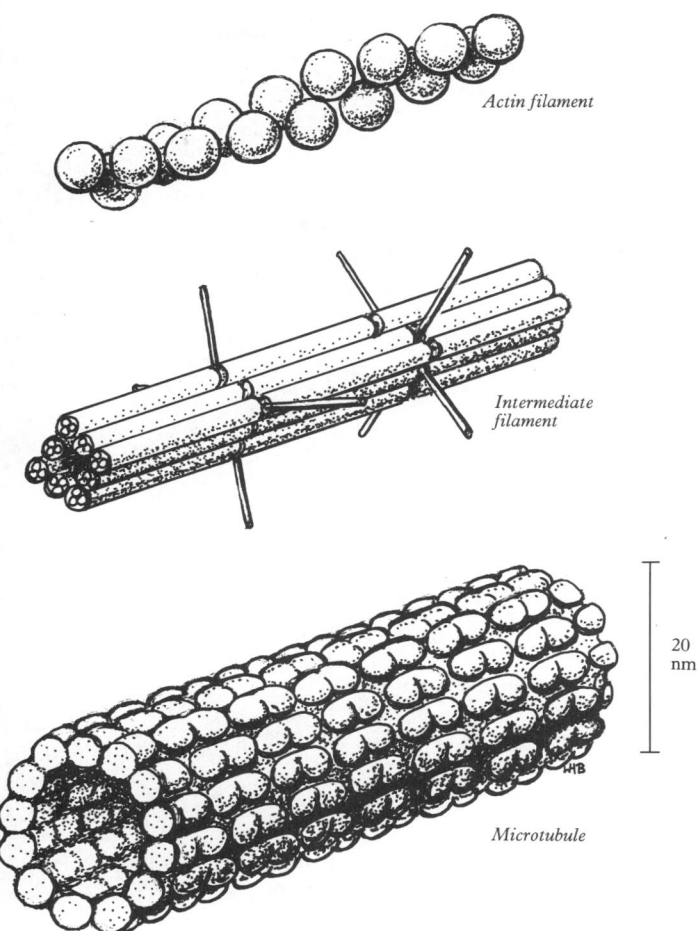

Actin filament

Intermediate filament

20 nm

Microtubule

1.22A Intermediate filaments, in this case composed of glial fibrillary acidic protein (GFAP), demonstrated in cultured retinal astrocytes with anti-GFAP antibodies coupled to rhodamine. Provided by Caroline Wigley, UMDS, Guy's Campus, London. Magnification × 5000.

1.22B The substructure of some major cytoskeletal components, including (top) an actin microfilament formed by polymerization of globular actin monomers; (middle) an intermediate filament composed in this example (e.g. vimentin) of eight sub-filaments, each containing two or three paired acidic and basic polypeptide chains, some extending laterally as side-arms; and (below) a microtubule composed of assemblies of tubulin dimers arranged in a tight helix.

kidney. In the small intestine absorptive cells bear microvilli which have a thick glycocalyx, probably of great importance in providing a site for activity of digestive enzymes. Several enzyme systems associated with active transport across membranes have been localized within striated borders.

Irregular microvilli, or *filopodia*, are also found on the surfaces of many types of cell, particularly free macrophages and fibroblasts. Again, these may be associated with transport processes, particularly phagocytosis, and with cell motility.

Large, regular microvilli (*stereocilia*) occur at some sensory surfaces, e.g. on cochlear and vestibular receptor cells and in the absorptive epithelium of the epididymis.

Transient microvillus-like structures found on developing or motile cells include elongate '*microspikes*' and *ruffled membranes* (see p. 49, also **1.48**).

CILIA AND FLAGELLA

These are, typically, motile hair-like (cilium = eyelash) and whip-like (flagellum = whip) projections of the cell surface, which create currents in the surrounding fluid or movements of the cell to which they are attached. Cilia occur on many internal surfaces, particularly the epithelia of most of the respiratory tract and parts of the male and female reproductive tracts, the ependyma lining the central canal of the spinal cord and ventricles of the brain and the mesothelia of the peritoneal and pleural cavities. They also occur at olfactory receptor and vestibular hair

cell endings and, modified, as portions of the rods and cones of the retina; cilia are, in addition, present on many dividing tissue cells. Many cilia may be present on a single cell, as in bronchial epithelium, or only one or two as with some mesothelial cells. Each male gamete possesses a single flagellum about 70 μm long.

By electron microscopy (**1.24,25**) each cilium or flagellum is seen to consist of a *shaft*, constituting most of its length, with a diameter of about 0.25 μm, a tapering *tip* and, within the surface cytoplasm of the cell, a *kinetosome (basal body, basal granule* or *blepharoplast)* about 1 μm long (Gibbons & Grimstone 1960). The whole structure is bounded, except at its base, by an extension of the plasma membrane. In freeze-fracture preparations chain-like groups of characteristic membrane particles, the *ciliary necklace,* surround the proximal end of the cilium (Gilula & Satir 1972). These particles may assist the control of ciliary beating. Within the membrane lies a cylinder of nine double microtubules (the *axoneme*), surrounding a central pair of single microtubules.

At the *base* of the cilium each microtubule doublet, the two parts of which are designated the A and B subfibres, are twisted through 40 degrees (**1.24**) and another microtubule, the C subfibre, is added. The central pair sometimes end above the cell surface in a dense sphere or *axosome*, beneath which lies a transverse partition or *basal plate*. Associated with the kinetosome are often one or more cross-banded filamentous *rootlets* and a plate-like *basal foot*, which probably help to anchor the cilium into the cytoplasm.

At the *apical end*, the various microtubular elements do not all continue to the tip; the A subfibres are shortest, then the B

31

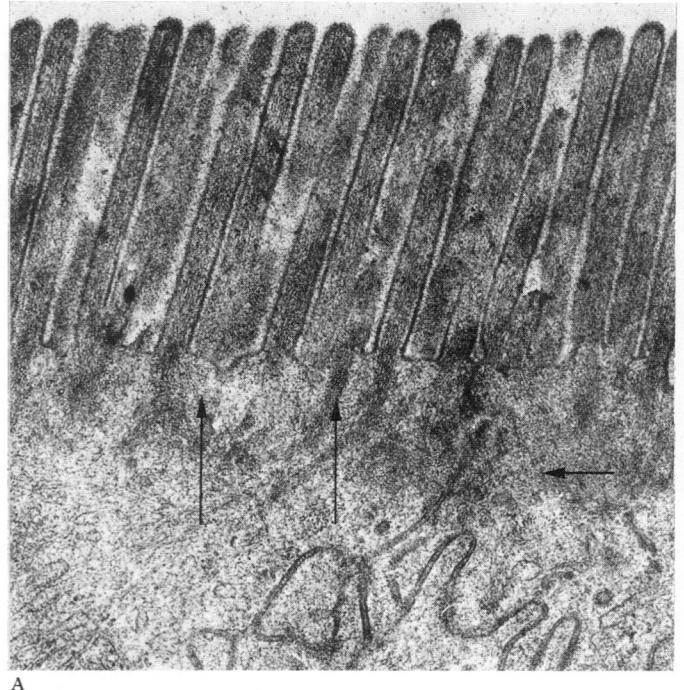

A

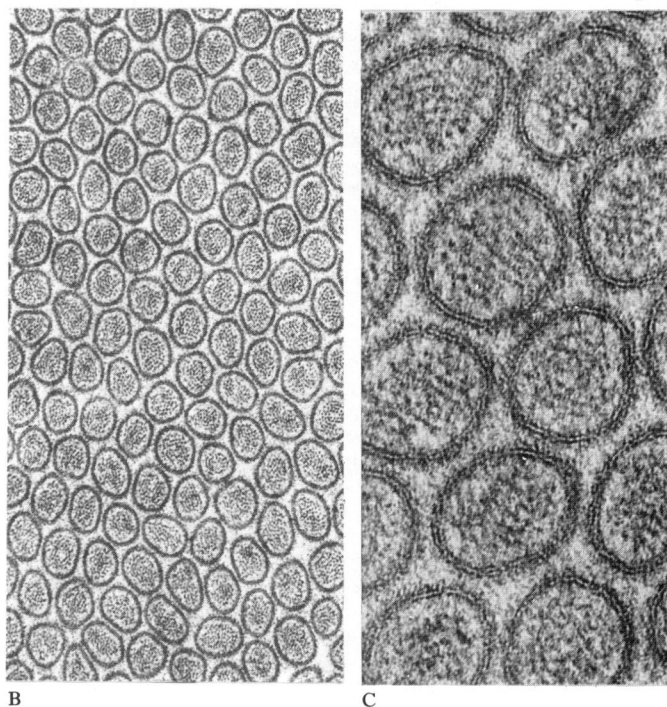

B

C

1.23 Electron micrograph of microvilli from the striated border of an absorptive columnar epithelial cell of the small intestine (compare with 1.5): A. vertical section showing microvilli with supporting microfilaments (long arrows) and terminal web (short arrow); B. horizontal section through cell surface cutting each microvillus transversely. Insert (C) shows the details of a few microvilli from B; note the bilayer structure of the plasma membrane and the fuzzy external cell coat. Magnifications: A and B × 20 000; C × 110 000.

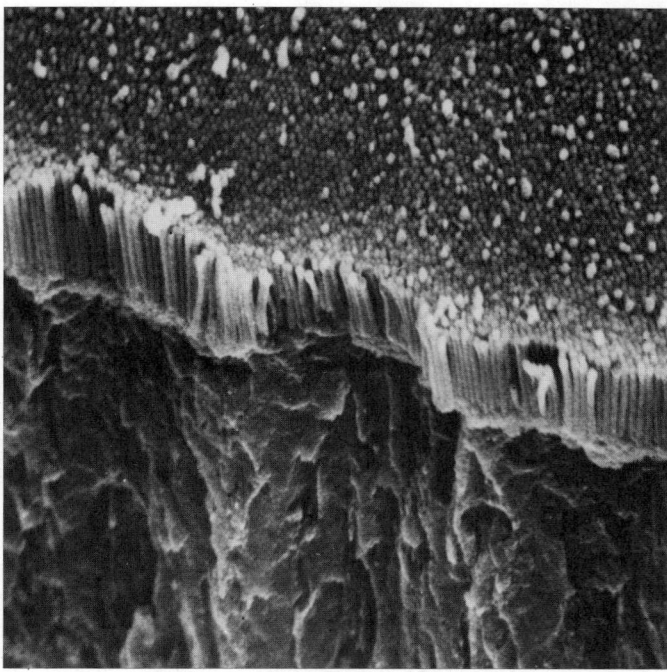

1.23D Scanning electron micrograph of a striated border from the small intestine showing numerous microvilli. The epithelium has been cut vertically to allow the microvilli to be viewed from the side as well as from above. Magnification × 6000.

pair and corresponds to the direction of bending.

Movements of cilia and flagella (see Sleigh 1974,1977), are similar in broad outline (1.26). Flagella move by rapid undulation passing from the attached to the free end. In human spermatozoa there is a helical component in this motion, although the precise form is hydrodynamically complex (see e.g. Brokaw 1975).

In cilia, the beating is planar, but asymmetrical. In the *effective stroke*, the cilium remains stiff except at the base where it bends to produce an oar-like stroke. Then follows the *recovery stroke* during which the bend passes from base to tip returning the cilium to its initial position for the next cycle. The activity of groups of cilia is usually coupled so that the bending of one is rapidly followed by the bending of the next and so on, resulting in long travelling or *metachronal* waves (see also p. 1276). These pass over the tissue surface in the same direction as the effective stroke. Mechanical coupling of adjacent cilia is caused by their viscous interaction, the bending of one initiating beating of the next.

The basic mechanism is still not clear; when a cilium bends, the microtubules do not change in length but slide on one another (Warner & Satir 1974). The 'arms' of peripheral doublets are made of a protein 'dynein' and slant towards the cilium base from their attached ends. Dynein has an ATPase activity, stimulated by magnesium ions, and appears to cause mutual sliding of adjacent doublets by first attaching to the next at its free end, then swinging upwards towards the tip of the cilium. The radial spokes and various other substructures appear to modify the form of the ciliary bending (Warner & Mitchell 1980).

The cilia of olfactory receptors (p. 1173) show the typical '9 + 2' arrangement of subfibres; the cilium segment of retinal rods and cones, however, lacks the central pair (p. 1199). Dividing cells of many tissues (e.g. the adrenal cortex) also show a small number of cilia which protrude from the cell surfaces into the intercellular spaces. Their function is unknown but they may represent some unavoidable, functionless by-product of cell division processes.

Cilia and flagella are formed by the polymerization of tubulin on centrioles, which are synthesized deep in the cell and then move to lie immediately beneath the cell membrane before the cilia begin to sprout (Sorokin 1968, Staprans & Dirksen 1974). Once initiated, they grow by the addition of tubulin and other materials to the distal end.

32 subfibres and finally the central pair terminate. Within the shaft lie several filamentous structures associated with the microtubules, such as radial spokes which extend inwards from the outer microtubules towards the central pair. The outer doublet microtubules bear two rows of tangential *dynein arms* attached to the A subfibre of one doublet, pointing towards the B subfibre of the adjacent doublet. Adjacent doublets are also linked by thin filaments. Other filaments partially encircle the central pair of microtubules, which are also united by ladder-like spokes. Because of the 9 + 2 pattern of tubules, there is a plane of symmetry which passes perpendicular to a line joining the central

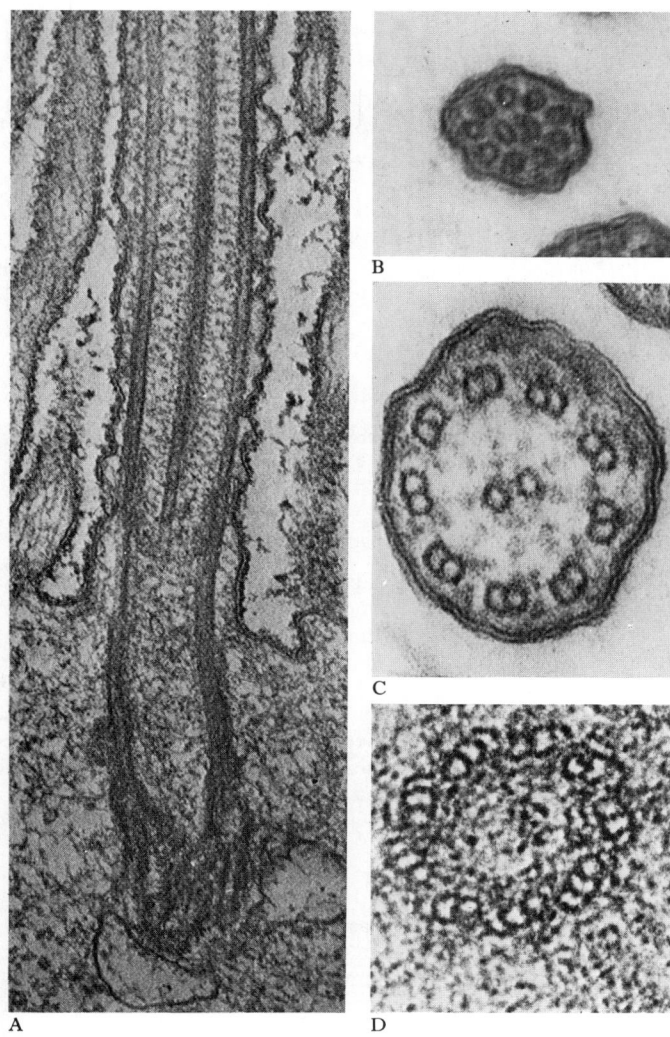

A   D

1.24 Electron micrographs of cilia cut (A) longitudinally, and (B, C, D) through the tip, shaft and base respectively. Magnifications: A × 50 000; B–D × 110 000.

CENTRIOLES

Centrioles are microtubular cylinders about 1μm long by 0.25 μm in diameter (Peterson 1980, Wheatley 1982), identical in structure with the bases of cilia (vide supra) but lying free in the cytoplasm (**1.5,27**). At least two centrioles occur in all cells capable of division, usually lying close and at right angles to each other, together often termed a *diplosome*, within a somewhat dense region of cytoplasm, the *centrosome*. Various filamentous or granular structures (*centriolar satellites* or *pericentriolar bodies*) surround them.

This centriolar complex is a centre for microtubule assembly, e.g. in the generation of the spindle and aster microtubules during cell division, the sprouting of cilia and the provision of axonal and dendritic microtubules in developing neurons. Prior to cell division a new centriole forms near each old one and the resulting pairs are passed on to the two new cells (Rattner & Phillips 1973).

Although the microtubules of cilia are assembled directly on the ends of centrioles, spindle microtubules grow from the surrounding centriolar satellites rather than centrioles themselves. The nature of this organizing capacity is not certain; in flowering plants spindle microtubules are formed without centrioles, so they may not be essential to this process.

The Nucleus

The nucleus is usually spherical or ellipsoidal, often indented, 4–10 μm in diameter, and separated from the cytoplasm by a membranous nuclear envelope. It stains deeply with basic dyes such as haematoxylin, by virtue of the nucleic acids in its chromosomes (deoxyribonucleic acid, DNA) and its nucleoli (ribonucleic acid, RNA). DNA is also responsible for the strong positive reaction with the Feulgen method. Basic proteins (histones) and some acidic proteins are also present in the nucleus.

Many experimental investigations of the functions of the nucleus have been carried out to determine its role in cellular life. These involved various approaches, e.g. transplantation of nuclei between different species of protozoa, removal of the nucleus and the fusion of dissimilar cells from different species after prior treatment with viruses or other agents (Harris 1970). Such experiments and many other observations, e.g. on the role of the gamete

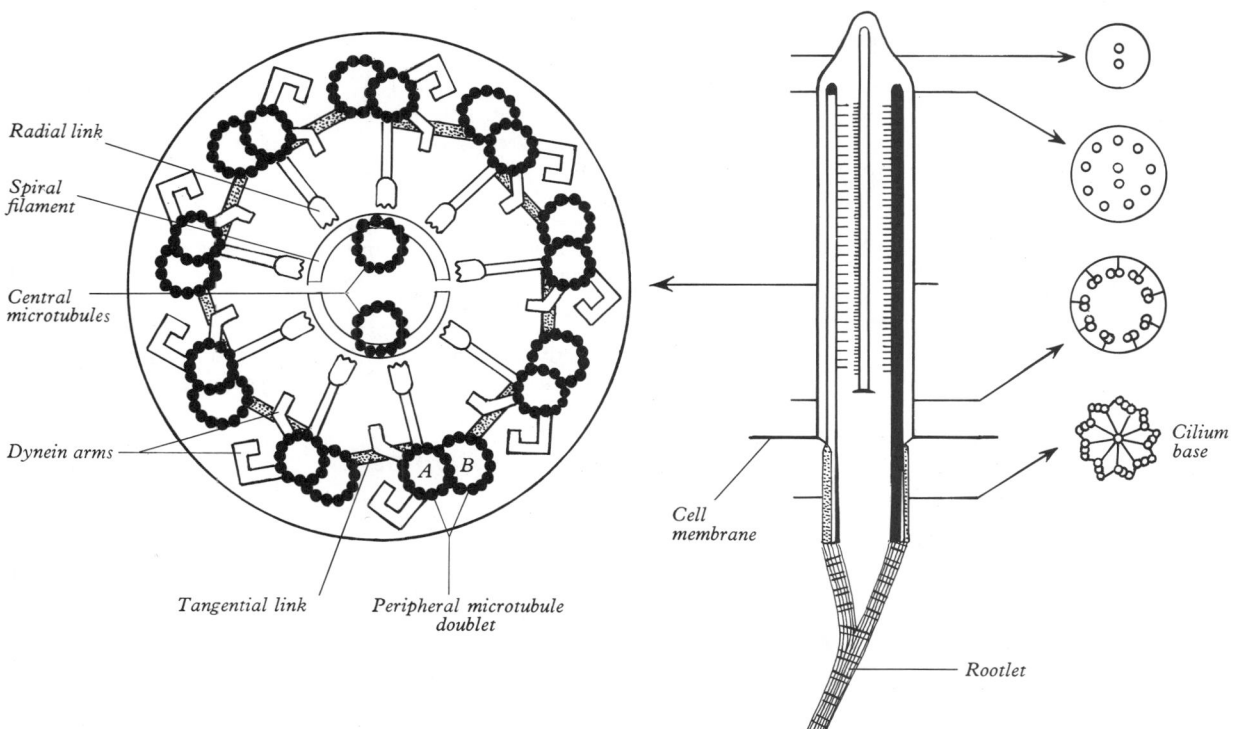

Radial link

Spiral filament

Central microtubules

Dynein arms

Cilium base

Cell membrane

Tangential link   Peripheral microtubule doublet

Rootlet

1.24E The internal structure of a cilium in longitudinal section (centre), the disposition of microtubules at various levels (right), and the detailed cross-sectional appearance of the shaft of the cilium (left). In the latter, hook-like dynein arms extend from the peripheral doublets and radial

links with expanded heads project towards the central pair of microtubules. A filamentous spiral encircles the central pair. Adjacent peripheral doublet microtubules are also connected by tangential links. Redrawn from Sleigh (1977).

1.25 Scanning electron micrograph of the ciliated surface in the trachea. Magnification × 10 000.

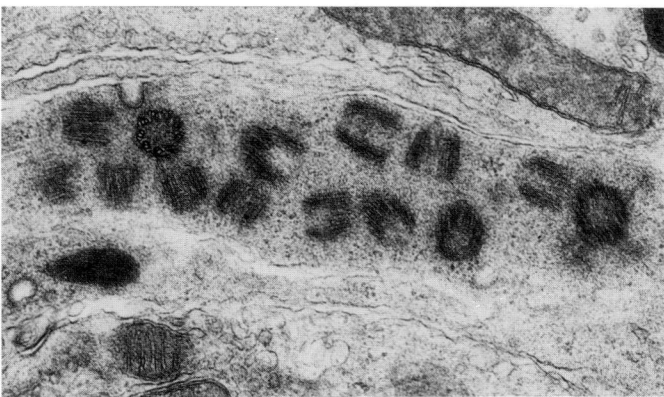

1.27 Transmission electron micrograph of a group of centrioles in a developing nerve cell dendrite. Provided by A Cuschieri. Magnification × 25 000.

nuclei in transmitting inheritable traits, show that the nucleus contains the inheritable instructions (genes) for the direction of the cell's activities so that, although the cytoplasm can survive for a limited time without a nucleus, protein synthesis ceases and the cell soon dies.

Many of the characteristics of intact cells in culture, such as division rate and motility, seem primarily determined by the nucleus. Isotopic labelling also shows a constant exchange of materials between nucleus and cytoplasm (see p. 39).

NUCLEAR ENVELOPE

This is a flattened sac formed by the membranes of the innermost two layers of the cytoplasmic membrane system, separated by a distance of about 20 nm (**1.5,28**). The outer membrane is part of the endoplasmic reticulum proper and often bears ribosomes; products of protein synthesis may accumulate between the two membranes. The inner membrane is quite distinct from other cell membranes, being the attachment site for the ends of chromosomes, which often adhere to this surface as a dense coating of chromatin during interphase. This membrane is also reinforced with a network of 10 nm filaments (p. 30) which, with other proteinaceous material, forms a dense *nuclear lamina* lining the inner surface of the nuclear envelope.

The envelope membranes show scattered zones of annular fusion to form *nuclear pores*. Each of these creates an aperture in the nuclear envelope, about 80 nm across and surrounded by a ring of eight groups of granules; a diaphragm partly covers the aperture and another granule occupies its central region, so that only particles less than 9 nm across can traverse the pore. The whole apparatus is the *pore complex* (Franke et al 1981). There are from 3000–4000 pores in a typical nucleus, providing diffusion channels for smaller proteins, messenger RNA, etc, while retaining the chromosomes within the nucleus and excluding potentially destructive organelles, e.g. lysosomes.

Chromosomes

Within the nucleus most of the nucleic acid is deoxyribonucleic acid (DNA) associated with protein as *chromatin*, in the form of elongated threads, the *chromosomes*. Between cell divisions, these are highly extended to create, collectively, a diffuse network which may be dispersed as weakly staining *euchromatin* or more folded as more densely staining *heterochromatin* (**1.29**). These two states are associated, respectively, with high and low degrees of RNA synthesis (vide infra). It has also been found that replication of DNA during S phase takes place later in heterochromatin than euchromatin. Euchromatic nuclei are known as *open-faced* and are larger than heterochromatic, *closed-face* nuclei. When heterochromatin takes the form of thick threads, as seen by light microscopy, it is said to be *leptochromatic* (e.g. in resting

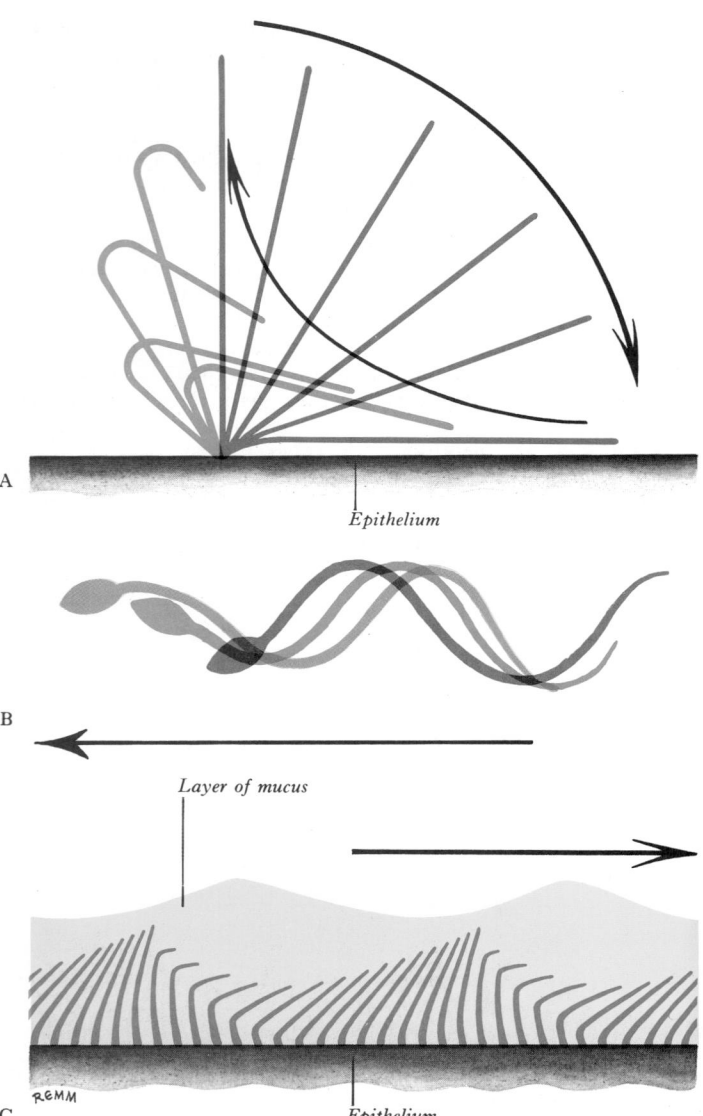

A

Epithelium

B

Layer of mucus

C

Epithelium

1.26 Diagram of ciliary (A) and flagellar (B) action. In A the effective stroke is shown in red and the recovery stroke in blue. In B successive movements of the sperm tail are indicated in red, blue, and green; C represents a number of cilia in the respiratory tract showing the metachronal wave and the direction of mucus movement.

lymphocytes). In moribund cells, the nuclei are often small and stain densely (*pyknotic nuclei*) or may disintegrate into vesicles (*karyorrhexis*). Pathological nuclei (e.g. in neoplastic cells) may be

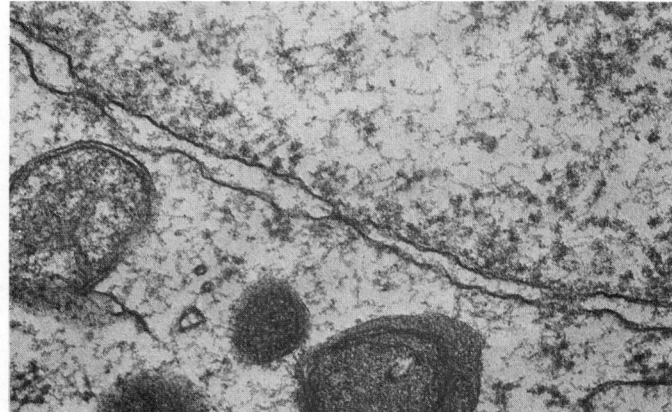

A

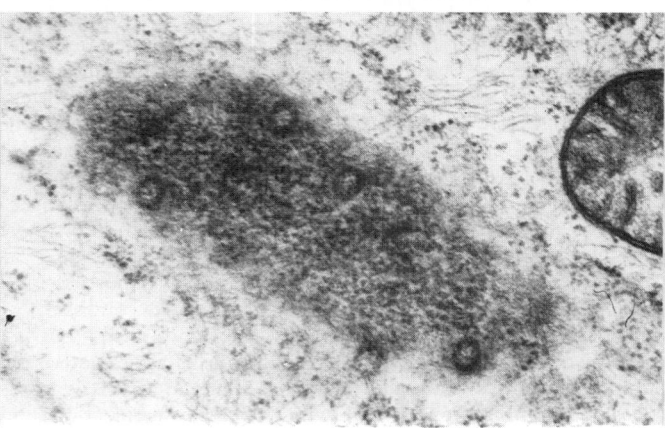

B

1.28 Electron micrograph of the nuclear envelope: A. in transverse section to show inner and outer membranes, and nuclear pores; B. in tangential section to show pore complexes (dense rings). Micrograph (B) provided by M Dyson of Guy's Hospital Medical School. Magnification × 25 000.

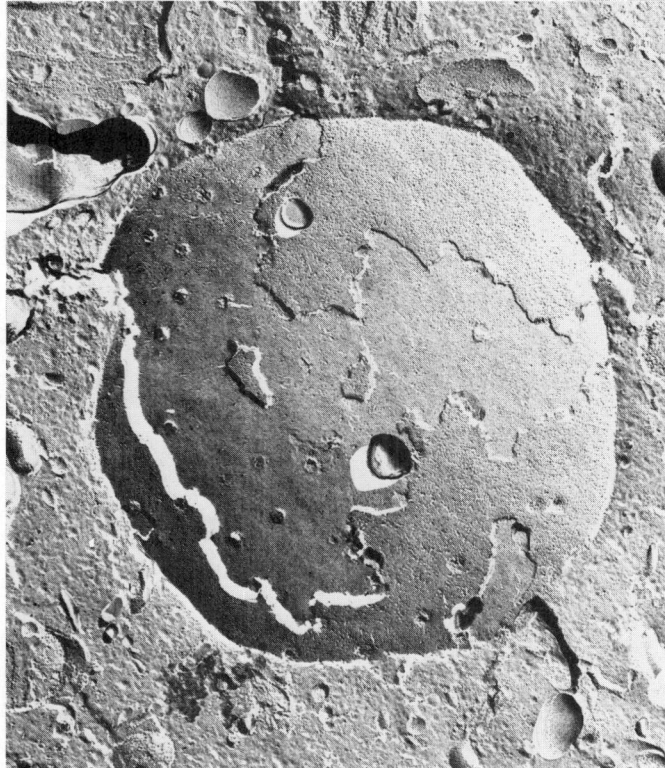

C

1.28C Freeze-fractured cell in which the fracture plane has passed through the nuclear envelope exposing portions of its two membranes and several nuclear pores. Magnification × 20 000.

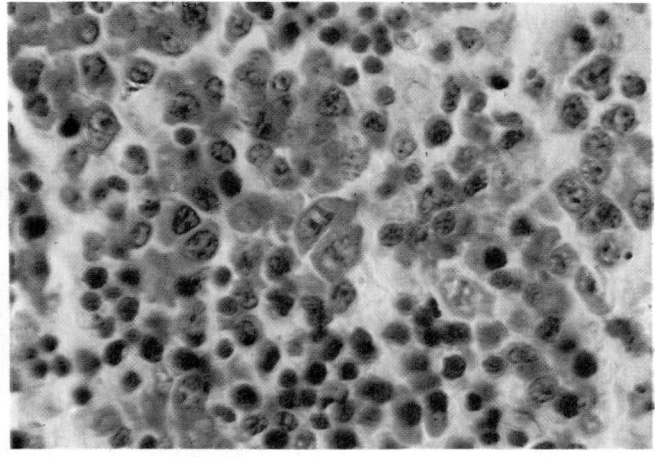

1.29 Photomicrograph showing the configuration of chromatin in 'closed-face', heterochromatic nuclei of lymphocytes (small cells) and 'open-face', euchromatic nuclei of larger reticular cells. The section is stained with methyl green-pyronin stain which also gives the characteristic red colour to the nucleolus and cytoplasm of the larger cells, denoting the presence of much RNA. Magnification × 1000.

highly irregular in shape and take on a variety of appearances (*pleomorphic nuclei*), reflecting disturbances in their chromosomal organization.

In most somatic cells both states of chromatin co-exist; heterochromatin typically lies mainly against the inside of the nuclear envelope, leaving gaps at the nuclear pores; euchromatin occupies the more central region of the nucleus. It must be added that the terms euchromatin and heterochromatin are used in a different sense in the discipline of cytogenetics, to denote specific configurations of chromatin in dividing cells (see p. 42).

During cell division chromosomes shorten very considerably; each cell is then seen to contain a fixed number of chromosomes characteristic for that species. In human somatic cells this number is 46, the *diploid* number; in gametes and their immediate precursors the number is half this (the *haploid* number, p. 42). When fertilization occurs, fusion of two haploid sets, one from each gamete, restores the diploid number. The number of sets is the *ploidy*. If more than two sets are present the cell is *polyploid*.

Each chromosome in a haploid set is unique in size and shape; in a diploid set there are two of each type of chromosome (*homologous* pairs), one each from the ovum and the spermatozoön. However, the chromosomes of one pair are not always identical with each other; these are the *sex chromosomes*, which in humans are distinct from the remaining 22 pairs of chromosomes (*autosomes*). In the female (the *homogametic sex*) the sex chromosomes are identical X *chromosomes*, so giving 44 autosomes + 2 X sex chromosomes. In the male (the *heterogametic sex*) there is one X and an unequal Y *chromosome* giving a diploid complement of 44 autosomes + X and Y sex chromosomes.

During gametogenesis homologous chromosomes separate, one to each gamete. Hence in the male, there are equal numbers of gametes with X or Y sex chromosomes whereas, in the female, all ova contain an X chromosome. At fertilization there is therefore an equal chance of a male (XY) or a female (XX) zygote being formed, although at birth there is a slight predominance of females because of a higher prenatal mortality in male embryos.

CHROMOSOMAL STRUCTURE

Chromosomes are most clearly observed during cell division, when they are shortest and thickest; then each pair of homologous chromosomes has a characteristic common basic structure. Each consists of two parallel and identical *chromatids* (1.30–32), joined together at a narrowed region, the *primary constriction*, within which is a pale-staining *centromere* or *kinetochore*; this is attached to the spindle fibres during cell division.

Each chromatid is divided by the centromere into two *arms*. The centromere may be close to the middle of the chromosome (*metacentric*) or placed somewhat away from the middle

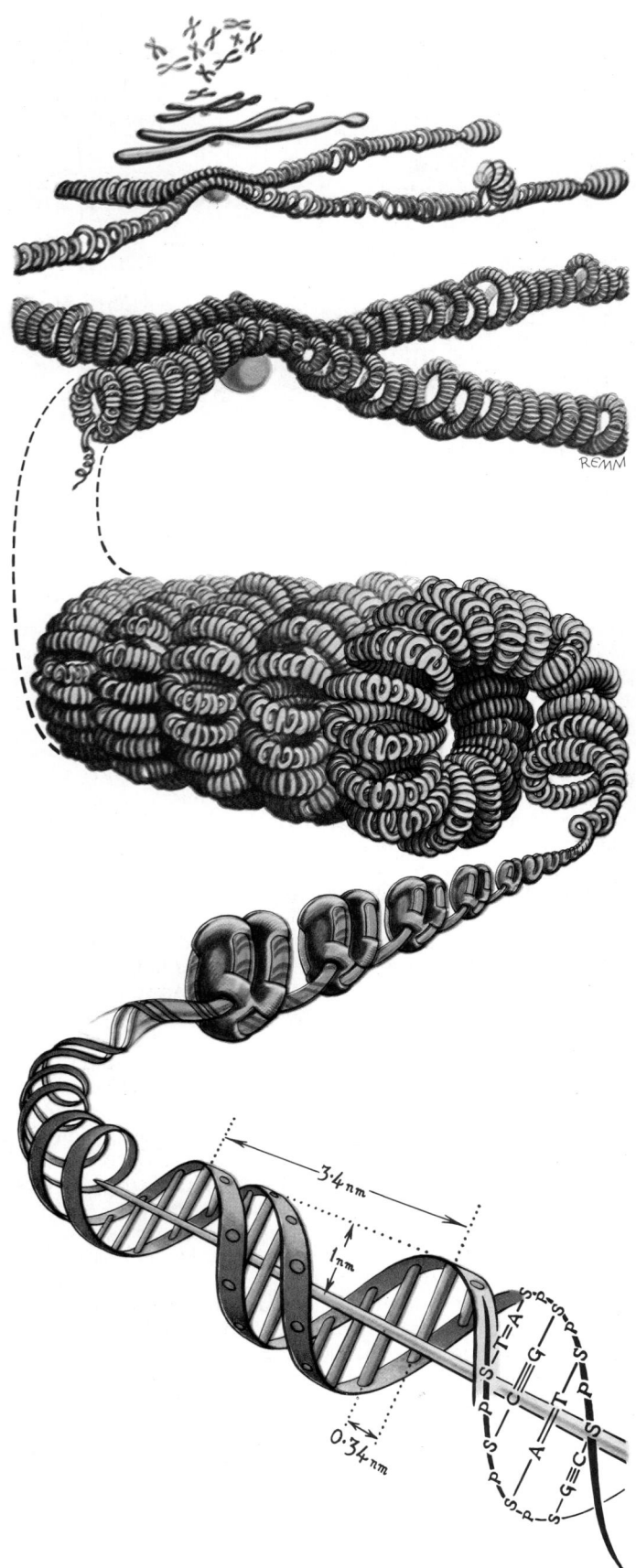

(*submetacentric*) or near one end (*acrocentric*). In some mammals the centromere may be placed at the extreme end (the *telocentric* condition). In certain chromosomes there is a *secondary constriction* near one end of each chromatid, dividing off terminal *satellite bodies*. These constrictions are believed to be associated with the formation of nucleoli (*nucleolar organizing centres*), which are attached to this region in interphase. The classification of human chromosomes on the basis of such characteristics will be considered later (p. 47).

Each complete diploid set of chromosomes contains the cell's hereditary instructions, or *genome*, and each cell in the body, apart from the germ cells, has an identical genetic complement. Chromosomal threads bearing these instructions are *chromonemata* (singular: *chromenema*). Each chromosome bears a linear sequence of *genes* (units of inheritance), each at a characteristic position (*gene locus*), which together determine the inheritable characters of the organism.

During interphase, each chromosome retains its integrity and the sequence of genes is thus preserved. Although this arrangement is highly stable, permanent, inheritable changes (*mutations*) can be caused by external agents such as ionizing radiation and exposure to certain chemicals. When the affected region is restricted these are called *gene mutations*. When whole arms or relatively large segments are involved, they are *chromosomal mutations*. Such changes are some of the principal causes of inheritable *biological variation* which all populations show (but see also meiotic cell division, p. 44). Within members of such a population, homologous chromosomes may determine a number of alternative characters by carrying mutant genes; such alternative genes are termed *alleles*.

In any genome homologous chromosomes may possess, at a particular locus, identical alleles (the *homozygous* condition) or non-identical ones (the *heterozygous* condition). During gametogenesis the homologous chromosomes are segregated into haploid sets and so their alleles are also segregated. The various types of recombination of these which occur at fertilization, and their expression in the resulting organism, are the substance of the basic laws of inheritance, first propounded by Gregor Mendel, forming the basis of classical genetics.

COMPOSITION AND ACTIVITIES OF CHROMOSOMES

Chromosomes are made of filaments of the nucleic acid DNA (**1**.30,33) each associated with various nucleoproteins. Human diploid somatic cell nuclei each contain about 5.6×10^{-12}g of DNA (Rudkin 1967). The identification of DNA as the hereditary

1.30 A possible model for the organization of DNA and protein in chromosomes during early metaphase in mitosis. Each chromosome consists of two chromatids united at the primary constriction where the centromere is also sited. In the exposed region of the DNA molecule, the chemical groups of the 'backbone' of each helix—S (sugar) P(phosphate)—and the bases—A(adenine), T(thymine), C(cytosine) and G(guanine)—are represented. Several nucleosomes are also depicted (see text).

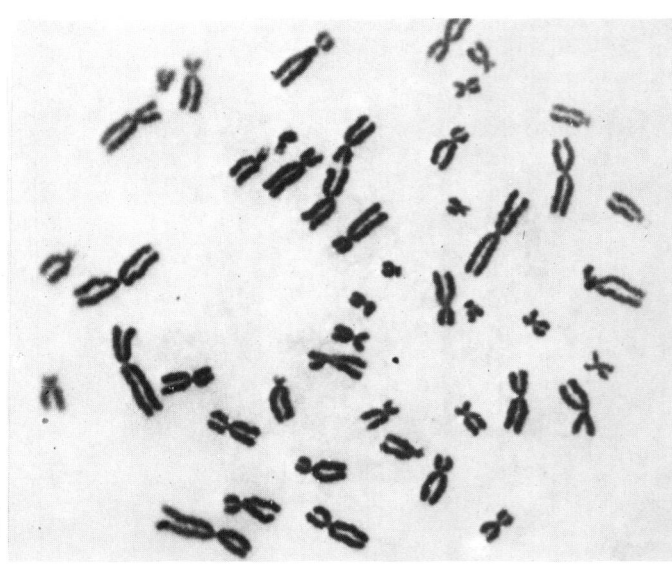

1.31 A spread preparation of metaphase chromosomes of a human fibroblast. Provided by the Paediatric Research Unit, Guy's Hospital Medical School.

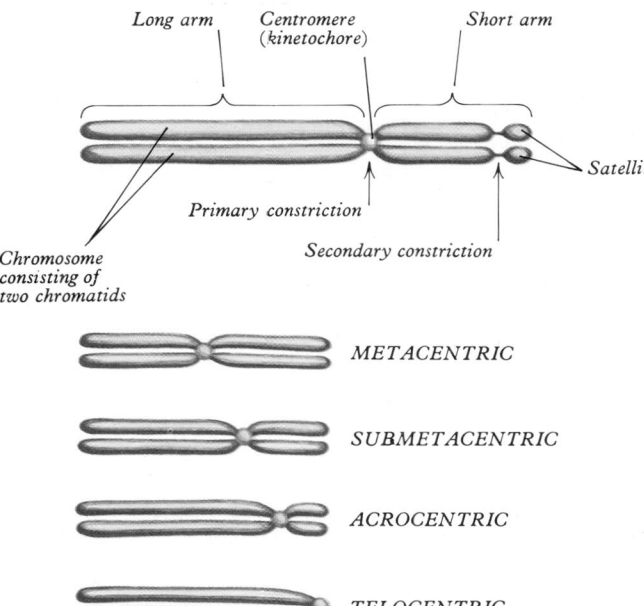

1.32 The major structural features of chromosomes seen in mitotic metaphase; the terminology of the different chromosomal regions is given together with the various major categories of chromosome, classified according to differences in the position of the primary constriction.

material has a long history, dating from the characterization of nucleic acids by Miescher in the 1860s, the recognition of the role of nuclei in fertilization and the cytological events in mitotic and meiotic divisions, and their correlation with the laws propounded by Mendel in 1865 and rediscovered in 1900. Further milestones were the analysis of the chemistry and distribution of nucleic acids in cells including bacteria, the investigation of viruses, which are largely composed of either RNA or DNA, and the resolution of DNA structure by X-ray diffraction crystallography. The latter led to the 'double helix' model of DNA (Watson & Crick 1953 a,b), which opened a new era of molecular biology. Since that time the nature of the genetic store and its mode of action have been studied more intensively, perhaps, than any other biological theme and, in the recent development of the science of molecular genetics, promises to transform many areas of medical research and clinical practice.

DNA and RNA are long molecules which bear coded information directing the activities of each cell and therefore the structure and metabolism of the whole organism (1.33). DNA is the central molecule, since its 'code' is handed on unaltered from parent cell to all of its descendants in the individual organism, and, by its germ cells, to subsequent generations. RNA is the transient intermediary, translating coded information from chromosomal DNA into the structure of enzymes and other proteins which determine the morphology and chemistry of the cell. This flow of information has been termed 'the central dogma' of molecular biology, i.e. DNA→RNA→protein. At each step chemical energy is expended and the mediation of previously existing enzymes is necessary.

All three of these compounds (DNA, RNA and proteins) have one characteristic in common: their molecules are all linear sequences of subunits. Proteins are long polymers of 20 types of amino acid arranged in different order. They vary in length, each arrangement determining the precise nature of a particular protein, e.g. the way in which its chain folds into more complex shapes, as in the macromolecules of globular proteins, and its catalytic activity in enzymes. Likewise, DNA is a long linear molecule, with a direct correspondence between the arrangement of its chemical groups and those of the protein for which it codes. Transfer of information from DNA to RNA by synthesis of the latter alongside the former is transcription and the final decoding of the RNA message into the amino acid sequence in a protein is translation. The other activity, the exact duplication of the DNA code to provide each derived cell with identical gene sequences, is replication.

THE CELL CYCLE

Synthetic activity in a cell is not uniformly continuous: e.g. appreciable protein or DNA synthesis does not occur during mitosis. Likewise in interphase, DNA synthesis is restricted mainly to a short period of the cell cycle, whereas protein synthesis occurs throughout it(Mitchison 1971). Accordingly, interphase can be divided into a G_1 (first gap), S (synthetic) and G_2 (second gap) stage (1.34). DNA replication occurs in the S phase and manufacture of proteins for mitosis takes place in G_2. The duration of the cycle and its stages varies greatly between cell types. G_1 may last from a few hours to many years or it may be absent, as in the rapidly dividing blastomeres of the early embryo. The S stage usually lasts about seven hours and the G_2 stage up to five hours. Mitosis in human cells is complete in approximately one hour.

Some cells, once formed, do not enter the S phase, being unable to divide and are then said to be in G_0, e.g. neurons.

DNA STRUCTURE

In the Watson–Crick model of DNA structure, now amply confirmed, DNA consists of two parallel molecular chains running in opposite directions as a double helix, 2.5 nm in diameter, repeating every 3.4 nm (see 1.30,33). Each chain is composed of a 'backbone' of pentose sugar (deoxyribose) groups, linked by phosphoric acid bridges in such a way that they confer directionality. Each sugar group bears a nitrogen-carbon ring base, directed sideways towards a corresponding group on the opposite helix, and is linked to it by hydrogen bonds. There are four bases: adenine and guanine (purines), and cytosine and thymine (pyrimidines). These are dissimilar in shape; the space between the two helices and the type of bonding available dictate that adenine on one side always pairs with thymine on the other, and guanine with cytosine. The rigid complementary nature of this pairing is the basis of the genetic code, written in terms of three-letter 'words', each made up of three bases in a row and corresponding to a single amino acid in the final protein. Multiples of such triplets encode a single polypeptide and such a sequence corresponds to a structural gene, or cistron, of classic genetics. Some triplets do not code for amino acids but initiate or end the formation of polypeptides, like punctuation in a sentence. Often several functionally related proteins, e.g. an interdependent system of enzymes, are encoded by a series of genes which may be controlled as a unit, a supergene or polycistron.

ORGANIZATION OF CHROMOSOMAL DNA

In all eukaryotes the DNA of each chromosome forms a single continuous duplex thread (i.e. consisting of one long double helical molecule) before the S phase, when DNA is duplicated and thereafter, or until cell division, two double helices, lying parallel to each other (vide supra). Each nucleus contains a total DNA length of about 170 cm and the longest human chromosomes are each as much as 7 cm if fully extended (Du Praw 1968).

Chromatin

In the nucleus, proteins are bound intimately to the DNA threads, in a 1:1 ratio by mass, to form a molecular chromatin. The proteins carry out various vital functions, including regulation of genetic expression and stabilizing the structure of the whole chromosome. The proteins include two major classes, histones and non-histone proteins. Histones carry net positive charges (i.e. they are basic) and form the great majority of nuclear proteins. In structure they vary little in animals and plants, being highly conserved, a fact pointing to a fundamental role in living processes (although they are absent in bacteria). Most of the histones are aggregated in spheroidal particles, each about 10 nm across, strung like beads along the DNA duplex strand, which is coiled around each particle to form a complex, the nucleosome, then passes to the next particle. The whole thread is then a chromonema. The intervening length of DNA (linker DNA) is considered to be part of the nucleosome complex (Olins et al 1976, Kornberg & Klug 1981). Nucleosome proteins are grouped

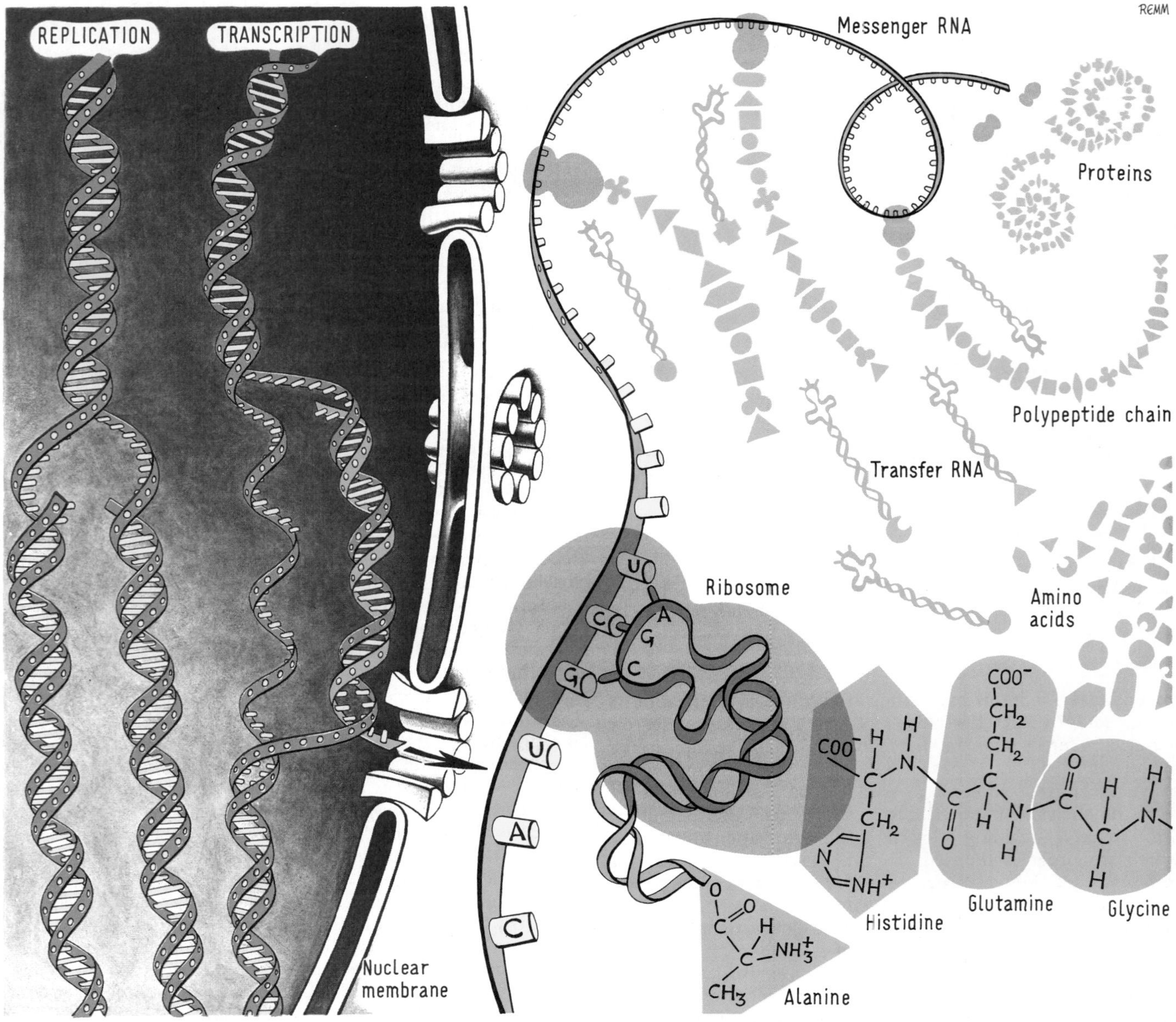

REPLICATION   TRANSCRIPTION

Messenger RNA

REMM

Proteins

Polypeptide chain

Transfer RNA

Amino acids

Ribosome

COO⁻
CH₂
CH₂
COO⁻
CH₂
C
N
C
N
C
Histidine

Glutamine

Glycine

Alanine

Nuclear membrane

1.33 The processes occurring during DNA replication and protein synthesis. The processes of DNA replication (on left) and transcription (middle, right) are indicated; abbreviations are the same as those used in **1.30**, with the addition of U = uridine.

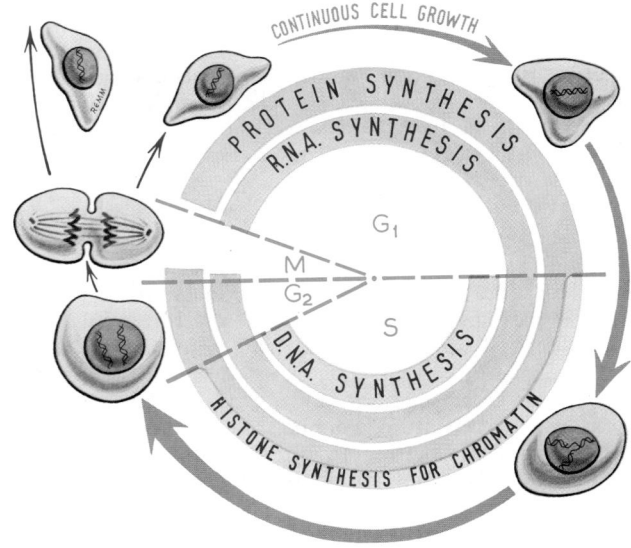

1.34 The major events occurring during the cell cycle of a growing and mitotically active cell to show the major stages: G_1 or 'first gap' phase; S or 'synthetic' phase; G_2 or 'second gap' phase; M or 'mitotic' phase.

eightfold, comprising pairs of four proteins, H2A, H2B, H3 and H4. Another histone, H1, acts as a link between adjacent nucleosomes and is responsible for aggregating these structures in more complex assemblies which e.g. generate the 30 nm diameter coiled form of chromatin. It has also been suggested that the nucleosomes may be able to move along the DNA or to become detached when the DNA is perhaps active in transcription. Non-histone proteins are present in much smaller amounts. They include important regulatory molecules, enzymes for replicating and transcribing and repairing DNA and many others needed for appropriate gene activity. Some of these (e.g. RNA polymerase) can be detected structurally as 5 nm granules attached to DNA threads.

Folding of Chromatin within Chromosomes

Much of the DNA exists in a highly folded state, a condition most extreme in heads of mature spermatozoa (p. 119). The narrowest chromatin threads are 10 nm wide (the diameter of the nucleosomes) but 30 nm threads are also abundant, consisting of spiral clusters of nucleosomes (the 'solenoid' configuration). Further folding and looping may also occur, depending on functional status; chromatin which is active in transcription (during protein synthesis) is most extended (euchromatin), while inactive chromatin is highly folded (heterochromatin). DNA synthesis during S phase (p. 37) is completed earlier in euchromatin than in heterochromatin. Chromatin is most folded during cell division (at metaphase) and, in active cells, least folded in interphase. However, some DNA is permanently folded and heterochromatic; this includes DNA which at some phase of early development was active but later becomes inactive (*facultative heterochromatin*, see Brown 1966), e.g. the inactive X chromosome in human females. Other chromosomal regions which appear permanently heterochromatic (*constitutive heterochromatin*) lie at the primary constrictions of some, and perhaps all, chromosomes and in the satellites of acrocentric chromosomes (p. 37) (Hsu & Arrighi 1971) amongst other sites. The significance of this chromatin is not clear although it may have some intrinsic role in maintaining chromosome stability. The coiled structure of chromatin can also vary locally with different sequences of nucleotides which alter the geometry of the duplex. Such regions may have special properties in causing specific types of folding or in promoting (or preventing) attachment of molecules involved in gene regulation.

Analyses of DNA show that a small proportion consists of highly *repetitive sequences of bases* and is therefore separable from the bulk of nuclear DNA by centrifugation; this is *satellite DNA*. Its functions are unknown, but generally constitutive heterochromatin seems to be identical with satellite DNA (Jones 1970). Satellite DNA is thought to be non-coding; it can be detected with special stains in mitotic chromosomes at their centromeric and terminal (telomeric) regions, including satellites (a coincidence in terminology) where these are present (p. 47). Amounts of repetitive DNA vary widely in different, even closely related, species so there is no obvious significance in its presence.

TRANSCRIPTION

In transcription the two helical strands of DNA separate locally to expose a particular cistron on one of them, the coding *strand* (the other being complementary). Upon this template, a single strand of RNA is constructed with the aid of an RNA polymerase and ATP. RNA is similar to DNA in composition but it is single stranded, contains ribose instead of deoxyribose and has the base uracil in place of thymine. The same type of base pairing now takes place so that a chain of RNA is formed alongside the DNA coding strand. The DNA triplets are thus transcribed into complementary triplets or *codons* of RNA. When the complete cistron has been transcribed, the codon-bearing RNA separates and moves to the cytoplasm, being therefore known as *messenger* RNA (mRNA). This eventually attaches to the small subunit of a ribosome which, with other chemical factors, acts as a complex enzyme. Another class of nucleic acid called *transfer* RNA (tRNA) recognizes specific cytoplasmic amino acids and matches them with the triplet codons of the mRNA. There exist a large number of specific tRNA molecules in the cytoplasm, each able to bind at the same time to a single mRNA codon and a corresponding amino acid. Recognition of codons again occurs by complementary base-pairing and the relevant part of the tRNA molecule involved is called the tRNA *anticodon*. Recognition of the specific amino acid is less well understood; it may not involve the amino acid directly but, rather, an intermediary specific enzyme involved in its linkage to the growing polypeptide chain.

Each ribosome moves along the mRNA strand, 'reading' its code, so that a corresponding sequence of amino acids is formed, each linked to the next by a peptide bond; this process is termed *translation*. Many ribosomes may be reading the mRNA strand at a given time, so forming polysome rosettes or helices. When decoding is complete, the ribosome detaches itself and both mRNA and protein are released.

Introns and Exons

During the last few years it has also been shown that the relatively simple mechanism described here has fundamental modifications in most cases so far studied. In the synthesis of many proteins, messenger RNA formed by DNA transcription is altered in the nucleus, being initially quite long (*heterogeneous RNA*, or *hnRNA*), and is later cut by enzymes, present in *small ribonucleoprotein particles* abundant in the nucleus, into shorter lengths, some of which are then spliced together, the others being destroyed. Thus the final sequence of bases in the final mRNA represents only a fraction of the total DNA used in encoding it.

DNA sequences which are ultimately translated into proteins are termed *exons* and those portions not expressed are *introns*. This mechanism allows a single length of DNA, consisting, in effect, of overlapping genes, to be used to synthesize more than one type of protein, by bringing together various sequences of triplet codons in their corresponding mRNAs (Darnell 1985). This arrangement is economical and also has the evolutionary potential of generating novel proteins after mutation (a rare event of course).

REGULATION OF GENE EXPRESSION

Since each nucleus contains the whole complement of genetic information for an individual, the expression of this must be subject to complex interlocking control mechanisms for any directed action to occur. These permit only a small fraction of the total to be transcribed at any moment, depending on demands made on the cell and its role in the economy of the total organism.

Jacob & Monod (1961) proposed a twofold mechanism of gene control on the basis of bacterial experiments, involving a hierarchial system of three gene types. In the first type of control, *regulation by induction*, each group of structural genes is closely linked to a controlling *operator gene*, together forming an *operon*. The operator gene is normally repressed by a protein, the *repressor*, produced by a *regulator gene* (not necessarily on the same chromosome) and so prevents RNA polymerase from transcribing the structural genes by binding tightly to a particular gene region, the operator. A metabolite of the cell which induces protein transcription does so by combining with the repressor, which then cannot repress the operator gene, so allowing transcription. Alternatively, a metabolite may inhibit protein synthesis—*regulation by repression*. This second scheme is similar to the first, except that the repressor can only inhibit the operator gene after combining with the inducer substance, which thus prevents transcription by the structural genes.

In higher organisms, gene regulation, although basically similar to that of prokaryotes, has important differences. In multicellular animals the cells of the body, while carrying identical genes, express them in a wide variety of different ways, so that in any cell most genes are inactive and only those few which are relevant to general maintenance and to its specialist functions are expressed. DNA appears to be maintained in the repressed state by its close association with *histones*, the major basic nucleoproteins of chromatin (von Holt 1985, and also p. 37). There are only a few classes of histones and these are very similar in a wide range of animals; they are hence unlikely to be related to any particular pattern of genes, unlike repressor substances in bacteria (vide supra, and p. 108).

Regulation of gene expression in eukaryotes is still not well understood but a number of different mechanisms appear to operate in this process. Genes in eukaryotes can be activated by *non-histone proteins* which interact with specific genes. Many such proteins are known and these vary widely in different cells and species. They probably operate by first disengaging the histones from their close association with a particular gene sequence and then attach to DNA to promote the co-attachment of a molecule of RNA polymerase which proceeds to transcribe the gene or group of genes. Such activities have been studied extensively in the interaction of hormones with specific 'target' cells. Sometimes, as in the action of progesterone (a steroid of low molecular weight) on the uterine epithelium, the hormone enters the cytoplasm and combines with specific receptor proteins; then, in this complexed form, it moves to the nucleus to combine further with an appropriate region of DNA through its associated histones, resulting in its transcription and the formation of proteins for cell growth (King & Mainwaring 1974, O'Malley & Schrader 1976). In other examples a hormone too large in molecular size to traverse the plasma membrane (e.g. adrenocorticotrophic hormone, ACTH) may instead stimulate an enzyme, adenyl cyclase, present in the plasma membrane, to release an intracellular 'second order messenger' which, by a complex train of events, eventually activates the relevant genes as do other hormones (Stein et al 1975). The sensitivity of a target cell to its stimulatory hormone presumably depends on its ability to synthesize appropriate receptor molecules and the relative susceptibility of its gene sequences to derepression. However, various other factors may affect transcription by DNA, including local chemical modification of its related histones, its degree of coiling and other variations in local geometry. Specific DNA sequences exist which are known to play important roles in the regulation of gene expression, including initiation of RNA transcription (e.g. the sequence of bases: TATA, where T is thymidine and A is adenine), frequency of transcription and enhancement of the rate of this process. The establishment of such patterns of gene regulation in a particular cell and its coordination are poorly understood at present, although this process, *cellular differentiation*, is known to be dependent on many factors both inside and outside the differentiating cell. In the Britten–Davidson model (1969), it is envisaged that small numbers of repetitive nucleotide sequences synthesize 'activator' molecules of some kind, when suitably stimulated, and that these switch on groups of many related genes ('batteries'). Although there is some evidence that repetitive DNA may in certain cases act in this way, this model remains unproved (see Davidson et al 1983, Klug & Cummings 1986).

CELLULAR DIFFERENTIATION

As development of an organism proceeds, its cells pass through a series of changes in gene expression, reflected in alterations of cell structure and behaviour (see also pp. 107–9). Initially, all cells possess rather similar properties but, as embryogenesis gathers momentum, they begin to diversify, first separating into broad categories (e.g. the principal germ layers, etc.) and then into narrower categories (tissue and subtypes of tissues) until finally they mature into the '*end cells*' of their particular lineage (Gurdon 1973). Some cells which are capable of giving rise to others throughout life *(stem cells)* never proceed to the ultimate point of this progression and retain some embryonic characteristics (p. 678) but in all cases there is a *sequential pattern of gene expression* which changes and limits the cell to a particular specialized range of activities. We can detect such changes as alterations in cell structure and biochemical properties, particularly in the types of proteins which are synthesized. At the level of the gene, differentiation of this kind must be based on a change in the pattern of repression and activation of the DNA sequences transcribing the specific proteins of that stage of development; in the lifetime of a particular cell lineage, many such changes will take place (see, e.g. the development of haemal cells, p. 678). It appears that the selection of a pattern of gene activity occurs some while before its expression in protein synthesis. Thus, a cell may be *committed* to a particular line of specialization without manifesting its commitment until later; once 'switched' in this way, cells are not usually able to revert to an earlier stage of development, so that an irreversible repression of some gene sequences must have occurred. *Stem cells* may remain permanently at a level of partial differentiation, although some of their offspring will be committed to full differentiation. In general, as the degree of differentiation progresses, cell division becomes less frequent (e.g. erythroblasts, p. 679), although some structurally specialized cells such as Schwann cells (p. 891) and certain glandular cells, when suitably stimulated, may undergo repeated mitotic divisions.

What constitutes the appropriate stimulus for a cell line to differentiate at successive stages in development is one of the most fascinating of biological problems and one which also lies at the heart of many pathological transformations including carcinogenesis. In the embryo, differentiation depends upon a wide variety of circumstances (see pp. 107–9), although in most cases chemical interactions between cells or between different regions of the same cell are thought to be of primary significance. The initiation of a particular pathway of cellular development may, however, also depend upon complex, competitive interactions between cells or upon the position which a cell occupies within a cell group (see discussion on 'positional information' on pp. 110, 111). In addition, differentiation may also depend upon some temporal sequence such as the number of previous cell divisions (pp. 104, 108). Even in mature tissues in which cell turnover occurs, similar mechanisms appear to ensure the final differentiation to a functional end cell. In some cases this is linked to the presence of a physiological stimulus as, e.g. in the case of the 'B'

lymphocytes which respond to exposure to an appropriate antigen by differentiating into plasma cells which secrete a neutralizing antibody. In other cases, particularly where a cell is part of a highly organized system, more subtle mechanisms must be present.

REPLICATION

Replication requires the separation of the two intertwined DNA chains (1.33) and the manufacture of complementary chains alongside each of the originals which act as templates for the exact copying of the sequence of bases. Thus, in each of the two resulting double helices, one chain of the original double helix is conserved, so that replication is *semi-conservative* (Taylor 1958). During replication DNA is synthesized by a DNA polymerase from nucleotide precursors, so that if radioactive thymidine, a precursor of the specific DNA base thymine, is added to a cell, it is incorporated into the chromosome and can be detected autoradiographically. Using this method it has been shown that DNA synthesis during the S phase occurs simultaneously along many short stretches of DNA, beginning at a series of *initiation points* where DNA polymerase molecules can attach and then move along the chromonema to synthesize new DNA chains. Synthesis spreads in both directions from the initiation points, forming *replication forks* which can also be visualized by autoradiography (Cairns 1966, Taylor & Hozier 1976), until the whole chromosome is replicated.

Each type of chromosome has its own pattern of DNA synthesis; permanently heterochromatic regions replicate later than the rest and the inactive heterochromatic X chromosome in females replicates later than the other chromosomes (German 1964).

While most of the chromosomal DNA synthesis is confined to the S phase, there is also a constant low level synthesis of DNA for the *repair* of minor chromosomal damage resulting, e.g. from ultraviolet radiation of epidermal cells. Various enzymes are needed for repair, including *endonucleases* which recognize damaged DNA strands and begin to cut them, *exonucleases* which excise damaged regions, a *DNA polymerase* which synthesizes new DNA alongside the unaffected chain and a *ligase* to join the new portion to cut ends of the old. These enzymes can be separated from bacterial cells, and are now widely employed to insert genes isolated from animal or human cells for their further analysis or to make specific proteins (i.e. for genetic engineering). In diseases where DNA repair mechanisms are faulty, the body may be very sensitive to mutagens, e.g. in *xeroderma pigmentosum*, where skin is easily damaged by ultraviolet light.

Limited DNA synthesis also occurs in the first meiotic prophase in gametogenesis (Stern & Hotta 1969) and this may be concerned with exchange of genetic material between homologous chromosomes (p. 44).

It is now known that a single DNA thread lies within a single chromatid. After DNA synthesis the two chromonemata separate into two distinct chromatids so that each cell now possesses the tetraploid amount of DNA (i.e. each *chromosome* contains two DNA strands and there are *two* of each type of chromosome in each somatic cell). At mitotic division (vide infra) the content in each cell is reduced to the diploid amount, to be doubled again in the following interphase.

It is interesting that wide variations of DNA content occur in the cells of different species, even though closely related, suggesting that segments of DNA may be repeated many times. It is possible that in some cases many copies of each gene exist so that, when activated, large-scale transcription can be initiated rapidly, as in the case of the nucleolar genes, of which there are multiple copies. However it appears that much of the repetitive DNA is not active in transcription and its function remains an enigma.

NUCLEOLI

Each nucleus possesses a number of spheroidal nucleoli, dense bodies which take up basic dyes strongly; the human number varies (one or several may occur) and collectively they are considered to consist of five pairs of nucleolar fragments related to

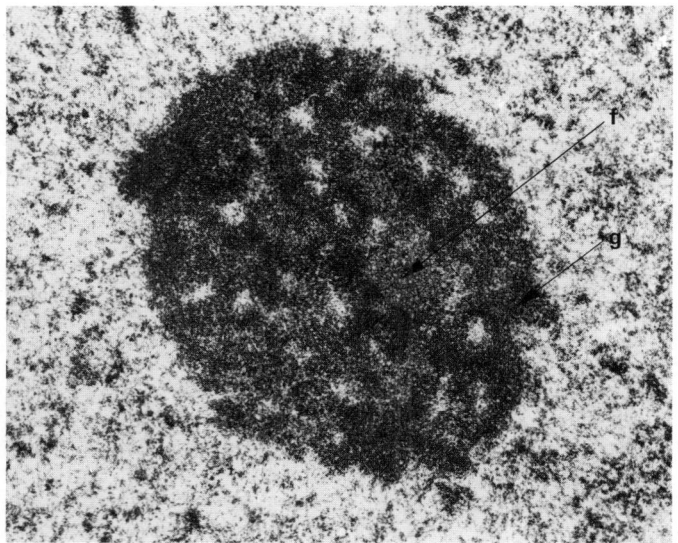

1.35 Electron micrograph of a nucleolus from a highly active cell, showing the pars filamentosa (f) enclosed in the darker pars granulosa (g). Magnification × 25 000.

particular chromosomes (vide infra). Nucleoli vary in size up to about 4 μm in diameter (1.5,6,35). Each is a heterogeneous mass of fine granules and fibrils set in a less dense matrix, the whole agglomeration often containing cavities or channels continuous with the surrounding nucleoplasm, so that with the light microscope it may have the appearance of a folded thread (nucleolonema). By electron microscopy, four structural elements can be detected: an inner region of dense filaments (pars filamentosa), an outer granular zone (pars granulosa), both embedded in an apparently structureless matrix, the pars amorpha; while centrally lies a series of somewhat less dense elongate DNA-rich regions, constituting the pars chromosoma. These parts can be correlated with the synthetic pathway of ribosomal RNA much of which is transcribed on the nucleolar organizing genes of certain chromosomes, in the pars chromosoma. Newly synthesized ribosomal RNA is exceptionally long, forming the basis of the fibrils in the pars filamentosa, but is later broken into fragments and combined with ribosomal proteins to form ribosome subunits. These accumulate in the pars granulosa, from which point they pass to the cytoplasm through the nuclear pores, eventually assembling into mature ribosomes.

In humans *nucleolar organizing centres* are present on those chromosomes with *secondary constrictions* (acrocentric chromosomes, with satellites; chromosomes numbers 13–15,21,22) and within a single nucleolus several such centres may be active. Each consists of many tandemly repeated DNA sequences which code for ribosomal RNA and which can therefore generate large quantities of this material very rapidly—a necessity if enough ribosomes are to be formed prior to active synthesis of other proteins. In some lower vertebrates there are large numbers of such organizing regions and in their oöcytes hundreds of nucleoli are formed; regions of such chromosomes synthesizing nucleolar RNA form lateral loops giving the whole chromosome a bristly appearance (*lampbrush chromosomes*) and these, under the electron microscope, are seen to be festooned with RNA strands in the process of being transcribed (Mott & Callan 1975).

Cells with a high rate of protein synthesis, and thus a large demand for ribosomes, have large nucleoli, e.g. large neurons with extensive axons and dendrites, active embryonic cells and the cells of malignant neoplasms.

During cell division, when protein synthesis is suspended, the nucleoli disintegrate and their chromosomal portions are drawn back into the main mass of their parent chromosomes; at the end of division nucleoli are rapidly reformed.

OTHER NUCLEAR INCLUSIONS

Other less prominent structures present in all nuclei (Wischnitzer 1973) including *granules*, some of which are associated with chromatin margins (*perichromatin granules*) about 40 nm across, and others lying between the chromatin masses (*interchromatin granules*) about 20 nm across; the significance of these is unknown. *Perichromatin fibres* also occur at the margins of the chromatin. In addition, many other nuclear inclusions have been described as constant features of some cells or occurring occasionally, including fibrillar material, membranous structures and larger granules. Many though not all of these have been shown to be nonpathogenic viruses which abound in the cells of the human body.

In the prophase of the first meiotic division, long, fibrillar *synaptonemal complexes* connect adjacent chromosomes (p. 44). Other structures which have been reported in different types of cell include membranous vesicles and lamellae, bundles of filaments, crystals and further varieties of granule. Some of these may represent symbiotic C viruses, common inhabitants of mammalian nuclei. The nature of the other inclusions is obscure. The nucleoplasm also contains numerous enzymes, nucleotide precursors and proteins, only identifiable by cytochemical or biochemical means; these are, of course, involved in nuclear metabolism.

REPRODUCTION OF CELLS

During prenatal development most cells undergo repeated division as the body grows in size and complexity. As cells mature they differentiate in structure and function and may finally lose the ability to divide, as do neurons, or may persist permanently as stem cells capable of dividing throughout life, as in the haemopoietic tissue of bone marrow.

Rates of cell division vary considerably in different tissues (see e.g. Potten et al 1979, Lewin 1980, Lloyd et al 1982); in many epithelia subject to mechanical stress, the replacement of damaged cells by division of stem cells may be rapid, as in crypts between intestinal villi; it may also vary according to demand, as in healing of wounded skin, where cell proliferation rises to a peak and then returns to the normal replacement level. The rate of cell division is therefore tightly coupled to the demand for growth and replacement; where this coupling is faulty, tissues either fail to grow, or replace their cells, or else they overgrow, giving rise to neoplasms.

The mechanism of normal control is poorly understood (see the reviews by Yanishevsky & Stein 1981, Alberts et al 1983, Brachet

1985). During early embryonic life, control appears to be local, involving the diffusion of metabolites from one cell group to another, so stimulating or inhibiting division of cells. At later stages, endocrine control is also involved. Several hormones affect cell division rates either generally (e.g. somatotrophic hormone) or locally (e.g. progesterone). Some hormones may act on division by affecting general metabolic rates (thyroid hormones) or protein synthesis (some corticosteroids and somatotrophic hormone). At least some of these effects may be mediated by the intracellular release of cyclic adenosine monophosphate (cyclic AMP) which inhibits synthesis of new DNA during the S phase.

In adult tissues, local control of cell division is an important factor in wound healing. A class of compounds termed *chalones*, which inhibit cell division, has been isolated from normal tissue cells. A possible model for their action is that normal cells produce chalones which inhibit division of the cells in their locality; when some cells are damaged or destroyed, the concentration of chalones drops allowing cell division to occur until the normal levels are restored, when division is again inhibited.

The Mechanism of Cell Division

Two distinct events occur in cell division: division of the nucleus (*karyokinesis*) and of the cytoplasm (*cytokinesis*); they are usually, but not always, coupled. For reviews of these processes, consult Prescott (1976), Lewin (1980), Inoué (1981), Zimmerman & Forer (1981), Alberts et al (1983) and Brachet (1985).

Nuclear division can occur in three ways. In the first, termed *amitotic* or *direct division*, nuclear material is distributed at random to the resultant cells. (This process, once thought to be common, is in fact restricted to pathological conditions.) In the other two modes of nuclear division, complex chromosomal manoeuvres take place (*indirect division*, comprising *mitosis* and *meiosis*).

Mitosis (**1**.36–8) occurs in most somatic cells and results in the distribution of identical copies of the parent cell's genome to the resulting cells. In *meiosis*, occurring in the divisions immediately before final production of gametes, the number of chromosomes is halved to the haploid number, so that at fertilization the diploid number is restored; some exchange of genetic material also occurs between homologous chromosomes, a *reassortment* of genes (see e.g. Klug & Cummings 1983). This leads to further genetic variability in a population, the essence of evolution (see also p. 44).

Mitosis and meiosis are alike in many respects, differing chiefly in chromosomal behaviour during early stages of cell division. In meiosis two divisions occur in quick succession; the first of them is unlike mitosis (*meiosis I, heterotypical* division) but the second more like mitosis (*meiosis II, homotypical* division).

MITOSIS

As already stated, new DNA is synthesized during the S phase of interphase, so that in diploid cells the *amount* of DNA has doubled by the onset of mitosis to the tetraploid value, although the chromosome *number* is still diploid. During mitosis, this amount is halved between the two resulting cells, so that DNA quantity and chromosome number are now diploid in both. The nuclear changes which achieve this distribution can be divided into four phases: prophase, metaphase, anaphase and telophase (**1**.36,37).

Prophase

The strands of chromatin, highly extended during interphase, begin to shorten, thicken and resolve themselves into recognizable chromosomes, each made up of two chromatids joined at the centromere. Outside the nucleus centrioles begin to separate, moving to opposite poles of the cell; parallel microtubules are synthesized between them to create the *central spindle* and others radiate to form the *astral rays*, collectively termed asters. The spindle and two asters together constitute the *achromatic figure* or *diaster* (amphiaster).

As prophase proceeds, the nucleoli disappear and finally the nuclear envelope suddenly disintegrates into small vesicles to release the chromosomes. This event marks the end of prophase.

Metaphase

As the nuclear envelope disappears, the spindle microtubules invade the central region of the cell and the chromosomes move towards the equator of the spindle (*prometaphase*). Once they have arrived at this imaginary plane (the *metaphase* or *equatorial plate*), the chromosomes attach by their centromeres to spindle microtubules and are so arranged in a star-like ring when viewed from either pole of the cell.

Anaphase

The centromere in metaphase is a double structure; its halves now separate both carrying an attached chromatid, so that the original chromosome has, in effect, split lengthwise into two new chromosomes. These move apart, one towards each pole. It is not yet known how this is achieved, but spindle microtubules are necessary for movement to occur. The two halves of each centromere may become attached to separate categories of microtubules (McIntosh et al 1979) which would then move towards their respective poles, possibly by sliding interaction with other stationary microtubules, pulling the chromosomes. Alternatively the progressive disintegration and reassembly of microtubules might generate similar movements by addition of tubulin at the equator and removal at the poles (Inoué & Sato 1967).

Telophase

At the end of anaphase the chromosomes are grouped at each end of the cell, both aggregations being diploid in number. The chromosomes now re-extend and the nuclear envelope reappears, beginning as membranous vesicles at the ends of the chromosomes. Nucleoli also reappear.

Meanwhile, cytoplasmic division, which usually begins in early anaphase, is progressing. This process is accompanied at late metaphase by cytoplasmic movements involving the equal distribution of mitochondria and other organelles around the cell periphery. In anaphase an infolding of the cell equator begins and deepens as the *cleavage furrow* (Schroeder 1973). Small vacuoles form in the cytoplasm along the plane of cleavage and eventually the furrow divides the cell into two. Where the constriction meets the remains of the spindle, a dense region of cytoplasm, the *midbody*, is visible but eventually the new cells separate, each with its derived nucleus. The spindle remnant now disintegrates. While the cleavage furrow is active, a peripheral band of actin and myosin appears in the constricting zone and the contraction of this is probably responsible for furrow formation. During telophase the filaments of the cleavage furrow contract down on the remaining spindle microtubules to form the dense mid-body which finally disappears.

Failure of *disjunction* and '*lagging*' of chromatids may sometimes occur, so that paired chromatids pass to the same pole. Of the two new cells, one will have more and the other less chromosomes than the diploid number. Exposure to ionizing radiation promotes such events and may, by chromosomal damage, inhibit mitosis altogether. A typical symptom of *radiation sickness* is the failure of epithelia to replace lost cells, with consequent ulceration of the skin and mucous membranes.

Mitosis can also be disrupted by several chemical agents, particularly colchicine and its derivatives colcemide, podophyllin and podophyllotoxin, and by viniblastine. These compounds inhibit or reverse spindle microtubule formation, so that mitosis is arrested in metaphase. This is useful in cytogenetic studies, since chromosomes are most easily examined in the metaphase condition, and also forms the rationale for attacking multiplying tumour cells with many types of cytotoxic drugs.

MEIOSIS

During meiosis (**1**.39–42) there are two cell divisions; in the interphase prior to the first division DNA is replicated in the usual manner, resulting in the tetraploid amount of DNA, the chromosomal number being diploid. During *meiosis I* the DNA is reduced to the diploid *amount* in each resultant cell, although the chromosome *number* is halved to the haploid value; in *meiosis II*, the DNA in each new cell formed is reduced to the haploid amount, the chromosome number remaining haploid.

Prophase I

This long and complex phase differs considerably from mitotic prophase and is customarily divided into five substages: the leptotene, zygotene, pachytene and diplotene stages and diakinesis (for details see Comings & Okada 1972, Whitehouse 1973, Moens 1974, Brachet 1985).

Leptotene stage. Chromosomes appear as individual threads attached at one end to the nuclear envelope and show characteristic beads (*chromomeres*) throughout their length.

Zygotene stage. Chromosomes have come together side by side in homologous pairs (a process which may already have occurred as early as telophase of the previous mitotic division). The homologous chromosomes pair point for point progressively,

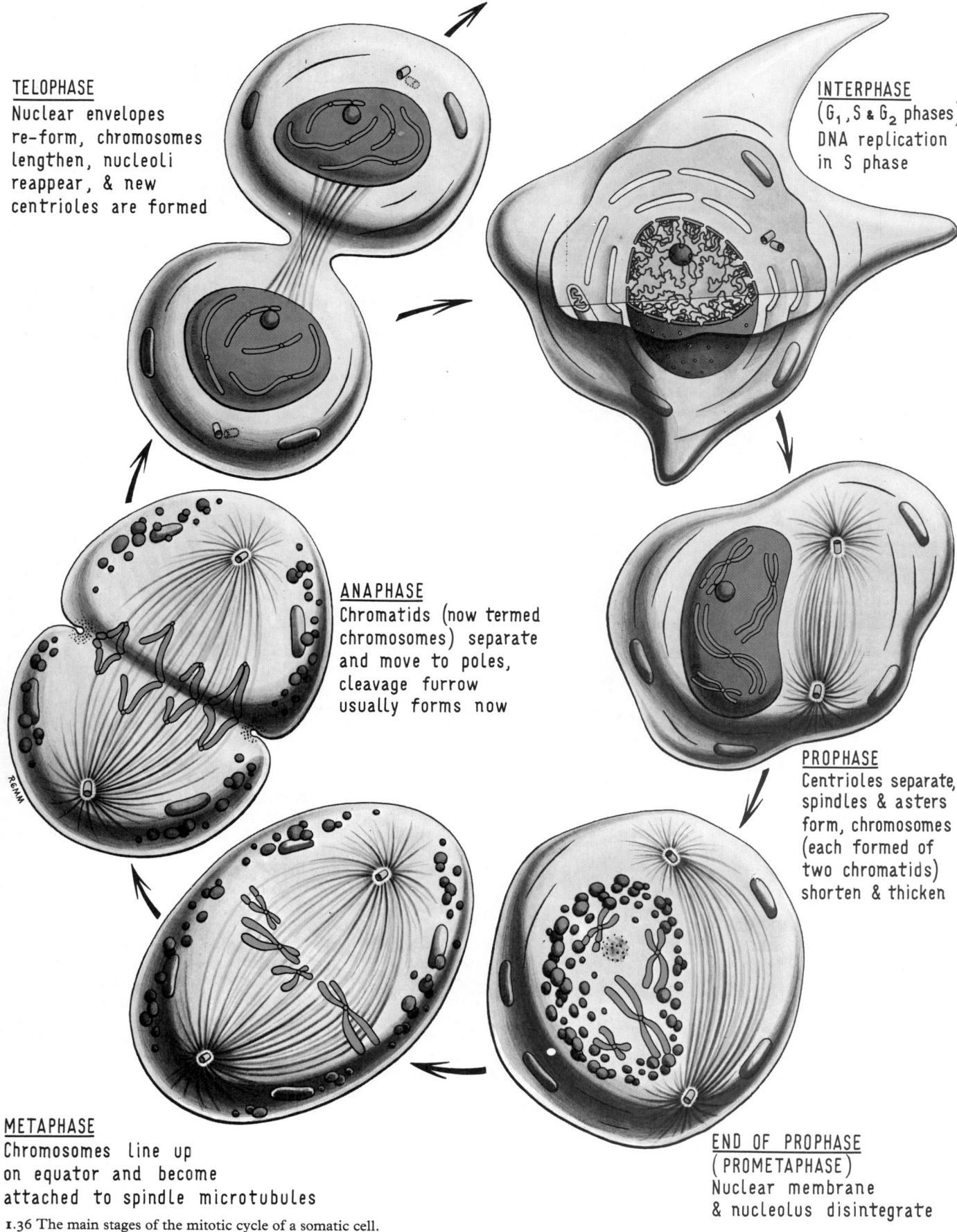

TELOPHASE
Nuclear envelopes
re-form, chromosomes
lengthen, nucleoli
reappear, & new
centrioles are formed

INTERPHASE
(G$_1$, S & G$_2$ phases)
DNA replication
in S phase

ANAPHASE
Chromatids (now termed
chromosomes) separate
and move to poles,
cleavage furrow
usually forms now

PROPHASE
Centrioles separate,
spindles & asters
form, chromosomes
(each formed of
two chromatids)
shorten & thicken

METAPHASE
Chromosomes line up
on equator and become
attached to spindle microtubules

END OF PROPHASE
(PROMETAPHASE)
Nuclear membrane
& nucleolus disintegrate

1.36 The main stages of the mitotic cycle of a somatic cell.

beginning at the attachment point to the nuclear envelope (Moens 1974), so that corresponding regions lie in contact. This process is *synapsis, conjugation* or *pairing*. Each pair is now a *bivalent*. In the case of the unequal X and Y sex chromosomes, which during zygotene and pachytene are sequestered in a secluded zone of the nucleus, the *sex vesicle* (Solari & Tres 1967), only limited *pairing segments* are homologous and these pair end to end (1.40); the remaining parts are *differential segments*. By electron microscopy,   43

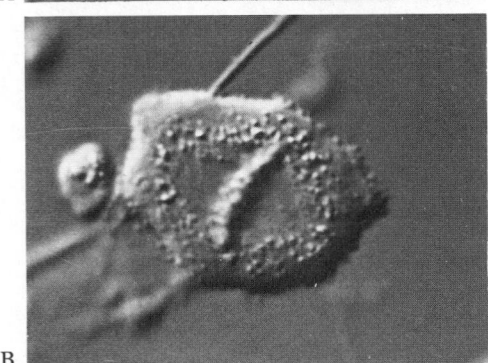

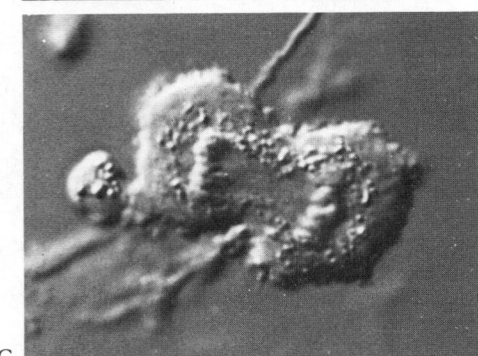

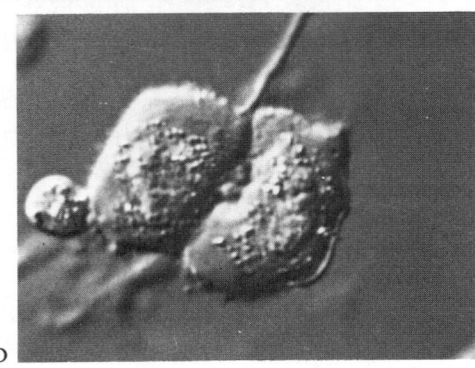

1.37 A series of micrographs taken at different stages of mitosis of a human fibroblast in tissue culture; compare with **1.36**. Nomarski interference microscopy. A. Early metaphase; B. metaphase; C. anaphase; D. telophase. Provided by the Paediatric Research Unit, Guy's Hospital Medical School. Magnification × 1500.

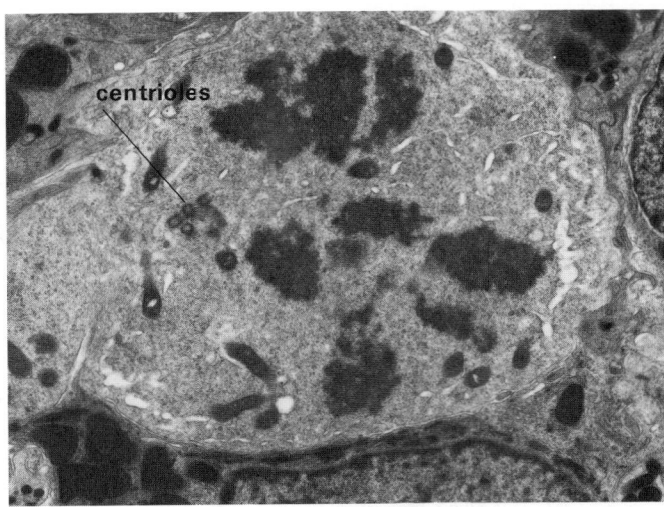

1.38 Electron micrograph of an epithelial cell in mitotic metaphase. The chromosomes are seen as blocks of dense material, and are sectioned approximately along the plane of the equator. Magnification × 10 000.

homologous chromosomes appear held together by a highly structured fibrillar band, the *synaptonemal complex* (**1.42**), which occupies the space (about 100 nm wide) between them (von Wettstein et al 1984, see also p. 44).

Pachytene stage. Spiralized shortening and thickening of each chromosome progresses and its two chromatids, joined at the centromere, become visible. Each bivalent pair therefore consists of four chromatids, forming a *tetrad*. Two chromatids, one from each bivalent chromosome, partially coil round each other, and during this stage it is probable that exchange of DNA (*crossing over* or *decussation*) occurs by breaking and rejoining, perhaps facilitated by the synaptonemal complex.

Within the latter structure, regions of DNA exchange are marked by the presence of dense proteinaceous masses about 90 nm in diameter (*recombination nodules*) in which the processes of cutting and rejoining the adjacent DNA strands may occur.

Diplotene stage. Homologous pairs, now much shortened, separate except where crossing over has occurred (*chiasmata*). Sometimes chiasmata appear to move towards the ends of the chromatids (*terminalization*); at least one chiasma forms between each homologous pair and up to five have been observed (even up to 10 in some species). In human ovaries, primary oöcytes become diplotene by the fifth month in utero and each remains in this stage until the period prior to ovulation (some for decades, and even up to 50 years).

Diakinesis. Remaining chiasmata finally resolve and the chromosomes, still as bivalents, become even shorter and thicker; they disperse, as bivalents, to lie against the nuclear envelope.

During prophase the nucleoli have disappeared and the spindle and asters have formed as in mitosis. At the end of prophase the nuclear envelope disappears and bivalent chromosomes move towards the equatorial plate (*prometaphase*).

Decussation. Exchange of genes between homologous chromosomes involves extraordinary precision whereby the DNA, at exactly corresponding positions on both, is severed and rejoined to the DNA of its corresponding partner. How this is achieved is not certain, although it has been proposed that the DNA double chains of the two exchanging chromatid segments are exchanged one at a time, perhaps with some remodelling of redundant DNA chains by partial dissolution and resynthesis in the correct position (Holliday 1964, Whitehouse & Hastings 1965). Such exchanges may occur during late zygotene and early pachytene (Whitehouse 1973), when there is some DNA synthesis; it is likely that the synaptonemal complex is an important mediator of genetic exchange.

Once segments of chromatid are exchanged and the chromosomes begin to separate, regions where crossing over has taken place become visible as *chiasmata*; these therefore represent a *past event* and chiasma formation is *not* synonymous with genetic recombination. After diakinesis the paired chromatids of a given chromosome adhere to each other more strongly than those of homologous chromosomes, even when exchange has occurred, producing the curious configurations which chromosomes adopt in metaphase of meiosis I (**1.39**).

Metaphase I

This resembles mitotic metaphase except that the bodies attaching to the spindle microtubules are bivalents, not single chromosomes. These become arranged so that the homologous pairs lie parallel to the equatorial plate with one member on either side.

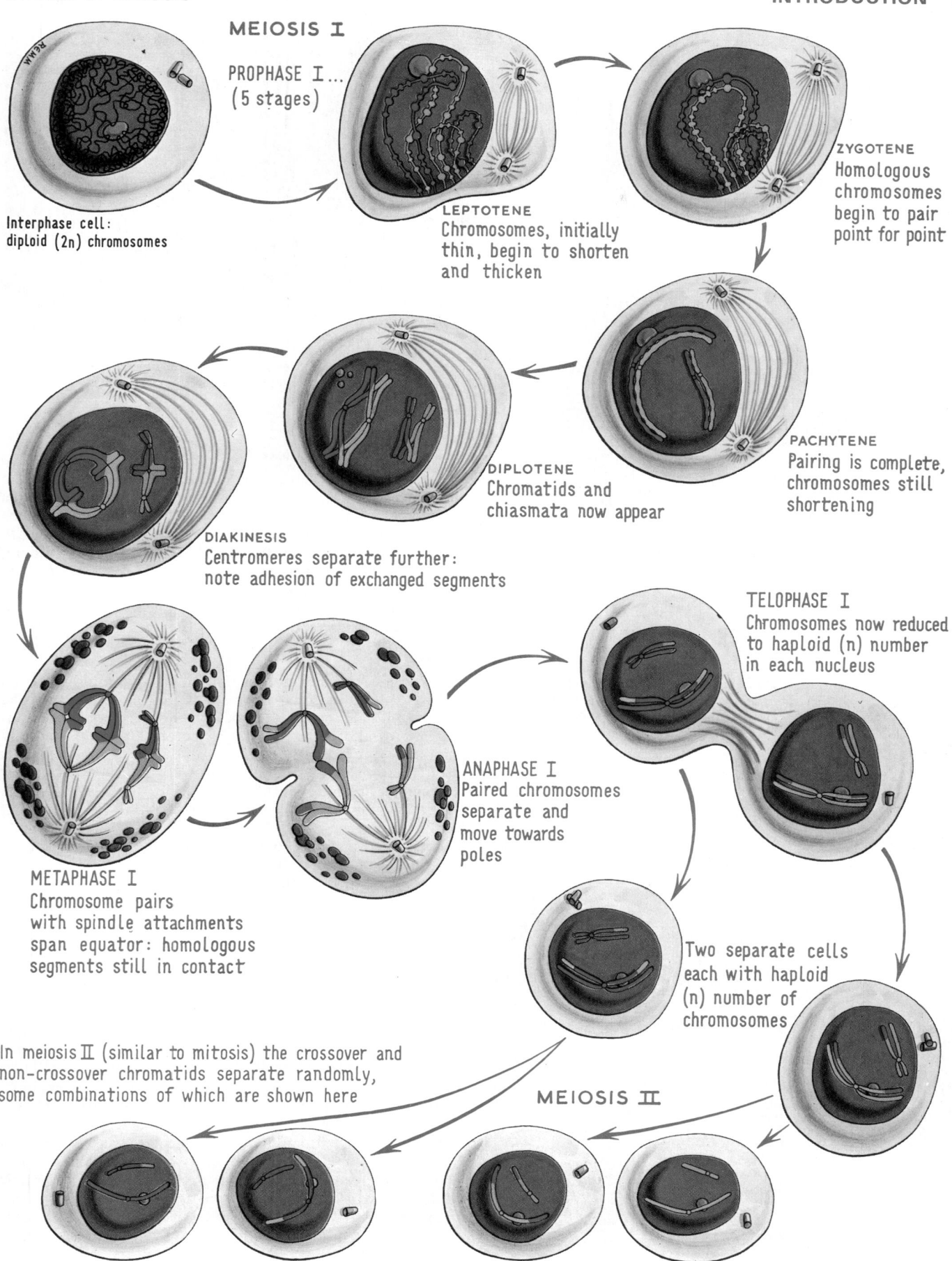

MEIOSIS I

PROPHASE I...
(5 stages)

Interphase cell:
diploid (2n) chromosomes

LEPTOTENE
Chromosomes, initially
thin, begin to shorten
and thicken

ZYGOTENE
Homologous
chromosomes
begin to pair
point for point

PACHYTENE
Pairing is complete,
chromosomes still
shortening

DIPLOTENE
Chromatids and
chiasmata now appear

DIAKINESIS
Centromeres separate further:
note adhesion of exchanged segments

TELOPHASE I
Chromosomes now reduced
to haploid (n) number
in each nucleus

ANAPHASE I
Paired chromosomes
separate and
move towards
poles

METAPHASE I
Chromosome pairs
with spindle attachments
span equator: homologous
segments still in contact

Two separate cells
each with haploid
(n) number of
chromosomes

In meiosis II (similar to mitosis) the crossover and
non-crossover chromatids separate randomly,
some combinations of which are shown here

MEIOSIS II

I.39 Chief stages in the meiotic cycle (male). For clarity only four
chromosomes out of the total 46 are shown. Please note that the detailed
movements of chromosomes during different stages of Meiosis II are not
shown in this figure.

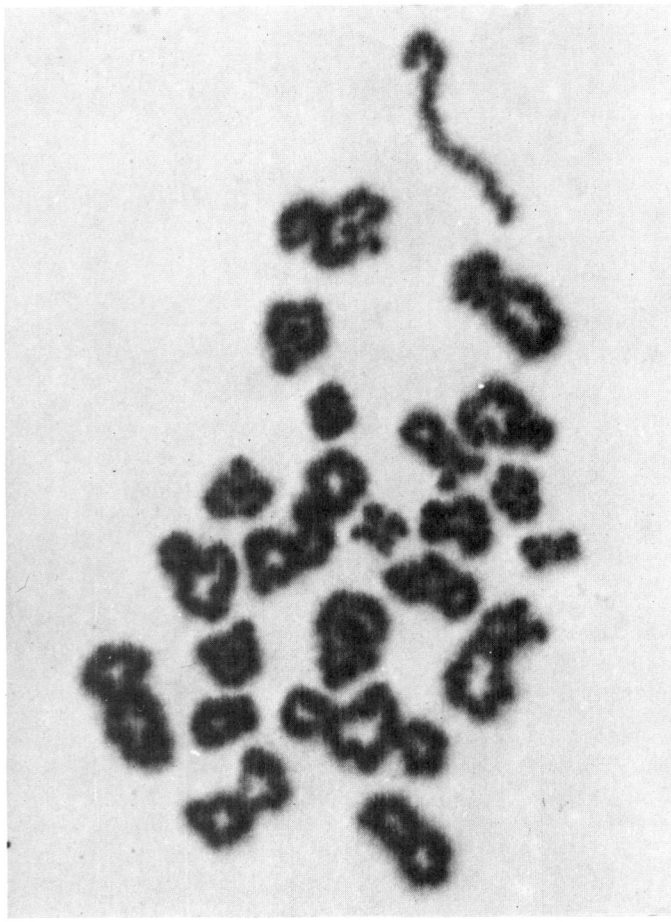

1.40 Photomicrograph of chromosomes from a human spermatocyte in late prophase (diakinesis) of the first meiotic division showing bivalents beginning to separate. Note the paired sex chromosomes which are joined end to end. Provided by the Paediatric Research Unit, Guy's Hospital Medical School.

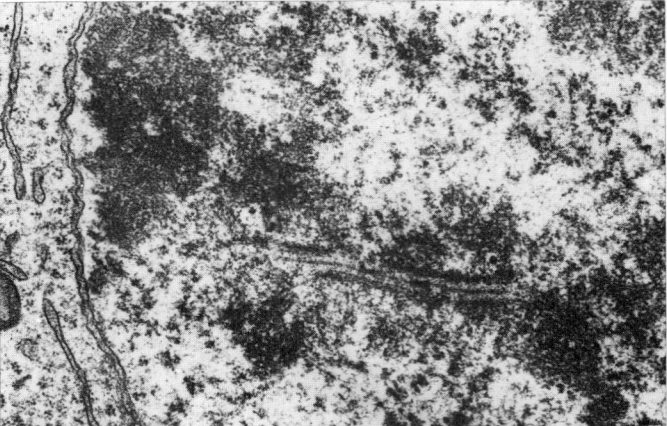

1.42 Detail of a synaptonemal complex in a cell similar to that shown in 1.41. Note the masses of heterochromatin and the fibrillar core of the synaptonemal complex. Magnification × 20 000.

Anaphase and Telophase I

These also occur as in mitosis, except that in anaphase the centromeres do not split; thus, instead of the paired chromatids separating to move towards the poles, whole homologous chromosomes made up of two joined chromatids depart to opposite poles. Since positioning of bivalent pairs is random, assortment of maternal and paternal chromosomes in each telophase nucleus is also random.

During meiosis I, cytoplasmic division occurs as in mitosis to produce two new cells.

Meiosis II

This commences after only a short interval during which no DNA synthesis occurs. This second division is more like mitosis, with separation of chromatids during anaphase; but in contrast to mitosis, the separating chromatids are genetically dissimilar. Cytoplasmic division also occurs and thus *four* cells result from meiosis I and II.

If cytoplasmic division fails during meiosis I, gametes with the diploid number of chromosomes result and thus give, at fertilization, a triploid zygote.

Other abnormalities, some of them viable, are produced by *non-disjunction* of chromosomes, or by the *lagging* of individual chromosomes, during anaphase. These will be discussed later (p. 48). There is much evidence that, as maternal age increases, so does the frequency of such abnormalities.

CLASSIFICATION OF HUMAN CHROMOSOMES

Since a number of genetic abnormalities can be directly related to the chromosomal pattern, the characterization or *karyotyping* of chromosomes is of considerable diagnostic importance. Their identifying features are most easily seen during metaphase, although recently prophase chromosomes have also been used for more detailed analysis (vide infra).

As a source of chromosomes, lymphocytes can be separated from blood samples or cells taken from other sources. For diagnosis of fetal chromosome patterns, samples of amniotic fluid, containing fetal cells, can be aspirated from the uterus (amniocentesis) or, as is now often preferred, a small piece of a chorionic villus (p. 145) removed from the placenta. Whatever their origin, the cells are cultured in vitro and stimulated to divide by treatment with phytohaemagglutinin or other chemicals. Mitosis is later interrupted at metaphase with spindle inhibitors (see p. 42) and chromosomes dispersed by swelling the cells in hypotonic solutions followed by air drying, or squashing, to rupture the cells and flatten the chromosomes. These can then be stained to allow the identification of individual chromosomes by their

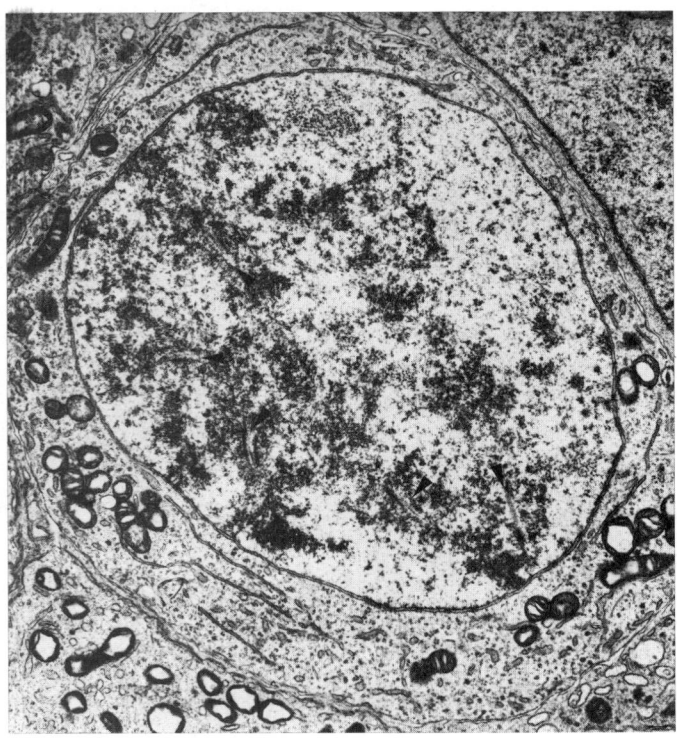

1.41 Electron micrograph of a primary spermatocyte (mouse) showing the nucleus in early prophase of the first meiotic division. Five synaptonemal complexes are visible (arrowheads). Magnification × 10 000.

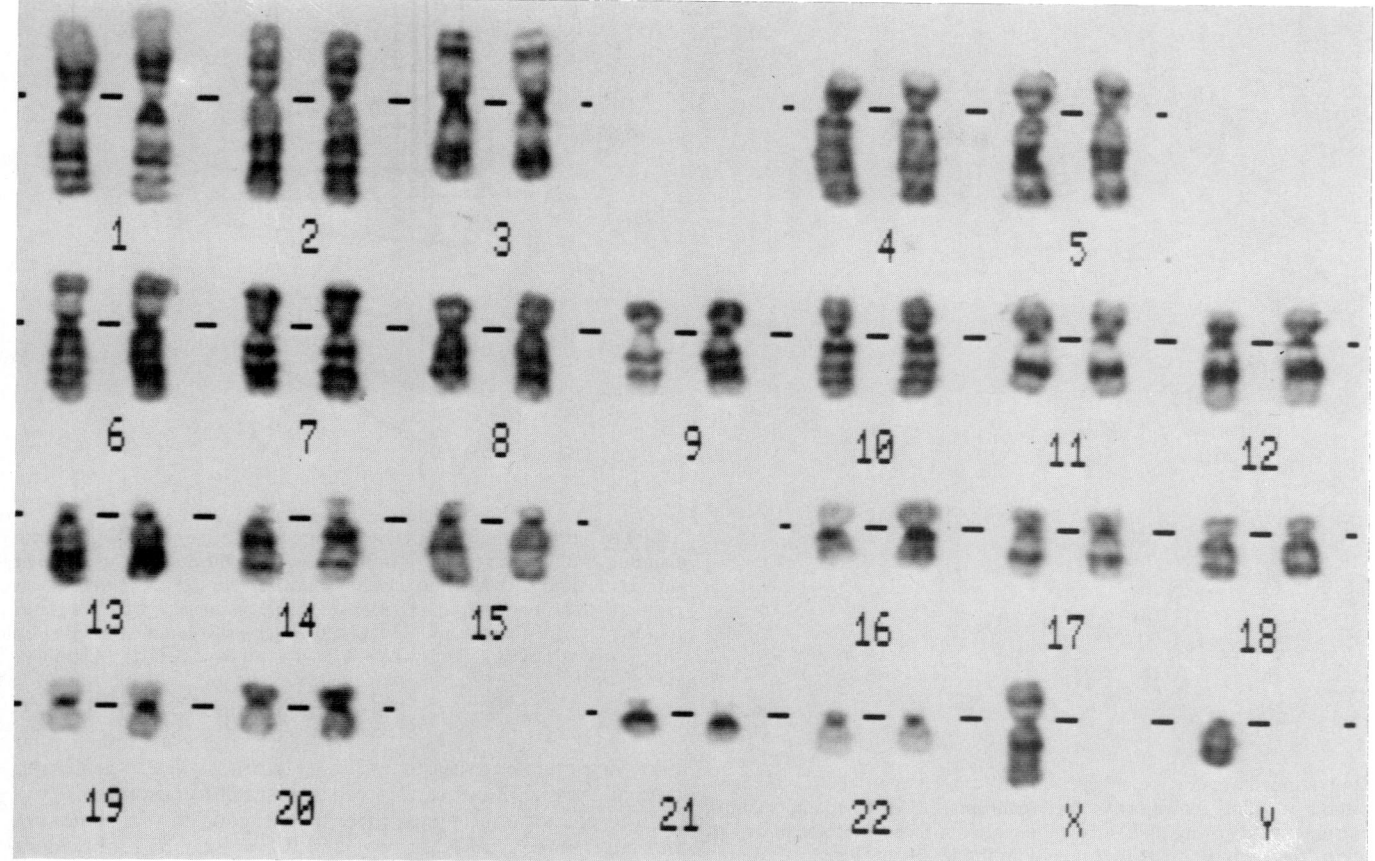

1.43A Human karyotype, stained with the Giemsa banding method. This micrograph was prepared with the assistance of computer analysis to assemble and enhance the appearance of the chromosomal banding.

Provided by the Paediatric Research Unit, Guy's Hospital Medical School. Magnification × 2500.

size, shape and distribution of stain. When large numbers of cells have to be examined for clinical purposes, automatic scanning light microscopy with computer control and analysis can greatly facilitate the identification of chromosomes and any abnormalities (1.43–45). Methods include: *general techniques* to show the obvious landmarks, e.g. length of arms, position of primary and secondary constrictions; and *banding techniques* to demonstrate differential staining patterns, characteristic for each chromosome type. Since the introduction of the first banding methods (Caspersson et al 1968), several others have been developed, such as fluorescence staining with quinacrine mustard and related compounds (Q bands) (1.46), Giemsa staining after treatment with alkali (G bands), 'reverse-Giemsa' staining in which the light and dark areas are reversed (R bands) and the staining of constitutive heterochromatin with silver salts (C-banding); T-banding methods stain the ends (telomeres) of chromosomes and other techniques can be used to demonstrate nucleolar organizing centres. In *high resolution banding*, chromosomes arrested in prophase are used, even greater detail being discernible in these more elongated forms (Yunis et al 1978).

By such methods, chromosomes can be classified by length, positions of primary and secondary constrictions, presence of satellites, number and positions of the transverse bands and the distribution of constitutive heterochromatin. A system of numbering autosomal pairs of chromosomes in order of decreasing size, from 1 to 22, has been adopted according to the conventions agreed in London (1960) and Chicago (1964). Before the introduction of banding, some chromosomes were difficult to distinguish and thus were grouped together in seven categories (A–G) according to size and detailed shape and updated more recently (ISCN 1985). However, the later methods allow unambiguous identification and a classification agreed first in Paris

(1972) is now generally accepted. A summary of the major classes of chromosomes is given in the table below.

Major Categories of Human Chromosomes

Group 1 – 3 (A) Large metacentric chromosomes
4 – 5 (B) Large submetacentric chromosomes
6 – 12 + X (C) Metacentrics of medium size
13 – 15 (D) Medium-sized acrocentrics with satellites (and nucleolar organizing centres)
16 – 18 (E) Shorter metacentrics (No. 16) or submetacentrics (Nos. 17 and 18)
19 – 20 (F) Shortest metacentrics
21 - 22 + Y (G) Short acrocentrics, 21 and 22 with satellites (and nucleolar organizing centres), Y without satellites.

These advances have also much improved the recognition of *abnormal chromosome patterns*, since the origins of extra chromosomes, or the fragments, can often be identified. Another means of identification exploits variations in the *time* of DNA synthesis during the S phase in different chromosomes, which is shown by pulse-labelling the replicating chromosomes with ^{3}H-thymidine at specific times during the S phase, followed by autoradiography of chromosomes during the ensuing metaphase (Gianelli 1970). They can also be pulse-labelled with an analogue of thymidine, bromo-deoxyuridine (BrDU), which stains differently from normal DNA (Pain et al 1976). For example, the inactive X chromosome replicates particularly late (p. 48) and extra X chromosomes can easily be detected by such techniques.

SEX CHROMATIN

The interphase nuclei of female mammals differ from those of males in that a heterochromatic body is usually present on the

47

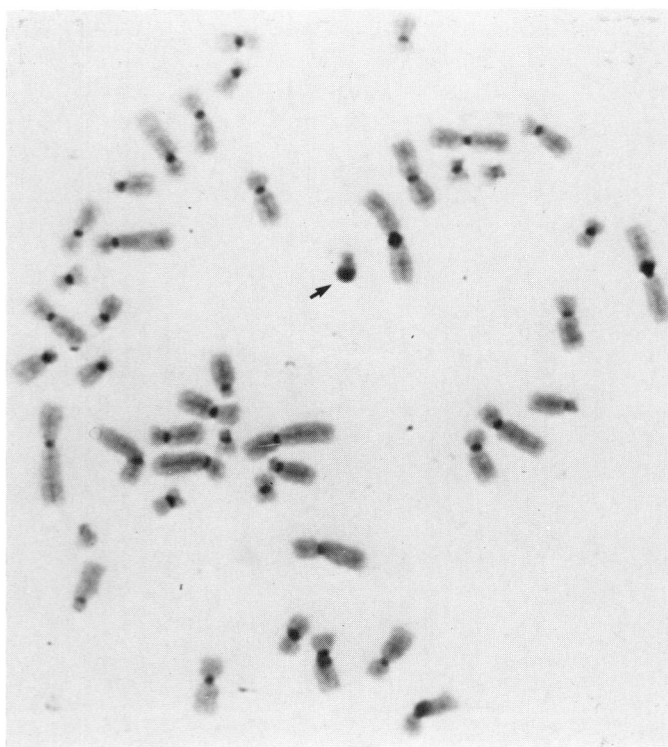

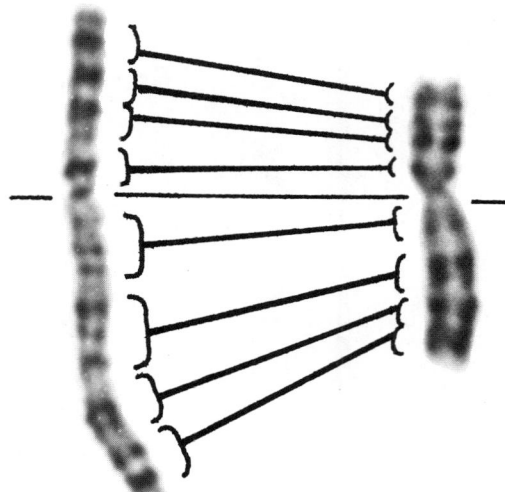

1.44 Two chromosomes (both chromosome number 2), showing Giemsa banding at different periods of mitosis. On the right the cell was arrested during metaphase when the chromosomes are shortest. On the left an earlier (late prophase–early metaphase) chromosome shows that the later bands are formed by the condensation of several sub-bands. Provided by the Paediatric Research Unit, Guy's Hospital Medical School. Magnification × 3700.

1.43B Human (male) metaphase spread stained with the C-banding method to show constitutive heterochromatin (dark spots) mainly centromeric in position. The arrow indicates the Y chromosome in which heterochromatin is situated terminally in the long arm. Provided by the Paediatric Research Unit, Guy's Hospital Medical School. Magnification × 1300.

inner face of the nuclear envelope in females (1.47). This structure, the *sex chromatin* or *Barr body* (Barr & Bertram 1949), may protrude from the surface of polymorphic nuclei (as a 'drumstick'). It is now regarded as one of the X chromosomes which is synthetically inert and thus heterochromatic, its euchromatic partner carrying out all necessary functions in RNA synthesis (Lyon 1962). The inert X chromosome also replicates later during interphase than its homologue.

CHROMOSOME ABNORMALITIES

As already stated, aberrations in mitotic and meiotic cell divisions cause chromosomal abnormalities. Extreme examples such as formation of more than the doublet (diploid) set of chromosomes to form *polyploid genomes* (triploid, tetraploid, octoploid, etc.) occur in some organisms, but are lethal in higher forms.

Failure of equal segregation of chromosomes in anaphase can result in more or less than the normal compliment. If this takes

place in gametogenesis, the resulting offspring will also possess an altered number of either autosomes or sex chromosomes.

Where, instead of a pair, three homologous chromosomes are present (*trisomy*), the additional chromosome can be readily identified as may unpaired chromosomes (*monosomy*). Trisomy of chromosome 21 (one of the smallest in the set) is found in Down's Syndrome ('mongolism'), giving a total of 47 chromosomes in the karyotype. However, the same syndrome is also found in individuals with 46 chromosomes, as a result of *translocation*, in which a segment of the extra chromosome 21 is attached to another chromosome (usually number 18). Other abnormalities of chromosome structure include *fragile sites*, seen as gaps in the staining patterns of some chromosomes. Many such examples have been described but, although inherited, they do not appear to result in any defect in their carriers. However, when X chromosomes are affected, mental retardation and various skeletal or other abnormalities may occur (*fragile X Syndrome* or *Martin-Bell Syndrome*; see Sutherland 1984, Pembrey et al 1986).

Anomalies of the sex chromosome complement are also well known. In the trisomy of Klinefelter's syndrome the cell contains an additional X chromosome (44 autosomes + XXY); these cases are male in character (phenotypically male) but the testes remain small and spermatogenesis fails to occur. On the other hand, in Turner's syndrome there is only a single X chromosome (44 autosomes + XO); the subject is feminine but the ovaries are rudimentary and, although the embryonic ovary may appear normal with plentiful oögonia, oögenesis is not completed. Higher

HUMAN KARYOTYPE CHROMOSOME NUMBER 2

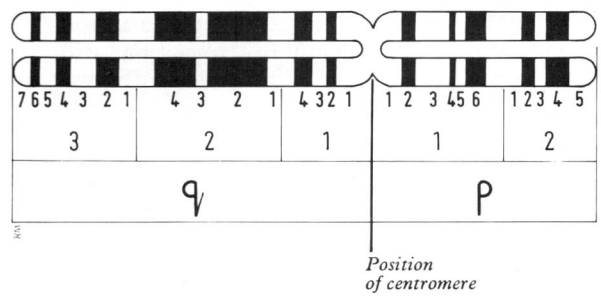

HUMAN KARYOTYPE CHROMOSOME NUMBER 15

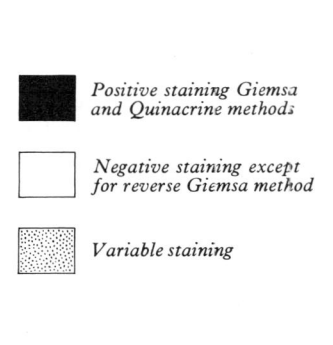

■ *Positive staining Giemsa and Quinacrine methods*

□ *Negative staining except for reverse Giemsa method*

▨ *Variable staining*

1.45 Nomenclature of the banding patterns in two human chromosome types, designated by the Paris Convention, for different staining techniques (For details see text.)

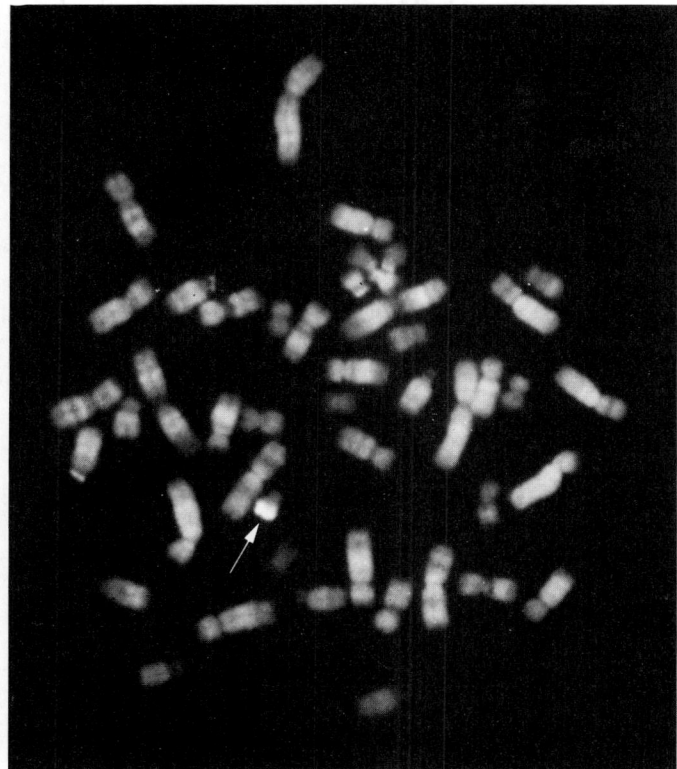

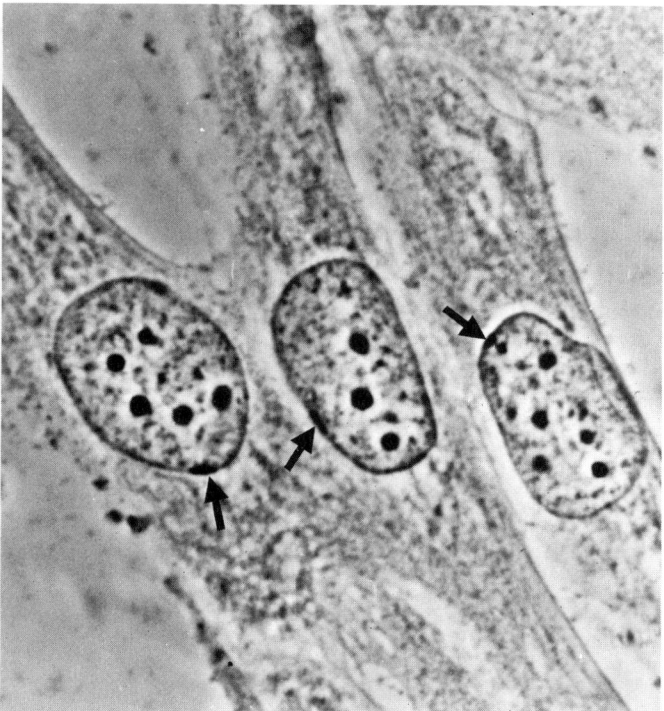

1.47 Fibroblasts from a human female in tissue culture, showing dense heterochromatic Barr bodies (arrows) lying close against the nuclear envelope. Kindly provided by the Paediatric Research Unit, Guy's Hospital Medical School. Magnification × 2000.

1.46 A chromosome spread from a normal human male cell, treated with a fluorescent dye (see text) and photographed using fluorescence microscopy. In addition to the terminal fluorescence of the long arm of the Y chromosome (arrow), bands which fluoresce less strongly are also present on the other chromosomes, providing a basis for their identification. Supplied by the Paediatric Research Unit, Guy's Hospital Medical School. Magnification × 1500.

multiples (e.g. 44 autosomes + XXXXY) have also been described.

Human chromosomal abnormalities and triploidy are quite common (about 0.6% of all conceptions) and sometimes are compatible with survival to birth. However, many such abnormalities are lethal in embryonic life, accounting for their rarity among survivors at birth; 20–30% of spontaneously aborted individuals are chromosomally abnormal.

Cell Motility

All cells, except perhaps erythrocytes, exhibit some form of motility ranging from the minute streaming of cytoplasm, characteristic of all nucleated cells, to the powerful contractions of muscle cells. The types of motility vary widely; in most tissues there are constant modulations of cell shape and detailed surface topography, including rounding prior to mitosis, flattening or elongation, e.g. in developing neuroblasts of the neural tube, and the formation of microvilli and pseudopodia at the cell surface (**1.48**). Such changes in shape are prominent in embryonic development and in the repair of damaged tissues; many morphogenetic movements of cell layers also result from such motility (p. 110), e.g. the rolling up of the neural tube, brought about by transformation of columnar cells of the neural plate into wedge-shaped elements or the formation of depressions by apical contraction of 'bottle cells' (**1.48**). Changes of shape caused by cytoplasmic movement occur in the locomotion of many cells. In amoeboid movement, typical of neutrophil and some other migratory cells, the central cytoplasm streams into pseudopodia at the advancing edge and is withdrawn at the rear. In other cells, e.g. macrophages, the leading edge shows delicate undulations, creating a much folded, thin *ruffled membrane* or, as in the growth cones of the neurites of nerve cells, long narrow *microspikes*, which are constantly formed and resorbed as the cone advances. Some cells carry motile cilia and flagella which generate fluid movements at their surfaces, either to move mucus on the surface

(e.g. of the respiratory tract) or to move the entire cell, in spermatozoa. Muscle cells and myoepithelial cells are extreme examples of motility, although their activities have much in common with those of less specialized cells (see Bretscher 1987).

Cytoplasmic streaming within cells moves metabolites and organelles much more rapidly than simple diffusion. Such streaming is often restricted to certain specific pathways, so that there is a rapid and directed flow of highly fluid cytoplasm through a landscape of relatively immobile organelles. Cytoplasmic streaming is well developed in extensive cells such as neurons, where metabolites may be conveyed for distances of a metre or more to the extremities of axons and dendrites, some materials then retracing the route back to the cell soma. Speeds of up to 25 mm/day have been recorded for the centrifugal *axoplasmic flow* in peripheral nerves.

Other internal movements include: passage of secretory vacuoles to the surface of glandular cells, with subsequent release of the secretory product; intake of materials at the cell surface (endocytosis, phagocytosis, and pinocytosis; see p. 29) and their movement into the cytoplasm; and the movements of chromosomes during mitosis and meiosis.

The mechanisms of many of these appear to be fundamentally similar, depending on the presence of actin, myosin and, depending on location, various other types of molecule, e.g. tropomyosin, tubulin. Motility has been investigated most thoroughly in the highly organized cytoplasm of striated muscle fibres whose orderliness has facilitated structural, biochemical and biophysical analyses (p. 556). Muscle contraction involves the energy-dependent sliding interaction of filaments of actin and myosin, and these two proteins are also present in a less organized form in nearly all other cells, including myoepitheliocytes, endothelial cells, fibroblasts, macrophages, platelets, etc. Actin is present in its polymerized form (*f-actin*) as microfilaments which are thought to exist in more than one state, some relatively stable but others highly labile, being assembled or dispersed in a few seconds (Wessels et al 1971). F-actin is in equilibrium with an intracellular pool of actin monomer (g-actin); agents which disturb this equilibrium either cause the formation, or the dismantling, of at least some classes of microfilament. Cell membranes often form sites of attachment for microfilaments by way of other proteins (see p. 30). Myosin, more difficult to demonstrate structurally, is probably present as small molecular groups of two or more in most cells; it is only in specialized contractile cells that larger assemblies are permanently present. In fibroblasts, small aggregates of myosin may form sliding cross-links between adjacent microfilaments, thus generating forces which cause the network of microfilaments to extend or contract. If the actin filaments were attached to the plasma membrane, such movements would affect the whole cell, flattening or rounding it according to the direction of sliding. Localized extensions of the actin network could form microvilli or, if directed by the presence of aligned bundles of microtubules or intermediate filaments to which actin may also be attached, could produce some types of locomotion. Contraction of bundles of actin filaments by sliding interaction with myosin probably also causes cytoplasmic cleavage during cell division (p. 42).

Certain very transient structures, including microspikes and ruffled membranes, are created by rapid conversion of g-actin into a network of filaments, then stabilized by interaction with myosin into finger-like processes or ridges. Such structures can be easily dismantled by depolymerization, their presence or absence depending upon local fluxes of certain ions or cell metabolites.

Actin-myosin interactions also probably accompany some types of endocytosis, such as pinocytosis and phagocytosis, where major cell surface movements occur. Cytochalasins, which depolymerize some though probably not all actin filaments, severely impair many cell movements so that, e.g. multinucleate cells are formed because the cleavage furrow fails to separate cells during mitosis.

A number of regulatory proteins accompany actin and myosin both in muscle and non-muscle cells, including tropomyosin, troponin and α-actinin; these can be demonstrated clearly, often in strikingly beautiful networks, by immunofluorescent labelling with antibodies evoked specifically against these proteins (see e.g. Lazarides 1976).

Another major protein associated with movement is tubulin, which polymerizes to form microtubules (p. 29). As the major component of mitotic spindles, these are vital for the anaphase movements of chromosomes during cell divisions. They are also involved in the movement of neurotransmitter vesicles in synaptic endings, in axoplasmic flow, in movements of cilia, flagella and in secretion by some glandular cells. Some microtubules can be disassembled by colchicine and other drugs which bind to the

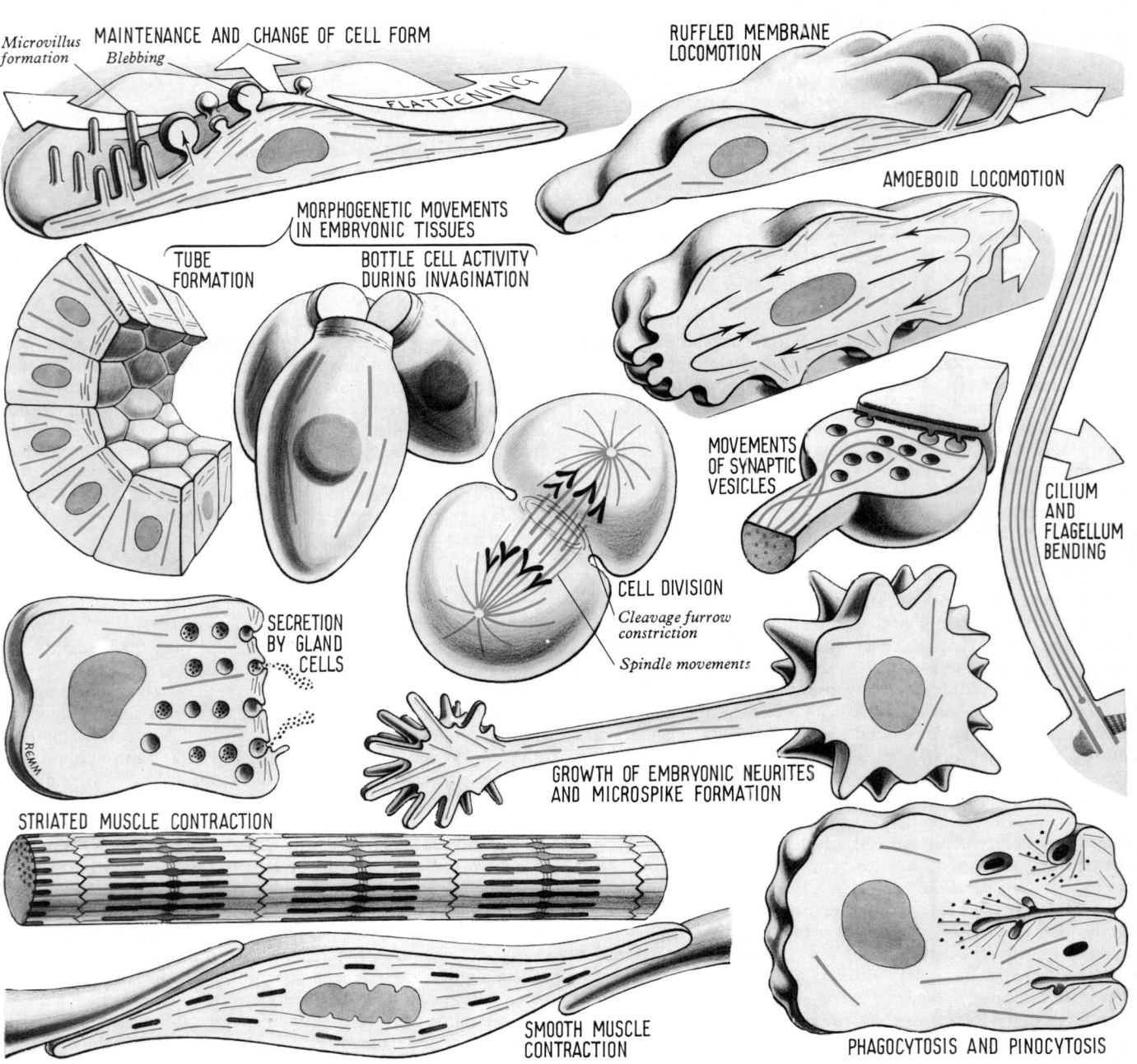

1.48 Diagram summarizing some major examples of cell motility associated with actin filaments (red) and microtubules (blue).

cytoplasmic pool of dimers and thus shift the equilibrium away from the polymeric form. Under these conditions motility is often affected (see e.g. Dustin 1980, Roberts & Hyams 1979). In cilia and flagella motility is achieved by the sliding interaction between adjacent microtubule pairs through the mediation of another protein, dynein (p. 32), which causes bending movements. In other sites it is probable that vesicles and other organelles or colloids may be able to slide along the outside of microtubules by means of interacting, temporary protein links by substances such as the protein *kinesin*, recently shown to be involved in vesicle movements along microtubule arrays. to produce directed movements and perhaps some types of cytoplasmic streaming (1.48). Other types of movement may involve growth by the extension of the ends of axonal microtubules as tubulin is added at that point

and also possibly the progressive assembly and disassembly of microtubules as in 'treadmilling'.

Finally, there are a number of other movements which probably operate by different mechanisms. These include the transport of various transport vesicles within the cytoplasm vesicles and the formation of 'coated' endocytotic vesicles (see p. 29).

The patterns of control of cell movements are also of great significance, particularly in the study of morphogenesis, teratogenesis and various pathological states in the adult. Ionic fluxes, perhaps associated in some cases with the release of such metabolites as cyclic adenosine monophosphate (cyclic AMP) and other 'second messengers' (p. 18) within the cell, are associated with many types of motile control, often involving the phosphorylation of 'motile' proteins.

THE TISSUES

EPITHELIUM

The term *epithelium* is applied to the layer or layers of cells covering body surfaces. Strangely, this term originally referred to the cellular covering of the nipple (*thelos*) but has been extended to include the cellular linings of all cavities as well as the exterior of the body. Embryologically, epithelia are derived from all three major cell layers; those lining or continuous with external surfaces (the integument, alimentary tract and the distal parts of the urogenital tracts) arise from ectoderm and endoderm. Those lining internal cavities and the proximal parts of the urogenital tract stem from mesoderm and are variously known as *mesothelia* (lining the pericardial, pleural and peritoneal cavities), *endothelia* (lining vascular channels) and *epithelia* in urogenital passages.

Most glands are epithelial, since they arise as diverticula of the body surfaces. The largest is the liver.

Epithelia function generally as selective barriers, aiding or preventing materials traversing the surfaces which they cover; some also protect underlying tissues against dehydration, chemical and mechanical damage; some elaborate and secrete materials into the spaces which they bound; others function as sensory surfaces. Indeed, many features of nervous tissue can be regarded as those of a modified epithelium.

Mesothelia and endothelia are often classified as distinct from epithelia, as they are derived from mesenchymal cells rather than ectodermal or endodermal elements, and have some characteristic mesenchyme-derived features, e.g. their cytoskeletal intermediate filaments (p. 30) are of the connective tissue type (vimentin) rather than the keratin filaments found in epithelia proper. Pathological changes are also more typical of mesenchyme-derived tissues.

Structure of Epithelia

Epithelia (1.49) are predominantly cellular, i.e. they possess little extracellular material. Their cells rest on a mucopolysaccharide *basement membrane* (see p. 67) and the 20 nm intercellular gaps between adjacent cells are occupied by glycoproteins. Intercellular junctions (p. 22) are usually numerous. Cell shape, typically polygonal, is partly determined by their cytoplasmic contents and partly by pressure from surrounding tissues, among other factors.

Epithelia can also regenerate when injured, as seen most dramatically in the liver, which can replace excised portions rapidly. Continuous replacement is also important where abrasion occurs and most epithelia covering external surfaces show a steady rate of cell division, which offsets loss of cells due to this cause. Perhaps because of this ability, tumours of epithelial origin (papillomata and carcinomata) are common.

Usually, blood vessels do not penetrate epithelia, diffusion from capillaries of neighbouring tissues providing nutrition. This limits the maximum thickness of living cell layers. Epithelia with their supporting connective tissue can often be removed surgically as one 'layer', collectively known as a *membrane*. Where the surface is moistened by mucous glands it is a *mucous membrane* and, if covered with a film of serum-like fluid, a *serous membrane*.

Classification of Epithelia (1.49)

They can be grouped into unilaminar (*simple*) epithelia—single layers of cells resting on a basement membrane; and multilaminar epithelia—more than one cell thick. The latter includes: *replacing* or *stratified squamous* epithelia, in which superficial cells are constantly replaced from the basal layers; *urothelium (transitional epithelium)* serving special functions in the genito-urinary tract; and other multilaminar epithelia which, like urothelium, are replaced only very slowly under normal conditions.

Unilaminar (Simple) Epithelia

This group may be further subdivided according to the shape of its cells into squamous, cuboidal, columnar and pseudostratified types. Cell shape is largely related to cell volume; where little cytoplasm is present, denoting few organelles and therefore a low metabolic activity, cells are squamous or low cuboidal. Highly active cells, e.g. ciliated and secretory forms, contain abundant mitochondria and endoplasmic reticulum and are typically tall cuboidal or columnar.

Unilaminar epithelia can also be subdivided into those which have special functions, i.e. those with cilia, microvilli (brush and striated borders), secretory vacuoles (in mucous and serous glandular cells) or sensory features. *Myoepitheliocytes*, which are contractile, occur as isolated cells associated with glandular structures, e.g. salivary glands (p. 1296).

SQUAMOUS (PAVEMENT) EPITHELIUM

This is composed of flattened, interlocking, polygonal cells (*squames*). The cytoplasm may in places be only 0.1 μm thick and the nucleus usually bulges into the overlying space (1.49,50). These cells line the alveoli of the lungs, renal corpuscles, the thin segments of the nephric tubules and various parts of the inner ear. The structurally similar *mesothelium* forms the surfaces of the pericardial, pleural and peritoneal cavities, each cell possessing one or more cilia. *Endothelium* lines the blood vascular and

E P I T H E L I A L   T I S S U E S

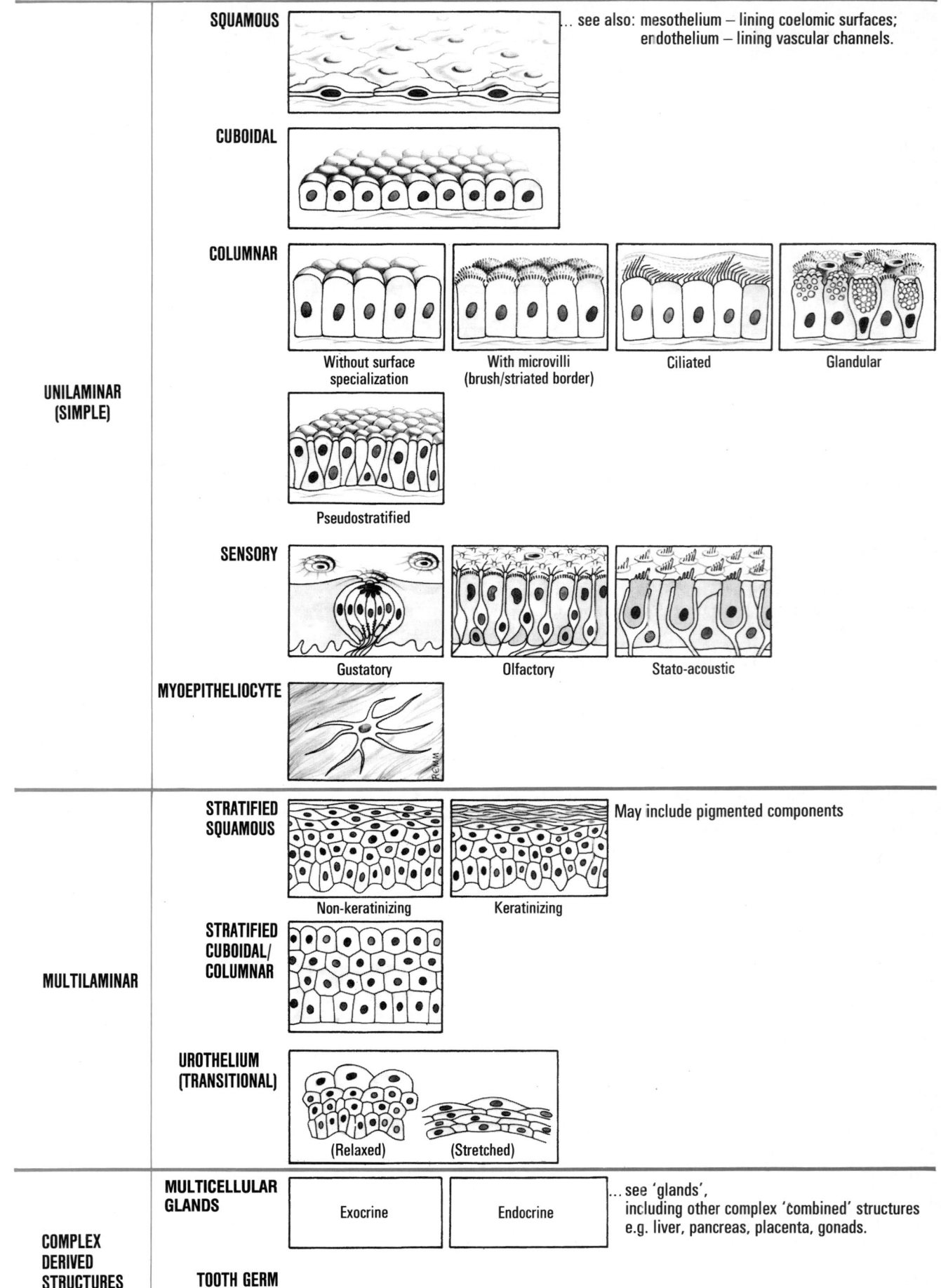

UNILAMINAR (SIMPLE)

SQUAMOUS

... see also: mesothelium – lining coelomic surfaces; endothelium – lining vascular channels.

CUBOIDAL

COLUMNAR

Without surface specialization

With microvilli (brush/striated border)

Ciliated

Glandular

Pseudostratified

SENSORY

Gustatory

Olfactory

Stato-acoustic

MYOEPITHELIOCYTE

MULTILAMINAR

STRATIFIED SQUAMOUS

May include pigmented components

Non-keratinizing

Keratinizing

STRATIFIED CUBOIDAL/ COLUMNAR

UROTHELIUM (TRANSITIONAL)

(Relaxed)

(Stretched)

COMPLEX DERIVED STRUCTURES

MULTICELLULAR GLANDS

Exocrine

Endocrine

... see 'glands', including other complex 'combined' structures e.g. liver, pancreas, placenta, gonads.

TOOTH GERM

NERVOUS TISSUE (often classified as a separate tissue, but retains many characteristics of its epithelial origins).
SEMINIFEROUS EPITHELIUM

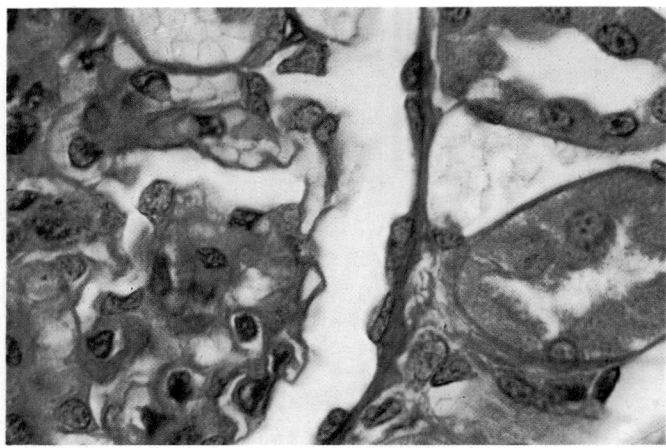

1.50 Part of a renal corpuscle showing vertically sectioned simple squamous epithelial cells in Bowman's capsule. Mallory's triple stain.

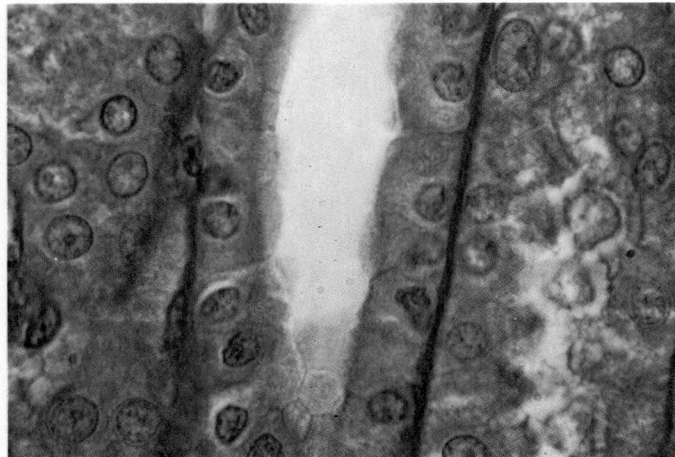

1.51 Simple cuboidal epithelium in renal collecting ducts. Masson's trichrome stain.

lymphatic channels (p. 684) and this term (i.e. endothelium) is also used for the layer of cells lining the inner surface of the cornea.

Simple squamous epithelia may also be *tessellated*, the cells having sinuous interlocking borders rather than straight boundaries as shown in **6.11**.

Although mesothelia and vascular endothelia resemble ordinary squamous epithelia structurally, there are some important differences in their composition, e.g. they contain vimentin filaments rather than epithelial keratin filaments; they also show characteristic pathological reactions and can form distinctive neoplasms.

Because it is so thin, squamous epithelium allows rapid diffusion of gases and water but may also engage in active transport, as indicated by the numerous endocytic vesicles often seen in such cells. Occluding junctions between adjacent cells (see Staehelin 1975) ensure that materials pass primarily *through* cells rather than between them, except where punctuated by holes, as in the fenestrated endothelia of the renal glomeruli and elsewhere (see p. 1401).

CUBOIDAL AND COLUMNAR EPITHELIA

These consist of regular rows of cylindrical cells (**1.51–56**). Cuboidal cells are square in vertical section, whereas columnar cells are taller than their diameter, and both are polygonal when sectioned horizontally. Commonly, microvilli are found on their

1.49 (*opposite*) Classification of the major types of epithelia and associated tissues described in the ensuing section.

free surfaces, providing a large absorptive area (p. 30), e.g. the epithelium of the small intestine (columnar cells with a striated border), the gallbladder (columnar cells with a brush border) and proximal and distal convoluted tubules of the kidney (large cuboidal cells with brush borders). Ciliated columnar epithelium lines most of the respiratory tract (**1.54–56**) as far as the respiratory bronchioles (excepting the lower pharynx and vocal folds), some of the tympanic cavity and auditory tube, the uterine tube and patches in the cavity of the infantile uterus and efferent ductules of the testis.

Mucous glands also line the respiratory tract and cilia sweep a layer of viscous fluid and trapped dust particles, etc. from the lung towards the pharynx (the *muco-ciliary rejection current*), clearing the respiratory passages of inhaled particles. In the uterine tube, cilia may assist the passage of ova from the peritoneal cavity to the uterus.

Some columnar cells are glandular, their apices containing mucus or proteinaceous vacuoles, e.g. mucin-secreting and chief cells of the gastric epithelium. Often mucous cells lie among nonsecretory ones, allowing expansion of the secretory apices to give a shape characteristic of these, the *goblet, chalice* or *calciform* cells (**1.52,53**) e.g. in the intestinal epithelium.

Mucus is a viscous suspension of complex glycoproteins of various kinds, as well as many other secretory products of adjacent epithelial cells. All mucous glycoproteins (*mucins*) consist of filamentous *core proteins* to which are attached carbohydrate chains, usually branched, standing out from the core protein like bristles in a bottle brush; up to 600 such chains occur in human salivary mucus, so that the entire molecule is quite huge. Furthermore, such glycoproteins typically are linked by sulphydryl groups into tetramers, with molecular weights of 2 million or more. Carbohydrate residues include glucose, fucose, galactose and the negatively charged N-acetylglucosamine (sialic acid), particularly near the free ends of the chains. The terminals of some carbohydrate chains are identical with the blood group antigens of the ABO (ABH) group in about 80% of the population ('secretors', bearing the *secretor gene* S^e) and can be detected in salivary mucus with appropriate clinical tests. There is an interesting correlation between the absence of these antigens in mucus and proneness to gastric ulcers and it is possible that they confer protective properties to gastric and perhaps other types of mucus. The long polymeric carbohydrate chains bind water and so protect surfaces against drying; when in dilute suspension, their molecules can slide over one another with ease, providing good lubricating properties, and their negative charges also bind cations such as Na^+. In more concentrated states, they form viscous layers which protect the underlying tissues against damage.

Mucus is synthesized in a stepwise fashion (Peterson & Leblond

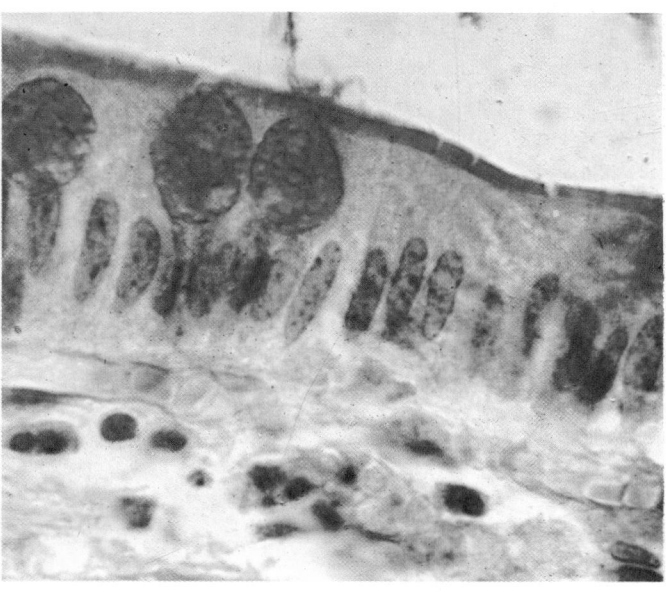

1.52 Simple columnar epithelium of the small intestine showing absorptive cells with a striated border and goblet cells. PAS–haematoxylin.

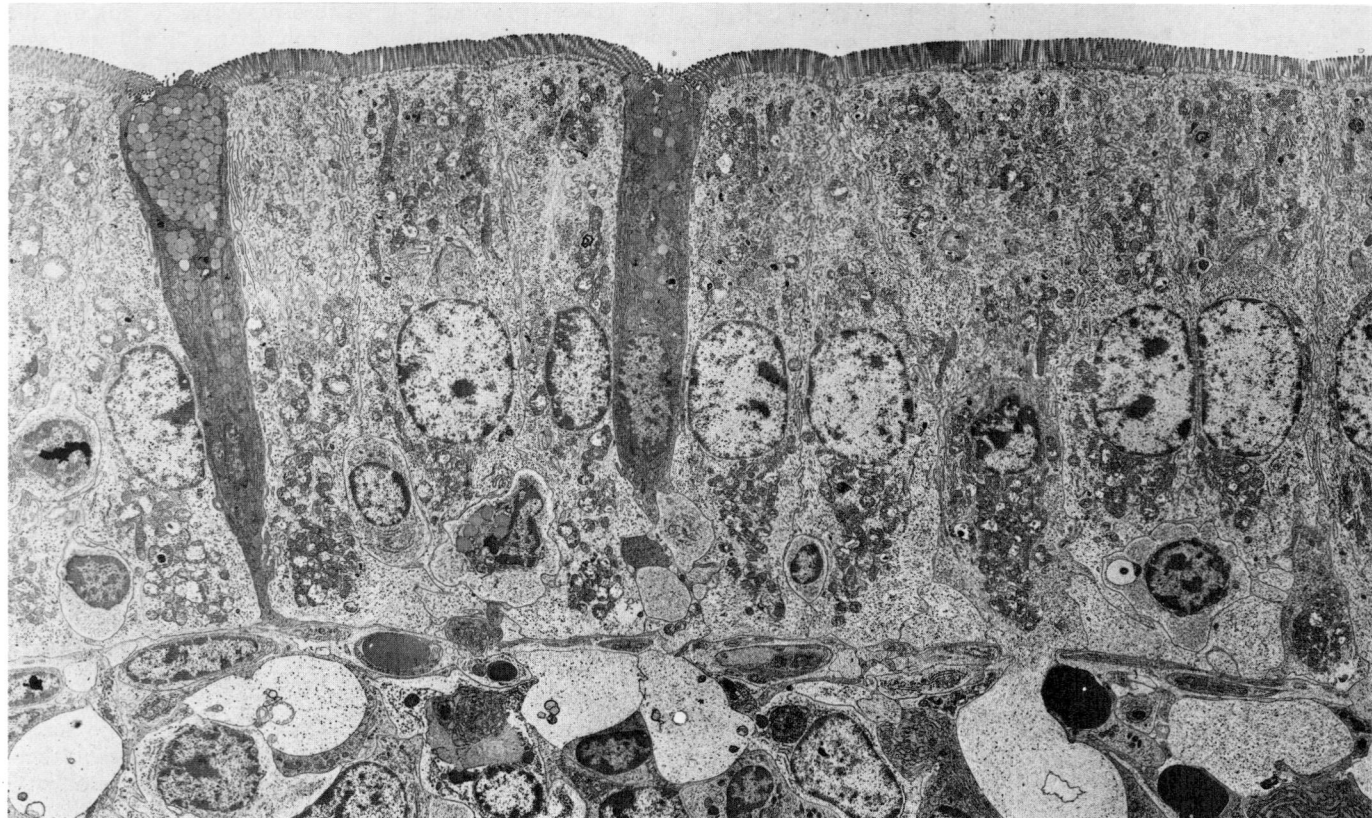

1.53 Low-power electron micrograph of a vertical section through simple columnar epithelium bearing microvilli; two goblet cells are also present. Note the presence of several small lymphocytes near the epithelial base. Small intestine. Provided by Derrick Lovell, Guy's Hospital Medical School. Magnification × 8000.

1964); protein synthesized in the granular endoplasmic reticulum is passed to the Golgi complex, where it is first conjugated with sulphated carbohydrates to form the glycoprotein, *mucinogen*, then exported in small dense vesicles which swell as they approach the cell surface and finally fuse with it to release the mucus.

PSEUDOSTRATIFIED EPITHELIUM

This is a simple columnar epithelium in which nuclei lie at different levels in a vertical section (**1.49,57**). Not all cells extend through the whole thickness of the epithelium, some constituting a basal cell layer, often mitotic and able to replace damaged mature cells. Migrating lymphocytes within columnar epithelia may also give a pseudostratified appearance because their nuclei are placed at different depths. Much of the ciliated lining of the respiratory tract is of the pseudostratified type.

SENSORY EPITHELIA

These are restricted to special sense organs of the olfactory, gustatory and vestibulo-cochlear receptor systems. All of these

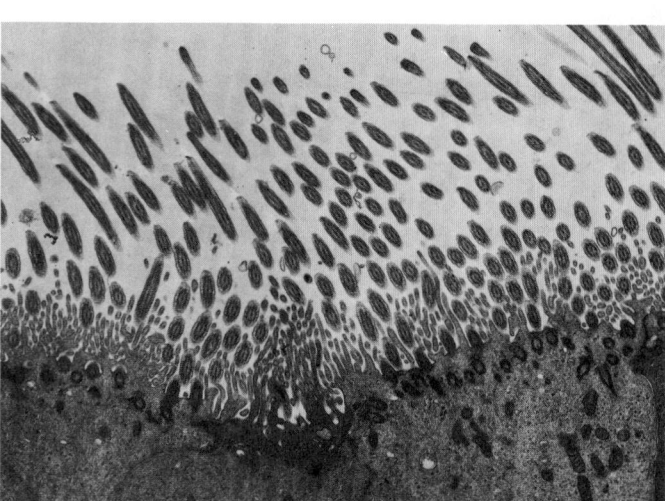

1.54 Detail of a group of cilia cut in various planes at the surface of a ciliated columnar epithelial cell from the upper respiratory tract. Note the presence of small microvilli interspersed between the cilia. Magnification × 7500.

1.55 Low-power scanning electron micrograph of the ciliated surface of the trachea. Magnification × 3000.

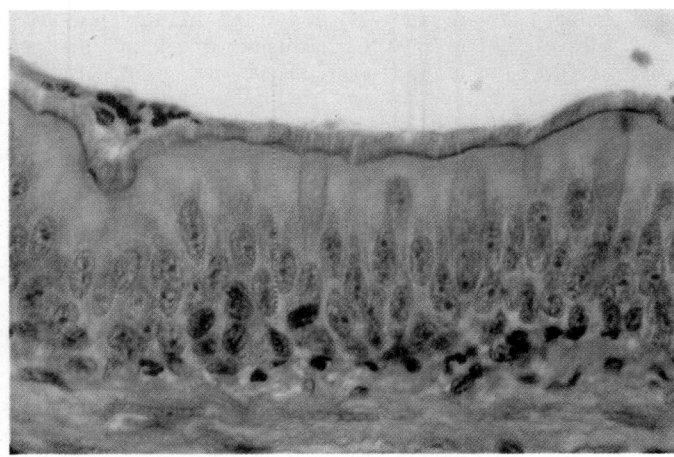

1.56 Simple ciliated columnar epithelium at the surface of a bronchus showing a superficial layer of mucus entrapping some particles, and several goblet cells. PAS–haematoxylin. Magnification × 850.

contain sensory cells surrounded by supportive, non-receptive cells. Olfactory receptors are modified neurons, their axons passing directly to the brain, but the other types are specialized epithelial cells synapsing with terminals of afferent (and sometimes efferent) nerve fibres (see Section 7 Neurology).

MYOEPITHELIOCYTES

Also sometimes termed *basket cells*, these are fusiform or stellate in shape, containing actin and myosin filaments; they contract when stimulated by nervous or neurohormonal signals (Abe et al 1981). They surround the secretory portions and ducts of some glands, e.g. mammary, lacrimal, salivary and sweat glands (Nagato et al 1980), lying within the basal laminae at the surface of these structures. Their contraction assists the initial flow of secretion into larger channels. Myoepitheliocytes are ultrastructurally similar to non-striated muscle cells (p. 559) in the arrangement of their actin and myosin but they originate from ecto- and endoderm.

Multilaminar Epithelia

These are found at surfaces subjected to mechanical wear and tear. They can be divided into those which continue to replace their surface cells from deeper layers (stratified squamous epithelia) and others where replacement is extremely slow except when injured.

STRATIFIED SQUAMOUS EPITHELIA

These are multilayered tissues in which there is a constant formation, maturation and loss of cells. Formed mitotically in the most basal layers, their cells move more superficially, changing from a cuboidal or columnar shape to a more flattened form, and eventually being shed from the surface. These epithelia occur in sites exposed to mechanical stresses and provide for a constant protection of the underlying tissues against mechanical, microbial and chemical damage. Typically, the cells are held together by numerous desmosomes to form strong continuous cellular sheets. The two major types of these epithelia are *keratinized* and *non-keratinized stratified squamous epithelia*.

Keratinized Epithelium

This is found at surfaces which are subject to drying as well as mechanical stresses (1.58). These include the entire epidermis, the mucocutaneous junctions of the lips, nostrils, distal anal canal and the outer surface of the tympanic membrane. Their cells, *keratinocytes*, are formed by the mitosis of basally-sited stem cells (the *stratum basale*) and gradually pass towards the surface in a column of moving cells. When newly formed, keratinocytes contain many ribosomes, mitochondria and other metabolically active organelles and have an extensive cytoskeleton of keratin filament bundles (tonofibrils), anchored into the plasma membrane at desmosomal contacts with neighbouring cells. The most basal layer of cells is also anchored to the basal lamina by hemidesmosomes (p. 24). As keratinocytes move away from the base, the cell surfaces become highly folded, tightly interlocking with those of adjacent cells to present a spiny appearance. These cells (*prickle cells*) together comprise the *stratum spinosum*. As they move more superficially they begin to flatten and synthesize other substances including a dense, basophilic protein, keratohyalin, which mingles with the keratin filaments and gives the cells of this layer (the *stratum granulosum*) a granular appearance. Simultaneously, the cells synthesize and secrete a glycolipid which coats the cell surface, forming a thick, adhesive, water-resistant lipidic cement between the flattened cells. At this stage the cells lose their nuclei, the keratin filaments become firmly embedded in the matrix protein and the cells are now dead, flattened plates (squames), forming the final layer, the *stratum corneum*. They eventually flake off from the surface.

This unusual combination of strongly coherent layers of living cells and more superficial strata made of plates of inert, mechanically robust protein complexes, interleaved with water-resistant lipid, makes this type of epithelium an excellent barrier against different types of injury.

Further details of this tissue are given on p. 73 with the skin.

Non-keratinizing Epithelium

This is present at surfaces subject to abrasion but protected from drying (1.59), including the buccal cavity, oro- and laryngopharynx, oesophagus, part of the anal canal, vagina, distal uterine cervix, distal urethra, the conjunctiva and cornea, inner surfaces of eyelids and vestibule of the nasal cavities. Characteristically,

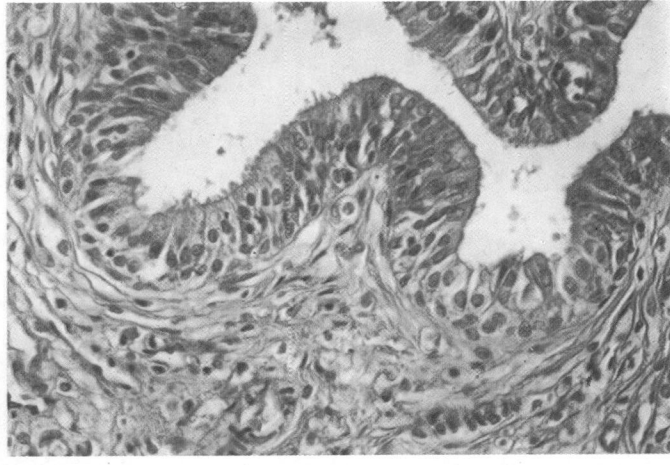

1.57 Pseudostratified epithelium from the male urethra. Haematoxylin and eosin. Magnification × 400.

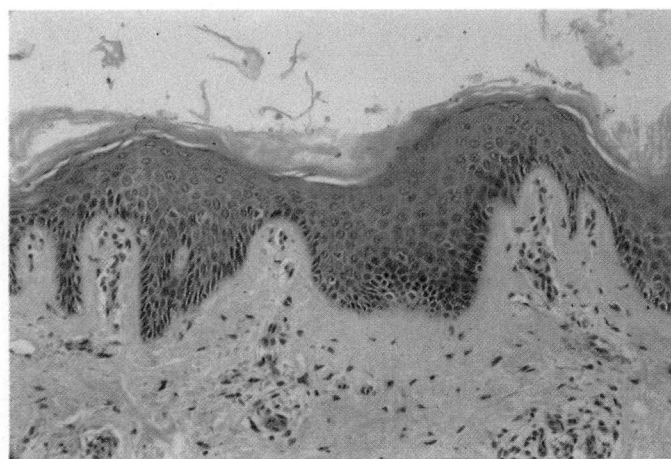

1.58 Section of keratinized stratified squamous epithelium forming the epidermis of the skin, in this example of the scalp. Note the presence of a superficial zone of anucleate keratinized cells. Connective tissue of the dermis underlies the epithelium. Haematoxylin and eosin. Magnification × 150.

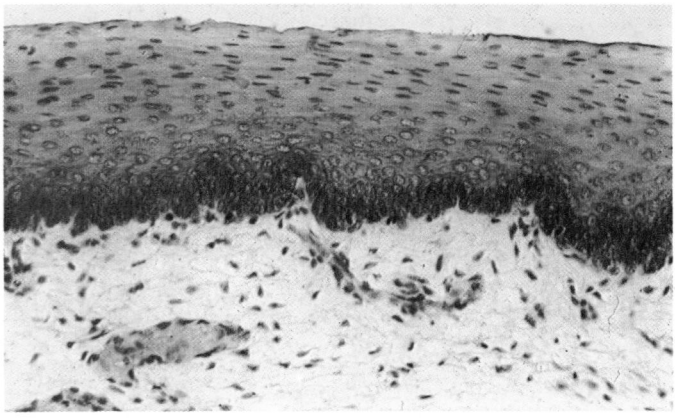

1.59 Non-keratinizing, stratified squamous epithelium from the human tongue. A vertical section stained with haematoxylin and eosin. Note the presence of nuclei in the surface cells. Magnification × 150.

its cells go through the same transitions in general shape as seen in the keratinized type but do not fill completely with keratin or secrete glycolipid, and retain their nuclei until they desquamate at the surface. In sites where considerable abrasion occurs the epithelium is thicker and its more superficial cells may synthesize some keratin (*parakeratinised* epithelia in contrast to the fully, or *orthokeratinized*, state of keratinized epithelium), as e.g. on the hard palate, gingivae and dorsum of the tongue. Diets deficient in vitamin A may induce keratinization of such epithelia and excessive doses may lead to its transformation into mucus-secreting epithelium.

Cells gently removed by swabbing epithelial surfaces can be collected, stained and examined for signs of pathological (e.g. cancerous) transformations (exfoliative cytology).

STRATIFIED CUBOIDAL AND COLUMNAR EPITHELIA

Two or more layers of cuboidal or columnar cells are typical of the walls of the larger ducts of some exocrine glands, such as the pancreas and salivary glands and the ducts of sweat glands, presumably affording more strength than a single layer. These are not continually replaced by basal mitoses and there is no progression of form from base to surface.

Urothelium (Urinary or Transitional Epithelium)

This (1.49,60,61) lines much of the urinary tract, extending from the ends of the collecting ducts of the kidneys, through ureters

and bladder to the proximal portion of the urethra. In males it covers the urethra as far as the ejaculatory ducts, then becomes intermittent, finally being replaced by pseudostratified epithelium in the membranous urethra (see also p. 1421). In females it also extends as far as the urogenital membrane. During development part of it is derived from mesoderm and part from ectoderm and endoderm. The epithelium appears to be 4–6 cells thick and lines organs which undergo considerable distension and contraction; it therefore can stretch greatly without losing its integrity. In stretching, the cells become flattened, without altering their positions relative to each other, since they are firmly connected by numerous desmosomes (see review by Hicks 1975). On close inspection, many cells are seen to be attached to the basal lamina of the epithelium by slender basal processes and it has been proposed that perhaps all cells, even the most superficial, are attached to the base; thus this tissue could be viewed as a type of specialized pseudostratified epithelium (see p. 54), in which case it should be reclassified as a unilaminar epithelium (see Scheidegger 1980, Walton et al 1982). When relaxed, the basally situated cells are cuboidal, uninucleate (diploid) and basophilic with many ribosomes. More apically the cells progressively fuse to form larger, binucleate or uninucleate but polyploid cells. The surface cells are largest and may even be octoploid; their luminal surfaces are covered by a plasma membrane bearing plates of glycoprotein particles embedded in its lipid bilayer. These arrays stiffen the membrane so that, when the epithelium is in the relaxed state and the surface area of the cells is reduced, the glycoprotein-lipid plates are partially taken into the cytoplasm within vacuoles or diverticula, re-emerging on to the surface when its area increases once more through stretching (see Minsky & Chlapowski 1978).

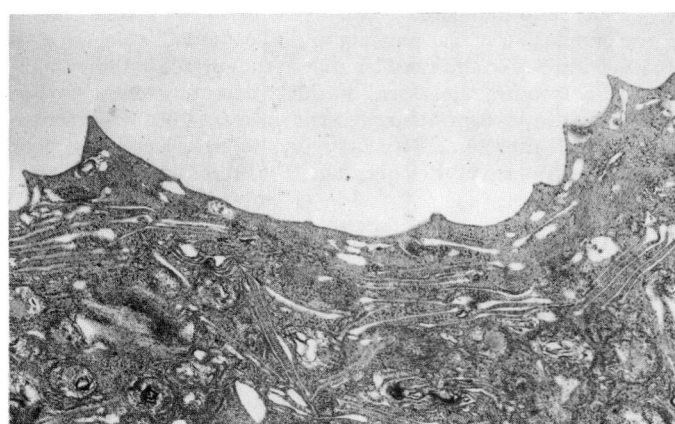

1.61A Transmission electron micrograph of the surface of the urothelium (transitional epithelium) lining the relaxed bladder. Note the angular profiles of the epithelial surface and the plate-like areas of membrane internalisation. Magnification × 15 000.

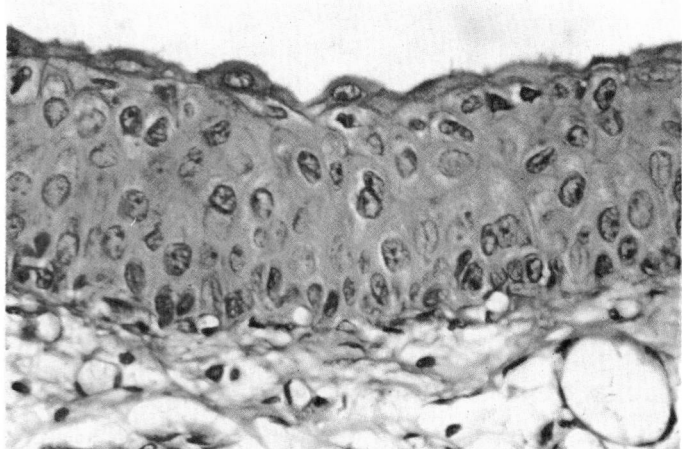

1.60 A vertical section through the surface of a ureter to show the urothelium lining its lumen, stained by Mallory's triple staining technique. Magnification × 600.

1.61B Scanning electron micrograph of the (relaxed) urothelial surface showing the plate-like arrangement of its plasma membrane. Magnification × 6000.

These unusual membranes, together with the occluding junctions of the surface cells, form an effective barrier preventing urine from passing into the epithelium or beyond into the adjacent tissues. The urothelium therefore creates a protective lining to the urinary system which prevents its rather toxic contents from damaging surrounding structures.

Normally, cell turnover is very slow, cell division being restricted to the basal layer and infrequent. When damaged, however, the epithelium regenerates quite rapidly (Annis 1962).

Pigmented epithelial cells are found in various parts of the body. As a single layer of cuboidal cells they form the external layer of the retina (p. 1195) and are present in the posterior epithelium of the iris (p. 1192). The pigment granules are densely crowded in the cytoplasm. *Melanocytes* within stratified squamous epithelium are not derived from ectoderm but are formed in the neural crest and migrate into epithelial sites during development (see p. 115).

Glands

Glands are epithelial derivatives in which cells elaborate a secretion (**1.**62,63, **8.**1A). Some, *exocrine* glands, discharge on to an epithelial surface in continuity with the exterior, e.g. the alimentary, respiratory and urino-genital tracts. Others, *endocrine* glands, become detached from the epithelial surface; no ducts exist (hence the term *ductless glands*) and the secretion is discharged into the blood and lymph streams instead.

Exocrine glands may be *unicellular* or *multicellular*. In the former the secretory cell is part of the epithelial surface on to which it secretes, e.g. goblet cells (**1.**53). In *multicellular glands*, secretory cells are carried in diverticula of the surface, usually divided into a *secretory portion* and a *duct*. *Simple* glands have a single duct, e.g. the tubular glands of the stomach wall (p. 1352), while *compound* glands such as the parotid have branched ducts. The secretory portion may be tubular (straight or coiled), flask-like (*acinar*) or rounded (*alveolar*). Intermediate types, such as tubulo-acinar and tubulo-alveolar, also occur. Glands with highly distended secretory portions (e.g. the prostate, p. 1434) are sometimes called *saccular* glands. These different types of organization appear to be in part related to the amount and destination of secretions; thus tubular glands are typically found where a large epithelial surface receives a modest flow of secretion over a wide area, e.g. the gastric glands. Where copious secretion into a cavity is required, glands are usually highly branched, with relatively large secretory portions secreting into a complex series of branched ducts, e.g. the pancreas. In some glands the ducts may modify the secretion by ionic exchange (e.g. the major salivary glands).

The larger glands are supported by a delicate connective tissue skeleton, mainly of reticulin fibres, which often groups secretory portions into lobules and lobules into larger lobes, blood vessels,

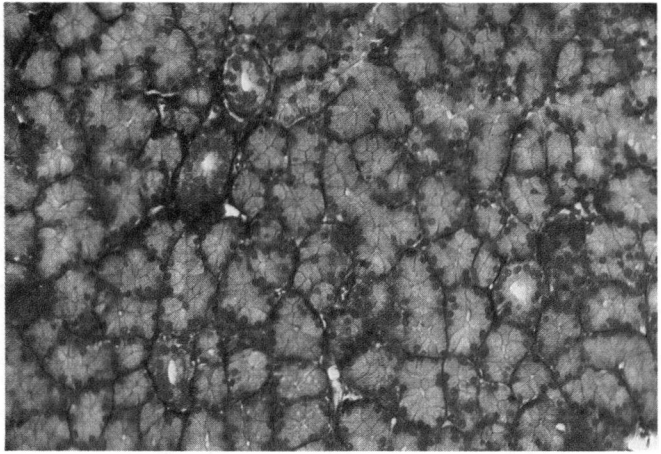

1.63 Section through a compound alveolar gland (nasopharynx), stained with periodic acid-Schiff to show mucus within secretory lobules. The connective tissue septa have been stained blue. Two small ducts are also visible. Magnification × 160.

nerve fibres and branched ducts traversing the larger connective tissue septa. Ducts may therefore lie *within* or *between* lobes (intralobar and interlobar ducts, respectively). The final, large and often multilaminar conduit is often referred to as an *excretory duct*.

The method whereby the epithelial cells of exocrine glands discharge their secretion varies. In *holocrine* glands, such as the sebaceous glands in the skin, the cells disintegrate to liberate the secretion. In *apocrine* glands, such as the mammary gland, a small amount of cytoplasm may be released with the secretory vesicles. In *merocrine* glands, which are the most widespread, the secretion is discharged by the fusion of the membranous walls of secretory vesicles with the plasma membrane, so that only the contents of the vesicle are released. This process involves a 'docking' mechanism by which proteins in the vesicle membrane recognize plasma membrane proteins, bind to them and bring the two membranes close enough for fusion to take place. Membrane added to the cell surface by repeated vesicle fusion can be taken back into the cell by endocytosis and re-cycled for further secretory vesicles (see Pearse & Bretscher 1981, Poisner & Trifaró 1982). The stimulus for secretion varies with the type of cell, but often appears to involve a classic 'second messenger' system such as the cyclic AMP pathway which causes internal calcium release in the cytoplasm (see Conn 1982, Novikoff & Novikoff 1979) e.g. mucous cells.

Glands are often classified by their secretory products. Exocrine glands of some regions, e.g. the buccal cavity, are divided into: *mucus-secreting* or *mucous* glands, whose cells possess frothy cytoplasm with basal flattened nuclei and stain with metachromatic stains and PAS methods, and *serous glands* where the cells have centrally placed nuclei, granular cytoplasm and secrete mainly proteins (e.g. lysozyme, a bactericide, or digestive enzymes).

Some glands are entirely mucous (e.g. the sublingual salivary glands), whilst others are mainly serous (e.g. the parotid salivary glands). The submandibular gland is mixed, some lobules being predominantly mucous and others serous. In some regions mucous acini are capped with crescents of serous cells (*serous demilunes*). The division into mucous and serous types is, however, a rather imprecise one, since their chemical nature is complex and a wide range of secretory products exists. Such terms are of limited use outside certain narrow contexts (e.g. salivary glands, p. 1290). Exocrine glands typically possess a rich blood supply, being metabolically active, and are also often under direct nervous or hormonal control.

Endocrine glands are composed of cell groups lying close to vascular channels; they mostly occur in *clumps* or *cords* of closely-knit cells supported by a network of fine reticular tissue; in the thyroid gland they form hollow balls of cells, or *follicles*, containing a secretory store. Endocrine glands have a particularly well

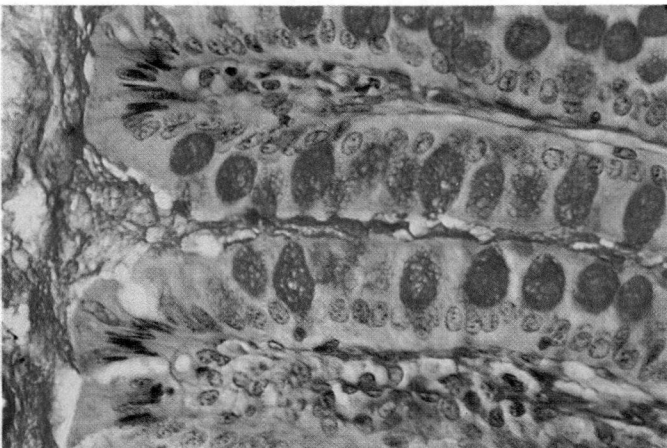

1.62 Section through part of a simple tubular gland (colon) stained with alcian blue to show mucous gland cells. Magnification × 500.

developed circulation and often the endothelium of capillaries is *fenestrated* (p. 689), facilitating the diffusion of large molecules of hormone into the blood. There are also numerous types of isolated endocrine cells, scattered in the epithelial lining of the alimentary tract and other sites, which have important activities in regulation of gut motility and many other processes (see also p. 1352).

The fine structure of gland cells is related to the type of secretion synthesized. In protein-synthesizing cells such as goblet and pancreatic acinar cells, the granular endoplasmic reticulum is extensive and Golgi complexes well developed, whereas in steroid-secreting cells (e.g. of the adrenal cortex) the elaborate endoplasmic reticulum is mainly smooth.

In salivary and mammary glands, *release* of secretion is accelerated by the neurally elicited contraction of *myoepithelial cells* which surround the alveoli and the initial segments of their ducts.

Patterns of synthesis and release of secretions by glandular cells vary considerably. In some, synthesis and release occur con-tinuously (but with quantitative fluctuations), e.g. the adrenal cortex. In others, secretory vacuoles may be stored for long periods, to be released when suitably stimulated (e.g. pancreatic islet β-cells). A third pattern is seen in goblet cells of bronchi, which repeatedly undergo cycles of synthesis followed by release of secretion.

COMPLEX STRUCTURES DERIVED FROM EPITHELIUM

These include those organs which are largely derived from epithelia and retain their highly cellular nature; they often possess typical epithelial features such as secretory, absorptive and transport functions and include the liver, placenta, and the early tooth germ; indeed the nervous system as a whole can be viewed as a highly elaborate epithelial structure.

THE CONNECTIVE TISSUES

Introduction

The connective tissues may be defined as that group of elements, derived largely from embryonic mesoderm, where the tissue is predominantly composed of intercellular material, secreted mainly by its cells which are consequently widely spaced. Many of the special properties of connective tissues are determined by the composition of the intercellular substance and their classification is also largely based on its characteristics.

Connective tissues play several essential roles in the body, both *structural*, since many of the extracellular elements possess special mechanical properties, and *defensive*, a role which has a cellular basis (p. 59). Structural connective tissues are conveniently divided into 'ordinary' or 'general' types, distributed widely, and special types, namely cartilage and bone which are described elsewhere (pp. 283–29). A third variety, the *haemolymphoid tissues*, are comprised of the cells of the blood and lymphoid tissue and their precursors; these are often considered to be akin to other types of connective tissue because of their similar mesenchymal origins and also because the various defensive cells of the blood form part of a typical connective tissue cell population. However, for convenience, these tissues will be considered separately in another section (p. 662) with the circulatory system.

Connective tissue is thus formed of *cells* and *extracellular matrix*. The matrix in turn is composed of *fibres* and a relatively amorphous, viscous *ground substance*. (Some authors apply the term 'matrix' to ground substance alone but it will not be used in this manner here.) It is noteworthy that a number of cell types of connective tissue are also found in the circulating blood and lymph and there is a dynamic equilibrium between the two.

The Cells of General Connective Tissue

In general connective tissue, these can be conveniently separated into: (1) the resident cell population (fibroblasts, adipocytes, persistent mesenchymal stem cells, etc.) and (2) a fluctuating population of immigrant, wandering cells with various defensive functions (macrophages, lymphocytes, mast cells, neutrophils and eosinophils) which may change their activities, structures and numbers according to defensive demand. A graphic summary of these cell types is presented in **1.64**.

Embryologically, fibroblasts and adipocytes arise from relatively undifferentiated *mesenchymal stem cells*, some of which may remain in the tissues, providing a postnatal source of new cellular elements. Most of the other cells migrate into the tissue from bone marrow, lymphoid tissue and other external sources.

FIBROBLASTS

These are usually the most numerous cells. They are flattened and irregular in outline, with branching processes; in profile they appear fusiform or spindle-shaped (**1.64,65**). Fibroblasts synthesize most of the extracellular matrix of connective tissue and accordingly have all the features typical of cells actively engaged in synthesis and secretion of proteins. Their nuclei are relatively large, active or euchromatic (open-faced) and possess prominent nucleoli. In young and active cells the cytoplasm is abundant and basophilic because of the high concentration of rough endoplasmic reticulum (Hall & Jackson 1968, Ross 1975, Hay 1981). In old and inactive fibroblasts (often termed *fibrocytes*) the cytoplasm is sparse, the endoplasmic reticulum scanty and the nucleus flattened and heterochromatic (close-faced). In active fibroblasts, mitochondria are abundant and several sets of Golgi complexes are present.

Fibroblasts are usually adherent to the fibres (collagen and elastin) which they lay down. In some highly cellular structures, e.g. glands and lymphoid tissue, fibroblasts and delicate collagenous fibres (reticulin fibres) form fibrocellular networks often called *reticular tissue*. The fibroblasts may then be termed reticular cells (see p. 62).

Fibroblasts are particularly active during wound repair, multiplying and laying down a fibrous matrix which becomes invaded by numerous blood vessels *(granulation tissue)* (see Ross 1968). Contraction of wounds (p. 87) is at least in part caused by the shortening of specialized contractile fibroblasts (myofibroblasts) which arise in such areas (Gabbiani et al 1973). Myofibroblast activity is influenced by various factors such as steroid hormone levels, dietary content and prevalent mechanical stresses. In vitamin C deficiency there is an impairment of collagen formation.

MACROPHAGES (HISTIOCYTES, CLASMATOCYTES)

These are also numerous in connective tissue where they may be either attached to matrix fibres (stationary, or fixed macrophages), when their shape is irregular with many filopodia (**1.66**); or they may be motile (nomadic macrophages) of a more rounded and regular form.

Macrophages are about 15–20 μm in diameter, their nuclei usually indented and somewhat heterochromatic, with a prominent nucleolus. Their cytoplasm is mildly basophilic and typically has a 'frothy' appearance under the light microscope. They are important phagocytes, forming part of the *mononuclear phagocyte system* (p. 669). They can engulf and digest particulate organic materials, such as bacteria and other foreign bodies, and are also able to dispose of dead or moribund cells prior to tissue regeneration.

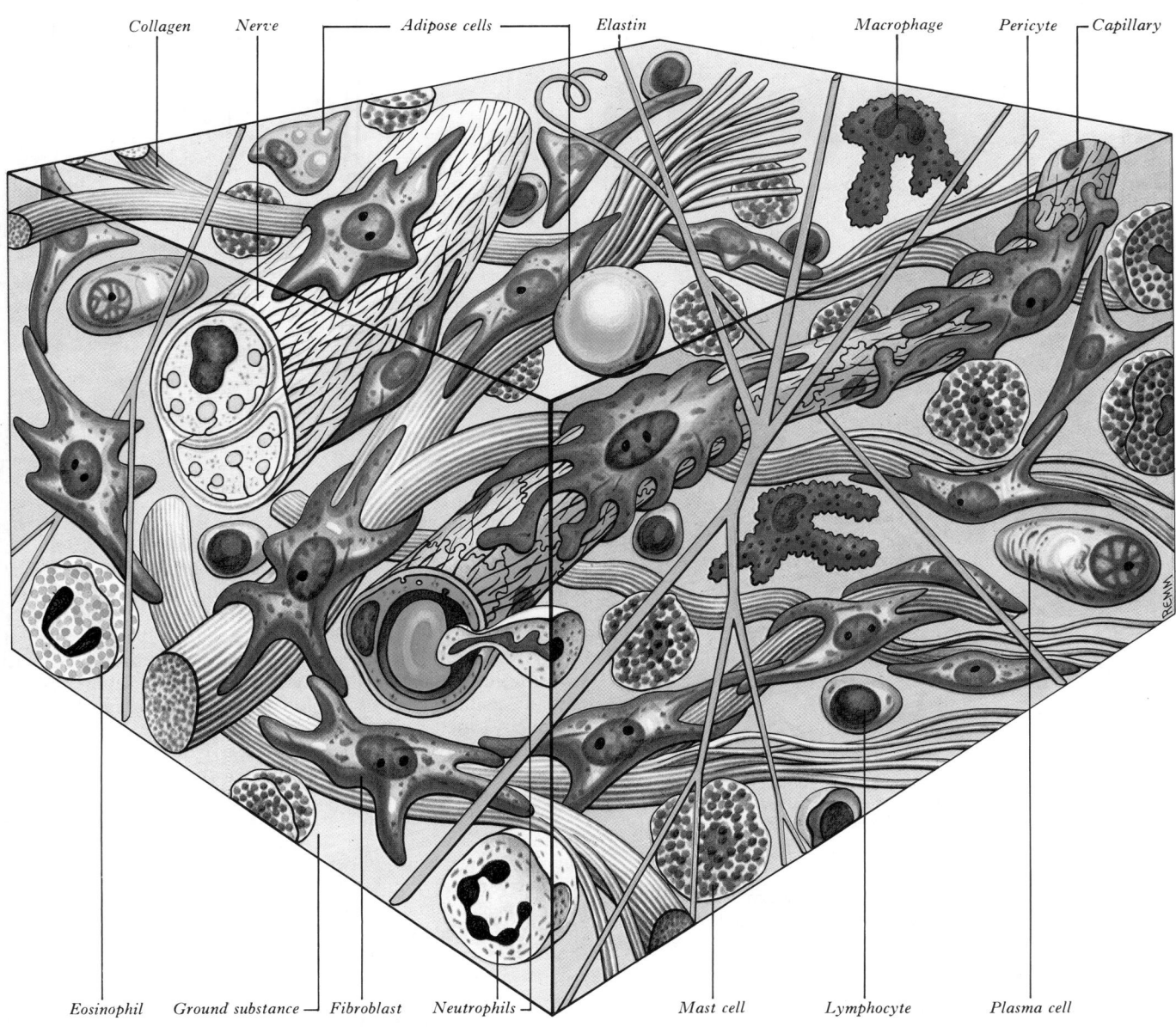

Collagen Nerve Adipose cells Elastin Macrophage Pericyte Capillary

Eosinophil Ground substance Fibroblast Neutrophils Mast cell Lymphocyte Plasma cell

1.64 Diagrammatic reconstruction of loose connective tissue showing the characteristic cell types, fibre and intercellular spaces.

Ultrastructurally, they contain numerous lysosomes which digest ingested materials (p. 27). Inert materials such as small particles of carbon or metals may also be taken up, a quality useful in demonstrating macrophages histologically, since their cytoplasm becomes filled with ingested particles if an experimental animal is previously injected with a suspension of India ink (1.67), trypan blue or lithium carmine (vital staining). Macrophages may also be separated magnetically from mixed cell samples by first treating them with iron carbamyl, which they ingest.

These cells are also of great importance in many immunological aspects of defence and their interactions with numerous defensive cells in the regulation of the immune system are widespread. Antigens adhering to macrophages can be (directly or after modification) passed to and thus stimulate lymphocytes (p. 671). They may selectively phagocytose particles previously coated by antibodies (opsonins, see p. 672) synthesized by lymphocytes and they are themselves sites for homocytotropic antibody attachment, which enables them to recognize and attack foreign substances (see also monocytes, p. 667).

Many properties of connective tissue macrophages are similar to those of a number of specialized cell types in other sites, particularly: circulating monocytes, alveolar phagocytes in the lungs, littoral phagocytes in the lymph nodes, spleen and bone marrow, Kupffer cells of the liver sinusoids, microglial cells of the brain, Langerhans cells of the epidermis and dendritic cells of lymphoid

tissue. Various cell-labelling experiments have shown that most precursors of these arise in the bone marrow as monocyte-like cells which then pass in the circulation to their final destinations, where, however, they may continue to divide. Recent evidence suggests strongly that these cells are specialized subclasses of a general macrophagic cell, having distinctive functions, preferred sites of action and possessing different antigenic markers as well as many common ones.

All macrophages are capable of motility, when suitably stimulated. When grouped around a large foreign body, macrophages may also fuse together to form syncytial giant cells and epithelioid cells. These and other aspects of macrophage biology will be considered further in Section 6 (p. 662).

LYMPHOCYTES

These cells are numerous in general connective tissue only in pathological states, migrating from adjacent lymphoid tissue or from the circulation. The majority are small cells (6–8 μm) with rounded, highly heterochromatic or often deeply indented nuclei (p. 671, and see 1.64,66). When appropriately stimulated they enlarge, developing numerous ribosomes. Two major functional classes exist, termed 'B' and 'T'-lymphocytes (see p. 672). B-lymphocytes originate in the bone marrow, then pass to various lymphoid tissue sites where they proliferate. When antigenically stimulated, they undergo further mitotic divisions, then enlarge

59

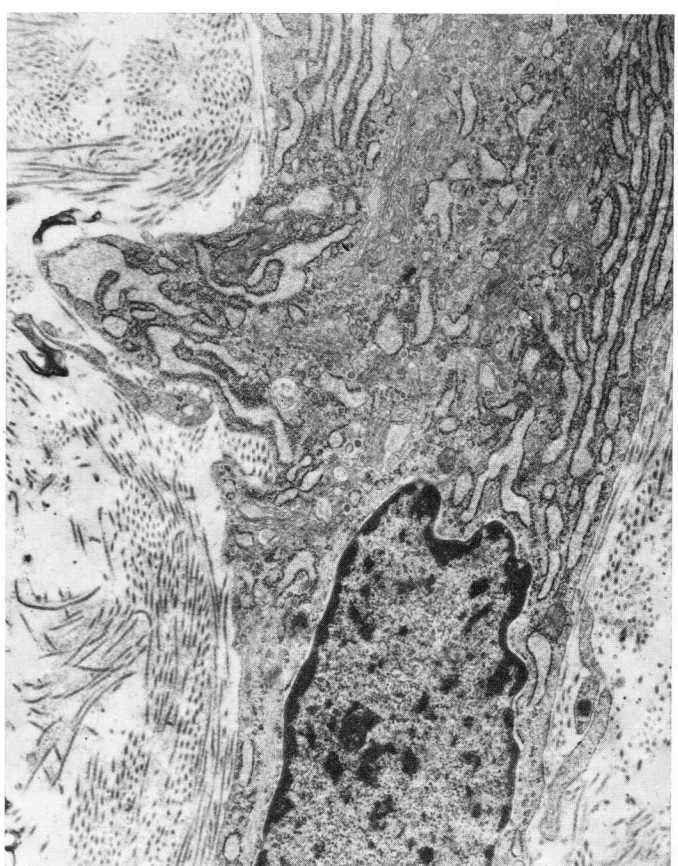

1.65 Transmission electron micrograph of a fibroblast. Note the abundant endoplasmic reticulum and extensive Golgi complex. Some extracellular collagen fibres are also visible. Magnification × 15 000.

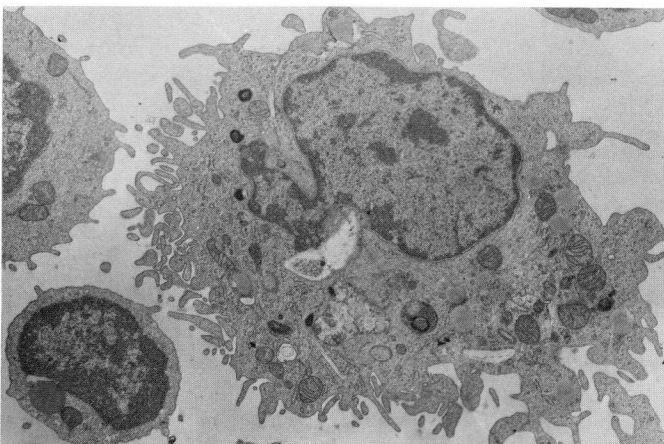

1.66 Electron micrograph of a macrophage (right) and a small lymphocyte (lower left). Magnification × 6000.

as they mature to form *plasmacytes* (plasma cells) which synthesize and secrete defensive proteins, the antibodies (immunoglobulins). Mature plasmacytes are rounded or ovoid, up to 15μm across, and have extensive arrays of granular endoplasmic reticulum. Their nuclei are spherical and have a characteristic 'cartwheel' configuration of heterochromatin which is regularly distributed in peripheral clumps. The prominent Golgi complex is also seen with a light microscope as a pale region to one side of the nucleus, whilst the remaining cytoplasm is deeply basophilic due to the abundant rough endoplasmic reticulum. Mature plasma cells are unable to divide.

T-lymphocytes stem from cell stocks initially formed by the bone marrow but later migrating to and multiplying within the thymus before passing into the peripheral lymphoid system where they continue to multiply. When antigenically stimulated, these cells enlarge and their cytoplasm becomes filled with 'free' polysome clusters. The functions of T cells are numerous and incompletely understood, but include the recognition and destruction of virus-infected cells, tumour cells, fungi, tissue and organ grafts, the modulation of B-lymphocytes and interactions with several other defensive cells. Different subsets of T-lymphocytes have distinctive roles in these activities.

Further details of the natural history of lymphocytes may be found on pp. 670–75.

MAST CELLS (MASTOCYTES, HISTAMINOCYTES)

Mast cells are important defensive cells which occur particularly in loose connective tissues and often in the fibrous capsules of certain organs such as the liver (see the review of this cell by Holgate 1983). They are characteristically numerous around blood vessels and nerves. Mast cells are round or oval, about 12 μm in diameter, with many filopodia extending from the cell surface (**1.68**). The nucleus is centrally placed and relatively small, being surrounded by large numbers of prominent vesicles, a well developed Golgi apparatus but scanty endoplasmic

reticulum. The vesicles show a strongly positive reaction with the periodic acid-Schiff (PAS) stain for carbohydrates and with toluidine blue, methylene blue, azure A and alcian blue; they show strong *metachromatic* staining reactions (staining red), also indicating an acid mucopolysaccharide content. Ultrastructurally the membrane-bounded vesicles (or 'granules') vary in size and shape (mean diameter about 0.5 μm) and have a rather heterogeneous content, differing with species. In man the vesicles usually contain dense osmiophilic material which may be finely granular, lamellar or in the form of membranous whorls; these variants may co-exist in the same vesicle, which may also present in places a crystalline substructure. For these reasons they have sometimes been termed *compound granules* (**1.68C**).

The available evidence points to the presence in mast cell granules of a wide variety of substances which can initiate and amplify a great range of defensive responses in the surrounding connective tissue, many of them associated with inflammation. The major granule components are the proteoglycan *heparin, histamine, tryptase, superoxide dismutase, aryl sulphatase, β-hexosaminidase* and various other enzymes; also present are *eosinophil chemotactic factors* (tetrapeptides) and *neutrophil chemotactic factors*. Mast cells may be disrupted to release some or all of their contents either by direct mechanical or chemical trauma, or following contact with particular antigens to which the body has previously been exposed (p. 672). The latter may result from interaction between the antigen and *homocytotropic antibodies* of the IgE and IgG₄ classes associated with the mast cell plasma membrane and may give rise to local responses (e.g.

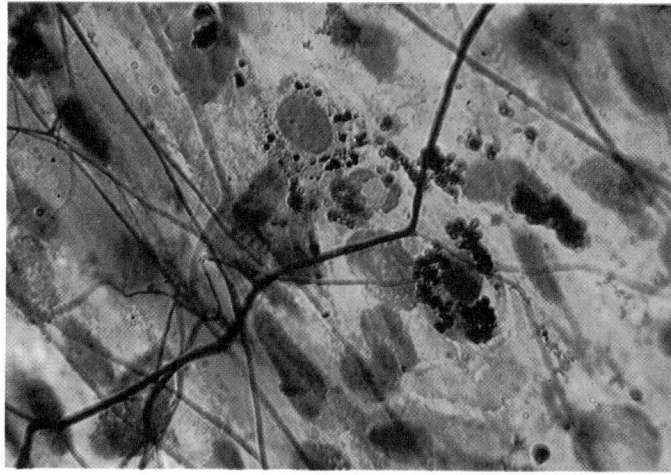

1.67 Loose connective tissue in the mesentery of a rabbit which had previously been injected intraperitoneally with India ink, showing fibroblasts and macrophages. The cytoplasm of the macrophages is full of phagocytosed particles, the collagen is stained pink and the elastin fibres black. Van Gieson's and Verhoeff's elastin stain. Magnification × 1000.

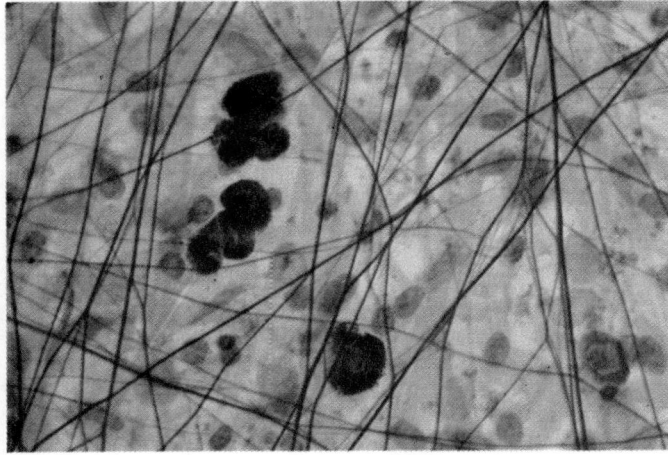

1.68A Mast cells (histaminocytes) in a whole mount spread of loose connective tissue (mesentery) stained with carmine-lithium carbonate to show mast cell granules. The network of elastin fibres has also been counterstained by the Verhoeff method. Magnification × 660.

urticaria) or generalized ones (anaphylactic shock) following the release of large amounts of histamine into the general circulation. They have thus been implicated in many of the phenomena occurring in inflammatory reactions, allergies and hypersensitivity states. The modes of action of their released mediators are exceedingly complex and involve many chemical reactions which alter capillary permeability, can cause smooth muscle contraction, activate and attract various other defensive cells, activate platelets, inhibit clotting and lead to many other local or systemic reactions.

Mast cells closely resemble basophil leucocytes of the general circulation and it is widely considered that they arise from the bone marrow and pass to the tissues perhaps as basophils, migrating through the capillary and venule walls to their final destination. However, there are minor differences between the basophil and the mast cell in terms of their cytochemistry, which suggest either a different lineage for the two, or perhaps that the basophil matures into a mast cell when it reaches its extravascular environment. In connective tissues at least two subclasses of mast cells have been recognized, one resident in the mucosae of the alimentary and respiratory tracts, and others situated elsewhere. They differ in their detailed chemistry and their reactions to drugs (Pearce 1983, Friedmann 1986).

ADIPOCYTES (LIPOCYTES, FAT CELLS)

These occur singly or in groups in the meshes of many but not all connective tissues, being specially numerous in *adipose tissue* (**1.69,70**, see also p. 69). When occurring singly the cells are oval or spherical in shape but when mutually compressed they are polygonal. They vary in diameter, averaging about 50 μm. Each cell consists of a peripheral rim of cytoplasm, in which the nucleus is embedded, surrounding a single large central globule of fat (see Slavin 1985). There is a slight accumulation of cytoplasm around the nucleus, which is oval in shape and appears compressed against the cell membrane by the lipid droplet, as does the Golgi complex. Ultrastructurally, lipid droplets are in direct contact with the surrounding cytoplasm without an enclosing membrane. Many microfilaments are also seen around the lipid vacuole as well as some endoplasmic reticulum and mitochondria.

In sections not specially prepared to preserve fat this is usually dissolved out by the solvents used, particularly xylol or benzene; only the nucleus and the peripheral rim of cytoplasm surrounding a central empty space are left, so that the cell has a signet-ring appearance. The fat is fixed and stained by osmium tetroxide and specially coloured by alcoholic solutions of certain dyes, notably Sudan III, Sudan black and Scharlach R, which are more soluble in fat than in the solvent; lipid is conveniently demonstrated in frozen or cryostat sections (**1.69A**). The fat consists of glycerol esters of oleic, palmitic and stearic acids.

Doubts exist as to whether fat cells are specifically and exclusively concerned with the storage, and perhaps the synthesis,

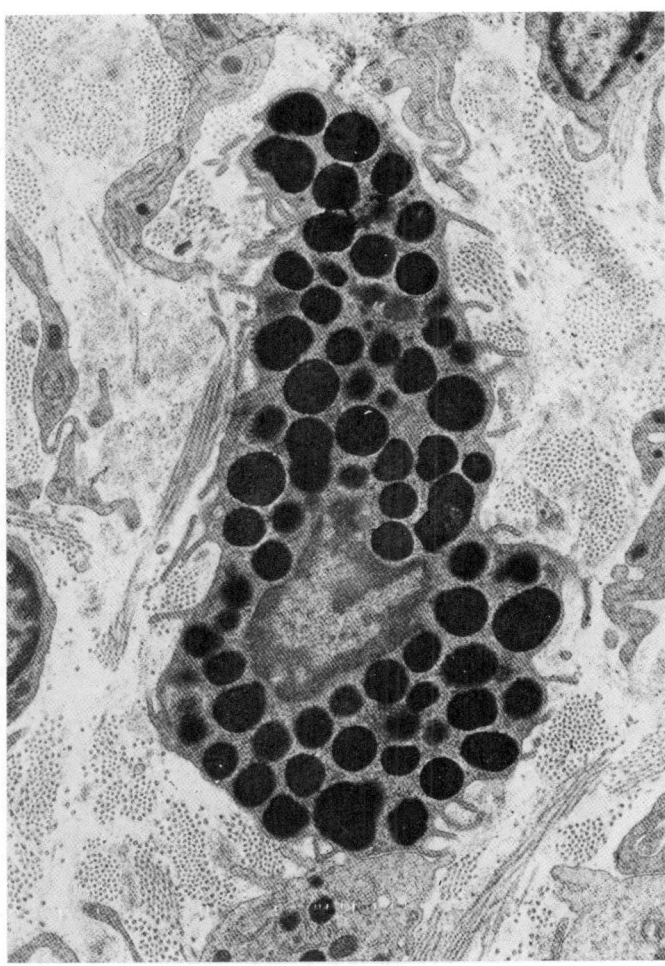

1.68B Electron micrograph of a mast cell showing the large densely staining membrane-bound cytoplasmic granules. Magnification × 6000.

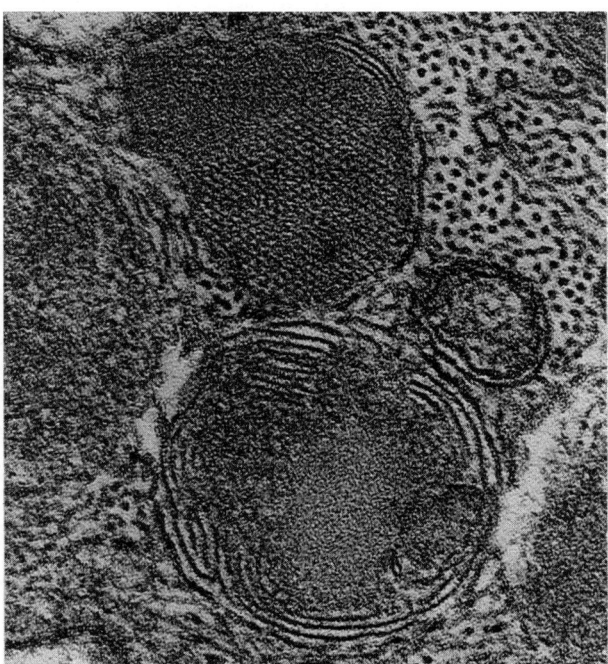

1.68C Higher magnification of part of a mast cell from human skin, showing the complex internal structure of its granules. Cross-sections of microtubules and intermediate filaments are also visible in this section. Magnification × 100 000.

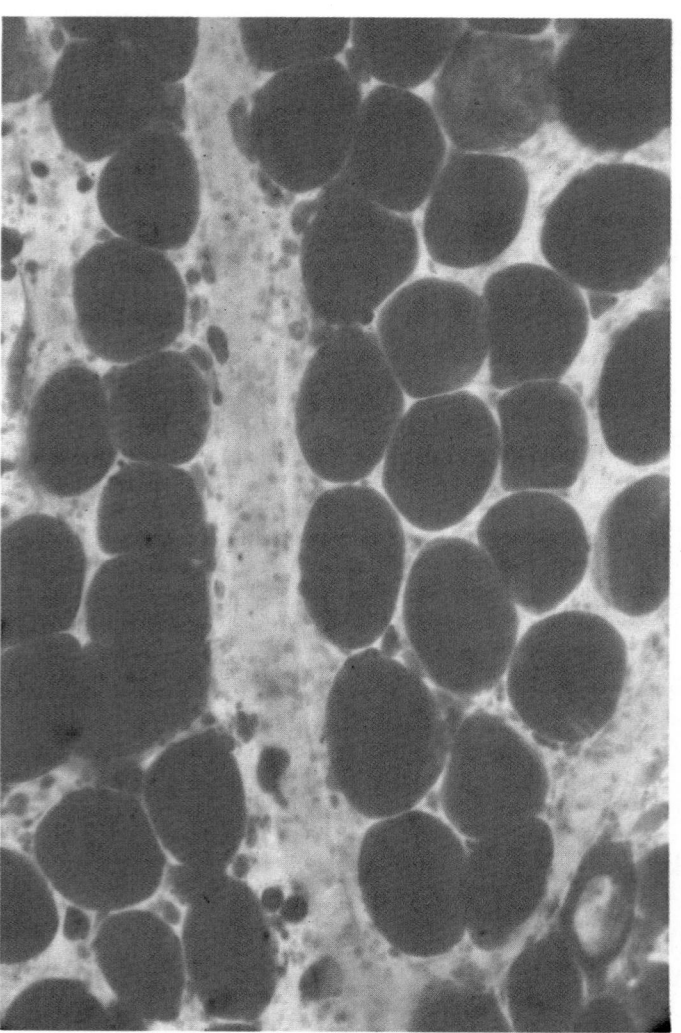

1.69 Adipocytes clustered around a blood vessel in a mesentery, stained as whole mount with Sudan red and mounted in an aqueous medium to avoid alcoholic extraction of lipid. Magnification × 175.

of fat. Prior to the storage of fat within them they are stellate in shape and difficult to distinguish from fibroblasts and, when depleted of fat, they revert to this appearance. As fat accumulates the cells enlarge and become rounded, the fat first appearing as isolated small globules which later coalesce to form a single large droplet. Conversely, during depletion the single large globule

1.70 A group of subcutaneous adipocytes in section after processing for routine histology. Note that the lipid has been totally extracted leaving the cells with only a narrow rim of cytoplasm containing (where sectioned) a nucleus. Compare with 1.69. Magnification × 125.

diminishes in size and then breaks up into droplets while the cells become stellate in shape. Fat cells, however, seem to have a well defined distribution within the body and there is evidence that they are indeed specific cells.

In most mammals which have been studied but especially in those which hibernate, certain deposits of fat are characterized by the presence of a large cell type in which the fat is present as several separate droplets and not as a single globule; their mitochondria also have unusually large and numerous cristae. These deposits of fat are often termed *brown* adipose tissue and are concerned with heat production, mediated by mitochondria. In many species they occur in the interscapular tissue, dorsally, in the subcutaneous tissue of the ventral thorax, axillae, renal capsule and posterior abdominal wall. In human neonates brown fat is present in the interscapular region and may be more widespread. Its importance in later years is uncertain. (See Cannon & Nedergaard 1985 for a review of this subject.)

The mobilization of fat is under nervous or hormonal control and noradrenalin released at sympathetic nerve endings in adipose tissue is particularly important in this respect.

PIGMENT CELLS

These occur in the corium (dermis) of skin, especially in dark races, and in the iris and choroid of the eye. Although some of these cells may be able to synthesize melanin (or related pigments), in which case they are described as *melanocytes* (derived from neural crest origins), the majority contain pigment which they have engulfed after its release from melanocytes, being unable to synthesize it themselves. Such cells are called *chromatophores* (or *melanophores*, when melanin is the pigment involved); they are typically stellate in form and may represent modified fibroblasts. They are generally stellate cells with long processes and numerous dark brown or black granules, believed to be melanin, in their cytoplasm. Their function is generally to prevent light from reaching adjacent cells.

RETICULAR TISSUE

In many situations, as in exocrine and endocrine glands, a fine meshwork of reticulin fibres supports the cellular elements (see p. 58 and Carr 1970). The cells responsible for forming this are indistinguishable from fibroblasts but are sometimes termed 'reticular cells'. This term is, however, also used for a variety of other cells in the reticulo-endothelial and haemopoietic tissues (p. 678), where reticulin is also laid down. The nomenclature and proposed relationships between these various elements is somewhat confused and will not be further explored here.

The Matrix of Connective Tissue

This includes all the extracellular materials of connective tissue, and may be divided into the *fibres* and the *ground substance*.

There are two major classes of fibre: collagen (including reticulin) and elastin. The *ground substance* is composed of hydrated networks of proteins, mainly associated with carbohydrates (glycoproteins and proteoglycans), the large spaces between which contain water, salts and other diffusible substances. These extracellular materials (fibres and ground substance) have complex mechanical properties, conferring varying degrees of strength as well as elasticity, depending on tissue type and site. They form the molecular environment of the contained cells and there are intimate biological interactions between them, governing many aspects of cell behaviour and chemistry. In the skeletal tissues (cartilage and bone), the matrix is stiff and can bear much greater stresses than other types of connective tissue. In cartilage, this is achieved mainly by a high concentration of collagen and proteoglycan complexes whereas, in bone, collagen and mineral crystals within and around the fibres are responsible for its strength.

COLLAGEN FIBRES

These form the predominant fibrous component of ordinary connective tissue and are also major components of cartilage and bone as well as special structures such as the cornea, sclera and vitreous body of the eye and nucleus pulposus of intervertebral discs and notochordal tissue. Tendons, aponeuroses, ligaments, fasciae, sheaths of muscles and nerves, the meninges, dermis and many other structures have a particularly high collagen content. Collagen fibres are flexible but also have a high tensile strength with only a little elastic recoil, and are vital to the mechanical properties of the body in general.

Seen with the light microscope, fresh collagen fibres are white and glistening and show faint longitudinal striations. They form bundles of various size, whose component fibres may leave one fascicle and interweave with others (1.64,67) but, unlike elastin fibres, do not branch. Their precise disposition varies with site (p. 69); in unstretched loose connective tissue they are sinuous, but may straighten under tension to limit the degree of tissue expansion. They stain lightly with eosin but can be dyed strongly with aniline blue (e.g. in Mallory's triple stain) and with aldehyde fuchsin, as in van Gieson's method. Generally, they give only a weak reaction with silver impregnation, except in the case of a fine fibred variety, *reticulin*, which is strongly argyrophilic. Because of their highly orientated fibrillar substructure, collagen fibres show form birefringence when viewed with a polarizing microscope, an effect which can be greatly enhanced by first treating them with chemicals such as Sirius red, whose long molecules bind parallel to those of collagen.

Under the electron microscope, collagen fibres are seen to be composites of narrower (20–200 nm) collagen *fibrils* and these in turn consist of fine (3.5 nm) *microfibrils*, which are aggregates of filamentous *tropocollagen molecules*.

Although there is some variation of detailed structure in different types of collagen (see p. 64), those which form fibres are mostly cross-banded with a characteristic regular repeating pattern of major and minor bands (1.71,72). X-ray diffraction measurements have shown the repeat distance of the major bands to be 67 nm, although in specimens prepared for electron microscopy, a process causing dehydration and slight shrinkage, a value of 64 nm is usually quoted. Longitudinal filaments are also visible in electron micrographs corresponding to linear arrangements of fibrillar tropocollagen subunits (1.71,72,73) whose ordered disposition is also responsible for the cross-striations, which are generated by alignments of chemical cross-links and other minutae of molecular composition (vide infra). The numerous minor crossbands of collagen are asymmetrically arranged within each repeat pattern, since the tropocollagen molecules all point in the same direction within one fibre, which therefore has polarity.

Some types of collagen exist which form only very narrow fibres, or no fibres at all, and in these cross bands are either not clear or are missing entirely (vide infra).

At every level of organization within collagen assemblies, there are cross-links between adjacent longitudinal elements and this is responsible for the great mechanical stability and tensile strength of collagen. Between microfibrils, fibrils and fibres, proteoglycans form regular connecting links (see 1.71), reflecting the periodic organization of the structures to which they are attached (Scott 1988). Individual tropocollagen molecules are also cross-linked by various chemical bonds between specific amino acid units along their length, as indeed are the polypeptide subunits of which they are composed (p. 65).

Tropocollagen

When appropriately treated with acids or alkalis, collagen fibres swell and disintegrate into the filamentous molecules of tropocollagen. In the fresh state these are 300 nm long (280 nm when dried) and 1.4 nm thick, with an Mr of 280 000–540 000 (see Table 1). *Gelatin*, obtained by boiling collagen, consists of disordered networks of such molecules.

Each tropocollagen molecule is composed of three polypeptide (*procollagen*) chains, arranged in a triple helix (1.73). Chemically, procollagen consists of about 100 repeating amino-acid triplets, each with glycine at the third position, the other two varying in

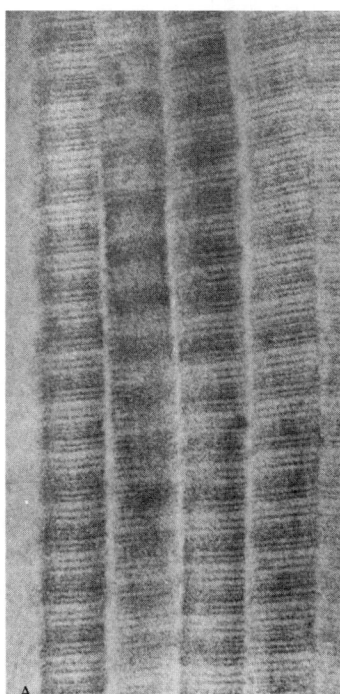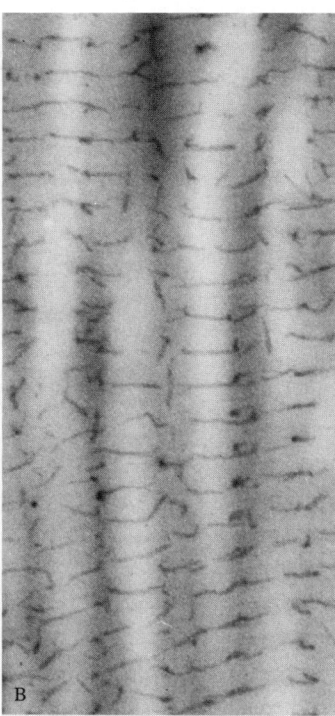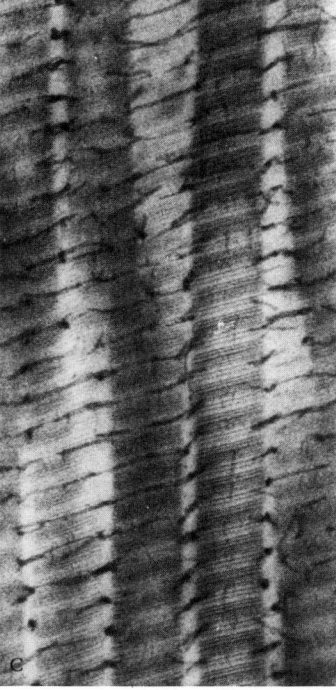

1.71 Transmission electron micrographs of skin collagen fibrils stained in different ways to demonstrate the banding pattern and the presence of associated proteoglycans. In A the fibres have been longitudinally sectioned and stained with uranyl acetate to show the characteristic major 64 nm repeat pattern with several sub-bands visible between. B has been stained with cupromeronic blue demonstrating the filamentous proteoglycans encircling and interconnecting collagen fibrils. In C both stains have been used to show that the proteoglycan connections correspond to the 64 nm cross bands. Micrographs provided by J Scott and Marion Haigh, Department of Chemical Morphology, University of Manchester. Magnifications: A × 100 000; B and C × 65 000.

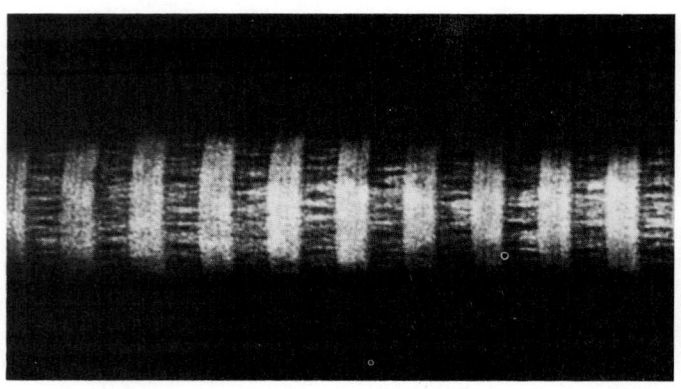

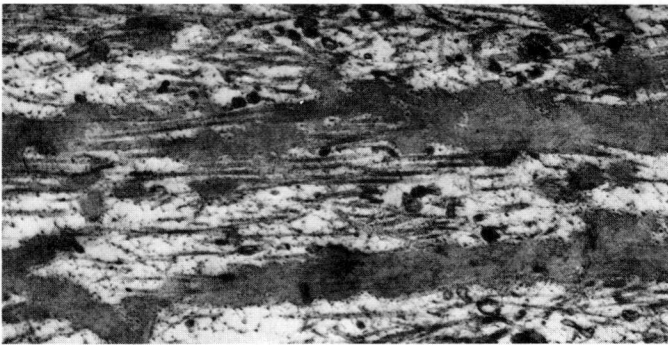

1.71D Portion of a collagen fibril, negatively stained with phosphotungstate to show the substructure of longitudinal tropocollagen filaments and the banding pattern produced by this method. Dark areas represent hydrophilic sites including spaces between filaments (see also **1.72**). Notice that the banding appearance is different from the positively stained fibrils shown in **1.71A**. Magnification × 140 000.

1.71E Developing elastin fibres and surrounding collagen microfibrils in a section stained with phosphotungstate. Notice the apparently amorphous appearance of the elastin. Provided by W Jayaratnam. Magnification × 20 000.

type but often being the amino acids hydroxyproline and hydroxylysine (substances rarely found outside collagen). These, and other amino acids, form covalent bonds between the polypeptide chains, maintaining their triple helical configuration. The hydroxylysine residues also form attachment sites for certain (D-glycosidically linked) carbohydrates, which appear to be necessary for the assembly of tropocollagen into fibrils when released into the extracellular space.

Several variations in amino acid sequences of the procollagen chains have been found, so that many combinations are possible. This variability has been exploited to create collagens with a wide variety of different functions.

Classes of Collagen

Biochemical analysis, and more recently immunohistochemical detection, has shown that there is a large family of chemically related molecular species of collagen (see Table, also the review by Miller 1984). Different tissues and different stages of development or tissue repair show characteristic types or mixtures whose recognized number is still growing. The particular combination of polypeptide chains affects many of the structural features and properties of collagen, including their ability to form fibrils, the

diameter of their fibrils, their binding to adjacent structures, degree of hydration and other aspects of collagen biology. Of these, types I–V are the best understood. The notation used to describe the composition of particular types of collagen is also shown in the Table.

Each tropocollagen molecule is a combination of three polypeptide chains with an α-helical chemical structure (α-chains), each major type having its own particular varieties (e.g. two possible chains, α1[I] or α2(I) for Type I collagen, etc.). There may be three identical chains in one tropocollagen (e.g. [α1(I)]$_3$ for one subclass of Type I), or a heterogeneous combination (e.g. α1(I)$_2$ α2(I).

Type I collagen is the most widely distributed, being the major collagen of ordinary connective tissue including tendons, ligaments, the dermis and bone. Its fibres tend to be quite flexible and are strongly cross-banded. In calcified tissues, regular gaps between the ends of tropocollagen molecules can provide sites for calcium salt deposition and this appears to be an important initiating mechanism for tissue calcification (see p. 302).

Type II occurs in cartilage (p. 285) and a few other sites—the cornea, vitreous humour, notochord and nucleus pulposus—and has a mixture of larger (100 nm diameter) and very narrow (20 nm), short fibrils with indistinct cross bands. The larger fibres are found mainly in articular cartilage, where mechanical reinforcement is required. The collagen of white fibrocartilage is a mixture of Type I, which forms the larger fasciculi, and Type II, whose

Table of collagen types and their distribution adapted from Miller and Gay (1987)

Type	Older terminology	Main source	Number of unique chains	Chain Mr	Molecular species	Form of aggregates
I	Collagen	Virtually all C.T.	2	95K	[α1(I)]$_2$α2(I) [α1(I)]$_3$	Fibres
II	Cartilage collagen	Hyaline cartilage	1	95K	[α1(II)]$_3$	Fibrils or fibres
III	Embryonic collagen	Distensible C.T.	1	110–95K	[α1(III)]$_3$	Reticular networks
IV	—	Basement membranes	2	185–170K	[α1(IV)]$_2$α2(IV) [α1(IV)]$_3$ [α2(IV)]$_3$ RC(IV)	Mesh-like, open, structures ?
V	A–B collagen	Virtually all C.T.	3	200–130K	[α1(V)$_2$α2(V) [α1(V)]$_3$ α1(V),α2(V),α3(V)	Pericellular and perifibrillar of unknown form
VI	Intimal or SC collagen	Placental villi	3	240–140K	α1(VI), α2(VI), α3(VI)	Microfibrils
VII	LC collagen	Placental membranes	1	>170K	[α1(VII)]$_3$	?
VIII	EC collagen	Endothelial cells	1	180–100K	?	?
IX	Type M or HMW–LMW collagen	Hyaline cartilage	1–3	85K	?	Perilacunar of unknown form
X	G or SC cartilage collagen	Hyaline cartilage	1	59K	?	Aggregates of unknown form in zones of hypertrophying cells
K	1α, 2α, 3α-collagen	Hyaline cartilage	3	?	1α, 2α, 3α	Perilacunar of unknown form

narrow fibres are not easily detectable by light microscopy. The finer fibres of the vitreous body (and incidentally of hyaline cartilage) are associated with unusual transparency; because of the larger amounts of attached carbohydrates, they bind much water and tend to form stiff, transparent gels of short, separated fibrils. The smaller fibrils also lack obvious cross-banding and do not join up to form fibres.

Type III forms the collagenous portion of reticulin (which is also rich in carbohydrate); these fibres are narrower than Type I and, seen with the light microscope, form extensive networks of delicate fibres providing support to many soft tissues, e.g. glands, lymphoid tissue, basement membranes, the papillary layer of the dermis, bone marrow etc. (see p. 66). Type III is chemically interesting because of the incomplete removal of terminal registration peptides (p. 66) which are normally cleaved away after secretion. This may in some way limit their growth in diameter. **Type IV** collagen is present in the lamina densa of basal laminae (p. 67); it consists of short filaments which do not form fibrils but aggregate by their ends into small groups of four, which then intermesh to form complex molecular feltworks. The individual filaments resemble procollagen molecules virtually unprocessed after secretion. Collagen of this type is important in forming the main skeleton of the basal lamina to which its other components are attached.

Type V collagen does not form fibrils either and is a relatively minor component of most connective tissues. It is highly glycosylated and tends to be present in pericellular zones, often associated with Type I fibres. Its functions are not certain.

Other minor collagen types. In addition to the foregoing classes, many other varieties of collagen have been described chemically. Some of these may merely be side-products of the synthesis or breakdown of other collagens, others more important but with as yet ill-understood functions. **Type VI**, like Type IV, forms tetramers but these assemble into microfibrils; it is often associated with elastin and may be related to the fine filaments which are a typical of elastic tissue (see p. 67). **Type VII** is a minor component of the total collagen of placental membranes, **Type VIII** is produced by endothelial cells and **Types IX, X** and **K** collagen are synthesized by chondrocytes, (X by hypertrophying cells) and are of unknown function.

Molecular Organization of Banded Collagen

The reason for the striated appearances of Type I and other fibrillar collagens lies in the ordered arrangement of tropocollagen molecules, a phenomenon which has been much studied. The essence of this organization is that, within each microfibril, the molecules, 300 nm long, are all aligned longitudinally in five longitudinal rows, with a gap of 40 nm between the end of each molecule and the next in longitudinal sequence. Thus each tropocollagen unit spans four 67 nm intervals, overlapping the fifth by a short distance of about 27 nm (see **1.72**). Each row is staggered by 67 nm on its lateral neighbour and cross-linked covalently by various amino acid residues which, because of alignments when the assembly is viewed from the side, generate differences in density and staining patterns which appear as complex cross-bands of different sizes. Within a fibril the microfibrillar subunits are aligned by glycoproteins to give a specific cross-striated pattern to the whole structure. In negatively-stained electron microscope preparations, the hydrophilic gaps between molecules are penetrated by stain and these include the 'gap' regions between the ends of tropocollagen units so that a somewhat different pattern of banding is seen relative to sectioned, stained specimens (**1.71**). This regularity allows the addition of new tropocollagen molecules to the ends of fibrils and their assembly into structures of indefinite length.

The Biosynthesis of Collagen

The involvement of fibroblasts in the biosynthesis of collagen (**1.73**) has been studied extensively biochemically and also by autoradiographic electron microscopy (see Prockop et al 1979, Alberts et al 1983). In many respects it appears to be similar to the synthesis of collagen by chondrocytes (p. 285). Amino acids are taken up by the cell and synthesized on the ribosomes of the rough endoplasmic reticulum to form long polypeptides, the *pro-α-*

chains (see review by Uitto & Lichtenstein 1976). These are longer than the final α-chains of tropocollagen because extra polypeptide segments (*extension peptides*), which are used to assemble the total molecule, are present at both ends, later to be removed. As chain synthesis proceeds and the polypeptide moves into the cisternae of the endoplasmic reticulum, various enzymes hydroxylate certain proline and lysine residues to hydroxyproline and hydroxylysine; ascorbic acid (vitamin C), molecular oxygen and other factors are needed for this step, so that vitamin C, iron deficiency and anoxia result in impaired collagen synthesis. Meanwhile, carbohydrate is attached to some hydroxylysine residues; then the three pro-α-chains associate at one end by means of their extension peptides, intertwining as cross-links are formed, to give the triple helix of the *procollagen molecule*. In this form, the molecules are transferred via the Golgi apparatus to the exterior in secretory vacuoles. Once outside, the extension peptides are removed by enzymes (procollagen peptidases) at the cell surface, so forming tropocollagen which aggregates spontaneously (given the correct ionic environment) to create collagen fibres, cross-linked at the lysine and hydroxylysine residues by the

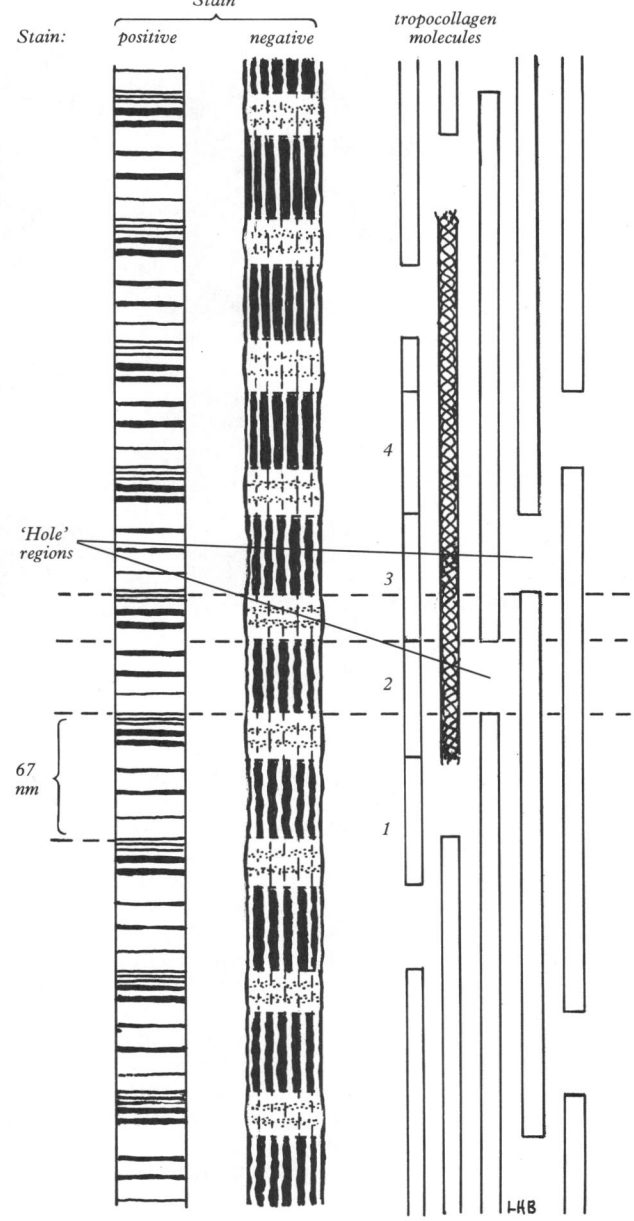

1.72 The organization of tropocollagen molecules in collagen (e.g. type I) and their spatial relationship to the appearance of banding patterns in a sectioned, positively stained fibril (left) and a negatively stained fibril (middle). The numbered segments of one tropocollagen unit indicate the degree of overlap.

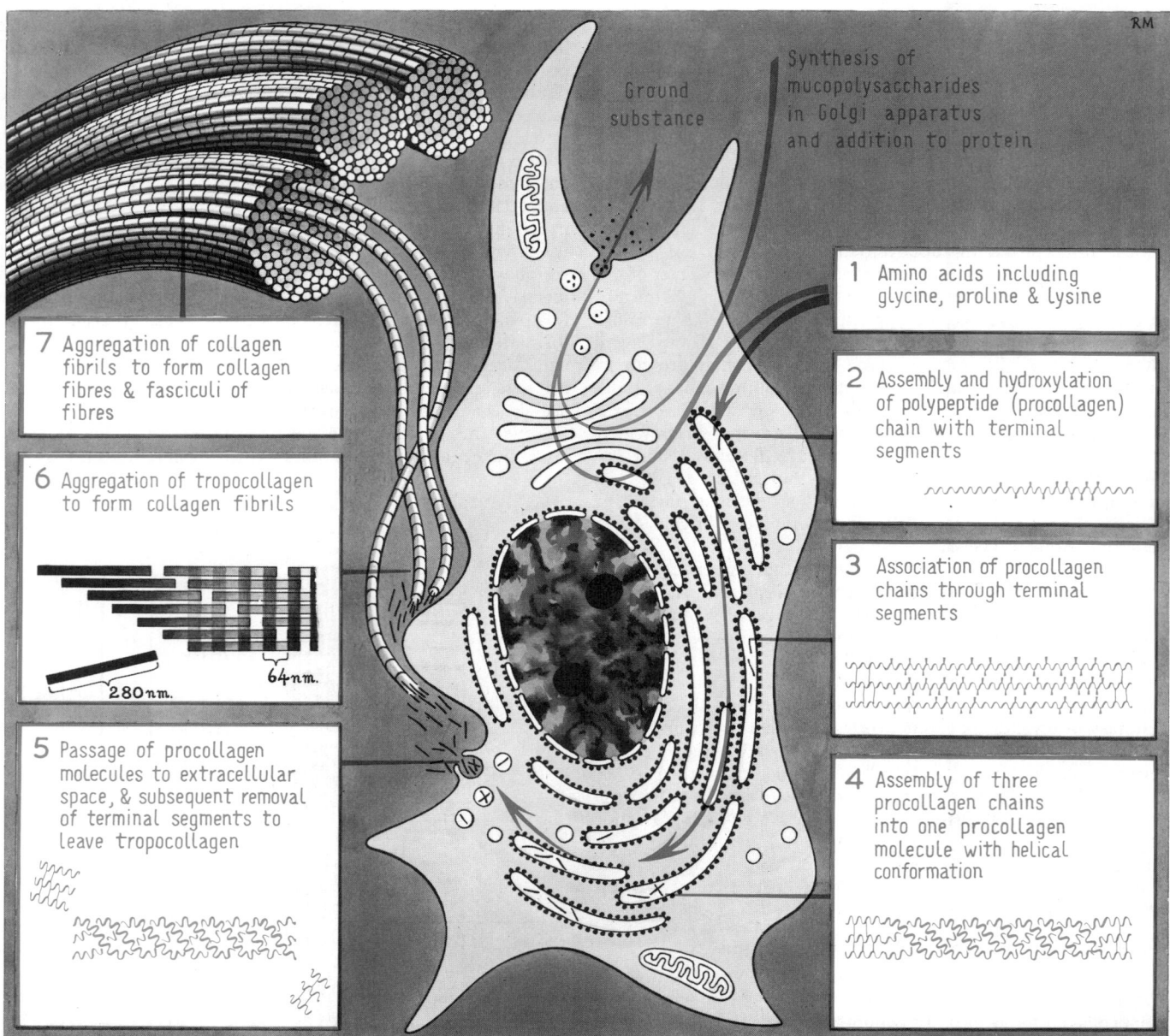

1.73 A schematic diagram showing the major steps in collagen synthesis by fibroblasts in connective tissue.

Within the figure:

Ground substance

Synthesis of mucopolysaccharides in Golgi apparatus and addition to protein

7 Aggregation of collagen fibrils to form collagen fibres & fasciculi of fibres

6 Aggregation of tropocollagen to form collagen fibrils

280 nm.   64 nm.

5 Passage of procollagen molecules to extracellular space, & subsequent removal of terminal segments to leave tropocollagen

1 Amino acids including glycine, proline & lysine

2 Assembly and hydroxylation of polypeptide (procollagen) chain with terminal segments

3 Association of procollagen chains through terminal segments

4 Assembly of three procollagen chains into one procollagen molecule with helical conformation

action of an enzyme, lysyl oxidase, released by the cell. Tropocollagen molecules can be re-aggregated from acid solutions, in their characteristic pattern, by raising the pH to 7. The 67 nm banding pattern can, however, be altered by changing the glycosamino-glycan concentration.

The *direction* of fibre formation appears to be dependent on the stresses acting in the tissue. The relation between stress and the rate and direction of fibre formation is uncertain but may involve movement of fibroblasts along lines determined by piezo-electric currents consequent upon the deformation of pre-formed collagen fibres.

It is remarkable that in many situations collagen is laid down in a precise geometrical pattern, with successive layers regularly alternating in direction. This occurs in the cornea, where successive layers lie at 90° to each other, helping to provide the cornea with its mechanical and unique optical properties. Regularity of pattern exists also in ligaments, tendons, aponeuroses and connective tissue capsules, etc. (**1.77**.)

RETICULIN FIBRES

Fine branching and anastomosing reticulin fibres form the supporting framework of many glands, the kidney and the lympho-reticular tissues (lymph nodes, spleen, etc; e.g. see **6.**175), also in association with basement membranes and often in the neighbourhood of collagen fibre bundles. Unlike other collagen fibres, reticulin takes up silver salts strongly but does not stain strongly with acid fuchsin.

Ultrastructurally reticulin shows a periodic banding which is identical with that of collagen, displays the same X-ray diffraction patterns and exhibits a similar chemical composition. It is now considered to be a type of collagen (Type III) with a particularly high degree of glycosylation (see p. 65). Fibroblasts are responsible for reticulin production in most or possibly all general connective tissues but are usually termed '*reticular cells*' when the amounts of reticulin are high.

ELASTIN FIBRES

These are less frequent than collagen fibres and are, in contrast, yellowish in colour and hence so are the tissues in which they abound. Elastin fibres (**1.**64,67,71E,74) branch and rejoin freely and are usually thinner (1.0–0.2 µm) than collagen fibres, although they can occur as thicker fibres, e.g. in the ligamenta flava, and sheets, as in the fenestrated elastic laminae of the aortic wall. Such fibres and laminae stretch easily, with an almost perfect recoil, Young's modulus of elasticity being 6×10^6 (Bergel 1961), although with advancing years they may calcify, losing elasticity.

66

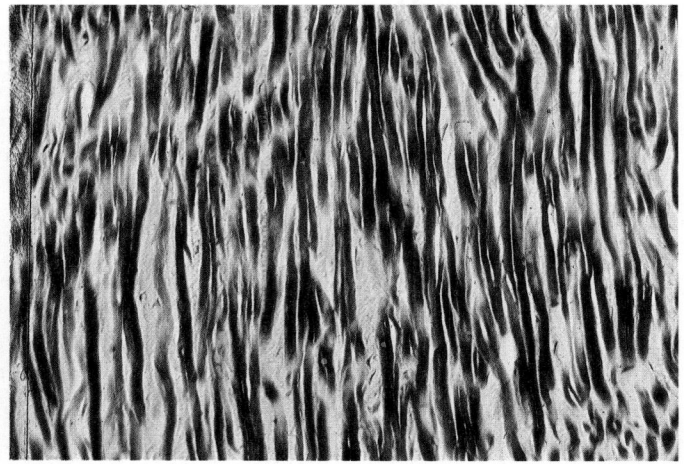

1.74 Longitudinal section through an elastic ligament of an ox showing elastin fibres (black) interspersed with collagen fibres. Verhoeff's and van Gieson's stains. Magnification × 200.

Elastin fibres are highly resistant to hydrolysis in aqueous solutions of high or low *p*H, or boiling water, and to treatment with organic solvents. However, crude preparations of trypsin which contain the enzyme *elastase* and bacterial elastases are capable of dissociating the fibre structure.

Elastin fibres can be stained with orcein, with Weigert's resorcin-fuchsin and with Verhoeff's stain. The fibres display only weak or no birefringence when unstretched, indicating a lack of regular orientation but become strongly birefringent when stretched, indicating the rearrangement of its molecules (strain birefringence).

Seen by electron microscopy, the fibres are composed chiefly of the relatively amorphous protein *elastin* surrounded by fibrils, about 10 nm thick, the 'elastic tissue microfibrillar component' (Sandberg et al 1981). Also frequently present are thin (10–12 nm) filaments of two related proteins (which can be considered as types of elastin): *oxytalan* and *elaunin*; these appear to have similar elastic properties but give some different staining reactions (see Braverman & Fonferko 1982). Traces of collagen, proteoglycans and lipid may also be associated with elastin. Elastin itself is composed of subunits of *tropo-elastin* which have an *M*r of 7000. These subunits are believed to consist of unfolded irregularly coiled polypeptide chains ('random coils'), strongly cross-linked in a three-dimensional network, allowing deformation with storage of mechanical energy when stretched, and recoil. Chemically, elastin contains a high content of the amino acids, valine and alanine, and another named desmosine, which is apparently unique to elastin and has a complex mode of synthesis (see Siegel 1979) involving the enzyme lysyl oxidase. In connective tissue, fibroblasts are responsible for elastin formation, although fibres and sheets of elastin can also be formed by smooth muscle cells. During their formation, the microfibrillar component is laid down first, providing a scaffolding on which the more amorphous elastin is deposited (see Serafini-Fracassini & Smith 1974, Bradamante & Svajger 1977).

THE GROUND SUBSTANCE

This is the non-fibrous part of the matrix, in which cells and fibres are embedded (1.64,65). Except in mineralized connective tissue, the ground substance is a viscous gel containing a high proportion of water, bound largely to long-chain carbohydrate molecules and protein–carbohydrate complexes. Two major classes of such molecules exist, the *proteoglycans* and the *structural glycoproteins*, both being combinations of carbohydrate and protein.

In **proteoglycans**, the carbohydrate is mainly in the form of glycosaminoglycan (GAG) molecules, long unbranched polymers of repeating disaccharides, each unit being a hexosamine residue bearing an acetyl group and a hexuronate or galactosyl residue (Hascall & Hascall 1981). The disaccharide may also bear either a carboxyl or a sulphuric ester group, conferring a strong negative charge on the whole polymer which acts as a poly-anion and therefore can be stained with cationic dyes such as Hale's colloidal iron, alcian blue and various metachromatic stains.

Such molecules range from 6000 to several million in relative molecular mass (*M*r). Because of their complex structure, wide variations in chemistry are possible and various combinations of these molecules can occur at any one site. In tissues, they are almost always found in combination with long filamentous proteins, these complexes being termed glycoproteins or proteoglycans depending on the ratio of the two compounds. A more detailed account of these and other ground substance components is given on p. 285.

The principal glycosaminoglycans are: chondroitin 4-sulphate and chondroitin 6-sulphate (chondroitin sulphates A and C), hyaluronate, dermatan sulphate (chondroitin sulphate B), keratan sulphate and heparan sulphate. All of these, except sometimes the hyaluronate, a huge randomly coiled molecule several million daltons in molecular weight, are normally attached to a core protein from which they extend like bristles on a brush, being repelled from each other by their negative charges. These GAG-protein units may be further grouped as even larger complexes by attachment to hyaluronate molecules; another *link* protein is necessary for this to occur. Such arrangements are usually highly viscous because of the formation of complex three-dimensional networks capable of binding water strongly. This water component is of great importance in facilitating diffusion of metabolites, electrolytes and gases between the capillaries and the cells embedded within the ground substance. The presence of negative charges determines their ion-binding power and they thus act as selective barriers to the passage of inorganic ions and charged molecules.

Structural glycoproteins (Yamada 1981) are filamentous proteins bearing small carbohydrate side-groups; they are important in the adhesion of cells to other matrix components and in general cell–matrix interactions of various kinds. They include *laminin* (*M*r: 220 000 – 800 000, depending on its aggregation of subunits), present in the lamina lucida of basal laminae of a wide variety of cells, and *fibronectin* (*M*r: 450 000) which is found both in the blood plasma (plasma fibronectin) and the connective tissue matrix and appears to act as a glue between different matrix components and between these and fibroblasts (see Hynes 1986); similarly *chondronectin* is a specific adhesive between chondroblasts and type II collagen (p. 286) and *osteonectin* has a similar role in bone (p. 296). Many other tissues may have similar specific adhesion molecules (see, e.g. p. 918).

Because of its high viscosity the ground substance also forms a mechanical barrier; some bacteria secrete the enzyme hyaluronidase which decreases its viscosity, so facilitating the spread of the organisms. The matrix is synthesized by fibroblasts (also by osteoblasts, chondroblasts and, in some sites, by smooth muscle cells) within the granular endoplasmic reticulum and Golgi complex. Its production may be affected by hormones, as shown e.g. in hypothyroidism, when much ground substance may be generated (myxoedema). The ground substance is difficult to demonstrate structurally since it is often highly soluble in water. At the ultrastructural level, fixation and dehydration usually cause the collapse of the large feathery proteoglycan complexes into dense granules, although isolated, negatively stained molecules show the highly extended condition well (see 3.16).

BASEMENT MEMBRANES

These are laminae of dense amorphous material (1.75) which vary in thickness and are associated with many types of cell embedded in or adjacent to connective tissue (e.g. Schwann cells, muscle cells, capillary endothelium and epithelia in general). In some situations they may be particularly thick and so, incidentally, easier to investigate, as in the glomerular membrane of the kidney, the lens capsule, the anterior limiting (Descemet's) membrane in the cornea and Reichert's membrane in the placenta of certain mammals (see Kefalides 1973, Heathcote & Grant 1982).

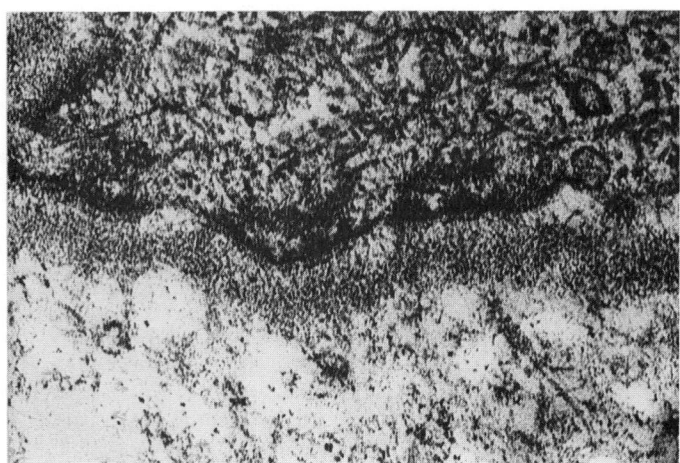

1.75A Section through the base of the epidermis (human) showing the basal lamina. Visible are the lamina lucida just below the dense plasmalemma of the overlying cell and the thick fuzzy lamina densa, below which is the paler area of the connective tissue matrix containing a few microfibrils in section. Also shown are several hemidesmosomes with filamentous connections crossing the lamina lucida. Compare with **1.75B**. Magnification × 60 000.

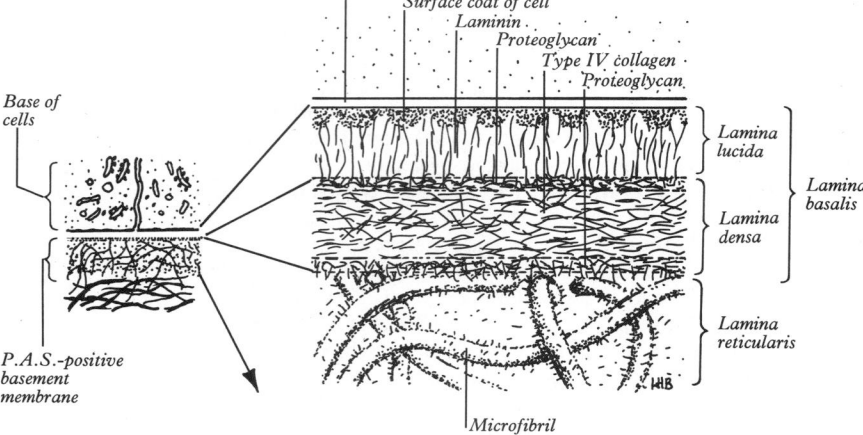

1.75B The arrangement of the components of a basal lamina, as currently envisaged. In this case the basal lamina of the epidermis is illustrated for comparison with **1.75A**. Note that some of the structures depicted in this diagram are based on cytochemical data and are not visible in routine electron micrographs.

Such membranes may form supporting layers for cell attachment, to restrict the passage of large molecules and to anchor the adjacent cells to connective tissue fibres. There is much current research on the more complex roles of this structure in determining metabolic and behavioural patterns of the cells contacting them, a function of great importance in directing the development, maintenance and regeneration of tissues. Changes in their thickness are also often associated with pathological conditions, as in the thickening of the glomerular membrane in glomerulonephritis and diabetes, although it is difficult to allot causes and effects in such cases.

Ultrastructurally, the basement membrane usually consists of two zones: a *basal lamina* lying close to the plasma membranes of the neighbouring cell layer, and a more diffuse *reticular lamina* which merges into the adjacent connective tissue matrix. The basal lamina is usually about 80 nm thick and consists of a fibrillar layer, the *lamina densa* (20–50 nm wide), with an electron-lucent zone, the *lamina lucida*, intervening between the adjacent cell membrane and the lamina densa (see, e.g., Stanley et al 1982). The lamina lucida shows granular or fibrillar features, regularly spaced. The lamina densa is composed chiefly of a delicate network of type IV collagen fibrils and heparan sulphate proteoglycan, whereas the lamina lucida has been shown to contain laminin, fibronectin and various proteoglycans; amongst

other activities these latter three groups of molecules may be important in the adhesion of cells to the basal lamina and adhesion of the basal lamina to the connective tissue matrix. These and perhaps other molecules may also carry information which can guide or otherwise affect the cells they adhere to during morphogenesis and subsequent life. Recently there has been much interest in their role in regeneration in the peripheral nervous system after injury (p. 903) when basal lamina components appear to be involved in guiding outgrowth of axons and the re-establishment of neuromuscular junctions.

Although all basal laminae have a similar form, their detailed organization varies with their site. At the epidermal-dermal junction, special glycoproteins are present in the lamina lucida (see p. 78) and special adhesive plaques (hemidesmosomes, p. 79) also anchor the neighbouring keratinocytes to the lamina densa. The kidney glomerular basal lamina is unusual in having a lucent zone on both sides of the lamina densa (laminae rarae interna and externa) and in various other features associated with the specialized filtering functions of this structure (p. 1401).

The *reticular lamina* is composed of condensed connective tissue matrix including a reticulum of fine collagen (not type IV) in the form of reticulin fibres; this layer may be quite thick and is the main component of the total basement membrane which can be demonstrated at the light microscope level with such staining techniques as the PAS method.

The basal lamina appears to be formed, in most cases, entirely by the adjacent cell layer (epithelium, endothelium, etc.), whereas the reticular lamina is, of course, of fibroblast origin.

Types of Connective Tissue

The connective tissues differ considerably in appearance, consistency and composition in different regions according to the local functional requirements. These differences are related to the predominance or otherwise of one or other of the cell types, the concentration, arrangement and types of fibre and the character of the ground substance. On these bases, *ordinary connective tissues* can be classified into *irregular* and *regular* types, distinguished by the absence or presence of a high degree of orientation in their fibrous elements.

IRREGULAR CONNECTIVE TISSUES

These can be further subdivided into *loose, dense* and *adipose*.

Loose (areolar) connective tissue (1.64,67) is the most generalized form. It is extensively distributed and its chief use is

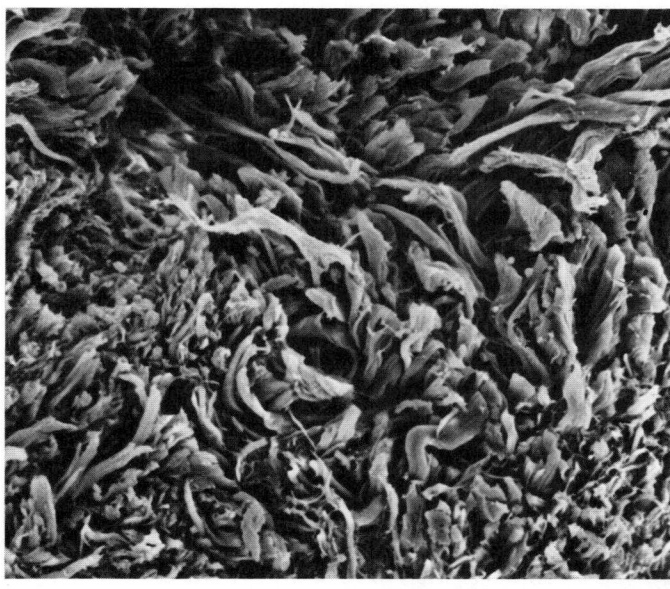

1.76 Scanning electron micrograph of dense irregular connective tissue (human dermis) showing the complex interwoven nature of its fibrous matrix. Magnification × 1000.

to bind parts together, though allowing a considerably amount of movement to take place because of its elasticity. It occurs as the submucous coat in the digestive tract and as subserous tissue. It forms the subcutaneous tissue in regions where this is devoid of fat as in the eyelids, penis, the scrotum and labia. It is also found between muscles, vessels and nerves, forming their investing sheaths and connecting them with surrounding structures. It is present in the interior of organs, binding together the lobes and lobules of the compound glands, the various coats of the hollow viscera and muscle and nerve fibres.

Loose connective tissue consists of a meshwork of thin collagen and elastin fibres interlacing in all directions to give a measure of both elasticity and tensile strength. The large meshes contain the soft, pliable semifluid ground substance and the different connective tissue cells, scattered along the fibres or in the meshes. Occasional adipocytes, usually in small groups, are seen, particularly around blood vessels.

Dense irregular connective tissue is found in regions which experience considerable mechanical stress and where protection is given to ensheathed organs. The matrix contains a high proportion of collagen fibres which form thick bundles (1.76,77,81,85) interweaving in three dimensions and giving considerable strength. Active fibroblasts are few in number and most are flattened with heterochromatic nuclei. The vascular supply is limited, as might be expected.

Examples may be found in the reticular layer of the dermis, the connective tissue sheaths of muscle and nerves and the adventitia of large blood vessels. The capsules of various glands, the coverings of various organs such as the penis and testes, the sclera of the eye and periostea and perichondria are all composed of dense irregular connective tissue.

Adipose tissue. A few fat cells occur in loose connective tissue in most parts of the body. However, in adipose tissue (1.69) these occur in great abundance and constitute the principal component. Adipose tissue occurs only in certain regions and this selective distribution suggests that the fat is deposited in genetically determined sites. It occurs in abundance in subcutaneous tissue, which is sometimes referred to as the *panniculus adiposus*, around the kidneys, in the mesenteries and omenta, in the female breast, in the orbit behind the eyeball, in the marrow of bones and as localized pads in the synovial membrane of many joints. Its distribution in subcutaneous tissue shows characteristic age and sex differences.

Adipose tissue consists of adipocytes (p. 61) embedded in a vascular areolar tissue, which is usually divided into lobules by stronger fibrous septa carrying the larger blood vessels, whence each lobule receives an independent blood supply. Within the lobules the fat cells are round or, when mutually compressed, polygonal. Areolar tissue and septa both contain the other cellular components of fibrous tissue. These fat deposits serve as energy stores, sources of metabolic lipids, thermal insulation (subcutaneous fat), mechanical shock-absorbers (e.g. soles of the feet, palms of hands, gluteal fat, synovial membranes etc.), and may also serve more complex, cosmetic or behavioural functions. In some regions the connective tissue framework of the fatty tissue often contains large amounts of elastic tissue and in emaciation these deposits tend to be spared until a late stage, where mechanically important, e.g. plantar and palmar superficial fasciae, perirenal fat. In the newborn human, as in many adult lower mammals, there are areas of specialized *brown fat*, which is capable of generating heat by the breakdown of nutrients in unusual mitochondria (see also pp. 26, 62).

REGULAR CONNECTIVE TISSUE

This includes those highly fibrous tissues with fibres regularly orientated, either to form sheets such as fasciae and aponeuroses, or thicker bundles as ligaments or tendons (1.77,78). The direction of the fibres within such structures is related to the stresses which they undergo but there is considerable interweaving of fibrous bundles, even within tendons, which increases their structural stability and resilience.

The fibroblasts which secrete the fibres may eventually become trapped within the fibrous structure, where they are compressed, and present highly angular outlines. Cross-sections of tendons show inactive fibroblasts with stellate profiles and small heterochromatic nuclei. Fibroblasts on the external surface may be active in continued fibre formation and they afford a pool of cells able to repair damage (see McMinn 1969).

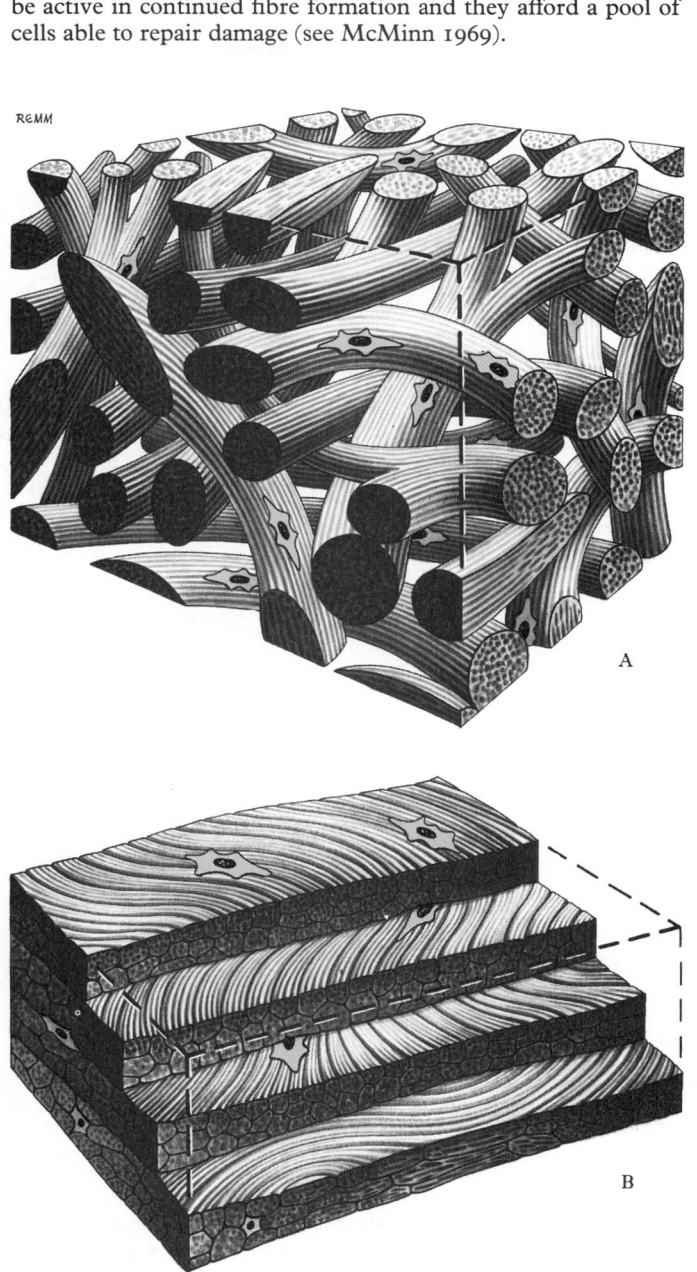

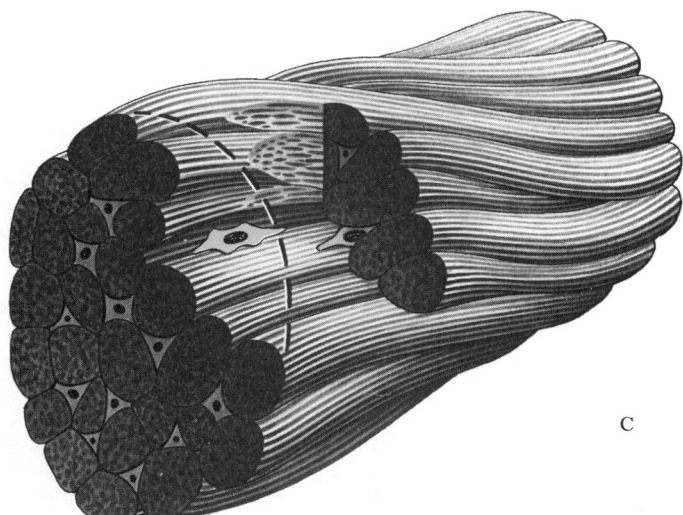

1.77 Three types of arrangement of collagen fibres in: (A) dense irregular connective tissue; (B) a ligament; (C) a tendon.

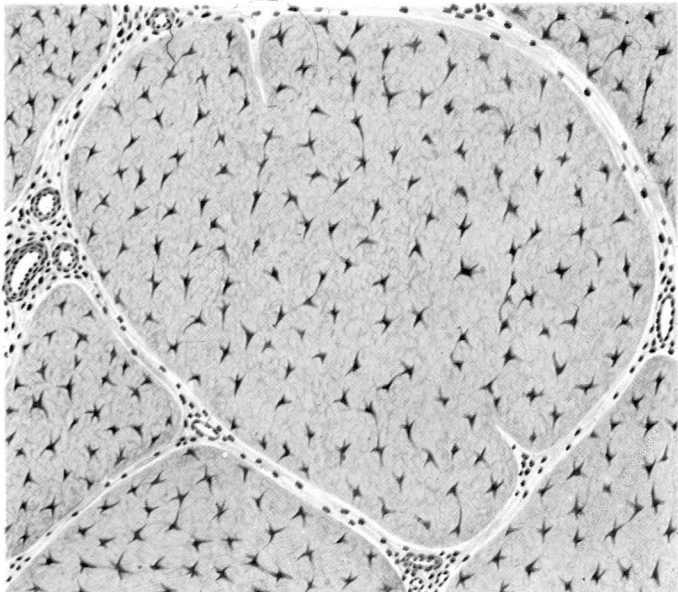

1.78A Transverse section of a tendon showing fibrocytes enclosed between bundles of collagen fibres.

1.78B Longitudinal section through a tendon, showing the parallel organization of its collagen fibres. A few long, flattened fibrocyte nuclei are also visible. Haematoxylin and eosin. Magnification × 200.

Regular connective tissue is predominantly collagenous but in some ligaments elastic components also occur, as in the ligamenta flava of the vertebral laminae and in the vocal folds. Some elastin is also present between the collagen lamellae of many other ligaments and fasciae. In other sites, the collagen fibres may form precise geometrical patterns, as in the cornea (see p. 1183).

Mucoid tissue is a fetal or embryonic type of connective tissue (1.79), found chiefly as a stage in the development of connective tissue from mesenchyme. It exists in the 'jelly of Wharton', which forms the bulk of the umbilical cord, and consists of a copious matrix, largely made up of hydrated 'mucosubstances' and a

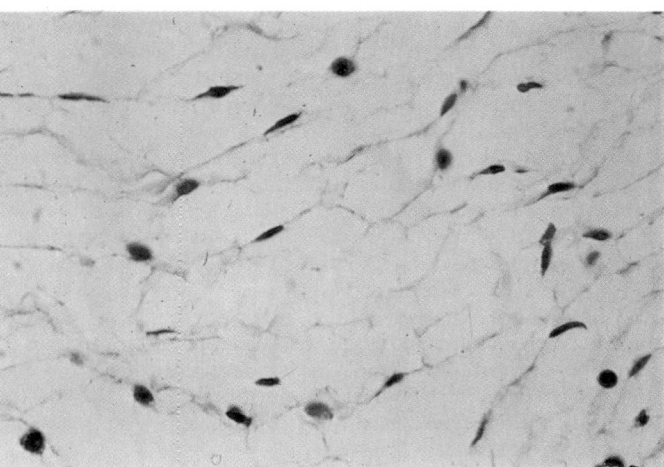

1.79 A section of fetal mesenchyme showing mucoid tissue sparsely populated with cells. Magnification × 500.

fine meshwork of collagen fibres, in which nucleated cells with branching process (probably fibroblasts) are found (Boyd & Hamilton 1970). However, the stellate appearance of the cells in Wharton's jelly probably results from fixation and staining of excised cords which have suffered a haemodynamic collapse (Parry 1970). When this is avoided the cells are aligned and strap-like. Such findings may well apply to other situations in which mucoid tissue is found. Usually few fibres occur in typical mucoid tissue, though at birth the umbilical cord shows a considerable development of perivascular collagen fibres; after birth it is still to be seen in the pulp of a developing tooth. In the adult the vitreous body of the eye is a persistent form of mucoid tissue in which the fibres and cells are very few in number, and the nucleus pulposus of the intervertebral disc is similar.

Pigmented connective tissue, such as occurs in the choroid and in the lamina fusca of the sclera of the eye, is composed of loose connective tissue, in which large numbers of pigment cells (melanocytes) are also present.

VESSELS AND NERVES OF CONNECTIVE TISSUE

Generally, few blood vessels supply the tissue itself, the numbers of cells and therefore the metabolic demand being relatively low, although many blood vessels may pass through *en route* for other tissues, as in loose connective tissue. In dense collagenous or elastic tissues the blood vessels usually run parallel to and between the longitudinal bundles, sending communicating branches across the bundles; in some of its forms, as in the periosteum and dura mater, they are fairly numerous. *Lymphatic vessels* are very numerous in most forms of connective tissue, especially in the loose tissue beneath the skin and the mucous and serous surfaces. They also occur abundantly in the sheaths of tendons, as well as in the tendons themselves. *Nerve* endings of various kinds also terminate in connective tissues and provide sensory innervation detecting mechanical stresses, painful stimuli and thermal changes (see p. 902).

THE INTEGUMENT

Introduction

The integument, or skin (cutis) is an anatomically and physiologically specialized boundary lamina essential to life. It is a major organ of the body, forming about 8% of its total mass and having an area of between 1.2–2.2 m². In total thickness it ranges from about 1.5–4.0 mm. Skin covers the entire external surface of the body, including the external auditory meatus and the lateral

aspect of the tympanic membrane. It is continuous with the mucosae of the alimentary, respiratory and urinogenital tracts at their respective orifices, where the specialized skin of mucocutaneous junctions occur; it also fuses with the conjunctiva at the margins of the eyelids and with the lining of the lacrimal canaliculi at the lacrimal puncta.

Structurally, skin is complex and highly specialized, as might be expected of the major interface between the body and its

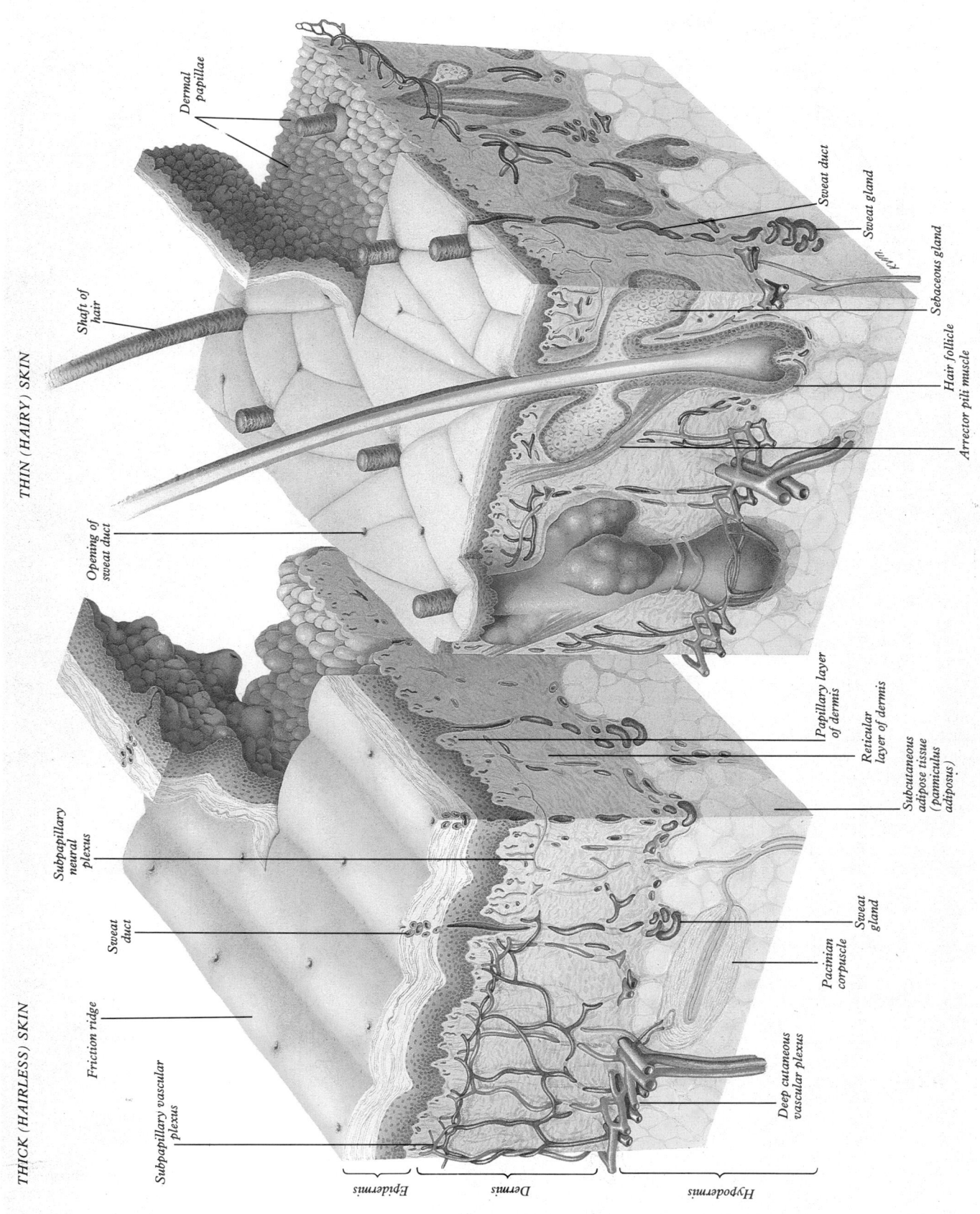

THIN (HAIRY) SKIN

THICK (HAIRLESS) SKIN

Dermal papillae

Shaft of hair

Opening of sweat duct

Sweat duct

Sweat gland

Sebaceous gland

Hair follicle

Arrector pili muscle

Papillary layer of dermis

Reticular layer of dermis

Subcutaneous adipose tissue (panniculus adiposus)

Subpapillary neural plexus

Sweat duct

Friction ridge

Subpapillary vascular plexus

Sweat gland

Pacinian corpuscle

Deep cutaneous vascular plexus

Epidermis

Dermis

Hypodermis

1.80 Schema showing the organization of skin, comparing the structures present in thick, hairless (plantar and palmar) skin and thin, hirsute skin. The epidermis has been partly peeled back in this picture, to show interdigitating dermal and epidermal papillae. For details of the innervation (yellow), see p. 912.

environment (see the reviews by Montagna 1968, Jakubovic & Ackerman 1985). Microscopically, it is formed as an intimate association between two distinct tissues: keratinized stratified squamous epithelium superficially, the *epidermis*, and a deeper layer of moderately dense connective tissue, the *dermis* (1.80,81). Because of this combination, it is within limits a most effective barrier against microbial invasion and dehydration and against mechanical, chemical, osmotic, thermal and photic damage. It limits and regulates heat loss, is a major sensory surface with elaborate systems of varied receptor types, is capable of limited excretion and absorption and carries out many specialized biochemical functions, including the formation of Vitamin D_3 from precursor steroids under the action of ultraviolet light. Skin also has good frictional properties, assisting locomotion and manipulation by its texture. Its sensory endings provide a means for subtle social communication and, in the case of facial skin, it can signal emotional states by means of muscular and vascular responses. Normal hormonal changes can affect the appearance and function of specific areas of the integument, e.g. the activity of its glands and growth of hair. The state of health, age and many other aspects of life are also reflected in the condition of the skin, and so clinically its appearance is important in the diagnosis of a wide variety of pathologies.

The outer surface of skin is covered by various markings, some of them large and conspicuous and others delicate to the point of being microscopic (see Montagna & Lobitz 1968, Millington & Wilkinson 1983) or only revealed after manipulation or incision of the skin. These are often referred to collectively as *skin lines* and include: externally visible grooves of the epidermis, i.e. *flexure lines*, positioned near or opposite synovial joints; *tension lines* which form a delicate pattern of geometric shapes over the surface of thin, hairy skin; and *papillary ridges*, forming fields of parallel lines on the thick, hairless skin of hands and feet. There are also lines which appear when tension is removed from skin (*relaxation lines*) and lines or bands which may appear after rapid local expansion of underlying structures, e.g. during pregnancy (*striae gravidarum*), etc. Pigmentation of the skin also varies in amount, type and distribution, with the occurrence in some races of abrupt boundary lines between darker and lighter areas on the upper limbs (Voigt lines) and, in young children, of blue-grey patches ('mongolian spots') on the trunk. Small, localized accumulations of pigment cells in the epidermis are *naevi* (moles). More hidden variations in structure are also revealed when skin is incised, a factor of importance in surgery (Langer's lines). These markings are considered in greater detail on pp. 80–82.

In addition to these variations, the appearance of skin is also affected by many other factors, e.g. size and shape and distribution of hairs and their follicles, of skin glands (sudorific, sebaceous and apocrine, see pp. 92–94), changes associated with maturation, ageing, metabolic changes, pregnancy and many other considerations.

Types of Skin

Although skin over the entire body is fundamentally of similar structure, there are many local variations in thickness, mechanical strength, degree of keratinization, sizes and numbers of hairs, frequency and types of glands, pigmentation, vascularity, innervation and other features. However, it is useful to distinguish between two major classes of skin which cover large areas of the body, but show important differences of detailed structure and functional properties; these are *thin, hairy (hirsute) skin*, which constitutes the great majority of the body's covering, and *thick, hairless (glabrous) skin* forming the surfaces of the palms of the hands, soles of the feet and flexor surfaces of the digits (1.80, 81,86–89). As mentioned above, these two classes differ in their surface markings, thick skin having complex patterns of friction ridges absent elsewhere on the body. As implied by their names, they also differ in thickness of both epidermal and dermal components, and in the presence or absence of hairs with attendant sebaceous glands and arrector pili muscles ('pilo-sebaceous units'). These dissimilarities reflect their distinctive functions. Thick hairless skin forms frictional surfaces for manipulation and

locomotion and requires extra strength and numerous sweat glands for cooling during sustained activity (a design feature perhaps more appreciated in warmer climates) and dense clusters of sensitive sensory endings with spatial discrimination unimpeded by the presence of hairs. Thin hairy skin is responsible for the general cutaneous functions over the remainder of the body.

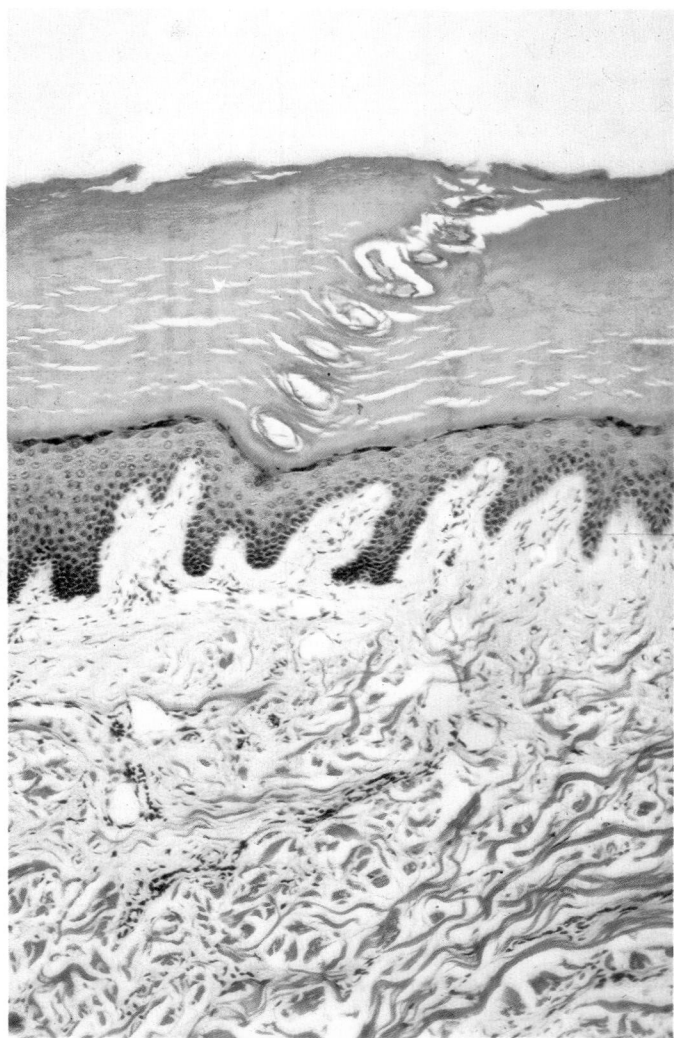

A

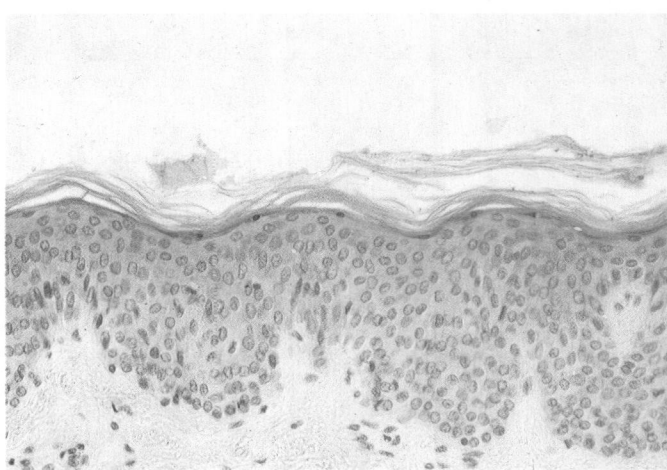

B

1.81 A. Low-power light micrograph of a vertical section through skin from the sole of the human foot: Mallory's triple stain. The finer collagen fibres of the papillary and reticular layers are stained blue and the coarser collagen of the reticular layer stains red. The epidermis is differentiated into the stratum corneum and stratum lucidum above (red-brown) and the strata basale, spinosum and granulosum, which stain blue-grey. A spiral duct from a sweat gland is also seen in vertical section. The base of the epidermis is irregularly ridged. B. Vertical section through the epidermis of the scalp between hair follicles. Note the reduced keratinized layer (compare with A). Haematoxylin and eosin. Magnifications: A × 150; B × 200

Minor areas of skin also have special features which do not fall into either of the major categories. The *mucocutaneous junctions* of the lips, outer rim of the anal canal and urethral openings each have a characteristic histology; e.g. the lips have a delicate epidermis lacking glands or hairs, as also does the covering of the glans penis and glans clitoridis.

Microscopic Structure of Skin

THE EPIDERMIS

The epidermis (1.80–85) is composed of keratinized stratified squamous epithelium (p. 55). In this tissue there is a continuous replacement of cells, a mitotic layer at the base replacing cells shed at the surface. As they move away from the base of the epidermis, they undergo progressive changes in shape and content, eventually transforming from polygonal living cells to dead, flattened squames full of the protein keratin. The cells which differentiate in this way are called *keratinocytes*, but there are also other less numerous cell types in the epidermis not engaging in constant replacement. These include pigment-forming *melanocytes*, phagocytic *Langerhans cells* and neurally-associated *Merkel cells*. It is usual to divide the epidermis into a number of strata, representing different stages in keratinocyte maturation from deep to superficial as follows: stratum basale, stratum spinosum, stratum granulosum, stratum lucidum (where present) and

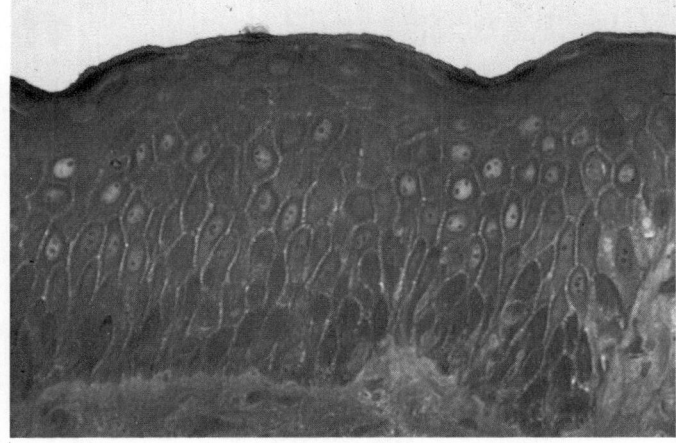

1.82A Section through human epidermis from the dorsum of the thorax. Note the cellular organization of its layers including the densely stained stratum corneum at the surface. Compare with 1.93. Epoxy resin section stained with toluidine blue. Magnification × 410.

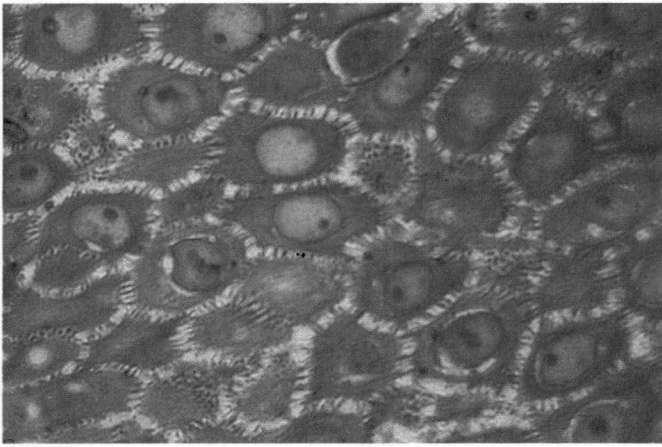

1.82B High magnification of the same specimen shown in 1.82A showing part of the stratum spinosum. The keratinocytes are held together by numerous desmosomal contacts giving the characteristic spiny appearance of cells in this layer (compare with 1.82A and 1.83). Magnification × 1650.

stratum corneum. The first three of these layers are metabolically active and are often grouped together as the *germinative zone* or *stratum Malpighii*. The more superficial strata of cells achieving terminal keratinization constitute the *cornified zone*. The term stratum germinativum is, however, quite commonly used as a synonym for the stratum basale, and stratum Mapighii for the stratum spinosum alone. We have not used these two collective names therefore in the present description and have treated each stratum separately. In addition to the different layers, various structures formed by ingrowth or other modifications of the epidermis (*epidermal appendages* or *adnexa*) include *glands* (sebaceous, sudorific, apocrine, ceruminous, tarsal, Meobomian glands, glands of Moll), *hair follicles, hairs* and *nails*. (For reviews of skin structure, see Montagna 1962, Breathnach 1971, Jakubovic & Ackerman 1985, Thody & Friedmann 1986.)

Epidermal Strata and Structural Maturation of Keratinocytes

The stratum basale includes the single layer of cells which lies in contact with the basal lamina. Keratinocytes are formed in this layer by division of a population of *stem cells*, then move apically, dividing again two or three times in the deeper regions of the stratum spinosum before ceasing mitotic activity (1.82–84). Structurally, the cells of the stratum basale are heterogeneous, since they comprise stem cells and keratinocytes in various stages of early maturation, as well as non-keratinocytes (vide infra). The majority of basal cells are columnar to cuboidal and are attached by hemidesmosomes to the basal lamina (p. 68). In most cases their basal surfaces are highly folded, interdigitating with projections of the basal lamina; some cells, however, have smoother bases and may constitute a separate group of keratinocytes with different turnover rates, present in epidermal papillae (Lavker & Sun 1982).

Ultrastructurally, basal cells have many free polyribosomes, many mitochondria and nuclei varying from euchromatic (in stem or young cells) to heterochromatic (older keratinocytes). They have many cytoskeletal filaments, including actin microfilaments and keratin filament bundles, some of them anchored to desmosomes or hemidesmosomes. Melanin granules are also present in many of these cells (vide infra).

The stratum spinosum (prickle cell layer) contains more mature keratinocytes several layers deep, the cells packed closely and interdigitating by means of numerous projections and indentations at their surfaces which are linked by many desmosomes. Internally, these cells possess large numbers of keratin filament bundles (formerly termed tonofibrils or tonofilaments), many of them attached to desmosomes at the cell surface. Because the cells are anchored together in this way, this stratum provides much tensile strength and coherence to the surface of the integument. When skin is prepared for routine histology, these cells tend to shrink away from each other except where joined by their desmosomes, so giving them a spiny appearance (hence 'prickle cells' and the name of this stratum). The nuclei of these cells are moderately euchromatic, with prominent nucleoli. In the cytoplasm there are numerous polyribosomes and typically vacuoles containing melanin granules (melanosomes). In heavily pigmented skin, these may be numerous and each vacuole may have several granules within it. The melanin is derived from epidermal melanocytes (vide infra); the granules are most numerous in the deeper parts of this stratum and are gradually degraded by the keratinocytes so that in the surface layers they are usually absent.

The stratum granulosum. In this layer various changes already begun in the upper stratum spinosum accelerate dramatically and a drastic change in keratinocyte structure occurs. The cells become flattened and accumulate many large (up to 5 μm), dense, basophilic granules (*keratohyalin*). Their nuclei become pycnotic and begin to disintegrate and other cellular organelles degenerate. Ultrastructurally, such cells contain feltworks of keratin filament bundles enclosed or enmeshed by the irregular masses of densely staining *keratohyalin granules*. These appear to contain some carbohydrate and lipid, but the predominant material is a highly phosphorylated protein rich in the amino acid, histidine (*histidine-rich protein*, also called *stratum corneum basic*

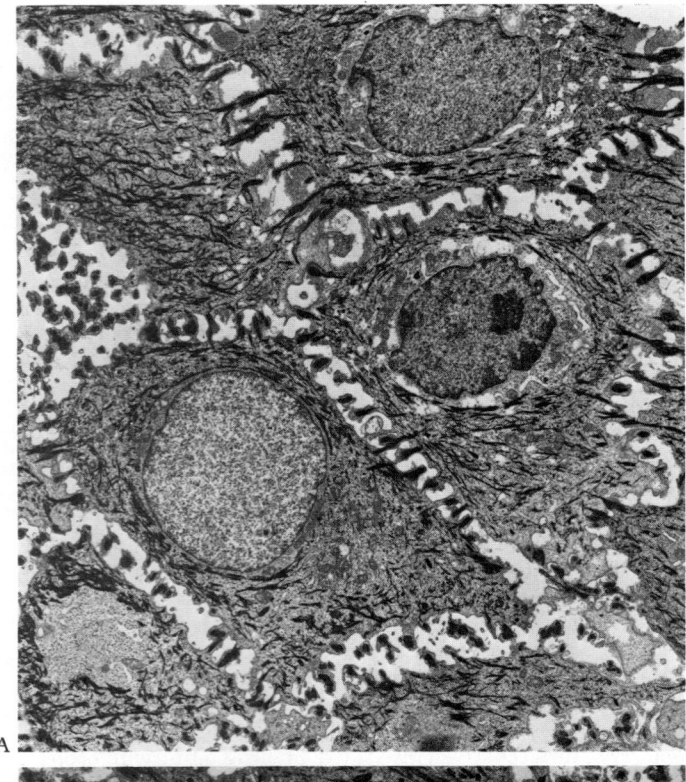

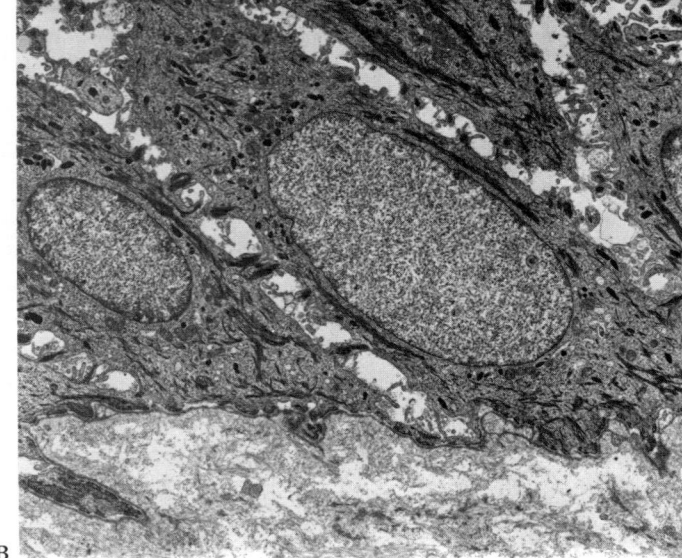

1.83 Electron micrographs of human epidermis showing (A) a section through the stratum spinosum with several keratinocytes interconnected by desmosomal contacts attached intracellularly to interweaving bundles of keratin filaments; (B) a section through the stratum basale where it abuts the basal lamina and papillary layer of the dermis (below). Note the presence of numerous melanin granules in the epidermal cells. Magnification: A × 50 000, B × 7000.

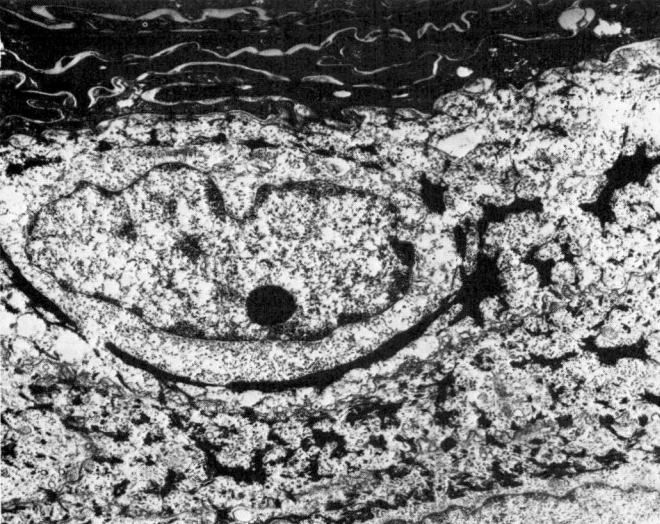

1.83C Electron micrograph of human skin in vertical section. This shows the transition between the stratum spinosum (below), stratum granulosum (middle) and stratum corneum (the dark laminae above). The microfilamentous bundles of the keratinocytes, below, become denser and more compact in the stratum granulosum; finally the cells flatten, becoming scale-like and electron dense. Magnification × 8000.

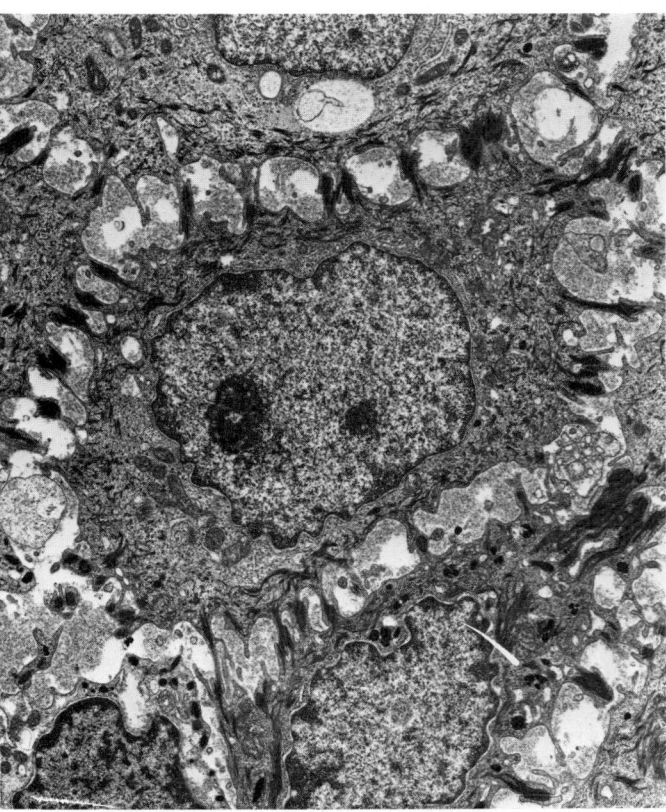

1.83D Electron micrograph of a keratinocyte from the stratum spinosum showing details of its cytoplasmic and nuclear structure and the desmosomal attachments between cells. Magnification × 14 000.

protein or *profilaggrin* which eventually matures into the protein filaggrin: (vide infra). Also visible are smaller, finely granular *dense homogenous bodies* and elliptical membrane-bounded vacuoles, 0.1 by 0.3 μm in diameter, known as *membrane-coating granules, lamellar bodies* or *Odland bodies* (see the review by Hayward 1979), containing carbohydrate, lipid and hydrolytic enzymes, in the form of transverse or obliquely orientated lamellae. These eventually discharge their contents into the intercellular spaces of the upper layers of this stratum, creating a thick, waterproof layer of lamellar cement, rich in neutral lipids, between the cells of the stratum corneum.

The stratum lucidum. This layer is only found in thick, glabrous skin and represents a rather poorly understood stage in keratinization; characteristically, it stains more strongly than the

stratum corneum with acidic dyes, is more refractile optically and often contains nuclear debris. Ultrastructurally, its cells closely resemble those of the stratum corneum.

The stratum corneum consists of closely packed layers of flattened, dead keratinocytes (*squames*). In thin skin, e.g. of the scalp, this stratum is only a few cells deep, but in thick skin it may be more than 50. The cells are compact and contain high concentrations of keratin filaments each about 8–10 nm thick, often lying parallel and 8 nm apart, and embedded in another protein, *filaggrin*, which is derived from histidine-rich protein. Disulphide bond formation appears to stabilize this protein complex.

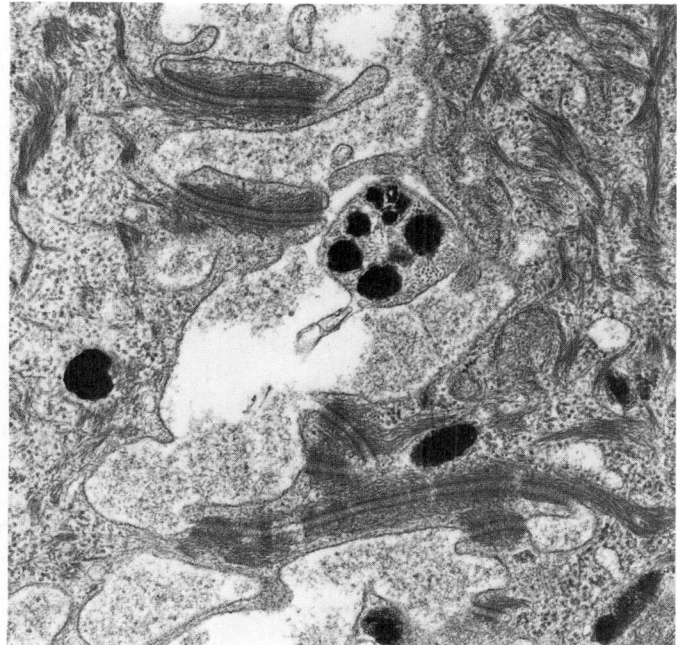

1.83E Higher magnification of desmosomal contacts between adjacent keratinocytes. Note the oblique arrangement of desmosomes and the intracellular keratin filaments. Also visible between keratinocytes is a circular profile of a melanocyte dendrite containing dense melanosomes, some of which have also been endocytosed by neighbouring keratinocytes. Magnification × 25 000.

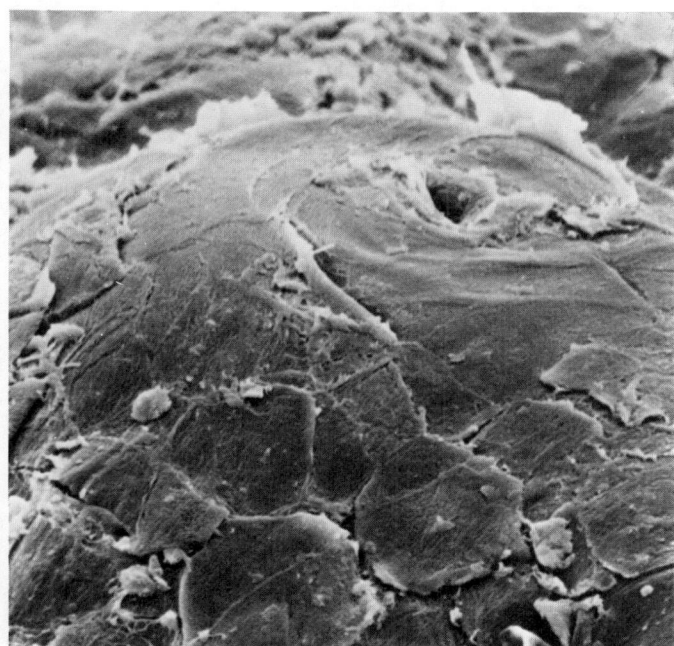

1.83F Scanning electron micrograph of a portion of the epidermal surface surrounding the aperture of a sweat duct. Several scale-like corneocytes, polygonal in form, are visible. Magnification × 2000.

This regularity can be detected in whole skin preparations by X-ray diffraction studies and was known before the advent of electron microscopy.

The internal surface of the cell membrane is also coated with a dense 12 nm layer, the *protein envelope*, composed of a number of highly cross-linked, chemically inert proteins including *keratolinin* and *involucrin* (Mr 36 000 and 140 000 respectively) which are probably derived from the dense homogeneous bodies of the stratum granulosum. The intercellular space is filled with a thick layer of lipidic cement (vide supra), crossed by occasional

remaining desmosomes which persist from the deeper layers. Eventually the cells disengage from each other and become detached from the epidermal surface (desquamation). The causes of this event are not certain but may involve the activation of lipases released earlier from the membrane-coating granules.

The cells of this layer, *corneocytes*, are quite large, with a surface area ranging from about 800–1100 μm². They can be stripped away with adhesive tape for histological or chemical analysis.

Cell Turnover in the Epidermis

Disturbances in division and maturation rates are common in various skin diseases and have received close scrutiny in both normal and pathological states. In animal models, epidermal cell population dynamics can be analysed by pulse-labelling with [³H]-thymidine, with subsequent microautoradiography of sections to determine the position and degree of nuclear labelling. Human skin samples can also be cultured for short periods and examined in the same way. These and other studies have shown that replacement rates (i.e. the time taken for a newly-formed keratinocyte to pass to the surface and be shed) vary with body position, epidermal thickness, degree of skin abrasion, ambient temperature, time of day, hormonal changes, age and many other factors. Typical replacement times have been variously estimated to range in human skin from 45 to 75 days (Halprin 1972, Bergstresser & Taylor 1977), although they may be much lower than this in thinner epidermal regions. In some pathologies of the skin, turnover rates can be exceedingly rapid; e.g. in psoriasis it may be as little as eight days and in such cases the surface cells do not have time to keratinize properly so that the normal barrier functions of the skin break down.

The reasons for the shedding of keratinocytes are not known but it still occurs even if replacement from below ceases, e.g. when mitotic activity is inhibited.

Epidermal proliferative units (1.84B). If the epidermis of some mammalian species (e.g. rats) is treated with dilute alkali before histological sectioning, the cells of the stratum corneum swell and can then be seen to be stacked in regular columns. From this observation, coupled with autoradiographic analysis of cell division and migration within the epidermis, the concept of the Epidermal Proliferative Unit (EPU) arose (Potten 1975, 1983). According to this view, the whole epidermis is a mosaic of prismatically-shaped territories, each resting on the basal lamina and extending to the epidermal surface. In the stratum basale at the base of each prism is a group of about eight cells capable of mitosis, surrounding a single *primary stem cell* from which they and all the keratinocytes in the overlying column are derived; a single Langerhans cell lies close to the primary stem cell, perhaps directing local mitotic activity. This cluster of basal cells and Langerhans cell together constitutes the Epidermal Proliferative Unit (EPU).

According to this theory, repeated divisions of the primary stem cell give rise to an encircling ring of basal cells which then migrate in sequence into the stratum spinosum, dividing two or three times as they do so ('amplification divisions'); they are thus viewed as *intermediate stem cells*. They then differentiate into non-mitotic keratinocytes and, after a variable period of transit within the stratum spinosum, eventually pass into the stratum granulosum. There, having become flattened and terminally keratinized, they move superficially as a single column of corneocytes until finally shed at the surface.

The distributions of melanocytes and Merkel cells do not correspond to the proliferative units and obey some other territorial rule.

Subsequent studies have in general provided support for this model where it applies to species and sites where epidermal cell turnover is regular and not too fast. However, regular columns of keratinized cells have not been found in normal human epidermis, perhaps due to abrasion from clothes, movement, washing, etc. which may have a stimulatory activity on cell renewal, disrupting any highly ordered migration of keratinocytes within it (Bergstresser & Chapman 1980). It has also been reported that in addition to the mitotic basal cells there is another type of basal cell, post-mitotic and with deep interdigitations at its base, which may have special anchoring functions (Lavker & Sun 1982).

Stratum
corneum

(Stratum
lucidum)

Stratum
granulosum

Stratum
spinosum

Stratum
basale

Dermis

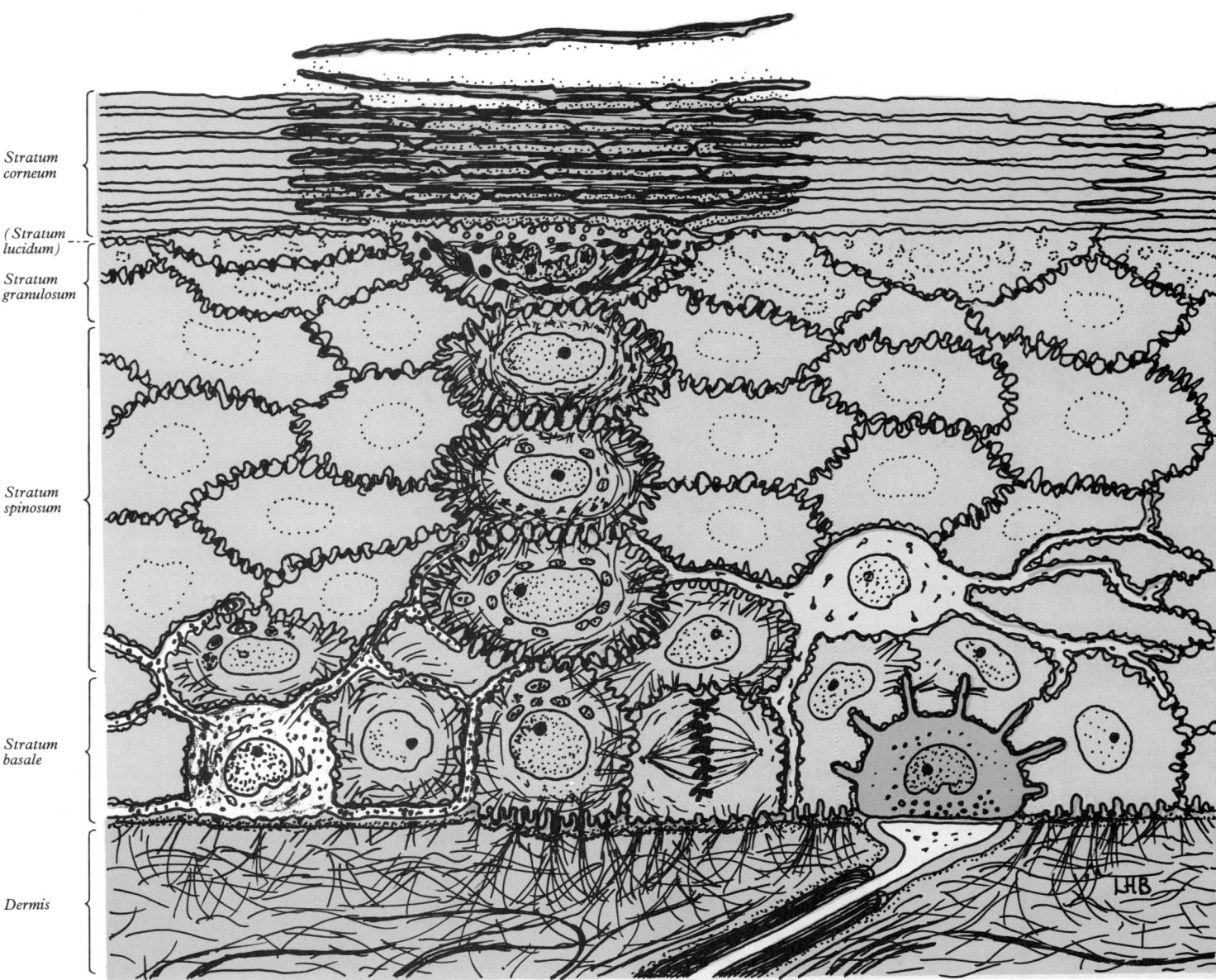

1.84A Diagram of the main features of the epidermis, including its cell layers and different types of cell, including: keratinocytes (pink); two dendritic varieties, the melanocyte (grey) and Langerhans cell (blue); and a Merkel cell (purple). Also depicted are the dermis (green) and the sensory axon (yellow) associated with the Merkel cell. In this picture the epidermis of thin skin is shown so that the position of the stratum lucidum as it would appear in thick (hairless) skin is only indicated. For clarity only a single column of keratinocytes arising from the mitosis of a basal cell is shown in any structural detail, although desmosomal contacts are not illustrated because of the level of magnification.

Epidermal Keratinization

Traditionally, this term applies to the final stages of keratinocyte maturation during which cells are converted into tough, cornified squames (*orthokeratinization*, in contrast with the incomplete *parakeratinization* of areas of the buccal mucosa, see p. 56). However, the keratin filaments of corneocytes are only one subtype of the general family of keratin filaments of similar size and broadly comparable chemistry which are characteristic of all varieties of epithelia (see p. 51). Indeed, recent immunocytochemical studies show that the cells of the epidermis have a varied and changeable pattern of keratin filament chemistry which changes as keratinocytes mature. Thus, keratins of basal stem cells are different from those of young keratinocytes and it is only in the upper layers of the stratum spinosum that the final type of keratin filament begins to emerge. During a keratinocyte's life span there must be sequential synthesis and probably some dissolution of different keratins as cells move from the base towards the surface. The classic process of epidermal keratinization therefore describes only the final stages of maturation in which dead squames filled with keratin and other proteins are formed.

The crucial steps in this process are, first, the formation of the appropriate keratin filament proteins by the free ribosomes of the keratinocytes, their assembly into 8–10 nm thick filaments, then their progessive aggregation on contact with keratohyalin in the stratum granulosum. Finally, chemical changes which involve the formation of disulphide bonds stabilize this filament–filaggrin complex as a mass of chemically inert, mechanically resistant parallel fibres embedded in a dense matrix.

Other types of terminal keratinization occur elsewhere, particularly in hairs and nails, where the keratin is chemically distinct and becomes much tougher than in the general epidermis ('hard keratin' as opposed to 'soft' keratin). These processes will be considered further below (p. 89). The *thickness* of the keratin layer can also be influenced by local environmental factors, particularly abrasion, which can lead to a considerable thickening of the whole epidermis including the stratum corneum, so that the soles of the feet become exceedingly resistant if shoes are dispensed with, and keratinized pads develop in areas of habitual pressure, e.g. corns from tight shoes, palmar callouses in manual workers, digital callouses in guitar players, etc. Exposure to high levels of sunshine or other stressful agents may also cause general epidermal thickening.

Epidermal Dendritic Cells, Langerhans Cells

These cells (**1.84**) are regularly scattered throughout the epidermis (vide supra) and the stratified squamous epithelia of the buccal mucosa. They are derived from bone marrow cells and are continually renewed (Katz et al 1979). The cell bodies of Langerhans cells are situated in the base of the stratum spinosum and their extensively branched dendrites are insinuated between the surrounding cells, although no desmosomal junctions occur with them. The nucleus is euchromatic and indented; the cytoplasm contains many mitochondria, granular endoplasmic reticulum and a well developed Golgi complex; lysosomes are also present. Characteristic of Langerhans cells are the membrane-lined *Langerhans bodies* (Birbeck bodies), a diagnostic feature of this cell type. They are elongated vacuoles approximately 0.5 μm long and 30 nm wide, usually with expanded spheroidal ends, and contain a central lamina marked with regular cross-striations at 9 nm intervals. These bodies are derived from the Golgi complex and discharge their striated contents at the cell surface into the intercellular space; their function is unknown.

Langerhans cells have many features in common with connective tissue macrophages, including immunochemical reactivity with macrophage-specific monoclonal antibodies, plasmalemmal receptors for the Fc fragment of Immunoglobulin G (IgG), the third component of Complement (C3) and the same class of histocompatibility antigens (HLA-D/DR in humans, I-A in mice, see Rowden 1981); they are also strongly positive for the enzyme ATPase and, like macrophages, they are derived from cells of the bone marrow. However, the unique Langerhans bodies show that they are not identical with macrophages and recent immunohistochemical findings suggest that the Langerhans cells may belong to a class of dendritic immune cell also present in other sites, including lymph nodes and the thymic medulla (Kashihara et al 1986).

Functionally there is much evidence that, like macrophages, they are important in cellular defence, particularly in detecting, binding and presenting antigens to local T-lymphocytes, as part of the immune mechanism of the skin. Such a system appears to be important in cell-mediated immunity to epidermal viral infections, in the elimination of epidermal cancers and in a diversity of other defensive responses, e.g. to ectoparasites such as ticks (Nithiuthi & Allen 1984). In addition, Langerhans cells may participate in the regulation of cell division, a role which has been postulated for other dendritic, macrophage-like cells elsewhere (see p. 669). The close relation of Langerhans cells and mitotic epidermal basal cells is certainly suggestive of such an activity, which could be either stimulatory or inhibitory, although this has not yet been demonstrated unequivocally.

Clinically, misdirected reactions of Langerhans cells are implicated in various skin pathologies, such as allergic contact dermatitis and in graft rejection. Their absence from the cornea may be an important factor in the ease with which grafts of this tissue can be accomplished. In the epidermis, they are adversely affected by ultraviolet light and various chemicals, which can inactivate them and deplete their numbers. This suppressive effect may contribute to the higher incidence of epidermal carcinomas in Caucasians exposed habitually to strong sunlight or artificial ultraviolet light and may play a part too in the chemical induction of neoplasms.

During development, Langerhans cells appear in the epidermis at six to seven weeks of gestation (Foster et al 1986) and are renewed continuously from a proliferative pool of myeloid precursors, as well as multiplying to some extent within the epidermis (Czernielewski & Demarchez 1987). In adult life they number 400–1000/mm². For further information regarding this cell type, see the reviews by Stingl et al (1980), Rowden (1981) and Friedmann (1986).

Recently another category of non-pigmented epidermal dendritic cell derived from myeloid tissue has been demonstrated in mice; instead of the typical macrophage-like surface markers (vide supra), they have an antigen, Thy-1, also found in various lymphocytic cell lines. They may represent a type of Natural Killer (NK) cell working cooperatively with Langerhans cells in the defence of the epidermis (see e.g. Romani et al 1986).

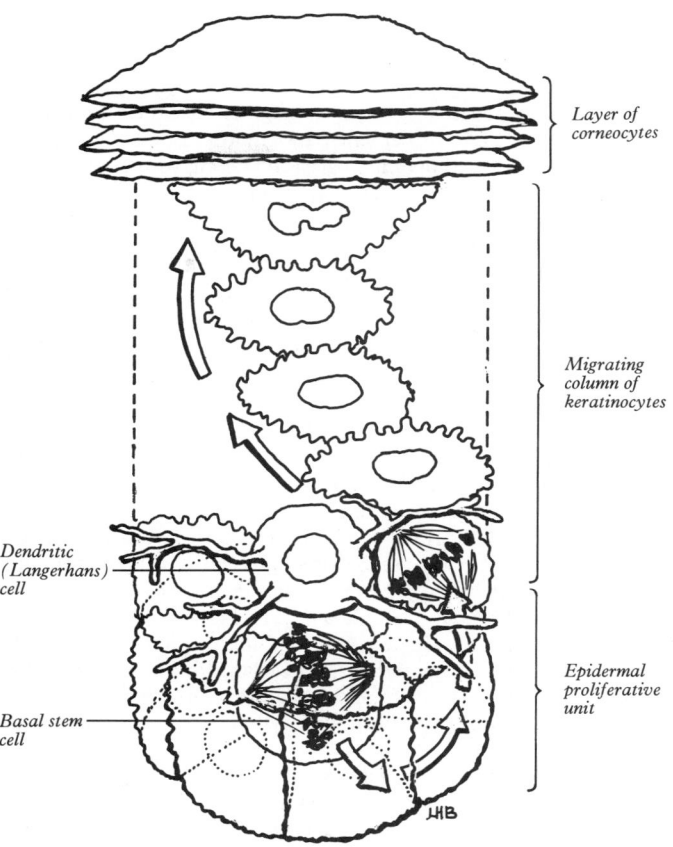

Layer of corneocytes

Migrating column of keratinocytes

Dendritic (Langerhans) cell

Epidermal proliferative unit

Basal stem cell

1.84B The concept of the epidermal proliferative unit and its relationship to the overlying column of differentiating keratinocytes, finally maturing through the stratum granulosum into the flattened corneocytes of the stratum corneum. In this model, keratinocytes arise by the repeated mitosis of a single basal stem cell, move laterally, then pass into the base of the stratum spinosum, where they may divide again one or more times before passing towards the surface. A single basal stem cell and its progeny within the stratum basale constitutes the epidermal proliferative unit as initially conceived. The dendritic Langerhans cell, situated above the basal stem cell, has also been suggested as a controlling influence on cell division. For further explanation see text.

Epidermal Melanocytes

The rounded cell bodies of melanocytes (**1.84A**) lie in the stratum basale in contact with the basal lamina; from their apices a number of cytoplasmic dendrites emerge and ramify extensively among the surrounding cells, reaching deep into the stratum spinosum. Developmentally, they originate from the neural crest and migrate as melanoblasts into the epidermis at 12 to 14 weeks' gestational age.

Melanocytes are attached to the basal lamina by hemidesmosomes but do not form desmosomal attachments to other cells. Their nuclei are rounded or elliptical and euchromatic; the perikaryal cytoplasm contains granular endoplasmic reticulum, a Golgi complex and numerous mitochondria. Microtubules, intermediate filaments and microfilaments extend peripherally within the dendrites. For a detailed account of their structure see Breathnach (1969, 1971), Fitzpatrick et al (1983).

The characteristic organelles of these cells are the *melanosomes*, membrane-bound bodies containing pigment, budded off from the Golgi complex then transported from the cell body to the ends of the dendrites. Terminal portions of dendrite cytoplasm containing melanosomes are periodically shed (*cytocrine secretion*) and are immediately engulfed by neighbouring keratinocytes and basal stem cells, and the melanosomes incorporated into their cytoplasm. In this way, the germinative cells of the epidermis are provided with a pigment shield against the damaging effects of ultraviolet light on their DNA. Of course, pigment is constantly being lost from the epithelium as the basal cells divide, migrate and are shed from the surface, so a continuous production of melanosomes is needed. Where protection is inadequate, intense ultraviolet light may cause acute damage and cell death in epidermal cells, as in sunburn, or after chronic exposure, epidermal neoplasms may occur because of damage to basal cell DNA; this is seen in the higher incidence of squamous cell carcinoma amongst pale-skinned people in areas of strong sunshine in comparison with the indigenous, melanized populations.

The darkness of the skin depends on a number of factors, including melanocyte frequencies and the number and composition of their melanosomes. These may vary depending on genetic, environmental and hormonal factors. Melanocytes are often most numerous on the face, mucosal orifices, external genitalia, nipple and areola and limbs, varying between 800 and 2000/mm³. Their numbers are similar in pale and dark races but their activity and the structure of their melanosomes differ (vide infra). Melanocytes are also present at the bases of hair follicles (see p. 91) and are responsible for hair colour. They are, however, absent from the matrix of the nail bed.

Melanosomes are oblong, membrane-bound, rounded bodies containing the dark brown pigment *eumelanin*. Similar, though more spheroidal, bodies enclosing a reddish-yellow pigment *phaeomelanin*, are called *phaeomelanosomes*; they occur mainly in reddish hair. In the more heavily melanized races, mature melanosomes are rod-shaped, (up to 3 μm long by 0.5 μm wide) and are intensely pigmented. In Caucasian skin they are smaller, elliptical (0.3–0.7 μm long) and have areas of differing density within each granule; in both cases their form changes with melanosome maturation.

Melanosomes are generated (see Fitzpatrick et al 1983) by budding from the granular endoplasmic reticulum as spherical vesicles containing granular material and a few peripheral cross-striated filaments (Type I melanosomes, promelanosomes). The filaments increase in number to fill the vacuole which changes to an elliptical shape (Type II melanosomes); then gradually more and more finely granular eumelanin is deposited amongst the filaments (Type III) until finally only the melanin is visible (Type IV), giving a homogeneous dense appearance punctuated by local spheroidal areas of lower density (vesiculo-globular bodies). Phaeomelanosomes remain spheroidal throughout their development and, instead of a homogeneous, dense interior, mature specimens have several rather granular masses with a distinctive structure (Breathnach 1965).

The filaments found in immature melanosomes appear to be molecular assemblies of the enzyme *tyrosinase* which take the form of helices 10μm in diameter, with a 10 nm periodicity and up to 500 nm long.

Eumelanin is a highly insoluble, proteinoid polymer of DOPA-quinone, formed by a series of reactions involving tyrosinase and other oxidative enzymes. The amino acid tyrosine is first converted by melanocytes into dihydroxyphenylalanine (DOPA). then into DOPA-quinone, which, after various chemical steps, polymerizes into eumelanin. These changes occur in the developing melanosomes, tyrosinase and other enzymes being added to promelanosomes by fusion of vesicles derived from the Golgi complex. Melanocytes can be demonstrated histochemically by incubating pieces of fresh epidermis with DOPA, which is then converted to pigment by their enzyme systems.

Phaeomelanin, reddish brown in colour, has a similar chemistry — but the synthetic pathway diverges from that of eumelanin, since it is a cysteine-rich polymer of DOPA and therefore contains disulphide bonds.

Control of Melanization

The extent and position of melanocyte activity is determined by various factors as already mentioned. Genetic differences are generally reflected in the numbers and types of melanosomes produced rather than the numbers of melanocytes. Various hormones can also alter these parameters, including melanocyte stimulating hormone (MSH) from the adenohypophysis. Oestrogens and progesterone can also increase melanization, particularly on the face, abdominal and genital skin and on the nipple and areola, which may remain permanently darker after the first pregnancy. Age and many pathological conditions can also affect the degree of local, or general, pigmentation.

Changes in colour caused by exposure to ultraviolet light (or other short-wavelength electromagnetic radiation including X-rays and gamma-rays) involves a number of different mechanisms. With brief exposure, individual melanosomes darken within minutes because of photo-oxidation of melanin, returning later to their original colour. With increased exposure duration there is evidence from animal experiments that melanocytes multiply mitotically (see e.g. Nordlund et al 1981) and increase their rates of both melanin synthesis and transferral to keratinocytes (which are also produced more rapidly) within two to three days.

The extent of these changes depends on genetic constitution as well as dosage and varies considerably within racial groups as well as between races. In some individuals, melanocytes with eumelanin are grouped together to give freckles or larger pigmented patches. In albinism, an autosomal recessive genetic condition, certain melanogenic enzymes (e.g. tyrosinase) are absent, so that melanin pigments are totally missing from all regions of the body, including skin and eyes. At least six biochemically distinct classes of oculocutaneous albinism are known.

Epidermal *melanocytic naevi* (moles) are regions where melanocytes are clustered in high densities and are considered to be small benign, quiescent neoplasms, melanomas, usually formed early in development. Although most such naevi remain inactive and innocent, some have the potential to become malignant, particularly if exposed to stresses such as excessive abrasion (e.g. on palms and soles and, because of chafing during walking, on the scrotum) or excessive sunlight. Malignant melanomas, which typically disrupt the surrounding skin and often give rise to smaller melanomas around them, are able to spread and grow rapidly, and become quickly fatal unless immediately checked.

In addition to melanin, there are other factors which determine epidermal colour, including the presence of carotenes in the epidermal keratinocytes, pigmented hairs inserted within it and various other normal and pathological factors. The causes of general skin colour will be considered later (p. 83).

Merkel Cells

These are present only in thick hairless skin (**1.84**). They lie at the base of the epidermis, usually protruding somewhat from it into the dermis, and form the terminal attachment of certain mechanoreceptive cutaneous nerve endings. The cells are elliptical in shape and from their apical surface several cytoplasmic spikes arise, inserted among the epidermal cells and connected to them by desmosomes. Within the cytoplasm are numerous dense-cored vesicles and intermediate filaments. They are thought to play some as yet uncertain role in sensory transduction and are considered in more detail on p. 910.

The Basement Membrane of the Epidermis

This, and basement membranes of other epithelia, consists of a basal lamina, about 80 nm thick (**1.75**), to which cells of the stratum basale are attached, and, on the dermal aspect, a reticular lamina which grades into the connective tissue of the dermis (see the reviews by Stanley et al 1982, Briggaman 1983). This structure and its role in the normal and pathological behaviour of skin are of great interest at present and it is known that it has a chemical composition peculiar to the integument.

Similar to other basal laminae (p. 68), that of the skin consists of a lamina lucida lying close to the epidermal cell bases and a deeper lamina densa. The lamina lucida is occupied by various macromolecules giving this layer a finely granular or filamentous appearance; here are laminin, heparan sulphate proteoglycan and pemphigoid antigen (a protein unique to skin and with an *Mr* of 220 000). The lamina densa includes a network of Type IV collagen molecules (p. 65), a glycoprotein named epidermolysis bullosa acquisita antigen (Woodley et al 1986), fibronectin and various proteoglycans. The lamina lucida is strongly adhesive to the epidermal cell membrane, while the meshwork of the lamina densa may limit the passage of macromolecules from the dermis to the epidermis. The basal lamina may also suppress keratinocyte differentiation in the stratum basale (Regnier et al 1986) and regulate various other cellular activities in the epidermis. However, a major function of the whole basement membrane is to stabilize the epidermis mechanically by its hemidesmosomal attachments, the fine filaments attached at one end to epidermal cell desmosomal plaques traversing the lamina lucida and inserting into the lamina densa, from which delicate microfibrils of elastin-like proteins (e.g. oxytalan, elaunin, see p. 67) are anchored into the dermis.

THE DERMIS

The dermis consists of irregular, moderately dense, soft connective tissue (1.80,85A, B). Its matrix consists of an interwoven collagenous meshwork, with a varying content of elastin fibres, proteoglycans (the glycosaminoglycans being predominantly hyaluronic acid and dermatan sulphate, with some chondroitin-6-sulphate, heparin sulphate), fibronectin and other matrix components, blood vessels, lymphatic vessels and nerves. Non-striated myocytes occur in the dermis as *arrector pili* (p. 92) muscles. The density of its fibre meshwork also varies, both within an area and in different sites, and its degree of elasticity also varies much, being particularly great in areas subject to much stretching (e.g. axillary skin). Mechanically, the dermis provides considerable strength to skin by virtue of the numbers and arrangement of its collagen fibres and it also has elastic recoil because of its elastin content. It also provides a compartment for blood vessels, lymphatics, nerves and defensive cells and so is vital to the survival of the epidermis and to much of the skin's physiology.

The dermis can be divided into two distinct zones, a narrow superficial *papillary layer* and a deeper *reticular layer*.

The papillary layer is immediately deep to the epidermis and is specialized to provide mechanical anchorage, metabolic support and trophic maintenance to the overlying tissue, as well as housing rich networks of sensory nerve endings and blood vessels. Its superficial surface is marked by numerous *papillae* which interdigitate with recesses in the base of the epidermis and form the dermo-epidermal junction at their interface. The papillae have round or blunt apices which may be divided into several cusps. In thin skin, especially in regions with little mechanical stress and minimal sensitivity, papillae are few and very small, while in the thick skin of the flexor aspects of the hands and feet they are much larger, closely aggregated and arranged in curved parallel lines following the pattern of ridges and grooves typical of these surfaces (1.86). Lying under each epidermal ridge are two longitudinal rows of papillae, one on either side of the midline through which sweat ducts pass within the epidermal rete pegs (see 1.80). Each papilla contains narrow, densely interwoven bundles of fine type I and III collagen fibres and some elastic fibres and microfibrils, many of them attached to the basal lamina of the epidermis. Also present is a capillary loop and in some sites, especially in thick hairless skin, Meissner's corpuscular nerve endings (see pp. 910–911).

The reticular layer merges with the deep aspect of the papillary layer. Its bundles of collagen fibres (mainly type I except early in development when type III predominate) are thicker than those in the papillary layer and interlace with them and with each other to form a strong yet deformable three-dimensional lattice, in which many of the fibres are parallel to each other and within which lies a variable number of elastic fibres. The predominant orientation of the collagen fibres may be related to the main direction of action of the mechanical forces to which the dermis is subjected locally and thus may be involved in the development of some of the surface lines which occur on the skin (p. 80). The deeper collagenous fasciculi are coarse and separated by loose connective and adipose tissue containing sweat ducts, some of which may extend into the underlying hypodermis.

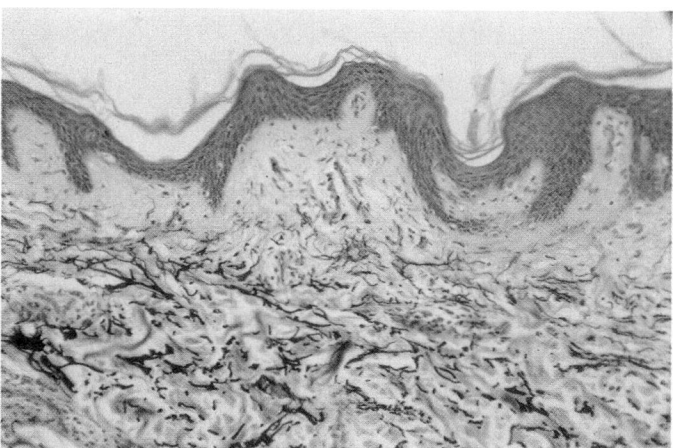

1.85B Section through thin skin, stained by the Verhoeff method to demonstrate elastin fibres. The fibres of the superficial layers of the dermis are thin while the deeper layers have more conspicuous coarser elastic fibres. Magnification × 120.

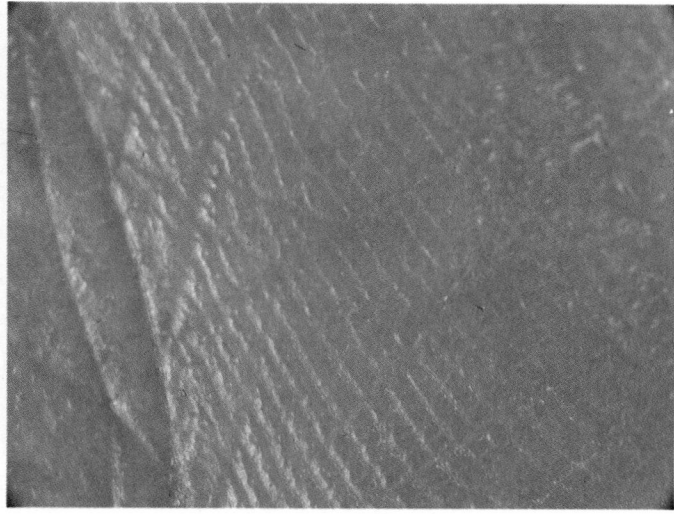

1.85A Scanning electron micrograph of the surface of a section through the skin showing the epidermis (above) and the layers of the dermis. The papillary layer close to the epidermis contains fine collagen fibres, while in the deeper reticular layer the fibres become increasingly more coarse. Magnification × 300.

1.86 Low-power light micrograph of hairless skin, in surface view, from the palm of the hand showing epidermal friction (papillary) ridges and larger flexure lines (left). Magnification × 6.

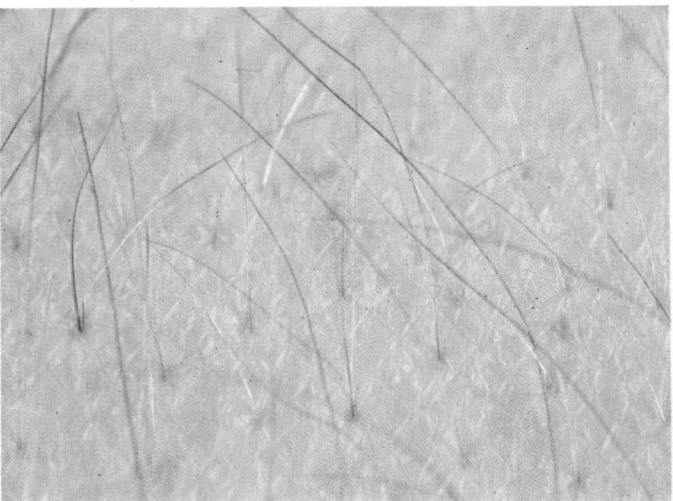

1.87 A light micrograph similar to that shown in 1.86 but taken from hairy skin on the extensor aspect of the forearm; note the pattern of surface grooves (tension lines) and hairs. The oblique direction of the emerging hair shafts points away from the pre-axial border of the limb. Magnification × 6.

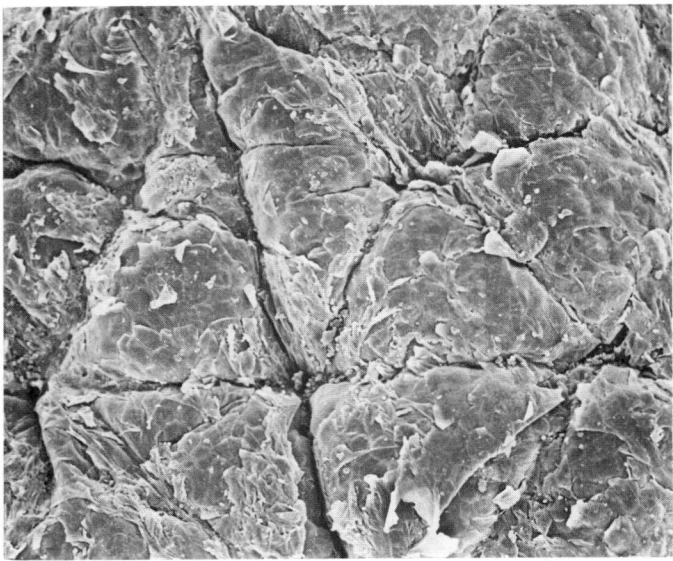

1.88 Scanning electron micrograph of the surface of thin skin (human: dorsum of thorax), showing the interlacing network of fine creases and predominantly triangular areas between them. Magnification × 400.

SKIN LINES

As already mentioned (p. 72), the surface of the skin and its deeper structures show various 'skin lines'. Over 35 different names, many of them synonymous, have been applied to such lines, relating to various systems of grooves, raised areas, preferred directions of stretching, lines of naevus occurrence and spread of infection. Some of these are clearly visible in intact skin, others are recognizable only after some sort of intervention, while yet others are at present only debatable postulates. Here, we will briefly describe some of the major varieties of lines, dividing them into externally visible markings (tension lines, flexure lines, friction ridges, striae gravidarum, pigmented markings or transitions) and deeper 'directional' lines of the dermis.

Externally Visible Skin Lines

These are related to various patterns of epidermal creasing, ridge formation, scarring and pigmentation.

Tension lines, skin creases (1.80,87,88). A simple lattice of skin creases or folds occurs on all major areas of the body other than the thick skin of volar and plantar surfaces. The lattice

pattern typically consists of polygons (generally parallelograms) formed by relatively deep primary creases visible to the naked eye, irregularly divided by finer secondary creases into triangular areas (Hashimoto 1974, Millington & Wilkinson 1983), which are further subdivided by tertiary creases limited to the stratum corneum of the epidermis and finally, at a microscopic level, by quaternary 'creases' which are simply the outlines of individual corneocytes. Apart from the quaternary creases, all the others increase the surface area of the skin, permitting considerable stretching and recoil and distributing stresses more evenly. Details of the pattern vary according to the region of the body; e.g. on the cheek the primary creases radiate from the hair follicles, on the scalp they form hexagons, while on the calf and thigh they form parallelograms, whose longer sides are inclined at about 30–40° to the vertical. Superimposed on this basic pattern are deeper creases associated with flexion (vide infra) (Bruner 1951).

Flexure lines (skin joints) are major markings found in the vicinity of synovial joints, where the skin is attached strongly to the underlying deep fascia (1.86). They are conspicuous on the flexor surfaces of the palms, soles and digits and, in combination with associated skin folds, facilitate joint movement. Their patterns on the palms and soles may vary and are to some extent genetically determined; in Down's syndrome, affected individuals tend to possess a prominent transverse palmar crease at the bases of the metacarpals, which is of some diagnostic importance.

Papillary ridges (friction ridges) are confined to the palms and soles and the flexor surfaces of digits, where they form narrow parallel and often curved arrays separated by narrow furrows (1.80,86,89). Along the midline of each ridge the apertures of sweat ducts open at regular intervals; each ridge corresponds to an underlying pattern of dermal papillae which help to anchor the epidermis and dermis together. The dermal papillae also have more subtle morphogenetic properties, as their pattern determines the early development and maintenance of the epidermal ridges throughout life. Functionally this arrangement increases the gripping ability of hands and feet and, because of the great density of tactile nerve endings beneath their epidermal lining, they are also important sensory structures (p. 914).

Since the patterns of ridges and furrows are formed early in the fetus and are thereafter stable throughout life and vary between individuals, they are of considerable interest as a means of identification and also clinically, because they are affected by certain abnormalities of early development, including genetic disorders.

The analysis of these patterns or *Dermatoglyphics* (skin carving, see Cummins 1926) has attracted much attention and has complex mathematical ramifications which are beyond the scope of this account.

Measurable parameters include the disposition of *tri-radii*, junctional areas where three sets of parallel ridges meet, and the frequency of ridges in particular patterns. Ridge configurations on the terminal segments of fingers can be separated into three major types: arches, loops and whorls (1.89), arches having no tri-radii, loops one, and whorls two or more. Arches may be simple or tented, loops may have a tri-radius towards the ulnar or radial side of the hand (ulnar and radial loops); whorls may be symmetrical, spiral, or double loop. Ridges within these patterns may branch or join, or may be discontinuous. Similar considerations apply to the toes. In any such configuration the number of ridges may also vary, the *ridge-count* being given by counting the total intersected when a line is drawn from the central point of a pattern to its nearest tri-radius. Such patterns can vary from digit to digit in any individual but, in the hand, loops are the most common and arches the least; with toes the converse is true. These variable features provide an astronomical number of possible combinations, so that each individual is almost certain to have a unique set of patterns.

Tri-radii also occur on the palms and soles, including the bases of each digit except the thumb; a characteristic tri-radius is also present on the proximal edge of the hand in the midline above the flexor retinaculum (the axial tri-radius). The precise positions, numbers and ridge-counts associated with tri-radii have an inherited basis but in general the genetics are multifactorial and too complex at present to be clinically useful. However, the total

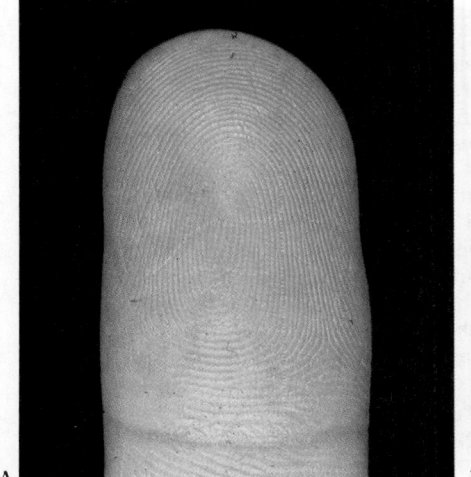 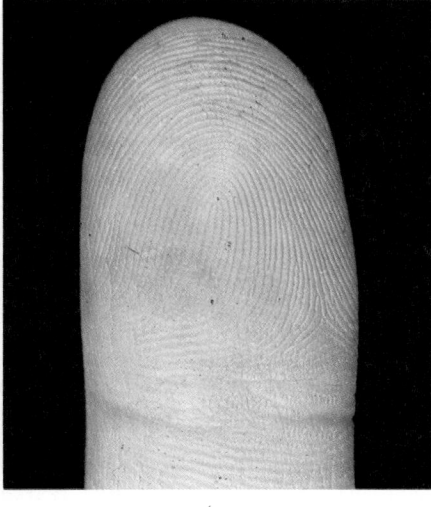 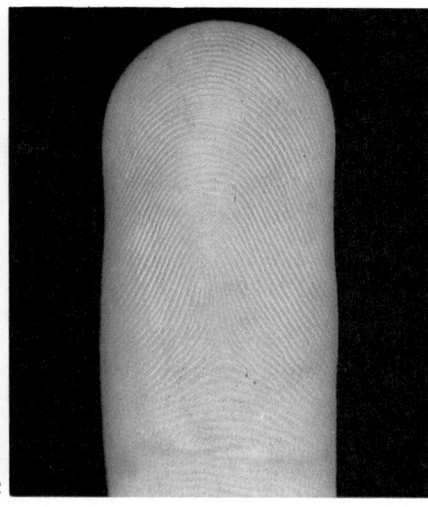

A B C

1.89 Photographs of the palmar aspect of a terminal phalanx in three different individuals to show the major types of pattern of the fingerprint ridges. The pattern in A is commonly termed a whorl: B is composed of loops; C is composed of arches. Note interphalangeal flexure lines.

ridge-count of all ten digits of the hand appear to have a simpler inheritance; in Down's syndrome, all ten finger-prints have ulnar loops in about 35% of affected persons, in contrast to about 8% of the normal population (Thompson & Bandler 1973). Other conditions affecting skeletal development, e.g. polydactyly, and teratogens such as thalidomide, also cause disturbances to the pattern of papillary ridges. For further details of this subject, see Cummins & Midlo (1961), Penrose & Loesch (1969) and Schaumann & Alter (1976).

Intrinsic scarring. If the mechanical demands placed on the skin are greater than the skin creases and the elasticity of the dermis can accommodate, the reticular layer of the dermis may rupture and be replaced first by highly vascular granulation tissue (p. 85) and finally by highly collagenous, poorly vascularized scar tissue. Such changes can be termed intrinsic, to distinguish them from scars formed by external wounding (see p. 84). Sites of dermal rupture are visible externally as lines, *striae* or 'stretch marks', which are initially usually pink in colour, later widen and become a vivid purple or red and eventually fade, becoming paler than the surrounding intact skin. They develop on the anterior abdominal wall of some women after pregnancy (Poidevin 1959), when they are termed *striae (lineae) gravidarum*. Although the development of striae may be due to excessive stretching of normal skin by the rapid growth of structures deep to it (e.g. a fetus,

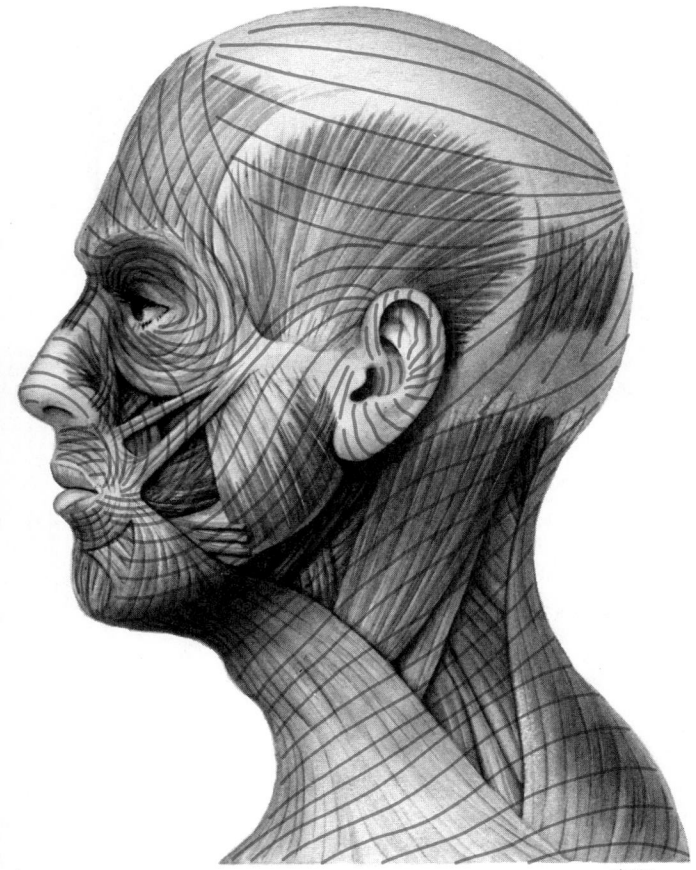

kvm

1.90A Distribution of Langer's cleavage lines on the face. For further details see text.

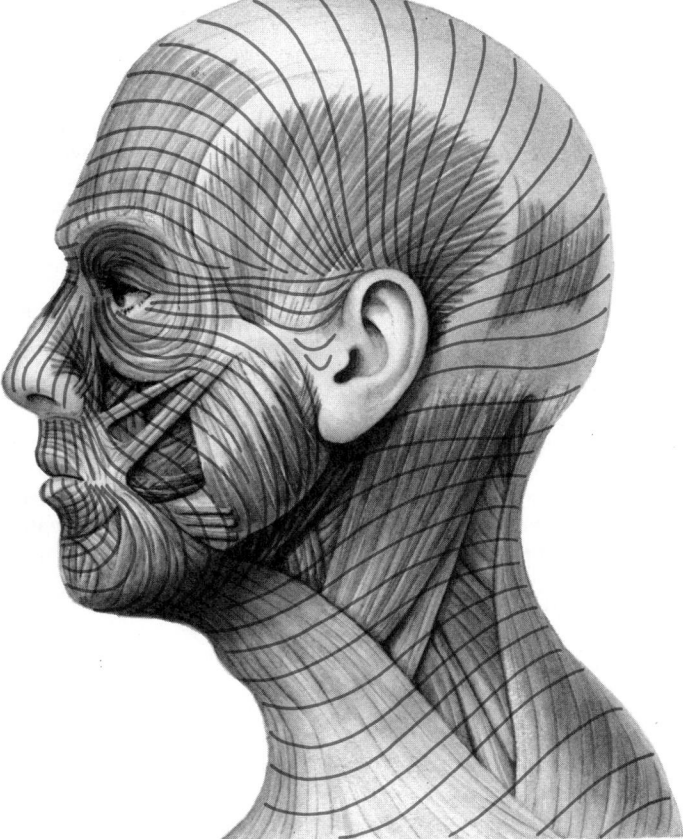

kvm

1.90B Distribution of the lines of Kraissl's (relaxed skin tension lines) on the face. It has been reported that incisions made along these lines heal with minimum scarring (see text).

adipose tissue, hypertrophying muscle, a tumour), more often they indicate pathological or hormonally-induced weakening of the dermis. Striae have been associated with many debilitating diseases, some chronic infections and changes in steroid hormone levels, and are possible side effects of the therapeutic administration of adrenocortical hormones (Geschwandtner 1973). Striae result from extensive dermal changes involving stressing of collagen fibre bundles (Arem & Kischer 1980), inducing a form of remodelling or dermal scarring, perhaps under the influence of steroid hormones.

Pigmentation. Variation in pigmentation can also produce externally visible lines, such as 'Voigt lines', on the surface of the skin. Found in some infants, Voigt lines mark differences in pigmentation between the darker extensor and paler flexor surfaces of the arms (Matsumoto 1913). They occur along the anterior axial lines (p. 1151), extending from the sternum to the wrist.

Lines Detectable after Manipulation or Incision

Lines of Langer and Kraissl (1.90A,B,C). Skin is normally under tension, the direction in which this is greatest varying regionally. The first consideration of skin tension was published in 1834 by Dupuytren, who observed in a post-mortem examination that a wound inflicted with an awl was elliptical although the weapon was round in cross-section. Malgaigne (1838) demonstrated that, in a particular area of the body, the long axes of such elliptical wounds were parallel to each other and formed a pattern of *cleavage lines*. Langer (1861) studied these patterns exhaustively in cadaveric material and proposed the hypothesis that the skin is always in a state of tension, due to the rhomboidal arrangement of its dermal fibres. When these fibres were disturbed, the tension exerted in the long axis of the rhomboid predominated and the wound became distorted. The surgical significance of Langer's cleavage lines was first recognized by Kocher (1892), who advised that surgical incisions should be made parallel with them to minimize postoperative scarring. The observation behind this advice was that wounds made at right angles to cleavage lines gaped widely and that linear scars widened and even hypertrophied, if the wounds extended into the dermis. Conversely, wounds made parallel to the lines remained narrow, as did any scar tissue. Unfortunately, the lines mapped out by Langer, using supine cadavers, do not always coincide with the lines of greatest tension (Conway 1938) and therefore are not the most appropriate lines for elective incision in the living (Kraissl 1951). In some locations Langer's lines and 'Kraissl's lines' (the latter being usually orthogonal to the line of action of any immediately subcutaneous muscle fibres and often coincident with wrinkle lines) are at right angles to each other. Examples of differences are shown diagrammatically in **1.90A** and **1.90B**. Lines of greatest tension similar to those described by Kraissl have been termed 'relaxed skin tension lines' by Borges & Alexander (1962). The position of the relaxed skin tension lines can be found, as their name suggests, by 'relaxing' the skin of a region by, e.g. joint mobilization or passive manipulation. The resulting skin creases follow the direction of relaxed skin tension lines. Incisions made in the direction of these lines result in minimal scarring.

Blaschko lines. These refer to the way in which patterns of naevi and related dermatological pathologies are distributed or develop (Blaschko 1901) along certain preferred cutaneous pathways. These do not appear to correspond to vascular or neural elements of the skin and may be related to early developmental boundaries of a 'mosaic' nature. (See Jackson 1976 for a review of this subject.)

AGE-RELATED CHANGES IN SKIN

The major features of skin are essentially formed well before birth and during the first two to three decades of life the main changes are an expansion of its surface area and thickening of the epidermis and dermis, as well as various changes in hair and gland patterns which occur at puberty. The arrangement and numbers of creases and friction ridges is essentially the same from their early fetal formation onwards, although their sizes increase until cessation of growth. However, from about the third decade onwards there is a gradual change in the appearance and mechanical properties of the skin which reflect natural ageing processes. In old age, or under adverse environmental conditions, these alterations may become very marked.

Normal human ageing is accompanied by epidermal and dermal atrophy, which result in some changes in the appearance microstructure and function of the skin; alterations include wrinkling, loss of elasticity and the development of purpura. Epidermal atrophy is accompanied by flattening of the dermal papillary ridges (Montagna & Carlisle 1979), the resulting reduction in interdigitation between epidermis and dermis decreasing resistance to shear. Dermal changes include thickening of elastic fibres and the localized appearance of material resembling degraded collagen. The latter is particularly marked in chronically sun-damaged skin (Mitchell 1967). Skin extensibility decreases with age (Leveque et al 1980) and so does elastic recoil. Although the amount of collagen in skin normally exposed to the sun decreases with age (Shuster et al 1975), no quantitative changes in total collagen have been observed in skin normally shielded from solar irradiation; there is, however, a change in the ratio of type I: type III collagen, the relative amount of type III collagen increasing after the age of about 65 years (Lovell et al 1987). Whether the increased proportion of type III collagen in older subjects is due to preferential synthesis of type III (possibly resulting from impaired synthesis of type I), preferential degradation of type I, or a combination of both, remains undetermined. Unlike the situation in fetal skin, where there is a temporary preferential synthesis of type III collagen, active increase in the synthesis of type III collagen is unlikely, since fibroblasts obtained from older skin synthesize less total collagen than those from younger skin, as well as having a shorter lifespan (Martin et al 1970). Changes which occur with age in the number, thickness and arrangement of the bundles of collagen fibres in the dermis

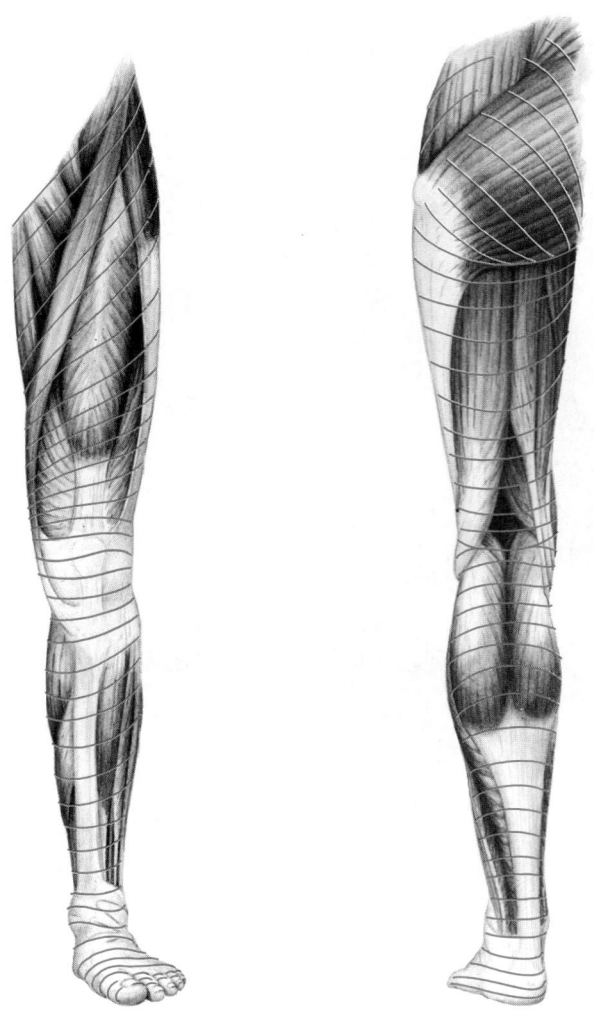

1.90C Lines of Kraissl associated with the anterior (left) and posterior (right) aspects of the lower limb.

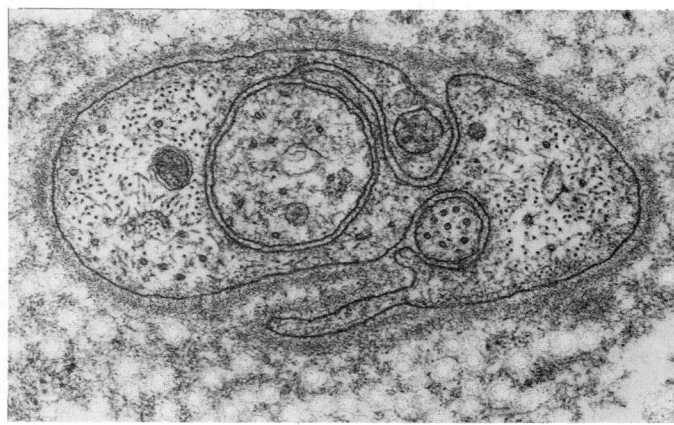

1.91 Electron micrograph of an example of an unmyelinated cutaneous afferent nerve bundle close to the surface of the dermis (human thin skin). Two axons are visible within the Schwann cell sheath. Note also the thick basal lamina and the pale profiles of sectioned collagen fibrils surrounding the nerve fibres. Magnification × 20000.

(Lovell et al 1987) may be correlated with the decrease in tensile strength reported in ageing human skin by Vogel (1983).

THE COLOUR OF SKIN

Cutaneous colour is determined by several factors, the most important being its intrinsic pigmentation, its vascular supply and thickness. Pigments of mature skin are usually present only in the epidermis and hair, the chief being the brown-black *melanin* and (mainly in hair) the closely related red-brown *phaeomelanin* (see p. 78). These vary greatly according to body site, genetic constitution, incident ultraviolet light and other factors. During fetal and early postnatal life pigment cells (*melanoblasts*) derived from the neural crest migrate into the epidermis from deeper sites and when still in the dermis may show as patches of grey-blue colour ('Mongolian spots') in young children of negroid and mongolian (and occasionally Caucasian) races. The boundaries between the darker dorsal and paler ventral aspects of the upper limbs may also be clearly defined (Voigt lines, see also p. 82).

Epidermal cells and underlying dermal and subcutaneous tissues may also contain yellowish carotenes giving poorly melanized skin a sallow hue, particularly when the superficial blood vessels are few or constricted, or when the epidermis is particularly thick. However, in the absence of large amounts of melanin, the main determinant of skin colour is the oxygenated haemoglobin of the dermal vascular beds, particularly the superficial (papillary) plexus (see p. 84). These are of course viewed through the overlying epidermis, whose surface layers scatter and reflect some of the light, and is somewhat opalescent, giving well-oxygenated skin a pink colour. Where less oxygenated blood is present, e.g. in congestive heart failure, a bluish hue results. Skin colour dependent on blood flow also varies with ambient temperature, exercise, emotional state, haemoglobin content of blood and various other features, often affected by general health.

INNERVATION OF SKIN

The integument has a rich nerve supply which is concerned mainly with two major functional roles of skin: as a major sensory organ and as a thermoregulator. These are mediated, respectively, by the regional sensory and efferent sympathetic fibres of segmental cranio-spinal nerves.

Cutaneous sense provides us with a wealth of information about the external environment and its interactions with the skin, through receptors tuned to mechanoreceptive stimuli of various kinds (rapid or sustained touch, pressure, vibration, stretching, bending of hairs, etc.), thermal stimuli (hot and cold), and damaging effects (moderate, perceived as discomfort, itching, etc. and more severe pain, associated with extremes of temperature, sharp penetration of the dermis and other overtly destructive stimuli).

These are sensed by a wide variety of specialized neurons whose cell bodies lie in the spinal and cranial ganglia and whose fibres terminate in the dermis and superficial fascia.

Efferent autonomic fibres are non-myelinated noradrenergic and cholinergic in type, innervating the arterioles, arrector pili muscles and myoepitheliocytes of sudorific and apocrine glands and, in the scrotum, labia majora, perineal skin and nipples, non-striated muscle fasciculi of the dermis and adjacent connective tissue. Except in the nipples and genital areas, activity of these nerves is concerned with regulation of heat loss by vasodilation and vasoconstriction, sweat production and (only incipiently in humans) pilo-erection.

Nerve fasciculi pass into the skin through the superficial fascia in which they often form a deep, subcutaneous plexus which serves sweat glands and the bases of hair follicles where these protrude into this layer. In thick, glabrous skin, fibres are given off here to Pacinian corpuscles (large encapsulated sensory endings) present in the subcutaneous tissue. On reaching the dermis, the fasciculi branch extensively to form a deep, *reticular plexus* which serves much of the dermis including most sweat glands, hair follicles and the larger arterioles. Many small fasciculi pass from this plexus to ramify in another superficial *papillary plexus* at the junction between the reticular and papillary layers of the dermis; twigs from this supply the dermal papillae or in some cases contact the base of the epidermis.

The detailed structure and behaviour of the sensory endings are described in detail in another section (p. 906) and will be considered here only briefly (see 1.80). They include the branched 'free' endings of fine non-myelinated (C) and narrow (Aδ) myelinated fibres which belong (in humans) to thermo- and noci-ceptors. Myelinated fibres of different sizes have mechanoreceptive functions: in thick skin many of these are encapsulated terminals including *Meissner's corpuscles*, rapidly adapting touch endings in dermal papillae, and Pacinian corpuscles (vide supra), already mentioned, which are vibration and flutter sensors. Genital skin also has numerous small encapsulated mechanoreceptive endings ('genital corpuscles'). The *Merkel endings* of thick skin are branched slowly adapting mechanoreceptors which are attached to the bases of a specialized type of epidermal cell (Merkel cell, see p. 78) which abound on the bases of friction ridges. They appear to be important in detecting pressure exerted on the skin surface. Also within the dermis are various types of terminal associated with hair follicles which signal deflections of hairs, and the thinly encapsulated *Ruffini endings*, slowly adapting stretch receptors in the dermis.

The density of cutaneous endings varies greatly over the body surface, as do the sizes of the receptive fields formed by the terminal

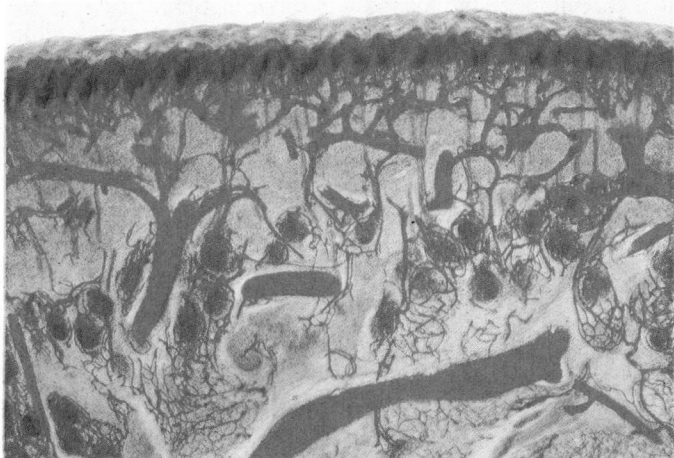

1.92 A thick vertical section through palmar skin, the arteries, arterioles and capillaries of which have been injected with red gelatin to demonstrate the pattern of dermal vascularization. At the base of the dermis a broad flat arterial plexus supplies a more superficial papillary plexus, which in turn gives off capillary loops which enter the dermal papillae. Sweat glands and their ducts are numerous in this specimen; they extend basally into the subcutaneous tissues. Magnification × 200.

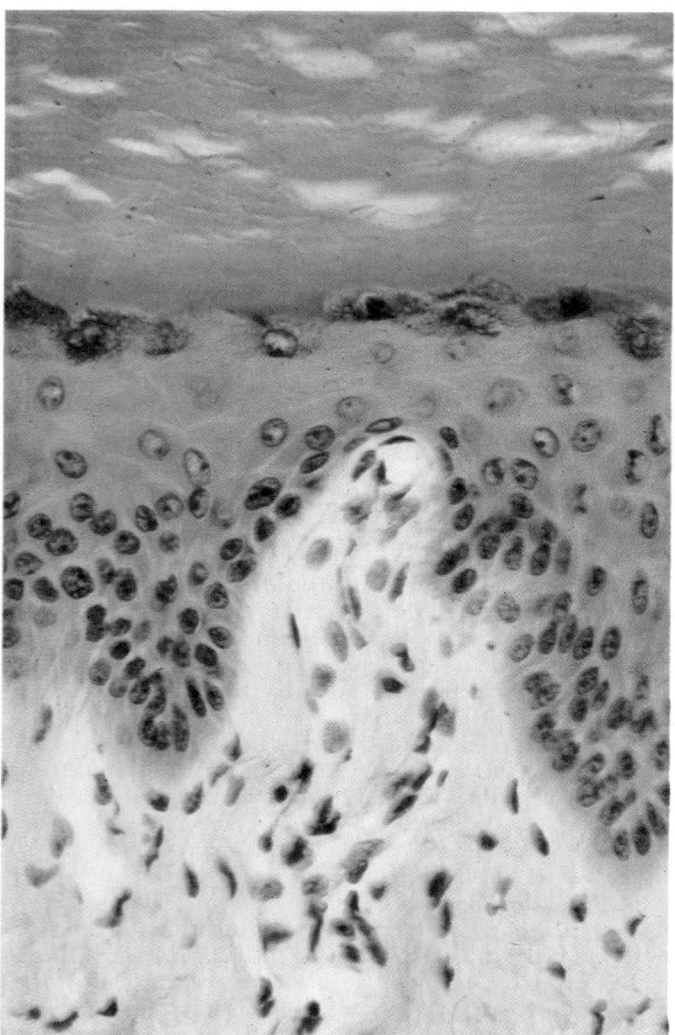

1.93 Vertical section through a dermal papilla and adjacent epidermis, showing a capillary loop. Notice the closeness of the vessel to the basal layer of the epidermis. Also visible are the layers of the epidermis, including a prominent stratum granulosum and above it the stratum lucidum stained dark orange and the paler orange stratum corneum. The section was taken from the thick skin of the foot. (Compare also 1.80 and 1.84A.) Mallory's triple stain. Magnification × 800.

branches of individual nerve fibres—and therefore their ability to locate stimuli and their point-to-point discrimination and also their thresholds to mechanical and thermal stimuli. In general, these properties vary in parallel with one another, being most marked in the extremities of the limbs, and progressively less so towards the abdomen, thorax and head, although some areas such as the skin around the mouth and ano-genital regions reverse this pattern. Discrimination is also to some extent asymmetrical, the right side of the body being generally better in this regard, at least in dextral individuals. Sex differences have also been reported, females generally having lower thresholds to pressure stimuli than males, although this is probably due to a thinner epidermis.

Sympathetic efferent fibres are typical non-myelinated axon groups which branch a number of times, and may have several synaptic zones where neurotransmission can occur. They end in close proximity to the structures they innervate.

For further reviews of these topics see, e.g. Sinclair (1973), Burgess & Perl (1973), Mountcastle (1980) and Brown (1986).

THE VASCULARIZATION OF SKIN

The metabolic demands of skin are not generally great, since the majority of it is dermal matrix containing few cells, and relatively few capillaries are present in the deeper, very fibrous layers of the dermis. However, the epidermis, glands, hair follicles and nerve

supply have many active cells and are closely related to rich dermal capillary beds (Montagna & Parakkal 1974, Keatinge & Harman 1980).

Blood enters the skin through small arteries penetrating the reticular layer from its deep aspect, ramifying and anastomosing in a sheet-like plexus, the *rete cutaneum (reticular plexus)*, at the interface between the dermis and the superficial fascia (1.80,92). From this plexus, some arterioles pass deeply to supply the adipose tissue and, where present at this depth, cutaneous sweat glands and hair follicles. Other arterioles pass superficially, some turning laterally to anastomose with other arterioles at this level, giving off capillaries to glands and hair follicles here while the majority form a second major plexus at the junction of the reticular and papillary layers of the dermis (*rete subpapillare, superficial plexus, papillary plexus*). Capillaries from this plexus loop into the dermal papillae (1.93) and so approach closely the base of the epidermis before passing back into a venous plexus present just beneath the arteriolar papillary plexus. This in turn drains into a flat intermediate plexus in the reticular layer and again into another venous plexus close to the deep arteriolar reticular plexus. Venous capillary beds surrounding glands and hair follicles drain into these plexuses at appropriate levels.

In the deeper layers of the dermis, arteriovenous anastomoses are common, particularly in the extremities subject to cooling (hands, feet, ears) where, as *glomera* (see p. 693), they are surrounded by thick muscle coats. Under autonomic vasomotor control, these vessels when relaxed divert blood away from the superficial plexus and so reduce heat loss, while at the same time ensuring some cutaneous circulation and preventing anoxia of structures such as nerves which might otherwise be at risk. Generally, muscle tone of the vascular bed is constantly adjusted by autonomic efferents according to the need for heat loss or retention or, in some areas of the body, also to emotional states. Various humoral agents (noradrenalin, angiotensin, histamine, etc.) also affect vascular muscle and therefore cutaneous blood flow. In very cold conditions, the peripheral circulation is greatly reduced but some blood flow is maintained even in the superficial layers by intermittent opening and closing of arterioles (the hunting reaction).

Many of the small arteries and large arterioles run closely parallel with veins, so that at low temperatures heat can be conserved by a countercurrent mechanism.

THE LYMPHATIC DRAINAGE OF SKIN

Blind-ending lymphatic capillaries are numerous in the reticular layer of the dermis and drain into plexuses within this stratum and into wider vessels of the superficial fascia. The lymphatics tend to be distended and the larger ones have prominent valves; lymph flows through these to the deeper lymphatics draining the region. Lymphatics of skin are quite profuse and there appears to be free anastomosis between vessels of all levels, so that there is ready interchange of lymph between adjacent areas of skin (Forbes 1938).

Wound Healing in Skin

The normal response of an organism to injury is either regeneration, the complete restoration of the damaged part, or repair, the reconstruction of the injured region in such a manner that scar tissue covered, where appropriate, by an epithelium is produced at the site of the injury. Evolutionary progress has been accompanied by an apparent loss of regenerative capacity; while regeneration of entire limbs is possible in some of the less advanced vertebrates (e.g. urodeles), it is restricted in man and other mammals to limited regeneration of some tissues (e.g. epithelia and endothelia, bone, peripheral nerves) and to partial regeneration of some organs, e.g. the liver. Whether or not this loss of regenerative capacity is irreversible remains to be discovered. When skin is injured the dermis responds by repair, while the epidermis responds by regeneration. The collective response of the skin to injury is termed *wound healing*.

DERMAL REPAIR

For descriptive convenience, dermal repair can be divided into three overlapping phases: *inflammation, proliferation* and *remodelling* (**1.94–98**).

Inflammation

Acute inflammation begins with the activation of platelets and mast cells as an immediate response to injury. During inflammation, haemostasis is achieved, removal of damaged tissue occurs and factors which start the formation of granulation tissue are released or deposited in the wound. Multiple interacting pathways of inflammation (**1.98**) are triggered when dermal blood vessels are injured and end automatically when the inflammatory stimuli dissipate; should these stimuli persist, then inflammation persists and wound healing is delayed.

Tissue injury and bleeding are followed by blood clotting. This involves many complex chemical interactions between components of the extravascular tissue and of blood, including activation of the Hageman factor (for intrinsic coagulation), factor VII (for extrinsic coagulation) and the activation of platelets. These are all responses to the surface adsorption and activation of specific coagulation pro-enzymes normally inhibited in intact tissues, but free to act in the protease inhibitor-free microenvironment temporarily provided by the wound. Blood clotting is a crucial part of inflammation, because activation of Hageman factor leads to bradykinin generation (Habal et al 1976), initiation of the classic complement cascade (Ghebrehiwet et al 1981) and possibly also to production of anaphylatoxins C3a and C5a (Clark 1985), among many other complex and as yet poorly understood reactions. These anaphylatoxins, together with bradykinin, increase local blood vessel permeability (Williams & Jose 1981), causing leakage of plasma proteins and formation of an extravascular clot. They also stimulate release of the vaso-active mediators, histamine and leukotrienes C4 and D4, from mast cells (Hugli & Muller-Eberhard 1978, Stimler et al 1982) and attract neutrophils and monocytes to the wound (Fernandez et al 1978).

The main function of the neutrophils while at the wound site is the phagocytosis of pathogenic bacteria. Once bacterial contamination has been controlled, neutrophil infiltration ceases and the early inflammatory phase of repair is at an end. In contrast, monocytes, which develop into macrophages on entering the wound bed, remain throughout the entire inflammatory phase. Macrophages are not only phagocytic but also release a host of biologically active materials, including substances essential for the initiation and propagation of granulation tissue during the next, proliferative, phase of repair.

Proliferation

During this stage, cells and intercellular substances increase greatly to form *granulation tissue*. This is a highly vascular material consisting largely of macrophages, pluripotent pericytes, fibroblasts and endothelial cells lining capillaries, all embedded in a matrix of fibronectin, proteoglycans rich in hyaluronic acid and collagen, which at first is mainly type III, changing later to type I. The profuse assemblies of capillaries, which are the main type

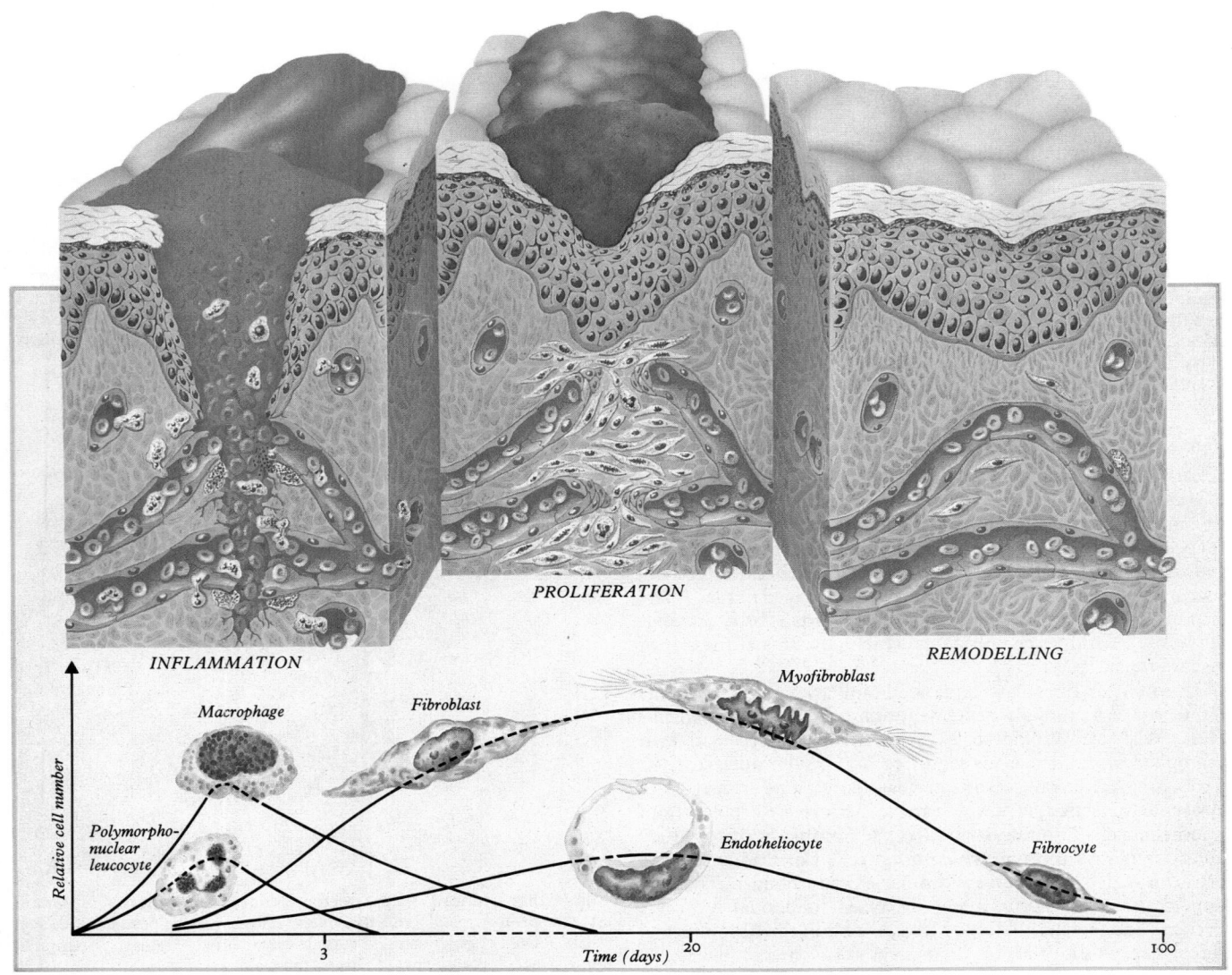

1.94 Diagrammatic representation of the normal response of skin to incision showing the changes in relative numbers of different cell types during inflammation, proliferation and remodelling. For further explanation see text.

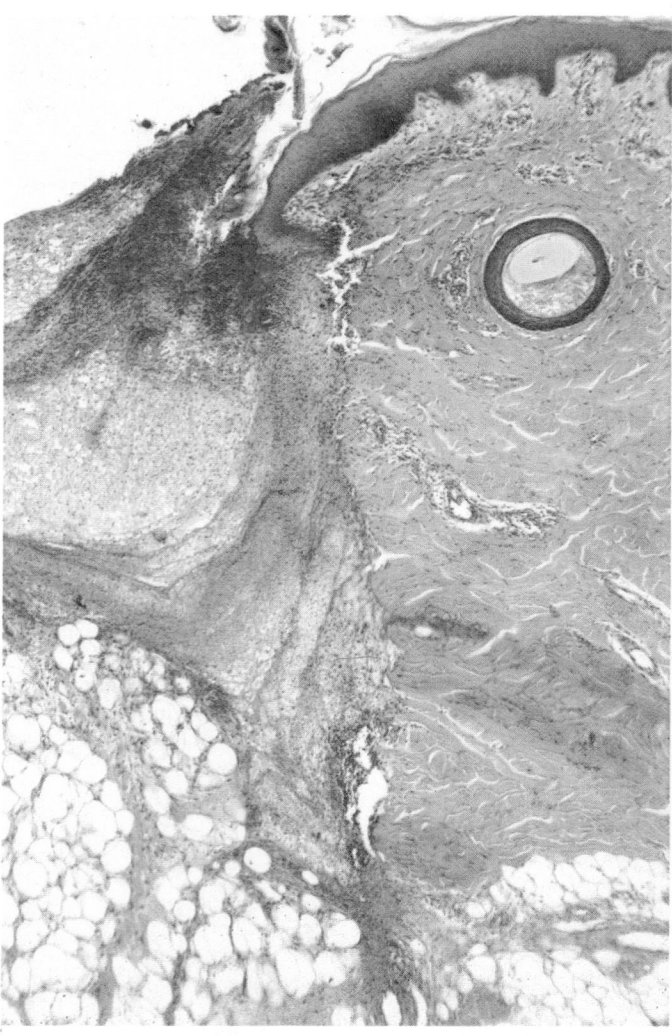

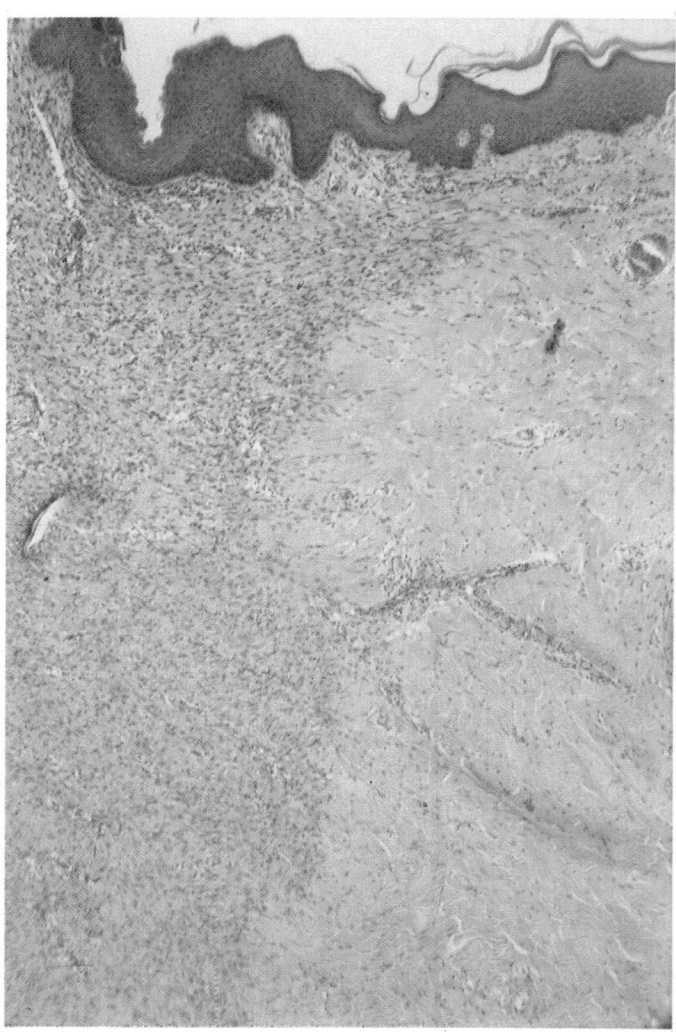

1.95 Low magnification light micrograph of part of the site of a full-thickness excised lesion produced in porcine skin, three days after injury, during the inflammatory phase of repair. Intact skin can be seen on the right. Polymorphonuclear leucocytes and macrophages are present beneath the exudate covering the wound bed, and epidermal migration from the intact skin has already commenced. Stained with haematoxylin and eosin. Material supplied by Steve Young and photographed by Kevin Fitzpatrick, Department of Anatomy, UMDS, Guy's Campus, London. Magnification × 135.

1.96 Low magnification light micrograph of part of the site of a full-thickness excised lesion produced in porcine skin, 10 days after injury, during the proliferative phase of repair. Granulation tissue, over which epidermal cells have migrated, fills the wound bed. Stained with haematoxylin and eosin. Material supplied by Steve Young and photographed by Kevin Fitzpatrick, Department of Anatomy, UMDS, Guy's Campus, London. Magnification × 135.

of blood vessel, give this tissue its 'granular' appearance when incised, hence its name (see **1**.94, 96, 97).

Granulation tissue forms in response to various signals, which may include chemotactic and growth factors, structural molecules and proteases which digest connective tissue matrix (Clark 1985). It forms a nutritive substrate over which the regenerating epidermis can migrate and is gradually replaced by scar tissue. Apart from the absence of osteogenic cells and chondroblasts, granulation tissue resembles the blastema which develops at the site of fracture repair (p. 314).

Macrophages, fibroblasts and blood capillaries migrate into the wound bed as a mutually dependent unit termed a *wound module* (Hunt & Van Winkle 1979). In the lead are activated macrophages, followed in sequence by newly differentiated fibroblasts, dividing fibroblasts and capillaries. The macrophages release chemotactic agents which attract pericytes, fibroblasts and endothelial cells into the wound. As the fibroblasts mature they produce a matrix through which other cells can readily migrate and from which delicate new capillaries can obtain mechanical support. As each capillary loop becomes functional it brings nutrients and oxygen to nearby cells, enabling the fibroblasts to secrete materials for the matrix, through which macrophages and other cells can migrate further. The above proliferative and migratory processes are repeated sequentially until the wound bed is filled with granulation tissue.

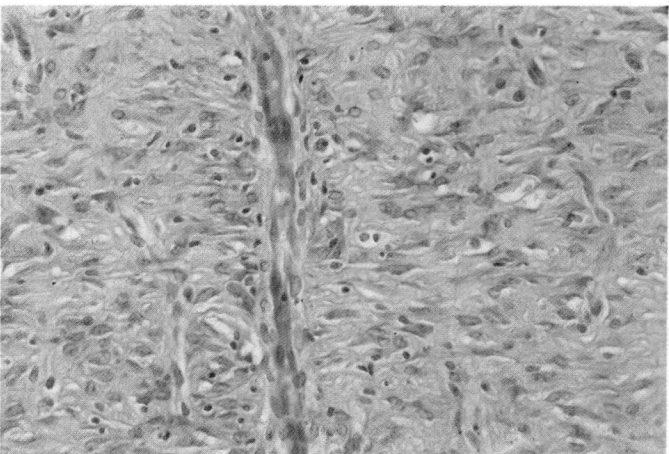

1.97 High magnification light micrograph of granulation tissue, 10 days after the production of a full-thickness excised lesion in porcine skin. The tissue is well vascularized and contains many myofibroblasts, orientated approximately parallel to the base of the wound and at right angles to the majority of the blood vessels. Material supplied by Steve Young and photographed by Kevin Fitzpatrick, Department of Anatomy, UMDS, Guy's Campus, London. Magnification × 300.

There is considerable experimental evidence to support the proposition that macrophages are key cells in dermal repair. They assist in tissue debridement, release chemotactic agents which attract fibroblasts and endothelial cells to the wound site and secrete lactate which stimulates collagen synthesis by fibroblasts (Comstock 1970), thus strengthening the tissue which develops within and adjacent to the wound (Silver 1974). Intercellular contacts between macrophages and fibroblasts (1.99) suggest that the cells may exert a direct effect on each other. If the migration of macrophages into the wound bed is prevented by anti-inflammatory steroids, or if they are eliminated from it by the application of antimacrophagic serum, the formation of granulation tissue is inhibited (Leibovich & Ross 1975). Inhibition due to anti-inflammatory steroids can be reversed by vitamin A administration, which permits migration of macrophages to the wound site (Hunt et al 1969).

During the proliferative phase of repair fibroblasts of the granulation tissue develop into cells termed *myofibroblasts* (Majno 1979). These cells, which are responsible for wound contraction, the centripetal movement of the wound margin and the consequent reduction of the size of the wound, are immunologically similar to non-striated myocytes, contain peripherally located myofilaments and become linked together by desmosomes and other intercellular contacts (1.99B). Links between the cells and their substrates have also been found (Ryan et al 1974). Intracytoplasmic filaments of actin and vinculin form co-linear assemblies, each termed a *fibronexus* (1.100), with extracellular matrix fibrils of fibronectin (Singer 1979) and types I and III procollagen (Furcht et al 1980). It has been suggested (Singer et al 1984) that a fibronexus is a cohesive complex which transmits the collective forces generated by contraction of all the myofibroblasts of the granulation tissue to the wound margins, thereby effecting wound contraction.

Angiogenesis is a vital part of the proliferative phase of dermal repair. Without it invasion of the wound bed by macrophages and fibroblasts would cease through lack of oxygen and nutrients. In vitro studies have shown that capillary endothelial cells release collagenase in response to angiogenic factors. This degrades the collagen of the basement membrane which later fragments, permitting migration of endothelial cells into the perivascular spaces, where they form buds which are added to by the proliferation of cells within and near the parent vessel (Kalebic et al 1983). During dermal repair these buds grow rapidly towards the free surface, where they branch at their tips and unite to form functional capillary loops. New buds then develop on these loops so that a superficial capillary plexus rapidly forms in the granulation tissue (1.101).

Although the factors responsible for angiogenesis during dermal repair remain unidentified, several candidates have been proposed. These include a macrophage-derived growth factor known to stimulate proliferation of endothelial cells in vitro (Martin et al 1981), low oxygen tension (Remensnyder & Majno 1968), lactic acid (Imre 1964) and biogenic amines (Zauberman et al 1969). Endothelial migration may be more significant than proliferation during angiogenesis after injury. If so, then chemotactic factors will play a key role in vivo. Such factors include platelet-derived substances (Wall et al 1978), heparin (Azizkhan et al 1980) and fibronectin (Bowersox & Sorgente 1982). Successful angiogenesis depends not only on chemotactic and mitogenic factors, but also on the presence of a suitable substrate over which migration of endothelial cells can occur. This may be produced, at least partly, by the endothelial cells themselves, since they have been shown to synthesize fibronectin (Birdwell et al 1978) and collagen (Madri & Stenn 1982).

Remodelling

Just as the proliferative overlaps the inflammatory phase, so remodelling overlaps proliferation. During remodelling the highly cellular and highly vascular granulation tissue is gradually replaced by scar tissue with few cells and blood vessels. During the process of remodelling, which may occupy months or even years, most of the fibronectin is removed from the matrix and there is a slow accumulation of large bundles of type I collagen fibres which, as they form cross-links, increase the tensile strength of the scar tissue. Changes in the arrangement of the collagen with time after injury have been studied by scanning electron microscopy (Forrester 1973). When it first appears in the granulation tissue of the wound bed it forms randomly arranged fibrils, which gradually develop into large irregular masses without evidence of any fibrillar substructure. The absence of the characteristic pattern found in uninjured dermis may be associated with the decrease in extensibility and tensile strength which are typical of scar tissue. During subsequent remodelling, the orientation of fibres becomes less random and its strength increases. This change may be caused by the action of mechanical forces exerted on the scar during normal usage to produce orientation of the collagen fibrils in the scar tissue and improve its mechanical function, so that it resembles uninjured dermis more closely.

Such forces may also produce a piezo-electric effect which affects the arrangement of collagen fibrils and fibres. There is now convincing evidence that the pattern of collagen within granulation and scar tissue can be altered by local forces, i.e. that remodelling occurs (Forrester 1973). As long ago as 1892 Wolff noted that bone responded structurally to functional demands. It is likely that scar tissue responds in a similar manner.

The cellular changes involved in dermal repair can be accelerated by treatments which improve the microenvironment of the wound (Dyson et al 1988) and by application of electrotherapy modalities which increase the rate of ingress of macrophages (Dyson 1987, Dyson & Young 1986), possibly by temporarily modifying their membrane structure. The use of such techniques has considerable clinical and surgical significance.

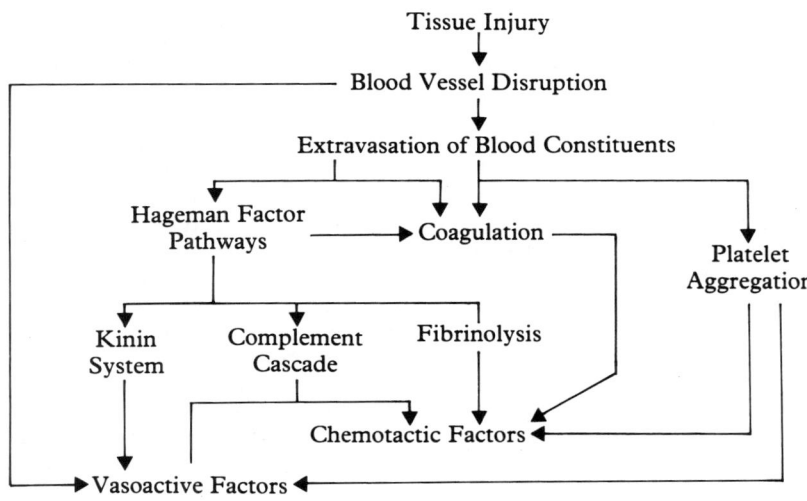

1.98 Mediator pathways of inflammation initiated by tissue injury. After Clark (1985).

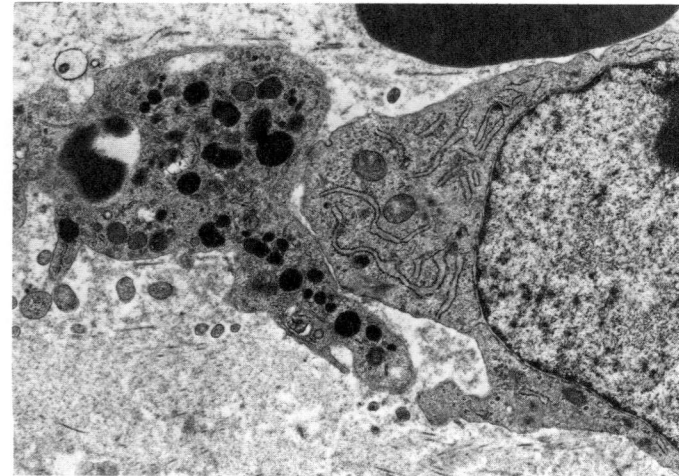

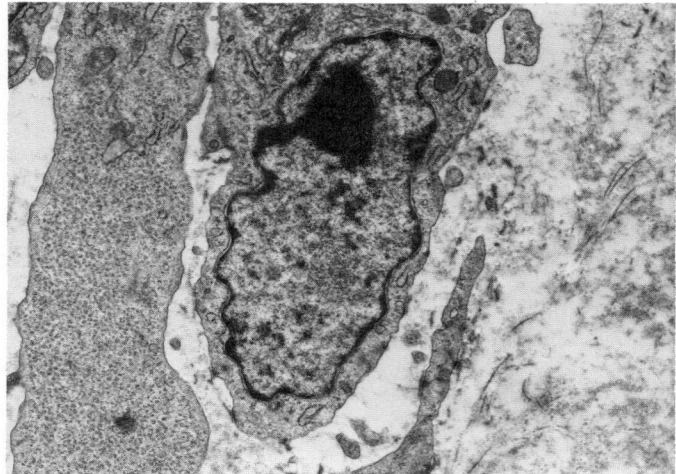

1.99 Electron micrograph showing intercellular contact between: (A) a macrophage and fibroblast in a healing, full thickness skin lesion, three days after trauma; (B) two myofibroblasts in a similar lesion, seven days after trauma. Supplied by Rachel Hickman, Department of Anatomy, UMDS, Guy's Campus, London. Magnification × 8000.

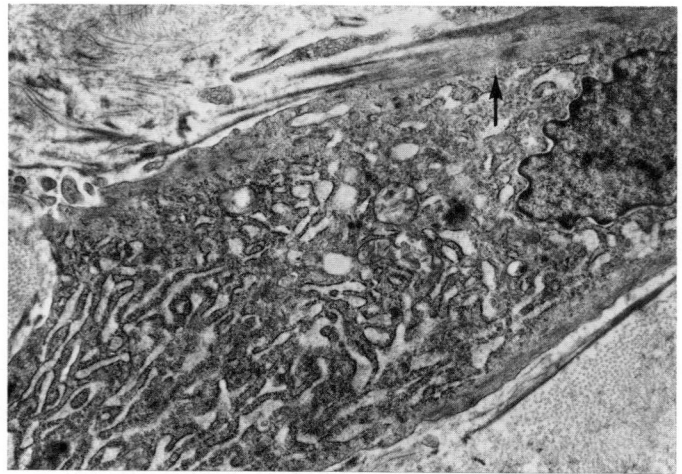

1.100 Electron micrograph showing a fibronexus, a region at the surface of a myofibroblast where there is an alignment between intracellular filaments and fibrils of the extracellular matrix (arrow). Supplied by Rachel Hickman, Department of Anatomy, UMDS, Guy's Campus, London. Magnification × 8000.

EPIDERMAL REGENERATION

Changes in the epidermis leading to re-epithelialization begin within a few hours of the formation of a cutaneous wound. Intact keratinocytes at the free edge of the cut epidermis begin to migrate across the defect (Winter 1962). Migration is made possible by a change in gene expression of the cells, involving temporary dissolution of hemidesmosomes and desmosomes (p. 24), freeing the cells to move, and the formation of peripherally located actin filaments enabling them to do so (Gabbiani et al 1978). According to the 'leap-frog' hypothesis of epidermal regeneration, (Winter 1962, 1964) cells superficial to the stratum basale at the edges of the wound elongate laterally and crawl over each other until they make contact with the wound bed; they then cease to move and begin to divide, producing a new supply of cells, some of which add to the thickness of the regenerating epidermis. Meanwhile other cells migrate over the first cells, reach the wound bed, divide and repeat the process in 'leap-frog' fashion until prevented from doing so by contact inhibition. According to this hypothesis no single keratinocyte moves more than about four or five cell diameters (approximately 40µm) from its original position during epidermal regeneration. Within 48 hours of injury the basal keratinocytes of the new epidermis begin to divide, generating more cells capable of migration (Hell & Cruickshank 1963). The stimulus for epidermal proliferation after injury is still unknown. It may be the removal of an inhibitory chalone (Bullough & Lawrence 1961) and/or the release of epidermal growth factors (Cohen 1965).

In shallow, partial thickness wounds of thin skin, each cut hair follicle acts as a source of reparative epidermal stem cells. These cells can be recognized by their ability to form a characteristic stem cell keratin (type 19). After migration and division, they produce cells which manufacture other keratins (types 9 and 16) which allow the cells to remain sufficiently flexible to migrate over the wound bed. Later products of keratinocyte division produce more rigid keratins (types 1 and 10), typical of mature epidermis.

If the injury is sufficient to disrupt the basement membrane, the keratinocytes migrate over a temporary matrix of fibronectin, fibrin and type V collagen (Clark et al 1982, Repesh et al 1982, Donaldson & Mahan 1983). Keratinocytes have been shown to secrete fibronectin in vitro (Kariniemi et al 1982, O'Keefe et al 1984, Kubo et al 1984) so it is possible that they may produce at least part of the matrix over which they migrate. Once migration ceases, the temporary matrix is replaced by basement membrane.

Regeneration of the basement membrane occurs in sequential stages. Bullous pemphigoid antigen (p. 78) is always present between the basal plasma membrane of the migrating cells and the temporary matrix and can thus be considered to be the first part of the basement membrane to regenerate (Stanley et al 1981, Clark et al 1982). Once the epidermal cells cease to migrate, first type IV collagen and then laminin become incorporated in the regenerating basement membrane (Clark et al 1982).

If the migrating keratinocytes make contact with small foreign particles of approximately 1µm in size, the cells may remove them by phagocytosis (Odland & Ross 1968), possibly after opsonization by fibronectin (Takashima & Grinnell 1984). The keratinocytes migrate deep to any larger particles and dead tissues which lie in their path. Secretion of plasminogen activators (Isseroff et al 1984), collagenases and neutral proteases (Donoff et al 1971) by the keratinocytes may help to clear the way for their migration.

Once re-epithelialization is complete, the keratinocytes revert to their original phenotype (Clark 1985).

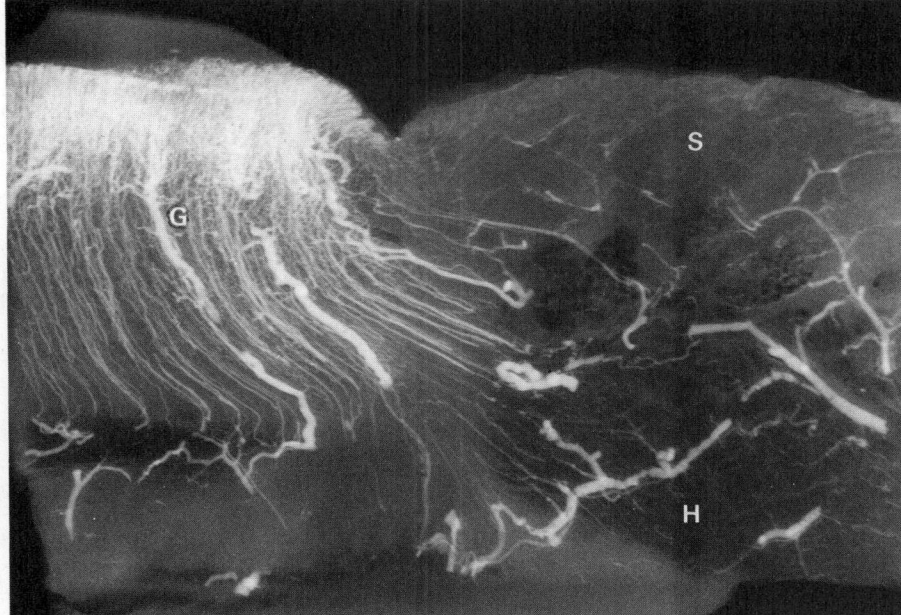

1.101 Microangiograph of a transverse section of granulation tissue (G) showing its invasion by new blood vessels; adjacent intact skin (S), and hypodermis (H) are also visible. The specimen was perfused with barium sulphate and gelatin 10 days after the production of a full-thickness excised lesion in porcine skin. The majority of the regenerating vessels lie at right angles to the surface and are linked by a superficially located capillary plexus. Supplied by Steve Young, Department of Anatomy, UMDS, Guy's Campus, London. Magnification × 130.

APPENDAGES OF THE SKIN

Nails (Ungues)

The nails are horny translucent plates of approximately rectangular shape lying on the extensor surface of the distal segment of each digit (1.102). They are mostly convex in both longitudinal and transverse axes, although there is much variation both between individuals and the different digits of one person. The thickness of mature nails varies from about 0.5 to 0.75 mm. During early development, nails are formed by invaginations of the embryonic epidermis at about nine weeks' gestation and are completely formed by 12 weeks (Zaias 1963).

The nail includes three major regions, the proximal *root (radix)*, the exposed *body* of the nail, and the free distal *border*. The root is inserted in a deep, curved cleft, about 5 mm long, with an overlying *proximal nail fold* whose stratum corneum is prolonged distally on to the body of the nail as the thin *cuticle* or *eponychium*. The sides of the body are bordered by the convex cutaneous *lateral nail folds* (nail walls), the *lateral nail grooves* marking this conjunction. The root of the nail rests on a thick plate of living epidermal cells, the *germinal matrix*, which is closely bound by fibrous dermal tissue to the dorsum of the underlying phalangeal bone (Forslind 1970, Hashimoto 1971). In most digits the matrix extends for some distance distally beneath the nail's body, where it contributes to the appearance of the pale, proximal crescentic area, the *lunule* (lunula); in the hand, the lunule is always visible on the pollex but on the other digits it is progressively covered by the proximal nail fold and on minimus it is usually hidden.

The majority of the body is firmly attached to the underlying epidermal *nail bed* or *sterile matrix* which is longitudinally fluted at its junction with the overlying nail and also at its dermal interface. Here the dermis is very vascular, accounting for the pink colour seen through the translucent nail. Under the distal border of the nail is a narrow crescentic zone of epidermis, the *hyponychium*, which is separated from the adjacent volar skin of the digital tip by a shallow *distal nail groove*.

Microscopically, nails are homologous with the stratum corneum, consisting of compacted, dead, anucleate keratin-filled squames, derived from the maturation of cells generated in the germinal matrix and underside of the proximal nail fold (Zaias & Alverez 1968). Both of these epithelial structures have mitotic basal zones which form typical keratinocytes. These undergo terminal keratinization without passing through a granular cell stage; the squames contain closely packed keratin filaments which lie transversely to the direction of growth and are embedded in a dense protein matrix. Marginal bands of dense protein, heavily cross-linked by disulphide bonds, lie beneath the cell membranes (Hashimoto 1971) and the desmosomes are numerous and persistent, giving nail its coherence; the cells are not shed from the surface as they are in the general epidermis. Very little lipid is present and there is usually a low water content, although nails are more than ten times more permeable to water than the epidermis. The ventral surface of the nail is formed by the matrix, while the dorsal strata are generated by the underside of the proximal nail fold; as new cells are formed, they keratinize and are then steadily extruded forward along the plane of the matrix and on to the nail bed. Growth occurs only in the root and lunule, so the thickness of the proximal end of the nail gradually increases as far as the lunule edge, after which it is constant.

The nail bed itself is lined by stratified squamous epithelium in which the surface layer is only parakeratinized, retaining its nuclei. The surface cells become firmly attached to the undersurface of the nail and move distally at the same rate, being shed imperceptibly at the distal border. This tight coupling, largely the consequence of the highly interlocking, longitudinally folded interface, is important, as it prevents the invasion of microbes and impaction of debris under the nail. Although there is a continual turnover of cells in the nail bed, they do not contribute to the nail substance, but provide a gliding surface for its growth. Beneath the epithelium is a richly vascularized dermis anchored to the periosteum of the distal phalanx and forming a distinct compartment. Because of this, infections of the nail bed or other sources

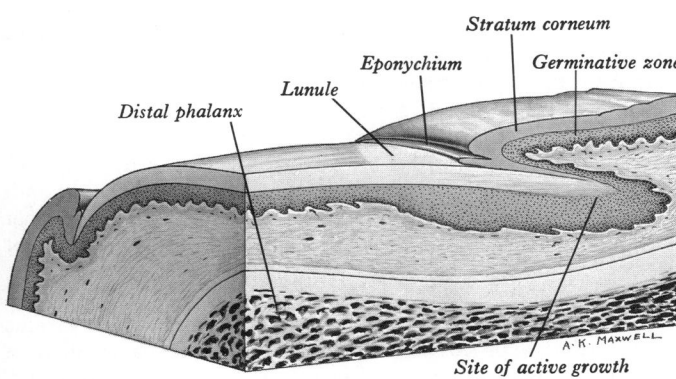

Stratum corneum

Eponychium   Germinative zone

Lunule

Distal phalanx

A·K·MAXWELL

Site of active growth

1.102 Longitudinal section through the root of a nail.

of local pressure rise (e.g. haematoma) may cause severe pain, relieved only by excision of part or all of the nail body.

As already stated, the lunule coincides with the externally visible part of the matrix lying beneath the nail body and obscuring the pink colour of the vascular dermis beneath. However, the nail itself is also cloudy in this area, probably because its keratin is not completely mature.

The rate of nail growth varies with digit, age, environmental temperature, time of day and other factors. Generally its speed is related to the length of the digit, being fastest (about 0.1 mm per day) in the third digit of the hand (medius) and slowest in the fifth (minimus). Fingernails grow up to four times faster than toenails. On average, nails grow quicker in the summer than the winter and faster in the young than the old. Disturbances of the growth pattern caused, e.g. by acute illness or by mechanical damage to its germinative cells, may result in transverse grooves (Beau's lines) or longitudinal furrows; psoriasis typically causes pits in the nail. These defects move forward with growth. More severe damage to the root may result in complete cessation of growth and the nail is then eventually shed, being replaced by a new one if germinative activity recovers. Minor disturbances of nail generation of uncertain origin result in other local defects, including minute air bubbles seen as white flecks.

The dermis beneath the nail is rich in blood vessels, including large arteriovenous shunts (glomera, p. 693). There are also numerous sensory nerve endings, including Merkel terminals (p. 910), and the nails subserve an important tactile function, aiding manipulation. Spatulate nails of this type are found in all the primate species, unlike other tetrapods where a simple keratinized claw typically terminates the digits, and appears to be an important aspect of the evolution of complex, manipulative hands and feet.

For further details of nail biology, see Forslind (1970) and Jakubovic & Ackerman (1985).

Hairs

Hairs (*pili*) are filamentous, keratinized structures present over almost all of the body surface (1.103–106) and are derivatives of the epidermis which assist in thermoregulation, provide some protection against injury, have sensory functions and subserve various subtle roles in social communication. Hairs are absent from a few areas of the body, including the thick skin of palms, soles and flexor surfaces of digits and certain other regions: umbilicus, nipples, glans penis and clitoris, the labia minora and the inner aspects of the labia majora and prepuce. Elsewhere, they vary from about 600 per cm² on the face to 60 per cm² on the rest of the body.

In length they range from less than a millimetre to more than a metre, in width from 0.005 to 0.6 mm. They vary in form, being straight, coiled, helical or wavy, and differ in colour depending on the degree and type of pigmentation. Such qualities differ with body region and between individuals of different ages, sexes and genetic constitution. In general, body hairs are longest and coarsest in Caucasians and least noticeable in Mongolian races.

Over most of the body surface hairs are short and narrow (*vellus* hairs) and in some areas such hairs do not project beyond their follicles, e.g. eyelid skin. In other region they are longer, thicker and often heavily pigmented (*terminal hairs*); these include the hairs of the scalp, the eyelashes, eyebrows and the post-pubertal skin of the axillae and pubis, and the moustache, beard and often the chest hair of males. The hairs of the nostrils and around the external ear canals are also coarser and of the terminal type. Curly hairs tend to have a flattened cross section and in general are weaker than straight hairs.

Each consists of a *shaft (scapus)* and a *root (radix)* lying within a tubular invagination of the epidermis, the *hair follicle (folliculus pili)*. At the proximal end of the root the hair is expanded to form the *hair bulb (bulbus)*, which is continuous basally with the epithelium of the hair follicle. The bulb is deeply indented on its deep surface by a conical vascular *dermal papilla*.

The hair shaft is composed of cells containing a particularly strong form of keratin in which many disulphide bonds cross-link the protein to form a resistant material with high tensile strength. In section, thick hairs show three concentric zones, the *cuticle, cortex and medulla* with different cellular constituents and types of keratin. In thinner hairs the medulla is usually absent. The *cuticle* forms the hair surface and consists of overlapping keratinized squames directed apically and slightly onwards (1.106). The *cortex* contains numerous closely packed, elongated cells, also filled with keratin and, except in white hairs, melanosomes (or phaeomelanin in red hair, see p. 78). In mature hair shafts these cells lack nuclei; between them are often air-filled spaces which may modify hair colour. Finally, the central *medulla* is composed of loosely aggregated and often discontinuous columns of rounded or discoidal cells containing vacuoles, keratin filaments, melanosomes and granular material. Air cavities lie between the cells, or even within them. The apical ends of uncut hair are usually tapered to a fine point.

The hair follicle may extend deeply into the hypodermis (1.103,104) with long hairs or may be more superficial within the dermis. Typically, the long axis of the follicle is oblique to the skin surface; with curly hairs, it is also curved. At its deeper end (*fundus*) it enlarges to accommodate the hair bulb and apically it receives the ducts of sebaceous (and sometimes apocrine) glands before expanding into a funnel-shaped aperture (1.103). The region above the sebaceous duct is the *infundibulum*, then basally as far as the attachment of the arrector pili muscle is the *isthmus* and deep to this, including the region enclosing the hair bulb, is the *inferior segment*. The follicle is surrounded by a thick perifollicular dermal coat containing type III collagen, elastin, sensory nerve fibres and blood vessels and into which blend the arrector pili muscles. Marking the interface with the dermis and follicular epithelium is a broad, PAS-positive basal lamina, the *glassy membrane*. Within this are arranged concentrically the epithelial *outer and inner root sheaths* surrounding the hair shaft.

The outer root sheath (1.105) is an invagination of the epidermis and has a similar although not identical stratification into basal and spiny cell layers which, in growing hairs, are rich in glycogen. Above the level of the inferior segment, its surface cells undergo keratinization, although no granular stratum is found. In deeper regions it is unkeratinized, forming a sheath a few cells thick. *The inner root sheath* occurs mainly in the inferior segment and is a three-layered structure formed basally in the hair bulb, moving apically with the growing hair and disintegrating in the isthmus of the follicle. Its three zones are, from outermost inwards: the *stratum epitheliale pallidum (Henle's layer)*, a single layer of keratinizing cells with pale, flat nuclei; then the *stratum epitheliale granuliferum (Huxley's layer)*, composed of two layers of partially keratinized cells containing strongly eosinophilic trichohyalin granules; and finally a single layer of flattened squames with atrophied nuclei, the *cuticle of the inner root sheath*, facing and loosely interlocking with the cells of the hair shaft cuticle. The keratin of the inner root sheath is soft and the cells of this structure disintegrate by the level of the sebaceous ducts, or before; they may play a part in protecting the incompletely keratinized basal regions of the hair root or in assisting its passage through the follicle.

The hair bulb (1.104) has at its base the *germinative matrix*, a zone of great mitotic activity which generates the hair and its surrounding inner root sheath, and more apically a *keratogenous zone* in which the cells undergo keratinization. The germinative matrix is composed of a mass of closely packed, pluripotent polygonal cells capping the dermal papilla. Cells arising mitotically from this group move apically and may differentiate along several different routes, depending on their position of origin. Those arising from the centre of the base may form the hair medulla; then, radially further out, successive concentric rings of cells will transform into the cortex and cuticle of the hair and, outside this, into the cells of the inner root sheath. The outer root sheath forms the walls of the whole structure but is apparently not derived from it.

This arrangement is of great theoretical interest, since the continuing activity of the whole root complex involves various morphogenetic processes in which different cell shapes, distinct chemical forms of keratin and migration patterns are produced, depending on which genes are being expressed. The entire unit

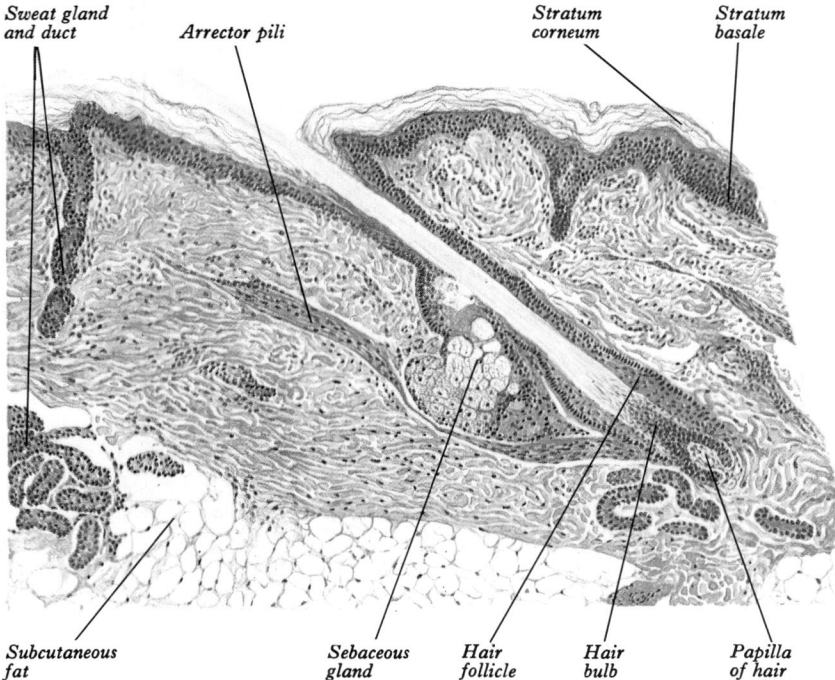

Sweat gland and duct · Arrector pili · Stratum corneum · Stratum basale

Subcutaneous fat · Sebaceous gland · Hair follicle · Hair bulb · Papilla of hair

1.103 A section through the skin, showing the epidermis and dermis (corium), a hair in its follicle, an arrector pili muscle and sebaceous glands opening into the hair follicle. Magnification × 100.

appears to depend on interactions with the connective tissue of the dermal papilla, whose presence is necessary for the hair to be formed and maintained; damage to the papilla results in abnormalities or loss of the hair.

Interspersed among the basal cells of the bulb are numerous melanocytes (**1.104**), their dendrites spreading and branching among the differentiating cells of this mass. Melanosomes (or in red hair, phaeosomes) are synthesized and passed on to the forming cells of the hair shaft, in the manner already described in the epidermis (p. 78), and so are responsible for hair pigmentation. Except in genetic albinism, white hairs have inactive melanocytes which do not contribute melanosomes to the hair root.

The growth rate of hairs varies with their site and thickness, ranging from about 1.5 mm (fine hair) to 2.2 mm (coarse hair) a week when actively growing. There are, however, cycles of growth and hair loss. Fast growing hairs are said to be in the *anagen phase*; this is followed by the involuting or *catagen phase*, when growth ceases, then the resting or *telogen phase*.

The anagen phase of large hairs such as those of the scalp lasts up to ten years, the catagen phase about three weeks and telogen up to four months, after which a new hair bulb is formed in the same follicle and a new hair also begins to grow. Hairs may be shed at any time after growth has ceased. In smaller hairs the growth cycle may be much shorter than this, as little as six months. During catagen, the germinative matrix ceases mitosis and the whole inferior segment of the follicle degenerates, so the expanded base of the hair moves upwards and is enclosed in an epithelial *sac*; such hairs are termed *club hairs*. During telogen the inferior segment of the follicle is totally absent and a new one has to be formed by interaction with a renewed dermal papilla before the next anagen can begin.

Cycles of growth and hair loss occur randomly in postnatal life, so that, e.g. in the scalp, about 85% of hairs may be in anagen phase. In some animals, hair growth and shedding is seasonally synchronized; this is also seen to a lesser extent in early development, as hairs of the fetus begin vigorous growth by the fifth month and tend to be shed in quantity during the third month of postnatal life.

Hair follicles are first seen as cord-like epidermal downgrowths into the mesenchyme at about the third month of gestation. Mesenchymal condensations then invade the base of these cords to form papillae; sebaceous glands bud off apically and arrector pili myocytes begin to differentiate to one side of the follicle. By

the fifth month the body is covered by fine, often deeply pigmented primary hairs (collectively the *lanugo*); on the back these hairs are more frequent than those of the gorilla and chimpanzee at a similar age. Lanugal hairs are mostly shed before birth and are replaced by the secondary, *vellus* hairs, except on the scalp, eyebrows and palpebral margins. Since no further follicles are formed after birth, hairs become more widely spaced as the area of skin increases with body growth.

With puberty, hair growth and generation of much thicker hairs occurs on the pubes and axillae in both sexes and, under the influence of androgens, on the face and trunk in males. The actions of hormones on hair growth are complex but in general androgens stimulate facial and general body hair formation and, after about the first 30 years, tend to cause the thick terminal hairs of the scalp to change to small vellus hairs and may eventually cause complete baldness. In females, oestrogens tend to maintain vellus hairs in their formation of minute hairs, and in post-

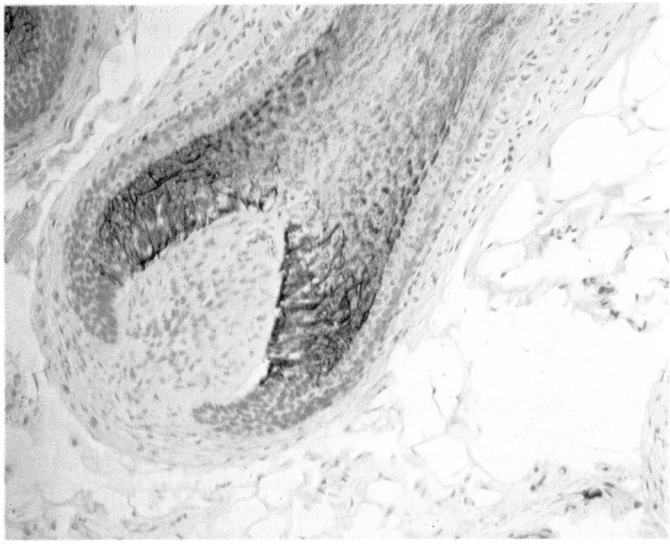

1.104 Vertical section through a hair root, showing the dermal papilla and numerous melanocyte processes extending into the matrix of the hair. Haematoxylin and eosin. Magnification × 250.

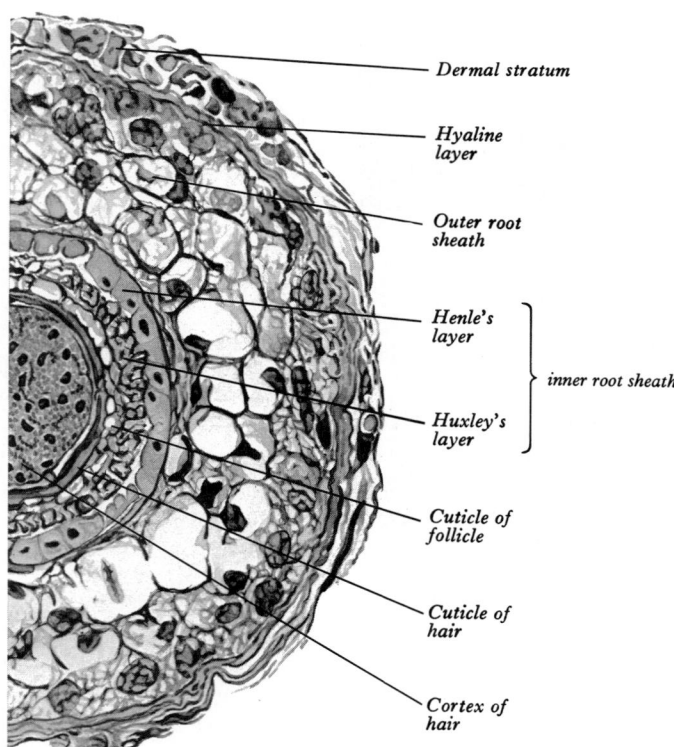

1.105A Transverse section of hair follicle from the scalp of a newborn infant. Magnification about × 600.

The labels for the figure read (from top):
Dermal stratum
Hyaline layer
Outer root sheath
Henle's layer ⎫
         ⎬ inner root sheath
Huxley's layer ⎭
Cuticle of follicle
Cuticle of hair
Cortex of hair

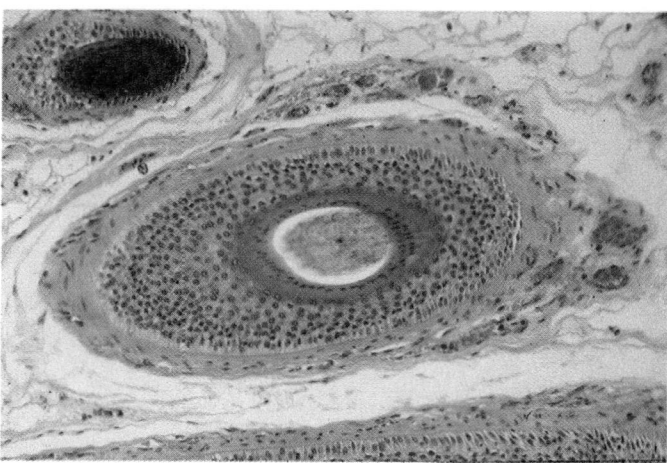

1.105B Section through a hair follicle a short distance below the level of the sebaceous gland ducts showing the different layers of its wall and a small hair. Visible are (from outside inwards); the dermal sheath, surrounding the epithelial layers. The hair which is obliquely sectioned lacks an obvious medulla in this example but streaks of melanin are present in its cortex. Compare with 1.105A. Magnification × 250.

menopausal life reduction in oestrogens may permit stronger facial and body hair growth. During midpregnancy, hair growth may be particularly active but some weeks later an unusually large number of hairs tend to enter the telogen phase and may be shed before the growth cycle recommences.

In older men, growth of hairs on the skin inside the eyebrows, nostrils and external ear canals tends to occur more rapidly, whereas elsewhere on the body growth slows and the hairs tend to become much finer.

Arrector Pili Muscles

These are small fasciculi of non-striated myocytes which form diagonal links between the dermal sheaths of hair follicles and the superficial (papillary) layer of the dermis (1.80,103). They are attached to the isthmus region of follicles, where the dermis contains many elastin fibres, and are directed obliquely and superficially towards the side to which the hairs slope. Contraction therefore tends to pull the hair more vertical and to elevate the epidermis surrounding it into a small hillock, while dimpling the surface where the muscle is inserted superficially, giving the appearance of 'goose flesh' on exposure to cold or emotional stimuli. The myocytes are innervated by noradrenergic terminals of cutaneous sympathetic nerves and may also respond to high levels of circulating catecholamines. Arrector pili muscles are absent from facial and axillary hairs and from eyelashes and eyebrows and the hairs around nostrils and external ear canals.

These muscles are believed to help express the secretions of sebaceous glands which occupy the angles between the arrector pili and the hair follicle. In many mammals, pilo-erection is a means of signalling aggression, fear and other social responses.

Sebaceous Glands

Sebaceous glands are small saccular structures (1.80,103) lying in the dermis and present over the whole body except the thick hairless skin of the palms, soles and flexor surfaces of digits; they secrete an oily substance, sebum (sebum cutaneum), over the skin surface and on to hairs.

Typically, sebaceous glands consist of a cluster of two to five (occasionally up to 20) secretory acini opening by a short com-

mon duct into the apical portion of a hair follicle and can be considered together with the follicle and arrector pili muscle as part of an anatomical unit, the pilo-sebaceous complex (see the extensive review by Wheatley 1986). In some areas of thin skin lacking hair follicles, their ducts open instead directly on to the skin surface, e.g. on the lips and corners of the mouth, nipples, female mammary areolae (Montgomery's tubercles), glans penis, inner surface of the prepuce (glands of Tyson), glans clitoridis and labia minora; at the margins of the eyelids, the large, complex palpebral tarsal glands (Meibomian glands) are of this type (p. 1215).

In general, numbers of sebaceous glands in any given area reflect the distribution of hair follicles, ranging from an average of about 100/cm² over most of the body to as many as 400–900/cm² on the face and scalp. They are also numerous in the midline of the back. Individual sebaceous glands are particularly large on the face, around the external auditory meatus, chest and shoulders and on the anogenital surfaces (sebaceous follicles), and may become distended by their accumulated secretions. Those of much of the face are related to very small vellus hairs (p. 90), whose investing follicles have particularly wide apertures, usually filled with sebum.

Microscopically, the glandular acini are seen to be invested by a basal lamina supported by a thin dermal capsule and a rich capillary network (1.107). Within this, each acinus is lined by a single layer of small, flat, polygonal epithelial cells (1.108). These possess euchromatic nuclei and large nucleoli, keratin filaments, agranular endoplasmic reticulum, free ribosomes and rounded mitochondria and are attached to each other by desomosomes. Functionally, they are mitotically active stem cells whose offspring move gradually towards the centre of the acinus, increasing in volume and accumulating increasingly swollen lipidic vacuoles. Their nuclei become pycnotic as the cells mature and finally the relatively huge, distended cells disintegrate, filling the central cavity and its effluent duct with a mass of fatty cellular debris. This mode of secretion, involving the total destruction of the glandular cells, is described as holocrine (p. 57). For further information see Strauss et al (1976) and the comprehensive review of sebaceous gland biology by Wheatley (1986).

The secretory products pass through a wide duct lined with keratinized stratified squamous epithelium into the apical part of the hair follicle (the infundibulum) and thence on to the surface of the hair and epidermal surface. When first formed, sebum is a complex mixture of which over 50% is di- and triglycerides, with smaller proportions of wax esters, squalene, cholesterol esters and cholesterol (Stewart et al 1983). As sebum moves along its duct its glycerides are partly hydrolyzed by bacterial action to free fatty acids. The normal functions of these components are not yet understood but it is likely that, at least in some cases, they form a

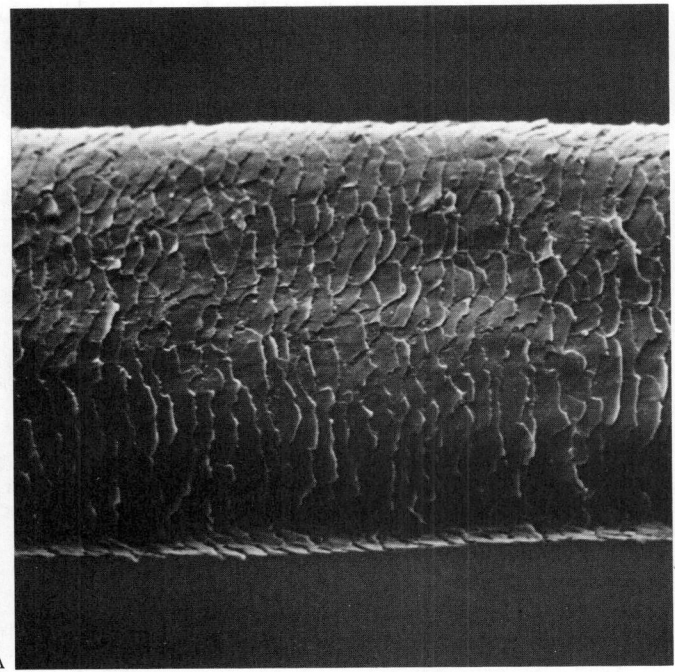

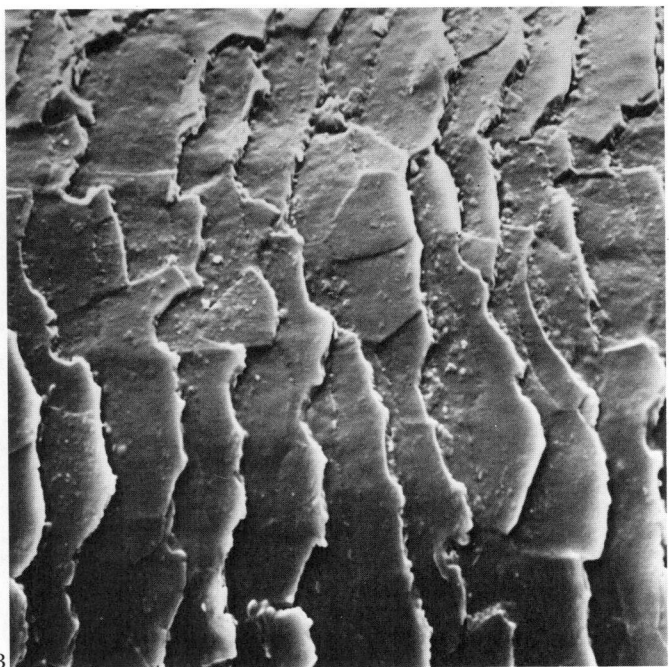

A

B

1.106A, B Scanning electron micrographs of a scalp hair showing details of surface structure. Note the manner in which the cuticular cells overlap each other; their free ends point towards the apex of the hair.

Prepared by Michael Crowder, Guy's Hospital Medical School, London. Magnifications: A × 370; B × 1850.

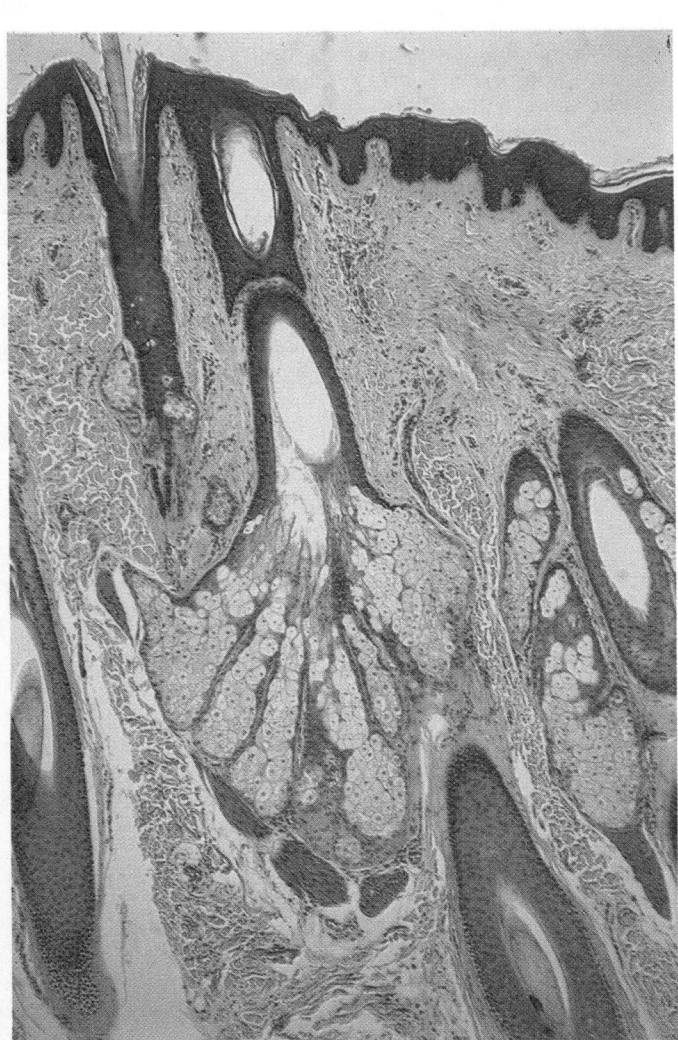

1.107 Section through a group of sebaceous gland saccules opening into a hair follicle. Note also the presence of arrector pili muscles in oblique section, stained red, adjacent to the saccule bases. Haematoxylin and eosin. Magnification × 140.

protective coating of lipid on hairs, possibly assist waterproofing of the epidermis, discourage blood-sucking ectoparasites and contribute to characteristic body odour, a feature which in our ancestral conditions may have possessed strong positive social connotations and in the new-born could play a part in the relationship between mother and child.

Prenatally, sebaceous glands arise by invagination from the epidermis between 13 and 15 weeks' gestation and at birth are quite large, regressing later until stimulated again at puberty. At that time, sebaceous gland growth and secretory activity increase greatly in both males and females, under the influence of androgens, and possibly growth hormone from the adenohypophysis amongst other factors. Excessive amounts of sebum may become impacted within hair follicles, causing damage and inflammation of the surrounding area and thus causing acne. Secretory activity does not appear to be under nervous control but contractions of the neighbouring arrector pili muscles attached to hair follicles may compress them and stimulate their discharge.

Sudorific (Sweat) Glands

These are divisible into two types: *eccrine* glands, numerous and present over almost all of the body surface and *apocrine* glands confined to a few, restricted areas (see the reviews by Zelickson 1971). *Ceruminous glands* of the ear canal and *ciliary glands* (of Moll) in the margins of the eyelids are also modified sweat glands.

Eccrine sweat glands (1.80) are long unbranched tubular structures, each with a highly coiled, wider secretory portion (the *body* or *fundus*, up to 0.4 μm in diameter) situated deep in the dermis or hypodermis and a narrower, straight or slightly helical *ductular portion*, which in the deeper layers of the dermis is convoluted or twisted. The walls of the duct fuse with the base of epidermal (rete) papillae and the lumen passes between the keratinocytes often, particularly in thick hairless skin, in a tight spiral (1.80), to open via a rounded aperture on to the cutaneous surface. In thick hairless skin, they discharge by a regular series of punctae along the centre lines of friction ridges, incidentally providing markers of fingerprint patterns for forensic purposes. Eccrine glands are absent from the tympanic membrane, margins of the lips, nail bed, nipple, inner preputial surface, labia minora, glans penis and glans clitoridis. Elsewhere they are numerous, their frequency ranging from 80 to over 600/cm², depending on

93

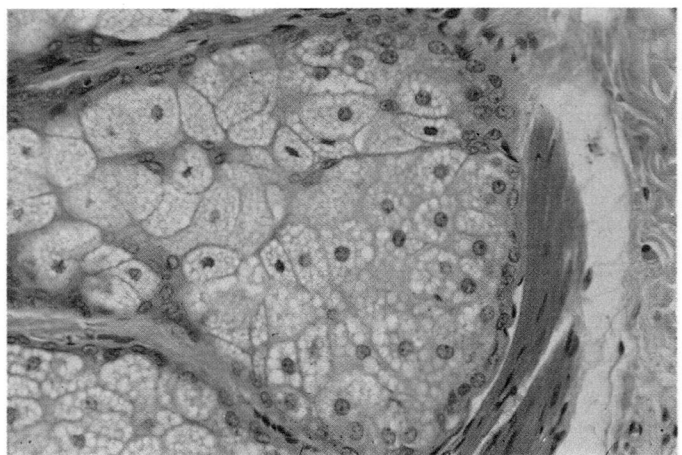

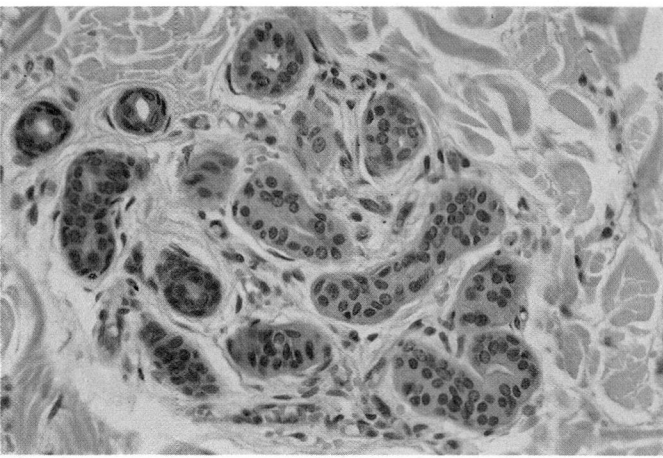

1.108 Higher magnification of a sebaceous gland showing the progression from small polygonal cells around its margins to the large, highly vacuolated cells in the interior of a saccule. The sebum has been extracted by histological processing, leaving an empty frothy appearance in cells about to undergo holocrine secretion. Haematoxylin and eosin. Magnification × 540.

1.109 Section through the basal coil of a sweat gland showing the wider secretory portion and the narrower initial region of the sweat duct composed of two layers of small cuboidal cells. Haematoxylin and eosin. Magnification × 540.

position and genetic variation (Montagna & Parakkal 1974, Millington & Wilkinson 1983). Numbers are greatest on the plantar skin of the feet, but there are also many on the face and flexor aspects of the hands, while the surfaces of the limbs generally have the fewest. Racial groups indigenous to warmer climates tend to have more than those of cooler geographical areas.

In microscopic structure, the whole length of the gland is surrounded by a thin fibrous dermal sheath and an investing basal lamina. The secretory portion consists of a pseudostratified epithelium enclosing a wide lumen (1.109). There are three types of cell: *clear cells*, from which most of the secretion derives, *dark (mucoid) cells* and *myoepitheliocytes*.

Clear cells are approximately pyramidal in shape, with their bases resting partly on the basal lamina and their apices not reaching the lumen of the gland, but instead facing one or more small intercellular canaliculi which are confluent with the lumen. Their apical aspects bear numerous irregular microvilli; their bases are also highly folded where they abut on to the basal lamina. The cytoplasm is rich in glycogen granules and mitochondria, and some granular endoplasmic reticulum and a small Golgi complex are present. The nucleus is rounded and moderately euchromatic.

Dark cells are also pyramidal, with their narrow ends on the basal lamina and their broad ends facing the lumen. Their cytoplasm contains organelles associated with secretory activity—granular endoplasmic reticulum, Golgi complex and numerous dense carbohydrate-rich secretory vacuoles which discharge into the lumen.

Myoepitheliocytes (see p. 55) are numerous; they are elongated and lie among the foot-processes of the other types of cell, just within the basal lamina of the gland. A rich sympathetic efferent innervation, mostly cholinergic, surrounds the glandular portion and is responsible for stimulating myoepitheliocyte contraction, resulting in an immediate secretory response as well as the more delayed increase in glandular activity due to increased blood flow in the area.

Eccrine dermal sweat ducts are lined by two layers of basophilic cuboidal epitheliocytes, the inner of which has short microvilli bordering the duct lumen. A strongly eosinophilic layer lines the apices of the inner cells. The ductular cells have many mitochondria and rounded nuclei. Ductular cells within the epidermis migrate and keratinize in the same way as surrounding keratinocytes.

Sweat glands secrete a clear, odourless fluid, hypotonic to tissue fluid and containing small quantities of many substances, predominantly sodium and chloride ions but also urea, lactate, amino acids, immunoglobulins and other proteins, bicarbonate, calcium ions etc. When initially secreted, the fluid is similar in composition to tissue fluid but is modified as it passes along the duct by the action of its lining cells, which resorb sodium and chloride and some water too. The hormone aldosterone enhances this activity. Secretion is stimulated chiefly by temperature rise but, in certain areas of the body (hands, face and axilla), the glands react most strongly to emotional stimuli. These effects are mediated by the cholinergic sympathetic innervation surrounding the glands.

Apocrine sweat glands are particularly large glands of the dermis or hypodermis but, unlike the eccrine variety, they typically discharge into the apical regions of hair follicles. They are present only in a few areas of the body, namely the axillae, perianal region, areolae, peri-umbilical skin, prepuce, scrotum, mons pubis, and labia minora. Ceruminous glands of the external auditory meatus and the ciliary glands of the palpebral margins (glands of Moll) are also usually included in this category. However, their secretions are quite different and these glands should be considered as distinct, specialized subtypes (for details see pp. 1222 and 1215 respectively).

Although apocrine glands resemble the eccrine category in having a coiled basal secretory body and a straighter duct, the secretory region may be as much as 2 mm wide, being greatly distended by accumulated secretions; its coils often anastomose with each other to form a labyrinthine network. Their walls are lined by a single layer of cuboidal epithelium with rounded euchromatic nuclei and complex, microvillous apices which tend to protrude into the lumen. Among the bases of these cells lie numerous longitudinally orientated myoepitheliocytes, the whole complex of cells resting on a thick basement and, outside this, a connective tissue capsule, rich in capillaries.

The cuboidal cells have typical secretory cell contents—granular and agranular endoplasmic reticula, Golgi complex, dense secretory vacuoles—their numbers varying with the cycle of synthesis and discharge. The mechanism of secretion is still not entirely clear but may involve a number of different processes (see Hashimoto 1978), including merocrine secretion of granules, detachment of large volumes of apical cytoplasm (*apocrine secretion*) and occasionally complete, holocrine disintegration of epithelial cells.

The secretions within these glands are a proteinaceous, thick milky fluid which at first is sterile and odourless but undergoes bacterial metabolism to generate potent odorous compounds, musky or urinous in smell, including short-chain fatty acids, etc. In many animals these are potent pheromonal signals important in courtship, parental and territorial behaviour, as well as in various other aspects of social life. Their role in modern humans is less certain, although there has been much speculation on the potency of, e.g. axillary odours in the more subtle aspects of human interactions (see Gower et al 1987).

Apocrine glands are formed at the same time as eccrine glands and are at first widely distributed over the fetal body surface but from five months' gestation onwards are progressively lost, except in the small areas where they are occur in adults. At puberty they enlarge and become actively secretory. For reviews of this subject, see Hashimoto (1978), Tani et al (1980) and Jakubovic & Ackerman (1985).

EMBRYOLOGY

2

Introduction

Distinguishing living organisms from inanimate objects intuitively seems simple and commonplace, but such distinction still defies concise scientific definition. This question is obviously central to any study of living forms, and biological texts commonly open by considering a series of attributes, such as irritability and other responses to environmental change, motility, ingestion followed by transformation (metabolism) of foodstuffs and excretion of end products, growth, reproduction and maturation.

However, the basic building blocks of living things, sub-atomic particles, atoms and molecules and interblock forces, are common to all materials known to man; the animate attributes listed above appear as inevitable consequences of particular arrays or *organizations* of the building units.

THE DYNAMISM OF LIVING ORGANISMS

Examination of both the physical universe and the biosphere has led to widely accepted fundamental generalizations: firstly the relativistic nature of all forms of observation, and secondly the interrelation between matter and energy. The latter, stated explicitly in the *laws of thermodynamics*, holds that the total energy content of the universe is constant but that the proportion of disorganized, randomized or dissipated energy (*entropy* level) is steadily *increasing*. Living systems, however, appear locally to reverse this trend by their growth, reproduction and increasing structural complexity, that is to say, their entropy level is *decreasing*. This apparent paradox is answered by considering life forms as interlocked structurally, functionally and energetically with their environment, from which they are supplied by a constant flow of energy essential for their maintenance (Schrödinger 1967, Lehninger 1965, Young 1964). The ultimate source of this energy is the structural degradation of materials in the sun with the emission of radiant energy partly as light. This is trapped by green plants and used in photosynthetic processes whereby simple substances are energetically transformed into complex molecules incorporated into the plant tissues and eventually provide a source of chemical energy, or food, for the animal kingdom. (A minority of elementary organisms have energy provided by degradation of materials in the earth's core.) This absolute dependency upon supplies of environmental energy reflects the *dynamic* nature of life processes, which constantly undergo change. Further, continued success of an organism in any habitat implies that its structural/functional organization is complementary to features of its environment, and they can only be studied intelligently together, that is, as one system. Indeed, some hold that the various control systems of an organism necessarily embody a 'model' of their environment for effective responses to occur (Young 1971, 1978).

That living organisms are dynamic, of course, denotes changing states with the passage of time, but the temporal scale chosen for study may vary from millions of years to fractions of milliseconds. Examples of the latter include brief biochemical and biophysical interactions such as the activation of visual pigments by photons, conformational changes in cell membranes with consequent alterations in permeability, as during the transmission of a nervous impulse or the release of a neurochemical transmitter, and the formation and breaking of linkages between actin and myosin subunits during muscle contraction. A relatively short time scale is also involved in the multitude of feedback systems whereby the state of the internal and external environment is constantly monitored and appropriate adjustments are made to preserve the internal environment within the fairly narrow range of values necessary to the continued life of cells. Such a preservation of internal constancy, a comparatively rapid variety of *homeostasis* (p. 860), despite *short-term* fluctuations in the surroundings, is a central feature of the operation of all the principal bodily systems, e.g. the regulation of temperature, blood pressure, hydration, osmolarity, electrolyte and hydrogen ion concentrations, glucose and oxygen levels, etc. To these may be added somewhat *slower homeostatic cycles* of endocrine secretory control systems, the varying rates of cell growth, death and replacement in different tissues, the behaviour patterns associated with mating, rearing of young, searching for and securing food, and also the ill-understood processes which lead to the laying down of short- and longer-term memory traces. (Long-term homeostasis is discussed below.)

REPRODUCTION AND EVOLUTION

Contrasting with such changes of short duration all extant organisms' ancestries stretch back over hundreds of millions of years, a period involving great changes in the natural environment. The latter were paralleled by a gradually increasing diversity of form and complexity of organisms, each more or less well adapted to life in some particular pocket of the changing environment. The *long-term homeostatic mechanism* which enabled this diversification is intimately connected to another aspect of the dynamism of living organisms. Thus, in addition to securing an adequate flow of solar or chemical energy to drive its cellular machinery and continually replace its substance, each individual organism also passes through stages of *growth*, increasing *maturity* and, in complex forms, eventual *senescence* and *death*. Continuing life of the species is therefore dependent upon some mode of *reproduction*. Elementary life forms increase their mass and maturity until, at a critical point, they reproduce by simple equating division of their substance. Each successive organism has the same essential physico-chemical constitution as the parent, and part of its structure operates as an information store, i.e. a chemical (or genetic) *memory system*, so that the offspring grow into identical replicas of the parent. Occasionally, however, apparently spontaneous changes occur in the organization of the genetic memory store and such *mutations* lead to the emergence of organisms with altered characteristics. Often such *mutants* are less well fitted for survival under environmental stresses and they succumb, but sporadically a *favourable mutation* occurs. The transmission of this to succeeding generations leads to the establishment of a species with an enhanced probability of survival and, occasionally, the ability to populate new or changing environments.

Such **asexual** methods of reproduction occur in many relatively simple extant forms of life, but the great diversity and complexity of the remainder followed the emergence of **sexual** methods of reproduction, each offspring the outcome of welding of half of the genetic memory of *two* parents. In essence each parent *segregates* certain specialized *reproductive cells* in which there is a *rearrangement* of the genetic material, and a *reduction* to half the quantity found in the general cells of the body (p. 42). Two reproductive cells, one from each parent, approach and *fuse*, and their genetic material intermingles in a specific manner (p. 123). The offspring, whilst in part possessing certain characteristics which resemble those of one or other parent, and of course those of the species in general, also possesses unique features due to the new genetic combination and the previous genetic rearrangement. When the effects of occasional favourable mutations are added to the infinitely varied genetic combinations resulting from sexual reproduction over countless generations, the opportunities for the initiation of altered species are vastly increased. On this wide range of variants the forces of *natural selection* operate and result in the endless array of life forms which populate all corners of the earth's surface, oceans and skies.

Any organism, therefore, including mankind, is not only interlocked structurally and energetically with its environment, exhibiting transient homeostatic responses and rhythmic variations, but is also at some point in its individually unfolding life history (*ontogeny*). It is progressing towards senescence, and possesses a unique genetic apparatus, the result of innumerable generations of diversifying, sexually reproducing ancestors (*phylogeny*). All these grades of dynamic change should be considered when any study of the fabric and behaviour of mankind is undertaken.

The present section attempts a summary of the ontogeny of the human body. Comprehensive texts and papers are cited as frequently as space restrictions permit.

EARLY DEVELOPMENT OF CHORDATA

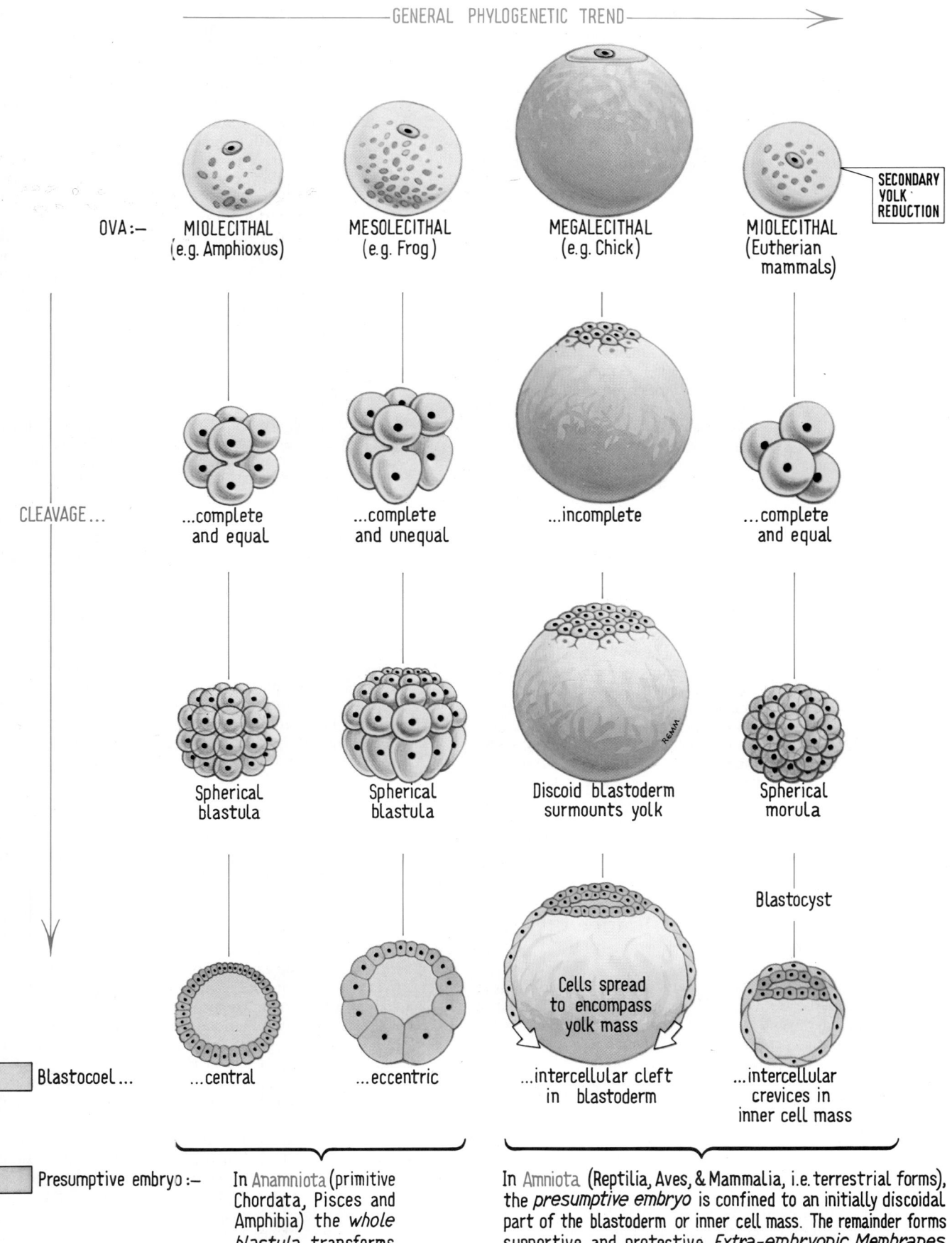

———————GENERAL PHYLOGENETIC TREND————————→

OVA:— MIOLECITHAL
(e.g. Amphioxus)

MESOLECITHAL
(e.g. Frog)

MEGALECITHAL
(e.g. Chick)

MIOLECITHAL
(Eutherian
mammals)

SECONDARY
YOLK
REDUCTION

CLEAVAGE...

...complete
and equal

...complete
and unequal

...incomplete

...complete
and equal

Spherical
blastula

Spherical
blastula

Discoid blastoderm
surmounts yolk

Spherical
morula

Blastocyst

Blastocoel... ...central ...eccentric

Cells spread
to encompass
yolk mass

...intercellular cleft
in blastoderm

...intercellular
crevices in
inner cell mass

Presumptive embryo:— In Anamniota (primitive Chordata, Pisces and Amphibia) the *whole blastula* transforms into an *embryo*.

In Amniota (Reptilia, Aves, & Mammalia, i.e. terrestrial forms), the *presumptive embryo* is confined to an initially discoidal part of the blastoderm or inner cell mass. The remainder forms supportive and protective *Extra-embryonic Membranes* –amnion, chorion, yolk sac, allantois, & placental structures.

2.1 Variations in the patterns of cleavage in *Chordata* associated with differences in oöcyte size, yolk volume and distribution. Note that the overall phylogenetic trend is from left to right, and the progression of cleavage from above downwards. Note also the morphological *differences* and often proposed '*homologies*' between the various groups. Not drawn to scale. Consult text for further comment.

Sexual Reproduction

Sexual reproduction can conveniently be divided into several chronological stages, namely *gametogenesis, fertilization, cleavage, gastrulation* and *organogenesis*, whilst to the latter three stages the general term *embryogenesis* may be applied. This essentially reflects orderly temporal and spatial variations in *growth, morphogenetic movement* and *patterned differentiation*, balanced against *regional degeneration* and *death* of cells and tissues. A brief survey of these phases of development will be given before proceeding to a more specific account of human ontogeny.

GAMETOGENESIS

A multicellular organism which reproduces sexually consists mainly of general body (*somatic*) cells which, although exhibiting extreme ranges of variation in size, shape and specific functional as well as morphological characteristics, all possess certain common features of their genetic apparatus. This contrasts with the highly distinctive differentiation process and genetic changes which occur within the *gonads* and result in the formation of mature *generative (germ)* cells or *gametes*. General somatic cells usually possess a full or *diploid number* of chromosomes, a half of which was originally derived from each parent. However, the *activity* and *expression* of the many different gene loci on the chromosomes varies in different cell types in adult tissues, and during development as differentiation occurs. The somatic cells which are capable of dividing do so by the process of mitosis (p. 42). Each chromosome, having first made a faithful replica of itself and thus of its genes, divides longitudinally and each resultant cell is again equipped with the diploid number of chromosomes which are replicas of those in the parent cell. During gametogenesis, however, a complex series of changes termed *meiosis* are set in train (p. 42). Essentially, during this, the original chromosome number is reduced to a half, the *haploid number*, in which a *recombination* of the genetic material has occurred. (Details are considered on p.44.) The male gonad or *testis* produces many small motile gametes or *spermatozoa* in which the cytoplasmic machinery is much reduced; each consists of a nucleus bearing the genetic apparatus, the chromosomes, with a closely applied acrosome derived from a Golgi apparatus and succeeded by a long flagellum which is partly ensheathed by an energy-transforming mitochondrial sheath (p. 119). In contrast the female gonad or *ovary* produces fewer, larger nonmotile *ova*, their cytoplasm containing a variable quantity of food reserves (*yolk* or *deutoplasm*) and exhibiting regional variations of the cortex (*oöplasmic segregation*) which are of importance in the initial stages of differentiation (p. 104 et seq).

Types of Ova and their Influence on Development

Ova vary enormously in their size, amount of yolk and number shed, throughout the animal kingdom, and within the Chordata a broad evolutionary trend can be recognized (**2.1**). Elementary aquatic forms usually produce many small *miolecithal* (yolk-poor) ova, which, after fertilization, rapidly develop into independent larvae, many of which succumb to predators and other inhospitable features of the environment. The larger ova of the amphibia are *mesolecithal* (with greater reserves of yolk), but again, those fertilized ova that survive are *wholly* transformed into aquatic larval stages which forage for food independently and exhibit long periods of growth, differentiation and metamorphosis before the mature state is reached.

Colonization of dry land by forms not restricted by a periodic return to the water for reproductive purposes was necessarily paralleled by drastic evolutionary modifications in their ontogeny. Reptiles, birds (and prototherian mammals) produce a small number of large, heavily yolked (*megalecithal*) eggs enclosed by membranes and a hard shell, which may be laid on land. They have no larval stage and much of the fertilized ovum does not contribute directly to the body of the embryo, but is concerned

with the elaboration of various *extraembryonic membranes*. The latter provide mechanisms for mobilization and transport of the massive yolk reserves to the developing embryo, deposition of waste end-products of metabolism, the provision of a local aquatic environment for the embryonic tissues and to allow exchanges of carbon dioxide and oxygen across the porous shell (**2.2**).

Eutherian mammals, including man, although derived from such ancestors with heavily yolked eggs, again produce *miolecithal* eggs, but these are few in number, possess no hard shell and development proceeds within the controlled environment of the female genital apparatus. Despite this drastic secondary reduction in yolk content, however, many features of mammalian development are only explicable by reference to such ancestry. Thus, although there are differences of detail between the various groups of mammals, they all develop the same array of extraembryonic membranes (yolk sac, allantois, amnion and chorion) with broadly similar functional associations. These membranes are, of course, modified in various ways for *viviparity*, i.e. development in utero, with nutritive, respiratory, excretory and immunological dependency upon the mother.

Recapitulation

Similarly, within the body of the mammalian embryo all the principal organ systems pass through stages which reflect fundamental patterns in the *ontogeny* of their ancestors. Such patterns are particularly clear in the basic metamerism of the trunk, the progression from a simple tubular heart to a complex, folded, double pump with intervening septa, the common plan of the early nervous system and special sense organs, the craniocaudal sequence of nephric tubules, the plan of the early chondrocranium and the succession of gill arches in the early pharyngeal wall. Such *recapitulation* during development was originally, and too simply, interpreted as if an embryo passed through sequential stages which mimicked the *mature* form of its main ancestral groups. Although this view was later modified to include only structural features during *embryonic life* of the ancestors, emphasis is now placed on the essential similarity of the *morphogenetic mechanisms* common to all chordates. Accordingly, the latter closely resemble each other in the early stages of development, and recognizable intergroup differences emerge gradually as development proceeds.

FERTILIZATION

Fertilization includes those mechanisms whereby a sperm approaches, attaches to, then penetrates the surface of an ovum, and the early changes which follow. Briefly, the egg surface becomes modified preventing entry of further sperms, the oöplasmic segregation of factors in the egg cortex, important in subsequent processes of differentiation, are modified and finalization of the meiotic cycle of the ovum occurs. The latter involves completion of the second meiotic division with the expulsion of one set of chromatids into the second *polar cell*. (The first division had halved the number of chromosomes; each chromosome, however, consisted of two chromatids.) The male and female nuclear derivatives (pronuclei) swell, approach each other and then coalesce (*karyogamy*) to form a single *segmentation nucleus*. Thus the diploid number of chromosomes is restored, the chromosomal sex of the zygote established and a series of mitotic *(cleavage)* divisions are initiated.

Much evidence from comparative and experimental embryology has accumulated concerning the forms of symmetry exhibited by ova, the mechanism of sperm attachment and entry, the site of sperm entry and its influence on the distribution of formative substances in the egg cortex and their relative fixity or lability. For some prominent earlier contributors to this important area of biology see Needham (1942), Waddington (1956, 1962), Brachet (1960, 1965, 1967), Raven (1958, 1961, 1963, 1966a,b). All ova are *polarized* with an axis extending between the largely cytoplasmic *animal pole*, near which the nucleus lies, and the more heavily yolked *vegetative pole*. Whilst ova of some species possess a fairly simple *radial symmetry* around this axis with experimentally

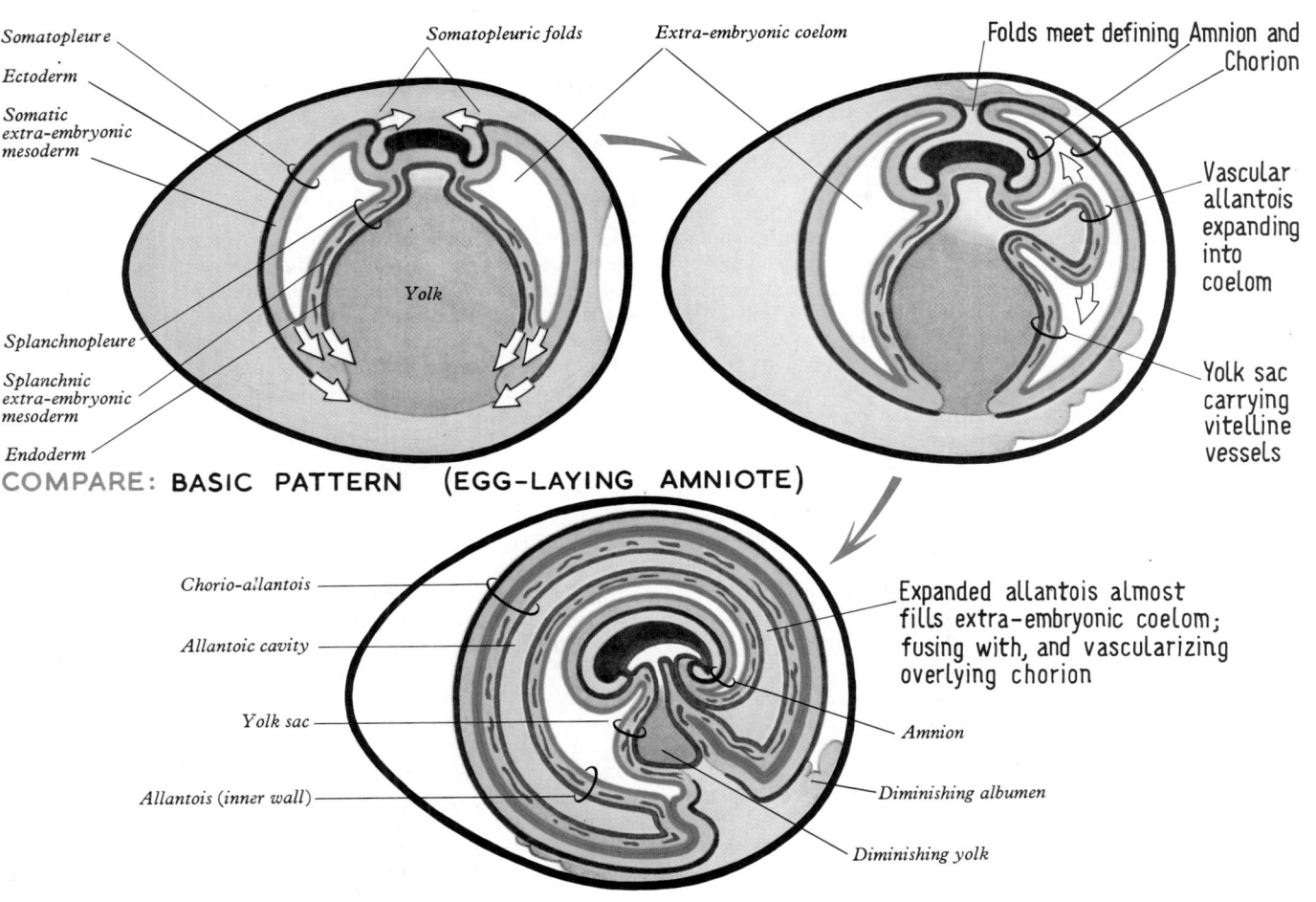

COMPARE: BASIC PATTERN (EGG-LAYING AMNIOTE)

WITH: EUTHERIAN MAMMALS —

SUPERFICIAL IMPLANTATION
e.g. PIG

INTERSTITIAL IMPLANTATION
e.g. MAN

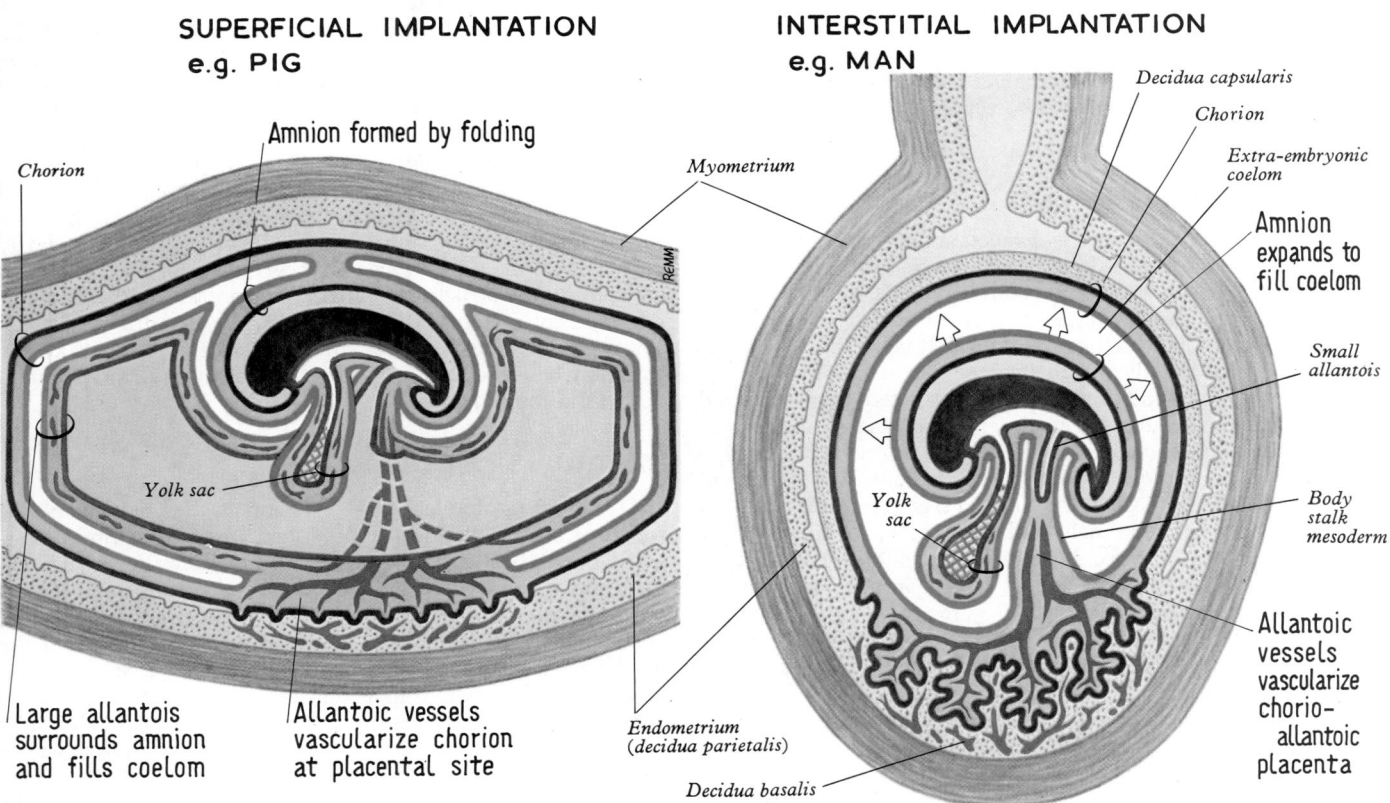

2.2 A generalized series of diagrams to allow comparison of the mode of formation of extraembryonic membranes in a megalecithal egg-laying amniote (e.g. chick) with that of a eutherian mammal exhibiting superficial implantation and a large allantois (e.g. pig), and a eutherian mammal showing interstitial implantation and a diminutive allantois (man). Note in each case the vascularization of the yolk sac, and the manner in which the expanded allantois (chick, pig) or its homologue in man, the body stalk mesoderm continuing from the small allantoic diverticulum, bear allantoic blood vessels which vascularize the overlying chorion. The latter is applied either to the shell, or to the uterine decidua.

demonstrable *gradients* of *developmental potential* extending between the poles, all others exhibit, in addition, further localizations of materials which confer a *bilateral symmetry* on the ovum. The time of appearance, degree and reversibility of this bilateral symmetry varies greatly between animal groups. Again, in occasional types of ova, the site of sperm entry is restricted by the egg coats to a small area on the surface whereas, in the majority, entry occurs at any point. Further, in some, the bilateral symmetry is reorganized in relation to the point of sperm entry whereas, in others, it is minimally affected. There is little evidence that sperms are specifically attracted towards ova by chemotropism, but in mammals it is often held that approach towards the egg surface, which initially is surrounded by granulosa cells of the corona radiata and the zona pellucida (p. 116), is facilitated by hyaluronidase and acrosin (acrosomase) produced by the acrosome of the sperm. However, the degree of development of an acrosome and its enzyme concentrations vary considerably—neither are prominent in man. In some instances the egg cortex and its coats have been shown to contain proteins (*fertilizins*) which can activate and agglutinate sperms, whilst the sperms contain *antifertilizins* which, when extracted, can agglutinate ova; penetration is assumed to involve an interaction of the antigen–antibody type. Finally, in some, the deeper part of the acrosome has been shown to extrude an *acrosomal filament* which, when it contacts the egg surface, becomes engulfed by a *fertilization cone* of cytoplasm which protrudes from the egg surface. The degree to which these postulated mechanisms, symmetries and rearrangements apply to human ova remains uncertain.

CLEAVAGE

Compared with the majority of somatic cells (excepting large neurons and certain other large cells), recently fertilized ova are unusual in possessing but a single diploid nucleus within a large volume of cytoplasm. The mitotic divisions of cleavage are, however, not accompanied by any substantial synthesis of cytoplasm and so the repeated replication and division of nuclear material, together with the partitioning of the existing cytoplasm, soon causes a return to the more usual *nucleo–cytoplasm ratio*. The resulting cells are called *blastomeres*, but their individual potentiality for further development varies considerably between animal groups; accordingly, a variety of terms have been used to describe cleavage 'types', which are briefly mentioned here.

TYPES OF CLEAVAGE (2.1)

Cleavage of miolecithal eggs is *complete* and *equal* since the whole zygote segments and resulting blastomeres are approximately equal in size. In contrast, mesolecithal eggs have a complete form of cleavage, which is *unequal*, with large yolk-laden blastomeres at the vegetative pole which grade into small yolk-free ones at the animal pole. In heavily yolked eggs the yolk mass does not segment (*incomplete*) and divisions are restricted to the discoid cytoplasmic accumulation at the animal pole. Such purely descriptive terms have been supplemented by others designed to summarize the *developmental potential* of the various regions of the zygote and the blastomeres derived from them. As has been intimated, experimental analysis of the *causal mechanisms* of early development has indicated that all zygotes exhibit some degree of segregation of cytoplasmic factors which, in their interactions with different blastomere nuclei, are fundamental in the subsequent emergence of different cell varieties. However, the degree, precision and lability of this segregation varies greatly between animal groups. In some invertebrate ova it is highly ordered and in such '*mosaic eggs*' cleavage often follows a fairly precise geometrical pattern (e.g. spirally cleaving forms) and is called *determinate*. The fate of each blastomere is relatively fixed and removal of blastomeres results in the development of a partial embryo. In contrast, in so-called '*regulative eggs*' with *indeterminate cleavage* the segregation, although present, is much more diffuse and flexible. In such forms, a blastomere when separated from its fellows often proceeds to develop into a complete, although small, embryo. However, such terms imply a distinction which is only one of degree. Many intermediate types exist and even the most mosaic ova show a small regulative capacity; conversely, the most regulative eggs are by no means equipotential systems; they possess some mosaicism, and only regions which include presumptive chorda-mesoderm (vide infra) continue to the formation of a complete embryo.

CLEAVAGE IN CHORDATA (2.1)

Cleavage of the miolecithal eggs of elementary chordates leads to the formation of a spherical *blastula*, i.e. a ball of cells with a central cavity or *blastocoele*, and the animal pole blastomeres are slightly smaller than those at the vegetative pole. The cleavage product of mesolecithal eggs is also a spherical blastula but, reflecting the large gradation in blastomere size from animal to vegetative pole, the blastocoele is eccentric and sited nearer the former. Cleavage in heavily yolked ova results in a circular *blastodisc* at the animal pole, several blastomeres thick and separated from the underlying yolk by a subgerminal cavity containing products of yolk degradation. The blastocoele is probably represented by a small cleft which develops between the superficial and deeper parts of the blastodisc (i.e. it is *not* homologous with the subgerminal cavity). Finally, cleavage of the miolecithal eggs of mammals leads first to a solid ball of blastomeres, the *morula* (Latin: *morum* = mulberry), soon to be transformed into a hollow *blastocyst* with an eccentric *blastocystic cavity*. The latter is surrounded on most aspects by a thin wall of cells (the early *trophoblast*) but at one point a localized group of cells, the *inner cell mass*, projects into the cavity. The mammalian blastocyst is to be carefully distinguished from the spherical blastula of the more elementary chordates. In the latter, the *whole blastula* undergoes a complex series of foldings, cell migrations and other *morphogenetic movements* and *interactions*, collectively termed *gastrulation*, and is then progressively transformed through early and late *gastrula*, then *neurula* stages, into the embryo. In contrast, only the *central cells* of the avian or reptilian blastodisc are engaged in these transformations. The peripheral blastomeres gradually grow to enclose the yolk, becoming arranged in several layers with an intervening cavity and, by complicated processes of differential growth, cellular changes and vascularization (2.2), form the various *extraembryonic membranes* (yolk sac, amnion, chorion and allantois) described more fully elsewhere. Similarly, in the early mammalian blastocyst, only a few *formative cells* in the centre of the *inner cell mass* are directly concerned in the future formation of the embryonic body. The remaining cells of the inner mass and the general walls of the blastocyst are destined to form the same extraembryonic membranes foreshadowed in their heavily yolked ancestors. The precocious segregation of blastomeres as extraembryonic structures in mammals contrasts with their later development in reptiles and birds and the detailed fate of the various membranes and their role in *placentation* shows a wide variation between the different mammalian orders (Mossman 1937, Amoroso 1952).

GASTRULATION

Gastrulation in simple forms (2.3) entails the conversion of a roughly spherical, hollow, single-walled *blastula* into a more elongated, double-walled *gastrula* containing a cavity, the *archenteron*, for reception and assimilation of food and with oral and blastoporic openings at opposite ends. This is achieved by the infolding of one half of the blastula into the other, the orifice formed by the invaginating material being the *blastopore*; the point where the first invaginated cells contact the inner aspect of the opposite hemisphere eventually breaks down to form an *oral orifice*. The addition of a third, intermediate, mesodermal layer in the wall is usual in most groups of animals and its origin is varied; in particular the fate of the walls of the primitive archenteron in the lower chordates is rather more complex (2.3). In these, the blastopore is regarded as presenting a dorsal, a ventral, and a pair of lateral *lips*. The material invaginated through the *dorsal lip* forms a midline strip in the roof of the archenteron which stretches from the blastopore to the margin of the future oral orifice; it underlies

EARLY DEVELOPMENT OF THE ELEMENTARY CHORDATE

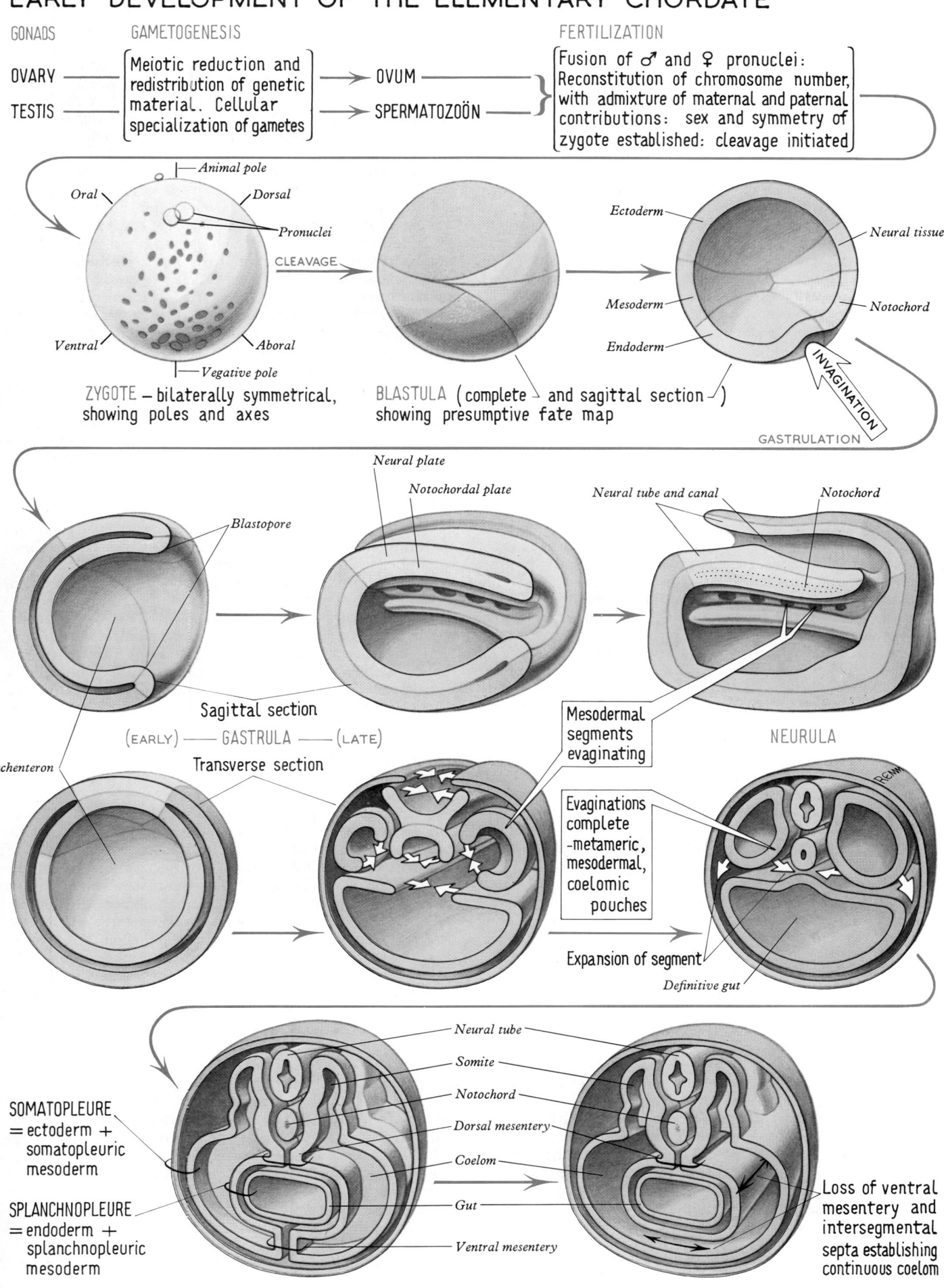

GONADS   GAMETOGENESIS   FERTILIZATION

OVARY ———— Meiotic reduction and   ——→ OVUM ————   Fusion of ♂ and ♀ pronuclei:
            redistribution of genetic   Reconstitution of chromosome number,
TESTIS ——— material. Cellular   ——→ SPERMATOZOÖN — with admixture of maternal and paternal
            specialization of gametes   contributions: sex and symmetry of
                                                          zygote established: cleavage initiated

ZYGOTE — bilaterally symmetrical, showing poles and axes

Animal pole
Oral
Dorsal
Pronuclei
Ventral
Aboral
Vegative pole

CLEAVAGE

BLASTULA (complete and sagittal section) showing presumptive fate map

Ectoderm
Neural tissue
Mesoderm
Notochord
Endoderm

INVAGINATION

GASTRULATION

Blastopore
Sagittal section

Neural plate
Notochordal plate

Neural tube and canal
Notochord

(EARLY) — GASTRULA — (LATE)

NEURULA

Archenteron
Transverse section

Mesodermal segments evaginating

Evaginations complete —metameric, mesodermal, coelomic pouches

Expansion of segment

Definitive gut

SOMATOPLEURE = ectoderm + somatopleuric mesoderm

SPLANCHNOPLEURE = endoderm + splanchnopleuric mesoderm

Neural tube
Somite
Notochord
Dorsal mesentery
Coelom
Gut
Ventral mesentery

Loss of ventral mesentery and intersegmental septa establishing continuous coelom

2.3 Principal stages in the early development of an elementary chordate. The tabbing and illustrations proceed from left to right in a series of rows from above downwards. Note the poles and axes of the zygote, the presumptive fate map of the blastula, and the progression of gastrulation, neurulation, and coelom formation. Consult text for further details.

101

the external cellular wall and eventually folds off dorsally to form the *axial notochord*. Laterally placed strips, which flank the future notochord, form a series of paired outpouchings, the *metameric coelomic pouches*. The ventrolateral walls of the archenteron grow across ventral to the pouches and notochord to re-form a gut tube. The external wall of the gastrula dorsal to the mesodermal pouches and notochord thickens and folds in to form the primitive nervous system, whilst the remainder of the external wall forms the epidermis and its derivatives. In many types the lateral lips of the blastopore approach and fuse to form a *primitive streak* as they continue to bud off generations of mesodermal cells into the walls of the elongating gastrula. The dorsal lip continues to contribute notochordal cells and, with the fusion of the lateral lips, it rounds off and thus temporarily connects the archenteron with the caudal end of the early nervous system (i.e. a transient *neurenteric canal*). Similarly the ventral lip closes around a primitive cloaca, continuous with the gut tube and later with the urogenital ducts.

Accordingly, during gastrulation, originally widely separated cell masses in the walls of the blastula come into close secondary apposition, as the *presumptive organ rudiments* and tissue layers assume their definitive positions. *Presumptive fate maps* of the various regions of blastular walls have been produced by staining small regions differentially with vital dyes and following the stained regions during subsequent development. Much work has also been directed towards an understanding of the mechanics of gastrulation itself and, as we shall see, the profound morphogenetic significance of the secondary association between formerly separated groups of blastomeres (2.4,5, p. 104). Gastrulation, although necessarily modified in the blastodisc of heavily yolked eggs and within the inner cell masses of mammalian blastocysts, remains an essentially similar process. Ectodermal and endodermal layers become defined and, within the former, a linear proliferative zone of uncommitted cells, the *primitive streak*, persists and produces generations of mesodermal cells which pass between the two layers. Near the headward end of the streak a zone corresponding to the dorsal blastopore lip forms a notochordal process, which invaginates and breaks through to form a neurenteric canal, while caudal to the streak a cloacal membrane forms (cf. the transformation of the human bilaminar embryonic disc, pp. 131–136). The basic structural correspondence in these stages throughout the Chordata and in forms as diverse as *Amphioxus* and Mammalia is assumed to imply similar fundamental *morphogenetic mechanisms*—where experimental analysis has proved feasible, this has been confirmed.

Some General Aspects of Development

The development of a large complex organism from a single diploid cell, the zygote, entails an increase in mass and cell number of thousands of millions and the emergence of a bewildering variety of cell types, each with a characteristic constitution, shape, specific activities and welded into the distinctive tissue masses, sheets, tubes, liquids and so forth, which cooperate in the functioning of the whole.

The *developmental biology* of such an organism, while endlessly complicated in detail, and with as yet rather rudimentary proposals concerning the control systems involved, can nevertheless, in principle, be reduced to a limited number of basic processes, namely: temporo-spatial variations in *growth, pattern formation, morphogenetic movement, differentiation*, together with *degeneration* and *death* of some cells and their aggregates.

The *description, definition* and *causal analysis* of these basic processes, and the degree to which they are independent or interdependent, still hold the centre of attention of modern developmental biologists, although many of the fundamental questions were first formulated, and some of the most dramatic experimental observations made, many years ago (vide infra). In the intervening years, as fresh experimental data have accumulated, the questions have often been rephrased; the definitions and terms used have been constantly revised and modified and there has been a gradual change of emphasis concerning the conceptual frameworks currently in fashion.

Thus, in addition to attempting to describe and define the various forms, combinations and differential rates of *positive growth* exhibited by organisms, there has been a more recent upsurge of interest in simultaneous *cell necrosis* and removal and the factors which may control regional differences in balance between these processes. Similarly, the detailed analysis of the *specific intracellular events* exhibited by cells undergoing *differentiation* has been paralleled by numerous hypotheses concerning the *spatial patterning* of differentiation. Concepts such as *oöplasmic segregation* of *morphogenetic factors*, *'organizer' phenomena*, *competence* of tissues, *induction, determination, morphogenetic gradients* and *developmental field phenomena* and, more latterly, ideas concerning *intercellular communication and positional information* available to cells in an aggregate, have emerged. In contrast, other authors have placed greater emphasis on the probable morphogenetic significance of variations in the *cell cycle* (p. 37) with the establishment of *specific cell lineages* and the emergence of *cell clones* (vide infra). Intense research is directed towards the *regional genomic control mechanisms* implicit in all the foregoing. Much interest has also recently been aroused by the possible morphogenetic implications of the mathematical formulation of the *catastrophe theory*. An immense scientific literature has accumulated concerning these fundamental fields of biological study, which of necessity can only be touched upon in a volume such as this, and the interested reader should consult any of the readily available reviews, monographs or original papers.

GROWTH

Growth is a term widely used in everyday conversation and applied to both living and inanimate objects; it may, of course, within biology be applied to entire communities, individuals or parts of an organism such as a whole body segment, an organ, tissue, group of cells or even subcellular features such as mitochondria, cell membrane and so on. This variety in the application of the term is paralleled by the number of definitions for it—evidence that none is wholly satisfactory. They range from the over-simplified 'increase in mass and size' to lengthy, complex and sometimes incomprehensible descriptions. In general, during ontogeny, growth implies an increase in size and mass which results from synthesis of protoplasm and extracellular materials which are specific tissue components.

Growth of a tissue is classically considered to be either multiplicative, auxetic, accretionary or, quite often, a combination of these.

Multiplicative growth, as the name suggests, involves an increase in the number of cells (or nuclei and associated cytoplasm in syncytia) by a succession of mitotic divisions, and all mammalian tissues exhibit this form of growth at some stage. Its persistence, however, varies greatly between tissues with increasing differentiation and maturity. In some, multiplicative growth continues throughout life with continual replacement of senescent or dead cells, as in the epidermis, intestinal epithelium, and myeloid precursors of circulating erythrocytes. At the other extreme, cell division during embryonic and fetal life may lead to the full complement of a particular cell, e.g. oöcytes and neurons, and thereafter multiplicative growth ceases.

Much work has thus been directed at establishing the ranges of mitotic rates in many tissues throughout development into maturity and senescence, and towards an analysis of any cyclic (e.g. diurnal) variation in rate and the extreme differences in potential for further growth during regeneration exhibited by different tissues. Despite adequacy of such quantitative investigations in tissues of different age, sex and type, and the recognition of many general factors which influence such growth, e.g. genetic, nutritional (p. 156), endocrine, thermal, photic, mechanical, etc., little is known of the intimate processes which initiate, maintain, modify and terminate complex three-dimensional arrays of mitoses. In certain tissues (e.g. the epidermis and the growing nervous system) specific growth-promoting factors have been demonstrated (Levi-Montalcini 1967). Additionally, substances of widespread occurrence in many tissues termed *chalones* have received increasing attention (Bullough 1967, Bullough et al 1967). They have been defined as internal secretions produced by a tissue which, by inhibition, control the mitotic activity of the cells of that tissue. It is considered that the chalone concentration is an important determinant of the balance which exists between mitotic rate, progressive cell differentiation and the assumption of a full functional role on the one hand, and the rate of cell ageing and progression towards senescence and death on the other. They are perhaps intermediaries between the general growth-influencing factors listed above and the differential activities of the genome (p. 109).

Auxetic growth involves an increase in the size of the specific individual cells characterizing a tissue. It is particularly prominent in certain invertebrate tissues such as the salivary glands of *Diptera*, but is also well shown by some mammalian tissues. For example, there is a vast post-natal increase of both surface area and cytoplasmic volume in many neurons and glial cells (p. 892); the growing striated muscle fibre (p. 553), the oöcyte, the myelinating Schwann cells, and the smooth muscle cells of the pregnant uterus furnish further obvious examples. The majority of other tissues, however, show some auxetic growth but of limited degree, while conversely, in some sites, continued multiplicative growth is accompanied by a *reduction* in cell volume (e.g. granule cells of the cerebellar cortex, and small lymphocytes in lymphoid tissue).

It is widely held that the general *nucleocytoplasmic ratio* to which most of the body's cells roughly approximate reflects the fixed quantity of DNA in their diploid nuclei which, in turn, imposes a rate limitation on the replacement of cytoplasmic proteins (each of which has a characteristic turnover rate). Thus, with continuing auxetic growth of a cell, its cytoplasmic volume eventually reaches a point beyond which the structural genes could not effectively replace the protein which is undergoing continual degradation. In some cases, growth ceases at this point, or nuclear replication with cell division occurs. The cases of auxetic growth cited above, however, often proceed far beyond the usual ratio of cytoplasmic volume to nuclear material and, in these, various methods of providing auxiliary nuclear support have emerged. The large dipteran salivary gland cells develop 'giant' *polytene chromosomes* containing some multiple of the diploid DNA content. The striated muscle fibre and other 'giant' cells such as megakaryocytes are, of course, *multinucleate syncytia*. Finally, the enlarging oöcyte and neuron (possessing but a single haploid and a single diploid nucleus, respectively) have their surfaces clothed by numerous *satellite cells* (follicular or glial cells). Such satellites probably provide auxiliary metabolic and nuclear support for the enlarged central cell, i.e. the two are functionally interlocked as a *cytophysiological unit*.

Accretionary growth denotes an increase in the amount of structural intercellular material between tissue cells—bone and cartilage are the most commonly cited examples. Further, perhaps less obvious, examples are the other fibrous connective tissues, tendons, joint capsules, aponeuroses and fasciae, the cornea, growing mesenchyme with its abundant intercellular matrix, and during the development of renal tissue and smooth muscle where the appearance of substantial basal laminae contributes significantly to the growing mass.

In practice, of course, these 'types' of growth as defined above are often welded together in various patterns, with differential growth rates and directions in different parts of the system; furthermore, they overlap with and merge into the phases of cell differentiation, full functional activity, ageing and senescence. These *differential growth patterns* with either *random* or *preferentially polarized* directions of mitotic division, together with alterations in cell size, shape and surface consistency, are central features of embryonic development and are responsible for the moulding of tissues into specific shapes whether solid masses, hollow balls, tubes, sheets and so forth. Equally important in some regions, however, is a process of tissue regression, with degeneration, cell death and tissue removal. Where multiplicative (and sometimes accretionary) growth continues throughout the thickness of a tissue mass and it grows as a whole expanding from within, its growth is termed *interstitial*. Experimental markers placed within such a tissue during growth are later found to have become increasingly separated—excellent examples are embryonic limb bud ectoderm (p. 174) and the interior of young masses of hyaline cartilage (p. 291). Alternative-

ly, in *appositional growth* new generations of cells and intercellular material are added to the *surface* of the tissue by the repeated division of the cells of a *cambial layer* which surrounds the tissue, e.g. periosteum and perichondrium (pp. 291, 300, 461).

Differential growth rates, as intimated, are an essential feature of many aspects of embryonic development. For example, if in an area of loose mesenchyme, a locus with a high proliferative rate appears surrounded by 'shells' of tissue with progressively lower rates, a mass of tightly packed cells which grades into the surrounding tissue is formed, i.e. a so-called *mesenchymal condensation*. Two adjacent loci in subepithelial mesenchyme with an intervening quiescent zone lead to neighbouring elevations of the epithelium, and an intermediate furrow (cf. the development of the facial prominences or 'processes', p. 166). In many situations, mesenchymal condensations appear and their *blastemal shape* foreshadows that of the structure, bone, cartilage or muscle mass, into which they subsequently differentiate. Many other examples of differential growth will be encountered as development of individual organ systems is considered. Particularly clear-cut examples are seen in descent of the gonads (p. 256), formation of the cauda equina (p. 182), neck elongation, septation of the heart (p. 209) and in the emergence of asymmetrical arterial, venous and lymphatic systems from initially bilaterally symmetrical patterns.

MORPHOGENETIC CELL DEATH

Cell death and tissue removal are as much an integral element in embryogenesis as the positive forms of growth discussed briefly in the foregoing paragraphs. As we have seen, it is currently held that mechanisms of *tissue homeostasis* exist, possibly operated, at least partly, by the local concentrations of tissue-specific chalones, whereby a balance is achieved between cell division, differentiation, functioning and senescence. Death of cells has been recognized as early as the blastocyst stage in rabbits (Daniel & Olson 1966) and demonstrated at many later stages of ontogeny throughout the vertebrates (Glücksmann 1951). An obvious example is the retrogression of the amphibian tail at metamorphosis, but similar degenerations occur during the development of the axial nervous system (Hughes 1968) and in association with the mammalian primitive streak. At subsequent and postnatal stages, bone growth and remodelling, in particular, are accompanied by surface deposition in some regions and erosion and removal in others (p. 309).

Particularly well studied have been the integrated patterns of growth and death of cells during the shaping of the tetrapod limb (Saunders & Fallon 1966). Massive zones of what has been dramatically termed *cataclysmic necrosis* sweep the length of the developing limb as its contours are moulded. The process being especially prominent in some localities and finally occurs in the interdigital clefts as digital outlines become sculpted. Neighbouring mesenchymal cells become intensely phagocytic and engulf whole dying cells or any resulting cellular débris. It should be emphasized that the *normal prospective fate* of such a cell mass is to degenerate and die, and since, in modern terminology, it has become fashionable to regard ontogeny as the enacting of a *developmental programme*, ultimately encoded in the genome, such cells are described as succumbing to *programmed cell death*. Numerous experiments have been devised in attempts to analyse the precise roles of such necrotic zones in normogenesis and to elucidate what factors affect the enactment and timing of the so-called '*death clock*'. The *posterior necrotic zone* of the developing chick wing bud has, because of its predictable localization, timing and susceptibility to microsurgical experiments, become a favourite object for study (Saunders & Gasseling 1968), and its relationship to the establishment of *polarity* in the wing bud is firmly recognized. (See extensive review and bibliography by Hinchliffe 1982.)

DIFFERENTIATION

The appearance of many highly distinctive cell types which become blended in various ways to form the functioning tissues and the control mechanisms involved, has been for decades, and remains today, one of the central problems in biology.

The differences between fully differentiated cells in terms of their size, shape, ultrastructural appearance, chemical constitution and functioning (e.g. mature muscle, nerve and glandular cells) are plain: no intermediaries exist, and there is no known experimental situation in which one form changes into another. As we shall see, however, the cells of early chordate embryos may, under suitable conditions, *ultimately* form any one of many different cell types, i.e. they are widely held by some developmental biologists to be *pluripotent* and, as development proceeds, there occurs a step by step reduction in their potential for forming a variety of cells so that finally the fate of most cells is limited to one type of fully differentiated cell only. Even in the mature body, however, some cells retain some degree of flexibility and can still transform into a number of types—this is seen most powerfully in the ability to regenerate a whole body segment, e.g. limb or tail, shown by certain amphibia.

It may be mentioned here, however, and outlined further below that other groups of cell biologists contest the view just given of most early blastomeres as undifferentiated, totipotent or pluripotent cells with *immediate* access to a wide range of possible phenotypical fates and advance instead hypotheses of much more strictly compartmentalized cell lineages during development.

PREFORMATION AND EPIGENESIS

For centuries, a controversy raged concerning the form taken by the differentiation which occurs during embryogenesis, whether there was a *preformation* in the early embryo, or whether *epigenesis* took place. The former view held that all the structures of the mature body, although so minute as to be invisible, were present in the zygote and simply enlarged and were thus revealed in subsequent development. The epigeneticists, however, held that the various organ primordia appearing during development were essentially new formations. In a modern context both views, somewhat modified, have their place. In one sense, *the total information content* of both the cytoplasm and the nuclear genetic apparatus of a zygote is a 'preformation' which through a series of complex interactions directs the 'epigenesis' of new structures.

The classical studies of *descriptive embryology* in which times of appearance, shape and size of the various embryonic layers and organ systems at different stages of development were meticulously described in serial histological sections, and similar *comparative studies* of different species, although providing an invaluable framework, could make no contribution to an *understanding* of differentiation. An analytical approach was necessary (**2.4, 5**), and the foundations of 'developmental mechanics' or an *experimental embryology*, which sought to understand the *causal interactions* of development, were laid by Wilhelm Roux late in the nineteenth century. His experiments on frog development were paralleled by other observations of fundamental importance by Hans Driesch on the early sea urchin embryo. Then, passing into the present century, contributions of outstanding penetrance were made by such investigators as Hans Spemann, Ross G. Harrison, Sven Horstadius, J. Holtfreter and many others. Only the most cursory mention of this field of biology can be contemplated, and those with wider interests should consult the original writings of these pioneers and, as an introduction to the equally important researches of modern developmental biologists, specific monographs (e.g. Spemann 1938, Willier et al 1955, Hamburger 1960, Balinsky 1981, and subsequent text references).

EXPERIMENTAL EMBRYOLOGY

The final blow to rigid preformationist concepts was delivered when Hans Driesch showed that if the first two blastomeres of a sea urchin egg were separated, each could develop into a complete, although small, embryo (**2.4**). Thus, any idea of strict compartmentalism in the fertilized ovum had to be abandoned, since in normal development each of the blastomeres would have formed half an embryo but, when separated, each could *regulate* sufficiently to form a whole. On this basis Driesch supposed (mistakenly) that there existed *no segregation whatever* in the egg

cytoplasm and that it constituted a '*harmonious equipotential system*' in all directions. Since subsequent differentiation occurred, he regarded this as evidence supporting revival of the Aristotelian *entelechy*, holding that such a phenomenon could not be explained by the laws of physics and chemistry and that a vital or spiritual '*internal perfecting principle*' was present.

However, as we have seen elsewhere (p. 100), much subsequent work showed that all ova and zygotes show some degree of *segregation of formative materials* in their cytoplasm, but its *degree* and *lability* varies between species. Ironically, had Driesch's separation been made in an equatorial plane rather than a meridional one, only partial embryos would have resulted!

The experimental methods employed by the earlier investigators (**2.4,5**) included the preparation of '*presumptive fate maps*' of the blastula or blastodisc, *ablation* of regions of the blastula, *autotransplantation* of cell masses from one place to another in the same embryo, *explanation* of such masses or combinations of these into salt solutions and nutrient media, and finally the powerful methods of *heteroplastic* and *xenoplastic grafting* between different species. In the latter, the tissues of both the graft and the host are so distinctive that their relative contributions and aspects of their interactions in any complex development that occurs after transplantation can be followed.

The production of maps (**2.3**) showing the *presumptive fate* in normal development of different regions of the blastula or blastodisc, by the application of spots of vital dyes and following their subsequent history, was a necessary prerequisite of experimentation using microsurgical methods.

Self and Dependent Differentiation

From the earlier autotransplantation experiments, and later reinforced by explanation of blastula fragments into salt solutions (**2.4**), the first concepts of two distinct tissue states emerged, i.e. those that exhibited *self-differentiation* and the remainder in which *dependent differentiation* was the rule. In the former, as the name suggests, a fragment, when transplanted to an apparently 'indifferent' region of the blastula or, more critically, into salt solutions, proceeded to develop into the tissues it would have formed if undisturbed. Regions with dependent differentiation, however, if transplanted early, failed to continue development and only did so if combined with some neighbouring mass of cells. If such a region, which at an early stage showed dependent differentiation only, was not transplanted until a later stage, it was often found to have become self-differentiating. Furthermore, when such an early region was combined with different regions of the embryo, development would often result in a changed final tissue (e.g. muscle rather than kidney).

From such work, it was realized that the *prospective significance* or *normal fate* of the various regions of the blastula did not reflect the *prospective potency* of these regions (i.e. the *various fates possible* under different experimental conditions). Further, it was found that from the earliest moments, certain *restricted regions only* exhibited *self-differentiation*, i.e. *determination* of their fate had occurred. The remaining regions with dependent differentiation, originally 'pluripotent', suffered a stepwise restriction in their possible fates, i.e. an increasing fixity of determination occurred as development proceeded.

Later, the features of dependent differentiation were more clearly defined. Classical experiments with many variations were performed showing the dependency of the development of the lens vesicle (which transforms into the ocular crystalline lens) on the approach and close apposition of an evagination from the primitive nervous system (the optic vesicle). These experiments showed that, in the absence of an optic vesicle, no lens formation occurred, that many different regions of 'indifferent' ectoderm, if exposed to the influence of the vesicle, could form lens tissue and similarly, if the same ectoderm was used to replace that covering the hindbrain, it would develop into a membranous labyrinth. In this manner, the concept of *embryonic induction* developed. This held that for a restricted period a particular region of pluripotent tissue was *competent* to react to an inductive influence exerted by an adjacent tissue.

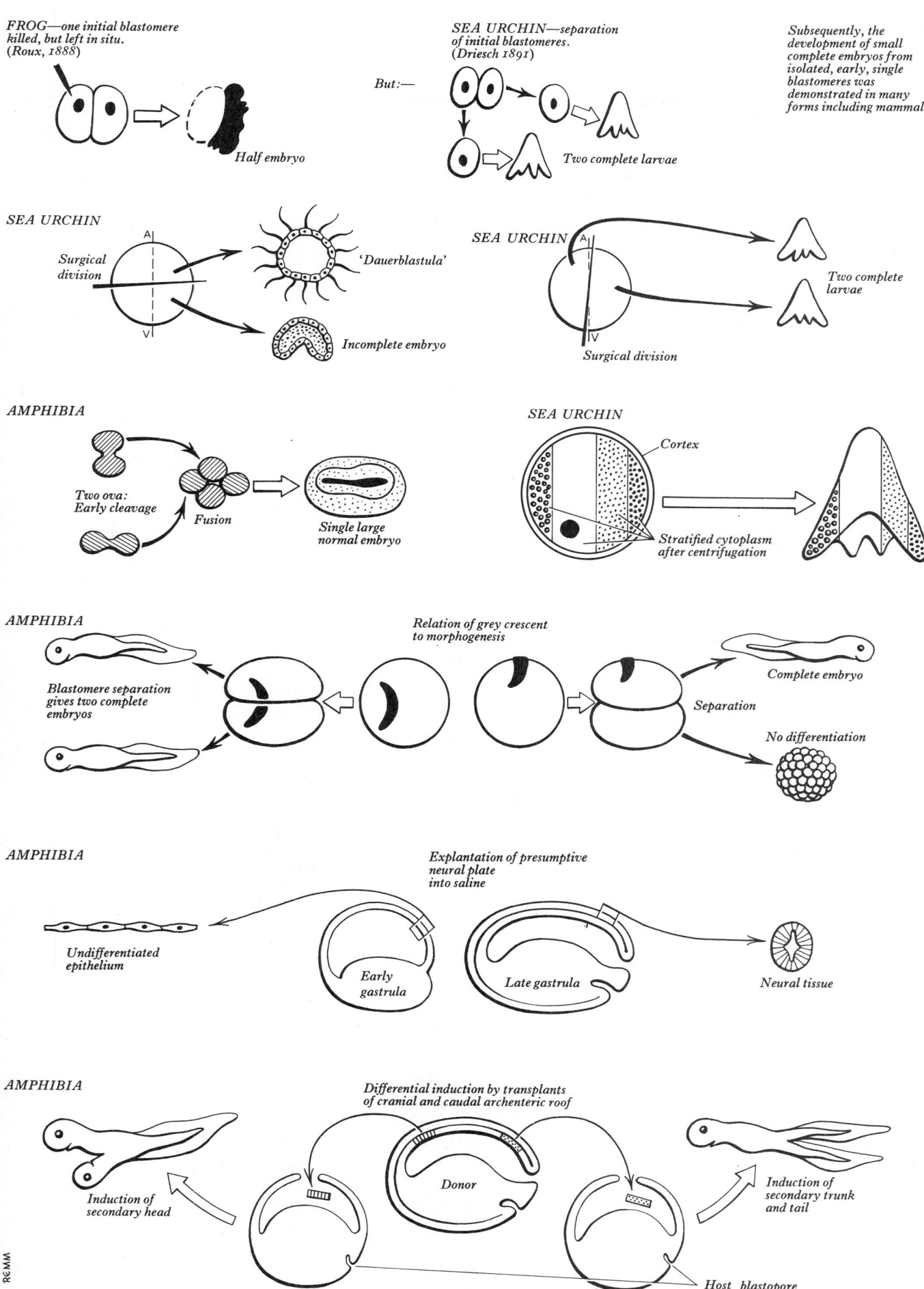

FROG—one initial blastomere killed, but left in situ. (Roux, 1888)

Half embryo

SEA URCHIN—separation of initial blastomeres. (Driesch 1891)

But:—

Two complete larvae

Subsequently, the development of small complete embryos from isolated, early, single blastomeres was demonstrated in many forms including mammals

SEA URCHIN

Surgical division

A

V

'Dauerblastula'

Incomplete embryo

SEA URCHIN

A

V

Two complete larvae

Surgical division

AMPHIBIA

Two ova: Early cleavage

Fusion

Single large normal embryo

SEA URCHIN

Cortex

Stratified cytoplasm after centrifugation

AMPHIBIA

Relation of grey crescent to morphogenesis

Blastomere separation gives two complete embryos

Complete embryo

Separation

No differentiation

AMPHIBIA

Explantation of presumptive neural plate into saline

Undifferentiated epithelium

Early gastrula

Late gastrula

Neural tissue

AMPHIBIA

Differential induction by transplants of cranial and caudal archenteric roof

Induction of secondary head

Donor

Induction of secondary trunk and tail

Host blastopore

REMM

2.4 A selection of the classic manoeuvres, which were milestones in the development of experimental embryology as a science. Consult text.

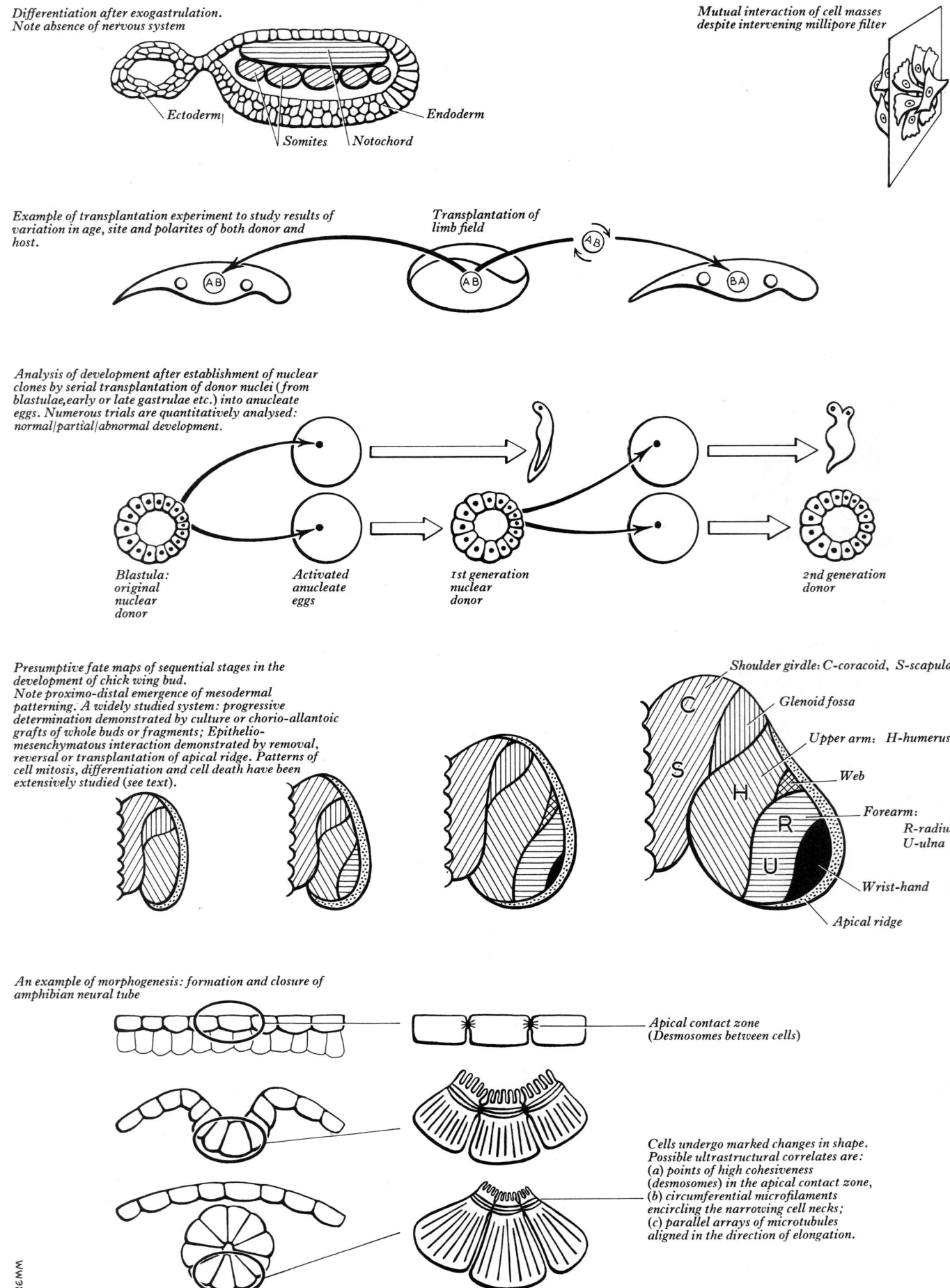

Differentiation after exogastrulation.
Note absence of nervous system

Ectoderm *Endoderm*

Somites *Notochord*

Mutual interaction of cell masses
despite intervening millipore filter

Example of transplantation experiment to study results of
variation in age, site and polarites of both donor and
host.

Transplantation of
limb field

Analysis of development after establishment of nuclear
clones by serial transplantation of donor nuclei (from
blastulae, early or late gastrulae etc.) into anucleate
eggs. Numerous trials are quantitatively analysed:
normal/partial/abnormal development.

Blastula:
original
nuclear
donor

Activated
anucleate
eggs

1st generation
nuclear
donor

2nd generation
donor

Presumptive fate maps of sequential stages in the
development of chick wing bud.
Note proximo-distal emergence of mesodermal
patterning. A widely studied system: progressive
determination demonstrated by culture or chorio-allantoic
grafts of whole buds or fragments; Epithelio-
mesenchymatous interaction demonstrated by removal,
reversal or transplantation of apical ridge. Patterns of
cell mitosis, differentiation and cell death have been
extensively studied (see text).

Shoulder girdle: C-coracoid, S-scapula

Glenoid fossa

Upper arm: H-humerus

Web

Forearm:
R-radius,
U-ulna

Wrist-hand

Apical ridge

An example of morphogenesis: formation and closure of
amphibian neural tube

Apical contact zone
(Desmosomes between cells)

Cells undergo marked changes in shape.
Possible ultrastructural correlates are:
(a) points of high cohesiveness
(desmosomes) in the apical contact zone,
(b) circumferential microfilaments
encircling the narrowing cell necks;
(c) parallel arrays of microtubules
aligned in the direction of elongation.

REMM

2.5 A further selection of classic experimental embryological procedures.
Consult text.

Primary Organization

Experiments by Hans Spemann involving the ablation, or the auto- or hetero-transplantation, of the future dorsal lip of the blastopore region, further clarified the process. In the absence of a dorsal lip, continued development of an organized nature ceased. When, however, such a region was transplanted to the flank of another gastrula a dramatic series of changes ensued. The *graft* proceeded to proliferate, invaginate and develop into a notochord, surrounding skeletal structures and adjacent muscle masses, showing that extremely early determination had occurred, and it followed its normal course of self-differentiation. In addition, the surrounding *host tissues* displayed an equally dramatic series of responses to the inductive influences of the graft. The superjacent ectoderm developed into a remarkably complete nervous system, with recognizable subdivisions of brain and spinal cord, laterally placed host mesoderm formed musculature and kidney tissues whilst, on occasions, a subjacent gut tube was induced. Host and graft tissues often continued to interlock in the development of what was virtually a complete second embryo, attached to the abdomen of the host. Because of the early determination and self-differentiating nature of the tissues of the dorsal blastopore region, and their profound inductive influence on surrounding tissues which lead to the development of an embryonic axis and adjacent tissues, it was called the *primary organizer region*. Then followed a series of experiments which demonstrated regional differences in the inductive capacity of the tissues sequentially invaginated through the dorsal lip (**2.4**); thus so-called head, trunk and tail 'organizer' regions were recognized. Subsequently, however, a wide variety of substances, including some inorganic compounds which do not exist in developing embryos, were shown to have inductive effects and similar effects followed the microsurgical killing of a few blastomeres or the introduction into the blastocoele of powdered glass. Interestingly, extracts of adult tissues produced varied responses, either a *mesodermalizing* effect or a *regional neuralizing* effect (with the formation of predominantly forebrain, hindbrain or spinal structures; see Toivonen 1967, Nieuwkoop 1967). Thereafter, many subordinate inductive systems were shown to operate in coordinated sequences throughout chordate development. In some cases competent tissue was separated from its relevant inductor tissue by a 'millipore' filter which prevented the passage of cells (**2.5**); inductive effects still occurred, presumably because of the passage of chemical messengers between the two. (It should be mentioned, in this context, that since the advent of the scanning electron microscope the full effectiveness of such filters has been subjected to a critical reappraisal.) Since the turn of the century much effort and ingenuity have been expended in attempts to isolate and identify specific intercellular chemical inducers but, despite many claims, none have been substantiated in the embryogenesis of metazoa (Grobstein 1967).

In contrast, considerable detail has now accumulated concerning the differential activation and repression of the genome in microorganisms associated with the *induction of enzyme synthesis* (vide infra) and also concerning the two-stage inductive events which firstly transform an uncommitted lymphoid stem cell to a committed cell (p. 670) and secondly express this committal by the production of specific antibody (Feldman & Globerson 1974). In the latter case the second stage inducer, the antigen, is of course often a chemically well-defined molecule. Whilst such studies may well have parallels with events in embryogenesis, direct extrapolation is premature, and it is perhaps unfortunate that the term induction has been applied to all these situations.

In recent years there has occurred a great change of emphasis when developmental biologists consider inductive phenomena. The dramatic responses demonstrated by Spemann following blastopore transplantation, coupled with his all-embracing term 'primary organizer', led for many years to the following view of early embryogenesis: namely, that the majority of early blastomeres were *undifferentiated, totipotent* cells, with highly labile control mechanisms and with *numerous* alternative developmental paths *immediately* available to them. A *master organizer* region, or regions, was assumed to produce complex organizer or inducer

molecules of *high information content* which passed to, and *directed*, the subsequent developmental path of the reactive cells. Naturally, great interest focused on the possibility of isolating and characterizing physico-chemically these *instructive molecules*. Gradually, however, as data have accumulated concerning other possible modes of interaction of cell aggregates (vide infra), the use of the term organizer has fallen into disuse, and the search for complex instructive molecules has slackened. Perhaps the inductive phenomena of embryogenesis are mediated by slight variations in ubiquitous molecules of relatively simple chemical constitution (e.g. metabolites) of low, or absent, intrinsic information content, and the developmental path followed is largely determined by the currently available metabolic options provided by the *cytoplasm* and *genome* of the reacting cells.

MODERN VIEWS OF DIFFERENTIATION

Differentiation has often been defined as the emergence of *irreversible, inheritable* differences between cells, in contrast to the many *reversible* changes shown by cells in response to changed local environments, termed *modulations*. However, the apparent irreversibility seems to be a feature only prominent in cells of vertebrates in the relatively few experimental situations so far studied. Further, many changes in the cells of plants and higher invertebrates closely resemble processes of differentiation, but are in fact reversible (see below). Perhaps the terms differentiation and modulation imply a rigid distinction between two states exhibited by cells, when in fact a range of states with varying degrees of reversibility exists.

It has been held (Sager 1965) that from the earliest times the simplest, non-nuclear (*prokaryote*) units of life consisted of '... a stablized tripartite system: nucleic acids for replication, a photosynthetic or chemosynthetic system for energy conversion and protein enzymes to catalyse the two processes'. To these essential ingredients common to all living forms were later added the enzyme systems which accompanied a dependency on environmental oxygen, and then a greater variety of cellular form paralleled the appearance of *eukaryote cells* with a well defined chromosome-bearing nucleus. It is in the *interactions* between these different phases of one cooperative system, the cell, that clues to the nature of differentiation are currently sought (De Reuck & Knight 1967, Bullough 1967, Toivonen 1967, Nieuwkoop 1967, Grobstein 1967, Wolstenholme & Knight 1970, Harris 1970, Ashworth 1973, et seq.).

As we have seen (p. 15 et seq.), proteins are of fundamental significance in the construction and operation of all prokaryote and eukaryote cells. In great variety they are essential constituents of cytoplasmic matrix, nucleoplasm, cytomembranes including the plasma membrane and walls of membrane-bound organelles and, with nucleic acids, as nucleoproteins of the genetic apparatus. Most significantly, they form the basis of the innumerable enzyme arrays which catalyse all the steps in the structural and energetic transformations which constitute 'life'.

Thus, it has been stated quite simply that when cells with the same genome synthesize different proteins, differentiation has occurred (Jacob & Monod 1961, 1963). This implies, therefore, that an understanding of the processes of differentiation lies not in the details of the protein synthetic pathways as such, but rather in the various control systems to which these are subjected. The latter are currently believed to involve varieties of chemical messengers which operate within and between cells.

The foregoing definition of differentiation, however, does not specify which gene products may be regarded as indices that differentiation has occurred and disregards the phases of the cell cycle. On such a view, early 'pluripotent' embryonic cells have a *wide repertoire* of possible responses, any one of which would *immediately* follow a local change in the operative control system, during any phase of the particular cell cycle occurring at the time. Thereafter, the early, labile control systems would become progressively stabilized in subsequent generations as the cells differentiate (**2.6A**).

Other groups of cell biologists, in contrast, propose an alternative set of hypotheses (Holtzer & Abbott 1968, Holtzer et al 1972, 1975, **2.6A, B**). These authors class materials synthesized by

THEORIES OF MITOSIS AND PROGRESSIVE DIFFERENTIATION

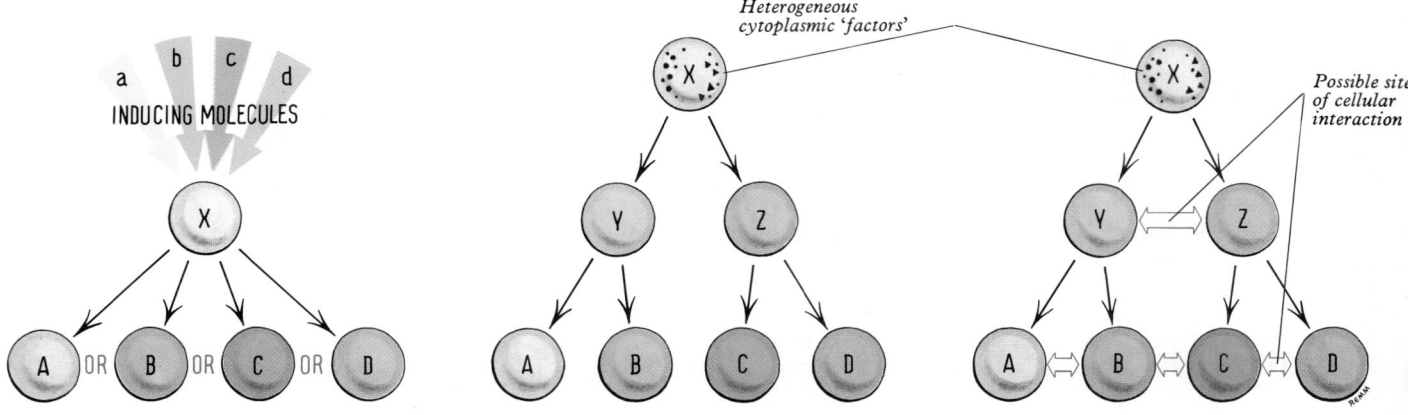

2.6A Theories of mitosis and progressive differentiation. The tabbing is further explained in the accompanying text.

cells as either *'essential molecules'*, i.e. those synthesized by most cells and required for their viability, or *'luxury molecules'* which are 'those cell-unique molecules responsible for the state of differentiation of the cell that either synthesizes them or has inherited them'. Secondly, cell mitotic cycles are regarded as of two fundamentally different types (**2.6B**): *proliferative cell cycles* with amplification of cell numbers, the progeny having precisely the same synthetic capacity as the parent cell; and *quantal cell cycles* which, at some point, make available for synthetic activities regions of the genome in subsequent cells which were *not available* in the mother cell. Such a view securely links essential steps in differentiation to particular phases of a *critical series* of quantal mitoses, perhaps interspersed by proliferative cycles and at each quantal cycle the parent cell is maximally *bipotential*. Thus, the division may be *symmetrical* giving derived cells with identical synthetic capacities or *asymmetrical* in which their capacities differ (**2.6B**). In one particular type of asymmetrical division, one of the progeny retains the synthetic capacity of the parent cell (**2.6B**), persisting as a *stem cell* (p. 40). A quantal cell division, therefore, is a *decision point* at which the progeny enter one of two new paths of differentiation. A series of such divisions would clearly lead to diversification of cell types, each ultimate *clonal type* being the product of a precise and strictly compartmentalized *cell lineage*. This hypothesis, presented here in the barest outline, rejects the conventional view of the pluripotentiality of early blastomeres, holding that all cells are optimally differentiated for the stage of development they have reached and, further, relegates the inductive capacities of exogenous molecules to a relatively minor role (Holtzer et al 1972, 1975). Such views are, however, not wholly in accord with those proposed by other groups in relation to the spatial patterning of differentiation (vide infra). These contrasting views are summarized in illustrations **2.6A**, B.

THE CONTROL OF GENE ACTION

The general structure and role of nucleic acids in the form of nuclear and extranuclear deoxyribonucleic acids (DNAs) and various ribonucleic acids (RNAs) in relation to protein synthesis has been discussed elsewhere (p. 39). Mention was also made of the concept of the *operon* which emerged from elegant investigations on microorganisms (Jacob & Monod 1961, 1963). Essentially, an operon is a functional grouping of DNA loci including several *structural genes*, an *operator segment (o)*, a *promotor region (p)* and a *regulator gene (i)*. Each structural gene carries the information necessary for the specification of a particular polypeptide (which forms part of either a structural or an enzymatic protein). In the case of the *lac operon* of *Escherichia coli* —the first to be thoroughly analysed by the study of the activities of mutant forms (Ashworth 1973)—the linear sequence of DNA loci was demonstrated to be: *i p o z y a (z, y* and *a* denoting structural genes which encode for the synthesis of β-

TYPES OF MITOTIC DIVISION

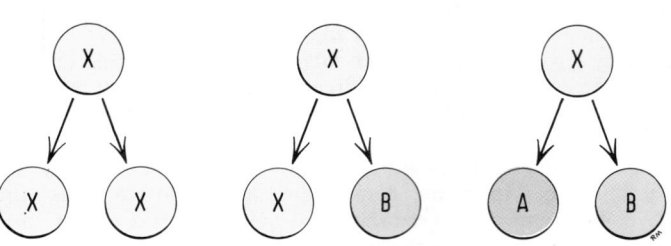

2.6B Varieties of mitotic cell division: these include a symmetrical proliferative mitosis, and two types of asymmetrical quantal mitosis. Consult text for discussion.

galactosidase, a permease and a trans-acetylase respectively). The prefix *lac* indicates that lactose is the physiologically important substrate for β-galactosidase. Subsequently, further examples of more complex operons have been analysed, with many features in common with the lac operon but with alternative controlling mechanisms. Such studies have led to rapid advances in our knowledge of the main variations in gene activity in *prokaryotes*.

In general the activity of the structural genes is dependent upon the state of their associated *operator*; they are active only when it is 'free' and they become inactive when it is 'blocked' by combination with a specific chemical group. The neighbouring *regulator gene* produces a *repressor protein* which, depending on local conditions, may exist in an active form capable of binding with the operator, or in an inactive form when it cannot do so. In this manner the activity or otherwise of a particular group of structural genes is dependent upon the production of a *first order* chemical messenger or repressor protein. In turn, specific *second order* messengers (*effectors*) determine the state of the repressor molecules. Some repressors can only block the operator when combined with effectors termed *co-repressor* molecules. In some instances it has been shown that low molecular weight end products of certain *biosynthetic pathways* act as co-repressors, thereby providing a negative feedback system controlling the synthesis of enzymes essential to the biosynthetic path. Thus, heightened activity in the biosynthetic path leads ultimately to gene *repression*, whereas diminished activity leads to *derepression* of the relevant genes.

Other effectors, such as *enzyme inducers*, act in a converse way —by combining with the repressor protein they so modify its structure that it is unable to combine with the operator and consequently the related structural genes become active. The *promotor region*, which lies adjacent to the operator, is now known to be the site where binding of DNA-dependent RNA polymerases (*transcriptases*) occurs, when the operator is not blocked by repressor protein. These determine the *loci* for the *initiation of transcription* and, further, variations in a number of factors entering into the constitution of the transcriptases (e.g. the σ-factor) result in *modulation of the rate* of transcription.

Thus, through the balanced activities of *negative* (repressor/operator) and *positive* (transcriptase/promotor) controls, a great flexibility of gene action may be achieved varying from *on/off* switches to delicate *adjustments of transcription rates*.

Certain gene loci have also been considered to control the *mitotic* behaviour of cells; on this view, when the loci are active, multiplicative growth persists and, when inactive, the cells are channelled into avenues of differentiation and ageing. The *tissue-specific chalones* mentioned previously have been proposed as co-repressors of such mitosis-initiating genes (Bullough 1967). The activities of the tissue chalones may be modified by the concentration of general circulating hormones and thus act as inter-mediaries between the endocrine system and the state of the genome in the tissue cells. (In certain invertebrates a further class of substances called *pheromones* even pass between different individual members of a community, and affect the reactions of their tissues and also their behaviour patterns.)

However, as indicated previously, some cell biologists would not regard the foregoing as an adequate view of the role of mitosis in cell differentiation. Firstly, many would prefer to consider the delicate on-going quantitative controls of protein synthesis as more the province of the cell physiologist, and regard true differentiation to occur only at certain definite phases of *quantal cell cycles* (p. 40). As stated, in the latter, regions of the genome of derived cells become available for synthetic activities which were *not available* in the parent cell. In this regard, mention must be made here of the elegant researches on the structure, distribution and possible roles of varieties of the basic proteins called *histones* (see reviews by Zubay 1968, Cooke 1975, Wasserman 1973, Borun et al 1975). Five principal classes of histone are present in eukaryote cells, and these form strong, stable complexes with each other, and with DNA, thereby being important structural and functional components of the chromosomes (see the discussion on the *nucleosome*, p. 37, and **1.30**). It has been proposed that in combination with certain 'operator-like' regions of the chromosomes the histones provide a powerful *non-specific repression* of these regions and their associated structural genes. Such regions, it is thought, become *unblocked* by the cooperation of *non-histone components* only at specific times during chromosome replication in quantal cell cycles. It is possible that such non-histone components, or their precursors, form the basis of the so-called *morphogenetic substances* which many workers have indirectly demonstrated to be differentially distributed in the cortex of the oöcyte, thereafter being unequally partitioned in the various blastomeres at cleavage and, further, they may form the substrate for cell–cell interactions in later development. Experimental analysis of the components that may interact with the histone complexes is still in its infancy. Particular interest has been centred upon the structure and metabolism of certain 'acidic' *non-histone chromosomal proteins* (Borun & Stein 1972, Stein & Borun 1972, Borun et al 1975) which may be responsible for the production of specific RNA transcripts and, accordingly, be closely related to aspects of the control of cellular differentiation. (Over 400 varieties of such proteins have been demonstrated in mammalian cells.)

Some authors have stressed that the genome of *prokaryotic cells* is permanently *unblocked* but can be repressed or derepressed (Tsanev & Sendov 1971, Tsanev 1975), while in metazoan *eukaryotic cells* the genome is initially *blocked* and during embryogenesis *sequential unblocking* of different gene combinations occurs. The *functional state* of the fully differentiated cell is thereafter determined by repression and derepression 'within the limits of the unblocked regions of the genome'. Computer simulation of model systems of metazoan development indicates that the final stages of cellular differentiation can be reached only through 'a chain of events in which both processes of repression-derepression (realized by embryonic induction) and of blocking-unblocking (realized by the mitotic cycle) are successively involved'.

It should be noted that many of the control systems described above seem to be mediated by that ubiquitous regulator molecule

cyclic adenosine monophosphate (3'5'cAMP), which therefore also appears to have a central role in cellular differentiation.

An abbreviated account such as the foregoing can do little justice to the numerous elegant and rapidly expanding researches in these fields. Nevertheless, enough has perhaps been said to emphasize that the earlier evidences stemmed largely from work with microorganisms, and although considerable advances have now been made (Lewin 1970), the positive identification of the specific roles of intra-cellular and intercellular chemical messengers in advanced metazoan tissues is still awaited. However, as a working hypothesis it is considered that the patterns of growth, differentiation and functional expression in metazoa result from the varied activation and repression, coupled with sequential unblocking, of gene groups which stem from an interlocked hierarchy of chemical messengers—repressors, corepressors, inducers, transcriptases, blocking/unblocking proteins, tissue chalones and circulating hormones (and in some communities pheromones), which act in coordinated sequences in space and time as integral parts of cell cycles. Many modifications of this scheme will undoubtedly prove necessary. Furthermore, the informational role of extrachromosomal nucleic acids and other self-replicating cytoplasmic substances (or *plasmagenes*) which may have major roles in normal development have as yet been little investigated in the higher metazoa.

CONCEPTS RELATING TO PATTERN FORMATION

The events outlined in previous sections dealt largely with hypotheses concerning *intracellular* mechanisms controlling cell differentiation, but neglecting two fundamental aspects of development, *pattern formation* and *morphogenesis*. The latter, acting in concert with cell differentiation, are the means by which the total genetic information available to the zygote is ultimately expressed in the mature functioning phenotype.

Pattern formation concerns the processes whereby the individual members of a mass of cells, initially *apparently* homogeneous, follow a number of different avenues of differentiation precisely related to each other in an orderly manner in space and time. The mechanisms by which such temporo-spatial differences are initiated and maintained currently engage the attention of many developmental biologists. The 'patterns' embraced by the term, of course, not only apply to regions of *regular* geometrical order such as the retina, crystalline lens, cerebellar cortex, etc., but all aspects of metazoan development including ordered but *asymmetrical* structures such as the tetrapod limb, mammalian liver, etc. Thus, e.g. in the case of the limb, the intimate *intracellular* events leading to the emergence of chondroblasts, myoblasts, osteoblasts and so forth will be identical in right and left sided limbs, forelimbs and hind-limbs, and their counterparts in different species; nevertheless, the *patterned arrangement* of these elemental processes differs in each case.

Morphogenesis, or the assumption of form by the whole, or part, of a developing embryo, is often for convenience treated as a separate field of study, involving differential rates of growth or degeneration of cell groups but, particularly in early embryogenesis, also the relative *morphogenetic movements* of cells or cell aggregates. The movements involved in gastrulation (p. 100) or the wide dispersal of neural crest cells (p. 183) are dramatic examples. Thus, the study involves an analysis of the *mechanisms* and *forces* responsible for cell movement, migration, changes in shape of individual cells and the differential adhesiveness of their surfaces (see pp. 22, 115). However, despite the conveniences for research groups, these facets of development are necessarily *interdependent*. Whether it is differential growth balanced against areas of degeneration, morphogenetic movements and migration or cellular differentiation under consideration, it is necessary first to *specify* within a group of cells *which particular ones* shall divide, degenerate, migrate or differentiate, and this is the province of *pattern formation*. As an example, when a cell group migrates to a distant location in normal development, mechanisms clearly exist to *locate* the group about to migrate, in relation to the remainder of the embryo. After a prelude of growth, the selected group enters the initial stages of differentiation, which lead to the development of those organelles, surface specializations and changes of cell shape responsible for the mechanics of migration; further, these events must follow in correct *sequence* and *orientation*. Following migration, the group is in a new cellular environment and further periods of growth and differentiation follow, with crucial interactions between the migrated group and the cells forming its new environment, often affecting the particular avenues of differentiation followed by *both*.

It should be noted that pattern formation is an essential ingredient not only of normal embryogenesis but of processes of *regeneration*, good examples being the regeneration of new heads and tails in transected flatworms, new apical and basal regions in transected *Hydra* or the replacement of a whole severed limb, or an appropriate part thereof, in adult Amphibia. In general, regeneration follows one of two main types: *morphallaxis,* in which the whole reforms by rearrangement and differentiation of the existing tissues without further growth, or *epimorphosis* in which there is new growth of blastemal tissue which subsequently matures into the full regenerate. Whatever control mechanisms are proposed for embryonic pattern formation, they must be sufficiently flexible to also account for morphallaxis and epimorphosis.

(For an extensive entry into the literature consult the chapters collated by Malacinski & Bryant 1984.)

It should be emphasized that the approach to the problems of pattern, form and differentiation contrasts sharply between workers. Some (Holtzer et al 1975) assume that finer and finer probing of the genome and the detailed control of protein synthesis, leading to a reasonably complete understanding of the *intracellular* molecular events of cytodifferentiation, would, in its turn and with little addition, provide an adequate explanation of changes in form and pattern. Others (Wolpert 1969, 1971 a,b, 1978, 1981, Wolpert and Stein 1984), whilst conceding the importance of the molecular aspects of cytodifferentiation, hold that the laws and principles of control of pattern and form must initially be sought at the *cellular* and *intercellular* level and that they will prove to be 'as general, universal, elegant and simple as those that now apply to molecular genetics'.

Since the closing decades of the last century, a series of concepts concerning pattern formation have emerged, but with time these have been rephrased and redefined, often resulting in differences of definition and terminological confusion. These concepts, mentioned earlier (p. 98), include ideas of polarity, axes, gradients and fields and suggestions concerning the coordinates necessary for their specification.

In the physical sciences, a notional system possessing two regions of opposite tendencies or qualities, the *poles,* which are considered joined by a linear *axis,* is commonplace, e.g. in magnetism and electrostatics. At each point in space along and around the axis and poles, a force exists having a specific intensity and direction, i.e. a *vectorial force field* is present in three-dimensional space and a *linear axial gradient* exists between the poles. Similar terms were adopted by developmental biologists, but whilst some authors (Wolpert 1971 a, b, 1978) stressed the analogy with the physical sciences, others (Waddington 1966) considered that, except for particular purposes, such approaches were grossly oversimplified. Descriptions of systems exhibiting forms of polarity, axes and gradients, abound in biology. Classic examples include regeneration following the transection of *platyhelminths* or *Hydra* at different levels. The details will not be followed here, but in general the presence of an axial gradient concerning the form and frequency of successful regeneration was demonstrated, together with gradients of metabolic activity and susceptibility to toxic materials. Further, the *apical region* (head in flatworms and hypostome in *Hydra*) regenerated first and subsequently inhibited the formation of additional apical structures, and appeared to coordinate the regeneration of succeeding parts (a phenomenon termed *apical dominance*). Similar poles, axes and gradients have been described in many *oöcytes,* e.g. the animal–vegetative, oral–aboral and dorso–ventral axes (see p. 98 and **2**.3). The term axis has, however, been used in rather different ways in different situations. Thus, the animal–vegetative axis of oöcytes is often an obvious structural and cytophysiological feature, reflected by the eccentricity of the nucleus and the differential distribution of yolk platelets, other organelles, pigment granules,

metabolic levels, etc. Further, in some forms which have been thoroughly analysed, e.g. the sea urchin (Gustafson & Wolpert 1963, Gustafson 1965), there exists an experimentally demonstrable 'morphogenetic gradient' along this axis, although it is as yet uncertain what are the relevant 'morphogenetic factors' which are distributed along the axis, although tentative suggestions have been made (see p. 109), and of course the mechanisms controlling their distribution are completely unknown. However, it is clear that the animal–vegetative axis is determined early and once formed is extremely stable; in contrast, the dorso–ventral axis is determined much later, is more difficult to demonstrate and is more labile. Similar concepts of polarity and axes have often been applied, e.g. to the developing tetrapod limb, and here again the proximo–distal, cranio–caudal and dorso–ventral 'axes' are determined at different times and vary in their lability.

Such concepts are central to many of the hypotheses relating to pattern formation and may be considered in association with the two forms of development mentioned earlier (p. 100) termed *mosaic* and *regulative*. In a mosaic system, removal of a part results in a specific defect in the final embryo, whereas removal of part of a regulative system is followed by a total proportionate reorganization of the remainder, with the formation of a small but complete embryo. Further, division of a regulative system often results in duplication, with two small complete embryos. Similar considerations also apply not only to whole zygotes, but also to presumptive organ rudiments. Although no animal groups are wholly mosaic or regulative, some, such as the annelids and molluscs, show a high degree of mosaicism, whereas the sea urchin and chordates are predominantly regulative.

During oögenesis, and variably modified at fertilization, it has been demonstrated (Raven 1958, 1963, 1966 a,b, Austin & Short 1984) that morphogenetically important substances are localized (oöplasmic segregation) in the cortical zone in geometric patterns related to the principal axes of the oöcyte. It is assumed that the distribution gradients of such substances are differentially parcelled between the blastomeres during cleavage. In *mosaic systems* it is thought that the cytoplasmic parcelling which results from their highly ordered and specifically orientated cleavage is almost wholly responsible for directing subsequent pattern formation, morphogenesis and cytodifferentiation, as a series of independent localized events with relatively little intercellular communication. In contrast, it has been proposed that while the blastomeres of regulative eggs also necessarily differ qualitatively and quantitatively, the phenomenon of *regulation* exhibits *global features* wherein the mass of cells reorganizes *as a whole*, and there must exist avenues of *intercellular communication* of crucial significance in development. Such ideas led to the early formulation of the concept of the *morphogenetic field* (Huxley & DeBeer 1934, Child 1927, 1941) defined as 'a region throughout which some agency is at work in a coordinated way, resulting in the establishment of an equilibrium within the area of the field'. Subsequent years saw the rather sporadic (and occasionally inappropriate) use of the term embryonic field, and then for a period its use was largely abandoned by many developmental anatomists. Only quite recently has the phrase been subjected to critical reappraisal and restatement; two provocative but contrasting views have emerged.

The first concerns the idea of *positional information* being available to individual members of a mass of cells by virtue of the presence of *reference points* at its borders, the whole constituting a *morphogenetic field*. Essential to the idea is that *gradients* of activity (information) extend from the reference points throughout the cell mass by *cell–cell communication:* accordingly, the unique position of each cell is specified, and subsequently each differentiates appropriately within the limits of the metabolic options provided at the time by the state of its genome. The picturesque model adopted to present this hypothesis is the so-called 'French Flag Problem' (**2**.7A,B). In this, a line of cells with channels of intercellular communication between them are considered to have three possibilities for molecular differentiation —blue, white or red—and they form a correctly proportioned French Flag whatever the number of cells in the line and even if parts of the original line are removed. It is assumed that each cell is assigned a positional value by appropriate signals with respect to reference sources at the ends of the line. Reference to

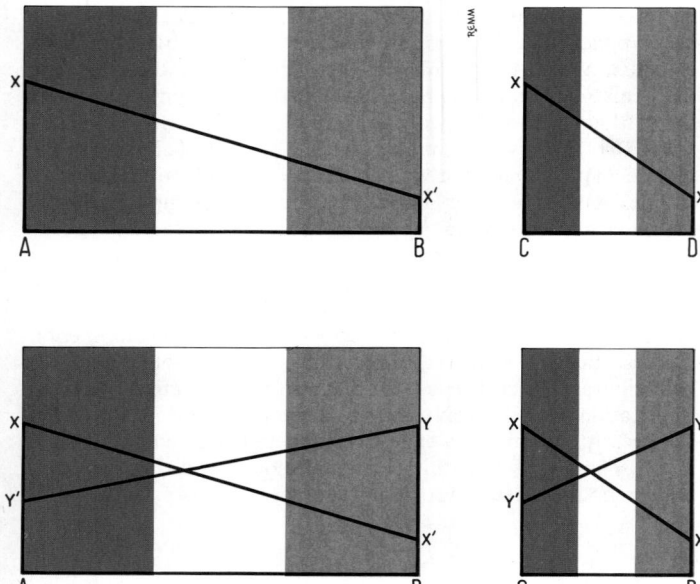

2.7A Diagram illustrating one hypothesis concerning positional information in relation to pattern formation, by reference to the so-called 'French Flag' problem. A-B and C-D indicate long and short rows of cells respectively, X, X' and Y, Y' the concentrations of morphogenetically active substances. Above: (X-X') shows the gradient of a single substance; below: (X-X' and Y-Y') the gradients of two substances. In each case a similar triple differentiation occurs. See text for discussion. (Modified from, and kindly provided by, Professor L Wolpert 1971 a,b, 1978.)

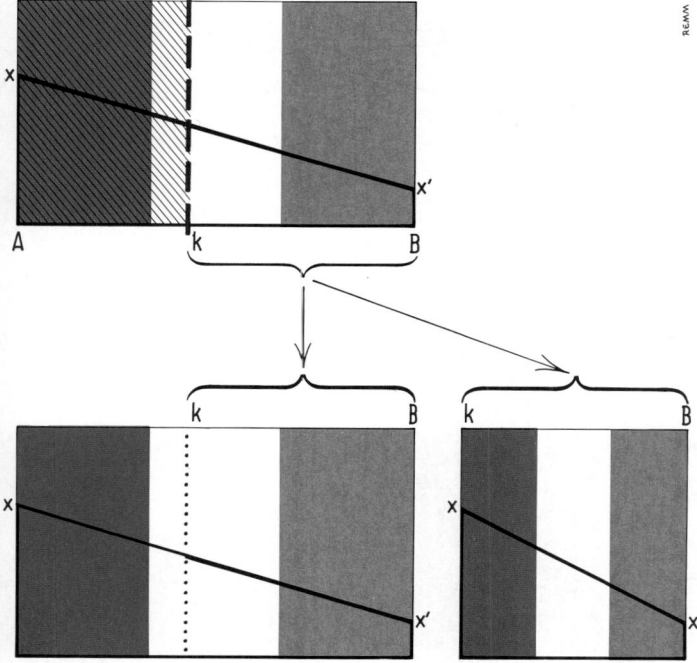

2.7B Diagram using the 'French Flag' analogy to illustrate application of the hypothesis of positional information advanced to explain regeneration, after ablation (cross-hatch), of part of an array of cells, either by morphallaxis (right), or by epimorphosis (left).

illustration **2.7A,B** will make it clear how such a proposal could successfully explain proportionate differentiation, not only with lines of different lengths but also to include epimorphosis and morphallaxis. A number of model systems proposing possible mechanisms for the mode of operation of the signalling system have been examined. Briefly these include, most simply, a source of a metabolic product at one end of the line, a sink at the other, with the establishment of a linear diffusion gradient between them. More complex models have proposed signalling gradients of two or more substances, or the reversible operation of multiple sources and sinks depending upon local conditions (the so-called homeostatic model), whilst others have invoked active transport

mechanisms or the presence of paths of selective permeability. An interesting alternative mechanism suggests (Cooke & Zeeman 1976) that a reference point may be provided by a *pacemaker cell* which emits two wave-like periodic propagations of biochemical activity which are transmitted from cell to cell throughout the morphogenetic field. Positional information would then be signalled by the *phase angle difference* between the arrival of the two waveforms at the cell in question. In a detailed analysis of amphibian gastrulation it was suggested that *two* such pacemaker regions were present, one at the animal pole and the other at the grey crescent. Their mutual positions change during gastrulation providing a dynamic *two-dimensional grid* of positional information in the mesodermal mantle, thereby specifying its subsequent pattern of cytodifferentiation. (For further discussion of these and related hypotheses consult Kitchin 1949, Muchmore 1951, Waddington 1956, Deuchar & Burgess 1967, Johnson 1970, Zeeman 1976b, Elsdale & Pearson 1979, Burgess 1981, 1983.)

Despite the fact that all the foregoing signalling mechanisms are relatively simple in principle, it has been stressed that such systems, when coupled with variations in the 'rules for interpretation of the signals' on the part of the genome of the reacting cells, could result in the production of highly complex, even asymmetrical patterns. Nevertheless, although universality has been claimed for the theory throughout biology (Wolpert 1971a, b, 1978), currently there is no direct evidence for the precise nature of the boundary zones (reference points or pacemaker cells), the detailed mechanisms of intercellular signalling and the manner of cellular interpretation of the signals. As yet, therefore, the hypothesis of a coordinate system which specifies positional information, whilst providing a powerful stimulus for new thought and experiment, must for the present remain the province of tentative theoretical biology.

During the period of emergence of theories of positional information in relation to morphogenetic fields, there has also occurred the convergence of ideas from two other sources. The first constituted a critical reappraisal of the definition of the term 'morphogenetic field' and associated terminology by the eminent developmental biologist C. H. Waddington (1940, 1966). The second was the evolution of a new set of descriptive and explanatory mathematical models which have been grouped under the dramatic name of *catastrophe theory*.

During attempts to redefine the term 'field' as applied to embryology, it was stressed that much confusion had arisen because of the original combination of the ideas of organization and induction (Waddington 1966); confusion was reinforced by the use of terms such as 'organization centre' and 'organizer substance'. Thus it was originally supposed, as indicated above, that such a centre, during an act of embryonic induction, produced organizer substances which transmitted the instructions necessary for the subsequent complex organization of a neighbouring mass of competent cells. In order to avoid this confusion it was proposed that *evocation* should be used for 'induction that does not *transmit* organization'. *Individuation* was suggested for organization of a mass of cells *per se*, whether preceded by an inductive *triggering mechanism* or not. In this sense it was proposed that a morphogenetic field was only appropriately named when applied to an *individuation field* and it was indicated that some authors had used the term too loosely, denoting merely the *locality* of a normal presumptive organ rudiment. It was also considered that to view a morphogenetic field as the manifestation of the presence of gradients of one, or at most a few, simple diffusable substances, was almost certainly a naive over-simplification. Emphasis was placed on the fact that when attempts are made to describe a system in developmental biology, cognizance must be given not only to the three dimensions of *space*, but also to the essential dimension of *time*, to embrace the dynamic progressive changes involved. Further, at successive intervals it would be necessary in such a *multi-dimensional function space* to establish coordinates allowing vectorial analysis of the *concentrations* and *directionality of interaction* of a large number of relevant chemical families. The mathematical concepts appropriate to a study of multidimensional space–surface transitions constitute *topology*. Those facets of topology which are of wide application in both the physical and biological sciences, including developmental biology, constitute the *catastrophe theory* which was invented, developed mathematically and provocatively propounded by the distinguished French mathematician René Thom (1969, 1975). Hitherto the great domains of the physical sciences, e.g. the laws of motion and gravitation, the theory of electromagnetism and the general theory of relativity, dealt with systems exhibiting gradual, smooth, continuous transitions which could be adequately described by the differential calculus invented by Newton and Leibniz. However, it has become clear that in many systems, both in physics and biology including the behavioural sciences, despite a gradual smooth increase or decrease of one or more parameters affecting the system, and although the response of the system is initially smooth and continuous, a point is reached when there occurs a sudden, dramatic, completely contrasting reaction. The science of topology is involved because the *resultant forces* in multidimensional space systems are best described by smooth, continuously curved *surfaces of equilibrium*. Abrupt changes in the form of the surface indicate the points at which the equilibrium breaks down, and at which equally abrupt changes in response

occur—such discontinuities are mathematically termed *catastrophes*. They are not amenable to analysis by differential calculus, and it is to embrace such phenomena that the new mathematical theory has been developed. The rigorous proof of the mathematical validity of the theory is highly complex (Thom 1969, 1975) and cannot concern us here, but a brief allusion to the manner of presentation of *results* of the proof are illustrated (2.8, 9).

The catastrophe theory initially had its greatest impact on other branches of mathematics; applications in physics include shock-wave propagation, non-linear oscillations and the detailed analysis of caustics and fluid flow. It was suggested, however, that in future years its potentially most powerful applications may lie in biology in relation to the behavioural sciences and in developmental studies. Illustration 2.8 is redrawn from a paper devoted largely to behavioural science, because the diagram's construction and some implications are relatively easy to understand. Space constraints in the present volume, however, preclude an extended description and discussion of even this elemental model, and the interested reader should consult the original sources and books cited (Thom

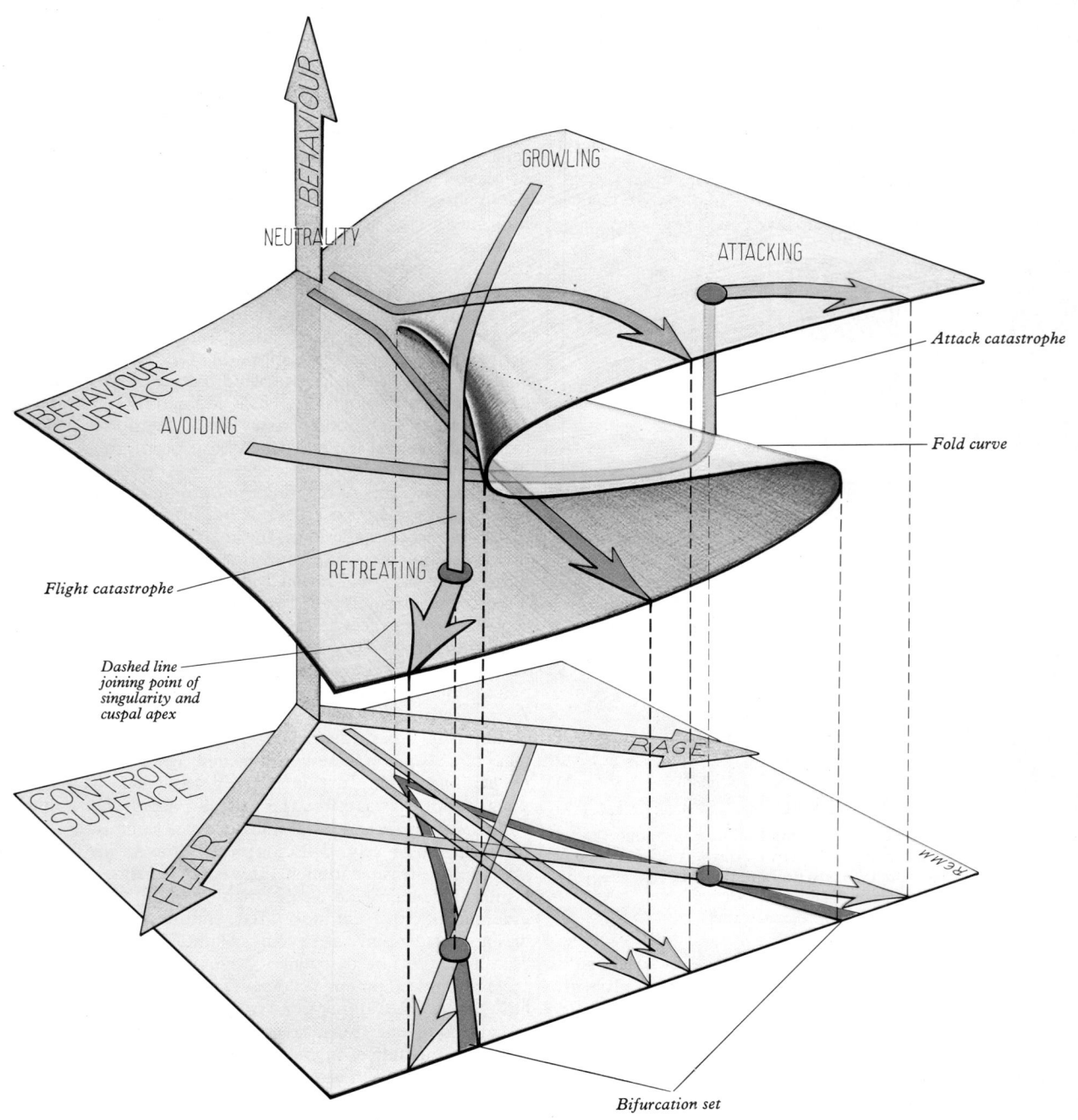

2.8 Diagram showing the use of a three-dimensional graph which embodies the single pleat of an elementary cusp-catastrophe model. In this, changes in the possible aggressive or submissive behavioural states resulting from the interactions of varying degrees of fear and rage in dogs have been illustrated. Consult adjoining text for a more detailed description, and explanation of the terms used. (Redrawn, and slightly modified, by kind permission from the author, Professor E C Zeeman, and the *Scientific American*.)

1969, 1975, Zeeman 1974, 1976a, b, Thom & Zeeman 1975, Poston & Stewart 1978). The illustration deals with aggressive behaviour in dogs and includes the following principal features. The assumption is made that aggressive behaviour is the resultant of two *measureable conflicting* factors, fear and rage, which are plotted as axes diverging from an origin on a horizontal *control surface*. The behaviour of the dog is plotted on a third, vertical axis, the most aggressive behaviour—growling leading to overt attack—being assigned the largest values, whilst submissive behaviour—avoidance reactions leading to frank retreat—being assigned progressively lower values. The total set of points on the vertical axis, plotted from every point on the control surface, together form the *behaviour surface*. It might be imagined that only *one* resultant behaviour would be predicted from each combination of fear and rage leading to a simple downward slope passing from attack, through growling, neutrality, avoidance and finally retreat. Over much of the behaviour surface there is, indeed, only one resultant behaviour point, but where rage and fear have roughly equal values there are *two* highly probable forms of behaviour, i.e. attack *or* retreat, but neutrality is the *least* probable form of behaviour. Application of catastrophe theory explains this *bimodality* in behaviour pattern. Expanding from a locus vertically above the origin along the distant margin and then including the right and left regions, the behaviour surface forms a single continuous slope which is high to the right and low to the left. Centrally, however, starting at a *point of singularity*, and expanding forwards, a continuous double fold or *pleat* is formed. The upper and lower layers denote *two* alternative behaviours each with a *high probability* corresponding to a *single* point on the control surface. (The intermediate layer of the pleat can be disregarded in terms of response because it represents the *least probable* form of behaviour.) The actual behaviour pattern followed can only be predicted from a knowledge of the particular *starting point* in the behaviour surface and then tracing its route as the relevant parameters change with time. Thus, an enraged but only slightly fearful dog shows aggressive behaviour: as fear increases, behaviour changes only slowly (gentle slope) remaining aggressive, but the point on the upper layer of the behaviour surface moves centrally. Eventually the edge of the pleat is reached, and the behaviour plot *suddenly* drops vertically to the lower sheet denoting a complete contrast in behaviour which is now meek, submissive and ultimately leads to retreat. The projection of the edges of the pleat back on to the control surface comprises a cusp-shaped pair of divergent curved lines, the *bifurcation set*, and the whole system predicting sudden changes in response is therefore termed a *cusp-catastrophe model*. In summary, this model predicts a single behaviour pattern for most parameter combinations, but *bimodality exists within the bifurcation set*. The points at which jumps from the upper to the lower sheet occur do not correspond with those points at which the converse direction of jumps occurs (*hysteresis*). It is also clear that starting at a point of neutrality, *slight* differences in the temporal change of parameters may lead to *widely divergent* responses (passing either on to the upper sheet or the lower sheet, the divergence occurring at the point of singularity). The forces which operate homeostatically within any particular system tending to maintain the behaviour point (or trajectory) in contact with either the upper or the lower sheet of the behaviour surface are collectively termed the *attractor* for that particular surface. The attractor effect is most powerful in unimodal regions which approach the horizontal and are of gentle slope. Its effect diminishes near the region of bimodality and increasing slope, and of course vanishes at the margin of the pleat where it is replaced by a new attractor.

When a maximum of four parameters are considered, Thom has shown mathematically that these can be satisfactorily described by one of seven *elementary catastrophe* models of which the cusp catastrophe is one of the simplest and, as yet, of widest application. The analysis of models embodying larger numbers of parameters and their possible applications is currently engaging leading mathematicians in many centres.

In view of the recent wide acceptance of the conceptual relevance of the catastrophe theory in such diverse fields as the physical and behavioural sciences, it is of considerable interest that the first formal scientific paper devoted to the theory should

have been preceded, by some years, by a brief statement in relation to *developmental biology* following conversations between Waddington, Thom and Zeeman (see Waddington 1966). Illustration **2**.9 summarizes some of the main points. For simplification, the developing three-dimensional embryo has been reduced to a pair of two-dimensional embryo maps (compare with the *control surfaces* described above) and these represent increasing developmental complexity at two successive times T1 and T2. From each point on the embryo maps, vectorial plots of the activities of relevant chemical groupings are made within multidimensional spaces termed *phenotype cubes*. These plots generate appropriate 'behaviour surfaces' the form of each indicating the homeostatic equilibria present. It will be noted that the almost horizontal featureless behaviour surface at T1, which embodies a single, powerful, effective *attractor*, has during development by time T2 been subdivided into two different attractor surfaces, separated by a breakdown in homeostatic equilibrium and indicated by the pleat of an *elementary cusp catastrophe*. The term *chreod* (derived from Greek roots meaning 'it is necessary or fated' and 'a path') was introduced by Waddington to refer to equilibrium surfaces showing attractor phenomena in multidimensional space systems. The region of strong effective attractor action was termed by Thom the 'support of the chreod', and the surrounding zone where its effectiveness diminishes, and catastrophes occur was named 'the cone of the chreod'. Accordingly 'fields' in embryogenesis were portrayed as subregions in a multi-dimensional space–time system, each being chreodic in character with its particular attractor system, and with progressive development divergence of chreods occurs by the interposition of catastrophe zones. In the central part of illustration **2**.9 each attractor surface has been plotted down to a simplified contour impression of the picturesque concept of an 'epigenetic landscape' which had been advanced by Waddington a quarter of a century earlier (Waddington 1940). This comprises a system of valleys, initially single but diverging into branch valleys with the passage of time. Each valley and its sloping walls constitutes a chreod. The homeostatic power of its attractor mechanisms is simulated by the steepness of the valley walls and it is this feature that confers on a developmental field, or chreod, its regulative capacity (p. 100). The points of divergence of valleys are the points at which the homeostatic attraction breaks down with the occurrence of catastrophes. Such a system may be thought of as analogous to the emergence of, e.g. a series of separate organ primordia during early development. The lowest part of illustration **2**.9 displays a modified cusp-catastrophe model which applies the catastrophe theory to two possible modes of change which may ensue when a group of 'undifferentiated' cells (*a*) are subjected to an array of physical and chemical parameters which change temporally and spatially. It will be noted that *modulation* of the cells to condition *A* is represented as a reversible process (remaining confined to the upper leaflet of the behaviour surface). In sharp contrast, *differentiation* of the cells to condition *B* can be considered as involving a sudden 'catastrophic' change (occurring, perhaps, when particular parameter combinations are present at the time of the *critical mitoses* discussed previously on p. 108). It should also be appreciated that in the model illustrated, the pleat in the behaviour surface is *incomplete*: i.e. there is no return path from *B* to *A* (which would involve a pleat edge). Thus it is assumed that, in this case, such a change (or dedifferentiation) back to condition *A* cannot occur.

It is hoped that the greatly abbreviated accounts given in the foregoing paragraphs relating to pattern formation and the concepts of positional information and catastrophe theory will be sufficient to convey the general conceptual framework employed by modern developmental biologists. It must be stressed, however, that whilst they have provided powerful stimuli for reappraisal of many outstanding problems, precise laboratory verification and clarification is still awaited. Thus, whilst the *phenomenon* of embryonic induction is widely accepted as adequately demonstrated, its *mechanism* is still unknown. Similar, perhaps identical, questions are posed, but remain unanswered, by the *intercellular signalling* mechanisms implicit in the theory of positional information. In like manner, *mathematical models* based upon the catastrophe theory have been constructed for the

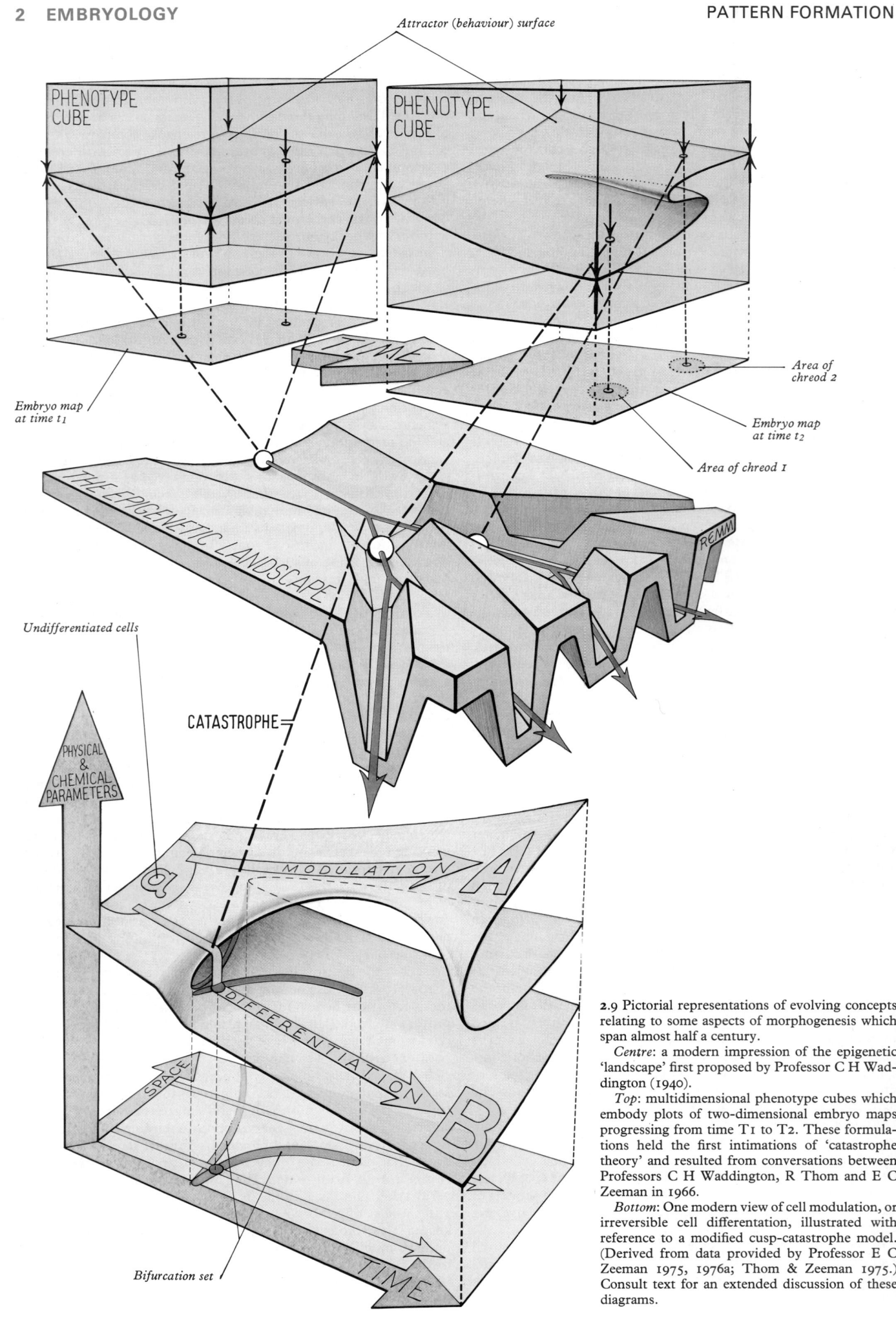

Attractor (behaviour) surface

PHENOTYPE CUBE

PHENOTYPE CUBE

Embryo map at time t₁

TIME

Area of chreod 2

Embryo map at time t₂

Area of chreod 1

THE EPIGENETIC LANDSCAPE

REMM

Undifferentiated cells

CATASTROPHE

PHYSICAL & CHEMICAL PARAMETERS

MODULATION   A

DIFFERENTIATION   B

SPACE

TIME

Bifurcation set

2.9 Pictorial representations of evolving concepts relating to some aspects of morphogenesis which span almost half a century.

Centre: a modern impression of the epigenetic 'landscape' first proposed by Professor C H Waddington (1940).

Top: multidimensional phenotype cubes which embody plots of two-dimensional embryo maps progressing from time T1 to T2. These formulations held the first intimations of 'catastrophe theory' and resulted from conversations between Professors C H Waddington, R Thom and E C Zeeman in 1966.

Bottom: One modern view of cell modulation, or irreversible cell differentation, illustrated with reference to a modified cusp-catastrophe model. (Derived from data provided by Professor E C Zeeman 1975, 1976a; Thom & Zeeman 1975.) Consult text for an extended discussion of these diagrams.

cardiac cycle, nerve impulse propagation, cell differentiation (2.9) and for amphibian gastrulation (Zeeman 1976a,b, Poston & Stewart 1978), but as yet experimental validation is lacking (and indeed, occasional authors view with scepticism the over-enthusiastic acceptance and application of the theory in the presence of incomplete data—see Croll 1976).

NUCLEAR CHANGES DURING DIFFERENTIATION

Two main lines of investigation over some decades have attempted to demonstrate changes in the nuclei of tissues which have undergone differentiation—firstly, experiments using nuclear transplantation techniques and secondly, an examination of the phenomenon of chromosome 'puffing' shown by some species. More recently the modern techniques of genetic manipulation ('engineering') have started to focus, with some success, on the regional variations in activity of short segments of chromosomes (see p. 39). Currently, immunofluorescence techniques are proving valuable methods of demonstrating the location and timing of first appearance of antigenic new gene controlled products during differentiation.

Many kinds of transplantation experiment have been carried out (King & Briggs 1965, Gurdon 1967, 1968) but in principle, the nucleus is removed from an activated egg and replaced by one removed from a particular site in, e.g. an early blastula, early or late gastrula or more highly differentiated post-gastrulation tissue, and then any subsequent development which proceeds is analysed. In practice more reliable quantitative results can be achieved by the *serial transplantation* of donor nuclei before proceeding to a developmental analysis (2.5). Although the results show considerable variation, in general, the transplantation of early nuclei is often followed by the formation of a complete embryo, whereas nuclei from the later stages are sometimes only able to support the development of incomplete embryos or cell masses of a restricted tissue type. However, in contrast, it has been shown that nuclei from advanced tissues such as amphibian neural plate cells and ciliated gut epithelial cells of the tadpole can still support the formation of complete embryos, although the proportion of successful experiments is reduced, whilst nuclei from fully adult tissues fail to do so (Gurdon 1967, 1968). Whilst such results may reflect some stable changes occurring in nuclei as differentiation proceeds, more work will be necessary before clear-cut interpretation is possible.

Direct visualization of differences between the chromosomes of various tissues and times is provided by the giant polytene chromosomes of the *Diptera* and the 'lampbrush' chromosomes of the newt. At certain loci such chromosomes exhibit hazy thickenings of their outlines ('*puffs*' or *Balbiani rings*). It is widely held that throughout inactive regions of chromosomes the DNA threads are closely coiled and supercoiled, but at sites of activity the threads become unwound and spread laterally from the axis of the chromosome and form the basis of the puffs. Each puff may include a number of active operons at which transcription and synthesis of messenger RNA is occurring. In short, although many details of the process of puff formation remain to be elucidated, the puff patterns of all the functional nuclei of a particular cell type in a tissue are identical but differ significantly from the patterns in other tissues. The patterns also change as development proceeds; e.g. there is a regularly repeating sequence of changes accompanying the 'moult' and 'intermoult' stages in the life of larvae. The latter system is particularly interesting since moulting is a response to a hormone, *ecdysone*, which has been isolated and which affects the activity of the genome—details of the process are under intense analysis.

It is important to emphasize the fundamental question posed by such investigations, namely—during the growth, patterning and differentiation of complete multicellular organisms, is the *essential sequence* of instructive molecules in the genome of the different emergent cell types *altered*, e.g. by deletion of segments?

Alternatively, is the overall sequence *preserved* and the *expression* of its various segments *modified* appropriately? The consensus supports the latter, with the reservation that the *degree of reversibility of expression control* varies widely with site and species.

Occasional examples of alteration of genomic sequences have been quoted, e.g. the early segregation of gonocyte and somatic cell nuclei in *Dipterans*, with the apparent deletion of sex-determining loci in the latter; another case involves the *amplification* of repetitive gene sequences during oögenesis at the other extreme; it has proved possible, by careful tissue culture techniques, to grow a complete carrot plant from a single isolated parenchymal cell from a mature carrot.

SURFACE INTERACTION BETWEEN CELLS

Further properties of developing cells, which reflect some aspect of their state of differentiation and have important implications in morphogenesis, concern the selective interactions between their surfaces when in contact. Such surface modifications are closely related to the formation of the cellular sheets, hollow balls, tubes, invaginations, solid masses and so on which characterize embryonic life. Following earlier tissue culture studies (Weiss 1941, 1961, 1970, Abercrombie 1961, Abercrombie & Heaysman 1953, 1954), two important principles were enunciated. Numerous experiments showed that some cells, when explanted on to a matrix with an heterogeneous, polarized substructure, often migrated along preferential paths determined by its architecture. This led to the concept of *contact guidance* of cells—particularly well shown in the preferred regeneration of nerve fibres along the surfaces of Schwann cells or, during ontogeny, the growth of later generations of nerve fibres along the surfaces of the earlier pioneer fibres.

In contrast, when similar cells which were migrating freely in tissue culture approached each other and made contact, all movement and often cell division in the plane of contact ceased and this phenomenon of *contact inhibition* was proposed as an important morphogenetic factor. For a brief account of the mechanisms of cell movement see p. 49, and for reviews on morphogenetic movements consult Ede (1976), Abercrombie & Heaysman (1953, 1954), Hinchliffe & Johnson (1980).

Since that time many experiments involving the *disaggregation* of tissue cells have been made (Townes & Holtfreter 1965). For example, if the cells of a neurula or those of a developing kidney are disassociated to form random cell suspensions, they often subsequently reaggregate and similar cells reassociate so that a recognizable neurula or a fragment of kidney tissue is reformed. Again, if cell suspensions from developing liver and kidney are intimately mixed, similar cells gradually reassociate and organize into kidney and liver tissue. In this manner, using different cell combinations, the potential of different cell types for mutual recognition, reorganization into tissues, sheet formation, invagination and so forth have been analysed. An adequate interpretation of the phenomena would, of course, necessitate an intimate understanding of the physico-chemical properties and control systems of cell surfaces. Until such become available, the investigators in this active field of research are pursuing quantitative variations in the *specific adhesiveness* between cells and proposing various models for further examination, such as *stereochemical surface configurations* which can only interlock with complementary surfaces. For reviews consult Curtis (1967, 1973), Karfunkel (1971).

In the subsequent sections a brief account of *human* development is given and, as such, is necessarily and solely *descriptive embryology*. When reading such an account of the emergence of primordial organs and subsequent changes in shape, the reader should constantly be aware of the many continuing fields of *experimental analysis* in other forms, which are daily adding to our comprehension of the *causal processes* of the almost incredible and beautiful phenomenon of development.

The Female Gamete: The Ovum

The precise origin of the primordial germ cells, from which both ova (Shettles 1960, Austin 1961, Austin & Short 1984) and spermatozoa are derived, remains uncertain (p. 254); but those which reach the genital ridge in the female, the *oögonia*, proliferate and differentiate into *primary oöcytes*. Their number maximizes at about 6 000 000 in each ovary by the fifth intra-uterine month (Baker 1963, 1966). Subsequently, before birth, widespread degeneration occurs, leaving a complement of about 2 000 000 primary oöcytes (Block 1953). However, many of these become atretic before puberty, when only some 40 000 oöcytes persist in each ovary (Pinkerton et al 1961). A mere 400 or so of these will be shed at ovulation with the potential of fertilization during the reproductive years of the average woman. (Each primary oöcyte passes through a series of changes, becoming a *secondary oöcyte* just before ovulation and mature *ovum* at, or just after, fertilization. The term ovum is nevertheless used with a confusing vagueness for all three and even for later stages, a practice which is best avoided.) Immediately after formation and still before birth, primary oöcytes enter the initial stages of prophase of the *first meiotic (reduction) division* in the process of *maturation*, which is not completed until shortly before ovulation. This reduction in the genetic apparatus and the recombination of genetic material are, of course, central features of maturation (p. 42). A primary oöcyte is distinguishable from other cells in the ovary by its large size, being about 35 µm in diameter, and by the fat granules in its cytoplasm. Its nucleus is relatively large, vesicular and usually eccentric, with a prominent nucleolus. The nuclear eccentricity confers a visible polarity on the cell which is emphasized to some degree during subsequent growth and maturation. The Golgi apparatus is well developed and consists of parallel cisternae and numerous small vesicles; the latter also appear throughout the cytoplasm. Juxtanuclear anulate lamellae occur interspersed between spherical or slightly elongate mitochondria. At first the endoplasmic reticulum is vesicular and displays few ribosomes. As the oöcyte grows all these organelles increase in number and usually in size, and they disperse from the juxtanuclear zone throughout the cytoplasm. (Although described as isolecithal—vide infra—this dispersal is not entirely uniform, varying with nuclear position.) Granular endoplasmic reticulum and free ribosomes, though not prominent, also increase. Lipid granules begin to appear; they correspond to the yolk platelets of earlier vertebrates but are smaller and less frequent in primates. By the secondary or antral stage of follicular growth (vide infra) the oöcyte has completed its increase in size. Usually each ovarian follicle contains one, but two or even more may be present. In some strains of certain animals polyovular follicles are more frequent. The oöcyte is surrounded by a single layer of cubical cells and at this stage electron microscopy shows that the relationship is one of close apposition; adjacent plasma membranes are smooth and separated by 12.5–14.0 nm.

THE OVARIAN FOLLICLE

During the earlier and subsequent stages of maturation considerable growth of the primary oöcyte and enlargement of the follicle occurs. By accumulation of cytoplasm, particularly its lipids, the oöcyte reaches a diameter of 117–142 µm—a volume increase of about a thousand times; this is achieved before follicular enlargement is complete. The follicular cells first multiply and form several layers; at, or even before, this stage electron microscopy reveals the accumulation of interrupted masses of amorphous material between the cells and the oöcyte. These soon fuse into a complete membrane around the oöcyte, seen with light microscopy as a PAS-positive, thick and radially striated envelope, the *zona pellucida* (z. striata). The zona contains considerable amounts of carbohydrate and neutral glycoprotein (Braden 1955). Sialic acid residues also occur and may contribute to its elasticity (Soupart & Noyes 1964, Soupart & Clewe 1965). Electron

microscopy reveals an intricate filamentous structure at the time of ovulation. The origin of the zonal membrane, whether from the oöcyte, follicular cells or both, is still undecided (Hope 1965); an outer layer, consisting of acid mucopolysaccharide, has however been attributed to the follicular cells (Wartenberg & Stegner 1960, 1961). Coincident with the development of the zona, the surface of the primary oöcyte presents numerous microvilli which project into the zona: less frequent processes from the follicular cells pass through the zona pellucida to make contact with the plasma membrane (oölemma) of the oöcyte, without actual cytoplasmic continuity. These intimate relationships may be concerned with transport of materials between the follicular cells and ovum (Shettles 1958, 1960, Chiquoine 1959, Sotelo & Porter 1959, Odor 1960).

The zona pellucida is semipermeable, is slowly lysed in alkaline media and degraded by the proteolytic enzyme acrosomase. It is best developed in placental mammals and persists until the blastocyst stage of development. It is probably the source of fertilizin in mammals (Bishop & Tyler 1956).

Dense extracellular granules, *Gall-Exner bodies*, develop between the maturing follicular cells. They are PAS-positive and display a complex ultrastructure. They may be derived from the Golgi apparatus but are of uncertain significance. They occur in human follicles but rarely in other mammals.

OVULATION

When the follicle is multi-laminar, a fluid-filled *antrum* appears among its proliferating cells, gradually dividing them into an internal stratum, the *cumulus ovaricus* (oöphorus), and an external *stratum granulosum*; but at one site the two strata maintain continuity (2.10). The primary oöcyte is in the cumulus; the granular stratum forming an envelope around the fluid gradually thins as the *liquor folliculi* accumulates. (Antral development has been analysed in particular detail by Zamboni et al 1972.) The primary oöcyte, which has remained in the prolonged prophase of the first meiotic division, retains its microvilli and its complex interrelationship with the follicular cells outside the zona pellucida (2.11A) until shortly before ovulation, when continued formation of the first polar cell is resumed (vide infra) and completed. Now a *circumvitelline space* appears between the zona and what is now a secondary oöcyte. The processes of the follicular cells are also withdrawn from the zona, which therefore loses the striation seen earlier with light microscopy and becomes more homogeneous. The follicular cells show an increasing abundance of mitochondria, granular endoplasmic reticulum and free ribosomes and their Golgi organelles become more prominent, indicating increasing activity. The stratum granulosum becomes enveloped in a cellular and fibrous sheath derived from the ovarian stroma (2.10), termed the *theca folliculi* (p. 1436). By enlargement, and perhaps also translation through the ovarian stroma, the ripening follicle projects from the surface of the ovary and finally ruptures, releasing the secondary oöcyte, surrounded by the zona and cells of the cumulus, into the peritoneal cavity, from which it rapidly enters the uterine tube; thereafter, slow transit to the uterus is aided by contractions of the fimbriae of the tube and of the tube itself, together with ciliary movements of its epithelium (Austin 1963). It has been noted that the secondary oöcyte cannot be strictly regarded as an 'ovum' until after completion of its second division and release of a second polar cell.

The follicular cells adherent to the liberated oöcyte are now termed the *corona radiata* and they form two or three irregular and loosely arranged layers (2.10, 11). Between them is a matrix containing hyaluronic acid, and in some species there is strong evidence that seminal hyaluronidase brings about the disintegration of the corona which soon occurs. As mentioned, the zona pellucida is partly degraded by seminal proteolytic enzymes derived from the acrosome of the sperm head, including acrosomase (acrosin). Unless fertilization occurs, the secondary oöcyte is discharged from the uterus in the debris of the next menstrual period; if it is fertilized, the zygote which results is

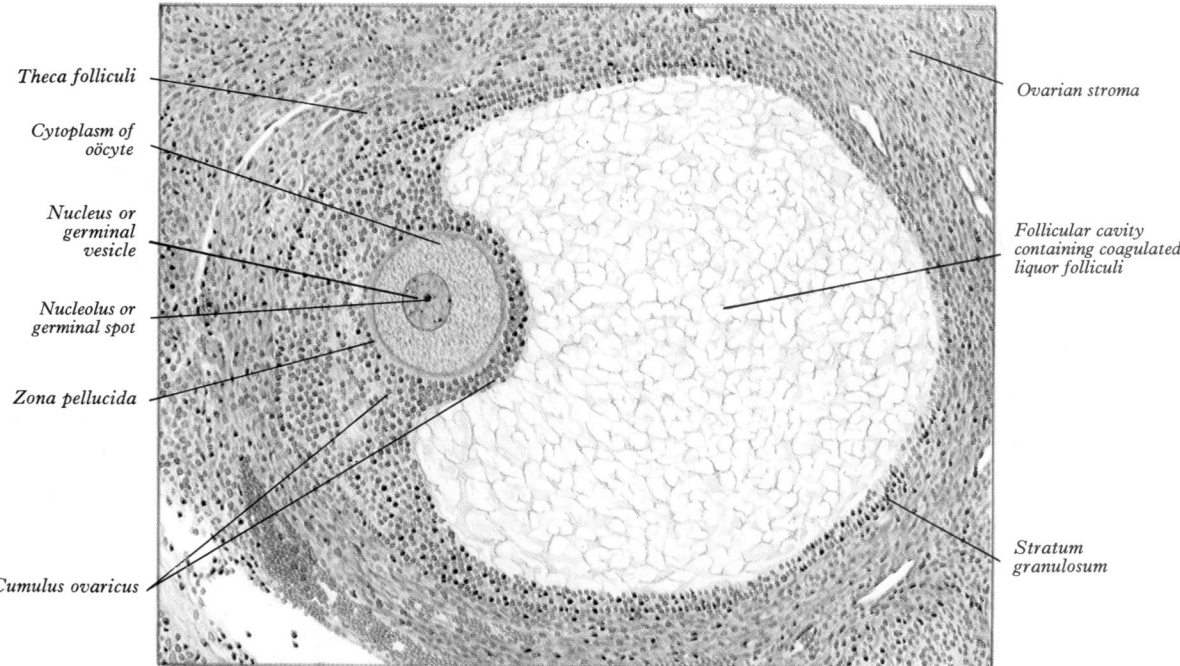

Theca folliculi

Cytoplasm of oöcyte

Nucleus or germinal vesicle

Nucleolus or germinal spot

Zona pellucida

Cumulus ovaricus

Ovarian stroma

Follicular cavity containing coagulated liquor folliculi

Stratum granulosum

2.10 Ovarian follicle from a woman aged 28 years. Haematoxylin and eosin. Magnification × c. 90.

retained and pregnancy begins. (For details of fertilization see p. 123.) Usually only one follicle matures fully and ruptures in each menstrual cycle in mankind, but multiple ovulation does occur (p. 125). Hence only a small fraction of the primary oöcytes which persist in the ovaries of the newborn female are destined to survive to ovulation, the remainder degenerating. Atresia occurs before birth, as stated above, and even at birth about half of the surviving oöcytes are abnormal. Several follicles usually develop normally to some extent in each reproductive cycle, but only one commonly completes the process, the others becoming

degenerate, being invaded by blood vessels and connective tissue which ultimately replaces them as *atretic follicles* (eventually contracting into small white fibrous *corpora albicantia*, see p. 1437).

Although an oöcyte is exceedingly large compared with other general body cells, it resembles them in general structure; some of its parts are given specific names. Thus, the cytoplasm is the *yolk* or *oöplasm*, the nucleus is termed the *germinal vesicle* and the nucleolus the *germinal spot*. The cytoplasm is also often known as the *vitellus* and hence the plasma membrane of the cell is the *vitelline membrane*. The yolk comprises two major components: (1) cytoplasm bearing organelles that characterize many other cells, termed *formative yolk*; (2) the *nutritive yolk* or *deutoplasm*, which consists largely of fatty droplets containing lecithin. In mammalian ova the deutoplasm is small in amount and is only partly sufficient for the initial stages of embryonic development prior to implantation in the uterus. Over this period the deutoplasm is supplemented by nutritive tubal secretions (p. 141) and 'uterine milk' (p. 146). In contrast, as we have seen (p. 98), the eggs of birds contain a supply which nourishes the

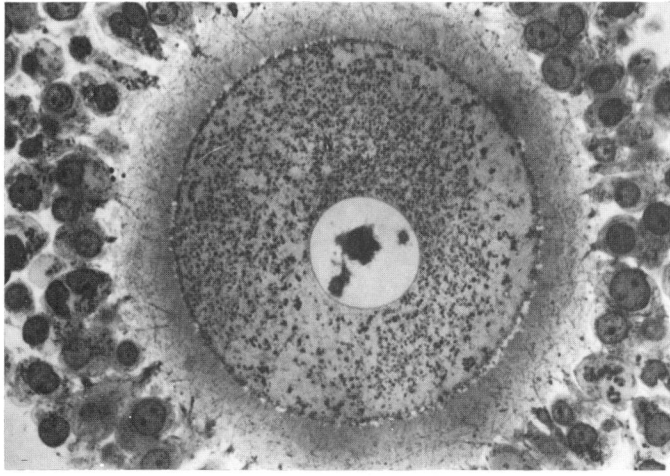

2.11A Drawing constructed from electron micrographs of a primary oöcyte surrounded by its zona pellucida, outside which are aggregated follicular cells of the corona radiata. Note extensive invasion of the zona by microvilli of the oöcyte interdigitating with cytoplasmic processes of the follicular cells. Where some of the latter approach the plasma membrane of the oöcyte, desmosomoid junctions are visible. Note the organelle-rich cytoplasm of the oöcyte. (Modified from Anderson & Beams 1960, with kind permission.)

2.11B Human oöcyte photographed in transverse section. Follicular cells surround it, the zona pellucida intervening. Epon-embedded, osmic-acid fixed, toluidine blue stained. Magnification × 550. (Baca & Zamboni, 1967)

117

chick throughout its entire development up to hatching. The distribution of the deutoplasm varies in different animal groups; the abundant deutoplasm of certain ova tends to accumulate in one part of the cell, which is therefore termed *telolecithal*. In human oöcytes the dispersion is rather more, but not completely, uniform or *isolecithal*. The formative yolk contains some of the organelles characteristic of cytoplasm, such as mitochondria, Golgi apparatus and centrosome, rough and smooth endoplasmic reticulum, varieties of lysosomes, microbodies granules, filaments and tubules; centrioles are sometimes visible in the neighbourhood of the nucleus, but they disappear when the female pronucleus is formed after fertilization (p. 124). (For a detailed analysis of the fine structure of the developing human oöcyte in vitro consult Zamboni et al 1972.) By analogy it is presumed that crucial, active, morphogenetic substances are differentially concentrated in various parts of the cytoplasm, particularly in the cortical zones.

MEIOTIC DIVISION OF THE OÖCYTE

The series of changes which produce a fertilizable 'ovum' are known as maturation (Monroy & Tyler 1967), some features of which have already been mentioned above. The primary oöcyte's initial period of growth by accumulation of deutoplasm is followed by two successive cell divisions (**2**.12), the first being *heterotypical, reduction division* or *meiosis*, during which the so-called *diploid* or full complement of chromosomes is halved, the second division being *homotypical* and involving no such reduction (p. 98). Unfortunately, both forms of division, intimately linked in the formation of mature ova (and spermatozoa) from their precursors, are, in current practice, grouped together under the term meiosis (which strictly implies a reduction or lessening), a custom which may be misleading. (The equally fundamental nuclear changes involving interchange of chromatid segments leading to *genetic reassortment* or *recombination* are detailed on pp. 42–46).

The primary oöcyte, which is already in the prophase of the first meiotic (true reduction) division, contains the diploid number of chromosomes. It should be emphasized that most, if not all, the primary oöcytes have completed the prophase stage of the first meiotic division before the individual's birth (Manotaya & Potter 1963, Ohno & Smith 1964). Instead of proceeding to metaphase, they enter the so-called *dictyotene* stage, a resting state between prophase and metaphase, which is characterized by a reticular arrangement of chromatin and lasts until the first meiotic division is resumed and completed shortly before ovulation. As this division recommences, the nucleolus disappears, the nuclear membrane disintegrates, the chromosomes appear and become

arranged in homologous or *bivalent* pairs at the equator of a spindle radially orientated at the animal pole of the oöcyte. The phenomena constituting *cross-over* of chromatid segments has now occurred. The chromosomes separate and one member of each pair passes centripetally into the central region of the cell, while the others move centrifugally, forming a projection at the animal pole of the oöcyte which becomes separated off as the *first polar cell* or 'body'. Unlike that of the nucleus, the division of the cytoplasm is highly *unequal*, the polar cell carrying with its numerically equal (but constitutionally altered) chromosomal complement an exiguous share of the cytoplasm. The larger cell resulting from this reduction division is a *secondary oöcyte* and it is not fully reconstituted before it divides again. Its haploid complement of chromosomes (22 autosomes + an X sex chromosome) is again rearranged around the equator of a second polar spindle at the animal pole of the cell. A nuclear membrane is not reformed before this second maturation division commences. A similar division may occur in the first polar cell but is abortive, since all polar cells degenerate. At the moment of spindle formation the secondary oöcyte is shed from its mature follicle to enter its uterine tube. The division is only completed after fertilization, as had long been suspected (Allen et al 1930, Hamilton 1944, 1949). *If fertilization ensues*, the haploid group of chromosomes split longitudinally so that each of the resultant two cells receives an equal amount of genetic material, although the chromosome content has been halved. The larger cell is the (sperm penetrated) ovum, which retains almost all the cytoplasm, a small amount only of which enters into the formation of the *second polar cell*. The first polar cell may not undergo division in every instance, but three, theoretically the full complement, have been observed in association with fertilized ova within the zona pellucida (Shettles 1955). Thus, in the human ovary, there is no true, free, non-sperm penetrated *ovum* stage: unfortunately, as mentioned elsewhere (p. 116) the term ovum is widely but imprecisely used for many stages of early development.

The chromosomes of the fertilized ovum now lose their visual identity and appear as a reticulum near the centre of the cell and constitute the *female pronucleus*, around which a nuclear membrane reappears. The cytoplasm shrinks a little so that a distinct circumvitelline space develops, in which the polar cells may be visible. At this stage the mature ovum is characterized by its chromosome content—half the typical number for the somatic cells of the species—and its large size, due to the great accumulation of deutoplasm. Its pronucleus is now ready for union with the male pronucleus, already in the cell, in the last act of fertilization, the formation of a zygote (p. 123). (The above remarks concentrate upon chromosomal and nuclear organization; for the equally essential organization of the *cytoplasm* of the oöcyte prior to fertilization and further development, see p. 98 et seq.).

The Male Gamete: The Spermatozoön

Gametogenesis in the male exhibits both marked similarities and differences in comparison with the development of ova. During maturation there is the same reduction of chromosomes to the haploid number and genetic recombination, but in the testis there is a continuous formation of spermatocytes and spermatozoa during reproductive life, linked with the enormous number of gametes which are formed. In each ejaculate there are many times more spermatozoa than there are germ cells in both ovaries at their peak content before birth; whereas the latter is of the order of 10 to 12 million, a single ejaculation may contain 300 million spermatozoa (**2**.13A–C), only one of which may fertilize an ovum. (The nomenclature of male gametes is not yet unified officially; spermatozoön, spermatoid, sperm and spermium are all used.)

THE MORPHOLOGY OF SPERMATOZOA

A spermatozoön, or sperm (Fawcett 1961a, 1975, Pikó 1969, Rothschild 1957), is a smaller cell than an oöcyte, highly specialized to reach the latter and to carry to it its own haploid chromosome complement. Its expanded *caput* or head contains little cytoplasm and is connected by a short constricted *cervix* or *neck* to the *cauda*

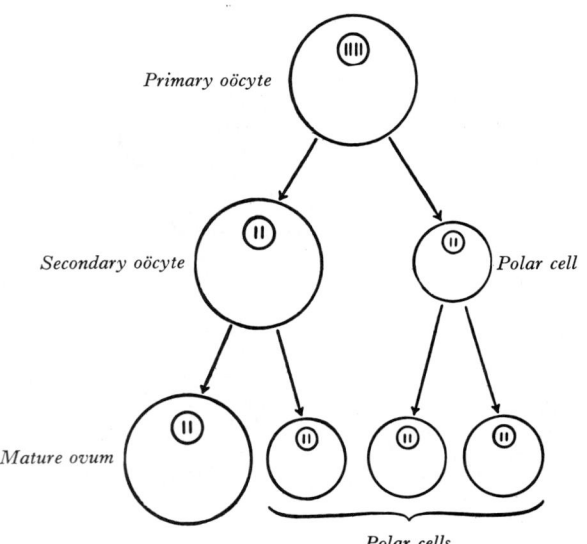

Primary oöcyte

Secondary oöcyte   *Polar cell*

Mature ovum

Polar cells

2.12 Diagram showing the reduction in number of the chromosomes during the maturation of the ovum. The first division is heterotypical, the second homotypical. The latter follow sperm penetration.

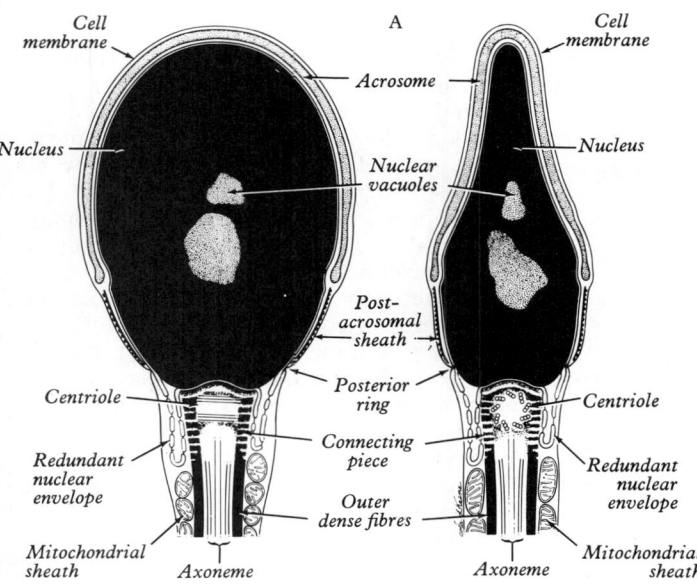

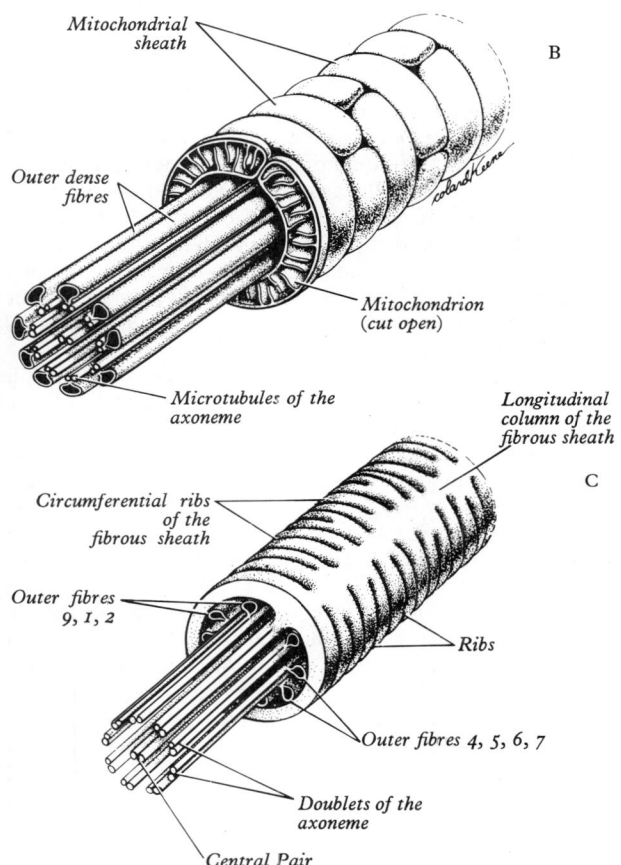

2.13 Diagrams of human spermatozoön showing; A. the head, viewed in its major (left) and minor (right) diameters; B. the middle part and C. the principal part or tail. (By courtesy of Fawcett & Hafez and C V Mosby Company)

or *tail*. The latter is a flagellum of complex structure, usually divided into *middle, principal* and *end parts* or *pieces*. Volumetrically the tail much exceeds the head, which varies greatly in different species (Rothschild 1957, Phillips 1975), being ovoid or piriform in man, somewhat flattened at the tip in lateral profile, with a maximum length of about 4 μm and a maximum diameter of 3 μm. The tail, about 45–50 μm in length, displays a greater uniformity between species.

The head (2.13A) is an extreme example of chromatin concentration, consisting largely of a dense and visually uniform nucleus, with a distinct bilaminar nuclear membrane and a bilaminar *acrosomal cap* (head cap), the latter covering the terminal two-thirds of the nucleus and partly derived from the spermatid Golgi apparatus. The acrosomal cap is thin in a human spermatozoön but in other species it is often large and more complex in shape. The acrosome has been shown to contain several enzymes including acid phosphatase, hyaluronidase and a protease (*acrosomase*), which are probably involved in penetration of the oöcyte. The nucleus and acrosome are enveloped in a continuous plasma membrane without intervening cytoplasm (Fawcett & Burgos 1956, Anberg 1957). The chromatin is stabilized by disulphide bonds, as if to protect its genetic content during the spermatozoön's journey (Fawcett 1975). So densely packed is the chromatin that it appears homogeneous even under electron microscopy. It has a strong affinity for basic stains, consisting of about 40% (dry weight) deoxyribonucleic acid and a protein rich in arginine (Daoust & Clermont 1955). It is also resistant to physical stress, e.g. ultrasonication (Henle et al 1938), and to mechanical shear (Mann 1949). Defects in condensation of nuclear material may be visible under light microscopy as relatively clear areas or *nuclear vacuoles*. Attempts to discover structural details in the nucleus in a variety of species, by polarization microscopy, X-ray diffraction and freeze-fracturing techniques, have shown a lamellar structure which cannot yet be equated with chromosomal content. The human Y chromosome has been identified using fluorescence microscopy (see 1.46).

Between the head and the middle part or body of the spermatozoön is a slight constriction, the *neck*, about 0.3 μm long. In its centre (2.13A), close to a shallow recess in the base of the nucleus, is a well-formed centriole, corresponding to the *proximal* centriole of the spermatid from which the spermatozoon differentiated (2.15). The axial filament complex (axoneme) is derived from the *distal* centriole, a funnel-shaped *connecting piece* or *basal body* from which the outer fibrils of the tail extend (vide infra). (The *nuclear recess*, or *implantation fossa*, is the region of attachment of the complex filamentary structure of the tail. It is continuous with the postacrosomal part of the nuclear envelope concerned in fusion with the ovum and its nucleus.) A small amount of cytoplasm exists in the neck, covered by a plasma membrane continuous with that of the head and tail.

The middle part or piece is a long cylinder, about 1 μm in diameter and 7 μm long. It consists of an *axial bundle of microtubules (fibrils)* or *axoneme* (the axial 'filament' of light microscopy), surrounded by a *mitochondrial sheath* in which the mitochondria of the spermatid have become arranged in a helical manner (2.13B), the whole being enveloped by cytoplasm and a plasma membrane, as in the neck. The axoneme consists of a central pair of microtubules within a symmetrical set of nine doublet microtubules, as in a typical cilium (p. 31), and outside this is a second ring of nine coarser fibres, less symmetrical in arrangement and unequal in size. These external fibres are also less regular in cross-sectional profile (vide infra), showing marked interspecific variations. They appear to be non-contractile despite showing a surface striation. Their function is obscure. The mitochondrial helix exhibits 10–14 turns (Reed & Reed 1948) but this sheath is subject to considerable variation in abnormal spermatozoa (Fujita et al 1970). The number of mitochondria seems excessive in some species, including *Homo sapiens*, when related to the energy requirements of the axonema. Their close relation to the external coarse fibres is suggestive but, as noted, these are apparently not contractile. At the caudal end of the middle part of the cell, immediately anterior to the tail, is an electron-dense body, the *annulus* (2.15). The mitochondria of the sheath are much compressed, but it is now certain that they retain their individuality (Fawcett & Ito 1965).

The principal part or tail of a spermatozoön is the motile part of the cell. Being about 40 μm long and 0.5 μm in diameter, it forms the greater part of the spermatozoön. The axial bundle of fibrils and the surrounding array of coarse fibres are continued uninterruptedly from the basal body through the mitochondrial sheath and through the whole length of the tail except for its terminal 5–7 μm, in which the axial bundle alone persists, the coarse fibres ceasing before them. It is only in this terminal *end part* or *piece* that the tail has the typical structure of a flagellum; the coarse fibres are peculiar to mammalian spermatozoa, which also display other specializations. External to the fibres and fibrils, coarse and fine, is a circumferentially orientated dense *fibrous sheath*, whose individual elements branch and re-unite to form a tight reticulum. A small amount of cytoplasm and a plasma membrane complete the major elements in the structure of the tail. The finer details of the structure have been studied intensively in mammals such as the guinea-pig (Fawcett 1965) and, while there is little doubt that the human spermatozoön is highly similar (Pedersen 1969), the following summary of these findings is based perforce on appearances in the guinea-pig.

MOTILITY OF SPERMATOZOA

In cross-section the tail is oval and tapers caudally and its central area is typical of a flagellum or cilium. The surrounding coarse fibres are obovate or petal-shaped and unequal in size, one being consistently the largest. This is given the number 1 and the rest are numbered from this in a clockwise manner. These fibres are separated into two unequal groups by slender *longitudinal* columns in the fibrous sheath which interrupt its circumferential fibres and extend inwards to meet the coarse fibres numbered 3 and 8. This divides the interior of the tail into *major* and *minor compartments*, containing respectively coarse fibres 4, 5, 6 and 7, and 9, 1 and 2. The plane through the two columns also passes through the central pair of the axial bundle of fibrils and can be used as a reference datum for other structural details. For example, the transverse diameter of the head has been considered to lie at right angles to the plane of the columns, but it has now been shown in the guinea-pig that the angle between the two planes is 20–30 degrees less than a right angle (Fawcett 1968). Such details may be instrumental in elucidating the motile activities of the tail. It is now generally accepted that the tail executes undulatory movements in one plane (p. 32), but it has also been suggested that a helical component is superimposed upon this, there being perhaps two separable mechanisms, one involving flat waves travelling along the tail, the other associated with torsional activity (Gray 1958, Bishop 1962, Lindahl & Drevius 1964).

The latter variety of movement has been linked with the unequal size and distribution of the coarse fibrils; it has also been suggested that the central pair of fine fibrils act as axial stiffeners. The asymmetry of the spermatozoan head has also been invoked to explain supposed helical movement. However, it has to be admitted that the full details of the mechanisms of spermatozoal motility are unknown.

Some further details deserve mention. The dense outer fibres have been shown to exhibit an oblique or helical striation in replicas of dried whole mounts of rodent spermatozoa (Phillips & Olson 1975). The central axoneme (2.13), consisting of the usual ciliary pattern of nine double fibrils or 'doublets', has been intensively studied (see p. 31 and Fawcett 1975 for survey of literature). Each fibril is a microtubule, itself constructed of a regular number of protofibrils. Protein bridges connect adjoining doublets at regular intervals and radial links extend centrally to the central doublet of the axoneme (p. 31).

Maturation of spermatozoa (2.14, 15) is a complex process which has received much attention. Spermatozoa show little independent motility while still in the male genital tract, though when removed from the epididymis they may display circular swimming movements or even directive movements if taken from the cauda epididymis near the beginning of the ductus deferens (Blandau & Rumery 1964). From the results of artificial insemination of rabbits with spermatozoa from the caput and cauda of the epididymis, it has been postulated that some form of maturation process takes place in this organ, during which the spermatozoön attains its specific pattern of motility (Gaddum 1968). Apart from these incomplete activities, spermatozoa are largely transported through the genital tract by ciliary action, by fluid currents set up by localized secretion and absorption and by muscular contractions. Associated with the maturation of spermatozoa in the genital tract in some mammals is the extrusion of a small mass of cytoplasm, the *kinoplasmic droplet* or *residual body*, which migrates backwards along the surface of the head and middle part of the spermatozoön before disappearing. It consists of membrane-enclosed cytoplasm containing fine tubules and vesicles (Bloom & Nicander 1961, Guraya 1963). Human spermatozoa have not been shown to undergo any demonstrable structural changes during passage through the epididymis, but there is evidence of an increase in sulphide cross-linking between proteins (Bedford & Calvin 1974). Moreover, restorative surgery after vasectomy indicates that at least part of the human epididymis is essential for motility (Bedford et al 1973).

As soon as they are ejaculated the spermatozoa display their full pattern of motility. The precise factors which trigger off these movements enabling human gametes to travel at a rate of 1.5–3 mm/min are not yet clear; but the other constituents of semen, derived from the epididymis and testis and from the seminal vesicle and prostate, are generally considered to exert an activating influence. The motility varies greatly in different species, disappearing in minutes in some fish but usually persisting in mammals for hours and even days when introduced into the female genital tract. Exact figures for its persistence in the human female are uncertain and are of doubtful value, since it is likely that, as in other mammals, human spermatozoa quickly lose their potency for fertilization, although still motile. They have been recovered in a motile state in human cervical mucus several days after insemination and will survive in this condition for as long as 7 days when implanted into such secretions in vitro. These survival periods may, however, be of little significance, in view of the speed with which spermatozoa reach the infundibulum of the uterine tube and the brevity of their fertilizing power. Spermatozoa have been shown to reach their tubal destination in a manner of minutes after ejaculation in some mammals and experiments on recently excised human uteri and tubes indicated a time of about 70 minutes (Brown 1944). The conclusion must be that factors other than their own motility are responsible for the transport of spermatozoa from their site of deposition in the vaginal fornix to the ovarian end of the uterine tube and there is considerable evidence that contraction of the uterine and tubal musculature is responsible (Bickers 1960).

It is not usually recognized that a spermatozoön must be adaptable to a wide range of environments in its long journey from the seminiferous tubule to the uterine tube, encountering major changes in the electrolyte and non-electrolyte constituents in the fluids with which it is successively surrounded. Nevertheless, a collectively vast amount of observation and experiment has been recorded in connection with the effects of the multitude of factors, both physical and chemical, in the natural media involved regarding the behaviour of these cells and particularly their motility and

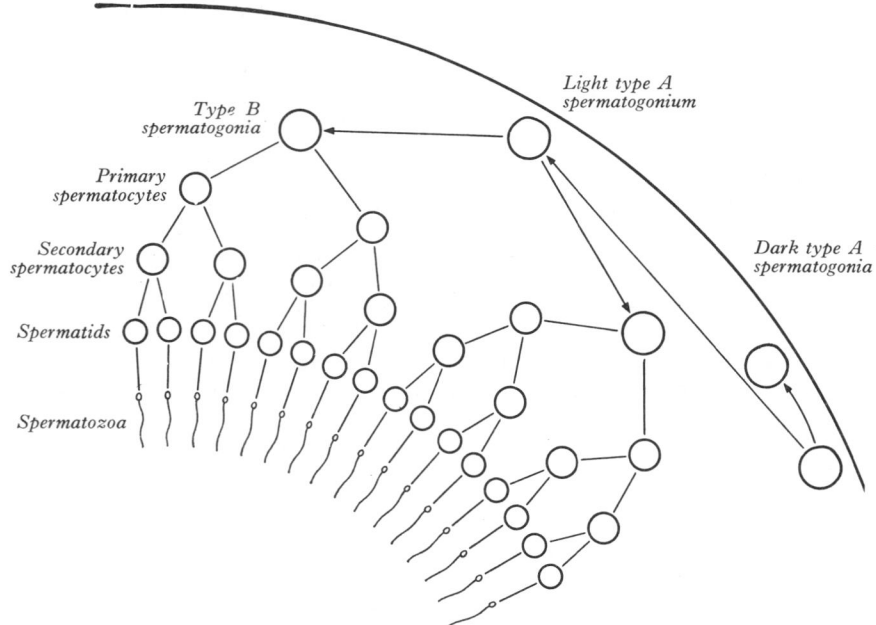

Light type A spermatogonium

Type B spermatogonia

Primary spermatocytes

Secondary spermatocytes

Spermatids

Spermatozoa

Dark type A spermatogonia

2.14 Diagram showing the stages in the maturation of the spermatozoön. The division of the primary spermatocyte is heterotypical; the remaining divisions are homotypical. Some have described a further division of spermatids in the human testis, but this is unconfirmed—see text for further details and compare with illustration 8.194.

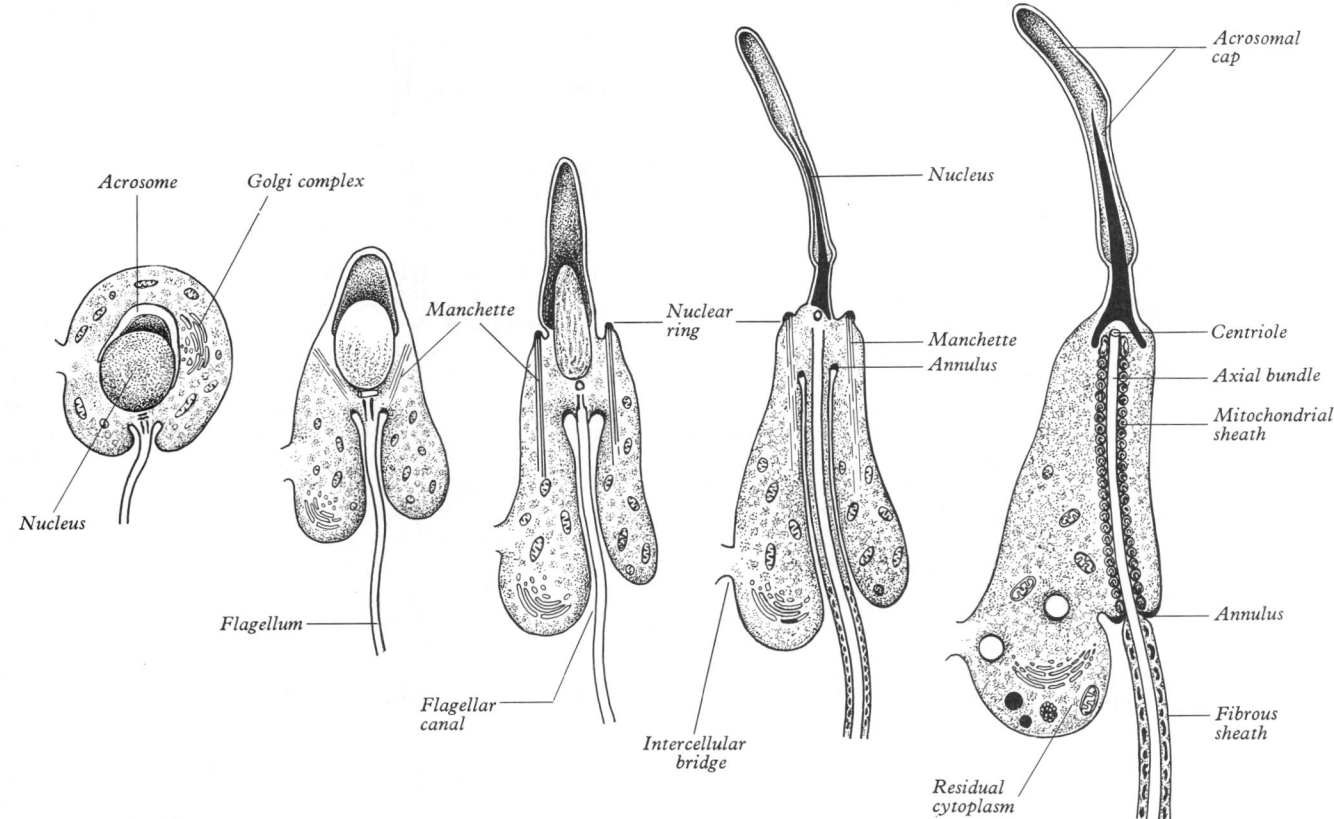

2.15 Differentiation of the spermatid (guinea pig), from left to right. Note nuclear condensation and elongation, appearance of the *manchette* and fibrous sheath, caudal movement of the annulus, and development of the mitochondrial sheath. (Modified after Fawcett in Bloom & Fawcett 1975)

fertility (Nelson 1967, Mann 1967); e.g. the effects of respiratory gas tensions, reaction, various ions, antibodies, vitamins, hormones, inhibitory substances, temperature, different forms of radiation and other factors have been studied in remarkable detail, for which monographs and original papers must be consulted. The effects of low temperatures in preserving spermatozoa and perhaps prolonging their vitality have attracted much research in connection with artificial insemination, both in stock-breeding and in infertile human marriage. Mammalian semen, including that of human beings (Parkes 1952), can be stored at temperatures of about − 70°C for weeks and even months, the motility and fertility of its suspended spermatozoa reappearing when the suspension is unfrozen. However, storage of human semen presents difficulties (Polge 1957).

Seminal plasma, the fluid component of *seminal fluid* or *semen*, contains a remarkable array of substances, including muco-proteins, a dozen or more identified proteolytic enzymes, the bases spermine, glycerylphosphorylcholine and ergothioneine, a group of organic acids called prostaglandins (which have pharmaco-dynamic actions on the uterus and smooth muscle in general), acids such as citric, ascorbic, uric, lactic and pyruvic, and the sugars sorbitol, inositol and fructose. The fructose added to the fluid by the secretion of the seminal vesicle is an essential substrate in the anaerobic glycolysis by which spermatozoa survive the low oxygen tensions existing in semen itself and in the female genital tract. Prostaglandins are now believed to play a role, perhaps by modulation of neurotransmitter release, in the contraction/relaxation activity of the non-striated muscle in the testicular capsule and interlobular septa adjacent to the seminiferous tubules (as suggested by von Euler 1936). Consult Ellis & Hargrove (1977) for literature.

CAPACITATION

After ejaculation into the female, the spermatozoa undergo the final step in their maturation, a process known as *capacitation*. It has been shown that spermatozoa are not able to fertilize ova until they have been within the genital tract of the female for a period of time, usually of hours but varying with the species (Austin 1951, Chang 1951, Austin & Walton 1960). The mechanism of capacitation, whereby the spermatozoön is activated to enter and fertilize the ovum is still uncertain. A confusing array of findings with regard to the interactions of the two gametes immediately prior to this event have been described, unfortunately in widely different vertebrates and invertebrates. It is probable that hyaluronidase hastens the separation of corona radiata cells from the ovum, and thus facilitates the spermatozoön's approach to the zona pellucida. The origin of hyaluronidase from the acrosomal cap is associated with subsequent loss of the cap (Leuchtenberger & Schrader 1950), at least in part (Austin & Bishop 1958). 'Capacitated' spermatozoa observed in the zona pellucida or perivitelline space have invariably lost most of their acrosomal material (Leuchtenberger & Schrader 1950, Pikó & Tyler 1964) and it is clear that capacitation is some process of activation which precedes penetration. Antigenic 'coating' substances on the surface of mammalian spermatozoa, including those of man (Weil 1965), have been recorded and it is possible that an immunological reaction may be involved. Interaction between a *fertilizin*, derived from the ovum or elsewhere in the female genital tract, and a spermatozoan *anti-fertilizin* has been associated with capacitation, but the interrelationship between the various events is still sub judice, as comprehensive reviews show (Metz & Monroy 1969, Chang & Hunter 1975). Capacitation may be regarded as the terminal event of maturation of the spermatozoön, prior to actual fertilization, for which it is preparation.

Spermatogenesis

This is the complex series of changes by which spermatogonia are transformed into spermatozoa, similar in some general features —particularly in reduction division and genetic recombination —to the evolution of ova from oögonia, but differing in the more profound morphological metamorphosis involved. Spermatogenesis may, for convenience, be divided into three phases. During the first, **spermatocytosis**, spermatogonia proliferate by

121

mitotic division to replace themselves and to produce primary spermatocytes. In the second phase, **meiosis**, two successive maturation divisions, the first a true reduction division as in the case of the oöcyte, produce *secondary spermatocytes* and then *spermatids*, all with the haploid number of chromosomes. In the third phase, **spermiogenesis** or **spermateliosis**, the spermatids become spermatozoa; it is during this period that the greatest visible transformation of structure occurs.

SPERMATOCYTOSIS

During embryonic, fetal and perhaps also early post-natal life, *primordial germ cells* (p. 254) in the tubules of the testes divide mitotically to produce spermatogonia (Witschi 1948, 1951, Mintz 1960), from which, at and subsequent to puberty, the development of spermatocytes and spermatozoa commences. It is the cyclic divisions of these cells, the details of which have attracted much attention in recent years, that form the starting place for production of the huge numbers of spermatozoa discharged into the seminal plasma to form the seminal fluid. The series of changes involved do not occur in a synchronous manner in all seminiferous tubules at the same time, although they do in considerable parts of an individual tube, with variations in different mammalian species, including mankind. As the cycle of change from spermatogonia to spermatids proceeds at any particular locus in a tubule, a succession of varying cell associations can be observed and measured; the process is termed the *cycle of the seminiferous epithelium* (Clermont & Leblond 1955) (p. 1427).

In man, three types of spermatogonia can be distinguished and are termed the *dark type A, light type A*, and the *type B* (Clermont 1963). Spermatogonia are large rounded cells, the three types showing little difference in size or in their cytoplasm, but the A series, light and dark, are distinguishable by their nucleoli which are eccentric and attached to the internal aspect of the nuclear membrane. The type B spermatogonia have a more constantly spherical nucleus, in which the nucleolus is central in position. The dark type A is distinguished from the pale type A by its dark nucleoplasm and a large pale-staining nuclear vacuole. The dark type A is now considered, largely on morphological grounds, to be the progenitor or stem spermatogonium (Clermont 1963). Such cells, peripherally situated in the tubule and often in pairs, divide mitotically at the beginning of a seminiferous cycle, some to produce two further dark type A spermatogonia, thus replenishing the complement of stem cells, others into two light type A spermatogonia. Mitotic division of a light type A spermatogonium furnishes two type B spermatogonia. On theoretical grounds, it is probable that in man a larger series of spermatogonial divisions may occur, so that the ultimate spermatocyte progeny of a stem cell may in fact be more numerous than is here indicated (Clermont 1966). Each type B spermatogonium then divides again mitotically into two *resting primary spermatocytes* or preleptotene spermatocytes. The dark type A cells remain arranged along the basement membrane of the tubule, whereas the pale type A, type B spermatocytes and the spermatids derived from them lie closer to the lumen, into which the free end-product, spermatozoa, will be discharged (8.191). These events constitute spermatocytosis, which is now followed by meiosis.

MEIOSIS OF SPERMATOCYTES

The primary spermatocytes soon enter the prophase of the *first maturation (reduction) division*, which is prolonged over several days through the successive stages of leptotene, zygotene, pachytene, diplotene and diakinesis (p. 44) (Clermont 1963, 1966). As the nuclear membrane now disappears in metaphase, the bivalent chromosomes are arranged on the equatorial plate, separating into two groups and moving to opposite poles in anaphase, followed in the usual manner by reformation of the nuclear membranes in telophase and division of the cell. These three phases occur much more rapidly than prophase and during the whole process there is a considerable increase of nuclear and cytoplasmic material, bringing the primary spermatocyte back to a size comparable with that of the stem spermatogonium. The two

secondary spermatocytes thus formed contain, of course, the haploid number of chromosomes, this *first* maturation division being the one which is strictly speaking *meiotic*. After a brief interphase each secondary spermatocyte now undergoes a *second maturation division*, which is by mitosis. The two resultant cells are spermatids and with their formation the phase of meiosis may be considered to end, being followed by their maturation into spermatozoa (spermiogenesis). Theoretically each primary spermatocyte may be expected to produce four spermatids but in mankind the yield is less than this, presumably because some spermatocytes degenerate during maturation.

Criticism of the traditional view that spermatids do not divide has been expressed (Roosen-Runge 1952) and tentative corroboration of this came from electron microscope and other studies (Fawcett et al 1959, Fawcett 1961b). These interpretations have subsequently been subjected to critical rescrutiny. Electron microscopy has also shown that the division of the cell body (cytokinesis) in spermatocytes may be delayed, so that fine cytoplasmic bridges remain interconnecting such cells even beyond the stage of the next nuclear division (Fawcett et al 1959). These bridges, which may remain in the case of spermatids until a late phase in their transformation into spermatozoa, are short, devoid of spindle fibres or other remnants and are enclosed in annular thickenings of the plasma membrane. They probably permit interchange of organelles, may be involved in synchronization of development and may contribute to the mechanical stability of the spermatid-Sertoli cell complexes. Such cytoplasmic interconnection may also help to explain the formation of multinucleated masses when the seminiferous epithelium is injured, as in making teased preparations. Except where connected together by bridges the developing spermatids are very closely associated with Sertoli cells, whose processes are insinuated between them.

During the differentiation of some rodent spermatids a fusiform conglomeration of microtubules, the 'spindle-shaped body', appears between the annulus and fibrous sheath. The same structure has been observed in human spermatids; its functional significance is uncertain but an association with the development of the fibrous sheath has been suggested (Pedersen 1969, Wartenburg & Holstein 1975).

SPERMIOGENESIS

During *spermiogenesis* (spermateliosis), spermatids go through a complex series of changes to become spermatozoa (2.15) and this metamorphosis has been studied in particular by light microscopy, using preparations stained by the periodic-acid Schiff technique (PAS) (Leblond & Clermont 1952, Clermont 1963). Electron microscopy has confirmed these observations (Fawcett & Burgos 1956). In the newly formed spermatid the Golgi apparatus (idiosome-Golgi complex) is large but otherwise typical, consisting of flattened membrane-enclosed vesicles, usually stacked in a parallel array, together with rounded minute vesicles which are possibly nipped off from the flattened variety, the whole complex being juxtanuclear. A few electron-dense homogeneous *paracrosomal granules*, which are intensely PAS-positive, develop in separate Golgi complex vesicles and the latter coalesce into a single large *acrosomal vesicle*, the separate granules fusing into a single spherical *acrosomal granule*. This vesicle, with its granule attached to its juxtanuclear wall, becomes adherent to the nuclear membrane over an area which will be anterior or 'leading' in the maturing spermatozoön. The granule flattens, but its central part bulges slightly into a shallow depression in the nucleus, which becomes progressively more ovoid (2.15). By successive absorption of further vesicles from the Golgi complex, the material in the acrosomal vesicle increases and the vesicle expands as a bilaminar cap over the anterior two-thirds of the nucleus. Coincident with these changes, the spermatid elongates and the Golgi complex and associated cytoplasm migrate to the posterior part of the cell, bringing the external wall of the acrosomal vesicle into contact with the plasma membrane at the anterior aspect of the cell. The acrosomal granule now spreads out between the layers of the vesicle until it is uniformly distributed and no longer a localized structure. When the centrioles begin to separate (vide infra), microtubules develop forming an inverted conical array, *the manchette*, perinuclear in position and expanding from the region of the acrosomal cap; its precise significance is not yet explained.

In the early spermatid the nucleus is of relatively low density, containing finely dispersed granules which aggregate into larger and denser masses as development proceeds. These finally agglomerate into a homogeneously dense mass, usually containing one or more regions of low electron-density and variable in size, position and shape—the *head vacuoles* (Fawcett & Burgos 1956). Correlated biochemical and ultrastructural studies indicate a considerable variation in the chromatin content of spermatozoa, this heterogeneity being more marked in mankind than other primates or rodents; it probably indicates a lower fertilizing power (Bedford et al 1973).

The two centrioles are near the posterior aspect of the nucleus from an early stage. One remains unmodified; the other becomes modified into the basal body of the spermatozoon (see p. 119). The anulus arises close to the latter (the distal centriole) but its origin from it is doubtful. The axial fibrils begin to develop from the basal body, extending 'caudally' into the cytoplasm of the cell as it becomes progressively more elongated. Only the proximal part of the bundle of fibrils remains surrounded by cytoplasm, to form the definitive middle part of the spermatozoön. In this, the mitochondria of the spermatid assemble to form the helical sheath. The detailed development of the fibrous sheath of the tail part is unknown. These changes complete what might be called the period of 'organogenesis' of the spermatid, whose further development into a spermatozoön is largely concerned with enlargement of the tail.

During the final maturation of spermatids into individual spermatozoa some of the cytoplasm is detached as a *residual body*. This contains some mitochondria, Golgi membranes and vesicles, RNA particles, lipid granules but of course no nucleus. Residual bodies are prominent when spermatozoa are being released into their tubule. They are engulfed by Sertoli cells, which accounts for the increase in lipid content of these cells at this period (Lacy 1960).

As already stated, there is a close relation between developing spermatids and Sertoli sustentacular cells. The spermatogonia are external or basal to the Sertoli cells in the tubule and the spermatocytes which develop from the former are embraced by Sertoli processes; the spermatids are even more deeply embedded in the supportive cells. (For further details see p. 1425.) These associations have been regarded as symbiotic, a single sustentacular cell being grouped with several spermatids. The Sertoli cells are phagocytic and absorb not only residual bodies but also degenerating germ cells. They have been attributed a metabolic role and may form, or at least transmit, hormones involved in the maturation of germ cells. Until released into the seminiferous tubule, spermatozoa are very firmly held by the Sertoli cells. Their release is sometimes termed *spermiation* and is followed by rapid translation of the spermatozoa to the epididymis.

Spermatogenesis is an orderly and complex sequence of events, with characteristic time constants and cell associations for each mammalian species. The details of these in the human testis will be discussed with that organ (p. 1427).

Fertilization—Union of the Gametes

It has become customary to speak of an ovum as being fertilized by a spermatozoön and, in view of the apparent passivity of the former and the extreme specialization of the latter for motility, it is tempting to consider the ovum as being merely stimulated to further development, in which the part played by the sperm nucleus may be consequently overlooked. It is of interest to note that in some of the earliest organisms to propagate by sexual reproduction, such as primitive algae, the gametes are all alike, except presumably in their genetic content, whether chromosomal or cytoplasmic. The profound differences which have nevertheless evolved in the gametes of the great majority of plants and animals—vertebrate and invertebrate—appear to depend on a conflict between adaptation for carriage of nutriment and improvement in motility. The effectiveness of *syngamy*, the

bringing together of two gametes to produce a new individual, has been furthered by the development of this marked dimorphism between them. This dimorphism has, in turn, entailed the evolution of equally profound differences between the individuals producing the two kinds of gametes, males and females, in regard to the organs concerned in bringing together these dissimilar gametes and ensuring the development of their fused product, the *zygote*, until able to undertake a separate existence. Strictly speaking, therefore, the expression 'fertilized ovum', insofar as it appears to assign a merely stimulatory role to the spermatozoön, is misleading and hence undesirable. However, it should be noted that almost all the cytoplasm of the zygote is derived from the ovum and, as indicated elsewhere (p. 98 et seq), its organization has a profound morphogenetic significance.

The occurrence of parthenogenesis in a very small fraction of animals, mainly invertebrates, does not invalidate the fact that *fusion* of two gametes, from separate parents, is the central feature of reproduction in most animals and plants (Tyler 1967b). There is also abundant evidence that the critical element of this fusion is the conjunction in one cell of two haploid complements of the particular species' chromosomes, these being sufficiently identical to entail the development of a viable new unit in the species. Differences involving a number of genes, arising during the phenomena of crossover and segregation of the chromosomal material or (very rarely) by sporadic mutation, are thus also able to express themselves in individual variation, with all its implications in evolutionary reactions between the species and its environment (p. 96 et seq).

Nevertheless, it would be equally misleading to regard the union of the ovum and spermatozoön as little more than restoration of the diploid number of chromosomes, because other events, of equal significance in the further development of the zygote, also occur. The ovum immediately resumes and completes its second maturation division; without fertilization this does not occur and the ovum begins to degenerate within 24 hours (Allen et al 1930, Hamilton 1949). A second polar cell is thus produced and, if the first has divided as it sometimes does in human beings (Shettles 1955), three may be present in the circumvitelline space. In many mammals, and probably in mankind, what is apparently a small excess of nutritive yolk or deutoplasm is extruded from the ovum into the same space after fertilization (*deutoplasmolysis*). This event is followed at once by segmentation, which is also dependent upon fertilization for its initiation.

(For overall reviews on mammalian reproduction see Austin & Short 1984.)

ACTIVATION AND SEX DETERMINATION

The phenomena which ensue upon the entry of the spermatozoön can thus be conveniently divided into those described in the preceding paragraph, which together constitute *activation*, and those concerned in the intermingling of the separate hereditary influences of the two parents and associated with the fusion of the nuclei of their gametes, a process described as *amphimixis*. The latter really falls within the scope of genetics, but a particular aspect—the determination of the sex of the zygote—may be mentioned here, because the actual determination is in a sense effected as soon as a spermatozoön has entered the ovum. In approximately equal numbers, spermatozoa of many animal groups, including mammals, contain either an X or a Y chromosome, whereas mature ova contain only an X chromosome. If both gametes contain an X chromosome the resultant individual is female; but when the spermatozoön contains a Y chromosome, the cells of the new individual will each contain an X and a Y chromosome, which is characteristic of the male. This may not, however, be the only factor involved in the determination of sex (p. 260).

UNION OF THE GAMETES

In mammals, generally only one spermatozoön pierces the vitelline membrane to enter the ovum, although numbers usually penetrate the corona radiata and even the zona pellucida. Enzymes of acrosomal origin cause dispersal of the cells of the corona, thus opening the route to the zona. Though some form of

123

chemotaxis, as exists in the syngamy of plant gametes, has been claimed for some animals, no clear evidence for this has been recorded (Tyler & Bishop 1963). The mechanism of union of the ovum and spermatozoön has been studied extensively in all its stages. The primary reaction between the gametes appears to be of an immunological nature, which would account for the exclusive specificity of fertilization. Metchnikoff (1900) and contemporaries first reported the antigenicity of spermatozoa and a species recognition system, dependent upon an agent in the ovum, a 'fertilizin', was identified surprisingly early (Lillie 1913). The sequence of the reactions of this with spermatozoan anti-fertilizins (associated with the plasma membrane) have attracted much attention (Metz & Monroy 1967, Bedford 1977). Such antibody–antigen interaction appears not only to be essential to contact between the gametes but also to initiate the subsequent entry of the spermatozoön by the lytic action of enzymes derived from the acrosome (see p. 119, Austin & Bishop 1958, Srivastava et al 1965). This *zona-lysin* enables the spermatozoön to pass through the zona pellucida into the circumvitelline space and to come into direct contact with the oöcyte. Penetration is followed immediately by permeability changes in the zona which effectively block the subsequent entry of other spermatozoa (Austin & Braden 1956). The *zona reaction*, as this phenomenon is called, is however not a complete block to polyspermy; two or more spermatozoa may pass *simultaneously* through the zona, a not uncommon event in the case of some rodents, seemingly also a possibility in man. The result is one of the types of *triploid* zygotes that may arise. Further, there is evidence of a second line of defence, the *vitelline block*, associated with the plasma membrane of the ovum. Both blocking mechanisms may depend upon discharge of the contents of *cortical granules*—electron-dense particles subjacent to the plasma membrane—into the circumvitelline space (Szollosi & Ris 1961, Szollosi 1967). However, recent evidence derived from in vitro fertilization of *human* ova suggests that the zona pellucida is, under these conditions, a complete block to polyspermy (Soupart & Strong 1975). Electron microscope studies

suggest that the actual entry of the spermatozoön into the ovum is effected by fusion of their plasma membranes, that of the former being left on the surface of the ovum (Szollosi & Ris 1961, Szollosi 1967). Once within the ovum, the head and neck of the spermatozoön become detached from the middle part and tail which soon disappear, their significance in fertilization, if any, and their ultimate disposal being obscure. For a review of the uncertainties and problems in regard to the mechanisms of fertilization consult Bedford (1977) and for ultrastructural details see Austin (1968). The occurrence of auto- and iso-immune anti-spermatozoal antibodies has been exhaustively surveyed.

The second maturation division of the ovum is now completed, with the extrusion of a second polar cell into the circumvitelline space, which has become opened up by shrinkage of the oöplasm. Failure of extrusion of the second polar cell is another mechanism of formation of a *triploid* zygote. Meanwhile, the head of the spermatozoön swells to become the *male pronucleus*, being at this juncture morphologically indistinguishable from the *female pronucleus* (2.16). Both nuclei move to a central position in the ovum and each duplicates its DNA content. Meanwhile, two centrioles appear, probably derived from the anterior centriole of the spermatozoön. The chromosomes become organized from each pronucleus and arranged on the spindle between the centrioles and, since 23 are contributed from each, the typical diploid human number is now re-established. The degree of fusion of the two pronuclei before the appearance of the chromosomes varies in different species but usually the chromosome groups remain separate during prophase. They now split longitudinally and the resultant halves segregate as in an ordinary mitosis. The two sets, each consisting of the diploid number of chromosomes with the normal amount of DNA, retreat to opposite poles of the cell, on whose surface a deepening groove develops as the zygote enters the next phase in its development, *cleavage* or *segmentation* (p. 100 and below).

The chronological elements in the events of the human fertilization are still only inexactly known.

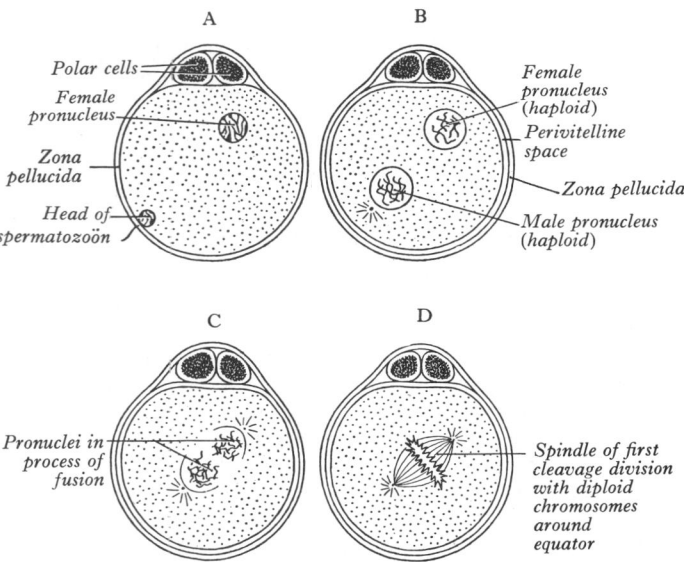

2.16 The process of fertilization in a mammalian ovum. Diagrammatic (After Sobotta). The female pronucleus and second polar cell, shown in A, are only formed after fertilization.

PARTHENOGENESIS

It is generally agreed that a mature ovum contains within itself all the material potentiality to form a new being. Apart from the *natural parthenogenesis* which occurs in some invertebrates, this potentiality of the ovum to develop further without fertilization can be released experimentally by a variety of mechanical, chemical and physico-chemical means, such as pricking, exposure to alteration of tonicity and reaction in the surrounding medium, treatment with various kinds of radiation, etc (Beatty 1957, 1967). Such *artificial parthenogenesis* has been studied in a wide range of invertebrate and vertebrate animals, including mammals. Such offspring rarely survive beyond the embryonic stage but viable young have occasionally been obtained, e.g. in the case of rabbits. The particular interest of these parthenogenetic phenomena in vertebrates is that they illustrate the fact that, apart from its normal necessary chromosomal contribution, the spermatozoön plays little part in embryogenesis. However, there is some evidence that the plane of bilateral symmetry is determined in some animals by the direction and place of entry of the spermatozoön. Nevertheless, both in this determination and the early polarization of the zygote, there must be a considerable flexibility, because the blastomeres resulting from its early segmentation are 'totipotent', in the sense that they may separate to form complete embryos, as in the formation of uniovular twins (but see p. 158).

EARLY DEVELOPMENT OF THE HUMAN EMBRYO

CLEAVAGE

As we have seen (p. 100), cleavage is the process whereby a *unicellular* fertilized ovum with an exceptionally large ratio of cytoplasm to nucleus is transformed into a *multicellular* mass of *blastomeres*, each of which approximates to the ratio found in

general somatic cells. For example, in the sea urchin, volumetric ratios are 550:1 at the onset of cleavage, reducing to 6:1 in individual blastomeres at its termination (Brachet 1960, 1965, 1967, et seq.). Comparative studies show that cleavage is essentially a series of mitotic divisions with little overall cytoplasmic growth but with a dramatic synthesis of nuclear DNA; also cytoplasmic factors important in subsequent development are differentially parcelled in the cytoplasm of the resultant blastomeres (p. 109).

As each generation divides, the nuclear population is, of course, doubled, preceded by replication of chromosomal DNA as in any other mitotic division (p. 42). The precursors for this replication are dispersed in the oöplasm of the zygote which is rich in various RNAs (p. 39); *cytoplasmic* DNA is contained in both mitochondria and yolk platelets, whilst the oöplasm also contains abundant low molecular weight nucleic acid precursors. The relative contributions from these different sources to the processes of replication, transcription and translation remains uncertain. Similarly, the interesting question of whether cytoplasmic nucleic acids are wholly broken into unitary molecules before incorporation into nuclear DNA, or are utilized in more substantial segments bearing coded genetic information, remains unresolved. (For earlier comparative reviews on the synthesis and transformations of nucleic acids during embryogenesis, consult Brachet & Quertier 1963, Brown 1966, Tyler 1967a, Gurdon 1968, Brachet 1969.)

Synthesis of *new* ribosomal RNA is low during cleavage; but some protein synthesis continues, particularly that related to the formation of microtubular spindles of mitoses, the production of increasing areas of cell membrane and additional DNA polymerases. The foregoing analyses, until recently, applied mainly to observations on non-mammalia and, for technical reasons, mammalian data were less plentiful and lacked the same precision (see Jones-Seaton 1950, Dalcq 1954, Austin 1961, 1965, 1968). However, detailed analyses on mammalian oöcytes and during early embryogenesis are now accumulating (Herbert & Graham 1974, Church & Schultz 1974).

Relatively few specimens of the earliest stages of human development occurring in the normal environment of the female genital tract were recovered and studied (Hertig et al 1954, 1956) —including one 2-cell stage, a few, possibly abnormal, up to the 12-cell stage, and one presumed normal 12-cell stage (vide infra). More recently a 7-cell stage has been recovered and analysed (Avendano et al 1975).

Successful attempts at the in vitro fertilization of mature mammalian ova and their subsequent culture up to the blastocyst stage have been made since the early 1930s (Lewis & Hartman 1933, Mulnard 1964, 1965). Early development was recorded using time-lapse cinematography, and over the last 25 years, similar techniques have been applied to human ova (Shettles 1953, 1955, 1958). Techniques of *in vitro fertilization* (IVF) of human ova have been greatly improved (Edwards et al 1965, 1969). Preovulatory oöcytes, recovered by laparoscopy in women under the control of gonadotrophins (Steptoe & Edwards 1970), have often been fertilized and cultured to the 8- and 16-cell stages and beyond (2.20). Reintroduction of cultured morulae into the uterine cavities of women with pathological uterine tubes and in whom pregnancy would otherwise be impossible (Steptoe & Edwards 1976), although initially unsuccessful, resulted in the first full-term birth late in 1978. This clinical approach is now being pursued in a number of centres with limited (< 20%) success. (Nevertheless, such techniques often result in multiple potentially viable zygotes and their fate, whether alternative clinical action, scientific observation, experimentation, or rejection is still provoking much ethical, philosophical and theological debate, despite the obvious clinical advantages. Recently (Asch et al 1984, 1985, Lim-Howe et al 1987) success with a new complementary/alternative procedure to IVF has provoked widespread interest; namely *gamete intrafallopian transfer* (GIFT). Partner selection is critical; female—under 38 years, confirmed tubal patency, regular ovarian cycles; male— sperm > 20 million per ml, > 40% motile, < 40% abnormal; both —post-coital test +ve, no antisperm antibodies in either. After hormonally induced superovulation, oöcytes are recovered and suitably prepared; similarly a sperm sample. Two oöcytes followed by a measured volume and concentration of sperm are then passed via a cannula into the ampullary lumen of each tube. Fertilization occurs (success rate > 40%) in this natural environment. (Mr J Erian, personal communication)

Fertilization normally occurs in the ampullary uterine tube, probably within 24 hours of ovulation. After completion of the second maturation division the zygote divides into two blastomeres of approximately equal size (2.17). (Human cleavage may

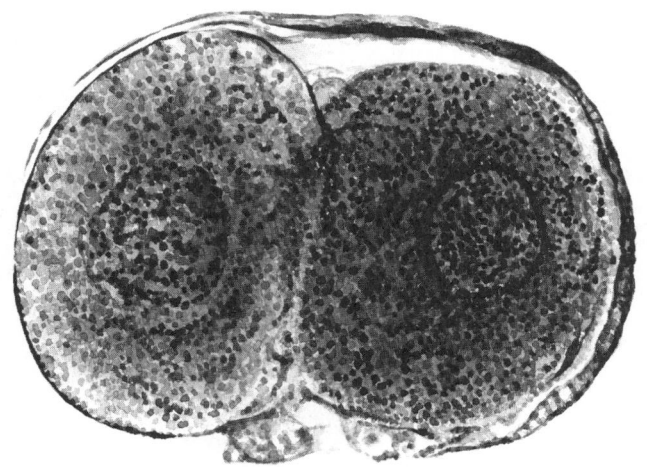

2.17 Early human cleavage. Two-blastomere stage recovered from the uterine tube. A polar cell is seen at each end of the cleavage plane. Magnification × c.700. (Hertig et al 1954.)

be designated *total, equal* and *indeterminate*, see p. 100). The human 2-cell stage recorded was recovered from the middle of the uterine tube 60 hours after a fertile mating (2.17), the menstrual pattern, however, indicating an age of 36 hours. By repeated division and subdivision of the blastomeres a mulberry-shaped mass of cells, hence termed the *morula*, is formed (2.20D–F, 21). Intense activity of their cytoplasm is evidenced by the constant streaming of contained granules, and the individual blastomeres are not motionless, but restlessly alter their positions relative to one another and reorientate themselves within the zona pellucida. The human morula is believed to enter the uterus at about the 8- to 12-cell stage and about 72 hours after fertilization. The four human morulae of 5–12 blastomeres which are recorded, however (all believed to have been abnormal), were recovered from the uterus on the fourth day after presumed ovulation and fertilization. The 12-cell stage, assumed to be normal, consisted of 11 small peripheral cells which surrounded a larger centrally placed one, indicating that during the earlier cleavage divisions two different groups of cells emerge and the earliest visible signs of such differentiation has now been claimed even at the 7-blastomere stage (Avendano et al 1975). Of these diverging cell lines one will form the embryo and the other will form the nourishing and protective membranes by which it is surrounded. The former is composed of centrally sited larger cells, few in number, which divide more slowly and retain to a high degree the developmental potential of the fertilized ovum, termed the *formative cells*; the latter consists of smaller and more numerous superficially placed

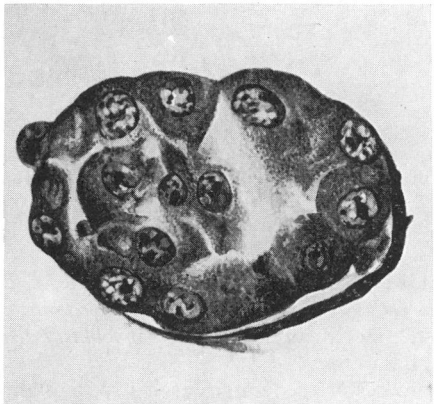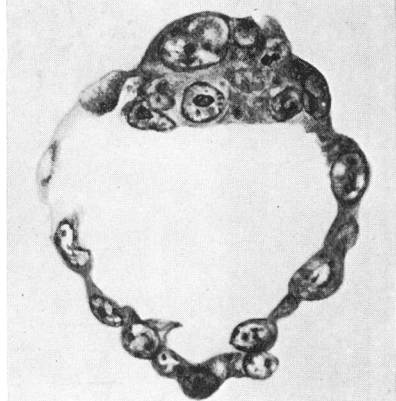

2.18 Section of a 58-cell human blastocyst recovered from the uterine cavity showing the zona pellucida, trophoblast and inner cell mass. Magnification × c.510. (A T Hertig et al, 1954.)

2.19 Section of a 107-cell human blastocyst recovered from the uterine cavity. The mural and polar trophoblastic cells and the inner cell mass can be distinguished. Magnification × c. 550. (A T Hertig et al, 1954.)

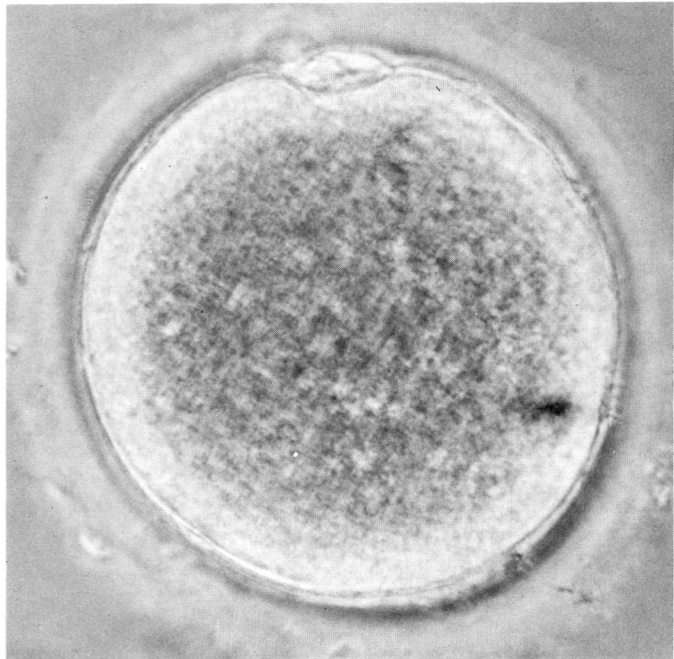

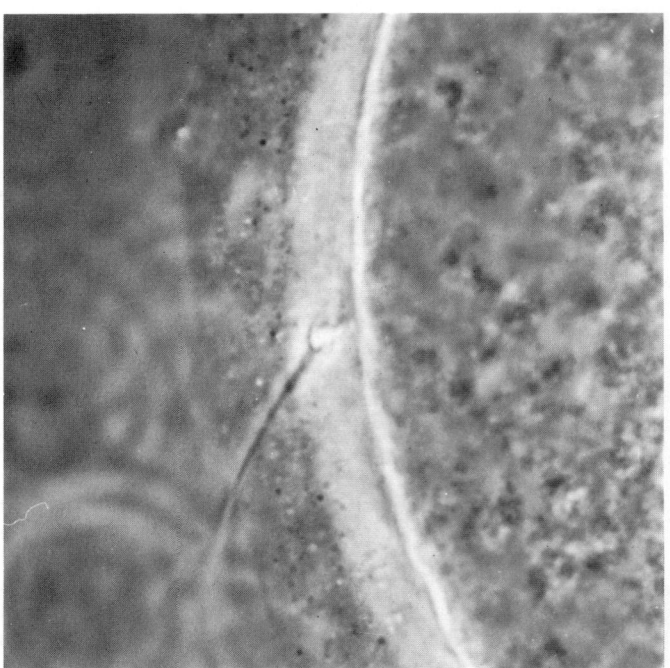

A An unfertilized oöcyte surrounded by the zona pellucida; the first polar cell can be seen.

B High magnification of the surface of an unfertilized living human oöcyte; the head of a spermatozoön is visible in the perivitelline space and its tail is beating outside the zona pellucida.

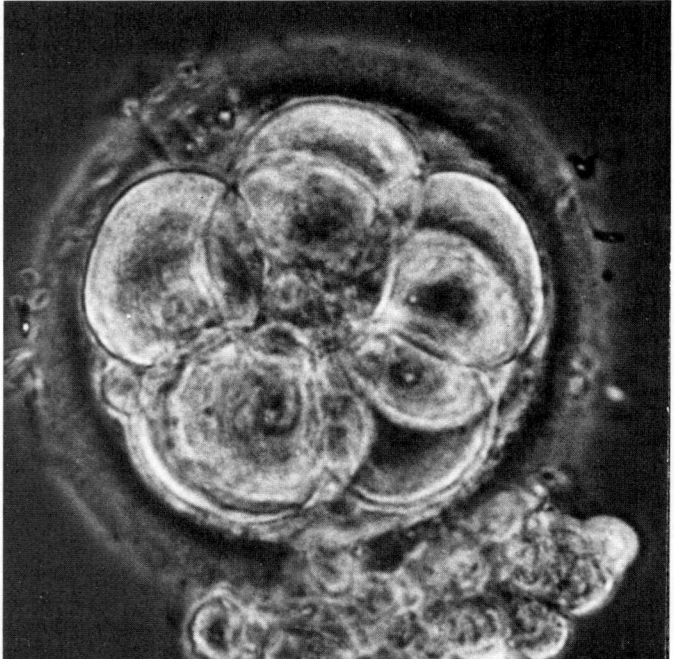

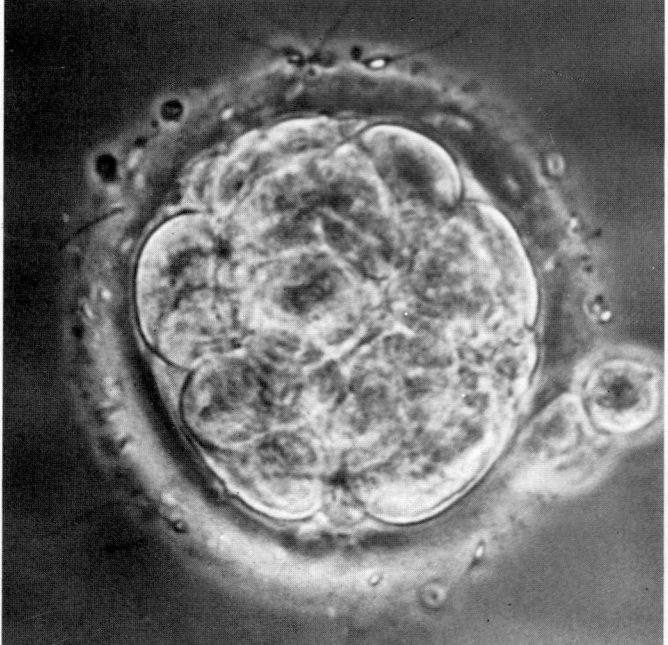

E 8-celled stage.

F 16-celled stage.

2.20A–H These are all photographs of living human specimens in tissue culture.

All these specimens were kindly provided by Dr R G Edwards of the

Anatomy Department, Cambridge University, B,E,F and G are reproduced by courtesy of *Nature* and C by courtesy of the *Journal of Reproduction and Fertility*.

cells, which divide more rapidly as they differentiate into progenitor *trophoblastic cells*. (For possible *causal mechanisms* leading to emergence of distinctive *formative* and *trophoblastic cell lines*, see Herbert & Graham 1974.)

If the first two blastomeres are separated, each is potentially capable of continuing development into a complete embryo—it has been claimed that 25–30% of instances of monozygotic twinning in man follow separation at this stage. It has been demonstrated experimentally in the mouse (Tarkowski 1961, 1963, Mintz 1962, 1964, 1965, 1970) that if one of the first two blastomeres is killed, the remaining one can continue to develop into a small but complete embryo. However, blastomeres isolated at the 4- or 8-cell stages often fail to continue normal development and are limited to the formation of simple *trophoblastic vesicles* which contain no inner cell mass and are therefore devoid of formative cells (Tarkowski 1965). Conversely, from the 2-cell to the morula stage, two cleaving zygotes may be fused, development proceeding to the formation of a single 'giant' embryo (2.4).

THE BLASTOCYST

Before the zona pellucida disappears, fluid, either secreted by the trophoblastic cells or derived from the uterine lumen, begins to accumulate within the morula. (Cell death of a number of deeply

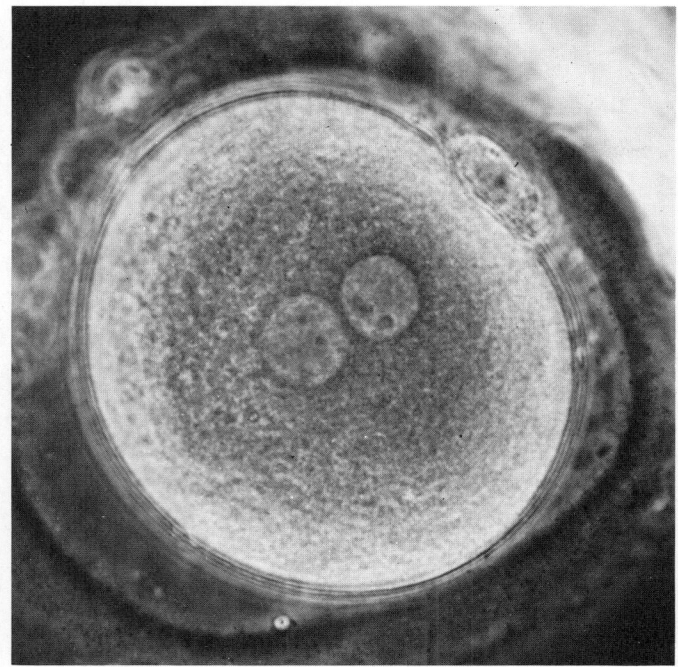

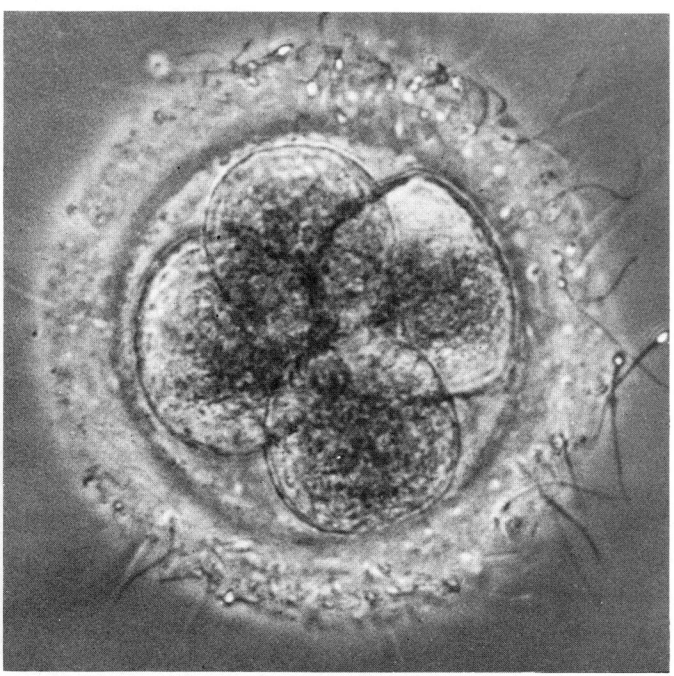

c After penetration by the spermatozoön, the male and female pronuclei can be seen near the centre of the 'ovum', whilst two polar cells lie beneath the zona pellucida.

D Human cleavage—4-celled stage.

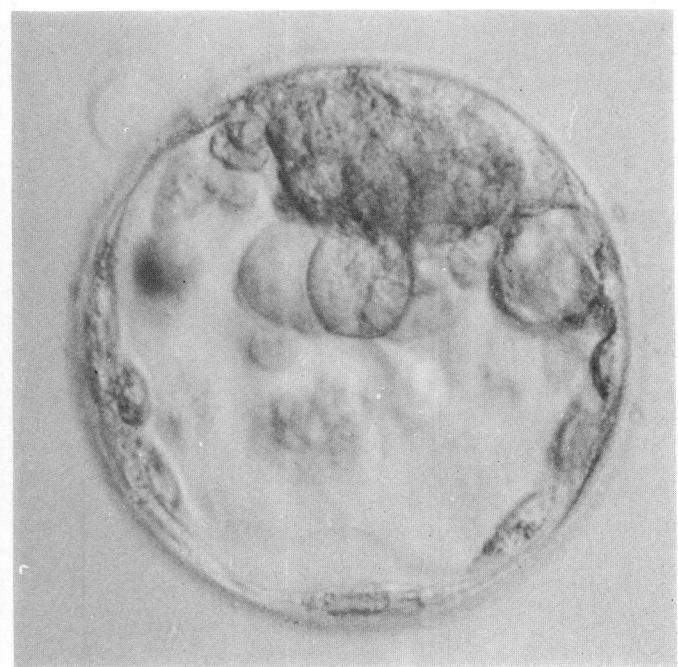

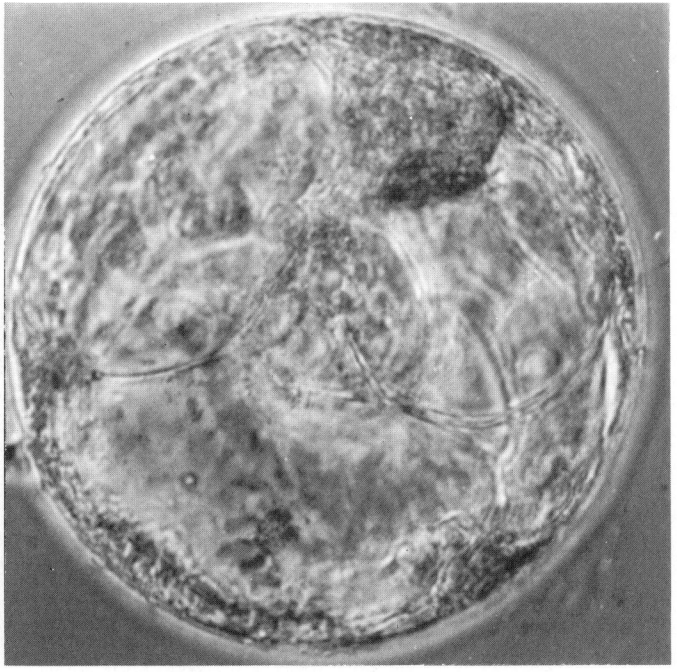

G An early living human blastocyst. Note the blastocyst cavity, small flattened mural trophoblastic cells, and the projecting clump of large cells constituting the inner cell mass.

H A blastocyst which has lost its zona pellucida and started to expand.

placed blastomeres has also been demonstrated in some mammals but, as yet, evidence for this phenomenon is lacking in man.) The intercellular spaces enlarge and coalesce to form a single fluid-filled *blastocyst cavity*. (It should be noted that the latter is not directly homologous with the *blastocoele* of primitive chordates, **2.1**, p. 100.) This cavity is largely bounded by *mural trophoblast cells*, except at one place where a clump of cells, termed the *inner cell mass (embryoblast* or *formative mass)* (**2.18,19,20,21**), projects into the cavity. This is widely considered to represent the residue of totipotential cells, some of which, the *embryogenic cells*, are destined to form the tissues of the embryo proper (see, however, p. 100). The segmenting zygote has now been converted into

a *unilaminar blastocyst*. Of some 60 cells composing the blastocyst at this stage, only about 5 are formative cells; the remainder constitute the flattened (*mural*) trophoblastic epithelium forming the wall of the vesicle (Hertig et al 1954). This *unilaminar blastocyst* is formed by the end of the fourth or the beginning of the fifth day after ovulation and lies free in the uterine cavity (**2.21**).

The zona pellucida disappears during the fifth day when the blastocyst is about 130–140 μm in diameter. The cells of the inner mass have multiplied and form an irregular clump insinuated into the wall of the blastocyst and flanked by trophoblast. One blastocyst described in detail at this stage (Hertig et al 1954) consists of 107 cells (**2.19**), of which 69 are *mural trophoblast cells*. Of the

127

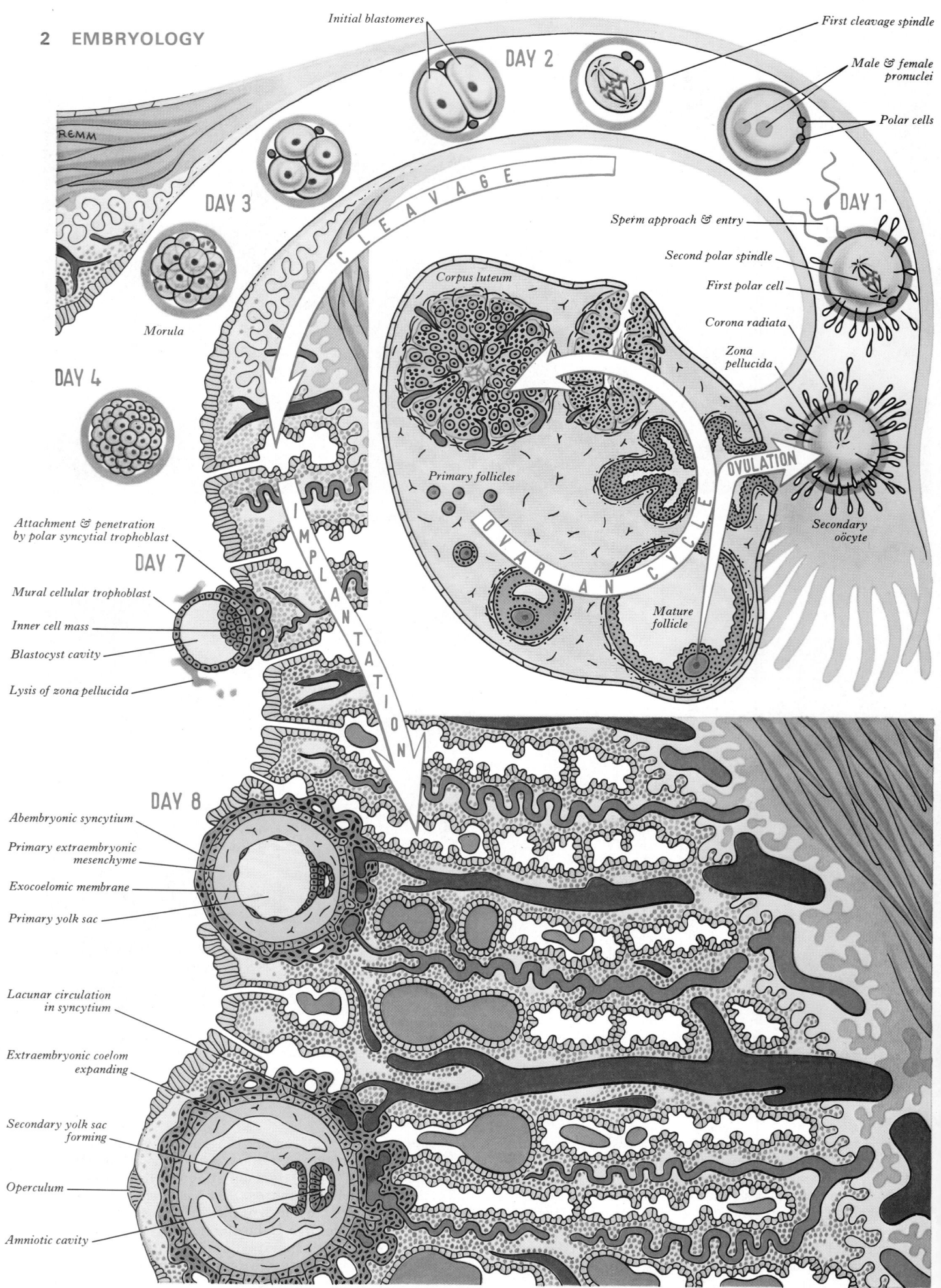

Initial blastomeres

First cleavage spindle

Male & female pronuclei

Polar cells

DAY 2

DAY 1

CLEAVAGE

DAY 3

Sperm approach & entry

Second polar spindle

First polar cell

Corona radiata

Zona pellucida

Corpus luteum

Morula

DAY 4

Primary follicles

OVULATION

OVARIAN CYCLE

Secondary oöcyte

Mature follicle

IMPLANTATION

Attachment & penetration by polar syncytial trophoblast

DAY 7

Mural cellular trophoblast

Inner cell mass

Blastocyst cavity

Lysis of zona pellucida

DAY 8

Abembryonic syncytium

Primary extraembryonic mesenchyme

Exocoelomic membrane

Primary yolk sac

Lacunar circulation in syncytium

Extraembryonic coelom expanding

Secondary yolk sac forming

Operculum

Amniotic cavity

2.21 A composite schema of the major events in the ovarian cycle, ovulation, fertilization, tubal transport and cleavage, differentiation of blastocyst, implantation, early embryogenesis, and incipient placentation.

Consult text and references for further details and alternative views of formation of amnion, yolk sac, and cytotrophoblast.

remainder, 8 are large *embryonic cells* constituting the *formative mass*, whilst the remaining 30 are future *polar trophoblast* cells and lie around its margins; they soon spread to cover the external aspect of the formative mass (at the *embryonic pole* of the blastocyst). At the same time, the inner aspect of the mass becomes lined by a single layer of polyhedral cells, the *embryonic endoderm* (*alternatively termed hypoblast or endoblast*).

Late in the sixth day after fertilization the polar trophoblast adheres to the uterine mucosa, exerting a histolytic action on its lining epithelium, then its underlying tissue. As they engulf the degenerating uterine tissue, the trophoblast cells in this situation divide with great rapidity and their individual progeny fuse with each other (Boyd & Hamilton 1966) so that all trace of cellular definition is lost. Accordingly the formative mass becomes covered with a multinucleated mass of cytoplasm, the *syncytial trophoblast* (*syntrophoblast*), which continues to burrow into the stratum compactum of the uterine mucosa. As the blastocyst embeds more deeply the remainder of its trophoblastic wall undergoes a similar change; but, at a slightly later stage, cell boundaries again appear on the inner or embryonic surface of the syncytial trophoblast. Thereafter the wall of the whole blastocyst consists of an inner germinative cellular layer, the *cellular trophoblast* (*cytotrophoblast*), covered on its outer surface with syncytial trophoblast, thickest over the formative mass, i.e. at the area of deepest penetration or *embryonic pole*, and thinnest over the area most recently embedded or *abembryonic pole*.

The *site of implantation* is normally on the posterior wall of the uterus, nearer to the fundus than to the cervix, and may be in the median plane or to one or other side (see, however, p. 131).

The youngest implanting human blastocyst hitherto recovered and described in detail (Hertig & Rock 1945) shows an early stage in the process. The polar trophoblast displays an extensive development of syncytium, which has destroyed a patch of uterine epithelium and underlying stroma (**2.22**). The blastocyst is not completely embedded and a portion of its wall at the abembryonic pole still projects into the uterine lumen. The fertilization age is believed to be $7\frac{1}{2}$ days (Hertig & Rock 1945).

In a slightly older specimen (Hertig & Rock 1945) ($9\frac{1}{2}$ days) the blastocyst is completely embedded. The syncytial trophoblast has undergone further proliferation and now covers the wall at the abembryonic pole as a thin layer. Irregular *lacunae* have developed in the syncytium which communicate with eroded maternal veins. The formative mass now consists externally of a thick plate of large, irregularly arranged cells, termed the *embryonic ectoderm (germ disc or embryonic epiblast)*, and internally of a single-layered sheet of *embryonic endoderm*, which intervenes between the disc and the blastocyst cavity. On its external surface the germ disc is in process of separation from the trophoblast by the formation of a cavity, which will soon become the *cavity of the amnion* (**2.**21,22,23,24). Re-examination of early rhesus monkey and human embryos has led to a modification of this long held view of amniogenesis (Luckett 1975). A *primordial amniotic cavity* develops by cavitation within the inner cell mass and initially has no direct relation with the overlying trophoblast. This occurs in 7-day human blastocysts. Subsequently the epiblastic roof thins and disappears forming a slightly cup-shaped germ disc which is surmounted by a transitory *tropho-epiblastic cavity*. The definitive amniotic epithelium forms by upfolding and

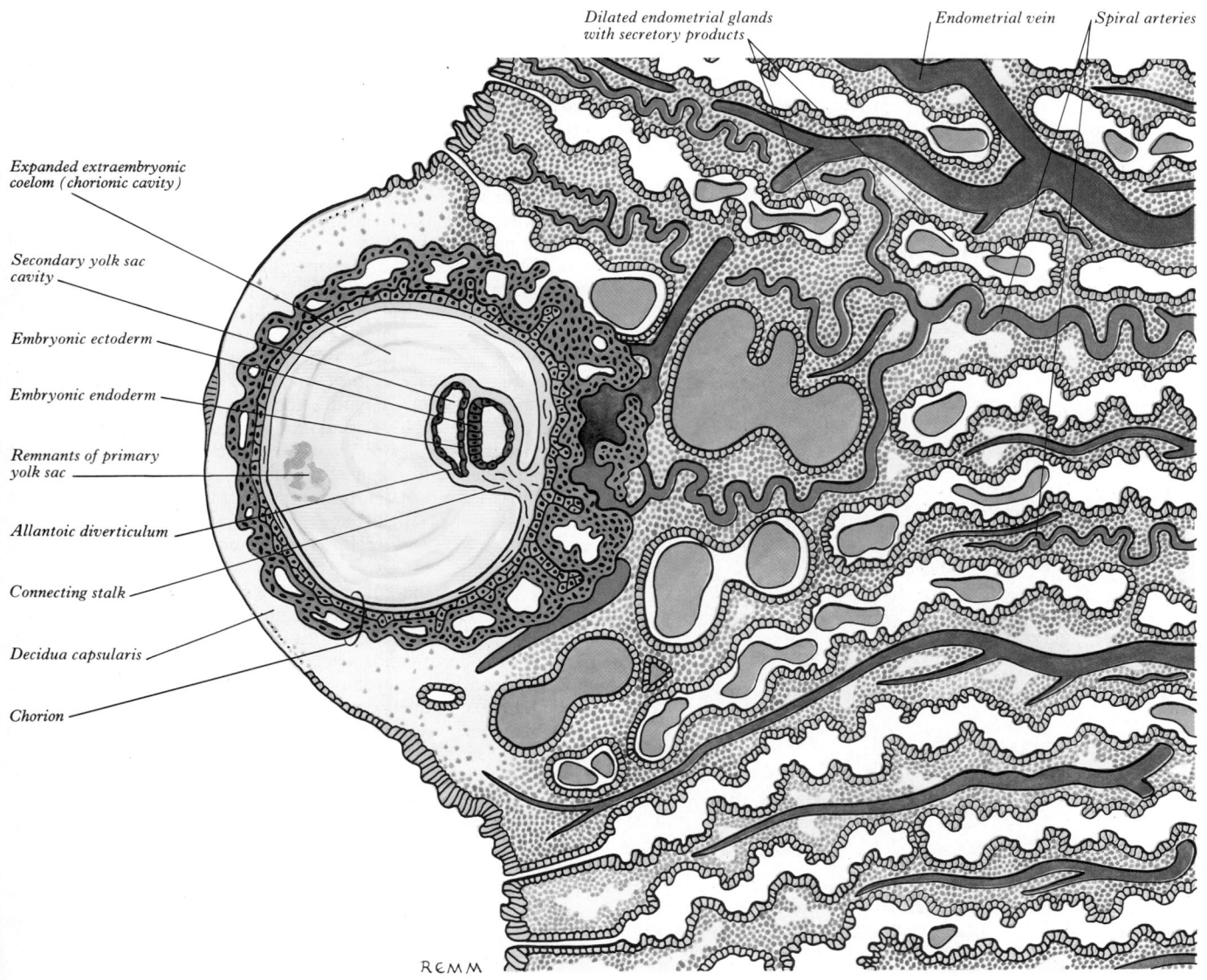

Dilated endometrial glands with secretory products   Endometrial vein   Spiral arteries

Expanded extraembryonic coelom (chorionic cavity)

Secondary yolk sac cavity

Embryonic ectoderm

Embryonic endoderm

Remnants of primary yolk sac

Allantoic diverticulum

Connecting stalk

Decidua capsularis

Chorion

REMM

2.21 (*contd*) Further expansion and differentiation of the blastocyst.

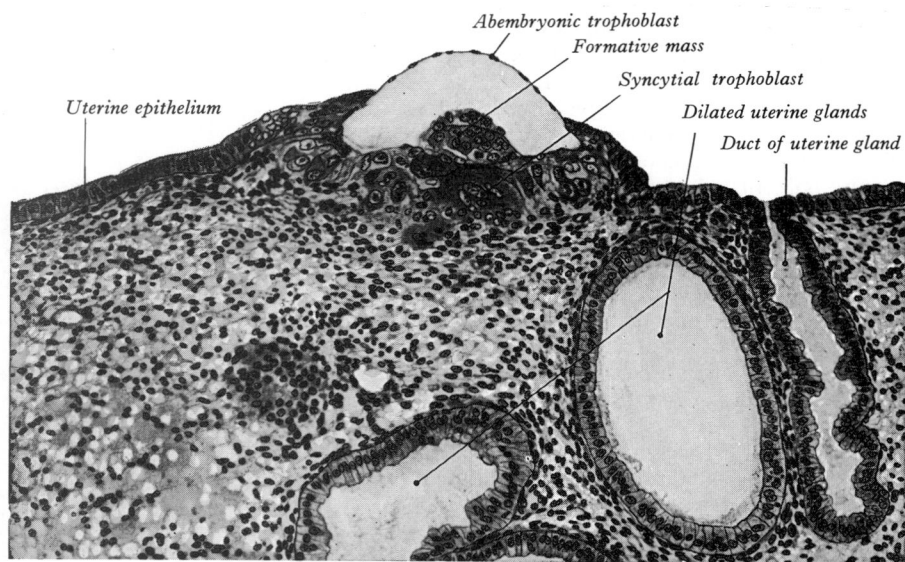

Uterine epithelium

Abembryonic trophoblast
Formative mass
Syncytial trophoblast
Dilated uterine glands
Duct of uterine gland

2.22 A human blastocyst (Carnegie 8020), fertilization age 7-7½ days, in process of embedding in the uterine mucosa. In the actual specimen the abembryonic trophoblast had collapsed on the formative mass but, for clarity, it has been shown projecting into the uterine cavity. (Rock & Hertig 1942, 1944). Magnification × c.150.

proliferation of the margins of the epiblastic disc, the process being completed by 9 days.

In the immediately succeeding stages the non-polar abembryonic walls of the blastocyst become *bilaminar* as its cavity becomes lined with a thin layer of flattened cells, which are in continuity with the embryonic endoderm around its margins and enclose a cavity now termed the *primary yolk sac*. It has been uncertain whether these are cells wholly derived from the cytotrophoblast, whether they are formed by an extension from

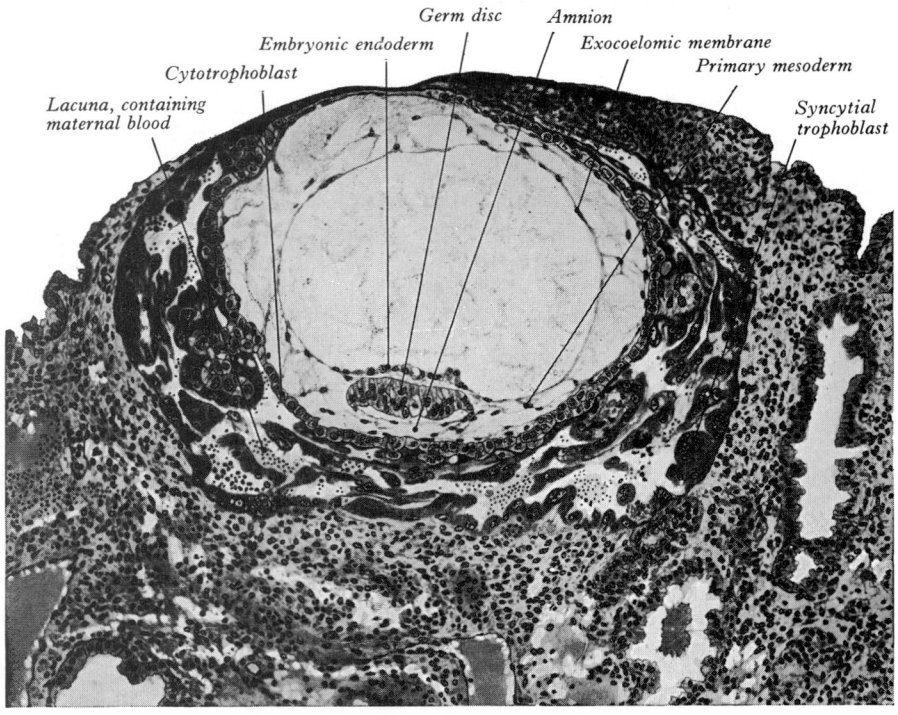

Germ disc *Amnion*
Embryonic endoderm *Exocoelomic membrane*
Cytotrophoblast *Primary mesoderm*
Lacuna, containing *Syncytial*
maternal blood *trophoblast*

2.23 A human blastocyst (Carnegie 7700), fertilization age 12-12½ days, embedded in the stratum compactum of the endometrium (Hertig & Rock 1941). Magnification × c.105. Compare with 2.22. Note lacunae in the syncytial trophoblast, many containing maternal blood; also that the primary yolk sac, surrounded by the exocoelomic membrane, does not fill the blastocyst cavity. The cells of the germ disc are now columnar and form an ectodermal plate. (Drawn from a photomicrograph by Dr A T Hertig.)

the endoderm (as in the rabbit, dog and some other mammals) or have a mixed origin. In a relatively recent review and analysis, Luckett (1978) is firmly convinced that in the rhesus monkey, and man, their derivation is exclusively hypoblastic (i.e. endodermal). Whatever their origin, a variety of terms have been applied to this tissue layer—*extraembryonic hypoblast* and later *extraembryonic endoderm* or the *exocoelomic (Heuser's) membrane*. The primary yolk sac, formed in this way (2.21, 23), appears to meet a functional need in the selective transport of the fluid which the trophoblast absorbs carrying cytolytic nutritive products from the uterine mucosa, making them available for the nourishment of the germ disc at this stage.

The cells of the germ disc, hitherto arranged in an irregular manner, now dispose themselves as a plate of columnar cells. At the same time extraembryonic mesodermal cells of hypoblastic origin and delamination from the inner surface of the cytotrophoblast intervene between the latter and the primary yolk sac. (The relative contributions to the mesenchymatous cells from these various sources remain controversial.) The wall of the sac becomes retracted from the wall of the blastocyst, although it often remains connected to it near the abembryonic pole by strands of mesoderm. Descriptions of several human blastocysts (Stieve 1936, Odgers 1937, Dible & West 1940, Hertig & Rock 1941, 1945, Davies 1944) which have reached this *trilaminar* stage and already are completely embedded in the uterine mucosa are now available e.g. see 2.21, 23, 24. The trophoblast is thickest over the embryonic pole and diminishes over the sides of the blastocyst but is exceedingly thin over the abembryonic pole, the last part to be embedded. It is lined over most of its extent by a single layer of cytotrophoblast and the syncytial trophoblast contains numerous lacunae. Maternal blood from invaded vessels occupies the lacunae and serves as a new source of nourishment.

Within the trophoblastic vesicle, the cells of the germ disc are now a columnar epithelium, the *embryonic ectodermal plate* or *disc*. Externally it is separated from the cytotrophoblast by the amniotic wall, cavity and contained fluid and some extraembryonic mesoderm; internally it is in close contact with the embryonic endoderm, although a distinct basement membrane intervenes. The embryonic endoderm consists of a single layer of low cuboidal epithelial cells, lining the ventral surface of the germ disc and forming the roof of the primary yolk sac, which is lined elsewhere by the exocoelomic membrane and which contains fluid and some apparent coagulum (this appearance resulting from the effects of histological fixation). The intervals between the trophoblast, on the one hand, and the primary yolk sac and developing amnion, on the other, are again occupied by fluid in which are found extraembryonic mesodermal cells, noted above, and mesenchymatous in nature, together with a certain amount of coagulum (2.24). This extraembryonic mesoderm soon forms a loose net, the *magma reticulare*, in which a series of enlarging clefts develop and begin to coalesce. As this proceeds the trophoblastic wall of the bastocyst with its lining of mesoderm becomes a separate membrane, the *chorion*, and the general cavity of the blastocyst can therefore now be termed the *chorionic cavity* or *extraembryonic coelom* (2.21, 24).

The cavity of the amnion with its roof of epiblastic 'amniogenic' cells and extraembryonic mesoderm, and its floor of embryonic ectoderm, now constitutes the *amnio-embryonic vesicle*. At first mesoderm connects chorion to amnion over a wide area but, with continued development and coalescence of clefts in the mesoderm extending the extraembryonic coelem, this attachment soon becomes increasingly circumscribed (2.21) and condenses to form a *connecting stalk* (body stalk), a permanent connection between (initially) the caudal rim of the embryonic disc and chorion, a pathway along which blood vessels of the embryo later establish communication with those of the chorion. Subsequently it becomes converted into part of the *umbilical cord* (vide infra); its embryonic attachment becoming transposed, later, to the ventral abdominal wall (p. 143).

While the connecting stalk is being defined, the primary yolk sac suffers a reduction in size, the mechanism of which is obscure, although it seems probable that part of its wall becomes drawn out by the mesodermal strands which anchor it to the abembryonic chorion, and which then loses its connection with the rest of the

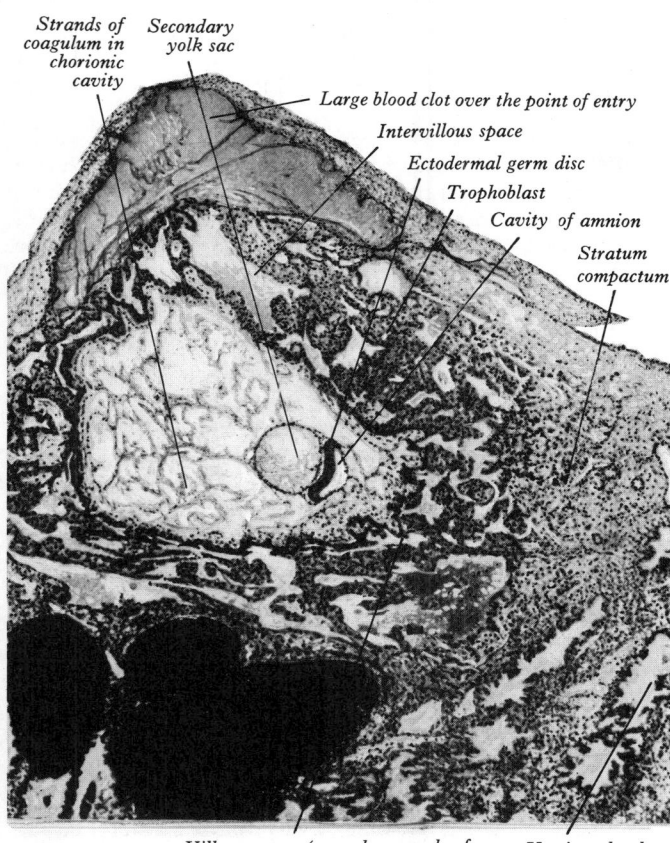

Strands of coagulum in chorionic cavity

Secondary yolk sac

Large blood clot over the point of entry

Intervillous space

Ectodermal germ disc

Trophoblast

Cavity of amnion

Stratum compactum

Villous stems (secondary grade of differentiation—see text)

Uterine gland

2.24 An advanced human blastocyst, embedded in the stratum compactum. Estimated age, 13.5 days (Rock & Hertig 1942).

sac. The resultant smaller cavity is the *definitive* (or *secondary*) yolk sac (**2.21, 24**) but is often referred to simply as *the yolk sac*. Some investigators (Heuser & Streeter 1941), however, maintain that the secondary yolk sac is a new formation by a rearrangement of the embryonic endodermal cells on the deep surface of the germ disc. (Consult Luckett 1978.)

At slightly older stages, of which many examples are available, the chorionic cavity contains two hollow vesicles (the amnion and the yolk sac), each covered with extraembryonic mesoderm, by which they are connected to the chorionic wall (**2.21, 24, 29**). Only the cells where the two vesicles are in contact with each other contribute to the formation of the body of the embryo and it is therefore this region which is termed the *bilaminar embryonic disc* (*embryonic shield* or *area*).

It is noteworthy that, of the multitude of cells derived from the fertilized ovum *at this stage*, only a relatively small number take part in the formation of the embryo, while the vast majority form its covering and nourishing membranes and certain other extraembryonic structures to be noted later. Hypotheses concerning the causal mechanisms of this early differentiation have been critically examined in relation to development in the mouse (Herbert & Graham 1974). The ectodermal cells of the germ disc are columnar and separated from the endoderm by a basement membrane. By the third week the ectoderm takes the form of three or four interlocking rows of cells continuous round its margins with the cells of the amnion which are flatter and more elongated and form only a single stratum, which is covered externally with a layer of extraembryonic mesoderm. The endodermal cells of the disc are flattened, but those lining the rest of the yolk sac vary in shape and patches of cubical or low columnar cells are found, especially on its caudal wall. Except in some localized situations, to be noted later (p. 135, 139), the endoderm seldom forms more than a single layer of cells.

An inconstant feature of the embryo is sometimes observed at this stage within the mesodermal connecting stalk between the amnion and disc to the trophoblast; this may contain an *amniotic duct*, extending from the amniotic cavity to end blindly near the

trophoblast. In reptiles, and in many mammals, the amnion arises by a process of folding around the periphery of the embryonic disc. The amniotic duct of the human embryo has been interpreted as the homologue of the point of closure of these folds (but it has also been suggested that its presence depends on the chance occurrence of a trophoblastic lacuna which breaks through into the amniotic cavity). See Luckett (1975), who describes human amniogenesis as a modified form of folding.

Concomitant with the changes which transform the zygote into a morula, the latter is slowly carried along the uterine tube by the action of the cilia of its lining epithelium and probably by gentle waves of muscular contraction which perhaps also generate currents in the tubal fluid. Passage along the tube covers a period of about 3 days, and the morula is then transformed into a blastocyst which remains *free* in the uterine cavity for a further 3 days before implantation. During this interval the uterine mucosa is preparing for the reception of the blastocyst (p. 146). The disappearance of the zona pellucida, after the formation of the blastocyst, allows the syncytial trophoblast to come into direct contact with the uterine mucosa and initiate implantation.

ECTOPIC IMPLANTATION

Implantation of the blastocyst normally occurs in the endometrium of the uterine body, more frequently on the posterior wall somewhere near the fundus, but may occur elsewhere in the uterus or in an *extrauterine* or *ectopic* site. Implantation near the internal os results in the condition of *placenta praevia* with its attendant risk of severe antepartum haemorrhage (p. 156).

Alternatively, the zygote or segmenting morula may be arrested at any point during its migration through the uterine tube and implant in its wall. Previous tubal inflammatory episodes may predispose to such tubal arrest and it has also been suggested that congenital abnormalities of the tube, tubal tumours, transperitoneal migration of an ovum from one ovary to the opposite tube, delayed ovulation and psychological stress, which may cause tubal spasm, are additional predisposing factors (Woodruff & Pauerstein 1969).

Nidation in the intramural part of the tube often results in early abortion of the conceptus whereas, if it occurs elsewhere in the tube, development often proceeds for about 2 months and is then usually followed by tubal rupture with death of the embryo and severe intraperitoneal haemorrhage—a grave surgical emergency. However, slow rupture of the tube may occur, accompanied by a further implantation of the conceptus into any adjacent peritonealized surface (*secondary abdominal pregnancy*), which again ultimately leads to rupture of the surface with similar consequences.

Primary ovarian or *abdominal* pregnancies have also been described, in which it has been presumed that the fertilization occurred in the vicinity of the ovary; most cases, however, are probably of the secondary type following a slow tubal rupture or a slow extrusion of the conceptus through the abdominal ostium of the tube.

Apart from their important clinical implications, such conditions emphasize the interesting fact that the conceptus can implant successfully into tissues other than a normal progestational endometrium. Further, prolonged development can occur in such sites and is usually terminated by a mechanical or vascular accident and not by a fundamental nutritive or endocrine insufficiency or by an immune maternal response.

Differentiation of the Embryonic Area

Near the end of the second week, the embryonic area thus consists of an almost *circular, bilaminar disc* which separates the cavity of the amnion from that of the secondary yolk sac. As we have seen, the amniotic aspect of the area consists of a thick plate of columnar embryonic ectodermal (epiblastic) cells, 2–4 cells thick, which is continuous around its margins with the flattened cells lining the amniotic cavity. A single layer of rather flattened polyhedral endodermal (hypoblastic) cells forms the roof of the secondary yolk sac; this roof is quite closely applied to the superjacent ectodermal plate, a basement membrane intervening, and it is continuous peripherally with the cuboidal or columnar cells lining the secondary yolk sac. At this stage few mesodermal cells are found between the ectoderm and endoderm of the embryonic area but thin strata of extraembryonic mesoderm clothe the outer surfaces of the amnion and secondary yolk sac and line the chorion (p. 145)—the space bounded by these various mesodermal layers constituting the *chorionic cavity* or *extra-embryonic coelom*.

The walls of the amnion on the one hand, and that of the yolk sac on the other, may now be defined as extraembryonic *somatopleure* and *splanchnopleure* respectively. The former consisting of an epithelial layer (trophoblast or amniotic lining) associated with a layer of *somatopleuric extraembryonic mesoderm*, whilst the latter comprises the endodermal epithelial lining of the yolk sac together with *splanchnopleuric extraembryonic mesoderm*. At a *junctional zone* surrounding the margins of the embryonic area, where the walls of the amnion and yolk sac converge, the somatopleuric and splanchnopleuric layers of mesoderm are continuous.

At an early stage the embryonic area is radially symmetrical, with no specialized features distinguishing its future cephalic, caudal and lateral borders; but with the extensions of the extraembryonic coelom noted above and the increasing separation of the amnion from the overlying chorion, the mesodermal connecting stalk becomes increasingly defined and its embryonic attachment is soon limited to a zone which marks the future *caudal edge* of the embryonic area.

The first change within the embryonic area itself occurs near the opposite (*cephalic*) edge, where, in a localized region, the endoderm in the yolk sac roof thickens, forming an oval plate of large vesicular cells, the *prechordal plate*, which at a later stage contributes cells to the general head mesenchyme (p. 136) and which, together with the closely applied superjacent ectoderm, forms the transient *oropharyngeal membrane*. The attachment of the body stalk and the position of the prechordal plate have now defined the *plane of bilateral symmetry* of the embryonic area.

During immediately succeeding stages the embryonic area changes from a circular *bilaminar* disc to first an oval, and then a pear-shaped *trilaminar* one as a third *mesodermal layer* becomes interposed between the embryonic ectoderm and endoderm. The long axis of the oval and pear lie in the future cephalo–caudal axis of the developing embryo, whilst the pear-shaped outline is narrow caudally and expanded at the cephalic end. These shape changes accompany the emergence of proliferative and cell migratory zones in the caudal midline of the primitive ectoderm of the embryonic area—a linear *primitive streak* with, at its cephalic end, a *primitive node* or *knot*. From the former generations of *intraembryonic mesodermal cells* and from the latter an axial *head*

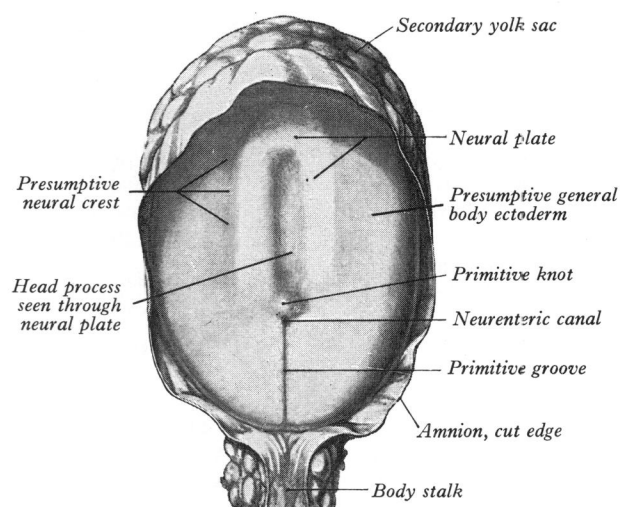

Presumptive neural crest

Head process seen through neural plate

Secondary yolk sac

Neural plate

Presumptive general body ectoderm

Primitive knot

Neurenteric canal

Primitive groove

Amnion, cut edge

Body stalk

2.25 A human embryo, 1.16 mm long. The amnio-embryonic vesicle has been laid open widely, most of the amnion having been removed. The embryonic area is exposed in almost its whole extent and shows an early stage of differentiation. Estimated age 19 days. (W C Gerge 1942.)

process (ultimately forming the notochord) pass between the ectoderm and endoderm. (For long these were considered the principal, if not the sole, derivatives of streak and node; however, vide infra.)

Presumably (based on the evidence of comparative studies, p. 107), under the primary inductive influences of the chordamesoderm, the central regions of the ectoderm which overlies it commences the formation of a *neural plate*, the primordium of the central nervous system (**2.25, 27**). The margins of the plate are surrounded by a strip of presumptive *neural crest* tissue (p. 137) which, again, is surrounded peripherally by the layer which will form the general body ectoderm. These various regions will now be considered further.

As noted elsewhere (p. 107) many experimental embryologists consider that the totipotency of the fertilized mammalian oöcyte and initial blastomeres suffers a stepwise reduction, as the determination and differentiation of the various embryonic regions proceeds. Alternative views have also been advanced (see p. 108). Despite the relative paucity of experimental analyses, because of technical difficulties on mammalia, it is widely held that by the blastocyst stage the vast majority of the cells have already suffered a severe restriction in their developmental potency, being destined to form extraembryonic structures (p. 98), whilst the pluripotent cells are restricted to the embryonic area. It was proposed (Streeter 1942) that, as the changes outlined above occur, these pluripotent cells become progressively circumscribed, first in the caudal part of the oval embryonic area and then as a narrowing midline strip in the caudal end of the increasingly pear-shaped area, i.e. the site of formation of the *primitive streak* which, as development proceeds, becomes more obvious as a linear opacity when viewed from the amniotic aspect. (For more recent *experimental* data on mammalia see Herbert & Graham 1974.)

In the early stages of development the living tissues are greyish and translucent, but any localized thickening due to cellular proliferation and patterns of cell migration naturally interferes with the translucency and causes a localized opacity as in the case of the primitive streak, indicating that rapid multiplicative growth is occurring throughout its length. Such growth also affects neighbouring epiblastic areas and both these and the streak are involved in complex cell migrations. Thus some cell progeny diverge from the streak and contribute to the neural plate, neural crest and general body ectoderm. Others converge on and enter the streak, these diverging dorsocaudally to form *extraembryonic mesenchyme*, including the body stalk. Further streams radiate in most directions as *intraembryonic mesenchyme* (vide infra). Finally, additions are made to the midline endodermal roof of the secondary yolk sac. At the headward end of the streak another area of exceptionally active growth and cell migration forms a knob-like thickening termed the *primitive knot* or *Hensen's node* (**2.25**) and here the ectoderm is fused with the endoderm.

From the primitive knot a rod-like column of cells grows headwards in the midline separating the presumptive neural plate from the subjacent roof of the yolk sac until its cranial tip reaches the caudal margin of the prechordal plate. This is termed the *head process (notochordal process)*—the forerunner of the skeletal axis of the body. This, initially a cylindrical solid rod of cells, now becomes canalized and caudally the canal breaks through on to the ectodermal surface at the posterior end of the primitive knot. More cranially the endodermal cells lying ventral to the canalized head process disappear and the head process then becomes, briefly, a constituent part of the roof of the yolk sac in the median plane. The cells forming the floor of the canal of the head process also break down so that the canal communicates freely with the yolk sac and, at its caudal end, a temporary communication is established between the yolk sac and the amniotic cavity. This connection, which pierces the embryonic area at the primitive knot (i.e. the caudal end of the neural groove), is the *neurenteric canal*. At this stage a transverse section across the embryonic area cranial to the primitive knot shows that, in the median plane, the roof of the yolk sac is, for a time, formed by the cells of the head process (**2.27A**), and this intercalation, which forms the *chordal or notochordal plate*, extends forwards to include the region which will subsequently form the roof, or dorsal wall, of the pharynx.

Later, these cells of the head process become separated from the endoderm and form the *notochord*, the roof of the yolk sac being repaired by the fusion of the adjoining endodermal cells. Subsequently the notochordal cells are encased by a homogeneous sheath and continued proliferation of cells within the sheath results in a solid but flexible rod, later surrounded by mesoderm to form the primitive or *blastemal vertebral column* (p. 159).

During the earlier part of this period another prominent change affects most of the embryonic area. From the sides and surroundings of the primitive streak occurs an intensely active multiplicative cell growth. The cellular progeny, having converged on the streak, sink into the substance of the germ disc and then spread laterally forwards and backwards until they extend over the whole embryonic area (*with the exception of most of the median plane*). They insinuate themselves between the ectoderm and the underlying endoderm and produce a third layer, the *intraembryonic mesoderm*. (**2.27,28**). For a review of experimental data of these stages of development in higher vertebrates consult Bellairs (1971).

The primitive streak is the principal but not the only source of intraembryonic mesoderm; the latter also receives accessions from the margins of the notochord and from the neural crest (p. 137), particularly in the head and branchial arches (Hörstadius 1950), and the latter also receive contributions from ectodermal placodes and from the thickened endoderm of the prechordal plate. As mentioned above, this plate forms an integral part of the future oropharyngeal membrane and it is probably involved in inductive processes of cranial structures (termed a *head organizer region* by earlier workers). Similarly, in the midline at the opposite end of the embryonic area, a patch of endoderm thickens and becomes closely applied to the overlying ectoderm, between the caudal tip of the primitive streak and the attachment of the body stalk, to form the future *cloacal membrane*.

As the streams of intraembryonic mesodermal cells radiate from the lateral margins of the primitive streak (**2.27,28**), they separate the ectoderm and endoderm of the embryonic area, then soon approach and become confluent with extraembryonic mesoderm around the margins of the area, i.e. at the *junctional zone* where the splanchnic and somatic strata of extraembryonic mesoderm merge. The streams passing in a *cephalic* direction flank the differentiating head process (later the notochord), skirt the margins of the prechordal plate and then converge medially to

fuse in the midline beyond its cephalic border. This transmedian mass is the *cardiogenic mesoderm* in which the heart and pericardium are to develop. Around the extreme *cephalic* margin of the embryonic area, the cardiogenic mesoderm fuses with the junctional zone of extraembryonic mesoderm mentioned above and this confluent mesodermal region will eventually form the *septum transversum* and *primitive ventral mesentery* of the foregut (p. 232).

The intraembryonic mesoderm which streams *caudally* from the primitive streak skirts the margins of the cloacal membrane and then converges towards the caudal midline extremity of the embryonic disc to become continuous with the extraembryonic mesoderm of the connecting stalk.

The ectodermal thickening of the *neural plate*, from which most of the central nervous system develops, initially corresponds precisely in length to the underlying notochord. Thus it extends from the cranial border of the primitive knot to the caudal border of the future oropharyngeal membrane but it also extends laterally beyond the notochord, to cover the medial strips of intraembryonic mesoderm. As the plate grows, its margins become raised (**2.26,27,28**) as the *neural folds*, with a longitudinal, midline *neural groove* between them.

The neural folds become particularly prominent at the anterior end of the area and the walls of the groove become expanded (**2.26**), the first sign of brain formation. It is not a continuous enlargement; two slight transverse constrictions indicate a division into three *parts* or *regions*, named craniocaudally: the *prosencephalon* or *forebrain*, the *mesencephalon* or *midbrain* and the *rhombencephalon* or *hindbrain* (see p. 185). As we have seen, between the cephalic margin of the forebrain and the cardiogenic mesoderm the ectoderm and the endoderm are in contact over a small area forming the *oropharyngeal membrane* (**2.28,29**). As the neural groove deepens, its dorsal edges come into contact with each other and fuse to convert the groove into a sagittal slit-like canal (**2.27A–E**). It should be noted that as the edges of the groove approximate they carry with them the adjoining general body ectoderm and, when the process of fusion occurs, like fuses with like, i.e. ectoderm with ectoderm and neural tissue with neural tissue to form a submerged *neural tube*. The process is effected first in the hindbrain or upper cervical region (**2.26**) in the latter part of the third week and extends both headwards and tailwards, until only a small opening is left at each end. These are the rostral and caudal *neuropores*; the former closes in the middle and the latter towards the end of the fourth week. The cells which lie in the line of fusion of the dorsal edges of the neural plate constitute the *neural crest* (**2.27**)—its fate is considered later (p. 137).

From initiation, the walls of the neural groove and tube are bathed by fluid in the amniotic cavity (*liquor amnii*) and are, in part, dependent on it for their nourishment whilst the neuropores remain open. It may be noted that closure of the neuropores coincides with the establishment of a blood vascular circulation for the neural tube; also the formation of choroid plexuses and the production of embryonic cerebrospinal fluid.

The fusion of the dorsal lips of the neural plate in the region of the brain results in the completion of the three zones mentioned above, and termed, somewhat inappropriately, the *primary cerebral vesicles* (see p. 185). Much of the walls of these 'vesicles' becomes thickened and develops into the nervous tissues including neuroglia of the brain and their cavities are modified to form the ventricles of the brain. In some important sites, however, the walls of the forebrain and hindbrain vesicles do not develop nervous tissue but, as mentioned, remain thin and modified into *choroid plexuses* (p. 192). At three loci in the roof of the hindbrain the thin tissue degenerates forming ventriculo-subarachnoid foramina (p. 186). The remainder of the tube forms the spinal cord, its cavity persisting as the central canal.

Meanwhile, throughout the extent of the notochord the intraembryonic mesoderm becomes arranged into three zones: (1) a thickened medial portion which lies immediately lateral to the grooved neural plate (later tube) and notochord, the *paraxial mesoderm*; (2) a narrower *intermediate mesoderm*, situated lateral but directly continuous with the paraxial mesoderm; and (3) a flattened *lateral plate* which extends from the intermediate mesoderm to the periphery of the embryonic area where it is continuous with the extraembryonic mesoderm on the outer

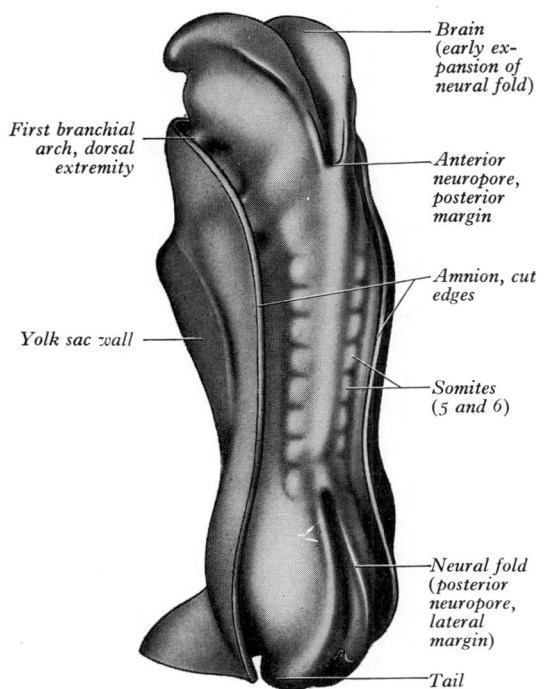

Brain (early expansion of neural fold)

First branchial arch, dorsal extremity

Anterior neuropore, posterior margin

Amnion, cut edges

Yolk sac wall

Somites (5 and 6)

Neural fold (posterior neuropore, lateral margin)

Tail

2.26 A human embryo, 2.1 mm long, with nine somites. Viewed from the left lateral and dorsal aspects. (From a model by Eternod.) Nearly all the yolk sac, and the caudal amnion have been excised to show the tail region.

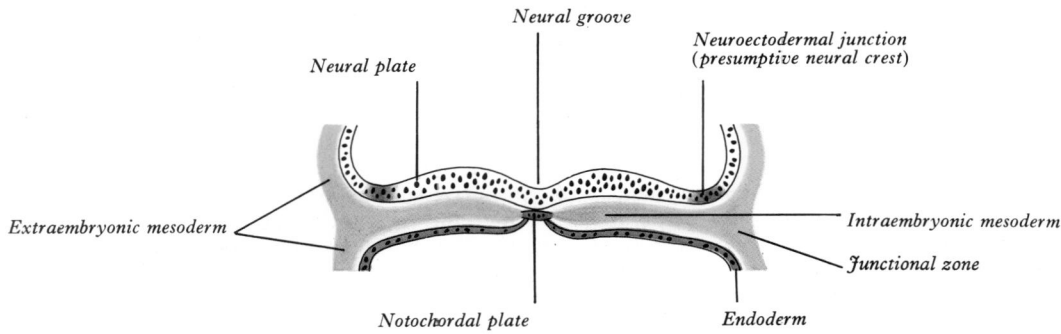

Neural groove

Neural plate

Neuroectodermal junction
(presumptive neural crest)

Extraembryonic mesoderm

Intraembryonic mesoderm

Junctional zone

Notochordal plate

Endoderm

A

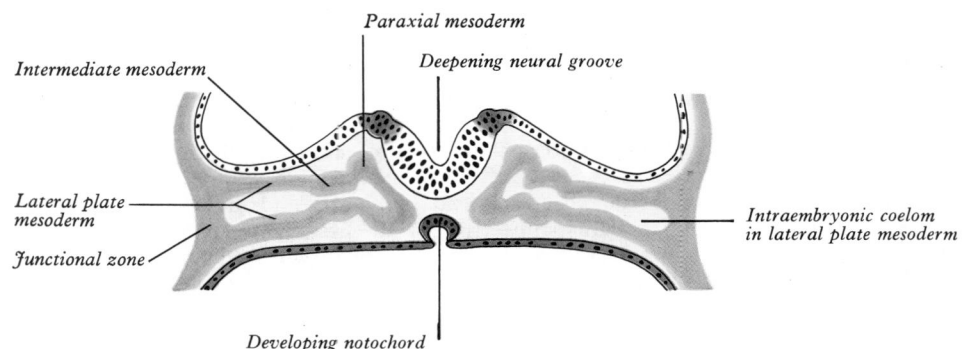

Paraxial mesoderm

Deepening neural groove

Intermediate mesoderm

Lateral plate
mesoderm

Junctional zone

Intraembryonic coelom
in lateral plate mesoderm

Developing notochord

B

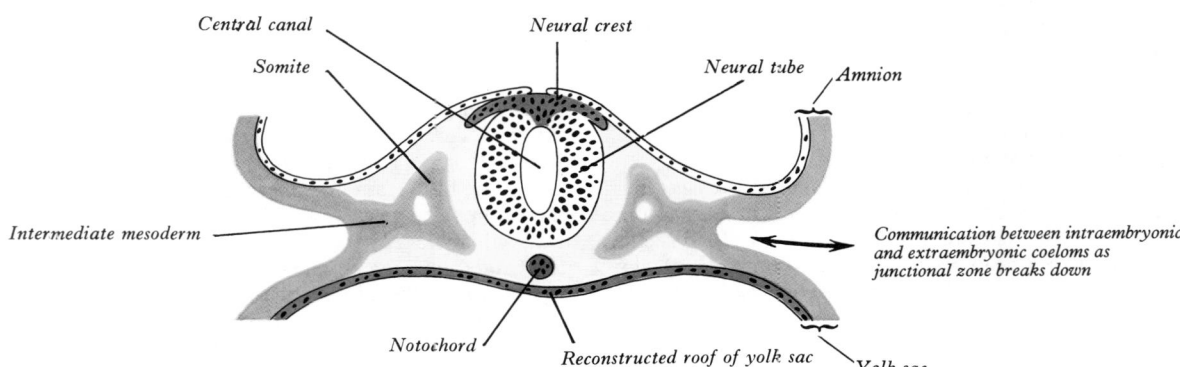

Central canal

Neural crest

Somite

Neural tube

Amnion

Intermediate mesoderm

Communication between intraembryonic
and extraembryonic coeloms as
junctional zone breaks down

Notochord

Reconstructed roof of yolk sac

Yolk sac

C

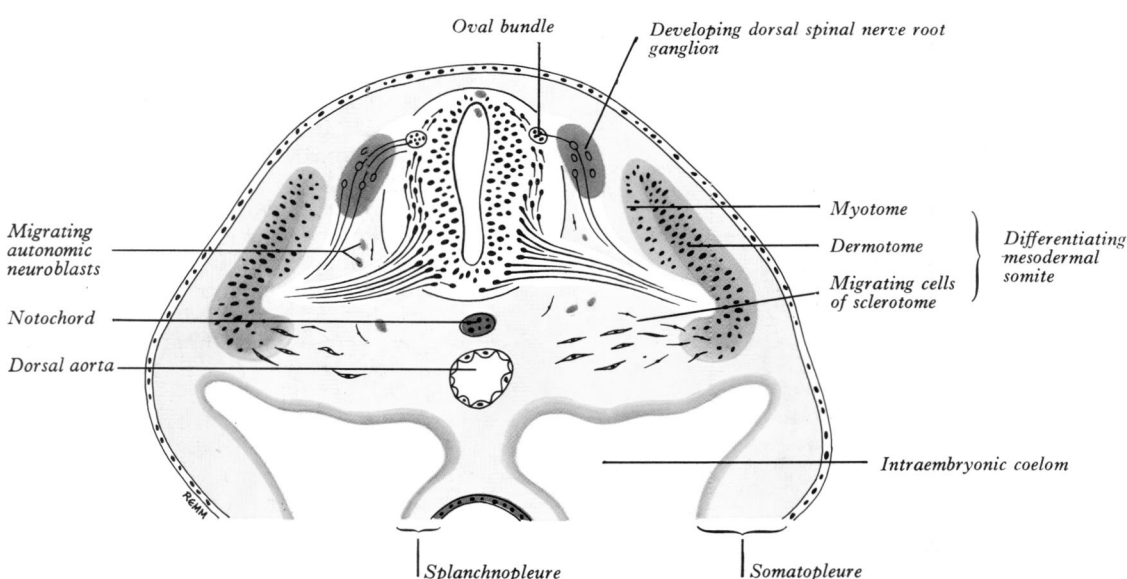

Oval bundle

Developing dorsal spinal nerve root
ganglion

Myotome

Dermotome

Differentiating
mesodermal
somite

Migrating
autonomic
neuroblasts

Migrating cells
of sclerotome

Notochord

Dorsal aorta

Intraembryonic coelom

Splanchnopleure

Somatopleure

D

2.27A–D A series of schematic diagrams showing important gross stages in the development and differentiation of the neural plate and tube, the neural crest, the notochord and the intraembryonic mesoderm and coelom. See text for a detailed description. Many histogenetic details have, intentionally, been omitted: the diagrams may serve, therefore, as an overall guide.

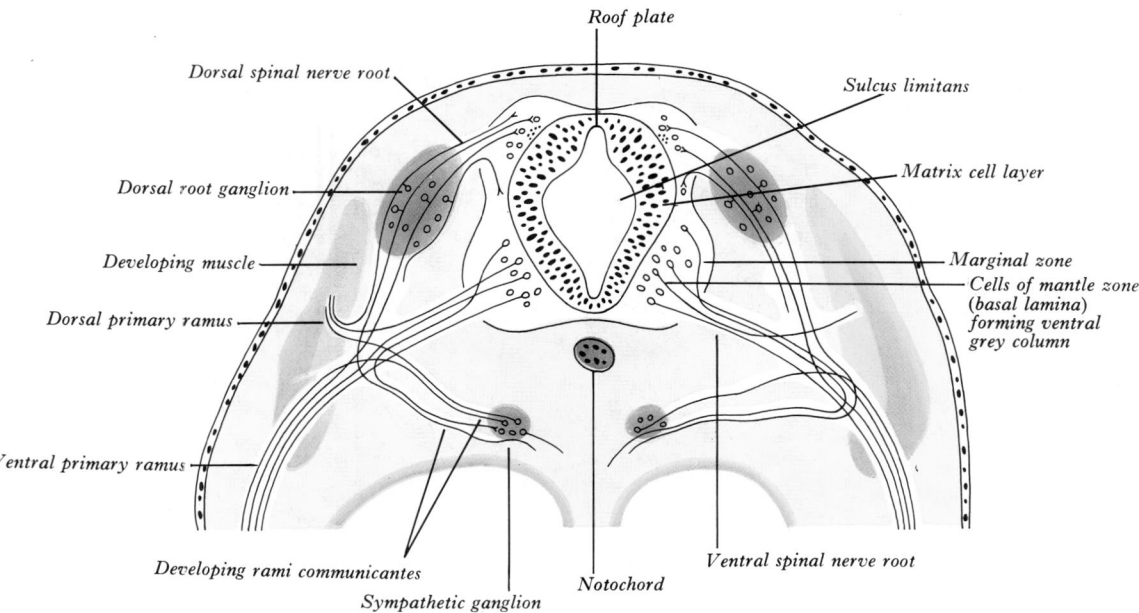

Roof plate

Dorsal spinal nerve root

Sulcus limitans

Dorsal root ganglion

Matrix cell layer

Developing muscle

Marginal zone

Cells of mantle zone (basal lamina) forming ventral grey column

Dorsal primary ramus

Ventral primary ramus

Developing rami communicantes

Ventral spinal nerve root

Sympathetic ganglion

Notochord

2.27E Further development of the neural tube, spinal and sympathetic ganglia. See notes in caption for **2.27A-D**.

surfaces of the amnion and the yolk sac, i.e. at the *junctional zone* of mesoderm.

THE EMBRYONIC LAYERS

The appearance of the intraembryonic mesoderm and its three major medio-lateral divisions on each side completes the first stage of differentiation of the embryonic area. The embryo now consists of an outer protective and neural layer, the ectoderm, an inner nutritive layer, the endoderm, and an intermediate layer, the intraembryonic mesoderm, which is available primarily as a muscle-forming layer. It is clear from their history and their early differentiation that these layers are of considerable phylogenetic and ontogenetic significance but too much stress must not be laid on their independence from one another. The differentiation at this stage does not leave the constituent cells of the three layers with potencies so limited that complete divergence occurs in subsequent development. The developmental potential of the individual cell layers is reduced from the pluripotent condition found at an earlier period, but not to such an extent that the potencies of the cells of one layer are entirely different from those of the two remaining layers. Further, during subsequent development of the various organ systems, derivatives of the different layers are often closely interlocked and interdependent in terms of fundamental *morphogenetic mechanisms* (pp. 104–115).

However, the contributions of the three layers in the human embryo to the formation of the different systems and organs may, for convenience, be summarized:

The primitive embryonic ectoderm consists of columnar cells, which become cubical towards the periphery of the embryonic area. It gives origin to: (1) the epidermis and the lining cells of the glands which open on it, and the appendages of the skin, the hair and nails; (2) practically the whole of the nervous system, including the cranial and spinal ganglia, the sympathetic ganglia, the posterior lobe of the hypophysis cerebri and the pineal gland; (3) the chromaffin organs; (4) the anterior lobe of the hypophysis cerebri; (5) the epithelium of the cornea, conjunctiva and lacrimal gland, canaliculi, nasolacrimal sac and duct; (6) the lens; (7) the plain muscle of the iris; (8) the neuro-epithelium of the sense organs; (9) the epithelium lining the nose and paranasal sinuses, the roof of the mouth, the gums and the cheeks, the external acoustic meatus and external stratum of the tympanic membrane; (10) some salivary glands and the enamel of the teeth; (11) the epithelium lining the lower part of the anal canal and that of the terminal male urethra, scrotal and penile epidermis, also that of the female vulval (pudendal) cleft as far as and including the labia minora.

The primitive embryonic endoderm consists at first of flattened cells, which subsequently become columnar. It gives origin to: (1) the epithelial lining of the whole of the alimentary canal, with the exception of those portions already ascribed to the ectoderm; (2) the lining cells of the glands which open into the alimentary canal, including the liver and the pancreas and their ducts, but excluding the parotid salivary glands; (3) the epithelium lining the auditory tube and tympanic cavity; (4) the epithelium of the thyroid and parathyroid glands and the thymus; (5) the lining epithelium of the larynx, trachea and the smaller air passages, including the alveoli and the air saccules; (6) the epithelium of most of the urinary bladder and much of the urethra; (7) the epithelium of the prostate and many other paraurethral glands.

The intraembryonic mesoderm gives origin to the remaining organs and tissues of the body. These include: (1) all the connective and sclerous tissues and their arthroses; (2) the teeth with the exception of the enamel; (3) the whole musculature of the body, both striated and unstriated, with the exception of the musculature of the iris; (4) the blood and the blood vascular, and lymph and the lymphatic systems; (5) the urogenital system, except most of the lining of the urinary bladder, prostate and urethra—derivation of the vaginal lining is controversial; (6) the cortex of the suprarenal glands and the mesothelial linings of the pericardial, pleural and peritoneal cavities.

THE EMBRYONIC TISSUES

The cells of the embryo become arranged at this early stage into two fundamental types of tissue, *epithelial* and *mesenchymatous* and, despite regional modifications introduced as the various tissues develop, these types persist in large measure throughout life.

Epithelial tissues are those in which the cells are closely packed, with narrow intercellular clefts containing minimal extracellular material; subsequently the cells usually show inter-cell surface specializations such as desmosomes, tight junctions, gap junctions, etc (pp. 22–25). Characteristically, they clothe internal and external surfaces as simple or compound cellular sheets which separate phases of differing composition (e.g. the external environment and the subepithelial tissue fluids, intravascular and extravascular fluids, etc). Traffic of materials in the intercellular clefts between cells is limited and passage occurs across the cells and their limiting membranes, which function as energy-dependent selective barriers, enhancing the passage of some materials and impeding the passage of others.

Clefts in lateral plate mesoderm

Pericardial cavity

Septum transversum

Oropharyngeal membrane

Cardiogenic mesoderm

Head process

Coelomic duct

A

B

C

Peritoneal part of coelom communicating with extraembryonic coelom

Primitive streak

Primitive knot

Notochord seen through floor of neural groove

Primitive knot

Primitive streak

Cloacal membrane

Allanto-enteric diverticulum

Connecting stalk

2.28 Diagrams to illustrate the formation of the intraembryonic coelom —the embryonic area is viewed from the dorsal aspect: structures are seen developing in, or through, the epiblast. Tissues dorsal to the coelom have been removed. Fold formation not shown.

The majority of the derivatives of the embryonic ectoderm and endoderm listed above retain their epithelial character throughout life. The third germ layer, mesoderm, also contributes many tissues of an epithelial type (although they may also be termed endothelia or mesothelia depending upon their topographic site). These include the visceral and parietal mesothelia of the coelomic cavities, many of the derivatives of the intermediate mesoderm such as the pro-, meso- and meta-nephroi and their ducts, the paramesonephric (Mullerian) duct system and the cortex of the suprarenal gland. Further, after first passing through a mesenchymatous stage (vide infra) many mesodermal cells contribute to the formation of the widespread endothelial cell layers which line the blood and lymphatic vascular systems. Still other mesothelia develop to line the various synovial cavities of the many joints, bursae and tendon sheaths, the anterior chamber of the eye and the periotic perilymphatic space of the internal ear. The myotome of mesodermal somite origin (p. 139) also maintains an epithelial arrangement throughout its early stages.

Mesenchyme, a name first introduced over a century ago (Hertwig 1881), forms the remaining tissue of the early embryo, occupying all the regions between the various epithelial layers described above. In contrast it consists of a loosely arranged tissue with wide spaces containing copious extracellular fluid which carries a variety of hydrated mucosubstances (p. 67), suspended in which are scattered amoeboid, primitive *mesenchymal cells*. The latter usually appear stellate in fixed and stained preparations, they often show some degree of phagocytic ability and they retain a potentiality for developing into a wide array of cell types. Many mesenchymal cells are derived from various regions of the early intraembryonic mesoderm, e.g. from the lateral plate where they form the core between the limiting epithelia of the splanchnopleure and somatopleure and from the dermotome and sclerotome of the mesodermal somites (p. 140).

Mesenchymal cells are, however, not exclusively of mesodermal origin; it is probable that the extraembryonic mesenchyme of the yolk sac and chorion receive contributions from the neighbouring endoderm and cytotrophoblast respectively. Within the embryonic body considerable accessions are also received, particularly in the head region from the neural crest, placodes and the endodermal prechordal plate.

In subsequent development mesenchymal cells differentiate into many different tissues including: the general and specialized connective tissues (cartilage, bone, dentine) and their many types of attendant cell (p. 300); smooth visceral, cardiac and, in some sites, striated muscle; the endothelium of the blood and lymphatic vascular systems and any attendant musculo-adventitial walls; the various blood-forming sites, lymphoid tissue, bone marrow,

spleen and the remaining elements of the reticulo-endothelial and macrophage systems; the tissues of joints including their synovial cells and those of bursae and tendon sheaths; finally the sheaths of nerves, muscles and the periostea of bones.

Subregions of the Embryonic Layers

In addition to the general summaries of the derivatives of the primitive embryonic layers given above, it may also prove convenient here to indicate the principal *subregions* of these layers and their main contributions to the developing embryo.

THE EMBRYONIC ECTODERM

This may be divided into (a) general body ectoderm; (b) neural plate; (c) neural crest; (d) ectodermal placodes.

General body ectoderm gives rise to: (1) most of the cutaneous epidermal cells; the secretory, duct-lining and myoepithelial cells of sweat, sebaceous and mammary glands; the hairs and nails; (2) the epithelia of the cornea and conjunctiva and the secretory and duct-lining cells of the lacrimal gland; (3) the respiratory nasal epithelium, the epithelia lining the paranasal sinuses, lips, cheeks, gums and palate, the secretory and duct-lining cells of the nasal, labial, palatine, general oral and salivary glands; (4) the embryonic enamel organ and dental enamel; (5) the epithelial lining of the external acoustic meatus and external epithelium of the tympanic membrane; (6) the epithelial lining of the lacrimal canaliculi, sac and nasolacrimal duct; (7) the epithelium of the lower anal canal and terminal male urethra.

Neural plate cells give rise to: (1) all the neurons of the *central* nervous system; these include not only those neurons wholly confined to the central nervous system but also those somatic motor and preganglionic efferent neurons with somata within the central system and axons which pass out into the peripheral nervous system; (2) the macroglial cells, i.e. varieties of astrocyte and oligodendrocyte; (3) generalized ependymal cells lining the ventricles, aqueduct and central canal of brain and spinal cord; (4) specialized ependymal cells, including tanycytes, those covering the choroid plexuses and distinctive regions of the third ventricle such as the subcommissural organ; (5) cells belonging to the APUD series. (The term is an acronym for cells showing characteristic amine handling properties—Amine Precursor Uptake and Decarboxylation. The evolution and modification of the concept is discussed more fully elsewhere, p. 1376 and at other appropriate points throughout the volume.) These include pinealocytes, magnocellular and parvocellular peptide-secreting neurons of the hypothalamus and reticular formation and certain

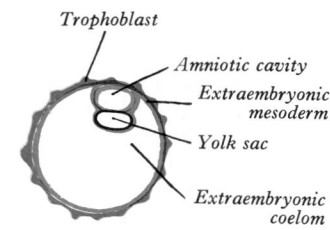

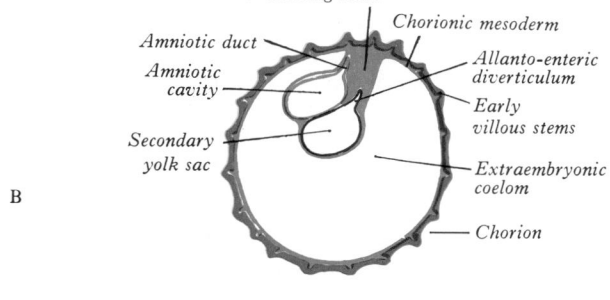

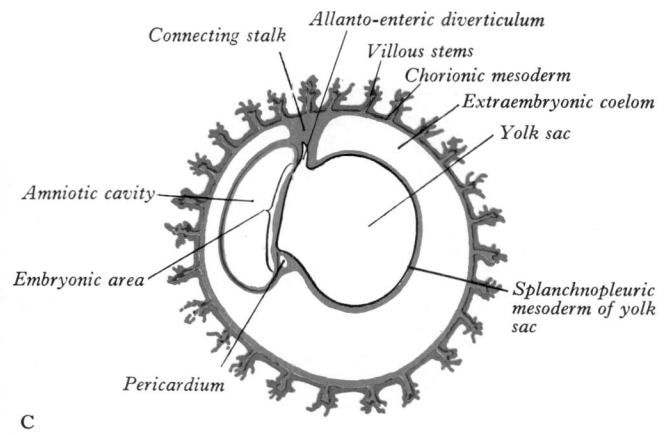

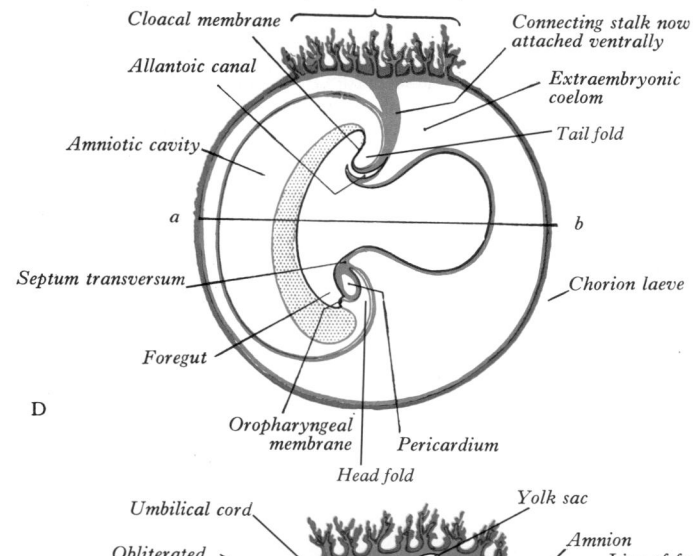

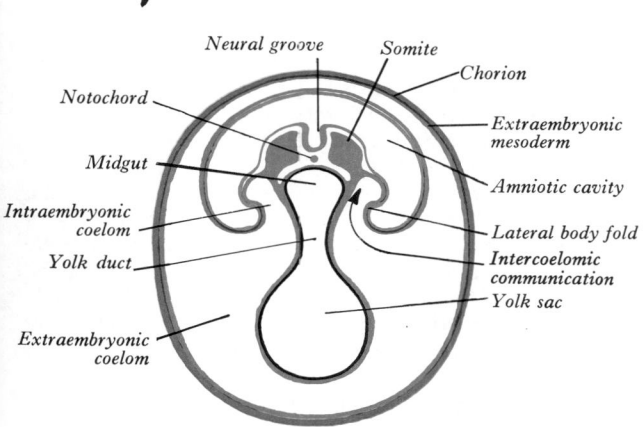

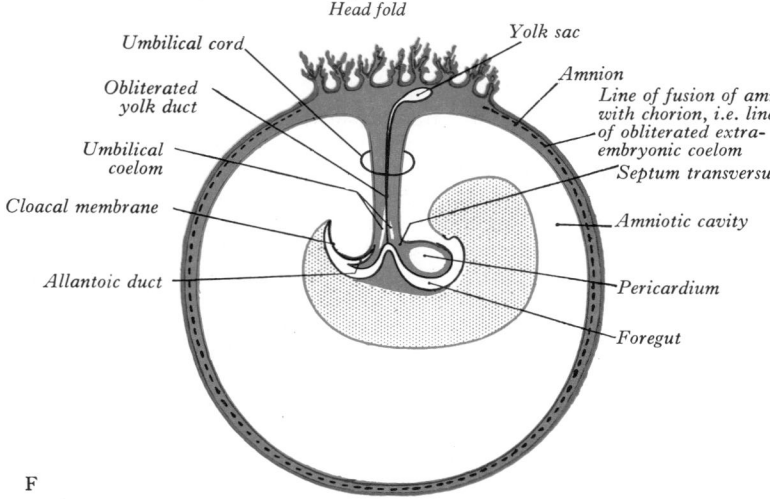

2.29A Diagram showing an early stage in development of the human blastocyst. B Diagram illustrating the early formation of the allanto-enteric diverticulum and the definition of the connecting stalk. C A later stage of the development. Observe that the heart occupies the most anterior part of the embryonic area and is separated from the prosencephalon by the oral membrane. D The formation of the head and tail folds, the expansion of the amnion, and the delimitation of the umbilicus. E A transverse section along the line *ab* in D. Observe that the intraembryonic coelom communicates freely with the extraembryonic coelom. F A later stage in the development of the umbilical cord.

It should be noted that whilst these diagrams, from an earlier edition, are useful for the general changes in disposition of the embryo and extraembryonic membranes, the modern view of development of the villous stems has been revised—consult text.

adenohypophyseal cells; (6) all retinal cells and the epithelia of the iris, ciliary body and processes.

It has been proposed, relating to (1) above, that neurons may be classified as either *central neurons* or *peripheral neurons* (Lieberman 1974); this highlights a particular mammalian distinction, but as yet has limited general usefulness.

Neural crest cells, first forming a strip of tissue flanking the neural plate then *migrating* widely, often in well-defined streams, and subsequently forming local aggregations or invading other tissues, give rise to: (1) the sensory neurons of the dorsal root ganglia of the spinal nerves; (2) many of the sensory neurons of the ganglia of the trigeminal, facial, vestibulocochlear, glossopharyngeal and vagal cranial nerves (other cranial nerves may have transient populations, see p. 200); (3) the principal postganglionic

autonomic neurons of both sympathetic and parasympathetic moieties, including the enteric plexuses, ganglionated sympathetic chains, and other ganglionic aggregates associated with, e.g. the cardiac, coeliac, mesenteric, renal and vesical plexuses; (4) many specialized peripheral sensory receptor cells (see also placodes below); Schwann cells, the perisomatic satellite cells of sensory ganglia, the 'glial' elements of autonomic ganglia and peripheral receptors; (5) the basal regions of the neurocranium and viscerocranium (p. 164), in particular the trabeculae cranii and cartilages of the branchial arches and possibly the osteogenic elements that become associated with these cartilages; (6) dentine production by odontoblasts of the tooth germs; probably admixed with mesenchyme from other sources, crest cells may contribute to the special visceral striated muscles derived from the branchial arches, and to the smooth muscle of the branchial vessels; (7) some, at least, of the meningeal pia-arachnoid; (8) cells belonging to the APUD series (vide supra and p. 1376). These include: thyroid parafollicular (C) cells derived from ultimobranchial tissue, classic chromaffin tissue (phaeochromocytes of the adrenal medulla, paraganglia and para-aortic bodies and other outlying masses, see 8.235), varieties of small intensely fluorescent (SIF) cells associated with sympathetic ganglionic neurons, the glomus (type I) cells of the carotid body (and probably many other functionally similar sites throughout the body) and finally both dermal and epidermal melanoblasts and other epidermal, non-melanoblastic dendritic cells; (9) contributions to connective tissues throughout the body. For a review of neural crest see Leikola (1976).

Ectodermal placodes are specialized patches of ectodermal cells (the named placodes being the olfactory, otic, dorsolateral, epibranchial, suprabranchial and lental) which invaginate, sometimes coming into close association with cells from other sources. They all have some characteristics in common with the neural crest and give rise to: (1) the receptor and probably sustentacular cells of the olfactory part of the nasal mucosa; (2) the epithelial walls of the membranous labyrinth and the specialized receptor cells of its utricular and saccular maculae, the ampullary crests of its semicircular ducts, and the cochlear organ of Corti; (3) *some* of the primary sensory neurons of the trigeminal, facial, vestibulocochlear, glossopharyngeal and vagal cranial nerves (although their precise contribution to these in the *human* embryo needs further clarification); (4) the crystalline lens of the eye; (5) some members of the APUD series, in particular the parathyroid chief cells. Certain authors prefer to regard the cells of the adenohypophysis as having a placodal origin through the medium of the ectodermal anlage of Rathke's pouch and similar considerations may apply to the parafollicular cells of the thyroid through the anlage of the ultimobranchial body. (The latter may therefore imply the existence of *endodermal* 'placodes'.)

It may be noted here that, despite cautious suggestions made about two decades ago (Pearse 1966), less than 10 of the over 40 cell varieties now accepted as belonging to the APUD series have definitely been shown to stem from the neural crest. Thus the large number of varieties of enterochromaffin cells present in the mucosae of the stomach and intestines, in the pancreatic islets and other varieties found in the bronchial mucosa, the lining epithelium of the urethra and prostatic parenchyma, have an uncertain origin. Their progenitors have been hypothecated as 'neuoendocrine-programmed ectoblastic cells' and a foregut endodermal origin has been postulated for the gastrointestinal/pancreatic varieties (Pearse 1977). For a further discussion of this topic and the introduction of the term 'paraneuron' see p. 1376 and refer to Fujita 1976.

THE EMBRYONIC MESODERM

This may be divided, as noted above, into *paraxial* mesoderm, *intermediate* mesoderm and the *lateral* plate. The paraxial mesoderm, as detailed below, segments into a longitudinal series of mesodermal somites and each somite is subdivided into a sclerotome, dermotome and myotome. The intermediate mesoderm forming a longitudinal, cord-like, cylindrical mass, which is continuous medially with the somites and laterally with the lateral plate, may best be considered as a cranio-caudal series of subregions—cervical, thoracic, lumbar and sacrococcygeal. Clefts appearing in the cardiogenic mesoderm and lateral plate coalesce to form the intraembryonic coelom (vide infra). The latter separates the lateral plate into somatopleuric (body wall) and splanchnopleuric (enteric) layers, which line or cover the ectoderm and endoderm and their derivatives respectively. As it enlarges the intermediate mesoderm projects into and also forms a partial boundary for the coelom, whilst the *zones of continuity* between the somatopleuric and splanchnopleuric mesoderm, after much differential growth, excavation, repositioning and in some places adhesion, become sculpted to form the definitive gastric, enteric, hepatic and splenic mesenteries. The epithelial or mesenchymatous nature of these various zones are discussed further below and in subsequent pages.

The paraxial mesoderm, through the medium of the somites, forms: (1) the vertebrae and their associated joints and ligaments, including much of the intervertebral discs; the ribs, costal cartilages, sternal plates (ultimately manubrium, sternebrae and xiphoid process), by migrations from the sclerotomes: cells from the latter probably contribute to the dura mater; (2) the dermotomic cells migrate ventrolaterally, admixing with neural crest cells and somatopleuric lateral plate mesoderm and, together with the costal elements, form the bulk of the tissues of the body wall (vide infra); (3) the cells of the myotomes, by migration, subdivision, rearrangement and sometimes fusion of cell masses from neighbouring myotomes, ultimately differentiate into the striated musculature of the trunk. (It appears that the limb muscles and lingual muscles, although segmentally innervated, develop in situ, with little obvious migration from the myotomes.) (4) perineural mesenchyme of uncertain, but possibly myotomic origin provides the microglia of the central nervous system.

The intermediate mesoderm through its principal longitudinal subdivisions, the nephrogenic cord, genital ridge and overlying medial and lateral coelomic mesothelia, forms: (1) the rudimentary pronephric and fully functional meso- and meta-nephric renal corpuscles and nephric tubules; (2) the mesonephric (Wolffian) duct and its derivatives—the vesical trigone, ureter and renal calices and collecting tubules in both sexes; male testicular fibrocellular framework and duct complexes, the epididymis, ductus deferens, seminal vesicle, ejaculatory duct and a small area of prostatic urethra; in the female merely rudimentary but homologous vesicular, tubular and duct systems are formed (p. 260); (3) the paramesonephric (Mullerian) duct and its derivatives—the female uterine tube and uterus, the development of the vagina being more conjectural; vestigial testicular structures and prostatic utricle in the male; (4) in both sexes, the whole of the gonadal tissues with the exception of the sex cells (gonocytes) themselves; (5) from the medial coelomic bay and encroaching on to the mesonephric ridge, the cortex of the adrenal gland; (6) the connective tissue framework, capsule, mesothelial covering and, where appropriate, the mesenteries of all the foregoing.

The lateral plate mesoderm becomes split, as noted, by the intraembryonic coelom, into somatic and splanchnic layers (p. 140), the somatic mesothelial lining of the coelom forming the serous parietal pericardium, pleura and peritoneum, whilst the splanchnic mesothelial lining forms the serous epicardium, pulmonary pleura and the visceral peritoneum that wholly or partly clothes the various abdominal and pelvic organs. The *somatopleuric mesenchyme*, as mentioned above and elsewhere, receives accessions from the neighbouring neural crest, and is also admixed with invading cells from adjacent dermotomes, myotomes and (in the thoracic region) sclerotomes. Thereafter mesenchymatous condensations of heterogeneous origins appear and growth and differentiation transform them into the osseous, cartilaginous, muscular, vascular, lymphatic and connective tissues of the body wall. As the appropriate spatio-temporal patterning of these tissues occurs, they gradually receive the terminals of ingrowing bundles of axonal processes (and accompanying Schwann cells) from the neuroblasts of the ventrolateral lamina of the neural tube and from the neuroblasts of the primordial spinal and autonomic ganglia. Similar tissues and transformations are involved in the formation of the limbs, the early limb buds consisting of an external covering of ectoderm and a mesenchymatous core of mixed origin.

The splanchnopleuric mesenchyme clothes the endodermal gut tube, including the endodermal cloaca, allanto-enteric diverticulum and their various subdivisions and glandular derivatives. Its cells grow and differentiate into the non-striated muscle, connective, vascular, lymphatic and adipose tissues of the walls of the stomach, small and large intestines, urinary bladder (excluding perhaps the trigone) and also the walls of (all but the finest terminal) biliary and pancreatic duct systems. Similar tissues are provided by splanchnopleuric mesenchymal cells in the walls of the tracheo-bronchial tree down to its finest pre-alveolar passages, but with the addition of cartilage as far as the commencement of the bronchiolar system. Most of these tissues receive the ingrowing terminals of post-ganglionic sympathetic axons and preganglionic parasympathetic terminals from the vagus (foregut and midgut and their derivatives) and the pelvic parasympathetic nerves (hindgut). Corresponding regions are invaded by postganglionic parasympathetic neuroblasts derived, it is presumed, from the vagal and sacral neural crest respectively.

It should be noted that vascular venous differentiation in the splanchnopleuric mesenchyme caudal to the developing diaphragm leads to numerous radicles draining virtually the whole of the subdiaphragmatic gut, liver, pancreas and spleen and these gradually combine to form the hepatic portal vein. In the primitive pericardial region, similar mesenchyme, including its covering mesothelium, forms all the non-nervous tissues of the heart. The spleen differentiates in the mesenchyme of the dorsal mesentery of the stomach (dorsal mesogastrium).

It should also be stressed that in the ventrolateral walls of the embryonic pharynx *no coelomic cavity* develops so that in the branchial arches there is *unsplit* lateral plate mesenchyme that has received substantial additions from the cranial neural crest and possibly also from adjacent sclerotomes and dermotomes. Thus, whilst it has become customary to loosely classify many branchial structures as 'visceral' or 'special visceral', it is by no means clear in mammals in general, and particularly in mankind, which of these sources of mesenchyme predominates in the anlage of any specific derivative.

Finally it may be mentioned that the splanchnopleuric mesenchyme of the yolk sac, liver and spleen are the earliest loci of red blood cell formation, preceding the appearance of progenitor cells in the red bone marrow.

THE EMBRYONIC ENDODERM

The main derivatives of the embryonic endoderm have already been listed above and will be further detailed in subsequent sections: to avoid excessive repetition, therefore, only some additional points or as yet unresolved questions will be mentioned here.

(1) Whilst in general it is usually stated that the endoderm forms the epithelium of the whole alimentary tract and its associated glands, from stomatodeum to proctodeum, its precise extent is less certain cranially and the origin of some of the epithelial cell types cannot be regarded as proven. The epithelia of the lips, gums, parotid glands, cheeks, dental enamel and anterior part of the hard palate are accepted as definitely of *ectodermal* origin; similarly, most authorities accept that the lingual epithelium and that of the lingual glands are endodermal; the exact site of the transition line between the two is not yet established. It must be stressed that the line does not necessarily correspond to any particular outstanding topographic landmark, such as the oropharyngeal isthmus or linguogingival sulcus, and authorities differ concerning the ectodermal or endodermal status they ascribe to the submandibular and sublingual salivary glands. In contrast, the generally accepted caudal limit of the hindgut endoderm is at the level of the anal 'valves'.

(2) The endoderm of the remainder of the post-oral foregut forms the great variety of epithelia that characterize the various zones, recesses, diverticula and glands of the pharynx and the general, glandular and duct-lining epithelia of the oesophagus, stomach and proximal duodenum. Some parts of the embryonic pharyngeal lining form endocrine glands that lose all connection with the gut, whilst other regions become associated with specialized localizations of lymphoid tissue.

(3) The origin of the hair cells of the lingual, palatal, pharyngeal and laryngeal taste buds (neural crest or 'neurally- programmed' foregut endoderm) is uncertain but, in addition to their sensory receptor function, they have been classified as members of the APUD neuroendocrine series and dubbed paraneurons (Fujita 1976).

(4) The lymphoid-associated endodermal derivatives constitute *Waldeyer's ring*: characteristically their epithelial surfaces are uneven (showing undulations, folds, crypts, fossulae, clefts or sinuses); they include the nasopharyngeal tonsil (adenoid), the (auditory) tubal tonsils, the palatine tonsils and the lingual tonsil.

(5) The pharyngeal epithelia range from respiratory, mucus-secreting, ciliated, columnar epithelium which lines the upper nasopharynx, the environs of the opening of the auditory tube and the superior part of the dorsal surface of the soft palate, together with the laryngeal vestibule, to the non-keratinized stratified squamous epithelium which lines the rest of the pharynx and palate.

(6) The epithelial lining of the auditory tube, tympanic cavity with its mesentery-like mucosal folds and internal lamina of the tympanic membrane, and of the tympanic antrum is derived from endoderm (p. 1228) and ranges from tall columnar and ciliated, through cuboidal to simple squamous in different locations.

(7) Other pharyngeal, endodermal derivatives include the general mucous glandular and duct-lining cells; the main follicular and parafollicular cells of the thyroid (from different locations); parathyroid secretory cells (debate persists concerning their origin—Fujita 1976, Pearse 1977); and the cytoreticulum and concentric corpuscles of the thymus.

(8) Little is to be added to the widely held view that the great majority of the epithelial cells, glandular and duct-lining cells of the oesophagus, stomach, jejunum, ileum, colon, rectum and upper anal canal are of endodermal origin: similar considerations applying to the hepatocytes and duct-lining epithelia of the biliary duct system and to the pancreatic acinar and duct-lining cells: again the majority of epithelial cells—ranging from tall columnar ciliated and mucus-producing to an extremely attenuated simple squamous type—that line the whole of the tracheobronchial-alveolar respiratory tree are endodermal in origin. A similar derivation is also accepted for most of the transitional epithelium lining the urinary bladder and urethra and for the parenchymal secretory cells of the prostate and lesser urethral glands. In almost all these locations, however, as intimated above, are numerous cell types loosely termed *enterochromaffin cells*: characteristically these show properties of *receptor mechanisms* on their surfaces and the production of *neuroendocrine secretions* and *neurotransmitter-like* substances. (Not all the cells postulated as enterochromaffin have, as yet, fulfilled all these criteria.) Particular interest has been focused upon the pancreatic islet cells and the so-called neuro-insular complex (p. 1384).

(9) Finally, reference must be made to the possible relationship between endodermal cells and the haemopoietic stem cells (p. 675) and to the origin of the primordial sex cells (p. 254).

SEGMENTATION OF THE MESODERM

In all vertebrate embryos the intraembryonic mesoderm becomes incompletely divided by a longitudinal groove into a *paraxial part* and a *lateral plate*, on each side of the midline (2.27). The mesoderm in the floor of the groove which connects these two parts is the *intermediate mesoderm*, most of which subsequently forms the *nephrogenic cord*. The latter progressively expands into the coelomic cavity and further longitudinal grooves develop delineating its medial and lateral limits and principal subdivisions. Soon after the appearance of the initial longitudinal groove the paraxial mesoderm becomes subdivided into prismatic blocks by a series of transverse grooves (2.26, 27). This is termed *segmentation of the mesoderm* and the blocks of paraxial mesoderm so formed are the *mesodermal somites (primitive segments or metameres)*. Commencing at the end of the third or the beginning of the fourth week in the region of the hindbrain, the process extends in a *cranio-caudal direction*, additional somites being added as the embryo grows in length until 42–45 pairs are present (2.125). The most cranial four or possibly five lying alongside the hindbrain participate in the formation of the skull and are named the *occipital somites*. However, other workers (de Beer 1937, Sensenig 1957) have suggested that originally nine somites were involved in the elaboration of part of the skull. The remainder flank the spinal cord and usually number: 8 cervical, 12 thoracic, 5 lumbar, 5 sacral and 8–10 coccygeal. Variations in this sequence may occur, usually involving either an additional or a missing segment in either the thoracic, lumbar or sacral regions, a variation in one region often being accompanied by a compensatory variation in a neighbouring region. The first occipital pair are often asymmetrical in their degree of development and one or both may be partly or completely suppressed. Some hold that mesodermal masses, serially homologous with the segmented paraxial mesoderm, arise beside the cranial tip of the notochord, the anlagen of the extrinsic ocular muscles (2.54). (For experimental data concerning mesodermal segmentation in higher vertebrates consult Bellairs 1971; for more recent analyses, hypotheses and related bibliographies see Burgess 1981, 1983.)

In the *human embryo* it is only the *paraxial* mesoderm alongside the notochord which is physically segmented but, in view of the obviously segmental arrangement of the nerves of the spinal cord and their areas of distribution to all the various embryonic layers, it is reasonable to suppose that forms of segmentation of other structures are present although more difficult to display.

This basic *metameric segmentation* is a characteristic feature of the whole phylum Chordata and is particularly well seen in the primitive miolecithal forms in which, after gastrulation (p. 102 and 2.3), the mesoderm arises as a series of evaginations from the archenteron, which form independent, paired, segmental, coelomic, mesodermal pouches. These expand, separating the ectoderm from the endoderm until they meet dorsal and ventral to the gut as permanent dorsal and (often temporary) ventral mesenteries. It is quite evident that in these forms the metameric segmentation involves not only paraxial structures but the intermediate mesoderm and both somatopleure and splanchnopleure (and the coelom itself).

THE SOMITE

A typical human *mesodermal somite* consists of tightly packed (epithelioid) cells and at first contains a transient central cavity which is, however, soon obliterated by cells proliferated from its walls. The ventromedial cells, the *sclerotome*, become mesenchymatous and migrate medially to provide the tissue from which the axial skeleton is ultimately derived (p. 159). The dorsolateral somite cells form a *dermomyotome*. Spindle-shaped cells proliferate from its margins to form a tightly packed cellular mass on its medial aspect, the *muscle plate* or *myotome* (p. 175). The remaining epithelially arranged cells constitute the *skin plate* or *dermotome* (2.27D). The myotome is the forerunner of much of the striated musculature of the body (p. 175) and, after it has commenced differentiation, the cells of the dermotome lose their epithelioid character and become mesenchymatous. They spread to mix with the somatopleuric mesenchyme and, probably with accessions from the neighbouring neural crest, lay down the foundation of the dermis.

THE EARLY INTRAEMBRYONIC COELOM

During early somite formation a midline cavity appears in the *cardiogenic mesoderm*, the first indication of the *intraembryonic coelom*. As segmentation of the mesoderm proceeds, a number of clefts are also formed on each side in the lateral plate mesoderm. The clefts expand, gradually coalesce and the lateral plate becomes divided into a *somatic* and a *splanchnic* layer (2.27, 20). The somatic layer, with its covering of ectoderm, constitutes the *intraembryonic somatopleure* whilst the splanchnic layer, with the underlying endoderm, constitutes the *intraembryonic splanchnopleure*.

The lateral extremities of the cavity in the cardiogenic mesoderm curve to extend caudally and link up with the coalescing clefts in the lateral plate and as a result the intraembryonic coelom is formed as an inverted U-shaped tube, from which the pericardial, pleural and peritoneal cavities are subsequently developed. Around the periphery of the embryonic area the somatopleuric and splanchnopleuric mesoderms are continuous, at first both with each other and with the similar extraembryonic layers, at the *junctional zone* mentioned previously. Soon, however, the process of cavitation in the lateral plate extends beyond the embryonic area and the continuity between these layers is broken; the intraembryonic coelom is now thrown into free communication with the extraembryonic coelom (2.27C, 28C, 29C). This process, however, is limited to the lateral regions; it does not affect the rostral rim of the pericardial area or the regions immediately adjoining it on either side, where the median *primitive*

pericardial cavity now opens into bilateral *pericardioperitoneal canals (coelomic ducts)* (2.28C). Accordingly, the junctional mesoderm around the rostrolateral border of the embryo which will form the *septum transversum* and *primitive ventral mesentery* of the foregut (p. 232) remains intact, as does the caudal junctional mesoderm which continues into the connecting stalk (2.28C).

The early formation of the intraembryonic coelom and its free communication with the extraembryonic coelom allows the fluid which fills the latter to gain access to the interior of the embryo. It thus provides a path for the passage of nutrients during the period which still has to elapse before the establishment of a blood vascular circulation (Streeter 1942). The walls of the coelom are formed of undifferentiated mesodermal cells (*mesoblasts*) which rapidly proliferate. From the mesoblasts of the somatopleure, together with those derived from the dermatome and some neural crest, the corium and subcutaneous tissues and numerous other mural connective tissues are formed. Splanchnopleuric mesenchymal cells, on the other hand, later differentiate into the muscles, blood vessels, lymphatics, adipose and connective tissues of the walls of the heart and gastrointestinal tract.

It is not until later that the mesoblasts which directly line the coelom itself become differentiated into the mesothelia which characterize the various serous membranes (pp. 51, 135).

The Formation of the Embryo

Hitherto, the embryonic area, initially bilaminar and later trilaminar, has been essentially a *flattened* disc which changes to an oval and then a piriform outline. Towards the end of the third week, however, the embryo begins to assume its definitive shape. The immediate cause is the difference in the rate at which adjoining areas grow. The periphery fails to keep pace with that within the embryonic area and, as the embryo is increasing more rapidly in its long axis, especially at its cranial end where the walls of the neural groove are expanding to form the forebrain, both extremities tend to project beyond the original limits of the area (2.29). In this way a *head fold* is developed cranially and a *tail fold* caudally. Simultaneously, and for similar reasons, less prominent right and left *lateral body folds* develop and the extension of these four folds (2.29) gradually constricts off the embryo from the yolk sac and imparts its characteristic shape. The latter has been described as 'comma-like', increasingly convex dorsally, expanded cranially and tapering to a curved cauda (tail) concave ventrally; the whole progressively projects into the ballooning amniotic cavity.

Resulting from head fold formation, the prosencephalon, which was hitherto separated from the cranial extremity of the embryonic area by the oropharyngeal membrane (p. 132) and pericardium, comes to lie at the extreme cranial tip of the embryo. This alteration in position of the forebrain is accompanied by a corresponding alteration in the relative positions of the membrane and pericardium (2.29 C–F). The former now lies on the *ventral surface* and forms the floor of a depression, the primitive mouth or *stomatodeum*. Cranially the stomatodeum is bounded by the projecting forebrain, and caudally by the pericardium. The latter has not only altered its position relative to the cephalic extremity of the embryo but has also undergone a reversal of its surfaces and cranio-caudal limits, as will be examined below.

In addition to these positional alterations and reversals, head folding results in the inclusion within the embryo of a portion of the yolk sac, now termed the *foregut (proenteron)*. The latter is thus placed with the oropharyngeal membrane and pericardium on its ventral aspect and with the hindbrain placed dorsally. It communicates at its caudal end with the *midgut (mesenteron)* through an opening, the *anterior intestinal portal*.

The ventral bend of the head fold is associated with a correspondingly pronounced *midbrain flexure* which is concave ventrally and which underlies a projecting dorsal convexity (*midbrain prominence*) of the overlying ectoderm, seen to best advantage when the embryo is viewed from the side (e.g. 2.55).

Before tail fold formation, the *caudal* end of the embryonic area is anchored to the trophoblast by the connecting stalk, which is covered on one aspect by the amnion (2.29C). The formation of the tail fold carries the attachment of the connecting stalk round on to the *ventral* aspect of the embryo so that it now assumes the permanent attachment site of the umbilical cord. It will be remembered that the stalk was originally connected to the embryo at the caudal end of the primitive streak (but separated from it by the cloacal membrane (vide infra) and in consequence the primitive streak also extends around and reverses towards the ventral aspect of the embryo to the region which later lies immediately adjacent to the anal orifice. Some, however, hold the view that the primitive streak is continued into the rudimentary tail and does not appear in the perineum or on the ventral surface of the embryo.

Just as a recess of the yolk sac is included within the head fold to form the foregut, so a corresponding part is included within the tail fold to form the *hindgut (metenteron)*. But the similarity between these two included recesses goes further. A portion of the endoderm (earlier the prechordal plate, p. 132) in the floor or ventral wall of the foregut is in direct contact with the ectoderm over an area termed the *oropharyngeal membrane*. This membrane soon disappears and the communication of the gut with the exterior through the mouth is thus established. In part of the hindgut a similar relationship exists. As we have seen, even before the tail fold is defined, the ectoderm and endoderm are in contact with each other at the caudal end of the embryonic area, forming the *cloacal membrane*. As will be described later, this membrane subsequently breaks down in two places to form the *urogenital* and *anal orifices*.

Before tail fold formation an endodermal diverticulum arises from the dorsocaudal wall of the yolk sac and grows into the mesoderm of the connecting stalk (2.21). This outgrowth is the *allanto-enteric diverticulum* (2.29B). As the tail fold is defined, the proximal part of the diverticulum becomes incorporated in the hindgut and its distal part persists as the *endodermal allantoic duct* (p. 143), which then communicates directly with the *ventral* surface of the hindgut. The region of hindgut caudal to this communication expands to form the *endodermal cloaca* (p. 233). As tail fold formation continues, the cloacal membrane itself also progressively becomes intercalated in the ventral wall of the hindgut, even extending on to the adjoining caudal aspect of the connecting stalk, where it is associated with the allantoic part of the allanto-enteric diverticulum. Even before the latter is incorporated in the hindgut, the cloacal membrane becomes interrupted and shortened by the interposition of mesoderm between the endoderm of the diverticulum and the covering epithelium of the body stalk (Florian 1930).

Between the head fold and the tail fold the embryo, as noted, additionally becomes constricted off by right and left *lateral body folds*. The intervening dorsal portion of the yolk sac, which these folds gradually include within the embryo, constitutes the *midgut (mesenteron)*. At first the midgut communicates freely on its ventral surface with the rest of the yolk sac but the continued growth of all the folds results in a narrowing of the connection, which becomes drawn out as the *splanchnopleuric yolk stalk* (vitello-intestinal duct) which contains the *endodermal yolk duct* (2.29E, F). The remainder (distal part) of the yolk sac remains extraembryonic (paraplacental) and is often termed the *umbilical vesicle*. The subsequent history of the duct and the vesicle will be dealt with later.

THE NUTRITION OF THE EMBRYO

In early development the blastomeres derive their nourishment in part from the store laid up within the cell body of the primary oöcyte. Such stores are possibly maintained at a high concentration and are subsequently liberated in a more dilute form for absorption from the cavity of the blastocyst and later from the primary and secondary yolk sac cavities (p. 130). In addition, it is assumed that the blastocyst derives nourishment from tubal and uterine secretions, subsequently, during the process of embedding, from products stemming from the lysed uterine tissues. Then follows a period of about 2 weeks during which the embryonic disc is dependent on the nutriment it can obtain from the fluids which fill the cavities of the amnion, the coelom and the yolk sac. These fluids contain material absorbed by the trophoblast from the lysed uterine tissues and extravasated maternal blood, selectively modified perhaps, as they pass through the cellular walls of these various cavities. However, at an early stage in development these sources of supply are much diminished. The lumen of the neural tube is isolated by closure of the neuropores, the extraembryonic coelom becomes greatly reduced (2.29D,F) and is later shut off from the intraembryonic coelom, and the obliteration of the yolk duct separates the yolk sac from the gut. It therefore becomes imperative that some other source should be available at an early stage. This involves the maternal circulation coming into close, although indirect, *apposition* with the developing embryonic circulation.

The differentiating mesenchyme in which the embryonic vessels and erythrocytes develop, *angioblastic tissue*, is probably first formed from the deepest layer of mesenchyme which clothes the endoderm of the yolk sac early in the third week (p. 206).

Slightly later, angioblastic tissue can also be recognized in the connecting stalk and mesenchyme of the chorion, and it then appears also within the embryonic area. Tissue spaces form in the angioblastic tissue and the cells which line them differentiate into typical, flattened endothelial cells whilst adjoining spaces join to form capillary plexuses. While the yolk sac spaces are forming, small, localized groups of mesodermal cells project into them and become cut off to form *blood islets*, their cells differentiating into embryonic erythrocytes (pp. 206, 675).

The vessels formed in the chorion soon establish an intimate relationship with maternal circulation (2.32, p. 145). Vessels develop in the embryo as two longitudinal channels which, at their headward ends, invade the wall of the pericardium; the position and direction of the invasion changes with the progress of head fold formation. They are the rudimentary right and left dorsal aortae and, after folding their cranial ends, curve ventrally in the lateral wall of the pharynx to reach the cranial end of the pericardium, where they fuse, becoming continuous with the developing primitive tubular heart. Caudally the aortae traverse the connecting stalk as the rudimentary umbilical arteries and break up into capillaries in the chorion. The venules from the chorion converge on the stalk where they form the right and left umbilical veins, which run headwards in the somatopleure, close to the margin of the embryonic area, to reach the caudal end of the tubular heart.

The pericardial cavity never communicates directly with the extraembryonic coelom, and (before head folding) at its craniolateral limits the somatopleure and splanchnopleure are continuous (2.81A). With formation of the head fold the mesodermal masses extending from surfaces of the pericardium are altered in disposition or even reversed and the original cranial mass comes into intimate relation with the *ventral* wall of the foregut as far as the cranial rim of the anterior intestinal portal (2.29C). After reversal the caudal wall of the pericardium deepens dorsoventrally; the mesenchyme between it, the gut and proximal yolk stalk forms a sheet, which is the *septum transversum*. (It has been noted that prior to reversal this mesoderm formed a U-shaped mass intervening between the pericardium and the extraembryonic coelom, p. 140). The post-reversal septum transversum plays an important part in the later development of the diaphragm (p. 239). At this stage it is bounded on its headward surface by the pericardium and on its caudodorsal surface by the foregut; on its dorsolateral surface it is limited by the bilateral coelomic ducts, which connect the pericardium with the peritoneal cavity, and on its caudolateral surface by the single crescentic opening of the peritoneal cavity into the extraembryonic coelom. The umbilical and body wall veins, which run in the somatopleure, and the vitelline veins, which run in the splanchopleure, meet in the junctional mesoderm of the septum transversum and so gain the venous end of the heart. Through these various channels the early embryonic circulation is established (p. 213, et seq.).

The Fetal Membranes and Placenta

THE ALLANTOIS

The allanto-enteric diverticulum (2.29B, C) arises early in the third week as a solid, endodermal outgrowth from the dorsocaudal part of the yolk sac into the mesoderm of the connecting stalk. It soon becomes canalized and, when the hindgut is developed, the proximal (enteric) part of the diverticulum is incorporated in its ventral wall and the distal (allantoic) part remains as the allantoic duct and is carried ventrally to open into the ventral aspect of the cloaca or terminal part of the hindgut (2.29A–F). The diverticulum, lined with endoderm, is surrounded by mesoderm of the connecting stalk, in which the umbilical vessels develop at a slightly later stage.

In reptiles, birds and many mammals the allantoic diverticulum develops into a stalked vesicle which continues expanding into the extraembryonic coelom and forms a vascular organ to which the term *allantois* should perhaps be restricted. In birds it projects to the right side of the embryo and gradually spreads over the dorsal surface of the embryo as a flattened sac between the amnion and the chorion (serosa), and ultimately also spreads to surround the yolk sac. Its outer wall becomes applied to, and fuses with, the chorion (forming an *allantochorion*) which lies immediately inside the shell membrane. Blood is carried to the allantoic sac by two *allantoic* (or *umbilical) arteries*, which are continuous with the primitive aortae, and, after circulating through the allantoic capillaries, is returned to the heart by (initially) two *umbilical veins*. In this way the chorio-allantoic circulation, which is of the utmost importance in connection with the respiration and nutrition of the chick, is established. Oxygen is taken from, and carbon dioxide is given up to, the atmosphere through the porous egg-shell. Simultaneously nutritive materials are absorbed from the yolk by the network of blood capillaries in the splanchnopleuric wall of the yolk sac. This net is fed and drained to the embryonic body via *vitelline arteries* and *veins*, constituents of the yolk stalk. With the formation of the amnion the embryo is, in most mammals, separated entirely from the chorion and is not united to the chorion again until the allantoic mesenchyme spreads to become applied to its inner surface. The human embryo, however, is never wholly separated from the chorion, its caudal end being from the first connected with the chorion by a thick band of mesoderm, the *connecting stalk*, which accordingly is regarded as precociously formed *allantoic mesoderm*.

THE AMNION

This is a membranous sac which surrounds the embryo; it is developed in reptiles, birds and mammals (Amniota), but not in amphibia or fishes (Anamniota).

In the human embryo the amnion appears as a cavity within the inner cell mass adjacent to the overlying trophoblast. (For details see p. 129 and Luckett 1975.) This cavity is roofed by a stratum of epithelial cells, and its floor is formed by the cells of the embryonic germ disc—continuity between the roof and floor being at the margin of the disc. In the first half of pregnancy the epithelial cells are flattened and contain abundant glycogen; in later stages they become cuboidal or columnar over the placenta, the glycogen diminishes in amount and lipid globules appear in the cytoplasm (Goto 1959). Ultrastructurally, two types of amniotic epithelial cell have been described (Thomas 1965): the 'Golgi type' and the 'fibrillar type'. The former has a pronounced Golgi apparatus and a considerable content of rough endoplasmic reticulum and membrane-bound vesicles; the fibrillar cytoplasm of the latter contains few organelles. Possibly they are physiological variants of the same cell. Their free surfaces are beset with irregular microvilli embedded in a surface coat, whilst their deep surfaces rest on a basement membrane. The intercellular clefts present scattered desmosomes, but elsewhere the clefts widen and contain interlacing microvilli between which complex tubules penetrate the cell cytoplasm. These features suggest an active role in selective transport across the membrane (Lister 1968, Wynn & French 1968).

Externally the amnion is covered with a thin layer of somatopleuric extraembryonic mesoderm, which is continuous at the margins of the disc both with the splanchnopleuric extraembryonic mesoderm covering the yolk sac and with the intraembryonic mesoderm. Through the connecting stalk it is continuous also with the extraembryonic chorionic mesoderm.

Fluid, the *liquor amnii*, occupies the amniotic cavity and increases steadily in volume as the sac gradually expands in the extraembryonic coelom (2.29); this continues until the coelom is obliterated, except for a small volume which is enclosed within the proximal part of the umbilical cord (the *umbilical coelom*). The liquor amnii increases in quantity up to the sixth or seventh month and then diminishes somewhat; at the end of pregnancy it is usually about a litre. It provides a buoyant medium which supports the delicate tissues of the young embryo and allows free movement of the fetus during the later stages of pregnancy. It also diminishes the risk to the fetus of injury from without. It contains less than 2% of solids, including urea, inorganic salts, a small amount of protein and frequently a trace of sugar.

The source of the liquor, whether fetal, maternal or both, is not finally settled. Its volume is regulated by a multiplicity of factors, both fetal and maternal, but it is suggested that normally it is predominantly fetal in origin (Davies 1960). In the early stages it resembles blood plasma in composition and is probably formed largely by transport across the amniotic membrane but as pregnancy advances it becomes progressively more dilute, partly by the addition of fetal urine. It has been shown experimentally that there is a considerable and rapid flux of water across the amniotic membrane. There is rapid exchange between the amniotic fluid and maternal and fetal circulations, probably via the placenta and fetal kidneys. By the end of the third month the expanding amnion has extensive contact with the chorion and only these thin membranes separate the amniotic fluid from the decidua parietalis, the tissues and vessels of which may provide another route for the exchange of water and dissolved substances (Plentl 1958). A volume of amniotic fluid in excess of 2 litres is generally considered to be abnormal and constitutes *hydramnios*. A deficiency is termed *oligamnios*. Both conditions may be associated with fetal abnormalities, e.g. fetuses with agenesis of the kidneys or atresia of the lower urinary tract are often associated with oligamnios.

It has been suggested that fetal swallowing of amniotic fluid is a normal occurrence, and also that respiratory movements aspirate some fluid into the fetal lungs. In either case the fluid may be absorbed into the fetal circulation and then pass the placental barrier into the maternal circulation. Cases of oesophageal atresia or anencephaly, in which swallowing is impossible or impaired, and open spina bifida are often associated with hydramnios. With these neural defects, impaired swallowing is accompanied by direct discharge of cerebrospinal fluid into the amniotic liquor.

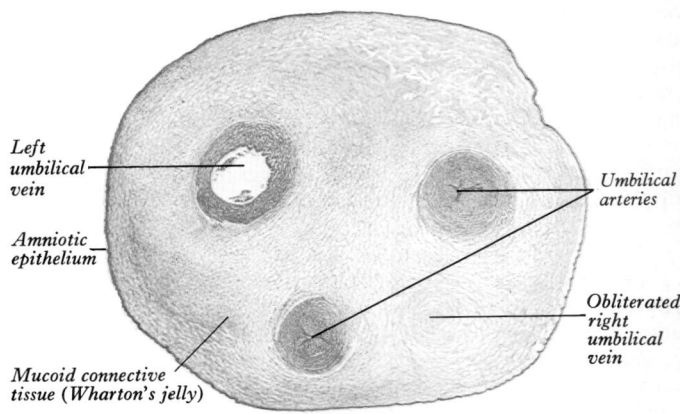

Left umbilical vein

Amniotic epithelium

Umbilical arteries

Obliterated right umbilical vein

Mucoid connective tissue (Wharton's jelly)

2.30 Transverse section through a human umbilical cord. Stained with haematoxylin and eosin. Magnification × c.8.

THE CONNECTING STALK AND UMBILICAL CORD

The connecting stalk (2.29) is, as described above, a mass of precociously formed allantoic mesoderm, which at first connects the caudal end of the embryonic area with the chorion. Proximally (its embryonic end) it surrounds the short allanto-enteric diverticulum but it is traversed throughout its length by the umbilical (allantoic) vessels. At first its dorsal surface is covered with the amnion and its ventral surface is bounded by the extraembryonic coelom. As a result of the folding of the embryo and distension of the amnion, the embryonic end of the connecting stalk comes to lie on the ventral surface of the embryo, and its mesoderm approaches that covering the yolk sac and its stalk. With continued expansion of the amnion, the extraembryonic coelom is largely obliterated (2.29) and its only remaining part surrounds the elongating yolk stalk; this part still communicates freely through the umbilicus with the intraembryonic coelom. The mesoderm-covered surfaces of the head, tail and folds of the expanding amnion now reach the chorion and converge on the connecting stalk and yolk stalk (and their vessels), and the umbilical cord is formed as they meet and their mesoderms fuse (2.29) thus almost completely closing off the intra-embryonic coelom. A limited exocoelomic recess persists in the embryonic end of the cord (umbilical coelom), retaining its communication with the intraembryonic coelom, and is involved in later enteric development (vide infra and p. 232).

The umbilical cord (2.30) thus consists of an outer covering of flattened amniotic epithelial cells, containing an interior mass of mesoderm of diverse origins (vide infra). Embedded in the latter are two endodermal tubes—the yolk and allantoic ducts—their associated vitelline and allantoic (umbilical) blood vessels and, near its fetal end, the remains of the extraembryonic coelom mentioned above.

The *mesodermal core* is derived from the somatopleuric extraembryonic mesoderm covering the amniotic folds: the splanchnopleuric extraembryonic mesoderm of the yolk stalk which carries the vitelline vessels and clothes the endodermal yolk duct; and similar allantoic mesoderm of the connecting stalk which clothes the allantoic duct and carries initially two umbilical arteries and two umbilical veins. These various mesoderms fuse and are gradually transformed into the viscid, mucoid connective tissue (*Wharton's jelly*) which characterizes the more mature cord. The tissue consists of widely spaced fibroblasts separated by an extensive intercellular space filled with a copious matrix consisting of a delicate three-dimensional meshwork of fine collagen fibres surrounded by a dilute ground substance containing a variety of hydrated mucopolysaccharides. In specimens which have been excised *before* fixation and staining, the fibroblasts present stellate profiles, as they do in other areas of mucoid tissue when similarly prepared. However, it has been shown that if the tissue is fixed *before* any haemodynamic collapse and consequent modification of the jelly has occurred, the cells are long and strap-like and present a regular orientation (Parry 1970).

The part of the extraembryonic coelom (the *umbilical coelom*) included in the base of the umbilical cord acts as a sac which receives the normal *umbilical hernia* of the midgut, developing in the embryo between the sixth and tenth weeks (p. 232). After the disappearance of this hernia the extraembryonic coelomic sac is normally obliterated.

The yolk sac becomes located between the amnion and chorion as they fuse near the placental attachment of the cord (2.29, 31, 40)—it continues to grow slowly and is sometimes found at term in this site, as a small vesicle usually less than 5 mm in diameter. The yolk stalk and its contained endodermal duct and accompanying vessels gradually elongate with growth in length of the umbilical cord. The duct and vessels slowly degenerate and they have usually disappeared by mid-pregnancy.

The endodermal allantoic duct, which is confined to the proximal end of the growing cord, also elongates and thins but may persist as an interrupted series of epithelial strands until term. At the umbilicus the proximal strand is often continuous with the median intra-abdominal *urachus*, which in turn continues into the apex of the bladder (p. 258).

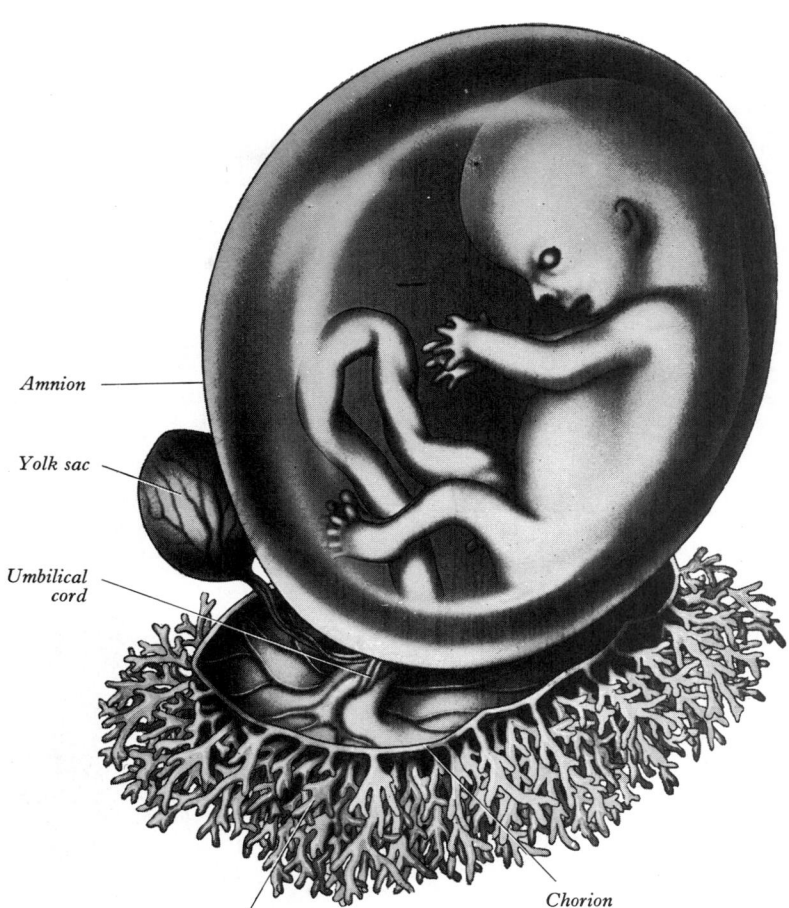

Amnion

Yolk sac

Umbilical cord

Chorion

Villous stems of chorion frondosum

2.31 A fetus of about 8 weeks, enclosed in the amnion, magnified about $2\frac{1}{2}$ diameters. A part of the chorion frondosum with its branching villous stems is shown in the lower part of the figure. The villous stems have been detached from the basal plate, which is not shown here.

Usually, the embryonic right umbilical vein disappears in the early months of pregnancy (and exceptionally only one artery may persist). The vessels of the umbilical cord are rarely straight but usually show a twisted conformation which may exist as either a right- or left-handed cylindrical helix. The number of turns involved may be relatively few or, at the other extreme, may even exceed 300. Their causation has been variously ascribed to unequal growth of the vessels, or to torsional forces imposed by fetal movements; their functional significance is obscure; perhaps their pulsations and contractions (vide infra) assist the venous return to the fetus in the umbilical vein. When fully developed the umbilical vessels, particularly the arteries, are provided with a strong muscular coat which contracts readily in response to mechanical stimuli. The outermost muscle bundles pursue an interlacing spiral course so that, when they contract, they produce shortening of the vessel and thickening of the media, with folding of the interna and considerable narrowing of the lumen. This action may account for the periodic sharp constrictions of contour—the so-called *valves of Hoboken* which often characterize these vessels.

When fully developed, the umbilical cord is on average some 50 cm long and 1–2 cm in diameter, but the length is subject to great variation (20–120 cm). Obstetrical complications during labour may accompany an exceptionally short or long cord.

Cords may also exhibit *knots* which are 'true' or 'false'. True knots presumably follow some exceptional form of fetal movement in utero, and may embarrass the circulatory flow in the umbilical vessels. False knots are sharp variations in contour which may reflect abnormally pronounced looping of one of the umbilical vessels, or a local accumulation of a mass of Wharton's jelly. For variations in the placental attachment of the cord see p. 156, and for further details of its morphology and an extensive review of the relevant literature see Boyd & Hamilton (1970).

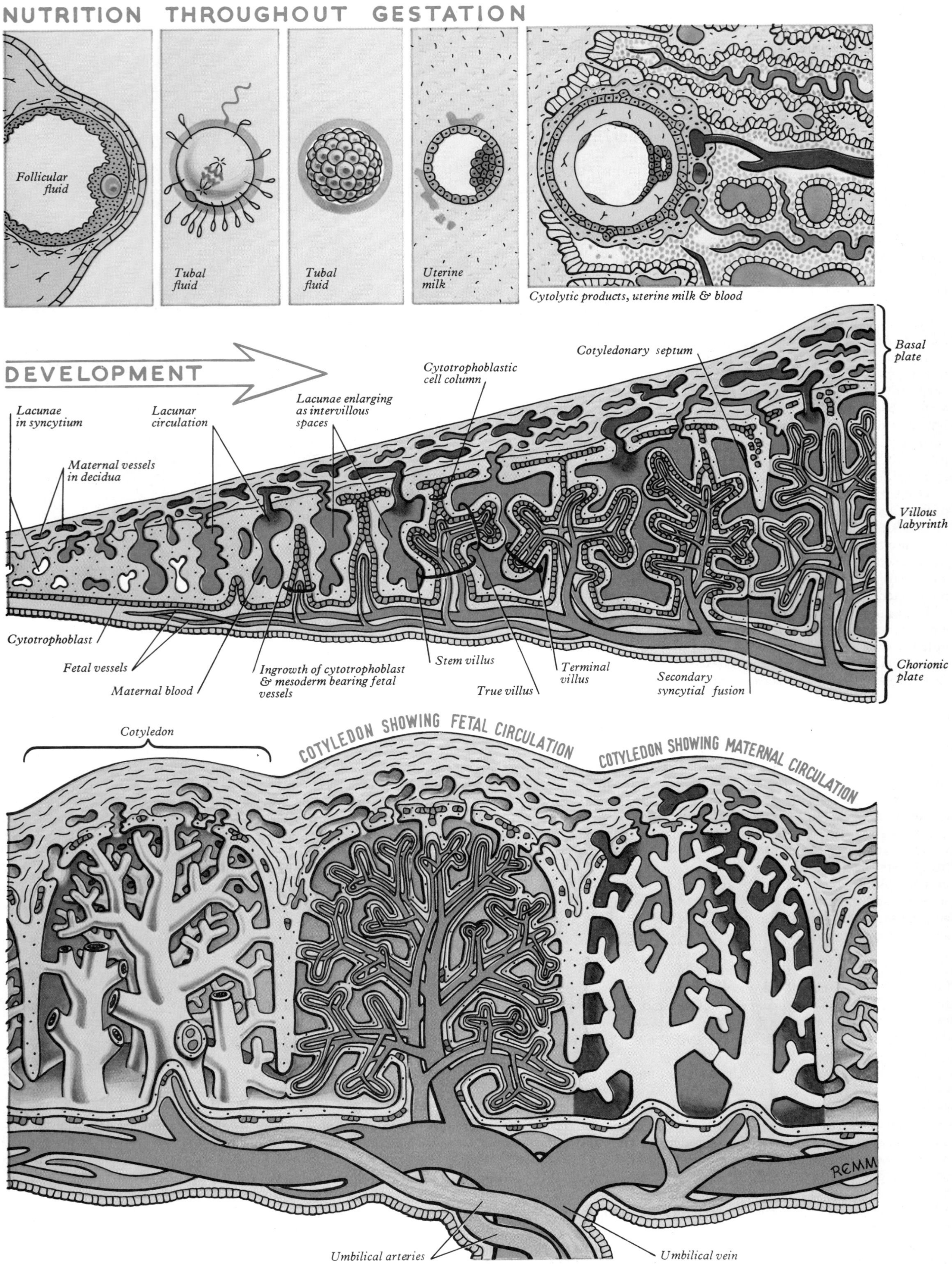

NUTRITION THROUGHOUT GESTATION
Follicular fluid
Tubal fluid
Tubal fluid
Uterine milk
Cytolytic products, uterine milk & blood
DEVELOPMENT
Lacunae in syncytium
Lacunar circulation
Lacunae enlarging as intervillous spaces
Cytotrophoblastic cell column
Cotyledonary septum
Basal plate
Maternal vessels in decidua
Villous labyrinth
Cytotrophoblast
Fetal vessels
Maternal blood
Ingrowth of cytotrophoblast & mesoderm bearing fetal vessels
Stem villus
True villus
Terminal villus
Secondary syncytial fusion
Chorionic plate
Cotyledon
COTYLEDON SHOWING FETAL CIRCULATION
COTYLEDON SHOWING MATERNAL CIRCULATION
RCMM
Umbilical arteries
Umbilical vein

2.32 Nutrition of oöcyte, zygote, morula, free and embedded blastocyst, embryo and fetus, throughout gestation. Blastocystic and placental development proceed from left to right. Aspects of mature placental structure and circulation are shown below.

As noted (p. 125), fertilization occurs in the lateral or ampullary part of the uterine tube, and is immediately followed by cleavage. The segmenting zygote is conveyed along the tube to the uterine cavity by ciliary action of the tube aided by muscular tubal contractions, the journey occupying about 3 days. Approaching the uterine lumen the morula becomes a blastocyst still surrounded by the zona pellucida. The lysis of the zona and the formation of polar syncytial trophoblast over the formative mass are prerequisites for its implantation in the uterine mucosa (see pp. 127, 129). In the interval between ovulation and blastocyst arrival in the uterine cavity, preimplantation changes also occur in the uterine mucosa. These progestational changes are detailed later (p. 148) but, having occurred, the syncytial trophoblast adheres to the uterine mucous membrane, destroys the epithelium over the area of contact and excavates a cavity in the membrane in which the blastocyst implants. In a conceptus described by Bryce & Teacher (1908) the point of entrance was visible as a small gap closed by a mass of fibrin and leucocytes; in another (Peters 1899), the opening was covered with a mushroom-shaped mass of fibrin and blood clot, the narrow stalk of which plugged the aperture in the mucous membrane. It is held that this *operculum* represents in part some syncytial trophoblast cut off by the decidua capsularis (Böving 1963).

THE TROPHOBLAST AND CHORION

Actively excavating the uterine mucous membrane is the syncytial trophoblast which increases rapidly in thickness over the embryonic pole and then forms a progressively thinner layer over the rest of the wall towards the abembryonic pole. As the blastocyst implants, it invades and digests the uterine tissues, including the walls of the uterine (maternal) blood vessels (see 2.22, 23 and Böving 1959, 1963). Lacunar spaces develop in the trophoblastic envelope and establish communications with each other. Early, many of them contain maternal blood (2.21, 23) derived from dilated uterine capillaries and veins, the walls of which have been partially destroyed. As the conceptus grows, the lacunar spaces enlarge, becoming confluent to form an initial blood-filled *intervillous space*; their trophoblastic walls are converted at first into an *irregular* spongework, or *labyrinth*. With further growth, however, the main strands of trophoblastic syncytium assume a *radial* arrangement. Thus, from early stages, the intervillous space is *completely spanned* by these radial strands which undergo an orderly sequence of three main histological changes (vide infra). Such strands extend from the syncytial layer of the chorion (on the embryonic aspect of which is a layer of cytotrophoblast, lined by vascularized fetal mesoderm) across the intervillous space to the layer of (peripheral) syncytium which is in close apposition to the excavated maternal tissues. Through spaces in the latter, extravasated maternal blood continues to enter the intervillous space.

The central core of each syncytial strand is now invaded sequentially by the following:

(1) A growing *column* of proliferating cells extends from the chorionic cytotrophoblast and grows throughout the length of the strand until it reaches the syncytium on the maternal aspect of the intervillous space, *within which* it 'mushrooms' tangentially to meet and fuse with neighbouring outgrowths to form a spherical *cytotrophoblastic shell* around the conceptus (2.21, 23, 32).

(2) Each *cytotrophoblastic cell column* now develops, centrally and throughout much of its length, a *core* of extraembryonic mesoderm. This never completely reaches the trophoblastic shell, and it is uncertain how much of this mesoderm is derived by *inflexion* and *invasion* from the chorionic mesoderm, or whether it arises by differentiation of the central cytotrophoblastic cells.

(3) Capillaries now differentiate within the mesodermal core (2.32) and soon establish connections with the radicles of the umbilical vessels in the general mesoderm of the chorion.

The radial trabeculae undergoing these transformations are termed the *villous stems* and they, and their subsequent side branches, may be regarded as passing through the three grades of *histological differentiation* (primary, secondary and tertiary) just

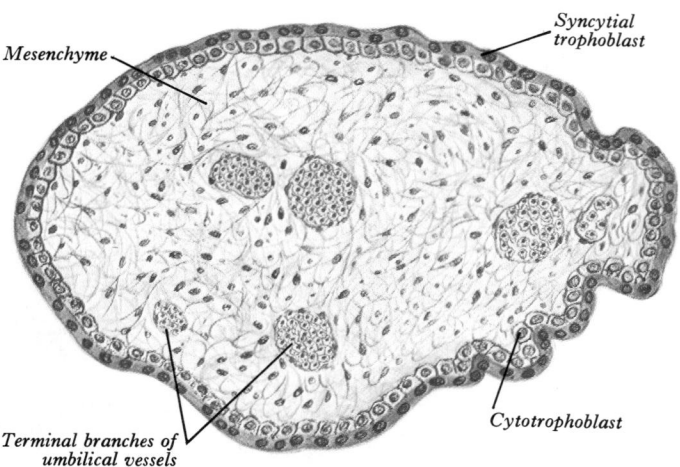

2.33 A transverse section of a terminal villus stained with haematoxylin and eosin.

described (2.32). The term 'villous stem' is used here in preference to the traditional primary, secondary and tertiary chorionic *villi* because it is evident that, *initially*, there are no true villi (i.e. finger-like projections with a free tip which grow across the intervillous space to reach the maternal tissues).

Each villous stem now consists, from its chorionic base and throughout much of its extent, of a vascularized mesodermal core, covered by a single (*Langhans*) layer of cytotrophoblast, which is again ensheathed by a layer of syncytium. Near the maternal end, however, it contains no mesodermal core but is formed of a solid *cytotrophoblastic cell column*, continuous peripherally with the trophoblastic shell and ensheathed by a layer of syncytium.

Two major changes now ensue (2.32); firstly, increasing complexity and subdivision of the maternal ends of the villous stems and, secondly, the development of free side branches. Expansion of the whole conceptus is accompanied by *radial growth* of the villous stems and, simultaneously, an integrated *tangential growth* with expansion of the trophoblastic shell. The attached maternal extremity of each villous stem undergoes a series of longitudinal divisions. Eventually each stem forms a complex consisting of a single *trunk (truncus)* attached by its base to the chorion, from which arise distally, second and third order branches (which have been named *rami* and *ramuli* respectively—Stieve 1926). Each of the latter remains attached at its maternal end to the trophoblastic shell.

Now, for the first time, 'true' villi with free tips make their appearance as outgrowths from the sides of the villous stems, and particularly from the rami and ramuli rather than from the truncus. These have been variously termed *free, terminal, absorption* or *fringing* villi and it is predominantly across their walls that exchanges between the fetal and maternal circulations occur. Each terminal villus commences as a syncytial outgrowth which, as it continues to grow, is again invaded successively by cytotrophoblastic cells which then develop a core of fetal mesoderm; this is finally vascularized by fetal capillaries (i.e. each villus passes through primary, secondary and tertiary grades of histological differentiation). The terminal villi continue to form and branch, within the confines of the definitive placenta (vide infra) throughout gestation, projecting in all directions into the intervillous space, many of them making contact with those growing from adjacent villous stems. When this occurs some form of secondary adhesion or fusion may occur between their tips. Some earlier workers suggested that this may proceed to mesodermal or even vascular fusion between adjacent villi, but a recent survey of available evidence strongly suggests that this does not occur to any appreciable extent and that connection between villi is limited to fibrinoid adhesion between their surfaces or, at most, fusion between their syncytial tips. As these changes proceed, the intervillous space, at first spanned by the early villous stems and their branches, and then increasingly permeated by growing free villi, is finally, once again, transformed into an exceedingly complex *labyrinth* of fine intercommunicating maternal vascular spaces.

2.34 Schematic diagram to show the arrangement of the placental tissues. Note the chorionic and basal plates, and the intervillous space spanned by a villous stem and its divisions (truncus, rami and ramuli). The sectioned surfaces show the disposition of the fetal and maternal blood vessels, the amniotic epithelium, the cellular and syncytial layers of trophoblast and the complex junctional zone between the fetal and maternal tissues in the basal plate containing deposits of fibrinoid material and isolated masses of peripheral syncytium. Note also the presence of surface syncytial sprouts, a stromal trophoblastic bud and Hofbauer cells associated with a terminal villus, and syncytial fusion occurring between the tips of two terminal villi. See text for further details. The region enclosed in the rectangle is shown greatly enlarged in 2.35.

The intervillous space, which contains the circulating maternal blood and is everywhere lined by syncytial trophoblast, is thus bounded:

(1) On its *fetal aspect* by a *chorionic plate*, consisting of syncytial, cytotrophoblastic and mesodermal layers of the chorion, the latter carrying radicles of the umbilical vessels, and soon to fuse with the mesoderm of the expanding amnion.

(2) On its *maternal aspect* by a *basal plate* consisting of the *peripheral syncytium* that has been divided by the growth of the cytotrophoblastic shell into an *inner layer* enclosing the intervillous space, and *external masses* which, together with the adjacent, modified and excavated uterine tissues, form a complex *junctional zone*.

(3) *Crossing it* from chorionic to basal plates, the main trunks of the *villous stems* dividing into their rami, ramuli and associated complex of free, or secondarily fused, terminal villi; the trunk and its branches may be regarded as the essential *structural, functional* and *growth unit* of the developing placenta.

Further consideration of the placenta must now be deferred until the preparation of the uterine tissues for the implantation and development of the blastocyst has been briefly described.

CYCLICAL CHANGES IN THE UTERUS

Throughout the period of reproductive life (i.e. from about the fifteenth to the forty-fifth year), except during pregnancy and lactation, a series of closely interrelated cyclical changes occur in the ovary, uterus and vagina. Each cycle extends over a period of about 28 days. In the *ovarian cycle*, which is described more fully elsewhere (pp. 116, 1436), one follicle usually reaches full maturity, ruptures and releases its secondary oöcyte during this period. The wall of the follicle is then transformed into an important endocrine gland, the *corpus luteum* (p. 1437). About 10 days after ovulation the corpus luteum begins to regress, then ceases to function and is replaced by fibrous tissue.

The changes of the *uterine cycle (menstrual cycle)* chiefly involve the lining endometrium of the body and fundus of the uterus and may, for convenience, be divided into four phases: (1) menstrual; (2) postmenstrual; (3) interval or proliferative; and (4) premenstrual or luteal (but see also below).

In the *menstrual (haemorrhagic) phase* the superficial part of the endometrium, next to the free surface, is shed piecemeal, leaving only the basal zone, adjacent to the uterine muscle (2.36A). Outwardly this phase is marked by a discharge of blood with necrotic epithelial and stromal debris from the uterus through the vagina. This discharge, the *menstrual flow*, lasts some 3–6 days.

In the *postmenstrual (reparative) phase*, and even before the menstrual flow ceases, the epithelium from the persisting basal parts of the uterine glands grows luminally over the denuded surface of the endometrium. The latter is now only 1–2 mm thick and lined by low cuboidal epithelium. The glands are straight and narrow. The stroma is dense and contains small numbers of lymphocytes amongst its more general population of rather ill-defined spindle-shaped cells (**2.36**B). This phase lasts about 4 days.

During the *proliferative (interval) phase* which lasts about 10–12 days, there is a growth of the endometrium associated with the presence in the bloodstream of oestrogenic hormones, internal secretions produced by the ovary (**2.38**A). The endometrium thickens to about 2–3 mm. Mitoses are present and the glands become distinctly tortuous. Their lining epithelium becomes tall columnar (**2.36**C). In the later part of the phase clear vacuoles

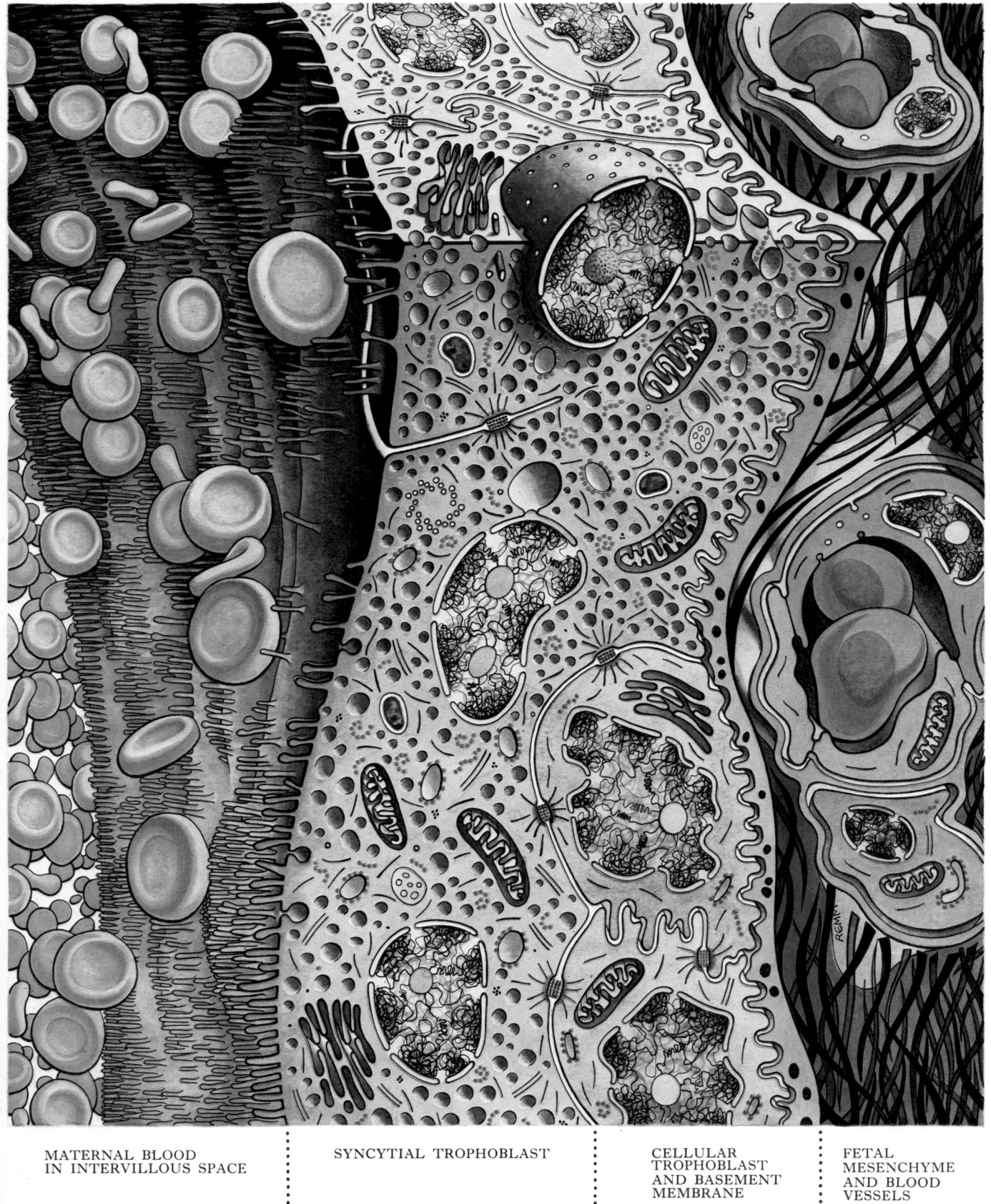

| MATERNAL BLOOD IN INTERVILLOUS SPACE | SYNCYTIAL TROPHOBLAST | CELLULAR TROPHOBLAST AND BASEMENT MEMBRANE | FETAL MESENCHYME AND BLOOD VESSELS |

2.35 Schematic diagram showing the detailed ultrastructural features of the tissues (enclosed in a rectangle in **2.34**) which intervene between the maternal and fetal blood streams. Note the contrasting architecture of the syncytial and the cellular trophoblasts, and the substantial basement membrane and delicate fetal mesenchyme which separate the trophoblast from the fetal vessels. Contrast the nucleate fetal, and anucleate maternal erythrocytes. For further description see text. (Based on data in Boyd & Hamilton 1970.)

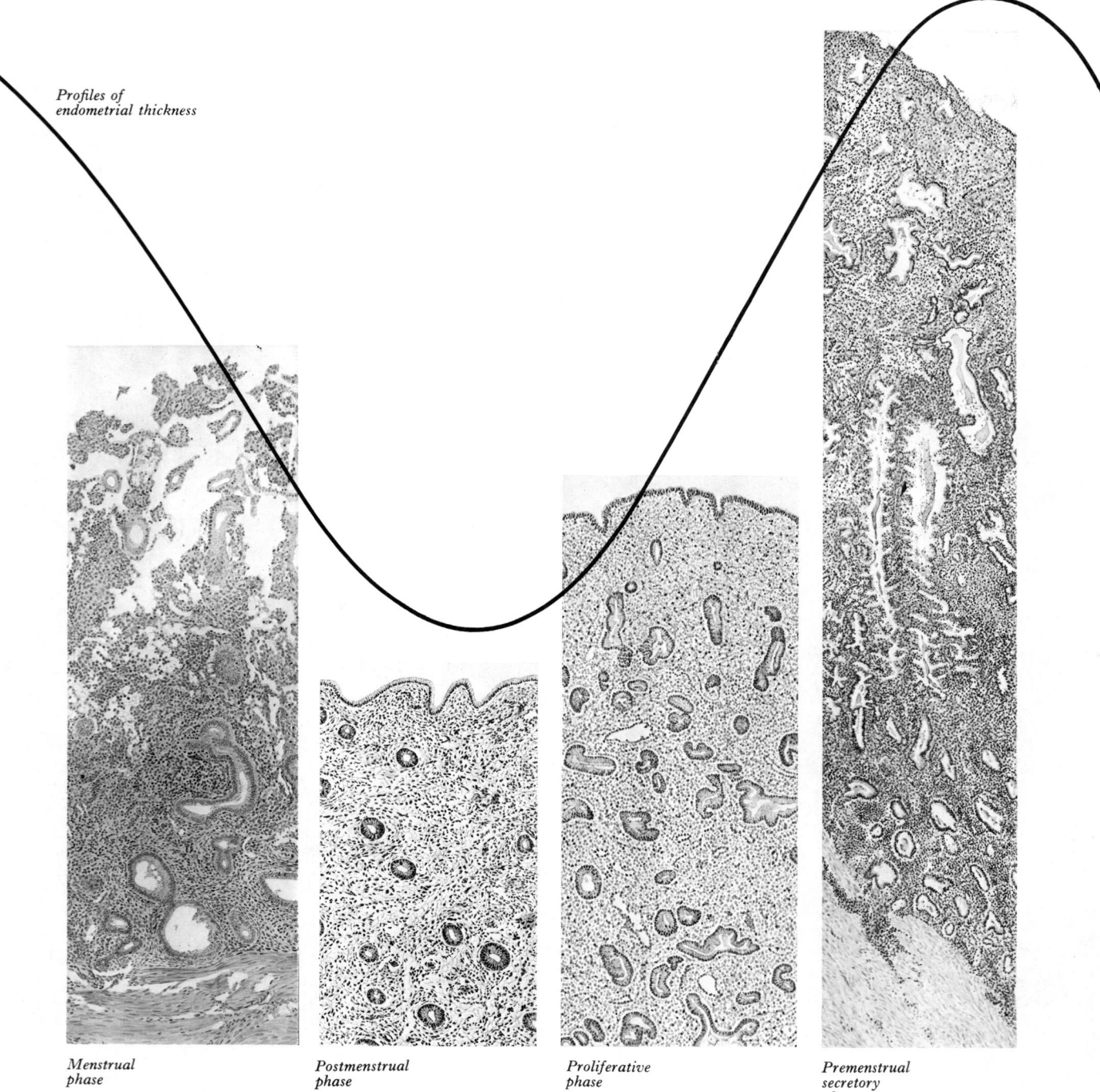

*Profiles of
endometrial thickness*

*Menstrual
phase*

*Postmenstrual
phase*

*Proliferative
phase*

*Premenstrual
secretory
phase*

2.36 Stages in the human menstrual cycle—the sections of the endometrium have all been reproduced at approximately the same magnification. The black line is a rough indication of the changes in endometrial thickness throughout the cycle. (Specimens kindly provided by the Shattock Museum, St Thomas's Hospital Medical School.)

appear in the basal parts of the epithelial cells lining the glands. These vacuoles enlarge, occupy more of the cytoplasm and the nuclei are displaced from the basal regions of the cells towards the lumina of the glands (**2.37**). Secretion is exuded into the lumen, and for a while the glands may be distended with secretion containing mucin and glycogen which can be demonstrated both in the cells and in the gland lumina by appropriate histochemical methods. Mitoses decrease and finally cease, the end of the proliferative phase.

Ovulation occurs about *14 days before the onset* of the next menstrual flow. The changes occurring in the premenstrual phase depend upon the presence in the bloodstream of the hormones, *progesterone* and *oestrogens*, both secreted by the corpus luteum (**2.38**).

The *premenstrual phase* occurs during the 7 days preceding the next menstrual flow. The mucous membrane thickens and terminally may be 7–8 mm deep. In their middle regions, the glands

become more voluminous and their walls folded upon themselves so that tuft-like processes project into the lumen. This gives the glandular wall a saw-toothed appearance in longitudinal section. Secretion is seen in the lumen of the active parts of the gland. The gland epithelial cells, which were tall and columnar at the commencement of the phase, become frayed and worn down on their luminal aspects and the nuclei resume their original basal position. The basal parts of the glands adjacent to the uterine muscle take little or no part in these changes. Later in the premenstrual phase characteristic changes appear in the interglandular stroma. These changes are most marked in the superficial part of the endometrium and around the blood vessels. Here the stromal cells, hitherto not clearly defined, become enlarged and swollen and for the first time definite cell outlines can be clearly seen. A few mitoses may be seen in the stroma at this stage. These changes constitute the *stromal premenstrual decidual reaction*. Three strata can now be clearly recognized in the endometrium (**2.36D**):

(1) *stratum compactum*, next to the free surface in which the necks of the gland are but slightly expanded and the stromal cells show a distinct decidual reaction, (2) *stratum spongiosum*, where the uterine glands are tortuous, dilated and ultimately only separated from one another by a small amount of interglandular tissue, (3) a thin *stratum basale*, next to the uterine muscle containing the tips of the uterine glands embedded in an unaltered stroma.

In the last days of the premenstrual phase lymphocytes appear in the endometrium in increasing numbers. They are found between and beneath the surface epithelial cells, amongst the stromal cells and around and between the gland cells. Towards the end of this period, as regression of the corpus luteum occurs, those parts of the stroma showing a decidual reaction and the glandular epithelium both undergo degenerative changes and the endometrium often diminishes in thickness. These degenerative changes precede the phase of bleeding.

During *menstruation* blood escapes from the superficial vessels of the endometrium forming small haematomata beneath the surface epithelium which raise it. Blood and necrotic endometrium then begin to appear in the uterine lumen. The shedding of the endometrium starts at the surface and extends into the deeper layers. The amount of tissue lost is variable, but usually the stratum compactum and most of the spongiosum are desquamated.

The endometrium is regenerated from the stratum basale and that part of the spongy layer which remains, the surface epithelium being reformed with remarkable rapidity.

The *vascular bed* of the endometrium undergoes significant changes during the menstrual cycle. The arteries to the endometrium arise from a *myometrial plexus* and consist of *short straight* vessels to the basal portion of the endometrium and more muscular *spiral arteries* to its superficial two-thirds. The venous drainage consists of narrow perpendicular vessels which anastomose by cross branches and is common to both the superficial and basal layers of the endometrium. The arterial supply to the basal part of the endometrium remains unchanged during the menstrual cycle. The spiral arteries to the superficial strata, however, lengthen disproportionately, become increasingly coiled and their tips approach more closely the uterine epithelium during the late proliferative and, particularly, in the premenstrual phases of the menstrual cycle. This leads to a slowing of the circulation in the superficial strata with some vasodilation. Immediately before the menstrual flow these vessels begin to constrict intermittently causing stasis of the blood and anaemia of the superficial strata. During the periods of relaxation of the vessels, blood escapes from the devitalized capillaries and veins, thus causing the *menstrual haemorrhage*.

The endometrial changes leading to premenstrual hypertrophy are a preparation for reception of the blastocyst and result from the action of oestrogen and progesterone (2.38). If fertilization of the ovum does not occur, the corpus luteum undergoes degeneration. The breakdown of the endometrium follows this cessation of function and is due to the drop in progesterone and oestrogen levels (p. 1437).

The uterine cycle may thus, alternatively, be considered as consisting of a *pre-ovulatory (follicular)* phase, which includes the postmenstrual and proliferative stages and a *post-ovulatory (progestational* or *secretory)* phase, which includes the second half of the menstrual cycle, i.e. the premenstrual and menstrual stages. These two *phases* are of approximately equal duration, although the pre-ovulatory phase is more variable than the post-ovulatory phase. The former is largely under the control of oestrogens alone and the epithelial cells in this phase possess large microvilli, whilst their cytoplasm contains much rough endoplasmic reticulum, lipid globules and lysosomes which become more numerous as the phase progresses. Histochemically the epithelium can be shown to contain large amounts of ribonucleoprotein and alkaline phosphatase, signifying the production and utilization of protein. In the progestational phase, which is under the control of both oestrogens and progesterone, the above activity declines and the surfaces of the cells become irregular with smaller and shedding microvilli and cytoplasmic processes; large vesicles appear in the apical cytoplasm and more numerous lysosomes are present. Acid

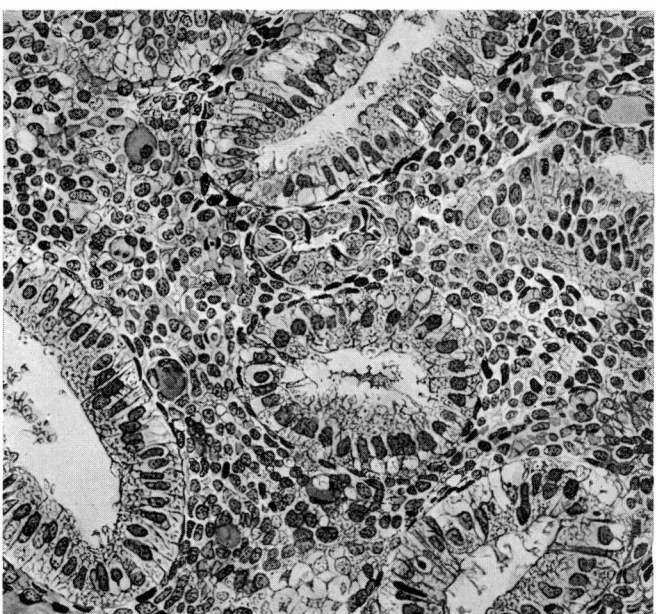

2.37 Section of human endometrium at about the seventeenth day of the menstrual cycle to show the accumulation of secretions in the basal parts of the epithelial cells lining the glands, resulting in displacement of the nuclei towards the lumen of the gland. Magnification × c.300. Stained with haematoxylin and eosin. (Kindly lent by the Shattock Museum, St Thomas's Hospital Medical School.)

phosphatases, malic and succinic dehydrogenases, cytochrome oxidase and adenosine triphosphatase are present with large amounts of glycogen and mucopolysaccharide (McKay et al 1956, Nilsson 1962).

A form of menstruation frequently occurs in the absence of ovulation (*anovulatory cycles*), particularly in post-pubertal girls and also in women approaching the menopause. Instead of liberating its ovum, the ripe follicle fails to rupture and undergoes degeneration. This is accompanied by a rapid reduction in oestrogen secretion with consequent breakdown of the uterine mucosa in the proliferative phase and in the absence of the changes evoked by the presence of progesterone in the bloodstream (an *oestrogen-withdrawal* bleeding—Corner 1938).

THE DECIDUA

If fertilization and successful implantation occur, a hormone secreted by the syncytiotrophoblast, *human chorionic gonadotrophin* (hCG), and controlled by cytotrophoblast secreted gonadotrophin-releasing hormone (GnRH), prolongs the life of the corpus luteum, which continues to secrete progesterone and oestrogens for the first trimester. (Thereafter these and other hormonal functions are the province of the definitive placenta, p. 156.) Menstruation does not occur and the vascular endometrium, now known as the *decidua of pregnancy*, thickens further to form a suitable nidus for the conceptus. (Chorionic gonadotrophin also appears early in the urine and its presence here is used as a basis for many of the tests for early pregnancy.) The interglandular tissue increases in quantity; it contains a number of leucocytes and is crowded with large round, oval or polygonal *decidual cells*. These are stromal cells which have accumulated glycogen and lipid in their distended cytoplasm. They continue to be conspicuous in the early stages of gestation but regress in the later months. Their precise significance is uncertain. They may offer a nutritive pabulum which is engulfed by the syncytial trophoblast but, on the other hand, they have been regarded as a defensive mechanism to protect the endometrium from excessive destruction. These changes are well advanced by the second month of pregnancy, when the three strata recognizable in the premenstrual phase, compactum, spongiosum and basale, are better differentiated and easily distinguished.

After the blastocyst is embedded, distinctive names are applied to different regions of the decidua: the part covering the

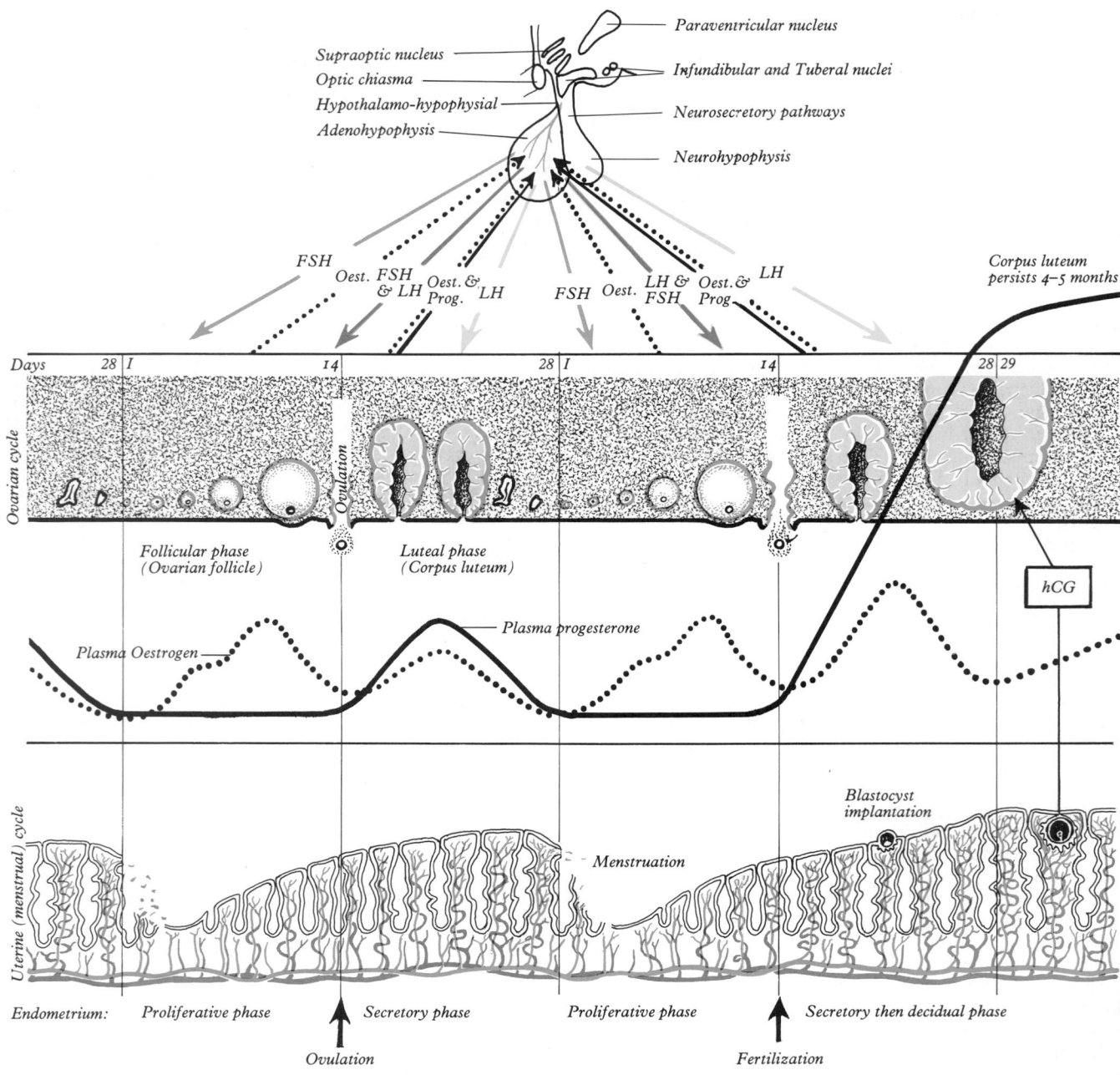

2.38 Composite diagram emphasizing some salient features of the female reproductive cycles. Note the (simplified) neuro-endocrine control loops, the periodic changes during the non-pregnant state of the ovarian cycle with concomitant endometrial changes of the menstrual (uterine) cycle, variations in circulating plasma hormone levels; finally, the dramatic sequelae when pregnancy supervenes. Neuro-endocrine controls are further detailed elsewhere (pp. 1436, 1453). FSH-follicle stimulating hormone; LH-luteinizing hormone; Oest.-oestrogens; Prog.-progesterone; hCG-human chorionic gonadotrophin. (Modified from Wendell-Smith et al 1984)

conceptus *decidua capsularis*; the part between the conceptus and the uterine muscular wall *decidua basalis*, and it is here that the placenta is subsequently developed; the part which lines the remainder of the body of the uterus *decidua parietalis* (**2.**40).

Coincidently with the growth of the embryo and the expansion of the amnion (p. 153), the decidua capsularis is thinned and distended (**2.**39, 40) and the space between it and the decidua parietalis gradually obliterated. By the beginning of the third month of pregnancy the capsularis and parietalis are in contact; by the fifth month the capsularis is greatly thinned, while during the succeeding months (**2.**41) it virtually disappears and the parietalis also largely atrophies, owing to the increased pressure. The glands of the compactum are obliterated and their epithelium is lost; in the spongiosum the glands are compressed, they appear as oblique slit-like fissures and their epithelium undergoes degeneration; in the limiting or boundary zone, however, the glandular epithelium remains cuboidal.

Implantation, the Chorion and Placenta

As we have seen (p. 125), during the early series of cleavage divisions of the zygote two distinctive groups of blastomeres emerge; one forms the *inner cell mass* including the *embryogenic cells*, the other consists of the outer, more numerous, smaller, rather flattened, polyhedral cells of the primitive *blastocystic trophoblast*. The latter spreads to cover the inner cell mass and consequently *polar* and *mural* regions of trophoblast may now be distinguished. Comparative studies have shown that such cells in the blastocysts of a number of different mammals possess many features in common. They are characterized by short, irregular microvilli on their external surfaces, numerous invaginations of the plasma membrane between the microvilli, with adjacent micropinocytotic vesicles in the cytoplasm, and junctional complexes between the cells at the outer end of the intercellular clefts (Enders & Schlafke 1965). Their cytoplasm carries many

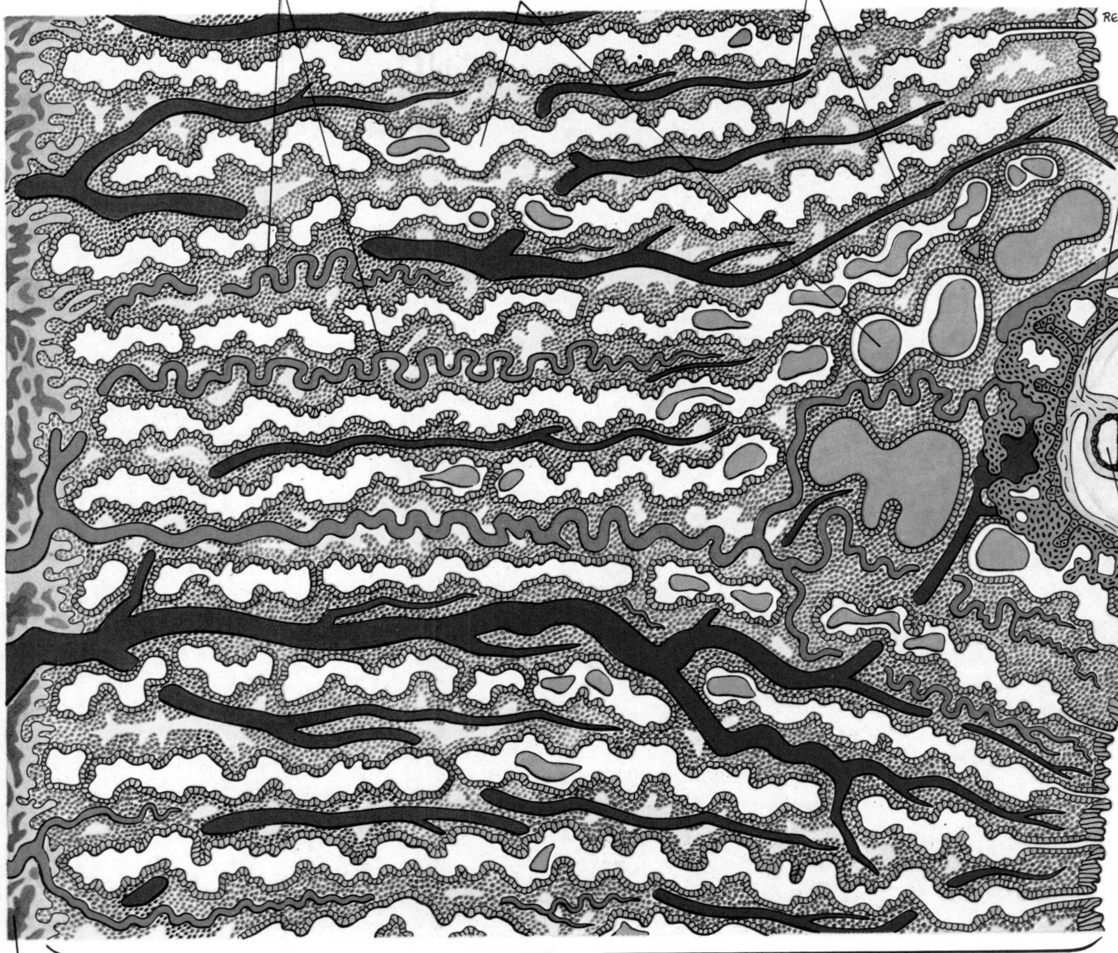

Spiral arteries   Dilated uterine glands   Endometrial veins
                   with secretory products

REMM   Connecting stalk

Syncytial trophoblast

Allantoic
diverticulum

Trophoblastic
lacuna

Cellular
trophoblast

Decidua
capsularis

Extra-
embryonic
mesoderm

Extraembryonic
coelom

Yolk sac cavity

Amniotic cavity

UTERINE CAVITY

Myometrium   Endometrium

2.39 A diagram showing the general structure of the implanting blastocyst and its relationship to the tissues of the endometrium on the 15th day after fertilization. Note the arrangement and gradation in thickness of the syncytial trophoblast which has eroded the maternal tissues. Some of the deeper trophoblastic lacunae already contain maternal blood. Note also the active dilated uterine glands, the endometrial venous plexus, spiral arteries, and the stage of development of the cellular trophoblast, amnion, yolk sac, allantoic diverticulum, connecting stalk and the extraembryonic mesodermal layers lining the extraembryonic coelom. For further details see text.

mitochondria, a variably developed Golgi complex, glycogen clusters, lipid inclusions, lysosomes, free ribosome rosettes, but scanty granular endoplasmic reticulum.

From the cells of this primitive layer some hold that all the elements of the chorion are derived, including the varieties of syncytium and also, at least in part, its mesodermal lining.

The origin of *syncytial trophoblast* was long debated, many workers holding that it arose by endomitosis of the primitive cells without cell division, but there is an increasing body of evidence (Boyd et al 1968) that it arises by proliferation of the cellular layer, with subsequent fusion of primarily independent cells on its external surface. During early implantation, and particularly in the polar trophoblast, it has been claimed that transiently nuclear division is so rapid that all traces of cellular definition are lost, a deep cellular layer only reappearing at a slightly later stage (Boyd & Hamilton 1966). The cellular layer itself persists, somewhat modified, as the germinal *cytotrophoblast* which, by multiplicative growth, can add additional cells which fuse with the overlying syncytium, may contribute cells to the underlying extraembryonic mesoderm and, by its continued growth, is an important factor in the expansion of the whole conceptus. It should be noted that proliferating and migrating epiblastic cells of the germ disc, converging on the primitive streak, entering the disc tissue and then diverging dorsocaudally, has been proposed as a prominent, if not exclusive, source of extraembryonic mesoderm (Luckett 1978).

6 or 7 days after ovulation the endometrium is well into the development of its secretory phase (p. 148) and, with the disappearance of the fibrillary zona pellucida and elaboration of polar

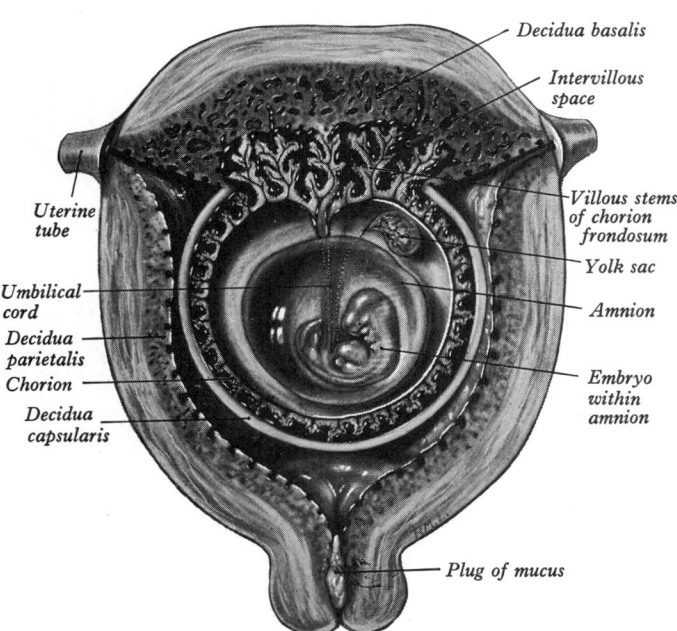

Decidua basalis

Intervillous
space

Uterine
tube

Villous stems
of chorion
frondosum

Yolk sac

Umbilical
cord

Amnion

Decidua
parietalis

Chorion

Embryo
within
amnion

Decidua
capsularis

Plug of mucus

2.40 A plan of the gravid uterus in the second month. A placental site precisely in the uterine fundus as indicated in the plan is, however, rather unusual. (The dorsal, ventral or lateral wall of the corpus uteri is more usual.)

2.41 Diagram showing a full-term human fetus *in utero*, including a sectional view of the placenta, the amnion (mauve), chorion (green), uterine wall (orange), the umbilical cord and its contained vessels, the cervix with a plug of mucus in the cervical canal, and the rugose vaginal wall. Note the characteristic flexed posture of the fetus and its limbs, and the overall position within the uterus which the fetus commonly occupies. (Other positions, although less fequent, are, however, also quite common.) Note also the single umbilical vein carrying oxygenated blood, the two umbilical arteries carrying deoxygenated blood, the arborization of these vessels in the chorionic plate (seen through the overlying amnion) and their branches which pass into the villous stems. The latter span the intervillous space and their trunci, rami, ramuli and terminal villi may be seen; incomplete placental septa project from the basal plate towards the chorionic plate. See text for further details.

synctium, the stage is set for the implantation of the blastocyst. From comparative studies (Böving 1959) the mechanisms of implantation have been described as *muscular,* involving blastocyst transport (and spacing in polyovular species), *adhesive,* whereby the blastocyst becomes attached to the uterine mucosa, and *invasive,* being the means by which it enters the uterine tissues. Whilst the earliest stages of human blastocyst attachment and penetration have not been available for study, a number of mammals, including primates (Wislocki & Streeter 1938), have been subjected to close scrutiny using light and electron microscopy (Reinius 1967, Nilsson 1967, Tachi et al 1970).

Early implantation is initiated by a close approach of the trophoblastic plasma membrane to the tips of the microvilli and irregular surface protrusions of the uterine epithelial cells. At these points the trophoblast surface often presents patches of 'bristle-coated' membrane, with adjacent cytoplasmic plaques having a crystalline substructure, with numerous neighbouring lysosomes. The microvilli shorten and disappear and for a period there is a close mutual adaptation between the contours of the trophoblast and the uterine epithelial cell surfaces. The syncytium now sends finger-like projections between adjacent epithelial cells towards the underlying basement membrane, the two layers becoming closely interlocked by the formation of numerous tight junctions between them. The epithelial cells then degenerate and are engulfed by the trophoblast, resultant cellular debris becoming enclosed in phagosomes within the trophoblastic cytoplasm. The subepithelial basement membrane is soon penetrated, with the development of oedema, hyperaemia and cell degeneration in the neighbouring stroma.

Subsequent implantation of the human blastocyst continues with erosion, degeneration and phagocytosis of maternal vascular endothelium, glandular epithelium with secretory products and decidual cells, until the blastocyst sinks into and occupies a ragged *implantation cavity* in the decidua (*interstitial implantation*).

Later stages of development, which include the definition of the layers of the chorion, the elaboration of syncytial trophoblast, which is thick at the embryonic pole and thins towards the abembryonic pole and carries a lacunar circulation, and the principal stages in the development of the intervillous space, the villous stems and their branches and the trophoblastic shell have already been described.

From the third week until about the second month of pregnancy the *entire chorion* is covered with villous stems which are thus continuous peripherally with the trophoblastic shell in close apposition with *both* the decidua capsularis and the decidua basalis. The latter, however, are stouter, longer and show a greater profusion of terminal villi. As the conceptus continues to expand, the decidua capsularis is progressively compressed and thinned, the circulation through it is gradually reduced and, accordingly, its related villous stems and villi slowly atrophy and disappear. This process starts at the abembryonic pole and by the end of the third month the abembryonic hemisphere of the conceptus is largely denuded. This continues until the whole chorion which is related to the capsularis is smooth (the *chorion laeve*); in contrast, the villous stems of the disc-shaped region of chorion related to the decidua basalis increase greatly in size and complexity (the *chorion frondosum*), and together with the basalis constitute the *definitive placental site*.

An exhaustive account of the growth, dimensional changes, vasculature and haemodynamics, cell varieties, ultrastructure and histochemistry of the placenta, and the physiological aspects of placental transfer and its status as a metabolic store and endocrine gland, lies beyond the scope of the present volume. What follows is necessarily an abbreviated account of selected topics and the interested reader should consult the profusion of original papers devoted to these subjects. The exhaustive classical volume by Boyd & Hamilton (1970) provides an unrivalled source of information and an extensive bibliography. A provocative survey of placental transfer mechanisms is to be found in Longo (1972).

Definition of the Human Placenta

The human placenta (Wislocki 1929, Hill 1932, Grosser 1936, Mossman 1937, Amoroso 1952) is defined as *discoidal* (in contrast

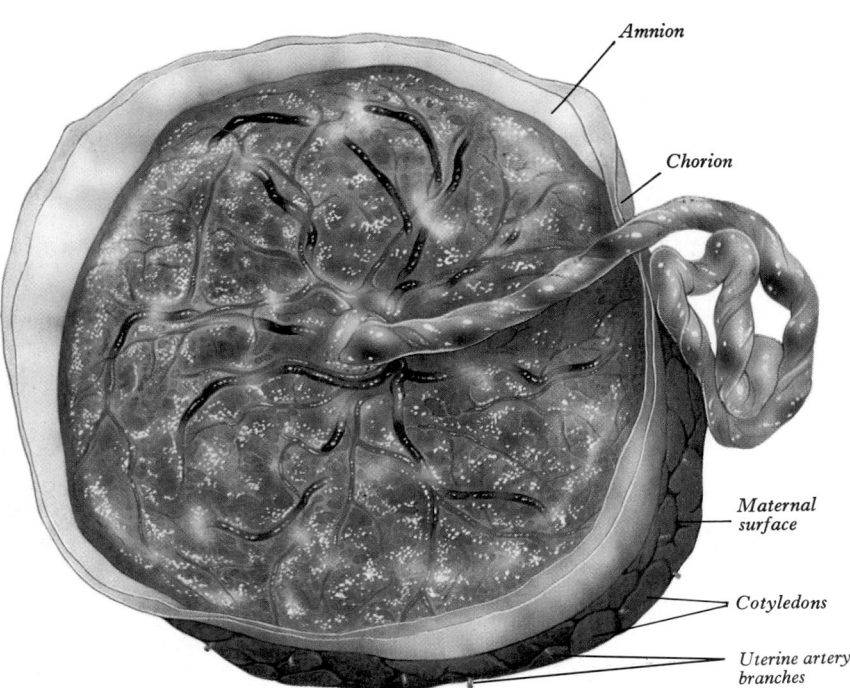

2.42 The fetal surface of a recently delivered placenta, drawn from a coloured photograph kindly provided by Mr E F Gibberd. The spiral umbilical vessels in the umbilical cord, and their radiating branches shine through the transparent amnion. The maternal surface is exposed in the lower and right corner of the figure. Note the small branches of the uterine artery entering a series of grooves. Note also the fringes of amnion and chorion, the the majority of which have been cut away near the placental margin.

to other shapes, e.g. *zonary, bidiscoidal, diffuse,* etc seen in other forms). It is initially *labyrinthine* as the early villous stems are formed, but becomes secondarily *villous* with the development of generations of terminal villi and finally there is a partial return to a *labyrinthine* state as superficial adhesion or partial fusion occurs between the tips of a number of villi (vide supra). Maternal blood bathes the surfaces of the chorion which bound the intervillous space and it is thus defined as *haemochorial*, distinguishing it from the different grades of fusion between the maternal and fetal tissues which exist in many other forms (*epitheliochorial syndesmochorial, endotheliochorial* and even an approach to a *haemoendothelial* condition but with the vascular endothelial basement membrane and a fine layer of perivascular connective tissue persisting). The chorion is vascularized by the allantoic blood vessels of the body stalk and the human placenta is termed *chorioallantoic* (whereas in some forms a *choriovitelline* placenta either exists alone or supplements the chorioallantoic variety). Finally, the human placenta is said to be *deciduate* because maternal tissue is shed with the placenta and membranes at term as part of the afterbirth (vide infra).

THE PLACENTA AT TERM

The expelled placenta (**2.42**) is a flattened discoidal mass with an approximately circular or oval outline, with an average volume of some 500 ml (range 200–950 ml), average weight about 500 g (range 200–800 g), average diameter 185 mm (range 150–200 mm), average thickness 23 mm (range 10–40 mm) and an average surface area of about 30 000 mm². Thickest at its centre (the original embryonic pole) it rapidly diminishes in thickness towards its periphery where it continues as the chorion laeve.

Macroscopically, its *fetal* or *inner surface*, covered by amnion, is smooth, shiny and transparent and the mottled appearance of the subjacent chorion, to which it is closely applied, can be seen through it. The umbilical cord, usually attached near the centre of the fetal surface, and branches of the umbilical vessels radiate out under the amnion from this point, the veins being deeper and larger than the arteries. Beneath the amnion and close to the attachment of the cord, the remains of the yolk sac can sometimes be identified as a minute vesicle, up to 5 mm in diameter, with a fine thread—a vestige of the yolk stalk—attached to it.

The *maternal surface* is finely granular and mapped into some 15–30 lobes by a series of fissures or grooves. The lobes are often, somewhat loosely, termed *cotyledons* (but see also below) and the grooves themselves correspond to the bases of incomplete *placental septa* which become increasingly prominent from the third month onwards, and extend from the maternal aspect of the intervillous space (the basal plate) towards, but do not quite reach, the chorionic plate. The septa form initially as ingrowths of the cytotrophoblastic shell, with a covering of syncytium, but from the fourth month they develop a complex core derived from the maternal tissues which includes reticulin fibres and associated cells, decidual cells, remnants of glandular epithelium, occasional blood vessels and, deep to the cytotrophoblast, a stratum of fibrinoid material. Later, cytotrophoblastic cells often penetrate into the fibrinoid at various points with a resultant admixture of the fetal and maternal tissues and, in the later months of pregnancy, patches of central degeneration occur in many septa, which, incidentally, increases the difficulty of histological interpretation. The nature of the maternal surface of the expelled placenta is of course determined by the tissue plane of separation of the placenta at parturition.

Studies of the human placenta include morphometric analysis, surface architecture using scanning electron microscopy, ultrastructural studies of angioarchitecture and possible mechanisms whereby the maternal placental circulation is controlled. The ultrastructure of biopsies taken from placental uterine beds which were presumed to be normal has been reviewed. A detailed investigation into the connective tissue 'skeleton' of the placenta, with a helpful bibliography, has been provided by Ockleford & Wakely (1982).

SEPARATION OF THE PLACENTA

After delivery of the fetus the placenta becomes separated from the uterine wall and, together with the so-called 'membranes', is expelled as the *afterbirth*. Separation takes place along the plane of the stratum spongiosum and extends beyond the placental area, detaching (1) almost the whole remaining thickness of the largely degenerate, confluent, decidua parietalis and decidua capsularis, (2) the chorion laeve and (3) the amnion. These three layers are partially fused together and are continuous with the placenta at its margin; they constitute the *membranes* familiar in obstetrics. The process of separation requires rupture of many uterine vessels but their torn ends are closed by the firm contraction of the muscular wall of the uterus after delivery of the placenta and membranes and thus, under normal circumstances, postpartum haemorrhage is limited in amount. When the placenta and membranes have been expelled, a thin layer of stratum spongiosum is left as a lining for the uterus, but it soon undergoes degeneration and is cast off in the early part of the puerperium. A new epithelial lining for the uterus is then regenerated from the remaining stratum basale.

PLACENTAL LOBES AND LOBULES

The placental *lobes* are demarcated by the grooves on its maternal surface, and they correspond in large measure to the major branches of distribution of umbilical vessels, particularly well seen in specimens X-rayed after intravascular injection of radio-opaque media. However, the application of the term cotyledon to these major lobes does not correspond directly to its usage in comparative placentology.

Particularly in those forms with a diffuse kind of placentation, the whole chorion does not bear villous structures evenly, for these are restricted to a large number of scattered discontinuous patches (*fetal cotyledons*) with non-villous chorion between them. These fetal cotyledons come into apposition with similarly distributed discrete areas of endometrium (the *maternal cotyledons* or *uterine caruncles*); the combination of one fetal unit with its associated maternal unit is termed a *placentome*.

The fetal cotyledon of the human placenta evidently corresponds to a major villous stem and its branches; early in pregnancy,

the chorion bears some 800–1000 of such stems but as pregnancy advances, with the formation of the chorion laeve and possibly some fusion between adjacent stems, the number is progressively reduced until only about 60 persist in the placental area in the last months of pregnancy. However, this number is distributed between the 15–30 lobes, i.e. each lobe (or 'cotyledon' of obstetrics) contains 2–4 major villous stems and is to be regarded as compounded of this number of *placentomes* or *lobules*.

THE PLACENTAL TISSUES

These are arranged as a chorionic plate, a basal plate and, between the two, the villous stems, their branches and the intervillous space (2.32–41).

The chorionic plate is covered on its fetal aspect by the amniotic epithelium, its cells described elsewhere, followed by a connective tissue layer carrying the main branches of the umbilical vessels, then a diminishing layer of cytotrophoblast and finally the inner syncytial wall of the intervillous space. The connective tissue layer derives from fusion between mesoderm-covered surfaces of amnion and chorion and is more fibrous and less cellular than Wharton's jelly of the umbilical cord, except near the larger vessels. The latter radiate and branch from the cord attachment (with variations in the branching pattern), until they reach the bases of the trunks of the villous stems, which the branches enter and then arborize within their rami, ramuli and terminal villi. There is no anastomosis between vascular trees of adjacent stems but, in contrast, the two umbilical arteries are normally joined by some form of substantial transverse (Hyrtl's) anastomosis at, or just before they enter, the chorionic plate.

The basal plate consists, from fetal to maternal aspect, of: (1) the syncytium (and perhaps in patches, maternal tissue) which forms the outer wall of the intervillous space; (2) Rohr's stria of fibrinoid; (3) what remains of the cytotrophoblastic shell; (4) Nitabuch's stria of fibrinoid and; (5) the remains of the decidua.

The striae of fibrinoid are irregularly interconnected and strands pass from Nitabuch's stria into the adjacent decidua. The latter contains basal remnants of the endometrial glands, large and small decidual cells scattered in a connective tissue framework which also supports an extensive venous plexus. These deeper layers of the basal plate also contain a variety of *giant cells* which may possess either multilobate nuclei or are in fact multinucleate. Their origin has been disputed, but the majority probably have a fetal origin, arising either by the isolation of masses of peripheral syncytium or by growth from the cells of the cytotrophoblastic shell. For some, however, a maternal origin cannot be excluded. Their role is uncertain but they may be one source of placental hormones. In this regard, the possession by the cytotrophoblastic cells in this region of prominent nucleoli, a particularly basophilic cytoplasm rich in RNA and varieties of cytoplasmic inclusions (Dallenbach-Hellweg & Nette 1964), both acidophil and basophil, may also be related to an endocrine secretory role.

Throughout the second half of pregnancy the basal plate is thinned and progressively modified, with a relative diminution of the decidual elements, and an increasing deposition of fibrinoid and admixture of fetal and maternal derivatives. It is, as we have seen, through the deeper decidual layers of the plate that separation of the placenta occurs at parturition and through its various layers the maternal blood vessels approach and reach the intervillous space. The spiral arteries of the endometrium do not open directly into the intervillous space until the fifth or sixth week, after which they pass through the plate and open through gaps in the cytotrophoblastic shell and peripheral syncytium. The terminal parts of the vessels are dilated, lose their muscular walls by degeneration and present a hypertrophy of their lining endothelium. Proliferations of cytotrophoblast often grow back for some distance along their lumina (Hamilton & Boyd 1960).

The veins which drain the blood away from the intervillous space pierce the basal plate and join tributaries of the uterine veins. The presence of a marginal venous sinus, which has hitherto been described as a constant feature, occupying the peripheral margin of the placenta and communicating freely with the intervillous space, has not been confirmed.

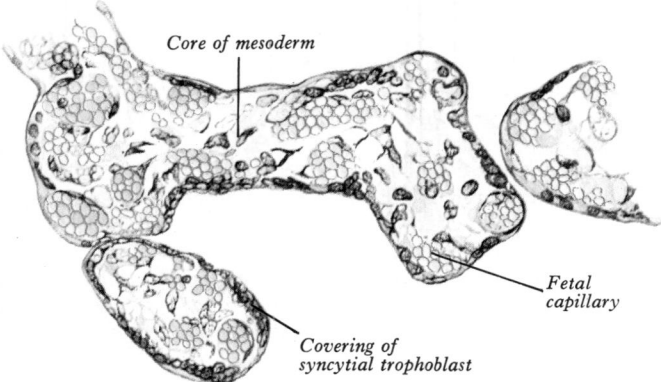

2.43 Part of a section of a branching terminal villus in a mature placenta (ninth month). Note the close approach of many of the fetal capillaries to the subepithelial basement membrane; see text for details.

In the macaque monkey, radio-opaque material injected into the aorta passes in spurts or jets to the intervillous space and at sufficient pressure to drive it towards the chorion, thus preventing a short circuit of arterial blood into the venous openings. The openings of the coiled arteries show intermittent activity, probably due to alternating constriction and relaxation of the arteries themselves. Myometrial contractions alter the pressure in the intervillous space and promote placental venous drainage (Ramsey et al 1963, Martin 1965).

THE STRUCTURE OF A VILLUS

Terminal villi are the essential structures involved in exchanges between mother and fetus and, accordingly, the villous tissues separating fetal and maternal blood are of crucial functional importance; numerous investigations have been instigated.

Each villus has a core of connective tissue bearing fetal capillaries and separated by a basement membrane from the ensheathing cyto- and syncytial trophoblast which is bathed by the maternal blood in the intervillous space (2.32,34,35,43). Cohesion between the cells of the cytotrophoblast and also between this layer and the syncytium is provided by numerous desmosomes between their apposed plasma membranes.

In earlier stages, the cytotrophoblast forms an almost continuous layer on the basement membrane, but after the fourth month it gradually expends itself producing syncytium (Midgley et al 1963). As cytotrophoblast reduces, syncytium becomes directly related to the basement membrane over an increasingly large area and simultaneously itself becomes progressively thinner, but a few cytotrophoblastic cells, usually disposed singly, persist until term.

The cells of the *villous* cytotrophoblast (*Langhans cells*) are pale-staining with only a slight basophilia. Ultrastructurally, they show a rather electron-translucent cytoplasm, with few free polysomes, little granular endoplasmic reticulum but a number of large mitochondria, a fairly extensive Golgi apparatus, microfilaments particularly in association with the desmosomes and occasional vacuoles and dense inclusions. Between the desmosomes, the cell membranes of adjacent cells may be smooth and straight, or they may undulate, and are separated by a featureless gap of some 20 nm. Sometimes the intercellular gap widens to accommodate microvillous projections from the cell surfaces; the gap occasionally contains patches of fibrinoid.

In contrast, the syncytial cytoplasm is more strongly basophilic and possesses many ultrastructural features which distinguish it from the cytoplasm of the Langhans cells. Where the plasma membrane adjoins basement membrane it is often complexly infolded into the cytoplasm, whereas the surface bordering the intervillous space is beset with numerous long microvilli, the cores of which show linear densities. These microvilli are responsible for the brush border of light microscopy.

The syncytial cytoplasm is exceedingly complex and more electron-dense than that of Langhans cells. It contains a wealth of free ribosomes, cisternae of granular endoplasmic reticulum, scattered representations of the Golgi complex, a cytoskeleton of microfilaments and a profusion of vesicles and vacuoles, some smooth and some coated, of a wide size range, numerous lysosomes, phagosomes and other electron-dense inclusions. It evidently is not, as was earlier claimed, a simple homogeneous semipermeable membrane but an intensely active tissue layer across which all transplacental transport must occur.

Glycogen is held to be present in both layers of the trophoblast at all stages but it is not always possible to demonstrate it by histochemical means. Lipid droplets are also present in both layers and free in the core of the villus. In the trophoblast they are found principally within the cytoplasm but also occur extracellularly between cytotrophoblast and syncytium, or between the individual cells of the cytotrophoblast and also in the basement membrane. These droplets may represent fat in transit from mother to fetus, their numbers diminishing with advancing age. Membrane-bound granular bodies of moderate electron-density also occur in the cytoplasm, particularly in the syncytium; some of these are probably secretion granules of hormones or hormone precursors and are believed to arise in the Golgi apparatus. The lysosomes and phagosomes are evidently concerned in the degradation of materials engulfed from the intervillous space.

The core of the villus contains fibroblasts, large phagocytic *Hofbauer cells*, more numerous in early pregnancy, many plasma cells and free lipid granules. Electron microscopy shows fine collagen fibres which are thicker and collected into bundles in the villous stems. The fetal vessels include arterioles and capillaries. Their endothelial cells contain fine cytoplasmic filaments and they may provide bulbous projections into the lumen (Rhodin & Terzakis 1962, Terzakis 1963, Iklé 1961, 1964). They are surrounded externally by the usual periendothelial basement membrane (p. 689).

The characteristics of both syncytium and cytotrophoblast, however, vary considerably from the brief account just given in sites other than a villus and at different stages of development and for such details original sources should be consulted. There are, however, certain specialized regions of the villous syncytium which should be mentioned.

Many villi show localized ingrowths of syncytium (with some associated cytotrophoblast) into their mesodermal cores, the *stromal trophoblastic buds*. They may remain attached by stems to the overlying trophoblast, or become detached to lie free, and they may even approach and project into the lumina of the fetal blood vessels. In the latter site they may be a source of the syncytial masses which have been described in the blood of the umbilical vein. The frequency and fate of such fetal cellular emboli is uncertain. On the free surface of the villus at least two other types of specialization occur. The first, termed syncytial *knots* or *clumps*, are localized thickenings in which there is a close aggregation of a number of nuclei. Their significance for long remained obscure but in some cases appears to be related to the maintenance of a neighbouring delicate, non-nucleated, *epithelial plate* of syncytium which comes into particularly close apposition with a fetal capillary. An ultrastructural study of syncytial knots (Jones & Fox 1977) has shown that their nuclear aggregates exhibit marked degenerative changes, and it was suggested that they represent a sequestration phenomenon involving removal of senescent nuclear material from adjacent metabolically active areas of syncytium. Fusion of knots from adjacent villi appears to be one basis of *syncytial bridge* formation, and for the latter a mechanical role as an internal *placental strut* system has been advanced. The second specialization is termed a *syncytial sprout* which continues to be formed throughout pregnancy. Some sprouts are unquestionably initial stages in the development of a new branch villus. Many others, however, project into the maternal bloodstream and become detached, forming *maternal syncytial emboli* which pass to the lungs. It has been computed that there is a passage of some 100 000 of such sprouts daily into the maternal circulation. In the lungs they provoke little local reaction and apparently disappear by lysis but they may, on occasion, form a locus for neoplastic growth.

MATURATION AND FUNCTIONS OF THE PLACENTA

In the early stages of placental development the blood in the fetal vessels is separated from the maternal blood in the intervillous space by the fetal vascular endothelial cells and their basement membranes, the relatively abundant connective tissue of the villus, the subepithelial basement membrane and its covering of cyto- and syncytial trophoblast. These constitute a *placental barrier* interposed between the bloodstreams, but it is a permeable barrier and allows water, oxygen and other nutritive substances and hormones to pass from mother to fetus, and some of the products of excretion to pass from fetus to mother.

During the first half of pregnancy the placenta not only increases its surface area but reaches its maximum thickness; this accompanies increases in the size, length and complexity of branching of the villous stems and is not accompanied by any further invasion of the uterine wall. In the latter half of pregnancy the placenta further increases its surface area, doubling its diameter, the overall thickness remaining static (Stieve 1948, Hamilton & Boyd 1951). Placental weight has also been correlated with parity (Boyd & Hamilton 1970). These gross measurements might indicate a placental failure to keep pace with the functional demands of the growing fetus but, in compensation, the *placental barrier* becomes *reduced* in thickness. After the fourth month the villous trophoblast becomes reduced; as it ages the syncytium becomes directly related to the subepithelial basement membrane over an increasing area and it also becomes thinner. The fetal capillaries approach the surface of the villus but are always separated from the basement membrane by delicate connective tissue.

The mechanism of transfer of substances across the placental barrier is complex. The surface area of the villi has been variously estimated; it may, however, be as large as 14 square metres (Wilkin & Bursztein 1958). The volume of maternal blood circulating through the intervillous space has been assessed at 500 ml per minute (Assali et al 1960). Simple diffusion suffices to explain gaseous exchange and a transfer of many dissolved substances of low molecular weight (e.g. sodium, potassium, chloride, iodide, phosphate); some electrolytes, glucose and fatty acids involve facilitated diffusion, and active transport mechanisms carry other electrolytes, amino acids and water soluble vitamins. In later pregnancy, water is interchanged between fetus and mother (in both directions) at about 3.5 litres per hour. The transfer of substances of high molecular weight such as complex sugars, some lipids, hormonal and non-hormonal proteins varies greatly in rate and degree, and is not so readily understood. Energy-dependent selective transport mechanisms, including micropinocytosis, are probably involved.

Lipids may be transported unchanged through and between the cells of the trophoblast to the core of the villus. The passage of maternal antibodies (immunoglobulins) across the placental barrier confers some degree of passive immunity on the fetus—in this instance transfer by micropinocytosis is widely accepted. The whole problem of investigations into transplacental mechanisms is complicated by the fact that the trophoblast itself is the site of synthesis and storage of certain substances, e.g. glycogen. For comprehensive reviews of placental transfer mechanisms consult Longo (1972), Boime & Boguslawski (1974).

The placenta is an important endocrine organ; some steroid hormones, various oestrogens, β endorphins, progesterone, human chorionic gonadotrophin (hCG), human chorionic somatomammotropin (hCS)—formerly named placental lactogen —are predominantly, perhaps exclusively, synthesized by the syncytium, whilst chorionic gonadotrophin releasing hormone (GnRH) is secreted by the cytotrophoblastic cells. It is of interest that steroid hormones are formed in isolated perfused placentae and also by placental explants in tissue culture. The trophoblast is rich in birefringent lipids and cytochemical methods show that it also contains enzyme systems which are associated with the synthesis of steroid hormones.

Leucocytes are more numerous in the blood of the umbilical vein than in that of the umbilical artery, suggesting that they may migrate from the maternal blood through the placental barrier into the fetal capillaries. It has also been shown that some fetal and maternal red blood cells may cross the barrier (Dancis 1959). The former may have important consequences, e.g. in rhesus incompatibility (p. 682). The vitamins pass the placental barrier with different degrees of facility; vitamins B, C and D pass readily, the remainder with greater difficulty.

The majority of drugs pass the barrier and many are apparently tolerated by the fetus, but some may exert grave teratogenic effects on the developing embryo (e.g. thalidomide, vide infra).

Finally, a wide variety of bacteria, spirochaetes, protozoa and viruses are known to pass the placental barrier from mother to fetus, although the mechanism of transfer is uncertain. The presence of maternal rubella in the early months of pregnancy is of especial importance in relation to the production of congenital anomalies (vide infra).

PLACENTAL VARIATIONS

As a rule the placenta is attached to the posterior wall of the uterus near the fundus, with its centre in or near the median plane. The site of attachment is determined by the point where the blastocyst becomes embedded but the factors on which this depends are not yet fully understood. The placenta, however, may be attached at any point on the uterine wall, but these variations offer no complications to a normal labour unless it is attached so low down that it overlies the internal os uteri, when it may give rise to serious antepartum haemorrhage, especially if it is nearly central in position. This condition, which occurs in about 0.5% of pregnancies, is known as *placenta praevia*. (*Extrauterine* sites of implantation are discussed on p. 131.)

The umbilical cord, although usually attached near the centre of the organ, may reach it at any point between its centre and margin, the latter known as a *battledore* placenta. Occasionally the cord fails to reach the placenta itself and ends in the membranes in its vicinity. With such a *velamentous insertion* of the cord, the larger branches of the umbilical vessels traverse the membranes before they reach and ramify on the placenta. A small *accessory* or *succenturiate* placental lobe is occasionally present, connected to the main organ by membranes and blood vessels; it may be retained in utero after delivery of the main placental mass and prolong postpartum haemorrhage. Occasionally other degrees of division occur (*bipartite* or *tripartite* placentae). Other variations include *placenta membranacea*, in which villous stems and their branches persist over the whole chorion, and *placenta circumvallata*, in which its margin is undercut by a deep groove. Pathological forms of adherence or penetration include *placenta accreta*, with exceptional adherence to the decidua basalis, *placenta increta*, in which the myometrium is invaded, and *placenta percreta*, when the invasion by placental tissue has passed completely through the uterine wall.

At birth, when ligature of the umbilical cord is delayed, the blood volume of the child is, on the average, appreciably greater than it is when the ligature is applied at the earliest possible moment (de Marsh et al 1942). It appears that in the former case much of the blood in the fetal placental vessels is transferred from the placenta to the fetus. The meaning of the phenomenon is far from clear, for in the first few days of life after late ligature of the cord the newly born suffers a loss both of plasma volume and of haemoglobin (Gotsev 1939).

Congenital Abnormalities—Teratology

The term teratology was at one time reserved for grosser examples of congenital abnormality—for the study of 'monsters'. Cleft palate or thalassaemia could scarcely be regarded as monstrosities —something to be pointed at with surprise or aversion. But the Greek τερας and the Latin *monstrum* both imply pointing of a different kind—a pointing to the future, a warning or portent. In primitive cultures the advent of abnormal offspring or prodigy, whether human or otherwise, was usually accepted as a portent of ill-omen. Hence records of human congenital malformations, in cave-paintings, sculptures and ultimately in writings, extend backwards into prehistory, as do also theories and beliefs in regard to their causes. Talipes, achondroplasia and conjoined twins were portrayed in most ancient times, together with centaurs, sirens, mermaids and other fanciful creatures (Barrow 1971). The Hippocratic School identified hydrocephalus. Aristotle described many major and minor malformations with accuracy but, unlike his contemporaries and his successors through many subsequent centuries, he denied the magical and divinatory aspects of teratology, ascribing 'monsters' to natural causes, a view which was not reawakened until the times of Harvey, Wolff, von Haller and the Hunters. They and their contemporaries of the seventeenth and eighteenth centuries, stimulated by the growing knowledge of embryology, initiated the *theory of embryonic arrest* to explain malformations. Saint-Hilaire experimented on developing chick embryos in the early nineteenth century, and with this approach the study of teratology was consolidated as a science. In parallel, however, superstition and prejudice persisted, producing, e.g. the 'hybrid theory' which attributed monstrous births to unnatural hybridization, a groundless hypothesis which cost many innocent men and women their lives. Another old theory, which is scarcely dead even today, was the concept that the experiences of the pregnant mother, usually visual, could influence her unborn offspring in an adverse manner. That factors in the *maternal environment* may influence the embryo has, of course, proved true, but in a different sense, for Watson, as long ago as 1749, suggested that fetal disease, contracted by a transplacental route, might be a cause of congenital abnormality, citing variola as an example. This theory slowly dwindled, and was almost extinguished a century later by the authority of Virchow and His, whose views dominated teratological opinion through most of the second half of the nineteenth century. A contemporary, working in obscurity, was Mendel; and as his work became known and genetics flowered into the twentieth century attention inevitably turned to the hereditary aspect of congenital defects (already foreshadowed by Paré and John Hunter, centuries earlier). In 1941 Gregg's observation that congenital cataract is associated with the infection of pregnant mothers by rubella revived interest in environmental factors. Subsequent to this, teratology has been largely concerned with *genetic* and *environmental factors*, particularly in attempts to explain the mechanisms of abnormal development.

The expansion of experimental embryology has revealed a wide array of *environmental agents* capable of affecting normal development, including temperature variations, mechanical insult, variation in substances such as lithium and magnesium in culture media, exposure to irradiation, hypoxia, hypo- and hypervitaminosis, hormonal effects (especially with oestrogens and androgens), nutritional defects and exposure to various drugs and other chemicals. These experiments have stimulated and illumined research into teratogenic agents. Although much of such work has been pursued on the embryos of fish (minnow), amphibians (frog) and birds (chick), experiments on mammals in producing cleft palate, anophthalmia and other abnormalities suggest that similar mechanisms must operate in human maldevelopment. The identification of human teratogens may be said to have commenced with Gregg's demonstration of the effects of rubella virus. Other viral and bacterial maternal infections have since been implicated. But with the multiplication of drugs, for such purposes as abortion (aminopterin) and sedation (thalidomide), some of which have proved tragically teratogenic, teratological research has been much concentrated upon drugs. In parallel with this, however, has been the recognition of the genetic conditioning of a growing list of abnormalities (McKusick 1986), including metabolic aberrations (Garrod 1963). These remain the two major fields of current teratology, pharmacological and genetic.

While the great volume of recorded research has led to the identification of a large number of causal agents, the detailed mechanisms by which these produce abnormalities are for the most part obscure. In some cases it is clear that normal embryonic processes are delayed, arrested or accelerated, leading to agenesis or to varying degrees of hypoplasia or hyperplasia. The recognition that cell death plays a normal role in prenatal growth and differentiation, leading to the disappearance, modelling or reduction of structures (p. 103), has been extended to explanations of congenital defects. The available evidence suggests that cell death

is a major factor in developmental aberrations associated with viruses, irradiation, nutritional defects and hypervitaminoses. Local hyper- or hypoplasia and disturbed morphogenetic movements of cells may obviously produce distortions of normal development. It cannot be doubted that such factors as these are concerned, but in most cases the explanations are partial and missing links in the causal chain persist (Saxén 1970, Johnston & Pratt 1975). This is not surprising, since the precise march of events in much of normal development is still imperfectly known.

A very large number of cytotoxic agents are now known, and many have been used as experimental teratogens, mostly on rodents. Some act as antimetabolites, amino-acid antagonists, anti-purines or spindle toxins and most are highly selective in their effects, which are produced only by *controlled dosage* at *specific periods* during development. Connors (1975) has collated two decades of reports in this field, emphasizing the complex array of variable parameters which help to determine whether an agent acts as a teratogen or not. These include embryonic age, the amount, route and mode of administration of the agent, placental and embryonic permeability, maternal or embryonic ability to inactivate the agent, the state of differentiation of target cells and their ability to recover. Sullivan (1975) has reviewed the literature concerning teratogenic drugs taken by pregnant women, classifying them on their sites of action, whether directly on the embryo (thalidomide, tetracycline antibiotics), on embryonic endocrine balance (oestrogens and androgens), on the placenta or on maternal tissues. These considerations are, of course, of intense clinical importance, but their contribution to *explanations* of teratogenic *mechanisms* is limited, except in so far as some drugs are known to be teratogenic at specific stages of development.

Genetic conditioning of congenital abnormalities, whether structural, metabolic or behavioural, has been the subject of very widespread research and observation. A long list of inheritable defects has accumulated (McKusick 1986). According to Polani (1973) about a third of recognized defects in the newborn are due to single gene abnormalities, a further twelfth to chromosomal errors and a further substantial, but unknown, fraction to polygenic interaction. About 2000 birth defects occur in mankind which are either known to have, or are presumed to have, a genetic background. Of these 1700 are probably autosomal and 150 sex-linked (McKusick 1986). Although much information is thus available which may prove of great value in genetic counselling, complete mechanisms from disturbed DNA coding to actual phenotypic expression of particular deformities or other aberrations are not likely to be definable. Such explanations are dependent upon a more complete knowledge of embryonic processes than is yet available. Presumably, different genetic errors may operate at different stages in the procession of events through multiplication, determination, aggregation, morphogenetic movement, differentiation, localized proliferation and cell death, each of which is itself a complex of biochemical activities. In some examples, however, it is possible to demonstrate or to hypothecate such events in detail, normal and abnormal metabolic processes in bacteria providing the most successful field. One of the earliest, if not the first of human abnormalities to be identified as genetic was alkaptonuria, now known to be due to deficiency of a single enzyme (homogentisic acid oxidase) and hence, presumably, to a single gene disturbance. Although the locus of the latter is not established as yet, the biochemical sequence during which 'alkapton' is produced is fully clarified. In sickle-cell anaemia, which is inherited as a mendelian recessive characteristic, there is demonstrably a defect in the production of the protein moiety of the haemoglobin molecule (in fact an amino-acid substitution). The actual 'sickling' of erythrocytes only occurs in certain conditions of local oxygen concentration, an interaction between genetic effect and environmental factor which appears to be involved in the development of many basically genetic abnormalities. In the case of more elaborate structural deformities it is obvious that complete explanations are dependent upon adequate clarification of embryogenetic processes as noted above.

A relatively common example of an autosomal genetic abnormality often cited is the complex termed Down's syndrome (mongolism), which occurs in 1 in 600 live births. It is due to a trisomy or tripling of chromosome 21. The incidence of Down's syndrome increases with maternal age, suggesting that the aetiology is complicated and may not be simply genetic. The discovery that sex chromosomes can be identified has expedited the recognition of several sex-linked conditions, although, of course, familial history alone had indicated the sex-linked nature of haemophilia at a much earlier date. In Klinefelter's syndrome, which is characterized by testicular atrophy and hence sterility and eunuchoidism, associated with a variable depression of intelligence, there is a condition of X-polyploidy, such as XXY, XXXY and even higher states of X. Development of the gonad into a testis is ensured by the Y chromosome and the additional X chromosomes presumably retard differentiation. In another congenital deviation, Turner's syndrome, which features dwarfism, rudimentary ovaries, amenorrhoea and failure of the appearance of secondary sexual characteristics, the Y chromosome is absent (XO) or imperfect. A similar sex inversion (XX) occurs in males. As in the case of autosomal abnormalities, sex-linked chromosomal defects are common in spontaneously aborted individuals. (XO is said to be the commonest, at about 1% of all conceptions.)

It is clear from the foregoing that although the exact causology of the great majority of human congenital malformations is uncertain, the role of genetic and environmental factors (including teratogenic drugs) cannot be doubted. Wherever adequate familial studies can be established there is frequently an indication of genetic mechanism. However, an increased frequency of certain abnormalities may be equally associated with depressed socio-economic status, climate or geographical regions and maternal age.

Epidemiological studies of congenital aberrations are of considerable interest and value, not only in genetic counselling but also in attempts to unravel the complex problems of causation. The incidence of congenital abnormality may be as high as 6% in surviving infants. With the decline in incidence of, and death from, infectious and nutritional diseases, congenital abnormality is assuming a proportionately greater importance in the medical care of many populations. Several large epidemiological surveys, carried out in many countries, are now available (e.g. Kennedy 1967, WHO 1972). These show not only the overall incidence of abnormality but also the comparative frequency of various malformations and the regional and racial variations which some display. For example, neurulation defects such as anencephaly and spina bifida, while being relatively frequent abnormalities, also show much variation in frequency. (Particularly high incidences have been recorded in Belfast and Bombay.) Galactokinase deficiency was identified about 15 times more often in a Canadian series (Manitoba) than in the USA (Massachusetts). Such variations are at present largely inexplicable; but the well-known coincidence of sickle-cell gene for haemoglobin S and of the thalassaemia trait with the distribution of malaria are interesting exceptions.

The prenatal identification of certain malformations and aberrations must be mentioned. Amniocentesis and examination of the amniotic fluid and its suspended cells permits the diagnosis of some metabolic disorders and potentially all chromosomal aberrations. It is of particular value in conditions of high incidence and early lethal effect such as Tay-Sachs disease, which occurs about 100 times more often in Ashkenazi Jews than in other races. Radiology alone may reveal anencephalic or hydrocephalic fetuses (Russell 1969); but the hazards of this technique have led to the substitution of ultrasonography, which is presumed to be safer and can discover similar conditions (Taylor 1978). Amniography consists in radiological examination after the injection of a water-soluble radio-opaque substance into the amniotic fluid; fetography is a variant of this technique which employs an oily radio-opaque injectant (Wiesenhaan 1972). By absorption of the radio-opaque substance in the vernix caseosa, the latter technique is claimed to yield clearer delineation of surface features. (For literature consult Persaud 1977.)

Twinning

Twinning occurs once in about every 80 births. Twins may be *dizygotic* (binovular or fraternal) or *monozygotic* (uniovular or identical), of which the former occurs more frequently. Twin boys are most common, less common are a boy and a girl and least common two girls.

Dizygotic twins result from the discharge and fertilization of two ova, either from two separate or a single ovarian follicle. Each embryo develops in its own chorionic sac, but occasionally synchorial fusion occurs. The twins are of different genetic constitution and may be of the same or different sex. They bear no greater resemblance to one another than other siblings of the same family. There is evidence that dizygotic twinning may be hereditary. The development of ovarian follicles and the release of ova at ovulation is under the hormonal control of circulating *gonadotrophins* produced by the β cells of the anterior lobe of the hypophysis cerebri. The control mechanisms are far from clear, but it has been demonstrated in a variety of mammals that raising the circulating gonadotrophic level by injections increases the number of ova discharged (Hammond 1952). Similar considerations apply to women who have had such treatment for long-standing amenorrhoea (Gemzell & Roos 1966). In one series of about 100 cases so treated there were 43 pregnancies including 20 singletons, 14 twins, 2 triplets, 3 quadruplets, 1 quintuplet, 2 sextuplets and 1 septuplet. At the time of writing, the clinical exhibition of gonadotrophins in cases of infertility has become commonplace.

Monozygotic twins arise from a single ovum fertilized by a single sperm. At some stage up to the establishment of the axis of the embryonic area and the development of the primitive streak the formative material separates into two parts, each of which gives rise to a complete embryo. The twins may share one or have independent chorionic sacs and possess an individual or a common amniotic sac. The twins are of the same sex, have the same blood groups and tissues of identical antigenic potencies. They resemble each other closely. Monozygotic twinning is a hereditary character.

Interestingly, a number of workers have suggested that in addition to the widely recognized two foregoing types, a *third type* of twin (with several variants) may exist. Essentially this type is *uniovular* but *dispermatic* following an irregular meiotic cycle during oögenesis. It is considered that either the ovum, or the secondary oöcyte, or the primary oöcyte may undergo an equal cytoplasmic division (instead of throwing off a diminutive polar cell). Evidently each of the three variants would have different degrees of genetic dissimilarity depending upon the stage at which such an equal division occurred.

Multiple births greater than twinning, such as triplets or quadruplets, can arise from multiple ovulations or a single ovum, the formative material of which later separates into several parts, or from a combination of these mechanisms.

Conjoined twins or double monsters arise from the incomplete division of the formative material of a single early ovum. There is, however, no evidence that the condition is hereditary or that it occurs more frequently in families given to uniovular twinning. The conjoined twins are usually united by corresponding regions and the degree of duplication of parts varies.

For an excellent review of the types, frequency, inheritance and embryology of twinning, the diagnosis of zygosity, the course and outcome of twin pregnancies, their possible evolutionary significance and some post-natal characteristics of twins, the interested reader should consult Bulmer (1970).

DEVELOPMENT OF INDIVIDUAL SYSTEMS

Embryonic development has so far been considered as a whole but, as the definition of its structures proceeds, overall description becomes so complicated as to actually impede the clarity of appreciation of the events occurring. It is hence customary and convenient to limit attention to individual systems in their further development; but it must never be overlooked that the analysis of a whole organism into such divisions—however attractive on morphological and functional grounds—is largely a product of the sequential nature of human perception. Not only do the several systems into which we divide the organism develop simultaneously, they also interact and modify each other. This necessary interdependence is not only supported by the evidence of experimental embryology but is also emphatically demonstrated by the phenomena of growth anomalies, which cut across the artificial boundaries of systems in most instances. For these reasons it is most desirable that the development of any one individual system should be frequently related to others, especially those most closely associated with it (both spatio-temporally and causally).

So far the development of the embryo has been taken to an age of between 3 and 4 weeks, the stage of early somite formation, equivalent to Horizons X or XI on the scale established by the studies of Streeter and others (1942–1949). (See also the analysis into 'Stages' by O'Rahilly 1973 and p. 261.) It is only partly constricted from the yolk sac, but the head and tail folds are well formed, with enclosure of the foregut and hindgut (proenteron and metenteron). The forebrain projection dominates the cranial end of the embryo, the oropharyngeal membrane and cardiac prominence being caudal and ventral to this. The intraembryonic mesoderm has begun to differentiate and its paraxial region is undergoing segmentation into somites. Neural groove closure is progressing to a neural tube and is separated from the dorsal aspect of the gut by the notochord. The earliest blood vessels have appeared and a primitive tubular heart occupies the pericardium. The chorionic circulation is soon to be established, after which the embryo rapidly becomes completely dependent for its requirements upon the maternal bloodstream. The intraembryonic part of the coelom consists of the transmedian pericardial cavity, leading dorsocaudally into right and left pericardioperitoneal canals (coelomic ducts). The canals occupy mesenchyme dorsal to the septum transversum, caudal to which they expand into the peritoneal cavity. This establishes, for a time, a free communication with the extraembryonic coelom (2.27, 28).

Development of the Skeletal System

The skeleton is a derivative of mesoderm (and in some parts neural crest), including not only its axial and appendicular divisions but also all accessory ossicles such as sesamoid bones and the osseous and cartilaginous elements of the branchial arches. Most of these parts initially enter a *blastemal stage* of mesenchymal condensation and this transforms into a *cartilaginous stage* before becoming ossified. In some bones, however, ossification follows immediately upon the blastemal stage, the intermediate phase of chondrification being omitted (p. 300). For a review of embryonic cartilage see Glenister (1976).

THE SKELETAL AXIS

Before reaching its final condition the skeletal axis passes through three preliminary states (Präder 1947, Sensenig 1949, Peacock 1951, 1952, Walmsley 1953, Tondury 1958, Verbout 1976, 1985 and 2.44). First it is formed by the *non-segmental notochord*, a flexible rod of cells enclosed by a thick, membranous sheath. The notochord is not, however, limited to the region in which the vertebrae will replace it, for it extends into the head as far as the caudal limit of the hypophysis cerebri, this cranial extension being subsequently incorporated into the basilar part of the occipital bone and the caudal part of the body of the sphenoid (2.45A).

In the second stage, the notochord provides a framework around which a *blastemal* or *mesenchymatous vertebral column* is formed. On each side fusiform cells of the sclerotomes multiply rapidly and migrate ventromedially enclosing the notochord in a mesenchymal sheath (2.44), which at first retains traces of its segmental origin. Sclerotomic cells later migrate dorsally around the spinal cord, passing between and lateral to the rudimentary spinal ganglia and ventrolaterally into intervals between myotomes.

Around the notochord each sclerotomic segment is divided into equal cranial and caudal parts by a transitory transverse split or loosening between the two groups of cells, the *sclerotomic fissure* (2.44). The parafissural mesoderm condenses to form a dense transverse plate, a *perichordal disc*, whilst the less dense, caudal part of one segment fuses with the slightly smaller also less dense cranial part of the adjoining segment. This fusion defines the blastemal centrum of a vertebra. In the dorsal extensions of the sclerotomes segmental condensations between the spinal ganglia define the neural arch and later its processes, while their ventrolateral extensions outline the costal processes, which are continuous dorsomedially with the perichordal disc.

In the third stage the vertebral components become more clearly defined and chondrification of these mesenchymatous models produces a *cartilaginous vertebral column*. Each centrum is chondrified from a pair of centres which appear during the sixth week and quickly coalesce. Each half of a neural arch is chondrified from a centre starting in its base and extending ventrally into the pedicles, to meet, expand and blend with the centrum, and dorsally into the laminae; but the latter do not meet in the midline until the fourth month. Thus the presumptive vertebral *body* is compound—a centrum with dorsolateral accessions, expansions from the pedicles of the neural arch. The transverse and articular processes are chondrified in continuity with the neural arches; intervening zones of mesenchyme which do not become cartilage mark the sites of their intervertebral and costovertebral joints, and synovial cavities appear later in these. The costal processes chondrify separately and in the thoracic region they extend ventrally, the more cranial members curving round in the body wall to reach the developing sternal plates. They are separated from the developing transverse processes by non-chondrified mesenchyme in which the costotransverse joints will appear. A similar process affects the sites of most of the capitular costovertebral joints involving the perichordal disc and adjacent parts of the neural arch-derived dorsolateral parts of the bodies. At other than thoracic levels the developing costal process (or pleurapophysis) becomes incorporated into the 'transverse process' of descriptive adult anatomy (2.44).

In both the blastemal and chondrifying stages, the cranial and caudal aspects of each perichordal disc proliferate, contributing to the growth of adjoining centra and eventually merging with them. The main, intermediate mass of each perichordal disc (which contains the sclerotomic fissure), together with an enclosed part of the notochord, becomes an *intervertebral disc*. (Vide infra for further reviews.)

HYPOCHORDAL ELEMENTS

A structure which, though present during the blastemal stage, only becomes clearly recognizable as an entity during chondrification is the *hypochordal arch* or bow. This connects the vertebral ends of two costal processes to each other and spans the ventral surface of a centrum; it is prominent only with the first three or four cervical vertebrae in man. Only in the case of the *atlas* does the hypochordal arch chondrify and become ossified during the first post-natal year; it forms the anterior arch and the ventral parts of the lateral masses of this vertebra. (The hypochordal arch may correspond to the paired intercentra, ventral elements in the development of vertebrae in many other animals. They may persist at any level, forming such arrangements as haemal or intercentral arches, which enclose the caudal vessels.) Some authorities, however, have tentatively suggested that certain

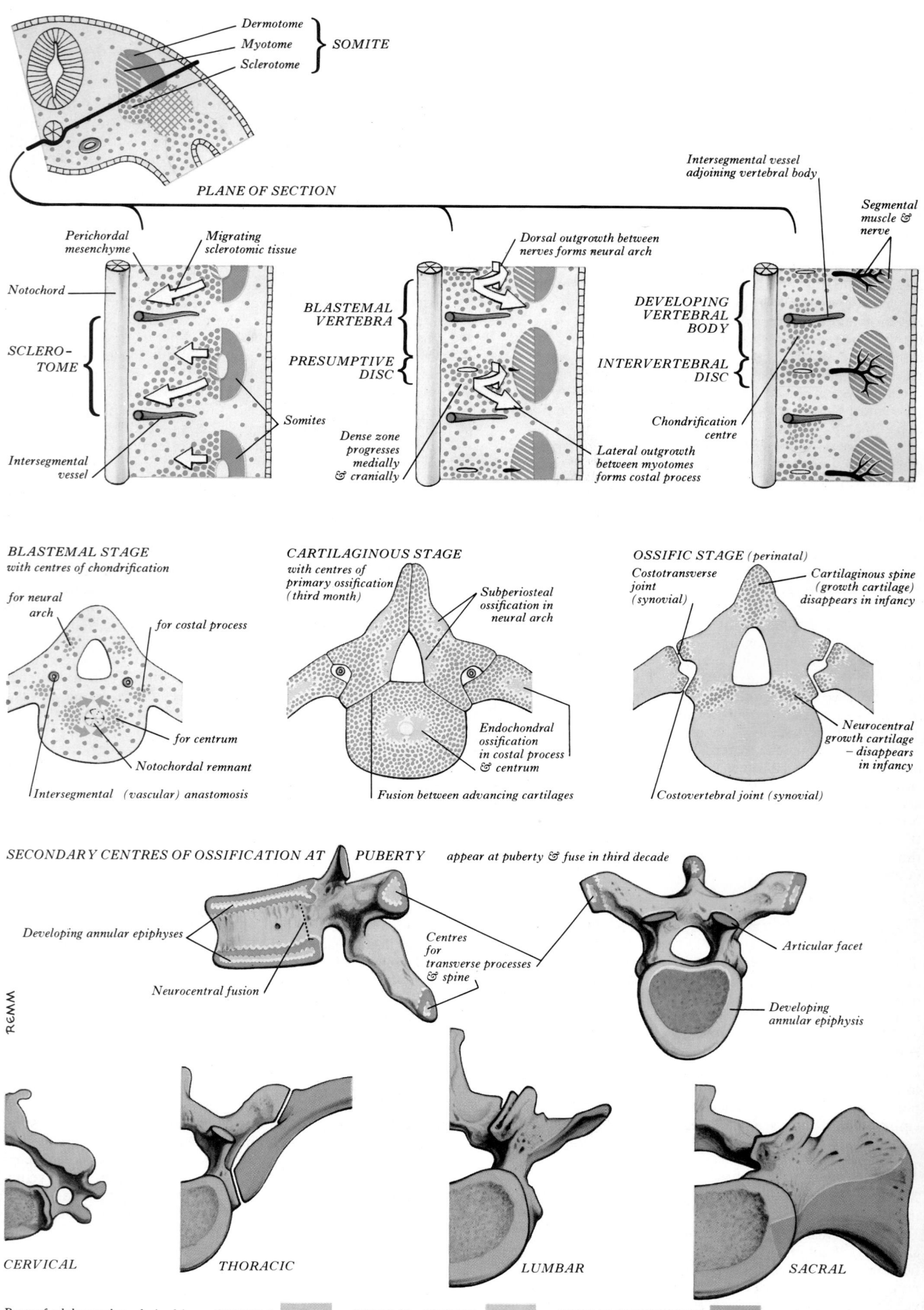

SOMITE
- Dermotome
- Myotome
- Sclerotome

PLANE OF SECTION

Intersegmental vessel adjoining vertebral body

Segmental muscle & nerve

Perichordal mesenchyme

Migrating sclerotomic tissue

Notochord

Dorsal outgrowth between nerves forms neural arch

BLASTEMAL VERTEBRA

DEVELOPING VERTEBRAL BODY

SCLERO-TOME

PRESUMPTIVE DISC

INTERVERTEBRAL DISC

Somites

Dense zone progresses medially & cranially

Lateral outgrowth between myotomes forms costal process

Chondrification centre

Intersegmental vessel

BLASTEMAL STAGE
with centres of chondrification

for neural arch

for costal process

for centrum

Notochordal remnant

Intersegmental (vascular) anastomosis

CARTILAGINOUS STAGE
with centres of primary ossification (third month)

Subperiosteal ossification in neural arch

Endochondral ossification in costal process & centrum

Fusion between advancing cartilages

OSSIFIC STAGE (perinatal)

Costotransverse joint (synovial)

Cartilaginous spine (growth cartilage) disappears in infancy

Neurocentral growth cartilage – disappears in infancy

Costovertebral joint (synovial)

SECONDARY CENTRES OF OSSIFICATION AT PUBERTY   appear at puberty & fuse in third decade

Developing annular epiphyses

Neurocentral fusion

Centres for transverse processes & spine

Articular facet

Developing annular epiphysis

REMM

CERVICAL

THORACIC

LUMBAR

SACRAL

Parts of adult vertebrae derived from: CENTRA   *, NEURAL ARCHES*   *, COSTAL PROCESSES*   *of embryonic vertebrae*

2.44 Sequential diagrams of vertebral development from early somitic and perichordal stages through blastemal, cartilaginous, and pre- and post-natal ossificatory stages. For views concerning segmentation see text. Bottom row indicates principal morphological parts of adult vertebrae.

collagenous, ventral, transcentral structures, e.g. part of the thoracic, costovertebral radiate ligaments, may have hypochordal origins.

The homologies of the various elements which fuse to form vertebrae are still subject to disagreements (Jollie 1962). It does, however, now appear certain that the original concept of Gadow (1933), that all tetrapod vertebrae can be derived from four paired elements (basidorsals, basiventrals, interdorsals and interventrals) must be abandoned. Subsequent authorities, provided with much more extensive evidence, especially embryological data, have not substantiated Gadow's somewhat theoretical views (Williams 1959, Jenkins 1969). Another widely accepted view—that the centrum of the atlas is united with that of the axis as its dens—has been thrown into doubt by observations on developmental stages in mammals (Jenkins 1969). In the case of the axis and succeeding vertebrae the hypochordal arches degenerate, either disappearing entirely or becoming incorporated into the ventral parts of the adjoining centra.

NOTOCHORDAL VESTIGES

The notochord may be identified for some time traversing the cartilaginous centra, but these parts of it ultimately atrophy and vanish. Between the developing vertebrae it expands as localized aggregates of cells with intervening mucoid matrix together forming the *nucleus pulposus* of the intervertebral disc. This nucleus is surrounded by the intermediate part of each perichordal disc which forms the *anulus fibrosus* and differentiates into an external laminated fibrous zone and an internal cuff around the nucleus pulposus. This inner zone contributes to the growth of the outer, and near the end of the second month of embryonic life it begins to merge with the notochordal tissue, being ultimately converted into fibrocartilage. After the sixth month of fetal life notochordal cells in the nucleus pulposus commence to degenerate, being replaced by cells from the internal zone of the anulus fibrosus. This degeneration continues until the second decade of life, by which time all the notochordal cells have disappeared (p. 490). Thus, in the adult, notochordal vestiges are limited, at the most, to non-cellular matrix.

SUMMARY

Each vertebra was thus usually described as formed from parts of two adjacent sclerotomes. The rearrangement of segmentation was said to result in an alteration in position of vertebrae with respect to myotomes, bringing the intersegmental vessels into line with the vertebral bodies, a relationship which persists in the lumbar and lower thoracic regions. However, an extensive study of the behaviour of the perichordal and sclerotomic mesenchyme in the embryos of sheep suggests that, in this species at least, 'resegmentation' of this kind does not occur, the primordia of vertebrae arising in their *definitive* positions, without any movement into new sites relative to surrounding structures (Verbout 1976, 1985). The essence of this view is that the vertebral *processes* and *centra* have different mesenchymal origins, the former from cells closely associated with myotomes (*sclerotomic tissue*), the latter from the non-segmented *perichordal mesenchyme*. The processes develop in a caudal condensation within their segment and are caudal to the numerically corresponding spinal nerve. It is suggested that a *gradient of segmentation* spreads medially from the myotomes to the precursors of the vertebral processes and from these to the perichordal tissue (2.40). For further data and an extensive review of these topics consult Verbout (1985).

Towards the end of the second month ossification commences in the cartilaginous vertebrae and the column then enters the fourth and last stage in its development. The further details of this are described elsewhere (p. 327).

RIBS, COSTAL CARTILAGES AND STERNUM

The ribs develop from the costal processes of the primitive vertebral arches, extending between the myotomic muscle plates. In the thoracic region (2.44) of the vertebral column these processes grow laterally to form a series of *precartilaginous ribs*. The transverse processes grow laterally behind the vertebral ends of the costal processes, at first connected by mesenchyme which later becomes differentiated into the ligaments and other tissues of the costotransverse joints. The capitular costovertebral joints are similarly formed from mesenchyme between the proximal end of the costal processes and the perichordal disc, and adjacent, neural arch derived parts of usually two (sometimes one) vertebral bodies. In *cervical vertebra* (2.44) the transverse process is dorsomedial to the foramen transversarium, while the costal process, corresponding to the head, neck and tubercle of a rib, limits the foramen ventrolaterally and dorsolaterally. The distal parts of these cervical costal processes do not develop, but occasionally they do so in the case of the seventh cervical vertebra, even developing costovertebral joints. Such *cervical ribs* may reach the sternum (p. 328). In *lumbar vertebrae* (2.44) the costal processes do not develop distally, but their proximal parts become the 'transverse processes' of these vertebrae, whose morphologically *true* transverse processes may be represented by their accessory processes (p. 322). Occasionally, movable ribs may develop in association with the first lumbar vertebra. Only the upper two or three *sacral costal processes* usually develop (2.44). They fuse into the lateral mass of the sacrum, forming its ventral part. The *coccygeal* vertebrae are apparently devoid of costal processes.

The sternum is formed from bilateral mesenchymatous condensations, sternal plates, which begin in the dorsolateral region of the body wall. They are immediately ventral to the rudiments of the clavicles and ribs, but are independent of them in their formation, in so far as experiments on their primordia in the mouse can testify to the human condition (Chen 1952). These plates chondrify, move ventrally towards each other from both sides as the costal processes lengthen, and they eventually fuse together across the midline in a craniocaudal direction. This forms a longitudinal cartilaginous bar, with which the clavicles and upper seven pairs of costal cartilages establish contact. The xiphoid process develops as a caudal extension of the sternal bar. Hypertrophy of the cartilage cells as a preliminary to ossification occurs opposite future intercostal spaces, as the first indication of segmentation of the sternum into manubrium, four sternebrae for the 'body', and xiphoid process. The ossification and further growth of the sternum and ribs is described later (pp. 332, 335).

Development of the Cranium (2.45)

The bones of the skull are developed in the mesenchyme which surrounds the cerebral vesicles but, before the osseous state is reached, the cranium passes through blastemal and cartilaginous stages like other parts of the skeleton. However, not all parts pass through a phase of chondrification; and hence the *chondrocranium* is incomplete, the remainder comprising the mesenchymatous, *blastemal desmocranium*. Most of the cranial vault and limited parts of its base are thus not preformed in cartilage. Though the mesenchymatous (membranous) and cartilaginous parts of the skull will, for convenience, be considered in sequence, they develop together and complement each other in forming the complete cranium, some of whose bones are composite structures derived from both sources. All elements, of course, pass first through a mesenchymatous phase (2.45A–F).

THE DESMOCRANIUM

The blastemal skull (desmocranium) begins to appear at the end of the first month as a condensation and thickening of the mesenchyme which surrounds the developing brain, forming localized masses which are the earliest distinguishable cranial elements. The first masses evident are in the occipital region, outlining the basilar (ventral) part of the occipital bone. These form an *occipital plate*, from which two extensions on each side grow laterally and spread to complete a foramen around each hypoglossal nerve. At the same time the mesenchymal condensation extends forwards, dorsal to the pharynx, to reach the primordium of the hypophysis, thus establishing the clivus of the cranial base and the dorsum

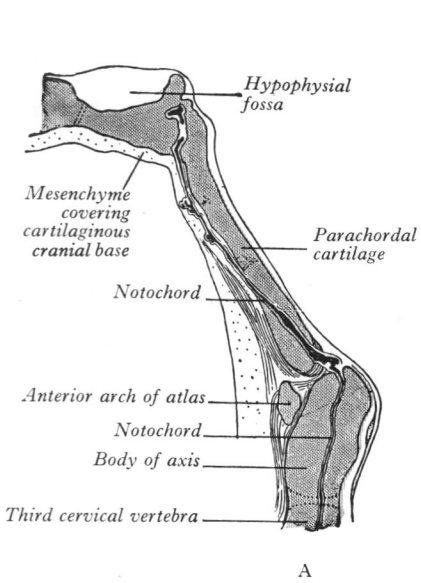

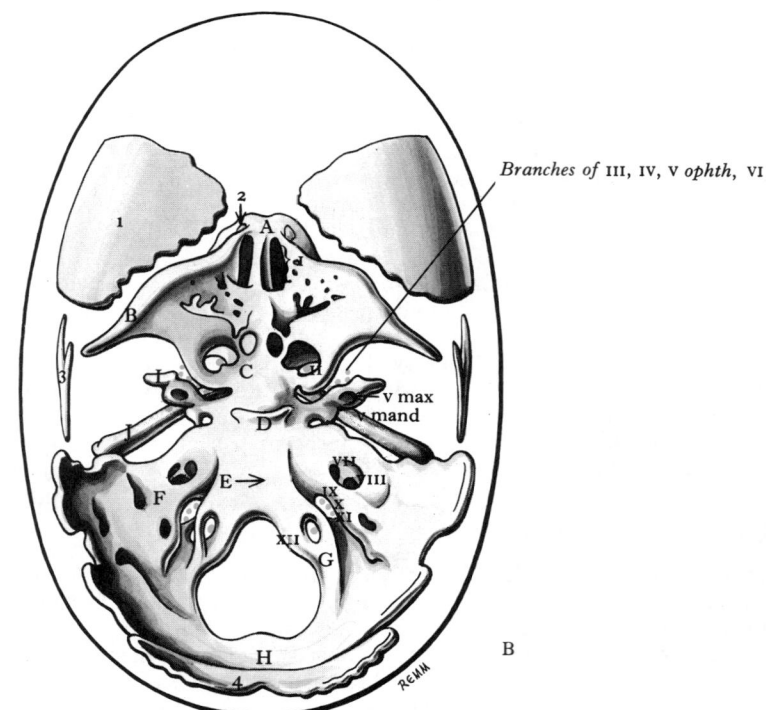

Key to chondral elements:

A   Nasal capsule
B   Orbitosphenoid
C   Presphenoid
D   Postsphenoid
E   Basi-occipital
F   Otic capsule
G   Exoccipital
H   Supra-occipital
I   Alisphenoid
J   Meckel's mandibular cartilage
K   Cartilage of malleus
L   Styloid cartilage
M   Hyoid cartilage
N   Thyroid cartilage
O   Cricoid cartilage
P   Arytenoid cartilage

Key to dermal (membrane) elements:

1   Frontal bone
2   Nasal bone
3   Squama of temporal bone
4   Squama of occipital bone
    (interparietal)
5   Parietal bone
6   Maxilla
7   Lacrimal bone
8   Zygomatic bone
9   Palatine bone
10 Vomer
11 Medial pterygoid plate
12 Tympanic ring
13 Mandible

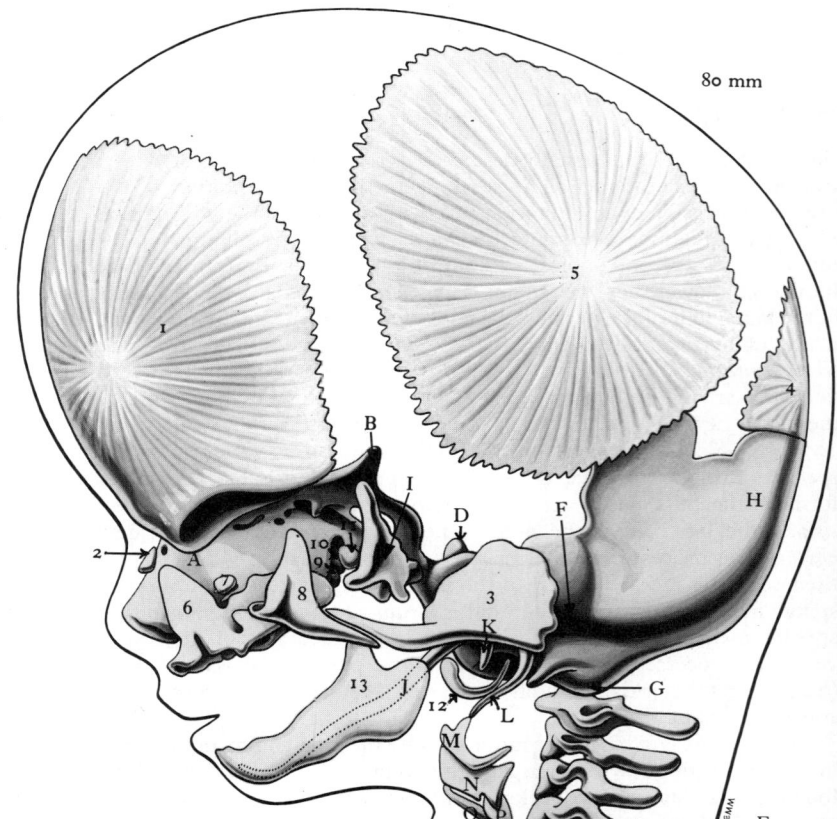

2.45 Representative stages in the development of the cranium. In all the diagrams the *chondrocranium* and cartilaginous stages of vertebrae are shown in blue, except where ossification is occurring and here the colour is green. The *desmocranium*, consisting of elements ossifying directly in mesenchyme is shown in yellow. Cranial nerves are indicated by the appropriate roman numeral.

A Sagittal section through the cranial end of the developing axial skeleton in an early human embryo of about 10 mm, showing the extent of the notochord. B Key for diagram C. C Superior aspect of cranium of human embryo at 40 mm. D Lateral aspect of C. E Key for diagram F. F Lateral aspect of cranium of human embryo at 80 mm.

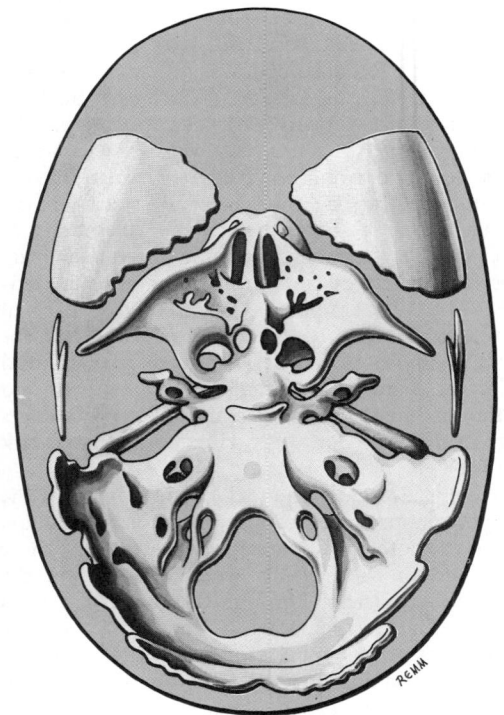

C

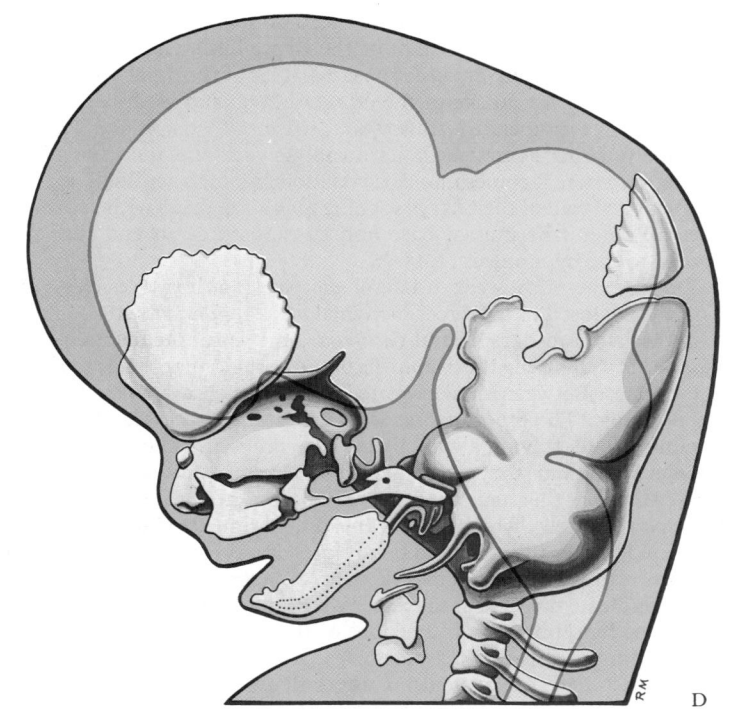

D

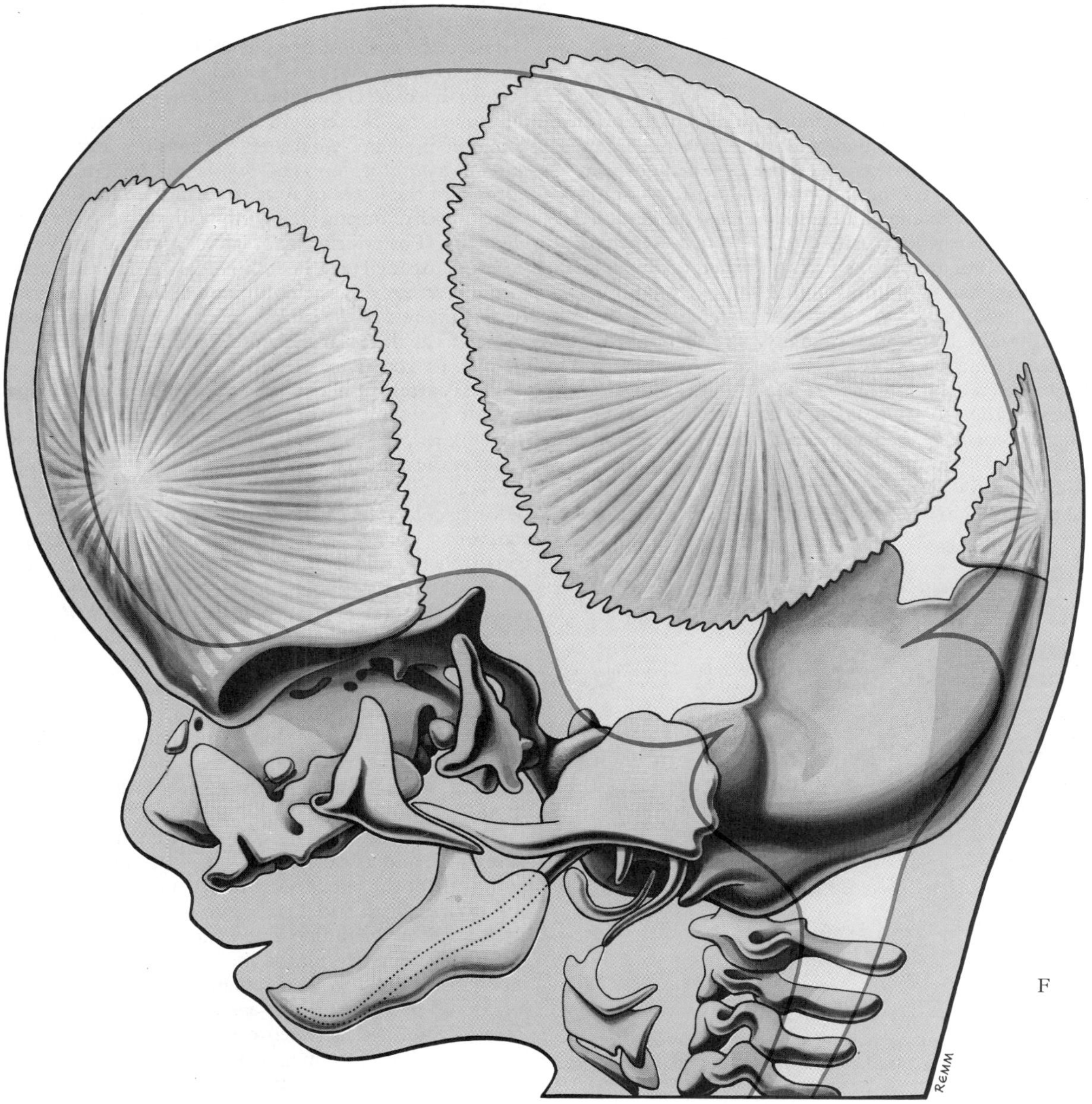

F

sellae of the future sphenoid bone. Early in the second month it surrounds the developing stalk of the hypophysis and extends ventrally and rostrally between the two halves of the nasal cavity, where it forms the anlage of the ethmoid bone and of the nasal septum. The notochord traverses the ventral occipital plate obliquely, being at first near its dorsal surface and then lying ventrally, where it comes into close relationship to the epithelium of the dorsal wall of the pharynx, being for a time fused with it. It then re-enters the cranial base and runs rostrally to end just caudal to the hypophysis (2.45A).

During the fifth week bilateral *otocysts* (auditory vesicles) become enclosed in their mesenchymal *otic capsules*, which soon differentiate into dorsolateral *vestibular* and ventromedial *cochlear* parts, enveloping the primordia of the semicircular canals and the cochlea. Between these two regions the facial nerve lies in a deep groove. The otocysts fuse with the lateral processes of the occipital plate, leaving a wide hiatus through which the internal jugular vein and the glossopharyngeal, vagus and accessory nerves pass. At this stage the mesenchyme around the developing hypophysial stalk, which is forming the rudiment of the postsphenoid part of the sphenoid bone, spreads out laterally to form the future greater wings of this element. Smaller processes rostral to this indicate the sites of the lesser wings of the sphenoid, while other condensations reach the sides of the nasal cavity and also blend with the still mesenchymatous septum.

The first signs of the vault or upper neurocranial part of the skull appear about the thirtieth day; they consist of curved plates of mesenchyme at the sides of the skull and gradually extend cranially to blend with each other; they also extend towards and reach the base of the skull, which will become part of the chondrocranium.

THE CHONDROCRANIUM

Chondrocranium (2.45) is a term applied to the parts of any vertebrate skull which pass through, and sometimes remain in, a cartilaginous stage. (In Chondrichthyes, the cartilaginous fishes, such as sharks, all the cranium chondrifies and persists in this state—see p. 340.) The crania of all land animals, the tetrapods, contain variable regions which ossify directly from mesenchyme. Such *dermal* (membranous) elements form much of the cranial vault, and in mammals *chondrification* is limited to the basal regions of the skull (2.45). This mammalian change occurs primarily in three regions: (1) caudally, in relation to the notochord; (2) intermediately, in relation to the hypophysis; and (3) rostrally, between the orbits and the nasal cavities. These may be named *parachordal, hypophysial* and *interorbitonasal* regions. The parachordal cartilage is developed from the mesenchyme related to the cranial end of the notochord; caudally it exhibits traces of four primitive segments separated by the roots of the hypoglossal nerves. The hypophysial cartilage ossifies to form the

postsphenoid part of the sphenoid bone, but its morphological status is uncertain. The interorbitonasal cartilage is perhaps to be equated with the trabeculae cranii of lower vertebrates and is usually known as the *trabecular cartilage*, which is a bilateral structure developing from two centres of chondrification. The trabeculae cranii may largely be derived from branchial arch (neural crest) mesoderm, i.e. from the **viscerocranium**, having been adapted into the cartilaginous or basal part of the **neurocranium** or 'brainbox'. From the evidence in embryos of earlier vertebrates most of the chondrocranium and the majority of the viscerocranium are derived from neural crest tissue, including almost all of the branchial skeleton. Only the caudal parts of the trabeculae, the parachordal bars, the otic capsules and the second basibranchial appear to be derived from general head mesenchyme. For these and other details of development and morphology of the mammalian chondrocranium, see references (Fawcett 1911, 1917, 1918, de Beer 1937, Hörstadius 1950, Stark 1965, Balinsky 1981).

In the human embryo cranial chondrification begins in the second month; cartilaginous foci first appear in the occipital plate, one on each side of the notochord (parachordal cartilages); these later fuse at the end of the seventh week surrounding the notochord, whose oblique transit through the region has been mentioned (2.45A). The cartilaginous posterior part of the sphenoid is formed from two hypophysial centres, flanking the stalk of the hypophysis and uniting at first behind, then in front, enclosing a *craniopharyngeal canal* containing the hypophysial diverticulum. The canal is usually obliterated by the third month; its association with the derivation of the anterior lobe of the hypophysis from the pharyngeal diverticulum of Rathke has been denied (p. 230).

The otic capsules, presphenoid, bases of the greater wings and lesser wings of the sphenoid, and finally the nasal capsules, in turn become chondrified. The whole nasal capsule is well developed by the end of the third month, consisting of a common median septal part (sometimes initially termed the *interorbitonasal septum*) and two lateral regions. The free caudal borders of the latter incurve to form the inferior nasal conchae, which ossify during the fifth month and become separate elements. Posteriorly each lateral part of the nasal capsule becomes ossified as the ethmoidal labyrinth, bearing on its medial surface ridges the future middle and superior conchae. Part of the rest of the capsule remains cartilaginous as the septal and alar cartilages of the nose; part is replaced by the mesenchymatous vomer and nasal bones.

The ventral surface of the chondrocranium is associated with the cartilages of the branchial arches, the *viscerocranium*, the development of which will be considered later (vide infra). The bones of the cranial base which are thus preformed in cartilage are the occipital (excepting the upper part of its squama), the petromastoid part of the temporal, the body, lesser wings and roots of the greater wings of the sphenoid, and the ethmoid. These constitute the cartilaginous part of the neurocranium, the rest of which, the mesenchymatous (membranous) neurocranium, corresponding to the cranial vault, is not preformed in cartilage. Its elements, frequently described as *dermal* bones because of their probable origin (p. 340), are the frontal bones, the parietals, the squamous parts of the temporal bones and the upper (interparietal) part of the occipital squama. To summarize, therefore, the base of the skull—except for the orbital plates of the frontal and the lateral parts of the greater sphenoidal wings—is preformed in cartilage, while the whole of the vault is ossified directly in mesenchyme (2.45B-F).

Ossification commences before the chondrocranium has fully developed, and as this change extends, bone overtakes cartilage until little of the chondrocranium remains. However, parts of it still exist at birth and small regions remain cartilaginous in the adult skull. At birth unossified chondrocranium still persists at: (1) the alae and septum of the nose, (2) the spheno-ethmoidal junction (p. 376), (3) the spheno-occipital and spheno-petrous junctions (p. 376), (4) the apex of the petrous bone (foramen lacerum), and (5) between ossifying elements of the sphenoid bone and between elements of the occipital bone. Most of these regions function as growth cartilages (p. 474). For further

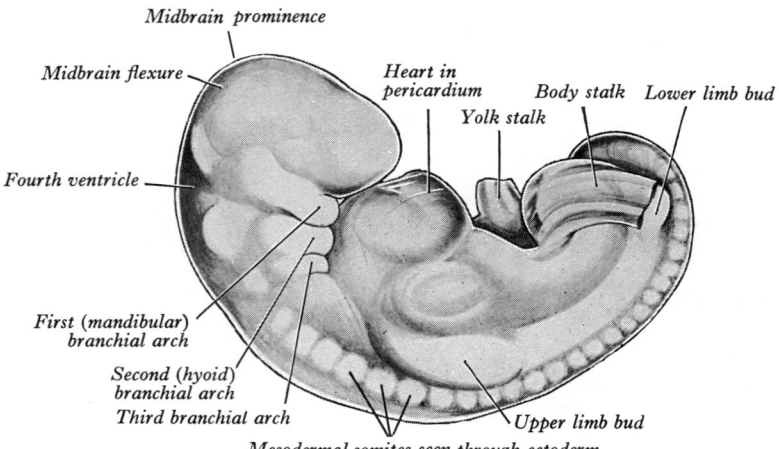

Midbrain prominence

Midbrain flexure

Heart in pericardium

Body stalk *Lower limb bud*

Yolk stalk

Fourth ventricle

First (mandibular) branchial arch

Second (hyoid) branchial arch

Third branchial arch

Upper limb bud

Mesodermal somites seen through ectoderm

2.46 A 26½-day macaque embryo, showing an early stage in the development of the branchial arches and the limb buds. Magnification × 12. (Heuser & Streeter 1941)

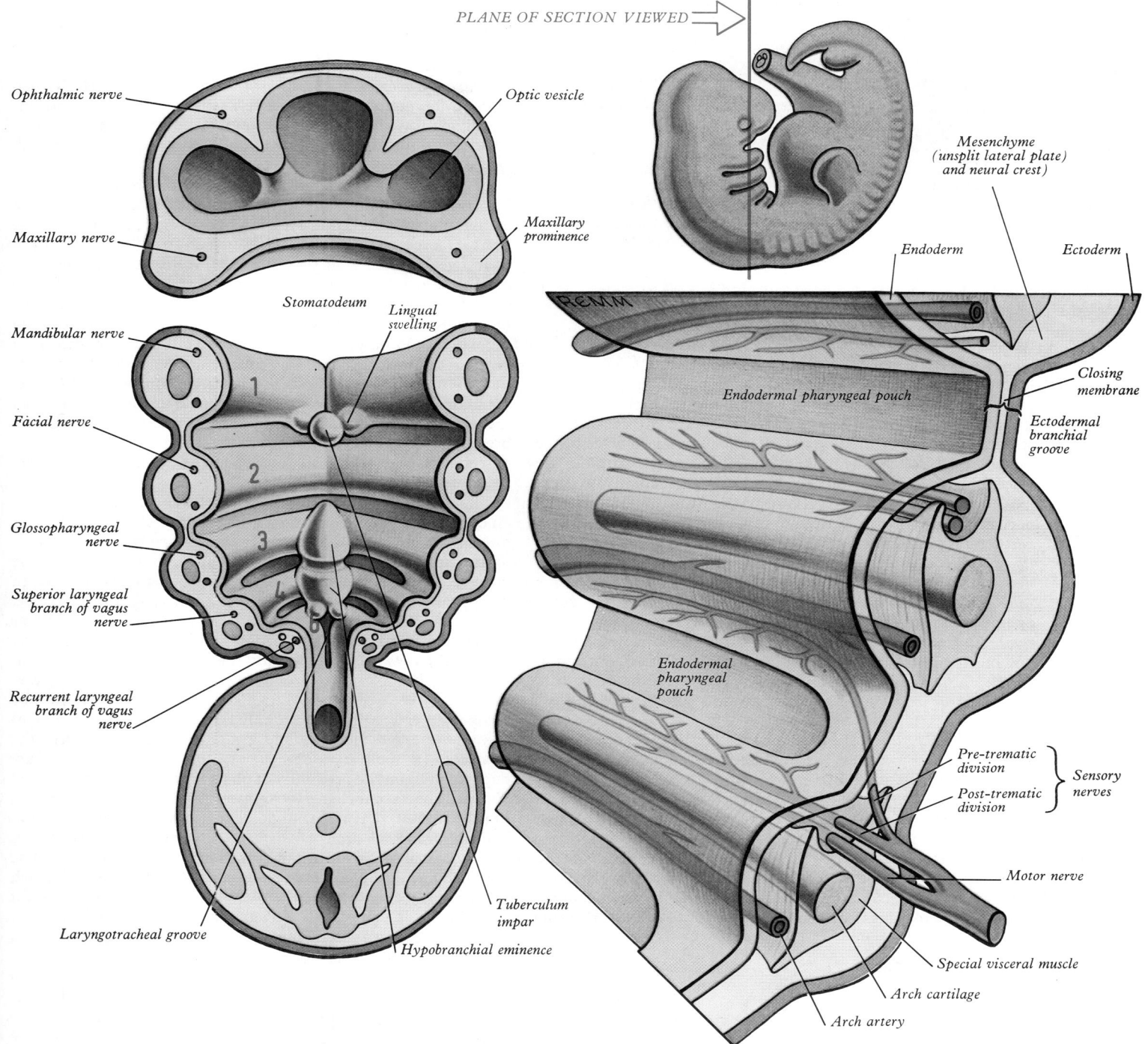

PLANE OF SECTION VIEWED

Ophthalmic nerve

Optic vesicle

Maxillary nerve

Maxillary prominence

Stomatodeum

Lingual swelling

Mandibular nerve

Mesenchyme (unsplit lateral plate) and neural crest)

Endoderm

Ectoderm

Facial nerve

Endodermal pharyngeal pouch

Closing membrane

Ectodermal branchial groove

Glossopharyngeal nerve

Superior laryngeal branch of vagus nerve

Endodermal pharyngeal pouch

Recurrent laryngeal branch of vagus nerve

Pre-trematic division

Post-trematic division

Sensory nerves

Motor nerve

Laryngotracheal groove

Tuberculum impar

Special visceral muscle

Arch cartilage

Hypobranchial eminence

Arch artery

2.47 Schema of developing branchial region showing (left) the pharyngeal floor and sectioned lateral walls, viewed from the dorsal aspect, and (right) details of generalized branchial constituents, including arches, endo- dermal pouches and ectodermal grooves. (Modified in part after Williams et al 1969.)

development of these areas and cranial bones in general see Section 3, Osteology and Section 4, Arthrology.

THE VISCEROCRANIUM

Certain cranial components are derived from the branchial or visceral arches, the *viscerocranium*, and therefore their skeletal development will be considered next, before that of the appendicular skeleton (p. 174). It is also convenient to include a general consideration of the branchial apparatus.

The Branchial Apparatus

After head fold formation, the *stomatodeum* or primitive mouth is bounded cranially by the projecting forebrain and caudally by the

cardiac prominence (**2.46**). The mandibular region and the whole of the neck, which will subsequently intervene between mouth and developing thorax, are absent, but will be formed by the appearance and modification of six paired **branchial arches**, which develop in the lateral aspects of the head adjacent to the hindbrain (**2.47, 48, 49**). In the earliest vertebrates, which were jawless *(Agnatha)*, the arches were a uniform series of bars between the gill clefts; but long before the evolution of the terrestrial vertebrates, remarkable adaptations had occurred in them. Structures commonly regarded as the first pair of arches became the jaws, upper and lower, of the jaw-bearing vertebrates *(Gnathostomata)*, including most fish; they are, therefore, usually named the *mandibular arches*. (The term 'mandibular arch' is widely used but not entirely appropriate because of the numerous maxillofacial, nasal, otic and palatopharyngeal derivatives from its dorsal end.) It should, however, be noted that since this early

165

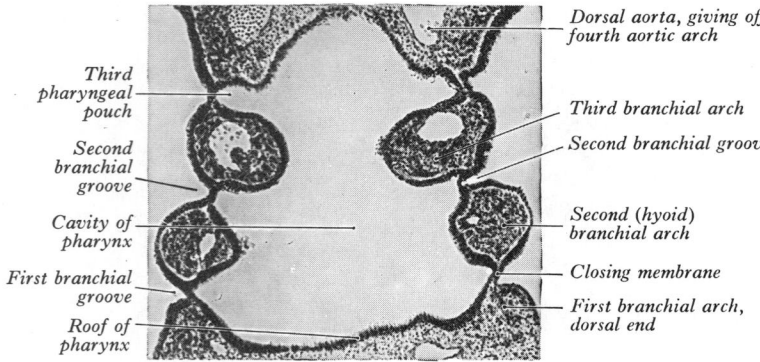

Third pharyngeal pouch

Second branchial groove

Cavity of pharynx

First branchial groove

Roof of pharynx

Dorsal aorta, giving off fourth aortic arch

Third branchial arch

Second branchial groove

Second (hyoid) branchial arch

Closing membrane

First branchial arch, dorsal end

2.48 An oblique section through the pharynx of a human embryo. C.R. length—2 mm. Magnification × c. 50. (Norris, 1938.)

identification, strong evidence has accumulated that, at least, a pair of *pre-mandibular arches* existed and have become adapted as the *trabeculae cranii* of subsequent vertebrate embryos. These are probably represented by the interorbitonasal cartilage of the human embryo, as mentioned above (p. 164), forming the visceral or branchial element in the chondrocranium. The next (*post-mandibular*) arch in the series is the *hyoid arch*; its skeletal derivatives form the varied hyoid elements present in all vertebrates with jaws. The most dorsal of the latter, the *hyomandibula*, is already present in cartilaginous fish as a strut between the skull and the primitive jaw joint, thereby reducing the cleft between the mandibular and hyoid arches to a small opening, the *spiracle*. The interesting further evolution of this region in land animals in connection with the auditory apparatus will be considered later (p. 204). The hyoid arch also contributes to the formation of a gill cover, or *operculum* in bony fish, and the remaining arches persist as the supports of the gill apparatus.

The mesenchyme of the branchial region (p. 138) at first exists as a thin sheet between the ectoderm and endoderm but, as growth proceeds, it proliferates to form a series of cylindrical masses which constitute the mesodermal cores of the *branchial arches*, covered, of course, by the two epithelia mentioned. At first the arches produce rounded ridge-like prominences both of the overlying ectoderm and of the endodermal lining of the lateral walls and floor of the pharynx. In the furrows between these prominences the ectoderm and endoderm are in virtual contact. The thin membranes so formed break down in gill-breathing

vertebrates but persist in the tetrapods, in which open channels or 'true clefts' are not formed. However, the external *branchial grooves* which correspond to them are frequently, less appropriately, called *branchial clefts* and their internal counterparts are the *pharyngeal sacs* or *pouches*. At this stage in the human embryo the pharynx is transversely wide at its cranial end but rapidly narrows in a caudal direction; it is also so dorsoventrally compressed that there is limited, or virtually no, true lateral wall (2.49).

In gill-breathing vertebrates the exchange of respiratory gases is directly from solution in water to solution in blood. From the cranial end (arterial, but carrying deoxygenated blood) of the heart emerge two *ventral aortae* which traverse the ventral pharyngeal wall, sending branches curving dorsally into the branchial arches, where they feed capillary plexuses in the gills. These are drained by corresponding arteries, which join two *dorsal aortae* supplying the general circulation. As water is taken in through the mouth and passed back through the gill clefts, its dissolved oxygen diffuses through the pharyngeal endoderm and endothelium of the gills to reach the blood, carbon dioxide diffusing out of the latter into the water. The intimate relationship between the developing mouth, branchial apparatus and heart in water-breathing vertebrates is repeated in the embryos of their tetrapod descendants, but with many modifications necessary to changed respiratory function.

The human circumoral *first branchial arch* (2.47, 50) consists, on each side, of two main regions — a *ventral part* or *mandibular prominence* and a *dorsal part* or *maxillary prominence*. Each mandibular prominence grows ventromedially in the floor of the pharynx to meet its fellow in the midline, being situated between the primitive mouth and the cardiac (pericardial) prominence. The *second* or *hyoid arches* are caudal to the maxillo-mandibular and similarly grow ventrally to meet and fuse in the midline. The *third* and *fourth arches*, especially the latter, are not prominent, being largely sunk in a depression produced by the caudal overlapping of the hyoid arch. The *fifth* and *sixth arches* cannot be recognized externally and can only be identified by the arrangement of the mesenchyme and by slight projections into the pharynx. In their development the branchial structures are dependent upon the proximity of pharyngeal endoderm, removal of which aborts branchial development. However, the first arch is, in part at least, dependent upon ectoderm of the stomatodeum.

Each branchial arch consists of an ectodermal exterior, a mesenchymal core and an endodermal interior (2.47, 48, 49). The mesenchyme produces a presumptive *skeletal element*, which subsequently chondrifies either wholly or in part of its length; if this change is complete the element extends dorsally until it comes into contact with the mesenchymatous cranial base lateral to the hindbrain. Much of the remainder of the core of mesenchyme becomes *striated muscle*, which usually migrates and may lose connection with the skeletal elements in arches which cease their original respiratory function. The identities of these muscle masses, where they assume new functions, can nevertheless be inferred by reference to their nerve supply. Motor nerves from the adjacent hindbrain (2.47, 49, 64) pass directly into the arches, which are ventral to it. However far muscle masses migrate from their sites of development, their original innervation almost always persists. The mandibular division of the trigeminal nerve innervates the musculature of the mandibular prominence, the facial nerve supplies the hyoid arch, the glossopharyngeal, the third arch, and the vagus and accessory nerves the rest of the arches. The recurrent laryngeal may be the nerve of the sixth arch and the superior laryngeal that of the fourth. The nerve of the fifth arch, itself difficult to identify in the human embryo, is uncertain.

The branchial arches play a large part in the formation of the face, oronasal cavities, neck, pharynx and larynx, but before the development and adaptations of each of the latter are detailed it is convenient to describe the general events which lead to the construction of the face and nasal cavity.

THE FACE, NASAL CAVITY AND PALATE

While the **mandibular prominence** is invading the floor of the pharynx, the mesenchyme between the central aspect of the

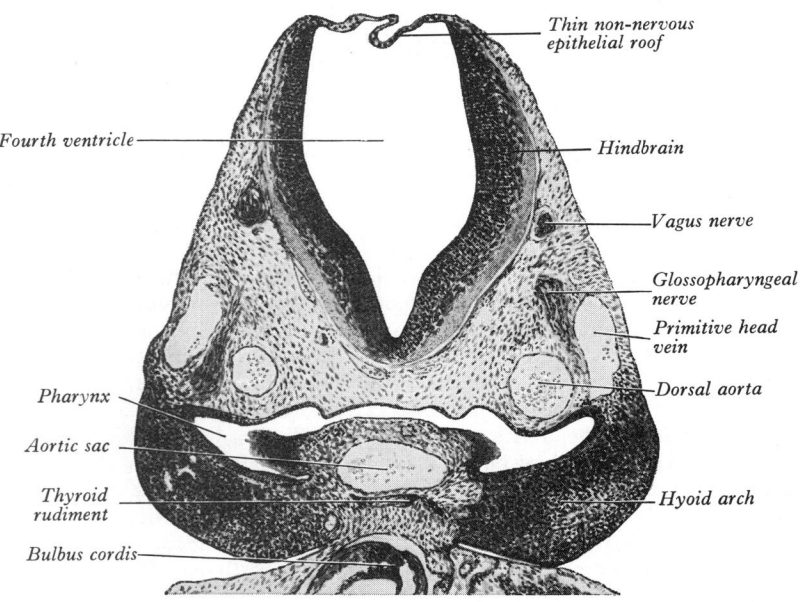

Fourth ventricle

Pharynx

Aortic sac

Thyroid rudiment

Bulbus cordis

Thin non-nervous epithelial roof

Hindbrain

Vagus nerve

Glossopharyngeal nerve

Primitive head vein

Dorsal aorta

Hyoid arch

2.49 Oblique section through the head of a mole embryo, 4.5 mm long. The section passes through the hindbrain, the pharynx, the second (hyoid) and a part of the third branchial arches.

forebrain and the epithelial roof of the mouth proliferates and bulges under the ectoderm to form the *frontonasal prominence*. During the fifth week a thickened plaque of ectoderm develops on each side ventrolateral to the frontonasal prominence, dividing the latter, on each side, into *medial* and *lateral nasal prominences* or folds; these thickenings are the *olfactory* or *nasal placodes*. The placodes are at first widely separated and coplanar with the surface ectoderm but, as the prominences develop, they soon become depressed to form the *olfactory pits* (nasal sacs). The olfactory placodes are the anlage of the olfactory and vomeronasal epithelia; the extensive epithelia of cavities yet to form are respiratory. The lateral nasal prominences are the more evident (2.50, 51B), but the medial nasal prominences, still separated by the median remainder of the frontonasal field, project caudally beyond the former. Extensions of mesenchyme from the medial prominence into the roof of the stomatodeum proliferate to form the *premaxillary* or globular fields—the globular prominences of His. Each nasal sac has a ventral fold from which develops an epithelial *nasal fin* passing caudally to fuse with the stomatodeal roof.

While these changes are progressing a somewhat triangular elevation swells ventrally from the cranial aspect of the dorsal region of the mandibular prominence. This is the **maxillary prominence**, and like the frontonasal prominence it consists of proliferating mesenchyme covered by ectoderm. The maxillary prominence grows in a ventral direction and fuses with the lateral nasal prominence, the two being at first separated by a *nasomaxillary groove* (naso-optic furrow) (2.50, Streeter 1948). The opposed margins of the lateral nasal and maxillary prominences growing together thus establish continuity between the side of the future nose and the cheek (2.50). The ectoderm along the boundary between them does not entirely disappear; it gives rise to a solid cellular rod, which at first develops as a linear surface elevation, the *nasolacrimal ridge*, and then sinks into the mesenchyme (Politzer 1952). Its caudal end proliferates to connect with the caudal part of the lateral nasal wall, while its cranial extremity later connects with the developing conjunctival sac. The solid rod becomes canalized to form the *nasolacrimal duct* (2.51B).

(It should be noted that the epithelial folds and elevations due to loci of proliferation of underlying mesenchyme were long termed processes. The International Nomenclature Committee felt that this was not entirely appropriate and their revised term 'prominence' has been adopted here.)

The relatively wide primitive mouth or *stomatodeal fissure* is progressively reduced, and the epithelial and connective tissues of the cheek enlarged, by fusion between the adjacent surfaces of the mandibular and maxillary prominences. This proceeds from the paraotic region to the angle of the definitive *oral fissure*.

THE NASAL CAVITY

The rounded apex of the triangular maxillary prominence extends beyond the lateral nasal prominence, crossing the caudal end of the olfactory pit to meet and fuse with the *premaxillary elevation* developing at the extremity of the frontonasal field. This closes off the lower or caudal edge of the olfactory pit, the upper part of the opening of which is thus defined as the primitive *external naris*. The growth of the surrounding mesenchyme leads to a deepening of the pit to become a primitive nasal cavity, or *nasal sac*, the epithelial wall of which, in the dorsocaudal part of its extent, the nasal fin, retains contiguity with the epithelium of the stomatodeal roof. This contact area becomes progressively greater as growth continues, and the nasal fin is eroded, ultimately forming a thin layer, the *oronasal membrane*, which also disappears later. Thereafter the primitive nasal cavity communicates with the stomatodeum through a primitive *internal naris (choana)*, which is at this stage still well forward or ventrally situated in the stomatodeal roof (Warbrick 1960). By these changes a new cranial boundary is set for the oral opening, consisting of the fused premaxillary and maxillary regions. This is the future upper lip, but it has not yet become separated from the deeper tissues which will form the maxillary alveolus. At the same time the nasal cavity acquires a floor through the fusion of the nasal

prominences and the maxillary prominences. At this stage the two external nares are still widely separated by an area derived from the frontonasal field, but this separation becomes reduced by the fusion of the premaxillary mesenchyme from the two sides. According to some investigators the mesenchyme of the maxillary prominences invades the premaxillary regions, the mesenchyme of which is said to become buried, to form later the premaxilla or os incisivum (p. 389, Boyd 1933, Baxter 1953). The maxillary mesenchyme is thus considered by some to contribute substantially to the formation of the philtrum of the upper lip, thus accounting for its maxillary innervation. Others, however, maintain that the philtrum is derived wholly from premaxillary tissue (Warbrick 1960, Keith 1948, King 1954, Wood et al 1967). The maxillary nerve primarily innervates the maxillary mesenchyme but apparently extends later into the territory of the frontonasal prominence. (See also p. 1103.) It should be added that some workers deny that sensory nerve distribution is a reliable guide to migration of mesenchyme in the case of the maxillary prominence.

THE PALATE

Once the primitive nasal cavities are defined the ventral part of the roof of the oral cavity can be regarded as the *primitive palate (median palatine prominence)* (2.51A). It is formed by the premaxillary regions and maxillary prominences, which become confluent and establish continuity with the thick median *nasal septal prominence* (primitive nasal septum). As the head grows in size, the region of mesenchyme between the forebrain and oral cavity increases greatly by proliferation and the nasal cavities deepen, extending towards the forebrain. Simultaneously they also extend dorsally from the primitive choanae as two narrow and deep grooves in the oral roof (2.51B,C) which are separated by a partition. The grooves and the partition deepen together, and the latter becomes the *nasal septum*, continuous rostrally with the *primitive nasal septum* (2.51B). The broad dorsocaudal border of the nasal septum is at first in contact with the dorsum of the developing tongue (2.51B), the right and left nasal cavities still communicating freely with the mouth except where the nasal floor is already established ventrally by the primitive palate.

During the sixth week the internal aspects of the maxillary prominences produce *palatine processes*, which grow towards the midline but are for some time separated from each other by the tongue. At this stage the roof of the oral cavity projects ventrally beyond its floor and the tip of the developing tongue actually lies in contact with the cranial (superior) surface of the primitive palate. A coronal section dorsal to this shows the maxillary palatine processes contiguous with the sides of the tongue and bent into a vertical position on each side of it (2.51B). With further growth, the mandibular region and the tongue are carried forwards (ventrally), and the lingual tip passes round to the caudal surface of the primitive palate. By some mechanism, still subject to much controversy (Kraus et al 1969), the palatine processes assume a horizontal position which allows them to grow towards each other and thus to fuse (2.51C); this occurs from before backwards. The change of position occurs very rapidly in some experimental animals and may be due to a sudden increase in turgor in the processes (Ferguson 1977). Other views, such as active contraction of neighbouring striated and non-striated muscle, have been advanced (e.g. Wee et al 1976). The change is also said to be rapid in the case of the human palate. These events obviously may have a direct bearing on maldevelopment of the palate (vide infra). Palatal elevation in mankind occurs during the eighth week; the development of the neck, allowing some descent of the tongue and floor of the mouth, is then also occurring and may be an additional factor in the elevation. This permits the palatine processes to grow medially along the inferior borders of the primitive choanae, uniting with them and with the margins of the median palatine prominence, except over a small area in the midline where a *nasopalatine canal* maintains connection between the nasal and oral cavities for some time and marks the future position of the incisive fossa. (The plates which form the early (primitive) palate are sometimes known as *median* palatine processes, the

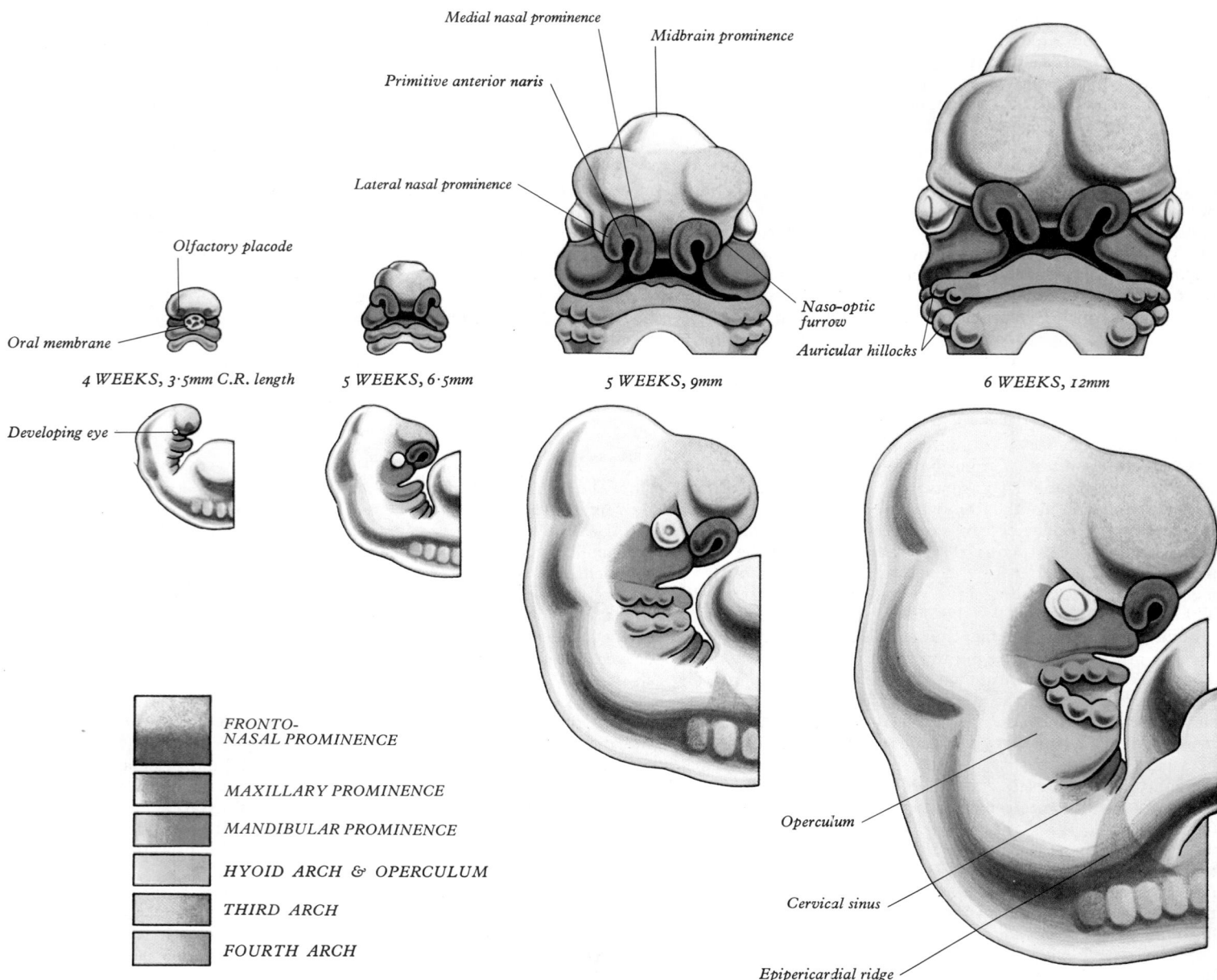

Medial nasal prominence

Midbrain prominence

Primitive anterior naris

Lateral nasal prominence

Olfactory placode

Oral membrane

Naso-optic furrow

Auricular hillocks

4 WEEKS, 3·5mm C.R. length

5 WEEKS, 6·5mm

5 WEEKS, 9mm

6 WEEKS, 12mm

Developing eye

FRONTO-NASAL PROMINENCE

MAXILLARY PROMINENCE

MANDIBULAR PROMINENCE

HYOID ARCH & OPERCULUM

THIRD ARCH

FOURTH ARCH

Operculum

Cervical sinus

Epipericardial ridge

2.50 A sequence of diagrams showing the superficial contributions of the facial prominences and branchial elements to the development of the face, including the external nose, circumorbital structures, external acoustic meatus and pinna, and neck. All diagrams are drawn to scale. Note changes in general proportions and relative positions.

maxillary contributions being then named the *lateral* palatine processes.) As the medial borders of the maxillary palatine processes fuse together, fusing also with the free border of the nasal septum, the nasal and oral cavities are progressively separated and the tongue is excluded from the former. The nasal cavities are thus extended dorsally and the choanae reach their final position, leaving the caudal edge of the nasal septum free in about its dorsal quarter as the partition between them. Slightly later the dorsomedial extremities of the palatine processes, which extend dorsally beyond the choanae, fuse together rostrocaudally to form the future epithelia and connective tissues of the soft palate (**2.**51C). This fusion, as is the case with most of the processes and prominences which form these regions, is more apparent than real, the epithelial cleft between the two processes being pushed to the surface by mesenchymal growth (Burdi & Faist 1967). There is later an upgrowth of myogenic mesenchyme from the third (and probably other) branchial arches into the palate and around the caudal margins of the auditory tubes, along a line corresponding in the final state to the palatopharyngeal arches (Baxter 1953).

On each side of the nasal septum, in a ventral or anterior position just above the primitive palate, placodal ectoderm is invaginated to form a pair of small diverticula, which extend dorsally and cranially into the septum. These are vestiges of the *vomeronasal organs* (**2.**51C), whose openings are close to the junction between the two premaxillae and the maxillae; they are always rudimentary in mankind, but are well-developed auxiliary olfactory organs in many vertebrates (pp. 1095, 1173, 1175). For bibliographies in the field of facial development consult Latham (1973).

Anomalies

Congenital malformations consequent upon arrest of development and failure of fusion of components in the formation of the face and palate are not uncommon. At the simplest, one maxillary prominence may fail completely to fuse with the corresponding premaxillary region (globular prominence), leading to a persistent fissure between the philtrum and lateral part of the upper lip on that side, *cleft lip* (less appropriately 'hare' lip). A similar but rare malformation follows failure of fusion between the maxillary prominence and the lateral nasal prominence, *facial cleft*, in which the nasolacrimal duct persists as an open furrow, a condition usually associated with a cleft lip on the same side. The palatine processes may fail to fuse with each other and the nasal septum to

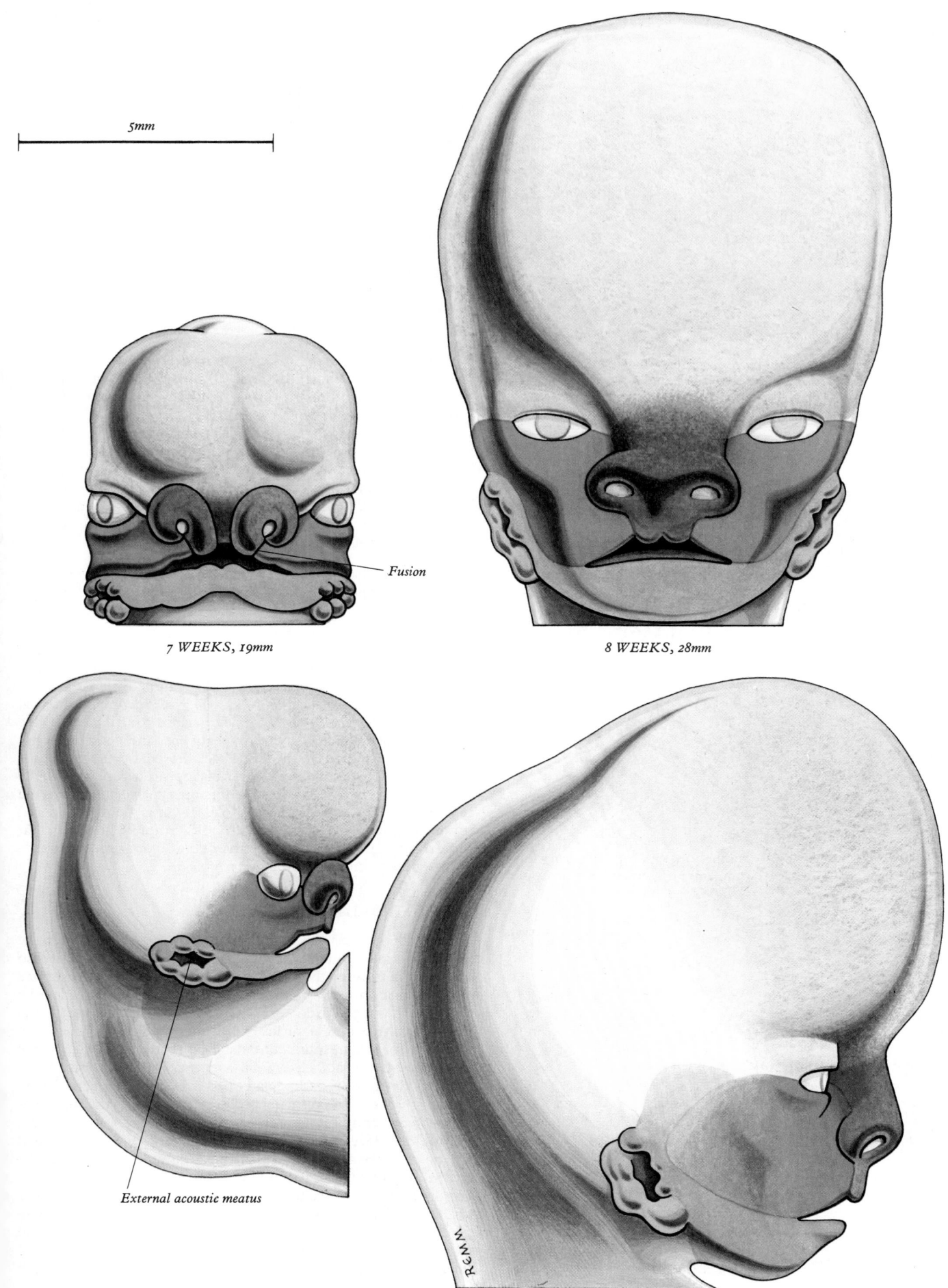

5mm

Fusion

7 WEEKS, 19mm

8 WEEKS, 28mm

External acoustic meatus

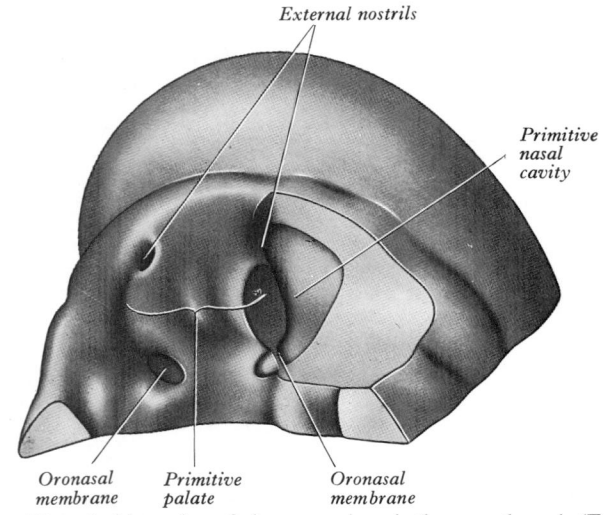

External nostrils

Primitive
nasal
cavity

Oronasal   Primitive   Oronasal
membrane   palate   membrane

2.51A The primitive palate of a human embryo in the seventh week. (From a model by K Peter.) The figure shows the anterior part of the roof of the mouth; large parts of the left lateral nasal prominence and the left maxillary prominence have been removed to expose the left primitive nasal cavity.

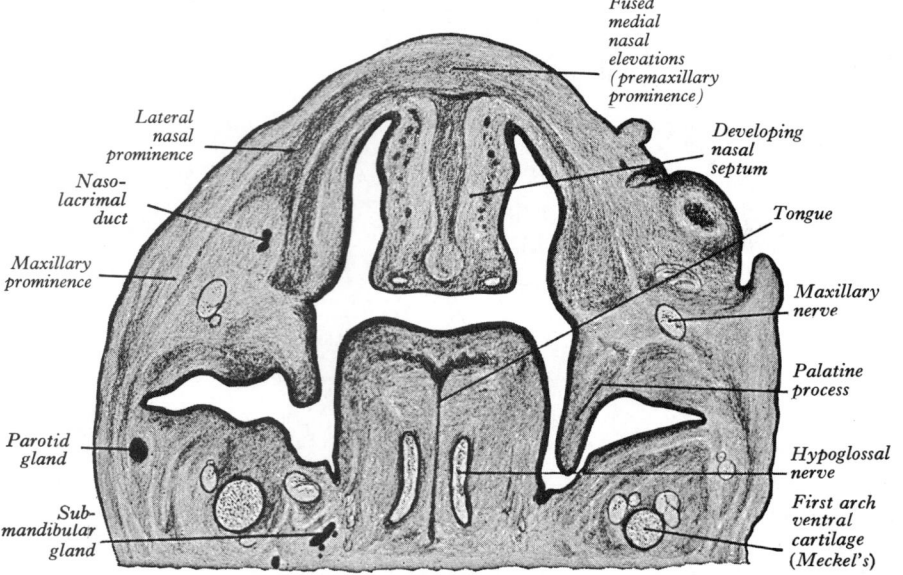

Fused
medial
nasal
elevations
(premaxillary
prominence)

Lateral
nasal
prominence

Developing
nasal
septum

Naso-
lacrimal
duct

Tongue

Maxillary
prominence

Maxillary
nerve

Palatine
process

Parotid
gland

Hypoglossal
nerve

Sub-
mandibular
gland

First arch
ventral
cartilage
(Meckel's)

2.51B Oblique coronal section through the head of a human embryo 23 mm long. The nasal cavities communicate freely with the cavity of the mouth.

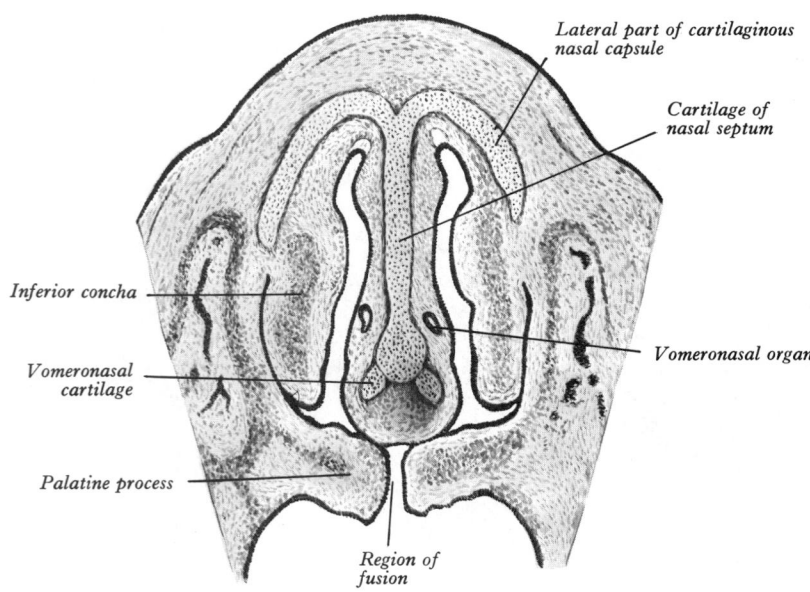

Lateral part of cartilaginous
nasal capsule

Cartilage of
nasal septum

Inferior concha

Vomeronasal
cartilage

Vomeronasal organ

Palatine process

Region of
fusion

2.51C Coronal section through the nasal cavity of a human embryo 28 mm long. (After Kollmann.)

variable degrees. In its severest form fusion is wholly lacking, leaving a wide fissure between the palatine processes through which the nasal septum is visible. On each side the premaxillary parts of the palate are separated from the maxillary palatine processes by clefts which are continuous ventrally with bilateral clefts in the upper lip. In such cases the philtrum is a separate entity, continuous cranially and dorsally with the nasal septum. The floor of the nasal cavity is deficient throughout its extent and the choanae are not completed. Many varieties of milder degrees of cleft palate have been observed; the commonest type is unilateral, only one side of the nasal cavity being in communication with the mouth, the extent of the cleft being variable. In the mildest forms only the soft palate is cleft, or even merely the uvula. Such examples of arrested development may be associated with disturbances in embryonic nutrition during the second and the third months of gestation (p. 156) and the grosser varieties are usually coupled with malformations in other regions of the body. In such cases the premaxillary region protrudes, with associated extension forwards of the nasal septum. For discussion see Latham (1973). Certain midline anomalies are rarely encountered, *median cleft lip* (true hare lip), *cleft nose* and *cleft lower jaw*. More common are minor degrees of *cleft chin* and *micrognathia* —under-development of the lower jaw.

The further growth of the face during the fetal period has received little attention, although this period is by no means characterized entirely by incremental growth. It is during fetal life that human facial proportions develop. The facial and cranial parts display different patterns of growth, though each influences the other. For an interesting analysis of the data observed from 280 fetuses consult Lavelle (1974).

Further Branchial Development

As has already been described to some extent, the branchial arches contribute extensively to the growth of the face, neck, nasal cavity and mouth, as they also do in the case of the palate, larynx and pharynx. Although there are never complete clefts between the arches, the external ectodermal *branchial grooves* and internal endodermal *pharyngeal pouches* make various contributions. The first branchial groove does partly persist as, and contributes to, the external acoustic meatus but it is closed by the tympanic membrane. The remaining grooves disappear, except in so far as such derivatives of their corresponding pharyngeal pouches as the tympanic cavity, auditory tube, tonsil, thymus, parathyroid and thyroid glands can be counted. It should be noted, however, that some contend that the surface course of the second groove persists as the curved submandibular *cervical flexure line*.

ECTODERMAL DERIVATIVES

The ectoderm over the mandibular prominences becomes the skin of the mandibular region of the face (2.50, 52), and it also takes part in forming the tragus of the auricle (p. 1220). Its surface facial contribution is roughly triangular; the apex includes the tragus, the upper border extending to the lateral angle of the mouth and free border of lower lip; its lower border curves to follow the principal submandibular flexure line of the neck. The ectoderm on the arched cranial aspects of these prominences thickens along a curve which later becomes the *labiogingival* (vestibular) *groove*. This epithelial proliferation invades the subjacent mesenchyme and subsequently breaks down to separate the lower lip from the developing gum. Before this, however, an epithelial lamina develops from its internal surface (8.78) from which the enamel organs of the teeth will eventually be derived (p. 1313). Similar phenomena occur in the caudal, arched circumoral border of the maxillary prominence. Its surface facial contribution extends from the supratragic point to the lateral angles of eye and mouth, includes the lower eyelid and follows the paranasal line of the nasolacrimal duct, finally including a controversial amount of the upper lip.

The first branchial groove is obliterated ventrally, as in all but the most ancient vertebrates. In man its dorsal end deepens to form the epithelium of the external acoustic meatus and the

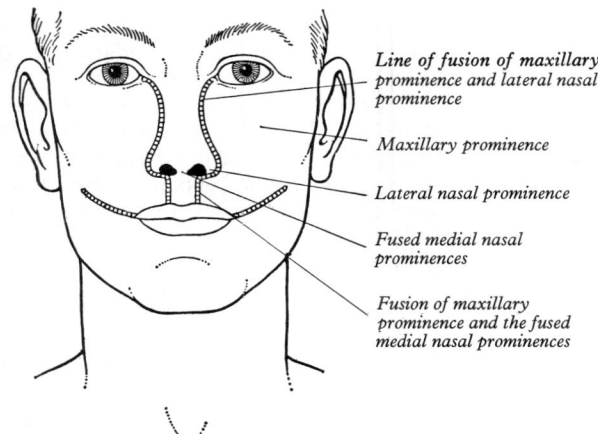

2.52 Diagram to show the parts of the adult face which are derived from the nasal elevations, and the maxillary and mandibular processes.

Line of fusion of maxillary prominence and lateral nasal prominence

Maxillary prominence

Lateral nasal prominence

Fused medial nasal prominences

Fusion of maxillary prominence and the fused medial nasal prominences

external surface of the tympanic membrane. (For details see p. 204.)

At the dorsal ends of the first, second and fourth branchial grooves thickened patches of ectoderm appear, the *epibranchial placodes*. These are closely related to the developing ganglia of the facial, glossopharyngeal and vagus nerves, to which they contribute (p. 199): these, and other placodal cells (*dorsolateral* and *suprabranchial*) also contribute to the trigeminal and vestibulocochlear ganglia. The interrelation, relative contribution and precise definition of placodal cells, and cranial neural crest remains speculative.

At the end of the fifth week the third and fourth arches are sunk in a retrohyoid depression, the *cervical sinus*. Cranially the sinus is bounded by the hyoid arch, dorsally by a ridge produced by ventral extensions from the occipital myotomes and by mesenchyme developing into sternocleidomastoid and trapezius. Caudally, the smaller *epipericardial ridge* separates the sinus from the pericardium and curves cranially near the midline, and then with its fellow reaches the lingual swelling of the mandibular prominence and the hypobranchial eminence. The muscle cells which are often held to migrate from the occipital myotomes to the tongue follow the epipericardial ridge together with the hypoglossal nerve (but see p. 175). The long held view that the sinus is obliterated by caudal growth of the hyoid arches to fuse with the cardiac elevation, excluding the succeeding arches from any part in the formation of the skin of the neck, has been criticized; an alternative view is that the sinus is reduced by gradual approximation of its walls from within outwards. During these changes the epibranchial placodes sink inwards as small vesicles which ultimately lose connection with the surface ectoderm, and are then more closely associated with possibly the trigeminal and with the acousticofacial, glossopharyngeal and vagus nerves. The vesicle derived from the fourth groove placode is also for a time associated with the rudiments of the thymus and parathyroid arising from the third endodermal pharyngeal pouch (p. 229, Garrett 1948), but there is no evidence that the human thymus receives any substantial contribution from this vesicle.

Experiment clearly shows that in lower vertebrates the trigeminal, acousticofacial, glossopharyngeal and vagus nerves derive elements from the overlying epibranchial placodes, or are dependent upon the associated placode for their full development. All the foregoing nerves also receive large accessions from the cranial neural crest, and the acousticofacial from the otocyst. In man, the placodal vesicles associated with the glossopharyngeal and vagus nerves are believed largely to regress, abnormal persistence giving rise to *branchial (lateral cervical) cysts*, which are lined by stratified squamous epithelium and are situated between the carotid sheath and sternocleidomastoid. It is claimed that branchial cysts, lined by columnar epithelium, may be derived from persistent remnants of pharyngeal pouches, and that *branchial (lateral cervical) fistulae* may result from the second branchial cleft or possibly the cervical sinus (Martins 1961). Blind ended

sinuses, with varying epithelia and either an external or internal opening, may have a similar origin. In contrast, *preauricular sinuses* may result from failed fusion between auricular hillocks, the forerunners of the pinna.

MESODERMAL DERIVATIVES

The mesoderm of the branchial arches (of mixed *unsplit lateral plate, neural crest and possibly placodal* origin) is still 'pluripotential'; the term is used here to indicate the ultimate variety of its derivatives (and does not imply the immediate premitotic potential of individual cells—p. 108). The derivatives include the following: (1) An endothelial tube, an *aortic arch*, develops in the substance of each arch, connecting the aortic sac, ventral to the pharynx (p. 214), to the dorsal aorta, which is dorsal to it. There are thus six pairs of aortic arches, but the appearance of the fifth pair in man is questioned; its position and connections are subject to disagreement (Shaner 1921, Barry 1951). For further details see p. 216. (2) Some of the mesodermal cells, many of neural crest origin, condense to form a skeletal element in each arch; and this is typically a bar of cartilage connected by its dorsal end to the caudal region of the chondrocranium and meeting its fellow ventrally in the midline. (3) Other mesodermal cells differentiate into 'special visceral' striated muscle tissue; and partly associated with this developing *branchial musculature* is to be noted the invasion of the mesoderm by a nerve supply. The proportion of branchial mesoderm from each of its various sources noted above that contributes to the different types of derivative is uncertain (except for comparative experimental results implying neural crest predominance in the cartilage).

Typically each arch is invaded by two nerves, derived from the hindbrain. One runs along the rostral border of the arch and is

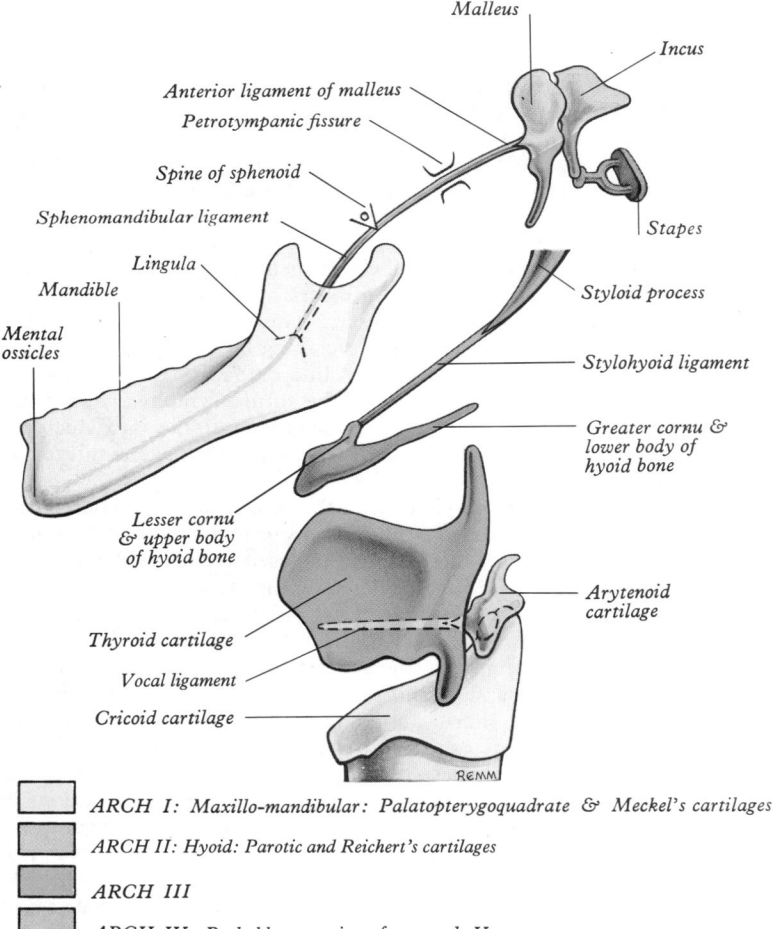

Malleus

Incus

Anterior ligament of malleus

Petrotympanic fissure

Spine of sphenoid

Sphenomandibular ligament

Stapes

Lingula

Mandible

Styloid process

Mental ossicles

Stylohyoid ligament

Greater cornu & lower body of hyoid bone

Lesser cornu & upper body of hyoid bone

Arytenoid cartilage

Thyroid cartilage

Vocal ligament

Cricoid cartilage

ARCH I: *Maxillo-mandibular: Palatopterygoquadrate & Meckel's cartilages*

ARCH II: *Hyoid: Parotic and Reichert's cartilages*

ARCH III

ARCH IV: *Probably accessions from arch V*

ARCH VI

2.53 Schema illustrating the skeletal derivatives (osseous and cartilaginous) of the branchial arches (viscerocranium). (Modified from Williams et al 1969)

hence described as *post-trematic*, because it is behind or caudal to the cleft or *trema* rostral to the arch. Similarly the other nerve, which is close to the caudal border, is *pretrematic*—with respect to the cleft caudal to it and its derivation as a branch from the principal post-trematic nerve of the immediately succeeding arch. In the human embryo the pre- and post-trematic nerves cannot be identified with certainty.

In arches 1–4 the aortic arch is caudomedial and the nerve rostromedial to the cartilage: in arch 6 the reverse obtains, with respect to the positions of artery and nerve.

The mesoderm of the branchial region also contributes to the fibro-areolar and other connective tissues of the face, tongue and neck. In the third arch it gives rise to some cellular elements of the *carotid glomus* or body (see p. 1473), which may also receive neuroblasts from the glossopharyngeal nerve and the developing superior cervical sympathetic ganglion (Boyd 1937, Adams 1958, Rogers 1965). Similarly *para-aortic* bodies are developed in association with the fourth and sixth arches, receiving accessions of neuroblasts from the ganglia of the vagus nerve (Hammond 1941).

DERIVATIVES OF THE BRANCHIAL SKELETON
(Throughout consult **2**.53)

The first arch contains on each side a dorsal and ventral element. The former represents the *palatopterygoquadrate bar*, a prominent element in earlier vertebrates forming part of the upper jaw, but much reduced in mammals. In human embryos its early appearance seems transient and its contribution to some permanent cranial structures, such as the maxilla, is uncertain (but vide infra for other derivatives). The *ventral cartilage* (of Meckel, **2**.53) extends from the developing otic capsule into the mandibular moiety of the first arch, meeting its fellow at its ventral end. The dorsal end of Meckel's cartilage becomes separated, and was often held to form the rudiments of both *malleus* and *incus*. However, there is strong paleontological (Romer 1970) and comparative anatomical (Shute 1956) evidence that the incus is, in part, to be regarded as a homologue of the *quadrate bone* of reptiles, and it is therefore probably more correctly regarded as a derivative of the palatopterygoquadrate cartilage. This cartilage may also contribute to the ala major of the sphenoid bone and the roots of its pterygoid plates. A more precise developmental analysis of this region was needed (vide infra). Beyond the rudiment of the malleus, the intermediate part of Meckel's cartilage disappears, but its sheath persists as the *anterior malleolar* and *sphenomandibular ligaments*. The ventral part, much the largest, is enveloped by the developing mesenchymatous mandible (p. 369); a small fraction of this, extending from the mental foramen almost to the site of the future symphysis, probably becomes ossified from invading mandibular tissue, into which it is incorporated, while the remainder of the cartilage is ultimately absorbed.

The cartilaginous element of the second arch (Reichert's cartilage) also extends from the otic capsule to the midline. Its dorsal end also separates and becomes enclosed in the developing tympanic cavity as the *stapes*. Thereafter the cartilage gives rise to the *styloid process*, *stylohyoid ligament*, the *lesser cornu* and probably the *cranial rim* of the body of the *hyoid bone* (**2**.53).

Early development of the auditory ossicles outlined above was based on comparative and paleontological evidence and studies of centres of chondrification and ossification in staged embryos and fetuses. This gained fairly wide acceptance, but is open to revision following close scrutiny of earlier embryos to establish the sources of mesenchyme that provided the condensations of the initial *blastemal ossicles* (Hanson et al 1962). The presumptive palatopterygoquadrate mesenchyme provided the body and short process of the incus; the dorsal moiety of Meckel's mesenchyme provided the head, neck, anterior and lateral processes of the malleus. Between them lay the interzone of the incudomalleolar joint. The dorsal parieties of the condensing core of the *second* arch contributed the long process of the incus and the handle of the malleus. The dorsal tip of the second arch condensation, after receiving accessions from the otic capsule, encircled the stapedial artery forming the anlage of the stapes.

Thus each ossicle was derived from at least *two* distinct main sources. Subsequent chondrification and ossification proceeded as detailed elsewhere.

Chondrification does not occur in the dorsal parts of the skeletal elements of the third to sixth arches. The ventral cartilage of the third arch becomes the *greater cornu* of the *hyoid bone* and the *caudal part* of its *body*. (The whole of the body may be formed from the third arch cartilage.) Alternatively, the hyoid body may be derived from cartilage formed in the base of the hypobranchial eminence (p. 228) and thus from third arch tissue alone (Frazer 1926), acquiring its connection with the second arch cartilage secondarily.

The final adaptations of the cartilages of the skeletal elements in the fourth, fifth and sixth arches are a source of disagreement, but the following represents a fairly general view. The thyroid cartilage develops from the fourth and fifth arches, which may also give rise to the arytenoid, corniculate and cuneiform cartilages. The cricoid cartilage may be derived from the sixth arch cartilage, or it may be a modified tracheal ring. The epiglottis is developed in the substance of the hypobranchial eminence and probably not from 'true' branchial cartilage.

DERIVATIVES OF BRANCHIAL MUSCULATURE
(Throughout consult **2**.54 and colour coded diagrams of skull muscle attachments in Osteology.)

The muscle mass of the mandibular part of the first arch forms the *tensor tympani*, *tensor veli palatini* and the *masticatory muscles*, including *mylohyoid* and the *anterior belly* of *digastric*, all being supplied by the mandibular nerve, the mixed 'post-trematic' nerve of the arch. The tensor tympani retains its connection with the skeletal element of the arch through its attachments to the malleus, and the tensor veli palatini to the base of the medial pterygoid process, which may be derived from the dorsal cartilage of the first arch, but the masticatory muscles transfer to the mandible, a dermal bone. The muscles derived from the hyoid arch mesoderm for the most part migrate widely but retain their original nerve supply from the facial. The *stapedius*, *stylohyoid* and *posterior belly of digastric* remain attached to the hyoid skeleton, but the *facial musculature*, *platysma*, *auricular muscles* and *epicranius* all lose connection with it. Their migration is facilitated by the early obliteration of some of the first cleft and pouch. (This cleft, the spiracle in fishes, is already much reduced in all but the earliest vertebrates.)

The muscle masses from the remaining arches are adapted to form the musculature of the *pharynx*, *larynx* and *soft palate*. The stylopharyngeus can be attributed to the third and the cricothyroid to the fourth arch; the rest of the laryngeal muscles are derived from the sixth arch, but the precise origin of the remaining palatal muscles and the pharyngeal constrictors is uncertain in man. A mixed origin, partly from branchial mesoderm and partly from adjacent myotomes, has been attributed to sternocleidomastoideus and trapezius (McKenzie 1955).

NEURAL ELEMENTS OF THE BRANCHIAL ARCHES
(Throughout consult **2**.47,64.)

The nerves of the branchial arches, derived from the hindbrain, immediately enter the dorsal ends of them (**2**.47, 64). They are typically mixed, their motor component supplying the muscles of the arch and their sensory fibres innervating the skin and mucous membrane derived from the region. In fish the trunks of the branchial nerves and their ganglia are close to the dorsal ends of the true clefts existing in these forms, each sending a post-trematic branch into its own arch and a pretrematic branch into the arch cranial to this. In mammals, some have claimed that both types of branch can be identified in the first arch, but only a single nerve can be identified with certainty in the second to sixth arches, with the exception of the fifth, the nerve of which is unknown and may have disappeared.

The trigeminal mandibular division is the post-trematic nerve of the first arch; the chorda tympani, or greater petrosal, have sometimes been regarded as its pretrematic nerve derived from the facial. The latter supplies the second arch, the glosso-

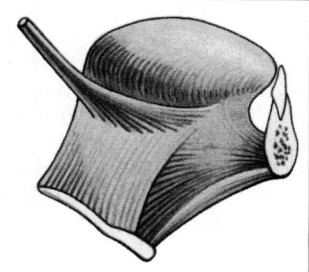

OCCIPITAL SOMITES

MUSCLES: *Extrinsic &
intrinsic lingual
muscles (except
palatoglossus).*

NERVE: *Hypoglossal.*

PRO-OTIC 'SOMITES'

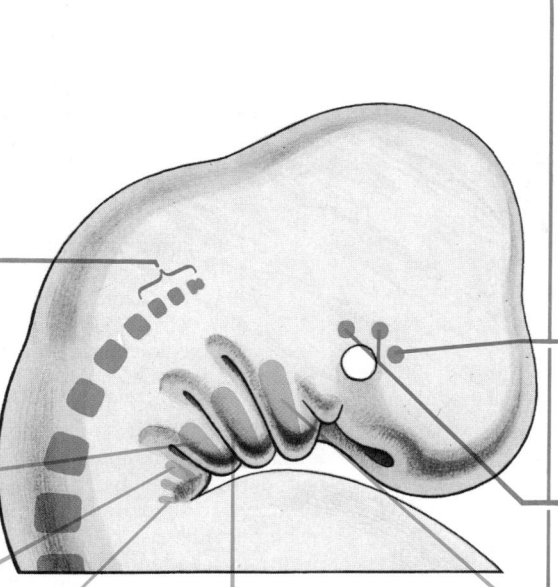

PREMANDIBULAR MESENCHYME

MUSCLES: *Levator palpebrae superioris;
superior, medial & inferior
recti; inferior oblique.*

NERVE: *Oculomotor.*

ARCH 3

MUSCLE: *Stylopharyngeus.*

NERVE: *Glossopharyngeal.*

ARCH 4

MUSCLE: *Cricothyroid.*

NERVE: *Superior laryngeal
branch of vagus.*

ARCH 6

MUSCLES: *Other intrinsic
laryngeal muscles.*

NERVE: *Recurrent laryngeal
branch of vagus.*

MAXILLOMANDIBULAR MESENCHYME

MUSCLES: *Superior oblique
& lateral rectus.*

NERVES: *Trochlear & Abducent.*

ARCH 2

MUSCLES OF FACIAL EXPRESSION
*including auricular muscles, occipitofrontalis,
posterior belly of digastric, stylohyoid,
stapedius & platysma.*

NERVE: *Facial.*

ARCH 1

MUSCLES OF MASTICATION
*Temporalis, masseter, pterygoids,
mylohyoid, anterior belly of
digastric, tensor veli palatini
& tensor tympani.*

NERVE: *Mandibular.*

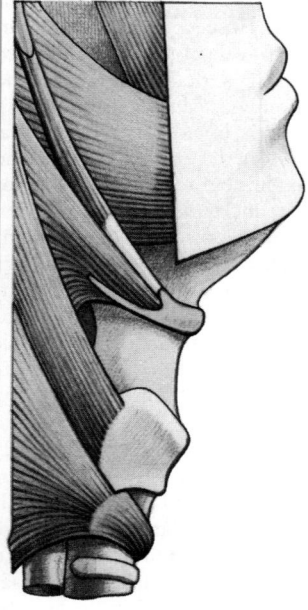

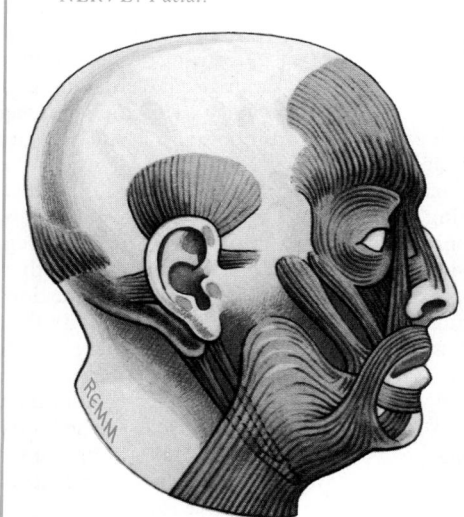

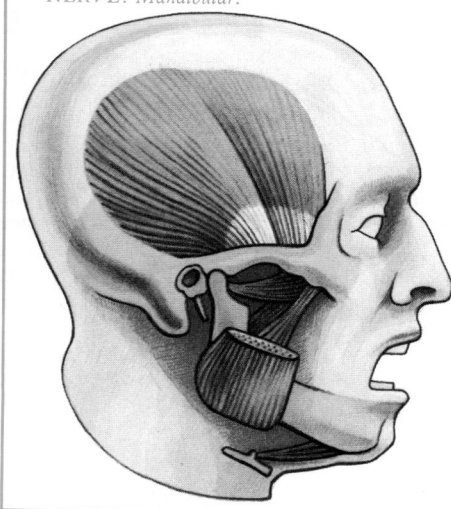

2.54 Schema illustrating the muscular derivatives of the branchial mesenchyme and the pre-otic and post-otic cranial 'somites'. (Modified from Williams et al 1969.)

CAUDAL ARCHES

*Remaining palatine muscles
& constrictors – precise
sources uncertain.*

*NERVE: Cranial accessory
via branches of vagus.*

pharyngeal the third, the superior laryngeal branch of the vagus the fourth and the latter's recurrent laryngeal branch the sixth. In lower vertebrates the fifth arch is also supplied by a vagal branch. Other branches that have, on occasion, been proposed as pretrematic are the tympanic branch of the glossopharyngeal and the auricular branch of the vagus. However, none of the foregoing fulfil sufficient criteria for them to be classified as pretrematic with confidence.

The difference in the courses of the recurrent laryngeal nerves can be explained by the development of the aortic arches. In arches 1–5 the arch nerve enters rostral to its aortic arch. However, the nerve enters its sixth arch *caudal* to the aortic arch, retaining this position on the left side and hence being caudal to and looping round the ligamentum arteriosum in its final disposition. However, on the right, owing to the disappearance of the dorsal part of the sixth aortic arch and the whole of the fifth, the nerve loops round the caudal aspect of the *fourth* aortic arch, i.e. the subclavian artery.

DERIVATIVES OF THE PHARYNGEAL POUCHES

The first four pharyngeal pouches appear in sequence cranio-caudally, and their endoderm approaches the ectoderm of the overlying branchial grooves to form thin *closing membranes* (2.47, 48). The blind recesses of the second, third and fourth pouches are prolonged dorsally and ventrally as angular, wing-like diverticula. From the fourth a diverticulum grows caudoventrally and is at first demarcated from the pouch by a groove in which may occur a transient fifth aortic arch artery. From this diverticulum a fifth pouch may develop and establish a connection with ectoderm. The remainder of this diverticulum is the *ultimobranchial body*. This, together with the fourth pouch and the transitory fifth, when present, constitute the *caudal pharyngeal complex*. Its communication with the cavity of the pharynx is the *common pharyngobranchial duct*. The ultimobranchial body is almost a constant feature of vertebrate development (Watzka 1955). Its form in the *human* embryo, however, has been a matter of controversy. Apparently it is incorporated into the rest of the caudal pharyngeal complex and contributes to the development of the lateral thyroid rudiment (p. 228). Ultimobranchial bodies exist in the adults of many lower vertebrates and *calcitonin* has been isolated from such tissue (Copp et al 1967). There is thus a strong presumption that the parafollicular cells of the human thyroid gland, which are a source of calcitonin, are derived from ultimobranchial tissue. (See proceedings of a symposium on: *Thyrocalcitonin and the C cells*, Taylor 1968, Bussolati & Pearse 1967, also pp. 304, 1463).

The further development of the endodermal derivatives of the pharyngeal pouches is intimately associated with that of the mouth, pharynx and larynx, and is considered with them (p. 228).

Development of Locomotor Structures

EARLY DEVELOPMENT OF THE LIMBS

Near the end of week four the limbs appear as small elevations, the **limb buds**, from slight lateral ridges extending the length of the trunk (2.46, 55). The forelimb bud shows first, opposite the more caudal cervical segments, the hind-limb bud being level with lumbar and upper sacral segments. Each bud contains a small proliferating mass of mesenchyme derived from the somatopleure, covered by ectoderm. At this stage the limb bud is not more than a slight accentuation of the lateral ridge, or *crest of Wolff*, which—though not identified in all embryos—is

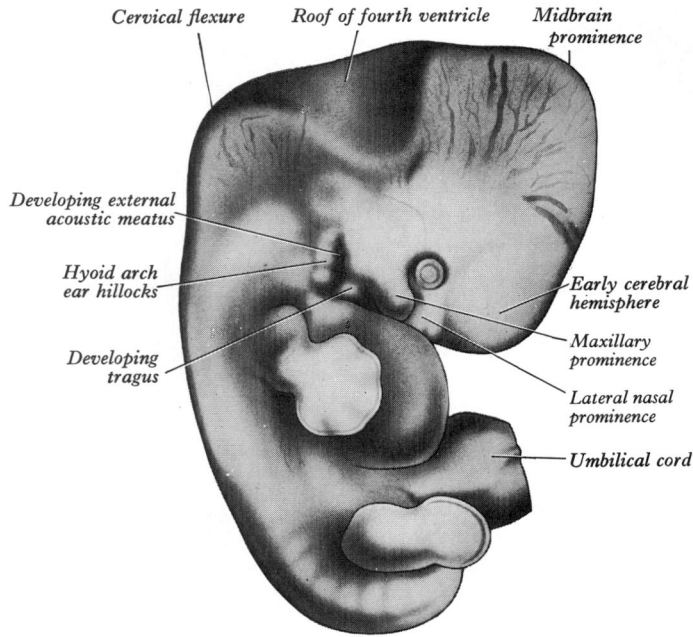

Cervical flexure *Roof of fourth ventricle* *Midbrain prominence*

Developing external acoustic meatus

Hyoid arch ear hillocks

Developing tragus

Early cerebral hemisphere

Maxillary prominence

Lateral nasal prominence

Umbilical cord

2.55 A human embryo, 15.5 mm long. Compare the upper and lower limb buds, concerning their stage of differentiation and degree of rotation. (Streeter 1948)

considered to correspond to the *finfold of Gegenbaur*, the hypothetical origin of vertebrate limbs. From their initial appearance, the limb buds display an external crest at their developing apices, the *apical ectodermal ridge* or cap; and despite some disagreement over details there is general concurrence that *interaction* between the mesenchymal core and the apical ectoderm is essential to limb formation. Most experimental evidence admittedly applies only to amphibian (Tschumi 1957), reptilian (Milaire 1957) and avian (Saunders 1948, Amprino & Camosso 1955, Amprino 1968, MacCabe et al 1974) embryos some confirmation in mammalian limb buds has also been recorded and observations on human embryos (Horizons XII to XVII), have indicated the occurrence of ectodermal ridges (O'Rahilly et al 1956). This evidence indicates that the ectoderm of the apical region of the limb bud, though itself relatively undifferentiated, is necessary to the development of successive accretions of mesenchyme as the bud grows. The potentiality of the initial mass of somatopleuric mesoderm is limited to the formation of proximal limb elements, i.e. the girdle region. The mesodermal development of the successive segments of the limb is induced in a proximo-distal sequence (cf. meristematic growth) by the ectodermal ridge, the cells of which apparently do not proliferate to add to the covering of the growing bud. The necessary increase in ectoderm for this appears to be derived by interstitial growth and from the body wall ectoderm. The apical ridge also appears to establish the polarity of the limb with respect to its pre- and post-axial borders and the arrangement of its basic vascular pattern. The evidence for these views has been surveyed by Zwilling (1961). Specific histochemical changes associated with morphogenetic events in the developing limb bud have been reviewed by Milaire (1965). Ultrastructural studies of the zone of epithelio–mesenchymal interaction at the apex of the bud have been reported (Kelley 1973). The evidence suggests that the sculpturing of the interdigital clefts is a process involving mesenchymal cell death with phagocytic removal (p. 103) accompanied by active epithelial invagination. For a detailed review and bibliography concerning these phenomena see Hinchliffe (1982).

The mesenchyme of the limb buds is invaded by the ventral rami of adjoining spinal nerves—fourth cervical to second thoracic for the fore-limb, twelfth thoracic to fourth sacral for hind-limb. The axial region of mesenchyme proliferates and condenses to produce the blastemal skeleton of the limb, chondrification and ossification following as the limb bud grows. The musculature of the limbs is developed in situ from the mesenchyme surrounding the skeletal elements. It is important to emphasize that there is no apparent migration of myotomic mesoderm into the limb bud, the so-called plurisegmental nature of the muscles in the limbs referring solely to their nerve supply. Such muscles as latissimus dorsi, with extensive attachments to the axial skeleton when fully developed, reach them secondarily by active migration.

The early stages of development of fore- and hind-limb buds are alike, except that the former precedes the latter in time by a few days in the human embryo (2.46, 262). Differentiation in general proceeds in a proximo-distal direction. Flexion creases appear indicating the sites of elbow, wrist, knee and ankle; and the hands and feet are at first mere plate-like expansions (2.55, 262). Near the end of the sixth week there is little difference in shape and posture between the arm and leg, the long axis of each being approximately orthogonal to the trunk with the prominences of elbow and knee both directed laterally (2.55, 2.262). As intimated the pre- and post-axial borders of the limbs are respectively cranial and caudal in orientation.

Preliminary observations on human limb buds, removed from embryos of 4–6 weeks and cultured for 4–18 days in vitro, have been reported (Yasuda 1973). There was a general retardation of development in vitro, and the effect more marked in fore- than in hind-limb buds but, amongst tissues, cartilage was the least retarded.

LIMB ROTATION

(Follow progress on pp. 262, 263 and fig. 2.125.)

During the seventh and eighth weeks differential growth rates are exhibited by the limbs which bring them into a position of adduction towards the ventral aspect of the trunk, and flexion at elbow and knee is increased. By the end of the eighth week the limbs have attained the *fetal position*, with the conspicuous differences that the elbow points caudally and the knee in a rostral direction. This divergence of final positioning is of the utmost importance in the future functioning of the limbs (p. 398); it is achieved by a kind of rotation in each limb around its axis, presumably in part due to growth changes, for it occurs before any joints are fully defined. The upper limb rotates laterally, so that its pre-axial border becomes lateral and its ventral surface rostral or anterior. The changes are the reverse of this in the hind-limb; medial rotation carries the pre-axial border into a medial position, turning the dorsal surface to face anteriorly. The same reorientation is illustrated by the radius and tibia, both pre-axial primitively in their limbs, the former becoming the lateral fore-limb bone, the tibia taking up a medial position in the foreleg. Inevitably this affects the hand and foot, whose pre-axial digits, thumb and great toe, take up lateral and medial positions. It should be noted that reverse *forearm* rotation at a later stage in most mammals, and the retention of joint mediated supination-pronation activity in the primate forearm, modify this arrangement (p. 398). The cutaneous nerve supply follows these movements and this explains the cutaneous innervation of the 'lateral' aspect (pre-axial border) of the human upper limb from the more cranial nerves of the brachial plexus (p. 1130) and the 'medial' border by its more caudal nerves. In the hind-limb the pattern is similar but less clear; but as a generalization, the upper spinal nerve elements in the lumbosacral plexus innervate the medial border, the lower ones the lateral (7.266, 267, see also p. 1152.)

The foot and hand resemble each other closely at first, as flattened expansions at the terminations of the limb buds. The mesenchyme in the periphery of these plates condenses to outline the pattern of the digits, and the thinner intervening regions break down from the circumference inwards. If this process is incomplete or becomes arrested varying degrees of webbing or *syndactyly* result. For a detailed study and review of phylogeny and ontogeny consult Čihák (1972).

Further development and ossification of the skeletal elements of the limbs is described in Section 3, Osteology. For the chronology of appearance of human embryonic ossific centres consult O'Rahilly & Gardner (1972).

DEVELOPMENT OF JOINTS

Except in the vertebral column and the caudal region of the chondrocranium, which derive from sclerotomes (pp. 140, 159), the mesenchyme from which other skeletal elements are derived *at first* shows no differentiation into primordia of individual bones. The skeletal mesenchyme appears as a continuous condensed mass not initially demarcated clearly from the surrounding myogenic tissue. However, centres of chondrification and ossification now develop in this mesenchymal core and rapidly extend to delineate individual skeletal elements, each containing its own centre or focus of change, the process of bone or cartilage formation spreading in an orderly and characteristic manner from it (Streeter 1949). Each element becomes limited by a compact layer of apparently undifferentiated cells from surrounding tissues; this layer proliferates and differentiates to produce chondroblasts and osteoblasts, which contribute to growth by surface accretion. This lamina gradually becomes more clearly demarcated from the underlying cartilage or bone as *perichondrium* or *periosteum*, continuing to generate chondroblasts and osteoblasts (p. 291). It appears to be an important factor in the determination of major bone growth characteristics in the formation of individual elements (Fell 1939, Barnett et al 1961, also p. 293 et seq).

Connecting adjacent skeletal elements as they become defined from continuous mesenchyme masses are regions which do not initially progress into cartilage or bone but persist as plates of *interzonal mesenchyme*. These, the sites of future joints, and their development varies according to the type of joint formed. In fibrous joints the interzone is converted into collagen, as the definitive connecting medium between the bones involved. In synchondroses it becomes cartilage of the modified hyaline type,

whereas in symphyses the tissue is predominantly fibrocartilage, but retaining narrow para-osseous laminae of hyaline cartilage. The interzonal mesenchyme of developing synovial joints becomes trilaminar, due to the appearance of a more tenuous intermediate zone between two dense strata next to the cartilaginous ends of the skeletal elements of the region. These latter are continous peripherally with the adjoining perichondrium and, like it, are chondrogenic and thus concerned with growth of cartilaginous epiphyses (p. 304). In some synovial joints, however, this trilaminar interzone may be modified (Gardner & Gray 1953). The intermediate stratum merges with the general mesenchyme of the limb, which is vascularized. From this, a cuff condenses as the fibrous capsule of the joint, in continuity with the perichondrium of the bones concerned. A thinner layer of vascular mesenchyme is enclosed within this as the precursor of the synovial capsule (Haines 1947, Gardner & Gray 1950, Gray & Gardner 1951, Gardner & O'Rahilly 1968).

As the skeletal elements chondrify and in part ossify, the dense strata of the interzonal mesenchyme also become cartilaginous and cavitation of the intermediate zone establishes the cavity or discontinuity of the joint. The synovial mesenchyme now forms the synovial membrane and probably also gives rise to all other intra-articular structures, such as tendons, ligaments, discs and menisci. In joints containing discs or menisci and in compound articulations more than one cavity may appear initially, sometimes merging later into a complex single one. As development proceeds thickenings in the fibrous capsule can be recognized as the specializations peculiar to a particular joint. In some, however, such accessions to the fibrous capsule are derived from neighbouring tendons, muscles or cartilaginous elements. For a review of the literature concerning the chronology of events in human embryonic limbs consult O'Rahilly & Gardner (1975).

DEVELOPMENT OF SKELETAL MUSCULATURE

With the exception of certain muscles of the head and neck, which are developed from branchial mesenchyme (p. 172), and the limb muscles, which develop in situ from the mesenchyme of the limb buds (p. 174), most somatic (striated or skeletal) muscles are derived from myotomes (p. 138). (The term 'voluntary muscle', although frequently encountered, is often misleading and its use is to be discouraged.) Typically, each myotome divides into dorsal (*epaxial*) and ventral (*hypaxial*) regions, the former sited dorsolateral to the vertebral column and innervated by the dorsal ramus of the corresponding spinal nerve, the latter migrating ventrally in the body wall or somatopleure and innervated by the corresponding ventral ramus. Primitively, in fishes, these two primary divisions of the myotomes are separated by vertebral transverse processes and a fibrous septum extends from these to the lateral body line. The myogenic masses derived from the myotomes may divide longitudinally or tangentially and the resultant parts may remain separate, e.g. intercostal muscles, or they may fuse with corresponding parts of adjacent myotomes to form sheets of muscle, such as the abdominal oblique and transverse layers. In mammals the derivatives of the ventral regions of the myotomes may subsequently migrate over the derivatives of their dorsal parts. Consequently, such muscles as the serrati posteriores, which reach the vertebral spines superficial to the erectores spinae, are nevertheless supplied by ventral rami of spinal nerves, indicating their origin from ventral myotomic mesoderm. It should be added that from experiments on chick embryos it is apparent that some parts of the intercostal and anterolateral abdominal musculature are not derived from myotomes but from the lateral plate of mesoderm (Straus & Rawles 1953). However, observations on amphibian (Detwiler 1955) and mammalian (Theiler 1957) embryos support their myotomic origin. Some muscles may also *migrate* in rostral or caudal directions, the latter as in the case of latissimus dorsi, which is derived from lower cervical myotomes and is innervated accordingly.

Although it is usually stated that the lingual muscles are derived from three or four pairs of occipital myotomes (Bates 1948, Deuchar 1958), with which the morphology of their supply from the hypoglossal nerve accords, the evidence for this in the human embryo is negligible, and earlier views that they arise in situ from mesenchyme associated with the developing tongue have not been controverted (Frazer 1926). The hypoglossal nerve is composite, representing the ventral roots of an uncertain number of segmental nerves cranial to the cervical series. It is considered to have acquired cranial status in the earliest land tetrapods, its exclusion from the modern amphibians being a secondary adaptation.

In lower vertebrates the extrinsic ocular muscles are developed from three paired head cavities which are considered to be three persistent pro-otic somites (Neal 1918). There is no direct evidence of this origin in man, although it is claimed that the most rostral of these cavities, the *premandibular*, is derived from mesoderm originating in the prechordal plate (see p. 136 and Gilbert 1952). This condensation is held to give rise to the muscles innervated by the oculomotor nerve. The remaining extra-ocular muscles are considered to develop from a mass of mesenchyme dorsal to the region of conjunction of the mandibular and maxillary prominences (which may correspond to the mandibular and hyoid head cavities identifiable in sharks, and which has been termed *maxillo-mandibular mesenchyme* by some authors). The dual origin of this mass is at least in accord with the innervation of these muscles by the trochlear and abducent nerves. The same mass is said to contribute to the formation of the scleral tunic of the eye (Gilbert 1957).

In the differentiation of striated muscle the primitive muscle cells, *myoblasts*, may arise from mesenchymatous *promyoblasts* of somite or non-somite (e.g. branchial, ventral pharyngeal, upper post-branchial, perianal or limb bud) origin. The myoblasts multiply by mitosis, but further recruitment from neighbouring mesenchyme also occurs. The cells elongate to form multinucleated *myocytes*, in which the nuclei move to the periphery as a myofibrillar structure and cross-striation begins to appear. Their maturation is described in more detail, with related bibliography, on p. 553. New fibres are said to form until mid-fetal stages of development, sometimes by splitting of pre-existing fibres. Thereafter growth of muscles is accomplished by enlargement of individual fibres (Cuajunco 1942). Muscle spindles can be identified as early as the twelfth week (Cuajunco 1940). The tendons of muscles develop independently in mesenchyme and their connection with their muscles is secondary.

In man and other higher vertebrates some derivatives of myotomes degenerate and some disappear entirely. Others may persist as fibrous tissue vestiges; in some instances regions formed from myotomes, and/or lateral plate mesoderm, which are muscular in ancestral forms, e.g. may become aponeurotic or ligamentous, the aponeuroses of the abdominal muscles and the sacrotuberous ligament.

SKIN AND APPENDAGES

The epidermis and its specialized appendages—hairs, nails, sudoriferous and sebaceous glands—are derivatives of the ectoderm, while the dermis (corium) is formed from the somatic layer of the mesoderm, with some contribution from dorsolateral aspects of the mesodermal somites (p. 140), and the neural crest. The ectoderm at first consists of a single stratum of cuboidal cells, but by the sixth week these have proliferated to form a double layer—a superficial *periderm*, or *epitrichium*, of flattened cells and a subjacent *stratum germinativum*. The cells of the latter are at first cuboidal but become columnar. By multiplication and differentiation of these cells the characteristic layers of the epidermis are developed. The periderm is merely a temporary covering stratum. By the sixth month most of the periderm, which becomes keratinized, has disappeared and the strata granulosum, lucidum and corneum of the epidermis are established. The superficial cornified cells, together with sebaceous secretion and the remains of the periderm, form a cheesy or caseous material, the *vernix caseosa*, which may exercise the function of protecting the underlying epidermis from maceration by amniotic fluid. Vernix caseosa is characteristic of the surface of the fetus during its final 3 months of intrauterine existence, finally desquamating soon after birth. The developing epidermis has a high content of glycogen which, apart from its metabolic role, may aid in repelling

the amniotic fluid, which also has a high concentration of glycogen.

Ultrastructural studies of the human periderm show that its cells possess microvilli, in contact with the amniotic fluid (Whittaker & Adams 1971). The functional significance of this transient surface layer remains obscure.

At an early, but still undetermined age, neural crest cells invade the developing skin and differentiate into both epithelial and connective tissue potential pigment cells, such as *melanoblasts*. These assume a branching form and hence are named *dendritic cells*. In negro fetuses many of these cells become truly melanoblastic developing granules of melanin (Rawles 1948, Billingham & Silvers 1960, Boyd 1960), which can be transferred to neighbouring non-dendritic cells in postnatal life. In less pigmented races similar cells exist in the epidermis and deeper tissues which, though normally containing little or no pigment, retain a melanogenic potentiality by which, given appropriate stimulation such as intense sunlight, they are able to form melanin granules.

For a summary of non-cutaneous ectodermal derivatives see p. 136.

The dermis is a mesodermal derivative, and it is widely assumed that at least some of its cells develop from the lateral wall of the somites—regions termed *dermatomes*. It is also held that dermal tissue is partly derived from mesenchyme associated with the somatopleuric sheet of lateral plate mesoderm and also the neural crest. Perhaps all such regions contribute; at present evidence is insufficient for quantification.

Towards the end of the third month the subepidermal mesoderm begins to condense and define the dermis, deep to which areolar connective tissue appears. A month later dermal papillae can be identified and the characteristic patterns of ridges on the ventral hairless skin of the extremities, best seen and most familiar as fingerprints, are quickly established and remain substantially unchanged in the individual apart from areal growth (Cummins & Midlo 1961, Hale 1952). The histological elements of the dermis differentiate at varying times; thus, collagenous fibres begin to appear and to aggregate during the third month, whereas elastic fibres become identifiable 2 or 3 months later.

Hairs initially appear in the third month as solid cylinders of epidermis which grow *obliquely* into the dermis (2.56), each *hair bud* having an external layer of columnar cells and a core of polygonal elements (Hale 1952). The obliquity of the buds (later hair follicles) varies in *degree* but they are roughly parallel in adjacent hairs; their corporate *direction* being specific for each region of the body. The deepest part of the bud

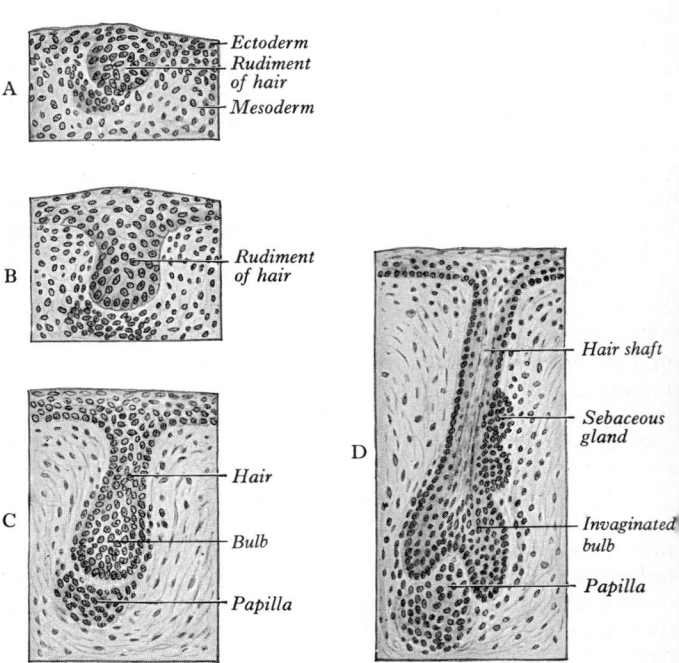

2.56 Successive stages in the development of hair.

expands into a *hair bulb*, which becomes invaginated by a mesodermal *hair papilla*. The central cells become keratinized to form the substance of the hair itself; and as further cells are added from the inner layer of the bulb these also contribute to the shaft of hair, which lengthens and begins to project above the skin surface. Further growth is due to continued proliferation of the epidermal cells around the mesodermal papilla (see also p. 90, 91). Hairs appear first on the face, and by the end of the third month there is a general covering over the body of somewhat scattered fine hairs, the *lanugo*. These are shed before birth and are replaced by thicker hairs arising in a new set of follicles. Melanoblasts become associated with developing hairs at an early stage and these produce much melanin in dark-haired individuals during the later months of gestation. For further details of hair structure, growth, associated glands and arrectores pili muscles see p. 90-92.

Sudoriferous glands first begin to develop in the skin of the palms and soles during the fourth month initially, like hairs, as solid downgrowths of epidermis into the dermis. These elongate and their deeper parts coil to form the main secreting part of the sweat glands. A lumen develops and opens superficially at about the seventh month. Both the duct and the coiled 'body' of each gland is double-layered, the inner epithelial, the outer, fine fibroareolar, with an intervening basal lamina. The paraluminal cells vary from cuboidal to polygonal (p. 93, 94) surrounded by contractile myoepitheliocytes (p. 55). External to the basal lamina the fibroareolar sheath invests the body of the gland, also its cylindrical duct, whilst near the general epidermis it merges with the superficial layers of the dermis. Some glands arise from the superficial parts of hair follicles, acquiring a surface opening later. Specialized sudoriferous glands are formed in some sites, such as in the skin of the axillae, eyelids and external acoustic meatuses (Montagna & Ellis 1961).

Sebaceous glands develop as solid epidermal buds from the cuboidal cells of the oblique hair follicles during the fifth month. They arise from that aspect of the follicle making an obtuse angle with the skin. The buds extend into the mesenchyme, branching into several oval alveoli, the lining cells of which are derived from the stratum germinativum. These proliferate, the older cells being pushed centrally, where they degenerate to form sebum, this being extruded into the hair follicle. Sebaceous glands develop independently of such follicles in the nostrils, eyelids (as tarsal glands) and in the anal region. The arrectores pilorum develop near the sebaceous glands and hair follicles, but at some little distance from them, arising independently from the mesenchyme; their attachment to follicles is a secondary development (Pinkus 1958). Some of the independent sebaceous glands continue forming after birth.

Nails are demonstrable as rudiments in the third month, appearing as *primary nail fields* of proliferative ectoderm on the *tips* of the terminal segments of the digits. The thickened ectoderm of these fields is in fact only transiently at the tips of the digits; but continued proliferation of the nail fields in a proximal direction progressively defines their final dorsal position. Due to their relatively slow rate of growth the fields become somewhat depressed, and the epidermis overlaps their sides and proximal ends to form *nail folds*. The distal ends of the nail fields are bounded by a shallow groove. The proximal part of each nail field proliferates to form the root of the nail, which becomes the *formative zone*. From the stratum germinativum of the root the actual substance of the nail, consisting of modified stratum lucidum cells, is continuously formed. The stratum at first covers the nail as the *eponychium*, but this disappears for the most part, remaining only as a narrow proximal crescentric fold overlying the *lunule* of the nail. The *hyponychium* is an accumulation of epidermal cells beneath the free edge of the nail (massive in claws and hooves). Fetal growth of the nails is gradual and their extremities have merely reached near the tips of the digits at birth, the fingers being rather more advanced than those of the toes.

Anomalous development of the epidermis and its derivatives is relatively common. Excessive or diminished growth, or even complete absence, may affect sebaceous or sudoriferous glands and hair, either locally or generally. Similarly, the epidermis may be excessively pigmented (*melanism*) or lack melanocytes

(*albinism*). Excessive keratinization leads to *ichthyosis*. A *naevus* or 'mole' is a locus of excessive pigmentation.

Mammary glands are considered to be much modified sudoriferous glands and as such they are basically ingrowths from the ectoderm, which forms their ducts and alveoli, supported by vascularized connective tissue derived from the mesenchyme. In embryos at about the fifth or sixth week two ventral bands of thickened ectoderm, the *mammary ridges*, extend from axilla to the inguinal region, and in many mammals paired mammary glands develop at intervals along these ridges. In the human embryo the ridges are not prominent features, and only a single pair of glands develops in the pectoral region. The ridges disappear later in embryonic life, but before this the cranial third of each begins to show proliferation to form the two glandular rudiments. Supernumerary rudiments may form anywhere along the path of the mammary ridges and may develop into actual mammae or merely accessory or supernumerary nipples.

As each mammary primordium develops, its ectodermal ingrowth branches into 15-20 solid buds of ectoderm which will become the lactiferous ducts and their associated lobes of alveoli in the fully formed gland. These are surrounded by mesenchyme which forms the connective tissue, fat, vasculature, and is invaded by the mammary nerves. By proliferation, elongation and further branching the alveoli are formed and the duct system defined. During the last 2 months of gestation the ducts become canalized and the epidermis at the point of original development of the gland forms a small *mammary pit*, into which the lactiferous tubules open. Perinatally the nipple is formed by mesenchymal proliferation. Should this fail the ducts open into shallow pits, a malformation known as inverted nipple. At birth the mammary glands are alike in their stage of development in both sexes, and in both some transient secretory activity may be observed, due presumably to circulating prolactin in the mother (Smith 1959). In males, thereafter, the mammary glands remain undeveloped, but in females at puberty, in late pregnancy and during the period of lactation they undergo further, hormone dependent, developmental changes (p. 1434 et seq). For reviews of the prenatal histogenesis and ultrastructural appearances of mammary tissues consult Tobon & Salazar (1974); for postnatal reviews, p. 1434 et seq.

The Nervous System and Special Sense Organs

The gross initial appearance of a neural plate and the development within it of a median neural groove bounded on each side by neural folds, which meet dorsally and fuse in the fourth week to form a neural tube, have been outlined (p. 133 and 2.27A–E).

Reference has also been made to the wealth of comparative experimental data, largely from submammalian forms, which indicate the fundamental role of *primary embryonic induction* in these early stages of development of the nervous system (p. 107).

Briefly, during gastrulation, the chorda-mesoderm, which is invaginated through the dorsal lip of the blastopore, forms a midline strip in the roof of the archenteron which is continuous cranially with the endoderm of the prechordal plate (p. 100) and becomes closely applied to the superjacent ectoderm. For a period the latter is competent to react to the inductive influence of the chorda-mesoderm and, in this manner, neural plate formation is initiated. The early midline neural tube is thus co-extensive with the notochord, stretching from the dorsal lip of the blastopore (primitive knot) to the orpharyngeal membrane. Further analyses also demonstrated a regional specificity in the inductive capacity of the various rostrocaudal regions of the archenteric roof. The region first invaginated, i.e. the endoderm of the prechordal plate and rostral tip of the notochord, induces the formation of forebrain and eyes, etc. (the so-called *archencephalon*); the intermediate part of the chorda-mesoderm induces hindbrain and associated structures (the *deuterencephalon*), whilst the remaining (caudal) roof of the archenteron operates as a *spino-caudal* inducer promoting the formation of spinal cord and neighbouring muscle masses.

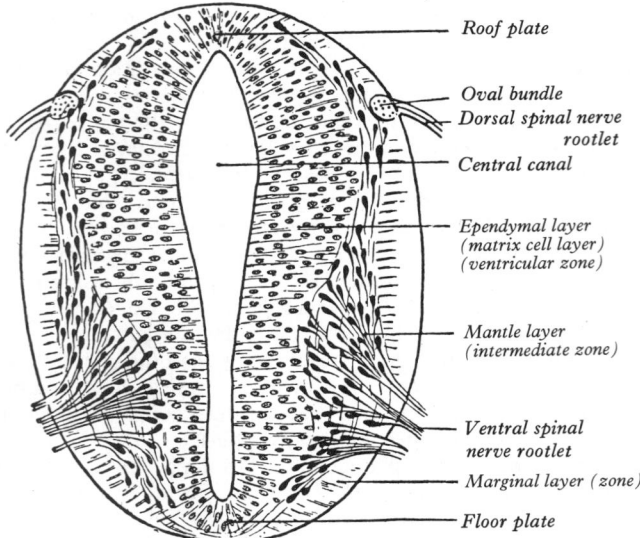

2.57 Transverse section through the developing spinal cord of a human embryo four weeks old. (After His.) Historically early terminology is retained; more recent terms in parenthesis.

Although the search for the chemical identity of inducing substances continues, as pointed out on p. 107, there has, in recent years, been a great change of emphasis on the part of developmental biologists. Nevertheless, despite the profound developmental modifications introduced by the secondary reduction of yolk in mammalian ova associated with viviparity (p. 98), it is widely assumed that similar primary mechanisms operate throughout the Chordata, including mankind. Thus, from the outset, the development of the nervous system is heavily dependent upon morphogenetic influences from neighbouring non-nervous structures. (For penetrative reviews of early neurogenesis consult Hughes 1968, Jacobson 1970, Gaze 1970, Bellairs 1971, 1974, Gottlieb 1974, Berry 1974, 1982, Berry et al 1980a,b,c,d, Schmitt & Worden 1979, Rakic & Goldman-Rakic 1982, Smart & McSherry 1982, Smart 1982, 1983.)

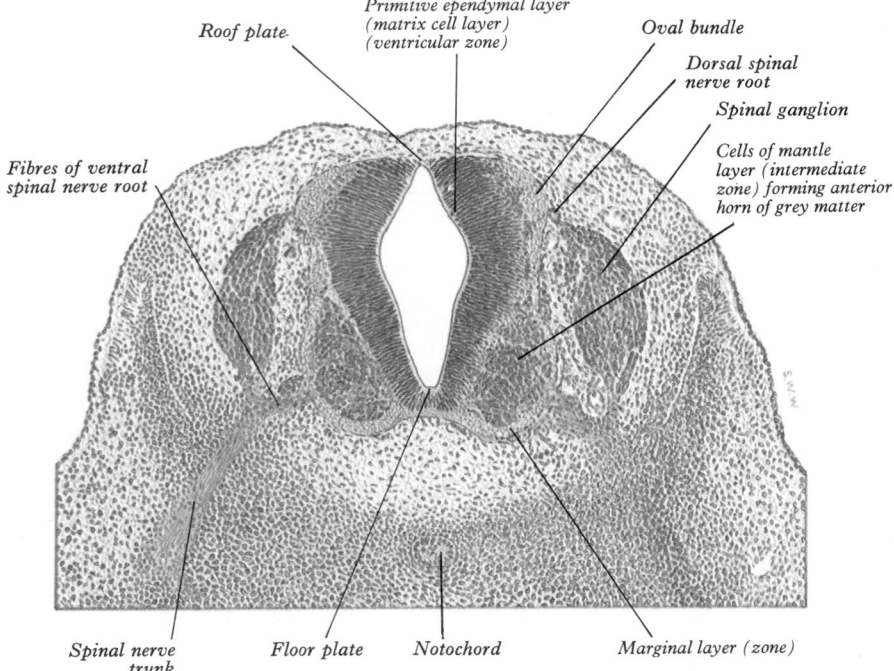

2.58 Transverse section of the developing spinal cord in the cervical region of a human embryo early in the sixth week. CR length = 8 mm.

THE NEURAL TUBE (NEURAXIS)

When the neural tube is closing, its walls consist of a single layer of columnar *neural epithelial cells*, the extremities of which abut on internal and external limiting membranes. The mechanism of rounding up of the neural plate into a neural tube has been studied particularly closely in amphibia (Burnside 1971). The columnar cells increase in length and develop numerous longitudinally disposed microtubules, whilst the borders of their luminal ends are firmly attached to adjacent cells by junctional complexes, the cytoplasmic aspect of the complexes being associated with a dense paraluminal web of microfilaments (see pp. 49, 106; **1**.13, **2**.5, also Watterson 1965). It is proposed that this disposition of organelles imparts a slight wedge conformation on the cells resulting in neural groove and eventually neural tube formation. As the cells elongate their nuclei become clustered at varying depths in the deeper part of the wall (i.e. nearer the internal limiting membrane) and, for a period, the epithelium is *pseudostratified* with an inner nucleated zone and an outer zone composed of the peripheral cytoplasmic processes of the cells. Soon, however, some of the peripheral cytoplasmic processes become detached from the (basal) external limiting membrane and rounded cells appear close to the inner membrane which, by their repeated mitotic division, form descendants which *migrate* outwards to take up an intermediate position in the wall of the tube. Histologically at this stage therefore, the wall of the tube presents three zones or layers (**2**.57, 58, 60). The internal **ventricular zone** (variously termed the *germinal, primitive ependymal* or *matrix layer*) consists of the nucleated parts of the columnar cells and the round cells undergoing mitosis. The **intermediate zone** or *mantle layer* consists of the migrant cells from the divisions occurring in the deeper layer just described. The outer **marginal zone** for a period consists of the external cytoplasmic processes of some of the original columnar cells, but they are soon invaded by tracts of axonal processes which grow from neuroblasts developing in the intermediate zone, together with varieties of non-nervous cells (glioblasts and later vascular endothelium and perivascular mesenchyme).

HISTOGENESIS OF THE NEURAL TUBE

The classical view of neural tube histogenesis was first propounded by Wilhelm His in 1890 and soon, sometimes with minor modifications, gained quite wide acceptance (e.g. Kölliker 1896, Ramón & Cajal 1911). His proposed that almost from the first the wall of the tube was *stratified* and contained a *variety* of distinct cell types—*spongioblasts, neuroblasts* and *germinal cells*. (Some contemporary cytologists claimed that the early neuroepithelium was syncytial, but this gained little acceptance.)

The primitive spongioblasts, originally elongate cells attached to both limiting membranes, with their nuclei in the ventricular zone, were considered to differentiate into a number of *sustentacular cell* varieties. By losing contact with one or both limiting membranes, they may differentiate into either astroblasts and astrocytes, oligodendroblasts and oligodendrocytes, or, retaining an internal attachment, into the definitive ependymal cells which line the central canal and regionally specialized tanycytes (pp. 894, 1454). The *neuroblasts*, one of the cell types found in the intermediate zone, differentiated into the wide array of neurons. The round deeply placed *germinal cells* or *medulloblasts* (Glees 1963) he regarded as undifferentiated stem cells, which by repeated division gave further generations of both spongioblasts (*glioblasts*) and neuroblasts. On this view, therefore, which constituted a *polyphyletic theory of neurogenesis*, the early neural epithelium was regarded as a *heterogeneous* grouping of cells and their various derivatives developed *simultaneously*.

The meticulous cytological studies of Sauer (1935a,b, 1936) led him to oppose the classical view, and he maintained that the early neural epithelium, including the deeply placed ventricular mitotic zone, consists of a *homogeneous* population of pluripotent cells, the varying appearances merely reflecting different phases in a *proliferative cycle*, the sequence being termed by Sauer *interkinetic migration*. This proposal was essentially a *monophyletic theory of neurogenesis*. In subsequent years much experimental

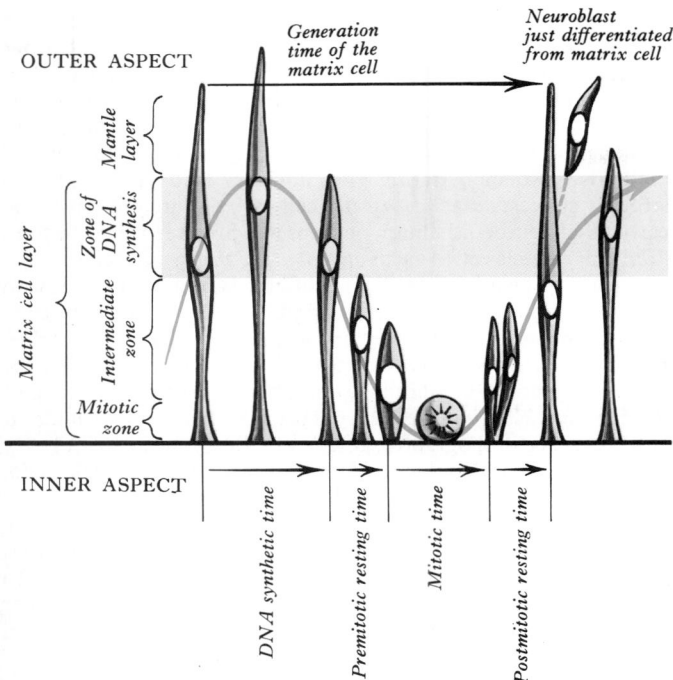

OUTER ASPECT

Generation
time of the
matrix cell

Neuroblast
just differentiated
from matrix cell

Mantle layer

Zone of DNA synthesis

Matrix cell layer

Intermediate zone

Mitotic zone

INNER ASPECT

DNA synthetic time

Premitotic resting time

Mitotic time

Postmitotic resting time

2.59 Diagram showing the cytogenetic cycle in the matrix cell layer and mantle layer in the wall of the developing neural tube. Note the various zones and their associated time scales, through which the matrix cell nuclei pass during their postulated 'elevator movement' or 'interkinetic migration' which is described in greater detail in the text. (From S Fujita, with kind permission of the author and the *Journal of Comparative Neurology*, 1963.)

evidence based upon colchicine studies (e.g. Watterson et al 1956), spectrophotometric nuclear analysis (e.g. Sauer & Chittenden 1959) and upon autoradiographic studies following the distribution of cells at varying times after labelling with tritiated thymidine (e.g. Sidman et al 1959, Sidman 1970, Fujita 1963, Fujita & Fujita 1963) and ultrastructural studies (e.g. Hinds 1971), in the main, supported the latter proposition. The scheme amplified by Fujita in a series of later publications is illustrated in

2.59. The ependymal layer of earlier workers (termed by Fujita the *matrix layer*) is considered to be populated by a single basic type of *matrix cell* and to exhibit three 'zones' (the M or mitotic, the I or intermediate, and the S or synthetic zones). As they pass through a complete inter-mitotic and mitotic cycle, the matrix cells show an 'elevator movement' progressively approaching and then receding from the internal limiting membrane. DNA replication occurs whilst the cells are extended and their nuclei occupy the S zone; they then enter a pre-mitotic resting period whilst the cells shorten and their nuclei pass through the I zone. The cells now become rounded close to the internal limiting membrane (in the M zone) and undergo mitosis; thereafter they elongate again, their nuclei passing through the I zone during the post-mitotic resting period, finally to enter the synthetic zone once again. The cells so formed may then start another *proliferative cycle* but others, presumably after a 'critical' or 'quantal' mitosis (see p. 108), rapidly migrate outwards (i.e. radially) and differentiate into *neuroblasts* as they approach and enter the adjacent stratum; possibly this differentiation is initiated as they pass through the I zone during the post-mitotic resting period. The *proliferative cycle*, with the production of clones of vast numbers of neuroblasts, continues for a time, but eventually neuroblast formation wanes; however, the mitotic activity of the remaining matrix cells continues for varying times, but then begins to decline as their progeny differentiate into *ependymal cell* and *macroglial cell* varieties. Thus the *simultaneous* appearance of cell varieties, implicit in the polyphyletic theory, contrasts with their partly successive nature in the monophyletic theory, in which maximal neuroblastosis *precedes* maximal glioblastosis.

It should be emphasized here that differences of opinion persist (and have increased in the last decade), concerning the precise details of the proliferative cycles, differentiating mitoses, migrations and timing, in different species, sites in the central nervous system and the possible mechanisms involved. Further, considerable confusion has obtained between different neuro-embryologists in relation to the *terminology* to be adopted for the various '*layers*' or '*zones*' at different times and places in the developing neural tube. An international group of neurocytologists (The Boulder Committee, 1970) proposed a less ambiguous nomenclature which is increasingly adopted. They termed the early pseudostratified neuroepithelium, in which the 'elevator movement' occurs, the *ventricular zone* (its

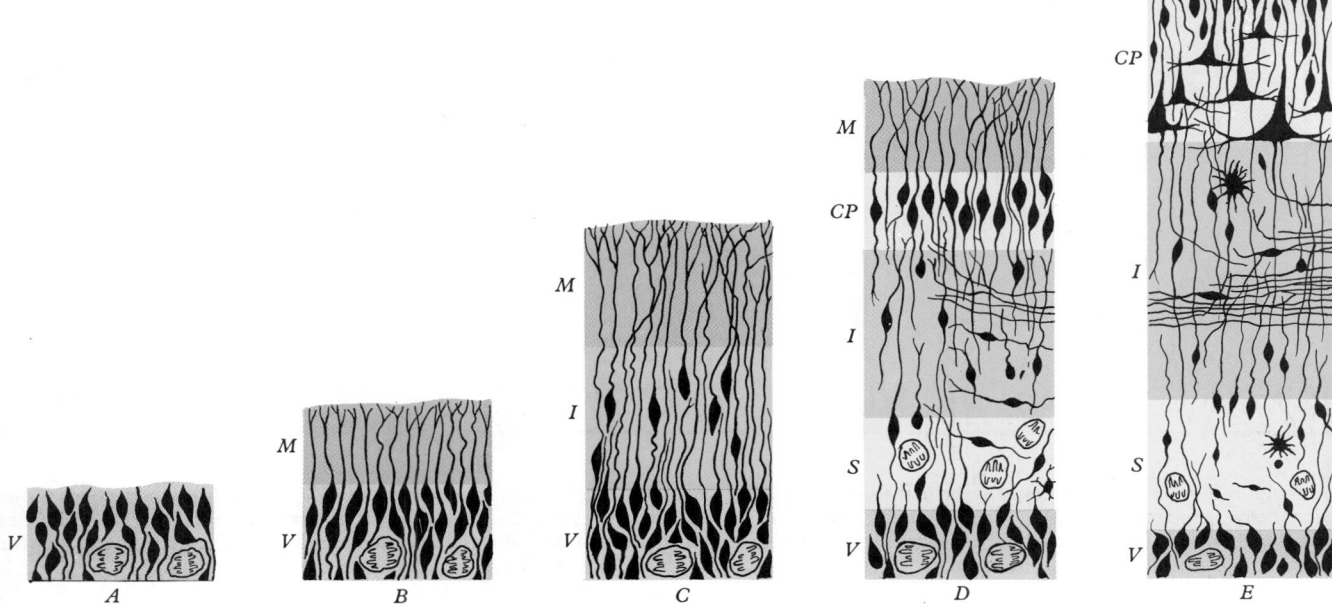

2.60 Schema of five sequential stages (A to E) in the development of part of the neural tube's wall in section to form cerebral cortex. Spinal cord development is limited to stages A to C. (Modified from Boulder Committee *initial* recommendations 1970.) V—ventricular zone;

M—marginal zone; I—intermediate zone; S—subventricular zone; CP—cortical plate. New research techniques led Rakic (1982) to advance many significant alterations and additions to these proposals: see p. 197 and **2.72**.

further development is summarized below, on pp. 188, 185, and in 2.60). For an excellent review of the biological and terminological problems posed, and an extensive bibliography to the date indicated, consult Berry (1974), and for further comments on the natural history of neurons see p. 917.

Illustration 2.60 summarizes the main stages of development of the neural tube and the nomenclature as *initially* proposed by the Boulder Committee (1970), the complete series of stages A–E applying to the cerebral neocortex: only stages A–C are relevant in the context of the spinal cord. As detailed in the caption, it will be seen that the early pseudostratified *ventricular* zone (V) is followed by the sequential appearance of *marginal* (M), *intermediate* (I), *subventricular* (S) zones and, concurrently with the latter, the early *cortical plate* (CP) (see p. 195). It is unfortunate that this use of the term *zone* for these *major ontogenetic strata* does *not* correspond to the M, I and S zones of Fujita detailed above (and in 2.59), the latter applying to subdivisions of the early neural tube only. More than a decade of research employing technical innovations, induced Rakic (1982) to suggest a series of additional features and modifications of the original Boulder proposals. These are mentioned at various points in the text, and summarized in relation to the neocortex on p. 195 and 2.72A–G.

Illustration 2.74 (which is based on the work of Berry & Rogers 1965, Berry 1974) may be noted here for comparison with 2.72; it stems from labelling studies on the *patterns of migration* of neuroblasts in the developing neocortex of the rat and will be briefly mentioned later on p. 195 et seq.

The following decade, and to the time of writing, has seen accelerating, enthusiastic application of newer, more refined techniques to these and related problems: in particular immunohistochemistry and computer analysis of neurocellular geometry. This has resulted in reappraisal of some foregoing conclusions. The cytokinetics of the early pseudostratified neuroepithelium ('elevator' movement, or interkinetic migration) has been substantiated, as has the establishment of the principal zones proposed by the Boulder Committee. However, the latter has been modified (at least in the forebrain) by the recognition of

an additional (though transient) stratum deep to the early cortical plate—the *subplate zone* (SP), see p. 195—and many further features of cortical and subcortical maturation. Throughout most of the neuraxis, proliferative cell cycles are limited to the ventricular (V) and subventricular (S) zones, but their homogeneous nature has been challenged. Immunochemical methods for demonstrating glial fibrillary acid protein (GFA) strongly suggests that *at least two* cell varieties are present in the early tubal neuroepithelium, perhaps in a precise mosaic pattern. (This may be classified as an *oligophyletic theory of neurogenesis*). The early disposition of radial glial fibres is proposed by some to have a profound morphogenetic significance. The latter concerns not only their spatially ordered array, but the probability that they provide migration paths, by contact guidance, for at least the earlier generations of migratory neuroblasts, thus specifying their destinations. Neuronal geometry is further considered on pp. 183, 188, 195. (See bibliographic reviews in: Berry 1974, 1982, Berry et al 1980a,b,c,d, Schmitt & Worden 1979, Levitt & Rakic 1980, Rakic & Goldman-Rakic 1982, Smart 1982, 1983.)

Illustration 2.61, based on and slightly modified from Rakic (1981, 1982), is a simplified analysis of earlier and current theories of some cell varieties and lineages in neural tube histogenesis. In a recent presentation (Levitt et al 1981) some aspects of all earlier theories have been incorporated, others rejected. See caption and text above for further comments.

THE SPINAL CORD

In the future spinal cord the median *dorsal lamina* (*roof plate*) and *ventral lamina* (*floor plate*) of the neural tube do not participate in the cellular proliferation affecting the lateral walls and hence remain thin. Their cells contribute largely to the formation of ependyma.

The neuroblasts of the *lateral* walls of the tube are large and at first round or oval (*apolar*). Soon they develop processes at opposite poles, becoming *bipolar neuroblasts*. One process is, however, withdrawn and the neuroblast becomes *unipolar* although

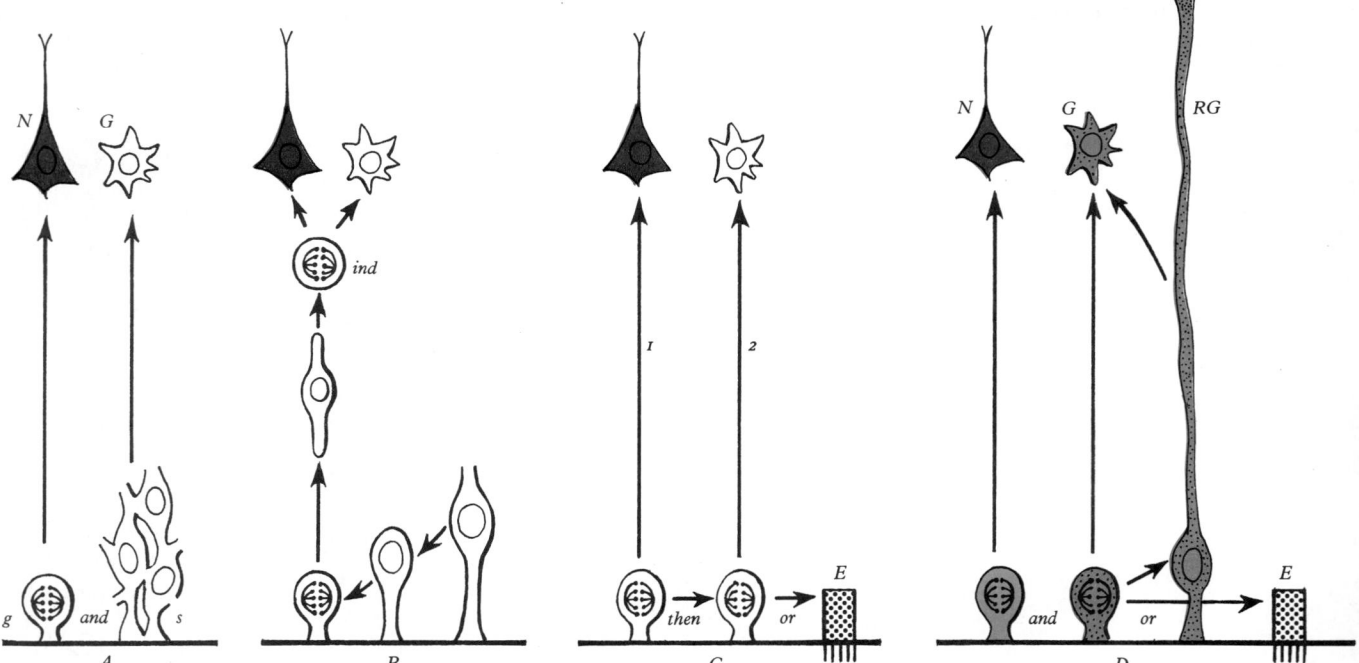

2.61 Simplified schemes of successively held theories of cell lineages during early neurogenesis. A His (1889) recognized two major cell varieties in the matrix layers (ventricular zone) lining the central canal and ventricles. Spheroidal, proliferative, germinal cells (g) their progeny migrating to form neuroblasts (N); neighbouring spongioblasts (s) had progeny forming varieties of glioblast (G). B Schaper (1897) held that ventricular zone mitoses gave outwardly migrating indifferent cells (ind), further division of which produced neuroblasts, glioblasts, or both. C Fujita (1963), using autoradiography, believed that a single proliferative type of matrix cell *firstly* gave migratory generations of neuroblasts,

secondarily (later) of glioblasts, finally, the persistent remainder formed ependymal cells (E). D Currently, electron immunohistochemistry has demonstrated at least two distinct cell varieties, — GFA (glial fibrillary acid protein) positive and negative, present in extremely early ventricular zones. The negative cells (blue) gave migratory neuroblasts (purple). The positive cells (magenta) gave, firstly, radial glioblasts (RG) that span the neural tube joining homologous internal and external points; later, directly or indirectly, generations of definitive glioblasts, and finally ependymal cells. The radial glial extensions probably provide contact guidance paths for echelons of neuroblast migration. (Modified after Rakic 1982.)

this is not invariably so in the case of the spinal cord. Further differentiation leads to the development of dendritic processes and they become typical *multipolar neuroblasts*. In the developing cord they occur in small clusters representing clones of neuroblasts.

At first the neural tube is oval in outline and its lumen is narrow and slit-like (**2.57**). As the lateral walls thicken, the lumen widens in its dorsal part and is somewhat diamond-shaped on cross-section (**2.58**). The widening of the canal is associated with the development of a longitudinal *sulcus limitans* on each side. This divides the ventricular and intermediate zones (ependymal and mantle layers) in each lateral wall into a *ventrolateral (basal) lamina* and a *dorsolateral (alar) lamina*. This separation indicates a fundamental functional difference, for neuroblasts in the ventrolateral lamina include the motor cells of the anterior and lateral grey columns, while those of the dorsolateral lamina exclusively form 'interneurons' (both short and long axoned), some of which receive the terminals of primary sensory neurons. (It should, perhaps, be mentioned at this point that the scientific utility of the term 'interneuron' has been seriously challenged—see Gray 1974.) Caudally the central canal of the cord exhibits a fusiform dilatation, the *terminal ventricle*.

(It will be apparent in the preceding paragraphs that the long held and widely used terms (1) roof plate, (2) floor plate, (3) alar lamina and (4) basal lamina have been, in the main, changed to: (1) dorsal lamina, (2) ventral lamina, (3) dorsolateral lamina and (4) ventrolateral lamina. These new terms were officially introduced in the *Nomina Embryologica* associated with *Nomina Anatomica* (Fourth Edition) by the International Anatomical Nomenclature Committee (Tokyo 1975). The modified names are more apposite *throughout* the *early* neural tube (i.e. spinal cord and brainstem).

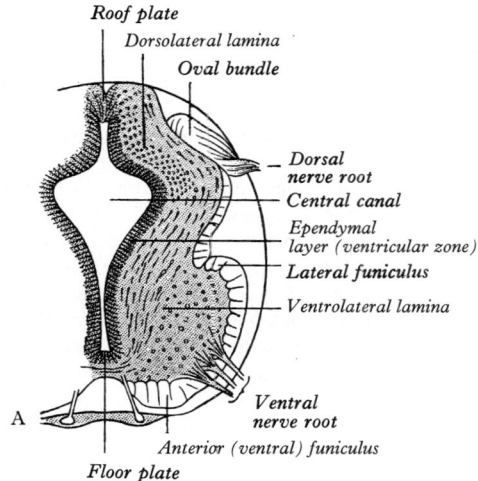

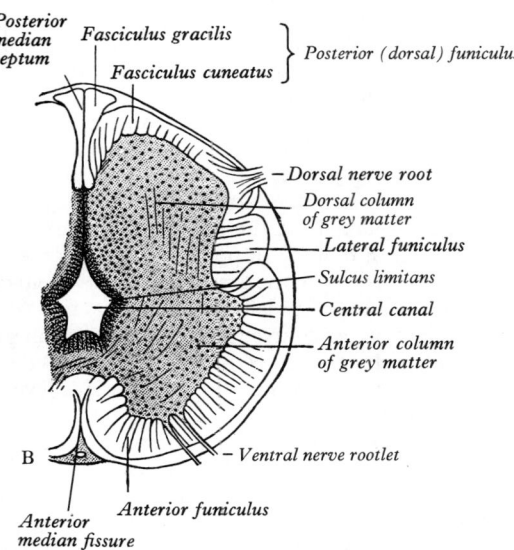

2.62 Transverse sections through the developing spinal cord of human embryos. A About 6 weeks old. B About 3 months old. (After His.)

However, these terms become much less appropriate in the rostral half of the medulla oblongata and caudal half of the pons, with the profound positional changes accompanying the formation of the pontine flexure—see p. 185. Nevertheless, for consistency they have been retained here.)

The cells of the ventricular zone are closely packed at this stage and arranged in radial columns (**2.58**). For experimental studies of radial migration patterns consult Berquist (1932, 1968), also bibliographies in Rakic (1981, 1982). Their disposition may be partly determined by contact guidance along the earliest *radial array* of glial fibres that traverse the *full thickness* of the early neuroepithelium. The cells of the intermediate zone are more loosely scattered, and they increase in number at first in the region of the ventrolateral (basal) lamina. This enlargement outlines the *anterior (ventral) column* of the grey matter and causes a ventral projection on each side of the median plane, the ventral lamina (floor plate) remaining at the bottom of the shallow groove so produced. As growth proceeds these enlargements, further increased by the development of the anterior funiculi, encroach on the groove until it becomes converted into the slit-like anterior median fissure of the adult spinal cord (**2.62A, B**). The axons of some of the neuroblasts in the anterior grey column traverse the marginal zone and emerge as bundles on the anterolateral aspect of the spinal cord as the *ventral spinal nerve rootlets*. These constitute, eventually, both the alpha-efferents which establish motor end plates on extrafusal striated muscle fibres and the gamma-efferents which innervate the contractile polar regions of the intrafusal muscle fibres of the muscle spindles (p. 915). (The histogenesis of beta-efferents is, of course, completely uncharted.)

In the thoracic and upper lumbar regions some intermediate zone neuroblasts in the dorsal part of the ventrolateral lamina outline a *lateral column*. Their axons join the emerging ventral nerve roots and pass as preganglionic fibres to the ganglia of the sympathetic trunk or related ganglia, the majority eventually myelinating to form *white rami communicantes*. The fibres constituting the rami establish synapses with the autonomic ganglionic neurons (pp. 199, 200), and the axons of some of the latter proceed as post-ganglionic fibres to innervate smooth muscle cells, adipose tissue or glandular cells. Some of the preganglionic sympathetic efferent axons pass to the cells of the suprarenal medulla (p. 1469). The innervation of other 'chromaffin' tissues is less certain (but see the carotid body, p. 1473). Similarly an autonomic lateral column is also laid down in the mid-sacral region and gives origin to the preganglionic parasympathetic fibres of the pelvic splanchnic nerves (p. 1166).

A number of investigations have been directed towards an understanding of the cell dynamics involved in the elaboration of the distinctive groupings of cells seen in the mature spinal cord of various species (Levi-Montalcini 1950, Hamburger 1952, Harris 1965, Romanes 1951, 1953, 1964).

The changes in cell number, position and density as seen in a longitudinal section of the chick spinal cord, based on investigations of Hamburger (1952), are illustrated in **2.63**. The anterior region of each ventrolateral (basal) lamina forms at first a continuous column of cells throughout the length of the developing cord. In many forms this soon develops into two columns (on each side)—one medially placed, concerned with innervation of axial musculature, and a lateral one innervating the limbs. At limb levels the latter column enlarges enormously but retrogresses at other levels.

Involved in transformations such as the foregoing, there is an interplay between a number of fundamental processes which vary in prominence at different times and at different levels—cell *proliferation, migration*, followed either by progressive cell *growth and differentiation* or, in complete contrast, by cell *degeneration and death*. In the example quoted, cell proliferation persists as a prominent feature at the levels concerned with limb innervation, whilst at thoracic levels a *dorsomedial* migration of neuroblasts occurs to lay the foundation of the visceral efferent column. Further, save at limb levels, massive cell degenerations occur in the lateral 'motor' columns, whereas the medial columns, which innervate axial musculature, persist throughout the cord. The phenomenon of cell death and removal on a large scale, balanced against local proliferation and migration rates, has only been

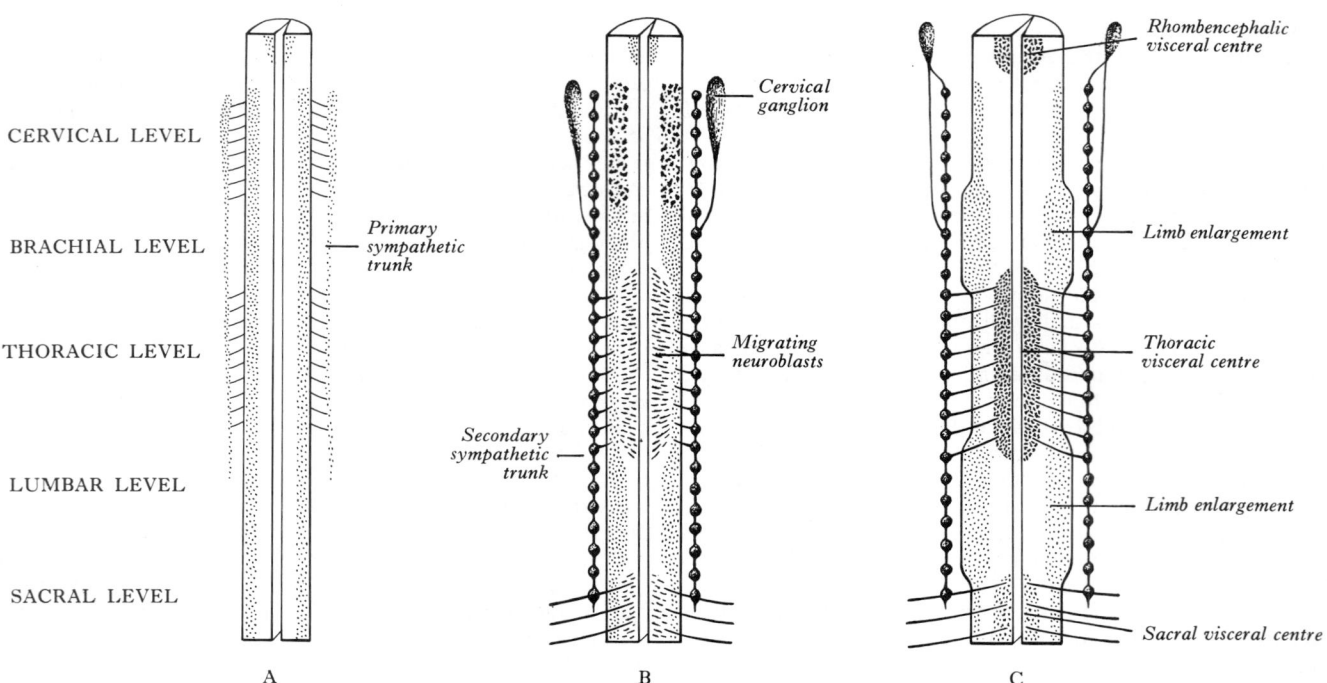

CERVICAL LEVEL

BRACHIAL LEVEL

THORACIC LEVEL

LUMBAR LEVEL

SACRAL LEVEL

Primary sympathetic trunk

Cervical ganglion

Migrating neuroblasts

Secondary sympathetic trunk

Rhombencephalic visceral centre

Limb enlargement

Thoracic visceral centre

Limb enlargement

Sacral visceral centre

A   B   C

2.63 Diagrams illustrating the changing morphogenetic patterns in the developing neural tube of the *chick* seen in longitudinal section. Note the progression from A to C as the initial simple homogeneous columnar organization (A) is transformed into the definitive pattern (C). These changes follow variations in the turnover rate and differentiation rate of neuroblasts in different regions of the tube, combined with migration of neuroblasts in some regions, and cell death in others. For further description see the text. (From V Hamburger, with kind permission of the author and the New York Academy of Sciences.)

recognized relatively recently (p. 103) as a fundamental feature in many morphogenetic situations. For an excellent review of this field in relation to the developing nervous system, the interested reader should consult Hughes (1968, 1974).

Thus, as we have seen, some ventrolateral laminal neuroblasts differentiate into the ventral horn neurons from which both alpha, beta and gamma efferent fibres arise, and these are accompanied at thoracic, upper lumbar and mid-sacral levels by preganglionic autonomic efferents from neuroblasts of the developing lateral horn. However, additionally, numerous interneurons develop in both these situations (including the well-studied Renshaw cells), but it is uncertain how many of these differentiate directly from ventrolateral lamina neuroblasts and how many migrate to their final positions from the dorsolateral lamina.

In the human embryo, the definitive grouping of the ventral column cells which characterizes the mature cord occurs early, and by the fourteenth week (the 80 mm stage) all the major groups can be recognized (Romanes 1941, 1942, 1946, 1953, 1964).

As the anterior and lateral grey columns assume their final form the germinal cells in the ventral part of the ventricular zone gradually cease to proliferate and the layer becomes reduced in thickness until it ultimately forms the single-layered ependyma which lines the ventral part of the central canal of the spinal cord. The *posterior (dorsal) column* is somewhat later in its development and, as a result, its ventricular zone is for a time much thicker in the dorsolateral lamina than it is in the ventrolateral lamina (**2.**58).

While the columns of grey matter are being defined, the dorsal region of the central canal becomes narrow and slit-like and its walls come into apposition and fuse with each other (**2.**62B). In this way the central canal becomes relatively reduced in size and somewhat triangular in outline.

About the end of the fourth week advancing axonal sprouts invade the marginal zone. The first to develop are those destined to become short *intersegmental* fibres from the neuroblasts in the intermediate (mantle) zone, and also fibres of *dorsal roots* of spinal nerves which pass into the spinal cord from neuroblasts of the early spinal ganglia. The earlier dorsal root fibres that invade the dorsal marginal zone stem from *small* dorsal root ganglionic neuroblasts. By the sixth week the latter form a well-defined *oval*

bundle near the peripheral part of the dorsolateral lamina (**2.**58, 62A); this bundle increases in size and, spreading towards the median plane, forms the *primitive posterior funiculus*; its constituent fibres are destined to be of fine calibre. Later, fibres derived from new populations of *large* dorsal root ganglionic neuroblasts join the dorsal root and invade the cord nearer the median plane; they are destined to become fibres of much larger calibre. As the posterior funiculi increase in thickness, their medial surfaces come into contact separated only by the *posterior median septum*, which is ependymal in origin, neuroglial in nature. A more detailed analysis of the temporal sequence of modifications of the dorsal lamina (roof plate), posterior median septum, and the lateral displacement of the *primitive posterior funiculus* with the later development of the fasciculus cuneatus followed by the fasciculus gracilis, based on a study of sectioned human embryos of 6–10 weeks and dissections of fetuses up to the end of the fourth month, has been provided by Hughes (1976). He proposed that the displaced primitive posterior funiculus may form the basis of the dorsolateral tract or fasciculus (of Lissauer), and also correlated the sequence, siting and calibre of the entrant dorsal root fibres with the changing size-distribution of the somata of the dorsal root ganglionic neuroblasts.

Long intersegmental fibres begin to appear about the third month and *corticospinal* fibres about the fifth month. All nerve fibres are at first without myelin sheaths and different groups *commence* to develop sheaths at different times, e.g. the ventral and dorsal nerve roots about the fifth month, the corticospinal fibres after the ninth month. In peripheral nerves the myelin is formed by Schwann cells; in the central nervous system oligodendrocytes are believed by the majority of workers to be concerned (p. 893). Myelination, of course, persists until overall growth of the central and peripheral nervous system has ceased. Thus, in many sites, slow growth continues for long periods, even into the postpubertal years.

The cervical and lumbar enlargements first appear simultaneously with the development of their respective limb buds.

In early embryonic life the spinal cord occupies the *entire length* of the vertebral canal and the spinal nerves pass at right angles to the cord. After the embryo has attained a length of 30 mm the vertebral column begins to grow more rapidly than the spinal cord, the caudal end of which gradually becomes more cranial in the vertebral canal. Most of this relative rostral migration occurs during the *first half of* intrauterine life. By the twenty-fifth week the terminal ventricle of the spinal cord has altered in level from the second coccygeal vertebra to the third lumbar, a distance of nine segments, and there remain but two segments before the adult position is reached (Streeter 1919). As the change in level begins rostrally the caudal end of the terminal ventricle, which has become adherent to the overlying ectoderm, remains *in situ* and the walls of the intermediate part of the ventricle and its covering pia mater become drawn out to form a delicate filament, the *filum terminale*. The separated portion of the terminal ventricle persists for a time but it usually disappears before birth. It does, however, occasionally give rise to congenital cysts in the neighbourhood of the coccyx. In the definitive state, the upper cervical spinal nerves retain their position roughly at right angles to the cord; proceeding caudally, however, the nerve roots lengthen and are progressively more oblique.

THE SPINAL NERVES AND NEURAL CREST

Each spinal nerve is connected to the spinal cord by a ventral root and a dorsal root. The fibres of the ventral roots grow out from neuroblasts in the anterior and lateral parts of the intermediate zone; these pass through the overlying marginal zone and external limiting membrane to enter the *myotomes* of the mesodermal somites, or penetrate the latter, reaching the adjacent somatopleure, and in both sites ultimately form the alpha-, beta- and gamma-efferents. At appropriate levels these are accompanied by the outgrowing axons of preganglionic sympathetic neuroblasts (segments T_1–L_2), or preganglionic parasympathetic neuroblasts (S_2–S_4).

The fibres of the dorsal roots are developed from the cells of the spinal ganglia. Before the neural groove is closed to form the neural tube (p. 133) a ridge of neurectodermal cells, the *neural crest (ganglion ridge)*, appears along the prominent margin of each neural fold (**2.27**A–E). When the folds meet in the median plane the two neural crests fuse into a wedge-shaped mass along the line of closure of the tube. Opposite each primitive mesodermal segment the neural crest cells proliferate rapidly to form a bilateral series of oval-shaped *primordial spinal ganglia*, and these migrate for a short distance in a lateral and ventral direction. From the ventral *region* of each a small part separates to form *sympathochromaffin* cells (pp. 200, 1465), while the remainder becomes a *definitive* spinal ganglion. The spinal ganglia are arranged symmetrically at the sides of the neural tube and, except in the caudal region, are equal in number to the primitive segments. The cells of the ganglia, like the cells of the intermediate zone of the early neural tube, are glioblasts and neuroblasts. The glioblasts develop into the satellite cells, which become closely applied to the ganglionic nerve cell somata (perikarya), into Schwann cells, and possibly other cells. The neuroblasts, at first round or oval, soon become fusiform, with extremities gradually enlongating into central and peripheral processes. The central processes grow into the neural tube, as the fibres of dorsal nerve roots, while the peripheral processes grow ventrolaterally to mingle with the fibres of the ventral root thus forming a *mixed spinal nerve*. As development proceeds the original bipolar form of the cells in the spinal ganglia changes; the two processes become approximated until they ultimately arise from a single stem in a T-shaped manner, to form a unipolar cell (sometimes, less appropriately, termed pseudo-unipolar). The bipolar form is, however, retained in the retina and in the ganglia of the vestibulocochlear nerve. Some observers hold that the T-form is derived from the branching of a single process which grows out from the cell.

It should be noted that the position of the early neural crest as a wedge-shaped mass along the line of tube closure noted above, and the identification of ganglionic cells in various positions in the wall of the early neural tube, and even within the central canal (Humphrey 1944, 1947), is strongly reminiscent of the developmental history of the *Rohon-Beard cells* in fish and amphibia (Rohon 1884, Beard 1896), which are thought to be important in the emergence of primitive locomotor patterns (Hughes 1968). In this regard, other investigators have claimed that in primitive chordates (Cyclostomes and Euselachians) the neural crest

develops as an *evagination* of the dorsal region of the dorsolateral lamina of the late neural folds and early neural tube (Conel 1942). For a review of the origin, widespread migration and differentiation of neural crest cells see, e.g. Weston (1970) Bellairs (1971) Leikola (1976) and p. 137.

For more than a century there have been many attempts to plan investigations which would aid an understanding of the *causal morphogenetic mechanisms* which operate during the development of the nervous system and to correlate the stage of development with emergent *behavioural patterns*. Whilst much has been established, answers to many fundamental questions still remain obscure. A review of this important and interesting area of biology lies outside the scope of the present volume and the reader should consult the pioneering works which have appeared over the intervening years. Consult references (His 1879, 1883, 1887, 1890, Ramón y Cajal 1928, 1955, Waldeyer 1891, Harrison 1906, 1907, 1910, 1914, 1924, Coghill 1929, Detwiler 1936, Spemann 1938, Weiss 1950, Hughes 1968, Gottlieb 1974).

Briefly, the nervous system is closely interlocked in terms of morphogenesis, with the 'periphery', i.e. surrounding non-nervous structures, and each is dependent upon the other for its effective structural and functional maturation.

Reference has already been made to the widely assumed initiating importance of a triggering system whereby the prechordal plate and chorda-mesoderm of the archenteron roof operate as a coordinated series of regional neural plate inducers—archencephalic, deuterencephalic and spinocaudal. (However, see discussions on pp. 107, 108–115 and consult Bellairs 1971.) Thereafter follows a period during which the main cell masses and subdivisions of the neural tube are established, and transplantation experiments have indicated that some of the earlier patterning is an expression of self-differentiation of the tube. (For a descriptive and experimental review of pattern formation in early neurogenesis in the chick consult Watterson 1965.) However, the polarization of ventrolateral lamina neuroblasts and those of the neural crest, and the formation of segmentally arranged ventral nerve roots and spinal ganglion primordia, is closely related to and dependent upon the orderly segmentation of the neighbouring mesodermal somites. Experimental removal or intercalation of somites results in corresponding disturbances of pattern in these nervous structures.

There are many other examples of the influence of peripheral structures on the development of the nervous system. Thus, if the normal peripheral innervation of the tetrapod limb is prevented there follows a gross disturbance in the balance between neuroblast proliferation, degeneration, migration and maturation in the relevant part of the neural tube and primitive spinal ganglia. Conversely, the denervated limb rudiment itself fails to complete its development in the absence of nerve-mediated influences. Other extensively investigated situations include the inductive influence of the optic vesicle on adjacent ectoderm with the formation of a lens vesicle (p. 202) and the reciprocal influences of the developing lens and peri-optic mesenchyme on the differentiation of the optic cup (p. 202); the profound influence of peripheral innervation on the regeneration of limbs in amphibia; the 'trophic' maintenance of striated muscle structure by peripheral nerves (p. 919); the dependence of regenerating peripheral nerves on the establishment of normal peripheral connections before a restoration of their structural and functional parameters ensues. Thus, whilst there are numerous experimentally demonstrable examples of the *interdependence* of the developing nervous system and the periphery, the precise causal mechanisms at work remain elusive. In this regard, however, mention should be made of a proposal (Prestige 1965, 1967, 1970) concerning the operation of *maintenance factors* thought to be produced by the tissues of developing tetrapod limbs, and which may enter the nervous system, perhaps to be stored in the cytoplasm of neurons, where their concentration influences the turnover of neuroblasts, i.e. the balance between the rate of proliferatration and degeneration. Similarly, other workers (Levi-Montalcini 1950, 1952, 1960, 1967, Levi-Montalcini & Chen 1971, Cohen 1958, Cohen & Levi-Montalcini 1956) have isolated various proteinaceous *nerve growth factors* from tissue and tumour extracts and snake venom which influence the in vitro form and extent of nerve cell growth. Although the operation of such factors in normal ontogeny has not been claimed by these workers, others have demonstrated the presence of similar substances in homogenates of the axial structures of chick embryos during early neurogenesis.

One problem which has attracted neuroembryologists for many years concerns the manner in which growing nerve fibres reach and make functional contact with their appropriate end organs (neuromuscular endings, secretomotor terminals or synapses with other neurons), during which they often pursue complex courses between the parent cell body and the sites of termination. If the cells concerned had been in contact from the first, the problem would have reduced to one concerning mechanisms of relative displacement and migration of both, with simple elongation of the nerve cell process between them. However, during the outgrowth of axonal processes from, e.g. neuroblasts within the ventrolateral lamina to reach presumptive myoblasts in the limb buds, the earliest nerve fibres are known to cross appreciable distances occupied solely by loose general mesenchyme.

The growing tips of the neurites have been studied in tissue culture (Harrison 1910, Lewis 1945, Speidel 1932, 1933, 1935, Nakai 1960, Nakai & Kawasaki 1959, Pomerat et al 1967, and vide infra) and in the tail of larval amphibia (Speidel 1932, 1933, 1935). Classically, the tip has been described as expanded into a *growth cone* which is constantly active, changing shape, extending and withdrawing processes which apparently 'explore' the local environment for a suitable surface along which extension may occur (see 'cell motility' p. 49). One process then enlarges at the expense of others and expands into a new growth cone, the exploratory behaviour recommencing. However, it has been suggested that the very rapid surface activity of growth cones is associated more with the micropinocytotic imbibition of water and solutes (perhaps in relation to a peripheral undulating cytoplasmic veil, or ruffled border, as in macrophages), forming cytoplasmic vacuoles which migrate centrally along the neurite. On this view the tip of the main trunk of the neurite moves in a series of discontinuous directional thrusts. Earlier ultrastructural studies also emphasized the presence within growth cones of microvesicles in the general cytoplasm (Estable et al 1957) and vacuoles within the cisternae of the endoplasmic reticulum (Bellairs 1959, but vide infra). In this context it is also widely held that products important in the growth mechanisms and synthesized within the cell bodies are passed outwards by some form of proximodistal *axoplasmic flow* along the neurites. Bulk axoplasmic flow was first postulated following the experimental constriction of nerves (Weiss & Hiscoe 1948), and since that time many intricate analyses of rapid and slow components of *bidirectional flow systems* within axons have been made (Lubińska 1964, and see *Neurosciences Research Programme Bulletin* **4**, vol 5, 1968, also p. 881).

However, in a series of ultrastructural studies on the in vitro outgrowth of neurites from explants of chick *spinal cord*, a number of established views concerning their form, structure and surface contacts have been challenged (Grainger et al 1968, James & Tresman 1969, Grainger & James 1970). In addition to describing the formation of neuromuscular junctions and varieties of synapse under these experimental conditions, the authors have shown that the neurites usually grow out in bundles rather than as single processes. The neurites contain arrays of longitudinal microtubules which are probably interconnected, microfilaments, mitochondria, linear densities in the cytoplasm and populations of vesicles. The expanded tip of the neurite bundle is not an aggregation of simple 'growth cones' but usually the neurite terminals end in association with a 'glial' cell. The latter has a relatively dense cytoplasm containing an array of organelles enabling it to be distinguished from the neighbouring neurites. Further, the glial cell is complex in form. It sends a major cytoplasmic process back towards the parent explant; it presents surface projections which partly enfold the terminal neurites, and distally, beyond the neurites, cytoplasmic sheets or finger-like processes extend to come into intimate relationship with the culture substrate. Thus, the essential contacts appear to be between glial cell membrane and substrate, and between neurite membrane and glial cell membrane, posing the question whether the glial cell, rather than the tip of the neurite, is basically involved in determining the direction

of outgrowth. Further investigations will be necessary to determine how far these appearances are reflected in the growth of both central and peripheral neurites in normal ontogeny, although strong supporting evidence has been adduced concerning the growth of *central* neuroblasts (Rakic 1971a,b, 1981, 1982). Further details cannot be pursued here, but an excellent review has appeared concerning the morphology, motility and directional growth of axons and their growth cones, including the in vitro establishment of synaptic contacts—for this consult James (1974). It should be pointed out that Rakic (1971a,b, 1981, 1982) suggested that within the central nervous system, the direction followed both by advancing neurites and by migrating neuroblast somata may stem from contact guidance (vide supra and pp. 115, 918) with previously established glioblast processes. On the other hand, studies of the growth cones of neurites from explanted dorsal root ganglia of the rat (including their endocytotic uptake and transport of horseradish peroxidase), using light, transmission and scanning electron microscopy, many surface features and ultrastructural details of the cones were provided; no 'leading' glial or Schwann cells were identified. For further discussion and bibliography consult Fischer (1977); and for a literature review concerning *dendritic* growth Berry (1974, 1982).

Over the years two principal theories emerged concerning the directional growth of nerve fibres—the *neurotropism* of Ramón y Cajal (1919) and the principle of *contact-guidance* of Weiss (1941). The former, based upon observations on the innervation of epithelia, proposed that growing fibres were guided by some form of attraction, presumably chemical, which emanated from the area to be innervated. The second view denied the existence of such attractive forces and, based upon many series of tissue culture experiments, held that pioneer neurites were guided to their destination by preferential growth along pathways provided by an orientation of the micellar substructure of the intercellular matrix in the intervening spaces. Despite the forceful arguments brought to bear on this question, it should be noted that in a critical review of this field (Hughes 1968), it was stated: 'Both contact guidance and neurotropism, however, associated or combined, are inadequate to explain the pattern of the peripheral nervous system and, *a fortiori*, of the enormously more intricate plan of the fully differentiated neural tube. No imaginable hypothesis is at present adequate to explain the phenomena of the regeneration of the optic nerve, which is the only central pathway which has so far been studied by experimental means. At present the only possible way of regarding the whole question is in terms of a succession of influences which operate in turn upon a growing nerve fibre, from its first emergence to its final link with an end organ. Such a sequence of controlling factors was postulated as far back as 1911.' Nevertheless, whilst few would disagree with the general tenets of these remarks, hypotheses abound, and there is currently a great upsurge of interest in neuroembryology. As an introduction to the extensive literature, the reader should consult, e.g. Sperry (1958, 1963, 1965), Schmitt et al (1970), Jacobson (1978), Gaze (1970), Gottlieb (1973, 1974), Mark (1974), Berry (1974, 1982), Rakic (1981, 1982). In addition to questions concerning the growth, guidance, and patterning of neurites and somata, the following kinds of problem confront neuroembryologists. What are the basic ontogenetic factors providing positional information (p. 109), controlling proliferative cycles, differentiating mitoses and the initial polarization and migration of neuroblasts? How much of the initial patterning is preprogrammed in the genome, and how much environmentally and functionally dependent? To what extent do regeneration experiments reflect ontogeny? Do neurons bear individually unique chemical markers? What are the structural bases of neuronal plasticity and the behavioural phenomena of instinct, learning and memory? Answers still elude us and we remain at the stage of attempting to plan meaningful experiments.

Development of the Brain

Prior to the closure of the neural tube the neural folds become expanded considerably in the head region as a first indication of a brain. Subsequent to the closure of the rostral neuropore

(p. 133) these regional expansions form the three *primary cerebral vesicles* (**2.**64, 67, 68). The customary use of the term 'vesicle', though followed here, is rather a misnomer as it suggests an exaggerated view of these localized accelerations of growth in the wall of the brain (O'Rahilly and Gardner 1971). In 'higher' vertebrates, and especially in primates, including mankind, the bulging is not initially marked, nor are the constrictions between them particularly narrow. The vesicles are more like gently fusiform tubes. The three regions are named (caudorostrally) the *rhombencephalon* or *hindbrain*, the *mesencephalon* or *midbrain* and the *prosencephalon* or *forebrain*—the first being continuous with the spinal cord. As a result of unequal growth of its different regions three flexures appear in the brain; two of these are concave ventrally and there are corresponding flexures of the head. The first is associated with the formation of the head fold and is a *midbrain flexure*; as a result the forebrain bends in a ventral direction around the cephalic end of the notochord and foregut and this progresses until its floor lies almost parallel with that of the hindbrain (**2.**64, 67, 68). The midbrain is thus, for a time, the most conspicuous part of the brain and underlies the marked *midbrain prominence* of the overlying ectoderm, seen to best advantage when viewed from the side. The second bend appears at the junction of the hindbrain and spinal cord, the *cervical* or *neck flexure* (**2.**64, 67, 68). This increases from the fifth to the end of the seventh week, by which time the hindbrain forms nearly a right angle with the spinal cord; after the seventh week, however, extension of the head takes place and the cervical flexure diminishes and eventually disappears. These bends are important factors in determining the shape of the cranial end of the embryo. The third bend, the *pontine flexure*, is at the level of the future pons. It differs from the other two in that its convexity is directed ventrally and it does not substantially affect the outline of the head. For more recent accounts of the early development of the brain consult Bartelmez & Dekaban (1962), Jacobson (1970), Gaze (1970), Eccles (1973), Gottlieb (1973, 1974), Mark (1974), Rakic (1981, 1982), Smart (1982, 1983).

The lateral walls of the hindbrain and midbrain, like those of the spinal cord, are divided into *dorsolateral (alar)* and *ventrolateral (basal) laminae* by the upward continuation of the limiting sulci of the spinal cord.

THE RHOMBENCEPHALON

By the time the midbrain flexure appears, the hindbrain exceeds in length the combined extent of the other two brain vesicles. Cranially it exhibits a constriction, the *isthmus rhombencephali* (**2.**67B), best viewed from the dorsal aspect. Ventrally the hindbrain is separated from the dorsal wall of the primitive pharynx only by the notochord, the two dorsal aortae and a small amount of mesenchyme; on each side it is closely related to the dorsal ends of the visceral arches (**2.**49).

The pontine flexure appears to 'stretch' the thin, epithelial roof plate (dorsal lamina) which becomes widened, the greatest increase in width corresponding to the region of maximum convexity, so that the outline of the roof plate becomes rhomboidal. By the same change the lateral walls become separated, particularly dorsally, and the cavity of the hindbrain, subsequently the fourth ventricle, becomes flattened and somewhat triangular on cross-section. The pontine flexure becomes increasingly acute until, at the end of the second month, the laminae of its cranial (metencephalic) and caudal (myelencephalic) slopes are opposed to each other (**2.**66C) and, at the same time, the lateral angles of the cavity extend to form the lateral recesses of the fourth ventricle.

About the end of the fourth week, when the pontine flexure is first discernible, a series of six transverse *rhombic grooves* appears in the ventrolateral lamina of the hindbrain. Between the grooves, the intervening masses of neural tissue have, at this stage, been termed *rhombomeres (neuromeres)*. These are closely associated with the pattern of the underlying motor nuclei of certain of the cranial nerves. The first two overlie the trigeminal nucleus, the third the facial nucleus, the fourth that of the abducent nerve, the fifth that of the glossopharyngeal and the sixth that of the vagus. These grooves, though transient, are constant in appearance, but their significance is uncertain. Some regard them as evidence of a

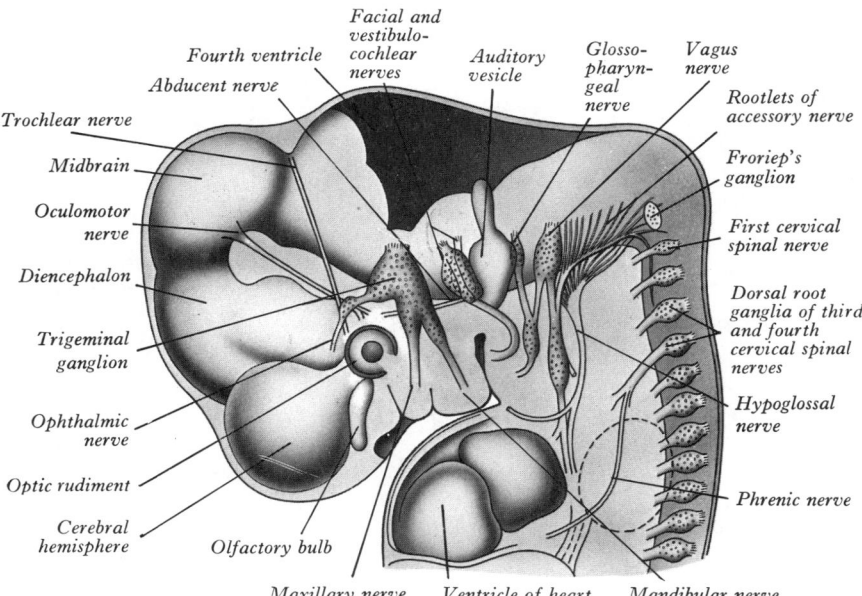

2.64 The brain and cranial nerves of a human embryo, 10.2 mm long. Note the midbrain, cervical and pontine flexures, and the expanding fourth ventricle. Note also the ganglia (stippled) associated with the trigeminal, facial vestibulocochlear, glossopharyngeal, vagus and spinal accessory nerves. Froriep's ganglion, an occipital dorsal root ganglion, is inconstant and soon disappears. (After His.)

segmental (neuromeric) origin for brain and cord (Berquist 1952).

The differentiation of the lateral walls of the hindbrain into ventrolateral and dorsolateral laminae has a similar significance to the corresponding differentiation in the lateral wall of the spinal cord (p. 181) and ventricular, intermediate and marginal zones are formed in the same way. Later extensive modifications occur.

The cells of the ventrolateral lamina are often, in elementary accounts, simply termed 'motor' (but vide infra), and they form *three* elongated, but interrupted, *columns*. The most ventral column is continuous with the anterior grey column of the spinal cord and will supply muscles considered 'myotomic' in origin. It is represented in the caudal part of the hindbrain by the hypoglossal nucleus, and it reappears at a higher level as the nuclei of the abducent, trochlear and oculomotor nerves, which are *somatic efferent nuclei*. The intermediate column is represented in the upper part of the spinal cord and caudal brain stem (medulla oblongata and pons) and is for the supply of branchial and post-branchial musculature. It is interrupted also, but the caudal brain-stem part, which gives fibres to the ninth, tenth and eleventh cranial nerves, forms the elongated *nucleus ambiguus*. The latter continues into the cervical spinal cord as the origin of the spinal accessory nerve. At higher levels parts of this column give origin to the motor nuclei of the facial and trigeminal nerves. These three nuclei are termed *branchial (special visceral) efferent nuclei*. The most dorsal column of the ventrolateral lamina (represented in the spinal cord by the lateral grey column) innervates viscera. It is interrupted also, its large caudal part forming some of the *dorsal nucleus of the vagus* and its cranial part the *salivatory nucleus*. These are termed *general visceral (general splanchnic) efferent nuclei* and their neurons give rise to preganglionic, parasympathetic nerve fibres.

It should be noted here that the neuroblasts of the ventrolateral (basal) lamina and their three columnar derivatives are only 'motor' in the sense that *some* of their number form either α, β or γ motor neurons, or preganglionic parasympathetic neurons. The remainder, which greatly outnumber the former, differentiate into functionally related interneurons and, in some loci, neuroendocrine cells.

The cell columns of the dorsolateral lamina are also interrupted and give rise to *general visceral (general splanchnic) afferent, special visceral (special splanchnic) afferent, general somatic afferent* and *special somatic afferent* nuclei (their relative positions, in simplified transverse section, are shown in **2.**65). The general

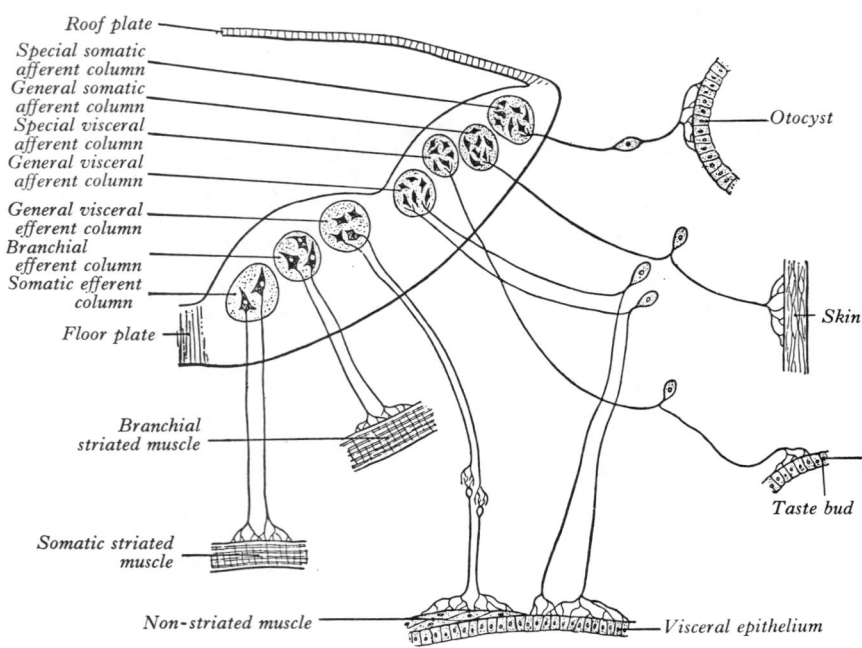

Roof plate
Special somatic afferent column
General somatic afferent column
Special visceral afferent column
General visceral afferent column
General visceral efferent column
Branchial efferent column
Somatic efferent column
Floor plate
Otocyst
Skin
Branchial striated muscle
Somatic striated muscle
Taste bud
Non-striated muscle
Visceral epithelium

2.65 Diagram of a transverse section through the developing hindbrain of a human embryo *c.* 10.5 mm long, to show the relative positions of the columns of grey matter from which the nuclei associated with the different varieties of nerve components are derived. Note the postganglionic neurons associated with the general visceral efferent column, the bipolar neurons associated with the otocyst and the unipolar afferent neurons associated with the other alar lamina columns.

visceral afferent column is represented by a part of the dorsal nucleus of the vagus (see also p. 953), the special visceral afferent column by the nucleus of the tractus solitarius, the general somatic afferent column by the afferent nuclei of the trigeminal nerve (Brown 1974) and the special somatic afferent column by the nuclei of the vestibulocochlear nerve. (Again it should be noted here that the relatively simple functional independence of these afferent columns implied by the foregoing classification is, in the main, an aid to elementary learning. The emergent neurobiological mechanisms are in fact much more complex and less well understood—see p. 1121.) Although they tend to retain their primitive positions, some of these nuclei are later displaced by differential growth patterns and by the appearance and growth of neighbouring fibre tracts, and possibly by active migration. It has been suggested that a nerve cell tends to remain as near as possible to its predominant source of stimulation and that when the danger of separation arises, owing to the development of neighbouring structures, it will migrate in the direction from which the greatest density of stimuli come. This phenomenon was termed neurobiotaxis (Kappers 1921, 1934). Cells can migrate in this way only by lengthening of their axons, which therefore trace the route taken by the cells on their transit. The curious courses of the fibres arising from the facial nucleus (p. 1108) and nucleus ambiguus (pp. 956, 1114) have been held to illustrate this. In the 10 mm embryo the facial nucleus lies in the floor of the fourth ventricle, occupying the position of the special visceral efferent column and it is placed at a higher level than the abducent nucleus. As growth proceeds the facial nucleus migrates at first caudally and dorsally, relative to the sixth nerve nucleus, and then ventrally to reach its adult position. As it migrates, the axons to which its cells give rise elongate and their subsequent course is assumed to map out the pathway along which the facial nucleus has travelled. Similarly the nucleus ambiguus arises initially immediately deep to the ventricular floor, but in the adult it is more deeply placed and its efferent fibres first pass dorsally and *medially* before curving laterally to emerge at the surface of the medulla oblongata. Attractive as the concept of neurobiotaxis may seem, it must be added that it is supported by little direct evidence; further, many other parameters affecting pattern formation must be considered.

THE MYELENCEPHALON

The caudal slope of the embryonic hindbrain constitutes the *myelencephalon*, which develops into the medulla oblongata. The nuclei of the ninth, tenth, eleventh and twelfth cranial nerves develop in the situations already indicated and afferent fibres from the ganglia of the ninth and tenth nerves form an oval marginal bundle in the region overlying the dorsolateral (alar) lamina. The dorsal edge of this lamina throughout the rhombencephalon gives attachment to the thin expanded roof plate (dorsal lamina) and is termed the *rhombic lip*. (The *inferior rhombic lip* is confined to the myelencephalon; the *superior rhombic lip* to the metencephalon.) As the walls of the rhombencephalon spread outwards, the rhombic lip protrudes as a lateral edge which becomes folded over the adjoining area. According to this widely accepted view it later becomes adherent to this area, and the cells of the rhombic lip migrate actively into the marginal zone of the ventrolateral *lamina*. In this way the oval bundle which forms the *tractus solitarius* becomes buried. Dorsolateral lamina cells which migrate from the rhombic lip are believed to give origin to the olivary and arcuate nuclei and the scattered grey matter of the nuclei pontis. While this migration is in progress the thin floor plate (ventral lamina) is invaded by fibres which cross the median plane (accompanied by neuroblasts which cluster in and near this plane), and it becomes thickened to form the *median raphe*. Some of the migrating cells from the rhombic lip in this region do not reach the ventrolateral lamina and form an oblique ridge across the dorsolateral aspect of the inferior cerebellar peduncle; the *corpus pontobulbare (nucleus of the circumolivary bundle)*.

The lower (caudal half) part of the myelencephalon takes no part in the formation of the fourth ventricle and, in its development, it closely resembles the spinal cord. The nuclei, gracilis and cuneatus, and some reticular nuclei, are derived from the dorsolateral lamina, and their efferent arcuate fibres and interspersed neuroblasts play a large part in the formation of the median raphe.

About the fourth month the descending *corticospinal* fibres invade the ventral part of the medulla oblongata to initiate the pyramids whilst dorsally, ascending fibres from the spinal cord, with olivocerebellar and parolivocerebellar fibres, external arcuate fibres, together with two-way reticulocerebellar and vestibulocerebellar interconnections, form the inferior cerebellar peduncle. (The reticular nuclei of the lower medulla probably have a dual origin from both laminae.)

THE METENCEPHALON

The rostral slope of the embryonic hindbrain is the *metencephalon*, from which both cerebellum and pons develop. Prior to formation of the pontine flexure the dorsolateral laminae of the metencephalon are parallel with one another. Subsequent to its formation the roof plate of the hindbrain becomes rhomboidal and the dorsal laminae of the metencephalon lie obliquely, being close at the cranial end of the fourth ventricle but widely separated at the level of its lateral angles. Accentuation of the flexure approximates the cranial angle of the ventricle to the caudal, and the dorsolateral laminae of the metencephalon now lie almost horizontally.

It should be noted that, caudal to the developing cerebellum (vide infra), the roof of the fourth ventricle remains epithelial, covering an approximately triangular zone from the lateral angles of the rhomboid fossa to the median obex (p. 980). Over this region nervous tissue fails to develop and vascular pia mater is closely applied to the subjacent ependyma. At each lateral angle and in the midline caudally the membranes break through forming the lateral and median apertures of the roof of the fourth ventricle. Subsequently, these are the principal routes by which cerebrospinal fluid, produced in the ventricles, escapes into the subarachnoid space. The vascular pia mater (tela choroidea), in an inverted V formation cranial to the apertures, invaginates the ependyma to form vascular fringes—the vertical and horizontal parts of the choroid plexuses of the fourth ventricle (p. 979).

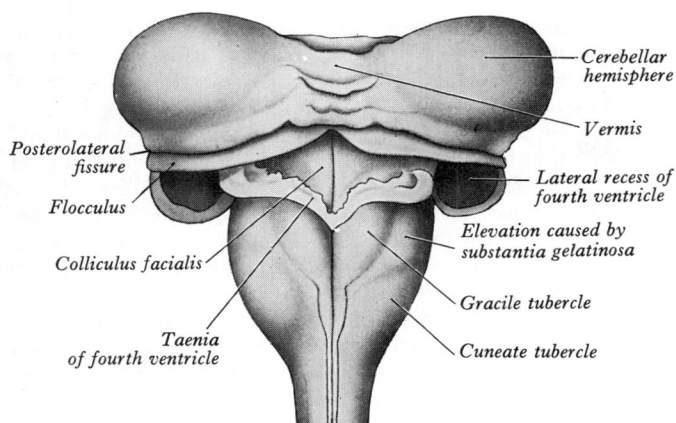

2.66A The cerebellum of a fetus in the fifth month. (After Kollmann.)

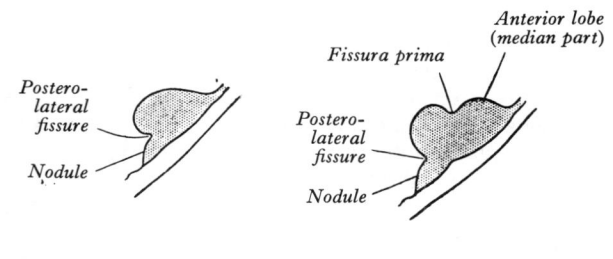

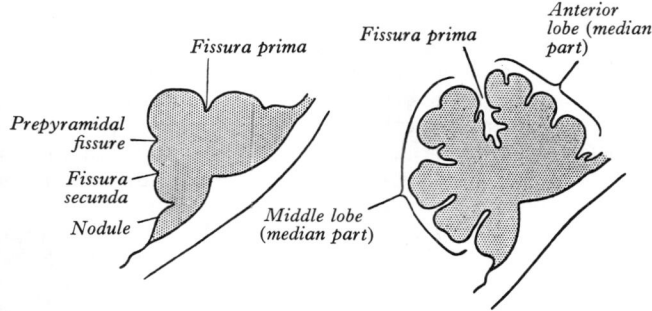

2.66B Median sagittal sections through the developing cerebellum, showing four different stages.

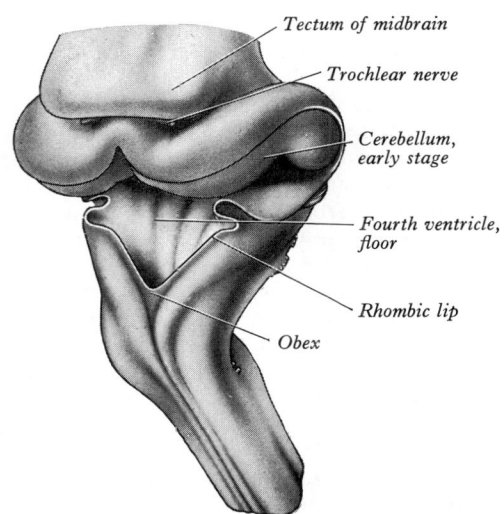

2.66C The dorsal aspect of the hindbrain of a human fetus about 3 months old. Viewed from behind and partly from the right side. (From a model by His.)

THE CEREBELLUM

While these changes are occurring the cells in the *superior rhombic lip* and adjacent dorsal part of the dorsolateral lamina of the metencephalon proliferate to form the *rudiment of the cerebellum*. Two rounded swellings are formed which at first project partly into the ventricle (2.66B,C), and they form the rudimentary cerebellar hemispheres. The most cranial part of the roof of the metencephalon originally separates the two swellings, but it becomes invaded by cells, which form the rudiment of the vermis. These cells were regarded as derivatives of both ventrolateral and dorsolateral laminae (Baxter 1953). At a later stage, *extroversion of the cerebellum* occurs, with reduction of its *intraventricular* projection and an increasing dorsal *extraventricular prominence*. The cerebellum now consists of a bilobar (dumb-bell shaped) swelling stretched across the rostral part of the fourth ventricle (2.66A), continuous rostrally with the anterior medullary velum, formed from the isthmus, and caudally with the epithelial roof of the myelencephalon. With growth a number of transverse grooves appear on the dorsal aspects of the cerebellar rudiment, as the precursors of the numerous fissures which characterize the surface of the mature cerebellum (2.66B, also p. 962).

The *posterolateral fissure*, in its lateral parts, appears first, demarcating the then most caudal area from the rest of the cerebellar rudiment, enabling the *flocculi* to be identified. The lateral parts of this fissure extend medially, meet in the median plane and demarcate the nodule. The *flocculonodular lobes* can now be recognized and are the most caudal part of the cerebellum at this stage but, owing to the growth of the adjoining areas, they progressively come to occupy the *anterior part* of the *inferior surface* in the adult. They are formed in close proximity to the line of attachment of the epithelial roof, i.e. to the rhombic lip (p. 963 and 2.66C).

At the end of the third month a transverse sulcus appears on the rostral slope of the cerebellar rudiment and deepens to form the *fissura prima*, which cuts into the vermis and both hemispheres, separating the most cranial region to form the anterior lobe.

About the same period two short transverse grooves appear on the inferior vermis; the first is the *fissura secunda*, which demarcates the uvula, and the second the *prepyramidal fissure*, demarcating the pyramid (2.66B). The whole cerebellum now grows dorsally and the caudoinferior aspects of the *hemispheres* expand much more than the inferior vermis, which therefore becomes buried at the bottom of a deep hollow—the *vallecula*. Meanwhile numerous additional fissures develop, which are approximately parallel to, and intervene between, the foregoing. They result in the relatively vast increase in surface area of the cerebellum which is occurring but their precise positions and systematic names have limited functional or morphological significance. The most extensive of these develops into the *horizontal fissure*.

In many mammals a part of the hemisphere immediately rostral to the posterolateral fissure becomes defined as an entity; in some it forms a very prominent part of the cerebellum, termed the *paraflocculus*, but the relationship is purely topographical and, in contrast to the flocculus, the paraflocculus receives afferent connections mainly, but not entirely, from the cerebral cortex. It is uncertain whether any homologue of the paraflocculus exists in the human cerebellum or whether, as has been proposed, it is represented by some small patches of grey matter which are frequently present on the inferior surface of the middle cerebellar peduncle (Larsell 1947).

The emergence of arrays of principal fissures ('fissuration') and folia ('foliation') has been studied experimentally, and with increasing refinement, morphometrically, by a number of workers. Pertinent results and bibliographies are given by Sievers et al (1981) and Allen et al (1981). It is apparent that *interactions* between developmental facets of many tissue components have an important bearing. These include changing spatio-temporal patterns of growth, migration and differentiation of the fibres constituting the central white core (p. 964), the Bergmann glioblasts, the neuroblasts of the external germinal layer, the pia/glial basal lamina and the pial fibroblasts. Selective experimental alteration of any of these elements alters or obliterates the folial pattern. (For these tissue components vide infra and pp. 968–972.)

Mammalian cerebellar histogenesis has been more completely described, both during normal ontogeny and after experimental intervention, than in any other part of the nervous system, due in large measure to its precisely ordered geometry and highly distinctive cell types. The connectivity, synaptology and electrophysiology of the latter have also been intensively studied, and some knowledge of these aspects are a necessary prerequisite of any account of histogenesis. (For this consult pp. 968–972.) Particularly valuable in histogenetic investigations have been nuclear labelling with tritiated thymidine, the analysis of genetic variants and the changes following surgical deafferentation, exposure to biochemically active agents, X-irradiation and virus diseases. Most studies have been confined to mice and rats, and whilst it seems probable that *qualitatively* similar cell migrations and contacts occur, *quantitative* findings and ontogenetic *timings* cannot, of course, be extrapolated to the human cerebellum. Only the briefest introduction can be entertained in this volume, and the interested reader should enter the literature by consulting such key references as Miale & Sidman (1961), Fujita (1963), Fujita et al (1966), Kornguth et al (1966, 1968), Mugnaini (1970), Eccles (1970), Hámori (1972), Altman (1972a,b,c), Eccles (1973), Swarz (1976), Swarz & del Cerro (1975, 1977), Berry (1982); Berry et al (1980a,b,c,d).

The early cerebellar rudiment consists of a pseudostratified epithelium showing interkinetic migration (p. 178) which soon develops the three basic zones—ventricular, intermediate, and marginal—as elsewhere in the neural tube. Originating in the ventricular zone, *proliferation* of as yet *uncommitted germinal* (matrix) *cells* proceeds, however, in *two* quite distinct sites, which become increasingly separated as the rudiment expands and thickens. Germinal cells continue proliferating in the deep (subventricular) part of the intermediate zone forming what, for convenience, may be termed the *internal germinal layer*. Slightly later, spreading lateromedially from each side to meet centrally in a subpial position (i.e. the most superficial part of the marginal zone), similarly uncommitted germinal cells proliferate forming an *external germinal layer* (sometimes, less appropriately, called the external granular layer). Both germinal layers continue *proliferative mitoses*, giving clonal progeny, until they are several cells thick. This *proliferative phase* gradually diminishes with the onset of the *phase of neurogenesis*, and this, in its turn, overlaps and *merges* into a *phase of gliogenesis*. Following a precisely ordered time sequence, sets of *critical (quantal) mitoses* occur (p. 108), the progeny becoming *committed neuroblasts* which do not divide again. Each set is destined to become one of the specific neuronal varieties that characterize the deep cerebellar nuclei and cortex, and they *migrate* to their definitive positions, develop their neurites and synaptic contacts, and mature, again in finely patterned spatio-temporal *sequences*, the latter often exhibiting some degree of overlap (Altman & Bayer 1978, Berry et al 1980d). Similarly, as neurogenesis wanes, sets of quantal mitoses of some remaining germinal cells continue to give generations of *committed glioblasts* which, although less well documented than the neuroblasts, also migrate to and mature in their definitive positions. (The specialized, radially disposed *Bergmann glioblasts*, however, are amongst the *first* cells to be committed in the early neuroepithelium and may provide contact guidance morphogenetic paths for some varieties of neuroblast.) The fate of each germinal layer will now briefly be reviewed.

Quantal mitoses of the uncommitted cells of the *internal germinal layer* first result in the emergence of two types, followed somewhat later by a third type, of definitive neuroblasts. Initially, *primitive Purkinje neuroblasts* and *primitive nuclear neuroblasts* emerge in approximately equal numbers, but whether this follows a series of symmetrical or asymmetrical mitoses (p. 108) is unknown. The *nuclear neuroblasts* remain embedded in the developing future white matter adjacent to the roof of the rostral part of the fourth ventricle. The main mass of nuclear neuroblasts then slowly subdivides into the primordial fastigial, emboliform, globose and dentate deep cerebellar nuclei, and the individual neuroblasts differentiate into either small intranuclear interneurons or the larger projection neurons. The axons of the latter invade the early cerebellar peduncles and pursue complex paths to their multiple destinations. The *Purkinje neuroblasts*, in contrast, *migrate* superficially towards their definitive position in the expanding cortex, where they slowly mature into their highly characteristic form of somata and dendritic trees. As they migrate,

the terminals of one neurite—the future axon—remain adjacent to, and ultimately in synaptic contact with, the nuclear neuroblasts, the remainder of the axon elongating as it 'trails' behind the advancing soma. The mature and developing Purkinje cell has been a favourite object for *quantitative* cytological and ultrastructural studies, both during normal ontogeny and after such experimental manipulations as suppression of granule cell (and therefore parallel fibre) development, or after prevention of climbing fibre growth (see e.g. Hámori 1972 and p. 971). Excellent illustrative examples of the maturation of normal rat cerebellar cortex are to be found in the writings of Altman (1972b), and after experimental manipulation, (Berry et al 1978, 1980a,b,c,d, Rakic 1981, 1982). When the formation of nuclear and Purkinje neuroblasts has proceeded for some time, a number of the remaining internal germinal cells give rise to generations of *Golgi neuroblasts* which also migrate superficially to gradually occupy, and mature in, their definitive position and morphology.

The first set of quantal mitoses in the *external germinal layer* gives generations of *basket neuroblasts* which migrate deeply to meet, and ultimately synapse with, the somata and preaxons of the ascending Purkinje neuroblasts, their axons passing transversely in the primordial cerebellar folia. Secondly, following intense and more prolonged proliferation of the external germinal cells, a second set of quantal mitoses results in vast numbers of *granule cell neuroblasts*. A widely held view is that these microneurons are at first bipolar, each presumptive axon growing in opposite directions along the long axis of the folium: bundles of such axons form the parallel fibres of the mature cortex (p. 971), and the more recently formed granule cell neuroblasts (with their parallel fibre axons) are placed in successive layers approaching the pial surface. The soma of the granule cell does not, however, remain bipolar and external to the Purkinje neuroblast layer. A neurite develops and grows deeply *through* the Purkinje layer, its tip ultimately dividing, meeting, and synapsing with the ingrowing mossy afferent terminals (p. 971), thus forming the primordial cerebellar glomeruli. As this centripetal neurite develops, the granule cell nucleus migrates along it to reach, and helps to form, the definitive granular layer of somata of the mature cortex. Rakic (1971a,b), however, reached a different conclusion from studies of migrating granule cell neuroblasts in the Rhesus monkey, holding that they migrated deeply by contact guidance along the surface of the long radial processes of the Bergmann glial cells (Mugnaini & Forstrønen 1967). He considered that the migrating soma 'trailed' behind it an elongating neurite which subsequently bifurcated to form the parallel fibre axons, whilst dividing neurites passing in advance of the soma met the mossy afferents to form cerebellar glomeruli.

The final generations of neuroblasts from the external germinal layer, relatively sparse and brief, occur whilst granule cell production is at its height; their progeny merely migrate locally and differentiate into outer stellate cells.

The origin of the cerebellar glial cells remains much more problematical and has been the subject of dispute for over a century. The view advanced by Obersteiner (1883) and Schaper (1897) was that the various macroglial cell varieties (Bergmann cells, astrocytes and oligodendrocytes) stemmed from final generations of germinal cells of *both* the internal and external germinal layers; this has received the more recent experimental support of such authorities as Fujita et al (1966), Fujita (1967), Meller et al (1969), Privat (1975). The second view proposed by Athias (1897), Ramón y Cajal (1911) held that the macroglia were formed exclusively from the internal germinal layer, and that the progeny of the external layer were solely the three varieties of neuroblast: this suggestion has been supported by the labelling studies of Swarz & del Cerro (1977). Clearly such diametrically opposed views must await further critical evaluation using alternative methods, sites and species. Further neither view embraces Rakic's analysis of the early emergence and possible roles of the pioneer mosaic of Bergmann glioblasts. All authorities seem agreed that the microglial elements are exogenous, invading the cerebellar rudiment from the surrounding mesenchyme.

The remainder of the metencephalon becomes the pons, but

little is known of the individual stages in the transformation. Ventricular, intermediate and marginal zones are formed in the usual way, and the nuclei of the trigeminal, abducent and facial nerves develop in the mantle layer. It is possible that the grey matter of the formatio reticularis is derived from the ventrolateral lamina and that of the nuclei pontis from the dorsolateral lamina by the active migration of cells from the rhombic lip. However, about the fourth month the pons is invaded by corticopontine, corticobulbar and corticospinal fibres, becomes proportionately thicker, and takes on its adult appearance.

The region of the *isthmus rhombencephali* undergoes a series of changes notoriously difficult to interpret. As a result, however, the greater part of the region apparently becomes absorbed into the caudal end of the midbrain, only the roof plate (dorsal lamina), in which the anterior medullary velum is formed, and the dorsal parts of the dorsolateral laminae, which become invaded by converging fibres of the superior cerebellar peduncles, remaining as recognizable derivatives in the adult. Note that originally the decussation of the trochlear nerves was *caudal* to the isthmus, but as the growth changes occur it is displaced in a cranial direction until it reaches its adult position. These changes are also responsible (1) for the rostral movement of the trochlear nucleus, whereby it comes to lie in the midbrain, and (2) for the position of the mesencephalic nucleus of the trigeminal nerve, also a derivative of the isthmus (Frazer 1928).

THE MESENCEPHALON

The mesencephalon or midbrain, derived from the intermediate primary cerebral vesicle, persists for a time as a thin-walled tube enclosing a cavity of some size, separated from that of the prosencephalon by a slight constriction and from the rhombencephalon by the isthmus rhombencephali. Later, its cavity becomes relatively reduced in diameter, and in the adult brain it forms the *cerebral aqueduct*. The ventrolateral laminae of the midbrain increase in thickness to form the *cerebral peduncles*, which are at first of small size, but enlarge rapidly after the fourth month, when their numerous fibre tracts begin to appear in the marginal zone (p. 917 et seq). The neuroblasts of the ventrolateral lamina give origin to the nuclei of the oculomotor nerve and some grey masses of the tegmentum, while the nucleus of the trochlear nerve, and also the mesencephalic nucleus of the trigeminal nerve, migrate rostrally into the midbrain owing to the developmental changes in the isthmus rhombencephali. It has been claimed (Chu-Wu & Wen-Kuei 1965) that some of the migrating neuroblasts of the mesencephalic nucleus reach the posterior commissure and its nucleus and also the interstitial nucleus. The cells of the dorsal part of the dorsolateral laminae proliferate and invade the roof plate, which therefore thickens and is later divided into corpora bigemina by a median groove. Caudally this groove becomes a median ridge, which persists in the adult as the frenulum veli. The corpora bigemina are later subdivided into the *superior* and *inferior colliculi* by a transverse furrow. The *red nucleus, substantia nigra* and *reticular nuclei* of the midbrain *tegmentum* may first be defined at the end of the third month. Their origins are probably mixed from neuroblasts of the ventrolateral and dorsolateral laminae.

The detailed histogenesis of the tectum, and its main derivatives, the colliculi, will not be followed here, but in general the principles outlined for the cerebellar cortex (p. 188) and the palaeopallium and neopallium (p. 195) also apply to this region. It may be noted, however, that there exists a high degree of geometric order in the developing retinotectal projection (p. 986), and also a precise somatotopy in tectospinal projection. These facts, coupled with the ability of the piscine and amphibian central nervous tracts to *regenerate* after severance, have led the retinotectal pathways to become classical sites for experimentation. A review of this field is beyond the scope of the present volume, but typical experiments included the electrophysiological and neuroanatomical mapping of the patterns of retinotectal connections that were established after, e.g. mere cutting of an optic nerve or, following nerve severance, the rotation of one or both eyes through 180°. Other experiments included bilateral optic nerve section together with median section of the optic chiasma; additionally the regeneration pattern was followed after deletion of different retinal sectors, or following various recombinations of half eye-cups. In general the results of such experiments led to the elaboration of a powerful theory of the *chemospecificity of neuroblasts*, the main proposition being that every neuroblast possessed a unique surface molecular configuration. It was further postulated that the growth pattern of a neurite followed a chemically coded path which led to a precise group of target cells bearing complementary surface configurations (see Sperry 1951a,b, 1958, 1963, 1965, 1971). While the *overall* results have received much subsequent experimental support, more detailed analyses of the re-established connectivity patterns, and in particular the *changing* patterns that occur with tectal growth (e.g. Chung et al 1972), and the modulation that occurs with altered environments, behaviour, experience and the phenomena of learning and memory, has cast doubt on the validity of a *rigid* theory of chemospecificity applying to *all* neuroblasts and neurons (e.g. see Szekely 1974, Hunt & Jacobson 1974, Mark 1974).

It may be pertinent to mention the interesting suggestion put forward by Jacobson (1969, 1970, 1974, 1978) that there are two principal classes of neuron (an elegant modern refinement of the old Golgi type I and type II varieties—see p. 976). Jacobson's *class I neurons* are phylogenetically ancient, develop early in ontogeny, possess prominent somata, long axons, specific target cells, and have invariant and unmodifiable connections and functions from their first establishment. It is presumed that their final form and activities are almost wholly encoded in the genome. In contrast, *class II neurons* are small, possessing a short (or no) axon (microneurons); they are phylogenetically recent and develop relatively late in ontogeny. They are considered to possess, initially, *multiple* functional potentialities that may persist for considerable intervals, and only become progressively restricted during subsequent fetal or post-natal growth, the microneuron *ultimately* becoming 'unipotential and unmodifiable'. It is also proposed that this restriction of functional potential occurs at varying times and rates in different locations in the developing nervous system. Such microneurons are, on this view, therefore regarded as one of the main structural bases of *plasticity* in nervous systems whereby structural/functional modifications result from environmental changes and the assessment of results of exploratory behaviour; they may thus be one of the sites where the elusive structural correlates of memory patterns should be sought (see essays by Young 1964, Gray 1974, Jacobson 1974).

THE PROSENCEPHALON

At an early stage, a transverse section through the forebrain shows the same parts as are displayed in similar sections of the spinal cord and medulla oblongata—thick lateral walls connected by thin floor and roof plates. Moreover, each lateral wall is divided into a dorsal area and a ventral area separated internally by the *hypothalamic sulcus*. This sulcus ends anteriorly at the medial end of the optic stalk (vide infra and p. 202); in the fully developed brain, it persists as a slight groove extending from the interventricular foramen to the cerebral aqueduct. It is analogous to, if not the homologue of, the sulcus limitans. The thin roof plate remains epithelial, but invaginated by vascular mesenchyme, the tela choroidea of the choroid plexuses of the third ventricle. Later, the lateral margins of the tela undergo a similar invagination into the medial walls of the cerebral hemispheres (vide infra). The floor plate thickens, developing the nuclear masses of the hypothalamus and subthalamus.

At a very early period, before the closure of the rostral neuropore (p. 133), two lateral diverticula, the *optic vesicles*, appear, one on each side, about the level of the forebrain; for a time they communicate with its cavity by relatively wide openings. The distal parts of the optic vesicles expand, while the proximal parts become the tubular *optic stalks*. (Their further development is given on pp. 201–204.) The forebrain next grows, its tip curving ventrally, and two further diverticula rapidly expand from it, one on each side. These diverticula are rostrolateral to the optic stalks and subsequently form the *cerebral hemispheres*; their cavities are the rudiments of the lateral ventricles; they

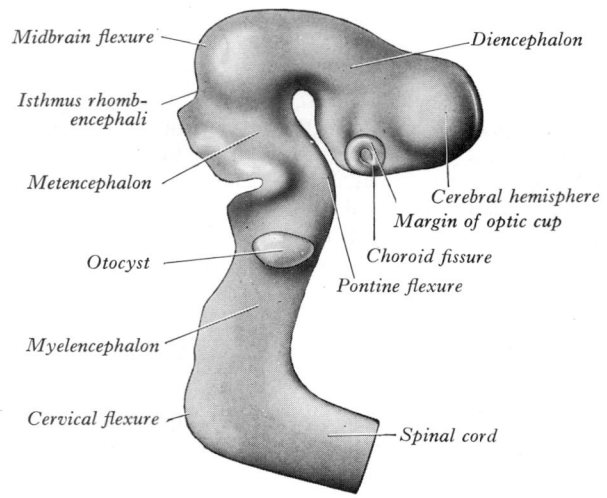

2.67A The right side of the brain of a human embryo, 9 mm long. (Drawn from a model by His.)

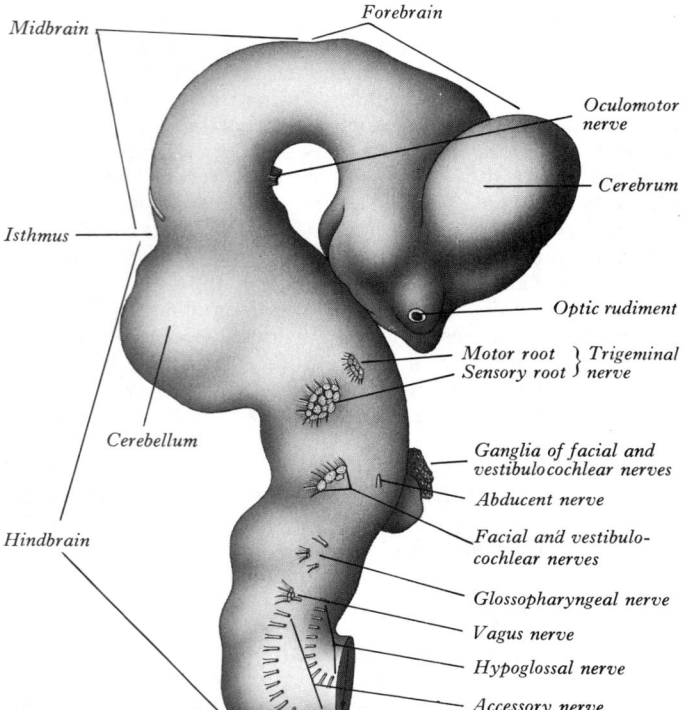

2.67B The brain of a human embryo about 10.2 mm long. Right lateral surface. (From a model by His.)

communicate with the median part of the forebrain cavity by relatively wide openings which ultimately become the inter-ventricular foramina. The anterior limit of the median part of the forebrain consists of a thin sheet, the *lamina terminalis* (2.68A, B, C), which stretches from the interventricular foramina to the recess at the base of the optic stalks. The anterior part of the forebrain, including the rudiments of the cerebral hemispheres, is the *telencephalon (end-brain)*, and the posterior part the *diencephalon (between-brain)*; both contribute to the formation of the third ventricle, although the latter predominates. The fate of the lamina terminalis is detailed below.

The diencephalon is broadly divided by the hypothalamic sulcus into the *pars dorsalis diencephali* and *pars ventralis diencephali*; these, however, are composite, each contributing to diverse neural structures. The *pars dorsalis* develops into the (dorsal) thalamus and metathalamus along the immediate suprasulcal area of its lateral wall, whilst the highest dorsocaudal lateral wall and roof form the epithalamus. The *thalamus* (2.68A–C) begins as a thickening which involves the anterior part of the

dorsal area (Cooper 1950). Caudal to the thalamus the lateral and medial geniculate bodies, or *metathalamus*, are recognizable at first as surface depressions on the internal aspect and as elevations on the external aspect of the lateral wall (Cooper 1945). As the thalami enlarge as smooth ovoid masses, they gradually narrow the wide interval between them into a vertically compressed cavity which forms the greater part of the third ventricle. After a time these medial surfaces may come into contact and become adherent over a variable area, the connection (single or multiple) constitut-ing the *interthalamic adhesion*. The caudal growth of the thalamus excludes the geniculate bodies from the lateral wall of the third ventricle.

At first the lateral aspect of the developing thalamus is separated from the medial aspect of the cerebral hemisphere by a cleft, but with growth the cleft becomes obliterated (2.69) as the thalamus fuses with the part of the hemisphere in which the cor-pus striatum is developing. Later, with the development of the projection fibres (corticofugal and corticopetal) of the neocortex (p. 194), the thalamus becomes related to the internal capsule, which intervenes between it and the lateral part of the corpus striatum (lentiform nucleus). Ventral to the hypothalamic sulcus the lateral wall of the diencephalon, in addition to median derivatives of its floor plate, forms a large part of the hypothalamus and subthalamus.

The *epithalamus*, which includes the pineal gland, the posterior and habenular commissures and the trigonum habenulae, develops in association with the caudal part of the roof plate and the adjoining regions of the lateral walls of the diencephalon. At an early period (12–20 mm CR length) the epithalamus in the lateral wall projects into the third ventricle as a smooth ellipsoid mass, larger than the adjacent mass of the (dorsal) thalamus and separated from it by a well defined *epithalamic sulcus*. In subsequent months growth of the thalamus rapidly overtakes that of the epithalamus; the intervening sulcus is obliterated. Thus, finally, structures of epithalamic origin are topographically relatively diminutive; in recent years, however, there has occurred a burgeoning of interest in, and understanding of their functional roles (p. 1457). The *pineal gland* arises as a hollow outgrowth from the roof plate, immediately adjoining the mesencephalon. Its distal part becomes solid by cellular prolifera-tion, but its proximal stalk remains hollow, containing the pineal recess of the third ventricle. In many reptiles the pineal out-growth is double. The anterior outgrowth (*parapineal organ*) develops into the pineal or parietal eye (p. 1456) while the posterior outgrowth is glandular in character. It is the *posterior*

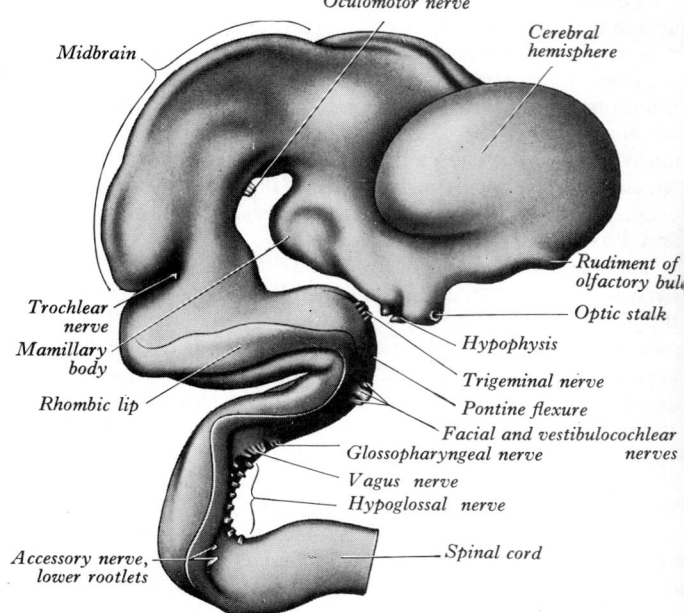

2.67C The right side of the brain of a human embryo, 13.6 mm long. The roof of the hindbrain has been removed. Compare with 2.68B. (From a model by His.)

outgrowth which is homologous with the pineal gland in man. The anterior outgrowth also develops in the human embryo but soon disappears entirely.

The *posterior commissure* is formed by fibres which invade the caudal wall of the pineal recess from both sides.

The *nucleus habenulae*, which is the most important constituent of the *trigonum habenulae*, is developed in the lateral wall of the diencephalon and is at first in close relationship with the geniculate bodies, from which it becomes separated by the dorsal growth of the thalamus. The *habenular commissure* develops in the cranial wall of the pineal recess.

It is the roof plate of the diencephalon, rostral to the pineal gland (and continuing over the median telencephalon), that remains thin and epithelial in character and is subsequently invaginated by the choroid plexuses of the third ventricle. Before the development of the corpus callosum and the fornix it lies at the bottom of the longitudinal fissure, between and reaching the two cerebral hemispheres, extending as far rostrally as the interventricular foramina and lamina terminalis. Here, and elsewhere, choroid plexuses develop by the close apposition of vascular pia

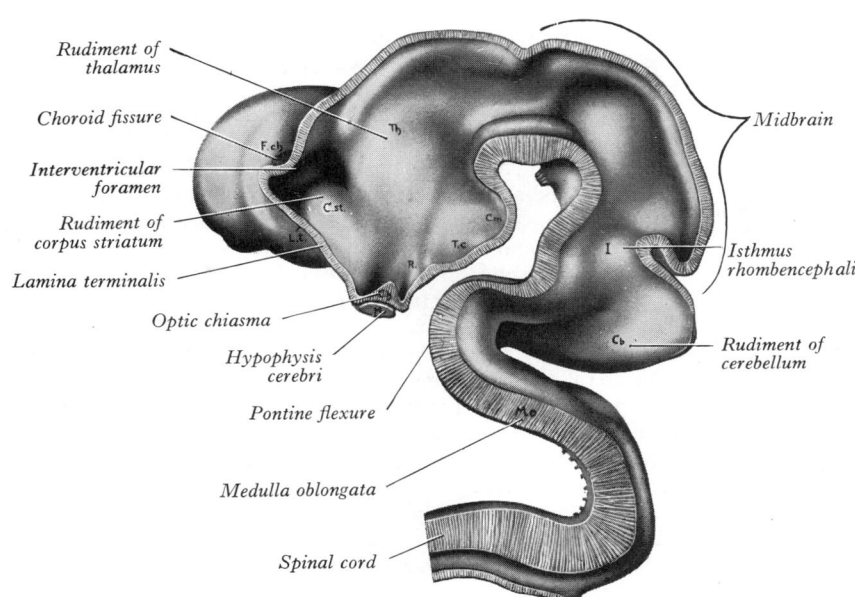

2.68B The brain of a human embryo, 13.6 mm long. Medial surface of right half. The roof of the hindbrain has been removed. (From a model by His.)

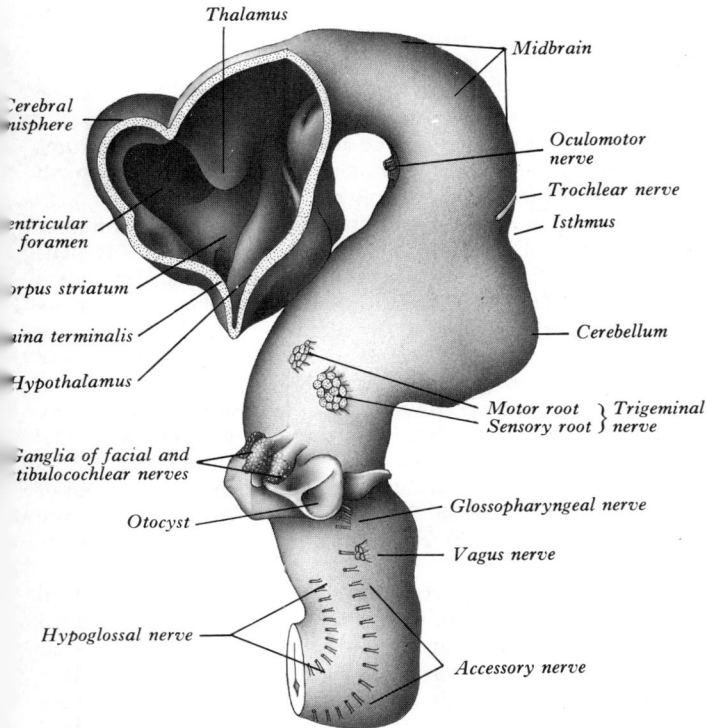

2.68A The brain of a human embryo, about 10.2 mm long. (From a model by His.)

mater and ependyma with no nervous tissue intervening. With development, the vascular layer is infolded into the ventricular cavity and develops a series of small villous (finger-like) projections, each covered by a cuboidal epithelium derived from the ependyma. The cuboidal cells carry numerous microvilli on their ventricular surfaces whilst basally their plasma membrane becomes complexly folded into the cell. The *early* choroid plexuses secrete a protein-rich cerebrospinal fluid into the ventricular system which may provide a nutritive medium for the primitive epithelial neural tissues. With increasing vascularity of the latter, however, the histochemical reactions of the cuboidal cells and the character of the fluid change to the adult type (Klosovskii 1963). It should also be noted that, in addition to choroid plexus formation, the remaining lining of the third ventricle does *not* simply form generalized ependymal cells. Many regions become highly specialized, developing concentrations of tanycytes or other modified cells, e.g. those of the *subfornical organ*, the *organum vasculosum (intercolumnar tubercle)* of the lamina terminalis, the *subcommissural organ* and those lining the *pineal*, *suprapineal*, and *infundibular recesses* (see Knigge et al 1975), collectively termed the *circumventricular organs*.

In addition to its subsulcal lateral walls, the floor of the pars ventralis diencephali takes part in the formation of the *hypothalamus*, including the mamillary bodies, the tuber cinereum and infundibulum of the hypophysis.

The *mamillary bodies* arise as a single thickening, which becomes divided by a median furrow during the third month. Anterior to them the *tuber cinereum* develops as a cellular proliferation which extends forwards as far as the infundibulum. In front of the tuber cinereum the floor of the diencephalon gives origin to a wide-mouthed diverticulum, which grows towards the stomatodeal roof and comes into contact with the posterior aspect of a dorsally directed ingrowth from the stomatodeum (Rathke's pouch, p. 229). These two diverticula together form the *hypophysis cerebri* (2.103). In the base of the neural outgrowth an extension of the third ventricle persists as the *infundibular recess*. The remaining caudolateral walls and floor of the ventral diencephalon are an extension of the midbrain tegmentum—the *subthalamus*. This forms the rostral limits of the red nucleus and

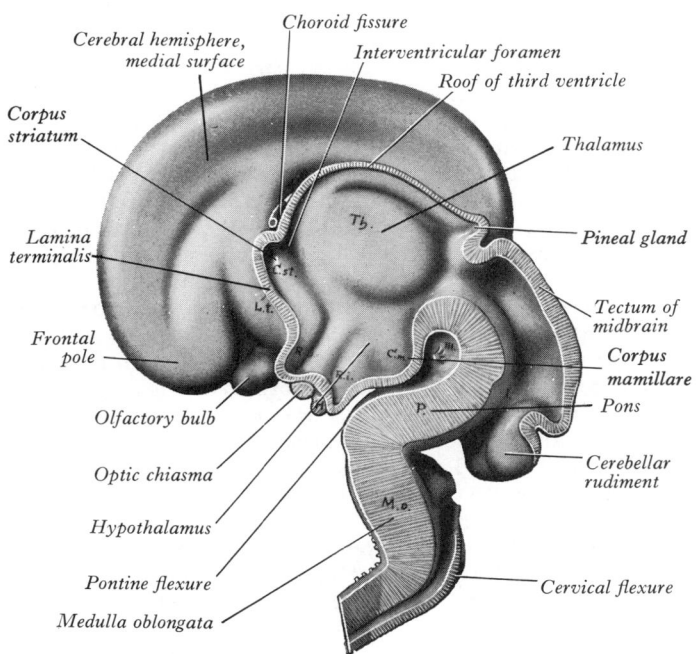

2.68C Medial surface of the right half of the brain of a human fetus, about 3 months old.

substantia nigra, numerous reticular nuclei and a wealth of inter-weaving nerve fibre bundles, ascending, descending and oblique, with many origins and destinations.

The optic vesicles, which are described with the development of the eye (p. 201), are derived from the lateral wall of the prosencephalon before the telencephalon can be identified. They are usually regarded as derivatives of the diencephalon and the optic chiasma is often regarded as the boundary between diencephalon and telencephalon.

The telencephalon (end-brain) consists of two lateral diverticula connected by a median region (the *telencephalon impar*). From the impar develops the anterior part of the cavity of the third ventricle, closed below and in front by the *lamina terminalis*. The lateral diverticula are outpouchings of the lateral walls of the telencephalon, which may correspond to the alar laminae, although this is uncertain; the cavities are the future lateral ventricles, and their walls the primordial nervous tissue of the *cerebral hemispheres*. The roof plate of the median part of the telencephalon remains thin and is, as noted, continuous behind with the roof plate of the diencephalon. In the floor plate and lateral walls of the prosencephalon, *ventral* to the primitive inter-ventricular foramina, the *anterior* parts of the *hypothalamus* are developed; these include the optic chiasma, optic recess and related nuclei. The chiasma is formed by the meeting and partial decussation of the optic nerves in the ventral part of the lamina terminalis, and from it the optic tracts subsequently grow back-wards to end in the diencephalon and midbrain.

The cerebral hemispheres arise as diverticula of the lateral walls of the telencephalon, with which they remain in continuity around the margins of the initially relatively large interventricular foramina, except caudally, where they are continuous with the anterior part of the lateral wall of the diencephalon (**2.68**A, B); as growth proceeds the hemisphere enlarges forwards, upwards and backwards and acquires an oval outline, with medial and superolateral walls and a floor. As a result the medial surfaces approach, but are separated from each other by a vascularized mesenchyme and pia mater that fills the median *longitudinal fissure*. At this stage the floor of the fissure is the epithelial roof plate of the telencephalon, which is directly continuous caudally with the epithelial roof plate of the diencephalon (**2.68**C), as already stated above.

At the early oval stage of hemispheric development, regional names are given in accordance with their future principal derivatives. The rostromedial and ventral *floor* becomes linked with the forming olfactory apparatus and may be termed the primitive *olfactory lobe* (*see* below and pp. 1029, 1033, 1175). The floor (ventral wall, or base) of the larger remainder of the hemisphere forms the anlage of the primitive corpus striatum and amygdaloid complex: hence this, including its associated rim of lateral and medial walls is the *striate part of the hemisphere*. The rest of the hemisphere, the largest in surface area but initially possessing rather thin walls, medial, lateral, dorsal and caudal, is thus the *suprastriate part of the hemisphere*. The whole of the latter (except the interventricular foramen and its extension, the choroidal fissure) together with the superficial (subpial) zone of the striate part are the areas where histogenesis of named apparent variants of cerebral cortex (or pallium) occur. Further details, and comments on their plethora of terminologies and (often unsatis-factory) classifications, are furnished below and on p. 1021. (See comments on neocortical *uniformity* p. 1039).

The rostral end of the oval hemisphere becomes the definitive *frontal pole* but, as the hemisphere expands, its *original* posterior pole moves relatively in a caudoventral and lateral direction, curving thence towards the orbit in association with the growth of the caudate nucleus and numerous other structures to form the definitive *temporal pole*, and a *new* posterior part becomes evident which persists as the definitive *occipital pole* of the mature brain. The great expansion of the cerebral hemispheres is characteristic of mammals and especially of man, and in their subsequent growth they overlap, successively, the diencephalon and the mesencephalon and then meet the rostral surface the cerebellum; the temporal lobes embrace the flanks of the brainstem.

The early diverticulum or anlage of the cerebral hemisphere contains initially a simple spheroidal *lateral ventricle* which is

continuous with the third ventricle via the interventricular foramen; the rim of the latter being, of course, the site of the original evagination. With expansion and the assumption of an oval outline by the hemisphere, the ventricle becomes firstly roughly ellipsoid and then a curved cylinder, convex dorsally. The ends of the cylinder expand towards (but do not reach) the frontal and (temporary) occipital poles—differentiating and thickening neural tissues separate the ventricular cavities and pial surfaces at all points, except along the line of the choroidal fissure (vide infra). Pronounced changes in ventricular form accompany the emergence of a temporal pole; the original caudal end of the curved cylinder expands within its substance. This temporal extension passes ventrolaterally to encircle its side of the upper brainstem (cf. choroidal fissure below). Finally, from the root of the temporal extension another may develop in the substance of the definitive occipital pole, passing caudomedially; this is quite variable in size, often asymmetrical on the two sides; one or both may be absent. Although a continuous system of cavities, specific parts of the lateral ventricle are now given regional names: the *central part (body)* extends from the interventricular foramen to the level of the posterior edge (splenium) of the corpus callosum. From the body three *cornua (horns)* diverge—*anterior* towards the frontal pole, *posterior* towards the occipital pole and *inferior* towards the temporal pole.

It may be noted that at these early stages of hemispheric development the term pole is preferred, in most instances, to lobe; the latter are defined by specific surface topographical features which will appear over several months, and differential growth patterns persist for a considerable period.

About the fifth week a longitudinal groove appears in the antero-medial part of the floor of each ventricle. This groove deepens and forms a hollow diverticulum continuous with the hemisphere by a short stalk. The diverticulum becomes connected on its ventral or inferior surface to the olfactory placode, the cells of which give rise to the afferent axons of its sensory cells. These terminate in the walls of the diverticulum. As the head increases in size the diverticulum grows forwards and, subsequently losing its cavity, becomes converted into the solid *olfactory bulb*. The forward growth of the bulb is accompanied by elongation of its stalk, which forms the *olfactory tract*, and the part of the floor of the hemisphere to which the tract is attached constitutes the *piriform area* (vide infra). For comments on the accessory olfactory bulb see p. 1033.

The pia mater which covers the epithelial roof of the third ventricle at this stage is itself covered with loosely arranged mesenchyme. In the meshes of this tissue numerous blood vessels develop and, as we have seen, on each side of the median plane these vessels subsequently invaginate the roof of the *third ventricle* to form its *choroid plexuses*. The lower part of the medial wall of the cerebral hemisphere, which immediately adjoins the epithelial roof of the interventricular foramen and the anterior extremity of the diencephalon, also remains epithelial, consisting of ependyma and pia mater, while elsewhere the walls of the hemisphere are thickening to form the *pallium*. The thin part of the medial wall of the hemisphere is invaginated by vascular tissue, continuous in front with the choroid plexus of the third ventricle and constituting the *choroid plexus* of the *lateral ventricle*. This invagination occurs along a line which arches upwards and backwards, parallel with and initially limited to the anterior and upper boundaries of the interventricular foramen; the curved indentation of the ventricular wall, where no nervous tissue develops between ependyma and pia mater, is termed the *choroid fissure* (**2.68**C,69A,B). The subsequent assumption of the complex, but exquisite, definitive form of the choroidal fissure naturally depends on related growth patterns in neighbouring structures: some are particularly relevant. These are the relatively slow growth of the interventricular foramen, the secondary 'fusion' between the lateral diencephalon and medial hemisphere walls, the encompassing of the upper brain stem by the forward growth of the temporal lobe and its pole towards the apex of the orbit and the massive expansion of two great commissures of the cerebrum—the fornix and corpus callosum. (Many of these features are detailed further below and in the Neurology section.) Nevertheless, the choroidal fissure is now clearly a caudal extension of the (much reduced) interventricular foramen, which arches above the thalamus and in this region is only a few millimetres from the median plane. Near the caudal end of the thalamus it diverges ventrolaterally, its curve reaching and continuing in the medial wall of the temporal lobe over much of its length (i.e. to the tip of the inferior horn of the lateral ventricle). The upper part of the arch is overhung by the corpus callosum and, throughout its convexity is bordered by the fornix and its derivatives (vide infra and p. 1039). Thus, the extensive and helicoid disposition of the choroid plexus of the lateral ventricle is explained.

At first growth proceeds more actively in the floor and the adjoining part of the lateral wall of the developing hemisphere, and elevations formed by the rudimentary *corpus striatum* (2.68A) encroach on the cavity of the lateral ventricle (Cooper 1946). The head of the *caudate nucleus* appears as three successive parts, medial, lateral and intermediate, which produce elevations in the floor of the lateral ventricle. Caudally these merge to form the *tail* of the *caudate nucleus* and the *amygdaloid complex*. From the outset the latter are close to the temporal pole of the hemisphere and, when the occipital pole grows backwards and the general enlargement of the hemisphere carries the temporal pole downwards and forwards, the tail is continued from the *floor* of the central part (*body*) of the ventricle curving into the *roof* of its temporal extension, the future *inferior horn*, and the amygdaloid complex encapsulates its tip. Anteriorly the *head* of the *caudate nucleus* extends forwards to the floor of the interventricular foramen, where it is separated from the developing anterior end of the thalamus by a groove (2.68B); later, the head expands in the floor of the anterior horn of the lateral ventricle. The *lentiform nucleus* is developed from two laminae of cells, medial and lateral, which are continuous with both the medial and lateral parts of the caudate nucleus. The internal capsule appears first in the medial lamina and extends laterally through the outer lamina to the cortex. It divides the laminae into two, the internal parts joining the caudate nucleus and the external parts forming the lentiform nucleus. In the latter, which consists of two main parts, the remaining medial lamina cells give rise predominantly to the (medially placed) globus pallidus and the lateral to the (laterally placed) putamen. Subsequently the putamen expands concurrently with the intermediate part of the caudate nucleus (Hewitt 1958, 1961).

As the hemisphere enlarges, the caudal part of its medial surface overlaps and hides the lateral surface of the diencephalon (thalamic part), being separated from it by a narrow cleft occupied by vascular connective tissue. At this stage (about the end of the second month) a transverse section made caudal to the interventricular foramen passes from the third ventricular cavity successively through (1) the developing thalamus, (2) the narrow cleft just mentioned, (3) the thin medial wall of the hemisphere and (4) the cavity of the lateral ventricle, with the corpus striatum in its floor and lateral wall (2.69A). As the thalamus increases in extent it acquires a superior in addition to medial and lateral surfaces, and the lateral part of its superior surface fuses with the thin medial wall of the hemisphere so that, finally, this part of the thalamus is covered with the ependyma of the lateral ventricle immediately ventral to the choroid fissure (2.69B). As a result the corpus striatum is approximated to the thalamus and separated from it only by a deep groove which becomes obliterated by increased growth along the line of contact. The lateral aspect of the thalamus is now in continuity with the medial aspect of the corpus striatum so that a secondary union between the diencephalon and the telencephalon is effected over a wide area, providing a route for the subsequent passage of projection fibres to and from the cortex.

Throughout the brain stem, as in the spinal cord, the *migration* and differentiation of neuroblasts to form nuclei is either minimal or limited, their progeny remaining immediately extraependymal or, partially displaced towards the pial exterior, being arrested deeply embedded in the myelinated fibre 'white matter' of the region. As noted, however, the 'roofbrain' of part of the fore-, mid- and hindbrains develops following an additional, fundamental pattern which results in a superficial layer of *grey matter*. The latter consists of neuronal somata, dendrites, the terminations of incoming (afferent) axons, the stems of (or the whole of)

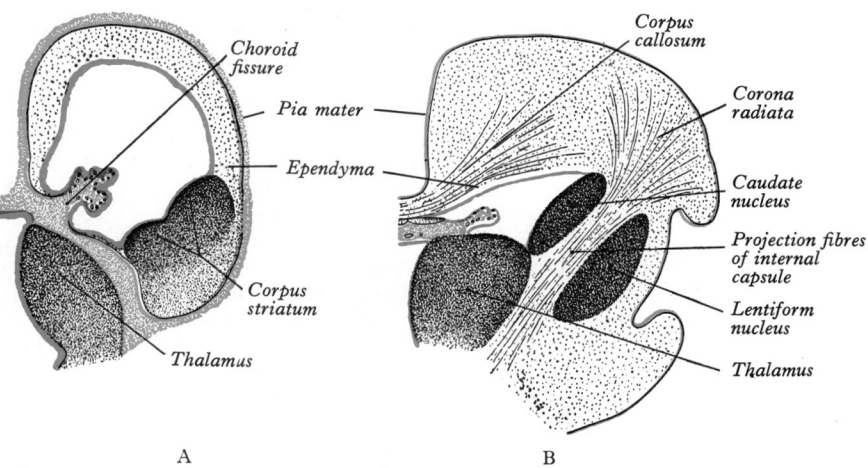

A   B

2.69 Diagrams illustrating transverse sections across the developing thalamus and cerebral hemisphere. Note that at the choroid fissure the vascular pia mater (blue) meets the ependyma (red) to form a choroid plexus. In A the lateral aspect of the thalamus is separated from the medial aspect of the hemisphere by an interval containing vascular mesenchyme. In B this interval has disappeared; the expanded upper surface of the thalamus is covered by the ependyma of the lateral ventricle, and the approximation of the thalamus and the corpus striatum has provided a pathway for the projection fibres of the internal capsule.

efferent axons, geometrically and functionally apposite glial cells and vasculature. Subsequent differentiation results in a highly organized subpial surface coat of grey matter termed the *cortex* (Latin–bark, e.g. of a tree) or *pallium* (Latin–pall, mantle or cloak). Pallium is used preferentially by neuroembryologists and some comparative zoologists; however, cortex is employed much more widely and (on occasion with its alternative in parenthesis) will be used here. It will be apparent that neither term is used (perhaps unfortunately) in the case of the mesencephalic tectum. In contrast cerebellar cortical transformations result in a virtually uniform neurohistogenesis, localized differences depending almost exclusively on its sources of 'input' and destinations of 'output' (pp. 967–968). In the cerebral hemisphere the superficial subpial regions of its wall, both striate and suprastriate (other than central areas of its medial wall, where secondary fusion with the diencephalon occurs and is encompassed by the lamina terminalis, interventricular foramen and curve of the choroid fissure), becomes invaded by migrating neuroblasts to form an elementary cerebral cortex. The cortical or pallial area which *borders* the lamina terminalis, the interventricular foramen, the convexity of the choroid fissure and continues into the diverging roots of the olfactory tract has been simply termed the *limbic lobe* (or bordering lobe), together with its numerous subdivisions and connections (p. 1028 et seq), the *limbic system*. Its cortical

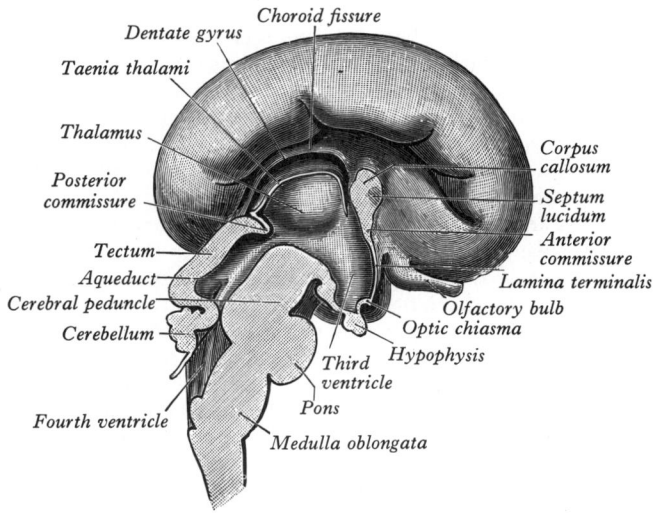

2.70 The brain of a human fetus, 4 months old. Medial aspect of left half.   193

derivatives possessing, some regard, a relatively elementary structure, have been grouped as the *allocortex* (other cortex).

It is the first part of the cortex to initiate differentiation (vide infra) and at first it forms a continuous, almost circular strip on the medial and inferior aspects of the hemisphere. Below and in front, where the stalk of the olfactory tract is attached, it constitutes a part of the *piriform area (palaeocortex* or *palaeopallium)*. The portion outside the curve of the choroid fissure (2.70) constitutes the *hippocampal formation (archaeocortex* or *archaeopallium)*. In this region the neuroblasts of the developing cortex proliferate and migrate (vide infra), and the wall of the hemisphere thickens and produces an elevation which projects into the medial side of the ventricle. This elevation is the *hippocampus* (Humphrey 1964, 1967). It appears first on the medial wall of the hemisphere in the area above and in front of the lamina terminalis (*paraterminal area*) and gradually extends backwards, curving into the region of the temporal pole where it adjoins the piriform area. The marginal zone in the neighbourhood of the hippocampus is invaded by neuroblasts forming the *dentate gyrus*. Both extend from the paraterminal area (see precommissural septum and prehippocampal rudiment, p. 1035) backwards above the choroid fissure and follow its curve downwards and forwards towards the temporal pole, where they continue into the piriform area. A shallow surface depression (which has been termed the *hippocampal sulcus*) grooves the medial surface of the hemisphere throughout the hippocampal formation.

The efferent fibres from the cells of the hippocampus collect along its medial edge and run forwards immediately above the choroid fissure. Anteriorly they turn ventrally and enter the lateral part of the lamina terminalis to gain the hypothalamus, where they end in and around the mamillary body and neighbouring nuclei. These efferent hippocampal fibres form the *fimbria hippocampi* and the *fornix*. For the sources of afferent fibres to the hippocampus, hippocampal commissures and multiple subdivisions of the fornix see pp. 1036–1039.

The terms archaeocortex and palaeocortex as the two principal divisions of the *allocortex* focus on the acceptance by earlier neuroanatomists concerning the phylogenetically ancient nature of these regions. The remainder of the hemispheric surface, particularly in mammalia, with a relatively vast expansion in primates, developed what was regarded as a six-layered and (in this respect) equal *isocortex*, the young cortex, *neocortex* or *neopallium*. Isocortex seemed less appropriate, as all subregions showed fine structural differences. Aspects of *numerical uniformity* have been demonstrated (p. 1043), however. Furthermore, all three varieties of cortex are present simultaneously in an initial form in extant reptilia. Their relationship to piscine or amphibian ancestry remains controversial.

The development of the commissures effects a very profound alteration on the medial wall of the hemisphere. At the time of their appearance the two hemispheres are connected to each other by the median part of the telencephalon. The roof plate of this area remains epithelial, whilst its floor becomes invaded by the decussating fibres of the optic nerves and developing hypothalamic nuclei. These two routes are thus not available for the passage of commissural fibres passing from hemisphere to hemisphere across the median plane, and these fibres therefore pass through the anterior wall of the interventricular foramen, i.e. the *lamina terminalis*. The first commissures to develop are those associated with the *palaeocortex* and *archaeocortex*. Fibres of the olfactory tracts cross in the ventral or lower part of the lamina terminalis and, together with fibres from the piriform and prepiriform areas and the amygdaloid bodies, form the *anterior part* of the *anterior commissure*. In addition the two hippocampi become interconnected by transverse fibres which cross from fornix to fornix in the upper part of the lamina terminalis as the *commissure of the fornix*. Various other decussating fibre bundles (known as the *supra-optic commissures*, although they are not true commissures) develop in the lamina terminalis immediately dorsal to the optic chiasma, between it and the anterior commissure.

The commissures of the neocortex develop later and follow the pathways already established by the commissures of the limbic system. Fibres from the tentorial surface of the hemisphere join the *anterior commissure* and constitute its larger *posterior part*. All the other commissural fibres of the neocortex associate themselves closely with the commissure of the fornix and lie on its dorsal surface. These fibres increase enormously in number and the bundle rapidly outgrows its neighbours to form the corpus callosum (2.70).

The *corpus callosum* commences in a thick mass connecting the two cerebral hemispheres around and above the anterior commissure. (This site was, for a period, called the precommissural area. However, this ontogenetic use has been rejected here because of increasing use of the adjective precommissural to denote the position of parts of the limbic lobe—prehippocampal rudiment, septal areas and nuclei, strands of the fornix in relation to the anterior commissure of the mature brain.) The upper end of this neocortical commissural area extends backwards to form the trunk of the corpus callosum. The *rostrum* of the corpus callosum develops later and separates some of the rostral end of the limbic area from the remainder of the cerebral hemisphere. Further backward growth of the trunk of the corpus callosum then results in the entrapped part of the limbic area becoming stretched out to form the bilateral septum pellucidum (Hewitt 1962). As the corpus callosum grows backwards it extends *above* the choroid fissure, carrying the commissure of the fornix on its under surface. In this way a new floor is formed for the longitudinal fissure, and additional structures come to lie above the epithelial roof of the third ventricle. In its backward growth the corpus callosum invades the area hitherto occupied by the *upper* part of the archaeocortical hippocampal formation, and the corresponding parts of the dentate gyrus (2.70) and hippocampus are reduced to vestiges—the *indusium griseum* and the *longitudinal striae*. However, the *posteroinferior* (temporal) archaeocortical regions of both dentate gyrus and hippocampus persist and *enlarge* because, with the forward growth of the temporal lobes, the brain stem presents a complete barrier to further extension of the corpus callosum in the median plane.

The growth of the *neocortex* and its enormous expansion are associated with the initial appearance of projection fibres (corticofugal and corticopetal) during the latter part of the third month. These fibres follow the pathway provided by the apposition of the lateral aspect of the thalamus with the medial aspect of the corpus striatum, and, as they do so, they (the *internal capsule*) divide the latter, almost completely, into a lateral part, the lentiform nucleus, and a medial part, the caudate nucleus, these two nuclei remaining confluent only in their antero-inferior regions. The corticospinal tracts begin to develop in the ninth week of fetal life and have reached their caudal limits by the twenty-ninth week. The fibres destined for the cervical and upper thoracic regions and implicated in the innervation of the upper limb are in advance of those concerned with the lower limbs, which in turn are in advance of those concerned with the face. The appearance of reflexes in these three parts of the body shows a comparable sequence (Humphrey 1960). For further analysis of the development of the projection fibres and corpus striatum see Hewitt (1961, 1962).

The preceding emphasis on corticospinal projection fibres is a reflexion of the limited information available to the earlier neuroanatomists. It should be emphasized, however, that the majority of subcortical nuclear masses receive terminals from descending fibres of cortical origin. Furthermore, the foregoing are joined by thalamocortical, hypothalamocortical and other afferent ascending bundles, the whole complex constituting the internal capsule that divides the early corpus striatum. It should also be noted that the internal capsular fibres pass *lateral* to the head and body of the caudate nucleus, the anterior cornu and central part of the lateral ventricle, the rostroventral extensions and body of the fornix, the dorsal thalamus and dorsal choroidal fissure; at similar levels they pass *medial* to the lentiform nucleus. With temporal lobe formation, the capsular fibres also lie *medial* to the inferior cornu of the lateral ventricle which has the amygdaloid complex capping its tip, the tail of the caudate nucleus in its roof, the hippocampus, dentate gyrus and fimbria of fornix in its floor, and temporal extension of the choroidal fissure in its medial wall.

At the end of the third month the superolateral surface of the cerebral hemisphere shows a slight depression anterosuperior to the temporal pole. This corresponds to the site of the corpus striatum in the floor and lateral wall of the ventricle, and its presence is due to the more rapid growth of the adjoining cortical regions. This *lateral cerebral fossa* gradually becomes overlapped and submerged, and is converted into the *lateral cerebral sulcus*; its floor becomes the *insula* (2.71A–G). The process, however, is not completed in its most anterior part until after birth. The presumptive neocortical areas that overlap the insula are termed the frontal, parietal and temporal *opercula*. The lentiform nucleus (lateral part of the corpus striatum) remains deep to and coextensive with the insula, the superficial zones of the latter transforming into varieties of cortex. Posteriorly the insula develops granular neocortex and an intermediate area forms agranular neocortex; finally the rostroventral area becomes similar to and continuous with the palaeocortex of the piriform area.

The growth changes in the temporal lobe which help to submerge the insula produce important changes in the olfactory and other neighbouring limbic areas. The olfactory tract, as it approaches the hemispheric floor, diverges into *lateral, medial* and (variable) *intermediate striae*. The medial stria is clothed with a thin archaeocortical *medial olfactory gyrus*; this curves up into further archaeocortical areas anterior to the lamina terminalis (paraterminal gyrus, prehippocampal rudiment, parolfactory gyrus, septal nuclei) and these continue into the indusium griseum. The lateral stria, clothed by the *lateral olfactory gyrus*, and, when present, the intermediate stria, terminate in the rostral parts of the *piriform area*. In brief, this includes the olfactory trigone and tubercle, anterior perforated substance and the uncus (hook) and entorhinal area of the anterior part of the future parahippocampal gyrus. Its lateral limit is indicated by the *rhinal sulcus*. (For details of the numerous subdivisions and putative interconnections of these areas, see p. 1029 et seq and 7.148–151.) The forward growth of the temporal pole and the general expansion of the neopallium cause the lateral olfactory gyrus to bend laterally, the summit of the convexity lying at the antero-inferior corner of the developing insula (2.71A–G). During the fourth and fifth months much of the *piriform area* becomes submerged by the adjoining neopallium and in the adult only a part of it remains visible on the inferior aspect of the cerebrum.

Apart from the shallow hippocampal sulcus and the lateral cerebral fossa the surfaces of the hemisphere remain smooth and uninterrupted until early in the fourth month (2.71A–G). The parieto-occipital sulcus also appears about that time on the *medial* aspect of the hemisphere and its appearance seems associated with the increase in the splenial fibres of the corpus callosum. Over the same period the posterior part of the *calcarine sulcus* appears as a shallow groove extending forwards from a region near the occipital pole. It is a true infolding of the cortex in the long axis of the *striate area*, producing an elevation, the *calcar avis*, on the medial wall of the posterior horn of the ventricle.

During the fifth month the *sulcus cinguli* appears on the medial aspect of the hemisphere, but not until the sixth month do sulci appear on the inferior and superolateral aspects. The *central, precentral* and *postcentral sulci* appear, each in two parts, upper and lower, which usually coalesce shortly afterwards although they may remain discontinuous. The *superior* and *inferior frontal*, the *intraparietal, occipital, superior* and *inferior temporal*, the *occipitotemporal, collateral* and *rhinal sulci* make their appearance during the same period, and by the end of the eighth month all the important sulci can be recognized (2.71A–G).

Histogenesis of the cortical (pallial) wall of the cerebral hemisphere has generated an impressive literature since the early 1930s. Nevertheless, because of the immense complexity, multiplicity of cell types and structural heterogeneity in different locations, descriptive and experimental analyses are less well understood and documented than those appertaining to the cerebellum, with its regular, geometrically ordered microstructure (p. 969). Only the briefest review of some basic principles, together with a few introductory key references, can be encompassed in this volume and the interested reader will find it apposite to constantly cross-refer to the sections devoted to mature neuronal and cortical architecture (pp. 1039–1043).

The wall of the earliest cerebral hemisphere, as elsewhere in the neural tube, consists of a pseudostratified epithelium, its cells exhibiting interkinetic migration (p. 178) as they proliferate to form clones of, it was assumed, as yet uncommitted germinal cells. The columnar cells elongate and (following the initial nomenclature proposed by the Boulder Committee, *Anat. Rec.*, 1970 and 2.60) their non-nucleated peripheral processes now constitute a *marginal zone*, whilst their nucleated, paraluminal and mitosing regions constitute the *ventricular zone*. Some of the mitotic progeny now leave the ventricular zone and migrate to occupy an *intermediate zone*. This *proliferative phase* continues for a considerable period of fetal (and in some species post-natal) life, but, as in the case of the cerebellar cortex, after a period, groups of germinal cells undergo quantal mitoses forming, at first, generations of *definitive neuroblasts* and, later, *definitive glioblasts*, which migrate to and mature in their final positions (see below for variant views). It must be appreciated, however, that these phases of proliferation, migration, differentiation and maturation are not precisely sequential for each cell variety but overlap each other in space and time. (While the foregoing was widely accepted, and was apposite in the 36th edition of this volume, there has been a burgeoning of investigation and publication on these topics; both prominent literature sources and classifications will receive brief mention.)

The earliest migration of definitive neuroblasts from the ventricular and intermediate zones occurs radially until they approach, but do not reach, the pial surface, their somata becoming arranged as a transient *cortical plate*. Subsequently, proliferation wanes in the ventricular zone but for considerable periods persists in the immediately subjacent *subventricular zone*. From the pial surface inwards, therefore, there may now be defined the following zones: *marginal, cortical plate, subplate, intermediate, subventricular* and *ventricular*. Briefly, whilst the foregoing *terminology* is relatively recent it has for long been accepted that the *marginal zone* forms the outermost layer of the cerebral cortex, the neuroblasts of the *cortical plate* and *subplate* form the neurons of the remaining cortical laminae (the complexity, of course, varying in different locations and with further additions of neuroblasts from the deeper zones), whilst the intermediate zone gradually transforms into the white matter of the hemisphere. Meanwhile other deep germinal cells have been producing generations of glioblasts which also migrate into the more superficial layers. As proliferation wanes and finally ceases in the ventricular and subventricular zones their remaining cells differentiate into general or specialized ependymal cells, tanycytes or subependymal glial cells. Illustration 2.72A–F summarizes the modifications of the Boulder Committee original proposals, suggested by Rakic (1982) in the light of more recent investigations.

As mentioned above, the phases of proliferation, neurogenesis and gliogenesis are by no means sequential as first envisaged but vary spatiotemporally with location and cell type. Further, the gliogenesis referred to was related to the (numerically largest) astrocyte and oligodendrocyte population of the mature tissue. However, as noted (p. 180), immunohistochemical studies of glial fibrillary acid protein (GFA) distribution showed a patterned array of GFA-positive columnar cells in the earliest pseudostratified neuroepithelium of the neural tube including the walls of the rudimentary cerebral hemisphere. The positive cellular elements are interspersed with GFA-negative columnar cells, both undergoing interkinetic migration or proliferation. The GFA-positive elements are presumptive glial cells which stretch radially across the full thickness of the wall of the telencephalon and provide contact guidance paths for the subsequent peripheral migration of groups of functionally related neuroblasts—the foundation of the cortical *columns* or *modules* of neurobiology. The first groups of neuroblasts to migrate are destined for the deep part of the column, later groups passing through them to more superficial regions (see below). The *subplate zone*, a transient feature most prominent during mid-gestation, contains neuroblasts surrounded by a dense neuropil: the neuroblasts appear to await the arrival of thalamocortical afferents before migrating to disperse in most cortical layers. The subplate neuropil is the site of the most intense *synaptogenesis* in the cortex.

Thus, with growth, both radial and tangential, there occurs a great increase in cortical thickness and a vast increase in surface area. Accordingly, the glial contact guidance paths, originally having a simple radial ventriculo-pial disposition, become

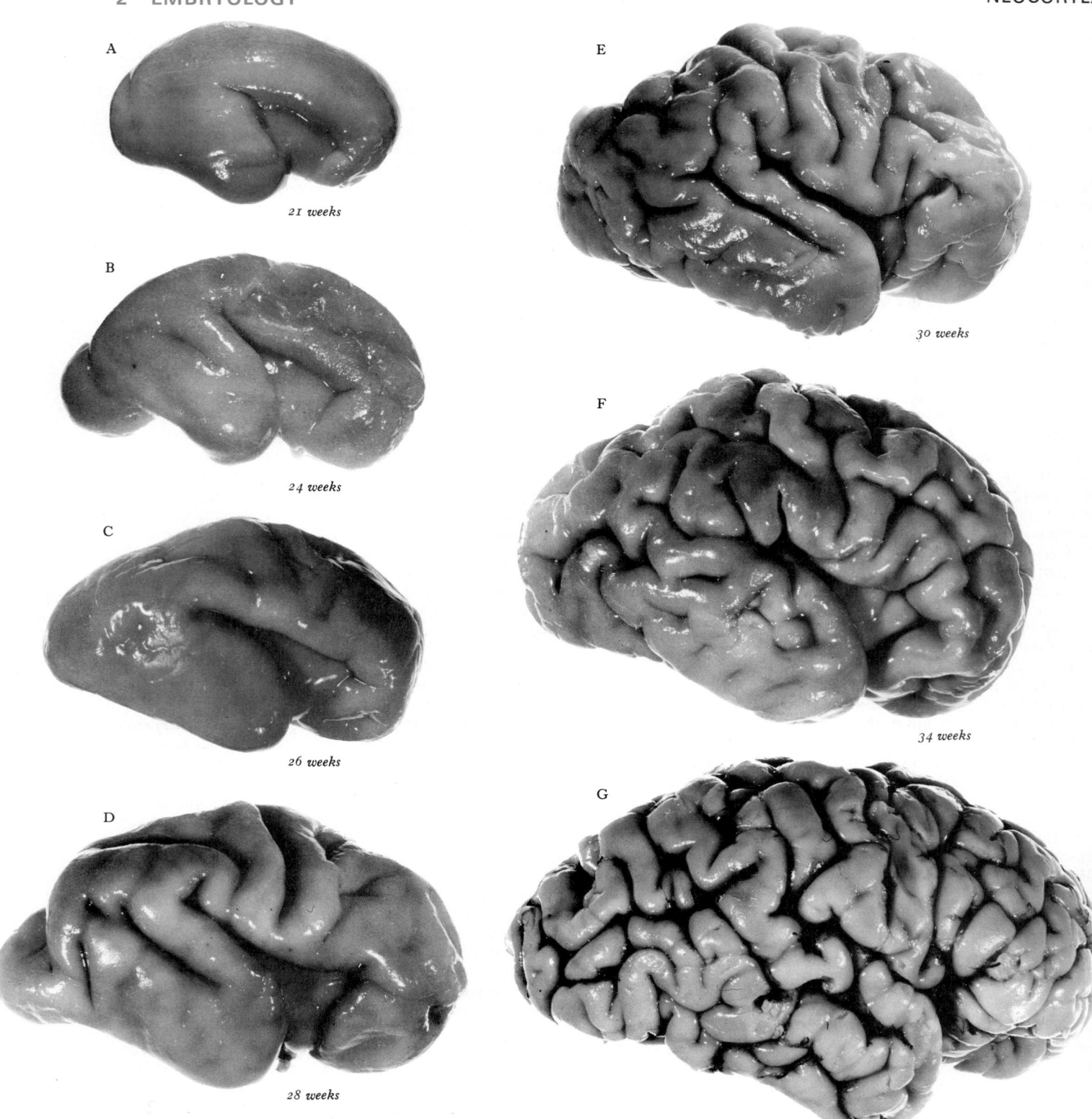

A

21 weeks

B

24 weeks

C

26 weeks

D

28 weeks

E

30 weeks

F

34 weeks

G

40 weeks

ALL SPECIMENS ARE ACTUAL SIZE

2.71 Series showing the superolateral surfaces of human fetal cerebral hemispheres at the ages indicated, demonstrating the changes in size, profile and the emerging pattern of cerebral sulci with increasing maturation. Note the changing prominence and relative positions of the frontal, occipital and particularly the temporal pole of the hemisphere. At the earliest stage (A) the lateral cerebral fossa is already obvious—its floor covers the developing corpus striatum in the depths of the hemisphere and progressively matures into the cortex of the insula. The fossa is bounded by overgrowing cortical regions, the frontal, temporal and parietal opercula, which gradually converge to bury the insula; their approximation forms the lateral cerebral sulcus. By the sixth month the central, pre- and post-central, superior temporal, intraparietal and parieto-occipital sulci are all clearly visible. In the subsequent stages shown all the remaining principal and subsidiary sulci rapidly appear and by 40 weeks all the features which characterize the adult hemisphere in terms of surface topography are already present in miniature.

The photographs were kindly supplied by Dr Sabina Strick of the Maudsley Hospital, London.

systematically deformed (2.73). Nevertheless, it is held that the *overall pattern* in the ventricular zone of the early hemisphere, which is probably genetically preprogrammed in the genome, is maintained in the array of definitive structural/functional cortical columns (modules).

Although the short account just given of the *general* histogenetic history of the cortex gained, and maintains, wide acceptance, it is of the greatest interest that, with the advent of modern nuclear labelling techniques, there has been a fundamental reappraisal of the *sequences* involved in the migratory patterns of the

196

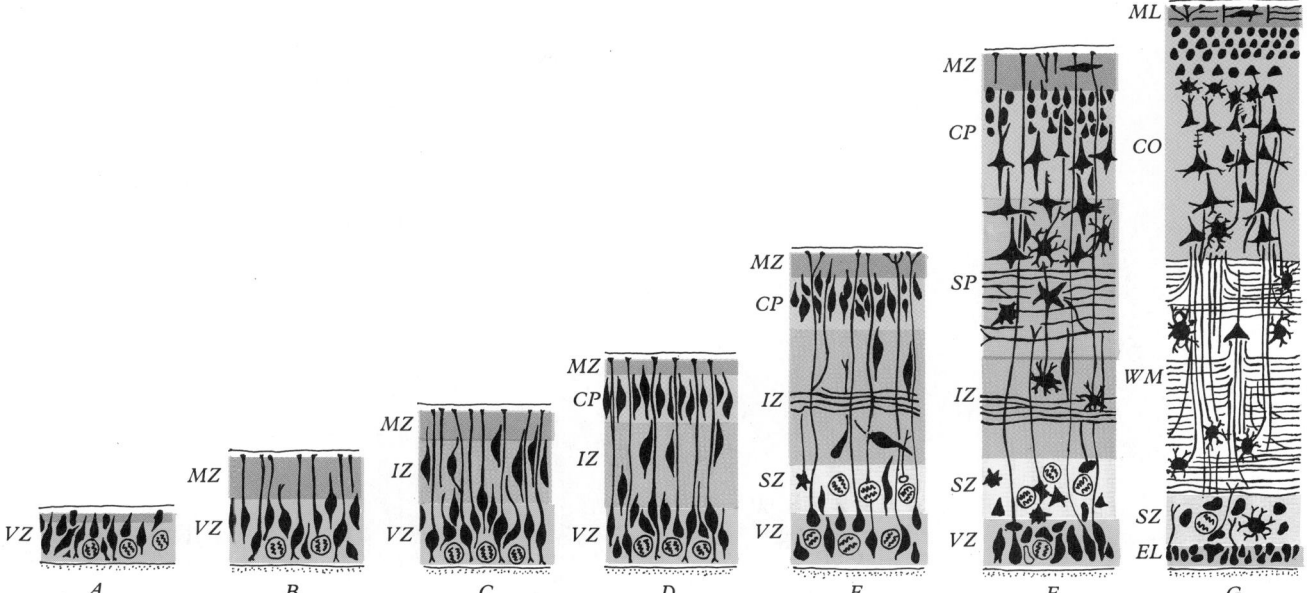

2.72 Formalized diagram of the laminar development of cerebral neocortex. The original proposals of the Boulder Committee (1970) have been revised, in the light of subsequent research, by Rakic (1982), with further additions and modifications in preparation for the present volume. See text for further comment. VZ — ventricular zone; MZ — marginal zone; IZ — intermediate zone; CP — cortical plate; SZ — subventricular zone; SP — subplate zone; CO — definitive neocortex; WM — white matter; EL — ependymal layer; ML — molecular layer.

neuroblasts, with consequently considerable ontogenetic and phylogenetic implications.

Pioneering studies into neuroblast migration in the developing mammalian *neocortex* were made by Tilney (1933), the technique available to him at this time being analysis of sections of Nissl stained tissue. Whilst it was clear that the subpial (marginal) zone formed the plexiform lamina (I), he considered that the remaining laminae stemmed from *three* quite distinct and *separate migrations* of neuroblasts up to the cortical plate. The first migration he thought differentiated into the external granular lamina (II) and the pyramidal lamina (III); the second migration forming the internal granular lamina IV; the third migration he held formed the ganglionic lamina V and the multiform lamina VI. On this view, therefore, the *outermost* layers were the *earliest* to be formed, with progressively deeper layers at successively later times.

The possibility that precisely the *reverse* sequence, progressing from *deep to superficial*, was first implied by the results of X-irradiation studies by Hicks et al (1959). Further irradiation studies (Berry & Eayrs 1966), and autoradiographic nuclear labelling studies (Berry & Rogers 1965, Berry 1974 and 2.74) supported this contention. These seminal investigations have, in the subsequent years, been amply confirmed in the rat (2.75, 76), mouse, opossum and golden hamster. (The individual references are too numerous to quote in this volume, but for an excellent overview of this and many other problems appertaining to cortical histogenesis the reader should refer to Berry 1974, 1982; see also comments and bibliographies in: Smart 1982, 1983, Rakic & Goldman-Rakic 1982, editors and contributors.) Thus in these various mammals, apart from the pre-existing anlage of Lamina I, the first laminae to be populated are VI and V, followed sequentially by laminae IV to II. Clearly, whilst the ontogenetic timings of migrations from these experimental sources are not directly appropriate to a volume on *human* anatomy, it is assumed from comparison of purely descriptive material that similar *patterns* of migration and elaboration occur in the human cortex. It should also be noted that, as yet only in the human cortex, a thin subpial lamina of densely staining cells, of unknown origin or destination, has been identified (Rabinowicz 1964, 1967, Brun 1965): they are not a prominent feature of cortical histogenesis but an analogy with the external germinal layer of the cerebellum has been suggested.

No attempt will be made here further to discuss neuroblast and glioblast differentiation, migration and maturation with the establishment of intercellular contacts; for these the reader is

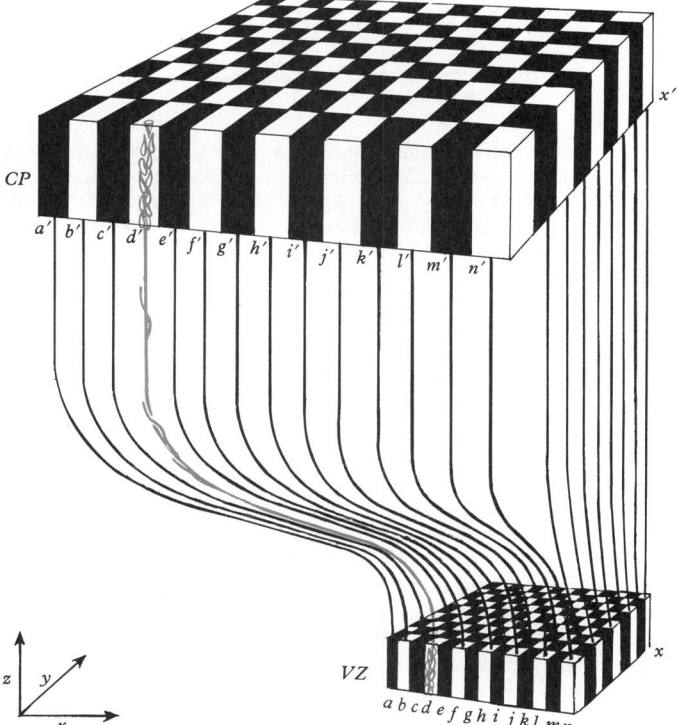

2.73 Development of the cerebral neocortex. Diagram of the spatial three dimensional (and implied temporal fourth-dimensional) relationship between a defined region of the proliferative ventricular zone (VZ) and its subsequent derivative, an enlarged subpial block of cortical plate (CP). The micro-architecture of each is represented as a chequer-board pattern on its upper surface each square extending as a miniature column, or module, through the full thickness of the lamina. The lines joining homologous modules are vertical distally but increasingly sigmoid proximally; they represent changes in the disposition of the primitive radial glial cell processes (contact guidance paths for generations of neuroblasts). Thus the tangential patterning of modules in the cortical plate reflects that already present in the ventricular zone. Vertical patterns of neuroblasts within a module, however, reflect the time of onset and rate of migration along a particular path. The first (i.e. oldest) to migrate reach and mature in the deeper part of the cortical plate; succeeding generations bypass these and occupy successively more superficial parts, the whole finally integrating into a radial functional cell column. (After Rakic 1982.)

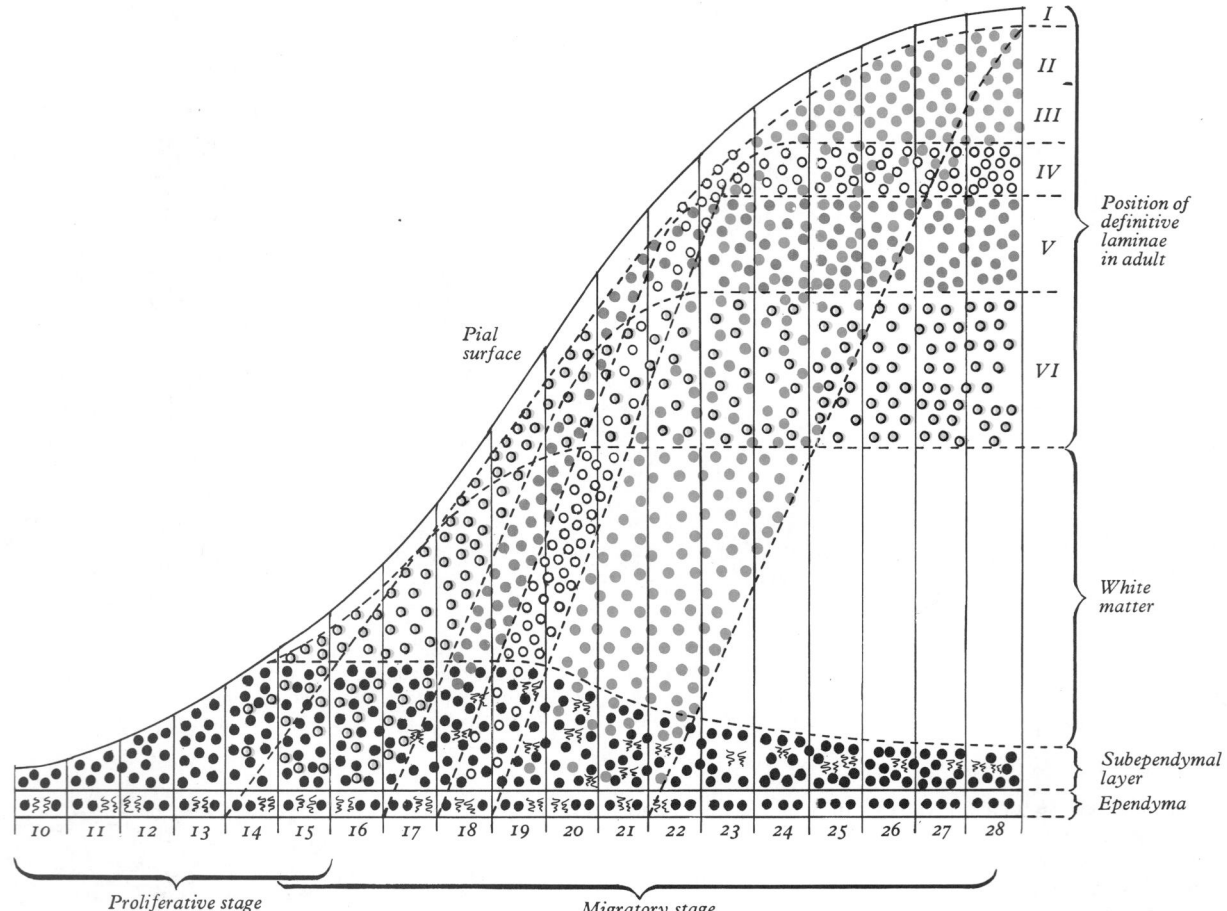

2.74 Schematic representation of the dynamics of neuroblast migrations during transformation of the early cranial neural tube to form the cerebral neocortex of the rat through days 10 to 28. Note the successive waves of migration. Coding: symbolic metaphase chromosomes—mitotic cells; *full black* discs—ventricular and subventricular zone neuroblasts; *full yellow discs*—infragranular neuroblasts destined for lamina VI; full *magenta discs* —infragranular neuroblasts destined for lamina V; *open black circles*— granular neuroblasts destined for lamina IV; *full blue discs*— supragranular neuroblasts destined for laminae III and II. (Redrawn and colour coded by kind permission from data provided by Professor M Berry (1974) of the Anatomy Department, Birmingham University, now Guy's Hospital Medical School, London.)

referred to pp. 178–180, 188, 197 (see also Molliver et al 1973, Kostović & Molliver 1974). However, some general hypotheses may be mentioned, involving a comparison of ontogeny and phylogeny. Firstly, all parts of the neural tube, from the presumptive spinal cord to presumptive neocortex, pass through the stage of a pseudostratified epithelium and a proliferative phase, followed by differentiation sets, with the emergence of ventricular, intermediate and marginal zones. Neuroblast migration, target cell contact and maturation (or, in some locations, *degeneration* and *cell death*), whilst *still confined to* the deeper reaches of the intermediate zone, are the principal events in spinal cord development. Throughout the encephalon, however, the primary difference is the *continued migration* of neuroblasts and their ultimate maturation far *beyond* the confines of the intermediate zone, forming either nuclear masses, variously displaced from the ventricular and aqueductal channels, or, in the 'roofbrain' regions, reaching the subpial marginal zones forming, initially, a simple cortical plate of neuroblasts. The latter then differentiates into subzones, showing a tangential *laminar* organization, whilst in some locations there emerges a well-defined *columnar (modular)* radial organization. Such cortical dispositions are evident in the hemispheric forebrain, tectal midbrain and cerebellar hindbrain. In the pallial walls of the mammalian cerebral hemisphere, the phylogenetically *oldest regions* and the *first* to differentiate during ontogeny are those that *border* the interventricular foramen, and its extension the choroidal fissure, the lamina terminalis and piriform lobe. There exists an increasingly complex level of organization from three to six tangential laminae, passing from the dentate gyrus and cornu ammonis through the subiculum until the general neocortex is reached. (It may be noted that many investigators find the simple progression from three to six major laminae a gross oversimplification, and numerous subdivisions have been proposed, e.g. see cornu ammonis p. 1037, and neocortex p. 1039.) The *deepest* and phylogenetically oldest *tangential laminae* are the first to be populated by migrating neuroblasts, more superficial layers being added in sequence, their neuroblasts migrating *through* the older layers; the number of superadded laminae depending upon the location with respect to the choroidal fissure. These broad

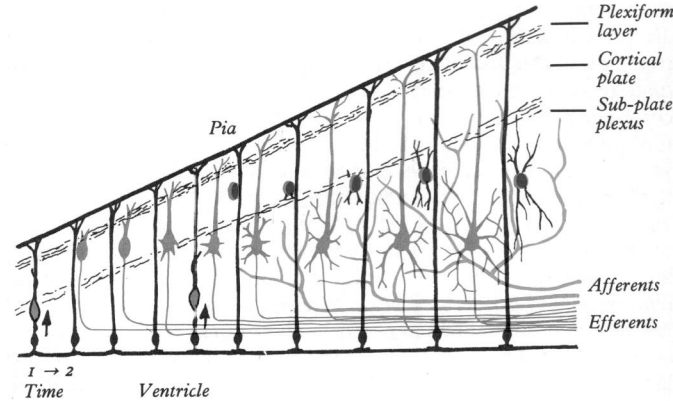

2.75 Diagram to show the manner of the initial stages of formation of apical and basal dendrites of pyramidal neurons, also of stellate neuron dendrites in the cortical plate. Note radial glial cells (black) extending from internal to external limiting membrane; these provide contact guidance paths for neuroblasts. 1. Migration of a presumptive pyramidal neuron (magenta). 2. Migration of a presumptive stellate neuron (purple). Time increments from left to right. (After Berry 1982.)

patterns have been demonstrated by nuclear labelling studies, not only in the neocortex but, with modifications, also in the dentate gyrus and hippocampus (see, e.g. Angevine 1975, Altman & Bayer 1975, Smart 1982).

The state of differentiation at birth and at various postnatal stages, as seen in Golgi (metal impregnation) preparations, has been described in considerable detail elsewhere (Conel—a series of publications 1939–1959). However, the *mechanisms* whereby variation in cell type, number, dendritic patterns and connectivity arise continue under intense investigation. Nevertheless, gross nutritional deficiencies, selective neural ablation, endocrine imbalances, sensory deprivation, neurotropic viruses, vascular abnormalities and perinatal anoxia may all disturb the normal pattern. (See bibliographies in Rakic & Goldman-Rakic 1982.)

At birth the volume of the brain is approximately 25% of its volume in adult life. The greater part of the increase occurs during the *first year*, at the end of which the volume of the brain has increased to 75% of its adult volume. The growth can be accounted for partly by increase in the size of nerve cell somata, the profusion and dimensions of their dendritic trees, axons and their collaterals and by growth of the neuroglial cells and cerebral blood vessels, but it is the acquisition of myelin sheaths by the axons which is principally responsible for it. The great sensory pathways, visual, auditory and somatic, myelinate first, the motor fibres later. During the second and subsequent years, growth proceeds much more slowly; the brain attains adult size by the seventeenth or eighteenth year. This is largely due to continued myelination of various groups of nerve fibres.

A summary of the parts derived from the cerebral vesicles is as follows:

Rhombencephalon (or hindbrain)	1. Myelencephalon	Medulla oblongata Caudal part of the 4th ventricle Inferior cerebellar peduncles
	2. Metencephalon	Pons Cerebellum Middle part of the 4th ventricle Middle cerebellar peduncles
	3. Isthmus rhombencephali	Anterior medullary velum Superior cerebellar peduncles Rostral part of the 4th ventricle
Mesencephalon (or midbrain)		Cerebral peduncles Tegmentum Tectum Aqueduct
Prosencephalon (or forebrain)	1. Diencephalon	Thalamus Metathalamus Subthalamus Epithalamus Caudal part of the hypothalamus Caudal part of the 3rd ventricle
	2. Telencephalon	Rostral part of the hypothalamus Rostral part of the 3rd ventricle Cerebral hemispheres Lateral ventricles Cortex (archaeocortex, palaeocortex, neocortex) Corpus striatum

THE CRANIAL NERVES AND THE NEURAL CREST

With the exception of the olfactory and optic nerves, which will be considered separately, the cranial nerves are developed in a manner similar in some respects to components of the spinal nerves.

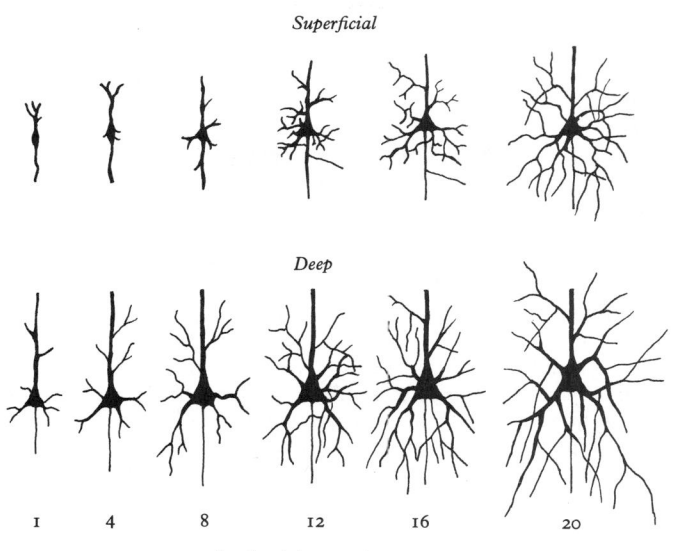

Superficial

Deep

| 1 | 4 | 8 | 12 | 16 | 20 |

Age (rat) days postpartum

2.76 The temporal sequence of the initial appearance, growth and maturation, of basal and oblique dendrites by pyramidal neurons in the superficial and deep neocortical laminae of the rat. In the first postpartum week the deep neurons are well differentiated relative to the superficial neurons; the latter have only recently terminated their migration from the ventricular zone. Maturation of the superficial neurons relatively hastens in the second and third weeks, and by the 20th postpartum day the degree of maturation of superficial and deep pyramidal neurons is the same. (After Berry 1982.)

The motor fibres of the cranial nerves to striated muscle are the axons of cells in the 'ventrolateral' lamina of the midbrain and hindbrain which grow outwards to their muscle fibres of distribution, but whereas the motor fibres of the spinal nerves form a single series, those of the cranial nerves form two, which are derived from the medial and intermediate parts of the ventrolateral lamina respectively. The first series (*somatic efferent*) comprises the oculomotor, trochlear, abducent and hypoglossal nerves, supplying somite-derived muscles or their homologues; the second (*branchial efferent*) comprises the accessory nerve and the motor parts of the trigeminal, facial, glossopharyngeal and vagus nerves, all of which supply the *striated* muscles derived from the branchial arches, or from postbranchial 'visceral' mesoderm. (It should be recalled, however, that some authorities regard the mesoderm of the branchial arches and immediate post-branchial region as neither somatic nor visceral but *mixed*, being *unsplit lateral plate* since it is devoid of a coelomic cavity; it also receives accessions from the cranial neural crest and varieties of 'head' placodes.) The dorsolateral part of the ventrolateral lamina contributes a variety of neuroblasts, both pre- and post-ganglionic during autonomic neurogenesis (vide infra and p. 200).

As the lips of the neural groove fuse in the region of the hindbrain and midbrain, a *neural crest* is formed which is homologous with the neural crest of the trunk, flanking the spinal cord (p. 137). (For a review of the origin, migration and differentiation of neural crest cells, consult Weston 1970, Bellairs 1971, Leikola 1976, Fujita 1976, Pearse 1977.) The ganglia of the vagal, glossopharyngeal, vestibulocochlear (in part), facial and trigeminal nerves are derived from the neural crest, but they migrate ventrally and soon lie on the ventrolateral aspect of the hindbrain. The vestibulocochlear nerve ganglion is believed to receive contributions from the otocyst (otic placode initially). There is also descriptive and experimental evidence, mainly from lower vertebrates, that overlying ectodermal thickenings or *placodes* (epibranchial, dorsolateral and suprabranchial) contribute to the ganglia of the trigeminal, facial, vestibulocochlear, vagus and glossopharyngeal nerves. It is claimed that these contributions give rise, in the main, to the *special somatic afferent* (acoustic and lateral line) and the *special visceral afferent* (chiefly gustatory) components of these nerves. The development of these cranial nerve ganglia and the role of the ectodermal placodes in the *human* embryo, however, still need clarifying (see p. 138). Caudal to the ganglion of the vagal nerve the occipital region of the neural crest is concerned with the ganglia of the accessory and hypoglossal nerves. Rudimentary ganglion cells may occur along the hypoglossal nerve in the human embryo; they undergo regression later. Ganglion cells are also found on the developing spinal root of the accessory nerve and these are believed to persist in the adult. The central processes of the cells of these various ganglia, where they persist, form some sensory roots of the cranial nerves and enter the dorsolateral lamina of the hindbrain; their peripheral processes join the efferent components of the nerve to be distributed to the various tissues innervated. Some incoming fibres from the facial, glossopharyngeal and vagal nerves collect to form an oval bundle, the *tractus solitarius* (p. 954), on the lateral aspect of the myelencephalon. This bundle is the homologue of the oval bundle of the spinal cord, but in the hindbrain it becomes more deeply placed by the overgrowth, folding and subsequent fusion of tissue derived from the rhombic lip on the external aspect of the bundle.

THE AUTONOMIC NERVOUS SYSTEM

The ganglion cells of the *sympathetic system* are derived from the neural crest through the medium of the *primitive spinal ganglia* (p. 183). Some cells in the ventral parts of the latter migrate towards the sides of the aorta, where they subsequently form the *ganglia of the sympathetic trunks* and certain other associated cells (vide infra). Others migrate still further and eventually form the subsidiary sympathetic ganglia such as the coeliac and renal (**2.77**). The original migration is limited to the thoracic and upper lumbar regions. Thereafter the chain grows rostrally and caudally until the whole trunk is laid down. The view has also been advanced that the sympathetic ganglion cells are, at least in part, derived from cells which migrate from the *ventrolateral lamina* along the ventral nerve roots. The results of destruction of the neural crest in chick embryos and excision of the neural crest and portions of the neural tube in frog embryos have been somewhat contradictory in the hands of different investigators. On the whole the balance of the evidence favours the earlier view. For experimental analysis of the neural crest regional mosaic of ganglionic neuroblast sources in quail-chick chimeras see Le Douarin (1980, 1981).

The ganglion cells of the cranial part of the *parasympathetic system* are probably derived from the neural crest through the medium of the primitive ganglia of the oculomotor, trigeminal, facial, glossopharyngeal and vagus nerves. The *ciliary ganglion* is formed by cells which migrate from the trigeminal ganglion along

the ophthalmic nerve, but it is almost certainly reinforced by cells migrating from the nucleus of the oculomotor nerve along which a few scattered cells are always demonstrable in post-natal life. The *pterygopalatine ganglion* receives contributions from the ganglia of the trigeminal and facial nerves, the *otic* from that of the glossopharyngeal and the *submandibular* from that of the facial cranial nerve.

The origin of the enteric ganglia in the walls of the gastrointestinal tract has been disputed, but a critical review of the available evidence (Andrew 1971), suggested that both the vagal neural crest and that of the trunk may provide sources. Further evidence was awaited. Le Douarin and Teillet (1974) studying explants of quail neural crest into the chick confirmed that avian enteric neuroblasts have both a vagal and lumbo-sacral neural crest origin. The sacral derivatives join the vagal and colonize 'postumbilical' hindgut (Le Douarin 1980).

Since the foregoing publications there has been a surge of interest in autonomic neurogenesis; e.g. in relation to neuroeffector junctions, the events occurring at 'prerecognition', 'recognition' and 'postrecognition' stages have been characterized. In general, cell sources, critical mitoses, migrations, neurochemical differentiation (cholinergic, noradrenergic, serotonergic, purinergic and an array of other peptidergic neuroblasts) have gained acceptance, together with numerous other facets. For details and extensive bibliographies see Burnstock (1981, editor and contributor). See also p. 891.

THE NEUROGLIA

The glial cells are partly neurectodermal and partly mesodermal in origin. The ventricular zone lining the early central canal of the spinal cord and the cavities of the brain give rise, as we have seen, to at least two types of cell (GFA positive and negative, see pp. 180, 188, 195); many of their proliferative progeny migrate into the intermediate zone. The negative cells form neuroblasts which differentiate into neurons, and the positive cells glioblasts which differentiate first into form *primitive radial* varieties, and then generations of *astroblasts* and *oligodendroblasts*; the latter mature into astrocytes and oligodendrocytes (for their mature morphology and hypothesized functional roles, see p. 892). The earliest glioblasts' radial processes extend both outwards to form the outer limiting membrane deep to the pia mater and inwards forming the inner limiting membrane around the central cavity. Their cell geometry may provide contact guidance paths for subsequent cell migrations, both neuroblastic and glioblastic. As the glioblasts differentiate into primitive neuroglia some lose their connections with both inner and outer limiting membranes. They may partially clothe the somata (between presumptive synaptic contacts) of neighbouring developing neuroblasts, or similarly enwrap intersynaptic surfaces of their neurites. (When many varieties of axon are involved, the encircling glial processes form internodal segments of myelin.) Further glial processes expand around intraneural capillaries as perivascular end-feet. Other glioblasts retain an attachment (or form new expansions) applied as pial end-feet to the innermost stratum of the meninges (pia mater)—this is the *pia intima* of some neurocytologists. Both strata may be termed *pia-glia*. Still other glioblasts remain lining the central canal and cavities of the brain as generalized or specialized ependymal cells, including tanycytes, but lose their peripheral attachments. In some situations, as in the anterior median fissure of the spinal cord, the ependymal cells retain their attachments to both the inner and outer limiting membranes. Thus, in addition to functioning as perineuronal satellites, the glia provide cellular channels interconnecting extracerebral and intraventricular cerebrospinal fluid, the cerebral vascular bed, the intercellular crevices of the neuropil and the cytoplasm of all neural cell varieties. (For further comments see pp. 892-896.)

In contrast to astrocytes and oligodendrocytes which are ectodermal in origin, the microglia are mesodermal derivatives. They appear in the central nervous system after this has been penetrated by blood vessels and invade it in large numbers from certain restricted regions, whence they spread in what have picturesquely been called 'fountains of microglia', to extend deeply

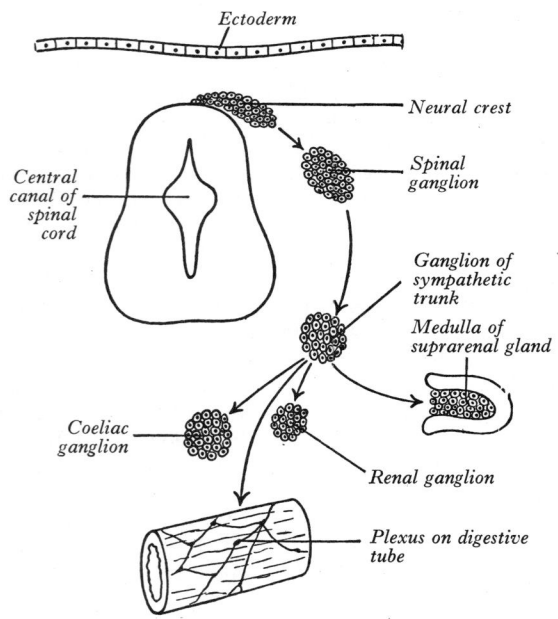

2.77 Diagram showing some derivatives of the neural crest.

Ectoderm

Neural crest

Central canal of spinal cord

Spinal ganglion

Ganglion of sympathetic trunk

Medulla of suprarenal gland

Coeliac ganglion

Renal ganglion

Plexus on digestive tube

amongst the nervous elements. Their origin has been variously ascribed to perivascular mesenchyme cells, vascular endothelial cells, and even to blood-borne leucocytes. See comments in Neurology, Section 7, p. 895.

THE MENINGES

The meninges may be divided in development into the *pachymeninx* (dura mater) and *leptomeninges* (arachnoid and pia mater). Experimental work in lower vertebrates indicates that the dura mater is mesodermal in origin (derived, at least in part, from the sclerotomes), whilst the arachnoid and pia mater are closely associated, being ectodermal in origin and largely derived from neural crest cells (Harvey & Burr 1926, Hörstadius 1950, Leikola 1976). However, morphological studies on the *spinal* meninges of human embryos suggest that all three membranes are initially derived from the loose mesenchyme of sclerotomic origin surrounding the spinal cord, hence termed the *meninx primitiva*. Neural crest cells mingle with this mesenchyme and, as mentioned above, appear to be predominantly involved in the formation of the pia and arachnoid (Sensenig 1951, Weston 1970, Leikola 1976). In the head, the mesoderm from which the *cranial* dura mater develops is closely associated at first with the mesenchyme which is chondrified and ossified, or ossified directly, to form the skull, and these layers are only clearly differentiated as the venous sinuses develop. For an interesting study of pre- and postnatal growth of the tentorium cerebelli, with a mathematical analysis, see Klintworth (1967).

The Chromaffin Organs

The tissue from which the sympathetic ganglia are formed is, at first, a mass of relatively undifferentiated *sympatho-chromaffin cells* but later, a number of cell varieties which may be roughly grouped into *small* and *intermediate-sized neuroblasts (sympathoblasts)* and *large*, initially rounded *phaeochromocytoblasts*. The intermediate-sized neuroblasts differentiate into the typical multipolar post-ganglionic sympathetic neurons (which secrete noradrenalin at their terminals) of classic autonomic neuroanatomy. The small 'neuroblasts' have only come into prominence relatively recently; they are currently the subject of intense scrutiny and they will unquestionably cause a radical reappraisal of autonomic terminology and views concerning ganglionic transmission. Ultrastructural and specialized light-microscopic fluorescence techniques led them to be termed, firstly small granulated cells and, latterly, *small intensely fluorescent* (SIF) cells, type I and II. Both have been shown (at least in some species and sites) to be dopamine-storing and secreting cells. It is postulated that type I function as true interneurons, synapsing with the principal post-ganglionic neurons. Type II are thought to operate as *local neuroendocrine cells*, secreting dopamine into the ganglionic microcirculation. Both types of SIF cells probably modulate the principal preganglionic/post-ganglionic synaptic transmission. The *large* cells differentiate into masses of columnar or polyhedral *phaeochromocytes* ('classic' chromaffin cells) which secrete either adrenalin or noradrenalin. These cell masses are termed *paraganglia* and may be situated near, on the surface of, or embedded in the capsules of the ganglia of the sympathetic chain, or in some of the large autonomic plexuses (see p. 1465 and **8.235**). The largest members of the latter are the *para-aortic bodies* which lie along the sides of the abdominal aorta in relation to the inferior mesenteric artery. During childhood the para-aortic bodies and the paraganglia of the sympathetic chain partly degenerate and can no longer be isolated by gross dissection, but even in the adult chromaffin tissue can still be recognized microscopically in these various sites (p. 1472). It may be noted here that both the phaeochromocytes and the SIF cells, using a wider and more recent classification, are regarded as chromaffin; they belong to the APUD series of cells and are paraneuronal in nature (see further discussions on pp. 136, 138, 1466, and Burnstock 1981, editor).

THE SUPRARENAL GLANDS

Each suprarenal gland consists of a cortex of mesodermal origin and a medulla of neurectodermal origin. The cortex is initiated during the second month as a proliferation of coelomic mesothelium into the underlying mesenchyme between the root of the dorsal mesogastrium and the mesonephros (Keene & Hewer 1927, Crowder 1957). (An older view that two separate proliferations form at first a fetal cortex before the definitive cortex has not been corroborated.) The proliferating tissue extends from the level of the sixth to the twelfth thoracic segments. It is soon disorganized dorsomedially by invasion by sympathochromaffin tissue from adjacent sympathetic ganglionic masses to form the medulla and also by the development of venous sinusoids. The latter are joined by capillaries which arise from adjacent mesonephric arteries and penetrate the cortex in a radial manner. When proliferation of the coelomic epithelium ceases the cortex is enveloped ventrally, later dorsally, by a mesodermal capsule derived from the mesonephros. The subcapsular nests of cortical cells are the rudiment of the *zona glomerulosa*. These nests proliferate cords of cells which pass deeply between the capillaries and sinusoids. The cells in these cords degenerate in an erratic fashion as they pass towards the medulla, becoming granular, eosinophilic and ultimately autolysed. These cords of degenerating cells constitute the *fetal cortex*, which undergoes a rapid degeneration during the first two weeks after birth with marked shrinkage of the gland. The *fascicular* and *reticular* zones of the adult cortex are proliferated from the glomerular zone after birth and are only fully differentiated by about the twelfth year.

Development of Special Sense Organs

THE NOSE

The early development of the olfactory placodes, external nose and nasal cavities have already been considered (pp. 166 et seq.).

The *olfactory nerve fibre bundles (fila olfactoria)* are developed from a proportion of the placodal cells which line the olfactory pits; these cells proliferate and give rise to *olfactory receptor cells*. Their central processes grow into the overlying olfactory bulb and thus form the axons of the olfactory nerves. It was claimed that the olfactory cells are from the first connected with the overlying brain by bridges of cytoplasm, within which the olfactory nerve fibres develop (Smith 1908, Ballantyne 1925). More recent accounts, however, suggest that the earliest pioneer neurites are naked cytoplasmic processes which cross a mesenchyme-filled gap between the placode and the superjacent brain. Later these and subsequent generations of centrally directed neurites become enclothed in Schwann cell processes, presumably derived from the rostral neural crest (Pearson 1941, Van Campenhout 1956, Dejean et al 1958). Within the olfactory bulb the terminals of the olfactory axons divide repeatedly, and establish complex synaptic contacts with a number of neuroblast types in rudimentary *olfactory glomeruli* (p. 1031). The single dendrite extends towards the nasal cavity surface of the olfactory epithelium where, in most regions, slight expansion with surface specializations occurs (p. 1173).

The remaining placodal cells, with probable accessions from neighbouring rostral neural crest and mixed head mesenchyme, differentiate into columnar *supporting (sustentacular) cells*, rounded *basal cells* and, by invagination, the flattened *duct-lining* and polyhedral *acinar cells* of the glands of Bowman. Later, basal infiltration by lymphocytes occurs.

For further details and references consult p. 1175.

THE EYES

The rudiments of the eyeballs appear as two hollow evaginations from the lateral aspects of the forebrain (Mann 1964, Duke-Elder 1963, O'Rahilly 1966, 1975). These rudiments are visible some time before the closure of the rostral neuropore; after its closure they are known as the *optic diverticula*. Their formation is dependent on the organizing influence of the mesoderm of prechordal

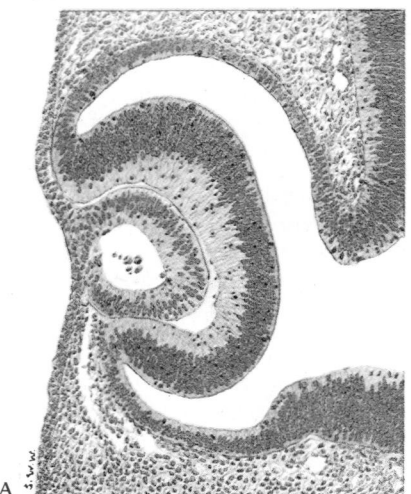

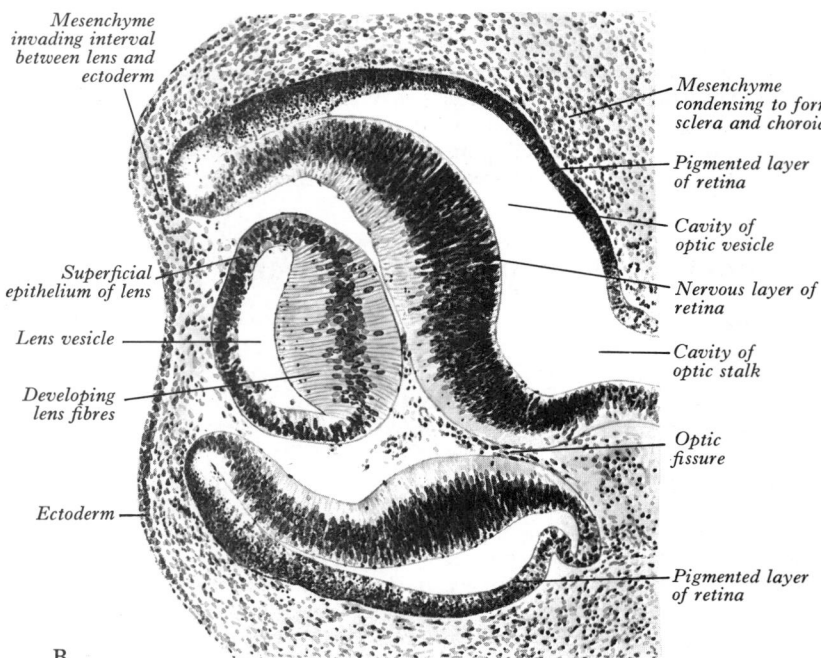

Mesenchyme invading interval between lens and ectoderm

Mesenchyme condensing to form sclera and choroid

Pigmented layer of retina

Cavity of optic vesicle

Nervous layer of retina

Superficial epithelium of lens

Lens vesicle

Developing lens fibres

Cavity of optic stalk

Ectoderm

Optic fissure

Pigmented layer of retina

B

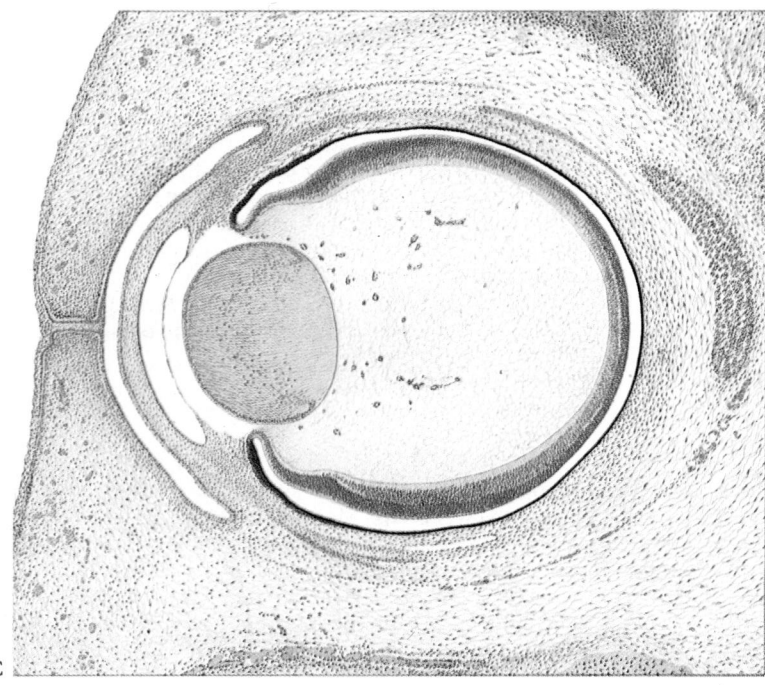

C

plate origin which underlies this part of the neural tube. They project *laterally* towards the sides of the head, and the distal part of each expands as an *optic vesicle (vesicula optica)* while the proximal part remains narrow as the *optic stalk* (2.78A, B). Each contains, initially, a correspondingly shaped part of the original diverticular *optic cavity*. Under the inductive influence of the subjacent optic vesicle the ectoderm overlying it becomes thickened and depressed in its centre. (The classical experiments of Spemann in this field were largely responsible for initiation of the concept of embryonic induction—see p. 107.) This *lens pit* deepens and its edges come together and fuse to enclose a hollow *lens vesicle (cupula optica)* (2.78A) which soon loses its connection with the surface ectoderm—the rudiment of the lens. The lateral, paralental wall of the optic vesicle increases in thickness and undergoes invagination into the optic cavity to form the *optic cup*, its wall consisting of two strata of cells (2.78A,B). These two strata are continuous with each other at the cup's margin, which grows forwards at the end of the third month, overlapping first the rim then the front of the lens, and finally converging to form the edge of the future pupil. The invagination is not limited to the lateral wall of the vesicle, but involves also its caudal surface and extends as a *groove* some distance along the optic stalk. Thus, for a time, a wide hiatus, the *optic* or *choroidal fissure and groove*, exists in the caudal part of the cup and distal stalk respectively. Through the groove and fissure mesenchyme extends into the optic stalk and cup, carrying the *hyaloid artery* with it; as growth proceeds, the edges of the groove and fissure become approximated and they close during the seventh week, including the artery in the distal part of the stalk. Failure of the optic fissure to close is a rare anomaly and there is always a corresponding deficiency in the choroid and the iris (*congenital coloboma*). It may be noted, however, that localized deficiencies of the choroid and iris can occur independently, and in such cases the retina is normal (Lopashov & Stroeva 1961). Although the cavity of the optic vesicle is largely obliterated by the invagination to form a cup, a potential space does persist, at the microscopic level, as the interval between rod and cone processes and the apical microvilli of the pigmented cells of the most external layer of the retina (p. 1194). The 'space' is of course originally continuous with that of the cerebral vesicle of the forebrain. It is also the site of pathological detachment of the retina.

At the stage of growth briefly described above, a few pertinent terminological points may be noted. Each eye has passed through the phases of a prosencephalon-derived optic diverticulum, conversion into a pedunculated vesicle which induces lens vesicle formation, invagination forming a partially incomplete optic cup and stalk and completion of these deficiencies. The walls of the cup, as mentioned, are two cellular strata thick—the *laminae externa et interna cupulae* separated by a rapidly diminishing *intraretinal space*; the strata are confluent at the margin of the future pupil. The cavity of the cup (*cavitas cupularis*), with growth of the cup rim, comes to engulf the developing lens, internal to which is the site of formation of the vitreous body and the fibro-elastic and membranous structures associated with these. Anteriorly develops the diminutive posterior chamber of the aqueous space. Because of the event of invagination, substrata and fine cellular zones derived from the *inner lamina* have become *reversed* with respect to their original position in the unaltered wall of the neural tube. Many structures in the ocular tunics are named in relation to their radial distance from the centre of the *mature eyeball*. Note in particular the plethora of names for

2.78 A Section through the developing eye of a human embryo. 8 mm CR length. The thick nervous and the thinner pigmented layers of the retina and the developing lens are shown. Stained with haematoxylin and eosin. Magnification × *c*. 114. (From material loaned by Professor R J Harrison.) B Section through the developing eye of a human embryo, 13.2 mm long. (Streeter 1948.) C Section through the eye of a human embryo. 40 mm CR length. Note the layers of the retina, developing lens, pupillary membrane, cornea, conjunctival sac, anterior and posterior aqueous chambers, the developing vitreous body, and condensing circumoptic mesenchyme, and the fused eyelids. Stained with haematoxylin and eosin. Magnification × *c*. 62.

sublayers of the neural retina and compare them with the line of the early (ventricular) intraretinal space. Another group of terms, occasionally causing confusion but approved by a number of developmental biologists, is, with qualification, to dub as 'retinal' (retinae, retinaris) almost *all* derivatives of the optic cup. Thus true photoreceptors and associated neuroblasts form in the internal lamina from a scalloped line a little posterior to the lens's rim (*ora serrata*) throughout the deeply placed remainder of the globe. This, with its associated external lamina, constitutes the *pars optica retinae*. The forward continuation of the (now non-nervous) double strata ends at the pupil's rim, the whole being termed the *pars caeca retinae*. Regionally this forms the columnar to cuboidal cells of the *pars ciliaris retinae* (p. 1186) and similar cells of the *pars iridica retinae* (p. 1194); double epithelia result in both sites.

The neural retina, as indicated, develops from the *pars optica* of the cup (2.78B,C). Its two strata are at first equipotential and mutually interchangeable. Depending on contact with surrounding mesoderm, however, the external stratum of the cup remains a single layer of cells, which assume a flattened to cuboidal shape, acquire pigment and form the *pigmented epithelium of the retina*, the pigment first appearing in the cells near the edge of the cup. The majority of the cells have a regular geometry, rectangular in radial section, hexagonal in tangential section. (For further structural details see p. 1195.) Under the influence of the lens the cells of the internal stratum proliferate and form a layer of considerable thickness from which are developed the nervous elements and the elongate sustentacular cells of the retina, together with a portion of the vitreous body.

The cells of the inner layer of the cup proliferate and form an outer *nuclear zone* and an inner *marginal zone*, devoid of nuclei. At 12 mm the cells of the nuclear zone invade the marginal zone, and at 17 mm the nervous stratum of the retina consists of inner and outer *neuroblastic layers*. The inner neuroblastic layer gives origin to the ganglion cells, the amacrine cells and the somata of the 'fibrous' sustentacular cells (of Muller); the outer neuroblastic layer is the source of the horizontal and rod- and cone-bipolar neurons and probably the rod and cone cells, which first appear in the central part of the retina. By the eighth month all the named layers of the retina can be identified. For bibliographies on retinal development, including ultrastructural studies, consult Spira & Hollenberg (1973), Fisher & Linberg (1975), Warwick et al (in press).

The deepest part of the optic fissure is at the centre of the floor of the optic cup. In this situation, which later is the site of the *optic disc*, the inner (neural) cell layer of the cup is continuous with the corresponding invaginated cell layer of the optic stalk and, as a result, the developing nerve fibres of the ganglion cells pass directly into the wall of the stalk, converting it into the *optic nerve*. The fibres of the optic nerve begin to acquire their myelin sheaths shortly before birth, but the process is not completed until some time after birth. The *optic chiasma* is formed by the meeting and partial decussation of the fibres of the two optic nerves in the ventral part of the lamina terminalis and it marks the junction of the telencephalon with the diencephalon in the floor of the third ventricle. Beyond the chiasma the fibres are continued backwards as the optic tracts principally to the lateral geniculate bodies and to the superior tectum.

The lens is developed from the lens vesicle (2.78A), which is overlapped by the margin of the cup and becomes separated from the overlying ectoderm by mesenchyme. The cells forming the posterior wall of the vesicle lengthen greatly and, although retaining their nuclei for many years, are usually described as converting into so-called lens fibres, which grow into and fill its cavity (2.78B,C). The cells forming the anterior wall retain their cellular character and form the epithelium on the anterior surface of the fully developed lens; at the equator of the lens the gradual transition of the cells into elongate lens fibres can be observed; lens fibre differentiation and growth continues throughout life. Characteristic ultrastructural changes have been described (Wulle & Lerche 1967). (For further structural, dimensional and age changes, see p. 1207.) By the second month the lens is invested by a vascular mesenchymal condensation, the *vascular capsule of the lens*, the ventral part of which is named the *pupillary membrane*; the blood vessels supplying the dorsal part of this capsule are derived from the *hyaloid artery*, those for the ventral part from the *anterior ciliary arteries*. By the sixth month all the vessels of the capsule are atrophied except the hyaloid artery, which becomes occluded during the eighth month of intrauterine life. Prior to this, during the fourth month, the hyaloid artery gives off retinal branches and its proximal part persists in the adult as the *central artery of the retina*, (together with its accompanying *central vein*). The *hyaloid canal*, which carries the vessels through the vitreous, persists after the vessels have become occluded. In the newly born child it extends more or less horizontally from the optic disc to the posterior aspect of the lens but when the adult eye is examined with a slit-lamp it can be seen to follow a wavy, curved course, sagging downwards as it passes forwards to the lens (Mann 1927). With the loss of its blood vessels the vascular capsule of the lens disappears, but sometimes the pupillary membrane persists at birth, giving rise to *congenital atresia of the pupil*.

The vitreous body is developed between the lens and the optic cup. The lens rudiment and the optic vesicle are at first in contact, but after closure of the lens vesicle and formation of the optic cup the former is withdrawn from the retinal layer of the cup; the two, however, remain connected by a network of delicate cytoplasmic processes. This network, derived partly from the cells of the lens and partly from those of the retinal layer of the cup, is the *primitive vitreous body*. At first these cytoplasmic processes spring from the whole of the neuroretinal area of the cup, but later are limited to the ciliary region where, by a process of condensation, they form the basis of the *ciliary zonule*. The mesenchyme which enters the cup through the choroidal fissure and around the equator of the lens becomes intimately united with this reticular tissue and also contributes to the formation of the vitreous body, which is therefore derived partly from the ectoderm and partly from the mesoderm. Considerable controversy regarding the precise derivation of the vitreous still persists. There are probably three sources: the local *mesenchyme*, the lens (*ectoderm*) and the retina (*neurectoderm*). The mesodermal component is probably represented only by the juxtazonular vitreous and the vitreous immediately adjacent to the hyaloid vestiges.

The aqueous chamber of the eye initially appears as a cleft in that part of the mesenchyme which intervenes between the lens and the ectoderm. The mesenchyme superfical to the cleft forms the *substantia propria* of the cornea that deep to the cleft the mesenchymal stroma of the *iris* and the *pupillary membrane*. Tangentially, this early cleft extends as far as the *iridocorneal angle* where communications are established with the sinus venosus sclerae. The cornea is induced by the lens and optic cup. The corneal epithelium is formed from the surface ectoderm and the endothelium of the anterior chamber from mesenchyme (O'Rahilly & Meyer 1959, Coulombre 1964). When the pupillary membrane disappears cavitation continues between the double epithelium on the deep surface of the iris and the anterior part of the lens capsule, until zonular lental suspensory fibres and ciliary processes are reached. Thus the aqueous chamber is now divided by the iris into *anterior* and *posterior chambers*, communicating through the pupil. Their walls furnish the sites of production, channels of circulation and reabsorption of the aqueous humour (p. 1190).

The sclera and choroid are derived from the mesenchyme surrounding the optic cup, and the anterior part of the choroid is modified to form the *ciliary body* and *ciliary processes*. The fibres of the *ciliary muscle* are derived from the mesoderm, but those of the *sphincter* and *dilatator pupillae* are unusual being of neurectodermal origin, developed from the cells of the pupillary part of the optic cup, as is the double epithelium on the posterior (lenticular) aspect of the iris.

The eyelids are formed as small cutaneous folds (2.78C). About the middle of the third month their edges come together and unite over the cornea; they are usually said to remain united until about the end of the sixth month. In a more recent study, however, it was concluded that the separation is slow but completed a month earlier (5.1 months, 155 mm stage). The same observers, from examination of 20 human embryos and fetuses, also stated that the tarsal plates and glands begin to develop respectively at 55 mm and 75 mm stages. For the chronology of development of the human eye at embryonic and fetal stages consult O'Rahilly (1966, 1975).

THE LACRIMAL APPARATUS

The epithelium of the alveoli and ducts of the *lacrimal gland* arise as a series of tubular buds from the ectoderm of the superior conjunctival fornix; these buds are arranged in two groups; one forming the gland proper, and the other its palpebral process. The *lacrimal sac* and *nasolacrimal duct* are considered to be derived from the ectoderm in the nasomaxillary groove between the lateral nasal prominence and the maxillary prominence (p. 1219). This thickens to form a solid cord of cells which sinks into the mesenchyme; during the third month the central cells of the cord break down and a lumen is acquired. In this way the *nasolacrimal duct* is established. The lacrimal canaliculi arise as buds from the upper part of the cord of cells and secondarily establish openings (*punctua lacrimalia*) on the margins of the lids; the inferior canaliculus cuts off a small part of the lower eyelid to form the *lacrimal caruncle* and *plica semilunaris*. The epithelium of the cornea and conjunctiva is of ectodermal origin, as are also the eyelashes and the lining cells of the tarsal, ciliary and other glands which open on the margins of the eyelids.

For general accounts of ocular developmental abnormalities consult Dejean et al (1958) and Mann (1964).

The Ears

The rudiments of the **internal ears** appear shortly after those of the eyes as two patches of thickened, surface epithelium, *otic placodes*, lateral to the hindbrain. These patches, though surrounded by general skin ectoderm, are probably neurectodermal in character. Each placode invaginates as an *otic pit* (**2**.125A). The mouth of the pit then closes and an *otocyst (auditory or otic vesicle)* is formed (**2**.79A), initially piriform in shape, and from it the epithelial lining of the *membranous labyrinth* is derived (**2**.79A-F). A vertical infolding of its wall progressively marks off a tubular diverticulum on the medial side, which differentiates into the *ductus* and *saccus endolymphaticus*, and they communicate via the ductus with the remainder of the vesicle—the *utriculosaccular chamber*—which is placed laterally. From the dorsal part of this chamber three compressed diverticula appear as disc-like evaginations; the central parts of the walls of the discs coalesce and disappear while the peripheral portions of the discs persist as *semicircular ducts*; the anterior duct is completed first, and the lateral last. From the ventral part of the utriculosaccular chamber arises a medially directed evagination which progressively coils as the *cochlear duct;* its proximal extremity constricts as the *ductus reuniens*. The central part of the chamber now represents the membranous vestibule, divided into a smaller ventral *saccule* and a larger *utricle* mainly by horizontal infolding which extends from the lateral wall towards the opening of the ductus endolymphaticus, leaving only a narrow *utriculo-saccular duct* between its divisions. This duct becomes acutely bent on itself, its apex being continuous with the ductus endolymphaticus (**7**.355). During this period the membranous labyrinth undergoes a rotation so that the long axis, originally vertical, becomes more or less horizontal (Bast & Anson 1949). Subsequently, otocyst derived cells (having contributed placodal cells to the vestibulocochlear ganglion) differentiate into the specialized paraneuronal hair cells of the utricle, saccule, ampullae of the semicircular ducts, and organ of Corti; they also differentiate into various specialized sustentacular cells and the unique epithelia of the stria vascularis and endolymphatic sac. The remainder form the general epithelial lining of the rest of the membranous labyrinth. (For structural details consult p. 1231 et seq.)

The mesenchyme surrounding the various parts of the epithelial labyrinth is converted into a *cartilaginous otic capsule*, and this is finally ossified to form most of the *bony labyrinth* of the internal ear. (Exceptions are the modiolus and osseous spiral lamina—vide infra.) For a time the cartilaginous capsule is incomplete and the cochlear, vestibular and facial ganglia are situated in the gap between its canalicular and cochlear parts. These ganglia

are soon covered by an outgrowth of cartilage and at the same time the facial nerve is covered in by a growth of cartilage from the cochlear to the canalicular part of the capsule. In the embryonic connective tissue between the cartilaginous capsule and the epithelial wall of the labyrinth the perilymphatic spaces are developed. The rudiment of the *periotic cistern* or vestibular perilymphatic space can be seen in an embryo of from 30 to 40 mm in length in the reticulum between the saccule and the fenestra vestibuli. The scala tympani is next developed and begins opposite the fenestra cochleae; the scala vestibuli is the last to appear (Streeter 1917). The two scalae gradually extend along each side of the ductus cochlearis, and when they reach the tip of the ductus a communication, the *helicotrema*, is developed between them. The modiolus and the osseous spiral lamina of the cochlea are not preformed in cartilage but ossified directly from connective tissue.

The auditory tube and tympanic cavity are developed from a hollow, termed the *tubotympanic recess* (Frazer 1914), between the first and third branchial arches, the floor of the recess consisting of the second arch and its limiting pouches. By the forward growth of the third arch the inner part of the recess is narrowed to form the tubal region, and the inner part of the second arch is excluded from this portion of the floor. The more lateral part of the recess subsequently develops into the *tympanic cavity* and the floor of this part forms the lateral wall of the tympanic cavity up to about the level of the chorda tympani nerve. From this it will be seen that the lateral wall of the tympanic cavity contains first and second arch elements, the first arch being limited to the part in front of the anterior process of the malleus. The second arch forms the outer wall behind this and turns on to the back wall to take in the tympanohyal region. Some observations, however, indicate that the tympanic cavity is derived wholly from the first pouch (Kanagasuntheram 1967). The tubotympanic recess is at first inferolateral to the cartilaginous otic capsule, but as the latter enlarges the relations become altered and the tympanic cavity becomes anterolateral. A cartilaginous process grows from the lateral part of the capsule to form the tegmen tympani and it curves caudally to form the lateral wall of the auditory tube. In this way, subsequent to ossification, the tympanic cavity and the proximal part of the auditory tube become included in the petrous region of the temporal bone. During the sixth or seventh month the mastoid antrum appears as a dorsal expansion of the tympanic cavity. Much of the cavity's basic development thus occurs during *fetal* life (Bok 1966). A study of the posterior part of the tympanic cavity and in particular of the sinus tympani, a recess between the promontory and pyramid (p. 1225), emphasizes the late fetal development of this region (Bollobas & Hajdu 1975).

The opinion long held as to the development of the auditory ossicles was that the *malleus* derived from the dorsal end of the ventral mandibular (Meckel's) cartilage (**2**.53) and the *incus* from the dorsal cartilage, probably corresponding to the quadrate bone of birds and reptiles. The *stapes* stems mainly from the dorsal end of the cartilage of the second (hyoid) arch, first as a ring (*annulus stapedis*) encircling the small stapedial artery (p. 214). The primordium of the stapedius muscle appears close to the artery and facial nerve at the end of the second month, and at almost the same time the tensor tympani begins to appear near the extremity of the tubotympanic recess (Candiollo & Levi 1969). Detailed analysis of early embryos concerning the *mesenchymal* origins of the blastemal ossicles, however, differs from the foregoing. Each ossicle has at least *two* distinct sources (see p. 172 and Hanson et al 1962).

At first the ossicles are embedded in the mesenchymal roof of the tympanic cavity and their extraneous origin is indicated in the adult by the covering which they receive from its mucous lining.

THE EXTERNAL EAR

The external acoustic meatus is developed from the dorsal end of the hyomandibular or first branchial groove. Close to its dorsal extremity this groove extends inwards as a funnel-shaped *primary meatus* from which the cartilaginous part and a small area of the roof of the osseous meatus are developed. From this funnel-

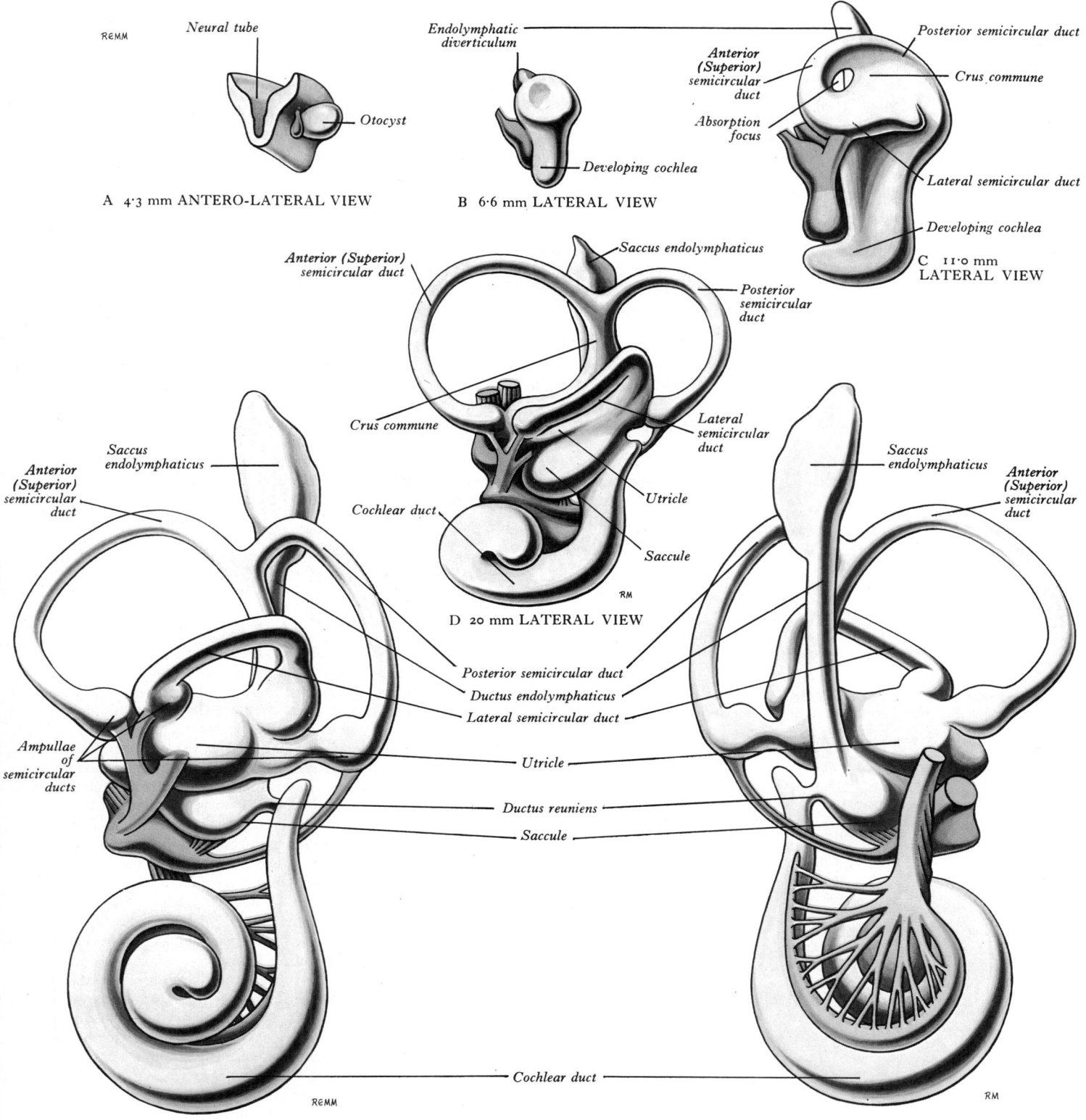

Neural tube

Otocyst

A 4·3 mm ANTERO-LATERAL VIEW

Endolymphatic diverticulum

Developing cochlea

B 6·6 mm LATERAL VIEW

Posterior semicircular duct

Anterior (Superior) semicircular duct

Absorption focus

Crus commune

Lateral semicircular duct

Developing cochlea

C 11·0 mm LATERAL VIEW

Anterior (Superior) semicircular duct

Saccus endolymphaticus

Posterior semicircular duct

Crus commune

Lateral semicircular duct

Cochlear duct

Utricle

Saccule

D 20 mm LATERAL VIEW

Saccus endolymphaticus

Anterior (Superior) semicircular duct

Anterior (Superior) semicircular duct

Saccus endolymphaticus

Anterior (Superior) semicircular duct

Ampullae of semicircular ducts

Posterior semicircular duct

Ductus endolymphaticus

Lateral semicircular duct

Utricle

Ductus reuniens

Saccule

Cochlear duct

E 30 mm LATERAL VIEW

F MEDIAL VIEW

2.79 A series of diagrams showing the stages in the development of the membranous labyrinth from the otocyst, at the embryonic stages and viewed from the aspects indicated. Note also the relationship of the vestibular (orange) and cochlear (yellow) parts of the vestibulocochlear nerve. (From a series of models prepared by His.)

shaped tube a solid epidermal plug extends inwards along the floor of the tubotympanic recess; by the breaking down of the central cells of this plug the inner part of the meatus (*secondary meatus*) is produced, while its deepest ectodermal cells form the epidermal stratum of the *tympanic membrane*. The fibrous stratum of the membrane is formed from the mesenchyme between the meatal plate and the endodermal floor of the tubotympanic recess.

The development of the *auricle* is initiated by the appearance of six hillocks which form round the margins of the dorsal portion of the hyomandibular groove. Of the six, three are on the caudal edge of the mandibular arch and three on the cranial edge of the hyoid arch (**2.125F**). These hillocks appear at the 4 mm stage, but they tend to become obscured as development proceeds and of those on the mandibular arch only the most ventral, which subsequently forms the *tragus*, can be identified throughout (**2.50**). The remainder of the auricle follows proliferation of the mesenchyme of the hyoid arch (Streeter 1922), which extends forwards round the dorsal end of the remains of the hyomandibular groove, forming a keel-like eleva-tion—the forerunner of the *helix*. The contribution made by the mandibular arch to the auricle is greatest at the end of the second month; as growth continues, it is relatively reduced; eventually the area of skin supplied by the mandibular nerve extends little above the tragus (7.340). The lobule is the last part of the auricle to develop.

The rudiment of the eighth nerve appears in the fourth week as the *vestibulocochlear ganglion*, which lies between the otocyst and the wall of the hindbrain. At first it is fused with the ganglion of the facial nerve (*acousticofacial ganglion*) but later the two separate. The cells of the vestibulocochlear ganglion are mainly derived from the neural crest, but with accessions from the neurectoderm of the otocyst; the ganglion divides into *vestibular* and *cochlear* parts, each associated with the corresponding division of the eighth nerve. The cells of these ganglia remain bipolar throughout life, each sending a proximal fibre into the brainstem, and a peripheral fibre to the internal ear. These neurons are also unusual in that many of their *somata* become enveloped in thin *myelin sheaths*.

The ganglionic fibres just described provide, of course, the afferent, sensory innervation of the labyrinthine hair cells. The latter soon become associated with the *outgrowing* axons from cells of the superior olivary complexes of the pons which provide an efferent innervation, the *olivocochlear bundle* (pp. 1112, 1241). Development details are, however, lacking in mankind.

Development of the Vascular System

The earliest blood vessels are derived from *angioblastic tissue*, which differentiates from extra-embryonic mesenchyme in three regions: (*a*) in splanchnopleure of the yolk sac; (*b*) in the body stalk (allantoic); and (*c*) somatopleure of the chorion (Hertig 1935, Bloom & Bartelmez 1940). In the yolk sac and *base* of the body

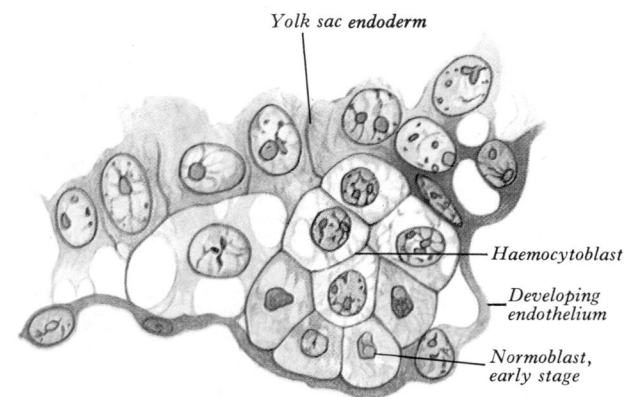

Yolk sac endoderm

Haemocytoblast

Developing endothelium

Normoblast, early stage

2.80A Part of a section through the wall of the yolk sac of an early human embryo to show an early stage in the differentiation of angioblastic tissue. (Hamilton et al 1962. Reproduced by permission of the authors and publishers.)

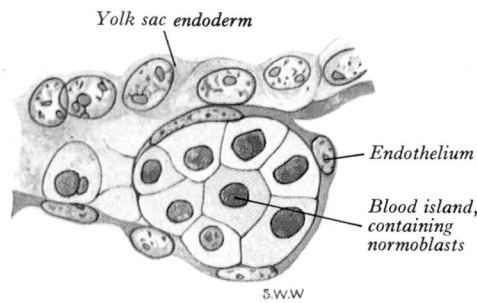

Yolk sac endoderm

Endothelium

Blood island, containing normoblasts

S.W.W

2.80B Part of a section through the wall of the yolk sac of an early human embryo, to show a developing blood vessel including a blood island. (Hamilton et al 1962. Reproduced by permission of the authors and publishers.)

stalk small, more or less spherical groups of cells are found early in the third week, termed *blood islands* (**2.80A,B**). Stages of trans-formation of islands into blood-containing *vessels* are controver-sial in detail, but it is widely believed that peripheral cells of the islands flatten as the vascular endothelium, while the central cells transform into primitive red blood corpuscles (**2.80B**). Later these small blood-containing spaces merge forming a continuous net-work of fine vessels. In the chorionic end of the body stalk and mesoderm lining the chorion typical blood islands are not found, but some mesodermal cells give rise to solid strands of angioblasts. Each strand contains two or three rod- shaped nuclei arranged in a single row, which soon develops a space occupied by one or more nucleated haemoglobin-containing cells. These spaces coalesce to form blood vessels which are lined by endo-thelial derivatives of the mesoderm; the precise source of their contained blood cells is uncertain. The earliest vessels, therefore, are formed at several separate centres; from the walls of these vessels buds grow out and become canalized, thus converted into new vessels which join with those of neighbouring areas to form a close meshwork. The heart and the blood vessels of the embryo arise from angioblastic tissue differentiated from the *intra-embryonic* mesoderm. Prior to the establishment of the circulation (p. 141), such new vessels (e.g. endothelial heart tubes, dorsal aortae, umbilical and early vitelline vessels) develop in situ, but thereafter they all take origin as outgrowths from pre-existing vessels. (Early vascular patterns are described on pp. 213 et seq. and 219 et seq., see also **2.88-97** and **2.87A,B**.) The subsequent development of the blood corpuscles is described on p. 675.

Development of the Heart

The precocious development of the heart has already been men-tioned (p. 141). In placental vertebrates it is the earliest major organ to function; for obvious nutritive reasons it must not only accommodate a stream of blood but also begin to propel it. These early *functional* demands on the heart represent an important factor in the dynamics of its development. The early appearance of cardiac activity in the tubular hearts of chick and rat embryos was noted many years ago (e.g. Sabin 1920 et seq., Goss 1942). First manifested by arrhythmic and sporadic ventricular contrac-tions, these are rapidly superseded by regular peristaltic activity propagated directionally along the atrioventricular tube. Earlier observers also described the emergence of valvular function by the accumulation, subjacent to the endothelium, of *cardiac jelly* or *subendocardial reticulum*. This occupies a large space between the primitive endocardium and myoepicardial mantle (vide infra and **2.81C,D**). These rapid and critical early events in cardiac mor-phogenesis have been re-examined with improved techniques, including electron microscopy, especially in the mouse (Challice & Viragh 1973). As the simple heart tube elongates and develops an asymmetrical twist, a succession of cavities, joined by more constricted regions, begin to define the sinus venosus, atrium, ventricle and bulbus cordis. Initially somewhat cylindrical, they rapidly become more spherical (within 24 hours in the mouse). Coincidentally, myoblastic strands invade the subendocardial reticulum from the cells of the myoepicardial mantle and form a complex network of intercommunicating *trabeculae*, external to but ultimately indenting and becoming clothed by endocardium. In many vertebrate groups the myocardium remains predominantly trabecular, but in birds and mammals compact layers of cardiac muscle develop external to the trabeculae. The latter also persist, however, but constitute a lesser volume of the propulsive tissue (vide infra).

These early events in cardiac development provide an approximate parallel between phylogeny and ontogeny. The evolution from a 'trabecular' heart to an organ with rounder chambers and a largely compacted myocardium is likely to be an expression of increased efficiency. An attempt to demonstrate this by mathematical analysis (Challice & Viragh 1973 et seq.) provides some corroboration and also an explanation for the persistence of the internal trabeculation in the mammalian heart. In the account which follows, attention is necessarily con-centrated upon the complex changes which transform a tube into

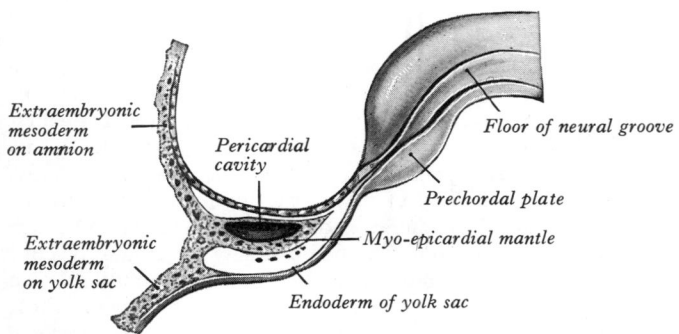

2.81A Median section through the cranial end of an early human embryo to show the position of the pericardium before the formation of the head fold. A few scattered angioblasts are seen between the cardiogenic plate and the yolk sac; they will ultimately form the endothelial heart tubes. (After C L Davis.)

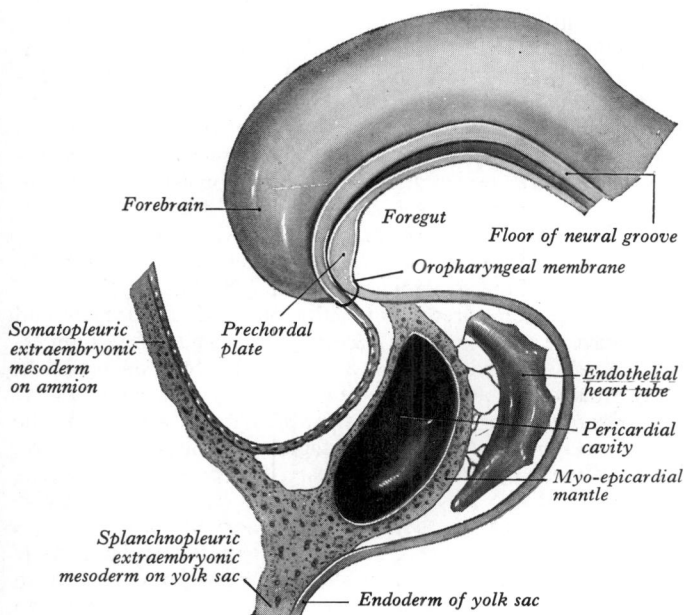

2.81B Median section through the cranial end of a young human embryo, showing the head fold in process of formation and its reversal effect on the position of the pericardium and endothelial heart; also intervening reticulum and myo-epicardium.

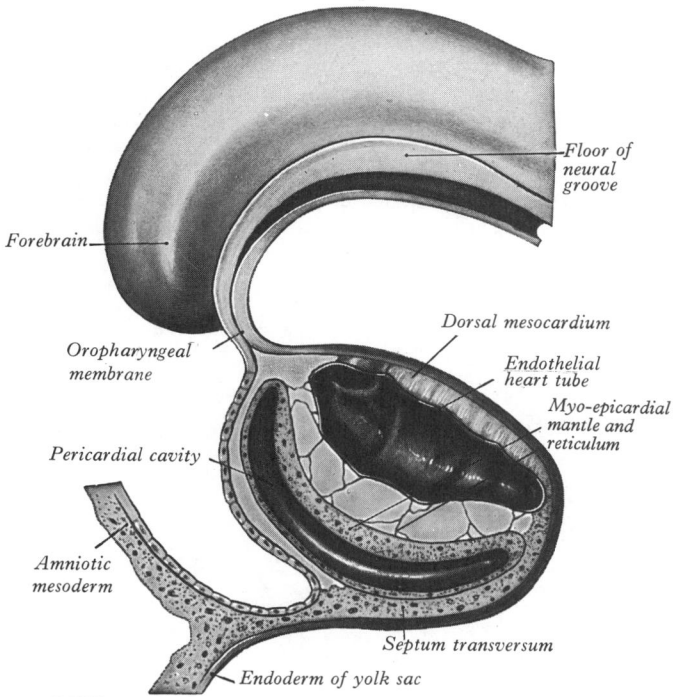

2.81C Median section through the cranial end of a young human embryo, after completion of the head fold and reversal of the pericardium.

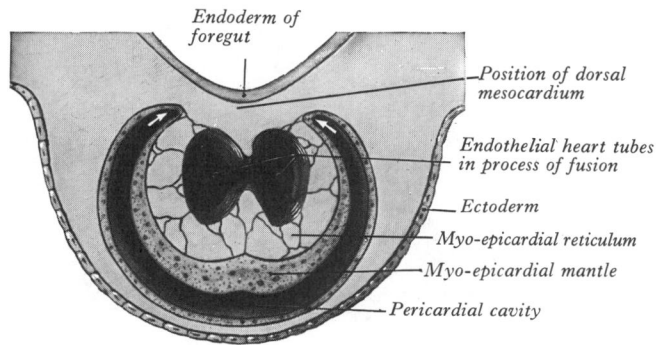

2.81D Horizontal section through the pericardium and developing heart of the embryo shown in **2.81C**. The arrows indicate the directions in which the dorsolateral recesses of the pericardium deepen so as to define the transient dorsal mesocardium. (After C L Davis.)

a chambered septate human heart. It is also necessary to keep in mind that at every stage the heart must be an effective circulatory pump.

The human heart, as in all vertebrates, is formed by the fusion of two symmetrically developing tubes (**2.81D**), but the fusion is progressive, commencing at the bulbar or arterial end and extending to the venous end (Tandler 1912, Davis 1927).

The pericardial cavity can be identified before the head fold is formed or during its early formation, when the embryo possesses only two somites. The heart is then represented by groups of angioblasts derived from the splanchnopleuric mesenchyme which lies between the pericardial cavity and the endoderm of the yolk sac (**2.81A**). At this stage the *ventral* (or yolk sac) pericardial wall, which will form both the epicardium and the myocardium, is thicker than the dorsal wall and is termed a *myo-epicardial mantle*. When the head fold is formed, the mantle becomes the *dorsal* wall of the pericardial *cavity* and lies *ventral* to the foregut. While this reversal of the pericardium is taking place (**2.81B,C**), the *cardiogenic mesenchyme* gives rise to two paramedian endothelial tubes which rapidly fuse to form a tubular heart. Except at its venous end the *tubular endothelial heart* is separated from the myo-epicardial mantle by a substantial accumulation of *cardiac jelly (gelatinoreticulum)*.

The proximity of the presumptive cardiac mesoderm and yolk sac endoderm at this stage of development is suggestive; in chick embryos, at least, experimental studies indicate that movements,

determination of polarity and regional differentiation of the cardiac mesoderm are in some way dependent upon the adjacent endoderm, and perhaps also ectoderm. For discussions and bibliographies concerning such experimental studies, and the importance of haemodynamic influences on regional cardiac morphogenesis, consult Stalsberg & De Haan (1968), Bellairs (1971), Balinsky (1981), Orts-Llorca et al (1982). Thus, in addition to such fundamental phenomena common to all organs, including genomic encoding and differentiation with pattern formation, discussed elsewhere, other aspects have received particular emphasis in cardiogenesis. As indicated the endothelial cardiac tube is separated from the proliferating myo-epicardial mantle by elastically deformable gelatinoreticulum. The latter is invaded by waves of mantle-derived fasicles of differentiating cardiac myoblasts and fibroblasts, whilst interiorly its endothelial lining receives the impact of changing directions of blood flow and pressure profiles. With the emergence of early cardiac chambers the interchamber protrusions of jelly provide a primitive valvular system. With development, these processes contribute to the sculpturing, not only of the subendocardial trabeculae, but also the papillary muscles, chordae tendinae and valve leaflets, the septal complexes between the heart chambers, the complex spiral bulbotruncal ridges and septum, and the fibrous 'skeleton' of the heart (vide infra, and pp. 696 et seq.).

The dorsal aortae arise in situ as paired endothelial vessels. They extend caudally into the body stalk, establishing continuity with the umbilical arteries, which precede them in time of appearance. At their cranial ends the dorsal aortae curve ventrally round the sides of the foregut to reach the pericardium and become continuous with the cranial end of the endothelial heart tube, thus forming the first pair of aortic arches (**2.82A**). In all vertebrates in which the heart and aortae are laid down *before* the formation of the head fold, the arteries communicate with the *caudal* end of the heart. When the head fold forms, as noted, the ends of the heart are *reversed* and the cranial ends of the dorsal

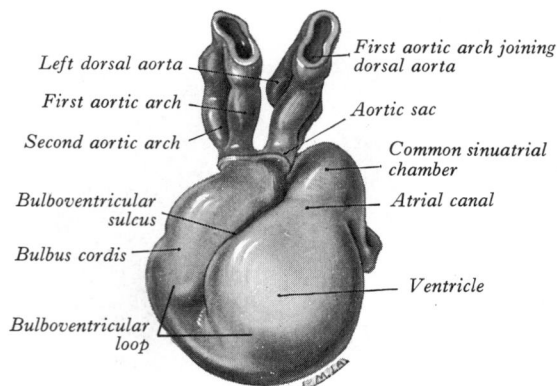

Left dorsal aorta

First aortic arch

Second aortic arch

Bulboventricular sulcus

Bulbus cordis

Bulboventricular loop

First aortic arch joining dorsal aorta

Aortic sac

Common sinuatrial chamber

Atrial canal

Ventricle

2.82A The heart of a 0.95 mm rabbit embryo, viewed from the ventral side. (Drawn from a model by G Born.)

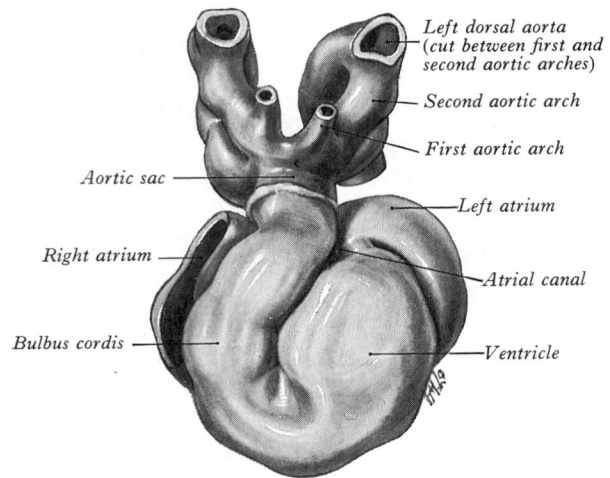

Left dorsal aorta (cut between first and second aortic arches)

Second aortic arch

First aortic arch

Aortic sac

Left atrium

Right atrium

Atrial canal

Bulbus cordis

Ventricle

2.82B The heart of a 1.7 mm rabbit embryo, viewed from the ventral side. (Drawn from a model by G Born.)

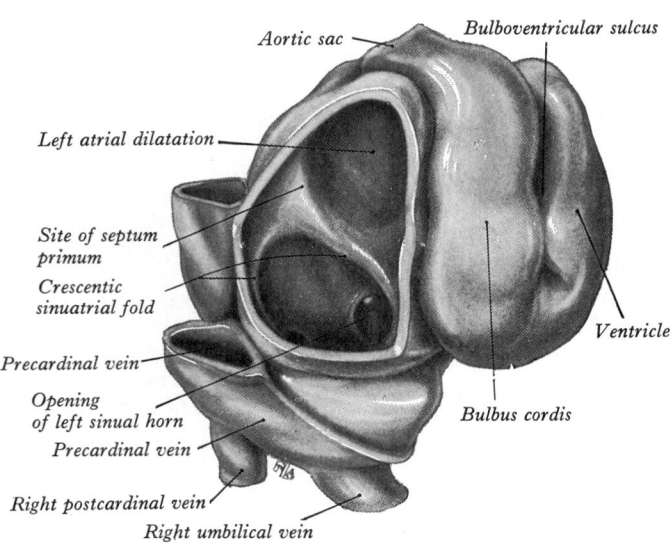

Aortic sac

Bulboventricular sulcus

Left atrial dilatation

Site of septum primum

Crescentic sinuatrial fold

Precardinal vein

Opening of left sinual horn

Precardinal vein

Right postcardinal vein

Right umbilical vein

Ventricle

Bulbus cordis

2.82C The heart shown in **2.82B**, viewed from the right side and slightly from the ventral aspect. The right wall of the common sinuatrial chamber has been removed to show the interior. (Drawn from a model by G Born.)

aortae are curved forwards round the sides of the foregut as the first aortic arches.

A transverse groove appears on the surface of the heart tube about its middle, the junction of the *bulbus cordis* with the *ventricle*. The bulbus is cranial to the groove and continues as the first pair of aortic arches. The ventricle shows a second groove at its caudal end where it opens into a *common atrium*, which, initially, is embedded in the floor of the pericardium (the future *septum transversum*) and the chamber is disposed transversely. On each side the common atrium is joined caudally by a short venous trunk, formed by the union of the corresponding umbilical vein with veins issuing from the *vitelline* (yolk sac) *plexus*. These trunks represent the right and left *horns* of the *sinus venosus (sinual horns)* so that the common atrium may justifiably be termed a *common sinuatrial chamber*. The umbilical and vitelline radicles of each sinual horn are soon joined laterally by a *common cardinal vein* (each the confluence of a *precardinal* and *postcardinal vein* —p. 223).

Early in the fourth week the heart tube undergoes a striking change. Hitherto the pericardium increased in length proportionately with the heart, but now the heart tube grows more rapidly than the pericardium and the bulboventricular tube bulges ventrally and caudally, forming a U-shaped loop; the bulb is the right limb and the ventricle the left. The loop is conspicuous throughout the fourth and fifth weeks, and seen as a deep *bulboventricular sulcus* externally (**2.82A**) and a corresponding *bulboventricular ridge* projects internally. Other factors often considered operative in determining the disposition of the loop are modifications in the tubular heart flow patterns, and co-ordinated ciliary action of the coelomic epithelium (Afzelius 1979).

The dorsolateral recesses of the pericardium deepen and approach one another (**2.81D**); their opposed walls fuse, completing the myo-epicardial covering of the heart thus converting its early broad dorsal attachment into a thinning cardiac mesentery. This *dorsal mesocardium* is transient breaking down early in the fourth week and establishing a passage across the pericardial cavity from side to side dorsal to the heart. This persists as the *transverse sinus* of the pericardium. While these bulboventricular changes are occurring, the atrial part is also affected; the atrioventricular opening moves cranially and to the left, and both parts of the common atrial or sinuatrial chamber 'rise' or emerge from the mesoderm of the septum transversum to grow cranially into the pericardial cavity dorsal to the ventricle. Owing to these changes the atrioventricular canal for a time connects the *left* part of the atrium to the ventricle and venous blood from the right side has to pass through both parts of the atrium.

At this stage (**2.87A**), about the middle of the fourth week, the bulbus cordis communicates with the dorsal aortae through the first pair of aortic arches, and both are connected with the capillary plexus associated with the developing cerebral vesicles. From this plexus the primitive head vein passes caudally, but ends blindly before it reaches the heart. The intersegmental arteries begin to grow out from the dorsal aorta on each side but have not yet established connections with their corresponding veins; and the postcardinal veins, which later drain the body wall caudal to the heart, are only in process of development. The umbilical arteries and veins are defined and, early in the fourth week, their terminals and radicles, respectively, link up with the capillaries which have developed in the chorionic villous stems, establishing the chorionic part of the circulation. Despite the fact that the channels for the remainder of the circulation are only partially established there is good ground for assuming, from observations made on living embryos by a variety of techniques, that the heart begins to contract about this time. Under the prevailing conditions, the effect can only be of an 'ebb and flow' nature, but this also serves to effect some movement in the nutritive fluid filling the pericardial cavity, coelomic ducts and exocoelom (p. 141), on which the embryo is still heavily dependent.

Towards the end of the fourth week the connection between the bulbus cordis and the first pair of aortic arches lengthens to form the *truncus arteriosus*, and the cranial end of this vessel becomes connected to the dorsal aortae by the remaining five pairs of aortic arches. By this time the *venous drainage* of the body wall and neural tube has been established. On each side a *precardinal vein*, from the cranial end of the embryo, unites with a *postcardinal vein* from the caudal region to form the *common cardinal vein (duct of Cuvier)*; the latter vessel opens close to the umbilical and vitelline veins into the dorsocaudal part of the common sinuatrial chamber (the three vessels forming the right or left *sinual horn*).

As the chorionic circulation already exists the embryo can now exchange materials with the maternal blood in the intervillous space. This is not effected suddenly; initially the blood volume and its cellular content in the heart and vessels of the embryo is insufficient to enable it to take full advantage of this new source of nourishment, and until this is rectified, the embryo continues to draw upon the coelomic fluid.

The separation of a definitive *sinus venosus* from the *common atrium* completes the definition of the primitive chambers of the heart. A crescentic groove appears on the left wall of the sinuatrial chamber and rapidly deepens to the right. Hence the left horn of the sinus venosus loses its connection with the left part of the atrium and becomes linked to the right sinual horn by separation of the caudal part of the sinuatrial chamber; the latter now constitutes the *body of the sinus venosus*. At the same time the right sinual horn becomes more clearly demarcated from the right part of the atrium by a shallow groove, and its wide connection with the atrium (**2.82**C) becomes relatively smaller (Foxon 1955). The right and left parts of the atrium grow cranially to occupy the dorsal part of the pericardial cavity, and later they bulge forwards, embracing the sides of the bulb (**2.83**A).

The embryo has now reached a length of nearly 4 mm (**2.87**B). It possesses 28 somites and has almost completed the fourth week of development. From this stage onwards it is more convenient to deal with the individual chambers with only occasional reference to the development of the heart as a whole.

It must be noted that the above account of early cardiogenesis, though widely subscribed to, is not without its critics. In considering the many factors governing cardiac development— phylogenetic, ontogenetic, and physiological— the last is usually underestimated and the first is perhaps stated too dogmatically (Foxon 1955). Ontogenetic mechanisms must conform to the early demand for a functioning heart, and cardiogenesis is not necessarily a mere repetition of phylogenetic steps, which are themselves uncertain, however plausible they may seem (De Vries & Saunders 1962).

THE SINUS VENOSUS

The right sinual horn increases rapidly in size at the expense of the left, due to the changes already outlined and to those occurring in the originally symmetrical arrangement of the umbilical and vitelline veins by the development of the liver (**2.93**-95, p. 219). As a result the vitello-umbilical blood flow enters the right horn through a wide but short vessel, the *common hepatic vein*, which becomes the cranial end of the inferior vena cava. In addition, the right horn receives the right common cardinal vein (from the body wall of the right side) and the body of the sinus, which conveys the blood from the left horn and left common cardinal vein. Later, when *transverse connections* are established between the cardinal veins (**2.96**-98), the blood from the body wall of the left side reaches the heart via the veins of the *right* side. The left common cardinal vein then becomes much reduced in size and forms the oblique vein of the left atrium and the fold of the left caval vein, while the left horn and the body of the sinus venosus persist as the *coronary sinus* (**2.97**).

The right sinual horn opens into the right atrium through its dorsal and caudal walls. The orifice, elongated and often slit-like, is guarded by two muscular folds, the *right* and *left sinuatrial (venous) valves (valvules)* (**2.83**A). These two valves meet cranially and become continuous with a fold which projects into the atrium from its roof, the *septum spurium*. Caudally the valves meet and fuse with the dorsal endocardial cushion of the atrial canal. The cranial part of the right sinuatrial valve loses its fold-like form, but its position is indicated in the adult heart by the crista terminalis of the right atrium; its caudal part forms the valve of the coronary sinus and most of the valve of the inferior vena cava. The medial (or left) end of the valve of the inferior vena cava is formed by a small fold continuous with the dorsal wall of the sinus

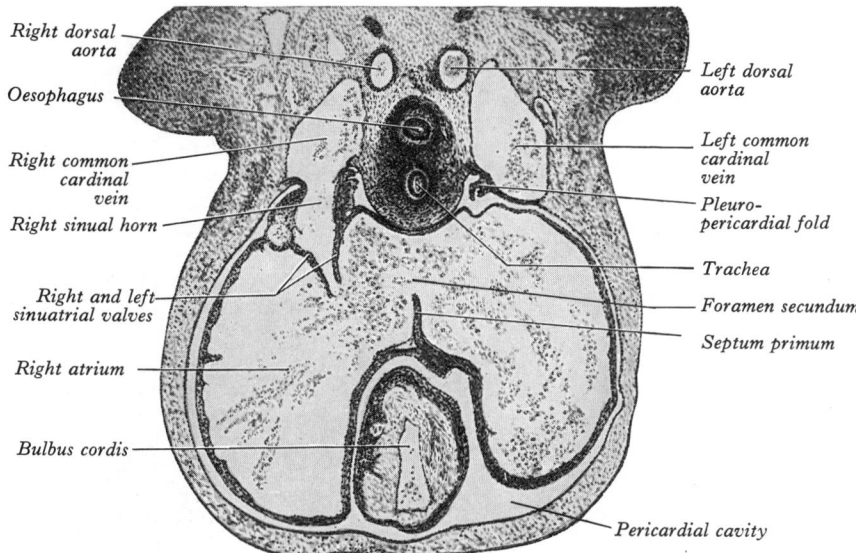

2.83A Transverse section of a human embryo, 8 mm long. Observe how the atria bulge forwards on each side of the bulbus cordis. The septum primum has broken down in its dorsal region and the two atria communicate through the ostium secundum or foramen ovale.

venosus, the *sinus septum*. The latter intervenes between the orifice of the common hepatic vein and the opening of the body of the sinus. (In the mature heart see the *tendon of Tondaro* and *triangle of Koch* pp. 703, 720.)

The left venous valve blends with the right side of the atrial septum and usually no trace of it can be seen in the adult heart.

As the sinuatrial valves undergo these changes the right sinual horn becomes incorporated in the right atrium and expands to form its smooth dorsal wall, medial to the crista terminalis. This part of the adult atrium is termed the *sinus venarum*, the receiving chamber of the large venous orifices. The right half of the primitive atrium forms the internally ridged, more muscular, wall anterior to the crista terminalis and the right auricular appendage.

THE RIGHT AND LEFT ATRIA

As stated, the common atrium is derived from the cranial part of the sinuatrial chamber. It receives the opening of the sinus venosus dorsocaudally and to the right of the median plane, while it communicates ventrally with the ventricle through the atrioventricular canal, which has resumed its median position by the middle of the fifth week, thus permitting both right and left parts of the atrium to communicate with the common ventricular cavity. Dorsal and ventral swellings appear in the walls of the atrioventricular canal between the endothelial tube and the myoepicardial mantle. These, the *atrioventricular endocardial cushions*, consist of a core of gelatinoreticular mesenchyme. They encroach on the canal and eventually fuse, leaving a relatively small orifice on each side. The fused tissue constitutes the *septum intermedium* (of His), which separates the two, small, right and left atrioventricular orifices and canals.

Internal separation into right and left atria is mainly effected by sequential growth of two septa (but with additional, less prominent structures). First the *septum primum* grows from the dorsocranial atrial wall as a crescentic fold (**2.83**B,A), separated from the left sinuatrial valve by the *interseptovalvular space*. The ventral horn of the crescent reaches the ventral atrioventricular cushion, the dorsal horn the dorsal cushion. Ventral and dorsal refer to the positions of the cushions after the atrium repositions to lie dorsal to the bulbus cordis. Strictly the cushions are ventrocranial and dorsocaudal in position, but ventral and dorsal are in general use and will be retained here. Ventral and caudal to the advancing edge of the septum the two atria communicate through the *foramen primum* (**2.83**B,A). Free passage of blood from right to left atrium is essential throughout fetal life, as oxygenated blood from the placenta reaches the heart via the inferior vena cava (p. 723); therefore, as the foramen primum diminishes, the septum primum breaks down dorsally and a new right-left shunt, the *foramen secundum*, is formed before the end of the fifth week. The foramen primum is finally occluded by fusion of the edge of the septum primum with the fused atrioventricular cushions, in the median plane. The foramen secundum enlarges, allowing sufficient free passage of blood from right to left atrium (2.83B), and it persists throughout intrauterine life as *part* of the valvular *foramen ovale* in the progressively changing *interatrial septal complex* (vide infra). At first the foramen secundum is sited craniodorsally in the septum primum but it becomes modified until it is cranioventral.

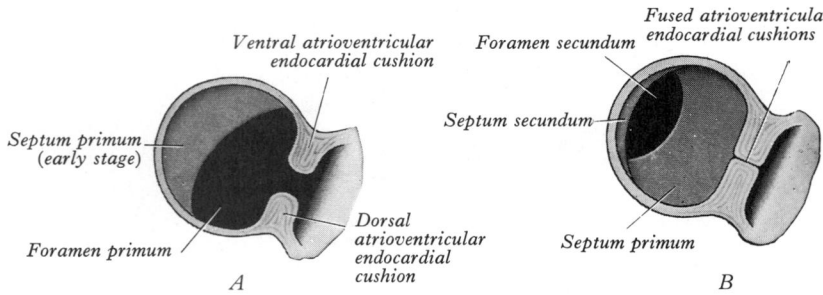

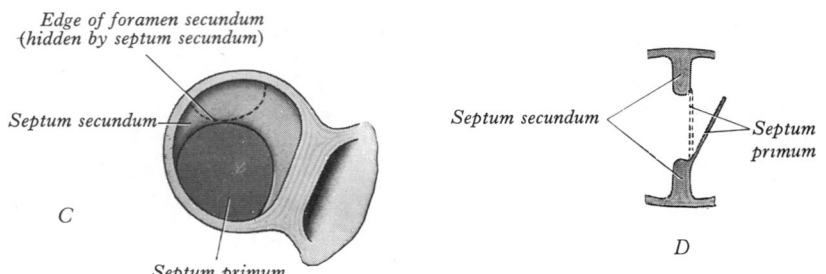

2.83B Diagrams representing three stages in the development of the atrial septum, viewed from the right side. The heart has been divided in its long axis to the right of its median plane and only the atria and the adjoining part of the ventricular cavity are depicted.

A The septum primum has not yet obliterated the original communication between the two atria and the atrioventricular endocardial cushions have not yet fused.

B The atrioventricular endocardial cushions have fused with each other and with the septum primum, which has broken down in its dorsal part. The foramen secundum, thus formed, subsequently moves to the position shown in C.

C The septum secundum has formed and hides the foramen secundum, the margins of which are indicated by the curved, dotted line.

D Section to show the valve-like character of the foramen secundum. When the pressure in the right atrium exceeds that in the left atrium, blood passes from the right to the left side of the heart, but when the two pressures are equal the septum primum assumes the position indicated by the dotted outline.

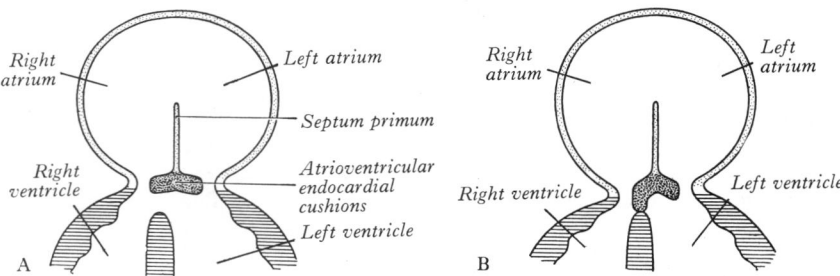

2.83C Diagram to show two stages in the formation of the adult ventricular septum. In A the right and left ventricles communicate with each other, but in B the interventricular communication has been closed by the fusion of the ventricular septum with the enlarged right extremity of the fused a.v. cushions. Note the position of the septum primum relative to the fused a.v. cushions, and observe that in B cushion tissue intervenes between the two ventricles (membranous part of ventricular septum), and also between the *right* atrium and the *left* ventricle (atrioventricular septum).

Towards the end of the second month the muscular wall of the atrium becomes invaginated as another crescentic septum on the right side of the septum primum (2.83B,*B,C*). This, the *septum secundum*, involves more than the whole width of the interseptovalvular space; thus the dorsal attachments of the septum primum and the left sinuatrial valve are carried into the interior of the atrium on its left and right surfaces respectively. The superior (ventrocranial) and inferior (dorsocaudal) horns of the septum secundum at first grow ventrally; the superior horn grows much more rapidly and fuses first with the septum intermedium; it is then continuous with the *sinus septum* (vide supra). Thus the free edge of the septum secundum (*crista dividens*) is at first directed caudoventrally and later caudally alone; it overlaps the foramen secundum (2.83B,*C,D*); thus the septum primum acts as a flap valve. Since the blood pressure is greater in the right atrium than in the left, the blood flows from right to left, but not conversely. The right-left flow occurs through the 'true', but somewhat misnamed *foramen ovale*, proceeding from the right atrium under the crescentic free border of the septum secundum, thence through the oblique cleft between the (parted) secondary and primary septal surfaces, to finally enter the left atrium through the foramen secundum. After birth the intra-atrial pressures are equalized and the free edge of the septum primum is therefore kept in contact with the left side of the septum secundum and fusion occurs. Not infrequently the fusion is incomplete, but the remaining cleft is usually small, valvular and has no functional significance. The initially free, crescentic margin of the septum secundum forms, after fusion, the *limbus fossae ovalis* and the septum primum the floor of the *fossa ovalis* of the adult heart. An alternative derivation of the septum secundum from a ridge developing to the right of the line of fusion between the confluent endocardial cushions and the septum primum has been advanced (Odgers 1934). The dorsal horn of the septum secundum is said to incorporate tissue derived from part of the left sinuatrial valve; its contribution, however, remains uncertain; sometimes, small vestigial remnants persist to maturity. Another view embodies the above suggestion, but regards the valve as of minor importance in this connection, describing, however, yet another ridge—the *septum accessorium*, contributing to the lower part of the dorsal border of the limbus (Christie 1963).

Early in the development of the septum primum a single, common pulmonary vein of controversial origin opens into the caudodorsal wall of the left atrium close to the septum. It is the union of a right and a left pulmonary vein, each formed by two small veins issuing in turn from each developing lung bud. Subsequently the common trunk and the two veins forming it expand and are incorporated in the left atrium to make up the greater part of its cavity. This expansion usually continues as far as the orifices of the four veins, which thus open separately into the left atrium; variations, however, are quite common. The left half of the primitive atrium is progressively restricted to the mature auricular appendage.

During the second month the two atria bulge ventrally one on each side of the bulbus cordis, which lies in a groove on their ventral surface (2.82B,83A). These projecting parts of the atria form the auricular appendages of the adult heart.

THE VENTRICLES BULBUS CORDIS AND THE TRUNCUS ARTERIOSUS

The process of separation of the ventricles is intimately related to that of the aortic and pulmonary orifices at the distal end of the bulbus (2.84A-C) and also to the division of the truncus arteriosus into pulmonary and aortic channels. Their interdependence is such that the history of the truncus arteriosus is dealt with here, although strictly it takes no part in the formation of the heart itself. Bulbotruncal separation is conveniently considered before final interventricular septation and valve modelling. As intimated these complex events are the result of mutual *interaction* between factors controlling pattern formation, differential growth and the continuously changing blood-flow paths, volumes and pressures. This interplay moulds the grooves, ridges, outpouchings, valve complexes and varying myocardial thickness and architecture. Although described sequentially, many of these events occur simultaneously. Blood enters the bulboventricular cavity through the right and left atrioventricular canals (ventricular *inflow tracts*) and is ejected through the proximal and distal bulbus (*outflow tracts*). That from the future left ventricle passes obliquely to the *dorsal* part of the bulbus, whereas right ventricular blood has a reverse inclination to the former and is expelled through the *ventral* part of the bulbus. These inclinations impose a mutually spiral flow on the two streams as they traverse the truncus.

Four endocardial cushions—ventral, dorsal, right and left—form in the distal part of the bulbus and the right and left cushions fuse to constitute a *distal bulbar septum*. This separates a ventral, *pulmonary orifice* from a dorsal, *aortic orifice*, and later the cushions divide and become modified to form the *semilunar valves* (vide infra).

The separation of the pulmonary trunk from the aorta is a more complicated process. Two ridge-like thickenings project into the interior of the truncus arteriosus between the entwined spiralized streams of blood. Proximally, the ridges project from the *lateral* walls of the vessel but, progressing distally, the right ridge passes obliquely on to the *ventral* and then the *left* wall, while the left ridge extends on to the *dorsal* wall and then the *right* wall (**2.84B**). The ridges are therefore spiral, and their fusion forms the *spiral aorticopulmonary septum*. Proximally this meets and fuses with the distal bulbar septum, and in accord with its spiral form the pulmonary trunk, which lies ventral to the aorta at its orifice, curves round to its left side as it ascends and finally lies dorsal to it (**2.84B**). Distally the aorticopulmonary septum meets the dorsal wall of the aortic sac (see p. 214) cranial to the point where it is joined by the sixth pair of aortic arches, and thus the latter become branches of the pulmonary trunk while the remaining arches retain communication with the aorta (**2.84B**).

The separation of the two ventricles from each other leaves the right ventricle in communication (inflow tract) with the right atrium and (outflow tract) the pulmonary artery, and the left ventricle in communication (inflow tract) with the left atrium and an outflow tract via the aorta. This involves a series of complex changes in which three distinct factors contribute to the formation of the *adult ventricular septum*—(*a*) the *fetal ventricular septum*, (*b*) the *proximal bulbar septum* and (*c*) the *atrioventricular endocardial cushions*.

(*a*) During the fifth week the right and left definitive ventricles appear as slight projections on the external surface of the primitive common ventricle. It is uncertain whether the right definitive ventricle is solely a derivative of the common ventricle, or of the caudal end of the primitive bulbus, or of both. In either event, the appearance of a caudal crescentic ridge in the inside of the heart indicates the separation between the two ventricles and, as the heart enlarges, this ridge deepens to form the early *ventricular septum*. The dorsal and ventral horns of the septum grow along the ventricular walls to meet and fuse with the corresponding endocardial cushions of the atrioventricular canal near their *right* extremities (**2.84A**). The septum has a free sickle-shaped margin which, with the endocardial cushions, bounds a circular *interventricular foramen* (**2.84B**).

At first the bulboventricular junction is marked by a distinct notch on the outside of the heart (**2.82A**) and inside is a corresponding *bulboventricular ridge*. The latter between the atrioventricular orifice and the caudal part of the bulb (**2.84A**) and its absorption is essential for the development of a four-chambered heart. Partly by absorption of the bulboventricular ridge and partly growth of the atrioventricular region, the *right* extremity of the atrioventricular canal comes to lie *caudal* to the orifice of the bulb (**2.84B**). This alteration in relative positions of the structures concerned occurs while the ventricular septum is forming, paving the way for completion of ventricular partition (Wenink 1971, 1976).

(*b*) The *proximal bulbar septum* separates the bulbus cordis into pulmonary and aortic channels (*ventricular outflow tracts*), and is formed by the *right* and *left bulbar ridges*, which are in continuity with the corresponding distal bulbar endocardial cushions (which form the *distal bulbar septum*—vide supra). The right bulbar ridge grows across the dorsal wall of the bulb and right extremity of the fused atrioventricular endocardial cushions to reach the dorsal

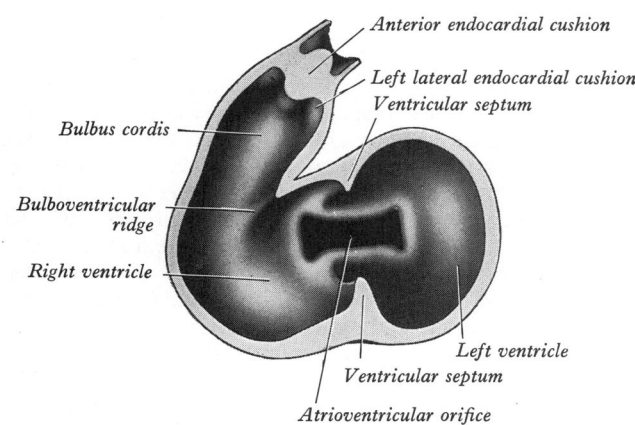

2.84A Diagram showing an early stage in the relations between the atrioventricular opening and ventricles, the cavity of the bulbus cordis, and the bulboventricular ridge. The endocardial cushions at the distal end of the bulb are shown in a more differentiated state than they really exhibit at this stage. (After J E Frazer.)

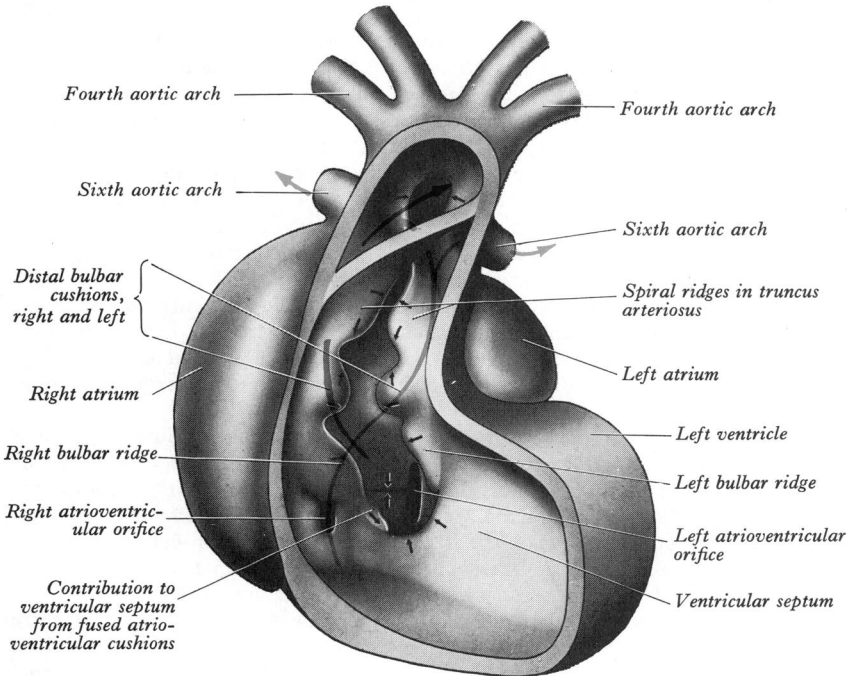

2.84B Diagram to show the mode of formation of the septa which separate the aortic and pulmonary channels in the embryonic heart. The red arrow indicates the aortic channel and the blue arrow the pulmonary. The small black arrows indicate the direction of growth. (From a model by James Whillis.)

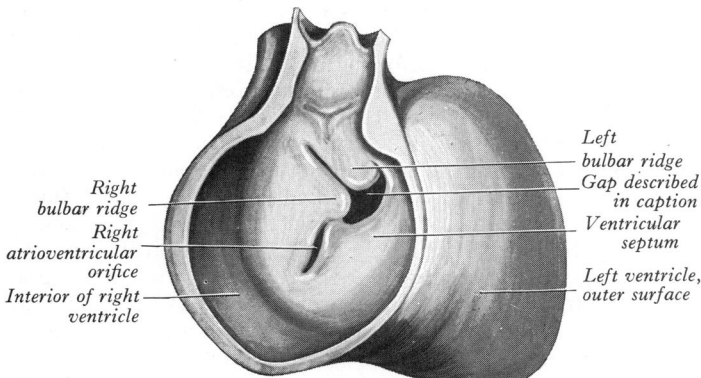

2.84C Diagram to show the part played by the fusion of the right and left bulbar ridges in the separation of the aortic and pulmonary channels. The darkly shaded area indicates the gap filled by the proliferation of cushion tissue (from the right extremity of the fused atrioventricular cushions) which establishes continuity between the proximal bulbar septum and the ventricular septum. Compare with **2.84B**.

horn of the free, crescentic edge of the ventricular septum and obliterates the ventral or cranial part of the right atrioventricular orifice (2.84B). The left bulbar ridge crosses the ventral wall of the bulb to reach the ventral or cranial horn of the ventricular septum. The bulbar ridges fuse thus separating the conus arteriosus of the right ventricle from the aortic vestibule; however, the caudal edge of the bulbar septum is still separated from the free crescentic edge of the ventricular septum by a diminishing interventricular channel. The latter is closed by growth of tissue from the right extremity of the fused atrioventricular cushions (Odgers 1938) and this fuses, on its one aspect with the caudal border of the proximal bulbar septum and on its other with the margin of the ventricular septum. The dorsal part of the bulb largely becomes absorbed, but its position is indicated by the dorsal wall of the aortic vestibule, which, however, is mainly formed by tissue extensions from the fused atrioventricular endocardial cushions.

(c) At their time of fusion the *atrioventricular endocardial cushions* are large relative to the size of the atrioventricular orifices. The atrial septal complex meets the approximate centre of the atrial surface of the cushions; the ventricular septum, however, meets them near their right margins. Thus a part of the fused cushions intervenes between the *right atrium* and the *left ventricle*, and it is this which forms the right wall of the aortic vestibule (*atrioventricular septum*, see 6.34). The *membranous* part of the *interventricular septum*, continuous dorsally with the membranous *atrioventricular septum* in the completed heart (2.83C), is also formed by proliferation of cushion tissue from the right extremity of the fused atrioventricular cushions. The persistence of an interventricular communication may follow anomalous development in this region. It may be noted that it is the craniodorsal part of the bulbar orifice, which lies above the ventricular septum (2.79B), and normally becomes incorporated in the aortic vestibule through which the left ventricle discharges into the truncal aortic channel. In some cases of persistent interventricular foramen the aortic orifice is described as 'overriding' the free upper border of the muscular interventricular septum (p. 725).

The valve complexes of the heart, four in number, arise in two main cardiac zones: the aortic and pulmonary valves at the distal bulbotruncal junction, the mitral and tricuspid complexes extending from their inception between the atrioventricular junctions and loci on the interior of the ventricular walls. Each commences as an internal endocardial (endothelial) projection of varying form enclosing cardiac mesenchyme (gelatinoreticulum). In some regions the mesenchymatous cells proliferate, transform into fibroblasts, and produce a geometrically organized collagenous framework that varies with site and functional demands. Elsewhere the core is invaded by differentiating cardiac myoblasts.

The *atrioventricular valves* develop as shelf-like projections from the margins of the atrioventricular orifices, directed as almost complete conical sheets towards the ventricles, their advancing edges continuing, initially as trabecular ridges, deep into the ventricular cavity. With continued differential growth and excavation on their ventricular aspects, each sheet develops two (mitral) or three (tricuspid) marginal indentations, defining the principal *valve leaflets*, minor marginal indentations (clefts) subdividing some leaflets into scallops. Each leaflet develops functionally significant regional variations in surface texture; its core condenses as a collagenous lamina fibrosa. The latter blends at its atrioventricular base with the inappropriately named fibroareolar valve 'annulus'—each a part of the complex, functionally crucial, fibrous 'skeleton' of the heart. The intermediate regions of the trabeculae pass between the margins of the leaflets, indentations, clefts and defined zones of the leaflet surface as the white, glistening, compacted, collagenous *chordae tendinae* and these converge towards the tip and sides of the single or grouped *papillary muscles* and blend with their connective tissue framework. The muscles are the ventricular ends of the original trabeculae and, whilst free throughout their length, their mural ends are confluent with mural ventricular musculature and receive a dense population of its nerves and specialized conducting tissues. (For details of the disposition, architecture and some functional implications of the cardiac valve complexes and fibrous skeleton see pp. 702–717.) The *aortic* and *pulmonary valves* are formed from the four endocardial cushions which appear at the distal end of the bulbus cordis. The completion of the distal bulbar septum results in division of each lateral cushion into two; thus the number of thickenings is increased to six: three associated with the pulmonary orifice and three with the aortic. These are the rudiments of the aortic and pulmonary valves. Each cushion-derived intrusion grows and is excavated on its truncal aspect to form a semilunar valve cusp. Similar events affect the adjacent truncal or septal wall. Thus the pouches between the valves and the walls of the vessels gradually enlarge and form their related *sinuses*. The core of each cusp forms a collagenous lamina fibrosa, delicate and thin in each crescentic lunule, thick and compact in the central nodule, with marginal radiate and basal bands. The latter blend with the complex, scalloped, mural valve ring (pp. 715–717). Initially, one cusp of the pulmonary valve lies anteriorly and the other two posterolaterally, whereas one cusp of the aortic valve lies posteriorly and the other two anterolaterally. However, a rotation of the heart to the left before birth changes the orientation of the cusps of the pulmonary and aortic valves and this is reflected in the various schemes for the designation of these cusps in the mature heart (see p. 707-712).

The development of the chambers of the heart has now been traced to a stage at which the main features of the adult heart are established. It is to be noted that the pattern has developed in such a way as to provide for the sudden establishment of the pulmonary circulation at birth (Dawes 1961, 1969), although it is adapted to the persistence of the placental circulation for the remainder of fetal life. The presence of the ductus venosus ensures that a substantial proportion of umbilical oxygenated blood gains the right atrium with a limited loss of oxygen to the liver. However, some umbilical blood and portal venous blood enter the hepatic sinusoids through venae advehentes; their drainage through venae revehentes (eventually the grouped hepatic veins) returns the blood to the hepatic segment of the inferior vena cava, some admixture occurring here. It was claimed that only minor further admixture of relatively oxygenated and deoxygenated blood occurs in the *right* atrium (Barclay et al 1939) and that nearly all the oxygenated blood passes through the foramen ovale into the left atrium, so gaining the *left* ventricle, aorta and systemic circulation. However, there is evidence for the opposing view that there is considerable mixing of the superior and inferior vena caval streams in the right atrium (Born et al 1954, Lind & Wegelius 1954). The inferior vena caval blood is directed by its valve towards the cleft ('foramen ovale') in the atrial septal complex. It has been estimated that about 75% passes through to the left atrium; the remainder mixes with the deoxygenated blood from the superior vena cava; separation occurs at the crista dividens of septum secundum. Because the transition from a placental to a pulmonary circulation occurs suddenly at birth, the right ventricle and the pulmonary trunk of the fetus are relatively large, although only a small amount of blood passes through the lung. Most of the blood expelled by the right ventricle to the pulmonary trunk passes through the ductus arteriosus to the descending aorta and therefore is under higher pressure than blood in the aorta. Thus the muscular wall of the *right* ventricle is *thicker* than the wall of the left, a condition which persists throughout fetal life but is progressively reversed postnatally. The origin of the carotid and subclavian arteries from the aorta above the junction with the ductus arteriosus may be correlated with the relatively rapid growth of the brain, demanding a copious blood supply, with the more advanced development of the upper limbs, relative to the lower, at birth, and the overall cephalocaudal gradient in the growth of the trunk.

In early development the arteries of the embryo are disproportionately large and their walls consist of little more than a single layer of endothelium. The cardiac orifices are also relatively large and the force of the cardiac contraction is weak. As a result, despite the rapid rate of contraction, the circulation is sluggish, but this is compensated for because the tissues are able to draw nourishment, not only from the capillaries but also from the large arteries. As the heart muscle thickens, compacts and strengthens, the cardiac orifices become both relatively and absolutely reduced in size, the valves increase their efficiency and the large arteries acquire their muscular walls and they too undergo a relative reduction in size. From this time onwards the embryo is dependent for its nourishment on the expanding capillary beds and henceforth the larger arteries' function becomes restricted to controllable distribution channels to keep its tissues constantly and appropriately supplied.

The conducting system of the heart is histologically identifiable towards the middle of the sixth week of embryonic life. Some claim that first to appear is the *atrioventricular node* developing as an outgrowth from the dorsal part of the muscular ring surrounding the atrioventricular canal. On this view the *atrioventricular bundle* is a subsequent extension which passes distally behind the dorsal endocardial cushion thence along the dorsal part of the curved free edge of the muscular ventricular septum, where it gives rise first to the left and then to the right limbs of the bundle (Shaner 1929, Walls 1947). Others maintain that the stem of the atrioventricular bundle appears first as a derivative of the musculature of the dorsal wall of the atrioventricular canal. From this, differentiation spreads both proximally to form the atrioventricular node and distally to form the right and left limbs of the bundle (Field 1951, Muir 1954). Whichever view is subscribed to, after completion of septation in the normal heart, the atrioventricular conducting tissue elements occupy precise sites or courses. The atrioventricular node lies deep to the septal endocardium of the right atrium in the triangle of Koch (for details see p. 703, 6.34). The main atrioventricular bundle proceeds through a channel traversing the *central fibrous body* (right fibrous trigone), then curves along the posterior and posteroinferior margins of the *pars membranacea septi* (atrioventricular and interventricular parts) to reach its locus of division over the crest of the muscular septum. From diverging fascicles of the main bundle branches a subendocardial terminal network of conducting cells (classic specialized Purkinje myocytes) spread to all myocardial parts of the ventricular walls; the net is particularly dense around the bases of the papillary muscles. Finally, fibres from the network enter the myocardial tissue of the ventricles; the patterned array of their ultimate contacts with general cardiac myocytes is, as yet, undetermined (p. 722).

The *sinuatrial node* is the last part of the conducting system to appear, but its site is indicated by an aggregation of nerve cells and fibres in the wall of the superior vena cava just above its junction with the atrium, before nodal tissue itself can be recognized late in the third month. A narrow band of muscle in the ventrolateral surface of the superior vena cava is modified into the characteristic adult histological structure (p. 720). Although at first caval in position, in late fetal life it becomes incorporated into the atrial wall at the upper end of the sulcus terminalis. Its cranial edge, however, remains in continuity with the muscular coat of the vessel, although elsewhere it makes multiple contacts with the muscular wall of the atrium. The node does not appear until the right horn of the sinus venosus has been incorporated in the right atrium. Sinuatrial nodal tissue appears at an early relative stage in some rodents and recently a *left* sinuatrial node has been described in the murine embryo (Heintzberger 1974) as a transient structure. The occurrence of such a bilateral symmetry has been hypothecated in mankind but not substantiated (Patten 1956).

Studies of conducting tissue development in human embryos emphasize the basic difficulty of identifying such tissue, particularly in its initial stages of differentiation (Wenink 1976). The earlier views, stated above, which derived elements of the conducting system by extensions from localized parts of only two of the muscular rings between successive chambers of the primitive heart tube, were given limited confirmation. However, Wenink (1976) claimed that parts of *all four interchamber rings* (sinuatrial, atrioventricular, ventriculobulbar and truncobulbar) were involved. Following the complex repositioning of the rings during folding and growth of the cardiac tube many parts of the rings retrogress, others variously invaginate, fused, subdivided or extended into surrounding myocardium. From these, it was proposed, the whole system developed (2.85,86). It may be noted that the 'four ring hypothesis' when followed in detail could provide a possible explanation of the less widely accepted tracts

mentioned below (and pp. 721–722), also the siting of nodal tissue in certain cardiac anomalies.

Controversy persists concerning the development, architecture, operation, or even the existence, of further specialized conductive tissue elements. The latter involve the mode of spreading of intermittent waves of excitation, generated by the rhythmic activity of the sinuatrial nodal tissue, to encompass *both* atria, to converge on the atrioventricular node and proximal bundle, thence to invade the ventricular walls through the ramifications of the main atrioventricular bundle and possibly other (occasional) atrioventricular connections. Some hold that sinuatrial excitation pervades the atrial walls through fascicles of ordinary 'working' cardiac myocytes and the nexuses (electrical couplings) between them. Thus variations in ionic current flow paths and velocities would result from the geometry of the atrial walls, e.g. the position and calibre of the large venous orifices, mural thickness differences over grooves, fossae, ridges, eminences etc. and presumably any rectilinear polarization of the fascicular cardiac myocytes and their nexuses. Others maintain, however, that excitatory dispersion occurs initially along tracts and networks of specialized conducting tissue. In general these may be classed as: *atrial*, to the walls of the right atrium; *interatrial*, crossing the interatrial septal complex from right to left atrium; *internodal* (anterior, middle and posterior *tracts*), converging on aspects of the atrioventricular node; *bypass fibres* joining the nodal end of the atrioventricular bundle directly; finally, slender, irregular, *accessory atrioventricular bundles* may occur at any point in the atrioventricular fibrous annuli (for further details see pp. 721–722.)

The *intrinsic nerves* of the heart, consisting of groups of ganglion cells associated with the great arteries, atria, and atrioventricular grooves, together with a developing epicardial plexus, are observable from the 15 mm stage (Navaratnam 1965, Smith 1970). Later, atrial and ventricular myocardia, coronary vessels and conducting tissues are involved. For a highly detailed description of the state of development of the *extrinsic nerves* of the heart in a human embryo of 8 weeks consult Smith (1971), and Gardner and O'Rahilly (1976). For a review of autonomic cardiac innervation see Gabella (1976); and for a bibliography on control of the fetal circulation see Mott (1982). Further details pp. 722, 723.

It will be noted that the heart commences to beat early prior to the development of the conducting system, and that a circulation is established before a competent valvular mechanism. The latter has been associated (Streeter 1945) with the incompressible but displaceable cardiac jelly occupying the myoendocardial interval.

FURTHER DEVELOPMENT OF THE ARTERIES

Apart from the aortae none of the main vessels of the adult arise as single trunks in the embryo. Along the course of each vessel a capillary network is first laid down and by selection and enlargement of definite paths in this the larger arteries and veins are defined. The branches of main arteries are not always simple modifications of the vessels of a capillary network but arise as outgrowths from the enlarged stem.

As mentioned, subsequent to head fold formation each primitive aorta consists of ventral and dorsal parts which are continuous through the first embryonic aortic arch. The dorsal aortae run caudally, one on each side of the notochord, but in the fourth week they fuse from about the level of the fourth thoracic to that of the fourth lumbar segment to form a single *definitive descending aorta*. Although in many animals paired ventral aortae arise from the truncus arteriosus and course headwards on the ventral surface of the pharynx, in the human embryo the ventral aortae are fused and form a dilated *aortic sac* (see 2.87A and consult Congdon 1922). The first aortic arches run through the mandibular arches, and caudal to them five additional pairs are developed within the corresponding branchial arches so that in all six pairs of aortic arches are formed (2.88A). The fifth arches are atypical and probably transient, at most, in mankind.

In fishes the aortic arches persist and give off branches to the gills, in which the blood is oxygenated. In mammals some of the

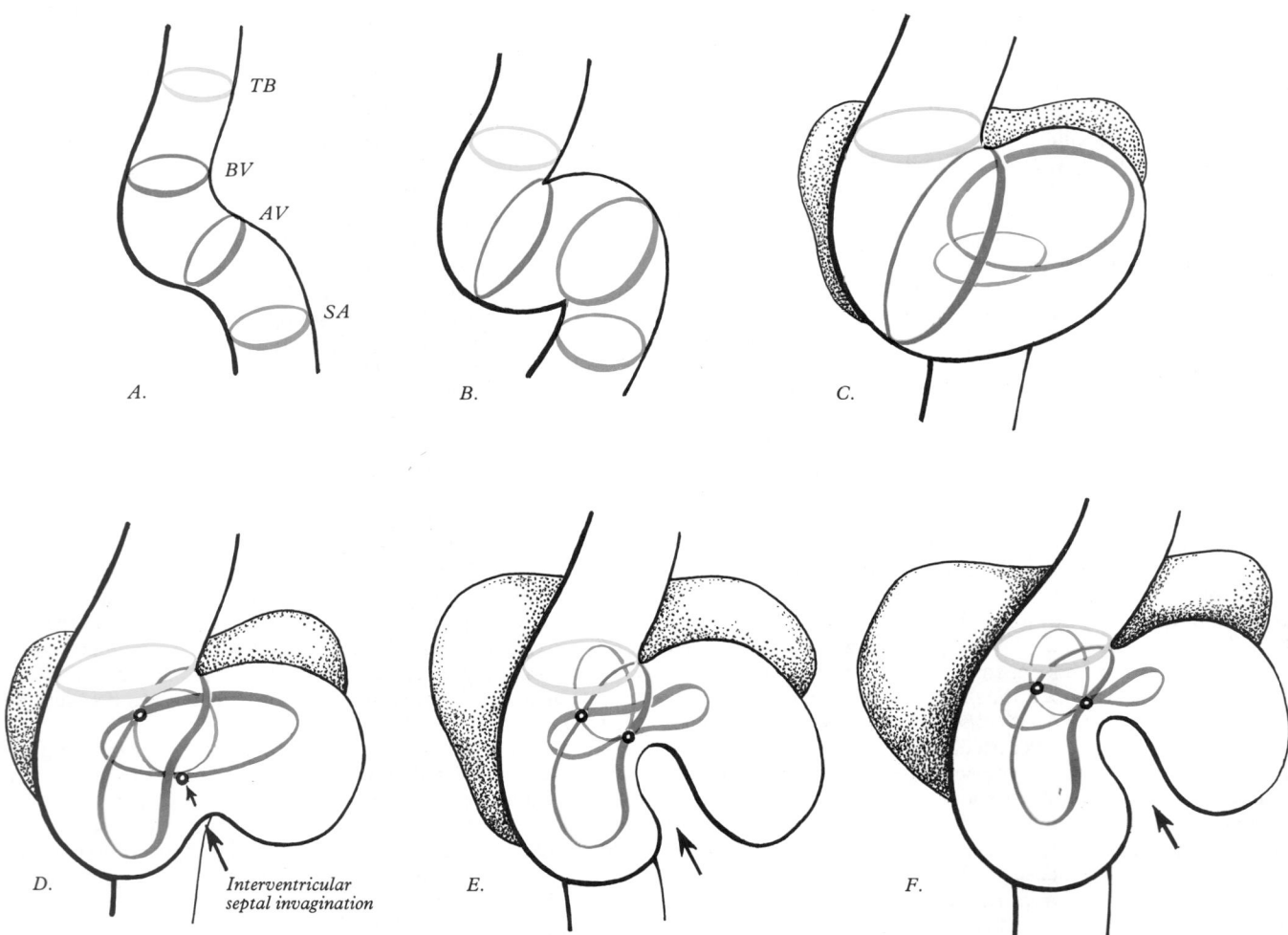

A.   *TB*   *BV*   *AV*   *SA*

B.

C.

D.   *Interventricular septal invagination*

E.

F.

2.85 Development of the cardiac conducting system: the *'four ring' theory*. The successive regions of the early cardiac tube are separated by four initially complete rings of interzonal specialized myocardium, coloured here to aid identification and named: SA — sinuatrial; AV — atrioventricular; BV — bulboventricular; and TB — truncobulbar. With progressive folding, differential expansion and mural changes, and the onset of septation, the positions, planes and shapes of the rings are profoundly modified. In figures A-F the alterations in the cardiac outline and rings are shown from the ventral aspect; in the mural tissue forming the floor of the bulboventricular and atrioventricular grooves loci where originally separated rings make secondary contact and become confluent are indicated.

arteries remain as permanent structures, while others disappear or are obliterated (**2.**88A-C).

Caution should moderate unqualified use of the term *aortic arch(es)*. The *embryonic aortic arches* are paired bilateral series joining the ventral aortae (or their fused expanded homologue) with the dorsal aorta of its side after traversing the core of a branchial arch. In contrast the *definitive aorta* consists of ascending aorta, aortic arch and descending (thoracic and abdominal) aorta—all parts of a single vessel in the mature state (but derived from multiple embryonic sources). For one detailed analysis see **2.**88A-C.

The aortic sac represents fused, paired ventral aortae (**2.**87A, B). As the embryo grows and the aorticopulmonary septum is formed part of the caudal end of the sac is incorporated in the pulmonary trunk. The cranial end of the sac becomes drawn out into *right* and *left limbs* as the neck lengthens. The right limb becomes the brachiocephalic trunk and the left limb forms that part of the definitive arch of the aorta which lies between the origin of the brachiocephalic trunk and the left common carotid artery. The remainder of the sac contributes to the formation of the ascending arch of the aorta.

The embryonic aortic arches (**2.**88A-C), with the exception of the fifth, are developed in a craniocaudal sequence, but the more cranial are in process of disappearing before the caudal ones are completed. The *first* and *second embryonic aortic arches* are already dwindling by the time the third is established. The first disappears entirely. The dorsal end of the second arch or *hyoid artery* remains as the stem of the *stapedial artery*, while the remainder of this arch artery also disappears (**2.**89). The external carotid artery first appears as a sprout which grows headward from the aortic sac close to the ventral end of the third arch artery. The common carotid arises from an elongation of the adjacent part of the aortic sac, and the third arch artery becomes the proximal part of the internal carotid artery. (Evidence against this view, however, has also been recorded: e.g. see Moffat 1959, Adams 1957.) The *fourth embryonic aortic arch* on the right forms the proximal part of the right subclavian artery, whilst the corresponding vessel on the left is believed to constitute the arch of the definitive aorta between the origins of the left common carotid and left subclavian arteries. It has, however, proved difficult to assess accurately the contributions of the *fourth embryonic aortic arches* and it has also been variously claimed that the left fourth aortic arch is subsequently drawn into the descending or ascending (or both) limbs of the *definitive aortic arch*, and the corresponding vessel on the right contributes to the brachiocephalic artery. The identity and status of the *fifth embryonic arch artery* is uncertain; it is usually incomplete and may connect the fourth aortic arch or subjacent aortic sac with the dorsal end of the sixth aortic arch (whereas the other embryonic aortic arches pass between sac and dorsal aorta). The fifth aortic arch eventually disappears on both sides. From its inception the *sixth embryonic arch vessel* is associated with a developing lung bud. Initially each bud is supplied by a capillary plexus from the aortic sac. Only later the plexus connects with the dorsal aorta and the sixth aortic arch is defined as a channel in the vascular connection between sac and dorsal aorta; however, this continues to supply the developing lung bud. When the aorticopulmonary septum divides the *truncus arteriosus* into pulmonary trunk and ascending aorta the sixth

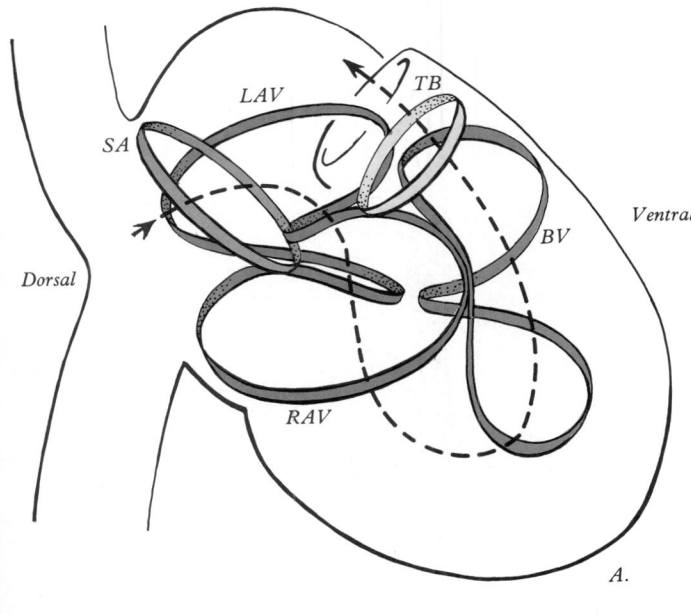

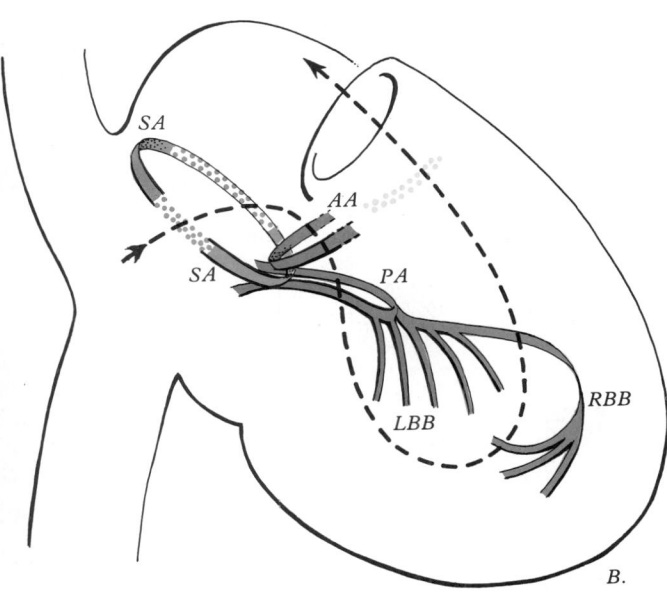

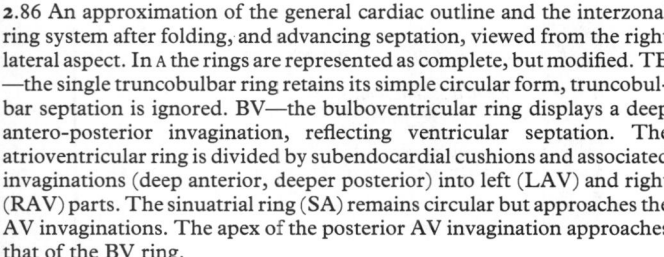

A.

B.

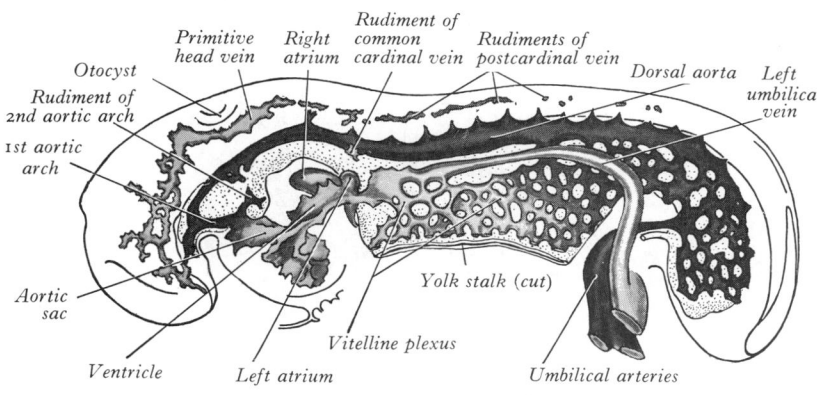

2.87A The blood vascular system of a human embryo with 14 paired somites. Estimated age, 23½ days. CR length 2.4 mm. Magnification × c. 45. The arteries and veins are only in process of development, so that no true circulation is possible at this stage. Only the endothelial lining of the heart tube is shown. (Streeter 1942.)

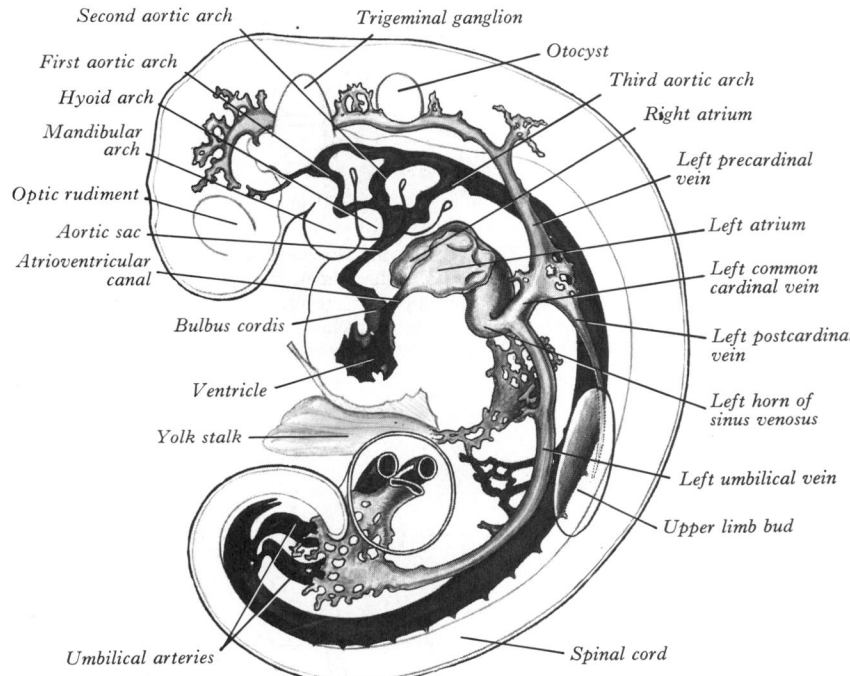

2.87B Profile reconstruction of the blood vascular system of a human embryo, having 28 somites. CR length, 4 mm. Estimated age, 26 days. *Note.* Only the endothelial lining of the heart chambers is shown and, as the muscular wall has been omitted, the pericardial cavity appears much larger than the contained heart. Observe that the atrioventricular canal still connects the left atrium with the single ventricle. (Streeter 1942.)

2.86 An approximation of the general cardiac outline and the interzonal ring system after folding, and advancing septation, viewed from the right lateral aspect. In A the rings are represented as complete, but modified. TB —the single truncobulbar ring retains its simple circular form, truncobulbar septation is ignored. BV—the bulboventricular ring displays a deep antero-posterior invagination, reflecting ventricular septation. The atrioventricular ring is divided by subendocardial cushions and associated invaginations (deep anterior, deeper posterior) into left (LAV) and right (RAV) parts. The sinuatrial ring (SA) remains circular but approaches the AV invaginations. The apex of the posterior AV invagination approaches that of the BV ring.

In B parts of the rings retrogress; the regions proposed as contributing to widely accepted parts of the conducting system are shown. SA — dorsocranially forms sinuatrial tissue. Ventrocaudally SA joins AA and PA (anterior and posterior atrioventricular invaginations), forming atrioventricular nodal tissue. PA continues as the common A-V bundle; then is continuous with BV ring invagination. LBB — 'fan' of left bundle branches of dual origin (PA and BV); RBB — compact right bundle branch and terminal divisions; single BV ring origin. TB — stipple, merges with aortic vestibule. SA — stipple, possible origin of more controversial intermodal and interatrial pathways. (Modified after Wenink 1976, with permission from *Journal of Anatomy*.)

aortic arches retain continuity with the former. On the right the ventral part of the sixth aortic arch persists as the stem of the right pulmonary artery, but its dorsal segment disappears, possibly due to a decreased blood flow resulting from partitioning of the aortic and pulmonary bloodstreams (Navaratnam 1963). On the left side

the ventral part of the sixth aortic arch is absorbed into the pulmonary trunk, while its dorsal segment persists as the *ductus arteriosus*, which is functional during intrauterine life but becomes obliterated after birth ultimately forming the fibrous *ligamentum arteriosum*. Postnatal *functional* closure nears completion within a few weeks but *structural* changes continue over many months (p. 724).

The transformation of the aortic arches described above is conditioned by environmental changes and results largely from changes in the pharynx and from the descent of the heart. The whole period of transformation can be divided into two phases, branchial and postbranchial. In the *branchial phase*, which lasts until about the 12 mm crown–rump length stage, the arrangement of the aortic arches resembles that in lower vertebrates. In this phase the course of the blood from the heart to the dorsal aorta follows a succession of different pathways—first arch, first and second arches, second and third arches, third and fourth aortic arches and finally third, fourth and sixth aortic arches. In the *postbranchial*

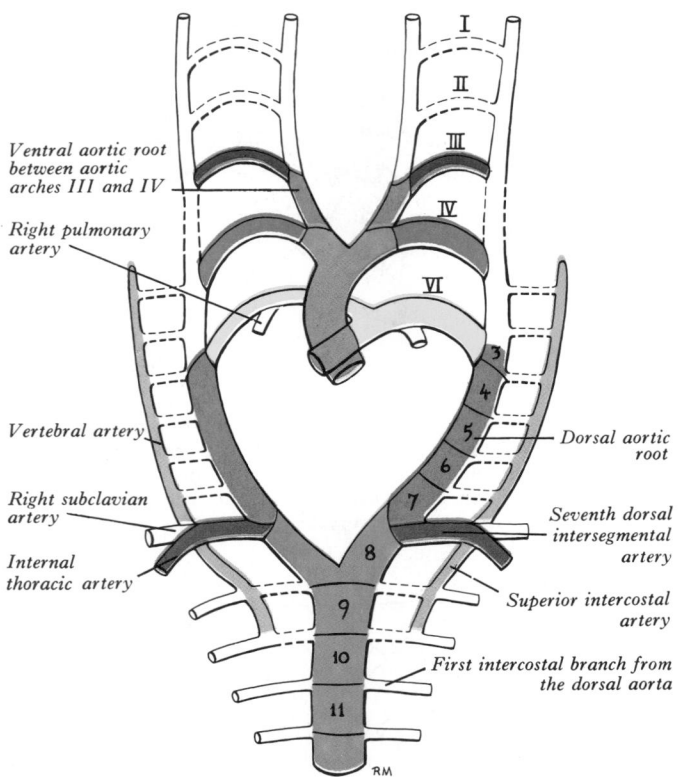

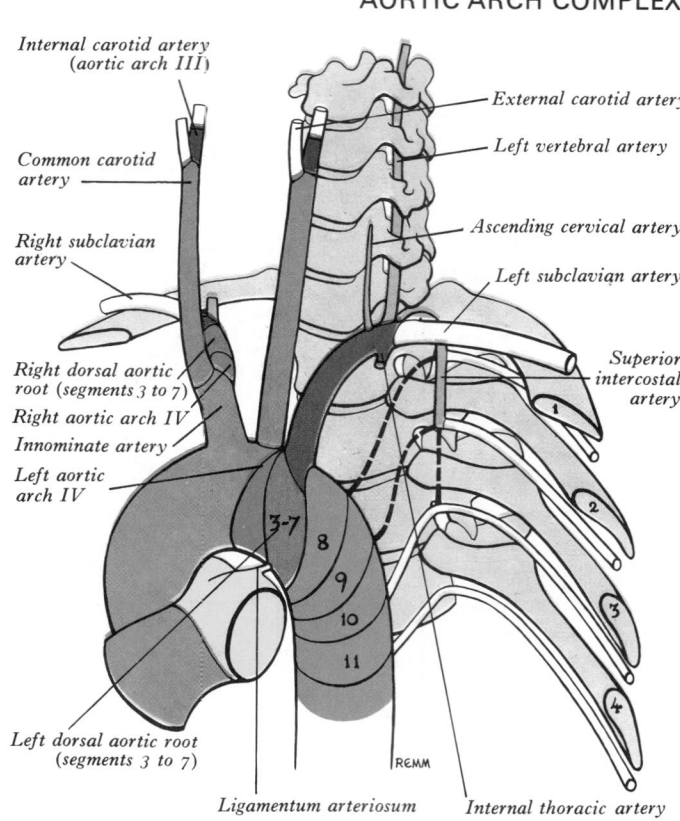

2.88A Schematic diagram showing the various components of the embryonic aortic arch complex in the human embryo. Structures which do not persist in normal development are indicated by interrupted lines. Roman numerals refer to the branchial arches concerned, and arabic numerals indicate the metameric body segments.

2.88c Diagram of the adult human aorta and its branches viewed from the left ventrolateral aspect, showing the position and relative sizes of the definitive contributions from the various embryonic components shown in **2.88** A and B.

Based on the work of A Barry, with kind permission of the author and *Anatomical Record*, **III**, 1951.)

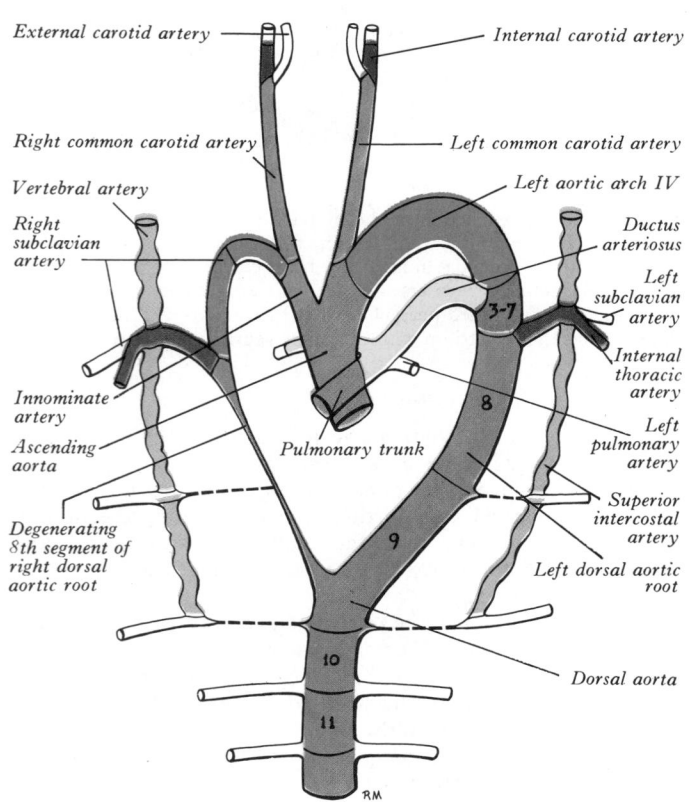

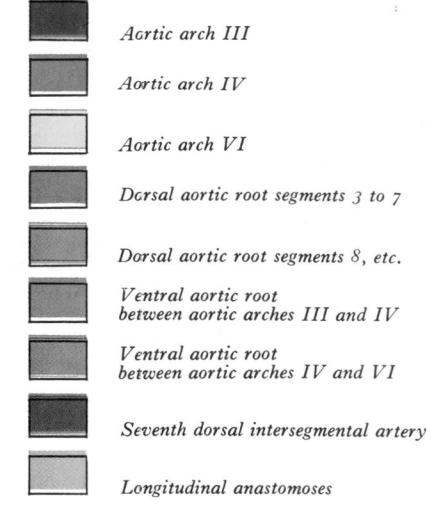

Aortic arch III

Aortic arch IV

Aortic arch VI

Dorsal aortic root segments 3 to 7

Dorsal aortic root segments 8, etc.

Ventral aortic root between aortic arches III and IV

Ventral aortic root between aortic arches IV and VI

Seventh dorsal intersegmental artery

Longitudinal anastomoses

2.88B Diagrammatic ventral view of the aortic arch complex of human embryo of 15 mm CR length. Note the asymmetry in the pattern that has developed by this stage. Compare with **2.88** A and C.

phase, which extends onwards into, and beyond, intrauterine life the definitive human pattern and disposition of the vessels is finally established.

The cranial arteries (2.89) develop in outline as follows (Padget 1948). The *internal carotid artery* is progressively formed from the third arch artery, the dorsal aorta cranial to this and a further forward continuation which differentiates, at the time of regression of the first and second aortic arches, from the capillary plexus extending to the walls of the forebrain and midbrain. At its anterior extremity this *primitive internal carotid artery* divides into *cranial* and *caudal* divisions, the former terminating as the *primitive olfactory artery*, supplying the developing regions implied, and the latter sweeping caudally to reach the ventral aspect of the midbrain, its terminal branches being the *primitive mesencephalic arteries.* Simultaneously bilateral longitudinal channels differentiate along the ventral surface of the hindbrain from a plexus fed by intersegmental and transitory presegmental branches of the dorsal aorta and its forward continuation. The most important of the presegmental branches is closely related to the fifth nerve—the *primitive trigeminal artery.* *Otic* and *hypoglossal* presegmental arteries also occur (Padget 1948); sometimes these persist. The longitudinal channels later connect, cranially, with the caudal divisions of the internal carotid arteries, each of which gives rise to an *anterior choroidal artery* supplying branches to diencephalon and midbrain, and caudally with the vertebral arteries through the first cervical intersegmental arteries. Fusion of the longitudinal channels results in the formation of the *basilar artery,* whilst the caudal division of the internal carotid artery becomes the *posterior communicating artery* and the stem of the *posterior cerebral artery.* The remainder of the latter develops comparatively late, probably from the stem of the posterior choroidal artery which is annexed by the caudally expanding cerebral hemisphere, its distal portion becoming a choroidal branch of the posterior cerebral artery. In the rat, where the vascular pattern is essentially similar to that in man, this artery is derived from the posterior communicating artery, the common stem of origin of the posterior choroidal, mesencephalic and diencephalic arteries, together with a new channel formed in the plexus on the medial wall of the cerebral hemisphere initially supplied by the anterior choroidal artery (Moffat 1961a). The cranial division of the internal carotid artery gives rise to *anterior choroidal, middle cerebral* and *anterior cerebral arteries,* the stem of the primitive olfactory artery remaining as a small medial striate branch of the anterior cerebral artery. In the rat the primitive olfactory artery and its recurrent branch form the anterior cerebral artery, the territory of which is initially supplied by the primitive maxillary and the cranial ramus of the internal carotid arteries (Moffat 1961b). The *cerebellar arteries,* of which the superior is the first to differentiate, emerge from the capillary plexus on the wall of the rhombencephalon.

Source of blood supply to the territory of the trigeminal nerve varies at different stages in development. When the first and second aortic arch arteries begin to regress the supply to the corresponding arches is derived from a transient *ventral pharyngeal artery,* which grows from the aortic sac. It terminates by dividing into *mandibular* and *maxillary* branches. Later the *stapedial artery* develops from the dorsal stem of the second arch artery and passes through the condensed mesenchymal site of the future ring of the stapes to anastomose with the cranial end of the ventral pharyngeal artery thereby annexing its terminal distribution. The fully developed stapedial artery possesses three branches, *mandibular,* *maxillary* and *supraorbital,* which follow the divisions of the trigeminal nerve (2.89). The mandibular and maxillary branches diverge from a common stem. When the external carotid artery emerges from the base of the third arch it incorporates the stem of the ventral pharyngeal artery, and its maxillary branch communicates with the common trunk of origin of the maxillary and mandibular branches of the stapedial artery and annexes these vessels. The proximal part of the common trunk persists as the root of the *middle meningeal artery.* More distally the meningeal artery is derived from the proximal part of the supraorbital artery. The maxillary branch becomes the infraorbital artery and the mandibular branch forms the inferior alveolar artery.

When the definitive *ophthalmic artery* differentiates as a branch from the terminal part of the internal carotid artery, it communicates with the supraorbital branch of the stapedial artery; distally this becomes the *lacrimal artery.* The latter retains an anastomotic connection with the middle meningeal artery. The dorsal stem of the original second arch artery remains as one or more *caroticotympanic branches* of the internal carotid artery.

The dorsal aortae (2.88A–C) persist on the cranial side of the third aortic arches as continuations of the internal carotid arteries. The dorsal aorta between the third and fourth aortic arches, the *ductus caroticus,* diminishes and finally disappears; but from fourth arch to the origin of the seventh intersegmental artery the right dorsal aorta becomes part of the right subclavian artery (2.88A–C). Caudal to the seventh intersegmental artery the right dorsal aorta disappears as far as the locus of fusion of thoracic

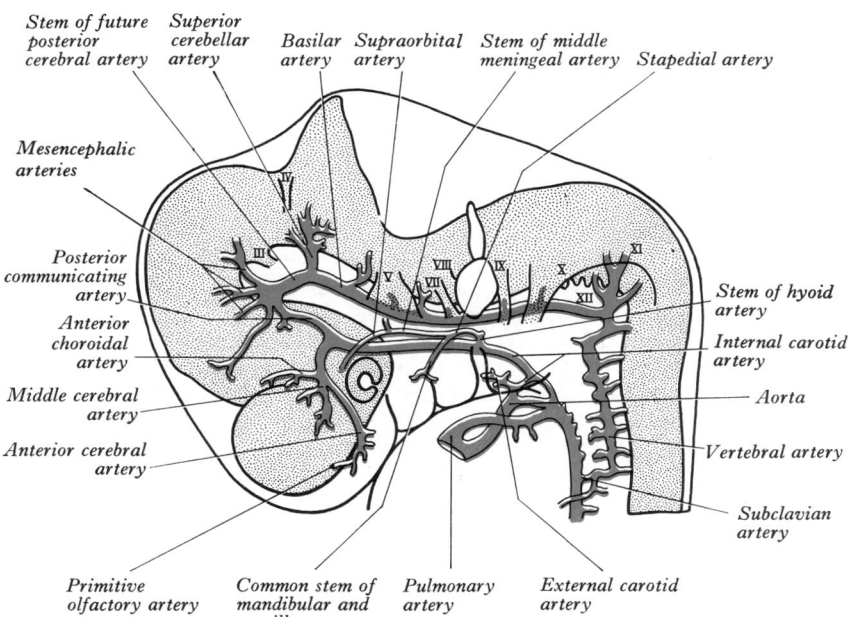

2.89 Diagram to show the origins of the main cranial arteries. (After Padget 1948.)

aorta. After disappearance of the left ductus caroticus. The remainder persists to form the descending part of the arch of the aorta. Thence the fused right and left embryonic dorsal aortae persist as the definitive descending thoracic and abdominal aorta. A constriction, the *aortic isthmus,* is sometimes present in the aorta between the final site of origin of the left subclavian artery and reception of the ductus arteriosus (see *coarctation* of the aorta, p. 766).

In the adult, the right subclavian artery occasionally arises from the arch of the aorta distal to the origin of the left subclavian and then passes upwards and to the right behind the trachea and oesophagus. This condition is possibly explained by the persistence of the embryonic right dorsal aorta and the obliteration of the fourth aortic arch of the right side.

In birds the right fourth aortic arch is transformed into the definitive arch of the aorta; in reptiles the fourth arches of both sides persist and give rise to their characteristic double aortic arch. In both these classes, development of the heart and aortic arches is probably along phylogenetic lines so divergent from the mammalian pattern that comparisons may be inappropriate.

The heart originally lies ventral to the pharynx, immediately caudal to the stomatodeum (2.87A); with the elongation of the neck and the development of the lungs it recedes within the thorax and, correspondingly, the vessels are drawn out and the original position of the fourth and sixth aortic arches is greatly modified. Thus, on the right the fourth aortic arch only recedes to the thoracic inlet, while on the left side it descends *into* the thorax. The recurrent laryngeal nerves (in contrast to the other arch nerves) originally pass to the larynx *caudal* to the sixth pair of aortic arches, and are therefore affected by the descent of these structures; thus in the adult the left nerve hooks round the ligamentum arteriosum *within* the thorax; on the right owing to the disappearance of the fifth and the dorsal part of the sixth aortic arch, the right recurrent laryngeal nerve hooks round the fourth aortic arch, i.e. the commencement of the right subclavian artery.

At first the aortae are the only longitudinal vessels present, for their branches all run at right angles to the long axis of the embryo. Later these transverse arteries become connected in certain situations by longitudinal anastomosing channels, which in part persist, forming such arteries as the internal thoracic, the superior and inferior epigastric, the gastro-epiploic, etc. Each primitive dorsal aorta gives off *segmental* branches to the digestive tube (*ventral splanchnic arteries*) and to the mesonephric ridge (*lateral splanchnic arteries*) and *intersegmental* branches to the body wall (*somatic arteries*).

The ventral splanchnic arteries are originally paired vessels which are distributed to the capillary plexus in the wall of the yolk sac, but after fusion of the dorsal aortae they merge as unpaired trunks distributed to the increasingly defined and lengthening primitive digestive tube. Longitudinal anastomotic channels connect these branches along the dorsal and ventral aspects of the tube, forming dorsal and ventral splanchnic anastomoses (2.90) (Ennabli & Niveiro 1967). These vessels obviate the need for so many 'subdiaphragmatic' ventral splanchnic arteries, and these are reduced to three: the coeliac trunk and superior and inferior mesenteric arteries. As the viscera supplied descend into the abdomen their origins migrate caudally, by differential growth; thus the origin of the coeliac artery is transferred from the level of the seventh cervical segment to the level of the twelfth thoracic, the superior mesenteric from the second thoracic to the first lumbar and the inferior mesenteric from the twelfth thoracic to the third lumbar. (Above the diaphragm, however, a variable number of ventral splanchnic arteries persist, usually four or five, supplying the thoracic oesophagus.) The dorsal splanchnic anastomosis persists in the gastro-epiploic, pancreaticoduodenal and the primary branches of the colic arteries, while the ventral splanchnic anastomosis forms the right and left gastric and the hepatic arteries. These arterial rearrangements have been investigated recently by angiography and explanatory haemodynamic hypotheses have been advanced (Barth et al 1976).

The lateral splanchnic arteries supply, on each side, the mesonephros, metanephros, the testis or ovary, and the suprarenal gland; all these structures develop, in whole or in part, from the intermediate mesoderm of the mesonephric ridge (p. 248). One testicular or ovarian artery and three suprarenal arteries persist on each side. The phrenic artery branches from the most cranial suprarenal artery, and the renal artery arises from the most caudal. Additional renal arteries are frequently present and may be looked on as branches of persistent lateral splanchnic arteries.

The somatic arteries are *intersegmental* in position and they persist, almost unchanged, in the thoracic and lumbar regions, as the posterior intercostal, subcostal and lumbar arteries. Each gives off a dorsal ramus which passes backwards in the intersegmental interval and divides into medial and lateral branches to supply the muscles and superficial tissues of the back (2.90). It also gives off a spinal branch, which enters the vertebral canal and divides into a series of branches to the tissues constituting the walls and joints of the osteoligamentous canal and neural branches to the spinal cord and spinal nerve roots (Somogyi et al 1973, Undi et al 1973). Having produced its dorsal branch the intersegmental artery runs ventrally in the body wall, gives off a lateral branch and terminates in muscular and cutaneous rami. Before their division, the stems of the somatic arteries, at thoracic and lumbar levels, provide small rami which enter the developing vertebral bodies.

Numerous longitudinal anastomoses link up the intersegmental arteries and their branches (2.90). On both sides a *postcostal anastomosis* connects their dorsal branches in the intervals between the necks of ribs and the vertebral transverse processes. This persists in the cervical region where it forms the greater part of the vertebral artery. A *post-transverse anastomosis* also connects the dorsal branches and forms the greater part of the deep cervical artery. A *precostal anastomosis* connects intersegmental arteries beyond the origins of their dorsal branches. The ascending cervical and the superior intercostal arteries are persistent parts of this vessel. Lastly, near the anterior median line intersegmental arteries are linked by a *ventral somatic anastomosis*. Most of these vessels persist bilaterally as the internal thoracic, the superior and inferior epigastric arteries.

The *umbilical arteries* at first are the direct caudal continuation of the primitive dorsal aortae and are present in the body stalk before any vitelline (yolk sac) or visceral branches emerge, indicating the dominance of the chorionic over the vitelline circulation in the human embryo. (On a comparative basis the umbilical vessels are chorioallantoic and therefore 'somatovisceral'.) After the fusion of the dorsal aortae the umbilical arteries arise from their ventrolateral aspects and pass medial to the primary excretory duct (Wolffian) to the umbilicus. Later the proximal part of each umbilical artery is joined by a new vessel which leaves the aorta at its termination and passes lateral to the primary excretory duct. This, possibly the fifth lumbar intersegmental artery, constitutes the *dorsal root* of the umbilical artery (the original stem, the *ventral root*). The dorsal root gives off the axial artery of the lower limb, branches to the pelvic viscera and, more proximally, the external iliac artery. The ventral root disappears entirely, the umbilical artery now arising from that part of its dorsal root distal to the external iliac artery, i.e. the internal iliac artery.

The arteries of the limbs: initially a number of vessels contribute to a primitive capillary plexus. In the upper limb bud, usually only one trunk—the subclavian—persists and it probably represents the lateral branch of the *seventh intersegmental artery*. Its main continuation (*axis artery*) to the upper limb (2.91), later the *axillary* and *branchial arteries*, passes into the forearm deep to the flexor muscle mass and terminates as a deep plexus in the developing hand. This vessel ultimately persists as the *anterior interosseous artery* and the *deep palmar arch*. A branch from the main trunk passes dorsally between the early radius and ulna as the *posterior interosseous artery*, while a second accompanies the median nerve into the hand, where it ends in a *superficial capillary plexus*. The *radial* and *ulnar arteries* are the latest arteries to appear in the forearm; at first the radial artery arises more proximally than the ulnar and crosses in front of the median nerve, supplying the biceps. Later, the radial artery establishes a new connection with the main trunk at or near the level of origin of the ulnar artery and the upper portion of its original stem usually disappears to a large extent (see also p. 759). On reaching the hand the ulnar artery becomes linked up with the superficial palmar plexus, from which the *superficial palmar arch* is derived, while the median artery commonly loses its distal connections and is reduced to a small vessel. The radial artery passes to the dorsal surface of the hand but, after giving off dorsal digital branches, it traverses the first intermetacarpal space and joins the deep palmar arch.

Because of their multiple and plexiform sources, the temporal succession of emergence of principal arteries, anastomoses and periarticular networks and functional dominance followed by regression of some paths, anomalies of the forelimb arterial tree are fairly common. In the main, such anomalous patterns present as divergences in the mode and proximodistal level of branching, the presence of unusual compound arterial segments, aberrant vessels connecting other principal vessels, arcades or plexuses and vessels occupying exceptional tissue planes (e.g. superficial fascia instead of the usual subfascial route) or having unexpected neural, myological or osteoligamentous relationships. For example, see variations on pp. 758, 759, 760, 762.

The *axis artery* of the lower limb (2.92) arises from the dorsal root of the umbilical artery, and courses along the *dorsal* surface of the thigh, knee and leg; below the knee it lies between tibia and popliteus and in the leg between the crural interosseous

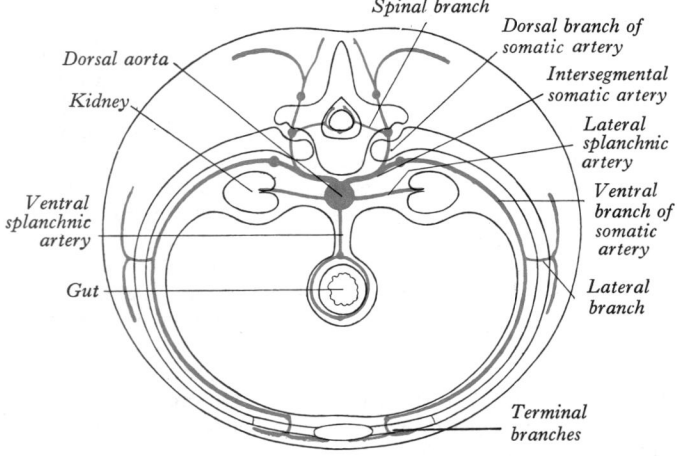

2.90 Diagram of the segmental and intersegmental arteries. Note the positions of the longitudinal anastomoses.

Spinal branch

Dorsal branch of somatic artery

Dorsal aorta

Intersegmental somatic artery

Kidney

Lateral splanchnic artery

Ventral branch of somatic artery

Ventral splanchnic artery

Gut

Lateral branch

Terminal branches

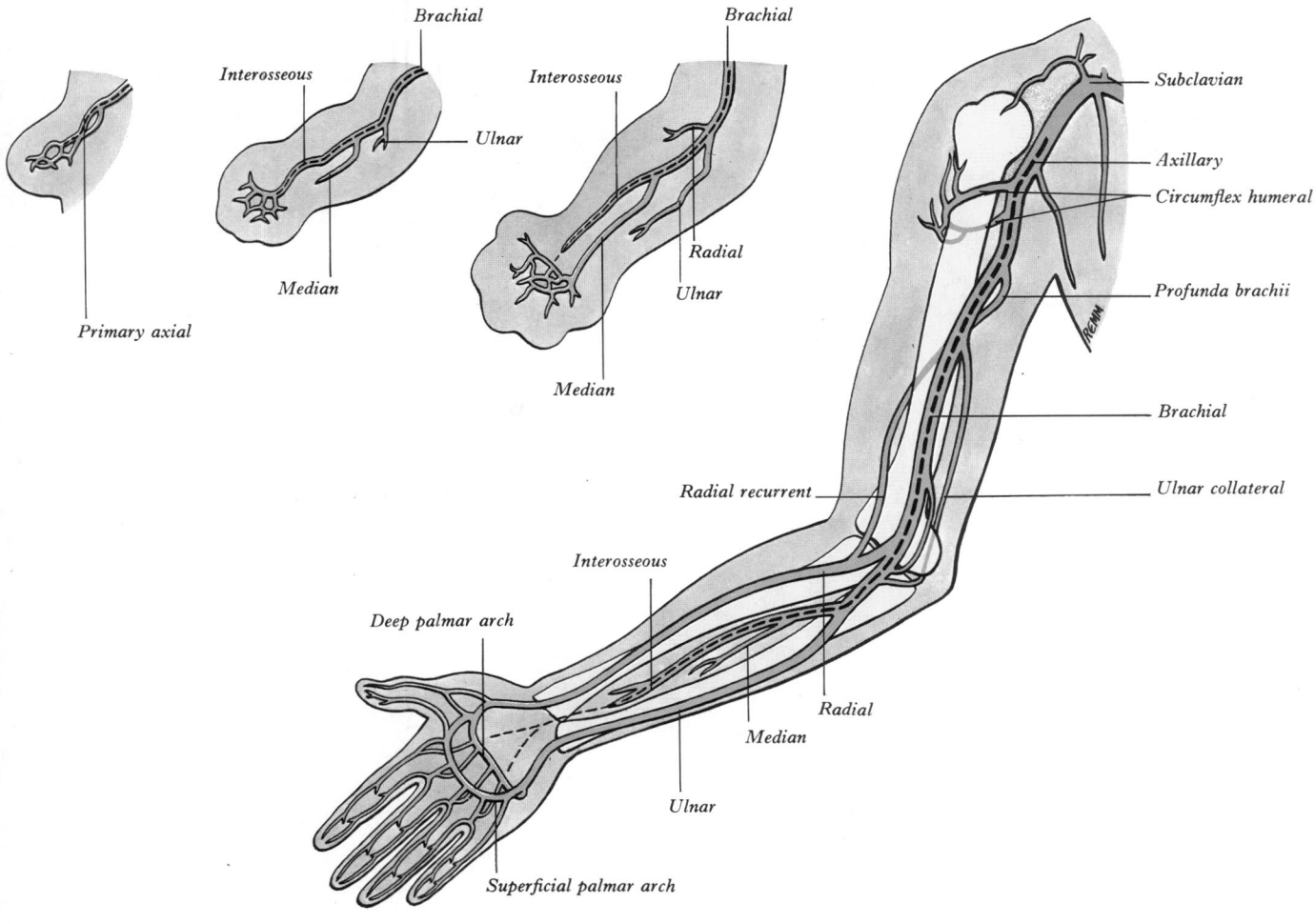

2.91 Stages in the development of the arteries of the arm. The original path of the axis artery is indicated by an interrupted line. (After Patten.)

membrane and tibialis posterior. Ending distally in a *plantar network*, it gives off a perforating artery traversing the sinus tarsi to form a *dorsal network*. The *femoral artery* passes along the *ventral* surface of the thigh, opening a new channel to the lower limb. It arises from a capillary plexus, connected proximally with the femoral branches of the external iliac artery and distally with the axis artery. At the proximal margin of the popliteus the axis artery provides a *primitive posterior tibial* and a *primitive peroneal branch*, which run distally on the dorsal surface of that muscle and on tibialis posterior to gain the sole of the foot. At the distal border of popliteus the axis artery gives off a *perforating branch*, which passes ventrally between the tibia and the fibula and then courses to the dorsum of the foot, forming the *anterior tibial artery* and *arteria dorsalis pedis*. The primitive peroneal artery communicates with the axis artery at the distal border of the popliteus and in its course in the leg (Senior 1919, 1920).

The femoral artery gradually increases in size and coincidentally most of the axis artery disappears; proximal to its communication with the femoral the root of the axis artery, however, persists as the *inferior gluteal artery* and the *arteria comitans nervi ischiadici*.

The proximal parts of the primitive posterior tibial and peroneal arteries fuse, but distally remain separate. Ultimately much of the primitive peroneal artery disappears, although a part of the axis artery is incorporated in the permanent peroneal artery. As in the forelimb (above) the same considerations apply to anomalies and variations.

FURTHER DEVELOPMENT OF THE VEINS

Often, for convenience and apparent simplicity, the early embryonic veins are segregated into two groups, *visceral* and *parietal*

(or *somatic*). The visceral group comprises the derivatives of the vitelline and umbilical veins, the somatic group includes all remaining veins. Such a classification is a potentially misleading oversimplification, based on inadequate criteria. Many embryonic veins, with time, change the principal tissues they drain; others have some radicles from patently parietal tissues that become confluent with drainage channels that are clearly visceral, thus forming a compound vessel; finally some veins differentiate from contrasting mesenchymal layers at different points along their course (each having its distinct phylogenetic history).

Less confusion attaches to the recognition of three main groups of veins — the *vitelline, umbilical* and *cardinal* vein complexes; nevertheless, even here, some imprecision persists where, for instance anastomoses or convergence of radicles from different groups occurs, where ontogenetically compound tissues are drained or when distinct developmental layers are traversed. The early embryonic veins develop initially with a symmetrical bilateral array of channels; the *principal* events correlated with changes in this are the craniocaudal and mediolateral gradients in growth and differentiation of the nervous system, skeleton and musculature; diversion of cardiac venous return to the right with concomitant cardiac asymmetry; 'descent' of the heart and lungs; gut rotation and repositioning; and venous involvement by the developing liver, pancreas, spleen and mesonephric ridges. (Clearly, the temporospatial developmental sequences of *all* vascularized tissues are involved to varying, but less obvious extents.) As noted, the primitive tubular symmetrical heart receives its venous return through the right and left sinual horns (horns of the sinus venous). The horns are initially embedded in the mesenchyme of the septum transversum and each receives, most medially, the termination of the principal vitelline vein (but vide infra), more laterally, the umbilical vein and, most laterally,

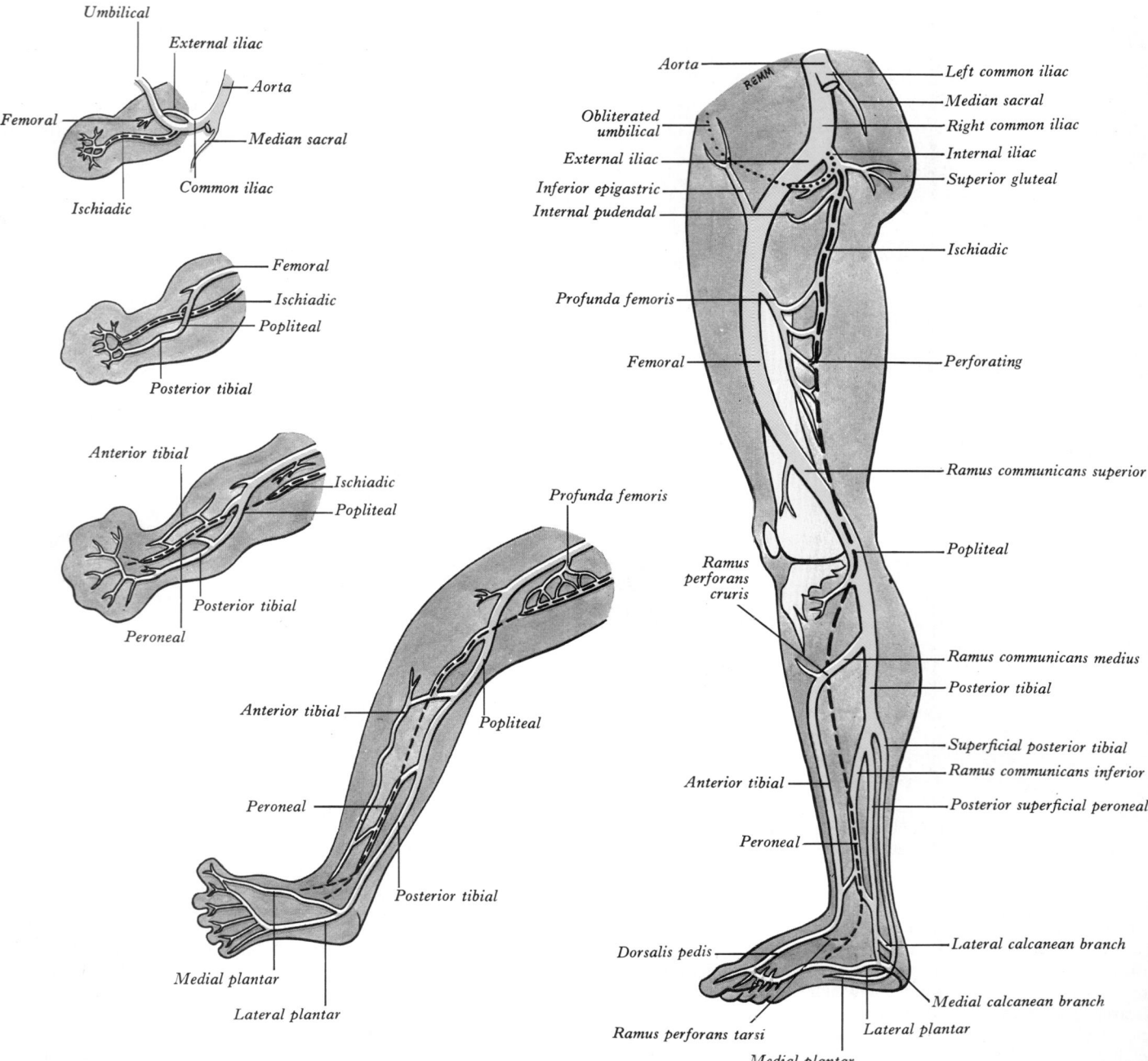

2.92 Stages in the development of the arteries of the leg. The original path of the axis artery is indicated by a dashed line. (After Senior.)

having encircled the coelomic duct, the common cardinal vein. These cardiac inputs correspond, in large measure, but not exclusively, with the groups of veins mentioned above.

The vitelline veins drain capillary plexuses developed in the splanchnopleuric mesenchyme of the secondary yolk sac. With head, tail and lateral fold formation, the upper recesses of the yolk sac are enclosed within the embryo as the splanchnopleuric gut tube extending from the stomatodeal oropharyngeal membrane to the proctodeal cloacal membrane. It may be emphasized that derivatives from *all* these levels possess a venous drainage, originally *vitelline in origin*, although many accounts are limited to the (mainly subdiaphragmatic-sacral) regions drained via the hepatic portal vein. The deep aspects of the maxillo-mandibular facial prominences, retrogingival oral cavity, the pharyngeal walls and their lymphoid and endocrine derivatives and the cervicothoracic oesophagus, all have drainage channels that connect with the *precardinal complex*, ultimately returning blood to the heart via the superior vena cava. Laryngeal and tracheobronchial veins also drain to the precardinal complex, whilst the capillary plexuses developed in the (splanchnopleuric) walls of the fine terminal respiratory passages and alveoli, converge on *pulmonary veins* of

increasing calibre, finally making secondary connections with the left atrium of the heart and may be grouped with the vitelline systems. Even the heart, itself, first differentiates in splanchnopleure that, after head fold formation, forms the dorsal wall of the primitive pericardial cavity (floor of the rostral foregut) and may therefore be considered a highly specialized vitelline vascular derivative. Similarly at the caudal extremity of the splanchnopleuric gut tube (the future lower rectum and upper anal canal—p. 233) the vitelline venous drainage makes connections with the internal iliac radicles of the *postcardinal complex*.

The increasingly extensive remainder of the gut tube, from the gastric terminal segment of the future oesophagus to the upper rectum, is, as elsewhere, clothed with splanchnopleuric mesoderm permeated by a capillary plexus; the latter drains into an anastomosing network of veins. The net is denser ventrally and in the central midgut region; for a while, it receives a leash of small veins from the definitive yolk sac that enter the embryo through the umbilicus, embedded in the yolk stalk. Later, in normal development, both stalk and vessels atrophy. Within the splanchnopleuric net, progressing rostrally, longitudinal channels anterolateral to the gut, become increasingly well

220

defined —the embryonic abdominal *vitelline veins*. Entering the septum transversum, the right and left vitelline veins incline slightly, becoming parallel to the lateral aspects of the gut; they establish connections with capillary plexuses in the septal mesenchyme, then continue, finally curving to enter the medial part of the cardiac sinual horn of their corresponding side. The parts of the gut closely related to the presinual segments of the vitelline veins, just described, are the future subdiaphragmatic end of the oesophagus, primitive stomach, the superior (first) and descending (second) regions of the duodenum, and the remainder of the duodenal tube. The early septum transversum is a thick mass of mesenchyme filling the interval between the median foregut and ventral body wall, and extending from the primitive pericardium to the rostral lip of the umbilicus; in it somatopleuric and splanchnopleuric mesoderms are confluent. When, later, the cardiac sinuatrium rises into the pericardium, the rostral part of the septum (*pars diaphragmatica*) is one of the multiple contributors to the framework of the definitive diaphragm; its caudal *pars mesenterica* provides ventral mesenteries, serous coats and mural tissues for the abdominal foregut, listed above. Where foregut continues into midgut (initially the rostral rim of the yolk stalk) the hepatic (and, transiently, ventral pancreatic) rudiments form as protrusions of the gut; the rapid and asymmetrical invasion of the septum cranioventrally is correlated with profound modifications of the transeptal parts of the vitelline and umbilical veins, and the splanchnopleuric mesothelium, framework and mesenteries of the liver, stomach and duodenum (pp. 240–248). The stem of the earliest hepatic rudiment projects from the ventral wall of the future second (descending) part of the duodenum; whilst expansion of the body of the liver continues as indicated, with gut rotation and differential growth, the ultimate derivative of the stem, the common bile duct, curves posterior to the proximal duodenum to reach its termination through the left posteromedial aspect of the second part of the duodenum.

The principal ascending vitelline veins flanking the sides of the abdominal part of the foregut receive venules from its splanchnopleuric capillaries, and those of the septal mesenchyme. Within these venular nets, enlarged (but still plexiform) anastomoses connect the two vitelline veins. (For clarity these are represented diagrammatically as simple transverse channels, 2.93A–C.) A *subdiaphragmatic intervitelline anastomosis* develops in the rostral septal mesenchyme, lying a little caudal to the cardiac sinuatrial chamber, connecting the veins near their sinual terminations. (Sometimes termed *suprahepatic* because of the position of the hepatic primordium, with expansion of the latter, the channel becomes partly *intrahepatic*.) The presumptive duodenum is crossed by three transverse *duodenal intervitelline anastomoses*; their relation to the gut tube alternates—most cranial, the *subhepatic* is ventral, the *intermediate* is dorsal and the *caudal* is ventral. It has become customary to describe the paraduodenal vitelline veins and their associated anastomoses as forming a figure 8. At this early stage when left and right embryonic veins are still symmetrical, the cranial duodenal anastomosis becomes connected with the subdiaphragmatic anastomosis by a *median* longitudinal channel, the *primitive ductus venosus*, which is dorsal to the expanding hepatic primordium, but ventral to the gut. The further development of the vitelline veins and anastomoses is, as indicated, closely interlocked with rapid hepatic expansion and gut changes; also umbilical vein disposition and modification is closely involved, and their early emergence and arrangement must be outlined before an account of later asymmetries is undertaken. The latter differs in part from long held descriptions, following the detailed analyses of goat and human embryos by Dickson (1957).

The umbilical veins form by the convergence of venules draining the splanchnopleure of the extraembryonic allantois. Throughout *Amniota* the allantois arises as a diverticulum from the caudal yolk sac wall (later, the ventrorostral wall of the cloacal part of the hindgut, or future bladder). Its degree of vesicular, then sometimes saccular, expansion is extremely variable; on occasion it virtually fills the extraembryonic coelom; the human endodermal allantois is diminutive, projecting merely into the embryonic end of the connecting stalk. The latter is regarded by many as precociously formed allantoic mesoderm, and the

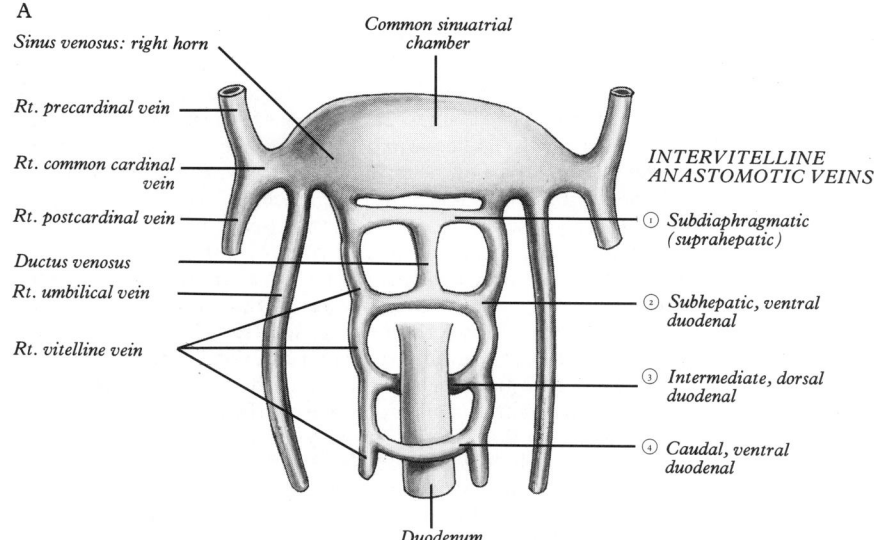

A

Common sinuatrial chamber

Sinus venosus: right horn

Rt. precardinal vein

Rt. common cardinal vein

Rt. postcardinal vein

Ductus venosus

Rt. umbilical vein

Rt. vitelline vein

Duodenum

INTERVITELLINE ANASTOMOTIC VEINS

① *Subdiaphragmatic (suprahepatic)*

② *Subhepatic, ventral duodenal*

③ *Intermediate, dorsal duodenal*

④ *Caudal, ventral duodenal*

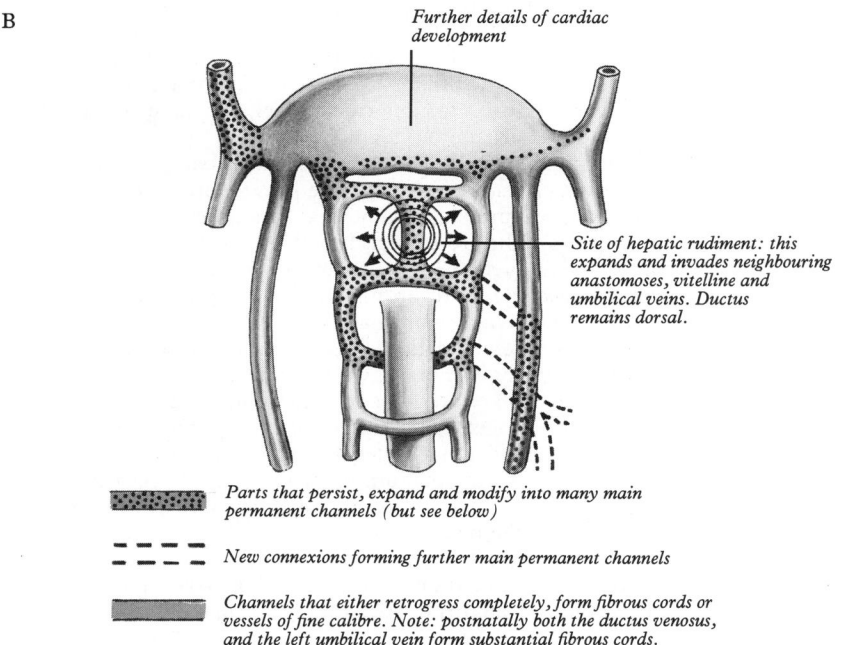

B

Further details of cardiac development

Common sinuatrial chamber

Site of hepatic rudiment: this expands and invades neighbouring anastomoses, vitelline and umbilical veins. Ductus remains dorsal.

Parts that persist, expand and modify into many main permanent channels (but see below)

New connexions forming further main permanent channels

Channels that either retrogress completely, form fibrous cords or vessels of fine calibre. Note: postnatally both the ductus venosus, and the left umbilical vein form substantial fibrous cords.

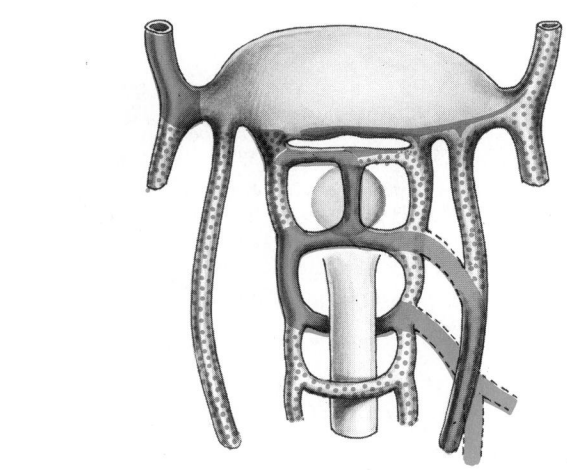

C

2.93 Development of the vitelline, umbilical, and terminal cardinal vein complexes: the early symmetrical condition. A The topography and nomenclature of the veins forming the right and left sinual horns; the intervitelline anastomoses; the median ductus venosus. B To assist understanding of later changes the symmetrical pattern is used to indicate which segments persist, retrogress, and the sites of formation of new channels and the intimately involved hepatic rudiment. C A simplified representation of the subsequent main flow paths of oxygenated and deoxygenated blood.

umbilical vessels as *allantoic* (with close affinities, because of their common phylogenetic history, with the vitelline vessels). In the human embryo the peripheral venules drain the mesodermal cores of the chorionic villous stems and terminal villi (extraembryonic *somatopleuric* structures); these are the radicles of a, usually single, *vena umbilicales impar* which traverses the compacting mixed mesenchyme of the umbilical cord, to reach the caudal rim of the umbilicus. Here, the single cordal vein divides into primitive right and left umbilical veins; each curves rostrally in the somatopleuric lateral border of the umbilicus (i.e. where intraembryonic and extraembryonic or amniotic somatopleure are continuous), where it lies *lateral* to the communication between both the intraembryonic and extraembryonic coeloms. Rostrolateral to the umbilicus the two umbilical veins reach, enter and traverse the junctional mesoderm of the septum transversum, connect with septal capillary plexuses, then continue, to enter their corresponding cardiac sinual horns (lateral to the terminations of the vitelline veins). This early symmetrical disposition of the vitelline veins and anastomoses, umbilical and common cardinal veins and locus of the hepatic primordial complex is summarized in **2.93**A–C.

Progressive changes in the vitello-umbilical veins are rapid, profound and closely linked with regional modifications of shape and position of the gut, expansion and invasion of venous channels by hepatic tissue, asymmetry of the heart and cardiac venous return. The principal events are summarized in **2.93–95**, only brief textual allusions can be included here; see Dickson (1957) for details and bibliography.

From the pars hepatica of the ventral hepatopancreatic rudiment (p. 237) interconnected sheets and 'cords' of endodermal cells, the presumptive hepatocytes, penetrate the mesenchyme-filled spaces of the pre-existing septal capillary plexuses. Possibly, under the inductive influence of the endodermal sheets, the plexuses become more profuse by the addition of angioblastic septal mesenchyme, which also forms masses of perivascular intrahepatic haemopoietic tissue. These processes extend along the plexiform connections of the vitelline, and later the umbilical veins until their intrahepatic (transeptal) zones themselves become largely plexiform—initially capillary in nature, but transform to a mass of rather wider, irregular, sinusoidal vessels with a discontinuous endothelium containing many phagocytic cells (p. 1392). The lengths of vitelline veins involved in these processes are the intermediate parts of the segments extending from the subhepatic (cranioventral duodenal) to the suprahepatic (subdiaphragmatic) transverse intervitelline anastomoses and the corresponding lengths of the umbilical veins. Thus at this early

stage the liver sinusoids are perfused by mixed blood reaching them through a series of branching vessels collectively called the *venae advehentes*, or *venae afferentes hepatis*; they are deoxygenated from the gut splanchnopleure via vitelline vein hepatic terminals and oxygenated from the placenta via hepatic terminals of the umbilical veins. Blood leaves the liver through four *venae revehentes (venae efferentes hepatis)*, two on each side reach and open into their respective cardiac sinual horns; this full complement of four *hepatocardiac veins* is only transient, becoming reduced to one dominant, rapidly enlarging channel. As detailed below the originally bilaterally symmetrical cardinal vein complexes, both rostral and caudal, develop transverse or oblique anastomoses whereby the cardiac venous return is restricted to the definitive right atrium. (The pulmonary veins are the only major ones returning to the left atrium.) The increasingly deep inflexion of the left wall of the common sinuatrial chamber, separation of the left sinual horn and 'body' of the sinus venosus, movement to the right of the sinuatrial orifice, and right atrial inclusion of the right sinual horn have been noted, (p. 209).

With these cardiac and concomitant hepatoenteric changes are accompanying events in supra-, intra- and sub-hepatic parts of the vitello-umbilical veins. Some vessels enlarge, persisting as definitive vessels to maturity and, in places, are joined later by other channels becoming defined in already established capillary plexuses. Other vessels retrogress, either disappearing completely, or remain as vestigial tags and occasionally vessels of fine calibre. Finally, some vessels of crucial importance in the circulatory patterns of embryonic and fetal life, *postnatally* become obliterated and transformed to substantial fibrous cords. Both right and left umbilical hepatocardiac and the left vitelline hepatocardiac veins continue, for a time, to discharge blood into their sinual horns; however they begin to retrogress (**2.93,94**). The right umbilical channel atrophies completely; the left channels also disappear, but their cardiac terminals may, on occasion, be found as conical fibrous tags attached to the inferior wall of the coronary sinus. The right vitelline hepatocardiac vein continues enlarging and ultimately forms the terminal segment of the inferior vena cava. The latter receives the right venae revehentes and new channels draining the territories of the left venae revehentes; these collectively form the upper and lower groups of right and (secondary) left hepatic veins. The terminal caval segment also shows the orifice of the right half of the intervitelline subdiaphragmatic anastomosis (p. 221) and a large new connection with the right subcardinal vein (vide infra).

The hepatic terminals of the right and left duodenal parts of the vitelline veins are destined to form the corresponding branches of

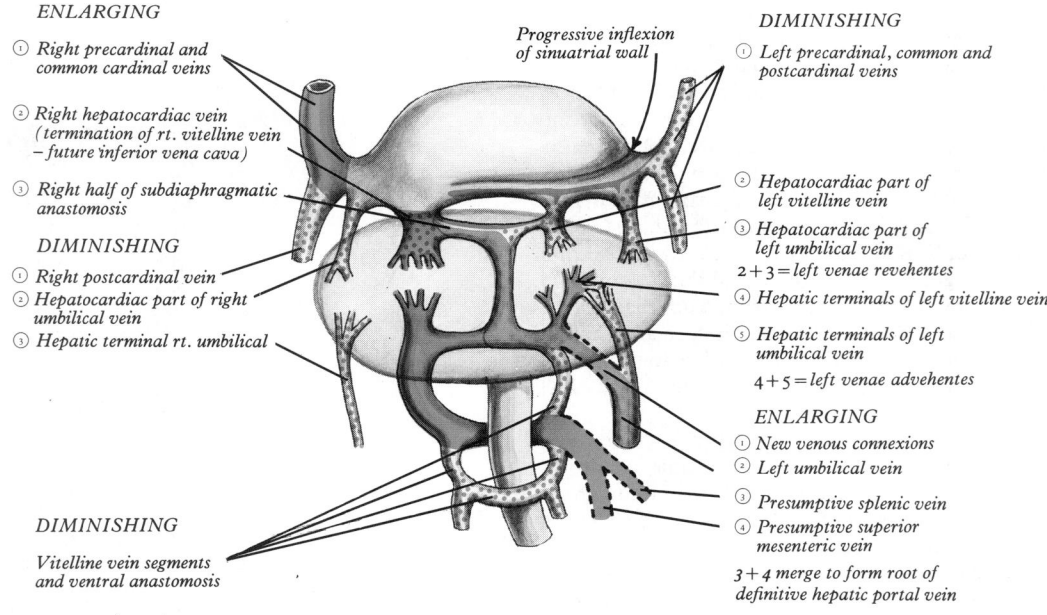

ENLARGING
① *Right precardinal and common cardinal veins*
② *Right hepatocardiac vein (termination of rt. vitelline vein – future inferior vena cava)*
③ *Right half of subdiaphragmatic anastomosis*

DIMINISHING
① *Right postcardinal vein*
② *Hepatocardiac part of right umbilical vein*
③ *Hepatic terminal rt. umbilical*

Progressive inflexion of sinuatrial wall

DIMINISHING
Vitelline vein segments and ventral anastomosis

DIMINISHING
① *Left precardinal, common and postcardinal veins*
② *Hepatocardiac part of left vitelline vein*
③ *Hepatocardiac part of left umbilical vein*
2 + 3 = left venae revehentes
④ *Hepatic terminals of left vitelline vein*
⑤ *Hepatic terminals of left umbilical vein*
4 + 5 = left venae advehentes

ENLARGING
① *New venous connexions*
② *Left umbilical vein*
③ *Presumptive splenic vein*
④ *Presumptive superior mesenteric vein*
3 + 4 merge to form root of definitive hepatic portal vein

2.94 Development of the vitelline, umbilical and terminal cardinal vein complexes: a mid-stage of asymmetry has been reached between the early symmetrical condition (**2.93**A–C) and the definitive late prenatal state.

Note the definition of the coronary sinus, the central role of the liver, and compare homologous left and right sided vessels.

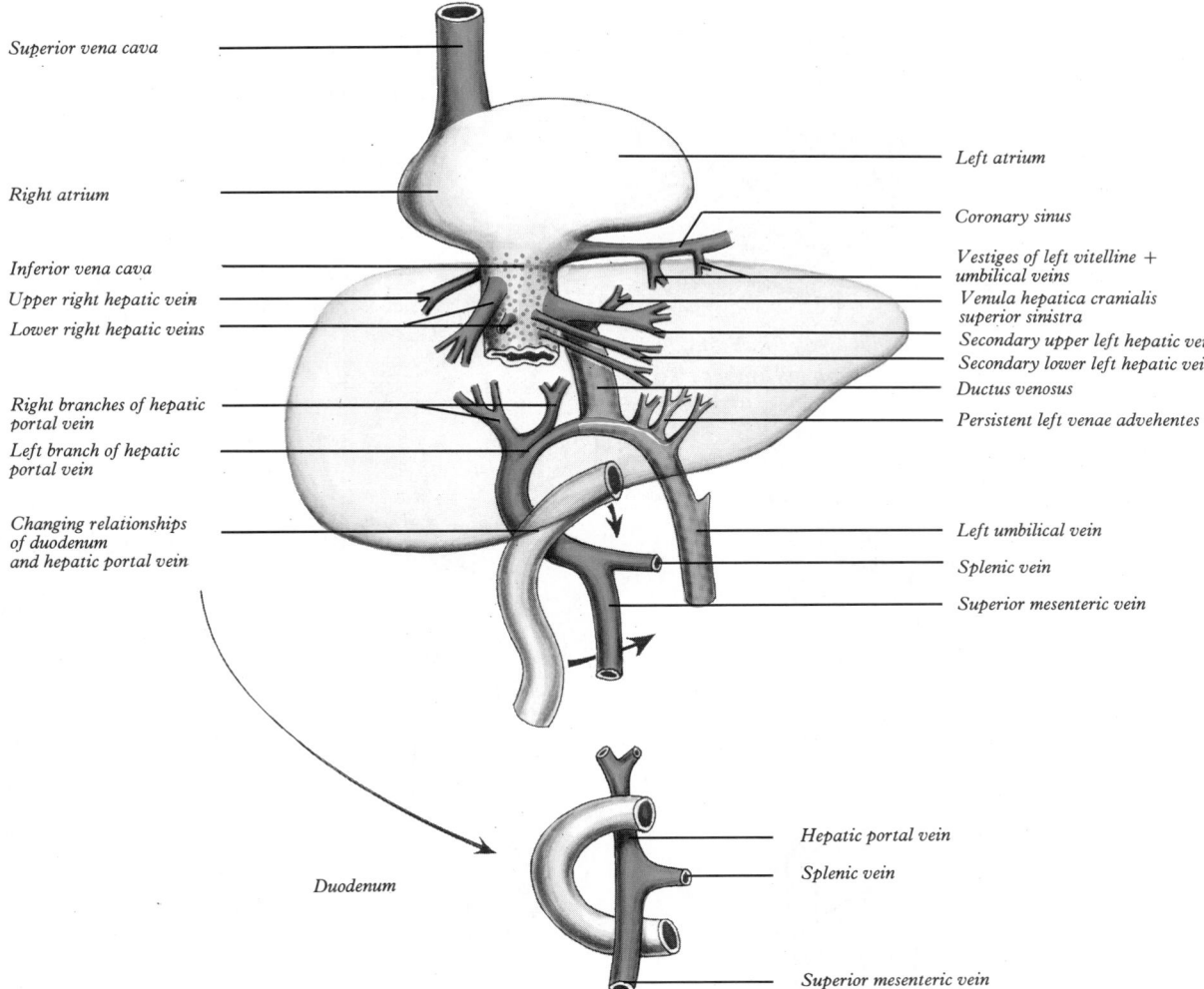

Superior vena cava

Right atrium

Inferior vena cava
Upper right hepatic vein
Lower right hepatic veins

Right branches of hepatic
portal vein

Left branch of hepatic
portal vein

Changing relationships
of duodenum
and hepatic portal vein

Left atrium

Coronary sinus

Vestiges of left vitelline +
umbilical veins
Venula hepatica cranialis
superior sinistra
Secondary upper left hepatic vein
Secondary lower left hepatic veins
Ductus venosus
Persistent left venae advehentes

Left umbilical vein

Splenic vein

Superior mesenteric vein

Duodenum

Hepatic portal vein

Splenic vein

Superior mesenteric vein

2.95 The condition of some main upper abdominal and intrathoracic right atrial terminal veins in the later prenatal months. Note the coronary sinus and attached vestiges, the terminations of the superior and inferior venae cavae; the grouped hepatic veins; the convergent formation and hepatic divarication of the hepatic portal vein, and the routes available for return of oxygenated left umbilical venous blood; also the changing disposition of the duodenal loop and its relationship to the definitive portal vein; the manner in which the left terminal branch of the latter establishes connections with the ductus venosus and the left umbilical vein. Return of placental blood to the fetal heart is along an (embryologically) complex path; illustrations **2.93-95** should assist clarification. See text for further details.

the *hepatic portal vein*, the left branch incorporating the cranial ventral intervitelline anastomosis. With rotation of the gut and formation of the duodenal loop, segments of the original vitelline veins and the caudal transverse anastomosis (indicated in **2.93-95**) atrophy, whilst new splanchnopleuric venous channels, the superior mesenteric and splenic veins, converge and join the left end of the dorsal intermediate anastomosis. The numerous other radicles of the portal vein and its principal branches, including the inferior mesenteric vein, are later formations.

For a period placental blood returns from the umbilicus via right and left umbilical veins, both discharging through venae advehentes into the hepatic sinusoids, where admixture with vitelline blood occurs. At approximately 7 mm crown-rump length, the right umbilical vein retrogresses completely; the left umbilical vein retains some vessels discharging directly into the sinusoids, but new enlarging connections with the left half of the subhepatic intervitelline anastomosis emerge. The latter is the commencement of a *by-pass channel* for the majority of the placental blood, that continues through the median ductus venosus, and finally the right half of the subdiaphragmatic anastomosis to reach the termination of the inferior vena cava. Postnatally these channels are obliterated, with the resulting ligamentum teres extending from the umbilicus to the porta hepatis, whence, having established connections with the left branch of the portal vein, it continues as the ligamentum venosum to join an upper left hepatic vein, and terminates in the suprahepatic inferior vena cava.

The cardinal venous complexes (**2.96,97**) are first represented by two large vessels on each side, the *precardinal* and *post-cardinal veins*; the former drain the rostral part of the embryo, the latter its caudal region. The two veins on each side unite to form a short *common cardinal vein*, which passes ventrally, lateral to the pleuropericardial canal (p. 239), to open into the corresponding horn of the sinus venosus (**2.87B**). (The cardinal complexes are often, less appropriately, called parietal or somatic. In addition to drainage of the latter, they receive many radicles from splanchnopleuric structures.)

Owing to the rapid development of the head and brain the precardinal veins become enlarged. They are further augmented by the *subclavian* veins from the upper limb buds, and so become the chief tributaries of the common cardinal veins; these gradually assume an almost vertical position in association with the descent of the heart into the thorax. That part of the original precardinal vein rostral to the subclavian is now the *internal jugular vein*, and their confluence the *brachiocephalic vein* of each side. The right and left common cardinal veins are originally of the same diameter. By the development of a large oblique transverse connection, the *left brachiocephalic vein* carries blood across from the left to the right (**2.97**). The part of the original right precardinal vein between the junction of the two brachiocephalics and the azygos veins forms the upper part of the **superior vena cava**; the caudal part of the latter vessel (below the entrance of the azygos vein) is formed by the right common cardinal. Caudal to the transverse branching of the left brachiocephalic vein the left precardinal and left common cardinal veins largely atrophy, the former constituting the terminal part of the left superior intercostal vein, while the latter is represented by the ligament of the left

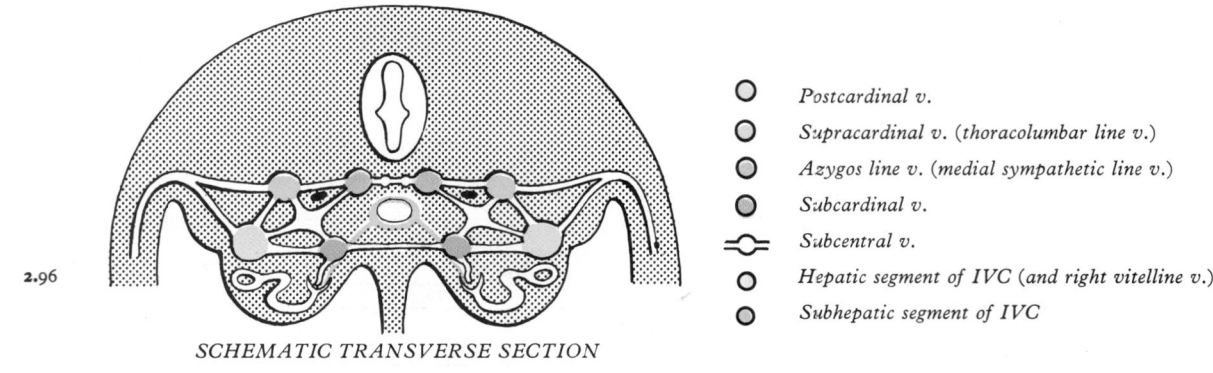

2.96

Postcardinal v.

Supracardinal v. (thoracolumbar line v.)

Azygos line v. (medial sympathetic line v.)

Subcardinal v.

Subcentral v.

Hepatic segment of IVC (and right vitelline v.)

Subhepatic segment of IVC

SCHEMATIC TRANSVERSE SECTION
THROUGH EMBRYONIC TRUNK

Precardinal vs.

Oblique interprecardinal anastomosis

Internal jugular vs.

Brachio-cephalic vs.

Sinus venosus

Common cardinal v.

Hepatocardiac vs.

Post-cardinal vs.

Sub-cardinal v.

Azygos line v.

Superior vena cava

Azygos v.

Superior intercostal v.

Oblique v. and ligt. of L. atrium

Hemiazygos vs.

IVC hepatic segment

IVC subhepatic segment

IVC subcardinal segment

L. suprarenal v.

L. renal v.

L. gonadal v.

IVC supracardinal segment

Supracardinal v.

Intersubcardinal anastomoses ('renal collar')

Interpostcardinal anastomosis

Common iliac vs.

2.97

J.A.H.

PROGRESSIVE ASYMMETRY:
RIGHT SIDED DOMINANCE:
SOME CHANNELS ENLARGE
OTHERS RETROGRESS

EARLY SYMMETRICAL
DISPOSITION OF VEINS

NOTE MATURATION AND
TRIBUTARIES OF SUPERIOR
VENA CAVA: SEGMENTS OF
DEFINITIVE INFERIOR VENA CAVA

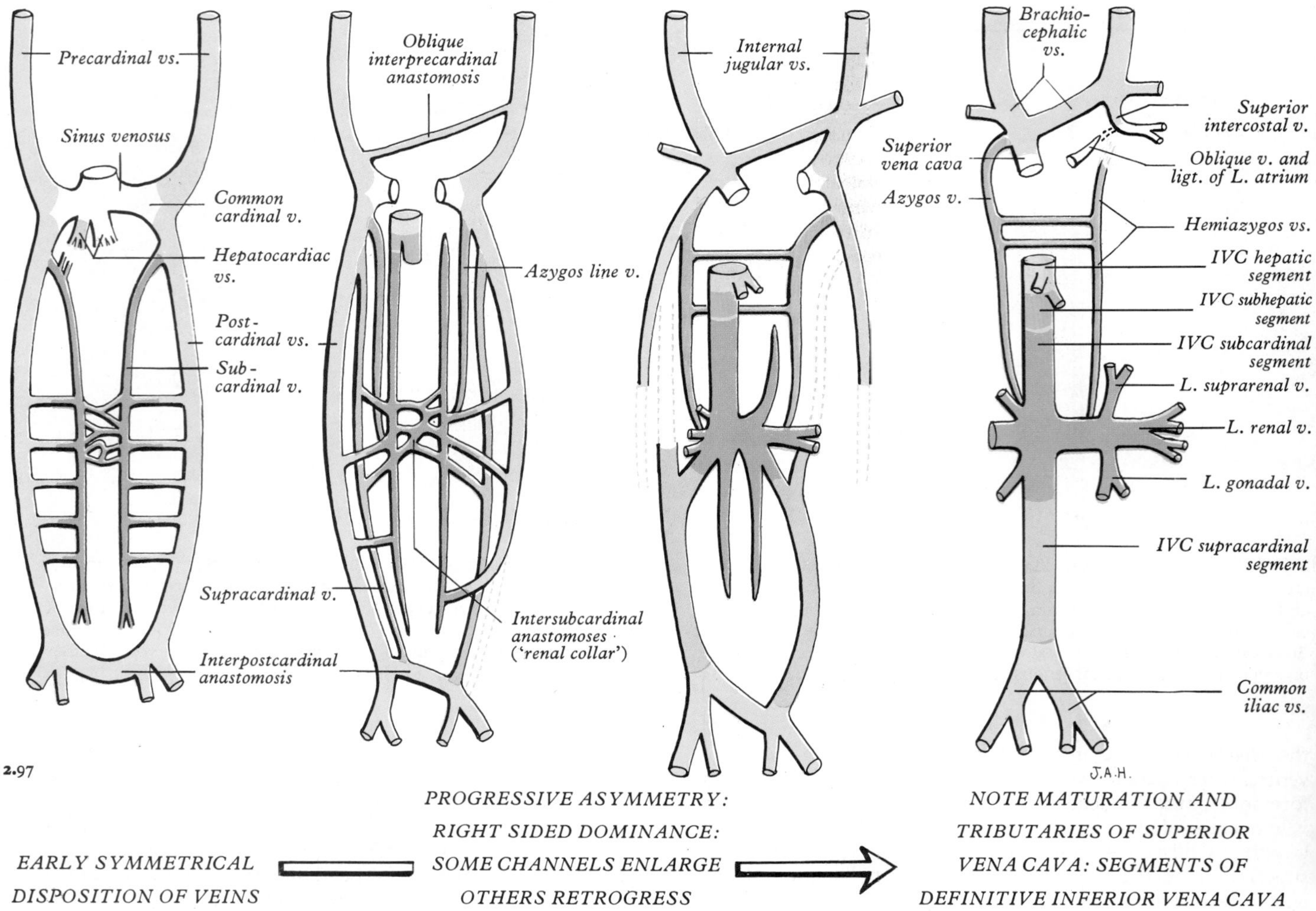

2.96 Somatic venous development: schematic section through the embryonic trunk. Principal longitudinal veins are colour-coded. Interconnections and intersegmental veins remain uncoloured. (Modified from Williams et al 1969.)

2.97 Plan of development of principal somatic veins from the early symmetrical state, through states of increasing asymmetry, to the definitive arrangement. (Modified from Williams et al 1969.)

vena cava and the oblique vein of the left atrium (2.97). The remainder of the left superior intercostal vein is developed from the cranial end of the postcardinal vein and drains the second, third and, on occasion, the fourth intercostal veins. The oblique vein passes downwards across the back of the left atrium to open into the coronary sinus, which, as already indicated, represents the persistent left horn of the sinus venosus. Right and left superior venae cavae are present in some animals, and occasionally persist in mankind.

The inferior vena cava (2.96,97) of the adult is a composite vessel, and the precise mode of development of its postrenal segment (caudal to the renal vein) still somewhat uncertain. Its function is initially carried out by the right and left *postcardinal veins*, which receive the venous drainage of the lower limb buds and pelvis and run in the dorsal part of the mesonephric ridges,

receiving tributaries from the body wall (*intersegmental veins*) and from the derivatives of the mesonephroi.

A second pair of longitudinal channels, the *subcardinal veins*, form in ventromedial parts of the mesonephric ridges and become connected to the postcardinal veins by a number of vessels traversing the medial part of the ridges. The subcardinal veins intercommunicate by a *pre-aortic anastomotic plexus*, which later constitutes the part of the *left renal vein* crossing anterior to the abdominal aorta.

The formation of an oblique transverse anastomosis between the iliac veins—which itself becomes the major part of the definitive *left common iliac vein*—diverts an increasing volume of blood into the right longitudinal veins, accounting for the ultimate disappearance of most of those on the left.

At its cranial end the subcardinal vein receives the *suprarenal vein* on each side, but on the right side it comes into intimate relationship with the liver. An extension of the vessel takes place in a cranial direction and meets and establishes continuity with a corresponding new formation which is growing caudally from the right vitelline hepatocardiac (common hepatic) vein. In this way on the right side a more direct route is established to the heart and the *prerenal* (cranial) segment of the inferior vena cava is defined.

The enlargement of the metanephros (p. 250) diverts the postcardinal vein from its course and the venous drainage of the mesonephric ridge is assumed by the subcardinal vein. At the same time new longitudinal channels appear and take over intersegmental venous drainage and the whole of the postcardinal vein disappears—except its extreme cranial and caudal ends. There are at least three such channels on each side, in addition to the *postcardinal* and *subcardinal* already mentioned, but, so far as is known, only two of them persist as large vessels in the adult: (1) A bilateral longitudinal channel forms dorsolateral to the aorta and *lateral* to the sympathetic trunk and takes over the intersegmental venous drainage from the posterior cardinal vein. This is the *supracardinal* or *thoracolumbar 'complex'*; alternatively *lateral sympathetic line vein*. (2) A second channel forms on each side, also dorsolateral to the aorta but *medial* to the sympathetic trunk. This, the *azygos line vein* or *medial sympathetic line vein*, gradually takes over the intersegmental venous drainage from the thoracolumbar line. The intersegmental veins now obviously reach their longitudinal channel by passing medial to the autonomic trunk, the relationship which the lumbar and intercostal veins maintain thenceforth. Cranially the azygos lines join the persistent cranial ends of the posterior cardinal veins. (3) Two *subcentral veins* are laid down directly dorsal to the aorta in the interval between the origins of the paired intersegmental arteries. These veins communicate freely with each other and with the azygos line veins, and these connections ultimately form the retro-aortic parts of the left lumbar veins and of the hemiazygos veins. (4) Some authorities also recognize a *precostal* or *lumbocostal* venous line, anterior to the vertebrocostal element, and posterior to the supracardinal. a possible derivative is the ascending lumbar vein (p. 818).

The thoracolumbar or supracardinal veins are, as indicated, lateral to the aorta and sympathetic trunks, which therefore intervene between them and the azygos lines. These veins communicate caudally with the iliac veins and cranially with the subcardinal veins in the neighbourhood of the pre-aortic intersubcardinal anastomosis. In addition, the supracardinal veins communicate freely with each other through the medium of the azygos lines and the subcentral veins. The most cranial of these connections, together with the supracardinal–subcardinal and the intersubcardinal anastomoses, complete a venous ring around the aorta below the origin of the superior mesenteric artery, termed the '*renal collar*' (Huntington 1920).

The right supracardinal vein persists and forms the greater part of the postrenal segment of the inferior vena cava, the continuity of the vessel being maintained by the persistence of the anastomosis between the right supracardinal and the right subcardinal in the 'renal collar'. The left supracardinal disappears, but some of the 'renal collar' formed by the left supracardinal-subcardinal anastomosis persists in the left renal vein. It must be added that much confusion and disagreement exists with regard to the disposition, homologies, and derivatives of the complicated array of longitudinal veins described above.

The inferior vena cava (2.97) is, therefore, formed from below upwards by (1) confluence of the common iliac veins, (2) short segment of the right postcardinal vein, (3) postcardinal-supracardinal anastomosis, (4) part of the right supracardinal vein, (5) right supracardinal–subcardinal anastomosis, (6) right subcardinal vein, (7) a new anastomotic channel of double origin —the hepatic segment of the inferior vena cava, and (8) the cardiac termination of the right vitelline hepatocardiac vein (common hepatic vein). It should be noted that only the supracardinal part of the inferior vena cava receives the intersegmental venous drainage, and that the postrenal (caudal) segment of the inferior vena cava is on a plane which lies dorsal to the plane of the prerenal (cranial) segment. Thus the right phrenic, suprarenal and renal arteries, which represent persistent mesonephric

arteries, pass behind the inferior vena cava, while the testicular or ovarian, which has a similar developmental origin, passes anterior to it.

In some animals the right postcardinal vein constitutes a large part of the postrenal segment of the inferior vena cava. In these cases the right ureter, on leaving the kidney, passes medially dorsal to the vessel and then, curving round its medial side, crosses its ventral aspect. Rarely, a similar condition is found in the human subject, and indicates the persistence of the right postcardinal vein and failure of the right supracardinal to play its normal part in the development of the vessel.

The ultimate arrangement of some of the embryonic abdominal and thoracic longitudinal cardinal veins may be summarized:

(1) The terminal part of the left postcardinal vein forms the distal part of the left superior intercostal vein; on the right side its cranial end persists as the terminal part of the vena azygos.

(2) The caudal part of the subcardinal vein is in part incorporated in the testicular or ovarian vein (McClure & Butler 1925) and partly disappears. The cranial end of the right subcardinal vein is incorporated into the inferior vena cava and also forms the right suprarenal vein. The left subcardinal vein, cranial to the intersubcardinal anastomosis, is incorporated into the left suprarenal vein. The renal and testicular or ovarian veins on both sides join the supracardinal-subcardinal anastomosis. On the left side this is connected directly to the part of the inferior vena cava which is of subcardinal status through an intersubcardinal anastomosis.

(3) The right supracardinal vein forms much of the postrenal (caudal) segment of the inferior vena cava. The left supracardinal vein disappears entirely.

(4) The right azygos line persists in its thoracic part to form all but the terminal part of the vena azygos. Its lumbar part can usually be identified as a small vessel which leaves the vena azygos on the body of the twelfth thoracic vertebra and descends on the vertebral column, deep to the right crus of the diaphragm, to join the posterior aspect of the inferior vena cava at the upper end of its postrenal segment. The left azygos line forms the hemiazygos veins.

(5) The subcentral veins give rise to the retro-aortic parts of the left lumbar veins and of the hemiazygos veins.

The veins of the head have a complicated developmental history (2.98A,B, and consult Markowski 1911, Streeter 1918, Padget 1957, Butler 1957, 1967, Browder & Kaplan 1976).

The primary vessels consist of a close-meshed capillary plexus drained on each side by the precardinal vein, which is at first continuous cranially with a transitory *primordial hindbrain channel*, lying on the neural tube medial to the cranial nerve roots. This is soon replaced by the *primary head vein* which runs caudally from the medial side of the trigeminal ganglion, lateral to the facial and vestibulocochlear nerves and otocyst, then medial to the vagus nerve, to become continuous with the precardinal vein. A lateral anastomosis subsequently brings it lateral to the vagus nerve. The cranial part of the precardinal vein forms the internal jugular vein; its caudal moiety has already been described on p. 223.

The primary capillary plexus of the head becomes separated into three fairly distinct strata by the differentiation of the skull and meninges. The superficial vessels, draining the integument and underlying soft parts, eventually discharge in large part into the *external jugular system*. They retain some connections with the deeper veins through so-called emissary veins. Deep to this is the *venous plexus of the dura mater*, from which the dural venous sinuses differentiate. This plexus converges on each side into *anterior, middle* and *posterior dural stems*. The anterior stem drains the prosencephalon and mesencephalon entering the primary head vein rostral to the trigeminal ganglion. The middle stem drains the metencephalon and empties into the primary head vein caudal to the trigeminal ganglion, while the posterior stem drains the myelencephalon into the commencement of the precardinal vein (2.98A,B). The deepest capillary stratum is the pial plexus from which the veins of the brain differentiate; it drains at the dorsolateral aspect of the neural tube into the adjacent dural venous plexus. In addition the primary head vein also receives, at its cranial end, the *primitive maxillary vein*, draining the maxillary prominence and region of the optic vesicle.

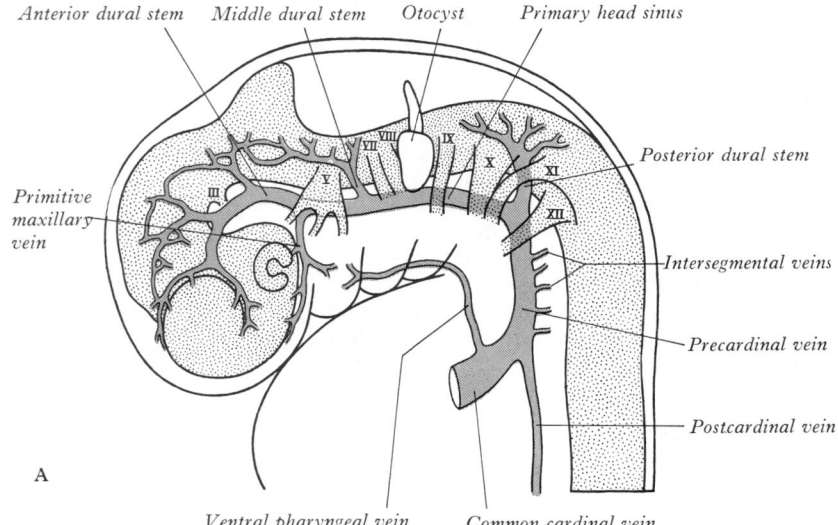

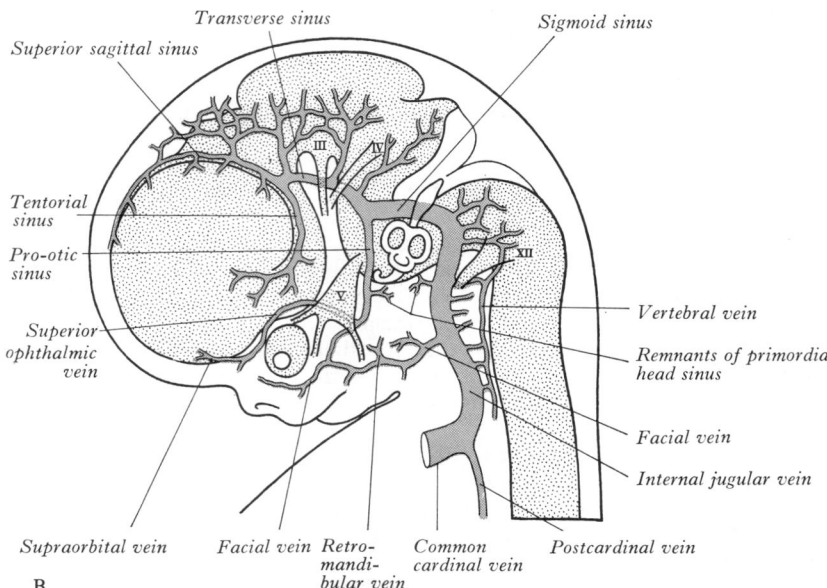

2.98 Diagrams illustrating successive stages in the development of the veins of the head and neck, (A) at approximately 8 mm and (B) at approximately 24 mm CR length.

The vessels of the dural plexus undergo profound changes, largely accommodating the growth of the cartilaginous otic capsule of the membranous labyrinth and expansion of the cerebral hemispheres. With growth of the otic capsule the primary head vein is gradually reduced and a new channel joining anterior, middle and posterior dural stems appears dorsal to the cranial nerve ganglia and the capsule (Butler 1967). Where this new vessel joins the middle and posterior stems, together with the posterior dural stem itself, forms the adult sigmoid sinus (**2.**98B).

Between the growing cerebral hemispheres and along the dorsal margins of the anterior and middle plexuses there forms a curtain of capillary veins, the *sagittal plexus*, in the position of the future falx cerebri. Rostrodorsally this plexus forms the *superior sagittal sinus* and is continuous behind with the anastomosis between the anterior and middle dural stems, which forms most of the *transverse sinus*. Ventrally the sagittal plexus differentiates into the *inferior sagittal* and *straight sinuses* and the *great cerebral vein*, and drains, more commonly, into the left transverse sinus. (For right/left variations in the terminations of sagittal and straight sinuses, and occurrence of a confluence see p. 800.)

The vessels along the ventrolateral edge of the developing cerebral hemisphere form the transitory *tentorial sinus*, which drains the convex surface of the cerebral hemisphere and basal

ganglia, and the ventral aspect of the diencephalon, to the transverse sinus. With expansions of the cerebral hemispheres, and in particular the emergence of the temporal lobe, the tentorial sinus becomes elongated, attenuated and eventually disappears, and its territory is drained by enlarging anastomoses of pial vessels which become the *basal veins*, radicles of the *great cerebral vein*.

The anterior dural stem disappears and the caudal part of the primary head vein dwindles; it is represented in the adult by the *inferior petrosal sinus*. The cranial part of the primary head vein, medial to the trigeminal ganglion, persists and still receives the stem of the primitive maxillary vein. The latter has now lost most of its tributaries to the anterior facial vein, its stem becoming the main trunk of the *primitive supra-orbital vein*, which will form the *superior ophthalmic vein* of the adult. Thus, the main venous drainage of the orbit and its contents is now carried via the augmented middle dural stem, the *pro-otic sinus*, into the transverse sinus and at a later stage into the cavernous sinus. The *cavernous sinus* is formed from a secondary plexus, derived from the primary head vein and lying between the otic and basioccipital cartilages. This forms the *inferior petrosal sinus* and drains through the primordial hindbrain channel into the internal jugular vein. The *superior petrosal sinus* arises later from a ventral metencephalic tributary of the pro-otic sinus; it communicates secondarily with the cavernous sinus (Butler 1957, 1967). The pro-otic sinus meanwhile has developed a new and more caudally situated stem draining into the sigmoid sinus; this new stem is the *petrosquamosal sinus* (p. 801) and the pro-otic sinus becomes, with progressive ossification of the skull, diploic in position. The development of the venous drainage and portal system of the hypophysis cerebri is closely associated with that of the venous sinuses (Wislocki 1937, Niemineva 1950, also p. 1455).

The venous drainage of the face, scalp and neck becomes established after the development of the skull. The first identifiable vessel is the *ventral pharyngeal vein*, draining the massive mandibular and hyoid arches into the common cardinal vein. With the elongation of the neck its termination is transferred to the cranial part of the precardinal vein which later becomes the *internal jugular*. The ventral pharyngeal vein, receiving tributaries from the face and tongue, becomes the *linguofacial* vein. With development of the face the primitive maxillary vein extends its drainage into the territories of supply of the ophthalmic and mandibular division of the fifth nerve, including the pterygoid and temporal muscles, and over the lower jaw it anastomoses with the linguofacial vein. This anastomosis becomes the *facial vein*; it receives a strong *retromandibular vein* from the temporal region, and drains through the linguofacial vein into the internal jugular. The stem of the linguofacial vein is now the lower part of the facial vein, whilst the dwindling connection of the facial with the primitive maxillary becomes the *deep facial vein*. The *external jugular vein* is developed from a tributary of the cephalic vein from the tissues of the neck and anastomoses secondarily with the anterior facial vein. At this stage the *cephalic vein* forms a *venous ring* around the clavicle from which it is connected with the caudal part of the precardinal. The deep segment of the venous ring forms the *subclavian vein* and receives the definitive external jugular vein. The superficial segment of the venous ring dwindles, but may persist in adult life (Padget 1957).

THE LYMPHATIC SYSTEM

Two different views are current as to the initial stages in the development of the lymphatic system (Rusznyák et al 1960). According to the first view (Huntington 1908, McClure & Butler 1925) lymphatic spaces commence as clefts in the mesenchyme, and their lining cells take on the characteristics of endothelium (Kampmeier 1969). These spaces form capillary plexuses from which certain *lymph sacs*, to be noted later, are derived. The connections of the lymphatic and venous systems are regarded as entirely secondary. In contrast, however, according to Sabin (1912), the earliest lymph vessels arise as capillary offshoots from the endothelium of the veins, as capillary plexuses. These plexuses lose their connections with the venous system and become confluent to form lymph sacs. The balance of the evidence suggests that all but the earliest lymphatic channels originate independently of the venous system and only acquire connections with it at a later stage (Kampmeier 1969).

In the human embryo the lymph sacs from which the lymph vessels are derived are six in number; two paired (the *jugular* and the *posterior lymph sacs*) and two unpaired (the *retroperitoneal* and the *cisterna chyli*). In lower mammals an additional pair (the *subclavian*) is present, but in the human embryo these are merely extensions of the jugular sacs.

The position of the sacs is as follows (**2**.99): (1) the *jugular*, the first to appear, at the junction of the subclavian vein with the precardinal, with later prolongations along the internal and external jugular veins; (2) the *posterior* encircling the left common iliac vein; (3) the *retroperitoneal*, in the root of the mesentery near the suprarenal glands; (4) the *cisterna chyli*, opposite the third and fourth lumbar vertebrae. From the lymph sacs the lymph vessels bud out along lines corresponding more or less closely with the course of embryonic blood vessels, most commonly veins, but many arise *de novo* in the mesenchyme and establish connections with existing vessels. In the body wall and that of the intestine the deeper plexuses are the first to be developed; by continued growth of these, the vessels in the superficial layers are gradually formed. The *thoracic duct* is, phylogenetically, a bilateral structure. In man it comprises the caudal part of the right vessel, a transverse anastomosis and the cranial part of the left vessel. According to the second view cited above it is formed from anastomosing outgrowths from the jugular sacs and cisterna chyli. At its connection with the cisterna it is at first double, but the vessels soon join. Numerous valves are laid down in the duct during the fifth month, but many of them disappear prior to birth. Those which persist are formed in situations where the duct may be subjected to pressure, e.g. where it is crossed by the oesophagus and the aortic arch.

All the lymph sacs except the cisterna chyli are, at a later stage, divided by a number of slender connective tissue bridges. Subsequently they are invaded by lymphocytes and transformed into groups of lymph nodes, the lymph sinuses representing portions of the original cavity of the sac. The caudal part of the cisterna chyli is similarly converted, but its rostral part remains as the definitive cisterna; in many cases the cisterna chyli is plexiform (p. 842). The siting of the major groups of lymph nodes follows a similar basic pattern amongst the mammals (Spira 1962).

Haemal lymph nodes are said to develop as mesenchymal condensations in close relation to blood vessels rather than lymphatics (Meyer 1917, Turner 1969).

The Development of the Alimentary and Respiratory Apparatus

THE BUCCAL CAVITY AND PHARYNX

The mouth is developed partly from the stomatodeum, and partly from the floor of the cranial portion of the foregut. By the rostral growth of the embryo and formation of the head fold, the pericardial area and oropharyngeal membrane come to lie on the ventral surface of the embryo (p. 140). With further expansion of the brain and bulging of the pericardium, the membrane is soon depressed between the two elevations. This depression is the *stomatodeum* or *primitive buccal cavity* (**2**.100). It is lined with ectoderm and separated from the cranial end of the foregut by the oropharyngeal membrane, which is formed by the apposition of stomatodeal ectoderm with foregut endoderm; at the end of the fourth week the membrane disappears and a communication is established between stomatodeum and cranial end of the foregut or future oropharynx. No vestige of the membrane is evident in the adult, and this embryonic communication must not be confused with the permanent oropharyngeal isthmus. The epithelium of the lips and gums and the enamel of the teeth are ectodermal in origin, from the stomatodeal walls, but the epithelium of the tongue and adnexa, developed in the floor of mouth and pharynx, is derived from endoderm. The development of the teeth and gums is described on p. 1310 et seq.

The branchial arches grow in a ventral direction and lie progressively between stomatodeum and pericardium; with the completion of the mandibular prominences and the development of the maxillary prominences (p. 167), the opening of the stomatodeum assumes a pentagonal form, bounded cranially by the frontonasal prominence, caudally by the mandibular prominences and laterally by the maxillary prominences (**2**.100). With the inward growth and fusion of the palatine processes (**2**.51B,C), the stomatodeum is divided into a nasal and a buccal part. Along the free margins of the prominences bounding the mouth cavity appears a shallow groove, and the ectoderm in its floor thickens and invades the underlying mesoderm; it divides into a medial *dental lamina* and a lateral *vestibular lamina*. The central cells of the latter degenerate and the furrow becomes deepened. It is now termed the *labiogingival groove* or *sulcus*; its inner wall contributes to the formation of the alveolar processes of the maxillae and the mandible, while its outer wall forms the lips and cheeks.

The salivary glands arise from the epithelial lining of the mouth. The *parotid gland* can be recognized in human embryos

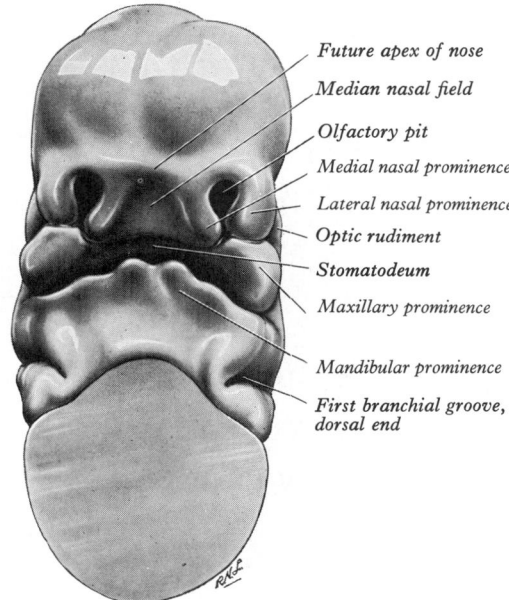

2.99 Scheme showing the relative positions of the primary lymph sacs. (After F R Sabin.)

Left brachiocephalic vein
Internal jugular vein
External jugular vein
Jugular lymph sac
Right brachiocephalic vein
Left common cardinal vein
Superior vena cava
Left postcardinal vein
Prerenal part of inferior vena cava
Left suprarenal vein
Left renal vein
Retroperitoneal lymph sac
Postrenal part of inferior vena cava
Cisterna chyli
Left common iliac vein
Posterior lymph sac
External iliac vein
Internal iliac vein

Future apex of nose
Median nasal field
Olfactory pit
Medial nasal prominence
Lateral nasal prominence
Optic rudiment
Stomatodeum
Maxillary prominence
Mandibular prominence
First branchial groove, dorsal end

2.100 The head of a human embryo in the sixth week. Ventral aspect. (From a model by K Peter.)

8 mm long as an elongated furrow running dorsally from the angle of the mouth between the mandibular and maxillary prominences. The groove, which is converted into a tube, loses its connection with the epithelium of the mouth, except at its ventral end, and grows dorsally into the substance of the cheek. The tube persists as the *parotid duct* and its blind end proliferates to form the gland. Subsequently the size of the oral fissure is reduced by partial fusion between the maxillary and mandibular prominences and the duct opens thereafter on the inside of the cheek at some distance from the angle of the mouth. The *submandibular gland* is identifiable in human embryos 13 mm long as an epithelial outgrowth from the floor of the *linguogingival groove* (vide infra). It increases rapidly in size, giving off numerous branching processes which later acquire lumina. At first the connection of the submandibular outgrowth with the floor of the mouth lies at the side of the tongue, but the edges of the groove in which it opens come together, from behind forwards, and form the tubular part of the *submandibular duct*. As a result, the orifice of the duct is shifted forwards till it is below the tip of the tongue, close to the median plane. The *sublingual gland* arises in embryos about 20 mm long as a number of small epithelial thickenings in the linguogingival groove and on the groove's lateral side, which later closes to form the submandibular duct. Each thickening canalizes separately; many of the multiple sublingual ducts open separately on the summit of the sublingual fold, others join the submandibular duct (p. 1293).

The tongue appears as a small median elevation, named the *median tongue bud (tuberculum impar)*, in the endodermal floor of the pharynx before the branchial arches meet ventrally; it subsequently becomes incorporated in the anterior part of the tongue. A little later two oval *distal tongue buds (lingual swellings)*

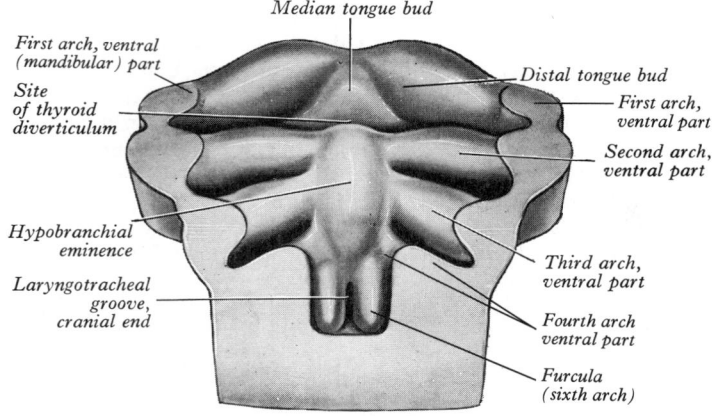

2.101A The floor of the pharynx of a human embryo at the beginning of the sixth week. (From a model by K Peter.)

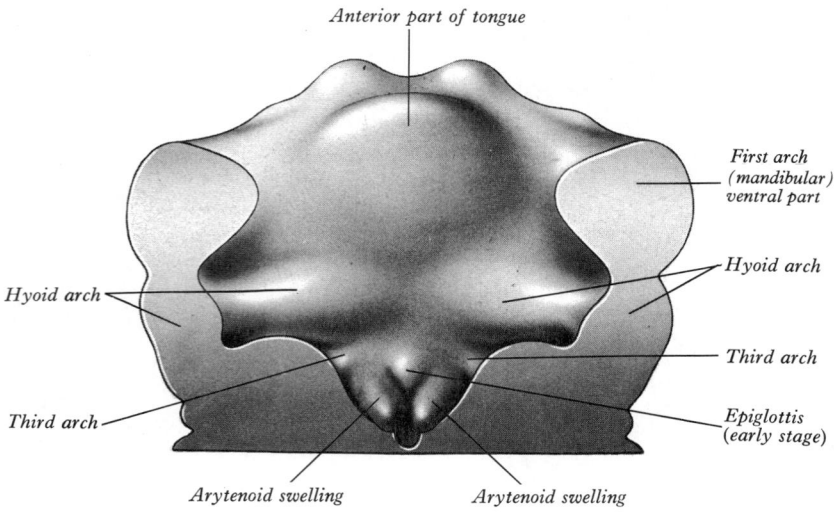

2.101B The floor of the pharynx of a human embryo, about 6 weeks old. (From a model by K Peter.)

appear on the endodermal aspect of the mandibular prominences. They meet each other in front, and caudally they converge on the median tongue bud, with which they fuse (2.101A,B). A sulcus forms along the ventral and lateral margins of this elevation and deepens, internal to the future alveolar process of the mandible, to form the *linguogingival groove*, while the elevation constitutes the anterior or buccal (presulcal) part of the tongue.

Caudal to the median tongue bud, a second median elevation, the *hypobranchial eminence* (copula of His), forms in the floor of the pharynx, and the ventral ends of the fourth, the third and, later, the second visceral arches converge into it. A transverse groove separates its caudal part to form the epiglottis, while ventrally it approaches the tongue rudiment, spreading in the form of a V, and forming the posterior or pharyngeal part of the tongue. In the process the third arch elements grow over and bury the elements of the second arch, excluding it from the tongue. As a result the mucous membrane of the pharyngeal part of the tongue receives its sensory supply from the glossopharyngeal, the nerve of the third arch. In the adult the union of the anterior and posterior parts of the tongue approximately corresponds to the angulated *sulcus terminalis*, its apex at the *foramen caecum*, a blind depression produced at the time of fusion of the constituent parts of the tongue, but also marking the site of ingrowth of the median rudiment of the thyroid gland.

At first the tongue consists of a mass of mesoderm covered on its surface by endoderm. During the second month occipital myotomes are often stated to migrate from the lateral aspects of the myelencephalon, then invading the tongue to form its musculature. They are held to pass ventrally round the pharynx to reach its floor accompanied by their nerve (the hypoglossal), which therefore crosses superficial to both the internal and external carotid arteries; however, see p. 175 for alternative views.

The composite character of the tongue is indicated by its innervation. Impulses from the anterior, buccal part are mediated by: (a) the lingual nerve, derived from the post-trematic nerve of the first arch (mandibular nerve); and (b) the chorda tympani, often held to be the pretrematic nerve to the first arch (but see p. 172). The posterior, pharyngeal part of the tongue is innervated by the glossopharyngeal, the nerve of the third arch. The muscles of the tongue being possibly 'myotomic' analogues, if not homologues, receive their nerve supply from the hypoglossal nerve, which is usually regarded as serially homologous with spinal ventral nerve roots.

The sulcus terminalis cannot be distinguished earlier than the 52 mm stage according to some observers. The vallate papillae appear at about the same time, increasing in number until the 170 mm stage. Serial reconstructions also suggest that the territory of the glossopharyngeal nerve extends considerably beyond these papillae.

The thyroid gland is first identifiable in embryos of about 20 somites, as a median thickening of endoderm in the floor of the pharynx between the first and second pharyngeal pouches and immediately dorsal to the aortic sac (Davis 1923). This area is later invaginated to form a median diverticulum which appears late in the fourth week in the furrow immediately caudal to the median tongue bud (2.101A). It grows caudally as a tubular duct the tip of which bifurcates and subsequently the whole mass divides into a series of double cellular plates, from which the isthmus and the lateral lobes of the thyroid gland are developed. The *primary thyroid follicles* differentiate by reorganization and proliferation of the cells of these plates. *Secondary follicles* subsequently arise by budding and subdivision (Norris 1916). These primary and secondary endodermal cells are the progenitors of the follicular parenchyma proper. The claim that the fourth pharyngeal pouches contribute thyroid tissue to the lateral lobes of the gland was long disputed and perhaps seemed unlikely on the grounds of comparative embryology. (But vide infra—the *ultimobranchial body* and the derivation of *parafollicular* or *C cells*.)

The original diverticulum, its bifurcation and generations of follicles invade mixed hypobranchial mesenchyme. From the latter are derived the thin connective tissue capsule, thinner 'interlobular' septa and delicate perifollicular investments. These carry the main vasculature, characteristic fenestrated capillaries, lymphatics and autonomic nerve supply.

The connection of the median diverticulum with the pharynx is termed the *thyroglossal duct*. The site of its initial continuation with the endodermal floor of the mouth is marked by the foramen caecum. From here it extends caudally in the median line ventral to the primordium of the hyoid bone, behind which it later forms a recurrent loop. The distal part of the duct commonly differentiates variably as the pyramidal lobe and levator muscle (or 'suspensory' fibrous band) of the thyroid. The remainder fragments and disappears, but the lingual part is often identifiable until late in fetal life and may branch and give rise to miniature salivary glands (Boyd 1964). Occasionally parts of the midline thyroglossal duct persist (occurring in lingual, suprahyoid, retrohyoid, or infrahyoid positions). They may form aberrant masses of thyroid tissue, cysts, fistulae or sinuses, usually in the midline (see p. 1459). A *lingual thyroid* situated at the junction of the buccal and pharyngeal parts of the tongue is not uncommon, but nodules of glandular tissue may also be found other than in the midline, e.g. laterally placed posterior to sternocleidomastoid, and, on occasion, below the level of the thyroid isthmus (see p. 1459).

The tonsils are developed from the parts of the second pharyngeal pouches which lie between the tongue and the soft palate. The endoderm lining these pouches grows into the surrounding mesenchyme as a number of solid buds. These buds are excavated by degeneration and shedding of their central cells, and thus the tonsillar *fossulae* and *crypts* are formed. Lymphoid cells accumulate around the crypts and become grouped as lymphoid follicles. A slit-like *intra-tonsillar cleft* (**8**.97) extends into the upper part of the tonsil and is a remnant of the second pharyngeal pouch.

The thymus is derived from the endoderm of the ventral part of the third pharyngeal pouch on each side (**2**.102). It cannot be recognized prior to the differentiation of the inferior parathyroid glands (vide infra), which occurs when the embryo is 10–12 mm long, but thereafter it is represented by two elongated diverticula which soon become solid cellular masses and grow caudally into the surrounding mesenchyme. Ventral to the aortic sac (p. 214) the two thymic rudiments meet and are subsequently united by connective tissue only; the rudiments themselves remain unfused. The connection with the third pouch is soon lost, but the stalk may persist for some time as a solid, cellular cord.

The development of thymic tissue from the ventral recess of the fourth pharyngeal pouch probably occurs in a proportion of embryos, although this has been denied by some authorities (Weller 1933, Norris 1938). Thymic tissue developing from this site is usually found near but outside the thyroid gland in close association with the superior parathyroid gland. An ectodermal contribution to the thymus, probably of placodal origin, occurs in some mammals but a similar contribution in man is conjectural (Garrett 1948).

Vascularized mesenchyme, including lymphoid stem cells, invades the cellular mass of the endodermal thymus and becomes partially lobulated. The cells of the cytoreticulum and the concentric corpuscles of the thymus are endodermal in origin. The

epithelial character of these cells is more obvious in fetal life; some are even ciliated (Sebuwufu 1968). Lymphoid cells enter and colonize the thymus from the haemopoietic tissue stem cells during the third month, according to observations from in vitro and animal experiments (Auerbach 1960, 1961, Taylor 1965, Moore & Owen 1967).

At birth the thymus is large relative to total body weight. Its absolute weight increases in the first two years after birth, but its relative weight decreases. There is little change thereafter until about the seventh year, when rapid growth again occurs to reach a maximum at about eleven years. After this it begins to decline to an adult weight which is very variable but averages 12–15 g. In old age the gland shrinks still further, especially after wasting diseases. For this and other reasons it is rarely identifiable in the preserved cadaver of the aged (Keynes 1954, Lasi 1959, and p. 833).

The parathyroid glands are also derivatives of the endoderm and adjacent mixed branchial mesenchyme. Prior to the appearance of the thymic rudiment from the third pharyngeal pouch, the epithelium on the dorsal aspect of the pouch and in the region of its duct-like connection with the cavity of the pharynx becomes differentiated as the primordium of the *inferior parathyroid gland*, recognizable by its cells, which stain more lightly than the other endodermal cells lining the pouch. Although the connection between the pouch and the pharynx is soon lost, the connection between the thymic and parathyroid rudiments persists for some time, and the latter passes caudally with the developing thymus. The *superior parathyroid glands* develop in a similar manner from the dorsal recess of the fourth pharyngeal pouches. They come into relation with, and appear anchored by, the lateral lobes of the thyroid gland and thus remain cranial to the parathyroid glands derived from the third pouch (see p. 1463). The mesenchyme provides the connective tissue envelopment, vasculature including fenestrated capillaries and lymphatics; also a route for vasomotor nerves.

The ultimobranchial body has already been noted as an endodermal diverticular part of the *caudal pharyngeal complex* (**2**.102). This separates from the ectoderm of the fourth branchial cleft and loses its connection with the pharynx by attenuation and rupture of the common pharyngobranchial duct. It becomes closely associated with the expanding lateral lobe of the thyroid gland, and the superior parathyroid (parathyroid IV) component of the complex lying dorsally and outside the thyroid gland. The remainder of the complex, which includes the ultimobranchial body and possibly some vestiges of the ventral recess of the fourth pharyngeal pouch and of the transitory fifth pharyngeal pouch, is enveloped by the thyroid gland. Although some controversy reigned, it is now strongly supported by evidence that the cells of the ultimobranchial body give rise to the 'C' or parafollicular cells producing calcitonin in the thyroid gland of many, if not all mammals (Halmi 1986). Calcitonin has been isolated from ultimobranchial tissue in vertebrates other than mammals (Copp et al 1967, Taylor 1968). The derivation of thyroid parafollicular cells has now been clearly demonstrated in embryonic sheep (Jordan et al 1973). For bibliographical sources see p. 1463 et seq.

The hypophysis cerebri consists of the adenohypophysis and the neurohypophysis; the former is derived from the placodal ectoderm of the stomatodeal roof (cf neural crest), the latter from the neurectoderm of the floor of the forebrain. (Consult p. 1451 for the varied usages of the older terms, anterior and posterior lobes, with the more satisfactory adenohypophysis and neurohypophysis, and their subdivisions.) From the outset the placodal roof and overlying forebrain are in close apposition and adherent, with no intervening mesenchyme; elsewhere, and up to the placodal margins, mesenchyme is interposed. Before rupture of the oropharyngeal membrane, proliferation of the periplacodal mesenchyme results in the placode forming the roof and walls of a saccular depression. This hypophysial recess (*pouch of Rathke*) (**2**.103A) is the rudiment of the adenohypophysis, lying immedately ventral to the dorsal border of the membrane, extending in front of the rostral tip of the notochord, and retaining contact with the ventral surface of the forebrain. It is constricted off by continued proliferation of the surrounding mesenchyme to form a closed vesicle, but remains for a time connected to the ectoderm

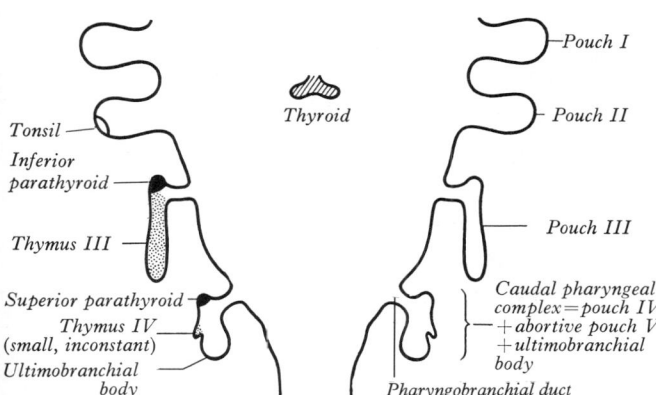

2.102 Scheme showing the development of the branchial epithelial bodies. The numbered sacs are the edodermal pharyngeal pouches.

Labels in figure:
Pouch I
Pouch II
Thyroid
Tonsil
Inferior parathyroid
Thymus III
Pouch III
Superior parathyroid
Thymus IV (small, inconstant)
Ultimobranchial body
Caudal pharyngeal complex = pouch IV + abortive pouch V + ultimobranchial body
Pharyngobranchial duct

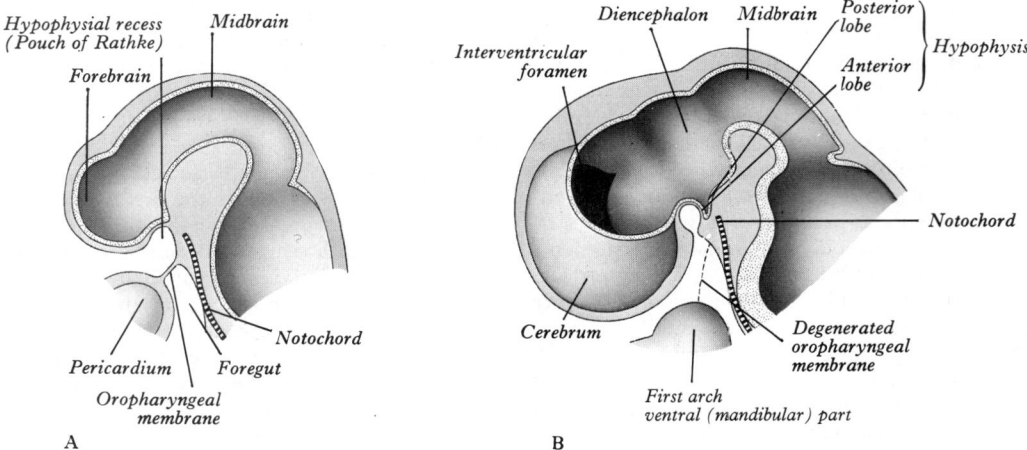

2.103 Schematic sagittal sections of heads of early embryos to show first stages in the development of the hypophysis.

of the stomatodeum by a solid cord of cells, which can be traced down the posterior edge of the nasal septum. Masses of epithelial cells form mainly on each side and in the ventral wall of the vesicle, and the development of the adenohypophysis progresses by the ingrowth of a mesenchymal stroma. Differentiation of epithelial cells into stem cells and three differentiating types is said to be apparent during the early months of fetal development (Dubois 1967). Recent work also suggests that different types of cells arise in succession, and that they may be derived in differing proportions from different parts of the hypophysial recess (Conklin 1968, see also p. 1451 et seq). A *craniopharyngeal canal*, which sometimes runs from the anterior part of the hypophysial fossa of the sphenoid bone to the exterior of the skull, is often said to mark the original position of the hypophysial recess (of

Rathke). Traces of the stomatodeal end of the recess are invariably present at the junction of the septum of the nose with the palate (see the *pharyngeal hypophysis* p. 1455). Others have claimed, however, that the craniopharyngeal canal itself is a secondary formation caused by the growth of blood vessels, and is quite unconnected with the stalk of the anterior lobe (Arey 1949). Just caudal to, but in contact with, the adenohypophysial recess a hollow diverticulum elongates towards the mouth from the floor of the diencephalon (**2.**103B). This neural outgrowth forms an *infundibular sac*, the walls of which increase in thickness until the contained cavity is obliterated except at its upper end, where it persists as the *infundibular recess* of the third ventricle. Formed in this way the neurohypophysis becomes invested by the adenohypophysis which extends dorsally on each side of it. In

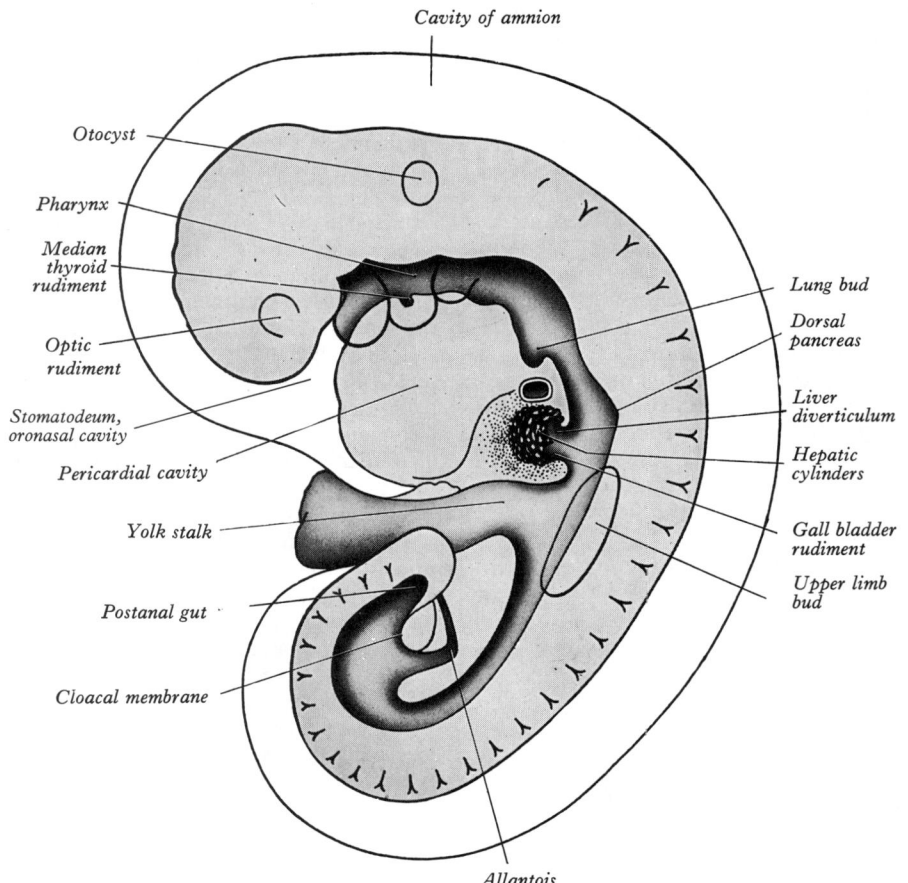

2.104A The digestive tube of a human embryo with 29 paired somites, a CR length of 3.4 mm and an estimated age of 27 days. Note branchial development. Magnification × c.25. (Streeter 1942.)

addition, the adenohypophysis gives off two processes from its ventral wall which grow along the infundibulum and fuse to surround it, coming into relation with the tuber cinereum and constituting the *tuberal portion* of the hypophysis. The original cavity of Rathke's pouch remains first as a cleft, and later scattered vesicles, and can be identified readily in sagittal sections through the mature gland. The dorsal wall of Rathke's pouch, which remains thin, fuses with the adjoining part of the neurohypophysis as the *pars intermedia*.

A small endodermal diverticulum, named *Seessel's pouch*, projects towards the brain from the cranial end of the foregut, immediately caudal to the oropharyngeal membrane. In some marsupials this pouch forms a part of the hypophysis, but, in man, it apparently disappears entirely; initially it may contribute some cells to the general head mesenchyme of mixed origin, and possibly functions as one of many rostral head organizers (together with the prechordal plate of endoderm).

The Pharynx

The pharynx is the cranial end of the foregut, and the branchial arches and pharyngeal pouches play an important part in its development (2.104A,B). The endodermal aspect of the first (maxillomandibular) arch in its dorsal part contributes to the formation of the lateral wall of the nasopharynx in front of the orifice of the auditory tube. The ventral end of the first pouch becomes obliterated, but its dorsal end persists and deepens as the head enlarges. It remains close to the ectoderm of the dorsal end of the first cleft (p. 170) and, together with the adjoining lateral part of the pharynx and dorsal part of the second pharyngeal pouch, constitutes the *tubotympanic recess*, which forms the tympanic cavity and the auditory tube (p. 204), and ultimately their extensions. The site of the second arch is partly indicated by the *palatoglossal arch*, but its dorsal end is separated from its ventral end by the forward growth of the third arch, which obliterates the

intermediate part. Some believe that the site of the second pharyngeal pouch is represented by the *intratonsillar cleft*, around which the tonsil is developed. The third arch forms the *lateral glosso-epiglottic fold*, and its dorsal end takes part in the formation of the floor of the auditory tube. The ventral ends of the fourth arches fuse with the caudal part of the hypobranchial eminence and so contribute to the formation of the *epiglottis* (p. 165). The adjoining portion becomes connected to the *arytenoid swelling* and may be identified in the *aryepiglottic fold*.

After the caudal part of the hypobranchial eminence has separated from the pharyngeal (posterior) part of the tongue (p. 238), it is in continuity with two linear ridges which appear in the ventral wall of the pharynx, the whole forming an inverted U, sometimes regarded as an independent formation, the *furcula* (of His). These vertical ridges have been identified as the sixth arches, placed very obliquely owing to the shortness of the pharyngeal floor compared with the greater extent of the roof. The ridges of the furcula are carried downwards on the ventral wall of the foregut and bound the median *laryngotracheal groove*, from which the lower part of the larynx, the trachea, bronchi and lungs are developed (p. 238). At the cranial end of the groove paired arytenoid swellings arise which convert the slit-like upper aperture of the respiratory system into a T-shaped opening. The aryepiglottic folds (fourth arch derivatives) can be recognized at this stage.

DEVELOPMENT OF THE POSTPHARYNGEAL GUT

The profound changes occurring late in the third and throughout the fourth weeks as, with head, tail and lateral folding, the trilaminar mass begins to assume its definitive embryonic form, have been outlined (p. 140). In the present context, the derivatives of the embryonic gut, a splanchnopleuric tube extending from oropharyngeal to cloacal membranes, are considered further; as a preliminary, they are summarized here. The central (intermediate, or 'middle') regions of the gut are initially still in

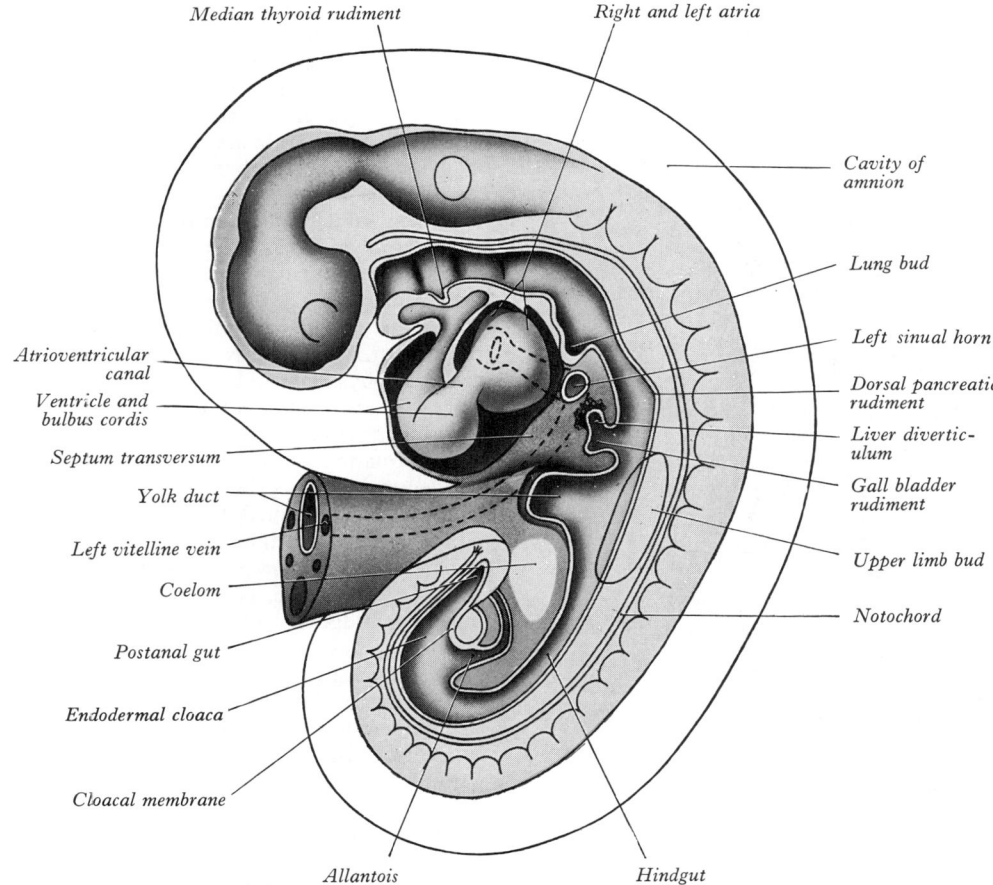

Median thyroid rudiment

Right and left atria

Cavity of amnion

Lung bud

Left sinual horn

Atrioventricular canal

Ventricle and bulbus cordis

Septum transversum

Yolk duct

Left vitelline vein

Coelom

Postanal gut

Endodermal cloaca

Cloacal membrane

Dorsal pancreatic rudiment

Liver diverticulum

Gall bladder rudiment

Upper limb bud

Notochord

Allantois

Hindgut

2.104B Composite diagram of a graphic reconstruction of a human embryo at the end of the fourth week. The alimentary canal and its outgrowths are shown in median section. The brain is shown in outline, but the spinal cord is omitted. The heart is shown in perspective, the left horn of the sinus venosus having been divided. The somites are indicated in outline. (After Streeter 1942.)

wide communication with the extraembryonic secondary yolk sac through the relatively large umbilicus (bounded by the advancing somatopleuric body folds). The *foregut* extends from the oropharyngeal membrane (floor of the stomatodeum) to its opening, or continuation, into the central midgut region through the anterior intestinal portal. Foregut derivatives are part of the buccal cavity, the pharynx (and their numerous sub-regions), the laryngo-tracheobronchio-pulmonary system, the oesophagus, stomach, the superior and about the proximal half of the descending parts of the duodenum, the liver, gallbladder and biliary duct systems, the pancreas and ducts; closely associated with these, but 'mesenteric' in origin, the spleen. The site of the original anterior intestinal portal is immediately caudal to the common hepatopancreatic ampulla and papilla. The *midgut*, between the intestinal portals, becomes tubular (with constriction and atrophy of the yolk duct), lengthens, being destined to form the remaining duodenum (postpapillary descending, transverse and finally ascending parts, then the jejunum, ileum, caecum and appendix, ascending colon, and much of the transverse colon. The site of the posterior portal remains controversial—from two-thirds across the transverse colon to the splenic flexure. The *hindgut* extends from the neighbourhood of the splenic flexure to (and a little beyond) the cloacal membrane. Its gut derivatives are the descending colon, sigmoid colon, rectum, and anal canal to the level of the anal valves. The caudoventral part of the hindgut is continuous with the allantoic duct; it separates from the alimentary hindgut, and contributes to the urinary bladder, urethra and associated glands.

The foregut immediately caudal to the pharynx remains as a splanchnopleuric tube which elongates to form the *oesophagus*. At five weeks its epithelium is two cells thick, the luminal columnar cells developing cilia at approximately 10 weeks, and only later begin to differentiate into stratified squamous epithelium, in the fifth month. Non-striated circular and longitudinal layers of muscle are identifiable at about the ninth week; striated fibres begin to appear at the tenth week. As the neck and thorax develop the oesophagus lengthens rapidly. At the end of the fourth and beginning of the fifth week the *stomach* can be recognized as a fusiform dilation (2.104–106), and a little beyond this is the wide opening of the midgut into the yolk sac; by the fifth week the opening has narrowed into a tubular *yolk stalk* (containing the endodermal *yolk duct*) (2.104–106), which soon loses its connection with the digestive tube (2.106). At this stage the stomach is median in position and separated cranially from the pericardium by the sep-

tum transversum (p. 140), which extends caudally on to the cranial side of the yolk duct and ventrally to the somatopleure. Dorsally, the stomach is related to the aorta and, reflecting the presence of the pleuroperitoneal canals on each side, is connected to the body wall by a short dorsal mesentery, the *dorsal mesogastrium* (2.114). The latter is directly continuous with the dorsal mesentery (*mesenteron*) of almost all of the remainder (except its caudal short segment) of the intestine. The liver develops as a hollow outgrowth from the ventral aspect of the foregut and grows cranially into the substance of the septum transversum (2.104,106), this part of the septum (*pars mesenterica*) now being termed the *ventral mesogastrium*. The rest of the intestine has no ventral mesentery.

In human embryos of 10 mm, the characteristic gastric curvatures are already recognizable. (What follows in this paragraph is largely a summary of long-held traditional views of gastric and omental development. More recently close reappraisal of human embryonic serial sections has led to many major descriptive changes and alternative proposals, see p. 240.) Growth is more active along the dorsal border of the viscus; its convexity markedly increases and the rudimentary fundus appears. Because of more rapid growth of the dorsal border, the pyloric end of the stomach turns ventrally and the concave lesser curvature becomes apparent (2.106). The stomach is now *displaced* to the left of the median plane and apparently becomes physically *rotated*, thus its originally right surface becomes dorsal and its left ventral. Accordingly the right vagus nerve is distributed mainly to the dorsal and the left mainly to the ventral surface of the organ. The dorsal mesogastrium increases in depth and becomes folded on itself; the ventral mesogastrium becomes more coronal than sagittal. The pancreatico-enteric recess (p. 240), hitherto usually described as a simple depression on the right side of the dorsal mesogastrium, becomes dorsal to the stomach and excavates downwards and to the left between the folded layers. It may now be termed the inferior recess of the *bursa omentalis*. The displacement, morphological changes and apparent 'rotation' of the stomach has been attributed variously to its own and surrounding differential growth changes, extension of the pancreatico-enteric recess with changes in its mesenchymal walls, and pressure, particularly by the rapidly growing liver (Kanagasuntheram 1957). (As intimated, the account just given prevails in basic courses and textbooks and perhaps suffices for some purposes. For a more complete treatment see p. 240 et seq.)

The midgut, while the gastric changes just outlined are occurring, increases in length more rapidly than the vertebral column, and forms a *midgut loop*, which acquires a dorsal mesentery as it lengthens, and projects into the coelomic cavity (2.106). The rapidly growing liver and the developing mesonephroi encroach on much of the available space in the coelom, and the midgut loop is extruded into the *umbilical coelom*, i.e. that part of the extra-embryonic coelom which lies in the proximal end of the umbilical cord (p. 143). This protrusion (often, quite inappropriately, called a 'physiological hernia') is a *normal* condition in human embryos between the 10 mm (end of fifth week) and the 40 mm (third month) stages; abnormally it may persist until term.

The rotation of the stomach reacts on the position of the duodenum, which prior to this stage is also a (smaller) right ventrolaterally directed *duodenal loop*, but is now carried further dorsally and to the right. At this stage the duodenum possesses a thick dorsal mesentery (*dorsal mesoduodenum*) which is continuous with the dorsal mesogastrium, rostrally, and the mesentery of the intestinal midgut loop, caudally. The extreme caudal free edge of the ventral mesogastrium extends on to the short initial segment of the duodenum (as a 'ventral mesoduodenum'): this also retains a coextensive part of the *dorsal mesoduodenum* (2.106). Later, the approximation of the duodenum to the dorsal abdominal wall leads first to the adhesion of the right side of its mesentery to the parietal peritoneum and later to the absorption of both layers. In this way almost the whole duodenum comes to be sessile or 'retroperitoneal' (except the short initial segment just mentioned). It may be restated that the second (descending) part of the duodenum includes the foregut-midgut junction. The lining epithelium of the duodenum proliferates and almost occludes the lumen by the sixth week; the channel is re-established by the third month when the epithelium reverts to a simple columnar form.

Facial and vestibulocochlear nerves
Otocyst
Glossopharyngeal nerve
Vagus nerve
Hyoid arch
Fourth ventricle
Third aortic arch
Hypoglossal nerve
Midbrain flexure
Vertebral artery
First cervical spinal nerve
Fourth aortic arch
Sixth aortic arch
Oesophagus
Left atrium
Trachea
Trigeminal ganglion
Left ventricle
Mandibular prominence
Upper limb bud
Umbilical cord
Maxillary prominence
Dorsal root ganglion
Olfactory placode
Allantois
Liver
Endodermal cloaca
Caecum
Ureter
Umbilical artery
Roots of dorsolateral intersegmental arteries
Notochord
Mesonephric duct

2.105 A human embryo of 7 mm greatest length. Fifth week. Left lateral aspect. (After Thompson.)

Early in the sixth week a small diverticulum appears on the antimesenteric border of the caudal limb of the midgut loop (2.105,106), and this later differentiates into *caecum* and *vermiform appendix*. Therafter it is possible to distinguish large from small intestine. Until the fifth month the diverticulum is conical, but thereafter its distal part remains rudimentary and forms the vermiform appendix, while its proximal part expands to form the caecum. At birth the vermiform appendix extends from the *apex* of the caecum; owing to unequal growth in the walls of the latter, it subsequently opens on the posteromedial caecal wall (p. 1366).

When the midgut loop enters the umbilical coelom it has already rotated through an angle of 90° so that the proximal limb (i.e. the limb nearer to the stomach) lies to the right and the distal limb to the left (2.106). This relative position is roughly maintained so long as the protrusion persists, but during this period the proximal part forming the small intestine becomes elongated and coiled, and the mesentery adapts itself to these changes in the gut. Thus its intestinal attachment also elongates and becomes folded, roughly along a 'horizontally' disposed zone, whereas its parietal attachment is much shorter, straight and for a while median and vertical. Overall, therefore, the mesenteric sheet is 'spiralized', with its contained vessels, through some 90°. The colic part of the loop elongates less rapidly and has no tendency to become coiled. By the time the fetus has attained a length of 40 mm (middle of third month), the peritoneal cavity has enlarged sufficiently for re-entry of the midgut to occur fairly rapidly, but in a definite sequence. The orderly progression is significant because, throughout, the gut *continues* the process of *rotation*, resulting ultimately, after differential growth, in the establishment of the definitive relationships which characterize the adult large intestine, including the relation of the transverse colon to the duodenum. The process has been analysed as follows: Frazer & Robbins (1915) may be consulted for the classic account of gut rotation: the present account, however, differs substantially from previous accounts in accord with the data presented by Harris & Jones (1976), from a radiological study of 64 fresh fetuses of 11 to 20 weeks of gestation. It should be noted, however, that *interpretation* of observations of the latter authors has not been followed in its entirety, and some common misconceptions stemming from the classic account will be pointed out.

While the midgut is still in the umbilical coelom the dorsal mesentery forms a dorsally median, but ventrally spiralized, partition from the dorsal wall to the umbilicus. As the gut re-enters the abdominal cavity, the coils of the small intestine, which return first, necessarily enter to the *right* of this partition, thrusting it over to the *left* and thus determining the position of the descending colon. They pass dorsal to the (oblique) superior mesenteric artery and determine its adult relationship to the horizontal part of the duodenum. This disposition of the duodenum is probably an important factor modulating midgut rotation. The caecum is the last to re-enter and at first lies on the surface of coils of ileum. Subsequently, with growth and some continued rotation the caecum is carried *dorsocaudally* and to the *right* where it lies *in*, or even a little *caudal* to the diminutive *right iliac fossa*. It must be appreciated that at this stage the fetal liver is still relatively grossly enlarged and the caecum can be termed *subhepatic* in position, but that the term does *not* imply (as it is held to do in most textbook accounts) that the caecum returns to the right *upper* quadrant of the abdominal cavity. Harris and Jones named midgut-derived colon *presplenic* and hindgut-derived colon *postsplenic*, the transitional zone, of course, corresponding to the presumptive splenic flexure. At an early stage, after the completion of gut rotation, the proximal part of the presplenic colon runs largely *transversely* across the *lower* abdominal cavity, its left extremity curving cranially to reach the *primitive splenic flexure*, from which point the postsplenic colon first descends vertically, and then curves to the right, finally descending in the midline—the primitive rectum.

The hindgut and colon: further development. The longitudinal muscle coat differentiates earlier in the small than in the large intestine, and it has been suggested that this may play a role in the return of the midgut into the abdomen from the umbilical coelom (Kanagasuntheram 1960). As the coils of small intestine re-enter, the mesentery of the *descending* postsplenic colon is thrust against the left dorsal abdominal wall and their opposed peritoneal sur-

faces become adherent and gradually absorbed. The descending colon thus loses its mesentery, becoming sessile. Since this change takes place towards the end of the third month, the position of the left colic vessels, on the posterior abdominal wall, is secondary and thus ventral to such structures as the left ureter and gonadal vessels, which are primarily associated with the posterior wall. The pelvic part of the postsplenic colon, however, does not become sessile but retains its dorsal mesentery; nevertheless, the original midline parietal attachment of the latter, by *partial* adherence and fusion, gradually assumes the inverted V shape characteristic of the adult.

As the visceral surface of the enlarged liver gradually recedes towards the costal margin, the presplenic colon changes its early disposition, becoming first oblique, then progressively develops its *ascending* and *transverse* parts. The former becomes sessile in the same manner as the descending colon, whilst the proximal part of the transverse colon is usually also sessile, and passes ventral to the descending (second) part of the duodenum, to which it adheres. The remainder of the transverse colon retains its dorsal mesentery, the *transverse mesocolon*, which now has a *horizontal* parietal attachment to the head, neck and body of the developing pancreas. Whilst this is occurring the omental bursa is expanding and the transverse colon passes dorsal to the bursal extension; a variable degree of fusion follows between the transverse mesocolon and the dorsal fold of that part of the dorsal mesogastrium destined to become the greater omentum (2.106). Meanwhile, as the hepatic flexure is gradually defined, there occurs a slow, short *ascent* of the ileo-caecal junction to the upper part of the right iliac fossa, and a more rapid ascent of the splenic flexure to its definitive position. (This account is in complete contrast to earlier descriptions, which maintained that the early caecum *descended* by slow growth from the subhepatic right upper abdominal quadrant caudally to the right iliac fossa.)

Rectum and *anal canal* development is associated with the growth of the tail fold, described earlier (pp. 140, 258). Further, it is intimately associated with that of the bladder and other elements of the urogenital system, hence associated illustrations appear later (see p. 258 and 2.122A–F). As growth proceeds, the gut lengthens, at first *pari passu* with the embryo, and an increasing segment of *hindgut* emerges between the caudal aspect of the yolk stalk and the root of the allantois (2.104B). The hindgut caudal to the latter level dilates forming a pouch, the *endodermal cloaca*, and in its ventral wall the *cloacal membrane* (p. 259) occupies the midline (2.122A). Proliferation of the surrounding mesoderm causes the cloacal membrane to lie as the floor of a shallow depression, the *ectodermal cloaca* (2.122A). The hindgut and allantois open into the endodermal cloaca from its inception, but in the fifth week the mesonephric ducts pierce its wall. When this occurs the ventral cloaca is wider than the dorsal. The latter remains very narrow, and it is into the dorsolateral corner (lateral horn) of the ventral part that the mesonephric duct opens (2.122B,C).

The mesenchyme outside the line of union of the dorsal and ventral parts of the cloaca grows rapidly and thrusts the endodermal epithelium inwards. As a result the two opposing walls come into apposition and fuse. This process commences opposite the connection of the allantoic canal with the cloaca and is continued caudally to form a *urorectal septum*, which separates the *dorsal segment* or presumptive rectum and upper anal canal from the *ventral segment*, which forms the urinary bladder and the urogenital sinus (2.122B). Caudally the urorectal septum reaches the cloacal membrane and divides it into *anal* and *urogenital membranes*. For a time a *cloacal duct* connects the two parts of the cloaca proximal to the growing urorectal septum (2.122C), and occasionally persists as an anomalous fistula between rectum and bladder or urethra. Anal tubercles (ectodermal elevations over loci of mesodermal proliferation) form round the margin of the anal part of the cloacal membrane, which thus comes to lie at the bottom of a depression, the ectodermal *proctodeum* (Tench 1936). With the absorption and disappearance of the anal membrane the anorectum communicates with the exterior (2.122E). The lower part of the anal canal is formed from the proctodeal ectoderm and surrounding mesoderm, but its upper part is lined by endoderm splanchnopleuric in origin and is derived from the caudal end of

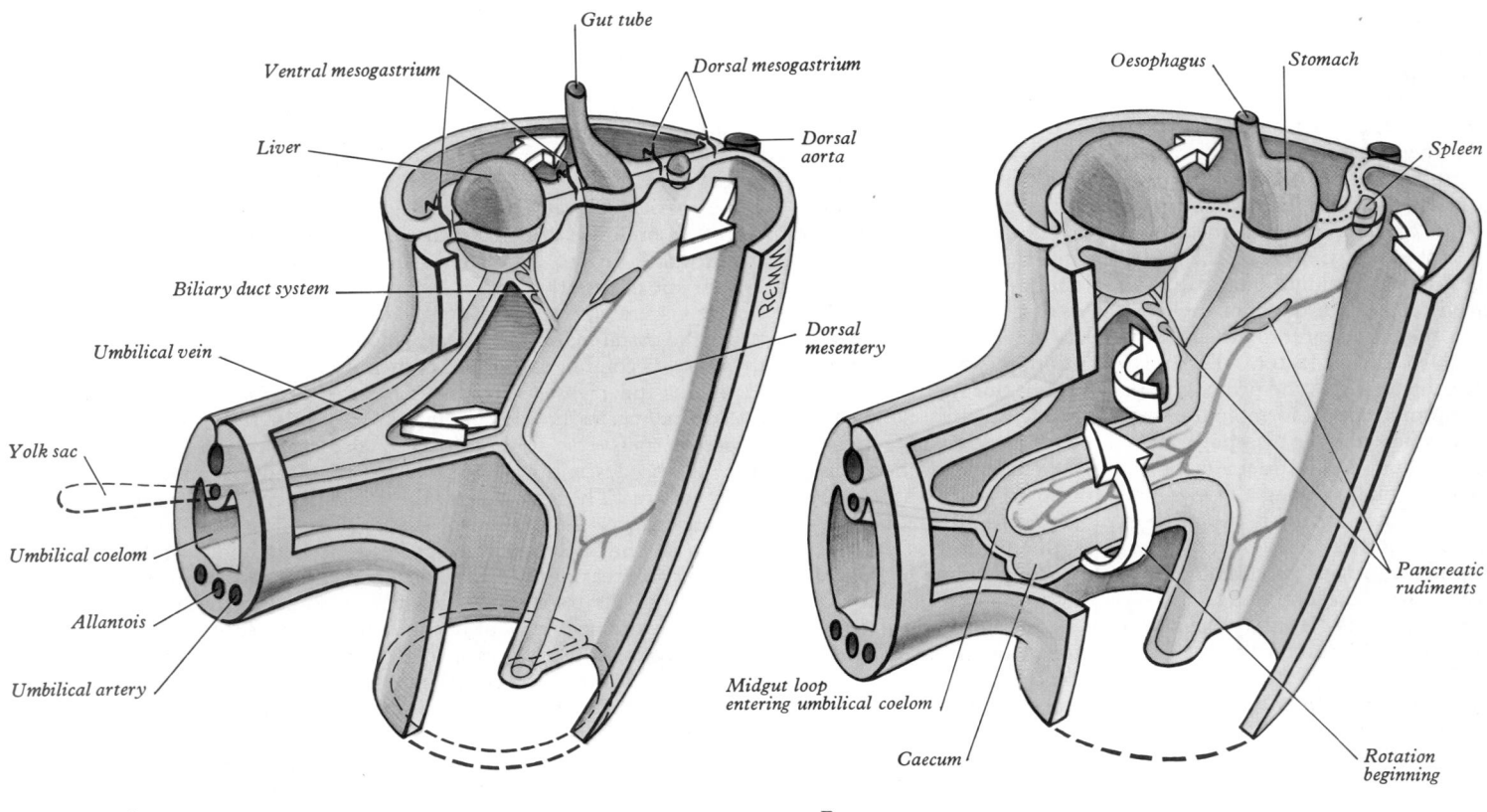

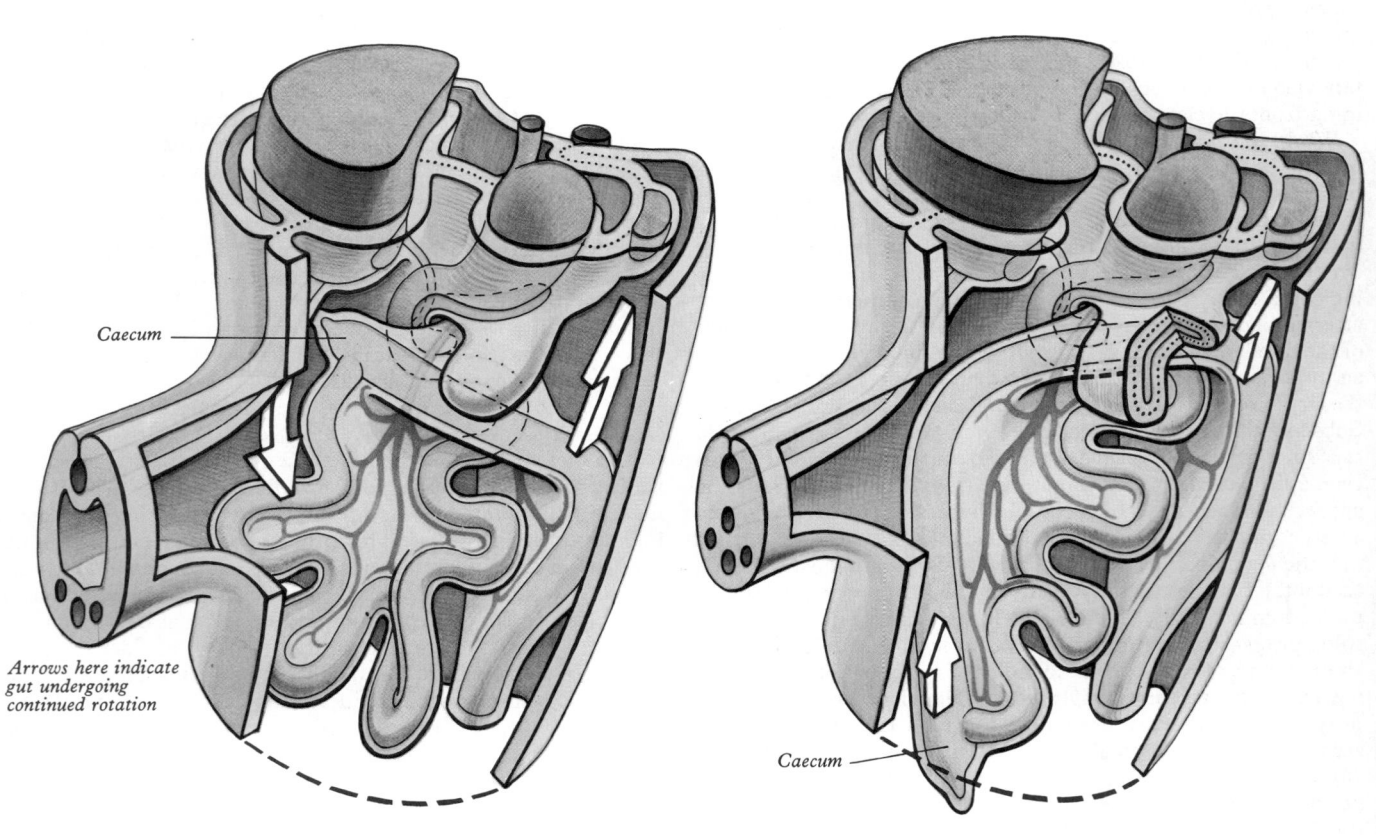

2.106 Three-dimensional schematization of the major developmental sequences of the subdiaphragmatic embryonic and fetal gut, together with its associated major glands, peritoneum and mesenteries, viewed from the left anterolateral aspect. The developmental sequence A–F spans 1½ months to the perinatal period. H denotes the general disposition of the

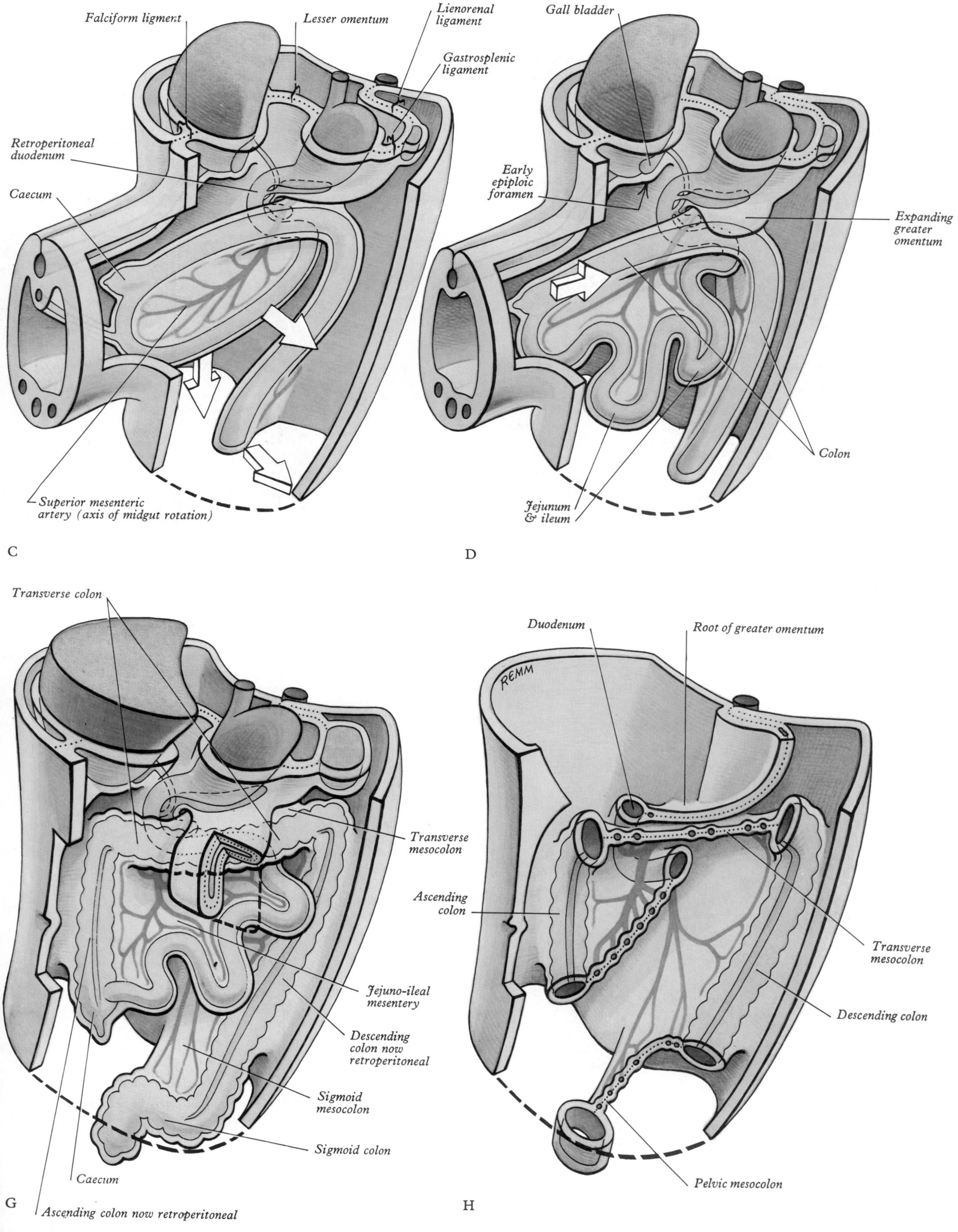

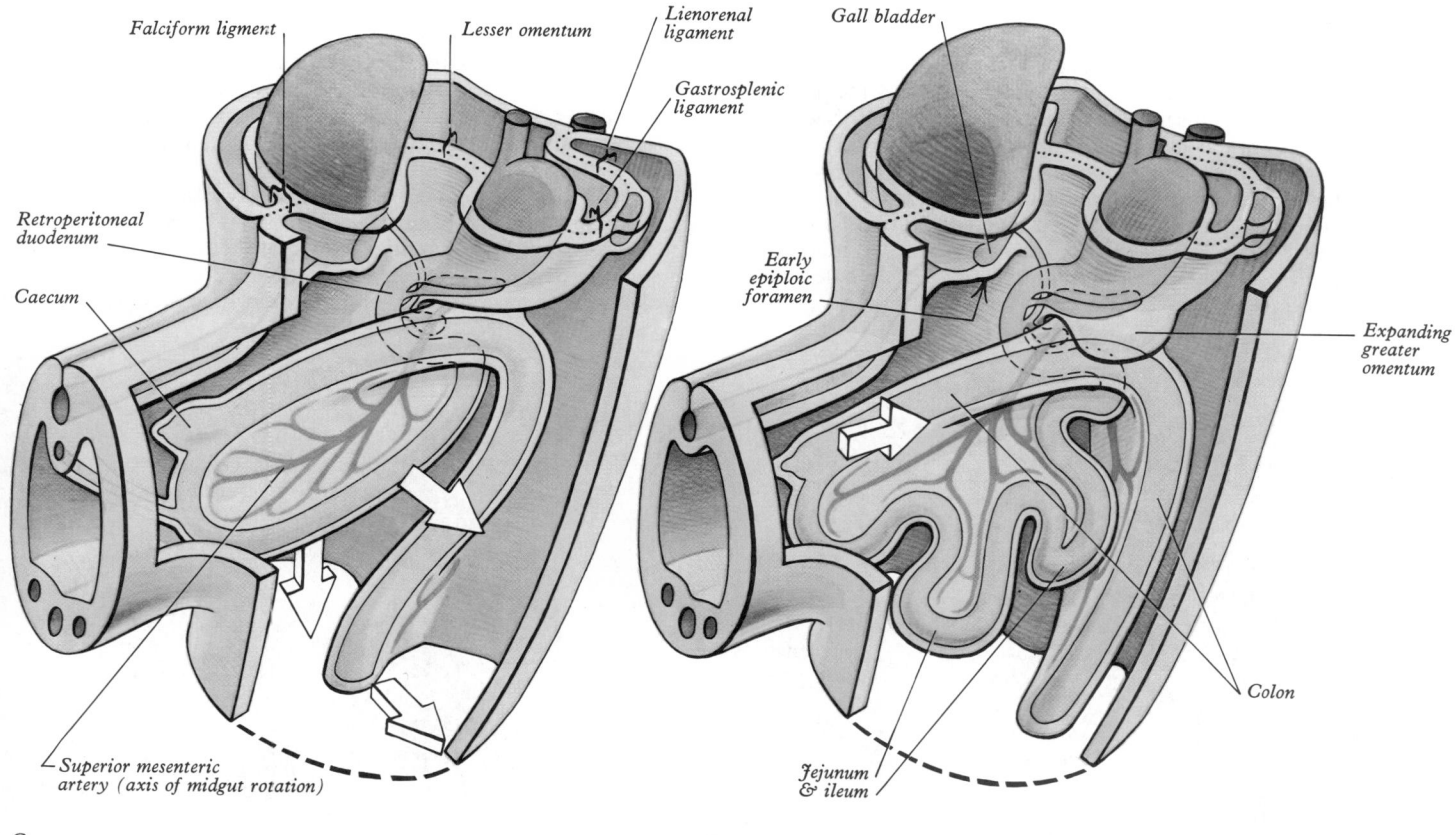

Falciform ligament

Lesser omentum

Lienorenal ligament

Gastrosplenic ligament

Retroperitoneal duodenum

Caecum

Superior mesenteric artery (axis of midgut rotation)

C

Gall bladder

Early epiploic foramen

Expanding greater omentum

Colon

Jejunum & ileum

D

Transverse colon

Transverse mesocolon

Jejuno-ileal mesentery

Descending colon now retroperitoneal

Sigmoid mesocolon

Sigmoid colon

Caecum

Ascending colon now retroperitoneal

G

Duodenum

Root of greater omentum

REMM

Ascending colon

Transverse mesocolon

Descending colon

Pelvic mesocolon

H

remaining viscera, mesenteric roots with their lines of attachment, and principal contained vessels, which approximate to the adult state for comparison. Some features of the sequence presented do not correspond to traditional descriptions—see text for further comment.

the dorsal division of the cloaca; the line of union corresponds with the edges of the anal valves in the adult (p. 1370). In the fourth and fifth weeks a small part of the hindgut, named the *postanal gut* (**2.122A**), projects caudally beyond the anal membrane; it usually disappears before the end of the fifth week. The dual origin of the anal canal is reflected by differences in arterial supply, venous and lymphatic drainage, innervation and epithelial specialization (summarized on p. 1373).

The development of the *nerve supply* of the gut has received relatively little attention in human material. A migration of neuroblasts to establish a vagal innervation has been described in human embryos between the eighth and twelfth weeks (Okamoto & Ueda 1967), which is distributed in a craniocaudal direction to establish the myenteric plexus; the vagal ramifications in the oesophageal wall have been identified at the 15 mm stage (Lecco & Balli 1968). Ultrastructural and histochemical observations suggest that, in the fetal small intestine and colon, the mucous membranes have already developed a potential for absorption (Lev & Orlic 1974).

The further development of the main body cavities, pericardial, thoracic and abdominal, their separation and serous membrane complexes and the differential growth of their contained viscera are so interdependent that any descriptive order chosen is defective, either because of over-simplification or omission in some

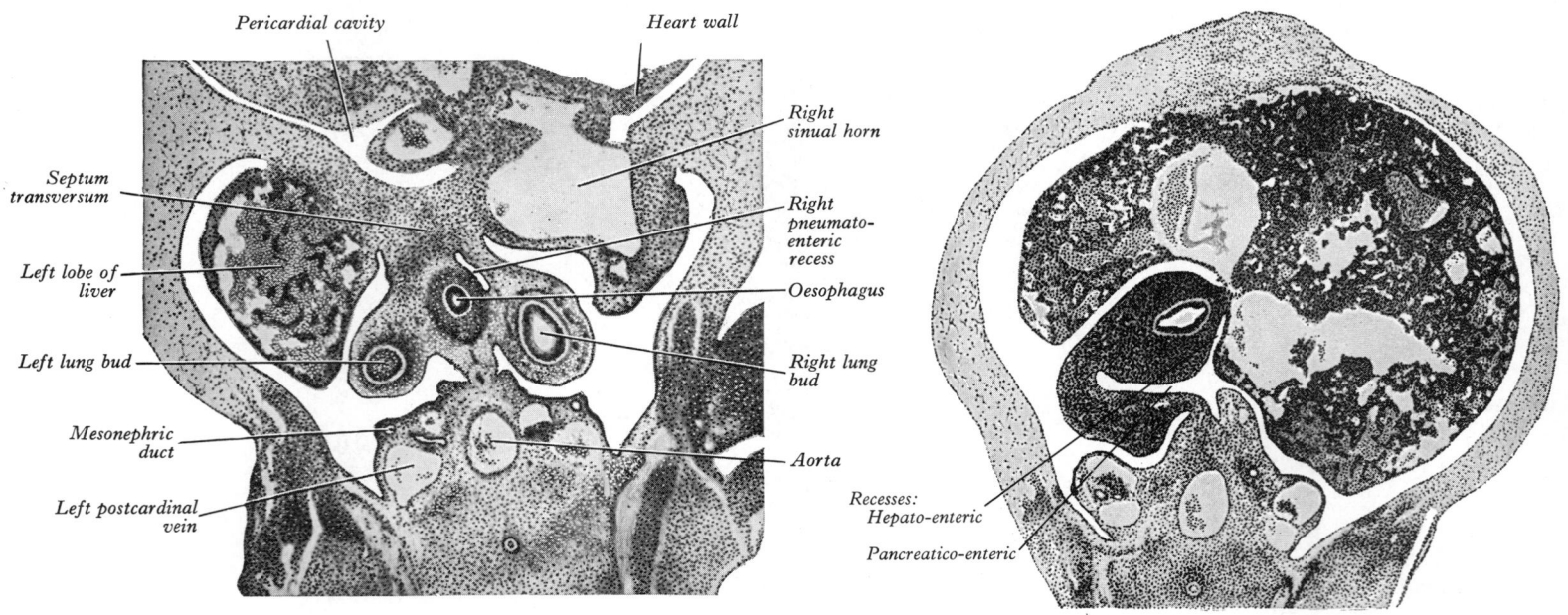

2.107A Transverse section of a human embryo, 8 mm long, showing the right pneumato-enteric recess.

2.107C Transverse section through the same embryo as **2.98B**, but 150 μm more caudally. Compare with the preceding figure and observe that the omental bursa (pancreatico-enteric recess) communicates with the general peritoneal cavity at this level.

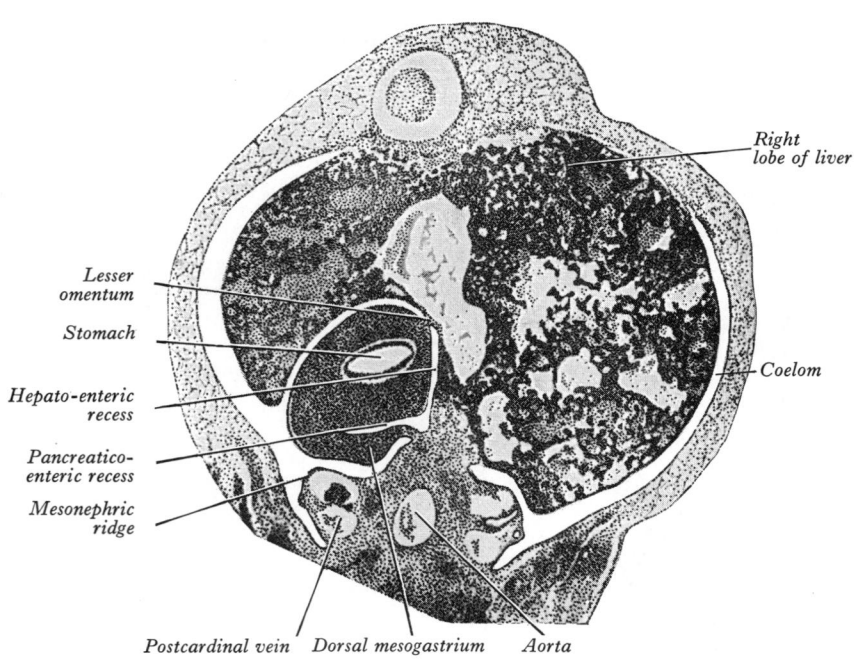

2.107B Transverse section through the same embryo as **2.98A** but 530 μm more caudally. Note that rotation of the stomach has taken place and that the sinusoidal spaces in the liver communicate freely with one another.

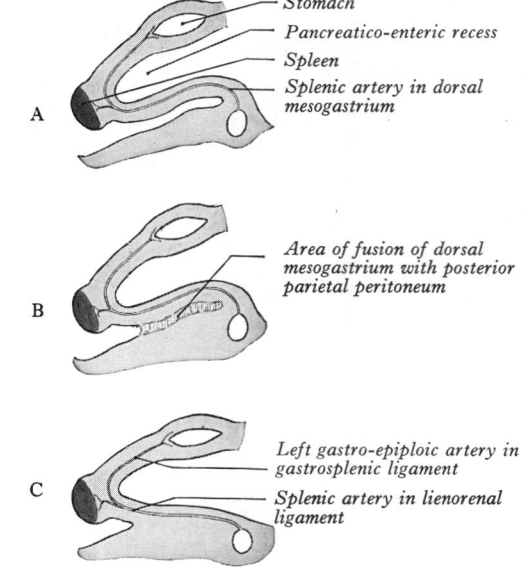

2.107D Diagram to show the fusion of the proximal part of the dorsal mesogastrium with the peritoneum on the posterior abdominal wall. Note also the conversion of the dorsal mesogastrium into the gastrosplenic and lienorenal ligaments. A represents a transverse section of an embryo in which the dorsal mesogastrium is still at the stage shown in **2.107**. B and C represent transverse sections of older embryos made at the same level (simplified, however, by retaining the shape and size of the stomach and spleen).

parts or, conversely, constant repetition. Cross references may help to minimize these problems. Here, brief accounts of the liver, pancreas and spleen, precede attempts to summarize the peritoneum and omental bursa, intestinal anomalies, then the respiratory system and separation of the body cavities.

The liver arises in the fourth week as a diverticulum from the ventral surface of the duodenal foregut, close to its junction with the midgut, where the latter is continuous with the yolk stalk (2.104A, B). This diverticulum, lined with endoderm, grows ventrally and cranially into the septum transversum; its tip diverges into two solid *hepatic buds* of cells, the future right and left lobes of the liver. The buds develop into epithelial trabeculae or sheets (so called *hepatic cylinders*), which branch and anastomose to form a close meshwork (*labyrinth* or *muralium*). The intervals of the meshwork become filled with blood *sinusoids*, and on section the organ has the appearance of a vascular sponge (2.107C). These vessels arise in situ possibly as the result of the influence exercised by the endodermal cells of the liver on the potentially angiogenic cells of the mesenchyme of the septum transversum (Streeter 1942); but also by dilatation of pre-existing septal capillaries that are plexiform tributaries of the transeptal vitello-umbilical veins (p. 220). The commonly described simple direct invasion of the vitelline veins by the epithelial trabeculae of the liver to form a sinusoidal system occurs only over a restricted area in a few mammals (Elias 1955). By the continued growth and ramification of the endodermal hepatic sheets the mass of the liver is gradually formed, but its connective tissue stroma and phagocytic sinusoid-lining cells are derived from the included mesenchymal cells of the septum transversum. Recent observations suggest that the formation of intrahepatic ducts is dependent upon contact between the developing embryonic liver mass and a preformed extrahepatic duct system (Elias 1967). The original diverticulum from the duodenum forms the *bile duct*, and from its distal part the cystic duct and gallbladder arise as an outgrowth, solid at first but later canalized. The bile duct first opens into the ventral wall of the duodenum; later, after gut rotation, it migrates to the left across the dorsal (originally right) surface of the duodenum to the position which it occupies in the adult on the medial border. The migration is ascribed to differing rates of growth in the duodenal walls and is aided by proliferation of the lining epithelium of the duodenum, which is most marked in this position (Kanagasuntheram 1960).

As the liver enlarges, it projects more and more into the abdominal cavity from the caudal surface of the septum transversum. In the process, mesenchyme of the septum becomes drawn out and cavitated ventral to the liver to form the falciform ligament, and craniodorsally to form the coronary right and left triangular ligaments, and the lesser omentum. These peritoneal folds are collectively the *ventral mesogastrium*. (Strictly, the lesser omentum, in and near its free border, also includes the less extensive *ventral mesoduodenum*.) At three months the liver almost fills the abdominal cavity and its left lobe is nearly as large as its right. Later the relative development of the liver is less active, more especially that of the left lobe, which actually undergoes some degeneration and becomes smaller than the right. The dominance of the right lobe is reflected in the large expanse of the coronary and right triangular ligaments, compared with the diminutive left triangular ligament. Until birth the liver remains relatively larger than in the adult.

The pancreas (2.108A,B) is developed in *dorsal* and *ventral* parts. The former arises in the latter half of the fourth week as a diverticulum from the dorsal wall of the duodenum a short distance cranial to the hepatic diverticulum; growing craniodorsally in the dorsal mesoduodenum it enters that part of the dorsal mesogastrium which is, before fusion, forming the dorsal wall of the inferior part of the bursa omentalis. (See pancreatico-enteric recess, p. 246.) It forms the whole of the neck, body and tail of the pancreas and a part of the head. The *ventral* part grows out from the primitive bile duct near the opening of the latter into the duodenum. This outgrowth is at first double, but the two fuse and the resulting mass, initially in the ventral mesoduodenum, soon grows round the gut into the dorsal mesoduodenum where it enlarges to form the remainder of the head of the gland (Odgers 1930). The duct of the dorsal part (the *accessory pancreatic duct*)

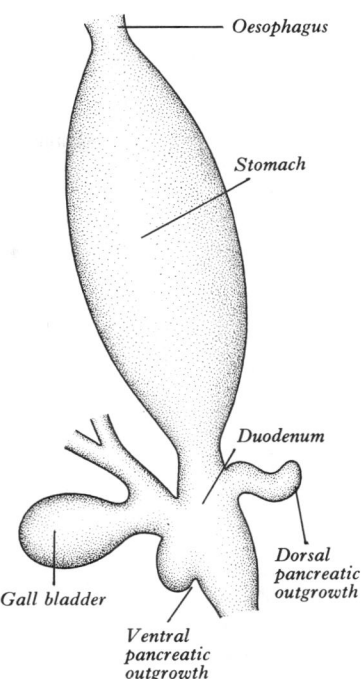

2.108A Diagram of an early stage in the development of the pancreas in a human embryo, 7.5 mm long; lateral view. (After Streeter.)

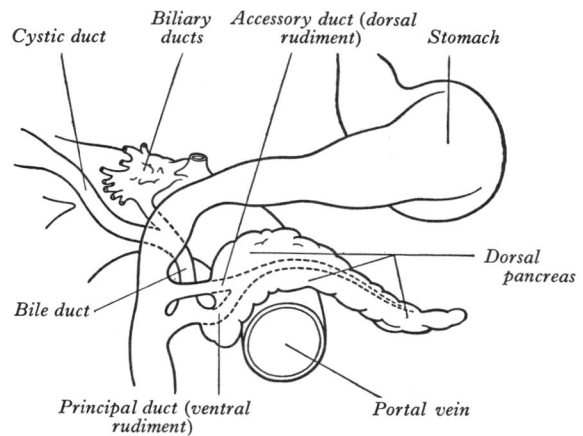

2.108B A later stage in the development of the pancreas in a human embryo, 14.5 mm long; ventral view. (After Streeter.)

therefore opens directly into the duodenum, while that of the ventral part (the *main pancreatic duct*) opens with the bile duct. Early in the seventh week the two parts of the pancreas meet and fuse, and an oblique communication is established between their ducts (2.108B). After this the accessory duct undergoes little or no enlargement, while the duct of the ventral part increases in size and forms the terminal of the main duct of the gland. The duodenal opening of the accessory duct is sometimes obliterated, and, even when patent, it is probable that most of the pancreatic secretion is conveyed through the main duct.

At first the body of the pancreas extends from the dorsal mesoduodenum, and progresses craniodorsally between the layers of the dorsal mesogastrium in the dorsal wall of the bursa omentalis (2.114). When the upper part of this wall fuses with the dorsal parietal peritoneum, the process of fusion continues caudally as far as the inferior border of the pancreas, and thus the gland finally becomes almost wholly sessile on the dorsal wall of the abdomen. The anterior border of the gland now provides the main line of attachment for the ascending (posterior two) layers of the greater omentum and the variably adherent transverse mesocolon. Having reached the anterior surface of the left kidney, the left end of the body ceases to be sessile and, as the glandular *tail*, passes between the (dorsal mesogastrial) layers of the lienorenal ligament to the hilum of the spleen.

For details of cytogenesis, exocrine and endocrine, of the human fetal pancreas consult Conklin (1962) and Laitio et al (1974).

The spleen (2.106,107D,114) is not a constituent part of the digestive system, but it is convenient to refer to its development here. It appears about the sixth week as a localized thickening of the coelomic epithelium of the dorsal mesogastrium near its cranial end, and the proliferating cells invade the underlying

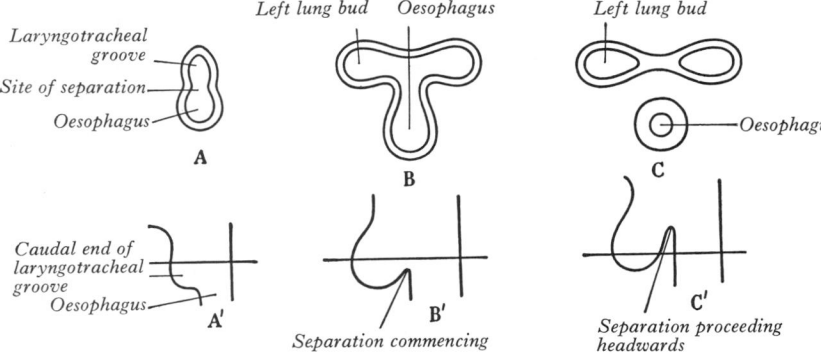

2.109A Diagrams to show the closure of the laryngotracheal groove and its separation from the oesophagus in the latter part of the fourth week. (After Streeter.) A, B and C represent transverse sections at the levels shown in A¹, B¹ and C¹, which are outline drawings of the oesophageal region in three closely following stages. Left lateral aspect.

In A and A¹ the laryngotracheal groove communicates freely with the oesophagus.

In B¹ the lower end of the laryngotracheal groove has begun to close and to form right and left evaginations, which represent the earliest rudiments of the lung buds, seen in B.

In C¹ the separation of the laryngotracheal groove from the oesophagus has proceeded further in a headward direction, and in C the primitive lung buds are now freed from the oesophagus.

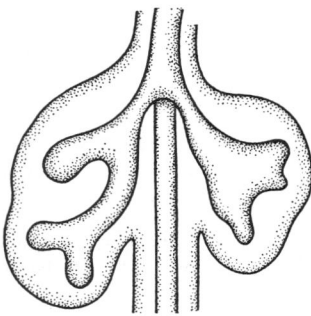

2.109B The lung buds from a human embryo, 11.8 mm long, showing commencing lobulation. Ventral aspect. (After Streeter 1948.)

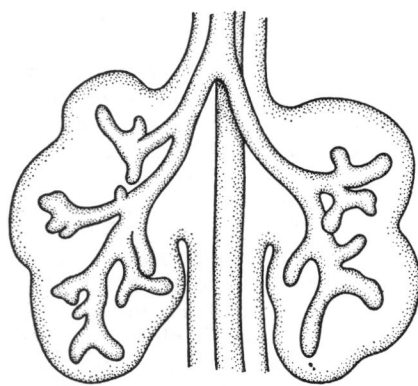

2.109C The lungs of a human embryo, 14.2 mm long, in the early part of the sixth week. (After Streeter.) Right/left differences in lobation are evident.

mesenchyme, which becomes condensed and vascularized. The process occurs simultaneously in several adjoining areas which soon fuse to form a lobulated spleen, of dual origin from coelomic epithelium and from mesenchyme of the dorsal mesogastrium. With enlargement, the spleen projects to the left so that its surfaces are covered by the peritoneum of the mesogastrium on its left aspect, thus forming a boundary of the general extrabursal (greater) sac. When fusion occurs between the dorsal wall of the bursa omentalis and the dorsal parietal peritoneum, fusion does not extend to the left as far as the spleen (2.107D, 114), which remains connected to the dorsal abdominal wall (left kidney and suprarenal) by a short lienorenal ligament, while its original connection with the stomach persists as the gastrosplenic ligament. The lienorenal ligament contains the tail of the pancreas. The earlier lobulated character of the spleen disappears, but is indicated by the presence of notches on its upper border in the adult.

The histogenesis of the spleen has attracted relatively little attention. For earlier accounts consult references (Sabin 1912, Thiel & Downey 1921, Lewis 1956, von Herrath 1958, Bloom & Fawcett 1975). The vascular reticulum is well developed at 8 to 9 weeks, with immature reticulocytes and numerous closely spaced thin-walled vascular loops. Differentiation of blood cells, macrophages, and of arteries, veins, capillaries and sinusoids has occurred by the eleventh to twelfth week. The capsule consists at first of cuboidal cells bearing cilia and microvilli (Weiss 1957).

The spleen is subject to various anomalies, including complete agenesis, multiple spleens or polysplenia, isolated small additional spleniculi and persistent lobulation. Attention has been directed to association of cardiac, pulmonary and other abnormalities with asplenia or polysplenia (Rose et al 1975).

DEVELOPMENT OF LARYNX, TRACHEA AND LUNGS

The rudiment of the respiratory tree appears in the fourth week as a median *laryngotracheal groove* in the ventral wall of the pharynx (i.e. caudal hypobranchial eminence, or furcula, p. 228). The groove deepens and its lips fuse to form a septum, converting it into a splanchnopleuric *laryngotracheal tube* (2.109). The process of fusion commences at the caudal end of the groove and extends cranially, but it does not involve the extreme cranial end of the groove, where the lips remain separate, bounding a slit-like aperture, where the tube opens into pharynx. The tube is lined with endoderm, from which the epithelial lining of the whole respiratory tract is developed. The cranial end of the tube forms the larynx, its succeeding part the trachea, while from its caudal end two lateral outgrowths arise and form the stem bronchi of the right and left *lung buds*. These grow into the pleural coeloms and are therefore covered with splanchnic mesenchyme (2.107A), from which the fibroelastic connective tissue, cartilage, nonstriated muscle and vasculature of the bronchi and lungs are developed.

The larynx, as an initial rudiment, is the cranial end of the laryngotracheal groove, bounded ventrally by the caudal part of the hypobranchial eminence (p. 228) and on each side by the ventral ends of the sixth arches. In the latter, two *arytenoid swellings* appear, one on each side of the groove (2.101A, B), and as they enlarge they approximate to each other and to the caudal part of the hypobranchial eminence (2.101A, B) from which the *epiglottis* is developed. The *opening* into the larynx, at first a vertical slit, is converted into a T-shaped cleft by the enlargement of the arytenoid swellings; the vertical limb of the T lies between the two swellings and its horizontal limb between them and the epiglottis. Soon after its appearance the epithelial walls of the cleft adhere to each other, and the primitive aperture of the larynx remains occluded until the third month when its lumen is regained. However, with rostral growth of the arytenoid swellings and the deepening of the *primitive aryepiglottic folds* forming the walls of the *vestibule*, a new *definitive aperture* is formed above the level of the primitive aperture, and the latter now corresponds to the level of the *glottis*. The arytenoid swellings differentiate into the *arytenoid* and *corniculate cartilages*, and the ridges joining them to the epiglottis become the definitive *aryepiglottic folds* in which the *cuneiform cartilages* are derived from the epiglottis. The *thyroid* *cartilage* is developed from the ventral ends of the cartilages of the fourth, or fourth and fifth branchial arches; it appears as two lateral plates, each chondrified from two centres and united in the mid-ventral line by a fibrous membrane in which an additional centre of chondrification develops. The *cricoid cartilage* arises from two cartilaginous centres, which soon unite ventrally, gradually extend and ultimately fuse on the dorsal surface of the tube (see also p. 172). For a recent compilation of literature on the early development of the larynx, consult O'Rahilly and Tucker (1973).

The right and left *lung buds* make their appearance before the laryngotracheal groove is converted into a tube. (An analysis of early pulmonary rudiment development in various mammals is provided by Hjortsjö 1950.) They grow out into the pleural passages caudal to the common cardinal veins where primary division of each lung bud into masses occurs. Three masses appear on the right and two on the left; these are the first indications of the corresponding lobes of the lung (2.109B,C). The buds undergo further subdivision and ramification, and ultimately end in minute expanded extremities—the *infundibula*. After the fifth to sixth month the *air sacs* begin to make their appearance as minute pouches of the infundibula. (For the detailed chronology of the branching, subdivision, phases and differentiation of the bronchial tree consult Reid 1976.) The splanchnopleuric mesenchyme surrounding the air passages and their terminals develops a dense capillary plexus. At first this receives blood directly from the aortic sac, and later from the sixth (branchial) aortic arches (future pulmonary arteries). The plexus is drained by radicles converging to form the pulmonary veins. The latter, of uncertain origin, open into, then contribute to, the walls of the cardiac left atrium (p. 210).

Three phases of differentiation—glandular, canalicular and alveolar—are described. In the *glandular phase*, the bronchial divisions are defined and differentiate; their epithelial cells have a prominent Golgi apparatus, scattered RNA granules and glycogen, but few organelles (Leeson & Leeson 1964). In the *canalicular phase* the respiratory parts are delineated and establish an intimate relation with the expanding blood vascular system. The *alveolar phase* extends from 6 months, but new bronchi and alveoli continue to be formed after birth (Wilson 1928). The mechanism of alveolar distension is uncertain; fetal respiratory movements, possibly involving aspiration of amniotic fluid into the lungs, may be involved (Duenhoelter & Pritchard 1973). The main expansion occurs with the onset of respiration at birth but is not complete for some days. During their development the lungs migrate caudally, so that by the time of birth the bifurcation of the trachea is opposite the fourth thoracic vertebra. As the lungs grow and project into the pleural passages the coelomic surface of the splanchnic mesenchyme enveloping each lung rudiment expands with it and becomes the visceral pleura. For bibliographies concerning pulmonary development consult Boyden (1972), O'Rahilly & Boyden (1973), Reid (1976).

DEVELOPMENT OF THE BODY CAVITIES

The formation of the intraembryonic coelom and its manner and route of communication with the extraembryonic coelom have already been considered (p. 132). After the formation of the head fold the median *pericardial coelom* connects dorsally with the right and left *coelomic ducts*, which open caudally into the *peritoneal coelom*. When the lung buds develop they project into the coelomic ducts, which may now be termed the *pleural coeloms*, and their communications with the pericardial and peritoneal coeloms become the *pleuropericardial* and *pleuroperitoneal canals* respectively (2.107,110). (When separation between these fluid-filled major coelomic regions is advancing towards completion, they are named the pericardial pleural and peritoneal *cavities*; the serous walls of the latter are often called *sacs*. In early embryos the cavities retain substantial volumes of fluid and their walls are separate; in later fetal and postnatal life cavities' walls are coapted, a mere microscopic film of serous fluid intervening.)

A curved elevation of tissue, the *pulmonary ridge*, develops on the lateral wall of the pleural coelom and partly encircles the pleuropericardial canal. The ridge is continuous with the

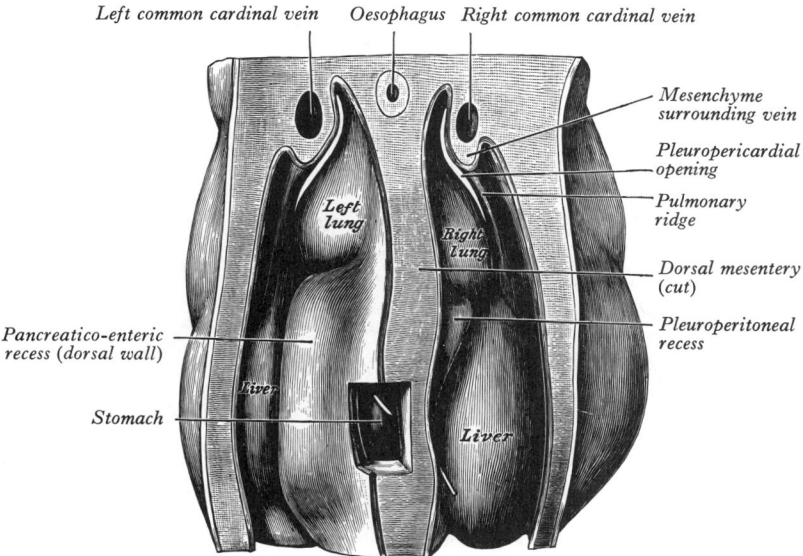

2.110 View, from the dorsal aspect, into the thoraco-abdominal part of the coelom of a human embryo 6.8 mm long, in the fifth week. Note that the dorsal body wall, including the spinal cord, developing vertebral column, the dorsal aorta and the mesonephroi, has been removed. A window has been made in the dorsal wall of the pancreatico-enteric recess to expose the posterior surface of the stomach and a wire has been passed through the epiploic foramen. (After Piper.)

dorsolateral edge of the septum transversum. The developing lung bud abuts on the ridge, which as a result divides into two diverging membranes meeting at the septum transversum. One is cranially placed and termed the *pleuropericardial membrane*; embedded within it the common cardinal vein and phrenic nerve reach the septum transversum by this route. The other membrane, caudally placed, is termed the *pleuroperitoneal membrane*. As the apical part of the lung forms it invades and *splits* the body wall and extends cranially on the *lateral aspect* of the common cardinal vein, carrying with it, or rather preceded by, an extension from the primitive pleural coelom to form part of the *secondary* or *definitive pleural sac*. In this way the common cardinal vein and the phrenic nerve come to lie medially in the mediastinum. The pleuropericardial canal, which lies medial to the vessel, is gradually narrowed to a slit, which is soon obliterated by the apposition and fusion of its margins (2.110). Its closure occurs early and is mainly effected by the growth and expansion of the surrounding viscera, heart and great vessels, lungs, trachea and oesophagus, and not by active growth of the pleuropericardial membrane across the opening to the root of the lung (2.110).

In addition to its extension in a cranial direction the lung and its associated vesceral and parietal pleura also enlarge ventromedially and caudodorsally (vide infra). With the ventromedial extension, the lung and pleurae therefore excavate and split the somatopleure over the pericardium, separating the latter from the ventral and lateral thoracic walls (2.111). Thus the ventrolateral fibrous pericardium, parietal serous pericardium and mediastinal parietal pleura, although topographically deep, are *somatopleuric in origin*.

Separation of pleural and peritoneal cavities is effected by development of the diaphragm. The *septum transversum* at first forms a block of mesoderm, caudal to the pericardial cavity and extending from the ventral and lateral regions of the body wall to the foregut. Dorsal to it on each side is the relatively narrow *pleuroperitoneal canal*. The liver grows into the septum transversum, which now can be seen to consist of two parts. One, the *pars diaphragmatica*, is disposed in the transverse plane and lies over the convex cranial surface of the liver. The other, the *pars mesenterica*, lies initially in the median sagittal plane and is expanded by the developing liver. At this stage the liver is widely attached to the pars diaphragmatica and to the ventral abdominal wall. These attachments are the forerunners of the coronary and triangular ligaments and of the falciform ligament respectively. Medial to the pleuroperitoneal canals are the oesophagus and stomach with

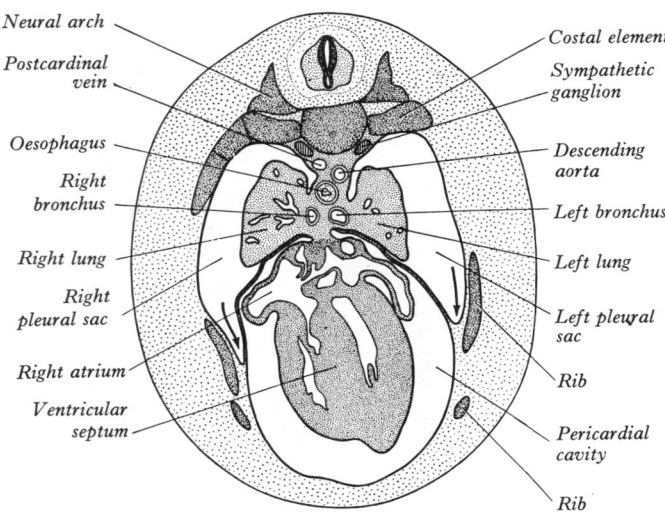

Neural arch
Postcardinal vein
Oesophagus
Right bronchus
Right lung
Right pleural sac
Right atrium
Ventricular septum

Costal element
Sympathetic ganglion
Descending aorta
Left bronchus
Left lung
Left pleural sac
Rib
Pericardial cavity
Rib

2.111 Transverse section of a 21 mm human embryo, showing how the pleural sacs extend ventrally on each side of the pericardium and split the body wall. The arrows indicate the directions of growth of the two secondary pleural sacs.

their dorsal mesentery, and at the root of the latter the dorsal aorta. Dorsolateral to the canals are the pleuroperitoneal membranes, which remain small; dorsally are the mesonephric ridges, suprarenals and gonads. Just as the enlargement of the pleural cavity cranially and ventrally is effected by a process of burrowing into the body wall, so its caudodorsal enlargement is effected in the same way. The expanding pleural cavities extend into the mesoderm *dorsal* to the suprarenal glands, the gonads and (degenerating) mesonephric ridges. Thus somatopleuric mesoderm is peeled off the dorsal body wall to form a substantial portion of the dorsolumbar part of the diaphragm. The pleuroperitoneal canal is closed by the fusion of its edges, which are carried together by growth of the organs surrounding it and, in particular, that of the suprarenal, which carries the dorsal margin of the canal ventrally to meet the pars diaphragmatica of the septum transversum (Wells 1954). The right pleuroperitoneal canal closes earlier than the left. Hence it is on the left that an abnormal communication persisting between the pleural and peritoneal cavities more frequently is encountered.

While these changes occur, the septum transversum undergoes a progressive alteration in relative position. In a 2 mm human embryo, the dorsal border of the septum transversum lies opposite the second cervical segment but, as the embryo grows and the heart enlarges, it migrates caudally. At first the ventral border moves more rapidly than the dorsal, but after the embryo has attained a length of 5 mm it is the dorsal border which migrates more rapidly (2.112). When the dorsal border of the septum transversum lies opposite the fourth cervical segment, the phrenic nerve (C3,4 and 5) and portions of the corresponding myotomes grow into it and accompany it in its later migrations. It is not until the end of the second month that the dorsal border of the septum transversum is opposite the last thoracic and first lumbar segments, the final position occupied by some of the dorsal attachments of the diaphragm and some derivatives of the pars mesenterica. However, the main derivatives of the pars diaphragmatica lie at considerably more cranial levels (vide infra and p. 241 et seq.).

THE DEVELOPMENT OF THE DIAPHRAGM

The closure of the pleuroperitoneal openings completes a mesodermal partition which thereafter separates thoracic from abdominal viscera and forms the framework for the future diaphragm. This has a composite origin. The sternal and costal parts are derived almost exclusively from the pars diaphragmatica of the septum transversum, with a small dorsolateral contribution from the pleuroperitoneal membranes and by delamination from the thoracic wall (costal part). Anterior to the oesophageal hiatus is a small contribution from the cranial oesophagophrenic con-

tinuation of the gastrohepatic part of the lesser omentum, both derived from the pars mesenterica of the septum transversum. Between the oesophageal and aortic hiatuses it is formed by the dorsal mesentery (strictly the dorsal mesoesophagus but often, less precisely, included as part of the dorsal mesogastrium). The remainder of the lumbar part of the diaphragm is formed from mesoderm around the abdominal aorta and more laterally from mesoderm of the dorsal body wall behind the suprarenal, mesonephric ridge and gonad (Wells 1954). Some authorities consider that much greater areas of the adult diaphragm are derived from the pleuroperitoneal membranes and from the chest wall. Gaps between the lumbar and costal parts of the diaphragm are usually due to under-development of the latter.

Premuscle tissue, derived principally from the fourth cervical myotomes, invades the septum transversum as already described and from there extends into the rest of the mesodermal partition, together giving rise to the muscular diaphragm (2.113) and its fibrous central tendon, or aponeurosis, with its trefoil shape, cruciform intersecting fibres and central nodal thickening.

THE PERITONEUM AND OMENTAL BURSA

Students addressing themselves de novo to topographical descriptions of the mature peritoneal cavity with its complex parietes, 'sessile' (or retroperitoneal) organs, peritonealized organs with their mesenteries (termed ligaments, folds or omenta in different locations), mesenteric contents and lines of reflexion and recesses, fossae, 'gutters', spaces, and bursae, face a formidable task. Comprehension is often lacking, and much of what is merely rote retention, quickly recedes. Preliminary study of a concise account of the development of an organ and its surroundings, or a survey of the whole region, is valuable; thus frequent cross reference should be made between any field of study of a mature organ, and its ontogeny.

The emergence of a bilaminar embryonic disc and cavitation of extraembryonic mesoderm, resulting in an exocoelom, has been reviewed (p. 131), as has proliferation, and spreading of intraembryonic mesoderm in the progressively pear-shaped, bilaterally symmetrical, trilaminar embryo (p. 132). The initial appearance of a median, rostrocaudally disposed protocardiac area, oropharyngeal membrane, notochord Hensens' node and primitive streak, is accompanied throughout the length of the notochord on both sides by intraembryonic mesodermal columns: mediolaterally—paraxial, intermediate and lateral plate. The plate is, at the disc margins, ringed by a band of extraembryonic mesoderm where somatopleure (amnion) and splanchnopleure are confluent; with growth, the band meets and fuses with the spreading lateral plate. The fusion of the three mesoderms occurs at the *junctional zone*; the latter completely encircles the early

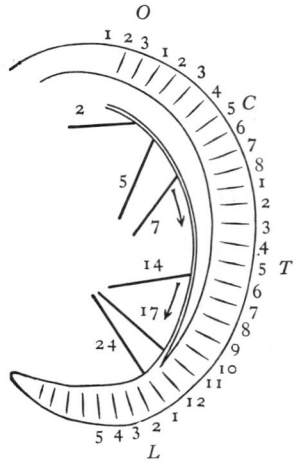

2.112 Schema showing stages in the descent of the dorsal attachment of the septum transversum. The numerals on the heavy lines indicate the length of the embryo in mm, and the position of the occipital, cervical, thoracic and lumbar segments is also shown. Note: straight lines ignore the true profile of the septum. (After Mall.)

embryonic plate periphery, but beyond the cranial tip of the notochord involves rostral streams of mesoderm which skirt the buccopharyngeal membrane, meet in the protocardiac area and continue into a substantial part of the junctional zone. Caudally, streams from the primitive streak skirt the cloacal membrane and blend with the connecting stalk around its contained allantoic duct (p. 140). Median cavitation in the protocardiac area and its confluence with multiple clefts in each lateral plate forming a ∩ shaped intraembryonic coelom which is bounded dorsally by somatopleure, ventrally by splanchnopleure, medially by intermediate mesoderm, and initially, rostrolaterally by the junctional zone (see pp. 131 et seq.). Loss of the junctional zone in the caudal half of the embryonic margins establishes communication between intraembryonic and extraembryonic coeloms. Head, tail and lateral folding imposes craniocaudal reversal of the headward and tailward structures, assumption of embryonic form, circumscription of an umbilicus, enfoldment of a splanchnopleuric gut tube and definition of the primary body cavities and coelomic regions. The manner of their separation into definitive pericardial, pleural and peritoneal cavities is described above: particular emphasis is placed on the invasion and splitting of the somatopleure with expansion of the secondary pleural cavities, submergence of the pericardium, also the positional changes and multiple derivations of the diaphragm. The intra-embryonic encompassing of a splanchnopleuric gut tube, its allocation into foregut, midgut and hindgut regions and a summary of their derivatives are given (pp. 231 et seq.). Also the period of extrusion of the midgut loop into the umbilical exocoelom, concomitant differential growth and varying degrees of rotation of almost all parts of the subdiaphragmatic gut, return of the midgut loop, and profound serous membrane modifications are traced (pp. 232 et seq).

Some repetition is unavoidable but, where apposite, brief summaries are appended to circumvent unacceptable degrees of cross-referencing; in other locations, further details and systematic names are given, when alternative morphogenetic events occur, but are infrequently mentioned in introductory accounts. Only those viscera developed in direct apposition to one of the primary coelomic regions, or a secondary extension of the latter, retain a partial or almost complete visceral serous cover. No coelomic cavitation occurs in the rostral and caudal ends of the embryo, i.e. the cranium, cervical vertebrae, buccal cavity, pharynx and their mural derivatives; similarly, the lower third of the rectum, anal canal, coextensive vagina and lower urinary tract. In all these cases, derived structures are embedded in unsplit *mixed somatovisceral mesenchyme*, often with accessions from the neural crest. (The possible classification of certain groups of *both* craniocervical *and* caudoperineal muscles and their nerves as 'special visceral' should be reviewed.) The cervicothoracic oesophagus is encased in prevertebral, retrotracheal and retrocardiac mesoderm and develops no true dorsal or ventral mesentery. In the lower thorax the oesophagus inclines ventrally anterior to the descending thoracic aorta; the dorsocaudally sloping midline diaphragm between oesophageal and aortic orifices may be homologized with part of a *dorsal meso-oesophagus*; in the same manner, a ventral midline diaphragmatic strip may be considered a derivative of a *ventral meso-oesophagus*. At superior and intermediate thoracic levels parts of the lateral aspects of the oesophagus come into closer relation to the secondary, mediastinal, *parietal* pleura (p. 1332).

The alimentary tube from the diaphragm to the commencement of the rectum possesses, throughout its length, initially, a sagittal dorsal mesentery; its line of continuity with the dorsal parietal peritoneum (i.e. its 'root' or 'line of reflexion') is, evidently, also midline. The abdominal foregut, from the diaphragm to the future hepatopancreatic duodenal papilla also has a ventral mesentery. The latter extends from the ventrolateral margins of the abdominal oesophagus and, as yet 'unrotated', primitive stomach and proximal duodenum, to cranially, the pars diaphragmatica of the septum transversum, anteriorly, to the ventral abdominal wall to the level of the cranial rim of the umbilicus, and caudally (between umbilicus and duodenum) presents a crescentic free border. The midgut and hindgut have no ventral mesentery; thus the pleural and supra-umbilical peritoneal cavities are

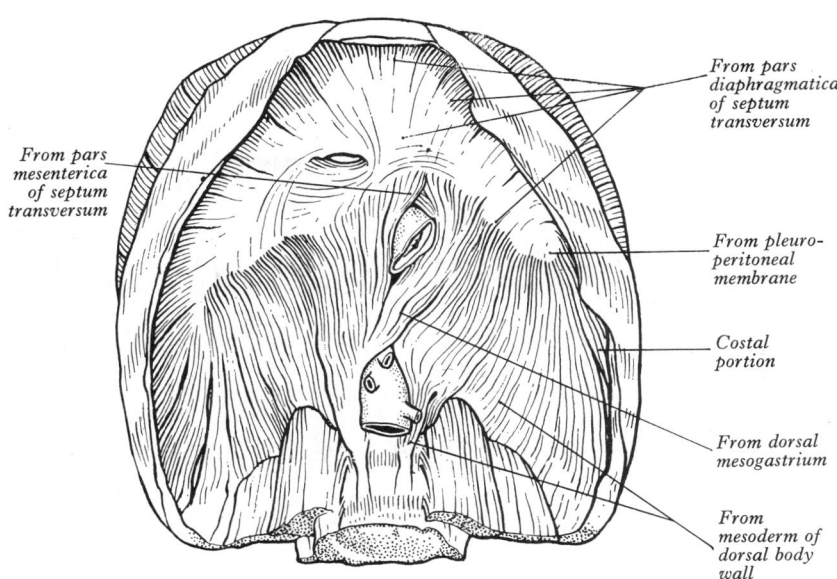

2.113 Inferior surface of the diaphragm showing the derivation of the different parts of its connective tissue framework. (After Wells 1954.)

initially (and transiently) bilaterally symmetrical above the umbilicus; below, the peritoneal cavity is freely continuous across the midline ventral to the gut.

A few general developmental points may be mentioned. Some organs, e.g. the lung, despite the extensive pulmonary growth and wide cavitation of a secondary pleural cavity, retain a relatively simple visceroparietal serous membrane disposition. The lung root, with its bronchial, neurovascular and lymphatic radicles is circumscribed by a comma-shaped line of reflexion, the dependent 'tail' or pulmonary ligament accommodating calibre variations in the contained tubes. Other organs with a single dorsal mesentery may undergo changes varying from slight to profound. Thus mesenteric changes are an integral accompaniment of alterations in their visceral contents including differential, sometimes asymmetrical growth, regional or progressive degeneration, repositioning such as sequential extrusion, retrusion, rotation, spiralization and relative ascent or descent. The *parietovisceral line of reflexion* of the originally midline dorsal mesentery may become oblique (see the jejuno-ileal mesentery) or transverse (see transverse mesocolon), angular (see sigmoid mesocolon) or complex and highly curved (see dorsal mesogastrium). In other locations, the events are more pronounced, the dorsal mesentery is lost entirely and the organ becomes 'sessile' or 'retroperitoneal', connective tissue only separating its external (usually posterior) surface from the abdominal parietes or the surface of another organ; the peritoneum is limited to some or all its (usually) ventral and perhaps ventromedial and also ventrolateral surfaces. The resulting parietovisceral or intervisceral depressions or grooves are called pouches, fossae or gutters in different locations. The term *recess* has two distinct connotations, *embryological* (vide infra), and in *mature topography* where it denotes a peritonealized, blind-ended channel, extending from a major region of the peritoneal cavity, continuing in a retrovisceral or paravisceral position, and often partly enclosed by one or two vascularized (or avascular) peritoneal folds.

The movements of lines of reflexion, assumption of a retroperitoneal site and other changes, are often held to affect relatively large endodermal 'organs' encased in a *fine* layer of splanchnic mesenchyme which is continued to the parietes by an equally tenuous, but vascularized, mesentery. With growth and also shape and positional changes, part or the whole of the mesentery lies against the parietal peritoneum; their apposed surfaces fuse and are absorbed. Thus the line of reflexion is altered, or the organ becomes retroperitoneal; further, mesenteric neurovascular bundles lie ventral to structures derived primarily from the intermediate mesoderm. Such mechanisms are significant throughout the subdiaphragmatic gut, but are predominant in the small and large intestine. However, such views fail to recognize the ability of all serous membranes to vary their thickness, lines of reflexion, disposition, 'space' enclosed and their channels of communication, by areal and thickness growth on one aspect combined with cavitation leading to expanding *embryonic recess* formation on the other. The ventral and dorsal foregut mesenterics are relatively large, thick blocks of mesoderm, compared with the slender endodermal tubes they encase. A complex series of recesses develop in these mesodermal masses, become confluent, and with foregut rotation, differential growth of stomach, liver, pancreas and spleen, and completion of the diaphragm, the territories of the greater sac and lesser sac (omental bursa) are delimited, and the mesenteric complexes of these organs (omenta and 'ligaments') are defined (vide infra).

It is convenient to first consider the mesenteries of the small and large intestine after rotation and the principal growth patterns have been achieved and the developing pancreas is becoming retroperitoneal. Most of the duodenal loop encircles the head of the pancreas and is retroperitoneal, the peritoneum covering principally its ventral and convex aspects. Exceptions are a short initial segment of the superior (first) part; this is more completely peritonealized having the attachments of the right margins of the greater and lesser omenta; peritoneum is lacking when there is close apposition of the transverse colon to the descending (second) part, or where the latter is crossed by the root of the transverse mesocolon; also where '*the*' mesentery crosses the transverse (third) part, and descends across the ascending (fourth) part from its upper extremity at the duodenojejunal flexure. In addition to the main peritoneal relations of the duodenum just mentioned, one or more of up to six different *duodenal recesses* may develop. Their variations in shape and size, their intestinal, mesenteric and vascular relations and, when adequately recorded, their frequencies and disposition of their orifices are given on p. 1344, and will not be repeated here.

The succeeding small intestine (jejunum and ileum) from the duodenojejunal flexure to the ileocaecal junction ('valve') undergoes, from a mesenteric standpoint, less modification of its embryonic form than other gut regions. Its early dorsal mesentery (no ventral component is present here, or caudally) is a continuous, single (but structurally bilaminar) sheet, with its parietal attachment—line of reflexion, or 'root'—in the midline. Usually, with development and the mechanisms mentioned above, the attachment of the root becomes an *oblique* narrow band from the left aspect of the second lumbar vertebra to the cranial aspect of the right sacro-iliac joint. Thus, from above downwards, it crosses the ascending and transverse parts of the duodenum, abdominal aorta, inferior vena cava, right psoas major and many structures related to these. For dimensions, contents and other details see pp. 233, 1344. Formal names—mesojejunum, meso-ileum—are seldom used; *the mesentery* is universal.

The caecum and vermiform appendix, as stated, arise as a diverticulum from the *antimesenteric* border of the caudal limb of the midgut loop and thus the caecum possesses no primitive mesocaecum. These regions of the gut undergo long periods of growth, often asymmetrical, and their final positions, dimensions and general topography show much variation (pp. 1365, 1366). The caecum is usually quite mobile with its lateral, medial, ventral, caudal and most of its dorsal surfaces clothed with visceral peritoneum. Peritoneum may be lacking dorsally near its continuation into the ascending colon. This arrangement is, however, usually complicated by the presence of two or three *caecal recesses*, bounded on one or more aspects by *local peritoneal folds* some of which carry vascular pedicles. For further details of the recesses (*superior* and *inferior ileocaecal* and *retrocaecal*) and folds (*lateral* and *medial parietocaecal, vascular caecal* and *ileocaecal*) see p. 1345. The vermiform appendix is also almost wholly clothed with visceral peritoneum, derived from the diverging layers of its rather diminutive *mesoappendix*. The latter, of triangular profile, carrying the appendicular vessels, lymphatics and nerves, is further detailed on p. 1344; it appears as a continuation of 'the' mesentery near its ileocaecal junction. Despite the opening remarks above, therefore, the mesoappendix should perhaps be regarded as a direct derivative of the primitive dorsal mesentery; on this view, a similar status for the vascular fold of the caecum

should be considered. With approaching completion of differential growth, rotation and circumabdominal displacement of the colonic gut, until the fourth month, this part of the gut retains its primitive dorsal mesentery (*mesocolon*) throughout. Its original root is still vertical, in the dorsal midline, from which it diverges widely, roughly as an incomplete, flattened pyramid to reach its colonic border (at the future taenia mesocolica). During the fourth and fifth months substantial areas of the primitive mesocolon adhere to, then fuse with, the parietal peritoneum; thus some colonic segments become sessile while others have a shorter mesocolon with an (often profoundly) altered parietal line of attachment (root). In one series of 100 specimens studied (p. 1367) the most common arrangement occurred in about 50%; in these a transverse and a sigmoid mesocolon persisted and the sessile state affected the ascending colon, right (hepatic) flexure and the descending colon. In the remainder, either the ascending or descending, or both, colonic segments also retained a mesocolon (varying from a localized 'fold' to a complete mesocolon). When sessile, the ventral, medial and lateral aspects of the ascending or descending colon are clothed with peritoneum, the protrusion of the viscus producing medial and lateral peritoneal *paracolic gutters* on each side. The multiple topographical relationships of these parts of the colon and particularly the dorsal structures, with a mere fibroareolar separation, are detailed on p. 1346. This form of apposition to underlying structures proceeds from the ascending colon to include the right colic (hepatic) flexure, and thence continuing anteroinferiorly to the left, thus involving the right-sided initial segment of the transverse colon. The right flexure is in contact with the caudal part of the ventrolateral surface of the right kidney; the colonic peritoneum passes cranially clothing this part of the kidney, sometimes reaching the rim of the right suprarenal. In either event, the peritoneum is next reflected ventrally as the lower layer of the coronary ligament to the dorsum of the right lobe of the liver: this renal peritoneum forms the floor of the *hepatorenal pouch* (of Morison). The pouch, in addition to its topographical relations with colon, kidney, suprarenal and liver, has clinically significant associations with the right lateral paracolic gutter, the epiploic foramen—its boundaries and their many contents—, the vestibule of the mature omental bursa, the gallbladder and biliary ducts. (Numerous references are made to all the foregoing in Section 8). The right extremity of the transverse colon, as mentioned, is also sessile, fibroareolar tissue separating it from the anterior aspect of the descending (second) part of the duodenum and the corresponding aspect of most of the head of the pancreas. The remainder of the transverse colon, up to and including the left (splenic) colic flexure, is almost completely peritonealized by the diverging layers of the *transverse mesocolon*. The root of the latter reaches the neck and whole extent of the anterior border of the body of the pancreas. The long axis of the definitive pancreas lies *obliquely*; also, the splenic colonic flexure is considerably more rostral than the hepatic flexure; in accord, the root of the mesocolon curves obliquely upwards as it crosses the upper abdomen from right to left. As it expands, the postero-inferior wall of the greater omental part of the bursa omentalis (vide infra) gradually covers, becoming closely applied to, the transverse mesocolon and its contained colon, finally projecting beyond the latter. Craniocaudal adherence now occurs between the omental wall and the pericolonic and mesocolonic layers. In the mature condition, therefore, a cursory examination suggests that the transverse colon is connected to the posterior abdominal structures by a single mesenteric sheet; close inspection, however shows this to be a compound structure, quadrilaminar in nature (p. 1339).

The left colic flexure receives much of its peritoneal covering from the left extremity of the transverse mesocolon; it also often is connected to the parietal peritoneum of the diaphragm over the tenth and eleventh ribs by a *phrenicocolic ligament*. The latter sometimes blends with a *presplenic fold* that radiates from the gastrosplenic ligament. The descending colon becomes sessile; it commences over the inferolateral border of the left kidney and first passes almost vertically in the depression between psoas major, quadratus lumborum and the aponeurotic origin of transversus abdominis, to the level of the iliac crest; here it inclines inferomedially crossing iliacus and psoas major to reach the brim of the true pelvis where it continues as the sigmoid colon. The numerous structures intervening between the posterior aspect of the descending colon and its fibromuscular 'bed' are detailed on p. 1368. Contrary to often stated (but unsupported) assumptions, it has now been clearly demonstrated (Kanagasuntheram 1957) that the process of fusion and obliteration of both ascending and descending mesocolons commences laterally and progresses medially.

The sigmoid colon is ultimately most variable in its length and disposition, with numerous topographical relationships (p. 1369). It retains its dorsal mesocolon, but the initial midline dorsal attachment of its root is considerably modified in its definitive state. The latter is commonly described as an inverted V which is, however, asymmetrical, and, other than the rectal termination of its right limb, lies wholly to the left of the midline. The apex of the V lies ventral to the bifurcation of the left common iliac artery; this separates it from the cranial part of the left sacro-iliac joint. The left limb of the V is shorter than the right and, bearing the inferior left colic (sigmoidal) vessels, is almost horizontal, inclining only slightly cranially as it is traced lateromedially from its origin to the apex. The right limb carries the superior rectal vessels; it is longer than the left and about 45° to the vertical, as it extends from the apex to its midline termination (at the cranial end of the rectum) on the ventral aspect of the third sacral vertebra. For details of the relationships and contents of the sigmoid mesocolon, also the presence, form and age dependency of its associated *intersigmoid apical recess*, see pp. 233, 1344, 1345.

The rectum continues from the ventral aspect of the third sacral vertebra to its anorectal (perineal) flexure antero-inferior to the tip of the coccyx, the distance, of course, changing with age. All aspects are encased by mesoderm; the early dorsally placed mass being named, by some authorities, the *dorsal mesorectum*. The latter does not form a true mesentery, however, but with progressive skeletal development it reduces to a woven fibro-areolar sheet with patterned variations in thickness and fibre orientation. The sheet is closely applied to the ventral concavity of the sacrum and coccyx, with numerous fibromuscular and neurovascular elements enclosed (p. 1369). Thus, the rectum becomes sessile, visceral peritoneum being restricted to its lateral and ventral surfaces. With the disappearance of the post-anal gut by the end of the fifth week, the ventrolateral peritoneum reaches the superior surface of the pelvic floor musculature, and this persists until late in the fourth month. In this period the ventral rectal peritoneum of the male is reflected to cover the posterior surface of the prostate and bladder trigone and associated structures; the female initially receives a reflexion covering almost the whole posterior aspect of the vagina, thence continuing over the uterus. Subsequently, the closely apposed walls of these deep peritoneal pouches fuse over much of their caudal extent, their mesothelia are lost, and the organs have an intervening, bilaminar, (surgically separable) fibrous stratum. The latter is the masculine recto-vesical fascia and posterior wall of the prostatic sheath; its female homologue is the posterior part of the fibrous envelope of the vagina, that intervenes between its middle two-fourths, and the rectum. Thus the proximal third of the rectum has a peritoneal tunic ventrolaterally; the lateral extensions are triangular—deep proximally and tapering to an acute angle when the middle third of the rectum is reached. Thereafter, the middle third has peritoneum restricted to its ventral surface where it forms the posterior wall of the shallower rectovesical or rectovagino-uterine pouch. The remaining rectum and anal canal are subperitoneal.

The many additional peritoneal eminences, ridges, folds, grooves, fossae and pouches, that on varying time scales characterize the true pelvis and lower abdomen, simply reflect the growth patterns of the subjacent viscera, vessels, some nerves or features of the parietes. They are mentioned either in the following pages, or with the appropriate organ under *Splanchnology*.

Subdiaphragmatic foregut peritoneum is complex in its definitive topography and, whilst having relatively simple symmetrical origins, becomes progressively more complicated during development. The latter involves rapidly changing three-dimensional arrays of serosal visceral surfaces, mesodermal masses and mesenteric sheets. Difficulties are compounded by the (rather surprising) paucity of structural ontogenetic accounts, the obligatory absence of human experimental data, some slackening of surgical interest in topographical minutiae and the adoption of simplified basic instructional courses. Here a few paragraphs offer scant justice to the topic and space limitations permit only brief allusion to the (often 'investigator specific') variations in terminology commonly encountered. Classic investigations, reviews and bilbiographies are found in Broman (1904, 1938); these may be contrasted with Kanagasuntheram (1957).

Some features have already been mentioned in other contexts, but will be summarized here to allow addition, deletion, criticism and suggested alternatives to the nomenclature and causal mechanisms invoked. (Throughout these and previous sections, reference should be made to 2.106A–H, and 2.114A–E.)

The subdiaphragmatic foregut includes, sequentially, the presumptive short terminal oesophagus, stomach and proximal duodenum (as far as, and including, the hepatopancreatic anlage). These subregions of the foregut initially (3–4 mm CR length) constitute a continuous endodermal tube of roughly uniform calibre (but slightly flattened laterally in the gastric region), encased in a *thick* stratum of visceral mesoderm. The latter is linked from *both* its *dorsal* and *ventral* aspects with substantial blocks of mesoderm that blend, respectively, with a wide dorsal midline strip of the body wall and diaphragm and the ventrolateral body wall and diaphragm including the caudal base of the pericardium. Thus, at this early stage, the abdominal foregut with its associated mesodermal masses comprises a thick sagittal septum dividing the peritoneal coelom into right and left symmetrical halves down to the level of the umbilicus. Until closure (p. 239), each half communicates with a pleural coelom via a small pleuroperitoneal canal. With the appearance of recognizable subregions of the foregut, and of the liver, pancreas and spleen, and the hepato-pancreatic duct systems, each undergoes asymmetrical differential growth (sometimes interpreted as physical 'rotation' of the whole organ—vide infra), expansion and relative displacement. Concurrently, the thick mesoderm that encases each organ forms most of its mural tissues, connective tissue framework and, where present, its visceral peritoneal surfaces (serosa). The mesoderm *between* organs (e.g. between stomach and liver, stomach and pancreas, stomach and spleen, spleen and left kidney) and the mesoderm extending from a viscus to the parietes increases in area and becomes relatively thinner, while its visceroparietal lines of attachment are altered, as are, necessarily, their topographical planes. These events follow the development of *multiple small clefts* within the thick mesoderm, their coalescence and secondary opening into the primitive peritoneal cavity rapidly follows and then further expansion of the *developmental recess* so formed. (The sequence of events closely mimics the formation of the primary intraembryonic coelom in the protocardiac area and lateral plate mesoderm. It is *not*, as commonly described (e.g. p. 232), a simple depression followed by excavation into one of the thick mesodermal strata from the primary peritoneal cavity. These morphogenetic changes result in a marked asymmetry of the upper abdominal viscera, their mesenteries and associated subregions of the peritoneal cavity. The latter, although increasingly complex topographically, remains a single cavity with numerous intercommunicating regions, pouches and recesses. (The only small peritoneal sacs to separate completely from the main cavity are the *infracardiac bursa*—vide infra and p. 1332—and the *tunica vaginalis testis*.)

The human definitive peritoneal cavity (there are extreme species specializations) is commonly described as comprising **lesser sac** and a **greater sac** of peritoneum.

The lesser sac (or *definitive bursa omentalis*) is compounded of a series of confluent recesses, mainly situated in the left, superior quadrant of the abdomen; however, it extends beyond the boundaries of the quadrant, crossing the median plane about 4 cm to the right, and a quite variable extent ventro-inferiorly. Briefly, it is intimately related to the retro-peritoneal organs of the 'stomach bed', the transverse colon and its mesocolon, the posteroinferior surface of the stomach and mesenteries (omenta and ligaments) attached to its greater and lesser curvatures, the spleen and its ligaments, the caudate lobe and process of the liver and the

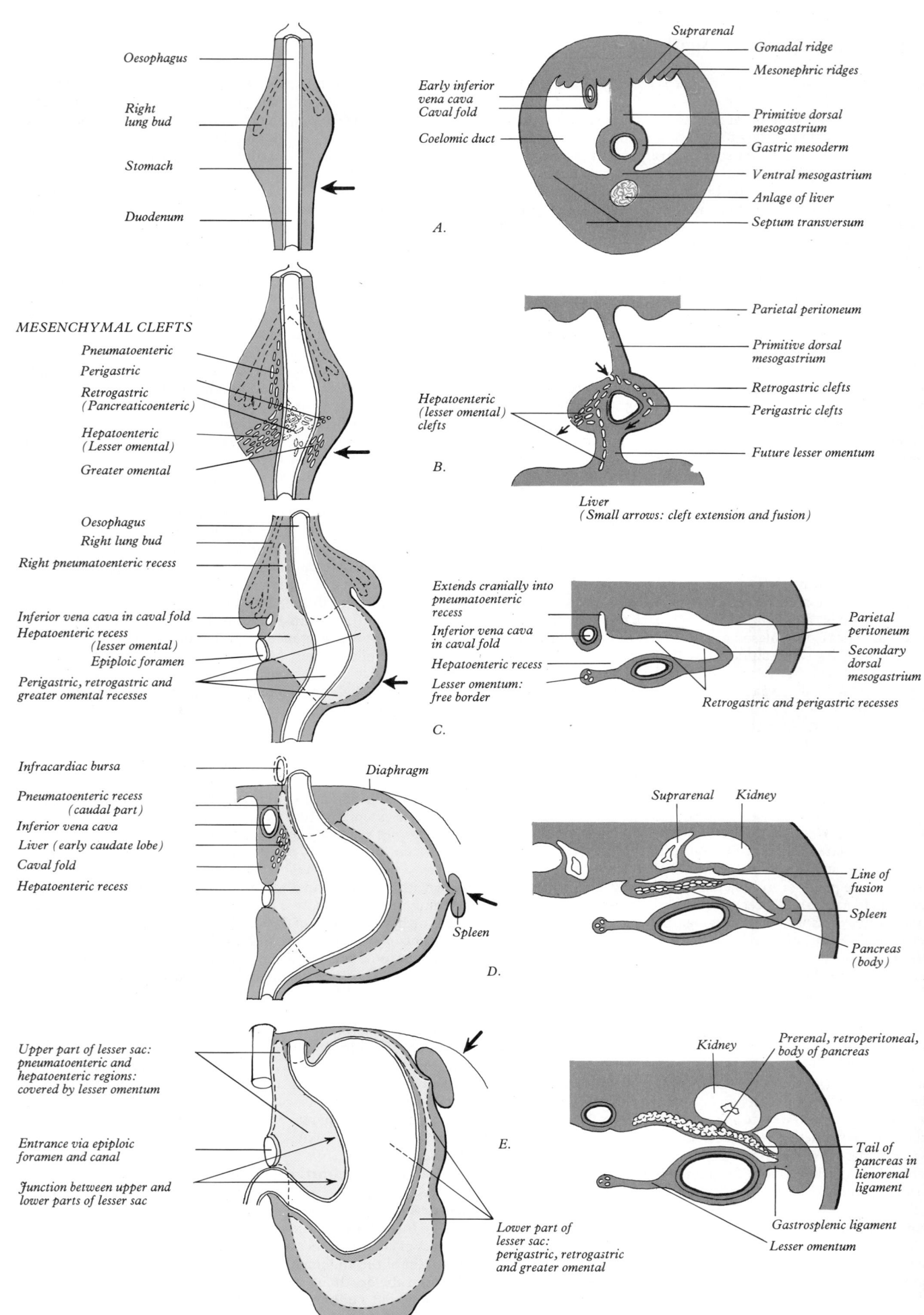

Oesophagus

Right
lung bud

Stomach

Duodenum

Suprarenal
Gonadal ridge
Mesonephric ridges

Early inferior
vena cava
Caval fold

Coelomic duct

Primitive dorsal
mesogastrium
Gastric mesoderm
Ventral mesogastrium
Anlage of liver
Septum transversum

A.

MESENCHYMAL CLEFTS

Pneumatoenteric
Perigastric
Retrogastric
(Pancreaticoenteric)
Hepatoenteric
(Lesser omental)
Greater omental

Parietal peritoneum

Primitive dorsal
mesogastrium

Retrogastric clefts

Perigastric clefts

Hepatoenteric
(lesser omental)
clefts

Future lesser omentum

B.

Liver
(Small arrows: cleft extension and fusion)

Oesophagus
Right lung bud
Right pneumatoenteric recess

Inferior vena cava in caval fold
Hepatoenteric recess
(lesser omental)
Epiploic foramen
Perigastric, retrogastric and
greater omental recesses

Extends cranially into
pneumatoenteric
recess
Inferior vena cava
in caval fold
Hepatoenteric recess
Lesser omentum:
free border

Parietal
peritoneum

Secondary
dorsal
mesogastrium

Retrogastric and perigastric recesses

C.

Infracardiac bursa

Pneumatoenteric recess
(caudal part)
Inferior vena cava
Liver (early caudate lobe)
Caval fold
Hepatoenteric recess

Diaphragm

Suprarenal   Kidney

Line of
fusion

Spleen

Spleen

Pancreas
(body)

D.

Upper part of lesser sac:
pneumatoenteric and
hepatoenteric regions:
covered by lesser omentum

Entrance via epiploic
foramen and canal

Junction between upper and
lower parts of lesser sac

Kidney

Prerenal, retroperitoneal,
body of pancreas

E.

Tail of
pancreas in
lienorenal
ligament

Gastrosplenic ligament
Lesser omentum

Lower part of
lesser sac:
perigastric, retrogastric
and greater omental

2.114 The development of the subdiaphragmatic foregut, its associated visceral and somatovisceral mesoderm and coelom, with particular reference to the terminal oesophagus, stomach, duodenum, spleen, the lesser sac of peritoneum and omenta. Seen in semicoronal section (left column) and transverse section at the levels indicated (right column). This differs in some important respects from traditional accounts. The tex[t] furnishes further definitions, details and discussion.

244

diaphragm. (Vide infra and p. 1339 for details, also p. 1340 for a discussion of the use of terms 'ligament' and 'omentum' in relation to peritoneal features.) A little above the centre of its right margin, the cavity of the lesser sac connects with that of the greater sac, i.e. the whole remainder of the peritoneal cavity, through a vertical slit, with apposed but easily separable walls, the *epiploic foramen* (p. 1339).

The greater sac is, for practical convenience, often allocated into a number of subregions, or 'spaces': all, except two, are peritonealized, and all connect with neighbouring spaces, either by a wide direct, or a more circuitous route. (In an overall peritoneal classification, as here, the cavity of the lesser sac and two extraperitoneal spaces are included.) The abdominal peritoneal cavity is (incompletely) divided by the transverse colon and its mesocolon into a *supracolic space* containing the foregut, its derivatives and mesenteries, and into an *infracolic space* containing the midgut and hindgut with derivatives and mesenteries, already described. The supracolic space is subdivided into six: *right* and *left subphrenic*, *subhepatic* and *extraperitoneal* spaces. (The left subhepatic space is a rather inappropriate name for the omental bursa; the right extraperitoneal space is the 'bare area' of the liver; the left extraperitoneal space surrounds the left suprarenal and upper pole of left kidney.) The infracolic space is divided by the mesentery of the small intestine into *right* and *left* moieties, while external to the ascending and descending colon lie the *right* and *left lateral paracolic gutters*. All the latter are more or less directly continuous with the fossae and pouches of the lesser pelvis. (For further details of these various spaces see p. 1346 and throughout the account of the peritoneum in *Splanchnology*.)

Abdominal foregut—associated mesoderm has an ontogenetic history too complex to receive more than a brief summary here; also the origin, admixture and final fate of some masses remain uncertain in the absence of experimental data. However, the nomenclature employed is, in some cases, a misleading simplification merely handed on in successive accounts over many decades. Sometimes, unproven mechanisms are held to operate which, on close examination, could not result in the well-known definitive topography. Some additional terms and divergences from current morphogenetic accounts are included.

Early embryonic nomenclature is often confined to the mesodermal masses lying dorsal and ventral to the foregut. However, the thick mesodermal strata of the wall of the gut itself are equally prominent morphogenetically. Although continuous with, they cannot be allocated quantitatively, as parts of the dorsal and ventral masses. It is useful, therefore, to refer to them simply as *oesophageal, gastric* and *duodenal mesoderms*. Some of the latter areas are *compound* in the sense that by cavitation and cleavage they may provide part of their thickness as *secondary extensions* of surrounding primary mesenteries with new lines of visceral attachment and areal expansion. These processes also result in modification of parietal lines of reflexion and related peritoneal cavity.

The septum transversum is the somewhat inappropriate name (but deeply ingrained in embryological literature) denoting all the mesoderm that extends from the ventral aspect of the caudal foregut to the ventrolateral aspects of the supra-umbilical abdominal and lower thoracic walls; its rostral surfaces reach, and contribute to the caudal 'bases' of the pericardium and secondary pleural cavities; its caudal free surface arches from the duodenum to the cranial rim of the umbilicus. Dorsally it borders the pleuroperitoneal canals, is continuous with the foregut mesoderms mentioned above, and contributes to the *caval fold* (vide infra). Some of these features have been encountered elsewhere (e.g. pp. 133, 140, 207, 239, 240) and will only receive a brief summary here. After head fold formation the most rostral part of the septum transversum has the common sinuatrial chamber of the heart embedded in it, each sinual horn of which receives the transeptal terminals of a vitelline, umbilical and common cardinal vein (for their fate see p. 222). As the pericardial cavity expands and the sinuatrial chamber 'rises' into it, septal mesenchyme follows (*pars pericardialis*) and condenses as the fibroserous diaphragmatic aspect of the pericardium. The intermediate stratum of septal mesenchyme (*pars diaphragmatica*) makes substantial contributions to the framework of the

sternocostal parts of the diaphragm (2.113). Additional extensions occur into the mesenchymatous bed in which the sub-hepatic-hepatic-hepatocardiac parts of the inferior vena cava develop (p. 225) and in relation to the expanding liver. Thus the pars diaphragmatica alone truly merits the name septum transversum and even this is both incomplete (until closure of the pleuroperitoneal canals) and also, with expansion of the lungs, secondary pleural cavities, liver and stomach, soon becomes less 'transverse' as it develops a marked convexity towards the thorax. Strong fibrous continuity between the perinodal area of the central tendon of the diaphragm and the overlying fibrous pericardium persists. Midline strips of diaphragm ventral and dorsal to the oesophageal hiatus are held by some to be mesenteric in origin (vide infra): in the writers' view this merely adds confusion to an already complex terminology. Since this is their normal fate it seems preferable to term these the ventral and dorsal pars diaphragmatica of the oesophageal mesoderm from the outset.

The remaining subdiapragmatic part of the septum 'transversum', as indicated, is designated the *pars mesenterica* and initially forms a sagittally placed ventral 'mesentery' for the foregut. Although a continuous sheet, thick subregions are recognized: a brief *ventral meso-oesophagus*, continuing into a more extensive *ventral mesogastrium*, the latter merging into a *ventral mesoduodenum* that has a caudal free border. Many (the majority of) accounts ignore these subregions and simply equate all derivatives of the ventral pars mesenterica as those of *the* ventral meso*gastrium*. Some confusion may occur, however, particularly with respect to the development and definitive courses of the main biliary ducts, hepatic artery, portal vein and attendent lymphatics and nerves, and to the hepatic and gastric serosae and mesenteries. As indicated, the mesenchymal modifications are correlated with the asymmetrical growth patterns of the stomach, duodenum and liver (see pp. 232, 237 and below).

The abdominal foregut dorsal mesoderm commences rostrally as an oesophageal *dorsal pars diaphragmatica* which, as indicated, contributes a narrow midline strip extending between the oesophageal hiatus and the aortic orifice. The remainder, the *dorsal pars mesenterica*, again initially forms a midline sagittal dorsal septum of thick mesenchyme; its subregions (corresponding to the ventral ones) and parts of one continuous mesenchymal block are: a brief *dorsal meso-oesophagus*, a more extensive *dorsal mesogastrium* and a *dorsal mesoduodenum*. Their modifications are even more extensive than those of the ventral mesenteries and also have the additional complications of the developing pancreas and spleen. Further, many accounts ignore the subregions and collectively dub them *the* dorsal meso*gastrium*. This may lead to (perhaps minor) misplacement of the anlage of certain organs; however, the siting of the initiation of the lesser sac of peritoneum in the dorsal mesogastrium is both incorrect and confusing.

The caval fold is a linear eminence with divergent rostral and caudal ends, that passes from the upper abdominal to the lower thoracic region, and protrudes from the dorsal wall of the pleuroperitoneal canal. Cranially it becomes continuous, lateromedially, with the root of the pulmonary anlage and pleural coelom, the pars pericardialis of the septum transversum and the retrocardiac mediastinal mesenchyme. Caudally it forms an arch with *dorsal* and *ventral horns*; the dorsal merges with the primitive dorsal mesentery and the mesonephric ridge (and associated gonad and suprarenal); the ventral horn is confluent with the dorsal surface of the septal mesenchyme. Thus the fold is a zone where intestinal, mesenteric, intermediate, hepatic, pericardial, pulmonary and mediastinal mesoderms meet and blend, and may justifiably be compared with the junctional zones of mesoderm described elsewhere (p. 131, 2.27). Contributions from septal mesenchyme are considered of particular importance by some authorities (Kanagasuntheram 1957). It provides a mesenchymal route for the upper abdominal, transdiaphragmatic and transpericardial parts of the inferior vena cava; it is also prominent in the development of parts of the liver, lesser sac of peritoneum and certain mesenteries. The left fold regresses whereas the right fold enlarges rapidly.

The lesser sac (*definitive bursa omentalis*) has already been mentioned in terms of its general disposition and communication with the greater sac. Unfortunately considerable variation exists between the terminologies adopted by different authors, both during ontogeny and in adult topography. Development of the lesser sac is so intimately interlocked with additional ontogenetic features, particularly of the liver and stomach, their mesenteries, and secondary extensions of the greater sac, that these must receive brief mention.

Invasion of the ventral pars mesenterica by the hepatic and (transiently) the ventral pancreatic rudiment, followed by trifurcation of the primary hepatic anlage into right and left submasses and a caudal pars cystica, has been described (p. 237). Hepatic growth on the two sides is approximately equal in early fetal months, and the liver mass almost fills the ventral mesoderm; but thick, short blocks of mesoderm intervene between the liver and the foregut, diaphragm and supra-umbilical ventral body wall. Removal of the viscera at these symmetrical stages reveals the cut edges of the mesothelium bordering the coelom and enclosing the wide areas of mesenchymal continuity. The latter has a large roughly circular central area with numerous tapering projections towards the abdomen: the obliquely cut mesenteries, caval folds, mesonephric and attendant ridges and the lateral rim of the pleuroperitoneal canals. The whole bears some resemblance to an inverted coronet, hence the name coronary ligament of the liver. (With growth and maturity the restricted use of the name is much less apposite.) Much of the mesodermal mass surrounding the liver develops cavities that coalesce and open into the general coelomic cavity as extensions of the greater (and lesser) sacs of peritoneum. Thus almost all the ventrosuperior, visceral and some of the posterior aspects of the liver become peritonealized. The process involving the greater sac continues over the right lobe and ceases when the future superior and inferior layers of the coronary ligament and the right triangular ligament are defined. Those, plus a medial boundary provided by an extension of the lesser sac, enclose the 'bare area' of the liver where loose areolar tissue of septal origin persists. Due to asymmetrical liver growth the same processes affecting the left lobe result in the smaller left triangular ligament. Where the superior layers of the coronary and left triangular ligaments meet they continue as a (bilaminar) ventral mesentery attached to the ventrosuperior aspects of the liver; its parietal attachment has a slight inclination to the right as it crosses the ventral diaphragm and then descends to the umbilicus. Its umbilico-hepatic free caudal border, somewhat arched, carries the left umbilical vein (or, postnatally, the fibrous obliterated vein or *ligamentum teres*, with fine calibre para-umbilical veins); the whole structure constitutes the *falciform ligament*. Most of the latter may be considered the final ventral part of the ventral mesogastrium; its free border, however, has a ventral mesoduodenal origin.

As stated, the endodermal subdiaphragmatic foregut of the early embryo is a tube of almost uniform calibre, but with a slight fusiform gastric dilatation with a minimal degree of side-to-side flattening, encased in thick gastric mesoderm. The latter is continuous with the primitive ventral and dorsal mesogastria and these indicate the sagittal positions of the *primitive* ventral and dorsal borders of the stomach. Many accounts state that the development of a ventrally directed concave *lesser curvature* and a more rapidly growing dorsocranially directed convex *greater curvature* and incipient *fundus* occur while the anlage is still sagittal. Thereafter, the whole miniature organ and attached mesogastria are supposed to physically rotate through almost 90° (see p. 232). On this view the original left surface becomes ventrosuperior and the primitive dorsal border is directed to the left as the greater curvature. The interpretation of various mesenteric and neurovascular sequelae were based on these assumptions. Inspection of closely-spaced series of embryos (Kanagasuntheram 1957, 1960) led to the suggestion that the morphogenetic changes in both stomach and duodenum and their attendant mesenteries were mainly, if not wholly, due to patterned differential growth rates and mesodermal cleft formation and cleavage. Thus the stomach, initially a tube with slight lateral flattening, developed a linear zone of high proliferation along the length of the central part of its (original) left lateral wall. At first it became triangular *in transverse section*, with rounded angles and apex pointing to the left; further growth resulted in an elliptical profile. The ends of

the sectional ellipse corresponded to the curvatures; the end directed to the right, and a little ventrally, the lesser curvature and justifiably regarded as a modified primitive ventral border; the left end, or greater curvature is an entirely *new formation*. For a while the two primitive mesogastria are *both* attached near the emerging lesser curvature (the sites of the primitive dorsal and ventral borders). The profound changes involved in reaching the definitive state only supervene with the development of the lesser sac.

The lesser sac is first indicated by the appearance of multiple clefts in the para-oesophageal mesenchyme on both left and right aspects of the oesophagus. Although they may become confluent, the left clefts are transitory and soon atrophy. The right clefts merge to form the *right pneumato-enteric recess* that extends from the oesophageal end of the lesser curvature as far as the caudal aspect of the right lung bud. At its gastric end it communicates with the general peritoneal cavity and lies *ventrolateral* to the gut; more rostrally it lies directly *lateral* to the oesophagus. It is *not*, as commonly stated, a simple progressive excavation of the right side of the *dorsal* mesogastrium. The right pneumato-enteric recess undergoes further extension, subdivision and modification mentioned below. From its caudal end a second process of cleft and cavity formation occurs producing the *hepato-enteric recess* that thins and expands the mesoderm between the liver and the stomach and proximal duodenum, and also reaches the diaphragm. The resulting, structurally bilaminar mesenteric sheet is the *lesser omentum* and is derived from the small mesooesophagus, the major part from the ventral mesogastrium, and the reduplicated strip including the free border from the ventral mesoduodenum. As differential growth of the duodenum occurs, the biliary duct is repositioned and most of the duodenum becomes sessile; the duodenal attachment of the free border and a continuous neighbouring strip of the lesser omentum becomes confined to the upper border of a short segment of its superior part. The free border, with contrasting growth and positioning of its attached viscera, gradually changes from the horizontal to the vertical. It carries the bile duct, portal vein and hepatic artery, and its hepatic end is reflected around the porta hepatis; an alternative name for this part of the lesser omentum is the *hepatoduodenal ligament*, and it forms the anterior wall of the epiploic foramen. The floor of the foramen is the initial segment of the superior part of the duodenum mentioned above, its posterior wall is the peritoneum covering the immediately subhepatic part of the inferior vena cava, and its roof the peritonealized caudate process of the liver. That major part of the lesser omentum from the lesser gastric curvature, passes in an approximately coronal plane to reach the floor of the increasingly deep groove for the ductus venosus (postnatally the ligamentum venosum) on the hepatic dorsum. This part is sometimes called the *hepatogastric ligament*.

The pneumato-enteric recess continues to expand to the right into the substance of the caval fold, ceasing near the left margin of the hepatic part of the inferior vena cava; the latter remains extraperitoneal, crossing the base of the now roughly triangular bare area of the liver (encompassed by coronary and right triangular ligaments) and this new expanded line of reflexion. With closure of the pleuroperitoneal canals the rostral part of the right pneumato-enteric recess is sequestered by the diaphragm but often persists as a small serous sac in the right pulmonary ligament. The remaining caval fold mesenchyme to the left of the inferior vena cava, and forming the right wall of the upper part of the lesser sac, becomes completely invaded by embryonic hepatic tissue and is transformed into the caudate lobe of the liver. This smooth, vertically elongate mass projects into the cavity of the lesser sac, and both its posterior and much of its anterior surface are peritonealized because of the increasing depth of the groove for the ductus venosus and the attachment of the lesser omentum to its *floor*. The narrow isthmus of the *caudate process* connects its right inferior angle to the rest of the right lobe (and is the roof of the epiploic foramen). The parts of the lesser sac thus far described above and to the right of the lesser curvature of the stomach have received a plethora of names by different investigators; sometimes these stem from changing or contrasting views of ontogeny. A few prominent ones will be mentioned. Some regard the right rim only of the structures already listed as constituting the boundaries of the epiploic foramen. The narrow channel, a few centimetres long, passing to the left, and coextensive with the caudate process and peritonealized part of the initial duodenal segment, is called the *vestibule* of the lesser sac (but vide infra). The remaining, much more extensive part above the lesser curvature is termed, variously, the *upper (superior) part* (recess) of the lesser sac, the *hepatic part* of the lesser sac or finally the *hepato-enteric part of the pneumato-enteric recess*. Some authorities group all the foregoing under the general term *vestibule*; they are intimately related to the diaphragm, caudate lobe of the liver, and coextensive with the lesser omentum. Varying use of the name vestibule partly reflects earlier, mistaken views of the ontogeny of the upper and lower parts of the lesser sac (about to be described). Much of the upper part was considered to be merely a part of the general peritoneal coelom 'captured' by differential growth and omental changes of plane, whereas fundamentally different mechanisms of excavation were considered to apply to the lower part, the *'true' omental bursa* of some authors. Current evidences suggest that the *same* array of developmental mechanisms, with only quantitative regional differences, apply to *all* coelomic regions. The general name omental bursa should be used with caution; both upper and lower parts have walls that are partly omental (lesser and greater omenta respectively); also neither part is functionally more nor less bursal than other coelomic regions —peritoneal, pleural or pericardial. Confusion stemming from the term 'recess' firmly established in embryology but, with different connotations, used sporadically in adult topography has been mentioned (p. 241). Thus, the upper (superior) and lower (inferior) parts of the lesser sac seem preferable terms. Ultimately the junction between upper and lower parts is oblique, curving upwards as it passes from right to left; ventrally lies the gastric lesser curvature, dorsally the body of the pancreas. The left limit is a curved ridge of mesenchyme (future *left gastropancreatic fold*) carrying the left gastric artery and the right a curved ridge (future *right gastropancreatic fold*) carrying the common hepatic artery.

The lower (inferior) part of the lesser sac commences development at about 8–9 mm CR length—at this stage the early pneumato-enteric and hepato-enteric recesses are well established. Differential gastric growth is progressing (p. 246) giving an elliptical transverse sectional profile, with a right sided lesser curvature, corresponding to the original ventral border of the gastric tube, and to which the lesser omental gastric part of the ventral mesogastrium remains attached. As indicated, the greater curvature is a new formation, rapidly expanding, with its convex profile projecting mainly to the left, but also rostrally and caudally. It was emphasized that the original dorsal border of the gastric tube now traverses the dorsal aspect of the expanding rudiment, curving along a line near the *lesser* curvature; here, transiently, the *primitive* dorsal mesogastrium is attached. The latter blends with the thick layer of compound gastric mesoderm clothing the posterior aspect and greater curvature of the miniature stomach; because of its thickness, the mesoderm projects rostrally, caudally, and particularly to the left, beyond the 'new' greater curvature of the endodermal lining of the stomach. (It may be noted that during these stages the geometries, and hence sectional profiles, of tissue strata and cavities are complex and change rapidly with varying levels of histological section and with time. Thus only a brief summary of some salient points can be attempted here. For details consult e.g. Kanagasuntheram 1957.) The processes already described in relation to the ventral mesenteries now supervene. Multiple clefts appear at various loci in the mesoderm; each group of clefts rapidly coalesce to form (transiently) isolated closed spaces. The latter, by a continuation of the mechanism, soon join with each other and with the preformed upper part of the lesser sac. The initial loci involve firstly the compound posterior gastric mesoderm nearer the lesser curvature and along its zone of blending with the primitive dorsal mesogastrium; secondarily in the dorsal mesoduodenum; thirdly, independently in the caudal rim where greater curvature mesoderm and dorsal mesogastrium blend. As these cavities become confluent and their 'reniform' expansion follows, matches and then exceeds that of the gastric greater curvature, there are some major sequelae. The primitive dorsal mesogastrium increases in area not only by intrinsic growth but, as cavitation proceeds, by substantial additions from the dorsal lamella

separated by cleavage of the posterior gastric mesoderm: together these may conveniently be called the *secondary dorsal mesogastrium*. The *gastric attachment* of the latter changes as a set of somewhat spiral lines, longitudinally disposed, that move, with time, from near the lesser curvature towards, and finally reaching, the definitive greater curvature. The *parietal* mesogastrial and (cleaving) mesoduodenal *attachment* remains, for a time, in the dorsal midline, but subsequently undergoes profound changes. With the confluence of the cavities that collectively form the lower part of the lesser sac, its communication with the upper part, corresponding to the lesser gastric curvature and right and left gastropancreatic folds, becomes better defined. *Ventral* to the lower part of the cavity lies the post-cleavage mesoderm covering the posteroinferior surface of the stomach and a short proximal segment of the duodenum. This ventral wall is continued beyond the greater curvature and duodenum as the mesodermal strip of visceral attachment of the secondary dorsal mesogastrium and mesoduodenum. The radial width of the strip is relatively short rostrally (gastric fundus) and gradually increases along the descending left part of the greater curvature; it is longest throughout the remaining perimeter of the greater curvature as far as the duodenum, and this prominent part shows continued marginal (caudoventral and lateral) growth with extended internal cavitation (its walls constituting the expanding *greater omentum*). The *margins* of the cavity of the inferior part of the lesser sac are limited by the reflexed edges of the ventrally placed strata derived from the secondary dorsal mesogastrium just described. These converge, forming the mesodermal *dorsal wall*, which is initially 'free' throughout except at its midline dorsal root. At roughly mid-gastric levels is where, encased in this dorsal wall, the pancreatic rudiment grows obliquely, its tail ultimately reaching the left limit of the lesser sac at the level of the junction between gastric fundus and body. (It may be mentioned that many earlier accounts held that the lower part of the lesser sac stemmed from a simple excavation of the dorsal mesogastrium —the *pancreatico-enteric recess*—with subsequent further extensions. This view has been discarded here.)

Centred on the dorsal mesogastrial region towards which growth of the pancreatic tail is directed, the anlage of the spleen appears in the coelomic epithelium and subjacent mesenchyme. As it expands into the upper left hypochondriac part of the *greater* sac, it retains its neurovascular pedicle and serous tunic, both of dorsal mesogastrial origin (p. 238, **2**.114).

The greater omentum and its contained extension of the lesser sac continue to grow both laterally, but particularly caudoventrally, covering and closely applied to the transverse mesocolon, transverse colon and inframesocolic and infracolic coils of small intestine. At this stage the quadrilaminar nature of the dependent part of the greater omentum is most easily appreciated. It will be recalled that 'simple' mesenteries, e.g. *the* mesentery of the jejuno-ileum, are bilaminar in that they possess two mesothelial surfaces enclosing a connective tissue core which bears blood and lymphatic vessels, lymph nodes, nerves, adipose tissue and other cell varieties. In the greater omentum, the gastric serosa covering its postero-inferior surface (single mesothelium) and the anterosuperior serosa (single mesothelium) converge, meeting at the greater curvature (and initial segment of the duodenum). The resulting bilaminar mesentery continues as the anterosuperior (or 'descending') stratum of the omentum which, on reaching the omental margins, is reflexed and 'returns' (or 'ascends') as its posterior bilaminar stratum to its parietal root. The two bilaminar strata are initially in fairly close contact caudally, but separated by a fine, fluid-containing, cleft-like extension of the lower part of the lesser sac. The posterior mesothelium of the posterior stratum makes equally close contact with the anterosuperior surface of the transverse colon (starting at its *taenia omentalis*) and with its transverse mesocolon.

At this stage, and subsequently, it is convenient to designate the lower part of the lesser sac as consisting of three *subregions*: *retrogastric, perigastric*, and *greater omental*. The names are self-explanatory but their confines are all modified by various factors. Two phenomena are particularly prominent—gastric 'descent' relative to the liver, and fusion of peritoneal layers with altered lines of reflexion, adhesion of surfaces and loss of parts of cavities.

After the third month the hepatic growth diminishes, particularly the left lobe, and the whole organ recedes into the upper abdomen; meanwhile the stomach elongates and some descent occurs, despite its relatively fixed cranial and caudal ends. This causes the angular flexure of the stomach that persists postnatally: the concavity of the lesser curvature is now directed more precisely to the right, the lesser omentum is more exactly coronal and its free border vertical ventral to the liver; the free border of the ligamentum falciformis passes steeply rostrodorsally from umbilicus to liver. The mesodermal dorsal wall of the lower part of the lesser sac, crossed obliquely by the growing pancreas, has hitherto remained free, with its original dorsal midline root. Substantial areas now fuse with adjacent peritonealized surfaces of retroperitoneal viscera, the parietes, or another mesenteric sheet or fold. In the latter case there occurs a variable loss of their apposed mesothelia and some continuity of their mesodermal cores, but they remain surgically separable and no vascular anastomosis develops across the interzone. Above the pancreas the posterior secondary dorsomesogastrial wall of the sac becomes closely applied to the peritoneum covering the posterior abdominal wall and its sessile organs—the diaphragm, much of the left suprarenal gland, the ventromedial part of the upper pole of the left kidney, the initial part of the abdominal aorta, the coeliac trunk and its branches, and other vessels, nerves, and lymphatics (pp. 1339, 1340). Their peritoneal surfaces fuse and, with some tissue loss, a single mesothelium covering these structures remains, intercalated as a new secondary dorsal wall for this part of the lesser sac. The pancreas, as indicated (p. 237), grows from the duodenal loop, penetrating the substance of the dorsal mesoduodenum and secondary dorsal mesogastrium, their mesoderms and mesothelia initially clothing its whole surface, except where there exist peritoneal lines of reflexion. Its posterior peritoneum becomes closely applied to that covering all the posterior abdominal wall structures it crosses (prominently—the inferior vena cava, abdominal aorta, splenic vein, superior mesenteric vessels, inferior mesenteric vein, portal vein, left renal vessels, the caudal pole of the left suprarenal, a broad ventral band on the left kidney and various muscles etc., see p. 1382). The intervening peritoneal mesothelia fuse and atrophy, the mesodermal cores forming fascial sheaths and septa. The pancreas is now sessile and the peritoneum covering the upper left part of its head, neck and anterosuperior part of its body forms the central part of the dorsal wall of the lesser sac. The pancreatic tail remains peritonealized by a persisting part of the secondary dorsal mesogastrium as it curves from the ventral aspect of the left kidney towards the hilum of the spleen. (The infracolic parts of the pancreas are covered with greater sac peritoneum.) In the greater omental subregion of the lower part of the lesser sac two contrasting forms of mesenteric adhesion occur. The posterior 'returning' bilaminar stratum of the omentum undergoes partial fusion with the peritoneum of the transverse colon (at the taenia omentalis) and with its mesocolon. (They provide a great increase in the functional area of the stomach 'bed'—p. 1349; also they remain surgically separable and no anastomosis occurs between omental and colic vessels.) In fetal life the greater omental cavity extends to the internal aspect of the lateral and caudal edges of the omentum. However, postnatally a slow but progressive fusion of the internal surfaces occurs with obliteration of the most dependent part of the cavity; this proceeds rostrally and, when mature, the cavity does not usually extend appreciably beyond the transverse colon. (The other fusions described above follow different patterns.) Transverse mesocolon-greater omental fusion commences early while the umbilical hernia of the midgut has not returned; it is initiated between the right margin of the early greater omentum and near the root of the *presumptive* mesocolon, later spreading to the left. The pancreas becomes sessile mediolaterally i.e. head, followed by neck and body from right to left. Paradoxically, in the suprapancreatic region, the direction of the process of fusion is reversed, progressing lateromedially from left to right. These events are wholly in accord with the numerous variations and anomalous visceral positioning and mesenteric arrangements that have been observed and recorded.

Correlated with the many facets of development of the lesser sac and its associated viscera, the original dorsal midline attachment to the parietes of the foregut dorsal mesentery is profoundly altered. However, despite the extensive areas of fusion, virtually the whole of the gastric greater curvature (other than a small suboesophageal area) and its topographical continuation, the inferior border of the first 2–3 cm of the duodenum, retain true mesenteric derivatives of the secondary dorsal mesogastrium and its continuation, the dorsal mesoduodenum. Thus, although regional names are given to assist identification and description, it must be emphasized that they are merely subregions of one continuous sheet; a short description of their 'root' or parietal line of reflexion is also given.

The upper (oesophagophrenic) part of the *lesser* omentum arches across the diaphragm and as this bilaminar mesentery approaches the oesophageal hiatus its laminae diverge, skirting the margins of the hiatus. They then descend for a limited distance and variable inclination, to enclose reciprocally-shaped areas on the dorsum of the gastric fundus and diaphragm. The area may be roughly triangular to quadrangular; it contains areolar tissue and constitutes the *bare area of the stomach* or, when large, the *left extraperitoneal space*. Its right lower angle is the base of the left gastropancreatic fold; its left lower angle reconstitutes the bilaminar mesentery. The root of the latter arches downwards and to the left across the diaphragm and suprarenal and gives the *gastrophrenic ligament* to the gastric fundus. Continuing to arch across the ventral surface of the upper part of the left kidney, its layers part to receive the pancreatic tail and initially extend to the hilum of the spleen as the *splenorenal (lienorenal) ligament*. The left half of this bilaminar 'ligament' provides an almost complete peritoneal tunic for the spleen (as indicated, projecting into the greater sac) then, reuniting with its fellow at the opposite rim of the splenic hilum, continues to the next part of the gastric greater curvature as the *gastrosplenic ligament*. The remaining major (perhaps two-thirds) of the gastric greater curvature and its short duodenal extension are the visceral attachment of the anterior, 'descending', bilaminar stratum of the greater omentum. Its returning, posterior, bilaminar stratum continues to its parietal root; the latter extends from the inferior limit assigned to the splenorenal ligament, continuing to curve caudally and to the right along the anterior border of the body of the pancreas, immediately cranial to the line of attachment of the transverse mesocolon. Crossing the neck of the pancreas, the same curve is followed for a few centimetres on to the glands' head; the omental root then is sharply recurved cranially and to the left, soon to reach the inferior border of the duodenum. Thus it reaches that part of the lesser sac provided by cleavage of the dorsal mesoduodenum from the greater sac, entering the epiploic foramen and traversing the epiploic canal between the caudate hepatic process and proximal duodenum, the lower right angle of the hepatoenteric recess, and crossing the right gastro-pancreatic fold, then descending behind the proximal duodenum to enter the right marginal strip enclosed by the greater omentum.

DEVELOPMENT OF UROGENITAL ORGANS

The urinary and reproductive organs are developed from the mesoderm of the *intermediate cell mass* (p. 133) and they are intimately associated with one another especially in the earlier stages of their development. Typically in lower vertebrates, the intermediate cell mass develops serial segmental *nephrotome* (2.115), each enclosing a cavity, the *nephrocoele*, which communicates with the coelom through a *peritoneal funnel*. The dorsal wall of each nephrotome evaginates as a *nephric tubule* communicating with the nephrocoele via a *nephrostome*. The dorsal tips of the (cranial) earlier developed nephric tubules bend caudally and fuse forming a longitudinal *primary excretory duct*, which grows caudally and curves ventrally to open into the cloaca. The more caudally placed and successively later developed tubules open secondarily into this duct or into tubular outgrowths from it. *Glomeruli* arise from the ventral wall of the nephrocoele (*internal glomeruli*) or the roof of the coelom adjacent to the peritoneal funnels (*coelomic* or *external glomeruli*), or in both situations (2.115). External glomeruli are usually transitory.

It has been customary to regard the renal excretory system as three organs, the *pronephros*, *mesonephros* and *metanephros*

PRIMITIVE VERTEBRATE EXCRETORY SYSTEM

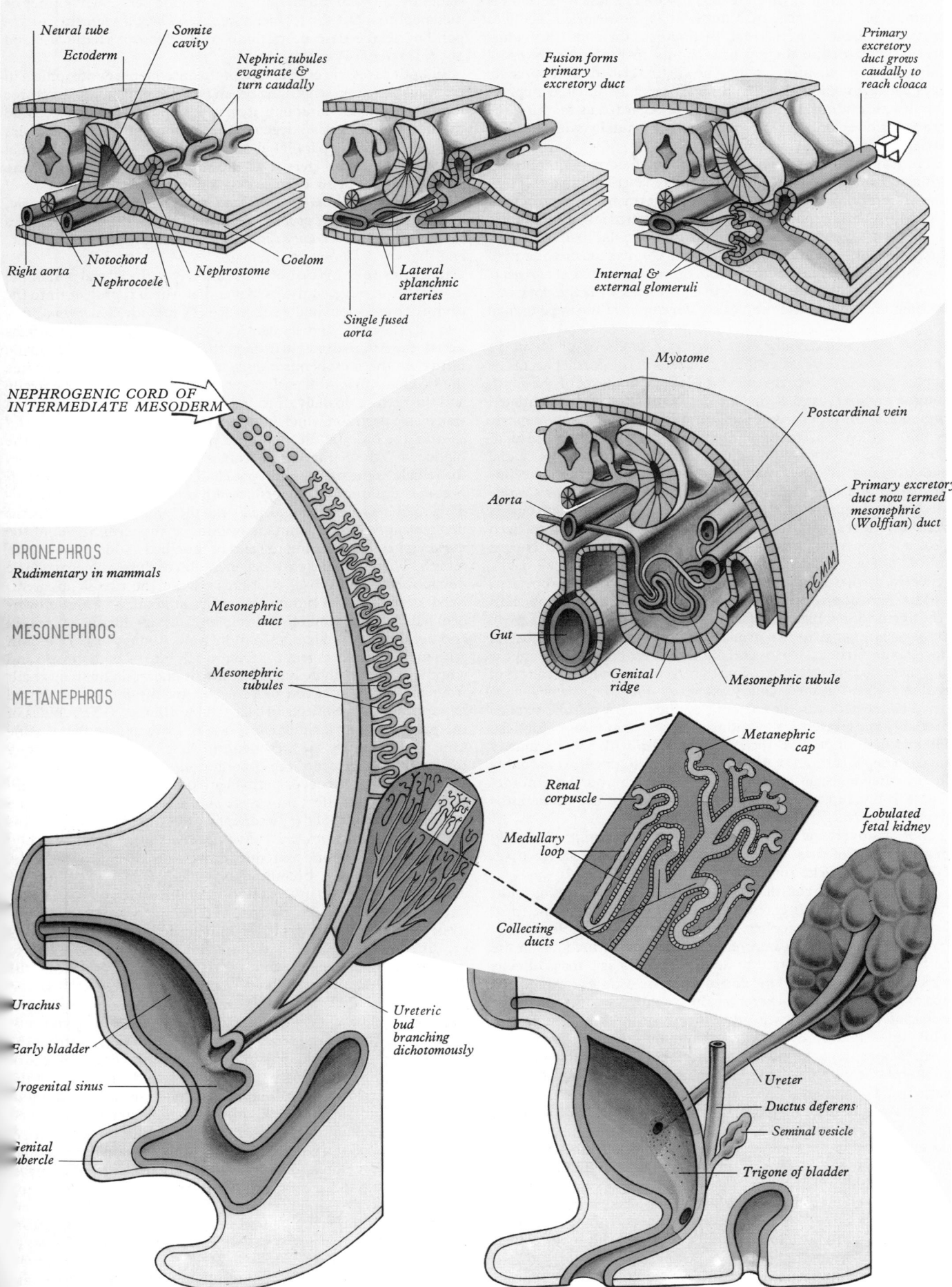

Neural tube
Ectoderm
Somite cavity
Nephric tubules evaginate & turn caudally
Fusion forms primary excretory duct
Primary excretory duct grows caudally to reach cloaca

Right aorta
Notochord
Nephrocoele
Nephrostome
Coelom
Lateral splanchnic arteries
Single fused aorta
Internal & external glomeruli

NEPHROGENIC CORD OF INTERMEDIATE MESODERM

Myotome
Postcardinal vein
Aorta
Primary excretory duct now termed mesonephric (Wolffian) duct

PRONEPHROS
Rudimentary in mammals

MESONEPHROS

METANEPHROS

Mesonephric duct

Mesonephric tubules

Gut
Genital ridge
Mesonephric tubule

Metanephric cap
Renal corpuscle
Medullary loop
Collecting ducts
Lobulated fetal kidney

Urachus
Early bladder
Urogenital sinus
Genital tubercle

Ureteric bud branching dichotomously

Ureter
Ductus deferens
Seminal vesicle
Trigone of bladder

115 Principal features of the primitive vertebrate nephric system for comparison with the development of the human nephric system. It should be appreciated that a considerable period of embryonic and fetal life has been necessarily compressed into a single, static diagram. (Modified from Williams et al 1969.)

249

succeeding each other in time and space. The last to develop is retained as the permanent kidney. It is, however, difficult to provide reliable criteria that distinguish them as individual organs or to define their precise limits in the embryos of all animals. A pronephros cannot be distinguished as a separate organ in man; the earliest and most cranially situated nephric tubules are rudimentary, transient and regarded as marking the pronephric region, and the latter merges caudally without clear demarcation into the mesonephros.

Human nephrotomes, with cavitation to form nephrocoeles which communicate with the coelom, are restricted to a few segments bordering the rostral limit of the intraembryonic coelom. Cranial to this (pronephric region) the intermediate mesoderm develops irregular, transient, solid or vesicular balls of cells. Caudal to the level of the eighth to tenth somites the mesoderm of the intermediate mass is fused into an unsegmented *nephrogenic cord*. This is connected at irregular intervals with the coelomic epithelium. No nephric tubules are developed at the most cranial levels.

The *primary excretory duct* begins in embryos of about 14 somites (23–24 days) as a solid rod of cells in the dorsal part of the nephrogenic cord. Its cranial end is about the level of the ninth somite and its caudal tip merges with the undifferentiated mesoderm of the cord. It differentiates before any nephric tubules; when the latter appear it is at first unconnected with them. In older embryos the duct has lengthened and its caudal end becomes detached from the nephrogenic cord to lie immediately beneath the ectoderm. From this level it grows caudally, independently of the nephrogenic cord, and then curves ventrally to reach the wall of the cloaca. It becomes canalized progressively from its caudal end to form a true duct which opens into the cloaca in embryos of 4–5 mm length (about 28 days). (Clearly, up to this stage, the name 'duct' is scarcely appropriate.)

The pronephros is first indicated as clusters of cells (rudimentary nephric tubules) in the nephrogenic cord (**2.115**). In regions cranial to the primary excretory duct these clusters develop no further. More caudally similar groups of cells appear and these become vesicular. The dorsal ends of the most caudal of these vesicles join the primary excretory duct, their central ends being connected with the coelomic epithelium by cellular strands probably representing rudimentary peritoneal funnels. Glomeruli are not developed in association with these cranially situated nephric tubules, which ultimately disappear. It is doubtful whether external glomeruli develop in the human embryo. These rudimentary nephric tubules constitute the pronephros described by earlier workers.

(For an overview and bibliographies concerning the early development of the human nephros consult Torrey 1954, O'Rahilly & Muecke 1972.)

The mesonephros develops caudal to the foregoing rudiments (**2.115–117**). It extends caudally to the level of the third lumbar segment, its nephric tubules are more completely differentiated and distinct internal glomeruli are formed. As in the more cranial region, each tubule first appears as a mass of cells which becomes hollow. One end of the vesicle grows towards and opens into the primary excretory duct, which is now termed the *mesonephric duct*, whilst the other forms the internal glomerulus. This dilates and invaginates; the outer stratum forms the *glomerular capsule*, the invaginated segment tissue from which the glomerular capillaries differentiate, with their enclothing cell varieties. These are supplied with blood through a lateral branch of the aorta (Streeter 1945). It is estimated that about 70–80 mesonephric tubules and a corresponding number of glomeruli develop, but these are *not* segmental in arrangement. These tubules, however, are not present at the same time and it is rare to find more than 30–40 in an individual embryo, for the cranial tubules and glomeruli develop then atrophy before the development of those situated more caudally. By the end of the sixth week each mesonephros is an elongated spindle-shaped organ which projects into the coelomic cavity, one on each side of the dorsal mesentery from the septum transversum to the third lumbar segment. This whole projection is the *mesonephric ridge* (Wolffian body); it develops subregions (vide infra), and the gonad is developed on its medial surface. There are striking similarities in structure between the mesonephros and the permanent kidney or metanephros, but the former's nephrons lack a segment corresponding to the descending limb of the loop of Henle (Leeson 1957, Davies & Routh 1957).

In both sexes the cranial end of the mesonephros atrophies; in embryos of 20 mm length the organ is found only in the first three lumbar segments, although it may still possess as many as 26 tubules. (This massive degeneration of the more cranial tubules appears to have no parallel in other mammals.) The most cranial one or two tubules persist as the *rostral aberrant ductules*; the succeeding five or six tubules develop into the *efferent ductules of the testis* and the *lobules of the head of the epididymis* in the male, and the tubules of the *epoöphoron* in the female; the caudal tubules form the *caudal aberrant ductules* and the *paradidymis* in the male, and the *paroöphoron* in the female.

The mesonephric duct runs caudally in the lateral part of the mesonephric ridge, at the caudal end of which it projects into the cavity of the coelom in the substance of a mesodermal *mesonephric fold* (**2.118–122**). (Subsequently, with the neighbouring formation of the paramesonephric duct, they form the *tubal fold* — vide infra.) As the mesonephric ducts approach the urogenital sinus the two folds fuse with each other, between the bladder ventrally and the rectum dorsally, forming across the cavity of the pelvis a transverse partition which is somewhat inaptly termed the *genital cord* (**2.120A,B**). In the male the peritoneal fossa between the bladder and the genital cord becomes obliterated, but it persists in the female as the uterovesical pouch. The mesonephric duct itself becomes the canal of the *epididymis, ductus deferens* and *ejaculatory duct*. The *seminal vesicle* and the ampulla of the ductus deferens appear as a common swelling at the termination of the mesonephric duct during the end of the third and into the fourth month. This coincides with degeneration of the paramesonephric ducts, though no causal relation between the two is apparent. Separation into two rudiments occurs at about 125 mm crownheel length. The seminal vesicle elongates, its duct is delineated and hollow diverticula bud from its wall. About the sixth month (300 mm crown-heel length) the growth rate of both vesicle and ampulla is greatly increased; the cause is uncertain but may result from increased secretion of prolactin by the fetal or maternal hypophysis, or the effects of placental hormones. The tubules of the prostate show a similar increase of growth rate at the same time. In the female the mesonephric duct is vestigial, becoming the longitudinal duct of the epoöphoron.

The metanephros, or permanent kidney, has a double origin (**2.115**). At about the 5 mm stage an outgrowth forms from the dorsomedial aspect of the mesonephric duct, near the cloaca. This is the *metanephric (ureteric) diverticulum* and it grows dorsally, then cranially. Its blind extremity grows into the caudal end of the nephrogenic cord, becomes expanded and the adjoining mesoderm condenses around it to form the *metanephrogenic mass* or *cap* (**2.115, 121**). The presence of the actively growing extremity of the diverticulum provides unidentified factors determining the differentiation of the mesoderm and, where the ureter fails to develop, no metanephrogenic cap is formed. The stalk of the diverticulum becomes the *ureter*, and its expanded cranial end gives origin not only to the *pelvis* of the kidney and its *calices* but also to the *collecting tubules* of the kidney. The secreting and convoluted tubules and the renal corpuscles are all derived from the metanephrogenic cap in a manner similar to the development of the mesonephric tubules and glomeruli. The blood vessels of the glomeruli develop in situ from mesenchyme in the concavity of the glomerular capsule (Lewis 1958a,b) and extend by repeated splitting of the lumina of the first-formed capillaries (Zamboni & De Martino 1968). (For details of the tubulo-vascular relationships in the developing kidney consult Speller & Moffat 1977.

From the expanded extremity of the ureter arise four collecting tubules of the first order and it itself forms the primitive renal pelvis. Each of the tubules ends in an ampullated extremity, from which collecting tubules of the second order bud out. These, in turn, give rise to collecting tubules of the third order and so on. In some animals the four collecting tubules of the first order are absorbed into the renal pelvis, which then presents a single renal papilla and no subdivision into calices. In man, however, the collecting tubules of the first and second order persist an

INDIFFERENT STAGE

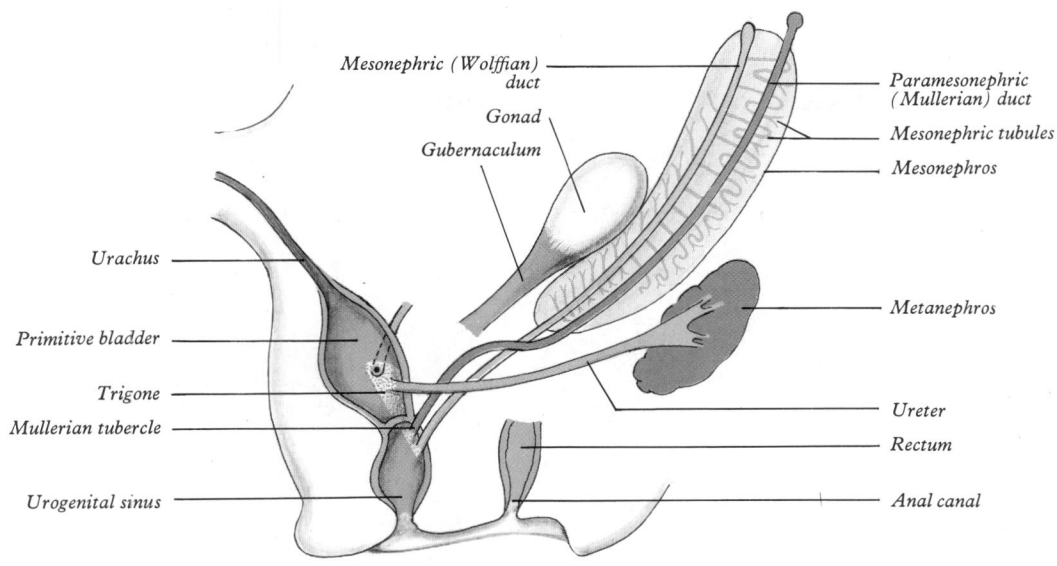

Mesonephric (Wolffian) duct
Gonad
Gubernaculum
Paramesonephric (Müllerian) duct
Mesonephric tubules
Mesonephros
Urachus
Metanephros
Primitive bladder
Trigone
Müllerian tubercle
Ureter
Rectum
Urogenital sinus
Anal canal

FEMALE

MALE

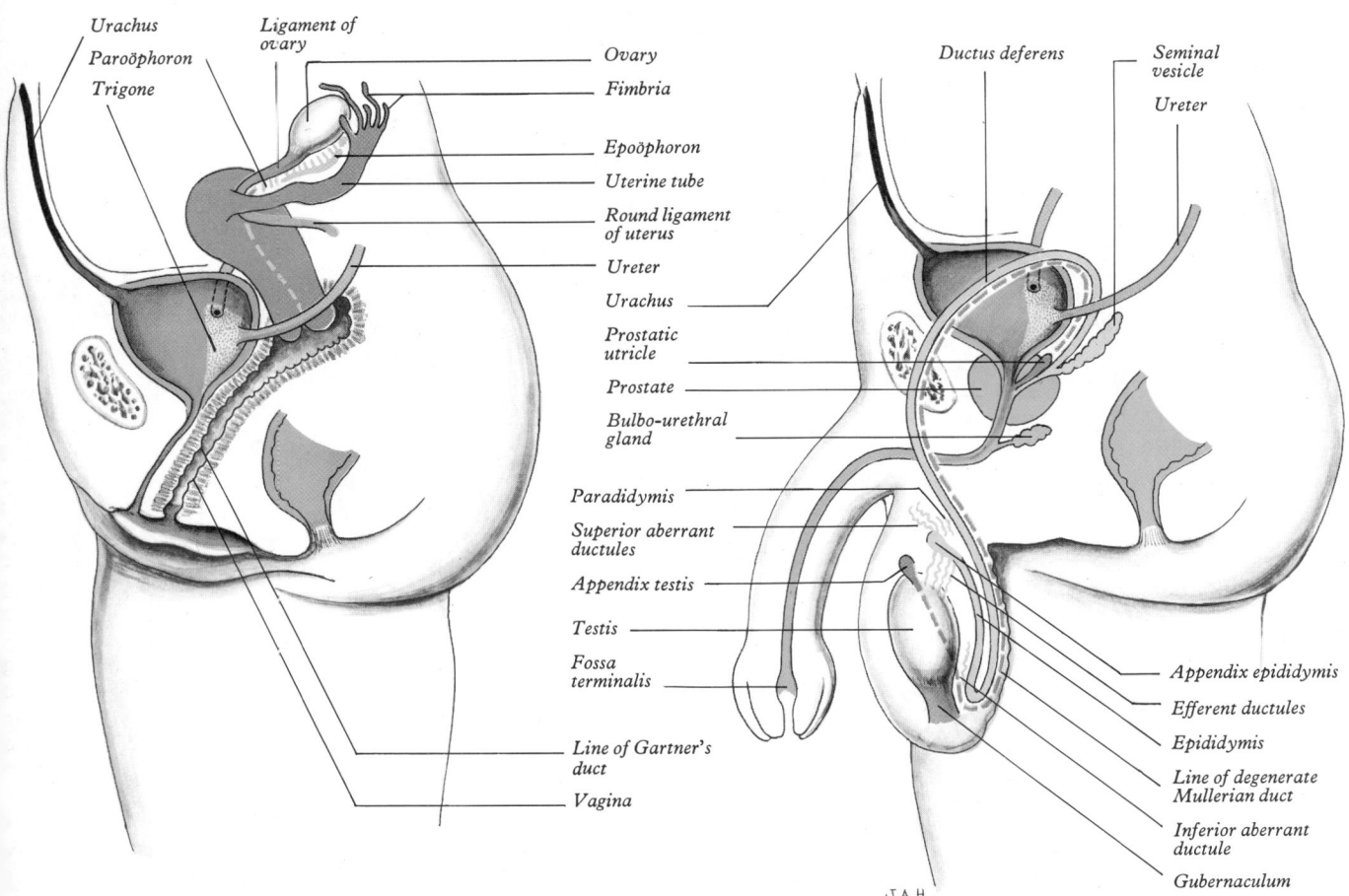

Urachus
Paroöphoron
Trigone
Ligament of ovary
Ovary
Fimbria
Epoöphoron
Uterine tube
Round ligament of uterus
Ureter
Urachus
Prostatic utricle
Prostate
Bulbo-urethral gland
Paradidymis
Superior aberrant ductules
Appendix testis
Testis
Fossa terminalis
Line of Gartner's duct
Vagina
Ductus deferens
Seminal vesicle
Ureter
Appendix epididymis
Efferent ductules
Epididymis
Line of degenerate Müllerian duct
Inferior aberrant ductule
Gubernaculum

J.A.H.

2.116 Disposition and fate of the mesonephric and paramesonephric ducts, mesonephric tubules, gonads and gubernaculum, primitive bladder and urogenital sinus, and ureters, as seen in the transformation from the indifferent stage to the definitive condition in the two sexes. (Modified from Williams et al 1969.)

constitute, respectively, the major and minor calices, while the tubules of the third and fourth orders are taken into the minor calices which, therefore, directly receive the openings of the collecting tubules of the fifth order.

When it first appears, the metanephric renal rudiment is sacral but, as the ureteric outgrowth lengthens, it grows cranially and when the embryo has a length of some 13 mm its expanded pelvis lies on a level with the second lumbar vertebra.

During this period the ascending kidney receives its blood supply *sequentially* from arteries in its immediate neighbourhood, the middle sacral and common iliac arteries; the definitive *renal artery* is not recognizable until the beginning of the third month. It arises from the most caudal of the three suprarenal arteries, all of which represent persistent mesonephric or lateral splanchnic arteries (p. 218). Additional renal arteries are by no means uncommon. They may enter at the hilum or at the upper or lower

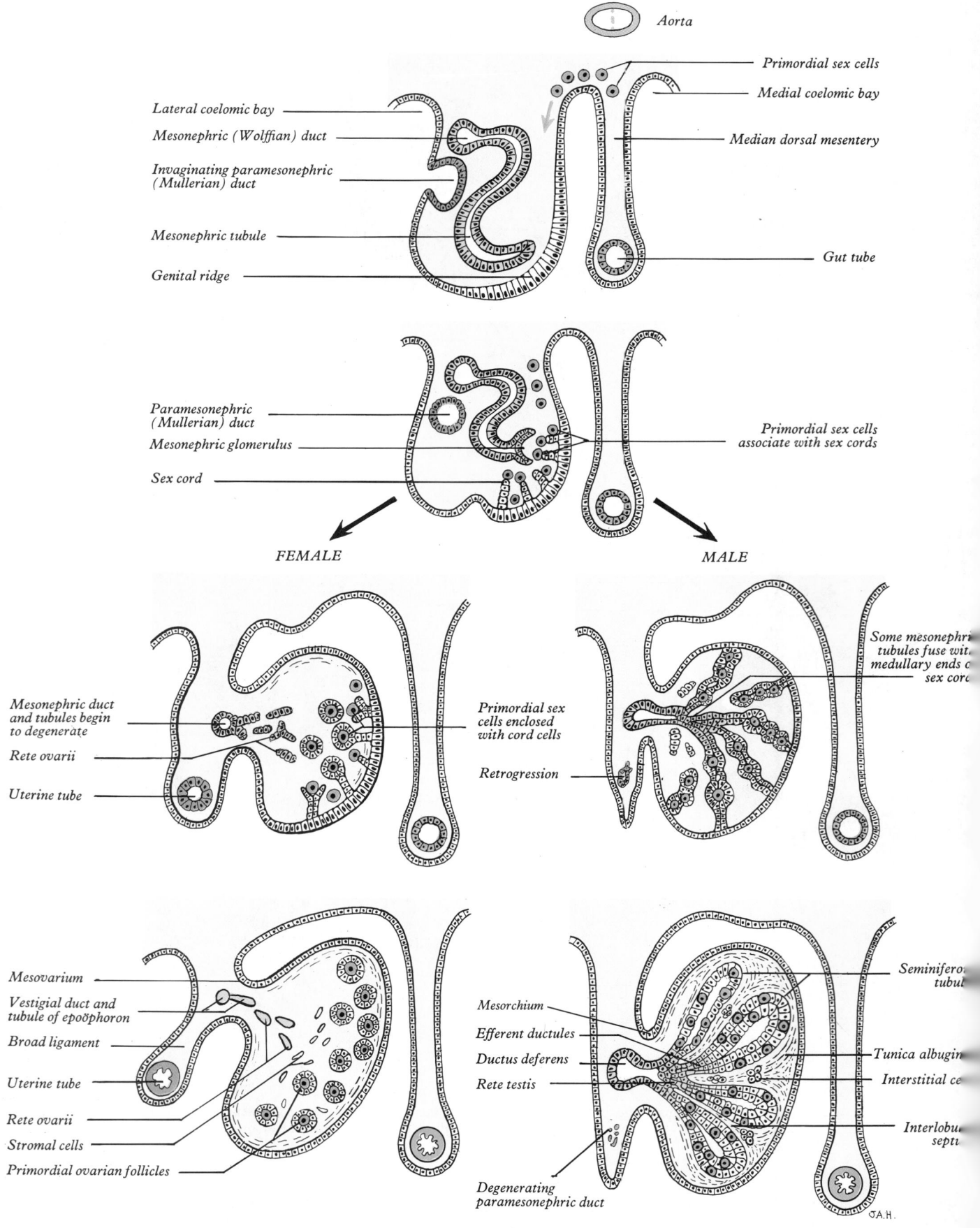

INDIFFERENT STAGES

Aorta

Primordial sex cells

Medial coelomic bay

Lateral coelomic bay

Median dorsal mesentery

Mesonephric (Wolffian) duct

Invaginating paramesonephric (Mullerian) duct

Mesonephric tubule

Genital ridge

Gut tube

Paramesonephric (Mullerian) duct

Primordial sex cells associate with sex cords

Mesonephric glomerulus

Sex cord

FEMALE

MALE

Mesonephric duct and tubules begin to degenerate

Some mesonephric tubules fuse with medullary ends of sex cords

Rete ovarii

Primordial sex cells enclosed with cord cells

Uterine tube

Retrogression

Mesovarium

Seminiferous tubules

Vestigial duct and tubule of epoöphoron

Mesorchium

Efferent ductules

Broad ligament

Ductus deferens

Tunica albuginea

Uterine tube

Rete testis

Interstitial cells

Rete ovarii

Stromal cells

Interlobular septum

Primordial ovarian follicles

Degenerating paramesonephric duct

2.117 Schema of the development of the gonads and associated ducts as seen in transverse section. Note fate of primordial sex cells, mesonephric duct and tubules and paramesonephric duct in the two sexes.

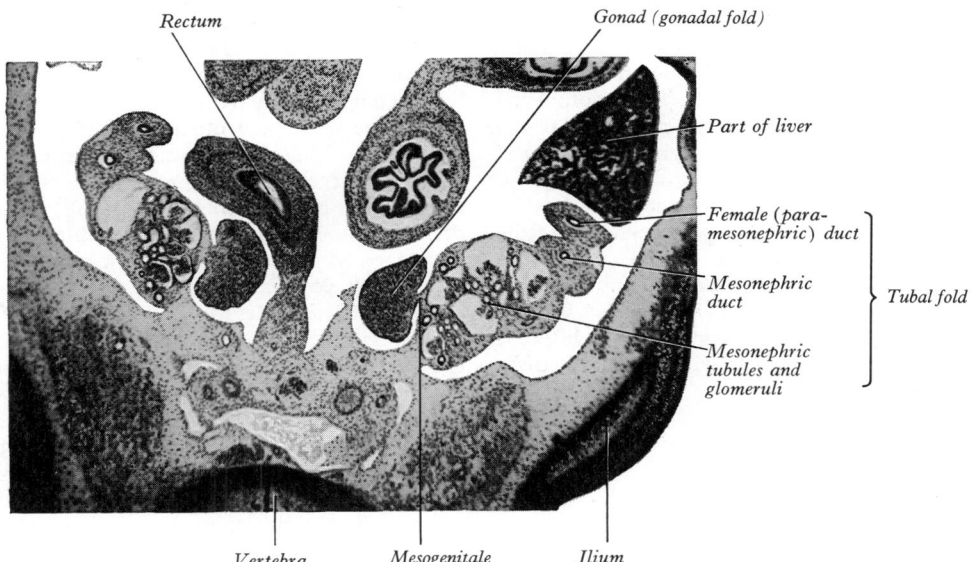

2.118 Transverse section through the lower part of the abdomen of a human fetus, nine weeks old, showing the connections and relative positions of the structures derived from the mesonephric ridge.

pole of the gland, and they also represent persistent mesonephric arteries.

At an early stage the metanephric kidney is lobulated, a condition which persists through fetal life but disappears during the first year after birth. Varying degress of lobulation, however, on occasion persist throughout life.

There occur many varieties of cystic renal disease, and a number of different classifications proposed (see Chisholm &

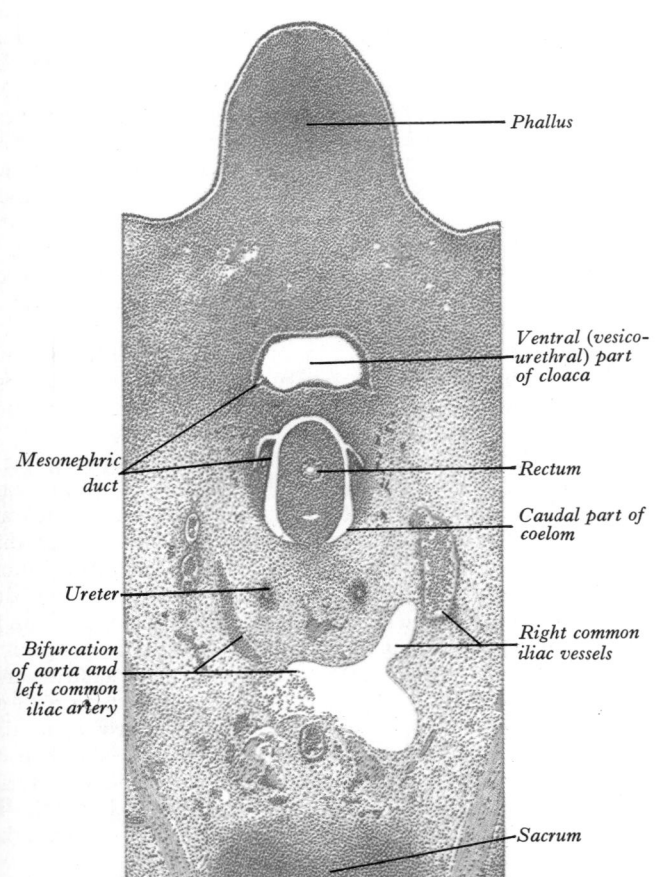

2.119 Transverse section of the tail end of a human embryo in the eighth week. CR length 22 mm. The projection from the dorsolateral aspect of the ventral portion of the cloaca marks the entry of the mesonephric duct. Stained with haematoxylin and eosin. Magnification × c.20.

Williams 1982). For long it has been held that renal cysts in most, if not all instances, result from vesicular cell clumps retained after failure of fusion between the tips of branches from the ureteric diverticulum on the one hand, and metanephrogenic cap tissue on the other. Such a view is no longer tenable. It has been demonstrated, convincingly, that the cyst-like formations are wide dilatations of a part of otherwise continuous nephrons (Moffat 1982). *Adult polycystic renal disease*, the commonest form, is inherited as an autosomal dominant; in this the dilatations may affect any part of the nephron, from Bowman's capsule to collecting tubule. Less common is *infantile cystic renal disease*, inherited as a recessive trait; the proximal and distal tubules are dilated to some degree but the collecting ducts are grossly affected.

Further development of the ureter has attracted less attention but one series included 45 human embryos (5–55 mm). It was claimed that the ureteric wall was highly permeable at an early stage (5 mm). Its lumen then becomes obliterated (13–22 mm), to be subsequently recanalized. Both processes were observed to begin at intermediate levels of the ureter and to proceed in both directions. The recanalization was not associated with metanephric function, but possibly with the rapid elongation of the ureter in conformity with embryonic growth. Two fusiform enlargements appear subsequently, affecting its lumbar and pelvic levels. The lumbar enlargement appears during the fifth month, the pelvic not until the ninth month, the latter being inconstant. As a result the ureter shows a constriction at its upper end (pelviureteric junction) and another as it crosses the pelvic brim. A third narrowing is always present at its lower end and is related to the growth of the bladder wall.

At first the caudal connection of the ureter is to the dorsomedial aspect of the mesonephric duct but, owing to differential growth, the connection becomes lateral to the duct. Thereafter the caudal end of the duct becomes incorporated in the developing bladder, and the orifice of the ureter opens separately into the bladder on the lateral side of the opening of the duct. Later the two orifices become separated still further and, although the ureter retains its point of entry into the bladder, the mesonephric duct opens into that part of the urogenital sinus which subsequently becomes the prostatic urethra.

The paramesonephric (Mullerian) ducts initially develop in embryos of both sexes, but become dominant in the development of the *female* reproductive system; they are not detectable, however, until the embryo reaches a length of 10–12 mm (early sixth week). Development in the 'indifferent' period is followed by further details of the female duct maturation and finally brief notes of the limited male derivatives. Each commences as a linear invagination of the coelomic epithelium (the *paramesonephric* *groove*) on the lateral aspect of the mesonephric ridge near its cranial end, and its blind caudal end continues to grow caudally into the substance of the ridge as a solid rod of cells which acquires a lumen as it lengthens. Throughout the extent of the mesonephros it is *lateral* to the mesonephric duct. At the caudal end of the mesonephros (which it reaches in the eighth week), the paramesonephric duct turns medially (**2.116**) and crosses *ventral* to the mesonephric duct to enter the genital cord (**2.120A,B**), where it bends caudally in close apposition with its fellow of the opposite side. The two ducts reach the dorsal wall of the urogenital sinus during the third month, and their blind ends produce an elevation on it termed the *Mullerian (sinus) tubercle* (**2.116,122F**). Each duct consists, at the end of the indifferent stage, of vertical cranial and caudal parts with an intermediate horizontal region.

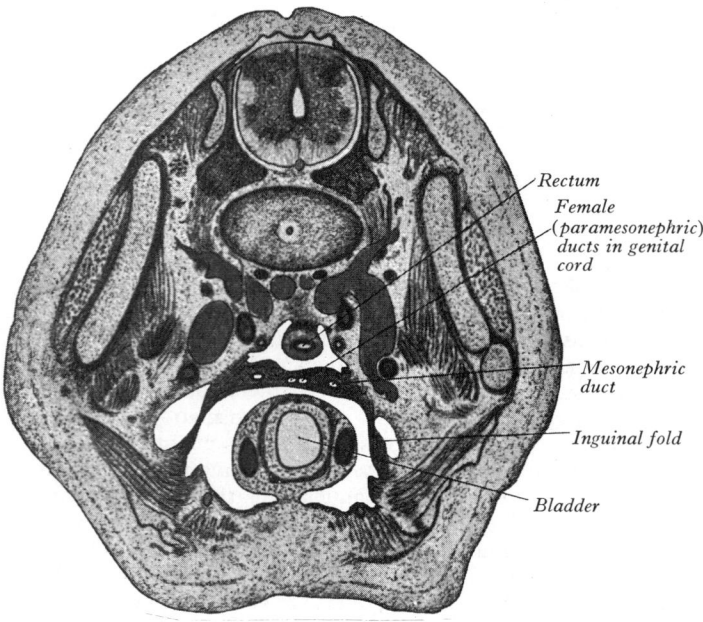

Rectum

Female (*paramesonephric*) ducts in genital cord

Mesonephric duct

Inguinal fold

Bladder

2.120A Transverse section through the pelvic part of a human fetus shows the formation of the genital cord and the inguinal folds.

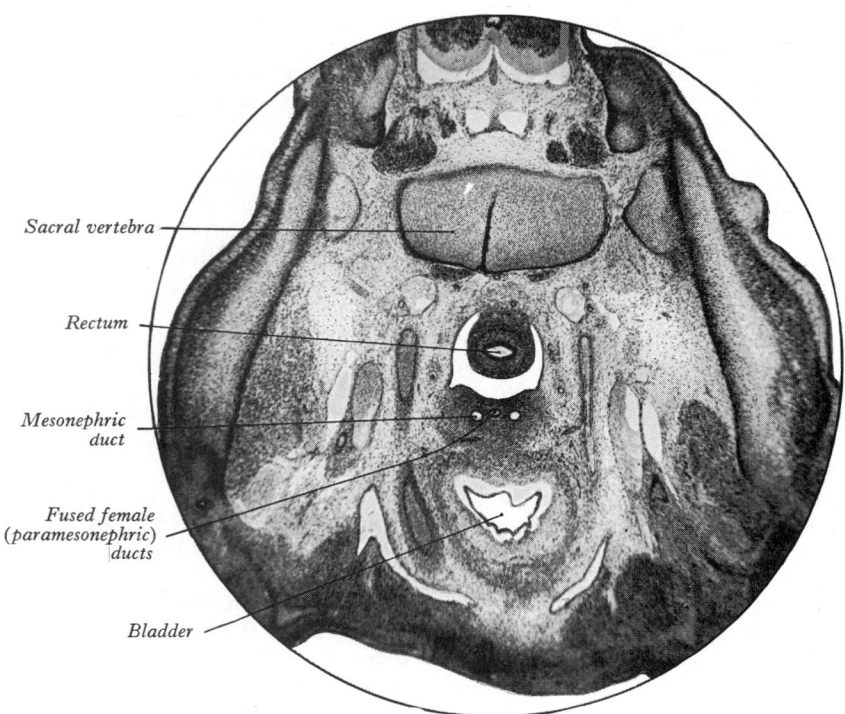

Sacral vertebra

Rectum

Mesonephric duct

Fused female (*paramesonephric*) ducts

Bladder

2.120B Transverse section through the pelvis of a nine-week-old male human fetus, showing the approximation of the genital cord to the dorsal wall of the urogenital sinus.

In the female the cranial part forms the uterine tube, and its original coelomic invagination remains as the pelvic opening of the tube, the fimbriae becoming defined as the cranial end of the mesonephros degenerates. The caudal vertical parts of the two ducts fuse with each other (**2.122F**) to form the *uterovaginal primordium*. This gives rise to the lower part of the uterus and, as it enlarges, it takes in the horizontal parts to form the fundus and most of the body of the adult uterus. The stroma of the endometrium and the uterine musculature (myometrium) are developed from the surrounding mesoderm of the genital cord.

At about 60 mm CR length an epithelial proliferation (the *sinuvaginal bulb*) arises from the dorsal wall of the urogenital sinus in the region of the sinus tubercle, and its origin marks the site of the future hymen. Whether the epithelium involved in the proliferation is from the sinus (Bulmer 1957) or epithelium of the mesonephric duct which has extended over the Mullerian tubercle (Vilas 1932, Meyer 1938, Forsberg 1963) is uncertain. The proliferation gradually extends cranially as a solid, antero-posteriorly flattened plate, inside the tubular mesodermal condensation of the uterovaginal primordium which will eventually become the fibromuscular vaginal wall. The caudal tip of the paramesonephric duct epithelium recedes until, at about the 140 mm stage, its junction with the sinus proliferation lies in the cervical canal.

Commencing from its caudal end, and gradually extending cranially through its whole extent, the solid plate formed by the sinus proliferation enlarges into a cylindrical structure; thereafter the central cells desquamate to establish the vaginal lumen. According to one view, the paramesonephric ducts do not directly contribute to the formation of the vagina (Frutiger 1969) but it has also been suggested that mesonephric and paramesonephric ducts are both concerned (Linkevich 1969). As the upper end of the vaginal plate enlarges it grows up into the mesoderm to embrace the cervix, and then is excavated to produce the vaginal fornices. As the lower end enlarges it invaginates the dorsal wall of the urogenital sinus around the hymeneal orifice, trapping a thin layer of mesoderm around the orifice to form the hymen (**2.123**). The hymen is lined on its superior surface by vaginal epithelium which, despite its origin, is clearly distinct from the sinus epithelium on the inferior surface. In the later stages, however, the lower surface of the hymen and a large part of the vestibule become lined by an epithelium indistinguishable from that of the vagina (Bulmer 1959). The urogenital sinus undergoes relative shortening craniocaudally to form the vestibule, which opens on the surface through the cleft between the genital folds.

During the later months of fetal life the vaginal epithelium is enormously hypertrophied, apparently under the influence of maternal hormones, but after birth it assumes the inactive form of childhood (Fraenkel & Papanicolaou 1938).

The differing embryonic origins of the vaginal epithelium and uterine epithelium have been correlated with their dissimilar responses in adult life to stimulation with oestrogenic hormones (Zuckerman 1940).

In the male the paramesonephric duct mostly atrophies, but a vestige of its cranial end persists as the *appendix testis* (p. 1424). The fused caudal ends of the two ducts are connected to the wall of the urogenital sinus by a solid *utricular cord* of cells. In this position it soon merges with a proliferation of sinus epithelium, the *sinu-utricular cord*, similar to, but less extensive than, the sinus proliferation in the female. This proliferating epithelium is claimed to be an intermingling of the endoderm of the urogenital sinus with the lining epithelia of the mesonephric and paramesonephric ducts, which have extended on to the surface of the sinus tubercle. As the sinu-utricular cord grows, so the utricular cord recedes from the tubercle. In the second half of fetal life the composite cord acquires a lumen and dilates to form the *prostatic utricle*, the lining of which consists of hyperplastic stratified squamous epithelium. The sinus tubercle becomes the *colliculus seminalis* (Vilas 1933, Glenister 1962).

The primordial germ cells, origin of the definitive germ cells, are segregated very early. This has been demonstrated at the 2-somite stage in the chick (Clawson & Domm 1969). They are large cells, in comparison with most somatic cells, being from 15 to 20 μm in diameter, and characterized by vesicular nuclei with well-defined nuclear membranes and by a tendency to retain yolk inclusions long after these have disappeared from somatic cells. (For the ultrastructure of human primordial germ cells consult Fukuda 1976.) It is not yet established whether the primordial germ cells are derived from particular blastomeres during cleavage, if they constitute a clonal line from a single blastomere or are the product of a progressive concentration of the germinal plasma of the fertilized ovum by unequal partition of this at successive mitoses (Bounoure 1939).

The germ cells can be identified in human embryos at the 13-somite stage (fourth week) in the endoderm at its caudal end and in the adjacent part of the yolk sac (Witschi 1948). At this stage the number of cells is probably not more than 20 to 30 (Hardisty 1967). When the tail fold has formed they appear in the endoderm and the splanchnic mesoderm of the hindgut as well as in the adjoining region of the wall of the yolk sac. By amoeboid movements and by growth displacement they migrate dorsocranially in the mesentery, passing around the dorsal angles of the coelom (medial coelomic bays) to reach the medial sides of the mesonephric ridges—the developmental sites of the gonads. In most vertebrates mitosis in the germ cells is arrested after their early segregation, to be renewed only when they reach the genital primordia. However, in mammals there is no such arrest and the cells proliferate both during and after migration to the mesonephric ridges; cells which do not complete this migration degenerate. After segregation the primordial germ cells are often termed *primary gonocytes*, which in turn divide to form *secondary gonocytes*. The distinction between the two generations is clear in most vertebrates, but the absence of mitotic arrest in mammals leads to a merging of the two stages.

The early association of primordial germ cells with the endoderm suggests that they are in fact derived from primitive endodermal cells or perhaps from common stem cells. Alternatively, the endoderm may exert an inductive effect on their site, proliferation and differentiation without being their actual source. For a discussion of this problem and citation of pertinent references consult Carlon & Stahl (1973).

While sexual differences in germ cell numbers occur in some vertebrate species, it is uncertain whether this represents an original difference at segregation or results from earlier and more rapid proliferation in one sex. No connection between numbers of primordial germ cells and fertility has been detected, but there is evidence that gross deficiency in number may affect individual fecundity (Hardisty 1967).

The formation of the gonads is first indicated by the appearance of an area of thickened epithelium on the medial side of the mesonephric ridge in the fifth week (2.117). Elsewhere on the surface of the ridge the coelomic epithelium is one or two cells thick, but over this genital area it becomes many-layered. The thickening rapidly extends in a longitudinal direction until it covers nearly the whole of the medial surface of the ridge. The thickened epithelium continues to proliferate, displacing the renal corpuscles of the mesonephros in a dorsolateral direction and forming a projection into the coelomic cavity, the *gonadal ridge*. Surface depressions form along the limits of the ridge which is thus connected to the mesonephros by an originally broad mesentery, the *mesogenitale*. In this way the mesonephric ridge becomes subdivided into a lateral part containing the mesonephric and paramesonephric ducts, which may be termed the *tubal fold*, and a medial part, termed the *gonadal fold*. The tubal fold also contains the nephric tubules and glomeruli at its base.

Up to the seventh week the gonad possesses no sexually differentiating feature. The proliferating epithelium now forms a number of cellular *gonadal cords*, separated by mesenchyme. These cords remain at the periphery of the primordium to form a cortex; more centrally a proliferation and labyrinthine cellular condensation of the mesenchyme of the mesonephros constitute a medulla. In the male all the progenitors of the definitive gonocytes become incorporated in the cords, but in the female a large number remain behind under the surface epithelium. At this stage in the male, an extension of the mesenchyme separates the gonadal cords from the surface and rapidly thickens to form the primitive *tunica albuginea*. This also develops in the female but to a lesser extent and at a later stage.

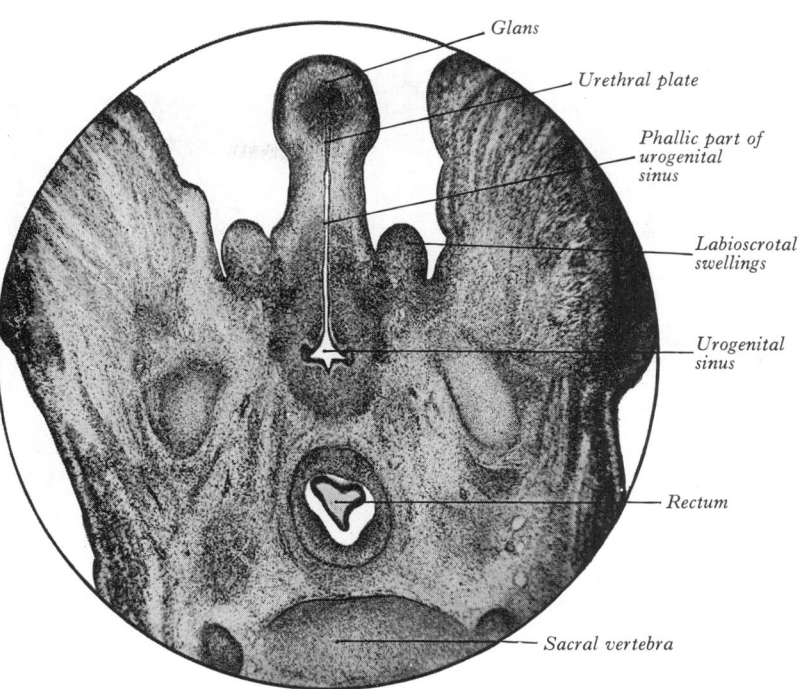

2.120C Transverse section through the lower part of the pelvis of a nine-week-old human fetus.

The testis (2.117). The cellular cords lengthen partly by additions from the coelomic epithelium and encroach on the medulla, where they unite with the network derived from the mesenchyme which ultimately becomes the *testicular rete*. The primordial germ cells are incorporated in the cords, which later become enlarged and canalized to form the *seminiferous tubules* (see Fukuda & Hedinger 1975). The cells derived from the surface of the early gonad form the *supporting cells* (of Sertoli). The *interstitial cells* of the testis are derived from the mesenchyme and possibly also from coelomic epithelial cells which do not become incorporated into the tubules. The cords of the testicular rete, which canalize later, become connected to the glomerular capsules in the cranial end of the (caudal) persisting part of the mesonephros and their associated glomerular tufts atrophy. The rete cords thus become connected to the mesonephric duct by the five to twelve most cranial *persisting* tubules and these become exceedingly convoluted and form the lobules of the head of the epididymis. The mesonephric duct, which was the primitive 'ureter' of the mesonephros, becomes the canal of the *epididymis* and the *ductus deferens* of the testis. The seminiferous tubules do not acquire lumina until the seventh month, but the tubules of the testicular rete somewhat earlier.

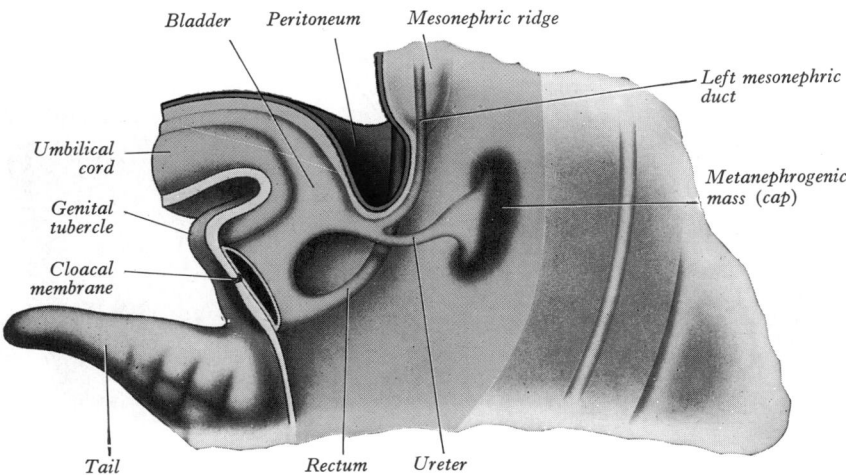

2.121 Schema, based on 2.122D, to show the formation of the pelvis of the kidney and the metanephrogenic mass or cap.

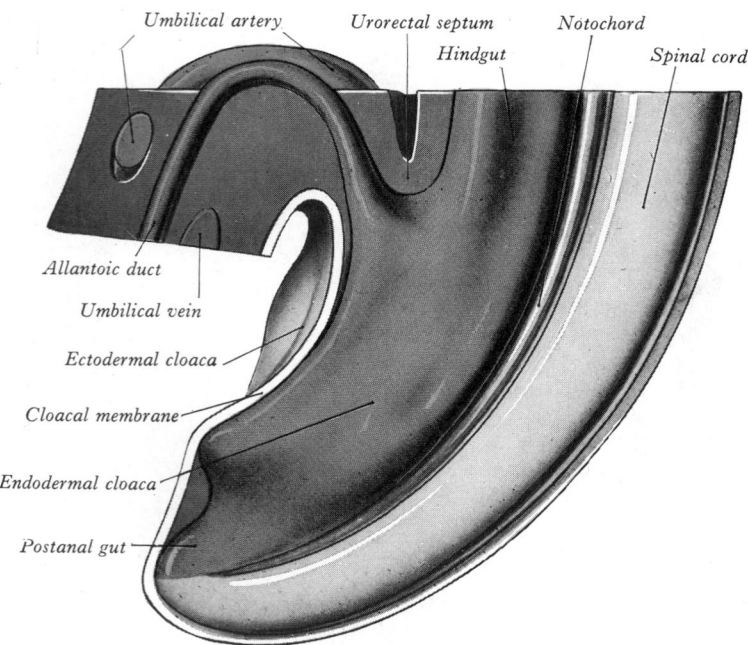

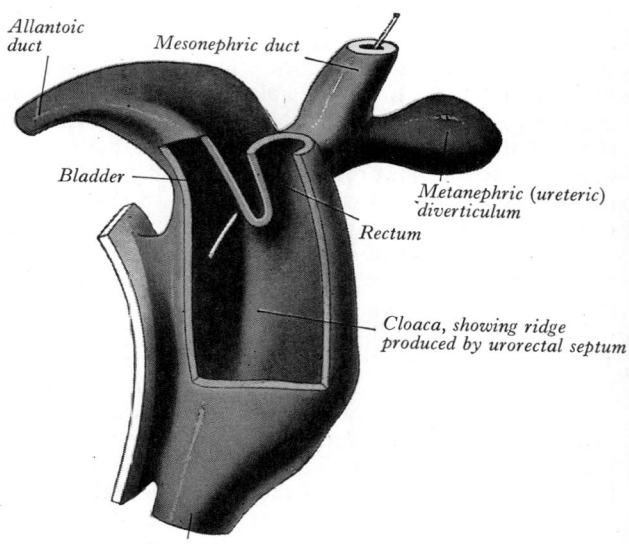

2.122A The tail end of a human embryo, about 4 weeks old. The model has been dissected to show the left lateral aspects of the spinal cord, notochord and endodermal cloaca. (After Keibel.)

2.122B The endodermal cloaca of a human embryo, near the end of the fifth week. A wire has been passed along the right mesonephric duct into the cloaca, and a part of the left wall of the cloaca, including the left mesonephric duct, has been removed, together with the adjoining portions of the walls of the developing bladder and rectum. A piece of the ectoderm around the cloacal membrane has been left *in situ* and is uncoloured. (After Keibel.)

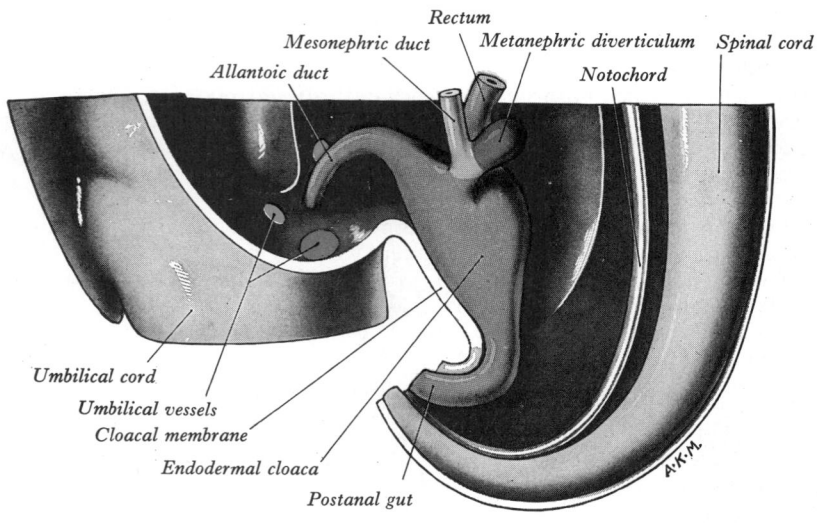

2.122C The caudal end of a human embryo, about 5 weeks old. (Drawn from a model by Keibel.)

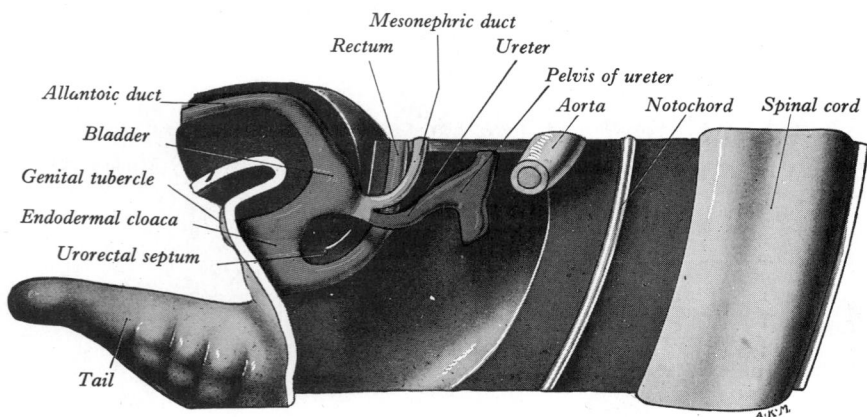

2.122D Part of the caudal end of a human embryo, about 6 weeks old. (Drawn from a model by Keibel. The model, which has been partly dissected, is seen from the left side.)

The ovary (2.117). In its earliest stages, the ovary closely resembles the testis, although it is slower to differentiate its characteristically female features. Few, if any, of the gonadal cords invade the medulla, the majority remaining in the cortex, where they may be joined by a second proliferation from the epithelium covering the gonad. In sections of the ovary in the third and subsequent months the cords appear as clusters of cells which may or may not contain primitive germ cells. These clusters are separated by fine septa of undifferentiated mesenchyme. An *ovarian rete* condenses in the medullary mesenchyme and some of its cords may form a junction with mesonephric glomeruli. The medulla subsequently regresses, and connective tissue and blood vessels from this region invade the cortex to form the stroma of the ovary. During this invasion the cell clusters break into individual groups which surround the *primordial 'ova'* now *primary oöcytes*, which have entered the prophase of the first meiotic division. These cells were derived from a mitotic division of the primordial germ cells (*naked oögonia*). Their epithelial capsules consist of flattened *pregranulosa cells* derived from proliferations of coelomic epithelium (Gillman 1948). The ovary now has its full complement of primary oöcytes. The majority undergo atresia at various stages during their development, but the remainder resume development by completing the first meiotic division shortly before ovulation (Zuckerman 1951, 1956). The capsular cells at the same time enlarge and multiply to form the stratum granulosum, and as they do so they become surrounded by thecal cells which differentiate from the stroma.

Organ culture of human fetal ovaries confirms the above, but it was considered that the primary oöcytes had progressed to the leptotene stage. Strands of epithelial cells containing oögonia developed during culture (Baker & Neal 1974).

Only the middle part of the gonadal ridge produces the ovary. Its cranial part is sterile and becomes the *suspensory ligament of the ovary* (infundibulopelvic fold of peritoneum). Its caudal region, also sterile, is incorporated in the *ovarian ligament*.

The descent of the testis is *not* merely a simple migration. At first it lies on the dorsal abdominal wall, but, as it enlarges, its cranial end degenerates and the remaining organ therefore occupies a more caudal position. It is attached to the mesonephric fold by a peritoneal fold, the *mesorchium* (2.118) (the *mesogenitale* of the undifferentiated gonad), which contains the testicular vessels and nerves and a quantity of undifferentiated mesenchyme. In addition, it acquires a secondary attachment to the ventral abdominal wall, which has a considerable influence on its

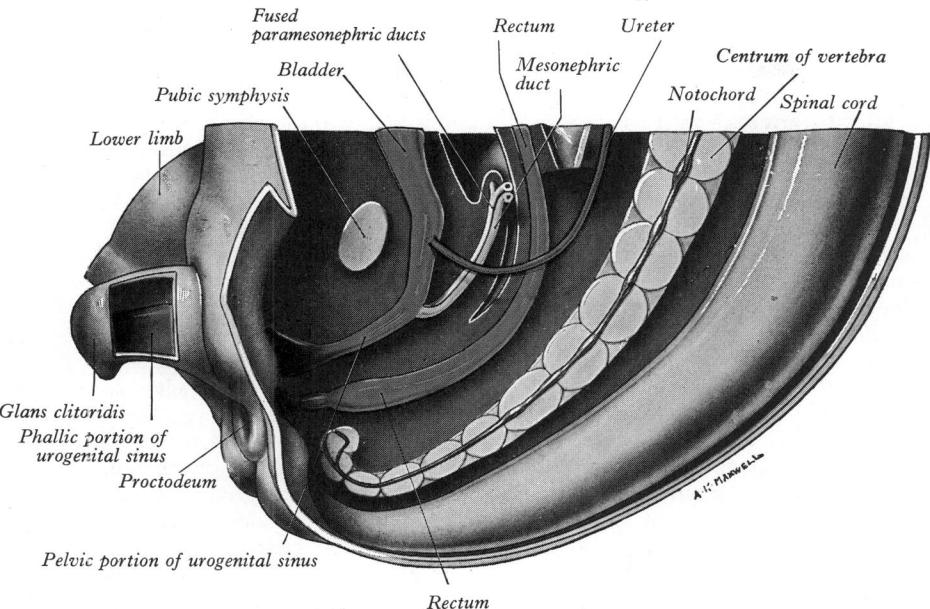

2.122E The tail end of a female human fetus, 8½–9 weeks old. The model has been dissected from the left side to show the structures in and near the median plane. Note that the cloaca has now been separated into urogenital and intestinal segments. (After Keibel.)

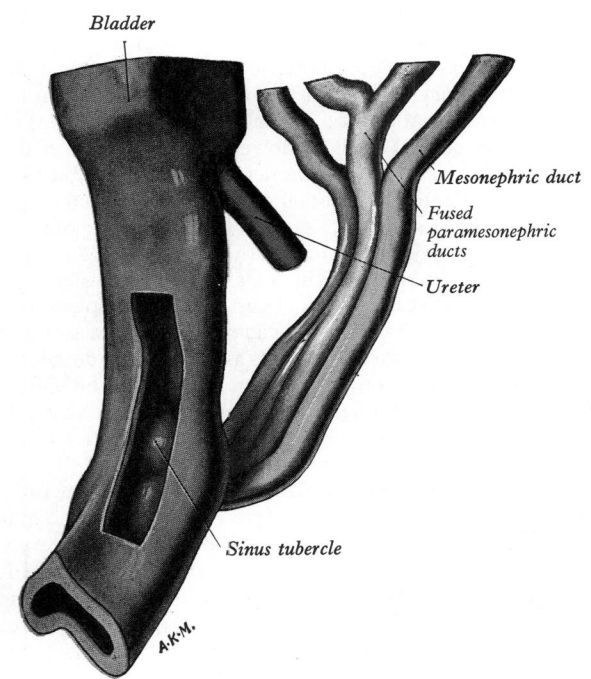

2.122F Part of the vesico-urethral portion of the endodermal cloaca of a female human fetus, 8½–9 weeks old. (Drawn from a model by Keibel.)

(p. 1339) and its lower end protrudes down the inguinal canal along the ventrosuperior aspect of the gubernaculum, as the *processus vaginalis*. The caudal pole of the testis is *retained* in apposition with the deep inguinal ring by the gubernaculum until the seventh month, when it abruptly and rapidly passes through the inguinal canal and gains the scrotum. As it descends it is necessarily accompanied by its peritoneal covering, and the adjoining peritoneum from the iliac fossa is drawn down into the processus vaginalis. The distal end of the processus vaginalis, into which the testis projects, forms the *tunica vaginalis testis* but the portion associated with the spermatic cord in the scrotum and in the inguinal canal normally becomes obliterated, usually leaving a fibrous remnant. Sometimes the remnant atrophies completely. Alternatively, its original cavity may persist, in whole or in part and any location. These variations may form the walls of hernial sacs or encysted fluid sites. The fascial coverings of the testis and spermatic cord, including the cremaster, are developed from the gubernaculum testis (Backhouse & Butler 1958).

The actual mechanism of the descent of the testis is still uncertain. It has been ascribed, by different investigators, to shortening and active contraction of the gubernaculum, to increased intra-abdominal pressure, to a simple growth process and to the effect on the convex surface of the gland of the active contraction of the lower fibres of the internal oblique muscle, squeezing it through the canal. None of these explanations is entirely convincing.

subsequent movements. At the point where the mesonephric fold bends medially to form the genital cord (p. 250), it becomes connected to the lower part of the ventral abdominal wall by an *inguinal fold* of peritoneum (2.120A). The mesenchymal cells occupying the core of the inguinal fold condense as another cord, the *gubernaculum*, extending from the skin which will later form the scrotum, through the inguinal fold and the mesorchium to the caudal pole of the testis. It traverses the site of the future inguinal canal, which is formed around it by the muscles of the abdominal wall as they differentiate. At the end of the second month the caudal part of the ventral abdominal wall is horizontal but, after the return of the intestines to the peritoneal cavity (p. 233), it grows in length and progressively becomes vertical. As a result the umbilical artery pulls up a falciform peritoneal fold, as it runs ventrally from the dorsal to the ventral wall, and this forms the medial boundary of a peritoneal fossa into which the testis projects. This fossa is the *saccus vaginalis* or *lateral inguinal fossa*

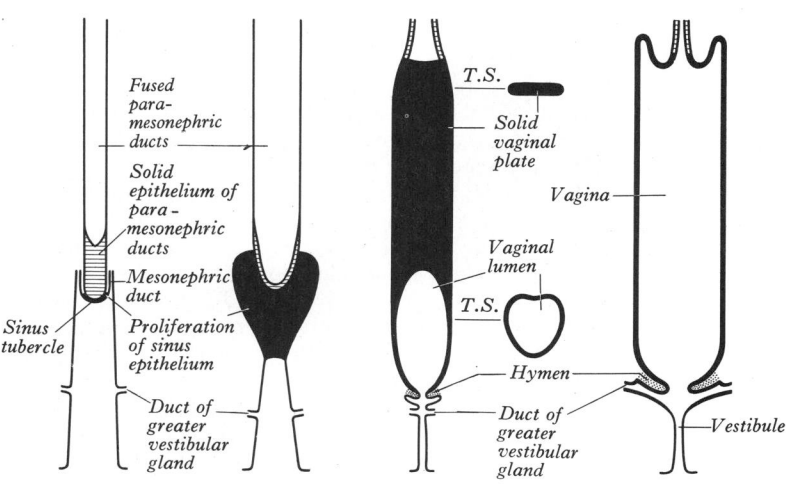

2.123 Diagrams to show the successive stages in the development of the uterus and vagina.

Whatever contribution the gubernaculum makes to the process, it cannot be by muscular contraction, because it contains no muscle, remaining a soft, mesenchymatous mass through its existence (Forssner 1928, Backhouse & Butler 1960). It precedes the testis both spatially and in rate of growth, forming a tapering column of soft tissue with the diminutive testis at its cranial pole. It continues to grow until the seventh month, by which time its caudal part has filled the future inguinal canal and has begun to expand the developing scrotum. In this it also precedes the processus vaginalis, but does not develop attachments to skin; nor is there any evidence that it produces the radiating extensions into the suprapubic, perineal and femoral sites, which are often used as explanations for the various forms of ectopia testis. By its soft consistency the gubernacular tissue may offer a route of low resistance to the descending testis, and the cessation of its growth in the last two months of gestation, coupled with an accelerating rate of growth in the testis and epididymis, may also be a factor in testicular descent as far as the inguinal canal. The mechanism of final, rapid descent into the scrotum remains unidentified. Endocrine effects seem certain, but this does not explain the actual agency. This account is based principally upon events as observed in porcine material by Backhouse & Butler (1960). A subsequent discussion of the problems of testicular descend and maldescent by Backhouse (1964) should be consulted. He has reviewed the literature since Hunter's original description (1762). Apparently the cremaster muscle develops in gubernacular mesenchyme and this may explain the development of the concept of the gubernaculum as a 'fibromuscular' ligament. The processus vaginalis also develops in close association with the soft, mucoid tissue of the gubernaculum and follows it into the genital swelling. Ultimately, as the testis and epididymis fill the scrotum, or genital swelling, the gubernacular mesenchyme is incorporated into the associated connective tissues.

Various abnormalities may occur in connection with testicular descent and obliteration of the processus vaginalis (p. 1424). The testis may remain in the abdomen, or it may fail to reach the scrotum and may then lie in any of the following situations: (1) in the perineum, (2) at the root of the penis, (3) at the superficial inguinal ring (p. 1377), (4) in the upper part of the thigh. These malpositions have been traditionally associated with certain additional extensions of gubernacular tissue. The largest extension normally passes to the scrotum while lesser extensions have been described as gaining attachment to the perineum, the root of the penis, the pubis, the inguinal ligament and the neighbourhood of the saphenous opening. The testis must follow the processus vaginalis and, should the latter for any reason follow any but the scrotal extension of the gubernaculum, malposition of the testis will result. It should be appreciated, however, that considerable doubt has now been expressed concerning these lesser expansions (previously the so-called 'tails of Lockwood'): possibly they reflect premature and abnormal fibrous partitioning of the gubernacular mesenchyme.

As noted, the processus vaginalis may remain completely patent, or its obliteration may be incomplete. When it retains a connection with the general peritoneal cavity it provides a preformed sac for a potential oblique inguinal hernia. It may be occluded at its upper end and may be shut off from the tunica vaginalis and yet remain patent in the intervening section. The patent portion may become distended with fluid, an *encysted hydrocoele* of the spermatic cord.

The descent of the ovary is less extensive. Like the testis, the ovary ultimately reaches a lower level than it occupies in the early months of fetal life but it does not leave the pelvis to enter the inguinal canal, except in certain anomalies. Connected to the medial aspect of the mesonephric fold by the *mesovarium* (homologous with the mesorchium), the ovary is also attached to the ventral abdominal wall through the medium of the inguinal fold. In this fold a mesenchymatous gubernaculum also develops but, as it traverses the mesonephric fold, it acquires an additional attachment to the lateral margin of the uterus near the entrance of the uterine tube. Its lower part, caudal to this uterine attachment, becomes the *round ligament of the uterus* and the part cranial to this, *the ovarian ligament*, these structures together being homologous with the gubernaculum testis in the male. This new

uterine attachment may be correlated with the restricted ovarian descent. At first the ovary is attached to the medial side of the mesonephric fold but, in accordance with the manner in which the two mesonephric folds form the genital cord (p. 250), its connection is finally to the posterior layer of the broad ligament of the uterus. The gubernaculum thus persists in the female, unlike the male, as *two* fibrous bands or ligaments on each side.

The *saccus vaginalis* also appears in the female; its prolongation into the inguinal canal (sometimes termed the *canal of Nuck*) normally undergoes complete obliteration, but may remain patent and form the sac of a potential oblique inguinal hernia (p. 1377). At birth the ovary and the lateral end of the corresponding uterine tube lie above the pelvic brim, and they do not sink into the lesser pelvis until the latter enlarges sufficiently to contain both of them and the other pelvic viscera, including the bladder.

The urinary bladder (2.122A-F) is derived partly from the so-called endodermal cloaca and partly from the caudal ends of the mesonephric ducts (2.116,119,122A-F). The walls of the cloaca involved are splanchnopleure, i.e. an endodermal lining encased in visceral mesenchyme. In contrast the mesonephric ducts are wholly mesodermal. After the separation of the rectum from the cloaca (p. 233), the ventral part of the latter becomes divided into three regions: (1) a cranial *vesico-urethral canal*, continuous with the allantoic duct—into this the mesonephric ducts open; (2) a middle, narrow channel, the *pelvic portion*; and (3) a caudal, deep, *phallic* section, closed externally by the *urogenital membrane* (2.122A-E). The second and third parts together constitute the *urogenital sinus*. The ureter and the mesonephric duct come to open separately into the vesico-urethral part (p. 250). The termination of the mesonephric duct then moves caudally to open into that part which will form the prostatic urethra. This occurs by the formation of a caudally directed loop of the duct behind the urogenital sinus, followed by absorption of the apposed walls. In this way the mesonephric duct contributes to the trigone of the bladder and dorsal wall of the proximal (superior) half of the prostatic urethra, i.e. as far as the opening of the prostatic utricle and ejaculatory ducts (or its homologue the *whole* female urethral dorsal wall). The remainder of the vesico-urethral part forms the body of the bladder and urethra; its apex is prolonged to the umbilicus as a narrow canal, the *urachus*. In postnatal life the urachus is drawn downwards as the bladder descends, but its upper end remains connected to one or both of the obliterated umbilical arteries. Its lumen persists throughout life and its lower end frequently communicates with the bladder near its apex (Begg 1930).

The prostate arises during the third month from the proximal part of the urethra. The earlier outgrowths, some 14–20 in number, arise from the endoderm around the whole circumference of the tube, but mainly on its lateral aspects and excluding the dorsal wall above the utricular plate. These outgrowths give rise to the *outer* glandular *zone* of the prostate (p. 1433). Later outgrowths from the dorsal wall above the mesonephric ducts arise from the epithelium of mixed urogenital, mesonephric and possibly paramesonephric origin covering the cranial end of the sinus tubercle. These produce the *internal zone* of glandular tissue. The outgrowths, which are at first solid, branch, become tubular and invade the surrounding mesenchyme, which is differentiating into non-striated muscles, associated blood and lymphatic vessels and connective tissues, and is invaded by autonomic nerves.

Similar outgrowths occur in the female but remain rudimentary. The urethral glands correspond to the mucosal glands around the upper part of the prostatic urethra and the paraurethral glands to the true prostatic glands of the external zone (Glenister 1962).

The *bulbo-urethral glands* in the male, and *greater vestibular glands* in the female, arise as diverticula from the epithelial lining of the urogenital sinus.

The external genital organs, like the gonads, pass through an indifferent state before distinguishing sexual characters appear (2.124). A surface elevation, the *genital tubercle*, appears at the cranial end of the cloacal membrane and lengthens to form the *phallus* (2.119). Within it is a longitudinal endodermal mass, the *urethral plate* (2.120,124), which grows forwards from the walls of the cloaca and urogenital sinus towards the tip of the organ. The

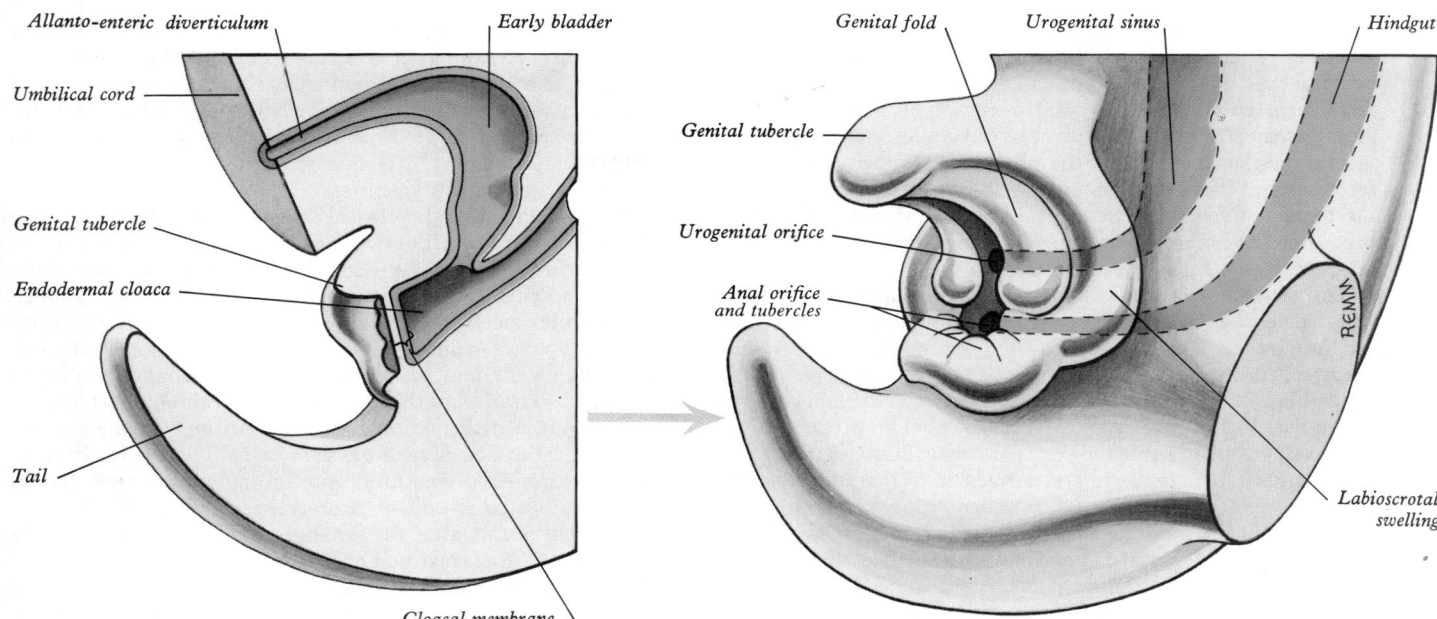

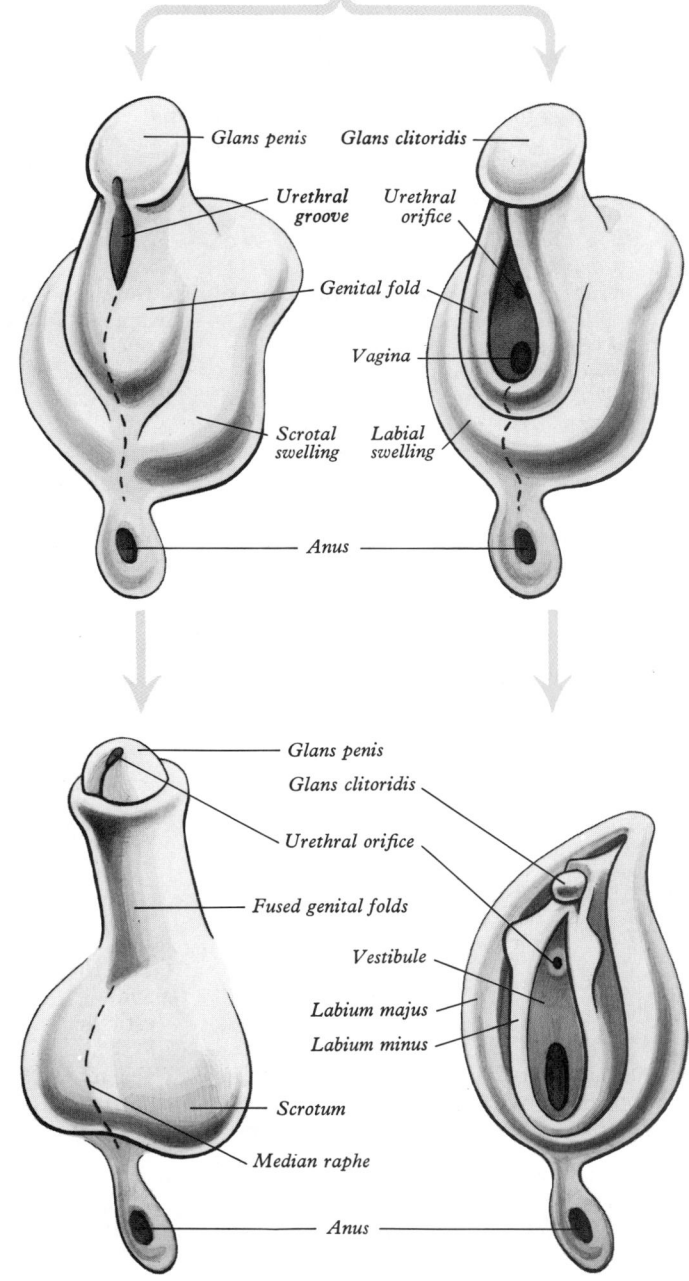

2.124 The development of the external genitalia from the indifferent stage to the definitive male and female conditions.

lower aspect of the plate is in contact with ectoderm lining a median groove, the *primary urethral groove*, which has meanwhile developed along the caudal surface of the phallus. The raised margins of the groove are the *genital folds* (**2.124**). Proximally they surround the urogenital membrane and terminate in a transverse ridge immediately ventral to the anus. The rupture of the urogenital membrane provides a common perineal orifice for the genital and urinary organs at the base of the phallus, bounded at the sides by the genital folds. Meanwhile disintegration of the cells of the urethral plate and contiguous ectoderm occurs, commencing at the base of the phallus and resulting in a deepening of what is now the *definitive urethral groove*.

While these changes are in progress two *labioscrotal (genital) swellings* have appeared, one on each side of the base of the phallus; these extend caudally, separated from the genital folds by distinct grooves (**2.124**).

The male phallus enlarges to form the penis, its apex being the glans. The genital swellings meet each other ventral to the anus and unite to form the scrotum. The genital folds fuse with each other from behind forwards enclosing the phallic part of the urogenital sinus behind to form the bulb of the urethra; similarly, the folds close the definitive urethral groove in front to form the greater part of the spongiose urethra. Fusion of the folds results in the formation of a median raphe and occurs in such a way that the lining of the postglandular urethra is mainly, perhaps wholly, *endodermal* in origin (Glenister 1954). Thus, as the phallus lengthens, the urogenital orifice is carried onwards until it reaches the glans. From the tip of the glans an ingrowth of a vertical plate of surface ectoderm has occurred to meet the anterior extremity of the urethral plate, and the disintegration of this gives rise to a glandular continuation of the urethral groove; closure of this completes the terminal part of the urethra contained within the glans, which is entirely *ectodermal* in origin.

The glans and shaft of the penis are hence recognizable by the third month. The prepuce also begins to develop in the third month, when the urethra still has its primary external orifice at the base of the glans. A ridge consisting of a mesodermal core covered by epithelium appears proximal to the neck of the penis and extends forwards over the glans. Deep to this ridge is a solid lamella of epithelium which extends backwards to the base of the glans. The ventral extremities of the ridge curve backwards to become continuous with the genital folds at the margins of the urethral orifice. As the urethral folds meet to form the terminal part of the urethra, the ventral horns of the ridge fuse to form the frenulum. Over the dorsum and sides of the glans, the epithelial lamella breaks down to form the preputial sac and thus free the prepuce

259

from the surface of the glans. Thereafter the prepuce grows as a free fold of skin covering the terminal part of the glans. The preputial sac may not be complete until 6 to 12 months or more *after birth* and, even then, the presence of some connecting strands may still interfere with the retractability of the prepuce.

The mesodermal core of the phallus is comparatively undifferentiated in the first two months, but during the third month the blastemata of the corpora cavernosa become defined. Nerves are present in the differentiating mesenchyme from the seventh week (Dail & Evan 1974).

The female phallus, which exceeds the male in length in the early stages, becomes the clitoris. The genital swellings remain separate as the labia majora and the genital folds also remain separate, forming the labia minora. The perineal orifice of the urogenital sinus is retained as the cleft between the labia minora, above which the urethra and vagina open. The prepuce of the clitoris develops in the same way as its male homologue.

The urethra, in the *female*, is derived entirely from the vesico-urethral region of the cloaca (p. 258), including the dorsal region derived from the mesonephric ducts. It is homologous with the part of the prostatic urethra proximal to the orifices of the prostatic utricle and the ejaculatory ducts.

In the *male*, the prostatic urethra proximal to the orifice of the prostatic utricle is derived from the vesico-urethral part of the cloaca and the incorporated caudal ends of the mesonephric ducts. The remainder of the prostatic part, the membranous part and probably the part within the bulb, are all derived from the urogenital sinus (p. 258). The succeeding section, as far as the glans, is formed by the fusion of the genital folds, while the section within the glans is formed from ectoderm (vide supra).

Homologies of the parts of the urogenital system in male and female (**2.**116,124) can be summarized:

Undifferentiated	Male	Female
Gonad	Testis	Ovary
Gubernacular cord	Gubernaculum testis	Ovarian and round ligaments
Mesonephros (Wolffian body)	Appendix of epididymis (?) Efferent ductules Lobules of epididymis	Appendices vesiculosae (?) Epoöphoron
	Paradidymis Aberrant ductules	Paroöphoron
Mesonephric duct (Wolffian duct)	Duct of epididymis Ductus deferens Ejaculatory duct	Duct of epoöphoron
	Part of bladder and prostatic urethra	Part of bladder and urethra
Paramesonephric (or Mullerian) duct	Appendix of testis	Uterine tube Uterus
	Prostatic utricle	Vagina (?)
Allantoic duct	Urachus	Urachus
Cloaca: dorsal part	Rectum and upper part of anal canal	Rectum and upper part of anal canal
ventral part	Most of bladder Part of prostatic urethra	Most of bladder and the urethra
urogenital sinus	Prostatic urethra distal to utricle	
	Bulbo-urethral glands	Greater vestibular glands
	Rest of urethra to glands	Vestibule
Genital folds	Ventral penis	Labia minora
Genital tubercle	Penis Urethra in glans	Clitoris

Congenital anomalies. Defects of the urethra, due to arrests of development, are not uncommon in the male. The urethra may open on the ventral (perineal) aspect of the penis at the base of the glans, and the part of the urethra which is normally within the glans is absent. This constitutes the simplest form of *hypospadias*. In more severe cases the genital folds fail to fuse, and the urethra

opens on the ventral aspect of a malformed penis just in front of the scrotum. A still greater degree of this malformation is accompanied by failure of the genital swellings to unite with each other. In these cases the scrotum is divided and, since the testes are also frequently undescended, the resemblance to the labia majora is very striking. Male children suffering from this deformity are often mistaken for girls.

Maldevelopment of the cloacal or urogenital membranes is a less common condition, but two varieties can be distinguished. (1) In *extroversion of the bladder* (ectopia vesicae) the lower part of the anterior abdominal wall is occupied by an irregularly oval area, covered with mucous membrane, on which the two ureters open (Wyburn 1937). Around its periphery this extroverted area, covered by urothelium, becomes continuous with the skin. (2) In *extroversion of the cloaca* the condition is very similar, but is complicated by the presence of intestinal openings in the median plane. In (2) the cloacal membrane is probably abnormally elongated and ruptures prematurely and throughout its whole extent, prior to the formation of the urorectal septum. In (1) the maldevelopment occurs after the separation of the ventral from the dorsal part of the cloaca. The urogenital membrane extends further cranially than it does in normal cases and the genital tubercle forms at its *caudal* limit. Rupture of the membrane throws the bladder open to the exterior.

In *epispadias* the urethra opens on the dorsal aspect of the penis at its junction with the anterior abdominal wall. No satisfactory explanation has yet been suggested for this anomaly. For a genetic and epidemiological study of urinary tract malformations consult Bois et al (1975).

Differentiation of the genital organs (Moore 1947, Burns 1955) is primarily directed by genetic factors many of which may mediate their effects through endocrine controls, notably the cortex of the suprarenal gland. The differentiation of the gonad may be modified by several factors including chromosomal constitution (p. 48), degree of ripeness of the ovum at fertilization, the number and distribution of the primordial germ cells and endocrine influences. A reduction in the number of primordial germ cells, believed to be associated with over-ripeness of the ovum at fertilization, exercises a masculinizing effect on genetic females. It is suggested that this is due to a failure of normal development of the superficial or cortical zone of the ovary, thus allowing the deep or medullary zone, which is the essentially male element, to dominate subsequent differentiation. The zones are not well marked in man but it is considered that they play an important role in the differentiation not only of the gonad but also of the genital ducts, mesonephric and paramesonephric, cloaca, external genitalia and eventually of the secondary sexual characters. During their development the accessory sexual organs also pass through an indifferent or potentially bisexual stage.

The part played by hormones in the differentiation of the reproductive organs is not clear, but both androgens and oestrogens profoundly modify this process, as instanced by work on the freemartin in cattle (Jost 1961). In twin pregnancy, where the fetuses are of opposite sex and their placental circulations anastomose, the male often exerts a masculinizing effect on its female twin. This has been attributed to the action of hormones elaborated in the interstitial tissue of the medullary zone in the testes of the male fetus. The gonad in turn affects the differentiation of the genital ducts, cloaca and external genitalia. It seems that the fate of the accessory organs, and particularly of the ducts, is determined largely by the presence or absence of a male factor, possibly hormonal, elaborated by the testes. The different parts of the accessory genital organs are most susceptible to this masculinizing effect over limited but differing periods.

Prenatal Growth in Form and Size

The absolute size of neither embryo nor fetus affords a reliable indication of either its true age or stage of structural organization even though graphs based on large numbers of observations have been constructed to provide averages. All such data suffer from the difficulty of equating dimensions and degree of differentiation with the actual time of conception, which can rarely, if ever, be established with complete exactness. The life of the individual really commences with fertilization, but the date of this cannot be exactly determined in mankind. It has long been customary to compute the age, whether in a normal birth or an abortion, from the last menstrual period of the mother but, since ovulation usually occurs near the fourteenth day of a period, this 'menstrual age' is about two weeks too much. Where a single coitus can be held to be responsible for conception, a 'coital age' can be established and the 'fertilization age' cannot be much less than this, because of the limited viability of both gametes; but it is usually held that the difference may be several days—a highly significant interval in the earlier stages of embryonic development. Even if the time of ovulation and coitus were known in instances of spontaneous abortion, not only would some uncertainty still persist with regard to the time of fertilization but there would also remain an indefinable period between the cessation of development and the actual recovery of the conceptus. With the legalization of abortion in some countries the latter source of inaccuracy may be expected to become less important.

To overcome these difficulties early embryos have been graded or classified, on the basis of both internal and external features, into developmental stages or 'horizons' (2.125A-H). Classic contributions in this field have been made by Lillie (1917), Streeter (1942-49), Hamilton (1944), Hertig et al (1956), Heuser & Corner (1957), and O'Rahilly et al (1966-75). Although it has become customary to describe these developmental levels as *stages*, the earlier descriptions are still accepted. However, Stages 1 to 9, covering the first three weeks of development, have been given a more reliable basis by O'Rahilly (1973), who has also gathered together the pertinent literature.

Stages 1 to 3 occupy the first 4–5 days after fertilization of the öocyte in the ampulla of the uterine tube. The initial 24 hours (Stage 1) are occupied by fertilization, the dominant feature of which is the fusion of the male and female pronuclei. This is followed by the first mitotic division, which is the onset of segmentation, or cleavage, and is arbitrarily regarded as the transition from Stage 1 to Stage 2. Stage 2 is characterized by the continuation of cleavage, starting with two blastomeres and ending with about 12. During this stage the developing morula moves along the uterine tube, by mechanisms still not wholly understood, a journey occupying about 4 days. During the fourth day a segmentation cavity appears within the morula and this is taken as initiating Stage 3, which thus corresponds to the establishment of a free blastocyst.

Although comparatively few human embryos representing Stages 1 to 3 have been recovered (see pp. 124–7 and consult O'Rahilly 1973), a considerable degree of agreement exists and, moreover, current observations of in vitro fertilization between human spermatozoa and ova (2.20A-H) support the earlier descriptions.

Stages 4 to 6 are concerned with the endometrial attachment of the blastocyst, trophoblastic development, implantation, further development of the blastocyst and the appearance of the primitive streak (2.20-24). Stage 4 corresponds to the fifth and sixth post-fertilization days. The blastocyst, now in the uterine cavity, loses its zona pellucida and it begins the rapid but complex activities of orientation in respect of the endometrium—adhesion, penetration and the cellular proliferation of trophoblastic growth. This establishment of dependence on the maternal circulation for nutritional requirements is far more rapid in primates than in other mammals. Stage 5 is reached when implantation has occurred, occupying the seventh to twelfth days; syncytiotrophoblastic and cytotrophoblastic strata have differentiated, the proamniotic cavity has appeared and a labyrinthine system of intercommunicating trophoblastic lacunae through which the maternal blood ebbs and flows, has developed. A little later in Stage 5 the exocoelomic membrane has been identified. In Stage 6 chorionic villous stems become defined and begin to develop side branches almost at once, producing an increasingly complex intervillous space. A little later the primitive streak becomes apparent and differentiation of the embryonic area has commenced and may now be distinguished from the various extraembryonic tissues.

Because of the complexity of the events in Stages 5 and 6, both have been subdivided by some authorities. A much greater number of human embryos have been recovered which represent Stages 4 to 6.

Stages 7 to 9 are characterized by basic embryogenic changes. During Stage 7 (sixteenth to eighteenth days) the primitive streak develops further and the notochordal ('head') process appears, together with the other mesodermal strata. The chorion and amnion continue to develop, villous stems being generally distributed over the former but more pronounced at the embryonic pole. Haemopoetic foci appear in the wall of the definitive yolk sac, and the cloacal membrane and allanto-enteric diverticulum are defined. Associated with the latter primordial germ cells have been noted. In Stage 8 (eighteenth to twentieth days) the prechordal plate, primitive pit, neural groove and notochordal and neurenteric canals are all definable. By Stage 9 (twentieth to twenty-second days) the neural groove is deepening and the first somites begin to appear about midway along it. The cranial half of the groove, representing developing brain, begins to develop a cephalic flexure, optic primordia become visible and early head and tail folds have appeared, as Stage 10 is approached. The foregut is becoming defined, and early pharyngeal pouches may be identified. The embryo is now 1.5–2.0 mm in length.

Stages 10 to 12, occupying days 22 to 26, feature continued formation of somites and, during this, the fourth week, head and tail folds are completed, the neural groove closes and primary cerebral vesicles appear. The cervical flexure can now be recognized, the optic vesicles form and the lental and otic placodes appear and become vesicular. The branchial arches are appearing, lateral folds are more clearly defined and the cloacal membrane and hindgut are becoming distinct. Rudimentary limb buds appear and the heart tubes fuse into a common loop in which contractile activity commences. The primordia of the thyroid gland, lungs, liver, pancreas and mesonephric tubules are all identifiable. The embryo is about 4.0 mm in length.

Stages 13 to 16 correspond approximately to the fifth week of development and the embryo grows from about 4 mm to 8 mm. It becomes markedly curved and its junction with the yolk sac relatively constricted. The cervical flexure is increased and the mesencephalic flexure is appearing. The dorsolateral (alar) and ventrolateral (basal) laminae are differentiating, and the emerging corpus striatum, thalamus epithalamus and hypothalamus are loci of proliferating cells. The cranial and spinal nerves are developing, together with their ganglia, from the associated placodes and neural crest elements. The limb buds are elongating, displaying joint flexures; the rudimentary hands and feet are differentiating. The olfactory placodes, maxillary, mandibular and frontonasal prominences, tongue primordia and the hypophysial pouch (of Rathke) are all appearing. The tubotympanic recesses are defined and the primordia of the thymus and parathyroid glands can be identified. In the developing heart the septum primum appears and its cornua define the foramen primum. The mesonephric ducts reach the cloaca and subsequently the metanephric (ureteric) buds appear and extend to the metanephrogenic masses of mesoderm in the sacral nephrogenic cord. The gonadal ridges, urorectal septum and genital tubercle are also developing.

Stages 16 to 20 are roughly equivalent to the sixth week, by the end of which the embryo has a length of about 13–15 mm. Its curvature has further increased and its head and relatively long tail are in contact with the developing umbilical stump. The pontine flexure, cerebral hemispheres and cerebellum are developing. The upper limbs and the facial region are growing and differentiating rapidly; the palatal processes and primitive nasal prominences are apparent and the oronasal membrane ruptures. The liver produces a surface prominence between the cardiac region and the umbilical cord. Into the latter the midgut loop herniates and the appendix and caecum become distinguishable in it. The spleen develops, as do the paramesonephric (Mullerian) ducts. The foramen primum in the heart closes, with simultaneous opening of the foramen secundum and septation of the bulbus cordis occurs. Cardiac muscle is differentiating. Haemopoiesis commences in the liver. Chondrification of many skeletal elements begins and ossification commences in mesenchymatous bones, the mandible and clavicles.

Stages 21 to 23 complete the series, extending to about day 47. This represents the major part of the seventh week and, with the

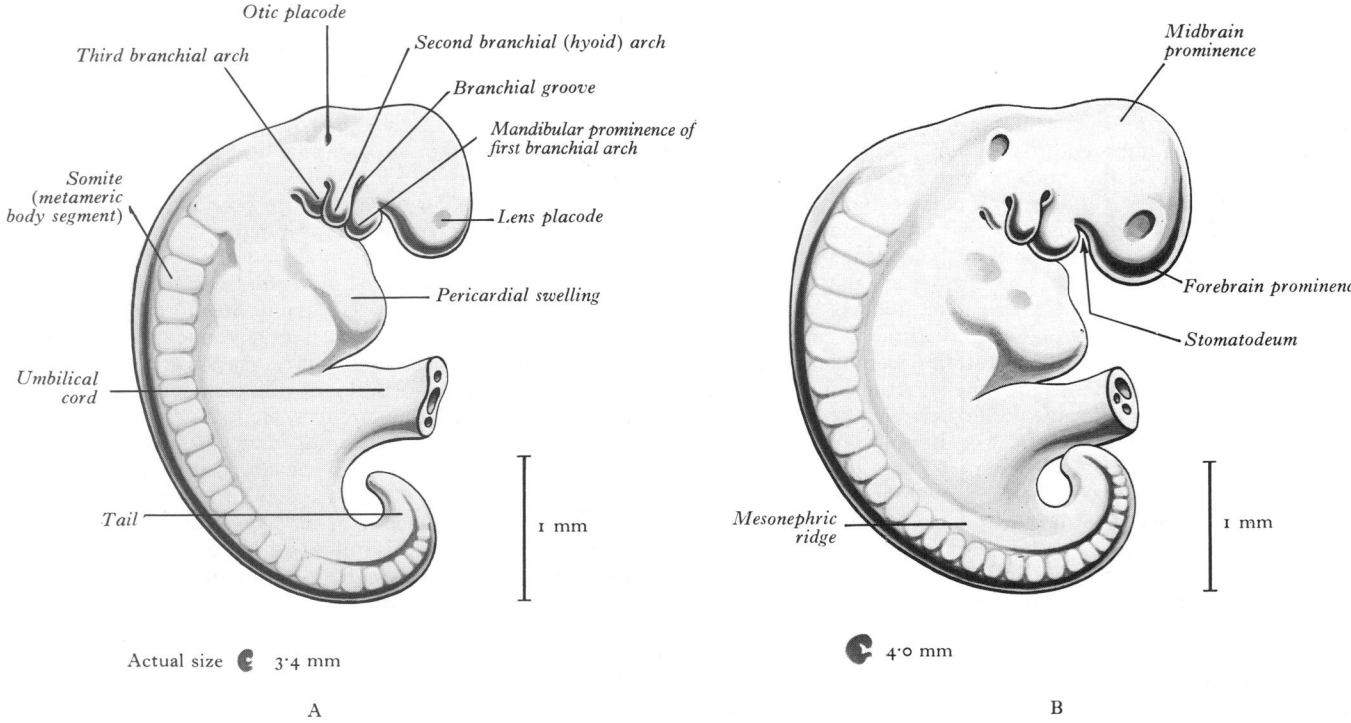

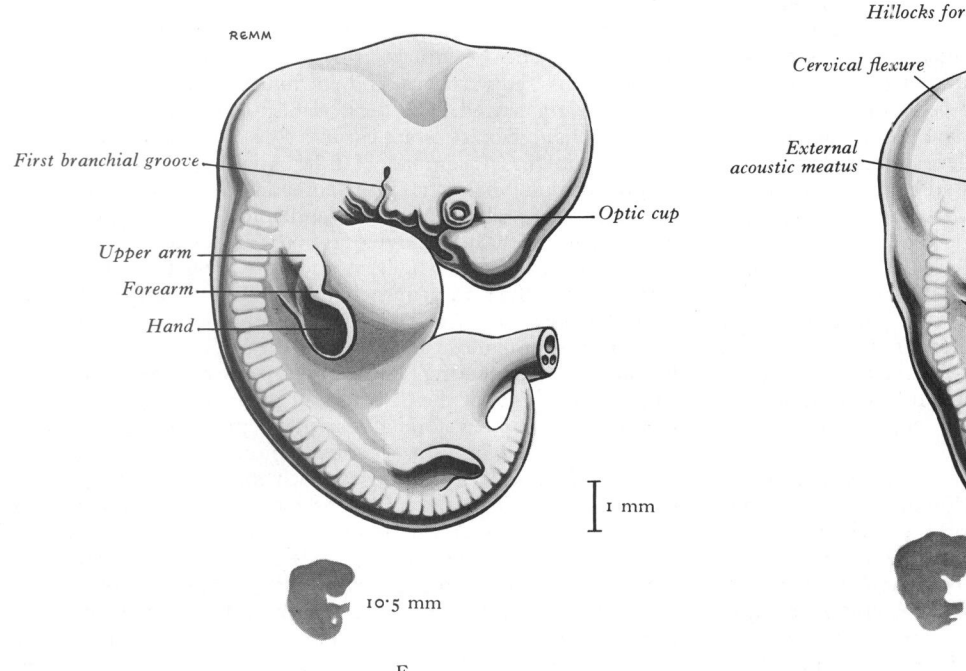

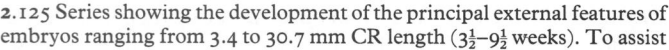

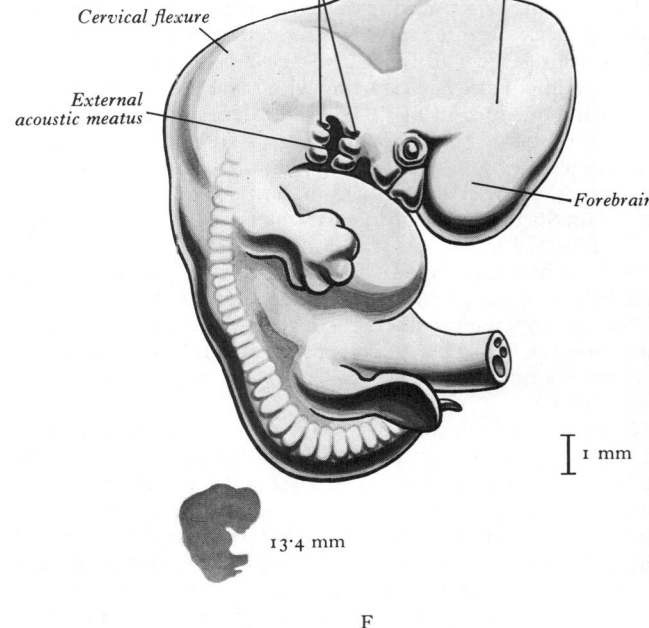

2.125 Series showing the development of the principal external features of embryos ranging from 3.4 to 30.7 mm CR length (3½–9½ weeks). To assist comparison a 1 mm scale is included in each case; the small silhouette is actual size.

eighth week, the *formative* or *embryonic period* is regarded as coming to an end (**2.125**A–H). This is a period during which *patterned differentiation* with consequent *organogenesis* tends to overshadow the growth which accompanies these events. Very considerable increase in size has, however, occurred; from a single cell about 0.14 mm in diameter the embryo has become a most complex and functioning creature, consisting of millions of cells and with a length of about 30 mm or more, and it has increased in weight many thousands of times. During the *fetal period*, which occupies the third to tenth lunar months (**2.126,127**), the accent is upon growth rather than differentiation but, of course, the latter

continues through this period and to a lesser degree after birth (and in some tissues, throughout life). During this period the overall rate of growth in length is greater, but not markedly so, from the fourth to sixth weeks the rate is about 1 mm per day, with a maximum of about 2 mm during the fourth month. The increase in length in the fetal period is from 30/40 to about 500 mm, and the increase in weight from perhaps 2 or 3 g to more than 3000 g.

During the *seventh* and *eighth weeks* there is a remarkable change in the external appearance of the embryo, for at the beginning of this period the individual still appears markedly 'embryonic', though clearly a primate, whereas at the end the form i

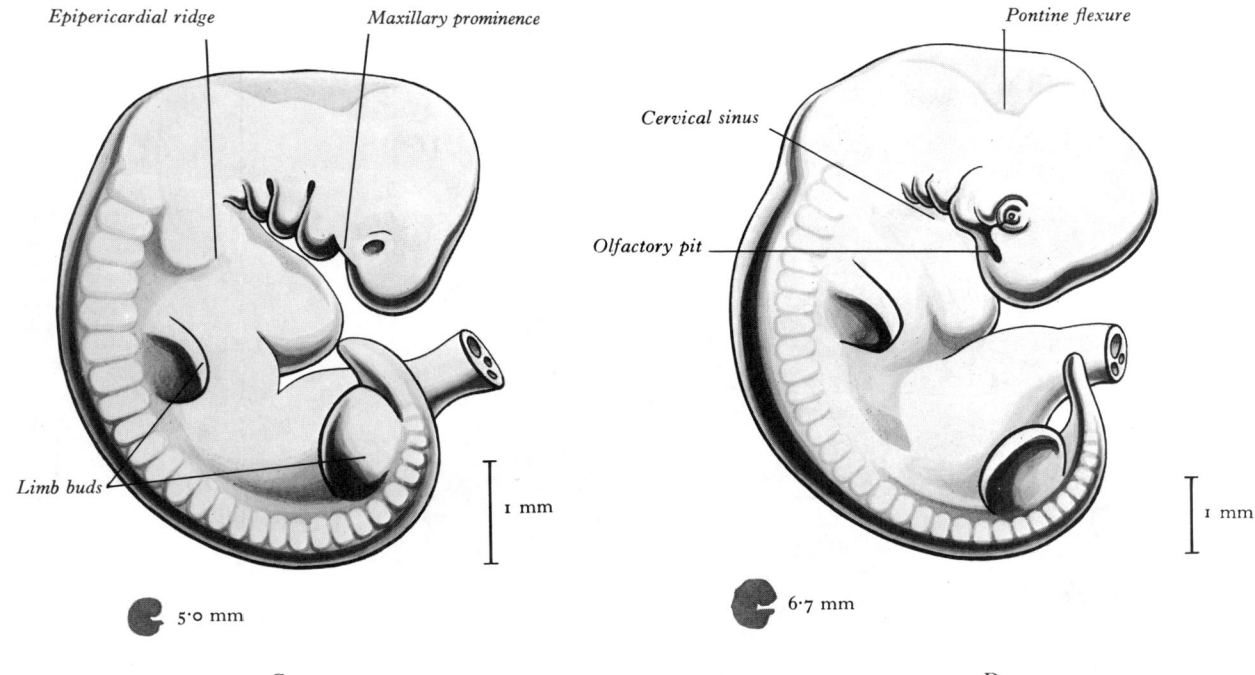

Epipericardial ridge   *Maxillary prominence*

Pontine flexure

Cervical sinus

Olfactory pit

Limb buds

1 mm

5·0 mm

C

1 mm

6·7 mm

D

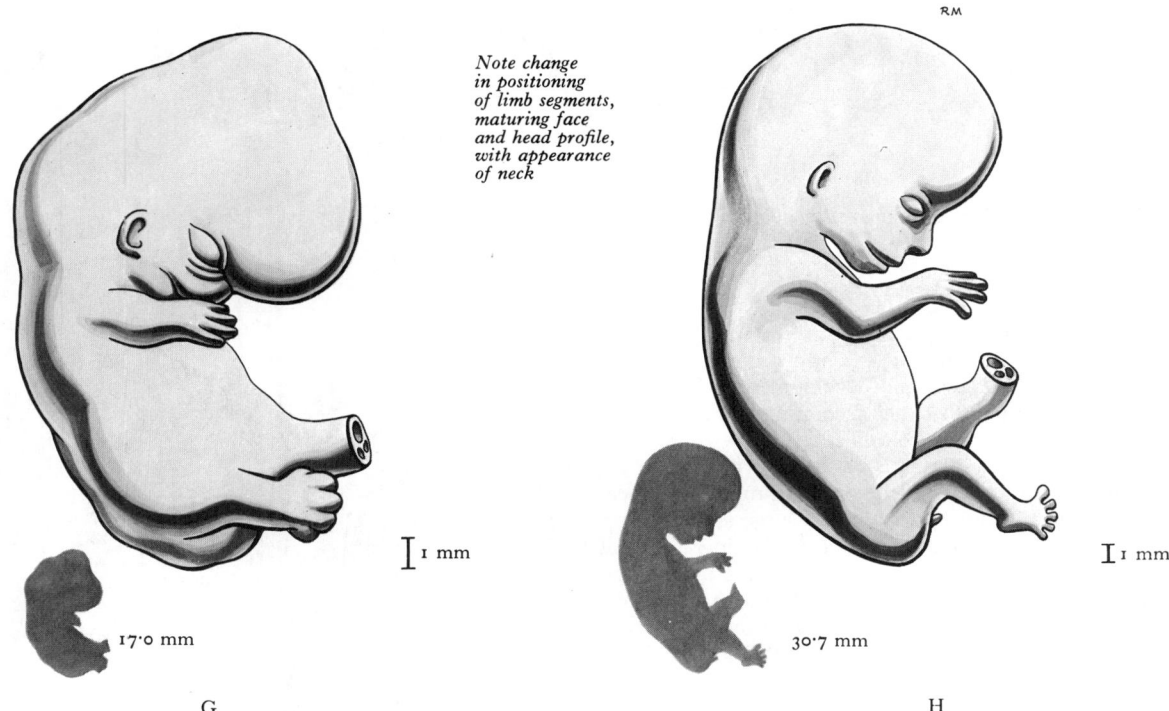

Note change in positioning of limb segments, maturing face and head profile, with appearance of neck

RM

1 mm

17·0 mm

G

1 mm

30·7 mm

H

most definitely human. The head is less flexed and the neck longer and clearly defined. The development of the face proceeds much further, with completion of the upper lip and nostrils, although the latter are plugged and the palate still incomplete. Enamel organs are developed from the dental laminae. The external ears and the eyelids are developing and the limbs elongate considerably, approaching much nearer their ultimate proportions and displaying well-formed hands and feet with separated digits. Early in this period the interventricular septum is completed. Skeletal and visceral muscle tissues begin to differentiate about this time, and generalized ossification occurs in enchondral bones. In the metanephrogenic mass vesicles appear and the remainder of the nephrons and the collecting tubules of the kidneys are defined. The

ovaries or testes are distinguishable and the paramesonephric (Mullerian) ducts are fusing to form the primordia of uterus and vagina. The external genitalia are further advanced and show sexual differentiation by the beginning of the eighth week. The cloacal membrane becomes perforate and the tail is retrogressing. By the end of the period the embryo possesses almost all the structural features, internal and external, characteristic of the human mammal and it now passes into the fetal period. During this there is much growth in the dimensions of the established organs, extensive changes in proportions brought about by variations of growth rate and widespread maturation in differentiating cells.

During the *third month* head flexion decreases further and the neck becomes proportionately longer. The eyelids meet and fuse

263

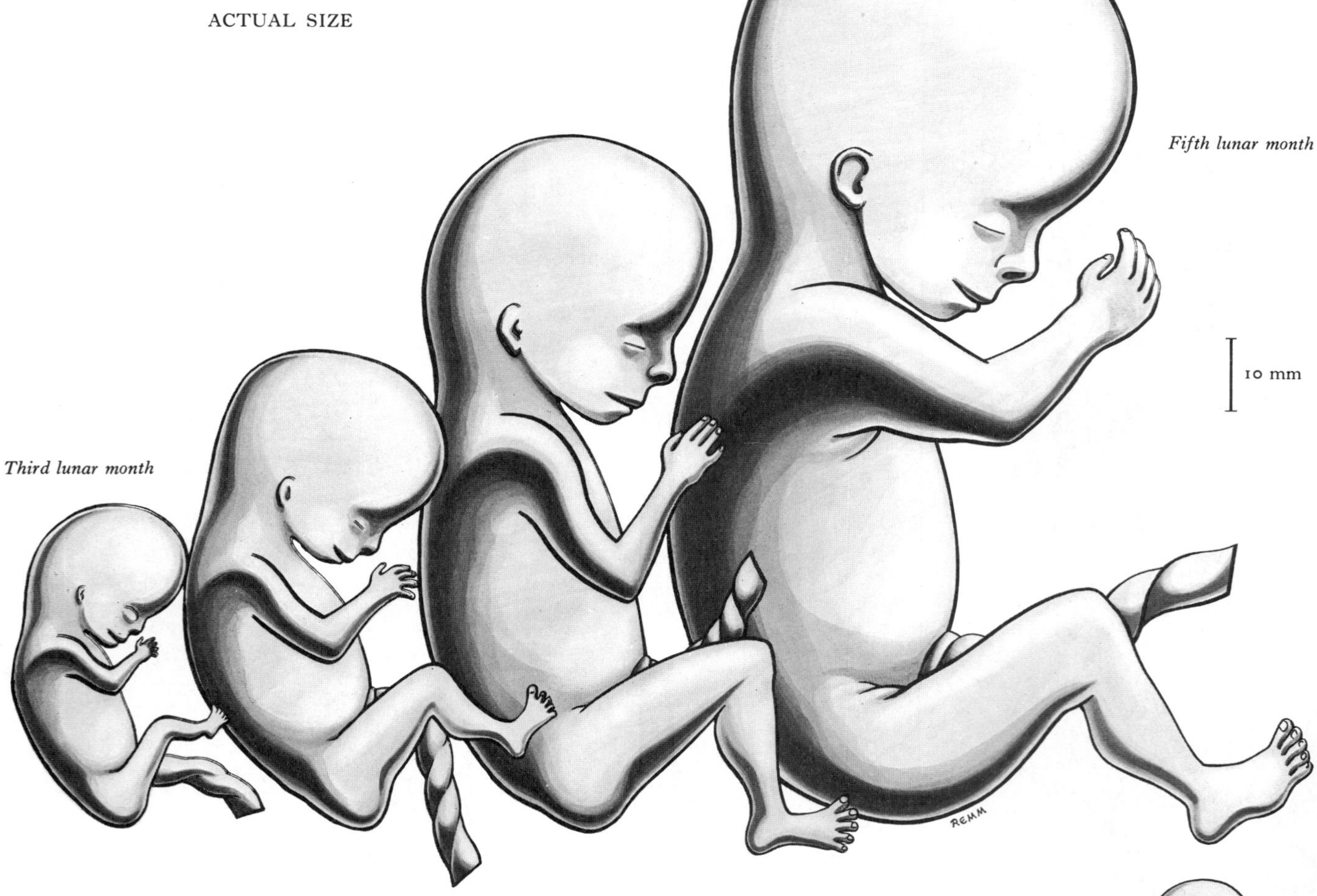

ACTUAL SIZE

Fifth lunar month

Third lunar month

10 mm

2.126 The progressive changes in fetal size and proportions during the third, fourth and fifth months.

and will remain temporarily united until the sixth lunar month. Nails appear on the digits and the upper limbs in general are comparatively accelerated in development. The umbilical protrusion of the gut is reduced accompanying a proportionate augmentation of abdominal volume.

During the *fourth month* the covering of primary hair appears —the *lanugo* (see p. 91). As in the previous month, the head and upper limbs are still disproportionately large and, although the trunk and lower limbs begin to catch up by increased rates of growth during the rest of uterine life, the same disproportion is present after birth and to a diminishing degree throughout childhood and on into the years of puberty. By the end of the fourth month the eyes have moved even further into an anteriorly directed position, but are still relatively wide apart. The external ear is approaching its characteristic form and is nearer its ultimate position, at the side of the head and no longer in the upper part of the neck. The total fetal length, including the lower limb, is now of the order of 230 mm. Its weight at the end of the *fifth month* is about 300 g, which will be increased more than tenfold during the second half of intrauterine life. Towards the end of this period sebaceous glands become active, and the sebum secreted blends with desquamated epidermal cells to form a cheesy covering to the skin, the *vernix caseosa*, usually considered to protect the former from maceration by the amniotic fluid. During this month the mother becomes conscious of fetal movement—so-called 'quickening'.

The *sixth month* witnesses a further general change of bodily proportions and facial appearance towards those of the infant at birth. The lanugo darkens and the skin becomes markedly wrink-led, presumably through a disparity in the growth rates of

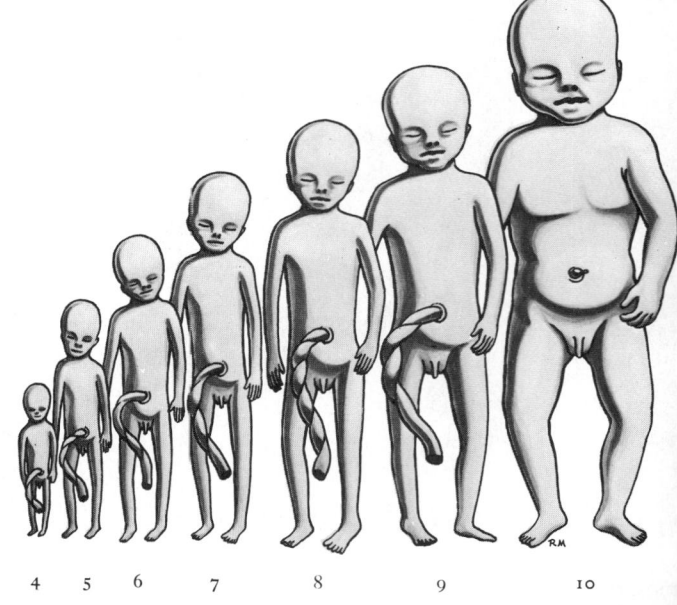

4 5 6 7 8 9 10

LUNAR MONTHS

2.127 Changes in *relative* size and bodily proportions at the fetal ages indicated.

cutaneous and subcutaneous tissues. The eyelids and eyebrows are now well developed. The vernix caseosa is more abundant. The length of the fetus is about 300 mm by the end of this month.

During the *seventh month* the hair of the scalp is lengthening and the eyebrow hairs and the eyelashes are well-developed. The eyelids themselves separate and the pupillary membrane disappears. The body becomes more plump and rounded in contour and the skin loses its wrinkled appearance due to increased deposition of subcutaneous fat. Towards the end of this month the fetus is viable and may in fact be successfully raised if born prematurely. Its length has increased to about 350 mm and it weighs about 1.5 kg.

Throughout the remaining three lunar months of normal gestation the covering of vernix caseosa is prominent. There is a progressive loss of lanugo, except for the hairs on the eyelids, eyebrows and scalp. The bodily shape is becoming more infantile, but despite some acceleration in its growth the leg has not quite equalled the arm in length proportionately even at the time of birth. The thorax broadens relative to the head, and the infra-umbilical abdominal wall shows a relative areal increase, so that the umbilicus gradually becomes more centrally situated. Average lengths and weights for the eighth, ninth and tenth months are 40, 45 and 50 cm and 2, 2.5 and 3 to 3.5 kg.

At the end of the *tenth lunar month*, just before birth, the lanugo has almost disappeared, the umbilicus is central and the testes, which begin to descend with the vaginal process of peritoneum during the seventh month and are approaching the scrotum in the ninth month, are usually scrotal in position. The ovaries are not yet in their final position at birth; although they have attained their final relationship to the uterine folds, they are still above the level of the pelvic brim.

The length of the *period of gestation* is regarded as 9 calendar months in obstetric practice—approximately 270 days. It is usually about 266 days—10 lunar months less 14 days. Legally, the infant is regarded as viable if born after the end of 7 calendar months but considerably more premature infants have been successfully reared.

OSTEOLOGY

3

MORPHOLOGY OF THE HUMAN SKELETON

Introduction

The human skeleton is bilaterally symmetrical (**3**.1,2,4), with the typical vertebrate pattern of an axis, divided into segments for flexibility, and of two pairs of limbs, pectoral and pelvic, also divided into jointed parts for locomotion, grasping, etc. The skull is the expanded and modified cranial end of the axis. Osseocartilaginous sesamoid bones develop in some tendons and ligaments. All these elements are collectively termed the 'skeleton'.

The human skeleton, as in other vertebrates, is internal to the muscles with which it has evolved. It is an *endoskeleton*, unlike the *exoskeleton* of many invertebrates, such as *Insecta*, whose muscles are attached to the internal aspects of jointed elements of chitin, a rigid material also offering protection. Many features of the human endoskeleton are also deemed defensive, with an excess of emphasis sometimes sufficient to obscure its primary association with muscle and hence with movement. Perhaps only in the vault of the skull and spinal column is the protective role paramount, but even here muscle attachments appear. Cranially, its superficial situation reflects its evolutionary origin, as generally accepted, from the dermal bony armour of earlier vertebrates, including fish, amphibians and reptiles, extinct and extant. The ossified dermal carapace of the tortoise is a familiar example. The maxilla, mandible, clavicle and dentine of teeth are also dermal derivatives; all are vestiges of more extensive assemblies of dermal bones from which they have been modified to form a human 'exoskeleton'.

Homologues of the branchial skeleton of fish also contribute a *visceral* component in higher vertebrates, including man. Much modified, these elements appear as ear ossicles, part of the mandible, maxilla, styloid process and hyoid bone. Some caudal branchial arches persist as the cartilaginous larynx, not customarily included in the human skeleton; supportive tracheal and bronchial cartilages maintain the larger respiratory lumina, much as do bones and cartilages of nasal cavities. In some ungulates bone develops in cardiac connective tissue as an *os cordis*, and an *os penis* occurs in many mammals. The human skeleton is thus a complex, derived from original endoskeleton, exoskeletal dermal elements, modified branchial arches, and ossification in structures such as tendons.

The Skeleton in Life

The living skeleton also includes other, non-osseous structures which are lost when bones are preserved for study since attached muscles, ligaments, periosteum and cartilages are all removed. Costal cartilages are of course grouped with the skeleton; but articular cartilages, functionally prominent parts of most living bones, are by custom excluded, as are ligaments, menisci and intervertebral discs. A macerated skeleton, therefore, is completely disjointed into elements which, while convenient for examination, have lost most of their functional implications. The bone marrow also disappears, and mechanical properties, such as elasticity, proper to living bone, are lost (Smith & Walmsley 1959).

The properties of living bone have been much studied (Bell 1956) at macroscopic, microscopic and ultrastructural levels, especially in relation to mechanical factors. Its intimate blend of hard inorganic and resilient organic components, resistant almost equally to compression and tension, differs from most materials used by man, which are usually better in one respect than the other. In tensile strength bone resembles cast iron, but with only a third of its weight, breaking stresses being respectively 15.5 and 18 metric tons per square inch: about 2400 to 3000 kg per cm² (Bell et al 1941, Tables on p. 278). In flexibility bone resembles steel more than iron, but only has half the strength of steel. In compression it has large margins of safety for weight-bearing (Koch 1917) and

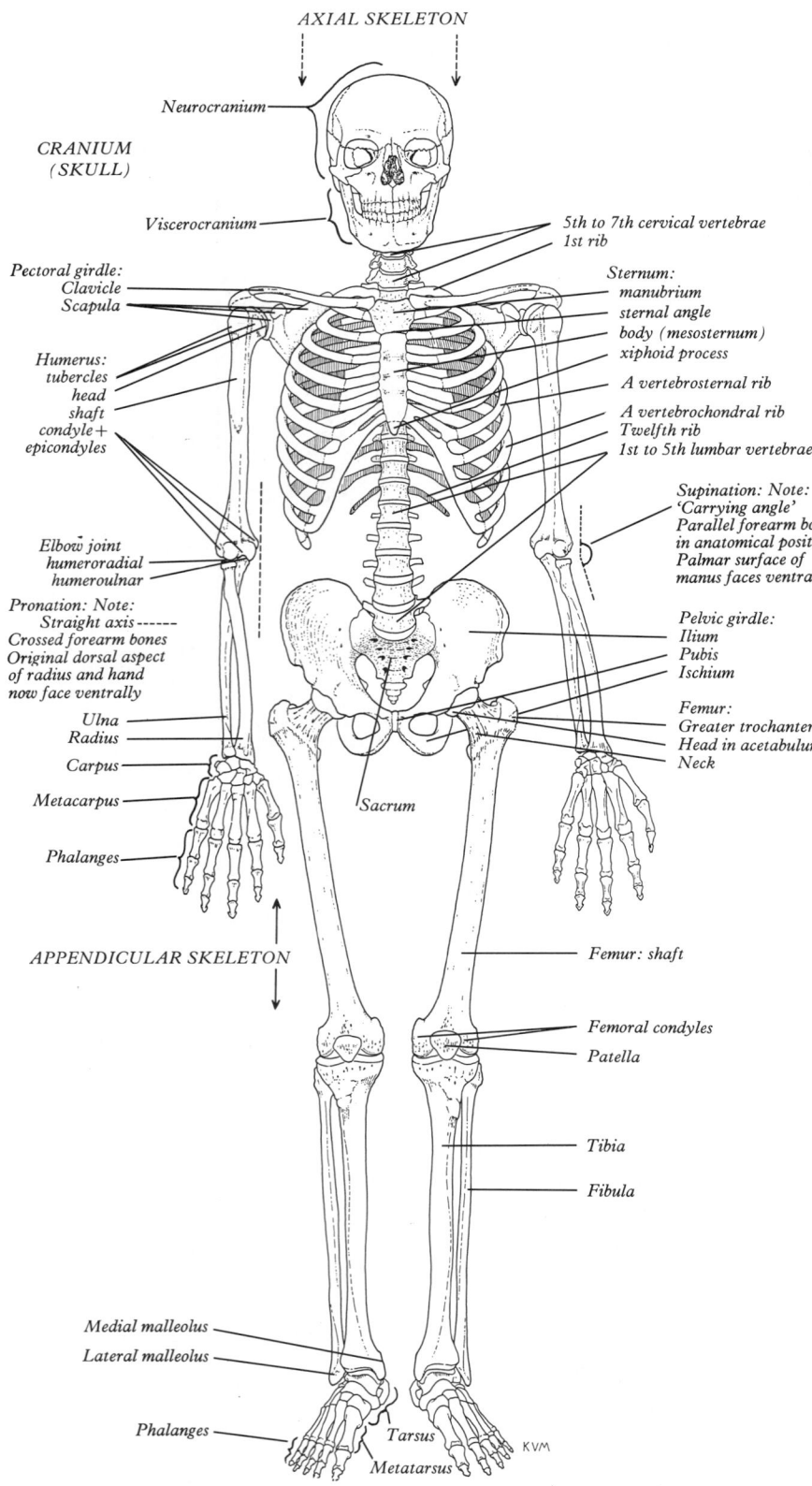

AXIAL SKELETON

Neurocranium

CRANIUM
(SKULL)

Viscerocranium

5th to 7th cervical vertebrae
1st rib

Pectoral girdle:
Clavicle
Scapula

Sternum:
manubrium
sternal angle
body (mesosternum)
xiphoid process

Humerus:
tubercles
head
shaft
condyle +
epicondyles

A vertebrosternal rib

A vertebrochondral rib
Twelfth rib
1st to 5th lumbar vertebrae

Supination: Note:
'Carrying angle'
Parallel forearm bones
in anatomical position
Palmar surface of
manus faces ventrally

Elbow joint
humeroradial
humeroulnar

Pronation: Note:
Straight axis
Crossed forearm bones
Original dorsal aspect
of radius and hand
now face ventrally

Ulna
Radius
Carpus

Pelvic girdle:
Ilium
Pubis
Ischium

Femur:
Greater trochanter
Head in acetabulum
Neck

Metacarpus

Phalanges

Sacrum

APPENDICULAR SKELETON

Femur: shaft

Femoral condyles
Patella

Tibia

Fibula

Medial malleolus
Lateral malleolus

Phalanges
Tarsus
Metatarsus

KVM

3.1 Adult male skeleton: anterior (ventral) view.

for impact. Contracting muscles are much the larger agent of pressure, even at weight-bearing joints, especially in active movement (Bell et al 1941, Williams & Svensson 1968). In hip joints a mere fraction of the pressure is due to body weight (vide infra).

A tubular structure, typical of shafts in many bones, is the strongest, lightest and hence most economical arrangement of material. This adaptation of form to habitual stresses is also seen in the predominant longitudinal orientation of osteons in elongate bones. Thickness of compact cortical bone is greatest at mid-shaft, where torsional and bending stresses are most severe and internal trabecular structure largely absent. At articular regions bones must chiefly withstand compression, often surprisingly large. In symmetrical standing each hip joint takes half the body weight, muscular pull multiplying this by a factor of about six; in walking or running, full body weight (except that of the weight-bearing leg) impinges alternately on each hip joint. Conservatively, in adult males, this yields a total load of 270 kg (600 lb), which may double in powerful exertion. In articular regions bone structure differs from that in shafts, the interior being entirely trabecular, supporting a thin shell of compact bone, an arrangement occurring also in smaller bones, e.g. carpal and tarsal, and vertebral bodies, all of which are mostly stressed by compression. In joints where weight-bearing is slight, pressure may still be large during action, due to the pull of muscles.

The trabeculae of cancellous bone, though individually small, collectively provide powerful support to the thin shells of compact bone, a form of construction widespread in vertebrate skeletons, only modified where bending, twisting and tensile forces demand larger masses of compact bone. Many examples of both arrangements will be noted in individual bones; most larger ones show both forms, and the intermediate arrays adapt to local mechanical needs. Mechanical forces influence growth and form in bones; and during protracted evolutionary time a most apt solution to mechanical demands has been reached, in compliance with nutrition, muscle power, and the best compromise between size and weight (p. 276).

Sections of trabecular bone (3.5,8,21,22) show patterns which resemble criss-crossing girderwork. Attempts to equate these patterns in particular bones, such as the femur and calcaneus, with lines of force in habitual stresses were made early (Ward 1838). Such architectural explanations of structure have been criticized (e.g. Murray 1936), but studies of individual bones, or even of the entire skeleton (Hall 1966), show an intimate correlation between stress and structure in trabecular *and* cortical bone. Wherever stresses are locally applied, as in attachments of tendons and ligaments, sections show that external features, e.g. ridges or facets, are not the only local changes. Compact cortical bone may also increase, with subjacent condensation of trabeculae, often pronounced. It is certain that patterns of stress and trabeculation are related, whatever the precise terms of the equation (p. 278). *Surface* features of bones are clear to the naked eye (3.5); but much more detail and variation is revealed by a hand lens, macrophotography, incident light microscopy and scanning electron microscopy (3.6).

The Shape and Proportions of Bones

Since bones vary in shape their gross appearances have led to a traditional grouping into long, short, flat and irregular bones. *Long* bones are typical of limbs, length reflecting degree of speed and power in movement. Their tubular shafts (diaphyses) have a central medullary cavity and they diverge (metaphyses) towards their expanded articular ends, which are epiphyses with separate centres of ossification, often multiple (e.g. humerus and femur). Metacarpals, metatarsals and phalanges are smaller examples, but they are proportionately greater in diameter of shaft and possess but a single epiphysis. So-called *short* bones occur in the carpus and tarsus, e.g. those appropriately termed cuboid, cuneiform, trapezoid or scaphoid. Being subject to pressure more than other stresses, they have typically a thin cortex of compact bone, supported by an interior which is wholly trabecular. *Flat* bones include curved laminar members of the cranial vault, in which trabecular bone, *diploë*, variable in thickness, is enclosed between

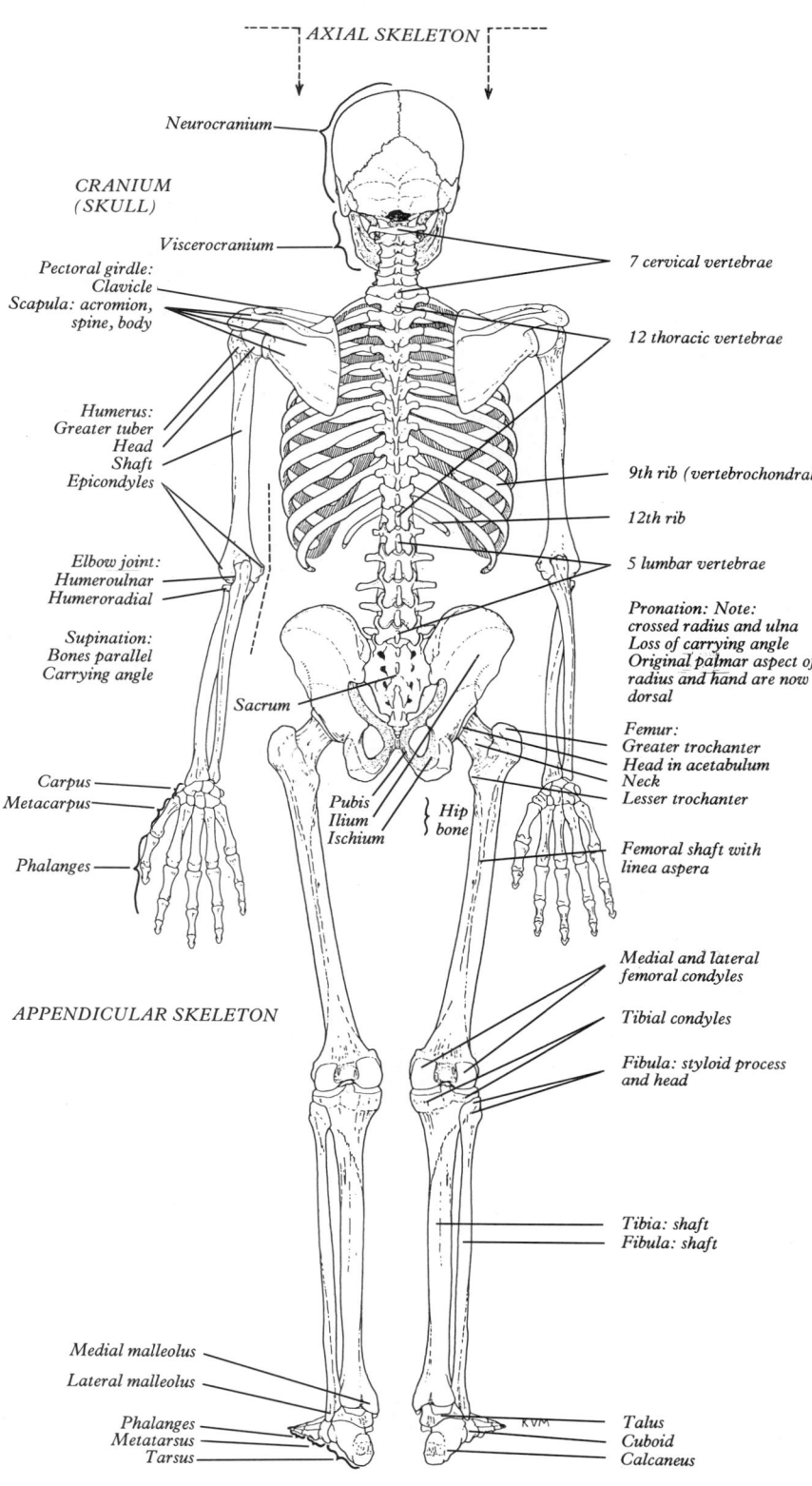

AXIAL SKELETON

Neurocranium

CRANIUM (SKULL)

Viscerocranium

Pectoral girdle:
Clavicle
Scapula: acromion, spine, body

Humerus:
Greater tuber
Head
Shaft
Epicondyles

Elbow joint:
Humeroulnar
Humeroradial

Supination:
Bones parallel
Carrying angle

Sacrum

Carpus
Metacarpus

Phalanges

Pubis
Ilium
Ischium

} Hip bone

7 cervical vertebrae

12 thoracic vertebrae

9th rib (vertebrochondral)

12th rib

5 lumbar vertebrae

Pronation: Note:
crossed radius and ulna
Loss of carrying angle
Original palmar aspect of
radius and hand are now
dorsal

Femur:
Greater trochanter
Head in acetabulum
Neck
Lesser trochanter

Femoral shaft with
linea aspera

Medial and lateral
femoral condyles

Tibial condyles

Fibula: styloid process
and head

APPENDICULAR SKELETON

Tibia: shaft
Fibula: shaft

Medial malleolus

Lateral malleolus

Phalanges
Metatarsus
Tarsus

Talus
Cuboid
Calcaneus

3.2 Adult male skeleton: posterior (dorsal) view.

269

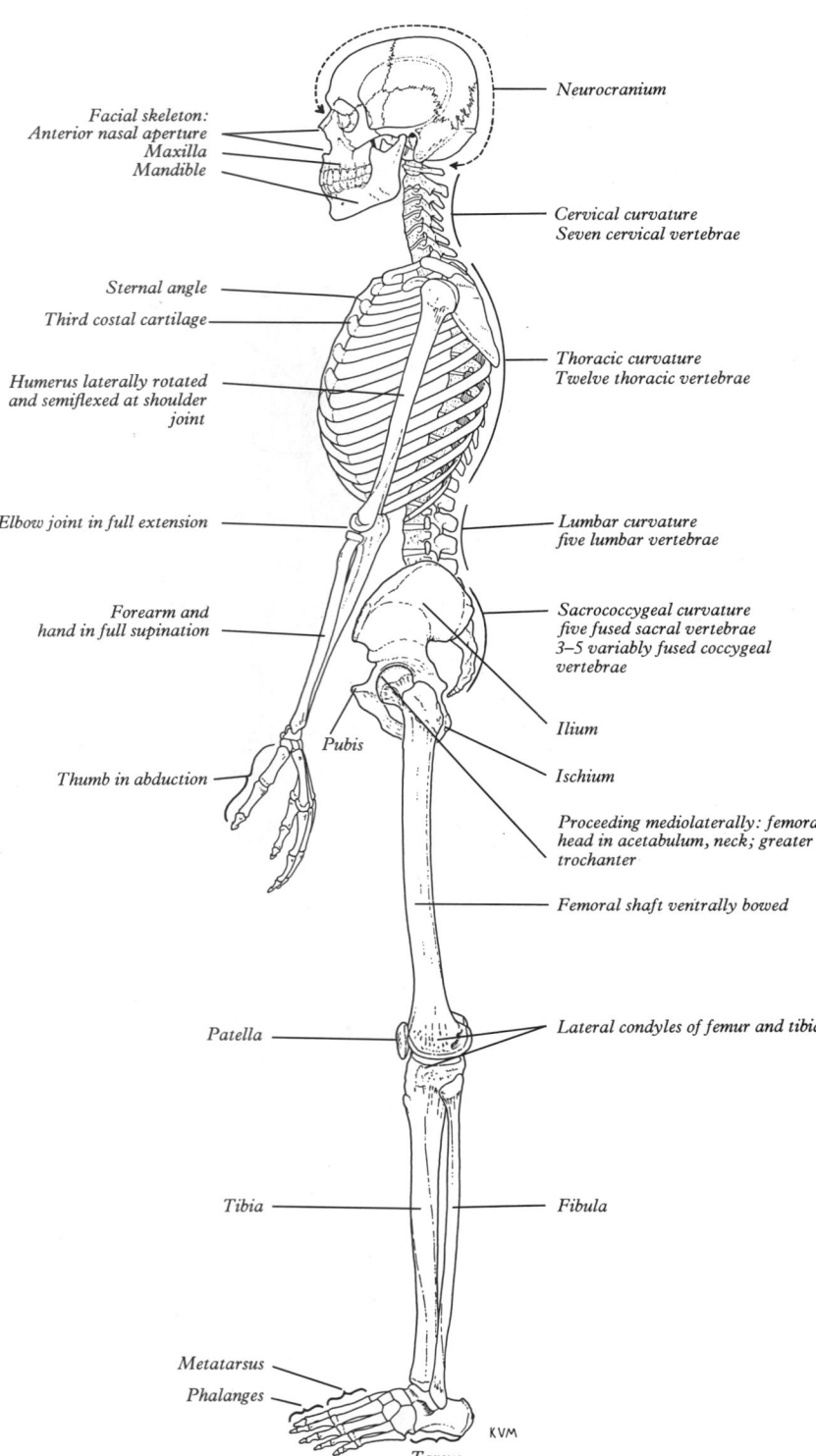

Facial skeleton:
Anterior nasal aperture
Maxilla
Mandible

Neurocranium

Cervical curvature
Seven cervical vertebrae

Sternal angle

Third costal cartilage

Thoracic curvature
Twelve thoracic vertebrae

Humerus laterally rotated
and semiflexed at shoulder
joint

Elbow joint in full extension

Lumbar curvature
five lumbar vertebrae

Forearm and
hand in full supination

Sacrococcygeal curvature
five fused sacral vertebrae
3–5 variably fused coccygeal
vertebrae

Ilium

Pubis

Ischium

Thumb in abduction

Proceeding mediolaterally: femoral
head in acetabulum, neck; greater
trochanter

Femoral shaft ventrally bowed

Patella

Lateral condyles of femur and tibia

Tibia

Fibula

Metatarsus

Phalanges

KVM

Tarsus

3.3 Adult male skeleton: lateral view.

laminae, or *tables*, of compact structure (**3**.5). The scapulae, despite their irregular form, are described as flat. *Irregular bones* include any element not easily assigned to foregoing groups. This time-honoured classification has no great merit. Bones must be studied individually, and in relation to functional demands.

In shape and structure, bones are affected by genetic, metabolic and mechanical factors, and each bone results from a long functional history through countless successive generations. Genetic determination of primary shape has been demonstrated by organ culture and transplantation of embryonic skeletal tissues (Murray & Huxley 1924, Fell & Canti 1934, Willis 1936), and all major characteristics appear to be self-determined, as perhaps is to be expected. Mechanical influences, such as moulding by muscular activity, often regarded as of first importance, are not operative when the primary form is being established. As muscles become active during prenatal life, they may influence bone growth, but to an extent difficult to estimate; after birth, up to adolescence, before all the epiphyses are fused, increased activity augments growth in length and girth. The reduced bone growth in limbs affected by paralyses, e.g. poliomyelitis, implies that activity is necessary to skeletal development. Experimental studies (Appleton 1934, Washburn 1947, Wolffson 1950), involving removal of muscles, point in the same direction. Again, increase in strength and stature show some temporal correlation in adolescents (Jones 1949).

Metabolic influences affect bone growth at all stages of development. The availability of calcium, phosphorus, vitamins A, C, and D, and secretions of the hypophysis, thyroid, parathyroid, adrenal glands and gonads, are all essential to osteogenesis (vide infra) and hence to skeletal form and dimensions. Disturbances in such factors entail recognized pathological states, but it is sometimes impossible to distinguish pathological from normal variation. Body height is an example: between the extremes of dwarfism and gigantism, resulting from hormonal dysfunction, much variation in height occurs. Variations in stature and other dimensions linked with age, sex and race are in part genetically conditioned; but racial variations in nutrition also have profound effects (Greulich 1951, Acheson 1960, Tanner 1962).

Bodily proportions, and absolute dimensions, vary widely in respect of age and sex (**3**.7) within racial groups, and between them. While partly due to variable muscularity and adiposity, such variations are chiefly skeletal; their study is *anthropometry* (Martin 1928, Hrdlička 1939). The data of anthropometry may be non-metrical, such as presence or absence of a feature (e.g. sagittal crest in Eskimo skulls, preauricular sulcus in female innominate bones), or persistence of an entity (e.g. interfrontal suture), or degree of development (e.g. frontal ridges, projection of chin). However, most anthropometric data are measurements, by internationally agreed techniques, in living subjects or skeletal material. These may involve the whole body, e.g. stature, or sections such as limbs or individual bones. The cranium attracted early attention, but metrical methods are now established for all skeletal elements. Proportions are expressed by indices; e.g. breadth of a skull as a percentage of length, the *cranial index*. Such indices may show ethnic variation. Mongoloid people display higher cranial indices than other races; they are 'broader' in head relatively and, in this instance, absolutely. A Mongolian child, of course, might have a cranial width less than that of a Negro adult, and yet be proportionately broader. Ratios between length of limbs and 'sitting height', or between arm and leg, upper arm and forearm, thigh and foreleg, are all used and show differences related to age, sex and race. Details of some indices are included later in accounts of particular bones.

Observations and measurements suggesting age, sex, size and race of an individual skeleton, or parts of it, are not only useful in anthropology and archaeology (Brothwell 1968, Warwick 1968) but sometimes essential to identification in forensic practice (Glaister & Brash 1937, Boyd & Trevor 1953, Harrison 1957, Stewart 1954).

Estimation of skeletal age involves many criteria, varying in value at different ages. Up to 25 years, including fetal life, dentition and ossification provide numerous data for assessment of age, with accuracy dependent on precision of observations, available statistics for sex and racial affinities of individuals under examination and their nutritional and endocrine history. The

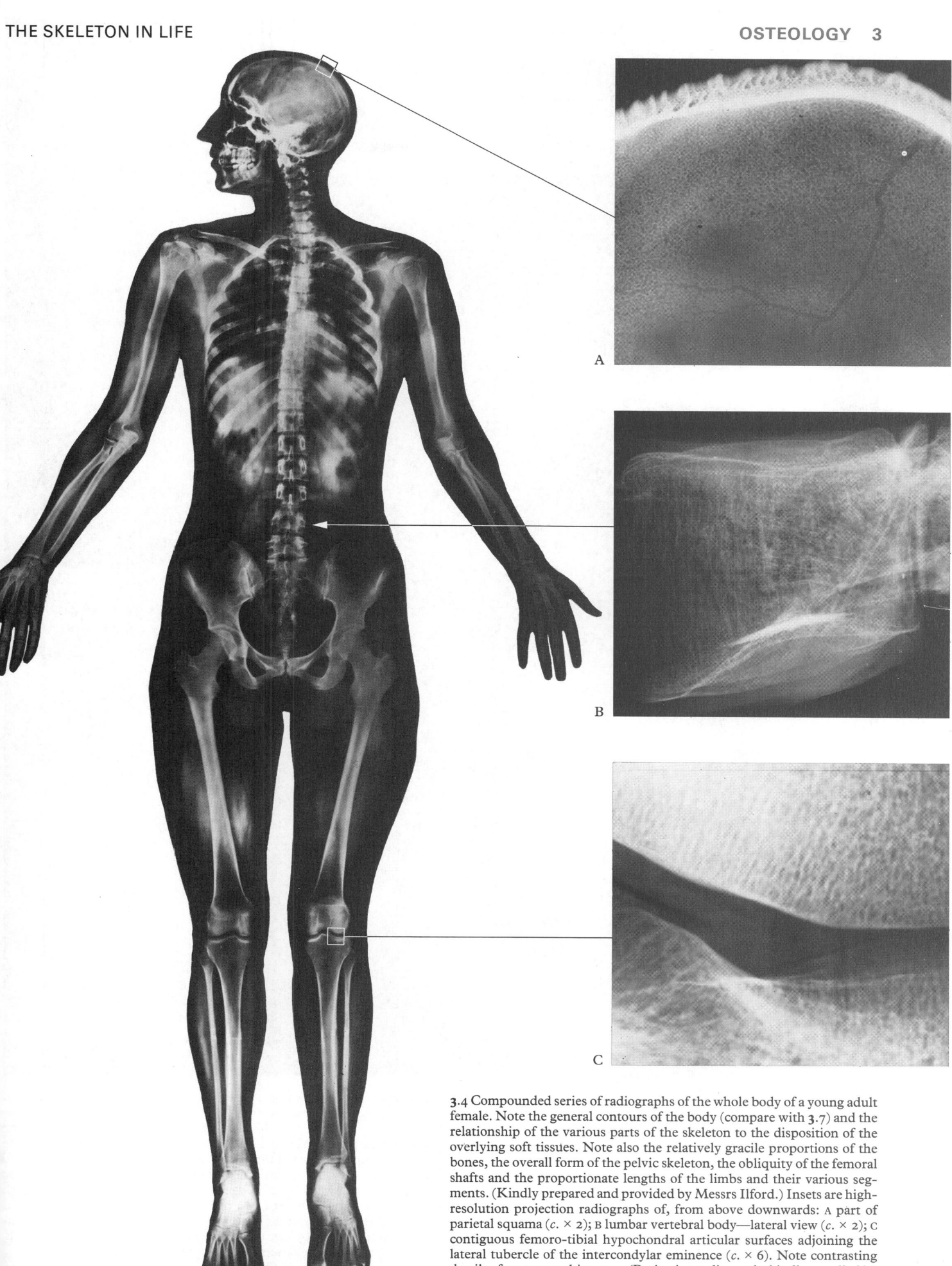

3.4 Compounded series of radiographs of the whole body of a young adult female. Note the general contours of the body (compare with **3**.7) and the relationship of the various parts of the skeleton to the disposition of the overlying soft tissues. Note also the relatively gracile proportions of the bones, the overall form of the pelvic skeleton, the obliquity of the femoral shafts and the proportionate lengths of the limbs and their various segments. (Kindly prepared and provided by Messrs Ilford.) Insets are high-resolution projection radiographs of, from above downwards: A part of parietal squama (*c.* × 2); B lumbar vertebral body—lateral view (*c.* × 2); C contiguous femoro-tibial hypochondral articular surfaces adjoining the lateral tubercle of the intercondylar eminence (*c.* × 6). Note contrasting details of osseous architecture. (Projection radiographs kindly supplied by C Buckland-Wright, Department of Anatomy, Guy's Hospital Medical School, London.)

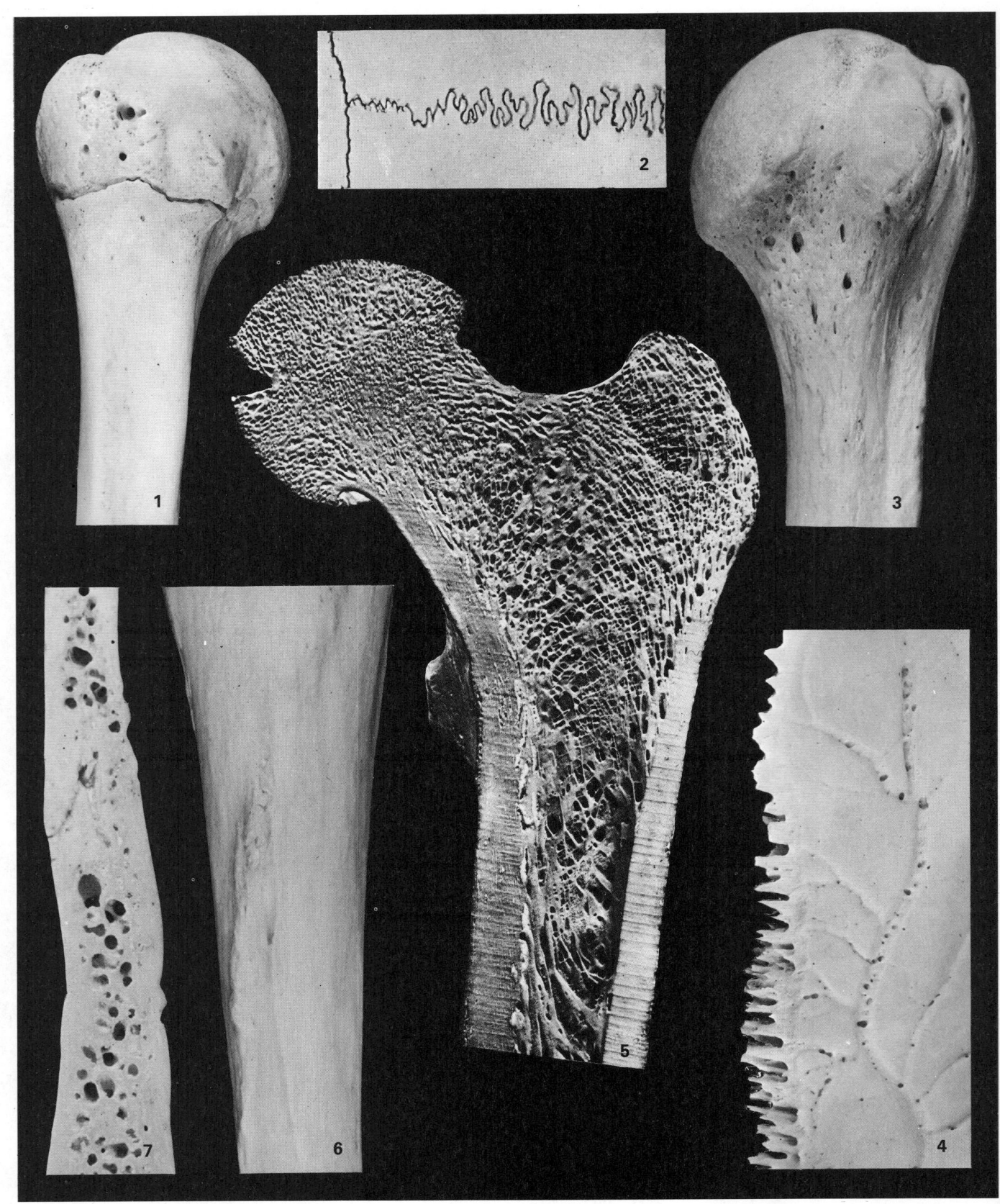

3.5 Some macroscopic features of bone structure. 1. The ventral aspect of the proximal end of an immature humerus. Note the contrasting surface characteristics of the smooth articular surface covered in life by articular cartilage, the smooth periosteal surface of the metaphysis, the vascular foramina at the base of the greater tubercle and the groove across which osseous fusion has not yet occurred, separating the compound epiphysis from the metaphysis. 2. The external aspect of part of the sagittal (horizontal) and coronal (vertical) cranial sutures. Note the variation in the form and degree of interlocking of the bones. 3. The ventral aspect of the proximal end of a mature humerus; complete osseous fusion has occurred between the compound epiphysis and the diaphysis. Note the variations in surface texture and the distribution and size of the vascular foramina. 4. The endocranial aspect of the sagittal margin of a parietal bone. Note the highly complex nature of the disarticulated sutural surface, the presence of vascular grooves and multiple vascular foramina on the endocranial surface. 5. A coronal section through the head, neck, greater trochanter and proximal shaft of an adult femur, showing the variation in thickness of the shell of compact cortical bone, and the geometrical patterning of the deeper intersecting bony trabeculae. 6. The posterior aspect of part of the shaft of a tibia. Note the relatively smooth areas which bear 'fleshy' attachments of muscles, the ridge where dense lamellae of collagen are attached and an oblique vascular foramen which transmits the main nutrient vascular bundle to the shaft. 7. A sectioned bone of the cranial vault showing the internal and external tables of compact bone separated by the trabecular diploë, the diploic spaces of which, in life, are filled with haemopoietic red bone marrow.

The sectioned femur (5) has been reproduced using a specialized photographic technique which produces a 'bas-relief' effect. (Photographs by Kevin Fitzpatrick, Guy's Hospital Medical School.)

latter data are rarely forthcoming; available tables of ossification usually apply to healthy children and adolescents in Caucasian communities of Europe or America. A few studies of other racial groups exist (Todd 1931, Modi 1957). Racial variations in the events of ossification would necessarily be genetic, but there is no clear evidence for this (Krogman 1962). Variations in ossification are affected by wide divergence in nutrition between, and even within, racial groups so far studied, in all of which data for females show earlier ossification and epiphysial fusion, a difference presumably genetic. Nevertheless the age of a complete skeleton up to 25 years can usually be assessed to within a year, or more accurately in earlier years, expecially if dental observations are available. (For details see Teeth and individual bones.)

Above 25 years skeletal age can be estimated within 5 years by the state of cranial sutures and bony surfaces of the symphysis pubis. From mid-twenties onwards, sutures exhibit progressive closure (p. 393, Todd & Lyon 1924–5). Closure begins internally, and without internal inspection observations are misleading (Singer 1953, Genovese & Messmacher 1959); complications due to racial variation have been recorded (Abbie 1950). Progressive changes occur in articular aspects at the pubic symphysis. Features typical of ages from late teens to fifties and beyond are established (Todd 1920–30, McKern & Stewart 1957), and this is now generally considered the best method for estimation of skeletal age in maturity. Sequences of age changes have also been described in other bones (scapula, sternum and costal cartilages), but provide less accurate estimates. Lipping of rims of vertebral bodies and at margins of other articular surfaces, exaggerated secondary markings and ossification into tendons and ligaments all suggest advancing age, but only vaguely indicate *actual* age.

Estimation of sex in complete human skeletons is usually easy, even without measurement, provided they are postpubertal. Sexual differences are marked in pelvis (p. 431) and skull (p. 395), but not equally in all populations. Thus 'sexing' of skeletons from one racial group is almost free from error, whereas assessment of an individual, of unknown extraction, is less certain. Postcranial bones other than the pelvis, especially larger limb bones, may provide clear evidence of sex, if others of like race and both sexes can be compared. Female bones are usually smaller than male equivalents, and also more slender, i.e. of lesser diameter in shaft relative to length. This is reflected in their comparative *weights*. In a study of Hindu femora, mean weights were 385 g in males, 279 g in females (Singh & Singh 1974).

Anatomists, anthropologists, and forensic experts have long judged the sex of skeletal material by non-metrical observations. Latterly, sexual divergence has been based upon measurements in many different bones (Montagu 1960, Krogman 1962). Of discriminant analyses, one example may be quoted: Rother et al (1977) analysed capital dimensions of 70 humeri, concluding that sex was not easy to establish (whereas approximate age *could* be assessed). Such studies emphasize the need for standards of sexual dimorphism in different populations. The pelvis remains, however, the most reliable region for assessing sex, even before puberty, in infancy (Reynolds 1947) and even fetal life (Boucher 1957).

In a recent study of thoracolumbar vertebrae in 1427 Australian subjects (aged 5 to 19 years), Taylor & Twomey (1984) detected sexual dimorphism, female vertebral bodies being more slender in proportions from the eighth year onwards, greater growth in transverse diameter occurring in males.

Estimation of size, particularly height, from measurements of limb bones has long been formulated (Rollet 1899), and with increasing accuracy as formulae have been refined (Trotter & Gleser 1958). All such calculation depends on the fact, familiar to artists (cf. Leonardo da Vinci), that major parts, trunk and limbs, exhibit consistent ratios among themselves and relative to total height, these ratios linked to age, sex and race. The relatively huge head, long trunk, short arms and shorter legs present a picture familiar in infants but grotesque even in older children, monstrous in adults. As an infant grows it also changes its proportions gradually towards adult shape, with divergence towards one sex at puberty. Limbs become relatively longer in adult males, shoulders broader and pelvis narrower, and there are other differences (3.7). Between major races, and even smaller ethnic groups, characteristic variations in proportions appear. Negroes are comparatively long in leg and arm; moreover, foreleg and forearm are long relative to upper arm and thigh. Consequently, formulae designed to estimate height from long bones in one population may not apply to another. Alternative formulae for the sexes must also be used, and immature bones must be recognized and subjected to suitable corrections.

Femoral length has commonly been used alone for estimates of stature by using a simple multiplier derived from comparison with known height in many individuals. The humerus, radius, ulna, tibia and fibula have all also been utilized; more complex formulae, based on several long bones, yield more accurate estimates. Whatever the methods used, estimates are no more than mean values with appropriate standard deviations; hence estimated stature of unidentified remains, however careful, may err by several centimetres.

Since estimates of stature (and other assessments) must sometimes be made from *fragments* of bones, attempts to establish reliable formulae have been made (Steele 1970).

Estimation of race from skeletal data has always been a central theme in anthropology. The skull has attracted most attention. An array of *non-metrical* features (Jones 1931) and of cranial, facial and mandibular indices are widely used (Martin & Saller 1958–61, Berry & Berry 1967, Berry 1975), with statistical analysis of *metrical* data (Giles & Elliot 1960). In the three major racial groups, Caucasian, Mongoloid and Negroid, perhaps 85–90% of skulls can be classified without elaborate measurement. However, with further racial division error rises steeply. Worldwide mingling of races renders 'pure' races difficult to indicate. In America, large collections of skeletons, both Caucasian and Negroid, well attested as to origin, have stimulated comparative studies. Critical survey of results (Krogman 1962) suggests that only the skull (but not the mandible) and pelvis (Todd & Lindàla 1928) have any value in estimating race. (For details of *cranial* features of racial significance see p. 395.) However, metrical peculiarities in Japanese limb bones have been claimed by Takahashi (1975, 1976). Racial significance of *postcranial* non-metrical characteristics has been re-examined by Finnegan (1978), with rigorous statistical analysis and summary of many findings. The value of non-metrical features in evaluating populations from skeletal data has been reviewed by Finnegan & Faust (1974).

Functions of Bone and Skeleton

Bone tissue itself, and struts and levers formed of it, are adapted exquisitely to resist all forms of stress, with suitable resilience, in various muscular activities. The skeleton is said to 'give shape and support' to the body; but shape is itself an expression of motor activities, which simply returns to a role in movement. In some lower vertebrates a cartilaginous skeleton, variably calcified, provides the levers of locomotion; but whether cartilage was ancestral to bone in this function remains uncertain (Romer 1942, 1970, Bassett 1962, Krompecher 1967). However, a bony skeleton is not merely a motor advantage; bone differs from cartilage in being intensely vascularized (p. 299) and it is possible that the calcium salts with which it is hardened could be as easily removed as deposited there (p. 303). Part of the osseous calcium can be mobilized into general circulation with little delay. Effects of depressed calcium levels in rickets and osteomalacia indicate that calcium in bone and the body at large are in constant balanced interchange (McLean & Urist 1969). The skeleton, thus integrated with other tissues in general metabolism of calcium, provides the major store (97%).

Some skeletal elements protect against extraneous forces; but the exoskeleton, so clearly protective in earlier vertebrates, is much reduced and modified in the human body. Even in the skull, much of whose cavity is enclosed by primitively protective dermal bones, this function is more than a barrier to *external* assault. Powerful muscles, masticatory and postural, have invaded cranial walls, demanding isolation of brain and its circulation from these intrinsic sources of disturbance. Again, ribs are commonly

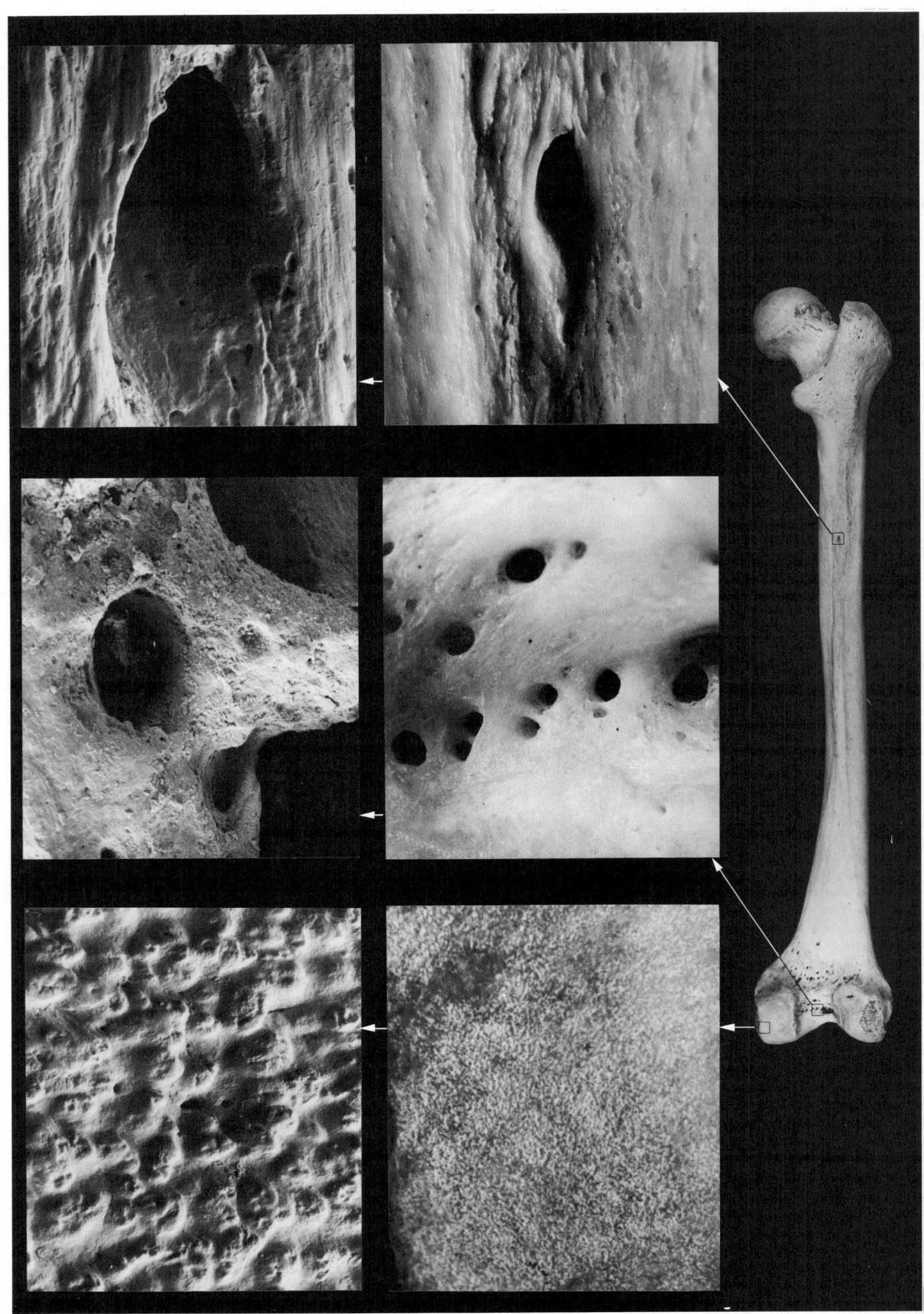

3.6 Surface details of a normal adult human femur photographed at the sites indicated, by macrophotography and scanning electron microscopy. The macrophotographs which are immediately adjacent to the whole bone photographs are at a magnification of × 5, except for that showing the diaphysial nutrient foramen which is × 4. All the scanning electron micrographs are × 80, with the exception of that showing the diaphysial

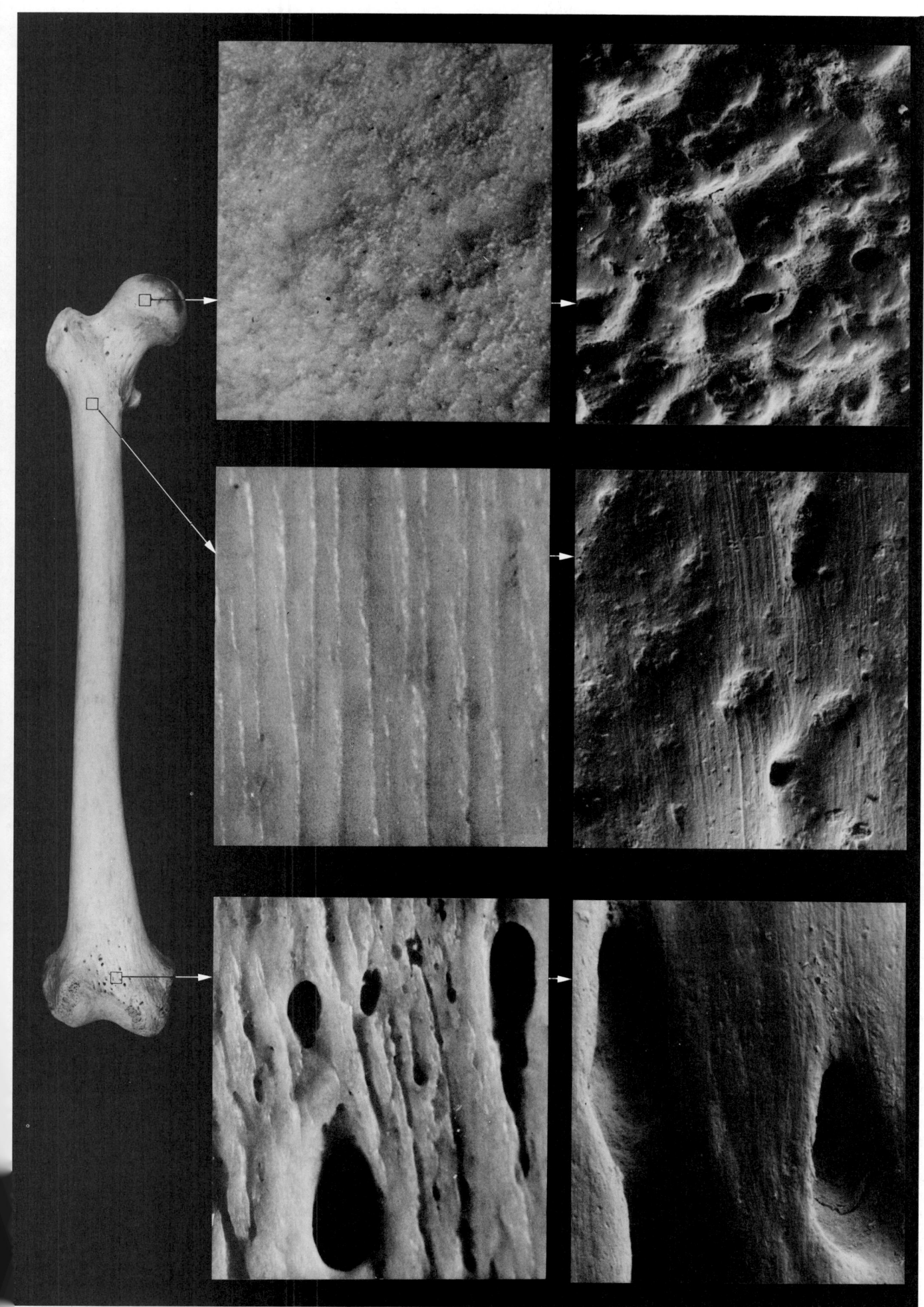

nutrient foramen, which is × 15. (Prepared and provided by Kevin Fitzpatrick, Derrick Lovell and Michael Crowder, Department of Anatomy, Guy's Hospital Medical School, London.)

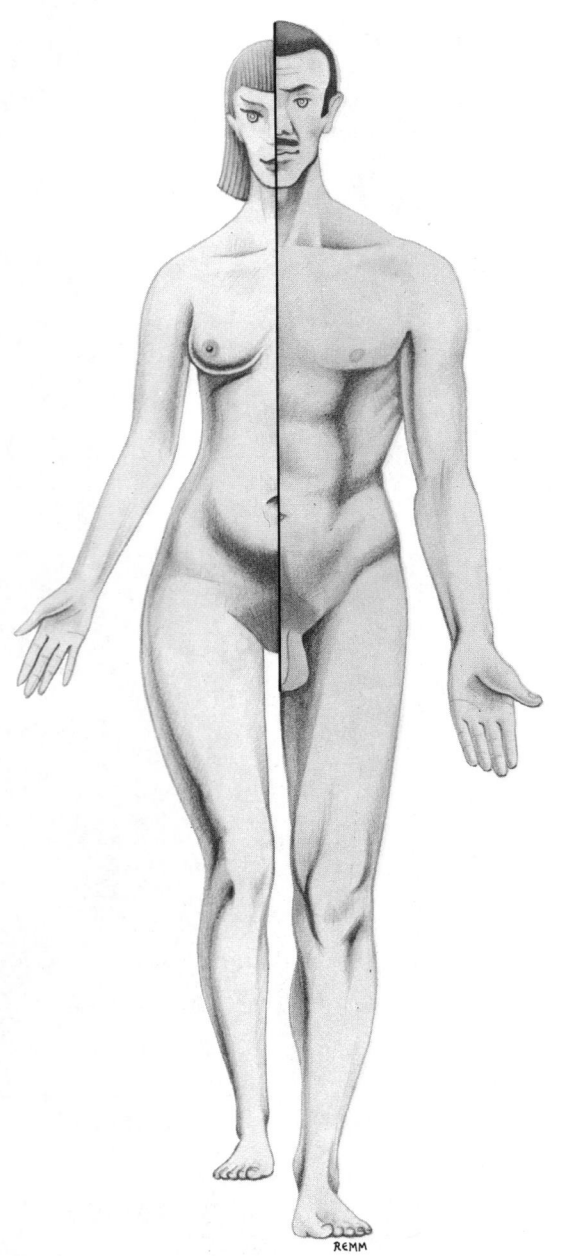

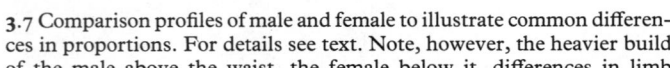

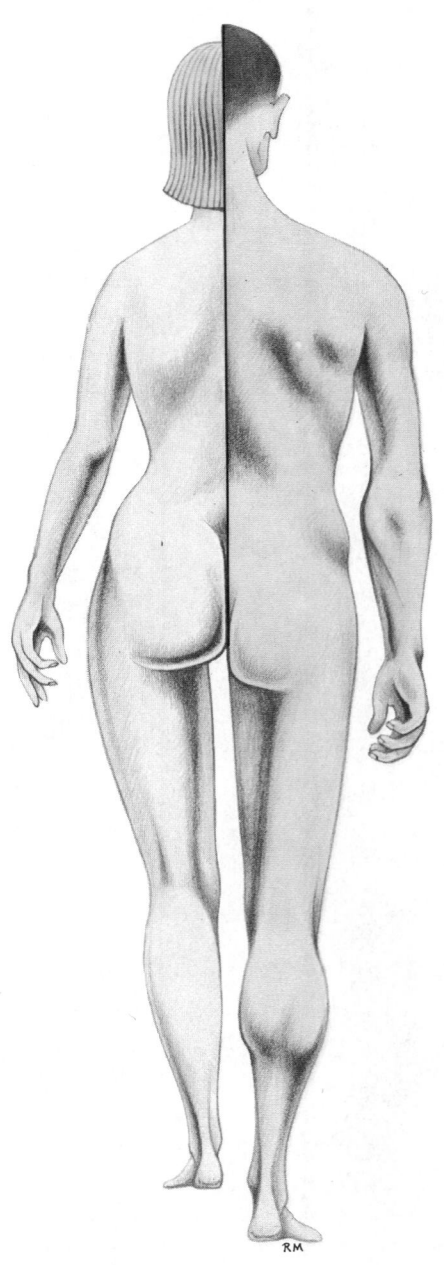

3.7 Comparison profiles of male and female to illustrate common differences in proportions. For details see text. Note, however, the heavier build of the male above the waist, the female below it, differences in limb proportions, carrying angle of forearm, muscularity and apparent length of neck.

dubbed protective. They certainly reduce risks of impact; but such dangers are occasional, whereas respiratory movements *depend* upon ribs.

All bones are, to some extent, internally trabecular, a pattern not only resistant to mechanical stresses but also for siting of bone marrow. Innumerable small spaces between trabeculae, with larger cavities in shafts of longer limb bones, are occupied by marrow, whether haemopoietic or adipose (p. 678). In some cranial bones air-filled cavities develop and they are then termed *pneumatic*. Pneumatization may consist of large hollows, such as the maxillary *sinus*, or multiple, small, communicating *air cells*, as in the temporal mastoid process (**3**.134,135). In human skulls, saving in weight thus effected can scarcely be as significant as it is in the pneumatized cranial vault of elephants; but the trabecular architecture of most bones is conducive to lightness without loss of strength, and with economy in use of materials. Some sinuses, particularly smaller air cells, appear to develop from trabecular tissue (diploë) in certain cranial bones, expressing the above principle. Larger sinuses affect the timbre of voice, but may be due merely to unusual growth patterns.

Mechanical Properties of Bone

The resemblance of bones to man-made levers, supporting columns, arches, struts and girders prompts application of concepts of mechanical engineering to them. Not only are bones like such structures in external shape and internal structure, but also in their uses. The apparently fragile but collectively strong lattices of struts and trusses of trabecular bone and skeletal forms such as tubes, H-girders, and ridges in cortical bone, predate human invention by aeons of time. Once human technology began to overtake natural biomechanics comparisons were inevitable. Galileo (1638) recognized the possible significance of trabeculation and also asserted that hollow cylinders are, weight for weight, *stronger* than solid rods. Havers described his 'systems' and their axial orientation (circa 1691); Ward (1838) likened trabecular patterns, with compressive and tensile elements, to supporting brackets and similar structures. Meyer (1867) and his mathematical collaborator, Culman, first clearly enunciated the *trajectorial theory*; and this, despite much argument, still survives. No more satisfactory hypothesis has replaced it. Basically it states that

276

trabecular pattern coincides with routes of stress; this has not yet been convincingly demonstrated nor disproved. Trabeculae must support shells of compact bone adjoining them, for these cortical strata are usually thin; but relations between stress and strain and trabecular pattern are often more complex than the simplified mathematical analyses applied to sections or models of bones (Kummer 1972).

In contrast to this largely theoretical approach is direct estimation of strength of bone as a material by engineering methods, to assess response in bone samples or entire skeletal elements. Wertheim (1847), anticipating Meyer, applied physical tests to bone tissue; subsequent workers have expanded such techniques. Developing interest in the mechanics of locomotion has stimulated the study of bone and a bewildering variety of values for various physical parameters of bone have been recorded. The problem is full of difficulties and fallacies. Obviously bone

material from various species differs, and sex and age are also factors. The form of specimen tested also varies; whole bones or blocks of cortical or trabecular bone, or mixtures of these behave differently, and source of specimens must be defined if results are to be comparable. Bones from preserved cadavers yield misleading values, especially as regards plastic deformation (Reilly & Burstein 1974), but also in elasticity, hardness, and compressive and tensile properties (Evans 1973). Dead bone, however, if kept wet, does not differ from living bone, tested in vivo, whereas dried bone is harder but less deformable. Sampling from a bone also presents problems, because varying distribution of Haversian systems or trabeculae may vitiate comparisons. Testing of specimens and intact bones, especially to rupture, involves attachment of tension devices, often incurring undesigned collapse at these points, especially in testing axial tension. Hence experimental data, recorded in profusion, present a complex picture, e.g. it is

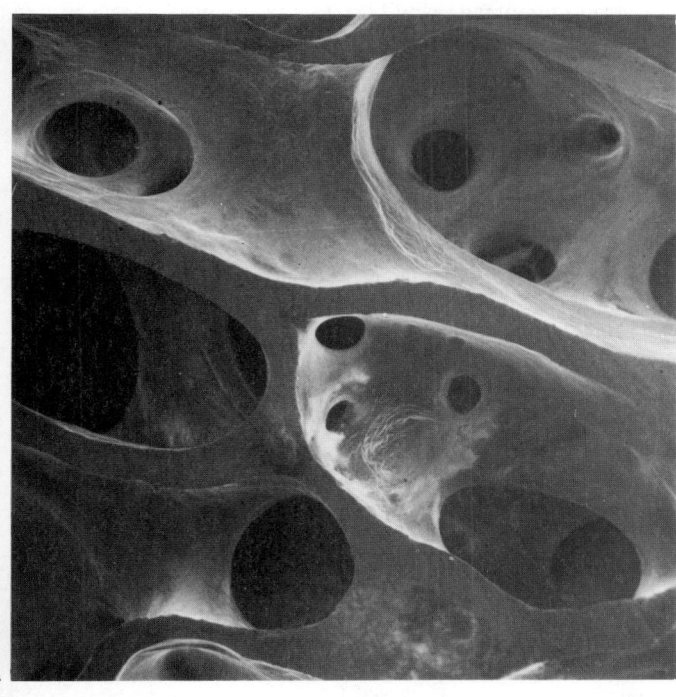
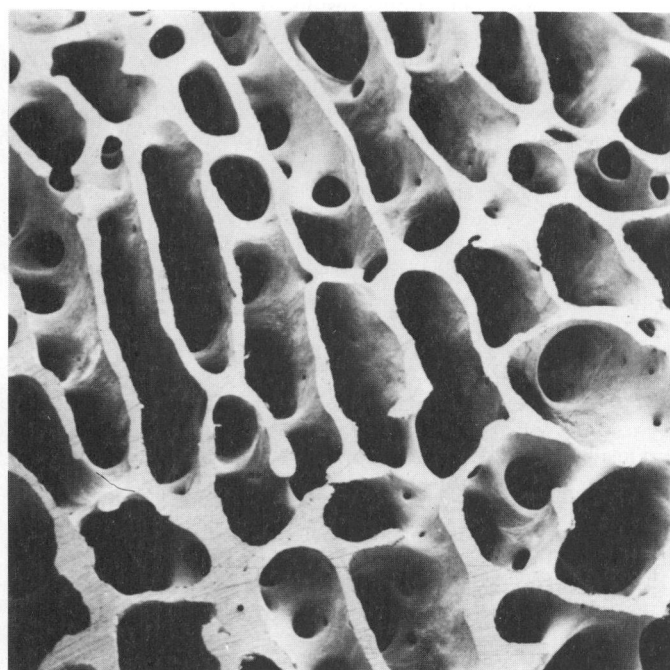
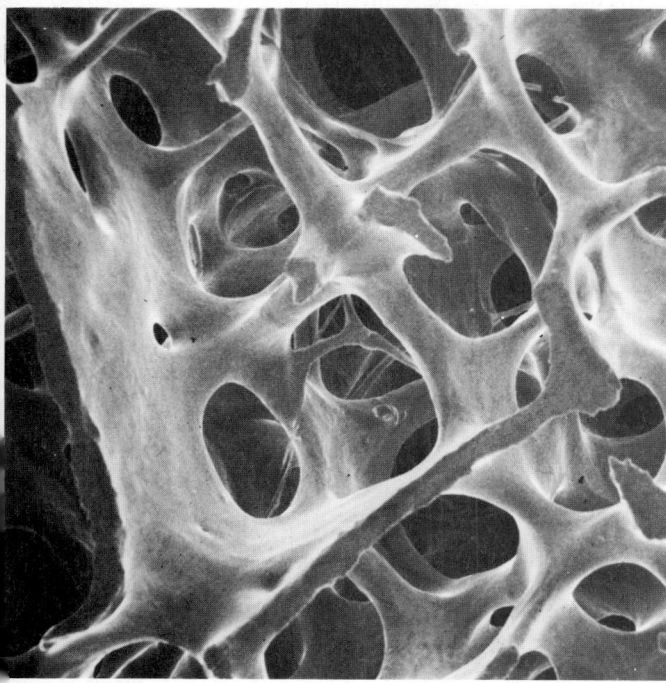
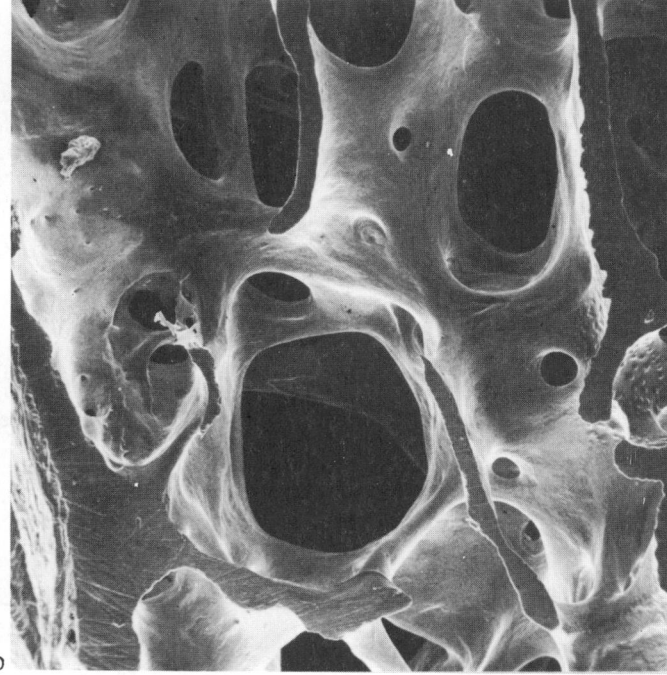

3.8 Scanning electron micrographs of trabecular bone at different sites in the proximal part of the same human femur. All fields are × 20. A is an area in the subcapital part of the neck, B and C are in the greater trochanter, and D in the rim of the articular surface of the head. Note wide variation in the thickness, orientation and spacing of the trabeculae. (Original photographs from Whitehouse & Dyson 1974, with permission from the authors, the *Journal of Anatomy* and Cambridge University Press.)

impossible to state even useful mean values for the physical properties of bone tissue, because resistance to stress and elasticity vary in different regions of the same bone. In view of the structural complexity in any skeletal element, including variations in thickness, density, cortical modelling and internal trabeculation (3.8A–D), some workers regard the definition of even an approximate picture of mechanical behaviour in any individual bone as improbable. The illustrations (3.8A–D) refer to the proximal end of the human femur (Whitehouse & Dyson 1974); similar regional variability exists in its distal end and in the proximal part of the tibia (Behrens et al 1974); in the tibia strength is not directly related to trabecular pattern or total density, indicating a multifactorial relation. Whitehouse (1975, 1977) has recorded data on the sternum and the iliac squama.

Various techniques, sometimes combined, have been used to assess the isolated physical properties of bone samples and whole bones, and to assess distribution of strains in them due to various forms of stress. Numerous results of tests by engineering techniques have been summarized by Ascenzi & Bell (1972), Kummer (1972) and Evans (1973). Although these values vary, greater disagreement exists in problems of transmission of forces. These have been investigated in bones, commonly the femur, and also plastic models, use of which is based on the false assumption that bone is isotropic and homogeneous. But bone is a viscoelastic, biphasic substance, analogous to fibre-glass, the tropocollagenous content corresponding to its 'fibres', and the crystalline hydroxyapatite to the 'glass'. Hence a plastic model can indicate only the *surface* behaviour of stresses in the bone it is supposed to imitate. In both, its stress pattern can be studied by inspection of cracks produced in a covering of colophonium resin (Kuntscher 1934) or special lacquers (Gurdjian & Lissner 1945); but these superficial cracks may indicate no more than a crude picture of strains in bone. Hallermann (1934) introduced a photoelastic technique, which reveals stress patterns in bones and plastic models by polarized light. Pauwels (1965) and Kummer (1966) exploited this technique, but Brekelmans et al (1972) emphasized its limitations, not only in models but also in oversimplified, two-dimensional mathematical analyses.

Benninghoff (1925) used the *split-line phenomenon* in decalcified bones, claiming that a surface pattern of split-lines or cracks, induced by puncturing with a round-bodied needle, follows distribution of osteons orientated in axes of compression or tension. Accepting the trajectorial theory, he considered it also directly related to behaviour of cortical osteons. Tappen (1954) has made similar claims, but admitted that patterns may follow 'immature' osteons. Isotupa (1972) considered that they represent distribution of points of weakness due to vascular spaces. Subsequently Buckland-Wright (1977) has attributed the cracks to weaker zones of bone (perhaps also concerned in genesis of fractures), and included cement-lines, interlamellar interfaces, osteocytic lacunae, and vascular canals. He considered the split-line technique as unreliable for analysis of structure and force transmission.

Strain gauges have been used to record the results of stresses in whole bones, either as dead, isolated elements, or during in vivo experiments, and also on bone samples. Gauges cannot be bonded to bone in large numbers, and the data obtained are limited; but interesting in vivo results have been obtained by Lanyon (1973) on ovine vertebrae, tibiae, and calcanei, recording deformation cycles during locomotion. Lanyon himself criticized the method, but it confirmed the elasticity of bone and showed that its elastic modulus varied with speed of movement.

Interference holography has been applied to the study of surface strains in human mandibles (Gupta & Knoell 1973); but the technique is difficult, especially in experiments on the living.

Mathematical analysis has been much developed in this field, but since basic data are limited or based on oversimplification inherent in use of models, usefulness is restricted. (Consult Koch 1917, Kummer 1966, Rybicki et al 1972.)

A combination of techniques perhaps offers most help in solving problems of bone mechanics, and this brief review of a complex and difficult field should hence end by reference to a study (Buckland-Wright 1978) of patterns of transmission of strains in the feline skull due to biting; it combined radiology and projection microradiography with use of strain gauges, colophonium resin experiments and histology, the limitations of each technique being complemented by the others. Projection microradiographs

Table of the approximate tensile strengths of bone and a few other materials (From Gordon 1968)

MATERIAL	TONS/INCH	MN/m^2
Traditional cast iron	5–10	75–150
Copper	10	150
Bone	10	150
Tendon	7	105
Wood, spruce (along grain)	7	105
Cotton	25	375

Table of Young's modulus of wet bone with the direction of load parallel to the long axis of the bone (From Reilly and Burnstein 1974)

BONE SPECIES AND AUTHORS	TYPE OF LOADING	YOUNG'S MODULUS ($\times 10^9$ N/m^2)	COMMENTS
HUMAN			
Dempster & Liddicoat (1952)	Tension, low strain rate	14·1	Dry, rewetted femur, tibia, humerus, mixed data, extensometer used.
Burnstein et al. (1972)	Tension, strain rate: 0·1 sec^{-1}	14·1	Femur; extensometer used for strain.
BOVINE			
Burnstein et al. (1972)	Tension, strain rate: 0·1 sec^{-1}	24·5 ± 5·10	Femur; extensometer used.
Simkin & Robin (1973)	Tension, low strain rate	23·8 ± 2·21	Tibia; unreported histology; extensometer used.
OTHER MATERIALS			
Methylmethacrylate: Plexiglass		8·6	
Wood: Douglas Fir		13·4	(68 per cent moisture.)
Steel		210·0	

Note: MN = 10^6 N 1 N = 1 kg × 1 m/s^2 = 10^5 dyn.

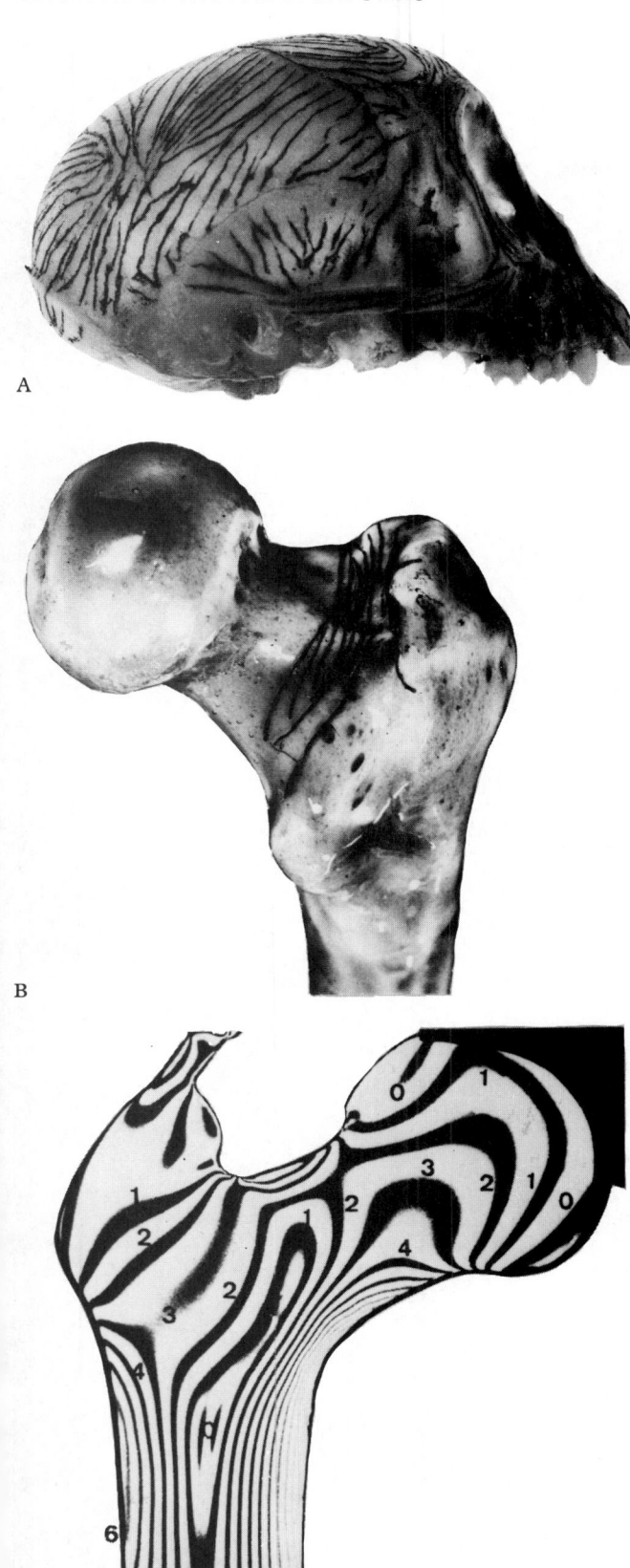

3.9 Photographs illustrating three methods used to investigate possible interrelations between structural organization and patterns of force transmission in bone. A split-line patterns produced in a decalcified skull (*Macaca mulatta*) by repeated puncture with an ink-tipped needle. B pattern of cracks in colophonium resin coating of a femur subjected to a compressive stress of the head, vertically applied. C photo-elastic pattern produced in a flat plastic sheet modelled upon the proximal end of the femur. A compressive force has been applied to the femoral 'head', whilst a tensile force has been applied to its 'greater trochanter'. C is from von Knief 1967, the others contributed by C Buckland-Wright, Department of Anatomy, Guy's Hospital Medical School, London.

were assessed by a stereoscopic technique, permitting three-dimensional correlation between bone structure and the impressed stresses. These observations led to a modified version of the trajectorial theory.

Alexander (1984d) has compared the optimum strengths of bone in terms of their habitual stresses and actual properties and has developed mathematical expressions for the balance between strength and weight of bones. He concludes that mammalian limb bones are often stronger than they need to be, but admits that the equation between stress and structure is difficult to predict with accuracy. Bacon & Griffiths (1985) have investigated texture, stress and ageing in human femora by neutron diffraction patterns in bone samples. They found a vertical orientation of hydroxyapatite crystals able to resist vertical stresses. The alignment was low at birth, maximal by about 13 years, decreasing steadily thereafter.

As stated above a profusion of estimates of physical parameters of bone tissue have been recorded. The following results are cited as representatives of such estimates, including comparison with other materials.

Growth of Individual Bones

As in other mammals human bones are mostly preformed in hyaline cartilage, some in condensed mesenchyme. Thus a model in soft tissue first appears and is gradually changed into bone by onset of osteogenesis, often at a centre from which it spreads, until the whole skeletal element is transformed. Such *centres of ossification* appear over a long period, many in embryonic life (3.10), some later in prenatal life and others well into postnatal growing period. The process is complex, and must be studied at different levels of organization. At microscopic, ultrastructural and molecular levels are the basic phenomena of histogenesis, growth and transformation of skeletal tissues, including primitive mesenchyme and its differentiation into other forms of connective tissue, cartilage and bone, all considered later (p. 304).

At first microscopic, the ossific centres soon become macroscopic and their growth then can be followed by plain inspection, dissection or radiological and other scanning techniques. It is the latter scale of events which will be considered here.

Many bones are ossified from a single centre, including carpal, tarsal, lacrimal, nasal, and zygomatic bones, inferior nasal conchae and auditory ossicles. Even in this limited group centres appear from the eighth intrauterine week to the tenth year, a wide sequence in studying growth or estimating age. However, most bones ossify from several foci, one appearing in late embryonic or early fetal life (seventh week to fourth month) centrally in the future bone (3.10), from which ossification progresses towards its ends, which are cartilaginous at birth (3.11,12), though typical in shape and articular congruities. These terminal regions are ossified from separate centres, sometimes multiple, appearing from birth into late teens; they are thus *secondary* to the earlier *primary* centre from which much of a bone is ossified. This is the mode in long bones, some shorter elements such as metacarpals and metatarsals, and in ribs and clavicles.

At birth a bone such as the tibia is typically ossified throughout its shaft, or *diaphysis*, by a primary centre appearing in the seventh prenatal week, whereas its cartilaginous *epiphyses* are transformed into bone by secondary centres. As these enlarge, replacing cartilage, they transform almost all the articular regions, but a layer of specialized hyaline cartilage persists at joint surfaces (p. 305) and a thicker zone between diaphysis and epiphysis. Persistence of this *epiphysial plate or disc (growth plate or growth cartilage)* allows increase in length until a bone's usual dimensions are attained, at which stage increase is abolished by discal ossification. The bone has then reached maturity. Coalescence of epiphysis and diaphysis is *fusion*, the amalgamation into one of separate osseous units, developed in the same cartilaginous or mesenchymal model, itself growing throughout the process.

Many bones have epiphyses at both ends, others at only one. Long limb bones illustrate the former, while metacarpals, metatarsals, phalanges, clavicles and ribs have only one epiphysis,

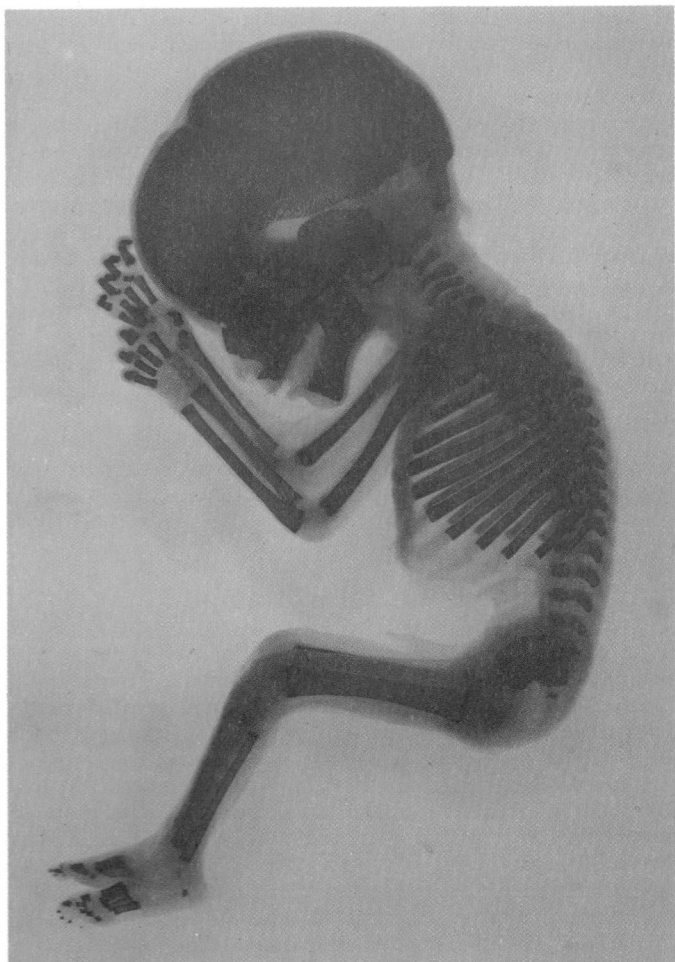

3.10 Alizarin stained and cleared human fetus of about 14 weeks in utero. Note the degree of progression of ossification from primary centres, which is endochondral in the appendicular and axial skeletons, excepting the clavicles, and the intramembranous centres for the majority of the cranial bones which are visible here. The carpus and tarsus are wholly cartilaginous except the primary centre for the calcaneus, as are, of course, the epiphyses of all the long bones. The centra and neural arches of the vertebrae are separate. The sternum is still unossified. The membranous anterolateral and posterolateral fontanelles are particularly obvious. (Photographed by Kevin Fitzpatrick, Department of Anatomy, Guy's Hospital Medical School, London, from a specimen prepared by Roslyn Holthouse, formerly of the same department.)

though costal cartilages may represent epiphyses normally devoid of ossification centres. (For discussion of a *pseudoepiphysis* at the first metacarpal's distal end consult Haines 1974.) Epiphysial ossification is sometimes more complex, e.g. the proximal end of the humerus, wholly cartilaginous at birth, develops three centres during childhood, which coalesce into a single mass before fusing to the diaphysis. Only one centre forms an articular surface, the others forming greater and lesser tubercles for muscular attachments. Similar composite epiphyses occur at the distal humeral ends and in the femora, ribs and vertebrae. Because some centres ossify in regions exposed to articular pressure, and others in regions subject to muscular traction, a classification into *pressure*, *traction* and *atavistic* epiphyses was proposed by Parsons (1903–5), the atavistic epiphyses being considered to represent skeletal elements separate at earlier evolutionary stages. Comparative morphology suggests that some mammalian bones are composites of separate reptilian or amphibian elements. The skull, clavicle, scapula and innominate bones afford examples; a small centre in the human coracoid process and an epiphysis at the medial end of the clavicle may be vestiges of skeletal elements separate in earlier vertebrates, repeated during development as transient features in subsequent mammalian forms. But the clavicle's medial end could equally be regarded a pressure epiphysis.

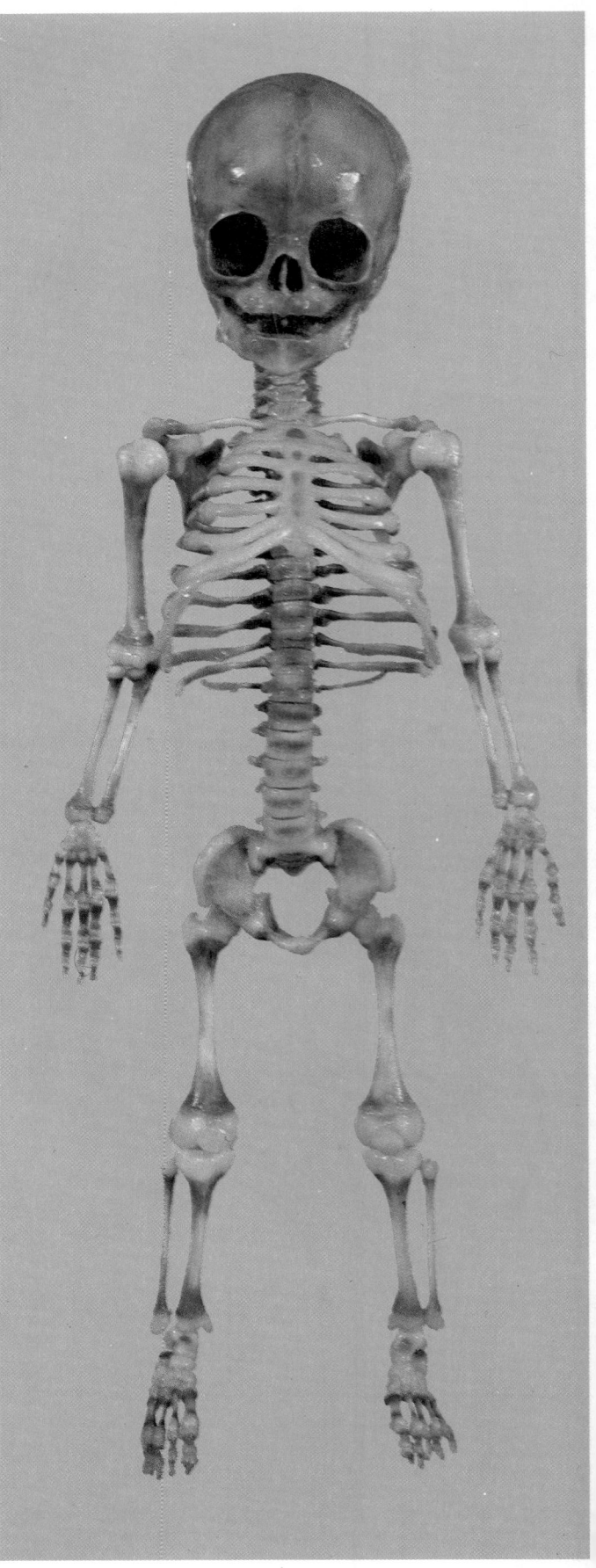

3.11 The skeleton of a neonatal infant, with all cartilages preserved. Note particularly proportions of neurocranium, orbital cavities and face, interfrontal (metopic) suture, sternebral ossification, extensive costal cartilages, large cartilaginous epiphyses and metaphysial flaring of long bones, especially the humeri, femora and tibiae. (Photographed by Kevin Fitzpatrick from a preparation by Michael C E Hutchinson, of the Department of Anatomy, Guy's Hospital Medical School, London.)

Many cranial bones ossify from multiple centres, and evidence suggests that some are atavistic. Marked reduction in number of cranial bones in mammals, compared with older vertebrate groups, is clear. The sphenoid, temporal and occipital bones are almost certainly composites of previously multiple elements, some represented by centres which are additional evidence of fusion in dermal (membranous) and cartilaginous derivatives, at first separate but united during growth to form a complex whole (p. 376). These events in cranial development do not precisely parallel atavistic epiphyses in postcranial bones, which are always *secondary*. In human composite cranial bones the result is coalescence of elements, each with a *primary* centre.

Classification of epiphyses, of little significance per se, does, however, direct attention to matters of interest in mammalian skeletal evolution. Epiphysial centres do not occur in other vertebrates, except sporadically in reptiles and birds (Haines 1937), and these are pressure (articular) epiphyses. Traction epiphyses are peculiar to mammals, but genetically established, for their appearance is not arrested by division of structures attached to them (Appleton 1922, Barnett & Lewis 1958); but experimental evidence in connexion with epiphysial development is limited.

The rate of growth in centres of ossification exceeds that of the cartilage in which they occur, though the latter is itself growing in concert with increase in size and change in proportions of the skeletal element for which it is a model. Thus, the epiphysial region, at first cartilaginous, gradually ossifies, with exceptions noted above. The process starts at a time and continues at a rate characteristic for each bone. (Individual, sexual and racial variations must be taken into account, as far as data permit, when assessing individual skeletal age.)

Rate of growth also varies both in bones with epiphysial plates and those without. Were the rate uniform, ossific centres would appear in a strict descending order of size. But primary centres for such different masses as phalangeal and femoral shafts are separated by a week of embryonic life at most. Those for carpal and tarsal bones show some correlation between size and order of ossification, from largest (calcaneus in fifth fetal month) to smallest (pisiform in ninth to twelfth postnatal year). In individual bones, succession of centres is related to the volume of bone which each produces. The largest epiphyses, e.g. the adjacent ends of femur and tibia, do ossify earliest (immediately before or after birth, points of forensic interest); but during comparison between other bones inconsistencies are numerous. Such dissimilar masses as a phalangeal epiphysis and lesser humeral tuberosity begin to ossify about the same time, and both long before the massive greater trochanter of the femur, showing widely disparate rates of ossification.

Experiments in mammals indicate that at epiphysial plates rate of growth is initially equal at both ends of bones possessing two epiphyses, but that after birth one grows faster (Brookes 1963); since this end also usually fuses later with the diaphysis, its contribution to length is greater. Though faster *rate* can only be presumed in human bones, *later* fusion is a radiological fact. Recurrent directions of nutrient arteries as they enter certain bones accords with the above facts (cf. p. 299).

The end of a long limb bone, which is more active in growth in length, is often termed the *growing end*, but this is obviously a misnomer. Variable rate of growth at epiphyses in general also shows in rate of increase of stature (Krogman 1941, Tanner 1962), which is rapid in infancy and again at puberty, but otherwise slower. The spurt at puberty, or slightly before, decreasing as epiphyses fuse in post-adolescent years, has been much studied. For general reviews of fetal and postnatal growth patterns, including the skeleton, consult Sinclair (1969), Tanner (1978).

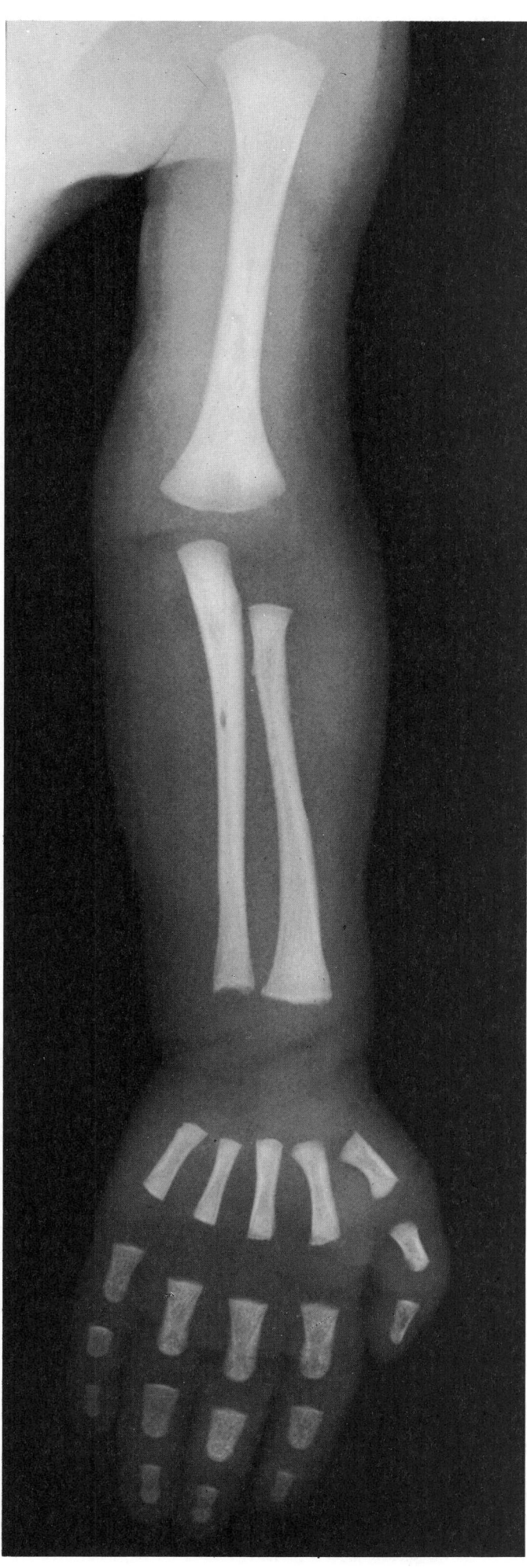

3.12A Radiograph of neonatal arm. Ossification from primary centres is well advanced in all bones except the carpals, which are still wholly cartilaginous. The gaps by which individual elements appear to be separated are, of course, filled by the radiolucent hyaline cartilage, in which epiphysial or carpal ossification will subsequently occur. In the long bones note the flaring contours, with narrow midshaft and relatively expanded metaphyses. Note also the proportions of the limb segments characteristic of this age and in particular the relatively large hand.

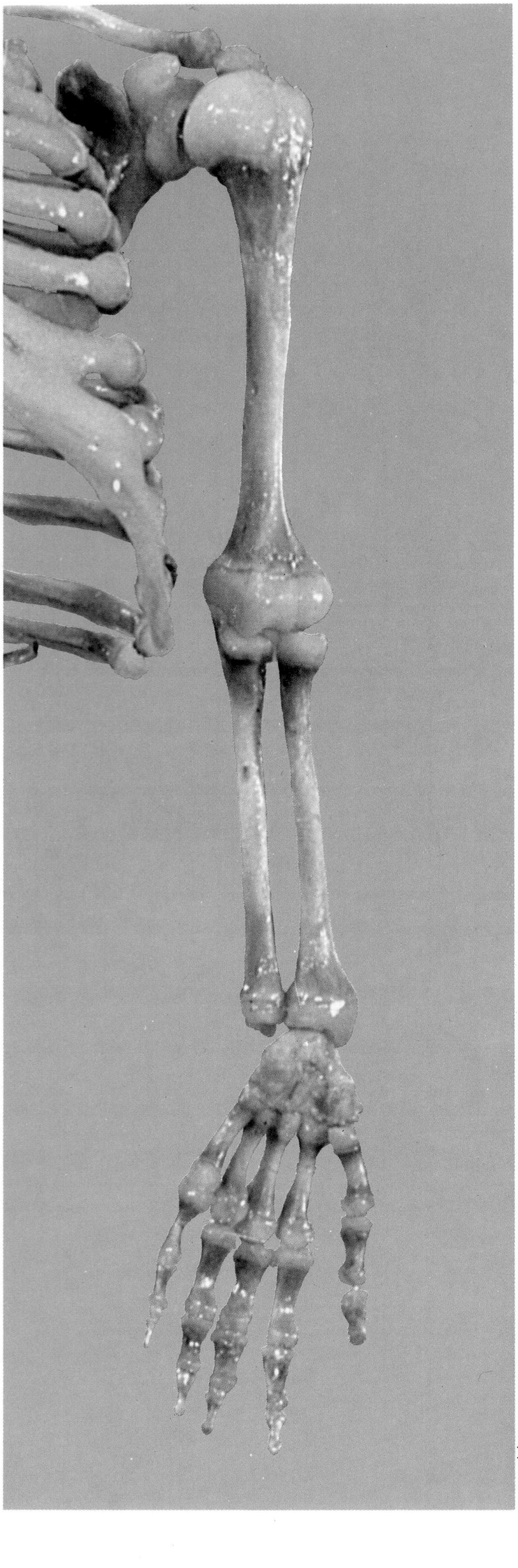

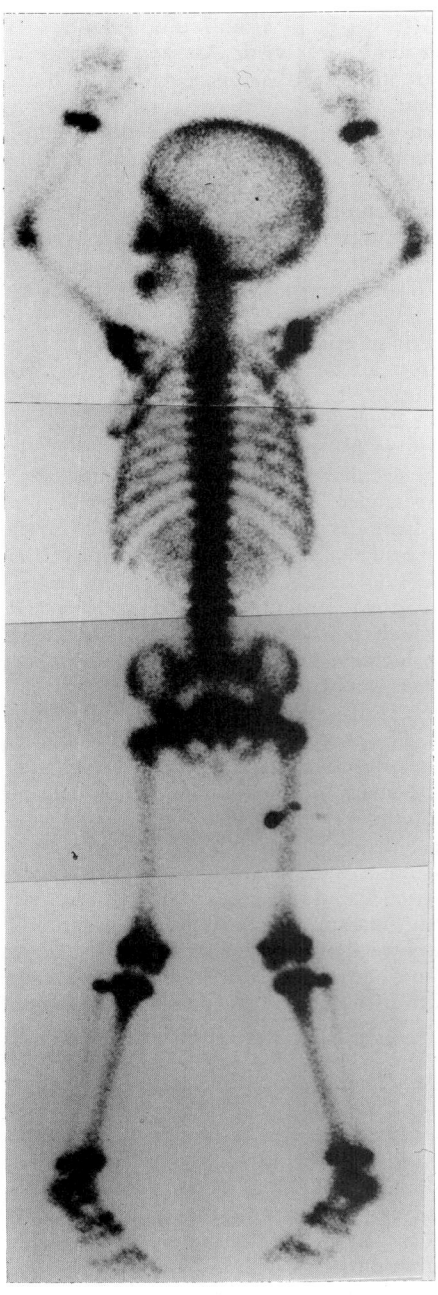

C

3.12B (*left*) Photograph of a preparation of a neonatal left arm (from the specimen shown in **3.11**). Compare the radiolucent areas in the radiograph (**3.12A**) with the preserved cartilaginous epiphyses and carpal elements in this specimen. (For acknowledgements see **3.11**.)

3.12C (*above*) Composite bone scan of a male child, aged approximately one year, showing sites of epiphysial growth activity. The tracer used (technetium-99m labelled methylene diphosphonate) also shows the kidneys and bladder. (Provided by Department of Nuclear Medicine, Guy's Hospital; photography by K Fitzpatrick and Sarah Smith.)

Growth cartilages do not grow uniformly at all points; this accounts for changes such as alteration in angle between the humeral shaft and its neck. On their diaphysial faces epiphyses do not have a uniformly flat junction with growth cartilages, nor indeed do the latter at their diaphysial junction. By differential growth osseous surfaces usually become reciprocally curved, the epiphysis forming a shallow cup over the convex end of the shaft (cartilage intervening). This form may resist shearing forces at this relatively weak region. Reciprocity of bone surfaces is augmented by small nodules and ridges, typical of such surfaces when denuded of cartilage. Such adaptations emphasize the formation of many immature bones from several elements held together by

B

epiphysial cartilages. Most human bones are such complexes, not only through the active years of childhood but also the even more vigorous adolescence; the bonding of bone to bone through cartilage is thus strengthened.

Forces at growth cartilages are largely compressive, but with an element of shear. At traction epiphyses, attached by cartilage prior to fusion, attached structures create tension, perhaps promoting development of abundant collagen fibres aligned along supposed routes of stress (Smith 1962a). Damage to epiphysial growth by violence may occur, but disturbance by constitutional disease, such as fevers, is more frequent, producing visible change in trabecular pattern of bone, visible radiographically as dense transverse *lines of arrested growth* (Harris 1933); several such lines may appear in limb bones of children afflicted by successive illnesses.

Details of fusion are described later (p. 307). The growing part of a diaphysis adjacent to epiphysial cartilage, the *metaphysis*, appears to overtake the cartilage, but epiphysial contribution is shown by the denticulated edges of both bone elements as they bridge the cartilaginous gap. These appearances, extensively described, provide criteria for estimations of times of fusion, by inspection *or* radiography (Stevenson 1924). Knowledge of sites, times of appearance, rates of growth and times of epiphysial fusion is clinically, forensically and anthropometrically valuable.

Reliable data depend upon adequate observations and standardized techniques. Variation in published figures is partly due to failure in these respects. Radiography needs to be more frequent than is usual. Angle of view must be carefully controlled; routine positions may be inadequate; e.g. the centre for lesser humeral tubercles is often missed, probably because its image is superimposed upon another centre.

Variation in skeletal development does, of course, occur between individuals, sexes and possibly races. The *sequence* of events, however, shows little variation; it is their timing which varies. *Females antedate males* in all groups studied, and differences, perhaps insignificant before birth, increase thereafter, rising to two years or so in later fusions of adolescence. Data are most reliable for Caucasians, but few of their skeletal parts have been compared in sufficient detail, numbers or over adequate periods. Perhaps the best studies available are for the hand (Todd 1937, Greulich & Pyle 1959, Mathiasen 1973), knee (Pyle & Hoerr 1955) and ankle and foot (Hoerr et al 1962).

Cartilage and bone are specialized connective tissues consisting of the same three elements: *cells* embedded in a *matrix*, permeated by arrays of *fibres*; but sclerous tissues differ from soft, pliant, connective tissues, their matrix being solidified. Cartilage and bone, nevertheless, differ in structure, physical properties, vascularization and modes of growth and regeneration.

THE SKELETAL CONNECTIVE TISSUES

The skeletal tissues, cartilage and bone, are essentially specialized connective tissues and consist of the same components—cells embedded in a matrix permeated by a system of fibres. Physically, however, matrices of skeletal tissues differ from those of general connective tissues in being solidified. Cartilage and bone are, nevertheless, quite distinct in their structure, physical properties, vascularization and in their patterns of growth and regeneration.

Structure of Cartilage

Cartilage is a phylogenetically ancient tissue, widespread in vertebrates as either a permanent or temporary skeletal component. During early fetal life the human skeleton is mostly cartilaginous, but is largely replaced by bone subsequently. In adults cartilage persists at the surfaces of synovial joints, in the walls of the larynx, trachea, bronchi, nose and external ears, in the epiglottis and as isolated small masses in the cranial base. Developmental replacement by bone is a complex process, and cartilaginous *growth plates* between ossifying epiphyses and diaphyses of long bones (and elsewhere) continue to proliferate, increasing the length of the bones concerned until they eventually ossify, when growth ceases. For reviews see Glenister (1976), Reid (1976, visceral cartilage), Stockwell (1978), Hall (1983, structure and biochemistry), for an alternative view of cartilaginous growth plate biology see Lutfi (1974).

Cartilage is essentially a type of stiff, load-bearing connective tissue. Its distinctive properties are a low metabolic rate and a vascular supply confined to its surface or to large, penetrating tunnels, a capacity for continued and often rapid interstitial and appositional growth (vide infra) and a high resistance to tension, compression and shearing, with some resilience and elasticity. Cartilage is covered by a fibrous *perichondrium* except at osseous junctions; at synovial surfaces, the latter are lubricated by secreted, nutrient fluid.

The *matrix*, containing *chondroblasts* and *chondrocytes*, varies in appearance in composition, and in the nature of its fibres. Hence there is *hyaline cartilage* (*hyalos* = glass), *white fibrocartilage* (with much collagen) or *yellow elastic fibrocartilage* (with an elastin network). A densely *cellular cartilage*, with thin septa of matrix between its cells, is a stage in development of early embryonic cartilage and a permanent tissue in many mammalian pinnae. White and yellow elastic fibrocartilage are specific, relatively unvarying tissues, whereas hyaline cartilage embraces a wide range of appearances varying much in composition and properties according to age and location as well as species.

Cartilage cells (3.13A,B,14) occupy small *lacunae* in the matrix which conform to their shape. Young cells (*chondroblasts*) are smaller, often flat, irregular in contour and bear many surface projections (*filopodia*) fitting complementary recesses in the matrix. Early postmitotic chondroblasts often have intercellular contacts which are necessarily transient, disappearing as matrix synthesis proceeds. As these cells mature, they lose the ability to divide and become metabolically less active. Some authors reserve the name *chondrocytes* for such cells; but this term is, by some, employed to denote cartilage cells of all degrees of activity, a practice which will largely be followed here. When appropriate, however, chondroblasts and chondrocytes are distinguished. Mature *chondrocytes* enlarge with age and, though rounder, still have a few filopodia (p. 31) and an occasional cilium. The internal structure of chondrocytes is typical of cells active in making and secreting proteins (Stockwell 1978, Sheldon 1983, Kühn 1986). Their nucleus is rounded or oval, euchromatic and possesses one or more prominent nucleoli. The cytoplasm is filled with granular endoplasmic reticulum, transport vesicles and Golgi complexes, and contains many mitochondria and frequent lysosomes. Also present are numerous glycogen granules, intermediate filaments (vimentin) and pigment granules. With final maturation of these cells to the relatively inactive chondrocyte stage, the nucleus becomes heterochromatic, the nucleoli smaller, and the apparatus of synthesis and secretion (endoplasmic reticulum, Golgi apparatus, etc.) much reduced; often such cells accumulate lipid vacuoles, quite large in diameter.

The matrix is composed of collagen and, in some cases, elastin fibres, embedded in a water-filled yet stiff ground substance (**3.17**). These components have various chemical features which are unique to cartilage, and confer upon it unusual mechanical properties. The *ground substance* is a firm gel, rich in carbohydrates and therefore stainable with the periodic acid-Schiff method; the carbohydrates are predominantly acidic, and are hence basophilic, strongly binding such dyes as haematoxylin, alcian blue and toluidine blue, and giving metachromatic coloration with the latter. The chemistry of the ground substance is complex, consisting mainly of water and dissolved salts, held in a meshwork of long interwoven proteoglycan molecules together with various other minor constituents, mainly proteins or glycoproteins and some lipid.

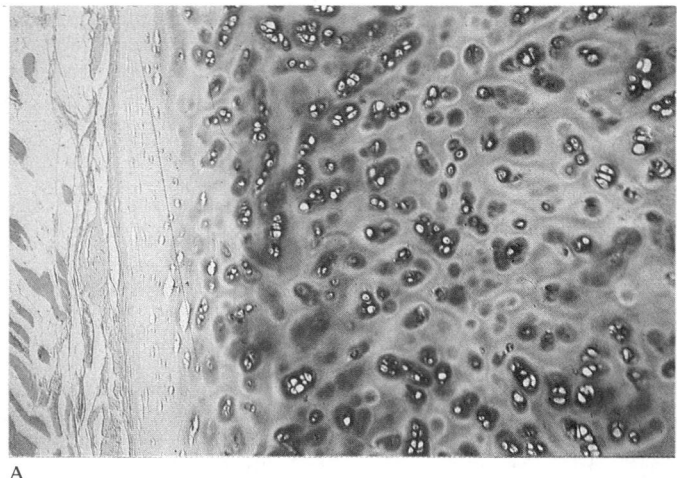

A

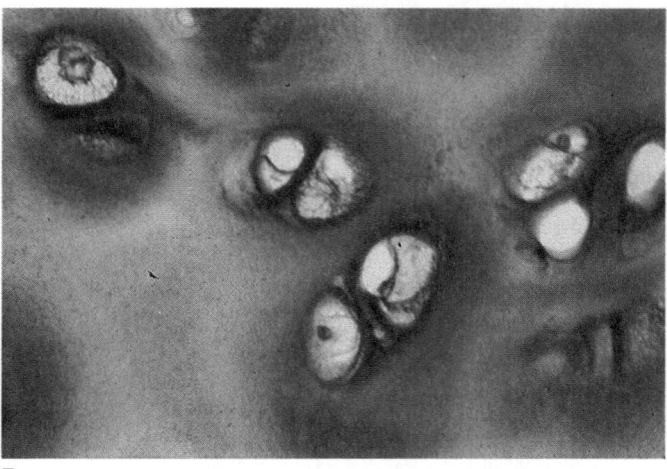

B

3.13 Sections through hyaline cartilage (human rib), stained with haematoxylin and eosin. A is a low-power view, showing perichondrium, chondroblasts and mature chondrocytes embedded in the basophilic matrix; magnification × 150. B higher magnification showing groups of chondrocytes within lacunae. Note the basophilic zones (rich in proteoglycans) around the cell clusters; magnification × 1000.

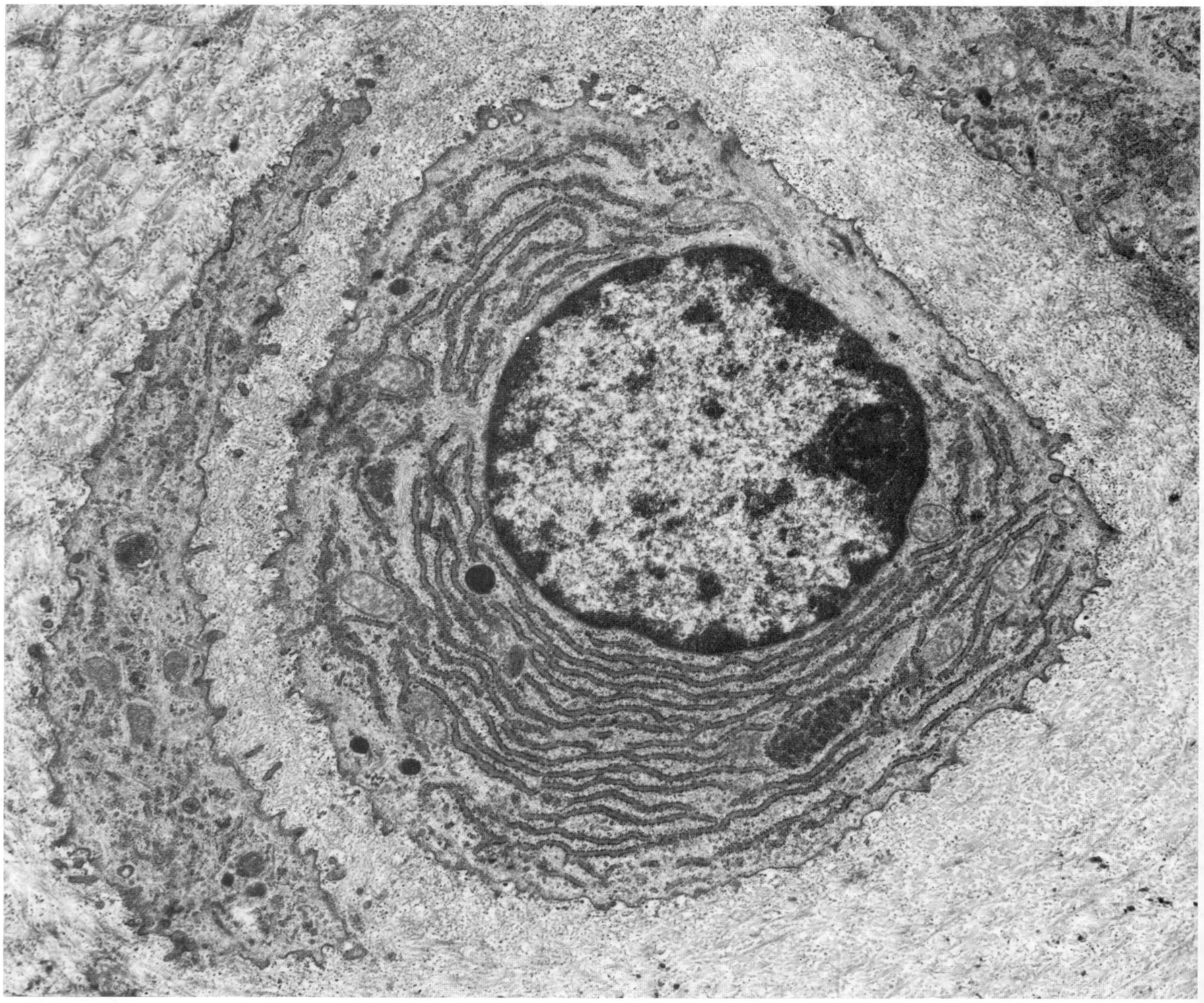

3.14 Transmission electron micrograph of an ultra-thin section of a perfusion-fixed specimen of a rabbit's femoral condylar cartilage. The centrally placed chondroblast contains an active euchromatic nucleus with a prominent nucleolus. Its cytoplasm contains a rich concentration of roughly parallel concentric flattened cisternae of rough endoplasmic reticulum, scattered mitochondria, lysosomes and aggregations of glycogen. Its plasma membrane bears numerous short filopodia which project into complementary recesses in the surrounding matrix. The latter shows a delicate fibrillary feltwork with a finely dispersed granular interfibrillary substance. No pericelluar 'lacuna' is present; the matrix separates the central chondroblast from the sectioned cytoplasmic periphery of two adjacent chondroblasts. The left profile, crescentic in outline, is characteristic of chondroblasts with an almost squamous form that are sited near the articular surface; magnification × 14 500. (Preparation by Susan Smith, Department of Anatomy, Guy's Hospital Medical School, London.)

The collagen of the cartilage matrix (**3.16**), forming up to 50% of its dry weight, is chemically distinct from that of most other tissues, being classed as Type II collagen; elsewhere this variety is only found in the notochord, the nucleus pulposus of the intervertebral disc, the vitreous body of the eye and in the primary corneal stroma. Its tropocollagen subunits are composed of triple helices of identical polypeptides (three α-1 chains; see p. 64), although collagen in the outer layers of the perichondrium and much of the collagen in white fibrocartilage belongs to the general connective tissue type I. The majority of the collagen fibres of cartilage are too small to be individually visible except by electron microscopy; they are relatively short, thin (mainly 10–20 nm diameter) structures with the characteristic cross banding (at 64 nm intervals in sections) and they are interwoven to create a three-dimensional meshwork linked by lateral projections of the proteoglycans associated with their surfaces. Various studies have shown that proteoglycans partially ensheath collagen fibres, and their concentrations often share the periodicity of the fibres. Proteoglycans and other organic molecules thus link collagen fibres with each other, with the interfibrillar material of the ground substance and also with the cells of cartilage (vide infra).

Collagen fibres vary greatly in their amounts, sizes and orientation in different types of cartilage, as also with maturity and position within the cartilage mass. In articular cartilage, collagen fibres close to the surfaces of cells are particularly narrow (4–6 nm across) and resemble fibres of Type II cartilage in non-cartilagenous sites (vide supra).

In addition to Type II collagen, minor quantities of other classes unique to cartilage are present (Mayne & Irwin 1986),

including Types IX and X, and isolated collagen polypeptide chains (1α, 2α, 3α). The significance of these is as yet unclear, but they may be involved in stabilizing Type II collagen networks (Type IX) and in hypertrophic changes in some types of cartilage (Type X). As described below, other types of fibre are also present in some classes of cartilage; these include elastin fibres in elastic fibrocartilage and, where ligamentous structures are inserted, as in fibrocartilage, Type I collagen, derived from fibroblasts rather than chondrocytes.

The proteoglycans of cartilage are similar in general outline to those of general connective tissue (see p. 67), although with features peculiar to cartilage. They consist of various long polymers, often branched, of carbohydrates termed glycosaminoglycans (GAGs; see Hascall & Hascall 1981). These are acidic, bearing anionic sulphate and carboxyl groups which give them a net negative charge. Groups of GAGs are covalently bound to a filament of protein (the 'core protein', with an Mr of 250 000, and a length of 300 nm) which may bear more than on hundred GAGs of different types sticking out sideways like the bristles of a bottle brush (**3.17**). In turn, several such proteoglycan assemblies can be bound along the length of a relatively huge (a million or more in relative molecular mass) hyaluronate molecule (another type of GAG) to form highly complex, filamentous aggregates. Other, smaller 'link proteins' are involved in this interaction. Because of the predominance of acidic groups there is a tendency for the chains to repel each other, thus standing out stiffly from the central core protein. Weak intermolecular forces hold these aggregates together as a three dimensional network with large water-filled spaces within. This arrangement allows the

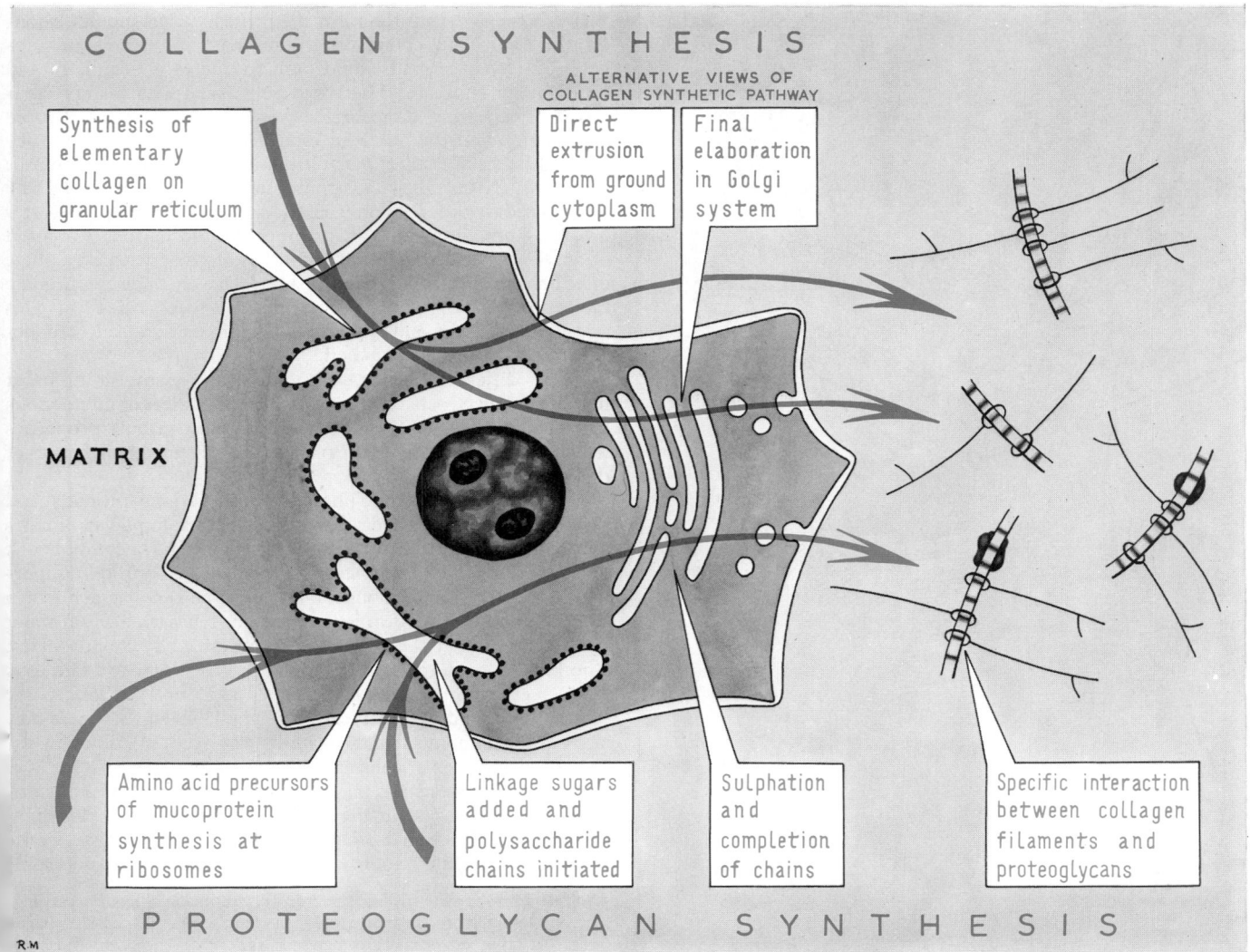

3.15 Summary of some of the important biosynthetic pathways of the chondroblast.

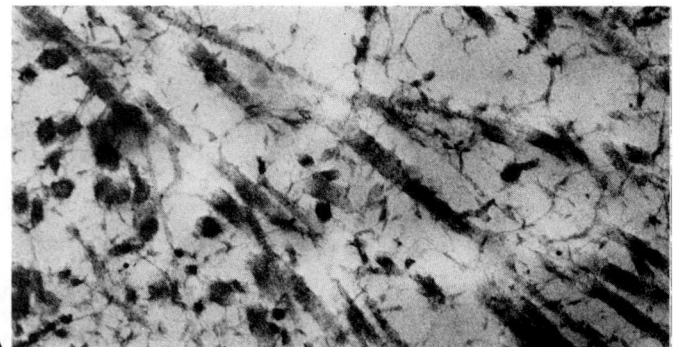

A

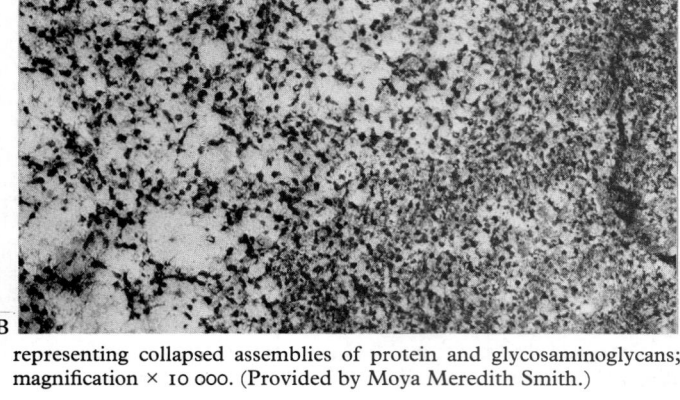

B

3.16 Electron micrography of hyaline cartilage matrix. The matrix in A shows thick, banded collagen fibrils and thinner, unbanded ones. Traces of finely filamentous proteoglycan complexes can also be seen adhering to and connecting collagen fibrils; magnification × 30 000. In B the matrix has been fixed in the presence of ruthenium red to preserve proteoglycan complexes, which are here visible as densely staining beaded structures representing collapsed assemblies of protein and glycosaminoglycans; magnification × 10 000. (Provided by Moya Meredith Smith.)

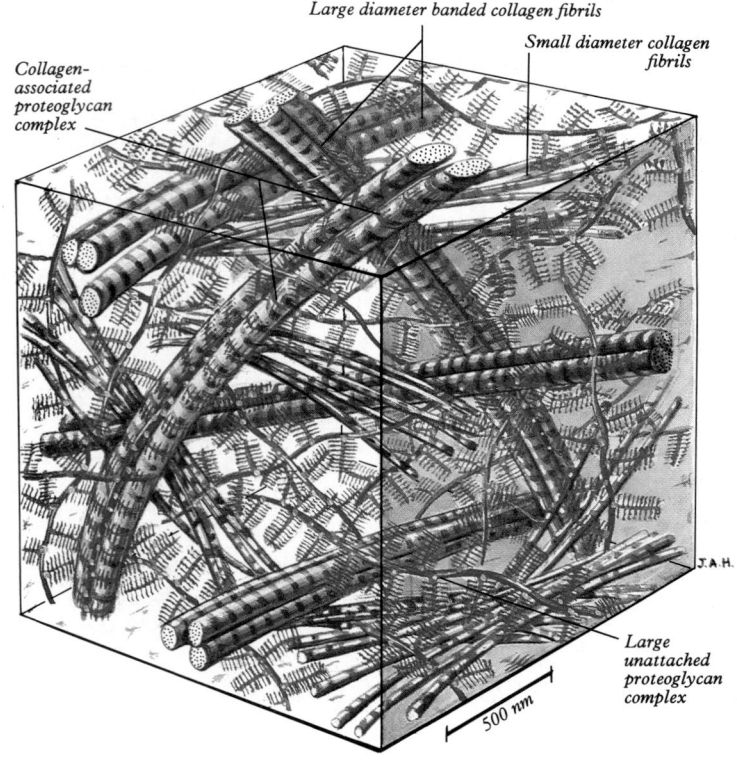

Large diameter banded collagen fibrils

Small diameter collagen fibrils

Collagen-associated proteoglycan complex

Large unattached proteoglycan complex

500 nm

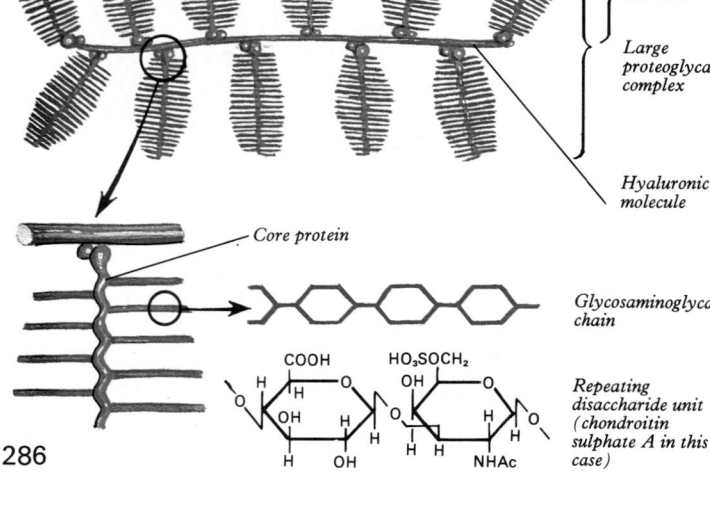

Proteoglycan side unit

Large proteoglycan complex

Hyaluronic acid molecule

Core protein

Glycosaminoglycan chain

Repeating disaccharide unit (chondroitin sulphate A in this case)

ready diffusion of water and dissolved materials through the ground substance, although water and other electrolytes are also loosely bound by the electrostatic forces on the surfaces of the charged macromolecules. In the living state the molecular aggregates appear to be compressed into a smaller volume than would be expected from their shape, the length of their chains and electrical repulsions between their subunits. It has been suggested that the proteoglycans may act as minute compressed springs, storing energy when further compacted then releasing it on recoil and so conferring elastic properties on the matrix.

In electron microscope sections, these large molecular arrays generally collapse to small dense granules about 20 nm across as their hydration sheaths are stripped away. However, when separated by centrifugation and negatively stained, their complex feathery nature is clearly visible.

Glycosaminoglycans found in the proteoglycans include chondroitin 4-sulphate, chondroitin 6-sulphate and dermatan sulphate (otherwise termed chondroitin sulphates A,C and B), and also keratan sulphate. These are polysaccharides mainly composed of repeating disaccharides, one invariably a substituted hexosamine and the other an esterified hexuronic acid or substituted hexose. Thus chondroitin 4- and 6-sulphates comprise repeating N-acetylgalactosamine and glucuronyl sulphate disaccharides, differing only in the siting of the sulphate ester linkage. Keratan sulphate is a polymer of disaccharides composed of an N-acetylgalactosamine and a D-galactose with a sulphate ester linkage of variable position. The relatively huge hyaluronate molecules have similar repeating disaccharides of N-acetylglucosamine and glucuronate composition, but are not sulphated and are unbranched.

These different chemical configurations result in different numbers of electrostatic charges, different degrees of interactions between adjacent chains and other variations in bulk properties which determine many of the physical and chemical properties of the matrix.

Cell adhesion proteins. These include various proteins and glycoproteins which have recently been shown to play an important role in the adhesion of chondrocytes to matrix components. The best known is *chondronectin*, a cartilage glycoprotein important in the adhesion of chondroblasts to Type II collagen fibres in the presence of chondroitin sulphates. It has an Mr of about 150 000 (Hewitt et al 1980); and is synthesized by chondrocytes. Another adhesion molecule is anchorin CII (von der Mark et al 1986), a glycoprotein with an Mr of 34 000 found at chondrocyte surfaces and binding specifically to Type II collagen. There may also be other similar substances which play a part in stabilizing the structural complex of cells and matrix components.

Lipid is also present as part of the matrix, in the form of small spheroidal masses stainable with lipid-soluble dyes for light microscopy (Meachim & Stockwell 1978). Some of this may be the result of cell disintegration, or shedding of cytoplasm, and some

3.17 Diagram showing the fine structural organization of hyaline cartilage matrix. Depicted are large proteoglycan complexes and type II collagen fibres of the coarser cross-banded and the narrower varieties. Proteoglycan complexes bind to the surface of these fibres and link them together. Detail shows the arrangement of glycosaminoglycans and core filaments. See text for further details.

is membranous material or portions of cells, e.g. the matrix vesicles found in epiphyseal plates (p. 302).

In summary, the matrix of cartilage is a labyrinth of protein and proteoglycan filaments, its interstices occupied by bound water, cations and various other small molecules. Reactions to mechanical stresses accordingly reflect interacting contributions from all these structural elements. The properties of the 'walls' and 'contents' of this molecular meshwork also affect the diffusion of nutrient and waste material throughout this remarkable tissue.

Synthesis of matrix by chondrocytes (3.15). It is now well established that all of the major components of the matrix are synthesized and secreted by chondrocytes. These processes have been studied extensively with a battery of biochemical, structural and autoradiographic methods, and more is known about the synthetic activity of these cells than any other cell type of connective tissue (see the reviews by Prockop et al 1979, Dorfman 1981, Mayne & von der Mark 1983). To summarize these findings briefly, collagen is synthesized within the granular endoplasmic reticulum in the same way as in fibroblasts (p. 65), except that Type II rather than Type I procollagen chains are made (Olsen 1981). These assemble into triple helices and some carbohydrate is added at this stage. After transport to the Golgi apparatus, where occurs further glycosylation, they are secreted as procollagen molecules into the extracellular space. Here, terminal registration peptides are cleaved from their ends, so forming tropocollagen molecules, and final assembly into collagen fibres takes place. In other regions of the granular endoplasmic reticulum, core proteins of the proteoglycan complexes are synthesized and addition of GAG chains is begun, completion of the process being carried out by the Golgi complex (Lohmander & Kimura 1986).

The large molecules of glucuronate appear to be synthesized enzymes inserted in the plasma membrane of the chondrocyte and are thought to be extruded directly into the matrix without passing through the endoplasmic reticulum (Prehm 1986).

NUTRITION OF CARTILAGE

Cartilage is often described as totally avascular. This is not wholly true but most cartilage cells are unusually distant from exchange vessels, which are mostly perichondrial. Between this perichondrial capillary network and chondrocytes, nutrient substances and metabolites diffuse along concentration gradients across the intervening matrix. This limitation is reflected in the restriction of most living cartilage masses to a few millimetres in thickness. Cartilage cells further than this from a nutrient vessel do not survive, and their surrounding matrix typically becomes calcified.

Cartilage Canals (Haines 1933–4, Hurrell 1934–5, Levene 1964, Blackwood 1965. Lutfi 1970, Brookes 1971, Stockwell 1978, Kuettner & Pauli 1983).

Nutrition is augmented, at least in cartilaginous epiphyses of large mammals (and their analogues in birds and reptiles) by branching of *cartilage canals* (Brookes 1971), each containing small vessels from centripetal rami of a perichondrial artery and vein. Perfusion (Brookes 1958) and histological examination (Wilsman & Van Sickle 1972) suggest that each canal contains a central small artery or arteriole surrounded by numerous venules and perivascular capillaries (3.18,19). The central vessel, at the blind end of the canal, forms several capillaries from which venules pass back to a parent perichondrial vein. Capillaries lined by fenestrated endothelium also branch off along the canal and, with terminal capillaries, mediate metabolic exchange with the cartilage mass (Wilsman & Van Sickle 1972). The canalicular vessels lie enclosed in loose connective tissue, abundant in fibroblasts and macrophages and continuous with the perichondrium.

Various theories have been proposed for their formation. The possibility that they are derived from perichondral vessels which have been engulfed by the appositional growth of cartilage around them is attractively simple, but does not explain many findings, e.g. that cartilage canals are formed more rapidly than the growing front of cartilage moves forwards. It appears

instead that canals actively invade cartilage, probably by the action of macrophages and other erosive cells (Wilsman & Van Sickle 1972); among the latter the endothelium of the central blood vessels may be important, as the ability of these cells to release chondrolytic agents has been demonstrated in tissue culture (Moscatelli et al 1981). However, chondrocytes themselves are also able to degrade cartilage matrix (*chondrocytic chondrolysis*, see Aydelotte et al 1986), and may also play a part in canal formation, either subsequently dying ahead of an advancing blood vessel to present it with a series of preformed cavities or, having freed themselves from the matrix, dedifferentiating to form other mesenchymally-derived cells of the cartilage canals.

Cartilage canals are not random, but have a characteristic pattern in each cartilage. Canals in a fetal epiphysis form two groups, the more conspicuous passing from the non-articular surfaces towards its centre, the less prominent from vessels in the ossification groove (of Ranvier, p. 299); capillary tufts at the ends of the latter are complex and glomerular, like those in postnatal

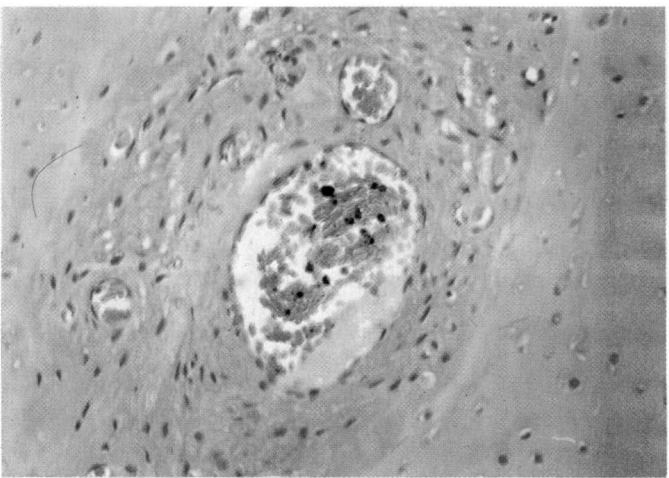

3.18 Section through the proximal cartilaginous end of the tibia of a young child showing a cartilage canal and its contained blood vessels, stained with haematoxylin and eosin.

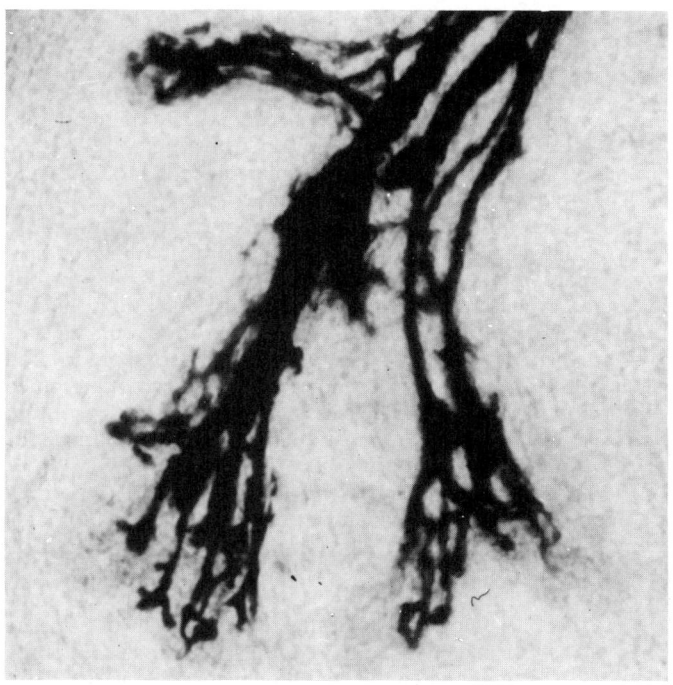

3.19 Capillary plexuses in the terminal arborization of a cartilage canal in the femoral condylar epiphysis of a seven-month human fetus. Specimen injected with Indian ink. (Provided by Murray Brookes, Department of Anatomy, Guy's Hospital Medical School, London.)

subchondral circulation (Trueta & Morgan 1960). According to Brookes (1971), the pattern of arteries in some adult bony epiphyses closely resembles the canalicular pattern in corresponding fetal epiphyses, concluding that vessels in the canals are forerunners of main epiphysial vessels in mature bone, and that vessels in the minor group of canals originating in Ranvier's groove are precursors of subchondral vascular networks on the epiphysial side of growth cartilages in postnatal life.

Cartilage canals are believed to assist nutrition of large cartilages, which would depend otherwise solely on perichondrial capillaries. This process appears to be faster than formerly considered both in young cartilage and in mature (vide supra); also labelled ions are concentrated by chondrocytes of mature articular cartilage with unexpected rapidity (Hall 1965), despite the absence of canals and perichondrium. Furthermore, the half-life of proteoglycans in rabbit articular cartilage is four days (Mankin & Lipiello 1969). While it might be expected that large cartilages, such as fetal limb epiphyses, possess canals, it is surprising that they also occur in carpal and sesamoid cartilages in early fetal hands, and even in the epiphyses of human fetal phalanges at 11 cm crown-rump length (Gray et al 1957).

While not denying a nutritional function, most workers consider that cartilage canals are involved in forming secondary centres of ossification (and primary centres in bones without epiphyses). But puzzling anomalies remain. Distal phalangeal and proximal metacarpal epiphyses develop elaborate cartilage canals, though lacking ossification centres. There is also no clear connection between onset of chondrification, the first appearance of canals and onset of ossification; all occur at times peculiar to each skeletal element (Brookes 1971). The contents of canals appear to provide osteogenic cells and capillaries for ossification when an ossification centre appears. The control of the timing of these events remains a complete enigma.

In the mandibular condylar cartilage, canals form until the second year, when they disappear. Canals in laryngeal and nasal cartilages first form in the seventh month in utero, persisting until old age; in costal cartilages they arise in the first year, reaching the centre of the shaft about the tenth year. In long-lived canals, marrow elements may appear after the twentieth year, persisting until the sixties, when atrophy supervenes, canals retaining a mucinous material.

VARIETIES OF CARTILAGE

Cartilage may be *hyaline*, *white* and *yellow fibrocartilage* and *cellular cartilage*. It may also be *calcified*.

Hyaline cartilage (3.20A) has a glassy, bluish, opalescent, homogeneous appearance, firm consistency and some elasticity. Costal, nasal, some laryngeal, tracheobronchial, all temporary and most articular cartilages are hyaline but there are marked differences in size, shape and arrangement of cells and fibres and proteoglycan composition at different sites and ages. Its cells vary in shape from quite flat near the perichondrium to rounded or bluntly angular deeper in the tissue. They are often in groups of two or more (*cell nests* or *isogenous cell groups*) which are the offspring of a common parent chondroblast; such cells have a straight outline where apposed to each other, but a rounded contour in general. The matrix is typically basophilic and metachromatic, both more pronounced where recently formed matrix bounds a lacuna. This distinctive zone is the *territorial matrix* or *lacunar capsule*, contrasting with the pale-staining *interterritorial matrix* between cell nests.

The fresh matrix is transparent and structureless, or dimly granular like ground glass when viewed by standard light microscopy. By polarized light or electron microscopy it appears permeated by fine fibrils and fibres of collagen, 10–20 nm in diameter (see p. 285). Electron microscopy shows that around each chondrocyte there is some zonation of the matrix. In well-fixed tissue where no cell shrinkage has occurred, the irregular surface of the chondrocyte is in direct contact with the matrix. However, there is often an encapsulating layer of proteoglycans containing fine filaments of uncertain composition, but no typical collagen, immediately surrounding the cell, probably corresponding to the limits of the lacuna seen by light microscopy. Outside this there

are typical collagen fibres, often arranged in a basket-like network around one or more cells in a group, grading into the parallel fibre domain between cells, the interterritorial zone. Chondroitin sulphates are particularly prevalent in the vicinity of chondrocytes, whereas keratan sulphate predominates in the interterritorial zone.

Articular hyaline cartilage (3.20E) covers articular surfaces in synovial joints, providing an extremely smooth, resistant surface bathed by synovial fluid, allowing almost frictionless movement (coefficient of friction 0.01–0.02, decreasing as loading increases). Its elasticity, with that of other articular structures, dissipates effects of concussions, giving the whole articulation some flexibility, particularly near extremes of movement. Articular cartilage is admirably constructed to resist the large compressive forces generated by weight transmission, especially during movement (p. 269).

It does not ossify, and varies from 1 to 7 mm in thickness; it is moulded to the shape of the underlying bone but its surface smooths, often accentuates and modifies the surface geometry. On convex osseous surfaces it is thickest centrally, the reverse being true of concave surfaces; its thickness decreases from maturity to old age. The surface of articular cartilage is devoid of perichondrium, but around its edges the synovial membrane overlaps then grades into its structure.

Adult articular cartilage shows a zonation in structure with increasing depth from the surface (Barnett et al 1961, Davies et al 1962, Palfrey & Davies 1966, Ghadially 1983). Except perhaps for a thin surface lamina, the matrix is pervaded by collagen fibres, those near the surface around lacunae being fine, plexiform and, at the most, faintly banded, whereas elsewhere they are coarser with the usual 64 nm banding. Their arrangement is variously described as plexiform, helical, or in the form of serial arcades radiating from the deepest zone to the surface, where they pursue a short tangential course before returning radially, as most workers confirm. It has long been known (Hultkrantz 1898, Amprino 1948) that if the surface of articular cartilage is pierced by a pin a longitudinal 'split-line' shows after withdrawal and that, for any given joint, the patterns of split-lines are constant and distinctive (cf. cleavage lines of skin). Bullough & Goodfellow (1968), using polarized light and electron microscopy, demonstrated that splits follow the *predominant* directions of collagen bundles in *tangential* zones of cartilage (vide infra). Such patterns probably reveal 'tension trajectories' set up in surrounding cartilage by local compression during habitual activities. Subsequent investigations have largely confirmed the views of Bullough & Goodfellow, e.g. split-line analysis by transmission electron microscopy (Meachim et al 1974), polarization microscopy (Ortmann 1975), and scanning microscopy (Minns & Steven 1977). Split-line arrangement, where superficial tangential bundles are not parallel, have been considered by Ortmann (1975); theoretical models for dynamics of articular cartilage have been proposed by Kempson et al (1968) and Mow et al (1974). It is pertinent to recall that some earlier workers considered split-line, and hence predominant collagen, orientation to reflect *direction* of *habitual movements*, the relation of this to tension trajectories being unknown. Others (e.g. McCall 1968) deny any significant collagen orientation in articular cartilage. Serafini-Fracassini & Smith (1974) propose that the basic function of articular cartilage collagen is not resistance to tensile stresses, but provision of *fixation anchors* for viscoelastic *domains* of proteoglycan molecules when subjected to deforming and displacing stresses. Clearly, no model is adequate unless it includes the dynamics of *all interacting* constituents of cartilage. (For an extensive review see Freeman 1979 and 3.20E.)

In the *superficial* or *tangential stratum* (*zone 1*) cells are small, oval or elongated, flat and parallel to the surface, and surrounded by fine tangential fibres; they have a few short projections, mainly from their lateral borders and deep surfaces, and display scattered mitochondria and small cisternae of granular endoplasmic reticulum; their Golgi apparatus is not prominent. Cytoplasm is free of filaments, unlike cells in deeper zones. The nature of the *articular surface*, the thin superficial layer of zone 1 (*lamina splendens*), remains uncertain. ('Lamina splendens' was introduced b

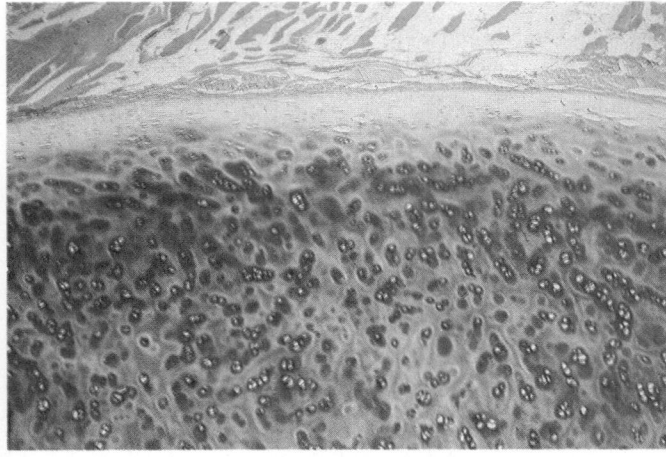

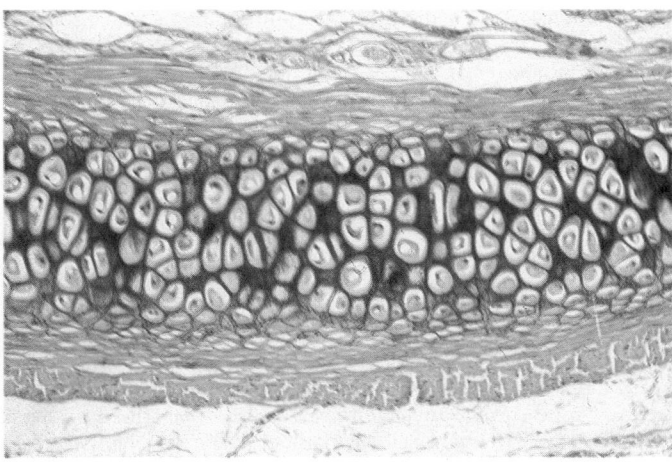

3.20 Different types of cartilage: A. Hyaline cartilage (chondral), (see also **3**.13 A,B) H & E, magnification × 150.

B. Elastic fibrocartilage, stained with Gomori's elastin stain (blue-black), and van Gieson's collagen stain (pink), which shows the fibrous perichondrium clearly. Pinna (rabbit); magnification × 150.

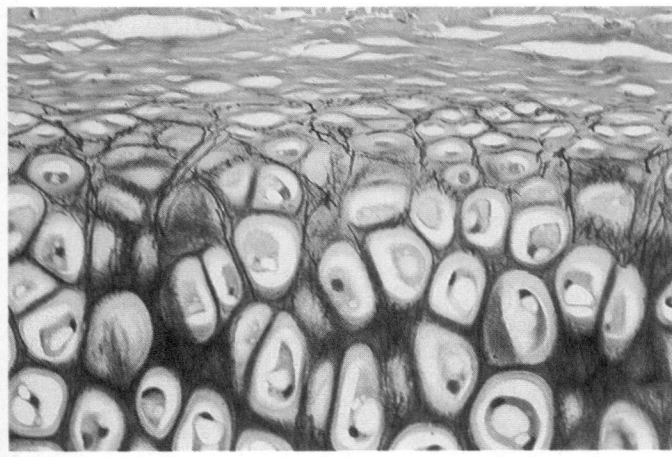

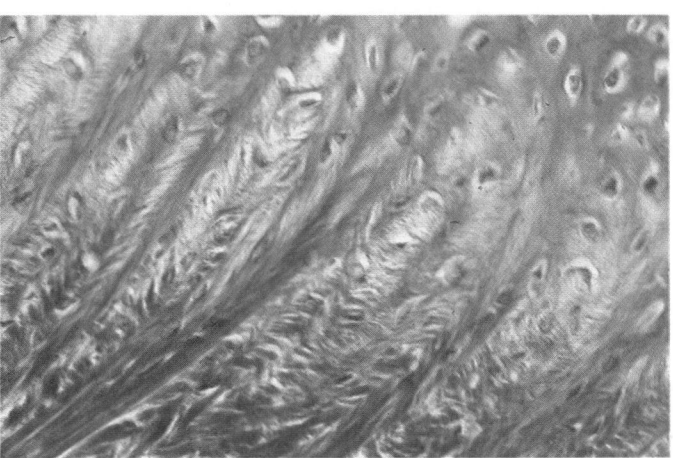

C. Higher magnification of B, showing fibrous perichondrium, chondroblasts and larger chondrocytes embedded in a matrix rich in elastin fibres, magnification × 400.

D. White fibrocartilage in late fetal intervertebral disc, showing chondroblasts between coarse collagen fibres derived from the annulus fibrosus; Mallory's triple stain, magnification × 150.

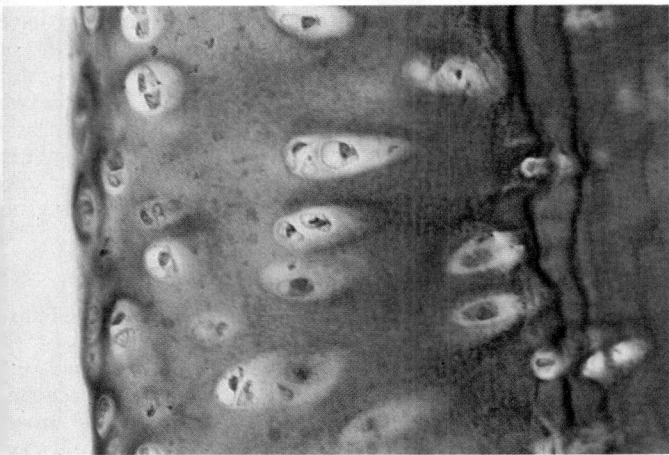

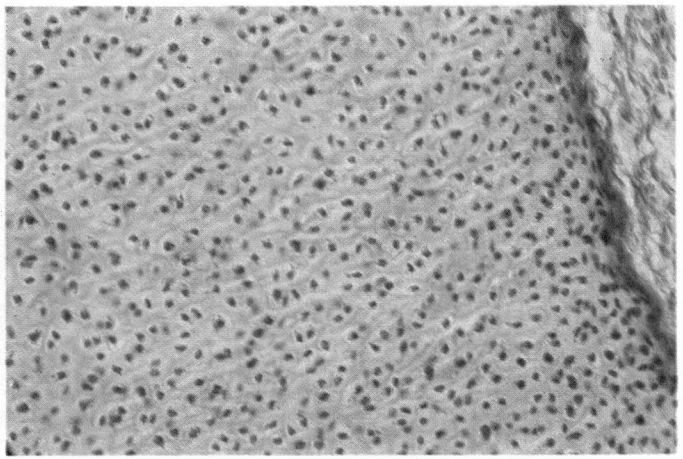

E. Articular cartilage stained with silver, showing cellular arrangement of the different layers. Note the absence of a periosteum, the superficial flattended cells of layers I and II, the more rounded chondrocytes of layer III, the lines of calcification in the deeper layer (IV) and the lamellar bone adjacent to the cartilage (see text for details). Interphalangeal joint (Rhesus monkey), magnification × 400.

F. Fetal cartilage, human phalanx, stained with Mallory's triple stain. Note the highly cellular appearance. The cells are small and almost uniform in distribution. Magnification × 150.

MacConaill 1951, for the brilliantly illuminated surface zone, in contrast to the dark main mass, visible in *oblique* sections by negative phase-contrast. Dimensions thus obtained cannot be compared with the data of other techniques, which limits the term's usefulness.) This surface zone is said to be a cell-free, 3 μm thick layer of compact, non-banded, randomly arranged, fine filaments 4 to 12 nm in diameter (Weiss et al 1968), which appear to branch frequently in a scanty proteoglycan ground substance. Deep to the lamina splendens are typical banded collagen fibres, regularly tangential, their diameters increasing with depth. These observers also noted that flat, oval cells deep in zone 1 closely resemble quiescent mature fibrocytes; they considered that synovial articular surfaces are probably formed by fine modified collagen filaments, their cells of origin more deeply placed. Others (Mow et al 1974) designated the surface layer of zone 1 the *superficial tangential zone*, described as sheets of tightly *woven* collagen fibrils parallel to the surface, from 5 to 20 nm in diameter and forming a layer from 1 to 200 μm in thickness. It is said to be poor in glycosaminoglycans but rich in hyaluronate.

The deeper cells of the *transitional* or *intermediate stratum* (*zone 2*) are larger, rounder, single or in isogenous groups. Most are largely typical active chondrocytes in ultrastructure (vide supra) and around them pass oblique collagen fibres.

Deeper still, in the *radiate stratum* (*zone 3*), cells are large, round, often in vertical columns, with intervening radial collagen fibres.

The deepest layer or *calcified stratum* (*zone 4*) adjoins the *hypochondral osseous lamina* of the epiphysis. These adjacent surfaces show reciprocal fine ridges, grooves and interdigitations, which, with the confluence of their fibrous arrays, resist shearing stresses due to postural changes and muscle action.

Concentrations of glycosaminoglycans vary: that of chondroitin sulphate increases fivefold in the intermediate zone, reducing to threefold in zone 4; but that of keratan sulphate increases linearly with depth; the latter is mainly in the interterritorial matrix, whereas chondroitin sulphates are largely circumlacunar.

This sequence of structural changes from normal hyaline cartilage through columnar arrays of cells, a zone of calcified cartilage and eventually epiphysial bone, is also typical of cartilaginous growth plates (p. 305). It follows radial epiphysial growth by the extension of endochondral ossification into overlying calcified cartilage. This ceases in maturity, but the zones persist throughout life. The same terminal mechanism also occurs in bones lacking epiphyses.

Cells of articular cartilage divide by mitosis, but mitoses are few except in young bones. Progressive loss of superficial cells from normal young joint surfaces, and their replacement by cells from deeper layers, is unconfirmed; but degenerating cells may occur in any of the four zones. This probably accounts for progressive reduction in cellularity of cartilage with advancing age, particularly in superficial layers (Stockwell 1967) and probably contributes to the variable lipid content of the matrix.

Articular cartilage may derive nutriment by diffusion from three sources: vessels of the synovial membrane, synovial fluid and hypochondral vessels of an adjacent medullary cavity, many capillaries from which penetrate and even traverse the calcified cartilage (Holmdahl & Ingelmark 1950, Barnett et al 1961), and have been estimated to contact 1–7% of the osseous aspect of the cartilage. For *relative* contributions from these sources evidence is lacking, but accurate determinations of permeability of articular cartilage are available (Maroudas et al 1975). Small molecules freely traverse articular cartilage, with diffusion coefficients about half those in aqueous solution. Larger molecules have diffusion coefficients inversely related to molecular size, e.g. glucose, inulin and haemoglobin (relative molecular masses: 180, 5000 and 68 000) have molal distribution coefficients in the ratio 85:11:1. Permeability of cartilage to large molecules is much affected by variations in its glycosaminoglycan content; e.g. a threefold increase multiplies the partition coefficient a hundredfold.

In costal cartilage cells and nuclei are large; the matrix is usually homogeneous and transparent, and tends to fibrous

striation, especially in old age. In their thickest parts a few large vascular channels and sometimes medullary elements may appear (p. 287). The xiphoid process and the cartilages of nose, larynx and trachea (except the elastic fibrocartilaginous epiglottis and corniculate cartilages) resemble costal cartilage in microstructure. The arytenoid cartilage changes from hyaline at its base to elastic cartilage at apex. Hyaline cartilages, after adolescence, are prone to calcification, especially in costal and laryngeal sites. Quantitative studies demonstrate diminishing cellularity throughout costal cartilages with advancing age (Stockwell 1979).

White fibrocartilage (3.20D) is dense, fasciculated, white fibrous tissue, with attendant fibroblasts and small interfascicular groups of chondrocytes; the cells are ovoid and surrounded by concentrically striated matrix. When amassed, as in intervertebral discs, fibrocartilage has great tensile strength with appreciable elasticity. In lesser amounts, as in articular discs (p. 470), glenoid and acetabular labra (pp. 501 and 519), the cartilaginous lining of bony grooves for tendons and some articular cartilages (vide infra), it is a tissue of strength and elasticity able to resist repeated pressure and friction. This tissue is unlike other types of cartilage in having much Type I (general connective tissue) collagen in its matrix; it is perhaps best regarded as a mingling of the two types of tissue, e.g. where a ligament or tendinous tissue inserts into hyaline cartilage, rather than a specific type of cartilage. The fibrocartilage of joints is often altogether lacking in Type II collagen and possibly represents a quite distinctive class of connective tissue, designated cartilage only for the convenience of histologists. The articular surfaces of bones which ossify in mesenchyme are covered by white fibrocartilage (e.g. squamous temporal, mandible and clavicle). Electron microscopy of such articular cartilages (Silva & Hart 1967) shows that *deep layers*, adjacent to hypochondral bone, resemble calcified, degenerative and hypertrophic regions of the radial zone of *hyaline* articular cartilage. The *superficial zone* contains dense parallel bundles of thick collagen fibres, interspersed with typical dense connective tissue fibroblasts and scanty ground substance. Fibre bundles in adjacent layers alternate in direction as in the cornea. A *transitional zone* of irregular bundles of coarse collagen and fibroblasts with prominent Golgi complexes separates the two. The fibroblasts are probably involved in elaboration of proteoglycans, or may be a germinal zone for deeper cartilage. For permeability of white fibrocartilage in intervertebral discs consult Maroudas et al (1975).

Yellow elastic fibrocartilage (3.20B,C) occurs in external ears, corniculate cartilages, epiglottis and apices of the arytenoids. It contains typical chondrocytes, but its matrix is pervaded by yellow elastic fibres except around lacunae, where it resembles typical hyaline matrix. As elsewhere its elastic fibres are unaffected by acetic acid, have an affinity for orcein and show no periodic banding (Sheldon & Robinson 1958). Cox & Peacock (1977) have reviewed the ultrastructure, histogenesis and growth of chondroblasts and matrix in the rabbit's pinna. Histogenesis of elastic fibres of cartilage (and probably other tissues) is in two stages; a glycoprotein microfibrillar framework of oxytalan first appears and is then impregnated by elastin (p. 67). Most sites in which elastic fibrocartilage occurs have *vibrational* functions, such as laryngeal sound-wave production and their collection and transmission in the ear.

Histogenesis, Growth and Regeneration

Normally cartilage is formed in embryonic mesenchyme (Glenister 1976, Serafini-Fracassini & Smith 1974, Cox & Peacock 1977). Mesenchymal cells proliferate and become tightly packed, the shape of their condensation foreshadowing that of a subsequent cartilage. The cells become rounded, with prominent rotund or oval nuclei but indistinct intercellular boundaries. Each cell soon begins to secrete a surrounding basophilic halo of matrix, composed of a delicate network of fine Type II collagen filaments and cartilage proteoglycans, indicating differentiation into chondroblasts, (3.20F). Continued secretion of matrix further separates the cells and so typical hyaline cartilage is now recognizable. In other sites collagen synthesis predominates, many cells becoming fibroblasts and chondroblastic activity appearing only in isolated groups or rows of cells. These are soon surrounded by dense bundles of collagen fibres to form white fibrocartilage. Similarly, elsewhere, the matrix of early cellular cartilage is permeated first by anastomosing oxytalan and, later, elastin fibres. In all cases, developing cartilage is surrounded by condensed mesenchyme differentiating into a bilaminar perichondrium in which the cells of the outer layer become fibroblasts secreting a dense collagenous matrix lined externally by vascular mesenchyme; the inner layer of perichondrium contains differentiated but mainly resting chondroblasts or pre-chondroblasts.

The growth of cartilage is both *interstitial* and *appositional*.

Interstitial growth (i.e. from within the cartilage) follows continued mitosis of early chondroblasts throughout the mass. When such a cell divides, its descendants temporarily occupy the same lacuna but are soon separated by thin septa of matrix, which thicken and further separate cells. Continuing division leads to isogenous groups. Interstitial growth is obvious only in young cartilage, where presumably plasticity of matrix permits continued expansion. However, the manner of matrix expansion and reorganization and indeed the factors determining total shape of a cartilage are as yet poorly understood.

Appositional growth at the cartilage surface results from the proliferation of cells of the internal, chondrogenic layer of perichondrium. Some resultant superficial chondroblasts secrete matrix around themselves, thus creating superficial lacunae. This process, continuing, adds additional surface while the entrapped cells can now participate in interstitial growth. Apposition is often stated to be prevalent in more mature cartilages, but interstitial growth must persist for long periods in epiphysial cartilages (pp. 279, 305).

Regeneration. In mammals regeneration of lost cartilage is poor, defects being slowly filled by vascularized connective (granulation) tissue, which may become less vascular and persist as fibrous tissue. Occasionally, cells in such tissue may become chondroblasts, secreting thin capsules of matrix.

Normal cartilage growth depends on adequate nutrition and hormonal environment; disturbances due to these will be considered with endochondral ossification (p. 304). With advancing years the matrix of many permanent cartilages becomes *calcified*, also an essential step in endochondral ossification during earlier development.

Interesting features of cartilage are the low antigenicity of its matrix, its relatively low vascularity and the isolation of its chondrocytes in lacunae, which permit successful homotransplantation without marked cellular or humoral immune reaction. The paucity of blood vessels is particularly striking; recent evidence suggests that their ingrowth is actively inhibited by factors released by chondrocytes (see Kuettner & Pauli 1983).

Bone as a Tissue

Bone is essentially a highly vascular, living, constantly changing mineralized connective tissue. It is remarkable for its hardness, resilience and regenerative capacity, as well as its characteristic growth mechanisms. Like all other connective tissues (see p. 58), bone consists of cells and an intercellular matrix, the great majority of its cells, osteocytes, lying embedded within it (3.21). The matrix is composed in part (about 40% dry weight in mature bone) of organic materials, mainly collagen fibres, and the rest consists of inorganic salts rich in calcium and phosphate. Together these give bone its unique mechanical properties (see, e.g. Hancox 1972, Currey 1984). Vascular canals ramify within bone, providing its cells with metabolic support and creating avenues of entry for other cells, including osteoclasts, capable of removing bone, and osteoblasts which can deposit it. While these features are found in all bone, their details differ widely with developmental state, site, prevailing mechanical forces and the metabolic state of the body. Bone's collagen framework, permeated with mineral salts, varies from almost randomly orientated coarse bundles (woven-fibred bone) when young, to a sys-

tem of highly ordered, parallel-fibred sheets or lamellae (lamellar bone) in the mature condition. Collagen fibres and mineralized matrix together usually form minute cylinders (osteons) arranged concentrically around blood vessels both in woven and lamellar bone while, in the mature state, the inner and outer surfaces of bones are lined by a few layers of continuous circumferential (outside) and endosteal (inside) lamellae. Bone may also be penetrated by bundles of collagen fibres derived from surface structures including tendons and ligaments (bundle bone). The outer surface of bone is always lined by a fibrocellular layer, the periosteum, and on the inner surface is a similar, though thinner, endosteum. In these layers lie osteoblasts, osteoclasts and other cells important in the biology of bone. The texture of mature bone also varies between dense (compact) and spongy (cancellous) osseous tissue which have distinctive mechanical and metabolic roles, often related to their positions within bones (diaphyseal, metaphyseal, epiphyseal, etc).

Developmentally, bone may form either by the direct transformation of condensed mesenchyme (intramembranous bone) or be preceded by a cartilage model which bone later replaces (endochondral bone). However, bones of different origins may show any of the features mentioned above, and can only be distinguished by a study of their genesis.

Because of these different types of classification, it may help to refer to the table set out below where the more commonly used terms have been grouped. The concepts which these reflect will be treated more fully in the following pages.

Classification and Terminology

1. **Macroscopic appearance of cut surfaces** (3.5).
   (a) *Compact bone*—the ivory surface layers of mature bone.
   (b) *Trabecular bone*—the interior of mature bones (also termed *cancellous* or *spongy* bone). Early embryonic bone is also spongiose—the *primary spongiosa* (*Os spongiosum primum*).
2. **Developmental origin**
   (a) *Intramembranous* (*mesenchymal* or *dermal* bone)—formed by direct transformation of condensed mesnchyme.
   (b) *Intracartilaginous* (*cartilage* or *endochondral* bone)—replacing a preformed cartilage model.
3. **Regions of long bones**
   (a) *Diaphysis*—intermediate region or shaft.
   (b) *Metaphysis*—developing, juxta-epiphysial regions of shaft.
   (c) *Epiphysis*—extremity with a separate centre of ossification.
4. **Organization of collagen fibres**
   (a) *Woven* bone (*coarse-bundled* bone), with an irregular collagen network—includes embryonic bone, isolated patches in adult bone, and repair tissue in fractures.
   (b) *Parallel-fibred* bone—includes all forms of lamellar bone and non-lamellar primary osteons.
5. **General microstructure**
   (a) *Non-lamellar* bone—includes early woven bone and primary osteons.
   (b) *Lamellar* bone—almost all mature bone.
6. **Disposition of lamellae**
   (a) *Circumferential* lamellae (*primary* lamellae)—parallel to both periosteal and endosteal surfaces.
   (b) *Osteonic* lamellae (*secondary* lamellae)—concentric lamellae around vascular canals of mature bone.
   (c) *Interstitial* lamellae between osteons.
7. **Types of osteon**
   (a) *Primary osteons* (*atypical Haversian systems*)—first formed, non-lamellar osteons.
   (b) *Secondary osteons* (*typical Haversian systems*)—concentric lamellae around vascular canals of mature bone.
8. **General terms**
   (a) *Surface bone*—usually circumferential lamellae but may include woven and bundle bone.
   (b) *Interstitial bone*—between osteons; often lamellar remnants of secondary osteons but may include woven or primary osteon fragments (vide infra).

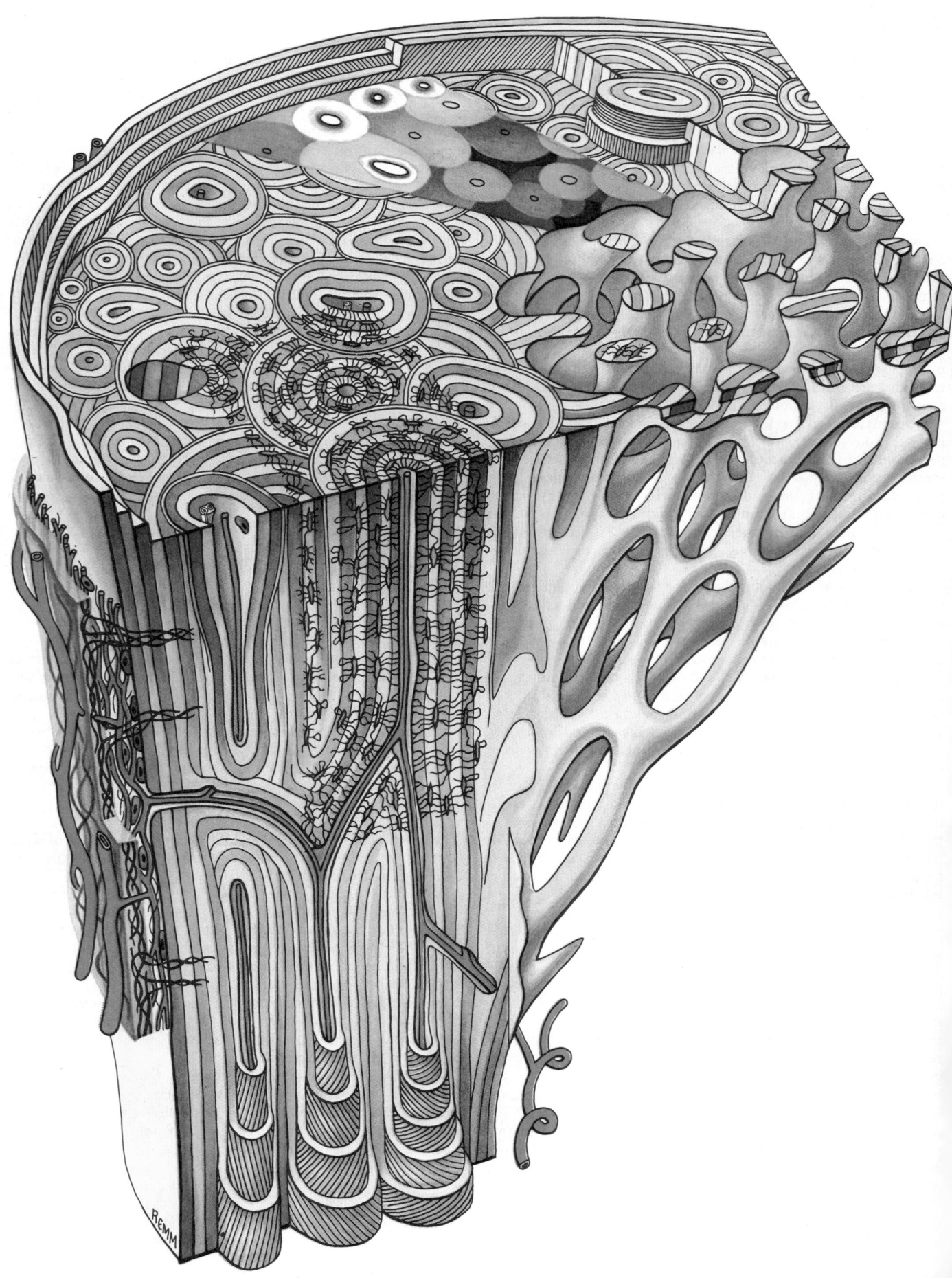

3.21 Schematic diagram of some of the main features of the microstructure of mature bone seen in both transverse section (top) and longitudinal section. Areas of compact and cancellous bone are included. The central area in transverse section simulates a microradiograph, the densities reflecting variations in mineralization. Note the general construction of the osteons, the distribution of the osteocyte lacunae, the Haversian canals and their contents, resorption spaces and the different views of the structural basis of bone lamellation. See text for more detail.

Structure of Mature Bone

MACROSCOPIC STRUCTURE

Macroscopically, living bone is white, with either a dense texture like ivory (*compact bone*), or honeycombed by large cavities, the bone being reduced to a latticework (*cancellus*) or bars and plates (*trabeculae*), in which case it is called *trabecular, cancellous* or *spongy bone* (3.21). Compact bone is usually limited to the cortices of mature bones (cortical bone) and is of supreme importance in providing their strength. Its thickness and architecture vary for different bones, reflecting their overall shape, position and functional roles. In contrast, cancellous bone lies chiefly in their interior (3.22), and particularly, in the case of long bones, within their expanded ends (metaphyses and epiphyses). Cancellous bone gives additional strength to bones and supports the tissues of their marrow. It also forms a reservoir of metabolic calcium and phosphate which can be readily added to or withdrawn by cellular action under hormonal control; this property is related to its relatively large surface area in comparison with a similar mass of compact bone, making bone resorption and deposition relatively easy.

The proportions of compact to cancellous bone vary greatly. In a long bone the thick cylinder of compact bone forming the shaft presents only a few trabeculae and spicules on its inner surface so that a large central medullary or marrow cavity is enclosed, communicating freely with the intratrabecular spaces of the expanded bone ends. In other bones, especially flat ones such as the ribs, the interior is uniformly cancellous, compact bone forming the surface. These cavities are filled with marrow, either red, haemopoietic or yellow, adipose, its character varying with age and site. In the mastoid bone of the skull, many of the cavities within are filled with air, a situation which is unique in the human body, although common in the wing-bones of most birds.

MICROSCOPIC STRUCTURE

As stated earlier, bone is composed of cells embedded in a stiff calcified matrix; these components will now be described in some detail, first individually then in terms of their overall organization.

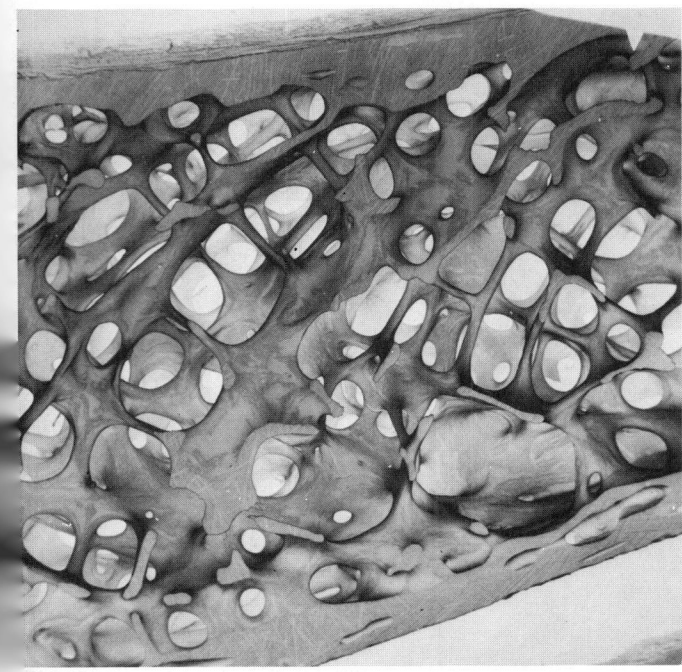

3.22 Low power scanning electron micrograph of a segment of a human rib. Note the cortical shell of compact bone and the delicate lattice of trabeculae, separated by labyrinthine marrow spaces. (Reproduced by permission of W J Whitehouse, E D Dyson and C K Jackson, Medical Research Council Radiology Unit, Harwell.) Such specimens are used currently for quantitative studies of bone architecture; magnification × 7.

The Cells of Bone

These consist of a number of types, including: (a) osteoprogenitor cells which give rise to various other bone cells; (b) osteoblasts which lay down bone; (c) osteocytes within bone; (d) bone lining cells on its surface; and (e) osteoclasts which erode it. In addition to these are the cells of its vascular and nervous system and other components of the periosteum, endosteum and marrow. Of these different cells, the first four are closely related, while the osteoclasts arise from a quite unrelated source. Vascular and nervous tissue are, of course, of external origins.

Osteoprogenitor cells. These can be defined as persistent, migratory stem cells which can proliferate and differentiate into osteoblasts prior to bone formation. They resemble young fibroblasts and are, like these, of mesenchymal origin. Such mesenchymally-derived cells are responsible for all bone formation during early development. In intramembranous bone they aggregate and undergo proliferation before transforming into osteoblasts while, during endochondral bone formation, similar cells migrate with the ingrowth of blood vessels from the perichondrium into areas of degenerating cartilage, then likewise differentiate into osteoblasts.

Cells with mesenchymal features, derived from bone, can be grown in culture and these give rise to new osseous tissue when implanted into animals. Ectopic bone formation can also be induced experimentally in various tissues of the body, e.g. skeletal muscle, by implantation of demineralized bone, dentine or urinary bladder epithelium; so it appears that there are two types of osteoprogenitor cell, one totally committed to bone formation (*committed osteoprogenitors*), found associated with bones, and the other (*inducible osteoprogenitors*) widely present in connective tissue and probably able to differentiate into various connective tissue cells (e.g. fibroblasts, adipose cells, chondroblasts, osteoblasts) depending on the nature of the inducer (Friedenstein 1976). It is also noteworthy that osteoblasts and osteoclasts were at one time thought to arise from the same progenitor cell, but it now appears that osteoclasts have a quite separate origin (Kahn et al 1983, see Osteoclasts p. 295; see also 3.40).

Osteoblasts (3.27,35,48). These are basophilic, roughly cuboidal cells about 15–30 μm across, found mainly on the surfaces of maturing or remodelling bone where they form a covering monolayer (see Friedenstein 1976, Kahn et al 1983). In relatively quiescent adult bones they appear to be present chiefly on the endosteal rather than periosteal surfaces, but also occur deep within compact bone where osteons are being remodelled. They are responsible for the synthesis, deposition and mineralization of bone matrix (Rodan & Rodan 1983) and, on becoming embedded in this, they finally transform into osteocytes (see also 3.38).

Ultrastructurally, osteoblasts have features typical of protein-secreting cells, i.e. a pale (euchromatic) oval nucleus placed away from the secreting surface, an extensive granular endoplasmic reticulum, large Golgi complex and numerous secretory vesicles. The cells are orientated so that their secretory surface abuts on to the adjacent bone. They have prominent bundles of actin, myosin and other cytoskeletal proteins associated with the maintenance of cell shape and motility. Their plasma membranes have many extensions, some contacting neighbouring osteoblasts and osteocytes at communicating junctions, by means of which the actions of quite large groups may be coordinated, as seen, e.g. in their concerted action to form large domains of parallel collagen fibres. There are irregular gaps between the cells which allow ready diffusion through the osteoblast monolayers to and from the underlying bone.

The surfaces of osteoblasts are rich in alkaline phosphatase activity, located at the plasma membrane. Some of this enzyme is shed and reaches the circulation where it can be detected in conditions of rapid bone formation or turnover. Osteoblasts of young bones have protrusions which appear to bud off vesicular structures (matrix vesicles) important in mineral deposition (vide infra p. 302). A major activity of osteoblasts is to synthesize and secrete the organic matrix of bone, i.e. Type I collagen (Wright & Leblond 1981) and various other macromolecules including the proteins osteocalcin (of uncertain function), osteonectin, which binds strongly to collagen, catalyzes the nucleation of calcium

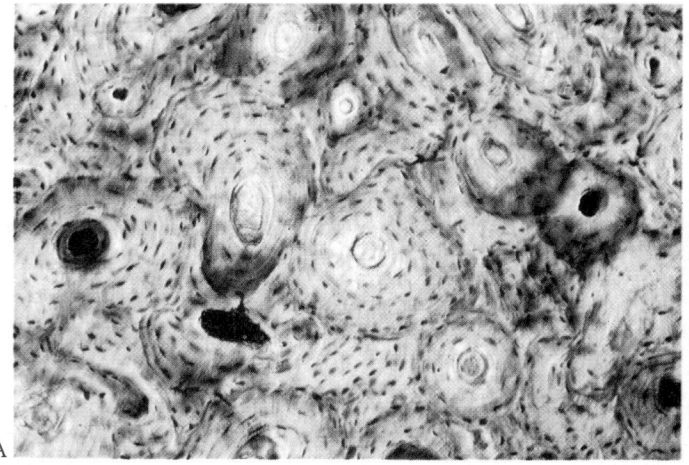

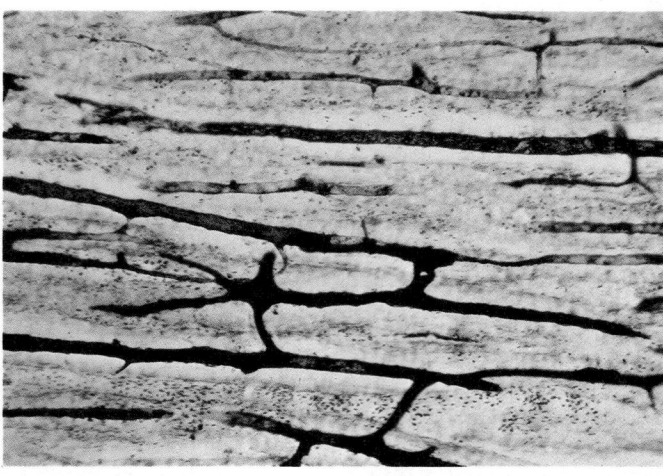

3.23 Transverse (A) and longitudinal (B) ground sections of compact bone from human femoral shaft. Note variation in shape and size of osteons and their canals and in distribution of lacunae in A. Haematoxylin staining in

B has emphasized the osteons, predominantly longitudinal but showing intercommunications. (Material for **3.23–25** prepared by David Ristow, Guy's Hospital Medical School.)

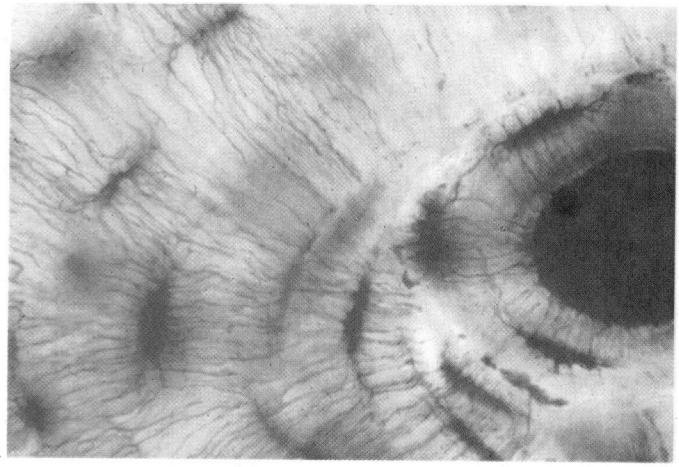

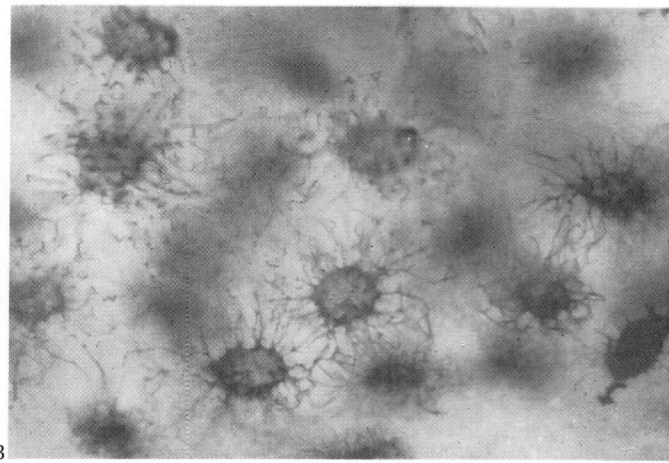

3.24A,B High-power view of part of an osteon in transverse section seen with transmitted light. Note the relation of the osteocyte lacunae and their canaliculi to each other, and to the central Haversian canal (black).

Tangential section of osteocyte lacunae and their associated canaliculi. Contrast with their appearance in A.

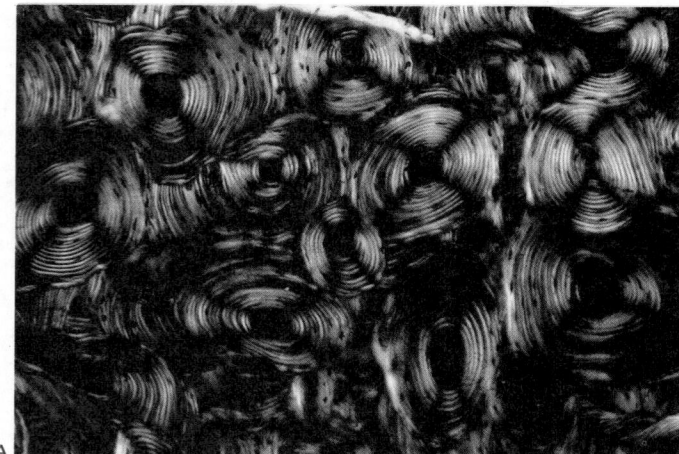

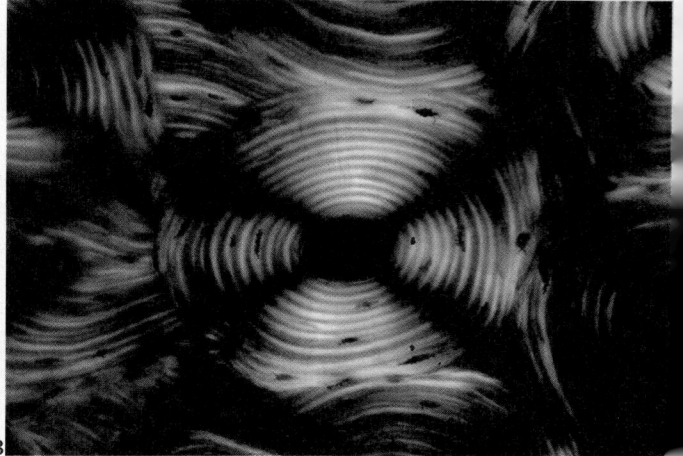

3.25A,B Ground transverse section of the compactum of the adult human femur photographed using polarization microscopy. Compare with **3.23A**.

A single secondary osteon viewed with high-power polarization optics to illustrate its lamellar architecture.

phosphate and may also be a cell adhesion factor, and some minor proteoglycans (see the review of this subject by Rodan & Rodan 1983). Prior to its mineralization this organic matrix is called osteoid or pre-bone (vide infra). An equally important osteoblast function is the mineralization of this matrix, alkaline phosphatase playing a vital role in this process. Collagen synthesis is carried out in much the same manner as in fibroblasts (see p. 65), initially

in the granular endoplasmic reticulum then in the Golgi apparatus, before being secreted to the exterior.

There is also much evidence that osteoblasts may play an important though indirect role in the hormonal regulation of bone resorption, since they bear receptors for parathyroid hormone and other stimulants of bone erosion. During bone deposition osteoblasts may inhibit osteoclast activity or may merely fail to stimulate them, but in the presence of bone resorption stimulators, e.g. parathyroid hormone (PTH), osteoblasts release various intermediaries which activate osteoclasts to remove osseous tissue. This subject, and others related to bone formation and turnover, will be considered more fully in the section on bone remodelling (p. 309).

Osteoblasts are differentiated, non-mitotic cells which arise from osteoprogenitor cells (vide supra) and also possibly by de-differentiation of osteocytes when these are released from bone during its resorption.

Osteoblasts may also themselves de-differentiate into osteoprogenitor cells under certain circumstances. When not active in bone resorption, they may also transform into the bone lining cells of endosteal surfaces, although this relationship is not certain.

Osteocytes (3.24,28). These constitute the major cell type of mature bone, lying scattered within its matrix, but interconnected by numerous cellular extensions to form a complex cellular network. They are derived from osteoblasts which have become enclosed in matrix, but retain contacts with each other and with cells at the surfaces of bone (i.e. osteoblasts and bone-lining cells) throughout their lifespan.

Structurally mature, relatively inactive osteocytes possess a lens-shaped soma, its long axis (about 25 μm) parallel to the surrounding bony lamella and its shortest axis transverse to it. The cytoplasm is faintly basophilic and contains relatively few organelles, including a little granular endoplasmic reticulum, a few free ribosomes and a small Golgi apparatus. The rather narrow rim of cytoplasm surrounds an oval nucleus. Younger, more active osteocytes are more rounded, with a well developed endoplasmic reticulum and larger Golgi complexes, both signs of greater synthetic and secretory activity.

Numerous dendrites emerge from the soma and branch a number of times to form an extensive tree. Such processes contain bundles of microfilaments and some smooth endoplasmic reticulum. At their distal tips they contact the dendrites of adjacent cells (other osteocytes and, at surfaces, osteoblasts and bone-lining cells), with which they form communicating ('gap') junctions and are thus in electrical and metabolic continuity.

The bone matrix surrounds the somata and dendrites but there appears to be a variable space between cell and cell wall containing extracellular fluid. From the cavity (lacuna) in which the soma lies extend up to 100 narrow branched tunnels or (canaliculi) about 0.25 μm wide containing the dendrites. Thus the bony matrix is riddled with minute branched canals and cavities which, beside housing living cells, provides a route for the diffusion of nutrients, gases and waste products for their maintenance (for a review see Atkinson & Hallsworth 1982). Canaliculi may infrequently extend through and beyond the reversal line surrounding an osteon or communicate with neighbouring systems and, usually at the outer limit of the osteon, loop back to their own lacuna. The walls of lacunae are lined with a variable (0.2–2 μm) layer of unmineralized organic matrix (osteoid).

The average lifespan of an osteocyte has been estimated as about 25 years; old osteocytes may retract their processes from the canaliculi and, when dead, their lacunae and canaliculi may become plugged with cell debris and minerals, so hindering diffusion through the bone. Dead osteocytes occur commonly in interstitial bone, becoming particularly noticeable by the second and third decades.

The exact functions of osteocytes are not yet clear but they must have an essential role in the maintenance of bone, as their death leads to the resorption of the matrix by osteoclast activity. They may be responsible for a small amount of matrix turnover, since the sizes of their lacunae can change and labelling of growing bone (e.g. with tetracycline) also indicates continuing matrix deposition. It is also possible that, by means of their communications with cells at the bone surface, they may act as local sensors of the mechanical and chemical state of the bone and initiate erosion or addition of matrix accordingly.

Bone lining cells. These are flattened epithelium-like cells found on the resting surfaces of bone, i.e. those not undergoing deposition or erosion (Menton et al 1982). They form continuous layers and are in contact with each other, and with neighbouring osteocytes, through communicating junctions. On the endosteal surface of narrow cavities they form the outer boundary of the marrow tissue and are present on the periosteal surface and line the system of vascular canals within osteons. Their role is not yet clear, but they may be inactive osteoblasts which can revert to the active state when suitably stimulated. They may also play an active role in regulating the differentiation of osteoprogenitor cells (Menton et al 1982).

Osteoclasts (3.29,30). These are large (20 μm or more) rounded cells with a variable number of oval, closely packed nuclei, often 15–20 (Gothlin & Ericsson 1976, Kahn et al 1983). They are found where there is active erosion of bone and lie in close contact with the bone surface in pits termed *resorption bays* or *lacunae of Howship*.

Osteoclasts contain numerous mitochondria and vacuoles, many being acid phosphatase-positive lysosomes. There is relatively little granular endoplasmic reticulum and only a small Golgi complex, unlike that in osteoblasts. The surface of the cell at the site of bone resorption is highly folded to form a ruffled membrane. Here, many lysosomes are concentrated at the bases of clefts between irregular cell extensions (*lobopodia*) and the plasma membrane has a dense, bristly undercoating similar to that of coated vesicles in other cells active in endocytosis (p. 29). Around the perimeter of the ruffled membrane is a zone of actin filaments, and here the osteoclast is closely attached to the bone surface, so forming a limit to the resorptive activities of the ruffled membrane. In tissue culture the ruffled membrane exhibits constant vigorous movement and active pinocytosis. Collagen fragments and other matrix debris have been found within its clefts and in cytoplasmic vesicles at their bases (Hancox & Boothroyd 1963). The size of the ruffled membrane varies with hormonal treatment, increasing when resorption of bone is stimulated and disappearing in the presence of agents favouring bone deposition, e.g. calcitonin.

Functionally, osteoclasts are responsible for the removal of bone, although exactly how is not known. Clearly, they cause demineralization, and there is also structural evidence of organic matrix destruction. Demineralization of the matrix occurs locally where the osteoclast's ruffled membrane approaches the bone surface; it has been proposed by Chambers et al (1984) that osteoclasts can only attack bone after its organic lining has been removed by osteoblasts or macrophages to expose the mineralized surface. However, there is also much evidence that osteoclasts can readily phagocytose collagen and other organic components of the matrix, so this question is not yet settled.

Agents responsible for stimulating osteoclasts to resorb bone appear to be multiple, including factors released by osteoblasts and probably various other cells such as macrophages and lymphocytes. Various humoral factors such as parathyroid hormone are also involved in bone resorption. This subject will be discussed more fully in a later section (see bone remodelling p. 309).

The cell lineage of osteoclasts has been much studied, and it is now established that they arise by fusion of mononuclear cells which originate in the bone marrow (Ash et al 1980, Ko & Bernard 1981) rather than from osteoblasts or osteoprogenitor cells as was previously believed. This conclusion is based on various types of labelling experiments with grafted cells. Several lines of evidence point to a close similarity between osteoclasts and mononuclear phagocytes (macrophages) which can also fuse to form giant multinuclear cells with phagocytic functions. Both cell types arise from the bone marrow, have an affinity for certain supravital dyes, possess many lysosomes and are motile. However, some monoclonal antibodies specific for macrophages fail to react with osteoclasts, so it is likely that they form a distinct class of cells, although possibly with a common myeloid precursor.

When bone resorption has finished, it is likely that these syncytial cells dissociate into mononuclear cells which may again fuse to form active osteoclasts when suitably stimulated.

The Bone Matrix

This is the extracellular material of bone and consists of a highly mineralized ground substance in which are embedded numerous collagen fibres, usually ordered in parallel arrays (3.26). In

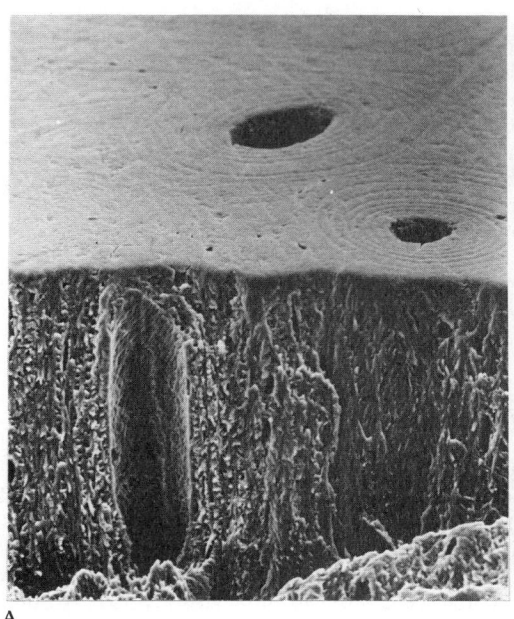 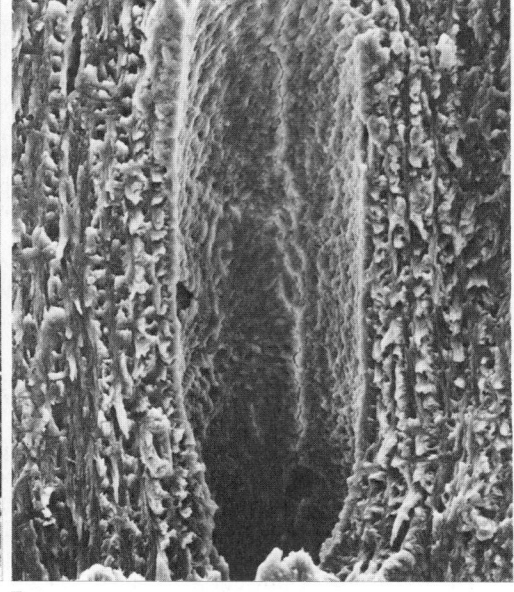 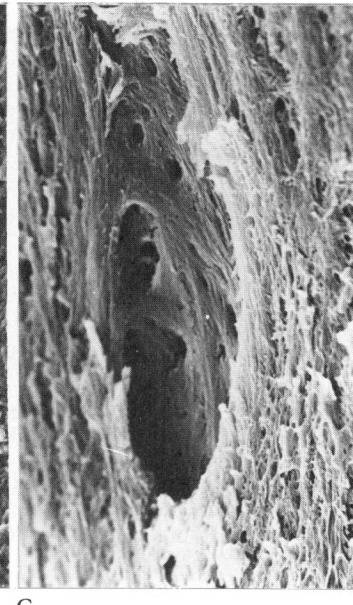

A   B   C

3.26 Scanning electron micrographs of osteonic bone from human femur. In A a piece of bone has been cut transversely to the osteons within it to show two vascular canals, concentric lamellae and some osteocyte lacunae; it has also been fractured parallel to the long axes of two osteons to expose the lamellar structure; magnification × 400. B shows a higher magnification of one of the osteons in A. The complex appearance of the lamellae reflect changing directions of collagen fibres and bone crystals in alternate layers. The surface of the central vascular canal is seen to be pitted with the minute orifices of canaliculi; magnification × 2000. C shows a more highly magnified osteocyte lacuna, broken open to show the origins of canaliculi on its internal surface; magnification × 8000.

mature bone, the matrix is moderately hydrated, 10–20% of its mass being water; of its dry weight, 60–70% is made up of inorganic, mineral salts (hydroxyapatite-like crystals and amorphous calcium phosphate), 30–40% is collagen and the remainder (about 5%) is protein and carbohydrate, mainly conjugated as glycoproteins (osteonectin, osteocalcin, α 2HS-glycoprotein, etc). The proportions of these various components vary with age, location and metabolic status. In the early stages of bone formation, before mineralization, the matrix is termed *osteoid*. In adult bones the amount of osteoid is very small, reflecting local remodelling of the bone in which mineralization rapidly follows the deposition of the organic matrix. In certain disease states where mineralization is defective, notably rickets, the amounts of osteoid are greatly increased.

Collagen in bone closely resembles that of many other connective tissues belonging to Type I (see p. 64). However, in some details it differs from collagen in loose connective tissue in that in bone its molecular structure is more strongly cross-linked internally, and the gaps within its fibres are somewhat larger. These features make it stronger and chemically more inert, and the internal gaps provide lodging for bone salt crystals—it has been estimated that up to 50% of the hydroxyapatite is located within collagen fibres. The crystals mostly lie in the gaps between the ends of tropocollagen subunits ('hole regions') which, in sectioned material, occur with a 64 nm repeat distance along the collagen (67 nm in hydrated tissue), emphasizing the regular banding pattern in electron microscope preparations. Various glycoproteins are associated with these collagen fibres, including *osteonectin* which acts as an important adhesive between collagen fibres, bone crystals and the cells of bone.

That collagen contributes much to the mechanical strength of this tissue is clearly demonstrated in bones heated to denature their collagen, when the matrix becomes brittle and easily fractured. The precise role of collagen in bone mechanics is still not understood in detail, however, since the direction of fibres, the association of mineral crystals within and outside the fibres and other local features all modify their behaviour under stresses of different kinds, making mathematical analysis a complex matter (see e.g. Frasca et al 1981, Currey 1984). But it is clear that besides contributing to the tensile, compressive and shearing strengths of bone, the small degree of elasticity shown by collagen imparts a measure of resilience to this tissue, helping to resist fracture when mechanically overloaded.

Collagen fibres are also vital to bone development by providing nucleation sites within their substructure for mineral deposition (see p. 302) and otherwise enhancing matrix mineralization.

Collagen fibres are synthesized by osteoblasts (vide supra), polymerizing from tropocollagen extracellularly and becoming progressively more cross-linked as they mature. In primary bone, they form a complex interwoven meshwork (non-lamellar or *woven bone*) which eventually is almost entirely replaced by regular laminar arrays of parallel collagen fibres (lamellar bone). Partially mineralized collagen networks can be seen on the internal surfaces of bone (endosteal faces and linings of vascular canals).

Collagen fibres from the periosteum penetrate cortical bone (*Sharpey's fibres*), anchoring this fibrocellular layer at its surface. The terminal fibres of tendons and ligaments are also inserted deeply into the matrix of cortical bone, traversing many osteons in their path.

Other organic components of the matrix. Various complex macromolecules also exist attached to collagen fibres and surrounding bone crystals in small amounts. *Osteonectin* is a phosphorylated glycoprotein with an Mr of 32 000, secreted by osteoblasts and unique to bone. It is important in adhesion of collagen to bone crystals and osteocytes (Termine et al 1981).

Osteocalcin is another glycoprotein, synthesized by osteoblasts and thought to play a part in calcium deposition, as is also the case with the *α-2HS-glycoprotein*, a substance made in the liver but concentrated by bone from blood plasma. Phospholipids glycosaminoglycans (including the chondroitin sulphates) phosphoprotein and various other organic substances are also present in small quantities within the matrix; apart from any other role in the ongoing activities of this tissue, these materials may be important in early mineralization and in the maintenance of calcium salts within mature bone.

The bone salts. These form the inorganic constituents of the bone matrix, conferring on bone its hardness and much of its rigidity. Major ions which compose the mineral part of bone include calcium, phosphate, hydroxyl and carbonate; less numerous ions are those of citrate, magnesium, sodium, potassium, fluoride, chloride, iron, zinc, copper, aluminium, lead, strontium, silicon and boron, many of these being present only in trace quantities. The mineral substances of bone are mostly soluble in dilute acids and can be removed by prolonged immersion, e.g. in 2% nitric acid, or in solutions of calcium chelators such as ethylene diamine tetracetic acid (EDTA). The bone then retains its shape but is highly flexible, so that a long thin bone (such as the fibula) can easily be tied in a knot. In histological sections the microstructure of the bone, including its cells, lamellae, osteons, etc., are well preserved after such gentle demineralization.

The mineral portion of mature bones is composed largely of *bone crystals* made of a substance generally known as *hydroxyapatite*, together with a small amount of *amorphous calcium phosphate*. Bone crystals take the form of elongated needles, thin plates or leaf-like structures about 10 nm long by 1.5–3 nm thick. They are often packed quite closely together, with long axes parallel to those of neighbouring collagen fibres, or lying within them (vide supra). The narrow gaps between the crystals contain water and organic macromolecules. X-ray crystallography shows that they have a hexagonal prismatic substructure which, although very similar to that of the mineral hydroxyapatite, $Ca_{10}(PO_4)_6(OH)_2$, differs in some respects, e.g. the presence of substantial amounts of other ions such as carbonate. Fluoride ions can substitute for hydroxyl ions, and various other ions can be held within crystal lattices. Likewise, different cations such as those of radium, strontium and lead all readily substitute for calcium and are therefore known as bone-seeking cations; these can be either radioactive or chemically toxic and so constitute a danger to health. When present in bone, they are particularly hazardous since their position in the body close to the haemopoietic tissues of bone marrow may induce various pathologies of that tissue.

Under the electron microscope, bone crystals are electron-dense and, where packed closely together, give the matrix a massive, solid appearance (3.28). Amorphous calcium phosphate is composed of more granular, semi-crystalline material and is thought to be the earliest form of mineralization in bone during its development, being largely replaced in mature bone by the crystalline form.

MICROSCOPIC ORGANIZATION OF BONE

While the mechanical properties of bone are dependent on the general composition of its matrix, the manner in which the different components are arranged is also of utmost importance in the strength and resilience of this tissue. Two quite distinct types of organization are found; in primary bone the most immature tissue of developing bones, collagen fibres and bone crystals are irregularly interwoven (*woven bone*). In all other situations, collagen fibres and crystals of bone salts are arranged in highly orientated flattened plates or *lamellae*, giving bone great strength.

Lamellar Bone

This type of bone therefore makes up almost all of the adult osseous skeleton. The precise arrangement of its lamellae varies from site to site, particularly between compact cortical bone and trabecular bone within.

Compact bone (3.23–25). As already mentioned, adult bone consists almost entirely of mineralized matrix and collagen fibres arranged in plates or lamellae, embedded in which are the osteocytes. In many bones a few such lamellae form continuous layers at the surface, termed *circumferential lamellae*. By far the greatest proportion, however, are arranged in concentric cylinders around neurovascular channels (*Haversian canals*), so forming the basic units of bone construction, the *Haversian systems* or *osteons* (more correctly called *secondary osteons* to distinguish them from the irregularly organized primary osteons formed during the initial generation of bone, as described below). Secondary osteons

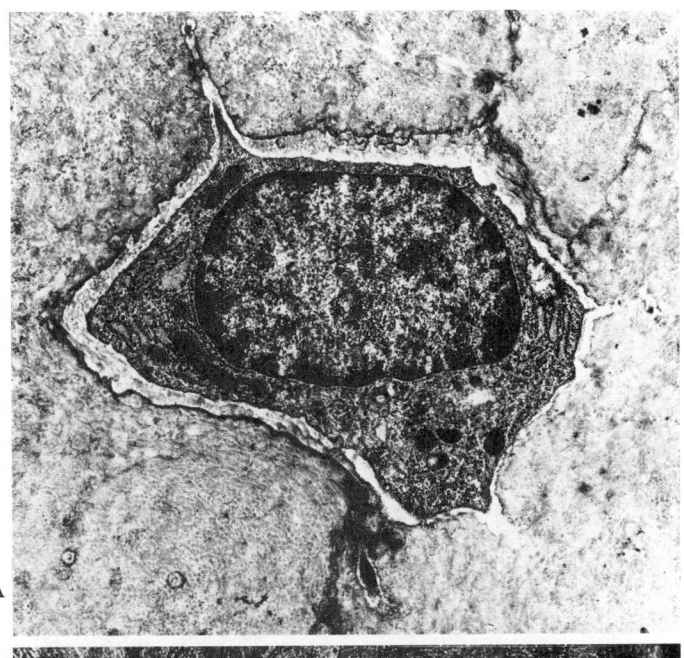

A

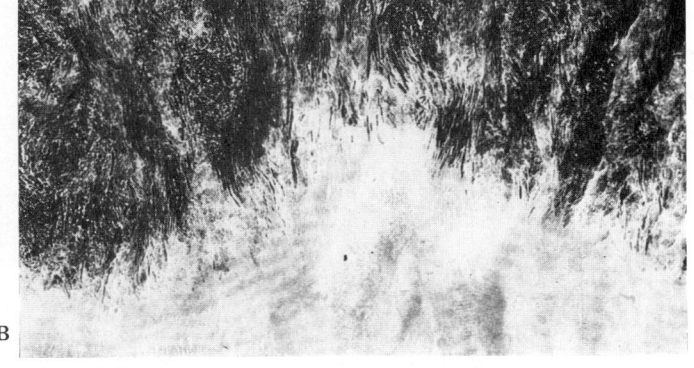

B

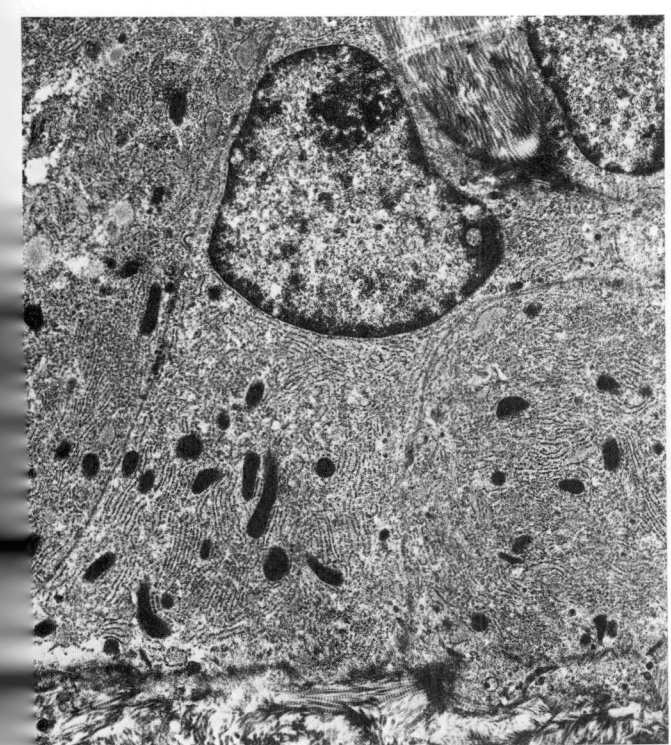

3.27 Electron micrograph of a series of osteoblast on a bone forming surface (below). The cytoplasm is filled with granular endoplasmic reticulum. Some banded collagen fibres are visible at the secretory front at the base of the picture; magnification × 4000. (Provided by A Hayward.)

3.28 Electron micrographs showing (A) a young osteocyte within bone matrix (decalcified during processing for microscopy). Visible are the bases of dendrites extending within the canaliculi, the cell body containing granular endoplasmic reticulum, the lacuna with less dense matrix surrounding the cell body, and many collagen fibres in the matrix, mainly sectioned transversely in this picture; magnification × 3000. (Provided by A Hayward.) B shows bone crystallites (above) and collagen fibres at a mineralizing front on the surface of bone; magnification × 3000. (Provided by Moya Meredith Smith.)

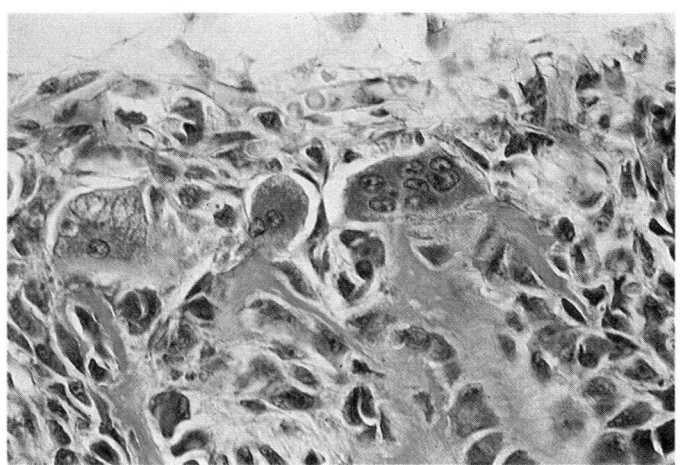

3.29 Section through developing membrane bone showing a group of large, multinucleate osteoclasts eroding spicules of bone (green). Many osteoblasts are also clustered at other surfaces of the spicules where bone is being formed. Masson's trichrome; magnification × 1000. (Provided by Moya Meredith Smith.)

usually lie parallel to each other and, in elongated bones such as those of the appendicular skeleton, with the long axis of the bone, but they frequently spiral, branch or intercommunicate and some end blindly. In transverse section they are round or ellipsoidal, varying from about 100–400 μm in diameter, and the number of lamellae is usually about six, rising to 15 in the largest osteons. Each osteon is permeated with canaliculi of its resident osteocytes which form the pathways for diffusion of nutrients, gases, etc. between the vascular system and the osteons. The rather small maximum diameter ensures that no osteocyte is more than about 200 μm from a blood vessel, a distance which may be a limiting

factor in cellular survival. In the intervening spaces between secondary osteons are the fragmentary remains of osteons of older bone which has been partially eroded before the new osteons were formed. This is termed *interstitial bone.*

The central osteonic canals vary in size, with a mean diameter of 50 μm; those near the marrow cavity are somewhat larger (Cohen & Harris 1958). Within each canal are one or two capillaries lined by fenestrated endothelium surrounded by a basal lamina which splits to enclose typical pericytes. Usually there are also some unmyelinated and occasional myelinated nerve fibres, 5–9 μm in diameter (Cooper 1969). The bony surfaces of osteonic canals are perforated by the openings of myriads of canaliculi and are also lined by collagen fibres (3.26).

Osteonic canals communicate directly or indirectly with the medullary cavity, channels which run obliquely or transversely in the direction of the osteons being known as Volkmann's canals. These are said not to be surrounded by concentric lamellae of bone. However, the majority of such channels appear to be simple interosteonic, anastomotic canals and the frequency of true vascular connections with the periosteum and endosteum may be rather low (see Cohen & Harris 1958).

All secondary osteons are demarcated from their neighbours by a *cement line* which has little or no collagen and is strongly basophilic due to a high content of glycoproteins and proteoglycans; this marks the limit of bone erosion prior to the formation of an osteon and is therefore also known as a *reversal line*. Similar basophilic lines also occur in the absence of erosion, where bony growth has been interrupted then resumed (*resting lines*). Canaliculi may sometimes pass through such cement lines, so providing a route for exchange between interstitial bone and vascular channels within osteons (Atkinson & Hallsworth 1982).

The lamellation of bone (3.20). While each lamella consists of a plate of mineralized matrix containing collagen fibres, the reasons for the lamellar appearance are not yet entirely understood. However, it appears to result from two features, namely the presence of a thin (0.1 μm) layer of matrix with a high mineral and low

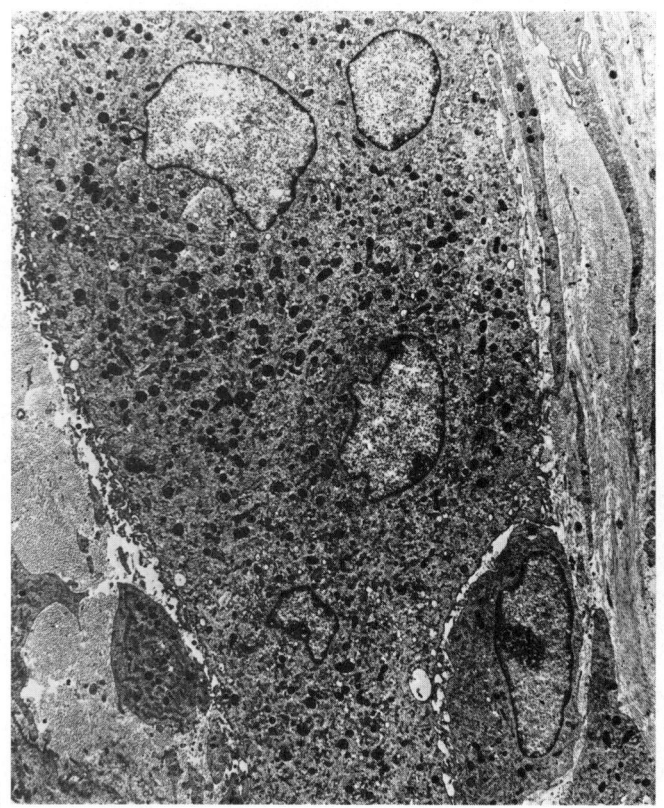

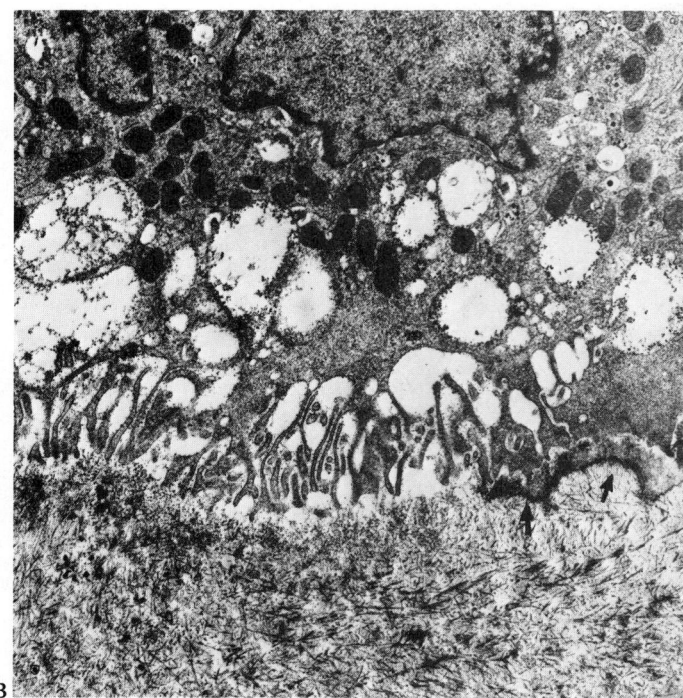

A   B

3.30 Electron micrograph of osteoclasts. A shows a large multinucleate cell close to the surface of bone, although not, in this case, within an erosion bay; magnification × 2500. B is a higher magnification of an osteoclast ruffled border, where active bone resorption is occurring. The cell has been stained cytochemically for acid phosphatase activity, which is shown

by the dense granular precipitate within the large (lysosomal) vacuoles in the cytoplasm. To the right is a dense junctional zone between the cell surface and the matrix arrows; magnification × 6000. (Both micrographs provided by Moya Meredith Smith.)

organic content between adjacent lamellae ('interlamellar bone') and a difference in orientation between the structures (i.e. collagen fibres and bone crystals) of adjacent lamellae, varying between 0° and 90°. Since crystallites are aligned parallel to collagen fibres, when viewed by polarized light microscopy (see **3**.25), the plane of polarization is rotated to different degrees by the different lamellae, emphasizing the alternate orientation of these structures.

The earliest descriptions of these fibres suggested that collagen fibres alternate between longitudinal and circumferential in successive lamellae, or that they had spiral paths of different pitches (Gebhardt 1905).

Although a degree of spiralization has been confirmed in investigations into the structure and compressive properties of isolated osteons (Ascenzi & Bonucci 1968, Ascenzi et al 1978), evidence from scanning electron microscopy shows that in each lamella there are small domains or areas of collagen fibres up to 100 µm across. Although fibres within a domain have the same direction, domains are not necessarily parallel with each other so that variation of orientation exists within each lamella (Boyde & Hobdell 1969, Frasca et al 1977). Collagen fibres run in branching fasciculi about 3 µm thick; they often run from one lamella to the next through discontinuities in the interlamellar zones.

Trabecular bone (**3**.22). The organization of this class of tissue is again basically lamellar, but the bone is in the form of branching and anastomosing bars and plates of various widths and lengths, limiting medullary spaces (see p. 277) and bounded by endosteal tissue. In general, bone lamellae are orientated parallel with the adjacent bone surface and there is the same arrangement of cells and matrix as found in circumferential and osteonic bone. However, except in thick trabeculae which may possess small osteons, blood vessels do not penetrate the bony tissue and osteocytes rely on canalicular diffusion from adjacent medullary vessels. In young bone, calcified cartilage may occur in the cores of trabeculae but this is generally replaced by bone during subsequent remodelling.

Variation in bone composition. Normal variation in composition of bone at different sites and ages is incompletely recorded, though valuable in radiological assessment of rarefaction (osteoporosis) due to disease or disuse. Whether this is actual reduction in total amount of tissue, rather than reduction in mineralization, is difficult to determine because of problems of separation of bone from other tissues. Furthermore, measurement of sizes of vascular, lacunar and medullary spaces is difficult. Minimization of difficulties by limiting estimates to cortical bone has been tried since trabecular is more labile than cortical bone. However, the difference in composition of trabecular and compact bone is considered normally slight. Overall, water content decreases and mineralization increases up to the sixth or seventh decade, but the organic content remains unchanged. Thereafter mineralization declines (Mueller et al 1966).

However, microradiography and interferometry of compact bone show uneven distributions of mineral salt. Inner and outer circumferential and interstitial lamellae are evenly and highly mineralized. Secondary osteons exhibit variable mineralization; in a single osteon concentration varies in different areas. Young osteons have a low but increasing concentration, initially highest centrally and least towards the periphery, a gradient which diminishes with age until in old, highly mineralized osteons mineral distribution is uniform. Bone resorption occurs in areas of high and low concentration (vide infra). In fetal bone mineralization reaches 70–80% in about three weeks, 100% much less rapidly. Events in subsequently formed osteons may have a similar time course.

Blood Vessels and Nerves of Bone

The osseous circulation supplies the living bone tissue, the marrow, perichondrium, epiphysial cartilages in young bones and in part the articular cartilages. Modern researches (Brookes 1964, 1967, 1971, Kelly 1968, Rhinelander 1968) have emphasized a *centrifugal* flow of blood through cortical bone in shafts of long bones, in contrast to an earlier concept of substantial *centripetal*

arterial flow into the cortex from periosteal vessels. The vascular supply of a long bone depends on several points of inflow, feeding complex and regionally variable sinusoidal networks within it, which in turn drain to venous channels leaving through all surfaces not covered by articular cartilage. These vascular patterns are summarized in **3**.31,32; see also **3**.33.

One or two main *diaphysial nutrient arteries* enter the shaft obliquely through *nutrient foramina* leading into *nutrient canals*. Long recognized (Havers 1691, Bernard 1835), their sites of entry and angulation are almost constant and characteristically directed *away* from the dominant growing epiphysis (p. 281). This is the basis of the *growing-end hypothesis* to explain the positions and orientations of nutrient foramina and canals (Clark 1965). But such a simple mechanism does not account for the exceptions to this pattern reported in various species and sites (Hughes 1952). However, in human long bones Mysorekar (1967) observed atypically directed canals only in the fibula, attributing this to its unique pattern of ossification. A subsequent study of 848 metacarpal and 811 metatarsal bones showed that, apart from a few with double or no foramina, over 90% had a single nutrient foramen in the middle third of the shaft. All foramina, single or double, slanted away from the epiphysis, supporting the above hypothesis. For a review of this and alternative hypotheses such as the *'periosteal slip'* theory of Schwalbe (1876), 'vascular theory' of Hughes (1952) and *'asymmetrical muscular development theory'* of Lacroix (1951), consult Patake & Mysorekar (1977). Finally, a comparative study of the contrasting growth patterns of foramen and canal in the rat femur and tibia has been contributed by Henderson (1978). Nutrient arteries do not branch in their canals, but divide into ascending and descending branches in the medullary cavity. These approach the epiphyses, dividing repeatedly into smaller rami which pursue helical courses in the juxta-endosteal medullary zone. Near the epiphyses they are joined by terminals of numerous *metaphysial* and *epiphysial arteries*: the former are direct branches of neighbouring systemic vessels, the latter from periarticular vascular arcades formed on non-articular bone surfaces (p. **3**.32). Numerous vascular foramina penetrate bones near their ends, often at fairly specific sites (see **3**.6); some are occupied by such arteries, but most contain thin-walled veins. Within bone these arteries are unusual in consisting of endothelium with only a thin layer of supportive connective tissue (Yoffey 1962). Epiphysial and metaphysial arteries quantitatively exceed the diaphysial supply, which they can complement, e.g. when the latter is experimentally destroyed.

Medullary arteries of the shaft (**3**.31,32) give off (a) centripetal branches to a hexagonal mesh of medullary sinusoids draining into a wide, thin-walled central venous sinus; (b) cortical branches passing through endosteal canals to feed fenestrated capillaries in Haversian systems. The central sinus drains veins retracing the paths of nutrient arteries, sometimes piercing the shaft elsewhere as independent emissary veins.

Cortical capillaries conform in their pattern to the Haversian canals, often described as longitudinal with oblique interosteonic connexions. However, an oblique, radial pattern has been demonstrated, vessels largely radiating from the primary ossification centre (Brookes 1964, 1967, 1971). At bone surfaces cortical capillaries make capillary and venular connections with the periosteal plexuses (**3**.31,32,34), which are formed by arteries from neighbouring muscles contributing vascular arcades with longitudinal links to the fibrous periosteum. From this external plexus a capillary network permeates the deeper, osteogenetic periosteum. At muscular attachments periosteal and muscular plexuses are confluent and the cortical capillaries then drain into interfascicular venules.

The concept of such an almost exclusively centrifugal supply to the cortex in the shafts of long bones has received increasing support, but some (Marneffe 1951, Morgan 1959) have described an appreciable centripetal arterial flow to outer cortical zones from periosteal vessels.

The large nutrient arteries of epiphyses form many intraosseous anastomoses, branches passing towards the articular surfaces within trabecular spaces of the bone. Near the articular cartilages these form serial anastomotic arcades (e.g. three or four in the femoral head) from which spring end-arterial loops often

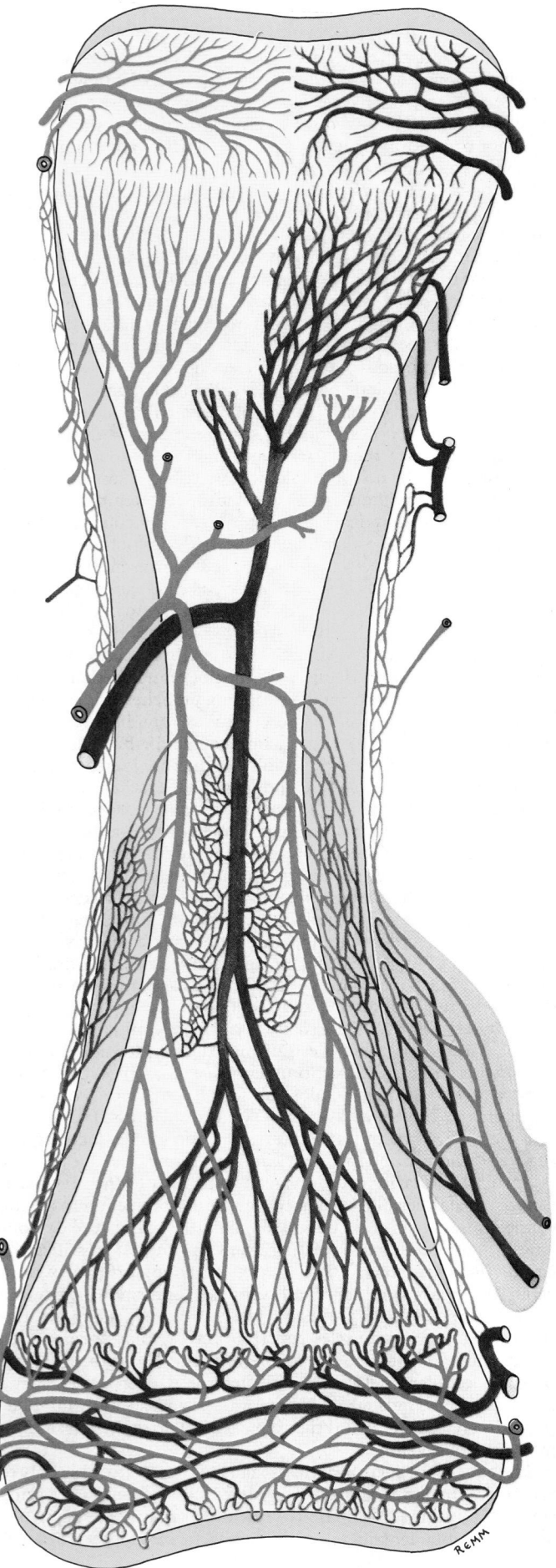

3.31 Scheme of the main features of the blood supply of a long bone based upon descriptions by M Brookes (Guy's Hospital Medical School). Note the contrasting supplies of the diaphysis, metaphysis and epiphysis, and their connections with periosteal, endosteal, muscular and periarticular vessels. Consult the text for a more extended description.

piercing the thin hypochondral compact bone (**3.6**) to enter, and sometimes traverse, the calcified zone of articular cartilage before returning to the epiphysial venous sinusoids.

In immature long bones the supply is similar, but the epiphysis is a discrete vascular zone; epiphysial and metaphysial arteries enter on both sides of the growth cartilage, anastomoses between them being few or absent. Growth cartilages probably receive a supply from both sources and also from an anastomotic collar in the adjoining periosteum, but how much from each is uncertain (Brookes 1964, 1967, 1971, Trueta & Morgan 1960, Rang 1969). Occasionally cartilage canals are incorporated into a growth plate (p. 287). Metaphysial bone is nourished by conjoined terminal branches of metaphysial arteries and primary nutrient arteries of the shaft. They form terminal sinusoidal loops in the zone of advancing ossification, the dilated bend of each separated slightly from adjacent partitions of calcified cartilage (p. 306). Such gaps are often packed with erythrocytes, possibly evidence of 'microrupture' at this point (Trueta & Morgan 1960). However, others have emphasized an intimate contact between sinusoidal endothelium and calcified cartilage, and to such loops have ascribed a nutritive *and* a chrondrolytic role (Cameron 1961, Brookes & Landon 1963). Young periosteum is more vascular; its vessels communicate more freely with those of the shaft than in adults, and have more metaphysial branches.

Large irregular bones like the scapula and innominate bone, receive a periosteal supply and often have large nutrient arteries penetrating directly into their cancellous bone. The two systems anastomose freely.

Short bones receive numerous fine vessels from the periosteum at non-articular surfaces, supplying their compact and cancellous bone and medulla. Arteries enter vertebrae close to the bases of transverse processes; their medulla drains to two large basivertebral veins converging to a foramen on the posterior surface of the vertebral body.

Flatter cranial bones are supplied by numerous periosteal or mucoperiosteal vessels. Large veins run tortuously in diploë (cancellous bone). Being thin-walled, they gape when cut.

Lymphatic vessels accompany periosteal vascular plexuses but have never been convincingly demonstrated in bone.

Nerves are most numerous in articular extremities of long bones, vertebrae and larger flat bones. They occur widely in periosteum, and fine myelinated and non-myelinate fibres accompany nutrient vessels into bone and even the perivascular spaces of Haversian canals.

The Histogenesis of Bone

Although there is no basic difference in the deposition of bone at various sites to form the skeleton, the precursors are of two distinct classes; bones such as those in the cranial vault are preceded by a fibrocellular membrane, whereas most bones are formed in rods or masses of cartilage. These two types of ossification are known as intramembranous and endochondral, respectively. They will now be considered in some detail, together with the general topic of the mineralization and subsequent fate of bone.

INTRAMEMBRANOUS (MESENCHYMAL) OSSIFICATION

This process, which is essentially the direct mineralization of a highly vascular connective tissue, spreads from regular *centres of ossification* (**3.**38A–D), at which differentiating mesenchymal (*osteoprogenitor*) cells (see p. 293) proliferate densely around a capillary network. Between cells and around vessels appears a fine mesh of collagen fibres and amorphous matrix (osteoid). The central cells enlarge and fine strips of eosinophilic matrix (earliest bone) appear between. These rapidly extend and fuse into a delicate labyrinth, enclosing vessels and transforming mesenchyme, whose cells enlarge, become polygonal, cuboidal, or low columnar and form an incomplete layer of osteoblasts (p. 293) in contact with the primitive, eosinophilic bony matrix. The earliest

3.32 Diagram of the circulatory arrangements in part of the diaphysis of a typical long bone. Note, in the marrow cavity, the large central venous sinus, the dense network of medullary sinusoids, longitudinal medullary arteries and their circumferential rami. From the latter, longitudinally oblique, transcortical capillaries emerge through minute 'comet-shaped' foramina (**3.34**) to become confluent with the periosteal capillaries and venules. Not to scale; obliquity of cortical capillaries is emphasized for clarity. (Constructed with the collaboration of M Brookes, Department of Anatomy, Guy's Hospital Medical School, London.)

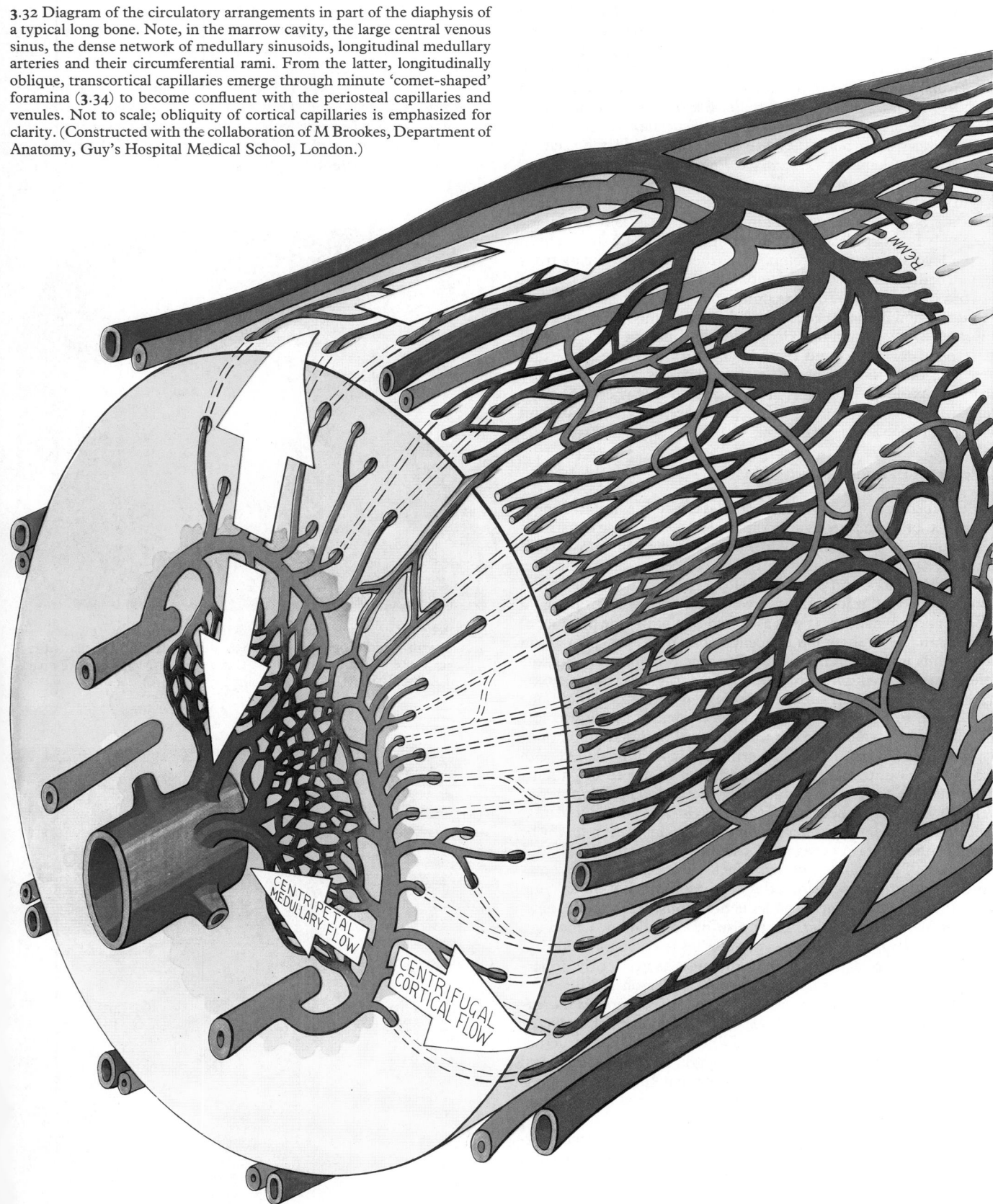

CENTRIPETAL MEDULLARY FLOW

CENTRIFUGAL CORTICAL FLOW

crystallites appear in minute extracellular *matrix vesicles* produced by these cells (see p. 302 and **3.**35,36).

These later change into typical hydroxyapatite crystallites which were, as noted, related first to cores of matrix vesicles and subsequently to collagen fibres which form increasingly wide-meshed reticula in walls of an early labyrinth of *woven bone* (p. 96), the *primary spongiosa*. As layers of calcifying matrix are added to these early trabeculae or *bone spicules*, some osteoblasts are enclosed by matrix in primitive lacunae. However, these

retain intercellular contact by means of their surface processes and, as these elongate, matrix condenses around them to form canaliculi. Further osteoblasts are added to trabecular surfaces by transformation of adjacent vascular mesenchyme cells.

As matrix secretion, calcification and enclosure of osteoblasts proceed, the trabeculae thicken and intervening vascular spaces become narrower. Where bone remains trabecular, the process slows and the spaces between them become occupied by haemopoietic tissue. Where compact bone is forming, trabeculae

301

continue to thicken and vascular spaces to narrow. Meanwhile the collagen fibres of the matrix, secreted on the walls of narrowing spaces between trabeculae, become organized as parallel, longitudinal or spiral bundles and enclosed cells occupy concentric sequential rows. These irregular, interconnected, sometimes cylindroid masses of compact parallel-fibred bone, with a central canal, are *primary osteons* or *atypical Haversian systems*. With intervening woven bone they are later eroded and replaced by generations of lamellar *secondary osteons* (see also pp. 297, 299, 309).

During these changes mesenchyme condenses on the surface to form a fibro-vascular periosteum, and bone is laid down increasingly by osteoblasts differentiating from mitotic stem cells in deeper layers of the periosteum. Thus there is an advancing front of matrix deposition which entraps further periosteal vessels and also osteoblasts which then become osteocytes. Further growth is a continuation of these processes with much remodelling by varying rates of resorption and deposition in different sites. Overall patterns of formation and remodelling vary with the shape and structure of particular bones, and these have been studied by many techniques. Examples are considered elsewhere (p. 309); here we will consider the general processes of calcification and bone resorption in further detail.

Calcification and the Osteoblast

As described above (p. 297) the essential crystallite unit in bone is composed of a complex hydroxyapatite-like substance consisting of calcium, phosphate and hydroxyl ions associated with various other species of ion. Initially the only ions needed for calcification are calcium (Ca^{2+}) and phosphate (PO_4^{3-}) and the mechanism of calcification was traditionally treated as simply a problem of precipitation dynamics. It was assumed that while the product of calcium and phosphate ion concentrations $[Ca^{2+}] \times [PO_4^{3-}]$ in body fluids was generally too low for precipitation to occur, the conditions in osteoid matrix were in some way especially favourable to this process. In support of this view, calcification of bone was known to be reduced when dietary calcium intake was low and, conversely, calcification of blood vessel walls and other non-bony sites could be induced experimentally in animals fed with excessive amounts of calcium or clinically occurred in patients given massive doses of vitamin D or where blood calcium levels were elevated in hyperparathyroidism. It was considered that conditions in the osteoid matrix favouring calcification might be alkaline phosphatase activity in osteoblasts (raising local concentration of phosphate), or some factor raising local pH to alkaline levels, with consequent reduction in solubility of the calcium salt. However, this simple model is now considered inadequate; relevant *activity coefficients* must also be considered, which are difficult to extrapolate from in vitro studies to osteoblasts in situ, the precise nature of whose environment is as yet poorly understood. For reviews consult Urist (1966), Anderson 1980).

An interesting development was the theory of *epitactic nucleation* (Neuman & Neuman 1958), based on the concept of *seeding* or *epitaxy*: a nucleus is somehow formed, with crystalline structure like hydroxyapatite, effective in aggregating calcium and phosphate ions. The hydroxyapatite crystal then grows spontaneously by addition of these from the saturated surrounding fluids.

General acceptance of the theory of epitaxy led to many attempts to determine the nature and distribution of nucleation sites, variously claimed to be points in periodic structure of collagen, ground substance links between collagen fibrils or structural aspects of proteoglycans (see the review by Bernard & Pease 1969). The latter authors (foreshadowed by Bonucci 1967) proposed a role for the osteoblast, considering initial nucleation sites to be cellular 'buds' or extrusions with a polysaccharide core and able, possibly with protein, to form an 'image' of crystal salt. Further aggregation of crystals around this initial locus was held to lead to formation of spherulitic bone nodules coalescing to form seams of bone, association with collagen fibres being secondary. Thus 'cellular seeds' initiating calcification would be specific cell products originating solely from osteoblasts and absent in general connective tissues.

Although epitactic nucleation has not been fully confirmed, the essential role of a particular cellular product extruded into the matrix, *the matrix vesicle*, has aroused intense interest and stimulated much investigation.

Matrix vesicles (3.36). These are membrane-bound spheres about 0.1 μm in diameter. They typically have an electron-dense core, often apposed eccentrically to the inner surface of the membrane. They are generally thought to be derived by budding (or fragmentation) of osteoblast processes, (or in other tissues, of the relevant cell). They are the loci of earliest hydroxyapatite

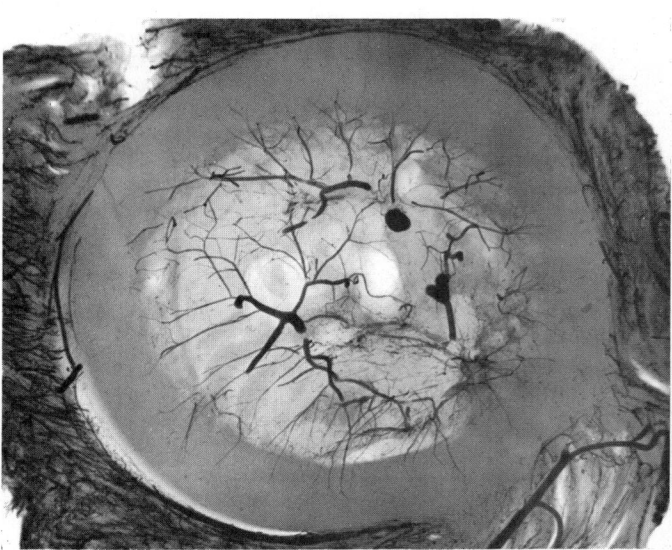

3.33 Microradiograph of long bone in which the arterial supply has been injected with a radio-opaque substance, demonstrating the centrifugal perfusion route of the blood supply. (Provided by M Brookes, Department of Anatomy, UMDS Guy's Campus, London.)

3.34 Scanning electron micrograph of the periosteal surface of the diaphysis of an adolescent tibia. Note the oblique 'comet-shaped' groove and foramen which in life transmitted a minute neurovascular bundle; numerous foramina of this general type are scattered over the surface of the diaphysis, but their pattern, direction and degree of obliquity vary with the overall subperiosteal deposition of bone that characterizes the various growth zones of the diaphysis; magnification × 80. (Preparation by Michael Crowder, Department of Anatomy, Guy's Hospital Medical School.)

crystallite formation in newly forming bone, and indeed in the initial mineralization of all mineralized tissues throughout vertebrates (see the review by Anderson 1985). They occur in calcifying cartilage (Anderson 1967, 1969, Bonucci 1967, 1970), woven bone (Bernard & Pease 1969), dentine (Sisca & Provenza 1972, Katchburian 1973), and subperiosteal bone (Anderson and Reynolds 1973, Reynolds 1976). Matrix vesicles with similar properties can hence be produced by chondroblasts, odontoblasts and osteoblasts. Their role in the initiation of calcification, however, remains speculative since they have not always been found in mineralizing fronts in osteoid of more mature bone. Being membrane-bound, vesicles separate internal and external environments. They carry high concentrations of alkaline phosphatase, ATPase, inorganic pyrophosphatase and, at some sites, high concentration of lipids. It is likely that matrix vesicles provide the enzymes and environment to concentrate calcium and phosphate sufficiently to initiate crystallization, which then spreads outside the vesicle. Other hypotheses include the siting in the vesicular membrane of a centripetal calcium binding by vesicular lipids.

While it is very probable that matrix vesicles are important in beginning the process of calcification, there are also various cal-cium-binding molecules among the proteins and glycoproteins of osteoid. These include collagen, as already mentioned (p. 296), osteocalcin and α 2HS-glycoprotein (p. 296). These are therefore likely candidates for substances able to increase local concentrations of ions to the point where precipitation occurs, and may be an important mechanism by which calcification continues once the matrix vesicle system has begun it.

Considerable controversy concerns the shape and size of the initial hydroxyapatite crystals which are variously described as plates, rods or needles. Most studies agree a smallest dimension of about 5 nm but estimates of length vary from 20–35 nm to values three times larger, reflecting a variety of techniques: X-ray diffraction and dark-field electron microscopy give low values for length, whereas transmission electron microscopy of ion-beam thinned section yields much larger values (see Boyde 1975, Jackson et al 1976).

Calcium Balance, Bone Resorption and the Osteoclast

A central theme of physiology is the dependence of many cellular activities on a relatively constant micro-environment (in terms of osmolality, types and concentrations of ions, pH, etc.); many finely balanced feedback systems operate to ensure this *homeostasis* (see Reynolds 1974). The level of circulating calcium ions is an example: it is preserved despite wide variation in diet, rates of bone growth and remodelling. Involved in this process but varying quantitatively at different times are both labile and stable areas of bone salts (vide infra), interacting types of cell (osteoblasts, osteocytes and osteoclasts) various hormones including parathyroid hormone and thyrocalcitonin as well as many other factors.

Calcium in blood and tissue fluids is constantly exchanging with calcium in bone to the extent of approximately one-quarter of ionic blood calcium each minute. The blood calcium is in equilibrium with that of tissue fluid, especially in the perivascular space in Haversian canals, lacunae and canaliculi—which present vast exposed areas of bone salt for such physico-chemical exchanges (p. 293). The mineral surface exposed to extracellular fluid in a 70 kilogram person is estimated at between 1500–5000 m² (Robinson 1964). The calcium content of newly formed bone is more labile than in older tissue and provides a ready reservoir of calcium ions. The gradual replacement of old osteons by new ones, which continues throughout life, can also be modulated to alter blood calcium levels. Bone salt release may depend on active erosion of matrix by osteocytes (Bélanger et al 1963) but how far such *active cellular osteolysis* contributes to rapid continuous turnover of calcium, or how much is simply a direct physico-chemical exchange, is uncertain.

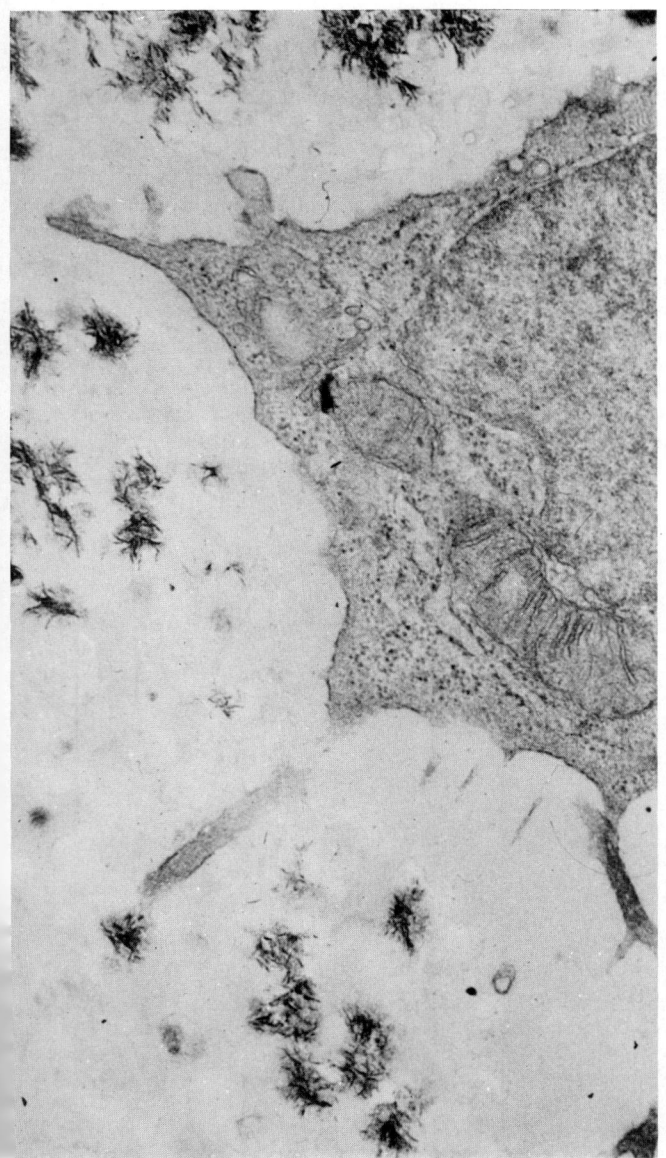

3.35 Transmission electron micrograph of part of an osteoblast, its surface bearing long filopodia which project into the surrounding intercellular matrix. Within the latter are clusters of 'matrix vesicles' which provide the initial nucleation sites for the formation of hydroxyapatite crystallites in the early mineralization of bone. The specimen is from the subperiosteal ossifying front of a chick's tibia in tissue culture.

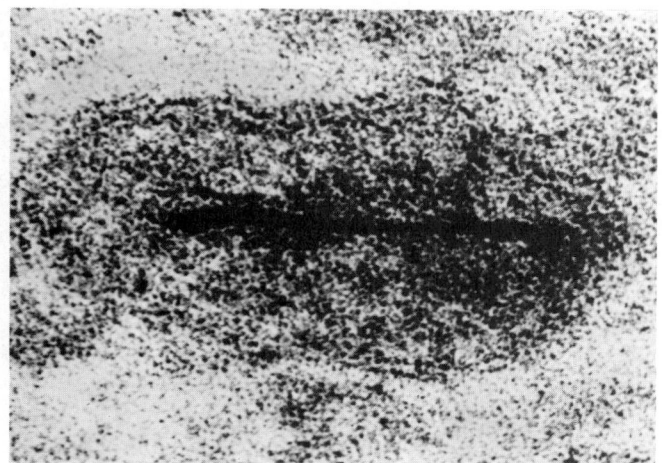

3.36 High power transmission electron micrograph of a single 'matrix vesicle' from the specimen shown in 3.35. The vesicle is membrane-bound, has a granular content and centrally, along its long axis, is a dense needle-shaped crystallite of hydroxyapatite; magnification × 200 000. (Specimens provided by J J Reynolds, Strangeways Laboratory, Cambridge.)

The depression of circulatory ionic calcium levels increases secretion of parathyroid hormone, which raises blood calcium level by several mechanisms: direct action on bone, increased renal tubular reabsorption of calcium and increased intestinal absorption of calcium—a minor effect. The actions of parathyroid hormone on bone are complex and obscure. They may directly modify the surface shells of ions associated with hydroxyapatite crystals, or may change the structure of collagen-associated proteoglycans and glycoproteins or increase osteocyte activity and bone resorption by osteoclast action (vide infra). Conversely, the rise in circulating calcium levels increases secretion of calcitonin (thyrocalcitonin) by thyroid parafollicular cells (p. 1460 and **8**.233), depressing circulating calcium by some direct action on bone. Calcium release from bone may be inhibited by the stabilization of collagen and by increased osteoblastic deposition of bone as well as specific inhibitors thought to be released locally by bone cells (including bone-lining cells).

During bone growth, remodelling occurs, involving deposition and removal at different sites. In mature bone also, continued remodelling of osteons is preceded by erosion. Surfaces undergoing massive localized resorption display large, multinuclear *osteoclasts* with foamy, lightly basophilic cytoplasm, often in contact with bone in small erosion cavities (*erosion bays* or *resorption lacunae of Howship*) formation; details of this cell type are given on p. 295.

Such appearances suggest bone removal, but osteoclastic action remains unclear. Lysosomes and high proteolytic enzyme activity may be significant, as may secretion of chelating agents to form soluble, weakly ionized complexes with bone salt constituents. When erosion ends osteoclasts disappear, perhaps degenerating or reverting to parental type.

INTRACARTILAGINOUS (ENDOCHONDRAL) OSSIFICATION (3.39–49)

Most human bones are preformed in cartilage: in early fetal life a 'long' bone is prefigured by a rod of hyaline cartilage (3.37,44), replacing a similar rod of condensed mesenchyme, both foreshadowing in shape the early bone. Smaller (e.g. carpal) bones are also preceded by appropriately shaped cartilaginous 'models' (see 3.37). In such models a sequence of orderly changes signals the appearance of *centres of ossification*. The significance of these, and of growth plates, has been considered elsewhere (p. 279) and only their microscopic changes will be described here.

The cartilaginous model is surrounded by vascular condensed mesenchyme or perichondrium, like that which precedes and surrounds intramembranous ossificatory centres (p. 300). Again, its deeper layers contain osteoprogenitor cells.

The first sign of a centre of primary ossification is seen when chondroblasts deep in the centre of the primitive shaft (3.44) enlarge greatly, their cytoplasm becoming vacuolated and accumulating glycogen. Their intervening matrix is compressed to thin and often perforated septa. As their lacunae enlarge, the cells begin to degenerate and finally die, leaving enlarged and sometimes confluent lacunae as *primary areolae*, whose thin walls have become calcified during these final stages. Simultaneously, cells in the deep layer of the perichondrium surrounding the centre of the model become osteoblasts and form a peripheral layer of fenestrated bone. This '*periosteal collar*', essentially formed by intramembranous ossification (3.44), is at first a thin-walled tube enclosing the central shaft, but as it increases in diameter it also extends towards the ends of the shaft (vide infra). It is noteworthy that both the calcifying cartilage and the matrix of the calcifying perichondral collar contain *matrix vesicles* (see p. 303).

Where the periosteal collar overlies the calcified cartilage walls of primary areolae, it is invaded from the deep periosteal layers by *osteogenic buds*—blind-ending capillaries and accompanying cells which continually divide to form osteoblasts, and osteoclasts. The latter excavate newly formed bone to pass into adjacent calcified cartilage and here continue to erode walls of primary areolae, leading to fusion of these into larger, irregular, communicating *secondary areolae* or *medullary spaces*. These fill with *embryonic medullary tissue* (vascular mesenchyme, osteoblasts and osteoclasts, haemopoietic cells, etc.); osteoblasts attach themselves to the delicate residual walls of calcified cartilage, laying down osteoid which rapidly changes firstly into patches then continuous linings of bone. Further layers of bone are added, enclosing young osteocytes in lacunae, and narrowing the perivascular spaces. As the formation of subperiosteal bone continues, bone deposition on the more central areolar walls ceases. Osteoclastic erosion of the early bone spicules then creates a primitive medullary cavity in which only a few trabeculae composed of bone with central

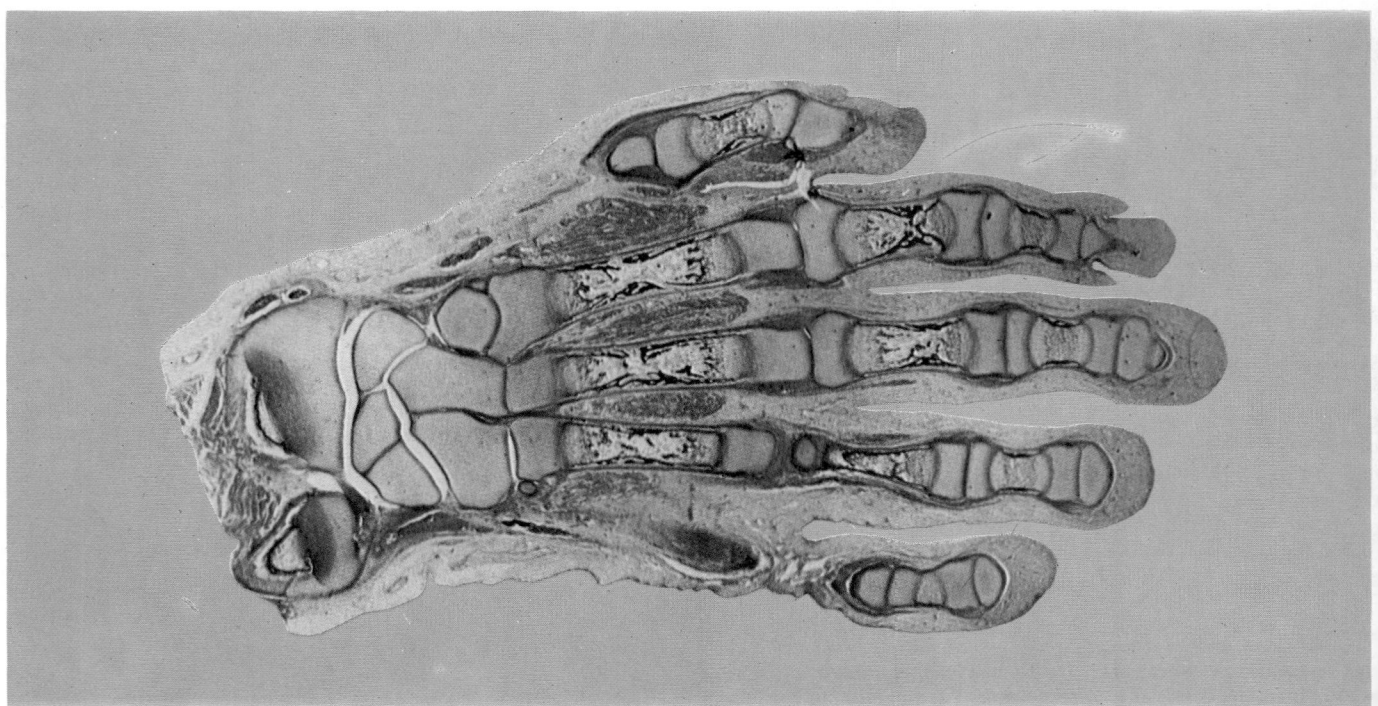

3.37 Survey photograph of a section of a fetal hand showing cartilaginous models of the carpal bones and various stages of development of primary ossification centres in the metacarpals and phalanges. Note that none of the carpal elements show any evidence of ossification. Photography by Kevin Fitzpatrick, Guy's Hospital Medical School, London.

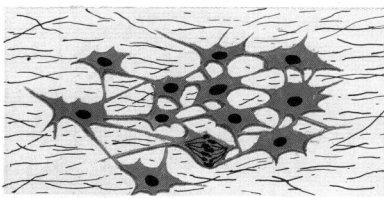

A. *Differentiation of osteoprogenitor cells in mesenchyme*

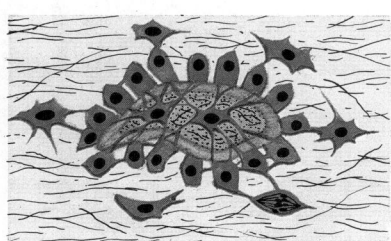

B. *Further differentiation of cells into osteoblasts and synthesis and partial calcification of a bone spicule*

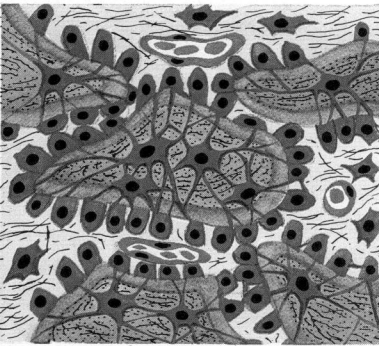

C. *Increase in sizes of bone spicules and invasion of intervening areas by blood vessels*

D. *Fusion of bone spicules to form a continuous plate of woven bone and beginning of remodelling by osteoclasts*

3.38 Schema of intramembranous ossification. Different stages in the formation of intramembranous bone are shown. Colour code: osteoblasts and osteocytes blue, osteoclasts pink, blood vessels red, surrounding mesenchyme matrix pale green, uncalcified matrix (pre-bone) dark green, and calcified bone orange-brown.

cores of calcified cartilage remain to support the developing marrow tissues. Such trabeculae soon become remodelled and replaced by more mature bone. Meanwhile cartilaginous regions near the shaft go through similar changes (3.37,43,46) and, when covered with bone, may become incorporated in the bone of the periosteal collar. Since these are most advanced centrally and the epiphyses remain cartilaginous, the intervening zones show a sequence of changes in growing bones, when viewed in longitudinal section (3.46A,B,47). This region, which persists until longitudinal growth of the bone ceases, is the *growth plate* or *epiphysial plate.*

The Growth Plate (3.47)

Growth of the cartilaginous extremity (usually an epiphysis) keeps pace with the rest of the bone both by apposition and interstitially (3.51). Where it meets the cartilaginous shaft an or-

ganized region of rapid growth appears, as the future growth plate between epiphysis and diaphysis (see e.g. Siffert 1956, Rang 1969), which grows in all dimensions. Transverse or latitudinal growth is due to occasional transverse mitoses and appositional growth due to matrix deposition by cells from the perichondrial collar (or ring) at this level. The future growth plate thus expands in concert with the shaft and adjacent future epiphysis. On the side of the plate closest to the epiphysis is a zone of relatively quiescent chondrocytes, but further towards the centre of the bone is an actively mitotic zone of cells. Here, the more frequent divisions in the long axis soon create numerous longitudinal columns (palisades) of discoidal or cuneiform chondrocytes, each in a flattened lacuna. This proliferation and column formation occupies the *zone of cartilage growth* and continued longitudinal interstitial expansion is the basic mode of elongation of a bone. Traced centrally in the shaft a column of cells shows increasing maturity, passing through the stages already described (vide supra). Thus they increase in size and accumulate glycogen and begin to show positive reactions for oxidative enzymes. Younger chondrocytes display surface projections into reciprocal recesses in lacunar walls, but as they hypertrophy and enter the *zone of cartilage transformation*, projections increase greatly, perhaps due to removal of cartilage matrix and progressive enlargement of lacunae. The largest cells withdraw their projections, then degenerate and die. The empty lacunae are now separated by transverse and longitudinal walls, impregnated with hydroxyapatite crystals (the zone of calcified cartilage). The calcified partitions enter the *zone of bone formation* and are invaded by vascular mesenchyme, with its osteoblasts, osteoclasts, chondroclasts, etc. from adjacent centres of primary ossification (3.48). Partitions, especially the transverse ones, are next partly eroded and on surfaces of the narrowing longitudinal walls occur osteoid deposition, bone formation and osteocyte enclosure as described above. As mentioned above, matrix vesicles occur in all sites undergoing calcification (see, e.g. Reinholt et al 1982). Chondrolysis of calcified partitions has been ascribed to osteoclastic (chondroclastic) action or to chondrolysis by endothelium in terminal loops of vascular sinusoids which occupy and come into close contact with each incomplete, columnar trabecular framework (3.39 and see Irving 1964). Whether these loops have intact endothelium or are temporarily fenestrated and are thus sites of microruptures (Brookes & Landon 1963, Trueta & Morgan 1960, Anderson & Parker 1966) is uncertain. Longitudinal and transverse partitions of matrix have also been considered to differ in structure and chondrolytic mechanisms, the transverse being lightly calcified, with sparse collagen and distinctive proteoglycans susceptible to lysosomes; the more resistant longitudinal ones succumb only to osteoclastic action (Anderson 1962, Dingle 1962).

Continuing cell division in the zone of growth adds to epiphysial ends of cell columns and, with the sequence of changes proceeding away from the diaphysial centre, the bone grows in length. Meanwhile, there is continued internal erosion and remodelling of the newly formed bone tissue and, as further subperiosteal bone deposition continues towards the epiphyses, the bone also grows in diameter, its medullary cavity enlarging transversely and longitudinally.

Growth continues thus for many months or years in different bones (p. 279) but eventually one or more secondary centres usually appears in the cartilaginous extremities. Such epiphysial centres (or ends of bones lacking epiphyses) do not at first display cell columns. Instead, isogenous cell groups hypertrophy and die, with matrix calcification and then invasion by osteogenic vascular mesenchyme (sometimes from cartilage canals, (see p. 287). Many irregular bone spicules, with intervening medullary spaces, are formed on calcified cartilage, as described above for the growth plate. As an epiphysis enlarges, its cartilaginous periphery forms a zone of proliferation and its organization becomes radial, with cell columns and zones of growth, hypertrophy, calcification, erosion and ossification at increasing depths from the surface (3.47). The early osseous epiphysis is thus surrounded by a superficial growth cartilage; but the growth plate next to the metaphysis soon becomes the most active region, and here enlarging cell columns are directed solely towards the metaphysial plate,

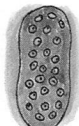

A.

Cartilage model

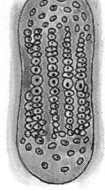

B.

Hypertrophy of central cells

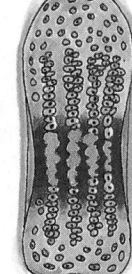

C.

Calcification of matrix in primary centre and formation of periosteal collar

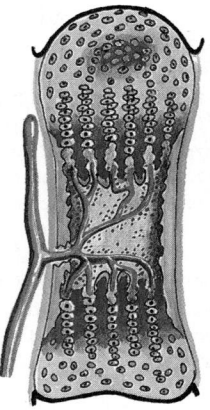

D.

Invasion of primary centre by vascular osteogenic buds

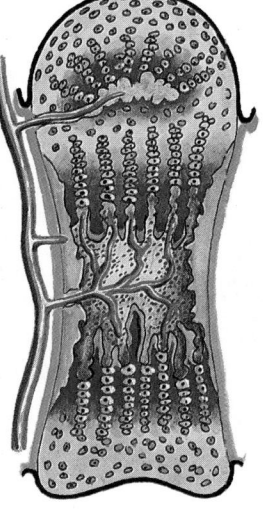

E.

Primary bone laid down on calcified cartilage remnants; secondary centre of ossification appears and becomes vascularised

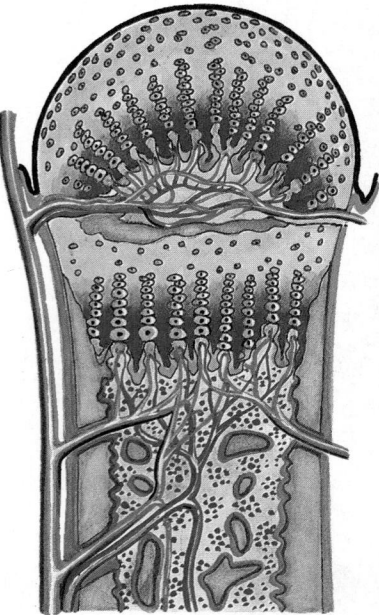

F.

Continued growth of cartilage of 'epiphyseal' plate and epiphysis; proliferation of red bone marrow

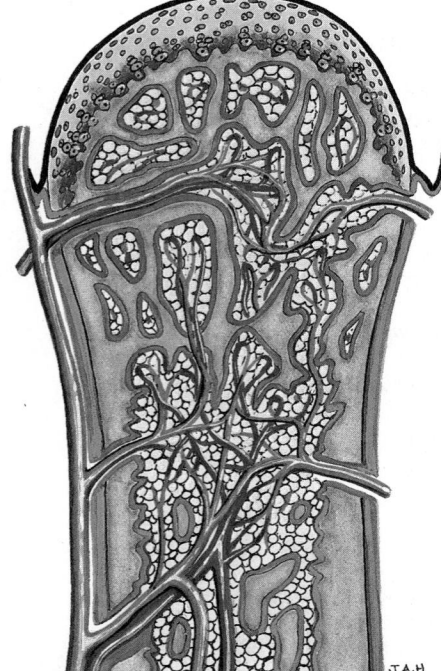

G.

Cessation of cartilage growth and complete ossification of 'epiphyseal' plate, (fusion of the epiphysis). Replacement of red bone marrow with yellow, adipose marrow (as in all long bones of the adult, except proximal humeral and femoral epiphyses)

3.39A–G Diagram depicting the stages of endochondral ossification in a long bone (e.g. a phalanx). For colour code see **3.38**; additionally, hyaline cartilage is light purple, calcified cartilage deep purple. For further details of postnatal endochondral ossification, see **3.47**.

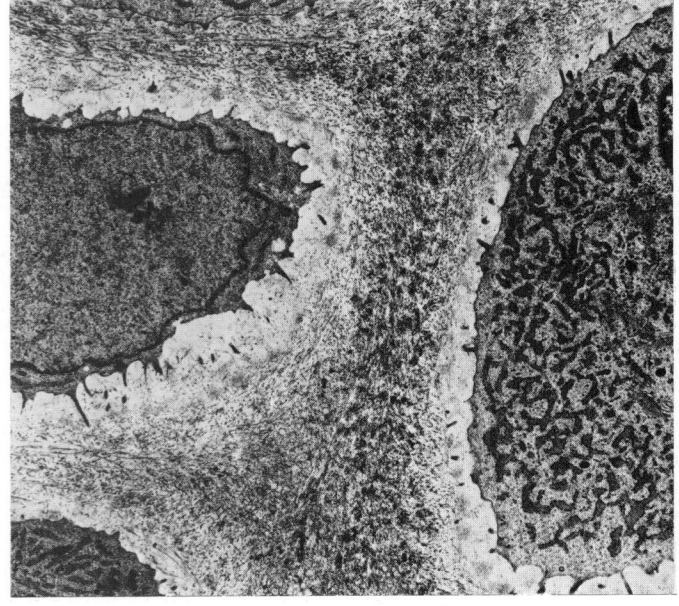

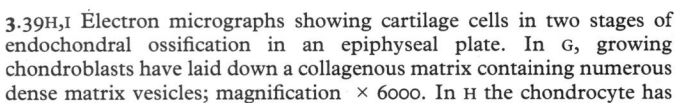

H

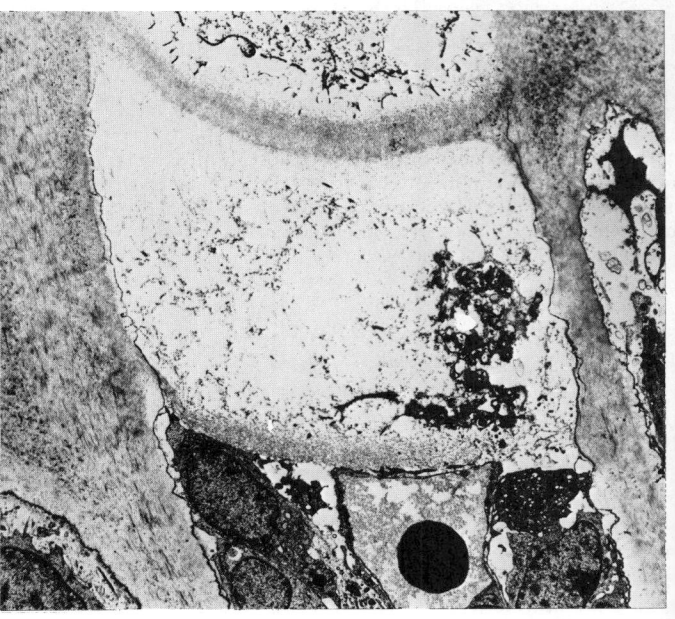

I

3.39H,I Electron micrographs showing cartilage cells in two stages of endochondral ossification in an epiphyseal plate. In G, growing chondroblasts have laid down a collagenous matrix containing numerous dense matrix vesicles; magnification × 6000. In H the chondrocyte has hypertrophied and disintegrated, while at its base is the tip of a ingrowing capillary surrounded by osteoblasts; magnification × 6000. (Micrographs taken by G Roberts.)

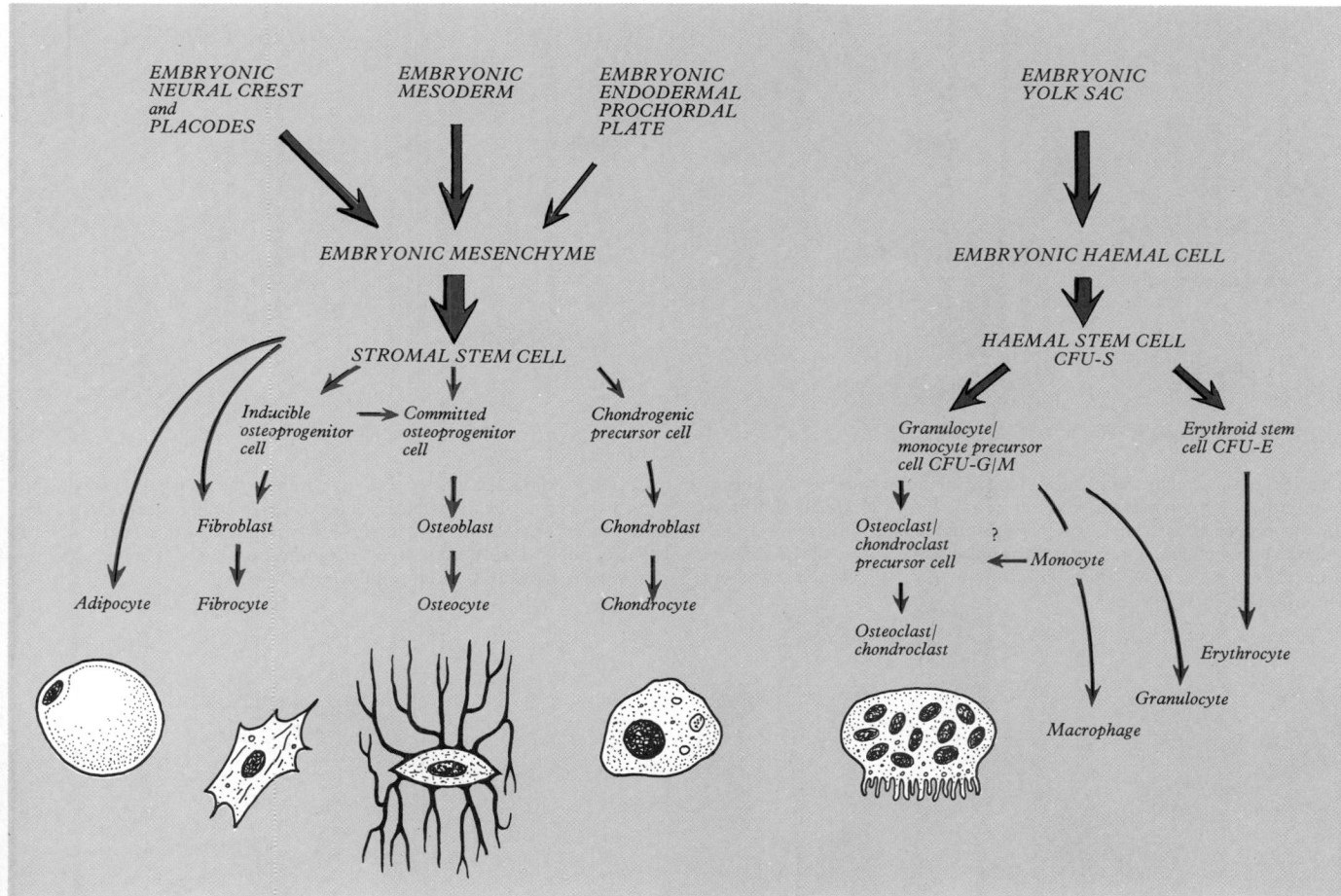

3.40 Diagram of origins and fates of the cells of mature bone, based on current evidence; the haemal series is not shown in detail. For further information see text.

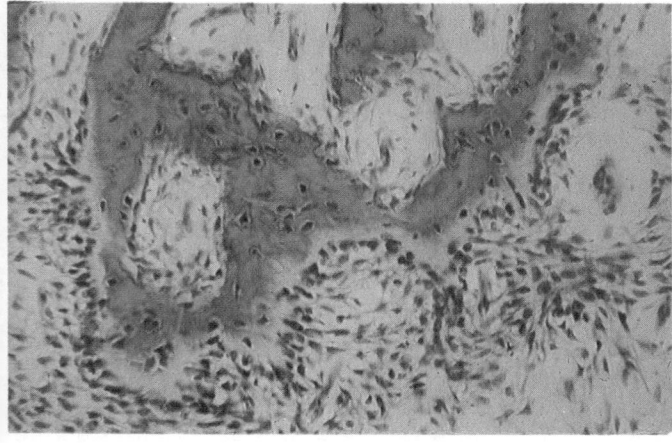

A

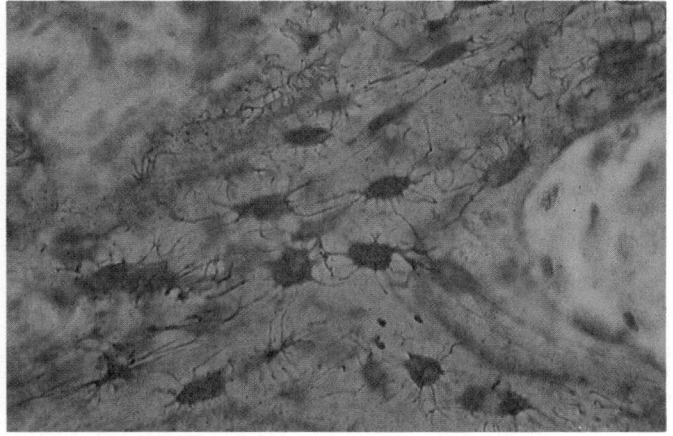

B

3.41 Micrographs of immature woven bone (fetal human). A is a section of part of the maxilla, stained with haematoxylin and eosin. The mineralized bone is eosinophilic, but shows a paler front of osteoid formation where many osteoblasts are clustered; magnification × 150. Osteocytes are also visible within the bone. B is a higher magnification of woven bone stained with picrothionin to show the typically large osteocytes and their branched processes; magnification × 800. (Both micrographs provided by Moya Meredith Smith, UMDS, Guy's Campus, London.)

whereas elsewhere they are directed to the underlying epiphysial ossification (Payton 1934, Rang 1969).

As the bone reaches maturity, epiphysial and metaphysial ossification processes gradually encroach upon this growth plate from either side, eventually meeting, when bony fusion of the epiphysis occurs and longitudinal growth of the bone ceases. Unlike changes in growth plates, events of fusion have attracted little interest (Rang 1969), but in the rat (Becks et al 1948) they are as follows. When growth ceases, the cartilaginous plate becomes quiescent and gradually thins. Proliferation, palisading and hypertrophy of

chondrocytes ceases; they become short, irregular conical masses. Patchy calcification is then accompanied by absorption of patches and some adjacent metaphysial bone. These erosion channels are invaded by vascular mesenchyme, some endothelial sprouts piercing the thin plate of cartilage, and here metaphysial and epiphysial vessels unite. Final bony fusion is by ossification around these vessels, spreading into the intervening zones. Such bone is very dense and visible in radiographs as an *epiphysial line* (a term also used for the level of the perichondral ring around the growth cartilage of immature bones, or the surface junction

3.42A Section through the surface of a mature bone showing the periosteum into which are inserted skeletal muscle fibres. Notice both fibrous and deeper cellular components of the periosteum. Viewed by half-polarized light which demonstrates the anisotropic nature of the lamellar bone visible here; H & E, magnification × 250. (Provided by Moya Meredith Smith.)

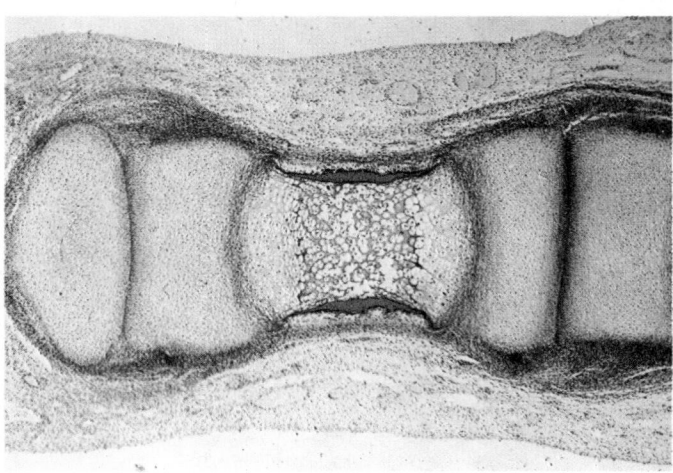

3.44 Longitudinal section of a phalanx (from the hand in 3.37) showing an early primary ossification centre. The cartilage cells in the shaft centre have hypertrophied and this region is surrounded by a delicate tube or collar of subperiosteal bone (red); magnification × 50. (Photography by Kevin Fitzpatrick, Guy's Hospital Medical School.)

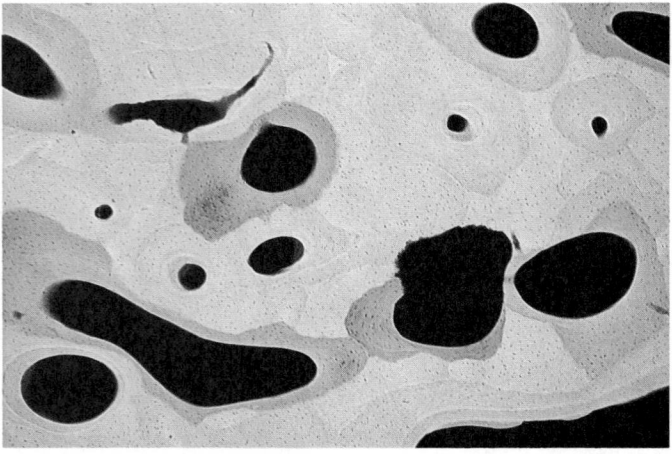

3.42B Microradiograph of a section through compact, osteonic bone, showing different degrees of mineralization of osteons which reflect progressive remodelling around large vascular canals. The most recent bone tissue appears denser; magnification × 150. (Provided by Moya Meredith Smith.)

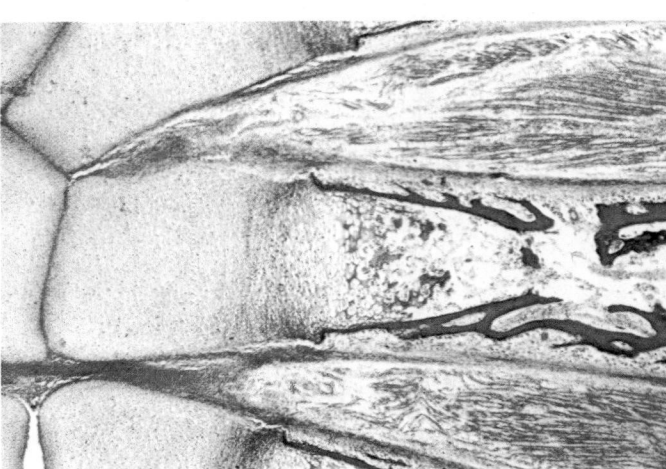

3.45 Longitudinal section of the proximal half of a fetal metacarpal bone (from 3.37) at a more advanced stage than the phalanx in 3.44. The periosteal collar of woven bone is thicker, contains radially-disposed vascular spaces; vascular invasion of the shaft centre has occurred and is proceeding towards its extremities; magnification × 40. (Photography by Kevin Fitzpatrick, Guy's Hospital Medical School.)

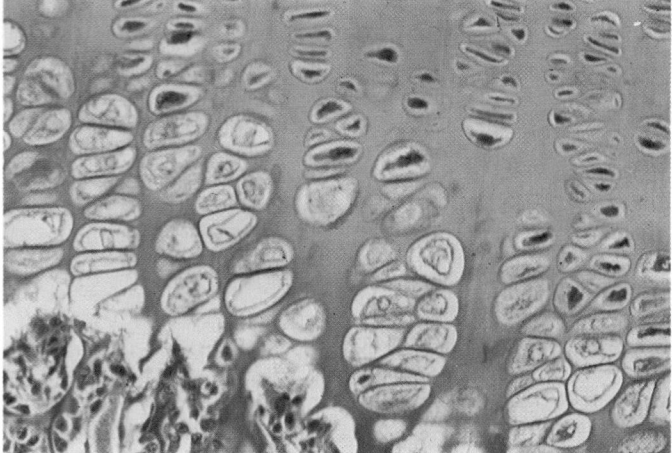

3.43 Section showing the transformation of cartilage cells and their lacunae as the ossifying front of an early primary centre of ossification is approached (below). Note the cell hypertrophy, lacunar enlargement with matrix partition reduction and increased density of the partitions following calcification; magnification × 600.

between epiphysis and metaphysis in a mature bone). For analyses of epiphysio-diaphysial union, studied in dog and man consult Haines (1975), who found an essentially similar sequence of events as seen in the rat. In smaller epiphyses, uniting earlier, there is usually one eccentric, initial area of fusion, with thinning of the residual cartilaginous plate. Subsequently the original sites of fusion, now mineralized cartilage ('metaplastic bone') with superficial lamellar bone and medulla inside, are eroded and replaced by new bone and medullary tissue extending into the whole cartilaginous plate until union is complete, so that no *epiphysial 'scars'* persist. In larger epiphyses, uniting later, similar processes also involve multiple perforations in growth plates, *islands* of epiphysial bone often persisting as epiphysial scars. (Special features of *metaplastic bone* subjacent to articular cartilage and attachments of tendons, ligaments, and other dense connective tissues, have been surveyed by Haines & Mohuiddin 1968).

The cartilaginous surfaces of epiphyses involved in synovial articulations remain unossified, but the typical sequence of cartilaginous zones in them persists throughout life.

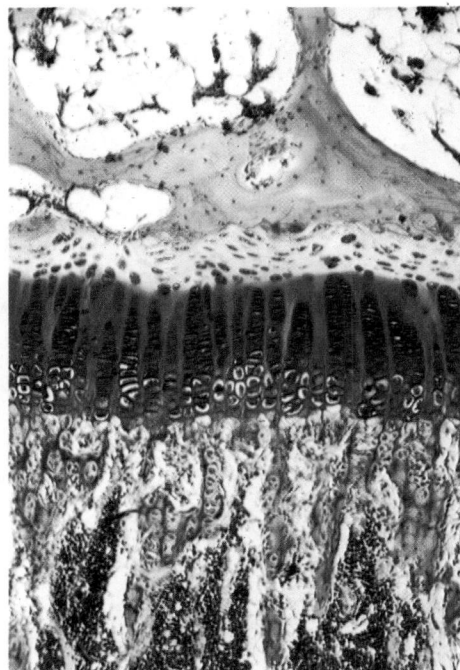

A

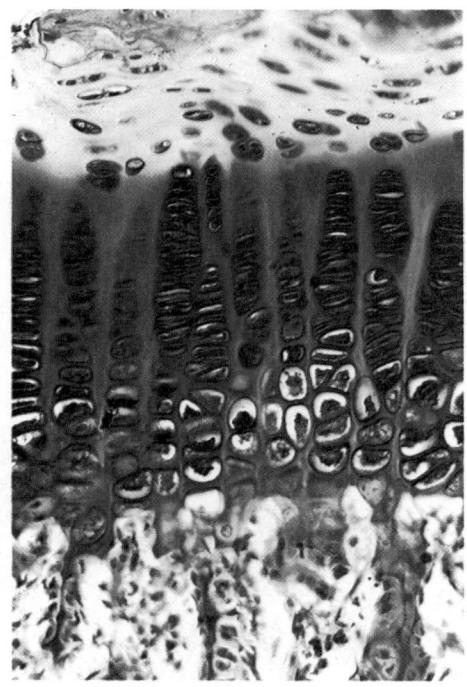

B

3.46 Low and intermediate magnifications of sections through the cartilaginous growth plate between the epiphysis and metaphysis (below) at the proximal end of a human tibia. Note the transition from hyaline cartilage, through zones of cell multiplication, hypertrophy, column formation, matrix calcification and partial chondrolysis to the ossifying front. Compare with **3**.47 and consult text for further details.

Further Development and Remodelling of Bone

In all situations so far described bone is initially trabecular, with a continuous labyrinth of large vascular spaces; osteocytes are irregularly scattered, there is no lamellation and collagen fascicles in the matrix form a random network. This *woven-fibred* tissue is typical of young fetal bones, but occurs also in adults (e.g. lining tooth sockets) and during repair of fractures. Later concentric tracts of non-lamellar, parallel-fibred bone, primary osteons or *atypical Haversian systems*, are deposited within the vascular spaces, narrowing or dividing them by osseous bridges. Collagen bundles in the matrix form parallel longitudinal spirals. No erosion precedes formation of such osteons; hence no cement lines separate them from surrounding woven bone. Bone deposited circumferentially on periosteal and endosteal surfaces is mostly lamellar with longitudinal or circumferential parallel fibres in successive lamellae; where these are thin, lamellation is indistinct. Circumferential bone may also be non-lamellar or woven, in some species, and the term *surface bone* is then preferable.

As bone matures, *typical Haversian systems (secondary osteons)* gradually replace primary osteons and woven bone. They are always preceded by erosion, usually eccentric, of the vascular channels within primary osteons or woven bone. Remodelling of the interiors of bone is marked by the balanced activities of osteoclasts and osteoblasts; osteoclasts first excavate a cylindrical tunnel by concerted action (a 'cutting cone'), and then they are followed by osteoblasts which fill in the space created by concentric deposition around a centrally ingrowing blood vessel (forming a 'closing cone'). Concentric deposition of lamellae of parallel-fibred bone follows on walls of *resorption cavities* and around vascular channels, which are thereby narrowed. A cement line always marks a site of reversal from erosion to deposition. Formation of secondary osteons does not end with growth but endures variably throughout life; some of the earliest secondary osteons are eventually eroded. Remnants of woven bone, primary osteons, circumferential lamellae and finally secondary osteons form *interstitial bone* occupying crevices between later osteons. The degree of remodelling is in general an indication of age, so that the number of osteons and osteon fragments can be used to estimate the age of skeletal material at death. (For quantitative data on the ageing of human bone see Courpron et al 1973, and for ageing effects of osteon remodelling in human tibia see Ortner 1975.)

Bone growth is solely *appositional*, i.e., new layers are added to pre-existing surfaces, osteocytes being enclosed in lacunae and ceasing to divide. The rigidity of mineralized bone matrix of course prevents internal expansion, so that interstitial growth, characteristic of most tissues, including cartilage, is absent in bone, though essential in growth cartilages, as discussed above.

Hence remodelling of bone, which involves either major readjustment after fracture or remodelling throughout life, depends upon delicate geometric balances between deposition and removal. Demonstrations of these main features have, with increasing refinement, spanned the last two centuries. Earlier experimenters used metal markers, such as wires encircling shafts of bones, or pellets embedded in them. An early observation that, in growing pigs fed with the plant *madder*, newly formed bone is identifiably pink initiated numerous studies of bone growth by alternating periods of feeding with madder and without it (Brash 1934). Intraperitoneal injection of alizarin red has been used with greater precision for the same purpose (Hoyte 1960). Such studies confirm that bone grows by accretion; later techniques have merely increased accuracy, making quantitative histological studies possible. These include administration of osseotropic isotopes followed by autoradiography (Leblond 1950), exposure to tetracycline at intervals followed by fluorescence microscopy, and historadiography (Amprino & Engström 1952, Amprino 1968).

Major remodelling, i.e. change in general shape, occurs in all growing bones; much studied examples are cranial and long bones with expanded extremities. A bone such as the parietal thickens and expands during growth, but decreases in curvature (**3**.50). Accretion continues at its edges by interstitial multiplication of osteoblasts at sutures. Periosteal bone is added externally and eroded internally, but not at uniform rates; the rate increases with radial distance from the centre of ossification (the future parietal eminence).

Long bones elongate mainly by extension of endochondral ossification into calcified zones of adjacent growth cartilages continually replaced by longitudinal interstitial growth of their proliferative zones, with minor additions by radial epiphysial growth (p. 304). Simultaneously diametric increases of growth cartilages and shafts occur by continuing subperiosteal deposition and endosteal erosion. In many bones, however, growth is at different rates, or even reversed, at different places. A bone

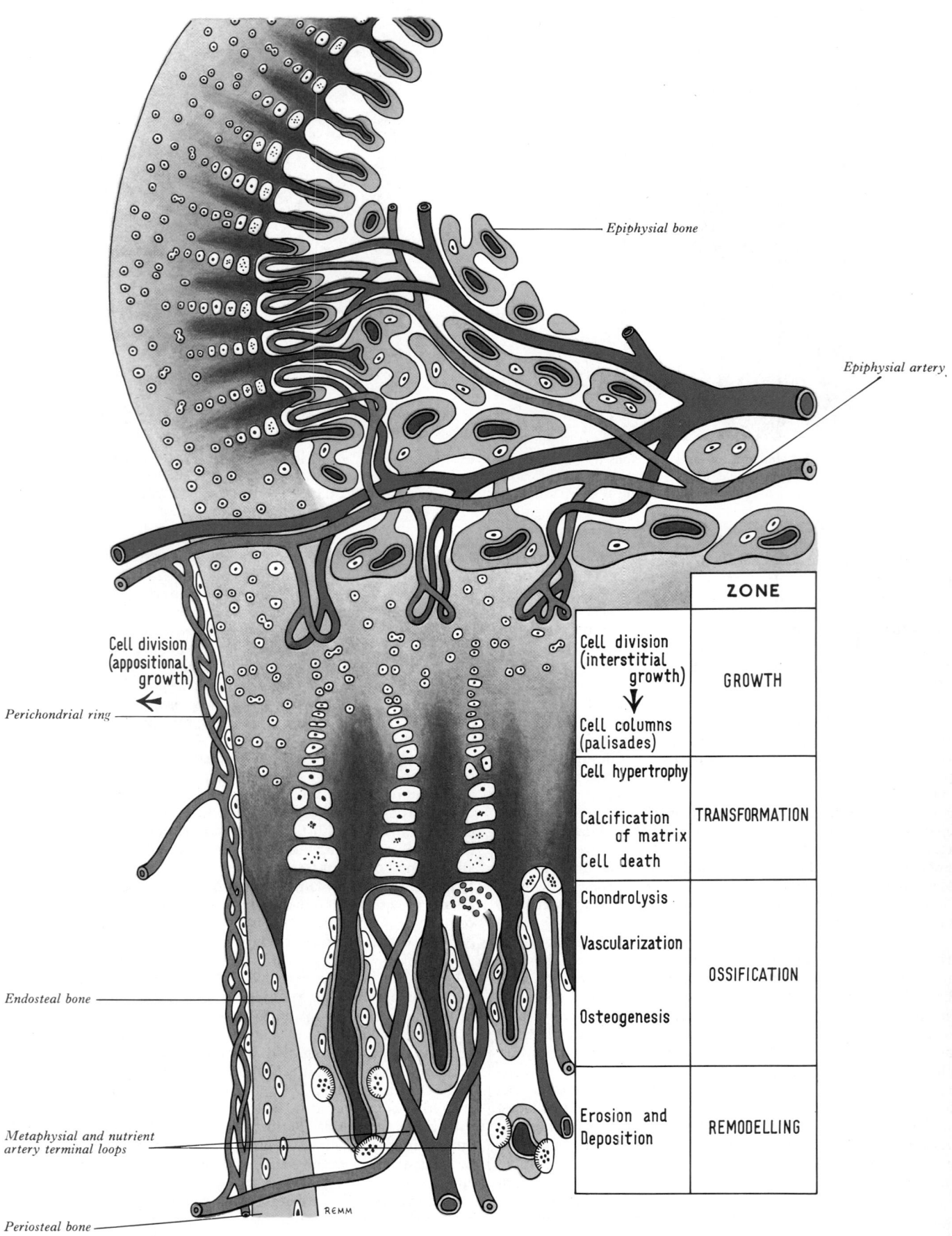

3.47 Scheme of the main features of an active growth cartilage and the adjacent metaphysis and epiphysis. The different views concerning the mechanism of chondrolysis described in the text are indicated.

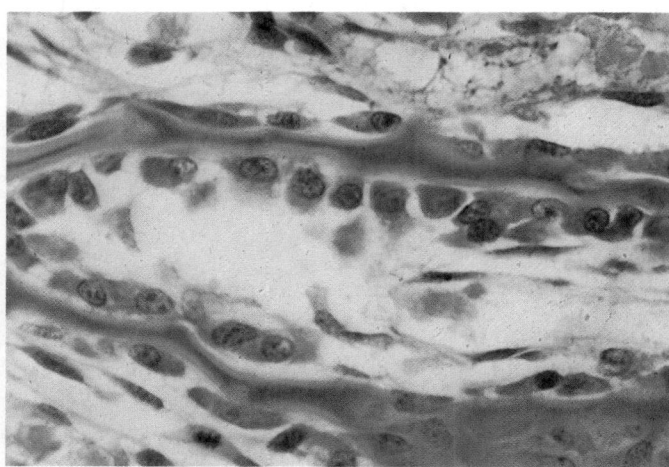

3.48 Section through the metaphysis of a fetal bone stained with haematoxylin and eosin showing developing spicules of early bone. Each spicule contains a deeply stained basophilic core of calcified cartilage, which is covered on both aspects by a lightly stained eosinophilic layer of young bone along which are ranged rows of active osteoblasts.

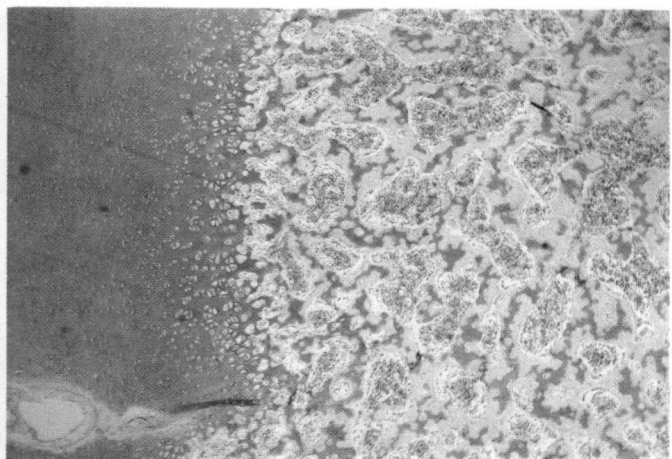

3.49 Transverse section through an epiphysis of a long bone at a more advanced stage of ossification. Note the gradation of cartilage cell lacunae (left) as the ossifying front is approached. The bony trabeculae (light pink) contain dark patches of basophilic calcified cartilage and the intertrabecular spaces are filled with red bone marrow.

initially tubular may thus become triangular in section, e.g. the tibia. Similarly, the waisted contours of metaphyses are preserved by differential rates of periosteal erosion and endosteal deposition (see **3.51**).

Tetracycline marking techniques have emphasized the mutability of even mature compact bone, where a single section may include mature quiescent osteons, and recently formed, forming or resorbing osteons. Lamellar bone forms at a variable rate; a resorption canal forms in 1–3 months and a new osteon in a similar period (Lee 1964). Internal remodelling continuously supplies young osteons with labile calcium reserves (p. 303), and a malleable osseous architecture responsive to altering patterns of stress (p. 276). Total morphogenetic control of shape is not clear, but suggestions largely involve responses to stress. Certainly bone resorption typically occurs when the gravitational or other mechanical stresses are reduced, as in bed rest, or in zero gravity conditions in space. Likewise bone subjected to constant pressure tends to resorb and, with constant tension, is deposited; this forms the basis of much orthodontic treatment, as teeth can be made to migrate slowly through alveolar bone by the application of steady lateral or medial pressure.

Like other crystalline substances, bone is piezo-electric, i.e. it generates electric currents when deformed, and distribution of potential differences reflects form of distortion. Such currents may affect cells responsible for osteolysis and bone deposition, promoting structural remodelling to resist predominating stresses (see Currey 1984 for a review of this subject). Data concerning sources, intensities and distributions of bioelectric phenomena in stressed connective tissues in general are not yet sufficient to show causal relationships between such events and to relate them to morphogenesis.

Metabolic and Endocrine Effects on Bone

The roles of bone salts, parathyroid hormone and calcitonin in regulating circulating calcium levels are considered elsewhere (p. 302). However, the normal development and maintenance of bone also require adequate intake and absorption of calcium, phosphorus, vitamins A, C and D and a balance between somatotrophic hormone, thyroid hormones, oestrogens and androgens. Various other factors, including different prostaglandins and glucocorticoids may also play important roles in the maintenance and turnover of osseous tissue (see Rodan & Rodan 1983).

Prolonged deficiency of calcium causes loss of bone mineral or *osteoporosis*, and consequent fragility of bones. Vitamin D influences intestinal absorption of calcium (and indirectly of phosphorus) and therefore affects circulatory calcium levels; prolonged deficiency (with or without low intake) leads, in adults, to *osteomalacia*, in which bones are poorly mineralized and often contain regions of deformable, uncalcified osteoid. Similar deficiencies, during growth, lead to classic rachitis (*rickets*), with severe disturbance of growth cartilages and ossification, including reductions of regular columnar organization in growth plates and failure of cartilage calcification (although chondrocytes proliferate); growth plates also become thicker and less regular than normal. In metaphysial regions cartilage is partially eroded, and osteoblasts secrete layers of osteoid, which fail to ossify on uncalcified or poorly calcified cartilage trabeculae. Ultimately gravity deforms such softened bones. (For reviews of the role of vitamin D, see Reynolds 1974, Rodan & Rodan 1983).

Deficiency of vitamin C causes *scurvy*, with widespread changes in all connective tissues, particularly growth cartilages and metaphyses of long bones. It appears essential for adequate

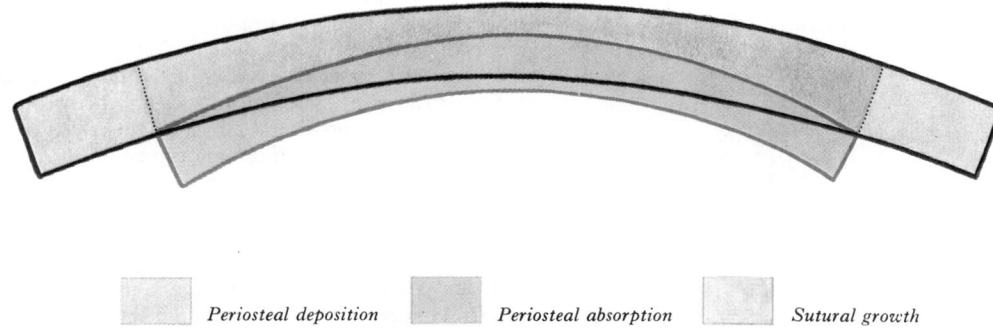

Periosteal deposition   Periosteal absorption   Sutural growth

3.50 Scheme illustrating the gross patterns of growth and remodelling of a bone of the cranial vault, in section. Note the changes in surface area, thickness and curvature. See text for further details.

synthesis of collagen *and* matrix proteoglycans in connective tissues and, when deficient growth plates become thin, ossification almost ceases and metaphysial trabeculae and cortical bone are reduced in thickness, causing fragility and delayed healing of fractures.

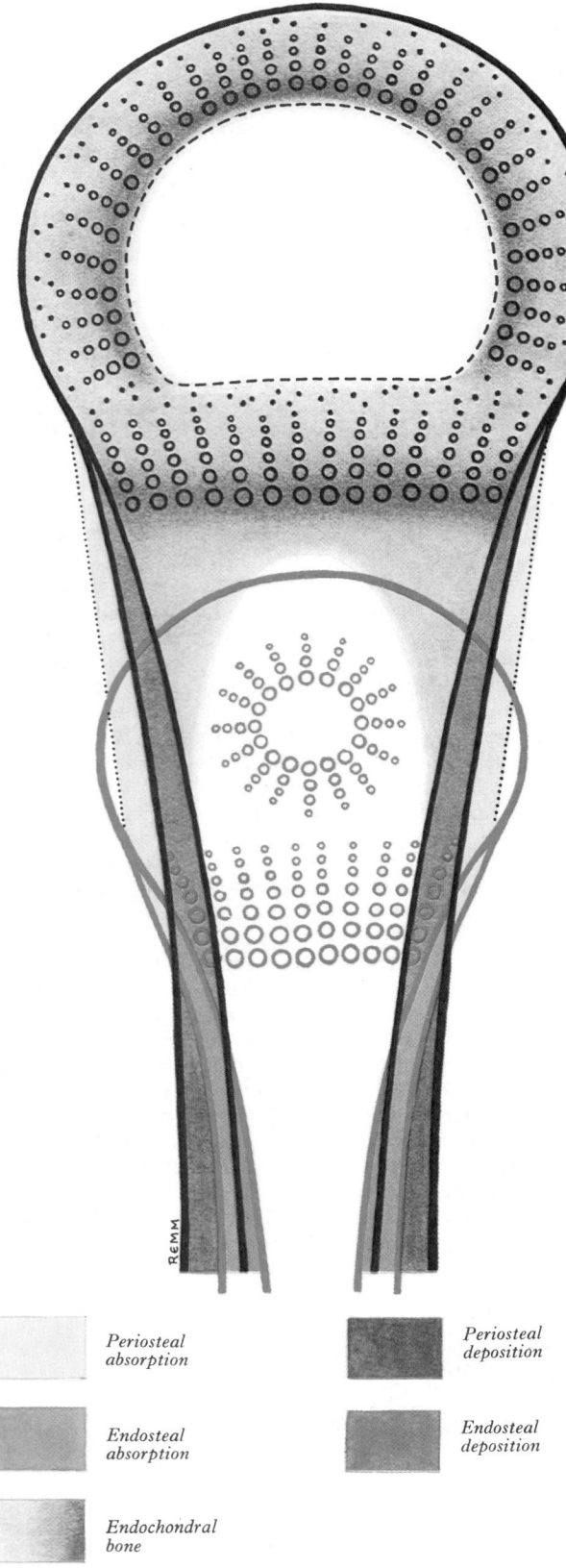

Vitamin A is also necessary for normal growth and for a correct balance of deposition and removal of bone. Deficiency retards growth due to the failure of internal erosion and remodelling, particularly in the cranial base. Foramina are narrowed, sometimes causing pressure atrophy of contained nerves; the cranial cavity and spinal canal may fail to expand with the central nervous system, impairing nervous function. Conversely, *excessive* vitamin A stimulates vascular erosion of growth cartilages, which become thin or totally lost; as a result, longitudinal growth ceases.

Balanced endocrine activities are essential to normal bone maturation, and disturbances may have profound effects. In addition to its role in calcium metabolism, parathyroid hormone in excess stimulates osteoclasts and perhaps osteocytes to widespread osteolysis and erosion, leaving a demineralized framework—a condition termed *osteitis fibrosa*.

Acidophil cells of the adenohypophysis cerebri secrete *growth hormone* (somatotropin) necessary for normal interstitial proliferation in growth cartilages and hence increase in stature. Termination of normal growth is imperfectly understood, but may involve a fall in hormone production or in the sensitivity of chondroblasts to hormones. Experimental hypophysectomy, or clinical reduction of growth hormone in the young, leads to quiescence and thinning of growth plates and hence *pituitary dwarfism*. Conversely, continued hypersecretion in the immature leads to *giantism*, but in maturity it results in thickening of bones by subperiosteal deposition, the mandible, hands and feet being most affected, an affliction known as *acromegaly*.

While continued longitudinal growth of bones depends on adequate levels of growth hormone, the effective remodelling to a mature shape also requires the action of the thyroid hormones, tri-iodothyronin and tetraiodothyronin (T_3 and T_4); growth and skeletal maturity are also closely related to endocrine activities of the ovaries, testes and adrenal cortices. High oestrogen levels increase deposition of endosteal and trabecular bone; conversely, osteoporosis in aged women may reflect reduced ovarian function. Fluctuations in the rate of growth and the timing of skeletal maturation reflect circulating levels of adrenal and testicular androgens (p. 1429). In hypogonadism, maturation (marked by growth plate obliteration) is late and the limbs therefore elongate excessively; conversely, in hypergonadism, premature fusion of the epiphyses results in diminished stature.

The normal growth and pathology of growth cartilages has a large literature, beyond the present scope, but several excellent accounts have appeared (Rang 1969, Serafini-Fracassini & Smith 1974, Kahn et al 1983).

General Features of Bones

Bones vary not only in their *primary* shape but also in lesser surface details, or *secondary* markings which appear mainly in postnatal life. Certain features—elevations and depressions, smooth areas and rough ridges—are found in many bones. For such, a repertoire of general terms is used, since the same form of marking or surface texture usually has the same functional significance wherever it occurs. For example, bones display *articular surfaces* at synovial joints with their neighbours; if small, these are *facets* or *foveae*. Knuckle-shaped surfaces are *condyles*, and a *trochlea* is grooved like a pulley. Adapted in shape to the movements of particular joints, such surfaces are smooth, though in life they are, of course, covered by articular cartilages forming the actual surfaces of synovial joints. The texture of such osseous surfaces is due to another feature, frequently overlooked; they are devoid of the vascular foramina typical of most other bone surfaces, articular cartilage being poorly vascularized. However, osseous articular surfaces are not completely impermeable, though the cartilage contiguous with subjacent bone is usually calcified; hence substantial interchange is improbable (the nutritive avenue to articular cartilage may be largely from synovial fluid). Scanning electron microscopy shows that *all* bony surfaces, including those appearing smooth to the naked eye or hand lens, display large numbers of minute foramina. The proportion of these occupied by vascular, nervous or collagenous elements is unknown. (For surface architecture consult 3.5,6.)

Periosteal absorption		*Periosteal deposition*
Endosteal absorption		*Endosteal deposition*
Endochondral bone		

3.51 Scheme of the patterns of remodelling which occur during the growth of a long bone. Note the changing shape of the epiphysial ossification centre, the altered organization of the cartilaginous growth plates and the varying zones of bone deposition and absorption. See text for further details.

Large tendons (e.g. adductor magnus, subscapularis) are attached to facets which lack the regular contours of articular surfaces but resemble them in texture, being poorly vascularized. Such tendon facets are sometimes depressed (the adductor 'tubercle' is often a small pit); alternatively they may surmount large elevations, e.g. the humeral tuberosities.

Depressions and elevations, varying in size and shape, interrupt otherwise featureless osseous surfaces. A depression is a *fossa*, and some articular surfaces are fossae (cf. temporomandibular joint). Lengthy depressions are *grooves* or *sulci* (e.g. humeral bicipital sulcus); a notch is an *incisura*, and an actual gap is a *hiatus*. A large projection is termed a *process* or, if elongated, slender or pointed, a *spine*. A curved process is a *hamulus* or *cornu* (cf. sphenoidal pterygoid hamuli and hyoid cornua). A rounded projection is a *tuberosity* or *tubercle*, occasionally a *trochanter*. Long elevations are *crests*, or *lines* if less developed; crests are wider and present boundary edges or *lips*; however, these terms are ill-defined, and one most substantial crest is the *linea aspera* (cf. femur). An *epicondyle* is a projection close to a condyle and usually an attachment for the collateral ligaments of the adjacent joint or common myotendinal attachments for superficial muscle groups (cf. humerus). The terms protuberance, prominence, eminence and torus are also less often applied to certain bony projections. The expanded proximal ends of many long bones are often termed the '*head*' or *caput* (cf. humerus, femur, radius, etc.).

A hole in bone is a *foramen*; foramina are *canals* when lengthy. Large holes may be called *apertures* or, if covered largely by connective tissue, *fenestrae*. Clefts in or between bones are *fissures*. A *lamina* is a thin plate; larger laminae may be called *squamae* (e.g. the temporal squama).

Large areas on many bones are featureless and indeed often smoother than articular surfaces, but they differ from these in possessing many visible vascular foramina. Such a texture occurs where muscle is directly attached; through the foramina pass small blood vessels from bone to muscle and perhaps vice versa. Areas covered only by periosteum are similar but vessels are less numerous.

Tendons are usually attached at roughened bone surfaces; and wherever any aggregation of collagen in a muscle reaches bone, surface irregularities correspond closely in form and extent to the pattern of such 'tendinous fibres'. Such markings are almost always elevated above the general surface, as if ossification advanced into the collagen bundles from periosteal bone. However, how such secondary markings are induced is uncertain. Evidence suggests that their prominence may be related to the power of muscles involved and they increase with advancing years, as if the pull of muscles and ligaments exercised an accumulative effect. Certainly, surface markings delineate the shape of attached connective tissue structures, whether an obvious tendon, intramuscular tendon or septum, aponeurosis, or tendinous fibres mediating otherwise direct muscular attachment. Hence markings may be facets, ridges, nodules, rough areas or complex mixtures, affording accurate data of junctions of bone with muscles, tendons, ligaments or articular capsules. Numerous examples will be described with individual bones.

Note that even in the so-called 'fleshy' or direct attachment of a muscle, its myocytes do not themselves adhere directly to periosteum or bone. The route of transmission of tension from contracting muscle to bone must be through the connective tissue which pervades all muscles as perimysium and endomysium. (See *myotendinal junction*, p. 556, although details of its mechanics are uncertain.) These two forms of attachment, at the extremes of a range of admixtures, differ in the density of collagen fibres between muscle and bone. Where collagen is visibly concentrated, markings appear on the bone surface. In contrast, the multitude of microscopic connective tissue ties of direct attachment, necessarily over a larger area, do not visibly mark it. Here the bone appears smooth to unaided vision and touch. (*See* 3.5,6 for magnified views of osseous surfaces.)

The Response of Bone to Injury

After an injury such as a fracture, bone generally undergoes spontaneous regeneration, which in favourable conditions can result in

virtually complete restoration of its anatomical structure before injury. Depending on the location of the injury, endochondral and/or intramembranous ossification (p. 300) may be involved. Exceptions to this occur in the calvarium, where fractures are usually repaired permanently by fibrous union and not by regeneration of new bone, and after rigid internal fixation, where bone can form directly by primary bone healing, without the preliminary development of either fibrous tissue or cartilage.

BONE HEALING WITHOUT OPERATIVE INTERVENTION

Bone healing consists of several overlapping phases: inflammation, soft callus formation, hard callus formation and remodelling. Soft and hard callus formation are collectively equivalent to the proliferative phase of wound healing (p. 84). The term 'soft callus' appears to be contradictory in that 'callus' implies hardness (Latin *callum* or *callus* = hard integument); it can be defined as the soft, collagenous, revascularizing, osteogenic blastema which unites the bone fragments and from which bone regenerates.

A

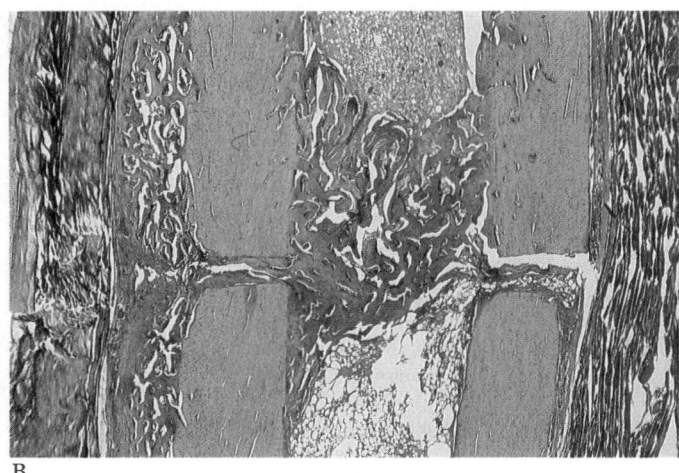

B

3.52 Midsagittal sections through the repairing fracture site of the lower tibia of a rabbit, held rigidly in an external fixator. Movat's hexachrome stain: undifferentiated blastemal tissue grey; cartilage turquoise; bone pale green. (Sections provided by R Brueton, Orthopaedic Research Department, Rayne Institute, St Thomas's Hospital, London; photographed by Kevin Fitzpatrick.)

A. 2 weeks after injury. A distinct fracture gap separates adjacent fragments of cortex. The mainly adipose marrow contains patches of blastemal tissue. External soft callus contains patches of cartilage; a few bone trabeculae and some cartilaginous areas are present endosteally.

B. 4 weeks after injury. A lamella of bone lies in the fracture gap. Endosteal soft callus unites the posterior cortical fragments. Periosteal soft callus is present posteriorly, but is not yet confluent across the gap, while anteriorly only fibrous tissue is present.

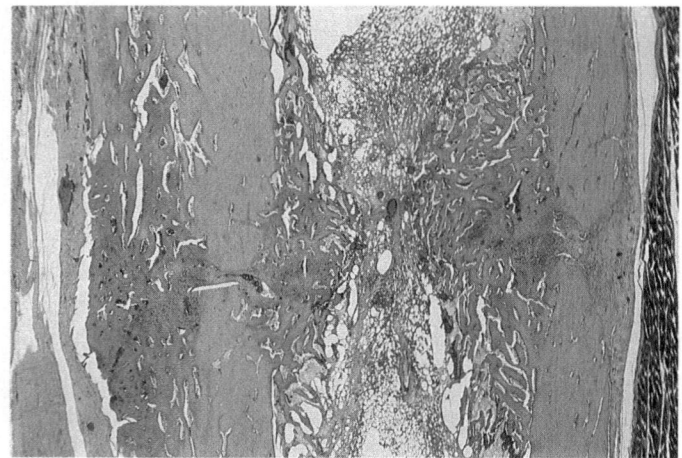

3.53 Midsagittal section through the remodelling fracture site of the lower tibia of a rabbit, held rigidly in an external fixator, six weeks after injury. A mass of periosteal hard callus unites the posterior cortices, while fibrous tissue is present anteriorly. Endosteal hard callus unites anterior and posterior cortices. Some soft callus lies between the fragments posteriorly. Remodelling of the marrow cavity has begun and there is a central axis of myelofibrotic material. Movat's hexachrome stain: undifferentiated blastemal tissue grey; cartilage turquoise; bone pale green. (Section provided by R Brueton, Orthopaedic Research Department, Rayne Institute, St Thomas's Hospital, London; photographed by Kevin Fitzpatrick.)

Inflammation

This begins immediately after fracture, overlaps with the phase of soft callus formation and typically lasts for about four days. It is characterized by local haemorrhage, haematoma formation, oedema and pain. Although there is some evidence that fibrin in the haematoma may stimulate the proliferation of local potentially osteogenic cells and assist in immobilizing the bone fragments, the haematoma is not generally considered to be a significant stimulator of bone healing (Heppenstall 1980).

Vascular damage causes the environment of the fracture to become hypoxic and acidic, resulting in osteocytic disruption and the release of lysosomal enzymes, followed by tissue necrosis at and close to the site of injury. Mast cells, polymorphonuclear leucocytes and macrophages appear at the site of injury and release mediators, some of which may, as in the repair of skin wounds (p. 84), stimulate the proliferation of reparative cells (here osteoblasts, endothelial cells and in some circumstances chondroblasts). The mast cells, some of the macrophages and the reparative cells are of local origin, while the polymorphonuclear leucocytes and the remainder of the macrophages are haematogenous.

During the inflammatory phase, osteoclasts and macrophages erode necrotic bone and remove tissue debris from the fracture site. Recent research suggests that macrophages also secrete factors or mediators, some of which stimulate collagen synthesis and angiogenesis in wound healing (Hunt et al 1984) and it is conceivable, though not yet demonstrated, that they act in a similar manner in the healing of bone.

As inflammation is overlapped by soft callus formation, cells near the fracture site are induced to develop into osteoblasts which produce new bone. These osteogenic cells either result from the modulation of osteocytes and osteogenic periosteal and endosteal cells, or from the division and differentiation of fibroblasts, endothelial cells, non-striated myocytes and mesenchymal cells. The inductive mechanism for modulation and differentiation is probably multifactorial and may involve a variety of interacting stimuli including mediators released from mast cells and macrophages, reduced oxygen availability and the level of bone morphogenic substance (Heppenstall 1980).

Soft Callus Formation

During this stage, which occupies approximately three or four weeks, a soft tissue blastema or soft callus develops around and between the fragments of bone, reducing their mobility (**3.**52A). The soft callus contains proliferating osteoblasts, fibroblasts and often chondroblasts, embedded in a matrix, rich in glycoproteins and collagen, into which new blood vessels grow. The soft callus can be subdivided into external and internal callus, the former derived from the proliferation of osteoblasts in the osteogenic layer of the periosteum, and the latter from endosteal cells. The enhanced proliferative activity of the osteogenic layer of the periosteum extends beyond the immediate fracture site, elevating the overlying fibrous component of the periosteum and producing a collar of soft external callus which unites the bone fragments. Periosteal cells located within this callus furthest from the fracture site (i.e. in a mechanically stable, well-oxygenated environment) form new osseous matrix directly but as the more mobile and poorly oxygenated fracture site is approached the cell population becomes mixed, containing chondrogenic as well as osteogenic cells, the former being the transformed progeny of the latter (Tonna & Pentel 1972). Although angiogenesis is in progress, proliferation is so rapid that cellularity outstrips vascularity, maintaining an oxygen gradient and keeping the fracture site relatively hypoxic (Wray 1963, Heppenstall et al 1975). Initially the periosteum supplies blood to the site of fracture, but later in the regenerative process the normal centrifugal flow of blood from the endosteal circulation is re-established (p. 299). The external and internal surfaces of the soft callus are electronegative throughout this stage of repair.

Hard Callus Formation

During this stage of fracture healing the external and internal soft callus are gradually converted into woven bone (**3.**52B) mainly by endochondral ossification, unless the bone fragments have been surgically immobilized with a compression plate, when intramembranous ossification predominates. Both cellularity and vascularity continue to increase, but in such a manner that the oxygen gradient referred to above is maintained. The pH of the matrix of the callus gradually increases to the neutral level but the external and internal surfaces of the callus remain electronegative. While both osteogenesis and chondrogenesis are in progress, osteoclasts continue to erode the matrix of any necrotic bone remaining at the fracture site. This stage of bone healing commences at about three or four weeks after injury and continues until attainment of firm bony union, about two or three months later for most adult long bones.

Remodelling

During this stage, which overlaps with the formation of hard callus and may continue for several years, the woven bone of the hard callus is gradually converted into lamellar bone. Osteoclasts remove excess bone from the exterior of the periosteal collar and remodel its endosteal aspect so that the medullary cavity is restored across the site of the fracture (**3.**53). As a result of changes in vascularity, the oxygen supply of the fracture site returns to normal. The surface charge of the fracture site also reverts to normal, being no longer electronegative. Remodelling can be considered to be complete, a state achieved more rapidly in children than in adults, when the site of the fracture can no longer be identified either structurally or functionally. The regenerative process is, however, extremely lengthy and, until it has been completed, the injured bone is functionally less efficient than it was before damage. The rate of regeneration can be accelerated by a number of measures, including surgical intervention and exposure to physical agents such as direct current (Friedenberg et al 1971, Brighton 1981), pulsed electromagnetic fields (Bassett et al 1974, reviewed by Barker & Lunt 1983) and ultrasound (Duarte 1983, Dyson & Brookes 1983). The mechanisms of induced acceleration of regeneration are currently under investigation.

PRIMARY BONE HEALING

In the primary bone healing which follows rigid internal fixation (Perren et al 1969), interruption of the blood supply at and near the fracture results in local destruction of osteons, and this stimulates the regeneration of new Haversian systems across the site of the fracture. Osteoclasts assemble at the ends of Haversian canals near to the fracture site forming *spearheads* or *cutting cones* which advance at a rate of 50 to 80 μm per day across the fracture, enlarging the canals as they advance; they are closely followed by osteoblasts which form new Haversian systems in the enlarged canals which now cross the fracture site and so link the fragments of bone. The entire process only lasts about five or six weeks but has the disadvantage that the major surgical intervention required subjects the tissues to further trauma.

THE AXIAL SKELETON

Introduction

Dividing the skeleton into axial and appendicular sections is not merely a convention. The axial structures, cranium and vertebral column and associated ribs and sternum, are primary; the appendicular elements in fins, limbs or wings were subsequent though early additions. Both primary and secondary elements are concerned in elaboration of locomotion. An axial endoskeleton, first a notochord and then a vertebral column, is the basic feature of *Chordata* and their subphylum, the *Vertebrata*, including mankind. A stiff but flexible axis, in bilaterally symmetrical animals that show an early tendency to elongation, prevents telescoping of the body during waves of contraction in successive segmental muscles to produce the sinuous movements, especially in the tail, which are the basic mode of locomotion in aquatic vertebrates. A chain of bones, flexibly connected by discs of deformable substance, developed around and largely replaced the notochord. However, notochordal vestiges occur in vertebrae of many fish, amphibians and reptiles, and centrally in mammalian intervertebral discs. This replacement is repeated in every vertebrate embryo. These new vertebral elements are complex and variable in pattern in earlier vertebrates but from reptiles onwards the most basic part is the *centrum*, forming most of the vertebral body, ventral to the spinal cord (spinal medulla). In a typical vertebra a *neural arch*, encircling the spinal cord, fuses ventrally with the centrum and usually bears a median dorsal *spinous process* and paired lateral *transverse processes* just dorsal to neurocentral junctions. This enclosure isolates the spinal cord from the play of axial musculature and protects it from external forces, thus insulating its vessels from extraneous compression.

The centrum and each half neural arch ossify from separate centres; and when these extend through cartilaginous precursors to meet and fuse, the dorsolateral parts of the vertebral *body* are formed from ventral ends of the neural arch. Centrum and body are therefore *not* synonyms, nor is the *vertebral arch* exactly equal to a neural arch. A centrum is somewhat less than a vertebral body, a neural somewhat more than a vertebral arch (p. 327).

Segmental muscles flexing the vertebral axis are only in part attached to vertebrae; in connective tissue septa (*myocommata*) between adjacent myotomes ribs evolve as levers for such attachments. Such costal struts first appeared *dorsally* in the axial musculature and extended ventrally into the body wall in early vertebrates. In fish *ventral* ribs also appeared, which enclosed caudal vessels in the tail. It is generally agreed that ribs of land tetrapods correspond to the dorsal piscine series (Romer 1970).

Ribs are thus intersegmental and segmental muscles, derived basically from myotomes, bend the vertebral column; vertebrae also become intersegmental, though their embryonic pattern (p. 159) is primarily segmental. In early vertebrates dorsal ribs adjoin most vertebrae, showing little regional adaptation except in the postanal tail. In land vertebrates, with elaboration of appendages for locomotion, the vertebral column is adapted to new patterns of force in the distribution of weight and muscular tensions.

In hind limbs the pelvic girdle articulates with several vertebrae, which fuse with each other and with costal elements to form a *sacrum*. Sacral vertebrae lose *individual* movement; in mammals there are three to five *fused sacral vertebrae*, and distal to these a variable number of *caudal vertebrae*, reduced to four degenerate elements fused into a *coccyx* in adult humans. Some movement persists, however, between the fused sacral mass and neighbouring vertebrae; also between sacrum and the pelvic girdle.

Mammalian presacral vertebrae vary in number and differentiation, distinguished, however, into *cervical* (neck), *thoracic* (where ribs persist) and *lumbar* vertebrae (devoid of mobile ribs). Cervical ribs are small or 'absent' (but *vide infra*); their disappearance in cervical *and* lumbar regions is linked to the change from breathing by gills to lungs, and to development of independent movements of the head. There is no neck in fish, the postcranial region being occupied by the branchial apparatus. Caudal shift of the respiratory apparatus in land vertebrates necessarily preceded development of a neck. Cervical vertebrae were early stabilized to seven in mammals (except tree sloths and manatees), even in such extremes as whales and giraffes. With respiratory adaptation of ribs in land animals and development of a diaphragm in mammals, well-formed ribs, articulating with, but separate from vertebrae, are limited to thoracic levels. But ribs do *not* disappear completely in *cervical* and post-thoracic regions: vestigial *costal elements* are combined with transverse processes of all such vertebrae (see **2.44**, and **3.57**).

The total number of vertebrae, excluding the tail, is reduced from the lemuroid and tarsioid to the anthropoid primates. Monkeys, apes and hominids (extinct and extant) show some uniformity. The cervical, thoracic, lumbar and sacral vertebrae number respectively 7, 11 to 15, 4 to 7, and 3 to 6, human values being 7, 12, 5 and 5. Caudal vertebrae vary much in number.

Such preliminary generalizations on the origin of the human vertebral column facilitate consideration of individual elements, beginning with the form and structure of a typical vertebra. Some reference to phylogeny of the ribs and sternum will be found in subsequent sections.

General Vertebral Features

A vertebra (**3.54**) has basically a ventral *body* and a dorsal *vertebral (neural) arch*, extended by lever-like processes, together enclosing a *vertebral foramen*, occupied by the spinal cord, meninges and their vessels. Opposed surfaces of adjacent bodies are bound together by *intervertebral discs* of fibrocartilage. The complete column of bodies and discs forms a strong but flexible central axis of the body supporting, in bipeds, the full weight of head and trunk. It also transmits even greater forces due to muscles attached to it directly and indirectly. The foramina form a *vertebral canal* for the spinal cord, and between adjoining neural arches, near their junctions with vertebral bodies, *intervertebral foramina* transmit mixed spinal nerves, smaller recurrent nerves and blood and lymphatic vessels (see also p. 1124).

The cylindroid *vertebral body* varies in size, shape and proportions in different regions and even more in different species. Its junctional aspects vary from approximately flat (but not parallel) to sellar, with a raised peripheral smooth zone formed from the 'annular' epiphysial disc (p. 328), within which the surface is rough. These differences in texture are due to variations in early structure of intervertebral discs (p. 159). In the horizontal plane profiles of most bodies are convex, but concave dorsally where they complete the vertebral foramen. Most vertical profiles are concave but dorsally flat. Small vascular foramina appear on the front and sides, but posteriorly there are small arterial foramina (Willis 1949) and a large irregular orifice (sometimes double) for exit of basivertebral veins (**3.55**). The adult vertebral *body* is *not* coextensive with the developmental *centrum* (p. 327) but includes, posterolaterally, parts of the neural arch, as already noted.

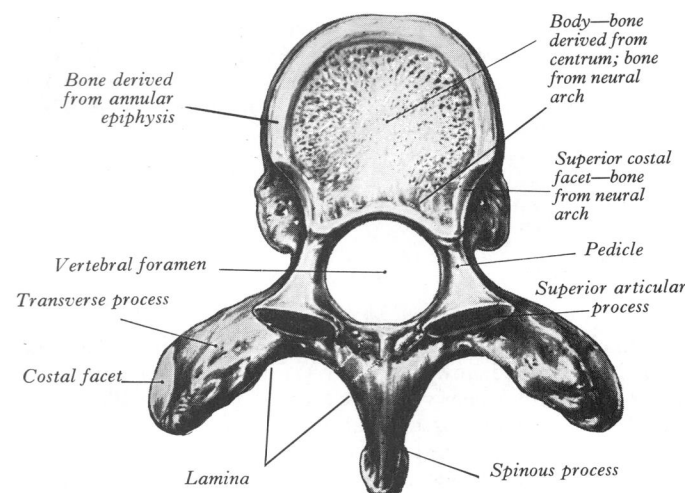

3.54 Typical thoracic vertebra: superior aspect.

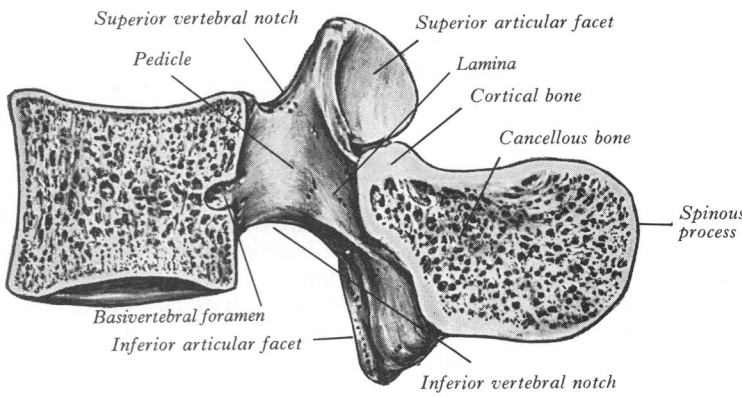

3.55 Median sagittal section through a lumbar vertebra.

Transverse processes project laterally from pediculolaminar junctions as levers for muscles and ligaments particularly concerned in rotation and lateral flexion. (The preceding comment is a simplification. In practice the activities of spinal musculature must be considered in terms of bilateral and surrounding groups; weight-bearing and initial posture are also crucial.) The thoracic transverse processes articulate with ribs. At other levels the mature transverse process is a composite of 'true' transverse process (*diapophysis*) and an incorporated *costal element*.

Costal elements (pleurapophyses) develop as basic parts of neural arches in mammalian embryos, but become independent only as thoracic ribs. Elsewhere they remain less developed and fuse with the 'transverse process' of descriptive anatomy (3.57).

Vertebrae are internally trabecular (3.55), with an external shell of compact bone perforated by vascular foramina. The shell is thin on discal surfaces but thicker in the arch and its processes. The trabecular interior contains red bone marrow and one or two large ventrodorsal canals for the basivertebral veins.

A most interesting technique of observing vertebral geometrical dimensions (14 cervical and 14 lumbar) in radiograms has been evolved by Gilad & Nassan (1985). This involves the analysis of nine dimensions for vertebral bodies and spines, affording an exact anthropometric technique.

Ratcliffe (1981) has described arterial patterns in bodies of thoracic vertebrae at ages from 29th prenatal week to the 15th year.

The Cervical Vertebrae

The seven cervical vertebrae (3.56–63), the smallest of the movable vertebrae, are typified by a foramen in each transverse process. The first, second and seventh have special features to be considered separately. A typical cervical vertebra has a small, but relatively broad *body*; the *vertebral foramen* is large and roughly triangular, because the *pedicles* project dorsolaterally (3.56A) and *laminae* curve dorsomedially from them; the vertebral canal here accomodates the cervical enlargement of the spinal cord. Superior and inferior vertebral notches are almost equal in depth, pedicles being midway between the discal aspects of the body (3.56B). *Laminae* are long and narrow, with a thin upper border, but roughened ventrally. The *spinous process* is short and bifid, with

The vertebral arch has on each side a vertically narrower ventral part, the *pedicle*, and dorsally a broader *lamina*. Projecting from their junctions are paired transverse, superior and inferior articular processes, and dorsally a median spinous process.

Pedicles are short, thick, rounded dorsal projections from the body at the junction of its lateral and dorsal surfaces and nearer the superior, so that the concavity formed by its curved superior border is shallower than the inferior one (3.55). Adjacent *vertebral notches* contribute to an *invertebral foramen* when vertebrae are articulated by the intervertebral disc and zygapophyseal joints. The complete perimeter of an intervertebral foramen consists, therefore, of the notches, the dorsolateral aspects of parts of adjacent vertebral bodies and intervening disc, and the capsule of the synovial zygapophyseal joint.

Laminae, directly continuous with pedicles, are vertically flattened and curve dorsomedially to complete, with the base of the spinous process, a vertebral foramen.

The *spinous process* or spine projects dorsally and often caudally from the laminal junction. Spines vary much in size, shape and direction. They are levers for muscles which control posture and active movements (flexion/extension, lateral flexion and rotation) of the vertebral column.

The *articular processes (zygapophyses)*, paired superior and inferior, project from the vertebral arch at pediculolaminar junctions. The *superior processes* jut cranially, bearing dorsal facets which may also have a lateral or medial inclination, depending on level. *Inferior processes* bulge caudally with articular facets directed ventrally, again with medial or lateral inclination dependent on level. Articular processes of adjoining vertebrae thus form small synovial zygapophysial joints (p. 490), forming the posterior aspect of the intervertebral foramina; these joints permit limited movement, restricted in range, between vertebrae. Mobility varies considerably with vertebral level.

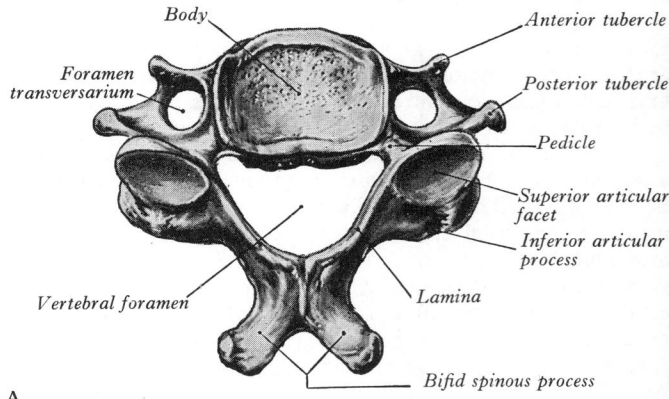

A

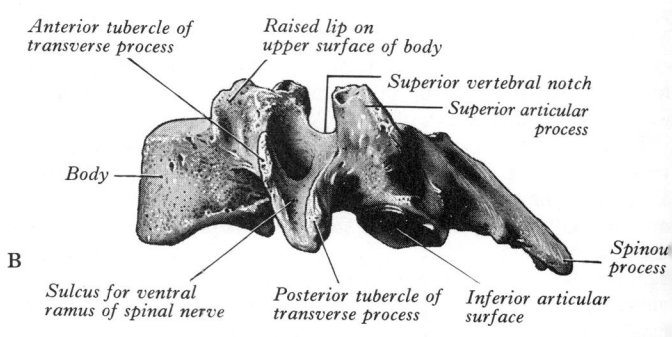

B

3.56 Typical cervical vertebra: A superior aspect. B Left lateral aspect.

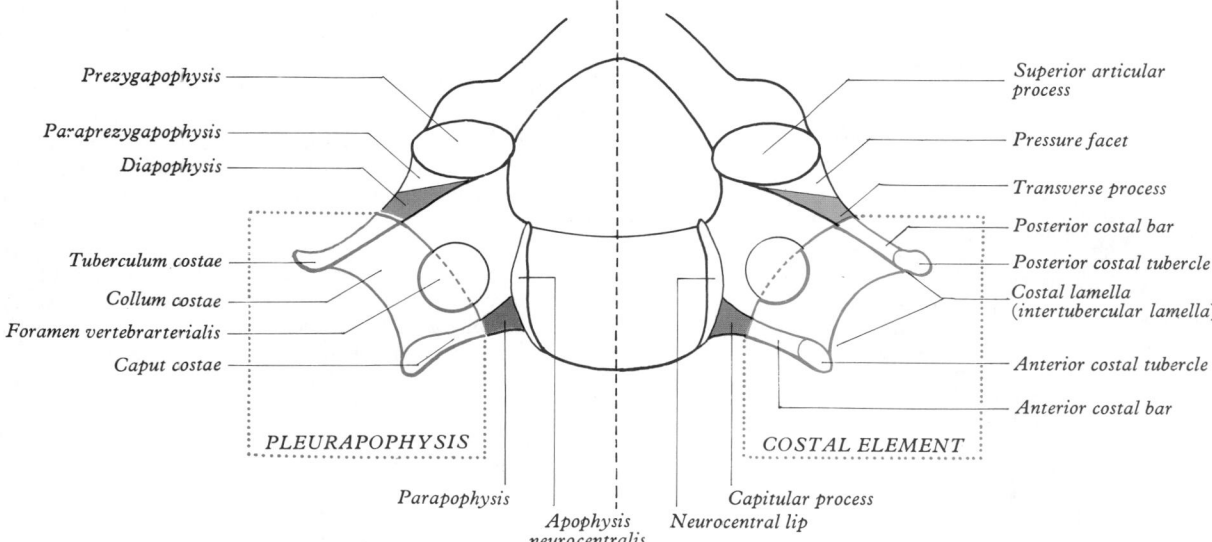

3.57 The morphology of a generalized cervical vertebra, with particular reference to the pleurapophyses. On the left the terms are zoological, on the right are alternatives for human anatomy suggested by Professor A J E Cave (1975). (Reproduced with permission from the author, the *Journal of Zoology* and Cambridge University Press.)

terminal tubercles often unequal in size. *Superior* and *inferior articular processes* form (when articulated) an *articular pillar*, each bulging laterally at the pediculolaminar junction. The mature 'transverse process' is morphologically composite around the *foramen transversarium*. Its dorsal and ventral *roots* or *bars* terminate laterally as corresponding *tubercles*. The roots are connected lateral to the foramen by an *intertubercular lamella* (often incorrectly called the costotransverse bar). Of these parts all except the medial end of the dorsal root are costal vestiges and this medial moiety is the sole representative of a true transverse process (*diapophysis*) such as exists in a thoracic vertebra (3.54, 64). Cave (1975), in a morphological study of the mammalian cervical 'costal element' or *pleurapophysis*, has defined it more precisely, especially in relation to cervical ribs (3.57, and p. 161). The foramen transversarium may be divided into two or more cervical vertebra.

The body's anterior surface, convex transversely, is marked at its discal margins by attachments of the anterior longitudinal ligament, and near the midline the bilateral depressions show the points of union of the vertical part of longus colli. The flat dorsal surface displays centrally two or more large *basivertebral*, venous foramina, and to its upper and lower borders the posterior longitudinal ligament is united. The cranial discal surface is transversely concave with marked bilateral lips, its anterior rim being slightly bevelled. The inferior surface is reciprocally convex transversely and dorsoventrally concave, i.e. both discal surfaces are sellar. The lateral edges form, with lateral lips of the subjacent vertebra, small synovial joints that comprise oblique clefts which develop at about 9–10 years and are limited medially by the intervertebral disc, laterally by capsular ligaments (Cave et al 1955). Some, however, site these clefts in the disc and deny them synovial status (Töndury 1943, Orofino et al 1960). The antero-inferior rim of the body overlaps the subjacent disc. Paired ligamenta flava extend from upper borders of laminae to quite extensive lower roughened anterior *surfaces* of those above. To spinous processes are attached the ligamentum nuchae and numerous deep extensors, including semispinalis thoracis and cervicis, multifidus, spinales and interspinales.

The dorsal rami of cervical spinal nerves curve dorsally close to the anterolateral aspects of the articular pillars, grooving the third and fourth pair, whose cranial facets are flat and almost oval and face craniodorsally; correspondingly similar inferior facets are anterocaudally directed. In all but the seventh cervical vertebra the transverse foramen transmits the vertebral artery and accompanying venous and sympathetic plexuses. The seventh foramen transmits accessory vertebral venous radicles and autonomic elements. Ventral parts of the fourth to sixth transverse processes end in rough tubercles for tendinous slips of scalenus anterior,

longus capitis and longus colli (5.36). The tubercle of the sixth vertebra is large and the common carotid artery can be effectively compressed here in the groove between tubercle and body, hence the term *carotid tubercle*. The intertubercular lamella is oblique in the third and fourth vertebrae, descending dorsolaterally; it is increasingly grooved from the fourth to sixth by emerging anterior spinal nerve rami. The lamella of the sixth vertebra is conspicuously wide and shallow. Dorsal parts of the transverse processes also end in rounded tubercles lateral to the anterior tubercles and slightly caudal in level except in the sixth vertebra, where they are approximatey level. Attached to dorsal tubercles are splenius, longissimus and iliocostalis cervicis, levator scapulae, and scalenus posterior and medius.

The atlas, first cervical vertebra (3.58), supports the 'globe' of the head, hence its name. It is unique in lacking a body, the position of which is occupied by the *dens* of the axis, a pivot around which the atlas turns (p. 493). Moreover, it has no true spine, and consists of two *lateral masses* connected by a short *anterior* and a longer curved *posterior arch*. It is thus irregularly annular. When atlas and axis are articulated, the *dens* of the latter is retained against the anterior arch by the *transverse ligament* (4.28). Morphological and developmental studies do not favour the tradition that the atlantal centrum is fused to the axis as its dens (Jenkins 1969).

The *anterior arch* is slightly convex ventrally, and carries a rough median *anterior tubercle*. Attached to this is the median cord-like (at this level) anterior longitudinal ligament and on each side the superior oblique part of longus colli. Its upper and lower

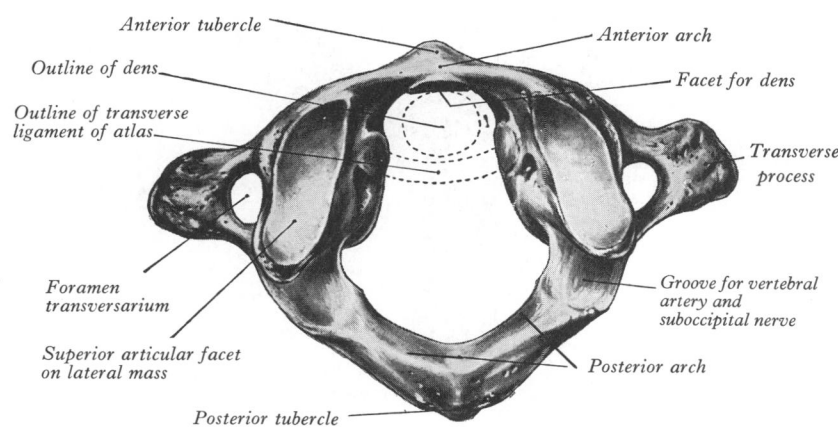

3.58 The first cervical vertebra, or atlas: superior aspect.

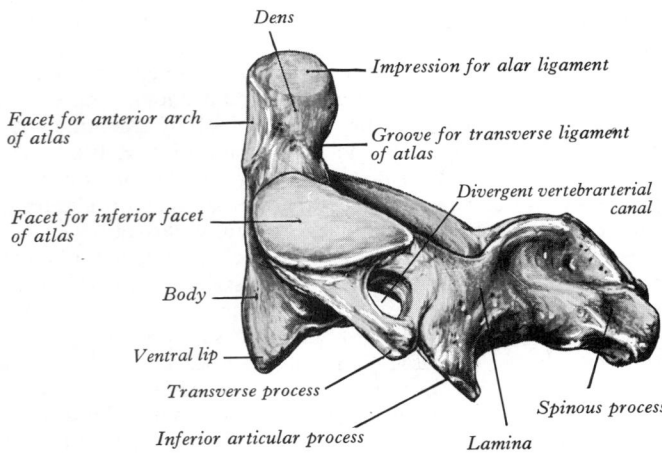

Dens (attachment of apical ligament)

Attachments of alar ligaments

*Groove for transverse ligament
of atlas*

*Superior articular
surface (facet)*

Foramen transversarium

Transverse process

Pedicle

*Inferior articular
process*

Lamina

Body

Spinous process

3.59 The second cervical vertebra, or axis: posterosuperior aspect.

Dens

Impression for alar ligament

*Facet for anterior arch
of atlas*

*Groove for transverse ligament
of atlas*

*Divergent vertebrarterial
canal*

*Facet for inferior facet
of atlas*

Body

Ventral lip

Transverse process

Spinous process

Inferior articular process

Lamina

3.60 The second veterbra, or axis; left lateral aspect.

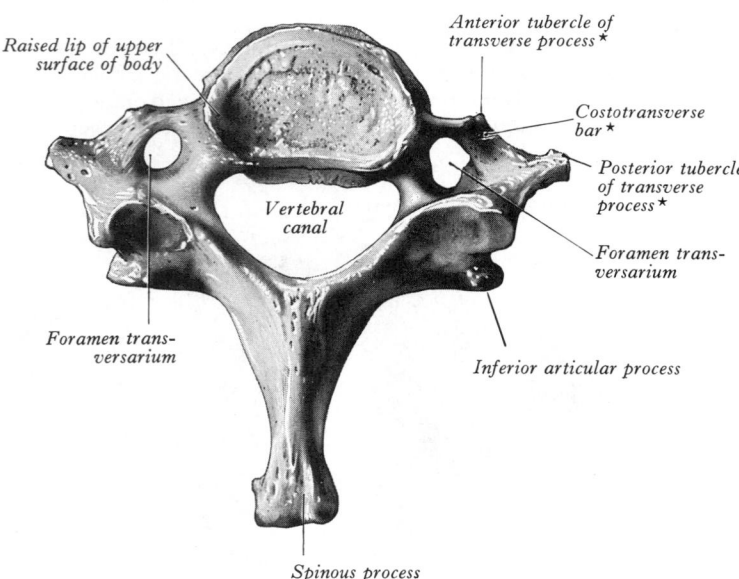

*Anterior tubercle of
transverse process* ★

*Raised lip of upper
surface of body*

*Costotransverse
bar* ★

*Posterior tubercle
of transverse
process* ★

*Vertebral
canal*

*Foramen trans-
versarium*

*Foramen trans-
versarium*

Inferior articular process

Spinous process

3.61 The seventh cervical vertebra: superior aspect. See text and **3.**57 for
alternative terms.

borders provide attachment respectively to the anterior atlanto-occipital membrane and diverging lateral parts of the anterior longitudinal ligament.

The roughly ellipsoidal *lateral masses* converge ventrally, each bearing an elongated, reniform, concave superior facet jointing with the corresponding occipital condyle. The facet, usually 'waisted' and sometimes divided into two (Singh 1965), faces superomedially in adaptation to nodding and lateral movements of the head at the atlanto-occipital joints. The inferior facet on each lateral mass is almost circular, flat or slightly concave, and articulates with a superior articular facet of the axis; these inferior facets face medially and slightly backwards. Attached to edges of the superior and inferior facets are articular capsules. Medially each lateral mass has a small tubercle for the transverse atlantal ligament (**3.**58), which is posterior to and retains the dens in place, also dividing the cavity of the atlas into a smaller, anterior part containing the dens and a larger, posterior region occupied by spinal cord, nerves, vessels and meninges. Anteriorly the lateral mass has rectus capitis anterior attached to it.

The *posterior arch* forms about two-fifths of the atlantal ring. Superiorly a wide neurovascular groove crosses behind each lateral mass, the latter overhanging the vertebral artery and venous plexus (**3.**58). Frequently (37% of 66 specimens) the groove is partly or wholly made a foramen by bone arching back from the superior surface of the lateral mass (Lamberty & Živanović 1973). The first cervical spinal nerve is also in the groove, between artery and bone. Posterior to the grooves the posterior atlanto-occipital membrane joins the posterior arch; its lateral border may be ossified, as mentioned, and curves to the lateral mass. The posterior arch is continuous at its lower margin with the highest pair of ligamenta flava. The posterior tubercle is a rudimentary spinous process, roughened by attachment of the ligamentum nuchae and lateral to this by the recti capitis posteriores minores.

The *transverse processes* are long (**3.**62): the atlas exceeds in width all cervical vertebrae save the seventh. They are strong levers for muscles rotating the head. (Maximal atlantal width ranges from 74 to 90 mm in males and 65 to 76 mm in females of European and American Caucasians—a useful criterion of sex in assessing human remains.) The apex of the transverse process is homologous with the *posterior* tubercle of typical cervical vertebrae; it is palpable between the mastoid process and mandibular angle. Anteriorly the lateral mass may bear a minute homologue of an *anterior* tubercle. An atlantal transverse process is thus equivalent to the posteror part and intertubercular lamella alone. Numerous muscles are attached to it: superiorly, rectus capitis lateralis in front of superior oblique; at the apex, inferior oblique; inferolaterally, slips of levator scapulae, splenius cervicis and scalenus medius. The first cervical ventral spinal ramus passes forwards on the lateral surface of the lateral mass and then down in front of the transverse process behind the internal jugular vein, crossed here anteriorly by the accessory nerve and occipital artery.

The axis, second cervical vertebra (**3.**59,60), is a pivot or axle for rotation of the atlas and head around its strong *dens* (odontoid process), projecting vertically from the body. A small oval anterior articular facet on the dens forms a synovial joint with a reciprocal facet on the back of the anterior atlantal arch; posteriorly the dens is grooved by the transverse atlantal ligament (**3.**59,60). The dens is conical, about 1.5 cm long and, as noted, waisted where grooved by the cartilage-covered ligament, a second, larger, synovial joint (sometimes termed a bursa) usually intervening. Its pointed apex is joined by the apical ligament (p. 495) and to its flat sides, above the groove, the alar ligaments are attached (**3.**59,60). The *body* contains relatively less compact bone than the dens. Above, the dens obscures the body and is flanked by two large, oval facets, extending laterally on to adjoining pedicles and articulating with the inferior atlantal facets. Unlike other superior facets, they do not form a pillar with the inferior facets, being considerably anterior to these. In front the body is hollowed on each side by attachment of the vertical part of longus colli. The anterior longitudinal ligament joins the lower discal border which projects downwards. Posteriorly the lower border affords attachment to the posterior longitudinal ligament and membrana tectoria (p. 494). The pedicles are thick, with deep inferior vertebral notches, the superior being shallow. The laminae, thicker than in other cervical vertebra, provide attachment to the ligamenta flava; they fuse posteriorly with the large spinous process which takes the pull of extensors, retractors and rotators of the head. The inferior obliques arise from rough areas on the lateral aspects of the spine, rectus capitis posterior major a little posterior to this. The ligamentum nuchae is attached in the apical notch in the spine with slips of semispinalis and spinalis cervicis, interspinalis and multifidus.

The *axial transverse processes* are small, blunt at their tips which are single tubercles and homologous with posterior tubercles at other levels. Small anterior tubercles are at or near the junctions of anterior parts of transverse processes and body, as in the atlas. To the tips of transverse processes are attached levator scapulae, between scalenus medius and splenius cervicis, and to their upper and lower surfaces the intertransverse muscles (p. 590). Each axial transverse foramen inclines superolaterally, because the vertebral arteries deviate laterally to traverse the more widely separated atlantal foramina. The inferior articular facets are at the pediculolaminar junctions, facing antero-inferiorly, as in typical cervical vertebrae. The vertebral foramen is large.

In a study of 60 cervical vertebral series (Krmpotić-Newmanić & Keros 1973) the longitudinal axis of the dens, relative to its vertebral body, varied with cervical curvature and also age. For discussion of homologies of transverse processes of atlas and axis and associated joints, muscles and nerves, consult Cave (1975).

The seventh cervical vertebra (3.61), *vertebra prominens*, has a long spinous process visible at the lower end of the nuchal furrow. The first thoracic spine is usually as prominent and sometimes more so; the seventh cervical is thick, almost horizontal and ends in a single tubercle for the ligamentum nuchae. Attached apically are trapezius, rhomboideus minor, serratus posterior superior, splenius capitis, spinalis cervicis, semispinalis thoracis, multifidus and interspinales. The transverse processes are large, particularly posterior to the transverse foramina (often reduplicated), the part anterior to them being slender and shorter; it is a costal element and may be separate as a *cervical rib*. The intertubercular lamella anterolateral to the foramen (3.61) is grooved superiorly by the ventral ramus of the seventh cervical nerve; it is often partly deficient. To the prominent posterior tubercle are attached scalenus minimus (pleuralis), when present, and also an aponeurosis covering the cervical pleural dome—the suprapleural membrane (p. 1268). The first pair of levatores costarum are attached to the transverse processes.

The Thoracic Vertebrae

The twelve thoracic vertebrae (3.54,64A–E,65A,B,66) increase in size caudally, like other vertebrae, due to increased loading from head to sacrum. All their bodies display lateral costal facets and all but the lowest two or three transverse processes have facets, articulating respectively with costal heads or tubercles. The first and ninth to twelfth have atypical features, to be considered separately; except for relatively minor details the rest are alike. (However, for comments concerning thoracic spinal curvatures and deviations of these, also individual asymmetries, see below and p. 284.)

The *body* is typically a waisted cylinder except where the vertebral foramen encroaches, transverse and anteroposterior dimensions being almost equal. On each side are *two costal facets* (really *demifacets*), the superior pair usually larger and at the upper border anterior to the pedicles, the inferior at the lower border anterior to the vertebral notches (3.64D). The *vertebral foramen* is small and circular; thus the *pedicles* do not diverge as in cervical vertebrae; also the thoracic spinal cord is smaller and more circular. The *laminae* are hence also short, thick, broad and overlapping from above downwards. The *spinous processes* slant downwards. The superior *articular processes*, thin plates of bone, project at pediculolaminar junctions, are almost flat and face dorsally, a little superolaterally. The *inferior processes* project down from the laminae, their facets directed forwards, a little superomedially. The large, club-like *transverse processes* project from the vertebral arch at pediculolaminar junctions. They point

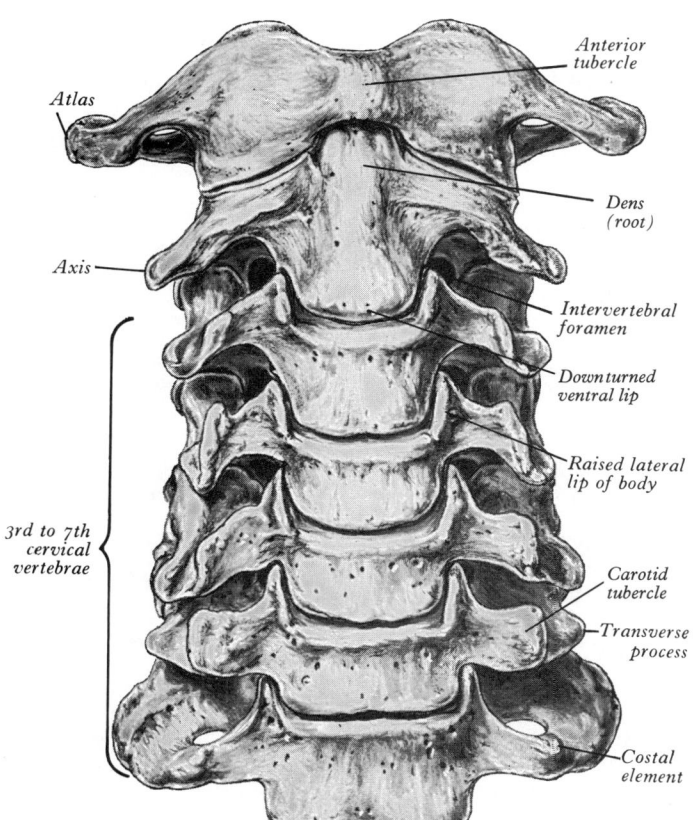

3.62 The cervical vertebrae: anterior aspect.

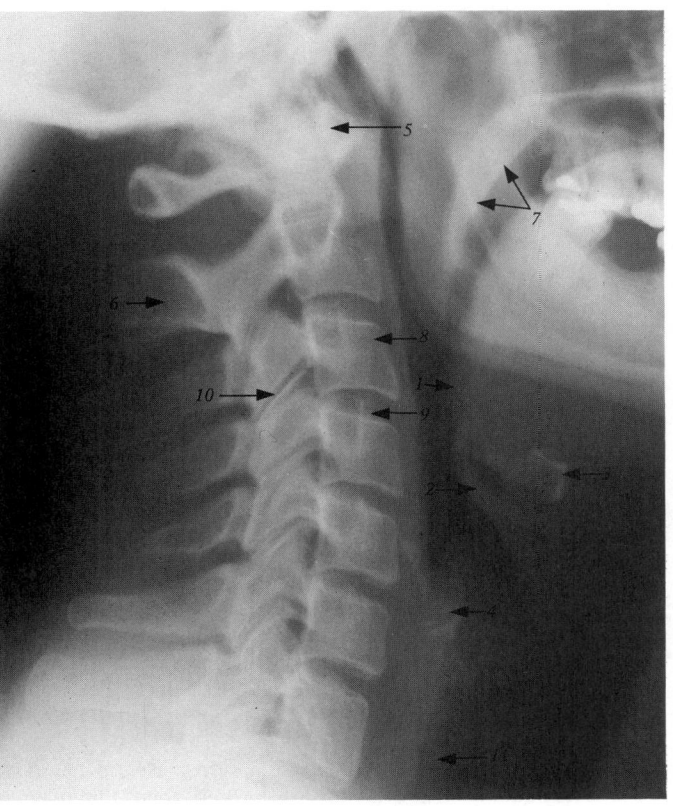

3.63 Lateral radiograph of the neck. The cervical curve of the vertebral column is well shown. The arrows point to 1. the pharyngeal part of the tongue; 2. the epiglottis; 3. the body of the hyoid bone; 4. the thyroid cartilage which is undergoing calcification; 5. the anterior tubercle of the atlas; 6. the spinous process of the axis; 7. the soft palate; 8. characteristic cervical body; 9. intervertebral disc; 10. zygapophysial joint; 11. air in trachea. (Provided by Shaun Gallagher; photography by Sarah Smith.)

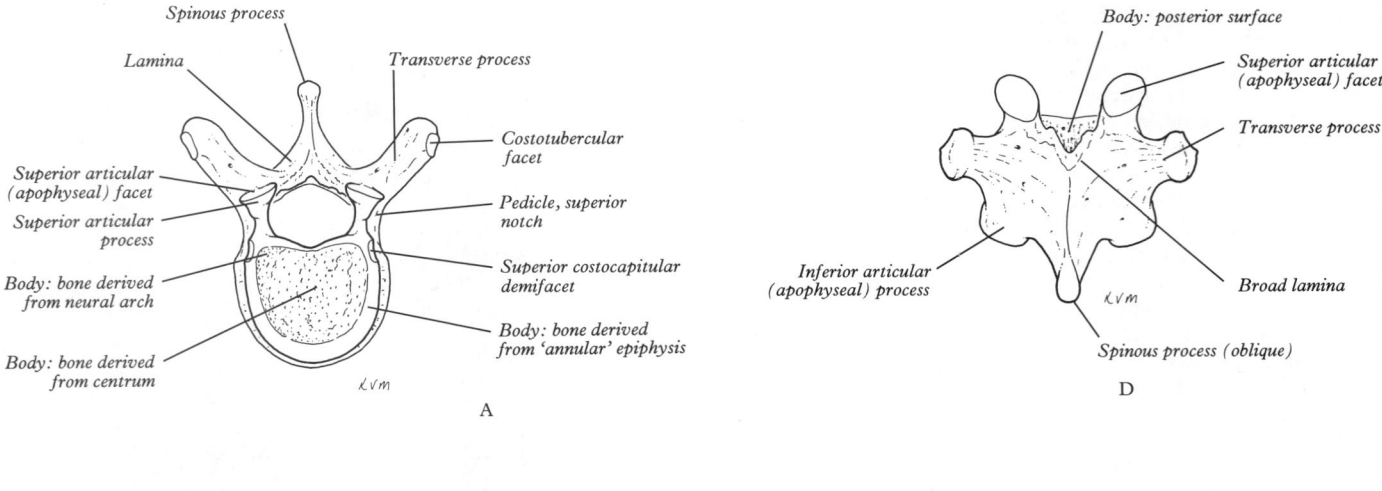

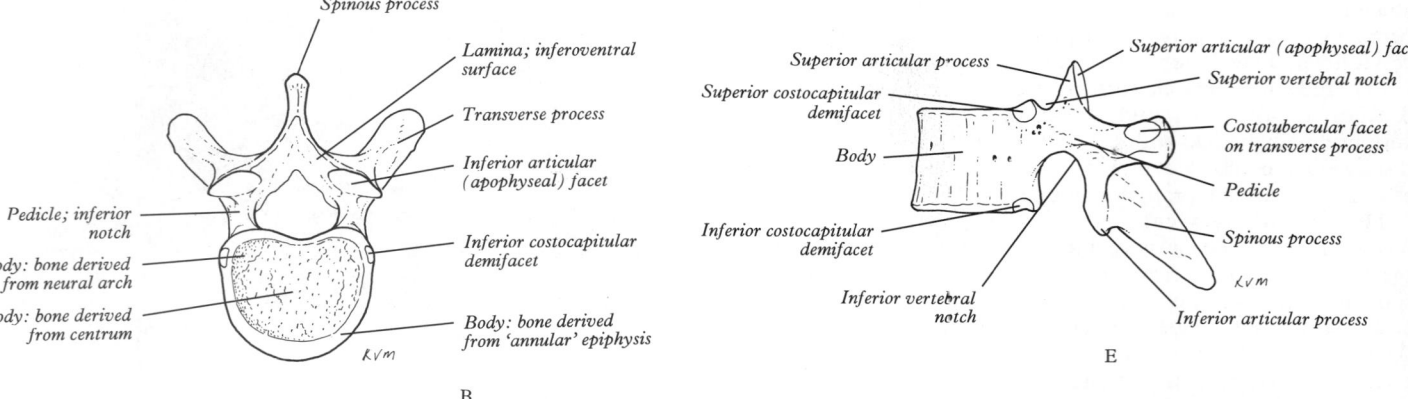

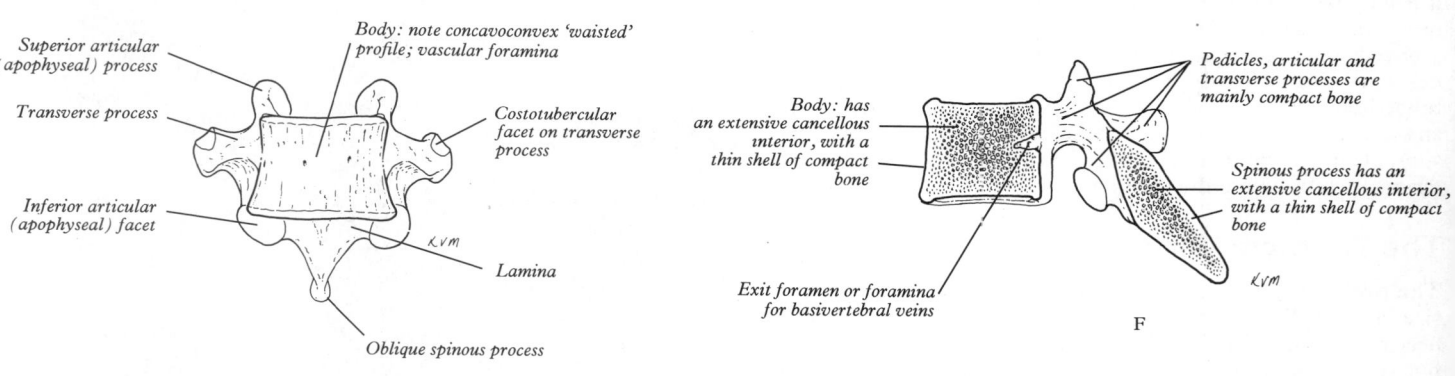

3.64 Typical thoracic vertebra. A. Superior aspect B. Inferior aspect C. Anterior aspect D. Posterior aspect E. Lateral aspect F. Longitudinal section.

posterolaterally and bear, near their tips, anterior oval facets articulating with tubercles of corresponding ribs.

The first thoracic vertebra (3.66) shows circular upper costal facets articulating with the whole facet on first costal heads, but the inferior are smaller and semilunar; they articulate, as do most thoracic vertebrae, with a demifacet on a costal head. The spine is thick, long, horizontal and commonly as prominent as the seventh cervical.

The ninth thoracic vertebra (3.66), otherwise typical, often fails to articulate with the tenth ribs and then the inferior demifacets are absent.

The tenth thoracic vertebra (3.66) articulates only with the tenth ribs and is excluded from the eleventh costovertebral joints. Therefore only superior facets appear and are usually large and semilunar or oval when the tenth ribs fail to joint with the ninth vertebrae and intervening disc. The transverse process may or may not be facetted for the tenth rib.

The eleventh thoracic vertebra (3.66) articulates only with heads of the eleventh ribs. The circular costal facets are close to the upper border of the body and extend on to pedicles. The small transverse processes hence lack articular facets.

The twelfth thoracic vertebra (3.66) articulates with the heads of the twelfth ribs by circular facets somewhat below the upper border, spreading on to the pedicles. The body is large and transverse processes are small; the vertebra has some lumbar features (vide infra).

Bodies of upper thoracic vertebrae gradually change from cervical to thoracic type, and the lower from thoracic to lumbar type. The body of the first is cervical, its transverse diameter almost twice the anteroposterior; the second retains a cervical shape, but its breadth is less and the two measurements differ less. The third body is the smallest; its anterior aspect is convex, unlike the flattened first and second. The remaining bodies increase in size and owing to its increased anteroposterior diameter, the fourth is

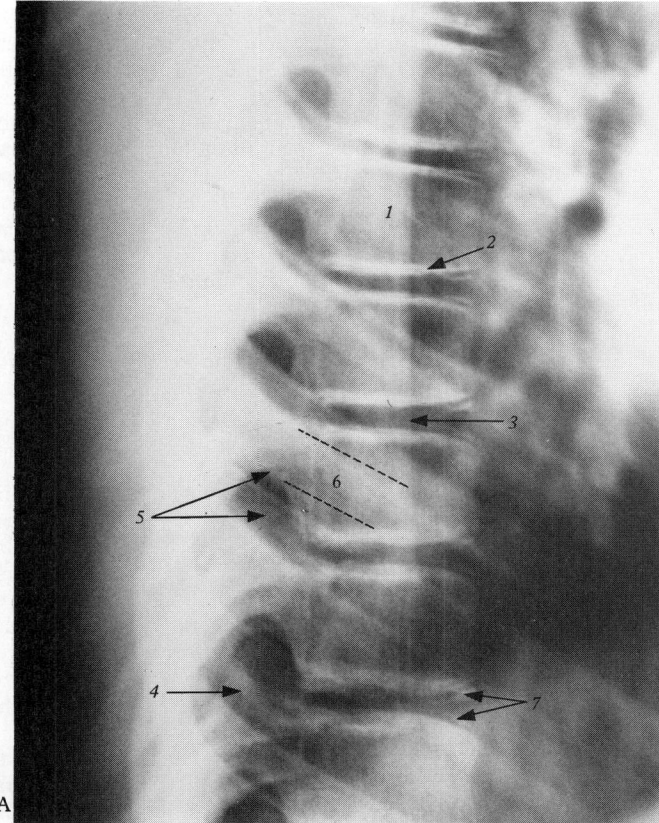

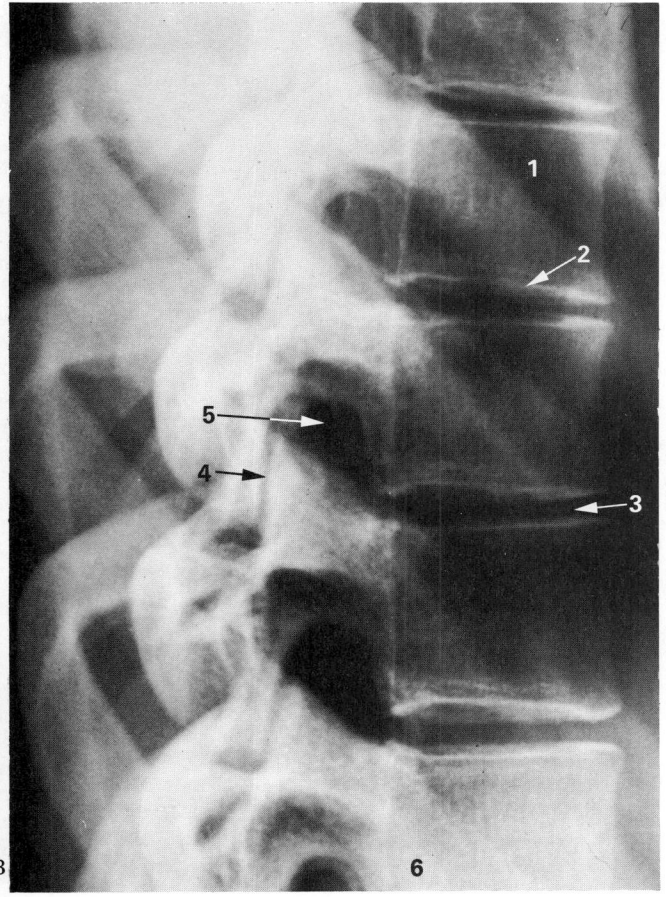

3.65 Lateral radiographs of mid-thoracic vertebral column in a child of 14 years (A) and an adult female of 22 years (B).
1. Cancellous bone of vertebral body. 2. Shell of compact bone; 3. Site of intervertebral disc. 4. Synovial joint between articular processes. 5. Intervertebral foramen. 6. Superimposed shadow of rib. 7. Ossification occurring in unfused annular epiphyses of vertebral bodies. (A supplied by Shaun Gallagher, Guy's Hospital; photography by Sarah Smith).

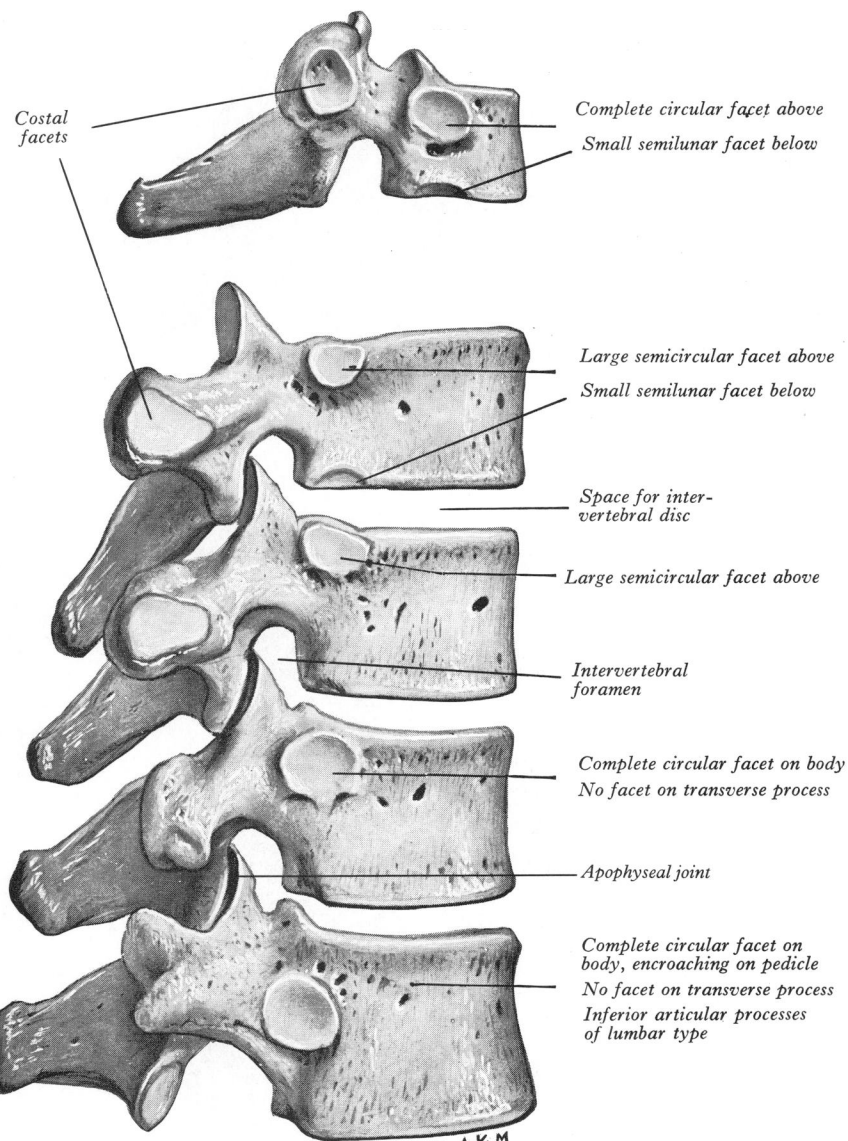

3.66 The first, ninth, tenth, eleventh and twelfth thoracic vertebra: right lateral aspect.

typically cordate ('heart-shaped'). The fifth to eighth increase in anteroposterior dimension but change little in transverse. These four, in transverse section, are *asymmetrical*, their left sides being flattened by pressure of the thoracic aorta. The rest increase more rapidly in all measurements, so that the twelfth body resembles that of a typical lumbar vertebra. Geometrical analysis suggests that these are also adaptations to a greater range of flexion-extension at the cervical and lumbar ends of the thoracic vertebral column (Veleanu et al 1972).

To borders of the bodies are attached the anterior and posterior longitudinal ligaments, and around the margins of costal facets are capsular and radiate ligaments of costovertebral joints. Longus colli arises from the upper three thoracic vertebral bodies, lateral to the anterior longitudinal ligament, and psoas major and minor from the side of the twelfth near its lower border.

Thoracic *pedicles* show a successive caudal increase in thickness. The superior vertebral notch is recognizable only in the first thoracic, but the inferior notch is deep in all. Ligamenta flava are attached at the upper borders and lower ventral surfaces of laminae, and rotatores to their dorsal aspects.

Thoracic *transverse processes* shorten in caudal succession. In the upper six (or five), the costal facets are concave and face anterolaterally; caudal to this facets are flatter and face superolaterally and slightly forwards. To the anterior surface medial to the facet is attached the costotransverse ligament, to its tuberculated apex the lateral costotransverse ligament and to its lower border the

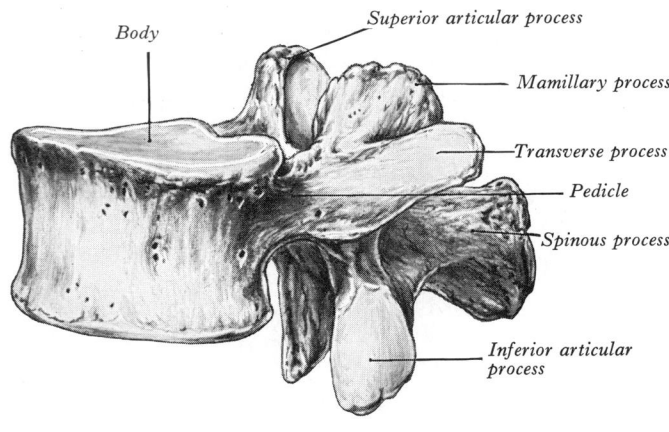

3.67 Lumbar vertebra: left lateral aspect.

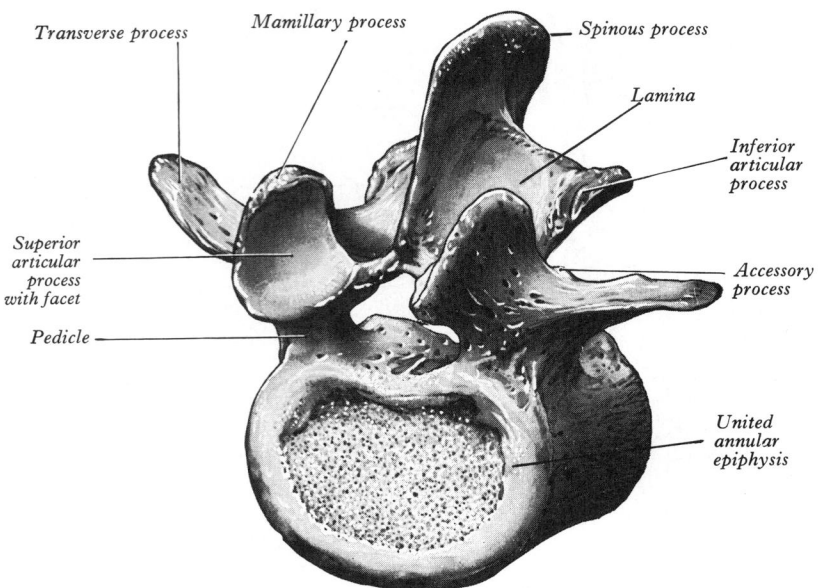

3.68 Lumbar vertebra: posterosuperior aspect, viewed obliquely from the left side.

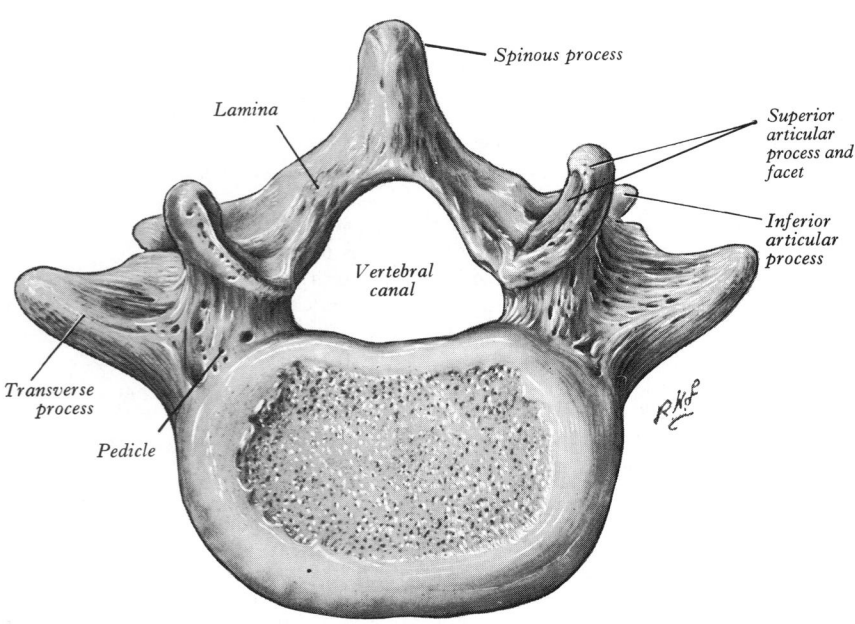

3.69 The fifth lumbar vertebra: superior aspect.

superior costotransverse ligament. Upper and lower borders also provide attachment for intertransverse muscles or fibrous vestiges and the posterior surface for deep dorsal muscles; dorsally on the apex is the levator costae.

Thoracic *spines* overlap from fifth to eighth, which is the longest and most oblique. (In quadrupeds most thoracic spines slope caudally and lumbar spines cranially, the change in inclination occuring at a lower thoracic *anticlinal* vertebra; its human equivalent is the eleventh.) Supraspinous and interspinous ligaments, trapezius, rhomboideus major and minor, latissimus dorsi, serrati posterior superior and inferior and many deep dorsal muscles are attached to thoracic spines.

The first thoracic resembles a cervical vertebra in its body, both in shape and posterolateral lipping, the latter forming the anterior border of the superior vertebral notch, a distinctive feature. The upper costal facet is often incomplete, the first rib then articulating with the seventh cervical and intervening disc. Below the facet a small, deep depression often occurs.

The eleventh and twelfth thoracic spinous processes are triangular, with blunt apices, a horizontal lower and an oblique upper border. The twelfth thoracic transverse process is replaced by three small tubercles; the superior is largest, juts upwards and corresponds to a lumbar mamillary process, though not so close to the superior articular process. The lateral tubercle is the homologue of a transverse process, the inferior the homologue of a lumbar accessory process. These two vertebrae can be distinguished by the size and shape of the transverse process and the distance between the costal facet and upper border (p. 320).

A change in orientation of articular processes from thoracic to lumbar type usually occurs at the eleventh thoracic vertebra, sometimes the twelfth or tenth. In the transitional vertebra the superior articular processes are thoracic, facing posterolaterally, while the inferior are transversely convex and face anterolaterally. This vertebra marks the site of a sudden change in degree from rotational to non-rotational function (p. 492) (Davis 1955).

Ratcliffe (1981) has described in much detail by microradiography the arterial supply of thoracic and lumbar vertebral bodies, with a review of literature.

The Lumbar Vertebrae

The five lumbar vertebrae (3.67,68,69,70A–D) are distinguished by large size and absence of costal facets and transverse foramina. The *body* is wider transversely and deeper in front. The *vertebral foramen* is triangular, larger than at thoracic but smaller than at cervical levels. The *pedicles* are short. The *spinous process* is almost horizontal, quadrangular and thickened along its posterior and inferior borders. The *superior articular processes* bear vertical concave articular facets facing posteromedially, with a rough *mamillary process* on their posterior borders. The *inferior articular processes* have vertical convex articular facets facing anterolaterally. The *transverse processes* are thin and long, except the more substantial fifth pair. A small *accessory process* marks the postero-inferior aspect of the root of each transverse process. Measurement of 338 third and fourth lumbar vertebrae, from both sexes aged 20–90 years, showed that *breadth* of body increases with age; in males, posterior height decreases relatively; in both sexes anterior height of the body decreases relative to breadth (Ericksen 1976). Twomey et al (1983), in a study of 97 adult vertebral columns, observed a reduction in bone density of lumbar vertebral bodies, principally due to a reduction in transverse trabeculae (more markedly in females), associated with increased diameter and increasing concavity in their juxtadiscal surfaces. Amonoo-Kuofi (1982, 1985) compared lumbar interpedicular distances in 150 males and 140 females by radiography, expressing these as a ratio of vertebral body width. Values varied much, but the ratio (in Nigerians) was most constant. He noted racial variation, in discussing the literature. The blood supply of developing lumbar vertebral bodies, from the 29th prenatal week to the 15th year, has been detailed by Ratcliffe (1981).

The fifth lumbar vertebra (3.69) has a massive *transverse process* continuous with the *whole of the pedicle and encroaching on the body*. The body is usually the largest and markedly deeper in front, contributing thus to the sacrovertebral angle.

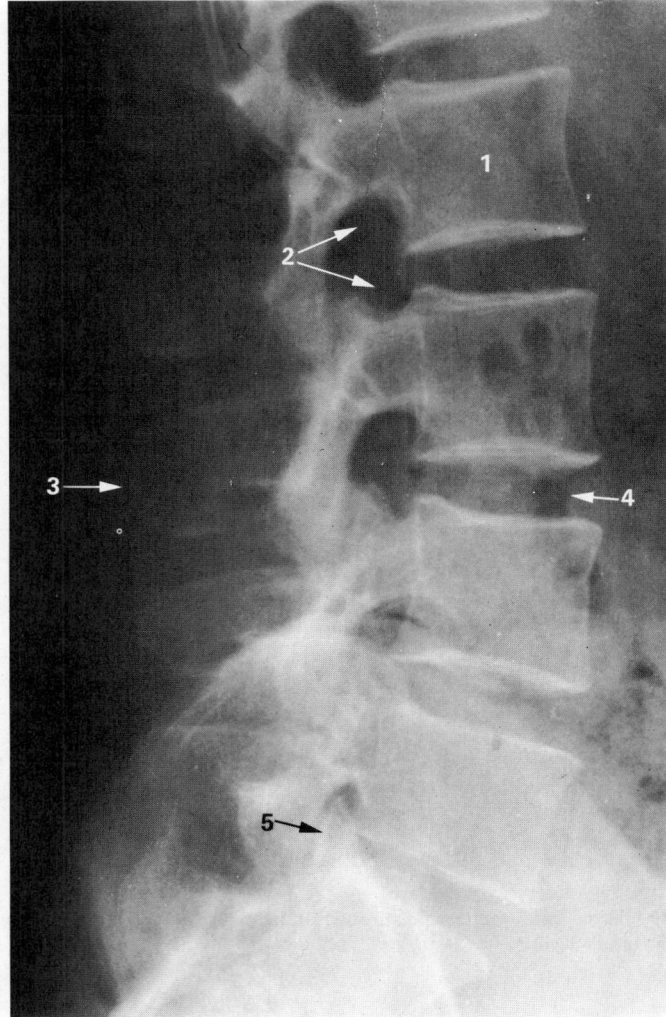

A

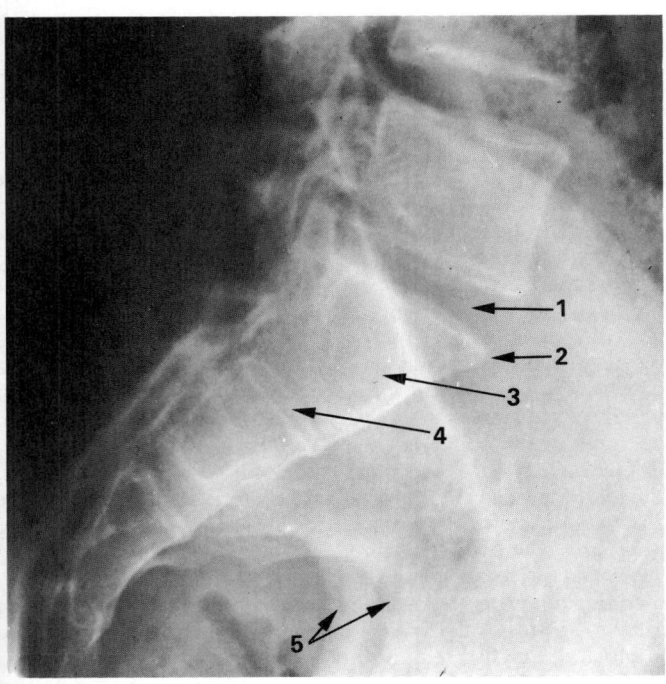

B

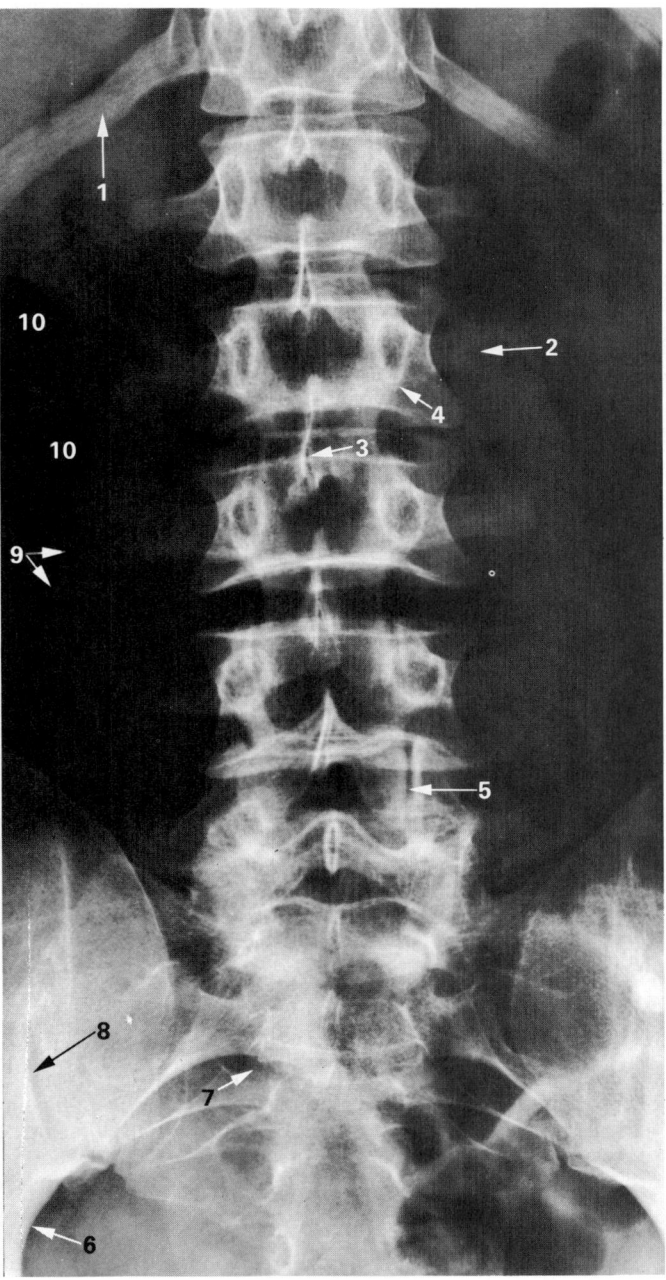

C

3.70C Anterosuperior radiograph of lumbosacral vertebral column in a young adult male aged 22 years. 1. Twelfth rib. 2. Transverse process. 3. Spinous process (2nd lumbar). 4. Compact bony shell of pedicle (2nd lumbar). 5. Joint between articular processes (4th and 5th lumbar). 6. Pelvic brim. 7. Anterior sacral foramen. 8. Sacro-iliac joint. 9. Lateral border of psoas major. 10. Gas in colon.

3.70A,B Lateral radiographs of lumbosacral vertebral column in an adult male aged 26 years.
A 1. Lumbar vertebral body. 2. Intervertebral foramen. 3. Spinous process. 4. Site of intervertebral disc. 5. Synovial joint between articular processes. Note slightly cuneiform profile of fifth lumbar vertebral body.
B 1. Site of lumbosacral disc. Note cuneiform shape. 2. Sacral promontory. 3. First sacral segment. 4. Remains of sacral intervertebral disc. 5. Profiles of greater sciatic notches.

Upper and lower borders of lumbar *bodies* give attachment to the anterior and posterior longitudinal ligaments (p. 489). Lateral to the anterior ligament the upper bodies (three on the right, two on the left) give attachments to the crura of the diaphragm. Posterolaterally, psoas major is attached to the upper and lower margins of all lumbar bodies; between these, tendinous arches carry its attachments across their concave sides. The first lumbar *vertebral foramen* contains the conus medullaris of the spinal cord, the lower foramina the cauda equina and spinal meninges. Strong paired *pedicles* spring posterolaterally from each body near its upper border. Superior vertebral notches are shallow, inferior ones deep. The laminae are broad and short but do not overlap as much as do the thoracic. To spinous processes are attached the posterior lamella of the thoracolumbar fascia, erectores spinae, spinales thoracis, multifidi, interspinal muscles and ligaments, and supraspinous ligaments. The fifth spine is smallest, its apex often rounded and down-turned. Upper lumbar superior articular processes

323

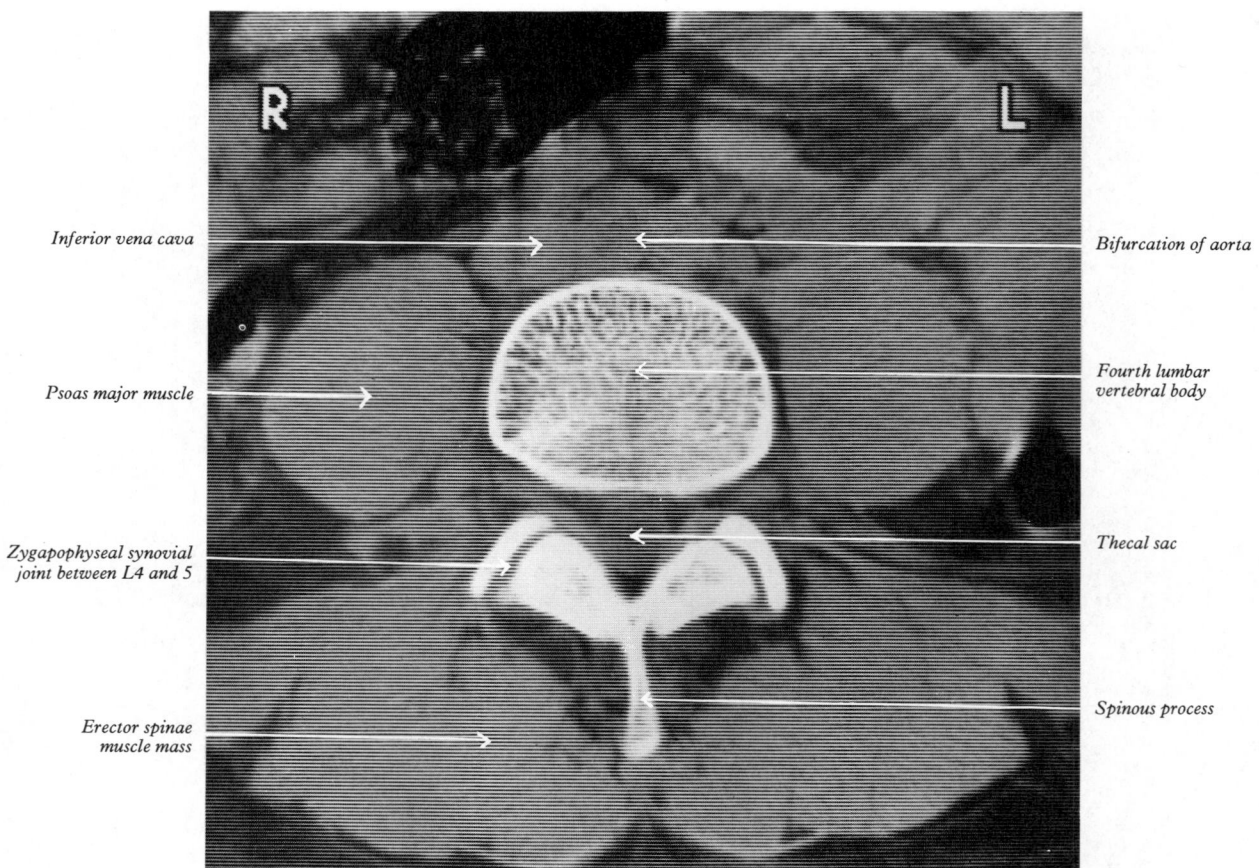

Inferior vena cava

Psoas major muscle

Zygapophyseal synovial
joint between L4 and 5

Erector spinae
muscle mass

Bifurcation of aorta

Fourth lumbar
vertebral body

Thecal sac

Spinous process

3.70D High resolution computed tomogram through posterior abdominal wall at the level of the body of the fourth lumbar vertebra, showing apophysial joints between fourth and fifth lumbar vertebrae. (Supplied by Shaun Gallagher, Guy's Hospital; photography by Sarah Smith).

are further apart than inferior ones, but the difference is slight in the fourth and negligible in the fifth. The articular facets are reciprocally concave (superior) and convex (inferior), allowing some degree of rotation as well as flexion-extension. Transverse processes, except the fifth, are dorsoventrally compressed and project dorsolaterally. The lower border of the fifth transverse process is angulated, passing laterally and then superolaterally to a blunt tip, the whole process presenting greater upward inclination than the fourth. The angle on the inferior border may represent the tip of the costal element and the lateral end the tip of the true transverse process. The lumbar transverse processes increase in length from first to third and then shorten. Thus, as noted, the fifth pair incline both *upwards* and posterolaterally. All lumbar transverse processes present a vertical ridge on the anterior surface nearer the tip which marks the attachment of the anterior layer of the thoracolumbar fascia, and separates the surface into a medial area for psoas major and a lateral for quadratus lumborum. The middle layer of the fascia is attached to the apices of the transverse processes; additionally to the first pair the medial and lateral arcuate ligaments (lumbocostal arches), and to the fifth the iliolumbar ligament, are attached. Posteriorly the transverse processes are covered by deep dorsal muscles and fibres of longissimus thoracis are attached to them. To their upper and lower borders lateral intertransverse muscle are attached. The mamillary process, homologous with the superior tubercle of the twelfth thoracic vertebra, gives attachment to multifidus and medial intertransverse muscle. To the accessory process, which is sometimes difficult to identify, is attached the medial intertransverse muscle. The costal element is incorporated in the mature transverse process. (For a discussion of homologies of all three processes consult Jones 1912.)

Applied Anatomy. Various methods of estimating the diameter of the lumbar vertebral canal, sometimes subject to stenosis, have been reviewed by Amonoo-Kuofi (1982), who considers radiographic estimation of interpedicular dimensions as a reliable

technique. The same observer (1985) considers that racial variation occurs in the sagittal dimension of the lumbar vertebral canal.

The Sacrum

The sacrum (**3**.71–5) is a large, triangular fusion of five vertebrae and forms the posterosuperior wall of the pelvic cavity, wedged between the two innominate bones. Its blunted, caudal *apex* articulates with the coccyx and its superior, wide *base* with the fifth lumbar vertebra at the *sacrovertebral angle*. The bone is set obliquely and curved longitudinally, its dorsal surface being convex, the pelvic concave (**3**.73). This ventral curvature increases pelvic capacity. Between base and apex are *dorsal, pelvic* and *lateral surfaces* and a *sacral canal*. In childhood, individual sacral vertebrae are connected by cartilage and separable by maceration; the adult bone also retains many vertebral features.

The base (**3**.74) is the upper surface of the *first sacral vertebra*, the least modified from typical vertebral plan. The *body* is large and wider transversely. Its anterior projecting edge is the *sacral promontory*. The vertebral foramen is triangular, its *pedicles* being short and posterolaterally divergent. The *laminae* are oblique, inclining down posteromedially to meet at a *spinous tubercle*. The *superior articular processes* project cranially, with concave articular facets directed dorsomedially to articulate with inferior articular processes of the fifth lumbar vertebra. The posterior part of each process projects, bearing laterally a rough area homologous with a lumbar mamillary process. The *transverse process* is much modified; a broad, sloping mass projects laterally from the body, pedicle and superior articular process (**3**.74A,B) —a unique feature, although foreshadowed in the fifth lumbar. It consists of transverse process and costal element fused together and to the rest of the vertebra, forming the superior part of the sacral *lateral mass* or *ala*.

The pelvic surface (3.71), is antero-inferior, and vertically and transversely concave but the second sacral body may produce a convexity. Four pairs of *pelvic sacral foramina* communicate through intervertebral foramina with the sacral canal, transmitting ventral rami of the upper four sacral spinal nerves. The large area between right and left foramina, formed by flat pelvic aspects of the sacral bodies, shows their fusion by four *transverse ridges*. The bars between foramina are *costal elements*, fused to the vertebrae. Lateral to the foramina the costal elements unite together and posteriorly with transverse processes to form the *lateral part* of the sacrum (which expands basally as the ala).

The dorsal surface (3.72) is convex and dorsosuperior. It has a raised, interrupted, *median sacral crest* with four (sometimes three) *spinous tubercles* representing fused sacral spines. Below the fourth (or third) an arched *sacral hiatus* in the posterior wall of the sacral canal, due to failure of the fifth pair of laminae to meet, exposes the dorsal surface of the fifth body. Flanking the median crest the posterior surface is formed by fused laminae and lateral to this are four pairs of *dorsal sacral foramina*. Like the pelvic foramina they lead into the sacral canal through intervertebral foramina; each transmits the dorsal ramus of a sacral spinal nerve. *Medial* to the foramina and vertically below each articular process of the first sacral, is a row of four small tubercles, collectively the *intermediate sacral crest*; these, sometimes termed *articular*, represent fused articular processes. The fifth inferior articular processes project caudally and flank the sacral hiatus as *sacral cornua*, connected to coccygeal cornua by intercornual ligaments. *Lateral* to the dorsal sacral foramina is a *lateral sacral crest* formed by fused transverse processes, whose apices appear as a row of *transverse tubercles*.

The lateral surface (3.73) is a fusion of transverse processes and costal elements. Wide above, it rapidly narrows in its lower part. The broad, upper part bears an *auricular surface* for articulation with the ilium, the area posterior to this being rough and deeply pitted by attachment of ligaments. The auricular surface, borne by costal elements, is like an inverted letter L. The shorter, cranial limb is restricted to the first sacral vertebra, the caudal descending to the middle of the third. Beyond this the lateral surface is non-articular and reduced in breadth. Caudally it curves medially to the body of the fifth sacral vertebra at the *inferior lateral angle*, beyond which the surface becomes a thin lateral border. A variable *accessory* sacral articular facet sometimes occurs.

The sacral apex, the inferior aspect of the fifth sacral vertebral body, bears an oval facet for articulation with the coccyx.

The sacral canal (3.75), formed by sacral vertebral foramina, is triangular in section. Its upper, basal opening is oblique but, owing to sacral inclination, is directed cranially in the standing position. Each lateral wall presents four intervertebral foramina (3.75), through which it is continuous with pelvic and dorsal sacral foramina. Its caudal opening is the *sacral hiatus*.

To the ventral and dorsal surfaces of the first sacral body are attached terminal fibres of the anterior and posterior longitudinal ligaments. Its upper laminar borders receive the lowest pair of ligamenta flava. Superiorly the ala is smooth, medially concave and laterally rough. It is covered almost entirely by psoas major. The smooth area is obliquely grooved by the lumbosacral trunk. The rough area is for the lower band of the iliolumbar ligament (p. 492), lateral to the fifth lumbar spinal nerve and to the ventral sacro-iliac ligament. Iliacus reaches the anterolateral part of this area (3.71).

The pelvic surface gives attachment to the piriformes (3.71). Emerging from the pelvic sacral foramina the first three sacral ventral rami pass anterior to piriformis. Medial to the foramina the sympathetic trunks descend in contact with bone, as do the median sacral vessels in the midline. Lateral to the foramina lateral sacral vessels are related to bone. Ventral surfaces of the first, second and partly third sacral bodies are covered by parietal peritoneum and crossed obliquely, left of the midline, by the attachment of the sigmoid mesocolon. The rectum is in contact with pelvic surfaces of the third to fifth sacral vertebrae and with the superior rectal artery's bifurcation between the rectum and third sacral vertebra.

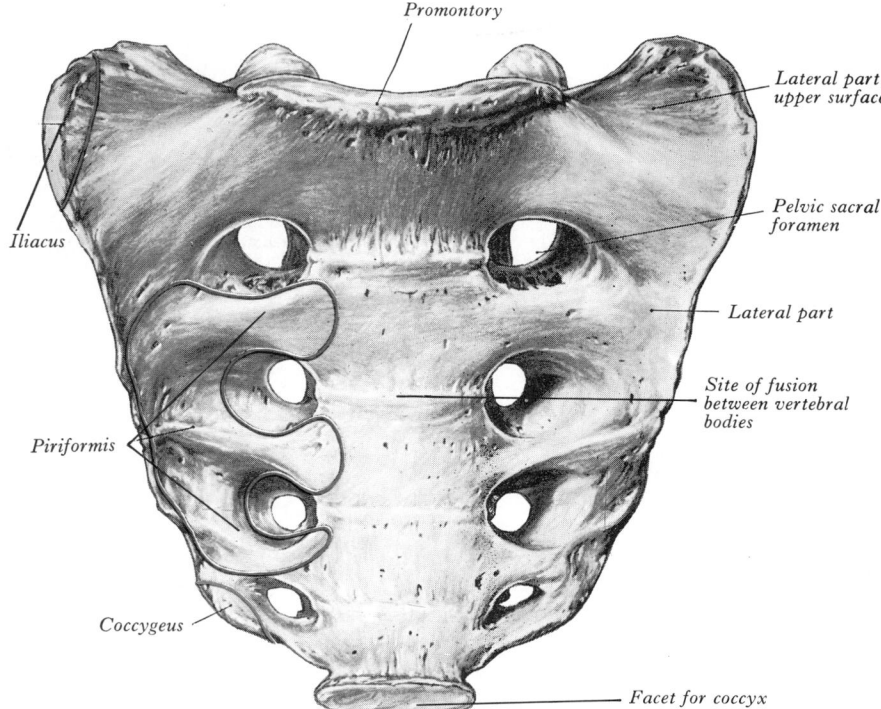

3.71 The sacrum: pelvic surface.

The dorsal sacral surface affords attachment to erector spinae by an elongated U-shaped area of spinous and transverse tubercles, covering multifidus which occupies the enclosed area (3.72). The upper three sacral spinal dorsal rami pierce these muscles as they emerge via dorsal foramina.

The auricular surface, covered by hyaline cartilage, is formed entirely by costal elements. It shows cranial and caudal elevations and an intermediate depression, dorsal to which, in the elderly, is a third elevation. The rough area behind the auricular surface shows two or three marked depressions for attachment of strong interosseous sacro-iliac ligaments. Below the auricular surface are attached gluteus maximus, sacrotuberous and sacrospinous ligaments and coccygeus, from behind forwards.

The sacral canal (3.71) contains the cauda equina (including its filum terminale) and spinal meninges. Near its mid-level subarachnoid and subdural spaces cease and lower sacral spinal roots and filum terminale pierce the arachnoid and dura maters. The filum terminale emerges at the sacral hiatus and traverses the dorsal surface of the fifth sacral vertebra and sacrococcygeal joint to reach the coccyx. The fifth sacral spinal nerves also emerge through the hiatus medial to the sacral cornua, grooving the lateral aspects of the fifth sacral vertebra.

Typical male or female sacra are easily identified but sexual differences are not always marked; identification is sometimes difficult. The female sacrum is shorter and wider, producing a

325

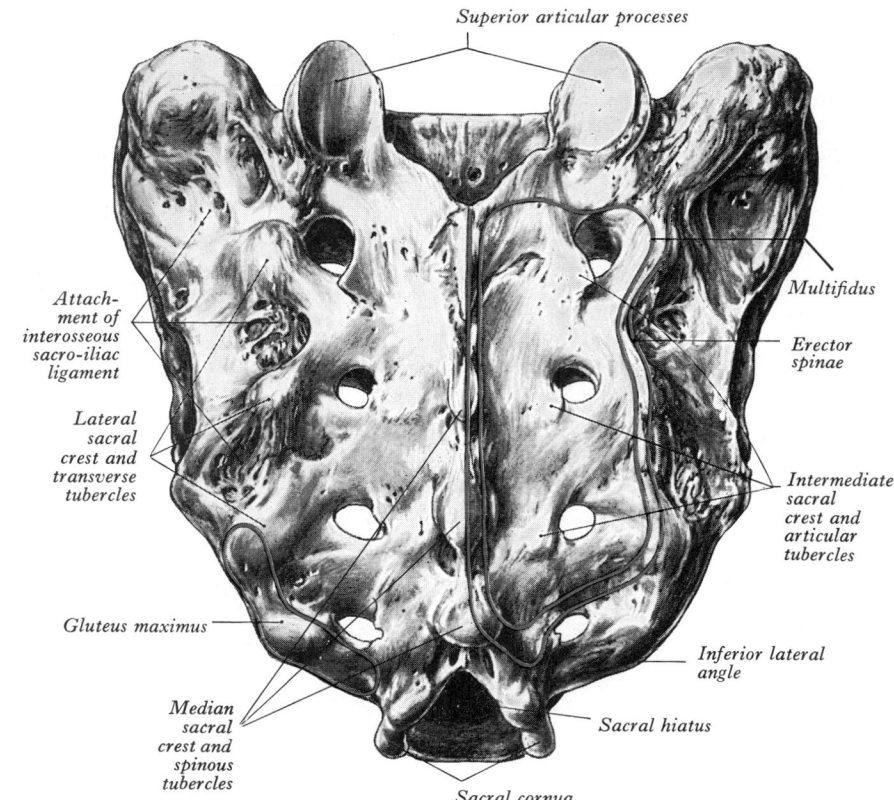

3.72 The sacrum: dorsal surface.

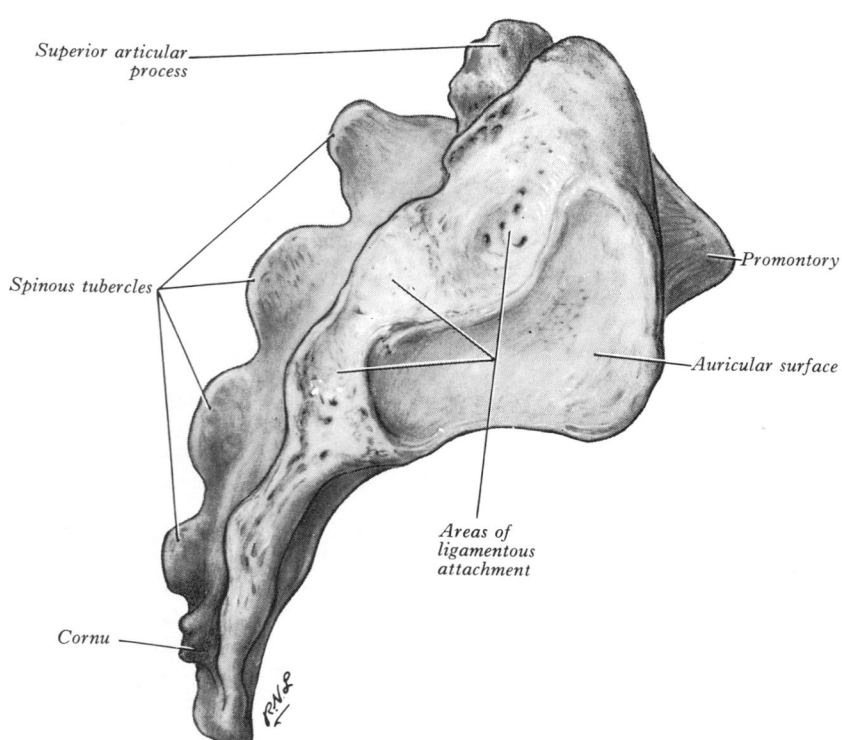

3.73 The sacrum: right lateral aspect.

wider pelvic cavity. Sacral width, as a percentage of length, yields a *sacral index*. The loci used in making these measurements are discussed on p. 430. The ventral concavity is deeper in females and its deepest point is usually higher than in males; curvature above this point is greater in the female. The dorsal protrusion of the second sacral vertebra (p. 432) is hence usually less prominent in males. In females the pelvic surface faces downwards more than in males, increasing the pelvic cavity and making the

sacrovertebral angle more prominent. The female auricular surface is shorter but in both sexes usually extends along the first three sacral vertebrae. Owing to the great size of the fifth *lumbar* body, the first *sacral* body occupies a larger proportion of the sacral base in the male, its transverse diameter exceeding the length of an ala; the female dimensions are roughly equal.

Structure. The sacrum consists of trabecular bone enveloped by a shell of compact bone variable in thickness.

Variations. The sacrum may contain six vertebrae, by development of an additional sacral element or by incorporation of fifth lumbar or first coccygeal vertebrae. Inclusion of the fifth lumbar (*sacralization*) is usually incomplete and limited to one side. In the most minor degree of the abnormality a fifth lumbar transverse process is large and articulates, sometimes by a synovial joint, with the sacrum at the dorsolateral angle of its base. Reduction of sacral constituents is less common but *lumbarization* of the first sacral vertebra occurs; it remains partially or complete-ly separate. The dorsal wall of the sacral canal may be variably deficient, due to imperfect development of laminae and spines. Orientation of the superior sacral articular facets (and hence the relation between the planes of the two zygapophysial lumbosacral joints) displays wide variation according to Čihák (1970): the chords of the concave sacral facets formed an angle with the sagittal plane varying from 20° to 90° in 132 sacra, the majority between 40° and 60°. Curvatures of the facets and their degree of asymmetry also varied.

The Coccyx

The coccyx (**3.**76A,B), a small triangular bone, usually consists of four fused rudimentary vertebrae but the number varies from five to three, the first being sometimes separate. The bone descends ventrally from the sacral apex, its pelvic surface being tilted upwards and forwards, its dorsum downwards and backwards. Orientation varies, of course, with its mobility.

The *base*, or upper surface of the *first coccygeal vertebral body* has an oval, articular facet for the sacral apex. Dorsolateral to this two *coccygeal cornua* project up to articulate with sacral cornua; they are homologues of pedicles and superior articular processes of other vertebrae. A rudimentary *transverse process* projects superolaterally from each side of the first coccygeal body and may articulate or fuse with the inferolateral sacral angle, completing the fifth sacral foramina.

The *second* to *fourth coccygeal vertebrae* diminish in size and are usually mere fused nodules, representing rudimentary vertebral bodies, though the second may show traces of transverse processes and pedicles.

Laterally on the *pelvic surface* and including the rudimentary transverse processes, the levatores ani and coccygei are attached, as is the ventral sacrococcygeal ligament to the front of the first and sometimes second coccygeal vertebral bodies (**4.**56), and to the *cornua* the intercornual ligaments. The gap between fifth sacral body and articulating cornua represents, on each side, an intervertebral foramen, transmitting the fifth sacral spinal nerve, whose dorsal ramus descends behind the rudimentary transverse process, its ventral ramus passing anterolaterally between the transverse process and sacrum with, laterally, the lateral sacrococcygeal ligament which connects the process to the inferiolateral sacral angle. To the *dorsal surface* are attached the glutei maximi, at its tip the sphincter ani externus, and in its median area the deep and superficial dorsal sacrococcygeal ligaments, the superficial descending from the margins of the sacral hiatus and sometimes closing the sacral canal. The filum terminale, between the two ligaments, blends with them on the dorsum of the first coccygeal vertebra.

Ossification of the Vertebral Column

For earlier stages of vertebral development consult p. 159 and illustration **2.**44.

A typical vertebra is ossified from three primary centres (**2.**44,**3.**77A), one in each half vertebral arch, one in the centrum. Centres in arches appear at the roots of transverse processes, ossification spreading back into laminae and spines, forwards into pedicles and posterolateral parts of the body, laterally into transverse processes and up and downwards into articular processes. Classically centres in *vertebral arches* are said to appear first in upper cervical vertebrae in the ninth to tenth week, and then in successively lower vertebrae, reaching lower lumbar levels at twelve weeks. However, in a radiographic study of unsexed human fetuses, of which 33 were of appropriate age (Bagnall et al 1977), a pattern was noted differing from such a simple craniocaudal sequence. A regular cervical progression was not observed. Centres first appeared in the lower cervical/upper thoracic region, quickly followed by others in the upper cervical region. After a short interval a third group appeared in the lower thoracic-lumbar region and remaining centres then appeared, spreading regularly and rapidly in craniocaudal directions. The body's major part, the *centrum*, was ossified from a primary centre dorsal to the notochord. (Centra are occasionally ossified from *bilateral* centres which may fail to unite. Suppression of one of these produces a *cuneiform vertebra*, a recognized cause of lateral spinal curvature. The condition is frequently multiple.) This centre first appears at lower thoracic levels in the ninth to tenth week, spreading craniocaudally and reaching the second cervical vertebra in the twelfth week. In the study by Bagnall et al (1977) these events were largely confirmed, but the earliest centres were in lower thoracic *and* upper lumbar regions; cranial progression was rather faster than caudal, the last centre being in the fifth sacral vertebra. (For the later and erratic ossification of the coccyx vide infra.) During early postnatal years the centrum is connected to each half neural arch by a synchondrosis or *neurocentral joint*. In thoracic vertebrae costal facets on bodies are *posterior* to neurocentral joints. At birth a vertebra consists of three ossifying elements, a centrum and two half arches, united by cartilage. During the first year the arches unite behind, first in the lumbar region and then through thoracic and cervical regions. In upper cervical vertebrae centra unite with arches about the third year, but in lower lumbar vertebrae union is not complete until the sixth. Until puberty the upper and lower surfaces of bodies and apices of transverse and spinous processes are cartilaginous but now *five secondary centres* appear, one in the apex of each transverse and spinous process and two annular epiphysial 'rings' for

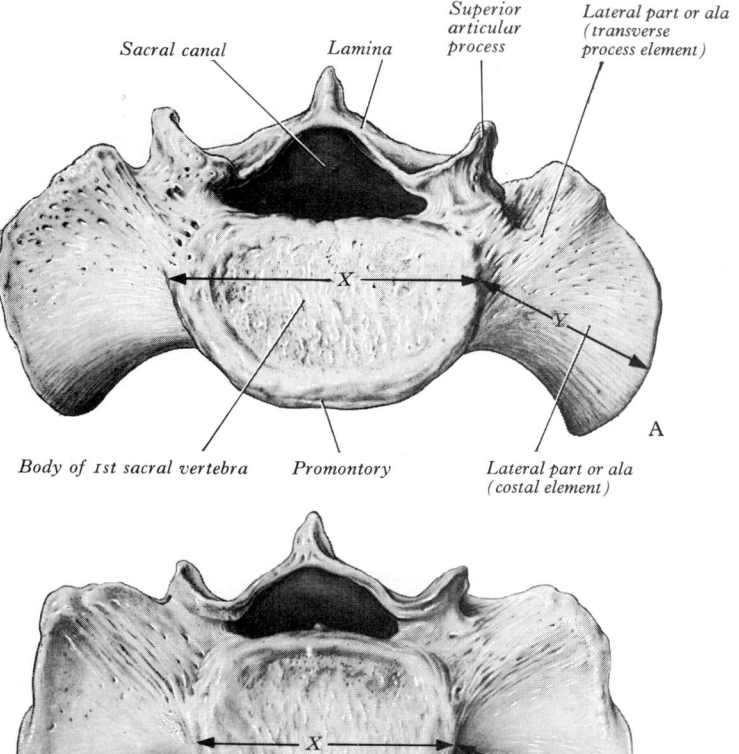

3.74 Base of the sacrum in the male (A) and in the female (B). The body of the first segment (X) forms a larger part of the breadth of the base in the male than it does in the female. In the latter the body is relatively smaller and the lateral, costal part (Y), the ala, is relatively broader.

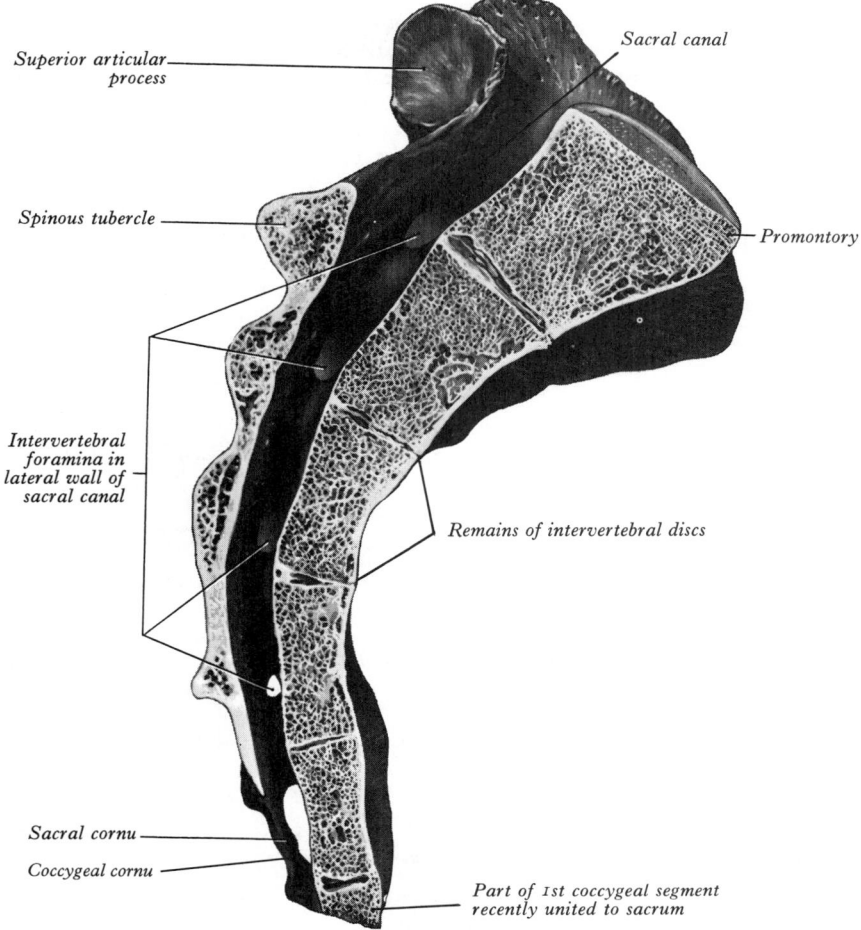

3.75 Median sagittal section through the sacrum.

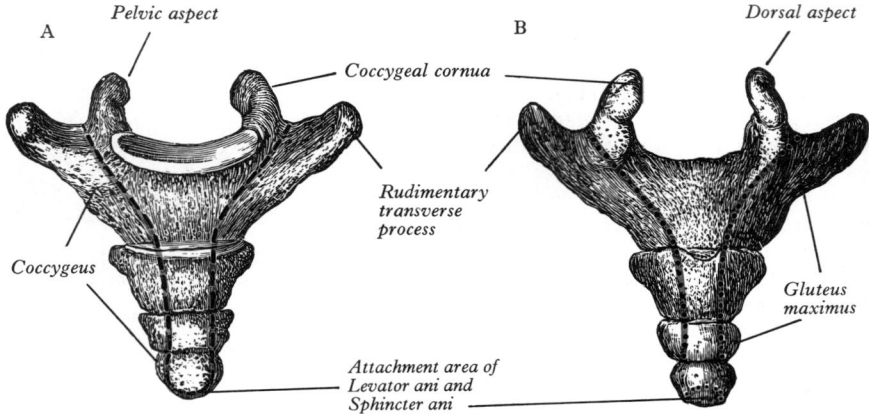

3.76 The coccyx: A. pelvic aspect; B. dorsal aspect.

circumferential parts of upper and lower surfaces of the body (2.44,3.77B,C). Costal articular facets are extensions of the annular epiphyses (Dixon 1920). These epiphyses fuse with the rest of the bone at about 25 years. In bifid cervical spinous processes there are two secondary centres. The annular 'epiphyses' of vertebrae probably cannot be equated with epiphyses of long bones. In most mammals they are complete osseous discs. For discussion consult François & Dhem (1974).

Sexual dimorphism in vertebrae has received little attention, but Taylor & Twomey (1984) have described radiological differences in adolescent humans, female vertebral bodies having a lower ratio of width to depth.

Exceptions to this pattern of ossification occur in the first, second and seventh cervical and in lumbar vertebrae.

The atlas is commonly ossified in three centres (3.77D): one in each lateral mass at about the seventh week, gradually extending into the posterior arch where they unite between the third and fourth years, directly or occasionally through a separate centre. At birth, the anterior arch is fibrocartilaginous; here a separate centre appears about the end of the first year, uniting with the lateral masses between the sixth and eighth, the lines of union extending across anterior parts of the superior articular facets. Occasionally the anterior arch is formed by extension and ultimate union of centres in the lateral masses, sometimes from two *lateral* centres in the arch itself.

The axis is ossified from five primary and two secondary centres (3.77E), the vertebral arch from two primary, the centrum from one, as in a typical vertebra. The former appear about the seventh or eighth week, that for the centrum about the fourth or fifth month. The dens is largely ossified from bilateral centres,

appearing about the sixth month and joining before birth to form a conical mass, deeply cleft above by cartilage. This cuneiform cartilage forms the apex of the odontoid process and in it a centre appears about the second year and unites with the main mass of the process about the twelfth. It has been widely regarded as a part of the cranial sclerotomal half of the first cervical segment or *pro-atlas* (Gadow 1933), but vide infra. The dens is separated from the body by a cartilaginous disc, the circumference of which ossifies while its centre remains cartilaginous until old age; in the disc possible rudiments of adjacent epiphyses of atlas and axis may occur. A thin epiphysial plate is formed inferior to the body around puberty. The dens has long been considered the atlantal centrum, secondarily fused with axis. This view was undermined by studies in a variety of mammals by Jenkins (1969), who regarded the dens as a new formation. Ganguly & Singh-Roy (1965) consider the apical centre for the dens as derived from the *pro-atlas*, which may also contribute to lateral atlantal masses.

Some observers regard the dens as the centrum of the pro-atlas.

In the seventh cervical vertebra separate centres for its costal processes appear about the sixth month and join body and transverse processes between the fifth and sixth years; they may remain separate and grow anterolaterally as cervical ribs (p. 319). Separate ossific centres may, on occasion, also occur in the costal processes of the fourth to sixth cervical vertebrae (p. 336).

The lumbar vertebrae (3.77F) have two additional centres for mamillary processes. In the fifth lumbar a pair of scale-like epiphyses usually appear on the tips of costal elements.

The sacrum (3.78A–E) resembles typical vertebrae in the ossification of its segments. Primary centres for the centrum and each half vertebral neural arch appear between tenth and twentieth

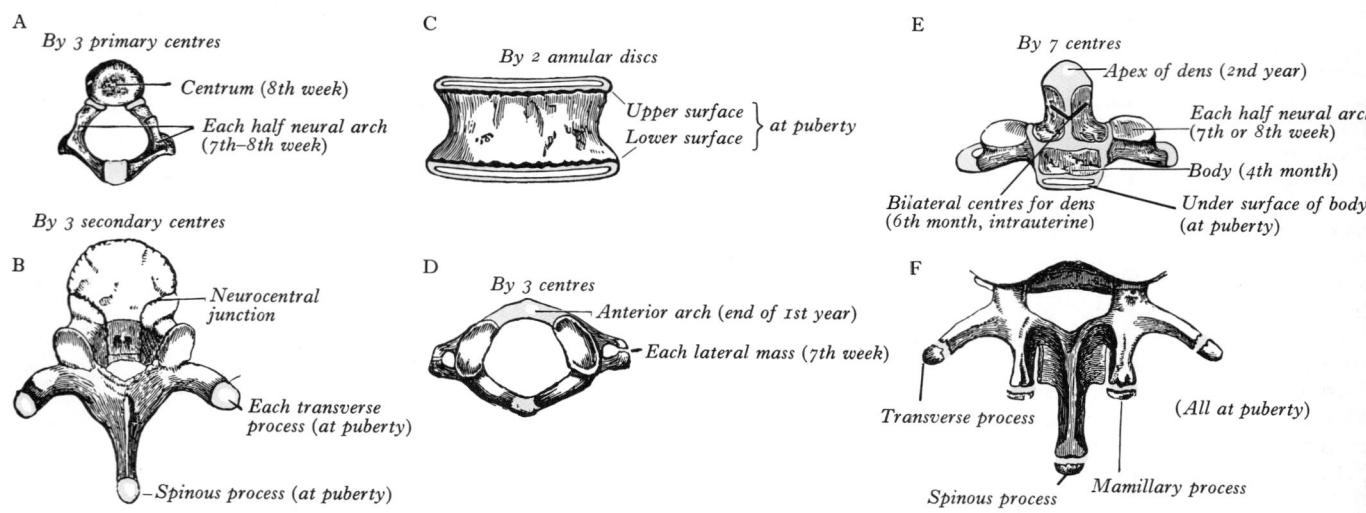

3.77 Ossification of the vertebral column. A. Typical vertebra. B. Typical vertebra at puberty. C. Body of a typical vertebra at puberty. D. The atlas. E. The axis. F. Lumbar vertebra. For further details consult text.

weeks. Primary centres for costal elements of the upper three or more segments appear superolateral to the pelvic sacral foramina, between sixth and eighth prenatal months. Each costal element unites with its half vertebral arch between the second and fifth years, and the conjoined element so formed unites ventrally with the centrum and dorsally with its opposite fellow at about the eighth year. Thereafter the upper and lower surfaces of each sacral body are covered by an epiphysial plate of hyaline cartilage separated by a fibrocartilaginous precursor of an intervertebral disc. Laterally, successive conjoined vertebral arches and costal elements are separated by hyaline cartilage; a cartilaginous epiphysis, sometimes divided into upper and lower parts, develops on each auricular and adjacent lateral surface. Soon after puberty the fused vertebral arches and costal elements of adjacent vertebrae begin to coalesce from below upwards. At the same time epiphysial centres develop for (1) upper and lower surfaces of bodies, (2) spinous tubercles, (3) transverse tubercles and (4) costal elements. The costal epiphysial centres appear at the lateral extremities of the hyaline cartilages *between* adjacent costal elements; two anterior and two posterior appear in each interval between first, second and third segments. Ossification spreads from these into the auricular epiphysial plates. One costal epiphysial centre appears anteriorly in each remaining interval; from these ossification spreads to the epiphysial plate covering the lower lateral surface. Sacral bodies unite at their adjacent margins after the twentieth year, but the central mass and most of each intervertebral disc remain unossified up to or beyond middle life. Available information is based on few specimens (Fawcett 1907, McKern & Stewart 1957).

The coccyx. Each segment is ossified from one primary centre but the incidence and timing are uncertain. A centre in the first segment appears about birth and its cornua may soon ossify from separate centres. Remaining segments ossify at wide intervals up to the twentieth year or later. Segments slowly unite; union between the first and second is frequently delayed until 30 years. Especially in females, the coccyx often fuses with the sacrum in later decades.

The Vertebral Curvatures

The vertebral column is median and dorsal in the body, the general vertebrate plan. In human males its length is about 70 cm, its cervical part approximately 12 cm, thoracic 28 cm, lumbar 18 cm and sacrum and coccyx 12 cm. In females the total length is about 60 cm.

The curves of the vertebral column viewed from the side (3.3,79B) are cervical, thoracic, lumbar and pelvic. The thoracic and pelvic curves are *primary*, concave ventrally during fetal life and thereafter. Cervical and lumbar curves are *secondary* or compensatory; the cervical appears before birth (vide infra) and is accentuated when the child begins to hold up its head (at 3 or 4 months) and to sit upright (about nine months). The lumbar curve appears at 12–18 months, when walking begins; it brings the trunk's centre of gravity over the legs. This change in mechanics of the lumbar region appears a potent factor in alterations in the proportions of vertebral bodies and intervertebral discs, especially in the fifth lumbar and the disc below it (Taylor 1975). The customary view that the secondary cervical curve appears around birth has been challenged (Bagnall et al 1977). Radiographic study of 195 human fetuses aged from 8 to 23 weeks, showed that 83% already possessed a cervical curve, visible as soon as the ossification permitted, i.e. at nine weeks. These observers also emphasized early onset of *fetal movements* as a factor in the emergence of an identifiable cervical curve.

In adults, the *cervical curve* is convex forwards and the least marked; it extends from atlas to second thoracic vertebra. The *thoracic curve*, concave forwards, reaching from second to twelfth thoracic vertebrae, is due to the greater posterior depth of the vertebral bodies. The *lumbar curve*, convex forwards, is greater in females; it reaches from the last thoracic vertebra to the lumbosacral angle and the convexity of the lower three segments is greater. It is mainly due to a greater anterior depth of intervertebral discs but the shape of vertebral bodies also contributes. Sexual differences in

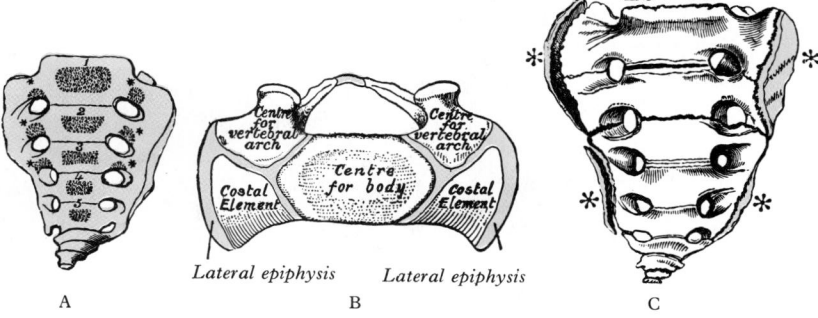

3.78A–C Ossification of the sacrum and coccyx: A at birth; B the base of the sacrum of a child of about four years of age; C at the twenty-fifth year. In C the epiphysial plates for each lateral surface are marked by asterisks.

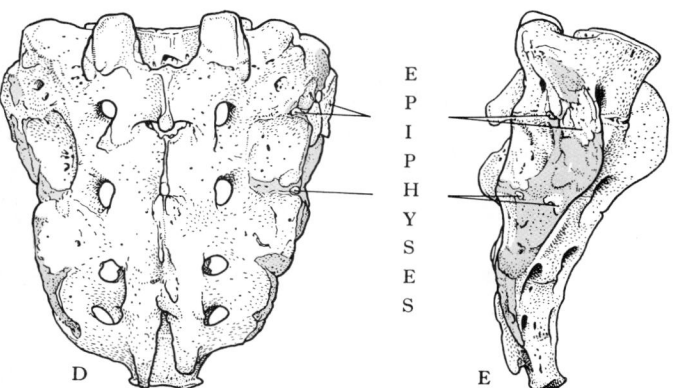

3.78D,E The epiphyses of the costal and transverse process of the sacrum at the eighteenth year.

lumbar and cervical curvatures were confirmed by Knussman & Finke (1977). The *pelvic curve* extends from lumbosacral joint to coccygeal apex; its concavity is antero-inferior. In the upper thoracic region there is often a slight *lateral* curvature, convex to the right in right-handed persons, and left in the left-handed.

Viewed from the front (3.1,79A), the width of vertebral bodies increases from second cervical to third lumbar, in association with increase in weight-bearing. There is some variation in size of the last two lumbar bodies but thereafter width diminishes rapidly to the coccygeal apex. In the two lowest lumbar vertebrae there is an inverse relation between areas of upper and lower surfaces of bodies, and size of pedicles and transverse processes, suggesting that the latter transmit some of the vertebral compressive forces from spine to pelvis (Davis 1961). From detailed dimensional study of primate lumbar vertebrae, Rose (1975) considered that the size of bodies is as much concerned with bending as compressive stresses, emphasizing that only the *fifth* human lumbar body *regularly* displays a greater ventral depth and concluding that other factors must contribute to lumbar curvature: as noted above this is principally the intervertebral discs.

In posterior view the approximately median spines projecting dorsally are the major feature (3.79C). In the cervical region (except for second and seventh vertebrae) these are short, nearly horizontal with bifid ends. In the upper thoracic region they slope obliquely down; intermediate thoracic spines are long and almost vertical; finally, ones in lower thoracic and lumbar regions are nearly horizontal. They are separated by intervals in cervical and lumbar regions but closely approximated in the midthoracic region. Occasionally a spinous process may deviate from the midline, an important clinical fact, since irregularities occur also in fractures or dislocations. The seventh cervical spine can be felt at the lower end of the nuchal furrow; first and second thoracic spines are also prominent. The third thoracic is level with the scapular spine, and the seventh with the inferior scapular angle, when the arm is pendent. The fourth lumbar spine is level with the summits of the iliac crests (a point useful in lumbar puncture), and the second sacral with the posterior superior iliac spines. Lateral to the vertebral spines *vertebral grooves* contain the deep dorsal muscles; at cervical and lumbar levels the grooves are shallow and

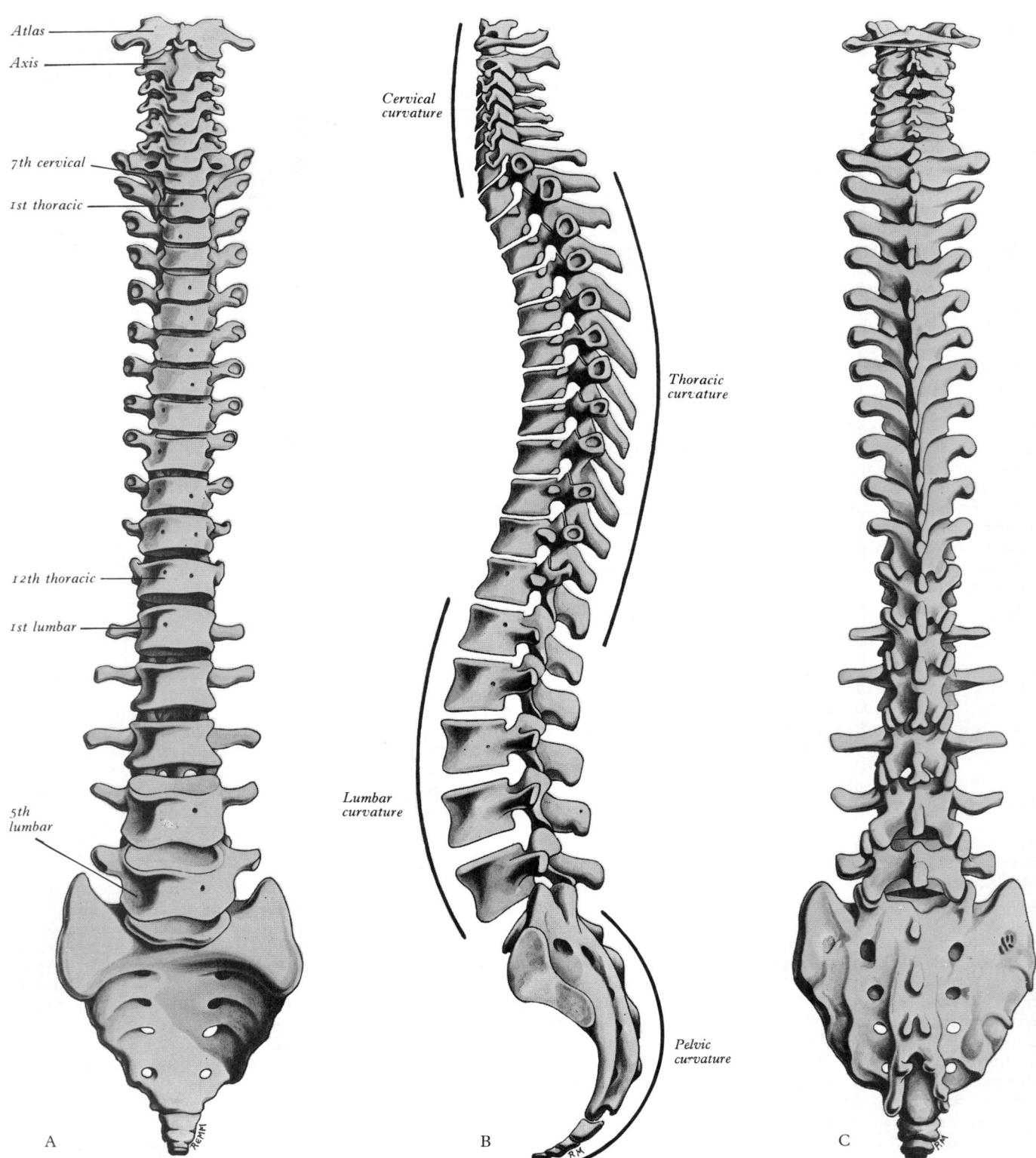

Atlas

Axis

7th cervical

1st thoracic

12th thoracic

1st lumbar

5th lumbar

A

Cervical
curvature

Thoracic
curvature

Lumbar
curvature

Pelvic
curvature

B

C

3.79 The vertebral column: A anterior aspect; B lateral aspect (Note curvatures); C dorsal aspect. Note slight sinuous, lateral, thoracolumbar curvature visible from both dorsal and anterior aspects.

mainly formed by laminae; in the thoracic region they are deep, wide and formed by laminae and transverse processes. Lateral to laminae are articular processes and still more lateral, transverse processes. In the thoracic region the transverse processes are in a plane considerably *behind* those of cervical and lumbar processes. Cervical transverse processes are anterior to articular processes, lateral to pedicles and between intervertebral foramina. In the thoracic region they are posterior to pedicles, intervertebral foramina and articular processes. In the lumbar region they are anterior to articular processes, but posterior to intervertebral foramina. The size of the atlantal transverse processes has been emphasized (p. 318) and transverse breadth of the atlas contrasted

with that of the axis. Breadth varies little from second to sixth cervical, but increases in the seventh. In thoracic vertebrae it is greatest in the first and diminishes to the twelfth, whose transverse elements are usually mere vestiges. The first lumbar vertebra is broader, the second more so, the third broadest of all, the fourth and fifth diminishing.

The *lateral aspect* of the vertebral column (3.3,79B) is arbitrarily separated from the posterior by articular processes in the cervical and lumbar regions and transverse processes in the thoracic. Anteriorly it is formed by the sides of vertebral bodies, with costal facets at thoracic level. The intervertebral foramina, behind bodies and between pedicles, are oval, smallest at cervical and upper thoracic levels and increasing in size to the last lumbar; they contain spinal and other nerves and various vessels.

The *vertebral canal* follows the vertebral curves, is large and triangular in cervical and lumbar regions, where movement is free, but small and circular in the thoracic, where motion is less. These differences also accommodate variations in diameter of the spinal cord and its enlargements.

Although movement between adjacent vertebrae is limited, these small ranges add up to considerable degrees of bending or rotation over the whole column (p. 492). Intervertebral discs tie vertebrae together but are also the principal sites of movement. By elastic deformability they permit tilting and torsion between vertebral bodies and also add compressibility to the column. This ability to absorb stresses is augmented by the column's sinuous curvature; forces transmitted with little loss by a straight column are largely expended against the pliancy of the spinal curves. Effects of weight, muscle traction and thrust from the feet, whether trivial as in walking, or large as in running and jumping, are smoothed out by the discs and curvatures.

Despite the prominent role of intervertebral discs in spinal dynamics, regional variations in mobility also depend on the disposition, properties and geometry of intervertebral synovial joints and ligamentous complexes attached to all parts of each vertebra (p. 492).

Fractures or fracture dislocations of the vertebral column result from (1) forced flexion, e.g. by a violent blow on the back, and (2) violence transmitted along its axis, e.g. by falls on to the feet or head. In the first type injury commonly occurs at the fifth or sixth thoracic; in the second, owing to normal curvatures, injury is also a flexion fracture, often between the ninth thoracic and second lumbar.

In spondylolisthesis, present in 5% of skeletons, the spine, laminae and inferior articular processes of the fifth (sometimes fourth) lumbar vertebra are united together but separate from the rest of the bone. The condition is probably congenital and it is suggested that each half vertebral arch ossifies from two primary centres which fail to fuse, but there is no clear evidence for this.

The Sternum

The sternum is confined to land vertebrates, and aquatic mammals such as seals; it is absent in fish. The elongate human form is typical of mammals, the sternum being a plate, often composite, in amphibians and reptiles. Though associated with shoulder girdles and ribs from its earliest appearance, it is often considered axial. Its embryonic development and ossification indicate an origin separate from ribs. The human sternum (**3**.80, 81A,B) consists of a cranial *manubrium* (prosternum), an intermediate *body* or mesosternum and a caudal *xiphoid process* (metasternum). Its total length in males is about 17 cm, less in females (Jit 1984). The ratio between manubrial and mesosternal lengths

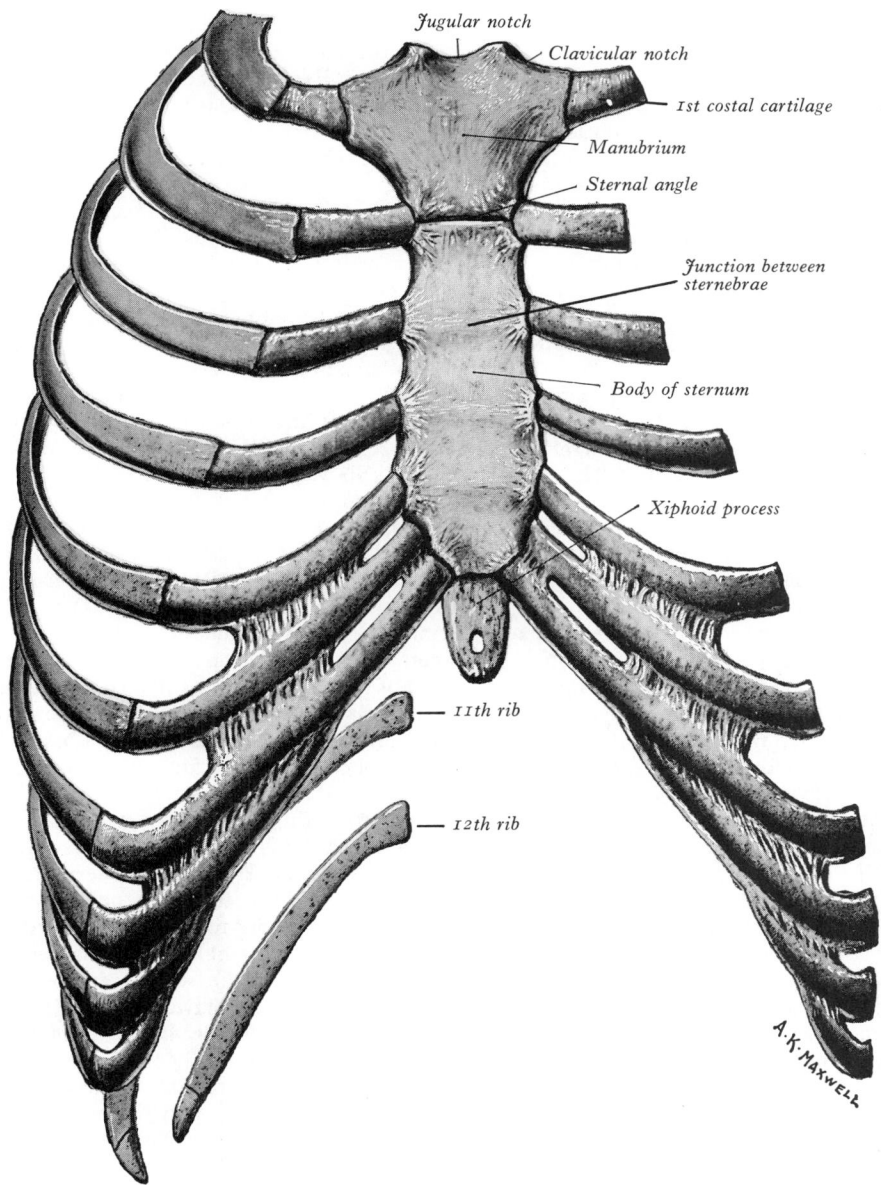

Jugular notch

Clavicular notch

1st costal cartilage

Manubrium

Sternal angle

Junction between sternebrae

Body of sternum

Xiphoid process

11th rib

12th rib

A.K. MAXWELL

.80 The sternum and costal cartilages: anterior aspect.

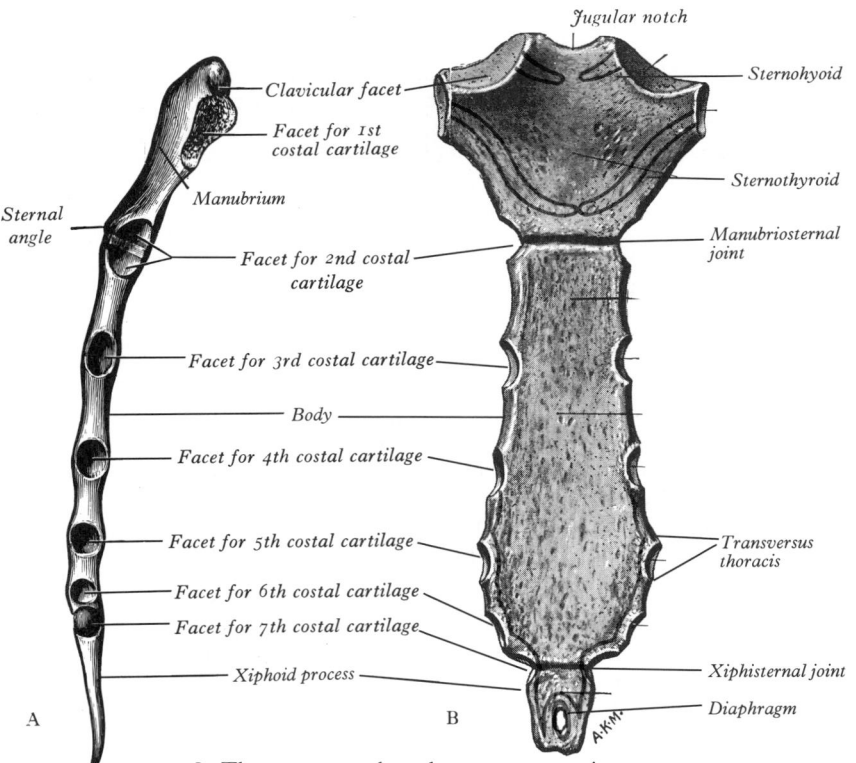

3.81 The sternum: A. lateral aspect; B. posterior aspect.

Labels on figure:
Jugular notch
Clavicular facet
Facet for 1st costal cartilage
Sternal angle
Manubrium
Sternohyoid
Sternothyroid
Facet for 2nd costal cartilage
Manubriosternal joint
Facet for 3rd costal cartilage
Body
Facet for 4th costal cartilage
Facet for 5th costal cartilage
Facet for 6th costal cartilage
Facet for 7th costal cartilage
Transversus thoracis
Xiphoid process
Xiphisternal joint
Diaphragm
A
B

differs in the sexes, but racial differences have not been established. Such data are complicated by possible continuation of growth beyond the third decade and even throughout life (Ashley 1956, Rother et al 1975). In addition to the three major sections, the mesosternum in early life consists of four *sternebrae*, which from costal relations appear to be intersegmental. In natural stance the sternum slopes down and slightly forwards. It is convex in front, concave behind and broadest at the junction with the first costal cartilages, narrow at the manubriosternal joint, below which it widens to its articulation with the fifth cartilages, narrowing again below this.

The manubrium sterni is broad and thick above, narrowing to its junction with the mesosternum. Its *anterior surface* is smooth, convex transversely, vertically concave. Its *posterior surface* is concave and smooth. The *superior border* is thick, with a central *jugular (suprasternal) notch* between two oval fossae directed up and posterolaterally for articulation with sternal ends of clavicles, hence termed *clavicular notches*. The *inferior border*, oval and rough, carries a thin layer of cartilage for articulation with the mesosternum. The *lateral borders* are marked above by a depression for the first costal cartilage and below by a small articular demifacet, which, with one on each superior mesosternal angle, joints with part of the second costal cartilage. Between these facets the narrow curved edge descends medially.

The mesosternum (body), longer, narrower and thinner than the manubrium, is broadest near its lower end. Its *anterior surface*, nearly flat, faces slightly upwards and has three variable transverse ridges at levels of fusion of its four *sternebrae* (vide infra). A *sternal foramen*, of varying size and form, may occur between the third and fourth sternebrae (p. 161). The *posterior surface*, slightly concave, also displays three less distinct transverse lines. The oval *upper end* articulates with the manubrium (*manubriosternal joint*) at the level of the *sternal angle*, an anterior ridge usually being palpable. A posterior transverse groove also marks the manubriosternal joint (3.81A). The *lower end* is narrow and continuous with the xiphoid process. On each *lateral border* (3.81B), at its superior angle, a small notch receives in part the second costal cartilage; below this, four *costal notches* articulate with the third to sixth cartilages; the inferior angle has a small facet, which, with the xiphoid process, receives the seventh costal cartilage. Between these articular depressions a series of curved edges diminish in length downwards, forming the anterior limits of the intercostal spaces.

The xiphoid process (*xiphisternum*), the smallest and most variable sternal element, may be broad and thin, pointed, bifid, perforated, curved or deflected. It is cartilaginous in youth, but more or less ossified in adults. It is continuous with the mesosternum's lower end at the *xiphisternal joint*. Anterior to its superolateral angles are demifacets for parts of the seventh costal cartilages (3.81).

The manubrium is level with the third and fourth thoracic vertebrae; the *sternal angle* is opposite the inferior border of the fourth vertebral body. To the manubrial *anterior surface* the sternal ends of the major pectoral and sternocleidomastoid muscles are attached and to its *posterior surface* the sternothyroids, opposite the first costal cartilages; above are the most medial fibres of the sternohyoids. This surface is the anterior wall of the superior mediastinum; inferiorly (approximately the vertically lower half) it is related to the aortic arch, above this to the left brachiocephalic vein and brachiocephalic, left common carotid and left subclavian arteries; laterally to lungs and pleurae. To the *jugular notch* are attached fibres of the interclavicular ligament. On the *lateral border* the manubriocostal joint (first costal cartilage) is a synchondrosis, unlike other sternocostal joints.

The mesosternum is level with the fifth to ninth thoracic vertebrae. To its *anterior surface* articular capsules of sternocostal joints and sternal fibres of pectoralis major are attached, and to its *posterior surface* the transversus thoracis (sternocostalis). Right of the median plane it is related to pleura and the thin, anterior border of the right lung, projecting between sternum and pericardium. The left half of the upper two sternebrae is related to left pleura and lung, the lower two directly to pericardium. To the *borders* the external intercostal membranes are attached between costal facets. Except for the first and sixth, the cartilages of 'true' ribs articulate with the sternum at junctions of its segments.

The xiphoid process is in the epigastrium. To its *anterior surface* are attached the most medial fibres of rectus abdominis and aponeuroses of external and internal obliques, to its *lower end* the linea alba, and to its *borders* the aponeuroses of internal oblique and transversus abdominis. To its *posterior aspect* slips of diaphragm are attached and it is here related to the liver.

The sternum contains highly vascular trabecular bone enclosed by a compact layer thickest in the manubrium between clavicular notches. The medulla contains *red bone marrow*. The internal trabecular pattern has been examined intensively by scanning electron microscopy and by computer analysis of metrical parameters (Whitehouse 1975). Centrally the bone is lightly constructed, the trabeculae being thicker and wider apart in lateral regions. No completely satisfactory explanation of these variations, or of those already noted in ribs and lumbar vertebrae, has been formulated. Pismenov & Zapetski (1977) have described the vascular supply of the sternum in detail.

Ossification (3.82A,B). The sternum is formed by fusion of two cartilaginous *sternal plates* (p. 161) flanking the median plane. Arrangement and number of ossificatory centres vary in relation to completeness and time of fusion of the sternal plates and to the width of the adult bone (Ashley 1956). Incomplete fusion leaves a *sternal foramen* (p. 161). (For incidence in Indian sterna, see Jit & Bakshi 1984.) The manubrium is ossified from one to three centres appearing in the fifth fetal month, and the first and second sternebrae usually from single centres about the same time. Centres in the third and fourth sternebrae are commonly paired and appear in the fifth and sixth months respectively, but one of either pair may be delayed until the seventh or even eighth month; the fourth sternebral centre may be absent (Paterson 1904). The xiphoid process begins to ossify in the third year or later. In some sterna all centres are single and median, in others the manubrial centre is single and the sternebral all paired, symmetrical or asymmetrical. Union between mesosternal centres begins at puberty and proceeds from below upwards; by 25 all are united (see also p. 498).

Little has been recorded of racial or sexual variation in the sternum. Jit et al (1980), and Jit and Hasjeet (1982) stated that, except in combined length of manubrium and mesosternum, sexing is not possible, the sternal angle being unreliable.

Suprasternal ossicles, paired or single, occur in about 7% of sterna. They may fuse to the manubrium or articulate posteriorly

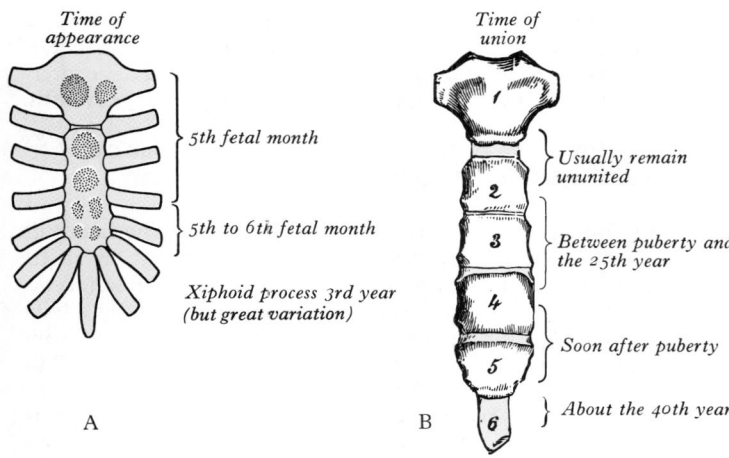

Time of appearance

5th fetal month

5th to 6th fetal month

Xiphoid process 3rd year (but great variation)

A

Time of union

Usually remain ununited

Between puberty and the 25th year

Soon after puberty

About the 40th year

B

3.82 The ossification of the sternum: A. before birth, B at puberty.

at the lateral border of the jugular notch. When well formed they are pyramidal, the base being articular. Cartilaginous at birth, they ossify during adolescence.

The Ribs

The ribs are elastic arches, connected posteriorly with the vertebral column, forming much of the thoracic skeleton. The twelve pairs may be increased by cervical or lumbar ribs or reduced to eleven by absence of the twelfth pair. The superior seven pairs are connected by costal cartilages to the sternum (**3.**80, 88), as *true* ribs. The remaining five are so-called *false* ribs, cartilages of the eighth to tenth joining the superjacent costal cartilage; the eleventh and twelfth, being free at their anterior ends, are sometimes termed *floating* ribs. (The tenth rib is also *usually* 'floating' in the Japanese, as also recorded in other races, see p. 336.)

Between ribs are *intercostal spaces*, which are deeper in front and between the upper ribs. Upper ribs are less oblique than the lower, obliquity being maximal at the ninth and decreasing to the twelfth. They increase in length from first to seventh, diminishing to the twelfth. In breadth they decrease downwards; in the upper ten the greatest breadth is anterior. The first two and last three present special features, the rest conforming to a common plan.

A typical rib has a shaft with anterior and posterior ends (**3.**83, 84). The *anterior, costal end* has a small concave depression for its cartilage's lateral end. The *shaft* has an external convexity and is grooved internally near its lower border, which is sharp, in contrast to its rounded upper border. The *posterior, vertebral end* has a head, neck and tubercle. The *head* presents two facets, separated by a transverse *crest*. The lower and larger articulates with the body of the corresponding vertebra, its crest attached to the intervertebral disc above it. The *neck* is the flat part beyond the head, anterior to the corresponding transverse process. It is oblique, facing anterosuperiorly. Its postero-inferior surface is rough and pierced by foramina. Its upper border is the sharp *crest of the neck*, its lower border rounded. The *tubercle* is postero-external at junction of neck with shaft; more prominent in upper ribs, it is divided into medial articular and lateral non-articular areas. The articular part bears a small, oval facet for the transverse process of the corresponding vertebra; the non-articular area is roughened by ligaments. The *shaft* is thin and flat with external and internal surfaces, superior and inferior borders. It is curved, bent at the *posterior angle* 5–6 cm from the tubercle, and also twisted in its long axis: the part behind the angle inclines superomedially, its *external surface* hence posteroinferior; in front of the angle it faces slightly up; it is convex and smooth, and crossed near the tubercle by a rough line, directed inferolaterally, at the *posterior angle*. The *internal surface* is smooth and marked by the *costal groove*, bounded below by the inferior border. The superior border of the groove continues behind the lower border of the neck, but terminates anteriorly at the junction of middle and anterior thirds of the shaft, anterior to which the groove is absent. (Vide infra concerning an *anterior angle*.)

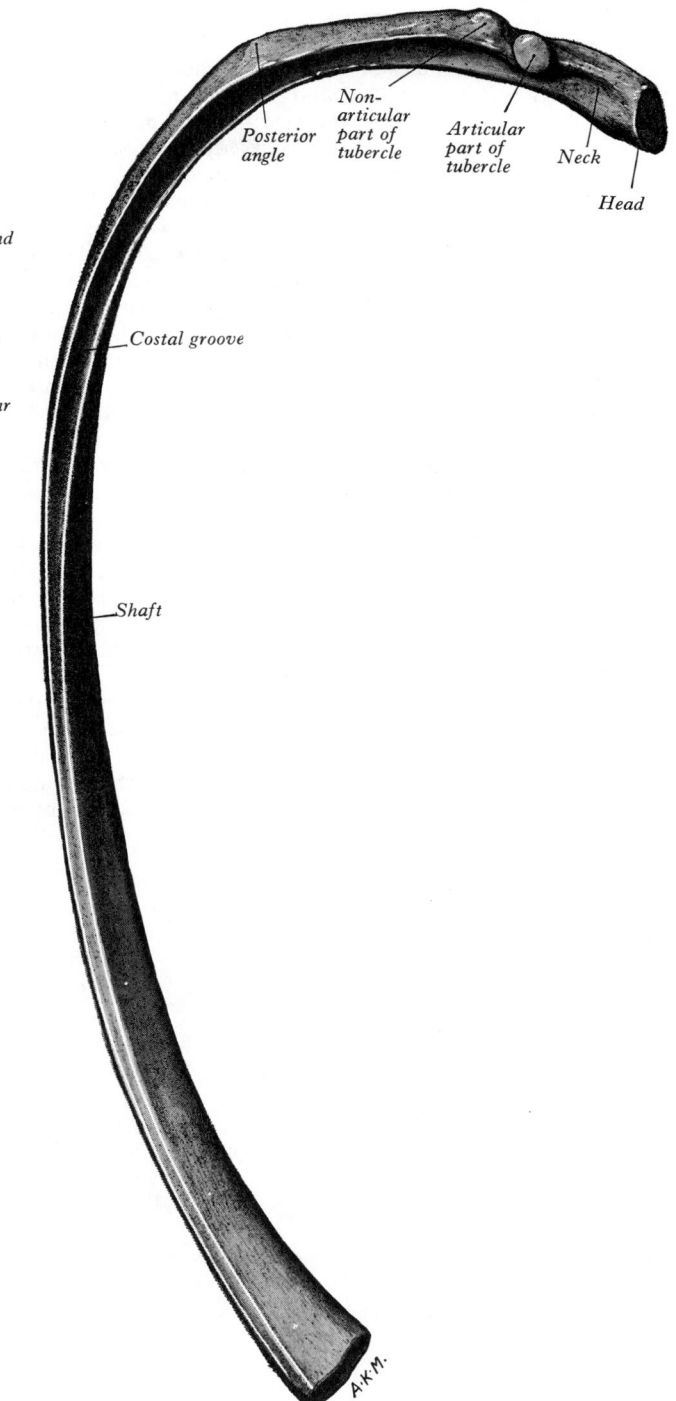

Posterior angle

Non-articular part of tubercle

Articular part of tubercle

Neck

Head

Costal groove

Shaft

A·K·M·

3.83 A typical rib of the left side: inferior aspect.

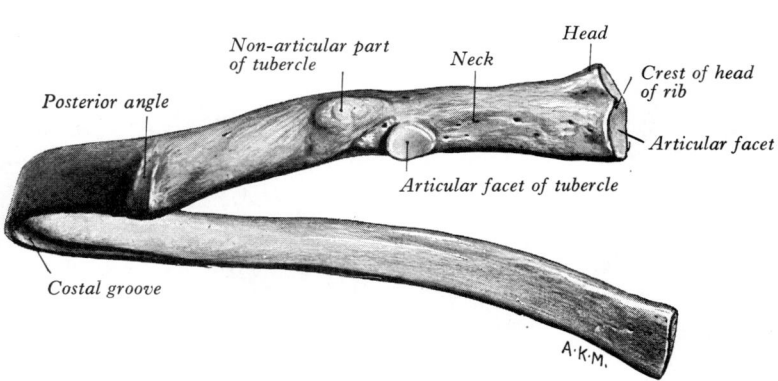

Non-articular part of tubercle

Neck

Head

Posterior angle

Crest of head of rib

Articular facet

Articular facet of tubercle

Costal groove

A·K·M·

3.84 A typical rib of the left side: posterior aspect.

333

The first rib (**3**.85), most acutely curved and usually shortest, is broad and flat, its surfaces superior and inferior, its borders internal and external. It slopes obliquely down and forwards to its sternal end. The *head* is small, round and bears an almost circular facet, articulating with the first thoracic vertebral body. The *neck* is rounded and ascends posterolaterally. The *tubercle*, wide and prominent, is directed up and backwards; medially an oval facet articulates with the first thoracic transverse process. At the tubercle the rib is bent, its head turned slightly down; angle and tubercle therefore coincide. The *superior surface* of the flattened shaft is crossed obliquely by two shallow grooves, separated by a slight ridge which ends at the internal border as a usually small, pointed, projection, the *scalene tubercle*. The *inferior surface* is smooth and ungrooved. The *external border* is convex, thick behind, thin in front. The internal *border* is concave and thin, with a scalene tubercle near its midpoint. The *anterior end* is larger than in any other rib.

The second rib (**3**.86) is twice the length of the first, with a similar curvature. The non-articular area of the *tubercle* is small. The *angle* is slight and near the tubercle. The *shaft* is not twisted, but at the tubercle is convex upwards, as in the first rib but less so. The *external surface* of the shaft is convex and superolaterally marked centrally by a rough, muscular impression; the latter continues posteromedially towards the tubercle as a narrow roughened ridge. The *internal surface*, smooth and concave, faces inferomedially with posteriorly a short costal groove.

The tenth rib has a single facet on its head which may articulate with the intervertebral disc above as well as the upper border of the tenth thoracic vertebra near its pedicle.

The eleventh and twelfth ribs (**3**.87) each have one large, articular facet on the head, but no necks or tubercles; their poin-

ted anterior ends are tipped with cartilage. The eleventh has a slight angle and shallow costal groove. The twelfth has neither, is much shorter and slopes cranially at its vertebral end. The internal surfaces of both face slightly *upwards*, more so in the twelfth.

Attachments and Relations of Ribs

To the *head of a typical rib* the radiate ligament is attached along its anterior border, and the intra-articular ligament to its crest. Its anterior surface is related to costal pleura and in lower ribs the sympathetic trunk. The *neck's anterior surface* is divided by a faint transverse ridge for the internal intercostal membrane and continuous with the *inner* lip of the superior border of the shaft. The area above the ridge, more or less triangular, is separated from the membrane by fatty tissue; the lower, smooth area is covered by costal pleura. The *neck's posterior surface* receives the costotransverse ligament and is pierced by vascular foramina. The *crest of the neck* is for attachment of the superior costotransverse ligament and extends laterally into the *outer* lip of the shaft's superior border. The neck's rounded *inferior border* continues laterally into the upper border of the costal groove, receiving the internal intercostal membrane. The *articular area of the tubercle* in the upper six ribs is convex and faces posteromedially; in the succeeding three or four it is almost flat and faces down, back and slightly medially. To the *non-articular area* the lateral costotransverse ligament is attached.

The ridge on the shaft's *external surface* (near its posterior angle) in a typical rib receives an upward continuation of thoracolumbar fascia and lateral fibres of iliocostalis thoracis. From second to tenth ribs the distance between angle and tubercle increases. Medial to the angle the external surface affords attachment to a levator costae and is covered by erector spinae. Near this surface's sternal end an indistinct, oblique line (*anterior 'angle'*) separates attachments of external oblique and serratus anterior (or latissimus dorsi, in ninth and tenth ribs). To the *costal groove* (internal surface) is attached the internal intercostal muscle, separating bone from intercostal vessels and nerve. At the vertebral end the groove faces down, its borders in the same plane. Near the posterior angle the shaft broadens and the groove reaches its internal surface; to the groove's superior rim intercostalis intimus is attached, rarely extending to the rib's anterior fourth. Posteriorly this rim meets the neck's lower border. To the sharp inferior costal border is attached an external intercostal muscle. The superior border has posteriorly two lips: to the inner are attached both intercostalis internus and intimus, to the outer only intercostalis externus.

The first rib (**3**.85) has a *tubercle for scalenus anterior*, which is also attached to adjoining parts of the *upper surface*. The groove anterior to this forms a bed for the subclavian vein; and the rough area between this and the first costal cartilage receives the costoclavicular ligament and, more anteriorly, subclavius. To the groove behind the tubercle is apposed the subclavian artery and usually the lower trunk of the brachial plexus. Behind this, as far as the costal tubercle, scalenus medius is attached. The obliquity of the first ribs accounts for the appearance of pulmonary and pleural apices in the neck.

The first rib's external border is covered behind by scalenus posterior descending to the second; the first digitation of serratus anterior is, in part, attached to it, behind the subclavian (arterial) groove. To the *internal border* is attached the suprapleural membrane, covering the pleura's cervical dome.

The second rib (**3**.86) is marked by *serratus anterior*, its rough prominence extending from just behind the midpoint of its external surface; to this tubercle the lower part of the first and the second digitation are attached. The second intercostal nerve is between the rib and pleura in most of its course. The distinct lips of the upper border are widely separated behind; in front of the angle the outer lip receives scalenus posterior and, below this, serratus posterior superior.

The twelfth rib (**3**.86A,B) has numerous attached muscles and ligaments. To the lower part of its *anterior surface*, in its medial half to two-thirds, quadratus lumborum and its anterior covering layer of thoracolumbar fascia are attached, the upper part being related to the costodiaphragmatic pleural recess. At or near the

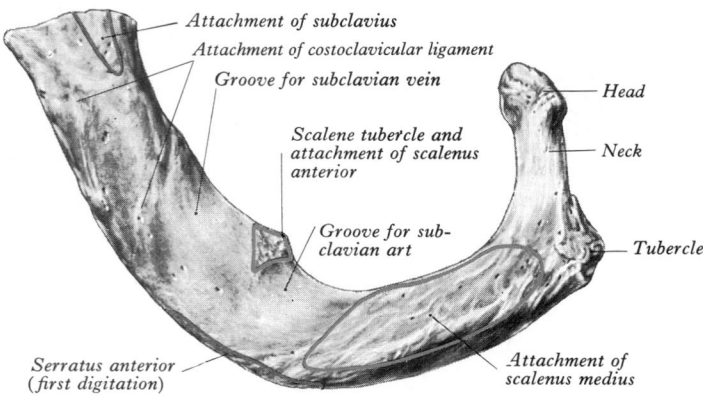

3.85 The first rib of the left side: superior aspect.

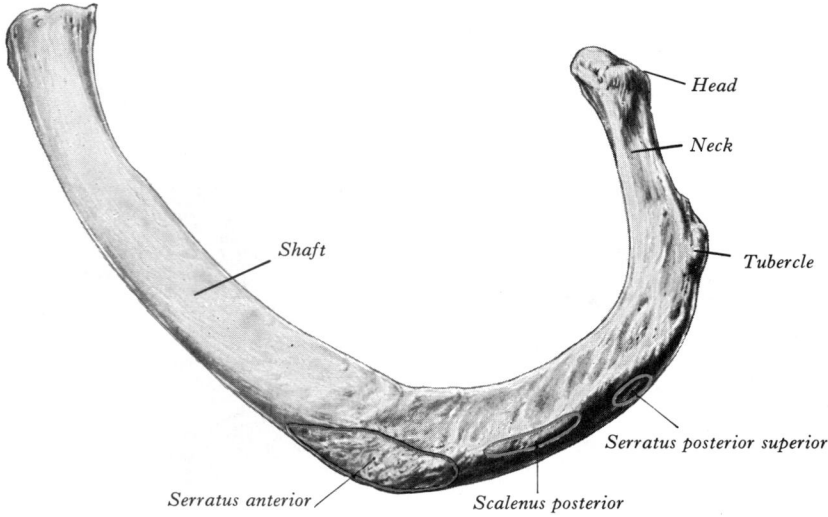

3.86 The second rib of the left side: superior aspect.

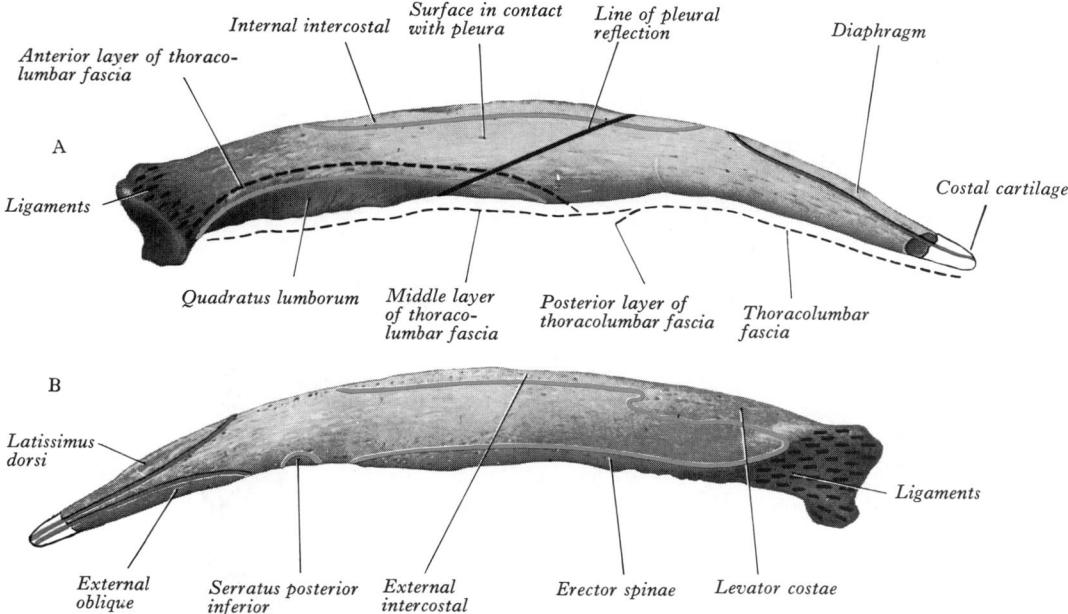

A

Anterior layer of thoraco-lumbar fascia
Internal intercostal
Surface in contact with pleura
Line of pleural reflection
Diaphragm
Costal cartilage
Ligaments
Quadratus lumborum
Middle layer of thoraco-lumbar fascia
Posterior layer of thoracolumbar fascia
Thoracolumbar fascia

B

Latissimus dorsi
External oblique
Serratus posterior inferior
External intercostal
Erector spinae
Levator costae
Ligaments

3.87 The twelfth rib of the left side. A. anterior aspect; B. posterior aspect.

upper border are attachments of the internal intercostal medially and laterally the diaphragm. The *lower border* affords attachment to the middle lamella of the thoracolumbar fascia and, lateral to quadratus lumborum, the lateral arcuate ligament (p. 592) and posterior lamella of the thoracolumbar fascia. Posteriorly, close to the head, is attached the lumbocostal ligament (p. 497), connecting it to the first lumbar transverse process. To the *external surface* are attached the lowest levator costae, longissimus thoracis and iliocostalis in its medial half and laterally serratus posterior inferior, latissimus dorsi and obliquus externus. Along the *upper border* the external intercostal is attached. These attachments vary; those of the internal intercostal, levator costae and erector spinae merge and those of latissimus dorsi, diaphragm and external oblique may reach the costal cartilage. The lower limit of the pleural sac crosses in front of the rib, approximately where it is crossed by the lateral border of iliocostalis. Its lateral end is usually below the line of costodiaphragmatic pleural reflexion and therefore not covered by pleura.

Ribs consist of highly vascular trabecular bone, enclosed in a thin layer of compact bone and containing large amounts of red marrow (see **3.22**).

Ossification. Each rib, except the first and last two, is ossified from a primary centre for the shaft and secondary centres for the head and articular and non-articular parts of the tubercle (Fawcett 1911) but not for non-articular parts of tubercles below the sixth or seventh. The primary centre appears near the angle, late in the second month, first in the sixth and seventh ribs. Secondary centres for head and tubercle appear about puberty, uniting to the shaft soon after 20. The first rib has a primary centre for the shaft, a secondary for the head but only one for the tubercle; the eleventh and twelfth, without tubercles, have each only two centres.

The Costal Cartilages

The costal cartilages (**3.88**), flat bars of hyaline cartilage, extend from the anterior ends of ribs, contributing much to thoracic mobility and elasticity. The upper seven pairs join the sternum; the eighth to tenth articulate with the lower border of the one above; the lowest two have free pointed ends in the abdominal wall. They increase in length from first to seventh, thence decreasing to the twelfth. They diminish in breadth from first to last, like the intercostal spaces. They are broad at their costal

continuity and taper therefrom; but the first and second are of even breadth and the sixth to eighth enlarge where their margins are in contact. The first descends a little, the second is horizontal and the third ascends slightly, the others being angulated and inclining up to the sternum or cartilage above, a little anterior to their ribs.

Each costal cartilage has two surfaces, borders and ends. The *anterior surface* is convex, facing anterosuperior; to it in the first cartilage are attached the sternoclavicular articular disc, costoclavicular ligament and subclavius and in the first six or seven medially is pectoralis major. The others are covered by the partial attachments of anterior abdominal muscles. The *posterior surface* is concave, and really posteroinferior; to the first cartilage is attached sternothyroid, to the second to sixth, transversus thoracis, and to the six lower ones transversus abdominis. To the concave *superior* and convex *inferior* borders are attached the internal intercostal muscles and external intercostal membranes. The inferior borders of the sixth to ninth cartilages project at points of greatest convexity and often also the fifth. Oblong facets on these projections articulate with facets on slight projections from the superior borders of subjacent cartilages. The *lateral end* of each cartilage is continuous with its rib; the *medial end* of the first is continuous with the sternum; those of the six succeeding cartilages are round and articulate with shallow costal notches on the lateral margins of the sternum; those of the eighth to tenth are pointed, each connected with the cartilage above; those of the eleventh and twelfth are pointed and free. Excepting the synchondrosis of the first rib and sternum, all these articulations are synovial (p. 497).

The tenth rib is usually united ventrally with the ninth by a fibrous joint (p. 498); but it frequently is free and pointed like the eleventh and twelfth. The incidence of a 'floating' tenth rib varies from 35% to 70% in different races (Shimaguchi 1974).

In old age costal cartilages tend to ossify superficially, losing pliability and becoming brittle. The histology (Amprino & Bairati 1933), histochemistry (Quintarelli & Dellovo 1966), and ultrastructure of such senile changes have been reviewed by Stockwell (1967), (p. 290). An increase in keratan sulphate is noted maximally in the subperichondrial zone.

The Thorax

The thoracic skeleton (3.88) is an osteocartilaginous frame around the principal organs of respiration and circulation. It is narrow

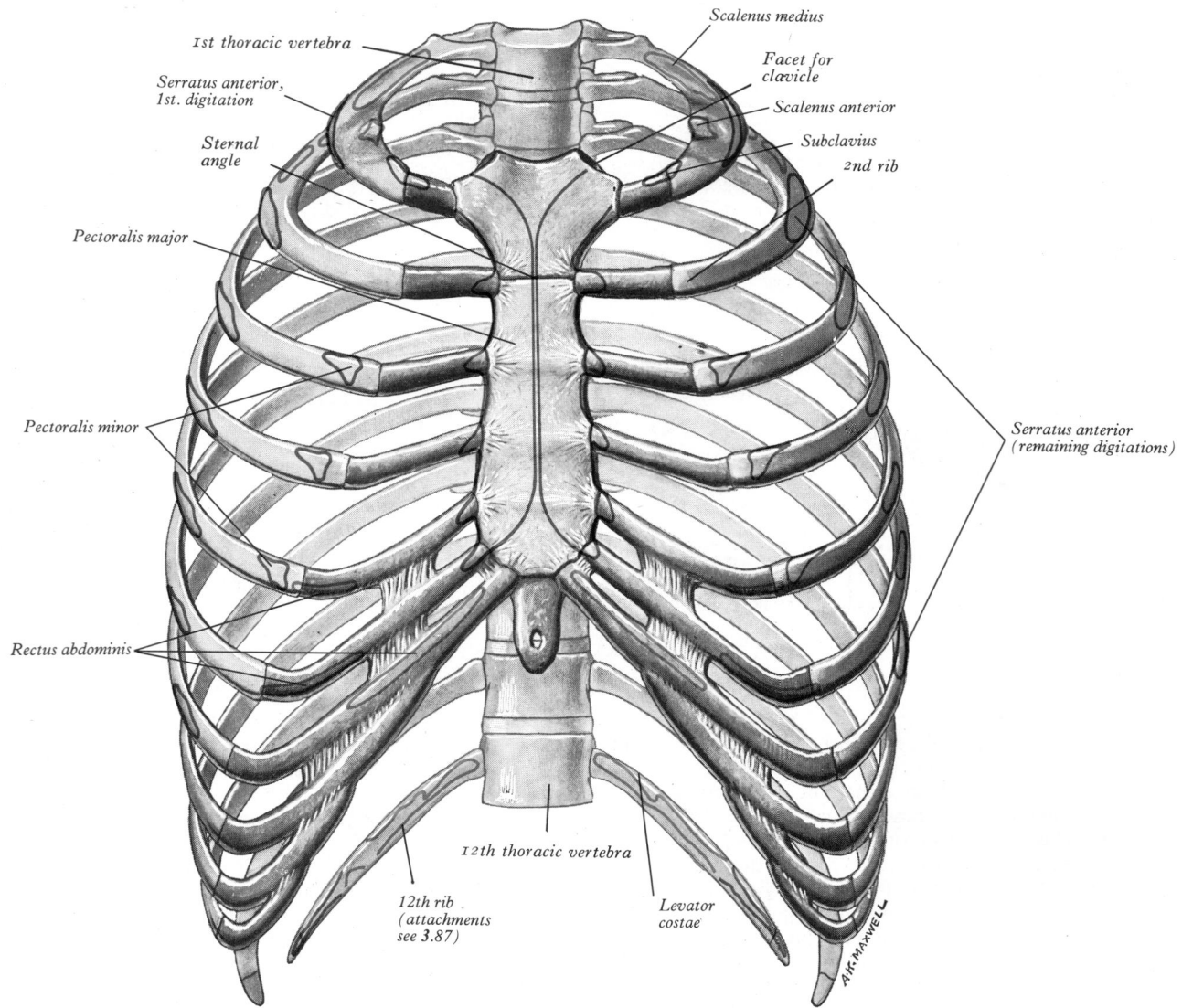

The skeleton of the thorax: anterior aspect.

1st thoracic vertebra

*Serratus anterior,
1st. digitation*

*Sternal
angle*

Pectoralis major

Pectoralis minor

Rectus abdominis

Scalenus medius

*Facet for
clavicle*

Scalenus anterior

Subclavius

2nd rib

*Serratus anterior
(remaining digitations)*

12th thoracic vertebra

*12th rib
(attachments
see 3.87)*

*Levator
costae*

3.88 The skeleton of the thorax: anterior aspect.

above, broad below, flattened anteroposteriorly and longer behind. It is reniform in horizontal section due to the forward projection of vertebral bodies.

Posteriorly the thorax includes thoracic vertebrae and posterior parts of the ribs. On both sides of the vertebral column is a large groove due to the posterolateral curvature of ribs from their vertebral ends to their angles. *Anteriorly* are the sternum, anterior parts of ribs and costal cartilages and this aspect is slightly convex. *Laterally* the thorax is convex and formed by ribs alone. Ribs and costal cartilages are separated by 11 intercostal spaces, occupied by intercostal muscles and membranes, neurovascular bundles and lymphatic channels.

The *thoracic inlet* is reniform, about 5 cm anteroposteriorly, and transversely about 10 cm. Its plane slopes down and forwards, bounded by the first thoracic vertebral body behind, the superior border of the manubrium sterni in front and first rib on each side. The *outlet* is limited behind by the twelfth thoracic vertebral body, the twelfth and eleventh ribs laterally, and in front by the tenth to seventh ribs, which ascend to form the *infrasternal angle*. The outlet is wider transversely and oblique, sloping down towards the back, and is closed by the diaphragm which forms a floor to the thoracic cavity.

Thoracic variations in dimensions and proportions are partly individual and also linked to age, sex and race. At birth the transverse diameter is relatively less but adult proportions develop as walking begins. In females capacity is less, absolutely and proportionately, the sternum being shorter, the thoracic inlet more oblique and the suprasternal notch level with the third thoracic

vertebra (second in males). The upper ribs are more mobile in females, allowing greater upper thoracic expansion. In tall thin individuals their thoraces usually show corresponding proportions, and a similar correspondence also occurs in the short and broad. Racial variations are also linked to stature and proportions in a like manner.

The prime function of the thorax is respiration (p. 498). The obvious protection afforded is fortuitous but many muscles are attached to it, not all primarily concerned with respiration, although they may assist it. Muscles of the arm, especially those acting on its girdle and humerus, those of the abdominal wall and spinal column, all have widespread thoracic attachments.

Elastic recoil of ribs, which suspend the sternum, may explain the rarity of sternal fractures. Despite their pliability, the ribs are much more frequently broken, middle ribs being most vulnerable. Since traumatic stress is usually due to compression of the thorax, the usual site of fracture is the rib's weakest point, just in front of the angle. Direct impact may fracture a rib anywhere and the broken ends of bone may be driven inwards, with possible injury to thoracic or upper abdominal viscera.

A *cervical rib*, the costal element of the seventh cervical vertebra, may be a mere epiphysis on its transverse process but more often has a head, neck and tubercle, with or without a shaft which, varying in length, extends anterolaterally into the posterior triangle of the neck, where it may end freely or join the first rib or costal cartilage, or even the sternum. It may be partly fibrous, but its effects are not related to the size of its osseous part. If it is long enough its relations are those of a first thoracic rib: the brachial plexus (usually lower trunk) and subclavian vessels are *superior* and apt to suffer compression in a narrow angle between rib and scalenus anterior. Hence cervical ribs may first be revealed by nervous and vascular symptoms, particularly due to pressure on the eighth cervical and first thoracic spinal nerves, with motor and sensory effects in structures supplied. According to Cave (1975) a cervical rib or pleurapophysis may show synostosis or diarthrosis with either the anterior (parapophysial) or posterior (diapophysial) 'roots' of the so-called seventh cervical transverse process or, more usually, with both (**3**.57).

THE SKULL

Introduction

The vertebrate skull is the most modified part of the axial skeleton. Whatever truth may reside in the hypothesis that it is partly derived from metameric elements such as modified vertebrae, palaeontology and comparative anatomy clearly indicate that from earliest times it has been a skeletal complex adapted to support the brain and organs of special sense and to secure food. Siting of special receptors, tactile, chemical and visual, and stresses due to play of powerful masticatory and axial musculature is less obvious but is continual. In addition to these extraneous forces, the rigid cranial walls provide continuous isolation for cerebral circulation. Moreover, the reputed buffering by meninges, subarachnoid space and contained fluid could only be effective within a rigid container.

The brain's extreme dependence on uninterrupted blood flow is well known; independence of cerebral arterial pressure from extracranial variations, due to some form of autoregulation as yet

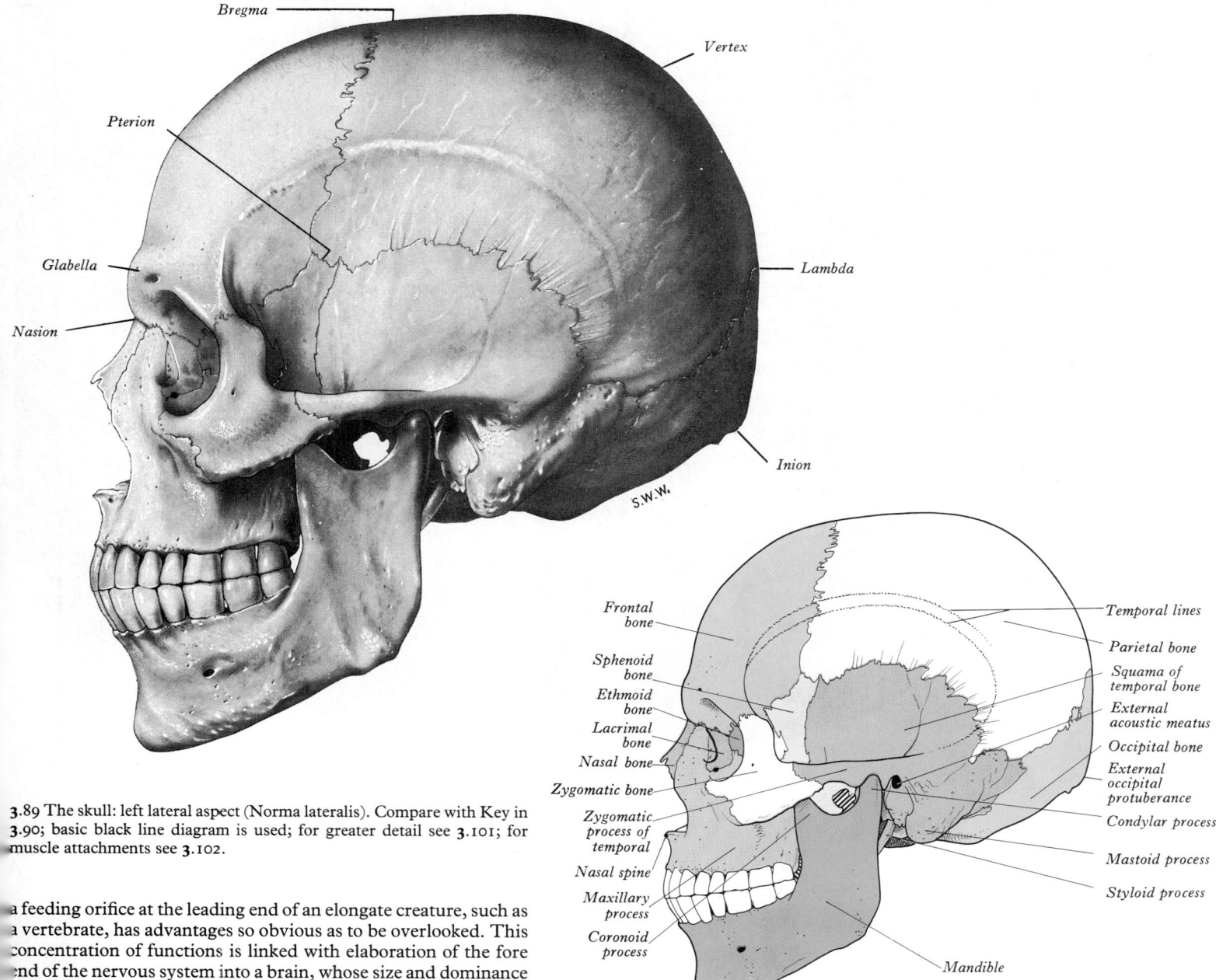

3.89 The skull: left lateral aspect (Norma lateralis). Compare with Key in **3**.90; basic black line diagram is used; for greater detail see **3**.101; for muscle attachments see **3**.102.

3.90 The skull (Norma lateralis). Key to **3**.89. Blue = frontal and occipital; brown = temporal, lacrimal and nasal; magenta = mandible and ethmoid; green = maxilla; yellow = sphenoid; white = zygomatic and parietal; oblique cross-hatch = bone is absent.

a feeding orifice at the leading end of an elongate creature, such as a vertebrate, has advantages so obvious as to be overlooked. This concentration of functions is linked with elaboration of the fore end of the nervous system into a brain, whose size and dominance have inceased throughout vertebrate evolution. The very size of the human brain emphasizes the skull's *cerebral* function, overshadowing others. Even in this limited role the cranium cannot be considered merely protective. Sporadic protection of brain from external impacts is of undoubted value; need of a barrier against

337

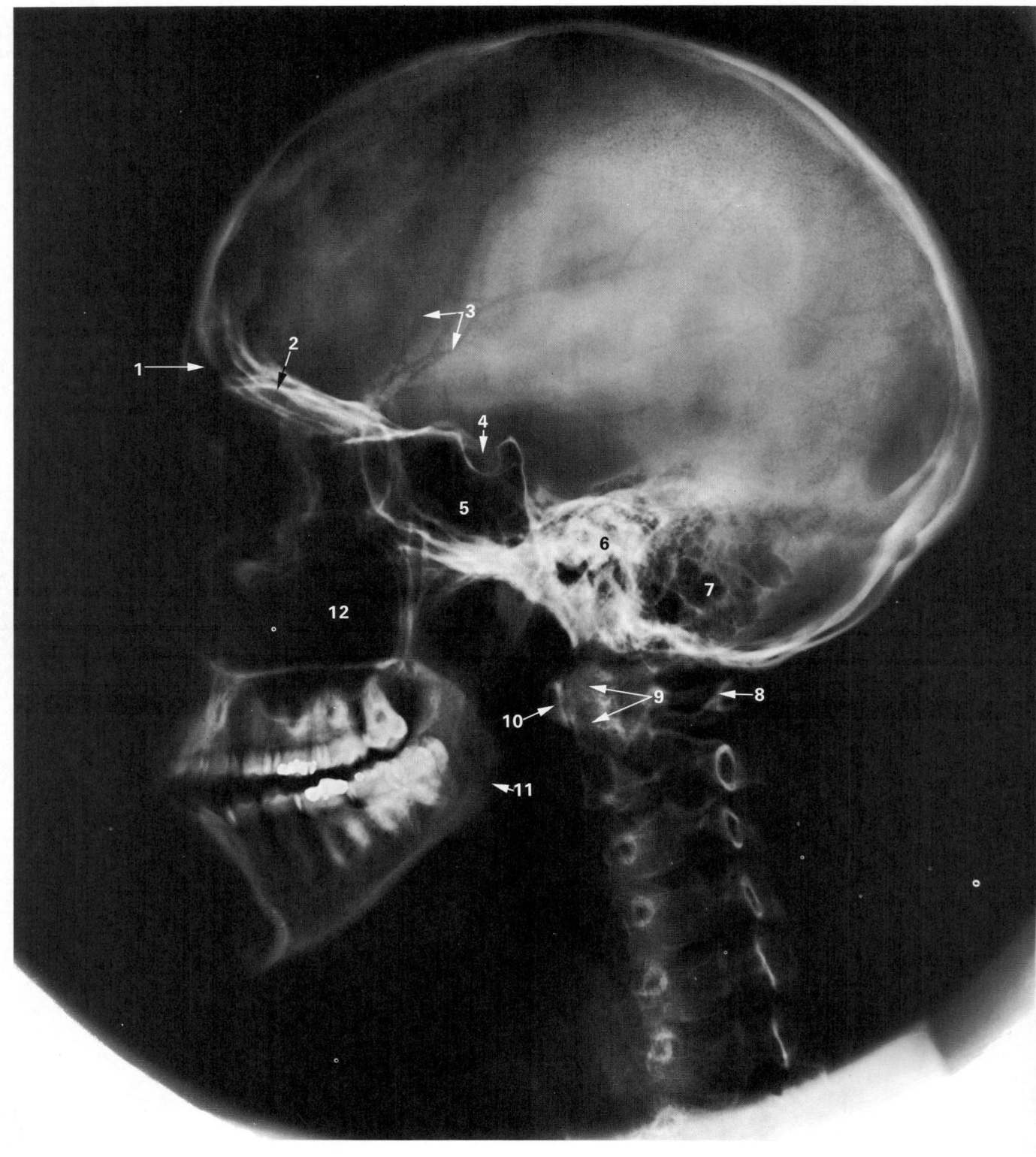

A

3.91 Lateral (A) and anteroposterior (B) radiographs of adult female skull. (Provided by R D Hoare, Neuroradiologist, Guy's Hospital.)

A 1. Frontal sinus. 2. Orbital roof. 3. Grooves for anterior branches of middle meningeal vessels. 4. Hypophysial fossa. 5. Sphenoidal sinus. 6. Dense shadow of petrous part of temporal bone. 7. Mastoid air cells. 8. Posterior arch of atlas. 9. Dens of axis. 10. Anterior arch of atlas. 11. Angle of mandible. 12. Maxillary sinus.

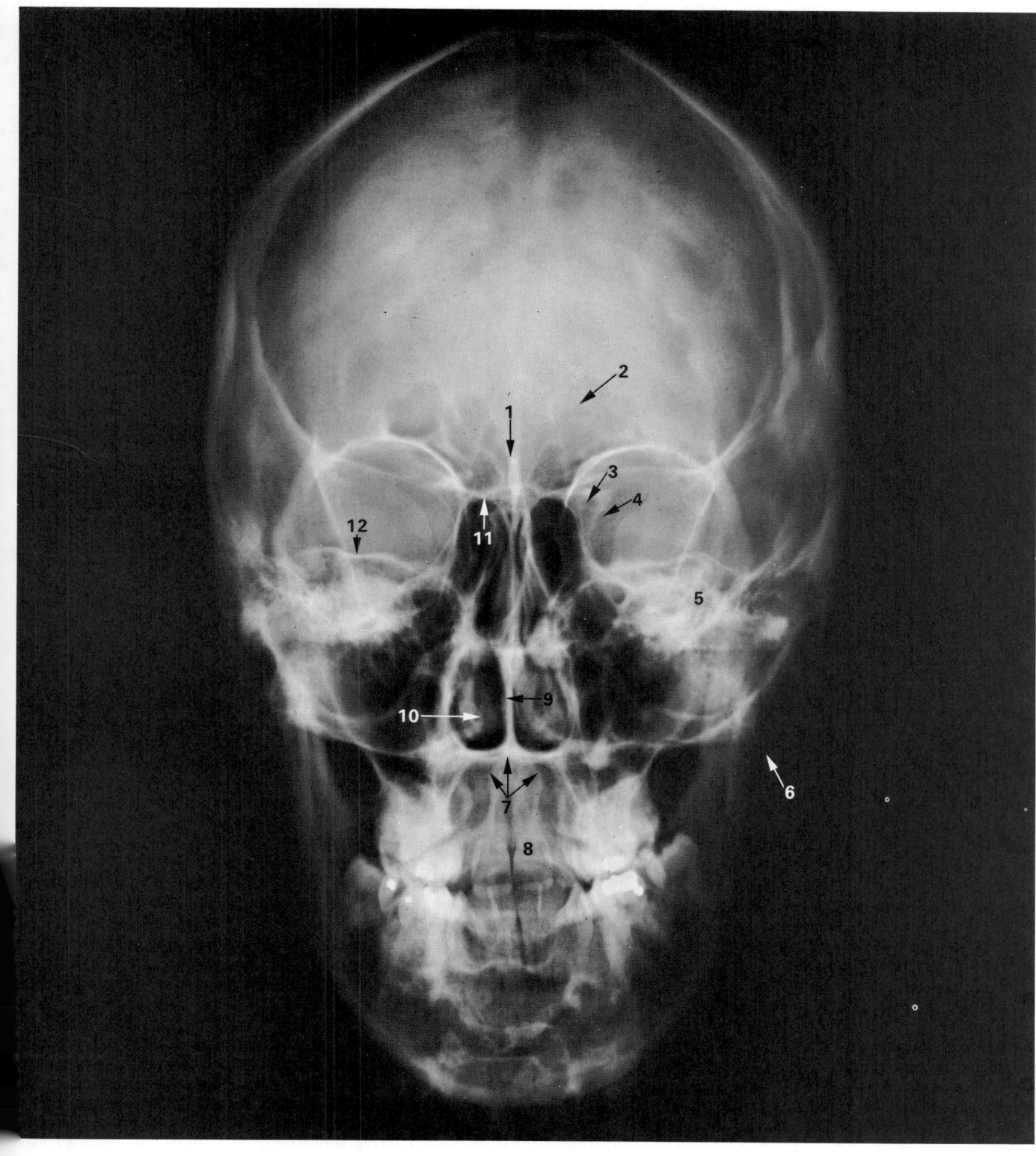

B

1. Crista galli. 2. Frontal sinus. 3. Optic canal. 4. Superior orbital fissure. 5. Dense shadow of petrous part of temporal bone. 6. Apex of mastoid process. 7. Dens of axis. 8. Upper central incisor tooth. 9. Nasal septum. 10. Inferior concha. 11. Cribriform plate. 12. Superior border of petrous part of temporal bone.

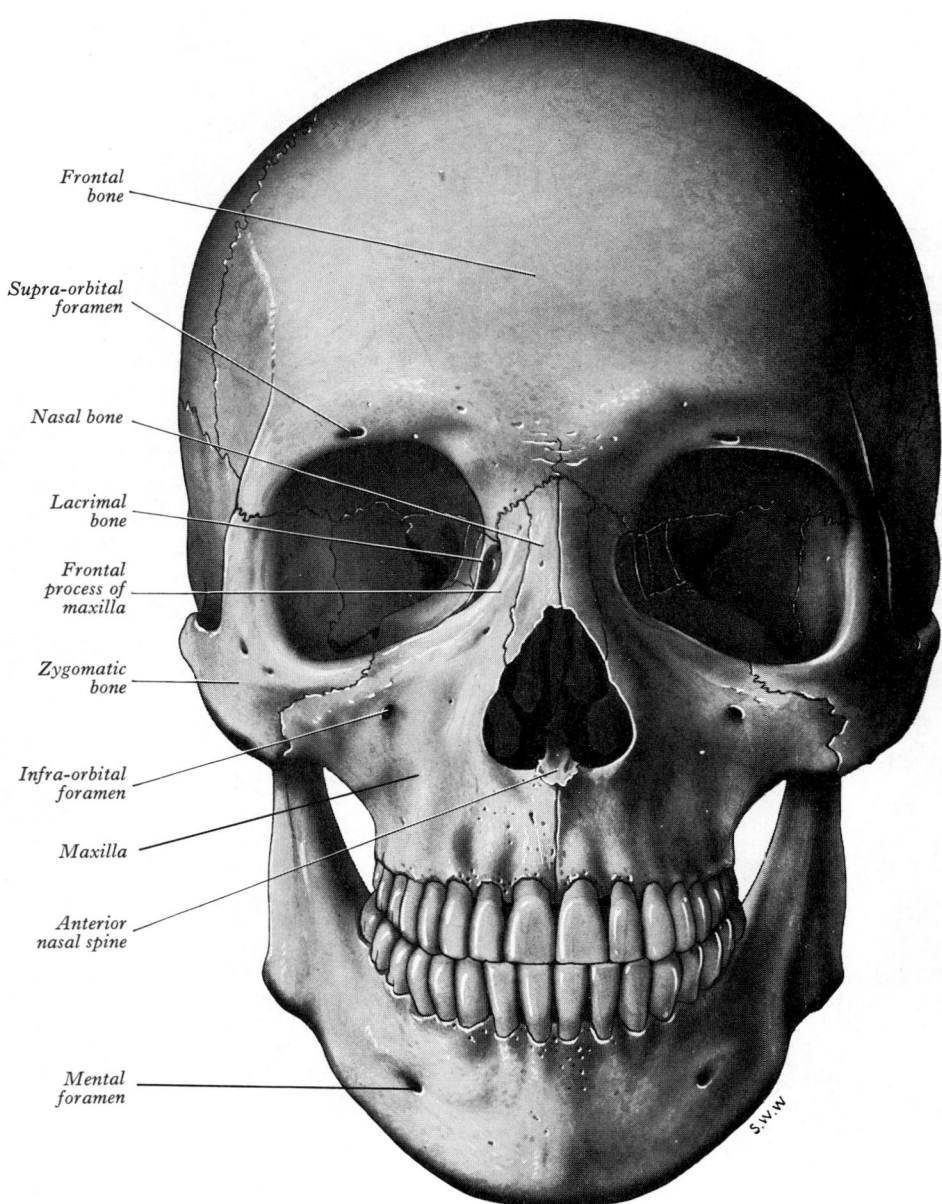

Frontal bone

Supra-orbital foramen

Nasal bone

Lacrimal bone

Frontal process of maxilla

Zygomatic bone

Infra-orbital foramen

Maxilla

Anterior nasal spine

Mental foramen

3.92A The skull: anterior aspect (Norma frontalis). Compare with Key in 3.92B, and also with 3.94 and 3.95.

unidentified, is also established. It appears likely that siting of a brain in a rigidly maintained space is a factor in such mechanisms, despite lack of exact evidence. Of course, the cranial cavity is not closed: cerebrospinal fluid passes freely through the foramen magnum, displaceable in either direction. Even the brainstem can be displaced through the same orifice by raised intracranial pressure. Variability in volume of fluid in the cerebral ventricular system and numerous connections between intra- and extracranial veins add to the complexity of the situation. Nevertheless, it appears undeniable that enclosure of the brain in an otherwise invariable space must be a factor in control of cerebral circulation. This peculiar location of the brain entails some penalties, however well-adapted to ordinary circumstances: lesions which occupy space can raise pressure within the cranium far more easily than elsewhere and with far more devastating effects.

Even in primitive vertebrates the *neurocranium*, derived from cartilages ventral to brain, as in human embryos (p. 164), has been joined by cartilaginous supports for external nares and olfactory receptors, eyeballs and labyrinths. With the addition of jaws, of branchial origin, and dermal or 'membrane' bones, arising in subcutaneous mesenchyme over the head and jaws and in the buccal roof, the vertebrate skull is defined in all its complexity.

Cartilaginous elements are usually ossified, except in minor groups such as sharks, where cartilage is considered degenerate rather than primitive. Moreover, some dermal elements are as ancient, judging by fossil records, as 'cartilage' bones, and may even have preceded them. It is misleading to regard cartilage as the primeval substance in this respect. Bone may arise by two major routes: by direct ossification in mesenchyme or indirectly in cartilage, itself derived from mesenchyme. Roof and sides of the neurocranium are developed in mesenchymal sheets, or membranes, in or subjacent to dermis; hence the terms 'membrane' or 'dermal' bone, the latter perhaps being more informative. Dermal parts of the cranial box are currently considered at least as primitive in origin as its chondrocranial base. In jaws, events appear more certain: primary cartilaginous elements, derived from the branchial apparatus, have been replaced by dermal bone.

The 'capsules', enclosing special sensory organs and integrated into the skull, are obviously protective but also confer less obvious but more basic advantages. Primitive nostrils in lower vertebrates are gustatory and olfactory; even when they lead into nasal cavities after development of a secondary palate in land vertebrates, olfactory function persists with new respiratory arrangements and both functions require open airways, whatever

340

sphincteric mechanisms are added. Circumocular skeletal elements have long provided sockets, more or less complete, containing not only eyeballs but also their muscles. The latter, in attachments to this optic 'capsule', have varied only in minor details since their appearance in earliest vertebrates—judging from extant forms. Their effects are largely rotatory, and it is difficult to envisage how such actions could be effective without a socket, which not only aids ocular location during rotation but also ensures stability of interocular distance, a necessary prelude to binocular vision.

The siting of labyrinths deep in cartilage or bone entails a fixed relationship between three semicircular canals; fusion of otic capsules into the cranial base determines the strict relation of all six canals both to each other and to the head itself, and without this no orderly correlation could evolve between these receptors and central nervous connections.

Axial muscles extending from the vertebral column to caudal aspects of the cranium have become more elaborate and massive in land animals with the development of necks and increasing problems of cranial suspension in quadrupeds. In all but jawless fishes (*Agnatha*, such as lampreys, a numerically insignificant class) the primitive skull is invaded by the ligamentous and muscular apparatus of the jaws, the maxillae being integrated into the skull at an early stage. These large muscles, both axial and mandibular, transmit strains to the skull which are sometimes very great and associated with extensive cranial modifications to resist and absorb stress. Even in humans, with modest masticatory and neck musculatures, the whole body can be suspended from the bite of the teeth. In other mammals, especially quadrupeds, greater relative size of jaws and weight of head lead to development of large plates, bars and buttresses, which owe little in origin to the protection which they fortuitously offer. A perfect example is the high cranial dome of elephants, which is associated with production of a large attachment for massive extensor muscles necessary to suspension of so heavy a head. Perhaps enough has now been said to emphasize the diversity of cranial function beyond its obvious protectiveness; exclusive emphasis upon protection, characteristic of anatomical texts, is purblind and belittles the skull's multifunctional adaptations.

General Cranial Features

The term *cranium* is sometimes reserved for the skull without its mandible but this strict usage is not adhered to here. Its upper part is a box enclosing the brain, often termed the *calvaria*, the remainder being the *facial skeleton*; its upper part is immovably fixed to the calvaria, the lower being the mobile mandible. The skull is clearly of greater practical interest, viewed as a whole, than its constituent bones. Nevertheless, the general positioning of these must first be considered (**3**.89A,B,90,91A,B,92A,B).

The skull may be viewed from above (*norma verticalis*), below (*norma basalis*), behind (*norma occipitalis*), the front (*norma frontalis*) and the side (*norma lateralis*). The calvarial roof or *calva* (skull cap) must be removed to examine its interior. In the erect attitude the lower margins of orbital openings and upper margins of external acoustic meatuses are near the same horizontal (Frankfurt plane), an important convention of orientation in description.

The forehead is formed by the *frontal bone* (**3**.89A,B,90,92A,B), passing back in the vault of the skull to the *coronal suture* to meet the anterior borders of the *parietal bones*, right and left, which form most of the cranial vault, articulating together at the median,

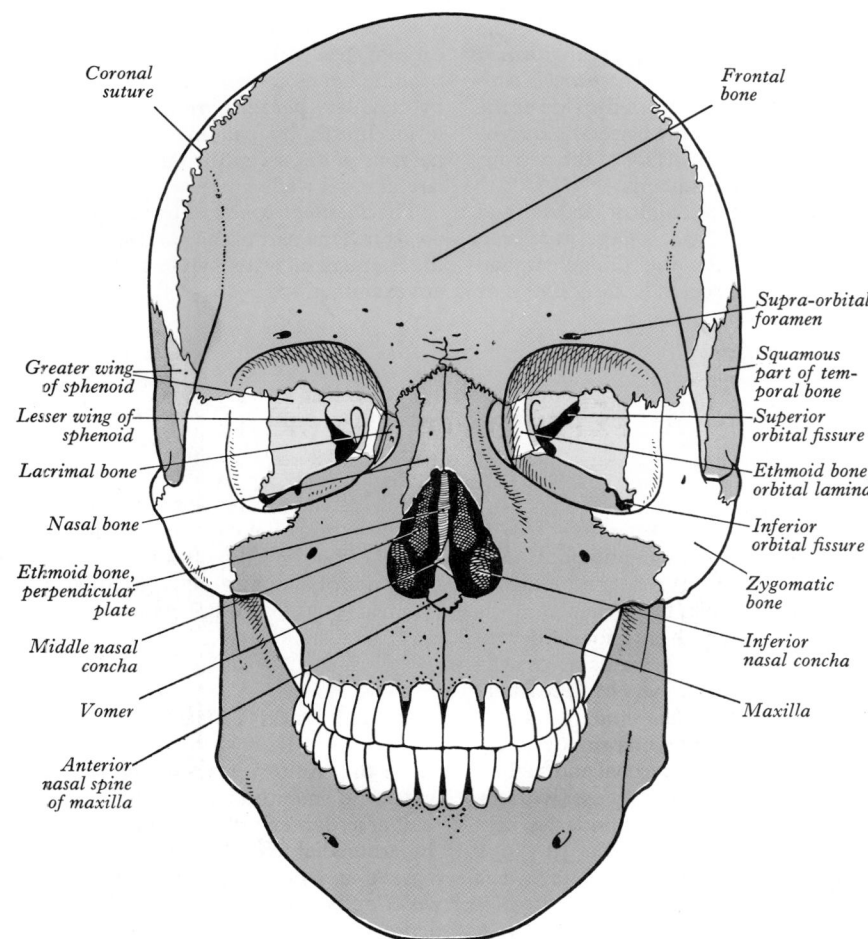

Coronal suture

Frontal bone

Greater wing of sphenoid

Lesser wing of sphenoid

Lacrimal bone

Nasal bone

Ethmoid bone, perpendicular plate

Middle nasal concha

Vomer

Anterior nasal spine of maxilla

Supra-orbital foramen

Squamous part of temporal bone

Superior orbital fissure

Ethmoid bone, orbital lamina

Inferior orbital fissure

Zygomatic bone

Inferior nasal concha

Maxilla

92B The skull: anterior aspect (Norma frontalis). Key to 3.92A. Blue = frontal bone; yellow = sphenoid bone; green = maxillae; brown = lacrimal, nasal, temporal bones and vomer; *magenta* = mandible; uncoloured = parietal, zygomatic and ethmoid bones.

serrated, *sagittal suture*. Posteriorly they meet the *occipital bone*, which forms the back of the skull (occiput). The suture between parietal and occipital bones is called *lambdoid*, after the Greek capital letter lambda, Λ, which it resembles in shape. Each parietal bone curves down as the side of the vault to the upper limit of the *greater wing* of the *sphenoid bone* in front, and the *squamous part* of the *temporal bone* behind. When the *calva* is removed, the saw cuts through the frontal and usually across the lower part of the parietal bones but may involve the temporal *squamae*; posteriorly the section cuts the occipital bone. The calva thus consists of a large part of the frontal bone, most of the two parietals, possibly small parts of temporal squamae and part of the occipital. When it is removed, the *calvarial floor or base of the skull* is revealed. It shows natural subdivision into three regions: anterior, middle and posterior *cranial fossae* (3.109A,B). These are considered in detail on pp. 361, 362, 364; here, as a preliminary, is a brief synopsis of their principal osseous boundaries.

The anterior cranial fossa, a little less than the base's anterior third, is limited behind by a sharp edge on each side. Note that it roofs the orbits and between them the nasal cavity. On each side an *orbital part* projects back from the *frontal bone* forming most of the orbital roof, these two plates being separated by a narrow interval occupied by a perforated strip, the *cribriform plate* of the *ethmoid bone*, forming much of the nasal roof; the rest of the ethmoid is in the lateral walls and septum of the nasal cavity. The cribriform plate bears a median *crista galli* on its upper surface. Posteriorly the floor of the anterior fossa is formed by parts of the *sphenoid bone*, the front of whose *body* meets the cribriform plate; on each side a narrow *lesser wing* projects laterally from it to the posterior margin of the orbital plate of the frontal, forming a sharp posterior border, the posterior limit of the floor of the anterior cranial fossa. These borders are adapted to the lateral cerebral fissures (p. 1023).

The middle cranial fossa (3.109A,B) immediately behind the anterior is of small median extent but expanded posteriolaterally on both sides. The narrow median region is formed by the sphenoid body and its cranial aspect presents a hollow for the hypophysis cerebri (pituitary gland). The lateral parts are formed by the *greater wings of the sphenoid* in front and the *petrous temporal bones* behind. Each greater wing curves from the side of the body in the base, then the side of the skull to the parietal's anteroinferior angle. Posterior to this the anterior surface of the petrous temporal bone continues laterally into the squama.

The posterior cranial fossa (3.109A,B), almost circular and occupying about two-fifths of the cranial base, is largely *occipital bone*. Through its large *foramen magnum* the brainstem and spinal cord are continuous. The anterior region is the *basilar part*

of the occipital bone, fused in front with the body of the sphenoid. On each side the fossa is formed by the posterior surface of the petrous temporal bone above and *lateral (condylar) part* of the occipital bone below. The temporal's *mastoid part*, posterolateral to the petrous, joins the occipital squama to complete the fossa.

In frontal view (*norma frontalis*) 3.92A,B,95) the *orbits* and *anterior nasal aperture* are apparent. Inferiorly is the *mandibular body*; above this are the *maxillae*, or upper jaws, separated by the teeth. The maxillae form much of the buccal roof and the inferolateral margin of the anterior nasal aperture; each forms the inferomedial orbital margin (completed laterally by the zygomatic bone) and a *frontal process* ascends to the frontal in the medial orbital margin. Between the bilateral maxillary frontal processes are two *nasal bones*, the upper boundary of the nasal aperture.

In lateral view (*norma lateralis*) (3.89A,B,91) the *mandibular ramus* ascends from the posterior end of its body to the cranial base. The *mandibular head*, surmounting the posterior border of the ramus, fits the *articular fossa* on the inferior aspect of the temporal squama. It is separated from the *external acoustic meatus* by the temporal *tympanic plate*. Anterosuperior to the meatus the temporal *zygomatic process* reaches to the zygomatic bone, forming the *zygomatic arch*, or *zygoma*, separated widely from the side of the skull. The *zygomatic bone* forms the prominence of the cheek and inferolateral orbital margin, ascending in the lateral orbital margin to the frontal bone.

With mandible removed (3.103A,B,C) it is easier to see behind the maxilla, the *pterygoid process* projecting down from the sphenoid at the root of its greater wing as a large *lateral pterygoid plate* reaching dental level, and a smaller *medial plate* ending as a hamulus. The plates are walls of the *pterygoid fossa*.

The inferior cranial aspect (*norma basalis*) (3.103A,B,104A,B), the exterior of its base, shows posteriorly the *occipital bone* and foramen magnum, lateral to which the occipital articulates with *mastoid parts* of the temporal bones and anterolaterally with their *petrous parts*, extending forwards almost to the roots of the pterygoid processes. Anteriorly the *osseous palate*, part of the buccal roof, lies within the maxillary dental arch; both maxillae and palatine bones contribute to it, its anterior three-fourths formed by maxillary *palatine processes*, meeting in the midline, the posterior fourth by palatine *horizontal plates*. The *perpendicular palatine plates* ascend from the horizontal plates as parts of the lateral nasal walls.

The *lacrimal bones,* anterior in the medial orbital walls, the *vomer*, a large part of the nasal septum, and the *inferior conchae*, in the lateral nasal walls, can be seen effectively when orbits and nose are examined (pp. 346, 365).

THE EXTERIOR OF THE SKULL

Norma Verticalis

Seen from above (3.93), cranial contour varies greatly but is usually ellipsoid (strictly, a modified ovoid), its greatest width nearer to its occipital pole; it displays three sutures: (1) the *coronal suture* is the junction of posterior frontal margin with anterior borders of the parietal bones, descending around and forwards across the cranial vault; (2) the *sagittal suture* is median between the interlocking medial parietal borders; (3) the *lambdoid suture* joins posterior parietal borders to the superior occipital margin, descending laterally and across the cranial vault. The coronal and sagittal sutures meet at the *bregma* and in the fetal skull (together with the temporary interfrontal suture) they form the boundaries of a diamond-shaped membrane-filled *anterior fontanelle* (p. 393). The latter persists until about 18 months after birth. The *lambda* is at the junction of sagittal and lambdoid sutures, the site of a similar *posterior fontanelle* (p. 393) which closes more rapidly.

Maximal parietal convexity is the palpable *parietal tuber (eminence)*, on each side, where norma verticalis passes into norma lateralis and occipitalis without distinct demarcation. A

parietal foramen, often absent, pierces each parietal bone near the sagittal suture about 3.5 cm anterior to the lambda. It transmits a small emissary vein from the superior sagittal sinus. An incidence of about 40–60% (in different races) is ascribed to this foramen (p. 331). Anteriorly the norma verticalis slopes into norma frontalis.

Norma Frontalis

From the front (3.92A,B,93,95) the skull appears roughly oval, wider and smooth above in its frontal region, but below this irregular and interrupted by the orbits and anterior nasal aperture. Superomedial to each orbit is a rounded *superciliary arch*, better marked in males, and between the two is a median elevation, the *glabella*, below which, where nasal bones meet frontal, is a depression at the root of the nose. The junction of internasal and frontonasal sutures is the *nasion*. Above each arch is a slightly elevated *frontal tuber* or tuberosity. These features are palpable and useful to both surgeon and anthropologist.

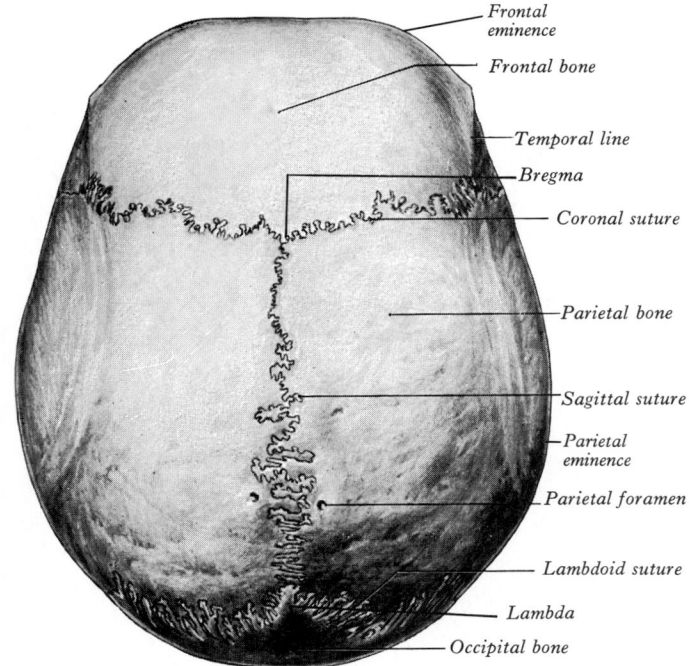

Frontal
eminence

Frontal bone

Temporal line

Bregma

Coronal suture

Parietal bone

Sagittal suture

Parietal
eminence

Parietal foramen

Lambdoid suture

Lambda

Occipital bone

3.93 The skull; superior aspect (Norma verticalis).

The orbital opening is somewhat quadrangular, its *supraorbital margin* formed entirely by the frontal bone, interrupted at the junction of its sharp lateral two-thirds and rounded medial third by the supraorbital notch (or foramen), which transmits the supraorbital vessels and nerve. The *lateral margin* is largely the frontal process of the zygomatic bone, completed above by the zygomatic process of the frontal bone; the suture between them is a palpable depression. The zygomatic bone laterally and maxilla medially form the *infraorbital margin*. Both these margins are sharp and palpable. The *medial margin*, not so obvious, is formed above by the frontal bone, below by the lacrimal crest of the maxillary frontal process, sharp and distinct only in its lower half.

The anterior nasal aperture is piriform, wider below, and bounded by nasal bones and maxillae; these articulate with each other, with their contralateral fellows and with the frontal bone above. Each nasal bone articulates behind with a maxillary frontal process; its lower border, to which the lateral nasal cartilage is attached, is the upper boundary of the nasal aperture (**3**.92A,B). The nasal cavities are, of course, *bilateral*; a *single* anterior nasal aperture presents in the macerated skull because various cartilages (septal, lateral nasal, major and minor alar) are lost during preparation.

The maxillae predominate in the facial skeleton, and their growth elongates the face between six and 12 years. Only the anterior surface is visible in norma frontalis, with a medial well-marked *nasal notch* (the lower and partly lateral border of the nasal aperture). The prominent *anterior nasal spine* marks the inter-maxillary junction in the aperture's lower boundary. It is palpable in the nasal septum. About 1 cm below the infraorbital margin is the *infraorbital foramen*, for infraorbital vessels and nerve; it is on or near a vertical through the supraorbital notch. The maxillary *alveolar process* contains sockets for upper teeth, best seen in basal view (p. 389). The short, thick *zygomatic process* from the superolateral region of the anterior surface has an oblique upper surface forming, with the zygomatic bone, a zygomaticomaxillary suture. The inferior border of the process meets the body above the first molar tooth and is palpable through cheek or buccal vestibule. The maxillary *frontal process* ascends posterolateral to the nasal bone to reach the frontal.

The *glabella* may show remains of the interfrontal suture, which ascends in about 9% of skulls to the coronal, indicating the frontal bone's formation by fusion of two halves, ossifying independently. To the medial part of the *superciliary arch* corrugator supercilii is attached; to the nasal part of the frontal bone and frontal process of the maxilla, the orbital part of orbicularis oculi. Between these the medial palpebral ligament is attached to the maxillary frontal process (**3**.95), and procerus to the nasal bone near the midline. The lower margin of the nasal bone usually bears a small notch, converted by the lateral nasal cartilage into a foramen for the external nasal nerve. In front of orbicularis oculi the levator labii superioris alaeque nasi is attached to the maxillary frontal process and, more laterally, levator labii superioris to the maxilla between infraorbital margin and foramen.

Norma Frontalis

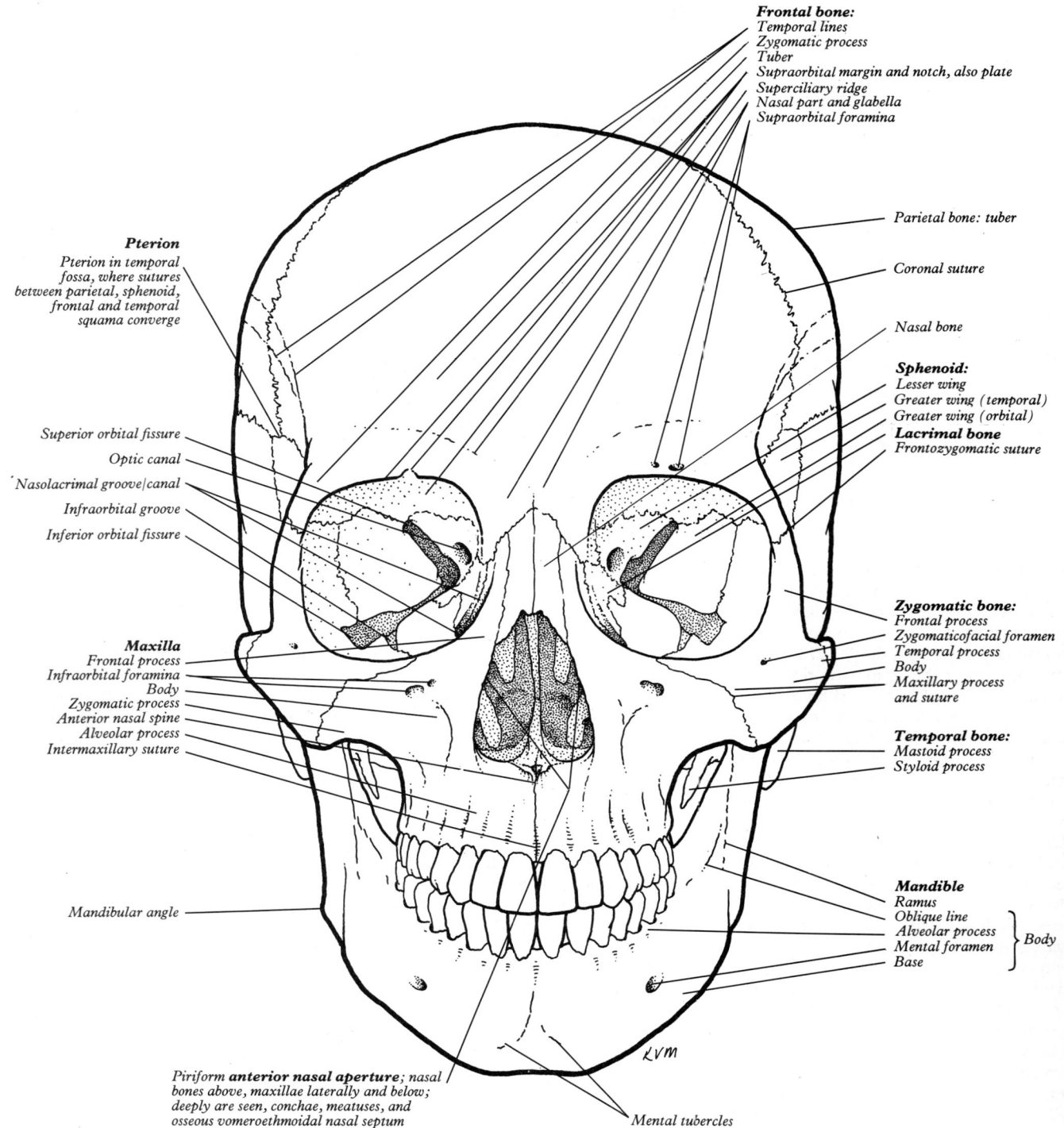

Frontal bone:
Temporal lines
Zygomatic process
Tuber
Supraorbital margin and notch, also plate
Superciliary ridge
Nasal part and glabella
Supraorbital foramina

Parietal bone: tuber

Coronal suture

Nasal bone

Sphenoid:
Lesser wing
Greater wing (temporal)
Greater wing (orbital)

Lacrimal bone
Frontozygomatic suture

Pterion
*Pterion in temporal
fossa, where sutures
between parietal, sphenoid,
frontal and temporal
squama converge*

Superior orbital fissure
Optic canal
Nasolacrimal groove/canal
Infraorbital groove
Inferior orbital fissure

Zygomatic bone:
Frontal process
Zygomaticofacial foramen
Temporal process
Body
*Maxillary process
and suture*

Maxilla
Frontal process
Infraorbital foramina
Body
Zygomatic process
Anterior nasal spine
Alveolar process
Intermaxillary suture

Temporal bone:
Mastoid process
Styloid process

Mandible
Ramus
Oblique line
Alveolar process   } *Body*
Mental foramen
Base

Mandibular angle

*Piriform **anterior nasal aperture**; nasal
bones above, maxillae laterally and below;
deeply are seen, conchae, meatuses, and
osseous vomeroethmoidal nasal septum*

KVM

Mental tubercles

3.94 (*above*) The skull: anterior aspect (Norma frontalis).

3.95 (*opposite*) The skull: anterior aspect (norma frontalis) showing muscle attachments, colour coded to indicate embryological origin and innervation. Refer to table; roman numerals indicate which cranial nerve is
involved. For spinal nerve sources consult individual muscles and
peripheral nerves.

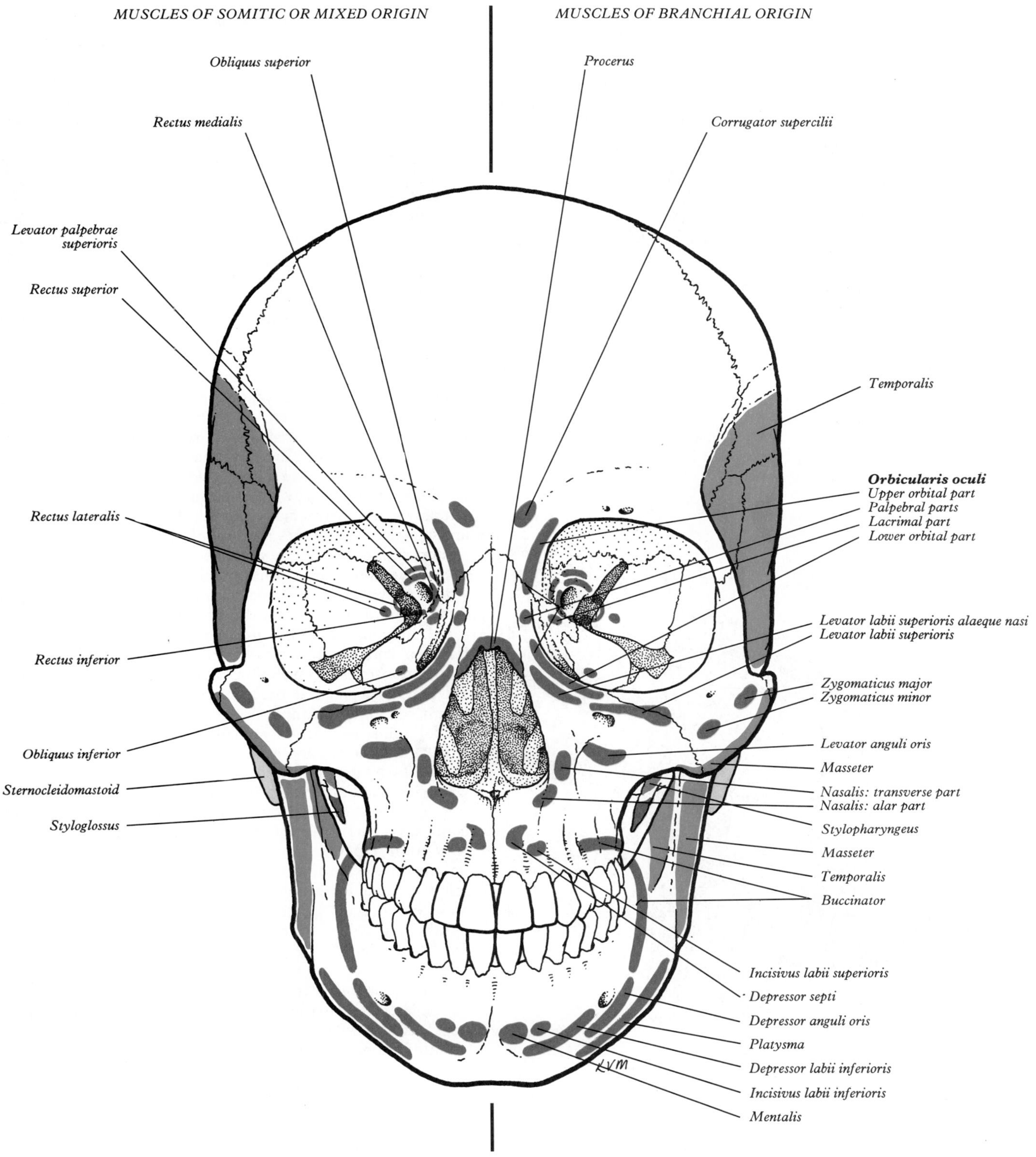

MUSCLES OF SOMITIC OR MIXED ORIGIN

MUSCLES OF BRANCHIAL ORIGIN

Obliquus superior

Rectus medialis

Levator palpebrae superioris

Rectus superior

Rectus lateralis

Rectus inferior

Obliquus inferior

Sternocleidomastoid

Styloglossus

Procerus

Corrugator supercilii

Temporalis

Orbicularis oculi
Upper orbital part
Palpebral parts
Lacrimal part
Lower orbital part

Levator labii superioris alaeque nasi
Levator labii superioris

Zygomaticus major
Zygomaticus minor

Levator anguli oris

Masseter

Nasalis: transverse part
Nasalis: alar part

Stylopharyngeus

Masseter

Temporalis

Buccinator

Incisivus labii superioris

Depressor septi

Depressor anguli oris

Platysma

Depressor labii inferioris

Incisivus labii inferioris

Mentalis

COLOUR CODING OF ATTACHMENTS OF MUSCLE GROUPS BASED ON DERIVATION AND INNERVATION

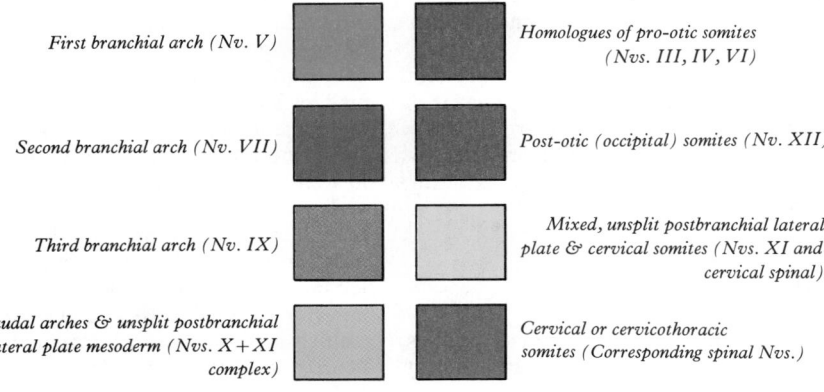

First branchial arch (Nv. V)	Homologues of pro-otic somites (Nvs. III, IV, VI)
Second branchial arch (Nv. VII)	Post-otic (occipital) somites (Nv. XII)
Third branchial arch (Nv. IX)	Mixed, unsplit postbranchial lateral plate & cervical somites (Nvs. XI and cervical spinal)
Caudal arches & unsplit postbranchial lateral plate mesoderm (Nvs. X + XI complex)	Cervical or cervicothoracic somites (Corresponding spinal Nvs.)

345

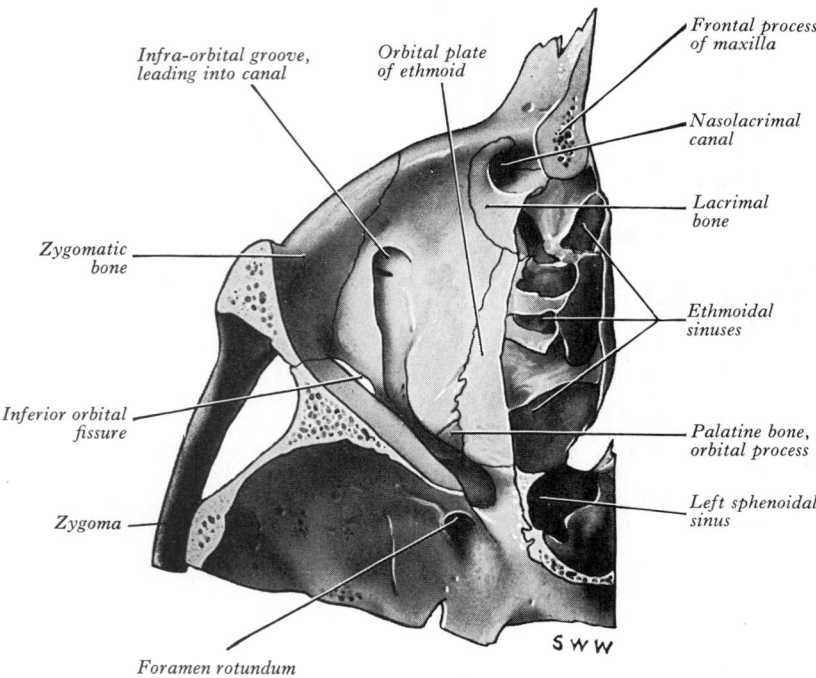

3.96 The right orbit: anterior aspect.

Supra-orbital notch
Lesser wing of sphenoid
Optic canal
Anterior ethmoidal foramen
Orbital plate of ethmoid
Nasal bone
Posterior lacrimal crest
Orbital process of palatine bone
Infra-orbital groove and foramen
Zygomatic bone
Zygoma
Inferior orbital fissure
Superior orbital fissure
Greater wing of sphenoid
Supra-orbital margin

Infra-orbital groove, leading into canal
Orbital plate of ethmoid
Frontal process of maxilla
Nasolacrimal canal
Lacrimal bone
Ethmoidal sinuses
Palatine bone, orbital process
Left sphenoidal sinus
Zygoma
Inferior orbital fissure
Zygomatic bone
Foramen rotundum

3.97 Horizontal section through the left orbit and nasal cavity viewed from above.

A *canine eminence*, due to the tooth's large root, appears between the lateral *canine* and medial *incisive fossae*. Levator anguli oris is attached in the canine fossa whilst, to the surface bordering the nasal notch, nasalis and depressor septi and below them the incisive muscle gain attachment.

In the zygomatic bone, near the junction of inferior and lateral orbital margins, is a small *zygomaticofacial foramen* (3.99B, 164), sometimes duplicated, for the so-named nerve and artery; below the foramen are attached zygomaticus minor and, more laterally, zygomaticus major. The foramen was absent in 12–30% of different populations in a series of 580 crania (p. 396).

The Orbital Cavity

The orbits (3.96,97,99A,B) contain the eyes, associated muscles, vessels and nerves, lacrimal apparatus, fascial strata and soft fat.

Each is pyramidal, with a base at the orbital opening and a long, posteromedially directed axis. It has a roof, floor, medial and lateral walls, a base and apex.

The superior wall or roof is a thin, frontal plate, gently concave on its orbital aspect which lies largely between orbital contents and brain in the anterior cranial fossa. Anteromedially it is bilaminated by the frontal sinus (q.v.); here it presents the *trochlear fovea* or spine where the annular pulley for obliquus oculi superior is attached (p. 383, 1208). Anterolaterally is a deep *lacrimal fossa* for the lacrimal gland's orbital part. Posteriorly, at the junction of its roof and medial wall, the *optic canal or foramen* connects orbit to the middle cranial fossa and transmits the optic nerve and ophthalmic artery. Near superior, medial and lower margins of the orbital opening of the canal the *common tendinous ring* of the four recti is attached to bone (7.329).

The medial wall (3.99A) is exceedingly thin except posteriorly and curves inferolaterally into the floor. Anterior is the vertical *lacrimal groove* for the lacrimal sac, opening below into the nasal cavity via the laterally inclining *nasolacrimal canal*, little more than 1 cm long. The floor of the groove separates orbital and nasal cavities, but more posteriorly ethmoidal sinuses intervene. The medial wall is related most posteriorly to the anterior region of the sphenoidal sinus, forming its lateral wall.

The inferior wall or floor (3.97) is thin and largely roofs the maxillary sinus (3.99A,B). Not quite horizontal, it ascends a little laterally. Anteriorly it curves into the lateral wall; posteriorly they are separated by the *inferior orbital fissure,* connecting the orbit posteriorly to the pterygopalatine fossa, and more anteriorly to the infratemporal fossa. The maxillary nerve traverses the fissure, whose medial lip is notched by the *infra-orbital groove*, passing forwards and sinking into the floor to become the *infra-orbital canal*, opening at an *infra-orbital foramen*. Groove, canal and foramen contain the infra-orbital nerve. The inferior orbital fissure transmits a connection between inferior ophthalmic vein and pterygoid plexus in the infratemporal fossa. The infra-orbital foramen is sometimes double, even multiple (Harris 1933), *accessory* foramina being usually smaller and recorded at incidences of 2–18% in various populations (p. 396).

The lateral wall (3.99B) is thickest, especially posteriorly where it separates the orbit from the middle cranial fossa. Anteriorly it separates the orbit and temporal fossa. Lateral wall and roof are continuous anteriorly but separated posteriorly by the *superior orbital fissure*, tapering laterally but widened at its medial end (3.96), its long axis descending posteromedially. It communicates with the middle cranial fossa and transmits the oculomotor, trochlear and abducent nerves, branches of the ophthalmic nerve and the ophthalmic veins. Where the fissure begins to widen, its inferolateral edge shows a projection, often a spine, for the lateral attachment of the annular tendon (p. 7.329). Royle (1973) has described an 'infra-orbital' sulcus, in 22 of 64 orbits, from the superolateral end of the superior orbital fissure towards the orbital floor, associated sometimes with an anastomosis between the middle meningeal and infra-orbital arteries; this feature has been confirmed in 45% of 100 orbits by Santo Neto et al (1984).

The boundaries of the *orbital opening* have already been described (p. 343). The *apex* of the orbit is near the medial end of the superior orbital fissure. Further details of the orbital walls are appended here.

The roof, almost entirely frontal orbital plate, includes posteriorly a part of the inferior aspect of the lesser sphenoidal wing. The suture between these is almost horizontal. The *optic canal* is between the roots of the lesser wing, bounded medially by the sphenoid body. As noted, near the junction of roof and medial wall, close to the orbital opening, a *trochlear fovea* or spine marks attachment of the fibrous loop for the superior oblique's tendon.

To the medial wall (3.99A), limited in front by the *anterior lacrimal crest* on the maxilla's frontal process, orbicularis oculi and lacrimal fascia are attached. Behind this crest is the maxillolacrimal suture in the *lacrimal groove's* floor. The *nasolacrimal canal's upper opening* is completed laterally by the *lacrimal hamulus*, curving anteromedially to the lower part of the anterior lacrimal crest. To the groove's *posterior lacrimal crest* (mostly lacrimal bone) are attached the lacrimal part of orbicularis oculi (p. 571) and lacrimal fascia, bridging the groove. Posterior the lacrimal's orbital surface is flat and articulates by a vertical suture with the the ethmoid labyrinth's orbital plate. The fronto-lacrimal and lacrimomaxillary sutures limit the medial wall in front; the ethmoid *orbital plate* contributes most to it. Almost rectangular, it is very thin, forming lateral walls to ethmoidal sinuses. Above, it articulates with the medial edge of the frontal orbital plate at a suture interrupted by *anterior* and *posterior ethmoidal foramina*, whose canals transmit their vessels and nerves (the posterior is often absent); they lead into the anterior cranial fossa at the lateral edge of the cribriform plate. Below, the ethmoid plate articulates with the medial edge of the maxilla's orbital surface and posteriorly with the palatine orbital process. Posteriorly it articulates with the sphenoid's body which forms the orbit's medial wall posteriorly, separated from the orbital roof by the optic canal. Analysis of racial and sexual variation in position and incidence of ethmoidal canals in 580 crania from several populations (p. 396), showed the anterior foramen lay outside the fronto-ethmoidal suture in 10–20% of several modern races and 62% out of 53 Peruvian crania.

The floor of the orbit (**3.97**), mostly formed by maxilla and zygomatic bone anterolaterally, contains posteromedially, adjoining the medial wall, a triangular area from the palatine orbital process. The *inferior orbital fissure* transmits the maxillary nerve, infra-orbital vessels, zygomatic nerve and rami of the pterygopalatine ganglion; it is bounded above by the greater sphenoidal wing, below by the maxilla and the palatine orbital process and laterally by the zygomatic bone or zygomaticomaxillary suture. In 35–40% of skulls, maxilla and sphenoid meet at the fissure's anterior end, excluding the zygomatic. Anteromedially, lateral to the lacrimal hamulus, a small maxillary depression may mark the attachment of the inferior oblique muscle.

The lateral wall (**3.99B**) is formed by orbital surfaces of the greater sphenoidal wing and anteriorly by the zygomatic bone's frontal process, meeting at the sphenozygomatic suture. This zygomatic surface has openings of minute canals for zygomatico-facial and zygomaticotemporal nerves, the former near the junction of floor and lateral wall, the latter at a slightly higher level, sometimes near the suture. The *superior orbital fissure* lies between the greater wing (below) and lesser wing (above) of the sphenoid, its body being medial. For contents see p. 346 and **7.347**.

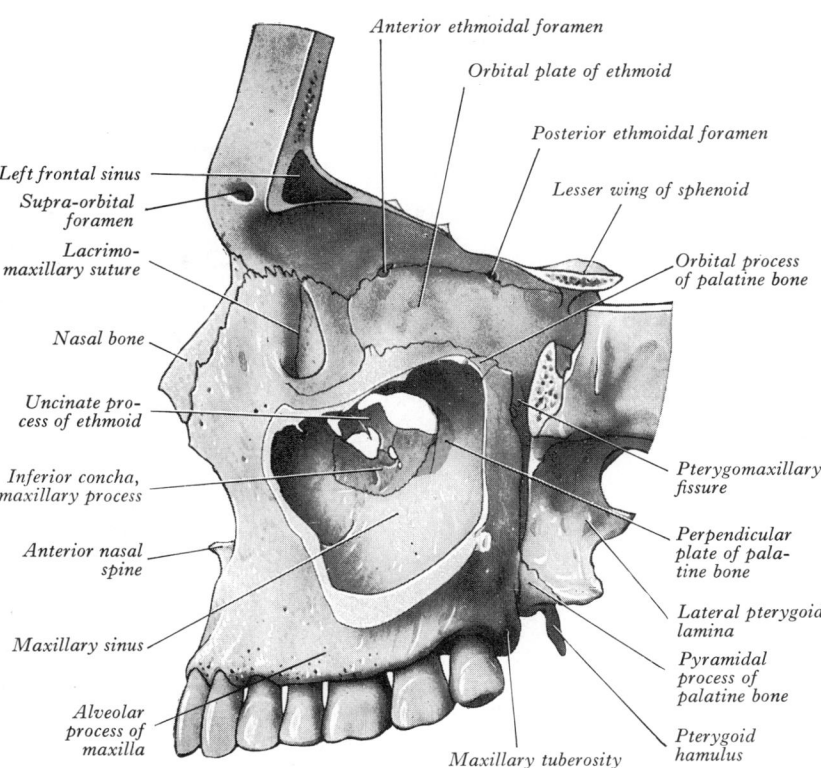

3.99A Oblique parasagittal section through the anterior part of the skull, showing the medial wall of the left orbit and the medial wall of the left maxillary sinus.

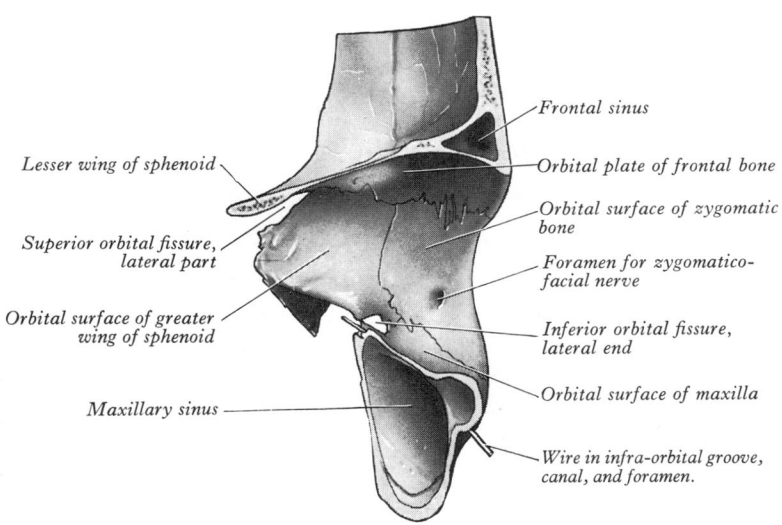

3.99B The lateral wall of the left orbit, viewed from the medial side. Compare with **3.99A**, which represents the opposite part of the same section of the skull.

Norma Occipitalis

The skull's posterior aspect is convex above and laterally flatter below. The *lambdoid suture*, entirely visible, is deeply serrated but less so inferolaterally. Inferiorly it meets *occipitomastoid* and *parietomastoid sutures* at the posteroinferior parietal angle (**3.89,90**). Sutural bones are common at the *lambda* (meeting of lambdoid and sagittal sutures) and along the lambdoid suture (pp. 373, 386). The main feature is the *external occipital protuberance* (**3.89**) and associated ridges. This is median and may overhang, being easily palpable at the upper end of the median posterior nuchal furrow. The *superior nuchal lines*, often sharp, pass laterally from the protuberance at the junction of scalp and neck; the

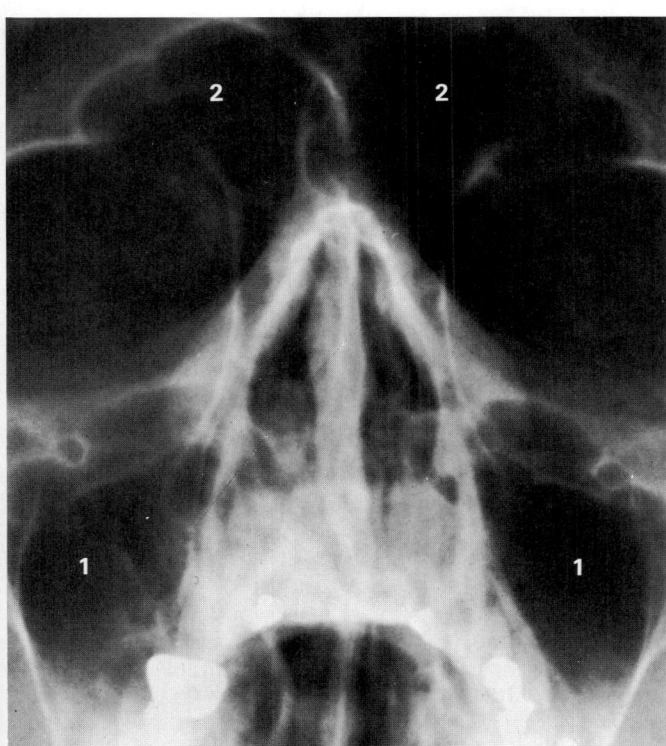

3.98 Radiograph of adult skull: frontal view. 1. Maxillary sinus. 2. Frontal sinus. Both infra-orbital canals are shown.

Norma Occipitalis

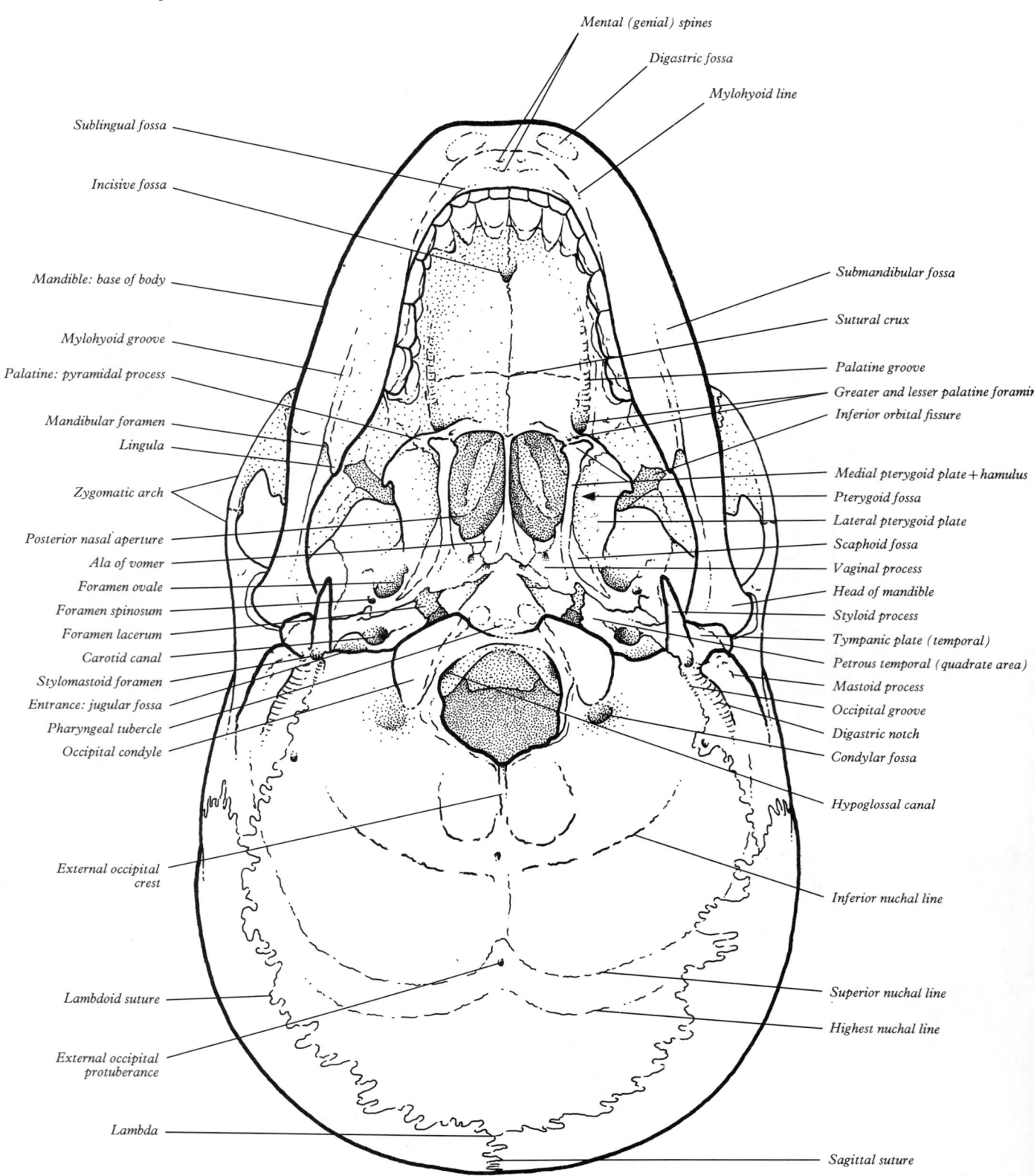

Mental (genial) spines

Digastric fossa

Mylohyoid line

Sublingual fossa

Incisive fossa

Mandible: base of body

Mylohyoid groove

Palatine: pyramidal process

Mandibular foramen

Lingula

Zygomatic arch

Posterior nasal aperture

Ala of vomer

Foramen ovale

Foramen spinosum

Foramen lacerum

Carotid canal

Stylomastoid foramen

Entrance: jugular fossa

Pharyngeal tubercle

Occipital condyle

External occipital crest

Lambdoid suture

External occipital protuberance

Lambda

Submandibular fossa

Sutural crux

Palatine groove

Greater and lesser palatine foramina

Inferior orbital fissure

Medial pterygoid plate + hamulus

Pterygoid fossa

Lateral pterygoid plate

Scaphoid fossa

Vaginal process

Head of mandible

Styloid process

Tympanic plate (temporal)

Petrous temporal (quadrate area)

Mastoid process

Occipital groove

Digastric notch

Condylar fossa

Hypoglossal canal

Inferior nuchal line

Superior nuchal line

Highest nuchal line

Sagittal suture

3.100A (*above*) The skull: norma occipitalis and norma basalis with the mandible in situ.

3.100B (*opposite*) The skull: norma occipitalis and norma basalis with the mandible in situ showing muscle attachments.

occipital region below them appears foreshortened and is seen better in norma basalis. The *highest nuchal lines*, when present, curve laterally from the protuberance, about 1 cm above the superior, and are more arched. In various ethnic groups their incidence varies from 3.6 to 40% (p. 395).

(Temporomastoid regions are inferolateral and examined more satisfactorily in norma lateralis, vide infra.)

The *inion* is the summit of the external occipital protuberance to whose lower part are attached the ligamentum nuchae and to it upper part, fibres of trapezius. The latter fibres spread to th *superior nuchal line*, to which laterally (3.104B,123) are attache posterior fibres of sternocleidomastoid and, below this, spleni capitis. To the *highest nuchal line* are attached the gale aponeurotica and laterally the occipital belly of occipitofrontali

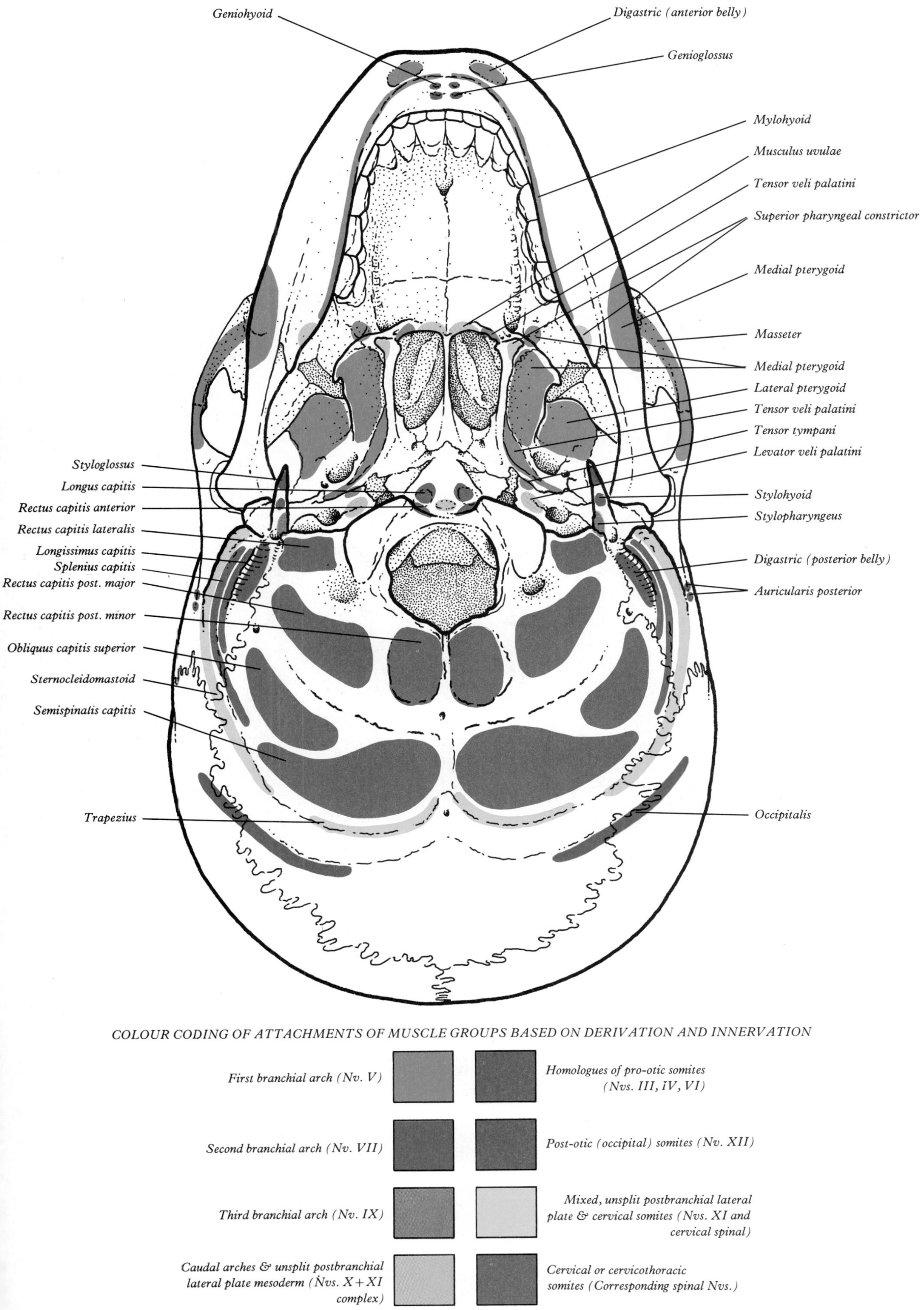

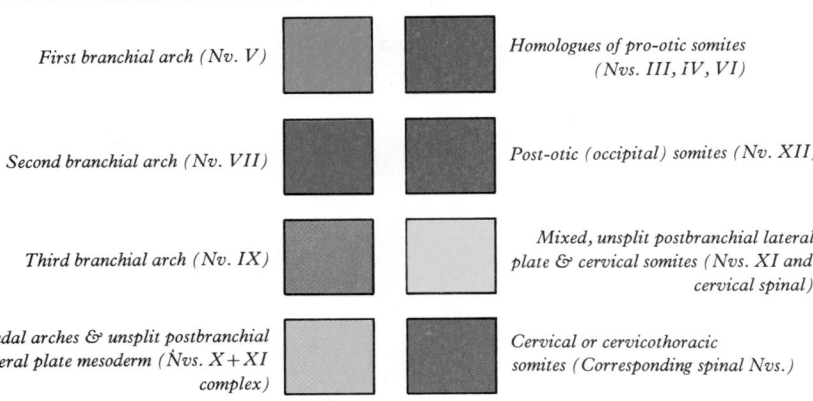

Geniohyoid

Digastric (anterior belly)

Genioglossus

Mylohyoid

Musculus uvulae

Tensor veli palatini

Superior pharyngeal constrictor

Medial pterygoid

Masseter

Medial pterygoid

Lateral pterygoid

Tensor veli palatini

Tensor tympani

Levator veli palatini

Styloglossus

Longus capitis

Rectus capitis anterior

Rectus capitis lateralis

Longissimus capitis

Splenius capitis

Rectus capitis post. major

Rectus capitis post. minor

Obliquus capitis superior

Sternocleidomastoid

Semispinalis capitis

Stylohyoid

Stylopharyngeus

Digastric (posterior belly)

Auricularis posterior

Trapezius

Occipitalis

COLOUR CODING OF ATTACHMENTS OF MUSCLE GROUPS BASED ON DERIVATION AND INNERVATION

First branchial arch (Nv. V)

Homologues of pro-otic somites (Nvs. III, IV, VI)

Second branchial arch (Nv. VII)

Post-otic (occipital) somites (Nv. XII)

Third branchial arch (Nv. IX)

Mixed, unsplit postbranchial lateral plate & cervical somites (Nvs. XI and cervical spinal)

Caudal arches & unsplit postbranchial lateral plate mesoderm (Nvs. X + XI complex)

Cervical or cervicothoracic somites (Corresponding spinal Nvs.)

Norma Lateralis

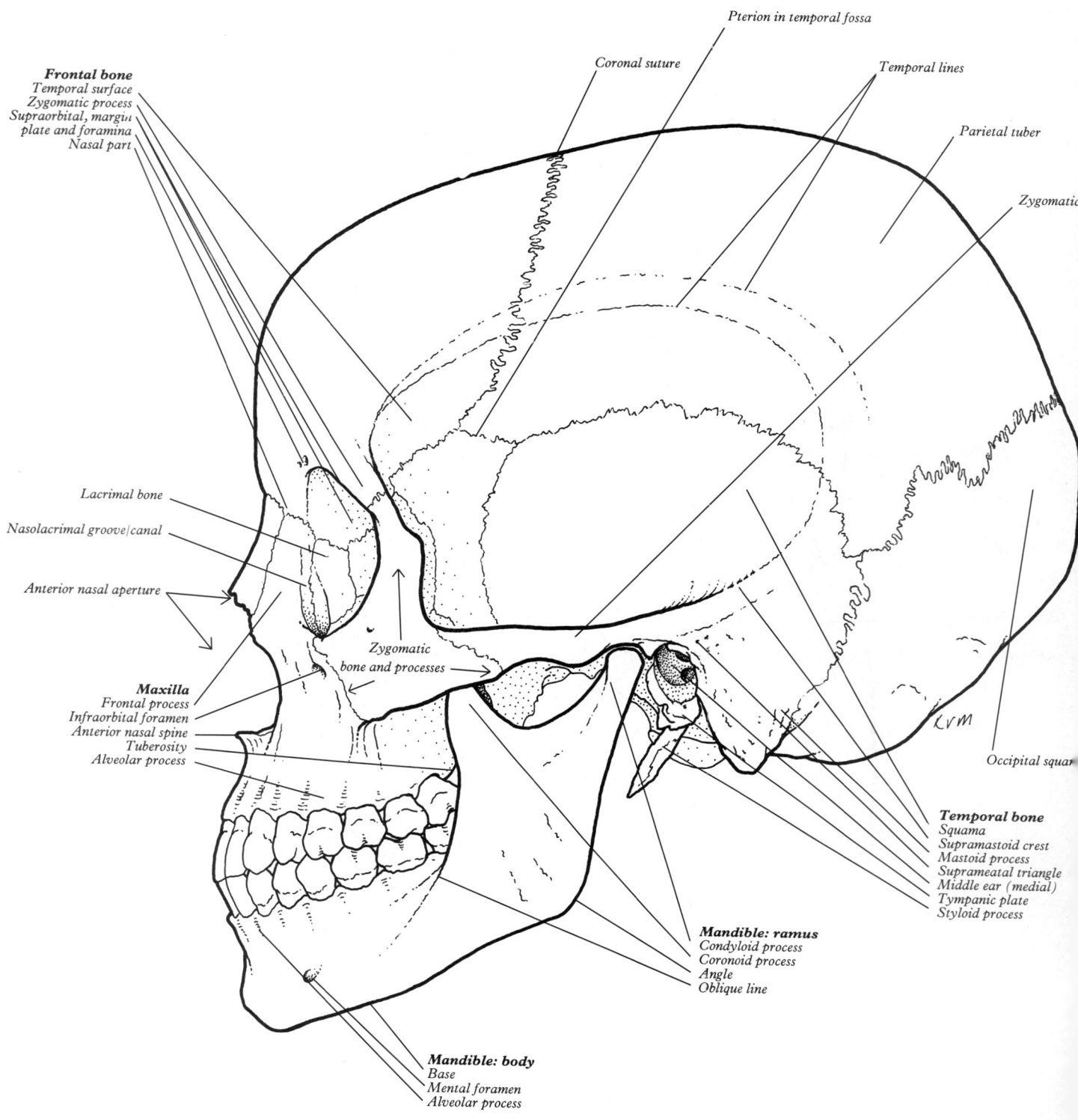

Frontal bone
Temporal surface
Zygomatic process
Supraorbital, margin
plate and foramina
Nasal part

Pterion in temporal fossa

Coronal suture

Temporal lines

Parietal tuber

Zygomatic (

Lacrimal bone

Nasolacrimal groove/canal

Anterior nasal aperture

Zygomatic
bone and processes

Maxilla
Frontal process
Infraorbital foramen
Anterior nasal spine
Tuberosity
Alveolar process

Occipital squa

Temporal bone
Squama
Supramastoid crest
Mastoid process
Suprameatal triangle
Middle ear (medial)
Tympanic plate
Styloid process

Mandible: ramus
Condyloid process
Coronoid process
Angle
Oblique line

Mandible: body
Base
Mental foramen
Alveolar process

3.102A (*opposite*) The skull: norma lateralis showing muscle attachmen

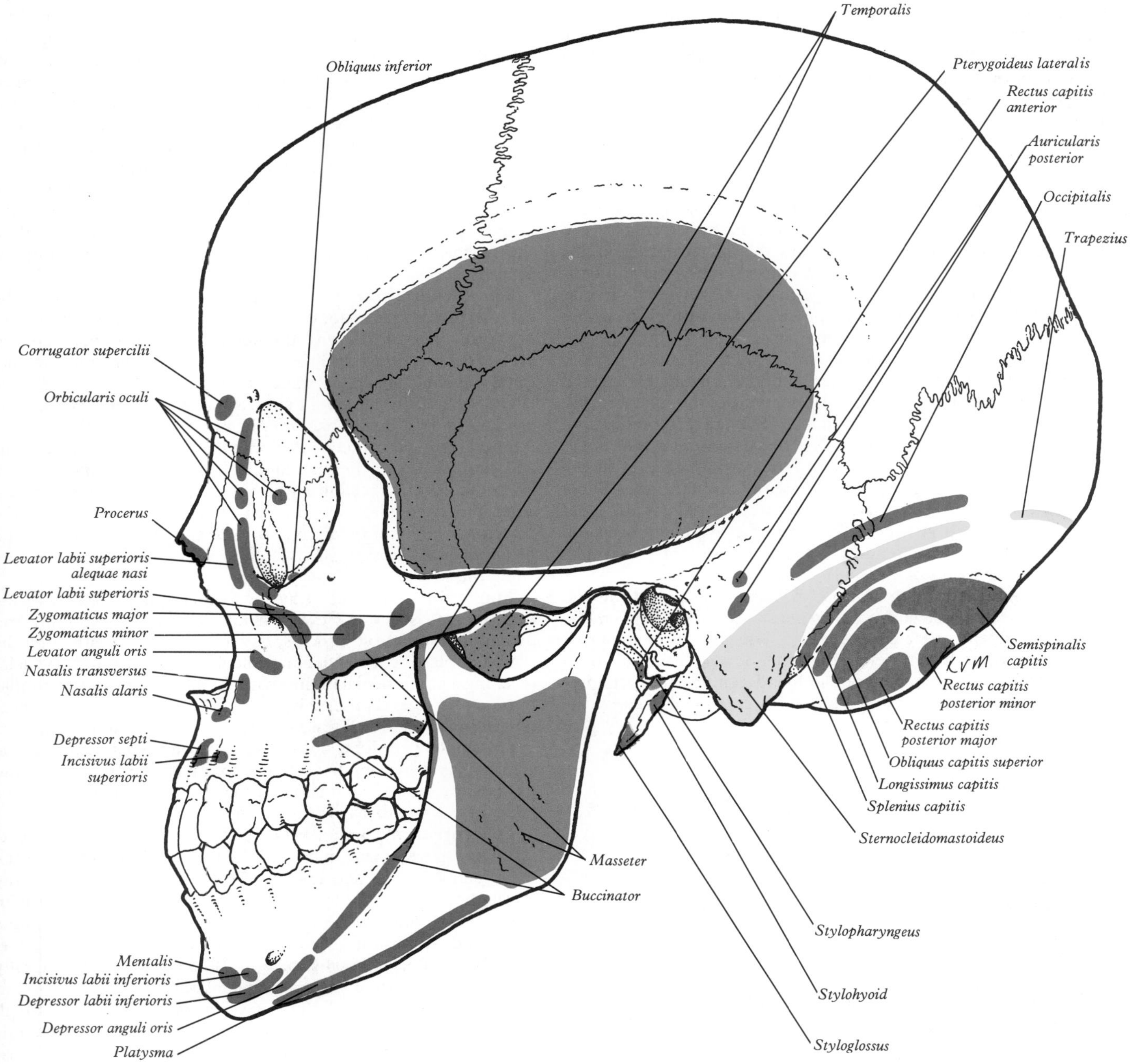

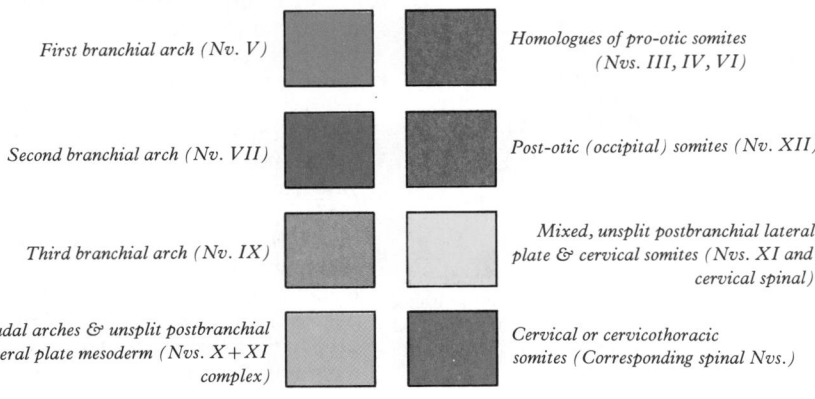

Obliquus inferior

Temporalis

Pterygoideus lateralis

Rectus capitis anterior

Auricularis posterior

Occipitalis

Trapezius

Corrugator supercilii

Orbicularis oculi

Procerus

Levator labii superioris
alequae nasi

Levator labii superioris

Zygomaticus major

Zygomaticus minor

Levator anguli oris

Nasalis transversus

Nasalis alaris

Depressor septi

Incisivus labii
superioris

Masseter

Buccinator

Semispinalis
capitis

Rectus capitis
posterior minor

Rectus capitis
posterior major

Obliquus capitis superior

Longissimus capitis

Splenius capitis

Sternocleidomastoideus

Stylopharyngeus

Stylohyoid

Styloglossus

Mentalis

Incisivus labii inferioris

Depressor labii inferioris

Depressor anguli oris

Platysma

COLOUR CODING OF ATTACHMENTS OF MUSCLE GROUPS BASED ON DERIVATION AND INNERVATION

First branchial arch (Nv. V)

Homologues of pro-otic somites
(Nvs. III, IV, VI)

Second branchial arch (Nv. VII)

Post-otic (occipital) somites (Nv. XII)

Third branchial arch (Nv. IX)

Mixed, unsplit postbranchial lateral
plate & cervical somites (Nvs. XI and
cervical spinal)

Caudal arches & unsplit postbranchial
lateral plate mesoderm (Nvs. X + XI
complex)

Cervical or cervicothoracic
somites (Corresponding spinal Nvs.)

351

Norma Lateralis

Much of the side of the skull (3.89,90,101,102) has already been described from other aspects, but not its central features. Above is the *temporal line*, arching up and back from the frontal's zygomatic process across the coronal suture to the parietal bone. Salient and palpable in front, it is less distinct on the parietal and usually becomes *two curved ridges*, enclosing a smooth strip. Posteriorly the superior line fades away, but the inferior again becomes more prominent as it curves down across the temporal squama above the mastoid process, where, as the *supramastoid crest*, it continues into the processes forming the zygomatic arch. The temporal line marks the periphery of the temporalis muscle and its fascia, the muscle attachment being limited by its inferior ridge, the temporal fascia by the superior ridge.

The temporal fossa is delineated by the zygomatic arch, temporal line, frontozygomatic processes and supramastoid crest; to its floor temporalis is attached. An irregularly H-shaped meeting of sutures in the fossa has as a horizontal limb the suture between the antero-inferior parietal angle and the apical border of the greater sphenoid wing. The frontal, sphenoid, parietal and temporal squama are here all close together (3.89A): a small circular area includes parts of all four and is termed the **pterion**, whose centre, an important surgical landmark, is on average 4.0 cm above the zygomatic arch and 3.5 cm behind the frontozygomatic suture (3.89A). It marks the anterior middle meningeal arterial ramus and the axial position of the lesser wing of the sphenoid; the latter is lodged in the stem of the lateral (Sylvian) cerebral fissure. Hence the term *Sylvian point*. The fossa's anterior wall is the temporal surface of the zygomatic bone, adjoining part of the greater sphenoid wing and a small area of frontal bone. These structures separate the fossa from the orbit; inferiorly the fossa merges with the infratemporal fossa between zygomatic arch and cranial wall; here the tendon and some fibres of temporalis descend to the mandible (p. 368).

The zygomatic arch (formed by temporal and zygomatic processes) is palpable and visible where cheek and temple meet. Its sharp upper border is obscured by attachment of temporal fascia, the lower by masseter; the latter is also attached to its medial aspect. The gap between arch and temple is deeper anteriorly; here the arch is crossed obliquely down and back by the zygomaticotemporal suture.

The *zygomatic process* of the temporal bone widens as it approaches the squama, dividing into: an *anterior root* passing medially in front of the *mandibular fossa* to the smooth *articular tubercle*, the anterior boundary of the fossa; and a *posterior root* passing back, lateral to the fossa, its upper border continuing into the supramastoid crest.

The external acoustic meatus, below the posterior zygomatic root, has rough margins, especially antero-inferiorly, for attachment of meatal cartilage. Posterosuperiorly the margin is formed by the temporal squama, the rest by the temporal's *tympanic plate*. The squamotympanic suture is anterosuperior, the posterior meatal (tympanomastoid) suture usually obliterated in adults, except for a canaliculus for the auricular branch of the vagus. Below the meatus the tympanic plate projects as a rough triangular area. Posterosuperior is often a small depression with a *suprameatal spine* in its anterior margin; within is a *suprameatal triangle*, bounded above by the supramastoid crest, in front by the posterosuperior meatal margin and behind by a posterior vertical tangent to the meatal margin. This triangle is the lateral wall of the mastoid (tympanic) antrum (p. 377) and is hence of surgical interest.

The mastoid part of the temporal bone, posterior to the meatus, is continuous above with squama in front. Its upper border forms behind this with the postero-inferior parietal angle a *parietomastoid suture* and, by its posterior border with the occipital squama, an *occipitomastoid suture*. These two meet the lateral end of the lambdoid suture at the *asterion*. The *mastoid process* (3.89), a breast-like inferior projection from the mastoid temporal bone, is postero-inferior to the external acoustic meatus and palpable under the ear's lobule. The *mastoid foramen* is near or in the occipitomastoid suture; it transmits an emissary vein from the sigmoid sinus. Sutural ossicles may appear in the parietomastoid suture, often at or near the asterion, the site of a *posterolateral fontanelle* (Le Double 1903), but also elsewhere along the suture.

The styloid process (3.89), a slender spike attached to the skull's base, is better viewed in norma lateralis. Anterior and medial to the mastoid process, its base partly ensheathed by the tympanic plate, it descends anteromedially medially, its tip usually reaching a point medial to the posterior margin of the mandibular ramus. The styloid process is, however, very variably developed, ranging in length from a few millimetres to a few centimetres, often approximately straight, but on occasion curved; a ventromedial concavity is more common, a dorsal concavity is infrequent. From its apex the stylohyoid ligament descends forward to the hyoid's lesser cornu (p. 370) as a cranial suspension.

The infratemporal fossa (3.103A) is an irregular, postmaxillary space, communicating with the temporal fossa between the zygomatic arch and lower temple. Medially its roof is the infratemporal surface of the sphenoid's greater wing and part of the temporal squama. Here the greater wing displays the foramina ovale and spinosum. Medial is the lateral pterygoid plate, described in norma basalis (p. 355). Behind, below and laterally the fossa is open. Its anterior and medial walls converge below but are separated above by the *pterygomaxillary fissure*, through which the infratemporal and pterygopalatine fossae connect. The fissure is continuous above with the *inferior orbital fissure's* posterior end, by which route the infratemporal and pterygopalatine fossae connect with the orbit (p. 347).

The pterygopalatine fossa is a small pyramidal space below the orbital apex; it communicates with the infratemporal fossa via the pterygomaxillary fissure, with the nasal cavity by the sphenopalatine foramen and the orbit by the medial end of the inferior orbital fissure. The foramen rotundum, in its posterior wall, is traversed by the maxillary nerve.

Further details. The floor of the *temporal fossa* bears a few vascular furrows, the most constant above the external acoustic meatus produced by middle temporal vessels. In its anterior wall the *zygomaticotemporal foramen* opens up and backwards from the zygomatic bone's posterior surface, and it transmits the zygomaticotemporal nerve and a minute artery. The tendon of temporalis descends and the deep temporal vessels and nerves ascend deep to the muscle, between zygomatic arch and cranial wall. The temporal bone's zygomatic process bears a small *tubercle*

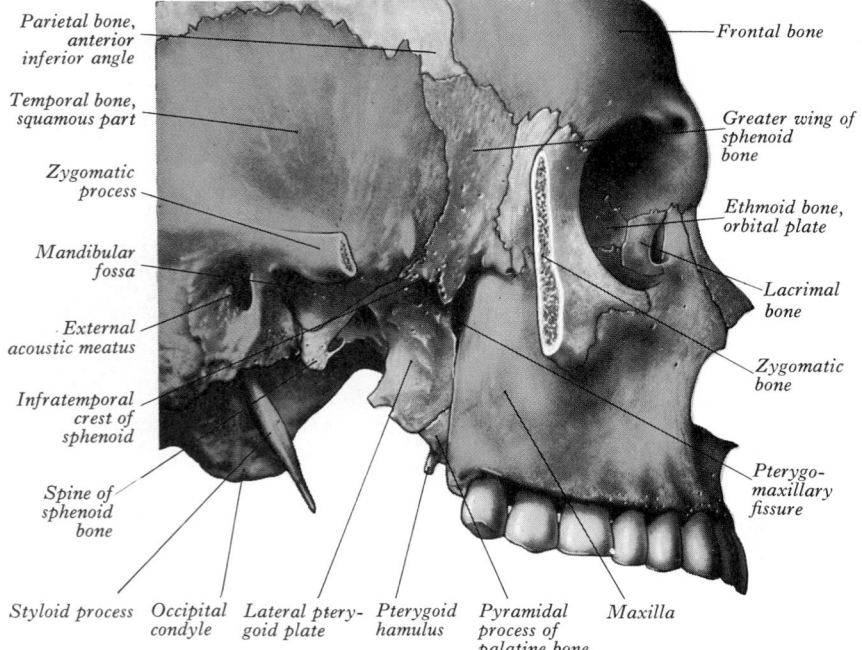

Parietal bone, anterior inferior angle

Temporal bone, squamous part

Zygomatic process

Mandibular fossa

External acoustic meatus

Infratemporal crest of sphenoid

Spine of sphenoid bone

Styloid process

Occipital condyle

Lateral ptery-goid plate

Pterygoid hamulus

Pyramidal process of palatine bone

Maxilla

Frontal bone

Greater wing of sphenoid bone

Ethmoid bone, orbital plate

Lacrimal bone

Zygomatic bone

Pterygo-maxillary fissure

3.102B The right infratemporal fossa: seen in norma lateralis after detachment of mandible and removal of zygomatic arch. Blue = frontal bone; yellow = sphenoid and lacrimal bones; brown = temporal and nasal bones; green = maxilla. The parts shown of the parietal, zygomatic, ethmoid and palatine bones are uncoloured.

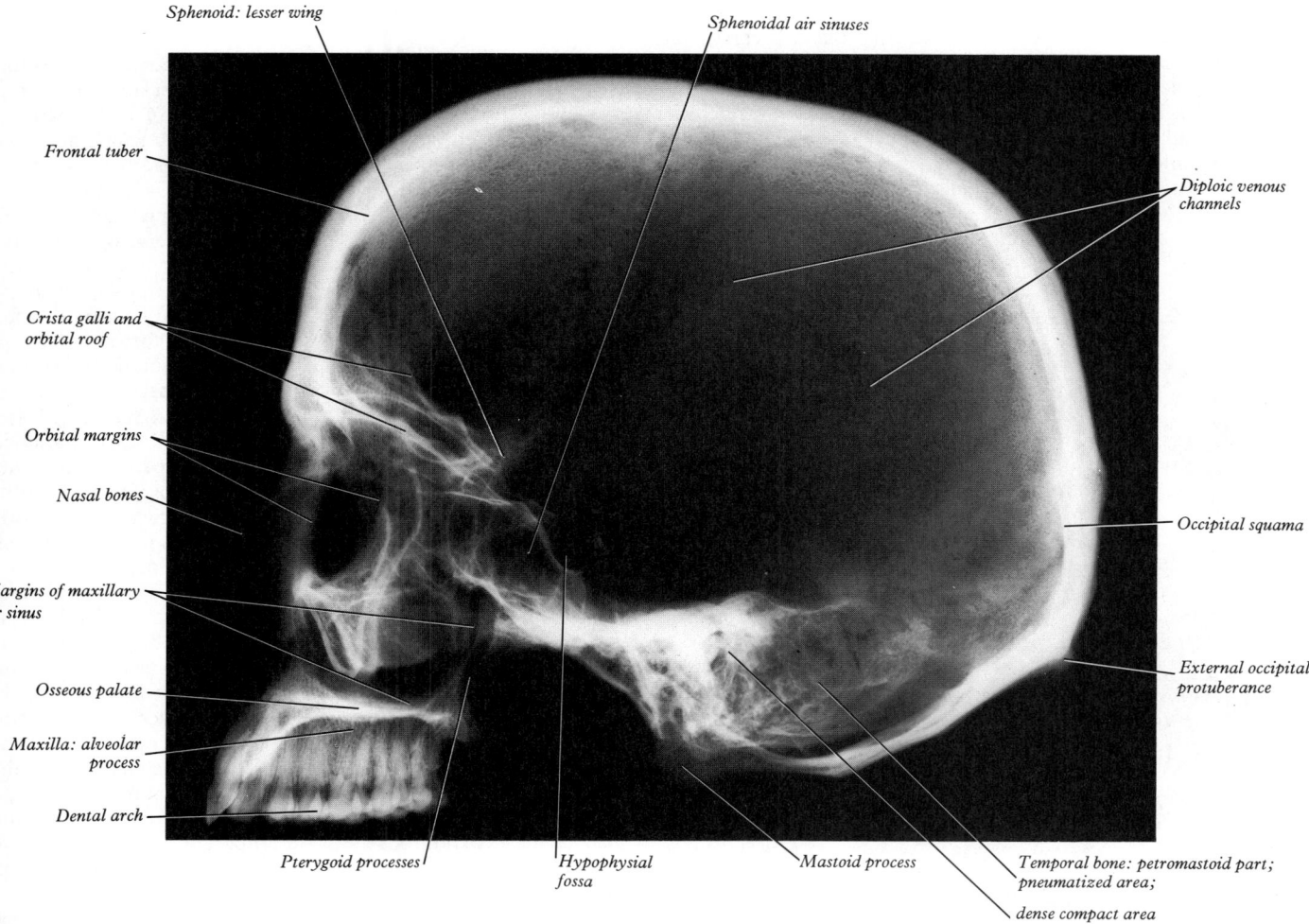

Sphenoid: lesser wing

Sphenoidal air sinuses

Frontal tuber

Diploic venous channels

Crista galli and orbital roof

Orbital margins

Nasal bones

Occipital squama

Margins of maxillary air sinus

Osseous palate

External occipital protuberance

Maxilla: alveolar process

Dental arch

Pterygoid processes

Hypophysial fossa

Mastoid process

Temporal bone: petromastoid part; pneumatized area;

dense compact area

3.102C Lateral radiograph of adult female skull, with mandible omitted.
(Supplied by P J Liepins; photography by Sarah Smith.)

of the anterior root of the zygoma, to which the lateral temporomandibular ligament is partly attached (**4.**20). It is palpable in front of the mandibular head. Behind the mandibular fossa the postglenoid tubercle descends from the zygoma's posterior root to meet the tympanic plate anterosuperiorly at the external acoustic meatus (**3.**132); its anterior aspect is a small part of the mandibular fossa.

The posterolateral surface of the *mastoid process* and its apex are attachments of sternocleidomastoid, splenius capitis and longissimus capitis, from before backwards (**3.**102,104B). Anterior and parallel to this area partially obliterated remains of the squamomastoid suture may be visible. Hence the suprameatal triangle's floor, and therefore the lateral wall of the mastoid antrum, is formed by temporal squama. The tympanomastoid fissure is on the base of the process; the mastoid canaliculus (**3.**136), which transmits the vagal auricular branch, opens in the fissure.

The *styloid process* is related laterally to the parotid gland, medially to the internal jugular vein. The stylohyoid is attached by a slender tendon to its posterior aspect near the base, styloglossus to the tip and adjacent anterior aspect, stylopharyngeus medially to its base and the stylomandibular ligament laterally near its tip. Behind its base the facial nerve emerges from the stylomastoid foramen crossing lateral to the process in the parotid gland.

The *infratemporal fossa* (**3.**103A) contains temporalis as it reaches the mandibular coronoid process and adjacent ramus; the maxillary artery and its rami, and the pterygoid venous plexus lie medial to the muscle and usually lateral to the lateral pterygoid. Deepest are the medial pterygoid, mandibular nerve and chorda tympani, the mandibular nerve entering the fossa through the foramen ovale in its roof to divide into terminal rami medial to the lateral pterygoid; these leave the fossa for other regions. The chorda tympani enters the fossa medial to the sphenoidal spine to join the lingual nerve. The maxillary nerve appears in its upper part between the pterygopalatine fossa and inferior orbital fissure. The anterior wall is pierced by small foramina for posterior superior alveolar vessels and nerves and is limited below by the maxilla's alveolar part behind the molar teeth. Here a strip of maxilla is covered by gingival mucosa and above this are attached upper fibres of buccinator, extending back on to the maxillary tuberosity. The fossa's medial wall, the lateral pterygoid plate, is completed below by the palatine bone's pyramidal process wedged between maxillary tuberosity and the plate. The superficial head of medial pterygoid is attached to the pyramidal process but also spreads to the maxillary tuberosity.

The *pterygomaxillary fissure*, between maxilla and pterygoid process, admits the maxillary artery to the pterygopalatine fossa; its uppermost part contains the maxillary nerve (p. 354).

The *pterygopalatine fossa* (**3.**103A) is bounded *behind* by the root of the pterygoid process and adjoining sphenoidal greater wing's anterior surface, *medially* by the palatine bone's perpendicular plate with its orbital and sphenoidal processes, *anteriorly* by the superomedial part of the maxilla's posterior surface. *Laterally* it connects with the infratemporal fossa via the pterygomaxillary fissure. Its main contents are the maxillary nerve, pterygopalatine ganglion and terminal rami of the maxillary artery. The pterygoid canal inferomedial to foramen rotundum transmits the pterygoid nerve and artery from the anterior wall of the foramen lacerum to the pterygopalatine ganglion; inferomedially the palatovaginal canal transmits the pharyngeal nerve and artery from the ganglion to the pharyngeal roof. In the medial wall the *sphenopalatine foramen* (**3.**115) is bounded above by the sphenoid's body, elsewhere by parts of the palatine bone, in front by its orbital

353

process, behind by its sphenoidal process, and below by the upper border of the palatine bone's perpendicular plate. It carries into the nasal cavity the nasopalatine nerve and vessels. A fifth foramen, placed inferiorly at the junction of anterior and posterior walls, leads into the *greater palatine canal* and descends between maxilla and palatine perpendicular plate. This canal transmits anterior, middle and posterior palatine nerves and greater and lesser palatine vessels, which emerge on the bony palate (vide infra).

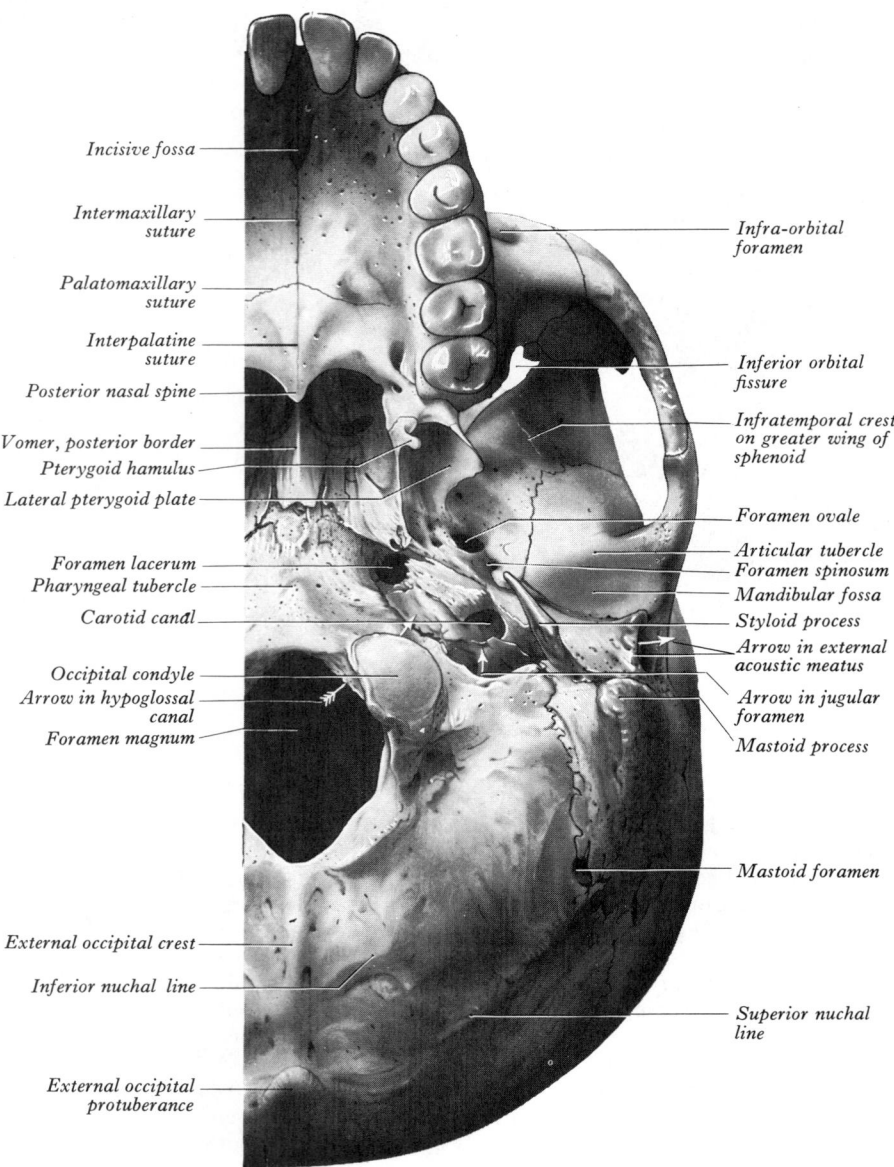

Incisive fossa

Intermaxillary suture

Palatomaxillary suture

Interpalatine suture

Posterior nasal spine

Vomer, posterior border
Pterygoid hamulus
Lateral pterygoid plate

Foramen lacerum
Pharyngeal tubercle
Carotid canal

Occipital condyle
Arrow in hypoglossal canal
Foramen magnum

External occipital crest

Inferior nuchal line

External occipital protuberance

Infra-orbital foramen

Inferior orbital fissure

Infratemporal crest on greater wing of sphenoid

Foramen ovale

Articular tubercle
Foramen spinosum
Mandibular fossa
Styloid process
Arrow in external acoustic meatus

Arrow in jugular foramen
Mastoid process

Mastoid foramen

Superior nuchal line

3.103A The inferior surface of the left half of the base of the skull with mandible removed (Norma basalis). Compare with Key, 3.103B.

Norma Basalis

The inferior cranial aspect is complex (3.103–107) extending from upper incisor teeth to superior nuchal lines of the occiput. Laterally are the post-incisor teeth, zygomatic arches and their posterior roots and mastoid processes (which reach the lateral limits of the superior nuchal lines). It is conveniently divided into anterior, middle and posterior parts, the anterior being the hard palate and alveolar arches, on a lower level than the rest, the rest being arbitrarily divided into middle and posterior parts by a transverse plane through the foramen magnum's anterior margin.

Anterior Part of Norma Basalis

The *bony palate* (3.106), within the superior dental arch, is formed by the maxillary palatine processes and palatine horizontal plates, meeting at a *cruciform suture* formed of intermaxillary, interpalatine and palatomaxillary sutures. The palate is arched sagittally and transversely, its depth and breadth variable but always greatest in the molar region. The *incisive fossa* is anterior and median; the *lateral incisive foramina*, via which incisive canals pass to the nasal cavity (p. 366), are in its lateral walls and the *median incisive foramina*, present in some skulls, open on its anterior and posterior walls. The *greater palatine foramen* is near the lateral palatal border behind the palatomaxillary suture (3.106), and a vascular groove, deep posteriorly, leads forwards from it. The *lesser palatine foramina*, usually two, behind the greater, pierce the *palatine pyramidal process* wedged between the lower ends of the medial and lateral pterygoid plates. (Accessory lesser foramina have a high incidence, with racial and sexual variation from 30–70%; see p. 396). The palate is pierced by many other small foramina and marked by pits for palatine glands. Near its sharp, gently arched bilaterally posterior border, also slightly curved, variably prominent *palatine crests* extend medially from behind the greater palatine formina. The posterior border projects back as a median *posterior nasal spine*. The *alveolar arch* has 16 sockets or *alveoli* for teeth, varying in size and depth, some single, some divided by septa in adaptation to dental roots. The lateral incisive foramen transmits terminal rami of the greater palatine vessels and nasopalatine nerve. When median incisive foramina occur, the left nasopalatine nerve traverses the anterior and the right posterior foramen. The lateral foramina are, some claim, in the line of fusion of an os incisivum (premaxilla) with the maxilla proper, as a primitive bucco-nasal communication. In young skulls a dubious bilateral suture between os incisivum and maxilla may extend from the posterior part of the incisive fossa to septa between roots of lateral incisor and canine teeth (but see p. 390).

The greater palatine nerve and vessels traverse their foramina, the vessels grooving the palate towards the incisive fossa. The lesser palatine foramina, usually two but sometimes three, contain lesser palatine nerves and vessels. To the palatine crest is attached part of the tendon of tensor veli palatini, to the posterior palatine border the palatine aponeurosis and to the posterior nasal spine musculus uvulae. These collagenous elements, although often described separately, are in fact blended as the osseous surface are approached. Margins of the median palatal intermaxillary suture are sometimes raised into a *palatine torus*, variably smooth, pitted or rough. A similar longitudinal *maxillary torus* may appear on the alveolar process, palatal to the upper molar roots.

Middle Part of Norma Basalis

This (3.103–107) extends posteriorly from the osseous palate to an arbitrary line through the anterior margin of the foramen magnum. Anteriorly the *posterior vomerine border* separates two posterior nasal apertures, behind which a posterior area of inferior sphenoid surface is continuous with that of the *basioccipital bone*, forming a broad bar sloping down to the foramen magnum. Convex transversely, wider behind, it bears, in front of the foramen a small midline *pharyngeal tubercle*, the highest attachment of the superior pharyngeal constrictor.

The pterygoid process, descending behind the third molar tooth from the junction of the sphenoid's greater wing and body, has medial and lateral pterygoid plates, separated by cuneiform *pterygoid fossa*, facing posterolaterally. Anteriorly the plates are fused, except below, where they are separated by the palatine's *pyramidal process*; sutures are usually discernible. Anteromedially the processes articulate with the posterior border of the palatine's perpendicular plate, forming a flat area in the posterior nasal aperture's lateral wall and nasopharynx. Laterally they are separated from the posterior maxillary surface by the pterygomaxillary fissure (3.103A). The *medial pterygoid plate*, narrower and projects directly back, its medial surface covered mucous membrane of the posterior nasal aperture's lateral wall and part of nasopharynx. Its posterior border is sharp, with small projection near midpoint, above which it is curved a attached to the pharyngeal end of the auditory tube; superiorly it divides to enclose the *scaphoid fossa* (**3**.104,127); below, it projects as a slender *pterygoid hamulus*, which curves laterally and is grooved anteriorly by the tendon of tensor veli palatini. The *lateral pterygoid plate* projects posterolaterally; its lateral surface is the medial wall of the infratemporal fossa. Superiorly it is continuous with the *infratemporal surface of the sphenoid's greater wing*, anterior in the roof of the infratemporal fossa. This surface, inferolateral, is almost pentagonal; anterior is the posterolateral border of the inferior orbital fissure, anterolateral the infratemporal crest. Laterally it articulates with the temporal squama; medially it is continuous with the pterygoid process and side of the sphenoid's body; posteromedially it articulates with the petrous part of the temporal bone.

The foramina ovale and spinosum, on the greater wing's infratemporal surface, transmit some large structures as follows (for further details vide infra): the *foramen ovale*, near the lateral pterygoid plate's posterior margin, transmits the trigeminal mandibular division. Posterolateral to it the *foramen spinosum* transmits the middle meningeal artery to the middle cranial fossa; it is much the smaller and circular. Posterolateral to it projects the irregular *spine of the sphenoid* whose medial surface is flat and forms, with the adjoining posterior border of the greater wing, the anterolateral wall of a groove, completed posteromedially by part of the petrous temporal bone. This groove contains the cartilaginous *pharyngotympanic (auditory) tube* and leads posterolaterally into an osseous canal for the tube in the petrous temporal bone and also anteromedially to the medial pterygoid plate's superior border. In the roof of the groove the posterior border of the greater wing and the anterior border of the petrous temporal meet at a petrosphenoidal suture. Occasionally the foramina ovale and spinosum are confluent, frequency varying from 0.7–10.4% in modern populations. Similarly, the foramen spinosum's posterior edge may be defective, with an incidence of 2–17% (p. 396). Posteromedial to this groove is the *inferior surface of the petrous temporal* between the sphenoid's greater wing and the basi-occipital bone. Anteriorly the surface is rough, its apex separated from the posterolateral aspect of the sphenoid's body by an irregular *foramen lacerum*. Behind this rough area and posterolateral to foramen lacerum, a large, almost circular foramen opens into the *carotid canal*, which ascends then turns anteromedially to reach the posterior wall of the foramen lacerum. Emerging from the canal the internal carotid artery turns up into the cranial cavity. Inferiorly the foramen lacerum is filled by fibrocartilage; this is not traversed by large structures.

Further details. From the sphenoidal spine's base the *squamotympanic fissure* runs posterolaterally between the tympanic plate and mandibular fossa, usually reaching the anterior margin of the external acoustic meatus but sometimes obliterated near its lateral end. The *mandibular fossa*, deeply concave sagittally, and transversely gently concave, is wider laterally. It contains the head of the mandible when the jaw is elevated. Anteriorly the articular surface invades a transverse rounded *articular tubercle*, continuous laterally with the zygoma's anterior root. In front is the part of the temporal squama in the roof of the infratemporal fossa. Behind the squamotympanic fissure the *temporal tympanic plate*, separating mandibular fossa and external acoustic meatus, is roughly triangular, its apex at the medial end of the fissure near the sphenoidal spine's root. Its lower border skirts the anterolateral margin of the inferior opening of the carotid canal, extending posterolaterally to the root of the styloid process. There it forms a *sheath of the styloid process*, longer and prominent laterally where the tympanic plate is fused with the rest of the temporal bone below and behind; it is free above, forming the anterior border of the external acoustic meatus.

The upper border of the *vomer*, applied to the inferior aspect of the sphenoid's body, expands into an *ala* on each side (**3**.104A,153), a median groove between them fitting the sphenoid's *rostrum*. The lateral border of each ala reaches a thin *vaginal process* projecting medially from the medial pterygoid plate. The two may merely touch or the alar edge may be overlapped inferiorly by the vaginal process, whose inferior surface bears an anteroposterior groove, converted into a canal anteriorly by the superior aspect of the sphenoidal process of the palatine

bone. This *palatovaginal canal*, opening anteriorly on the pterygopalatine fossa's anterior wall, transmits a pharyngeal branch of the pterygopalatine ganglion and a pharyngeal ramus from the third part of the maxillary artery. A second, *vomerovaginal canal*, medial to the palatovaginal, may exist between ala and vaginal process, leading into the anterior end of the palatovaginal canal.

Anterior to the pharyngeal tubercle the basi-occipital bone is related to the roof of the nasal part of the pharynx and the

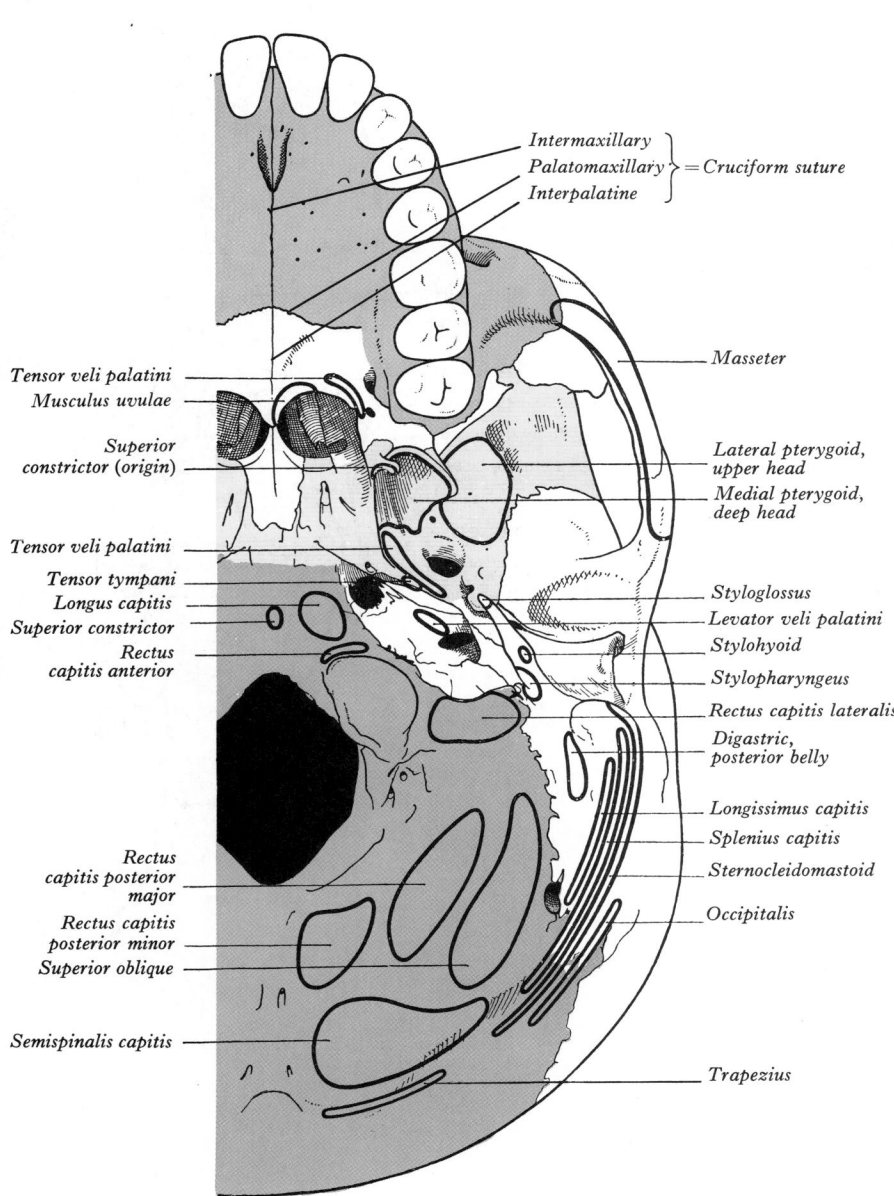

Intermaxillary
Palatomaxillary ⎫ = *Cruciform suture*
Interpalatine ⎬

Masseter

Tensor veli palatini
Musculus uvulae

Superior constrictor (origin)

Lateral pterygoid, upper head
Medial pterygoid, deep head

Tensor veli palatini
Tensor tympani
Longus capitis
Superior constrictor
Rectus capitis anterior

Styloglossus
Levator veli palatini
Stylohyoid
Stylopharyngeus
Rectus capitis lateralis
Digastric, posterior belly

Longissimus capitis
Splenius capitis
Sternocleidomastoid

Rectus capitis posterior major
Rectus capitis posterior minor
Superior oblique

Occipitalis

Semispinalis capitis

Trapezius

3.103B Outline drawing of norma basalis showing the attachments of named but unclassified muscles. Key to **3**.103A. Blue = occipital bone; yellow = sphenoid bone; green = maxilla. For muscle groups see **3**.104A,B.

pharyngeal tonsil. Anterolateral to the tubercle, longus capitis is attached and posterior to this rectus capitis anterior, immediately ventral to the occipital condyle and medial to the hypoglossal canal. At the root of the posterior border of **the medial pterygoid plate** the *scaphoid fossa* receives anterior fibres of tensor veli palatini, which descend along the plate's lateral surface and posterior border to reach the *hamulus*, around the anterolateral aspect of which the muscle's tendon twists medially into the soft palate. To the plate's posterior border, notched above by the auditory tube (p. 376), is attached the pharyngobasilar fascia, to its lower part the ventral highest fibres of the superior pharyngeal constrictor; these curve up and medially to the pharyngeal

Norma Basalis

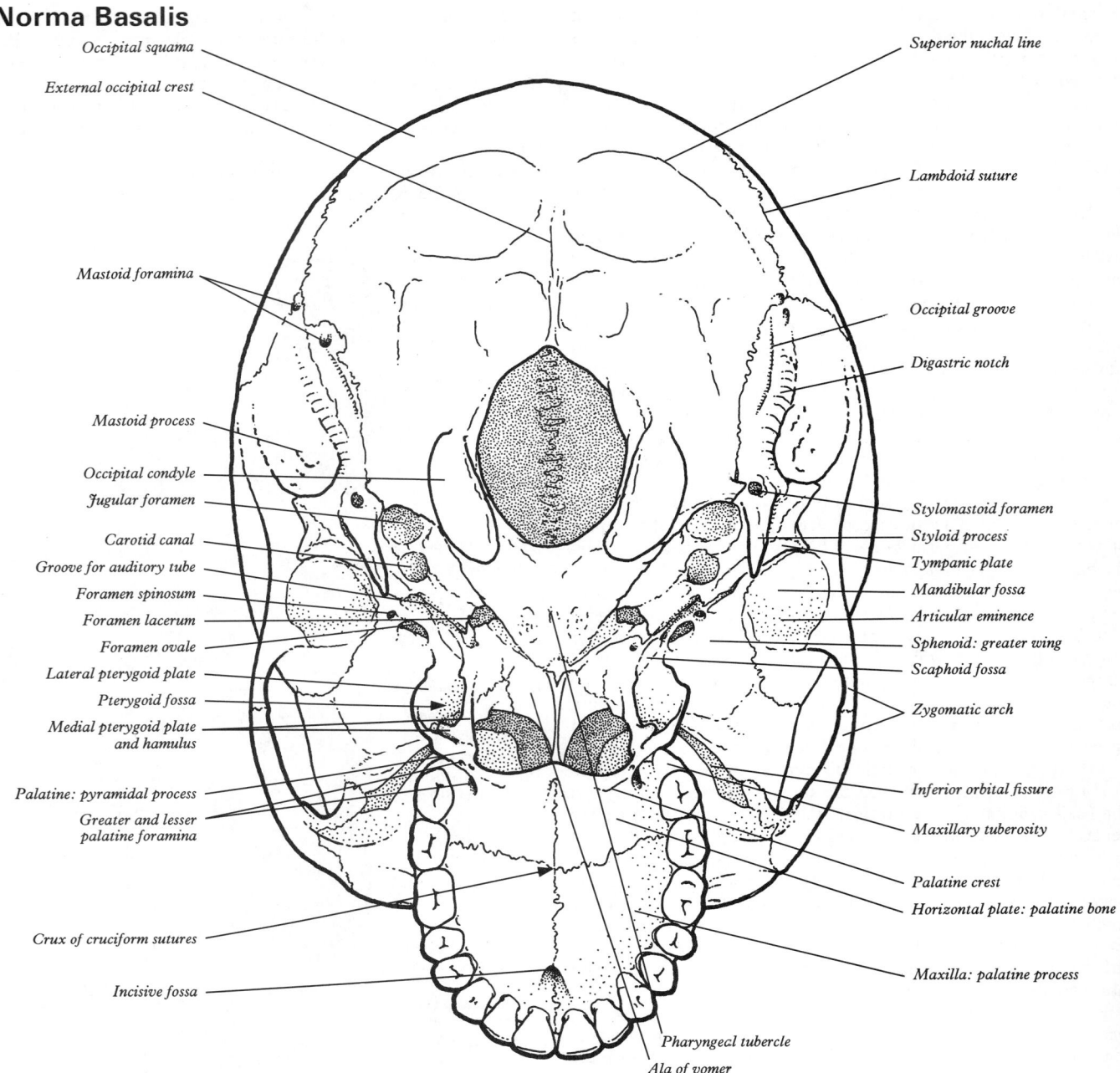

Occipital squama

External occipital crest

Mastoid foramina

Mastoid process

Occipital condyle

Jugular foramen

Carotid canal

Groove for auditory tube

Foramen spinosum

Foramen lacerum

Foramen ovale

Lateral pterygoid plate

Pterygoid fossa

Medial pterygoid plate
and hamulus

Palatine: pyramidal process

Greater and lesser
palatine foramina

Crux of cruciform sutures

Incisive fossa

Superior nuchal line

Lambdoid suture

Occipital groove

Digastric notch

Stylomastoid foramen

Styloid process

Tympanic plate

Mandibular fossa

Articular eminence

Sphenoid: greater wing

Scaphoid fossa

Zygomatic arch

Inferior orbital fissure

Maxillary tuberosity

Palatine crest

Horizontal plate: palatine bone

Maxilla: palatine process

Pharyngeal tubercle

Ala of vomer

3.104A (*above*) The skull: norma basalis without mandible.

3.104B (*opposite*) The skull: norma basalis without mandible, showing muscle attachments.

tubercle. To the tip of the hamulus the pterygomandibular raphe is attached. High on the plate's posterior border a small tubercle, medial to scaphoid fossa, projects back below the posterior opening of the *pterygoid canal*, which opens anteriorly on the pterygopalatine fossa's posterior wall. The canal transmits its nerve and vessels; it is in the line of fusion of the pterygoid process and greater wing with the sphenoidal body. The *pterygoid fossa*, between the pterygoid plates, is completed antero-inferiorly by the palatal pyramidal process.

The lateral pterygoid plate (3.103) is wider than the medial. A variable *pterygospinous process* on its irregular posterior border is connected by a ligament (sometimes ossified) to the sphenoid spine. To its rougher, lateral surface is attached the lower head of lateral pterygoid and to the medial surface most of medial pterygoid. Attached to the lateral aspect of the *palatine pyramidal process* are some fibres of the superficial slip of medial pterygoid.

The greater wing's infratemporal surface is an attachment of the upper head of lateral pterygoid and is crossed by the deep temporal and masseteric nerves, between muscle and bone. **The**

foramen ovale contains the accessory meningeal artery in additio to the mandibular nerve; to its sharp posterior border fibres o tensor veli palatine are attached between nerve and auditory tub **The foramen spinosum**, in addition to the middle meningea artery, transmits a meningeal branch of the mandibular nerv Between foramen ovale and scaphoid fossa a small *sphenoid emissary foramen* may contain an emissary vein from the cave nous sinus. Attached to the *spine of the sphenoid*, variably sharp c blunt, is the sphenomandibular ligament. The spine is relate laterally to the auriculotemporal nerve, medially to chorda tyr pani, by which it may be grooved, and to the auditory tub Posterior fibres of tensor veli palatini are attached posteriorly it. The groove for the tube varies in width and depth, its ro occasionally completed by fibrous tissue; its lateral, sphenoid wall gives attachment posteriorly to fibres of tensor tympani.

Laterally on the petrous temporal bone's posterior surfa levator veli palatini is attached. **The foramen lacerum** is bou ded in front by the sphenoid body and adjoining roots of t pterygoid process and greater wing, posterolaterally by the ap

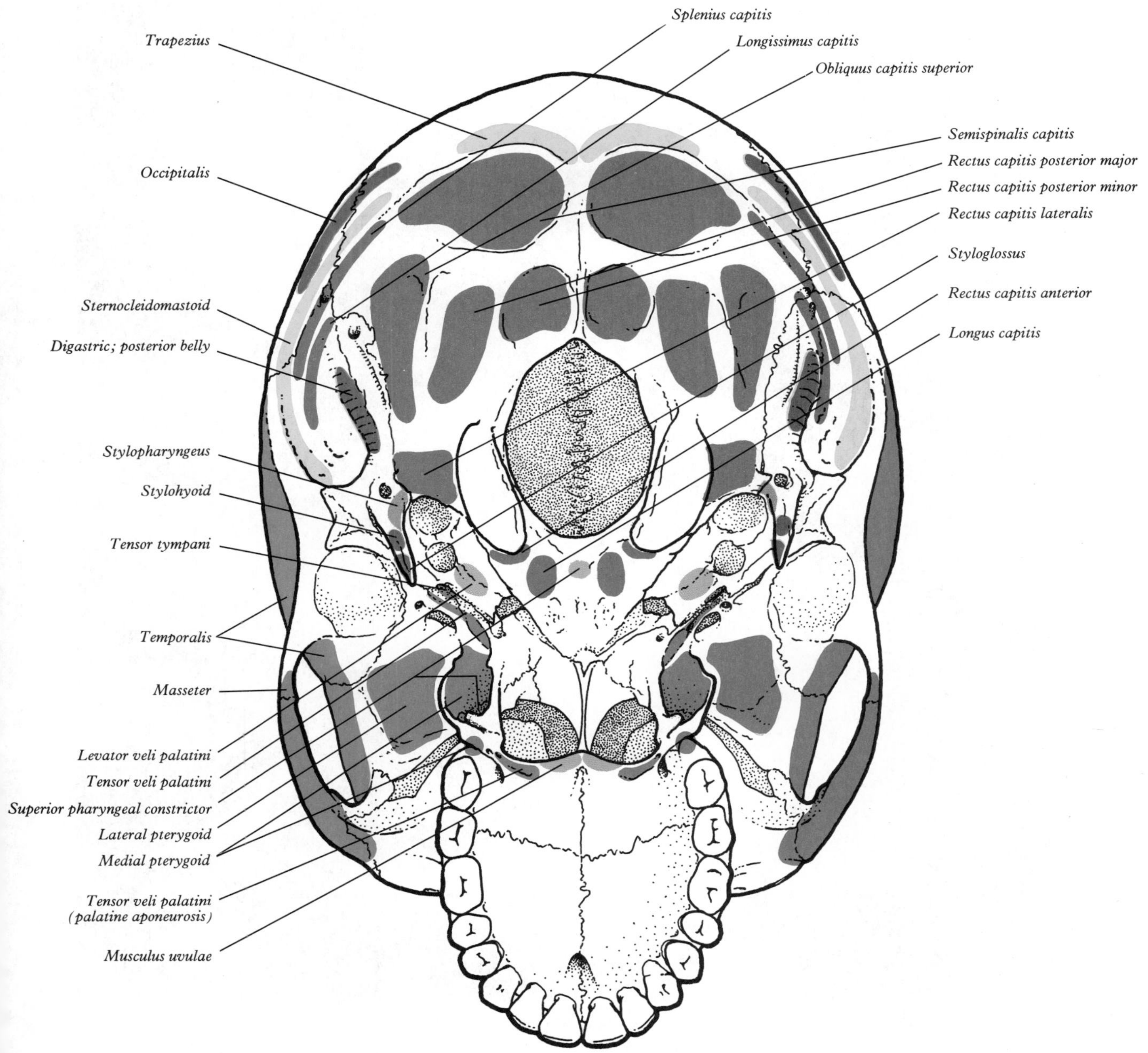

Splenius capitis

Longissimus capitis

Obliquus capitis superior

Trapezius

Semispinalis capitis

Rectus capitis posterior major

Rectus capitis posterior minor

Rectus capitis lateralis

Occipitalis

Styloglossus

Rectus capitis anterior

Longus capitis

Sternocleidomastoid

Digastric; posterior belly

Stylopharyngeus

Stylohyoid

Tensor tympani

Temporalis

Masseter

Levator veli palatini

Tensor veli palatini

Superior pharyngeal constrictor

Lateral pterygoid

Medial pterygoid

Tensor veli palatini
(palatine aponeurosis)

Musculus uvulae

COLOUR CODING OF ATTACHMENTS OF MUSCLE GROUPS BASED ON DERIVATION AND INNERVATION

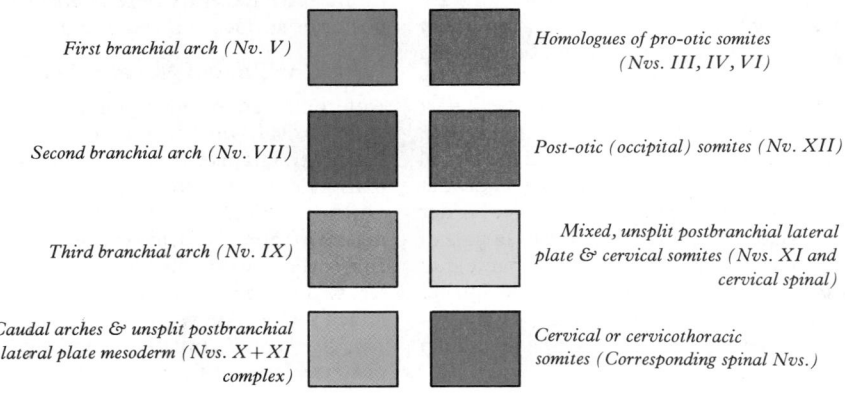

First branchial arch (Nv. V)

*Homologues of pro-otic somites
(Nvs. III, IV, VI)*

Second branchial arch (Nv. VII)

Post-otic (occipital) somites (Nv. XII)

Third branchial arch (Nv. IX)

*Mixed, unsplit postbranchial lateral
plate & cervical somites (Nvs. XI and
cervical spinal)*

*Caudal arches & unsplit postbranchial
lateral plate mesoderm (Nvs. X + XI
complex)*

*Cervical or cervicothoracic
somites (Corresponding spinal Nvs.)*

357

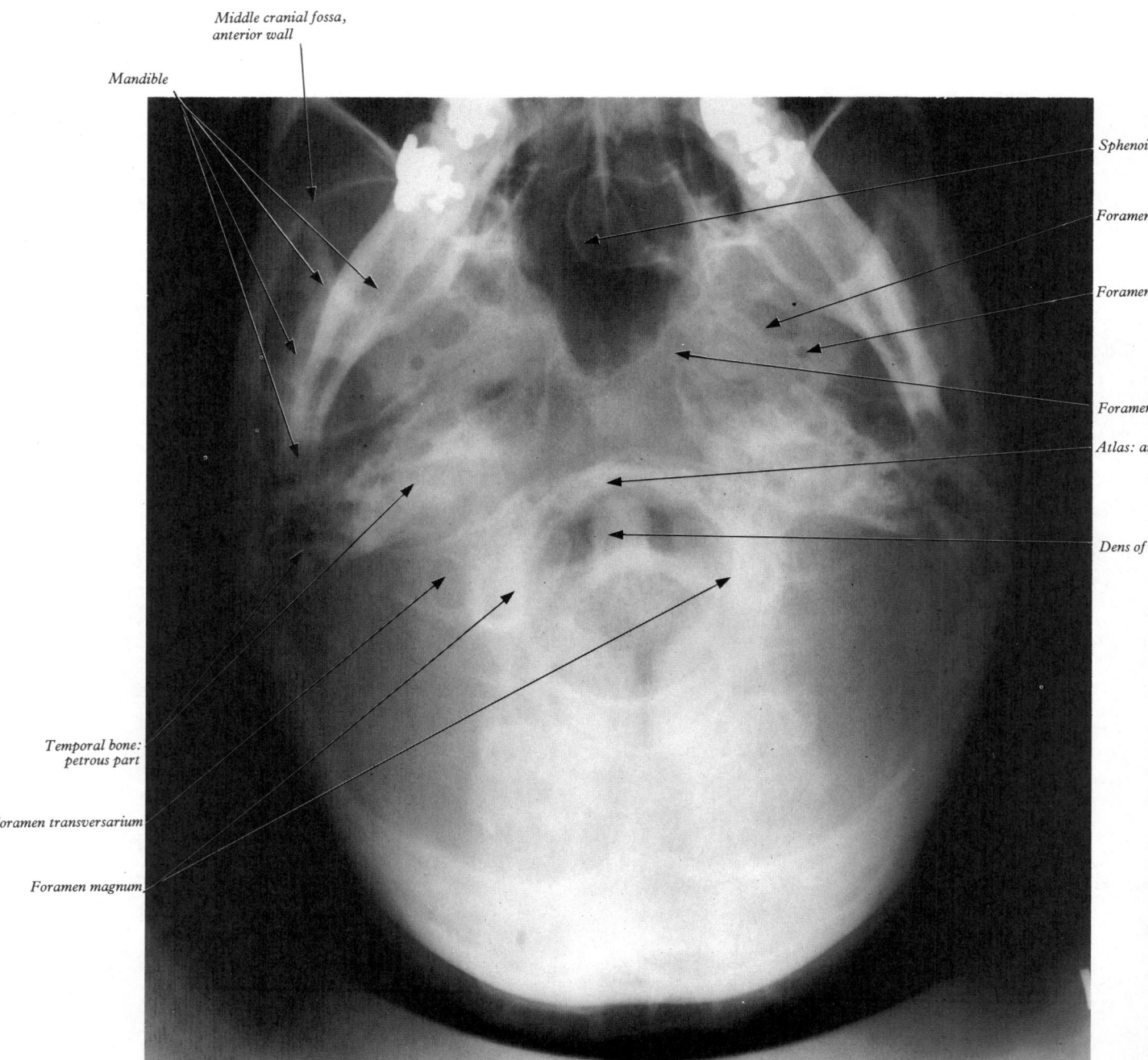

*Middle cranial fossa,
anterior wall*

Mandible

Sphenoidal a

Foramen ova

Foramen spir

Foramen lac

Atlas: anter

Dens of axi

*Temporal bone:
petrous part*

Foramen transversarium

Foramen magnum

3.105 Radiograph of base of skull (submento-vertical view) to show
foramina. (Supplied by Shaun Gallagher, Guy's Hospital; photography
by Kevin Fitzpatrick.)

of the petrous temporal, medially by the basi-occipital; it is nearly 1
cm long, but no large structure completely traverses it. The carotid
canal's anterior orifice opens posteriorly; this artery and its venous
and sympathetic plexuses ascend through its upper end. In the
foramen the deep and the greater petrosal nerves join to form the
nerve of the pterygoid canal, which opens low on the anterior wall.
Meningeal rami of the ascending pharyngeal artery, and emissary
veins from the cavernous sinus traverse the foramen. Cartilage in
its lower part is a remnant of the chondrocranium.

The *mandibular fossa's* thin floor is below the most lateral part
of the middle cranial fossa and covered by white fibrocartilage (p.
486). To the *tubercle of the root of the zygoma* is attached the lateral
temporomandibular ligament. A thin edge of bone may appear at
the medial end of the *squamotympanic fissure*; it is the lower border
of the down-curved lateral edge of the tegmen tympani, a part of
the petrous temporal. It partly divides the squamotympanic into
petrotympanic and *petrosquamous* fissures. Through the former
the chorda tympani, in its anterior canaliculus, escapes antero-
inferiorly from the tympanic cavity; the anterior tympanic branch
of the maxillary artery traverses the same fissure.

The temporal bone's tympanic part (**3.**107) is separate
from the temporomandibular joint by the parotid gland, usuall
containing the auriculotemporal nerve. It is thinnest centrally an
occasionally deficient (p. 380). Its grooved upper aspect forms th
anterior wall, floor and lower posterior wall of the external acou
tic meatus. Except where it ensheathes the styloid process, i
posterior surface is fused with the petromastoid bone.

Posterior Part of Norma Basalis

Anteromedian in this region (**3.**103,104) is the *foramen magnu*
which is oval, wider behind, its greatest diameter being anteropo
terior. It contains the medulla oblongata's lower end. Anteriorly
margin is slightly overlapped by the *occipital condyles*, projecti
down to articulate with superior articular facets on the late
masses of the atlas. Oval in outline, each condyle is oblique, its an
rior end nearer the midline, markedly convex anteroposterior
less so transversely. Its medial aspect is roughened by ligamento
attachments. Above it, anteriorly, is the *hypoglossal (anter
condylar) canal*, directed laterally and slightly forwards from
posterior cranial fossa and containing the hypoglossal nerve.

358

condylar fossa of variable depth, posterior to the condyle, is sometimes pierced by a *condylar canal* for an emissary vein from the sigmoid sinus. Incidence of condylar canals shows racial variation—from 13.3% in modern Palestinians to 70% in Peruvian crania; it also illustrates sexual variation, occurring in 58% of male Burmese, but only 31% of females (p. 395). Lateral to each condyle a *jugular process* joins the petrous temporal; its anterior border is the posterior boundary of the *jugular foramen*.

The jugular foramen is a large irregular hiatus *between* occipital and petrous temporal bones at the posterior end of the petro-occipital suture. Anteriorly it is separated from the inferior carotid orifice by a ridge, related laterally to the medial aspect of the styloid sheath and separated from the hypoglossal canal by a thin osseous bar. Its long axis is directed anteromedially, the right foramen usually being larger. Its anterior part contains the inferior petrosal sinus, its intermediate part the glossopharyngeal, vagus and accessory nerves and its posterior part the internal jugular vein. When the vein's superior bulb is well developed the jugular fossa of the petrous temporal is expanded superolaterally.

Posterior to the styloid process a *stylomastoid foramen* contains the facial nerve. Posterolateral to the foramen the *mastoid process* projects down and forwards as the lateral wall of the *mastoid notch*, in which the posterior belly of digastric is attached. Medial to the notch the temporal bone may be grooved by the occipital artery (3.107).

In the midline, posterior to foramen magnum, the occipital squama has a median *external occipital crest* to which is attached the ligamentum nuchae. The crest ends in the external occipital protuberance; *inferior nuchal lines* curve posterolaterally from its midpoint, almost parallel to the *superior nuchal lines*, which also extend from the protuberance and may form a distinct crest medially.

Further details. The foramen magnum is a wide communication between the posterior cranial fossa and vertebral canal. Anteriorly the apical ligament of dens and membrana tectoria are in it, both attached to the upper surface of the basi-occipital bone. Its wider, posterior part contains medulla oblongata and meninges. In the subarachnoid space spinal rami of the accessory nerves and vertebral arteries, with their sympathetic plexuses, ascend into the cranium; the posterior spinal arteries descend, posterolateral to the brainstem, as does the anterior spinal artery anteromedian to it. The cerebellar tonsils may project into the foramen. To its *anterior margin* is attached the anterior atlanto-occipital membrane, continuous on each side with capsular ligaments of atlanto-occipital joints, to its *posterior margin* the posterior atlanto-occipital membrane, and to rough medial condylar areas, the alar ligaments.

The hypoglossal canal also contains a meningeal branch of the ascending pharyngeal artery and an emissary vein from the basilar plexus. Sometimes it is divided by a spicule of bone (p. 373). To the inferior surface of the occipital jugular process rectus capitis lateralis is attached.

The jugular foramen (3.103B) ascends posteromedially; externally its apparent size is increased laterally by the jugular fossa, its floor separating the superior jugular bulb from the tympanic cavity. A minute *mastoid canaliculus* on the fossa's lateral wall transmits the vagal auricular branch. Passing laterally, this nerve is very near the facial canal, finally emerging in the tympanomastoid suture; it is extracranial at birth but surrounded by bone as tympanic plate and mastoid process develop. On or near the ridge between the jugular fossa and carotid canal is the *tympanic canaliculus*, transmitting to the middle ear the tympanic branch of the glossopharyngeal nerve. Medial on the upper boundary of the jugular foramen a small notch, more easily identified internally, contains the inferior glossopharyngeal ganglion. The orifice of the cochlear canaliculus (p. 379) is at the notch's apex; its projecting edges may reach the occipital bone to trisect the foramen.

The stylomastoid foramen is posterior to the styloid root at the anterior end of the mastoid notch. The facial nerve emerges from it near the digastric's posterior belly, supplying it before entering the parotid gland. The foramen also contains the stylomastoid artery. A groove across the inferior aspect of the temporal bone, medial to the mastoid notch, is related to the occipital artery; it is absent when the vessel is lower than usual, between splenius and longissimus capitis instead of medial to

both. To the area below the *inferior nuchal line* are attached medially rectus capitis posterior minor, laterally rectus capitis posterior major (3.104B). In the interval between inferior and superior nuchal lines is attached, medially, semispinalis capitis and, laterally, obliquus superior. Medially to the *superior nuchal line* the highest fibres of trapezius are attached, laterally fibres of sternocleidomastoid and, more anteriorly, splenius capitis.

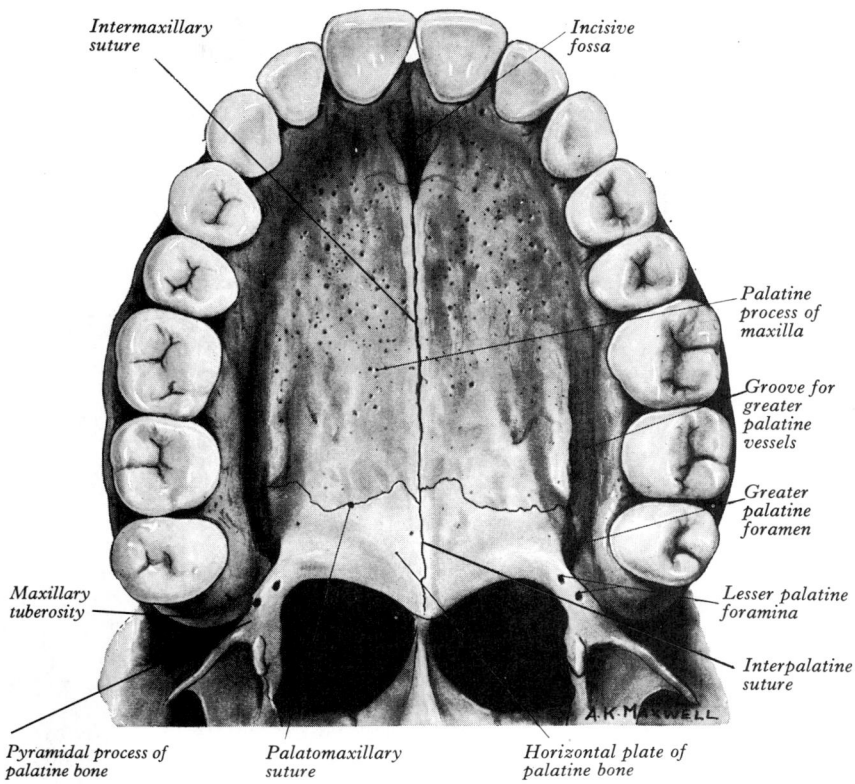

3.106 The bony palate and the alveolar arch: inferior aspect.

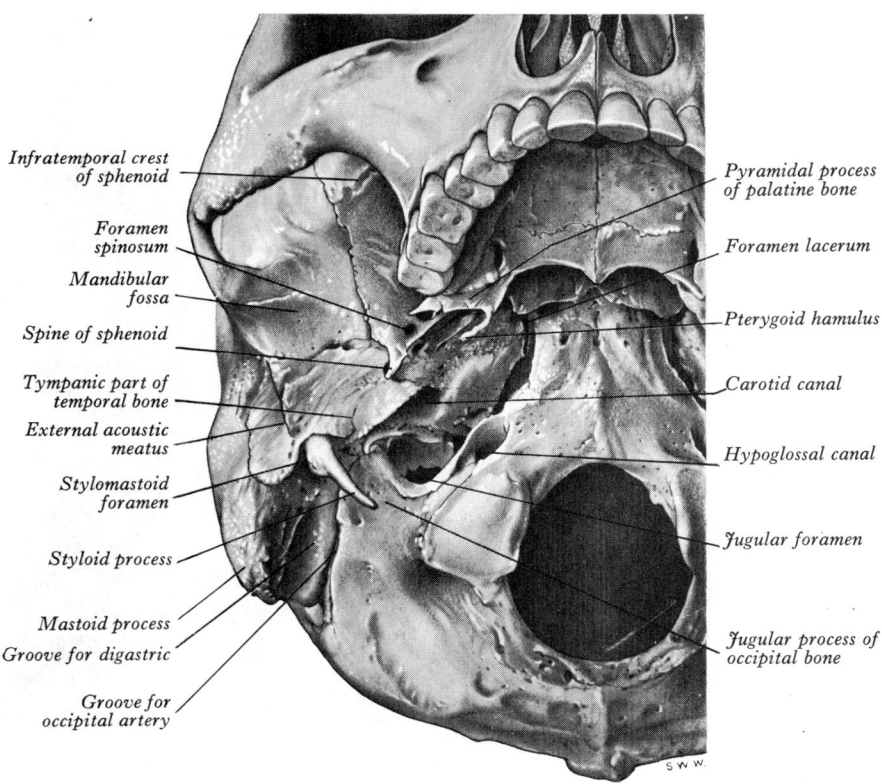

3.107 The right and part of the left side of the norma basalis of the skull. To show some features to better advantage, the anterior end of the skull has been elevated so that the Frankfurt plane is tilted to an angle of about 45° to the horizontal.

INTERIOR OF THE CRANIUM

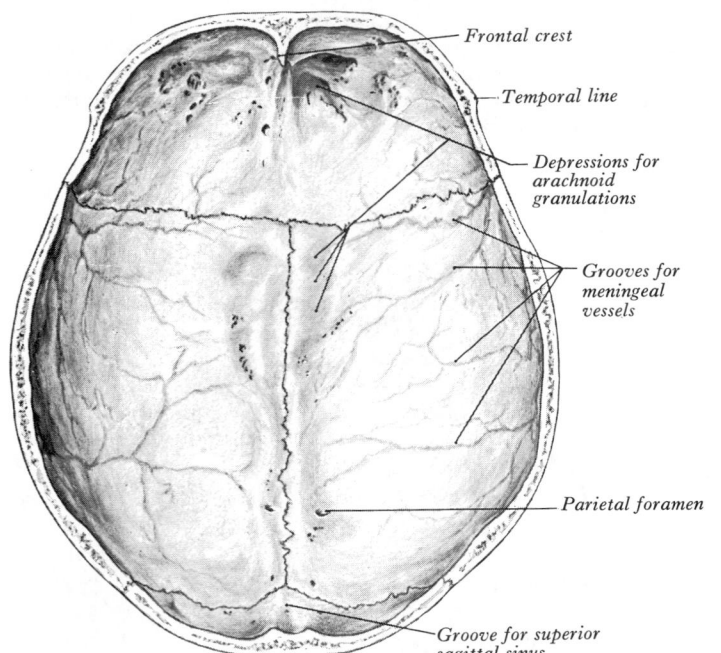

Frontal crest

Temporal line

Depressions for arachnoid granulations

Grooves for meningeal vessels

Parietal foramen

Groove for superior sagittal sinus

The cranial cavity contains the brain, pineal and hypophysis cerebri, parts of cranial and spinal nerves, blood vessels, meninges and cerebrospinal fluid. It is contained by frontal, parietal, sphenoid, temporal and occipital bones and, in part, the ethmoid, all lined by fibrous *endocranium*, the external zone of the dura mater, which traverses various foramina to join the external periosteum, the *pericranium*. Both membranes are blended with sutural ligaments or cartilages in the narrow interosseous intervals.

The cavity's walls vary in thickness in different regions and individuals, but tend to be thinner where covered by muscles, e.g. in temporal and posterior cranial fossae. Most cranial bones display *outer* and *inner tables* of compact bone, separated by *diploë*, trabecular bone containing red bone marrow. The inner table is thinner and more brittle, the outer generally very resilient. Many bones are so thin that the tables are fused, e.g. vomer, pterygoid plates. The skull is thicker in some races and individuals, but no relation exists between this and cranial capacity; in all races, it is thinner in women and children. The interior may

3.108 (*left*) The internal (endocranial) surface of the skull cap or calva.

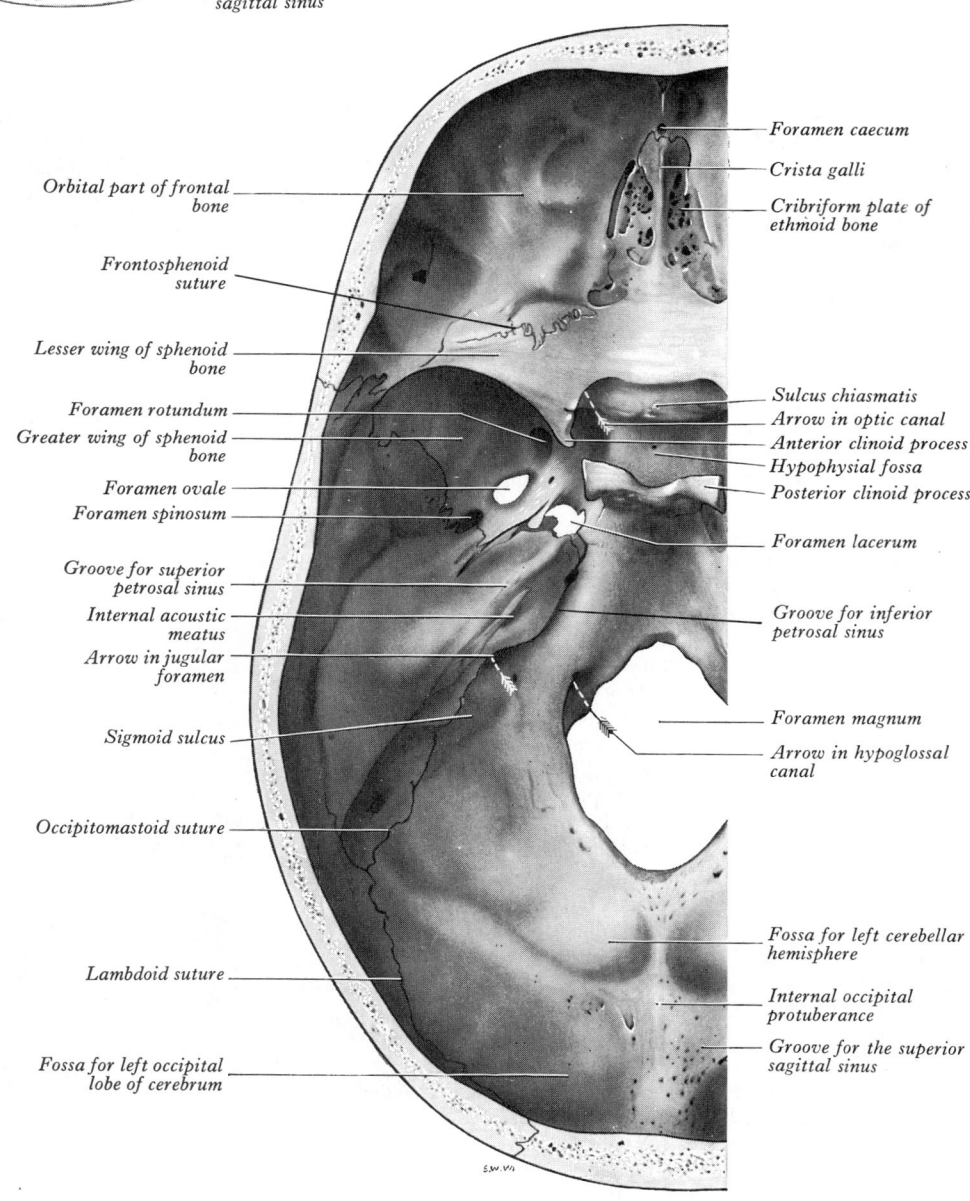

Orbital part of frontal bone

Frontosphenoid suture

Lesser wing of sphenoid bone

Foramen rotundum

Greater wing of sphenoid bone

Foramen ovale

Foramen spinosum

Groove for superior petrosal sinus

Internal acoustic meatus

Arrow in jugular foramen

Sigmoid sulcus

Occipitomastoid suture

Lambdoid suture

Fossa for left occipital lobe of cerebrum

Foramen caecum

Crista galli

Cribriform plate of ethmoid bone

Sulcus chiasmatis

Arrow in optic canal

Anterior clinoid process

Hypophysial fossa

Posterior clinoid process

Foramen lacerum

Groove for inferior petrosal sinus

Foramen magnum

Arrow in hypoglossal canal

Fossa for left cerebellar hemisphere

Internal occipital protuberance

Groove for the superior sagittal sinus

3.109A The internal (endocranial) surface of the left half of the base of the skull. Compare with Key, 3.109B.

be described in two parts: the internal surface of the skull 'cap' or calva and that of the cranial base.

Internal Surface of Cranial Vault (Calva)

The *calva* (3.108) includes most of the frontal and parietal bones and the upper occipital squama and hence the coronal, sagittal and lambdoid sutures unless fusion, which begins internally, has obliterated them. The cranial vault, deeply concave, presents numerous vascular furrows and cerebral grooves.

The anteromedian *frontal crest* projects back for attachment of the falx cerebri, grooved by the commencement of the *sagittal sulcus*, accommodating the superior sagittal sinus, the groove widening as it progresses back below the sagittal suture. On each side of it are irregular depressions, *granular foveolae*, which become larger and more numerous as skulls age; they are adapted to arachnoid granulations.

The middle meningeal vein's frontal branch, and sometimes the artery's, groove bone deeply just behind the coronal suture, corresponding to the precentral cerebral sulcus. Rami of both and their parietal branches ascend backwards, grooving the internal

parietal surface. Smaller grooves may mark the frontal and occipital bones. *Parietal foramina* may occur near the sagittal sulcus, about 3.5 cm anterior to the lambdoid suture, for emissary veins of the superior sagittal sinus. *Impressions for cerebral gyri* are less distinct on vault than cranial base, and most obvious near the latter.

Internal Surface of Cranial Base

This shows clear division into anterior, middle and posterior cranial fossae (3.109–112). It is irregular, partly due to impressions for cerebral gyri, especially conspicuous in the anterior and middle fossae, reflecting the pattern of corresponding cerebral surfaces. Dura mater is firmly adherent to the whole area and through all foramina and fissures its outer layer, endocranium, is continuous with the external periosteum.

Anterior Cranial Fossa

This fossa (3.109A,B,110) is formed at front and sides by the frontal bone, its floor by the frontal's orbital plate, the ethmoid's

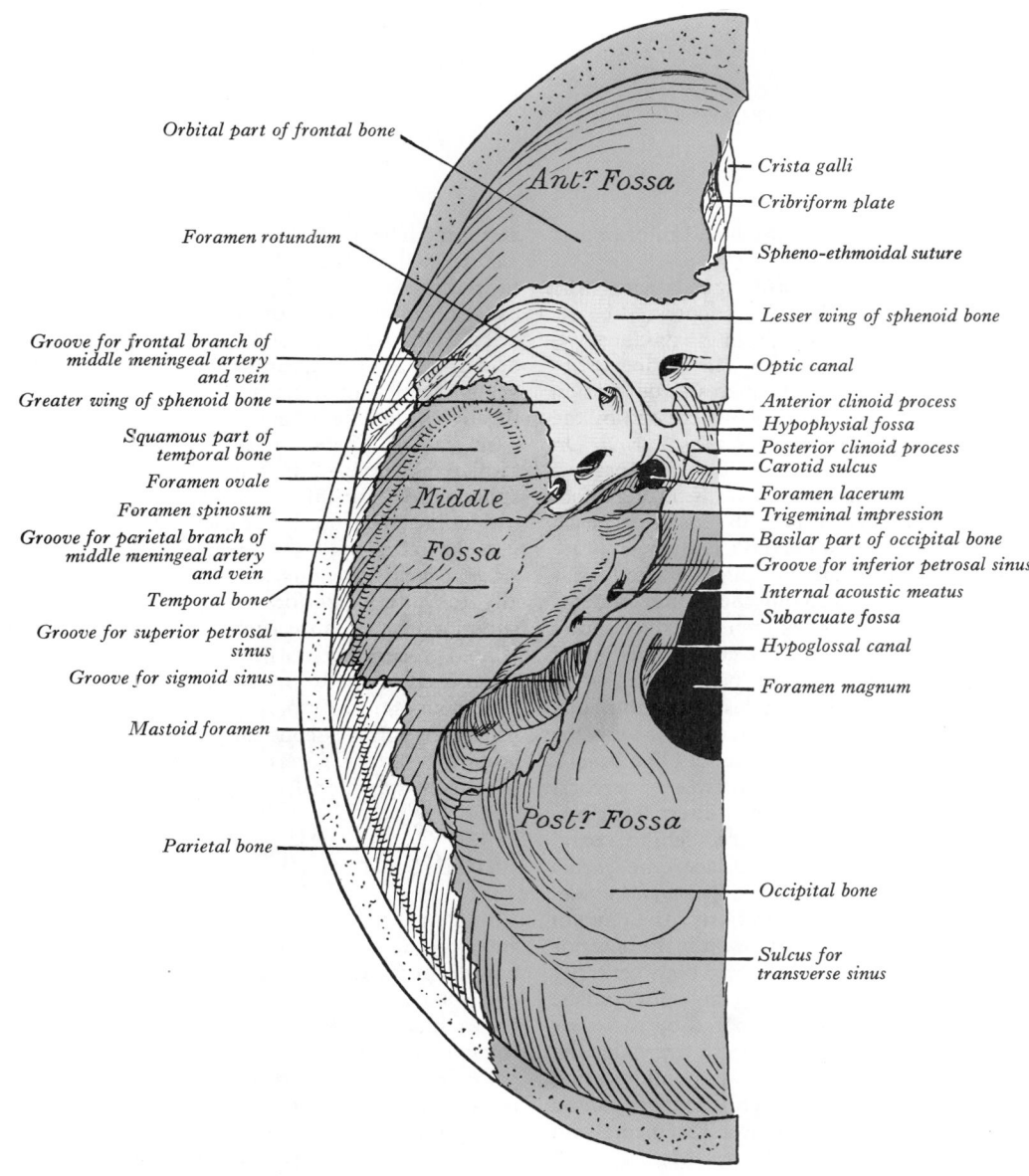

109B The internal surface of the left half of the base of the skull (Basis cranii interna). Key to 3.109A. Blue = frontal and occipital bones; yellow = sphenoid bones; brown = temporal bone; white = parietal and ethmoid bone.

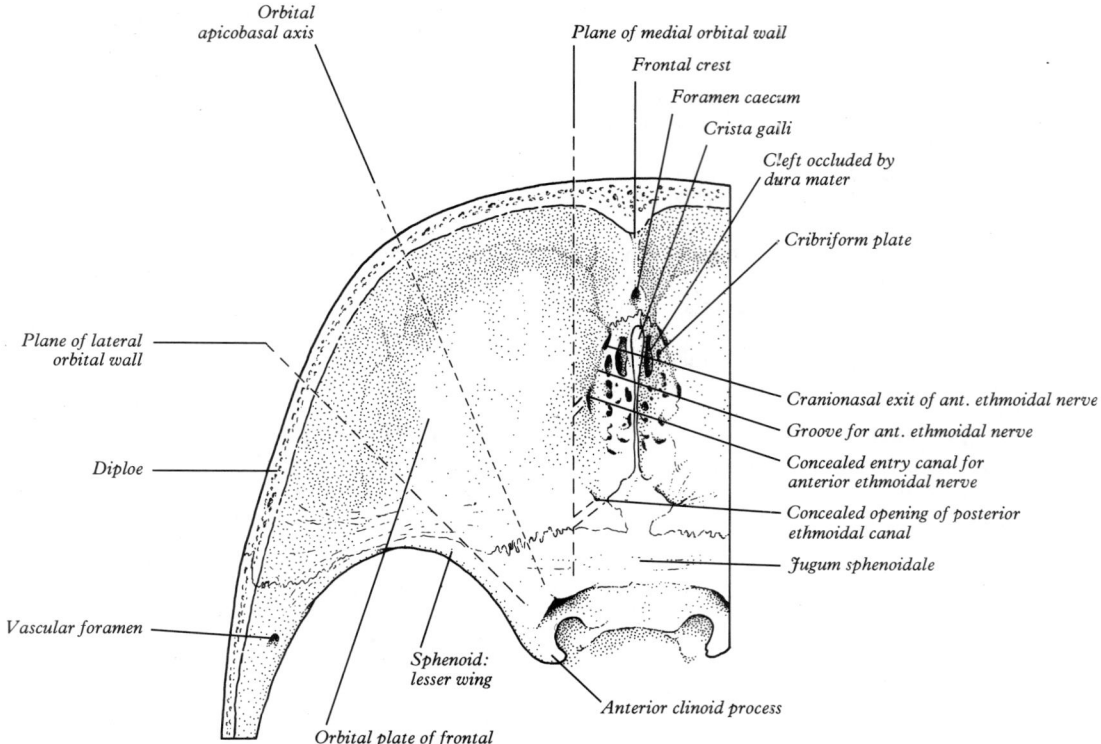

3.110 Anterior cranial fossa.

cribriform plate, and the sphenoid's lesser wings and anterior part of its body.

The cribriform plate of the ethmoid spreads across the midline between the frontal orbital plates, but depressed below them, separating the anterior cranial fossa from the nasal cavity, whose roof it helps to form (3.144). Anteriorly its median *crista galli* projects up between the cerebral hemispheres. A depression between crista galli and *crest of the frontal bone* is crossed by the fronto-ethmoidal suture and displays the *foramen caecum*. On each side the crista galli is separated from the frontal bone by a narrow region with numerous small foramina transmitting olfactory nerves from nasal mucosa to olfactory bulb. Posteriorly each cribriform plate articulates with the sphenoid's body.

The orbital plates of the frontal form most of the fossa's floor on each side of the ethmoid, separating the orbital contents from the cerebral frontal lobe. Its convex cranial surface shows impressions for cerebral gyri and small grooves for meningeal vessels. Anteromedially it is split to contain part of the *frontal sinus*. Medially each orbital plate excludes the ethmoidal labyrinth from the anterior cranial fossa. Posteriorly it joins the *sphenoid's lesser wing's* anterior border. The frontal bone bears a median *frontal crest*, which projects between cerebral hemispheres and narrows upwards on the bone's internal surface.

The sphenoid bone completes the fossa's floor behind, centrally by the anterior part of its upper surface, the *jugum sphenoidale*; this separates the fossa from paired *sphenoidal sinuses* in the sphenoid body (3.114,115). Anteriorly the jugum articulates with the cribriform plate; posteriorly it is the anterior bank of the *sulcus chiasmatis* crossing the sphenoid body centrally in the middle cranial fossa and connecting the optic canals. Lateral to the jugum, the anterior fossa is floored by the *lesser wings* of the sphenoid, whose posterior margins overhang the middle fossa. Laterally each lesser wing tapers to a point, meeting the suture between the frontal bone and greater wing at or near the superolateral end of the superior orbital fissure. The medial end of its posterior border is the *anterior clinoid process*. Medially each lesser wing joins the sphenoid body by two roots, separated by the *optic canal*; the anterior, broad and flat, is continuous with the jugum sphenoidale, the posterior, smaller and thicker, joins the sphenoid body near the posterior bank of the chiasmal sulcus.

To the crista galli and frontal crest the falx cerebri is attached. The foramen caecum between them, usually blind, occasionally accommodates a vein from nasal mucosa to superior sagittal sinus. Lateral to the crista is the gyrus rectus; the olfactory bulb lies on the medial edge of the frontal orbital plate. The *anterior ethmoidal canal* opens in the cribrofrontal suture (3.98A,144) behind the crista galli and is difficult to identify, being overlapped by the orbital plate. It transmits the anterior ethmoidal nerve and vessels, which then run forward under dura mater to descend through a slit-like foramen at the side of the crista galli into the nasal cavity. The *posterior ethmoidal canal* opens at the posterolateral corner of the cribriform plate and is overhung by the sphenoid. It transmits the posterior ethmoidal vessels.

The posterior border of each lesser wing fits the stem of the lateral cerebral sulcus and may be grooved by the sphenoparietal sinus. Above is the inferior surface of the frontal lobe and, medially, the anterior perforated substance. Inferiorly, it bounds the superior orbital fissure, completing the orbital roof. Each anterior clinoid process receives the free border of tentorium cerebelli and is grooved medially by the internal carotid artery leaving the cavernous sinus. It may be connected to the middle clinoid process by a thin osseous bar, completing a *caroticoclinoid foramen* around the artery. Parts of gyri recti and olfactory tracts are above the jugum sphenoidale.

Middle Cranial Fossa

This fossa (3.109A,B,111), deeper and more extensive than the anterior, particularly laterally, is bounded in front by the posterior borders of the lesser wings, anterior clinoid processes and sulcus chiasmatis, behind by the superior borders of the petrous temporal bones and sphenoid's dorsum sellae, laterally by the temporal squamae, parietal bones and sphenoidal greater wings.

Centrally the floor is narrower and formed by the sphenoid body. The chiasmal sulcus, connecting the optic canals, is rarely in contact with the optic chiasma, which is usually posterosuperior. Each **optic canal**, between the roots of a lesser wing and, medially, the sphenoid body, descends a little anterolaterally, containing optic nerve, ophthalmic artery and meninges. Behind the sulcus the upper sphenoid surface is the *sella turcica*, whose anterior slope bears a median *tuberculum sellae*, behind which is the *hypophysial fossa* (3.109A). The fossa's floor part of the roof of the sphenoidal sinuses (3.91,114,115); posterior

362

to it the *dorsum sellae* projects up and forwards. The superolateral angles of the dorsum are expanded as the *posterior clinoid processes*. Lateral to the sella turcica the sphenoid has a shallow *groove for the internal carotid artery*, here turning anteriorly from the foramen lacerum. A small elevation on the groove's medial edge is the *middle clinoid process*; it may be joined to the anterior process, as noted above. Posterolaterally the groove may be deepened by a small projecting *lingula* (**3.126**).

Laterally the middle fossa is deep and supports the temporal lobes. In front are the cerebral surfaces of sphenoid's greater wings, behind are anterior surfaces of petrous temporal bones and, laterally, cerebral surfaces of temporal squamae between the former two. Anteriorly are the orbits, laterally the temporal fossae, inferiorly the infratemporal fossae. The middle cranial fossa communicates with the orbits by **the superior orbital fissures**, each bounded above by a lesser wing, below by a greater wing, and medially by the sphenoidal body; each fissure is wider medially, with a long axis sloping inferomedially and forwards. Each transmits the terminal rami of the ophthalmic nerve, the ophthalmic veins, oculomotor, trochlear and abducent nerves and smaller vessels (vide infra).

The foramen rotundum, in the greater sphenoidal wing, is just below and behind the medial end of the superior orbital fissure. It leads forwards into the pterygopalatine fossa, to which it conducts the maxillary nerve (vide infra). **The foramen ovale** is posterior to the foramen rotundum, lateral to the lingula and posterior end of the carotid groove. It opens into the infratemporal fossa and transmits the mandibular nerve. **The foramen spinosum**, posterolateral to foramen ovale, transmits the middle meningeal artery which, with companion veins, ascends lateral to the temporal squama, turns anterolaterally across the spheno-squamosal suture to the greater wing to divide into frontal and parietal branches. The frontal ascends across the pterion (p. 352) to the anterior part of the parietal bone; at or near the pterion it is

often in a bony canal. The parietal branch runs back and up on to the temporal squama, crossing the squamosal suture to gain the parietal bone. These arteries and veins groove the floor and lateral wall of the middle cranial fossa (vide infra).

At the posterior end of the carotid groove, posteromedial to the foramen ovale, is the foramen lacerum, bounded behind by the petrous apex, in front by the body and posterior border of the sphenoid's greater wing, where it contains the internal carotid artery and accompanying sympathetic and venous plexuses (vide infra). Posterior to the foramen the anterior surface of the petrous temporal has near its apex a shallow *trigeminal impression* adapted to the trigeminal ganglion. Posterolateral to this is a shallow pit, limited posteriorly by a rounded *arcuate eminence*, and produced by the anterior semicircular canal, where it is closely related to the floor of the fossa. Lateral to the impression a narrow groove passes posterolaterally into the *hiatus for the greater petrosal nerve*, lateral to which is the *hiatus for the lesser petrosal nerve*. Anterolateral to the arcuate eminence the anterior petrous surface is formed by the *tegmen tympani*, a thin osseous lamina in the roof of the tympanic cavity, extending anteromedially above the auditory tube. Lateral to the eminence the posterior part of the tegmen tympani roofs the mastoid antrum (which is continuous anteriorly with the tympanic cavity). The superior border of the petrous temporal separates middle from posterior cranial fossae. Behind the trigeminal impression it is grooved by the superior petrosal sinus.

On each side of the sphenoid body a cavernous sinus extends from the superior orbital fissure to the apex of petrous temporal bone. It contains the internal carotid artery, its sympathetic plexus, oculomotor, trochlear, abducent and ophthalmic nerves, but only the artery contacts bone. An anterior intercavernous sinus crosses the tuberculum sellae and a posterior sinus crosses the dorsum sellae; they connect the two cavernous sinuses. The diaphragma sellae surrounds the infundibulum, spreading from the tuberculum in front to the dorsum sellae behind. To the

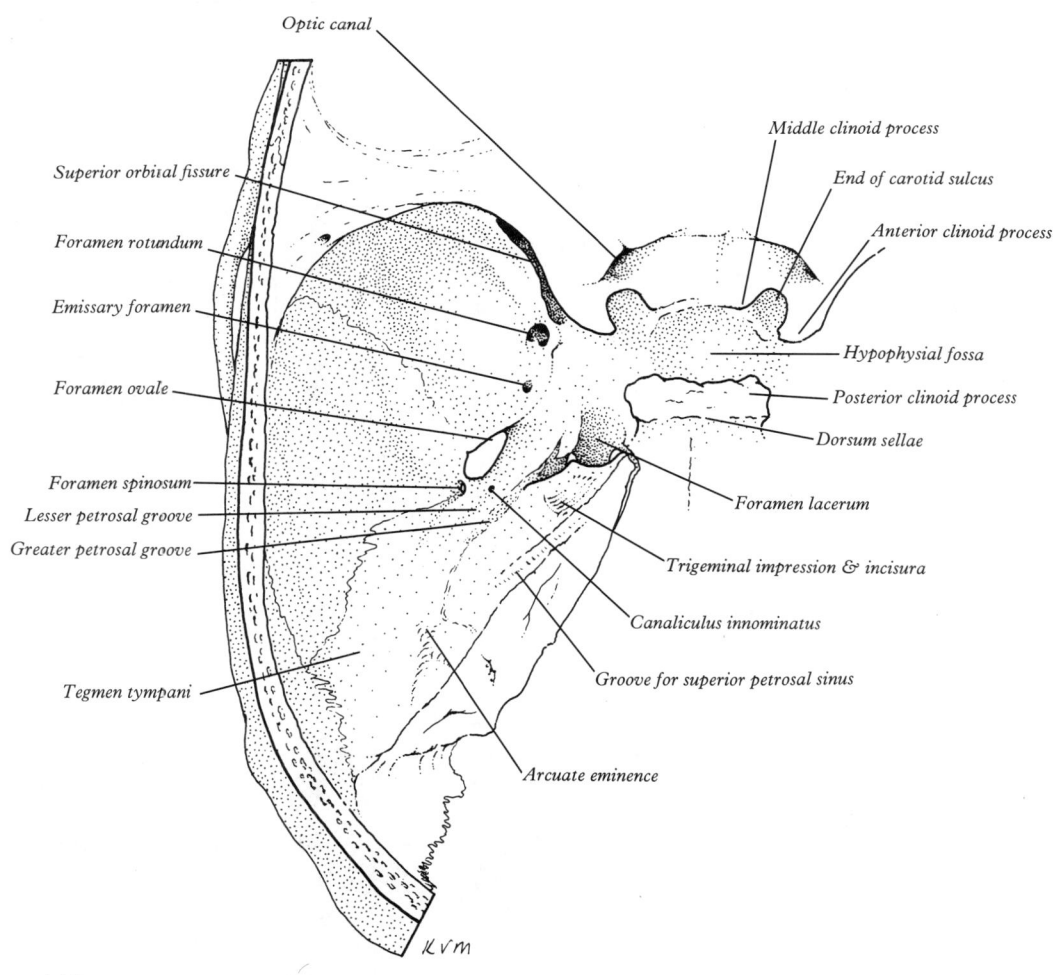

Optic canal

Middle clinoid process

End of carotid sulcus

Superior orbital fissure

Anterior clinoid process

Foramen rotundum

Emissary foramen

Hypophysial fossa

Foramen ovale

Posterior clinoid process

Dorsum sellae

Foramen spinosum

Lesser petrosal groove

Foramen lacerum

Greater petrosal groove

Trigeminal impression & incisura

Canaliculus innominatus

Groove for superior petrosal sinus

Tegmen tympani

Arcuate eminence

11 Middle cranial fossa.

posterior clinoid process are attached the anterior limit of the fixed margin of the tentorium cerebelli and also the petrosphenoidal ligament (p. 11077).

Further details. The superior orbital fissure (3.96) is between the orbit's roof and lateral wall. To its lower border is attached the common tendinous ring. At its superolateral end the greater wing articulates with the frontal orbital plate (p. 347). **The foramen rotundum,** like the superior orbital fissure's medial end, is near the lateral wall of the sphenoidal sinus. Originally part of the fissure, the latter is secondarily separated. Medial to it a small foramen may occur at the root of the greater wing; this *emissary sphenoidal foramen* transmits a vein from the cavernous sinus. **The foramen ovale** transmits the mandibular nerve, accessory meningeal artery and, sometimes, the lesser petrosal nerve. **The foramen spinosum** contains the middle meningeal artery and meningeal branch of the mandibular nerve. Both foramina are at first notches on the greater wing's margin and later converted into foramina. **The foramen lacerum** (p. 361) is a short canal completely *traversed* only by meningeal branches of the ascending pharyngeal artery and small veins. The internal carotid artery pierces its posterior wall and curves to ascend through its upper end. The greater petrosal nerve, leaving its hiatus, runs in its groove on the petrous temporal bone, then turns down to partly traverse the foramen lateral to the artery and joins the deep petrosal nerve. The nerve of the pterygoid canal, thus formed, leaves the foramen lacerum by a *pterygoid canal* in its anterior wall.

In young skulls a petrosquamous suture may be visible at the lateral limit of the *tegmen tympani* but is obliterated in adults; the tegmen then turns down as a lateral wall of the osseous auditory tube; its lower border may appear in the squamotympanic fissure (p. 359). Lateral to the tegmen's anterior part, the temporal squama is thin over a small area near the mandibular fossa's deepest part.

Anteromedial to the groove for the superior petrosal sinus, on upper border of petrous temporal, a smooth *trigeminal notch* leads into the *trigeminal impression*; here the trigeminal nerve separates sinus from bone. To a tiny spicule, directed anteromedially at the notch's anterior end, is attached the petrosphenoidal ligament; anterior to it the abducent nerve bends sharply across the upper petrous border between this ligament and the dorsum sellae.

Posterior Cranial Fossa

The posterior cranial fossa (3.109A,B,112), the largest and deepest of the cranial fossae, is bounded in front by the dorsum sellae and posterior aspects of sphenoidal body and basi-occipital bone, behind by the lower part of the occipital squama, laterally by petrous and mastoid parts of the temporal bone and lateral parts of the occipital and, above and behind, by the mastoid angles of the parietal bones. It contains the cerebellum, pons and medulla oblongata.

The foramen magnum (p. 359) is in the fossa's floor, surrounded by the basilar part of the occipital bone in front, its lateral parts on each side and a small part of its squama behind. Anterior to its transverse diameter it is narrowed by the two occipital condyles, being hence somewhat ovoid and wider behind. Its anterior part is above the dens of the axis; its posterior part communicates with the vertebral canal and here medulla oblongata and spinal cord become continuous.

Anterior to the foramen, in sequence, the basi-occipital, the posterior part of the sphenoid's body and then the dorsum sellae form a sloping *clivus*, which is concave transversely and placed antero-inferior to the pons and medulla oblongata. On each side the clivus is separated from the petrous temporal bone by a petro-occipital fissure, filled by a thin plate of cartilage and limited behind by the jugular foramen. Its margins are grooved by the inferior petrosal sinus.

The jugular foramen (p. 359), sited at the posterior end of the petro-occipital fissure, descends anterolaterally to the exterior. Its upper border is sharp and irregular, with a *notch for the glossopharyngeal nerve*; its lower border is smooth. Posteriorly is the sigmoid sinus, continuous below with the internal jugular vein, and anteriorly are the accessory, vagus and glossopharyngeal nerves from behind forwards. Medial to its lower border is

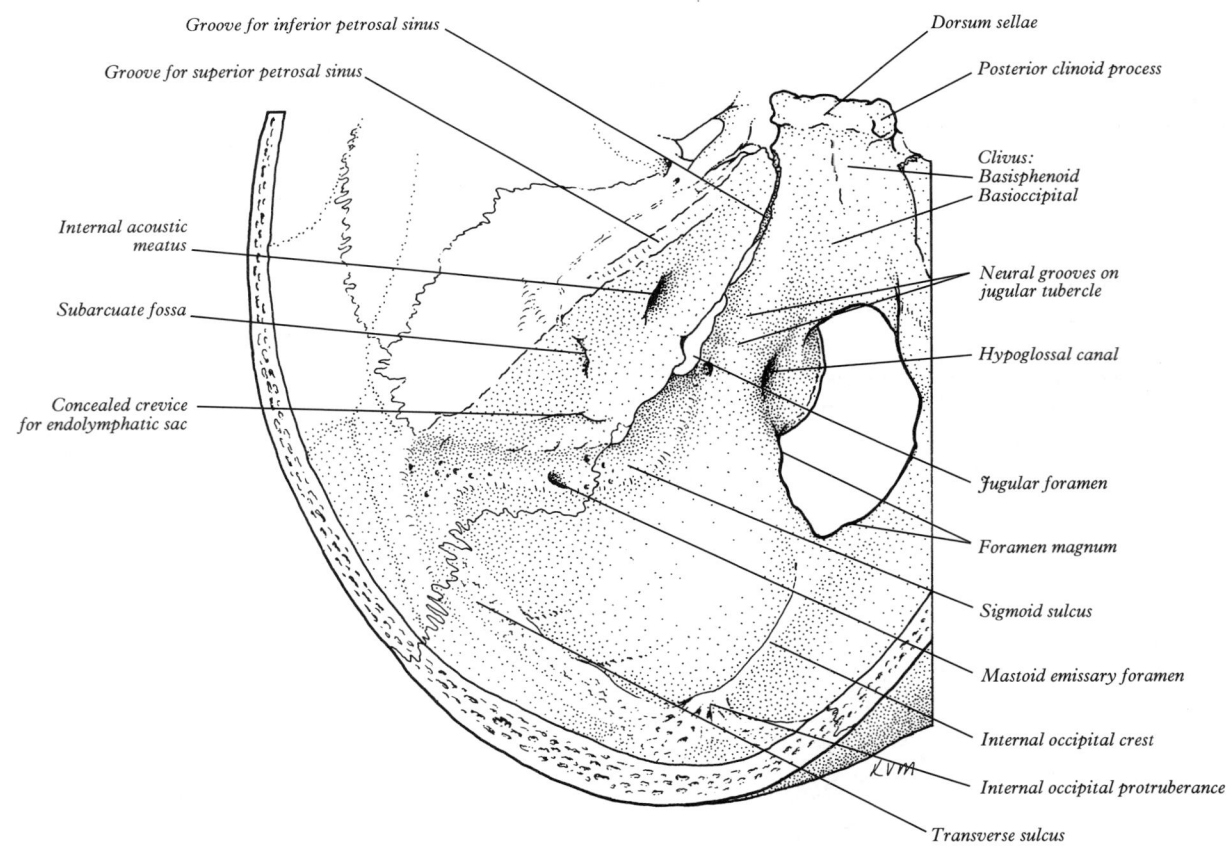

Groove for inferior petrosal sinus

Groove for superior petrosal sinus

Internal acoustic meatus

Subarcuate fossa

Concealed crevice for endolymphatic sac

Dorsum sellae

Posterior clinoid process

Clivus:
Basisphenoid
Basioccipital

Neural grooves on jugular tubercle

Hypoglossal canal

Jugular foramen

Foramen magnum

Sigmoid sulcus

Mastoid emissary foramen

Internal occipital crest

Internal occipital protruberance

Transverse sulcus

3.112 Left posterior cranial fossa. Viewed obliquely from a position above, slightly behind, and to the right.

the rounded *jugular tubercle*, anterosuperior to the internal opening of *the hypoglossal canal*, at the junction of basilar and lateral parts of the occipital bone.

The petrous temporal bone's posterior surface forms much of the posterior cranial fossa's anterolateral wall. Anterosuperior to the jugular foramen the *internal acoustic meatus* (**3.**109,111,112) runs laterally. It is about 1 cm long, closed laterally by a plate of bone separating it from the internal ear and perforated by facial and vestibulocochlear nerves, nervus intermedius and labyrinthine vessels.

Behind the petrous temporal, the posterior fossa's lateral wall, is the mastoid part of the temporal bone. Anteriorly is a wide *sigmoid sulcus* running forwards and downwards, then downwards and medially and finally forwards to the jugular foramen. It contains the sigmoid sinus, (**3.**109,133). Superiorly, where it touches the parietal bone's mastoid angle, the groove is continuous with one for the transverse sinus, and crosses the parietomastoid suture. It descends *behind the mastoid antrum* and here is sited a *mastoid foramen* for an emissary vein from the sinus. The lowest part of the sulcus crosses the occipitomastoid suture and grooves the occipital bone's jugular process. The right sulcus is usually larger.

Behind the foramen magnum the occipital squama has a median *internal occipital crest*, ending above in an *internal occipital protuberance*, on each side of which a wide sulcus curves laterally with an upward convexity, to the parietal bone's mastoid angle. Produced by the transverse sinus, it is usually deeper on the right and, on both sides, continuous with its sigmoid sulcus. Below this *transverse sulcus* the internal occipital crest separates two shallow fossae, adapted to the cerebellar hemispheres.

When a *condylar canal* is present (**3.**122–124), its internal orifice is posterolateral to that of the hypoglossal canal. It contains a sigmoid emissary vein.

The clivus (anterior wall of the posterior fossa) is related to the plexus of basilar sinuses connecting inferior petrosal sinuses and joining below the internal vertebral venous plexus. Anterior to the foramen magnum the membrana tectoria is attached to the basi-occipital (**4.**29), behind attachment of the apical ligament. The jugular tubercle is often grooved by glossopharyngeal, vagus and accessory nerves as they enter the jugular foramen. In addition to its nerve the hypoglossal canal, often subdivided (p. 373), transmits a meningeal branch of the ascending pharyngeal artery. To the rough medial aspect of the occipital condyle (**3.**123) the alar ligament is attached.

The lower, posterior borders of the jugular foramen (pp. 359, 364) are smooth, its upper border being sharp and notched; sometimes the margins of the notch extend to divide the foramen into two or three compartments. In the deepest part of the notch appears the *cochlear canaliculus*, containing the perilymphatic 'duct' (p. 1230).

The internal acoustic meatus is separated at its lateral *fundus* from the internal ear by a vertical plate divided unequally by a *transverse crest* (**3.**113), above which anteriorly is a *facial canal* conducting its nerve through the petrous temporal to the stylomastoid foramen. Posterior to this a small, depressed *superior vestibular area* presents openings for nerves to the utricle and anterior and lateral semicircular ducts. Below the crest is an anterior *cochlear area* in which a spiral of small holes, the *tractus spiralis foraminosus*, encircles the central cochlear canal. Behind this the *inferior vestibular area* bears openings for saccular nerves. Most postero-inferior is a *foramen singulare* for the nerve to the posterior semicircular duct.

Behind the internal acoustic meatus a thin plate with an irregularly curved margin projects back, bounding a slit containing the *vestibular aqueduct's opening* (**3.**133); within, it carries the saccus and ductus endolymphaticus (p. 1233) and a small artery and vein. Between internal acoustic meatus and aqueductual opening a small *subarcuate fossa*, containing (**3.**109B,133) dura mater; near the superior petrous border the fossa is pierced by a small vein. In infants the fossa is a relatively large blind tunnel under the anterior semicircular duct, it corresponds to the floccular fossa of some animals.

In addition to its emissary vein the *mastoid foramen* transmits a meningeal branch of the occipital artery, sometimes large enough to groove the occipital squama. The internal occipital crest, to

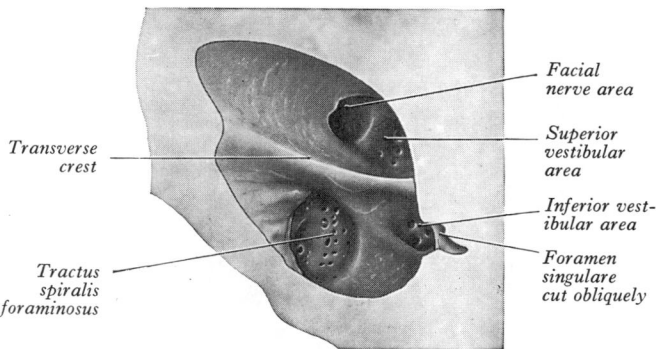

3.113 The fundus of the right internal acoustic meatus, exposed by a section through the petrous part of the right temporal bone nearly parallel to the line of its superior border.

which falx cerebelli is attached, may be grooved by the the occipital sinus. Its lower end adjoins the inferior vermis. The internal occipital protuberance is close to the confluence of sinuses and bilaterally grooved by the transverse sinuses, to the margins of which are attached the two layers of the tentorium cerebelli. Bone flanking the internal occipital crest is thin, even translucent, contrasting with the occipital's thick crest and protuberance.

The Nasal Cavity

The nasal cavity, an irregular space between buccal roof and cranial base, is divided by a vertical *septum* (**3.**114), approximately median. In macerated skulls the septum is deficient anteriorly, leaving a single *anterior nasal aperture* in norma frontalis, but it reaches the posterior limit of the cavity, leading into the nasopharynx through a pair of *posterior nasal apertures*, above the posterior osseous (hard) palatal border. The cavity is wider below than above, but widest and vertically deepest in its central region. It communicates with frontal, ethmoidal, maxillary and sphenoidal sinuses. Each half cavity has a roof, floor, lateral and medial walls, the medial being the nasal septum.

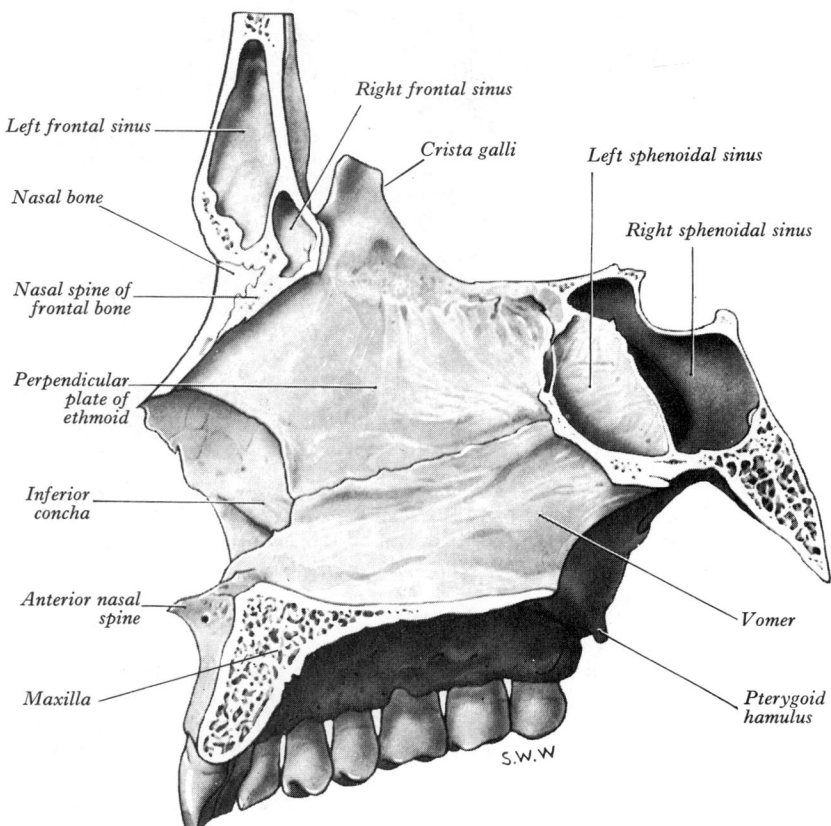

3.114 The bony nasal septum: left side. An enlarged part of **3.**117.

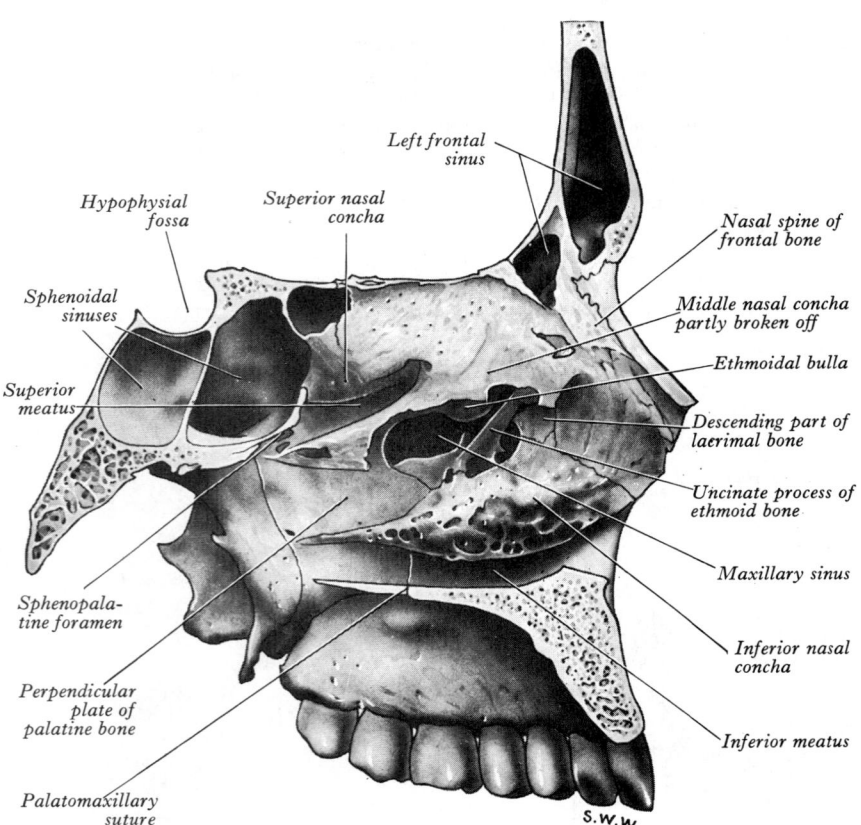

Lacrimal bone

Uncinate process of ethmoid

Middle concha

Superior concha

Left frontal sinus

Right frontal sinus

Inferior concha

Incisive fossa and canal

Sphenoidal sinus

Medial pterygoid plate

Sphenopalatine foramen

Perpendicular plate of palatine bone

3.115 The roof, floor and right lateral wall of the right nasal cavity.

Left frontal sinus

Hypophysial fossa

Superior nasal concha

Sphenoidal sinuses

Superior meatus

Sphenopalatine foramen

Perpendicular plate of palatine bone

Palatomaxillary suture

Nasal spine of frontal bone

Middle nasal concha partly broken off

Ethmoidal bulla

Descending part of lacrimal bone

Uncinate process of ethmoid bone

Maxillary sinus

Inferior nasal concha

Inferior meatus

S.W.W.

3.116 The lateral wall of the left nasal cavity, with an irregularly shaped portion removed from the lower part of the middle concha, exposing much of the middle meatus.

The roof (3.114–117) is centrally horizontal but descends in front and behind. The anterior slope is formed by the frontal's nasal spine and nasal bones, contributing to the external nose. The horizontal region is the ethmoid's cribriform plate separating the nasal cavity and median region of anterior cranial fossa's floor. Its numerous small perforations contain the olfactory nerves. The posterior slope is formed by the sphenoid's body and is interrupted, on each side, by an orifice of a sphenoidal sinus.

The floor is smooth, concave transversely, and slopes up from

anterior to posterior aperture. It is the upper surface of the osseous palate. Anteriorly the maxillary palatine processes and, behind them, palatine horizontal plates articulate in the midline and with each other. Anteriorly, near the septum, a small infundibular opening in the nasal floor leads into the *incisive canals* (p. 389).

The medial wall or *nasal septum* (3.114,117), between roof and floor, is a thin sheet of bone with a wide anterior deficiency occupied by septal cartilage; its bony part is largely vomer and perpendicular plate of the ethmoid. The *vomer* extends from the sphenoidal body to bony palate, forming the postero-inferior region, including the posterior border; it is furrowed by vessels and nerves. The *perpendicular plate of the ethmoid* forms the septum's anterosuperior part (3.114), continuous above with cribriform plate (p. 384). The septum is often deviated, most commonly at the vomero-ethmoidal suture.

The lateral wall (3.115,116,148) is irregular, due to three projections: inferior, middle and superior nasal *conchae*. It is formed largely by the maxilla antero-inferiorly, by the palatine's perpendicular plate posteriorly and superiorly by the ethmoidal labyrinth, separating nasal cavity from orbit. The conchae curve inferomedially, each roofing a groove, or meatus, open to the nasal cavity.

The *inferior concha*, thin, curved and an independent bone, articulates with the nasal surface of maxilla and palatine perpendicular plate; its free lower border is gently curved; the subjacent *inferior meatus* reaches the nasal floor. It is the largest meatus, extending along almost all the lateral nasal wall. It is deepest at the junction of anterior and middle thirds, where the inferior orifice of the nasolacrimal canal appears.

The *middle* and *superior conchae* are medial processes of the ethmoidal labyrinth. The *middle concha*, much the larger, extends back to articulate with the palatine bone's perpendicular plate above the *middle meatus*, whose lateral wall can be examined only after the removal of the concha (3.116); its upper part is the rounded *ethmoidal bulla*, which contains middle ethmoidal air cells; antero-inferior to this a thin, curved *uncinate process of the ethmoid* descends back across the large opening of the maxillary sinus. The curved gap (3.116) between process and bulla is the *hiatus semilunaris*; at its upper end it is continuous with the *ethmoidal infundibulum*, a short canal receiving anterior ethmoidal air cells and then ascending in the labyrinth to the frontal sinus. But the infundibulum very often ends blindly; the frontal sinus then opens directly into the middle meatus. The middle ethmoidal air cells open above, or near, the bulla.

The *superior concha*, a small curved lamina, posterosuperior to the middle, roofs the *superior meatus*, shortest and shallowest of the three; into it open the posterior ethmoidal air cells. Posterior to it the sphenopalatine foramen (occluded by mucosa) leads into the pterygopalatine fossa. A narrow *spheno-ethmoidal recess* separates superior concha and anterior aspect of the sphenoid's body, through which the sphenoidal sinus connects with the nasal cavity.

The posterior nasal apertures, or *choanae*, are separated by the posterior vomerine border, each being limited below by the posterior border of the palatine bone's horizontal plate, above by the sphenoid and laterally, on each side, by its medial pterygoid plates.

Further details. The anterior nasal aperture is described on page 343. In addition to the cribriform plates the horizontal part of the roof has a separate, anterior foramen for the anterior ethmoidal nerve and vessels. The roof's posterior slope is the anterior aspect of the sphenoid's body, with which the sphenoidal conchae (p. 325) are fused; below this are the vomerine alae and palatine bones' sphenoidal processes.

The nasal floor (3.115) is crossed at the junction of middle and posterior thirds by the palatomaxillary suture. Anteromedian are the incisive canals (p. 389), descending to the palatine incisive fossa; they may traverse the union of os incisivum (premaxilla) with maxillae; they represent a primitive bucconasal communication (p. 389).

At upper and lower limits of **the medial wall** (3.116) other bones make minor contributions to the septum: antero-superior are the nasal bones and frontal's nasal spine, posterosuperior are the sphenoid's rostrum and crest, and inferiorly, the maxillary and palatine nasal crests. The vomer is grooved by nasopalatine nerves and vessels.

The lateral wall (3.115,116,148), additionally, is formed anterosuperiorly by the nasal bone and maxillary's frontal process. Behind the latter, and articulating with its posterior border, lies the lacrimal bone which reaches the middle meatus to articulate with the lacrimal process of the inferior concha, thus forming the nasolacrimal canal's medial wall (3.116,151), which conveys the nasolacrimal duct to the inferior meatus. Posteriorly the lacrimal bone articulates with the ethmoidal labyrinth to complete some ethmoidal air cells. The *uncinate process* from this part of the labyrinth curves postero-inferiorly in the lateral middle meatal wall. It is thin, fragile, about 3 mm wide, curving across the maxillary hiatus to the inferior conchal ethmoidal process. Its concave, posterior border forms the medial edge of the hiatus semilunaris; it helps to form a medial wall of the maxillary sinus; its convex anterior border is free only above. The *maxillary hiatus*, a wide defect in the maxillary's nasal surface (3.160), is much reduced by neighbouring bones; it is covered below by the inferior concha and its maxillary process, above by ethmoid's uncinate process, behind by palatine's perpendicular plate and antero-superiorly by small parts of ethmoidal labyrinth and lacrimal bone (3.116). Thus the hiatus is sometimes reduced to a single orifice in the floor of the posterior part of the hiatus semilunaris, but usually additional openings exist both between the inferior concha and the uncinate process and also behind the latter. The *ethmoidal bulla*, variable in size and shape, may be fused with the uncinate process, and then the duct of the frontal sinus opens into middle meatus medial to the blind end of the infundibulum. A *concha suprema* often exists on the medial surface of the ethmoidal labyrinth posterosuperior to the superior concha's posterior end (spheno-ethmoidal recess); it is merely a ridge, separated from the superior concha by a depression. The *sphenopalatine foramen* (3.115) is posterior to the superior meatus and transmits the sphenopalatine artery and nasopalatine and superior nasal nerves from the pterygopalatine fossa. The foramen is bounded above by sphenoidal body and concha, below by the superior border of palatine perpendicular plate, in front and behind by the palatine's orbital and sphenoidal processes (3.160–162).

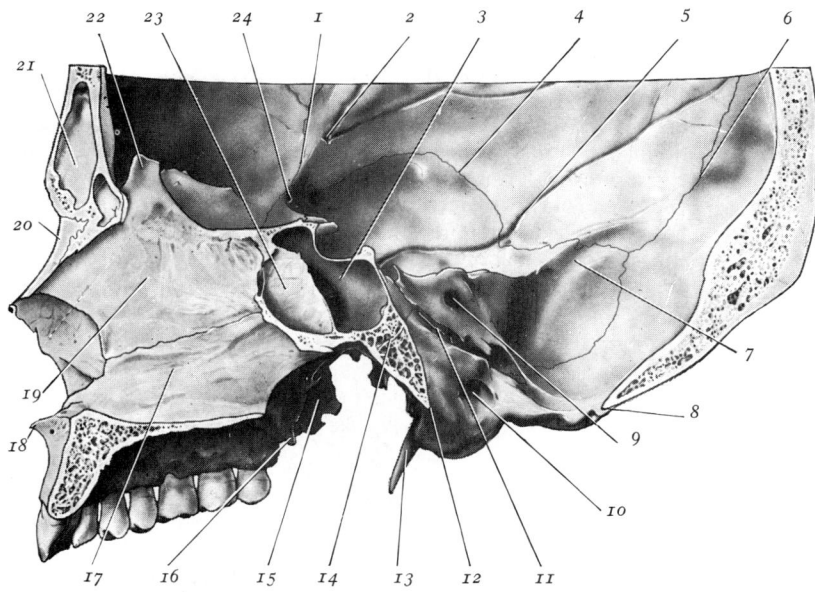

3.117 Sagittal section through the lower part of the skull, slightly to the left of the median plane. 1. Frontosphenoidal suture. 2. Bony canal for middle meningeal vessels, frontal branches, upper orifice. 3. Right sphenoidal sinus. 4. Squamosal suture. 5. Groove for parietal branches of middle meningeal vessels. 6. Lambdoid suture. 7. Groove for transverse sinus. 8. Posterior margin of foramen magnum. 9. Internal acoustic meatus. 10. Hypoglossal canal. 11. Petro-occipital suture in floor of groove for inferior petrosal sinus. 12. Anterior margin of foramen magnum. 13. Styloid process. 14. Line of occipitosphenoidal junction. 15. Lateral pterygoid plate. 16. Pterygoid hamulus. 17. Vomer. 18. Anterior nasal spine. 19. Perpendicular plate of ethmoid. 20. Nasal bone. 21. Frontal sinus. 22. Crista galli. 23. Left sphenoidal sinus. 24. Bony canal for frontal divisions of middle meningeal vessels, lower orifice.

THE INDIVIDUAL CRANIAL BONES

The Mandible

The mandible (3.118A,B) the largest, strongest and lowest bone in the face, has a horizontally curved *body*, convex forwards, and two broad *rami*, ascending posteriorly.

The mandibular body, somewhat U-shaped, has external and internal surfaces, separated by upper and lower borders. Anteriorly, the upper *external surface* shows a faint median ridge, often absent, indicating fusion of the halves of the fetal bone (*symphysis menti*); inferiorly this ridge divides to enclose a triangular *mental protuberance*, its base centrally depressed but raised on each side as a *mental tubercle*. Below the interval between the premolar teeth, or the second premolar, is the *mental foramen*, from which emerge the mental nerve and vessels; its posterior border is smooth accommodating the dorsolaterally emerging nerve (Warwick 1950). A faint *oblique line* ascends backwards from each mental tubercle, sweeps below the foramen, then becomes more marked as it continues into the anterior border of the ramus.

The body's lower border, the *base*, extends posterolaterally from the symphysis into that of the ramus behind the third molar tooth. Near the midline, on each side, is a rough *digastric fossa*, behind which the base is thick, rounded, with a slight anteroposterior convexity. As the ramus is approached, this changes to a gentle concavity; thus, in profile, the whole base is sinuous.

The upper border, the *alveolar part*, contains 16 *alveoli* for roots of teeth, varying in size and depth, some being multiple.

The *internal surface* is divided by an oblique *mylohyoid line*, sharp and distinct near the molars, faint in front and extending from behind the third molar, a centimetre from the upper border,

to the mental symphysis between the digastric fossae. Below this line is the slightly concave *submandibular fossa*; the area above it widens anteriorly into a triangular *sublingual fossa*. Above the latter and extending back to the third molar, the bone is covered by oral mucosa. Above the anterior ends of the mylohyoid lines, the posterior symphysial aspect has a small elevation, often divided into upper and lower parts, the *mental spines* (genial tubercles). Posteriorly the *mylohyoid groove* extends down and forwards from the ramus below the mylohyoid line's posterior part. Superior to the mental spines most mandibles display a median pit opening into a canal (Ingram, personal communication). As yet its development and contents are uncertain but it is a useful radiological landmark (p. 395): the name *genial foramen* has been proposed. (See also *accessory mandibular foramina*, p. 1304.) Above the mylohyoid line, medial to the molar roots, a rounded torus mandibularis sometimes appears: for its incidence consult Mayhall et al (1970) and Berry (1975).

The mandibular ramus (3.118A,B) is quadrilateral, with two surfaces, four borders and two processes. The flat *lateral surface* has oblique ridges in its lower part; the *medial* presents, a little above centre, an irregular *mandibular foramen*, leading into the *mandibular canal*, curving down and forwards into the body to its mental foramen (but vide infra and p. 1304). Anteromedially the foramen is overlapped by a thin, triangular *lingula*. The *mylohyoid groove* descends forwards from behind the lingula; the surface behind it is marked by short ridges. The *inferior border*, continuous with mandibular base, meets the posterior border at the *angle*. This is typically everted, but in females frequently incurved. The thin *upper border* bounds the *mandibular incisure*, 367 surmounted in front by the somewhat triangular, flat, *coronoid process*, behind by a strong *condylar process*. The *posterior border*, thick and rounded, extends from condyle to angle, being gently convex backwards above, and concave below; it is in contact with the parotid gland. The *anterior border* is thin above and continuous with that of the coronoid process, and thicker below and continuous with the oblique line.

The coronoid process projects up and slightly forwards. Its posterior border bounds the mandibular incisure, its anterior continues into that of the ramus. Its margins and medial surface are attachments for most of the temporalis.

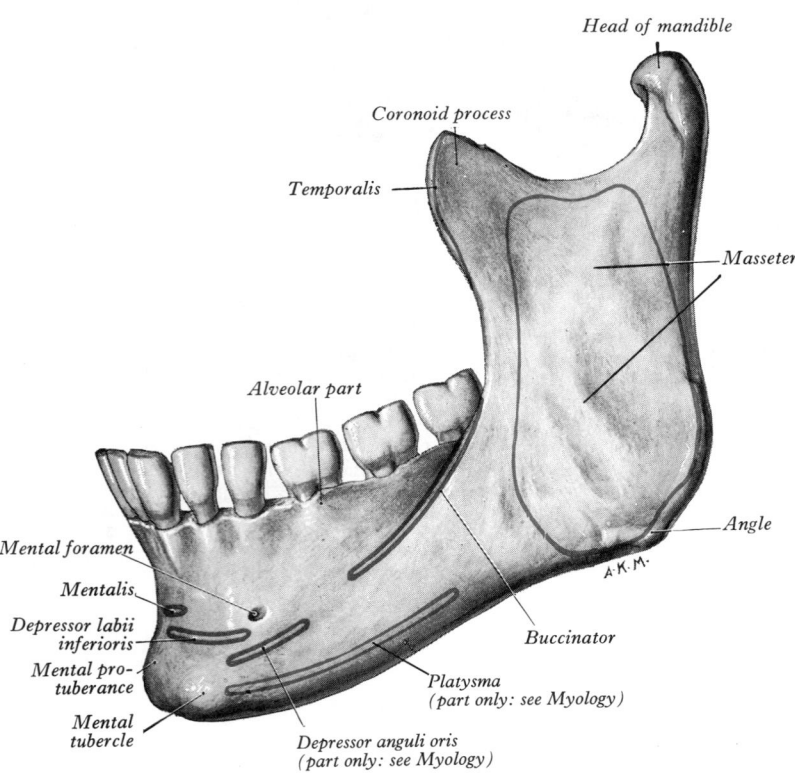

3.118A The left half of the mandible: lateral (external) aspect.

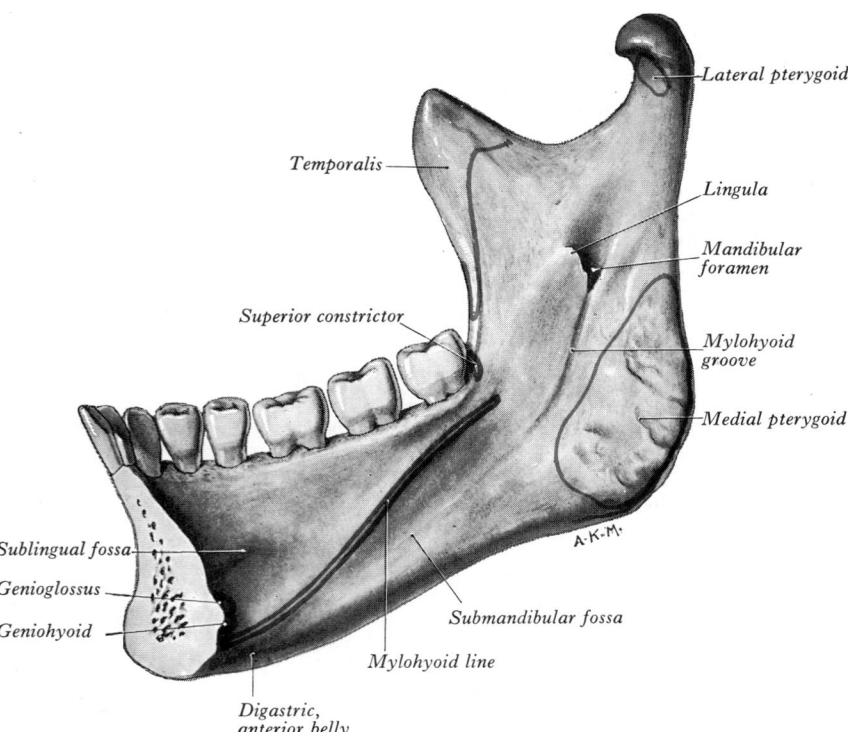

3.118B The right half of the mandible: medial (internal) aspect.

The condylar process is apically enlarged as a *head* or *condyle*, covered by fibrocartilage. It articulates with the temporal bone's mandibular fossa, with an articular disc between. It is convex in all directions, its transverse dimension greater. Its lateral aspect is a blunt projection, palpable in front of the auricular tragus. As the mouth opens the condyle descends forwards, admitting a finger-tip towards its vacated fossa. Below the head is the narrower *neck*, slightly flattened from before backwards, its anterior aspect overlapped laterally by the mandibular incisure's margin, medial to which the neck's anterior surface bears a rough *pterygoid fovea*.

The mandibular canal descends obliquely forwards in the ramus from the mandibular foramen, then horizontally forwards in the body below the alveoli, with which it communicates by small canals. It contains the inferior alveolar nerve and vessels, from which branches enter dental roots, periodontal sockets and septa. Between the roots of the first and second premolars, or below the second, the canal divides into *mental* and *incisive* parts; the mental canal swerves up, back and laterally to the mental foramen; the incisive canal continues below the incisor teeth. (See p. 1304 for variations.)

The mandibular body bears a small shallow *incisive fossa* below the incisors, an attachment for mentalis and part of orbicularis oris. To the anterior ends of the oblique lines are attached depressores labii inferioris and anguli oris, and platysma to bone below and backwards beyond them. Adjoining the alveolar border bone is covered by oral mucosa and, below this in the molar region, the buccinator has a linear attachment extending medially behind the last molar to the pterygomandibular raphe.

To the mylohyoid line is attached the mylohyoid muscle and, above its posterior end the superior pharyngeal constrictor, some retromolar fasicles of the buccinator, and the *pterygomandibular raphe* behind the third molar. Although usually described separately, here the constrictor, buccinator and the raphe are blended and jointly attach to the mandibular periosteum. The lingual nerve reaches the tongue above the mylohyoid line, closely related to bone near its posterior end (p. 1105); often the nerve is accommodated in a shallow, curved groove. To the superior mental spines are attached the genioglossi, to the inferior the geniohyoids. The *submandibular fossa* adjoins submandibular lymph nodes as well as the salivary gland; the facial artery usually descends here to curl round the base of the mandible, sometimes making a shallow groove. The *digastric fossa* is for attachment of the muscle's anterior belly.

The mandibular ramus and its processes provide attachment for muscles of mastication, much of its *lateral surface* to masseter, except posterosuperiorly, where it is covered by the parotid gland; the *medial surface* receives the medial pterygoid on the roughened area postero-inferior to the mylohyoid groove. To the *lingula* the sphenomandibular ligament is attached, posterior to which the mylohyoid nerve and vessels enter the *mylohyoid groove*, reaching the mandibular body below the mylohyoid line; they then pass superficial to mylohyoid. Below the lingula, but above the roughened attachment mentioned above, the medial surface of the ramus is related to the medial pterygoid, the lingual nerve being between muscle and bone. The lowest attachment of temporalis descends beyond the coronoid process to the anterior ramal border and particularly its adjoining medial surface. To the area posterosuperior to the mandibular foramen the maxillary artery and its inferior alveolar branch are related, and lateral pterygoid to the area near the mandibular incisure. The *mandibular incisure* transmits the masseteric nerve and vessels from the infratemporal fossa.

The coronoid process is covered laterally by the anterior par of masseter descending to its attachment on the ramus. Its anterior border is palpable below the zygoma; this is most eviden during mouth opening. **The condylar process** projects more a its medial pole. Its *articular surface* descends only a little on it anterior surface, covers the whole of its superior aspect an descends 5 mm posteriorly. Its projecting lateral part is separate from the cartilaginous external acoustic meatus by the parotic Laterally on its *neck* the joint's lateral ligament is attached (4.18 covered by parotid gland. The pterygoid fovea, anterior on th neck, receives the lateral pterygoid. The neck's medial surface is related to the auriculotemporal nerve above and maxillary artery below.

The *parotid gland* is below the external acoustic meatus and lies between the ramus and mastoid process, with the styloid process medial; but it extends forwards lateral to the temporomandibular joint and to the exposed ramus behind the masseter. It also curls round the posterior border to the medial aspect of the ramus above the attachment of the medial pterygoid.

Accessory foramina of the mandible are usually unnamed and infrequently described. Yet a study of 300 mandibles yielded a count of 2449 accessory foramina (Sutton 1974). Since many transmit auxiliary nerves to teeth (from facial, mylohyoid, buccal, transverse cervical cutaneous and other nerves), their occurrence is significant in dental blocking techniques. **Further mandibular variants** include *lingual depressions, molar or canine, variable position of mental foramen, multiple mental foramina, lingual fenestrations* of molar sockets, *retromolar foramina* and *condylar defects*. For incidences in 125 mandibles see Azas & Lustmann (1973).

Aspects of mandibular structure. In addition to variable mandibular canals (Fawcett 1895, Carter & Keen 1971), numerous analyses have been made of the structure of surface tables and buttresses of compact bone and the geometry of trabeculation in attempts to relate these to habitual functional stresses (Beltrani 1945, Seipel 1948, Dal Pont 1960, Scott & Symons 1977, Mercier et al 1970 a,b). Holographic interferometry has been used to study surface strains induced by orthodontic forces (Hewitt 1977).

Ossification. The mandible forms in dense fibromembranous tissue lateral to the inferior alveolar nerve and its incisive branch and also in the lower part of Meckel's cartilage (3.119A,B). Each half is ossified from a centre appearing near the mental foramen about the sixth week, i.e. just after the clavicle's primary centre (p. 405). From this, ossification spreads medially and dorsocranially to form body and ramus, first below, then around the inferior alveolar nerve and incisive branch and upwards, initially forming a trough and later crypts for developing teeth. By the tenth week Meckel's cartilage below the incisor rudiments is surrounded and invaded by bone (p. 172). Secondary cartilages appear later (3.119A,B): a conical mass, the *condylar cartilage*, extends from mandibular head down and forwards in the ramus, contributing to its growth in height; though it is largely replaced by bone by mid-fetal life, its proximal end persists as proliferating cartilage under articular fibrocartilage until the third decade. The orientation and growth patterns in the condylar cartilage are one (of many) important determinants of co-ordinated craniofacial growth. Another secondary cartilage, which soon ossifies, appears along the anterior coronoid border, disappearing before birth. One or two cartilaginous nodules also occur at the symphysis menti; about the seventh month these may ossify as variable *mental ossicles* in symphysial fibrous tissue, uniting to adjacent bone before the end of the first postnatal year (Lebourg & Champagne 1951, Sicher 1962, Scott & Symons 1970, and 3.120A). It may be mentioned here that ^{3}H-thymidine labelling in the *rat* showed no interstitial growth in condylar cartilage, which expanded by surface accretion only (Kvinnsland & Kvinnsland 1975).

Age Changes in the Mandible

At birth (3.120A,B) the two halves of the mandible are united by a fibrous *symphysis menti*. Anterior ends of both rudiments are covered by cartilage, separated only by a symphysis. Until fusion occurs new cells are added to each cartilage from symphysial fibrous tissue, ossification on its mandibular side proceeding towards the midline; when the latter process overtakes the former, extending into median fibrous tissue, the symphysis fuses but details are uncertain. At this stage the body is a mere shell, enclosing imperfectly separated sockets of deciduous teeth. The mandibular canal is near the lower border, the mental foramen opens below the first deciduous molar and is directed forwards (Warwick 1950). The coronoid process projects above the condyle.

In the first to third postnatal years (3.120) the two halves join at their symphysis from below upwards; separation near the alveolar margin may persist into the second year. The body elongates,

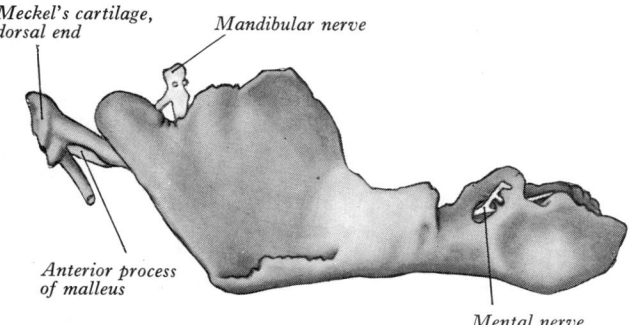

Meckel's cartilage, dorsal end

Mandibular nerve

Anterior process of malleus

Mental nerve

3.119A The right half of the mandible of a human embryo, 95 mm long: lateral aspect. Blue = cartilage; yellow = bone. (Reconstruction by A Low.)

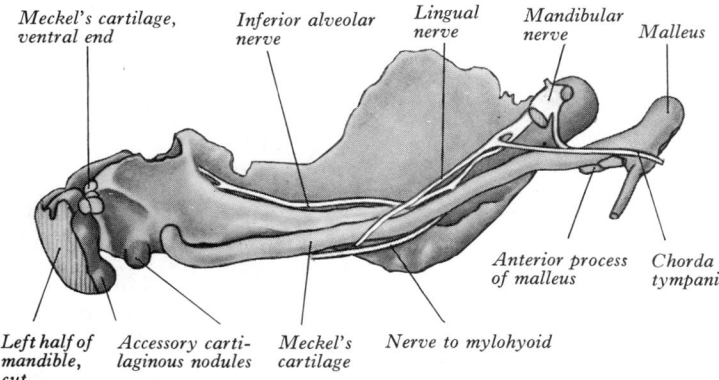

Meckel's cartilage, ventral end

Inferior alveolar nerve

Lingual nerve

Mandibular nerve

Malleus

Anterior process of malleus

Chorda tympani

Left half of mandible, cut

Accessory cartilaginous nodules

Meckel's cartilage

Nerve to mylohyoid

3.119B The right half of the mandible of a human embryo, 95 mm long: medial aspect. The developing bone is shown in yellow, and the cartilage in blue. (Reconstruction by A Low.)

especially behind the mental foramen, providing space for three additional teeth (pp. 1317–1318). During the first and second years, as a chin develops, the mental foramen alters direction from anterior to posterosuperior, and then almost horizontally posterior, as in adults (Warwick 1950), accommodating a changing direction of the emerging mental nerve. The proximal zone of the conical condylar cartilage persists as an epiphysial plate. Its proliferation contributes to vertical increase in the ramus and to general mandibular growth which is essentially *downwards and forwards*. With its antimere it also adapts the intercondylar distance to the widening cranial base. The condylar cartilage is covered on its articular aspect by self-perpetuating fibrous tissue, deep to which a proliferating intermediate zone is responsible for ramal growth (and thus displacement of the whole mandible). Beneath this are hypertrophic chondrocytes and then bone (Blackwood 1959). As depth of the body increases, alveolar growth makes room for roots of teeth, the subalveolar region becoming thicker and deeper. After eruption of permanent teeth (3.120D,E) the mandibular canal is a little above the mylohyoid line, and the mental foramen occupies its adult position. As the mandible increases in size, bone is added at posterior borders of ramus and coronoid process, absorption occuring at their anterior borders. This remodelling is continuous until adult size is reached, allowing alveolar parts to accommodate permanent molar teeth (Enlow & Harris 1964). Organ culture of condylar cartilage in rats shows no growth potential, but in vivo all mandibular growth cartilages respond to mechanical forces (Petrovic 1972).

In adults (3.120E) alveolar and subalveolar regions are about equal in depth, the mental foramen midway between upper and lower borders; the mandibular canal nearly parallels the mylohyoid line. The *angle* between the lower border of body and a plane touching the posterior surface of condyle above, and ramus below, diminishes as ramal height increases with age; but X-ray photographs at different ages (Brodie 1941) show that its *(gonial) contour* remains unaltered.

In old age (**3**.120F) the bone is reduced in size as teeth are lost and the alveolar region absorbed; the mandibular canal and mental foramen are nearer the superior border. Both may even disappear, exposing the inferior alveolar nerve (Gabriel 1958). The ramus becomes oblique, the angle about 140°, and the neck inclined

Symphysis menti

Mental ossicles

A

B

Birth

C

3 years

D

6 years

E

Adult

F

Old age

3.120 The mandible at different periods of life.

backwards (Fawcett et al 1924). Absorption affects chiefly the thinner alveolar wall (lingual or labial) and, after completion, a linear *alveolar ridge* is left at the superior border of the mandible (and inferior border of the maxilla). In the mandible the labial wall is thinner in incisor and canine regions, the lingual wall in the molar. The mandibular alveolar ridge hence is within the line of teeth in the former but outside it in molar regions, forming a curve wider posteriorly than that of the line of the teeth, but intersecting it near the premolars. In the maxilla, however, the labial wall is everywhere thinner and after absorption its maxillary alveolar ridge is entirely within the line of teeth.

The Hyoid Bone

The U-shaped or *hyoid* bone (**3**.121A,B) is suspended from the tips of the (bilateral) styloid processes by the stylohyoid ligaments. It has a body, two greater and two lesser cornua. **The body** is irregular, elongated and quadrilateral. Its *anterior surface* is convex, facing anterosuperiorly, and is crossed by a transverse ridge with a slight downward convexity; often a vertical median ridge bisects the body; the upper part of this vertical ridge is usually present, the lower part is rare. The *posterior surface* is smooth, concave, faces postero-inferiorly, and is separated from the epiglottis by the thyrohyoid membrane and loose areolar tissue, with a bursa between bone and membrane. In early life the **greater cornua** are connected to the body by cartilage, but after middle age they are usually united by bone. They project back (curving dorsolaterally) from the lateral ends of the body; horizontally flattened, they taper posteriorly, each ending in a tubercle. When the throat is gripped between finger and thumb above the thyroid cartilage, the greater cornua can be identified and the bone can be moved from side to side.

The lesser cornua are two small, conical projections at the junctions of the body and greater cornua, connected basally to the body by fibrous tissue and occasionally to the greater cornua by synovial joints, which occasionally become ankylosed (vide infra).

Attached to most of the body's *anterior surface* is geniohyoid, above and below the transverse ridge; but the medial part of hyoglossus invades the lateral geniohyoid area (**3**.121B). The lower anterior surface gives attachment to mylohyoid, above the sternohyoid medially and omohyoid laterally. To the rounded *superior border* the lowest fibres of the genioglossi, the hyo-epiglottic ligament and (most posteriorly) the thyrohyoid membrane are attached; to the *inferior border* are attached sternohyoid medially and omohyoid laterally, sometimes with the medial fibres of thyrohyoid and levator glandulae thyroideae, when present. The oblique postero-inferior surface was mentioned above.

To the *upper surface of each greater cornu* the middle pharyngeal constrictor and, more laterally (superficially), hyoglossus, are attached along its whole length. Near the junction of cornu with body the stylohyoid muscle is attached, lateral to hyoglossus, and a little posterior to this, the fibrous loop for the digastric tendon. To the *medial border* is attached the thyrohyoid membrane, to the *lateral border* anteriorly the thyrohyoid. The oblique *inferior surface* is separated from the thyrohyoid membrane by fibro-areolar tissue.

At the *posterior* and *lateral aspects of lesser cornua* are the attachments of the middle pharyngeal constrictors. To their apices are attached the stylohyoid ligaments, often partly ossified, and to the medial aspects of their bases, the chondroglossi.

Ossification. The hyoid bone is evolved from cartilages of second and third visceral arches, the lesser cornua from second, the greater from the third and the body from fused ventral ends of both (p. 172). Chondrification begins in the fifth fetal week in these elements, completed in the third and fourth months. Ossification proceeds from six centres—a pair for the body, one for each cornu—commencing in the greater cornua towards the end of intrauterine life, in the body shortly before or after birth, and in lesser cornua around puberty. The greater cornual apices remain cartilaginous until the third decade and epiphyses may occur here. They fuse with the body. Synovial joints between the greater and lesser cornua may be obliterated by ossification in later decades.

The Occipital Bone

The occipital bone (3.122–124), forming much of the cranium's back and base, is trapezoid and internally concave. It encloses 'basally' the *foramen magnum*; the expanded plate posterosuperior to this is the *squama (squamous part)*, the massive quadrilateral part anterior to it being the *basilar part (basi-occipital)*; on each side of the foramen is a *lateral part (exoccipital)*.

The squama is convex externally and concave internally. The *external surface* presents, midway between summit and foramen magnum, the *external occipital protuberance*. On each side two curved lines extend laterally from this; the upper, faintly marked and often almost imperceptible, is the *highest nuchal line* to which the epicranial aponeurosis is attached; the lower is the *superior nuchal line*. The surface above the highest nuchal lines is smooth and covered by the occipital part of occipitofrontalis; below this it is rough and irregular for attachment of muscles. From the external occipital protuberance the median *external occipital crest*, often faint, descends to the foramen magnum and is an attachment of the ligamentum nuchae; on each side an *inferior nuchal line* spreads laterally from the crest's midpoint. Areas of muscular attachment are shown in 3.100,104,123. The posterior atlanto-occipital membrane is attached posterolaterally just outside the foramen magnum's margin.

The squama's *internal surface* is divided into four deep fossae by an irregular *internal occipital protuberance* and by ridged sagittal and horizontal extensions from it. The two superior fossae are triangular and adapted to the cerebral occipital poles, the inferior are quadrilateral and shaped to accommodate the cerebellar hemispheres. A wide groove, with raised banks, ascends from the protuberance to the squama's superior angle—the *superior sagittal sulcus*. The posterior part of the falx cerebri is attached to its margins. A prominent *internal occipital crest* descends from the protuberance, for attachment of the falx cerebelli, and bifurcates near the foramen magnum; the occipital sinus, sometimes double, lies in this attachment. At the crest's lower end a small *vermian fossa* may exist, occupied by part of the inferior cerebellar vermis. On each side a wide *sulcus for the transverse sinus* extends laterally

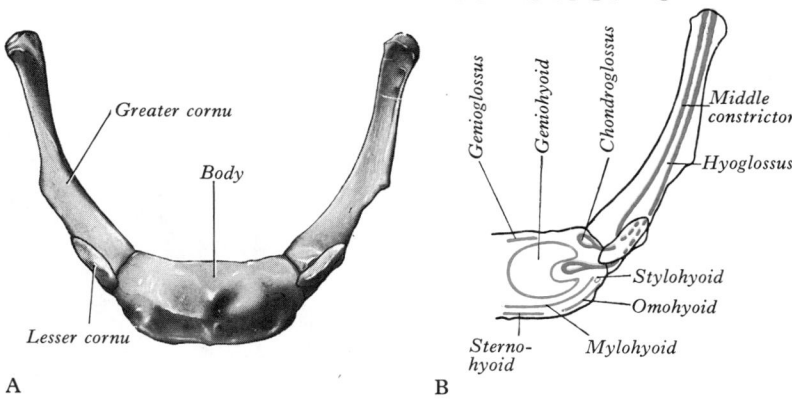

3.121 A The hyoid bone: anterosuperior aspect. B Drawing of the left half of the hyoid bone to show the muscular attachments: superior aspect.

from the protuberance; to the margins of these sulci the tentorium cerebelli is attached. The right sulcus is usually larger, passing into the sulcus for the superior sagittal sinus, but the left may be larger or both almost equal in size. The position of this *confluence of sinuses* is indicated by a depression on one side of the protuberance.

The *superior angle* at the squama's summit meets the occipital angles of the parietal bones, the position of the fetal *posterior fontanelle*. The *lateral angles* of the squama, marked internally by ends of the transverse sulci, project between parietal and temporal bones. The *lambdoid borders* extend from superior to lateral angles, serrated for articulation with the occipital borders of the parietals at the *lambdoid suture*. The *mastoid borders* extend from lateral angles to jugular processes, articulating with the mastoid parts of temporal bone. A variety of ossicles may occur at or near the lambda (p. 386 and Srivastava 1977), e.g. the 'interparietal' (Inca bone or ossicle of Goethe).

The occipital's basilar part extends anterosuperiorly from the foramen magnum, fusing with the sphenoid in adults. In

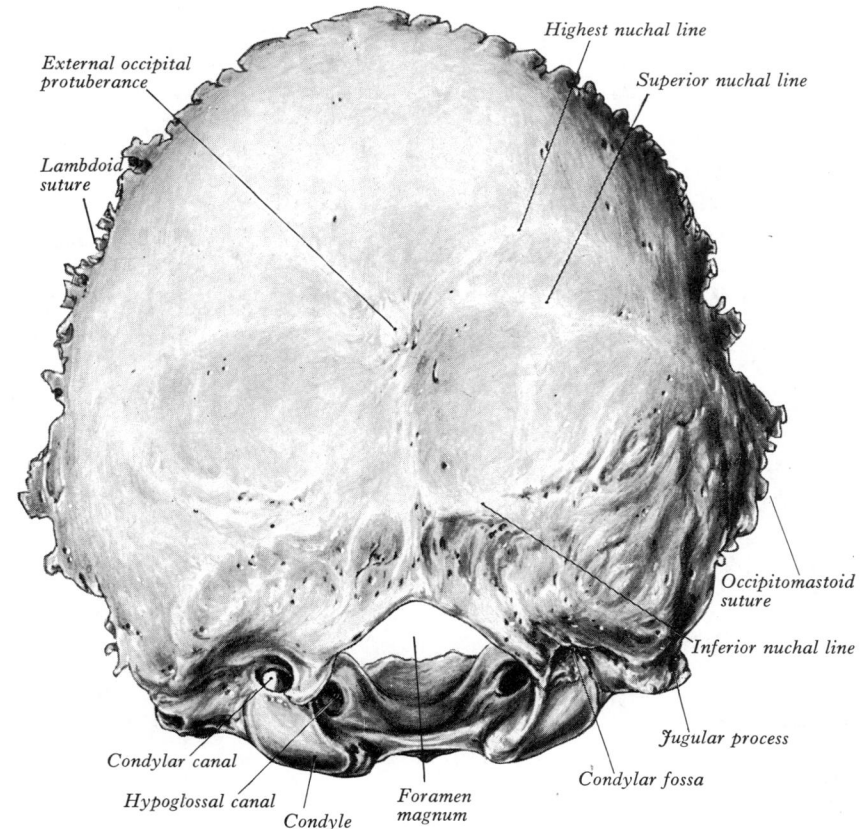

3.122 The occipital bone: posterior aspect. The condylar canal was present on the left side only in this specimen.

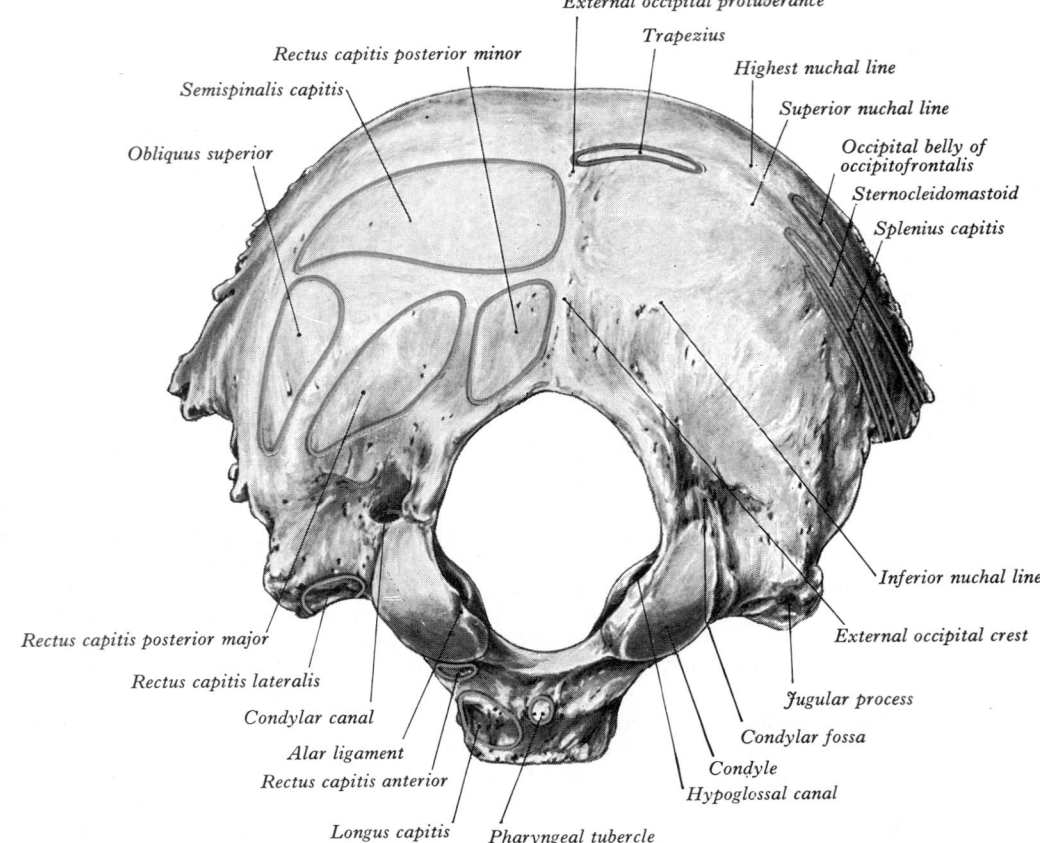

3.123 The occipital bone: inferior aspect. Drawn from the same specimen as **3.**122.

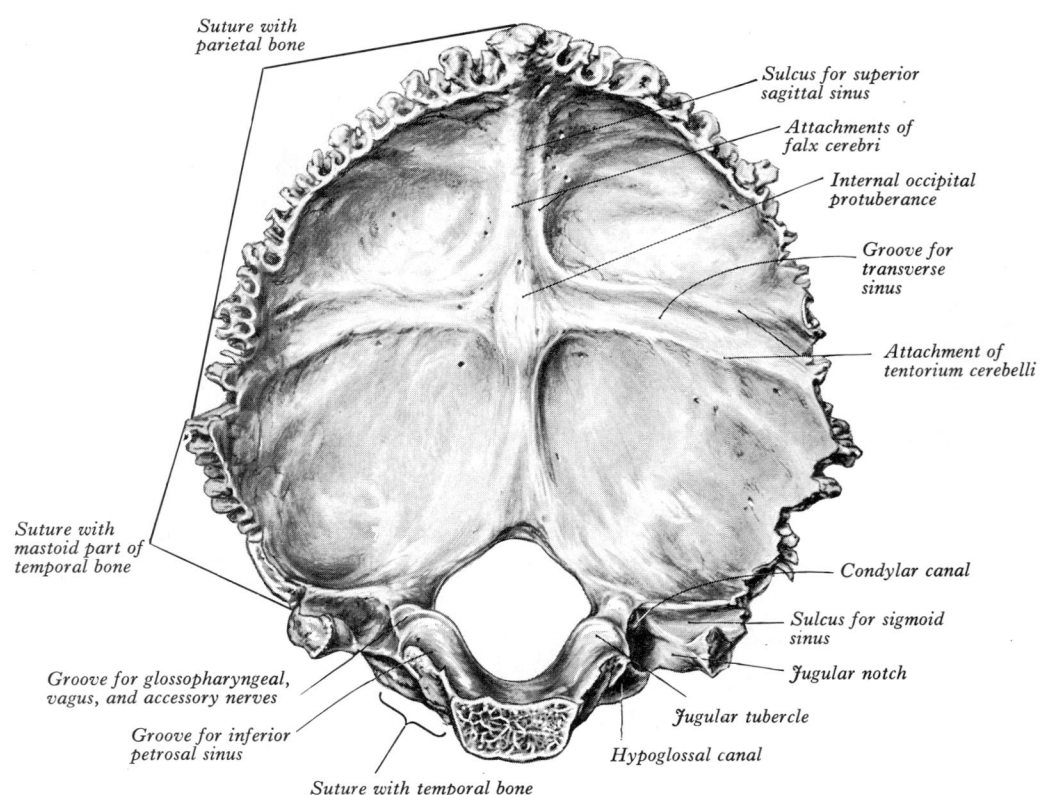

3.124 The occipital bone: internal aspect.

young skulls a rough and uneven surface is joined to the body of the sphenoid by a growth cartilage. By the twenty-fifth year this plate has ossified and occipital and sphenoid bones are fused.

The inferior surface of the basi-occipital, about 1 cm anterior to foramen magnum, bears a small *pharyngeal tubercle* for attachment of the fibrous pharyngeal raphe. Longus capitis is attached anterolateral to the tubercle and rectus capitis anterior to a small depression immediately anterior to the occipital condyle which, on occasion, may be replaced by a small *precondylar tubercle*, of low incidence, but relatively frequent in Mexican and Burmese crania (p. 396). To the anterior margin of the foramen the anterior atlanto-occipital membrane is attached.

The *superior basi-occipital surface* is a broad groove, part of the *clivus*, which ascends anteriorly from foramen magnum; above it are the medulla oblongata and lower pons. Near the foramen the membrana tectoria and axial apical ligament are attached to the clivus. On its lateral margins are *sulci of the inferior petrosal sinuses*, below which the lateral margins articulate with petrous temporal bones.

The lateral (condylar) parts of the occipital bone flank the foramen magnum; on their *inferior surfaces* are *occipital condyles* for articulation with superior atlantal facets. They are oval or reniform, their long axes converging anteromedially. Their anterior ends invade the basi-occipital, level with the foramen's centre. The articular surfaces, wholly convex, face inferolaterally; they are occasionally constricted and a condyle may be in two parts (p. 396). The atlantal superior articular facets are also frequently constricted and occasionally 'double' (Singh 1965). Medial to each facet a tubercle gives attachment to an alar ligament. Anteriorly above each condyle is a *hypoglossal canal*, which starts internally a little above the anterolateral part of the foramen magnum and continues anterolaterally. It may be partly or wholly divided by a spicule of bone and transmits the hypoglossal nerve and a meningeal branch of the ascending pharyngeal artery. A *condylar fossa*, behind each condyle, fits the posterior margin of the superior atlantal facet in full extension; its floor is sometimes perforated by a *condylar canal* for a sigmoid emissary vein. The *jugular process*, jutting laterally from the posterior half of each condyle, is a quadrilateral plate, indented in front by a *jugular notch*, the posterior part of the jugular foramen. This notch is sometimes partly divided by a small *intrajugular process*, projecting anterolaterally. The jugular process' inferior surface is roughened by attachment of rectus capitis lateralis; a *paramastoid process* sometimes projects down and may even articulate with the atlantal transverse process. Laterally the jugular process has a rough quadrilateral or triangular area joined to the jugular surface of the temporal bone by a growth plate of cartilage, which begins to ossify at about 25 years. For variations of jugular foramen, processes, etc., consult Solter & Paljan (1973). Though rare or absent in many populations, a paramastoid process was recorded in 44% of 149 male and 31% of 137 female American Indians (Finnegan 1972).

On the superior condylar surface an oval *jugular tubercle* overlies the hypoglossal canal, its posterior part often bearing a shallow furrow for the glossopharyngeal, vagus and accessory nerves. On the jugular process' superior surface a deep groove, curving anteromedially around a hook-shaped process, ends at the jugular notch; it contains the end of the sigmoid sinus. Near the groove's medial end the condylar canal opens into the posterior cranial fossa.

The *foramen magnum* is described on page 359.

Structure. The occipital, in common with many parts of other cranial bones, consists of two compact lamellae, *outer* and *inner plates*, enclosing trabecular bone or *diploë*; the bone is thick at ridges, protuberances and condyles, and in the anterior basi-occipital; in lower parts of cerebellar fossae, however, it is thin, semitransparent and devoid of diploë. As noted, the intertrabecular spaces (cancelli) of the diploë are occupied by haemopoietic red bone marrow (p. 360), the latter being drained by radicles of the wide diploic veins (p. 798), of which the occipital diploic vein is usually the largest.

Ossification (3.125). A common but *oversimplified* account of occipital ossification states that above the highest nuchal lines the squama is developed in a fibrous membrane and ossified from two

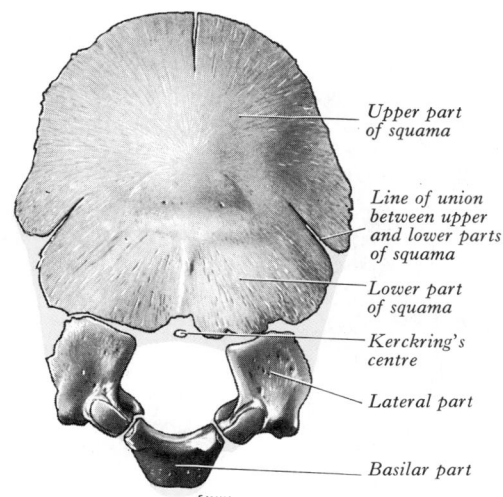

Upper part of squama

Line of union between upper and lower parts of squama

Lower part of squama

Kerckring's centre

Lateral part

Basilar part

S.W.W

3.125 The occipital bone of a newborn child: external surface. Parts of the chondrocranium still unossified are shown in blue.

centres, one on each side from about the second fetal month; this part may remain separate as the *interparietal bone*; the rest is preformed in cartilage. Below the highest nuchal lines, the squama ossifies from two centres, appearing in about the seventh week and soon uniting. These two regions of the squama unite in the third postnatal month but the line of union is recognizable at birth.

An occasional centre appears in the posterior margin of the foramen magnum in about the sixteenth week (Kerckring); it unites with the rest of the squama before birth. From a survey of literature and examination of 620 human skulls for anomalies of the occipital squama, however, Srivastava (1977) has proposed a more complex developmental history. He regards the *membranous (dermal)* part above the nuchal lines as compounded of *interparietal* and *preinterparietal* parts, the interparietal consisting of *two lateral plates* and a *central piece*. The intramembranous centres proposed for these are a pair for each lateral plate and two for the central piece of the interparietal, additionally a pair of centres for the pre-interparietal. Pal et al (1984) have recently criticized the views of Srivastava (1977); they consider the so-called pre-interparietal elements as probably sutural; but it has to be stated that much uncertainty persists regarding the status of these accessory ossicles. Fusion may fail, partly or completely, between any of these elements. To the *cartilaginous supra-occipital* Srivastava allots five endochondral centres, a pair each for right and left *lateral segments* and one for the *central segment* (the latter may correspond to Kerckring's centre). Each lateral (condylar or exoccipital) part ossifies from one centre, appearing during the eighth prenatal week. The basi-occipital is ossified from one centre appearing about the sixth week. Near the second year's end the squama unites with condylar parts and by the sixth year the bone is one entity. Between the eighteenth and twenty-fifth years the occipital and sphenoid bones unite. Metrical study of the occipital bone (Olivier 1975) suggests that squamous and basilar parts have independent parameters of growth, and that sexual differences are chiefly evident in condylar regions. Routal et al (1984) have described sexual dimorphism, chiefly difference in size, in the foramen magnum.

The Sphenoid Bone

The sphenoid bone (3.126–131) is in the base of the skull, 'wedged' (as its name implies) between the frontal and temporal and occipital bones. It has a central body, paired greater and lesser wings spreading laterally from it and two pterygoid processes, descending from junctions of the body and greater wings.

The body is cuboidal; it contains two large air sinuses, separated by a septum. The *cerebral* or *superior surface* (3.111,126) articulates in front with the ethmoidal cribriform

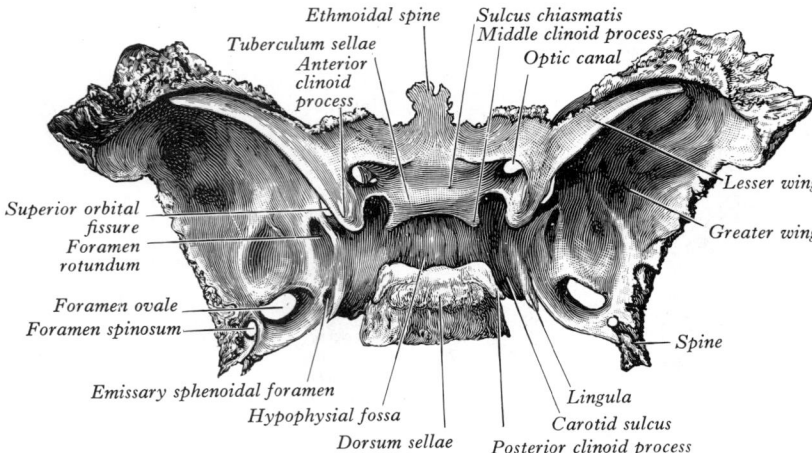

3.126 The sphenoid bone: superior aspect.

Labels on figure 3.126:
Ethmoidal spine
Sulcus chiasmatis
Middle clinoid process
Tuberculum sellae
Anterior clinoid process
Optic canal
Superior orbital fissure
Foramen rotundum
Lesser wing
Greater wing
Foramen ovale
Foramen spinosum
Spine
Emissary sphenoidal foramen
Hypophysial fossa
Dorsum sellae
Lingula
Carotid sulcus
Posterior clinoid process

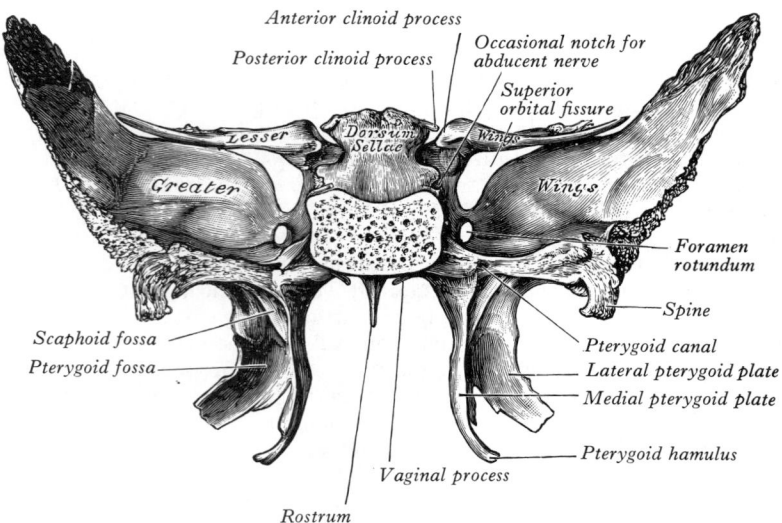

3.127 The sphenoid bone: posterior aspect.

Labels on figure 3.127:
Anterior clinoid process
Posterior clinoid process
Occasional notch for abducent nerve
Superior orbital fissure
Lesser
Dorsum Sellae
Wings
Greater
Wings
Foramen rotundum
Spine
Scaphoid fossa
Pterygoid fossa
Pterygoid canal
Lateral pterygoid plate
Medial pterygoid plate
Pterygoid hamulus
Vaginal process
Rostrum

plates; anteriorly it is the smooth *jugum sphenoidale*, related to gyri recti and olfactory tracts. The jugum is bounded behind by the anterior border of the *sulcus chiasmatis*, leading laterally to the *optic canals*. Posterior to this is the *tuberculum sellae*, behind which the deeply concave *sella turcica* contains the hypophysis cerebri in the *hypophysial fossa*. The sella's anterior edge is completed laterally by two *middle clinoid processes* (3.126), posteriorly by a square *dorsum sellae*, the superior angles of which bear variable *posterior clinoid processes*, for attachment of tentorium cerebelli. On each side, below the dorsum sellae, a small *petrosal process* articulates with the apex of the petrous temporal bone. (For variations in sella turcica see Kinman 1977, Lang 1977.) Posterior to the dorsum sellae the sphenoid body slopes directly into basi-occipital bone in adults, together forming the *clivus*. The latter's sphenoidal part lies beneath the upper pons.

The body's *lateral surfaces* are united both with the greater wings and the medial pterygoid plates. Above the root of each wing a broad *carotid sulcus*, curved like a tilde or 'swung dash' (∼), it accommodates the internal carotid artery and cavernous sinus and a series of closely related nerves. It is deepest posteriorly, overhung medially by the petrosal process and has a sharp lateral margin, the *lingula*, which continues back over the posterior opening of the pterygoid canal.

On the *anterior surface* (3.128) a median triangular, structurally bilaminar *sphenoidal crest* forms a small part of the nasal septum; its anterior border joins the ethmoid's perpendicular plate; on each side of it opens a *sphenoidal sinus*. The sphenoidal sinuses (p. 1179), two large, irregular cavities in the body, are separated by a usually asymmetrical septum. Varying in form and size, each

sinus is partially divided by bony laminae. A lateral recess may extend into the greater wing and lingula (Cope 1917) and may even invade the basi-occipital bone almost to the foramen magnum. Trans-sphenoidal surgical approach to the hypophysis cerebri has prompted classification into types: *conchal*—a small sinus separated from the sella turcica by about 10 mm of trabecular bone; *presellar*—a sinus not extended dorsally to the tuberculum sellae; and *sellar*—the sinus extended, as described above, at variable distances beyond the tuberculum (Hammer & Rådberg 1961, Kinman 1977, Lang 1977). In the articulated state they are closed antero-inferiorly by the *sphenoidal conchae* (p. 376), leaving openings by which each communicates with its spheno-ethmoidal recess. Each half anterior surface of the body consists of: (*a*) a superolateral depressed area joined to the ethmoid labyrinth, completing the posterior ethmoidal sinuses; its lateral margin articulates with the ethmoid's orbital plate above and palatine's orbital process below; (*b*) an inferomedial, smooth, triangular area, forming the posterior nasal roof; near its superior angle is the orifice of a sphenoidal sinus.

The body's *inferior surface* bears a median triangular *sphenoidal rostrum*, embraced above by the diverging lower margins of the crest, its narrow ventral end fitting into a fissure between the anterior parts of the vomerine alae. Posterior ends of the sphenoidal conchae flank the rostrum, articulating with vomerine alae. On each side of the posterior part of the rostrum, behind the sphenoidal concha's apex (vide infra), a thin *vaginal process* projects medially from the base of the medial pterygoid plate (3.128).

The greater wings, strong processes, curve broadly superolaterally from the body. Posteriorly each is triangular, fitting the angle between petrous and squamous parts of the temporal bone (3.103B.C) at a sphenosquamosal suture. The *cerebral surface* (3.126) is an anterior part of the middle cranial fossa. Deeply concave, its undulating surface is adapted to the anterior gyri of the temporal lobe. Anteromedial is the *foramen rotundum* for the maxillary nerve, posterolateral to this the *foramen ovale* for the mandibular nerve, accessory meningeal artery and sometimes the lesser petrosal nerve, which may have a special *canaliculus innominatus* medial to the foramen spinosum. A small *emissary sphenoidal foramen* exists on one or both sides in 40% of skulls; it opens below, lateral to the scaphoid fossa, and transmits a small vein from the cavernous sinus. Anteromedial to the sphenoidal spine the *foramen spinosum* transmits the middle meningeal artery and meningeal branch of the mandibular nerve.

The *lateral surface* (3.103A) is vertically convex and divided by a transverse *infratemporal crest* into temporal (upper) and infratemporal (lower) surfaces; to the upper surface the temporalis is attached; the lower is directed downwards and, with the infratemporal crest, it is the attachment of upper fibres of the lateral pterygoid. It presents the *foramen ovale* and *foramen spinosum* and posteriorly the *spine of the sphenoid* (p. 355)—a small, sometimes pointed process, projecting downwards; its medial side shows a faint antero-inferior groove for the chorda tympani and also appears in the lateral wall of the sulcus for the auditory tube (p. 1226). To its tip is attached the sphenomandibular ligament. Medial to the anterior end of the infratemporal crest a triangular process is part of the lateral pterygoid's attachment; a ridge, descending medially from this to the front of the lateral pterygoid plate, is a posterior boundary of the pterygomaxillary fissure.

The quadrilateral *orbital surface* (3.128) faces anteromedially as a posterior part of the lateral orbital wall. Its serrated upper edge articulates with the frontal's orbital plate, its serrated lateral margin with the zygomatic bone. Its smooth inferior border is the posterolateral edge of the inferior orbital fissure. Its sharp medial margin is the inferolateral edge of the superior orbital fissure; on it a small tubercle affords partial attachment of the common annular ocular tendon (p. 1208). Below the superior fissure's medial end a grooved area forms the posterior wall of the pterygopalatine fossa, pierced by the foramen rotundum.

The irregular *margin of the greater wing* (3.127), from sphenoid body to spine, is in its medial half an anterior limit of the *foramen lacerum* and displays the posterior aperture of the pterygoid canal. Its lateral half articulates with the petrous temporal bone at a *sphenopetrosal synchondrosis*. Inferior to this the *sulcus tubae*

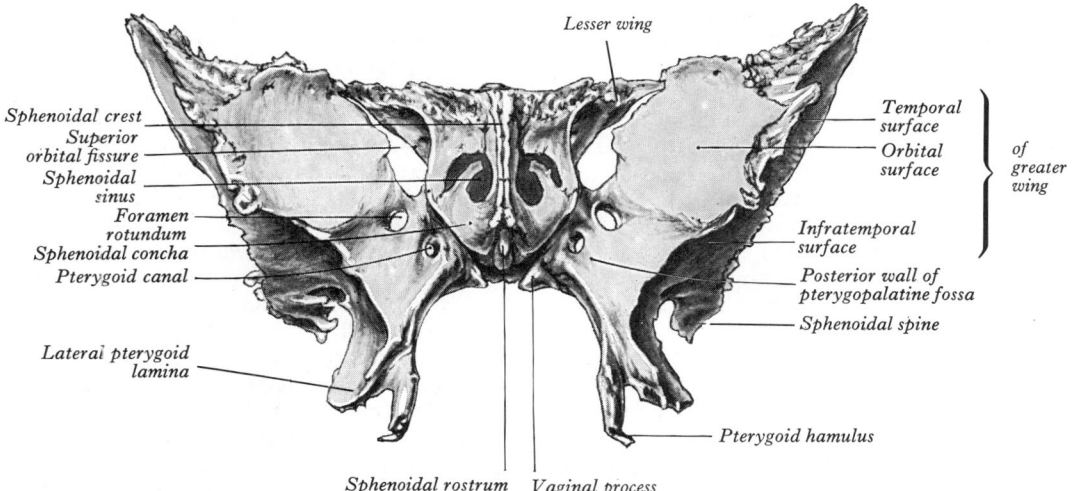

Sphenoidal crest
Superior
orbital fissure
Sphenoidal
sinus
Foramen
rotundum
Sphenoidal concha
Pterygoid canal

Lesser wing

Temporal
surface
Orbital
surface

of
greater
wing

Infratemporal
surface

Posterior wall of
pterygopalatine fossa

Sphenoidal spine

Lateral pterygoid
lamina

Pterygoid hamulus

Sphenoidal rostrum   Vaginal process

3.128 The sphenoid bone of an eight-year-old child: anterior aspect.

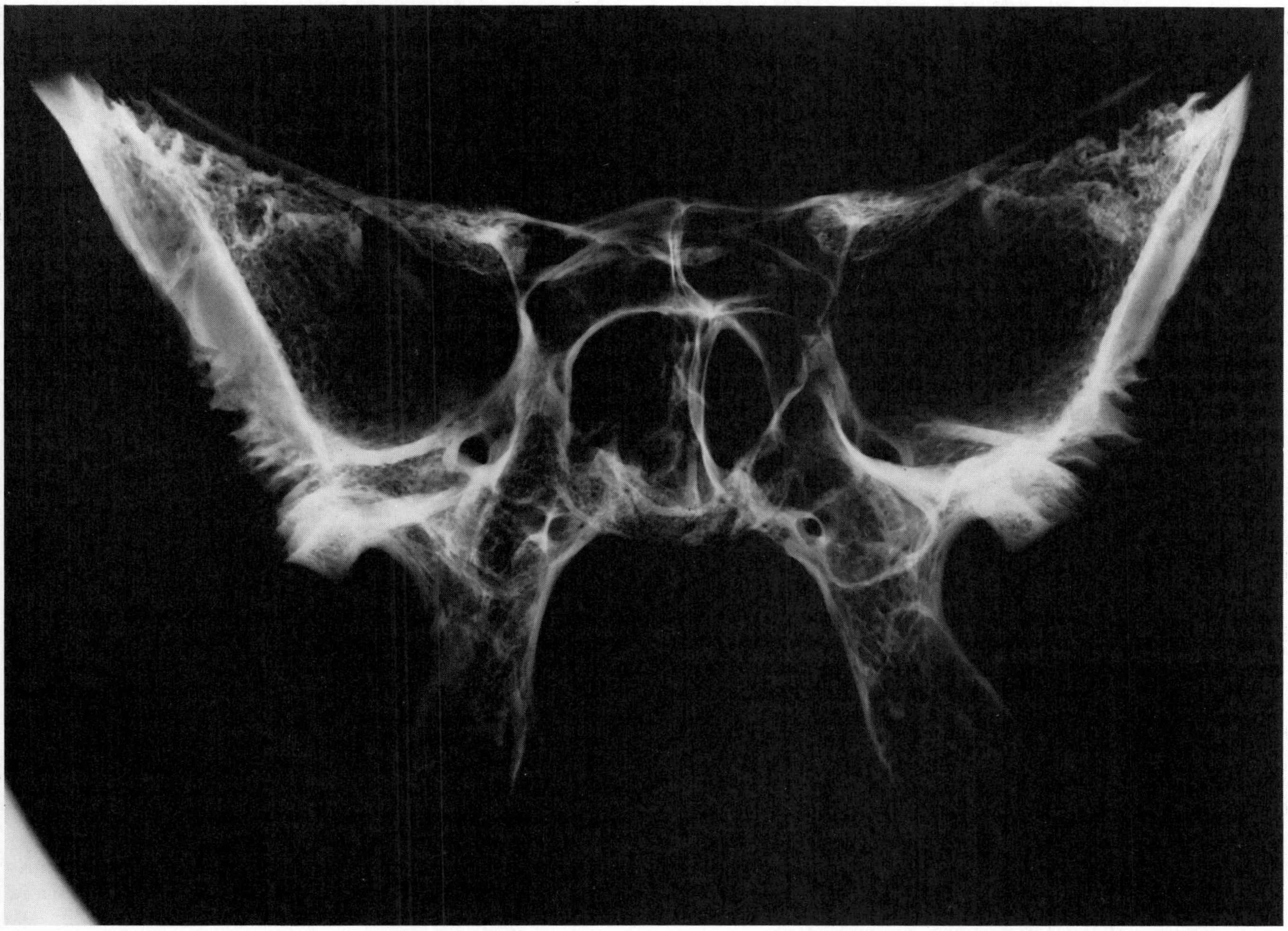

3.129 High resolution projection microradiograph of the sphenoid bone (anteroposterior projection). Note in particular the sphenoidal body and its contained sinuses, the architecture of the greater and lesser wings, and pterygoid processes and the disposition of the optic canal, superior orbital fissure, foramen rotundum and pterygoid canal. (Contributed by C Buckland-Wright, Department of Anatomy, Guy's Hospital Medical School, London.)

contains the cartilaginous auditory tube. Anterior to the sphenoidal spine the concave *squamosal margin* is serrated (bevelled internally below, externally above) for articulation with he temporal squama. The tip of the greater wing, bevelled nternally, articulates with the parietal's sphenoidal angle at the terion. Medial to this, a triangular rough area articulates with the rontal bone, its medial angle continuous with the inferior boundary of the superior orbital fissure, its anterior angle by a serrated rticulation with the zygomatic bone.

The lesser wings are triangular, pointed plates protruding laterally from the anterosuperior regions of the body (3.126,127). The *superior surface* of each is smooth and related to the cerebral frontal lobe. The *inferior surface* is a posterior part of the orbital roof and upper boundary of the *superior orbital fissure*; it overhangs the middle cranial fossa. The *posterior border* projects into the lateral cerebral fissure; its medial end is the *anterior clinoid process*, for attachment of the anterior end of the free tentorial border. The anterior and middle clinoid processes are sometimes

375

united to form a *caroticoclinoid foramen*. The lesser wing is connected to the body by an anterior root, thin and flat, and a posterior, thick and triangular root; between them the *optic canal* contains the optic nerve and ophthalmic artery. Growth of the posterior root is closely associated with variations in the canal (Kier 1966), whose cranial opening may be duplicated (Warwick 1951). More often the division is incomplete.

The superior orbital fissure, triangular and connecting the cranial cavity and orbit, is bounded medially by sphenoid body, above by the lesser wing, below by the medial margin of the orbital surface of the greater wing; it is completed laterally, between greater and lesser wings, by the frontal bone. For its contents see page 1209 and 7.329.

The pterygoid processes (3.127–129) descend perpendicularly from the junctions of greater wings and body. Each consists of a medial and lateral plate, their upper parts fused anteriorly. They are separated below by the angular *pterygoid fissure*, whose margins articulate with the palatine's pyramidal process. They diverge behind and the cuneiform *pterygoid fossa* between them contains the medial pterygoid and tensor veli palatini. Above is the small, oval, shallow *scaphoid fossa*, formed by division of the upper posterior border of the medial plate; to it part of the tensor veli palatini is attached. The anterior surface of the root of the pterygoid process is broad and triangular and is the pterygopalatine fossa's posterior wall, pierced by the anterior orifice of the *pterygoid canal*.

The lateral pterygoid plate is broad, thin and everted, its *lateral surface* part of the medial wall of the infratemporal fossa and an attachment of lower part of the lateral pterygoid; its *medial surface* is the pterygoid fossa's lateral wall and to it most of the medial pterygoid is attached. The upper part of its *anterior border* is a posterior boundary of the pterygomaxillary fissure; the lower part articulates with the palatine bone. Its *posterior border* is free.

The medial pterygoid plate is narrower and longer; its lower end curves into the lateral, unciform *pterygoid hamulus*, which deflects the tendon of the tensor veli palatini; the pterygomandibular raphe is attached to it. The *lateral surface* is the pterygoid fossa's medial wall; the tensor veli palatini adjoins it; the *medial surface* is a lateral boundary of the posterior nasal aperture. The medial plate is prolonged above on the sphenoid body's inferior aspect as the thin *vaginal process*, articulating anteriorly with the palatine's sphenoidal process and medially with the vomerine ala. Inferiorly it has a furrow, anteriorly made into a canal by the palatine sphenoidal process; this *palatovaginal canal* transmits pharyngeal branches of the maxillary artery and pterygopalatine

ganglion. To the whole of the medial plate's posterior margin the pharyngobasilar fascia is attached and to its lower end the superior pharyngeal constrictor. At its upper end is a small *pterygoid tubercle*, just below the pterygoid canal's posterior opening. Projecting back near the margin's midpoint is the *processus tubarius*, supporting the auditory tube's pharyngeal end. The plate's anterior margin, in its lower part, articulates with the posterior border of the palatine's perpendicular plate.

The sphenoidal conchae (3.128) are two thin, curved platelets, attached antero-inferiorly to the sphenoid body; the superior concave surface of each is the anterior wall and partly floor of a sphenoidal sinus. They are largely destroyed in disarticulating a skull; in situ each has anterior vertical, quadrilateral and posterior horizontal, triangular parts. The anterior part consists of (*a*) a superolateral depressed area, completing the posterior ethmoidal sinuses and joining below with a palatine's orbital process, and (*b*) an inferomedial area, smooth and triangular and part of the nasal roof, perforated above by the round opening connecting sphenoidal sinus and spheno-ethmoidal recess. Anterior parts of the two bones meet in the midline, protruding as the sphenoidal crest. The horizontal part appears in the nasal roof and completes the sphenopalatine foramen; its medial edge articulates with the sphenoid's rostrum and vomerine ala; its apex, directed back, is superomedial to the vaginal process of the medial pterygoid plate and joins the posterior part of the ala. A small conchal part sometimes appears in the medial orbital wall between the ethmoid's orbital plate in front, the palatine's orbital process below and the frontal bone above.

Ossification. Until the seventh or eighth month in utero the sphenoid body has a *presphenoidal part*, anterior to tuberculum sellae, with which the lesser wings are continuous, and a postsphenoidal part, comprising sella turcica and dorsum sellae, and integral with the greater wings and pterygoid processes. Much of the bone is preformed in cartilage. There are six ossificatory centres for the presphenoidal and eight for postsphenoidal parts. This multiplicity accords with the sphenoid's evolution from a number of elements, such as the median presphenoid and basisphenoid homologous with the human parts defined above. The lesser wings, primitively separate orbitosphenoids, show a tendency to fusion with the body in mammals.

Presphenoidal part. About the ninth fetal week a centre appears in each wing, lateral to the optic canal; a little later two bilateral centres appear in the presphenoidal body. Each sphenoidal concha has a centre, appearing superoposteriorly in the nasal capsule in the fifth month in utero; as this enlarges it partly surrounds a posterosuperior expansion of nasal cavity, which becomes the sphenoidal sinus. The posterior conchal wall is absorbed and the sinus invades the presphenoid. In the fourth year the concha fuses with the ethmoidal labyrinth and before puberty with the sphenoid and palatine bones. Its anterior deficiency persists as an orifice of a sphenoidal sinus.

Postsphenoidal part. First centres appear in the greater wings about the eighth fetal week, one in each below the foramen rotundum in the wing's basal cartilage; this centre forms only the wing's root, near the foramen rotundum and pterygoid canal; the remainder is ossified in mesenchyme, spreading also into the lateral pterygoid plate. About the fourth fetal month two centres appear, flanking the sella turcica, and soon fuse. The medial pterygoid plates are also ossified in 'membrane', a centre in each probably appearing about the ninth or tenth week; the hamulus is *chondrified* during the third fetal month and at once begins to ossify (Fawcett 1905). Medial and lateral pterygoid plates join about the sixth fetal month; during the fourth a centre appears for each lingula, soon joining the body.

Presphenoidal and postsphenoidal parts fuse about the eighth month in utero, but an unciform cartilage persists after birth in lower parts of the junction. At birth the bone is tripartite (3.131): a central part, body and lesser wings, and lateral parts each comprising a greater wing and pterygoid process. During the first year the greater wings and body unite around the pterygoid canals and the lesser extend medially above the body's anterior part, meeting to form the smooth, elevated *jugum sphenoidale*. By the twenty-fifth year sphenoid and occipital bones are completely fused. Anterior in the hypophysial fossa is an occasional vascular

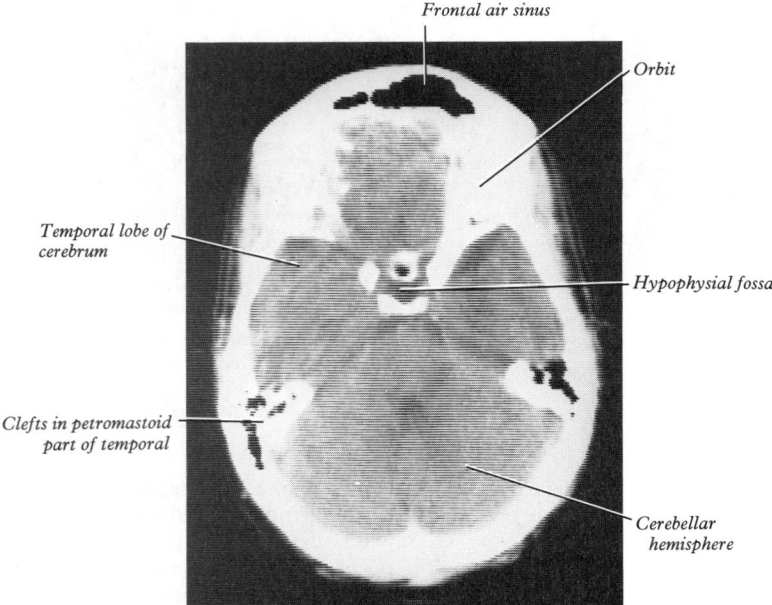

Frontal air sinus

Orbit

Temporal lobe of cerebrum

Hypophysial fossa

Clefts in petromastoid part of temporal

Cerebellar hemisphere

3.130 Computed tomogram through the head at the level of the pituitary fossa of the sphenoid. (Supplied by Shaun Gallagher, Guy's Hospital; photography by Sarah Smith).

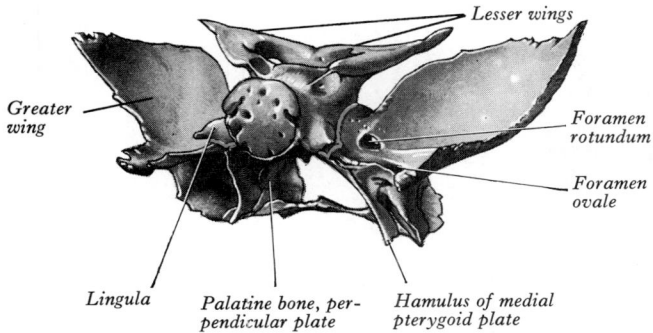

3.131 The sphenoid bone at birth, viewed from behind and from the right side. The blue strip indicates the cartilage between the central and the lateral parts on the right side; this is only partly visible on the left. Note that the two palatine bones are in situ.

foramen, often erroneously termed the *craniopharyngeal canal* (p. 230). (For discussion of cartilage canals in fetal spheno-occipital synchondrosis consult Moss-Salentijn 1975.)

The sphenoidal sinus before birth is an extension of the nasal cavity into the sphenoidal concha. In the second or third year it spreads into the presphenoid and later invades the postsphenoid, reaching full size in adolescence. As age advances it often enlarges further by absorption of its walls.

Certain sphenoidal parts are connected by ligaments which occasionally ossify, such as the *pterygospinous*, between sphenoid spine and upper part of lateral pterygoid plate (p. 376); the *interclinoid*, joining anterior to posterior clinoid process; and the *caroticoclinoid*, connecting anterior to middle clinoid process. For details of fetal and perinatal development of optic foramen and canal consult Kier (1966). Lang (1977) has surveyed ossificatory variations of the sella turcica.

Premature synostosis of the junction between pre- and postsphenoidal parts, or of the spheno-occipital suture, produces a characteristic appearance, obvious in profile. It is an abnormal depression of the nasal bridge, often observed in achondroplasia. Anomalous development of presphenoidal elements may lead to excessive separation of orbits and an abnormally broad nasal bridge (*hypertelorism*).

The Temporal Bones

The temporal bones (**3.**132–136) in the sides and base of the skull are developmentally divisible into *squamous, petromastoid, tympanic* and *styloid parts*. These morphologically distinct elements have fused during the evolution of higher vertebrates. The *squamous part* is a dermal bone evolved to help enclosure of brain. The *petromastoid part* is preformed in cartilage; it preserves precise orientation of the membranous labyrinth. The *tympanic part*, formed in mesenchyme, is homologous with the os angulare, part of the composite lower jaw of many reptiles and osseous fishes, integrated into the skull and adapted to form part of the tympanic cavity and external acoustic meatus and to support the tympanic membrane, all concerned in sound transmission. The *styloid process* is the dorsal element of the hyoid arch. Fusion of these parts and inclusion of tympanic cavity and auditory ossicles are discussed on page 172.

The squamous part or squama, anterosuperior in the bone, is thin and partly translucent. Its *temporal surface* (**3.**132) is smooth, slightly convex, and part of the temporal fossa for attachment of temporalis; above the external acoustic meatus it is grooved vertically by the middle temporal artery. The *supramastoid crest* curves back and up across its posterior part; it is an attachment of temporal fascia and muscle. Junction between squamous and mastoid parts is about 1.5 cm below this crest; traces of the squamo-mastoid suture may persist. Between the anterior end of the crest and posterosuperior quadrant of the external acoustic meatus is the *suprameatal triangle*, a depression marking the mastoid antrum, medial to it at a depth of about 1.25 cm (p. 380); in it anteriorly is usually a small *suprameatal spine*.

The *temporal's zygomatic process*, part of the **zygoma**, juts forwards from the squama's lower region. Its triangular posterior part has a broad base directed laterally, its surfaces superior and inferior. The process then twists anteromedially, its surfaces becoming medial and lateral. The posterior part's superior surface is concave and continuous with that of the squama; the inferior surface is bounded by *anterior* and *posterior roots*, converging into the anterior part of the process. At the junction of the roots the *tubercle of the zygomatic root* gives attachment to the lateral temporomandibular ligament. The posterior root is prolonged forwards above the external acoustic meatus, its upper border continuing into the supramastoid crest. The anterior root juts almost horizontally from the squama; its inferior surface, with an anteroposterior convexity covered by cartilage, contacts the joint's articular disc, forming a short semicylindrical *articular tubercle*, the anterior limit of the mandibular fossa. Very rarely the squama is perforated above the posterior root by a *squamosal foramen*, transmitting the petrosquamous sinus (p. 801).

The zygomatic process' anterior part is thin and flat. To its superior border the temporal fascia is attached, to the inferior, short and arched, some fibres of masseter. The convex lateral surface is subcutaneous; the medial is concave and is an attachment for part of the masseter. The anterior end is deeply serrated and slopes obliquely back and down to articulate with the zygomatic bone's temporal process. Anterior to the articular tubercle a small triangular area is part of the roof of the infratemporal fossa, separated from the squama's temporal surface by a ridge, continuous behind with the zygomatic process' anterior root, in front with the infratemporal crest of the sphenoid's greater wing.

The mandibular fossa, limited in front by the articular tubercle, has an anterior articular area, formed by temporal squama, and a posterior non-articular area, formed by the tympanic element. The *articular surface*, smooth, oval and deeply hollow, articulates with the temporomandibular disc; the non-articular area sometimes contains part of the parotid gland. A small, conical *postglenoid tubercle* separates the articular surface laterally from the tympanic plate; it is prominent in some mammals, descending behind the mandibular condyle to prevent backward displacement; it is sometimes described as a third root of the zygomatic process. (For a multivariate analysis of this

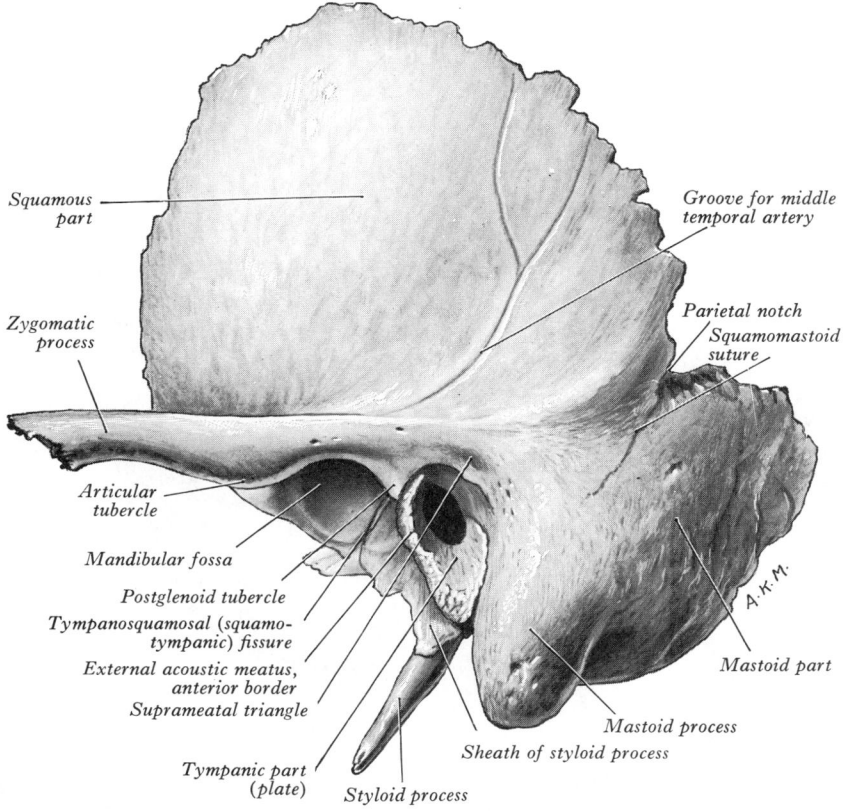

3.132 The left temporal bone: external aspect.

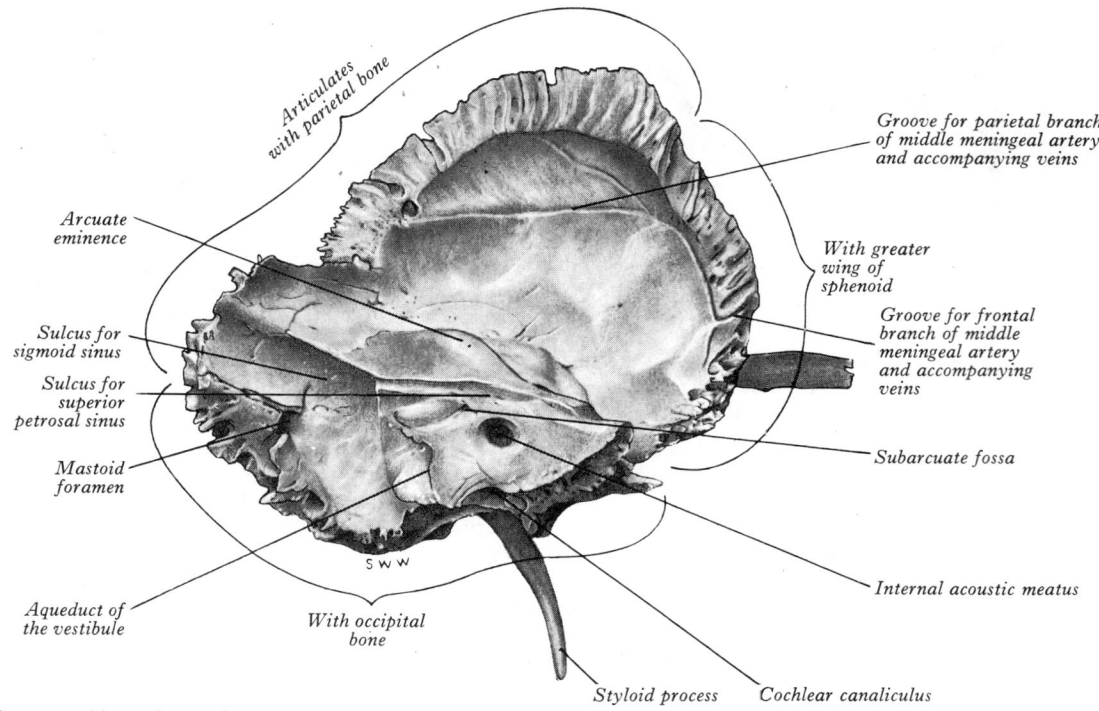

Articulates
with parietal bone

Groove for parietal branch
of middle meningeal artery
and accompanying veins

Arcuate
eminence

With greater
wing of
sphenoid

Groove for frontal
branch of middle
meningeal artery
and accompanying
veins

Sulcus for
sigmoid sinus

Sulcus for
superior
petrosal sinus

Subarcuate fossa

Mastoid
foramen

Internal acoustic meatus

Aqueduct of
the vestibule

With occipital
bone

Styloid process   Cochlear canaliculus

3.133 The left temporal bone: internal aspect. Note **3.**132 and **3.**133 were painted at different times by different artists and from different specimens. Contrast curvatures of the styloid processes (see text).

articular surface (and many other cranial parameters) in modern and fossil mankind and apes, and evaluation of functional deductions, consult Ashton et al 1976, Ashton & Moore 1980). Between the medial part of the articular fossa and the tympanic plate is the *squamotympanic fissure*, into which the anterolateral edge of tegmen tympani turns down; the *petrotympanic fissure* is between this plate and the tympanic part; it leads into the tympanic cavity and contains an anterior malleolar ligament and anterior tympanic branch of the maxillary artery. At the fissure's medial end is the anterior opening of the *anterior canaliculus for the chorda tympani*. Rarely, a *postglenoid foramen* exists anterior to the external acoustic meatus in the line of fusion of the squama and tympanic part; it replaces the squamosal foramen noted above and transmits the petrosquamous sinus (p. 801).

The squama's *cerebral surface* (**3.**133) is concave; its depressions correspond to convolutions of the temporal lobe, its grooves to middle meningeal vessels; its lower border is fused to the anterior petrous surface, but traces of a petrosquamosal suture often appear in adult bones. The *superior border* is thin, bevelled internally and overlaps the parietal's inferior border at the squamosal suture. Posteriorly it forms an angle with the mastoid element. The *antero-inferior border*, thin above and thick below, joints with the greater wing; above it is bevelled internally, below externally.

The petromastoid part of the temporal bone, morphologically one element (p. 377), is conveniently described in mastoid and petrous parts.

The mastoid part, posterior region of the temporal bone, has an *outer surface* (**3.**132) roughened by attachments of occipital belly of occipitofrontalis and auricularis posterior. Frequently near its posterior border is a *mastoid foramen*, traversed by a vein from sigmoid sinus and a small dural ramus of the occipital artery. Its position and size vary; it may be in the occipital or occipitotemporal suture. It is parasutural in 40–50% of crania (p. 395) but may be absent. The mastoid part projects down as the conical **mastoid process**, larger in adult males. To its lateral surface sternocleidomastoid, splenius capitis and longissimus capitis are attached and, medially, in a deep *mastoid notch*, the posterior belly of digastric. Medial to this a shallow *occipital groove* contains the occipital artery. The *internal* mastoid surface (**3.**133) bears a deep, curved *sigmoid sulcus* for the venous sinus and posteriorly the mastoid foramen. The sulcus is separated from innermost mastoid air cells merely by a thin lamina of bone. The air cells (**3.**134,135) and *mastoid antrum* are described on page 1225. The mastoid bone's *superior border* is thick and serrated for articulation with the mastoid angle of the parietal. Its serrated *posterior border* articulates with the inferior border of the occipital between its lateral angle and jugular process. The mastoid element is fused with the descending process of the squamous part: below, it appears in the posterior wall of the tympanic cavity.

The petrous part of the temporal bone, wedged between sphenoid and occipital in the cranial base (**3.**109A,B), is inclined up and anteromedially; it has a base, apex, three surfaces and margins. The acoustic labyrinth is within it.

The *base*, an artificial concept, corresponds to the suture between petrous and squamous elements, though this disappears soon after birth. Since a mastoid process is a postnatal petrous

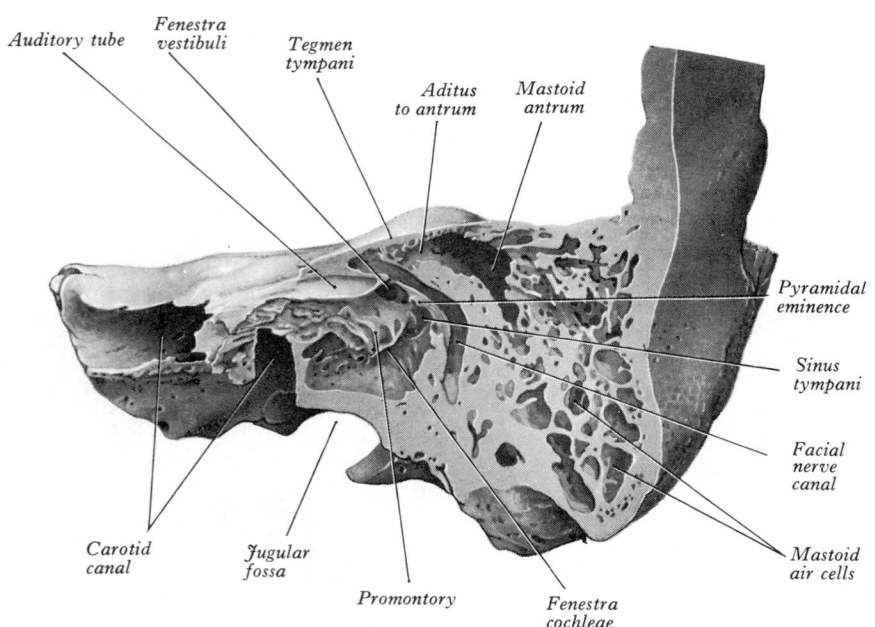

Auditory tube

Fenestra
vestibuli

Tegmen
tympani

Aditus
to antrum

Mastoid
antrum

Pyramidal
eminence

Sinus
tympani

Facial
nerve
canal

Mastoid
air cells

Carotid
canal

Jugular
fossa

Promontory

Fenestra
cochleae

3.134 Oblique vertical section through the left temporal bone in the long axis of the tympanic cavity. The lateral surface of the medial half of the bone is shown.

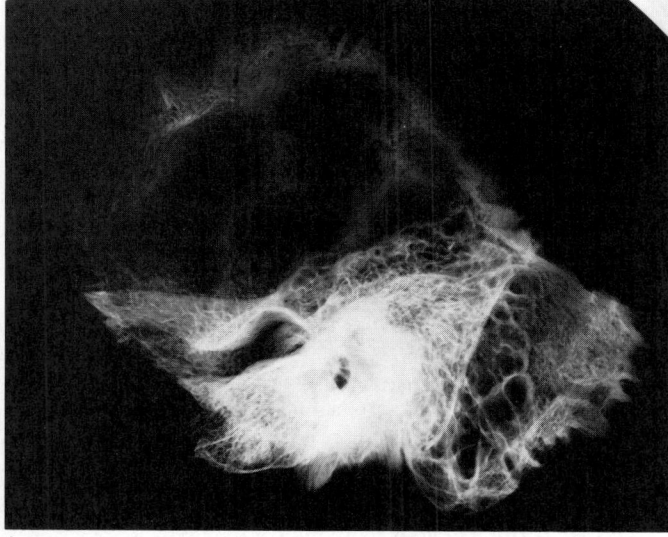

A

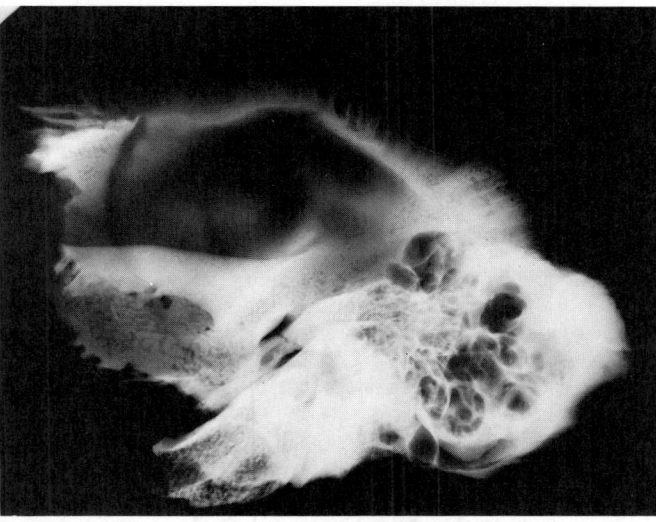

B

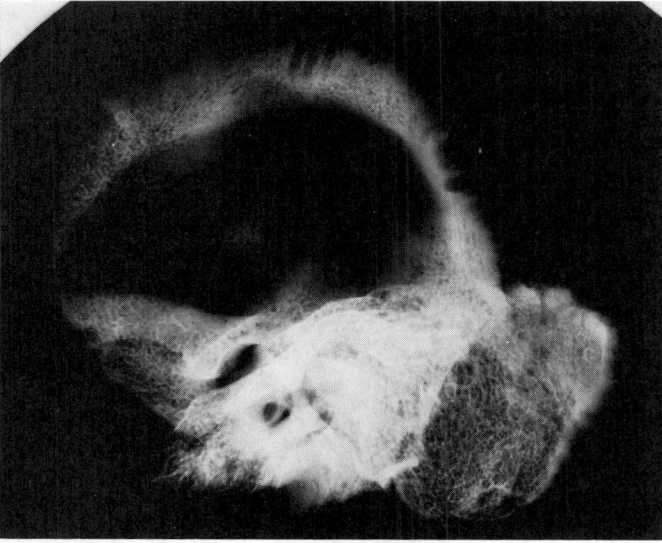

C

3.135 High resolution projection microradiographs of the petromastoid, tympanic and squamous parts of three temporal bones (mediolateral projection), showing marked variation in extent and mode of pneumatization. Above (A), the air-cells are comparatively large and extend beyond the mastoid process as far as the mandibular fossa. In the middle photograph (B) the cells are large but restricted to the post-otic and mastoid regions of the bone. Below (C), obvious pneumatization is absent; a fine meshwork of trabecular bone occupies the core of the mastoid process. (Prepared and contributed by C Buckland-Wright, Department of Anatomy, Guy's Hospital Medical School, London.)

development, the base is arbitrary but indicated by partial separation due to the mastoid antrum.

The *apex*, blunt and irregular, is angled between the posterior border of the greater wing and the basi-occipital bone; it contains the carotid canal's anterior orifice and limits posterolaterally the foramen lacerum.

The *anterior surface* partly floors the middle cranial fossa and is continuous with the cerebral surface of the squamous part, although the petrosquamosal suture often persists late in life. The whole surface is adapted to the inferior temporal gyri. Behind the apex is a *trigeminal impression* for the ganglion. Bone anterolateral to this roofs the anterior part of the carotid canal, but is often deficient. A ridge separates the trigeminal impression from another hollow behind it which partly roofs the internal acoustic meatus and cochlea. This is limited behind by the *arcuate eminence* (3.133) and raised by the anterior semicircular canal. Laterally it roofs the vestibule and, partly, the facial canal. Between the squamous part laterally and medially, the arcuate eminence and the hollows just described, the surface is formed by the *tegmen tympani*. This thin plate of bone, roof of the mastoid antrum, extends forwards above the tympanic cavity (3.134) and the canal for the tensor tympani. Its lateral margin meets the squama at the petrosquamosal suture, turning down in front as the lateral wall of the canal for the tensor tympani and the osseous part of the auditory tube; its lower edge is in the squamotympanic fissure (p. 358). Anteriorly the tegmen bears a narrow groove, passing posterolaterally to enter bone anterior to the arcuate eminence by a hiatus for the greater petrosal nerve, passing forwards to the foramen lacerum. A smaller, more lateral hiatus transmits the lesser petrosal nerve from the tympanic plexus; the bone in front of this hiatus may be grooved. The posterior slope of the arcuate eminence overlies the posterior and lateral semicircular canals; lateral to it the posterior part of the tegmen tympani roofs the mastoid antrum.

The *posterior surface* of the petrous (3.133) is an anterior part of the posterior cranial fossa, continuous with the internal mastoid surface. Near its centre is the **internal acoustic meatus** (p. 365), behind which a small slit, almost hidden by a thin plate of bone, leads to the *vestibular aqueduct*, containing the saccus and ductus endolymphaticus with a small artery and vein. The terminal half of the saccus endolymphaticus protrudes through the slit between periosteum and dura mater. Above these openings is the *subarcuate fossa* (p. 365).

The irregular *inferior surface* (3.136) is part of the cranial base's exterior. Near the petrous apex a quadrilateral area is partly attachment for the levator veli palatini and cartilaginous auditory tube, and partly connected to the basi-occipital bone by dense fibrocartilage. Behind this is the large, circular aperture of the **carotid canal** (p. 355), behind which is the **jugular fossa** of variable depth and size and containing the superior jugular bulb. Anteromedial to this, below the internal acoustic meatus, is a triangular depression for the inferior glossopharyngeal ganglion; at its apex is a small opening into the *cochlear canaliculus*, occupied by the perilymphatic duct, a tube of dura mater and a vein from the cochlea to the internal jugular vein. On the ridge between the carotid canal and jugular fossa is a *canaliculus for the tympanic nerve* from the glossopharyngeal (p. 356). Lateral in the fossa is the *mastoid canaliculus* for the vagal auricular branch. Behind the fossa the *jugular surface*, a rough quadrilateral, is covered by cartilage joining it to the occipital's jugular process.

The *superior border*, the longest, is grooved by the superior petrosal sinus, the tentorium cerebelli being attached to the groove's edges except at its medial end, where it is crossed by trigeminal roots. The *posterior border*, intermediate in length, bears medially a sulcus which forms, with one on the occipital bone, a gutter for the inferior petrosal sinus. Behind this the **jugular fossa** forms, with the occipital jugular notch, the jugular foramen and is also notched by the glossopharyngeal nerve. Bone on either or both sides of the jugular notch may meet the occipital bone and divide the foramen into two or three. The *anterior border* is joined laterally to the temporal squama at the *petrosquamosal suture*; medially it articulates with the sphenoid's greater wing.

At the junction of petrous and squamous parts two canals exist, one above the other, separated by a thin osseous plate. Both lead

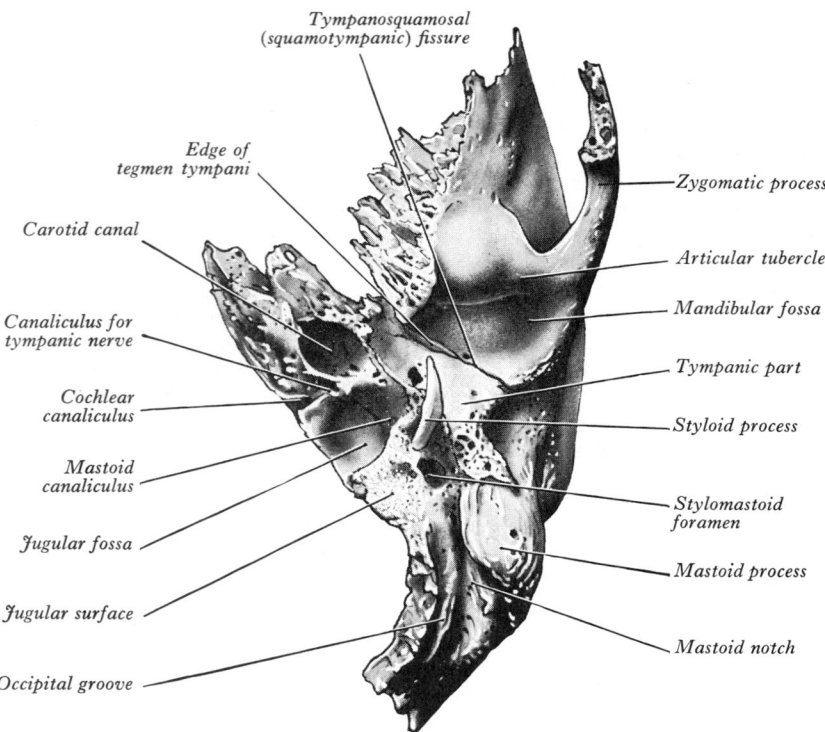

3.136 The left temporal bone: inferior aspect.

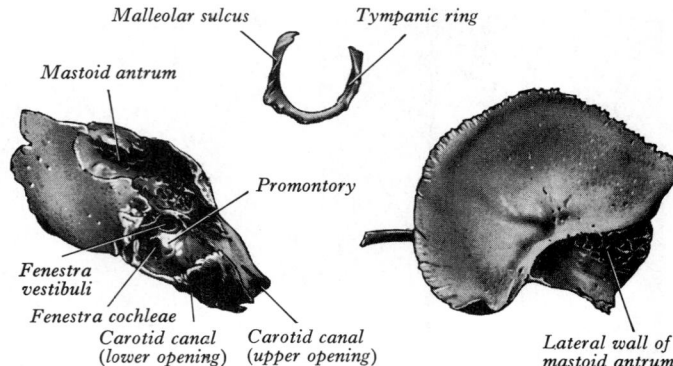

3.137 The three principal parts of the right temporal bone at birth, from left to right: lateral aspect of petromastoid part; medial aspect of tympanic ring; medial aspect of squama.

to the tympanic cavity, the upper containing the tensor tympani, the lower the auditory tube.

The tympanic part of the temporal bone (3.136) is a curved plate below the squama, anterior to the mastoid process. Internally it fuses with the petrous part and appears between this and the squama, where it is inferolateral to the auditory orifice. Behind, it fuses with the squama and mastoid process and is anterior limit of the tympanomastoid fissure. Its concave *posterior surface* forms anterior wall, floor and partly posterior wall of the external acoustic meatus. (Its postero-inferior wall may display a longitudinal *auditory torus*.) Medial on this surface is a narrow *tympanic sulcus* for attachment of the tympanic membrane. The *anterior surface*, quadrilateral and concave, is the posterior wall of the mandibular fossa and may contact the parotid gland. Its rough *lateral border* forms most of the external acoustic meatus' margin and is continuous with its cartilaginous part. Laterally the *upper border* is fused with the back of the postglenoid tubercle; medially it is the posterior edge of the petrotympanic fissure. The *inferior border* is sharp, splitting laterally to form, at its root, a *sheath of the styloid process* (vaginal process). Centrally the tympanic part is thin, often perforated. Between the styloid and mastoid processes the *stylomastoid foramen*, the external end of the facial canal, transmits the facial nerve and stylomastoid artery.

The external acoustic meatus, about 16 mm long, slopes down anteromedially; its floor is convex upwards. In sagittal section it is oval or elliptical with a long axis directed down and slightly back. Its anterior wall, floor and lower posterior wall are formed by the tympanic plate, its roof and upper posterior wall by the temporal squama. Its medial end is closed by tympanic membrane, the outer bounded above by the posterior zygomatic root, below which a *suprameatal spine* may exist.

The styloid process, slender, pointed, about 2.5 cm in length, projects down and forwards from the temporal bone's inferior aspect. Its curvature is somewhat variable (3.133). Its proximal part (*tympanohyal*) is sheathed by the tympanic plate, especially antero-laterally; to its distal part (*stylohyal*) are attached muscles and ligaments (p. 352). The process is covered laterally by the parotid gland; the facial nerve crosses its base, the external carotid artery its tip, embedded in the gland. Medially the process is separated from the commencement of the internal jugular vein by the attachment of the stylopharyngeus.

In structure the temporal squama is like other cranial bones: the mastoid part is trabecular and variably pneumatized, the petrous

part compact.

Ossification. The four temporal components ossify independently (3.137,138). The *squama* is ossified in a sheet of condensed mesenchyme from a single centre near the zygomatic roots, appearing in the seventh or eighth week in utero. The *petromastoid part* has several centres appearing in the cartilaginous *otic capsule* (p. 204) during the fifth month. As many as 14 have been described, variable in order of appearance. Several are small and inconstant, soon fusing with others. This multiplicity accords with numerous otic bones in earlier forms. The otic capsule is almost fully ossified by the end of the sixth month. The *tympanic part* is also ossified in mesenchyme from a centre identifiable about the third month; at birth it is an incomplete *tympanic ring*, deficient above (3.137,138), its concavity grooved by a tympanic sulcus for the tympanic membrane. Inclined obliquely down and forwards across the medial aspect of the anterior part of the ring

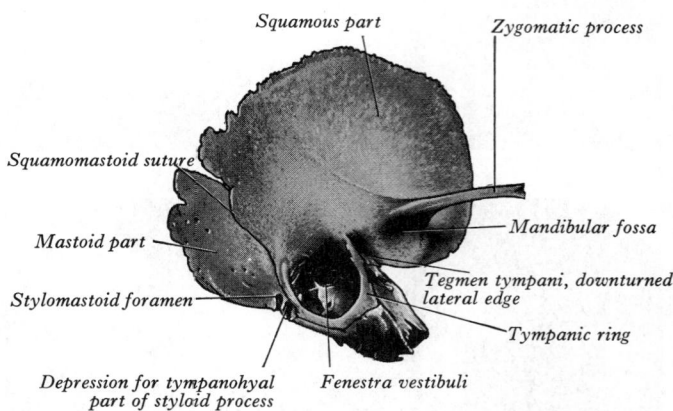

3.138A The right temporal bone at birth: lateral aspect. Note that the petromastoid part has not been coloured; the tympanic part is coloured yellow and the squamous part is coloured brown. The rudimentary base of the styloid process has been removed.

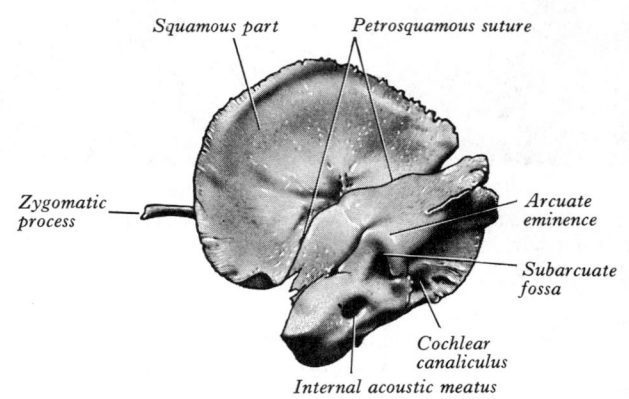

3.138B The right temporal bone at birth: medial aspect.

is the *malleolar sulcus*, (3.137) for the anterior malleolar process, chorda tympani and anterior tympanic artery. The *styloid process* is developed at the cranial end of cartilage in the second visceral or hyoid arch (p. 172) by two centres: a proximal, for the tympanohyal, appearing before birth; the other, for the distal stylohyal, after birth. The tympanic ring unites with the squama shortly before birth, the petromastoid fusing with it and the tympanohyal during the first year. The stylohyal does not unite with the rest of the process until after puberty and may never do so.

During ossification, the tympanic cavity, mastoid antrum and posterior end of the auditory tube are all in bone. The petrous part forms the roof, floor and medial wall of the cavity; the squama and tympanic part, with the membrana tympani, form its lateral wall. *At birth* the tympanic cavity, mastoid antrum and tympanic membrane are all almost *adult size*, as are the auditory ossicles; but its anterior process does not join the malleus until six months later. The internal acoustic meatus is about 6 mm in horizontal diameter, 4 mm vertically and 7 mm in length at birth, adult diameters being 7.7 mm and 11 mm.

After birth and apart from general growth, changes in the temporal bone are: (1) The tympanic *ring* extends posterolaterally to become cylindrical, growing into a fibrocartilaginous *tympanic plate*, which forms the adjacent part of external acoustic meatus at this stage. This growth is not equal but is rapid in the anterior and posterior regions, which meet and blend; thus, for a time, there is in the floor an opening (*foramen of Huschke*), usually closed at about the fifth year, but sometimes permanent (in 5–46% of adult crania from ancient and modern populations, p. 395); in Burmese crania a marked sexual difference occurs (25% in males, 40% in females). By posterior extension the tympanic plate ensheathes the styloid process and extends medially over the petrous bone to the carotid canal.

(2) The mandibular fossa is first shallow, facing more laterally, then deepening and ultimately facing downwards. Posteroinferiorly the squama grows down behind the tympanic ring to form the lateral wall of the mastoid antrum.

(3) The mastoid part is at first flat, the stylomastoid foramen and rudimentary styloid process being immediately behind the tympanic ring. As mastoid air cells develop the lateral mastoid region grows down and forwards to form the mastoid process, styloid process and stylomastoid foramen, becoming inferior. Descent of the foramen lengthens the facial canal. Not until late in the second year is the mastoid process perceptible.

(4) The subarcuate fossa is gradually filled and almost obliterated.

The external acoustic meatus is relatively as long in children as in adults, but the canal is fibrocartilaginous, whereas its medial two-thirds are osseous in adults. Surgical access to the tympanic cavity is via the mastoid antrum; in children only a thin scale of bone must be removed in the suprameatal triangle to reach antrum (p. 1225).

The Parietal Bones

Parietal bones (3.139,140) form most of the cranial roof and sides. Irregularly quadrilateral, each has two surfaces, four borders and four angles.

The *external surface* (3.139) is convex and smooth, with a central *parietal tuber (tuberosity)*. Curved *superior and inferior temporal lines* cross it, forming posterosuperior arches; to the superior arch is attached the temporal fascia; the inferior indicates the upper limit of attachment of the temporalis. Above these lines is the *epicranial aponeurosis* (galea aponeurotica), below them part of the temporal fossa. Posteriorly, close to the sagittal (superior) border, a *parietal foramen* transmits a vein from the superior sagittal sinus and sometimes a branch of occipital artery; the foramen is sometimes absent.

The concave *internal surface* (3.140) is marked by cerebral gyri and grooves for the middle meningeal vessels which ascend, inclining backwards, from the *sphenoidal (antero-inferior) angle* and posterior half or more of its inferior border. Along the sagittal border is a *groove for the superior sagittal sinus*, completed by the

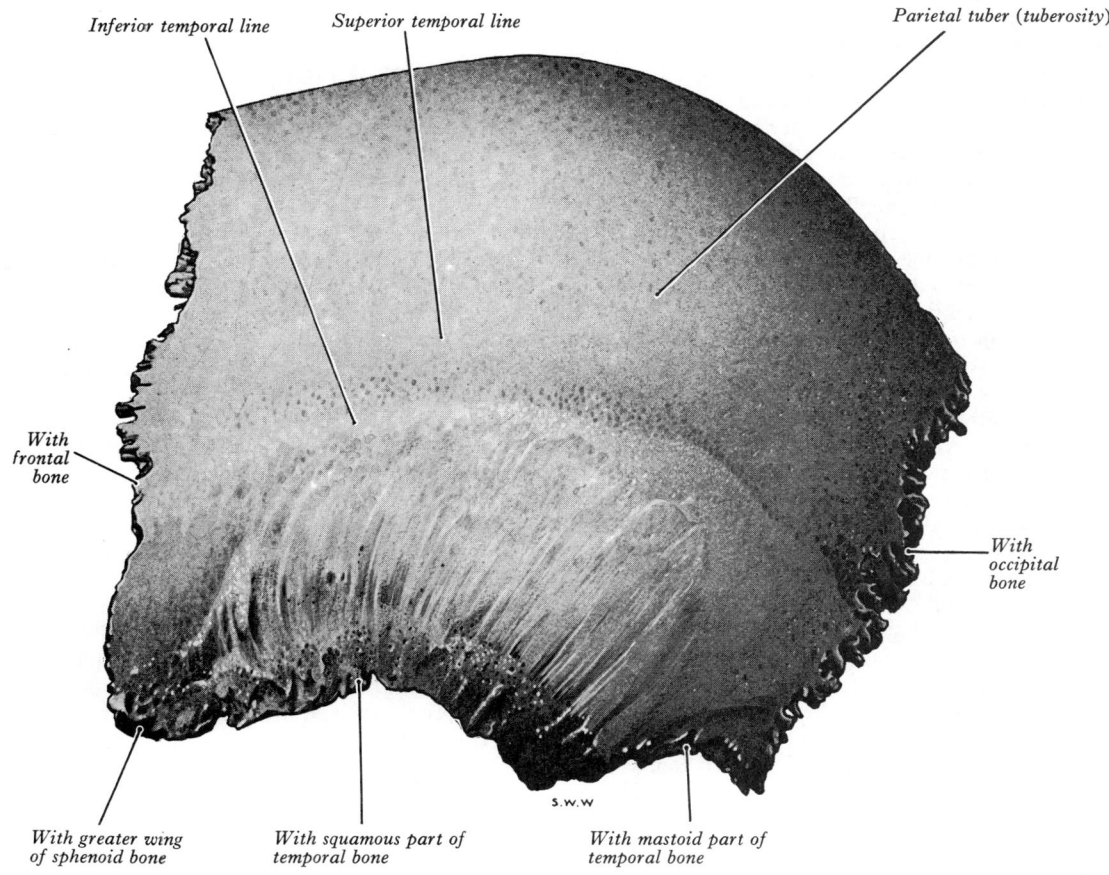

Inferior temporal line   Superior temporal line   Parietal tuber (tuberosity)

With frontal bone

With occipital bone

With greater wing of sphenoid bone   With squamous part of temporal bone   With mastoid part of temporal bone

S.W.W

3.139 The left parietal bone: external surface.

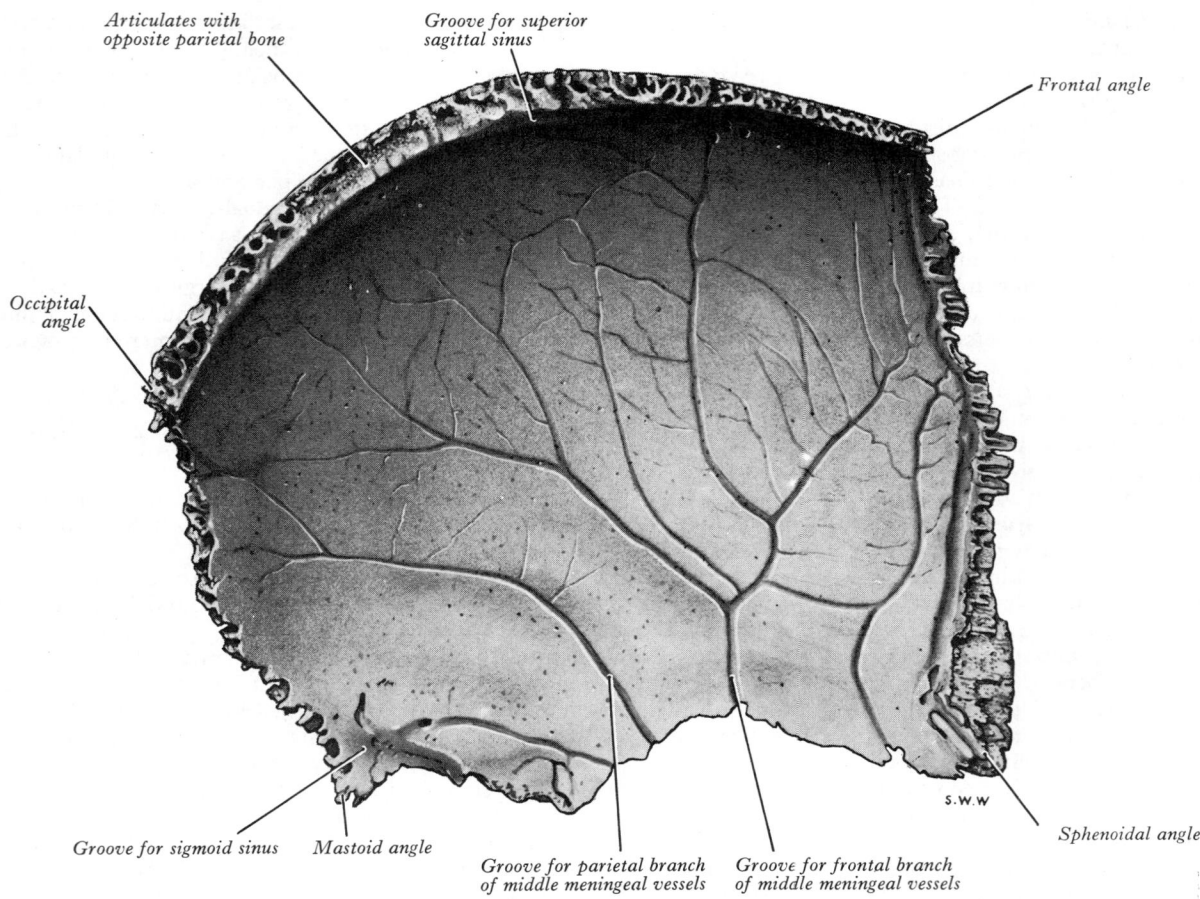

*Articulates with
opposite parietal bone*   *Groove for superior
sagittal sinus*   *Frontal angle*

*Occipital
angle*

Groove for sigmoid sinus   *Mastoid angle*   *Groove for parietal branch
of middle meningeal vessels*   *Groove for frontal branch
of middle meningeal vessels*   *Sphenoidal angle*

3.140 The left parietal bone: internal aspect.

opposite parietal; the falx cerebri is attached to its edges. *Granular foveolae* for arachnoid granulations flank the sagittal sulcus, most pronounced in old age.

The dentated *sagittal border*, longest and thickest, articulates with the opposite parietal at the sagittal suture. In the *squamosal (inferior) border* the anterior part is short, thin and truncated, bevelled externally and overlapped by the greater sphenoid wing; the middle is arched, bevelled externally and overlapped by temporal squama; the posterior part is short, thick and serrated for articulation with the mastoid bone. The *frontal border* is deeply serrated, bevelled externally above, internally below, and articulates with the frontal bone to form half the coronal suture. The *occipital border*, deeply dentated, articulates with the occipital, forming half the lambdoid suture.

The *frontal (anterosuperior) angle*, almost 90°, is at the *bregma* (meeting of the sagittal and coronal sutures). The *sphenoidal (anteroinferior) angle* is between the frontal bone and greater sphenoid wing, its internal surface marked by a deep groove, sometimes a canal, for the frontal branches of the middle meningeal vessels. The frontal sometimes meets the temporal squama, the parietal then failing to reach the greater wing. Where these four bones meet is the *pterion* (p. 352). (Frontotemporal articulation shows racial variation, probably epigenetic; its incidence varied from almost zero in a British seventeenth-century cemetery to 9.8% in Nigerian crania.) The rounded *occipital (postero-superior) angle* is at the *lambda*, the meeting of sagittal and lambdoid sutures. The blunt *mastoid (postero-inferior) angle* articulates with the occipital and mastoid temporal bones. Internally it bears a broad, shallow groove for the junction of transverse and sigmoid sinuses.

Ossification. Each parietal is ossified from two centres, appearing one above the other near its tuberosity at about the seventh week in utero in dense mesenchyme. These unite early; ossification radiates towards the margins; the angles are consequently last to be ossified and hence fontanelles occur at these sites (p. 393). At birth the temporal lines are low down,

reaching their final position after the eruption of molar teeth. Occasionally the parietal is divided by an anteroposterior suture.

The Frontal Bone

The frontal bone is like half a shallow, irregular cap forming the forehead or *frons*; on each side a horizontal *orbital part* roofs most of an orbital cavity.

The *external surface* (**3.**141) has a rounded *frontal tuber (tuberosity)* about 3 cm above midpoint of each supra-orbital margin. These tubera vary, but are especially prominent in young skulls and more so in adult females than males. Below them, separated by a shallow groove, are two curved *superciliary arches*, medially prominent and joined by a smooth median elevated *glabella*; they are more prominent in males, depending partly on size of frontal sinuses; but prominence is occasionally associated with small sinuses. Inferior to the arches are curved *supra-orbital margins* of the orbital openings, their lateral two-thirds sharp, the medial third rounded, and at each junction is a *supra-orbital notch* (or *foramen*) containing supra-orbital vessels and nerve. Medial to it a small *frontal notch* or *foramen* occurs in 50% of skulls. Berry (1975) has recorded incidences of frontal and supra-orbital *foramina* and *notches* (incisures). A supra-orbital notch or foramen occurs equally in some populations (e.g. 51% of Mexican crania). A frontal *foramen* occurs in 15–87% in various ethnic groups. Both features show sexual dimorphism (Kimura 1977).

The supra-orbital margin ends laterally in a *zygomatic process*, strong, prominent and meeting the zygomatic bone. From this a line curves posterosuperiorly, dividing into *superior* and *inferior temporal lines*, continued on the temporal squama.

The region between supra-orbital margins is the *nasal part*. Inferiorly a serrated *nasal notch* articulates with the nasal bones and, laterally, with the maxillary frontal processes and lacrimal bones. From the centre of the notch posteriorly the bone projects antero-inferiorly behind nasal bones (**3.**114) and maxillary frontal processes, supporting the nasal bridge. The region ends in a sharp *nasal spine*, on each side of which a small grooved surface partly roofs the ipsilateral nasal cavity. The nasal spine also contributes slightly to the nasal septum; in front it articulates with the crest of the nasal bones and behind with the perpendicular ethmoidal plate (3.114).

The *temporal surface*, postero-inferior to temporal lines, forms the anterior part of the temporal fossa and an attachment of temporalis (3.102). The *internal surface* (3.142) is concave. Its upper, median part has a vertical *sulcus for the sagittal sinus*, the edges of which unite below as the *frontal crest*; the sulcus contains the superior sagittal sinus (anterior part); to its margins and frontal crest part of falx cerebri is attached. The crest ends in a small notch, completed to form a *foramen caecum* (p. 362) by the ethmoid bone. The internal surface shows impressions of cerebral gyri and small furrows for meningeal vessels. Several *granular foveolae* usually exist near the sagittal sulcus, for arachnoid granulations.

The *parietal (posterior) margin* is thick, deeply serrated, bevelled internally above and externally below; inferiorly it becomes a rough, triangular surface for the greater sphenoid wing.

Orbital parts of the frontal bone are two thin, curved, triangular laminae, largely forming the orbital roofs and separated by a wide *ethmoidal notch*.

The *orbital surface* (3.142) of each plate is smooth and concave, with a shallow anterolateral *fossa for the lacrimal gland*. Below and behind the medial end of the supra-orbital margin, midway between supra-orbital notch and frontolacrimal suture, is the *trochlear fovea* (or *spine*) for attachment of a fibrocartilaginous trochlea for the superior oblique muscle. The convex *cerebral surface* is marked by frontal gyri of and faint grooves for meningeal vessels.

The quadrilateral *ethmoidal notch* (3.142) is occupied by the ethmoid's cribriform plate. Inferior to its lateral margins parts of several ethmoidal air cells are visible, completed when the ethmoid is in position. Two transverse grooves across each margin

are converted into *anterior* and *posterior ethmoidal canals* by the ethmoid; these open on the medial orbital wall, transmitting anterior and posterior ethmoidal nerves and vessels.

Openings of *frontal sinuses* (3.142) are anterior to the ethmoidal notch, lateral to the nasal spine. These two irregular cavities ascend posterolaterally for a variable distance between the frontal laminae, separated by a thin septum, usually deflected from the median plane; the sinuses are hence rarely symmetrical. Rudimentary at birth, usually well-developed by the seventh or eighth year, they reach full size after puberty, being most variable in size and larger in males. Each communicates with the middle meatus in the ipsilateral nasal cavity by a *frontonasal canal*. Degree of development is linked to prominence of superciliary arches, which are a response to masticatory stresses (Weinmann & Sicher 1955). The sinuses show a primary expansion with eruption of the first deciduous molars and again when permanent molars begin to appear in the sixth year. With advancing age osseous absorption may lead to further enlargement.

Posterior borders of orbital plates are thin and serrated to articulate with lesser sphenoid wings; their lateral parts usually appear in the middle cranial fossa between greater and lesser wings.

Structure. The frontal bone is thick and has trabecular tissue between two compact laminae, trabeculae being absent near frontal sinuses. The orbital plates, entirely of compact bone, are thin and even translucent posteriorly, and may be partly absorbed in old age.

Ossification (3.143). The frontal bone is ossified in fibrous mesenchyme from two primary centres appearing in the eighth week in utero, one near each frontal tuber. Ossification extends up to form half the main part of the bone, back to form orbital and down to form nasal parts. No secondary centres occur, except two for the nasal spine described as appearing about the tenth year (Inman & Saunders 1937). At birth the bone consists of two halves which may remain separate, a *metopic suture* persisting (Montagu 1951, Tongerson 1951, Linc & Fleischmann 1968). Such

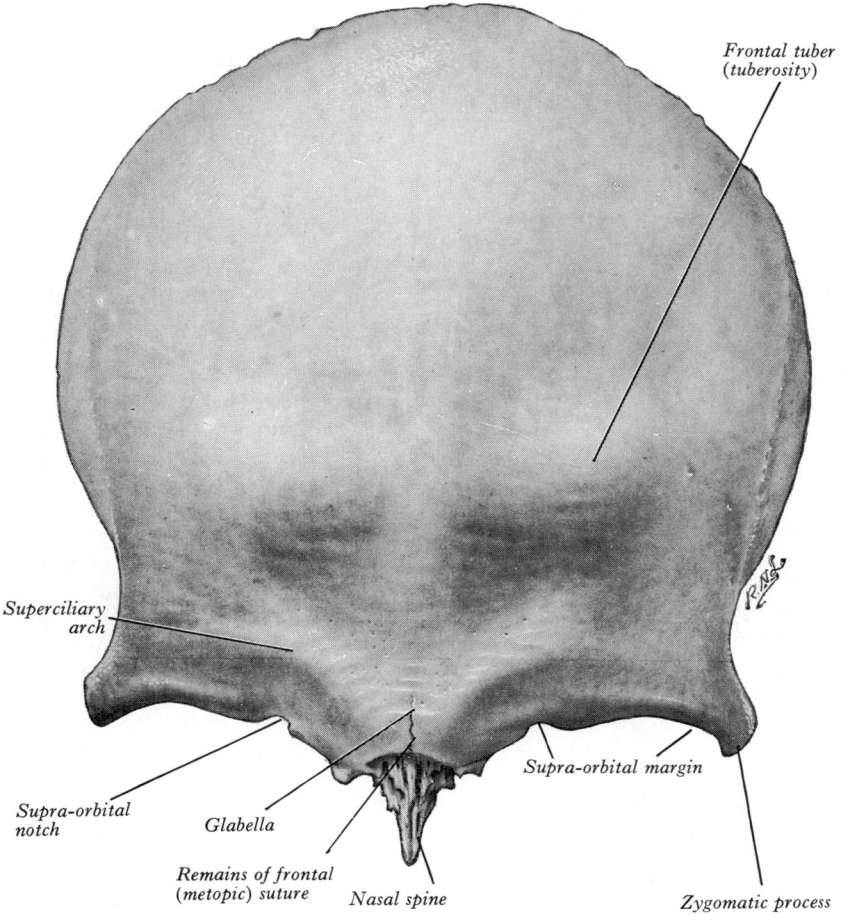

Frontal tuber (tuberosity)

Superciliary arch

Supra-orbital margin

Supra-orbital notch

Glabella

Remains of frontal (metopic) suture

Nasal spine

Zygomatic process

3.141 The frontal bone: external aspect.

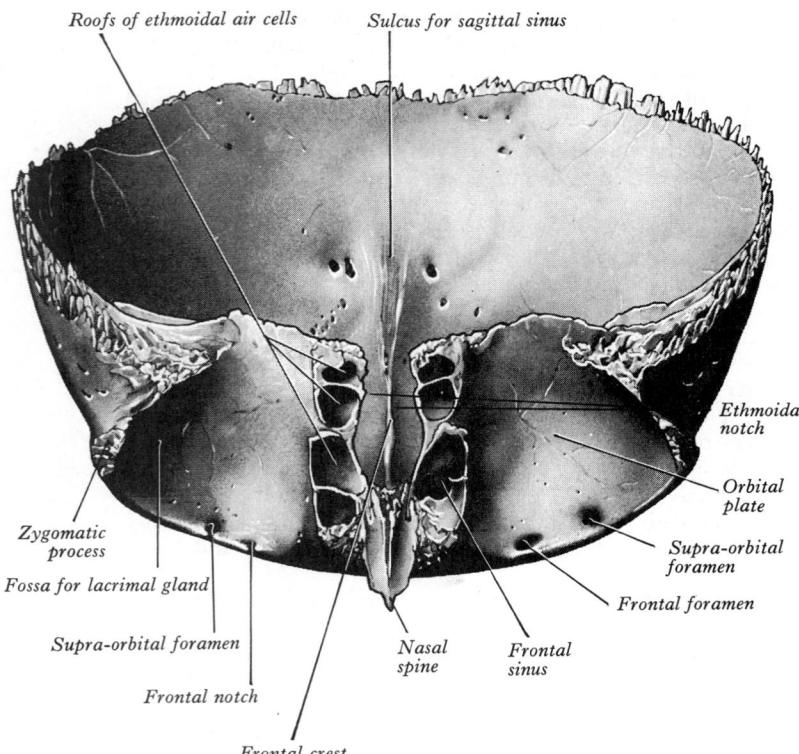

3.142 The frontal bone: inferior aspect.

Roofs of ethmoidal air cells

Sulcus for sagittal sinus

Ethmoidal notch

Orbital plate

Supra-orbital foramen

Frontal foramen

Zygomatic process

Fossa for lacrimal gland

Supra-orbital foramen

Nasal spine

Frontal sinus

Frontal notch

Frontal crest

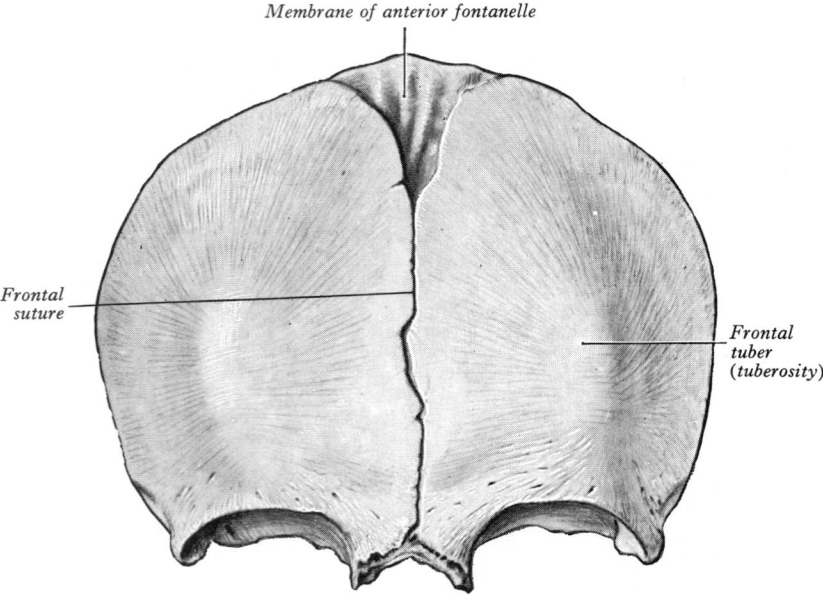

3.143 The frontal bone at birth: anterior aspect. Note that at this stage the bone consists of right and left halves connected by the frontal suture.

Membrane of anterior fontanelle

Frontal suture

Frontal tuber (tuberosity)

metopism has been assessed at 0–7.4% of individuals in various ethnic groups (Berry 1975). In a group of 206 Nigerian skulls (Ajmani et al 1983) incidence was 3.4%. It is high in mongoloids —10% according to Woo (1949).

The Ethmoid Bone

The ethmoid bone, cuboidal and fragile, lies anteriorly in the cranial base, and is involved in the structure of medial orbital walls, nasal septum, the roof and lateral walls of the nasal cavity. It is described as having a horizontal, perforated *cribriform plate*, a median *perpendicular plate* and two lateral *labyrinths*.

The cribriform plate (3.144) fills the frontal ethmoidal notch

as a large part of the nasal roof. A thick, smooth, triangular, median *crista galli* projects up from this lamina; to its posterior border, thin and curved, the falx cerebri is attached. Its shorter, thick, anterior border joints with frontal bone by two small *alae*, completing the foramen caecum (p. 362). Its sides are smooth but sometimes bulged by air cells. On both sides of crista galli the cribriform plate is narrow, depressed and related to the gyrus rectus and olfactory bulb above it; its numerous foramina transmit olfactory nerves. Anteriorly on each side of the crista is a small slit occupied by dura mater. A foramen carrying anterior ethmoidal nerve and vessels to the nasal cavity is anterolateral to this slit; a groove runs forwards to it from the anterior ethmoidal canal.

The perpendicular plate (3.145,146), thin, flat, quadrilateral and median, descends from the cribriform plate to form the upper part of the nasal septum, usually a little deflected. Its *anterior border* meets the frontal's nasal spine and crest of the nasal bones, its *posterior border* joining the sphenoidal crest above and vomer below. The thick *inferior border* is attached to the nasal septal cartilage. Its surfaces are smooth, except above, where numerous grooves and canals lead to medial foramina in the cribriform plate for filaments of olfactory nerves.

The ethmoidal labyrinths consist of thin-walled *ethmoidal air cells*, in *anterior*, *middle* and *posterior groups* between two vertical plates (p. 1178); the lateral, *orbital plate* is part of the medial orbital wall, the medial being part of lateral nasal wall. In the disarticulated bone many air cells are open, but closed when articulated with adjoining bones, except where they open into the nasal cavity. The *superior surface* (3.144) shows open air cells, completed by edges of the frontal ethmoidal notch (3.142). It is crossed by two grooves completing *anterior* and *posterior ethmoidal canals* with the frontal. On the *posterior surface* (3.146) large air cells are completed by sphenoidal conchae and the palatine's orbital process. The *lateral surface*, (3.147) thin, smooth and oblong, is the *orbital plate*, which covers middle and posterior ethmoidal sinuses. It articulates superiorly with the frontal's orbital plate, inferiorly with the maxilla and palatine's orbital process, anteriorly with the lacrimal and posteriorly with the sphenoid bone (3.98).

A few air cells lie anterior to the orbital plate, their walls completed by the lacrimal bone and frontal process of the maxilla. A thin, curved *uncinate process*, variable in size, projects postero-inferiorly from the labyrinth, appearing in the medial wall of the maxillary sinus (3.98A) as it crosses the hiatus maxillaris to join the inferior nasal concha's ethmoidal process. The upper edge of this process is a medial boundary of the hiatus semilunaris in the middle meatus.

The *medial surface* of the labyrinth (3.148) forms part of the lateral nasal wall as a thin lamella descending from the inferior surface of cribriform plate to end as the convoluted *middle nasal concha*. Above this surface shows many vertical grooves for olfactory nerves. Posteriorly it is divided by the narrow, oblique *superior meatus*, bounded above by the thin, curved *superior nasal concha*; posterior ethmoidal air cells open into this meatus. Antero-inferior to the superior meatus the middle concha's convex surface extends along all the medial surface of the labyrinth, its lower edge thick, its lateral surface concave and part of the *middle meatus*. Middle ethmoidal air cells produce a swelling, *bulla ethmoidalis*, on the lateral wall of the middle meatus (3.116); on the bulla, or above it, these cells open into the meatus. A curved *infundibulum* extends up and forwards from the middle meatus, communicating with anterior ethmoidal sinuses; in more than 50% of crania it continues up as the frontonasal duct.

Ossification. The ethmoid bone ossifies in the cartilaginous *nasal capsule* from three centres: one in the perpendicular plate, one in each labyrinth. The latter two appear in the orbital plates between the fourth and fifth months in utero, extending into the ethmoid conchae. At birth, the labyrinths, although ill-developed, are partially ossified, the remainder cartilaginous. During the first year the perpendicular plate and crista galli begin to ossify from the median centre, fusing with labyrinths early in the second year. The cribriform plate is ossified partly from the perpendicular plate, partly from labyrinths. Ethmoidal air cells begin to develop in utero; in the newborn they are narrow pouches.

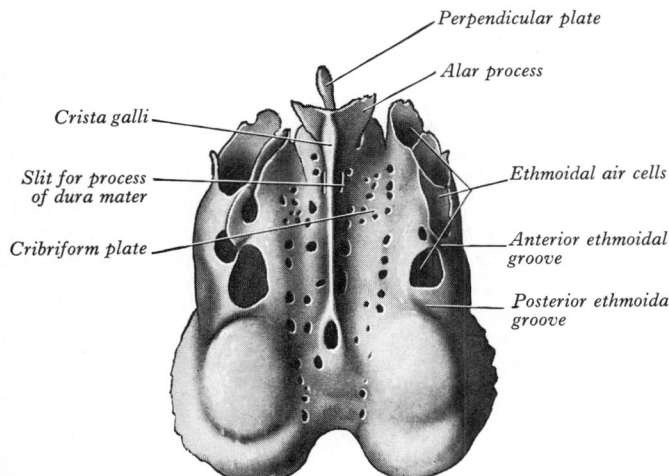

3.144 The ethmoid bone: superior aspect.

Perpendicular plate
Alar process
Crista galli
Slit for process of dura mater
Cribriform plate
Ethmoidal air cells
Anterior ethmoidal groove
Posterior ethmoidal groove

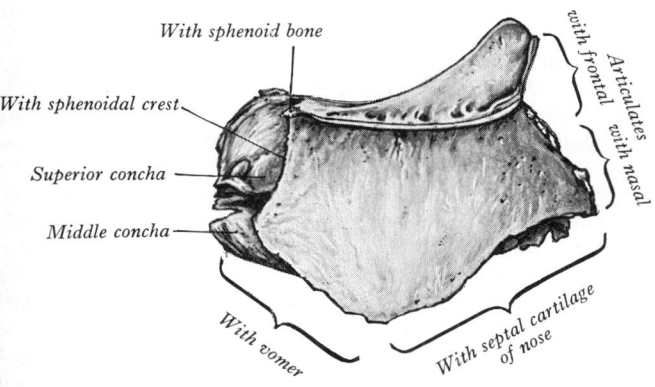

3.145 The perpendicular plate of the ethmoid bone: right lateral aspect, shown after removal of the right ethmoidal labyrinth.

With sphenoid bone
With sphenoidal crest
Superior concha
Middle concha
With vomer
With septal cartilage of nose
Articulates with frontal with nasal

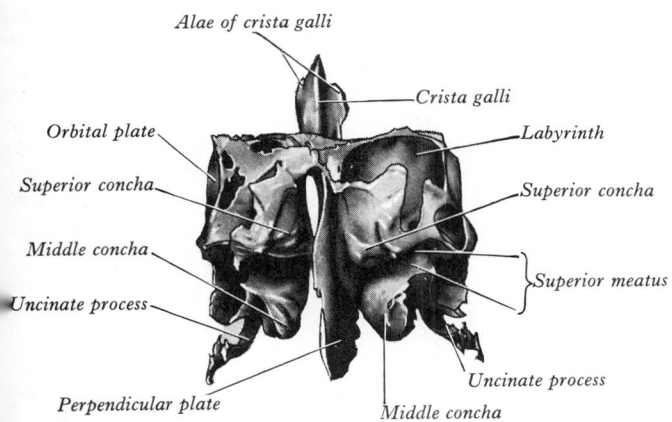

3.146 The ethmoid bone: posterior aspect.

Alae of crista galli
Crista galli
Orbital plate
Labyrinth
Superior concha
Superior concha
Middle concha
Superior meatus
Uncinate process
Uncinate process
Perpendicular plate
Middle concha

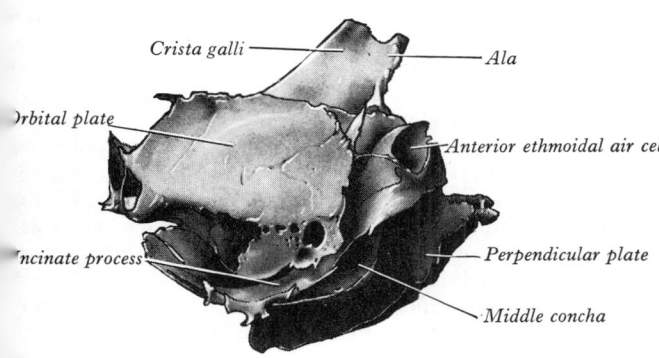

.147 The ethmoid bone: right lateral aspect.

Crista galli
Ala
Orbital plate
Anterior ethmoidal air cell
Uncinate process
Perpendicular plate
Middle concha

The Inferior Nasal Conchae

These bones are curved horizontal laminae in the lateral nasal walls (3.148). Each has two surfaces, borders and ends. The *medial surface* (3.149A) is convex, much perforated, and longitudinally grooved by vessels. The *lateral surface* is concave (3.149B) and part of the inferior meatus. The *superior border*, thin and irregular, may be divided into three regions: an anterior articulating with the maxillary conchal crest, a posterior with the palatine conchal crest and a middle with *three processes*, variable in size and form. Of these the *lacrimal process*, small and pointed, at the junction of anterior with posterior three-fourths of the border, articulates apically with a descending lacrimal process (3.150) and, by its margins, with edges of the nasolacrimal groove on the medial maxillary surface, thus completing the nasolacrimal canal. Most posterior, a thin *ethmoidal process* ascends to the ethmoid uncinate process (3.116). An intermediate thin *maxillary process* curves inferolaterally to articulate with the maxilla and maxillary process of the bone as part of the maxillary sinus' medial wall (p. 387). The *inferior conchal border* is thick and spongiose, especially in its midpart. Both *conchal ends* are more or less tapered, the posterior more so.

Ossification is from one centre, appearing at about the fifth month in utero in the incurved lower border of the cartilaginous nasal capsule's lateral wall. It loses continuity with the capsule during ossification.

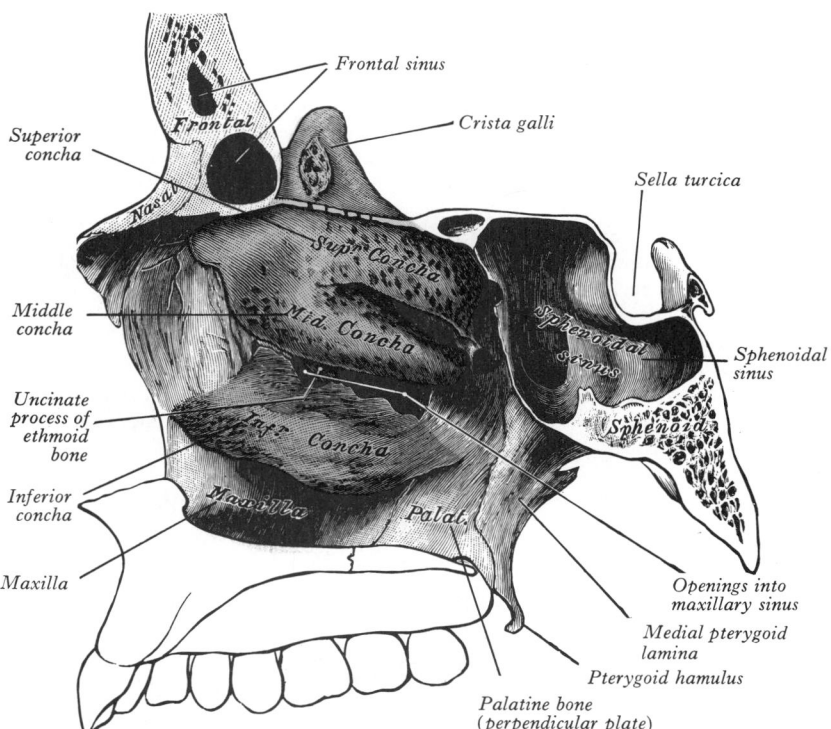

Frontal sinus
Superior concha
Crista galli
Sella turcica
Middle concha
Sphenoidal sinus
Uncinate process of ethmoid bone
Inferior concha
Maxilla
Openings into maxillary sinus
Medial pterygoid lamina
Pterygoid hamulus
Palatine bone (perpendicular plate)

3.148 The lateral wall of the right half of the nasal cavity, showing the ethmoid bone (coloured brown) and the inferior nasal concha (coloured blue) in position.

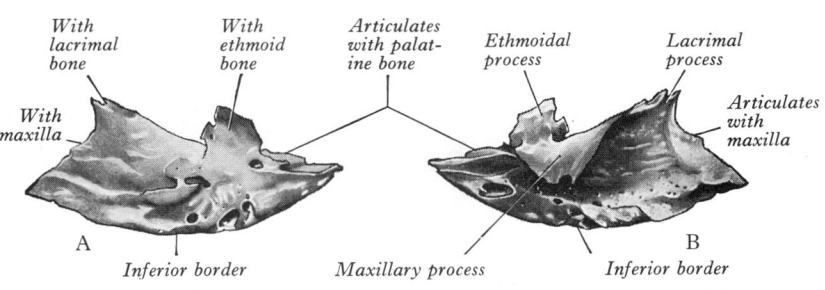

With lacrimal bone
With ethmoid bone
Articulates with palatine bone
Ethmoidal process
Lacrimal process
With maxilla
Articulates with maxilla
A
B
Inferior border
Maxillary process
Inferior border

3.149 The right inferior nasal concha. A. medial aspect; B. lateral aspect.

The Lacrimal Bones

These smallest and most fragile of cranial bones lie anteriorly in the medial orbital walls (3.98A). Each has two surfaces and four borders. The *lateral, orbital surface* (3.150) is divided by a vertical *posterior lacrimal crest*, anterior to which is a vertical groove, its anterior edge meeting the posterior border of the frontal process of the maxilla to complete the *fossa for the lacrimal sac*. The groove's medial wall is prolonged by a descending process (3.151)

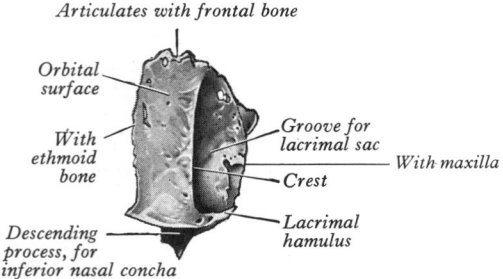

Articulates with frontal bone

Orbital surface

With ethmoid bone

Groove for lacrimal sac

With maxilla

Crest

Descending process, for inferior nasal concha

Lacrimal hamulus

3.150 The right lacrimal bone: lateral aspect.

helping to form the nasolacrimal canal by joining the lips of the maxillary nasolacrimal groove and lacrimal process of the inferior nasal concha. Behind the crest is a smooth part of the medial orbital wall. To this surface and crest the lacrimal part of orbicularis oculi is attached; the surface ends below in the *lacrimal hamulus* which, with the maxilla, completes the upper orifice of the nasolacrimal canal (3.97); the hamulus may be a separate *lesser lacrimal bone*. The antero-inferior region of the *medial (nasal) surface* is part of the middle meatus; its posterosuperior part meets the ethmoid to complete some anterior ethmoidal air cells. The *anterior lacrimal border* articulates with the frontal process of the maxilla, the *posterior* with the ethmoid's orbital plate, the *superior* with the frontal bone, and *inferior* with the maxillary orbital surface (3.97).

Ossification is from a centre appearing at about the twelfth week in mesenchyme around the nasal capsule. In later life the lacrimal is subject to patchy erosion.

The Nasal Bones

These are small, oblong, variable in size and form, placed side by side between the frontal processes of the maxillae; they jointly form the nasal bridge (3.91A,B,116).

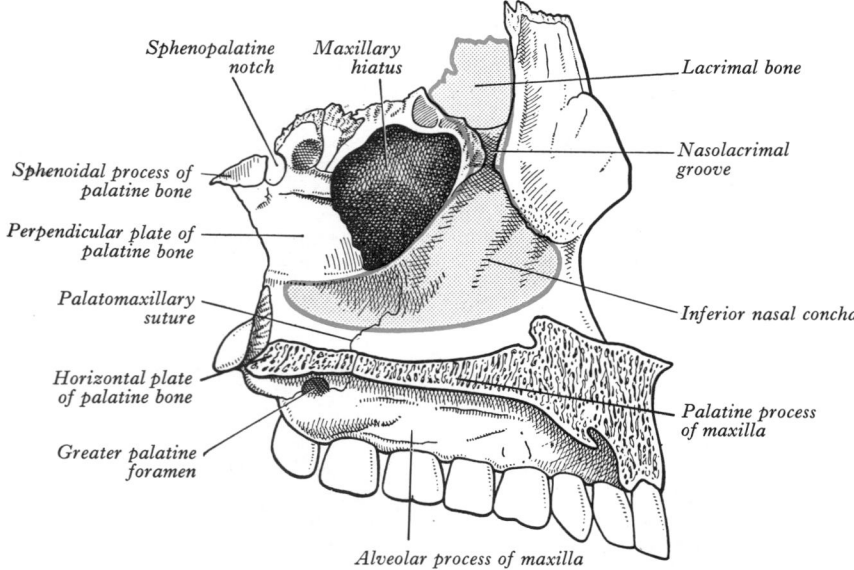

Sphenopalatine notch

Maxillary hiatus

Lacrimal bone

Sphenoidal process of palatine bone

Nasolacrimal groove

Perpendicular plate of palatine bone

Palatomaxillary suture

Horizontal plate of palatine bone

Inferior nasal concha

Greater palatine foramen

Palatine process of maxilla

Alveolar process of maxilla

3.151 Drawing to show how the medial wall of the nasolacrimal canal is formed by the articulation of the descending process of the lacrimal bone with the lacrimal process of the inferior nasal concha.

Each has two surfaces and four borders. The *external surface* (3.152A) has a descending concavoconvex profile and is transversely convex. It is covered by the procerus and nasalis muscles and perforated centrally by a small venous foramen. The *internal surface* (3.152B), transversely concave, is traversed by a longitudinal groove for the anterior ethmoidal nerve. The *superior border*, thick and serrated, articulates with the frontal. The *inferior border*, thin and notched, is continuous with the lateral nasal cartilage. The *lateral border* joins the frontal process of the maxilla. The *medial border*, thicker above, meets its fellow and projects behind as a vertical crest, a small part of the nasal septum, articulating from above with the nasal spine of the frontal, the ethmoid's perpendicular plate and the nasal septal cartilage. **Ossification** is from a centre which appears early in the third month in mesenchyme overlying the cartilaginous anterior part of the nasal capsule.

The Vomer

The vomer is thin, flat, and almost trapezoid, forming the postero-inferior part of the nasal septum (3.114); it has lateral surfaces and four borders. Both *surfaces* (3.153A,B) are marked by grooves for nerves and vessels, one obliquely antero-inferior for nasopalatine nerve and vessels. The *superior border* is thickest, with a deep furrow between projecting *alae* which fits the sphenoidal rostrum; the alae articulate with the sphenoidal conchae, sphenoidal processes of the palatine bones and vaginal processes of medial pterygoid plates. Where each ala is between the sphenoid's body and the vaginal process its inferior surface helps to form the vomerovaginal canal (p. 355). The *inferior border* articulates with the median maxillary and palatine nasal crests. The *anterior border*, the longest, articulates in its upper half with the ethmoid's perpendicular plate; the lower half is cleft to receive the inferior margin of nasal septal cartilage. The *posterior border* is concave, separating the posterior nasal apertures; it is thick and bifid above, thin below. The anterior end articulates with the posterior margin of the maxillary incisor crest and descends between the incisive canals.

Ossification. The nasal septum is at first a plate of cartilage part of which is ossified above to form the ethmoid's perpendicular plate; its antero-inferior region persists as septal cartilage the vomer being ossified in strata of connective tissue covering it on each aspect in its postero-inferior part. About the eighth week two centres appear flanking the midline and in the twelfth these unite below the cartilage, forming a deep groove (3.153B) for the nasal septal cartilage. Union of the bony lamellae progresses anterosuperiorly, while intervening cartilage is absorbed; by puberty they are almost united, but the bilaminar origin remains in the everted alae and anterior marginal groove (3.153A).

Sutural Bones

Additional ossificatory centres may occur in or near sutures giving rise to isolated *sutural bones* (3.154). Usually irregular in size and shape, and most frequent in the lambdoid suture, they sometimes occur at fontanelles, especially the posterior, perhaps representing a *pre-interparietal* element, a true *interparietal* or some composite. An isolated bone at the lambda is sometimes dubbed the *Inca bone* or *Goethe's ossicle* (p. 373). One or more *pterion ossicles* or *epipteric bones* may appear between the parietal sphenoidal angle and sphenoid's greater wing, varying much in size, but more or less symmetrical. Sutural bones usually have little morphological significance, with notable exceptions (vide infra and p. 373). There are often only two or three, but they appear in great numbers in hydrocephalic skulls. They have therefore been linked with rapid cranial expansion, but this is unproven. For a detailed analysis of these and other *epigenetic variations* in 585 adult crania, consult Berry & Berry (1967), who discuss genetic identification of populations by such criteria subsequently providing further data (Berry 1975). A report of their occurrence in fetal skulls (El-Najjar & Dawson 1977) supports a genetic factor.

The Maxillae

The maxillae, largest of the facial bones excepting the mandible, jointly form the whole upper jaw (3.92A,B,93), most of the buccal roof, floor and lateral wall of nasal cavity, orbital floors, in part the infratemporal and pterygopalatine fossae and inferior orbital and pterygomaxillary fissures. Each has a body and zygomatic, frontal, alveolar and palatine processes (3.155–158).

The maxillary body, roughly pyramidal, has anterior, infratemporal (posterior), orbital and nasal surfaces, enclosing the maxillary sinus. The *anterior surface* (3.95,102,155,156), facing anterolaterally, displays inferior elevations overlying the roots of teeth. Above the incisors is a shallow *incisive fossa* in which is attached the depressor septi; to the alveolar border below this a slip of orbicularis oris is attached, and superolateral to it is the nasalis. Lateral to this fossa is a larger, deeper *canine fossa*, separated from it by the *canine eminence*, over the canine socket. In the canine fossa the levator anguli oris is attached. Above it is the *infra-orbital foramen*, the anterior end of its canal, transmitting infra-orbital vessels and nerve. Above the foramen a sharp border, dividing anterior and orbital surfaces, is part of the orbital opening's rim; attached near it is the levator labii superioris. Medially the anterior surface ends at a deeply concave *nasal notch*, ending in a pointed process which, with its fellow, forms the *anterior nasal spine*. To the anterior surface near the notch, the nasalis and depressor septi are attached.

The convex *infratemporal surface* (3.155,156), facing posterolaterally, forms the anterior wall of the infratemporal fossa and is separated from the anterior surface by the maxilla's zygomatic process and a ridge ascending to it from the first molar socket. Near its centre are apertures of two or three *alveolar canals*, containing posterior superior alveolar vessels and nerves. Posteroinferior is the *maxillary tuberosity*, rough superomedially where it meets the palatine bone's pyramidal process (3.155,158); a few fibres of medial pterygoid are attached to it, and sometimes it articulates with the lateral pterygoid plate. Above this is the smooth anterior boundary of the pterygopalatine fossa, grooved by the maxillary nerve as it passes laterally and slightly upwards into the infra-orbital groove on the orbital surface.

The *orbital surface* (3.97,155), smooth and triangular, forms most of the orbital floor. Anteriorly its *medial border* bears a *lacrimal notch*, behind which it joints with the lacrimal bone, the ethmoid's orbital plate and, posteriorly, the palatine's orbital process (3.158). Its *posterior border* is smoothly rounded, forming most the anterior edge of the inferior orbital fissure; central is the infra-orbital groove. The *anterior border* is part of the orbital margin, continuous medially with the lacrimal crest of the maxilla's frontal process (p. 389). The *infra-orbital groove*, for so-named vessels and nerve, begins midway on the posterior border, continuous with a groove on the posterior surface and passes forwards into the *infra-orbital canal*, which opens on the anterior surface below the infra-orbital margin. Near its midpoint the canal has a small lateral branch for the anterior superior alveolar nerve and vessels; this *canalis sinuosus* (Wood Jones; see Jones 1939) descends in the orbital floor lateral to the infra-orbital canal and curves medially in the anterior wall of the maxillary sinus. It passes below the infra-orbital foramen to the margin of the anterior nasal aperture in front of the anterior end of the inferior concha; it then follows the aperture's lower margin to open near the nasal septum in front of the incisive canal. Anteromedial in the orbital surface and lateral to the lacrimal groove, the attachment of the inferior oblique muscle may make a small depression.

The *nasal surface* (3.158) displays posterosuperiorly the large, irregular *maxillary hiatus* leading into the sinus. At the upper hiatal border are parts of air sinuses completed by the ethmoid and lacrimal bones; the smooth concave surface below the hiatus is part of the inferior meatus, and behind it a rough surface meets the palatine's perpendicular plate; this surface is traversed by a groove, descending forwards from the mid posterior border; it is converted into a *greater palatine canal* by the palatine's perpendicular plate. Anterior to the hiatus a deep groove, continuous above with the lacrimal groove (p. 386), makes about two-thirds of the circumference of the nasolacrimal canal, the rest being the lacrimal's descending part, and the lacrimal process of the inferior

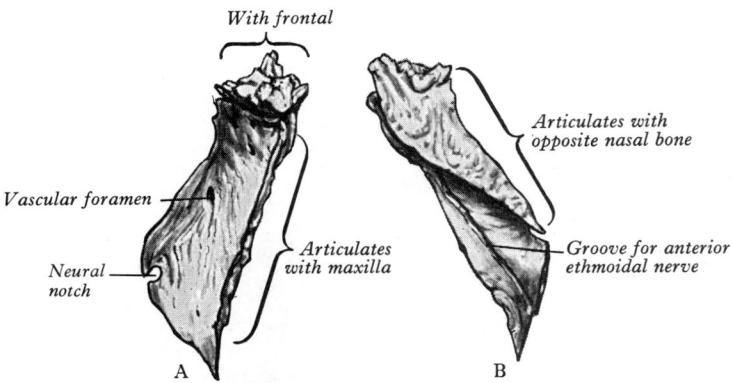

With frontal
Articulates with opposite nasal bone
Vascular foramen
Groove for anterior ethmoidal nerve
Neural notch
Articulates with maxilla

A   B

3.152 The left nasal bone: A. external aspect; B. internal aspect.

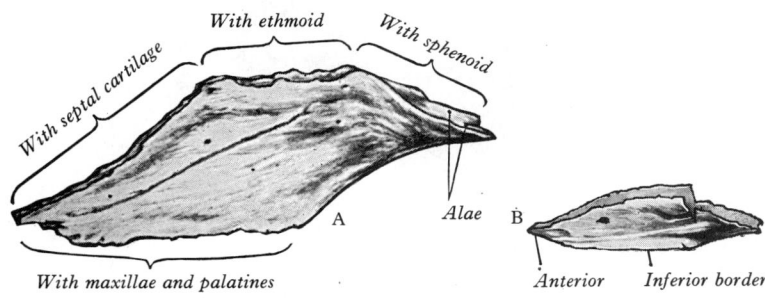

With ethmoid   *With sphenoid*
With septal cartilage
Alae
With maxillae and palatines
Anterior   *Inferior border*

3.153 The vomer: A. left lateral aspect; B. at birth.

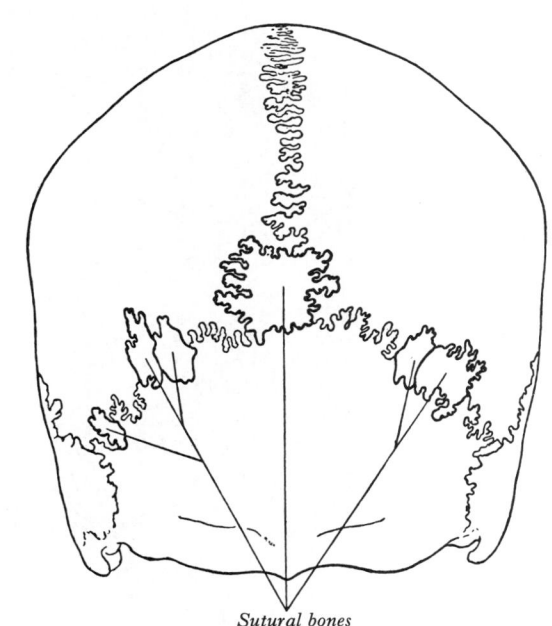

Sutural bones

3.154 Sutural bones in the lambdoid and sagittal sutures.

nasal concha (3.151). This canal leads the nasolacrimal duct to the inferior meatus (3.116). More anterior is an oblique *conchal crest* for the inferior nasal concha; the concavity below it is part of the inferior meatus; above it the surface is part of the atrium of the middle meatus.

The maxillary sinus (3.157,158), a large pyramidal cavity, has thin walls corresponding to orbital, alveolar, facial and infratemporal aspects of the maxilla. Its lateral, truncated *apex* extends into the zygomatic process, sometimes into the zygomatic bone; its *base* is medial and is the lateral wall of the nasal cavity, displaying in the maxillary hiatus when seen in a disarticulated bone. This hiatal aperture is reduced by the ethmoid uncinate process and descending part of the lacrimal bone above, the maxillary process of the inferior nasal concha below, and the

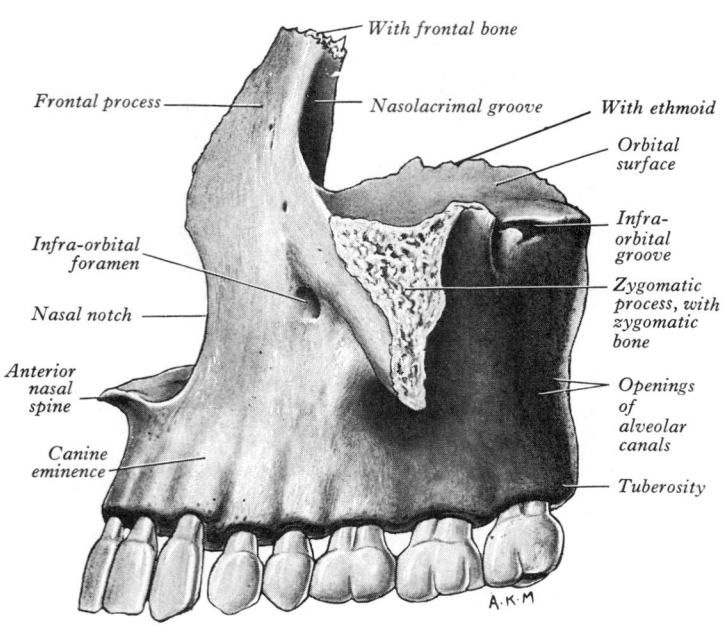

3.155 The left maxilla: lateral aspect.

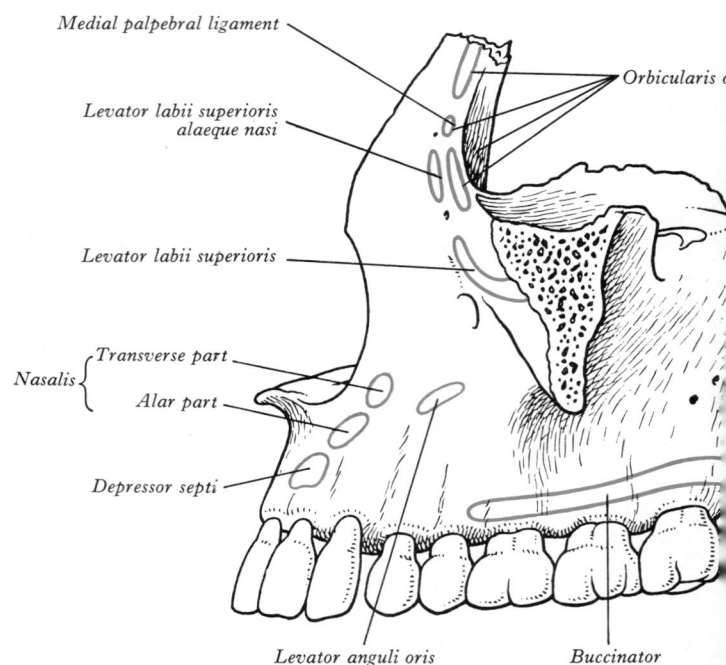

3.156 Outline of left maxilla, showing muscular attachments.

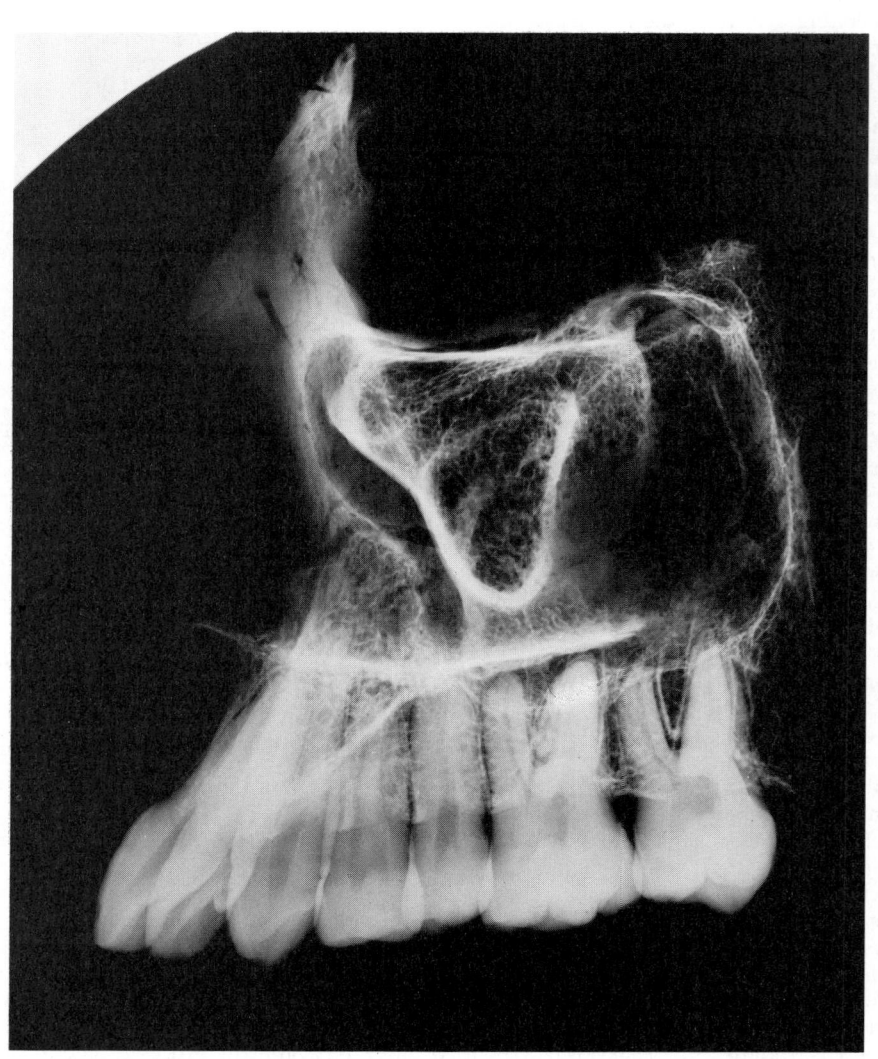

3.157A High resolution projection microradiograph of a maxilla (mediolateral view). Note particularly the distribution of compact and trabecular bone, the extent of the profile of the maxillary sinus and its relation to the dental alveoli, the compact bone of the *laminae durae* of the latter, the hard tissues, pulp cavities, root canals and foramina of the teeth. Contrast the dense shells of compact bone in the frontal, zygomatic and palatine processes with the fine filigree of trabecular bone in much of the alveolar process, interior of the zygomatic process and the posterior wall of the sinus. (Contributed by C Buckland-Wright, Guy's Hospital Medical School.)

palatine's perpendicular plate behind (3.116,158). The maxillary sinus hence connects only with the middle meatus, usually by two small holes, one usually closed by mucous membrane. Its *posterior wall* contains *alveolar canals*, conducting posterior superior alveolar vessels and nerves to molar teeth; these canals may ridge the sinus, whose *floor* is formed by the alveolar process, its lowest part about 1.25 cm below the nasal floor. Radiating septa of varying size usually spring from the sinual floor between adjacent dental roots; sometimes it is perforated by molar roots (p. 1179). The infra-orbital canal usually ridges the sinus from *roof* to *anterior wall*. The cavity's size varies, even between sides of a single skull (p. 1179).

Applied anatomy. Because of the extreme thinness of sinual walls, a tumour may push up the orbital floor and displace the eyeball, project into the nasal cavity, protrude on the cheek or spread back into the infratemporal fossa or down into the mouth. Extraction of molar teeth may damage the floor, and impact may fracture its walls.

The zygomatic process is a pyramidal projection where anterior, infratemporal and orbital surfaces converge. *In front* it merges into the maxillary body's facial surface; *behind* it is concave and continuous with the infratemporal surface; *above* it is roughly serrated for articulation with the zygomatic bone; *below* an arched border separates facial (anterior) and infratemporal surfaces.

The frontal process projects posterosuperiorly between nasal and lacrimal bones (3.89,158). Its *lateral surface* (3.101,158) is divided by a vertical *anterior lacrimal crest* for attachment of medial palpebral ligament and continuous below with the infraorbital margin. At junction of the crest and orbital surface a small palpable tubercle is a guide to the lacrimal sac. The smooth area anterior to the lacrimal crest merges below with the body's anterior surface; part of orbicularis oculi and levator labii superioris alaeque nasi are attached here (3.95,102). Behind the crest a vertical groove combines with one on the lacrimal bone to complete the lacrimal fossa. The *medial surface* (3.160) is part of the lateral nasal wall. A rough subapical area joints with the ethmoid and closes anterior ethmoidal air cells. Below this an oblique *ethmoidal crest* articulates posteriorly with the middle nasal concha and anteriorly underlies the *agger nasi*, a ridge anterior to the concha on the lateral nasal wall; the crest is the upper limit of the atrium of the middle meatus. The frontal process joints apically with the frontal's nasal part, its *anterior border* with the nasal bone, its *posterior* with the lacrimal.

The alveolar process is thick and arched, wide behind, and socketed for dental roots. Eight sockets on each side vary according to contained teeth. That for the canine tooth is deepest, those for molars widest and subdivided into three by septa, those for incisors and second premolar single, that for first premolar sometimes double. The buccinator is attached to the external alveolar aspect, as far forwards as the first molar. In articulated maxillae the processes form the *alveolar arch*. Occasionally a longitudinal *maxillary torus*, variably prominent, appears on the palatal aspect of the process near upper molar sockets.

The palatine process, thick, strong and horizontal, projects medially from the lowest part of the medial maxillary aspect, forming a large part of the nasal floor and palate; it is much thicker in front. Its *inferior surface* (3.106) is concave, uneven and forms with its fellow about the anterior three-fourths of the osseous palate. It displays numerous vascular foramina and depressions for palatine glands and, posterolaterally, two grooves containing greater palatine vessels and nerves. Between the maxillae the infundibular *incisive fossa* appears behind incisor teeth, posterior to which is the median intermaxillary palatal suture—a little uneven but relatively flat on its oral aspect. However, its bony margins are sometimes raised into a prominent longitudinal *palatine torus* (p. 354), incidence of which varies (p. 396), e.g. it is rare in Burmese, but frequent in English crania (29% in males, 48% in females among nineteenth-century Londoners). In the *incisive fossa* are orifices of two lateral *incisive canals*; each ascends into its half of the nasal cavity, transmitting terminations of greater palatine artery and nasopalatine nerve. Occasionally two additional median apertures exist: *anterior* and *posterior incisive foramina*, transmitting the nasopalatine nerves, the left passing anterior. On the inferior palatine surface a fine groove, sometimes termed *incisive*

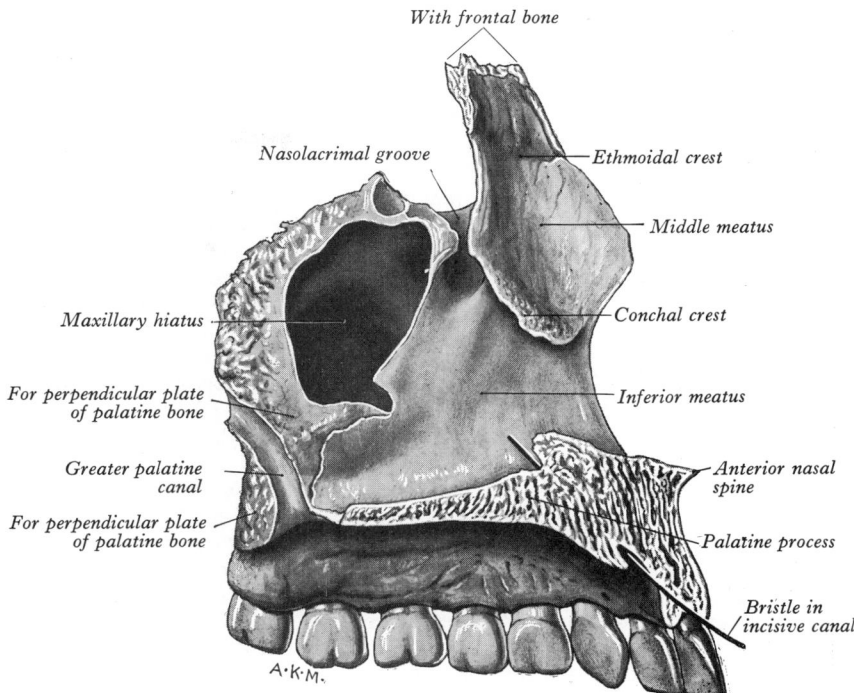

3.157B The left maxilla: medial aspect.

With frontal bone
Nasolacrimal groove
Ethmoidal crest
Middle meatus
Conchal crest
Maxillary hiatus
Inferior meatus
For perpendicular plate of palatine bone
Greater palatine canal
Anterior nasal spine
For perpendicular plate of palatine bone
Palatine process
Bristle in incisive canal
A·K·M·

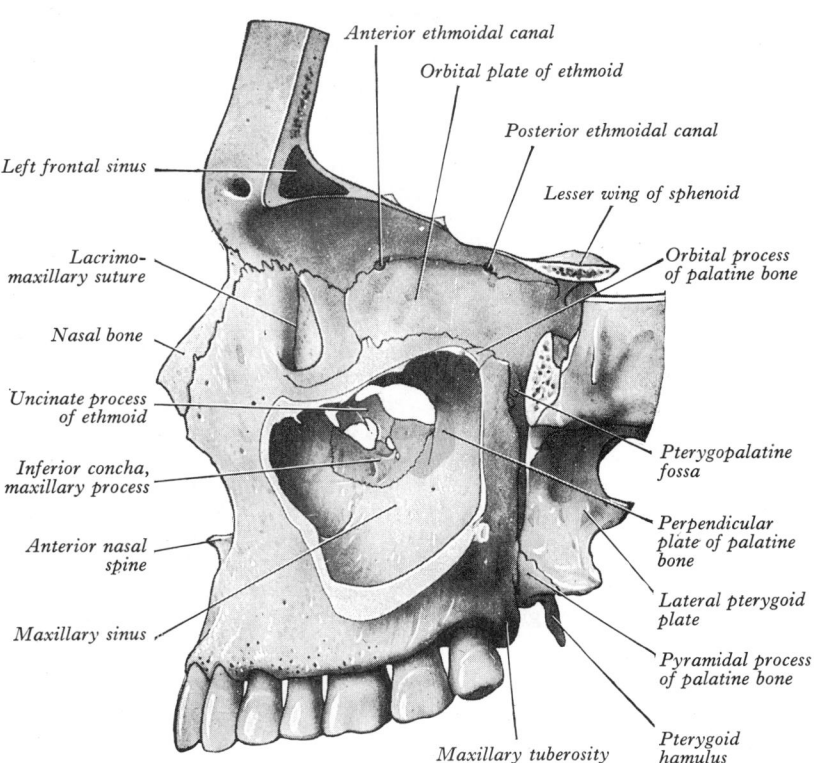

3.158 Oblique sagittal section through the anterior part of the skull, showing the medial wall of the left orbit and the medial wall of the left maxillary sinus. The inferior concha is shown in yellow and the perpendicular plate of the palatine bone in blue.

Anterior ethmoidal canal
Orbital plate of ethmoid
Posterior ethmoidal canal
Left frontal sinus
Lesser wing of sphenoid
Lacrimo-maxillary suture
Orbital process of palatine bone
Nasal bone
Uncinate process of ethmoid
Pterygopalatine fossa
Inferior concha, maxillary process
Perpendicular plate of palatine bone
Anterior nasal spine
Lateral pterygoid plate
Maxillary sinus
Pyramidal process of palatine bone
Maxillary tuberosity
Pterygoid hamulus

suture and prominent in young skulls, may be observed in adults where it extends anterolaterally from the incisive fossa to the interval between lateral incisor and canine teeth. Anterior to this supposed suture is the *os incisivum*, long considered to represent the premaxilla, a separate element in most vertebrates including primates. This view has been criticized (vide infra). The *superior surface* of the palatine process is concave transversely, smooth and forms most of the nasal floor; anteriorly, near its median margin, is the incisive canal. The *lateral border* is continuous with the maxillary body. The *medial border*, thicker in front, is raised into a *nasal crest* which, with its fellow, forms a groove for the vomer. The front of this ridge rises higher as an *incisor crest*, prolonged forwards into a sharp process which, with its fellow, forms an *anterior nasal spine*. The *posterior border* is serrated for the palatine's horizontal plate.

Ossification. The maxilla ossifies in a sheet of mesenchyme superficial to the nasal capsule. Three centres are described: one for the main maxillary mass appears above the canine fossa at about the sixth intra-uterine week; two others are ascribed to the *premaxillary* region, the so-called *os incisivum*, which corresponds in position with a true premaxilla. Of these two 'premaxillary' centres, reputed to occur in human embryonic upper jaw (Woo 1949, Noback & Moss 1953), the principal one is said to appear above the incisor tooth germs in the seventh week; a second, sometimes called *paraseptal* or *prevomerine*, is considered to begin in the medial wall of a paraseptal cartilage at the ventral margin of nasal septum, fusing almost at once with the maxillary's palatine process. Bone formed by the principal premaxillary centre is reputedly overgrown by bone from the main maxillary mass, fusing along its anterior limit with the maxilla's alveolar process. (This junction may be discernible as the *interalveolar suture of Farmer*.) This would explain the absence, after the third month in utero, of any *facial* indication of a premaxilla (*os incisivum*). However a suture, or what appears to be one, may occur on the nasal floor behind its anterior margin. When identifiable postnatally, this suture passes medially on each side to the incisive canal. Moreover, a suture or cleft is always visible at birth anterior in the palate; diverging on each side from the incisive fossa it runs to the septum between lateral incisor and canine teeth (rarely between canine and first premolar). This *palatal* sign of separation between the os incisivum and the rest of the maxilla may persist until the middle decades. Considered together, these features delineate a separate element, anterior to incisive fossa and canals, forming parts of incisor alveoli. Being fused anteriorly with an overlap from the maxillary centre, this 'os incisivum' has no facial representation.

Regular occurrence of a premaxilla in other primates, as a separate bone, has naturally prompted attempts to identify a human homologue. Despite early disagreements (Fawcett 1911a), the above description of maxillary development has become standard, conveniently equating the os incisivum with mammalian premaxilla. This argument depends primarily on supposed ossificatory centres in the os incisivum. One study of serial sections in human embryos strongly suggested that ossification in it is merely an extension from the maxillary centre (Wood et al 1969). The osseous lamina developing from this centre, as in the mandible, is complex, appearing in more than one place in some sections and suggesting multiple centres. It was claimed that their continuity is clear in serial sections, providing a welcome simplification for complexities of maxillary development. Extension from a supposed initial 'premaxillary' centre to form the frontal and palatine processes can thus be ascribed instead to the main maxillary centre. But this simplification does not explain various sutures, clefts and grooves held to delineate the human os incisivum.

Whether patterns of pre-ossificatory mesenchymal condensation in the region have any phylogenetic significance remains uncertain; the status of a human premaxilla is thus problematical. There is strong evidence, at least, that it has no specific centre of ossification.

The maxillary sinus appears as a shallow groove (3.159C) on the nasal aspect at about the fourth month in utero. Infraorbital vessels and nerve are for a time in an open groove in the orbital floor; its anterior part is converted into a canal by a lamina growing in from the lateral side.

Age Changes in the Maxilla

At birth transverse and sagittal maxillary dimensions are greater than vertical. The frontal process is prominent, but the body little more than an alveolar process, its alveoli reaching almost to the orbital floor, the maxillary sinus a mere furrow on the lateral nasal wall. In adults the vertical dimension is greatest, owing to development of the alveolar process and enlargement of the sinus. If all teeth are lost, in old age or earlier, the bone reverts towards infantile shape: its height diminishes, the alveolar process is absorbed (p. 393) and lower parts of the bone contracted and reduced in thickness at the expense of the labial wall (p. 369). Differences in mode of alveolar absorption in maxilla and mandible are of practical importance in fitting dentures.

The Palatine Bones

The palatine bones are posteriorly placed in the nasal cavity between the maxillae and sphenoid's pterygoid processes (3.116); they contribute to the nasal floor and lateral walls, to the palate and orbital floors, and to pterygopalatine and pterygoid fossae and inferior orbital fissures. Each resembles a letter L in its *horizontal* and *perpendicular plates*, with three processes: *pyramidal*, inclining down and posterolaterally from the junction of these plates; *orbital* and *sphenoidal*, surmounting each perpendicular plate and separated by the deep sphenopalatine notch.

The palatine's horizontal plate (3.106,162) is quadrilateral with two surfaces and four borders. The *nasal surface*, transversely concave, forms the posterior nasal floor; the *palatine surface* forms, with its fellow, a posterior quarter of the bony palate; near its posterior margin a curved *palatine crest* often exists. The *posterior border* is thin and concave; to it and its adjacent surface behind the palatine crest the expanded tendon of tensor veli palatini is attached. Medially the posterior border forms, with its fellow, a median *posterior nasal spine* for attachment of the uvulae muscle. The serrated *anterior border* articulates with the maxillary palatine process. The *lateral border* is continuous with the perpendicular plate and marked by a *greater palatine groove*. The *medial border*, thick and serrated, articulates with its fellow at midline, forming the posterior part of the *nasal crest*, which articulates with the posterior part of the vomer's lower edge and the latter is continuous anteriorly with the maxillary nasal crest

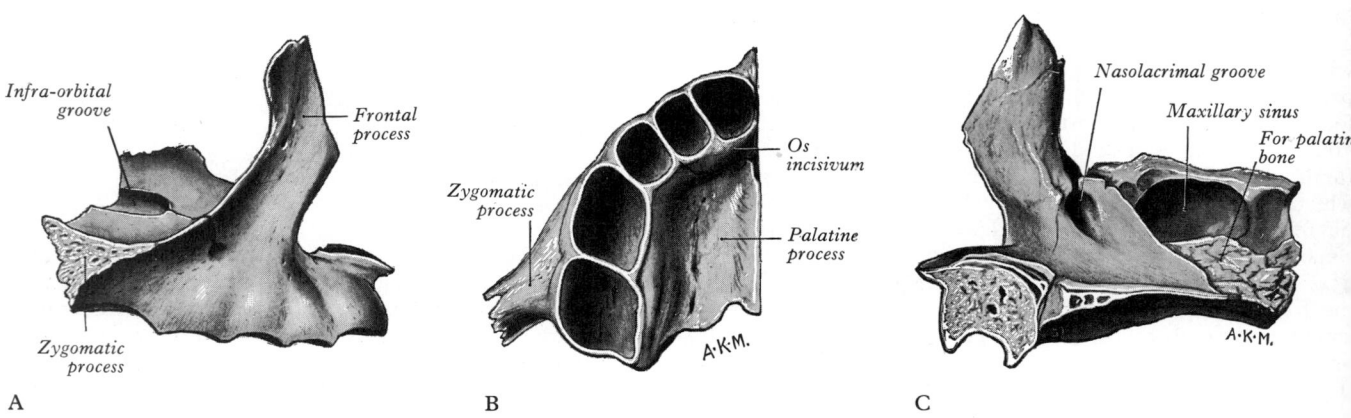

Infra-orbital groove — Frontal process — Zygomatic process — Zygomatic process

Os incisivum — Palatine process — Zygomatic process

Nasolacrimal groove — Maxillary sinus — For palatine bone

A.K.M.   A.K.M.

A   B   C

3.159 The right maxilla at birth. A. Lateral aspect; B. Inferior aspect; C. Medial aspect. Note alveoli for deciduous teeth (B) and rough articular area for opposite maxilla (C).

The perpendicular plate (3.161,162), thin and oblong, has two surfaces and four borders. The *nasal surface* bears inferiorly a broad depression, part of the inferior meatus. Above this the horizontal *conchal crest* articulates with the inferior concha; above this again a shallow depression forms part of middle meatus. This last depression is limited above by an *ethmoidal crest* for the middle nasal concha, above which a narrow, horizontal groove forms part of the superior meatus. The *maxillary surface*, largely rough and irregular, articulates with the maxilla's nasal surface; posterosuperiorly it is a smooth medial wall to the pterygopalatine fossa; its anterior area, also smooth, overlaps the maxillary hiatus from behind to form a posterior part of the medial wall of the maxillary sinus (3.158). Posteriorly on this maxillary surface is the deep, obliquely descending *greater palatine groove*, converted into a *canal* by the maxilla; it transmits the greater palatine vessels and nerve.

The *anterior border* is thin and irregular; level with the conchal crest a pointed lamina projects below and behind the maxillary process of the inferior concha, articulating with it and so appearing in the medial wall of the maxillary sinus (3.158). The *posterior border* (3.162) has a serrated suture with the medial pterygoid plate and is continuous above with the palatine's sphenoidal process, and expanding below into its pyramidal process. From the *superior border* project both the orbital and sphenoidal processes separated by the *sphenopalatine notch* (made into a *foramen* by the sphenoid body). This foramen connects the pterygopalatine fossa to the posterior part of the superior meatus, transmitting sphenopalatine vessels and posterior superior nasal nerves. The *inferior border*, continuous with the lateral border of the horizontal plate, bears the lower end of the greater palatine groove in front of the pyramidal process.

The pyramidal process slopes down posterolaterally from the junction of the horizontal and perpendicular palatine plates into the angle between the pterygoid plates. On its *posterior surface* a smooth, grooved triangular area, limited on each side by rough articular furrows articulating with pterygoid plates, completes the lower part of the pterygoid fossa, and is a partial attachment of the medial pterygoid muscle. Anteriorly the *lateral surface* articulates with maxillary tuberosity; posteriorly a smooth triangular area appears low in the infratemporal fossa between the tuberosity and lateral pterygoid plate (3.103A). The *inferior surface*, near its union with the horizontal plate, shows the *lesser palatine foramina* for the corresponding nerves and arteries (3.106).

The orbital process (3.161,162), directed superolerally from in front of the perpendicular plate with a constricted 'neck', encloses an air sinus and presents three articular and two non-articular surfaces. Of the former (1) the oblong *anterior* or *maxillary* faces down and anterolaterally to articulate with the maxilla; (2) the *posterior* or *sphenoidal*, directed up and posteromedially, bears the opening of an air sinus, usually connecting with the sphenoidal sinus and completed by a sphenoidal concha; (3) the *medial* or *ethmoidal*, facing anteromedially, articulates with the ethmoid's labyrinth. The sinus sometimes opens on this surface, connecting with posterior ethmoidal air cells; more rarely it opens on ethmoidal *and* sphenoidal surfaces, communicating with posterior ethmoidal air cells *and* the sphenoidal sinus. Of the non-articular surfaces, (1) the triangular *superior* or *orbital* is directed superolaterally to the posterior part of the orbital floor (3.97); (2) the *lateral* is oblong, faces the pterygopalatine fossa and is separated from the orbital surface by a rounded border, a medial part of the lower margin of the inferior orbital fissure; this surface may present a groove, directed superolaterally, for the maxillary nerve and is continuous with the groove on the upper maxilla's posterior surface (p. 389). The border between the lateral and posterior surfaces descends anterior to the sphenopalatine notch.

The sphenoidal process (3.161,162), a thin plate smaller and lower than the orbital, is directed superomedially. Its *superior surface* articulates with the sphenoidal concha and, above it, the root of the medial pterygoid plate; it carries a groove helping to form the palatovaginal canal. The concave *inferomedial surface* is part of the nasal roof and lateral wall. Posteriorly the *lateral surface* articulates with the medial pterygoid plate; its smooth anterior region is part of the medial wall of the pterygopalatine fossa.

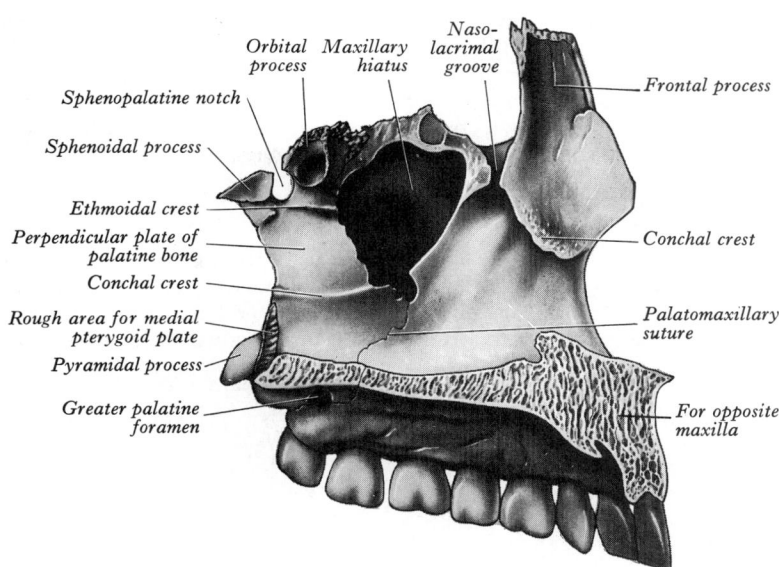

3.160 The left palatine bone in articulation with the left maxilla: medial aspect.

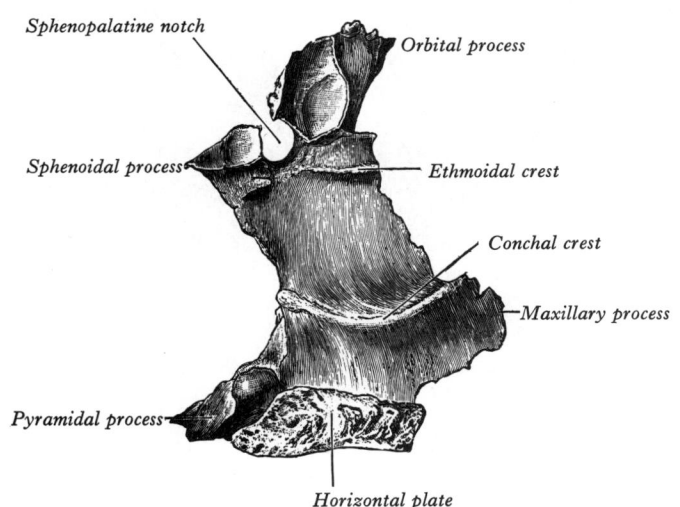

3.161 The left palatine bone: medial aspect (enlarged).

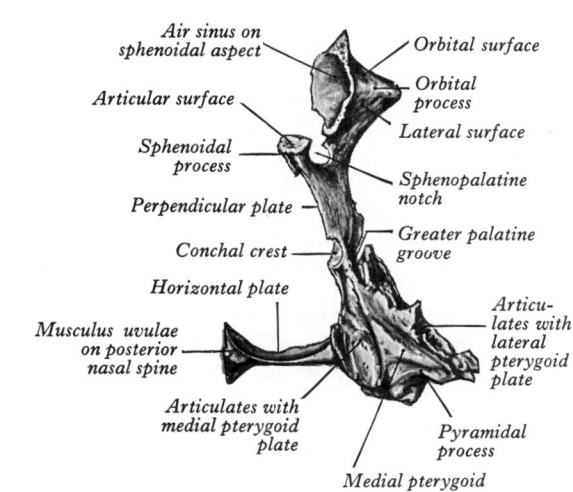

3.162 The right palatine bone: posterior aspect.

The *posterior border* articulates with the vaginal process of the medial pterygoid plate. The *anterior border* is the posterior edge of the sphenopalatine notch. The *medial border* articulates with the vomerine ala. The *sphenopalatine notch*, between the two processes, becomes a foramen by articulation with the sphenoidal body; sometimes the processes themselves unite.

Ossification is in mesenchyme from one centre, appearing during the eighth week in the perpendicular plate. From this, ossification spreads into all parts. At birth height of the perpendicular plate equals width of the horizontal, but in adults is almost twice as much, a change in proportions according with those in the maxilla.

The Zygomatic Bones

Each zygomatic bone forms the prominence of a cheek, contributes to the lateral orbital wall and floor, parts of the walls of temporal and infratemporal fossae and completes the zygomatic arch (3.163). It is roughly quadrangular with anteromedial

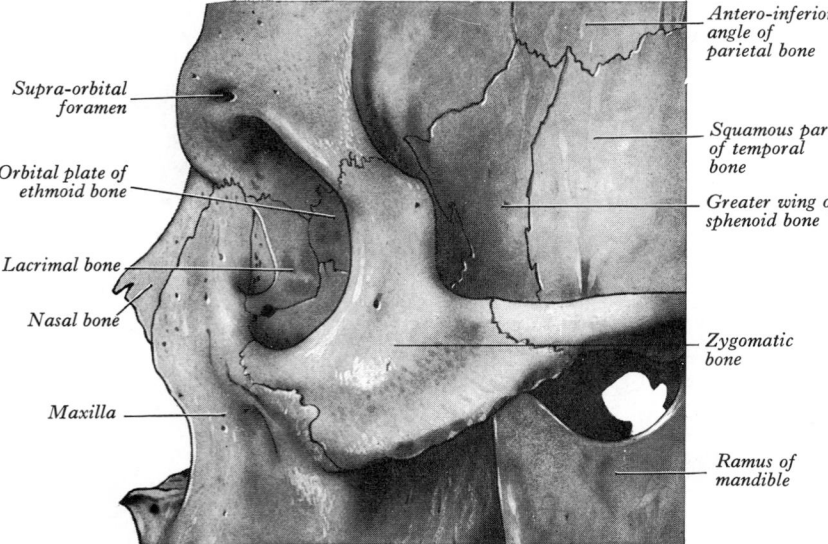

Antero-inferior angle of parietal bone

Supra-orbital foramen

Squamous part of temporal bone

Orbital plate of ethmoid bone

Greater wing of sphenoid bone

Lacrimal bone

Nasal bone

Zygomatic bone

Maxilla

Ramus of mandible

3.163 The left zygomatic bone in situ.

and frontal processes. It can be described as having three surfaces, five borders and two processes.

The *lateral surface* (3.163,164A), really anterolateral, is convex and pierced near its orbital border by the *zygomaticofacial foramen* (often double) for the zygomaticofacial nerve and vessels; below this the zygomaticus minor and, posteriorly, zygomaticus major are attached. The zygomaticofacial foramen is sometimes absent (p. 396). The postero-medial *temporal surface* (3.164B) has a rough anterior area for articulation with the maxilla and a smooth, concave posterior area extending up posteriorly on its frontal process as the anterior aspect of the temporal fossa. It also extends back on the medial aspect of the temporal process as an incomplete lateral wall for the infratemporal fossa. The *zygomaticotemporal foramen* pierces this surface near the base of the frontal process. The *orbital surface* (3.164A,B), smooth and concave, is the anterolateral part of the orbital floor and adjoining lateral wall, extending up on the medial aspect of its frontal process. It usually bears *zygomatico-orbital foramina*, openings of canals leading to zygomaticofacial and temporal foramina.

The smoothly concave *anterosuperior* or *orbital border* forms the inferolateral circumference of the orbital opening, separating orbital and lateral surfaces. The *antero-inferior* or *maxillary border* articulates with the maxilla; its medial end tapers to a point above the infraorbital foramen; near the orbital margin a part of the levator labii superioris is attached. The *posterosuperior* or *temporal border* is sinuous, convex above, concave below and continuous with the posterior border of the frontal process and upper border of the zygomatic arch; attached to it is the temporal fascia. Below

the frontozygomatic suture is often a small *marginal tubercle*, easily palpable. The *postero-inferior border* is roughened by attachment of the masseter. The serrated *posteromedial border* articulates with the sphenoid's greater wing above, and orbital surface of the maxilla below. Between these serrated regions a short, concave, non-articular part usually forms the lateral edge of the inferior orbital fissure; it is sometimes absent, the fissure then being completed by junction of the maxilla and sphenoid or a small sutural bone between them.

The *frontal process*, thick and serrated, articulates above with the the frontal's zygomatic process and behind with sphenoid's greater wing. On its orbital aspect, within the orbital opening and about 1 cm below the frontozygomatic suture, a tubercle of varying size and form, observed in 95% of skulls (Whitnall 1932), is the attachment of the lateral palpebrae ligament (p. 1215), suspensory ligament and part of the aponeurosis of levator palpebrae superioris (p. 1207). The *temporal process*, directed backwards, has an oblique, serrated end articulating with the temporal zygomatic process to complete the zygomatic arch.

Ossification is from one centre, appearing in fibrous tissue about the eighth week. The bone is sometimes divided by a horizontal suture into a larger upper and lower divisions.

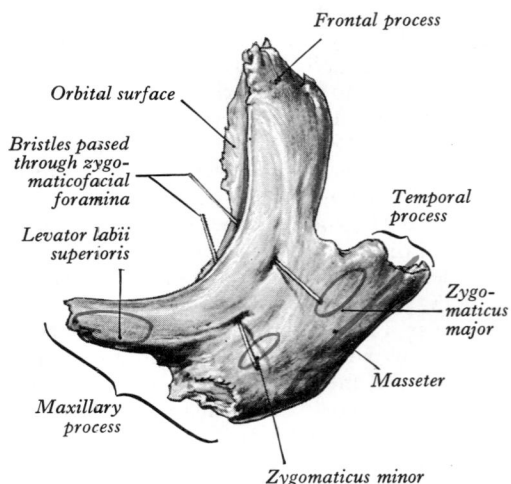

Frontal process

Orbital surface

Bristles passed through zygomaticofacial foramina

Temporal process

Levator labii superioris

Zygomaticus major

Masseter

Maxillary process

Zygomaticus minor

A

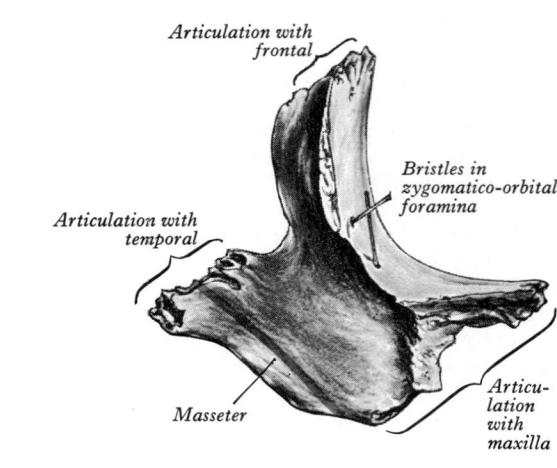

Articulation with frontal

Bristles in zygomatico-orbital foramina

Articulation with temporal

Articulation with maxilla

Masseter

B

3.164 The left zygomatic bone, showing muscular attachments. A. Lateral aspect. B. Medial aspect.

CRANIAL CHARACTERISTICS AT DIFFERENT AGES

The skull at birth is large in proportion to other skeletal parts, but the facial region is relatively small, only about one-eighth of the neonatal cranium, compared with half in adult life. Smallness of face at birth is due to the rudimentary state of mandible and maxillae, non-eruption of teeth and the small size of maxillary sinuses and nasal cavity. The latter is almost entirely between the orbits, the lower border of the piriform nasal aperture being only slightly below the orbital floors. The large size of the calvaria, especially the cranial vault, is related to precocious cerebral growth. The cranial base is relatively short and narrow and, although middle and internal auditory parts are almost adult in size, petrous temporal bones are generally far from adult dimensions. Bones of the cranial vault are unilaminar and without diploë. Frontal and parietal tuberosities are prominent and in norma verticalis the greatest width is between parietal tuberosities (p. 396). The glabella, superciliary arches and mastoid processes are not developed.

Ossification is incomplete, many bones being still in several elements united by fibrous tissue or cartilage. The 'os incisivum' is continuous with the maxilla (p. 390); pre- and postsphenoids have just united, but halves of frontal bone and mandible, and squamous, lateral and basilar parts of occipital bone are all separate. A second styloid centre (*stylohyal*) has not appeared and parts of the temporal bones are separate except for the commencing fusion of the tympanic with petrous and squamous parts. The fibrous membrane, forming the cranial vault before ossification, is unossified at the angles of parietal bones, leaving six *fonticuli* (*fontanelles*), two median (anterior and posterior) and two lateral pairs (sphenoidal and mastoid). The *anterior fontanelle* (3.165), the largest, is at the junction of sagittal, coronal and frontal sutures, hence rhomboid, and about 4 cm in anteroposterior and 2.5 cm in transverse dimensions. The *posterior fontanelle* (3.165), at the junction of sagittal and lambdoid sutures, is hence triangular. The *sphenoidal* (anterolateral) and *mastoid* (posterolateral) *fontanelles* (3.165) are small, irregular and at sphenoidal and mastoid angles of the parietal bones.

At birth the orbits are large and the germs of developing teeth are near their orbital floors. Temporal bones differ greatly from their adult form. Internal ear, tympanic cavity, auditory ossicles and mastoid antrum are almost adult, the tympanic plate is an incomplete ring and the mastoid process absent. Hence the external acoustic meatus is short, straight, unossified and wholly fibrocartilaginous. The external aspect of the tympanic membrane faces *down* rather than laterally, in accord with the basal cranial contour. The stylomastoid foramen is exposed on the lateral surface of the skull; the styloid process has not fused with the temporal bone; the mandibular fossa is flat and more lateral and its articular tubercle undeveloped. Paranasal sinuses are rudimentary or absent; only the maxillary sinuses are usually identifiable.

During birth the skull is moulded by slow compression; that part of the scalp which is more central in the birth canal is often temporarily oedematous due to interference with venous return and is called the *caput succedaneum*. Fontanelles and width of sutures allow bones of the cranial vault some overlap. The skull is compressed in one plane with compensatory elongation orthogonal to this. These effects disappear within a week.

Postnatal growth. Co-ordinated postnatal growth of the calvarial and facial skeleton proceeds at different rates and periods, the cranial cavity being related to cerebral growth, facial skeleton to the development of teeth, muscles of mastication and tongue; but growth of the cranial base is not at the same rate as that of the vault. Therefore the three regions must be considered separately. The anterior part of the cranial base is a *zone of interaction* between facial and cerebral growth (Brash 1924, Scott 1967).

Growth of the vault is rapid during the first year and slower to the seventh, by which time it has reached almost adult dimensions. For most of this period expansion is largely concentric; form is determined early in the first year, remaining thereafter largely unaltered (Weinmann & Sicher 1955, Scott & Symons 1977, Brodie 1941, Hoyte 1966). That *shape* of the vault is not directly related to cerebral growth but to genetic factors is supported by the great range of cranial indices and shapes in racial groups. During the first and early second years growth of the vault is mainly by ossification at apposed margins of bones, which possess an osteogenic layer (p. 468), accompanied by some accretion and absorption of bone at surfaces to adapt to continually altering curvatures. Growth in *breadth* occurs at sagittal, sphenofrontal, spheno-temporal and occipitomastoid sutures and petro-occipital cartilaginous joints, growth in *height* at frontozygomatic and squamosal sutures, pterion (p. 352) and asterion (p. 396). During this period fontanelles are closed by ossification of bones around them, but separate centres may convert them into sutural bones. Sphenoidal and posterior fontanelles 'fill in' within two or three months of birth, mastoid fontanelles usually near the end of the first year and the anterior fontanelle at about the middle of the second, by which time calvarial bones have interlocked at sutures, commencing early in the first year. Further expansion is chiefly by accretion and absorption on external and internal surfaces respectively (Ford 1956). Meanwhile bones thicken, but not uniformly. At birth the vault is unilaminar, tables and intervening diploë appearing about the fourth year, with maximal differentiation at about 35 years, when diploic veins are prominent in radiograms. Thickening of the vault and development of external muscular markings are related to the development of masticatory and neck muscles. The mastoid process is a visible bulge in the second year and invaded by air cells in the sixth; Galli et al (1976) have recorded metrical studies of it.

Growth of the base is responsible for much of the cranial lengthening, mostly at cartilaginous joints between the sphenoid and ethmoid, and especially between sphenoid and occipital. It is largely independent of cerebral growth, and continues at the occipitosphenoid synchondrosis until the eighteenth to twenty-fifth year, a period prolonged by continued expansion of jaws to accommodate erupting teeth, and by growth in muscles of mastication and those of the nasopharynx. However there is some evidence that growth may cease at about 15 years (Latham 1966). A pubertal growth spurt has been ascribed to both sexes, about two years earlier in females; considerable postpubertal growth, up to 17.5 years in males (Roche & Lewis 1974), has been described. For a review of studies on basal growth consult Hoyte (1975), and for labelling studies of growth in craniofacial cartilages (in rats) see Kvinnsland & Kvinnsland (1975). Buranaruga & Houghton (1981) have applied multivariate analysis to Cartesian coordinates of cranial landmarks in Polynesian crania; they claim considerable independence in the growth and positioning of segments noted above. Trenouth (1984) has contributed an extensive study of 60 fetal cranial profiles and he draws attention to the effects of disproportionate growth of brain and face and the result of this in changing the developing cranial base. It is to be hoped that this series of observations can be linked with pre- and postnatal changes.

Growth of the face occupies a longer period than the calvaria. Much information is derived from serial radiography (Salzmann 1961). Ethmoid bone, orbital and upper nasal cavities have almost completed growth by the seventh year. Orbital and upper nasal growth is achieved by sutural accretion with deposition of bone on the facial aspects of margins. The maxilla is carried *down and forwards* by expansion of orbits and nasal septum and sutural growth, especially at fontanelles and zygomaticomaxillary and pterygomaxillary sutures. In the first year growth in width occurs at symphysis menti and midpalatal, internasal and frontal sutures; but such growth is diminished or even ended when the symphysis menti and frontal suture close during the first few years, even though the midpalatal suture persists until mature years. Facial growth in this period continues to puberty and later, linked with the eruption of permanent teeth. After sutural growth, near the end of the second year, expansion of the facial skeleton is by surface accretion on the face, alveolar processes and palate, with resorption in the walls of the maxillary sinuses, the upper surface of hard palate and the labial aspect of the alveolar process. Co-ordinated growth and divergence of the pterygoid processes is due

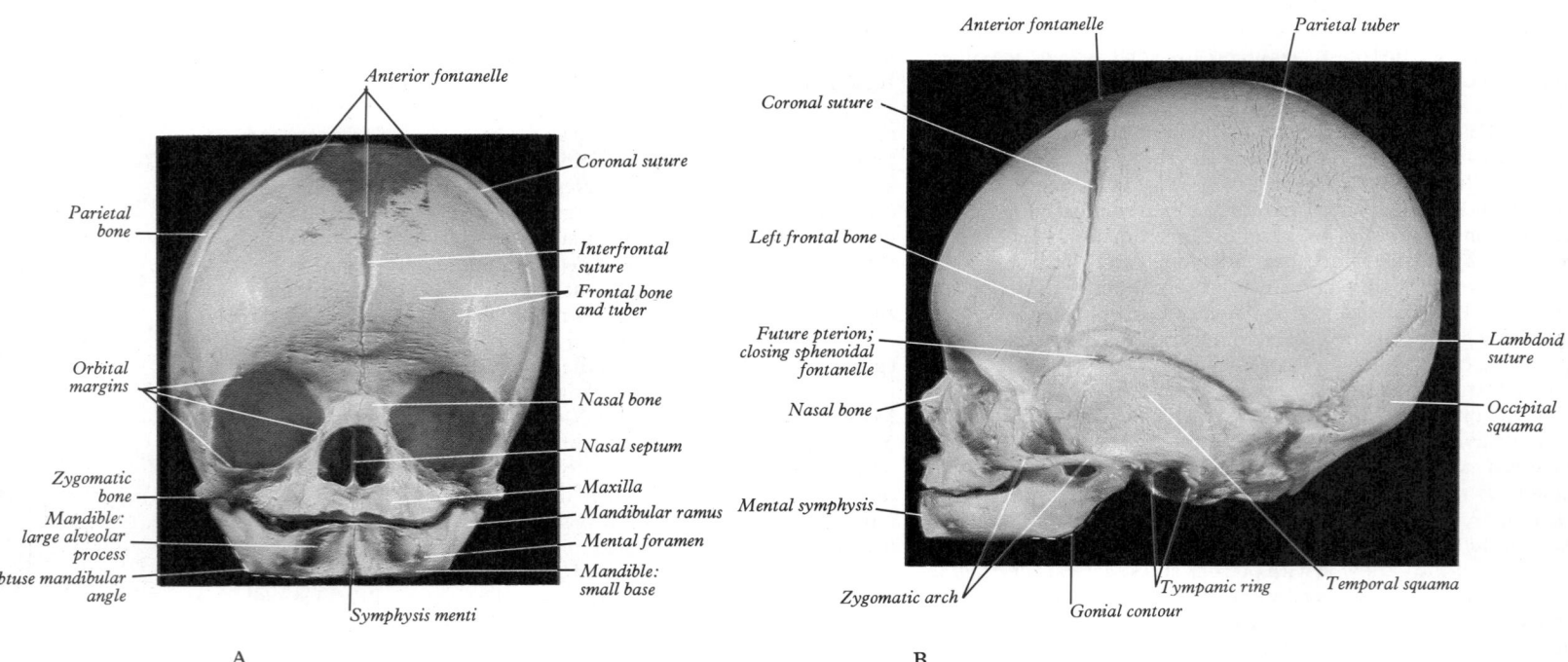

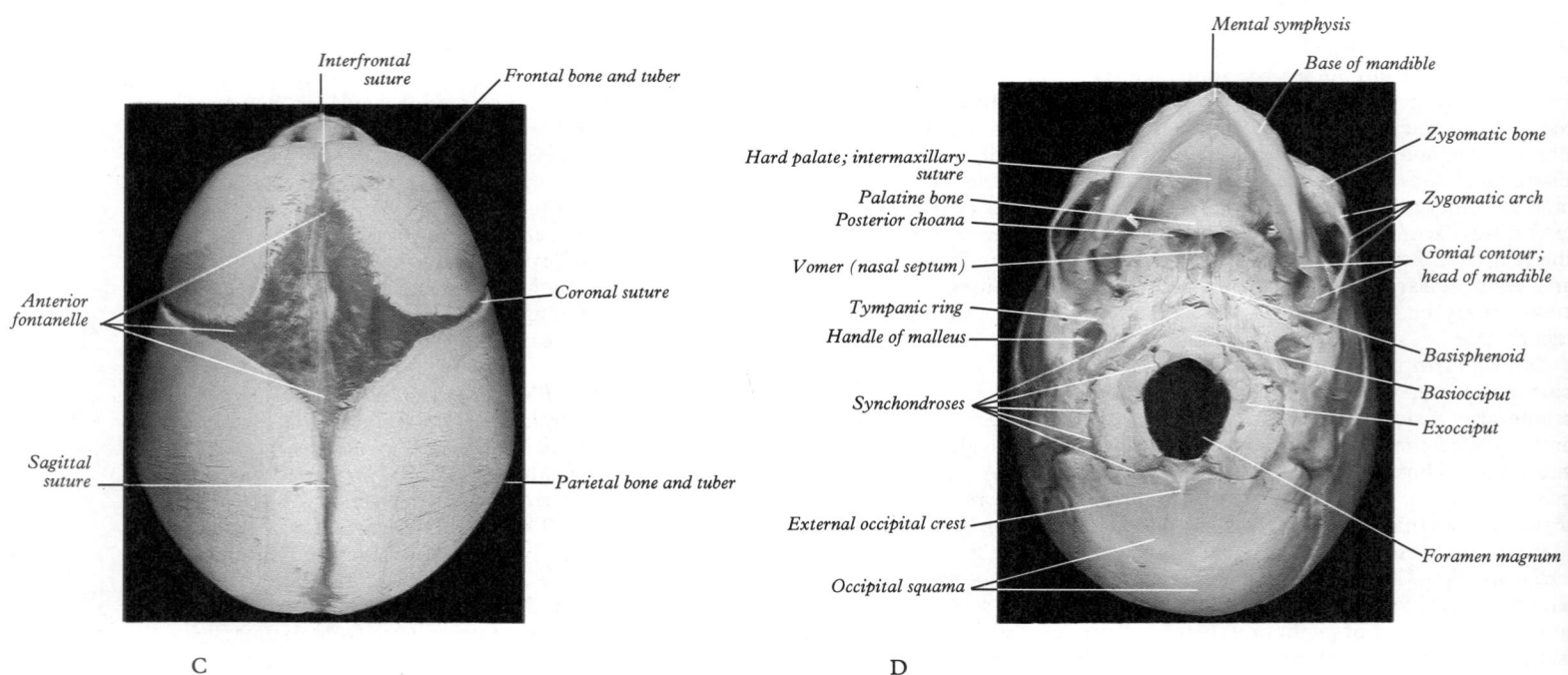

3.165 Skull of newborn infant: A. Anterior aspect; B. Lateral aspect; C. Superior aspect; D. Basal aspect. (Photograph by K. Fitzpatrick).

to deposition and resorption of bone on appropriate surfaces. Mandibular growth is described on pages 369-370.

Obliteration of the calvarial sutures progresses with age, commencing between 30 and 40 years internally, about 10 years later on the exterior, but closure times vary greatly (Todd & Lyon 1924, 1925, Abbie 1950, Singer 1953). Obliteration usually begins at the bregma, extending into sagittal, coronal and lambdoid sutures, in that order. In old age the skull becomes thinner and lighter, but occasionally the reverse. The most striking senile feature is diminution in size of the mandible and maxillae following loss of teeth and absorption of alveolar bone. This reduces the vertical depth of the face and increases the mandibular angles (**3**.120) (pp. 369-370).

Sexual and Racial Differences in the Cranium

Until puberty there is little sexual difference in skulls; the adult female's is a little lighter and smaller, its capacity about 10% less; its walls are thinner and muscular ridges less marked; the glabella, superciliary arches, and mastoid processes are less prominent; air sinuses are smaller; tympanic plates are smaller and their margins less rough; the upper orbital margins are sharper, the forehead vertical, frontal and parietal tuberosities prominent and the vault somewhat flattened; the facial contour is rounder, facial bones smoother and mandible and maxillae and contained teeth are smaller. Thus, more childhood characteristics are retained in females. Typical male or female skulls are easily recognized, but in some characteristics are so indeterminate that diagnosis of sex is difficult or impossible.

Sexual dimorphism is generally less marked in humans than other primates. This is associated with the *paedomorphic* tendency of human stock, females and even males being less divergent in adult development from their own juvenile form, a tendency less apparent in other primates, especially in males (Abbie 1952, Schultze 1956). Exaggerated sexual differences occur in most anthropoid apes and some monkeys in the form of enlarged canines in males, with associated jaw development, accentuated muscular ridges and total size of skull. These excesses are typified in the gorilla, orang utan, and many species of baboon, in which males also much exceed females in stature and physique. Even in extinct human forms sexual dimorphism, on available fossil evidence, appears to have been less than in other primates. In modern humans differences are further reduced but degree of sexual divergence in cranial size and proportions varies in racial groups. Such sexual differences have been less exactly assessed than general ethnic differences. In both, assessment depends on observation of two kinds of feature: (*a*) those which cannot be measured, for which no satisfactory quantitative technique has, as yet, been devised (e.g. size of mastoid process, prominence of chin); and (*b*) those expressed as actual measurements or indices (e.g. cranial capacity, orbital index). Examples of the former category in distinguishing sex have been cited above; more exhaustive lists exist (Keen 1950). Use of such features in judging sex or race in isolated crania is dependent on observers' experience but, where many are available from a single ethnic group, both types of assessment are more certain. Distinction of three major races by non-metrical traits can be effected with some confidence (Todd & Tracey 1930). Incidences of many variations have been observed on a racial basis, e.g. metopism and other sutural variants (Berry & Berry 1967), and are of considerable ethnic but lesser forensic interest. Berry (1975) made a special study of *non-metrical* human cranial variations; these include:

Highest nuchal line (p. 348).
Os suturale at lambda (pp. 373, 378).
Ossa suturalia in lambdoid suture (pp. 347,371).
Parietal foramen (p. 342).
Os suturale at bregma (p. 386).
Frontal suture (metopism) (p. 383).
Ossa suturalia in coronal suture (p. 383).
Epipteric os suturale (p. 386).
Frontotemporal articulation (p. 352).
Parietomastoid os suturale (p. 352).
Os suturale at asterion (p. 352).
Tympanic foramen of Huschke (p. 381).
Extrasutural mastoid foramen (p. 378).

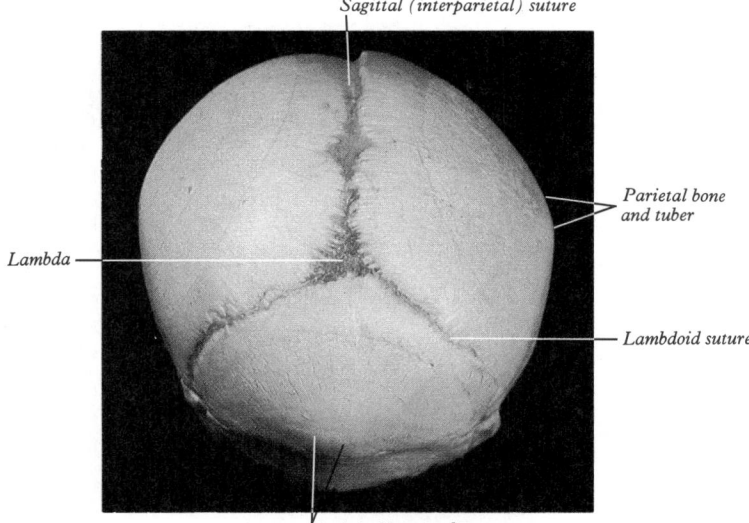

3.165E Skull of newborn infant: posterior (occipital) aspect. (Photograph by K. Fitzpatrick)

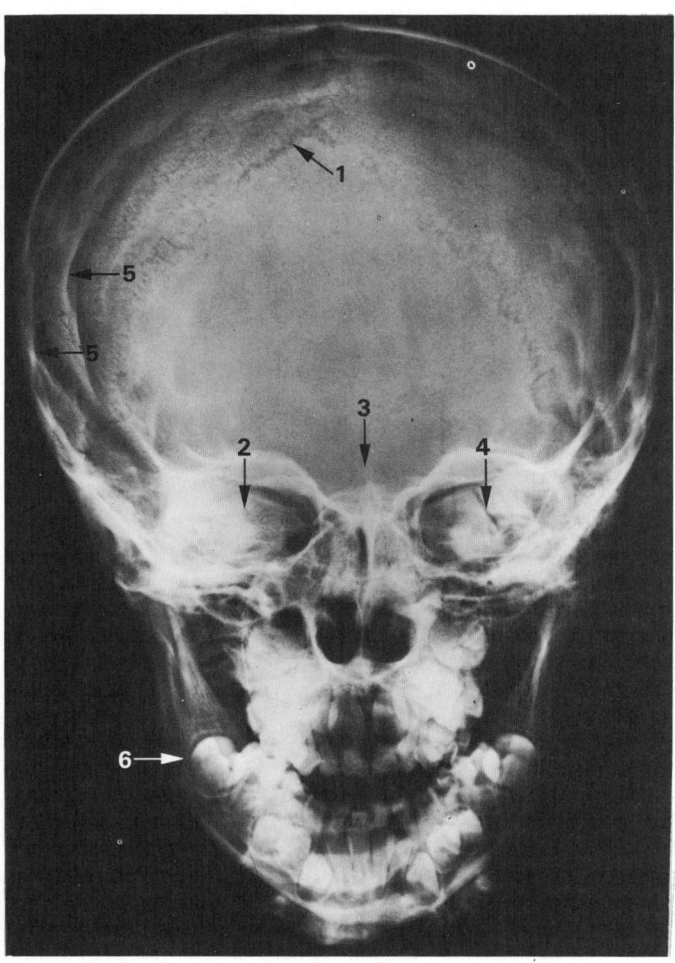

3.166 Radiograph of child's skull, aged seven: occipitofrontal view. 1. Lambdoid suture. 2. Petrous portion of right temporal bone, seen through the cavity of the orbit. 3. Crista galli of the ethmoid bone. 4. Fracture through petrous portion of left temporal bone, seen through the cavity of the orbit. 5. Impressions for cerebral gyri. 6. Second permanent molar tooth, not yet erupted.

14. Absence of mastoid foramen (p. 378).
15. Patent condylar canal (p. 359).
16. Double condylar facet (p. 373).
17. Precondylar tubercle (p. 373).
18. Double hypoglossal canal (p. 373).
19. Incomplete foramen ovale (p. 355).

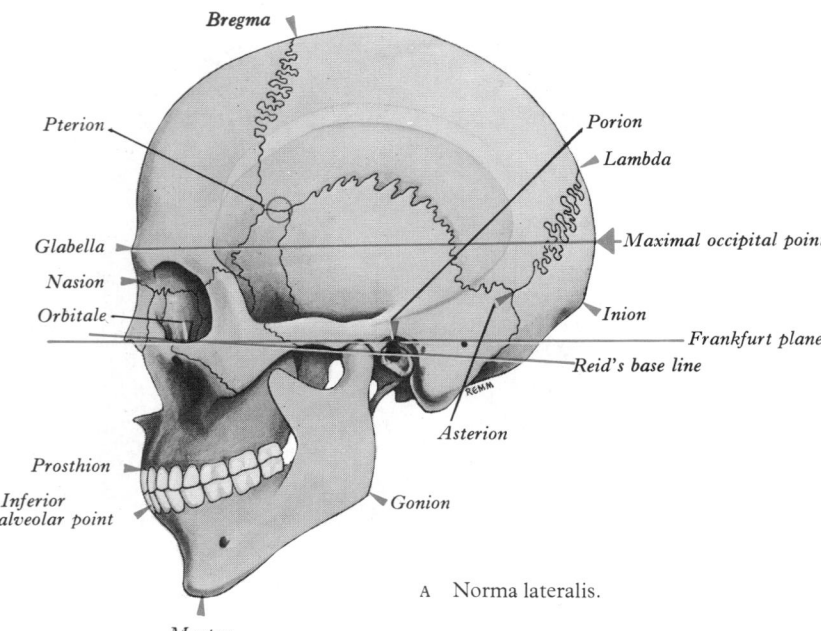

A   Norma lateralis.

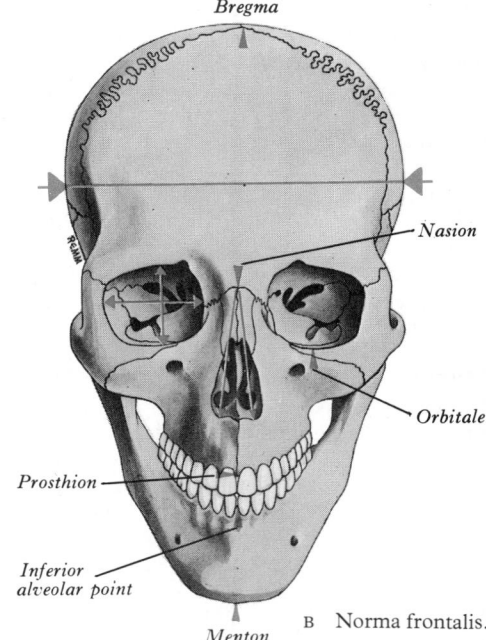

B   Norma frontalis.

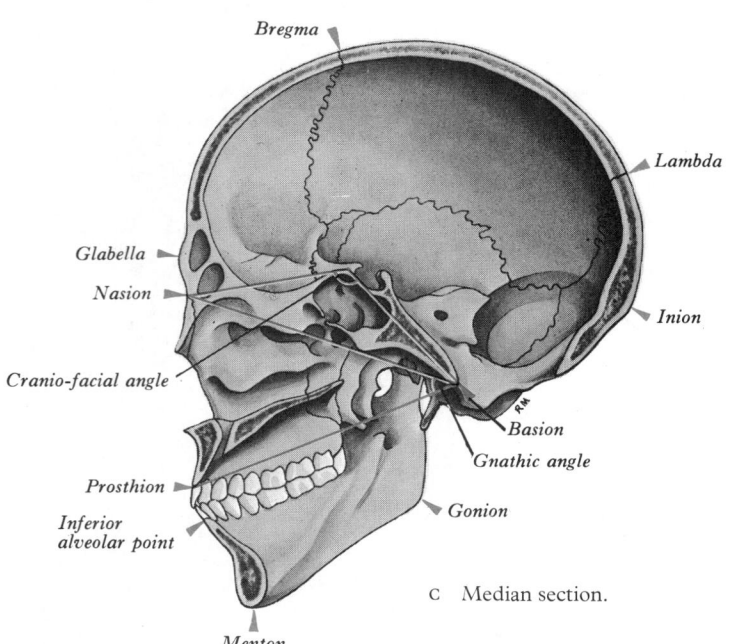

C   Median section.

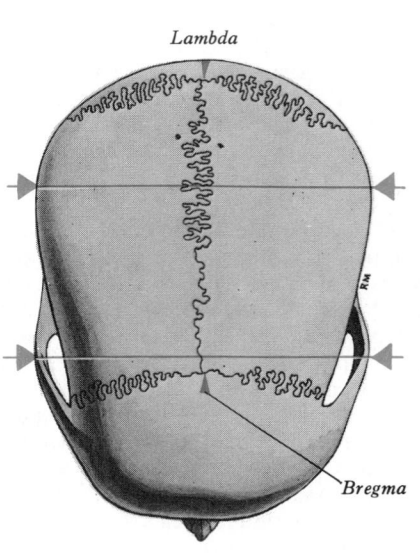

D   Norma verticalis.

3.167 These diagrams illustrate the cranial points used, by international agreement, in making linear and certain angular measurements in anthropometry. In all four views the skull is in the standard orientation, that is, with the Frankfurt plane as a horizontal. The point at which the cranio-facial angle is measured is not named; it corresponds to the midpoint of the chiasmal groove.

20. Incomplete foramen spinosum (p. 355).
21. Accessory palatine foramina (p. 354).
22. Maxillary torus (p. 389).
23. Palatine torus (pp. 354, 389).
24. Occurrence of zygomaticofacial foramen (pp. 343, 392).
25. Supraorbital foramen or incisure (p. 382).
26. Frontal foramen or incisure (p. 382).
27. Extrasutural anterior ethmoid foramen (p. 347).
28. Absence of posterior ethmoid foramen (p. 347).
29. Accessory infraorbital foramen (p. 346).
30. Interparietal bone (pp. 373, 386).
31. Paramastoid process (p. 373).
32. Auditory torus (p. 380).

Although many have studied these and other cranial variants (e.g. the mandibular torus, p. 367), few studies have been quantitative. They have been reviewed by Berry & Berry (1967) and Berry (1975), with extensive statistical data; they have assessed which cranial variations exhibit racial (and sometimes sexual) correlation. The interested reader should also consult Czarnetzki (1971) and Hjarnø et al (1974).

To achieve a more objective racial assessment, metrical studies have long been practised; internationally accepted techniques of *craniometry* have promoted a large corpus of comparable ethnic data for males and, to a lesser extent, females. Many standard measurements are used; only major dimensions and examples of indices derived from them are included here (**3.**167,168). The calvarial part of the skull is measured as follows:

A. Maximal cranial length   summit of glabella to furthest occipital point.

B. Maximal cranial breadth   greatest breadth, at right angles to median plane.

C. Cranial height   from basion (median point on anterior rim of foramen magnum) to bregma.

All measurements are to the nearest millimetre. From these three dimensions, three indices are calculated: B/A, C/A and C/B and expressed as percentages.

The breadth/length ratio is the *cranial index* (*cephalic index* in the living). Its recorded range of variation is high, as with other skull or head indices. All ranges are arbitrarily divided into steps usually covering 5% sections of the total range; to each step a specific term is applied. For example:

(*Maximal breadth/maximal length*) × 100 = cranial or cephalic index

Up to 74.9	= dolichocranic or dolichocephalic
75.0 to 79.9	= mesocranic or mesocephalic
80.0 to 84.9	= brachycranic or brachycephalic

Distinction between measurement of dried crania and living heads, by the two sets of terms, is not rigidly applied. The 'cephalic' terms are more common, often used for both purposes. Other indices in common use are:

(a) Total facial index = (nasion-gnathion height/bizygomatic breadth) × 100. The nasion is the point where the internasal suture meets the frontal bone. The gnathion is the midpoint of the lower mandibular border.

(b) Upper facial index = (nasion-prosthion length/bizygomatic breadth) × 100. The prosthion is the midpoint of the maxillary alveolar rim, between the central incisors. The bizygomatic breadth is the greatest distance measured by trial between zygomatic arches on external aspects.

(c) Nasal index = (nasal breadth/nasal height) × 100. Breadth is the horizontal maximum across the nasal aperture, and height is from nasion to the mean between the two lowest points on the aperture's lower border.

(d) Orbital index = (maximal orbital height/maximal orbital breadth) × 100.

(e) Palatal index = (maximal palatal breadth/maximal palatal length) × 100.

(f) Gnathic index = (basion-prosthion/basion-nasion) × 100.

Indices provide a system for *metrical* record of sizes and proportions of cranial features in place of subjective impressions. If a group is said to be dolichocephalic, this does not (or should not) imply a vague comparison, but that the index is within known numerical limits. It can be observed that the orbital opening appears rounder in females, but this statement can be quantified and

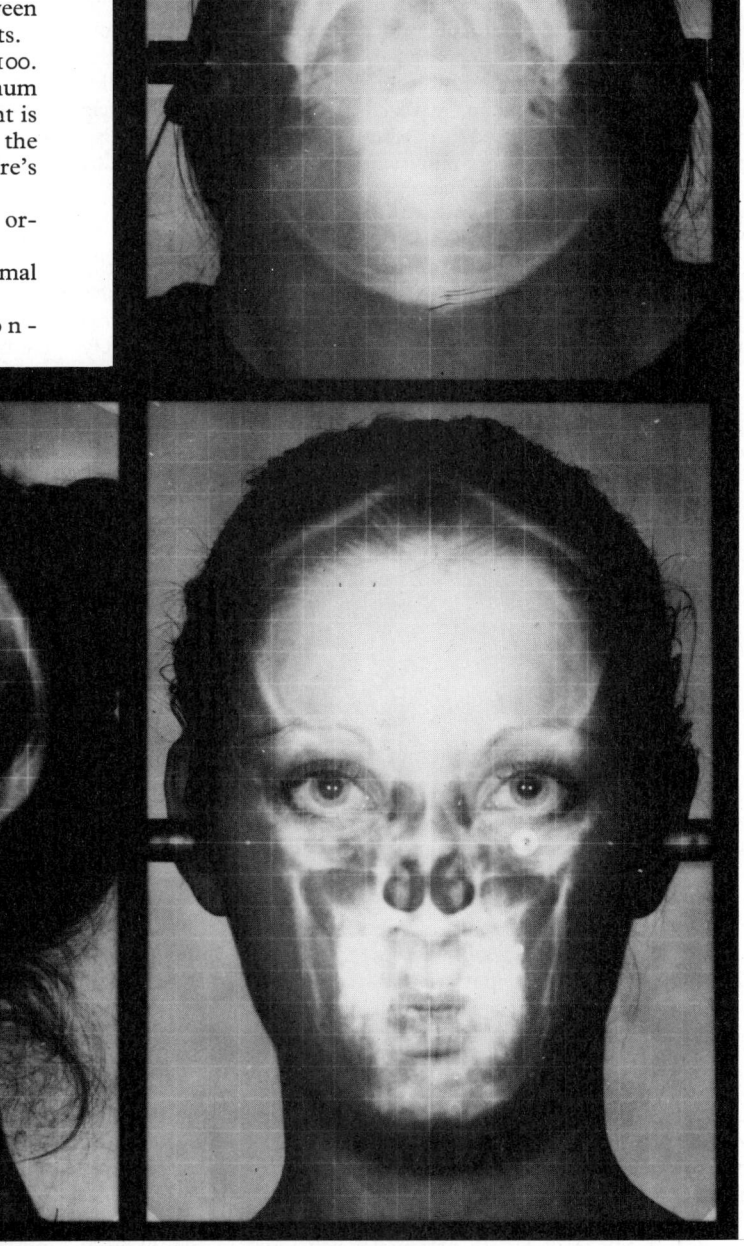

168 Combined, accurately superimposed photographs and radiographs of the kind shown here are currently being used to produce a computerized analysis of metrical relations between a selection of datum points in the skeletal and soft tissue components of human heads and faces. The technique, known as morphanalysis, has been elaborated by Dr G P Rabey, to whom we are indebted for these records. Both photographs and radiographs are produced in the three primary planes and hence the morphometric analysis is three-dimensional. For further information see Rabey (1968, 1971, 1980).

is then far more valuable. Again, jaws project more in some races, a fact determinable by inspection alone when the condition is pronounced; the gnathic index yields a numerical expression and comparisons are then permissible.

The following measurements are applied to mandibles: *(a)* height of symphysis, *(b)* length of body, *(c)* length of ramus, *(d)* bigonial width (between angles), *(e)* bicondylar width, and so on. The angle between body and ramus is also often estimated. Since the mandible is often missing from skulls and moreover is considered less reliable in assessing racial affinity (Morant 1936), its measurements are less widely used.

Other forms of cranial measurement are employed. Horizontal circumference and various arcs and contours are sometimes measured. Radiographic studies make it possible to extend classic craniometry and to measure directly certain angles, e.g. the *gnathic angle*, between basion–nasion and basion–prosthion lines. The *craniofacial* or *cranial base angle* is of special interest, representing the degree of cranial flexure. The human angle has a value of about 130°, probably achieved by birth, little further change occurring. Its high human value has been linked with translation of tongue towards pharynx, a change sometimes regarded as essential to speech (Schulter 1976). Radiographic technique for the estimation of endocranial volume has been evolved (Haas 1952). *Cranial capacity*, an indication of brain volume, can be assessed directly by filling the cavity with lead shot, millet seed or other particulate materials suitable for volumetric measurement when poured out again. Several formulae have also been constructed to give cranial capacity from length, breadth and height of the cranium. Examples are:

Males: $0.000337 \, (L-11) \, (B-11) \, (H-11) + 406.01$ cc.

Females: $0.000400 \, (L-11) \, (B-11) \, (H-11) + 206.60$ cc.

In these formulae, L and B are length and breadth, and H is the *auricular height*, measured to the vertex from the external acoustic meatus. All measurements are in millimetres, but such methods involve some inaccuracy, and various corrections do not entirely remove this (Hrdlička 1947, Montagu 1960).

Attempts to establish reliable craniometric differentiation between races are as old as craniometry. Though mandibular data and cranial capacity are less dependable, satisfactory differentiation is possible in some groups, especially between Caucasians and Negroes. When several measurements are treated by discriminant analysis, an accuracy of over 90% can be expected (Giles & Elliot 1960). Multivariate analysis may also elucidate cranial growth (Liebgott 1977). Craniometric methods are useful in forensic identification when cranial remains can be compared with existing photographic and radiographic records (Glaister & Brash 1937), and in attempts to reconstruct appearance in life of individuals represented only by skeletal remains (Stewart 1954). Cephalometry has also been applied in plastic and oral surgery concerned with craniofacial deformity. A development in this field is *morphanalysis*, which uses gridded radiographs and photographs universally related in three dimensions (3.168). Such standardized records enable diagnostic comparison between patients and the normal population (Rabey 1968, 1971).

Applied anatomy. Factors preventing cranial fracture are elasticity, round shape and construction from secondary elastic arches, each derived from a single bone. Where bone is thin, overlying muscles may cushion blows, e.g. temporal squama and inferior occipital fossa. Thickness of calvaria in 28 Caucasians (from 15 to 82 years) averaged 5.8 mm. In males thickness increased with age, in females the reverse is true (Lippert & Käfer 1974).

The commonest site of basal fracture is through the middle fossa; fracture starts from the point struck, usually near the parietal tuberosity, descends through parietal and temporal squamata and across the petrous region, often traversing the internal acoustic meatus, to the foramen lacerum. This explains various sequelae: thus, if the internal acoustic meatus is damaged, injury to facial and vestibulocochlear nerves may result, with consequent facial paralysis and deafness; if it extends through semicircular ducts, vertigo ensues; if the arachnoid around these nerves in the meatus is torn and when the internal ear and tympanic cavity communicate and the tympanic membrane is torn, as it often is, cerebrospinal fluid may escape.

Facial bones are sometimes fractured by direct violence, most commonly the nasal bones and mandible. Nasal fracture is usually transverse, about 1.25 cm from the free margin, broken edges being displaced backwards or more often to one side by direct blows. The commonest site of mandibular fracture is near the canine tooth, whose deep socket weakens the bone; next most frequent is the angle. Occasionally a double fracture may occur, in both halves of the bone, and is usually compound, with laceration of oral mucosa.

THE APPENDICULAR SKELETON

Phylogeny and Functions

The ancestral vertebrates, *Agnatha*, were not only jawless but limbless, like surviving representatives such as lampreys (Young 1962, Romer 1970). Two continuous ventrolateral finfolds probably preceded the separated fins which appear in fossil remains of the earliest fish possessing jaws, the *Placoderms*. These sometimes displayed more than two pairs of appendages, but in their descendant bony and cartilaginous classes, the *Osteicthyes* and *Chondricthyes*, the vertebrate pattern of two pairs, pectoral and pelvic, was stabilized, except in such forms as snakes, which have subsequently lost their limbs. Appendicular skeletons of fossil and extant fish vary but show a common plan of 'girdle' bones, embedded in body musculature, and rodlike elements radiating into the mobile fin, with an intermediate joint of the ball and socket type. In many orders of *Osteicthyes* homologues of the pectoral and pelvic components of terrestrial limbs can be identified, as well as those of the proximal segments, arm and forearm or thigh and foreleg (Bolk et al 1931–9). Distal equivalence is less clear; but the multiplicity of small bony units in fins at least resembles the extremities, hand and foot, of land tetrapods.

Terrestrial vertebrates are grouped as *tetrapods* because their four limbs are an outstanding common feature, although the pectoral pair may be modified for flight. They are probably derived from a primitive order, the *Crossopterygii* ('fringe-finned'), related to another ancient group, *Dipnoi* or lungfish, species of which still survive. The crossopterygians were known only as fossils until 1938, when a living specimen of a suborder, *Coelacanthus*, was captured off the coast of South Africa; further coelacanths have been secured near the Comoro Islands and elsewhere in the Indian Ocean. These possess lobed fins with a narrowed attachment, and from such appendages tetrapod limbs can most plausibly be derived.

In the change to terrestrial environment, during evolution of primitive amphibians from crossopterygians, two major adaptions began. The first, already foreshadowed in fins, was a repositioning of these limb-like paddles for more effective support and propulsion on land. The *forelimbs* rotated *laterally* at joints with pectoral girdles so that the hinging of the elbow, at first pointing away from the body as it still does in amphibians and reptiles, finally projected backwards. In contrast, the *hindlimbs* rotated *medially* at their pelvic joints into opposite positions, knees projecting forwards. These opposite changes rotated the *preaxial* or leading *border*, corresponding with the 'first' digit, pollex or hallux, from an anterior to lateral position in the pectoral limb. (As described later on p. 399, this changed orientation of the forelimb necessitates a complementary *pronation* of its radioulnar segment to achieve a plantigrade appendage, a compensation unnecessary in the hindlimb.) Coupled with these growing distinctions between fore and hind extremities was a second modification—articulation of pelvic girdles with the spinal column, a condition unknown in fish but uniform in tetrapods, except some fossil amphibians.

With evolution of amphibians, a general skeletal pattern of tetrapod limb was defined and, apart from minor modifications, has persisted through subsequent vertebrates. This *primitive pentadactyl limb* terminates typically in five digits, reduction in which is frequent, especially in mammals. But the primates, including humans, have retained the full number and are in this respect unspecialized.

Of the girdles the *pectoral* always comprises a dorsal *scapula*, but varies in ventral components, commonly a mixture of endochondral and dermal bones. Of these the *coracoid* and *clavicle* persist in mammals, the former reduced in size and fused to the scapula. The clavicle, sole mammalian survivor of a variety of dermal elements in lower vertebrates (e.g. the *cleithrum* and *interclavicle*), is also often reduced and even lost in some mammals, especially fast-running or bounding types of quadruped. Its persistence in primates is linked with the development of forelimbs for grasping and climbing; the clavicle acts as a mobile strut, at the lateral end of which the rest of the limb can be variably positioned (Watson 1917). The *pelvic girdle*, in contrast, contains only endochondral elements; the primitive dermal armour only extended far enough caudally to permit incorporation of dermal elements into the pectoral but not the pelvic girdle. The latter developed a ventral *symphysis* in fish, which in tetrapods contains distinguishable pubic and ischial elements. The dorsal ilium, though present in some fish, is not well developed. In land animals the need of larger muscles to support the body's weight and to impart thrust through hindlimbs, leads to greater iliac size and articulation with the sacral spinal column. Throughout tetrapods these pelvic entities, *pubis, ischium* and *ilium*, are more uniformly arranged than in fish. They also show an increasing tendency to fuse into an *innominate* bone, especially in adult mammals. The two innominate bones are joined ventrally at a symphysis pubis and to the sacrum dorsally at sacroiliac articulations. Thus is established a *pelvis*, primarily locomotor, conducting stresses between axial skeleton and limbs. The term pelvis, a basin, is misleading; it is more of an irregular ring which, being at the caudal end of the body cavity, inevitably surrounds certain viscera, a relation which in human anatomy obscures its essentially appendicular origin and function. Since the cloaca is immediately caudal to it, not only the hindgut but other tubes, urinary and genital, must traverse the pelvis, even though separate external openings evolve in mammals. During birth, the offspring must also traverse the pelvis, and their comparatively large size in human females entails obstetric considerations (p. 431), which again may temporarily overshadow the somatic locomotor nature of hindlimb girdles.

In mammals functional differences between fore- and hindlimbs, accentuated in the human primate, are associated with structural divergence in their girdles. Major differences, (ignoring the possible status of e.g. the coracoid complex) are as follows:

Pectoral Girdle	Pelvic Girdle
Dermal and endochondral	Entirely endochondral.
Two principal components, clavicle and scapula, which remain separate. (However, see above and p. 406.)	Three components, pubis, ischium and ilium, which fuse into a single innominate bone.
No articulation with vertebral column.	Articulates with sacral vertebrae.
No direct ventral articulation (clavicles connected only by interclavicular ligament).	Direct ventral articulation at symphysis pubis.
Articulations of clavicles with axial skeleton (sternum) are relatively small, mobile and ventral.	Articulations of innominate bones with axial skeleton (sacrum) are relatively large, capable of limited movement only, and dorsal.
Comparatively lightly built for mobility.	Massively constructed for resistance to stress, rather than for mobility.

7. Resilient to thrust. — Transmits thrust between vertebral column and leg.

8. Shallow joint with limb, allowing wide range of movement. — Deep joint with limb, limiting range of movement.

Both girdles articulate with the distal, freely movable parts of limbs by 'ball and socket' joints, the first segment of each limb containing a single *humerus* or *femur*. Shoulder and hip joints are the same in mechanism, allowing some movement in all directions, though more limited in the hip. Even in quadrupedal mammals, where limbs are walking props mainly adapted to fore and aft movement, greater mobility is usual at the shoulder, particularly in abduction, permitting splaying of the forelimbs. Throughout tetrapods the shoulder region shows a greater range of structural adaptation to changing function. Scapular orientation in relation to the thorax and humerus varies far more than hindlimb equivalents, even in quadrupeds, but the difference is most extreme in bipedal primates (Ashton et al 1976), whose adoption of an arboreal habitat, in which they climb, walk, run and swing, accords with the development of highly mobile and prehensile forelimbs and similar but lesser changes in hindlimbs. Equally to be emphasized is the habit of *sitting* upright, even in primates such as baboons which have returned to predominantly quadrupedal life on the ground; arms are thus released for manipulative use, activities of the profoundest significance to human evolution.

In modern mammals, *Eutheria*, the articular head of the humerus is usually directed dorsocaudally and the scapular glenoid socket reciprocally. While this may favour quadrupedal use, it prevents movement of the humerus into the vertical position so easily assumed by primates in reaching up above the head. This entails reorientation of the humeral head relative to its shaft, the head being directed *medially* rather than caudally. Coincident with reorganization of the shoulder is a change in maximal thoracic diameter from dorsoventral to transverse, the former being typical of quadrupeds, the latter of bipedal primates. This is coupled with the retention of clavicles in primate mammals and altered scapular orientation. The scapulae come to face ventrally, rather than medially, and their glenoid fossae more laterally. These changes and correlated muscular alterations are linked with a wider range of movement at the shoulder.

The tubercles and bicipital groove of the humerus partly share in the rotation of its head, but the distal end retains its transverse orientation (Martin 1932). Hence a radial axis through the centre of the head is approximately at right angles to the axis of the lower end in most *Eutheria*, because the head is directed caudally. As it becomes more medial this angle increases above 90°. It is still 95° in carnivora, rises to about 100° in monkeys, 120° in apes, and from 135° to 165° or more in man. This change in the primate humerus has been considered torsional and the angle termed *angle of torsion*. The term is misleading; there is no evidence of any true 'twisting' of the humeral shaft; spiralling of the radial nerve and its groove predates this change. Nor is there anything specially human in the torsion angle. It reaches 180° (where the axes are parallel) in birds and was apparently nearly as high in mammal-like reptiles, returning to about 90° in *Metatheria* (marsupials). A word of caution is needed concerning methods of recording angles of humeral torsion. As defined here, the angle is measured between articular 'axes' of the shoulder and elbow joints. However, Krahl (1944, 1976), sometimes quoted in textbooks, measured the angle between the axis of the elbow joint and a line orthogonal to the axis of the shoulder joint; his values, therefore, differ by 90° from those cited here. Neither method is conceptually satisfactory, but very similar figures are obtained (*mutatis mutandis*). Krahl's work thus supports the view put forward here, that change in orientation of the humeral head has involved alterations at the proximal end of the bone and has not been accomplished by an evolutionary twisting of the shaft. Racial and sexual differences have been recorded (Kate 1968).

In the lower limb the state is like that of mammal-like reptiles (and birds), with the two axes almost parallel. (The human femoral head and long axis of its neck are anteverted about 16° relative to the transcondylar femoral axis.) This condition is said not to have varied in the lower limbs, unlike the upper. Lability of the humeral 'torsion' angle may be coupled with greater variability in forelimb activity, especially in bipedal forms, with the consequent release of the forelimbs.

Intermediate segments of limbs, and their joints with humerus and femur, show similarities and differences. The differences are greater in bipeds; in quadrupeds similarities are more evident. Both segments initially contained *parallel* elements, radius and ulna, tibia and fibula (but vide infra). Primitively both pairs articulate with the single proximal element but in most mammals, as in man, fibulae withdraw from knee joints, leaving a *preaxial* tibia as major lever and prop. In the forearm the ulna, a *postaxial* bone, becomes the main force-transmitting strut at the elbow.

Both elbow and knee are hinge joints in quadruped mammals; but, while this was the primitive tetrapod state in knees, elbows had an initial potentiality for *rotation*. The hindlimb turns medially, bringing its extensor surface forwards, so that the foot, hinged at the ankle, projects forwards with its plantar surface on the ground. This typical *plantigrade* habit of many quadrupeds brings the hind-limbs under the body as serially hinged levers adapted for propulsion and support. Note that tibia and fibula remain *parallel*; but the *lateral* rotation of fore-limbs which turns the extensor aspects backwards, entails *pronation* at the elbow to bring the 'hand' forwards, palm downwards, for plantigrade locomotion. This brings the *preaxial* radius anteromedially across the ulna, and both are stabilized in this relation in a large majority of quadrupeds, including all four-*footed* mammals. Both forearm and foreleg bones tend to fuse in such mammals, particularly in those with a reduced number of digits and usually *digitigrade* in habit (e.g. Carnivora), contact with the ground being restricted to flexor aspects of the remaining digits.

A rotatory potential at the elbow, retained and refined in primates, allows free *pronation* and *supination*, essential to the development of manual skills. These movements are impossible in legs. At primate elbows, the ulna forms a massive hinge with the humerus but tapers to a small distal end; the radius is reciprocal: its upper end is small and adapted for rotation, the lower enlarged to carry the hand. Thus, when the radius revolves it takes the hand with it, turning round the distal end of the ulna.

Both wrist and ankle joints are hinges, but wrists can also adduct and abduct. The *carpus* and *tarsus* contain small bones between forearm or foreleg and digits. They are much modified throughout tetrapods, and the two groups show some divergence. But, however modified, both are derivable from the same primary arrangement in the primitive *pentadactyl limb*. This consisted of a proximal row of three, a distal row of five, articulating with five digits, and a central intermediate group, probably four, wedged between these rows. By fusion, loss and modification in size, shape and articulation, carpal and tarsal bones display much variation (particularly reduction). Tracing homologies between primitive elements and subsequent transformations (structural and functional) has aroused much controversy (Broom 1901, Wood Jones 1949, Lewis 1964a), but agreement has largely been reached. Authorities disagree in minor details, but generally accepted homologies are shown in the accompanying table.

Main divergencies between mammalian carpus and tarsus, apart from the latter's greater size in primates, are three. Firstly, the carpus usually remains more primitive, retaining a full proximal row, all articulating with the forearm bones; in the tarsus only one (talus) retains this role in most mammals, including the primates. Secondly, the single proximal tarsal bone (calcaneus) projects back as a lever behind the tibia and fibula; this does not occur in the carpus. Thirdly, tarsal bones at intermediate joints permit some rotation (inversion/eversion and pronation/supination), an adaptation to uneven surfaces, which may be met in some forelimbs by pronation and supination between *forearm* bones and some abduction and adduction at the wrist and intercarpal joints. In all three modifications the tarsus is more advanced. Evolution of the calcanean lever is a *mammalian*, not a solely primitive or a human modification. Its high development in the human foot and the resemblance of other primate feet to hands, may superficially suggest that it is a *human* characteristic. (Consult Lewis 1983 for a detailed account, with extensive bibliography, of the evolution of the architecture of the mammalian foot.)

Occurrence of small supernumerary bones, pre- or postaxial in position in both carpus and tarsus, has been taken as evidence of additional digits (Wood Jones 1941), suggesting an ancestral

HOMOLOGIZATION OF THE PRIMITIVE TETRAPOD AND HUMAN CARPUS AND TARSUS

	TETRAPOD CARPUS	HUMAN CARPUS		HUMAN TARSUS		TETRAPOD TARSUS	
PROXIMAL ROW	Os Radiale	Scaphoid	Scaphoid minus tubercle	Talus	Navicular minus tubercle	Os Tibiale	PRE-AXIAL
	Os Intermedium	Lunate			Talus	Os Intermedium	
	Os Ulnare	Triquetral		Calcaneus		Os Fibulare	POST-AXIAL
CENTRAL ELEMENTS	Os Centrale	Absent	Tubercle of Scaphoid	Navicular	Tubercle of Navicular	Os Centrale	
DISTAL ROW	Os Carpale 1	Trapezium		Medial Cuneiform		Os Tarsale 1	PRE-AXIAL
	Os Carpale 2	Trapezoid		Intermediate Cuneiform		Os Tarsale 2	
	Os Carpale 3	Capitate		Lateral Cuneiform		Os Tarsale 3	
	Ossa Carpalia 4 and 5	Hamate		Cuboid		Ossa Tarsalia 4 and 5	POST-AXIAL

N.B. Alternative views have been stated for the carpal scaphoid and the tarsal navicular. In each case the most widely accepted view is on the left. In the case of the talus the anomalous persistence of its posterior process as a separate element, an *os trigonum*, may have influenced the orthodox view of the talus as an amalgam of ossa tibiale et intermedium. If it becomes accepted that the talus is merely the os intermedium, the main residual controversies concern the scaphoid and navicular, centred on the possible derivation of each bone's tubercle from the corresponding os centrale. It should be added that the *centrale* element has been shown as singular in this table, though there were probably a variable number of centralia in the ancestral form.

pattern of more than five digits, but all known tetrapods show the orthodox five and often less. The five metacarpal and metatarsal bones originally articulated each with a separate carpal or tarsal, which are reduced to four by fusion of fourth and fifth even in the pentadactyl primates. In mammals with reduced or lost digits, metacarpals and metatarsals as well as phalanges are involved. An example exists in Carnivora, whose first digit, pollex or hallux, is degenerate and useless, though often still clawed; another more extreme example is the horse, which retains only the middle digit, with a greatly enlarged metapodial bone.

The *metapodial* (metacarpal or metatarsal) *formula* puts these bones in order of length and number. In humans it is usually 2>3>4>5>1 in hands, and 1>2>3>4>5 in feet. The *phalangeal formula*, starting from the preaxial digit, pollex or hallux, numbers the phalanges in each digit. In many primitive reptiles it is 2.3.4.5.3, in primitive mammals 2.3.3.3.3.for hand and foot, as in humans. In horses it is 0.0.3.0.0. A *digital formula* denoting relative lengths of digits is sometimes used, especially in comparing primate extremities.

In all tetrapods pollex and hallux tend to diverge from other digits, perhaps as a prop. Some early mammals were probably arboreal, using hands as primates do, with a grasp between the preaxial digit and the rest. Primate prehension is developed in hands and feet, though absent from human feet and not always effective in primate hands. In gibbons hands are used as hooks in brachiating from branch to branch; thumbs are not effectively opposable to other digits, though the feet have an excellent grasp. In most primates the distal end of the first metatarsal is not tied by ligaments to the second, allowing the hallux marked freedom. But the human first and second metatarsals *are* so connected, all five metatarsal heads being tied by ligaments. Human thumbs, on the contrary, are even better developed for *opposition* than in other primates, in which the joint between the pollicial metacarpal and *trapezium* (os carpale I of the primitive tetrapod carpus) permits

thumbs to flex and rotate towards fingers, either in a 'power' grip, when thumb and fingers wrap around opposite sides of an object such as a branch, weapon or tool, or in a 'precision' grip, in which the thumb is opposed to a single finger, often the index, in more accurate manipulations (Napier 1966).

Bones in human limbs, when compared functionally, are excellent examples of adaptation. In legs, the massive, almost immobile pelvic girdle is directly articulated to the vertebral column; the limited ball-and-socket mechanism at the hip is adapted for fore and aft swinging but allows some abduction to effect a steady stance in bipeds; the hinging knee and ankle joints give resilient propulsion in walking, running and springing; and the powerful lever of the foot is arched for resilience in the final thrust of take-off and impact of landing. All these features produce a limb highly developed for *locomotion*.

Arms present different adaptions, divisible into various components, but integrated in action. The scapula floats in a muscular suspension, strutted by a slender mobile clavicle, its only axial connexion. A *shallow* ball-and-socket joint allows a far wider range for the reaching, grasping, 'inquisitive' hand. Flexion *and* rotation at the elbow bring the hands within the arena of vision, providing a steady but adjustable carpal base for fine movements of opposable thumb and highly mobile fingers. Endlessly variable manipulations, entailing delicate muscular adjustments, are thus developed; with the constant feedback of stereoscopic vision and tactile sensation, a limitless horizon for manual dexterity in invention and cerebral development has been opened.

In ourselves we see the arm and hand at the end of several hundred million years of evolution. Perhaps it is not the 'end' but, even were the human upper limb to evolve no further, it has, with growing excellence of cerebral control, accomplished such a mastery over the environment as to deflect the whole course of evolution into a new channel (Huxley 1942).

THE SKELETON OF THE UPPER LIMB

The Scapula

The scapula (3.169,170,171), a large, flat, triangular bone, overlaps in part the second to seventh ribs on the posterolateral thoracic aspect. It has costal and dorsal surfaces, superior, lateral and medial borders, inferior, superior and lateral angles, and three processes—spinous, acromial and coracoid. The lateral angle is truncated by a glenoid cavity for articulation with the humerus, sometimes regarded as the *head*, connected to the body by an inconspicuous neck. The long axis of the scapula (from 'head' to inferior angle) is nearly vertical and much thickened. The relatively featureless costal surface contrasts with the dorsal, which is divided by the spine (3.170).

The costal surface (3.169), anteromedial when the arm is pendent, is slightly hollow especially above, and presents near the lateral border a rounded ridge, prominent near the neck, less so below and separated from the border by a narrow groove. **The dorsal surface** (3.170) is divided by the shelf-like *spine of the scapula* into a small upper and a large lower area, which are confluent at a *spinoglenoid notch* between the spine's lateral border and the neck's dorsal aspect. A flat dorsal strip, for muscular attachments, adjoins the **lateral border**, which is a sharp, rough and sinuous ridge, from the inferior angle to the glenoid cavity. Superiorly it widens into a rough, triangular *infraglenoid tubercle* (3.169,170). It is covered by muscles and cannot clearly be palpated. It is described as thick because the grooved part of the costal surface, the narrow flat lateral strip of the dorsal surface and adjacent thickened ridge (3.169), are often included in it during clinical examination. The **medial border**, from the inferior to the superior angle, is easily felt in its inferior two-thirds but its upper third is too deep. The **superior border**, thin, sharp and shortest, is separated laterally from the coracoid process by a *suprascapular notch*.

The scapular angles: the inferior angle overlies the seventh rib or intercostal space. Palpable through the skin and covering muscles, it is also visible as it advances round the thoracic wall when the arm is raised. The *superior angle*, at the junction of superior and medial borders, is obscured by muscles. The *lateral angle*, truncated and broad, is the 'head', bearing the *glenoid cavity* and forming a glenohumeral joint with the humerus. It provides a shallow, and a limited socket for the humeral head; its outline is piriform, narrower above (3.171). Just above it a small rough *supraglenoid tubercle* encroaches on the root of the coracoid process. The *anatomical neck*, the constriction adjoining the rim of the glenoid cavity, is most distinct at its inferior and dorsal aspects. Ventrally and dorsally it extends between the infraglenoid and supraglenoid tubercles, passing lateral to the root of the coracoid process. Some authorities describe a *surgical neck*, best distinguished dorsally. Commencing inferiorly, near the glenoidal rim, it passes upwards and laterally through the deepest part of the spinoglenoid (great scapular) notch continuing to the anterior border of the suprascapular notch, thence medial to the coracoid and is completed by an ill-defined, but corresponding ventral line.

The scapular spine (3.170) projects from the upper dorsal surface and is triangular. Its lateral border, thick and rounded, helps to form the *spinoglenoid (great scapular) notch*, between it and the neck's dorsal aspect. Anteriorly it joins the dorsal surface along a line running laterally and slightly up from the junction of upper and middle thirds of the medial border. The fairly flat scapular body is, however, bent along this junction, accentuating the concavity of the upper costal surface. The dorsal border is the *crest of the spine*, mainly subcutaneous; this expands medially into a smooth, triangular area. Elsewhere its upper and lower borders and surface are roughened by muscular attachments. The concave superior surface of the spine widens laterally, forming with the

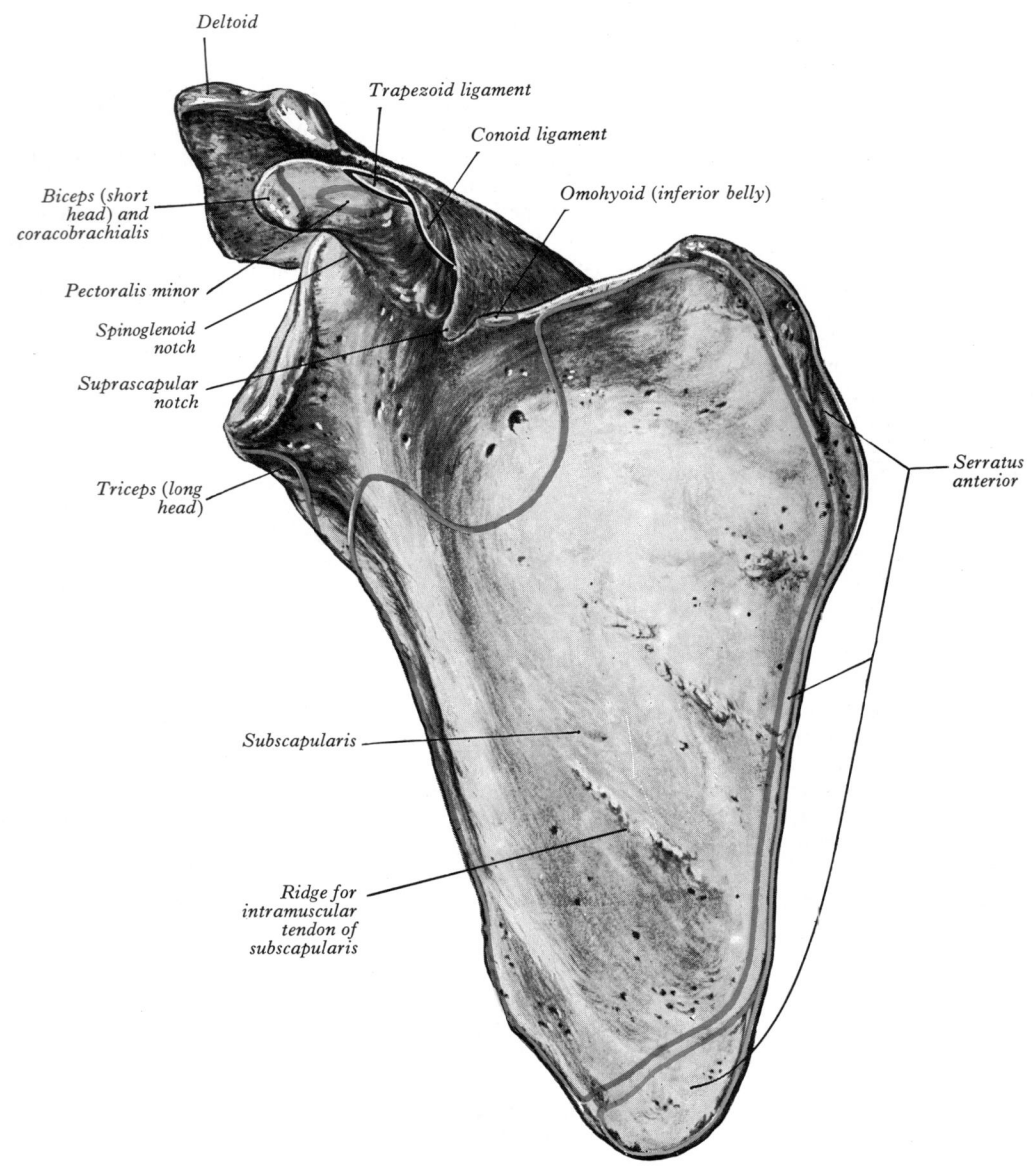

Deltoid

Trapezoid ligament

Conoid ligament

Omohyoid (inferior belly)

Biceps (short head) and coracobrachialis

Pectoralis minor

Spinoglenoid notch

Suprascapular notch

Triceps (long head)

Serratus anterior

Subscapularis

Ridge for intramuscular tendon of subscapularis

3.169 The right scapula: anterior (costal) aspect.

dorsal surface of the upper body the *supraspinous fossa*. The spine's inferior surface is overhung by the crest's medial, narrow end, but is gently convex in its wider lateral part; with the lower dorsal surface it forms the *infraspinous fossa*, continuous with the supraspinous at the spinoglenoid notch.

The acromion projects forwards, almost at right angles, from the lateral end of the spine. The inferior border of the crest and lateral border of the acromion are continuous at the *acromial angle*, a visible, subcutaneous landmark. The medial acromial border is short, bearing anteriorly a small, oval facet directed superomedially for articulation with the clavicle's lateral end. The lateral border, tip and superior surface are easily palpable. An accessory articular facet may occur on the acromion's inferior surface.

The coracoid process (**3.**169,170) springs from the summit of the scapular head and hooks slightly laterally and forwards. With the arm pendent, it points almost straight forwards. Its enlarged tip is palpable, though covered by the anterior part of the deltoid. It is about 2.5 cm below the junction of lateral fourth and the rest of the clavicle. The supraglenoid tubercle at the root of the process adjoins the upper part of the glenoid cavity. On the coracoid's dorsal aspect, where it changes direction, is an impression for the conoid part of the coracoclavicular ligament.

Further details. On the costal surface the subscapularis (**3.**169) is attached to almost the whole area, including much of the lateral groove but excluding the vicinity of the neck. Intramuscular

tendons are attached to radiating rough ridges dividing this surface incompletely into smooth areas. The neck's anterior aspect is separated from the tendon of subscapularis by a synovial protrusion of the joint (subscapular 'bursa'). To an almost oval area near the inferior angle, the lower five or six digitations of serratus anterior are attached. The upper (**3.**167) fibres are attached to a narrow ventral strip along the medial border, which is wider above for the large first digitation. Longitudinal thickening near the lateral border provides a lever to withstand the pull of the serratus anterior on the inferior angle during anterior scapular rotation, by which the glenoid cavity is turned up as the arm is raised against gravity.

On the dorsal surface, in the medial two-thirds of the supraspinous fossa, is attached the supraspinatus, and the fascia over it attaches to the fossa's margins. The flat strip near the lateral border has teres minor attached to it in its upper two-thirds. It is grooved near midpoint by the circumflex scapular vessels, which pass between muscle and bone into the infraspinous fossa. The lower limit of the muscle's attachment is an oblique ridge, which runs from the lateral border towards the inferior angle, separating the attachment of teres minor from a somewhat oval area for teres major. Except for the area near the neck, the remaining infraspinous fossa is the attachment of infraspinatus, the aponeurosis of which passes on to teres minor and major and sends the fascial septa between them marking the bone at the limits of their attachments.

402

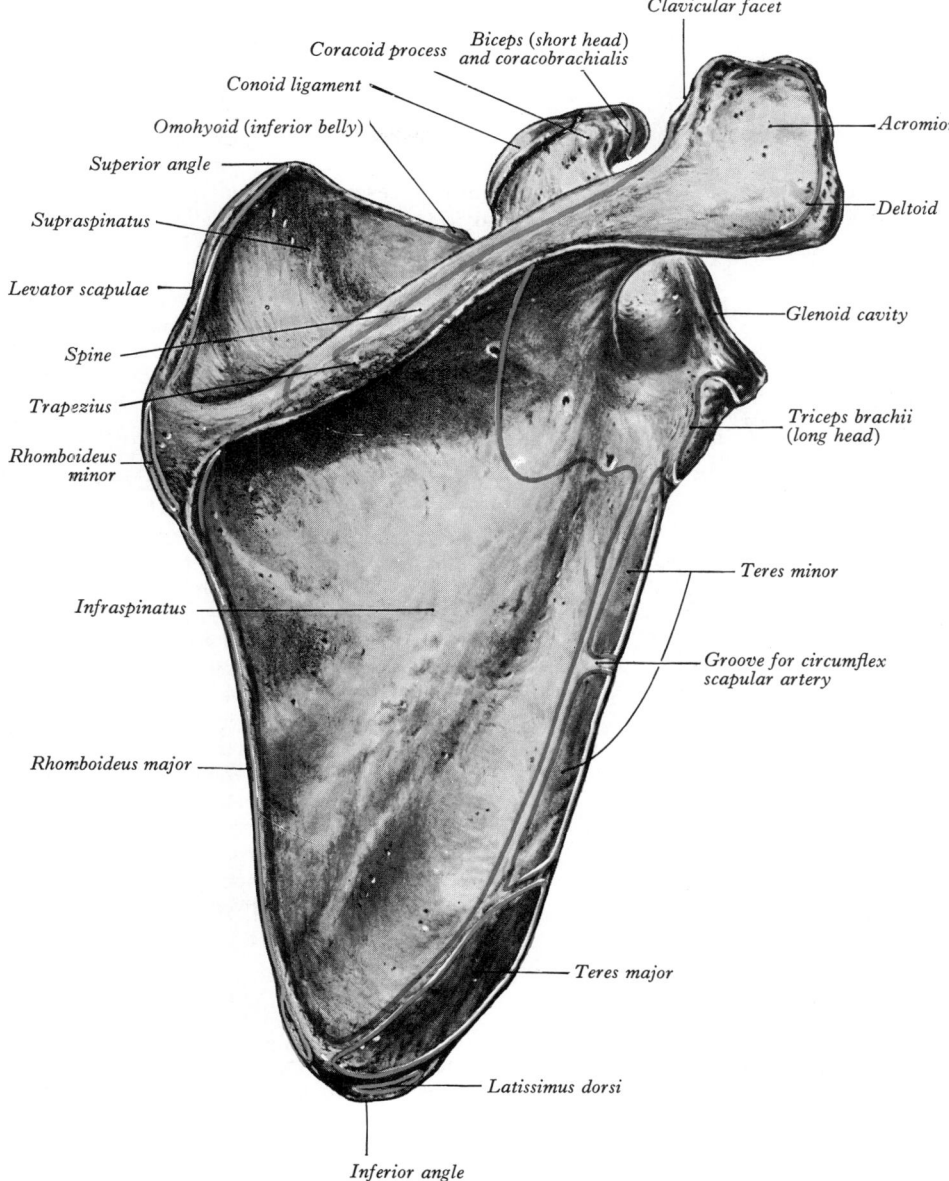

Clavicular facet
Coracoid process
Biceps (short head) and coracobrachialis
Conoid ligament
Omohyoid (inferior belly)
Acromion
Superior angle
Supraspinatus
Deltoid
Levator scapulae
Glenoid cavity
Spine
Trapezius
Triceps brachii (long head)
Rhomboideus minor
Teres minor
Infraspinatus
Groove for circumflex scapular artery
Rhomboideus major
Teres major
Latissimus dorsi
Inferior angle

3.170 The right scapula: dorsal aspect.

The *lateral border* separates attachments of subscapularis and the teretes. These muscles project beyond bone and, with latissimus dorsi, make the border impalpable. To the infraglenoid tubercle the long head of the triceps brachii is attached. The *medial border* is thin and angled opposite the spine; to a narrow strip, from the superior angle to this point, is attached levator scapulae and, below this opposite the spine, rhomboideus minor. The rest of the border is taken up by rhomboideus major (p. 610). The *superior border* is thin; near the suprascapular notch the inferior belly of the omohyoid is attached. Crossing the notch is the superior transverse ligament, attached laterally to the root of the coracoid process, medially to the notch; it is sometimes ossified. The foramen, thus completed, conducts the suprascapular nerve to the supraspinous fossa, the suprascapular vessels passing above the ligament.

The *inferior angle* is covered dorsally by the upper border of latissimus dorsi, which frequently receives a small slip from it. The *superior angle* is covered by the upper part of trapezius. The *lateral angle* bears the *glenoid cavity* covered by hyaline articular cartilage; to its anterior margin are attached glenohumeral ligaments (p. 501), to the *supraglenoid tubercle* the long head of biceps, to the *infraglenoid tubercle* the long head of triceps brachii.

To the *scapular spine's* upper and lower surfaces the supra- and infraspinatus muscles are attached. The triangular area at its root, opposite the third thoracic spine, is played over by the tendon of the trapezius, a bursa intervening. The crest's lower border is occupied by posterior fibres of the deltoid, its upper border by middle fibres of trapezius, the lowest part of which ends as a flat triangular tendon gliding over the area noted above and attached to the so-called *deltoid tubercle* on the spine's dorsum lateral to the area.

The *acromion* is dorsally subcutaneous, covered only by skin and superficial fascia. Its lateral border, thick and irregular, and its tip as far round as the clavicular facet are the attachment of the mid part of the deltoid. To its summit medially, and below the deltoid, the lateral end of the coracoacromial ligament is attached. The articular capsule of the acromioclavicular joint is attached to the rim of the clavicular facet. Behind the facet, the acromion's medial border receives the horizontal fibres of trapezius. The smooth inferior aspect of the acromion, together with the coracoacromial ligament and coracoid process, forms a *coracoacromial arch* over the shoulder joint. The tendon of supraspinatus, below the overhanging acromion, is separated from both it and the deltoid by the subacromial bursa.

The *coracoid process*, below the junction of the lateral fourth with the rest of the clavicle, is connected to it by the coracoclavicular ligament. For attachment of its *conoid* part see page 402; the *trapezoid* part is attached to the superior aspect of the horizontal part of the process (3.169), which also receives the pectoralis minor. To its lateral border is attached the wider, medial end of the coracoacromial ligament and, below this, the coracohumeral ligament. To the coracoid apex is attached the

403

coracobrachialis medially and the short head of biceps laterally. The apex is covered by the anterior edge of the deltoid and is palpable under the lateral border of the infraclavicular fossa.

The human coracro-acromial ligament is a trait shared only with other hominoid primates, according to Ciochon & Corruccini (1977), who devised a *coracoacromial projection index*: *projection height* (i.e. vertical distance from supraglenoid tubercle to a line between most lateral points on acromial and coracoid apices) divided by *height of the glenoid cavity*. They claim that their observations reflect specialized locomotor and feeding adaptations in *Hominoidea*.

Structure. The main processes, and thicker parts of the scapula, contain trabecular bone; the rest consists of a thin compact layer. The central supraspinous fossa and most of the infraspinous are thin and even translucent; occasionally the bone is deficient in them, gaps being filled by aponeurotic tissue.

Ossification (3.172A,B). The cartilaginous scapula is ossified from eight or more centres: one in the body, two each in the coracoid process and the acromion, one each in the medial border, inferior angle and lower glenoid rim. The centre for the body appears in the eighth intrauterine week. Ossification begins centrally in the coracoid process in the first year, occasionally before birth; it joins the rest of the bone in about the fifteenth year. At or soon after puberty, centres appear in the coracoid root (subcoracoid centre), in the lower glenoid rim, often in the coracoid apex, in the acromion, inferior angle and contiguous medial border and in the medial border. A variable part of the glenoid cavity, usually the upper third, is ossified from the subcoracoid centre, uniting in the fourteenth (females) to seventeenth (males) year. A crescentic epiphysis for the lower glenoid rim, thicker peripherally, converts the child's flat cavity into the gently concave adult fossa. The acromial base is an extension from the spine; the remaining acromion is ossified from two centres which unite and join the spinous extension. The various scapular epiphyses have all fused by about the twentieth year.

The Clavicle

The clavicle (3.173,174) extends laterally and almost horizontally across the neck from manubrium to acromion, being wholly subcutaneous. It struts the shoulder and enables the limb to swing clear of the trunk, transmitting part of its weight to the axial skeleton. Its flat *lateral, acromial end* articulates with the medial aspect of the acromion; the enlarged *medial, sternal end* articulates

Acromion

Coracoid process

Acromial angle

Glenoid cavity

Infraglenoid tubercle

For subscapularis

Ventral aspect

Lateral border

Inferior angle

3.171 The left scapula: lateral aspect.

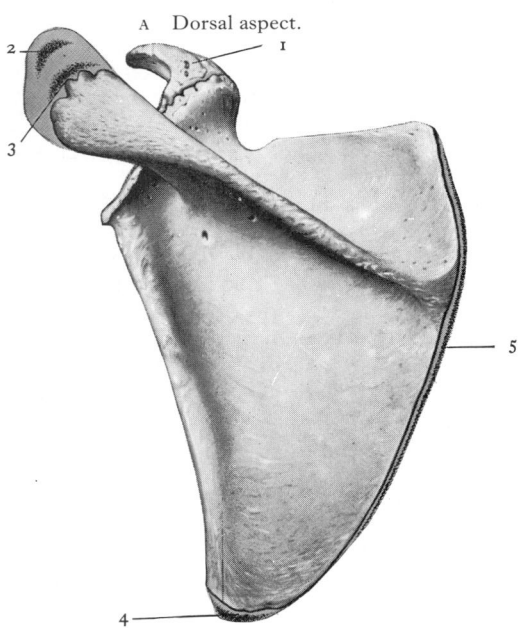

A   Dorsal aspect.

A

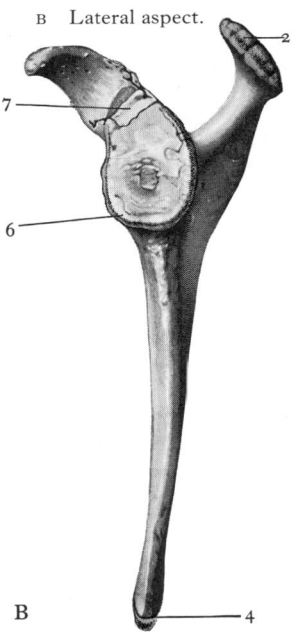

B   Lateral aspect.

B

3.172 The ossification of the scapula. A. Dorsal aspect; B. Lateral aspect. 1. Coracoid centre. 2. Distal acromial centre. 3. Proximal acromial centre. 4. Centre at inferior angle. 5. Centre for medial border. 6. Glenoid centre. 7. Subcoracoid centre.

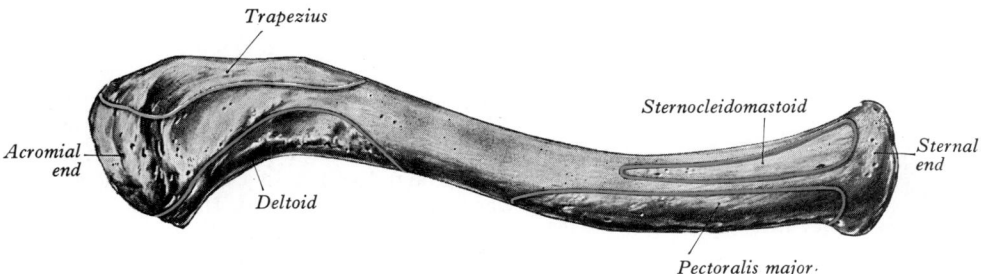

3.173 The right clavicle: superior aspect.

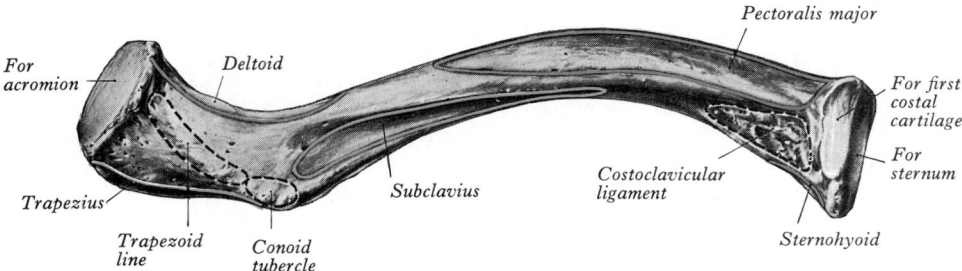

3.174 The right clavicle: inferior aspect.

with the manubrial clavicular facet and first costal cartilage. The *shaft* is sinuous, being convex forwards in its medial two-thirds, and concave lateral to this. The *inferior* aspect of its intermediate part is grooved in its long axis.

The clavicle's lateral third is flat, with superior and inferior surfaces, and anterior and posterior borders. The *anterior border* is concave, thin, rough and may show a small *deltoid tubercle*. The *posterior border*, also roughened by muscular attachments, is convex. The palpable *superior surface* is rough near its margins but otherwise smooth. The *inferior surface* presents, near its posterior border and at the junction of the lateral fourth with the rest of the shaft, a prominent *conoid tubercle* for the conoid part of the coracoclavicular ligament; from the lateral side of this a narrow, rough strip (*trapezoid line*) runs anterolaterally, almost to the acromial apex (**3.174**), for the ligament's trapezoid part. **The medial clavicular two-thirds** is roughly cylindrical or prismatic with four surfaces, the inferior often a mere ridge. *Anterior* and *superior* surfaces are largely rough but laterally smooth and rounded above the infraclavicular fossa (p. 611). The *posterior surface* is smooth, the *inferior* marked near its sternal end by a rough oval area, often depressed, for the costoclavicular ligament. Rarely, this area is smooth or even raised and may form a synovial joint with the first rib (Cave 1961). The lateral half the inferior surface is a groove for attachment of the subclavius.

Further details. The *acromial end* is flat, with a small oval articular facet for the medial aspect of the acromion which faces laterally and slightly down. **The sternal end**, directed medially and a little down and forwards, articulates with the manubrial clavicular notch. The sternal surface is quadrangular (sometimes triangular) and slightly rough above for ligamentous attachments. The articular surface is otherwise smooth and extends on to the inferior surface to articulate with the first costal cartilage. The sternal end projects above the manubrium and can be felt and usually seen (a prominent clinical landmark) in the lateral wall of the jugular fossa.

The shaft's lateral third has attached to it anteriorly the deltoid, posteriorly the trapezius, both reaching the superior surface. The coracoclavicular ligament, attached to the conoid tubercle and trapezoid line (**3.174**), transmits the weight of the upper limb to the clavicle, counteracted by the trapezius supporting its lateral part (p. 609). From the conoid tubercle weight is transmitted medially by the shaft to the axial skeleton; a fracture medial to it interrupts this transmission, almost all weight then being supported by the trapezius which is unable to meet the demand, and the limb therefore droops.

The clavicle's medial two-thirds provides attachment, anteriorly, to the clavicular part of pectoralis major, the area being usually visibly marked. The clavicular part of sternocleidomastoid is attached to the medial half of the superior surface but marks it little. The smooth *posterior surface* is devoid of attachments except near its sternal end, where part of the sternohyoid is attached. Medially this surface is related to the internal jugular vein's lower end (separated by sternohyoid), the end of the subclavian and beginning of the brachiocephalic vein. More laterally, the clavicle curves in front of the trunks of the brachial plexus and the third part of the subclavian artery. The suprascapular vessels are related to the upper part of this surface. On the *inferior surface* is the subclavius in its groove (**3.174**). To the edges of this groove is attached the clavipectoral fascia, which encloses the muscle; the posterior edge of the groove runs to the conoid tubercle, where fascia and conoid ligament merge. A nutrient foramen, lateral in the groove, is inclined laterally. Medially an impression for the costoclavicular ligament varies greatly in texture and outline (vide supra and Cave 1961).

To the margins of the *acromial* articular facet is attached the acromioclavicular capsule, and to the rough upper part of the *sternal end* the interclavicular ligament, sternoclavicular capsule and articular disc. The sternal (osseous) articular surface is usually irregular and pitted.

The female clavicle is shorter, thinner, less curved and smoother, its acromial end carried lower than the sternal; in males it is level with, or slightly above, the sternal end when the arm is pendent. Midshaft circumference is the most reliable single indicator of sex, but a combination of this with weight and length yields better results (Olivier 1951, Jit & Singh 1966); however consult Jit & Sahni (1983) for statistical analysis and literature. The clavicle is thicker and more curved in manual workers, its muscular attachments marked. The clavicle is trabecular internally, with a shell of compact bone much thicker in its shaft. Although elongate, the clavicle is unlike typical long bones; it usually has no medullary cavity.

Ossification (**3.175**). The clavicle begins to ossify before any other bone, the shaft from medial and lateral primary centres, appearing in condensed mesenchyme between fifth and sixth weeks and fusing about the forty-fifth day. A secondary centre for the sternal end appears in late teens, or even early twenties, usually two years earlier in females. Fusion is probably rapid but reliable data are lacking. An acromial secondary centre sometimes develops at about 18 to 20, but this epiphysis is always small and rapidly joins the shaft (Todd & D'Erico 1928). In a study of the

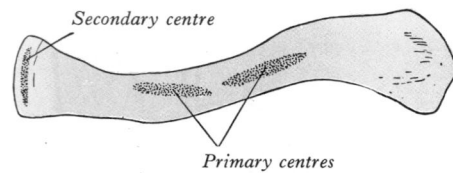

Secondary centre

Primary centres

3.175 Diagram showing the three constant centres of ossification of the clavicle.

sternal epiphysis in 684 Punjabis, ossification occurred from 11 to 19 years (both sexes) and fusion at 18 to 25 years (Jit & Kulkaria 1976). Comparison suggests racial differences. Such assessment by radiological appearances is, of course, not devoid of fallacy.

However, clavicular development is not only by intramembranous ossification. In 14 mm embryos the clavicle is a band of condensed mesenchyme between the acromion and apex of the first rib, and continuous with the sternal rudiment. In this band medial and lateral zones of early cartilage transformation ('precartilage') occur, and in mesenchyme between them intramembranous centres of ossification appear and soon fuse. Sternal and acromial zones soon become true cartilage, into which ossification extends from the shaft. Length increases by interstitial growth of these terminal cartilages, which develop zones of hypertrophy, calcification and advancing endochondral ossification as in other growth cartilages. Diameter is increased by subperichondrial deposition in the extremities and subperiosteal deposition in the shaft. Epiphyses are endochondral and probably fuse as in long bones which are *primarily* cartilaginous. Defects of ossification in clavicle and primarily intramembranous cranial bones occasionally coincide as the condition of cleidocranial dysostosis.

The primitive reptilian pectoral girdle comprises a dorsal scapula and two ventral elements, an anterior (cranial) precoracoid and posterior (caudal) coracoid. Of the three pelvic elements, the ilium is homologous with the scapula, the pubis with the precoracoid, the ischium with the coracoid. The clavicle, dermal in origin and therefore morphologically distinct, is an addition to the pectoral girdle with no pelvic equivalent. It is doubtful whether any trace of precoracoid persists in the human skeleton, but the double primary clavicular centre may indicate that human clavicles contain reptilian precoracoid and clavicle. Some believe that the first coracoid centre represents the precoracoid element and subcoracoid centre the caudal ventral element of reptilian girdles. The clavicle is absent from forelimbs used principally or entirely for progression, e.g. in ungulates and carnivores; but it is present and well-developed in prehensile limbs, e.g. in many rodents, primates and man.

The clavicle is often fractured, commonly by indirect forces, due to violent impacts to hand or shoulder, the break being at the junction of lateral and intermediate thirds, where its curvature changes, for this is its weakest part. Consequent deformity is caused by the weight of the arm, which depresses the lateral fragment, forces being transmitted through the coracoclavicular ligament; the medial fragment is usually little displaced.

The Humerus

The humerus (3.176A,B,177A,B et seq), the longest and largest bone in the arm, has expanded ends and a shaft. Proximally a round 'head' forms with the scapular glenoid cavity an enarthrodial articulation (cf p. 502). The distal end, loosely termed 'condylar', is adapted to the forearm bones at the cubital joint.

The head (3.176,177), at the proximal end, is slightly less than half a spheroid; it has an articular surface covered by hyaline cartilage, directed posteromedially and upwards to the glenoid cavity in the pendent arm. The humeral articular surface much exceeds that of the glenoid cavity, only part of it being in glenoid contact in any position of the joint.

The anatomical neck, directly adjoining the articular head's margin, is a slight constriction, least apparent near the greater tubercle, and superiorly, and for some distance continuing

anteriorly and posteriorly, indicates the line of capsular attachment of the glenohumeral joint. However, inferiorly, the capsular attachment diverges markedly from the anatomical neck (vide infra).

The lesser tubercle (tuberosity) is anterior and just beyond the anatomical neck; it has a smooth, muscular impression palpable through the deltoid 3 cm below the acromial apex, moving when the humerus rotates. Its sharp lateral edge is part of the medial border of the intertubercular sulcus.

The greater tubercle (tuberosity), is the most lateral part of the humeral proximal end and in the shoulder region projects beyond the acromion and, covered by the deltoid, produces the shoulder's round contour. Its posterosuperior aspect, near the anatomical neck, bears three smooth impressions for tendons. Between the tubercles is the **intertubercular sulcus**. The humeral proximal end tapers into the shaft as an ill-defined 'surgical neck', medial to which are the axillary nerve and posterior humeral circumflex artery (3.177A).

The humeral shaft, in its proximal half almost cylindrical, is distally prismatic (in section) and anteroposteriorly compressed. It is not directly palpable due to the large muscles around it. Its three surfaces and borders are not equally obvious.

The *anterior border* descends from the front of the greater tubercle almost to the distal end of the bone. Its proximal third is the lateral edge of the intertubercular sulcus, marked by muscular attachments. Beyond this, as the anterior edge of the deltoid tuberosity, it is also rough but smoothly rounded in its distal half.

The *lateral border* is distally sharp and rough along its anterior aspect; its proximal two-thirds are barely discernible, although sometimes traceable to the posterior aspect of the greater tubercle. Centrally it is interrupted by a wide, shallow groove, descending obliquely laterally and forwards, the *sulcus for the radial nerve*.

The *medial border*, rounded in its distal half, bears a roughened strip just below its midpoint; proximal to this it is indistinct but reappears as the medial lip of the intertubercular sulcus, where it is again rough and reaches the lesser tubercle.

The *anterolateral surface* is between the anterior and lateral borders; just proximal to its midpoint is the rough *deltoid tuberosity*, tapering to a distal apex. Behind this the radial sulcus descends, fading distally when it reaches the lateral border a little beyond the tuberosity. The *anteromedial surface*, between the anterior and medial borders, forms in its proximal third the rough floor of the intertubercular sulcus, but the rest is smooth. Near its midpoint a nutrient foramen opens, its canal directed distally. The *posterior surface*, between medial and lateral borders, is the most extensive. A ridge, sometimes rough, descends laterally across its proximal third. The middle third is crossed by the commencement of the radial sulcus. The distal third is an extensive, flat surface, which widens distally.

The distal end of the humerus (3.176–178), basically a modified *condyle*, is wide transversely and has articular and non-articular parts. The *articular part* joints with the radius and ulna at the elbow and is divided into a lateral, convex *capitulum* and a medial, pulley-shaped *trochlea*.

The capitulum, less than half a sphere, includes anterior and inferior surfaces of the condyle laterally, but not its posterior surface. It articulates with the discoid radial head, which abuts the inferior surface in full extension but slides on to the anterior surface during flexion.

The trochlea is like part of a pulley, occupying anterior, inferior and posterior surfaces of the humeral condyle medially; it is separated laterally from the capitulum by a faint groove; all aspects of its medial margin project. It articulates with the trochlear notch of the ulna. In extension the inferoposterior trochlear circumference contacts the ulna, but in flexion the trochlear notch slides on to the anterior aspect, the posterior being uncovered. The projecting medial trochlear edge is a main determinant of angulation between the long axes of humerus and ulna when the forearm is extended and supinated (p. 506).

The *non-articular* condyle includes medial and lateral epicondyles, olecranon, coronoid and radial fossae. *The medial epicondyle*, a blunt projection on the condyle's medial, is subcutaneous and usually visible in passive flexion. Its smooth posterior surface is crossed by the ulnar nerve in a shallow sulcus as it enters the forearm, and here the nerve can be rolled against bone. If it is jarred against the epicondyle, characteristic tingling sensations result. Distally the anterior epicondylar surface is marked by the attachment of superficial forearm flexors.

The lateral epicondyle, lateral non-articular part of the condyle, does not project beyond the lateral border. It has an anterolateral impression for superficial forearm extensors. Its posterior surface, slightly convex, is easily felt in a depression visible behind the extended elbow. The lateral humeral border ends at the *lateral epicondyle*, from which, extending proximally,

is its distal part, the *lateral supracondylar ridge*. The medial border ends at the *medial epicondyle* and is, distally, the *medial supracondylar ridge*. A deep hollow, the *olecranon fossa*, on the condyle's posterior surface, proximal to the trochlea, contains the apex of the olecranon in the extended elbow. Its floor is always thin and may be deficient. A smaller *coronoid fossa*, immediately proximal to the trochlea on the anterior surface, accommodates the margin of the ulnar coronoid process in full flexion. A shallow *radial fossa*, proximal to the capitulum and lateral to the coronoid fossa, is related to the margin of the radial head in full flexion.

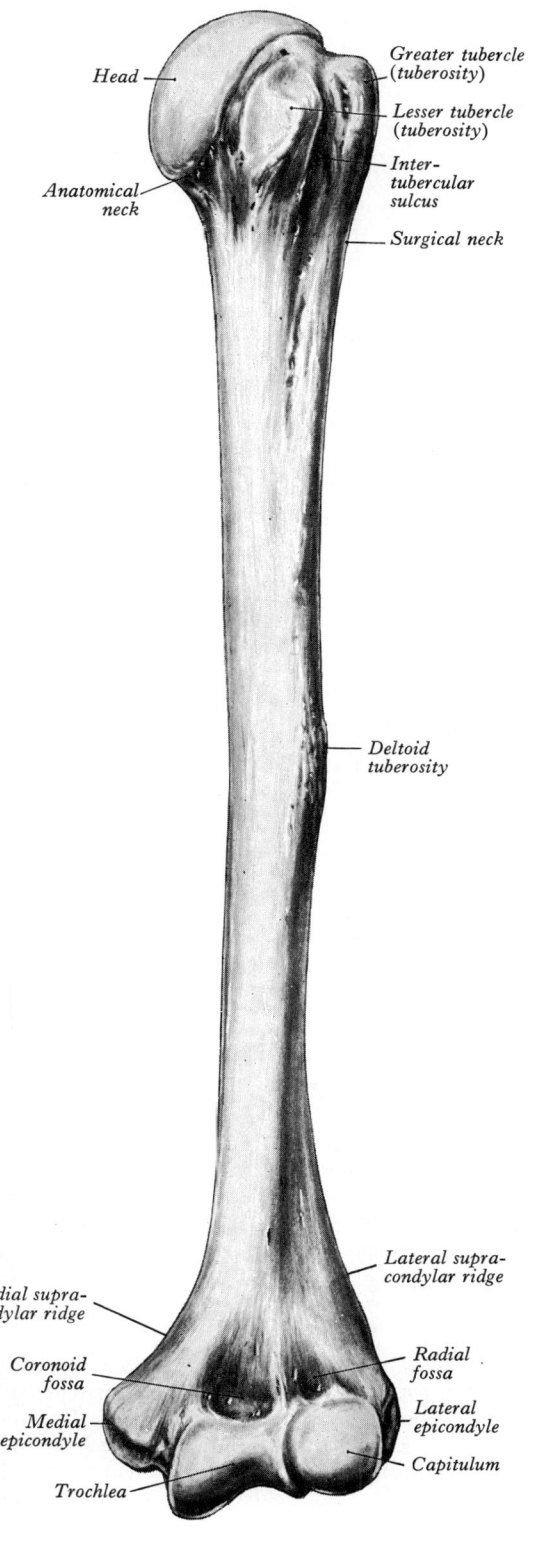

176A The left humerus: anterior aspect. Compare with Key, 3.176B.

3.176B Key to 3.176A. The interrupted lines indicate the attachment of the capsular ligaments; the stippled lines mark the position of the epiphysial lines.

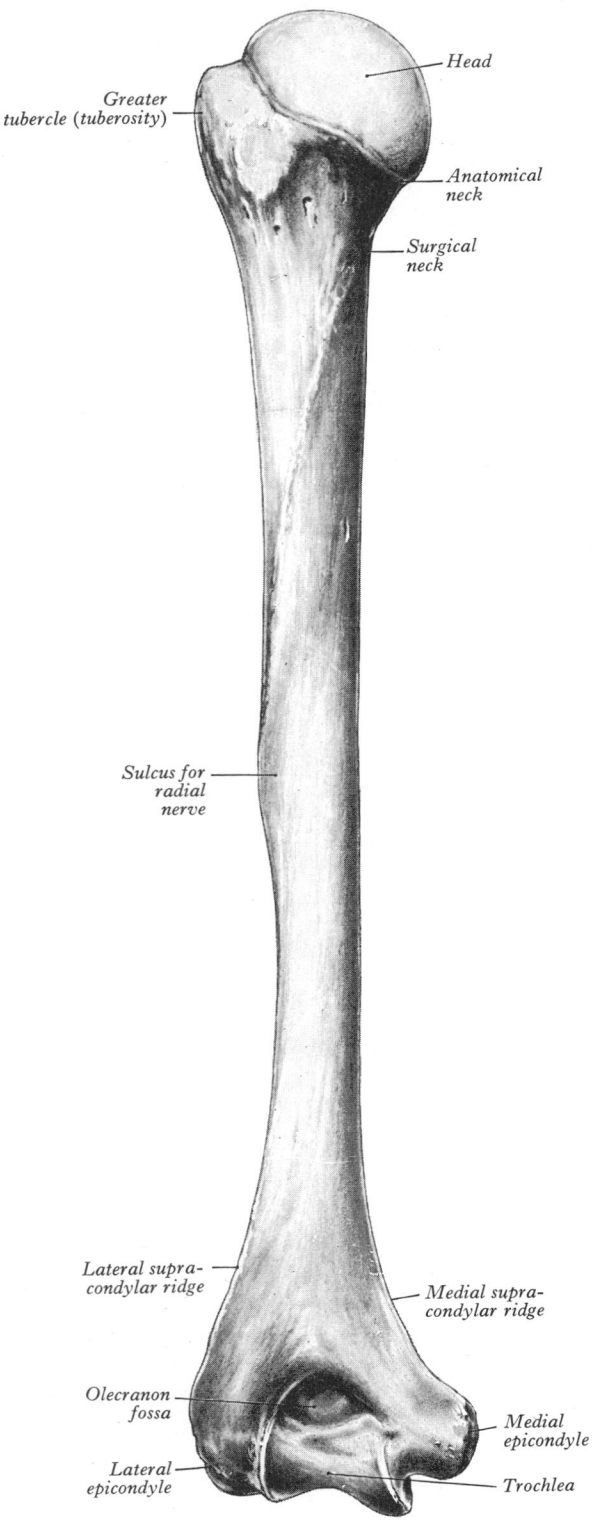

3.177A The left humerus: posterior aspect.

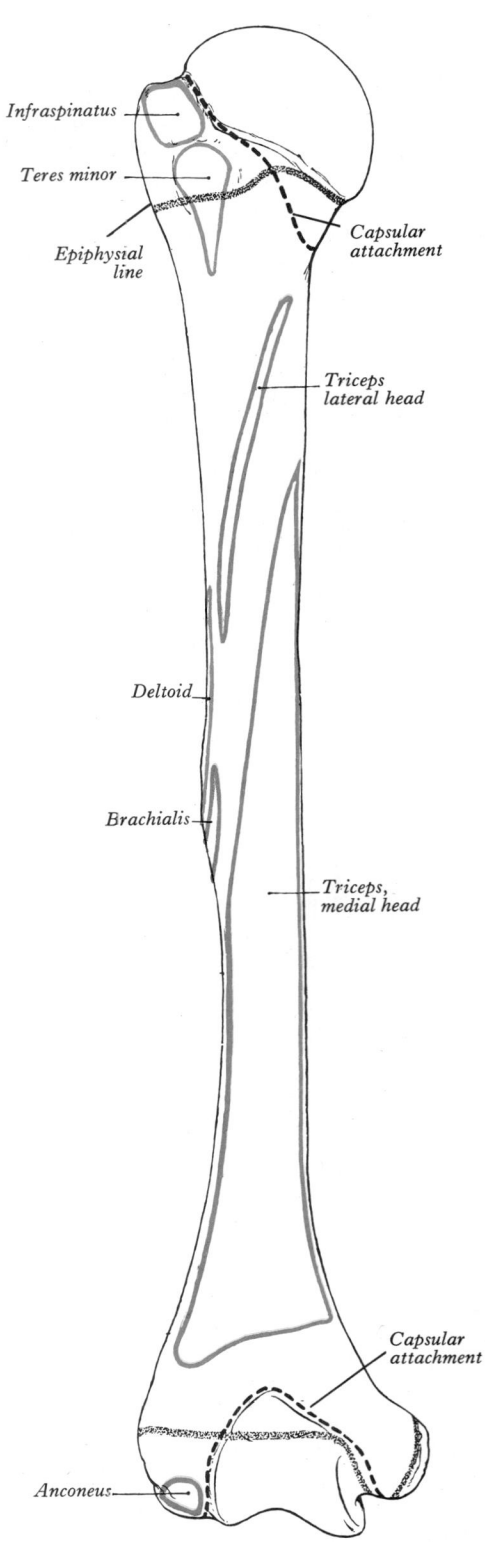

3.177B Key to **3.177A**. The interrupted lines indicate the attachment of the capsular ligaments; the stippled lines mark the position of the epiphysial lines.

In lower mammals the longest axes of proximal and distal humeral articular surfaces make an angle with each other of a little more than 90°. But the proximal human humerus appears to have rotated laterally, so that the angle has been increased to about 164° (**3.**179), the *angle of 'humeral torsion'*. It is greater in males and adults, and less in anthropoid apes. Krahl (1976) claims that the angle increases with age until fusion of the epiphysis (p. 399).

The head's articular cartilage is thick centrally, and thins peripherally; in sectional profile it is spheroidal (strictly ovoidal, see p. 501).

To the *anatomical neck* is attached the *capsular ligament* of the shoulder joint (**3.**176,177), except at the intertubercular sulcus, where the long tendon of biceps emerges. But medially the attachment descends 1 cm or more on the shaft. To the *lesser tubercle* the subscapularis is attached (**3.**176A) and to its lateral margin the transverse ligament of the shoulder joint. The *greater tubercle* shows three impressions for tendons, the uppermost for supraspinatus, the middle for infraspinatus and the lowest for teres minor (**3.**177B). Its projecting lateral surface has many vascular foramina and is covered by the deltoid; a subacromial bursa may

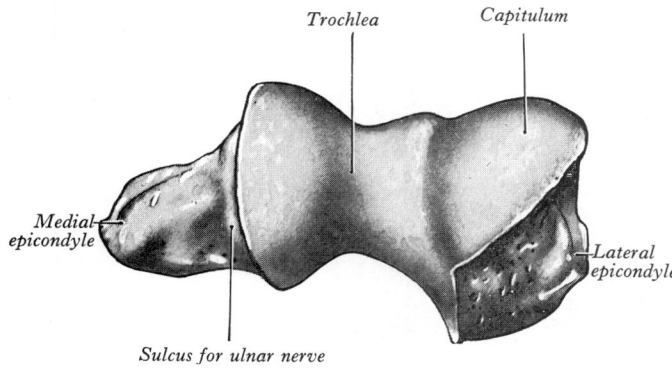

Trochlea   *Capitulum*

Medial epicondyle

Lateral epicondyle

Sulcus for ulnar nerve

3.178 The lower end of the left humerus: distal (inferior) aspect.

separate the deltoid from the tubercle. The attachments of subscapularis and teres minor are not confined to their respective tubercles, but extend for varying distances on to adjacent metaphysis.

The *intertubercular sulcus* (bicipital groove) contains the long tendon of biceps, its synovial sheath, and an ascending branch from the anterior circumflex humeral artery. The groove's lateral lip is marked by the tendon of pectoralis major, its floor by that of the latissimus dorsi and its medial lip by that of the teres major. Attachment of the pectoralis major extends beyond teres major's, that of latissimus dorsi being least extensive. Distal to the sulcus a small area of anteromedial surface is devoid of muscular attachment, but its distal half is occupied by the medial part of the brachialis (**3.176B**). The rough strip on the medial border is for the coracobrachialis. Near its distal end the medial supracondylar ridge has a narrow area for the humeral part of the pronator teres, and the ridge itself for the medial intermuscular septum.

To the oblique ridge across the *posterior surface* the lateral head of triceps is attached; proximal to this axillary (circumflex) nerve and posterior circumflex humeral vessels curve round the 'surgical neck' deep to the deltoid; distally the radial groove, containing the nerve and profunda vessels, descends laterally to the anterolateral surface. The medial head of the triceps occupies much of the posterior surface in a long triangular area, its apex medial on this surface and *proximal* to the distal limit of the teres major, widening distally over the whole dorsal surface almost to its distal end (**3.177B**).

Proximally the *anterolateral surface* is smooth and covered by the deltoid, which is attached to its tuberosity; distally it is occupied by part of the brachialis, ascending into the floor of the radial groove (**3.177B**). The lateral supracondylar ridge, in its proximal two-thirds, is for the brachioradialis and in its distal third for the extensor carpi radialis longus. Behind these, the lateral intermuscular septum is attached to the ridge.

The articular region of the humeral condyle is so curved that anterior and posterior surfaces are anterior in plane to corresponding surfaces of the shaft. The trochlear groove spirals posterolaterally from the anterior to the posterior surface; posteriorly it is wider, deeper and more symmetrical; anteriorly its medial flange is longer and the convex surface adjoining its projecting medial margin is adapted to the medial arc of the coronoid articular surface. These asymmetries entail varying angulation between the humeral and ulnar axes, together with some conjunct rotation (p. 502).

The *cubital capsular ligament* (**3.176B,177B**) is attached to the proximal limits of the radial and coronoid fossae—which are intracapsular and covered by synovial membrane—and also to the medial aspect of the trochlear rim and root of the medial epicondyle. Posteriorly it ascends almost to the upper margin of the olecranon fossa, which is also intracapsular and covered by synovial membrane. Laterally it skirts the borders of the trochlea and capitulum, lying medial to the lateral epicondyle.

The common superficial flexor tendon springs from the medial epicondylar epiphysis, being wholly extracapsular. The common superficial extensor tendon is attached to the lateral epicondyle, also outside the articular capsule; to its posterior surface the

anconeus is attached (**3.177B**). The medial epicondyle turns slightly backwards, the lateral slightly forwards.

With the humerus pendent, the medial epicondyle is posterior in plane to the lateral; the humeral head is directed almost equally *backwards* and *medially*, the posterior surface of the shaft facing posterolaterally. Since the glenoid cavity faces anterolaterally, the humerus is not rotated medially *relative to the scapula* in this position of rest; but it *is* so rotated relative to the *conventional anatomical position*. This must be remembered when movements of arm and forearm are considered (pp. 503 and 509; see also *Introduction*).

A hook-shaped *supracondylar process*, from 2 to 20 mm in length, occasionally projects from the shaft's anteromedial surface, about 5 cm proximal to the medial epicondyle. It curves distally and forwards, its apex connected to the medial border, proximal to the epicondyle, by a fibrous band, to which part of the pronator teres is attached. The foramen so formed usually encloses the median nerve and brachial artery, but sometimes only the nerve or perhaps nerve plus the ulnar artery in a high division of the brachial artery. A groove for the artery and nerve usually exists behind the process. It is the homologue of the *entepicondylar foramen* of many animals and may protect the nerve and artery from compression by muscles. These skeletal features may aid in assessing racial affinities (p. 273).

Ossification (**3.180**). The humerus is ossified from eight centres, in shaft, head, greater and lesser tubercles, capitulum and lateral trochlea, medial trochlea and both epicondyles. The shaft begins to ossify centrally in the eighth week with gradual extension to its ends, and the humeral head usually during the first six months after birth, but just before it in 20% of individuals. The greater and lesser tubercles begin to ossify in the second and fifth years in males, about a year earlier in females. By the sixth year head and tubercles have become a single epiphysis, *hollow* inferiorly (**3.182**) to fit the conical end of the diaphysis and fusing with it about the twentieth year in males, two years earlier in females. Some deny a centre in the lesser tubercle, perhaps because it is often obscured in the usual anteroposterior

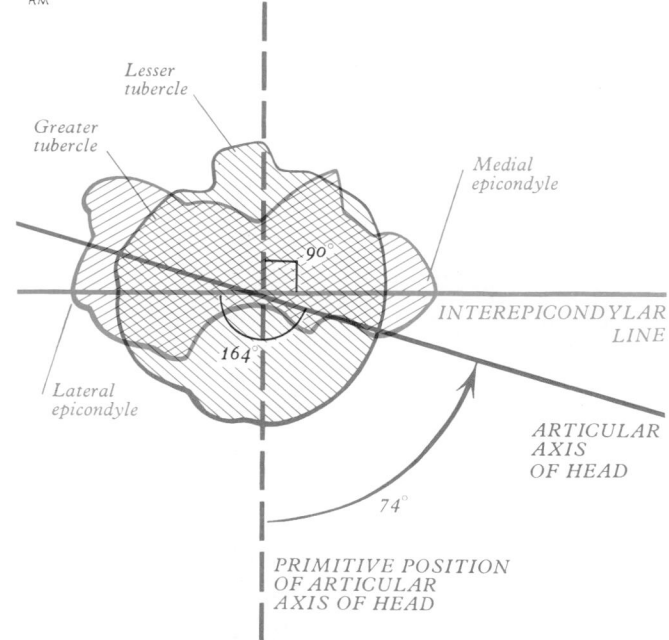

RM

Lesser tubercle

Greater tubercle

Medial epicondyle

90

INTEREPICONDYLAR LINE

164

Lateral epicondyle

ARTICULAR AXIS OF HEAD

74°

PRIMITIVE POSITION OF ARTICULAR AXIS OF HEAD

3.179 This diagram illustrates the concept of humeral 'torsion'. The left humerus is viewed distally along its length, with its proximal end (magenta) superimposed upon the distal (blue). The 'axes' of the two extremities are shown at right angles, as in early tetrapods and most quadrupedal mammals. In primates, including mankind, the axis of the head is directed medially, and the angle which this now makes with the 'primitive' anteroposterior position of the axis is the so-called angle of torsion. This is on average 74° in man: some authorities add to this the original angle of 90°, giving a total of 164°. See text for more detailed explanation.

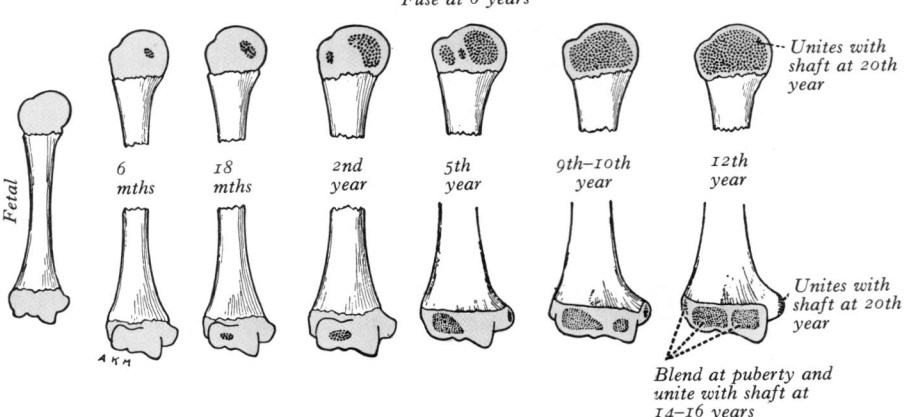

Fuse at 6 years

Fetal *6 mths* *18 mths* *2nd year* *5th year* *9th–10th year* *12th year*

Unites with shaft at 20th year

Unites with shaft at 20th year

Blend at puberty and unite with shaft at 14–16 years

3.180 Stages in the ossification of the humerus (not to scale).

radiological views (**3**.181). In the distal end, during the first year, ossification begins in the capitulum and extends to form most of its articular surface; a centre for the medial region of the trochlea appears in the ninth year in females and tenth in males. Ossification begins in the medial epicondyle in the fourth year in females, sixth in males, and in the lateral epicondyle about the twelfth year. Centres for lateral epicondyle, capitulum and trochlea fuse

around puberty and the composite epiphysis unites with the shaft in the fourteenth year in females, sixteenth in males. The medial epicondyle is a separate epiphysis, entirely extracapsular (**3**.176B,177B), from a centre posteromedial in it. It is separated from the distal epiphysis by a downgrowth of shaft, with which it unites about the twentieth year.

The proximal epiphysis joins the shaft later than the distal, and growth in length is predominantly due to the proximal growth cartilage. Hence, after amputation through the arm in youth, the humerus continues to grow, its lower end progressively moulding the soft tissues and making the stump conical. Fractures are comparatively common and at almost any level. Muscular action is more often the cause than in any other long bone; it is usually the shaft, below attachment of the deltoid, which is broken. Hence the radial nerve may be injured in its groove and is sometimes involved later in the growth of callus. Non-union is commoner than in any other bone except the tibia. Fractures at the proximal end may damage the axillary nerve, and at the medial epicondyle the ulnar nerve. Late fusion of this epicondyle may be misdiagnosed as a fracture.

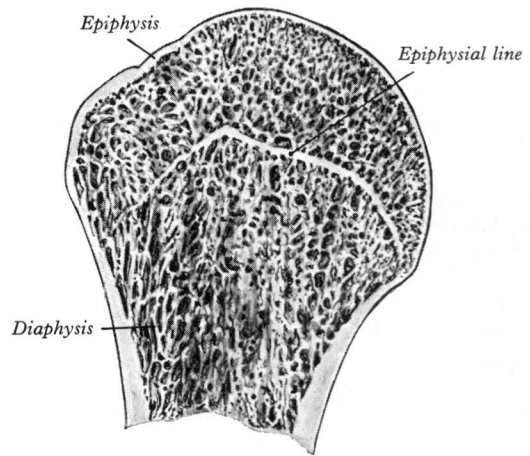

Epiphysis

Epiphysial line

Diaphysis

3.182 Longitudinal section through the head of the right humerus.

The Radius

The radius (**3**.183–186), lateral in the forearm, has expande[d] proximal and distal ends, the distal much the broader. The sha[ft] widens rapidly towards its distal end, is convex laterally and co[n]cave anteriorly in its distal part. **The proximal end** includes [a] head, neck and tuberosity. *The head* is discoid, its proximal su[r]face a shallow cup for the humeral capitulum. Its smooth *articul*[ar]

3.181 Anteroposterior radiograph of the right shoulder in a boy aged 11. 1. Coracoid process. 2. Growth plate of cartilage at upper end of humeral diaphysis. 3. Acromion. 4. Lateral end of clavicle, not yet completely ossified. 5. Proximal humeral epiphysis. Note its conical junction with the diaphysis.

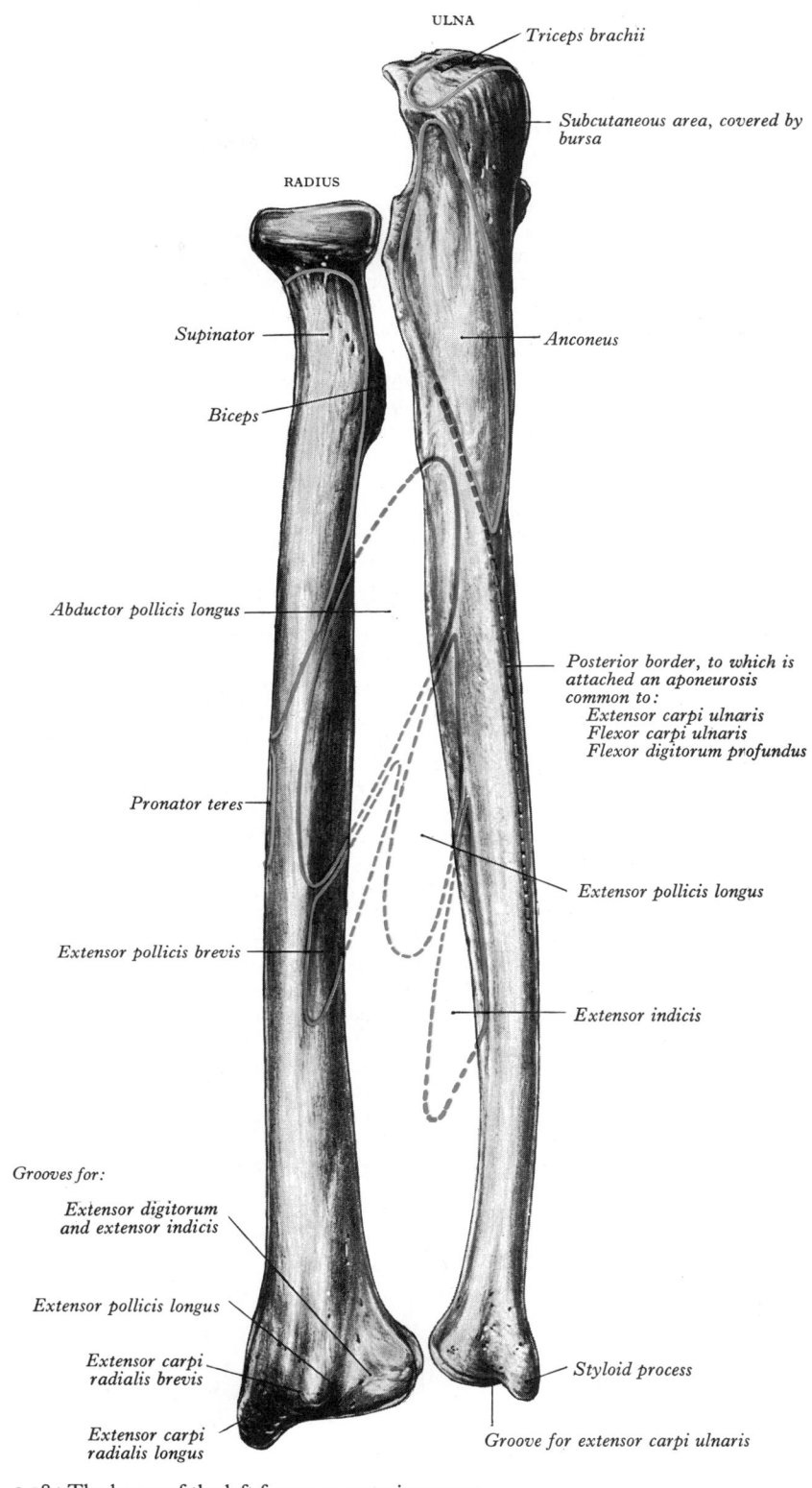

ULNA

Olecranon

Trochlear notch

Coronoid process

Flexor digitorum superficialis

Pronator teres, ulnar head

Brachialis

Flexor pollicis longus, occasional head

Supinator

Flexor digitorum profundus

Extension of muscular attachments on to interosseous membrane

Head

Styloid process

RADIUS

Head

Neck

Biceps (radial tuberosity)

Supinator

Flexor digitorum superficialis (oblique line)

Pronator teres

Flexor pollicis longus

Pronator quadratus

Brachioradialis

Styloid process

3.183 The bones of the left forearm: anterior aspect.

ULNA

Triceps brachii

Subcutaneous area, covered by bursa

Anconeus

Posterior border, to which is attached an aponeurosis common to:
Extensor carpi ulnaris
Flexor carpi ulnaris
Flexor digitorum profundus

Extensor pollicis longus

Extensor indicis

Styloid process

Groove for extensor carpi ulnaris

RADIUS

Supinator

Biceps

Abductor pollicis longus

Pronator teres

Extensor pollicis brevis

Grooves for:

Extensor digitorum and extensor indicis

Extensor pollicis longus

Extensor carpi radialis brevis

Extensor carpi radialis longus

3.184 The bones of the left forearm: posterior aspect.

periphery is vertically deepest medially, where it contacts the ulnar radial notch. Its posterior surface is palpable in a small depression on the lateral side of the back of the extended elbow. The *neck* is the constriction distal to the head, which overhangs it, especially on the lateral side. *The tuberosity* is distal to the medial part of the neck; it is posteriorly rough but anteriorly usually smooth.

The shaft has a lateral convexity, in section triangular (3.188), but only its *interosseous border* is sharp, except proximally, near the tuberosity; distally it is the posterior margin of a small, elongated, triangular area, proximal to the ulnar notch (vide infra), the two areas forming a so-called *medial* surface. To its distal three-fourths the interosseous membrane is attached, connecting radius

to ulna. The *anterior border* is obvious at both ends but rounded and indefinite between them. It descends laterally from the anterolateral part of the tuberosity as the *anterior oblique line*, distally a sharp, palpable crest along the lateral margin of the anterior surface. The *posterior* border is well-defined only in its middle third; proximally it ascends medially towards the postero-inferior part of the tuberosity; distally it is merely a rounded ridge. The *anterior surface*, between anterior and interosseous borders, is concave transversely and shows a distal forward curvature. Near its midpoint is a proximally directed nutrient foramen and canal. The *posterior surface*, between interosseous and posterior borders, is largely flat but may be slightly hollow in the proximal area. The *lateral surface* is gently convex; proximally, due to the 411 obliquity of anterior and posterior borders, it encroaches on the anterior and posterior aspects and is here slightly rough. A finely irregular oval area occurs near mid-shaft; beyond this the surface is smooth.

The distal end, the widest part, is four-sided in section. Its *lateral surface* is slightly rough, projecting distally as a *styloid process* palpable when tendons around it are slack. Distal (**3.189**) is the smooth *carpal articular surface*, divided by a ridge into medial and lateral areas, the medial quadrangular, the lateral triangular and curving on to the styloid process. The *anterior surface* is a thick, prominent ridge, palpable even through overlying tendons, 2 cm proximal to the thenar eminence. The *medial surface* is the *ulnar notch*, smooth, antero-posteriorly concave for articulation with the ulna's head. The *posterior surface* displays a palpable dorsal tubercle, limited medially by an oblique groove and in line with the cleft between the index and middle fingers. A wide, shallow groove, lateral to it, is divided by a faint vertical ridge. A similar but undivided groove is medial to the tubercle.

The *proximal articular surface* of the radial head and its *circumference* are covered by hyaline cartilage, the upper rim (margin) of the latter fitting the groove between capitulum and trochlea and invading the radial fossa in flexion. The articular circumference joints with the ulnar radial notch and annular ligament, within which it rotates in pronation and supination. The radial *neck* is enclosed by the ligament's narrower, distal part, separated by a synovial protrusion from the superior radio-ulnar joint.

The posterior area of the *tuberosity* is marked by the biceps tendon, but the latter is separated from a smooth anterior area by a bursa, distal to which the oblique cord is attached.

Proximally on the *anterior border* the thin, wide radial head of the flexor digitorum superificialis is attached, and to its conspicuous distal part the lateral edge of the extensor retinaculum. The small, triangular area proximal to the ulnar notch is for the deepest part of the pronator quadratus.

The proximal two-thirds of the *anterior surface* is an extensive area for the flexor pollicis longus, which conceals the nutrient foramen, the distal quarter is for the pronator quadratus. A rough area on the *lateral surface*, midway at its maximal curvature, is for the pronator teres. Proximally, the lateral surface widens, encroaching on the anterior and posterior, into a long V-shaped area (**3.**183,184) for the supinator. Distal to the pronator teres, the lateral surface is covered by tendons of radial extensors. Proximally on the *posterior surface* the abductor pollicis longus is attached and, more distally, the extensor pollicis brevis. The remaining surface is devoid of attachments but covered by the long and short pollicial extensors.

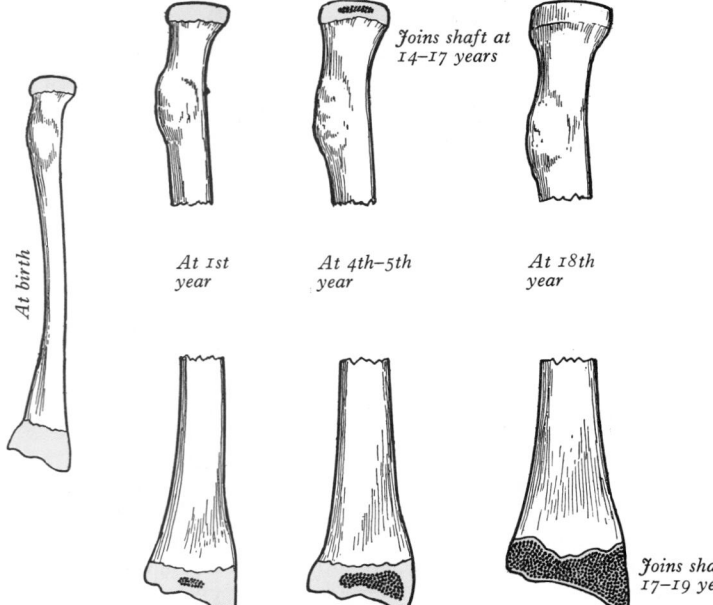

412   3.185 Stages in the ossification of the radius (not to scale).

Joins shaft at 14–17 years

At birth

At 1st year   *At 4th–5th year*   *At 18th year*

Joins shaft at 17–19 years

Epiphysial lines

Attachment of capsule of wrist radiocarpal joint

3.186 The epiphysial lines of the left radius in adolescence: anterior aspect. The line of attachment of the articular capsule of the wrist joint is in blue.

The radial *styloid process* projects beyond that of the ulna, its apex concealed by tendons of the abductor pollicis longus and extensor pollicis brevis. The carpal lateral ligament is attached to its tip. The *lateral surface*, near the styloid, receives the brachioradialis and is crossed obliquely, down and forwards, by tendons of the abductor pollicis longus and extensor pollicis brevis. The terminal ridge on the *anterior surface* of the lower end is an attachment for the palmar radiocarpal ligament. To a smooth ridge distal to the ulnar notch the base of the triangular articular disc of the inferior radio-ulnar joint is attached. From the latter, a narrow protrusion of synovial membrane extends proximally anterior to the lower end of the interosseous membrane (p. 508). The lateral part of the *carpal articular surface* joints with the scaphoid and the medial part of the surface with the lateral part of the lunate bone; in full adduction the latter's proximal surface is wholly in contact with the radius.

The radial *dorsal tubercle* receives a slip from the extensor retinaculum and is grooved medially by the tendon of the extensor pollicis longus. The wide groove lateral to the tubercle contains tendons of the extensor carpi radialis longus laterally, extensor carpi radialis brevis medially and their synovial sheaths. Medially the dorsal surface is grooved by the tendons of extensor digitorum, but extensor indicis and the posterior interosseous nerve separate these from the bone. Attached to the distal margin of this surface is the dorsal radiocarpal ligament.

Ossification (**3.**185,186). The radius ossifies at three centres, in the shaft, appearing centrally in the eighth week, and in each end. Near the end of the first postnatal year ossification begins in the distal epiphysis, and in the proximal at the fourth year in females, fifth in males. The proximal fuses in the fourteenth year in females, seventeenth in males, the distal in the seventeenth and nineteenth years respectively. A fourth centre sometimes appears in the tuberosity about the fourteenth or fifteenth year.

The Ulna

The ulna (**3.**183–192) is medial to the radius in the supinated forearm. Its proximal end is a massive hook (**3.**187), concave forwards. The shaft's lateral border is a sharp (interosseous) crest. The bone diminishes progressively from its proximal mass throughout almost its whole length, but at its distal end expands into a small rounded head and styloid process. The shaft is triangular in section and has an appreciable double curve. In its whole length it is slightly convex dorsally; but mediolaterally its profile is sinuous; the proximal half has a slight laterally concave curvature, the distal half a medially concave curvature.

The **proximal end** (3.187) has large olecranon and coronoid processes and trochlear and radial notches articulating with the humerus and radius. The *olecranon*, more proximal, is bent forwards at its summit like a beak, which enters the humeral olecranon fossa in extension. Its posterior surface is smooth, triangular and subcutaneous, its proximal border being the elbow's 'point'. In extension it can be felt near a line joining the humeral epicondyles, but in flexion it descends, the three osseous points forming an isosceles triangle. Its anterior, articular surface forms the proximal area of the trochlear notch. Its base is slightly constricted where it joins the shaft and narrowest part of the proximal ulna. The *coronoid process* projects anteriorly distal to the olecranon, its proximal aspect forming the distal part of the trochlear notch, distal to which, on the lateral surface, is a shallow smooth, oval *radial notch* for articulation with the radial head. Distal to this the surface is hollow to accommodate the radial tuberosity during pronation and supination. The coronoid's anterior surface is triangular, its distal part the *tuberosity of the ulna*. Its medial border is sharp and proximally bears a small tubercle.

The *trochlear notch* articulates with the humeral trochlea. It is constricted at the junction of the olecranon and coronoid processes, where their articular surfaces may be separated by a narrow, rough non-articular strip. A smooth ridge, adapted to the humeral trochlea's groove, divides the notch into medial and lateral parts, the medial fitting to the trochlear flange. The *radial notch* (3.187), an oval or oblong proximal depression on the lateral aspect of the coronoid, joints with the radial head's periphery, separated from the trochlear notch by a smooth ridge.

The distal end, a little expanded, has a head and styloid process. The *ulnar head* is visible in pronation on the posteromedial carpal aspect and can be gripped when the supinated hand is flexed. Its lateral convex articular surface fits the radial ulnar notch. Its smooth distal surface (3.189) is separated from the carpus by an articular disc, the apex of which is attached to a rough area between the articular surface and *styloid process*. The latter, a short, round, posterolateral projection of the ulna's distal end, is palpable (most readily in supination) about 1 cm proximal to the plane of the radial styloid. A dorsal vertical groove is present between the head and styloid process.

The shaft is triangular in section (3.188) in its proximal three-fourths, but distally almost cylindrical. It has anterior, posterior and medial surfaces and interosseous, posterior and anterior borders. The *interosseous border* is a conspicuous lateral crest in its middle two-fourths. Proximally continuous with the posterior border of a depression distal to the radial notch as the *supinator crest*, it disappears distally. The rounded *anterior border* commences medial to the ulnar tuberosity, descending backwards and

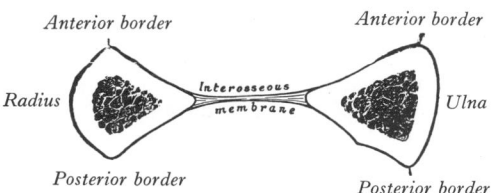

3.188 Transverse section through the left radius and ulna, showing the attachment of the antebrachial interosseous membrane: superior aspect.

usually traceable to the styloid process's base. The *posterior border*, also rounded, descends from the apex of the olecranon's posterior aspect, curving laterally to reach the styloid process. It is *palpable throughout* in a longitudinal furrow most obvious with the elbow in full flexion.

The *anterior surface* (3.183), between interosseous and anterior borders, is longitudinally grooved, sometimes deeply. Proximal to its midpoint is a nutrient foramen, directed proximally and containing a branch of the anterior interosseous artery. Distally, it is crossed obliquely by a rough, variable prominence, descending from the interosseous to the anterior border. The *medial surface*, between anterior and posterior borders, is transversely convex and smooth. The *posterior surface* (3.184), between posterior and interosseous borders, is divided into three areas, the most proximal limited by a sometimes faint oblique line ascending laterally from the junction of the middle and upper thirds of the posterior border to the posterior end of the radial notch (3.187). The region distal to this line is divided into a larger medial and narrower lateral strip by a vertical ridge, usually distinct only in its proximal three-fourths.

Further details. To the proximal *olecranon* surface, anteriorly, is attached the cubital capsular ligament, and to its rough posterior two-thirds the tendon of triceps; these may be separated by a smooth bursal area. Its medial surface is marked proximally by attachment of the posterior and oblique bands of the the ulnar collateral ligament and ulnar part of the flexor carpi ulnaris. The smooth area distal to this is the most proximal attachment of flexor digitorum profundus. To the lateral olecranon surface, and the adjoining posterior surface of the ulnar shaft as far as its oblique line (3.184,187), the anconeus is attached. Its posterior surface is separated from the skin by a subcutaneous bursa.

To the anterior surface of the *coronoid process*, including the ulnar tuberosity, the brachialis is attached. Its medial border is sharp; to a small tubercle at its proximal end are attached oblique and anterior bands of the ulnar collateral ligament and the distal part of the humero-ulnar slip of flexor digitorum superficialis. Distal to this the margin is for the ulnar part of the pronator teres. An ulnar part of the flexor pollicis longus may be attached to the

Olecranon —

Non-articular strip in trochlear notch —

Coronoid process —

Radial notch —

Tuberosity —

Supinator crest —

Supinator surface —

Oblique line on posterior surface —

Vertical ridge —

Interosseous border —

Posterior border —

A·K·M·

3.187 The proximal part of the left ulna: lateral aspect.

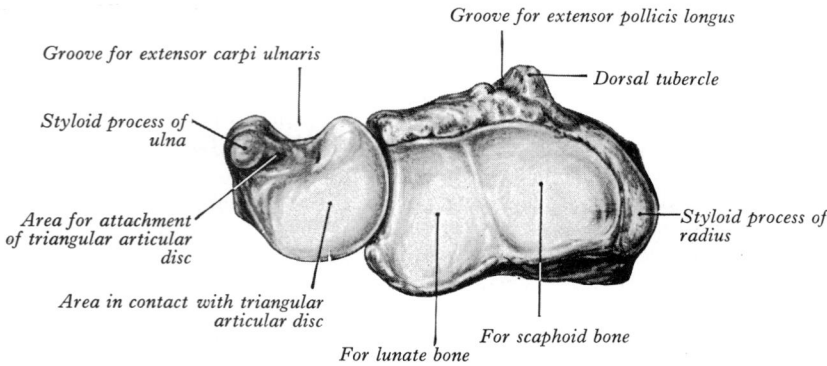

Groove for extensor carpi ulnaris

Groove for extensor pollicis longus

Dorsal tubercle

Styloid process of ulna

Area for attachment of triangular articular disc

Styloid process of radius

Area in contact with triangular articular disc

For lunate bone

For scaphoid bone

3.189 The distal ends of the right radius and ulna: distal (inferior) aspect.

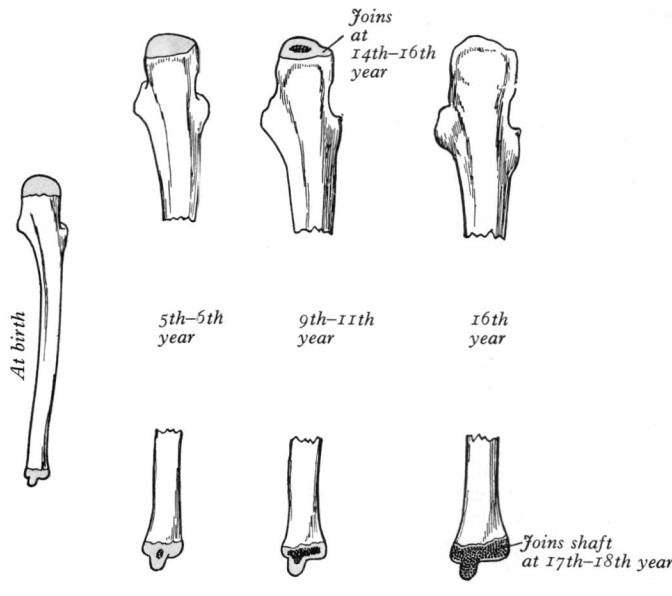

Joins at 14th–16th year

At birth

5th–5th year

9th–11th year

16th year

Joins shaft at 17th–18th year

3.191 Stages in the ossification of the ulna. The diagram is simplified in relation to the epiphysial ossification of the olecranon, which is held to have two centres (see Porteous 1960 and the description in the text).

lateral or, more rarely, the medial border of the coronoid process (Martin 1958); to its medial surface are attached fibres of the flexor digitorum profundus. To the anterior rim of the radial notch the annular ligament is attached and posteriorly to a ridge at or just behind the notch's posterior margin. The depressed area distal to the notch is limited behind by the *supinator crest*, both providing attachment for the supinator.

The olecranonic area of the *trochlear notch* is usually divided into three areas: the most medial faces ventromedially, and is grooved to fit the medial flange of the humeral trochlea with which it makes increasing contact during flexion; a flat intermediate area fits the lateral flange and the most lateral, a narrow strip, abuts the trochlea but only in extension (p. 504). The articular surface is narrower than the olecranon's base; resulting non-articular parts are related to the synovial processes (**4.**44). The coronoid area of the trochlear notch is also divided, its medial and lateral areas corresponding to medial and intermediate olecranonic areas. The medial is hollower, conforming to the convex medial trochlear flange (p. 505); to its medial and anterior borders, medial and anterior parts of the capsular ligament are attached.

To the subcutaneous *posterior border* is attached the forearm's deep fascia, which is also, in its proximal three-quarters, an aponeurotic attachment for the flexor digitorum profundus; in its

proximal half it is another attachment for the flexor carpi ulnaris and, in its middle third, for the extensor carpi ulnaris. These muscles thus connect with the posterior border through a common blended aponeurosis. To the *interosseous border*, except proximally, is attached the interosseous membrane. To the proximal three-fourths of the *anterior border* the flexor digitorum profundus is attached.

The *anterior surface*, in its proximal three-fourths, is for the flexor digitorum profundus, as are the anterior border and *medial surface*, ascending medial to the coronoid process and olecranon. The rough strip across the distal fourth of the anterior surface is an attachment of the pronator quadratus. The anconeus is attached to the *posterior surface* proximal to the oblique line and lateral to the olecranon. The narrow strip between interosseous border and vertical ridge is for three deep muscles: the abductor pollicis longus from the proximal fourth, and (with a ridge sometimes interposed) from the succeeding fourth for the extensor pollicis longus; the extensor indicis is attached to the third quarter. The broad strip medial to the vertical ridge is merely covered by the extensor carpi ulnaris, whose tendon grooves the posterior aspect of the ulna's distal end. The ulnar collateral ligament is attached to the apex of the *styloid process*.

For an account of ulnar vascularization consult Fischer et al (1974).

Ossification (**3.**190,191). The ulna ossifies from four main centres, one each in the shaft and distal end and two in the olecranon. Ossification begins in mid-shaft about the eighth fetal week, extending rapidly. In the fifth (females) and sixth (males) years a centre appears in the distal end, extending into the styloid process. The distal olecranon is ossified as an extension from the shaft, the remainder from two centres, one for the proximal trochlear surface, and a thin scale-like proximal epiphysis on its summit (Porteous 1960). The latter appears in the ninth year in females, eleventh in males; the whole proximal epiphysis has joined the shaft by the fourteenth year in females, sixteenth in males. The distal epiphysis unites with the shaft in the seventeenth year in females, eighteenth in males.

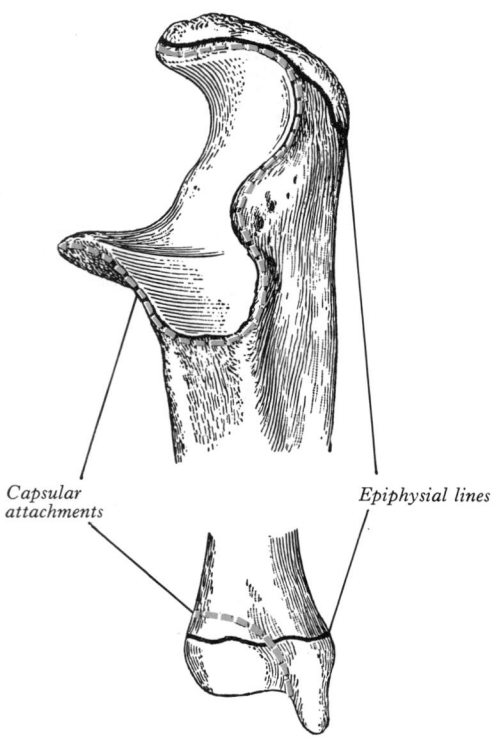

Capsular attachments

Epiphysial lines

3.190 The epiphysial lines of the left ulna in adolescence: lateral aspect. The lines of attachment of the articular capsules are in blue.

414

3.192 (*opposite left*) Anteroposterior radiograph of the forearm of a girl aged 11. 1. Proximal radial epiphysis. 2. Conjoined epiphyses of capitulum and lateral epicondyle. 3. Epiphysis of medial epicondyle. Diaphysial bone. 5. Trochlear epiphysis. 6. Cartilaginous growth plate 7. Distal ulnar epiphysis. 8. Distal radial epiphysis.

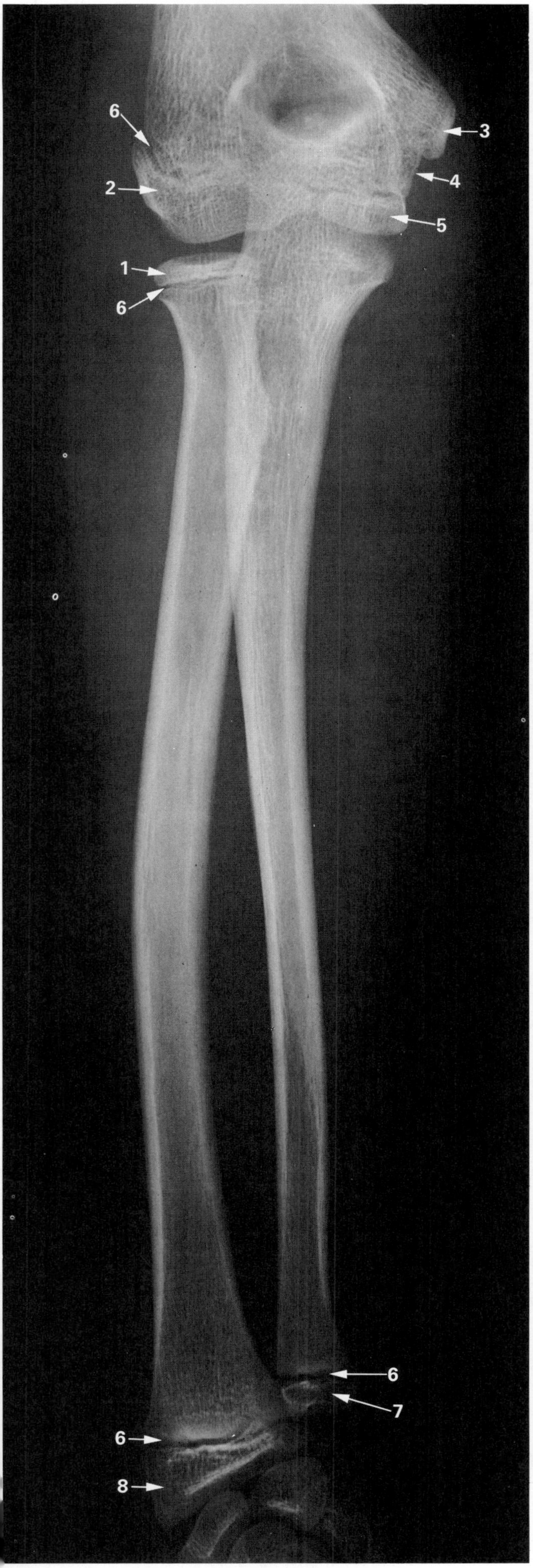

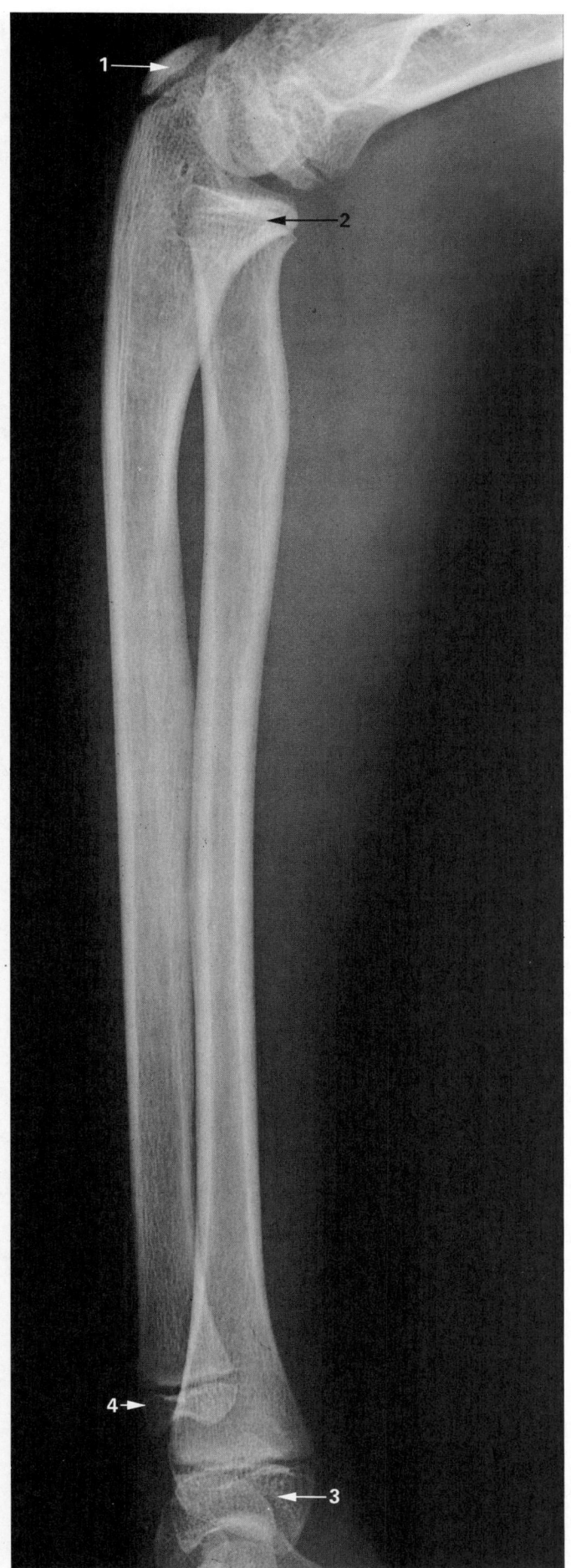

3.193 (*above*) Lateral radiograph of forearm of a girl of 11 years, semiflexed at the elbow. Note following epiphyses and adjacent radiotranslucent growth cartilages: 1. Olecranon. 2. Proximal radial. 3. Distal radial. 4. Distal ulnar.

THE SKELETON OF THE HAND

The hand's skeleton has three regions: (1) the carpus, (2) the metacarpus and (3) the phalanges. In the following description *proximal* and *distal* are used in preference to *superior* and *inferior*, and *palmar* and *dorsal*, which are self-explanatory, rather than *anterior* and *posterior*.

The Carpus

General features. The carpus (**3.194A,B,195A,B**) contains eight bones in proximal and distal rows of four. Proximally, in lateral to medial order, are the *scaphoid*, *lunate*, *triquetral* and *pisiform*

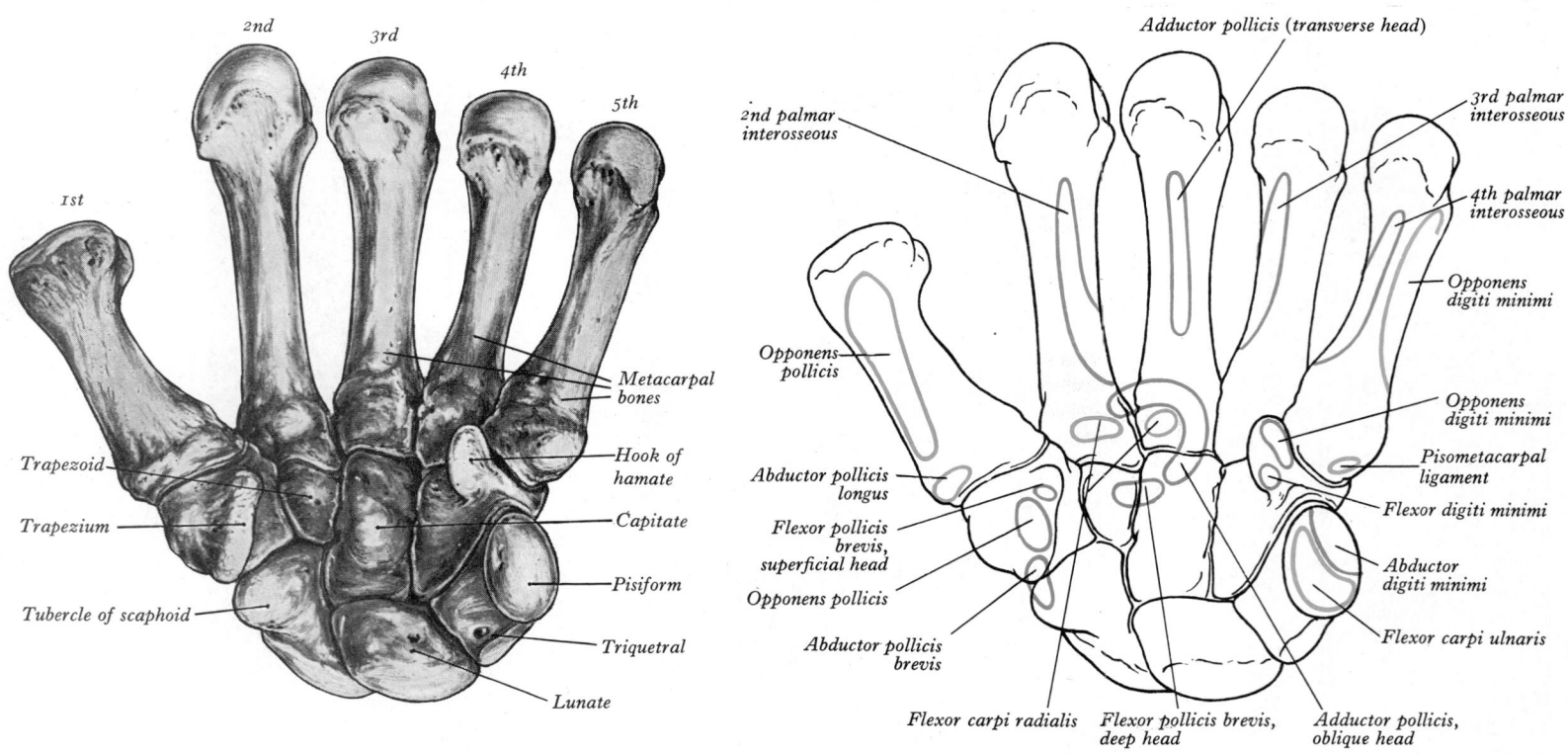

3.194A The carpal and metacarpal bones of the left hand: palmar aspect.

3.194B Muscle attachments to the carpal and metacarpal bones of the left hand: palmar aspect. Dorsal interossei not shown.

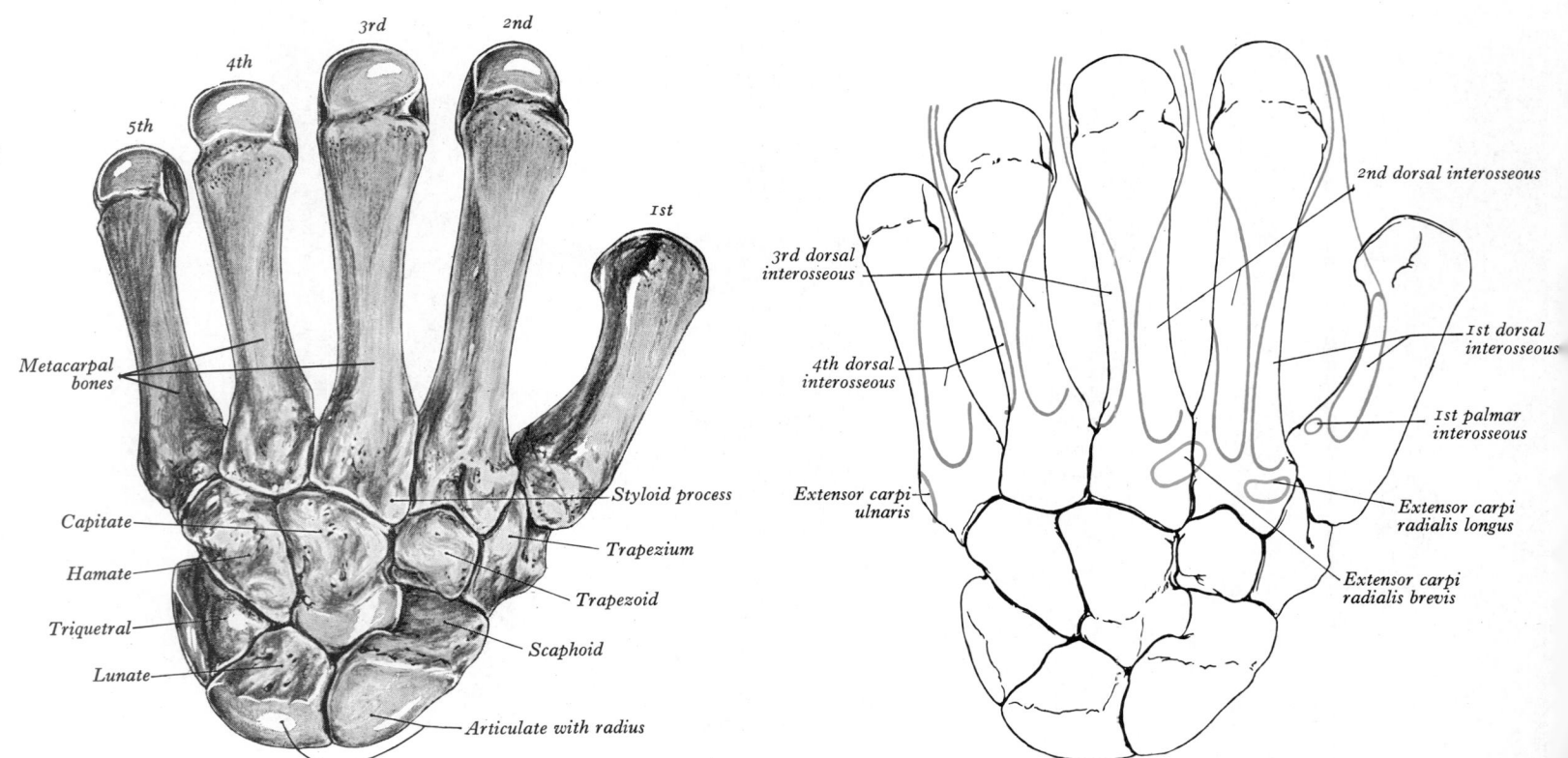

3.195A The carpal and metacarpal bones of the left hand: dorsal aspect.

3.195B Muscle attachments to the carpal and metacarpal bones of the left hand: dorsal aspect.

bones; in the distal row are the *trapezium, trapezoid, capitate* and *hamate bones*. The pisiform articulates with the palmar surface of the triquetral, thus separated from other carpal bones, all of which articulate with their neighbours. The other three proximal bones form an arch proximally convex, articulating with the radius and articular disc of the inferior radio-ulnar joint. The arch's concavity is a distal recess embracing, proximally, the projecting aspects of capitate and hamate bones; the two rows are thus mutually and firmly adapted without any loss of movement.

The dorsal carpal surface is convex and the palmar forms a deeply concave *carpal groove*, accentuated by the palmar projection of the lateral and medial borders. The medial projection is formed by the *pisiform bone* and the *hamulus*, an unciform palmar process of the *hamate* bone. The pisiform is at the proximal border of the hypothenar eminence, medial in the palm; it is easily felt in front of the triquetral. The hamulus is concave laterally, its tip palpable 2.5 cm distal to the pisiform, in line with the radial border of the ring finger. The ulnar nerve's superficial division can be rolled on it. The lateral border of the carpal groove is formed by the *tubercles of the scaphoid and trapezium*. The former is distal on the anterior scaphoid surface and palpable, sometimes also visible, as a small medial knob at the proximal border of the palmar thenar eminence, lateral to the tendon of the flexor carpi radialis. The trapezial tubercle is a vertically rounded ridge on the bone's anterior surface, slightly hollow medially and just distal and lateral to the scaphoid tubercle; it is difficult to palpate. (Both the scaphoid and trapezium may be grasped individually and moved passively, by firm pressure between an opposed index finger and thumb applied to the palmar surface and anatomical snuff-box simultaneously.) The carpal groove is made into an osseofibrous *carpal tunnel* by a fibrous retinaculum attached to its margins; the tunnel carries flexor tendons and the median nerve into the hand (see p. 1133). The retinaculum strengthens the carpus and augments flexor efficiency. Palmar and dorsal surfaces of carpal bones, apart from the triquetral and pisiform, are attachments of the radiocarpal, intercarpal and carpometacarpal ligaments.

The Individual Carpal Bones

The Scaphoid Bone

The scaphoid (**3.**196A,B), largest element in the proximal carpal row, has a long axis which is distal, lateral and slightly palmar in direction. Its round tubercle on the distolateral part of its *palmar surface* is directed anterolaterally (**3.**194A), and is an attachment of the flexor retinaculum and abductor pollicis brevis; it is crossed by the tendon of the flexor carpi radialis. The rough *dorsal surface* is slightly grooved, narrower than the palmar, and pierced by small nutrient foramina, often restricted to the distal half (13%) (Obletz & Halbstein 1938), an observation of clinical significance.

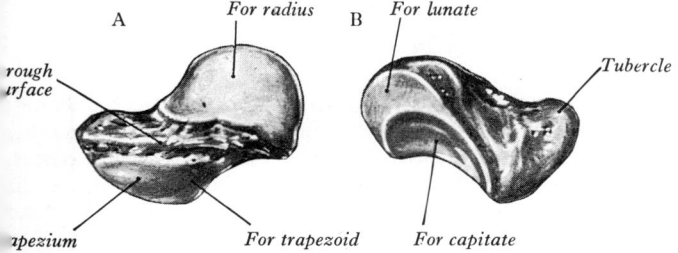

3.196 The left scaphoid bone: (A) dorsal, (B) palmar aspects.

The *lateral surface*, also narrow and rough, has the radial collateral ligament attached to it. The remaining surfaces are articular: the *radial surface* is convex, proximal and directed proximolaterally; the *lunate* surface is flat, semilunar, facing medially; the *capitate* surface is large, concave and distal, directed distomedially. The surface for *trapezium and trapezoid bones* is continuous, convex and distal.

The Lunate Bone

The lunate (**3.**197A,B), approximately semilunar, articulates between the scaphoid and triquetal in the proximal carpal row. Its

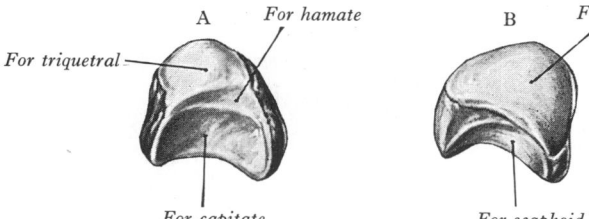

3.197 The left lunate bone: A. distomedial; B. proximolateral aspects.

rough *palmar surface*, almost triangular, is larger and wider than the rough *dorsal surface*. Its smooth convex *proximal surface* articulates with the radius and articular disc of the distal radio-ulnar joint. Its narrow *lateral surface* bears a flat semilunar facet for the scaphoid. The *medial surface*, almost square, articulates with the triquetral and is separated from the distal surface by a curved ridge, usually somewhat concave (**3.**197A) for articulation with the edge of the hamate bone in adduction. The *distal surface* is deeply concave to fit the medial part of the capitate's head.

The Triquetral Bone

The triquetral (**3.**198), somewhat pyramidal, bears an oval isolated facet for articulation with the pisiform on its distal *palmar* surface. Its *medial* and *dorsal surfaces* are confluent, marked distally by attachment of the ulnar collateral ligament but smooth

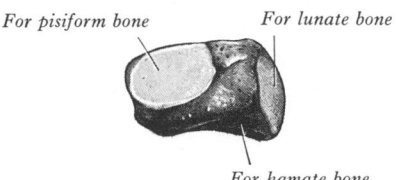

3.198 The left triquetral bone: palmar aspect.

proximally for the articular disc of the distal radio-ulnar joint in full adduction. The *hamate surface*, lateral and distal, is concavoconvex, broad proximally, narrow distally. The *lunate surface*, almost square, is proximal and lateral.

The Pisiform Bone

The pisiform (**3.**199), as its name implies, is shaped like a pea, with a dorsal flat articular facet for the triquetral; it has a distolateral long axis. The palmar non-articular area, surrounding and projecting distal to the articular surface, has the tendon of the

3.199 The left pisiform bone: dorsal aspect.

flexor carpi ulnaris attached to it and also the tendon's distal continuations, the pisometacarpal and pisohamate ligaments. Hence the pisiform has attributes of a sesamoid bone (see p. 457), which it almost certainly is, being unrepresented in primitive carpalia (p. 400).

The Os Trapezium

The os trapezium (**3.**200A,B) has a tubercle and groove on its rough *palmar surface*. The groove is medial and contains the tendon of the flexor carpi radialis; to its margins are attached two layers of the flexor retinaculum (**5.**73 and p. 625). The *tubercle* is obscured by attached thenar muscles, opponens pollicis between flexor pollicis brevis distally and abductor pollicis brevis proximally (**3.**194B). The elongate, rough *dorsal surface* is related to the radial artery. The large *lateral surface* is rough for attachment of the radial collateral ligament and capsular ligament of the pollicial carpometacarpal joint. A large *sellar surface* faces disto-laterally

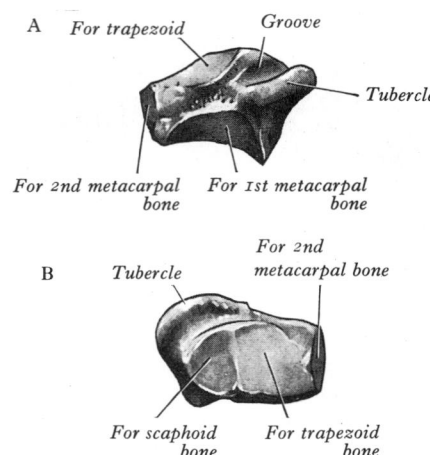

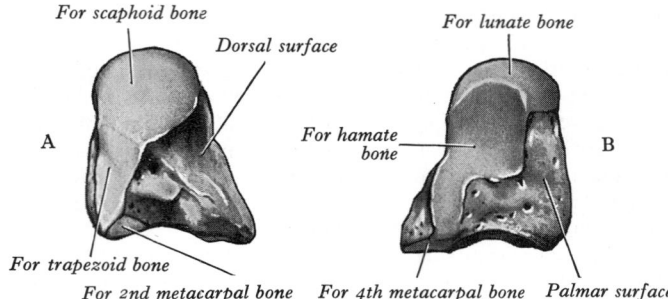

3.202 The left capitate bone: A. lateral; B. medial aspects.

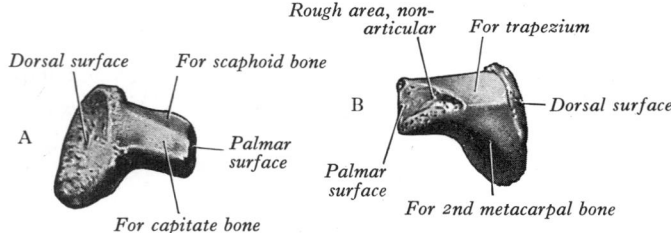

3.200 The left trapezium: A. palmar; B. proximomedial aspect.

for the hamate, deeper proximally where it is partly non-articular. *Palmar* and *dorsal surfaces* are roughened for carpal ligaments; the dorsal is larger.

The Hamate Bone

The hamate (**3**.203A,B) is cuneiform with an unciform *hamulus* projecting from the distal part of its rough *palmar surface;* the hamulus is curved with a lateral concavity and its tip inclines laterally contributing to the medial wall of the carpal tunnel. To the hamular apex is attached the flexor retinaculum. Distally, on the hamular base, a slight transverse groove may be in contact with the ulnar nerve's terminal deep branch. The remaining *palmar surface*, like the *dorsal*, is rough for ligaments. A faint ridge divides the *distal surface* into a smaller lateral facet articulating

for the base of the pollicial metacarpal bone. Most distally it projects between bases of the first and second metacarpal bones and carries a small, quadrilateral, distomedially directed facet articulating with the second metacarpal base. The large medial surface is gently concave for articulation with the trapezoid. The proximal surface is a small, slightly concave facet for the scaphoid. Due to its contribution to mobility of the pollex, the metacarpal articular surface has attracted much attention. Its ridge or 'summit', fitting the concavity of the first metacarpal base, extends in a palmar and lateral direction, at an angle of about 60° with the plane of the second and third metacarpals (Kuczynski 1974, p. 514 and 634). Abduction and adduction occur in the plane of the ridge, which is shorter than the corresponding metacarpal groove; contours of both vary reciprocally, being more curved near the second metacarpal base, with a longer radius of curvature away from this. The two surfaces are not completely congruent, the area of close contact probably moving towards the palm in adduction and dorsally in abduction (see p. 514).

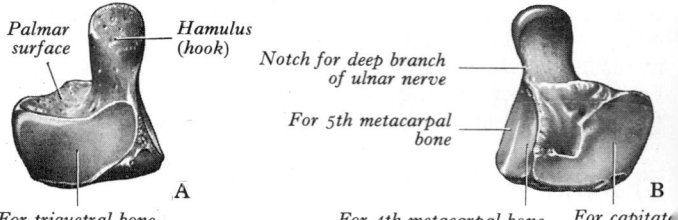

3.203 The left hamate bone: A. medial; B. lateral aspects.

The Trapezoid Bone

The trapezoid (**3**.201A,B), small and irregular, has a rough *palmar surface* narrower and smaller than its rough *dorsal surface*, the former invading the lateral aspect. The *distal surface*, articulating

with the fourth metacarpal base and a medial facet for the fifth. The *proximal surface*, the thin margin of the wedge, usually bears a narrow facet contacting the lunate in adduction. The *medial surface* is a broad strip, convex proximally, concave distally, articulating with the triquetral; distally a narrow medial strip is non-articular. The *lateral surface* articulates with the capitate by a facet covering all but its distal palmar angle.

The Metacarpus

The metacarpus (Singh 1959) consists of five metacarpal bones, conventionally numbered in lateromedial order. These are miniature long bones, with a distal head, shaft and expanded base. The rounded *heads* articulate with the proximal phalanges. Their articular surfaces are convex, less so transversely, and extend further on palmar surfaces, especially at their margins. The familiar knuckles are produced by metacarpal heads. The metacarpal bases articulate with the distal carpal row and with each other, excepting the first and second. The *shafts* have longitudinally concave palmar surfaces, forming hollows for palmar muscles. Their dorsal surfaces have a distal triangular area, continued proximally as a round ridge. These flat areas are palpable proximal to the knuckles.

The *medial four metacarpal bones* are sometimes described as parallel; strictly they diverge somewhat, radiating gently proximodistally. However, *the first metacarpal*, relative to the others, is more anterior and *rotated medially* on its axis through a right angle, so that its morphologically *dorsal* surface is *lateral*, its radial border palmar, its palmar surface medial and its ulnar border dorsal. Hence the pollex flexes medially across the palm and

3.201 The left trapezoid bone: A. proximomedial; B. distolateral aspects.

with the grooved second metacarpal base, is triangular, convex transversely and concave at right angles to this. The *medial surface* articulates, by a concave facet, with the distal part of the capitate, the *lateral surface* with the trapezium and the *proximal* with the scaphoid bone.

The Capitate Bone

The capitate (**3**.202A,B), the central and largest carpal bone, articulates with the third metacarpal base, its triangular *distal surface* concavoconvex for this. Its lateral border is a concave strip for the medial side of the second metacarpal base; its dorsomedial angle usually bears a facet for the fourth metacarpal base. The *head* projects into the concavity formed by the lunate and scaphoid bones, its *proximal surface* articulating with the lunate and the *lateral* with the scaphoid. The scaphoid and trapezoidal facets, usually continuous on the distolateral surface, may be separated by a rough interval. The *medial surface* has a large facet can be rotated into opposition with each finger. Such opposition depends on medial rotation and is the prime factor in manual dexterity; when an object is grasped, fingers and thumb encircle it from opposite sides, greatly increasing the power and skill of the grip. Lewis (1977) has made an extended study of human metacarpals and their articulations with the carpus and phalanges, in an explanation of evolving human skills.

Individual Metacarpal Bones

The first metacarpal bone (3.204) is short and thick. (Caution should be exercised when considering descriptions of this bone; here, morphological terms are used supplemented, in places, by their topographical equivalents.) Its dorsal (lateral) surface can be felt to face laterally; its long axis diverges distolaterally from its neighbour. The *shaft* is flattened, dorsally broad and transversely convex. The palmar (medial) surface is concave in long axis and divided by a ridge into a larger lateral (ventral) and smaller medial

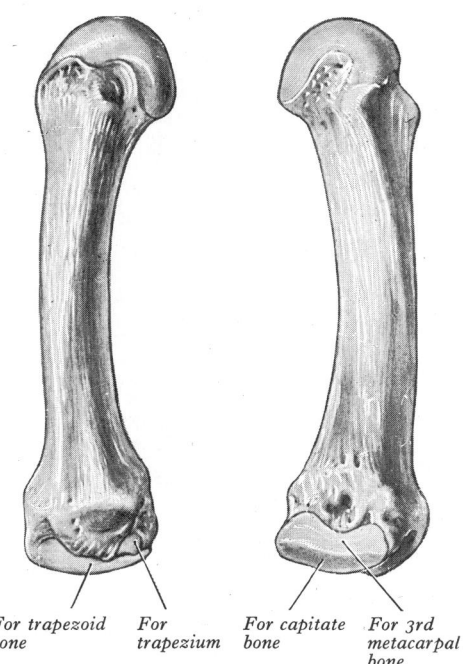

For trapezoid   For   For capitate   For 3rd
bone   trapezium   bone   metacarpal
                                            bone

3.205 The left second metacarpal bone: dorsolateral and medial aspects.

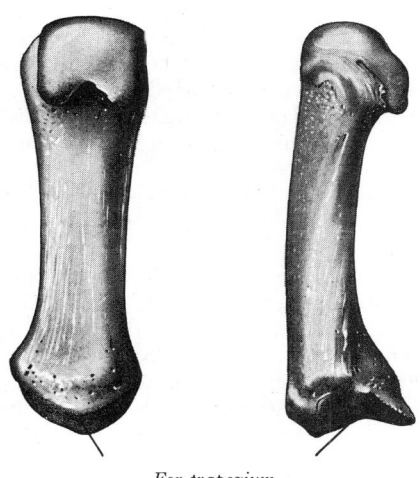

For trapezium

3.204 The first right metacarpal bone: palmar and lateral aspects.

(dorsal) part. Opponens pollicis is attached to the radial border and adjoining palmar surface, the first dorsal interosseous muscle (radial head) is attached to its ulnar border and adjacent palmar surface. The *base* is concavoconvex and articulates with the trapezium (p. 514). On its lateral (palmar) side abductor pollicis longus is attached, to its ulnar side the first palmar interosseous muscle (p. 632). The *head* is less convex than in other metacarpals and contrasts in being transversely broad. On its palmar aspect ulnar and radial angles form two articular eminences, on which glide sesamoid bones.

The second metacarpal bone (3.205) has the longest shaft and largest *base*, which is grooved in dorsopalmar direction for the trapezoid bone; medial to the groove a deep ridge articulates with the capitate bone; laterally, nearer the base's dorsal surface, is a quadrilateral trapezial facet, and just dorsal to this a rough impression marks attachment of the extensor carpi radialis longus. On the palmar surface a small tubercle or ridge receives flexor carpi radialis. The base's medial side articulates with the third metacarpal base by a long facet, centrally narrowed. The *shaft* is prismatic in section and longitudinally curved, convex dorsally, concave towards the palm. Its dorsal surface is distally broad but proximally narrows to a ridge and is covered by extensor tendons of the index finger; its converging borders begin at the tubercles, one on each side of its head for the attachment of collateral ligaments (p. 515). Proximally the *lateral surface* inclines dorsally for the ulnar head of the first dorsal interosseous muscle. The *medial surface*, inclining similarly, is divided by a faint ridge into a palmar strip for the second palmar interosseous and a dorsal for the radial head of the second dorsal interosseous muscle.

The third metacarpal bone (3.206) has a short *styloid process*, projecting proximally from the radial side of the dorsal

surface. Its *base* articulates with the capitate bone by a facet ventrally convex but dorsally concave where it invades the styloid process on the lateral aspect of its base. A strip-like facet, constricted centrally, articulates with the second metacarpal base (laterally) and the fourth metacarpal base (medially), the latter by two oval facets. The palmar facet may be absent; less frequently the facets are connected proximally by a narrow bridge. The base's palmar surface receives a slip from the flexor carpi radialis tendon; to its dorsal surface, beyond the styloid process, extensor carpi radialis brevis is attached. The *shaft* resembles that of the second metacarpal. To its lateral surface the ulnar head of the second dorsal interosseous muscle is attached, to its medial surface the radial head of the third dorsal interosseous and to the intervening palmar ridge in its distal two-thirds the transverse

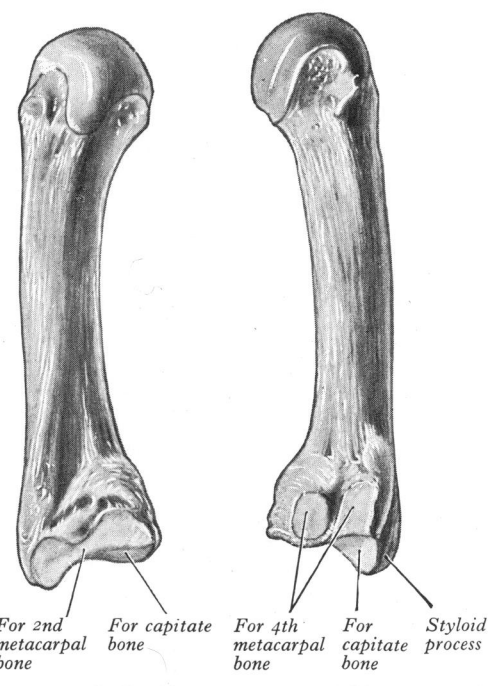

For 2nd   For capitate   For 4th   For   Styloid
metacarpal   bone   metacarpal   capitate   process
bone   bone   bone

3.206 The left third metacarpal bone: lateral and medial aspects.

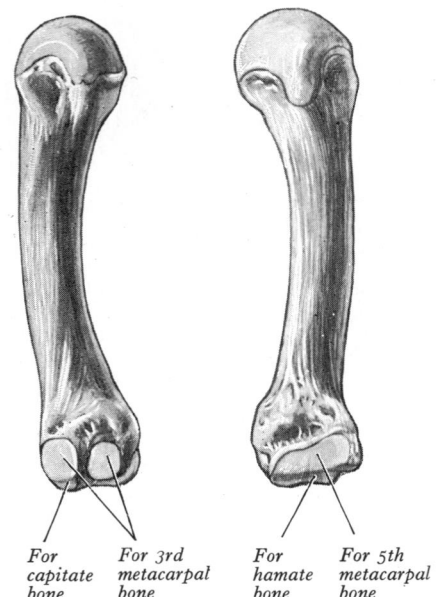

3.207 The left fourth metacarpal bone: lateral and medial aspects.

For capitate bone   For 3rd metacarpal bone   For hamate bone   For 5th metacarpal bone

head of adductor pollicis. Its dorsal surface is covered by the extensor tendon.

The fourth metacarpal bone (3.207), shorter and thinner than second and third, displays on its *base* two lateral oval facets for the third metacarpal base (vide supra), the dorsal usually larger and proximally in contact with the capitate. A single medial elongate facet is for the fifth metacarpal base. The proximal, quadrangular surface articulates with the hamate bone, being ventrally convex, dorsally concave. The *shaft* is like the second, but a faint ridge on its lateral surface separates attachments of the third palmar interosseous and the ulnar head of the third dorsal interosseous muscles. To the medial surface the radial head of the fourth dorsal interosseous is attached.

The fifth metacarpal bone (3.208) differs in its medial basal surface, which is non-articular and bears a tubercle for extensor carpi ulnaris. The lateral basal surface is a facet, transversely concave, convex from palm to dorsum, for articulation with the hamate. A lateral strip articulates with the fourth metacarpal base. The *shaft* has a triangular dorsal area almost reaching the base; the lateral surface inclines dorsally only at its proximal end. To the

medial surface opponens digiti minimi is attached; the lateral is divided by a ridge, sometimes sharp, into a palmar strip for the fourth palmar interosseous and a dorsal strip for the ulnar part of the fourth dorsal interosseous.

The Phalanges of the Hand

There are 14 phalanges, three in each finger, two in the pollex. Each has a head, shaft and proximal base. The *shaft* tapers distally, its dorsal surface transversely convex. The palmar surface is transversely flat but gently concave ventrally in its long axis. *Bases* of *proximal phalanges* carry concave, oval facets adapted to metacarpal heads, their own heads smoothly grooved like pulleys and encroaching more on palmar surfaces. Conforming to this, *bases* of *middle phalanges* carry *two* concave facets separated by a smooth ridge. Middle phalangeal *heads* are also pulley-like, to which the bases of distal phalanges are adapted; the latter's own heads are non-articular but carry a rough, crescentric *palmar tuberosity*, to which the pulps of finger tips are attached.

In addition to articular ligaments, phalanges afford attachment to numerous muscles: to the base of a distal phalanx on its palmar surface a corresponding tendon of flexor digitorum profundus and, on its dorsal surface, extensor digitorum; to the sides of a middle phalanx a tendon of flexor digitorum superficialis (p. 617) and its fibrous sheath; to the base dorsally, a part of extensor digitorum. To the sides of a proximal phalanx a fibrous flexor sheath is attached; to its base, laterally, part of the corresponding dorsal interosseous, and, medially, another dorsal interosseous muscle are attached (p. 631).

Phalanges of the minimus and pollex differ: attached to the medial side of the base of the former's proximal phalanx are abductor and flexor digiti minimi; to the base of the proximal pollicial phalanx are attached, dorsally, the tendon of extensor pollicis brevis and the oblique head of adductor pollicis and, medially, oblique and transverse heads of the adductor, sometimes conjoined with the first palmar interosseous (p. 631).

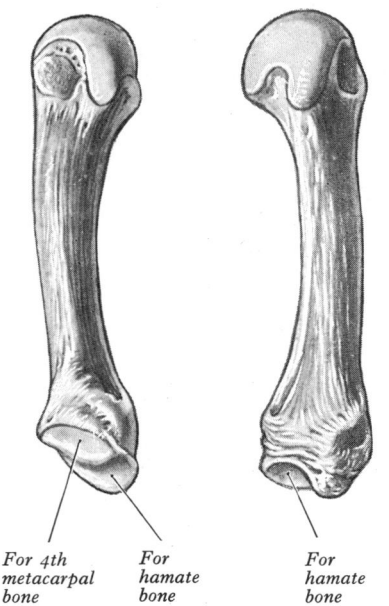

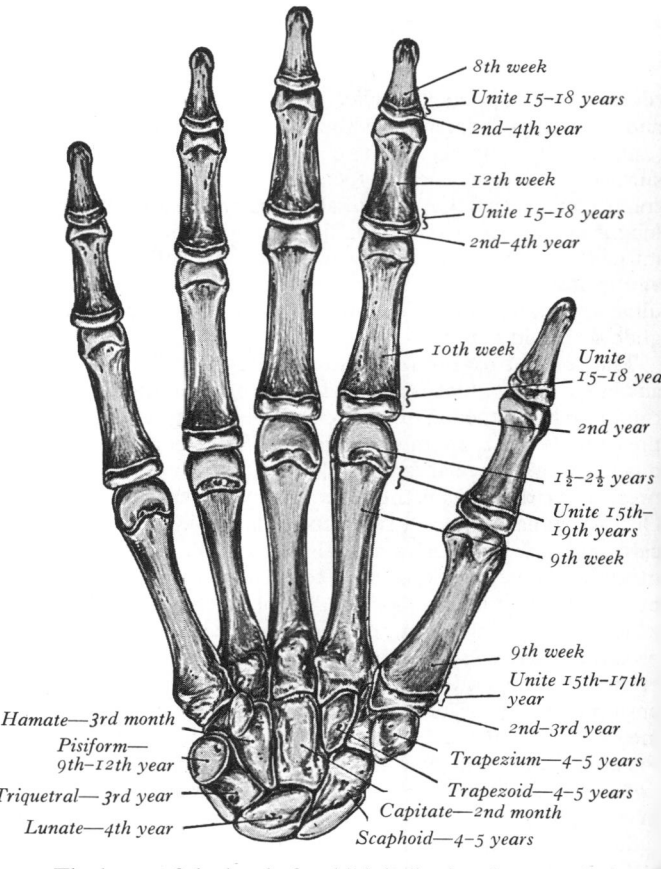

For 4th metacarpal bone   For hamate bone   For hamate bone

420   3.208 The left fifth metacarpal bone: lateral and medial aspects.

8th week
Unite 15–18 years
2nd–4th year

12th week
Unite 15–18 years
2nd–4th year

10th week
Unite 15–18 years

2nd year

1½–2½ years
Unite 15th–19th years
9th week

9th week
Unite 15th–17th year

2nd–3rd year

Trapezium—4–5 years
Trapezoid—4–5 years
Capitate—2nd month
Scaphoid—4–5 years

Hamate—3rd month
Pisiform—9th–12th year
Triquetral—3rd year
Lunate—4th year

3.209 The bones of the hand of a child, indicating the general plan of ossification.

Margins of the proximal pollicial phalanx are not sharp, the fibrous sheath being less strongly developed than in other digits.

Ossification of the Bones in the Hand (3.209–213)

Carpal bones are cartilaginous at birth, but ossification may have started by then in the capitate and hamate. Each carpal is ossified from one centre, capitate first, pisiform last; but the order in others varies. The capitate begins to ossify in the second *month*, the hamate at the end of the third, triquetral in the third *year*, lunate during the fourth and scaphoid, trapezium and trapezoid in the fourth in females, fifth in males; the pisiform begins in the ninth or tenth year in females, twelfth in males. The order varies according to sex (Garn & Rohmann 1960), nutrition and, possibly, race (Shakir & Zaini 1974, Wingerd et al 1974). Occasionally an *os centrale* (p. 400) occurs between the scaphoid, trapezoid and capitate bones; during the second prenatal month it is a cartilaginous nodule usually fusing with the scaphoid. Sometimes the styloid process of the third metacarpal is a separate ossicle. Occasionally, lunate and triquetral elements may fuse; other fusions and accessory ossicles have also been described (O'Rahilly 1953, 1956).

Each **metacarpal bone** ossifies from a primary centre for the shaft and a secondary in the *base* of the first and *heads* of the other four. Ossification begins in mid-shaft about the ninth week. Centres for second to fifth metacarpal heads appear in that order in the second year in females, and between one and a half to two and a half years in males. They unite with shafts about the fifteenth or sixteenth year in females, eighteenth or nineteenth in males. The first metacarpal base begins late in the second year in females, early in the third in males; it unites before the fifteenth year in females, seventeenth in males (Joseph 1951). For wider assessments of range of variability consult Modi (1957) and Krogman (1962).

The pollicial metacarpal bone ossifies like a phalanx; some therefore consider the pollicial skeleton to consist of three *phalanges*; others believe the distal to represent fused middle and distal phalanges, a condition occasionally observed in the fifth toe (Broom 1930). (When the pollex has *three* phalanges, the metacarpal has a distal *and* a proximal epiphysis. It occasionally bifurcates distally, the medial branch, with no distal epiphysis, bearing *two* phalanges, the lateral showing a distal epiphysis, and *three* phalanges—Nicholson 1937.) The existence of only a *distal* metacarpal epiphysis may be associated with a greater range of movement at the metacarpophalangeal joint. In the pollex, it is the carpometacarpal joint which has the wider range, and a *basal* epiphysis in the first metacarpal may be attributable to this. However, a distal epiphysis may appear in the first, and a proximal in the second metacarpal.

Growth studies of 1700 children, aged one to 18 years, favour the view of the pollicial metacarpal as a proximal phalanx (as first suggested by Vesalius in 1543). But this ignores the observations that the so-called distal epiphysis of the first metacarpal is usually (if not always) a diaphysial prolongation into terminal cartilage, rather than a separate centre. A review of the data concerning such 'pseudo-epiphyses' corroborates this description (Haines 1974).

Phalanges are ossified from a primary centre for the shaft and a proximal epiphysial centre. Ossification begins prenatally in shafts as follows: distal phalanges in the eighth or ninth week, proximal phalanges in the tenth, middle phalanges in the eleventh

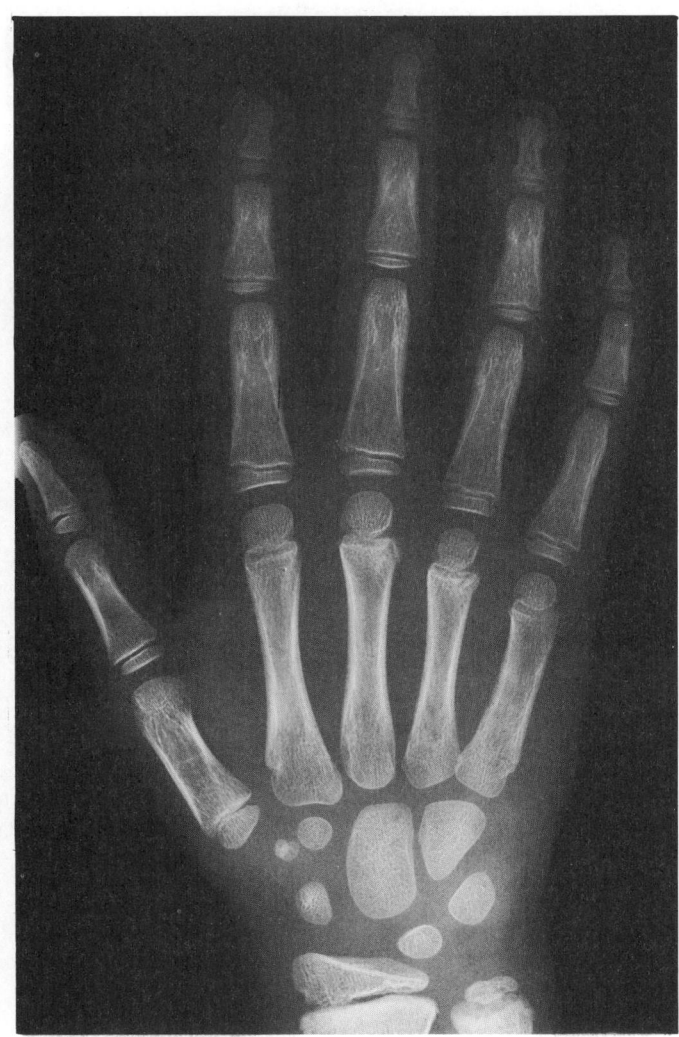

3.210 Radiograph of a hand at 2½ years (male). Note early stages of ossification in epiphyses at proximal ends of phalanges and first metacarpal, at distal ends of remaining metacarpals and radius and in the capitate, hamate and lunate bones. The last is more usually preceded by the centre for the triquetral. Compare with 3.211-213.

3.211 Radiograph of a hand at 6½ years (male). Note the more advanced state of the centres of ossification already visible in **3.210**, and additional centres in the distal ulnar epiphysis and in the triquetral, scaphoid, trapezium and trapezoid carpal bones.

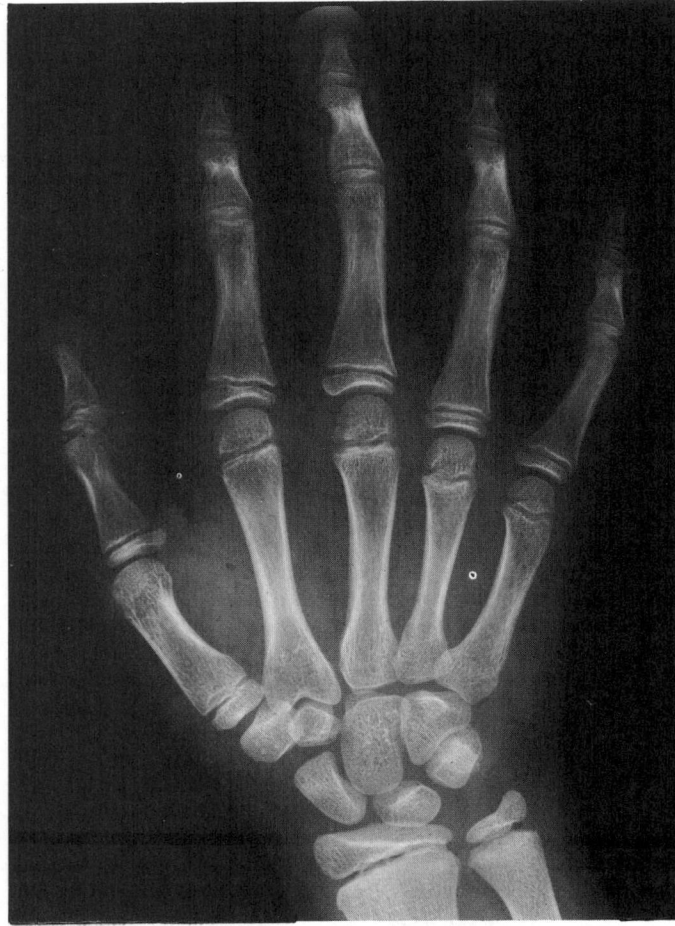

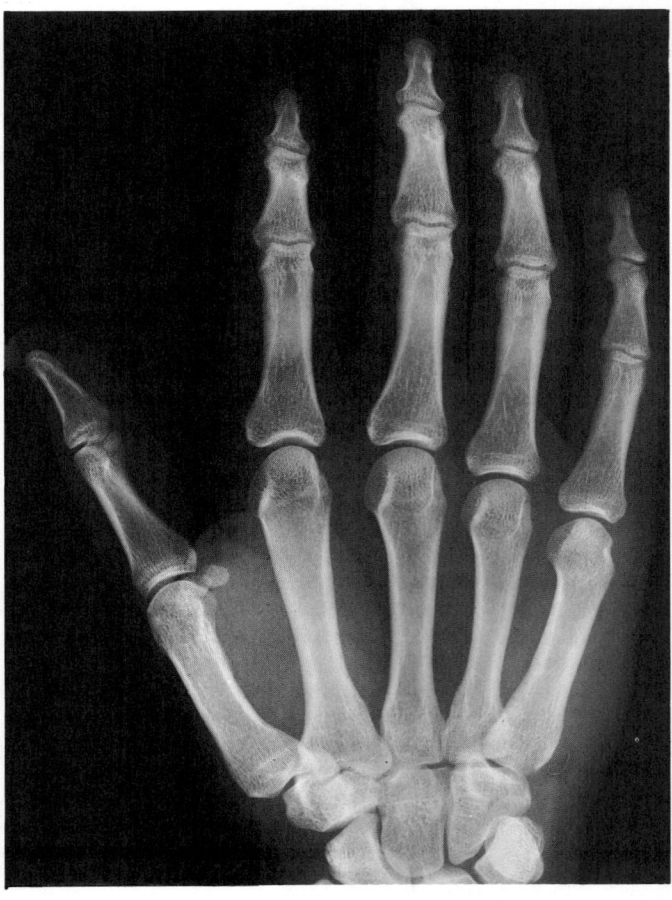

3.213 Radiograph of adult hand for comparison (male of 19 years). Note additional ossification in the sesamoid bones of the thumb.

3.212 Radiograph of a hand at 11 years (female). Note the maturing shapes of all the ossifications previously seen in **3.210** and **3.211**, with the addition of the pisiform bone.

week or later. Epiphysial centres appear in proximal phalanges early in the second year (females), and in its later months (males); in middle and distal phalanges in the second year (females), third or fourth (males). All epiphyses unite about the fifteenth to sixteenth year in females, seventeenth to eighteenth in males.

Applied anatomy. The scaphoid is the most frequently fractured carpal bone, the fracture usually crossing the long axis. Fractures of its proximal part or its 'waist' may fail to unite, possibly because the proximal fragment, devoid of nutrient foramina in about 13% (p. 417), has lost its blood supply. Ventral dislocation of the lunate bone is often associated with scaphoid fracture. The displaced lunate may compress the median nerve against the flexor retinaculum.

THE SKELETON OF THE LOWER LIMB

The Innominate Bone

The innominate or hip bone (**3.214**A,B,**215**A,B) is large, irregular, constricted centrally and expanded above and below. Its *lateral surface* has a deep, cup-shaped *acetabulum*, articulating with the femoral head. Distal (antero-inferior) to this is the large, oval or triangular *obturator foramen*. Above the acetabulum the bone widens into a plate with a sinuously curved *iliac crest*.

The bone articulates in front with its fellow, to form the pelvic girdle (p. 399). Each has three parts, **ilium, ischium** and **pubis**, connected by cartilage in youth but united as one bone in adults, the principal union being in the acetabulum (**3.214**B,**215**B). The *ilium* includes the upper acetabulum and expanded area above it; the *ischium* includes the lower acetabulum and bone postero-inferior to it; the *pubis* forms the anterior acetabulum, separating ilium from ischium, and the anterior median region where the pubes meet.

The Ilium

The ilium, so named because it supports the *flank*, may be described as having upper and lower parts and three surfaces. The smaller, lower part forms a little less than the upper two-fifths of the acetabulum, the upper part being much expanded, with gluteal, sacropelvic and iliac (internal) surfaces. The *gluteal surface*, posterolateral, is an extensive rough area; the anteromedial *iliac fossa* is smooth and concave; the *sacropelvic surface* is medial and is postero-inferior to the fossa, separated from it by the *medial border*.

The iliac crest, the ilium's superior border, is convex upwards but sinuously curved, internally concave in front, the reverse behind. Its ends project as anterior and posterior superior iliac spines. The *anterior superior iliac spine* is palpable at the lateral end of the inguinal fold; the *posterior superior iliac spine* is not palpable but often indicated by a dimple, about 4 cm lateral to the second sacral spine above the medial gluteal region (buttock). The crest has ventral and dorsal segments: the ventral is slightly more than the anterior two-thirds of the crest and its prominence is associated with changes in iliac form due to the emergence of the upright habit; the dorsal segment, about the posterior third in mankind, exists in all land vertebrates. The iliac crest's ventral segment has internal and external lips, the rough intermediate zone being narrowest centrally. The *tubercle of the crest* (**3.214**A) projects on the outer lip about 5 cm dorsosuperior to the anterior

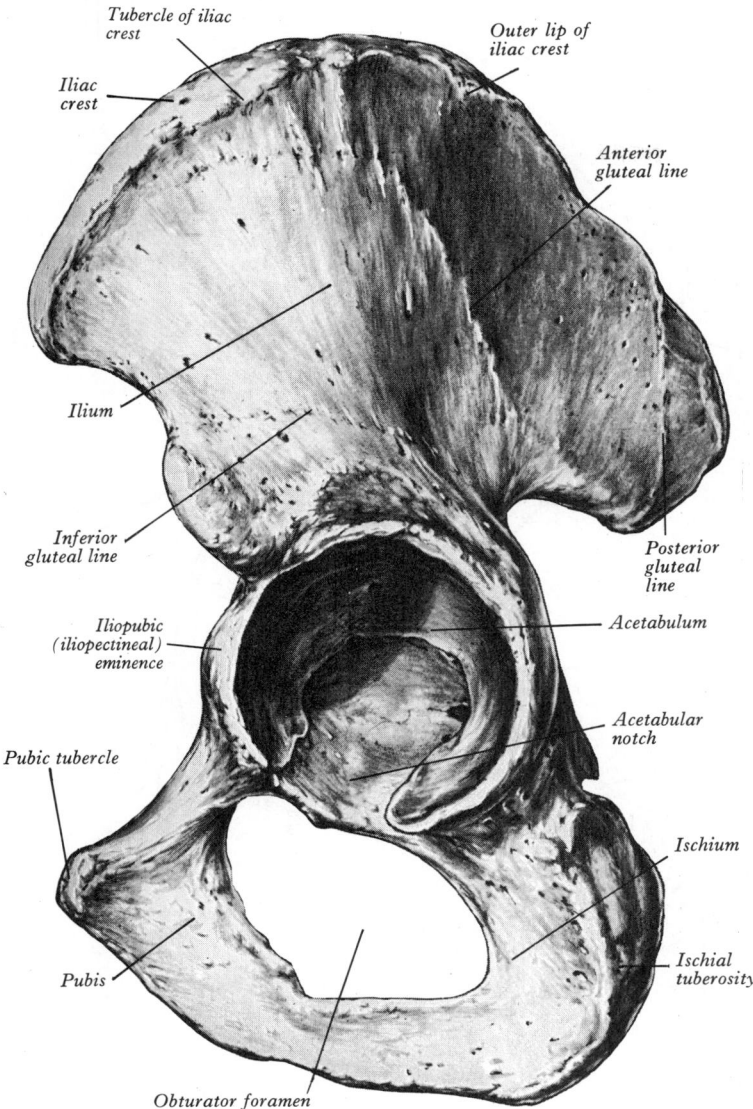

Tubercle of iliac crest

Iliac crest

Outer lip of iliac crest

Anterior gluteal line

Ilium

Inferior gluteal line

Posterior gluteal line

Iliopubic (iliopectineal) eminence

Acetabulum

Pubic tubercle

Acetabular notch

Ischium

Pubis

Ischial tuberosity

Obturator foramen

3.214A The left innominate bone: lateral (external) aspect. Compare with Key, **3.214B**.

superior spine. The dorsal segment has two sloping surfaces separated by a longitudinal ridge ending at the posterior superior spine. The crest's *summit*, a little behind its midpoint, is level with the interval between the third and fourth lumbar spines. The lower part of the ilium will be described with the acetabulum (p. 428).

The *iliac anterior border* descends to the acetabulum from the anterior superior spine. Superiorly it is concave forwards; inferiorly is a rough *anterior inferior iliac spine*, immediately above the acetabulum.

The *iliac posterior border*, irregularly curved (**3.215A,B**), descends from the posterior superior spine at first forwards, with a dorsal concavity forming a small notch. At the lower end of the notch is a wide, low projection, the *posterior inferior iliac spine*; here the border turns almost horizontally forwards for about 3 cm and finally down and back to join the posterior ischial border. Together they form a deep *greater sciatic notch*, bounded above by ilium, below by ilium and ischium (**3.215B**).

The *iliac medial border*, separating the iliac fossa and the sacropelvic surface, is indistinct near the crest, rough in its upper part, then sharp where it bounds an articular surface for the sacrum, and finally rounded. The latter part is the *arcuate line*, which inferiorly reaches the posterior part of the *iliopubic (iliopectineal) eminence*, marking the union of ilium and pubis.

The gluteal surface (**3.214A,B**), facing inferiorly in its dorsal part, laterally and slightly downwards in front, is bounded above by the iliac crest, below by the upper acetabular border and by anterior and posterior borders. It is rough and curved, convex in

front, concave behind, and marked by three gluteal lines. The *posterior gluteal line* is shortest, descending from the external lip of the crest about 5 cm in front of its posterior limit and ending in front of the posterior inferior spine. Above, it is usually distinct, but inferiorly ill-defined and frequently absent. The *anterior gluteal line*, the longest, begins near the midpoint of the superior margin of the greater sciatic notch and ascends forwards into the crest's outer lip, a little anterior to its tubercle. The *inferior gluteal line*, rarely well-marked, begins posterosuperior to the anterior inferior spine, curving back and downwards to end near the apex of the greater sciatic notch. Between the inferior gluteal line and the acetabular margin is a rough, shallow groove. Behind the acetabulum the lower gluteal surface is continuous with the posterior ischial surface, the union marked by a low elevation.

The iliac fossa, the internal concavity of the ilium, facing anterosuperior, is limited above by iliac crest, in front by the anterior border, behind by the medial border, separating it from the sacropelvic surface. It forms the smooth and gently concave posterolateral wall of the greater pelvis. Below it is continuous with a wide shallow groove (**3.215A**), which is bounded laterally by the anterior inferior spine and medially by the iliopubic eminence.

The sacropelvic surface (**3.215A**), the postero-inferior part of the medial iliac aspect, is bounded postero-inferiorly by the posterior border, anterosuperiorly by the medial border, posterosuperiorly by the iliac crest and antero-inferiorly by the line of fusion of ilium to ischium. It is divided into iliac tuberosity,

423

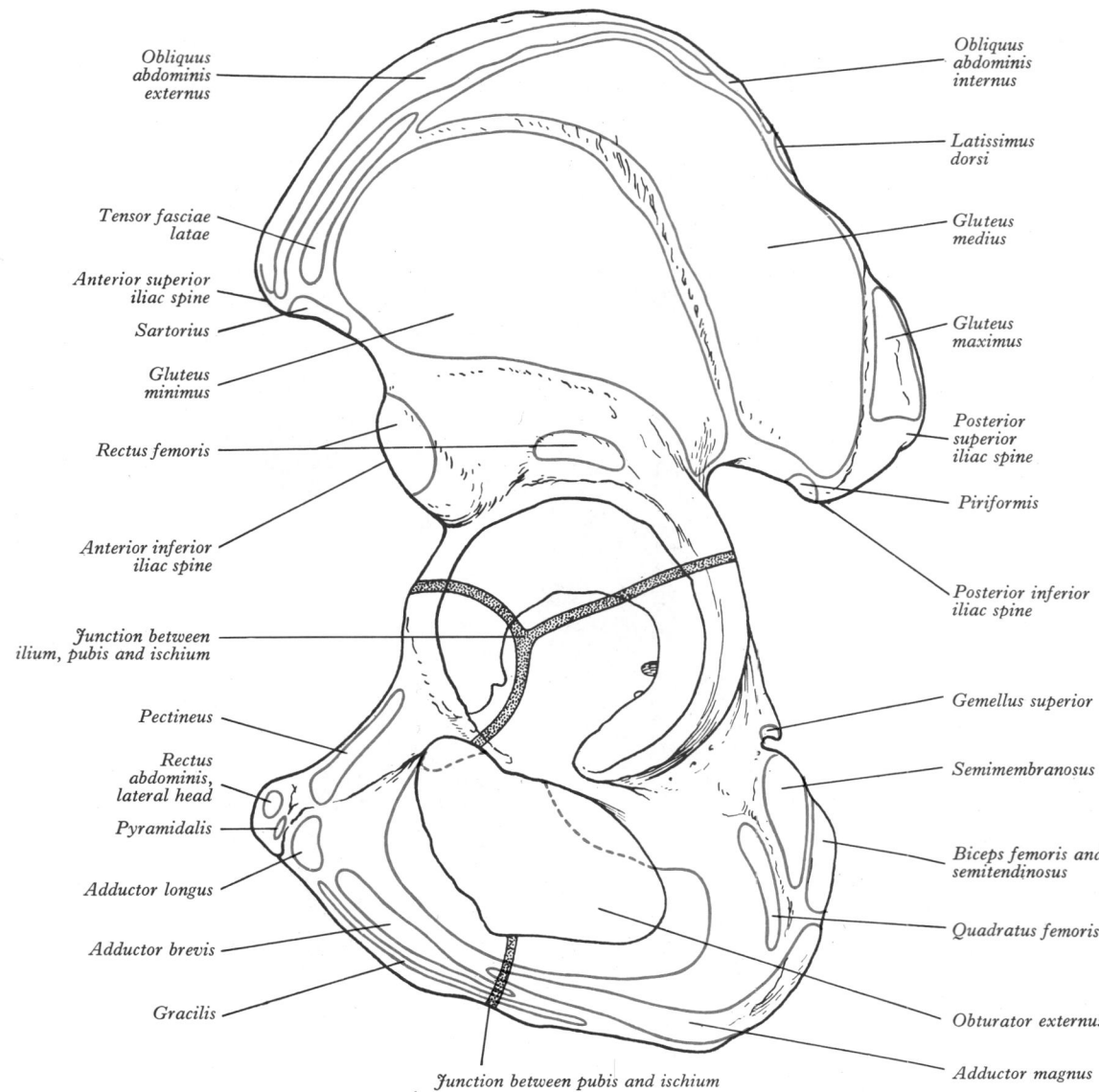

Obliquus abdominis externus

Obliquus abdominis internus

Latissimus dorsi

Tensor fasciae latae

Gluteus medius

Anterior superior iliac spine

Sartorius

Gluteus maximus

Gluteus minimus

Posterior superior iliac spine

Rectus femoris

Piriformis

Anterior inferior iliac spine

Posterior inferior iliac spine

Junction between ilium, pubis and ischium

Gemellus superior

Pectineus

Semimembranosus

Rectus abdominis, lateral head

Pyramidalis

Biceps femoris and semitendinosus

Adductor longus

Quadratus femoris

Adductor brevis

Gracilis

Obturator externus

Adductor magnus

Junction between pubis and ischium

3.214B Key to 3.214A the stippled bands indicate the limits of the iliac, pubic and ischial parts of the bone.

auricular and pelvic surfaces. The *iliac tuberosity*, a large, rough area below the dorsal segment of the iliac crest, shows cranial and caudal areas separated by an oblique ridge and connected to the sacrum by the interosseous sacroiliac ligament. The *auricular surface* (3.215A,B), immediately antero-inferior to the tuberosity, articulates with the lateral sacral mass. Shaped like an ear, its widest part is anterosuperior, its 'lobule' postero-inferior and on the medial aspect of the posterior inferior spine. Its edges are well-defined, but the surface, though articular, is rough and irregular. The *pelvic surface* is antero-inferior to the acutely recurved part of the auricular surface, contributing to the lateral wall of the lesser pelvis. Its upper part, facing down, is between the auricular surface and the upper limb of the greater sciatic notch; its lower region faces medially and is separated from iliac fossa by the arcuate line. Antero-inferiorly it extends to the line of union of ilium with ischium; this is usually obliterated, but passes from the depth of the acetabulum to, roughly, the middle of the inferior limb of the greater sciatic notch.

Further details. The iliac crest, approximating to the lower limit of the waist, is an attachment for lateral abdominal and dorsal muscles, fasciae and muscles of the lower limb (3.214B,215B). To the *outer lip* and *tubercle* of its ventral segment (p. 422) the fascia lata and iliotibial tract are attached, anterior to its tubercle tensor fasciae latae; to its anterior two-thirds, the lower fibres of the external oblique and, just behind its summit, lowest fibres of latissimus dorsi. A variable interval exists between

the most posterior attachment of external oblique and most anterior attachment of latissimus dorsi, and here the crest is the base of the lumbar triangle. The crest's *intermediate area* receives the internal oblique. To the anterior two-thirds of its *inner lip* transversus abdominis is attached, and behind this the lumbodorsal fascia and quadratus lumborum. To the crest's dorsal segment (p. 422), on its lateral slope, the highest fibres of gluteus maximus are attached; from its medial slope erector spinae and, along its medial margin, interosseous and dorsal sacro-iliac ligaments arise.

To the **anterior superior spine** is attached the lateral end of the inguinal ligament and, below it, the sartorius extends down the *anterior border*. The **anterior inferior spine** is divided indistinctly into an upper area for the straight part of rectus femoris and a lower area extending laterally along the upper acetabular margin to form a triangular impression for the iliofemoral ligament.

To the upper part of the **posterior border** upper fibres of the sacrotuberous ligament are attached and in front of the posterior inferior spine (i.e. on the upper border of the greater sciatic notch) fibres of piriformis. The superior rim of the notch is related to superior gluteal vessels and nerve. The lower part of the border (i.e. the lower margin of the greater sciatic notch) is covered by piriformis and related to the sciatic nerve, which, however, largely adjoins the ischium.

The gluteal surface is divided by the three gluteal lines into four areas (3.214A,B): (a) behind the posterior line, its upper

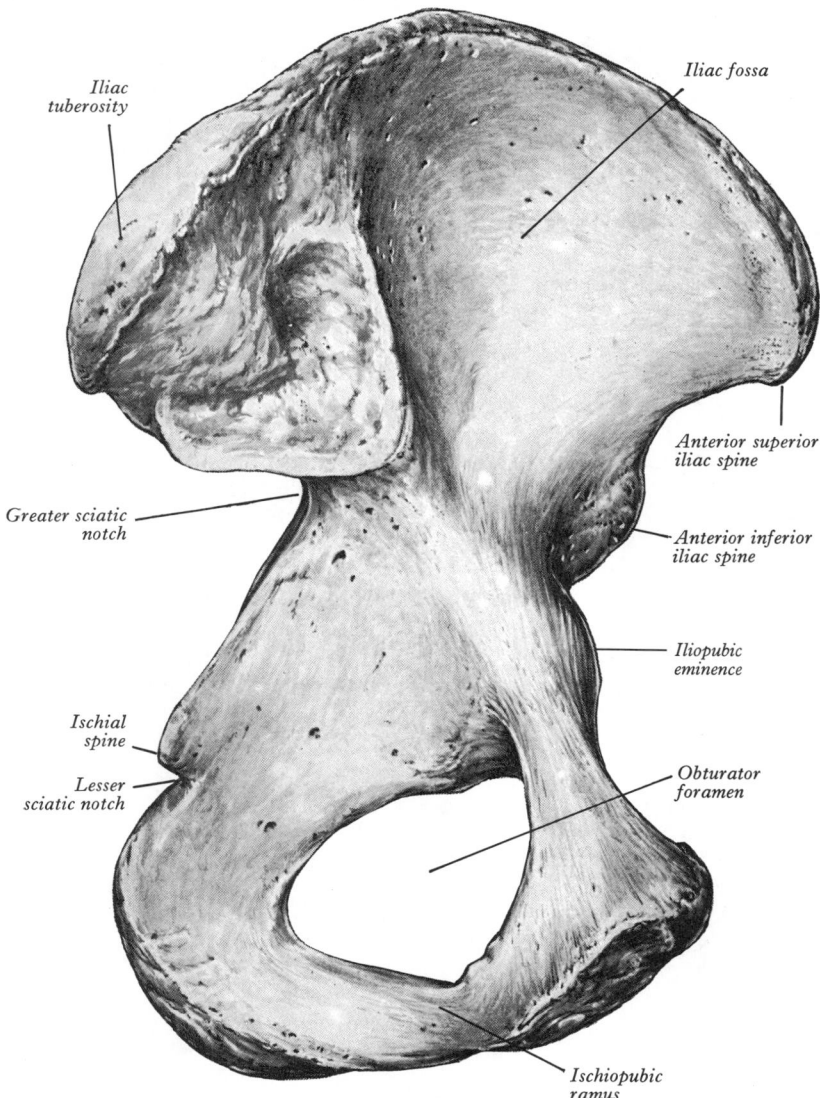

Iliac tuberosity

Iliac fossa

Greater sciatic notch

Anterior superior iliac spine

Anterior inferior iliac spine

Iliopubic eminence

Ischial spine

Lesser sciatic notch

Obturator foramen

Ischiopubic ramus

3.215A The left innominate bone: medial (internal) surface. Compare with Key, **3**.215B.

rough part being for upper fibres of the gluteus maximus and its lower, smooth region for part of the sacrotuberous ligament and iliac head of the piriformis; (*b*) between posterior and anterior lines, below the iliac crest, for the gluteus medius; (*c*) between anterior and inferior lines for the gluteus minimus; (*d*) below the inferior line, where there are many vascular foramina. Attached to a curved groove above the acetabulum is attached the reflected head of the rectus femoris and to an area adjoining the acetabular rim is the articular capsule; most of this area is covered by gluteus minimus. Postero-inferiorly, near the union of ilium and ischium, the bone is related to the piriformis. The vascular foramina on the iliac gluteal aspect may lead into large vascular canals in the bone (Sirang 1973).

The **iliac fossa**, in its upper two-thirds, provides attachment for the iliacus (**3**.215B), also related to the lower third; branches of the iliolumbar artery run between muscle and bone, one entering a large nutrient foramen often postero-inferior in the fossa. The wide groove between the anterior inferior spine and iliopubic eminence is occupied by converging fibres of iliacus laterally and the tendon of psoas major medially, the tendon separated from bone by a bursa. The right iliac fossa contains the caecum, and often the vermiform appendix and terminal ileum; the left one houses the end of the descending colon.

The **iliac tuberosity** (sacropelvic surface) gives attachment to dorsal sacroiliac ligaments and, behind the auricular surface, the interosseous sacroiliac ligament. To its anterior part is attached the iliolumbar ligament and above this is the medial part of the

quadratus lumborum. **The auricular surface** articulates with the sacrum and is reciprocally shaped (p. 516). To its sharp anterior and inferior borders the ventral sacroiliac ligament is attached. The narrow part of the **pelvic surface**, between the auricular surface and the upper rim of the greater sciatic notch, often shows a rough *preauricular sulcus* for the lower fibres of the ventral sacro-iliac ligament, more apparent in females. However, its unreliability in determination of sex was shown in a study of 237 Indian pelves (Jit & Gandhi 1966), but consult also Finnegan & Faust (1974) and Finnegan (1978). Lateral to the sulcus, piriformis is sometimes, in part, attached and, to the more extensive remainder of the pelvic surface, is attached part of the obturator internus (**3**.215B).

The Pubis

The pubis is the ventral part of the innominate bone and forms a median cartilaginous *pubic symphysis* with its fellow. From its anteromedial body a superior ramus passes up and back to the acetabulum and an inferior ramus passes back, down and laterally to join the ischial ramus inferomedial to the obturator foramen.

The body, dorsoventrally compressed, has anterior, posterior and symphysial (medial) surfaces and an upper border, the *pubic crest*. The *anterior surface* also faces *inferolaterally*; it is rough superomedially and elsewhere smooth, affording attachment for medial femoral muscles. The smooth *posterior surface* faces   425

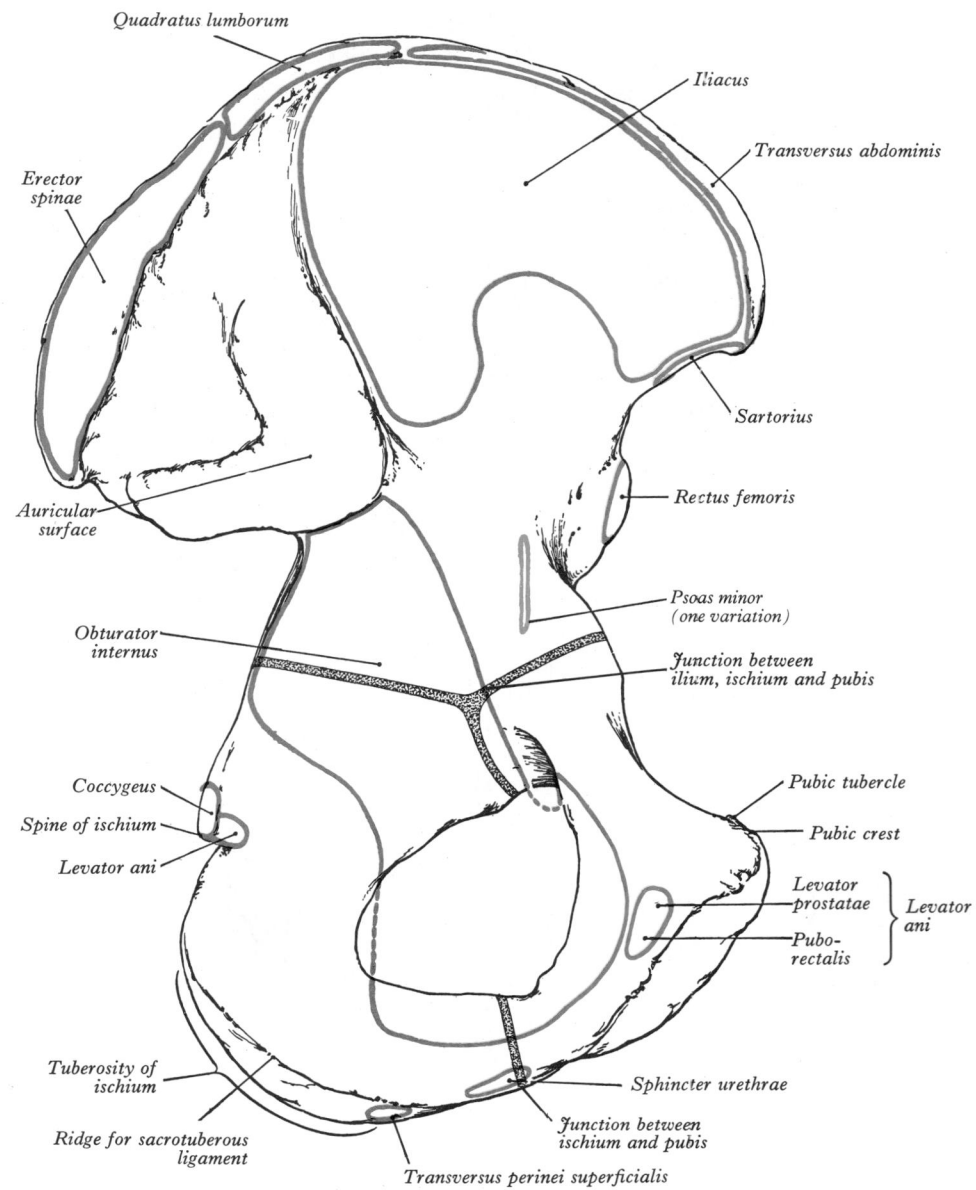

Quadratus lumborum

Iliacus

Transversus abdominis

Erector
spinae

Sartorius

Rectus femoris

Auricular
surface

Psoas minor
(one variation)

Obturator
internus

Junction between
ilium, ischium and pubis

Coccygeus

Pubic tubercle

Spine of ischium

Pubic crest

Levator ani

Levator
prostatae

Levator
ani

Pubo-
rectalis

Tuberosity of
ischium

Sphincter urethrae

Ridge for sacrotuberous
ligament

Junction between
ischium and pubis

Transversus perinei superficialis

3.215B Key to **3.215A** the stippled bands indicate the limits of the iliac, pubic and ischial parts of the bone.

upwards and backwards as the oblique anterior wall of the minor pelvis and is related to the urinary bladder. The *symphysial surface* is elongate and oval, united by cartilage to its fellow at the pubic symphysis. Denuded of cartilage it has, in the elderly, an irregular surface of small ridges and furrows or nodular elevations, varying much with age (Todd 1920, 1921, Brooks 1955, McKern & Stewart 1957, Suchey et al 1979), features of obvious forensic value. Pal & Tamankar (1983), in a series of 60 male and 60 female pubes of Northern Indians, aged from 14 to 60 years, found that accepted data gave inaccurate values, some 'over-ageing' their specimens, some the reverse. The *pubic crest* is the rounded upper border of the body, which overhangs the anterior surface (3.214A), its lateral end the rounded *pubic tubercle*. Both crest and tubercle are palpable, the latter partly obscured in males by the spermatic cord, crossing above it from scrotum to abdomen. The *pubic rami* diverge posterolaterally from the lateral corners of the body.

The superior pubic ramus passes upwards, backwards and laterally from the body, superolateral to the obturator foramen to reach the acetabulum. Triangular in section, it has three surfaces and borders. Its anterior, *pectineal surface*, tilted slightly up, is triangular in outline and extends from the pubic tubercle to the iliopubic eminence (3.215A). It is bounded in front by the rounded *obturator crest* and behind by the sharp *pecten pubis (pectineal*

line*)* which, with the crest, is the pubic part of the *linea terminalis* (i.e. anterior part of pelvic brim). The dorsosuperior, *pelvic surface*, medially inclined, is smooth and narrows into the posterior surface of the body, bounded above by the pecten pubis and below by a sharp *inferior border*. The *obturator surface*, directed down and back, is crossed by the *obturator groove* sloping down and forwards; its limit in front is the *obturator crest* and, behind, the inferior border.

The inferior pubic ramus, an inferolateral process of the body, descends inferolaterally to join the ischial ramus medial to, and below, the obturator foramen. The union may be locally thickened, but not obviously so in adults. The ramus has two surfaces and borders. The *antero-external surface*, continuous above with that of the pubic body, faces the thigh and is marked by muscles, its lateral limit being the margin of the obturator foramen and, medially, the rough anterior border. The *postero-internal surface* is continuous above with that of the body and transversely convex, its medial part often everted in males (3.223) and connected to the crus penis. This surface faces the perineum medially, its smooth lateral part tilted up towards pelvic cavity.

The pubic tubercle is a medial attachment of the inguinal ligament in the floor of the superficial inguinal ring and is crossed by the spermatic cord; ascending loops of cremaster are also attached to it. Lateral on the **pubic crest** are attached the lateral

426

part of the rectus abdominis and, below it, the pyramidalis. Medially the crest is crossed by the medial part of the rectus abdominis, ascending from ligamentous fibres interlacing in front of the pubic symphysis.

The *anterior surface* of the pubic body faces the femoral adductor region; on its medial part and to a rough strip, which is wider in females, the ventral pubic ligament is attached. In the angle between its upper end and the pubic crest the tendon of the adductor longus is attached and below this, to a line near the medial border extending down to the inferior ramus, is attached the gracilis. Lateral to the gracilis the adductor brevis is attached to the body and inferior ramus. Obturator externus is attached laterally to the anterior surface, spreading to both rami (3.214B).

The *posterior pubic surface* is separated from the urinary bladder by the retropubic fat; near its centre anterior fibres of the levator ani are attached, more laterally the obturator internus, and this extends to both rami. Medial to the levator ani the puboprostatic ligaments are attached.

To the *pectineal surface of the superior ramus* along its upper part the pectineus is attached, covering the rest of the surface (3.214B).

The pecten pubis is the sharp, superior edge of the pectineal surface. Attached at its medial end are the conjoint tendon and lacunar ligament, along the rest of it a strong fibrous *pectineal ligament* (p. 597) and, near its centre, psoas minor when present. The smooth *pelvic surface* is separated from parietal peritoneum only by areolar tissue, in which the lateral umbilical ligament descends forwards across the ramus and, laterally, the ductus deferens passes backwards. The *obturator groove*, converted to a canal by the upper borders of the obturator membrane and obturator muscles, transmits obturator vessels and nerve from pelvis to thigh. To the lateral end of the *obturator crest* (3.215B) some fibres of the pubofemoral ligament are attached.

To the *external surface of the inferior ramus* are attached gracilis, adductor brevis and obturator externus, in medio-lateral order. Also, the adductor magnus usually extends from the ischial ramus on to the lower part of the inferior pubic ramus between adductor brevis and obturator externus. The *internal surface* is indistinctly divided into medial, intermediate and lateral areas. The medial area faces inferomedially in direct contact with the crus penis, limited above and behind by an indistinct ridge for attachment of the inferior fascia of the urogenital diaphragm (p. 606). To the intermediate area, related to the dorsal penile nerve, internal pudendal vessels and their fascial sheath, may be attached some inner fibres of sphincter urethrae, and to the lateral area, fibres of obturator internus. The *medial margin* of the ramus, strongly everted in males, is an attachment of the fascia lata and the membranous layer of the superficial perineal fascia.

The Ischium

The ischium, inferoposterior part of the innominate bone, has a body and ramus, the body having upper and lower ends and femoral, dorsal and pelvic surfaces (3.214–216). Above, it forms the inferoposterior part of the acetabulum; below, its ramus ascends anteromedially at an acute angle to meet the descending pubic ramus, completing the obturator foramen.

The *femoral surface*, facing down, forwards and laterally towards the thigh, is bounded in front by the margin of the obturator foramen and a lateral border, indistinct above but well-defined below, forming the lateral limit of the ischial tuberosity. The *dorsal surface*, facing superolaterally, is continuous above with the iliac gluteal surface, and here a low convexity follows the acetabular curvature. This surface is inferiorly the upper part of the ischial tuberosity, above which is a wide, shallow groove on its lateral and medial aspects. The *ischial tuberosity* (3.216) is a large, rough area on the ischium's lower dorsal surface and inferior extremity. Though obscured by gluteus maximus in extension, it is palpable in flexion. It is 5 cm from midline and about the same distance above the gluteal fold (p. 637). Elongate, widest above, tapering inferiorly, it is the attachment of the posterior femoral muscles. The ischial dorsal aspect is between the lateral and posterior borders. The *posterior border* blends above with that of the ilium, helping to complete the inferior rim of the *greater sciatic*

notch, the posterior end of which has a conspicuous *ischial spine*. Below this the rounded border is the floor of the *lesser sciatic notch*, between ischial spine and tuberosity. The *pelvic surface* is smooth and faces the pelvic cavity; below this is part of the lateral wall of the ischiorectal fossa.

The ischial ramus has anterior and posterior surfaces continuous with those of the inferior pubic ramus; the former is really anteroinferior and roughened by the attachment of medial femoral muscles. The smooth *posterior surface* is partly divided into perineal and pelvic areas, like the inferior pubic ramus. The *upper border* completes the obturator foramen; the rough *lower border*, together with the medial border of the inferior pubic ramus, bounds the subpubic angle and pubic arch.

Further details. Attached below to the *femoral surface* of the **ischial body** is part of the obturator externus (3.214B) and along the upper part of the lateral border of the ischial tuberosity is the quadratus femoris. Below the acetabulum, to the lateral border, the ischiofemoral ligament is attached. Above the ischial tuberosity the *dorsal surface* is crossed by the tendon of obturator internus and the gemelli; between them and bone is the nerve to the quadratus femoris. At a higher level the femoral surface is covered by piriformis, partially separated by the sciatic nerve and the nerve to the quadratus femoris.

The ischial tuberosity is divided nearly transversely into upper and lower areas (3.216), the upper subdivided by an oblique line into a superolateral part for semimembranosus and an inferomedial for the long head of biceps femoris and semitendinosus; the lower region narrowing as it curves on to the inferior ischial aspect is subdivided by an irregular vertical ridge into lateral and medial areas. The larger lateral area is for part of the adductor magnus, the medial is covered by fibro-adipose tissue, usually containing the ischial bursa of gluteus maximus and supporting the body in sitting. Medially the tuberosity is limited by a curved ridge passing on to the ramus, to which the sacrotuberous ligament and its falciform process are attached (3.215B). Many fibres of biceps femoris pass into the ligament, an interesting fact, since the sacrum and posterior part of the ilium are primitive mammalian attachments of biceps femoris—the tuberosity being a secondary attachment, the ligament representing, at least in part, remains of primitive tendon.

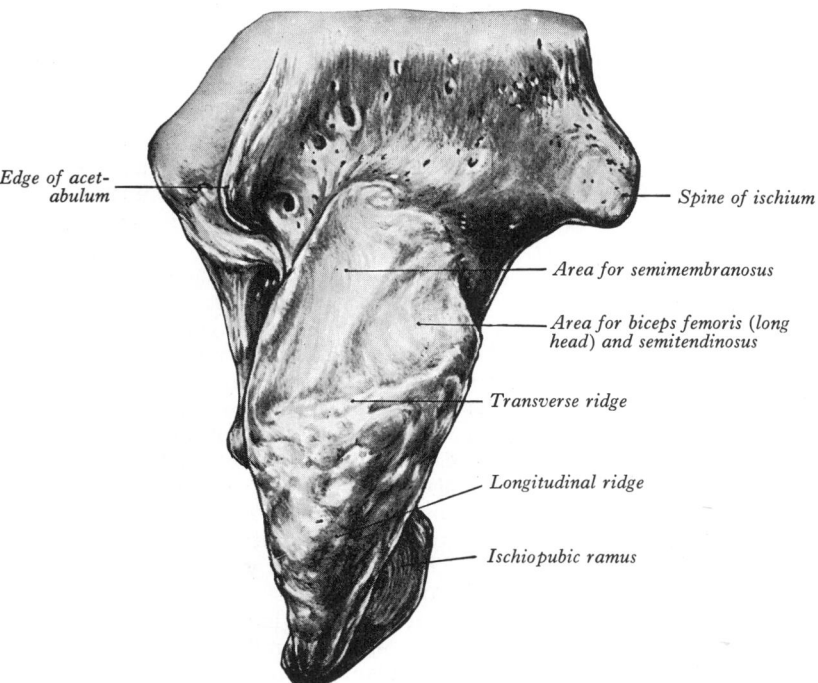

Edge of acet-
abulum — Spine of ischium

— Area for semimembranosus

— Area for biceps femoris (long head) and semitendinosus

— Transverse ridge

— Longitudinal ridge

— Ischiopubic ramus

3.216 The left ischial tuberosity: posterior aspect. The *transverse ridge* forms the lower boundary of the area for the hamstring muscles and separates it from the lower half of the tuberosity, which is divided into lateral and medial areas by the *longitudinal ridge*. To the lateral area is attached the adductor magnus; the medial area is covered with fibroadipose tissue and supports the body in the sitting posture.

Superomedial to the tuberosity the posterior surface has a wide, shallow groove, usually covered by hyaline cartilage, with a bursa between it and the tendon of obturator internus. To the lower margin of the groove, near the tuberosity, the gemellus inferior is attached, and to the upper margin, near the ischial spine, gemellus superior.

The ischial spine projects down and a little medially. To its margins is attached the sacrospinous ligament, separating the greater from the lesser sciatic foramen (4.56). It is crossed dorsally by internal pudendal vessels and nerve to the obturator internus. Its pelvic surface is an attachment of coccygeus (coextensive with the sacrospinous ligament) and the most posterior fibres of the levator ani. Various structures traverse these foramina (p. 517).

To the smooth *pelvic ischial surface*, in its upper part, is attached obturator internus, converging on the lesser sciatic notch (foramen) and covering the rest of this surface, except the pelvic aspect of the ischial spine. The muscle and its fascia separate bone from the ischiorectal fossa.

The *anterior surface* of the *ischial ramus* faces the adductor region. The obturator externus above, anterior fibres of the adductor magnus and, near the lower border, gracilis are all attached here. Between adductor magnus and gracilis the attachment of adductor brevis may descend from the inferior pubic ramus. The *posterior surface* is divided into pelvic and perineal areas; the former, facing back, has part of the obturator internus attached to it; the perineal area faces medially, its upper part related to crus penis or clitoridis, with sphincter urethrae attached to it, and below this ischiocavernosus and transversus superficialis perinei. The inferior fascia of the urogenital diaphragm is attached below the ridge between perineal and pelvic areas and above the areas for the penile crus and sphincter urethrae. The *lower border* of the ramus is an attachment of fascia lata and a membranous layer of the superficial perineal fascia.

The acetabulum (3.214A), an approximately hemispherical cavity central on the lateral aspect of the innominate bone, faces antero-inferiorly; it is surrounded by an irregular margin deficient *inferiorly* at the *acetabular notch*. The *acetabular fossa* is the cavity's central floor, which is rough and non-articular and has an articular *lunate surface*, widest above, where weight is transmitted to the femur. On this crescentic surface, covered with cartilage, the head of the femur slides. All three innominate elements contribute to the acetabulum in man, but unequally. The pubis forms the anterosuperior fifth of the articular surface, the ischium forms the fossa's floor and rather more than the postero-inferior two-fifths of articular surface, and the ilium the remainder. A linear defect may cross the acetabular surface from *superior* border to acetabular fossa, but does *not* follow any junction between the main morphologic parts of the innominate bone.

The obturator foramen, below and slightly anterior to the acetabulum, is between the pubis and ischium. It is rimmed above by the grooved obturator surface of the superior pubic ramus, medially by the pubic body and its inferior ramus, below by the ischial ramus, laterally by the anterior border of the ischial body, including the margin of the acetabular notch. The foramen is almost closed by the *obturator membrane* which is attached to its margins, except above, where a communication remains between pelvis and thigh; this free edge is attached to an *anterior obturator tubercle* at the anterior end of the the inferior border of superior pubic ramus, and a *posterior obturator tubercle* on the anterior border of the acetabular notch; these tubercles are sometimes indistinct. Since the tubercles lie in different planes and the obturator groove crosses the foramen's upper border, the acetabular rim is in fact a spiral. The foramen is large and oval in males, but smaller and nearly triangular in females.

Structure. The thicker parts of the innominate are trabecular, encased by two layers of compact bone; the thinner parts, e.g. in the acetabulum and central iliac fossa, are often translucent, consisting of one lamina of compact bone. In the upper acetabulum and along the arcuate line, i.e. the route of weight transmission from sacrum to femur, compact bone is increased and subjacent trabecular bone displays two sets of pressure lamellae. They start together near the upper auricular surface and diverge to impinge on two strong buttresses of compact bone, whence two similar sets

of lamellar arches start and converge on the acetabulum (Wakeley 1929). Because of frequent use for biopsy puncture the ventral iliac crest has been much studied as regards distribution of cortical and trabecular bone. Whitehouse (1977) has surveyed these studies; his own observations, by scanning electron micrography, indicate that the cortical bone is very porous, being only 75% bone, and merely 35% near the anterior superior iliac spine; denser cortical bone commences at margins of the crest, thickening rapidly below it on both aspects of the iliac squama.

Ossification. This is by three primary centres, for ilium, ischium and pubis; the iliac appears above the greater sciatic notch prenatally about the eighth week, the ischial in its body in the fourth month, and the pubic in its superior ramus between the fourth and fifth months. At birth some parts are still cartilaginous: the whole iliac crest, the acetabular floor and inferior margin (3.217); the acetabulum is still a cartilaginous cup with a triradiate stem extending medially to the pelvic surface as a Y-shaped epiphysial plate between ilium, ischium and pubis (Harrison 1957), including also the anterior inferior iliac spine. Cartilage along the inferior margin also covers the ischial tuberosity, forms (temporarily) conjoined ischial and pubic rami and continues to the pubic symphysial surface and along the pubic crest to its tubercle.

The ossifying ischium and pubis fuse to form a continuous ramus at the seventh or eighth year (3.218). Secondary centres appear about puberty and join between fifteenth and twenty-fifth years. There are usually two for the iliac crest, but these rapidly fuse, and two for the acetabular cartilage, occurring between ilium and pubis, and ilium and ischium. These two centres fuse (*os acetabuli*) as ossification spreads to include the whole stem of triradiate cartilage, forming a substantial part of the articular surface along the junction of the triradiate stem, cartilage cup and around its periphery. The anterior inferior iliac spine may be ossified from the triradiate centre or a separate one. Cartilage at the inferior acetabular margin begins to ossify over the ischial tuberosity, spreading forwards. The part ossifying first fuses almost at once, before more anterior parts have ossified, and fusion also extends forwards. The pubic tubercle, crest and symphysial surface may have separate centres.

The Skeletal Pelvis

The term *pelvis*, 'a basin', is vaguely applied to the skeletal ring formed by innominate bones and sacrum, the cavity within them and even the entire region where trunk and lower limbs meet. It

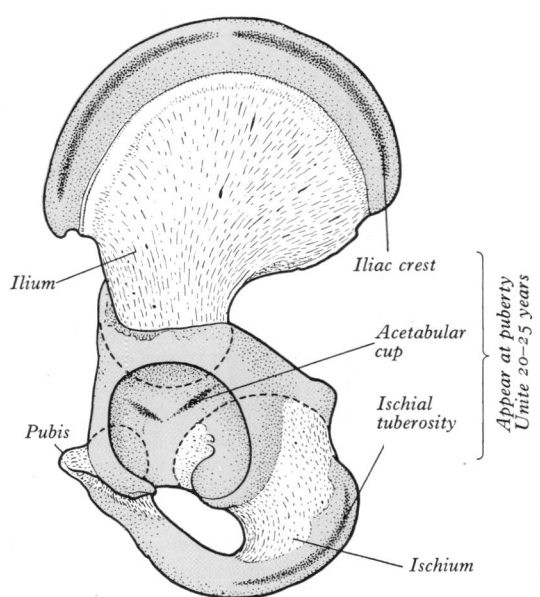

Ilium

Iliac crest

Acetabular cup

Appear at puberty
Unite 20–25 years

Pubis

Ischial tuberosity

Ischium

3.217 The innominate bone at birth. More heavily stippled areas indicate the secondary centres of ossification and the Y-shaped medial extension of the cartilaginous acetabular cup is indicated by interrupted lines. The ossified part of the ischium is shown in the floor of the acetabulum.

is used here in the skeletal sense, for the irregular osseous girdle between femoral heads and fifth lumbar vertebra. It is massive, because its primary function is to withstand compression and other forces due to body weight and powerful musculature, mechanisms considered elsewhere (p. 518). Here, we are concerned with metrical and other features of sexual significance, and hence of obstetric, forensic and anthropological application.

The pelvis can be regarded as having greater and lesser segments, the true and false pelves, arbitrarily divided by an oblique plane, passing through the sacral promontory behind and *lineae terminales* elsewhere. Each linea terminalis includes the iliac arcuate line (p. 423), pecten (iliopectineal line) and pubic crest. But the segments are continuous, and parts of the body cavity which they enclose are also continuous through the superior pelvic aperture or pelvic inlet (3.219).

The greater pelvis consists of iliac flanges above the lineae terminales and the sacral base. Structure of this junctional zone is massive, forming powerful arches from acetabular fossae to vertebral column around the visceral cavity which is, of course,

part of the abdomen; because of the pelvic inclination (vide infra) it has little anterior wall.

The lesser pelvis encloses a true basin when soft tissues of the pelvic floor are in place. Skeletally it is a narrower continuation of greater pelvis, with irregular but more complete walls around its cavity. Naturally of great obstetric importance, it has a *median curved axis*. It has superior and inferior openings, the superior occupied in life by viscera, the inferior largely closed by the pelvic floor and its sphincters.

The superior pelvic aperture (pelvic inlet), variable in contour—round or oval—is encroached upon by the sacral promontory. Its boundary, described as *pelvic brim*, is obstetrically important (3.219 and vide infra) and has also long been measured for anthropological reasons, as has the pelvic cavity; much information is available, especially for females. Data of different observers vary, being founded on differing racial and economic groups. Data cited here are merely approximate samples of values in Europeans (Martin 1928). Three dimensions of the *pelvic inlet* have become conventional:

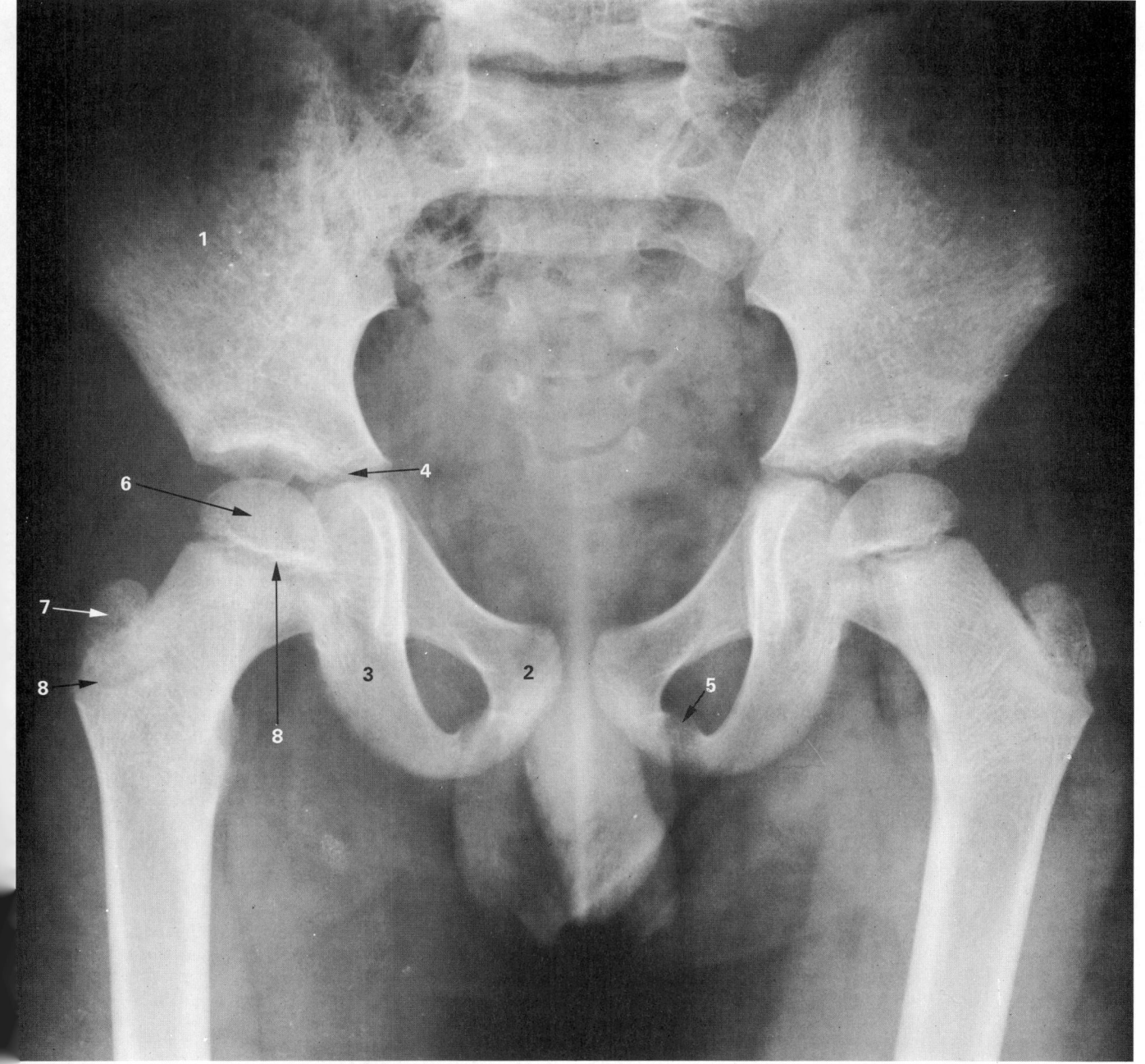

3.218 Anteroposterior radiograph of the pelvis of a boy aged seven. 1. Ilium. 2. Pubis. 3. Ischium. 4. Part of triradiate growth cartilage. 5. Cartilage between pubic and ischial rami. 6. Superior femoral epiphysis. 7. Ossifying greater trochanter. 8. Cartilaginous growth plates.

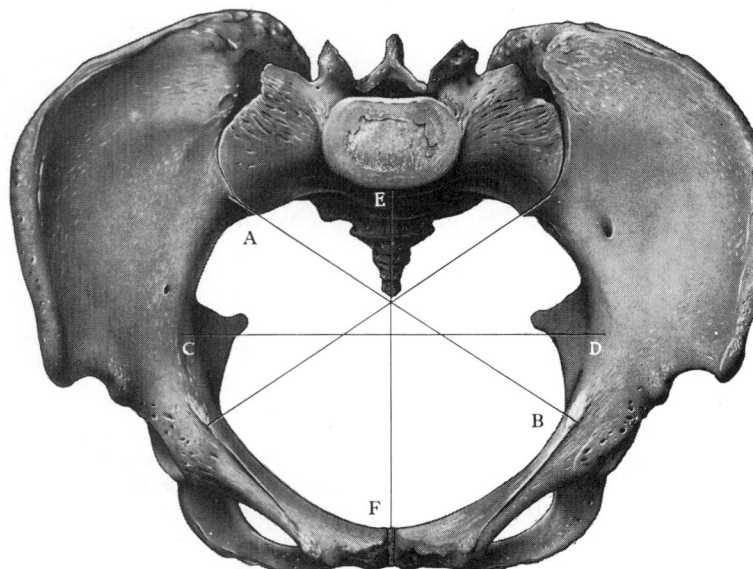

3.219 The diameters of the superior aperture of the lesser pelvis (female). A. sacroiliac joint; B. iliopubic eminence; C,D. middle of pelvic brim; E. sacral promontory; F. pubic symphysis.

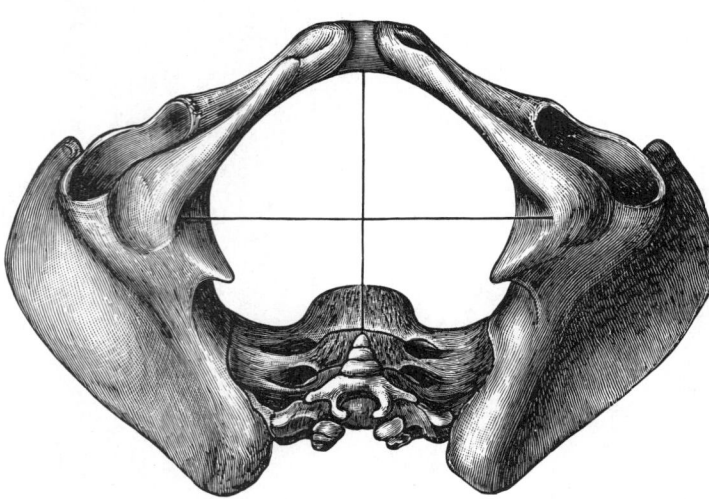

3.220 The diameters of the inferior aperture of the lesser pelvis (female). The oblique diameter is not shown.

A. *Anteroposterior diameter* (true conjugate) is measured between midpoints of the sacral promontory and upper border of symphysis pubis (male 100 mm, female 112 mm).
B. *Transverse diameter* is the maximum distance between similar points (assessed by eye) on opposite sides of the pelvic brim (male 125 mm, female 131 mm).
C. *Oblique diameter* is measured from iliopubic eminence to opposite sacroiliac joint (male 120 mm, female 125 mm).

The cavity of the lesser pelvis, short and curved, is markedly longer in its posterior wall. Antero-inferiorly it is bounded by pubic bones, their rami and symphysis; posteriorly by the concave anterior sacral surface and coccyx; and on each side by the smooth quadrangular pelvic aspect of fused ilium and ischium. The region thus enclosed is the pelvic cavity proper, through which pass, in all land vertebrates, the ends of the alimentary canal and urogenital ducts. Thus in humans its contents are rectum, bladder and parts of the reproductive organs. The rectum is posterior, bladder anterior, uterus intermediate. The cavity must also permit passage of the fetal head.

Pelvic diameters are often measured and while this can be done at many levels, measurements are usually chosen at approximately midlevel.

A. *Anteroposterior diameter* is measured between midpoints of the third sacral segment and posterior surface of symphysis pubis (male 105 mm, female 130 mm).

B. *Transverse diameter* is the widest transverse distance between side walls of the cavity, and often the greatest transverse dimension in the whole cavity (male 120 mm, female 125 mm).
C. *Oblique diameter* is the distance from lowest point of one sacroiliac joint to midpoint of contralateral obturator membrane (male 110 mm, female 131 mm).

The inferior pelvic aperture (pelvic outlet) (3.220), less regular in outline than the superior, is indented behind by coccyx and sacrum and bilaterally by ischial tuberosities. Its perimeter thus consists of three wide arcs; anterior is the *pubic arch*, between the converging ischiopubic rami; between the sacrum and coccyx posteriorly and the ischial tuberosities laterally are two large sciatic notches, divided, on both sides, by sacrotuberous and sacrospinous ligaments into *greater* and *lesser sciatic foramina* (p. 517). With ligaments included, the inferior aperture is rhomboidal, its anterior limbs being ischiopubic rami (joined by the inferior pubic ligament), its posterior the sacrotuberous ligaments, with the coccyx median. The outlet is thus not rigid in its posterior half, being limited by ligaments and coccyx, all slightly yielding. Even with the sacrum taken as the posterior midline limit (more reliable for measurement), there remains slight mobility at the sacroiliac joints. Note also that a plane of the inferior aperture is merely conceptual: the anterior, ischiopubic part has a plane inclined down and back to a transverse line between the lower limits of the ischial tuberosities, and the posterior half has a plane approximating to sacrotuberous ligaments, sloping down and forwards to the same line. Dimensions of the inferior aperture are measured as follows:

A. *Anteroposterior diameter* is usually measured from coccygeal apex to the midpoint of the lower rim of the symphysis. The lowest sacral point may also be used (male 80 mm, female 125 mm).
B. *Transverse diameter* is measured between ischial tuberosities at the lower borders of their medial surfaces. Hence the term *bituberous diameter* (male 85 mm, female 118 mm).
C. *Oblique diameter* extends from the midpoint of sacrotuberous ligament on one side to the contralateral ischiopubic junction (male 100 mm, female 118 mm).

Apart from these main measurements, by consensus the basis of pelvic osteometry, other planes and measurements are used in obstetric practice. *Plane of greatest pelvic dimensions*, an obstetrical concept, represents the most capacious pelvic level, between pelvic brim and midlevel plane, corresponding with the latter anteriorly (midsymphysis) but ascending slightly to the disc between second and third sacral segments. *Plane of least pelvic dimensions* is said to be at about midpelvic level; its transverse diameter is between apices of ischial spines; most difficulty in parturition occurs here. *Posterior median diameter* is the part of any anteroposterior diameter behind its intersection with transverse diameter, and thus assessable at any level or plane as an indicator of capacity of the posterior pelvic segment. Because direct measurement of osseous dimensions is impracticable in living patients, except by radiological methods, indirect measurements may be used.

The *diagonal (oblique) conjugate* is the distance from the sacral promontory to the lower symphysial border, measured *per vaginam*, and a guide to the 'true' conjugate. This, like all pelvic diameters described anatomically, largely disregards soft tissues. The *intercristal* and *interspinous diameters*, respectively the greatest widths between iliac crests and anterior superior iliac spines, are sometimes compared in females. Average values are (intercristal) 250 mm and (interspinous) 275 mm, the difference between them being an indication of width of pelvic cavity; iliac crests do not turn medially at their anterior ends to the same extent in females as males. The method's value is dubious and correlation between such sexual characteristics in the pelvis is uncertain.

Morphological Classification of Pelves

Interest in the above dimensions is primarily obstetric and, less frequently, forensic. All pelvic measurements display, as other elsewhere, individual variation; values quoted are means from limited surveys. Sexual and racial differences also occur. The range in any group, for both sexes, is large when adequately assessed. Measurements are on skeletonized pelves in the most extensive studies; such data are not only of anthropological interest but also the basis of attempts to define clinically identifiable forms of the pelvis without resort to the full range of measurements, which are not all possible in the living, although radiological pelvimetry has become a refined technique (Borell & Fernström 1960). In the female patient it is *her* measurements which are significant, not average values; comparison between her dimensions and a fetal head are the principal concern. These values can satisfactorily be assessed only by radiographic techniques. Hence modern pelvimetry and fetal cephalometry have replaced obstetric measurement by calipers whenever requisite expertise is available (Clyne 1963, Lewis 1964).

Pelvic measurements, however obtained, have been analysed by many anatomists, anthropologists, obstetricians and radiologists in attempts to classify human pelves, especially the female. *Pelvic brim index*—(ant-post./trans. diameters) × 100—was an early attempt to define the shape of the pelvic cavity (Turner 1886). Like the cephalic index (p. 396) its range is divided into steps, by which pelves are classified as *platypellic* (transversely flat), *mesatipellic* (intermediate) and *dolichopellic* (anteroposteriorly long); on anatomical and radiological data a *brachypellic* form is added, also known as *android* and responsible for most cases of severe obstruction in childbirth (Greulich & Thoms 1938, 1939, Thoms 1940). Such classification is equally applicable to both sexes and various ages, e.g. in one large series, children and males were predominantly dolichopellic, females mostly mesati- and brachypellic. Platypellic pelves are comparatively rare in all series. Another classification depends jointly on anatomical and radiological data, the latter including qualitative and metrical observations (Caldwell & Moloy 1933, Caldwell et al 1940). This divides the superior pelvic aperture by its transverse diameter into an anterior *forepelvis* and a *hindpelvis*, the latter more variable in shape and capacity. The slope of pelvic walls, whether radiographically straight, convergent or divergent, pelvic depth and shape and size of the greater sciatic notch (lateral radiographs) and shape of the subpubic arch are all considered. This study led to wide acceptance of four pelvic types (shown in the table below and **3.225**) depending fundamentally on skeletal measurement, although frequently applied somewhat subjectively in clinical practice. Though intermediate forms exist, the classification is claimed to embrace all but a very small percentage. Basic skeletal mean values in the original study were as follows:

PELVIC TYPE	CON-JUGATE DIA. (mm)	TRANS-VERSE DIA. (mm)	WHITE FEMALES (%)	NEGRO FEMALES (%)
Gynaecoid (=mesatipellic)	108·5	137·6	41·4	42·1
Android (=brachypellic)	105·9	135·6	32·5	15·7
Anthropoid (=dolichopellic)	117·5	129·4	23·5	40·5
Platypelloid (=platypellic)	85·5	144·5	2·6	1·7

Apart from racial differences (e.g. high rate of anthropoid pelves in Negro women), such observations emphasize the low incidence of the platypelloid form. Children (under nine years) and adult males show a high incidence of dolichopelly, perhaps indicating a paedomorphic trait in males. The platypelloid pelvis has been described as 'ultrahuman'.

Correlation of pelvic type with general bodily physique is uncertain (Smout et al 1969). It has been claimed, e.g. that women with the dangerous android pelvis are stout, short-necked and broad-shouldered (Kenny 1944), views regarded as unreliable by others (Ince & Young 1940).

Pelvic Axes and Inclination

The axis of the superior pelvic aperture traverses its centre at right angles to its plane, directed down and backwards; when prolonged (projected) it passes through the *umbilicus* and *mid-coccyx*. An axis is similarly established for the inferior aperture; projected up it impinges on the *sacral promontory*. Axes can likewise be constructed for any plane, and one for the whole cavity is a concatenation of an infinite series of such lines. It follows the cavity's curvature, indicated by the profile of sacrum and coccyx in lateral views (**3.221**). The form of this *pelvis axis* and disparity in depth between anterior and posterior contours of the cavity are prime factors in the mechanism of fetal transit in the *pelvic canal*.

In the standing position the pelvic canal curves obliquely back relative to trunk and abdominal cavity (**3.221**). The whole pelvis is tilted forwards, the plane of the pelvic brim making an angle of 50 to 60° with the horizontal. The plane of the inferior aperture is tilted likewise to about 15°, dorsal parts of both planes being thus above ventral. Strictly, the pelvic outlet has two planes, an anterior passing back from pubic symphysis and a posterior passing forwards from the coccyx, both descending to meet at the intertuberous line. In standing, the pelvic aspect of the symphysis pubis faces as much up as back and the sacral concavity is directed antero-inferiorly. The front of the symphysis and anterior superior iliac spines are in the same vertical plane. In sitting, body weight is transmitted through inferomedial parts of the ischial tuberosities, with variable soft tissues intervening, and anterior superior spines are in a vertical plane through the acetabular centres, the whole pelvis being tilted back and the lumbosacral angle somewhat diminished at the sacral promontory.

SEXUAL DIFFERENCES IN THE PELVIS

The pelvis obviously provides the most marked skeletal differences between male and female (**3.222–225**); surprisingly, distinction can be made even during fetal life, particularly in the subpubic arch (Boucher 1957). Radiographic pelvic studies in American children during their first postnatal year show that male infants exceed females in dimensions of the whole pelvis, but usually females exceed males in the size of pelvic cavity (Reynolds 1945). From two to nine years a similar distinction prevails; but the difference is said to be maximal at 22 months, decreasing in later childhood (Reynolds 1947). Sexual differences in adults, which have been much assessed (Genovese 1959), are divisible

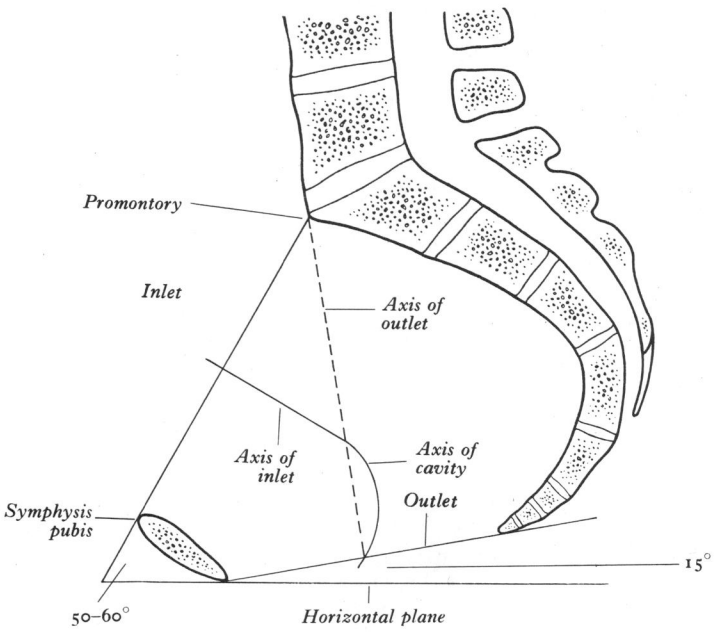

3.221 Median sagittal section through the female pelvis to show the planes of inlet and outlet and their relation to each other and the horizontal plane. The curved axis of the pelvic cavity is also shown. It should be observed that, as depicted in this section, the curve of the sacrum affects the lower part of the third and the upper part of the fourth sacral vertebra. In many cases the curve is restricted to the fourth vertebra only. Difficulties in defining the plane of the outlet are mentioned in the text.

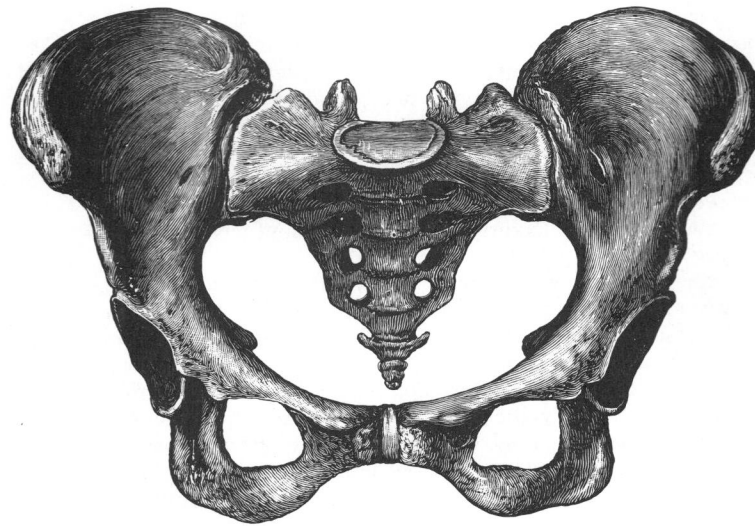

3.222 The female pelvis: anterior aspect. (From a specimen in the museum of the Royal College of Surgeons of England.)

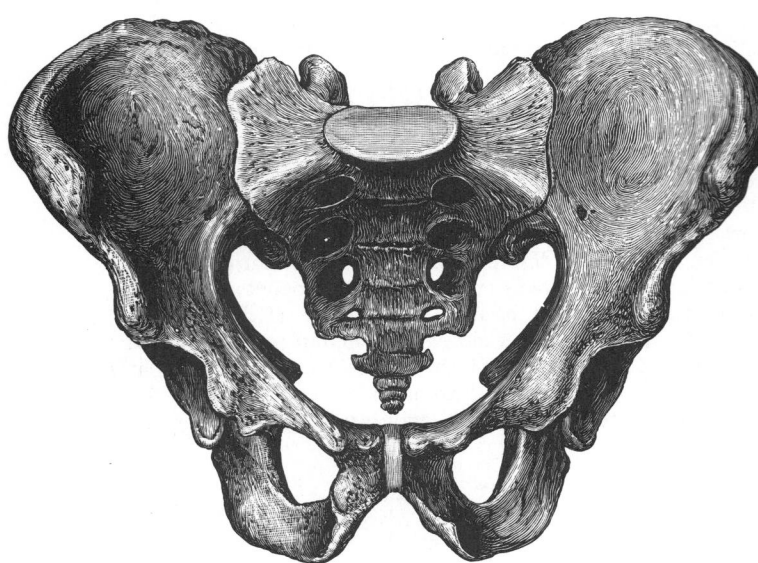

3.223 The male pelvis: anterior aspect.

into metrical and non-metrical features, the range of most, overlapping between sexes.

Differences are inevitably linked to function. While primary pelvic function in *both* sexes is locomotor, the pelvis is adapted to parturition in females, particularly the lesser pelvis, changes in which variably affect proportions and dimensions of the greater pelvis. Since males are distinctly more muscular and therefore more heavily built, overall pelvic dimensions, such as intercristal measurement, are greater, markings for muscles and ligaments more pronounced and general architecture heavier. The male iliac crest is more rugged and more medially inclined at its anterior end; in females crests are less curved in all parts. The iliac blades are more vertical in females, but do not ascend so far; iliac fossae are therefore shallower and each iliopectineal line more vertical. These iliac peculiarities probably account for the greater prominence of female hips.

The male is relatively and absolutely more heavily built *above* the pelvis, with consequent differences at lumbosacral and hip joints. The sacral basal articular facet for the fifth lumbar vertebra and intervening disc is more than a third of the total sacral basal width in males but less than a third in females, whose sacrum is also relatively broader, accentuating this difference. Thus the female has relatively broader alae. The male acetabulum is absolutely larger and its diameter about equal to the distance be-

tween its anterior rim and symphysis pubis. But in females acetabular diameter is usually less than this distance, not only because it is absolutely smaller but also because the anterolateral wall of the cavity is comparatively and often absolutely wider. Female symphysis and adjoining parts of pubis and ischium, forming the anterior pelvic wall, are also absolutely less in height, producing a somewhat triangular obturator foramen, which is more ovoid in males. Differing pubic growth is also expressed in the subpubic arch below symphysis and between inferior pubic rami. It is more angular in males, being 50 to 60°; in females it is rounded, less easy to measure and usually 80 to 85°. Associated with pubic width in females is greater separation of pubic tubercles. Ischiopubic rami are also much more lightly built and narrowed near the symphysis. In males they bear a distinctly rough, everted area for attachment of penile crura, the corresponding attachment for the clitoris being poorly developed. Ischial spines are closer in males, being more inturned. The greater sciatic notch is usually wider in females; comparisons of its width and angle yield mean values for males and females of 50.4° and 74.4° (Hanna & Washburn 1953); greater female values for angle and width are associated with increased backward sacral tilt and greater anteroposterior pelvic diameter, especially at lower levels. A method for comparing *depth* of the notch is reported by Jovanović & Živanović (1965). More recently, detailed analysis of dimensions and derived indices in relation to sex determination at the greater sciatic notch has been made by Singh & Potturi (1978), based on a study of 200 adult hip bones (120 males and 80 females). Using defined points (Jit & Singh 1966) they assessed maximal width and depth, posterior segment of width, total angle, and posterior angle. Two indices were also assessed (**3.224**). Width and depth of the notch per se were found valueless for determining sex: the posterior angle was the best single parameter, the length of the posterior segment and index II were also highly effective, especially in females.

The sacrum displays metrical sexual differences, in addition to orientation. Female sacra are less curved, curvature being most marked between the first and second segments and the third and fifth, with an intervening flatter region. Male sacra are more evenly curved, relatively long and narrow and more often exceeding five segments (by addition of a lumbar or coccygeal vertebra). The *sacral index* compares sacral breadth (between most anterior

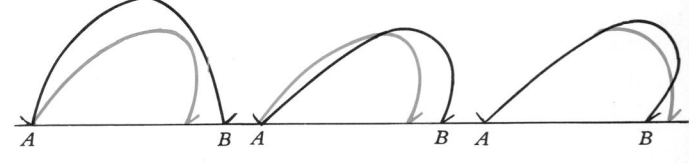

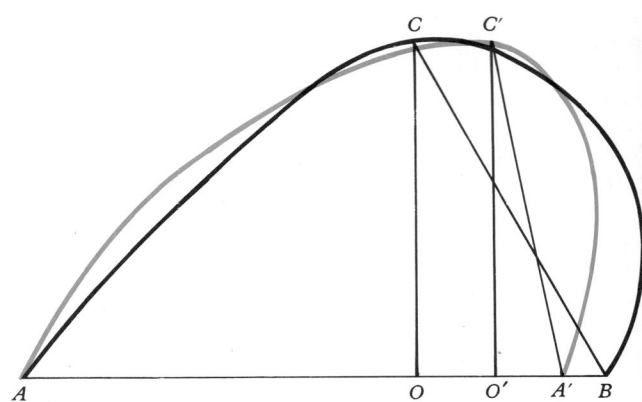

3.224 (*Above*) three common profiles of the greater sciatic notch in males (blue) and females (black). (*Below*) points, axes and angles utilized in mensuration of the greater sciatic notch. AB = *maximal width*, i.e. distance between the tip of the ischial spine and a tubercle marking attachment of the piriformis muscle. OC = *maximal depth* (perpendicular to AB). OB = *posterior segment* of width. Index I = OC × 100/AB. Index II = OB × 100/AB. ACB = *total angle*. BCO = *posterior angle* (Modified, with permission, from Singh & Potturi 1978.)

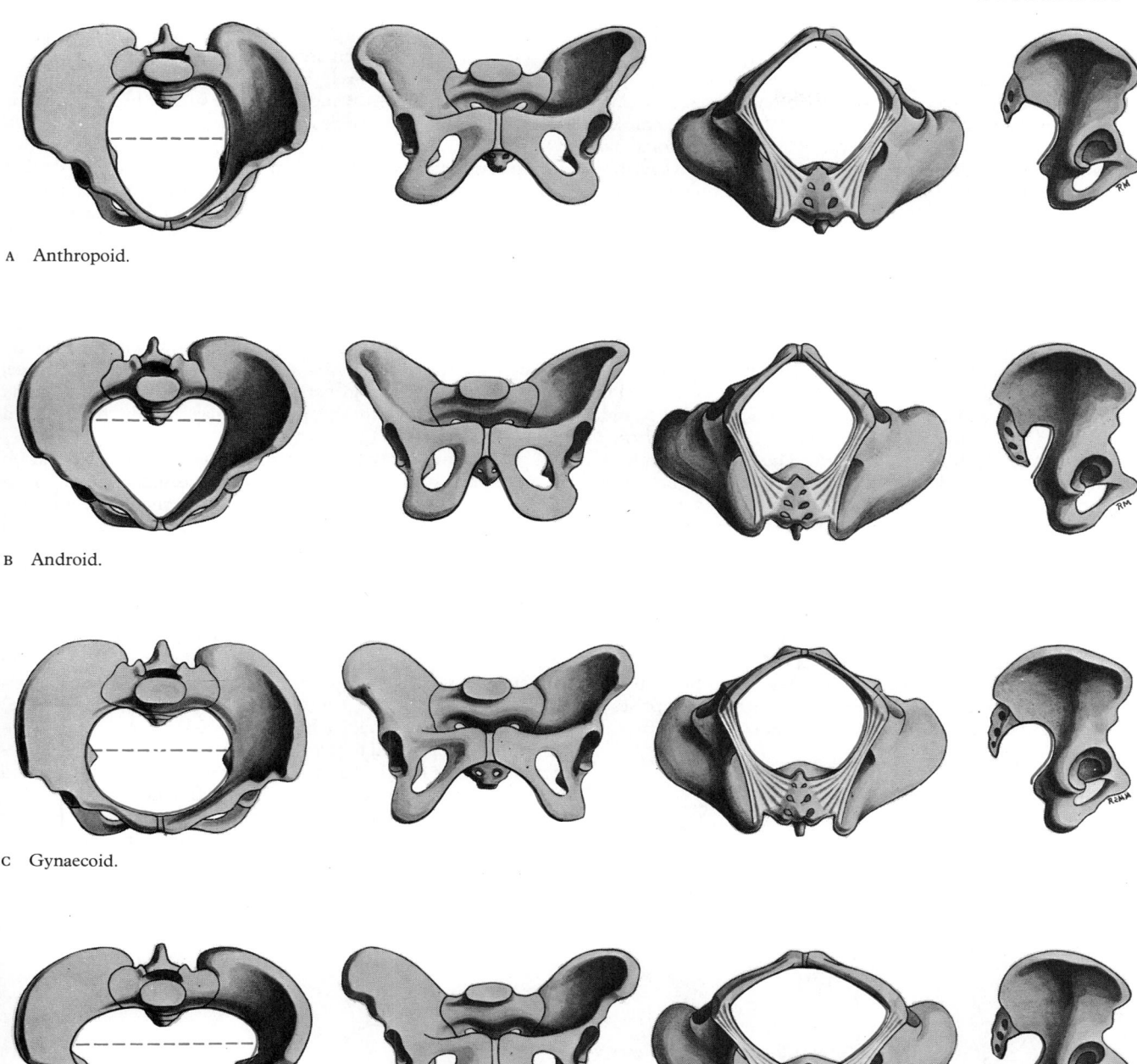

A   Anthropoid.

B   Android.

C   Gynaecoid.

D   Platypelloid.

3.225 The major differences between the four types of pelvis in the most widely accepted classification. Types shown in A and B are commonest in males, B and C in females, while D is rare, even in females. Note the variations in superior and inferior apertures, greater sciatic notch, and subpubic arch. Note varying proportion between fore-and hind-pelvis, anterior and posterior to transverse diameter of the inlet. (Caldwell and Moloy's Classification, after Clyne 1963.)

points on the auricular surfaces) with length (between midpoints on the anterior margins of promontory and apex); average values for males and females are 105 and 115%. Auricular surfaces are relatively smaller and more oblique in females; contrary to a common statement, they extend on the upper *three* sacral vertebrae in *both* sexes. The dorsal auricular border is more concave in females (Weisl 1954). Many differences are summarized in the generalization that the pelvic cavity is longer and more conical in males, shorter and more cylindrical in females; but in both the axis is curved. Differences are greater at the inferior aperture than the rim, where in absolute measurements males are not as different from females as sometimes stated. But the superior aperture is more likely to be anthropoid or android in males, gynaecoid or android in females, the sexes overlapping to this extent.

In forensic practice identification of human skeletal remains (sometimes fragmentary) usually involves diagnosis of sex, and this is most certainly established from the pelvis. Even parts of the pelvis may be useful. Several studies of metrical characteristics in various pelvic regions have been made, leading to various indices. The ilium has received particular attention, e.g., one index essentially compares the pelvic and sacro-iliac parts of the bone (Derry 1923). A line is extended back from the iliopectineal eminence to the nearest point on the anterior auricular margin and thence to the iliac crest. The auricular point divides this *chilotic line* into anterior (pelvic) and posterior (sacral) segments, each expressed as a percentage of the other. Such *chilotic indices* display reciprocal values in the sexes, the pelvic part of the chilotic line being predominant in females, the sacral part in males. The most detailed metrical study of the ilium, involving several indices, indicates its limited reliability in 'sexing' pelves (Straus 1927). However, the higher incidence and definition of the female preauricular sulcus was confirmed (p. 425). A *puboischial index*, based on maximum lengths of ischium and pubis, measured from their acetabular junction, produced values of 83.7% and 100.0%

for American males and females (Washburn 1949); when this was correlated with the angle of the sciatic notch, it was claimed that sex of 98% of pelves could be deduced. The desirability of correlation of all available metrical data is to be emphasized; when a range of pelvic data can be combined, especially if they are metrical, 95% accuracy should be achieved. Complete accuracy has been claimed when the rest of the skeleton is available. Nevertheless, assessment of sex in isolated and often incomplete human remains cannot always be absolutely certain.

The Femur

The femur is, of course, the longest and strongest bone (3.226–233). Its length is associated with striding gait, its strength with weight and muscular forces. Its *shaft*, almost cylindrical in most of its length and bowed forward, has a proximal round, *articular head* projecting mainly medially on its short *neck*, a medial curvature of the proximal shaft. The distal extremity is more massive, being a double 'knuckle' or *condyle* articulating with the tibia. In standing, the femora are oblique (3.1–3), their heads separated by the pelvic width, their shafts converging down and medially to where the knees almost touch. Since the foreleg bones descend vertically from the knees, the femoral obliquity approximates the feet, bringing them under the line of body weight in standing or progression. The narrowness of this base detracts from stability but facilitates forward movement by increasing speed and smoothness. Femoral obliquity varies but is greater in women, due to relatively greater pelvic breadth and shorter femora. Proximally the femur (3.226,227) comprises a head, neck, and greater and lesser trochanters.

The femoral head, slightly more than half a 'sphere' (but see pp. 478, 518), faces up and anteromedially to articulate with the acetabulum. (More precisely, the head is not part of a true sphere, but rather it is sphenoidal and part of the surface of an ovoid.) Its smoothness is interrupted postero-inferior to its centre by a small, rough *fovea*.

The femoral neck, about 5 cm long, connects head to shaft at an angle of about 125°. This facilitates movement at the hip joint, enabling the limb to swing clear of the pelvis. The neck is narrowest in its midpart, widest laterally. Its contours are rounded, the upper almost horizontal and slightly concave above, the lower straighter but oblique, directed inferolaterally and backwards to the shaft near the lesser trochanter. On all aspects the neck diverges as the capitular articular surface is approached. The neck's anterior surface is flat and marked at the junction with the shaft by a rough *intertrochanteric line* (3.228A). The posterior surface, facing back and up, is transversely convex, and concave in its long axis; its junction with the shaft is marked by the rounded *intertrochanteric crest* (3.226,229A).

The greater trochanter, large and quadrangular, projects up from the junction of the neck and shaft. Its posterosuperior region projects superomedially (3.226,229A) to overhang the adjacent posterior surface of the neck and here its medial surface presents the rough *trochanteric fossa*. The trochanter's proximal border is a hand's breadth below the iliac tubercle, level with the centre of the femoral head. It has an anterior rough impression, its lateral surface being divided by an oblique, flat strip, wider above, crossing it down and forwards; this aspect is palpable and, with the muscles (3.227) relaxed, can be gripped. The trochanteric fossa occasionally presents a tubercle or exostosis, a useful non-metrical racial characteristic (p. 273). **The lesser trochanter** (3.226) is a conical posteromedial projection of shaft at the postero-inferior aspect of its junction with the neck. Its summit and anterior surface are rough, but its posterior surface, at the distal end of the intertrochanteric crest, is smooth. It is not palpable.

The intertrochanteric line, at the junction of the anterior surfaces of neck and shaft (3.228), is a prominent ridge, descending medially from a tubercle, superomedial on the anterior aspect of the greater trochanter to the neck's lower border, level with but

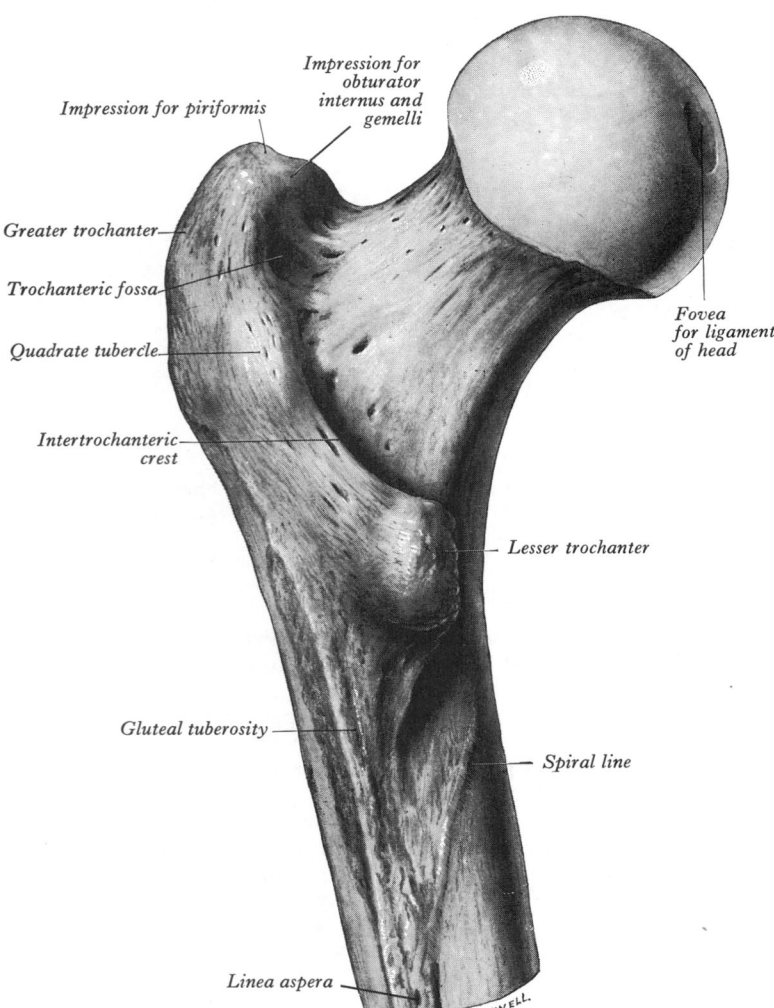

Impression for obturator internus and gemelli

Impression for piriformis

Greater trochanter

Trochanteric fossa

Quadrate tubercle

Intertrochanteric crest

Fovea for ligament of head

Lesser trochanter

Gluteal tuberosity

Spiral line

Linea aspera

A. K. MAXWELL.

3.226 The proximal part of the left femur: posterior aspect.

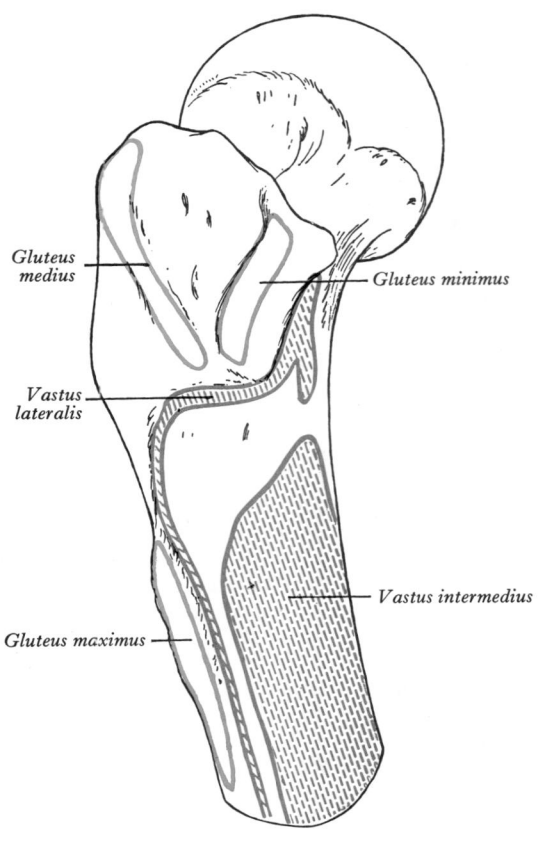

Gluteus medius

Gluteus minimus

Vastus lateralis

Vastus intermedius

Gluteus maximus

3.227 The proximal part of the right femur: lateral aspect.

in front of the lesser trochanter, often ending in a second tubercle. Distally it is continuous with the *spiral line* (vide infra). **The intertrochanteric crest** (3.226,229A), at the junction of the *posterior* surface of neck with shaft, is a smooth ridge, descending medially from the posterosuperior angle of the greater to the lesser trochanter. A little above its centre is a low, rounded *quadrate tubercle*.

Relations and Attachments

The femoral head is intracapsular, encircled lateral to its equator by the acetabular labrum. Its periphery is distinct, except anteriorly, where the articular surface extends to the neck. To its fovea (3.226) is attached the ligament of the head; the head's anterior surface is separated inferomedially by the tendon of psoas major, the subpsoas bursa, and the articular capsule from the femoral artery. Its blood supply is from a vascular circle around the neck, near the attachment of the fibrous capsule, supplied by medial and lateral circumflex arteries; from this rami pierce the capsule (under its zona orbicularis, p. 519) to ascend the neck beneath the reflected synovial membrane. These divide into metaphysial branches entering the neck and epiphysial rami to peripheral non-articular parts of the head to supply the epiphysis. During growth, territories of the two groups are separated by the epiphysial plate; after osseous union of head and neck, they anastomose freely. A small supply reaches the head along its ligament (pp. 439 and 783) by the acetabular branches of obturator and medial femoral circumflex arteries; these anastomose with the epiphysial vessels (Crock 1965). Observations on developmental patterns of this supply in late fetal and early postnatal periods modify the above description: though medial and lateral circumflex arteries at first contribute equally, two major branches of the medial are the final supply, both related posterior to the neck (Ogden 1974). Supply from the lateral circumflex diminishes and the arterial circle is interrupted. As the femoral neck elongates, the extracapsular circle becomes more distant from the epiphysial part of the head.

The neck has numerous vascular foramina, especially anterior and posterosuperior. Its angle with the shaft is widest at birth and diminishes until adolescence; it is less in females. Its *anterior surface* is intracapsular; the capsular ligament extends laterally attaching to the intertrochanteric line. Facets, often covered by extensions of articular cartilage, and various imprints frequently occur here; though attributed to squatting habits, they are not always so associated; their cause is uncertain (Kostick 1963). One such feature, the *cervical fossa* (of Allen), may be racial (p. 273). On the *posterior surface* the capsule does not reach the intertrochanteric crest (3.229A,B); little more than the medial half of the neck is intracapsular. The anterior surface adjoining the head and covered by cartilage is related to the iliofemoral ligament. A groove spirals across the neck's posterior surface in a proximolateral direction, produced by the tendon of obturator externus approaching the trochanteric fossa. The neck is *not* in the plane of the shaft, but inclined forwards. Hence the head's transverse axis makes an angle with the transverse condylar axis, the *angle of femoral torsion* (approximately 15°).

The greater trochanter (3.227,228B,229B) provides attachment for most gluteal muscles: gluteus minimus to its rough anterior impression, gluteus medius to its lateral oblique strip, the area anterior to this separated from its tendon by a trochanteric bursa; the area behind is covered by deep fibres of gluteus maximus, with part of its trochanteric bursa interposed. To the trochanter's upper border is attached the tendon of piriformis, to its medial surface the common tendon of obturator internus and the gemelli; at their attachment these two tendons are often variably fused. The trochanteric fossa receives the tendon of obturator externus.

The lesser trochanter has psoas major attached to its summit and anteromedial surface. To the medial or anterior surface of its base iliacus is attached, descending a little behind the spiral line. Adductor magnus (upper part) plays over its posterior surface, with sometimes an interposed bursa.

To the *intertrochanteric line*, lateral limit of the coxal capsular ligament, at its proximal and distal parts with their tubercles, are attached the upper and lower bands of the iliofemoral ligament.

The most proximal fibres of the vastus lateralis are attached to its proximal end, those of vastus medialis distally. The *intertrochanteric crest*, above the *quadrate tubercle*, is covered by gluteus maximus, separated distally from the muscle by quadratus femoris and the upper border of adductor magnus. To the tubercle and the bone distal to it quadratus femoris is attached (3.229B).

The shaft (3.228,229), narrowest centrally, expands a little upwards, but more so towards its distal end. Its long axis diverges about 10° from the tibia's. Its *middle third* has three surfaces and borders. The extensive *anterior surface*, smooth and gently convex, is between lateral and medial borders, both round and indistinct. The *lateral surface*, really posterolateral, is bounded behind by the posterior border, the broad, rough **linea aspera**, usually a crest with lateral and medial edges; its subjacent compact bone is augmented to withstand compressive forces concentrated here by the shaft's anterior curvature. The *medial surface* is posteromedial, smooth like the others, bounded in front by the indistinct medial border and behind by linea aspera. In its *proximal third* the shaft has a fourth, *posterior surface*, bounded medially by a narrow, rough *spiral line*, continuous proximally with the intertrochanteric line, distally with the medial edge of linea aspera. Laterally this surface is limited by the broad, rough, *gluteal tuberosity*, ascending a little laterally to the greater trochanter and descending to the lateral edge of linea aspera; this surface is triangular. In its *distal third* the shaft also has a fourth, *posterior surface*, between *medial* and *lateral supracondylar ridges*, continuous above with corresponding edges of the linea aspera; the lateral is the more distinct. Proximally the medial ridge is partly obliterated where the femoral artery is close to bone, slanting back to the popliteal fossa. This *popliteal surface* (3.229A) is also triangular, in its distal medial part rough and slightly elevated.

Further details. The shaft (3.228B,229B) is surrounded by muscles and is impalpable. Its *anterior* and *lateral surfaces* are attachments in its proximal three-fourths for vastus intermedius and distal to this for slips of articularis genus. The distal anterior surface, for 5 to 6 cm above the patellar articular surface, is covered by a suprapatellar bursa, between bone and muscle. The distal lateral surface is covered by vastus intermedius. The *medial surface*, devoid of attachments, is covered by vastus medialis.

Vastus lateralis has a linear attachment from in front of the greater trochanter's base to the proximal end of the gluteal tuberosity, and along the latter's lateral margin to the proximal half of the lateral edge of the linea aspera. Similarly, the vastus medialis is attached from the distal end of the intertrochanteric line along the spiral line to the medial edge of the linea aspera and thence to the medial supracondylar line; it receives many fibres from the aponeurotic attachments of adductor magnus (vide infra).

The gluteal tuberosity may be an elongate depression or a ridge. It may in part be prominent enough to be dubbed a *third trochanter*. It receives the deeper fibres of the distal half of the gluteus maximus and, at its medial edge, pubic fibres of adductor magnus, distal to which the adductor magnus is attached to the linea aspera and aponeurotically to the proximal part of the medial supracondylar ridge, remaining fibres forming a large tendon attached to the adductor tubercle (p. 438), with an aponeurotic expansion to the distal part of the medial supracondylar ridge. Lozanoff et al (1985) have devised a metrical technique associating various femoral dimensions with the incidence of a third trochanter in human femora, finding that increased incidence is associated with short femora bearing large epiphyses. Though this series is limited, the metrical technique applied is commendable.

Between gluteal tuberosity and spiral line, the posterior femoral surface receives pectineus and adductor brevis, the former at a line, sometimes slightly rough, from the lesser trochanter's base to the linea aspera. The adductor brevis is attached lateral to the pectineus and beyond this to the proximal part of the linea aspera, medial to the adductor magnus.

In addition, the **linea aspera** receives adductor longus, intermuscular septa and the short head of the biceps femoris, inseparably blended at their attachment. The perforating arteries cross the linea laterally under tendinous arches in the adductor magnus and biceps. Nutrient foramina appear in the linea aspera, varying

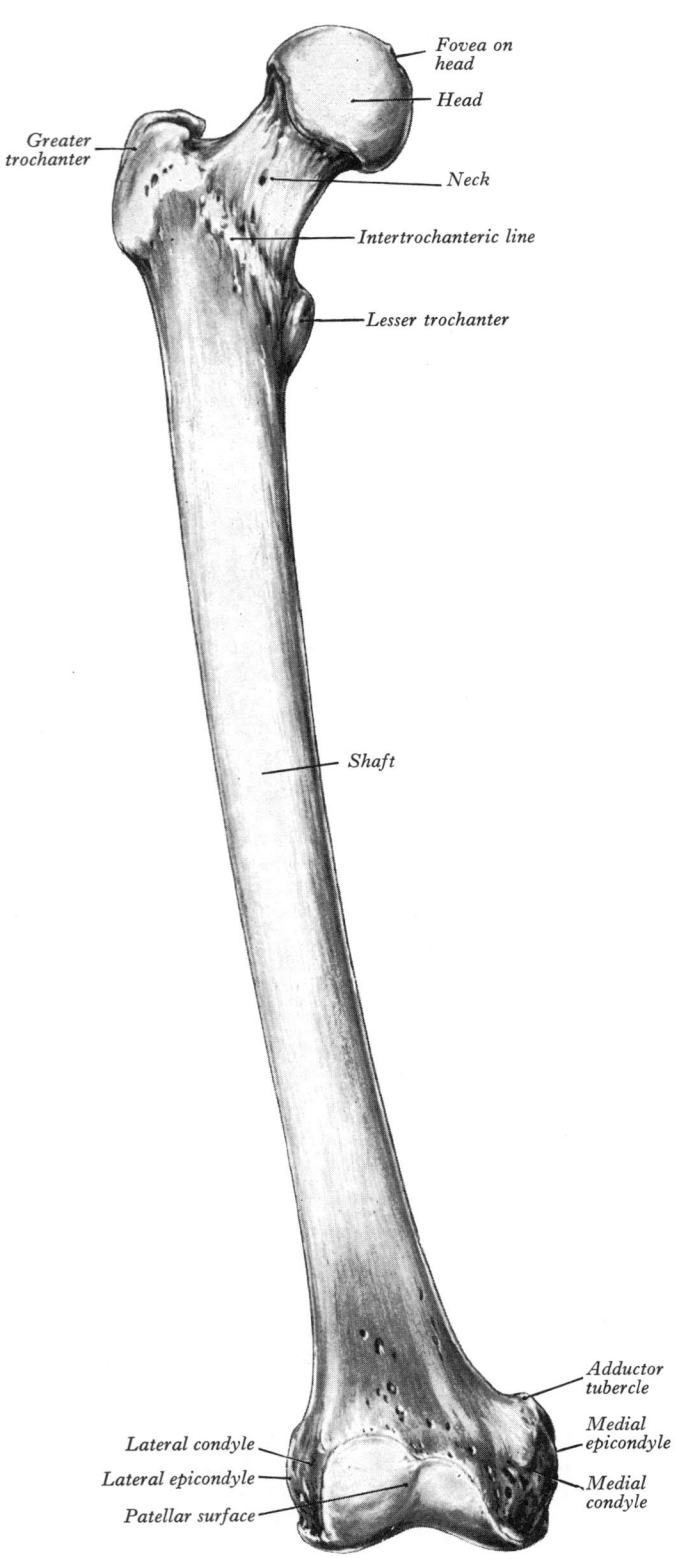

3.228A The right femur: anterior aspect. Compare with Key, 3.228B.

3.228B Key to 3.228A. The epiphysial lines are stippled. The interrupted lines correspond to the attachments of the capsular ligaments. The dotted part of the interrupted line indicates the site of communication between the cavity of the knee joint and the suprapatellar synovial bursa.

in number and site, one usually near its proximal end, usually a second near its distal end. They are directed proximally.

The popliteal surface, proximal floor of the popliteal fossa, is covered by variable amounts of fat separating the popliteal artery from bone. The superior medial genicular artery, from the popliteal artery in the intercondylar notch, arches medially above the medial condyle. It is separated from bone by the medial head of gastrocnemius, attached a little above the condyle, distal to which may be a smooth facet underlying a bursa for the medial

head of the gastrocnemius. More medially, proximal to the articular surface, is often an imprint which in flexion is close to a rough tubercle on the medial tibial condyle for attachment of semimembranosus (Kostick 1963). The superior lateral genicular artery arches up laterally proximal to the lateral condyle but separated from the bone by attachment of the plantaris to the distal part of the lateral supracondylar line.

The lateral supracondylar line is most distinct in its proximal two-thirds, where the short head of the biceps femoris

436

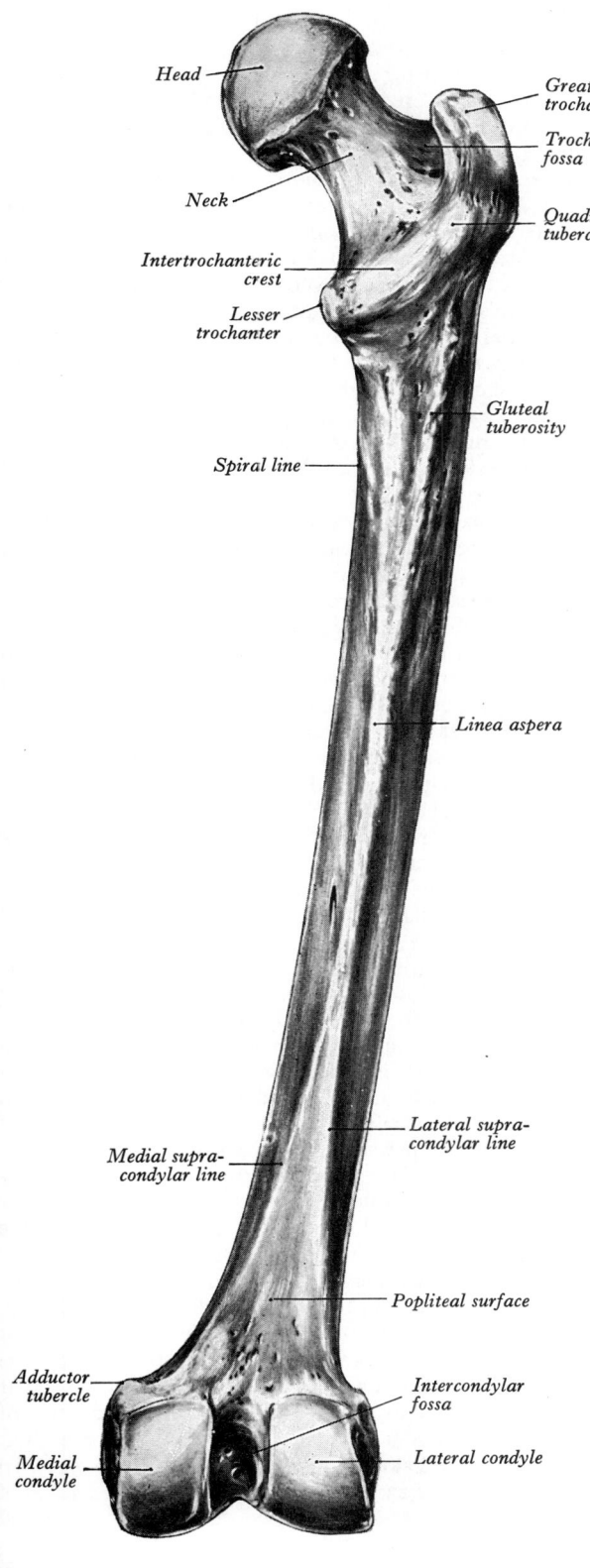

Head
Greater
trochanter
Trochanteric
fossa
Neck
Quadrate
tubercle
Intertrochanteric
crest
Lesser
trochanter
Gluteal
tuberosity
Spiral line
Linea aspera
Lateral supra-
condylar line
Medial supra-
condylar line
Popliteal surface
Adductor
tubercle
Intercondylar
fossa
Medial
condyle
Lateral condyle

3.229A The right femur: posterior aspect. Compare with Key, 3.229B.

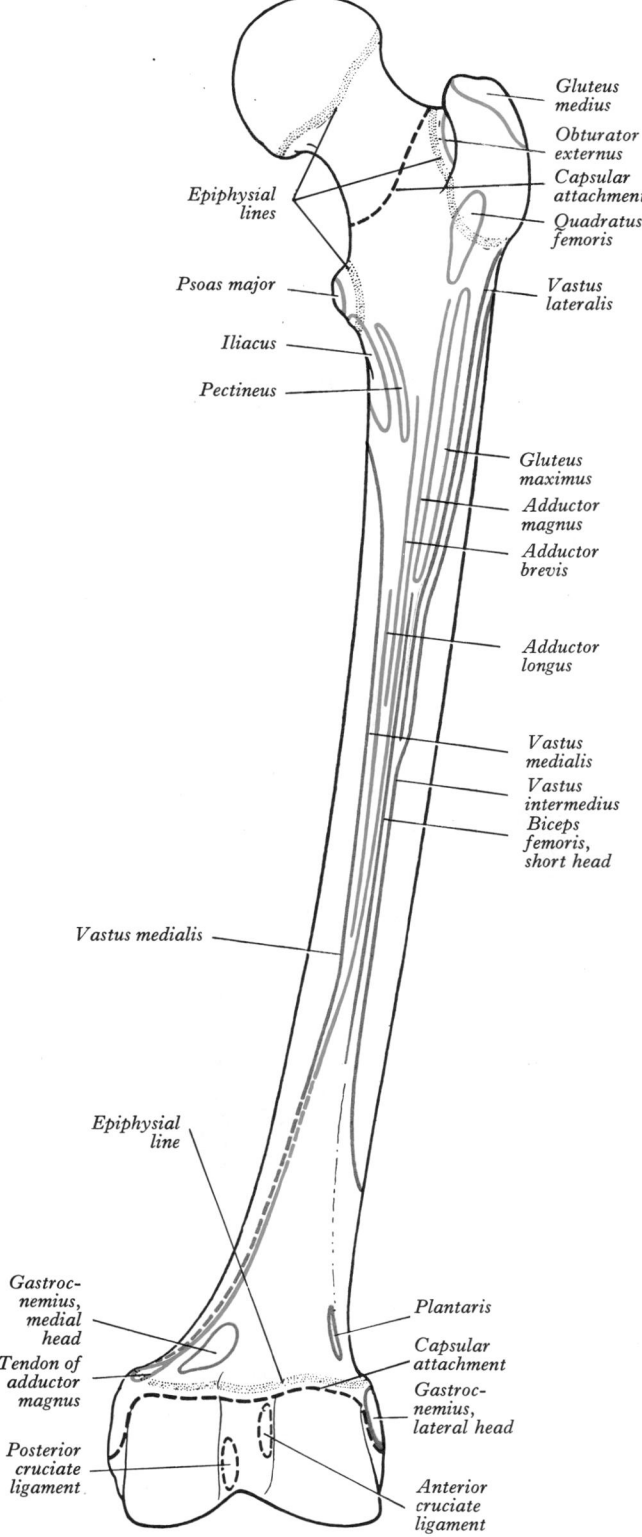

Gluteus
medius
Obturator
externus
Capsular
attachment
Epiphysial
lines
Quadratus
femoris
Psoas major
Vastus
lateralis
Iliacus
Pectineus
Gluteus
maximus
Adductor
magnus
Adductor
brevis
Adductor
longus
Vastus
medialis
Vastus
intermedius
Biceps
femoris,
short head
Vastus medialis
Epiphysial
line
Gastroc-
nemius,
medial
head
Plantaris
Capsular
attachment
Tendon of
adductor
magnus
Gastroc-
nemius,
lateral head
Posterior
cruciate
ligament
Anterior
cruciate
ligament

3.229B Key to 3.229A. The epiphysial lines are stippled; the interrupted lines indicate the attachments of the capsular ligaments.

and lateral intermuscular septum are attached. Its distal third has a small rough area for the plantaris, often encroaching on the popliteal surface. The **medial supracondylar line** is indistinct in its proximal two-thirds, where there is attachment of the vastus medialis. Proximally it is crossed by femoral vessels entering the popliteal fossa from the adductor canal. It is often sharp for 3 or 4 cm proximal to the adductor tubercle, and here the lower membranous expansion from the tendon of adductor magnus is attached.

The femur's distal end, widely expanded as a bearing surface for transmission of weight to the tibia, has two massive *condyles*,

partly *articular*. Anteriorly the condyles unite and continue into the shaft, posteriorly separated by a deep *intercondylar fossa* and projecting beyond the plane of the popliteal surface. The *articular surface* is a broad area, like an inverted U, for the patella above and the tibia (3.230). The *patellar surface* extends anteriorly on both condyles, but largely the lateral; transversely concave, it is vertically convex and grooved for the posterior patellar surface. The *tibial surface* is divided by the intercondylar fossa but is anteriorly continuous with the patellar surface; its medial part is a broad strip on the convex inferoposterior surface of the medial condyle, gently curved with a medial convexity; its lateral part covers

437

similar aspects of the lateral condyle but is broader and passes straight back. Heiple & Lovejoy (1971) regarded a deep patellar groove as a hominid feature, concerned with bipedal gait; others disagreed. The role of the groove in preventing lateral patellar subluxation is often repeated, but remains uncertain.

The lateral condyle (3.230,231), laterally flat and less prominent than the medial, is more massive and more directly in line with the femoral shaft, hence transmits more weight to the tibia. Its most prominent point is the *lateral epicondyle*; its whole lateral aspect is palpable. A short groove, deeper in front, separates lateral epicondyle posteriorly from the articular margin. The medial surface is the lateral wall of the intercondylar fossa. Its lateral surface projects beyond the shaft. To its lateral epicondyle is attached the fibular collateral ligament and, to an impression posterosuperior to this, part of the lateral head of the gastrocnemius. Attached anteriorly in the groove is the popliteus (3.231), whose tendon is in it in full flexion; in extension it crosses the articular margin and may groove it. Adjoining the margin a strip of condyle, 1 cm broad, is intracapsular, covered by synovial membrane except for attachment of the popliteus.

The medial condyle has a bulging, convex medial aspect, easily palpable. Proximally its *adductor tubercle* (3.229A) receives the tendon of adductor magnus, an important surgical landmark most readily identified from above, but often a facet. The condyle's summit, the *medial epicondyle*, is antero-inferior to the tubercle. The medial condyle's lateral surface is the medial wall of

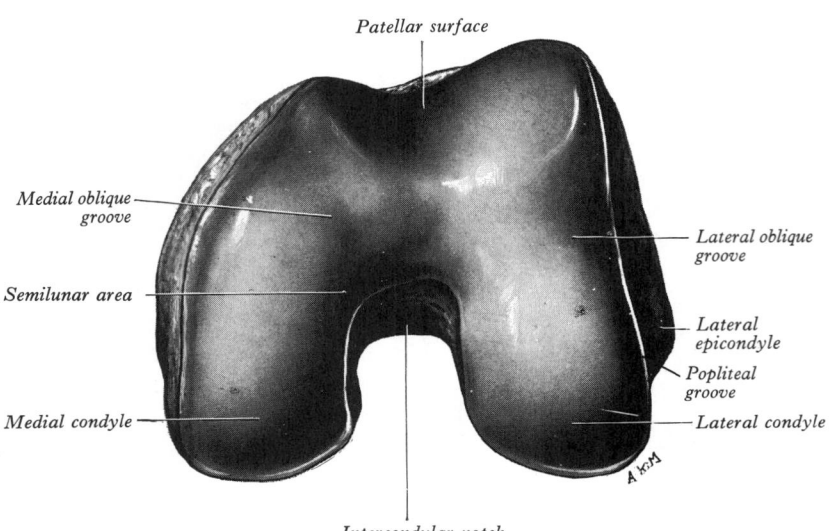

Patellar surface

Medial oblique groove

Semilunar area

Medial condyle

Lateral oblique groove

Lateral epicondyle

Popliteal groove

Lateral condyle

Intercondylar notch

3.230 The distal end of the left femur: inferior aspect.

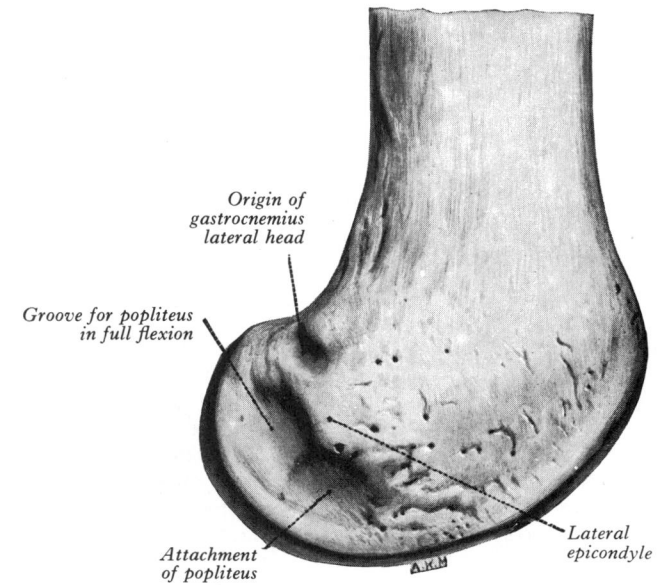

Origin of gastrocnemius lateral head

Groove for popliteus in full flexion

Attachment of popliteus

Lateral epicondyle

3.231 The distal end of the right femur: lateral aspect.

the intercondylar fossa. The condyle projects distally so that, despite the shaft's obliquity, the profile of the distal end is almost horizontal. A curved strip, about 1 cm wide, adjoining the medial articular margin, is covered by synovial membrane and is inside the joint's capsule. The medial epicondyle, proximal to this, receives the tibial collateral ligament.

The intercondylar fossa separates the two condyles distally and behind. In front it is limited by the distal border of the patellar surface, behind by an *intercondylar line*, separating it from the popliteal surface. It is within the capsular ligament but covered by synovial membrane over a very limited area.

The patellar surface extends proximally more on the lateral side; its proximal border is hence oblique and runs distally and medially (3.228A,230), separated from tibial surfaces by two faint grooves, which cross the condyles obliquely. The lateral groove is more distinct (3.230); it runs laterally and slightly forwards from the front of the intercondylar fossa and expands to form a faint triangular depression, resting on the anterior edge of the lateral meniscus with the knee fully extended. The medial groove is restricted to the medial part of the medial condyle and rests on the anterior edge of the medial meniscus in full extension; where it ceases, the patellar surface continues back to the lateral part of the medial condyle as a semilunar area adjoining the anterior region of the intercondylar fossa; this area articulates with the patella's medial vertical facet in full flexion; it is not distinct in outline in most femora. In habitual squatters articular cartilage may extend to the lateral aspect of the lateral condyle under vastus lateralis.

The tibial surfaces are transversely convex in all directions. Anteroposterior curvature of both surfaces is not uniform, with a shorter radius posteriorly; the medial surface is longer anteroposteriorly, and has a slight laterally concave profile. These differences are important determinants of rotatory movements, adjunct and conjunct (p. 481), at the knee joint (p. 530).

The intercondylar fossa, between projecting parts of condyles, is intracapsular but largely extrasynovial. Its lateral wall, the medial surface of the lateral condyle, has a flat, posterosuperior impression spreading to the fossa's floor near the intercondylar line for proximal attachment of the anterior cruciate ligament. The fossa's medial wall, lateral surface of the medial condyle, bears a similar larger area for proximal attachment of the posterior cruciate ligament, sited anteriorly and extending to the fossa's anterior floor. Both impressions are smooth, and largely devoid of vascular foramina like most attachments of tendons or ligaments; the rest of the fossa is rough and pitted by vascular foramina, but a bursal recess between the ligaments may ascend to it. To the *intercondylar line* is attached the capsular ligament and, laterally, the oblique popliteal ligament, and to the fossa's anterior border, the infrapatellar synovial fold (p. 527).

Structure. The femoral shaft is a cylinder of compact bone, with a large medullary cavity. The wall is thick in its middle third, where the femur is narrowest and the medullary cavity most capacious; but proximally and distally the compact wall is progressively thinner, the cavity gradually filling with trabecular bone; the extremities, especially where articular, consist of trabecular bone within a thin shell of compact bone; their trabeculae are disposed along lines of greatest stress (p. 524). In the proximal end (3.232,233) the main trabeculae form a series of plates orthogonal to the articular surface, converging to a central dense wedge, which is supported by strong trabeculae passing to the sides of the neck, especially along its upper and lower profiles. Force applied to the femoral head is hence transmitted to the wedge and thence to the junction of neck and shaft. This junction is strengthened by dense trabeculae extending laterally from the lesser trochanter to the end of the neck's superior aspect, thus resisting tensile or shearing forces applied to the neck through the head. A smaller bar across the junction of the greater trochanter with the neck and shaft resists shearing due to muscles attached to it. These two bars are proximal layers of arches between the sides of the shaft, transmitting to it forces applied to the proximal end. A thin vertical plate, *calcar femorale* (3.233), ascends from the compact wall near the linea aspera into the trabeculae of the neck. Medially it joins the posterior wall of the neck, laterally it continues into the greater trochanter, dispersing into general trabecular bone. It is thus in a plane anterior to the trochanteri crest and base of the lesser trochanter. Scanning electron microscopy of the proximal femoral end, in a small series of specimens, has partly confirmed these details (Whitehouse & Dyson 1974). Wide variation in metrical parameters of trabecular pattern in the femoral head and neck was nevertheless observed. Tensile and compressive tests of 40 human femora also indicate that axial trabecular tissue of the femoral head withstands much greater stresses than do peripheral trabeculae. In the distal femoral end trabeculae spring from the entire internal surface of compact bone, descending perpendicular to the articular surface; proximal to condyles these are strongest and most accurately perpendicular. Horizontal planes of trabecular bone, arranged like crossed girders, form a series of cubical compartments.

Ossification (3.228B,229B,234) of the femur is from five centres: in the shaft, head, greater and lesser trochanters and the distal end. Excepting the clavicle, it is the first long bone to ossify, this beginning in midshaft in the seventh prenatal week and extending to produce a miniature shaft largely ossified at birth. Secondary centres appear in the distal end during the ninth month (from which condyles and epicondyles are formed), in the head during the first six months after birth, in the greater trochanter during the fourth year and in the lesser between twelfth and fourteenth. Note that the centre in the cartilaginous head is restricted to it until the tenth year, so that the epiphysial line (3.218) is horizontal and the inferomedial part of the articular surface is on the neck (3.234). The medial epiphysial margin later grows over this part of the articular surface. Thus, the mature epiphysis is a *hollow cup* on the neck's summit. The epiphysial line follows the articular margin except where separated superiorly from the articular surface by a non-articular area; here blood vessels enter the head (Trueta 1957, Smith 1962a). The epiphyses fuse independently, the lesser trochanter soon after puberty, then the greater, the head at the fourteenth year in females, seventeenth in males, and the distal end at the sixteenth year in females, eighteenth in males. Note that the distal epiphysial plate traverses the adductor tubercle (3.228B).

Applied Anatomy. The distal femoral end is the only epiphysis in which ossification constantly starts just before birth, a most reliable indicator that a dead newborn child was viable. Since the epiphysial plate is level with the adductor tubercle, the epiphysis does not form all the distal end covered by synovial membrane; hence operations here may damage the distal epiphysial cartilage in children and entail subsequent shortening of the leg. Fractures of the femoral neck are usually due to transmitted stress, as in tripping over obstructions. The trunk continues to advance and, overbalancing, twists and imposes excessive medial rotation on thigh and leg. Before 16 years the usual injury is a spiral fracture of the shaft, but between 16 and 40 a crescentic tear of the medial meniscus is frequent. Between 40 and 60, a common result is fracture of the foreleg, but over 60, fracture of the femoral neck is common, because of osteoporotic changes in ageing bones. Women are more liable, their bones being lightly built.

Normally, with lower limb aligned with trunk, the greater trochanter's apex is on the line joining the anterior superior iliac spine and the most prominent part of the ischial tuberosity (Nelaton's line). In displacements due to fracture of the neck the trochanter may be above this line.

The Patella

The patella (3.235–237), the largest sesamoid bone, is embedded in the tendon of quadriceps femoris anterior to the knee joint. Flat, distally triangular, proximally curved, it has anterior and posterior surfaces, three borders and an apex. In the living, when standing, its distal apex is a little proximal to the line of the knee joint.

The convex *anterior surface*, subcutaneous, is perforated by nutrient vessels. It is longitudinally striated, separated from skin by a prepatellar *bursa* and covered by an expansion from the tendon of quadriceps femoris, which blends distally with superficial fibres of the so-called patellar *ligament*, strictly the continued *tendon* of quadriceps. The *posterior surface* has a proximal smooth, oval, articular area, crossed by a smooth vertical ridge, which fits the

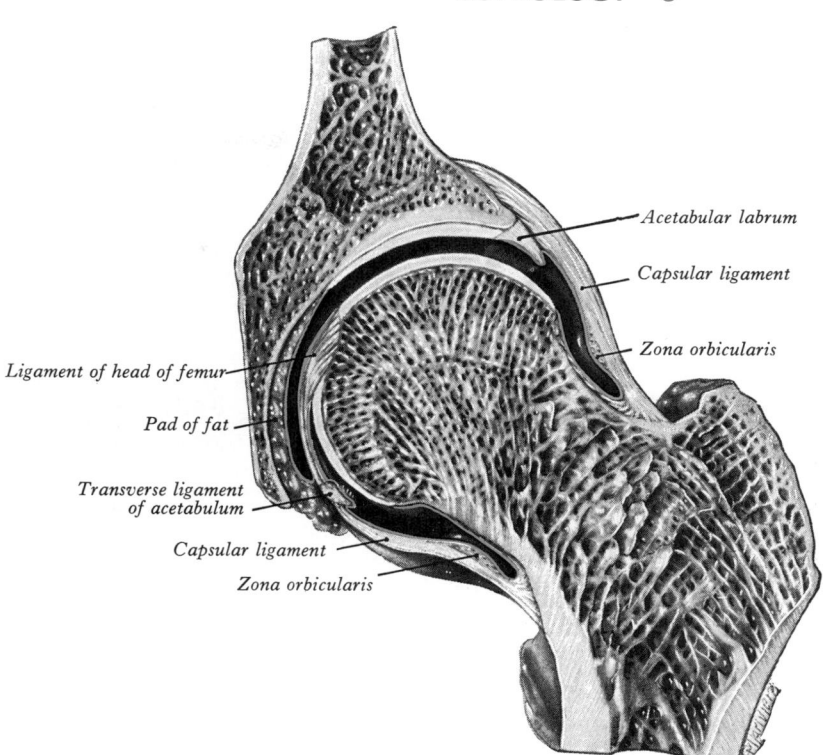

Labels: *Acetabular labrum*, *Capsular ligament*, *Zona orbicularis*, *Ligament of head of femur*, *Pad of fat*, *Transverse ligament of acetabulum*, *Capsular ligament*, *Zona orbicularis*

3.232 Section through the hip joint.

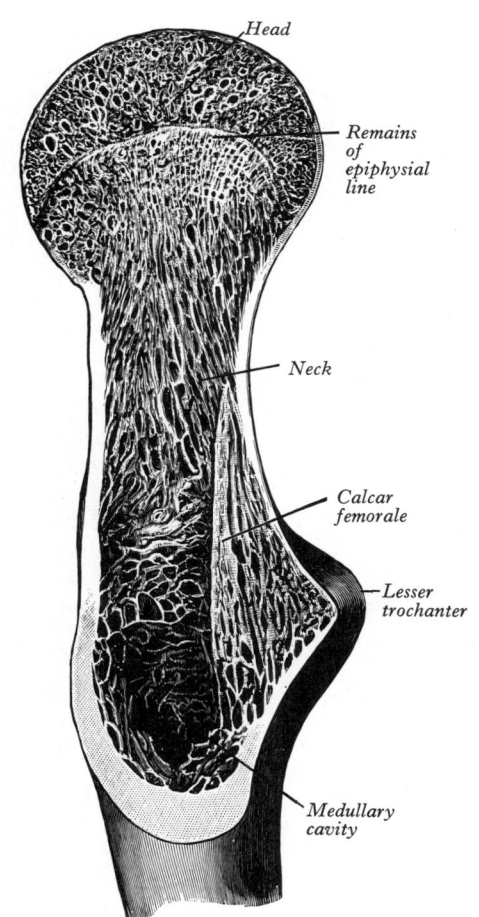

Labels: *Head*, *Remains of epiphysial line*, *Neck*, *Calcar femorale*, *Lesser trochanter*, *Medullary cavity*

3.233 Oblique section through the proximal end of the left femur showing the trabecular architecture, calcar femorale, medullary cavity and variations in cortical thickness. Compare with 3.5 and 3.232.

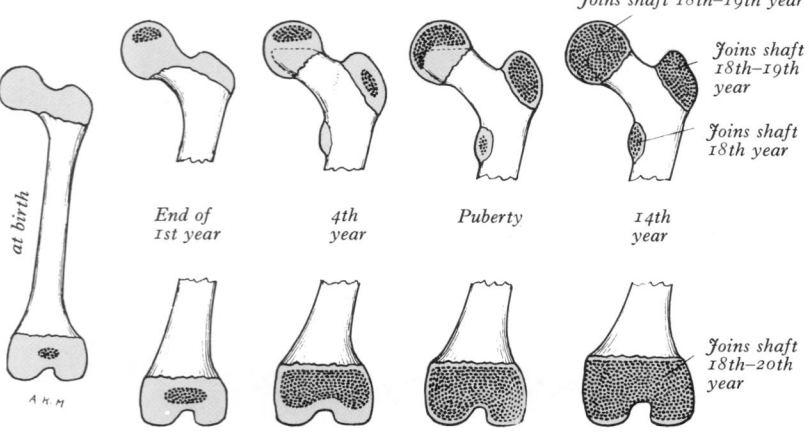

3.234 Stages in the ossification of the femur. Note how the neck, which is ossified as an extension from the shaft, invades, and is partly enclosed by, the cartilaginous head (not to scale).

to this, the area between the roughened apex and articular surface is covered by an infrapatellar pad of fat.

The thick *superior border* slopes down and forwards; except near its posterior margin, it is an attachment for quadriceps femoris (rectus femoris and vastus intermedius). The *medial* and *lateral borders* are thinner and converge distally, and to them are attached expansions of the tendons of vastus medialis and lateralis termed the *medial* and *lateral patellar retinacula* (p. 527). The lateral retinaculum receives accessions from the iliotibial tract. Near the superolateral angle is a shallow, circular depression for a distinct part of the tendon of vastus lateralis (vasti medialis and lateralis).

Structure. The patella consists of almost uniformly dense trabecular bone, covered by a thin compact lamina. Trabeculae beneath the anterior surface are parallel with it; elsewhere they radiate from the articular surface to other parts.

Ossification. Several centres appear during the third to sixth years and quickly coalesce. Accessory marginal centres appear later and fuse with the central mass (Hellmer 1935, Prakash et al 1979).

The Tibia

groove on the femoral patellar surface and divides the patellar articular area into medial and lateral facets, the lateral being larger. The 'ridge' and flanking facets are, of course, continuously covered by articular cartilage. (Degenerative changes in this cartilage, associated with ageing, have been reported by Meachim 1982.) A narrow strip, proximally broader, is marked off medially from the medial facet; this contacts the medial femoral condyle in extreme flexion. Distal to the articular surface the *apex* is roughened by attachment of ligamentum patellae and, proximal

The tibia (3.238–240), medial to and much stronger than the fibula, is exceeded in length only by the femur. Its shaft is prismoid in section, with expanded ends, the smaller distal end having a strong *medial malleolus* projecting distally. The anterior border is sharp, curving medially towards this malleolus; the anterior, together with medial and lateral borders, defines three surfaces.

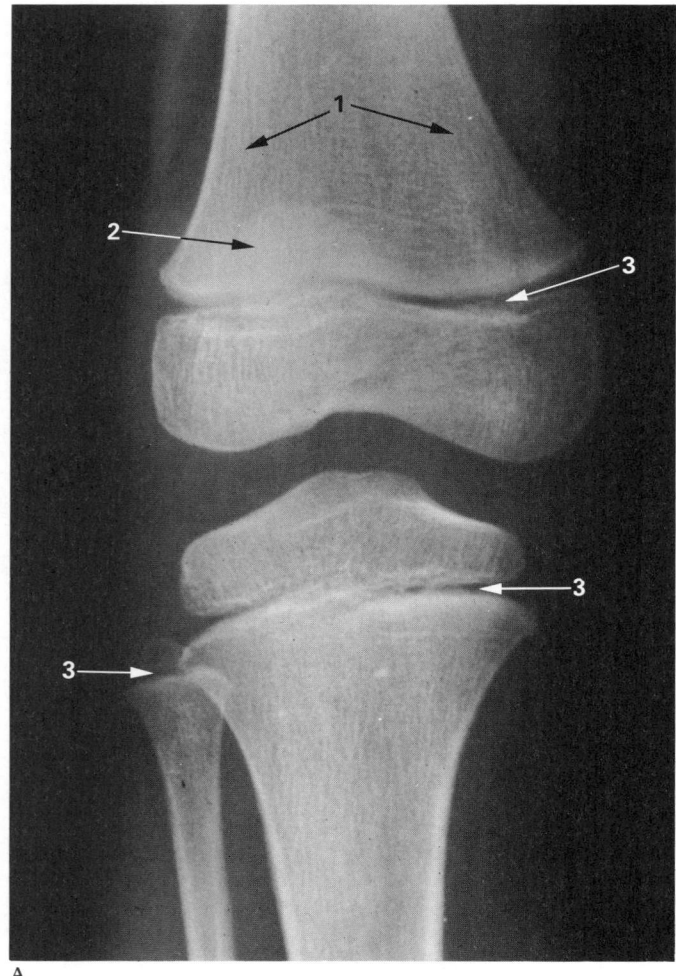

A

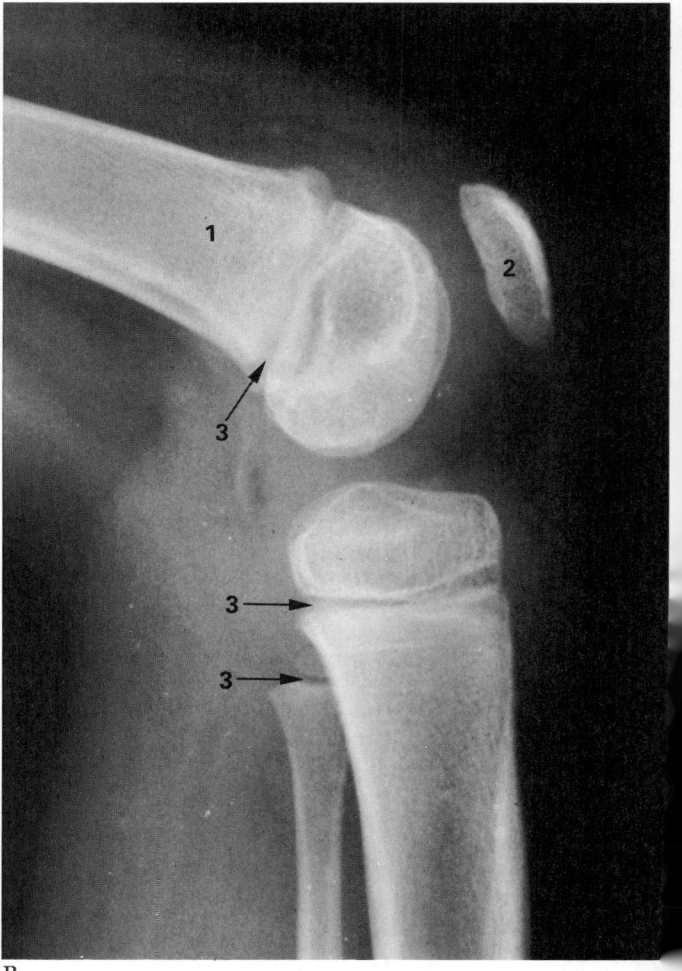

B

3.235 Anteroposterior (A) and lateral (B) radiographs of the knee in a girl aged six. 1. Flared femoral metaphysis. 2. Patella. 3. Cartilaginous growth plates with adjacent epiphyses. Note early stage of ossification i[n] fibular epiphysis.

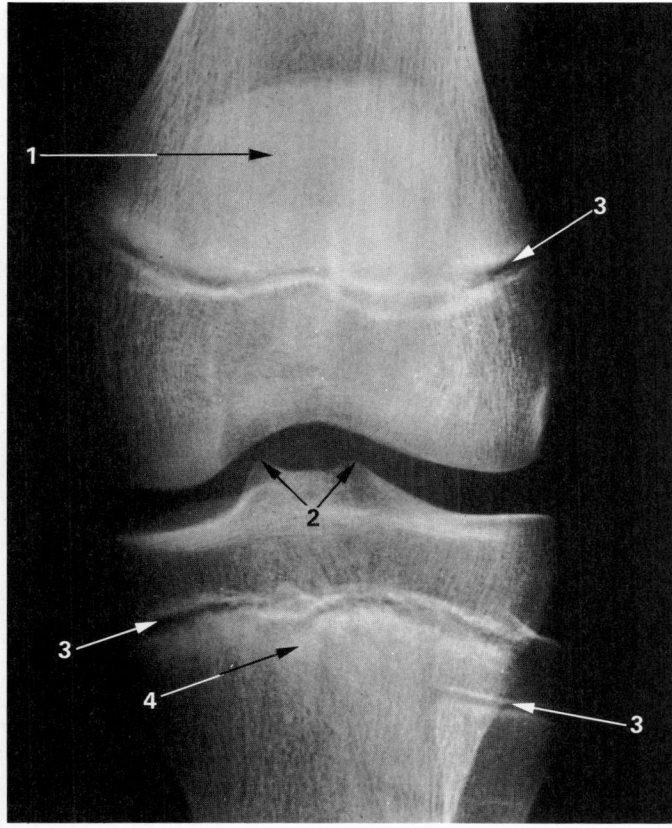

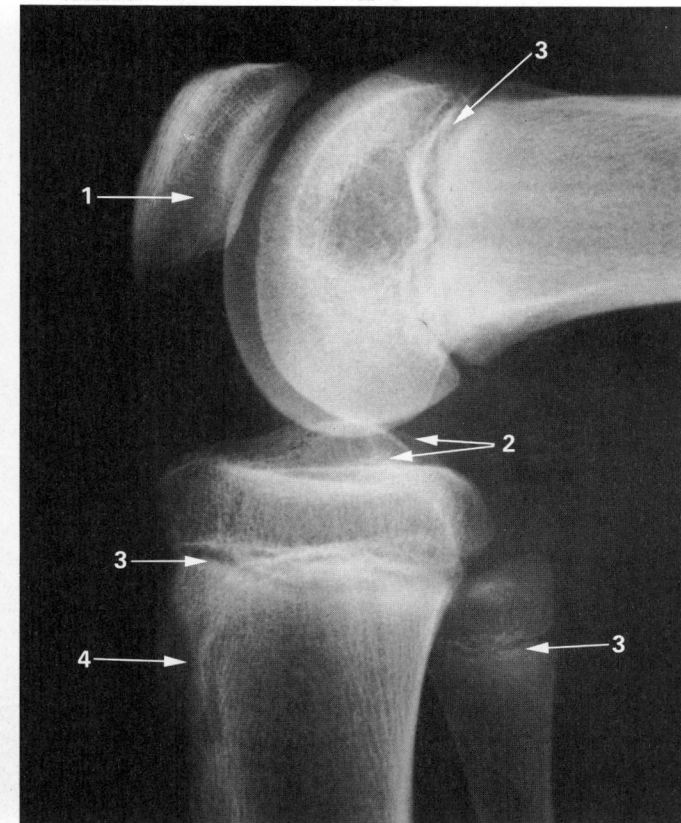

A

B

3.236 Anteroposterior (A) and lateral (B) radiographs of the knee in a boy aged 14. 1. Patella. 2. Intercondylar eminences. 3. Cartilaginous growth plates with adjacent epiphyses. 4. Prolongation of proximal tibial epiphysis and growth plate forming the tibial tuberosity.

The proximal end, expanded especially transversely, is a bearing surface for body weight transmitted through the femur; it has massive *medial* and *lateral condyles*, an intercondylar area and tibial tuberosity. The condyles overhang the shaft's proximal posterior surface, and both have proximal articular surfaces separated by the irregular intercondylar area. The condyles are visible and palpable at the sides of the ligamentum patellae, the lateral being more prominent. In the passively flexed knee the anterior *margins* of the condyles are palpable in fossae flanking the patellar ligament.

The medial condyle is larger but projects less. Its *oval* articular surface (**3.238**A–C) is concave and its lateral border, deepening the concavity, extends on to a *medial intercondylar tubercle*. Its posterior surface, distal to the articular margin, has a horizontal, rough groove; its anteromedial surface is a rough strip, separated from the shaft's medial surface by an inconspicuous ridge.

The lateral condyle overhangs the shaft posterolaterally above a small circular facet for articulation with the fibula. The proximal, articular surface (**3.238**A–C) for the lateral femoral condyle is almost *circular*, centrally concave, its medial border extending on a *lateral intercondylar tubercle*. Its posterior, lateral and anterior edges are rough.

Anterior condylar surfaces are continuous with a large triangular area; its apex is distal and formed by the tibial tuberosity, whose lateral edge is a sharp ridge between the lateral condyle and lateral surface of the shaft.

The intercondylar area (**3.238**A–C), between the condylar articular surfaces, is rough, narrowest centrally and here forms an *intercondylar eminence*, edges of which project slightly proximally as *lateral* and *medial intercondylar tubercles*. At the back and front of the eminence the intercondylar area widens as the articular surfaces diverge.

The tibial tuberosity, at the proximal end of the anterior border, is the truncated apex of a triangular area, mentioned above, where anterior condylar surfaces merge. It projects little, and is divided into a distal rough and a proximal smooth region. The former is palpable, separated from skin merely by a sub-

cutaneous *infrapatellar bursa*; to the proximal part the ligamentum patellae is attached.

The proximal tibial surface slopes posteriorly and downwards relative to the shaft's long axis; the tilt, maximal at birth, decreases with age; it is more marked in habitual squatters (Kate & Robert 1965). The *medial articular surface* is *oval* (long axis anteroposterior) and perceptibly longer in conformity with differences between the femoral condyles (p. 438). Around its anterior, medial and posterior margins it is related to the medial meniscus, the area of contact being flat. The meniscal imprint, wider behind, narrower anteromedially, is often discernible. The surface is otherwise concave; its raised lateral margin partly invades the medial intercondylar tubercle. The *lateral articular surface* is more *circular*; like the medial it is adapted to its meniscus. Elsewhere the surface, slightly concave to fit the femoral condyle, has a raised medial margin spreading to the lateral intercondylar tubercle. Its articular margins are sharp, except posterolaterally, where the edge is round and smooth; here the tendon of popliteus is in contact with bone.

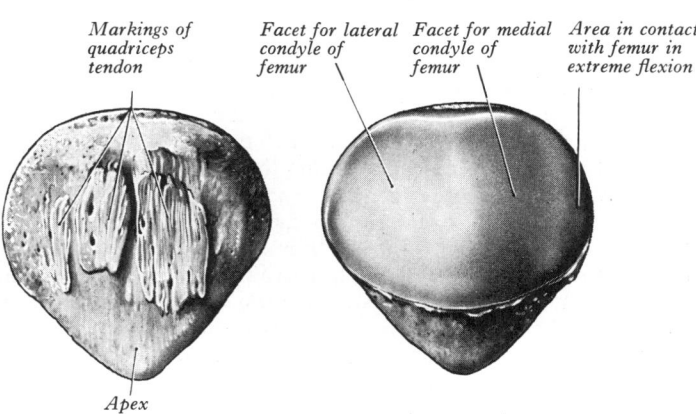

Markings of quadriceps tendon

Facet for lateral condyle of femur

Facet for medial condyle of femur

Area in contact with femur in extreme flexion

Apex

3.237 The left patella: anterior and posterior aspects, drawn from a fresh, unmacerated specimen, with the articular cartilage preserved.

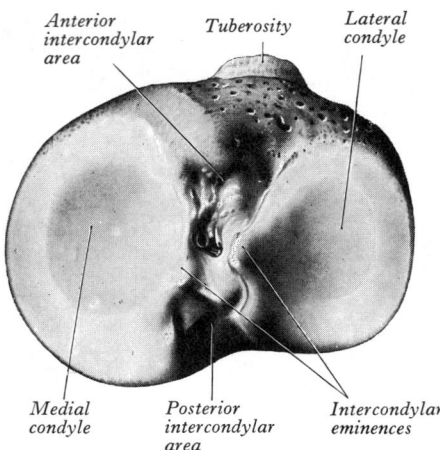

Anterior intercondylar area   Tuberosity   Lateral condyle

Medial condyle   Posterior intercondylar area   Intercondylar eminences

3.238A The proximal articular surface of the right tibia. The imprints of the menisci were very conspicuous in this specimen.

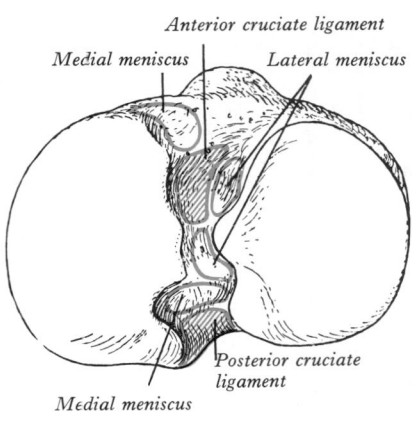

Anterior cruciate ligament

Medial meniscus   Lateral meniscus

Posterior cruciate ligament

Medial meniscus

3.238B Outline of A showing the attachments of the menisci and cruciate ligaments.

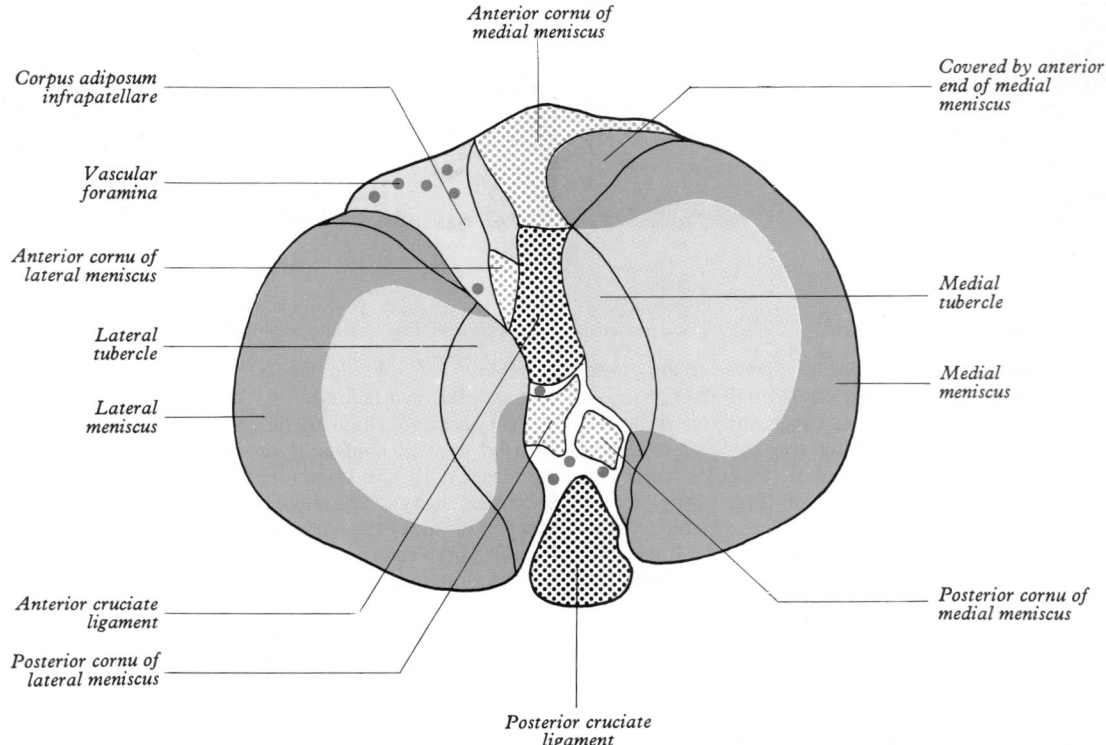

Anterior cornu of medial meniscus

Corpus adiposum infrapatellare

Covered by anterior end of medial meniscus

Vascular foramina

Anterior cornu of lateral meniscus

Medial tubercle

Lateral tubercle

Medial meniscus

Lateral meniscus

Anterior cruciate ligament

Posterior cornu of medial meniscus

Posterior cornu of lateral meniscus

Posterior cruciate ligament

3.238C Detailed analysis of the surface features of the proximal aspect of the human tibia (left). The condylar areas are shown in full blue, the parts in contact with the menisci being in deep blue, while the remaining condylar regions are in light blue. The attachments of the meniscal cornua are in blue stipple, and those of the cruciate ligaments are in black stipple. The yellow area indicates the extent of contact with the corpus adiposum and the red dots signify vascular foramina. (Adapted with permission from Klaus Jacobsen, Department of Orthopaedic Surgery, The Gentofte Hospital, Copenhagen, *Journal of Anatomy* and Cambridge University Press.)

The anterior intercondylar area (3.238B,C), widest anteriorly, bears on its anteromedial area, anterior to the medial articular surface, a depression in which the anterior cornu of the medial meniscus is attached. Behind this a smooth area receives the anterior cruciate ligament. The anterior cornu of the lateral meniscus is attached anterior to the intercondylar eminence, lateral to the anterior cruciate ligament. The eminence, with medial and lateral tubercles, is the narrow, central part of the area. To its posterior slope the posterior cornu of the lateral meniscus is attached and behind this a **posterior intercondylar area** inclines down and back. A depression behind the base of the medial intercondylar tubercle is for the posterior cornu of the medial meniscus. The rest of the area is smooth and for attachment of the posterior cruciate ligament, spreading back to a ridge for the capsular ligament.

To the proximal edge of the groove on the medial condyle's posterior surface are attached the capsular and posterior part of the tibial collateral ligaments; its distal edge receives the semimembranosus. At the lateral end of the groove a tubercle is the main attachment of the tendon of semimembranosus (Cave & Porteous 1958). To medial and anterior condylar surfaces, marked by vascular foramina, the medial patellar retinaculum is attached.

Measurement of 13 healthy knee joints, with observations on 7 macerated specimens, yielded a highly detailed plan of ligamentous and other attachments (Jacobsen 1974 and 3.238C); much variation was noted; dense fibrous attachments produced facets separated by more porous areas (p. 444).

The tibia has been investigated stereometrically by Ljungre (1976); analysis confirmed functional adaptation of curvature

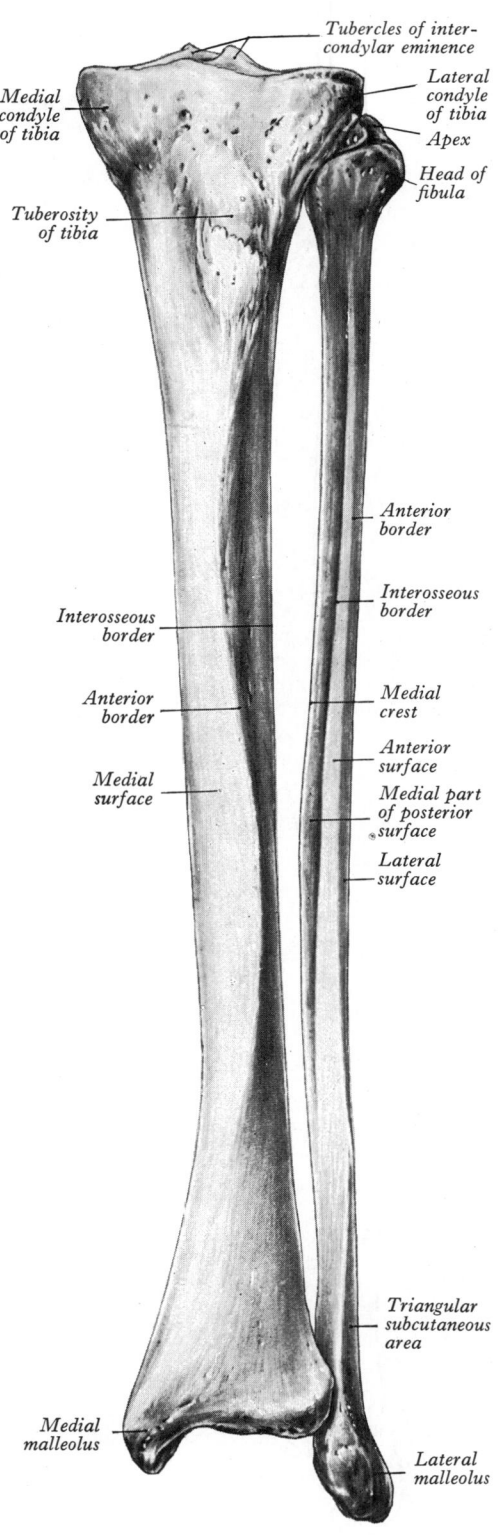

Tubercles of inter-condylar eminence

Medial condyle of tibia

Lateral condyle of tibia

Apex

Head of fibula

Tuberosity of tibia

Anterior border

Interosseous border

Interosseous border

Anterior border

Medial crest

Anterior surface

Medial surface

Medial part of posterior surface

Lateral surface

Triangular subcutaneous area

Medial malleolus

Lateral malleolus

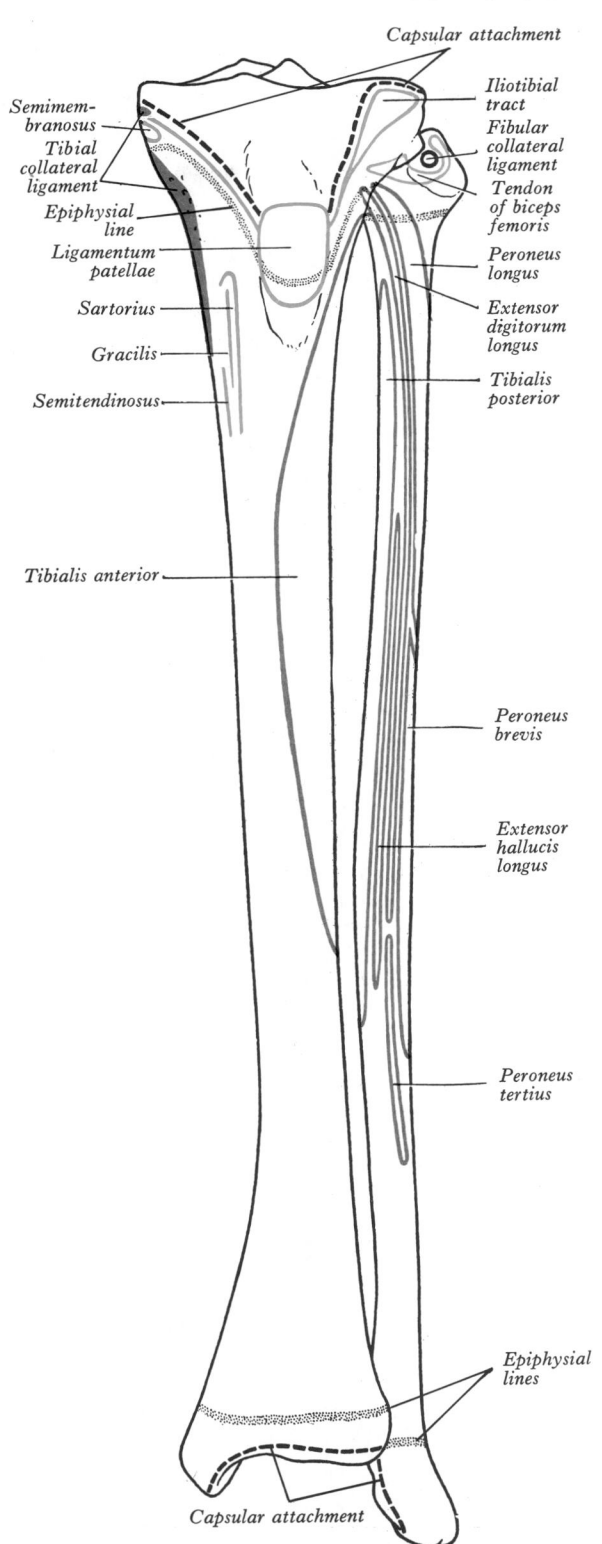

Capsular attachment

Semimem-branosus

Iliotibial tract

Tibial collateral ligament

Fibular collateral ligament

Epiphysial line

Tendon of biceps femoris

Ligamentum patellae

Peroneus longus

Sartorius

Extensor digitorum longus

Gracilis

Semitendinosus

Tibialis posterior

Tibialis anterior

Peroneus brevis

Extensor hallucis longus

Peroneus tertius

Epiphysial lines

Capsular attachment

3.239A Left tibia and fibula: anterior aspect. Compare with key, **3.239B**.

3.239B Key to **3.239A**. The epiphysial lines are stippled; the interrupted lines correspond to the attachments of the capsular ligaments.

and inclinations of tibial condyles and tuberosity, and these were correlated with habits of locomotion in different racial groups.

The fibular facet on the lateral condyle faces distally and posterolaterally. Superomedial to it the condyle is grooved posteriorly by the tendon of popliteus, with a synovial recess between tendon and bone. The condyle's anterolateral aspect is separated from the shaft's lateral surface by a sharp margin for attachment of deep fascia. The iliotibial tract makes a flat but definite marking (triangular and facet-like) on its anterior aspect. Slips from the tendon of biceps femoris are attached anteroproximal to the fibular facet, distal to which proximal fibres of extensor digitorum longus and occasionally peroneus longus are attached.

A line across the **tibial tuberosity** marks the distal limit of the epiphysial line (**3.239B** and p. 445); ligamentum patellae is attached to the smooth bone proximal to this, its superficial fibres reaching a rough area distal to the line; the attachment may be marked distally by a slight ridge (Lewis 1958a). Distally the tuberosity is subcutaneous; proximal to it bone is related to the deep aspect of the ligamentum patellae, but a *deep infrapatellar bursa* and fibroadipose tissue intervene. In habitual squatters a vertical groove on the anterior surface of the lateral condyle is occupied by the ligament's edge in flexion.

The shaft (**3.239A,B,240A,B**), triangular in section, has medial, lateral and posterior surfaces, separated by anterior, lateral (interosseous) and medial borders. It is thinnest at the junction of middle and distal thirds, expanding towards both ends.

The *anterior border* descends from the tuberosity to the anterior margin of the medial malleolus and is subcutaneous throughout;

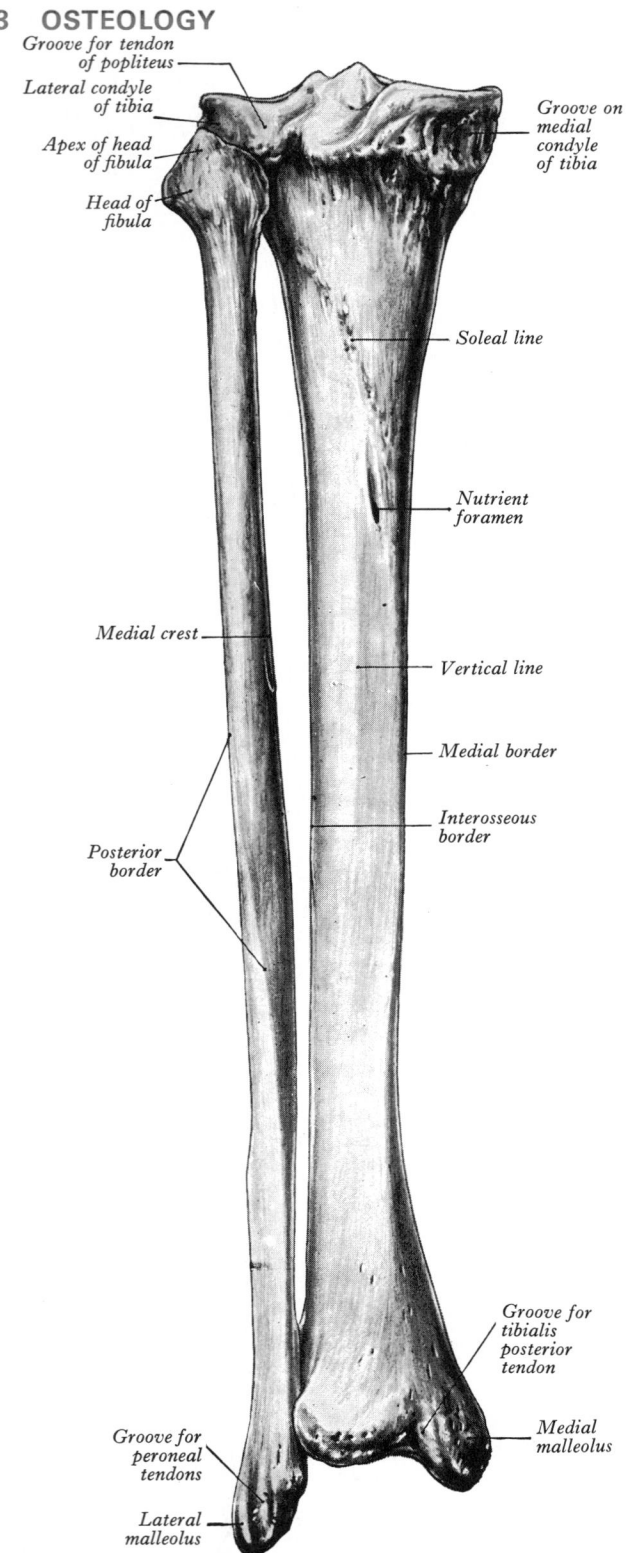

Groove for tendon
of popliteus

Lateral condyle
of tibia

Apex of head
of fibula

Head of
fibula

Groove on
medial
condyle
of tibia

Soleal line

Nutrient
foramen

Medial crest

Vertical line

Medial border

Interosseous
border

Posterior
border

Groove for
tibialis
posterior
tendon

Medial
malleolus

Groove for
peroneal
tendons

Lateral
malleolus

3.240A Left tibia and fibula: posterior aspect. Compare with Key, 3.240B.

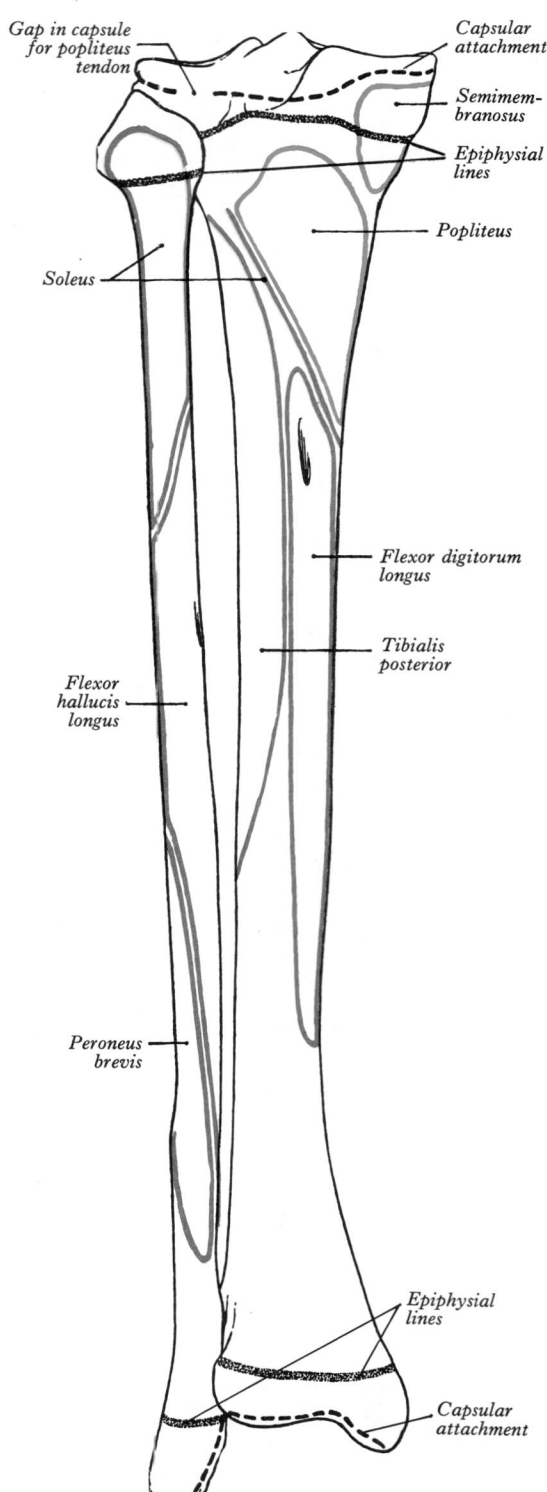

Gap in capsule
for popliteus
tendon

Capsular
attachment

Semimem-
branosus

Epiphysial
lines

Popliteus

Soleus

Flexor digitorum
longus

Tibialis
posterior

Flexor
hallucis
longus

Peroneus
brevis

Epiphysial
lines

Capsular
attachment

3.240B Key to 3.240A. The epiphysial lines are stippled; the interrupted lines correspond to the attachments of the capsular ligaments.

except in its distal fourth, where it is indistinct, it is a sharp crest, slightly sinuous and turning medially in its distal fourth. The *interosseous border* commences distal and anterior to the fibular facet and descends to the anterior border of the fibular notch; to most of it is attached the interosseous membrane, connecting tibia to fibula; above it is indistinct. The *medial border* descends from the anterior end of the groove on the medial condyle to the posterior margin of the medial malleolus. Its proximal and distal fourths are ill-defined, its middle region sharp.

The *medial surface* (really *antero-medial*), between anterior and medial borders, is broad, smooth and almost entirely subcutaneous. The broad, smooth *lateral surface*, between anterior and interosseous borders, faces laterally in its proximal three-fourths and is transversely concave. Its distal quarter swerves

anteriorly, due to medial deviation of the anterior and distal interosseous borders. This part of the surface is somewhat convex. The *posterior surface*, between interosseous and medial borders, is widest above, where it is crossed distomedially by an oblique, rough *soleal line* (3.240A), from the centre of which a faint *vertical line* descends but soon fades. A large vascular groove adjoins the line's end, descending distally into bone; it may be lateral or medial to the vertical line. To the *anterior border* is attached crural deep fascia and, proximal to the medial malleolus, the medial end of the superior extensor retinaculum. Proximal to the soleal line the *medial border* is an attachment of popliteal fascia, the posterior fibres of the tibial collateral ligament and slips of semimembranous; distal to it some fibres of soleus and fascia covering deep crural muscles are attached. The distal media border merges into the medial lip of a groove for the tendon of tibialis posterior. To the *lateral border* interosseous membrane is attached, except at its extremes. It is proximally indistinct where a large gap in the membrane transmits anterior tibial vessels; distally the border is the anterior boundary of the fibular notch; the anterior tibiofibular ligament is attached to it.

The *medial surface* bears proximally, near the medial border, an area about 5 cm long and 1 cm wide for the anterior part of the tibial collateral ligament and, behind this, semimembranosus; anterior to this are from before, backwards, linear attachments of the tendons of sartorius, gracilis and semitendinosus, which rarely mark the bone (**3.239B**). The remaining surface is subcutaneous but crossed obliquely by the great saphenous vein ascending from *in front* of the medial malleolus.

The *lateral surface* is, in its proximal two-thirds, an attachment for the tibialis anterior. Its distal third, devoid of attachments, is crossed by the tendon of tibialis anterior (lateral to the anterior border), extensor hallucis longus, anterior tibial vessels and nerve, extensor digitorum longus and peroneus tertius, in mediolateral order.

To the *posterior surface* popliteus is attached in a triangular area proximal to the soleal line, except near the fibular facet, and to the *soleal line* the popliteal aponeurosis, soleus and its fascia and deep transverse crural fascia. Proximally the line does not reach the interosseous border and has a tubercle for the medial end of the tendinous soleal arch; lateral to it the posterior tibial vessels and nerve descend on tibialis posterior. Distal to the soleal line, the *vertical line* separates attachments of flexor digitorum longus (medial) and tibialis posterior (**3.240B**). The surface's distal quarter has no attachments, but is crossed medially by the tendon of tibialis posterior travelling to a posterior groove on the medial malleolus. Flexor digitorum longus crosses obliquely behind the tibialis posterior; but posterior tibial vessels and nerve and flexor hallucis longus contact only the lateral part of the distal posterior surface.

The tibia's distal end, slightly expanded, has anterior, medial, posterior, lateral and distal surfaces. It projects inferomedially as the medial malleolus.

Its smooth *anterior surface* bulges beyond the distal surface, separated from it by a narrow groove, continuing the shaft's lateral surface (p. 443). The *medial surface*, smooth and continuous above and below with medial surfaces of shaft and malleolus, is subcutaneous and visible. The *posterior surface* is crossed near its medial end by a nearly vertical, but slightly oblique groove, usually conspicuous, extending to the posterior surface of the malleolus; elsewhere it is smooth and continuous with the shaft's posterior surface. The *lateral surface* is the triangular

fibular notch, bound by ligaments to the fibula; its anterior and posterior edges project and converge proximally to the interosseous border. The floor of the notch is roughened proximally by a substantial interosseous ligament, but is smooth distally and sometimes covered by articular cartilage. The *distal tibial surface*, articulating with the talus, is wider in front, concave sagittally and transversely slightly convex (i.e. it is *sellar*). Medially it continues into the malleolar articular surface. This articular surface may extend into the groove (vide supra) separating it from the shaft's anterior surface. Such extensions, medial or lateral or both, are *squatting facets*, articulating with reciprocal talar facets (p. 450) in extreme dorsiflexion. These features have been used in racial evaluation (p. 273).

The medial malleolus, short and thick, has a smooth lateral surface with a crescentic facet articulating with the medial talar surface. Its anterior aspect is rough and its posterior continues the groove on the shaft's posterior surface. The distal border is pointed anteriorly, posteriorly depressed.

The distal end is related anteriorly to tendons, vessels and nerves lateral to the tibial shaft (vide supra). To an anterior groove near the articular surface, the articular capsule of the ankle joint is attached. The posterior groove is adapted to the tendon of tibialis posterior, which usually separates that of flexor digitorum longus from bone. More laterally, posterior tibial vessels and nerve and flexor hallucis longus are in contact with this surface. To the edges of the *fibular notch* are attached anterior and posterior tibiofibular ligaments. The *medial malleolus* ends proximal to the lateral malleolus, which is also more posterior in plane. To its anterior surface is attached the ankle joint's articular capsule. To the groove for the posterior tibial tendon, on its prominent medial border, the flexor retinaculum is attached. The deltoid ligament, proximal to the distal malleolar border, is attached to its apex and depression.

Ossification of the tibia is by three centres (**3.239B, 240B, 241**) in the shaft and both epiphyses. It begins in midshaft about the seventh intrauterine week. The proximal epiphysial centre is usually present at birth; at about 10 years a thin anterior process from it descends to form the smooth part of the tibial tuberosity (**3.241**), but a separate centre for this may appear about the twelfth year, soon fusing with the epiphysis. Distal strata of the epiphysial plate are of dense collagenous tissue, the fibres of which are aligned with the ligamentum patellae. This peculiar structure is attributed to large tensile stresses transmitted via the ligament (Lewis 1958a, Smith 1962b). The distal epiphysial centre appears early in the first year, joining the shaft about the fifteenth in females, seventeenth in males. The proximal epiphysis fuses in the sixteenth year in females, eighteenth in males. The medial malleolus is an extension from the distal

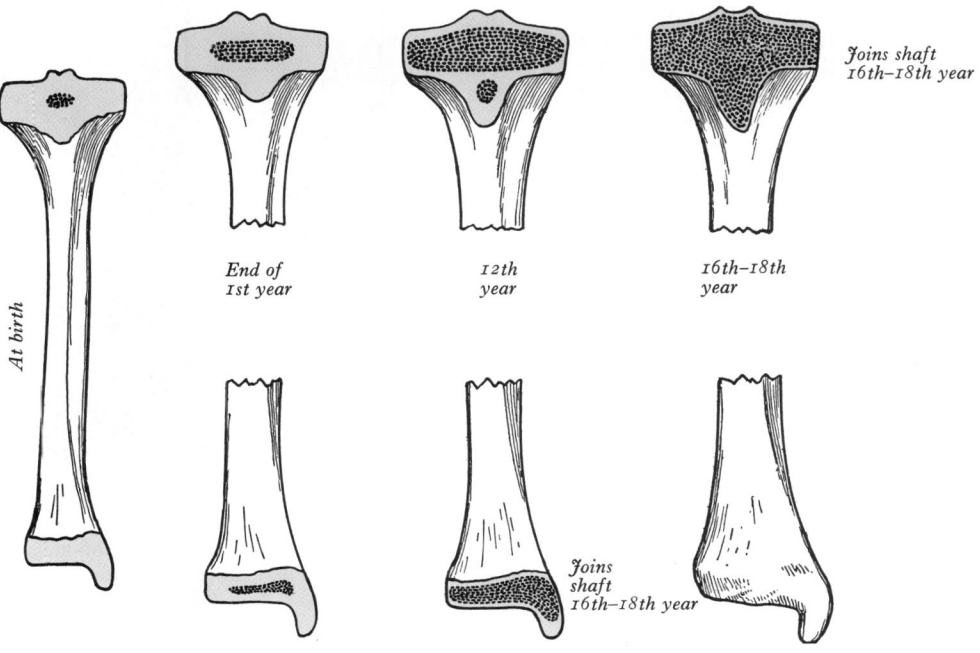

Joins shaft 16th–18th year

End of 1st year

12th year

16th–18th year

At birth

Joins shaft 16th–18th year

3.241 Stages in the ossification of the tibia (not to scale).

epiphysis, commencing to ossify in the seventh year; it may have a separate centre.

The Fibula

The fibula (3.239–244) is much more slender than the tibia, not being directly involved in transmission of weight. It has a proximal *head*, a long *shaft* and a distal *lateral malleolus*. A thinner part near the head is the *neck*. The shaft varies in form, being variably moulded by attached muscles; these variations may be confusing.

The head, slightly expanded, projects in front, behind and laterally. A round facet, on its proximomedial aspect, articulates with a facet on the inferolateral surface of the lateral tibial condyle; it faces proximally and anteromedially. A blunt *apex (styloid process)* projects proximally from its posterolateral aspect, palpable about 2 cm distal to the knee joint. The common peroneal nerve crosses posterolateral to the neck and can be rolled against bone, causing tingling sensations on the dorsum of foot and toes, especially on the medial side of the hallux.

The distal end or **lateral malleolus** projects distally and posteriorly. Its *lateral aspect* is subcutaneous; its *posterior aspect* has a broad groove with a prominent lateral border. Its *anterior aspect* is rough, round and continuous with the tibial inferior border. The *medial surface* has a triangular articular facet, vertically convex, its apex distal (3.244); it articulates with the lateral talar surface. Behind the facet is a rough *malleolar fossa*.

The shaft (3.239,240) has three borders and surfaces, each associated with a particular group of muscles. The *anterior border* ascends proximally from the apex of an elongated, triangular area continuous with the lateral malleolar surface (vide infra), to the anterior aspect of the fibular head. The *posterior border*, continuous with the medial margin of the posterior groove on the lateral malleolus, is usually distinct distally, but often rounded in its proximal half. The *interosseous border* is medial to the anterior border and usually more posterior (3.242); but in the proximal two-thirds of the bone they approximate, the 'surface' being narrowed to 1 mm or less.

The *lateral surface*, between anterior and posterior borders and associated with peroneal muscles, faces laterally in its proximal three-fourths, the distal quarter twists (spirals) to become continuous with the *posterior groove* of the lateral malleolus. The *anteromedial* (sometimes simply termed *anterior*, or *medial* unqualified) *surface*, between anterior and interosseous borders, usually faces anteromedially but often forwards. Distally wide, it narrows in its proximal half and may be a mere ridge; it is associated with extensor muscles. The *posterior surface*, the largest, between interosseous and posterior borders, is associated with flexor muscles. Its proximal two-thirds is divided by a longitudinal *medial crest*, separated from the interosseous border by a grooved surface, directed medially; the rest of the surface faces back in its proximal half, but its distal half curves on to the medial aspect; distally this area occupies the tibia's fibular notch, roughened by the attachment of the principal interosseous tibiofibular ligament.

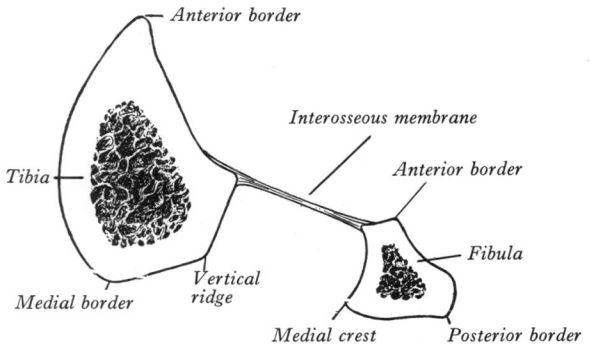

3.242 Transverse section through the right tibia and fibula, showing the attachment of the crural interosseous membrane: proximal aspect.

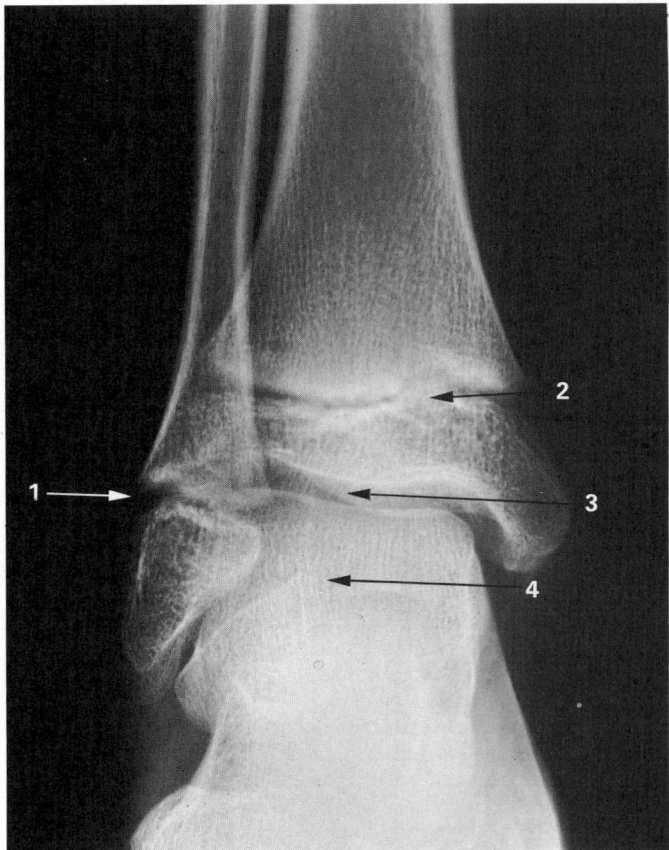

A

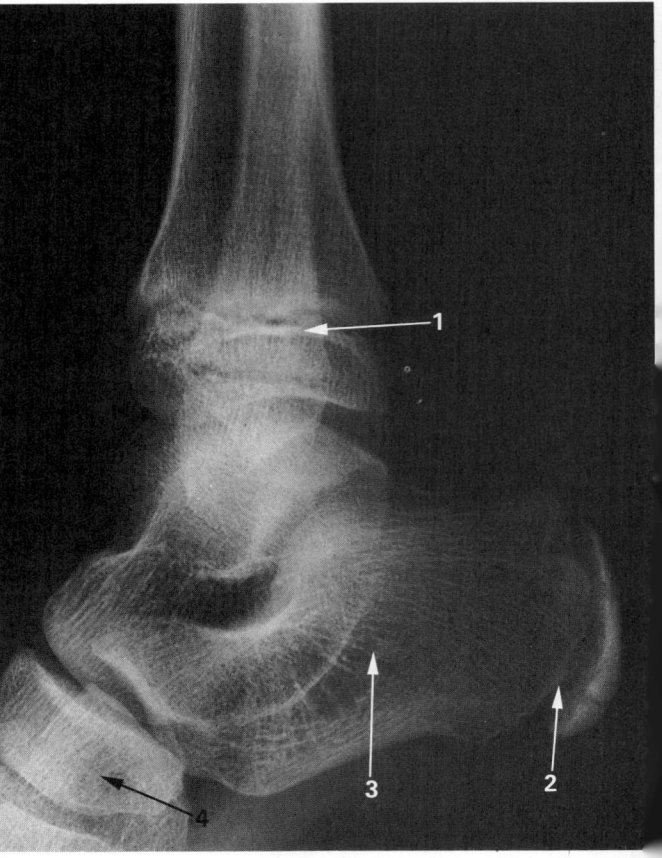

B

3.243 Ankle region of a child of 10 in plantar flexion. A. Obliquely anteroposterior. B. Lateral.

A 1. Inferior growth cartilage of fibula. 2. Inferior growth cartilage of tibia. 3. Talocrural joint. 4. Talus. Note that the fibular growth cartilage is approximately at the level of the talocrural joint.

B 1. Tibial growth cartilage. 2. Growth cartilage of epiphysis of posterior surface of calcaneus. 3. Note trabecular pattern in calcaneus. 4. Shadow of navicular bone superimposed on that of cuboid.

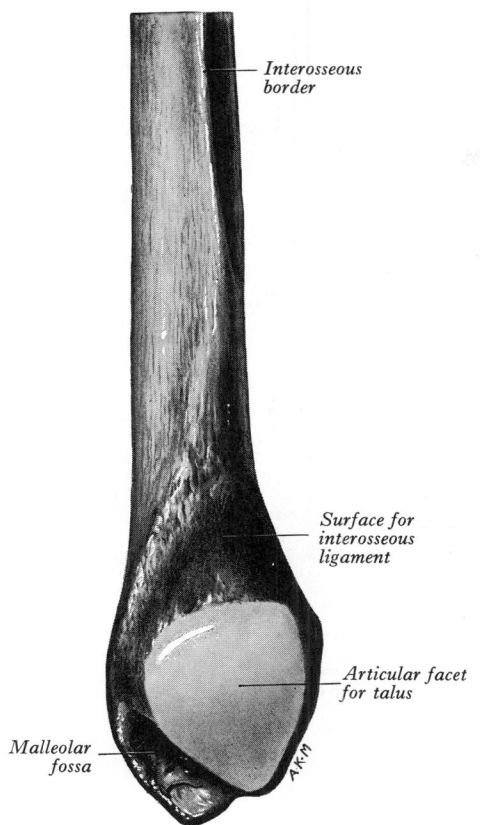

3.244 The lower end of the left fibula: medial aspect.

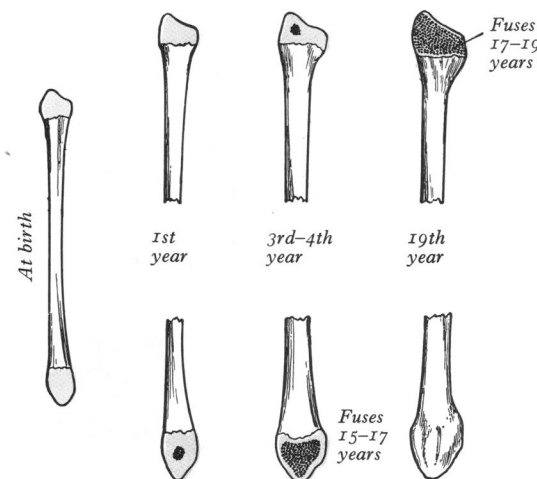

3.245 Stages in the ossification of the fibula.

The triangular area proximal to the lateral surface of the lateral malleolus (**3.239**A) *is subcutaneous*; the rest of the shaft is obscured by muscles. To the *fibular head* are attached extensor digitorum longus in front, peroneus longus anterolaterally and soleus behind; the fibular collateral ligament is attached in front of its apex and embraced by the main attachments of biceps femoris. To the margins of the articular facet is attached the tibiofibular capsular ligament.

The *anterior border* divides distally into two ridges enclosing the subcutaneous triangular surface (**3.239**A). The anterior intermuscular septum is attached to its proximal three-fourths, the lateral end of the superior extensor retinaculum distally on the anterior border of the triangular area. Distally on the triangular area's posterior margin is attached the lateral end of the superior peroneal retinaculum. The *interosseous border* ends at the proximal limit of the rough area for the interosseous ligament. The interosseous membrane attached to it does not reach the fibular head, leaving space for anterior tibial vessels. The *posterior border* is proximally indistinct, the posterior intermuscular septum attached to all but its distal end. The *medial crest* is related to the peroneal artery, with a nutrient foramen on or near it close to mid-shaft. Also attached to it is a layer of deep fascia separating tibialis posterior from flexor hallucis longus and flexor digitorum longus.

The *anteromedial (extensor) surface* has attached to it extensor digitorum longus, extensor hallucis longus and peroneus tertius (p. 646). To the *lateral (peroneal) surface* peroneus longus is attached to its whole width in its proximal third but in its middle third only from its posterior part, behind the peroneus brevis (p. 647). The latter continues its attachment almost to the distal end of the fibula's shaft. The *posterior surface*, divided longitudinally by the medial crest, has complex attachments: between crest and interosseous border it is concave and often crossed by an oblique ridge for an intramuscular tendon of tibialis posterior, which is attached throughout much of this region—usually confined to the proximal three-fourths. Between crest and posterior border in the proximal fourth of the posterior surface the soleus is attached, its tendinous arch being attached to it proximally. Distal to the soleus on this surface the flexor hallucis longus is attached, almost to the bone's distal end. A little proximal to its midpoint it is pierced by a nutrient foramen, directed distally, for a branch of the peroneal artery.

To the anterior surface of the *lateral malleolus* the anterior talofibular ligament is attached and, to the notch anterior to its apex, the calcaneofibular ligament. Tendons of peroneus brevis and longus groove its posterior aspect, the latter superficial and covered by the superior peroneal retinaculum. In the *malleolar fossa* (**3.244**), pitted by vascular foramina, the posterior tibiofibular ligament is attached posteriorly and, distal to this, the posterior talofibular ligament.

Ossification of the fibula is by three centres (**3.243,245**) in the shaft and extremities. It begins in the shaft about the eighth intrauterine week, in its distal end in the first year and in its proximal about the third in females and fourth in males. The distal epiphysis unites with the shaft about the fifteenth year in females, seventeenth in males, whereas the proximal does not unite until about the seventeenth year in females, nineteenth in males. In this respect the fibula reverses the ossificatory pattern in other long bones.

THE SKELETON OF THE FOOT

Functionally the pedal skeleton may be divided into tarsus, metatarsus and phalanges or digital bones. In this description the terms *plantar* and *dorsal*, which are self-explanatory, are used, anterior and posterior being inapposite. The terms proximal and distal are also usually employed, with the same significance as in limbs generally. Rotation occurring in early stages of development of limbs (p. 398) explains that, whereas the pollex is the most lateral in the hand, the hallux is the most medial in the foot (p. 400). **Note**. It must be emphasized, as noted elsewhere (p. 12), that these remarks are only valid with reference to the widely adopted descriptive Standard Anatomical Position. Perhaps of assistance to embalmers and prosectors, this has little to recommend it concerning habitual functioning, comparative studies and energetics.

The Tarsus

The seven tarsal bones occupy the proximal half of the foot (**3.246,247**). The tarsus and carpus are homologous, but tarsal

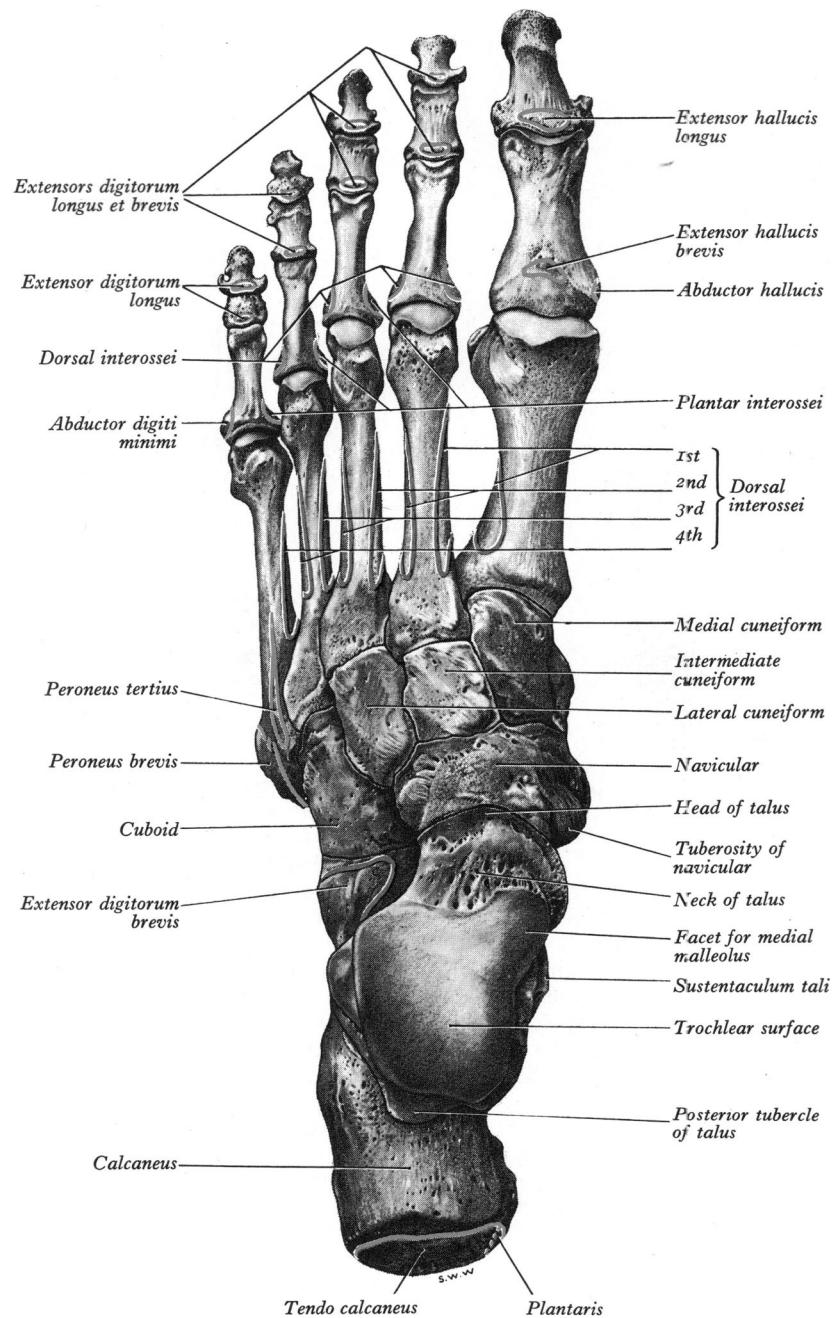

Extensor hallucis
longus

Extensors digitorum
longus et brevis

Extensor digitorum
longus

Extensor hallucis
brevis

Dorsal interossei

Abductor hallucis

Abductor digiti
minimi

Plantar interossei

1st
2nd } Dorsal
3rd } interossei
4th

Medial cuneiform

Intermediate
cuneiform

Peroneus tertius

Lateral cuneiform

Peroneus brevis

Navicular

Head of talus

Cuboid

Tuberosity of
navicular

Extensor digitorum
brevis

Neck of talus

Facet for medial
malleolus

Sustentaculum tali

Trochlear surface

Posterior tubercle
of talus

Calcaneus

Tendo calcaneus   Plantaris

3.246 The skeleton of the left foot: dorsal aspect.

elements are larger to support and distribute weight. As in the carpus, tarsal bones are arranged in proximal and distal rows, but medially is a single intermediate tarsal element. The proximal row comprises the *talus* and *calcaneus*, the former's long axis inclined anteromedially and down, its distal head being medial to the calcaneus and at a higher level. The distal row contains, mediolaterally, *medial, intermediate* and *lateral cuneiforms* and the *cuboid*, which are roughly in parallel and form a transverse arch dorsally convex. Medially, the *navicular bone* is interposed between the talus and cuneiforms. Laterally, the calcaneus articulates with the cuboid.

The foot is at right angles to the leg in standing, tarsus and metatarsus arranged to form intersecting longitudinal and transverse arches. Hence thrust and weight are not transmitted from the tibia to the ground (or vice versa) directly through the tarsus, but are distributed through tarsal and metatarsal bones to the ends of the longitudinal arches (p. 457). In this connection, the cancellous structure of tarsal bones has been reviewed by Sinha (1985). For description each tarsal bone is arbitrarily considered

to be cuboidal in form with six surfaces.

Structure of Tarsal Bones

The internal structure of tarsal bones has attracted relatively littl[e] attention (Wood Jones 1949, Le Gros Clark 1975). Sinha (1985) in a small series of 10 cadavers, has described the cancellou[s] architecture of all seven bones. His observations emphasize th[e] variation, from the mixed fine and coarse pattern of lamellae in th[e] calcaneus, to the more widely spaced, thicker lamellae of th[e] navicular, and he attempts to equate these appearances wit[h] functional stresses acting on these bones. The subject is of con[-] siderable kinematic and clinical interest and deserves furthe[r] study. One such investigation (Bacon et al 1984), using a neutro[n] beam technique on thin sections, shows how the orientation [of] hydroxyapatite crystals may be used to analyse the cancellou[s] architecture and to associate this pattern with effects of nor[-] osseous structures, such as the plantar aponeurosis.

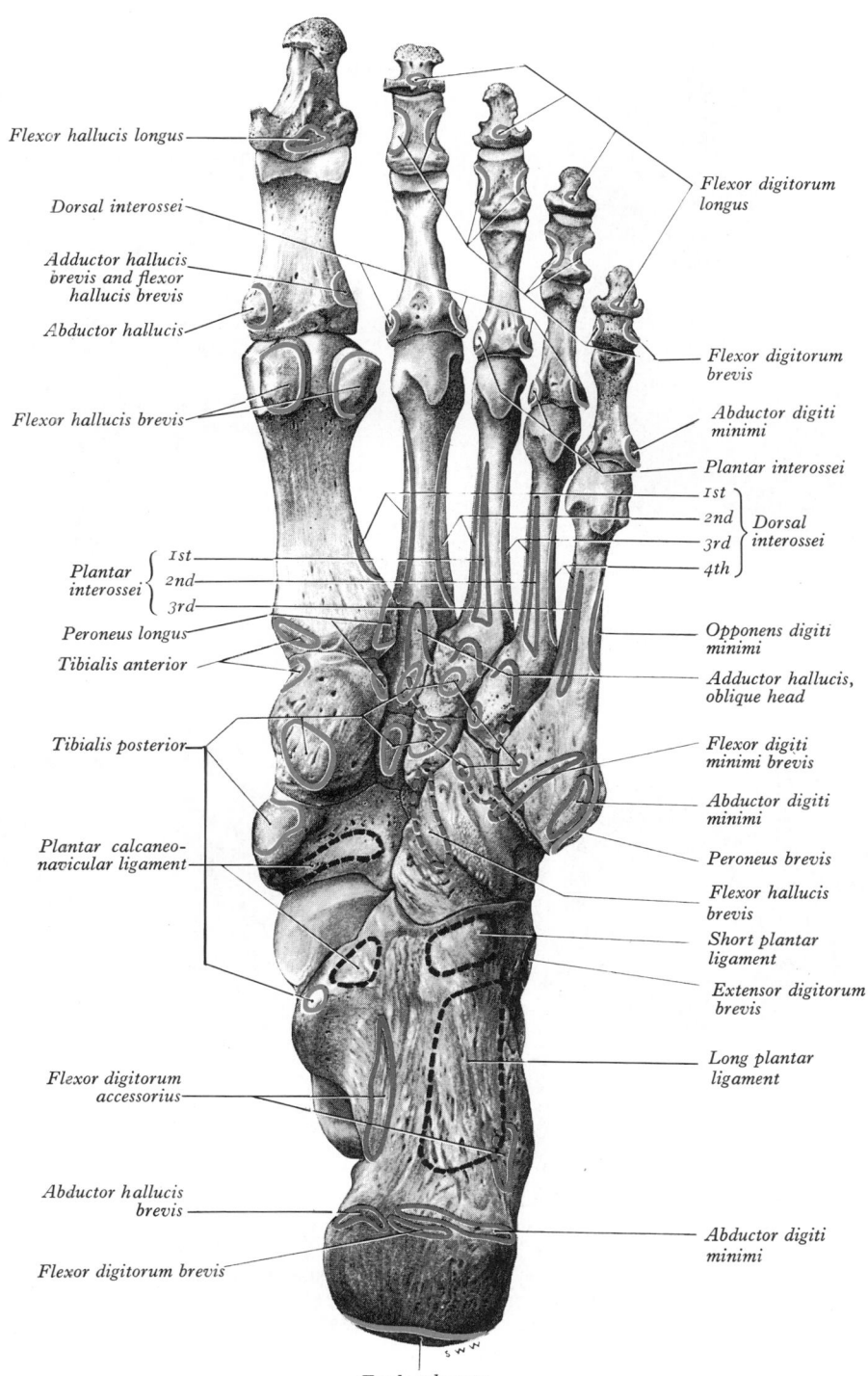

Flexor hallucis longus

Dorsal interossei

Adductor hallucis
brevis and flexor
hallucis brevis

Abductor hallucis

Flexor hallucis brevis

Plantar { 1st
interossei { 2nd
{ 3rd

Peroneus longus

Tibialis anterior

Tibialis posterior

Plantar calcaneo-
navicular ligament

Flexor digitorum
accessorius

Abductor hallucis
brevis

Flexor digitorum brevis

Flexor digitorum
longus

Flexor digitorum
brevis

Abductor digiti
minimi

Plantar interossei
1st
2nd } Dorsal
3rd } interossei
4th

Opponens digiti
minimi

Adductor hallucis,
oblique head

Flexor digiti
minimi brevis

Abductor digiti
minimi

Peroneus brevis

Flexor hallucis
brevis

Short plantar
ligament

Extensor digitorum
brevis

Long plantar
ligament

Abductor digiti
minimi

Tendo calcaneus

3.247 The skeleton of the left foot: plantar aspect. The attachments of the tibialis posterior to the metatarsals vary; those to the third and fifth may be absent. The adductor hallucis (oblique head) and the flexor hallucis brevis arise in large part from the ligaments and tendinous extensions in the sole of the foot and not directly from bone; these attachments are shown by interrupted lines.

The Individual Tarsal Bones

The Talus

The talus (**3.**248A–D) is the link between the foot and leg, through the ankle joint. Its rounded distal head, proximal trochlear surface for the tibia, facet for the lateral malleolus, and its neck and body are its main features.

The talar head, directed distally and somewhat inferomedially, has a *distal surface*, oval and convex, with its long axis also inclined inferomedially to articulate with the proximal navicular surface. The head's *plantar surface* has three articular areas, separated by smooth ridges; the most posterior and largest is oval, slightly convex and rests on a shelf-like medial calcanean projection, the *sustentaculum tali*. Anterolateral to this and usually continuous with it, a flat articular facet rests on the anteromedial part of the dorsal (proximal) calcanean surface; distally it continues into the navicular surface. Medial to these two calcanean facets a part of the talar head is covered with articular cartilage, continuous with the calcanean navicular areas (**3.**248B) and in contact with the *plantar calcaneonavicular ligament* (p. 537) which is covered here, superiorly, by a plaque of fibrocartilage. When the foot is inverted passively, the dorsolateral aspect of the head is visible and palpable about 3 cm distal to the tibia, but is hidden by extensor tendons when the toes are dorsiflexed.

The talar neck, the narrow region between head and body, is medially inclined. Its rough surfaces are for ligaments; the medial plantar surface has a deep *sulcus tali* which, when talus and calcaneus are articulated, roofs the *sinus tarsi*, occupied by interosseous talocalcanean and cervical ligaments.

449

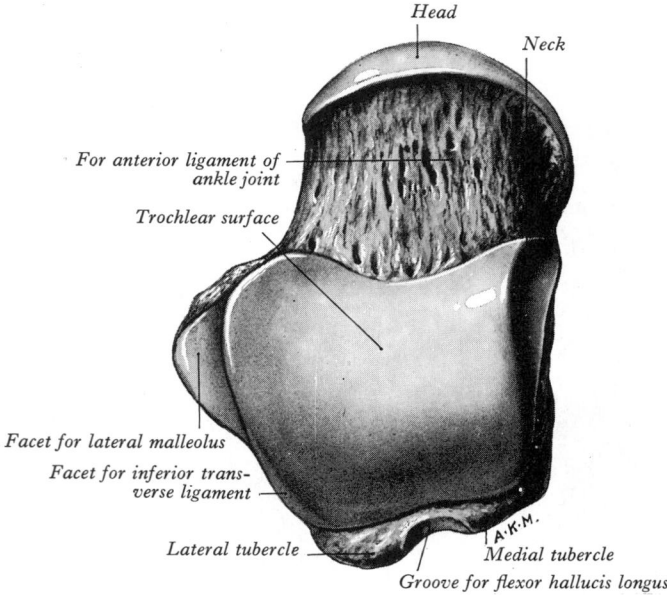

A

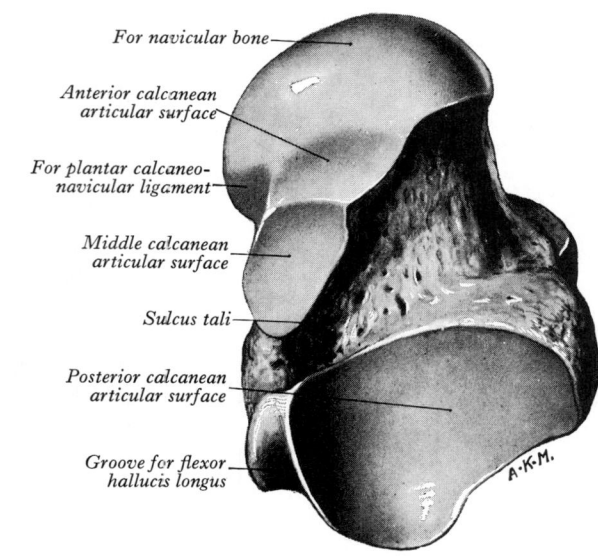

B

C

D

3.248 The left talus: A. dorsal (superior) aspect; B. plantar (inferior) aspect; C. medial aspect; D. lateral aspect.

The talar body is cuboidal, covered dorsally by a *trochlear* surface articulating with the tibia's distal end. It is anteroposteriorly convex, transversely gently concave, widest anteriorly and, therefore, sellar. The triangular *lateral surface* is smooth and vertically concave for articulation with the lateral malleolus. Superiorly it is continuous with the trochlear surface; inferiorly its apex is a *lateral process*. The *medial surface* is proximally (posterosuperiorly) covered by a comma-shaped facet, deeper in front and articulating with the medial malleolus, distal to which the surface is rough with numerous vascular foramina. The small *posterior surface* is the rough projecting *posterior process*, marked by an oblique groove between two tubercles; the *lateral tubercle* is usually larger, the *medial* less prominent and immediately behind the sustentaculum tali (**3.246**). The *plantar surface* articulates with the middle third of the dorsal calcanean surface by an oval concave facet, its long axis directed distolaterally at an angle of about 45° with the median plane.

No muscles but many ligaments (**4.246,247**) are attached to the talus, since it is involved in talocrural, subtalar and talocalcaneonavicular joints.

The long axis of the **neck**, inclined down, distally and medially, makes an angle of about 150° with that of the body; it is smaller (130°–140°) at birth (**3.249**), accounting in part for the inverted foot in young children. To its dorsal surface the dorsal talonavicular ligament and talocrural articular capsule are

attached distally, leaving its proximal part intracapsular. The medial articular facet and part of the trochlear surface may extend on the neck, especially in childhood (**3.249**). A dorsolateral *squatting facet* commonly occurs on the neck in Indians, but seldom in adult Europeans, articulating with the anterior tibial margin in extreme dorsiflexion (Barnett 1954, Singh 1959); the facet may be

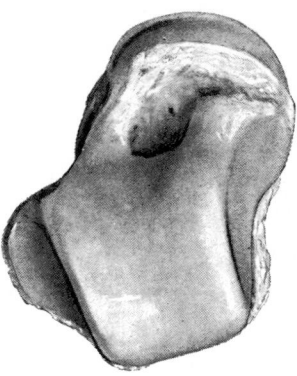

3.249 The left talus of a newborn infant: superior aspect. Compare wit **3.248**A, and note the angle which the axis of the neck makes with the lo axis of the body of the bone.

double. Laterally on the neck is attached the anterior talofibular ligament, spreading along the adjacent anterior border of the lateral surface. To the neck's inferior surface the interosseous talocalcanean and cervical ligaments are attached (p. 537).

The medial edge of the **trochlear surface** is straight but its lateral edge inclines medially in its posterior part, often broadened into a small elongated triangular area, in contact with the posterior tibiofibular ligament in dorsiflexion.

To the *posterior process* is attached the posterior talofibular ligament, which extends up to the groove, or depression, between the process and posterior trochlear border. To its plantar border the posterior talocalcanean ligament is attached. The *groove* between the tubercles of the process contains the tendon of flexor hallucis longus, continuing distally into the groove on the plantar aspect of the sustentaculum tali. To the *medial tubercle* the medial talocalcanean ligament is attached below, whilst above the tubercle are attached the most posterior superficial fibres of the deltoid ligament, its deep fibres are attached still higher to the rough area immediately below the comma-shaped articular facet on the medial surface (**3.**248C).

Sex differences in talar dry weight have been recorded (Singh & Singh 1975); the ranges in adult Indian males and females were 15.1 to 36.8 and 6.0 to 20.5 g, respective means being 23.50 and 15.32 g. Estimation of sex by dimensions has been claimed by Steele (1976).

The Calcaneus

The calcaneus (**3.**250A–D), the largest tarsal bone, projects posterior to the tibia and fibula as a short lever for muscles of the calf attached to its posterior surface. It is irregularly cuboidal, its long axis inclined distally up and laterally. Its smooth, articular anterior end contrasts with its larger, rough posterior aspect. The dorsal surface bears centrally a large articular facet and the plantar is rough; the lateral surface is flat, the medial hollowed.

The superior or **proximal surface** is divisible into three: the posterior third is rough, concavoconvex, the convexity transverse; it supports fibroadipose tissue between the calcanean tendon and ankle joint; the middle third carries the *posterior talar facet*, oval and convex anteroposteriorly; the anterior third is partly articular; distal (anterior) to the posterior articular facet a rough

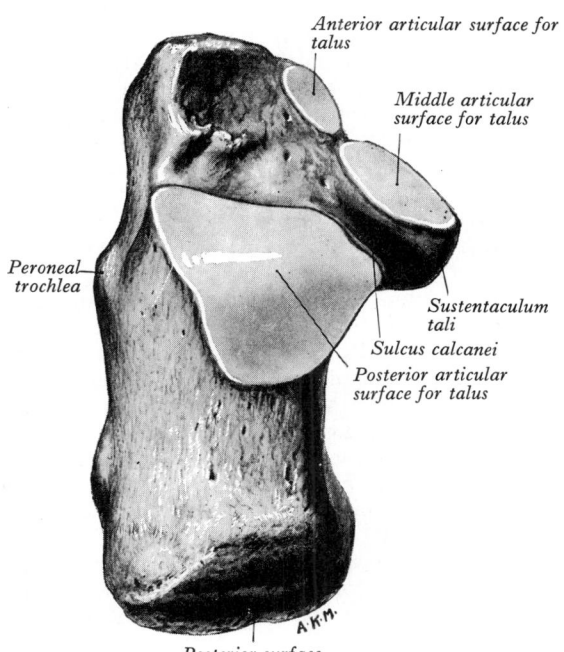

Anterior articular surface for talus
Middle articular surface for talus
Peroneal trochlea
Sustentaculum tali
Sulcus calcanei
Posterior articular surface for talus
Posterior surface

A

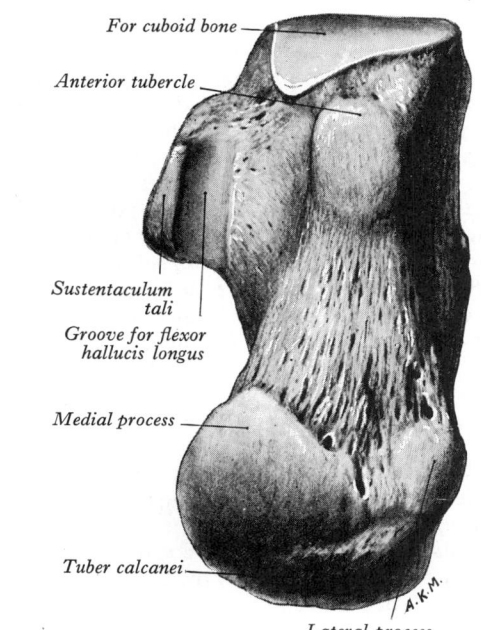

For cuboid bone
Anterior tubercle
Sustentaculum tali
Groove for flexor hallucis longus
Medial process
Tuber calcanei
Lateral process

B

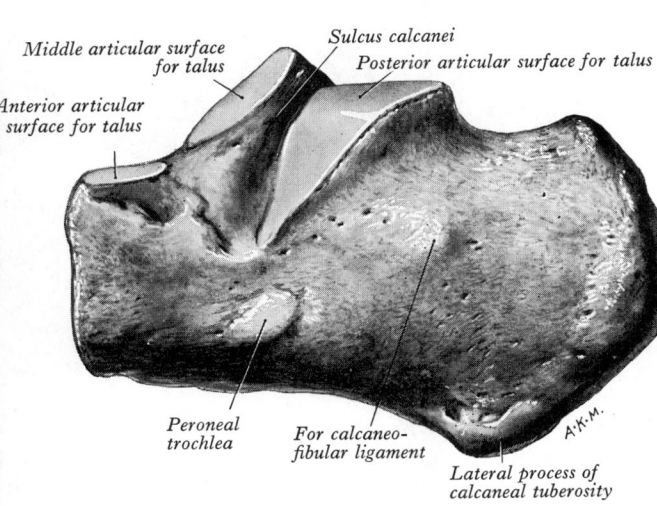

Middle articular surface for talus
Sulcus calcanei
Posterior articular surface for talus
Anterior articular surface for talus
Peroneal trochlea
For calcaneo-fibular ligament
Lateral process of calcaneal tuberosity

C

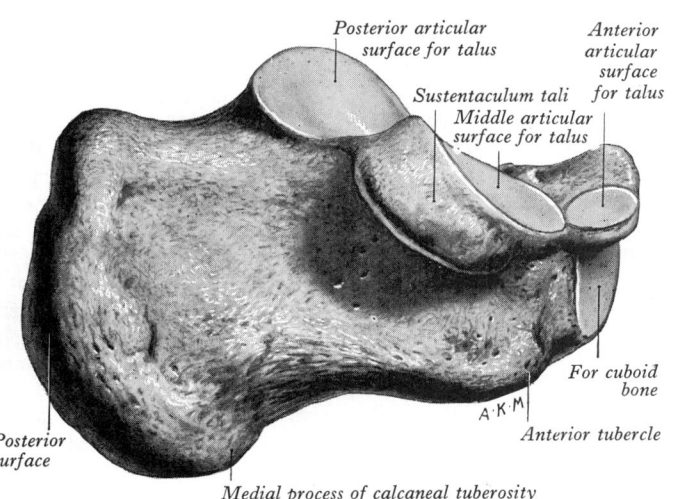

Posterior articular surface for talus
Anterior articular surface for talus
Sustentaculum tali
Middle articular surface for talus
For cuboid bone
Anterior tubercle
Posterior surface
Medial process of calcaneal tuberosity

D

3.250 The left calcaneus: A. dorsal aspect; B. plantar aspect; C. lateral aspect; D. medial aspect.

depression narrows into a groove on the medial side, the *sulcus calcanei*, which completes the sinus tarsi with the talus. Distal and medial to this groove an elongated articular area covers the sustentaculum tali, extending distolaterally on the bone's body. This facet is often divided by a non-articular interval at the anterior limit of the sustentaculum tali, forming *middle* and *anterior talar facets*, the incidence of which varies with sex and race. Rarely, all three facets on the upper surface of the calcaneus are fused into one irregular area (Bunning & Barnett 1963, 1965). A detailed analysis of patterns of *anterior* talar articular facets in a series of 401 Indian calcanei revealed four types. Type I (67%) showed one continuous facet on the sustentaculum extending to the distomedial calcanean corner; type II (26%) presented two facets, one sustentacular, one distal calcanean; type III (5%) possessed only a single sustentacular facet; and type IV (2%) showed all anterior and posterior facets confluent.

The anterior surface, the smallest, is an obliquely set concavoconvex articular facet for the cuboid bone.

The posterior surface is divided into three: a smooth proximal (superior) area separated from tendo calcaneus by a bursa and adipose tissue; a middle area, the largest, limited above by a groove, below by a rough ridge, for the calcanean tendon; and a distal (inferior) area inclined down and forwards, vertically striated, which is the subcutaneous weight-bearing surface.

The plantar surface is rough, especially proximally, as the *calcanean tuberosity*, the *lateral* and *medial processes* of which extend distally, separated by a notch. The medial is longer and broader (3.250B). Further distally, an *anterior tubercle* marks the distal limit of the attachment of the long plantar ligament.

The lateral surface is almost flat, proximally deeper and palpable on the lateral aspect of the heel distal to the lateral malleolus. Distally it presents the *peroneal trochlea (tubercle)* (3.250C), exceedingly variable in size and, when well developed, palpable 2 cm distal to the lateral malleolus. It has an oblique groove for the tendon of peroneus longus and a shallower proximal one for the tendon of peroneus brevis. About 1 cm or more behind and above the peroneal trochlea a second elevation may exist for attachment of the calcaneofibular part of the lateral ligament.

The medial surface is vertically concave, its concavity accentuated by the *sustentaculum tali* projecting medially from the distal part of its upper border (3.250D). Superiorly the process bears the middle talar facets and inferiorly a groove continuous with that on the talar posterior surface for the tendon of flexor hallucis longus (3.250B). The medial aspect of the sustentaculum tali can be felt immediately distal to the tip of the medial malleolus; occasionally it is also grooved by the tendon of flexor digitorum longus.

In the *calcanean sulcus* are attached the interosseous talocalcanean and cervical ligaments (p. 537) and the medial root of the inferior extensor retinaculum; the non-articular area distal to the posterior talar facet is the attachment of the extensor digitorum brevis (in part), the principal band of the inferior extensor retinaculum and the stem of the bifurcated ligament.

To the *medial process of the tuberosity*, at its prominent medial margin, abductor hallucis and the superficial part of flexor retinaculum and, distally, the plantar aponeurosis and flexor digitorum brevis are all attached. To the *lateral process* is attached the abductor digiti minimi, extending medially to the medial process. The rough region between the processes proximally and extending to the anterior tubercle distally is the attachment of the long plantar ligament, whilst the tubercle and area distal to it is the attachment of the short plantar ligament. The lateral tendinous head of the flexor digitorum accessorius is attached distal to the lateral process near the lateral margin of the long plantar ligament. To the *posterior surface*, which is wider below, and near the medial side of the tendo calcaneus, the plantaris is attached. The anterior part of the *lateral surface* is crossed by the peroneal tendons, but is largely subcutaneous (p. 651). The calcaneofibular ligament is attached about 1–2 cm proximal to the peroneal trochlea, usually to a low rounded elevation.

The *sustentaculum tali* is dorsally part of the talocalcaneonavicular joint; its plantar surface is grooved by the tendon of flexor hallucis longus, margins of the groove giving attachment to

the deep part of the flexor retinaculum. To the medial margin of the sustentaculum, which is narrow, rough and convex, the plantar calcaneonavicular ligament is attached distally (p. 537), proximally are attached a slip from the tendon of tibialis posterior, and superficial fibres of the deltoid ligament and medial talocalcanean ligament. Distal to the attachment of the deltoid ligament the tendon of flexor digitorum longus is related to the margin of the sustentaculum and may groove it. Distal to the groove for the flexor hallucis longus the large medial head of flexor accessorius is attached.

Singh & Singh (1975) studied variations in calcanean dry weight in both sexes in adult Indians. The male range was 23.3 to 56.8 g (mean = 36.64), and 13.0 to 33.0 g (mean = 24.75) in females. Steele (1976) has claimed sex differentiation on dimensional data. Comparison of right and left calcanei in 52 adults (Webber & Garnett 1976) showed greater density on the side of hand preference; the evidence suggested that the concordance was determined before birth.

The Navicular Bone

The navicular bone (3.251A,B) articulates between the talar head proximally and cuneiform bones distally. Its *distal surface* is transversely convex and divided into three facets (the medial is largest) for articulation with cuneiform bones. The *proximal surface*, oval and concave, articulates with the talar head. The *dorsal surface* is rough and convex; the *medial*, also rough, continues into a prominent *tuberosity*, palpable about 2.5 cm distal and plantar to the medial malleolus. The *plantar surface*, rough and concave, is separated from the tuberosity medially by a groove. The *lateral surface* is rough, irregular and often bears a facet for articulation with the cuboid.

The facet for the medial cuneiform is roughly triangular, its rounded apex is medial and its 'base' often markedly curved; those for intermediate and lateral cuneiform bones are also triangular with plantar apices. The facet for the lateral cuneiform may approach a wide crescent or a semicircle rather than a triangle (3.251A). To the dorsal navicular surface are attached dorsal talonavicular, cuneonavicular and cubonavicular ligaments. The **tuberosity** is the main attachment of tibialis posterior; a groove lateral to it transmits part of the tendon distally to the cuneiform bones and middle three metatarsal bases. To a slight projection lateral to the groove and adjacent to the proximal surface the plantar calcaneonavicular ligament is attached. To the rough part of the lateral surface is attached the calcaneonavicular part of the bifurcated ligament.

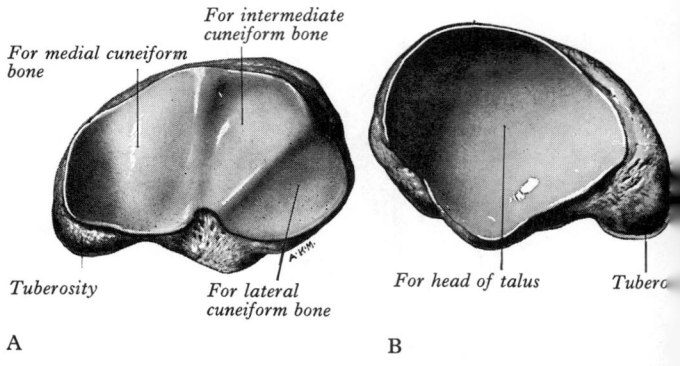

For intermediate cuneiform bone

For medial cuneiform bone

For head of talus

Tubero

Tuberosity

For lateral cuneiform bone

A

B

3.251 The left navicular bone: A. distal aspect; B. proximal aspect.

The Cuneiform Bones

The wedge-like cuneiform bones (3.252,253,254) articulate with the navicular proximally and with the bases of the first to third metatarsal bones distally; the medial is the largest, the intermediate smallest. In intermediate and lateral cuneiforms the dorsal surface is the base of the wedge but in the medial the wedge is reversed, a prime factor in shaping the transverse arch. Proxim surfaces of all three form a concavity for the navicular, but medial and lateral cuneiforms project distally beyond the intermediate, forming a recess for the second metatarsal base.

The medial cuneiform bone (3.252A,B), articulating with the navicular and first metatarsal base, has a rough, narrow *dorsal surface*. The *plantar surface* receives a slip from the tendon of tibialis posterior. The *distal surface* is a reniform facet for the first metatarsal base, its 'hilum' being lateral. The *proximal surface* bears a piriform facet for the navicular, concave vertically, dorsally narrowed. The *medial surface*, rough and subcutaneous, is vertically convex; its distal–plantar angle carries a large impression for most of the tendon of tibialis anterior (3.252A). The *lateral surface* is partly non-articular; along its proximal and dorsal margins is a smooth right-angled strip for the intermediate cuneiform; its distal dorsal area is separated by a vertical ridge from a small, almost square facet for the dorsal part of the medial surface of the second metatarsal base. Plantar to this the medial cuneiform is attached to the medial side of the second metatarsal base by a strong ligament. Proximally an interosseous intercuneiform ligament connects this surface to the intermediate

cuneiform. The surface's distal and plantar area is roughened by attachment of part of the peroneus longus tendon (3.252B).

The intermediate cuneiform bone (3.253A,B), articulating with the navicular and distally with the second metatarsal base, has a narrow, *plantar surface* receiving a slip from the tendon of tibialis posterior. The *distal* and *proximal surfaces*, both triangular articular facets, articulate with the second metatarsal base and navicular respectively. The *medial surface* is partly articular; along its proximal and dorsal margins a smooth, angled strip, occasionally double, articulates with the medial cuneiform. The *lateral surface* also is partly articular; along its proximal margin a vertical strip, usually indented, abuts with the lateral cuneiform. Strong interosseous ligaments connect non-articular parts of both surfaces to adjacent cuneiforms.

The lateral cuneiform bone (3.254A,B) is between the intermediate cuneiform and cuboid, articulating also with the navicular and, distally, the third metatarsal base. Like the intermediate, its *dorsal surface*, rough and almost rectangular, is the base of the wedge. Its narrow *plantar* surface receives a slip from the tibialis posterior and sometimes part of the flexor hallucis brevis. The *distal surface* is a triangular articular facet for the third metatarsal base. The *proximal surface* is rough on its plantar aspect, but its dorsal two-thirds articulates with the navicular by a triangular facet. The *medial surface*, partly non-articular, has on its proximal margin a vertical strip, indented for the intermediate cuneiform; on its distal margin a narrower strip (often two small facets) articulates with the lateral side of the second metatarsal base. The *lateral surface*, also partly non-articular, has a triangular or oval proximal facet for the cuboid; a semilunar facet on its dorsal and distal margin articulates with the dorsal part of the medial side of the fourth metatarsal base. Non-articular areas of medial and lateral surfaces receive intercuneiform and cuneocuboid ligaments, important in the transverse arch.

The Cuboid Bone

The cuboid bone (3.255A,B), most lateral in the distal tarsal row, is between the calcaneus proximally and the fourth and fifth metatarsals distally. Its *dorsal surface*, really dorsolateral, is rough for the attachment of ligaments. The *plantar surface* is crossed distally by an oblique *groove for the tendon of peroneus longus*, bounded proximally by a ridge ending laterally in the *cuboid's tuberosity*, the lateral aspect of which is faceted for a sesamoid bone or cartilage frequent in the peroneal tendon. Proximal to its ridge the rough plantar surface, due to obliquity of the calcaneocuboid joint, extends proximally and medially, making its medial border much longer than the lateral. The *lateral surface* is rough; from a deep notch on its plantar edge extends the groove for the peroneus longus. The *medial surface*, much more extensive and partly non-articular, bears an oval facet for the lateral cuneiform and proximal to this another (sometimes absent) for the navicular, the two forming a continuous surface separated by a smooth vertical ridge. The *distal surface* is divided vertically into a medial

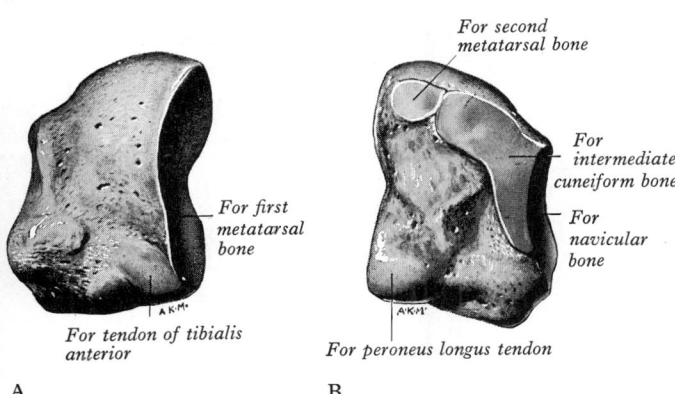

For second metatarsal bone

For first metatarsal bone

For intermediate cuneiform bone

For navicular bone

For tendon of tibialis anterior

For peroneus longus tendon

A

B

3.252 The left medial cuneiform bone: A. medial aspect; B. lateral aspect.

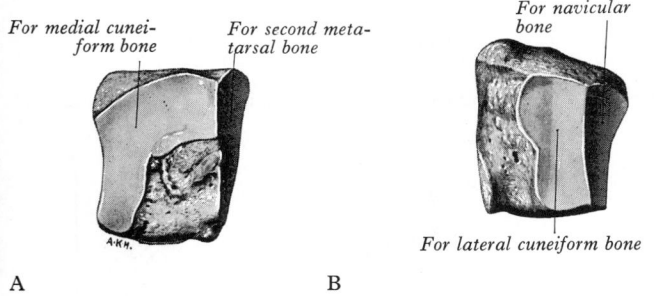

For medial cuneiform bone

For second metatarsal bone

For navicular bone

For lateral cuneiform bone

A

B

3.253 The left intermediate cuneiform bone: A. distal and medial aspect; B. proximal and lateral aspect.

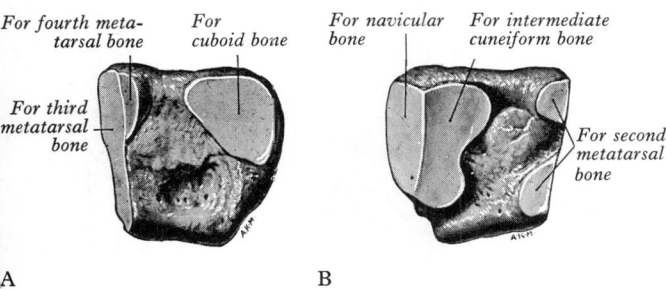

For fourth metatarsal bone

For cuboid bone

For navicular bone

For intermediate cuneiform bone

For third metatarsal bone

For second metatarsal bone

A

B

3.254 The left lateral cuneiform bone: A. distal and lateral aspect; B. proximal and medial aspect.

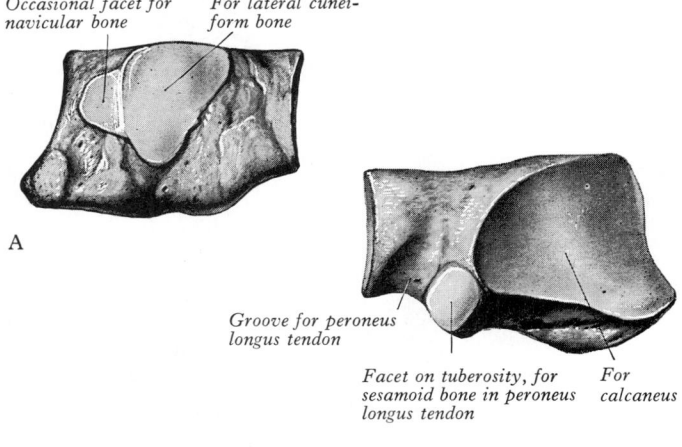

Occasional facet for navicular bone

For lateral cuneiform bone

A

Groove for peroneus longus tendon

Facet on tuberosity, for sesamoid bone in peroneus longus tendon

For calcaneus

B

3.255 The left cuboid bone: A. medial aspect; B. proximal and lateral aspect. 453 quadrilateral articular area for the fourth metatarsal base and a lateral triangular area, its apex lateral, for the fifth metatarsal base. The *proximal surface*, triangular and concavoconvex, articulates with the distal calcanean surface; its medial–plantar angle projects proximally and inferior to the distal end of the calcaneus.

To the dorsal surface are attached dorsal calcaneocuboid, cubonavicular, cuneocuboid and cubometatarsal ligaments, and, to the proximal edge of the plantar ridge, deep fibres of the long plantar ligament. To the projecting proximomedial part of the plantar surface are attached a slip of the tendon of tibialis posterior and the flexor hallucis brevis. To the rough part of the medial cuboidal surface are attached interosseous, cuneocuboid and cubonavicular ligaments, and proximally the medial calcaneocuboid, which is the lateral limb of the bifurcated ligament.

The Metatarsus

The five metatarsal bones, distal in the foot, connect tarsus and phalanges. Like metacarpals, they are miniature long bones, with a shaft, proximal base and distal head. Excepting the first and fifth, the *shafts* are long and slender, longitudinally convex on dorsal, concave on plantar aspects. Prismatic in section, they taper distally. Their *bases* articulate with the distal tarsal row and with each other. The line of each tarsometatarsal joint, except the first, inclines proximally and laterally, metatarsal bases being oblique relative to their shafts. The *heads* articulate with proximal phalanges, each by a convex surface passing farther on to its plantar surface, where it ends on the summits of two eminences. The sides of the heads are flat, with a depression surmounted by a dorsal tubercle for a collateral ligament of the metatarsophalangeal joint.

The Individual Metatarsal Bones

The first metatarsal bone (3.256), shortest and thickest, has a strong *shaft*, of marked prismatic form. The *base* sometimes has a lateral facet or ill-defined smooth area due to contact with the second metatarsal. Its large proximal surface, usually indented on medial and lateral margins, articulates with the medial cuneiform; its circumference is grooved for tarsometatarsal ligaments and, medially, part of the tendon of tibialis anterior is attached; its plantar angle has a rough, oval, lateral prominence for the tendon of peroneus longus. To the shaft's flat lateral surface the medial head of the first dorsal interosseous muscle is attached. The large *head* has a plantar elevation separating two grooved facets, the

medial larger, on which sesamoid bones glide. (For the arterial supply of the first metatarsal consult Jaworek 1973.)

The second metatarsal bone (3.257) is the longest. Its cuneiform *base* bears four articular facets: a proximal one, concave and triangular, for the intermediate cuneiform; a dorsomedial one for the medial cuneiform, variable in size and usually continuous with that for the intermediate cuneiform; two lateral, dorsal and plantar, separated by non-articular bone, each divided by a ridge into distal demifacets articulating with the third metatarsal base and a proximal pair (sometimes continuous) for the lateral cuneiform; these areas vary, particularly the plantar facet, which may be absent (Singh 1960). An oval pressure facet, due to contact with the first metatarsal, may appear on the medial side of the base, plantar to that for the medial cuneiform. To the shaft's medial and lateral surfaces are respectively attached the lateral head of the first dorsal interosseous muscle and medial head of the second.

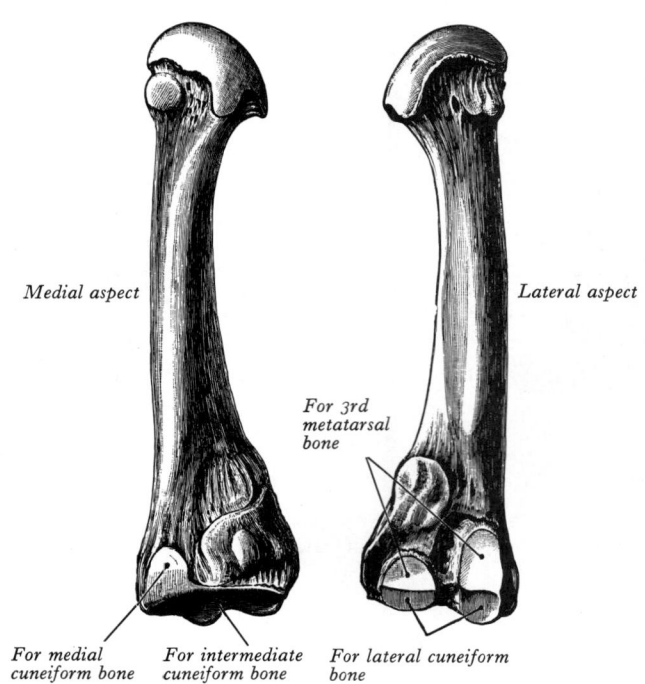

Medial aspect   *Lateral aspect*

For 3rd metatarsal bone

For medial cuneiform bone   *For intermediate cuneiform bone*   *For lateral cuneiform bone*

3.257 The left second metatarsal bone: medial and lateral aspects.

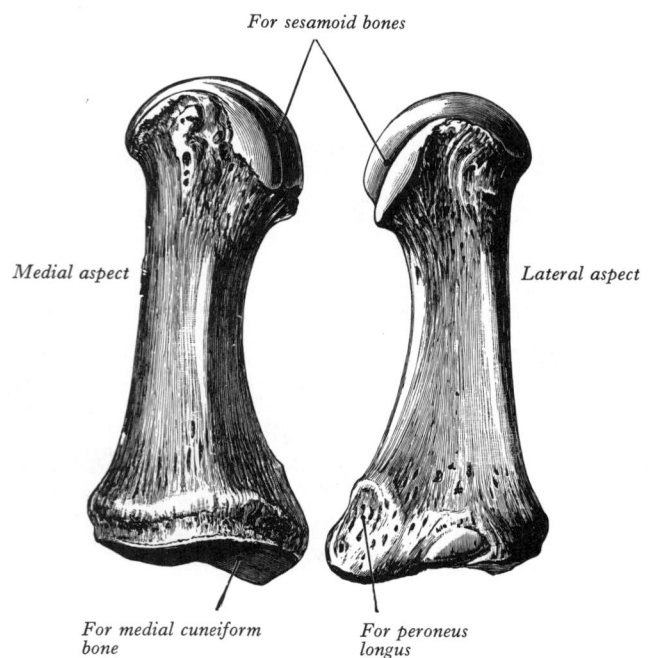

For sesamoid bones

Medial aspect   *Lateral aspect*

For medial cuneiform bone   *For peroneus longus*

3.256 The left first metatarsal bone: medial and lateral aspects.

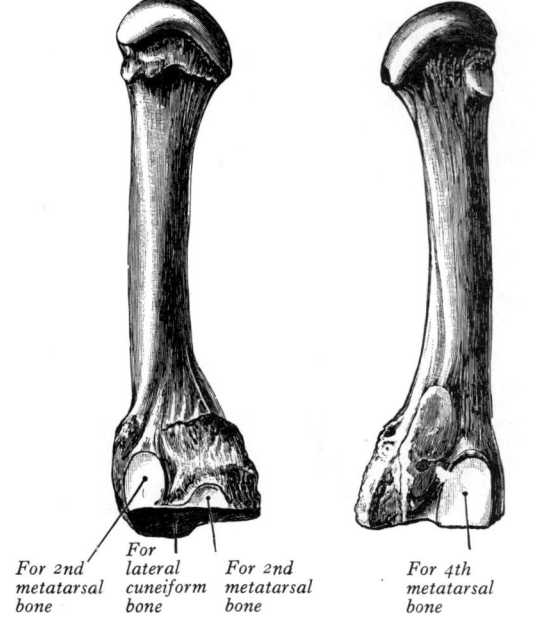

For 2nd metatarsal bone   *For lateral cuneiform bone*   *For 2nd metatarsal bone*   *For 4th metatarsal bone*

3.258 The left third metatarsal bone: medial and lateral aspects.

The **third metatarsal bone** (3.258) has a flat triangular *base*, articulating proximally with the lateral cuneiform; medially, dorsal and plantar facets articulate with the second metatarsal and, laterally, a single facet with the dorsal angle of the fourth. The medial plantar facet is frequently absent. To the shaft's medial surface the lateral head of the second dorsal interosseous muscle and first plantar are attached, to its lateral surface the medial head of third dorsal interosseous muscle.

The **fourth metatarsal bone** (3.259) is smaller than the third. Proximally its *base* has an oblique quadrilateral facet for articulation with the cuboid, laterally a single facet for the fifth metatarsal and medially an oval facet for the third, sometimes divided by a ridge, the proximal part then articulating with the lateral cuneiform (3.259). To the medial surface the lateral head of the third dorsal and second plantar interosseous muscles are attached; to the lateral surface is the medial head of the fourth dorsal interosseous muscle.

The **fifth metatarsal bone** (3.260) has a *tuberosity (styloid*

process) on the lateral side of its base. The base articulates proximally with the cuboid by a triangular, oblique surface, medially with the fourth metatarsal. The tendon of peroneus tertius is attached to the medial part of the dorsal surface and medial border of the shaft, that of the peroneus brevis to the dorsal surface of the tuberosity. A strong band of the plantar aponeurosis, sometimes containing muscle, connects the apex of the tuberosity to the lateral process of the calcanean tuberosity. The plantar surface of the base is grooved by the tendon of the abductor digiti minimi and the flexor digiti minimi brevis is attached here. To the shaft's medial side are attached the lateral head of the fourth dorsal and the third plantar interosseous muscles. The tuberosity can be seen and felt midway along the foot's lateral border; in acute inversion it may be fractured.

The Phalanges of the Foot

The phalanges (3.246,247) in general resemble those in the hand; there are two in the hallux, and three in each of the other toes. They are, however, much shorter and their shafts, especially those of the proximal set, are compressed from side to side. In proximal phalanges the compressed *shaft* is convex dorsally, with a plantar concavity. The *base* is concave for articulation with a metatarsal head and the *head* is a trochlea for a middle phalanx. Middle phalanges are small and short, but broader than the proximal. Distal phalanges resemble those in fingers, but are smaller and flatter; each has a broad base for a middle phalanx and an expanded distal end; a rough *tuberosity* on the plantar aspect of the latter is an attachment for the pulp, and a wider area for weight-bearing.

Tendons of the digital long flexor and extensores are attached to plantar and dorsal aspects of bases of the lateral four distal phalanges. In the hallux, flexor hallucis longus and extensor hallucis are likewise attached. The bases of middle phalanges also receive tendons of flexor digitorum brevis and extensores digitorum. The second, third and fourth proximal phalanges each receive a lumbrical on the medial side and an interosseous on each side (Wood Jones 1949). For further details of muscular, capsular and ligamentous arrangements in the toes see 3.246,247 and pp. 541 and 653. The terminal phalanx of the hallux normally shows a small degree of valgus (lateral) deviation, as may the proximal, even in persons who have never worn shoes (Barnett 1962). Comparison of the angulation in 30 students (18–21 years) and 35 aged individuals showed no increase in the latter. The deviation has also been observed in fetal specimens (Wilkinson 1954).

Ossification of the Bones of the Foot

Tarsal bones each have one centre, except the calcaneus, which has a scale-like posterior epiphysis (but vide infra). Centres for calcaneus and talus appear prenatally in the third and sixth month respectively. A detailed study of the calcaneus in 177 human fetuses between 49 and 150 mm (crown-rump length) showed 16% with a lateral perichondral mesenchymatous centre and 11% with an endochondral centre, while only 2% possessed both (Meyer & O'Rahilly 1976.) The cuboid frequently begins to ossify before birth; its centre has usually appeared by six months after birth. The medial cuneiform may have two centres and begins in the second year, the intermediate cuneiform and navicular in the third and lateral cuneiform in the first. The calcanean epiphysis covers most of the posterior and a little of the plantar surface; it begins thus to ossify in the sixth year in females, eighth in males, uniting in the fourteenth or sixteenth respectively. The posterior talar process sometimes has its own centre, and may remain separate or connected to the rest of the bone by cartilage as an *os trigonum*. Other accessory bones occur and may lead to radiological misinterpretation (Trolle 1947 and 3.262).

Metatarsal bones ossify from a primary centre for the shaft and a secondary epiphysis for the base of the first and for the head in the other four. In the second to fourth ossification begins in midshaft in the ninth prenatal week, in the first and fifth about the tenth week. The centre in the first metatarsal *base* appears about the third year, those for second to fifth metatarsal *heads* between the third and fourth years; all unite between seventeenth and

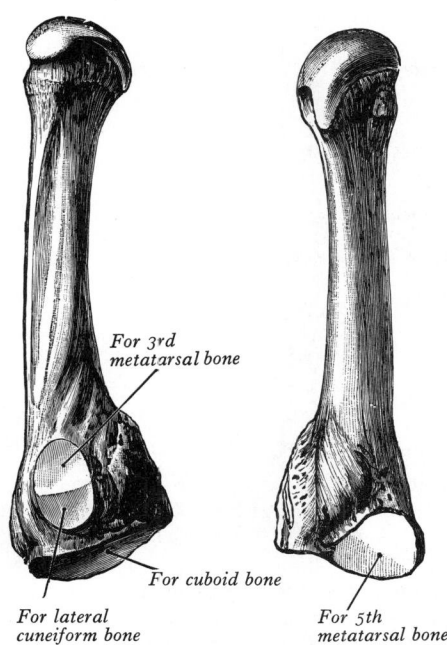

For 3rd
metatarsal bone

For cuboid bone

For lateral
cuneiform bone

For 5th
metatarsal bone

3.259 The left fourth metatarsal bone: medial and lateral aspects.

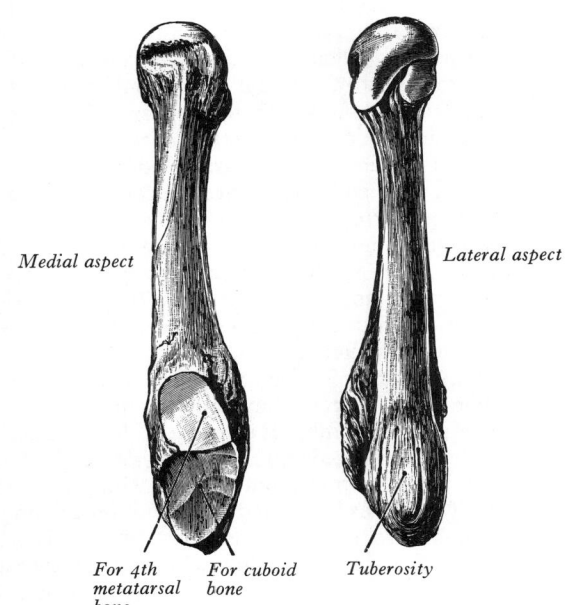

Medial aspect

Lateral aspect

For 4th   For cuboid   Tuberosity
metatarsal   bone
bone

3.260 The left fifth metatarsal bone: medial and lateral aspects.

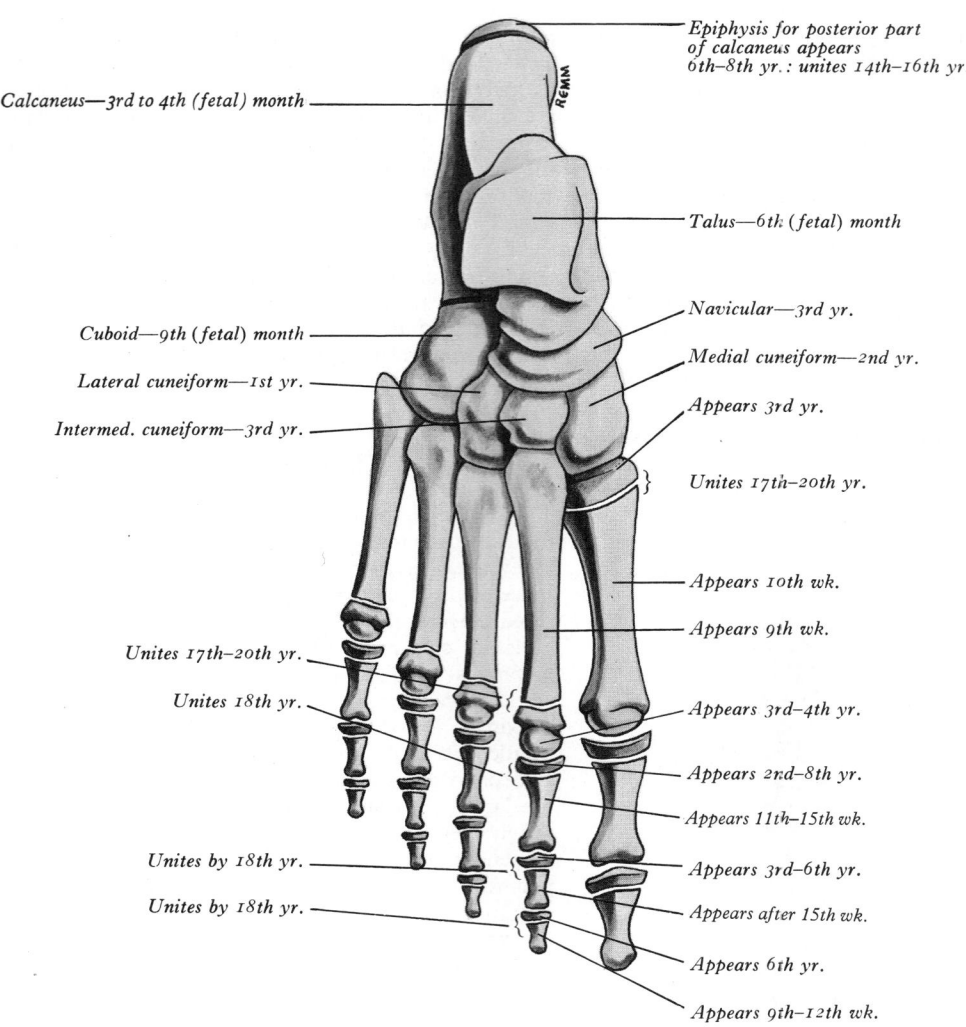

Epiphysis for posterior part
of calcaneus appears
6th–8th yr.: unites 14th–16th yr.

Calcaneus—3rd to 4th (fetal) month

Talus—6th (fetal) month

Cuboid—9th (fetal) month

Navicular—3rd yr.

Lateral cuneiform—1st yr.

Medial cuneiform—2nd yr.

Intermed. cuneiform—3rd yr.

Appears 3rd yr.

Unites 17th–20th yr.

Appears 10th wk.

Appears 9th wk.

Unites 17th–20th yr.

Unites 18th yr.

Appears 3rd–4th yr.

Appears 2nd–8th yr.

Appears 11th–15th wk.

Unites by 18th yr.

Appears 3rd–6th yr.

Unites by 18th yr.

Appears after 15th wk.

Appears 6th yr.

Appears 9th–12th wk.

3.261 Plan of the ossification of the bones of the foot.

twentieth years. Like the first *metacarpal* (p. 421), the first metatarsal may also have an epiphysis for its head. The basal tubercle of the fifth metatarsal bone is often an epiphysis.

Phalanges are ossified from a primary centre for shaft and a basal epiphysis. Primary centres for distal phalanges appear between the ninth and twelfth prenatal weeks, and even later in the fifth digit, those for proximal phalanges appear between the eleventh and fifteenth weeks, and later for intermediate phalanges, with wide variation. Basal centres appear between the second and eighth years (usually second or third in the hallux), uniting by the eighteenth year (Venning 1956a,b). Much variation appears in different reports; racial differences probably exist (Kraus 1961). For a summary of the chronology of pedal ossification see **3.261**.

Comparison of Bones of Hand and Foot

The origin of skeletal elements in the hand and foot from terminal segments of primitive pentadactyl limbs has already been considered (p. 400); similarities and differences can now be further examined. Both hand and foot consist of proximal, intermediate and distal components, i.e. carpus/tarsus, metacarpus/metatarsus and phalanges. Carpus and tarsus both contain seven elements (omitting the sesamoid pisiform); although more or less cubical, they vary much in shape and size. Carpal bones are smaller, are concerned in transmission of smaller forces and have retained primitive alignment in rows more than the tarsal bones. Articulation of the tarsus with the foreleg is reduced to one bone (talus); the whole tarsus is divisible into two groups, corresponding to their location in the principal longitudinal arches, with a degree of independent movement absent from the carpus. The calcaneal lever has no equivalent in the *primate* carpus, although similar *carpal* arrangements exist in some quadripedal mammals.

Digital components, metacarpus/metatarsus and phalanges, are more alike. Metacarpals and metatarsals are similar in form, both bound by soft parts into a structural entity. Mobility is very limited at their bases, where they articulate with the carpals or tarsals and with each other. Between both are interosseous muscles, and first and fifth digits also have short intrinsic muscles. On the flexor aspects in hand and foot are large arterial anastomotic arches, superficial and deep to the long flexor tendons crossing them to enter the free parts of the digits. The distal ends of metacarpals and metatarsals are slightly more mobile than the bases; but the third metacarpal and second metatarsal are relatively fixed as axes about which neighbouring elements exercise limited movements concerned with adaptation of the hand's grip to uneven objects and the plantigrade foot to inequalities of surface. Digits can flex and extend through considerable arcs at metacarpo- and metatarsophalangeal joints, with a lesser range of abduction and adduction, all better developed in the hand. Phalanges of the foot are much shorter; and though the phalangeal formula is usually the same, reduction of phalanges by fusion or loss, especially in the fifth digit, is more frequent in the foot.

Functions of the human hand and foot differ greatly and general skeletal similarity is much modified in details. Greater length and mobility of fingers, and especially preservation of a free-ranging opposable thumb, are essential to the grasping habit of a prehensile limb. Divergence of the pollicial metacarpal and its specialized joint with the trapezium contrast with arrangement in the hallux. The pollex is rotated about 90° on its long axis relative to the fingers so that, in the resting position, it is halfway to opposition with them. But the hallux is in the same plane as the other toes, its flexor surface also facing the ground. Moreover, it is tied by transverse metatarsal ligaments to the second metatarsal, thus integrated with its fellows and adapted for propulsion

456

This has led to a different digital formula in the foot: in the hand the middle digit projects most, in the foot it is the hallux, taking the major stress transmitted through the lever of the foot in locomotion. Dorsiflexion of the foot to a right angle with the leg, an obvious adaptation to plantigrade habit, is sometimes said to be peculiar to man; but it is the primitive plantigrade adaptation in ancestral mammal-like reptiles. While most mammals become varyingly digitigrade, bringing foot and hand into line with the rest of their limbs, most primates and some carnivores, such as bears, retain plantigrade adaptation in the foot. The outstanding human peculiarity is loss of opposition in the hallux, distinguishing man even from other primates. The foot is, indeed, in general far more specialized for the bipedal habit in man than in any other primate animal. The strong build of the hallux in its metatarsal and phalanges, and the elongation and strength of tarsal bones, together make a powerful lever, on which the whole body can be elevated to add impetus and spring in running and jumping. The arched foot, with a lateral arch to steady it on the ground and a medial for thrust in propulsion, are specializations absent from the hand (pp. 632-635, 4.54 and 4.94). The rotational element in the foot, movements of inversion and eversion and *pedal* pronation and supination (p. 400), are also absent from the hand, which rotates *with the forearm* in supination and pronation. Lewis (1983) has surveyed *in extenso* the evolution of the mammalian foot.

The Sesamoid Bones

Sesamoid bones, like the seeds of sesame, are usually ovoid nodules a few millimetres in diameter, but vary in shape and size, some being large, e.g. the patella (Bizarro 1921). They are not always fully ossified and may be dense fibrous tissue, cartilage and bone in varying proportions, but most are partly ossified. They are usually embedded in tendons closely related to articular surfaces or where tendons angle sharply round bony surfaces. In both sites the sesamoid surface related to another bone is covered by articular cartilage and slides over it, itself often part of an articular surface, as in the pollicial metacarpophalangeal joint. This entails that the tendons involved are partly fused with an articular capsule, as is exemplified by the relation between patella, quadriceps femoris tendon and knee joint (p. 527). In other sites, e.g. the tendon of peroneus longus (p. 647), the sesamoid articular aspect glides over a cartilage-covered osseous surface, here a facet on the cuboid's tuberosity, all enclosed within a so-called bursa, which could well be termed a joint capsule, for the arrangement has all the essentials of a synovial joint. Some consider the sesamoids as *primarily* articular, i.e. embedded in articular capsules, their association with tendons being *secondary* (Patterson 1946). Despite association with articulations, the precise functions of sesamoids are not clear. They may modify pressure, diminish friction and sometimes alter a tendon's direction of pull, e.g. the patella. Where a tendon is acutely deflected close to bone, a sesamoid, if ossified, may aid local circulation, bone being able to resist pressures which could compress the vessels in a tendon; these are, however, sparse and prolonged pressure is unlikely, except perhaps in the foot.

Since sesamoid elements appear during fetal life and in greater numbers than later, they are phylogenetic parts of the skeleton, not merely results of local physical factors; but this possibility cannot be dismissed in every instance. They are much more numerous in extremities of many other mammals. In reptiles, intratendinous ossification near the ends of some limb bones produces nodules like sesamoids, but some fuse later and may be traction epiphyses (Haines 1942a). The proximal epiphysis of the olecranon is an example. Experiments on the patella suggest that phylogenetic and other factors act in the formation of sesamoid bones. Transplantation of limb bud fragments from chick embryos to chorioallantoic membrane has led to the development of a patella, indicating a self-differentiating mechanism (Murray & Huxley 1924). Conversely, removal of the patella in young dogs has been followed by regeneration, provided that normal activity of quadriceps femoris was allowed (Carey et al 1927), suggesting that local mechanical factors are necessary to the formation of sesamoids and perhaps also a parallel between sesamoids and

bursae, since the latter also appear regularly before birth in certain sites, but can also develop in others as adventitious structures, apparently in response to local conditions.

The incidence of sesamoid bones has received much attention, but numerical data for individual sites are not always available, and onset and progress of ossification has been little studied.

In the upper limb, sesamoid nodules associated with joints are limited to the palm, two being almost constant in the tendons of adductor pollicis and flexor pollicis brevis, the medial being larger. They articulate with the palmar facets on the joint surface of the first metacarpal head; since they are firmly attached in the articular capsule, and in the tendons involved, they probably conduct some muscular pull to articular ligaments. A sesamoid is

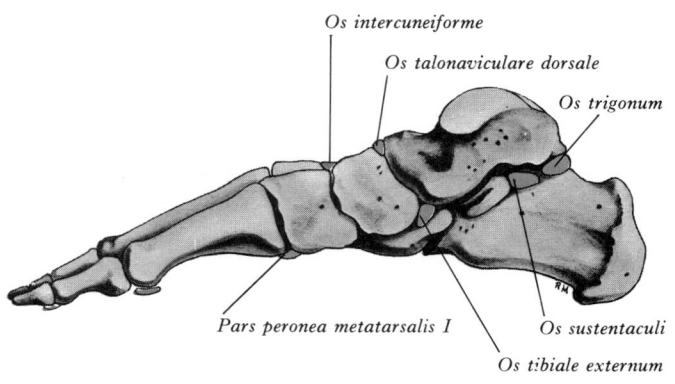

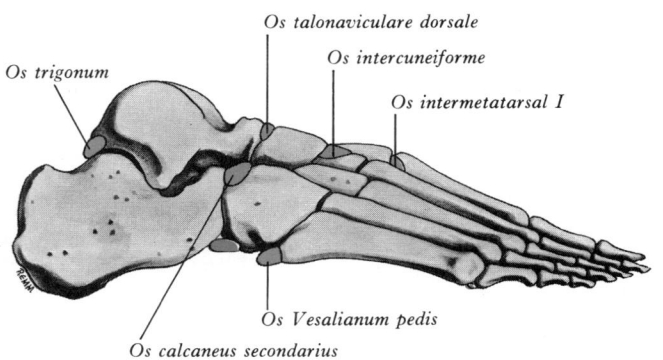

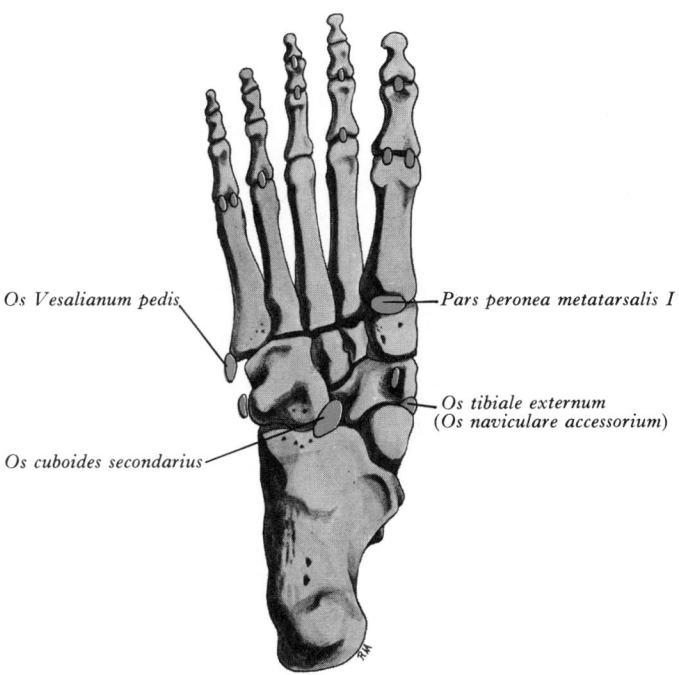

3.262 The sites of sesamoid (red) and accessory bones (blue) in the human foot.

embedded in tendons anterior to the metacarpophalangeal joint of the index digit (35% of hands) and of the fifth (70%), where it is occasionally double. Less frequently, similar sesamoids occur in the third and fourth digits and, in a majority of hands (73%), one exists on the palmar aspect of the pollicial interphalangeal joint (Gray et al 1957). Sesamoid elements in non-articular sites are rare in the upper limb, but one occasionally appears in the biceps tendon near the radial tuberosity.

In the lower limb, the patella is the largest articular sesamoid bone; its relation to the tendon of quadriceps femoris and the capsule of the knee joint exemplifies the arrangements around all sesamoids integrated into synovial articulations. In the foot, sesamoid distribution is like that in the hand. Two, both sometimes double (prompting fallacious diagnosis of fracture), always exist in the tendons of flexor hallucis brevis, plantar to the hallucial metatarsophalangeal joint; they are bound to adjacent ligamentous structures, including the medial part of the plantar aponeurosis. Single sesamoids may occur in the plantar capsule of this joint in all the toes and in interphalangeal joints.

Non-articular sesamoid bones or cartilages are more frequent in the lower limb; that in peroneus longus has been mentioned above. Another, usually appearing later and therefore perhaps adventitious, occurs in the tendon of tibialis anterior where it contacts the distal part of the medial surface of the medial cuneiform bone; the tendon of tibialis posterior may contain a sesamoid gliding on the medial side of the talar head. Sesamoids may also occur in the lateral part of gastrocnemius, behind the lateral femoral condyle, in the tendon of psoas major where it is in contact with the ilium, in the tendon of gluteus maximus as it passes over the greater trochanter and in tendons deflected by the malleoli. In many such sites bursal arrangements occur, and opposing bony surfaces are covered by articular cartilage, as in a synovial articulation. The sesamoid occurring in the gastrocnemius is a *fabella*, articulating with the lateral femoral condyle, and is hence an 'articular' sesamoid.

Sesamoid bones ossify relatively late, except the patella (p. 440), e.g. the hand in early teens. In a group of Caucasians, those sesamoids associated with the thumb began to ossify in males between 12 and 15 years, and in other digits between 15 and 18 years, dates for females being three years earlier (Joseph 1951). No such data appear available for large mixed or other racial groups. Figures for the lower limb are scant and vague.

GRAY'S
ANATOMY

ARTHROLOGY

4

The Scope of Arthrology

The rigid nature and mode of growth of the skeletal tissue, bone, make it imperative that the skeleton consists of *multiple* osseous elements, each joined to its neighbours by a variety of structural arrangements. All such co-aptations are grouped as **arthroses** (synonyms are: *articulationes, juncturae*, anglicized: joints, articulations and junctions). Arthroses are concerned with *differential growth*, with *transmission of forces* (tensile, compressive, shear and torsion) and with *movement* (from consolidation and complete rigidity at one extreme, through grades of restricted movement to, in great variety, relatively free but controlled movement). Which of these attributes predominates varies with site and age, in many instances changing markedly with the latter; scientific study of their functional topography and temporal variation constitutes **Arthrology**. Arthroses are classified in a number of ways, dif-

ferent criteria and degrees of quantitative accuracy being adopted as adequate for different groups of workers. Hence, sources limited to a single classification should be read with due regard to the intended audience, and vary from a simplified introductory grouping, through a somewhat more detailed vocational grouping, to greater mensural information for orthopaedicians and finally the most comprehensive classifications—used by specialist kinesiologists. In the present section *all* these approaches to classification are given, first as a synopsis of principal headings and terms; in later pages and (where indicated) elsewhere, these are defined and described in greater detail. Where the morphology is mixed, or changes radically with time, and in some exceptional topographical situations (e.g. the costal cartilages and the larynx), additional classificatory groups have been included. Although unusual, this confers a more complete logic to the frameworks employed.

INTRODUCTORY CLASSIFICATION OF ARTHROSES

EITHER (CLASSIC NOMENCLATURE)

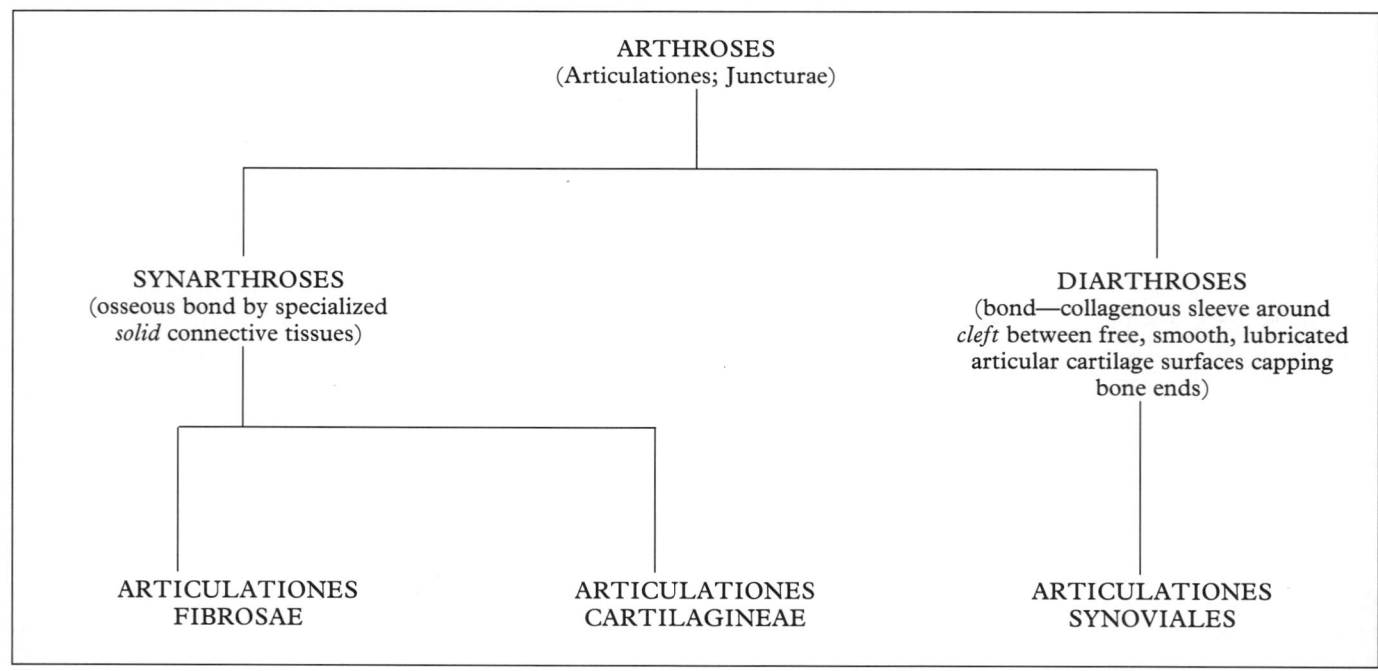

OR (ANGLICIZED NOMENCLATURE)

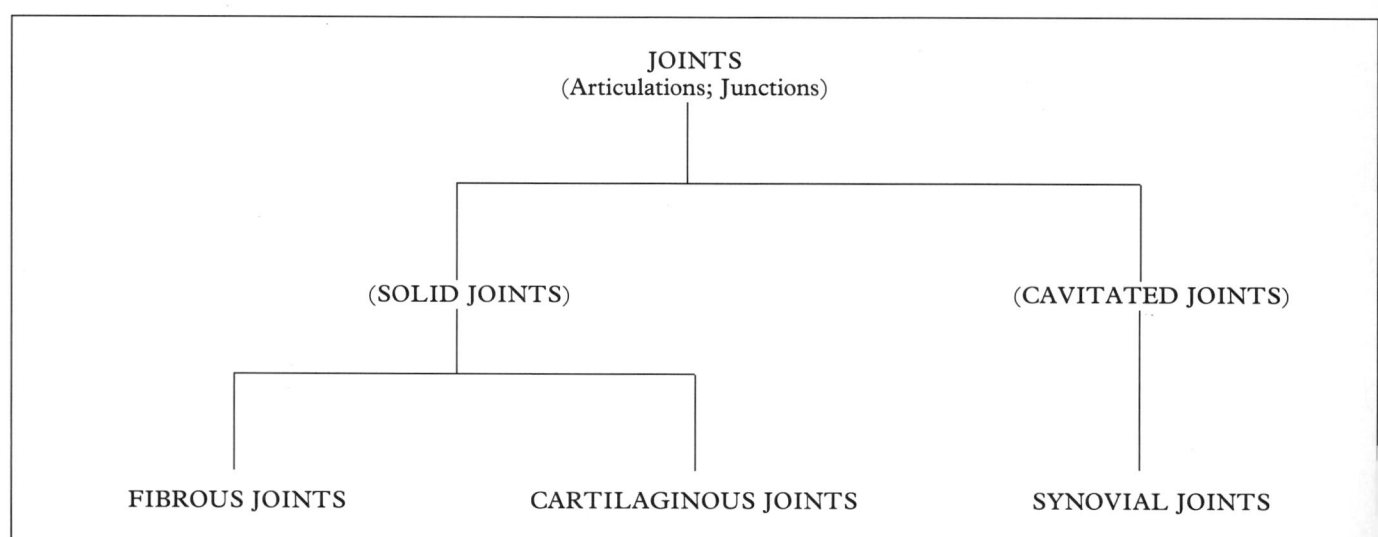

ARTHROSES: MAIN VARIETIES

The anlagen of all skeletal units (expanding, initially separate centres of ossification and ultimately consolidated whole bones) are derived from condensations of mesenchyme. In many sites the mesenchyme becomes increasingly fibrous and bone formation occurs directly in this tissue (*intramembranous ossification*, p. 300); elsewhere the mesenchyme differentiates into a hyaline cartilaginous model and, within this, an orderly sequence of transformations leads to *endochondral ossification*. The biological nature, mechanical properties, changes and fate of the specialized connective tissues interposed between adjacent osseous surfaces form the foundation for the broad classification of **arthroses**, firstly into *two* fundamentally contrasting groups. The specialized connective tissues may remain *solid*, often changing character with time, or may develop a narrow but extensive, enclosed, fluid-containing *cavity*. *Solid (non-synovial) joints* are termed **synarthroses** and are commonly grouped according to the principal type of interosseous connective tissue into **fibrous joints** and **cartilaginous joints**. These reflect the two modes of ossification already mentioned.

Synarthroses, of both types, characterize almost all the cranial junctions; their respective derivations, clearly, are the desmocranium and chondrocranium (p. 164); both types are also present in large numbers throughout the post-cranial skeleton which, other than the clavicle, is mainly endochondral in origin.

Cavitated (synovial) joints are formally termed **diathroses** and, with few exceptions (the temporomandibular joint and those involving the clavicle), are between the ends or other circumscribed surfaces of endochondral bones. Here each articular surface is of specialized hyaline cartilage strongly adherent to its bone on one aspect; on the other, it is a free, lubricated, macroscopically smooth, wear-resistant surface which can glide over its fellow with minimal friction. The fine cleft between apposed surfaces contains lubricant and nutritive synovial fluid (*synovia*) and this *synovial cavity* extends to the *synovial membrane* where the fluid is produced and which lines the fibrous capsule and other non-articular surfaces of the joint.

INTRODUCTORY CLASSIFICATION OF ARTHROSES (continued)

I. SYNARTHROSES   bone—solid connective tissue—bone

1. FIBROUS JOINTS (**Articulationes fibrosae**)

   a. Sutures: bone—collagenous sutural ligament—bone

   b. Syndesmoses: bone—collagenous interosseous ligament, membrane or cord—bone (elastic fibrous tissue is occasionally prominent)

   c. Gomphoses: bone—complex collagenous periodontium—dental cement

2. CARTILAGINOUS JOINTS (**Articulationes cartilagineae**)

   a. Synchondroses: bone—hyaline growth cartilage—bone
   (*Primary cartilaginous joints*)

   b. Symphyses: bone—hyaline growth cartilage—fibrocartilaginous disc—hyaline growth cartilage—bone
   (*Secondary cartilaginous joints*)

SYNOSTOSES: rigid bony union; after growth has ceased this is the normal fate of synchondroses, ultimately most sutures, and some symphyses

II. DIARTHROSES   bone—cavitated connective tissue—bone

SYNOVIAL JOINTS: bone—articular cartilage—synovial fluid in cavity—articular cartilage—bone

(*Articulationes synoviales*)   Bond: surrounding sleeve of collagenous fibrous capsule lined by synovial membrane; extrinsic and intrinsic ligaments. Occasional intracapsular ligaments, tendons, fat pads, fibrocartilaginous discs or menisci

---

MORE DETAILED NOMENCLATURE (see text for definitions)

(A)

SYNARTHROSES

FIBROUS JOINTS (ARTICULATIONES FIBROSAE)

Sutures (Suturae)
Morphological terms:
serrate
denticulate
squamous
limbous
plane
schindylesis (wedge and groove)
Suturae cranii (33 officially recognized and systematically named)

Gomphoses
('peg and socket';
articulationes dentoalveolares)

Syndemoses
Bonding by collagenous (sometimes elastic) interosseous ligament, cord or membrane sheet. There are 12 officially recognized, named, syndesmoses (see text). Criteria for inclusion (or exclusion) remain arbitrary.

SYNARTHROSES: MIXED

Many large synchondroses have localized bundles of dense collagen. Many sutures possess islands of fibrocartilage (see text).

CARTILAGINOUS JOINTS (ARTICULATIONES CARTILAGINEAE)

Synchondroses (Primary cartilaginous joints; hyaline growth cartilages)
Varieties and sites:
Synchondroses cranii (between ossific centres in the chondrocranium, within and between named bones. See text for details and names.)
Synchondroses postcranii
epiphysiodiaphyseal
epiphysiocorporeal
intra-epiphysial
multiplex (in compound, e.g. hip, bones)
sternales (between sternebrae)
manubriosternalis (young)
xiphosternalis
sacrales (young)
costal cartilages are atypical (see text)

Symphyses (Secondary cartilaginous joints)
Bone surfaces encased in hyaline growth cartilage are bonded by fibrocartilaginous disc. All are median.
S. *manubriosternalis*
S. *intervertebrales*
S. *sacrales*
S. *pubis*
S. *menti* (atypical and temporary)

SYNARTHROSES: PERSISTENCE OR CONSOLIDATION (Synostoses)

Syndesmoses, gomphoses and some symphyses grow with their surrounding tissues, the latter contributing to osseous growth; when growth ceases, however, they *persist* throughout life with only minimal age changes. Some symphyses ossify wholly or in part. Synchondroses and sutures possess both functionally important mechanical properties and are most prominently *growth mechanisms*. When growth ceases the intervening cartilage or fibrous tissue is progressively ossified, bony union occurs and they become **synostoses**. It is usually stated that 'the joint is obliterated' and uncommon to group **synostoses** with **arthroses**. Total function, however, makes it quite logical.

THE SITES OF ARTHROSES

The human skeleton, often simply classed as an endoskeleton, is actually compound; the numerous elements which were endoskeletal from their inception have been blended with exoskeletal derivatives, e.g. the clavicles and bones of the cranial vault, sides, nasofacial region, jaws, palate and tympanic plates. In the middle decades of a full normal human lifespan the mature, 'complete' (but far from quiescent) *named bones* are the constituent structural units of the osseous skeleton. The detailed topography of individual bones including the fleshy/tendinous/aponeurotic attachments of muscles, fasciae, ligaments, articular capsules and other relationships, e.g. neurovascular, visceral, coelomic or bursal, together with a summary of their patterns of ossification are provided in Section 3, *Osteology*. The latter section includes well-defined areas on the surface of a bone which are in *intimate structural/functional juxtaposition* with complementary areas on a neighbouring bone or bones; specialized connective tissues blend strongly with the osseous surfaces, *solid* or *cleft*, and occupy all the intervening territory. Their developmental history, progressive changes (some continuing to senescence) and distinctive functional attributes provide these **arthroses** with an acceptable framework for classification in the present section. Much information, from many viewpoints, is presented concerning the

multiple varieties of the sclerous connective tissues, cartilage and bone, also the presently relevant dual modes of ossification, *intramembranous* and *endochondral*, in the opening chapters of *Osteology*. Collagen, prominent in the present context, is considered in terms of varieties, dispositions, mechanical properties and biosynthesis in *The Matrix of Connective Tissue* in Section 1, *Introduction*; here also the less prominent elastic components are reviewed. Brief accounts of the development of skeletal, junctional and related locomotor structures may be found in Section 2, *Embryology*. (Consult tables for nomenclature of varieties of arthroses.)

The foregoing comments have, however, been restricted to surface areas, or borders, of *whole mature* bones. In the present context, throughout late embryonic, fetal and mainly the first two postnatal decades of life, other relevant structural skeletal units are also the separate *centres of ossification*, either endochondral or intramembranous. Each ossific centre and its surrounding specialized tissue are interdependent and co-operate with neighbours in integrated patterns of differential *growth*. As noted (p 279 et seq) a number of bones, some endochondral (carpals, most tarsals), some intramembranous (nasals, lacrimals, palatines, zygomatics), develop from a single centre of ossification and are classed as a *primary* centre. The remaining (endochondral) postcranial bones also possess, commonly, *one primary centre, bu*

MORE DETAILED NOMENCLATURE (see text for definitions)

(B)

DIARTHROSES
(ATICULATIONES SYNOVIALES)
SYNOVIAL JOINTS
(To avoid repetition Joint or Articulation is omitted below)
General Morphology
Simple (one pair of articulating surfaces; male and female)
Compound (more than one pair of surfaces)
Complex (with intracapsular meniscus or disc)

Widely used terms
(Useful approximations)
Surface shape
  plane
  spheroid (**enarthrosis**; 'ball and socket')
  ellipsoid
  ginglymus ('hinge')
  bicondylar
  trochoid ('pivot')
  sellar ('saddle-shaped')
Axes of movement
  uniaxial
  biaxial
  triaxial
  polyaxial
Types of movement
  translation
  rotation (*conjunct* or *adjunct*)
  angulation
  flexion/extension
  abduction/adduction
  circumduction
  elevation/depression
  protraction/retraction
  fixation (neuromuscular)
  fixation (mechanical: close-packing)
  During locomotion, postural adjustment and exploratory and
  manipulative tasks, the above analytical unitary movements are
  combined in complex patterned arrays at numerous joints.
Fundamental Joint Positions
  Loose packed—controlled free *mobility*
  Close packed—position of functional *rigidity*

Precise analytical terms
(Less frequently encountered)
Surface topology
  ovoid, simple; male and female
  ovoid, compound; male and female
  sellar; male and female
Almost all synovial articular surfaces are quantitative variants of
the above.
Joint mechanics
Movements are related to the concept of the *mechanical axis* of a
bone (see text). Movements are all resolvable as rotations around
one, two or three orthogonal axes, i.e. possessing 1–3 *degrees of
freedom.*
Types of movement
Terms refer to one mobile articular surface moving relative to its
fixed partner.
Spin: pure rotation of surface around its mechanical axis.
Swing: all movements other than pure spin. There are two
  varieties of swing—*pure* or *impure* (the latter with associated
  spin):
  Cardinal swing: tip of mechanical axis traces a chordal path.
  There is no spin.
  Arcuate swing: tip of mechanical axis traces an arcuate path.
  Joint topology necessitates some associated spin (i.e. conjunct
  rotation)

All the above terms, and others, are defined and amplified in the
following pages.

---

phylogenetically compound bones (e.g. vertebrae, the scapula, innominate or hip bone, sternum and sacrum) possess a number of *separate primary centres*; the majority appear between the seventh week of embryonic life and the fifth month of intrauterine fetal life. Some, however, appear at intervals throughout the first 10–12 postnatal years. *Secondary centres of ossification (epiphyses)* appear at times, varying considerably with location, over an even longer period, the earliest at full term and the last after adolescence, about the fifteenth year. Miniature long bones develop a simple, single, terminal epiphysis (see metacarpals, metatarsals, phalanges); other long bones (radius, ulna, fibula, and often tibia) develop similar simple epiphyses at *both* ends; the humerus (both ends) and femur (proximal end), however, have *multiple* secondary loci of ossification. Many epiphyses also characterize the compound post-cranial bones mentioned above. The remaining cranial bones each possess two or more, sometimes many, primary centres of ossification; some are confined to the desmocranium and are almost exclusively intramembranous in nature (frontal and parietals), whereas the centres for the ethmoid are endochondral, being wholly confined to the rostral chondrocranium. The four bones forming the cranial base, and the lateral and posterior walls of its vault—the sphenoid, paired temporals and the occipital—are complex in phylogenetic

origin and development; each possesses many centres of ossification, some intramembranous, the remainder endochondral.

Cartilaginous Joints

As indicated in the previous paragraphs and tables, synarthroses, bone junctions bonded by solid connective tissue, are generally divided into two main groups: cartilaginous joints and fibrousjoints. In most instances the distinction is quite obvious and apposite; in a number of sites however some admixture occurs, where a predominantly fibrous articulation contains occasional islands of cartilage or conversely a predominantly cartilaginous articulation contains aligned dense bundles of collagen (vide infra).

Articulationes cartilagineae (cartilaginous joints) are themselves classified into two groups: *synchondroses (primary cartilaginous joints)* and *symphyses (secondary cartilaginous joints).* When fully formed each group has some distinctive structural and functional features; other features however, although varying quantitatively, they share in common. The terms primary and secondary should, perhaps, be abandoned; they are only apposite in the instances of certain symphyses (vide infra) that, developmentally, are preceded by synchondroses within which further differentiation occurs.

SYNCHONDROSES

These junctions occur where originally separate, but adjacent, centres of ossification appear within a continuous mass of hyaline cartilage. As ossification spreads it invades the *actively growing* zone of cartilage which occupies the interval between contiguous osseous surfaces. Thus a synchondrosis consists of two ossifying fronts closely bonded by a specialized *hyaline growth cartilage*. Structurally, passing towards an ossifying surface the growth cartilage has successive recognizable zones of: relative *quiescence*; proliferative and interstitial *growth*; *transformation* into columns, with cell hypertrophy, matrix calcification and cell death; *ossification* involving chondrolysis, vascularization and osteogenesis. (For details see p. 300.) Where the rate of ossification of both surfaces forming a synchondrosis is approximately equal (e.g. in the cranial base), the growth cartilage has a central quiescent zone equidistant from the surfaces and the above-mentioned zones proceed symmetrically towards *both* surfaces. Where the growth rates are unequal, the growth cartilage structure is correspondingly asymmetrical; extreme examples are synchondroses between the diaphysis and terminal epiphyses of long bones where growth and progressive ossification are mainly (but not exclusively) diaphyseal. In the latter the quiescent stratum lies near the epiphyseal bone and the zonation extends towards the diaphysis. The cartilaginous growth plate grows in girth by the interstitial and subperichondrial methods described elsewhere (p. 309); also activity varies at different radial distances from its centre. In this manner the overall shape of the cartilage and its attendant bones are changed, sometimes profoundly. Thus, the early epiphysis of the femoral head is separated from (bound to) the femoral neck, an extension of the diaphysis, by a relatively simple horizontal growth plate of cartilage. With differential growth the capitular tip of the neck becomes increasingly conical, clothed and tightly bound to a conical growth cartilage and both received by a deep conical recess in the inferolateral part of the otherwise spheroidal ossifying femoral head.

Functionally, synchondroses are primarily *growth mechanisms* and, although contributing slightly to the more flexible skeleton of youth, their growth potential is combined with a considerable ability to successfully resist forces, whether of compression tension, shear or torsion. Thus they permit growth to continue, while there are free but controlled, often powerful, movements executed at neighbouring synovial joints, or more restricted movements at symphyses and some fibrous joints. These features are characteristic of the numerous **post-cranial synchondroses**. Since, with the exception of the clavicle and areas of subperiosteal bone accretion, all post-cranial centres of ossification are endochondral, synchondroses are present in all post-cranial bones derived from two or more centres, i.e. the majority of post-cranial bones. The carpal and most tarsal bones however each develop from a single endochondral centre; their ossifying surfaces advance invading a *complete encasement* of *growing cartilage*. Parts of the latter persist as the specialized articular cartilaginous surfaces of carpal and tarsal synovial joints; elsewhere the bone approaches the perichondrium (now periosteum) and provides attachment for fibrous joint capsules, tendon sheaths, interosseous ligaments, fascial septa and muscles. Clearly, these relatively small bones possess no synchondroses but their essential growth mechanisms have much in common with that of epiphyses. (For details see p. 421.)

Most **post-cranial synchondroses** have not been given specific topographical names, but the general morphological group to which each belongs should be clear after reference to previous paragraphs and Table A of 'More Detailed Nomenclature'. They may be *epiphysio-diaphyseal* (or more precisely epi-*metaphyseal*, or epi-*corporeal*), *intra-epiphyseal* in compound epiphyses, *multiplex* in compound bones with multiple primary centres. Examples of the last having topographical names are: the *triradiate acetabular cartilage* of the developing hip bone, *sternales* between growing sternebrae, *manubriosternalis* (young, before transformation to a symphysis), *xiphosternalis*, *vertebrales* confined to individual vertebrae, *sacrales centrales* (young, soon to be temporary symphyses) and *laterales* (between centra, neural arches and costal processes). Structurally, the pattern of zonation

mentioned above and elsewhere (p. 279) requires some amplification. Growth cartilages vary considerably with age, site and species. For reviews of the extensive literature see Knese (1979), Hall (1983) and for recent quantitative studies, Moss-Salentijn et al (1987), and Taylor et al (1987). Most intensively studied are certain synchondroses in small laboratory mammals, mouse, rat and rabbit, and in particular the proximal tibial synchondrosis. In the embryonic stage the cartilaginous 'models' of future endochondral bones consist of generalized hyaline cartilage, its chondrocytes being randomly distributed within copious matrix. With formation and longitudinal advance of the primary endochondral centre of ossification the site of the *presumptive growth plate* of cartilage is indicated by ovoid aggregates (or clones) of chondrocytes. The full zonal pattern is only seen in the *definitive growth plate* between the advancing osseous surface of the metaphysis and that of the epiphysis. Human growth plates are relatively much thicker than those of the laboratory mammals studied, due in large measure to the depth of the 'quiescent' or small cell zone adjacent to the epiphysial bone. This small cell zone is narrow in small mammals but in man it may account for 60–70% of total plate thickness; its significance is controversial. As implied by its earlier name, some regard it as relatively inert; others, however (see reviews and quantitative references), consider it a pool of stem progenitor cells. From the latter, throughout the life of the growth plate (the phase of growth of the bone), successive generations of chondrocyte columns are formed, transformed, partially eroded then form a scaffold for advancing ossification. Thus, when fully active, a growth cartilage contains chondrocyte columns that vary in length, maturity and position. Young columns extend limited distances from the small cell zone (*epiphysial* position); mature forms are *full length* columns; old columns are again short, partly eroded and *metaphysial* in position. (It should be emphasized that these names are purely positional; they were devised for quantitative analyses of growth cartilages in appendicular long bones and as these grow and retreat they all contribute to the advancing surface of *metaphysial* ossification.) The dual views concerning the small cell zone may, reasonably, both be correct; the cells near the epiphysial bone have a low growth potential while the potential to function as a pool of chondrocyte progenitor cells increases markedly as the tips of the chondrocyte columns are approached.

The chondrocytes and matrix of young growth cartilages have the three-dimensional architecture described for hyaline cartilage (p. 288) and the changes in the partitions enclosing chondrocyte columns have been mentioned (p. 305). In the latter the collagen fibrils have been termed 'ladderlike' (Eggli et al 1985). Some large growth cartilages that are habitually subjected to unusually large stresses develop, in addition, dense bundles of collagen, their fibro-osseous disposition reflecting the stress. These *fibrous epiphysial (growth) plates* have been studied by Smith (1962 a,b). In the proximal growth plate of the human tibia the approximately flat condylar parts of the plate are traversed by regularly aligned horizontal bundles (i.e. at right angles to the bone's long axis) and they suffer lateral compression during weight bearing. Where the epiphysis descends anteriorly to form the tibial tuberosity it receives shearing stress due to the oblique pull of the quadriceps femoris and here the cartilage of the growth plate is almost completely replaced by oblique dense collagen. (For other examples of structurally mixed growth cartilages consult references quoted above; and for a general review of the epiphysial vascularization of growth cartilages see Moss-Salentijn 1976.) The fates of post-cranial and cranial synchondroses are considered together below; also mentioned are some unclassified or inappropriately classified junctions involving cartilage.

Cranial synchondroses, as indicated, occur between neighbouring endochondral centres of ossification that develop in the chondrocranium; for its origin, chondrification, and topographical regions, see p. 309; and for a general developmental review Bosma (1976). Some regions of the chondrocranium remain unossified: the lateral, alar and septal nasal cartilages and remnants occupying and near the foramen lacerum. In contrast the largely endochondral auditory ossicles interconnect by specialized synovial joints, while the malleus and stapes are connected to the temporal bone by equally specialized fibrou

A SYNCHONDROSES

CARTILAGINOUS JOINTS

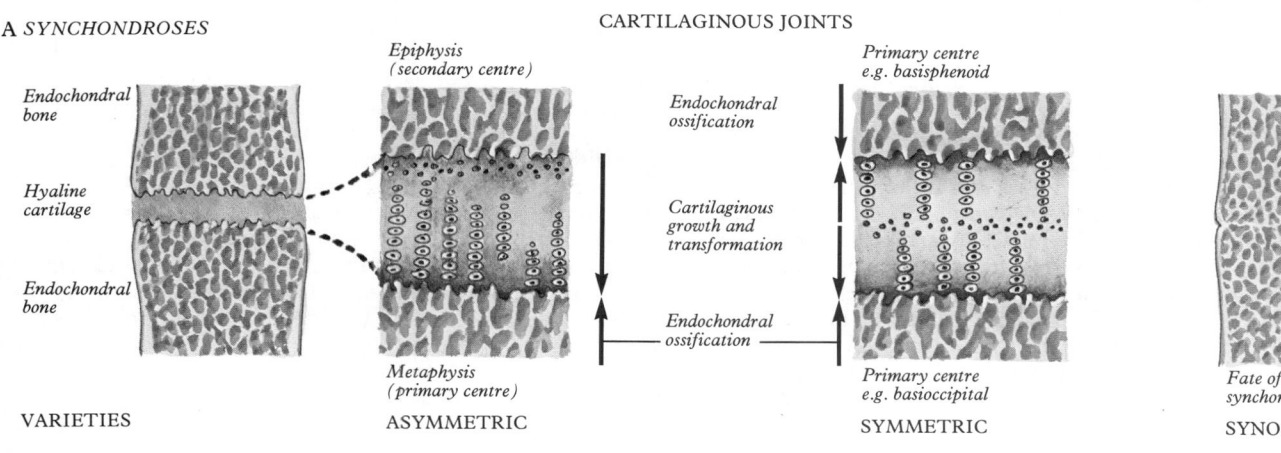

Endochondral bone

Hyaline cartilage

Endochondral bone

Epiphysis (secondary centre)

Metaphysis (primary centre)

Endochondral ossification

Cartilaginous growth and transformation

Endochondral ossification

Primary centre e.g. basisphenoid

Primary centre e.g. basioccipital

Fate of synchondrosis

VARIETIES   ASYMMETRIC   SYMMETRIC   SYNOSTOSIS

B SYMPHYSES

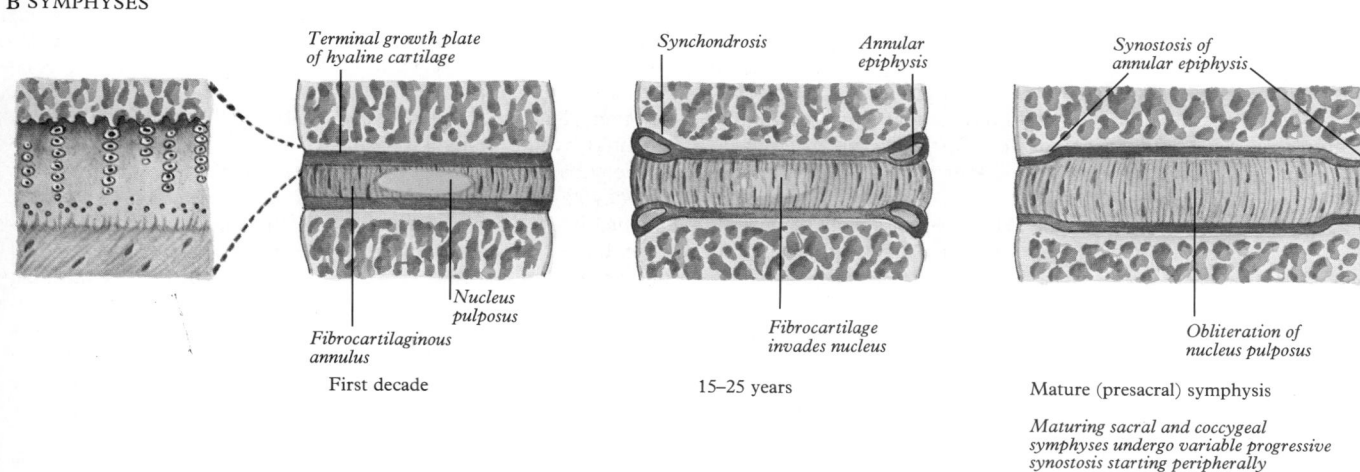

Terminal growth plate of hyaline cartilage

Synchondrosis   *Annular epiphysis*

Synostosis of annular epiphysis

Fibrocartilaginous annulus

Nucleus pulposus

Fibrocartilage invades nucleus

Obliteration of nucleus pulposus

First decade   15–25 years   Mature (presacral) symphysis

Maturing sacral and coccygeal symphyses undergo variable progressive synostosis starting peripherally

C LESS COMMON CHONDRAL JOINTS

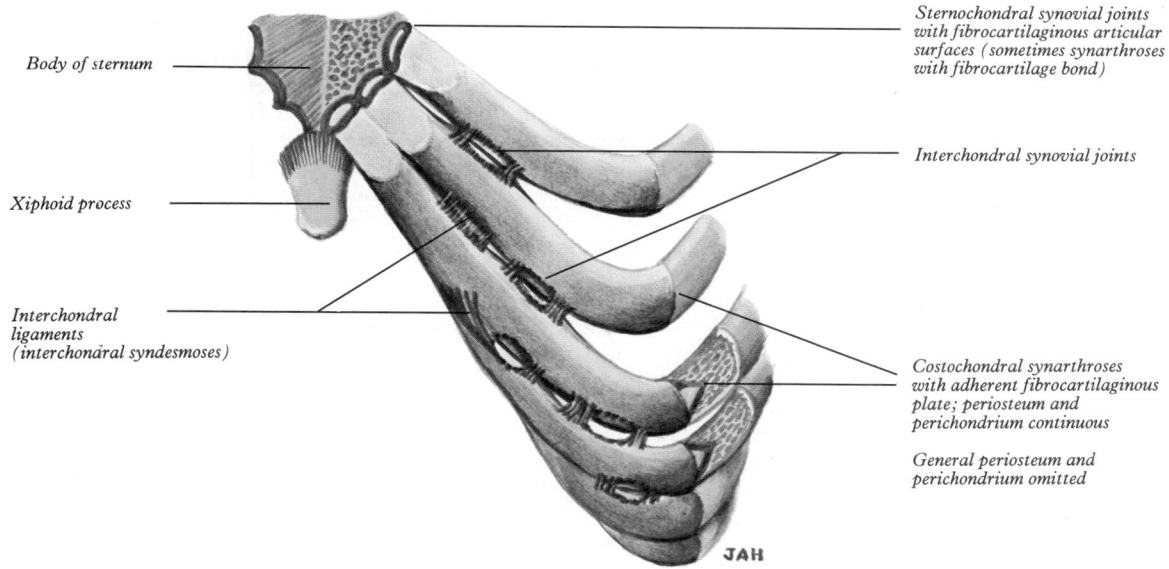

Body of sternum

Xiphoid process

Interchondral ligaments (interchondral syndesmoses)

Sternochondral synovial joints with fibrocartilaginous articular surfaces (sometimes synarthroses with fibrocartilage bond)

Interchondral synovial joints

Costochondral synarthroses with adherent fibrocartilaginous plate; periosteum and perichondrium continuous

General periosteum and perichondrium omitted

JAH

.1 Varieties of cartilaginous joints (articulationes cartilagineae). A. Synchondroses in section view. The principal tissues involved, more detailed architecture and main growth patterns of symmetrical and symmetrical synchondroses. In some locations lesser degrees of asymmetry occur; synostosis is the normal fate of almost all synchondroses when *endochondral* growth has ceased. B. Intervertebral symphyses (presacral) in sectional view, displaying changes with age. Note that partial or complete synostosis is the normal fate of sacral and coccygeal symphyses. For discussion and for manubriosternal and pubic symphyses see text and individual joints. C. Less common interchondral and osseochondral junctions; see text for other locations.

arrangements, the fibrous stratum of the tympanic membrane (p. 380) and the annular ligament respectively. The endochondral centres of ossification are for the inferior nasal concha (single), the ethmoid (three), the sphenoid (multiple pre- and post-sphenoidal centres for the body, conchae, lesser wings and roots of greater wings, see p. 376), the temporal bones (multiple centres for each petromastoid part and styloid process, see p. 380), the occipital bone (multiple centres for basilar, condylar and squamous parts up to the highest nuchal lines, see p. 373). Clearly, the sites of basicranial synchondroses are numerous, some within the individual bones, others between them. All contribute in some measure to basicranial growth but those that persist for limited periods, and coalesce early, are unnamed; official names are reserved for those making substantial contributions. Avoiding repetition of the prefix synchondrosis in each case, these are: *spheno-ethmoidalis*, *spheno-occipitalis*, *sphenopetrosa*, *petro-occipitalis*, *intraoccipitalis anterior* and *intraoccipitalis posterior*. Although they are not officially named, presumably because of their relatively early fusion, it seems appropriate to include the main intrinsic synchondroses of the sphenoid. Suggested names are: *intrasphenoidalis transversus* between pre- and post-sphenoidal parts of the body and *intrasphenoidalis lateralis*, bilaterally separating the body from the conjoined greater wings and pterygoid processes (see p. 376). As mentioned, the cranial synchondroses are approximately symmetrical in structure, their zones of growth and differentiation passing from the centre towards both ossifying surfaces.

The fate of synchondroses. When those aspects of growth of parts of a bone, or between bones, joined by a synchondrosis near completion, classic cartilaginous growth, transformation and endochondral ossification ceases. For a period the cartilaginous plate is relatively quiescent, but becomes irregularly thinned; complex histological changes now supervene and involve the cartilage from both osseous surfaces. (For histological details and references see p. 307.) Finally, the synchondrosis is wholly replaced by complete bony union between the originally separate osseous surfaces, i.e. it has formed a **synostosis**. The latter has lost its cartilaginous growth potential and mechanical properties, but has assumed the maximal rigidity of bone. (For mechanical properties see p. 276 et seq.) This continues and enhances the effectiveness of post-cranial bones in carrying a variety of powerful stresses, either in almost static postures or during movement patterns involving synovial and (to more limited degrees) median symphysial or fibrous joints. Rigidity of the neurocranium, both during and after completion of growth, is functionally imperative, affording geometric and protective support and encasement of the brain, special sense organs and their vasculature and associated fluids (blood and cerebrospinal fluid).

Terminology: Some Brief Comments

The fate of synchondroses raises a number of points, some directly, others indirectly, which require amplification; they also provide examples of the kind of limitations inherent in many biological classifications. Thus, it is commonly stated that 'synchondroses are temporary junctions, and with cessation of growth, the joint is obliterated as bony union occurs'. Apparently useful and self-explanatory, all parts of this statement need comment. *Temporary*: while strictly correct in relation to all true synchondroses (vide infra), it must be emphasized that their periods of activity and times of synostosis vary greatly in different sites. Some fuse early, in fetal or perinatal life, others throughout the first decade, many in the latter half of the second decade, while the spheno-occipital basicranial synchondrosis may continue well into the third decade (p. 376, see also under ossification of individual bones). Many are therefore fully active through many years of intense patterns of locomotion and their essential properties are a unique *combination* of growth potential and strength of interosseous bonding. *Cessation of growth*: this clearly does not refer to overall growth, e.g. of the whole body, skeleton, body segment or even a whole bone. Patterns of osseous fusion are complex and, as mentioned above, extend over lengthy periods; even some long bones that develop simple

466

terminal epiphyses at both ends (e.g. fibula, ulna) have asynchronous fusion times. The growth referred to specifically is solely the equilibrium between endochondral ossification, cartilaginous growth plate proliferation and transformation of the growth plate in question. When fusion *has* occurred, however, subperiosteal and subendosteal bone accretion and resorption and trabecular remodelling may *continue*; finally, at the microscopic level, osteon removal and replacement at varying rates continues throughout life. *The joint is obliterated*: this clearly implies acceptance of synarthroses as a major classificatory group, with synchondroses (also, symphyses, sutures, syndesmoses, gomphoses) as well-defined subgroups. Osseous fusion is not included in the official Nomenclature or by the majority of authors. Some authorities, however, find it quite logical to include **synostoses** as extreme forms of synarthroses. This becomes increasingly so when growth potential and rigidity are regarded as equally relevant arthrological ingredients as active movement.

Unusual Junctions Involving Cartilage

Hitherto in this section, and in general use, arthroses have been defined as co-aptations of *osseous* surfaces, and their classification stemmed from the variety of specialized intervening connective tissue and its solid or cleft state. A number of junctions, however, have many arthrodial features but are not accommodated by the general classification of arthroses; these occur where substantial masses of cartilage remain unossified in normal development. (Some masses develop irregular patches of calcification in later decades.) Thus the junctions may be between bone and cartilage, or between two adjacent cartilages. The bone involved may be endochondral or intramembranous; or finally a cartilage may join with a fibrous suture or a synchondrosis. The junctions are similar to either fibrous synarthroses or to synovial joints. Particular reference can be made to the nasal, laryngeal and costal cartilages, also those of the auditory tube and auricle (pinna). Fibrous junctions occur between the borders of contiguous nasal cartilages; the septal cartilage is, in part, continuous with the lateral nasal cartilage but elsewhere fibrous bonds extend to the internasal suture, perpendicular plate of the ethmoid, the vomer, the nasal crest (and intermaxillary suture) and anterior nasal spine of the maxillae and the septal process of the lateral nasal cartilage. At these junctions perichondria fuse or perichondrium and periosteum (and in some sites sutural ligament) blend.

A fibrous junction joins the perichondrium of the elastic auricular fibrocartilage and the periosteum of the roughened outer rim of the external acoustic meatus. A similar fibrous union suspends the perichondrium of the superior recurved border of the cartilage of the auditory tube and the inferior perichondrium of the sphenopetrosal synchondrosis (periosteum after synostosis, the upper attachment spreading on to the petrous quadrate area). The larynx and hyoid bone furnish further examples of specialized and uncommon varieties of connection and junction. Most extensively the laryngeal cartilages are connected to each other, and the thyroid and epiglottic cartilages to the hyoid bone, by three-dimensional complexes of fibro-elastic membranes and their ligamentous thickenings. These *interchondral* and *osseo-chondral* structures are detailed on pp. 462, 468; although not officially listed as synarthroses (subgroup *syndesmoses*) it seems quite logical to regard them thus. Additionally, the cricoid cartilage bears four discrete articular surfaces, one bilateral pair on the sloping superior border of its lamina, another lateral pair at each junction of lamina and arch. These articulate respectively with the articular surface on the base of each arytenoid cartilage and those on the medial aspects of the tips of the inferior cornua of the thyroid cartilage. The crico-arytenoid and cricothyroid articulations to which these surfaces contribute (p. 1252) are *interchondral* synovial joints.

The costal cartilages are often, usefully, considered as unossified cartilaginous extensions of the ribs; first to seventh reaching the sternum, eighth to tenth only reaching the cartilage above, eleventh and twelfth being short blunt cones with free intermuscular tips. The cartilages are, however, traditionally described as separate topographical entities (p. 335) and depending on its level, each engages in from one to five different types of junction (p. 498). The costochondral junctions are unusual synarthroses where the convex tip of the cartilage is received by a complementary recess in the tip of the rib; perichondrium and periosteum are continuous, as are the collagenous elements of bone and cartilage. The sternochondral joints (often called sternocostal joints) vary; the second to seventh are often synovial, with *fibrocartilaginous* articular surfaces on both chondral and sternal aspects. (For the disposition of the intra-articular, capsular and other ligaments see p. 498.) Quite frequently synovial cavities are absent in some or all these joints; in these cases a thin dense lamina of tightly adherent fibrocartilage is interposed between cartilage and bone—an unclassified variety of synarthrosis. This type of junction is usually present at the manubrial end of the first costal cartilage. (It may be noted that it is commonplace for the *whole* first costal cartilage and attached rib and sternum, or the manubriochondral junction alone, to be classified quite incorrectly as a synchondrosis.) The fifth to ninth costal cartilages carry a variable number of simple interchondral synovial joints and also irregular syndesmoses in the form of short interchondral ligaments between borders or apical (particularly cartilages eight to ten), anchoring their tips to the superjacent border.

SYMPHYSES

A symphysis is another variety of cartilaginous synarthrosis; distinctive attributes although prominent are few; they share many features with other arthroses and only differ quantitatively. Topographically, all symphyses are median and, with one exception, they are confined to the axial skeleton. The latter include the following named symphyses: the *manubriosternalis* between manubrium and sternal body; *intervertebralis* between successive vertebral bodies (regionally grouped—cervical, thoracic, lumbar, sacral and coccygeal); also axial is the *symphysis menti* between the bilateral halves of the fetal mandible and continuing only into the first postnatal year when synostosis supervenes. Histologically the mental junction is quite unlike other symphyses (see p. 367), but the use of its name is so widespread that retention seems inevitable. Finally bilateral appendicular bones, the medial surfaces of the bodies of the pubes, are conjoined at the *symphysis pubis*.

Ignoring regional specializations (mentioned below) the general architecture of a symphysis consists of two well-defined surface areas of articulating endochondral bones; the osseous surfaces are from a few millimetres to over a centimetre apart and bonding them are strong, tightly adherent, solid connective tissues. Each bony surface is firmly attached to a thin lamina of hyaline cartilage and, in turn, these blend with the surfaces of a thick, strong, but deformable pad (or disc) of fibrocartilage. Collagenous ligaments extend from the periostea across the symphysis and blend with the hyaline and fibrocartilaginous perichondria; they do not form a complete capsule (as is the case in synovial joints), but they are similar in containing plexuses of afferent nerve terminals, which also penetrate the periphery of the fibrocartilage. The combined strength of the hyaline and fibrocartilage with the ligaments exceeds even that of the bones to which they connect; thus they can easily withstand the habitual range of stresses (compression, tension, shear and torsion) encountered without disruption. Tears are usually the result of sudden, unexpected massive stresses occurring when the body is in an inappropriate posture. The architecture of the fibrocartilaginous tissues is such that it combines the *strength* mentioned with limited degrees of elastic deformability, thus allowing the restricted but appreciable ranges of *movement* that characterize symphyses. Fibrocartilaginous disc compression narrows the interosseous interval, while tension lengthens it; thus when these occur simultaneously in opposite sectors of the disc, the latter becomes cuneiform (wedge-shaped) and its attached bones are relatively inclined (angulated). Torsional stress increases or decreases the spiralization of collagenous discal microarchitecture and permits slight relative rotation. Although the ranges of movements possible at a symphysis are not great and primarily determined by its intrinsic anatomy, this is, however, further

limited by the additional articulations and ligaments that involve the bones but are extrasymphyseal; e.g. the numerous syndesmoses and synovial zygapophysial joints between adjacent vertebrae, articulations of the clavicles and costal cartilages with the sternum, and the sacro-iliac joints, interosseous sacro-iliac, sacrotuberous and sacrospinous ligaments. All these features profoundly affect movement patterns at their related symphyses.

Further comments on symphyseal structure, functional attributes and nomenclature. Thick fibrocartilaginous pads or discs that most characterize symphyses are not simple deformable masses of feltwork, but present precise zonal variations in microstructure, particularly in their arrangement of collagenous and elastic fibres adapted to their habitual stresses. Best investigated and most complex are the intervertebral discs. (See p. 489 for their main subregions, fibre architecture, cellularity, physicochemical properties and profound age changes.)

Solid specialized connective tissues, by definition, characterize synarthroses and, while this obtains in thoracolumbar and sacrococcygeal series of intervertebral symphyses, significant proportions of the remainder develop a single central or a bilateral pair of narrow fluid-filled clefts in (or near) their fibrocartilaginous pads. Over 30% of manubriosternal symphyses develop a central horizontally elongate cleft (p. 498) and over 50% of interpubic discs develop a median elongated cleft in the oblique long axis of the disc. (For variations in size and position see p. 517.) The second to sixth (sometimes seventh) cervical intervertebral discs develop laterally placed small fluid-filled cavities in the fibrous tissue between the inferolateral bevelled edge of the vertebral body above and the superolateral lip of the vertebral body below. Some regard these cavities as sited in the peripheral parts of the discs and non-synovial in character; others regard them as small synovial joints near, but external to, the discs (p. 490). The synovial or non-synovial nature of all the above clefts remains controversial; they perhaps reflect a stage in the evolution of synovial joints. Caution should also prevail concerning the dogmatic assertion that *no* movements of translation occur between their free moist surfaces until adequate objective evidence has accumulated. Distinct possibilities seem to be translations at the cervical cavities during spinal movements and translation with distraction at the interpubic cleft, particularly during the later stages of pregnancy.

The common statement that synchondroses are temporary and concerned with growth, whereas symphyses are permanent and concerned with movement, is an oversimplification, only partly correct and with much omitted. *Both* are concerned with *strength* and the ability to withstand and transmit considerable stresses and with *growth*; and, in contrasting ways, both contribute either directly or indirectly to the total *movement patterns* of the parts involved. The strength and mechanical properties of cartilaginous joints need little further comment, but it may be re-emphasized that the rigidity of synchondroses increases the efficiency of positive movements at related syndesmoses, symphyses and particularly synovial joints. It may also be reiterated that the movements at symphyses are not a simple extrapolation of the mechanical properties of a fibrocartilaginous pad or disc assessed in experimental isolation. For example, movement of a vertebra relative to its neighbour is a three-dimensional summation of the properties of *all* its intervertebral arthroses (syndesmoses and synovial joints, in addition to the complex symphysis) acting in concert, but each with its particular array of stresses. (During their growth phases the mechanical properties of the *intravertebral synchondroses* are also significant.)

The prominent role of synchondroses in skeletal *growth* is widely recognized, investigated and discussed, whereas growth of and involving symphyses is often ignored or receives but scant mention. *Symphysial growth* may, for convenience, be considered from two aspects (although interrelated); these are intrinsic growth of the fibrocartilaginous disc and growth of the hyaline cartilaginous plates into which (e.g. vertebral) endochondral ossification progresses and later 'annular' epiphyses form (vide infra). At the manubriosternal symphysis its limiting plates of proliferative hyaline cartilage are progressively invaded by the ossifying fronts of the manubrium and first sternebra; normally no epiphyses are formed. At the pubic symphysis, however, the growth plates of hyaline cartilage that are involved with endochondral ossification on the symphyseal aspects of the bodies of the pubes after puberty commonly develop scale-like epiphyses, usually anterosuperiorly. The occurrence of intervertebral and pubic epiphyses adds some complexity to the traditional description of symphyses. At these sites, as indicated, the epiphyses appear at, or soon after, puberty and continue slow growth for about a decade before synostosis with the main bone. During this period the epiphysis is completely encased by proliferative hyaline cartilage, its interosseous lamina being a narrow (epiphysiocorporeal) synchondrosis and its symphyseal lamina a simple extension of the terminal plate of hyaline growth cartilage. The latter was, of course, originally derived from the early embryonic cartilaginous vertebral model (p. 159). These growth plates are the proliferative source of cartilage into which all endochondral ossification of the vertebral *bodies* occurs cranially and caudally (and constituting about 80% of the length of the mature vertebral column, see p. 315). The remaining 20% is provided by growth of the fibrocartilaginous discs; their regional variations in thickness and their cuneiform sagittal profile (vertically thicker ventrally) in the cervical and lumbar secondary spinal curvatures are described on p. 315. Where the fibrocartilage blends with the lamina of hyaline cartilage some bundles of collagen pass from the former and, without interruption, interlace within the matrix of the latter. The hyaline laminae show the zonal variation in structure previously described and common to all growth cartilages.

Terminology. It may be noted that clinically and radiologically the term intervertebral symphysis is seldom used; instead '*the intervertebral disc*', or simply '*the disc*', and radiologically '*the disc space*' are often used to indicate *all* tissues between the vertebral bodies. Formally, however, the fibrocartilaginous intervertebral disc, with its age-dependent nucleus pulposus (p. 490), is regarded as part of a symphysis, completed by paired laminae of terminal growth cartilages and their associated ossifying or mature osseous surfaces.

Fates of symphyses. The widely quoted generalization that 'synchondroses are temporary while symphyses are permanent' is only partly correct; there are prominent exceptions. Normally permanent, but exhibiting age changes, are the cervical, thoracic, lumbar and lumbosacral intervertebral symphyses, the pubic symphysis and about 90% of manubriosternal symphyses. The age changes in intervertebral discs are detailed on p. 490; the development of narrow fluid-filled clefts are mentioned above and under individual joints; also allusions to changes in the pubic symphysis in the later stages of pregnancy are made above and on p. 518. In contrast, the joints between successive sacral bodies, between sacrum and coccyx and between coccygeal segments, after preliminary stages, form well-developed *synostosis*; however, their normal fate is partial or complete *synostosis*; the process is slow and lengthy (pp. 329, 332). After 30 years of age about 10% of manubriosternal symphyses develop partial or complete synostosis; again the phenomenon may be slow and lengthy (p. 393).

Articulationes Fibrosae

Fibrous joints in most instances consist of predominantly collagenous junctions between bones but in a minority of situations fibro-elastic tissue predominates. Three main groups of fibrous articulation are generally recognized, namely, sutures, gomphoses and syndesmoses (4.2).

SUTURES

Sutures, limited to the skull, occur wherever margins or broader surfaces of bones are separated only by connective tissue, the *sutural ligament* or membrane, which is a surviving unossified part of mesenchymatous sheets in which dermal bones develop (p. 340). Sutural ligaments display regions of differentiation concerned in growth and binding of apposed bone surfaces (Pritchard et al 1956). Each bone is covered on its sutural aspect by a layer of osteogenic cells (the 'cambial' layer), itself overlaid by a capsular

lamella of fibrous tissue (4.2), both collectively corresponding to periosteum and continuous with this at the margins of the sutural surfaces, both inside and outside the skull. Between these two layers of sutural periosteum is a central stratum of loose fibrous connective tissue, varying in width according to age and the interval between the particular bones involved. This central stratum contains thin-walled blood vessels, the veins communicating with diploic vessels, intracranial venous sinuses and external veins in the scalp. The fibrous periosteum adherent to the bones crosses the interval between them, as two uniting layers (external and internal) which enclose the sutural ligament and add to its strength. Experimental evidence suggests that during active growth the orientation of collagen fibres in sutural membranes is adaptable to several factors, in particular to the direction of growth of minute bone spicules (Koskinen et al 1976).

In view of the occurrence of fibrous tissue in synchondroses (p. 463), it is interesting to find that, during growth, secondary cartilage formation often occurs in sutural ligaments, suggesting a relation between fibrous and cartilaginous joints.

When cranial growth, including in its earlier stages growth at sutures, ends, osteogenic cells generally bring about complete ossification of sutural ligaments; but this is a slow process (p. 393), ultimately leading to obliteration and rigid synostosis. Sutural fusion, however, does not even begin until the late twenties, proceeding slowly thereafter; yet it is clearly necessary that sutures should cease to be slightly flexible joints as soon as possible after birth. Sutural ligaments may effect an almost immovable bond between large areas of bone, especially where these show reciprocally adapted irregularities, even if fine as in the intermaxillary junction; but such immobility cannot be effected at narrow edges of bones in the cranial vault. At these, however, their margins develop spikes and recesses which interlock so well that the bones are difficult to separate even when denuded of all fibrous connective tissue. Where the edges are saw-like, the junction is a *serrate suture*. A *denticulate suture* has small toothlike projections, often widening towards their ends to provide an even more effective interlocking. When such sutures are tied by sutural ligament and periosteum, almost complete immobility results. The sagittal suture is serrated and much of the lambdoid denticulate. Where bones overlap, as at the temporoparietal suture, this is a *squamous suture*; the adjacent bone surfaces are reciprocally bevelled and, if mutually ridged or serrated, the junction is sometimes termed a *limbous suture*. Simple apposition of contiguous surfaces, usually rough and reciprocally irregular, is inappropriately named a *plane suture*, instances being sutures between the palatine bones, between the maxillae and at the palatomaxillary sutures. Although surface demarcations between such bones show none of the interlocking evident at serrate or denticulate sutures, irregular surfaces of contact, tied by wide expanses of sutural ligament, provide much resistance to shearing or torsion; like other sutures they are, for all practical purposes, immovable. In summary therefore sutures, although in some locations providing slight perinatal flexibility, are essential structural mechanisms for sutural bone growth, combined with the necessary rigidity and geometry in the upper neurocranium and in the nasofacial and palatine skeleton.

A schindylesis is a specialized suture where a ridged bone fits into a groove on a neighbouring element, e.g. the cleft between the alae of the vomer that receives the rostrum of the sphenoid.

GOMPHOSES

A gomphosis or **peg-and-socket joint** (*articulatio dentoalveolaris*) is a specialized fibrous articulation restricted to the fixation of teeth in alveolar sockets in the mandible and maxillae. The collagen of the periodontium connects dental cement with alveolar bone (see pp. 1309, 1310, for details).

SYNDESMOSES

A syndesmosis is a fibrous articulation in which bony surfaces are bound together by an interosseous ligament, by a slender fibrous cord or an aponeurotic membrane, usually affording slight but occasionally more extensive movement between them. Fo

FIBROUS JOINTS

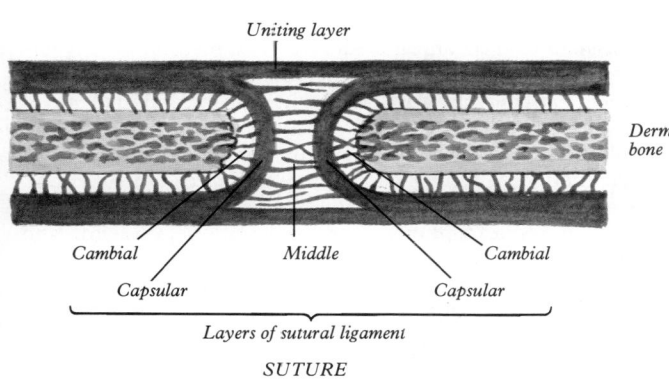

Uniting layer

Dermal bone

Cambial   *Middle*   *Cambial*
Capsular   *Capsular*

Layers of sutural ligament

SUTURE

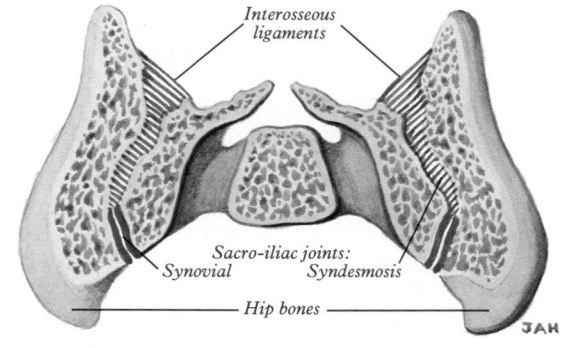

Interosseous ligaments

Sacro-iliac joints:
Synovial   *Syndesmosis*

Hip bones

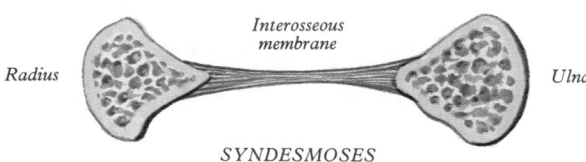

Interosseous membrane

Radius   *Ulna*

SYNDESMOSES

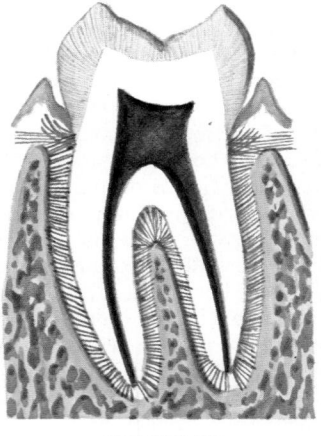

GOMPHOSIS
(*Dentoalveolar joint*)

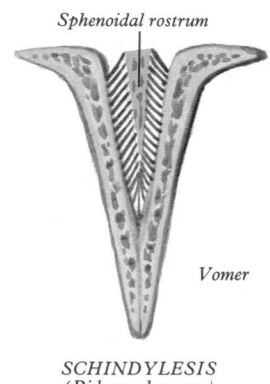

Sphenoidal rostrum

Vomer

SCHINDYLESIS
(*Ridge and groove*)

4.2 Some examples of the principal varieties of fibrous joints (articulationes fibrosae). A diagram of each variety in section is clearly indicated. Note some do not regard schindyleses as sufficiently distinct from sutures to merit a separate group. Syndesmoses are difficult to define because of the large numbers and great variety of fibrous interosseous structures. The officially (internationally) recognized list has changed a number of times over recent decades (the list increasing with welcome additions). It should, perhaps, be indicated that the massive interosseous sacro-iliac ligament illustrated is not, as yet, in the official list.

long considered rare in mammals, the term was until quite recently restricted to the inferior tibiofibular joint *alone* (but vide infra). Clearly this was unsatisfactory; nevertheless interosseous fibrous connections occur in such profusion and great variety of form that, despite repeated attempts, no consensus concerning a precise definition of *syndesmosis* has emerged. The following remarks may serve to illustrate some inconsistencies and difficulties and perhaps provide a background for the curious assortment of structures officially recognized as syndesmoses given below.

Though not usually so described, the dorsal part of the sacro-iliac junction, through its massive interosseous sacro-iliac ligaments, *is*, rather than 'closely resembles' a syndesmosis (4.2). The sacro-iliac joints proper, primarily synovial, also are often invaded by fibrous tissue late in life and may become entirely fibrous articulations, differing little from syndesmoses (p. 466). The inferior tibiofibular joint, for long accepted as a typical *syndesmosis*, alternatively could well be considered no more than an interosseous ligament, adjacent to the ankle joint or to a synovial extension of it when this exists, as it occasionally does in man and regularly in other primates. If this is accepted, the term syndesmosis could be extended to many other interosseous ligaments, as in the carpus and tarsus, and also to include interosseous membranes of the forearm and foreleg, especially since these are already described as 'intermediate' joints in the radio-ulnar and tibiofibular series (but vide infra). Since ligaments are almost all

'interosseous', it becomes difficult to restrict use of the term 'syndesmosis' at all, unless it be insisted that only very short *extrinsic* ligaments close to a synovial joint are so designated. (Plainly intrinsic and intracapsular ligaments of synovial joints must be excluded.) These appear to be the kind of criteria which limited use of the term to a single human joint, a restricted and irrational usage rendering all the more unsatisfactory the occasional substitution of the term 'syndesmology' for arthrology. The former is a word of unfamiliar derivation and hence obscure in meaning, whereas the stem **arthron**, a joint, is widely familiar in medical and biological application, e.g. arthritis, arthroplasty, and the name of the largest animal phylum, the *Arthropoda*. Official re-adoption of the internationally accepted term **arthrology** is most welcome.

In *Nomina Anatomica* (1977) the term **syndesmosis** was extended to include the following ligaments: pterygospinous, stylohyoid, interspinous, supraspinous, intertransverse, ligamenta flava and ligamentum nuchae. The *syndesmosis radio-ulnaris* includes the antebrachial interosseous membrane and oblique cord, the *syndesmosis tibiofibularis*, the crural interosseous membrane and ligament and the anterior and posterior tibiofibular ligaments. This particular choice of ligaments is, however, quite arbitrary; many others could as reasonably have been included as syndesmoses. In the opinion of the present writers the list should be greatly expanded within a strictly defined framework.

Synovial Joints

Synovial articulations (4.3,4) operate differently from *non-synovial* fibrous and cartilaginous joints. Although the bones involved are linked by a fibrous capsule usually having intrinsic ligamentous thickenings, and often by accessory ligaments internal or external to it, the osseous surfaces concerned are *not* in continuity. They are covered by articular cartilage, a stratum of specialized hyaline cartilage (occasionally fibrocartilage) of varying thickness and precise topology, and contact is strictly between these cartilaginous surfaces, which have a very low coefficient of friction (Charnley 1959). Sliding contact is facilitated by viscous *synovial fluid* (synovia), like a lubricant in some respects but also concerned in the maintenance of living cells in the articular cartilages. Its properties are considered later (p. 475). The coefficient of friction in synovial joints has been picturesquely described as equal to 'ice on ice'.

A fibrous capsule encloses such joints completely, with some exceptions, e.g. the hip joint (p. 520), where it is interrupted by synovial protrusions; such exceptions are described with individual joints. The capsule is lined by *synovial membrane*, which also covers all intra-articular surfaces except those in compression contact during activity; these include non-articulating osseous surfaces, tendons and ligaments partly or wholly within the fibrous capsule, as at the shoulder and knee. Where a tendon is attached inside a joint and issues from it, a synovial prolongation usually accompanies it beyond the capsule. Some extracapsular tendons are separated from it by a synovial bursa (p. 502) continuous with the joint's interior. Such protrusions are potential avenues for spread of infection into joints.

A third type of intra-articular structure, *not* covered by synovial membrane, is an *articular disc* or *meniscus*. This occurs between articular surfaces where congruity is low and consists of fibrocartilage, its fibrous element usually predominant. A disc may extend across a synovial joint, dividing it structurally and functionally into *two* synovial cavities. Peripherally discs are connected to fibrous capsules, usually by vascularized connective tissue; but the union is sometimes closer and stronger, as in the knee and jaw joints. Discal peripheries are invaded by vessels and afferent and motor (sympathetic) nerves. Their main part contains few cells, but surfaces may have an incomplete stratum of flat cells, continuous at attached perimeters with adjacent synovial membrane; however the two groups of cells are not alike (p. 474). The term *meniscus* should be reserved for incomplete discs, like those in the knee joint and occasionally the acromioclavicular joint. Complete *discs* occur in the sternoclavicular and inferior radio-ulnar joints; that in the temporomandibular joint may be complete or incomplete. Discs often have small perforations; where menisci are usual, complete discs may occur or may be

slightly perforated. The function of intra-articular fibrocartilages is uncertain, experiments being difficult to devise or effect. Views advanced are deductions from structural or phylogenetic data, with the help of mechanical analogies. The plethora of suggestions is hence not surprising; it includes shock absorption, improvement of fit between surfaces, facilitation of combined movements, checking of translation at joints such as the knee, deployment of weight over larger surfaces, protection of articular margins, facilitation of rolling movements and spread of lubricant. The temporomandibular disc has attracted particular attention because of its exceptional, perhaps unique, design and biomechanical properties. Osborn (1985) has surveyed these properties and examined some factors relating to the *stabilization* of the disc and its possible role as a *destabilizer* of the mandibular condyle, permitting trauma-free complex movements (see p. 487).

In the study of evolution of articular discs attention has been focused on equating them with skeletal vestiges, such as the quadrate bone (jaw joint) or the os intermedium carpi (inferior radio-ulnar joint), not upon comparative function. Such contentions are usually based on tenuous evidence but nodules of bone, or lunules, do occur in some primate discs but rarely in mankind (Lewis et al 1970). A more interesting comparative functional observation is that joints containing discs usually display translatory movements, combined of course with others (vide infra, p. 477). For example, in the temporomandibular joint, translation of the condyle forwards and backwards accompanies angulation or hinging. However, the disc, though usually thin, also occurs in carnivores where such translation is negligible. It is apposite to note (Moffat 1957) that, during evolution of mammalian from primitive reptilian jaw joints, the tendon of the lateral pterygoid muscle may have been partially incorporated into the temporomandibular disc; but it is absent in earlier mammals, monotremes and marsupials. Such conflicting observations illustrate how indecisive attempts to elucidate function from morphology often prove. It is more plausible that the pliant and sometimes elastic qualities of discs and menisci provide adaptable surfaces in joints where they occur, facilitating simultaneous movements of different kinds in the compartments of a joint thus divided (p. 486), perhaps also ensuring uninterrupted lubrication (MacConaill 1932). Where translation occurs, complete congruity is impossible except in a (hypothetical) entirely plane articulation, the nearest approximation to this being some sesamoid joints. All articular surfaces exhibit some curvature and translation is always compounded with some other movement (p. 477). But translation does occur and intra-articular cartilages may aid in controlling shift from one position to another, as in rolling at the knee joint (p. 530). The tapering profile of menisci enables them to fit between incongruent surfaces, obviating large

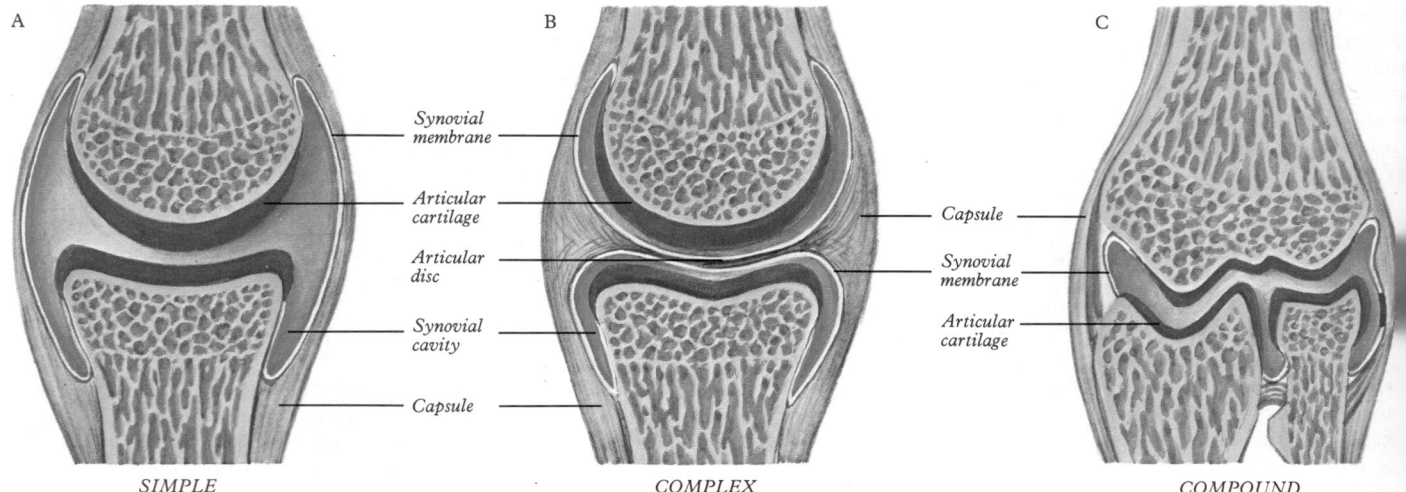

SIMPLE   COMPLEX   COMPOUND

4.3 Synovial joints, some main structural features and one elementary type of classification: A. simple; B. complex; C. compound joints (see text). For clarity, the articular surfaces are artificially separated. A and B are purely diagrammatic and not related to particular joints. C, however, is a simplified representation of some features of an elbow joint but the complicated contours due to the olecranon, coronoid and radial fossae and profiles of articular fat pads present in a true section have been omitted.

collections of synovial fluid between them and preserving a filmy distribution over articular surfaces. Advantages of this in damping turbulence and drag accord with the lubrication theory. This argument provides an attractive hypothesis but its experimental demonstration is at present difficult to envisage.

A complete articular disc creates in effect two joints in series, comparable with concatenations of multiple joints, whose individual small ranges of movement nevertheless summate. This arrangement occurs in the carpus and tarsus but is carried further in multiplication of interphalangeal joints in flippers of extinct aquatic reptiles and extant aquatic mammals such as some whales. By such analogy one may liken the talus of some mammals to a meniscus, in terms of function; it is equally tempting to equate some menisci with degenerate skeletal elements. The various views propounded are not reciprocally exclusive; discs may have evolved by more than one route, just as they may assist in more than one way the smooth performance of synovial joints.

Functions of two other quite common types of intra-articular structure, *labra* and *fat pads*, are also uncertain. A labrum is a fibrocartilaginous annular *lip*, usually triangular in cross-section like a meniscus and attached to an articular margin, e.g. the glenoid and acetabular labra, which deepen their respective sockets and add to areas of contact. They may act as lubricant spreaders and, like menisci, may reduce synovial space to capillary dimensions, thus limiting drag. However, unlike menisci, labra are not compressed between articular surfaces. Small *fibrous labra* (*connective tissue rims*) have been described along the ventral or dorsal margins of the zygapophysial joints at lumbar levels and also meniscus-shaped *fibro-adipose meniscoids* at the superior or inferior poles of the same joints. (For brief comments and reference see p. 490.) Fat pads are closely associated with synovial membrane, with which they are therefore described (p. 474).

EVOLUTION OF SYNOVIAL JOINTS

Much speculation, observation and experiment have been directed to the mechanics of synovial joints by anatomists, orthopaedic specialists, physicists and engineers. The basic problem is to account convincingly, in terms of movement, loading and lubrication, for the efficiency by which joints preserve smoothness of action without jamming under most variable conditions, except when diseased. It has proved impossible even to postulate, much less to construct, a single mechanical joint meeting all requirements of living arrangments. Lubrication theory has not yet explained the outstanding effectiveness of synovial joints, although progress has been made (p. 475).

What are the advantages of sliding articulations? That these are considerable is shown by the increasing dominance of synovial joints in vertebrate phylogeny. Evolution of joints has attracted little attention; few joints or vertebrate groups have been systematically studied. Synovial articulation occurs as far down the scale as lungfish (Dipnoi), especially in joints of jaws (Haines 1942b). Hence, the synovial joint is not particularly novel, in general features at least. Nevertheless, most movable bony junctions in piscine ancestors of land vertebrates were simpler, as in surviving species. These simpler joints may illustrate stages in the development of synovial joints. Fibrous and cartilaginous joints may be assumed to be the simplest and probably primary form, the next step being appearance of multiple fluid-filled cavities in the deformable tissue. A further advance is held to be union of these into a single joint cavity, surrounded by a substantial cuff of tissue uniting the skeletal components involved. This pattern resembles cavitated symphyses (p. 468). Finally complete dissolution of continuity between bones would markedly increase amplitude of movement. With subsequent development of a synovial stratum and a fibrous capsule, the fluid-containing cavity confined and the fluid reduced to a mere film between surfaces approximated in smooth, sliding contact.

Examples of all such stages occur in living vertebrates; interestingly synovial joints in more primitive land animals, although they replace other arrangements in limb articulations, are in some aspects inferior to piscine mandibular joints, which may have been the first vertebrate articulations refined to synovial status.

The stages hypothecated above accord with major events in prenatal development of human synovial joints. Perhaps most significant is breakdown of interzonal mesenchyme (p. 175) to form a presumptive synovial cavity. Potentiality of mesoderm for cleavage (forming considerable cavities or clefts between surfaces in contact and hence mobilized) is clearly significant to the evolution of synovial joints. In addition to established mesodermal discontinuities, including the whole range of coelomic and synovial arrangements, fully differentiated connective tissue retains this potentiality, revealed in the occurrence of such acquired structures as adventitious bursae and pseudarthroses. It is reasonable to suppose that, once a region of pliant connective tissue is established between rigid skeletal elements, the ultimate appearance of a synovial cavity is an evolutional and mechanically significant probability.

It is considered likely that synovial joints have developed not only as discontinuities in skeletal bars but also by approximation of separate elements. Mammalian tibiofibular joints may have thus evolved, the bones being out of contact in reptiles; but these synovial arrangements are probably extensions of the knee and ankle joints and not new formations. A better example, perhaps, is the occasional synovial joint between clavicle and coracoid process (p. 500, Lewis 1959), which may consist merely of a bursa or, more frequently than either arrangement, a fibrocartilaginous junction, an excellent example of lability of joints in occurrence and structure. A synovial joint such as the sacro-iliac may retrogress to a simpler form and may, like others such as interphalangeal, carpal and tarsal articulations, disappear entirely either by synostosis in individuals or as an evolutionary phenomenon. The avian tibiotarsus and tarsometatarsus illustrate reunion of once separate bones; the opposite tendency, towards multiplication of elements by new joint formation, is apparent in phalangeal patterns in the paddle-like extremities of some aquatic mammals and extinct reptiles.

Evolution of synovial joints at mammalian level shows two tendencies. Firstly, their number increases by replacement of non-synovial joints, particularly in limbs, eventually reaching their smallest terminal articulations. Secondly, synovial joints increasingly specialize, especially by limitation of movement to that habitually required. A typical non-synovial joint, such as a symphysis, is essentially multi-axial, however limited in range; its movements are basically angulation, torsion or rotation ('swing' and 'spin', see p. 480) and 'translation'; all others are combinations of these. From the restricted data available, the earliest synovial joints are likely to have had the same repertoire, with improvement in range and smoothness. The ancestral vertebrate limb probably had joints approximating to a 'ball-and-socket' form; palaeontological data and synovial joints in extant primitive amphibians and reptiles support this supposition. This view also entails that proximal joints have remained less specialized than the more distal; they certainly are closer to multi-axial activity. A multi-axial joint requires more complex muscular control than a bi-axial, in which more reliance can be placed on ligaments, an advantage even greater in uni-axial joints. This advantage obtains both in dynamic and static situations: for when a multi- or bi-axial articulation is in substantially uni-axial movement, muscle must be used to prevent unwanted movement in other axes. Similarly, when a joint is to be maintained static in a particular posture, or as an adjunct to some other phase of movement, this is effected more economically if joint surfaces and disposition of adjoining ligaments limit movement in some directions. Unless this interferes with overall activity, the trend is towards uni-axial function and this appears to be the evolutionary tendency in limbs. This is tantamount to saying that joints are adapted to motions required in the behaviour of a particular animal form. To take the argument further, a joint may even disappear if its activity ceases; examples have occurred in many mammals; in the human skeleton the process appears as synostosis of sternal and sacral segments, conferring rigidity on joints having a limited degree of mobility in some other mammals.

Refinement of a joint, in limitation of direction and range of movement to whatever is habitually required, favours skilled control and the most advantageous distribution of available muscle power. Bulky muscles around a joint may impede it and it is an

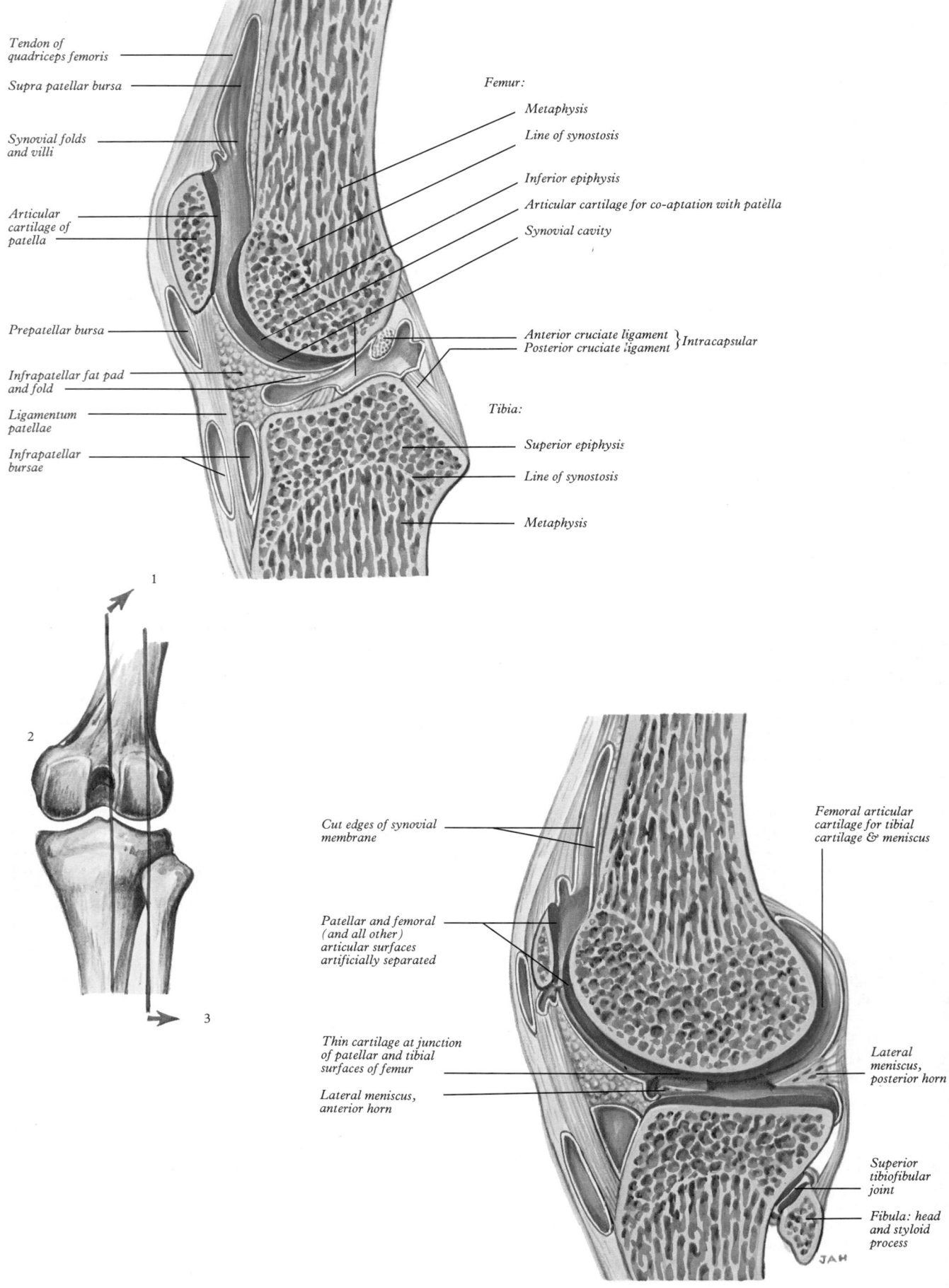

Tendon of quadriceps femoris

Supra patellar bursa

Synovial folds and villi

Articular cartilage of patella

Prepatellar bursa

Infrapatellar fat pad and fold

Ligamentum patellae

Infrapatellar bursae

Femur:

Metaphysis

Line of synostosis

Inferior epiphysis

Articular cartilage for co-aptation with patèlla

Synovial cavity

Anterior cruciate ligament
Posterior cruciate ligament } Intracapsular

Tibia:

Superior epiphysis

Line of synostosis

Metaphysis

Cut edges of synovial membrane

Patellar and femoral (and all other) articular surfaces artificially separated

Thin cartilage at junction of patellar and tibial surfaces of femur

Lateral meniscus, anterior horn

Femoral articular cartilage for tibial cartilage & meniscus

Lateral meniscus, posterior horn

Superior tibiofibular joint

Fibula: head and styloid process

JAH

4.4A, 1–3. Principal structural features displayed in two parasagittal sections of the left knee joint, which is synovial and both compound and complex. 2. Line diagram of the posterior aspect showing the planes of section in 1 and 3. The synovial cavity and bursae are, for clarity, shown as slightly distended; the articular surfaces of the patella, femur and tibia and the meniscal surfaces are thus artificially separated. These illustrations should be studied both with the accompanying introductory text and also with the detailed account of the knee joint (p. 526 et seq).

advance if restraining activities can be transferred from muscles to ligaments and articular geometry. Joints may be arrested in temporary but prolonged static positions, as required of human hip and knee in standing. If joints can be maintained in a nearly close-packed position (p. 483) by gravity, muscular effort can be reduced, e.g. the usual human mechanism of standing (p. 657). This commonly cited example of joint control prompts a concept of *stability* at joints, a now somewhat overworked term with *static* implications. Joints are, however, primarily *dynamic*; during movements qualities stabilizing them in one position are equally important in controlling transit from one attitude to another. In the medical context it is natural to view these factors as preserving articular *integrity*—a clinical approach deviating more towards factors preventing dislocation than towards functional interpretation of structure. Synovial joints of higher vertebrates are precisely engineered to accomplish habitual movements with greatest efficiency commensurate with the resources available; but, before analysis of individual joints, the intimate structure of the tissues of synovial joints must be considered.

STRUCTURE OF SYNOVIAL JOINTS

Articular surfaces are mostly formed by a special variety of *hyaline cartilage* reflecting their preformation as parts of cartilaginous models in embryonic life (Barnett et al 1961, Ghadially & Roy 1969); but, as exceptions to this, surfaces of the sternoclavicular and acromioclavicular joints and both temporomandibular surfaces are of dense fibrous tissue, with isolated groups of chondrocytes and little surrounding matrix, again reflecting their formation by *mesenchymatous* ossification. However, to regard cartilage of most articular surfaces as unmodified hyaline cartilage and merely the unossified surface sectors of growing cartilaginous models obscures its special features. In long bones articular cartilage is a specialized tissue long before the subjacent bone has ossified (Davies & Edwards 1948); the organization of mature articular cartilage is distinctly radial, with variation in cellular type and arrangement, fibrous architecture and calcification at varying levels from its surface (p. 288). (Split-line phenomena in articular cartilage and their possible relations to architectural and mechanical properties are discussed on p. 290. Examples of split-line patterns in a talocrural joint are shown in 4.4B.)

Articular cartilage has a wear-resistant, low-frictional, lubricated surface, slightly compressible and elastic and thus ideally constructed for easy movement over a similar surface but also able to absorb large forces of compression and shear generated by gravity and muscular power, qualities specially important at one extreme of a joint's most habitual movement, the so-called 'close-packed position' (p. 483).

The thickness of articular cartilage is said to range from 1–2 mm, but this is more typical of small bones in aged individuals; in youth it may reach 5–7 mm in larger joints; such young cartilages are typically white, smooth, glistening and compressible. Ageing cartilages are thinner, less cellular, firmer and more brittle, with a less regular surface and a yellowish opacity. In contrast to the 'glassy' moistness of fresh, wet articular cartilage and to the measurements held to support this early impression of smoothness, more refined techniques have emphasized the relative 'roughness' of its free surface. Dowson et al (1969) and Longfield et al (1969) observed a 'centre-line average' of undulations varying from 30×10^{-6} to 200×10^{-6} inch, much inferior to mechanical bearings (5×10^{-6} to 15×10^{-6} inch); but when covered by synovial fluid the surface has a very low coefficient of friction (< 0.002). Under load the 'troughs' between the 'crests' have been considered to trap pools of synovial fluid. Others (McCutchen 1959) have proposed a porous nature (with a pore size of 6 nm) like a sponge, saturated at rest with synovial fluid; increasing load and compression is presumed to 'weep' fluid from this porous surface. Scanning electron microscopy has renewed interest in these surfaces. When *detached* blocks of articular cartilage were first viewed many described patterns of undulation as constant features (e.g. McCall 1968, Gardner & Woodward 1969, Walker et al 1969, Redler & Zimny 1970, Redler 1974). But this view has

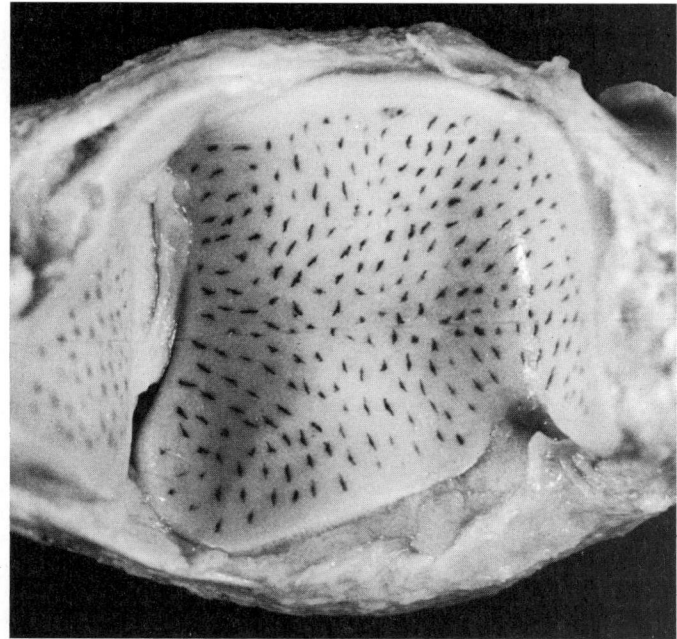

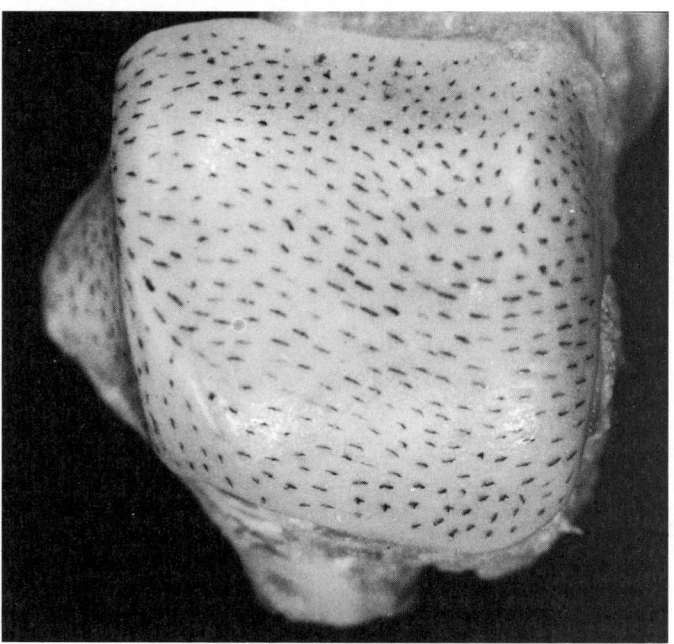

4.4B Articular surfaces of the left talocrural joint demonstrating the patterns of split-lines in the articular cartilages produced by multiple insertions of a round-bodied needle, previously immersed in Indian ink. *Above*, tibiofibular mortice; note interosseous and inferior transverse tibiofibular ligaments. *Below*, superior (tibial) and lateral (fibular) surfaces of the talus. For possible significances of split-line configuration, see p. 278. Photography by Kevin Fitzpatrick, Guy's Hospital Medical School, London.

been criticized; undulation was not apparent in cartilage *still attached* to bone and was therefore regarded as artefactual, atypical or pathological (Clarke 1973a, Ghadially et al 1976, Ghadially 1983). The only constant features revealed in 'normal', mature articular cartilage were numerous oval, fusiform or round *shallow pits*, considered to overlie groups of chondrocytes (Clarke 1973a,b, Ghadially et al 1974,1975,1976). In juvenile cartilage pits are replaced by small 'humps' of similar size. Doubt was thus cast on the 'enrichment theory' of joint lubrication (p. 476 and Ghadially et al 1977). Nevertheless, despite impressive micrographs of attached cartilage, the tissue was of course subjected to all the rigors of preparation for scanning electron microscopy. To what extent 'pits and humps' are due to this remains uncertain.

Such difficulties have been circumvented by studying fresh necropsy specimens of weight-bearing areas of lateral femoral condyles by reflected light interference microscopy (Longmore 1976, Longmore & Gardner 1978). Observed curvatures, undulations and irregularities were classified into groups of decreasing dimension: (1) *primary anatomical contours*, e.g. ovoid or sellar surfaces; (2) *secondary undulations* of 100–500 μm crest to crest; (3) *tertiary hollows* 20–50 μm in diameter and 0.5–2 μm deep. (4) *quaternary ridges* 1–4 μm in diameter and 130–275 μm deep. Longmore & Gardner (1978) emphasized the tertiary hollows, whose size, shape and distribution were shown to be age-dependent (a phase of *maturation* from 0–21 years being followed by *ageing* from 22 to 50 years). In the youngest surfaces tertiary hollows were largely circular and tightly packed; with increasing age they were separated, becoming circular or oval, sometimes in clusters or 'figure of eight' pairs and the areas between them, although 'granular and fuzzy', were relatively smooth. From 21 years onwards quaternary ridges, appearing sporadically between 10 and 20 years, occurred more often in focal patches in some 'inter-hollow areas'. Tertiary hollows appeared to correspond to underlying chondrocytic lacunae, quaternary ridges perhaps representing incipient cartilage fibrillation associated with ageing, though Bloebaum & Wilson (1980) concluded that they were simply shrinkage 'wrinkles' attributable to drying of the surface when exposed to air.

Articular cartilages are moulded to bone, but variations in thickness often accentuate subjacent osseous surface shape. Typically, convex surfaces are thickest centrally, thinning peripherally; concave surfaces are the reverse of this. Precise configuration, degree of congruence in various positions and dispositions of surrounding capsule and ligaments are all related to the types and ranges of movement permitted at a joint (vide infra). (For a simplified classification based on articular geometry see pp. 477 and 479).

Articular cartilage has no nerves or blood vessels (except occasional vascular loops reaching and even penetrating the calcified zone (p. 290). Nutrition is considered to depend on (1) a peripheral vascular plexus in synovial membrane (*circulus vasculosus articuli*), (2) synovial fluid and (3) blood vessels in adjacent marrow spaces; the relative importance of these is disputed.

The free surface of articular cartilage, devoid of perichondrium, was described by MacConaill (1951) as consisting of a special layer, the *lamina splendens* (p. 288), appearing as a bright line when oblique sections are examined by negative phase contrast microscopy. This is now known to be a diffraction effect which occurs at any edge between regions of different refractive index (e.g. the cut edges of a cartilage section) and provides no evidence of an anatomically distinct surface layer (Aspden & Hukins 1979); it is now considered that the term lamina splendens is inappropriate and should be abandoned. The zone of articular cartilage adjacent to the joint cavity is mainly a layer of collagen fibres arranged in various planes with small, oval chondrocytes lying in the matrix deep to it. An interrupted electron-dense surface coat of a particulate or filamentous appearance, generally 0.03 – 0.1 μm thick, covers it, with a layer of proteoglycan particles and associated filaments occasionally intervening (Ghadially et al 1982). Synovial fluid and matrical lipidic debris, the product of chondrocytic necrosis, may contribute to the surface coat which is ephemeral in nature, the stable, permanent articular surface being that bounded by the most superficial collagen fibres (Ghadially 1983).

With advancing age, undulations of articular surfaces deepen and develop minute, ragged projections perhaps due to wear and tear. Erosion occurs in pathologically 'dry' joints and where synovial viscosity is altered; but in healthy joints changes are extremely slow. Replacement of eroded surface by proliferation of deeper layers is uncertain; mitoses are absent from adult articular cartilage and amitotic cell division has not been substantiated.

Ultrastructural observations (Broom & Marra 1984, 1986) reveal a highly complex, three-dimensional reticulum of interconnected fibrils in articular cartilage, with obvious functional implications.

The fibrous capsule has parallel but interlacing bundles of white collagen fibres, forming a cuff with its ends attached continuously round the articular ends of the bones concerned, usually in small bones near the peripheries of articular surfaces, but this varies considerably in long bones; in these part or all of the attachment may be a significant distance from the surface. It is perforated by vessels and nerves and may have apertures through which synovial membrane protrudes as bursae. The capsule usually has local thickenings of parallel fibre bundles; such *capsular (intrinsic) ligaments* are named by their attachments. Some capsules are reinforced or replaced by tendons of nearby muscles or expansions from them. *Accessory ligaments* are separate from capsules and may be *extracapsular* or *intracapsular* in position.

All ligaments, although yielding little to tension, are pliant and do not resist normal actions, being designed to check excessive or abnormal movements. All ligaments are taut at the normal limit of a particular movement (p. 485); they are, however, *slightly elastic* and protected from excessive tension by reflex contraction of appropriate muscles (Smith 1954).

Synovial membrane (Barnett et al 1961) derived from embryonic mesenchyme lines non-articular areas in synovial joints, bursae and tendon sheaths, all regions where movement occurs between apposed surfaces, which are lubricated by a fluid superficially like egg-albumin (hence named *synovia*) secreted and absorbed by the membrane. In joints it lines fibrous capsules and covers exposed osseous surfaces, intracapsular ligaments and tendons. It is absent from intra-articular discs or menisci and ceases at the margins of articular cartilages, the peripheral few millimetres of which are a structural *transitional zone* between synovial membrane and articular cartilage.

Pink, smooth and shining, the internal synovial surface has a few small *synovial villi* (Ghadially 1983) which increase in size and number with age (Curtis 1964). Elsewhere folds and fringes may project into the joint cavity, some constant enough to be named, e.g. alar folds and ligamentum mucosum of the knee. Accumulations of adipose tissue occur in the synovial membrane in many joints; these *articular fat pads* were once misnamed 'mucilaginous glands'. Such pads, folds and fringes are flexible, elastic and displaceable cushions occupying potential spaces and irregularities in joints which are not wholly filled by synovial fluid; they accommodate to the changing shape and volume of these irregularities during movement. They increase synovial area and may promote distribution of lubricant over articular surfaces (cf. intra-articular discs and menisci). Synovial villi are normally few but more numerous where the membrane rests on areolar tissue near articular margins and on surfaces of folds and fringes. They increase with age and become prominent in some pathological states.

Synovial membrane varies in local structure. Basically it has a cellular *intima* on a fibrovascular *subintimal lamina* (*subsynovial tissue*) often loose and areolar, but sometimes of organized lamellae of collagen and elastin fibres lying parallel to the membrane's surface, between which are fibroblasts, macrophages, mast cells and fat cells (Davies 1950, Shaw & Martin 1962). The elastic component may prevent redundant fold formation during joint movement, folds which might be compressed between articular surfaces. Subintimal adipose cells form compact lobules surrounded by fibro-elastic interlobular septa, which are very vascular and impart firmness, deformability and elastic recoil. Where synovial membrane covers intracapsular ligaments or tendons the subintima is scarcely a separate zone, being merged with the adjacent capsule, ligament or tendon.

The synovial intima (*lamina propria synovialis*), or synovial lining layer, consists of pleomorphic *synoviocytes* embedded in granular, amorphous, fibre-free intercellular matrix. There is considerable regional variation in synoviocyte morphology and numbers, which appears to be dependent on the underlying subintimal tissue (Barnett et al 1962). The synoviocytes of normal human joints form an interlacing, discontinuous layer, one to three cells and 20–40 μm deep, between the subintima and the joint cavity (Castor 1962). They are not separated from the subintima by a basement membrane and are distinguished from the subintimal cells only by their association to form a superficial layer (Henderson & Pettipher 1985). In many locations, but particularly over areolar subintimal tissue, areas free from synoviocytes are commonly found, while conversely over fibrosubintimal tissue the synoviocytes may be flattened and closely packed, forming endothelioid sheets. Neighbouring cells are often separated by distinct gaps but where they approach more closely their processes may interdigitate. The latter situation is common in compact areas of rat's synovial intima, where cells may be linked by tight junctions and desmosomes (Roy & Ghadially 1967), although these junctions have not been identified in human synovial tissue (Henderson & Pettipher 1985).

Human synoviocytes are generally elliptical, with numerous cytoplasmic processes (Castor 1962) but can vary considerably in form (Key 1932). They consist of at least two morphologically distinct populations, termed *type A* and *type B* (Barland et al 1962, Wynne-Roberts & Anderson 1978, Ghadially 1983).

Type A synoviocytes are macrophage-like cells characterized by surface ruffles or lamellipodia (often described as filopodia, since they resemble these when sectioned), plasma membrane invaginations and associated micropinocytotic vesicles, a prominent Golgi apparatus but little granular endoplasmic reticulum. There is immunohistochemical evidence for the presence of surface receptors characteristic of macrophages on what are believed to be type A synoviocytes (Theofilopoulos et al 1980).

In contrast, *type B* synoviocytes, which predominate in the intima of most species (Ghadially 1983) and resemble fibroblasts, have abundant granular endoplasmic reticulum but contain fewer vacuoles and vesicles and have a less ruffled and branched plasma membrane than type A synoviocytes.

Cells intermediate in appearance between types A and B have been described and variously named as *intermediate, type C* or *type AB* (Ghadially 1983). Using ultrathin serial sections of rat synovial membrane, Graabaek (1984) has demonstrated conclusively that there are two distinct synoviocyte types, equivalent to A and B, and that cells which in a single section seemed intermediate in appearance could be shown by examination of other sections to be invariably of a type equivalent to B (termed type S by Graabaek). It has been suggested that the intermediate type of synoviocyte might be a precursor type A and B (Kinsella et al 1970) but the findings of Graabaek cast doubt on this.

The possiblity that precursors of type A synoviocytes may originate in bone marrow and be part of the mononuclear phagocyte system was proposed by Fell (1978) and demonstrated by Edwards (1982), using genetic markers for tracing cell lines related to mononuclear macrophages in radiation chimeras. The derivation of type B synoviocytes is still uncertain. Edwards & Willoughby (1982) have shown that in normal synovial membranes type B cells are derived neither from the bone marrow nor from type A cells. They could be of local origin from within the intima and be replaced by cell division or, as suggested by Ghadially (1983), derive from a stem cell population in the local subintima. Synoviocytes are not, however, an actively dividing cell population in normal synovial membranes (Key 1932, Coulton et al 1980), although their division rate increases dramatically in response to acute trauma and acute haemarthrosis (Ghadially 1983). It is interesting to speculate that in such conditions it is the type B cells which divide in situ while the type A cell population is increased by migration of bone-marrow derived precursors.

The *functions* of the cells of the synovial intima include the removal of debris from the joint cavity (mainly by type A cells) and the synthesis of some of the components of the synovial fluid (by both types of cells).

1. *Removal of debris*. Once a synoviocyte has ingested particulate matter it becomes capable of migrating into the subintima (Vernon-Roberts et al 1976), and it is probable that migrating synoviocytes carry ingested material to the lymphatic channels of the subintima which remove it from the joint. Fluid containing protein may be taken into the synoviocytes by micropinocytosis; there is evidence of the removal of injected peroxidase by type A cells but not by type B in this manner (Linck & Porte 1981).

2. *Synthesis*. It is generally assumed that some of the hyaluronic acid of synovial fluid is synthesized by the synoviocytes, although it could be manufactured by subintimal fibroblasts. While there is evidence that, like most mammalian cells, synoviocytes synthesize some forms of glycosaminoglycans, the evidence that these include hyaluronic acid is inconclusive (Henderson & Pettipher

1985). The presence of hyaluronic acid within the vesicles of synoviocytes could be due to micropinocytotic uptake from synovial fluid instead of, or as well as, synthesis by the cells. The ultrastructure of the intimal synovial cells suggests strongly, however, that they are involved in some form of synthesis and secretion, although the nature of the secretory products is so far only speculative. One or both of the synoviocyte types could be involved in the secretion of the proteoglycan-like material, lubricin, which acts as an articular cartilage lubricant; it has been shown to be produced by cells isolated from the synovial membrane and is not produced by fibroblasts from other sources (Swann 1982). Synoviocytes could also secrete some of the constituents of the extracellular matrix which surrounds them, e.g. collagen, proteoglycans and fibronectin (Henderson & Pettipher 1985). It has been proposed by Glynn (1977) that type A synoviocytes synthesize and release lytic enzymes during the phagocytosis of joint debris, damage to joint tissues being limited by the secretion of enzyme inhibitors by type B cells.

3. *Other functions*. Some intimal synoviocytes can present antigens to lymphocytes and can therefore rapidly stimulate an immune response to foreign material appearing in the joint cavity (Klareskog et al 1982). They may also have some control over the functioning of chondrocytes; it is known that mononuclear cells secrete factors which stimulate chondrocytes to degrade their matrix (Jasin & Dingle 1981) and it is possible that type A cells have a similar function. They may also control blood flow in the synovial membrane by the release of prostanoids such as the vasoconstrictor thromboxane A_2 and the vasodilator prostacyclin (Blotman et al 1983, Salmon et al 1983).

Synovial fluid occupies synovial joints, bursae and tendon sheaths. In synovial joints it is clear or pale yellow, viscous, slightly alkaline at rest (diminishing in activity) and has a small mixed population of cells and metachromatic amorphous particles. Viscosity, volume and colour vary in different joints and species; it is difficult to correlate these variations with particular joints or with size, weight or exertion. In human joints volume is low; usually less than 0.5 ml can be aspirated even from a large joint such as the knee.

The physical properties of synovial fluid include viscous, elastic and plastic components. Disagreements and inconsistencies in earlier investigations were often due to inadequate technique and ignorance of its non-Newtonian properties; i.e. with low rates of shear, the fluid is highly viscous but viscosity decreases with increased rates of shear, suggesting that in slow movement weight-bearing capacity would be maximal and in fast movement impedance due to fluid drag would be reduced. It was subsequently shown that the *product* of viscosity and shear rate is almost constant and hence also the weight-bearing capacity. Viscosity is very sensitive to changes in dilution and falls with increasing temperature and pH. Elasticity is similarly affected by changes in dilution, pH and temperature but it increases with higher rates of shear, unlike viscosity in similar conditions. For earlier reviews consult McCutchen (1959), Barnett et al (1961), Dintenfass (1963), Negami (1964), MacConaill (1966), Dowson et al (1969), Longfield et al (1969), Freeman (1973).

The composition of synovial fluid suggests a dialysate of blood plasma, containing protein (about 0.9 mg/100 ml) mainly derived from blood (Swann 1978), and added mucin, mostly hyaluronate, a sulphate-free glycosaminoglycan containing equimolar concentrations of glucuronic acid and N-acetyl-glucosamine; much evidence shows its visco-elastic and thixotropic (plastic) properties as largely due to hyaluronate content. Synovial mucin has been variously considered an attrition product of articular cartilage matrix, a secretion of subintimal mast cells, or a product of type A synovial cells. Light microscopy, histochemistry and tissue culture have proved inadequate but colloidal-iron techniques combined with electron-microscopy show a positive reaction for glycosaminoglycan in plasma membrane, filopodia, cytoplasmic vesicles and Golgi apparatus of type A synovial cells. This, with comparable results in other cells synthesizing glycosaminoglycans, is considered to confirm type A cells as sites of synovial hyaluronate synthesis (Ghadially & Roy 1969).

Synovial fluid protein is partly free and partly bound to glycosaminoglycans, including hyaluronate. Much appears to be of haematogenous origin, the relative concentration of the individual proteins probably depending on their molecular weights and on the permeability of the synovial vessels (Swann 1978). The protein associated with hyaluronate may be either loosely or firmly bound. About 10–30% of the loosely bound protein is derived from blood plasma (Fraser et al 1977), as is most of that firmly bound (termed hyaluronate-protein). The remainder of the latter, some 2%, differs from plasma protein and is probably produced by synovial intimal cells (How et al 1969). Approximately 0.5% of synovial fluid protein can be separated as a lubricating glycoprotein fraction (LGP–1); this lubricates articular cartilage like whole synovial fluid. Although hyaluronic acid is an efficient lubricator of soft connective tissues, including synovial membrane, tests have shown that it is ineffective on articular cartilage, where LGP–1 appears to be uniquely involved. (See Swann 1978 and Swann et al 1981 for further information.)

The origin of non-haematogenous proteins of synovial fluid is uncertain but ultrastructure of the type B cells of the synovial intima is typical of that of cells manufacturing and secreting protein, suggesting that they may be sources of at least some of them.

The small tally of cells (about 60 per ml in resting human joints) includes monocytes, lymphocytes, macrophages, synovial intimal cells and polymorphonuclear leucocytes (Bauer et al 1940). Higher counts, characteristic of youth and of other species, may be due to more active movements before sampling. The amorphous, metachromatic particles and fragments of cells and fibrous tissue in synovial fluid are considered due to wear and tear.

Functions of synovial fluid include provision of a liquid environment, small in range of pH, for joint surfaces, nutrition of articular cartilages, discs and menisci, lubrication and reduction of erosion. Its nutritive role, compared with direct diffusion from vascular plexuses, is uncertain; despite general agreement on its lubricant value details of its actions are still disputed.

Models proposed for lubrication in joints have largely paralleled current advances in engineering physics. First proposed was 'fluid film' or 'hydrodynamic' lubrication, familiar in engineering, bearing surfaces being separated by a substantial layer of lubricant, its effectiveness depending on the rheological properties of the fluid in bulk. Later theoretical refinement included consequences of fluid elasticity, the 'elastohydrodynamic model'; but criticism of suitability of the articular environment for simple fluid-film lubrication arose and 'boundary lubrication' was proposed, in which properties of the solid surfaces are combined with those of extremely thin layers of lubricant molecules. 'Weeping lubrication' followed, in which the porous, fluid-filled deformable nature of the articular surface was emphasized; it was proposed that surfaces under load were lubricated by a film of fluid expressed from the 'pores'. Then 'boosted lubrication' came forward suggesting that compression of articular cartilages traps fluid pools in irregularities of their surfaces; increasing compression is supposed to force a small-moleculed, mobile fraction of synovial fluid into the cartilage contact area, fluid left in the 'valleys' becoming increasingly enriched in hyaluronate and hence more viscous and thus a more effective and protective lubricant. But major ridges and valleys are probably artefacts (p. 473) and 'pits' and 'humps' in vivo are not proven. Considering the wide variation in joint geometry, structure and activity, it seems likely that multiple mechanisms may operate under different conditions. Despite much research, theories of joint lubrication remain speculative (Freeman 1973).

Synovial membrane not only produces fluid but also removes materials from articular cavities. Small molecules of crystalloids and soluble dyes can cross it directly into subintimal capillaries and venules, the former fenestrated according to Kos (1970). Particles pass into subintimal lymphatic capillaries for transport to regional lymph nodes. Intra-articular introductions of tracers, such as thorotrast, colloidal carbon and ferritin, followed by electron microscopy, has revealed the marked phagocytic powers of type A synovial cells, which rapidly enclose particles in phagocytic vesicles. The source of subintimal macrophages is uncertain; some may enter joints from blood but many are probably type A cells of intimal origin which enter subintimal tissues (Ghadially & Roy 1969).

CLASSIFICATION AMD MOVEMENTS OF SYNOVIAL JOINTS

Several criteria have been used in classifying synovial joints and their movements; but these differ in universality, scientific accuracy and utility. They include complexity and number of articulating surfaces, number and position of principal axes of movement, general geometry of surfaces and major movements, and association of more precise geometry with analysis of movements as bases for *human kinesiology*. This can be considered only briefly here (p. 477); detailed treatises are available (Steindler 1955, Barnett et al 1961, MacConaill & Basmajian 1977).

Complexity of form. Most synovial joints have *two* surfaces (*male* and *female*) and are *simple* articulations. In some one surface is wholly convex (male) and greater in area than its opposing concave (female) surface; in a few, both are concavoconvex, the larger being considered male. A joint with more than two surfaces is *compound*, e.g. the elbow. (The humerus presents two male surfaces, capitulum and trochlea, which 'mate' with the radial and ulnar female surfaces. The convex, male, circumference of the radial head also 'mates' with the female radial notch of the ulna.) In all compound joints articulating territories remain distinct; male surfaces never pass on to female surfaces of an adjoining pair. Thirdly, when it contains an intracapsular disc or meniscus a joint is *complex*.

Degrees of freedom. Analysis of positional changes of one of a pair of articulating bones is often best effected by considering rotations around three mutually perpendicular axes, whose directions may, for convenience, vary with the joint concerned. Often they correspond to the main bodily planes, i.e. vertical, transverse and ventrodorsal axes; but alternatives may be more appropriate, e.g. many prefer to consider the movements of humerus on scapula around a vertical axis, an oblique transverse axis in the plane of the scapula and a third at right angles to both (4.5).

When movement is practically limited to rotation at one axis, a joint is termed *uni-axial*; it has *one degree of freedom*. If *independent* movements can occur around two axes, it is *bi-axial*, with *two degrees of freedom*. Since there are three axes for independent rotation, joints may have up to *three degrees of freedom*. This apparently simple classification requires amplification.

Firstly a bone can rotate at a ball-and-socket joint on three main axes *and* any others in intermediate positions. Such a *multi-axial* joint, however, still has only three degrees of freedom; all intermediate rotations can be resolved into components involving *three* orthogonal axes.

Secondly if movement referred to one plane is examined in its *full range*, a single relevant axis cannot be fixed in space but is a succession, changing continuously as movement progresses; because articular surfaces are not simple, their radius of curvature changes across any profile (p. 480) to a variable extent in different forms of joint (vide infra). But a *mean position* of such a moving axis is often 'the axis' of usual description. For many purposes such approximate axes are useful particularly when gross movements only are considered.

Thirdly simple movements apparently limited to one plane are often in fact compound. For example, what appears simple angulation is often accompanied by rotation of one bone about its long axis, a direct consequence of the complex articular curves (p. 479). When, as is usual, such rotation is not independent, i.e. never isolated, it is not considered an additional degree of freedom (e.g. movements of ulna, p. 509).

Fourthly many movements involve *translation*: one surface slides bodily across its partner with minimal rotation or angulation (cf. 'plane' joints, infra). Translation, however, often accompanies large angulations (e.g. shoulder joint). Note that translation is not included in the above classification.

General Classification of Synovial Joints

General shape of synovial joints has been widely used to classify them, usually into seven varieties. While this has some practical value, it affords no exclusive basis for separation because they are merely variations, sometimes extreme, of two basic forms (p. 478).

Plane joints are appositions of almost flat surfaces (e.g. intermetatarsal and some intercarpal joints); slight curvature is usual although often disregarded, movements being considered pure translations or slides between bones. Nevertheless, in precise dynamics even slight curvatures cannot be ignored (vide infra).

Ginglymi (hinge joints) resemble hinges and are shaped to restrict movement to one plane, i.e. they are uni-axial. They have strong collateral ligaments, interphalangeal and humero-ulnar joints being examples. But surfaces of such biological hinges differ from regular mechanical cylinders: their profiles are not arcs but varyingly spiral and the main to-and-fro swing must include some (conjunct) rotation (vide infra).

Trochoid (pivot) joints, also uni-axial, have an osseous pivot in an osteoligamentous ring, allowing rotation only around the pivot's axis. Pivots may rotate in rings, as the radial head rotates within the annular ligament and ulnar radial notch; or rings may rotate, as the atlas (with its transverse ligament and carrying the head) rotates around the dens of the axis.

Bicondylar joints are largely uni-axial, with a main movement in one plane but also with limited rotation about a second axis orthogonal to the first. The rotation is of two varieties: *conjunct*, an integral and inevitable accompaniment of the main movement; *adjunct* which can occur independently and may or may not accompany the principal movement. They have two convex *condyles* (knuckles) articulating with concave surfaces (sometimes also inappropriately named condyles). Condyles may be almost parallel (e.g. knee) with a common fibrous capsule or well apart in separate capsules (e.g. temporomandibular joints) which necessarily co-operate in all movements as a *condylar pair*.

Ellipsoid joints are bi-axial, with an oval, convex (male) surface apposed to an elliptical (female) concavity, as in radiocarpal and metacarpophalangeal joints. *Primary* movements are about two orthogonal axes (e.g. flexion–extension, abduction–adduction), which may be combined as circumduction (vide infra); rotation around the third axis is largely prevented by general articular shape.

Sellar (saddle) joints, also bi-axial, have concavoconvex surfaces; each is most convex in a particular direction but at *right angles* to this they are maximally *concave* (p. 478). Convexity of the larger is apposed to concavity of the smaller surface and vice versa. Primary movements occur in two orthogonal planes but articular shape causes axial rotation of the moving bone. Such *conjunct rotation*, as mentioned above, is never independent and is not simply a by-product of 'imperfect' mechanics but is functionally significant in habitual positioning and limitation of movement (pp. 483, 485). The most familiar sellar joint is the pollicial carpometacarpal; others include the talocrural (ankle) and calcaneocuboid joints.

Spheroidal joints ('ball-and-socket') are formed by reception of a globoid 'head' into an opposing cup, e.g. hip and shoulder joints. These are multi-axial, with three degrees of freedom. Their surfaces, although resembling parts of spheres, are not strictly spherical but slightly ovoid. (*Articulatio ovoidalis* is an accepted alternative.) Consequently, in most positions congruence is not perfect, occurring only in *one* position, at the end of the commonest movement (vide infra).

Articular Movements and Mechanisms

Articular movement will first be considered in terms and concepts usual in standard textbooks, followed by a review of the more advanced concepts and theories of modern kinesiology. Articular movements are usually considered to be gliding and angulation, circumduction and rotation. Almost always these four are combined in infinite variety. Where movement is slight, reciprocal surfaces are near in size; where it is wide, the bone habitually more mobile has the larger articular surface.

Translation, the simplest motion, is sliding without appreciable angulation or rotation. Often combined with other movements, in some carpal and tarsal articulations it is often considered the only motion permitted; but even here cineradiography reveals considerable rotation and angulation of small carpal and tarsal bones during their movements.

Angulation implies change in angle between (topographical) axes of articulating bones. Two types are common, especially in limbs, around orthogonal axes: (1) *flexion* or bending and *extension* or straightening and (2) *abduction* and *adduction*.

Flexion, a widely used term, is difficult to define. It often means approximation of two *ventral* surfaces around a *transverse* axis. But the thumb is almost at right angles to the fingers; its 'dorsal' surface is thus lateral and flexion and extension at its joints is around anteroposterior axes. At the shoulder flexion is more reasonably referred to an oblique axis through the centre of the humeral head in the plane of the scapular body, the arm moving anteromedially forwards and hence nearer to the trunk's ventral aspect. Again, flexion at the hip, which has a transverse axis, brings the thigh's morphologically *dorsal* (but topographically ventral) surface to the trunk's *ventral* aspect reflecting rotation of the hind limb bud in early embryonic life (p. 174). Description at the talocrural (ankle) joint is again complicated by the foot's posture at a right angle to the leg. Elevation diminishes this angle and is sometimes termed flexion; but it is approximation of two dorsal surfaces and might equally be called extension. Flexion has also been defined as fetal posture, implying that elevation of the foot is flexion, a view supported by withdrawal reflexes in which elevation is always associated with flexion at knee and hip, the converse holding in crossed extensor reflexes. Definitions based on morphological and physiological grounds are thus contradictory. To avoid confusion in this instance *dorsiflexion* and *plantar flexion* are used for talocrural movements.

Throughout this section and elsewhere, therefore, it is evident that care must be exercised, when using positional terms, to indicate whether they stem from phylogenetic, ontogenetic or physiological data and how these relate to descriptions of mankind in 'the anatomical position'.

Abduction and adduction occur around anteroposterior axes except at the pollicial carpometacarpal and shoulder joints, for reasons stated. The terms generally imply lateral or medial angulation, except in digits, where arbitrary planes chosen are midlines of the middle manual digit and second pedal digit, because these are least mobile in this respect. Abduction of the thumb is around a transverse axis and away from the palm. Similarly, abduction of humerus on scapula occurs in the scapular plane around an oblique axis at right angles to it (4.5).

Circumduction occurs when the distal end of a long bone circumscribes the base of a cone, its apex at the joint, e.g. shoulder and hip joints. It combines successive flexion, abduction, extension and adduction.

Rotation is another term used widely but often imprecisely. Its *restricted sense* denotes movement around some notional 'longitudinal' axis, which may even be in a separate bone, e.g. the dens of the second cervical ('axis') vertebra on which the atlas rotates. An axis may be approximately the centre of the shaft of a long bone, as in medial and lateral humeral rotation (4.5). Again, it may be at an angle to the bone's topographical axis, as in movement of radius on ulna in pronation and supination, where it joins the centre of the radial head to the base of the ulnar styloid process; and in medial and lateral femoral rotation, where the axis joins the centre of the femoral head to the lateral femoral condyle (p. 523). In these examples rotations can be independent, as *adjunct rotations*, providing an additional degree of freedom. They must be distinguished from *conjunct rotations* which always accompany some other main movement due to articular geometry (p. 481). However, in some articulations *conjunct* may be combined with some *adjunct* rotation and the latter may either increase or nullify the former at different times. Furthermore, with the exception of slight simple translation, *all* movements are in fact rotations. For example (p. 503) medial and lateral humeral rotations occur around the bone's long axis (vertical, in the anatomical position); swinging, 'angular' movements (flexion–extension and abduction–adduction) are rotations around the other two axes.

Kinesiology

The preceding account summarizes views and terminologies evolved over many years by simple observation of articular

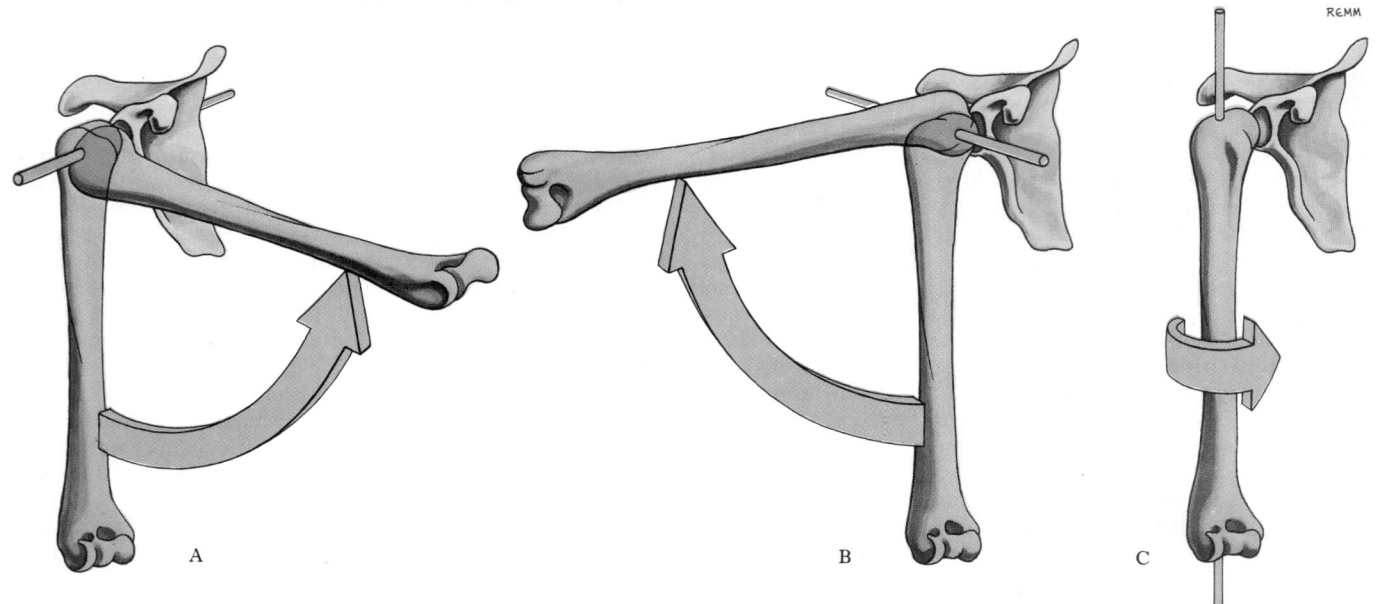

A   B   C

4.5 The shoulder joint is polyaxial and possesses three degrees of freedom. The three mutually perpendicular axes around which the principal movements of flexion–extension (A), abduction–adduction (B) and medial and lateral rotation (C) occur are shown. Note that in the case of the shoulder the axes are referred to the plane of the scapula and not to the coronal and sagittal planes of the erect body as a whole. An infinite variety of additional movements may, of course, occur at such a joint, e.g. those involving intermediate planes, or there may be movement combinations or sequences. However, these can always be resolved mathemetically into components related to the three axes which are illustrated.

anatomy and obvious geometrical changes. For many purposes this is adequate. More recently closer attention has amassed more objective observations, experimental results and theory to form an emerging science of *kinesiology*, a branch of biomechanics requiring new terminology, revision of older concepts and unfamiliar mathematical treatments. These developments are considered briefly here; useful expositions are given by Mac-Conaill (1946, 1950, 1953, 1964, 1966a, 1977), Steindler (1955), Barnett et al (1961), Kapandji (1970). Two fields of study are relevant here: (*a*) *osteokinematics* dealing primarily with overall bone movements with little reference to joints, and (*b*) *arthrokinematics*, concerned with articular mechanics. Certain generalizations about shapes of articular surfaces must first be considered and then replacement of an irregular bone by a single *mechanical axis*—concepts which necessarily precede study of surfaces and movements of particular joints.

The Shape of Articular Surfaces

Although the usual classification of synovial joints into seven types is based on shape, this does *not* entail a like number of fundamentally different surface shapes. Articular surfaces are never truly flat, nor exactly parts of spheres, cylinders, cones or ellipsoids (4.6). They are more nearly parts of surfaces of *ovoids* (4.6B) or compounded of more than one such surface. When these are either convex or concave, they may be termed male or female *ovoid articular surfaces*. The other well-recognized type is *sellar*, convex in one plane and concave at approximately right angles to this; even here curvatures are convex or concave *ovoid* profiles. Articular profiles vary from nearly flat to nearly spheroidal; but evidence increasingly indicates all to be *ovoid* or *sellar* (4.7). However, quantitative topology of conarticular surfaces has, for most joints (especially human), not yet been detailed. Palfrey & Ziemer (4.6A) have attempted such analyses on fresh talocrural joints (amputation specimens) and a more extensive analysis of tarsal joints has been contributed by Langelaan (1983). In the former study physical damage and desiccation were minimized (but not eliminated) and multiple casts in dental wax were made of talar and tibial trochlear surfaces; coronal and sagittal slices (3 mm or less) were prepared. Multiple points, marked along curved cartilage profiles, were assigned spatial co-ordinates determined by projection and tracing upon grid paper. Similar data were assessed for subchondral osseous surfaces. Subsequent analysis yielded two-dimensional contour maps and three-dimensional reconstructions. Much more exact data of shape, congruence, variation in thickness of cartilage, conarticular surface areas and differences between profiles of cartilage and subchondral bone were obtained.

Certain terms applied to ovoid surfaces and their properties are relevant both to movements of articular surfaces and of whole bones. Two points on an ovoid may be joined by a curved line which is the shortest distance between them, this being a *chord* distinct from any longer line joining them which is an *arc* (4.8) (MacConaill & Basmajian 1977). Any point moving on an ovoid surface traces a chordal or an arcuate line (or a succession of these), as will be considered further below. This usage of *chord* and *arc* is unfamiliar (they are usually applied to *circles* or plane surfaces). To clarify the concept imagine two points on a plane but deformable surface; these may be joined by the shortest route, a straight line, a *chord (geodesic)*; *any* longer route may be called an *arc (non-geodesic)*. Even if the surface is now formed into a sphere, cylinder, ovoid, sellar surface etc., relation between chord (geodesic) and arc (non-geodesic) is of course preserved.

A figure enclosed by three chords is a *triangle* (4.8), the sum of its angles exceeding 180° on an ovoid surface but being less than this on a sellar surface and precisely 180° when it is flat. Deviation from 180° depends on degree of curvature, an important fact in arthrokinematics (vide infra). Any three-sided figure in which at least one side is an arc is a *trigone*.

Sectional profiles of ovoid surfaces (4.9A) reveal two properties firstly radius of curvature varies continuously, a profile being considered a series of short segments of circles of different radii and a line joining their centres being the *evolute* of the profile. Rotation of a bone by sliding on such a profile is referrable not to one axis but to successive points on the evolute. Secondly if one convex ovoid surface a smaller segment of a similarly curved concave surface slides (4.9B), the two will fit perfectly only in one position (cf. close-packed position, described below); in all other positions surfaces are not congruent, the area of contact being much reduced and cuneiform intervals separating them elsewhere. (Theoretically, in a 'perfect', incompressible ovoid contact would be linear; it is an *area* because of superficial undulations (p. 473) and compressibility of cartilage.)

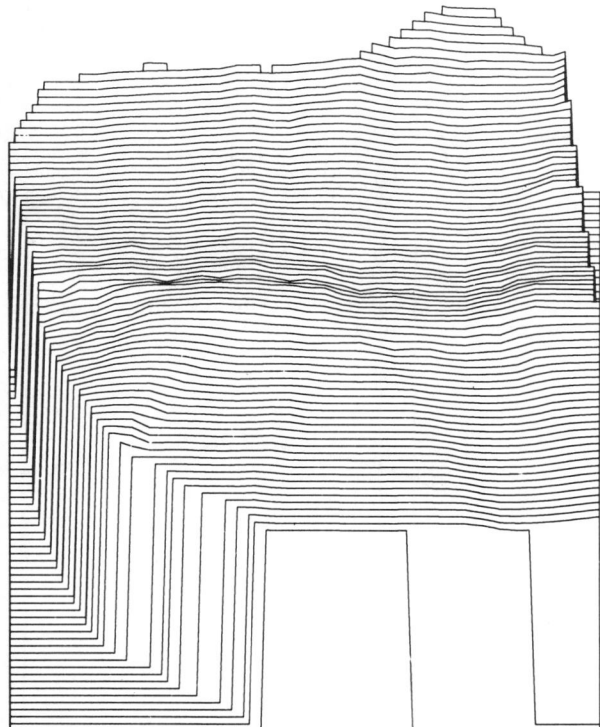

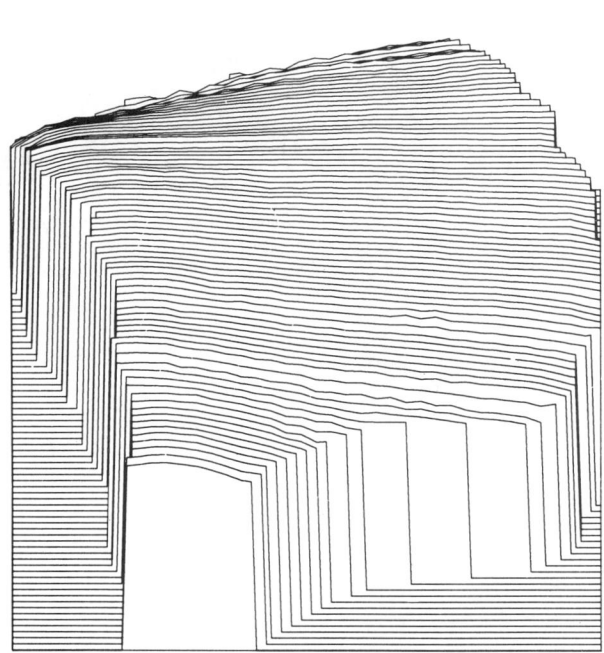

4.6A *Left*, a computer reconstruction of a sagittally sectioned cast of the trochlear surface of a fresh human tibia (left side, female, aged 63) viewed posteriorly. The medial malleolar surface is not included. It is a modified female sellar surface and reciprocally curved to its mating surface on the talus. Anteroposteriorly the surface is concave; lateromedially (left to right) there is a central convexity flanked by anterior and posterior concavities.

Right, a computer reconstruction of a coronally sectioned cast of the trochlear surface of a fresh human talus (left side, male, aged 62) viewed anteriorly. The triangular facet for the 'inferior transverse ligament' part of the posterior tibiofibular ligament has been excluded. It is a modified male sellar surface, which is convex anteroposteriorly; mediolaterally (left to right) it presents a medial convexity, a central concavity and a rather flattened lateral convexity. Reconstructions provided by A J Palfrey and Linda K Ziemer of the Department of Anatomy, Charing Cross Hospital Medical School, London.

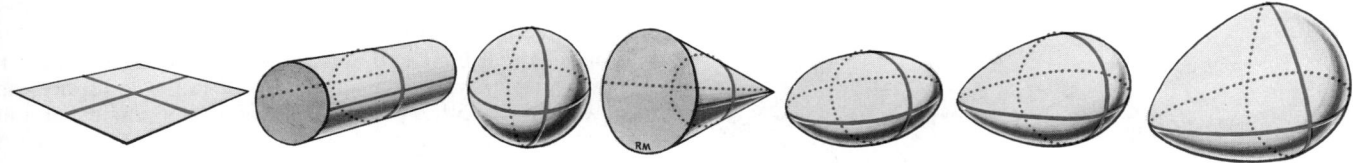

4.6B A variety of geometric figures to which reference is commonly made in simple classifications of synovial joints. However, such comparisons are rough approximations only—no synovial joint possesses articular surfaces which are truly plane, cylindrical, spherical, conical or ellipsoid. Commonplace simple and compounded ovoids are included for comparison.

Mechanical Axes of Bones

Bones are irregular, their articular surfaces bearing no simple or symmetrical relation to their form. Hence comparison of movements of individual *bones* may be misleading in terms of *joint mechanics*. To aid accurate comparison the concept of the *mechanical axis* has been introduced. This is made clearer in **4.10A** and B. In A a hypothetical, symmetrical long bone is shown with a terminal joint in mid-range of movement. A rod through the bone, ending perpendicular to the articular centre, is its mechanical axis and can represent the bone when *articular movements* are considered; in this special case of symmetry axis coincides with shaft. In **4.10B**, however, the bone is not symmetrical, the joint again is in mid-range and its mechanical axis is as shown; mechanical axis and shaft diverge widely. Corresponding *articular movements* hence result in similar displacements relative to the mechanical axis, but consequent movements of the shaft appear dissimilar. (See below for varieties of 'spin' and 'swing'.) Despite the usefulness of a mechanical axis in some analyses, in others it is less so because of arbitrary choice of a 'central articular point', difficult to define with precision. Mechanical axis selection might be better defined as that of *most habitual conjunct rotation*, the limit of which is the *close-packed* position of the joint concerned (vide infra). In

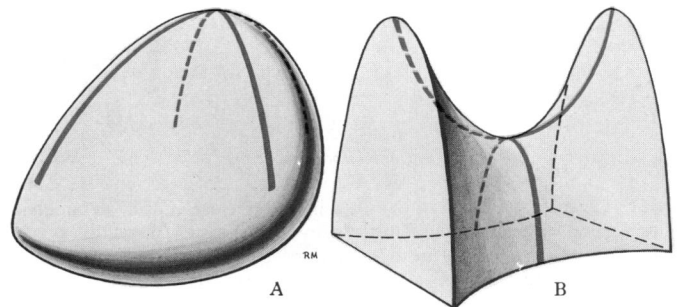

4.7 The two fundamental geometric types of articular surfaces. A. Ovoid, which may be convex (male) or concave (female). Note that the compound solid body illustrated presents different ovoid profiles in two planes at right angles and that the curvatures of the two may be different. B. Sellar or saddle-shaped surfaces, which are concavoconvex. In practice both types of surface may vary from only slightly to highly curved. Thus ovoid surfaces may be 'almost flat' or 'almost spherical' but the majority show intermediate grades of curvature and much variation in change of radius from place to place. Sellar surfaces may also be compound and have a marked asymmetry, e.g. at the talocrural and patellofemoral joints.

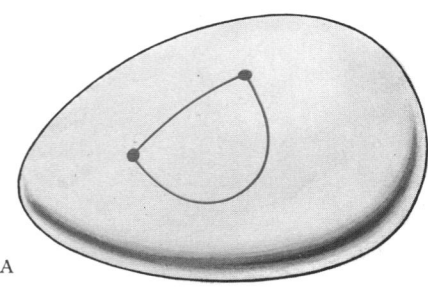 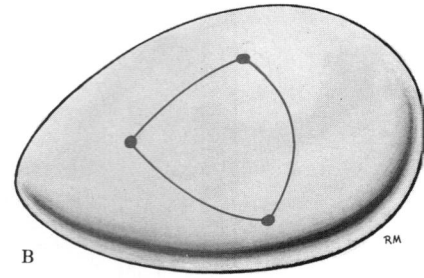 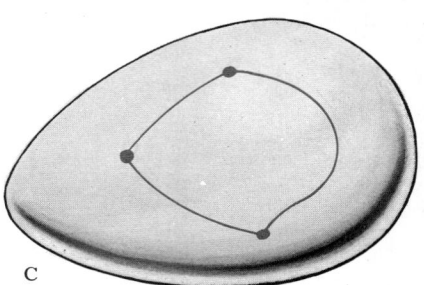

A   B   C

4.8 Geometric paths across ovoid surfaces, and three-sided figures enclosed by such paths. A. A chord (the shortest path between two points) and one example of an arc (any longer path between the points). B. A triangle enclosed by three chords. c. A trigone, a three-sided figure in which one or more sides is an arc.

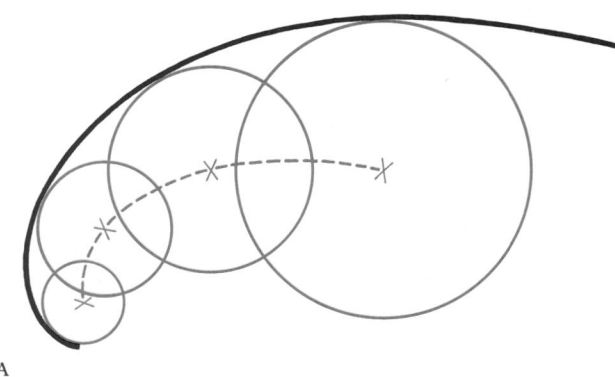

A

4.9A Profile of a section through an ovoid surface showing that it may be considered as a series of segments of circles of changing radius. The (dashed) line joining the centres of the circles is the evolute of the profile.

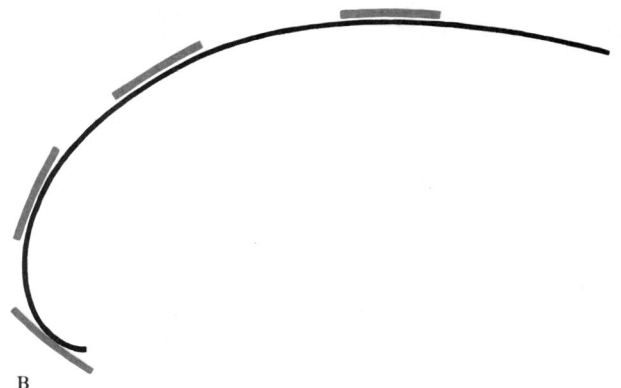

B

4.9B A small section of an ovoid profile (red) in various positions, in relation to a more extensive profile (black). The two fit perfectly (i.e. are fully congruent) in only one position.

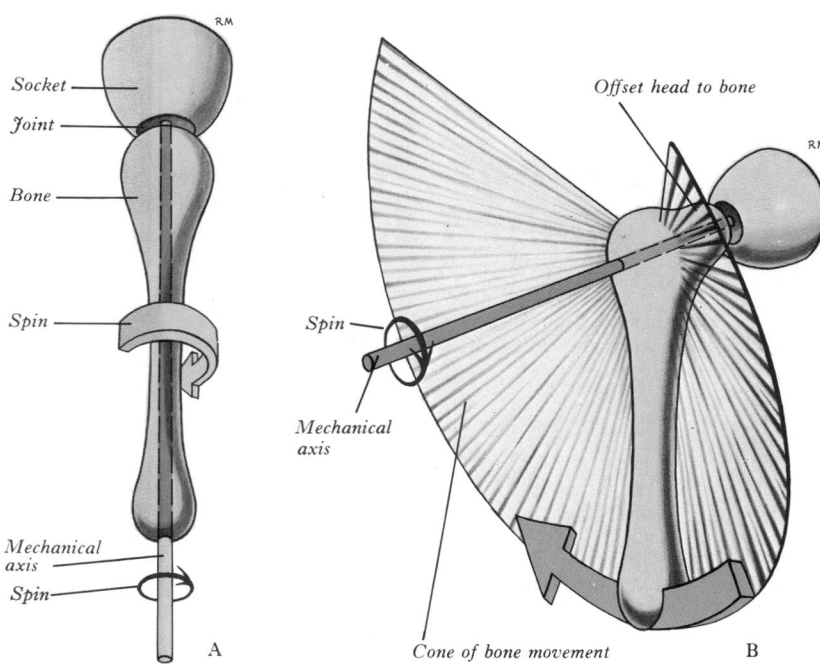

Socket

Joint

Bone

Spin

Mechanical axis

Spin

A

Offset head to bone

Spin

Mechanical axis

Cone of bone movement   B

4.10 The mechanical axis: (A). in a hypothetically simple, symmetrical long bone, with a terminal joint; a 'spin' is occurring around this axis; (B) in a bone with an offset head; a similar 'spin' is occurring between the articular surfaces, but the shaft of the bone traces part of the surface of a cone.

spheroidal, multi-axial joints definition of a *single* mechanical axis may also be an insufficient datum for the many combinations possible.

Movements of Bones

Apart from passive accessory movements due to external forces (p. 481), all others are rotations; since this term is often used in a

restricted sense (p. 477), new terms become necessary. Any bone solely rotating around its stationary mechanical axis is said to show pure *spin*; *any* point on the bone, outside the axis, describes an arc (of a circle), with the axis as its centre (**4.**10). All other displacements of bone and axis are *swings*, which may be *pure* (**4.**11A) or *impure* when spin also occurs (**4.**11B). (In a spheroidal joint what is virtually pure spin is theoretically possible around many alternative 'mechanical axes'.)

Another useful concept is the so-called *ovoid of motion*: during any swing a point on the mechanical axis, distant from its related joint, describes a curved path in space and all such possible paths (**4.**12) are on part of an ovoid's surface (as would be expected, since they are generated from such an articular surface). Area and shape of such *ovoids of motion* vary for individual bones and articular surfaces. During any particular swing, therefore, a point on the mechanical axis moves from X to Y on the ovoid of motion either along a chord (a *cardinal swing*) or by a longer route from X to Y (an *arcuate swing*). Synchronously the articular end of the axis traces a reciprocal chord or arc across its opposing articular surface. A consequence of such swings is related to presence or absence of associated spin: during a cardinal swing along a chord (an unusual movement) there is *no associated spin*; in an arcuate swing there must be some spin which is functionally significant (vide infra). Similarly, spin occurs if a movement involves two chords in series at an angle to each other (one form of *diadochal movement*), e.g. X to Y and Y to Z in **4.**12.

In some apparently simple swings of bones, spin was long undetected; but its occurrence can be predicted mathematically as a consequence of articular curves and is in most instances verifiable by careful inspection or cineradiography. Diagrams may assist comprehension: in Fig.**4.**13A three points, A, B and C, are on the surface of a sphere, AB, BC, and CD being shortest distances (chords) between them. AB and AC are 'lines of longitude', meeting the equator at 90° at B and C. A model limb moves successively along chords A→B, B→C, C→A; between initial and final positions it has 'spun' 90° (by which the sum of angles of 'triangle' ABC *exceeds* 180°). Fig. **4.**13B shows intermediate 'lines

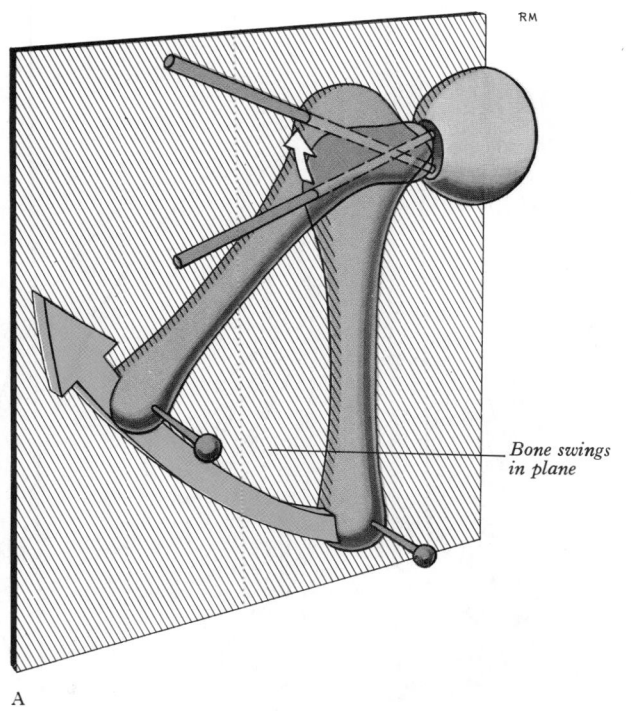

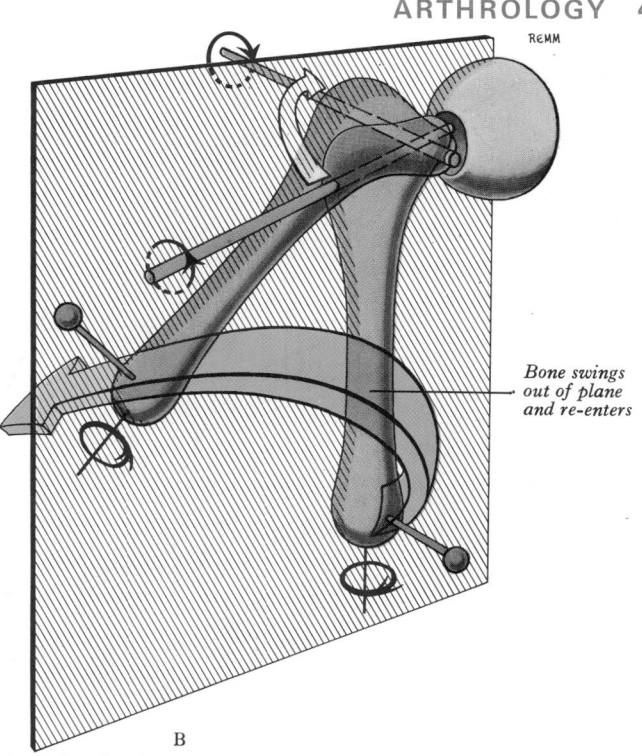

A   B

4.11 Further movements of the bone and joint shown in **4.10B**. A. A cardinal swing. Note that the mechanical axis moves in one plane. (Its proximal end traces a chordal path at the joint and its distal end a chordal path on the ovoid of motion, see **4.12**). There is no spin. B. An arcuate swing. Note that the bone moves out of the plane illustrated and then returns to it. The ends of the mechanical axis trace arcuate paths at the joint and on the ovoid motion. There is an associated spin or conjunct rotation around the *mechanical axis* (and a rotation around the *long axis* of the *bone shaft*).

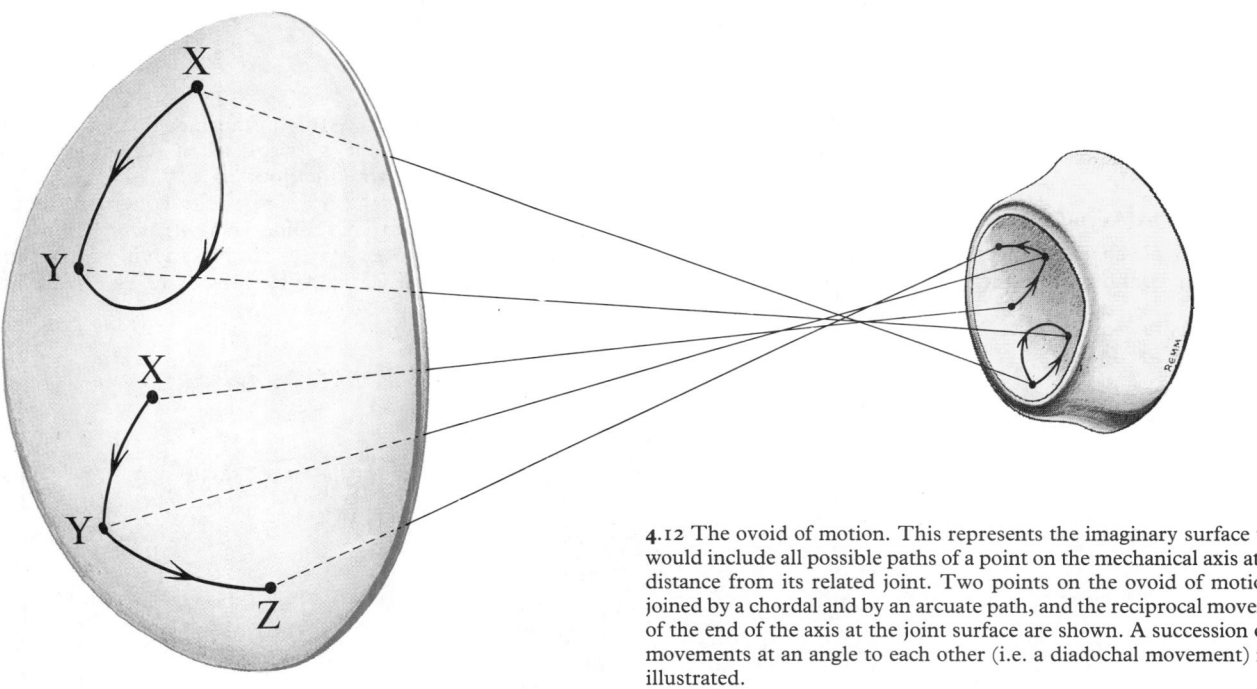

4.12 The ovoid of motion. This represents the imaginary surface which would include all possible paths of a point on the mechanical axis at some distance from its related joint. Two points on the ovoid of motion are joined by a chordal and by an arcuate path, and the reciprocal movements of the end of the axis at the joint surface are shown. A succession of two movements at an angle to each other (i.e. a diadochal movement) is also illustrated.

of longitude'; in each case spin equals the sum of the three angles minus 180°. This relation holds mathematically for all ovoid or sellar surfaces. Spin is imparted during the B→C movement (i.e. along a second chord at an angle to the first). An arcuate path may be analysed as a series of chords which change angle. These spins, inevitable in certain swings, are *conjunct rotations* and characteristic of sellar and most ovoid joints. *Habitual* movements at *all* joints always involve some demonstrable conjunct rotation; an important consequence will be best appreciated after consideration of related movements and 'fit' of articular surfaces (vide infra).

Beyond these basic arthrokinematics of *conjunct rotations* it must be added that other rotations (due to interplay of gravity, external forces and muscle action) are *adjunct*. Adjunct rotations occur only in joints with two or more degrees of freedom, involving either a bone undergoing *no* conjunct rotation or one in which it must be considered. In the first, factors causing adjunct rotation may generate *pure spin*; in the second, effects of adjunct rotation may be in the *same sense* (e.g. both 'clockwise'), i.e. additive, increasing rotation and hence termed a *cospin*. Alternatively, adjunct rotations may be in *opposite sense*, reducing or nullifying effects of conjunct rotation and hence termed *antispins*. Thus a bone starting one or a succession of arcuate swings which, if unmodified, would involve some conjunct rotation may also be subjected to *nullifying antispin*: the latter may be *gradual*, applied throughout movement or *sudden*, near its end. Therefore, while

481

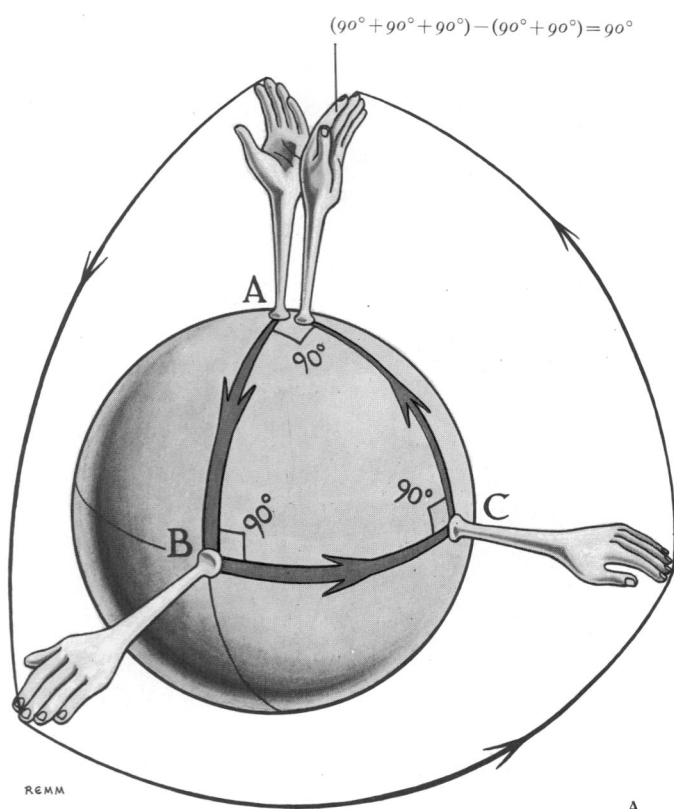

$$(90° + 90° + 90°) - (90° + 90°) = 90°$$

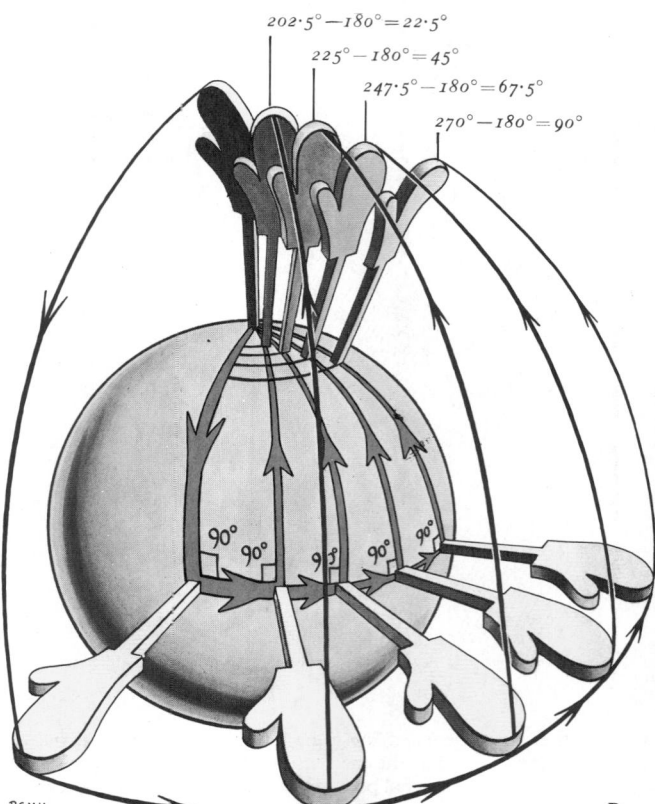

$$202·5° - 180° = 22·5°$$
$$225° - 180° = 45°$$
$$247·5° - 180° = 67·5°$$
$$270° - 180° = 90°$$

4.13 The 'spin' or conjunct rotation which accompanies a diadochal movement. A model hand and arm moves over the surface of a sphere, first along a line of longitude, then one of latitude to return along a line of longitude. A. The arm traverses three chordal paths which enclose a 'triangle' with three right angles (i.e. the sum of the angles = 270°). Note the *path* of the bone is a more or less complex *arcuate* swing, it is modified to become *quasichordal* because there is no net spin.

that as it passes along the line of latitude from B to C, a spin of 90° is imparted. B. This illustrates how the amount of spin imparted during a diadochal movement varies with the length and angulation of the second stage of the movement.

Movements at Articular Surfaces

The 'fit' of ovoid surfaces (p. 478) is precise only at one end of the joint's most common excursion, a feature of the close-packed position discussed below. In all other positions surfaces are not fully congruent, the joint is 'loose-packed'. Changes in congruity are illustrated in the shoulder joint in **4.17**.

Where articular surfaces are not fully congruent movements can be analysed into spin, slide and roll, which must first be considered at a convex (male) surface moving on a stationary concave (female) surface, and vice versa (**4.14,15**). In most natural

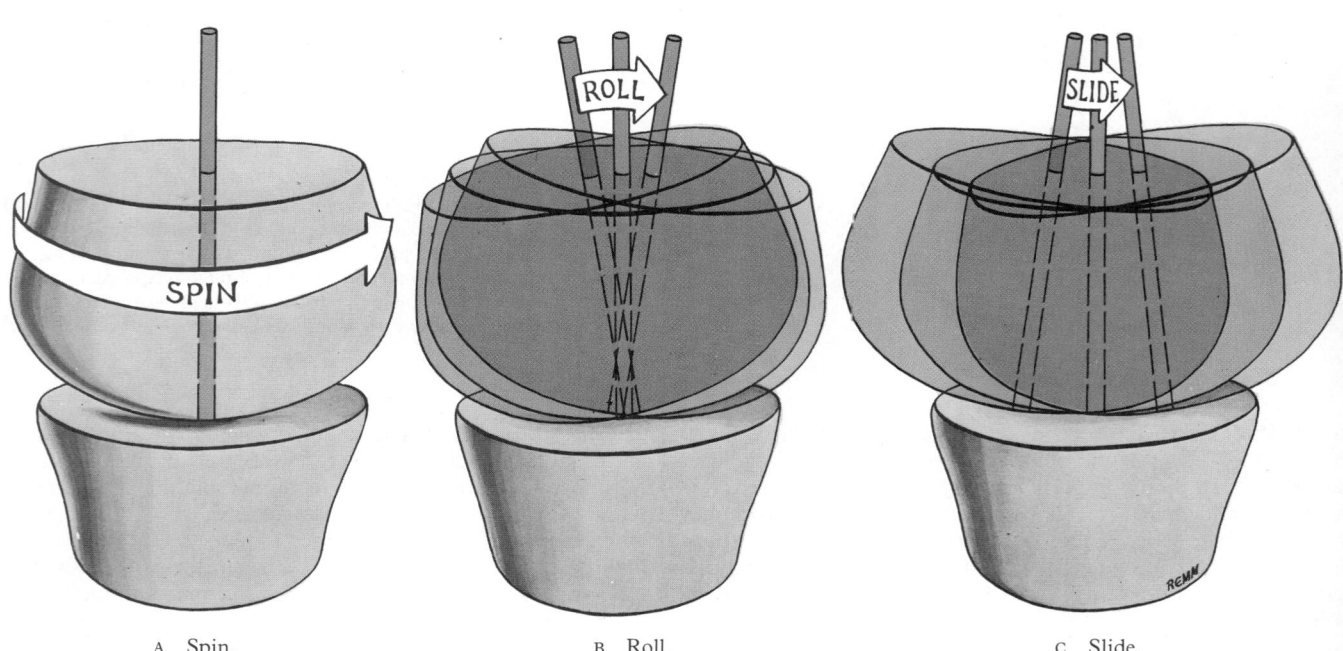

A Spin.

B Roll.

C Slide.

4.14 An analysis of the types of movement which occur (usually in combination) between articular surfaces when a male surface moves over a stationary female surface.

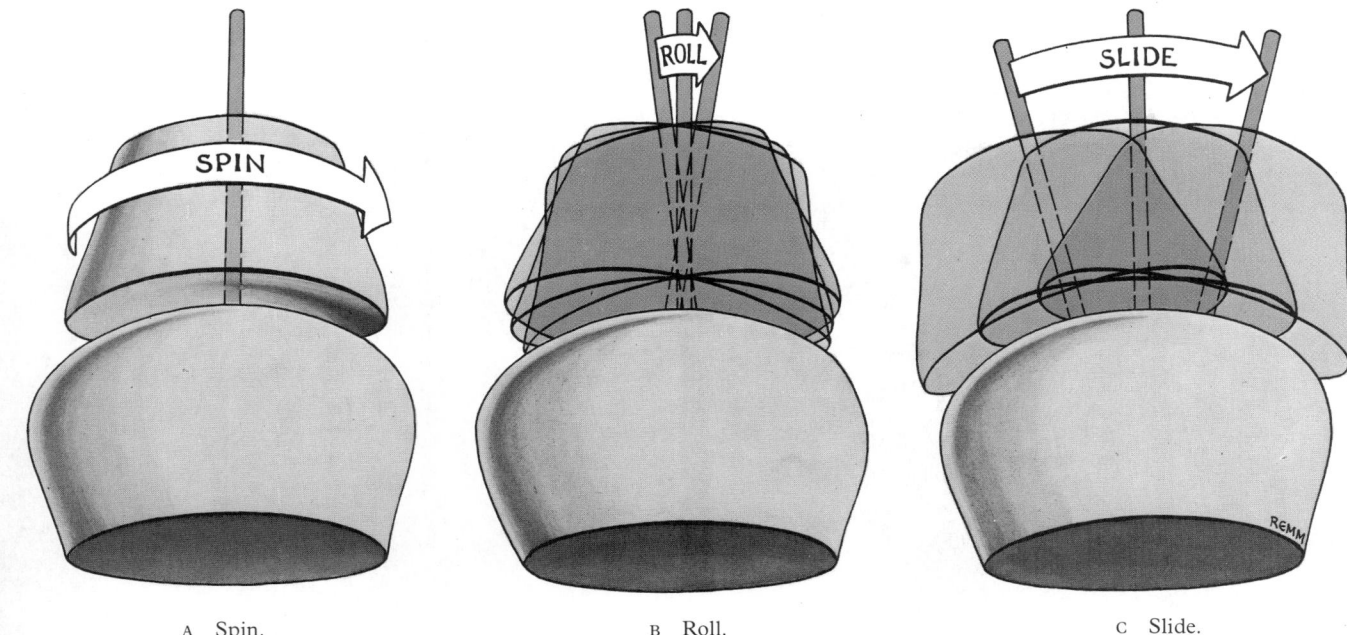

| A Spin. | B Roll. | C Slide. |

4.15 Articular movements. A female surface moves on a stationary male surface.

movement these are combined (4.16). With a moving convex surface, slide and roll are simultaneous but *opposite*; with a moving concave surface they are in the *same* direction; both combinations increase possible angulation, without a comparable increase in articular surface area. The significance of these movements and variations in congruity is best considered with the basic states of 'loose' and 'close-pack' (vide infra).

Joint Positions

Ovoid or sellar surfaces are fully congruent in only one position, at one extreme of *most habitual articular movement* (e.g. full extension at knee, wrist and interphalangeal joints, dorsiflexion at ankle, abduction with lateral rotation at shoulder). As full congruence is approached, other changes occur: attachments of fibrous capsule and ligaments increasingly separate and tense. The conjunct rotation of all natural movements adds to this by imposing spiral twist in them. In final **close-packing**, therefore, surfaces are fully congruent, in maximal contact and tightly compressed or 'screwed-home', fibrous capsule and ligaments being maximally spiralized and tensed; no further movement is possible. Close-packed surfaces cannot normally be separated by external force (as they may in other positions); bones can be regarded as temporarily locked, as if no joint existed. This extreme is only assumed when special effort is required; the rigidity and stresses generated expose articular structures maximally to trauma. Close-packing is a final, limiting position; any force which tends to further change is resisted by reflex contraction of appropriate muscles. Movement *just short* of close-packing is physiologically most important. Ligaments and articular cartilage are to a small degree elastically deformable and in the final stages the articular position is an equilibrium between the external torque applied (often gravity) and resistance to deformation of tissues by the tense, twisted capsule and compressed cartilages. In symmetrical standing, knee and hip joints approach close-packed positions sufficiently to maintain an erect posture with minimal muscular energy (Joseph 1960, 1975).

In all other positions articular surfaces are not congruent and parts of the capsule are lax; the joint is **loose-packed**. Capsules are lax enough near mid-range of many movements to allow separation of surfaces by external forces. Congruence in which a convex surface has a smaller radius than a concave is advantageous: firstly it allows combined spin, roll and slide, a feature of such joints; secondly contact area is greatly reduced and variable, which may diminish friction and erosion; thirdly small cuneiform intervals between surfaces around contact areas contain synovial fluid, their shape perhaps a factor in maintaining efficient lubrication and nutrition of avascular articular cartilages; finally, combination of sliding and rolling increases effective range of movement. With slide or roll alone, this could be achieved only by more extensive surfaces (4.16 and p. 433).

According to MacConaill & Basmajian (1977) positions of close- and loose (least)-pack of joints are as follows:

Joint	Close-packed position	Least-packed position
Shoulder	Abduction + lateral rotation	Semiabduction
Ulnohumeral	Extension	Semiflexion
Radiohumeral	Semiflexion + semipronation	Extension + supination
Wrist	Dorsiflexion	Semiflexion
Metacarpophalangeal (2–5)	Full flexion	Semiflexion + ulnar deviation
Interphalangeal (fingers)	Extension	Semiflexion
First carpo-metacarpal	Full opposition	Neutral position of thumb
Hip	Extension + medial rotation	Semiflexion
Knee	Full extension	Semiflexion
Ankle	Dorsiflexion	Neutral position
Tarsal joints	Full supination	Semipronation
Metatarsophalangeal	Dorsiflexion	Neutral position
Interphalangeal (toes)	Dorsiflexion	Semiflexion
Intervertebral	Dorsiflexion	Neutral position

Opinions may vary in connection with a few of the statements in the above table. For example a 'closed-packed' position may possibly occur in occasional joints at *both* extremes of the range of movement. It is also most difficult to assess the situation in small tarsal, carpal and the first carpometacarpal joints. Intervertebral movements are, of course, the result of integrated simultaneous changes at *all* elements constituting the intervertebral articular complex; i.e. massive symphyses (including the discs), equally massive syndesmoses and the relatively small synovial zygapophysial joints. It is doubtful whether 'intervertebral' should appear in the table.

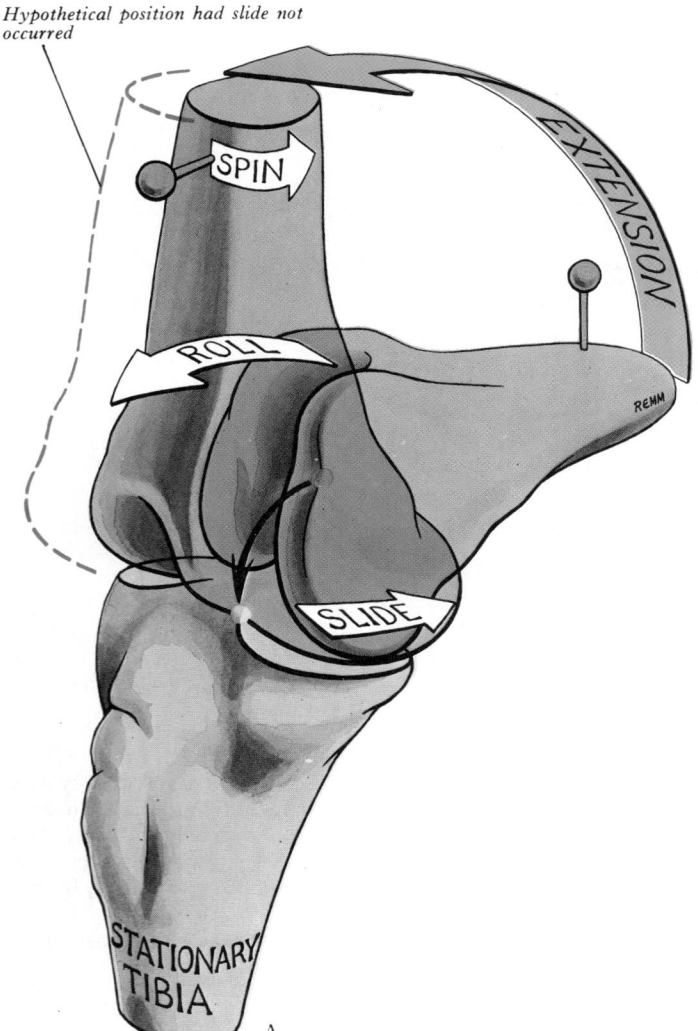

Hypothetical position had slide not occurred

4.16 Analysis of the articular movements which are combined during extension at the knee joint. (Right knee viewed from medial aspect; patella, menisci and other structural features omitted.) A. With a stationary tibia, i.e. moving male femoral condylar surfaces. B. With a stationary femur, i.e. moving female tibial condylar surfaces. Notice that in each case elements of slide, roll and spin occur together. In A the roll and slide are in opposite directions, whereas in B they are in the same direction. See text for further description.

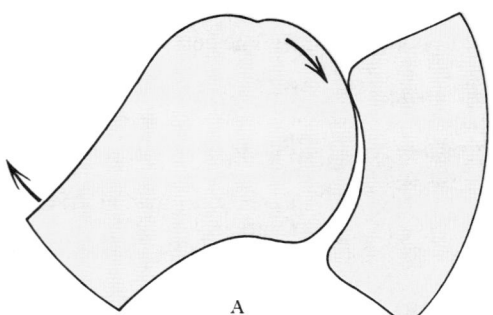

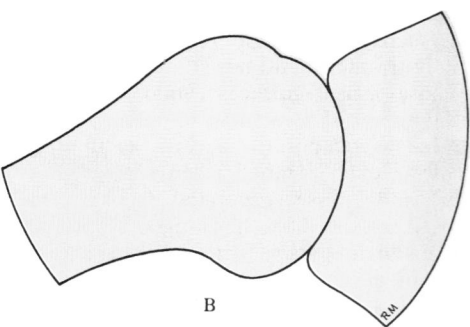

4.17 Congruence of articular surfaces. A. In loose-packed positions of a joint, e.g. the shoulder, the surfaces are not congruent (this has been over-emphasized for clarity). B. The close-packed position of the joint with close-fitting or full congruence of the surfaces.

However, most of the positions cited here do, it is true, correspond with postures adopted when maximal stress is encountered.

General concepts as set out above have much clarified joint mechanics but aspects of joint lubrication, maintenance of articular cartilage and detailed mechanics of many joints await resolution (Freeman 1973).

Accessory movements. Movements actively performed at any joint do not always include all that its structure permits. Certain movements, voluntarily impossible or limited in range, can only be produced maximally when resistance to active movements occurs (accessory movements, first type), e.g. only when a solid object is grasped in the hand can fingers be *maximally* rotated at metacarpophalangeal joints (see p. 515). (However, a moderate degree of rotation of the *non-grasping* fingers toward the centre of the palm during flexion and its converse during extension is commonplace.) Some movements can be produced only passively (accessory movements, second type), their widest range obtained when the muscles of a joint are relaxed, e.g. when the supported arm is passively abducted at the shoulder, the humerus can be distracted from the glenoid cavity, a feature of many joints when loose-packed. Such movements are commonly termed 'passive' but all movements, active or not, can be made passively when the muscles concerned are relaxed; the term 'accessory' will be used for all movements impossible in the absence of resistance (Salter 1955).

Limitation of movements is due to several factors, of which tension in ligaments is prominent, as is obvious in attempted hyperextension of the unfixed cadaveric knee or hip. Increasing ligamentous tension, balanced by increased compression between opposed articular surfaces, are integral factors in producing close-packing, limiting most habitual movements. But tension of antagonistic muscles is equally important, involving both *passive elastic components* of muscles (and other structures around the joint) and *reflex contraction* when stimulation of mechanoreceptors in articular and periarticular tissues reaches a critical level. Muscles as limiting factors are exemplified in flexion at the hip; with knee extended it is much more limited in range; with knee flexed hamstring muscles are relaxed allowing flexion of thigh to the abdominal wall, such *approximation of the soft parts* being a third factor in some movements, e.g. flexion at elbow and knee. Contact (occlusion) of teeth obviously limits mandibular elevation.

In synovial joints, where bones are connected only by ligaments and muscles, parts of articular surfaces are in constant apposition in all positions. (Some maintain that 'apposition' implies a fine film of synovial fluid, of 10 μm or *less*, between adjacent surfaces.) Apposition is assisted by atmospheric pressure and cohesion between surfaces; but these are subsidiary to balanced contraction of muscle groups around the joint. When these contract, the force generated is vectorially resolvable into components (p. 567). Some maintain or alter positions of bones ('swing' and 'spin' components) and oppose internal and external resistances, including gravity. Another component is *transarticular* ('shunt' component), which increases compression between articular surfaces and helps apposition in various postures and movements (p. 567). Effects of external compressive or tensile forces, including gravity, vary with posture of the body and its members in space and the direction of applied force. Thus gravity or load-bearing may sometimes provide a *distractive force*, tending to separate conarticular surfaces, or may exert a largely *translatory/swing* force between them. They often exert a considerable *compressive force* at surfaces.

The preceding remarks are merely an introduction to the main concepts of current kinesiology. Brief references to myokinetics are on p. 567. For further analyses the references quoted should be consulted. Readers with greater facility in mathematics and physics will be interested in attempts to present a 'generalized mechanics of articular swing', ranging from Aristotelian and Newtonian physics to the relativity theory and quantum mechanics (MacConaill 1978 a,b,c)!

Blood supply and lymphatics. Joints receive blood from peri-articular arterial *plexuses* whose numerous rami pierce capsules to form subsynovial vascular plexuses. Some synovial vessels end near articular margins in an anastomotic fringe, the *circulus articularis vasculosus* (p. 299). A lymphatic plexus in the synovial subintima drains along blood vessels to the regional deep lymph nodes.

Nerve supply. Movable joints are innervated in general by nerves supplying their muscles, probably establishing local reflex loops involved in movement and posture. Although the branches concerned vary, each innervates a specific capsular region but their territories freely overlap. The region tautened by muscular contraction is usually innervated by nerves supplying antagonists (Gardner 1948). For example, the hip joint's capsule, on stretch inferiorly in abduction, is here supplied by the obturator nerve, tension in it thus producing reflex contraction of adductors, usually enough to prevent damage. However this is not so at the shoulder, where the axillary nerve innervates the antero-inferior capsular region.

Myelinated fibres in articular nerves have Ruffini endings, lamellated articular corpuscles and some like the neurotendinous organs of Golgi. Simple endings are numerous at attachments of capsule and ligaments; they are terminals of non-myelinated and finely myelinated fibres believed to mediate pain (Gardner 1950). Ruffini end organs are variably orientated in the knee joint's capsule, principally in its flexor region; they respond to stretch and adapt slowly. Lamellated corpuscles, less numerous than Ruffini endings, are sited laterally and adapt rapidly; they respond to rapid movement and vibration; both register speed and direction of movement. Golgi end organs, with the largest myelinated nerve fibres (10–15 μm diameter), are like those at neuromuscular junctions and slow to adapt (Boyd & Roberts 1953, Boyd 1954, Skoglund 1956); they mediate position sense (Stopford 1921, Mountcastle & Powell 1959, Gardner 1967) and are concerned in stereognosis, i.e. recognition of shape in objects held (Renfrew & Melville 1960). Many non-myelinated fibres are sympathetic, ending near vascular non-striated muscle and believed to be vasomotor or vasosensory, although evidence is sparse. In synovial membrane no special end organs or even simple endings occur, except near blood vessels. The membrane is relatively insensitive to pain (Kellgren & Samuel 1950, Barnett et al 1961). For a review concerned with receptors and sensation see Wyke (1981); for histological and functional details and classification of articular nerve endings see p. 914.

THE INDIVIDUAL ARTICULATIONS

Temporomandibular Joints

Each joint involves the temporal articular tubercle and anterior part of the mandibular fossa above and mandibular condyle below. Articular surfaces are covered by white fibrocartilage in which collagen fibres predominate and cartilage cells are few. An articular disc divides the joint, usually completely, into upper and lower parts. (Sometimes, however, the disc is perforated.) Commonly described as 'condylar', they are preferably termed *ellipsoid*, right and left joints forming a *bicondylar* articulation.

The fibrous capsule is attached *above* anteriorly to the articular tubercle, posteriorly to the lips of the squamotympanic fissure and between these to the edges of the mandibular fossa; *below* to the mandibular neck. *Above* the articular disc the capsule is loose and it is taut *below* it. **Synovial membrane** lines the capsule, above and below the disc (but does not cover the disc); thus, on each side it lines the nonarticular surfaces of superior and inferior synovial compartments. Below the disc the capsular synovial membrane is reflected upwards along the mandible's neck and lateral pterygoid tendon to reach the condylar articular cartilage.

The lateral temporomandibular ligament (4.18), close to the capsule, is attached above to the tubercle on the zygoma's root, below to the lateral surface and posterior border of the mandibular neck, its fibres sloping *down* and *back* deep to the parotid gland.

The sphenomandibular ligament (4.19), medial to and separate from the capsule, is a flat, thin band descending from the sphenoidal spine and widening to reach the lingula of the mandibular foramen. Superolateral to it are the lateral pterygoid muscle and auriculotemporal nerve; inferior to this it is separated from the mandibular neck by maxillary vessels, below which the inferior alveolar vessels and nerve and a parotid lobule separate it from the mandibular ramus; here the vessels and nerve to the

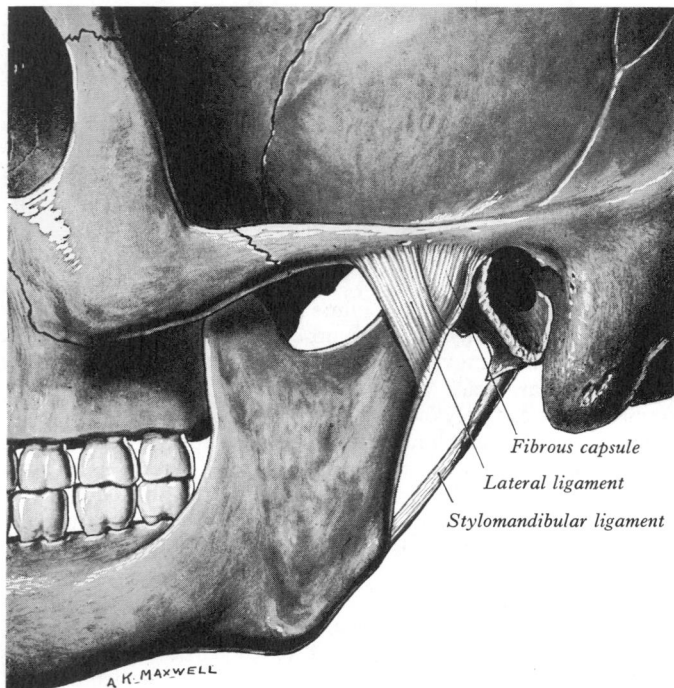

4.18 The left temporomandibular joint: lateral aspect.

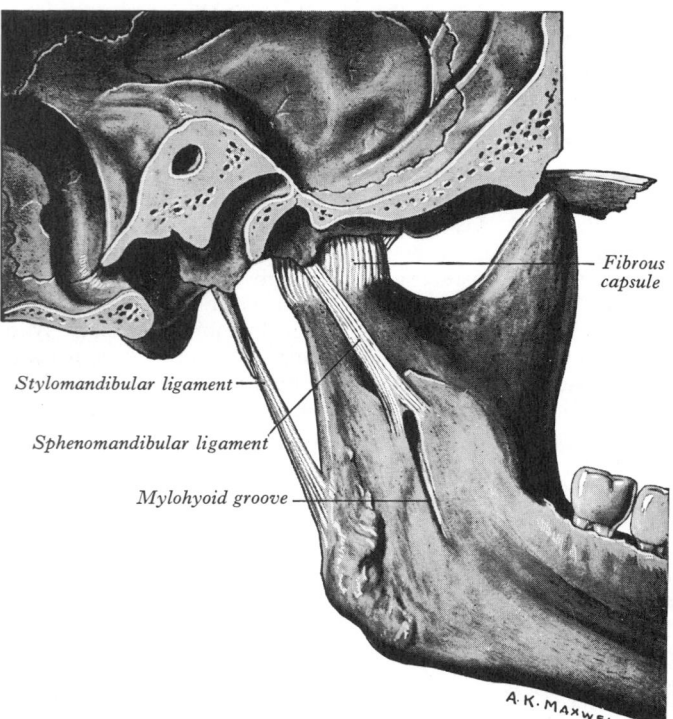

4.19 The left temporomandibular joint: medial aspect.

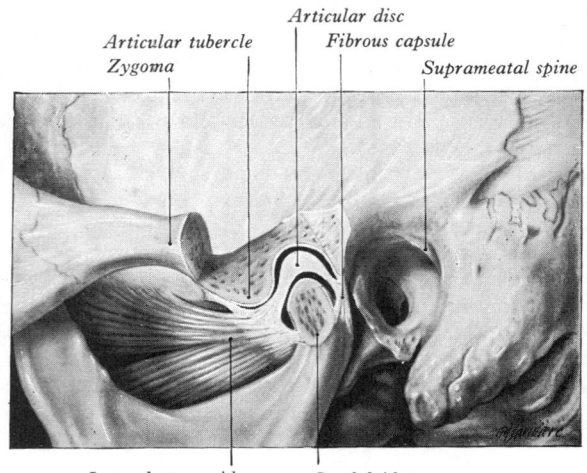

Articular tubercle
Zygoma
Articular disc
Fibrous capsule
Suprameatal spine

Lateral pterygoid   *Condyloid process*

4.20 Sagittal section through the left temporomandibular joint.

mylohyoid pierce the ligament with the medial pterygoid muscle inferomedial. It is separated from the pharynx by fat and pharyngeal veins and, near its upper end, it is crossed by the chorda tympani. Some fibres traverse the medial end of the petrotympanic fissure to the anterior malleolar process as a vestige (the anterior ligament of the malleus) of the dorsal end of Meckel's cartilage (p. 1227). The role of this vestigial ligament in *mandibular* mechanics is negligible.

The articular disc (4.20,21A,B,C), an oval plate of fibrous tissue shaped like a peaked cap, completely divides the joint. Its upper surface is sagittally concavoconvex to fit the articular tubercle and fossa, its inferior concave surface adapted to the man-

dibular head. Its circumference blends with the fibrous capsule and, in front, the tendon of lateral pterygoid (p. 581). Medial and lateral, short, strong bands pass from its margins to the medial and lateral condylar poles, ensuring that disc and condyle move together in protraction and retraction. Posteriorly in the disc a venous plexus separates upper and lower layers, the *upper* of fibro-elastic tissue attached to the fossa's posterior margin, the *lower* of non-elastic fibrous tissue attached to the back of the condyle. Though variable, the disc is thickest behind its centre, in the deepest part of the mandibular fossa. Rees (1954) has described two thick regions (anterior and posterior bands) with thinner zones between, subdivisions being from before backwards: anterior extension, anterior band, intermediate zone, posterior band and the bilaminar region mentioned above (4.21A,B,C). The description of the articular disc given by Rees and followed above (also in 4.21,22) is widely quoted; however it was derived from fixed cadaveric material and perhaps undue emphasis is placed on anteroposterior zonal organization alone. Mediolateral variations are equally relevant. An interesting analysis with some alternative proposals concerning the design, formation, possible functional roles and derangements of the disc have been provided by Osborn (1985). He emphasizes the largely fibrous nature of the highly non-congruent articular surfaces and views the interposed disc as a visco-elastically deformable pad co-extensive with the cranial articular surface (i.e. much more extensive than the mandibular condyle). Compression forces (increased during chewing, biting, grinding) *thin* an area of disc between condyle and the posterior slope of the articular eminence, thereby 'squeezing out' material to form a thickened zone, the *annulus* of Osborn, which surrounds the thin area—a recess for the mandibular condyle. (It may be noted that the anterior and posterior thickened bands shown in 4.21, after Rees, are in fact continuous medial and lateral to the mandibular condyle.) The question of whether the shape of the disc is genetically determined, produced mainly by biomechanical constraints or both is also discussed. The supposedly non-elastic nature of the inferoposterior (mandibular) lamina of the disc is also questioned (see also below). Postnatal development of the disc up to 2 years, described by Wright & Moffet (1974), showed at no stage any chondrocytes in the disc, which is flat at birth and develops its sigmoid profile as the articular eminence enlarges. From the fifth decade it so often shows macroscopic evidence of degeneration (fraying, thinning and perforation) that this can be regarded as 'normal' ageing. Weisengreen (1975) found such degeneration in 40% of 183 individuals between 40 and 90 years. The disc, often incomplete, is variably perforated.

The stylomandibular ligament (4.18), a specialized band of deep cervical fascia (p. 582), stretches from the apex and adjacent anterior aspect of the styloid process to the mandible's angle and posterior border. It can be considered only accessory to the joint and of uncertain function.

Innervation is from the auriculotemporal and masseteri

486

branches of the mandibular nerve, **arteries** from superficial temporal and maxillary arteries. Klineberg & Wyke (1973) have detailed the distribution of mechanoreceptors in the joint and suggested their role in mastication.

Movements. The mandible can be depressed or elevated, protruded or retracted and, since both joints always act together but may differ in actual movement, some rotation (around a vertical axis, vide infra) may occur. These actions involve gliding, spin, roll and angulation. (The last is another variety of rotation around a dynamic transverse axis, vide infra.)

In the *position of rest* upper and lower teeth are slightly apart; in closure they are apposed, the *occlusal position*. When the mouth opens the mandibular 'condyles' rotate on a common horizontal axis and also glide forwards and downwards on the inferior surfaces of their articular discs, which slide in the same direction on the temporal bones due to their attachments to the mandibular heads and to contraction of the lateral pterygoids drawing heads and discs on to the articular tubercles. Discal sliding ceases when their posterior fibro-elastic attachments to the temporal bones are stretched to their limit. Further hinging and gliding of the condyles brings them into articulation with the most anterior parts of the discs as the mouth opens fully. In *closure* movements are reversed: each head glides back and hinges on its disc, still held by the lateral pterygoid, which relaxes to allow the disc to glide back and up into the mandibular fossa. The full cycle is shown in 4.22, derived from observations by Rees (1954). Osborn (1985) presents alternative interpretations; he regards the fibrous articular surfaces as particularly susceptible to trauma during the arthrokinematic roll, spin and glide executed when the joint is maximally loaded. The deformable visco-elastic articular disc's primary function is to permit these activities while reducing the risk of trauma. The *disc* is thought to be stabilized by its inherent visco-elasticity; increasing loads thicken its annulus (vide supra) and prolapse is prevented. Activity of the lateral pterygoid and presence of elastic tissue in the posterior bilaminar zone are regarded as secondary features in disc stabilization. An unusual proposition is that while the disc is *self-stabilizing* its other principal role is to *destabilize* the mandibular condyle and allow complex free movement under load. The elastic tissues may act to withdraw tissues and thus prevent entrapment between the articular surfaces during mouth closure. In *protrusion* the teeth are parallel to the occlusal plane but variably separated, the lower carried forwards by both lateral pterygoid muscles. In *retraction* the mandible is returned to the position of rest. In rotatory movements of mastication (in the occlusal *plane*, but clearly not in *occlusion*), one head with its disc glides forwards, rotating around a vertical axis immediately behind the opposite head, then glides backwards rotating in the opposite direction, as the opposite head comes forward in turn. This alternation swings the mandible from side to side (Sarnat 1951).

Measurement of mandibular movements is of clinical interest; in adults incisors may separate by 50–60 mm, maximal lateral displacement and protrusion being about 10 mm, with much individual variation. Adult range is reached earlier in females (*c.* 10 years) than in males (*c.* 15 years) in accord with postnatal development of the articular profiles (Wright & Moffet 1974), adult form being reached between six and 12 years.

Muscles Producing Movements

Depression: lateral pterygoids, aided when the mouth is widely open or against resistance, by digastric, geniohyoid and mylohyoid muscles.

Elevation: temporalis, masseter, medial pterygoid, of both sides. In closure each mandibular head is retracted by the posterior fibres of the temporalis before elevation. The temporales maintain the position of rest (Latif 1957).

Protrusion: lateral and medial pterygoid muscles.

Retraction: temporales (posterior fibres), assisted by middle and deep parts of the masseters, digastric and geniohyoid muscles.

Lateral movements: medial and lateral pterygoid of each side, acting alternately.

Such a list, though useful, obscures the complex integrations of simultaneous contraction and lengthening of many muscles.

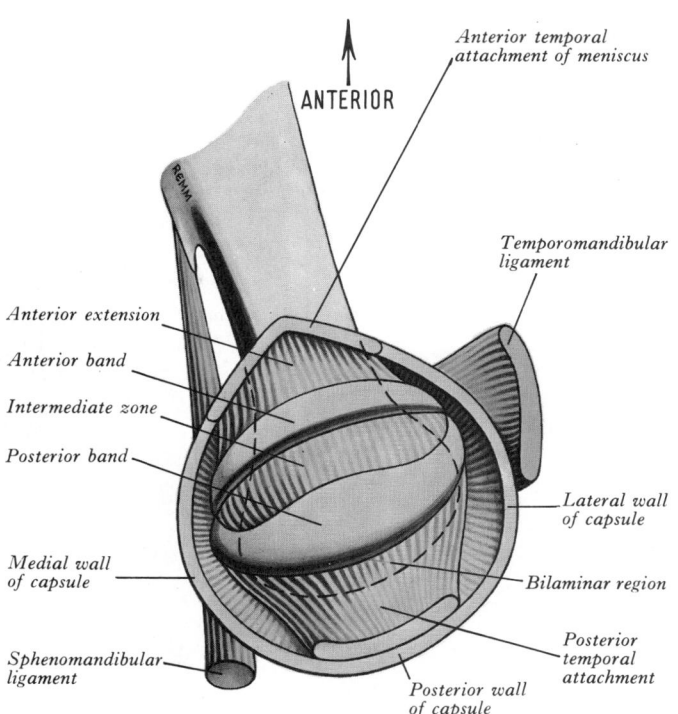

A Superior aspect.

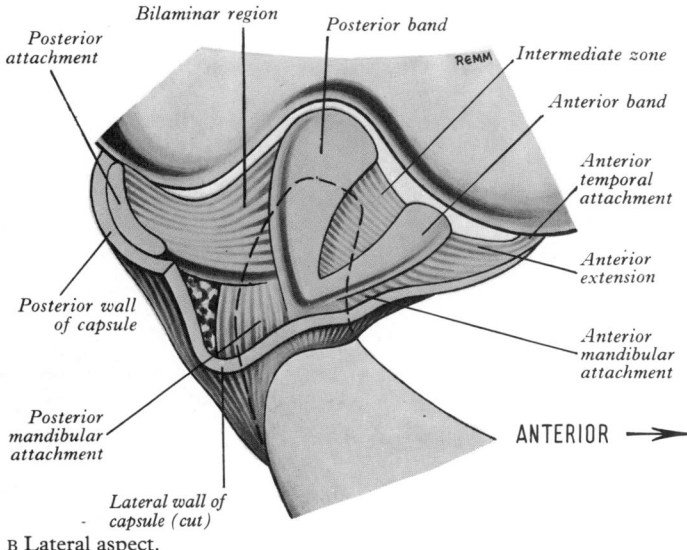

B Lateral aspect.

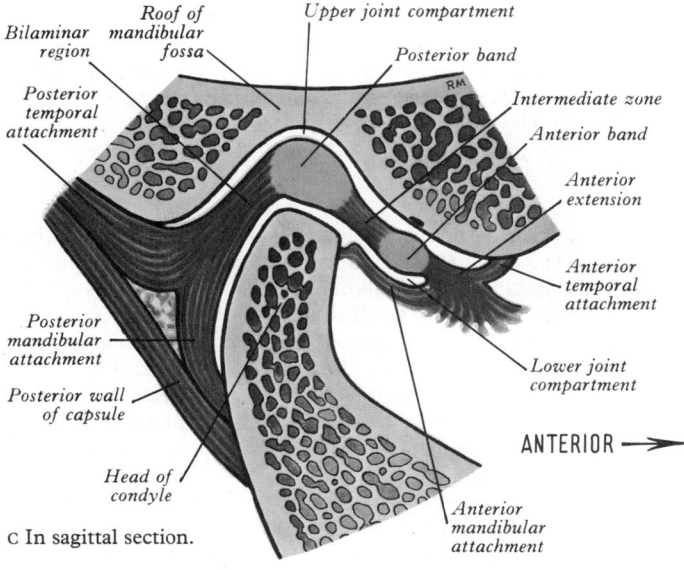

C In sagittal section.

4.21 Form, subdivisions and thickness variations of the intra-articular disc in the temporomandibular joint. See text for details and for alternative interpretations.

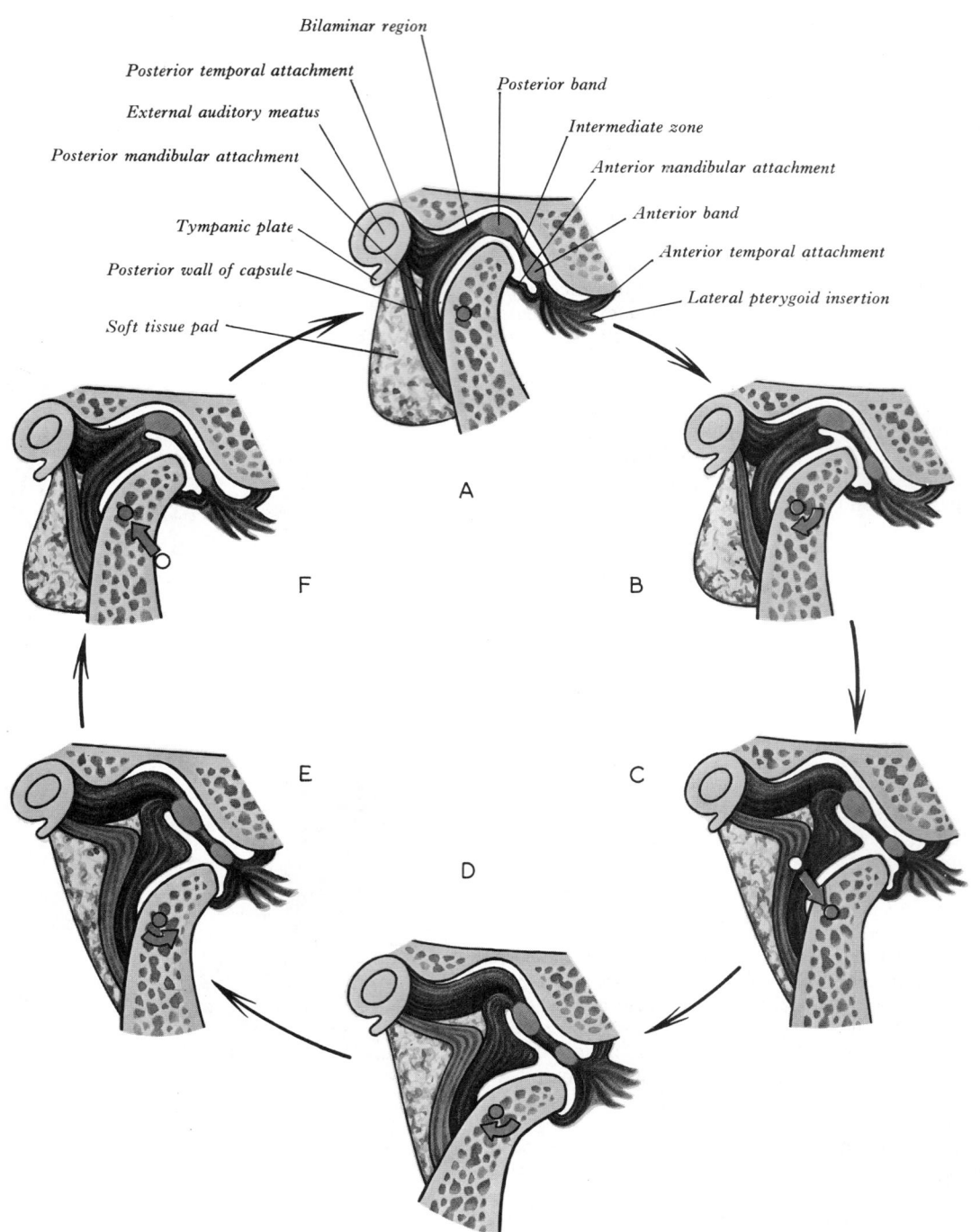

Bilaminar region
Posterior temporal attachment
External auditory meatus
Posterior mandibular attachment
Posterior band
Intermediate zone
Anterior mandibular attachment
Tympanic plate
Posterior wall of capsule
Anterior band
Soft tissue pad
Anterior temporal attachment
Lateral pterygoid insertion

4.22 Changing relationships of the condyle of the mandible, the articular disc and the articular surface of the temporal bone during one complete opening (A→D) and closing (D→A) cycle of the mouth.

4.21 and **4.22** are based upon Rees (1954) with permission of the late Professor J J Pritchard, on behalf of the late Leonard A Rees and the *British Dental Journal*. See also alternative hypotheses proposed by Osborn (1985).

Alternative descriptions of mandibular movements have been proposed (Kraus et al 1969). Successive positions of an indicator point on the mandible form a three-dimensional *envelope of movement; border movements* are those at its surface, *free movements* when the point moves within it and *contact movements* when maxillary and mandibular teeth remain apposed. For a more detailed analysis of biting, chewing and deglutition see Kraus et al (1969); and for a mathematical model relating patterns of human mandibular movement to biomechanical constraints see Baragar & Osborn (1984).

An unresolved controversy concerning forces transmitted at temporomandibular joints must be noted. A proposal that they carry no load, or are subject to little transarticular compression, has been supported by Wilson (1920), Scott (1955), Steinhardt (1958) and others; while later views (e.g. Smith & Savage 1959, Turnbull 1970) oppose this. Most observations involve

vertebrates other than mammals; few reliable experimental data are available for man according to Gingerich (1971). Analysi both theoretical and involving electromyography (integrated an quantitative) in man, however, favours a force-transmittin function (Barbenel 1972, 1974), supported by observations b Crompton & Hiiemae (opossum, 1969) and by Buckland Wright (cat, 1978). In view of the complex activities at thes joints, both views may be correct at times. It is only in 'hing movements of limited extent, as in biting, that controversy nee exist.

Applied Anatomy. The mandible is dislocated only forward With the mouth open, the condyles are on the articular eminenc and sudden violence, even muscular spasm (a convulsive yawn may displace one or both into the infratemporal fossa. Reductic involves depressing the jaw posteriorly, at the same time elevatin the chin. Downward pressure overcomes spasm in masseteri temporal and pterygoid muscles; elevation of the chin forces the condyles backwards. Derangement of a disc may follow trauma, by overclosure with backward displacement of a condyle or malocclusion; clicking and pain on movement result. In operations, rami of the facial nerve overlying the joint must be preserved. Changes in occlusion may lead to remodelling of temporomandibular articular surfaces (Moffet et al 1964).

JOINTS OF THE VERTEBRAL COLUMN AND THORAX

All vertebrae, from second cervical to first sacral, articulate by cartilaginous joints between their bodies, synovial joints between their articular processes (zygapophyses) and fibrous joints between their laminae and also between their transverse and spinous processes.

Joints of Vertebral Bodies

Vertebral bodies are united by anterior and posterior longitudinal ligaments and by fibrocartilaginous intervertebral discs between laminae of hyaline cartilage, together forming symphyses.

The anterior longitudinal ligament (4.23A), a strong band, extends along the anterior surfaces of the vertebral bodies. It is broader caudally, thicker and narrower in thoracic than in cervical and lumbar regions; also relatively thicker and narrower opposite vertebral bodies than at the levels of intervertebral symphyses. Attached to the basilar occipital bone, it extends to the atlantal anterior tubercle, thence to the front of the body of the axis and thereafter continues caudally to the front of the upper sacrum. Its longitudinal fibres, strongly adherent to intervertebral discs, hyaline cartilage laminae and margins of adjacent vertebral bodies, are but loosely attached at intermediate levels of the bodies, where the ligament fills their anterior concavities, flattening the vertebral profile (4.23A). At these various levels ligamentous fibres blend with the subjacent periosteum, perichondrium and periphery of annulus fibrosus. It has several layers, most superficial fibres being longest and extending over three or four vertebrae, intermediate between two or three, the deepest from one body to the next. Laterally short fibres connect adjacent vertebrae.

The posterior longitudinal ligament (4.23A,24A) is in the vertebral canal on the posterior surfaces of the vertebral bodies, attached to the body of the axis and continued to the sacrum; above it is continuous with the membrana tectoria (p. 494). Its smooth, glistening fibres, attached to intervertebral discs, laminae of hyaline cartilage and adjacent margins of vertebral bodies, are separated between attachments by basivertebral veins and the venous rami draining them into anterior internal vertebral plexuses. At cervical and upper thoracic levels the ligament is broad and uniform in width, but in lower thoracic and lumbar regions (4.24A) it is denticulated, narrow over vertebral bodies and broad over discs (strictly symphyses). Its superficial fibres bridge three or four vertebrae, the deeper extending between adjacent vertebrae as *periivertebral ligaments* close to and, in adults, fused with annuli fibrosi of the intervertebral discs. These layers are more distinct in the immediate postnatal years.

Intervertebral discs (4.23A,B), between adjacent surfaces of vertebral bodies from axis to sacrum, are their chief bonds. Discal outlines correspond with the bodies which they connect, thickness varying in different regions and parts of the same disc. They are thicker anteriorly in cervical and lumbar regions, contributing here to anterior convexity, but nearly uniform in the thoracic region, anterior concavity here being largely due to vertebral bodies. Discs are thinnest in upper thoracic, thickest in lumbar regions and adherent to thin layers of hyaline cartilage on superior and inferior vertebral surfaces (p. 315); together disc and hyaline cartilages form an *intervertebral symphysis*. Except for their peripheries, supplied from adjacent blood vessels, discs are avascular and supported by diffusion through the trabecular bone of adjacent vertebrae. Vascular and avascular parts differ in

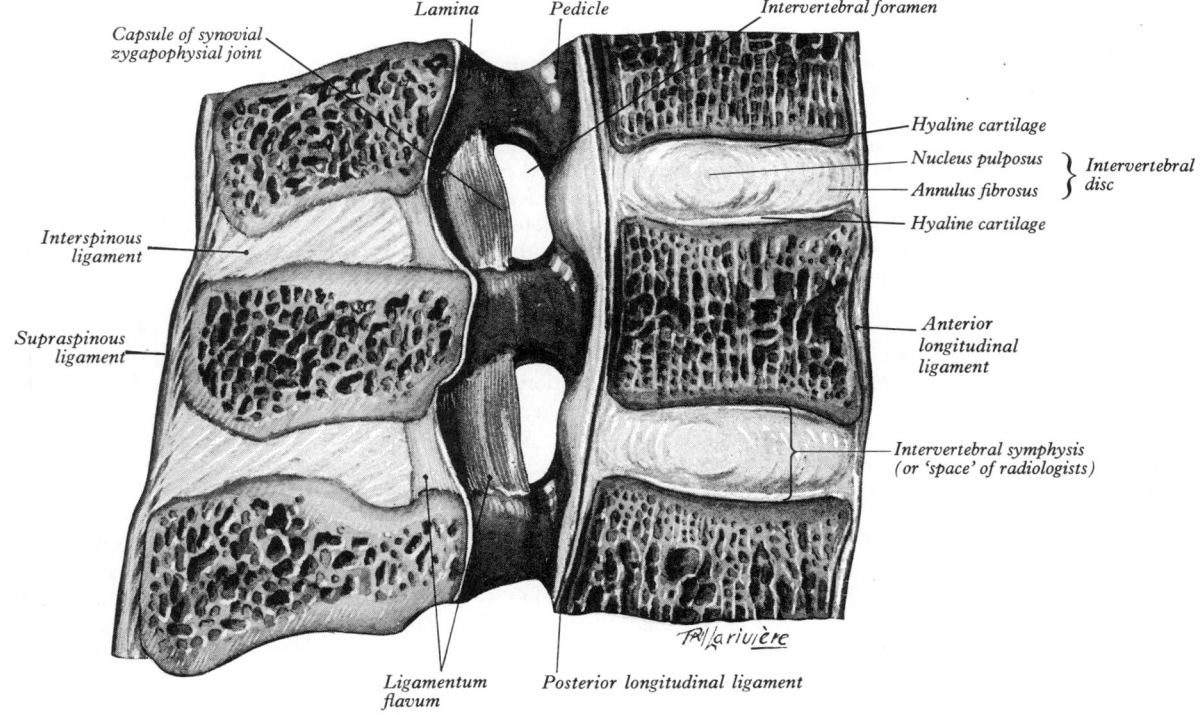

Capsule of synovial zygapophysial joint

Lamina

Pedicle

Intervertebral foramen

Hyaline cartilage

Nucleus pulposus

Annulus fibrosus

Intervertebral disc

Hyaline cartilage

Interspinous ligament

Supraspinous ligament

Anterior longitudinal ligament

Intervertebral symphysis (or 'space' of radiologists)

Ligamentum flavum

Posterior longitudinal ligament

23A Median sagittal section through part of the lumbar region of the vertebral column. Note the boundaries of intervertebral foramina. For contrasting details concerning the direction of fibre bundles in the interspinous ligaments see text and Heylings (1978).

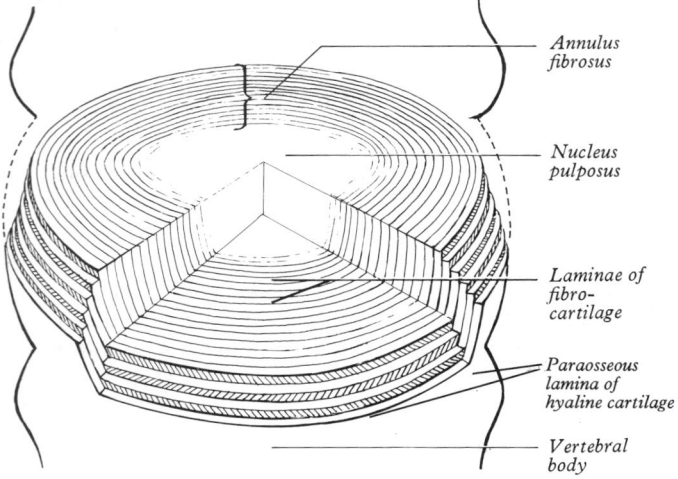

Annulus fibrosus

Nucleus pulposus

Laminae of fibro-cartilage

Paraosseous lamina of hyaline cartilage

Vertebral body

4.23B Simplified schema of the main structural features of an intervertebral disc. The fibrocellular structure of the nucleus pulposus is omitted. For clarity the number of fibrocartilaginous laminae has been greatly reduced, since they are in fact of microscopic dimensions. Note alternating obliquity of collagen fascicles in adjacent laminae. (Modified after Inoue 1973.)

reaction to injury (Smith & Walmsley 1951). Discs are connected to anterior and posterior longitudinal ligaments; at the thoracic level they are tied laterally by intra-articular ligaments to the heads of ribs articulating with adjacent vertebrae. Intervertebral discs (excluding the first two vertebrae) form a fifth of the postaxial vertebral column, cervical and lumbar regions having, in proportion to length, more discs than the thoracic and hence being more pliant (Harris 1939). Each disc consists of an outer laminated *annulus fibrosus* and an inner *nucleus pulposus* (**4.23A,B**).

The *annulus fibrosus* has a narrow outer collagenous zone and a wider inner fibrocartilaginous zone. Its laminae, convex peripherally seen in vertical section, are incomplete collars connected by fibrous bands overlapping one another. (The internal vertical concavity of the laminae conforms to the surface profile of the nucleus pulposus.) Posteriorly, laminae join in a complex manner; fibres in the rest of each lamina are parallel and run obliquely between vertebrae; fibres in contiguous laminae crisscross, (**4.23B**) thus limiting rotation in both directions. Predominantly *vertical* posterior fibres have been described (Zaki 1973), predisposing to herniation. Obliquity of fibres in deeper zones varies in different laminae (Inoue 1973). Johnson et al (1985) have described elastic fibres in a small number of human lumbar annuli fibrosi. Hickey et al (1981) have described fetal collagen fibril diameters in these structures; they also include elastic fibres.

The *nucleus pulposus*, better developed in cervical and lumbar regions, is nearer the disc's posterior surface. At birth it is large, soft, gelatinous and of mucoid material with a few multinucleated notochordal cells, invaded also by cells and fibres from the inner zone of adjacent annulus fibrosus. Notochordal cells disappear in the first decade, followed by gradual replacement of mucoid material by fibrocartilage (Sylven 1951), derived mainly from the annulus fibrosus and the hyaline cartilaginous plates adjoining vertebral bodies. The nucleus pulposus, hitherto distinct, now becomes less differentiated from the remainder of the disc (Walmsley 1953, Peacock 1952, Tondury 1958). In lumbar discs cellularity (6000 cells/mm³ overall) is highest in peripheral annulus fibrosus and in hyaline cartilage nearest to the vertebral bodies, with a glucose diffusion coefficient of 2.5 cm² per second, comparable with values for cartilage elsewhere (Maroudas et al 1975). However, nutritional conditions may be more critical, especially in large lumbar discs. With these changes the nucleus pulposus becomes amorphous and sometimes discoloured. Its water-binding capacity and elasticity diminish (Puschel 1930), because these properties are due to its mucopolysaccharide and protein component (Hendry 1958). When the disc is not loaded, pressure in the nucleus pulposus is low at all ages (Nachemson

1960). Pech & Haughton (1985) have recently used cadaveric intervertebral discs to show the high correlation between appearances in X-rays and by magnetic resonance (MR).

Applied Anatomy. In young adults intervertebral discs are so strong that violence first damages bones. It is impossible to damage a healthy disc except by forcible flexion. After the second decade, however, degenerative changes in discs may result in necrosis, sequestration of the nucleus pulposus, softening and weakening of the annulus fibrosus. Then comparatively minor strains may cause *either* internal derangement with eccentric displacement of the nucleus pulposus *or* external derangement; it then bulges or bursts through the annulus fibrosus, usually posterolaterally. In the former unequal tension in the joint causes muscle spasm and sudden violent pain, acute *lumbago*; in the latter a herniated nucleus may press on adjacent nerve roots with resultant referred pain, *sciatica*. Such derangements are usually in the lower lumbar region, especially at the lumbosacral ·joint, and sometimes at the levels of C.5–7. Motor effects, with loss of power and reflexes may ensue.

Joints of Vertebral Arches

Joints between vertebral articular processes (zygapophyses) are synovial and vary in shape with vertebral level (p. 316); the laminae, spines and transverse processes are connected through syndesmoses constituted by ligamenta flava, interspinous, supraspinous and intertransverse ligaments and the ligamentum nuchae. Some authorities prefer to group these various fibrous structures as accessory ligaments of the zygapophysial joints; others, as here, classify them as officially recognized, named syndesmoses.

ZYGAPOPHYSIAL JOINTS

These joints are of the simple (cervical and thoracic) or complex (lumbar) synovial variety. Their mating surfaces are carried on mutually adapted zygapophyses and are clothed with hyaline articular cartilage. Their size, shape and topology varies with spinal level and are described with individual vertebrae in *Osteology*.

Articular capsules are thin and loose and attached peripheral to the articular facets of adjacent zygapophyses; they are longer and looser in the cervical region. (See Intervertebral Foramen below.)

Zygapophysial lumbar specializations. Engel & Bogduk (1982) studied 82 lumbar zyapophysial joints and identified three types of intracapsular structure. These were: (1) *adipose tissue fat pads* either superoventral, inferodorsal or both; (2) *fibro-adipose 'menisci' (meniscoids)* at the superior or inferior pole, or both; (3) *connective tissue rims* either ventral, dorsal or both. The rims are inflexions of fibrous capsule; the fat pads are similar to those in many other joints; the meniscoids are remarkable in possessing an expanded, vascularized, fibro-adipose base, attached to and sometimes perforating the capsule, an adipose core, poorly vascularized, and a firm flattened fibrous apex. They project into the crevices between non-congruent articular surfaces but their function is conjectural; they are possibly significant clinically. *All* specimens presented at least one of these features and more than 50% presented two or more features.

INTERVERTEBRAL SYNDESMOSES

Ligamenta flava (**4.23A,24B**) connect laminae of adjacent vertebrae in the vertebral canal. Their attachments extend from zygapophysial capsules to where laminae fuse to form spines; here their posterior margins meet and are partially united, intervals being left for veins connecting internal to posterior external vertebral venous plexuses. Their predominant tissue is yellow elastic tissue, whose almost perpendicular fibres descend from the lower anterior surface of one lamina to the posterior surface and upper margin of the one below it. The ligaments are thin, broad and long in the neck, thicker in the thoracic region, thickest at lumbar levels. They brake separation of the laminae in spinal flexión, obviating abrupt limitation, and assist restoration to an erect attitude after flexion, perhaps protecting discs from injury.

The **supraspinous ligament** (4.23A), a strong fibrous cord connecting spinous process apices from the seventh cervical to the sacrum, is thicker and broader at lumbar levels and intimately blended with neighbouring fascia. Its most superficial fibres extend over three or four vertebrae, the deeper spanning two or three, the deepest connecting adjacent spines and here continuous with interspinous ligaments. Between the seventh cervical spine and external occipital protuberance it is expanded as the ligamentum nuchae. Heylings (1978) considers supraspinous ligaments to cease at the fifth lumbar spine.

The **ligamentum nuchae**, a bilaminar fibro-elastic intermuscular septum, is often considered homologous with, but structurally distinct from, supraspinous and interspinous ligaments in the neck. In structure, its dense bilateral fibro-elastic laminae are separated by a tenuous layer of areolar tissue; the laminae are blended at its dorsal free border. The latter is superficial and extends from the external occipital protuberance to the seventh cervical spine. From this the fibro-elastic laminae are attached to the median part of the external occipital crest, the posterior atlantal tubercle and the medial aspects of the bifid spines of cervical vertebrae, as a septum for bilateral attachment of cervical muscles and their sheaths. In bipeds it is the reduced representative of a much thicker, complex elastic ligament which, in quadrupedal mammals, aids suspension of the head and modifies its flexion, functioning like ligamenta flava.

Interspinous ligaments (4.23A), thin and almost membranous, connect adjoining spines, their attachments extending from root to the apex of each. They meet ligamenta flava in front and the supraspinous ligament behind. Narrow and elongated in the thoracic region, broader, thicker and quadrilateral at lumbar levels, they are poorly developed in the neck. (Some observers allocate all cervical bundles as part of the ligamentum nuchae; others regard them, although tenuous, as distinct interspinous fasicles.) Their fibres are usually described as obliquely dorsocaudal. In a study of 28 cadavers Heylings (1978) found them obliquely *dorsocranial* at lumbar levels.

Intertransverse ligaments between transverse processes consist at cervical levels of a few, irregular fibres, largely replaced by intertransverse muscles; in the thoracic region they are cords intimately blended with adjacent muscles; in the lumbar region they are thin and membranous.

Intervertebral joints (symphyses, syndesmoses and synovial) are all innervated by adjoining spinal (and sympathetic) nerves, particularly by their dorsal divisions.

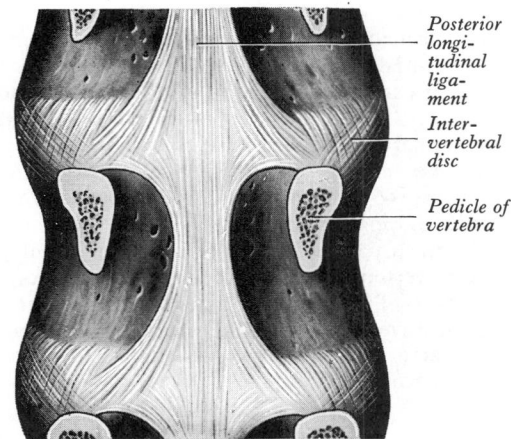

4.24A The lumbar region of the posterior longitudinal vertebral ligament.

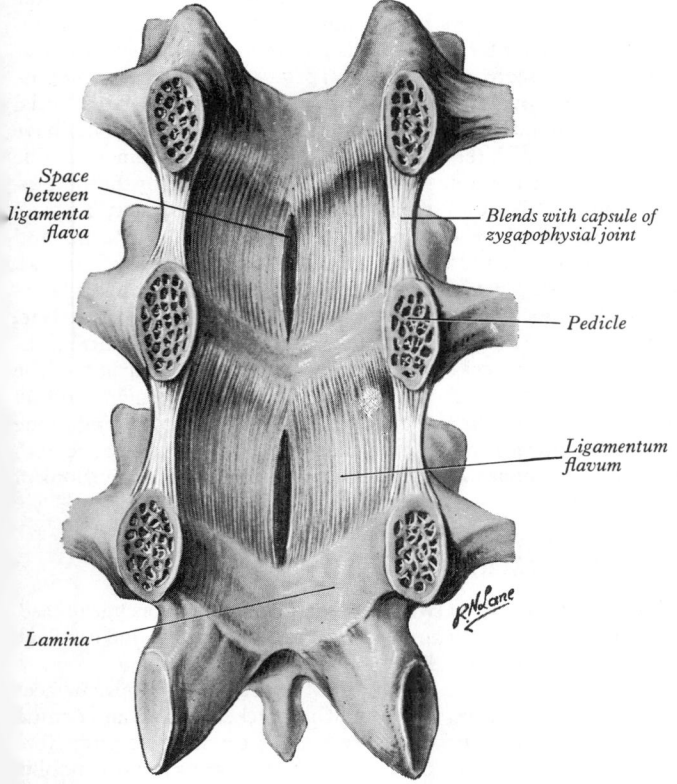

4.24B The ligamenta flava of the lumbar region: anterior aspect.

Posterior longitudinal ligament

Intervertebral disc

Pedicle of vertebra

Space between ligamenta flava

Blends with capsule of zygapophysial joint

Pedicle

Ligamentum flavum

Lamina

Intervertebral Foramina

Closely related to some of the main intervertebral articulations are the principal routes of entry and exit to and from the vertebral canal, the **intervertebral foramina**. (Minor routes occur between the median, often partly fused, margins of the ligamenta flava, vide supra.) Between the axis and sacrum, despite some quantitative and structural regional variations, essentially they conform to the same general plan; because of their construction, contents and susceptibilities to multiple disorders, they are loci of great biomechanical, functional and clinical significance. The specializations cranial to the axis and at sacral levels are described with the individual bones and articulations. **The boundaries** of a generalized intervertebral foramen are: *anteriorly*, from above downwards, periosteum of the posterolateral aspect of the superior vertebral body (thin compact osseous shell over red bone marrow containing cancellous bone); posterolateral aspect of the intervertebral symphysis (including the disc), the curved collagen fascicles here may be regarded as the outer lamina of the annulus fibrosus or as extensions of perivertebral ligaments (distinctions of little importance), finally a small (variable) periosteum-covered posterolateral part of the body of the inferior vertebra (structure as above); *superiorly*, the compact bone of the deep arched inferior vertebral notch of the vertebra above; *inferiorly*, the compact bone of the shallow superior vertebral notch of the vertebra below; *posteriorly* a part of the ventral aspect of the fibrous capsule of the zygapophysial synovial joint. Cervical intervertebral foramina are distinct in having superior and inferior vertebral notches of almost equal depth; also in accord with the direction of the pedicles they face *anterolaterally* and external to them, in the same direction, is the complex transverse process and foramen transversarium (p. 317). The thoracic and lumbar intervertebral foramina face *laterally* and their transverse processes are posterior. The first to tenth thoracic foramina have additionally as anteroinferior boundaries the articulations of the head of a rib, the capsules of double synovial joints with the demifacets on adjacent vertebrae and the intra-articular ligament between the costocapitular ridge and the intervertebral symphysis. Note that the lumbar foramina lie *between* the two principal lines of vertebral attachment of the psoas major muscle. The walls of each foramen are, as noted, clothed throughout by collagen, either periosteal, perichondrial, annular or capsular. **Contents** are: a segmental mixed spinal nerve and its sheaths, from two to four recurrent meningeal (sinu-vertebral) nerves (pp. 1124, 1125), variable spinal arteries (for origins, branches and distribution see p. 765), and plexiform venous connections between internal and external vertebral venous plexuses (p. 811). These structures, particularly the nerves, may be affected by trauma or one of the numerous disorders of the tissues bordering the foramen: i.e. fibrocartilage of the annulus fibrosus, in earlier decades the nucleus pulposus, the highly vascular red bone marrow occupying the cancellous bone of the vertebral bodies; the compact bone of the pedicles; the capsules, synovial membranes, articular cartilages (and at lumbar levels fibro-adipose meniscoids, fat pads, fibrous labra, p. 490) of the synovial zygapophysial joints; and additionally at thoracic levels tissues of the complex synovial costocapitular joints.

Movements of the Vertebral Column

Range of movement between vertebrae is restricted by the limited deformation of intervertebral symphyses (particularly the discs), whose greater thickness at cervical and lumbar levels increases individual ranges. It is also limited by the topography of the zygapophysial joints and by concomitant changes in tension of the ligamentous syndesmoses. Although movements between individual vertebrae are small, their summation gives a wide total range to the vertebral column in flexion, extension, lateral flexion, rotation and circumduction.

In *flexion* the anterior longitudinal ligaments become relaxed and the anterior parts of intervertebral discs are compressed; at its limit the posterior longitudinal ligament, ligamenta flava, interspinous and supraspinous ligaments and posterior fibres of intervertebral discs are tensed; interlaminar intervals widen, inferior articular processes glide on superior processes of subjacent vertebrae and their capsules become taut. Tension of extensor muscles is also important in limiting flexion, e.g. when carrying a load on the shoulders. Flexion is most extensive at the cervical level.

In *extension* the opposite events occur; it is limited by tension of the anterior longitudinal ligament, anterior discal fibres and approximation of spines, zygapophyses and compression of posterior discal fibres. Marked in cervical and lumbar regions, it is much less at thoracic level, partly because discs are thinner and also due to effects of thoracic skeleton and musculature. In full extension the axis of movement has been described as behind the articular processes, moving forwards as the column straightens and passes into flexion, reaching the centre of the vertebral bodies in full flexion (Wiles 1935).

In *lateral flexion* intervertebral discs are laterally compressed and contralaterally tensed and lengthened, motion being limited by tension of antagonist muscles and ligaments. It is always combined with rotation. Lateral movements occur in any part of the column but are greatest in cervical and lumbar regions.

Rotation involves twisting of vertebrae relative to each other, with torsional deformation of intervening discs. Although slight between individual vertebrae, it summates along the column, its upper part being turned to one or other side. Movement is slight at cervical level, greater in the upper thoracic and least in the lumbar region. *Circumduction* is limited and merely a succession of preceding movements.

Extent and direction of vertebral movements are guided by articular facets. Although often described as *plane*, they are never truly flat but *ovoid*, opposing surfaces being reciprocally concave and convex. In the *cervical* region upward inclination of the superior articular facets allows free flexion and extension; the latter is usually greater and checked above by locking of the posterior edges of the superior atlantal facets in the occipital condylar fossae and below by slipping of the seventh cervical inferior processes into grooves inferoposterior to the first thoracic superior articular processes. Flexion stops where the cervical convexity is straightened, checked by apposition of the projecting lower lips of vertebral bodies with shelving surfaces on subjacent bodies. Cervical lateral flexion and rotation are always combined; superomedial inclination of superior articular facets imparts rotation during lateral flexion. In the *thoracic* region, especially above, all movements are limited, reducing interference with respiration; lack of upward inclination of superior articular facets prohibits much flexion, extension being checked by contact of the inferior articular margins with laminae and of spines with each other. Thoracic rotation is freer; its axis is in the vertebral bodies in the

mid-thoracic region, in front of them elsewhere, so that rotation involves some lateral displacement (Davis 1959, Davis et al 1965). Direction of articular facets would allow free lateral flexion but this is limited in the upper thoracic region by resistance of ribs and sternum. Rotation is usually combined with slight lateral flexion to the same side. Lumbar extension is wider in range than flexion and some lateral flexion and rotation can also occur. Range of rotation is limited by absence of a common centre of curvature for right and left articular facets (Putz 1976). Functional transition between thoracic and lumbar regions is usually between the eleventh and twelfth thoracic vertebrae (p. 322), where zygapophysial joints of the vertebral arches usually fit tightly, slight compression locking them and preventing all but flexion.

Muscles producing vertebral movements. The spinal column is moved both by intrinsic muscles attached to it and by muscles attached to other bones, acting indirectly. Gravity also always plays a part.

Flexion: longus cervicis, scaleni, sternocleidomastoid and rectus abdominis of both sides.

Extension: the erector spinae complex, splenius, semispinalis capitis and trapezius of both sides.

Lateral flexion: longissimus and iliocostocervicalis, oblique abdominal muscles and flexors on the side of *lateral* flexion.

Rotation: rotatores, multifidus, splenius cervicis and oblique abdominal muscles.

Extension, principally lumbar, occurs commonly from a stooping position. Initial extension is mainly at hips and knees; continuous lumbar extension is delayed, with little or no activity in erector spinae. In lifting heavy weights there is considerable compression of lumbar intervertebral discs initially, with large rises in thoracic and abdominal pressure, which may resist flexion (Davis 1963, Davis et al 1965). Pal & Rontal (1986) have recently studied the mechanics of weight transmission via the vertebral arches. In contrast with the usual view, that the major, almost only, factor is contributed by vertebral bodies and intervertebral discs, they find that, on the basis of areal and other measurements, the vertebral arches and their zygapophysial joints are, at cervical levels, also a considerable factor in weight transmission. For an extensive view of lumbar backache see Wyke (1980).

Lumbosacral Joints

Articulations between the fifth lumbar and first sacral vertebrae resemble those of others. Their bodies are united by a symphysis including a large intervertebral disc, deeper ventrally (its anterior margin at the sacrovertebral angle, p. 329) with anterior and posterior longitudinal ligaments adherent to it. Their zygapophysial joints are separated by a wider interval than those above. They have ligamenta flava, interspinous and supraspinous ligaments. The fifth lumbar vertebra is attached to ilium and sacrum by the iliolumbar ligament. These joints vary much in geometry (p. 325, 4.25).

The iliolumbar ligament (4.56), attached to the tip and antero-inferior aspect of the fifth lumbar transverse process, sometimes has a weak attachment to the fourth. It radiates laterally and is attached by two main bands to the pelvis: a lower one, the *lumbosacral ligament*, from the inferior aspect of the fifth lumbar transverse process to the anterosuperior lateral surface of the sacrum, blending with the ventral sacro-iliac ligament; and an upper, partial attachment of quadratus lumborum, passing to the iliac crest anterior to the sacro-iliac joint, continuous above with the thoracolumbar fascia. Innervation is from dorsal divisions of spinal nerves.

THE SACROCOCCYGEAL JOINT

This is a symphysis between the sacral apex and coccygeal base united by a fibrocartilaginous disc, remnants of hyaline cartilage and ventral, dorsal and lateral ligaments.

The fibrocartilage, a thin disc between contiguous surfaces of the sacrum and coccyx, is somewhat thicker in front and behind than laterally. Its surfaces carry hyaline cartilage varying from thin veils to small islands. Occasionally the coccyx is more mobile, the joint being synovial.

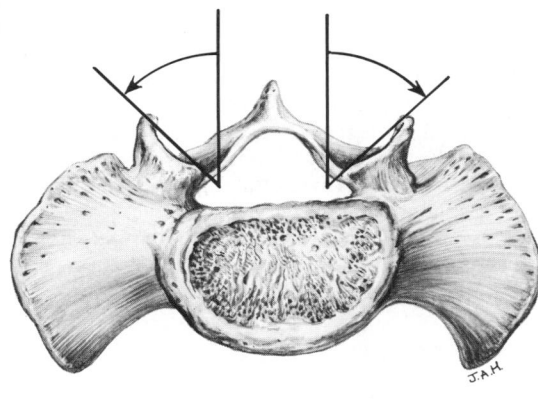

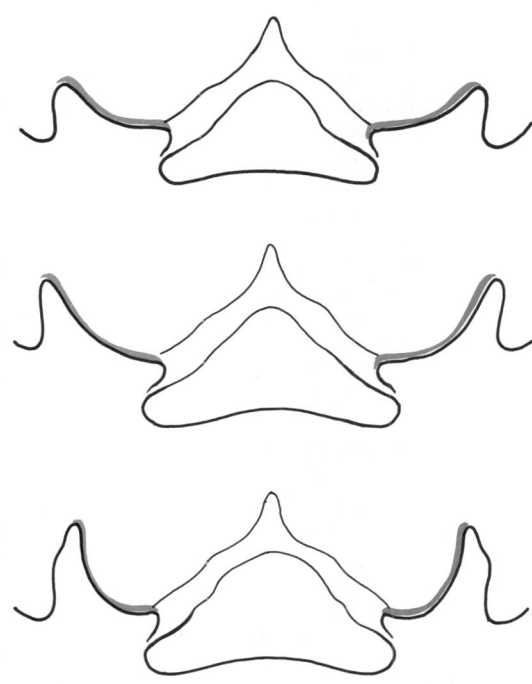

4.25 The method employed by Čihák (1970) to measure the inclinations of articular surfaces of the lumbosacral zygapophysial joints (above). The lower profiles depict three degrees of increasing curvature and inclination. For quantitative data derived from 132 human sacra consult the original paper.

The ventral sacrococcygeal ligament (4.56) consists of irregular fibres descending on the pelvic surfaces of both sacrum and coccyx, attached like the anterior longitudinal ligament.

The superficial dorsal sacrococcygeal ligament is flat and passes from the margin of the sacral hiatus to the dorsal coccygeal surface (4.57). It roofs the lower sacral canal.

The deep dorsal sacrococcygeal ligament passes from the back of the fifth sacral vertebral body to the dorsum of the coccyx and corresponds to the posterior longitudinal ligament.

A lateral sacrococcygeal ligament on each side, like an intertransverse ligament, connects a coccygeal transverse process to an inferolateral sacral angle, completing a foramen for the fifth sacral spinal nerve.

The intercornual ligaments connect sacral and coccygeal cornua on each side. A fasciculus also connects the sacral cornua to the coccygeal transverse processes.

Intercoccygeal Joints

These are symphyses, with thin discs of fibrocartilage, between coccygeal segments in the young. Segments are also connected by extension of the ventral and dorsal sacrococcygeal ligaments. In adult males all segments are united comparatively early but in females union is later. In advanced age the sacrococcygeal joint is

obliterated. Occasionally the joint between the first and second segments is synovial; the apex of the terminal segment is connected to overlying skin by white fibrous tissue.

Craniovertebral Joints

Articulation of the cranium and vertebral column is by means of paired atlanto-occipital joints and the ligaments between the axis and occipital bone; it is appropriate that the joints between atlas and axis at which the head rotates are also included here.

ATLANTO-AXIAL JOINTS

Articulation of atlas to axis is at three synovial joints, a pair between inferior atlantal and superior axial facets and a median complex between the axial dens and atlantal anterior arch and transverse ligament.

The lateral atlanto-axial joints are often classified as planar but the cartilaginous *articular surfaces* are ovoid, the atlantal slightly concave, the axial reciprocally convex (pp. 317, 318). According to Kapandji (1970) both facets are transversely cylindrical, engaging like wheels. Fibrous capsules are thin, loose, attached at their articular margins and lined by synovial membrane. Each has a posteromedial *accessory ligament* attached below to the axial body near the base of its dens, above to the lateral atlantal mass near the transverse ligament. The vertebrae are connected in front by the anterior longitudinal ligament (4.26), here a strong and wide band fixed above to the lower border of the anterior atlantal arch, below to the front of the axial body. It thickens into a median cord connecting the anterior atlantal tubercle to the axial body. They are joined by a broad, thin membrane (4.27) attached above to the lower border of the atlantal arch, below to upper edges of the axial laminae in series with ligamenta flava; it is pierced laterally by the second cervical spinal nerves.

The median atlanto-axial joint is a pivot between the axial dens and a ring formed by the atlantal anterior arch and transverse ligament; it has *two* synovial cavities which act in concert. A facet on the anterior dental surface articulates with one on the posterior aspect of the anterior atlantal arch, having a weak, loose fibrous capsule lined by synovial membrane. Posteriorly is a larger

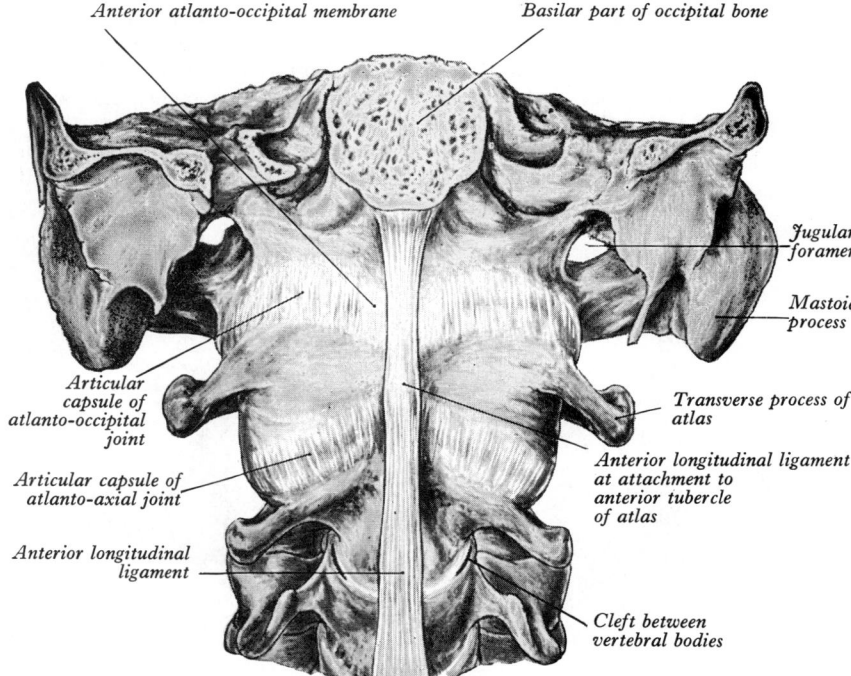

4.26 Atlanto-occipital and atlanto-axial joints: anterior aspect. On each side a small cleft has been opened between the lateral part of the upper surface of the body of the third cervical vertebra and the bevelled, inferior surface of the body of the axis. (See also p. 317.)

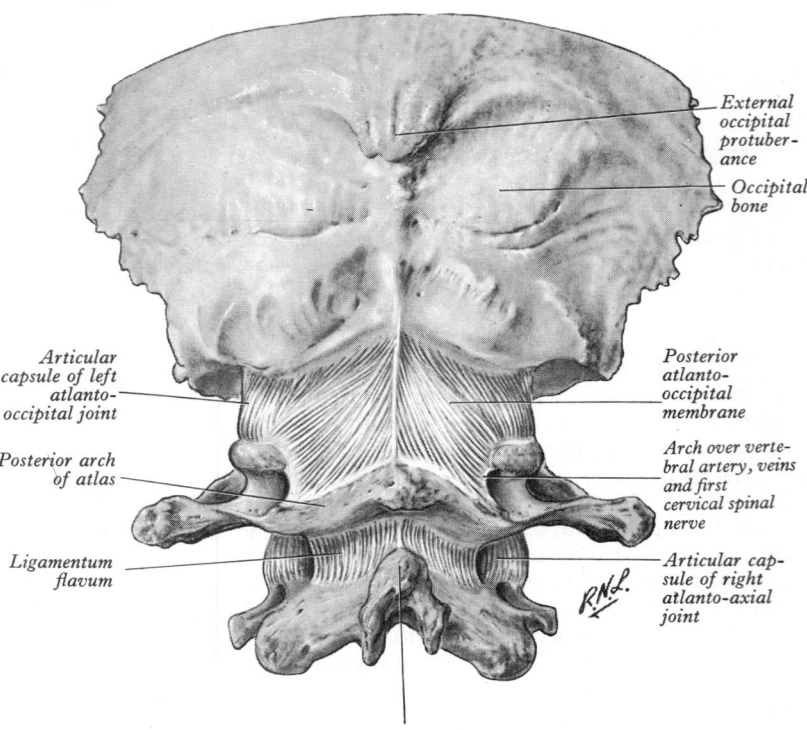

Atlanto-occipital and atlanto-axial joints: posterior aspect.

External occipital protuberance

Occipital bone

Articular capsule of left atlanto-occipital joint

Posterior atlanto-occipital membrane

Posterior arch of atlas

Arch over vertebral artery, veins and first cervical spinal nerve

Ligamentum flavum

Articular capsule of right atlanto-axial joint

Spine of axis

4.27 Atlanto-occipital and atlanto-axial joints: posterior aspect.

both slightly convex in anteroposterior axes. Therefore, when gliding back or forwards on the lower facet, the upper also descends, stretch of capsule being diminished by descent of its upper attachments, making capsular laxity unnecessary. The lateral atlanto-axial joints support the head via the atlas, the pivot guides rotation. Kapandji (1974) has analysed these and other cervical movements, emphasizing the slight helical component in lateral atlanto-axial motion, as described above.

Muscles producing rotation: obliquus capitis inferior, rectus capitis posterior major and splenius capitis of one side, and contralateral sternocleidomastoid.

ATLANTO-OCCIPITAL JOINTS

Each joint has a facet on its corresponding lateral atlantal mass which is adapted to an opposed occipital condyle: it is ellipsoid in type and *articular surfaces* are reciprocally curved. Atlantal facets are concave, tilted medially and centrally constricted, the surface partially, sometimes completely, divided (Singh 1965). The bones are connected by articular capsules and anterior and posterior atlanto-occipital membranes.

Fibrous capsules surround the occipital condyles and superior atlantal articular facets, thickened posterolaterally but thin and sometimes deficient medially, where the synovial cavities often connect with the bursa between dens and transverse atlantal ligament (Cave 1934).

The anterior atlanto-occipital membrane (4.26), broad and of densely woven fibres, connects the anterior margin of the foramen magnum to the upper border of the anterior atlantal arch; laterally it blends with the capsular ligaments; it is strengthened by the anterior longitudinal ligament to form a median cord between the basilar occipital bone and anterior atlantal tubercle (4.26).

The posterior atlanto-occipital membrane (4.27), broad but thin, connects the posterior margin of the foramen magnum to the upper border of the posterior atlantal arch. It arches over the grooves for the vertebral arteries, completing openings for entrance of the arteries and exit of venous plexuses and the first cervical spinal nerves. The ligamentous border arching over artery, veins and nerve is sometimes ossified.

In movements at the atlanto-occipital joints their long (topographical) axes run anteromedially; because of this and their articular curvatures, the profiles of *both* joints are parts of the surface of *one* ellipsoid, its long axis transverse. Hence joints act as one around transverse and anteroposterior axes of movement but *not* around the vertical, therefore permitting flexion/extension and slight lateral flexion.

Muscles Producing the Movements

Flexion: longus capitis and rectus capitis anterior.

Extension: recti capitis posteriores major et minor, obliquus capitis superior, semispinalis capitis, splenius capitis and trapezius (cervical part).

Lateral flexion: rectus capitis lateralis, semispinalis capitis, splenius capitis, sternocleidomastoid and trapezius (cervical part).

Ligaments Connecting Axis and Occipital Bone

The membrana tectoria, paired alar and median apical ligaments extend between axis and occipital bone.

The membrana tectoria (4.29,30), inside the vertebral canal, is a broad strong band covering the dens and its ligaments as a prolongation of the posterior longitudinal ligament (p. 489). Its superficial and deep laminae are both attached to the posterior surface of the axial body, the superficial lamina expanding as it ascends to the *upper* surface of the basilar occipital bone, and is attached in front of the foramen magnum, blending with the cranial dura mater; the deep lamina has a median band, extending also to the basilar occipital, and two lateral bands which ascend medial to the atlanto-occipital joints to the margins of the foramen magnum.

The alar ligaments (4.29), strong rounded cords, start on each side of the apex of the dens and ascend laterally to rough impressions on the medial sides of the occipital condyles. They

synovial cavity (or bursa) between the *cartilaginous* anterior surface of the transverse ligament and the posterior grooved aspect of the dens (4.30), often continuous with one of the atlanto-occipital joints.

The transverse atlantal ligament (4.28,29,30), a thick strong band across the atlantal ring, holds the dens against the anterior arch. Attached to a small, medial tubercle on each atlantal lateral mass, it is broader centrally and bears an anterior thin layer of articular cartilage. From its superficial fibres median longitudinal bands ascend and descend, the upper to the superior surface of the basilar occipital bone between the apical ligament of the dens and membrana tectoria; the lower band, which may be absent, reaches the posterior axial surface. The whole complex forms the *atlantal cruciform ligament*. The transverse ligament divides the ring into unequal parts (4.28), the posterior and larger surrounding spinal cord and meninges, the anterior containing the dens, whose neck is embraced posteriorly by the ligament, retaining it in position even when all other ligaments are divided.

Movement at the atlanto-axial joints are simultaneous at all three, consisting basically of atlantal rotation on the axis and limited by alar ligaments (p. 495). The opposed lateral facets are

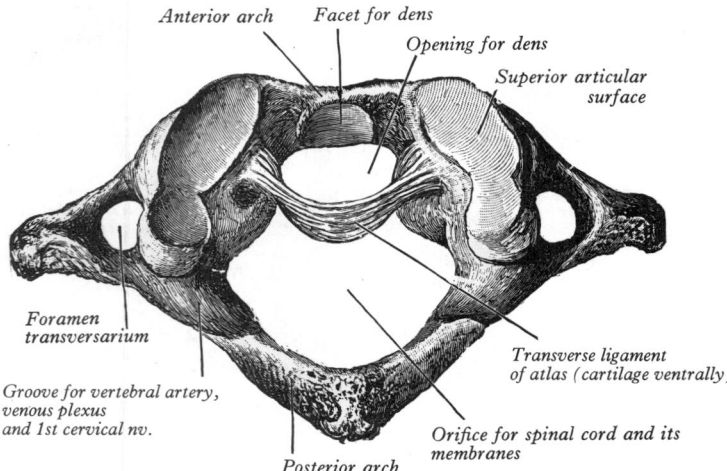

Anterior arch

Facet for dens

Opening for dens

Superior articular surface

Foramen transversarium

Groove for vertebral artery, venous plexus and 1st cervical nv.

Transverse ligament of atlas (cartilage ventrally)

Orifice for spinal cord and its membranes

Posterior arch

4.28 The atlas vertebra, with the transverse ligament.

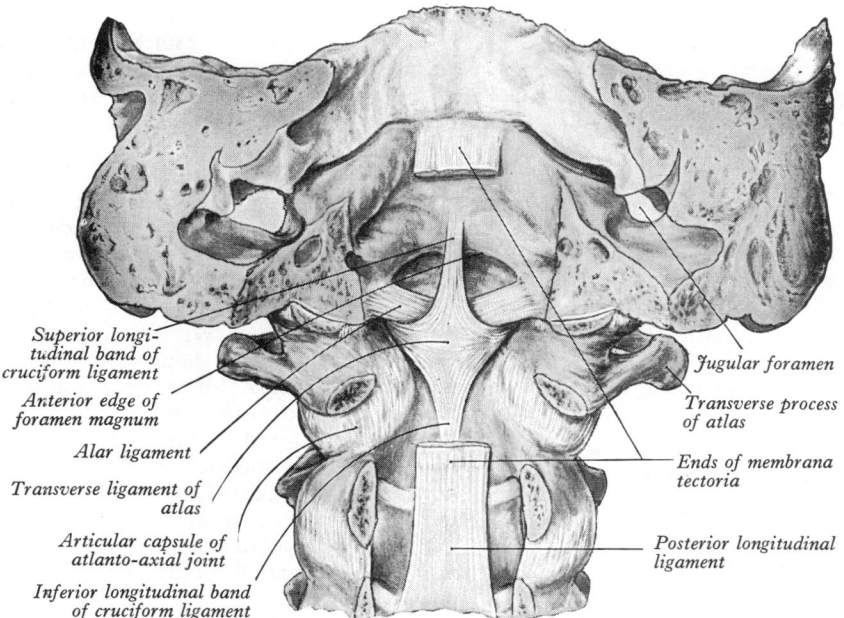

Superior longi-
tudinal band of
cruciform ligament

Anterior edge of
foramen magnum

Alar ligament

Transverse ligament of
atlas

Articular capsule of
atlanto-axial joint

Inferior longitudinal band
of cruciform ligament

Jugular foramen

Transverse process
of atlas

Ends of membrana
tectoria

Posterior longitudinal
ligament

4.29 Posterior aspect of the atlanto-occipital and atlanto-axial joints, after removal of the posterior part of the occipital bone and the laminae of the upper cervical vertebra. The atlanto-occipital joint cavities have been opened.

relax in extension, tauten in flexion, thus limiting movement. They would prevent rotation of the head, were it not for a slight descent of the atlas which relaxes the ligaments enough to compensate for rotational tension. Rotation to the right is checked by tension of those fibres of the right ligament attached to the dens in front of the axis of movement and fibres of the left ligament attached behind the axis, and vice versa in left rotation.

The apical ligament of the dens (4.30) extends from its apex to the anterior margin of the foramen magnum between the alar ligaments; it blends with deep fibres of the anterior atlanto-occipital membrane and the upper vertical band of the cruciform

ligament. It is said to be the core of the centrum of the pro-atlas and contains traces of notochord (Ganguly & Roy 1964). The ligamentum nuchae (p. 491) also connects cervical vertebrae with the cranium.

Applied Anatomy. Dislocation of atlas from axis, with rupture of the transverse ligament and consequent spinal injury, occurs in death by hanging which may, however, fracture the axis or tear the disc between it and the third cervical vertebra. Dislocation of atlas on axis may follow pathological softening of the cruciform ligament or disease in the lateral atlanto-axial joints (e.g. rheumatoid arthritis, Bogduk et al 1984).

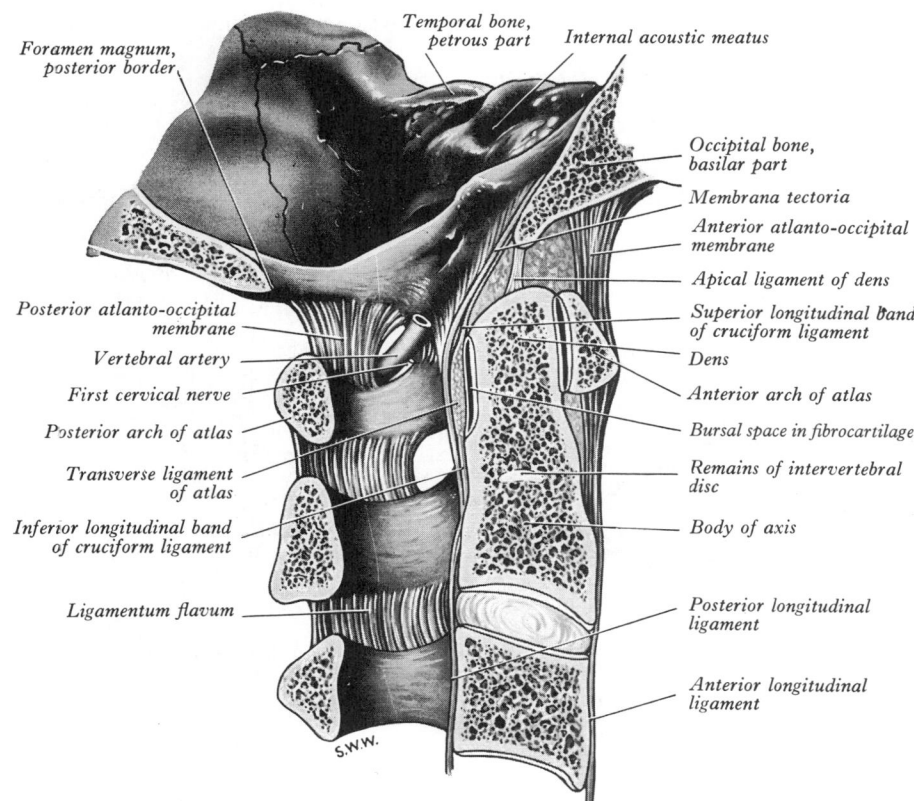

Foramen magnum,
posterior border

Posterior atlanto-occipital
membrane

Vertebral artery

First cervical nerve

Posterior arch of atlas

Transverse ligament
of atlas

Inferior longitudinal band
of cruciform ligament

Ligamentum flavum

Temporal bone,
petrous part

Internal acoustic meatus

Occipital bone,
basilar part

Membrana tectoria

Anterior atlanto-occipital
membrane

Apical ligament of dens

Superior longitudinal band
of cruciform ligament

Dens

Anterior arch of atlas

Bursal space in fibrocartilage

Remains of intervertebral
disc

Body of axis

Posterior longitudinal
ligament

Anterior longitudinal
ligament

S.W.W.

30 Median sagittal section through the occipital bone and the first to
third cervical vertebrae.

Costovertebral Joints

Articulations of the ribs with the vertebral column connect costal heads to vertebral bodies (costocorporeal) and costal necks and tubercles to transverse processes (costotransverse).

JOINTS OF COSTAL HEADS

Heads of typical ribs articulate with facets (often termed demifacets) on the margins of adjacent thoracic vertebral bodies and with intervertebral discs between them (4.31). The first and tenth to twelfth ribs articulate with a single vertebra by a simple synovial joint; in the others an intra-articular ligament bisects the joint. Double synovial compartments result, the joint being classified as both compound and complex. Often inaccurately described as plane, their articular surfaces are slightly ovoid and the upper and lower synovial articulations are obtusely angled to each other (4.31). Ligaments are capsular, radiate and intra-articular.

The fibrous capsules connect costal heads to the circumference of articular surfaces formed by intervertebral discs and demifacets of adjacent vertebrae; some of their upper fibres traverse their intervertebral foramina to blend with the backs of intervertebral discs (strictly symphyses); posterior fibres are continuous with costotransverse ligaments.

Radiate ligaments connect the anterior parts of each costal head to the bodies of two vertebrae and their intervertebral disc. Each is attached to the head just beyond its articular surface. Superior fibres ascend to the vertebral body above, inferior to the body below; intermediate fibres, shortest and least distinct, are horizontal and attached to the disc. In the first rib's joint the radiate ligament is attached to the seventh cervical and first thoracic vertebrae. In joints of the tenth to twelfth ribs, articulating with single vertebrae, the ligament is attached to this and the one above.

The intra-articular ligament is a short, flat band, attached laterally to the crest between the costal articular facets and medially to the intervertebral disc, dividing the joint; from the first and tenth to twelfth joints the ligament is absent.

COSTOTRANSVERSE JOINTS

The facet of a costal tubercle articulates reciprocally with the transverse process of its corresponding vertebra (4.32). The eleventh and twelfth ribs lack this articulation; in the upper five or six joints articular surfaces are reciprocally curved but below this flatter (4.33). Their ligaments are capsular, costotransverse, superior and lateral costotransverse. **The fibrous capsule** is thin and attached to articular peripheries; it has a synovial lining.

The superior costotransverse ligament has an anterior layer attached between the crest of the costal neck and lower aspect of the transverse process above (4.31); laterally it blends with the internal intercostal membrane and it is crossed by intercostal vessels and nerve; its posterior layer is attached dorsally on the costal neck, ascending posteromedially to the transverse

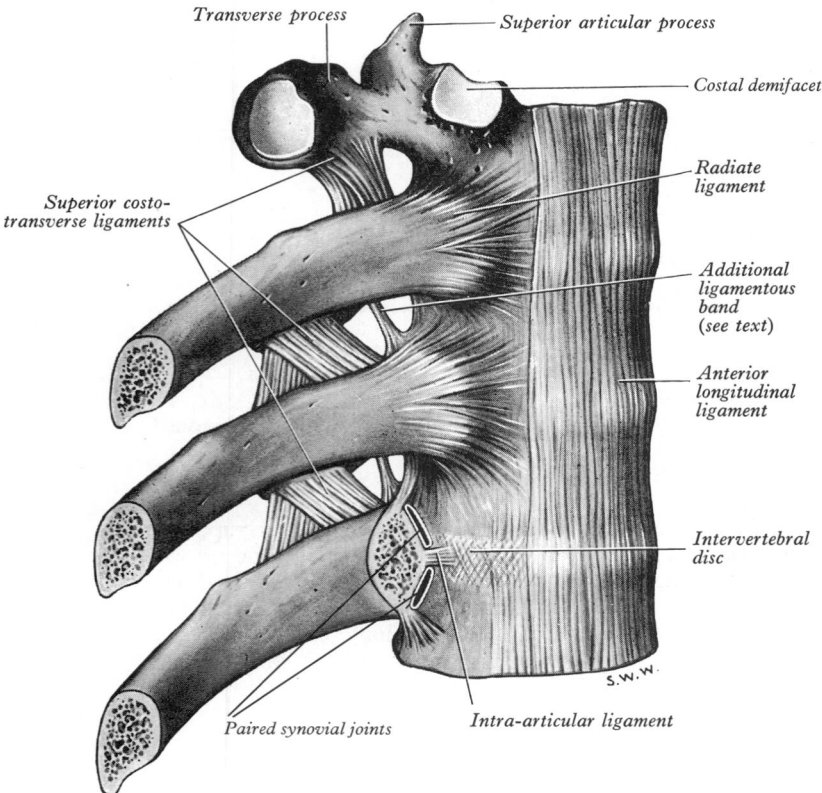

Transverse process — *Superior articular process*

Costal demifacet

Superior costo-transverse ligaments

Radiate ligament

Additional ligamentous band (see text)

Anterior longitudinal ligament

Intervertebral disc

S.W.W.

Paired synovial joints — *Intra-articular ligament*

4.31 Costovertebral joints: right anterolateral aspect. In the lowest joint shown most of the radiate ligament and the anterior part of the head of the rib have been excised to show the two joint cavities and the intra-articular ligament between them.

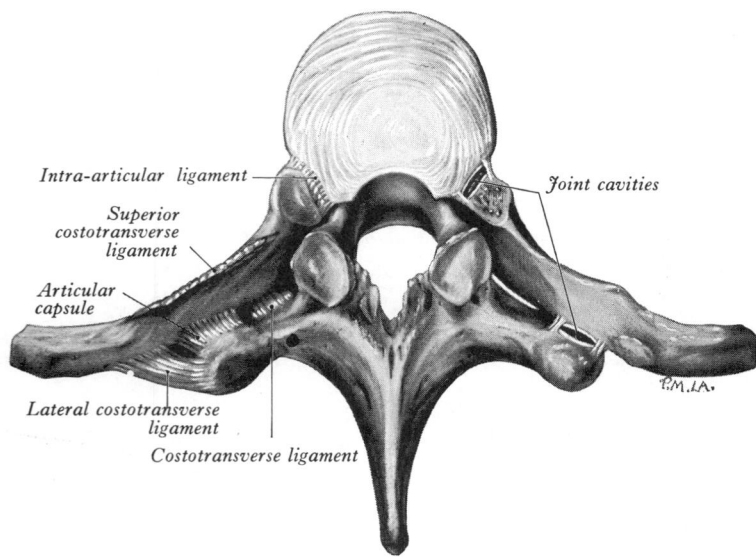

Intra-articular ligament

Superior costotransverse ligament

Articular capsule

Joint cavities

P.M.IA.

Lateral costotransverse ligament

Costotransverse ligament

4.32 Costovertebral joints: superior aspect. On the left the free demifacet superior to the intra-articular ligament is apparent on the rib's head. On the right the inferior costovertebral and costotransverse synovial cavities have been opened.

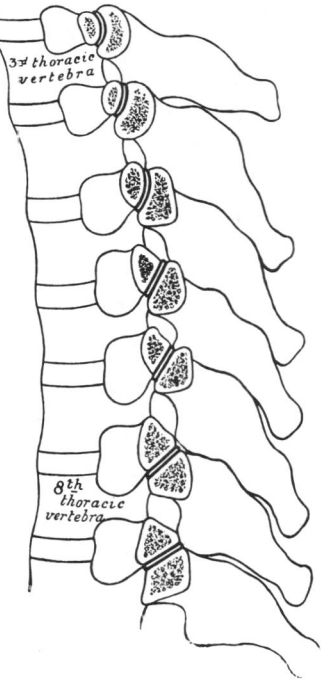

3rd thoracic vertebra

8th thoracic vertebra

4.33 Section through the costotransverse joints from the third to the ninth inclusive. Contrast the concave facets on the upper with less curved facets on the lower transverse processes.

process above; laterally it blends with the external intercostal muscle. The first rib has no such ligament; the shaft of the twelfth, near its head, is connected to the base of the first lumbar transverse process by a *lumbocostal ligament* in series with the superior costotransverse ligaments. An **accessory ligament**, usually present and medial to the superior costotransverse, is separated from it by the dorsal ramus of a thoracic spinal nerve and accompanying vessels (4.31). Variable in attachments, such bands usually pass from a depression medial to a costal tubercle to the inferior articular process immediately above; some fibres also pass to the base of the transverse process.

The costotransverse ligament fills the *costotransverse foramen* between costal neck and its adjacent corresponding transverse process. Its numerous short fibres extend back from the posterior rough surface on the costal neck to the anterior surface of the transverse process. In the eleventh and twelfth ribs it is rudimentary or absent.

The lateral costotransverse ligament, short, thick and strong, passes obliquely from the apex of a transverse process to the rough nonarticular part of the adjacent costal tubercle. The ligaments of upper ribs *ascend* from their transverse processes; they are shorter and more oblique than those of the lower ribs, which *descend*.

Movements at costotransverse joints. Costal heads are so firmly tied to vertebral bodies by radiate and intra-articular ligaments that only slight gliding can occur; strong ligaments binding costal necks and tubercles to transverse processes also limit movements at costotransverse joints to slight gliding, guided by the shape and direction of articular surfaces (4.33); those on tubercles of the upper six ribs are oval and vertically convex, fitting corresponding concavities on the anterior surfaces of transverse processes; hence up and down movements of tubercles involve rotation of costal necks on their long axes.

Articular surfaces of the seventh to tenth tubercles are almost flat, facing down, medially and backwards; opposing surfaces are on the upper aspects of transverse processes. Hence when these tubercles ascend they also move posteromedially. Both sets of joints move simultaneously and in the same directions; the costal neck moves as if at a single joint, the two articulations forming its ends. In the upper six ribs the neck moves slightly up and down but its chief movement is one of rotation on its long axis, downward rotation of its ventral aspect being associated with depression and upward rotation with elevation of the shaft and anterior end of the rib. In the seventh to tenth ribs the neck ascends posteromedially or descends anterolaterally, increasing or diminishing the infrasternal angle; slight rotation accompanies these movements. The muscles involved in respiration are considered on p. 594.

Sternocostal Joints

Costal cartilages join small concavities on the lateral sternal borders (*chondrosternal articulations*, 4.34). Perichondrium and periosteum are continuous. The first costal cartilage joins by an unusual variety of synarthrosis (see p. 467; this is often inaccurately called a synchondrosis) while the second to seventh articulate by synovial joints; articular cavities are often absent, particularly in lower joints. Articular surfaces are *fibrocartilaginous* and this tissue also unites costal cartilages to sternum where cavities are absent (Gray & Gardner 1943). Ligaments involved are capsular, radiate sternocostal, intra-articular and costoxiphoid. According to Sick & Koritke (1976) the seventh costosternal joint may be synovial or 'symphysial'.

Fibrous capsules surround the second to seventh sternocostal joints. They are thin, blended with sternocostal ligaments and

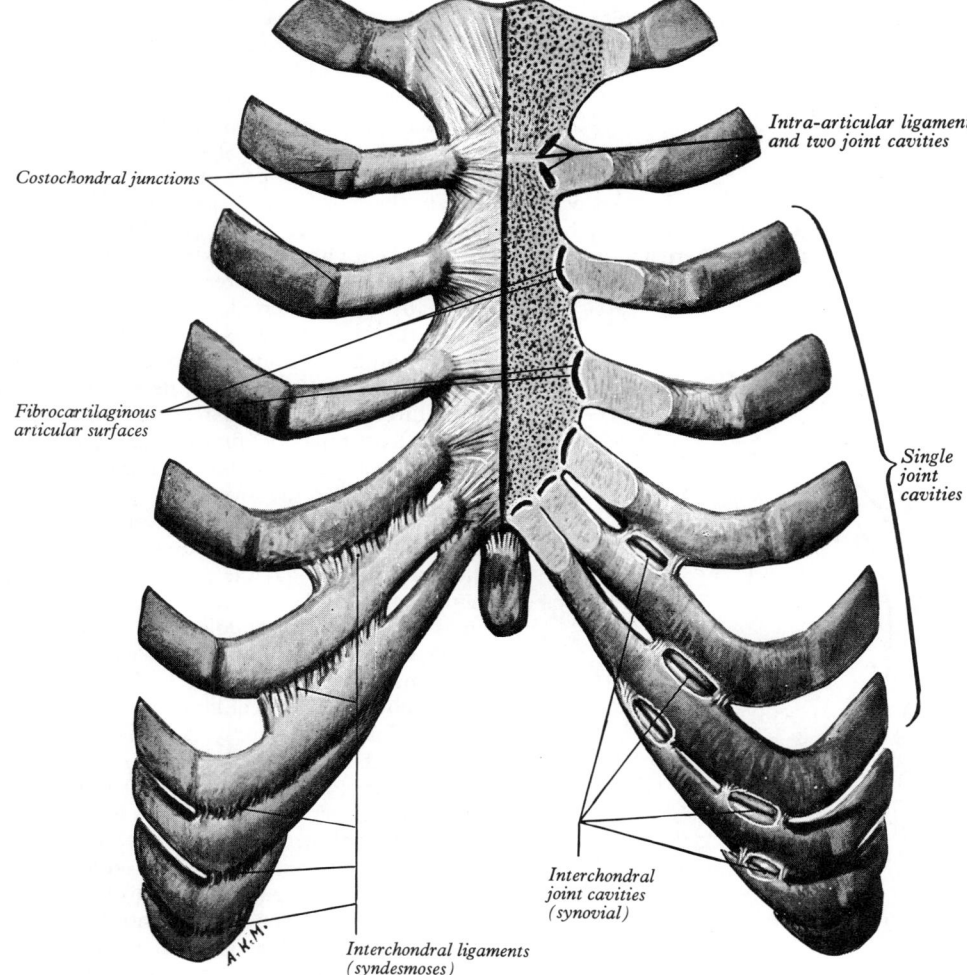

Costochondral junctions

Fibrocartilaginous articular surfaces

Intra-articular ligament and two joint cavities

Single joint cavities

Interchondral joint cavities (synovial)

Interchondral ligaments (syndesmoses)

34 Sternocostal and interchondral joints: anterior aspect.

strengthened above and below by fibres connecting cartilages to sternum.

Radiate sternocostal ligaments are broad, thin bands radiating from the front and back of sternal ends of cartilages of true ribs to corresponding sternal surfaces. Their superficial fibres intermingle with adjacent ligaments above and below, with those of the opposite side and with tendinous fibres of pectoralis major, forming a thick fibrous membrane around the sternum, more markedly in its lower part.

Intra-articular ligaments are constant only between second costal cartilages and sternum. The second costal cartilage's ligament extends from the costal cartilage to the fibrocartilage uniting manubrium and sternal body and is therefore intra-articular. Occasionally the third sternal cartilage is connected with the first and second sternal segments by a similar ligament. Fibrocartilaginous strands may occur in the third and lower joints. Articular cavities may be absent at all ages.

Costoxiphoid ligaments connect anterior and posterior surfaces of the seventh costal cartilage (and sometimes sixth) to the same surfaces of the xiphoid process. They vary in length and breadth, the posterior being less distinct.

Movements. Slight gliding movements occur at sternocostal joints, sufficient for respiration (p. 594).

Interchondral Joints

Contiguous borders of the sixth to ninth costal cartilages articulate by apposition of small oblong facets, each articulation being enclosed in a thin *fibrous capsule*, lined by synovial membrane with lateral and medial *interchondral ligaments* (4.34). Sometimes the fifth cartilage, more rarely the ninth, articulate at their inferior borders with adjoining cartilages, this connection being more often by ligamentous fibres. Articulation between the ninth and tenth cartilages is never synovial and sometimes absent (p. 467). For further comments on chondral arthroses see p. 335.

Costochondral Junctions

The costal cartilages are persistent, unossified anterior parts of cartilaginous models preceding fully developed ribs. Artificially separated from its rib a costal cartilage has a rounded end, the rib a depression. Across junctions periosteum and perichondrium are continuous and the collagen of the osseous and cartilaginous matrices blend. No movement occurs there.

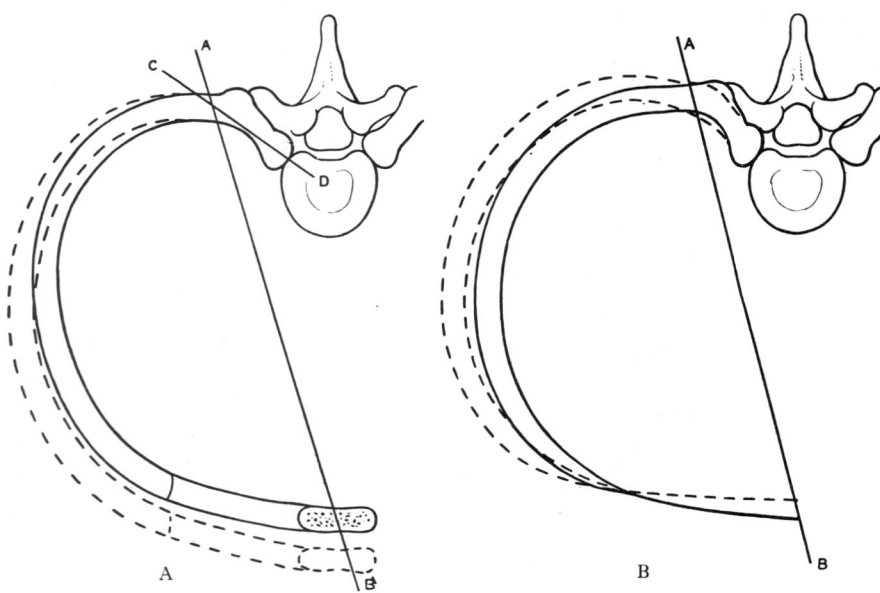

4.35 A. Diagram showing the axes of movement (AB and CD) of a vertebrosternal rib. The interrupted lines indicate the position of the rib in inspiration. B. Diagram showing the axis of movement (AB) of a vertebrochondral rib. The interrupted lines indicate the position of the rib in inspiration.

498

Sternal Joints

The manubriosternal joint between the manubrium and sternal body is usually a symphysis, the bony surfaces covered by hyaline cartilage and connected by a fibrocartilage which may ossify in the aged. In more than 30% the central part of the disc is absorbed and the joint *appears* synovial; similar cavitation may occur in the symphysis pubis (see p. 467, 517). The two are also connected by a fibrous membrane enveloping the whole bone. In 10% of all over 30 years the manubrium is joined to the sternal body by bone but the intervening cartilage may be only superficially ossified; it is in the aged that this is complete. Early synostosis has been attributed to a persistent synchondrosis in place of a symphysis (Ashley 1954). In the newborn union is by collagenous and elastic fibres without chondrocytes.

Movements. The symphysis permits a small range of angulation between longitudinal axes of manubrium and corpus sterni and also limited anteroposterior displacement. A study of these movements in 62 male athletes yielded (standing position) mean values of 162.7° (full inspiration) and 164.7° (full expiration) for the manubriosternal angle (Constantinescu 1974). Both movements contribute to respiratory excursions of the sternum.

The xiphisternal joint between xiphoid process and corpus sterni is also a symphysis, usually transformed to a synostosis by the fortieth year but it sometimes remains unchanged even in old age.

Mechanism of the Thorax

Each rib has its range and direction of movement contributing to thoracic respiratory excursions. Each acts as a lever, its fulcrum immediately lateral to its costotransverse articulation; hence, when the shaft is elevated, the neck is depressed and vice versa; since the lever's arms differ much in length, slight movement at the vertebral end is much magnified at the anterior end.

Anterior costal ends are lower than posterior; therefore when shafts are raised they move forwards. The mid-shaft is below the ends so that when the shaft is raised it also spreads laterally. Further, each rib is part of a curve greater than that of the rib above; therefore costal elevation increases transverse thoracic diameter at higher levels. (Modifications of rib movements at vertebral ends are described on p. 497.) Modifications also result from attachments of anterior costal ends with different movements of vertebrosternal, vertebrochondral and vertebral ribs.

Vertebrosternal ribs (4.35A). The first moves little, except in deep respiration about an oblique axis through the neck; the shaft rises laterally in inspiration, its inferomedial surface becoming more directly inferior, an impossible movement if the cartilage is calcified, as it may be; first ribs and manubrium then move as a unit about a transverse axis through their costotransverse joints. The second rib also moves little in quiet respiration; elevation of third to sixth ribs thrusts their anterior ends up and forwards, mostly by backward rotation at their necks; this also moves the sternum similarly at the manubriosternal joint, increasing the anteroposterior thoracic diameter. As this action ceases, an elevating force raises the middle parts of their shafts, everting their lower borders, opening the costochondral angle and increasing the transverse thoracic diameter. Measurements of sternal respiratory movements reveal large excursions, especially in fit adult males (Constantinescu 1974). Between full inspiration and expiration the suprasternal notch may move 31 mm, excursions at the superior (34 mm) and inferior (37 mm) ends of the corpus sterni being increased by changes in the sternal angle.

Vertebrochondral ribs (4.35B), including the seventh, assist thoracic respiratory enlargement and they also increase upper abdominal space for the viscera displaced by diaphragmatic descent, although abdominal relaxation accounts in part for this. Costal cartilages, through their joints, push each other up, the final thrust forcing the lower sternum up and forwards. Elevation of their anterior ends is limited by the very slight rotation possible at their necks. Elevation of shafts is accompanied by outward and backward movement, the former everting their anterior ends and opening the infrasternal angle, while backward movement retracts anterior ends and counteracts the forward thrust of elevation, most noticeably in the lower ribs which are shortest. The result is an increase in transverse and diminution in median anteroposterior diameters of the upper abdomen, while its lateral anteroposterior diameters are increased.

Vertebral ribs have free anterior ends and costovertebral joints without intra-articular ligaments and may move a little in all directions. As other ribs rise they are depressed and fixed by the quadrati lumborum to aid diaphragmatic action of the diaphragm. The muscles which produce these movements are discussed on p. 594.

JOINTS OF THE UPPER LIMB

The Sternoclavicular Joint

Involved in the sternoclavicular joint are the sternal end of the clavicle and the sternal clavicular notch, together with the adjacent superior surface of the first costal cartilage (4.36). The clavicular *articular surface*, much the larger, is covered by *fibrocartilage*, thicker than the fibrocartilaginous lamina on the sternum. Convex vertically but anteroposteriorly slightly concave, it is therefore sellar; the sternal clavicular notch is reciprocally curved but the two surfaces are not fully congruent. (For their variations see p. 405.) The joint is divided by an articular disc. Its ligaments are capsular, anterior and posterior sternoclavicular, interclavicular and costoclavicular.

The fibrous capsule is thickened in front and behind; but above and especially below it is little more than loose areolar tissue.

The anterior sternoclavicular ligament is broad, attached above to the anterosuperior aspect of the clavicle's sternal end, and passes inferomedially to the upper anterior aspect of the manubrium, spreading to the first costal cartilage.

The posterior sternoclavicular ligament, a weaker band posterior to the joint, descends inferomedially from the back of the clavicle's sternal end to the back of the upper manubrium.

The interclavicular ligament, continuous above with deep cervical fascia, unites the superior aspects of the sternal ends of both clavicles; some fibres are attached to the superior manubrial margin. When present, *suprasternal ossicles* (p. 332) are in this ligament.

The costoclavicular ligament is like an inverted cone but short and flattened, with anterior and posterior laminae attached to the upper surface of the first rib and costal cartilage, ascending to the margins of an impression on the inferior clavicular surface at its medial end. Fibres of the anterior lamina ascend laterally and shorter fibres of the posterior lamina ascend medially (4.36); they fuse laterally and merge medially with the capsule. Between them a bursa suggests separate functions (Cave 1961). Probably each is tensed at opposite extremes of clavicular axial rotation.

The articular disc, flat and almost circular, between the sternal and clavicular surfaces, is attached above to the superoposterior border of the clavicular articular surface, below to the first costal cartilage near its sternal junction and by the rest of its circumference to the capsule. Thicker peripherally, especially in its superoposterior part, it divides the joint into a larger superolateral part and a smaller inferomedial one. The capsule around the former is more lax, movements between clavicles and discs being more extensive than those between discs and sternum. Movement is considered below (p. 500). Sellar shape of the surfaces permits *angulation* in approximately anteroposterior and vertical planes and some rotation (about the clavicular long axis) or *spin* (30° according to Kapandji 1970), which is conjunct. Close-packing probably coincides with maximum posterior spin associated with full scapular rotation (vide infra). Some anteroposterior translation also occurs.

Arteries to the joint are branches from the internal thoracic and suprascapular arteries; *nerves* are from the anterior supraclavicular nerve and the nerve to the subclavius. For details of vascularization consult Sick & Ring (1976).

Applied Anatomy. The joint's strength depends on ligaments, especially its disc. This and the usual transmission of forces *along* the clavicle make dislocation rare and fracture far more common.

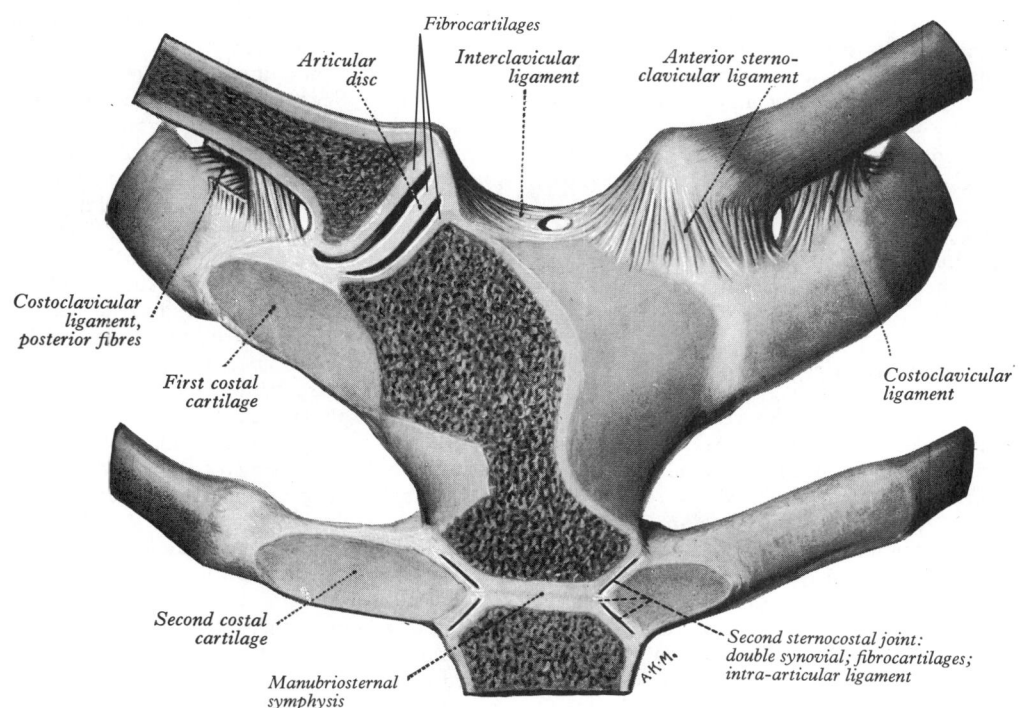

Fibrocartilages

Articular disc — *Interclavicular ligament* — *Anterior sterno-clavicular ligament*

Costoclavicular ligament, posterior fibres

First costal cartilage

Costoclavicular ligament

Second costal cartilage

Second sternocostal joint: double synovial; fibrocartilages; intra-articular ligament

Manubriosternal symphysis

4.36 Sternoclavicular joints: anterior aspect; left joint intact and right in coronal section.

The Acromioclavicular Joint

This articulation (4.40,42), between the clavicle's acromial end and the medial acromial margin, is approximately plane; but either surface may be slightly convex, the other reciprocally concave. *Both* are covered by *fibrocartilage*, the clavicular being a narrow, oval area facing inferolaterally and overlapping a corresponding facet on the medial acromial border. The long axis is anteroposterior. Ligaments are capsular, acromioclavicular and coracoclavicular.

The fibrous capsule completely surrounds the articular margins and is strengthened above by the acromioclavicular ligament.

The acromioclavicular ligament, quadrilateral and above the joint, extends between upper aspects of the clavicle's acromial end and the adjoining acromion. Its parallel fibres interlace with the trapezial and deltoid aponeuroses.

An articular disc often occurs in the joint's upper part, partially separating articular surfaces (de Palma 1957). More rarely it completely divides the joint.

Movements at this joint (vide infra) are like those of the sternoclavicular. Axial rotation of the clavicle is said to be about 30° (Kapandji 1970). The two joints together permit about 60° of scapular rotation. Angulation with the scapula occurs in any direction. Spin and angulation both tense the coracoclavicular ligament at their extremes; spin tightens the capsule by spiralization. According to MacConaill & Basmajian (1977) close-packing occurs when the angle between the superior scapular border and clavicular shaft reaches about 90°. This 'opening' (*ouverture*) of the angle tenses the conoid part of the coracoclavicular ligament (Kapandji 1970); further scapular rotation is due to rotation at the sternoclavicular joint.

The coracoclavicular ligament (4.37) interconnects clavicle and scapular coracoid process. Though separate from the acromioclavicular joint it is a most efficient accessory ligament, maintaining apposition of clavicle to acromion. Its *trapezoid* and *conoid parts*, usually separated by fat or, frequently, a bursa, connect the medial horizontal part of the coracoid process and lateral end of the subclavian groove of the clavicle; these adjacent areas may even be covered by cartilage to form a *coracoclavicular joint* (Lewis 1959).

The *trapezoid part*, the anterolateral fasciculus, is broad, thin and quadrilateral, ascending slightly from the upper coracoid surface to the trapezoid line on the inferior clavicular surface. It is almost *horizontal*, its anterior border free, its posterior joined to the conoid part, forming an angle projecting backwards.

The *conoid part*, the posteromedial fasciculus, a dense almost *vertical* triangular band, has its base attached to the clavicle's conoid tubercle and its inferior apex attached posteromedially to the conoid tubercle at the root of the coracoid process in front of the scapular notch.

Arterial supply is from suprascapular and thoraco-acromial arteries, *nerve supply* from suprascapular and lateral pectoral nerves.

Applied Anatomy. In acromioclavicular dislocation the coracoclavicular ligament is torn and the scapula falls away from the clavicle. Owing to the flatness and orientation of the joint surfaces, dislocation readily recurs.

MOVEMENTS OF THE SHOULDER GIRDLE

Clavicular movements at their sternoclavicular and acromioclavicular joints are always associated with those of the scapula, which, in turn, are usually accompanied by humeral movements. Therefore this account should be amplified by reference to pp. 502 and 503. The acromioclavicular joint allows the acromion, and hence the scapula, anteroposterior gliding and rotation on the clavicle; but scapular range is much increased by movements at the sternoclavicular joint. Analysis is clarified by considering *scapular movements*, primarily analysed as (1) elevation and depression, (2) protraction and retraction round the thorax, (3) rotation forwards or upwards, with its inferior angle as reference point, and its reverse.

1. *Scapular elevation and depression*, picturesquely exemplified by 'shoulder' shrugging, do not necessarily imply movement at

the shoulder joint. In elevation slight angulation or swing occurs at the acromioclavicular joint but the clavicle's sternal end, rotating on an anteroposterior axis through the bone above the medial attachment of the costoclavicular ligament, slides down over the articular disc (*translation*); this is checked by antagonist muscles and tension in the costoclavicular ligament and lower capsule. It is effected by trapezius (upper part) and levator scapulae; since these tend to rotate the scapula in opposite directions, pure elevation can occur.

In the reverse movement slight angulation occurs at the acromioclavicular joint, but at the sternoclavicular joint the clavicle slides up on the disc, checked by antagonist muscles, the interclavicular and sternoclavicular ligaments and articular disc. This role of the interclavicular ligament, suspension of the depressed clavicle, has been confirmed by experiment (Bearn 1967). Usually gravity alone is sufficient, but serratus anterior (lower part) and pectoralis minor are, when necessary, active effectors.

2. *Forward movement (protraction)* round the thoracic wall occurs in pushing, thrusting and reaching movements, usually with some forward rotation. The acromion advances over the clavicular facet to the limit and the shoulder is simultaneously advanced by forward movement of the lateral end of the clavicle and posterior translation of its sternal end over the sternal facet, carrying the disc with it. Antagonist muscles and the anterior sternoclavicular and costoclavicular (posterior lamina) ligaments check backward slide of the sternal end. Serratus anterior and pectoralis minor are prime movers and maintain continuous apposition of the scapula, especially its medial border, in smooth gliding on the thoracic wall. The upper part of latissimus dorsi also acts like a strap across the inferior scapular angle in protraction and forward rotation.

In *scapular retraction*, bracing back the shoulders, movements are reversed and checked at the sternoclavicular joints by posterior sternoclavicular and costoclavicular (anterior lamina) ligaments. Trapezius and the rhomboids are prime movers, but gravity may also produce retraction when the weight of the trunk is taken by the arms in leaning forwards, to a degree controlled by protractive musculature.

When force is applied at the end of an outstretched arm, e.g. in a fall on the hand, pressure transmitted to the glenoid cavity tends to drive the sloping acromial facet below the clavicle's acromial end but also tenses the trapezoid ligament, which resists the displacement; and, unless the fall is unexpected, even greater forces are available to resist dislocation.

3. *Forward rotation of the scapula* increases the range of *humeral elevation* by turning its glenoid cavity to face almost directly up (as in raising an arm above the head), a movement always associated with some humeral elevation and with protraction of the scapula. Scapular rotation entails movements at the sternoclavicular and the acromioclavicular joints: the sternoclavicular permits elevation of the clavicle's lateral end, a movement almost complete when the arm is abducted to 90°. The acromioclavicular joint moves in the first 30° of abduction when the conoid ligament becomes taut and thereafter is accompanied by clavicular rotation at the sternoclavicular joint around the bone's longitudinal axis, also with further depression of its medial end as the lateral end continues to rise. Some acromioclavicular movement also occurs in the final stages of humeral abduction (Inman et al 1944). Trapezius (upper part) and the serratus anterior (lower part) are prime movers.

Return rotation is usually by gravity, gradual active lengthening of trapezius and serratus anterior being sufficient to control it. When more force is needed the levator scapulae, rhomboids and in the initial stages pectoralis minor are prime movers in returning the scapula to a position of rest.

Muscles which are antagonists in one movement may combine as prime movers in another. Movements, not muscles, are represented in cerebral motor areas; muscles are not grouped unalterably in nervous control but can be variably combined as demands dictate. Thus serratus anterior and trapezius are opposed in scapular movement round the thorax but combine as prime movers in its forward rotation.

In all scapular movements subclavius probably steadies the clavicle by drawing it medially and downwards, although it inaccessibility makes its role uncertain. Scapular movements on the thoracic wall are facilitated by areolar tissue between subscapularis and serratus anterior and the costal wall. With the arm pendant the shoulder girdle's normal posture relative to the trunk involves moderate activity in trapezius and serratus anterior (Basmajian 1967), which obviously must increase when the limb is loaded.

LIGAMENTS OF THE SCAPULA

The main scapular ligaments (4.37) are the coraco-acromial and superior transverse scapular; there may also be a weaker, variable inferior transverse (spinoglenoid) ligament.

The coraco-acromial ligament, a strong triangular band between coracoid process and acromion, is attached apically to the acromion anterior to its clavicular articular surface and by its base to the whole lateral coracoid border. With coracoid process and acromion it completes an *arch* above the humeral head. It may have two strong marginal bands with a thinner centre; when occasionally the pectoralis minor is inserted into the humeral capsule instead of the coracoid process, its tendon passes between the bands (p. 611). The subacromial bursa (p. 502 and 4.39) facilitates movement between the coraco-acromial arch and the subjacent supraspinatus muscle and shoulder joint, functioning as a secondary synovial articulation.

The superior transverse scapular (suprascapular) ligament makes the scapular notch into a foramen and is sometimes ossified. A flat fasciculus, it narrows towards its ends, attached to the base of the coracoid process and medial side of the scapular notch. The suprascapular nerve traverses the foramen; suprascapular vessels cross above the ligament.

A membranous *inferior transverse (spinoglenoid) ligament* may stretch from the scapular spine's lateral border to the glenoid margin forming an arch over the suprascapular nerve and vessels entering the infraspinous fossa; it is often absent.

The Humeral (Shoulder) Joint

This is a multi-axial spheroidal joint (3.181,4.37,38A,B,C,41,42) possessing three degrees of freedom between the roughly hemispherical humeral head and shallow scapular glenoid cavity, a construction allowing much movement but reducing security. Skeletally the joint is weak, and depends for support on surrounding muscles more than on its shape and ligaments. However, the coraco-acromial arch overhangs it (vide supra).

The *articular surfaces*, reciprocally curved, are really *ovoids* (see p. 480). Here, as in the hip, where ovoid surfaces are almost spherical they are often termed *spheroidal*. The humeral convexity much exceeds in area that of the glenoid concavity (4.38A,B) and only a minor part opposes the glenoid in any position, the remaining capitular articular surface being in contact with the capsule. The glenoid cavity (4.38A,C,39,40) is deepened by a fibrocartilaginous rim, the *glenoidal labrum*. Both articular surfaces are covered by hyaline cartilage: on the humerus it is centrally thickest, thinner peripherally; the reverse in the glenoid cavity. In most positions, their curvatures are not fully congruent, the joint being loose-packed (Saha 1961). Full congruence (close-packing) is reached with the humerus abducted and laterally rotated (see pp. 502, 503; and 4.5,17). When the arm is dependent, the anterior glenoid edge can be represented by a laterally concave line descending 3 cm from a point just lateral to the coracoid apex; it lies over the joint's lower half. The ligaments are the *glenoidal labrum, fibrous capsule, glenohumeral, coracohumeral* and *transverse humeral*.

The fibrous capsule (4.37,38) envelops the joint, attached medially to the glenoid margin outside the glenoidal labrum, and encroaching on the coracoid process to include the attachment of the long head of biceps. Laterally, it is attached to the humeral anatomical neck, i.e. near the articular margin, except inferomedially where it descends more than 1 cm on the humeral shaft. It is so lax that the bones can be distracted for 2 or 3 cm. This accords with a very wide range of movement. However, such unnatural separation requires relaxation of the upper capsule by abduction.

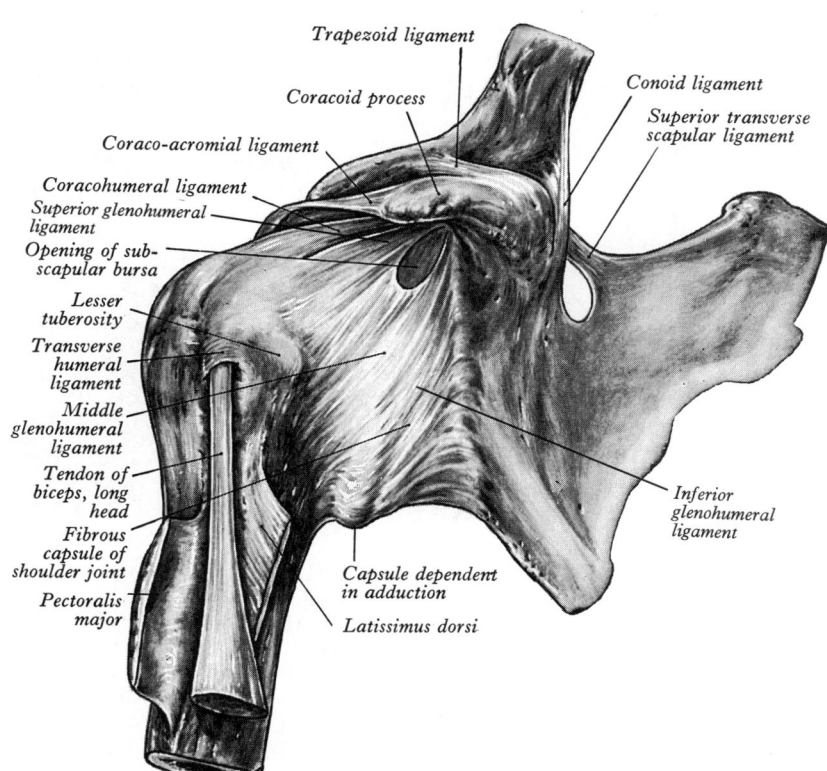

4.37 The right shoulder joint: anterior aspect.

The fibrous capsule is supported by the tendons of supraspinatus (above), infraspinatus and teres minor (behind), subscapularis (in front) and by the long head of triceps (below). All but triceps blend with the capsule as the *rotator cuff* which reinforces the capsule and actively supports it unless the muscles are fully relaxed. The triceps is separated from the capsule by the axillary nerve and posterior circumflex humeral vessels as they pass back from the axilla (4.41). Inferiorly the capsule is hence least supported and also subjected to the greatest strain, being stretched tightly across the humeral head in full abduction.

The capsule usually has two or three openings: an anterior, below the coracoid process, connects the joint to a bursa behind the subscapular tendon; another, between the humeral tubercles (tuberosities), transmits the long bicipital tendon and its synovial sheath; a third, inconstant and posterior, connects the joint to a bursa under the infraspinatous tendon.

Three *glenohumeral ligaments*, best visible from within the joint (4.40), reinforce the capsule. At their scapular ends all are attached to the superomedial glenoid margin, blended with the glenoidal labrum. The *superior* passes along the medial edge of the bicipital tendon to attachment above the lesser humeral tubercle; the *middle* reaches the lower part of this tubercle; the *inferior* extends to the lower part of the humeral anatomical neck. The capsule is also strengthened in front by extensions from the tendons of pectoralis major and teres major.

Synovial membrane lines the capsule and covers parts of the anatomical neck. The long bicipital tendon traverses the joint in a synovial sheath which continues into the intertubercular sulcus as far as the humeral surgical neck (4.37,38).

The coracohumeral ligament (4.37), a broad thickening of the upper capsular region, descends laterally from the lateral border of the coracoid root to the front of the greater tubercle, blending with the supraspinatous tendon; its inferoposterior border blends with the capsule, the anterosuperior does not.

The transverse humeral ligament (4.37) is a broad band passing between the humeral tubercles, converting the intertubercular sulcus into a canal; its attachment is superior to the epiphysial line. It is simply a retinaculum for the long tendon of biceps.

The glenoidal labrum (4.39), a fibrocartilaginous rim round the glenoid cavity, is triangular in section, its base attached to the cavity's margin, its thin margin projecting as a continuation of the

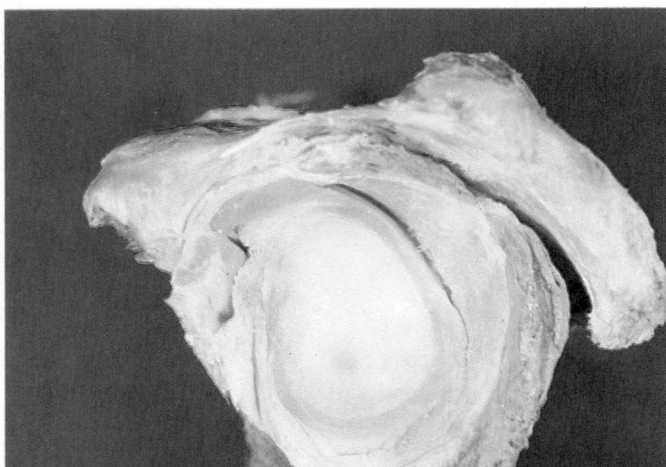

A

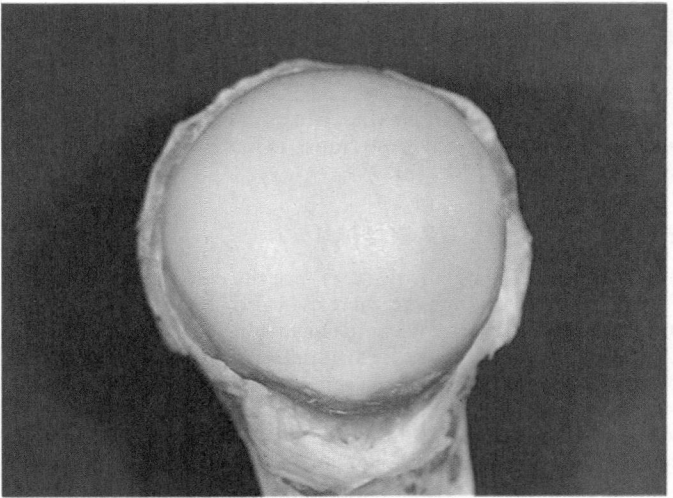

B

4.38 A,B Dissections from a preserved cadaver showing the articular surfaces and periarticular structures of the glenohumeral joint. A shows the glenoid cavity covered by articular cartilage and the glenoidal labrum; note the intracapsular part of the tendon of the long head of the biceps sectioned near the supraglenoid tubercle, the fibrous capsule intimately blended with 'rotator cuff' musculature, the subacromial bursa, and the

superjacent coraco-acromial arch complex. B displays the much more extensive spheroidal articular surface of the head of the humerus and the pendant inferior part of the fibrous articular capsule. (Preparation by M C E Hutchinson and photography by Kevin Fitzpatrick, Guy's Hospital Medical School, London.)

curve of the glenoid. It blends above with two fasciculi from the long tendon of biceps. It deepens the cavity, may protect bone and probably assists lubrication (p. 471). Its attachment is sometimes partly deficient; synovial membrane may protrude through such gaps.

Many bursae adjoin the shoulder joint: (1) between the subscapular tendon and articular capsule (4.40), communicating with the joint between the superior and middle glenohumeral ligaments; (2) sometimes between the infraspinatous tendon and capsule, occasionally opening into the joint; (3) the *subacromial bursa* (4.39) between deltoid muscle and capsule, non-communicating but prolonged under the acromion and coraco-acromial ligament, and between them and supraspinatus; (4) on the superior acromial aspect; (5) frequently between the coracoid process and capsule; (6) sometimes behind coracobrachialis; (7) between teres major and the long head of triceps; (8) and (9) anterior and posterior to the tendon of latissimus dorsi.

Related *muscles* are: supraspinatus (above), long head of triceps (below), subscapularis (in front), infraspinatus and teres minor (behind), long bicipital tendon (intracapsular). Deltoid covers the joint in front, behind and laterally (4.39).

Arteries are from the anterior and posterior circumflex humeral and suprascapular vessels. *Nerves* are mainly from the posterior brachial cord and from suprascapular, axillary and the lateral pectoral nerves. The suprascapular supplies the posterior and superior, axillary antero-inferior and the lateral pectoral antero-superior parts of the capsule (Gardner 1948B).

Movements. The shoulder, as a multi-axial spheroidal joint, is capable of any combination of swing and spin over a very wide range (p. 482), all movements analysable as rotations around three orthogonal axes, i.e. it has three degrees of freedom (p. 476). Classically, flexion–extension, abduction–adduction, circumduction and medial and lateral rotation (p. 476) are assigned to it. Laxity of the capsule, and a humeral head which is large relative to the shallow glenoid cavity, afford a wider range of movement than at any other joint. However, with the arm dependent, even when moderately loaded, the supraspinatus and tension in the upper capsule prevent downward displacement of the humerus (Basmajian 1962).

In analysis of shoulder movements it is preferable to refer humeral movement to the scapula, rather than to conventional anatomical planes, relevant axes being shown in 4.5A,B,C. When the arm hangs at rest the glenoid cavity faces almost equally forwards and laterally, and humeral capitular and scapular (topographical) axes correspond, although the humerus, relative to the anatomical position, is medially rotated (p. 477). *Flexion* carries the arm anteromedially on an axis through the humeral head orthogonal to the glenoid cavity at its centre. *Abduction and adduction* occur in a vertical plane orthogonal to that of flexion–extension, the axis being horizontal, through the humeral head, parallel to the glenoid plane (Flecker 1929). Pure abduction raises the arm *anterolaterally in the plane of the scapula*. However, when referred to the *trunk*, flexion and extension are in the paramedian plane, abduction and adduction in the coronal plane. Raising the arm vertically from flexion (in this sense) or raising it from abduction (in this sense) are both accompanied by humeral rotation in opposite directions. Whether 'scapular' or any other plane of abduction is described, these are selections from an infinite series. In scapular abduction points on the humeral surface pursue vertical chords but in rotation they are horizontal. In 'pure' flexion–extension, in a plane orthogonal to the scapular, the axis of movement (and the notional 'mechanical axis', p. 480) are regarded as projected from the centre of the glenoid cavity.

Glenohumeral abduction is about 90° (Kapandji 1970) but angles up to 120° have been stated (Inman et al 1944). About 60°

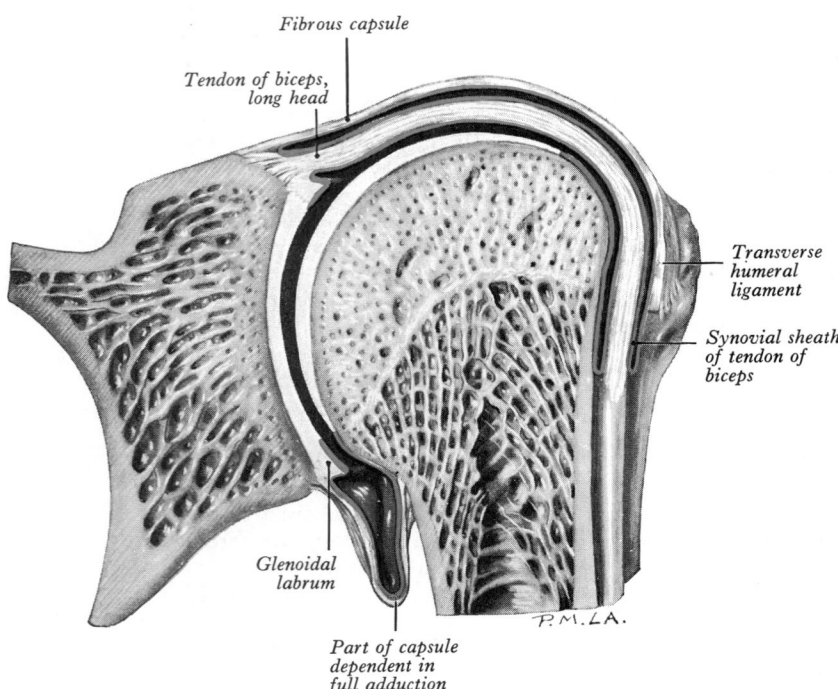

Fibrous capsule

Tendon of biceps, long head

Transverse humeral ligament

Synovial sheath of tendon of biceps

Glenoidal labrum

P.M.LA.

Part of capsule dependent in full adduction

4.38C Section through the shoulder joint. The synovial membrane is in blue.

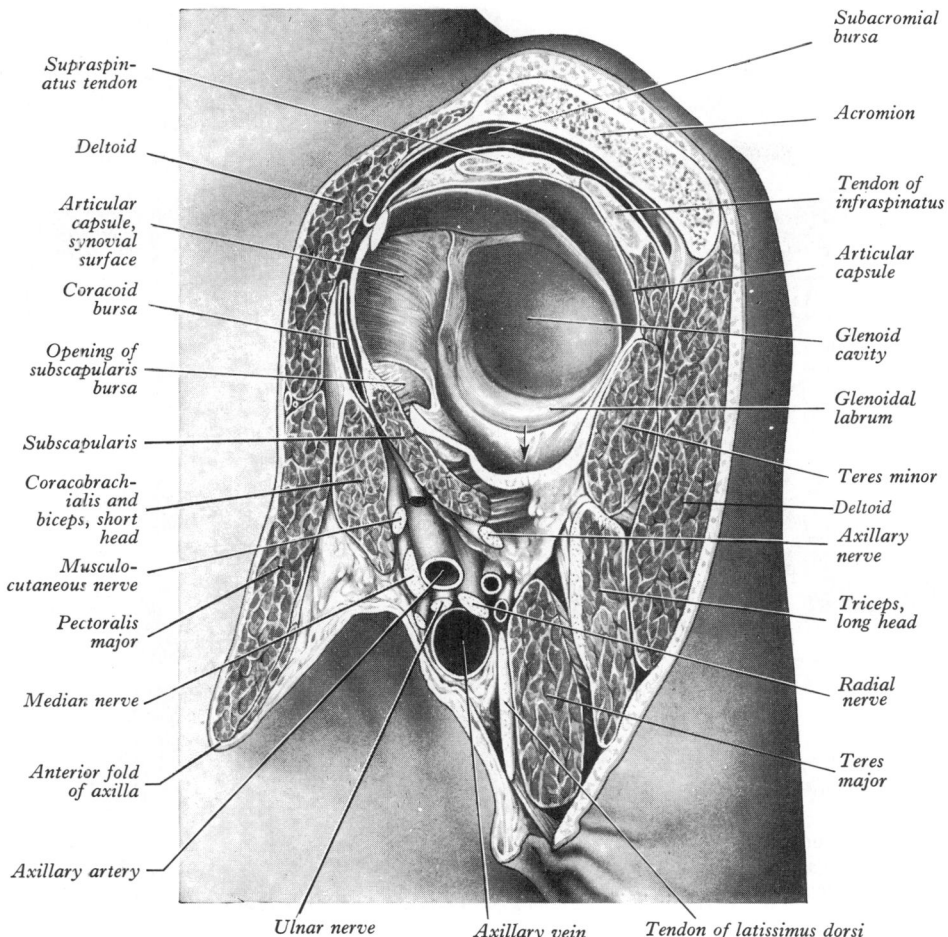

Supraspinatus tendon
Deltoid
Articular capsule, synovial surface
Coracoid bursa
Opening of subscapularis bursa
Subscapularis
Coracobrachialis and biceps, short head
Musculocutaneous nerve
Pectoralis major
Median nerve
Anterior fold of axilla
Axillary artery
Ulnar nerve
Axillary vein
Subacromial bursa
Acromion
Tendon of infraspinatus
Articular capsule
Glenoid cavity
Glenoidal labrum
Teres minor
Deltoid
Axillary nerve
Triceps, long head
Radial nerve
Teres major
Tendon of latissimus dorsi

4.39 An obliquely coronal section through the left shoulder and shoulder joint, in the plane of the glenoidal labrum, dissected after removal of the upper limb. The arrow points into the dependent part of the articular capsule. Note: the relations of the axillary vessels to each other and to the branches of the brachial plexus as displayed in this dissected section may appear unfamiliar and the reader is referred to **6.92** for a more usual view. Compare, also, with **4.40**.

further abduction occurs at the sterno- and acromioclavicular articulations. Contralateral vertebral flexion also aids in bringing the arm vertical. However, during active elevation movements at the glenohumeral and scapuloclavicular joints are simultaneous, except in the initial 25°–30°, when most and often all movement is glenohumeral. For every 15° of elevation, glenohumeral is said to be 10° and scapular 5°.

In flexion, however, the humerus swings at right angles to the scapular plane and scapular rotation cannot increase elevation (120°) obtainable in full flexion. If the fully flexed humerus is also abducted, elevation increases pro rata until, when the humerus reaches the scapular plane, i.e. when true abduction is reached, 180° of elevation becomes possible. In *rotation*, medial or lateral, the humerus revolves about one-quarter of a circle around a vertical axis; range is greatest when the arm is pendent, least when it is vertical. When assessing rotational range at the glenohumeral joint, the forearm should be flexed to a right angle at the elbow, preventing misinterpretation due to superadded pronation or supination in the pendent limb. In *circumduction*, a succession of the foregoing movements, the distal end of the humerus describes the base of a cone, its apex at the humeral head, but this glenohumeral movement can be much increased by scapular movements; the combination is exemplified in acts of slinging objects with force.

Because of the offset position of the humeral head relative to the shaft's longitudinal topographical axis, flexion and extension involve almost pure 'spins' in the glenoid cavity (p. 480). Other movements are an infinite variety of cardinal or arcuate swings or successions of these (pp. 478, 480).

The peculiar relation of the long bicipital tendon to the shoulder joint may serve several purposes. By its connection with both shoulder and elbow the muscle harmonizes their actions as an elastic ligament during all their movements. It helps to prevent

the humeral head impinging on the acromion when the deltoid contracts (vide infra) and to steady it in movements of the arm.

A wide range of *accessory movements* (p. 484) occurs at the shoulder joint. The humeral head can be translated in any direction relative to the glenoid cavity and, in abduction when accessory movements are most free, the articular surfaces can be separated by traction.

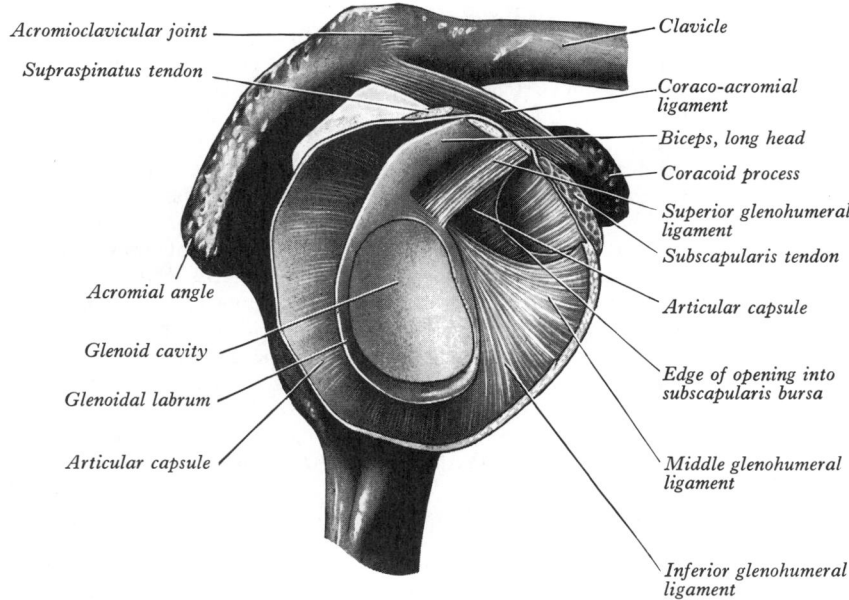

Acromioclavicular joint
Supraspinatus tendon
Acromial angle
Glenoid cavity
Glenoidal labrum
Articular capsule
Clavicle
Coraco-acromial ligament
Biceps, long head
Coracoid process
Superior glenohumeral ligament
Subscapularis tendon
Articular capsule
Edge of opening into subscapularis bursa
Middle glenohumeral ligament
Inferior glenohumeral ligament

4.40 Interior of the right shoulder joint: anterolateral aspect.

503

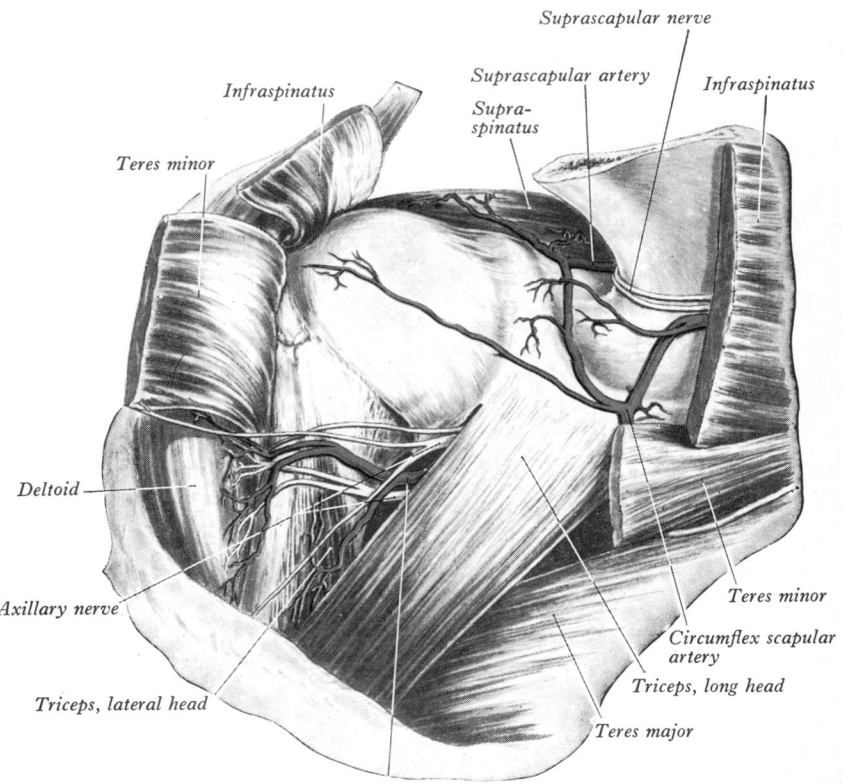

Infraspinatus

Teres minor

Suprascapular nerve

Suprascapular artery

Supra-
spinatus

Infraspinatus

Deltoid

Axillary nerve

Triceps, lateral head

Triceps, long head

Teres minor

Circumflex scapular
artery

Teres major

Posterior circumflex humeral artery

4.41 Posterior aspect of the left shoulder joint. An oblique section has been made through the spine of the scapula and the acromion has been removed. Parts of the infraspinatus and teres minor have been excised and their tendons have been turned forwards.

Muscles producing the movements may be divided into (1) those acting on the pectoral girdle and (2) those acting on the glenohumeral joint.

(1) **Muscles acting on the pectoral girdle** have been considered on p. 500.

(2) **Muscles acting at the glenohumeral joint** are principally deltoid, pectoralis major, latissimus dorsi and teres major. All converge on the humerus, acting at mechanical advantage on a joint which, owing to glenoid shallowness and capsular laxity, is relatively unstable, a condition counteracted by the short muscles attached nearer to it, viz. subscapularis, supraspinatus, infraspinatus and teres minor (the 'rotator cuff'), which function as postural muscles retaining the humeral head and glenoid in correct alignment and resisting skid during active movements.

Flexion: pectoralis major (clavicular part, p. 611), deltoid (anterior fibres) and coracobrachialis assisted by biceps. The sternocostal part of pectoralis major is a major force in flexion forwards to the coronal plane *from the full extension*.

Extension: deltoid (posterior fibres) and teres major, from the pendent position. When the fully flexed arm is extended against resistance, latissimus dorsi and the sternocostal part of pectoralis major act powerfully until the arm reaches the coronal plane.

Abduction: deltoid, but initially its effect is mainly upward and, unless opposed, would so displace the humerus. Subscapularis, infraspinatus and teres minor are the opposing force exerting downward traction; these three and deltoid constitute a 'couple' to produce abduction in the scapular plane. Supraspinatus assists in effecting and maintaining this movement but its precise role is controversial.

Medial rotation: pectoralis major, deltoid (anterior fibres), latissimus dorsi, teres major and, with the arm pendent, subscapularis.

Lateral rotation: infraspinatus, deltoid (posterior fibres) and teres minor.

Applied Anatomy. Owing to the joint's shallowness it is the most frequently dislocated, usually with the arm abducted, the humeral head pressing against the antero-inferior aspect of the

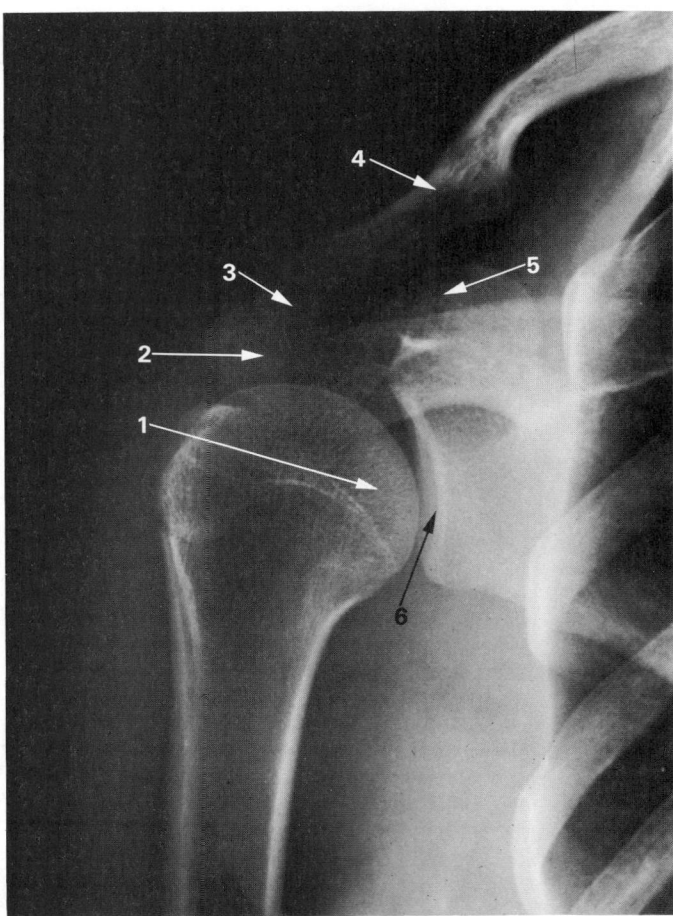

A

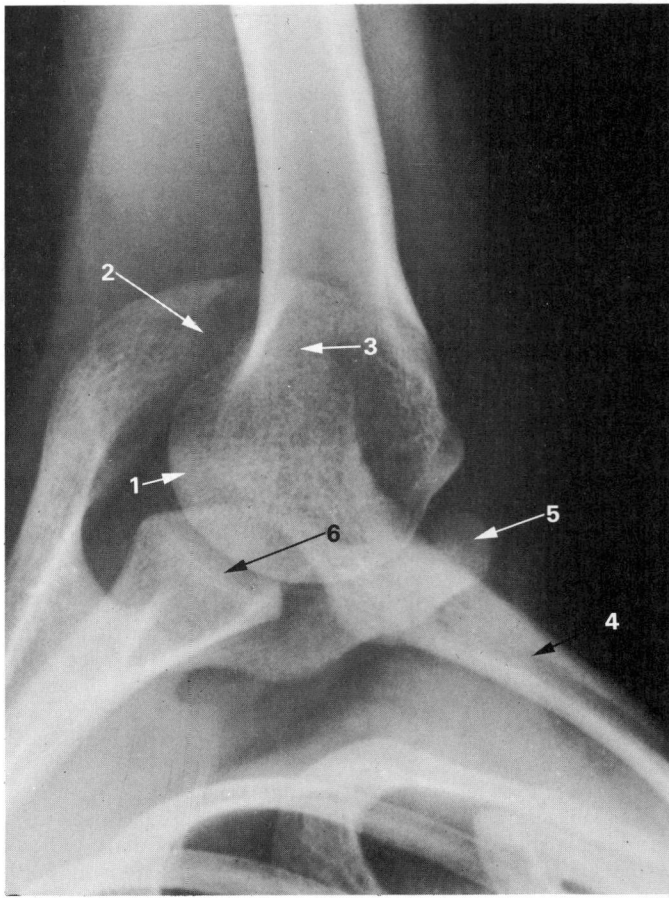

B

4.42 Radiograph of shoulder in a young female of 18 years in anteroposterior view (A) and axillary view with the arm abducted (B). 1. Head of humerus. 2. Acromion. 3. Acromioclavicular joint. 4. Clavicle. 5. Coracoid process. 6. Glenoid (osseous, subchondral) articular surface.

capsule, where it is thinnest and least supported. The tear is almost always here, dislocation being primarily subglenoid. (Further degrees are subcoracoid and subclavicular.) If, after reduction, abduction is prevented, dislocation cannot recur.

If the shoulder joint ankyloses, loss of movement is partly compensated by increased scapular mobility. When ankylosis is likely the humerus should be positioned as if the palm were to be placed on the back of the neck, i.e. abducted, to make full use of scapular mobility.

The Cubital (Elbow) Joint

This (4.43–48) includes two articulations: (1) *humero-ulnar*, between humeral trochlea and ulnar trochlear notch, and (2) *humeroradial*, between humeral capitulum and radial head. It is hence a *compound* synovial joint. Its complexity is increased by continuity with the superior radio-ulnar joint, this complex being the *cubital articulation*. The superior radio-ulnar joint will, however, be considered separately.

Articular surfaces are the humeral trochlea and capitulum, and the ulnar trochlear notch and radial head. The trochlea is not a simple pulley as its medial flange exceeds its lateral, thus projecting to a lower level so that the plane of the joint, about 2 cm distal to the inter-epicondylar line, is tilted inferomedially; the trochlea is also widest posteriorly and here its lateral edge is sharp. The trochlear notch is not wholly congruent with it; in full extension the medial part of its upper (olecranon) half is not in contact with the trochlea and a corresponding lateral strip loses contact in flexion. The trochlea has an asymmetrical sellar surface, largely concave transversely, convex anteroposteriorly; sections show that these profiles are compounded spirals. Consequently swing is accompanied (as in all hinge joints) by screwing and conjunct rotation (pp. 480, 482). Olecranon and coronoid parts of the trochlear notch are usually separated by a rough strip, devoid of articular cartilage and covered by fibro-adipose tissue and synovial membrane. The capitulum and the radial head are reciprocally curved; closest contact occurs with a semiflexed radius in mid-pronation. The rim of the head, more prominent medially, fits the groove between humeral capitulum and trochlea.

Since the humero-ulnar and humeroradial articulations form a largely uni-axial joint, ligaments are capsular and ulnar and radial collateral.

The articular capsule (4.44,45) is anteriorly broad and thin, attached proximally to the front of the medial epicondyle and humerus above the coronoid and radial fossae, and distally to the edge of the ulnar coronoid process and annular ligament (p. 507), being continuous at its sides with the ulnar and radial collateral ligaments. Anteriorly it receives numerous fibres from brachialis. Posteriorly the capsule is thin and attached to the humerus behind its capitulum and near its lateral trochlear margin, to all but the lower part of the olecranon fossa's edge, and to the back of the medial epicondyle. Inferomedially it reaches the olecranon's superior and lateral margins and is laterally continuous with the superior radio-ulnar capsule deep to the annular ligament (p. 507). It is related posteriorly to the tendon of triceps and to anconeus.

The synovial membrane (4.43,44) extends from the humeral articular margins, lines the coronoid, radial and olecranon fossae, the flat medial trochlear surface (3.136), the capsule's deep surface and the lower part of the annular ligament. Projecting between radius and ulna from behind is a crescentic synovial fold, partly dividing the joint into humeroradial and humero-ulnar parts; irregularly triangular, it contains extrasynovial fat (4.46). Between capsule and synovial membrane are three other pads of fat; the largest, at the olecranon fossa, is pressed into it by triceps during flexion; the other two, at the coronoid and radial fossae, are pressed in by brachialis during extension. They are all slightly displaced in contrary movements. Smaller synovial-covered tags of fat project into the joint near constrictions flanking the trochlear notch (p. 413), covering small non-articular areas of bone.

The ulnar collateral (medial cubital) ligament (4.45A), a triangular band, has thick anterior, posterior and inferior parts united by a thin region. The *anterior part* is attached by its apex

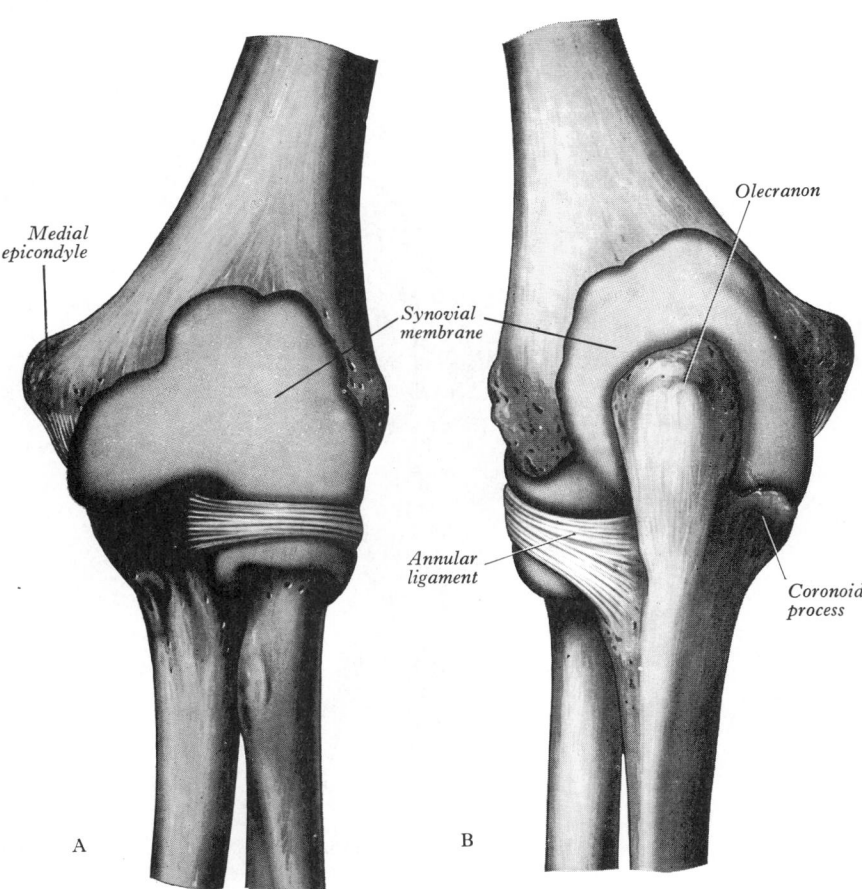

4.43 A. Synovial cavity of the left elbow joint, partially distended: anterior aspect. (Originally drawn from a specimen prepared by J C B Grant.) The fibrous capsule of the elbow joint has been removed but the thick part of the annular ligament has been left in situ. Note that the synovial membrane descends below the lower border of the annular ligament. B. Synovial cavity of the left elbow joint, partially distended: posterior aspect of the specimen represented in A. In A and B a small part of the ulnar (medial) collateral ligament may be seen.

to the front of the medial epicondyle and by its broad distal base to a proximal tubercle on the medial coronoid margin. The *posterior part*, also triangular, is attached low on the back of the medial epicondyle and to the olecranon's medial margin. Between these two bands intermediate fibres descend from the medial epicondyle to an inferior, *oblique band*, often weak, between olecranon and coronoid processes, converting a depression on the medial margin of the trochlear notch into a foramen, by which the intracapsular fat pad is continuous with extracapsular fat medial to the joint. The ulnar collateral ligament is related to triceps, flexor carpi ulnaris and the *ulnar nerve*. Along it, anteriorly, the attachment of flexor digitorum superficialis extends from the medial epicondyle to the medial coronoid border.

The radial collateral (lateral cubital) ligament (4.45B) is attached low on the lateral epicondyle and to the annular ligament, some of its posterior fibres crossing the ligament to the proximal end of the ulna's supinator crest. It is intimately blended with attachments of supinator and extensor carpi radialis brevis.

Muscles related to the joint are: (in front) brachialis, (behind) triceps and anconeus, (laterally) supinator and the common extensor tendon, and (medially) the common flexor tendon and flexor carpi ulnaris.

Articular arteries are from the numerous peri-articular anastomoses (6.97). *Articular nerves* are mainly from the musculocutaneous and radial, but the ulnar, median and sometimes anterior interosseous nerves also contribute. The musculocutaneous branch is from the nerve to brachialis and innervates an anterior part of the capsule; branches of the radial supply its posterior and anterolateral regions and come from the nerve to anconeus and the ulnar collateral branch to the medial head of triceps. The ulnar nerve supplies the ulnar collateral

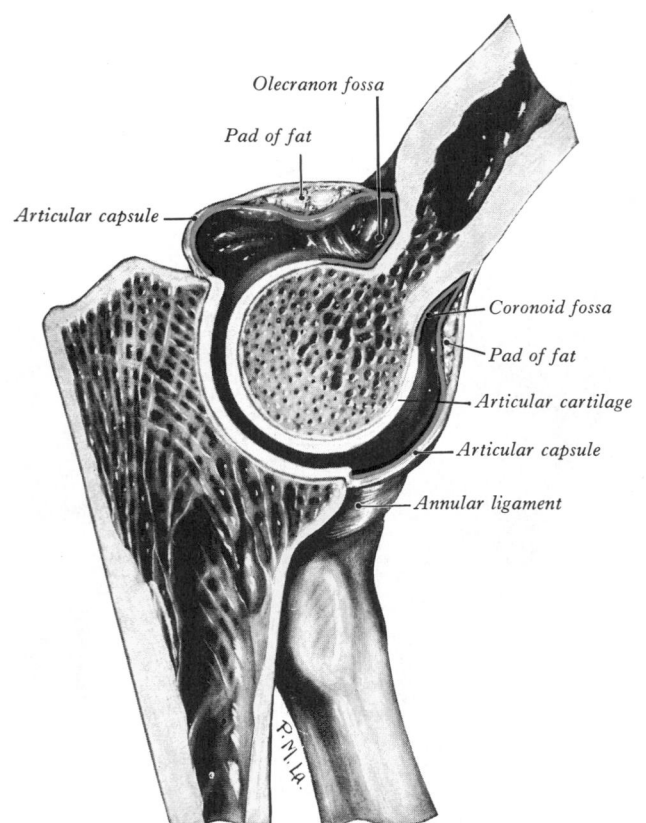

Olecranon fossa

Pad of fat

Articular capsule

Coronoid fossa

Pad of fat

Articular cartilage

Articular capsule

Annular ligament

4.44 Sagittal section through the left elbow joint: medial aspect. The synovial membrane is shown in blue.

However, ulnar flexion–extension is not a pure swing but accompanied by slight conjunct rotation, the ulna being slightly pronated in extension, supinated in flexion. Since the capitulum is smaller than the radial facet, the head of the radius can be felt at the back of the joint in full extension, which is limited by tension in the capsule and muscles anterior to the joint (extension being the close-packed position) and the entry of the tip of the olecranon into the olecranon fossa; flexion is limited chiefly by apposition of soft parts; in full flexion the rim of the radial head and the tip of the ulnar coronoid process enter the radial and coronoid humeral fossae respectively.

When the forearm is fully extended and supinated, it diverges laterally forming with the upper arm a 'carrying angle' of about 163° (Steel & Tomlinson 1958); its ulnar border cannot contact the lateral surface of the thigh. The 'carrying angle' is caused partly by projection of the medial trochlear edge about 6 mm beyond its lateral edge and partly by obliquity of the coronoid's superior articular surface, which is not orthogonal to the ulna's shaft. Tilt of humeral and ulnar articular surfaces is approximately equal; hence the carrying angle disappears in full flexion, the two bones reaching the same plane. When the adducted arm is flexed the little finger meets the clavicle, due to the position of the resting humerus (p. 502); with the humerus rotated laterally, the hand reaches the front of the shoulder. The carrying angle is also masked by pronation of the extended forearm, which brings the upper arm, semipronated forearm and hand into line, increasing manual precision in full extension of the elbow or during extension.

Accessory movements are limited to slight ulnar screwing, abduction and adduction, and anteroposterior translation of the radial head on the humeral capitulum. In translation the radial head moves on the ulnar radial notch and the annular ligaments are slewed backwards and forwards, more so when the elbow is semiflexed.

Muscles Producing the Movements

Flexion: brachialis, biceps and brachioradialis. In slow flexion or its maintenance against gravity, brachialis and biceps are principally involved, even for light loads. With increasing speed, activity in brachioradialis is increasingly prominent (MacConaill 1949, Basmajian 1959); due to its attachments it acts most effectively in mid-pronation. (See shunt muscle activity pp. 567, 568.) Against resistance pronator teres and flexor carpi radialis may act.

ligament behind the medial epicondyle (Gardner 1948b). These articular nerves accompany blood vessels supplying the synovial membrane, fat pads and epiphyses; they presumably contain vasomotor fibres as well as afferent fibres serving pain and proprioception.

Movements. Being a uni-axial joint the elbow allows flexion and extension, ulna moving on trochlea, radial head on capitulum.

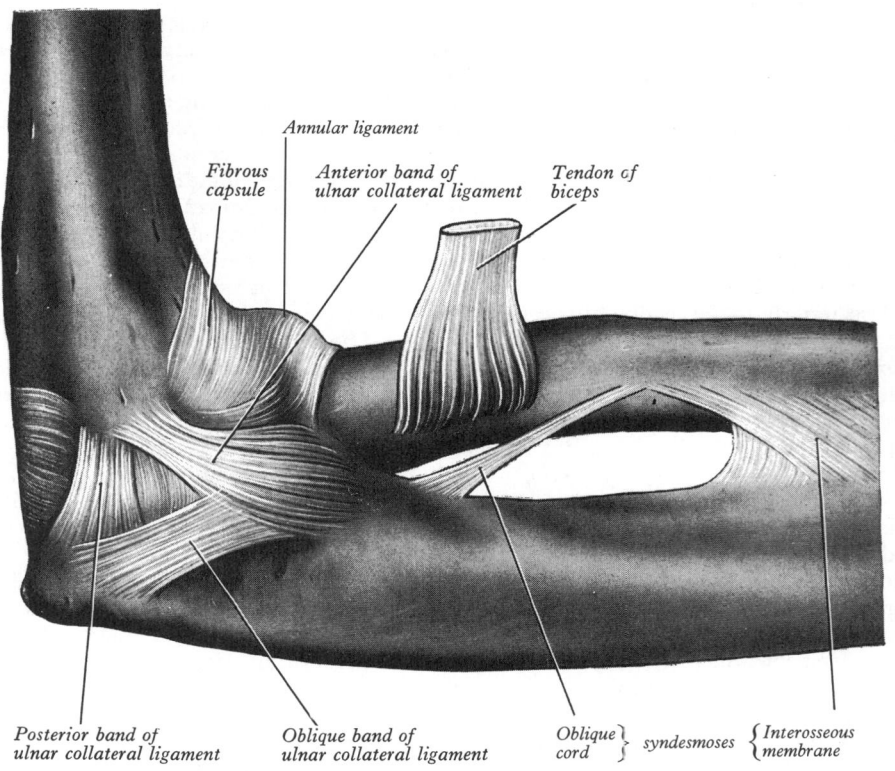

Annular ligament

Fibrous capsule

Anterior band of ulnar collateral ligament

Tendon of biceps

Posterior band of ulnar collateral ligament

Oblique band of ulnar collateral ligament

Oblique cord } *syndesmoses* { *Interosseous membrane*

4.45A Left elbow joint: medial aspect.

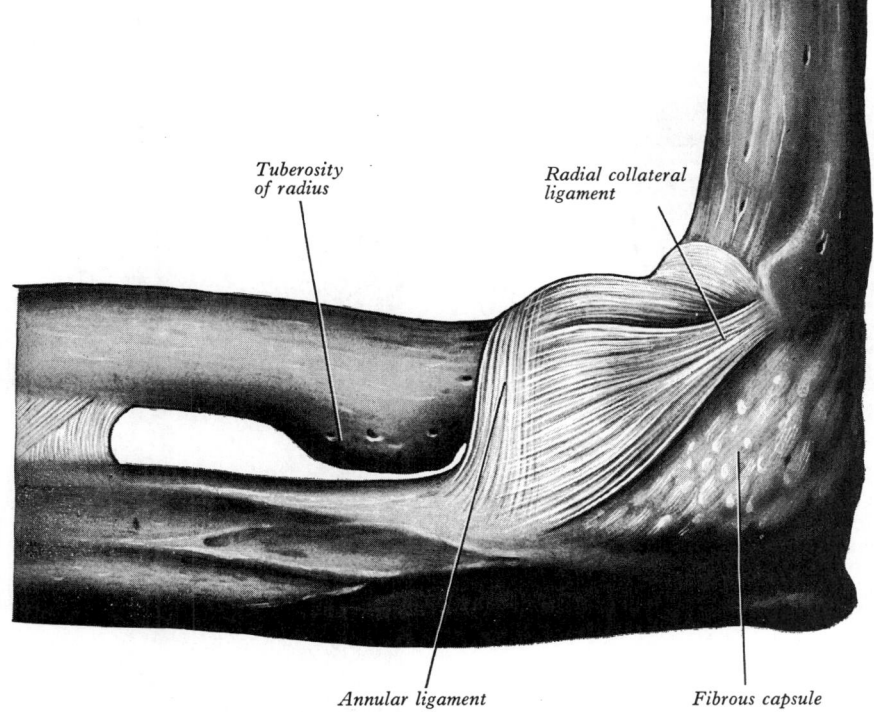

Tuberosity
of radius

Radial collateral
ligament

Annular ligament

Fibrous capsule

4.45B Left elbow joint: lateral aspect.

Extension: triceps, anconeus and gravity. In rapid extension brachioradialis again acts as a shunt muscle (pp. 567, 568).

Radio-ulnar Joints

The radius and ulna are connected by *proximal* and *distal radio-ulnar synovial joints* and by an intermediate interosseous membrane and ligament, a non-synovial *middle radio-ulnar joint*.

THE PROXIMAL RADIO-ULNAR JOINT

This a is uni-axial pivot between the circumference of the radial head and the osseofibrous ring made by the ulnar radial notch and annular ligament.

The annular ligament (4.45–47,49), a strong band, encircles the radial head, holding it against the ulna's radial notch. Forming about four-fifths of the ring, it is attached to the notch's anterior margin, broadens posteriorly and may divide into several bands and is attached to a rough ridge at or behind the notch's posterior margin; diverging bands may also reach the lateral margin of the trochlear notch above and proximal end of the supinator crest below. The proximal annular border blends with the cubital capsule, except posteriorly where the capsule passes deep to the ligament to reach the posterior and inferior margins of the radial notch. From the distal annular border a few fibres pass over reflected synovial membrane to

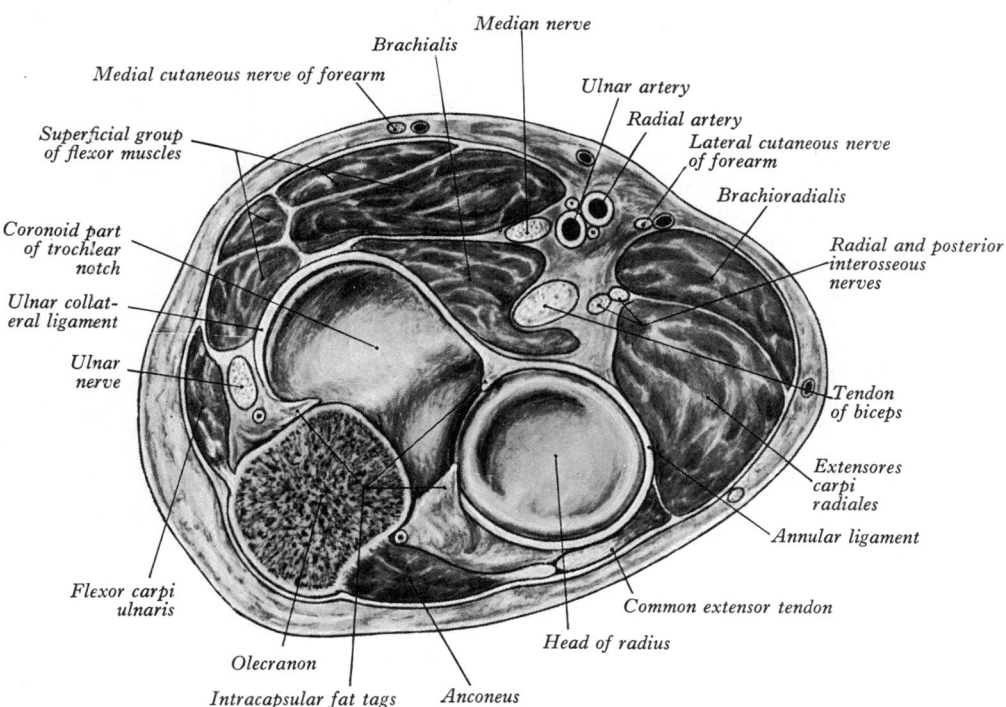

Median nerve

Brachialis

Medial cutaneous nerve of forearm

Superficial group
of flexor muscles

Ulnar artery

Radial artery

Lateral cutaneous nerve
of forearm

Brachioradialis

Coronoid part
of trochlear
notch

Ulnar collat-
eral ligament

Ulnar
nerve

Radial and posterior
interosseous
nerves

Tendon
of biceps

Extensores
carpi
radiales

Flexor carpi
ulnaris

Annular ligament

Common extensor tendon

Head of radius

Olecranon

Intracapsular fat tags

Anconeus

4.46 Transverse section of the right elbow joint to show the relations of the joint: superior aspect. Note the intracapsular fat 'tags' with meniscoid transverse-sectional profiles.

Coronoid part of trochlear notch

Neck of radius

Annular ligament (cartilaginous surface)

Quadrate 'ligament'

Olecranon   *Radial notch*

4.47 Distal socket of the superior radio-ulnar joint. The radial neck is transected and its head withdrawn. The socket consists of the chondrified aspect of the annular ligament, the articular cartilage of the radial notch of the ulna and the lax almost retiform quadrate 'ligament'. Section at the olecranon's base reveals, ventrally, the sellar articular surface of the coronoid part of the trochlear notch (humero-ulnar joint).

attach loosely on the radial neck. A thin, fibrous *quadrate ligament* covers the synovial membrane on the joint's distal surface (Martin 1958). The annular ligament's external surface blends with the radial collateral ligament and is an attachment of part of supinator. Posterior to it are anconeus and the interosseous recurrent artery. Internally the ligament is thinly covered by *cartilage* where it is in contact with the radial head; distally it is covered by synovial membrane, reflected up onto the radial neck.

THE MIDDLE RADIO-ULNAR UNION

The radial and ulnar shafts are connected by syndesmoses—an oblique cord and an interosseous membrane.

The oblique cord (4.49) is a small, inconstant, flat fascial band on the deep head of supinator, extending from the lateral side of the ulnar tuberosity to the radius a little distal to its tuberosity. Its fibres are at right angles to those in the interosseous membrane. Its functional significance is dubious.

The interosseous membrane (4.49) is a broad, thin, collagenous sheet, its fibres slanting distomedially between the radial and ulnar interosseous borders, its distal part attached to the posterior division of the radial border. Two or three posterior bands occasionally descend distolaterally across the other fibres. The membrane is proximally deficient, starting about 2 or 3 cm distal to the radial tuberosity. It is broader at mid-level and has an oval aperture near its distal margin conducting the anterior interosseous vessels to the back of the forearm. Between its proximal border and the oblique cord is a gap for the posterior interosseous vessels. The membrane augments attachments of the deep forearm muscles and connects the two bones. Its fibres appear to transmit to ulna and humerus forces acting proximally from hand to radius; however, it is only tense when the hand is midway between prone and supine positions and is relaxed in complete pronation and supination; the hand is usually pronated when subject to such forces. Moreover, the radius can transmit substantial forces directly to the humerus (Travill 1964). *Anteriorly* the membrane is related, in its proximal three-quarters, laterally to flexor pollicis longus and medially to flexor digitorum profundus, between them to the anterior interosseous vessels and nerve, and in its distal quarter to pronator quadratus; *posteriorly* are supinator, abductor pollicis longus, extensores pollicis brevis, pollicis longus and indicis and, near the carpus, the anterior interosseous artery and posterior interosseous nerve.

THE DISTAL RADIO-ULNAR JOINT

This is a uni-axial pivot between the ulna's convex distal end (head) and the concave ulnar notch of the radius; these surfaces

are enclosed by a capsule and connected by an articular disc. The fibrous capsule is thicker in front and behind, proximally lax and lined by synovial membrane projecting proximally between radius and ulna as a *recessus sacciformis* in front of the distal part of the interosseous membrane (4.52).

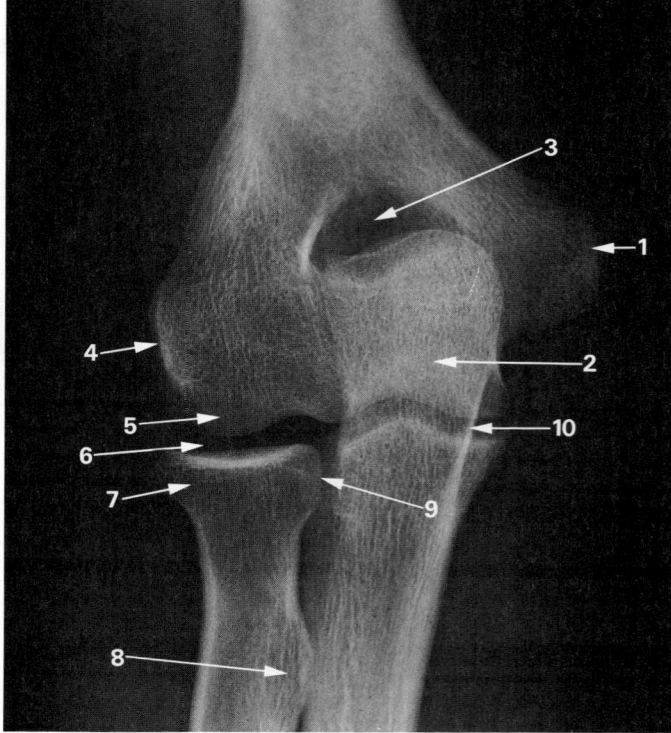

A

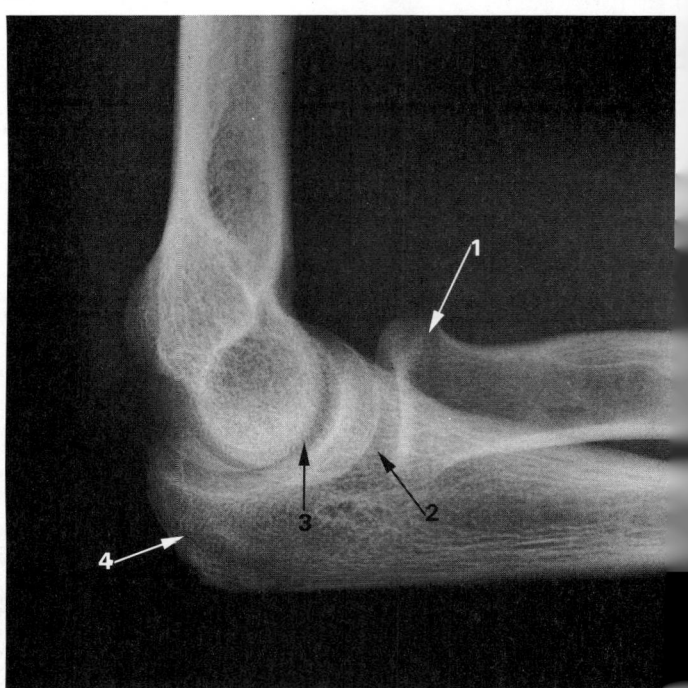

B

4.48 Anteroposterior (A) and lateral (B) radiographs of an adult elbow joint. The joint is semiflexed in B.

A. 1. Medial humeral epicondyle. 2. Shadow of olecranon superimposed on trochlea. 3. Olecranon fossa. 4. Lateral epicondyle. 5. Capitulum 6. Humeroradial joint. 7. Head of radius. 8. Radial tuberosity. 9. Radial head articulating with radial notch of ulna. 10. Humero-ulnar joint.

B. 1. Head of radius. 2. Profile of capitulum. 3. Profile of trochlea 4. Olecranon.

The articular disc (4.52) is fibrocartilaginous (collagen and elastic fibres in the young) and is triangular, binding the distal ends of ulna and radius. Its periphery is thicker, its centre sometimes perforated. It is attached by a blunt, thick apex to a depression between the ulnar styloid process and distal articular surface and by its wider thin base to the prominent edge between the ulnar notch and carpal articular surface of the radius. Its margins are united to adjacent carpal ligaments, its surfaces smooth and concave; the proximal articulates with the ulnar head, the distal is part of the radiocarpal joint, articulating with the lunate bone and, when the hand is adducted, the triquetral. Mikić (1978) studied 180 wrist joints from 100 cadavers, ranging from fetuses to 94 years, and found that articular discs show age-dependent degeneration: progressively reduced cellularity, loss of elastic fibres, mucoid degeneration, exposure of collagen fibres, fibrillation, ulceration, abnormal thinning and ultimate perforation. No perforations appeared in the first two decades, 7.6% in the third, 18.1% in the fourth, 40% in the fifth, 42.8% in the sixth and 53.1% in those over 60. Comparable degenerative changes frequently occur in the 'discal' surfaces of ulna and lunate bone.

Movements. Movements at the radio-ulnar joint complex pronate and supinate the hand. In *pronation* the radius carrying the hand, turns anteromedially obliquely across the ulna, its proximal end remaining lateral, its distal becoming medial. During this action the interosseous membrane is *spiralized*. In *supination* the radius returns to a position lateral and parallel to the ulna (and the interosseous membrane is *despiralized*). The hand can be turned thus through 140°–150° and, with the elbow extended, this can be increased to nearly 360° by humeral rotation and scapular movements. Power is greater in supination, affecting the design of nuts, bolts and screws, which are tightened by supination in right-handed persons. Moreover, supination is an *anti-gravity* movement (with a pendant upper arm, and semiflexed forearm); in seizing objects for examination or manipulation, pronation is merely a preliminary and is *aided* by gravity.

The axis for pronation and supination is often represented as a line through the centre of the radial head (proximal) and the ulnar attachment of the articular disc (distal). More correctly this is the axis of movement *radius relative to ulna* and it *does not remain stationary*. The radial head rotates in the annular ligament and ulnar radial notch, its distal lower end and articular disc swing on the ulnar head. During rotation of the radial head in this ring its proximal surface spins on the humeral capitulum. The distal end of the ulna *is not stationary* during this; it moves a variable amount along a curved course, posterolaterally in pronation, anteromedially in supination. This entails that the axis, as defined above, is displaced laterally in pronation, medially in supination. Hence the axis for *supination and pronation of the whole forearm and hand* passes *between* the bones at both the proximal and distal radio-ulnar joints, when ulnar movement is marked, but through the centres of the radial head and ulnar styloid when it is minimal. The axis may be prolonged through any digit, depending on medial or lateral displacement of the distal end of the ulna (Ray et al 1951). Note that this displacement is not conjunct rotation; it can be varied at will. For example, the index finger and thumb, in a precision grip (p. 632), can describe arcs of varying radius according to functional demands. Greatest radius of arc involves minimal ulnar deviation, whereas rotation of opposed digits around a virtually fixed point (often necessary in manipulation of tools and instruments) involves marked and even maximal ulnar displacement. Ulnar movements are possible because of the incongruence of the trochlea and trochlear notch, therefore occurring *without* humeral rotation, whether the elbow is flexed, semiflexed or extended. Since this incongruence persists in all cubital positions, the 'close-packed position' of the humero-ulnar articulation at terminal elbow extension depends on ligaments rather than articular geometry.

Accessory movements include backward and forward translation of the radial head on the ulnar radial notch (p. 413) and the ulnar head likewise on the radial ulnar notch.

Muscles Producing the Movements

Pronation: pronator quadratus (p. 621), aided in rapid movement and against resistance by pronator teres (Basmajian & Travill 1961). Gravity also assists.

Supination: supinator, in slow unresisted movement and extension, assisted by biceps in fast movements in flexion, especially when resisted.

Electromyographic studies have not confirmed activity in brachioradialis during pronation and supination.

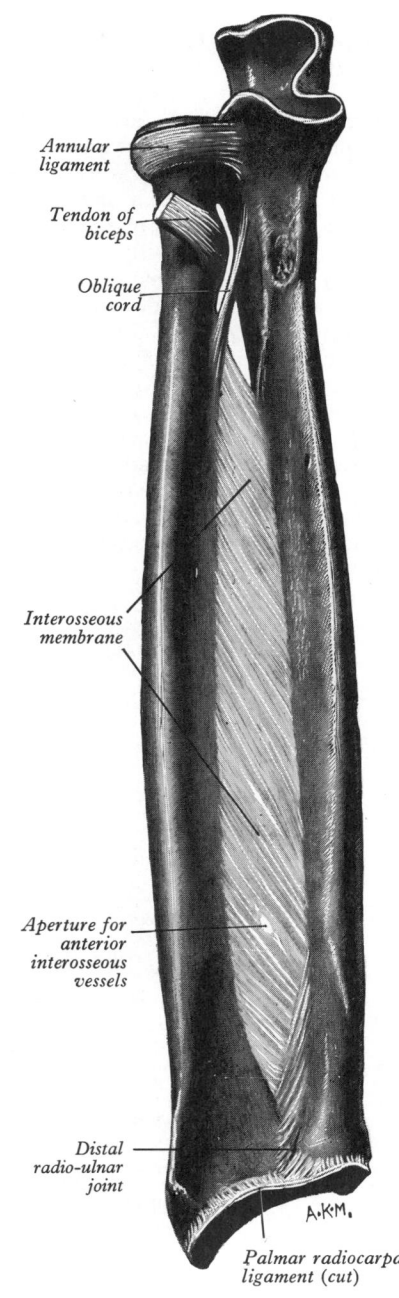

4.49 Interosseous membrane and oblique cord of the forearm: anterior aspect. Membrane and cord are syndesmoses.

Applied Anatomy. At the elbow joint backward dislocation with ulnar abduction are commonest, the former often complicated by fracture of the coronoid process. Owing to the strength of collateral ligaments, the medial epicondyle is frequently torn off in lateral dislocations.

In acute synovitis the joint cavity becomes distended, bulging most around the olecranon, due to capsular laxity here. There is often some swelling above the radial head or the whole elbow may become fusiform.

Dislocation of the radial head alone is not uncommon, most frequently in youth from falls on the hand with forearm extended and supinated, the head being displaced forwards, with rupture of the annular ligament. Occasionally a peculiar injury, supposedly a subluxation, occurs in children. The radial head appears to be displaced distally in the annular ligament; its upper border is folded over the head between it and the capitulum; the small size of the head in children predisposes to this. The forearm becomes fixed in semiflexion, midway between supination and pronation.

The Radiocarpal or Wrist Joint

The radiocarpal joint (4.50,51,52), bi-axial and ellipsoid (vide infra), is formed by articulation of the distal end of the radius and the triangular articular disc with the scaphoid, lunate and triquetral bones. The radial articular surface and distal discal surface form an almost elliptical, concave surface with a transverse long axis; but the radial surface is bisected by a low ridge into two concavities. A similar ridge usually appears between the medial radial concavity and the concave distal discal surface. Proximal articular surfaces of scaphoid, lunate and triquetral bones and their interosseous ligaments form a smooth convex surface, received into the proximal concavity. Surface projection of the joint is a line, convex upwards, joining radial and ulnar styloid processes (4.52–54).

The articular capsule is lined by *synovial membrane* which is usually separate from that of the inferior radio-ulnar and intercarpal joints; but a protruding *prestyloid recess*, ventral to the articular disc, is present and ascends close to the styloid process. The recess is bounded distally by a *fibrocartilaginous meniscus*,

projecting from the ulnar collateral ligament between the tip of the ulnar styloid process and the triquetral (4.52); both are clothed with hyaline articular cartilage. The meniscus may ossify (Lewis et al 1970). The capsule is strengthened by palmar radiocarpal and ulnocarpal, dorsal radiocarpal and radial and ulnar collateral ligaments.

The palmar radiocarpal ligament (4.50), a broad membranous band, is attached to the anterior margin of the distal end of the radius and its styloid process, its fibres passing distomedially to the anterior surfaces of scaphoid, lunate and triquetral bones, some reaching the capitate. It is partly intracapsular.

The palmar ulnocarpal ligament is a rounded fasciculus from the base of the ulnar styloid process and anterior margin of the articular disc to the lunate and triquetral bones. Lewis et al (1970) and Mayfield et al (1976) regard both palmar carpal ligaments as intracapsular; the latter also divide the radiocarpal ligament into three and the ulnocarpal ligament into two parts, each named by its attachments e.g. radiocapitate, ulnolunate, etc. It is not clear whether the ligaments are totally intracapsular, i.e. separate from the overlying articular capsule. (See also pp. 470 and 511.) The palmar carpal ligaments are perforated by vessels and are related anteriorly to tendons of flexores digitorum profundus and pollicis longus.

The dorsal radiocarpal ligament (4.51), thinner than the palmar, is attached to the posterior border of the distal end of the radius, its fibres descending distomedially to the dorsal surfaces of scaphoid, lunate and triquetral bones and continuing into the dorsal intercarpal ligaments. It is related posteriorly to the carpal and digital extensor tendons, their synovial sheaths and the posterior

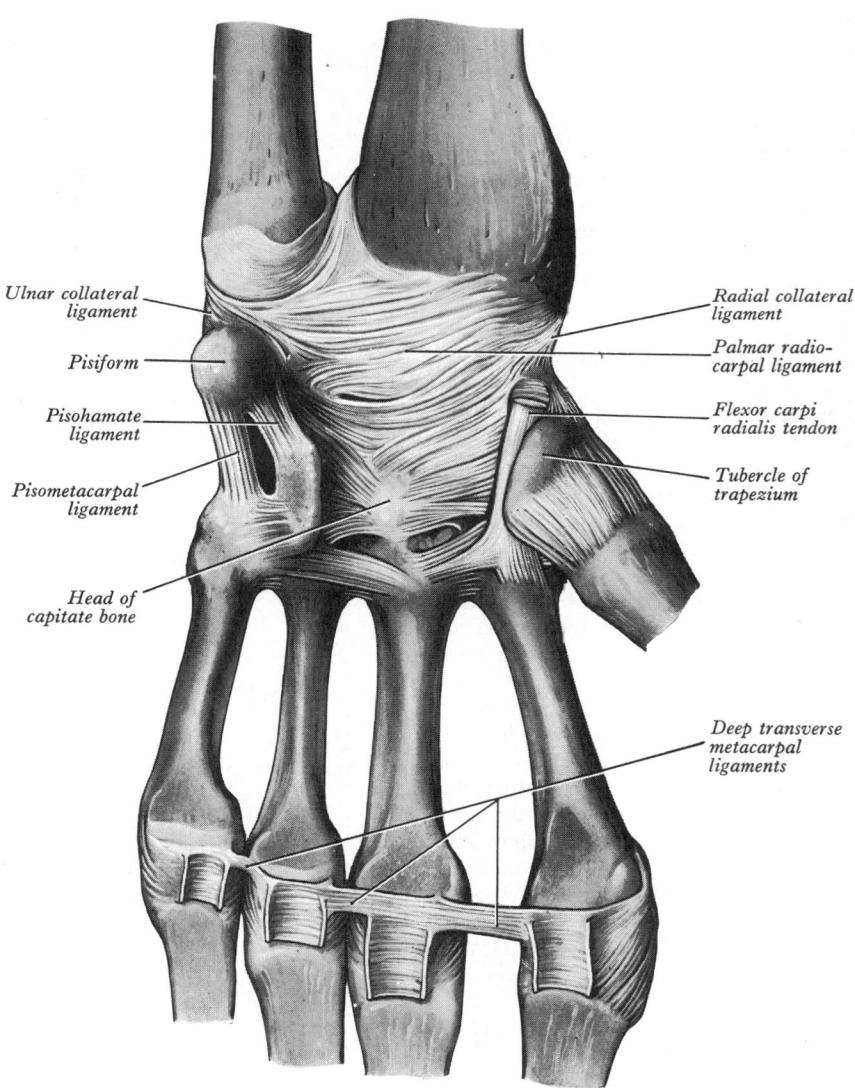

Ulnar collateral ligament

Pisiform

Pisohamate ligament

Pisometacarpal ligament

Head of capitate bone

Radial collateral ligament

Palmar radio-carpal ligament

Flexor carpi radialis tendon

Tubercle of trapezium

Deep transverse metacarpal ligaments

interosseous nerve; anteriorly it is blended with the disc of the distal radio-ulnar articulation.

The ulnar collateral carpal ligament (4.50,51) is attached to the apex of the ulnar styloid process; it divides into two fasciculi, one attached to the medial side of the triquetral, the other to the pisiform bone. Attached to the ligament's deep (lateral) aspect is a small fibrocartilaginous *meniscus* which is the distal boundary of the synovial *prestyloid recess* (4.52, p. 510).

The radial collateral carpal ligament (4.50,51) extends from the tip of the radial styloid process to the radial side of the scaphoid bone; some fibres are prolonged to the trapezium. It is related to the radial artery, curving round laterally between the ligament and tendons of abductor pollicis longus and extensor pollicis brevis. Both collateral ligaments are relatively weak. *Arteries* supplying the joint are the anterior interosseous, anterior and posterior carpal branches of radial and ulnar, palmar and dorsal metacarpals and recurrent rami of the deep palmar arch. *Nerves* are from the anterior and posterior interosseous.

Movements accompany those of the intercarpal and midcarpal joints, described together on p. 512. The close-packed position is full extension.

Intercarpal Joints

These interconnect carpal bones and may be summarized as (1) joints between the proximal carpal bones, (2) joints between the distal row, (3) a complex joint between the rows, the midcarpal joint. Carpal bones are connected by an extensive array of ligaments, not all specifically named. In addition to those described, the flexor retinaculum (p. 625) is an accessory intercarpal ligament. Articular surfaces are either sellar, ellipsoid or spheroidal.

JOINTS OF THE PROXIMAL CARPAL ROW

(*a*) The scaphoid, lunate and triquetral bones are connected by dorsal, palmar and interosseous ligaments.

The dorsal and palmar ligaments are transverse, connecting scaphoid to lunate and lunate to triquetral. The palmar ligaments are weaker. Mayfield et al (1976) described a distinct ligament (in 28 dissections) between palmar aspects of capitate and triquetrum which crosses the hamate bone.

The interosseous ligaments (4.52), two narrow bundles, one connecting lunate and scaphoid, the other lunate and triquetral, are attached near their proximal surfaces, forming part of the convex articular radiocarpal surface.

(*b*) The pisiform articulates with the palmar surface of the triquetral bone at a small, oval, almost flat, synovial *pisotriquetral joint*; ligaments are the capsular, pisohamate and pisometacarpal. The thin capsule surrounds the joint; the synovial cavity is usually separate but may communicate with the radiocarpal joint.

The pisohamate and pisometacarpal ligaments connect the pisiform to the hamate's hook and the base of the fifth metacarpal respectively (4.50). Both are continuations of the tendon of flexor carpi ulnaris and are misnamed.

JOINTS OF THE DISTAL CARPAL ROW

Bones of the distal carpal row are connected by dorsal, palmar and interosseous ligaments.

The dorsal and palmar ligaments extend transversely between trapezium and trapezoid, trapezoid and capitate, capitate and hamate bones. **The interosseous ligaments** are usually thicker than in the proximal row; three unite the four bones, that between capitate and hamate being strongest; one or both of the other two are often absent.

THE MIDCARPAL JOINT

This joint, between scaphoid, lunate and triquetral bones (proximally) and trapezium, trapezoid, capitate and hamate (distally) is a compound articulation divided descriptively into medial

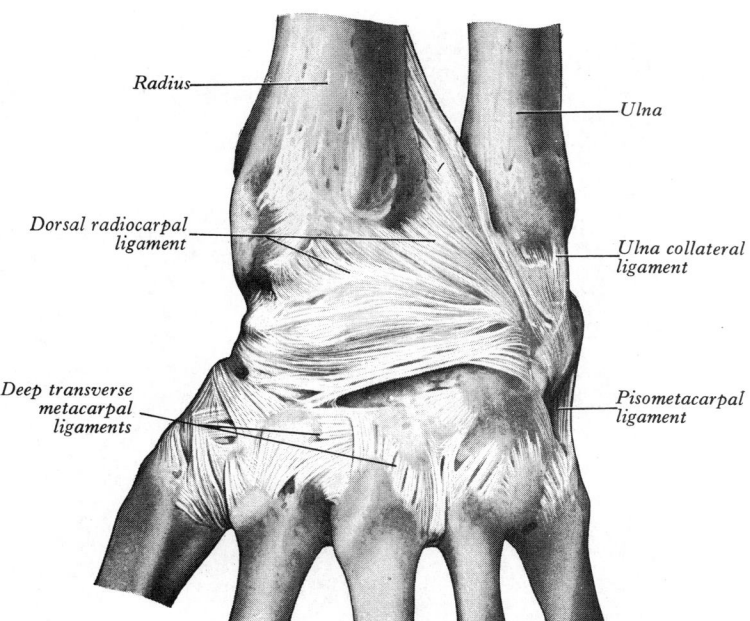

4.51 Ligaments of the left wrist: dorsal aspect.

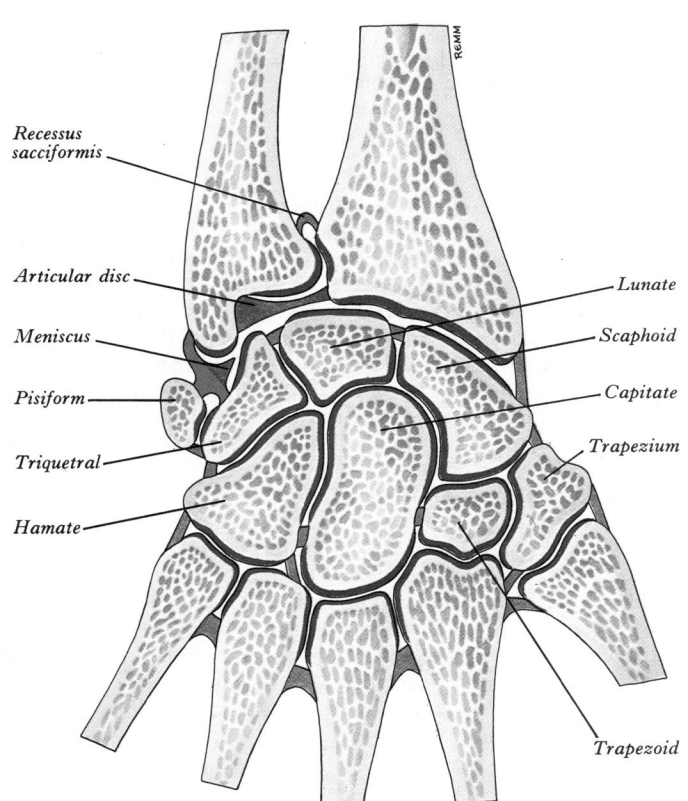

4.52 Coronal section through the distal ends of the radius and ulna, the carpus and the proximal ends of the metacarpals, showing the general form of the articular surfaces (blue), synovial cavities, interosseous ligaments (green) and fibrocartilages (purple). Partly after Lewis et al (1970).

and lateral parts. Throughout most of the *medial compartment* the convexity of the capitate's head and hamate articulate with a reciprocal concavity formed by scaphoid, lunate and much of the triquetral bones; however, most medially the curvatures are reversed, thus forming a compound sellar joint. In the *lateral compartment* the trapezium and trapezoid articulate with the scaphoid, forming a second compound articulation, often said to be plane but which is also sellar. The ligaments are dorsal, palmar and collateral.

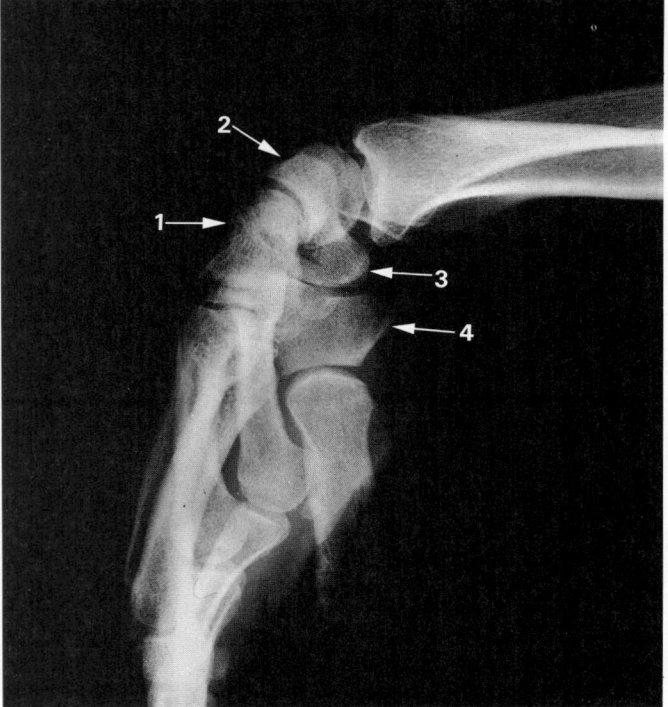

4.53A Radiograph of the hand and wrist in full flexion: lateral aspect. The arrows point to: (1) the capitate bone; (2) the lunate bone; (3) the tubercle of the scaphoid bone; (4) the tubercle of the trapezium. Compare with 4.53B and note the relative positions of the capitate and lunate, and the lunate and radius.

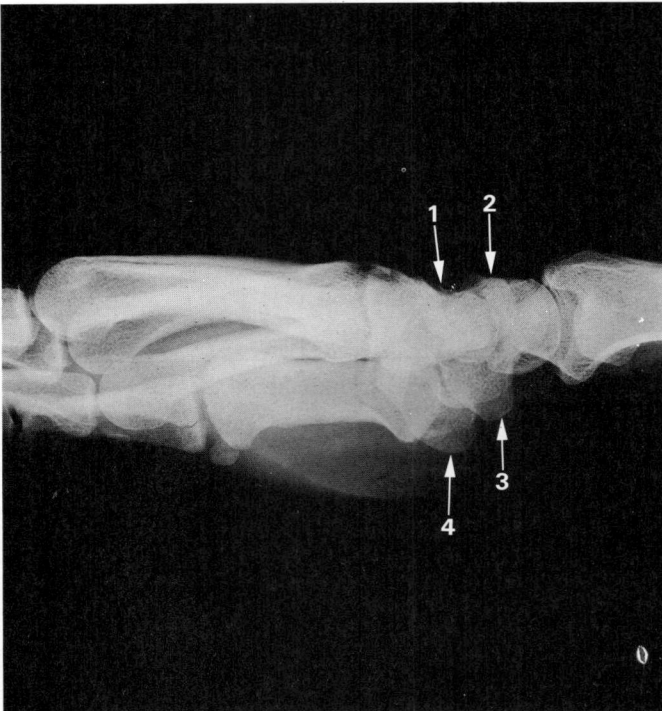

4.53B Radiograph of the hand and wrist: lateral aspect. The long axes of the third metacarpal bone, the capitate and the lunate are, approximately, in line with the long axis of the radius. The arrows point to the same structures as in 4.53A. Note the relative positions of the capitate and lunate, and the lunate and radius.

The dorsal and palmar ligaments are short, irregular bundles between the bones of the first and second carpal rows. Palmar fascicles radiate from the head of the capitate to surrounding bones as the *radiate carpal ligament*.

The collateral ligaments, radial and ulnar, are very short. The radial, stronger and more distinct, connects the scaphoid and trapezium, the ulnar connects triquetral and hamate; they are continuous with corresponding collateral ligaments of the wrist joint. A slender interosseous band sometimes connects capitate and scaphoid bones but does not completely divide the midcarpal synovial cavity.

The carpal synovial membrane is most extensive (4.52), lining a most irregular articular cavity, its proximal part between distal surfaces of the scaphoid, lunate and triquetral bones and the proximal surfaces of the second carpal row. It has proximal prolongations between scaphoid and lunate, lunate and triquetral and three distal prolongations between the four bones of the second row. The prolongation between trapezium and trapezoid or between trapezoid and capitate is, due to absence of an interosseous ligament, often continuous with corresponding carpometacarpal joints, variably from the second to fifth or often second and third only. If the latter, the joint between the hamate and fourth and fifth metacarpal bones has a separate synovial membrane; interposed (4.52) is the carpometacarpal interosseous ligament (p. 514). Synovial cavities of carpometacarpal joints are prolonged slightly between the metacarpal bases. The synovial joint between pisiform and triquetral is usually isolated.

RADIOCARPAL AND INTERCARPAL MOVEMENTS

The movements of radiocarpal and intercarpal joints are considered together as a common mechanism acted on by the same muscles. Active movements are flexion (c. 85°), extension (c. 85°), adduction (ulnar deviation), abduction (radial deviation) and circumduction.

In *carpal flexion* radiocarpal and midcarpal joints are involved, the range being greater at the latter; in *extension* the reverse occurs, with more movement at the radiocarpal (4.53). Hence the proximal surfaces extend further posteriorly on the lunate and scaphoid bones. These movements are limited chiefly by

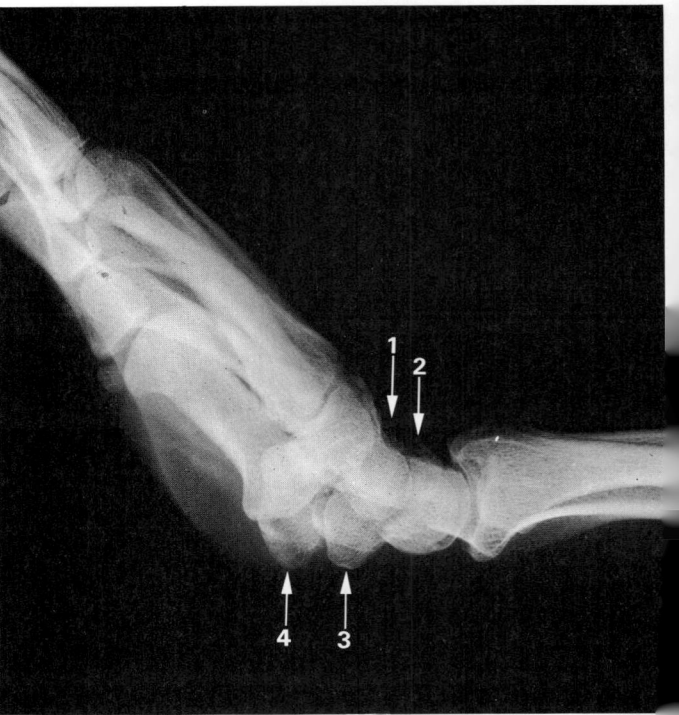

4.53C Radiograph of the hand and wrist in full extension: lateral aspect. The arrows point to the same structures as in 4.53A. Compare with 4.53 and note the alterations in the relative positions of the capitate and lunate bones, and the lunate bone and the radius.

antagonistic muscles; therefore the range of flexion is perceptibl diminished when fingers are flexed, due to increased tension i the extensors. Only when the joints are forced to the limits flexion or extension are dorsal or palmar ligaments fully stretche (but vide infra).

Adduction of the hand (c. 45°) is considerably greater tha abduction (c. 15°), perhaps due to the more proximal site of t ulnar styloid process. Most adduction occurs at the radiocarpal joint; the lunate bone, articulating with both radius and articular disc when the hand is in midposition (4.52), in adduction passes off the disc to articulate solely with the radius (4.54A). Much of the proximal articular surface of the scaphoid becomes subcapsular beneath the radial collateral ligament and forms a smooth, convex, palpable prominence in the floor of the anatomical 'snuff box'.

Abduction occurs almost entirely at the midcarpal joint; radiographs of abducted hands show that the capitate rotates round an anteroposterior axis so that its head passes medially and the hamate conforms to this; the distance between lunate and the apex of hamate is increased (4.54B). The scaphoid rotates around a transverse axis; its proximal articular surface has retreated from the capsule and is now in sole articulation with the radius. Movements are limited by antagonistic muscles and, at extremes, by the collateral carpal ligaments.

Circumduction of the hand is not rotatory but successive flexion, adduction, extension and abduction or the reverse.

Abduction–adduction movements are of special functional value. The hand is commonly used with the carpus slightly extended and the forearm in midposition. Skilled abduction–adduction movements manipulate a large variety of precision tools, from fine needles to hammers.

Accessory movements are largely radiocarpal and more easily observed in flexion than extension. The carpus can be moved bodily backwards and forwards on the radius and articular disc and *rotated axially to a considerable extent*. Some side-to-side movement is also possible.

ADDITIONAL ANALYSIS OF RADIOCARPAL AND CARPAL MOVEMENTS

A 'link' joint has been suggested as the mechanical equivalent of the radiocarpal and medial midcarpal complex, in simple form illustrated by a bicycle chain. This is stable only when under tension and 'on centre' with links in line; unless strengthened by a 'stop' device it buckles under longitudinal compression, especially when 'off centre'. But some advantages are inherent: since the range of movement at each unit is less than the total, articular surfaces can be flatter than in a single joint of the same total range and are better adapted to pressure. Further, overlying tissues are less disturbed by squeezing at the extremes of movement.

Other views contrast sharply with the foregoing (MacConaill 1941). In most manual positions carpal bones are loose-packed and relatively mobile; they become a rigid block only in full extension, i.e. the close-packed position for radiocarpal and most carpal joints; close-packing is achieved in *two* stages (vide infra). The carpus is envisaged in four functional units: (1) trapezium, (2) scaphoid, (3) hamate, capitate and trapezoid, and (4) triquetrum and lunate. From full flexion to full extension the stages considered to occur are: firstly the distal row (3) moves on the proximal (2 and 4) until hand and forearm are in line, hamate, capitate, trapezoid and scaphoid coming into mutual close-pack to form a rigid mass (i.e. 2 + 3). Secondly this mass moves upon the triquetrum and lunate, which move at the radiocarpal joint until full extension is reached, with close-packing of the radiocarpal and most carpal joints (i.e. except articulations of pisiform and trapezium). In this position, adopted only for special effort, very large forces can be transmitted through the articular structures. A similar position is often adopted to resist falls on the outstretched hand, and commonly results in damage, e.g. a 'supination' (Colles') fracture of radius, fracture of scaphoid or dislocation of lunate. Mayfield et al (1976), studying the behaviour of carpal ligaments in experimental carpal fractures, have corroborated this dorsal 'locking' of the carpus and made further analysis of the roles of carpal ligaments in progressive intercarpal slides ending in close-packing. During initial stages of the movement, the scaphoid acts with the proximal carpal row but in later stages with the distal row. The trapezioscaphoid joint is not considered involved in general close-packing of the carpus, perhaps due to the thumb's functional independence when the remainder of the carpus becomes a rigid mass.

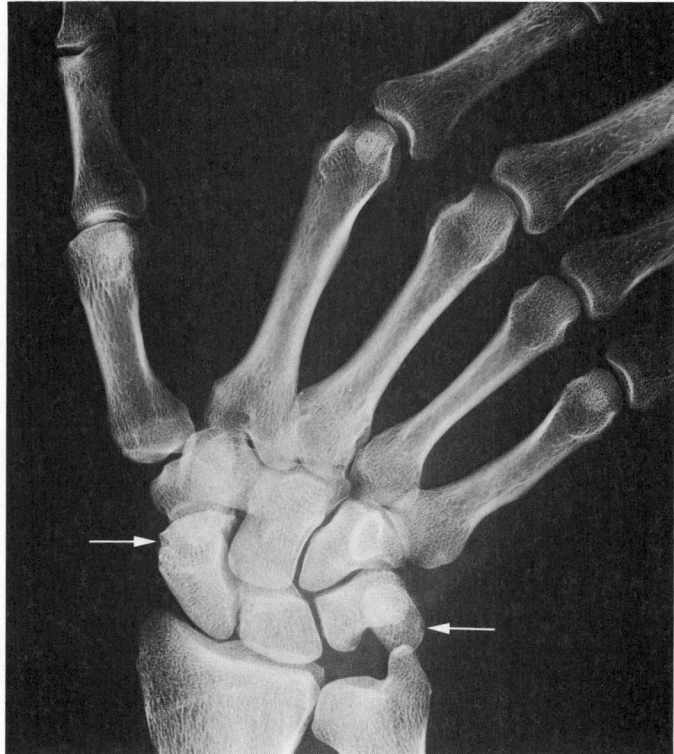

4.54A Radiograph of the hand in full adduction. The arrows point to the scaphoid bone on the lateral side and to the pisiform bone on the medial side. Note that the shadow of the pisiform bone overlaps the shadow of the tip of the styloid process of the ulna. Compare with 4.54B and observe that the movements occur at both the radiocarpal and intercarpal joints.

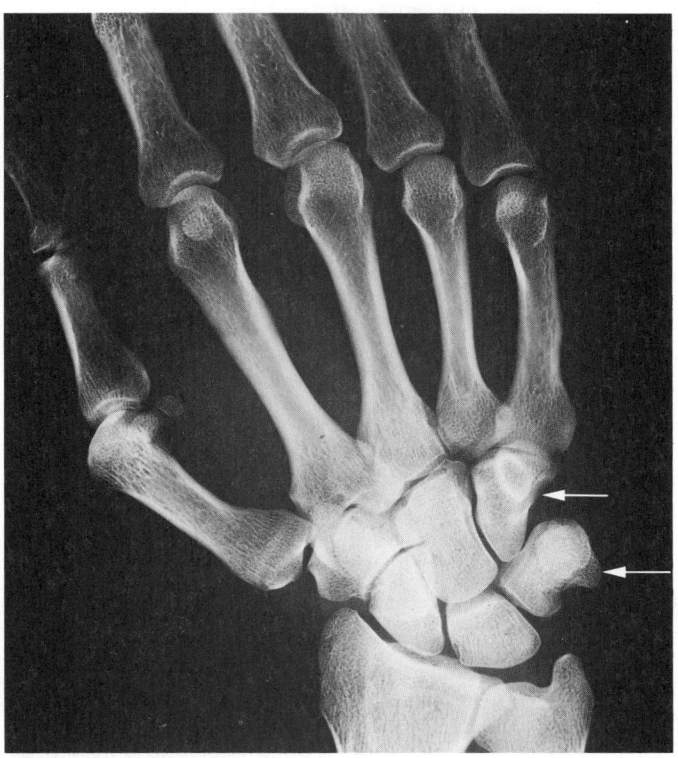

4.54B Radiograph of the same hand in full abduction. The arrows point to the hamate and pisiform bones. Compare with 4.54A and note that: (1) the scaphoid and lunate bones have passed medially so that the latter articulates to a large extent with the articular disc of the inferior radio-ulnar joint; (2) the pisiform is now widely separated from the styloid process of the ulna; (3) the scaphoid, having rotated round a transverse axis, is much foreshortened; (4) the apex of the hamate bone has been thrust away from the lunate by the rotation of the capitate around an anteroposterior axis; (5) a gap has opened up between the distal portions of the hamate and triquetral bones; and (6) the long axes of the capitate and lunate bones are now almost in the same straight line.

Further analysis by Kauer (1974) continues the concept of articular links but emphasizes *rotations* of lunate and scaphoid relative to radius and disc; during flexion–extension both rotate relative to each other and to the capitate. Proximal curvatures of lunate and scaphoid are different, dictating some independence of movement, although this is limited, their interdependence being determined by the biomechanical properties of their interosseous ligament.

Muscles Producing Carpal Movements

Flexion: flexores carpi radialis and ulnaris and palmaris longus, assisted by flexores digitorum superficialis and profundus, flexor pollicis longus and abductor pollicis longus.

Extension: extensores carpi radiales longus, brevis and ulnaris, assisted by extensores digitorum, digiti minimi, indicis and pollicis longus.

Adduction: flexor and extensor carpi ulnaris in unison.

Abduction: flexor carpi radialis, extensores carpi radiales longus and brevis in unison, with abductor pollicis longus and extensor pollicis brevis.

Carpometacarpal Joints

THE CARPOMETACARPAL JOINT OF THE POLLEX (THUMB)

This *sellar* joint, between the first metacarpal base and the trapezium, enjoys wide mobility due to its extensive articular surfaces and their topology. It has a fibrous capsule, thick but loose, extending from the circumference of the metacarpal base to the rough rim of the distal trapezial articular facet; it is thickest laterally and dorsally. The synovial membrane is separate (4.52). The first metacarpal and trapezium are connected by lateral, anterior and posterior ligaments and the capsule.

The broad *lateral ligament* runs from the lateral trapezial surface to the radial side of the metacarpal base. The *palmar* and *dorsal ligaments* are oblique bands converging to the ulnar side of the metacarpal base from the palmar and dorsal trapezial surfaces respectively.

Relations. The joint's *palmar surface* is covered by the thenar muscles, its *dorsal surface* by long and short extensors. *Medial* is the first dorsal interosseous muscle and radial artery, the latter passing through the first interosseous space, and the tendon of flexor pollicis longus; *lateral* is the tendon of abductor pollicis longus and that of extensor pollicis brevis.

Movements. *Active* movements are flexion–extension, abduction–adduction, rotation, and circumduction. In the resting position of the first metacarpal bone (p. 634) flexion and extension are parallel to the palmar plane, abduction and adduction at right angles to this. Except at initiation, flexion is accompanied by medial rotation; conversely, medial rotation involves flexion. Linkage of movements is largely due to the articular sellar shape imposing some conjunct rotation and to obliquity of the dorsal ligament, which when taut anchors the ulnar side of the metacarpal base while its radial side continues to move. Contraction of flexor pollicis brevis, assisted by opponens pollicis, hence produces medial rotation with flexion (Napier 1955); combined with abduction this brings the pollicial pulp into contact with those of slightly flexed fingers, a movement termed *opposition*. (The flexed fingers have varying degrees of *lateral metacarpophalangeal rotation*, minimal in the index but maximal in minimus.) Conversely, full extension of the pollicial metacarpal entails slight lateral rotation (Kuczynski 1974), attributable also to the sellar form of joint and to action of the *palmar* ligament, similar to the dorsal ligament in flexion (vide supra).

These rotations are conjunct, inevitably accompanying flexion–extension, with spiralization and tautening of the joint capsule and balanced conarticular compression at extremes of these movements. Close-packing is said to occur in powerful opposition, when greatest forces are transmitted to the joint. (For analyses consult Haines 1944, Kapandji 1963, 1970, Kuczynski 1974, MacConaill & Basmajian 1977.)

Accessory movements are axial rotation in the position of rest, and distraction.

Muscles Producing the Movements

Flexion: flexor pollicis brevis and opponens pollicis, aided by flexor pollicis longus when the other joints of the thumb are flexed. Flexion entails medial rotation.

Extension: abductor pollicis longus and extensores pollicis brevis and longus. In full extension extensor pollicis longus, owing to its oblique pull and the disposition of the palmar ligament, rotates the thumb laterally and draws it dorsally, i.e. slightly adducts it.

Abduction: abductores pollicis brevis and longus. When abduction is maximal the digit and metacarpal are not in line, the pollex being abducted at both metacarpophalangeal and carpometacarpal joints.

Adduction: adductor pollicis alone.

Opposition: opponens pollicis and flexor brevis pollicis simultaneously flex and medially rotate the abducted pollex. Interpulpal pressure, or that generated by digital grasping, is increased by adductor pollicis and flexor pollicis longus.

Circumduction: the above muscle groups acting consecutively, extensors, abductors, flexors and adductors in this or reverse order.

THE SECOND TO FIFTH CARPOMETACARPAL JOINTS

Joints between the carpus and second to fifth metacarpal bones, although widely classed as plane, have curved articular surfaces often of complex sellar shape (pp. 477–480). The bones are united by articular capsules, dorsal, palmar and interosseous ligaments.

The dorsal ligaments are strongest, connecting the dorsal surfaces of the carpal and metacarpal bones. The second metacarpal has two, from trapezium and trapezoid; the third also has two, from trapezoid and capitate; the fourth two, from capitate and hamate; the fifth a single band from the hamate continuous with a similar palmar ligament, forming an incomplete capsule.

The palmar ligaments are similar, except for the third metacarpal, which has three: a lateral from the trapezium, superficial to the tendon sheath of flexor carpi radialis, an intermediate from the capitate and a medial from the hamate bone.

The interosseous ligaments are two short, thick, fibrous bands and limited to one part of the carpometacarpal articulation; they connect contiguous distal margins of capitate and hamate with adjacent surfaces of the third and fourth metacarpal bones and proximally they may be united.

The synovial membranes are often continuous with those of the intercarpal joints. Occasionally, the joint between the hamate and fourth and fifth metacarpal bones has a separate synovial cavity, bounded laterally by the medial interosseous ligament and its extensions to the palmar and dorsal parts of the capsule (4.52).

Intermetacarpal Joints

The second to fifth metacarpal bases articulate reciprocally by small cartilage-covered facets connected by dorsal, palmar and interosseous ligaments.

The dorsal and palmar ligaments pass transversely from bone to bone, **interosseous ligaments** connecting contiguous surfaces just distal to their articular facets. **The synovial membranes** are continuous with those of the carpometacarpal articulations.

Movements at carpometacarpal and intermetacarpal articulations are limited to slight gliding, sufficient, however, to permit some flexion-extension and adjunct rotation, ranges varying in different joints. They are *partly* accessory movements of the first type (p. 484), occurring when the palm is 'cupped', as in grasping an object. Active movements also occur and are familiar, e.g. to the pianist or violinist. The fifth metacarpal is most movable, then the fourth, the second and third being least mobile. These variations are easily demonstrated by opposing each digit to the thumb over the palmar centre. About two-thirds of the movements are pollicial, as described above, but during opposition to minimus the latter is flexed, abducted and laterally rotated, accounting for the remaining third. These actions occur at both the carpometacarpal and metocarpophalangeal joints. (For a more extensive description of digital, including pollicial movement see below.) The close-packed position probably coincides with carpal extension, as in gripping. Further accessory movements are spiral twisting of the whole metacarpus on the carpus.

Metacarpophalangeal Joints

These (4.55A,B) are usually considered ellipsoid, but metacarpal heads, adapted to shallow concavities on the phalangeal bases, are not regularly convex but partially divided on their palmar aspects and thus almost bicondylar (pp. 418, 420). Each joint has a palmar and two collateral ligaments.

The **palmar ligaments** are unusual, being thick, dense and fibrocartilaginous, sited between the collateral ligaments and connected to them. They are attached loosely to metacarpal bones but firmly to the phalangeal bases. Their palmar aspects are blended with deep transverse palmar ligaments and are grooved for the flexor tendons, whose fibrous sheaths connect with the sides of the grooves. Their deep surfaces increase articular areas for the metacarpal heads.

The **deep transverse metacarpal ligaments** are three short, wide, flat bands connecting the palmar ligaments of the second to fifth metacarpophalangeal joints (4.55A) and related anteriorly to the lumbricals and digital vessels and nerves, posteriorly to interossei. Bands from the digital slips of the central palmar aponeurosis join their palmar surfaces (p. 627). On both sides of the third and fourth metacarpophalangeal joints (but only the ulnar side of the second and radial side of the fifth) transverse bands of the dorsal digital expansions (p. 632) join the deep transverse metacarpal ligaments; *ventral* to this band are the lumbricals and phalangeal attachments of dorsal interossei and *dorsal* to it are the remaining attachments of dorsal interossei and palmar interossei (p. 622 and 5.70,71).

The **collateral ligaments**, strong, round cords, flank the joints, each attached to the posterior tubercle and adjacent pit on the side of its metacarpal head and each passing distoventrally to the side of the ventral aspect of its phalangeal base (4.55B).

On dorsal surfaces of these joints the fibrous capsules are thin and often separated from extensor tendons by bursae (5.70). In a proportion of instances, however, part of the long extensor tendon is intercalated into the dorsal capsule and some authorities then regard the tendon as forming the capsule on this aspect. Their sensory nerve supply has been detailed by Sathian & Devanandan (1983), who found abundant Paciniform corpuscles but few Ruffini endings and no Golgi type receptors in association with these joints.

Movements are flexion, extension, adduction, abduction, circumduction, and limited rotation, the last of which cannot occur in isolation but may accompany flexion–extension; this, however, can be initiated voluntarily in the free hand (e.g. each finger flexing and rotating to place its tip near the palmar centre); the *range* of rotation is frequently increased due to the resistance of a grasped object. Extension ranges a few degrees, flexion is almost 90°; both are limited chiefly by antagonistic muscles but flexion is commonly terminated by resistance of a grasped object. Abduction and adduction are also relatively small and in flexion negligible. The pollicial metacarpophalangeal joint has a flexion–extension range of about 60° (almost entirely flexion); other movements are adduction–abduction (maximal range 25°) which invariably accompanies the corresponding carpometacarpal movements and increases their combined range; also some slight conjunct rotation, but greater adjunct rotation, accompanying flexion–extension. Of the other metacarpophalangeal joints the second is most mobile in adduction–abduction (c.30°), followed by fifth, fourth and third. (For further analyses of these movements see p. 632 et seq.)

Accessory movements are further rotation (marked in the thumb), anteroposterior and lateral translation of a phalanx or metacarpal, and distraction.

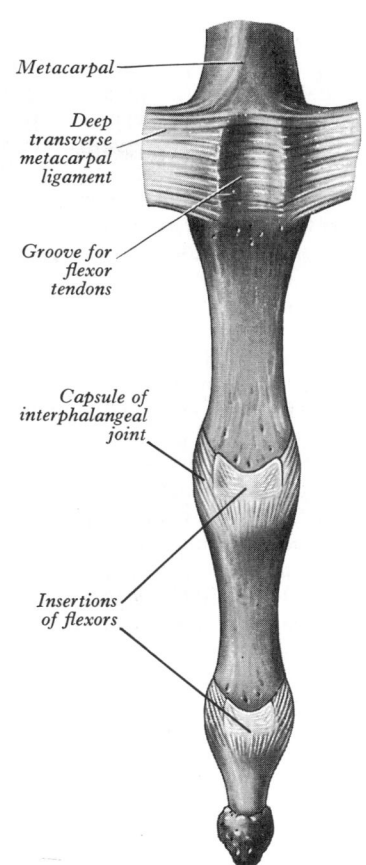

4.55A Metacarpophalangeal and digital joints of the middle finger: palmar aspect.

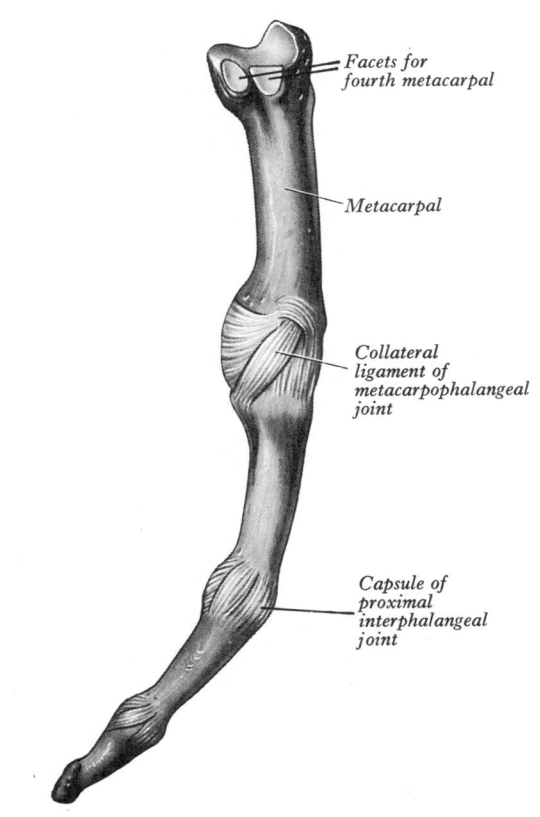

4.55B Metacarpophalangeal and digital joints of the right third finger: medial aspect.

Muscles Producing the Movements

Flexion: flexores digitorum superficialis and profundus, assisted by lumbricals and interossei (Long et al 1961) and, in the minimus, by flexor digiti minimi brevis; in the pollex, by flexores pollicis longus and brevis and the first palmar interosseous. Slight lateral rotation accompanies digital flexion of digits 3–5. Flexion of the index may be accompanied by minimal lateral or no rotation; frequently a small degree of *medial* rotation can be observed.

Extension: in the third and fourth digits extensor digitorum, assisted respectively in second and fifth by extensor indicis and extensor digiti minimi; in the thumb, by extensores pollicis longus and brevis.

Adduction: in extended fingers, palmar interossei; during flexion long flexors are predominant. In the pollex the limited metacarpophalangeal adduction possible is slightly attributable to adductor pollicis and first palmar interosseous.

Abduction: in extended fingers, dorsal interossei assisted by long extensors except in the middle finger; in minimus the abductor digiti minimi. In the thumb, abductor pollicis brevis produces the slight abduction possible, most often one factor contributing to opposition. When fingers are flexed, active abduction is impossible; but, if the long digital flexors are inactive, passive abduction is free. Inability to abduct actively in this position may be due to shortening of the dorsal interossei and abductor digiti minimi by flexion, but the altered line of pull of the interossei relative to the axis of movement is probably the determining factor; while in digital extension the axis of lateral movements is anteroposterior, in flexion it is proximodistal, the line of pull of the interossei being then nearly *parallel* to the axis.

Interphalangeal Joints

These are uni-axial hinge joints (4.55A,B), each with a fibrous capsule, palmar and two collateral ligaments, arranged as in the metacarpophalangeal joints (p. 514). Long extensor tendons take the place of the dorsal capsular ligaments.

Movements (active) at interphalangeal joints are flexion and extension, greater in range at proximal joints. Flexion is considerable, extension limited by tension of digital flexors and terminated by tension in palmar ligaments and conarticular compression. Full extension is the close-packed position, assumed whenever fingers are used as props to transmit body weight or powerful thrust. (In contrast alternatively the fully clenched fist may be used.) Flexion and extension are accompanied by slight conjunct rotation; during flexion this turns the digital pulps slightly laterally, i.e. to face the opposed thumb. An opposite rotation occurs during extension (vide supra.)

Accessory movements are limited rotation, abduction, adduction and anteroposterior translation. These permit the fingers to adapt to shapes of objects gripped and provide against stresses and strains.

Muscles Producing Movements

Flexion: at proximal interphalangeal joints, flexores digitorum superficialis and profundus; at the distal, flexor digitorum profundus; at the pollicial interphalangeal joint, flexor pollicis longus.

Extension: extensores digitorum, digiti minimi and pollicis longus, in association with abductor pollicis longus and extensor pollicis brevis. Extension occurs simultaneously in both joints in digits 2–5.

Simultaneous *flexion* at metacarpophalangeal and *extension* at interphalangeal joints of a digit are essential in the fine movements of writing, drawing, threading a needle etc. Lumbricals and interossei have long been accepted as not only primary agents in flexing metacarpophalangeal joints but also in extending interphalangeal joints via their attachments to the dorsal digital expansions (p. 630). Also, when lumbricals and interossei flex metacarpophalangeal joints, it has been claimed that balance between the tension of digital flexors and extensors alters in favour of extensors and that this alone is responsible for the extension of interphalangeal joints (Braithwaite et al 1948). However, lumbricals and interossei alone can both extend these joints (Sunderland 1945b, see also p. 632).

Flexure Lines of Wrist and Hand

Flexure lines commonly crease the skin across flexor surfaces of wrist and hand (6.105). Though not all directly over their functionally related subjacent skeletal joints they result from adhesion of skin to subjacent deep fascia and are sites of folding of skin during movement. Such flexures have often hence been termed 'skin joints'. They are useful landmarks. Less often mentioned, but quite prominent, although less regular, there are crease-line complexes, mainly transverse but with varying curvatures, which are centred over the *dorsal (extensor)* aspects of the radiocarpal, carpal, metacarpophalangeal and interphalangeal joints. During flexion the dorsal skin is stretched and the lines become less prominent (but can still be identified). During extension the now redundant skin becomes increasingly puckered and the lines are finally maximally prominent. (For a general review of 'skin lines' see p. 80.)

Near the junction of carpus and forearm there are usually three *ventral* transverse lines; the proximal marks the proximal limit of flexor synovial sheaths, an intermediate line overlies the wrist joint and a distal line is at the proximal border of the flexor retinaculum.

In the palm a curved *radial longitudinal line* encircles the thenar eminence, ending at the palm's lateral margin; medial and roughly parallel to it are several less constant longitudinal lines. *Proximal* and *distal transverse lines* ascend medially across the palm. The proximal begins at the distal end of the radial longitudinal line and runs obliquely to the middle of the hypothenar eminence across the shafts of metacarpal bones; the distal begins at or near the cleft between the index and medius and traverses the palm with a proximal convexity over the second to fourth metacarpal heads, near the proximal ends of fibrous flexor sheaths.

The second to fifth digits show proximal, middle and distal sets of transverse lines. The *proximal*, often double, are at the digital roots, about 2 cm *distal* to the metacarpophalangeal joints; the *middle* are typically double, proximal members being *directly over* the proximal interphalangeal joints; the *distal* lines, usually single, are *proximal* to the distal interphalangeal joints, their level sometimes marked by a fainter, more distal line. The free pollicial base is partly encircled by a line starting on the radial side and crossing distally over the metacarpophalangeal joint to end between pollex and index level with the base of the proximal pollicial phalanx. A second, shorter crease is about 1 cm distal to this. At the interphalangeal joint of the thumb are two lines comparable to middle digital lines in other digits.

JOINTS OF THE LOWER LIMB

The Sacro-iliac Joint

This synovial articulation between sacral and iliac auricular surfaces (4.59), although often termed plane, is nearly flat only in infants; in adults surfaces are irregular, often markedly, and sometimes sinuous. Their curvatures and irregularities, greater in males, are reciprocal; they restrict movements and contribute to the joint's considerable strength in transmitting weight from the vertebral column to the lower limbs. According to Putschar (1931) and Schunke (1938) the sacral surface is covered by hyaline cartilage and the iliac by fibrocartilage, an unusual arrangement (but not subsequently confirmed). Fibrous adhesions and gradual obliteration occur in both sexes, earlier in males, after the menopause in females. A radiological study of 94 healthy individuals (Cohen et al 1967) showed such changes in 6% before 50 years, in 24% thereafter. In old age the joint may be completely fibrosed and occasionally even ossified. The *articular capsule* is attached close to both articular margins. Ligaments are ventral, interosseous and dorsal sacro-iliac.

The ventral sacro-iliac ligament (4.56) is an antero-inferior capsular thickening, particularly well-developed near the arcuate line and posterior inferior iliac spine, where it connects the third sacral segment to the lateral side of the pre-auricular sulcus. It is thin elsewhere.

The interosseous sacro-iliac ligament (not officially recognized but the largest typical *syndesmosis*, p. 468), is the massive chief bond between the bones and fills the irregular space posterosuperior to the joint (4.60); it is covered superficially by the dorsal sacro-iliac ligament. Its deeper part has superior and inferior bands passing from depressions posterior to the sacral auricular surface to those on the iliac tuberosity. These bands are covered by, and blend with, a more superficial fibrous sheet connecting the dorsosuperior margin of a rough area posterior to the sacral auricular surface to the corresponding margins of the iliac tuberosity. This sheet is often partially divided into superior and inferior parts, the former uniting the superior articular process and lateral crest on the first two sacral segments to the neighbouring ilium as a *short posterior iliac ligament* (4.57).

The dorsal sacro-iliac ligament lies over the interosseous, but intervening are the dorsal rami of sacral spinal nerves and vessels. It has several weak fasciculi connecting the intermediate and, below this, the lateral sacral crests to the posterior superior iliac spine and internal lip of the iliac crest at its dorsal end. Inferior fibres, from the third and fourth sacral segments, ascending to the posterior superior iliac spine, may form a separate *long posterior sacro-iliac ligament* (4.57); the latter is continuous laterally with part of the sacrotuberous ligament and medially with the thoracolumbar fascia's posterior lamina (Weisl 1954a).

Accessory synovial articulations between lateral sacral crest and posterior superior iliac spine and iliac tuberosity are not uncommon (Trotter 1937).

Movements. A little anteroposterior rotation occurs around a transverse axis about 5–10 cm vertically below the sacral promontory (Weisl 1955). This accompanies flexion and extension of the trunk; its range is the same in males and non-pregnant females; it is increased in pregnancy. The greatest sacral movement relative to the iliac bones is in rising from a recumbent to a standing position (4.61). The sacral promontory advances as much as 5–6 mm as body weight impinges on the sacrum; backward movement of the lower end of the sacrum is less. Movement is not simple rotation, the axis being dynamic (Weisl 1953); some translation is associated with it.

VERTEBROPELVIC LIGAMENTS

Each ilium is connected to the fifth lumbar vertebra by an *iliolumbar ligament* (p. 324, 4.56) and the sacrum to the ischium by *sacrotuberous* and *sacrospinous ligaments*. (Further attachments are detailed below.)

The sacrotuberous ligament (4.56,57) is broadly attached by its base to the posterior iliac spine, partly blended with the dorsal sacro-iliac ligaments, to the lower transverse sacral tubercles and the lateral margins of the lower sacrum and upper coccyx. Its oblique fibres descend laterally, converging to form a thick, narrow band which widens again below and is attached to the ischial tuberosity's medial margin. It then spreads along the ischial ramus as the *falciform process*, whose concave edge blends with the fascial sheath of internal pudendal vessels (p. 778) and pudendal nerve. To the sacrotuberous ligament's posterior surface are attached the lowest fibres of gluteus maximus; superficial fibres of the ligament's lower part continue into the tendon of biceps femoris. The ligament is pierced by coccygeal branches of the inferior gluteal artery, the perforating cutaneous nerve and filaments of the coccygeal plexus.

The sacrospinous ligament (4.56), thin and triangular, extends from the ischial spine to the lateral margins of sacrum and coccyx anterior to the sacrotuberous ligament with which it blends. Its anterior surface is *muscular* and constitutes the coccygeus; the ligament is often regarded as a degenerate part of the muscle. (When pelvic and gluteal topographical relations are under consideration, it must be emphasized that muscle and ligament are co-extensive and are the ventral and dorsal aspects of the *same structure*.)

The sacrotuberous and sacrospinous ligaments oppose upward tilting of the lower part of the sacrum under downward thrust at its upper end (vide infra). They also convert the sciatic notches into foramina.

The greater sciatic foramen is bounded anterosuperiorly by the greater sciatic notch, posteriorly by the sacrotuberous and inferiorly by the sacrospinous ligament and ischial spine. It is partly filled by the emerging piriformis, above which the superior gluteal vessels and nerve leave the pelvis; below it, the inferior gluteal vessels and nerve, internal pudendal vessels and pudendal nerve, sciatic and posterior femoral cutaneous nerves and nerves to obturator internus and quadratus femoris also leave the pelvis.

The lesser sciatic foramen is bounded anteriorly by the ischial body, superiorly by its spine and sacrospinous ligament, posteriorly by the sacrotuberous ligament. It transmits the tendon of obturator internus, the muscle's nerve and the internal pudendal vessels and pudendal nerve.

The Pubic Symphysis

The pubic bones meet in the midline at a fibrocartilaginous *pubis symphysis* (4.58), connected by superior and arcuate pubic ligaments.

The superior pubic ligament connects the bones above, extending to the pubic tubercles. **The arcuate pubic ligament**, a thick arch of fibres, connects the lower borders of the symphysial pubic surfaces, bounding the pubic arch. Superiorly it blends with the interpubic disc and extends laterally attached to the inferior pubic rami; its base is separated from the ventral border of the urogenital diaphragm by an opening for the deep dorsal vein of penis or clitoris.

The interpubic disc connects the medial pubic surfaces, each covered by a thin layer of tightly adherent hyaline cartilage (surface growth cartilage in the young). The junction is not flat (p. 426) but marked by reciprocal crests and papillae. Theoretically this would resist shearing. The surfaces of hyaline cartilage are connected by fibrocartilage, varying in thickness. It often contains a cavity, probably due to absorption, rarely appearing before the tenth year and non-synovial. Better developed in females, it is usually posterosuperior but may reach the front or even occupy most of the cartilage (4.58). The disc is strengthened anteriorly by several interlacing collagenous fibrous layers, passing obliquely from bone to bone, decussating and interweaving with fibres of the external oblique aponeuroses and the medial tendons of the recti abdominis.

Movements have been little described. Angulation, rotation and displacement are possible but slight and are likely in activities at the sacro-iliac and hip joints. Some separation is held to occur late in gestation and during childbirth.

The Pelvic Mechanism

The skeletal pelvis undoubtedly supports and protects contained viscera, but is primarily part of the lower limbs, affording wide attachment for leg and trunk muscles. It is the major mechanism in transmitting the weight of head, trunk and upper limbs to the legs. It may be considered as two arches divided by a vertical transacetabular plane. The posterior arch, chiefly concerned in transmitting weight, consists of the upper three sacral vertebrae and strong pillars of bone from the sacro-iliac joints to the acetabular fossae (p. 428). The anterior arch, formed by the pubic bones and their superior rami, connects these lateral pillars as a tie-beam preventing separation; it also acts as a compression strut against medial femoral thrust. The sacrum, as summit of the posterior arch, is loaded at the lumbosacral joint; theoretically this force has two components, one thrusting the sacrum down

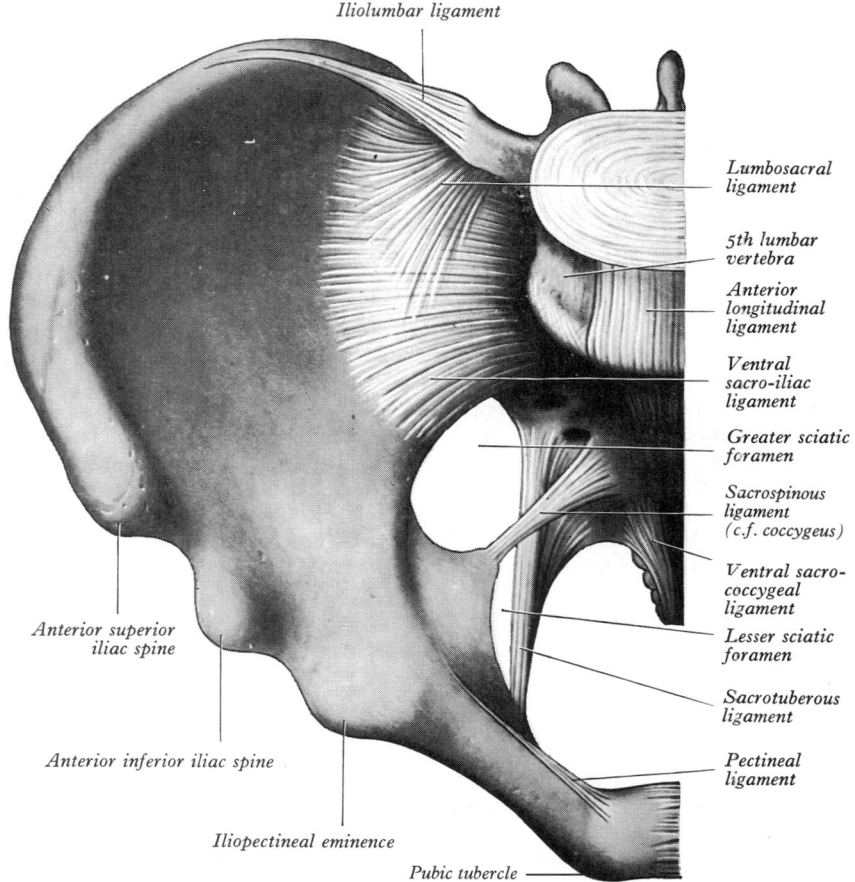

Iliolumbar ligament

Lumbosacral ligament

5th lumbar vertebra

Anterior longitudinal ligament

Ventral sacro-iliac ligament

Greater sciatic foramen

Sacrospinous ligament (c.f. coccygeus)

Ventral sacro-coccygeal ligament

Lesser sciatic foramen

Sacrotuberous ligament

Pectineal ligament

Anterior superior iliac spine

Anterior inferior iliac spine

Iliopectineal eminence

Pubic tubercle

4.56 Joints and ligaments of the right half of the pelvis: ventral aspect. Anterior superior iliac spine and pubic tubercle are in the same coronal plane. Note the inclination of the 'brim' (inlet) of the lesser (true) pelvis, the boundaries of the sciatic foramina and the (partly obscured) pelvic outlet.

and back between the iliac bones, the other thrusting its upper end down and forwards. Sacral movements are regulated by osseous shape and massive ligaments. The sacrum is a tilted wedge, its base anterosuperior. The first component therefore acts against the wedge, its tendency to separate iliac bones resisted by the sacro-iliac and iliolumbar ligaments and symphysis pubis.

Vertical coronal sections through the sacro-iliac joints suggest division of the (synovial) articular region of the sacrum into three segments. In the *anterosuperior segment*, involving the first sacral vertebra, articular surfaces are slightly sinuous and almost parallel. In the *middle segment* (4.60) the dorsal width between articular margins is greater than the ventral, and centrally a sacral concavity fits a corresponding iliac convexity, an interlocking mechanism relieving strain on the ligaments due to body weight. In the *postero-inferior segment* the ventral sacral width is greater than the dorsal and here its sacral surfaces are slightly concave. Antero-inferior sacral dislocation by the second component (of force) is prevented therefore mainly by the middle segment, owing to its cuneiform shape and interlocking mechanism. However, some rotation occurs, in which the anterosuperior segment tilts down and the postero-inferior segment up (4.61). 'Superior' segmental movement is to a small degree limited by wedging but chiefly by tension in the massive dorsal and interosseous sacro-iliac ligaments; movement of the 'inferior' segment is also slightly checked by wedging but chiefly by tension in the sacrotuberous and sacrospinous ligaments. In all movements sacro-iliac and iliolumbar ligaments and symphysis pubis resist separation of the iliac bones.

Locomotor significance of structural data has been more extensively studied on primate, including human, pelves than on any other skeletal element, except perhaps the femur. As an introduction to such metrical analysis consult Zuckerman et al (1973). Muscular, gravitational and inertial forces have been analysed in a mathematical model by Goel & Svensson (1977).

Applied Anatomy. During pregnancy pelvic joints and ligaments relax, while movements increase. Relaxation renders the sacro-iliac locking mechanism less effective, permitting greater rotation and perhaps allowing alterations in pelvic diameters at childbirth, although the effect is probably small (Young 1940). The impaired locking mechanism diverts the strain of weight-bearing to ligaments, with frequent sacro-iliac strain after pregnancy. After childbirth ligaments tighten and the locking mechanism improves; but this may occur in a position adopted during pregnancy. Such sacro-iliac 'subluxation' causes pain by unusual ligamentous tension; reduction by forcible manipulation may be attempted. The most common position in this condition of subluxation is believed to be backward rotation of the innominate bone relative to the sacrum; usually unilateral, it is on occasion bilateral.

The Coxal (Hip) Joint

This joint is multi-axial and of ball-and-socket (spheroidal, cotyloid) type. The femoral head articulates with the cup-shaped (cotyloid) acetabulum, its centre lying a little below the middle third of the inguinal ligament (4.59). (The profile of the joint's ventral margin is also *parallel to* the middle third of the inguinal ligament.) Articular surfaces are reciprocally curved but neither co-extensive nor completely congruent. Close-pack position is full extension, with slight abduction and medial rotation. As in the shoulder joint, surfaces are considered ovoid or spheroid rather than spherical but this is controversial (Hammond & Charnley 1967, Bullough & Goodfellow 1968, Greenwald & Haynes 1972). Evidence favours spheroid and slightly ovoid (p. 478) surfaces becoming almost spherical with advancing age. The femoral head is covered by articular cartilage, except for a rough pit for the *ligament of the head* (4.62C); in front the cartilage extends laterally over a small area on the adjoining neck; it is thickest centrally. Kurrat & Oberländer (1978) measured acetabular and femoral articular cartilages in 10 human joints, finding maximal thickness

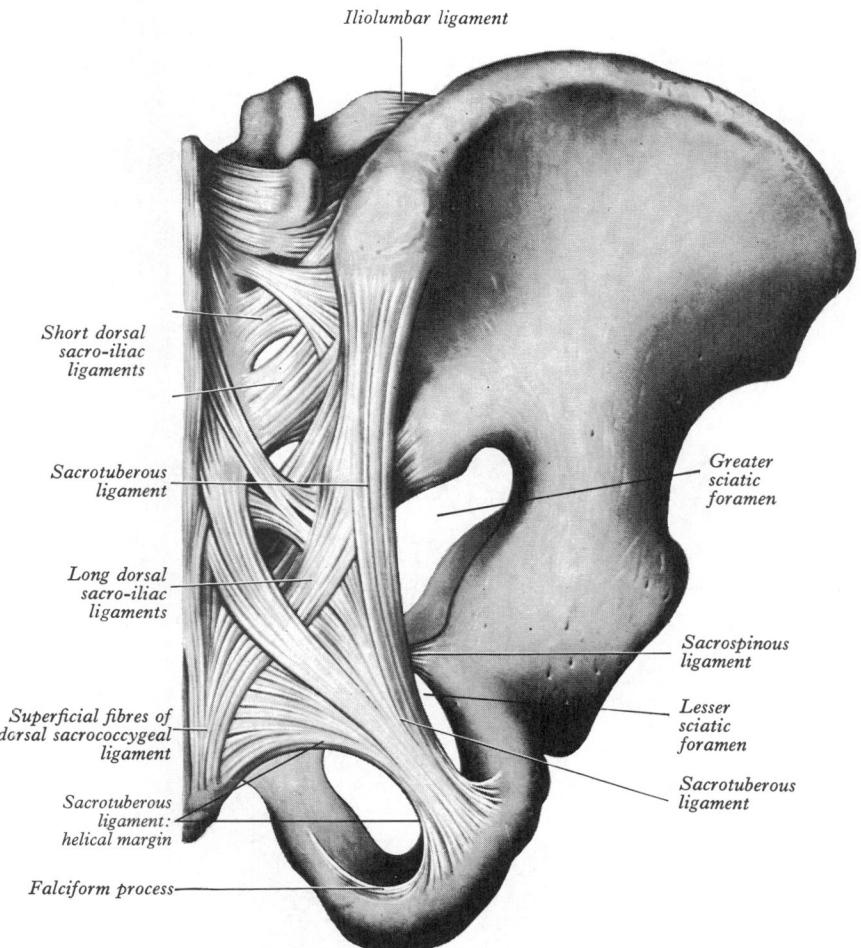

4.57 Joints and ligaments of the right half of the pelvis and fifth lumbar vertebra: posterior aspect.

in the acetabulum's anterosuperior quadrant and anterolateral on the femoral head (**4.62A**). Meachin & Allibone (1984) have described topographical variation in a calcified zone in this cartilage. The acetabular articular surface (**4.62B**) is an incomplete ring, the *lunate surface*, broadest above where pressure of the body weight falls in the erect posture, narrowest in its pubic region. It is deficient below opposite the acetabular notch and covered by articular cartilage, thickest where the surface is broadest. The acetabular fossa within it is devoid of cartilage but contains fibro-elastic fat largely covered by synovial membrane. Acetabular depth is increased by a fibrocartilaginous *acetabular labrum*. Ligaments include the fibrous capsule, labrum, ligament of the head, and iliofemoral, ischiofemoral, pubofemoral and transverse acetabular ligaments.

The fibrous capsule (**4.64,65**), strong and dense, is attached above to acetabular margin 5–6 mm beyond its labrum, in front to the outer labral aspect and, near the acetabular notch, to its transverse acetabular ligament and the adjacent rim of the obturator foramen. It surrounds the femoral neck and is attached *in front* to the trochanteric line; *above* to the base of the femoral neck; *behind* about 1 cm above the trochanteric crest; *below* to the femoral neck near the lesser trochanter (**4.64**). Anteriorly many fibres ascend along the neck as longitudinal *retinacula*, containing blood vessels for both femoral head and neck (p. 783). The capsule is thicker anterosuperiorly, where maximal stress occurs, particularly in standing; postero-inferiorly it is thin and loosely attached. It has two sets of fibres, circular and longitudinal. The circular fibres (*zona orbicularis*) are internal (**4.63A,B**), forming a collar round the femoral neck; though partly blended with pubofemoral and ischiofemoral ligaments, these fibres are not directly attached to bone. Externally, longitudinal fibres are most numerous in the anterosuperior region, reinforced by the *iliofemoral ligament*. The capsule is also strengthened by *pubofemoral* and *ischiofemoral ligaments*; externally it is rough, covered by muscles and separated from psoas major and iliacus by a bursa.

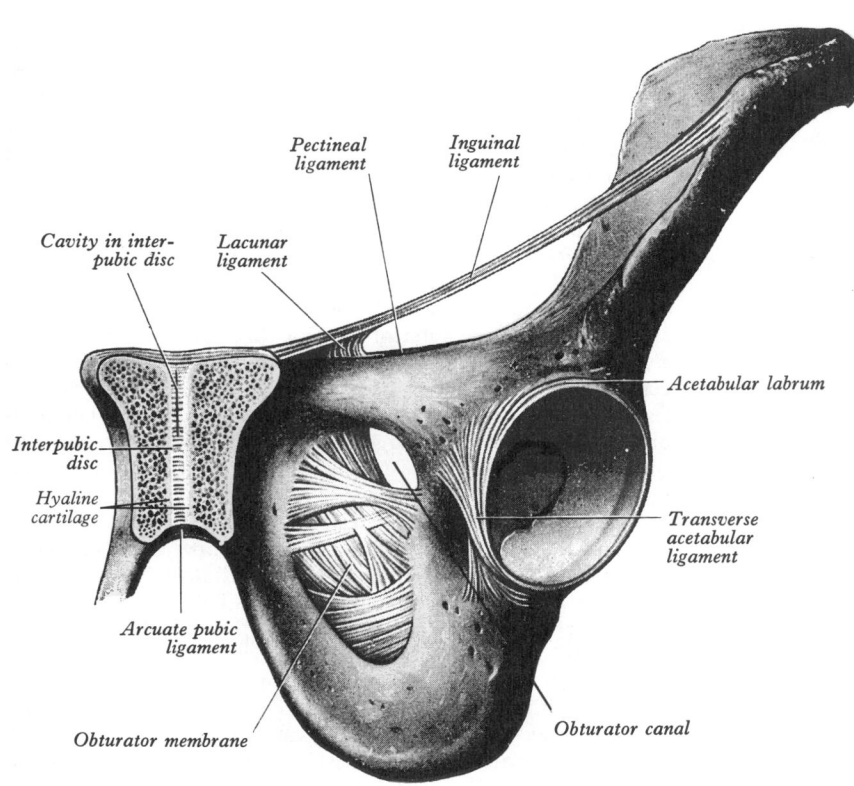

4.58 An obliquely coronal section through the pubic symphysis: antero-inferior aspect.

519

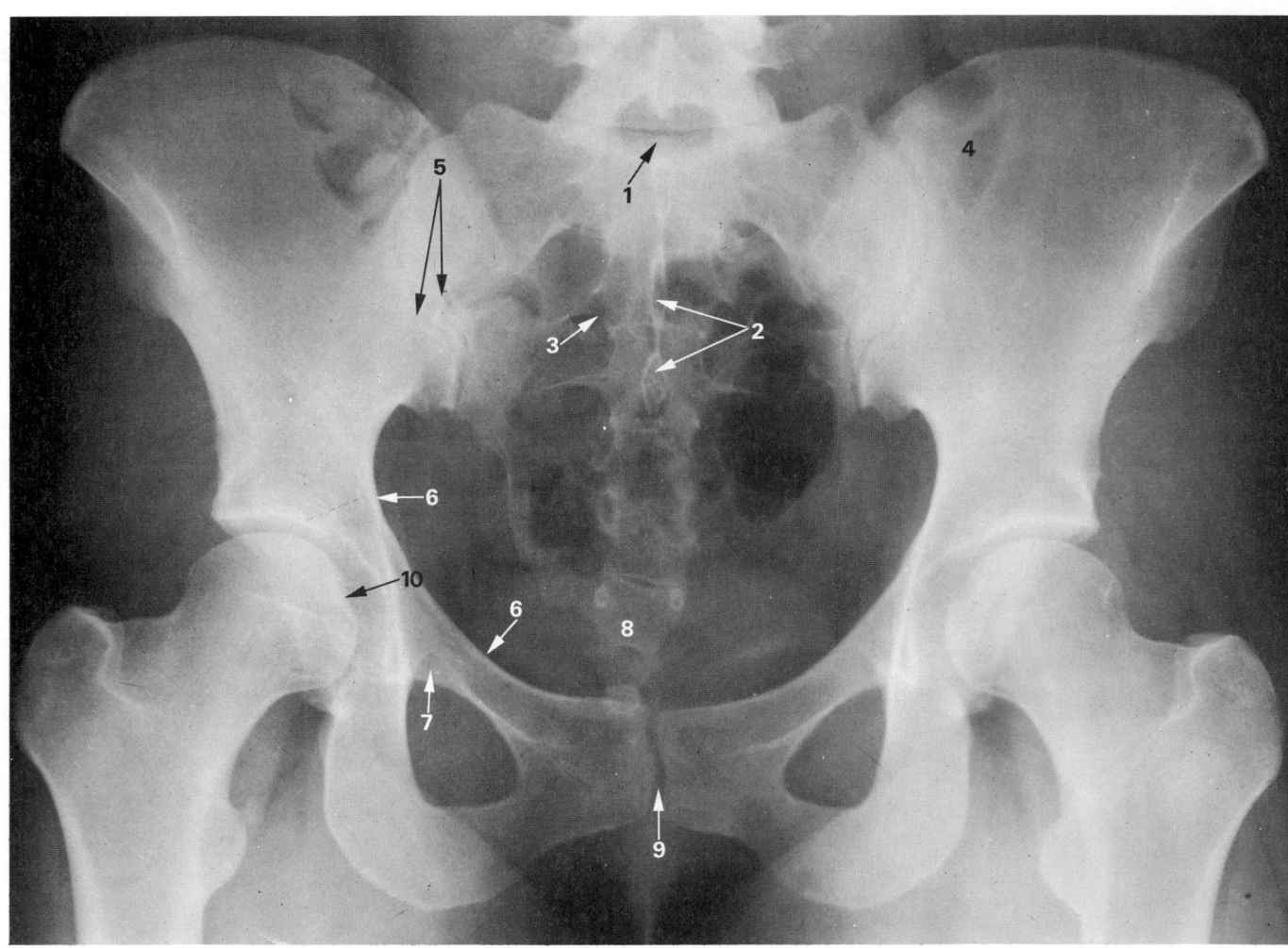

4.59A Anteroposterior radiograph of adult female pelvis. 1. Sacral promontory. 2. Sacral spinous crest. 3. Margin of anterior sacral foramen. 4. Gas in pelvic colon. 5. Sacro-iliac joint. 6. Pelvic brim. 7. Obturator groove. 8. Coccyx. 9. Symphysis pubis. 10. Fovea of femoral head.

The synovial membrane, starting from the femoral articular margin, covers the intracapsular part of the femoral neck, then passes to the capsule's internal surface to cover the acetabular labrum, ligament of the head and fat in the acetabular fossa. It is thin on the deep surface of the iliofemoral ligament where it is compressed against the femoral head and sometimes is even absent here. The joint may communicate with the subtendinous iliac (psoas) bursa (p. 635) by a circular aperture between the pubofemoral and the vertical band of the iliofemoral ligament.

The iliofemoral ligament (4.64), triangular and very strong, is anterior and intimately blended with the capsule, its apex attached low on the anterior inferior iliac spine, its base to the trochanteric line. Its medial and lateral bands are strong, its central part weaker; the medial, vertical band reaches the inferomedial (distal) end of the trochanteric line; the lateral is oblique and attached to a tubercle at the line's superolateral (proximal) end. Hence it is often named the Y-shaped ligament, its lateral band the *iliotrochanteric ligament*.

The pubofemoral ligament (4.64), also triangular, has a base attached to the iliopectineal eminence, superior pubic ramus, obturator crest and membrane; it blends distally with the capsule and deep surface of the medial iliofemoral band.

The ischiofemoral ligament (4.65) thickens the back of the capsule. From the ischium, postero-inferior to the acetabulum, it *spirals* superolaterally behind the femoral neck, some fibres blending with the zona orbicularis, others attached to the greater trochanter deep to the iliofemoral ligament.

The ligament of the femoral head (4.66) is a triangular flat band, its apex attached anterosuperiorly in the pit on the femoral head; its base is principally attached on both sides of the acetabular notch, between which it blends with the transverse

ligament. (However, it receives weaker accessions from the margins of the acetabular fossa.) Ensheathed by flattened tubular synovial membrane, it varies in strength; occasionally its synovial sheath alone exists, without a core; rarely both ligament and sheath are absent. The ligament appears to tense when the thigh is semiflexed and adducted and to relax in abduction.

The acetabular labrum (4.58,62) is a fibrocartilaginous rim attached to the acetabular margin, deepening the cup. It bridges the acetabular notch as the *transverse acetabular ligament*, completing the circle. Triangular in section, it is attached by its base to the acetabular rim, the apex being its free margin. The acetabular cavity is constricted by the labral rim which embraces the femoral head.

The transverse acetabular ligament (4.58) is part of the labrum but has no cartilage cells. Its strong, flat fibres cross the notch forming a foramen through which vessels and nerves enter the joint.

Relations. The coxal capsule is surrounded by muscles (4.67) *Anteriorly*, lateral fibres of pectineus separate its most medial part from the femoral vein; lateral to this the tendon of psoas major, with iliacus lateral to it, descends across it, partly separated by a bursa The femoral artery is anterior to the tendon; the femoral nerve deep in a groove between tendon and iliacus. More laterally the straight head of rectus femoris crosses the joint with a deep layer of the fascial iliotibial tract, which blends with the capsule under the muscle's lateral border. *Superiorly*, the reflected head of rectus femoris contacts the capsule medially, while gluteus minimus covers it laterally, closely adherent. *Inferiorly*, lateral fibres of pectineus adjoin the capsule and, more posteriorly, obturator externus spiralizes obliquely to its posterior aspect. *Posteriorly*, the lower capsule is covered by the tendon of obturator externus, separating

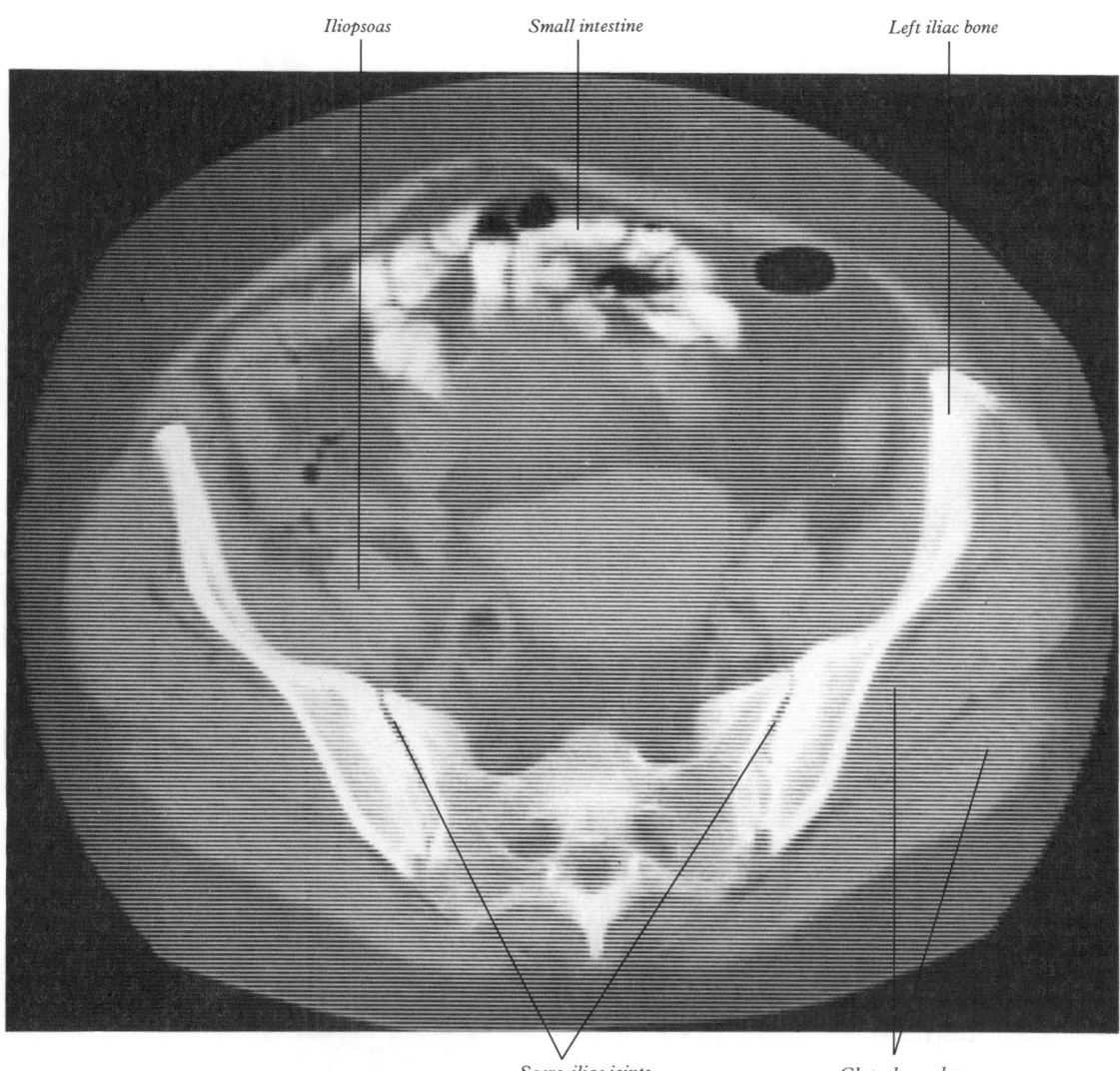

Iliopsoas   *Small intestine*   *Left iliac bone*

Sacro-iliac joints   *Gluteal muscles*

4.59B Computed tomogram in the transverse plane showing the sacro-iliac joints. (Supplied by Shaun Gallagher; photography by Sarah Smith.)

it from quadratus femoris and accompanied by an ascending branch of the medial circumflex femoral artery; above this the tendon of obturator internus and the gemelli contact the joint, separating it from the sciatic nerve; the nerve to quadratus femoris is deep to the obturator internus tendon, the nerve descending most medially on the capsule. Above this, the joint's posterior surface is crossed by piriformis.

Articular arteries are branches from the obturator, medial circumflex femoral, and superior and inferior gluteal arteries. *Nerves*

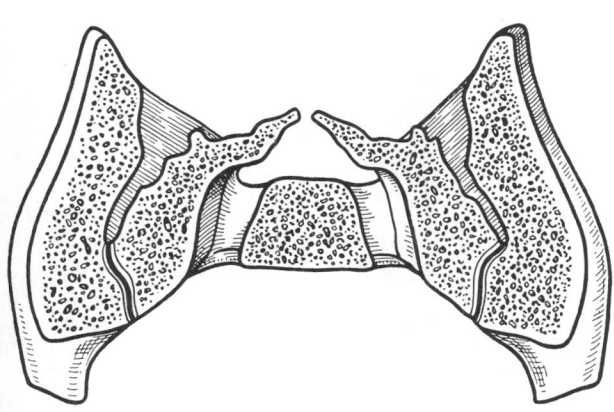

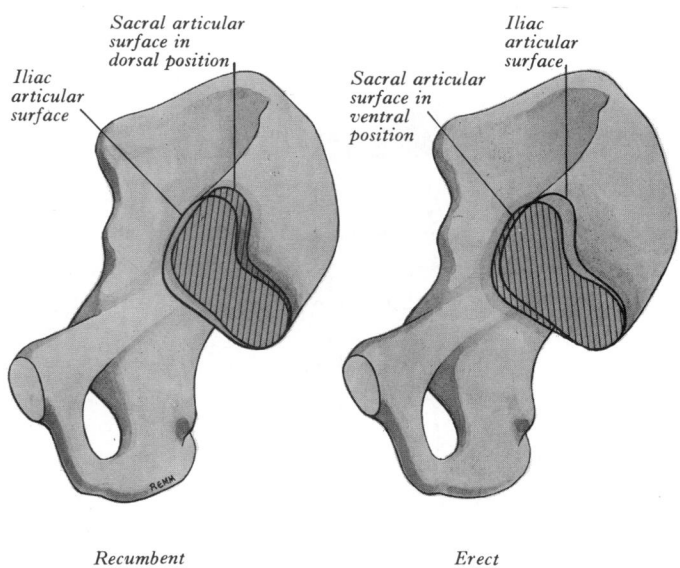

Sacral articular surface in dorsal position   *Iliac articular surface*

Iliac articular surface   *Sacral articular surface in ventral position*

Recumbent   *Erect*

60 Coronal section passing through the middle segment (see text) of the cro-iliac joints. The sacrum is in its normal position with the body in the ect attitude.

4.61 The changing relation (rotation) of the auricular surface of the sacrum and that of the ilium when changing from a recumbent to an erect posture. (Based upon work by H Weisl 1953, with permission of the author.)

521

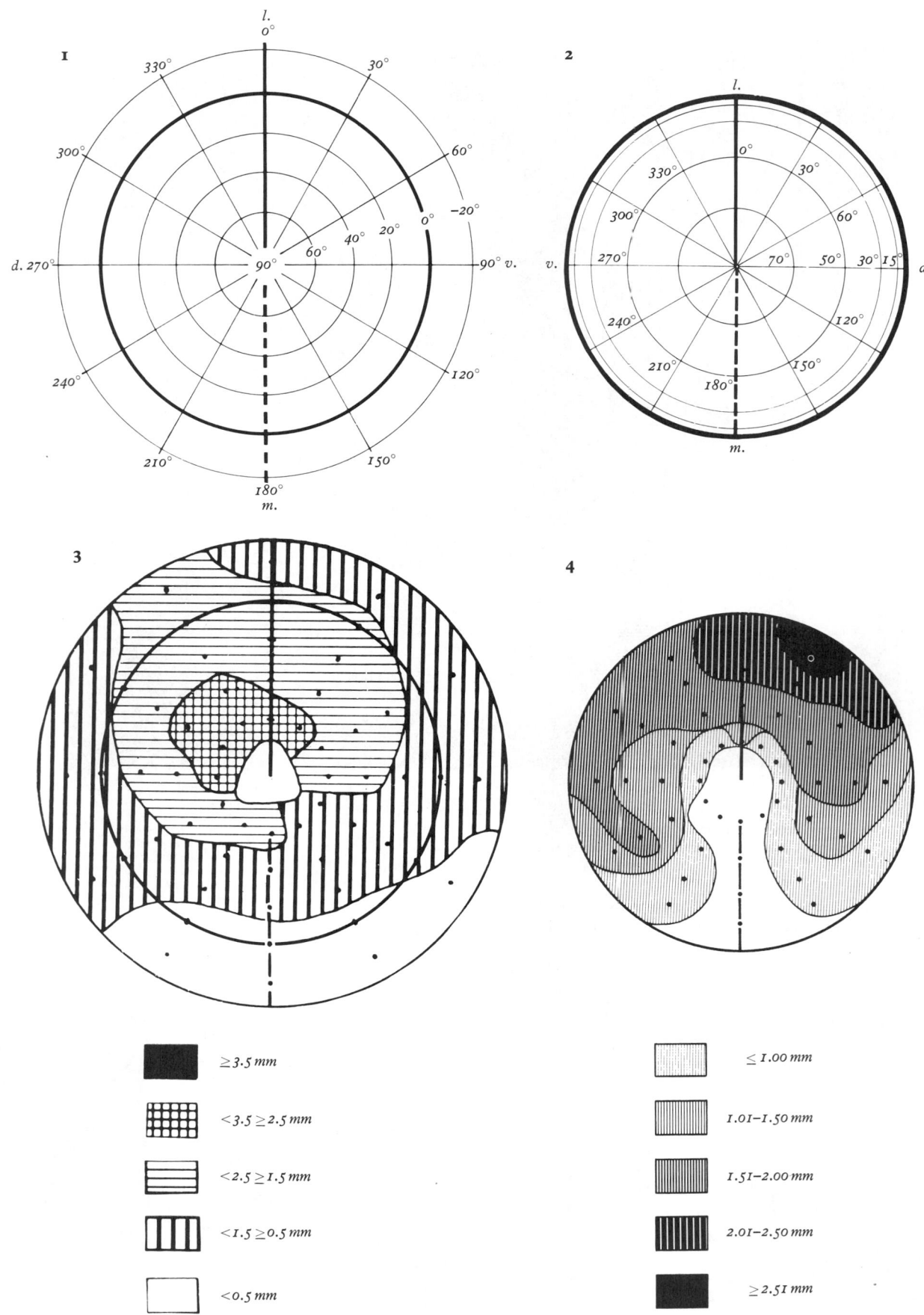

4.62A Variations in the thickness of the articular cartilages of the femoral head (1 and 3) and the lunate surface of the acetabulum (2 and 4). Diagrams 1 and 2 are reference grids by which the distance and angular direction of sampling points are measured from the centres of the femoral head and of the acetabulum. Diagrams 3 and 4 show the average contours of the thickness ranges indicated in the shading codes included below. Dorsal and ventral aspects are denoted as 'd' and 'v'; 'l' and 'm' denote superolateral and inferomedial points on the circumference of the femoral articular surface. The black dots indicate the intersections of lines of 'longitude' and 'latitude' and also represent sampling points. It is immediately apparent that this method does not provide a detaile representation of the graded topology of the articular surfaces. (See p. 47 for an alternative methodology, more recently introduced.) (Provide from the paper by H J Kurrat & W Oberländer, 1978 by courtesy Cambridge University Press.)

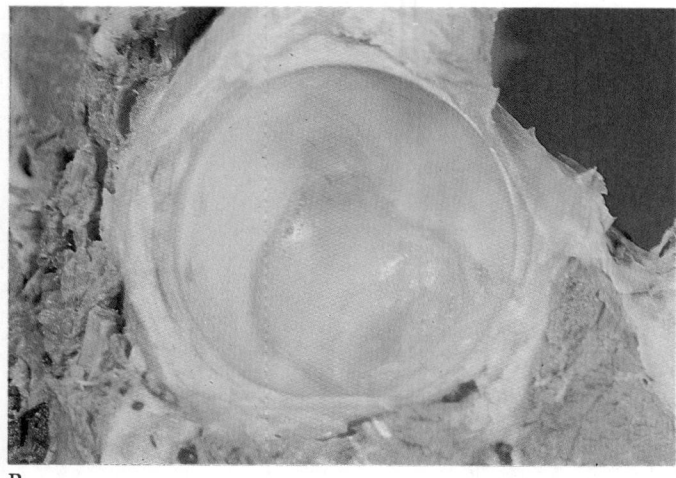

B

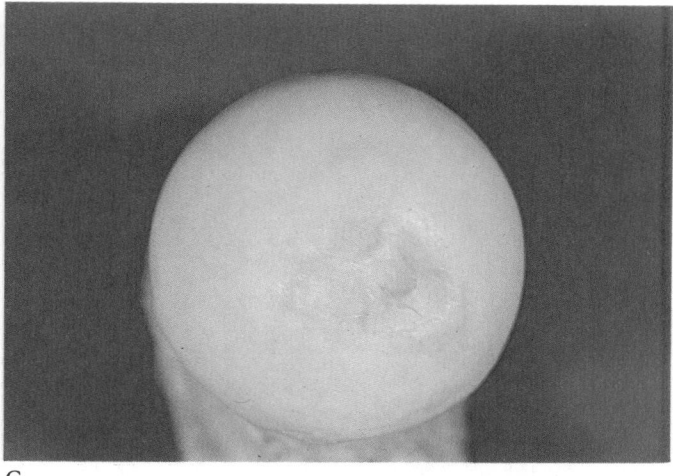

C

4.62 B,C Articular surfaces of a cadaveric hip joint. B shows the acetabulum, comprising the articular cartilage-covered lunate surface, the acetabular fossa with its articular pad of fat, the acetabular labrum, the transverse acetabular ligament and portions of the peri-articular musculature in-

cluded in the field of view. C displays the spheroidal articular cartilage covering the femoral head. Note the severed ligament of the head of the femur. (Preparation by M C E Hutchinson and photography by Kevin Fitzpatrick, Guy's Hospital Medical School, London.)

are from the femoral or its muscular branches, the obturator, accessory obturator, the nerve to quadratus femoris and the superior gluteal nerves (Gardner 1948a).

Movements can be categorized as flexion–extension, adduction–abduction, circumduction, medial and lateral rotation, conveniently considered as rotations around three orthogonal axes (p. 476); but when femoral movements are considered in relation to articular surfaces, length and angulation of the neck in relation to shaft must be remembered. When the thigh is flexed or extended, the femoral head 'spins' in the acetabulum on an approximately transverse axis; conversely, the acetabula rotate around similar axes in flexion and extension of the trunk on stationary femoral heads. Medial and lateral femoral rotation have a vertical axis through the femoral head's centre and lateral condyle, with the

foot *stationary* on the ground. Such rotations are the inevitable conjunct rotations accompanying terminal extension or initial flexion at the knee joint (pp. 530–533). Because of the relation of this axis to the whole femur (with its angulated neck and oblique shaft), during medial rotation the medial condyle moves in an arc *backwards* on the medial tibial condyle and the greater trochanter simultaneously in a *forward* arc: converse movements occur during lateral rotation. With the foot in loose contact or free, medial and lateral adjunct rotation of the whole lower limb occurs around variable axes, all through the femoral head and any part of the foot. Conversely, with one foot stationary and the other free, the whole trunk may rotate on one femoral head—as in cross-kicking. Abduction and adduction are around an anteroposterior axis through the femoral head; but since this is not truly spherical

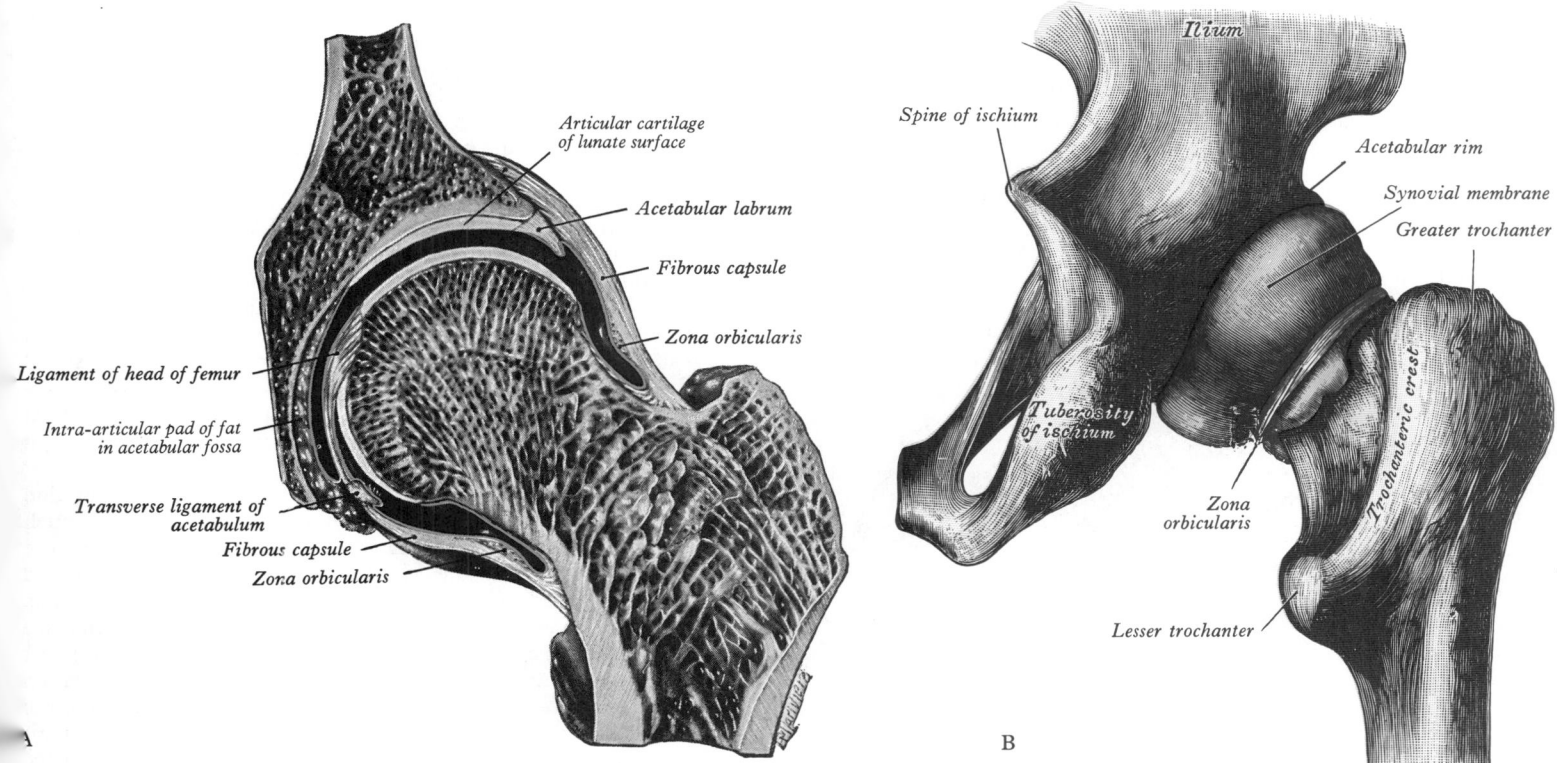

Articular cartilage
of lunate surface

Acetabular labrum

Fibrous capsule

Zona orbicularis

Ligament of head of femur

Intra-articular pad of fat
in acetabular fossa

Transverse ligament of
acetabulum

Fibrous capsule

Zona orbicularis

A

Ilium

Spine of ischium

Acetabular rim

Synovial membrane

Greater trochanter

Tuberosity
of ischium

Zona
orbicularis

Trochanteric crest

Lesser trochanter

B

4.63 A. Section through the hip joint. The synovial membrane is shown in blue. B. Synovial cavity of the right hip joint (distended); posterior aspect.

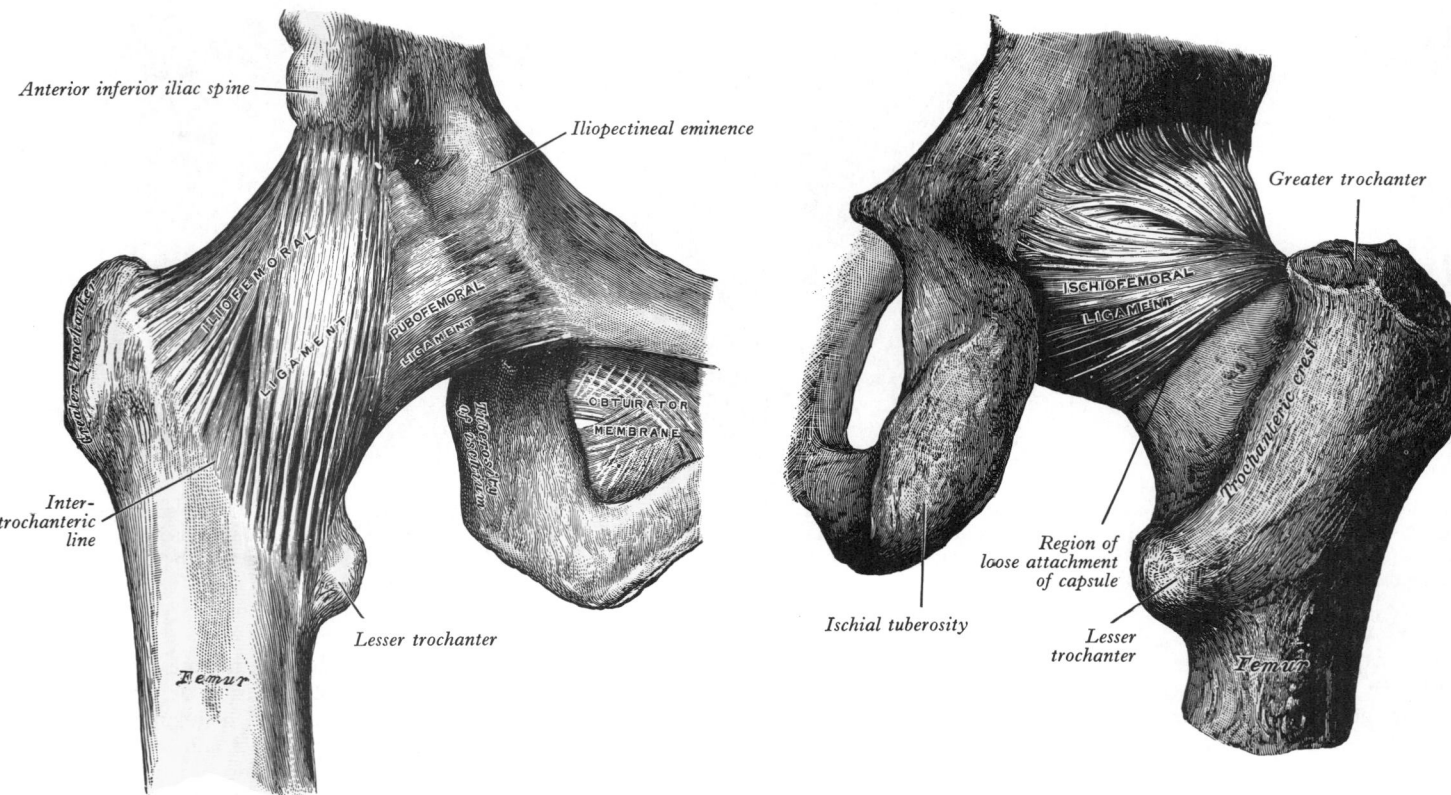

4.64 Right hip joint: anterior aspect.

4.65 Right hip joint: posterior aspect.

no axis is stationary (Walmsley 1928). Some kinesiologists refer to a *mechanical axis* coincident with the topographical long axis of the femoral neck, impinging on the approximate centre of its head's articular surface (p. 435), making extension and flexion of the thigh relatively pure 'spins' at the hip joint and most effective in spiralizing and tautening or straightening and relaxing the cap-

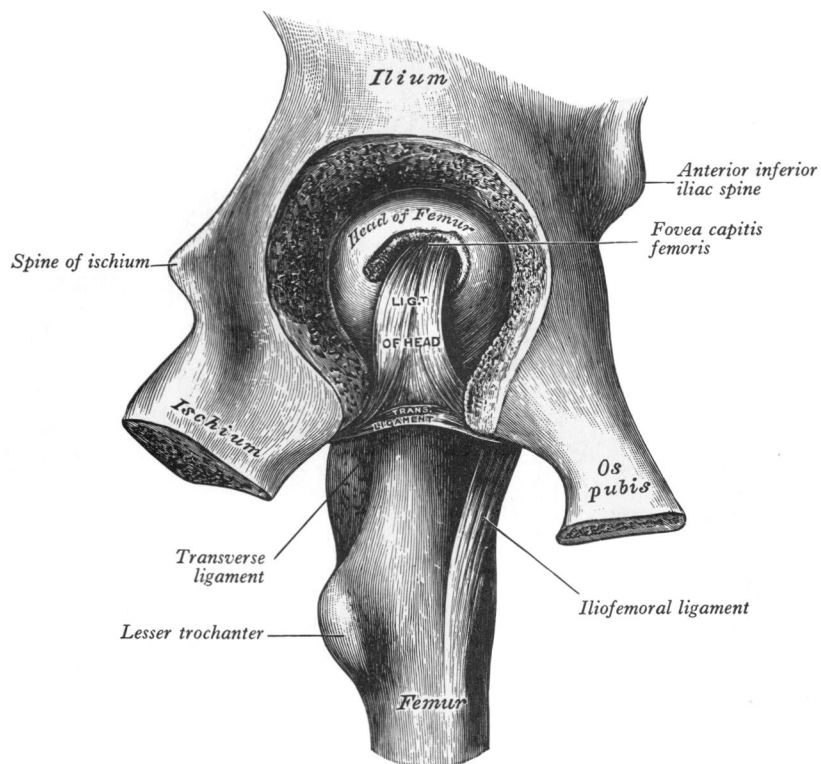

4.66 Left hip joint, opened by the removal of the floor of the acetabulum from within the pelvis.

524

sular ligaments (vide infra). All other movements are regarded as pure or impure swings (p. 480). While this permits analysis closer to actual articular function than a cardinal tri-axial system, related arbitrarily to the 'anatomical' position, the mechanical axis is itself (incorrectly) regarded by some as dynamic, because forms of apparent dynamic 'spin' may occur in many positions; this view stems from an imprecise definition of the varieties of axes employed in arthrokinematics. The notional mechanical axis of the kinesiologist makes an angle of about 125° with the long topographical axis of the femoral shaft, the angle varying with age and sex. The mechanical axis is not dynamic *relative to the femur*. It is stationary during pure spins; it moves relative to its con-articular surface in chordal or arcuate paths during pure or impure swings respectively.

Simple flexion is possible to 90°–100° from the *vertical*; extension beyond the *vertical* is limited (perhaps 10°–20°). Both movements are augmented by adjustments of spinal column and pelvis, flexion of the knee and concomitant medial or lateral coxal rotation. For example, genual flexion (lessening tension in the posterior femoral muscles) increases coxal flexion to 120°; the thigh can be drawn passively to the trunk, though with some spinal flexion. Extension in walking, running, etc., is increased by forward inclination of the body, pelvic tilting and rotation and lateral coxal rotation. (For analyses consult Kapandji 1965, 1974, Joseph 1975.) Abduction and adduction can be similarly increased.

The hip joint differs much from the shoulder joint in limitation of its movements. The humeral head's articular surface much exceeds that of the glenoid cavity and movements are little restrained by the capsule. In the hip, in contrast, the femoral head is closely fitted to the acetabulum in an area of almost half a sphere, embraced more closely by its labrum, being held in place even when the capsule is divided. The iliofemoral is the strongest of all ligaments and is progressively tightened by spiralization when the femur extends to the line of the trunk. The pubofemoral and ischiofemoral ligaments also tauten and, as the joint approaches close-packing, resistance to an extending torque rapidly increases.

No accessory movements occur, except for very slight separation effected by strong traction.

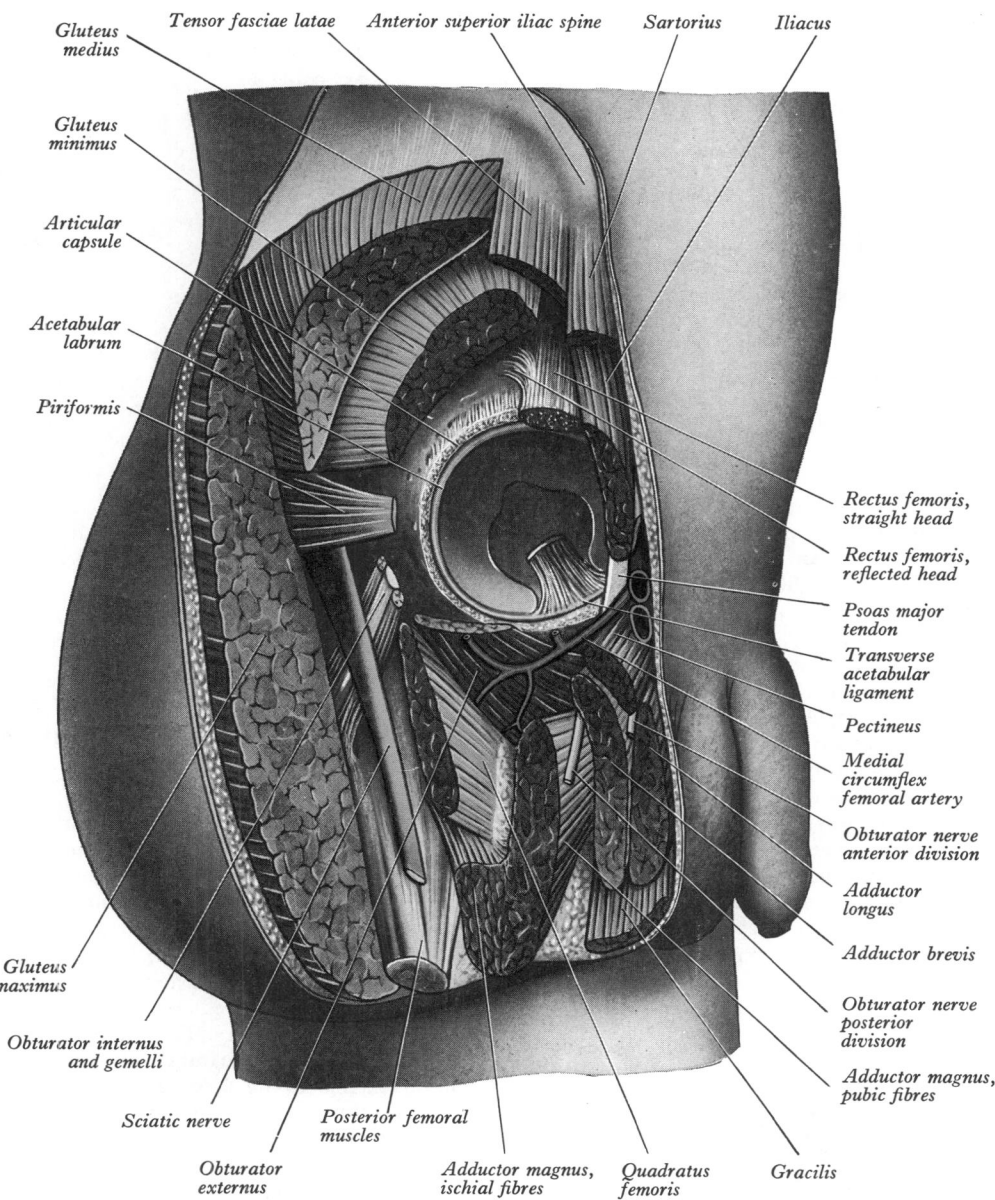

Gluteus medius　　*Tensor fasciae latae*　*Anterior superior iliac spine*　*Sartorius*　*Iliacus*

Gluteus minimus

Articular capsule

Acetabular labrum

Piriformis

Rectus femoris, straight head

Rectus femoris, reflected head

Psoas major tendon

Transverse acetabular ligament

Pectineus

Medial circumflex femoral artery

Obturator nerve anterior division

Adductor longus

Adductor brevis

Obturator nerve posterior division

Adductor magnus, pubic fibres

Gluteus maximus

Obturator internus and gemelli

Sciatic nerve　　*Posterior femoral muscles*

Obturator externus　　*Adductor magnus, ischial fibres*　*Quadratus femoris*　*Gracilis*

4.67 Dissection to display the structures surrounding the right hip joint. The head of the femur has been disarticulated and removed.

Muscles Producing Coxal Movements

Flexion: psoas major and iliacus, assisted by pectineus, rectus femoris and sartorius. Adductors, particularly adductor longus, also assist, especially in early flexion from full extension.

Extension: gluteus maximus and posterior femoral muscles. In full erect posture with pendant arms a vertical through the body's centre of gravity is behind a line joining the centres of the femoral heads; therefore the body tends to incline backwards but is counterbalanced by ligamentous tension and congruence and compression of articular surfaces with the hip joints in the close-packed position. Under increased loading of the trunk or leaning backwards, these resistive but passive factors are assisted by active contraction of the joint's flexors. In swaying forwards at the ankles, or when the arms are stretched forward, and also in forward bending at the hip, the line of body weight moves in front of the transverse axis and the posture adopted, or the rate of change of posture, is largely controlled by the posterior femoral muscles ('hamstrings') which, although powerful *flexors* of the knee, are equally strong *extensors* of the hip. Gluteus maximus only becomes active when the thigh is extended against resistance, as in rising from a bending position or climbing.

Abduction: glutei medius and minimus (p. 642), assisted by tensor fasciae latae and sartorius. Abduction is limited by adduc-

tor tension, the pubofemoral ligament and medial band of the iliofemoral ligament. These muscles are consistently involved in walking or running (p. 658), contracting periodically at precise phases of the walking or running cycle.

Adduction: adductores longus, brevis and magnus, assisted by pectineus and gracilis. Adduction is limited by contact with the opposite limb but its range is wider with the thigh flexed when it is limited by the abductor muscles, the lateral band of the iliofemoral ligament and ligament of the femoral head.

Medial rotation: tensor fasciae latae and anterior fibres of glutei minimus and medius. It is relatively weak and is limited by the lateral rotators, ischiofemoral ligament and posterior part of the capsule. Electromyographic data suggest that adductors usually assist in *medial* rather than lateral rotation but this is, of course, dependent on the primary position.

Lateral rotation: obturator muscles, gemelli and quadratus femoris, assisted by piriformis, gluteus maximus and sartorius. It is a powerful action and limited by tension in the medial rotators and the lateral band of the iliofemoral ligament.

Applied Anatomy. The iliofemoral ligament is rarely torn during dislocation, an advantage in reduction because it can act as the fulcrum to a lever, its long arm the femoral shaft, the short being the femoral neck. Congenital dislocation is more common at the hip than elsewhere. Displacement is usually to the gluteal surface of the ilium, the acetabular rim being deficient above. In manipulation of the sacro-iliac joints the surgeon takes advantage of the tautness of the iliofemoral and ischiofemoral ligaments in coxal extension. So strong are they that forcible coxal hyperextension with forward pressure on the iliac crest produces sacro-iliac movement.

The Genual (Knee) Joint

The knee (4.4), largest of human joints, is *compound* (4.69–78) and sometimes derived from a primitive double condylar articulation, although contrary evidence exists (Haines 1942a). Despite its single cavity in man, it is convenient to describe it as two condylar joints between femur and tibia and a sellar joint between patella and femur. The former are partly divided by menisci between corresponding articular surfaces. (Strictly, the joint is therefore also classified as *complex*.) The level of the joint is at the (palpable) proximal margins of the tibial condyles (4.77).

The articular surfaces (pp. 437, 439, 441) are most incongruent. Femoral condyles, bearing articular cartilage, are almost wholly convex; in lateral profile both are spiral with a curvature increasing posteriorly, the lateral condyle more rapidly. The tibial surfaces are also cartilage-covered areas, separated by the intercondylar region; each is gently hollow centrally and flattened peripherally where a meniscus rests. The lateral tibial articular surface is almost circular and smaller, the medial oval with a longer anteroposterior axis. Both tibial surfaces slope up near the intercondylar area on to its eminences; posteriorly the lateral is prolonged on the back of its condyle, in relation to the popliteal tendon; anteriorly, near the anterior cornu of the lateral meniscus, it descends onto the anterior sloping condylar surface (vide infra). Femorotibial congruence is improved by the menisci, shaped to increase the concavity of tibial surfaces, the combined lateral tibiomeniscal surface being deeper. The lateral femoral condyle has in front a faint groove, resting on the peripheral edge

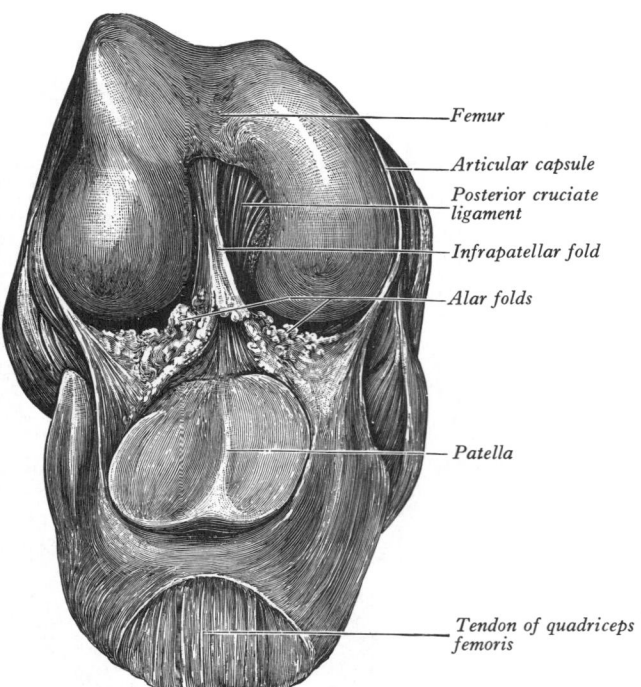

4.69 Right knee joint in flexion: anterior exposure.

of the lateral meniscus in full extension. A similar groove appears on the medial condyle but does not reach its lateral border, where a narrow strip contacts the medial patellar articular surface in full flexion. These grooves demarcate the femoral patellar and condylar surfaces, the latter articulating only with tibia and menisci. The distal outlines (4.69) of femoral surfaces conform to tibiomeniscal articular surfaces. The lateral femoral surface is almost circular, the medial is larger, somewhat oval but curved with a lateral concavity. These differences correlate with movements of the joint (vide infra). The surfaces approach congruence in full extension, the close-packed position, as confirmed by casting techniques and radiological studies of excised human joints. (Vide infra for detailed discussion of close-packing and consult Kettlekamp et al 1972 for relevant bibliography.)

The patella's articular surface is adapted to the femoral surface (p. 439), which extends on the anterior surfaces of both condyles like an inverted U. An oblique groove, descending a little laterally, divides the femoral patellar surface into a larger lateral and a medial area; the lateral is wider, passes more steeply on to the prominent anterior boss of the lateral condyle and ascends higher on its anterior surface. Since the whole area is concave transversely and parasagittally convex, it is an asymmetrical sellar surface (p. 478). A rounded, almost vertical ridge, dividing the articular surface of the patella also into larger lateral and medial areas, fits the corresponding femoral groove; but the two areas are not fully congruent with those of the femur. The patellar surface may be subdivided by two faint, horizontal ridges which, with the vertical ridge, map out three pairs of facets. But in many patellae there is only *one* horizontal ridge, better marked laterally, and the upper lateral facet is more deeply hollowed. Medially, a second vertical ridge separates a narrow semilunar strip from the medial border which contacts the lateral anterior end of the medial femoral condyle in full flexion, when the highest lateral patellar facet contacts the anterior part of the lateral condyle. As the knee extends, the middle patellar facets contact the lower half of the femoral surface; in full extension only the lowest patellar facets are in contact with the femur (4.79). Seedham & Tsubuku (1977) have applied a special technique for study of contact areas between visco-elastic bodies to the patella. They emphasize temporal factors in compressive contacts.

Ligaments of the joint are its fibrous capsule, ligamentum (tendo) patellae, tibial and fibular collateral, oblique and arcuate popliteal, anterior and posterior cruciate and transverse genual ligaments.

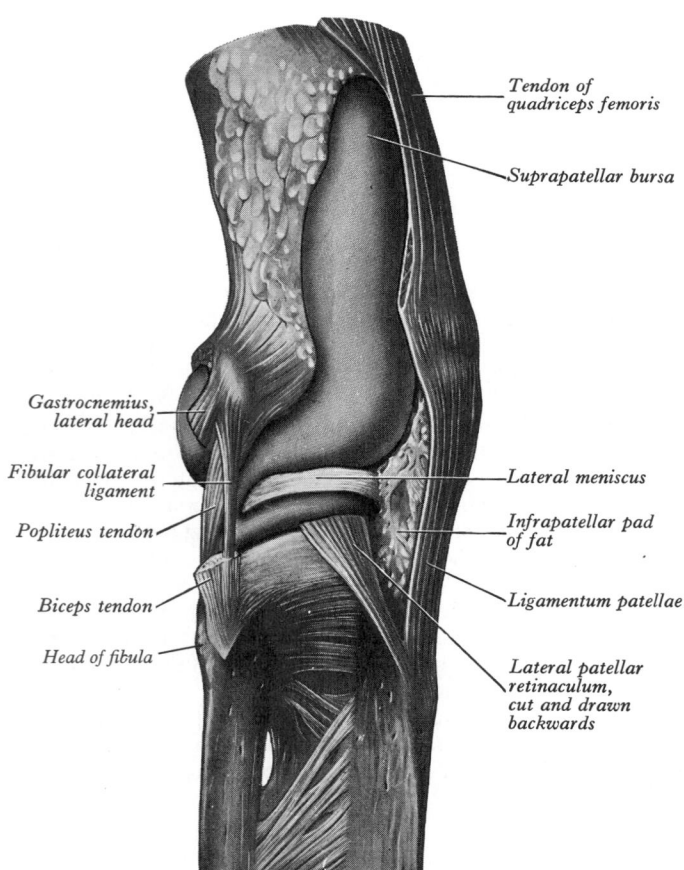

4.68 Dissection of the right knee joint: lateral aspect. The joint cavity has been injected and the synovial membrane is coloured blue.

The fibrous capsule is complex, partly deficient and partly augmented by expansions from adjacent tendons. Its *posterior, vertical fibres* are attached proximally to the posterior margins of the femoral condyles and intercondylar fossa and distally to the posterior margins of the tibial condyles and intercondylar area; proximally on each side it is blended with attachments of gastrocnemius and centrally strengthened by the oblique popliteal ligament. *Medial capsular fibres* are attached to femoral and tibial condyles just beyond their articular margins; here the capsule blends with the tibial collateral ligament. Between the medial epicondyle and the convex border of the medial meniscus is a capsular thickening, a deep component of the tibial collateral ligament (Last 1948). *Lateral capsular fibres*, attached to the femur above popliteus, descend over its tendon to the tibial condyle and fibular head. The fibular collateral ligament is separated from the capsule by fat and the inferior lateral genicular vessels and nerve. *Anteriorly* the capsule does not pass proximal to the patella or over the patellar area; elsewhere it blends with expansions from the vasti medialis and lateralis; these are attached to the patellar margins and ligamentum patellae, extending back to the corresponding collateral ligaments and distally to the tibial condyles. They form *medial* and *lateral patellar retinacula*, the lateral being augmented by the iliotibial tract. Proximal to the patella an absence of capsule allows continuity of the suprapatellar bursa (p. 640) with the joint. Posterior capsular attachment to the lateral tibial condyle is interrupted where popliteus emerges (4.72) but elsewhere, as mentioned, the oblique popliteal ligament, derived from the tendon of semimembranosus, thickens the posterior capsule. A prolongation of the iliotibial tract fills the interval between oblique popliteal and fibular collateral ligaments, partly covering the latter; its expansions also reach the lateral patellar retinaculum and patellar ligament. Medially, expansions from sartorius and semimembranosus ascend to the tibial collateral ligament. Internally the capsule is attached to the meniscal rims, connecting them to the tibia by short *coronary ligaments*.

The synovial membrane of the knee is the most extensive and complex in the body. At the proximal patellar border, it forms a large *suprapatellar bursa* between quadriceps femoris and lower femoral shaft (4.68,75). This is, in practice, an extension of the genual cavity, sustained by the *articularis genus* which is attached to it. Alongside the patella the membrane extends beneath the aponeuroses of the vasti, more extensively under the medial. Distal to the patella the synovial membrane is separated from the ligamentum patellae by the *infrapatellar pad of fat*, covering which the membrane projects into the joint as two fringes, or *alar folds*; these bear villi and then converge posteriorly to form the single *infrapatellar fold (ligamentum mucosum)* which curves posteriorly to its attachment in the femoral intercondylar fossa (4.69); this may be a vestige of the inferior boundary of an originally separate femoropatellar joint (p. 526). At the sides of the joint the synovial membrane descends from the femur, lining the capsule as far as the menisci whose surfaces have no synovial covering. Posterior to the lateral meniscus the membrane forms a *subpopliteal recess* between a groove on the meniscal surface and the tendon of popliteus (4.70); this may connect with the superior tibiofibular joint (p. 533). The relation of synovial membrane to cruciate ligaments is described on p. 528.

The ligamentum patella (4.71) is the central band of the tendon of quadriceps femoris, continued distally from patella to tibial tuberosity. It is strong, flat, about 8 cm in length, attached proximally to the apex, adjoining margins and to rough areas on the anterior surface and on the depression on the distal posterior patellar surface, and distally to the superior smooth area of the tibial tuberosity. Its superficial fibres are continuous over the patella with the tendon of quadriceps femoris, medial and lateral parts of which descend, flanking the patella, to the sides of the tibial tuberosity, merging into the fibrous capsule as *medial* and *lateral patellar retinacula*. The ligamentum patellae is separated from synovial membrane by a large infrapatellar fat pad and from the tibia by a bursa (4.75).

The oblique popliteal ligament (4.75) expands from the tendon of semimembranosus, blends partly with the capsule and ascends laterally to the lateral part of the intercondylar line and lateral femoral condyle. Its fasciculi are separated by apertures for

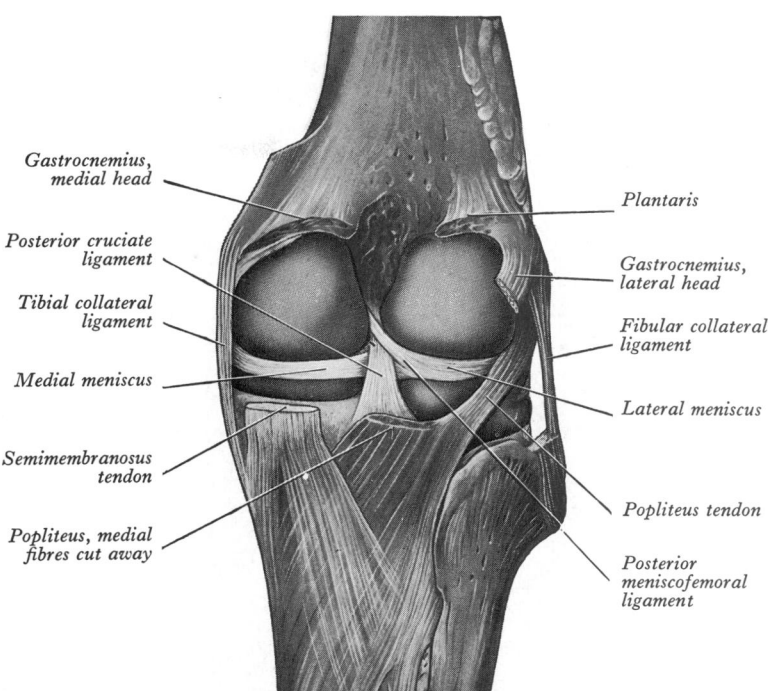

4.70 Posterior dissection of the right knee joint. The fibrous capsule has been removed, exposing the unopened synovial membrane which is coloured blue. The cavity of the joint had been partially distended by injection.

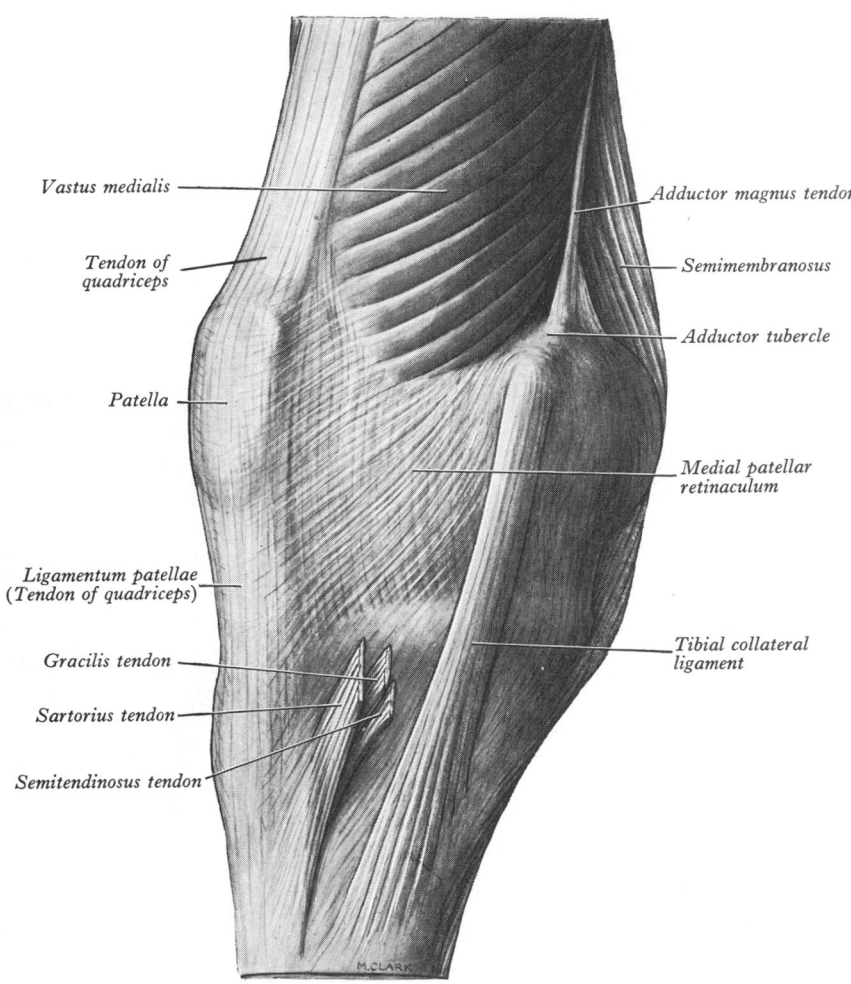

4.71 Right knee joint: anteromedial aspect.

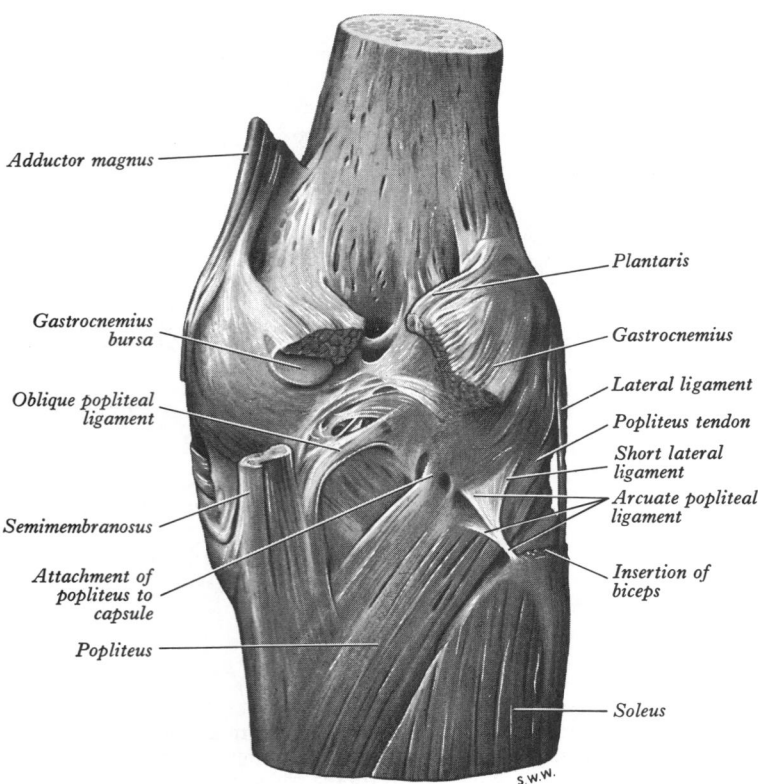

Adductor magnus

Gastrocnemius
bursa

Oblique popliteal
ligament

Semimembranosus

Attachment of
popliteus to
capsule

Popliteus

Plantaris

Gastrocnemius

Lateral ligament

Popliteus tendon

Short lateral
ligament

Arcuate popliteal
ligament

Insertion of
biceps

Soleus

4.72 Right knee joint: posterior aspect.

popliteus to the posterior border of the tibial intercondylar area; the anterior limb, sometimes absent, extends to the lateral femoral epicondyle, is connected with the lateral head of gastrocnemius and is often termed the *short lateral genual ligament*.

The tibial collateral ligament (4.71,76), a broad flat band nearer the back of the joint, extends from the medial femoral epicondyle, immediately distal to the adductor tubercle, to the medial meniscus, tibial condyle and adjacent shaft. Its *anterior part* is flat, about 10 cm long, and may be separated from the capsule and medial meniscus by one or more bursae (Brantigan & Voshell 1943). It slopes anteriorly as it descends to the medial margin and posterior medial surface of the tibial shaft where it is crossed by tendons of sartorius, gracilis and semitendinosus, with a bursa interposed. It lies over the medial inferior genicular vessels and nerve and anterior part of the tendon of semimembranosus, and may have tenuous connections with the latter. Its *posterior part* fans cut to blend with the back of the capsule; it is short and descends posteriorly to the medial tibial condyle proximal to the groove for semimembranosus.

The fibular collateral ligament (4.73), a strong cord, is attached to the lateral femoral epicondyle, proximal to the popliteal groove, and extends to the fibular head in front of its apex. It is largely overlapped by the tendon of biceps femoris, which embraces and partly blends with it; deep to the ligament are the tendon of popliteus and the inferior lateral genicular vessels and nerve. The ligament is *not* attached to the lateral meniscus.

The cruciate ligaments are very strong and sited a little posterior to the articular centre. They are termed *cruciate* because they cross and *anterior* and *posterior* from their *tibial* attachments. Their disposition suggests identification with collateral ligaments of originally separate medial and lateral femorotibial joints (pp. 526, 527).

The anterior cruciate ligament (4.74) is attached medially to the tibial *anterior* intercondylar area and partly blended with the anterior cornu of the lateral meniscus; it ascends *posterolaterally*, twisting on itself and fanning out to attach to the posteromedial aspect of the *lateral* femoral condyle (Last 1951). It is anterolateral to the posterior cruciate.

The posterior cruciate ligament (4.73,74) is stronger, less

vessels and nerves; it is in the floor of the popliteal fossa with the popliteal artery in contact.

The arcuate popliteal ligament (4.72), a Y-shaped mass of capsular fibres, has a stem attached to the head of the fibula; its posterior limb arches medially over the emerging tendon of

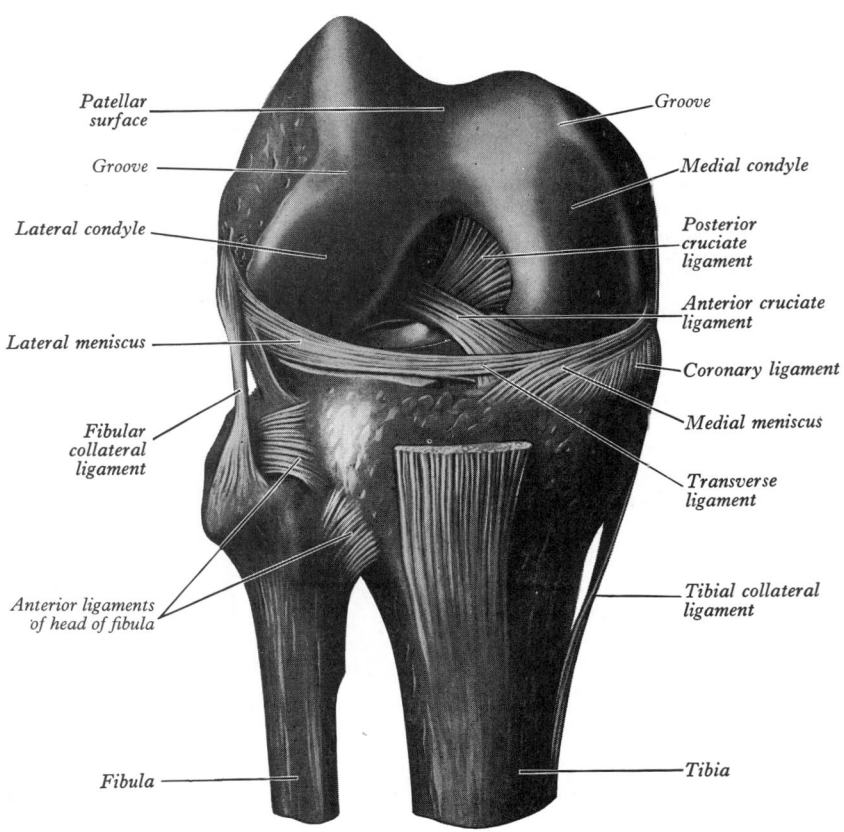

Patellar
surface

Groove

Lateral condyle

Lateral meniscus

Fibular
collateral
ligament

Anterior ligaments
of head of fibula

Fibula

Groove

Medial condyle

Posterior
cruciate
ligament

Anterior cruciate
ligament

Coronary ligament

Medial meniscus

Transverse
ligament

Tibial collateral
ligament

Tibia

4.73 Right knee joint in full flexion: dissection from anterior aspect.

oblique, with some shorter fibres and it is attached to the *posterior* intercondylar area and posterior cornu of the lateral meniscus; it ascends *anteromedially*, broadening out to its attachment on the lateral surface of the *medial* femoral condyle.

Observations on 24 fresh joints (Girgis et al 1975) suggested that each ligament has a main posterolateral and a smaller antero- medial band, which behave differently in movement (vide infra). Spiralization is apparent and more pronounced in certain positions. Average dimensions were: anterior cruciate length 38 mm, width 11 mm; posterior cruciate length 38 mm, width 13 mm. Odensten et al (1985) in a functional and surgical study gave a mean length of 31.3 mm in 33 adult cadavers (Swedish).

It should be noted that these ligaments have substantial cross- sectional areas and attachments that are not only in orthogonal planes but also displaced dorsoventrally; all these factors introduce difficulties during mensuration, particularly deter- minations of 'length'.

Synovial membrane almost surrounds the ligaments but is reflected posteriorly from the posterior cruciate to adjoining parts of the capsule. Thus the intercondylar part of the posterior region of the *fibrous capsule* has no synovial covering. A synovial bursal recess intrudes between the ligaments from their lateral aspect (4.75) and may reach the medial wall of the femoral intercondylar fossa.

The menisci (semilunar cartilages) (4.74) are crescentic laminae deepening the articulation of the tibia which receives the femur. Their peripheral attached borders are thick and convex; their free borders are thin and concave. Their peripheral zone is vascularized by capillary loops from the fibrous capsule and synovial membrane, their inner regions avascular (Davies & Edwards 1948). Their proximal surfaces are smooth and concave, in contact with articular cartilage on femoral condyles, their distal surfaces smooth and flat, resting on tibial articular cartilage. Each covers about two-thirds of its tibial articular surface. Ghadially et al (1983) have detailed meniscal ultrastructure.

The medial meniscus, almost a *semicircle* and broader behind (4.74), is attached at its *anterior end* (cornu) to the anterior tibial intercondylar area in front of the anterior cruciate ligament, its posterior fibres continuous with the transverse genual ligament. This anterior cornu is in the floor of a depression medial to the upper part of the ligamentum patellae. The *posterior cornu* is fixed to the posterior tibial intercondylar area, between attachments of the lateral meniscus and posterior cruciate ligament. Its peripheral border is attached to the fibrous capsule and the deep surface of the tibial collateral ligament (4.78).

The lateral meniscus, about four-fifths of a *ring* (4.74), covers a larger area than the medial. Its breadth is uniform, except for the short tapering cornua; it is grooved posterolaterally by the popliteal tendon, separating it from the fibular collateral liga- ment. Its *anterior cornu* is attached in front of the tibial inter- condylar eminence, posterolateral to the anterior cruciate liga- ment, with which it partly blends. Its *posterior cornu* is attached behind this eminence, in front of the posterior cornu of the medial meniscus. Its anterior attachment is twisted, its free margin facing posterosuperiorly, the anterior cornu resting on the anterior slope of the lateral intercondylar tubercle. Near its posterior attach- ment it commonly sends a *posterior meniscofemoral ligament* (4.74) superomedially behind the posterior cruciate ligament to the medial femoral condyle. An *anterior meniscofemoral ligament* may also connect the posterior cornu to the medial femoral condyle anterior to the posterior cruciate ligament. The meniscofemoral ligaments are often the sole attachments of the posterior cornu of the lateral meniscus. More medially, however, part of the popliteal tendon is attached to the lateral meniscus and mobility of its posterior cornu may thus be controlled by the meniscofemoral ligaments and popliteus (Last 1948). A *meniscofibular ligament* has been described in about 80% of knee joints (Živanović 1973).

That menisci meet a functional need is shown by their reforma- tion after full excision, regeneration being from peripheral vascular tissue. Prior to such reformation the knee joint shows no instability; but if it is subjected to continued violent exercise, subsequent history indicates that articular cartilage suffers per- manent damage, perhaps due to inefficient lubrication.

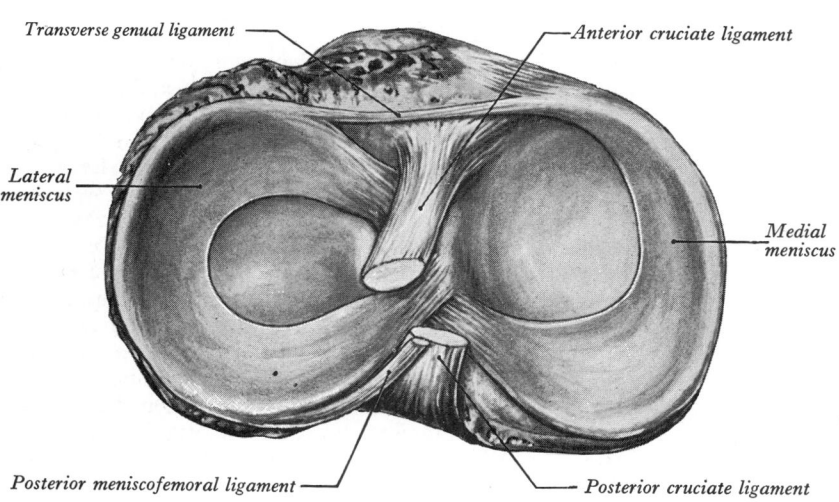

4.74 Superior aspect of the left tibia, showing the menisci and the tibial attachments of the cruciate ligaments.

The transverse genual ligament (4.74) connects the ante- rior convex margin of the lateral to the anterior cornu of medial meniscus; it varies in thickness and may be absent. According to Živanović (1974) this *menisco-meniscal* ligament varies much and is absent from about 40% of joints. A posterior menisco-meniscal ligament was observed in 20% of 300 knee joints; the same obser- ver described other inconstant ligaments.

Bursae associated with the knee are numerous. **Anteriorly** are: (1) a large *subcutaneous prepatellar bursa* between the lower

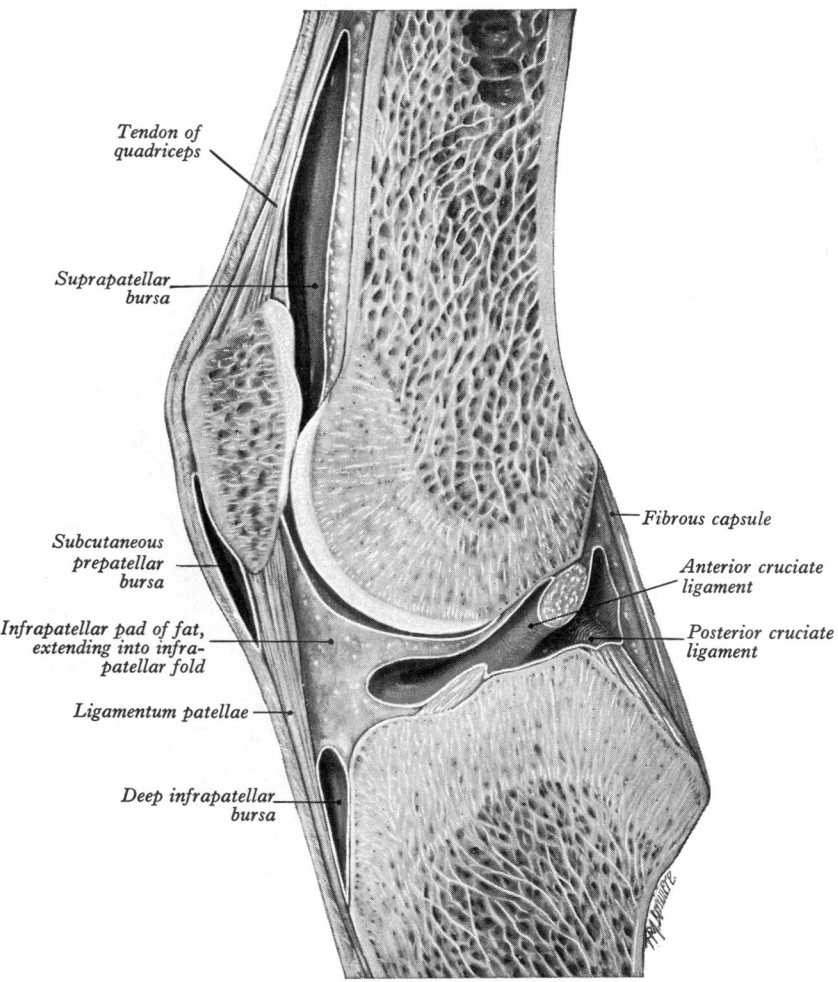

4.75 Sagittal section through the left knee joint: lateral aspect. The synovial membrane is shown in colour. See also 4.4.

patella and skin; (2) a small *deep infrapatellar bursa* between the tibia and ligamentum patellae; (3) a *subcutaneous infrapatellar bursa* between the distal part of the tibial tuberosity and skin; and (4) a large *suprapatellar bursa* between the femur and quadriceps femoris (4.75), developing separately although (4) is later continuous with the joint and better regarded as part of it. **Laterally** there are: (1) a bursa between the *lateral head of gastrocnemius* and joint capsule (sometimes continuous with the joint); (2) one between the *fibular collateral ligament* and *tendon of biceps femoris*; (3) one also between the *fibular collateral ligament* and *tendon of popliteus* (sometimes an expansion from 4); (4) one between the *tendon of popliteus* and *lateral femoral condyle*, usually an extension from the joint. **Medially** are bursae: (1) between the *medial head of gastrocnemius* and fibrous capsule, with a prolongation between the medial tendon of gastrocnemius and the tendon of semimembranosus, often communicating with the joint; (2) superficial to the *tibial collateral ligament*, between this and *tendons of sartorius, gracilis and semitendinosus*; (3) *deep* to the *tibial collateral ligament* in variable numbers and positions between this ligament and the femur, capsule, medial meniscus, tibia or tendon of semimembranosus; (4) between the *tendon of semimembranosus* and the *medial tibial condyle* and also the medial head of gastrocnemius, named the *semimembranosus bursa* which may communicate with (1); (5) occasionally between the tendons of semimembranosus and semitendinosus. Posteriorly, bursal extensions vary.

Relations. *Anterior* are the tendon of quadriceps femoris enclosing and attached to nonarticular surfaces of the patella, the tendon's continuation, the ligamentum patellae, and tendinous expansions from vasti medialis and lateralis extending over *anteromedial* and *anterolateral* aspects of the capsule respectively, as patellar retinacula. *Posteromedial* is sartorius, with the tendon of gracilis along its posterior border, both descending across the joint. *Posterolaterally* biceps tendon, with the common peroneal nerve medial to it, is in contact with capsule, separating it from popliteus (4.76). *Posteriorly* the popliteal artery and associated lymph nodes are on the oblique popliteal ligament, with the popliteal vein posteromedial or medial and tibial nerve posterior to both. The nerve and vessels are overlapped by both heads of gastrocnemius and laterally by plantaris. Around the vessels gastrocnemius contacts the capsule and medial to its medial head semimembranosus is between capsule and semitendinosus.

Arteries supplying the joint are the descending genicular branches of the femoral, superior, middle and inferior genicular branches of the popliteal, anterior and posterior recurrent branches of anterior tibial, the circumflex fibular artery and the descending branch of lateral circumflex femoral. (For genicular anastomoses see p. 786.) Scapinelli (1968) and Wladmirow (1968) showed a penetrative ligamentous supply, even the cruciate ligaments receiving small vessels. *Nerves* are from the obturator, femoral, tibial and common peroneal nerves (Gardner 1948a, Freeman & Wyke 1967).

Movements are customarily described as flexion, extension, medial and lateral rotation. Flexion and extension differ from true hinging: firstly, the spiral profiles of femoral condyles shift the axis up and forwards during extension, back and downwards during flexion; secondly, with the foot fixed, the last 30° of extension involves conjunct medial femoral rotation and early flexion entails corresponding lateral rotation. These conjunct rotations are due to the geometry of articular surfaces and the disposition of ligaments. Conversely, with the foot free to move, extension involves lateral rotation of tibia and flexion a medial rotation. In full flexion posterior parts of tibial surfaces contact posterior parts of femoral articular surfaces. During extension tibia and menisci glide forwards on the femoral condyles, the area of contact increasing and also moving forwards, carrying menisci with it. As movement progresses flatter parts of the femoral condyles contact the tibia and menisci are opened out, their anterior ends moving forwards, the posterior moving little. In flexion the reverse occurs so that menisci, moving with tibia, adapt to the curves of the femoral condyles where contact occurs.

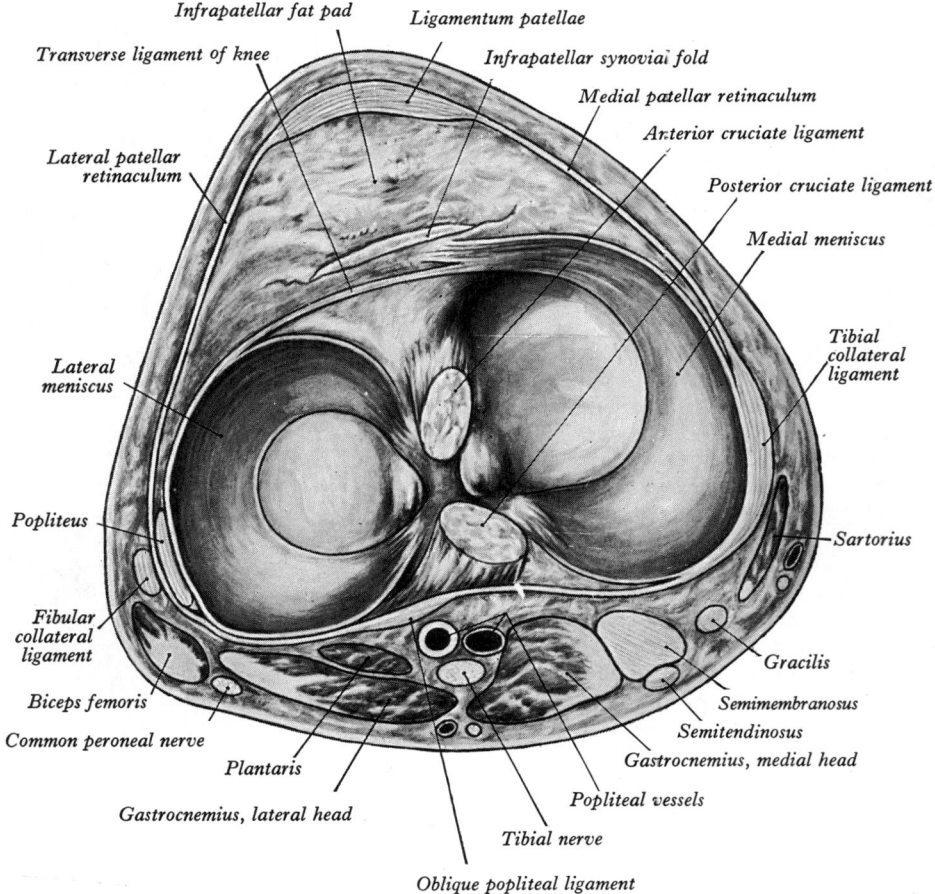

4.76 Transverse section of the left knee joint; superior aspect, to show the relations of the joint.

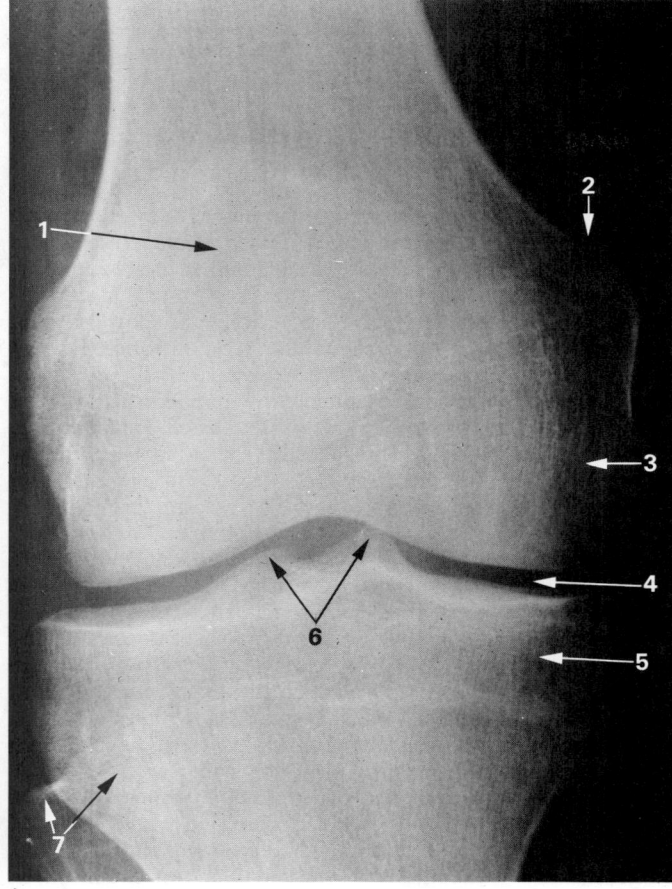

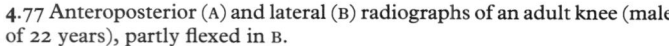

A

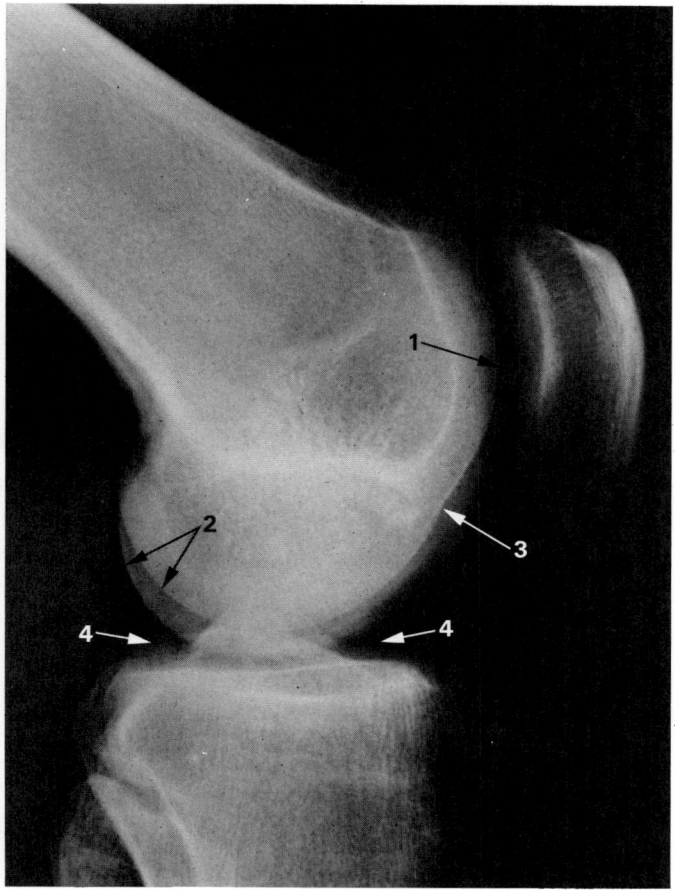

B

4.77 Anteroposterior (A) and lateral (B) radiographs of an adult knee (male of 22 years), partly flexed in B.

A 1. Shadow of patella superimposed on femur. 2. Adductor tubercle. 3. Medial femoral condyle. 4. Radiotranslucent space occupied by medial meniscus and articular cartilages. 5. Medial tibial condyle. 6. Intercondylar eminences. 7. Head of fibula.

B 1. Patellar surface of femur. 2. Spiral profiles of femoral condyles. 3. Groove impinging on anterior end of meniscus in full extension. 4. Note marked incongruity of femorotibial joint surfaces.

Rotations have a smaller range than flexion and extension, and involve translation of menisci *with* femoral condyles on the tibia. These rotations are *conjunct*, integral with flexion and extension but can be *adjunct* and independent, best demonstrated with the knee semiflexed.

Range of extension is about 5–10° beyond a vertical femorotibial axis, flexion about 120° with the hip extended, or 140° when it is flexed, and 160° when a passive element such as sitting on the heels aids it. *Passive* rotation is about 60–70° but *conjunct* rotation only about 20°.

Conjunct medial rotation of femur on tibia in later stages of extension is part of a 'locking' mechanism, an asset when fully extended knees are subjected to strain. Full extension is the position of close-packing, with maximal congruence, compression and contact area, maximal spiralization and tautening of ligaments. The roles of articular surfaces, musculature and ligaments in generating conjunct rotations have been much disputed (Barnett 1952, Kapandji 1974, Girgis et al 1975, Rajendran 1985) but the following points can be made. The *lateral* combined meniscotibial 'receiving surface' is smaller, more circular, more deeply concave. The lateral femoral articular surface is also smaller, rounder and flattens more rapidly. Consequently, the lateral femoral condyle approaches full congruence with the opposed surface about 30° before full extension (well before the medial condyle). Simple extension cannot continue but medial rotation of the femur occurs on a vertical axis through its head and lateral condyle, the medial femoral condyle moving backwards in an arc; while rotation of the lateral femoral condyle and meniscus brings the latter's anterior cornu on to the anterior slope of the lateral tibial condyle. Full congruence of lateral condyle is thus delayed by this deformation of its 'receiving surface' and extension continues. Rotation and extension follow simultaneously and

smoothly until final close-packing of both condyles coincides. At the beginning of flexion from full extension (with foot fixed) lateral femoral rotation occurs, 'unlocking' the joint. While joint surfaces and many ligaments are again similarly involved, electromyographic evidence shows that contraction of popliteus is important, pulling down and backwards on the lateral femoral condyle, *lateral* to the axis of femoral rotation. Through its attachment to the lateral meniscus, it also retracts its posterior cornu during lateral rotation and continuing flexion, obviating traumatic compression (Last 1950).

During extension femoral rotation is observable about 30° before full extension; it first progresses slowly, but more rapidly in the last 5°, when there is a progressive increase in passive mechanisms resisting further extension, i.e. spiralizing and tautening of ligaments, increasing congruence and compression of articular surfaces and gradually increasing tension in all extra-articular tissues crossing the joint's posterior aspect (Smith 1956). Any actual position of extension adopted is a balance between forces (torque) extending the joint and passive mechanisms resisting this. The range *near* to close-packing is functionally important. In symmetrical standing, the line of body weight is anterior to the transverse axes of the knee joints but the *passive* mechanisms noted above preserve posture with minimal muscular effort (Joseph 1960). Active contraction of extensors and a close-packed position only occur in asymmetrical postures, in leaning forward, heavy loading or when powerful thrust is needed.

In extension parts of both cruciate ligaments, tibial and fibular collateral ligaments, the posterior capsular region, oblique posterior ligament, skin and fasciae are all taut; passive and sometimes active tension exists in the hamstrings and gastrocnemius, and the anterior parts of menisci are compressed between femoral and tibial condyles. (Kaplan 1958 and Evans 1977 have suggested

531

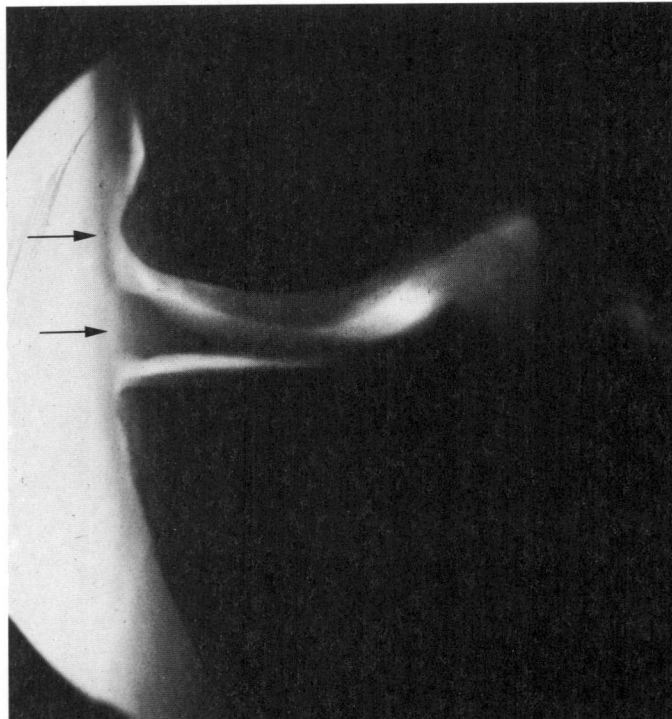

4.78A Radiograph taken after injection of air into the knee joint, showing the shadow thrown by the medial meniscus. The upper arrow points to the medial ligament and the lower to the medial meniscus. (Aerogram by A A Butler, RAF.)

the iliotibial tract as a strong tibial collateral ligament, see p. 528.) During extension the ligamentum patellae is tightened by quadriceps femoris but is relaxed in the erect attitude. When the knee flexes the fibular collateral and *posterior* part of the tibial collateral ligament relax but the cruciate ligaments and *anterior* part of the tibial collateral remain taut; posterior parts of the menisci are compressed between femoral and tibial condyles. Flexion is checked by the quadriceps extensor, anterior parts of the capsule, posterior cruciate ligament, and compression of soft tissues behind the knee. In extreme *passive* flexion, contact of calf with thigh may be the limiting factor; parts of both cruciate ligaments also tense. In addition to conjunct rotation with terminal extension or initial flexion, relaxed collateral ligaments allow independent medial and lateral rotation (adjunct rotation) when the joint is flexed. One cruciate ligament at least is taut in all positions; it acts as a direct bond between tibia and femur, limiting the former's translation back or forwards. In extension collateral ligaments assist the cruciate ligaments in this role. Forward gliding of tibia on femur is prevented by the anterior cruciate, backward gliding of tibia on femur by the posterior cruciate ligament.

During femoral extension on a fixed tibia, femoral articular surfaces simultaneously roll forward, slide back and spin medially. However, tibial extension on a relatively fixed femur involves simultaneous forward tibial roll, slide and lateral spin. This analysis is illustrated in 4.16A,B. For geometrical analysis consult Low & Lewis (1977).

Menisci probably assist lubrication, facilitate combined sliding, rolling and spinning, and may cushion extremes of flexion and extension (p. 529). Seedham et al (1974,1979) have demonstrated their role in cushioning compression forces.

Accessory movements. Wider rotation can be obtained by passive movements when the knee is semiflexed; the tibia can also, to a limited extent, be translated back and forwards on the femur. Abduction and adduction are prevented in full extension by

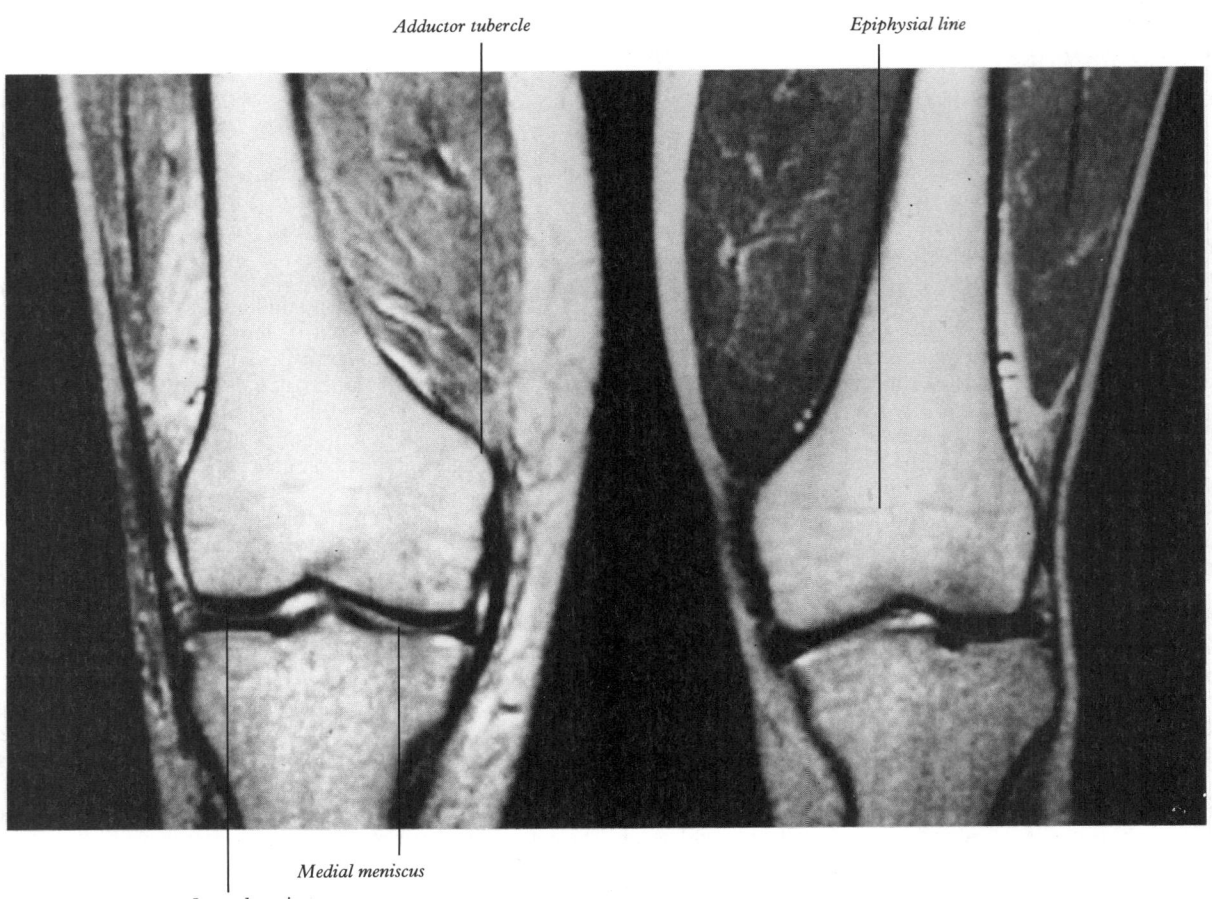

Adductor tubercle

Epiphysial line

Medial meniscus

Lateral meniscus

4.78B Magnetic resonance imaging scan of the knee joints of an adult male, showing the menisci. (Supplied by Philips Medical Systems; photographed by Sarah Smith.)

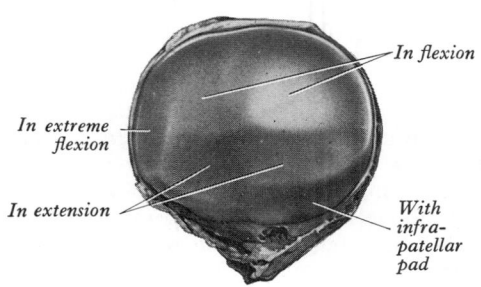

4.79 Posterior surface of the right patella, showing areas of contact with the femur and infrapatellar fat pad in different positions of the knee joint as indicated.

collateral and cruciate ligaments. With the knee slightly flexed limited adduction and abduction are possible, passive and active. Slight separation of femur and tibia results from strong traction.

Muscles Producing Movements

Flexion: biceps femoris, semitendinosus and semimembranosus, assisted by gracilis, sartorius and popliteus. With the foot stationary, gastrocnemius and plantaris also assist.

Extension: quadriceps femoris, assisted by tensor fasciae latae.

Medial rotation of flexed leg: popliteus, semimembranosus and semitendinosus, assisted by sartorius and gracilis.

Lateral rotation of flexed leg: biceps femoris.

Applied Anatomy. The knee appears to be an insecure joint. Between the two longest bones, it is subject to much leverage; articular surfaces are poorly congruent and range of motion is great. Nevertheless, the powerful ligaments and strong muscles concerned make it one of the strongest joints. Many muscles have direct attachments to the fibrous capsule. Traumatic dislocation is rare. Injuries of a meniscus are, however, common, resulting from twisting strains applied to the knee when slightly or fully flexed. Damage may be a complete tear across the full width, a partial split extending from the free border or a longitudinal tear within the cartilage; occasionally its periphery is detached from the capsule. The torn or detached part may be displaced centrally and jam between surfaces of femur and tibia, arresting all movement. Menisci are largely avascular; a tear cannot heal unless close to the capsule (p. 529). The medial meniscus is more commonly injured, probably because it is more securely attached to the medial capsule and ligaments and less adaptable to sudden change of position. During rotation of a flexed or partially flexed knee the medial cartilage moves more than the lateral. On the other hand, medial fibres of popliteus may draw the posterior part of the lateral meniscus back onto the groove behind the lateral tibial condyle, preventing trapping between articular surfaces.

Injuries to cruciate ligaments are also common, ranging from sprain to rupture. The anterior is more commonly affected. Damage to the tibial collateral ligament is less common, since excessive strain is likely only in full extension.

Osteoligamentous preparations have aided analysis of excessive movements due to ligamentous injuries. In a flexed knee the fibular collateral and posterior part of the tibial collateral ligaments relax. If either cruciate ligament is cut, tibial gliding is increased. Rotation is increased if either cruciate or tibial collateral ligaments are cut. Division of both collateral ligaments allows excessive lateral rotation but medial rotation is unchanged. Abduction and adduction are excessive if both cruciate ligaments are cut.

Acute traumatic synovitis is frequent in the knee. When its cavity is distended with fluid, the swelling shows above and alongside the patella, 5 cm or more above the femur's patellar surface, and extends a little more under vastus medialis than under the lateralis; the synovial membrane descends just below the tibial plateau.

Genual bursae are sometimes distended. Subcutaneous prepatellar or infrapatellar bursae are frequently caused by excessive kneeling. The semimembranosus bursa is occasionally enlarged, forming a fluctuant swelling at the back of the knee. During extension it is tense but in flexion soft, if it communicates with the joint.

Tibiofibular Articulations

Tibia and fibula are connected at both ends by a proximal synovial and distal fibrous joint. Their shafts are also tied by the crural interosseous membrane. The distal fibrous joint and interosseous membrane are syndesmoses.

THE PROXIMAL TIBIOFIBULAR JOINT

This (4.80) is an almost plane joint between the lateral tibial condyle and fibular head. The *articular surfaces* vary in size, form and inclination. The fibular facet is usually elliptical or circular, almost flat or slightly grooved. Surfaces are hyaline cartilage, with anterior and posterior ligaments.

The fibrous capsule is attached to the margins of rims of facets on tibia and fibula and anteriorly and posteriorly thickened. In about 10% the synovial membrane is continuous with that of

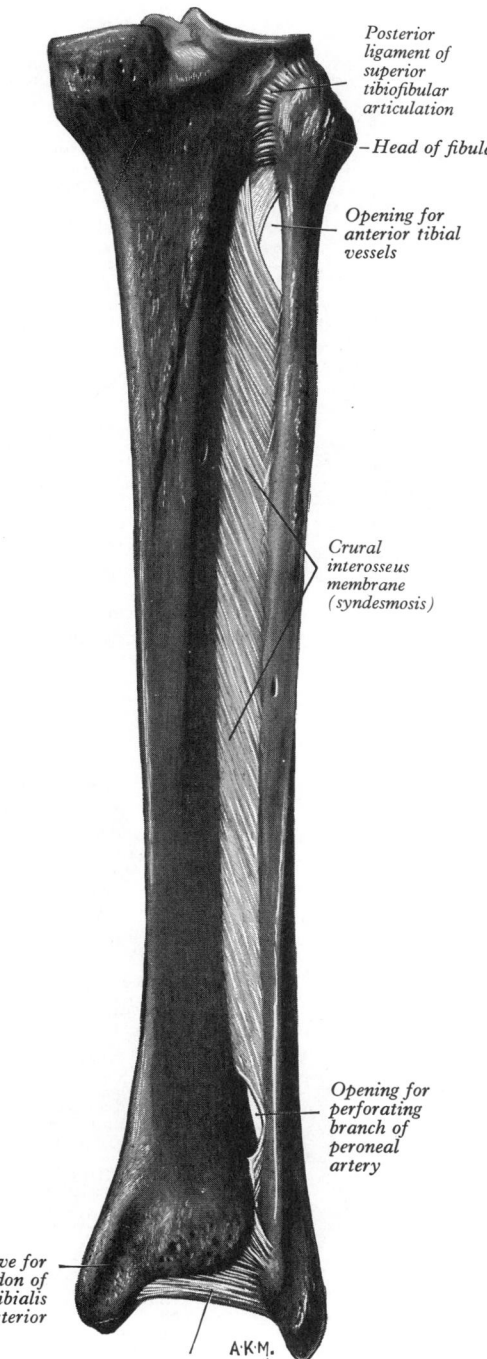

Posterior ligament of superior tibiofibular articulation

– Head of fibula

Opening for anterior tibial vessels

Crural interosseus membrane (syndesmosis)

Opening for perforating branch of peroneal artery

Groove for tendon of tibialis posterior

A·K·M·

Inferior transverse ligament

4.80 Crural interosseous membrane: posterior aspect. Note contrasting directions of collagen fibre bundles around both vascular apertures.

533

the knee joint via the subpopliteal recess (p. 527). **The anterior ligament** has two or three flat bands, passing obliquely up from the fibular head to the front of the lateral tibial condyle. **The posterior ligament**, a thick band, ascends obliquely between posterior aspects of the fibular head and the lateral tibial condyle, covered by the popliteal tendon. These ligaments are not entirely separate from the capsule.

Arteries are from anterior and posterior tibial recurrent rami of the anterior tibial, *nerves* from the common peroneal nerve and from the nerve to the popliteus.

THE CRURAL INTEROSSEOUS MEMBRANE

This (4.80) connects the interosseous borders of tibia and fibula, separating the anterior and posterior muscles in the leg, some being attached to it. The anterior tibial artery passes forwards through a large oval opening near the membrane's proximal end; distally the perforating branch of the peroneal artery pierces it. Its fibres, largely oblique, mainly descend laterally but a few descend medially including a bundle at the proximal border of the proximal opening. The membrane is distally continuous with the interosseous ligament of the distal tibiofibular joint. Anterior to it are tibialis anterior, extensors digitorum longus and hallucis longus, peroneus tertius, the anterior tibial vessels and deep peroneal nerve; posterior to it, tibialis posterior and flexor hallucis longus.

THE DISTAL TIBIOFIBULAR JOINT

This is between the rough, medial *convex* surface on the fibula's distal end and the rough *concave* surface of the tibia's fibular notch, which are separated distally for about 4 mm by a synovial prolongation from the talocrural joint and may be covered by articular cartilage in the lowest millimetre. The joint is usually considered a syndesmosis (p. 468).

The anterior tibiofibular ligament (4.87), a flat band, descends laterally between adjacent margins of tibia and fibula, anterior to the syndesmosis.

The posterior tibiofibular ligament (4.81) is stronger and disposed similarly on the posterior aspect of the syndesmosis. Its distal, deep part is the *inferior transverse ligament*, a thick band of yellow fibres crossing from the proximal end of the lateral malleolar fossa to the posterior border of the tibial articular surface almost to the medial malleolus. The ligament projects distal to the bones, in contact with the talus. Its colour reflects its content of yellow elastic fibres.

The interosseous ligament, continuous with the crural interosseous membrane, contains many short bands between rough adjacent tibial and fibular surfaces and is the strongest bond between the bones.

Movements

The tibiofibular joints permit only slight movement. Due to the varying slope of the talar lateral malleolar surface, the fibula rotates laterally a little during dorsiflexion at the ankle (Barnett & Napier 1952), the bones being also slightly separated. Although slight bending or torsion of the fibular shaft might permit

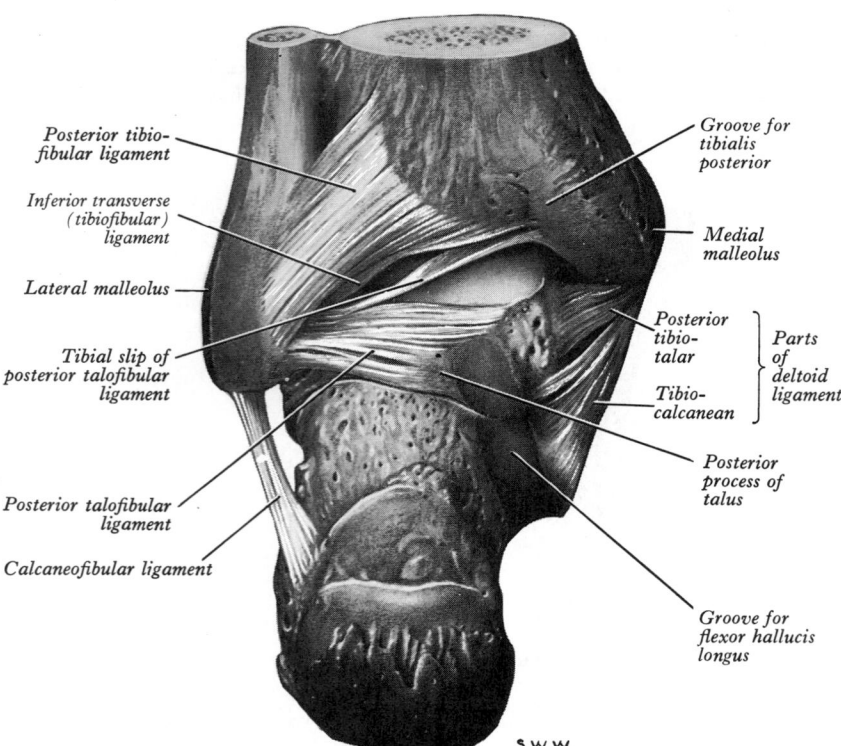

Posterior tibio-
fibular ligament

Inferior transverse
(tibiofibular)
ligament

Lateral malleolus

Tibial slip of
posterior talofibular
ligament

Posterior talofibular
ligament

Calcaneofibular ligament

Groove for
tibialis
posterior

Medial
malleolus

Posterior
tibio-
talar

Tibio-
calcanean

Parts
of
deltoid
ligament

Posterior
process of
talus

Groove for
flexor hallucis
longus

S.W.W.

4.81 Left talocrural joint: posterior aspect. (Drawn from a specimen in the Museum of the Royal College of Surgeons of England, with permission from the Council.)

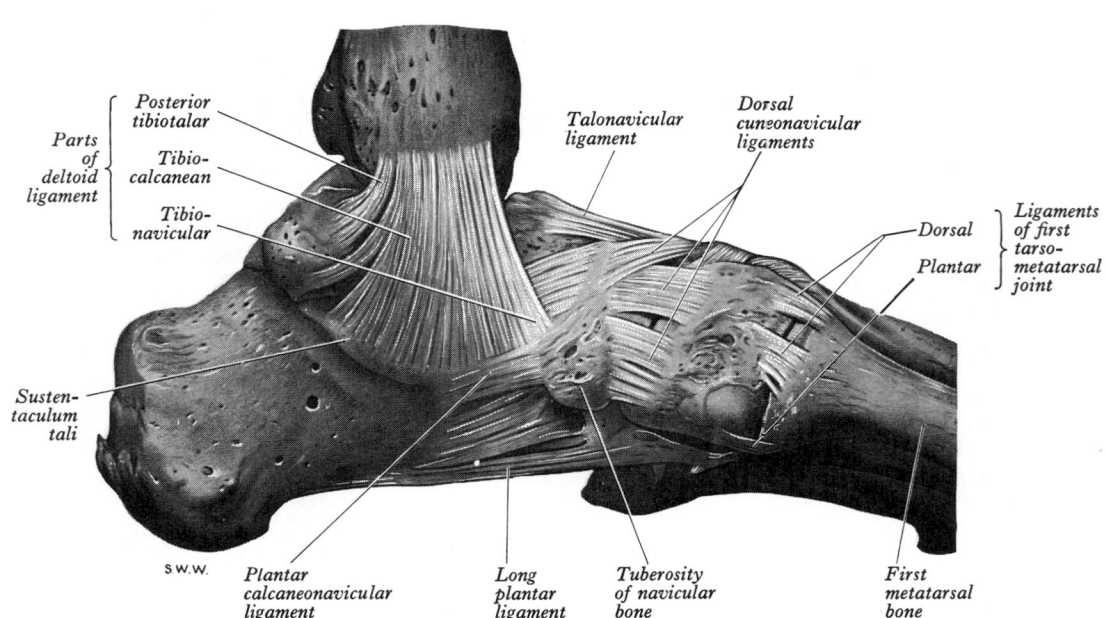

Parts
of
deltoid
ligament

Posterior
tibiotalar

Tibio-
calcanean

Tibio-
navicular

Talonavicular
ligament

Dorsal
cuneonavicular
ligaments

Dorsal

Plantar

Ligaments
of first
tarso-
metatarsal
joint

Susten-
taculum
tali

S.W.W.

Plantar
calcaneonavicular
ligament

Long
plantar
ligament

Tuberosity
of navicular
bone

First
metatarsal
bone

4.82 Ligaments of the left talocrural and tarsal joints: medial aspect. (Drawn from a specimen in the Museum of the Royal College of Surgeons of England, with permission from the Council.)

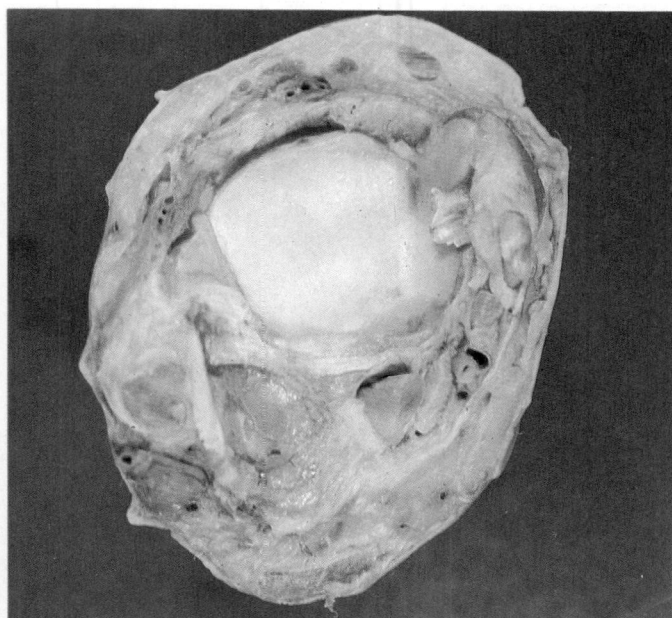

A

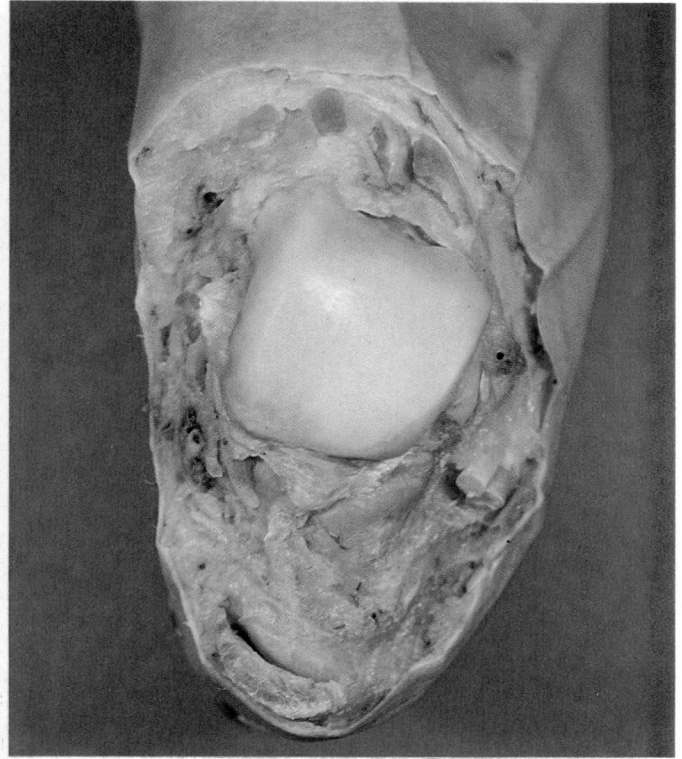

B

4.83 Articular surfaces of a human cadaveric talocrural joint. A displays the tibiofibular mortice, which includes the principal talar surface of the tibia with its medial malleolar facet, the interosseous and inferior transverse tibiofibular ligaments and the articular surface of the fibular (lateral) malleolus. B shows the characteristically sellar articular surface of the talus; note the obliquity of its principal axes relative to the sagittal axis of the foot. (Preparation by M C E Hutchinson; photographed by Kevin Fitzpatrick, Guy's Hospital Medical School, London.)

movements at the distal tibiofibular joint, the proximal probably also helps (p. 536).

Arteries are from the peroneal perforating branch and medial malleolar rami of anterior and posterior tibial arteries. *Nerves* are from the deep peroneal, tibial and saphenous nerves.

The Talocrural Joint

The talocrural (ankle) joint is approximately uni-axial. The tibia's lower end and its medial malleolus, with the fibular lateral malleolus and inferior transverse tibiofibular ligament, form a deep recess for the body of the talus. Its line is judged by the anterior margin of the tibia's distal end, which is palpable when the overlying tendons are relaxed. Although it appears a simple hinge, usually styled 'uni-axial', its axis of rotation is dynamic, shifting during dorsi- and plantar flexion (Sammarco 1977).

Articular surfaces are covered by hyaline cartilage. The talar trochlear surface, convex parasagittally, transversely gently concave, is wider in front; the distal tibial articular surface is reciprocally curved. The talar articular surface for the medial malleolus is a proximal area on the medial talar surface, being fairly flat, comma-shaped and anteriorly deeper. The larger lateral talar articular surface is triangular and vertically concave, that on the lateral malleolus is reciprocally curved. Posteriorly the edge between trochlear and fibular articular surfaces of the talus is bevelled to a narrow, flat triangular area articulating with the inferior transverse tibiofibular ligament (4.85,86). All these talar surfaces are continuous. The bones are connected by a fibrous capsule, medial (deltoid), anterior and posterior talofibular and calcaneofibular ligaments.

The fibrous capsule around the joint is thin in front and behind, attached proximally to the borders of tibial and malleolar articular surfaces and distally to the talus near the margins of its trochlear surface, except in front where it reaches the dorsum of the talar neck. It is strengthened by strong collateral ligaments. Its posterior part is mainly of transverse fibres. It blends with the inferior transverse ligament and is thickened laterally where it reaches the fibular malleolar fossa.

Synovial membrane lining the capsule ascends as a short vertical recess between tibia and fibula (4.88).

The medial ligament (deltoid collateral 4.82,88) is a strong, triangular band, attached to the apex and to the anterior and posterior borders of the medial malleolus. Of its *superficial fibres* the anterior (*tibionavicular*) pass forwards to the navicular tuberosity and behind this blend with the medial margin of the plantar calcaneonavicular ligament; intermediate (*tibiocalcaneal*) fibres descend almost vertically to the whole length of the sustentaculum tali; posterior fibres (*posterior tibiotalar*) pass

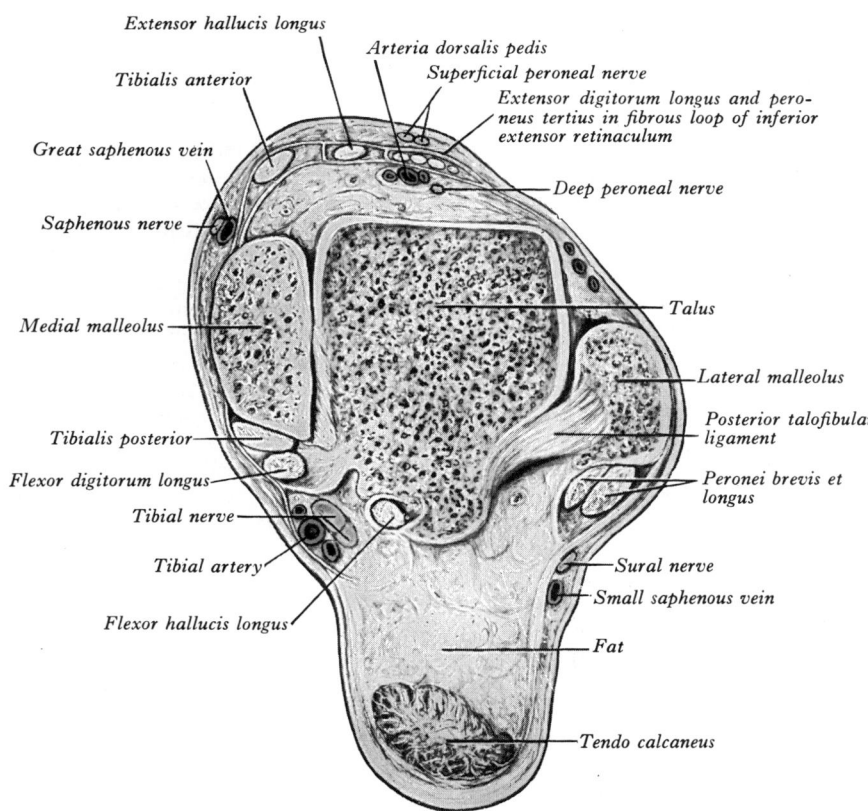

4.84 Transverse section through the lower part of the right talocrural joint: superior aspect.

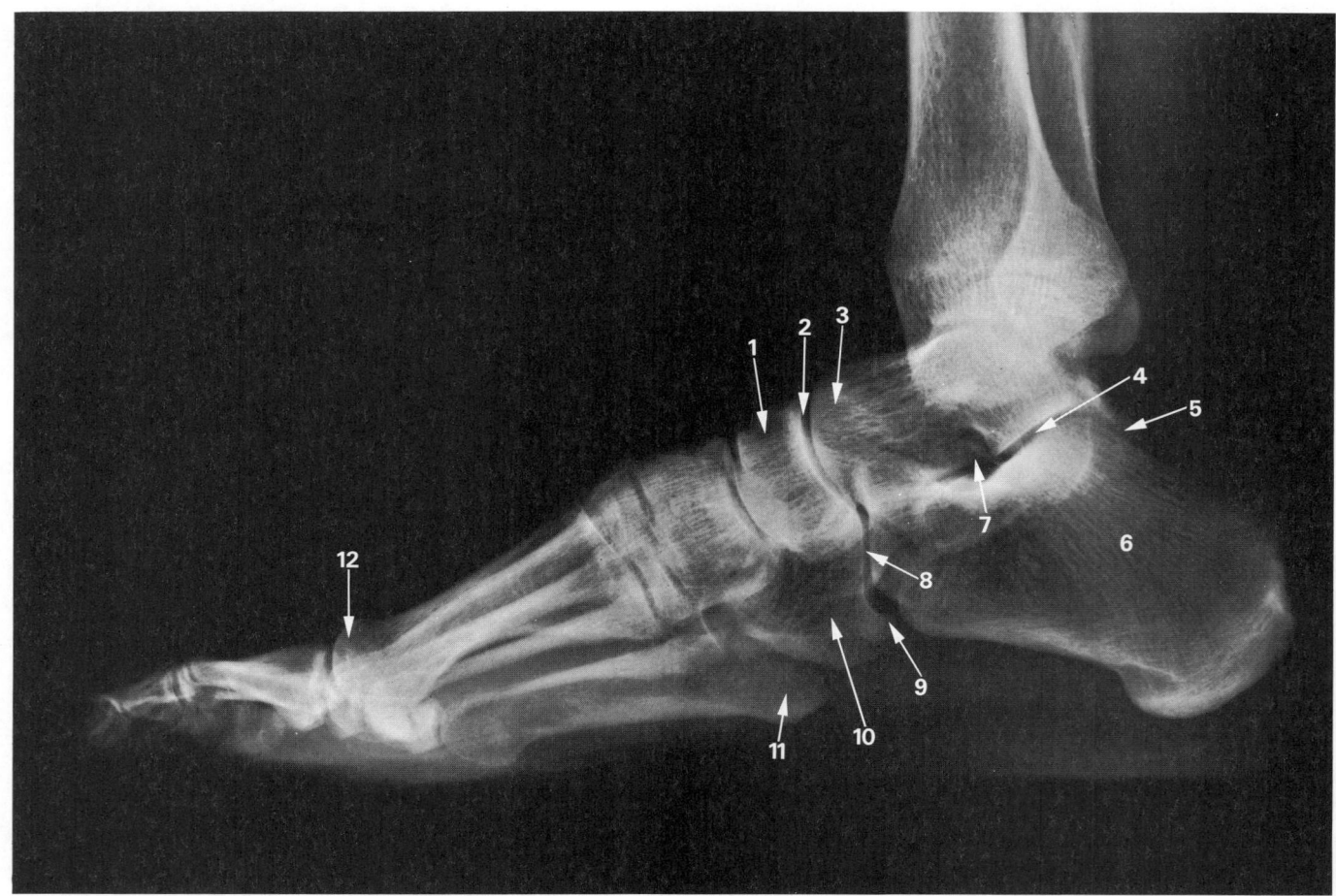

4.85 Lateral radiograph of ankle and foot in full plantigrade contact with the ground, during symmetrical standing, of a man aged 44 years. 1. Navicular. 2. Talonavicular joint. 3. Head of talus. 4. Subtalar joint. 5. Os trigonum. 6. Calcaneus. Note trabecular pattern. 7. Sinus tarsi. 8. Calcaneocuboid joint. 9. Sesamoid in tendon of peroneus longus. 10. cuboid. 11. Tuberosity on base of fifth metatarsal. 12. Head of first metatarsal.

posterolaterally to the medial side of talus and its medial tubercle. The *deep fibres* (*anterior tibiotalar*) pass from the tip of medial malleolus to the non-articular part of the medial talar surface. The ligament is crossed by tendons of tibialis posterior and flexor digitorum longus.

The anterior talofibular ligament (4.87) extends anteromedially from the anterior margin of the fibular malleolus to the talus, attached in front of its lateral articular facet and to the lateral aspect of its neck. **The posterior talofibular ligament** (4.81) runs almost horizontally from the distal part of the lateral malleolar fossa to the lateral tubercle of the posterior talar process. A 'tibial slip' of fibres connects it to the medial malleolus. **The calcaneofibular ligament** (4.87), a long cord, runs from a depression anterior to the apex of fibular malleolus to a tubercle on the lateral calcaneal surface and is crossed by tendons of peroneus longus and brevis. The three ligaments above constitute the **lateral ligament** of the talocrural joint.

Relations. *Anteriorly*, from the medial side, tibialis anterior, extensor hallucis longus, anterior tibial vessels, deep peroneal nerve, extensor digitorum longus and peroneus tertius; posteriorly from the medial side, tibialis posterior, flexor digitorum longus, posterior tibial vessels, tibial nerve, flexor hallucis longus; in the groove behind the fibular malleolus, tendons of peroneus longus and brevis (4.83).

Arteries are from malleolar rami of anterior tibial and peroneal arteries. *Nerves* are from deep peroneal and tibial nerves.

Movements. When the body is upright and the foot at right angles to the leg, *active movements* of the joint are dorsiflexion (*c.*10°) and plantar flexion (*c.* 20°), the former decreasing the anterior angle between leg and foot, the latter increasing it. Dorsiflexion is the position of 'close-pack', with maximal congruence and ligamentous tension; from this position all major thrusting movements are exerted, in walking, running and jump-

ing. The malleoli embrace the talus; even in relaxation no appreciable lateral movement can occur without stretch of the inferior tibiofibular syndesmosis and slight bending of the fibula. The superior talar surface is broader in front, and in dorsiflexion the malleolar gap is increased by slight lateral rotation of the fibula by 'give' at the inferior tibiofibular syndesmosis and gliding at the superior tibiofibular joint. The medial (deltoid) ligament is very strong and is even able to resist forces which tear the bone to which it is attached. Its middle part, with the calcaneofibular ligament, binds leg firmly to foot, resisting displacement in all directions. The posterior talofibular ligament assists the calcaneofibular in resisting posterior displacement, deepening the joint for the talus. The anterior talofibular ligament limits anterior displacement. Plantar flexion is limited by the opposing muscles, anterior fibres of the medial (deltoid) and by the anterior talofibular ligaments. Dorsiflexion is limited by the calcaneal tendon and contraction of triceps surae, posterior fibres of medial (deltoid) and calcaneofibular ligaments (4.87). Dorsi- and plantar flexion are increased by intertarsal movements, adding about 10° to the former, 20° to the latter.

Accessory movements. Slight amounts of side-to-side gliding, rotation, abduction and adduction are permitted, when the foot is plantarflexed.

Muscles Producing the Movements
(see also pp. 645–651, 656–657)

Dorsiflexion: tibialis anterior, assisted by extensores digitorum longus and hallucis longus, and peroneus tertius.

Plantar flexion: gastrocnemius and soleus, assisted by plantaris, tibialis posterior, flexores hallucis longus and digitorum longus.

In symmetrical standing the line of body weight is anterior to the ankle joints, which are not even near their close-packed position. Stability requires continuous action by the solei increasing (often with gastrocnemii involved) with leaning forward and vice versa. If backward sway takes the 'weight line' posterior to the transverse axes of the talocrural joints, plantar flexors relax and dorsiflexors contract.

Applied Anatomy. Owing to its depth the talocrural joint is rarely dislocated without malleolar fracture. So-called sprains of the talocrural joint are almost always actually abduction sprains of *subtalar joints*, although some medial (deltoid) fibres may also be torn. True sprains are usually due to forcible plantar flexion, resulting in capsular tears in front (most commonly of the anterior talofibular ligament) and bruising by impaction of structures behind the joint. In ankylosis the optimal position is slight plantar flexion.

Intertarsal Joints

THE SUBTALAR (TALOCALCANEAL) JOINT

There are anterior and posterior articulations between calcaneus and talus, forming a functional unit often termed the 'subtalar joint'; this term is used here only for the posterior, the anterior being treated as part of the talocalcaneonavicular joint (vide infra).

The subtalar joint proper involves the concave posterior calcaneal facet on the posterior part of the inferior surface of the talus and the convex posterior facet on the superior surface of the calcaneus. The joint is modified multi-axial and its permitted movements are considered together with those at other tarsal joints. The bones are connected by a fibrous capsule, lateral, medial, interosseous talocalcaneal and cervical ligaments.

The fibrous capsule (4.89) envelops the joint, its fibres being short and attached to its articular margins; it is split into slips, between which it is thin; its synovial membrane is separate from other tarsal joints.

The lateral talocalcaneal ligament, a short flat fasciculus, descends obliquely back from the lateral talar process to the lateral calcaneal surface and is attached anterosuperior to the calcaneofibular ligament.

The medial talocalcaneal ligament connects the medial talar tubercle to the back of the sustentaculum tali and adjacent medial surface of the calcaneus. Its fibres blend with the medial (deltoid) ligament, the most posterior fibres lining the groove for flexor hallucis longus between the talus and calcaneus.

The interosseous talocalcaneal ligament (4.88,89) is a broad, flat, bilaminar transverse band in the sinus tarsi, descending obliquely and laterally from sulcus tali to sulcus calcanei. Its medial fibres are taut in eversion. The ligament's posterior lamina is associated with the talocalcanean joint, its anterior lamina with the talocalcaneonavicular joint (vide infra).

The cervical ligament is just lateral to the sinus tarsi and attached to the superior calcaneal surface, medial to the attachment of extensor digitorum brevis, whence it ascends medially to an inferolateral tubercle on the talar neck (Barclay-Smith 1896, Smith 1958). It is considered taut in inversion.

Movements between talus and calcaneus accompany those at the talocalcaneonavicular joint and will be so described.

THE TALOCALCANEONAVICULAR JOINT

This compound articulation is multi-axial; the ovoid talar head is continuous with the triple-faceted anterior area of its inferior surface (p. 450) and the whole fits the concavity formed by the posterior surface of the navicular, the middle and anterior talar facets of the calcaneus and the superior fibrocartilaginous surface of the plantar calcaneonavicular ligament. The reverse of the subtalar (talocalcaneal) joint, the proximal talocalcaneonavicular surface is convex. The joint's position is judged by the talar head, to be seen and palpated about 3 cm anterior to the lower end of the tibia, with the foot passively inverted. The bones are connected by a fibrous capsule and three ligaments—talonavicular, plantar calcaneonavicular and calcaneonavicular part of the bifurcated ligaments.

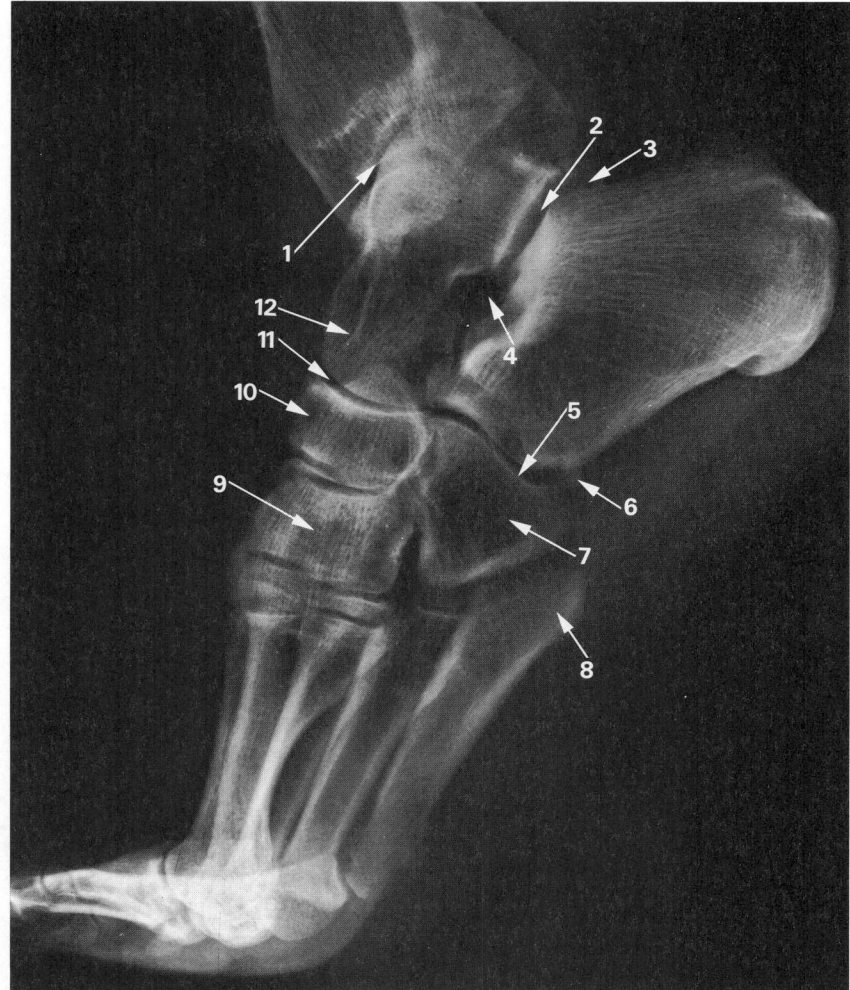

4.86 The same foot as in 4.85 with heel raised from the ground, as near the end of the 'stance phase' in walking. Note the forward angulation of the leg bones, plantar flexion at the talocrural joint and the position of the metatarsals and phalanges. 1. Talocrural joint. 2. Subtalar joint. 3. Os trigonum. 4. Sinus tarsi. 5. Calcaneocuboid joint. 6. Sesamoid in peroneus longus tendon. 7. Cuboid. 8. Tuberosity on base of fifth metatarsal. 9. Cuneiforms. 10. Navicular. 11. Talonavicular joint. 12. Head of talus.

The fibrous capsule is poorly developed, except posteriorly, where it is thick and is the anterior part of the interosseous ligament filling the sinus tarsi.

The talonavicular ligament (4.82,87), a broad, thin band, connects the dorsal surfaces of the neck of the talus and the navicular; it is covered by extensor tendons. The *plantar calcaneonavicular ligament* (p. 539) and the *calcaneonavicular part of the bifurcated ligament* (4.87) are the joint's plantar and lateral ligaments respectively.

Movements. Gliding and rotation occur at the talocalcaneal and talocalcaneonavicular joints, by which the calcaneus and navicular, carrying the (*free, non-weight-bearing*) foot, rotate medially on the talus, thus elevating the medial and depressing the lateral border of the foot, bringing its plantar aspect medial. This is *inversion*, occurring mainly at the subtalar joint and the complex articulation between the talar head, with its plantar extensions, and the sustentaculum tali, calcaneus, plantar calcaneonavicular ligament and posterior navicular surface. Joint surfaces are reciprocally curved but in opposite directions in the two joints, and movements at them have been likened to those between radius and ulna at the superior and inferior radio-ulnar joints. The axis joins their approximate centres of curvature, ascending anteromedially from the back of the calcaneus, crossing the sinus tarsi obliquely to reach the superomedial aspect of the talar neck (Shepard 1951). Axial obliquity explains, in part, the adduction and slight plantar flexion of the foot (but to a greater degree the forefoot) in inversion. Movement of calcaneus around talus,

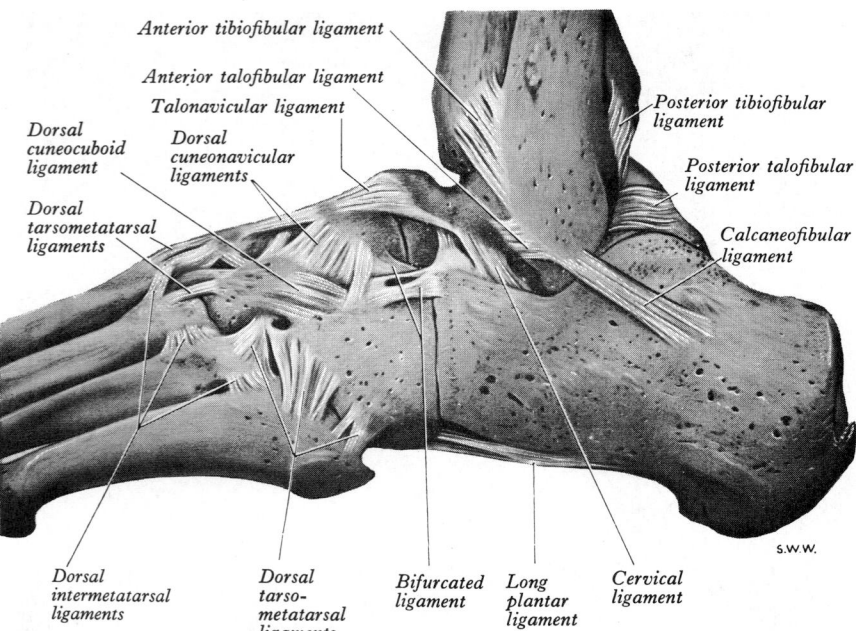

Anterior tibiofibular ligament

Anterior talofibular ligament

Talonavicular ligament

Dorsal cuneocuboid ligament

Dorsal cuneonavicular ligaments

Dorsal tarsometatarsal ligaments

Posterior tibiofibular ligament

Posterior talofibular ligament

Calcaneofibular ligament

S.W.W.

Dorsal intermetatarsal ligaments

Dorsal tarso-metatarsal ligaments

Bifurcated ligament

Long plantar ligament

Cervical ligament

4.87 Ligaments of the left talocrural and tarsal joints: lateral aspect. (Drawn from a specimen in the Museum of the Royal College of Surgeons of England, by permission of the Council.)

however, also involves movement at the *transverse tarsal articulation* (talonavicular and calcaneocuboid joints); forefoot range is thereby increased. These joints are almost in one transverse plane; during inversion the navicular rotates on the talar head, while the cuboid glides down and rotates at the sellar calcaneocuboid joint. Range of inversion is further increased in plantar flexion; the narrow part of the talar trochlear surface now occupies the tibiofibular mortice and slight talar movement increases adduction with inversion. The opposite movement, more limited in range, is *eversion*.

Inversion is chiefly checked by peroneal muscles and the lateral part of the interosseous talocalcaneal ligament (p. 537). The other

tarsal interosseous ligaments and the calcaneofibular are less powerful factors. Eversion is arrested by tibiales anterior and the posterior and medial (deltoid) ligament.

The complex actions of inversion and eversion described above refer to changes in the whole foot (with minor movements of the talus), *when it is off the ground*. The fully inverted foot is also plantar flexed; conversely, eversion is linked with dorsiflexion. *When the foot is transmitting weight or thrust* these movements are modified to maintain plantigrade contact. The distal tarsus and metatarsus are *pronated* or *supinated* relative to calcaneus and talus, pronation involving a downward rotation of the medial border and hallux; supination being the reverse, both bring the lateral border into plantigrade contact. Further, when there is *angulation* of tibia and fibula in the coronal plane, either laterally but particularly medially, these bones and their distal mortice carry the talus with them in their movements. The talus tilts (rotates) around the oblique transarticular intercentroid axis (described above) at first, relative to the rest of the subtalar foot and moving at the talocalcaneal and talocalcaneonavicular joints. When close-packing occurs between talus and calcaneus, further angulation is accompanied by tilting of the functional unit of the combined *talus and calcaneus*, the latter involving the calcaneocuboid joint. Unfortunately, the terms inversion and eversion are often confused with pronation and supination, the latter being *components* of the former, like the slight adduction and abduction which may accompany inversion and eversion. Supination and pronation can be dissociated from inversion and eversion in adaptation of the forefoot in plantigrade stance or progression (pp. 542–544, 656–659).

Terminology of pedal movements is somewhat confused amongst orthopaedists, kinesiologists and others. The above account represents widely accepted usage (MacConaill 1950, Kapandji 1970, MacConaill & Basmajian 1977, Keller 1977). According to Kapandji (1970), inversion is a combination of adduction and supination, eversion involving abduction and pronation.

Muscles Producing the Movements

Inversion: tibiales anterior and posterior.
Eversion: peroneus longus and brevis.

THE CALCANEOCUBOID JOINT

Articular surfaces of the calcaneocuboid joint, which is 2 cm proximal to the tubercle, on the fifth metatarsal base, are sellar. Its ligaments are: the fibrous capsule, calcaneocuboid part of the bifurcated ligament, long plantar and plantar calcaneocuboid ligaments.

The fibrous capsule is thickened dorsally as the *dorsal calcaneocuboid ligament*. The synovial membrane is distinct from other tarsal articulations (4.92).

The bifurcated ligament (4.87), a strong Y-shaped band, is attached by its stem proximally to the anterior part of the upper calcaneal surface; distally it divides into calcaneocuboid and calcaneonavicular parts. The (*medial*) *calcaneocuboid ligament* extends to the dorsomedial aspect of the cuboid bone, forming a main bond between the two rows of tarsal bones; the (*lateral*) *calcaneonavicular ligament* is attached to the dorsolateral aspect of the navicular bone.

The long plantar ligament (4.90), the longest ligament in the tarsus, extends from the plantar surface of the calcaneus (anterior to its tuberosity's processes) and from its anterior tubercle, to the ridge and tuberosity on the cuboid's plantar surface, to which deep fibres are attached, more superficial fibres continuing to the bases of the second to fourth and sometimes fifth metatarsal bones. This ligament, with the groove on the cuboid's plantar surface, makes a tunnel for the tendon of peroneus longus. It is a most powerful factor limiting depression of the lateral longitudinal arch (p. 543).

The plantar calcaneocuboid ligament (*short plantar ligament*) is nearer to bone than the preceding, from which it is separated by areolar tissue. It is a short, wide band of great strength, stretching from the anterior calcaneal tubercle and the depression anterior to it to the adjoining part of the plantar surface of the cuboid; it also sustains the lateral longitudinal arch

Interosseous ligament of inferior tibiofibular syndesmosis

Medial malleolus

Body of talus

Deltoid ligament

Tendon of tibialis posterior

Sustentaculum tali

Tendon of flexor digitorum longus

Tendon of flexor hallucis longus

Lateral malleolus

Posterior talofibular ligament

Interosseous talo-calcanean ligament

Body of calcaneus

Tendon of peroneus brevis

Tendon of peroneus longus

4.88 Coronal section through the left talocrural and talocalcaneal joints.

CUNEONAVICULAR JOINT

Movements. Movements between the calcaneus and cuboid are gliding with conjunct rotation upon each other during inversion and eversion of the whole foot and during pronative or supinative changes between fore- and hindfoot (pp. 542–544).

LIGAMENTS CONNECTING CALCANEUS AND NAVICULAR BONE

Though calcaneus and navicular do not articulate directly, they are connected by (lateral) calcaneonavicular and plantar calcaneonavicular ligaments. **The (lateral) calcaneonavicular ligament** has been described above as the medial band of the bifurcated ligament.

The plantar calcaneonavicular (spring) ligament (4.82,89,91), a broad, thick band, connects the anterior margin of the sustentaculum tali to the navicular's plantar surface. It ties the calcaneus to the navicular below the head of the talus as part of its articular cavity; it sustains the foot's medial longitudinal arch (p. 543). The ligament's *dorsal surface* has a triangular fibrocartilaginous facet on which part of the talar head rests (4.89). Its *plantar surface* is supported medially by the tendon of tibialis posterior and laterally by tendons of flexores hallucis longus and digitorum longus; its *medial border* is blended with anterior superficial fibres of the lateral (deltoid) ligament. There is no real evidence that the 'spring' ligament is particularly resilient.

THE CUNEONAVICULAR JOINT

The navicular articulates distally with the cuneiform bones at a compound joint often described as plane although the distal navicular surface is transversely convex and divided into three facets by low ridges, adapted to the proximal, slightly curved cuneiform surfaces. Its capsule is continuous with those of intercuneiform and cuneocuboid joints, as is its synovial cavity, and it is connected also to the second and third cuneometatarsal joints and inter-metatarsal joints between the second to fourth metatarsal bones (p. 540).

Dorsal and plantar ligaments connect the navicular to each cuneiform bone; of the three dorsal ligaments, one is attached to each cuneiform. The fasciculus from navicular to medial cuneiform is continued as the joint's capsule around its medial aspect, and then blends medially with the plantar ligament. Plantar

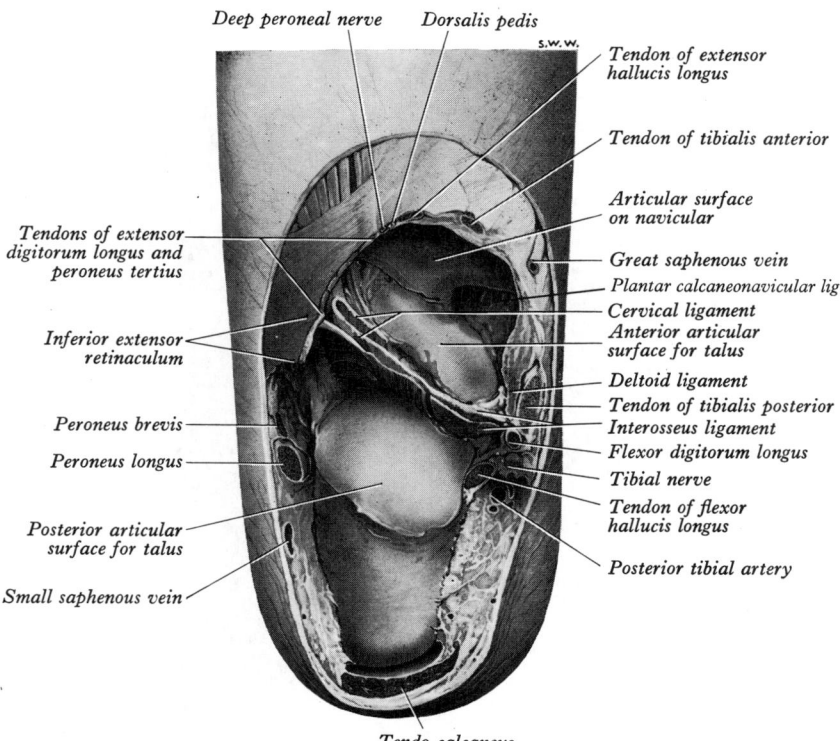

Deep peroneal nerve — *Dorsalis pedis*
s.w.w.
Tendon of extensor hallucis longus
Tendon of tibialis anterior
Articular surface on navicular
Tendons of extensor digitorum longus and peroneus tertius
Great saphenous vein
Plantar calcaneonavicular ligt
Cervical ligament
Anterior articular surface for talus
Inferior extensor retinaculum
Deltoid ligament
Tendon of tibialis posterior
Interosseus ligament
Flexor digitorum longus
Peroneus brevis
Tibial nerve
Peroneus longus
Tendon of flexor hallucis longus
Posterior tibial artery
Posterior articular surface for talus
Small saphenous vein
Tendo calcaneus

4.89A Left talocalcaneal and talocalcaneonavicular joints. Exposed from above by removal of the talus.

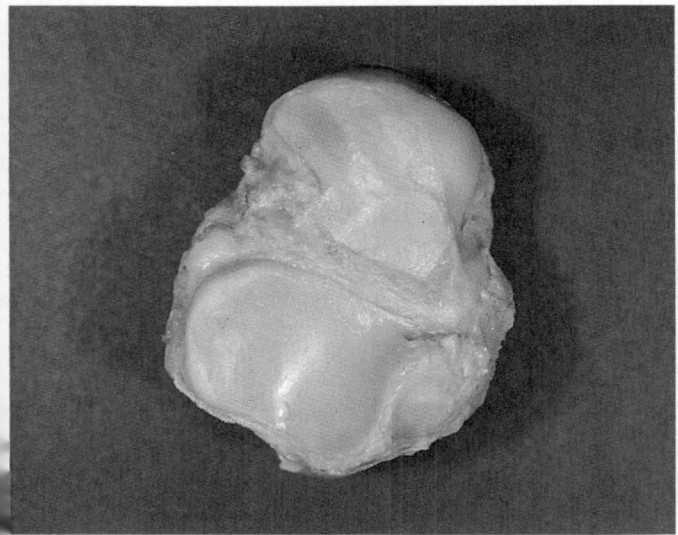

B

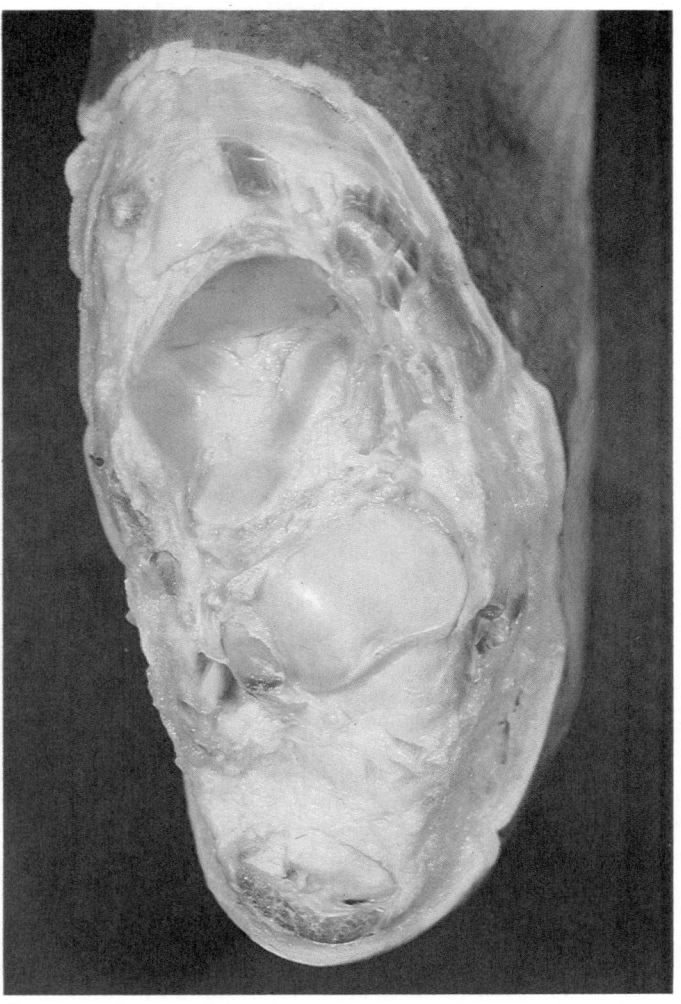

C

4.89B,C. A dissection in which the talus has been removed from the subjacent osseous and ligamentous tarsal elements. Compare the reciprocal articular curvatures of the inferior surface of the talus (B) and those of the superior aspects of their conarticular tarsal elements (C). Note the ovoid surfaces of the (posterior) talocalcaneal articulation (the *subtalar* joint of some authorities). Anteriorly, the complex convexity of the talar head is received into the '*acetabulum pedis*' which is formed by the calcaneus, navicular, the plantar calcaneonavicular ligament and part of the deltoid ligament. The detailed ligamentous architecture between these articulations is described in the text, and illustrated diagramatically (4.89A).

(Preparation by M C E Hutchinson; photographed by Kevin Fitzpatrick, Guy's Hospital Medical School, London.)

Tarsometatarsal Articulations

These are approximately plane synovial joints. The first metatarsal articulates with the medial cuneiform; the second is recessed between medial and lateral, articulating with the intermediate cuneiform; the third articulates with lateral cuneiform; the fourth with the lateral cuneiform and cuboid and the fifth with the cuboid. The joints are approximately on a line from the fifth metatarsal's tubercle to the hallucial tarsometatarsal joint, except for that between the second metatarsal and intermediate cuneiform, which is 2–3 mm proximal (4.92). The hallucial joint has its own capsule; articular capsules and cavities of the second and third are continuous with intercuneiform and cuneonavicular joints but separated from the fourth and fifth joints by an interosseous ligament between the lateral cuneiform and fourth metatarsal base. The bones are connected by dorsal and plantar tarsometatarsal and interosseous cuneometatarsal ligaments.

The dorsal ligaments are strong and flat. The first metatarsal is joined to the medial cuneiform by an articular capsule; the other tarsometatarsal capsules are blended with dorsal and plantar ligaments. The second metatarsal receives a band from each cuneiform, the third from the lateral cuneiform, fourth from lateral cuneiform and cuboid, fifth from the cuboid alone. **The plantar ligaments** are longitudinal and oblique bands, less regular than the dorsal. Those for the first and second metatarsal bones are strongest; the second and third metatarsals are joined by oblique bands to the medial cuneiform, the fourth and fifth by a few fibres to the cuboid.

The interosseous cuneometatarsal ligaments comprise: (1) the strongest from the lateral surface of the medial cuneiform to the adjacent angle of second metatarsal (4.92); (2) one connecting the lateral cuneiform to the adjacent angle of second metatarsal; this does not divide the joint between the second metatarsal and lateral cuneiform and is inconstant; (3) one connecting the lateral angle of the lateral cuneiform to the adjacent fourth metatarsal base.

Movements between tarsal and metatarsal bones are limited to gliding, very limited in range except between medial cuneiform and first metatarsal, where appreciable *passive* metatarsal flexion, extension and rotation are possible with muscles relaxed but all occurring *actively* in standing and walking to maintain plantigrade contact. Being recessed between medial and lateral cuneiform bones, the second is the least mobile metatarsal. Proximal articular profiles of cuneiform bones bring them into a close-packed state in plantar flexion, splaying them out in dorsiflexion.

Pronation and **supination** keep the feet in plantigrade apposition in a range of stances, from wide straddling to crossed legs. This is usually but inaccurately ascribed to inversion or eversion, because these involve the whole unsupported foot beyond the talus and even including the talus, if the slight adduction and abduction possible at the plantar-flexed ankle are added.

Feet planted far apart are wholly 'inverted' but relative positions of the forefoot and hindfoot must also change to maintain plantigrade contact, because each talus tilts medially with the tibiofibular extremities and its inferior mortice and talar tilt is followed by the calcaneus when limited movement between them is used up. If the rest of the foot followed this tilting (inversion) weight would be largely taken from the lateral to the medial border. To correct this and distribute weight by plantigrade contact, the *forefoot rotates* in the opposite sense, i.e. it *supinates* or *untwists*. This largely occurs at the transverse tarsal joint but others contribute, especially the hallucial cuneometatarsal. Conversely when feet are together or even crossed, contact with the ground can only be maintained by maximal *pronation* of the forefoot relative to the calcaneus and talus (pp. 542–544).

INTERMETATARSAL JOINTS

The first metatarsal *base* is not connected to the second by ligaments; in this respect hallux resembles pollex (but differs in being distally connected). A small bursa often occurs between first and second metatarsal bases (4.92). Second to fifth metatarsal bases are connected by dorsal, plantar and interosseous ligaments. All metatarsal *heads* are connected indirectly by deep transverse

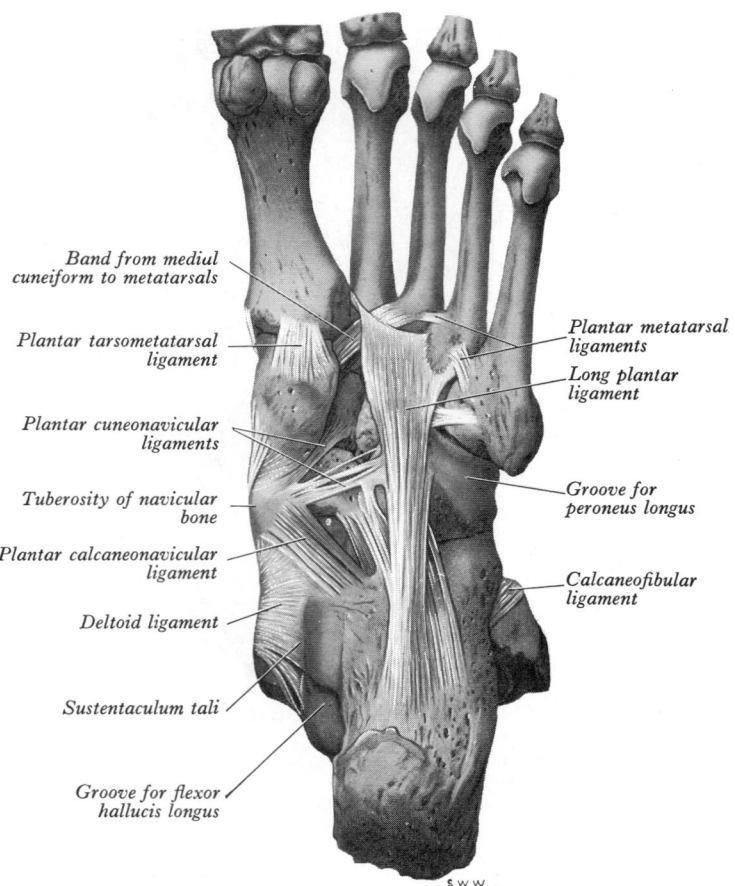

Band from mediul cuneiform to metatarsals

Plantar tarsometatarsal ligament

Plantar cuneonavicular ligaments

Tuberosity of navicular bone

Plantar calcaneonavicular ligament

Deltoid ligament

Sustentaculum tali

Groove for flexor hallucis longus

Plantar metatarsal ligaments

Long plantar ligament

Groove for peroneus longus

Calcaneofibular ligament

S.W.W.

4.90 Ligaments of the plantar surface of the left foot. Some of the fibres of the long plantar ligament which arise in front of the medial tubercle of the calcaneus have been removed. (Drawn from a specimen in the Museum of the Royal College of Surgeons of England, by permission of the Council.)

ligaments have similar attachments and receive slips from the tendon of tibialis posterior.

THE CUBOIDEONAVICULAR JOINT

This is usually a fibrous joint, its bones connected by dorsal, plantar and interosseous ligaments. **The dorsal ligament** extends distolaterally, **the plantar** nearly transversely from cuboid to navicular. **The interosseous ligament**, of strong transverse fibres, connects non-articular parts of adjacent surfaces of the two bones (4.92). This syndesmosis is often replaced by a synovial joint, almost plane, its articular capsule and synovial lining continuous with the cuneonavicular.

INTERCUNEIFORM AND CUNEOCUBOID JOINTS

These are all synovial and approximately plane or slightly curved. Their articular capsules and synovial linings are continuous with the cuneonavicular joint. The bones are connected by dorsal, plantar and interosseous ligaments.

The dorsal and plantar ligaments each have three transverse bands, between medial and intermediate cuneiform, intermediate and lateral cuneiform and between lateral cuneiform and cuboid bones. The plantar ligaments receive slips from the tendon of tibialis posterior. **The interosseous ligaments** connect nonarticular areas of adjacent surfaces and are strong agents in the maintaining transverse arch (p. 543).

Movements at cuneonavicular, cuboideonavicular, intercuneiform and cuneocuboid joints are merely slight gliding and rotation during pedal pronation or supination (p. 542), i.e. in alterations of a loaded foot in contact with the ground. For example, they increase suppleness when the forefoot is stressed, as in the initial thrust of running and jumping.

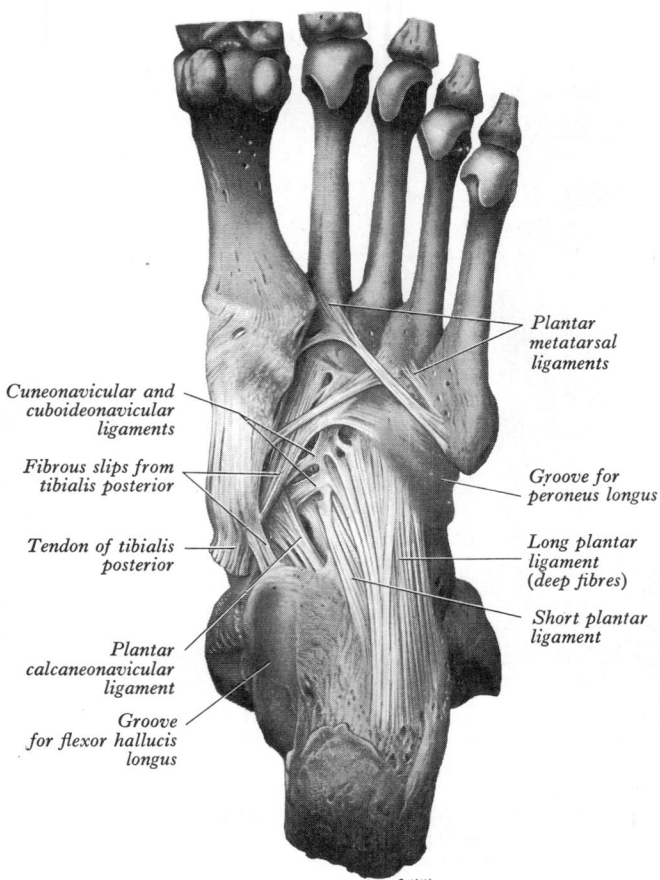

Plantar
metatarsal
ligaments

Cuneonavicular and
cuboideonavicular
ligaments

Fibrous slips from
tibialis posterior

Tendon of tibialis
posterior

Groove for
peroneus longus

Long plantar
ligament
(deep fibres)

Short plantar
ligament

Plantar
calcaneonavicular
ligament

Groove
for flexor hallucis
longus

S.W.W

4.91 Ligaments on the plantar surface of the left foot. The long plantar
ligament has been removed. (Drawn from a specimen in the Museum of
the Royal College of Surgeons of England, by permission of the Council.)

metatarsal ligaments (p. 515). **Dorsal and plantar ligaments**
pass transversely between adjacent *bases*. **Interosseous ligaments** are strong transverse bands connecting non-articular
parts of the adjacent surfaces (4.92).

Movements between the tarsal ends of metatarsal bones are
slight gliding when the forefoot is working under load (cf. intercuneiform joints, p. 540).

Synovial Arrangements of Tarsus and Metatarsus

Synovial cavities (4.92) in tarsus and metatarsus are: (1) subtalar;
(2) talocalcaneonavicular; (3) calcaneocuboid; (4) a complex of
cuneonavicular, intercuneiform, cuneocuboid; (5) intermediate
and lateral cuneiform respectively with second and third metatarsals; (6) second to fourth intermetatarsals; (7) medial cuneiform
with first metatarsal; and (8) cuboid with fourth and fifth metatarsals. A small synovial cavity sometimes occurs between the
navicular and cuboid, usually communicating with that between
cuboid and lateral cuneiform.

Metatarsophalangeal Articulations

These joints are ovoid or ellipsoid between rounded metatarsal
heads and shallow cavities on proximal phalangeal bases. They
are 2.5 cm proximal to the webs of the toes.

Articular surfaces cover the distal and plantar aspects of
metatarsal heads but not the dorsal. The plantar aspect of the first
metatarsal head has two longitudinal grooves separated by a ridge;
each articulates with a sesamoid bone embedded in the joint's
capsule, formed here by tendons of intrinsic hallucial muscles.
Articular areas on the proximal phalangeal bases are concave.
Ligaments are capsular, plantar, deep transverse metatarsal and
collateral.

The fibrous capsules are attached to their articular margins.
Thin dorsally, they may be separated from the long extensor
tendons by small bursae or they may be replaced by the tendons:
they are inseparable from plantar and collateral ligaments.

The plantar ligaments, thick and dense, are between and
blend with the collateral ligaments; their attachment to metatarsal
bones is loose but firm to phalangeal bases. Their margins blend
with the deep transverse metatarsal ligaments. Their plantar surfaces are grooved for the flexor tendons, whose fibrous sheaths
connect with the edges of the grooves. Their deep surfaces extend
articular areas for metatarsal heads.

The deep transverse metatarsal ligaments are four short,
wide, flat bands uniting plantar ligaments of adjoining metatarsophalangeal joints. Dorsal to them are the interosseous
muscles, whereas lumbricals and digital vessels and nerves are
plantar. They resemble the deep transverse metacarpal ligaments
(p. 515) *but* connect with the plantar ligament of the hallucial
metatarsophalangeal joint.

The collateral ligaments are strong cords flanking each
joint, attached to the dorsal tubercles on the metatarsal heads and
the corresponding sides of phalangeal bases; they slope down and
forwards.

Movements are like those at the hand's corresponding joints
but differ in range. Unlike that of the hand, the range of active
extension (50–60°) is greater than flexion (30–40°); an adaptation
to the needs of walking and a difference even more marked in the
hallucial joint where flexion is a few degrees while extension may
reach 90°. When the foot is on the ground metatarsophalangeal
joints are already extended to at least 25°, because metatarsal
bones slope up in the foot's longitudinal arches (4.93A,B). *Passive*
range in these joints is 90° (extension) and 45° (flexion), according
to Kapandji (1974). Adduction is linked to flexion, abduction to
extension; but abduction of the fifth toe is always linked with
slight flexion. As in the hand, *accessory movements* are gliding and
rotation of phalanges on their long axes.

Muscles Producing the Movements

Flexion: flexor digitorum brevis, lumbricals and interossei, assisted by flexores digitorum longus and accessorius. In the fifth toe
flexor digiti minimi brevis assists; in the hallux flexores hallucis
longus and brevis are the only flexors.

Extension: extensores digitorum longus and brevis, extensor
hallucis longus.

Adduction: adductor hallucis; in the third to fifth toes, the first,
second and third plantar interossei respectively.

Abduction: abductor hallucis; in the second toe, first and second
dorsal interossei; in the third and fourth, corresponding dorsal
interossei; in the fifth, abductor digiti minimi.

Note. The line of reference for adduction and abduction is
along the *second digit*, whose metatarsal is least mobile. This is
hence 'abducted' by both its interossei medially or laterally.

Interphalangeal Articulations

These are almost pure hinge joints in which trochlear surfaces on
phalangeal heads articulate with reciprocally curved surfaces on
adjacent phalangeal bases. Each has an articular capsule and two
collateral ligaments, as in the metatarsophalangeal joints (vide
supra). The capsule's plantar surface is a thickened fibrous plate,
like the plantar metatarsophalangeal ligaments, and is often termed the *plantar ligament*.

Movements are flexion and extension, greater in amplitude
between the proximal and middle phalanges than the middle and
distal. Flexion is marked, but extension is limited by tension of
the flexor muscles and plantar ligaments. Flexion and extension
are accompanied by slight conjunct rotation. *Accessory movements*
are abduction, adduction and rotation.

Muscles Producing the Movements

Flexion: flexores digitorum longus, brevis and accessorius, flexor
hallucis longus.

Extension: extensores digitorum longus and brevis, extensor
hallucis longus.

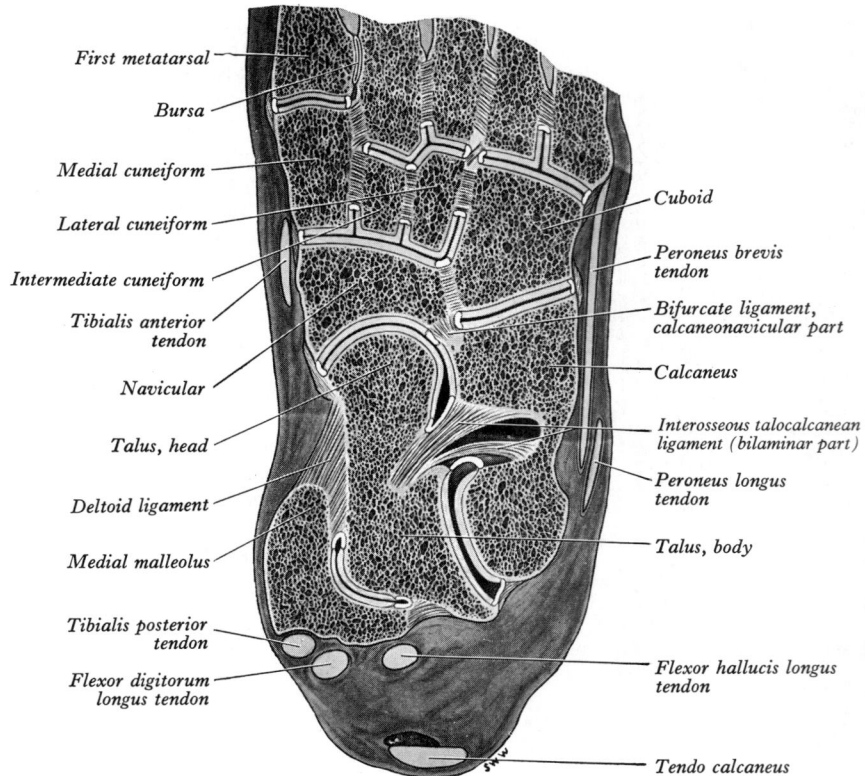

First metatarsal

Bursa

Medial cuneiform

Lateral cuneiform

Intermediate cuneiform

Tibialis anterior tendon

Navicular

Talus, head

Deltoid ligament

Medial malleolus

Tibialis posterior tendon

Flexor digitorum longus tendon

Cuboid

Peroneus brevis tendon

Bifurcate ligament, calcaneonavicular part

Calcaneus

Interosseous talocalcanean ligament (bilaminar part)

Peroneus longus tendon

Talus, body

Flexor hallucis longus tendon

Tendo calcaneus

4.92 Oblique section that descends mediolaterally through the right foot, showing the synovial cavities of the intertarsal and tarsometatarsal joints; also the medial malleolar part of the talocrural joint: Superior aspect. Note: the section passed below the joint between the medial cuneiform and the base of the second metatarsal bone; no synovial joint was present between the navicular and cuboid bones. The laminae of the interosseous talocalcaneal ligament form accessory ligaments to the fibrous capsules of the talocalcaneonavicular and subtalar (posterior talocalcaneal) joints. The (diagrammatic) apparently uniform thickness of the articular cartilages obscures the variations in thickness and curvature that obtain in life.

Movements of the Foot

The foot may move both *off the ground*, freely and relatively only to the leg, or while *on the ground*, bearing weight or transmitting thrust. The latter movements are more limited and partly imposed from the leg by body weight but are also the result of muscular contraction (pp. 656–659.)

Active movements occur at talocrural, talocalcaneonavicular and subtalar joints. At the talocrural joint movements are almost restricted to dorsi- and plantar flexion, but slight rotation may occur in plantar flexion; at the talocalcaneonavicular joint and subtalar joint ranges of movement are greater and here inversion and eversion largely occur.

With the foot grounded, body weight causes some supination, with flattening of the longitudinal arches; about one-third of the weight borne by the forefoot is taken by the first metatarsal's head. When a resting position becomes active, on starting to walk, the foot is *pronated* by muscular effort; the first metatarsal head (the second less so) is *depressed*, accentuating the medial longitudinal arch to its maximum height (Hicks 1953). Similar changes can be imposed on a weight-bearing foot by active lateral femoral rotation, which is transmitted through the tibia to the talus. This entails passive inversion of the foot at the subtalar joints. Medial femoral rotation has an opposite effect. When the foot is grounded and immobile, muscles which move it when freely suspended may exert effects *on the leg*, e.g. dorsiflexors can then pull the *leg forwards* at the talocrural joint.

The foot has two major functions: (1) to support the body in standing and progression; (2) to lever it forwards in walking, running and jumping. To fulfil the first the pedal platform must be able to spread stresses of standing and moving and be pliable enough for uneven and sloping surfaces. To fulfil its second function it must be transformable into a strong adjustable lever to resist inertia and powerful thrust. A segmented lever can best meet such stresses if arched in form.

In infants and young children plantar amassed, fatty, connective

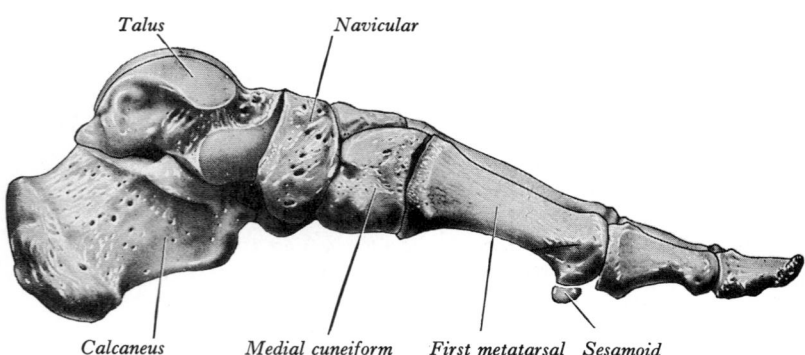

Talus

Navicular

Calcaneus

Medial cuneiform

First metatarsal

Sesamoid

4.93A Skeleton of the left foot: medial aspect. Note the height and number of bones constituting the medial part of the longitudinal arch and compare with the lateral part in 4.93B.

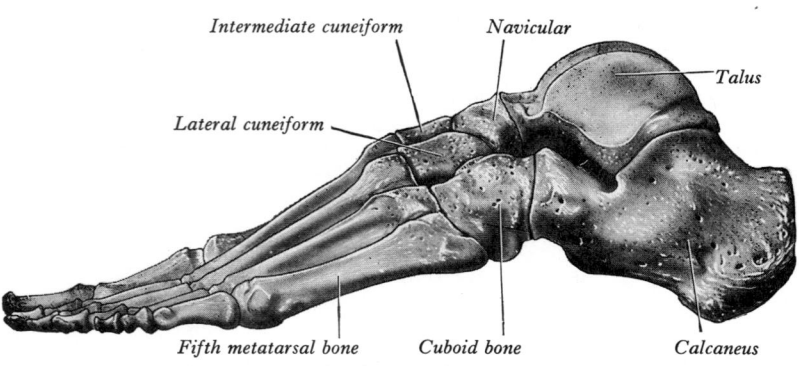

Intermediate cuneiform

Navicular

Lateral cuneiform

Talus

Fifth metatarsal bone

Cuboid bone

Calcaneus

4.93B Skeleton of the left foot: lateral aspect.

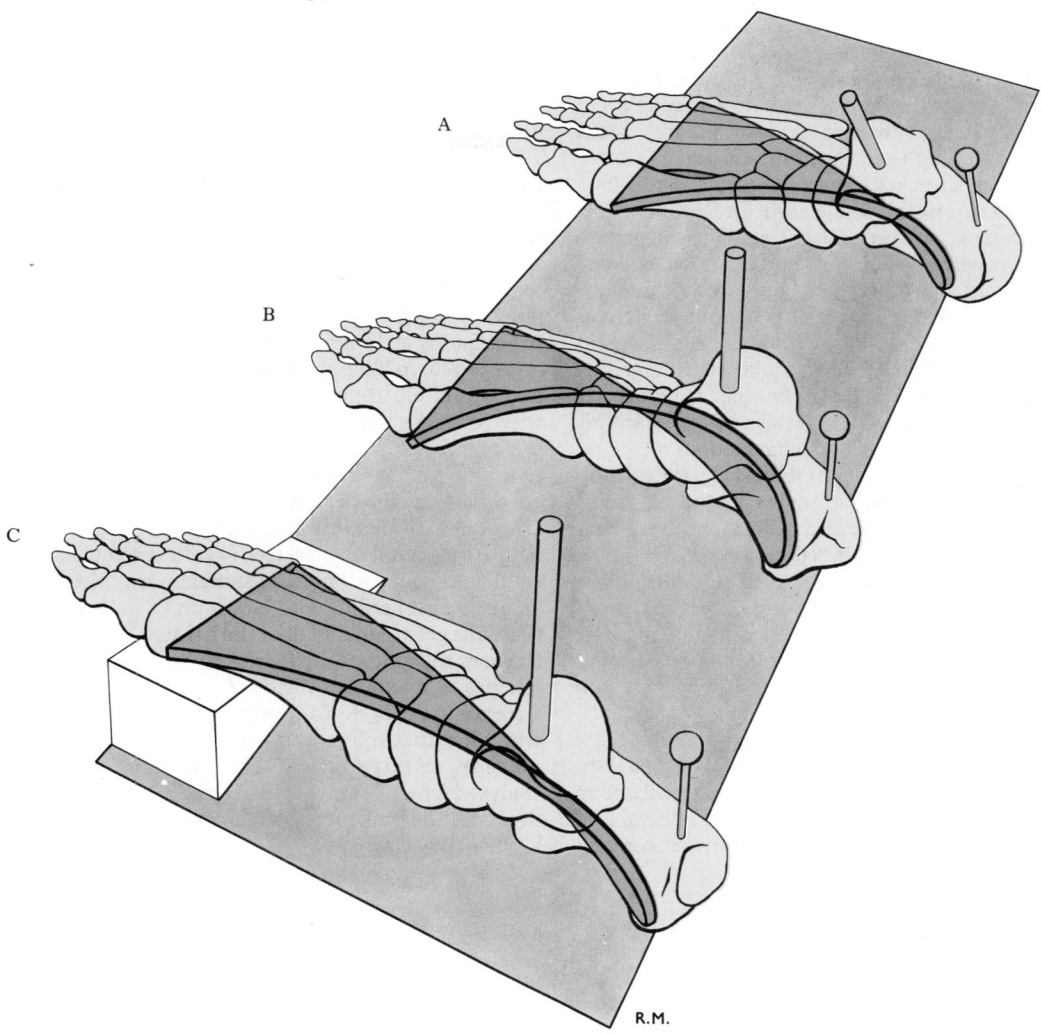

4.94 The concept of the pedal skeleton as a twisted plate which may be untwisted (supination) or further twisted (pronation) during the maintenance of a plantigrade stance in various positions of the foot. (Based upon MacConaill 1945, 1950.) A. The pedal skeleton in supination, as in standing with the feet widely separated. Note the marked medial tilting of the talus and, to a lesser degree, of the calcaneus and the depression of the medial longitudinal arch. B. Relative pronation of the foot, as in standing with the feet close together. C. Supination of the foot when standing on an inclined surface; if the position of the wedge had been reversed, the pedal skeleton would, of course, approach maximal pronation.

tissue may give the foot a flat appearance; soft tissues variably modify its appearance at all ages. But the human foot, alone among primates, is normally arched in its *skeletal* basis, usually with visible concavity in the sole. The word 'arch', so applied, has perhaps become too architectural, imposing rigidity on classical descriptions of curved pedal form and differences of interpretation more linguistic than factual. The word has several meanings and doubtless 'arches of the foot' has various implications. As a start, at simplest, the term implies little more than a curved form, concave on the plantar aspect. Such an arch should not be compared with static masonry, with pediments on terra firma and an intermediate keystone structure. The pedal arch is dynamic; muscles and ligaments are functionally inseparable. Moreover, its heel is often *off* the ground. In this account, therefore, 'arch' denotes no more than curved form, just as the back is merely curved when 'arched'.

This curvature is customarily analysed into longitudinal and transverse arches, of some value in initial analysis. They vary individually in height, especially the longitudinal in its medial part, and, being dynamic, in different phases of activity. Arches are often said to be mainly dependent on osseous shapes and ligamentous ties, the associated muscles playing a secondary role (Jones 1941, Hicks 1955). But clinical experience points to muscular insufficiency as the commonest cause of flat foot, in which ligaments elongate and bones in consequence alter in shape. It is perhaps unwise to give any factor major emphasis; all function together in the living foot. Loading experiments, on amputated legs and by electromyography in the living, show that

in *standing* ligaments play a major role; but as soon as movement occurs muscles predominate. Longitudinal curvature is usually described as composed of medial and lateral arches, a division partly justified by differences in osseous arrangement and function in the foot's medial and lateral regions.

The medial arch contains the calcaneus, talus, navicular, cuneiform and medial three metatarsal bones. Its summit is at the superior talar articular surface, taking full thrust from the tibia and passing it backwards to the calcaneus, forwards through the navicular and cuneiforms to the metatarsals. When the foot is grounded these forces are transmitted through the three metatarsal heads and calcaneus (especially its tuberosity). This arch is higher, more mobile and resilient than the lateral; its flattening progressively tightens plantar ligaments of all joints involved, including the plantar calcaneonavicular (p. 539) and plantar fascia. Tibialis posterior, flexor digitorum longus, flexor hallucis longus and intrinsic muscles all act on the curvature of the medial arch; they are adjustable, unlike ligaments. Nevertheless, they are often considered secondary and claimed to be usually inactive in standing but markedly active in any movement elevating the arch. This view, as regards static posture, does not accord with clinical experience in the aetiology of flat foot.

The lateral arch contains the calcaneus, cuboid and lateral two metatarsal bones; its summit is considered to be the subtalar articulation. It is hence skeletally lower than the medial. Its main joint is the calcaneocuboid part of the 'transverse tarsal joint' (p. 538), of very limited range. The lateral arch, being lower and less mobile, is adapted to transmit weight and thrust rather than to absorb such forces. As it flattens, long plantar and plantar calcaneocuboid ligaments tighten; its special muscles are peroneus longus and the intrinsic muscles of the fifth toe. It makes contact with the ground more extensively than the medial arch. With the foot flat its anterior and posterior ends or 'pillars' (heads of the lateral two metatarsals and calcanean tuberosity) of course transmit forces; but, as it flattens, an increasing fraction of load traverses soft tissues inferior to the whole arch. In fact the whole lateral border usually touches the ground; the medial border does not but is visibly concave, usually even in standing. This explains the familiar outline of human footprints (though this varies with position of feet, apart or together (4.94), development of soft tissues and the nature of the surface). As soon as the heel rises, in any activity, the toes are extended and muscular structures (including plantar aponeurosis) tighten up in the sole, accentuating the longitudinal arches. (It is suggested by Hicks 1955 that in this phase tension diminishes in deeper plantar ligaments.)

Since the sole is transversely concave, both in skeletal form and usually external appearance, serial transverse arches, most developed inferior to the metatarsus and adjoining tarsus, are usually described. Apart from metatarsal heads and to some degree along the lateral border, transverse arches cannot transmit forces though subjacent soft tissues; medially, only metatarsals heads can do so. Hence the foot has been likened to a half-dome, its plantar concavity directed medially, a complete dome forming when feet are close together. Peroneus longus is a strong agent in maintaining transverse curvature.

The foregoing structural analysis is a necessary prelude to the study of activities in the living. Direct observation, coupled with electromyography, kinesiology and clinical data all contribute to understanding the dynamic events in the foot during natural use. Unfortunately, disagreements persist with consequent uncertain-

ties. The following remarks are hence not to be taken as dogmatic.

In standing, with only body weight to support, intrinsic and extrinsic muscles appear to relax, plantar ligaments being relied on to tie bones into an arched form. This they can do only if longitudinal arches are allowed to sink by muscular relaxation. The medial arch is more elevated when feet are together than when they are apart; i.e. inversion with supination increases as feet are separated. This medial sag can be countered by voluntary contraction of muscles such as the anterior tibial. Pronation and supination ensure that in standing, whatever the position of the feet, a maximal weight-bearing area is grounded, from metatarsal heads along the lateral border to calcaneus. Twist imparted by pronation (partly undone in supination) prompted MacConaill (1945, 1950) to liken the foot to a twisted but resilient plate (4.94); this would ensure adequate ground contact whatever the angle between foot and leg, also imparting adaptable resilience in standing and progression. Such a form would strengthen the foot in leverage, perhaps by spiralization of ligaments.

In walking, the suspended foot, arches accentuated by muscles, is swung forward until the heel meets the ground. As it plantar-flexes, the lateral border, from heel to metatarsal heads, rolls into contact, the foot supinated for maximum contact. As it now rises, heel first, pronation (with elevation of the medial arch) transfers thrust largely to the 'ball' of the hallux. The dorsiflexed ankle and elevated metatarsus, being now near 'close-pack', are adapted to stresses of maximum thrust. As in repetitive actions at hip and knee joints, a cycle from a position of near close-pack coupled with maximal effort and return to a mobile range of joints, prepares re-entry into a phase of thrust. Kapandji (1970) has analysed foot movements in extenso. For a scholarly discussion of the evolution of the mammalian *pes*, in structure and function, see Lewis (1983).

MYOLOGY

5

Introduction

The ability to generate movement is a feature of all animals, enabling them to transport materials within their bodies and to react appropriately to external and internal stimuli. As outlined in Section 1 (p. 49), almost all cells of the body possess organelles with motile functions, prominent among which are complex interactive systems of actin and myosin filaments (see e.g. Lackie 1986). In many cells these serve to maintain shape or cause slow movements such as those seen in cellular locomotion or morphogenetic transformations in embryonic tissues. In other cells, specialized for more powerful and rapid movement, actin, myosin and other associated proteins are present in high concentrations and are deployed so as to produce efficient, linear contraction. Their plasma membranes are excitable and control the activities of the contractile apparatus they enclose. Such cells are the *myocytes* (muscle cells) which, when grouped into distinct tracts, form *muscles*. They are associated with other tissues such as those of the skeleton and, co-ordinated directly or indirectly by the nervous system, produce effector systems of great variety, capable of a wide range of highly complex actions.

Myocytes are formed in various regions of the body from myoblasts, cells of mesenchymal origin which may differentiate along one of three possible lineages to form the myocytes of *skeletal, cardiac* or *non-striated muscle* (5.1). Other contractile cells also exist as isolated individuals or small groups in various parts of the body, including *myofibroblasts* in regenerating connective tissue and *myoepitheliocytes*, cells arising from embryonic ectoderm which are associated with glands. In skeletal and cardiac muscle, the myosin and actin filaments are organized in a highly regular pattern to give myocytes a cross-striated appearance under the microscope, a cellular feature which suits them for rapid contraction.

Skeletal muscle, so named by virtue of its widespread (though not exclusive) attachment to the skeleton, is also known less appropriately as *voluntary muscle*, because in many sites it is on occasion under voluntary control, or as *striated* or *striped muscle* (an ambiguous term, of course, because cardiac muscle is also striated). Skeletal muscle is innervated by somatic motor nerves and forms the bulk of the body's muscular tissue.

Cardiac muscle is confined to the heart and is rhythmically contractile; its rate of beating is under nervous control but many individual cells do not receive direct innervation. In **non-striated muscle**, known also as *smooth, plain* or *involuntary muscle*, the actin and myosin are less ordered and the contraction is slower though of greater extent and may be sustained for long periods. Non-striated muscle is innervated by the autonomic system and is therefore not under voluntary control. It occurs in: the walls of the viscera including most of the alimentary, respiratory and urinogenital tracts; in blood vessels; in the dermis (as arrector pili muscles); in the intrinsic muscles of the eye; and the dartos muscle of the scrotum. Elsewhere there are several sites where smooth muscle fasciculi intermingle with those of skeletal muscle, particularly that classified on embryological grounds as 'special visceral muscle' (p. 172): e.g. anal and vesical sphincters, the tarsi of the upper and lower eyelids and the orbitalis muscle, the suspensory ligament of the duodenum (but see p. 1356), the intermediate zones of the oesophagus and the many fasciculi which are admixed with the fasciae and ligaments on the pelvic aspect of the pelvic diaphragm (termed by some authors the *smooth muscle pelvic diaphragm*). The mode of contraction of non-striated muscle is in general particularly suited to the regulation of the internal environment and disturbances in its function often result in impaired homeostasis.

These three types arise in different regions of the embryo (see p. 175). Skeletal muscles are largely derived from the myotomes of paraxial mesodermal somites but some differentiate in situ, e.g. the extrinsic ocular muscles in condensations of premandibular and maxillo-mandibular mesenchyme of the pharyngeal floor (often considered to be a derivative of occipital myotomes, p. 228), limb muscles from condensed limb-bud mesenchyme of somatopleuric origin (p. 174) and the various striated 'special visceral' muscles from condensations of unsplit lateral plate mesenchyme (of mixed origin) in the branchial arches and the

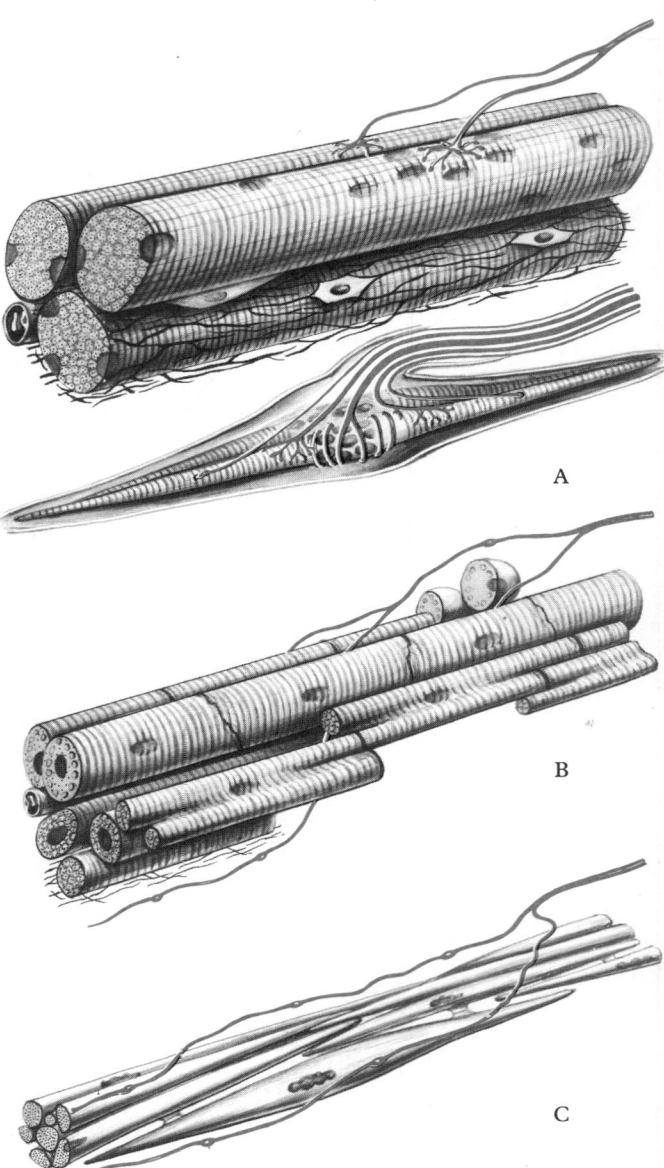

5.1 A diagrammatic representation of the major classes of muscle fibres, illustrating their chief microscopic features, including their innervation (red). A. Skeletal muscle fibres, showing a group of three fibres with satellite cells (above), and a muscle spindle containing intrafusal muscle fibres (below). B. Cardiac muscle, with smaller myocardial myocytes, larger Purkinje cells of the conducting tissue and rounded nodal cells. C. Non-striated myocytes.

post-branchial region (p. 174). Cardiac muscle stems from splanchnopleuric mesenchyme of the primitive pericardium, while the non-striated muscle of the viscera also arises from splanchnopleuric cells elsewhere or from those derived from intermediate mesoderm. Vascular non-striated muscle may, however, develop in any region from uncommitted mesenchyme cells. The non-striated muscle of the iris is thought to be derived from ectodermal cells near the margin of the optic cup; similarly, myoepitheliocytes appear to be ectodermal, whereas the non-striated ciliaris oculi and arrector pili muscles are mesodermal in origin.

Skeletal Muscle

The cellular units of skeletal muscle (5.1–3) are the *muscle fibres*, each a long, cylindrical structure bounded by a plasma membrane (the *sarcolemma*) enclosing numerous nuclei (muscle fibres are syncytial, see p. 15) and a relatively large amount of cytoplasm (*sarcoplasm*). Many muscle fibres are grouped into bundles

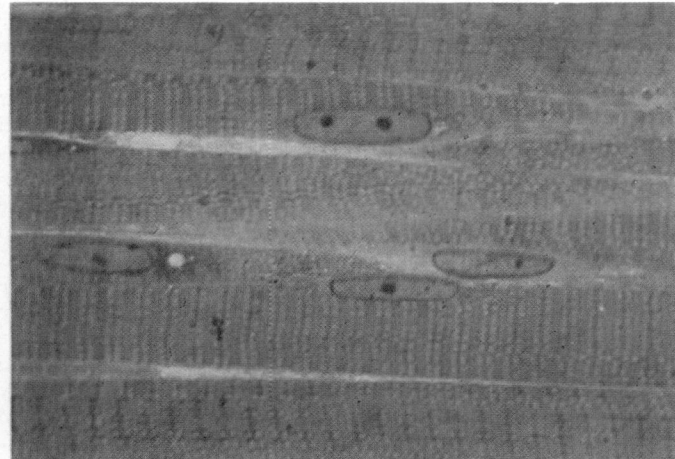

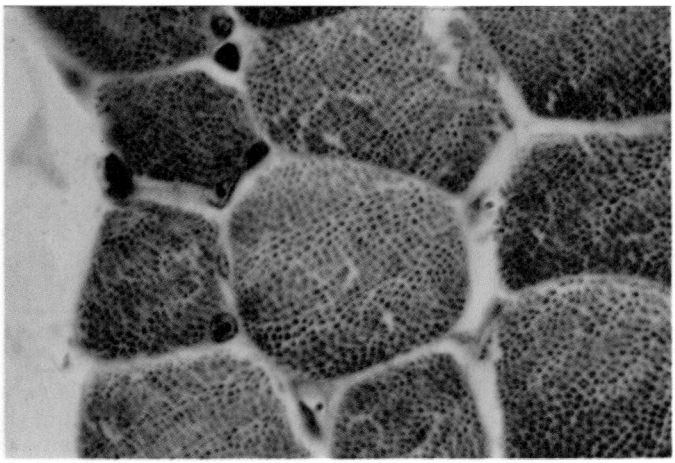

5.2A Longitudinal section of uncontracted human skeletal muscle show-ing characteristic banding pattern of large A bands (stained dark purple) and I bands (light pink). The Z bands are thin and transect the I bands. Araldite section stained with methyl green PAS. (Material supplied by R O Weller.)

5.2B Transverse section of skeletal muscle showing muscle fibres contain-ing myofibrils and muscle cell nuclei; endomysial sheaths lie between the muscle fibres. Silver stain. Magnification × 800.

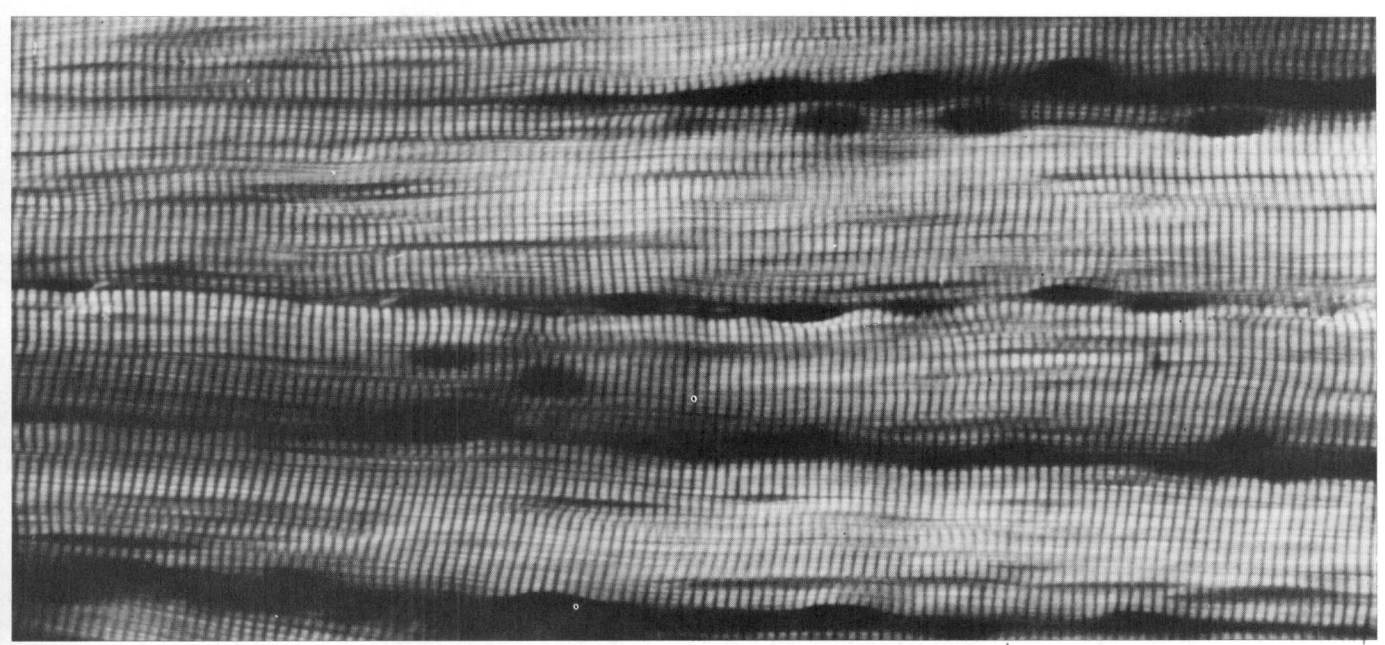

5.3 Longitudinal section of skeletal muscle viewed microscopically using crossed polaroid filters to show the birefringence of the A bands. Note also the dark muscle cell nuclei and longitudinal striations corresponding to the myofibrils. (Photography by Kevin Fitzpatrick, Department of Ana-tomy, UMDS, Guy's Campus, London). Magnification × 500.

(*fasciculi*) of varying size and pattern and an individual muscle may consist of many fasciculi. Connective tissue sheaths enclose the different muscle components: the delicate networks sur-rounding and infilling the spaces between muscle fibres are collec-tively termed *endomysium*; the stronger *perimysium* ensheaths in-dividual fasciculi; and the more substantial *epimysium* bounds the whole muscle but is also continuous with the perimysial septa within and the connective tissue external to it.

Microscopically, muscle fibres are rounded or polygonal in cross section, varying from 10–100 μm in diameter in mature muscle. Each fibre is elongated and may extend for many cen-timetres in long muscles. Usually they traverse only part of a muscle, ending in tendinous extensions or intersections within it. In longitudinal section, muscle fibres appear as long ribbons, but their cylindrical or prismatic shape can be readily seen in teased preparations. The nuclei are flattened, moderately euchromatic with one or more nucleoli and lie peripherally in the fibre im-mediately beneath the sarcolemma. They are especially numerous in the region of the neuromuscular junction (p. 556). Within the sarcoplasm are numerous longitudinal cylindrical *myofibrils*, of-

ten apparently aggregated in small groups (fields of Cohnheim; see however **5.2B**) when viewed in transverse section. The myofibrils are, in turn, composed of numerous *myofilaments* of actin and myosin, but these are visible only with the electron microscope. Myofibrils are crossed at regular intervals by trans-verse bands which have distinctive staining and optical proper-ties; in general the bands of all myofibrils are in close register, giving the whole fibre its characteristic cross-striated appearance. The myofibrils extend for long distances within muscle fibres, which therefore also display longitudinally striations.

In longitudinal sections stained with haematoxylin, there is an alternation of light and dark-staining bands; the lighter also rotate the plane of polarized light only slightly and are known as *Isotrop-ic (I or J bands)*, while the darker bands rotate the plane of polarized light strongly (*Anisotropic* or *A bands*, **5.3**, sometimes named *Q bands*). Each I band is bisected by a thin, densely stain-ing *Z band* (Zwischenscheibe, or inter-band, also termed Krause's membrane). The A bands are bisected by a paler *H* or *Hensen's band*. In relaxed muscle these bands are all clearly visible in appropriately stained sections and in unstained sections with polarizing microscopy but, when muscle fibres shorten, the I and H bands narrow to extinction, while A bands are unaffected and eventually *contraction bands* appear on either side of Z bands. Other, less well defined bands have also been described in the I bands (N lines).

Although these details have been known for over a hundred years, their significance in muscle contraction was first appreciated only relatively recently when two groups of investigators came independently to the same conclusion; A F Huxley & Neidergerke (1954) chiefly from polarized light studies; and H E Huxley & Hanson (1954) from electron microscopic observations. Their results and subsequent studies by many other investigators using a variety of cytological, biophysical, biochemical and physiological methods have provided a wealth of information about the structure and function of skeletal muscle, which will now be briefly outlined. For further reviews of ultrastructure, the reader is referred to the recent comprehensive accounts by H E Huxley (1983) and Ishikawa (1983).

THE ULTRASTRUCTURE OF SKELETAL MUSCLE

With the electron microscope, the myofibrils of skeletal muscle are seen to be divided transversely into serial units, *sarcomeres*, each about 2.5 μm long in resting muscle (5.4–8). These are composed of two major filament types, myosin filaments about 12 nm thick and actin filaments 6 nm in diameter. The myosin filaments correspond to the A band of light microscopy; the actin filaments are attached at one end to a Z band and interdigitate at the other with the myosin filaments; the I bands correspond to the adjacent regions in two sarcomeres where actin filaments do not overlap myosin (5.7). H bands are the middle regions of myosin filaments into which actin filaments have not penetrated.

Adjacent myosin filaments are connected to each other in their central regions by transversely running filaments which together constitute the *M band*. Along the sides of myosin filaments are also regularly spaced projections which, during contraction, can move to form cross-bridges with the adjacent actin filaments; however, these are absent from a short zone around the M band, sometimes called the pseudo-H band. Various other faint cross-striations have also been described, including several *C bands* across the myosin filaments and at least two *N bands* crossing the exposed regions of actin filaments. The strongest of these is the N_2 band close to the Z band. These minor bands will be discussed again in more detail later (p. 551).

The arrangement of actin and myosin in transverse sections is highly regular (H E Huxley 1972), the myosin filaments forming triangular arrays with each myosin surrounded by six actin filaments (5.5C, 6, 7). However, in the I band region, the actin filaments have a quadrilateral spacing, attaching in this configuration to the Z band, which has a corresponding square lattice substructure.

Sliding Filament Mechanism of Contraction

Electron microscope studies of the lengths of the filaments in muscle fixed in different degrees of contraction show that the actin and myosin filaments do not change their lengths during muscle shortening but slide over one another to draw the Z bands towards the middle of each sarcomere (5.4, 7). This behaviour has also been confirmed in X-ray diffraction studies of living muscle. This sliding movement is generated by the activities of the projections from the sides of each myosin filament, which can sequentially attach to and briefly pull on the adjacent actin filament before letting go and swinging back to repeat the process (5.8B). As many such cross-bridges are active but not synchronously during muscle contraction, the net effect is to exert a continuous pull on the actin filaments. Interestingly, because of the cyclical nature of cross-bridge activity, they can exert a tractive force not only when the muscle is shortening but also when it is contracting isometrically, i.e. without shortening, or even when it is being extended, e.g. by the action of opposing muscles, to decelerate a limb movement (p. 567), or opposing gravity in postural readjustments and manipulations.

Measurements of the contractile force developed by skeletal muscles stimulated to generate tension at fixed lengths (isometric contraction) have shown that the power generated by a muscle varies with its degree of stretch. Highly extended muscles are relatively weak but, as they progressively shorten, their contractile force increases to a maximum, then falls again. This behaviour is correlated closely with the amount of overlap between actin and myosin filaments, as might be expected if the contractile force is dependent on the number of cross-bridges which can be formed between them; in a muscle clamped so that its sarcomeres are stretched more than about 3.5 μm, there is very little overlap between myofilaments and little tension developed. At shorter sarcomere lengths, tension rises to a maximum at a sarcomere length of 2.0 μm, when maximum overlap and cross-bridge formation occur. At shorter lengths, tension drops again as the actin filaments slide across the mid-point of the sarcomere and begin to mutually interfere with cross-bridge formation, finally dropping to zero tension at about 1.9 μm; the Z bands are then drawn into contact with the ends of the myosin filaments, distorting the latter to convey the impression of 'contraction bands' when observed with the light microscope.

In view of these findings, it is interesting that most skeletal muscles are arranged in the body so that, relative to their overall lengths, they undergo only limited shortening under normal physiological conditions and are thus generally maintained near the right length for most efficient actin–myosin overlap and hence maximal force generation.

Detailed Structure of Myofilaments

Myosin filaments are about 1.8 μm long and are composed of about 180 myosin molecules each. Myosin molecules have a high molecular weight ($Mr = 500\ 000$) and consist of a rod-like tail region, which lies mostly along the axis of the myofilament, and a head region which projects at an angle to it. The myosin molecule has six distinct subunit polypeptides consisting of two heavy chains stretching the whole length of the molecule and four light chains confined to the head. The light chains constitute a pair of 'regulatory light chains' (LC-2) which appear to control myosin ATPase activity and a pair of so-called 'essential light chains' (LC-1 or LC-3, of uncertain function). The heavy chains can be cleaved by proteolytic enzymes into two parts: *light meromyosin* consisting of most of the rod-like portions of the two heavy chains; and *heavy meromyosin*, composed of the cross-bridge forming parts of the myosin molecule, namely the light chains and the remainder of the two heavy chains which can be further cleaved proteolytically into two fragments (S1 and S2). Under the electron microscope, isolate myosin molecules show that the head region has a deep longitudinal cleft, corresponding to the symmetrical arrangement of the heavy and light chains in this region (5.8A). The points of enzyme attack are thought to represent regions where the molecule flexes during muscle contraction. In the presence of ATP and magnesium ions, heavy meromyosin can bind to actin filaments and this property can be used experimentally to demonstrate the presence of actin in electron micrographs and to determine the 'direction' in which the actin filaments are organized at the molecular level, as the myosin heads bind at an acute angle with the actin in an arrowhead configuration, according to the polarity of the filament.

Myosin molecules have a complex chemistry which varies to some extent with the subtype of skeletal muscle fibre they belong to, and this can be exploited immunohistochemically to identify them experimentally or for clinical purposes (see also p. 552).

Myosin molecules can aggregate by their tails to form long filaments. The head, and part of the attached tail of each molecule projects at an angle from the main filament, except in its middle zone where the tails reverse direction. Heads are arranged with great precision, a factor which probably enhances the rapidity and efficiency of muscle contraction; although precise details are not certain, it is likely that the heads emerge in radially symmetric triplets which are arranged in a spiral pattern, each triplet being rotated by 40° on the adjacent triplet.

The heads are able to swing out from the main filament to attach to actin filaments, as seen in electron micrographs and X-ray diffraction images of contracting muscle, or muscle in rigor.

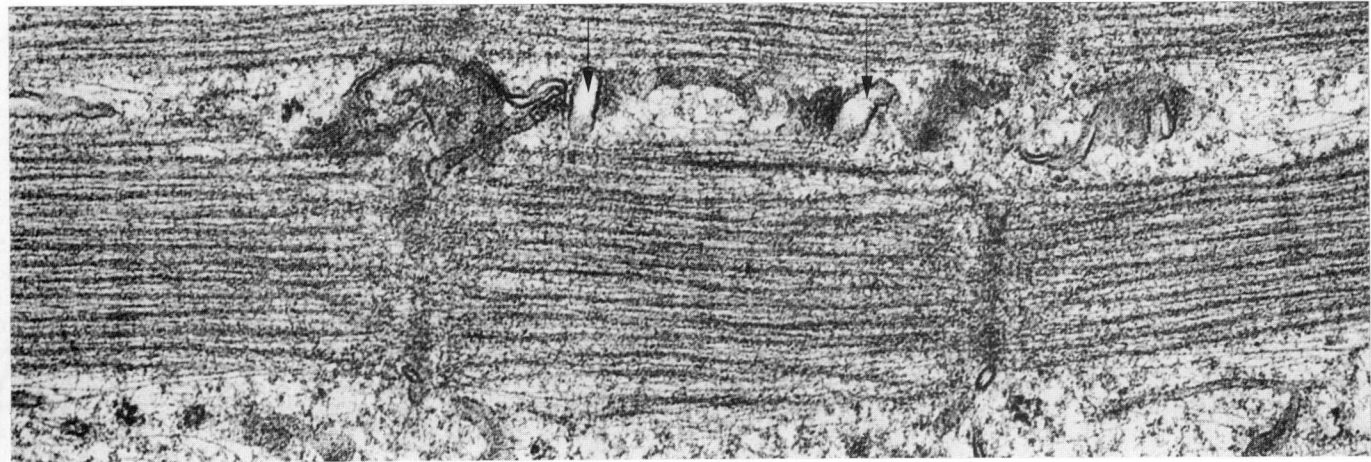

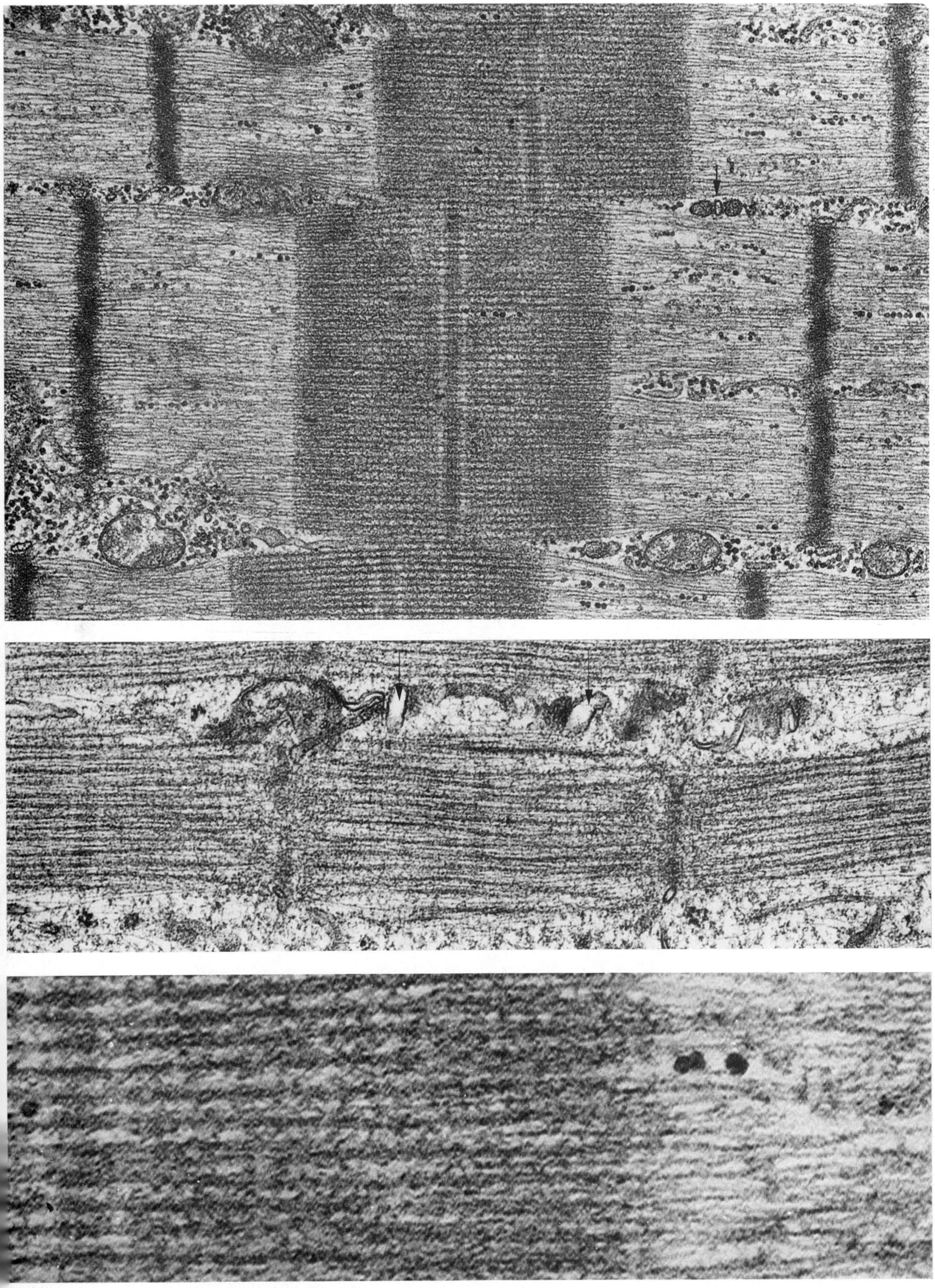

.4 Electron micrographs of skeletal muscle in longitudinal section. ᴀ. Uncontracted myofibrils. ʙ. Contraction has occurred, shortening the ᴀrcomeres. ᴄ. Detail of unshortened muscle showing cross-bridges between actin and myosin filaments. Arrows indicate the position of the T-system components. Compare with 5.6,7. (Material for ᴀ provided by Professor R O Weller; cf. 5.1.). Magnifications: ᴀ and ʙ × 15000, ᴄ × 100000.

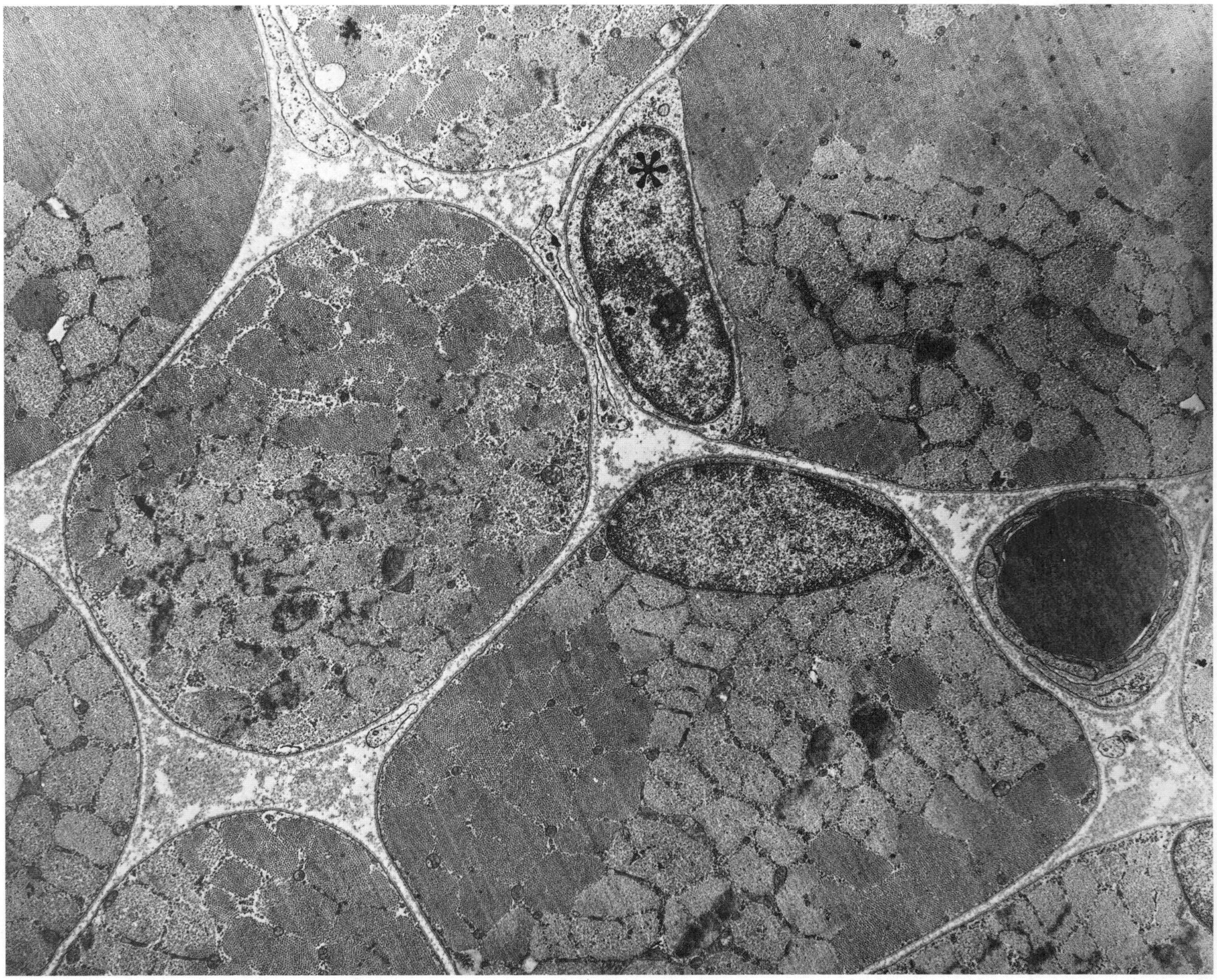

5.5A Transverse section of skeletal muscle from a child showing muscle fibres, myofibrils and mitochondria. A satellite cell(⋆) is also present, lying beneath the basal lamina. Immediately below this is a muscle fibre nucleus. Note the transversely cut endomysial capillary. Magnification × 4000.

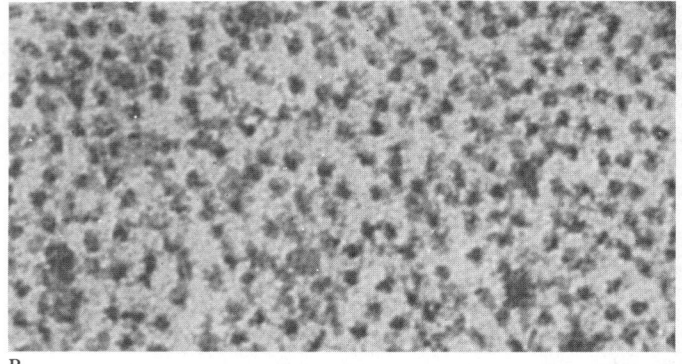

B

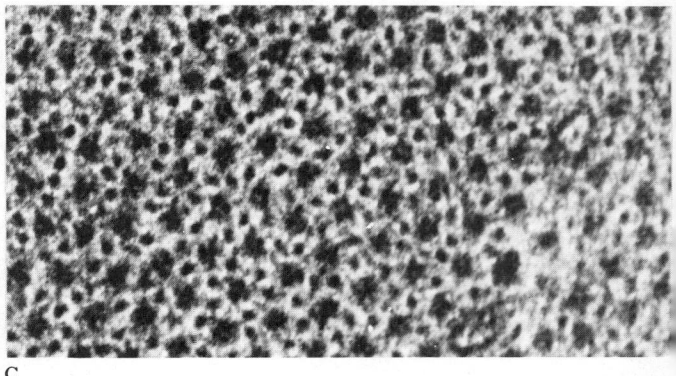

C

5.5B, C Details of the configuration of actin and myosin filaments in transverse section. B is through the I band and shows only the actin filaments. C is through the A band and shows thin actin filaments arranged round thick myosin filaments, with cross bridges stretching between them. Magnification × 200 000.

Actin filaments (see p. 30) are composed of a staggered helical array of globular actin subunits ($Mr = 43\,000$), 13 in each turn of the helix (5.8A), and are each about 1.0 μm long. In the grooves between the helices (one on each side of the filament) lie the filamentous molecules of another protein, *tropomyosin*, each molecule spanning 7 actin subunits. At 40 nm intervals there is a fourth protein, *troponin*, formed by four polypeptides *(troponins T, I and C)* which are bound to actin. When muscle is resting, the position of tropomyosin prevents the myosin heads from binding to actin but, when muscle is activated, raised levels of calcium cause the troponin to change its shape, displacing tropomyosin and allowing myosin to attach to actin.

The precise mechanism by which force is generated when myosin and actin interact is still uncertain, but it is likely that it involves first a binding of myosin heads to actin, then a change of angle in the hinge region of the stalk connecting the myosin head to the main filament, followed by detachment of the head from the actin and a swinging back of the stalk to its original angle (5.8B).

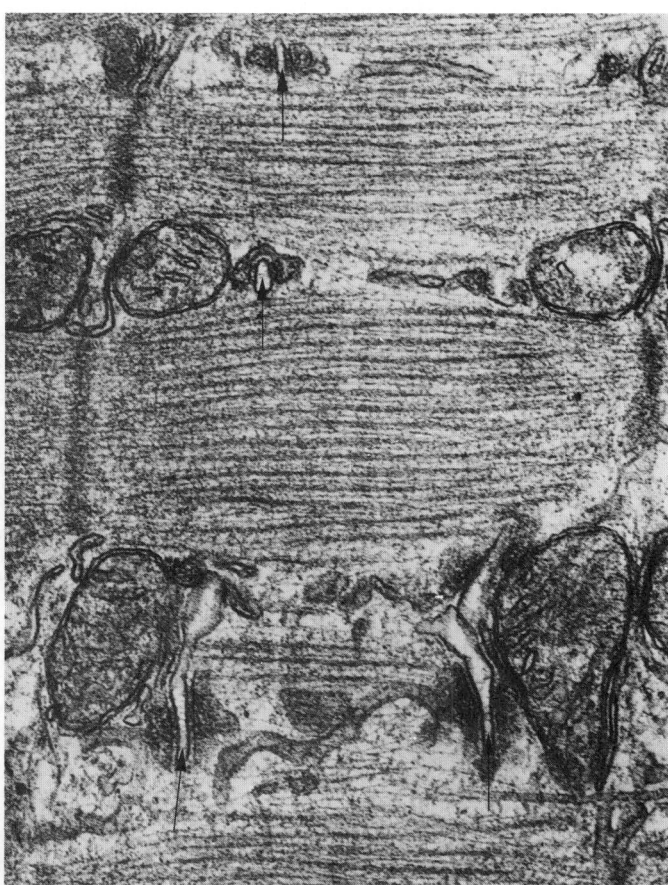

5.5D Longitudinal section through a group of myofibrils, showing the sarcoplasmic reticulum and associated transverse tubules (arrow). Note the darkly granular terminal cisternae of the sarcoplasmic reticulum where it lies adjacent to the transverse tubules. Magnification × 40 000.

As mentioned, this cyclical process is dependent on the presence of magnesium ions, ATP and sufficient calcium ions to unblock the myosin binding sites on the actin filament and also to activate certain other enzyme systems of the myosin heads which regulate their ATPase activity.

The direction in which active sliding occurs is also dependent on the orientation of actin filaments, the helices of which spiral in opposite directions on either side of each Z band.

Other components of myofibrils. The Z bands are complex structures which in section present a zig-zag appearance, to the points of which the actin filaments are attached. Viewed face on, the Z bands appear as a square lattice of fine filaments, embedded in a densely staining granular matrix. The precise organization of this array is not entirely certain but may be in the form of numerous regularly linked tetrahedra, from the apices of which actin filaments emerge (see e.g. Landon 1970, Kelly & Cahill 1972, see also the review by Wang 1985). The predominant protein of Z bands is α-actinin, which may form the dense matrix component, but various other proteins are also present.

In addition there are many other proteins associated with the contractile apparatus, some soluble and therefore not localized to any particular structure, others bound to the myofilaments, including: the 'C proteins' (Offer 1973) bound at regular positions on the myosin filaments; a creatine kinase associated with the filaments of the M band; and the proteins *titin (connectin)* which are thought to form an additional set of myofilaments within the sarcomeres, providing a degree of *elasticity* to muscle fibres (see Wang 1985). The existence of such elastic filaments has long been postulated and there have been various reports of thin (3–4 nm) filaments in addition to actin and myosin which may be responsible for the known elastic properties of muscle fibres and for helping to maintain their coherence during actin–myosin sliding. The presence of the rather mysterious N bands in the I band region may reflect the architecture and possible interlinking of these fine filaments.

The Non-myofibrillar Structures of the Sarcoplasm

These include the essential organelles required for cellular metabolism, such as ribosomes, mitochondria, Golgi apparatus, lysosomes, transport vesicles, lipid droplets; the majority of these organelles are located around the nuclei near the cell periphery and to a lesser extent scattered between the myofibrils. During active growth either during normal development or as a response to exercise (p. 554), these organelles are particularly numerous. Glycogen is also distributed in small granular clusters between myofibrils and among the actin filaments (5.4A); it is an important energy store and can be rapidly mobilized close to where it is needed. In addition to these usual cellular organelles there are some specialized features, as follows.

The sarcolemma has the ultrastructure typical of a normal plasma membrane, with an external basal lamina to which adhere the collagen fibres and other components of the endomysium. Narrow tubular invaginations of the sarcolemma penetrate deeply into the sarcoplasm, branching and anastomosing around the myofibrils; these are the *transverse tubules* or *centrotubules* (Franzini-Armstrong 1975, Ishikawa 1983). In most mammalian skeletal muscle, as also in reptiles and birds, these systems lie perpendicular to the sarcolemma, penetrating among myofibrils transversely at the planes of the A–I band junctions. Commonly in lower vertebrates, including amphibians and fishes, they lie at the level of the Z bands. The lumina of these tubules are of course in continuity with the extracellular space, as shown by the infusion of colloidal tracer substances into the endomysium; such markers can diffuse within the T-tubules, reaching into the deepest parts of muscle fibres, but do not pass into any other of the cell's membranous channels. Physiologically the sarcolemma has special properties in that it can generate and conduct action potentials rapidly to all parts of the muscle fibres, both along its external surface and into the interior via the T-tubules, as these are invaginations of the sarcolemma. In this way all parts of the muscle fibre can be activated rapidly and nearly synchronously to produce a characteristic twitch. The specialized region of the fibre which initiates this event is the *sole-plate*, part of the neuromuscular junction, which will be considered in detail elsewhere (p. 847).

The sarcoplasmic reticulum is a specialized form of agranular endoplasmic reticulum which plays a vital role in muscle contraction. It is a plexus of anastomosing membranous channels filling much of the space between myofibrils (5.5D, 6), expanding into larger sacs (terminal cisternae) where it comes into close approximation to the T-tubules. The combination of a T-tubule and the terminal cisternae on either side is known as a *muscle triad*. The membranes of terminal cisternae and the centrotubule are connected by rows of regular, hollow filaments ('feet') about 8 nm long, which may have a role in coupling the electrical excitation of the sarcolemma and the intracellular release of ions, leading to muscle contraction (Franzini-Armstrong 1970, Somlyo 1979).

Electron micrographs of the sarcoplasmic reticulum often show a densely staining granular interior (5.5D). This system can concentrate calcium ions to much higher level than found elsewhere in the sarcoplasm; when an action potential sweeps down the membrane of the T-tubules, the sarcoplasm releases free calcium ions into the myofibrils, initiating actin–myosin cross-bridge activity and causing muscle contraction. As the sarcolemma becomes electrically stable once more, calcium ions are re-bound to the sarcoplasmic reticulum and the muscle fibre relaxes.

Mitochondria (5.5D, 6) are usually elongate and have closely packed cristae. Their number varies according to muscle type, being much greater in slow twitch than in fast twitch muscle (see p. 552).

The cytoskeleton of skeletal muscle fibres has recently been investigated in detail (see Wang 1985) and it is now known that the class of intermediate filaments termed in muscle *desmin* filaments (p. 30) plays an important part in maintaining the relative positions of myofibrils and other organelles within the muscle cell, connecting adjacent myofibrils and also lying parallel to them in a manner suggesting that they may participate in transmitting contractile forces to the sarcolemma and throughout the muscle cell.

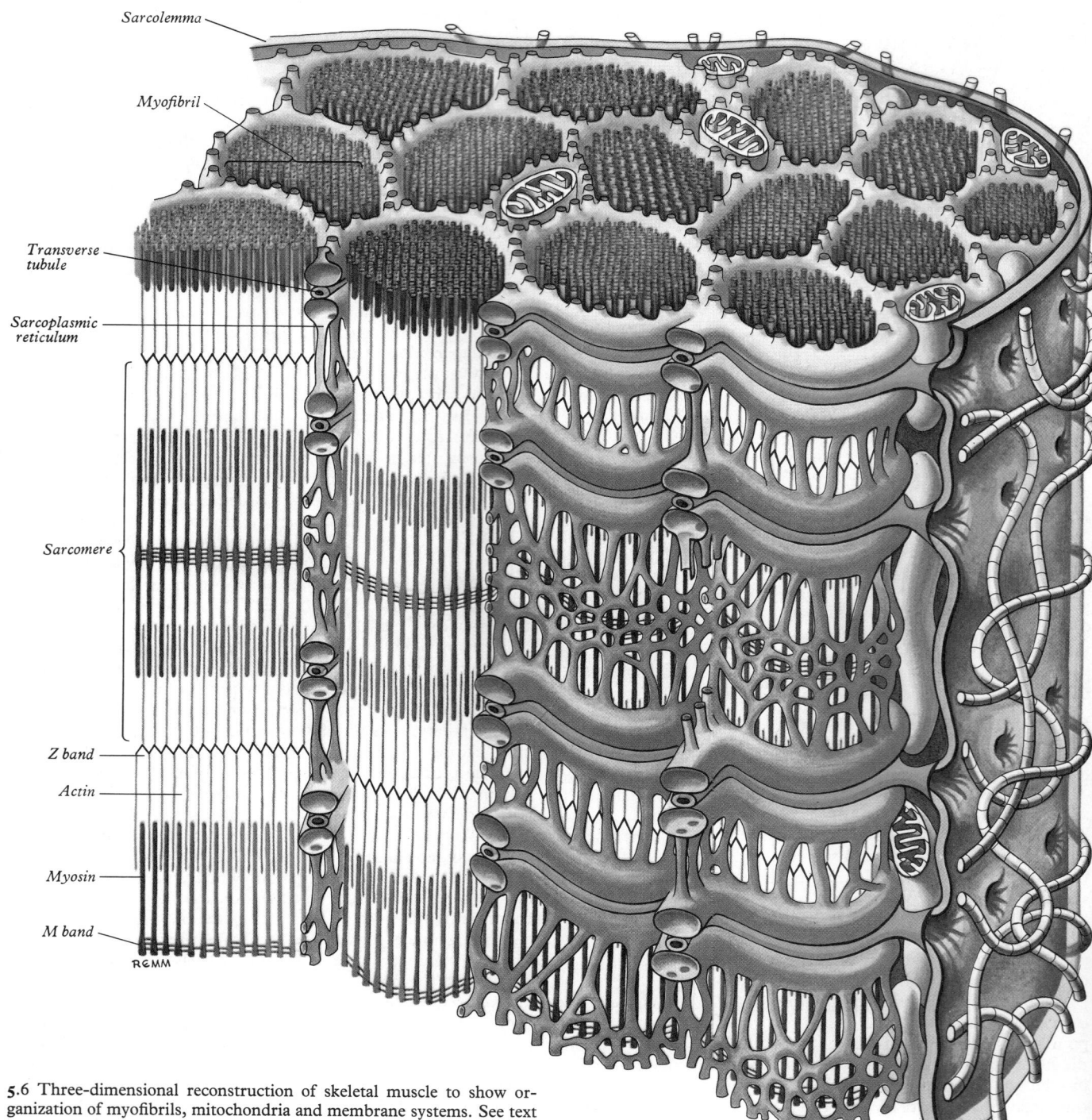

Sarcolemma

Myofibril

Transverse
tubule

Sarcoplasmic
reticulum

Sarcomere

Z band

Actin

Myosin

M band

RGMM

5.6 Three-dimensional reconstruction of skeletal muscle to show organization of myofibrils, mitochondria and membrane systems. See text for further details. (Modified from a diagram by D Fawcett.)

CLASSIFICATION OF SKELETAL MUSCLE FIBRE TYPES

Although all skeletal muscle fibres have the same basic sarcomeric organization, a number of distinct types have been described according to structural, physiological and biochemical criteria. In all mammals the great majority can be divided into three main types (Brooke & Kaiser 1970): red slow twitch (Type I), white fast twitch (Type IIA) and intermediate (Type IIB). Beside these are some highly specialised fibres, including slow tonic (non-twitch) components of extrinsic ocular muscles, the stapedius muscle of the middle ear and also the intrafusal (bag and chain) fibres of the neuromuscular spindle (see p. 858). (For a review of slow tonic muscle, see Morgan & Proske 1984.) In most muscles, Types I and II fibres are intermingled, although there is usually a predominance of one or the other; their ratio is closely related to the contractile properties of the muscle as a whole. There is also some evidence that other types of muscle fibre may exist, e.g.

'visceral' striated muscle in the pharynx and upper oesophagus where the motor innervation is from the vagus nerve (Whitmore & Notman 1987). Many vertebrates other than mammals show a wide range of muscle fibre types, including those of conspicuously red and white muscles, e.g. of birds, but there is no simple relationship between these and mammalian fibre classes.

Type I, red or slow twitch muscle fibres (slow oxidative fibres), take about 75 ms to complete their contraction and such fibres have other distinctive properties; they are red even when exsanguinated, because of a large concentration of myoglobin, an oxygen-storing pigment similar to haemoglobin, and can undergo extensive repetitive contraction without fatigue. Their sarcoplasm is rich in mitochondria and they are strongly positive for the oxidative enzymes of the Krebs cycle and cytochrome chain. Structurally their fibres tend to be narrower than Type II fibres, have thicker Z and M bands, more glycogen (Eisenberg & Kuda 1976) and tend to show a staggered arrangement of sarcomeres (5.2,4A). Their neuromuscular junctions also tend to be less extensive (p. 904) and their vascular supply is denser, their oxygen demands being more marked. When stretched during contraction, their fibres are also much stiffer than fast muscle, due to a slower rate of actin–myosin cross-bridge cycling, and are thus well suited for resisting or decelerating stretch, as seen in a particularly marked form in postural muscles, but also as a variable property of most skeletal muscles.

Type II, fast twitch muscle fibres, in contrast, can contract in as little as 25 ms. Two subclasses are usually distinguished: *IIA (fast oxidative and glycolytic or intermediate fibres)* and *type IIB (fast glycolytic fibres)*. Type IIA fibres obtain their energy from glycolysis and oxidative phosphorylation, having more mitochondria and being more resistant to fatigue than type IIB. The latter is dependent mainly on glycolysis for energy. There are also many other features which separate these types, for instance their myofibrillar ATPases are optimally active at different pHs and their myosin molecules are chemically distinct, so they can be readily separated histochemically. In general, Type II fibres appear to dominate in non-postural limb muscles, being able to generate powerful forces over a relatively short period (e.g. biceps brachii); while type I fibres form a high proportion of some postural muscles, e.g. thoracic erector spinae muscles and soleus (Brooke & Engel 1969, Johnson et al 1973, Širca & Kostevc 1985). However, these ratios are to some extent variable within a population and may change considerably during development and ageing (Larsson et al 1978).

Variations in enzyme content of muscle fibres appear to depend partly on their activity, as shown by crossing nerves supplying fast and slow muscles, which induces a partial change in enzyme patterns (Buller 1970, Close 1972). This might be caused by secretion of a neural 'trophic factor', but subsequent work indicates that type of muscle fibre is determined primarily by frequency of neural activity in its axon (Riley & Allin 1973). Muscle fibre ratios are also affected by training, the proportion of IIA fibres increasing with exercise, giving the muscle a better ability to resist fatigue (Maxwell et al 1973).

DEVELOPMENT, GROWTH AND REGENERATION OF SKELETAL MUSCLE

The precursors of myocytes are embryonic mesenchymal cells of myotomes or various other regions (pp. 136, 138, 175). The different stages of differentiation into mature muscle fibres have been the subject of much controversy and several schemes of nomenclature have been proposed (see e.g. Boyd 1960, Fischman 1972, Murray 1972). It is generally agreed that specific mesenchymal cells (*promyoblasts*) grow into fusiform elements up to 400 µm long (*myoblasts*), which cease to divide and fuse end to end (**5.8c**), forming long multinucleate *myotubes*, their nuclei initially lining up in a central row with a surrounding 'tube' of cytoplasm. Myotubes elongate by further fusion of myoblasts at their ends and sides, to form fetal muscle fibres. After this, assembly of myofilaments begins, forming myofibrils with organized sarcomeres. Myofibrils first appear near the cell's centre; more myofilaments are added outside existing ones and nuclei are thus displaced to the periphery. Later, the sarcoplasmic

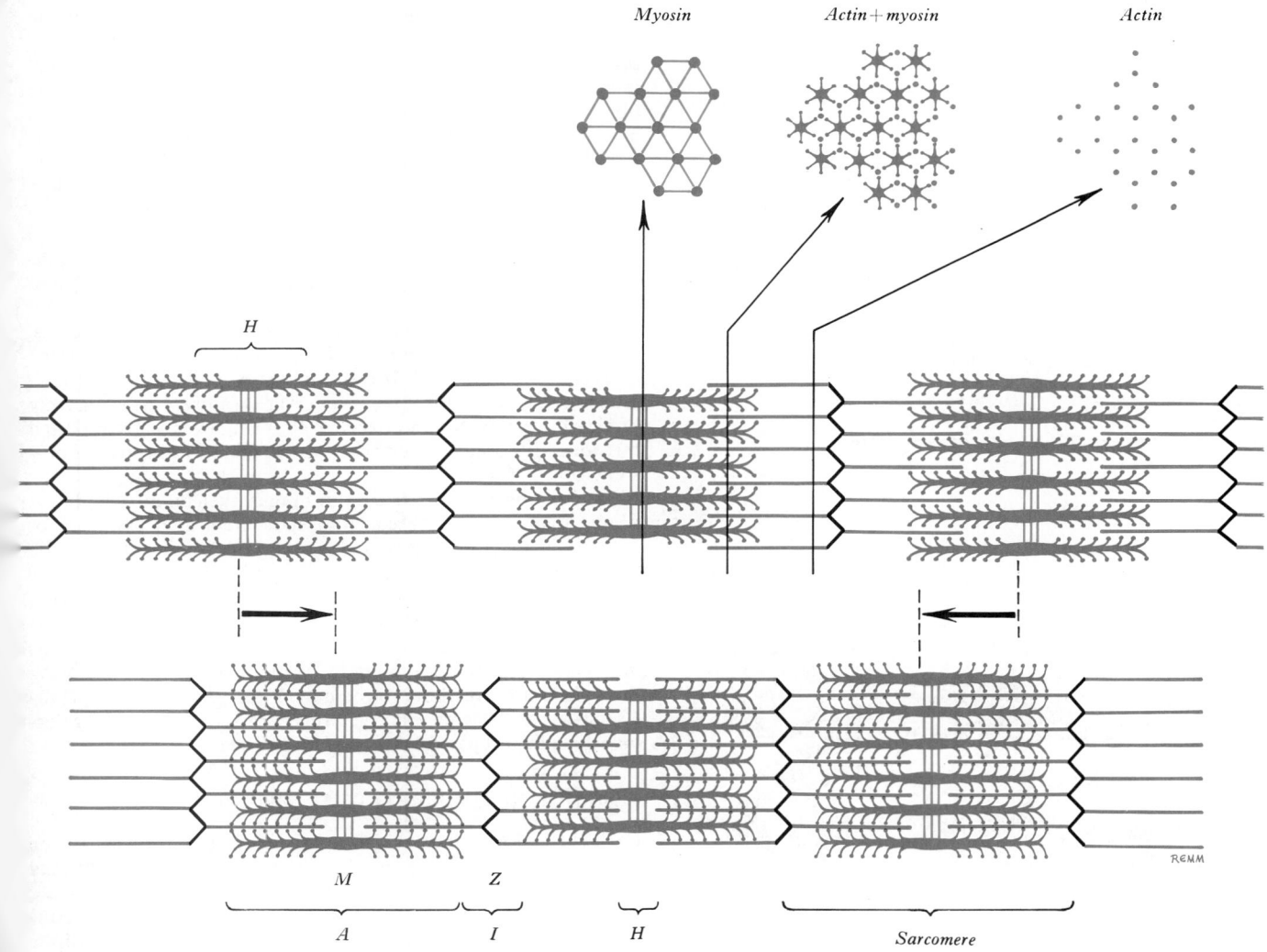

Myosin   *Actin + myosin*   *Actin*

7 Diagram showing the organization of sarcomeres in skeletal and cardiac muscle and the changes occurring during shortening. Transverse sections are shown at various levels and indicate the packing of actin and myosin filaments. Compare with 5.4 and see text for a detailed description.

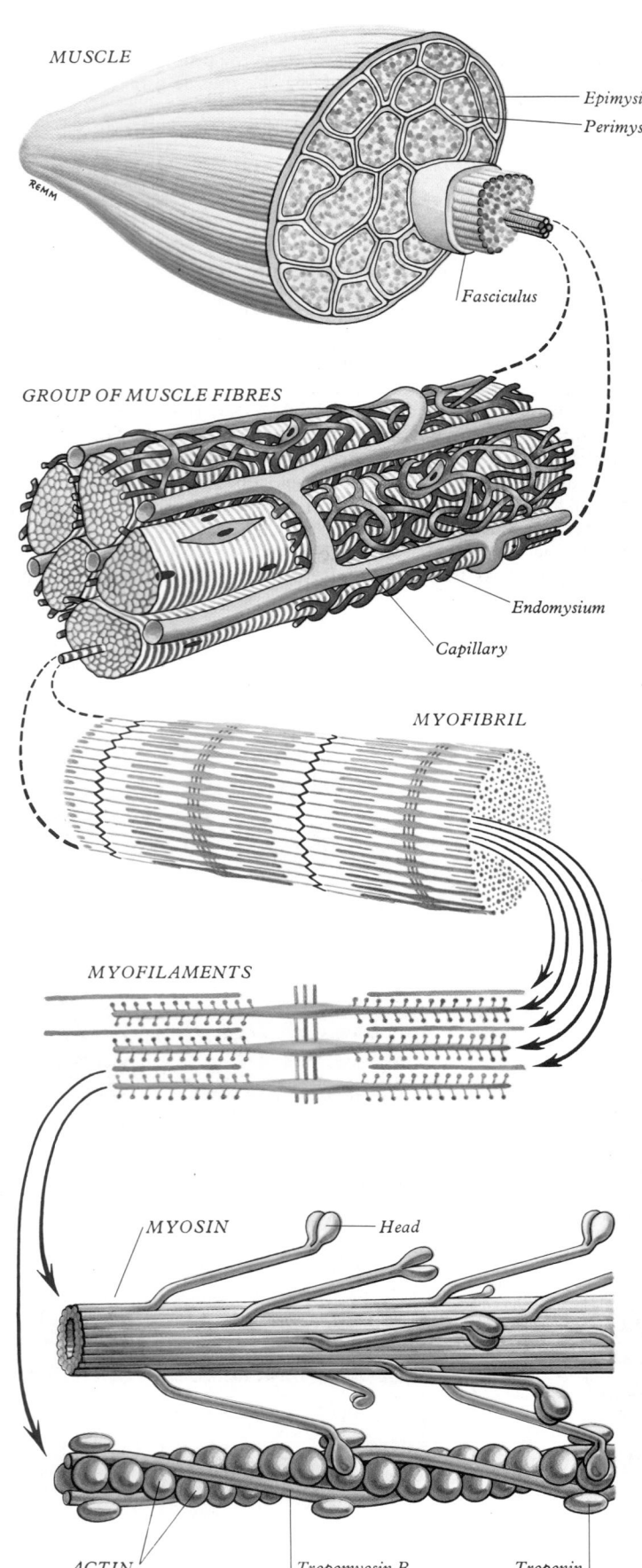

MUSCLE

Epimysium

Perimysium

Fasciculus

GROUP OF MUSCLE FIBRES

Endomysium

Capillary

MYOFIBRIL

MYOFILAMENTS

MYOSIN   Head

ACTIN   Tropomyosin B   Troponin

5.8A Diagram showing successive levels of organization within a skeletal muscle, from whole muscle, through fasciculi, fibres, myofibrils, myofilaments, down to molecular dimensions.

554

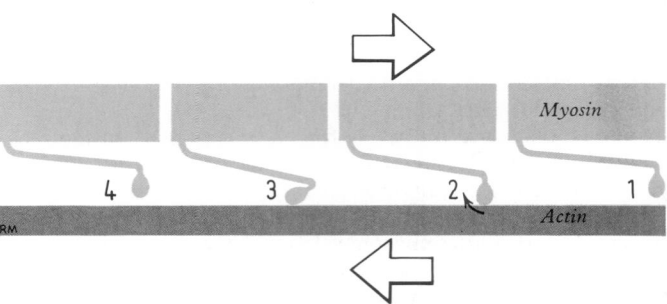

Myosin

4   3   2   1   Actin

5.8B Diagram showing the changes in myosin–actin interactions during the generation of shearing forces between the two types of filament. The cycle of attachment, flexing and detachment of the myosin heads is indicated in the sequence 1–4.

reticulum begins to form, followed by ingrowth of the transverse tubules.

Myotubes show three grades of increasing maturity. For a period areas of their surfaces are enveloped by a multinucleate syncytium with numerous pseudopodial projections invaginating the surface, especially opposite myotube nuclei (Kelly & Zacks 1969, Landon 1970, Yamashiro et al 1984). This superficial syncytium has been interpreted as a stage in fusion with the myotube or a close conjunction of morphogenetic significance without fusion, the surface syncytium eventually disappearing. The final number of fibres in a muscle is attained before birth (Jaubert 1955). Subsequent longitudinal growth is primarily due to an increase in sarcomeres added at the ends of fibres by progressive fusion of myoblasts. Elongation is also due to small increases in the resting lengths of sarcomeres. Growth in muscle fibre diameter is by addition of the myofilaments around myofibrils which, reaching a critical width, appear to split longitudinally. Growth continues into postnatal life, myoblasts probably fusing with established fibres to increase their nuclear number. Some myoblasts may persist as *satellite cells* which lie in contact with fibres, enclosed within their basal laminae (5.5A) (Schmalbruch & Hellhammer 1976). Fibre diameter is subsequently affected by use, reacting to exercise by hypertrophy and to disuse by atrophy (Rowe & Goldspink 1969). The postnatal growth of skeletal muscle is also much affected by hormone levels, particularly anabolic steroids such as testosterone, responsible for greater development of fibre size in males, and by thyroid hormones and somatotropin. Denervated fibres progressively diminish in diameter and degenerate, being eventually replaced by connective tissue, which becomes increasingly fibrous and may show contracture.

Limited regeneration occurs in human skeletal muscle; fibres near a damaged zone break into short nucleated cylinders of cytoplasm; macrophages arrive and engulf debris but leave the basal laminae intact. The cylinders now fuse and grow back inside original basal laminae to form myotubes, until the growing undamaged ends finally fuse to fill the gap (Carlson 1973). Varying degrees of maturation follow. Satellite cells (*myosatellitocytes*) may participate in regeneration, fusing at the ends of existing sarcoplasm to contribute to the new fibres (Bischoff 1975, Hall-Craggs 1974). If damage is great, regeneration may not occur; lost muscle is then replaced by connective tissue.

INNERVATION OF SKELETAL MUSCLE

Each skeletal muscle is supplied by one or more nerves. In the limbs, face and neck, the nerve supply is often single (but may stem from several neural segments), but where they originate from successive embryonic segments, e.g. muscles of the abdominal wall, their nerves of supply are correspondingly multiple. Neurovascular bundles usually enter the deeper surface of a muscle near its less habitually mobile attachment, at a position which is approximately constant for each muscle (see Brash 1955, Coërs & Woolf 1959).

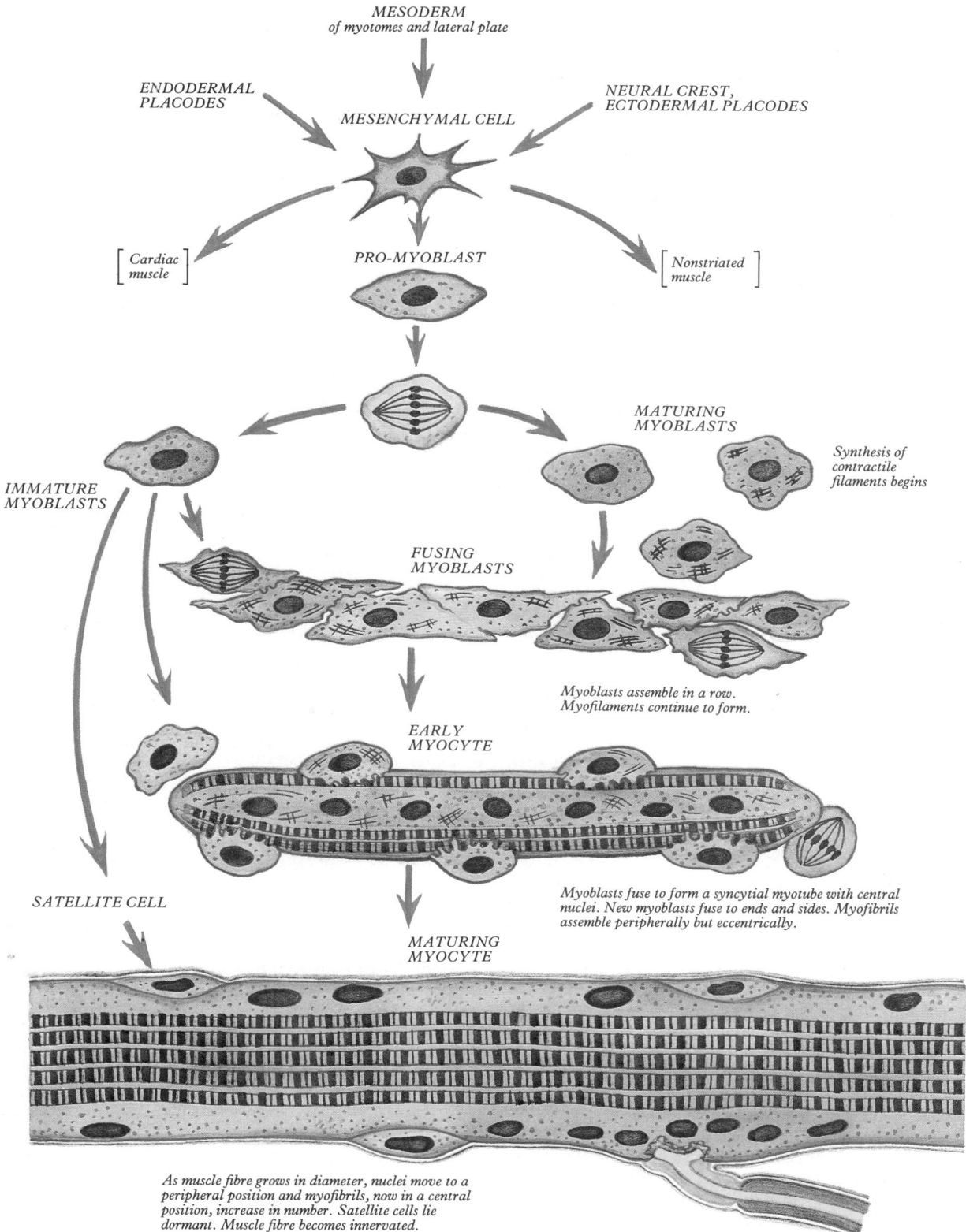

MESODERM
of myotomes and lateral plate

ENDODERMAL
PLACODES

NEURAL CREST,
ECTODERMAL PLACODES

MESENCHYMAL CELL

[*Cardiac*
muscle]

PRO-MYOBLAST

[*Nonstriated*
muscle]

MATURING
MYOBLASTS

Synthesis of
contractile
filaments begins

IMMATURE
MYOBLASTS

FUSING
MYOBLASTS

Myoblasts assemble in a row.
Myofilaments continue to form.

EARLY
MYOCYTE

SATELLITE CELL

Myoblasts fuse to form a syncytial myotube with central
nuclei. New myoblasts fuse to ends and sides. Myofibrils
assemble peripherally but eccentrically.

MATURING
MYOCYTE

As muscle fibre grows in diameter, nuclei move to a
peripheral position and myofibrils, now in a central
position, increase in number. Satellite cells lie
dormant. Muscle fibre becomes innervated.

5.8c Early development of skeletal muscle, showing different stages including early differentiation of myoblasts from mesenchymal cells and fusion of myoblasts to form a myotube maturation into a striated myocyte.

Muscle nerves have a wide variety of motor and sensory components. **Motor fibres** include large myelinated alpha-efferents, which supply most muscle fibres, small gamma-efferents to muscle spindles and fine non-myelinated autonomic efferents (C fibres) to vascular non-striated muscle. **Sensory fibres** comprise myelinated afferents of a range of diameters, from the large spindle and tendon organ fibres to the fine non-myelinated fibres supplying pain and other sensory terminals in the connective tissue of the muscle sheaths. For a fuller consideration of this topic, the reader is referred to pp. 898 and 915.

Within muscles, nerves form plexuses in the epi- and perimysial septa before entering the endomysial spaces around the muscle fibres. Alpha-efferents branch repeatedly before losing their myelinated sheaths and terminating on muscle fibres. The position of such motor terminals is usually to be found in a narrow band across the central zone of the muscle (*the motor band*). Many gamma-efferents to muscle spindles end in the same region, although muscle spindles may also be present elsewhere. Autonomic fibres ramify in the endomysium throughout the muscle, supplying its vessels. For the innervation of tendon, see pp. 563, 913.

Each muscle fibre receives a single terminal branch of an alpha-efferent axon which ends at a *neuromuscular junction*. This is composed of the axon terminal, the *motor end plate*, giving off a few short branches over a discoidal patch of sarcolemma, the *sole plate*. In a slow tonic fibre (vide supra) the terminals branch to form several small neuromuscular junctions with a muscle fibre over an extended distance (*en grappe endings*). Some or perhaps most of such muscle fibres may receive the terminal branches of more than one motor neuron; intrafusal (muscle spindle) fibres also have in some cases multiple innervation and their neuromuscular terminals take a number of different forms, e.g. *en plaque* endings of a discoidal form and long *trail* endings (see p. 916). Neuromuscular junctions are considered in more detail elsewhere (p. 904).

A motor unit is essentially a functional entity, defined as the combination of a single alpha motor neuron with all of the muscle fibres which it innervates. (The term *myone* for the scattered group of muscle fibres forming the *structural* equivalent of a motor unit has also been proposed by MacConaill & Basmajian 1977 and the muscle fascicle or *myoneme* is the *kinematic unit*. However, these terms are not in widespread use but are mentioned below.) Motor unit size varies greatly, units being smaller where precise control is required, e.g. in extraocular muscles where each motor neuron innervates only 6–12 muscle fibres (Bors 1926); in large limb muscles the ratio may reach 1 to 2000. The total force which can be developed by each unit is therefore inversely related to precision of control.

Muscle fibres innervated by one motor neuron are often widely spread throughout a muscle and do not correspond to fascicles or any other obvious structural division (Burke & Tsairis 1973). Even when only a few motor units are active, the force they generate is transmitted diffusely within the muscle. Each axon probably gives motor terminals to a rounded territory, with overlap between adjacent zones of innervation (Sissons 1969).

CONTROL OF MUSCLE CONTRACTION

Excitation of muscle fibre occurs when a volley of action potentials passes along the motor neuron axon to reach its motor end plate. This causes the secretion of acetylcholine into the gap between nerve ending and sarcolemma. This neurotransmitter is rapidly inactivated by the enzyme acetylcholinesterase at the surface of the sarcolemma but, in the brief period before this happens, acetylcholine is bound by receptor molecules in the membrane of the sole-plate, causing a change of conductance in the membrane and a local graded depolarization (the *end-plate potential*) from the normal resting potential of about 80 mV (inside negative). The size of the end-plate potential depends on the rate of acetylcholine release, which in turn is related to the frequency of action potential conduction in the motor axon. If the sole-plate membrane is depolarized beyond a certain critical level, the surrounding sarcolemma undergoes an abrupt but transient reversal of potential due to an influx of sodium ions, becoming about 20 mV, inside positive, before returning to the resting potential as potassium ions carry positive charges out of the cell. Such action potentials are propagated along the sarcolemma at about five metres per second and penetrate to the deepest parts of the muscle fibre along the transverse tubule system (p. 551). This event elicits a twitch of contraction lasting 25–75 ms. Repeated action potentials and corresponding twitches can be evoked by continuing neural activity, although there is a limit to twitch frequency, as a *refractory period* of about 10 ms follows each muscle action potential.

Since action potentials are all the same size, they are termed 'all or none' responses. Similarly, muscle twitches are maximal if they occur at all but the degree of contraction also depends on the initial length, loading and metabolic state of the muscle fibre. Contraction in whole muscles is graded by the differential activity of motor units (p. 939). Individual motor units can vary much in twitch frequency and the number of units active also fluctuates. In small contractions only a few units are active but, as contractions increase, more motor units are recruited, summating as a steady contraction of a whole muscle, the single muscle fibres twitching repetitively but asynchronously. Under experimental conditions above a critical stimulation frequency, the relaxation of individual muscle fibres cannot occur before the next contraction starts, resulting in partial or complete fusion of twitches or *tetanus*. The threshold for tetanic stimulation varies with muscle type: for slow twitch muscles it is about 30 and for fast twitch, 100 per second, or more. Since all muscle contractions are summations of individual twitches, even those which appear 'steady' show fine oscillation on closer examination. Like energetic states of subatomic particles, release of neurotransmitters is *quantal* and muscle contraction may also be so regarded; for some purposes quantum mechanics may thus prove an appropriate means of analysis.

BLOOD VESSELS OF SKELETAL MUSCLE

The blood supply of muscles arises from muscular branches of adjacent arteries. As already stated, in many cases branches of a principal artery, vein and nerve enter together along a strip, the neurovascular hilum, while subsidiary arteries often enter elsewhere. All branch into smaller arteries and arterioles which ramify in perimysial septa, giving off capillaries to the endomysium. These lie mainly parallel with the muscle fibres which they serve and are interconnected by numerous transverse anastomoses (5.8A). The density of capillary beds varies; in predominantly 'red' muscles it is higher than in those with a majority of 'white' (fast twitch) fibres, corresponding to the greater oxygen demands of red muscle (see p. 552).

On the basis of various experiments including the use of isotopic sodium as a tracer (Barlow et al 1961) and microscopic examination (Grant & Payling Wright 1968), two distinct circulations have been reported in skeletal muscle. The *nutritive circulation* is fed from arteriolar branches of the hilar arteries which reach the endomysium, where all blood passes through capillaries to collecting venules and veins, leaving by the hilum. Alternatively, some blood passes into arterioles of epi- and perimysium, which contain few capillaries; arteriovenous anastomoses are abundant here, most blood returning directly to veins without traversing the capillaries. This *non-nutritive* pathway allows blood to pass when flow in endomysial capillary beds is impeded, e.g. during muscle contraction (see also p. 693).

The *lymphatic drainage* commences as capillaries in epi- and perimysial sheaths which apparently do not enter the endomysium. They then converge to form larger lymphatic vessels accompanying the veins to the hilum, draining to the regional lymph nodes.

CONNECTIVE TISSUES OF SKELETAL MUSCLE

The general arrangement of connective tissue in muscle has already been described (p. 547). Connective tissue performs important roles in the functional organization of muscles, providing mechanical coherence, allowing independent movement of individual fibres and fasciculi and transmitting contractile forces to adjacent structures such as tendons, aponeuroses and bones. Connective tissue planes are also routes for nerves and blood vessels. The endomysium is of special significance, since it forms the external environment of muscle fibres and is the site of metabolic exchange between fibres and capillaries. Its proteoglycan matrix also mediates the flow of ions during excitation of muscle fibres and acts as an ionic reservoir.

The fibres of collagen (chiefly Type I, see p. 64), reticulin and elastin and their associated proteoglycans, also affect the mechanical characteristics of muscle in contraction and passive extension (Hill 1970). Transmission of contractile force to adjacent structures requires attachments between muscle fibres and their associated connective tissue. These include the junctional regions between the ends of fibres and their tendinous attachments, *myotendinal junctions* (Schippel & Reissig 1968, Hanak & Bock 1971). Here, the various orders of connective tissue sheaths fuse to form the fibrous fasciculi of the tendon. Muscle fibres may

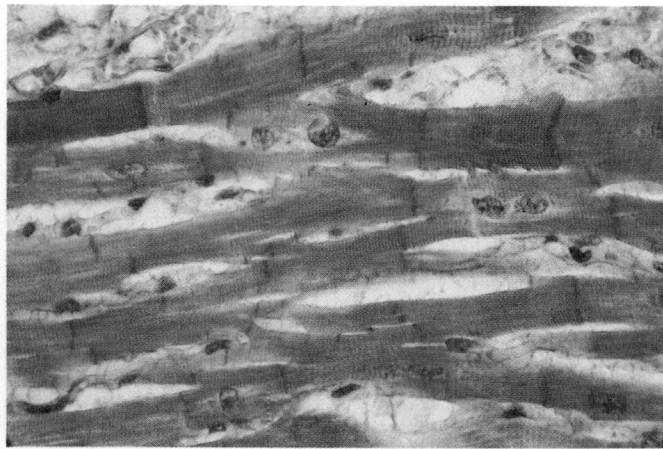

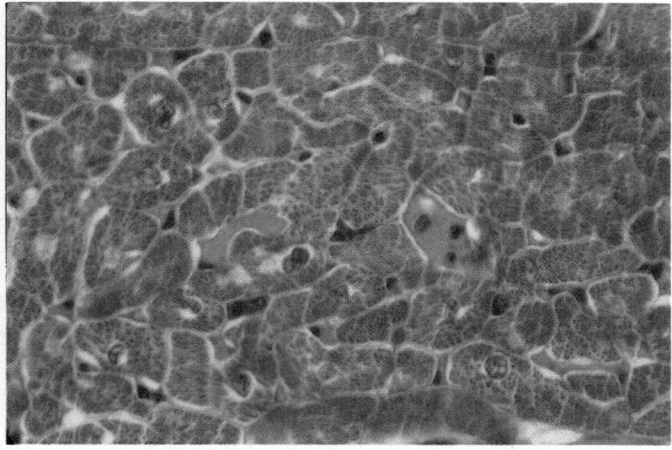

5.9A Longitudinal section of cardiac muscle showing branched cardiac myocytes with cross striations and centrally placed nuclei. Darkly staining transverse intercalated discs are also visible. Magnification × 800

5.9B Transverse section of cardiac muscle showing cardiac myocytes with central nuclei and peripheral myofibrils. Note the grouping of myocytes into fasciculi, with connective tissue septa between. Magnification × 800

taper or have flattened ends or show a terminal expansion. Electron microscopy shows a highly indented sarcolemma in such regions, with a dense sub-sarcolemmal layer of cytoplasm into which actin filaments of adjacent sarcomeres penetrate. Externally the basal lamina is prominent, with collagen and reticulin fibres in close contact. There are however no desmosomal attachments and it is probable that both here and along the whole fibre contractile forces are transmitted from cell surface to the connective tissue matrix primarily by viscous adhesive forces.

Cardiac Muscle

Heart muscle has distinctive structural and physiological properties which differ in several respects from those of skeletal and non-striated muscle. It is composed of tracts of uninucleate cells (5.9), each up to about 80 µm long and 15 µm in diameter in man, with a central nucleus; each cell may divide into two or more branches at its ends, which abut against other cells, giving the appearance of a network of branching and anastomosing cylinders. Indeed, cardiac muscle was long considered to be syncytial. In the intercellular spaces is an endomysium similar to that of skeletal muscle, containing capillaries and nerve fibres. The larger tracts of myocytes, varying from a few hundred to many thousands, are parcelled into fasciculi by perimysium, as also in skeletal muscle.

Each *cardiac myocyte* shows striations like those of skeletal muscle, with A,I,Z and H bands, although these are not as obvious in routine histological preparations. At the ends of the myocytes are conspicuous densely staining cross-striations, *intercalated discs*. The same pattern of sarcomeres with actin and myosin filaments also exists (Fawcett & McNutt 1969, Page & Fozzard 1973) but these are not arranged in discrete myofibrils. Mitochondria are large and numerous and are interspersed between myofilament clusters (5.11). T-systems are also similar to those of skeletal muscle (Forssmann & Girardier 1970) but penetrate the sarcoplasm opposite the Z bands rather than at the I–A band junction (5.11), possibly reflecting the slower contraction time of cardiac muscle; the sarcoplasmic reticulum is not as abundant. Glycogen and myoglobin are present abundantly and histochemical preparations are strongly positive for oxidative enzymes. These features, like those of mammalian slow twitch skeletal muscle, reflect the high oxygen demand and constant rhythmic expenditure of energy typical of cardiac muscle, as does the richness of the myocardial vascular bed and consequent blood flow. Cardiac myocytes have a spontaneous rhythm of contraction and relaxation, even when isolated. This *myogenic rhythm* has a parallel oscillation in membrane potential dependent on a slow inward leakage of sodium ions. The rhythm of isolated cells is slower than that of the whole heart but, in the intact heart, myocytes are synchronized by the more rapid

rhythm of the pacemakers and conducting system (vide infra and p. 720).

Intercalated Discs

Ultrastructurally these are seen to be special regions of the sarcolemma and underlying cytoplasm, present at junctions between myocytes. They may lie transverse to the cell's long axis but are often stepped with alternating transverse and longitudinally orientated planes (5.9,10). At these junctions, adjacent myocytes are anchored together by wide belts of desmosomes alternating with gap junctions (see p. 24), the latter being the means by which cardiac myocytes are electrically coupled. Hence, under normal conditions, excitation and contraction spread rapidly along tracts of interconnected cells and the heart behaves *electrically* as a syncytium. Immediately beneath these membranous junctions, the cytoplasm is dense and appears to anchor actin filaments of neighbouring sarcomeres as well as other cytoskeletal structures, including intermediate filaments, a feature which allows contractile forces to be transmitted effectively from one cell to the next.

Conducting Tissues of the Heart

These tissues are also described with the heart (p. 720). Three major types of cell form the different parts of the conducting system: *nodal myocytes, transitional myocytes* and *Purkinje myocytes* (Sommer & Johnson 1968, James & Sherf 1974).

Nodal myocytes, responsible for pacemaker activities, are mononucleate, rounded, cylindrical or polygonal cells, existing in groups in the sinu-atrial and atrioventricular nodes (5.1). They have a few randomly orientated myofibrils, lack a regular sarcotubular system and show a pale zone free of organelles around a large central nucleus.

Transitional myocytes also occur in nodes, extending into the stem and principal rami of the conducting system as connecting pathways between nodal and Purkinje myocytes. Transitional myocytes are narrower than cardiac myocytes but possess the same type of contractile apparatus. They may conduct more slowly than the larger fibres and this appears to be the basis of the conduction delay in the atrioventricular node and its principal rami.

Purkinje myocytes are large mononucleate cells, wider and shorter (30 µm in diameter and 20–50 µm long) than general cardiac myocytes. They also have larger gap junctions at their ends and also laterally. They contain fewer myofibrils, many mitochondria, abundant glycogen and a well-developed sarcotubular system, all features correlated with rapid conduction in Purkinje fibres (about 2–3 metres per second, compared with 0.6 metres per second in myocardium). It should be noted at this point that Purkinje *fibres* are composed of rows of Purkinje *cells* and it therefore seems inappropriate that the term 'Purkinje fibre' has often been used to denote single Purkinje myocytes.

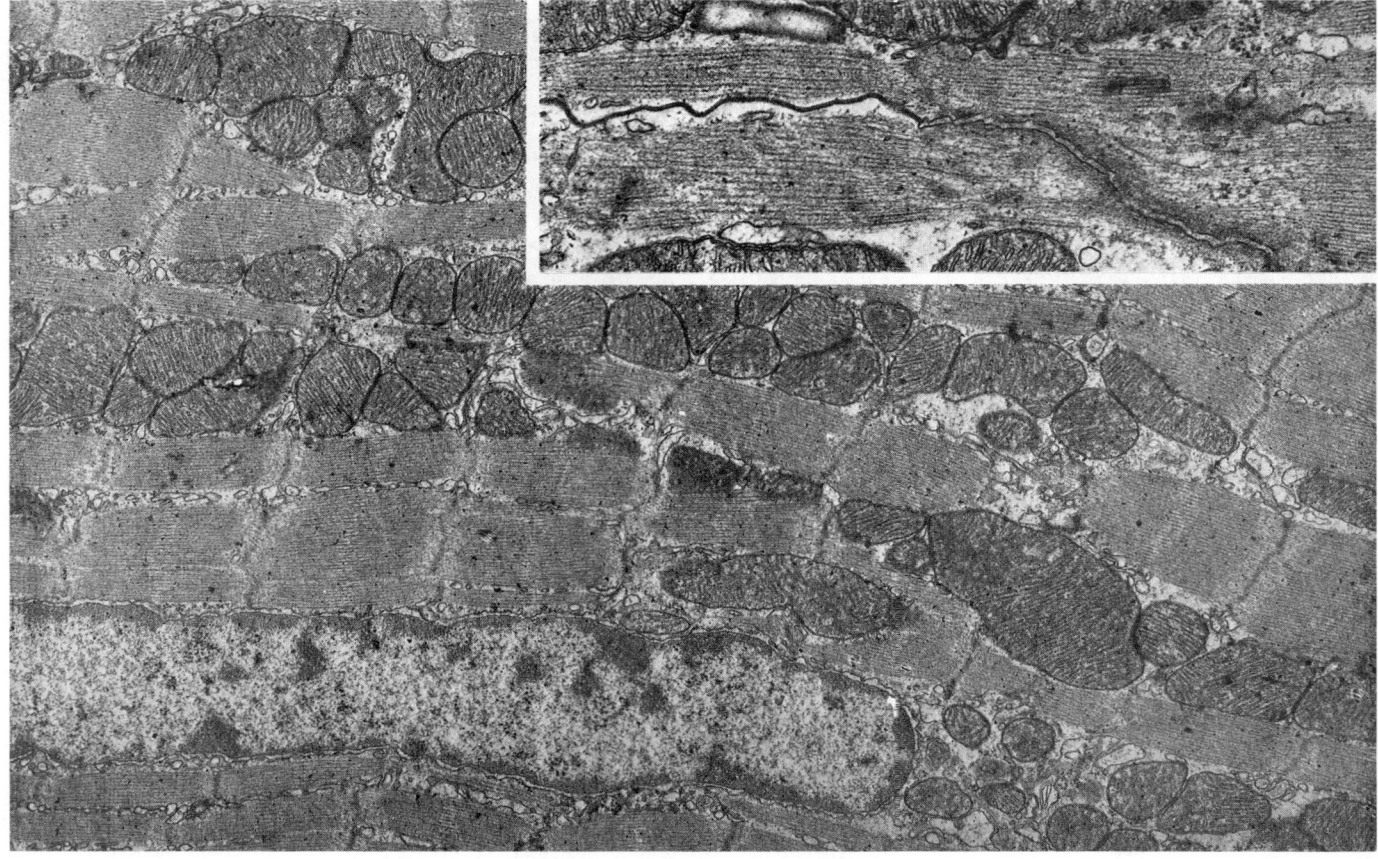

5.10 Electron micrograph of cardiac muscle in longitudinal section showing abundant mitochondria and intercalated discs. Magnification × 4000.

Detail shows part of intercalated disc with gap junction in left half of field and a desmosomal belt on the right. Magnification × 15 000.

Although these cell types and their arrangement within the major conducting pathways of the heart are well established, there is less certainty about some conducting routes which have been adduced from electrophysiological evidence. These include the proposed pathways from right to left atrium (interatrial), between the sinu-atrial and atrioventricular nodes (internodal) and the accessory atrioventricular bundles. For a further discussion of this topic, see p. 722.

Innervation of Cardiac Muscle

Non-myelinated, post-ganglionic, sympathetic and parasympathetic axons permeate cardiac connective tissue septa and end close to but not in contact with individual myocytes. The quantitative aspects of distribution of autonomic efferent and afferent nerve terminals in all the complex localities of the heart and its vessels are virtually an uncharted field.

Blood Vessels of the Myocardium

The myocardium has a dense vascular bed derived from the coronary vessels (p. 731), which branch to feed a rich endomysial plexus of anastomosing vessels, each cardiac myocyte being no more than about 8 μm from a capillary (Wearn 1941). Vascular channels occupy about 60% of the total interstitial space in rabbits' myocardium (Frank & Langer 1974). Lymphatic capillaries also penetrate the endomysium, unlike those of skeletal muscle.

Development, Growth and Regeneration

Cardiac muscle develops from mesenchyme of the myo-epicardial mantle in the embryonic heart (p. 207), a source confirmed by Morris (1976) in human embryos. Cardiac promyoblasts are at first stellate, adhering by intercellular junctions which later become intercalated discs (Challice & Viragh 1973, Morris 1976). They show repeated mitoses and synthesize myofilaments, thus becoming cardiac myoblasts, which begin rhythmic contraction early in fetal life. Cell proliferation continues into postnatal life but gradually yields to growth by hypertrophy of individual cells,

a process to some extent dependent upon their work load. Regeneration of damaged muscle does not seem to occur and it is replaced by fibrous scar tissue (Hudson & Field 1973).

Non-striated Muscle

Non-striated muscle (smooth, involuntary or plain muscle) differs considerably from other muscle tissues in structure and behaviour (5.1,12–15). It is composed of mononucleate, fusiform myocytes, varying in length from 15 μm in small arterioles to 500 μm or more in the myometrium in pregnancy (Csapo 1962, Huddart & Hunt 1975). Disposed with long axes parallel to the direction of contraction, they are usually arranged in small fasciculi, separated by connective tissue septa containing afferent and efferent vessels and nerves. In regions where concerted contraction in one direction occurs, fasciculi lie parallel, as in the muscularis externa of the intestines. Where contraction reduces the surface area, fasciculi interweave in all directions, as in the urinary bladder. In muscular arteries non-striated muscle forms thick sheets devoid of capillaries. Within fasciculi much of each cell is covered almost entirely by a prominent basal lamina; attached to the exterior of this structure (basement membrane), are numerous fine elastin, reticulin, and collagen fibres, forming complex networks around cells. Cells are separated by a space of 40–80 nm except at intercellular junctions (vide infra). Fibroblasts and other connective tissue cells are usually lacking in fasciculi, and the extracellular matrix is characteristically their own product. At their boundaries connective tissue fibres interweave with those of interfascicular septa, presumably transmitting forces generated by contraction of individual cells.

Two types of non-striated muscle are distinguished by innervation and behaviour: *multi-unit* and *unitary* muscles (Bozler 1948, Burnstock 1970). In *multi-unit muscles*, rich nerve plexuses provide many myocytes with motor terminals; contraction is usually initiated by neural action, i.e. is not spontaneous or due

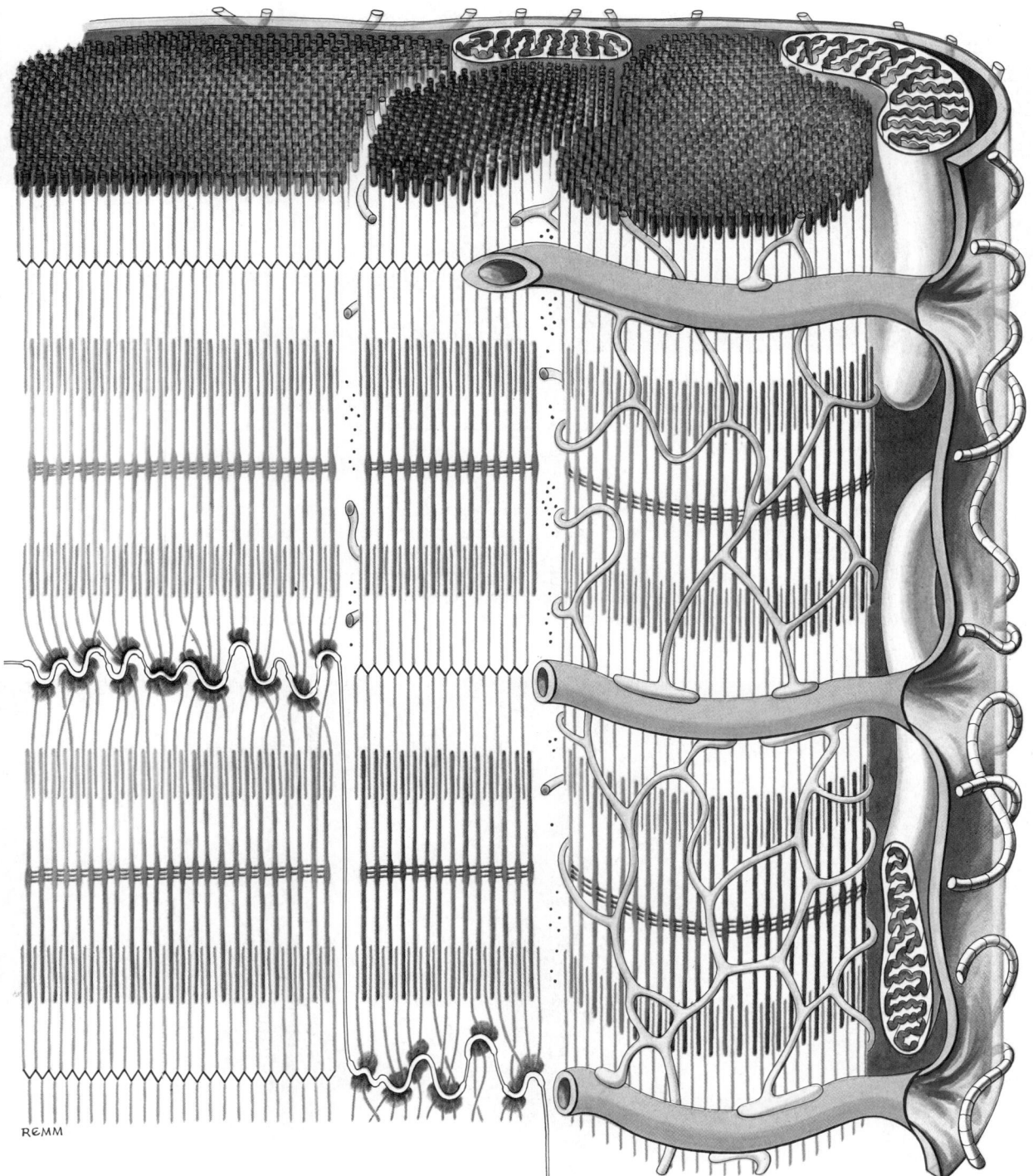

5.11 Three-dimensional reconstruction of cardiac muscle showing organization of myofibrils and membrane systems. As in previous diagrams, actin is shown red and myosin green; mitochondria are coloured blue; the sarcolemma and luminal surface of the invaginated transverse tubules are purple, the intracellular surface of the transverse tubule is orange, and the sarcoplasmic reticulum is coloured yellow. An intercalated disc with intermittent desmosomes and communicating 'gap' junctions is depicted on the lower left of the picture. Compare with **5**.6, and see text for further details. (Modified from a diagram by D Fawcett.)

stretch, and small groups of cells may act independently. Muscle in larger arteries, the iris and ductus deferens is of this type. *Unitary muscles* have few motor nerves; rhythmic contractions are spontaneous, i.e. *myogenic*, and may be governed by *pacemaker regions* in the muscle or elicited by tension. Examples are muscle layers of the stomach, intestines, uterus, ureter and some smaller blood vessels. Here, neural control enhances or depresses the rate and force of spontaneous, myogenic rhythmic contractions. Intermediate types of muscle exist and the divisions outlined may merely represent the extremes of a wide spectrum of activities.

There are also various structural differences which indicate other types of subdivision, e.g. into visceral and vascular non-striated muscle, although it is likely that many variations in detailed structure and physiology may occur in different parts of the body.

THE STRUCTURE OF NON-STRIATED MYOCYTES

Non-striated myocytes have a central nucleus, much elongated in stretched muscle and ovoid, crenated or sinuous in contraction. The myocytes are weakly birefringent, indicating a longitudinal orientation of filamentous components. Ultrastructurally their cytoplasm consists mainly of closely-packed fine filaments lying parallel to the long axis (5.13,14,15B). Golgi complexes,

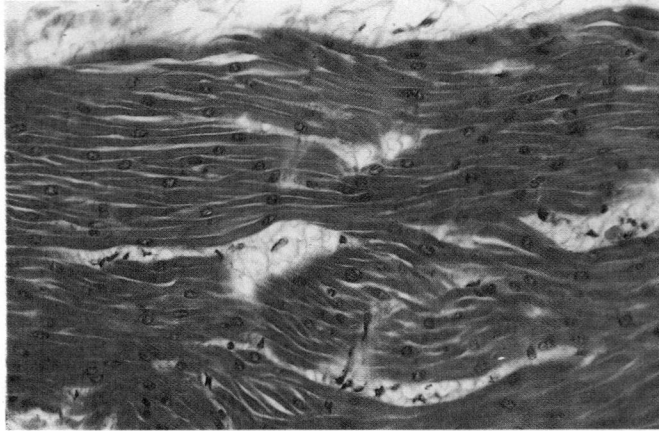

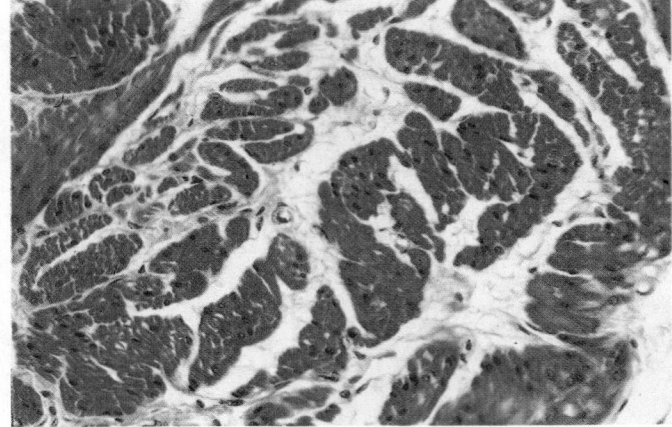

5.12A Longitudinal section of non-striated muscle from the small intestine showing groups of non-striated myocytes arranged in a fasciculus. Haematoxylin and eosin. Magnification × 500.

5.12B Transverse section of non-striated muscle from the small intestine. Haematoxylin and eosin. Magnification × 500.

agranular and granular endoplasmic reticulum, free ribosomes and lyosomes are positioned near the nuclear poles but groups of mitochondria are located either peripherally or scattered at deeper levels, varying in different types of muscle (Gabella 1973, Huddart & Hunt 1975). An irregular sarcoplasmic reticulum, able to sequester calcium ions, ramifies between the myofilaments (Popescu & Diculescu 1975), its terminal sacs lying under the plasma membrane. The plasma membrane has great numbers of small cup-like invaginations (caveolae), superficially resembling endocytic vesicles but not engaged in uptake of particles. Some of them at least may represent reservoirs of plasma membrane which allow the cell to stretch considerably without rupture. The

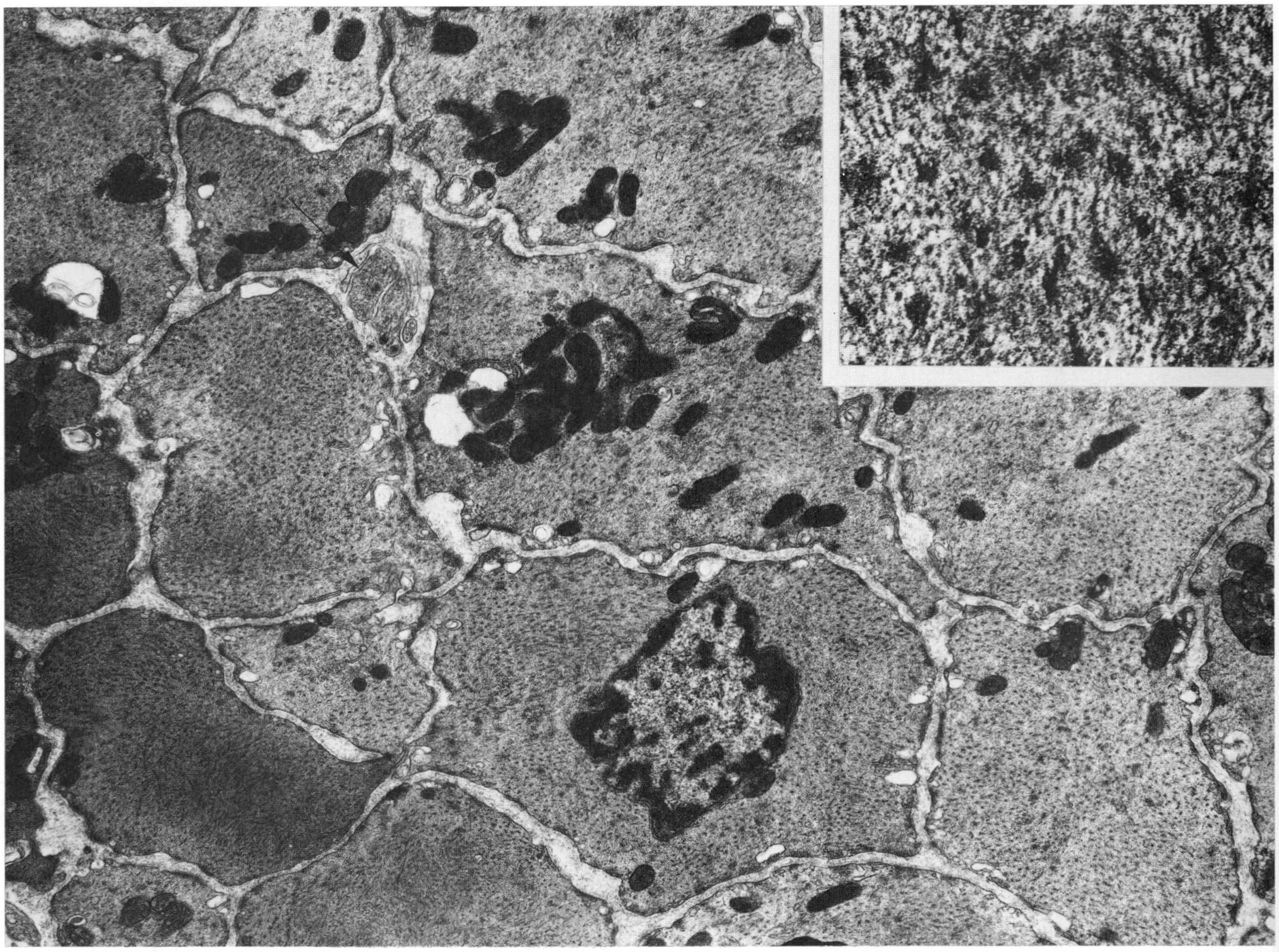

5.13 Low-power electron micrograph of a group of non-striated myocytes in transverse section, showing a centrally placed nucleus in one of the cells, and numerous darkly staining mitochondria. An autonomic nerve fibre (arrow) is visible among the myocytes. Specimen taken from the ileum. Magnification × 6000. The inset shows a highly magnified sample of cytoplasm from one myocyte demonstrating thin actin and thick myosin filaments irregularly arranged. Magnification × 100 000.

5.14A (*right*) A low-power electron micrograph of a longitudinal section through two non-striated myocytes, one showing a nucleus. Note the irregular outlines of the cells and the endoplasmic reticulum and mitochondria within. Fine collagen fibrils lie in the intercellular spaces. Specimen from the ileal wall. Magnification × 8000.

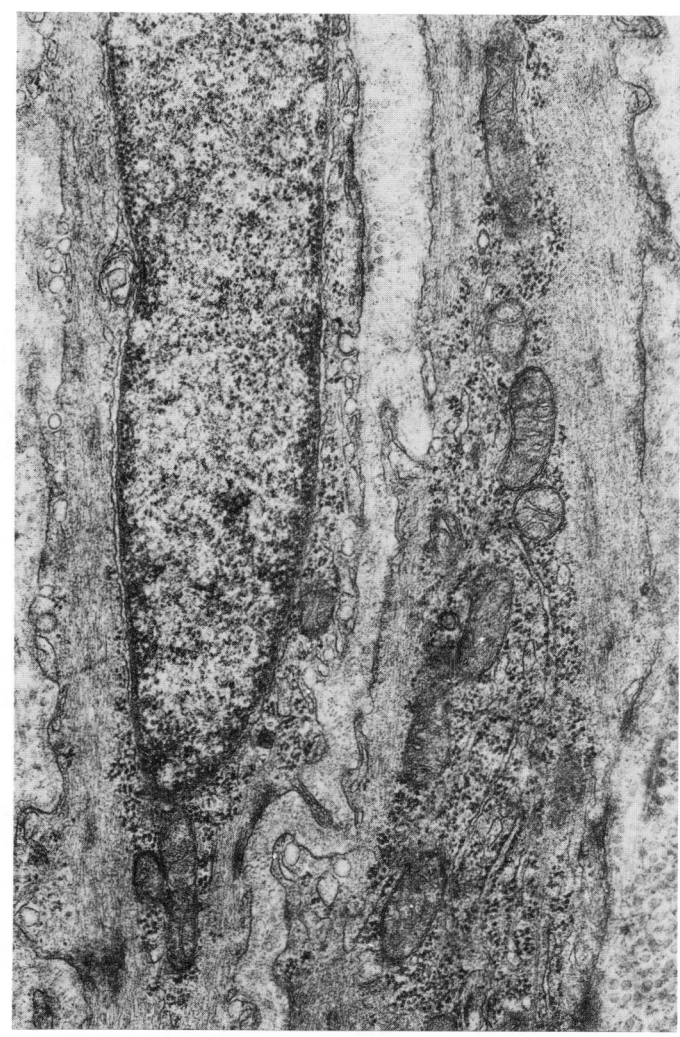

surface also shows occasional projections (5.13) and various junctions with other cells. For further details of cell structure see Gabella (1981).

Actin and myosin myofilaments are present in non-striated muscle and their mutual sliding brings about contraction, as in other types of muscle. Actin filaments have long been described in electron microscope studies but it was only with the advent of improved fixation methods that the myosin filaments were regularly seen, as they appear to be much less stable than those of the striated muscles (Somlyo et al 1973). Myosin filaments are thick ribbons or rod-like aggregates of myosin molecules up to 15 μm long and 0.1 μm wide. How the myosin is arranged in these large filaments is still uncertain but they may point in opposite directions on either side of each filament so that interactions with actin filaments may also be in opposite directions on the two sides. It has been suggested that this type of organization is necessary to produce directional contraction, in the absence of any obvious sarcomeric structure (Hinssen et al 1978). Each myosin filament is surrounded by up to 15 actin filaments, forming a type of contractile unit, which may be obliquely orientated in the myocyte (Small 1974). Actin filaments are also attached to cytoplasmic densities within the cell and also to projections of dense material lying under the plasma membranes (Cooke 1976), which may transmit contractile forces to the cell surface and to the

5.14B (*below*) An electron micrograph of the cytoplasm of a non-striated myocyte in longitudinal section showing the irregularly distributed thick myosin and thin filaments. Note also the numerous caveolae and sarcoplasmic reticulum at the cell boundaries. Specimen from the ileal wall. Magnification × 20 000.

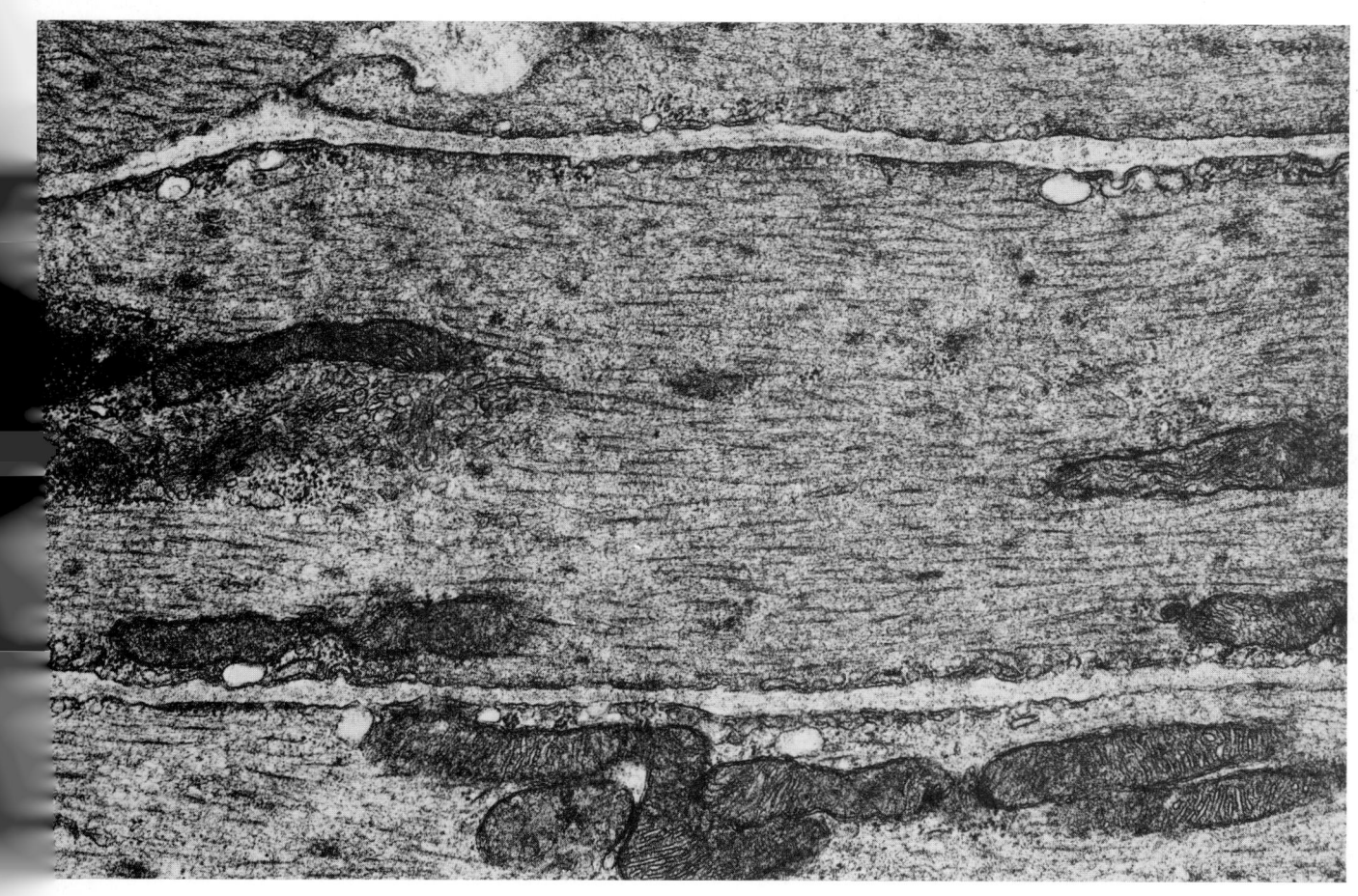

extracellular matrix (Gabella 1984). The chemistry of the myofilaments is also distinct from that of striated muscles; troponin is not present but instead a protein termed Leitonin is attached to actin filaments and appears to have a regulatory function in contraction. Unlike skeletal muscle, non-striated myosin is activated to bind to actin only when its light chain is phosphorylated by a calcium-dependent kinase (Ebashi 1983).

In addition to myofilaments, non-striated muscle contains numerous intermediate filaments (desmin filaments) which appear to provide coherence to the muscle fibre, uniting the various contractile units and transmitting forces throughout the cell.

Excitation of non-striated myocytes involves depolarization of their plasma membranes, caused by a transient influx of calcium ions activating actin-myosin filament sliding systems. Myocytes also excite each other through gap junctions of *maculae communicantes* (**5.**15A,B) (Cobb & Bennett 1969, Lowry & Small 1970, Watanabe & Yamamoto 1974). In some locations a few score occur on each cell. In the cytoplasm beneath such junctions are often cisternae of smooth endoplasmic reticulum, which may correspond in function to the sarcoplasmic reticulum of striated muscles (Fry et al 1977), perhaps mobilizing calcium stores during excitation. Electrical coupling ensures that myocytes act in concert, enabling one nerve fibre to control many myocytes. Myocytes are also linked to each other by desmosomes and to surrounding matrix by specialized surface features resembling zonulae adherentes of epithelial cells (p. 24).

Other Activities of Non-striated Muscle

Extracellular fibres always surround non-striated muscle fasciculi; in vitro non-striated myocytes secrete elastin, collagen and other matrix components (Scott et al 1977); the ability of non-striated myocytes to synthesize and secrete matrix components suggests strong affinities to fibroblasts and indeed some of the latter cells (e.g. myofibroblasts, p. 58) show pronounced contractile behaviour. In elastic arteries, non-striated myocytes are typically contractile during early years but, as they progressively lay down large quantities of elastin, they appear to lose this ability and come to resemble normal fibroblasts in this respect. However, myocytes are typically present in organized well-defined tracts and are under regular neural control, indicating that they are a distinctive cell type and not merely highly contractile fibroblasts (see Ross 1971).

Development and Regeneration of Non-striated Muscle

As already stated (p. 546), non-striated muscle arises from mesenchymal cells in various parts of the body. In visceral smooth

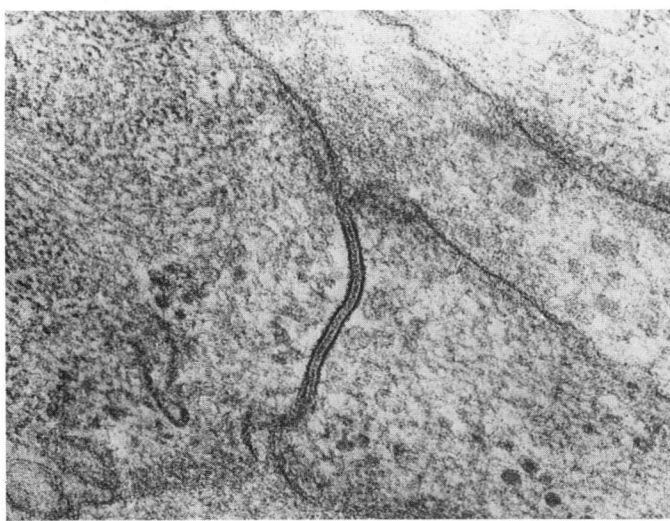

5.15A A gap (communicating) junction between two non-striated myocytes. Magnification × 70 000.

muscle, the first sign of differentiation is the appearance of a cluster of stellate mesenchymal cells which come together in a confluent mass, then differentiate into myoblasts, then into elongate myocytes which may continue to grow. These form fasciculi and individual cells come into contact at gap (communicating) junctions. Finally, fasciculi may associate to form well-defined muscular layers. The later stages of differentiation appear to be accelerated by the ingrowth of autonomic nerves (Burnstock 1981). Non-striated myocytes can also undergo mitotic division and this may be stimulated by hormones and by damage; in the latter case, regeneration of the muscle may be completely effected by myocyte proliferation (Burnstock 1981). New myocytes can originate, even in adults, from persisting mesenchymal cells, e.g. in the walls of regenerating blood vessels; non-striated myocytes are also able to multiply mitotically, e.g. in the myometrium after treatment with progesterone (Crandall 1938); to what extent this occurs naturally is uncertain (Ross & Klebanoff 1967).

Innervation and Blood Supply of Non-striated Muscle

Autonomic motor and sensory nerves from non-myelinated plexuses send branches into muscle fasciculi. Neuromuscular

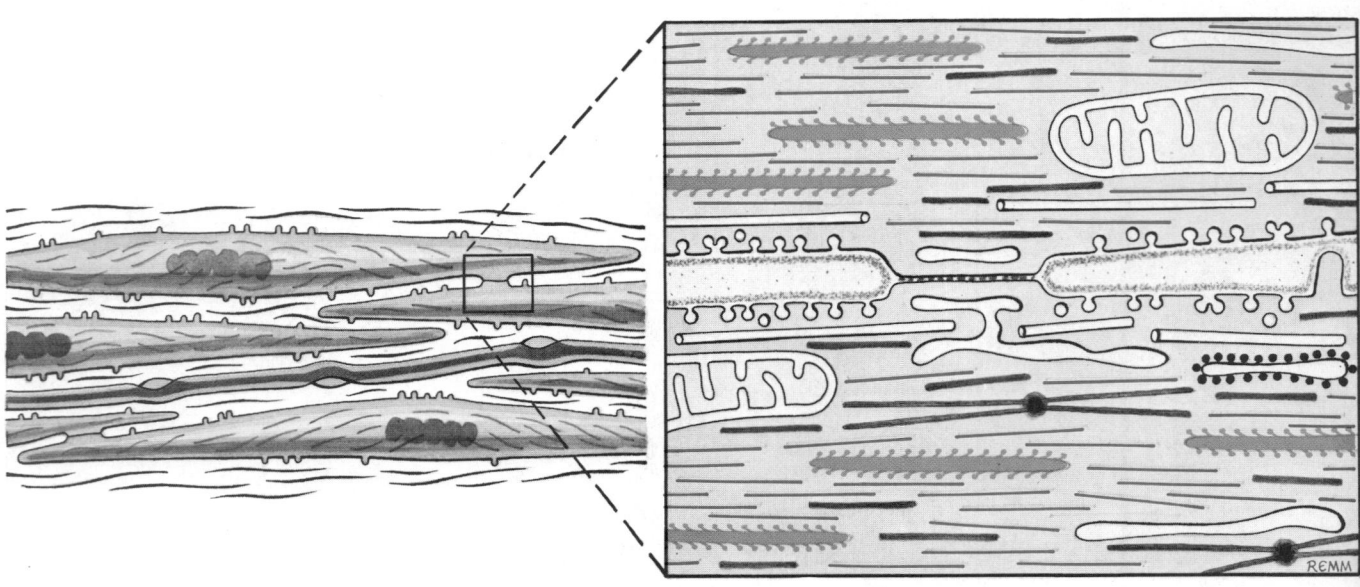

5.15B Diagram showing the principal features of organization of non-striated myocytes. On the left a group of myocytes shows the fusiform shape of the contracted cells and the arrangement of the autonomic efferent innervation; on the right the main features of the cytoplasmic filaments and membrane systems are depicted. Thick myosin filaments (green) are surrounded by irregularly arranged actin filaments (red) microtubules and intermediate filaments, some with associated dense bodies, are also illustrated. The elongated membranous cisternae correspond to the sarcoplasmic reticulum; endocytic vesicles and a communicating junction between two adjacent myocytes are also visible.

terminals of sympathetic and parasympathetic efferents are considered in detail elsewhere (pp. 905, 906); briefly, they consist generally of long, branched terminals which bear numerous varicosities, each swelling being a possible site of neurotransmitter release. In the slower visceral muscle the nerve terminals are at a much greater distance from the myocyte surface than in the quick-reacting muscle of the iris and neural control is less precise as well as slower. The myocyte plasma membrane has various types of response to a range of neurotransmitters, which vary with location in the body (see also p. 889) and the types of control may be quite complex, involving excitation, inhibition and other types of modulation of contractile behaviour (see Gabella 1974, Bülbring & Shuba 1976, Jones 1981) (pp. 890, 1363).

Capillary networks exist in connective tissue sheaths of smaller fasciculi in most locations, but vascularization is not as dense as in skeletal muscle. In some sites, e.g. tunica media of muscular arteries, capillaries may not penetrate muscle at all, reflecting the low energy requirements in such regions.

MYOEPITHELIOCYTES

These contractile cells, whose embryonic origin is believed to be ectodermal, exist in glandular tissue, close to the secretory parts and ducts. They are either *stellate ('basket cells')* with long dendritic extensions clasping an adjacent glandular acinus, as in salivary and lacrimal glands, or *fusiform*, e.g. in intercalated ducts of salivary glands (p. 55). They also exist around mammary acini and the secretory parts of sweat glands. They are often innervated by the autonomic nervous system; when they contract they initiate a rapid flow of secretion into the glandular ducts. Ultrastructurally (Ellis 1965, Tandler et al 1970) these cells show extensive desmosomal attachments to surrounding tissues and internally large numbers of actin and myosin filaments, similar to those of nonstriated myocytes. They also contain numerous intermediate filaments which immunohistochemistry has shown to be of the epithelial class (cytokeratin), strong evidence of an ectodermal origin (Franke et al 1980). They are also rich in alkaline phosphatase, ATPase and adenyl cyclase (Han et al 1976).

ATTACHMENTS, GENERAL FORM AND ACTIONS OF SKELETAL MUSCLES

Tendons, Aponeuroses and Fasciae

Tendons are integral parts of muscles, forming a component which can be considered for practical purposes to be of unvarying length (but vide infra). Being largely made of collagen fibres, they are resistant to stretch, but are also relatively flexible and can be angled round osseous surfaces or deflected under retinacula (vide infra) to change the final angle of pull. Since they are composed of collagen and their vascular supply is sparse, they appear white. Taking the form of cords or straps, they are round, oval or elongate in section and consist of fascicles of collagen fibres, largely parallel to the long axis but also partly interwoven. The fasciculi may be conspicuous enough to give tendons a striated appearance to the unaided eye, or even to form duplicated or multiple cords. Some tendons are commonly double and some have minor strands connecting with adjacent tendons. Their surfaces are smooth but longitudinal ridging by coarse fasciculi is common in large tendons, e.g. the osseous aspect of the angulated tendon of obturator internus (p. 643). Loose connective tissue permeates between fascicles, providing routes for vessels and nerves and condensing on surfaces as a so-called sheath or *epitendineum*, which may contain elastic and irregular collagen fibres. This has no real surface, being continuous with surrounding tissue, whose flexibility imposes minimal drag on tendon movements. Some tendons are, however, separated from their surroundings by synovial membrane, permitting much more extensive movement (vide infra).

Tendons (and aponeuroses, vide infra) are strongly attached to bones, not only via their periostea but also often by tendon fasciculi which may penetrate deeply into cortical osseous tissue. Sections of fresh bone show that at the sites of tendonal attachment there is a plate of white fibrocartilage (p. 290); where this is thin or absent the bone surface is rough but, when fibrocartilage is present, the bone surface is quite smooth in cleaned skeletal material (Benjamin et al 1986). The functions of such layers of fibrocartilage are not clear; where tendons insert into bones which replace a cartilage precursor relatively late in development, e.g. an epiphysis, fibrocartilage may be particularly thick and may represent a remnant of the earlier state; however, cartilage is also found at other sites and may therefore have a continuing functional importance, as yet unknown.

The tensile strength of tendons is similar to that of bone, i.e. half that of steel, a property much in excess of ordinary demand. A tendon 1 cm² in cross section will support 600–1000 kilograms. Though only a little extensile, tendons *are* slightly elastic and some contractile energy is stored in them when muscles contract. They also exhibit viscosity. Arnold (1974) has reviewed their biomechanical and rheological properties, with special reference to human feet. Certain avian wing tendons are highly elastic and have complex, ultrastructural interactions between their collagen and elastin fibres (see Oakes & Bialkower 1977).

The blood supply of tendons is relatively scant; small arterioles from adjacent muscle ramify longitudinally between their fascicles, anastomosing freely, accompanied by venae comitantes and lymphatic vessels. This longitudinal plexus is augmented by small vessels from adjacent loose connective tissue or synovial sheaths (Edwards 1946). At osseous attachments, their vessels are restricted to transverse capillaries and no vessels pass between bone and tendon at these sites; such osseous surfaces are usually devoid of foramina (p. 269). The metabolic rate of tendons is very low but increases in reaction to infection or injury. Repair involves the initial proliferation of connective tissue cells associated with collagen fibres, followed by interstitial deposition of new fibres (Potenza 1963).

The nerve supply is largely or perhaps solely sensory; evidence of any vasomotor control is lacking. Specialized *neurotendinous sensory endings* (Golgi tendon organs, p. 913) exist near musculotendinous junctions; their large, myelinated afferent fibres course centrally within branches of muscular nerves or in small rami of adjacent peripheral nerves (Stillwell 1957).

Postnatally, tendons grow interstitially, particularly at musculotendinous junctions, where high concentrations of fibroblasts occur performing rapid elaboration of collagen. There is a decreasing growth gradient towards osseous attachments.

Synovial sheaths and bursae exist where moving structures are in tight apposition, particularly where tendons are deflected around bone or under retinacula near joints. The simplest arrangement occurs at a few sites, such as over the olecranon and the patella, where skin moves freely over subcutaneous bone, often under pressure. Such *bursae* are simply flat sacs of synovial membrane, supported by dense irregular connective tissue in the loose areolar tissue between skin and bone, by their position termed *subcutaneous bursae*. The extremely flattened form is not perhaps indicated clearly by the name 'sac', for in each the opposed walls are separated by a mere film of fluid; it is an enclosed cleft, its walls tethered to periosteum and dermis and hence sliding over each other as the adjacent tissues move. Although they are often considered devices to reduce friction, they are basically absolute discontinuities between tissues, yielding total freedom of movement over a limited range. The capillary film of lubricant synovial fluid also provides the lining synovial cells with a wet environment and a source of nutrients.

Most synovial bursae lie between tendons and bones or ligaments, or between two tendons (i.e. they are *subtendinous*). Some occur between muscle and a bone, tendon or ligament (*submuscular synovial bursae*). Some separate aponeuroses from bone (*subfascial bursae*), some are *interligamentous*. Many situated near joints communicate with them, with continuity of their synovial membranes and fluid. Such *communicating bursae* may develop separately, connecting only later with the articular cavity; sometimes they ensheath tendons as they emerge from intra-capsular attachments.

Synovial tendon sheaths surround tendons where they pass under ligamentous bands and retinacula or through fascial slings and osseofibrous tunnels. They are composed of an outer and an inner synovial membrane separated by a thin film of synovial fluid, the two layers being continuous at both ends of the tunnel they form to create a closed cylindrical envelope around the tendon. The internal ('*visceral*') layer is attached by loose connective tissue to the underlying tendon and the external (*parietal*) layer is similarly tethered with adjacent periosteum or other connective tissue structures. It should be added that the precise shape of synovial tendon sheaths varies much, from simple cylindrical to highly complex, depending on the shape of the enclosed tendon and the surrounding tissues and on other mechanical factors (e.g. p. 619). Since the tendon is invaginated into the sheath, the visceral and parietal layers are often connected by a longitudinal *mesotendon*, providing a degree of mechanical stability and a route for neurovascular supply. Some mesotendons are reduced to localized cords, often multiple, e.g. the vincula tendinum of digital tendon sheaths (p. 621).

Where skin undergoes frequent displacement and pressure, as in the forearm or elbow in writing, or the buttock in some sedentary occupations such as hand-weaving, *adventitious bursae* may appear, providing the overlying skin with greater freedom, a phenomenon perhaps indicating that mechanical factors may be important in the evolution of bursae in general. However, the established series of bursae develop in utero and therefore must be determined genetically.

Some bursal mechanisms are more elaborate and share some of the features of synovial articulations (p. 471). Osseous or cartilaginous sesamoids occur in some tendons, closely applied to the osseous surface. In such conjunctions, the apposed surfaces are cartilage-covered and enveloped within a localized bursa, e.g. the sesamoid in the tendon of peroneus longus where it articulates with the cuboid bone (p. 647), in effect forming a synovial joint.

Aponeuroses are sheets of white compacted collagen fibres attached directly or indirectly to skeletal muscles and distributing their contractile forces to fasciae, other muscles, cartilage or bone. They are frequently striated, their large fascicles separated by loose connective tissue, and when the fibres are thin and regularly parallel they may give a general iridescence to the aponeurosis in the fresh stage, since they diffract light like an optical grating. They usually have several layers, fasciculi being parallel within a layer but differently inclined in adjacent laminae. The term 'aponeurosis' was originally applied to any sheet of connective tissue spreading from ('apo-') the edge of a tendon ('neuron', a term once indiscriminately applied to tendons and nerves) and providing additional attachment, e.g. levator palpebrae superioris and the retinacula of the quadriceps extensor tendon. But the term has come to denote any sheet of dense connective tissue associated with muscle. Sometimes a whole attachment is aponeurotic, e.g. obliquus externus abdominis; elsewhere aponeuroses provide large auxiliary attachment areas for muscle fibres (e.g. supra- and infraspinous aponeuroses). Some major aponeuroses interlace at linear (sometimes multiple) *decussations* or *raphes* (see e.g. recent accounts of the ventrolateral abdominal aponeuroses pp. 600–603).

Fascia is a term so vague in usage that it signifies little more than collections of connective tissue large enough to be visible to the unaided eye. The advent of tissue fixatives, especially formalin, which preserve and accentuate areolar loose connective tissue, greatly encouraged regional naming of fasciae; the habit of attaching a local term to any such aggregation sizeable enough to dissect is still current, though perhaps diminishing. During development mesodermal cells differentiate into bone, muscle, vessels, renal and splenic tissue, etc.; but large numbers persist in connective tissues permeating all regions, not only as a loose microscopic component between, e.g. the fibres of muscle, nerve and tendon, but also in macroscopic accumulations between whole muscles and viscera. The arrangement of such connective tissue is highly variable. In dissection it appears as condensations on the surfaces of muscles, etc. and is here named *investing fascia*, although its function is by no means so simple. Between muscles which move extensively it takes the form of loose areolar connective tissue, facilitating movement. Peripheral nerves, blood and lymph vessels lie in loose 'fascia' between other structures, often bound together into *neurovascular bundles*. Sometimes large vessels, such as the common carotid and femoral arteries, have denser sheaths of connective tissue. Their precise functions are conjectural: they may aid venous return by approximating large veins to pulsating arteries but equally they may be generated primarily as tissue responses to the pulsatile forces created by neighbouring arteries.

Lipid may accumulate in cells of loose connective tissue, forming adipose tissue, which is often well-developed between muscles and skin. The dermis is composed of dense irregular connective tissue; deep to and continuous with it is a layer of loose connective tissue, often adipose and of variable thickness, the **superficial fascia**. It increases the mobility of skin and is a thermal insulator as well as a store of energy for metabolic use. Subcutaneous nerves, vessels and lymphatics travel in the superficial fascia, their main trunks lying in its deepest part where adipose tissue is scant. Superficial fascia also contains skin muscles, such as platysma, a remnant of more extensive sheets of skin-twitching musculature found in other mammals. The quantity and distribution of subcutaneous fat differs in the sexes. It is generally more abundant and widely distributed in females; in males it diminishes from trunk to the extremities, and this distribution is more obvious in middle age, when the total amount increases in both sexes. There is an association with climate (rather than race), superficial fat being more abundant in colder geographical regions. The superficial fascia is most distinct on the lower anterior abdominal wall, where it contains much elastic tissue and appears many layered as it passes through inguinal regions into the thighs. It is similarly differentiated in limbs and the perineum but is thinnest over the dorsal aspects of hands and feet, sides of neck and face, peri-anally, and over the penis and scrotum. It is almost absent from the external ears but is particularly dense in scalp, palms and soles, permeated in these regions by numerous strong connective tissue bands binding the superficial fascia and skin to underlying structures; these come under the general term, *deep fascia*, but are known regionally as aponeuroses of the scalp, palm and sole.

Deep fascia is also composed chiefly of collagenous fibres but these are compacted and often so regularly arranged that the deep fascia may be indistinguishable from aponeurotic tissue as, in the latter, parallel fibres in one layer are usually inclined to those in the next. In limbs, where deep fascia is well developed, their fibres are longitudinal or transverse, condensing into a tough, inelastic sheath around the musculature. In both upper and lower limbs, some muscles are attached to the internal aspects of deep fascia, which functions as an aponeurosis. Wherever deep fascia contacts the periosteal coverings of bone it fuses with it and thus transmits to bone the pull of its attached muscles. A powerful band of deep femoral fascia transmits the pull of tensor fasciae latae and most of gluteus maximus to the femur and tibia (p. 637), as the iliotibial tract. Biceps brachii is partly attached to deep forearm fascia; the palmar aponeurosis is the deep fascia of the palm, receiving the pull of a number of muscles (palmaris longus, when present, and in part the intrinsic muscles of pollex and minimus).

In the neck and limbs laminae of the deep fascia pass between groups of muscles and connect extensively with bone. Such *intermuscular septa* may incidentally separate muscles or groups with different actions, developmental histories and innervations. Except as phenomena of growth, however, most of these 'compartments' are of little except clinical significance. Septa often connect rather than separate muscles which may be attached to both aspects of what is in fact an *intermuscular aponeurosis*.

In some sites, near the wrist and ankle being prime examples, localized transverse thickenings of the deep fascia are attached at both of their ends to local osseous prominences. These are *retinacula*, so termed because they *retain* tendons deep to them from 'bowing' out of position during activity. In such osseofibrous channels, tendons are deflected to work effectively, smooth action being aided by synovial sheaths. Similar tunnels of deep fascia and bone form fibrous sheaths in which digital flexor tendons of the hand and foot are retained, surrounded by synovial sheaths (pp. 618, 654).

In the lower limb particularly, deep fascia contributes to the efficiency of venous return, containing and compressing distended deeper veins, while regurgitation into superficial veins is largely prevented by numerous valves.

Form of Muscles

Muscles vary much in size, shape, fascicular architecture and form of attachment; each is adapted to provide a range, direction, velocity and force of contraction appropriate to the habitual needs at articulations on which it acts. Before such features are detailed in individual muscles, general properties must be considered, such as overall geometry, variation in number, size and direction of fibres relative to their line of traction and the aspects of muscle action which such structural variants serve.

Muscle fibres are not uniform in diameter, ranging from 10–60 μm, and they vary in length from a few millimetres to many centimetres (15–30 cm in sartorius). The statistical distribution of fibre diameters varies between whole muscles, or sometimes parts of a single muscle, as does their grouping into *fasciculi* by perimysial septa (p. 547). Visible texture reflects the proportions and distribution of connective tissue; muscles with fine fasciculi are often concerned in precision movement (e.g. extraocular muscles), whereas coarse fasciculation characterizes many larger limb muscles (e.g. gluteus maximus). But biomechanical sig-

nificance, morphogenesis and vascular and innervation patterns in such *fascicular units (myonemes)* of muscles have been little studied. These must be distinguished from *motor units (myones)*, the basic neuromuscular functional units, which have a different spatial distribution (p. 556); their interrelations are currently under investigation.

Muscles may be classified by fascicular orientation, which may be *parallel*, *oblique* or *spiral* relative to the muscle's direction of pull (5.16). Where fasciculi are largely parallel muscles vary from flat, short and *quadrilateral* (e.g. thyrohyoid) to long and *straplike* (e.g. sternohyoid, sartorius); individual fibres often traverse almost the whole muscle or shorter segments between transverse, *tendinous intersections* at intervals along the muscle (e.g. rectus abdominis). Fasciculi also approach the parallel in 'bellies' of *fusiform* muscles, which may be short, converging to a tendon, sometimes lengthy at one or both ends. Tendons concentrate the force of contraction to a small osseous area and may alter its direction of pull; the relative lengths of belly and tendon are closely related to the resultant *range* of contraction, often very limited, a tendon being then a device that causes relatively small excursions at some distance from the muscular attachment.

However, the maximal *force* generated by a muscle finally depends on its effective *mass* of contractile tissue, i.e. the number and dimensions of contained fibres, whereas the maximal *range* of unrestrained contraction depends on the *length* of its fibres. Where fasciculi are parallel to the line of traction, available force and range act at full advantage (5.17A). However, in strap muscles, fibres are relatively few in number but long, making them particularly effective where wide range is coupled with moderate power. Bellies of fusiform muscles may be slender or massive, long or short, with like variation in length and strength of the tendons, yielding various grades of power and range.

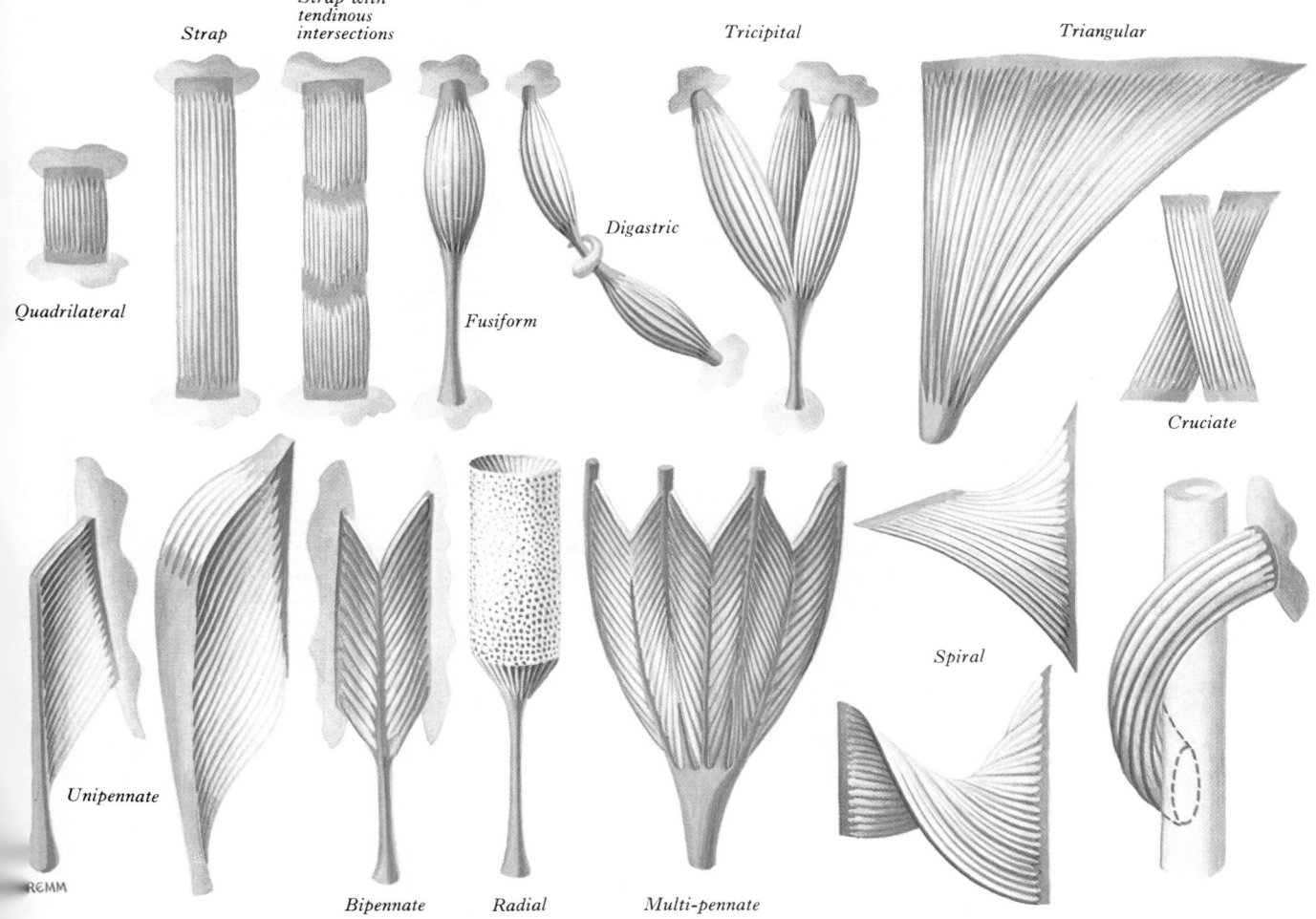

16 The morphological 'types' of muscle based upon their general form and fascicular architecture.

Where fasciculi are *oblique* to the line of traction (5.17B), muscles may be *triangular* (e.g. temporalis, adductor longus) or *pennate* (feather-like) in construction. The latter vary in complexity (see 5.16) from *unipennate* (flexor pollicis longus), *bipennate* (rectus femoris and dorsal interossei) to *multipennate* (deltoid). Similar general architecture also obtains in other muscles, e.g. in the main part of soleus, where fasciculi pass obliquely between deep and superficial aponeuroses, giving in sagittal section a kind of 'unipennate' form. In other sites muscle fibres start from walls of osteofascial compartments, converging obliquely to a central tendon in *circumpennate* form, e.g. tibialis anterior. (Unipennate and bipennate are unfortunate terms; feathers or *pennae* rarely if ever consist of a one-sided set of plumules attached to a midrib, like a 'unipennate' muscle. They are almost invariably 'bipennate'. More logical would be 'semipennate' and 'pennate'.) In all such forms, only part of the available force and range are effective along the tendon's axis. As shown in 5.17B, force generated by oblique fibres can be resolved into components, one along the tendon, one at 90° to this; the latter is largely counteracted in bi- or multipennate forms by opposed fibres but tends to deviate the tendon's line in unipennate muscles. Similarly, range is restricted and proportional to length of muscle fibre × cosine of its angle of attachment to tendon. Loss of efficiency in transmission of force to tendon is in natural use outweighed by the large numbers of short fibres in such muscles, which typically occur where limited range and considerable power are habitually required.

Some muscles are *spiralized* or 'twisted' in arrangement; the form and degree of spiralling varies with the particular muscular geometries, the state of contraction and posture adopted, e.g. the large triangular trapezius has a long, linear, *vertical*, median attachment and a *horizontal*, curved, lateral attachment, with a 90° twist between, in erect posture with arms pendent. Sternocostal fibres of pectoralis major and latissimus dorsi undergo a 180° twist between their median and lateral attachments. When such muscles contract they approximate their attachments but also tend to bring them into the same plane, undergoing some '*despiralization*' (5.20). Some muscles are spiralled around a bone, forming one attachment, e.g. supinator, which winds obliquely round the proximal radial shaft. Their contraction again reduces the spiral, imparting rotational force to a bone (vide infra). Allied to spiral muscles are those with two or more planes of fasciculi arranged in differing directions, sometimes termed *cruciate* (5.16); examples are sternocleidomastoid, masseter and adductor magnus, all partially spiral and cruciate. Many muscles exhibit regional variations in structure, compounded of more than one of the foregoing and adapted to several contrasting functional roles; in some actions one region only may be active, the remainder quiescent (p. 609).

Actions of Muscles

Analysis of sarcomeric construction and its relation to contraction has made rapid progress (p. 547). Muscle twitch, fusion of mechanical responses to repetitive stimulation, and functional neuromuscular territories or motor units (*myones*) have been discussed (pp. 556, 894). Isotonic and isometric contraction, varying relations between sarcomeric length and tension developed under isometric and other conditions have also been considered (p. 548). These are essential stages in understanding the biology of muscle; but the temporospatial dynamics of such events in a whole muscle, though difficult to analyse by experiment, provide greater range and flexibility of response than would first appear in such cellular analyses.

Repetitive, synchronous stimulation of all or most motor units does not occur in vivo; habitual demands are met by *asynchronous* contraction of a *proportion* of units available. During natural activities, at any instant, a muscle contains a mixture of motor units in quiescent, stimulated, contracting and relaxing states, the overall state being a *summation*. An ever-changing mosaic of active, relaxing and quiescent units thus occurs. During rotation of activities among available motor units the *active proportion* may increase, decrease or remain constant to meet functional demands. Whether a muscle lengthens, shortens or maintains its length and what tensions ensue depends on many factors,

·A

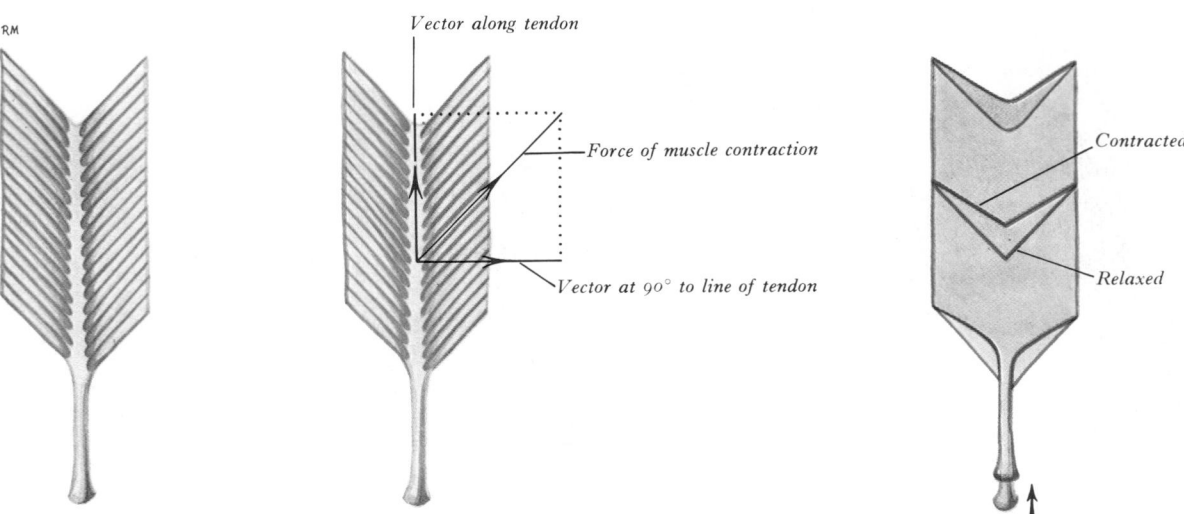

B

5.17 Simple mechanical considerations related to the actions of (A) strap, and (B) bipennate muscles. (For details of the actions of these different muscle types consult text.)

including initial length, proportion of active motor units and rate of change of this, and the sum of various extrinsic and intrinsic forces to which it is subjected, partly through its attachments. This allows an infinite variety of quantitative responses in any muscle.

At one extreme, a muscle may be fully relaxed, when much electromyographic evidence shows no activity; the muscle is electrically silent (Joseph et al 1955, Joseph 1960). This contrasts with earlier accounts by electrophysiologists, who held that 'full relaxation' was still accompanied by 'background' activity rotating among a few units. Misconceptions concerning such 'muscle tonus' were based on these views. But even a fully relaxed muscle is a complex structure endowed with *passive viscous, elastic* and other properties providing some resistance to deformation (e.g. by stretching). During activity also, such properties of inactive motor units and connective tissues provide a basal resistance, to be overcome by active units before external forces can be opposed and work performed. Such passive attributes contribute greatly to the space–time–force muscular equation during initiation, increase and maintenance of activity and equally during its diminution and/or cessation.

When a muscle fibre or motor unit contracts, it can be considered as attempting to approximate its attachments; whether it does so depends on the force it generates (a function of its geometry, mass and initial average sarcomeric length) balanced against *extrinsic* and *intrinsic* forces transmitted through its connective tissues and opposing contraction. In an active muscle, therefore, the following alternatives are possible: its original length may be *maintained* (with slight fluctuations) and the tension generated may persist nearly unchanged or may increase or decrease, depending on external factors and number and state of active motor units; similarly, the muscle may *shorten*, with tension increasing, persisting or even decreasing with varying conditions; again, a muscle may *lengthen*, its tension decreasing, persisting or increasing with varying demands. Hence, in considering whole muscles it is naive to correlate 'contraction' merely with shortening or increase of tension and 'relaxation' with mere lengthening or decrease of tension. Many more subtle grades of combination are not only possible but are the foundation of the infinitely varied patterns of activity.

When a muscle contracts to initiate a movement, e.g. a limb muscle across a single joint, factors opposing its shortening are many, including: the passive mechanical properties of muscle mentioned above; similar properties of articular tissues, opposing musculature, and other soft tissues in the region; the contractile state of opposing musculature; the inertia of the part moved and its external load; and finally the effects of gravity on all the foregoing. Sufficient motor units must be activated to overcome these resistances and impart an angular acceleration to a limb segment until the required velocity is reached, beyond which sufficient units continue contracting to achieve the rest of the movement. (Termination of the movement is considered below.)

In natural use, movements at different joints are carried out by *orderly patterns of activity* in muscle groups, which may be classified according to their role.

Apart from movements assisted by gravity (vide infra), one or more muscles are constantly active in initiation and maintenance of any movement and they are its *prime movers* (e.g. brachialis in elbow flexion). Muscle or muscles which wholly oppose this, or initiate and maintain its converse, are *antagonists* (e.g. triceps in elbow flexion). However, views concerning such interrelations have been much revised. A long-held view was that contraction of prime movers to start a movement is accompanied by partial contraction of antagonists, smooth continuation depending upon progressive contraction of prime movers with simultaneous lengthening and 'relaxation' of antagonists, the latter finally increasing contraction to decelerate and end the movement. The neural connections presumed responsible for this co-ordination were incorporated as a '*principle of reciprocal innervation*'. But apart from the variable transient activity in antagonists at a movement's start, they remain electrically quiescent until final deceleration. The active prime movers are not unrestrained, being balanced against the *passive, inertial* and *gravitational* effects mentioned above, the last varying, of course, with altering relation of moving parts to the direction of gravity.

Sometimes prime movers and antagonists *contract together as fixators*, increasing *transarticular compression* (vide infra) which thus *stabilizes* a joint, often to make an immobile base for other prime movers to act. Sometimes stability can be achieved by gravity alone at a joint in or near its close-packed position (cf. knee and hip joints in erect posture, pp. 483, 613), or by prime movers balancing passive forces (cf. stabilization of glenohumeral joint by supraspinatus with a pendent arm, pp. 502, 612). Otherwise, and when powerful external forces are encountered, both groups contract, holding the joint in any required position.

It must be emphasized that all movements (except the theoretically possible but improbable case of a uniaxial joint acting precisely in a horizontal plane) are opposed or assisted by gravity, which profoundly influences the play of musculature in certain movements. Prime-mover activity is always replaced by gravity whenever appropriate, the roles of muscles then often being reversed. For example, the prime mover extending the forearm at the elbow is triceps (e.g. when thrusting). When a loaded hand is lowered, however, extension (by gravity) is controlled by *active lengthening* of *flexors*; similarly, *increasing the length* and *force* of contraction in extensors of the hip, controls flexion of the trunk at hip joints. (Such considerations apply to numerous commonplace activities. Thus the names given to many individual muscles or muscle groups are oversimplifications. They emphasize only one of a number of habitual actions; the functional roles implied by many names are incomplete and should be extrapolated with caution.)

In some cases contraction of a prime mover, across a *uniaxial* joint, produces a simple movement; but at *multiaxial* joints or when muscles cross more than one joint, prime movers may produce additional movements, elimination of which is effected by contraction of *synergic* muscles, often partial antagonists of the prime mover. For example, prime flexors of the fingers at interphalangeal and metacarpophalangeal joints are the long flexors, superficial and deep; but these also cross intercarpal and radiocarpal joints and, unrestrained, they also flex these, with loss of efficiency. But simultaneous synergic contraction of the carpal extensors eliminates this, often producing some carpal extension and increasing efficiency of digital *flexion*. Electromyography shows that movements, wherever feasible, are effected without energy-consuming dependence on synergists; but most complex movements result from graded interplay between external forces, including gravity, inertia, the passive properties of various tissues and integrated patterns of tension–length variations in prime movers, antagonists, synergists and fixators. These patterns and positions of joints are constantly surveyed via receptors in various connective, periarticular and muscular tissues and thus linked to integration and control at all levels in the central nervous system.

Kinesiology

In recent decades interesting concepts and a terminology have emerged from kinesiological studies of articulo-muscular mechanics (MacConaill & Basmajian 1977). These are considered briefly here; for comprehensive treatment, the references quoted should be consulted.

Illustration 5.18A shows two articulating bones, one considered mobile and one fixed. A muscle attached eccentrically to the mobile bone contracts, exerting a force along the line of its tendon which is, relative to the mobile bone, vectorially resolvable into three components: (1) *transaxial*, at right angles to the bone, tending to change angulation at the joint (i.e. *swing or spurt component*); (2) *paraxial*, acting along the shaft towards the joint, tending to compress articular surfaces (i.e. *transarticular* or *shunt component*); and (3) *tangential*, acting around the bone's long axis, tending to rotate it about this axis (i.e. *spin component*). The terms *spurt* and *shunt* may be unfamiliar, although they were used by nineteenth-century British engineers (MacConaill 1949).

Actual movement will depend upon constraints due to articular shape, ligaments, adjacent musculature and extrinsic and intrinsic forces opposing it. Nevertheless, each component has its functional import, enhanced in some muscles, suppressed in others

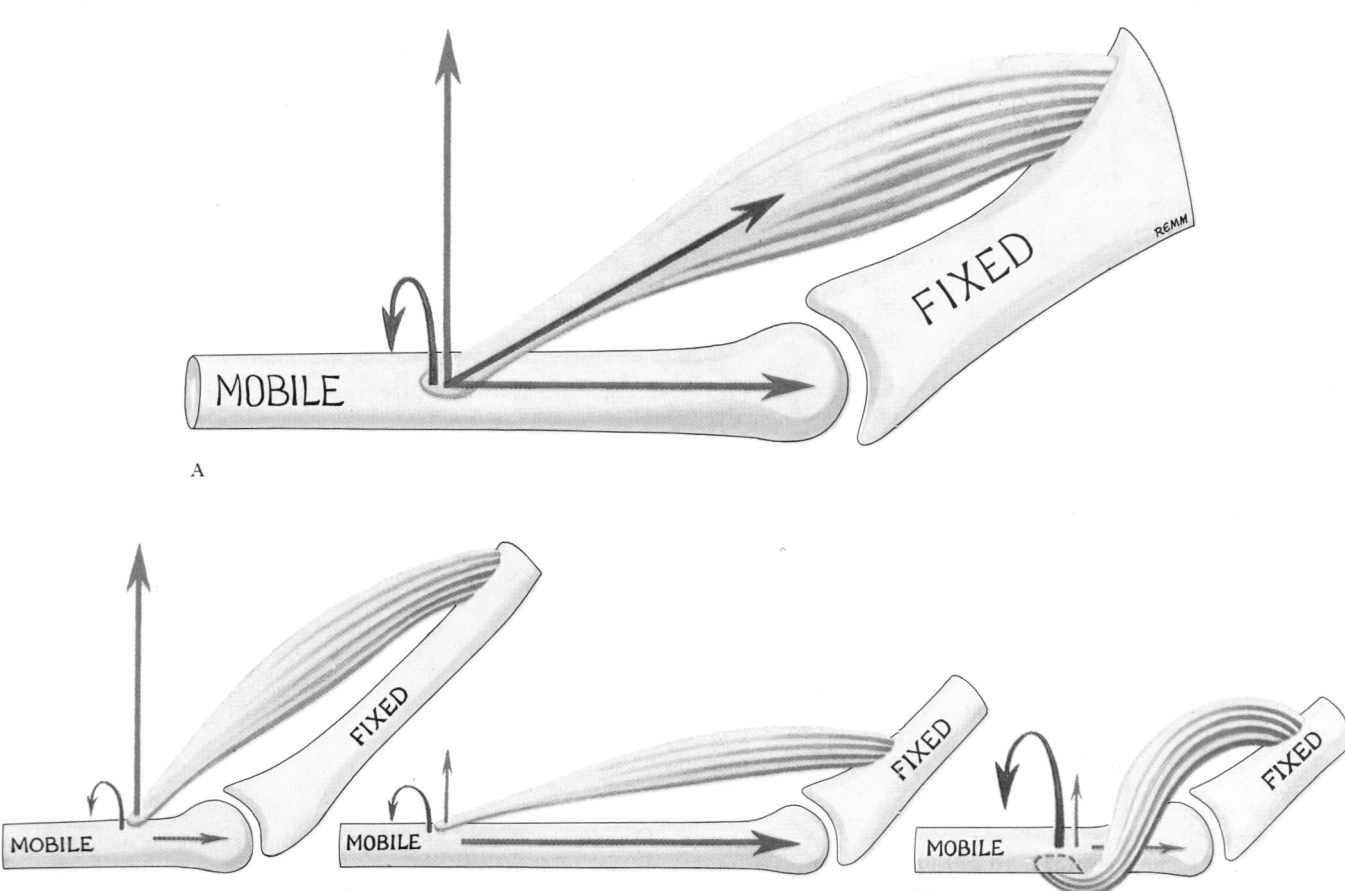

5.18 A. Vectorial analysis of the *force* of contraction generated by a muscle which is attached to a fixed base, crosses a single multi-axial joint and is attached eccentrically to the shaft of a mobile bone. Colour code of arrows: purple = muscle force; red = transaxial 'swing' vector; green = paraxial 'shunt' vector; blue = tangential 'spin' vector. Similar analyses relating to muscles which are *predominantly* (B) 'spurt', (C) 'shunt' and (D) 'spin' in

their actions. It should be appreciated that these diagrams have been constructed to illustrate certain principles discussed in the text; they do *not* represent any specific human bone or muscle. It is also clear that (in B and C), if the fixity and mobility of the two bones were interchanged, B would now operate as a 'shunt' muscle and C as a 'spurt' muscle.

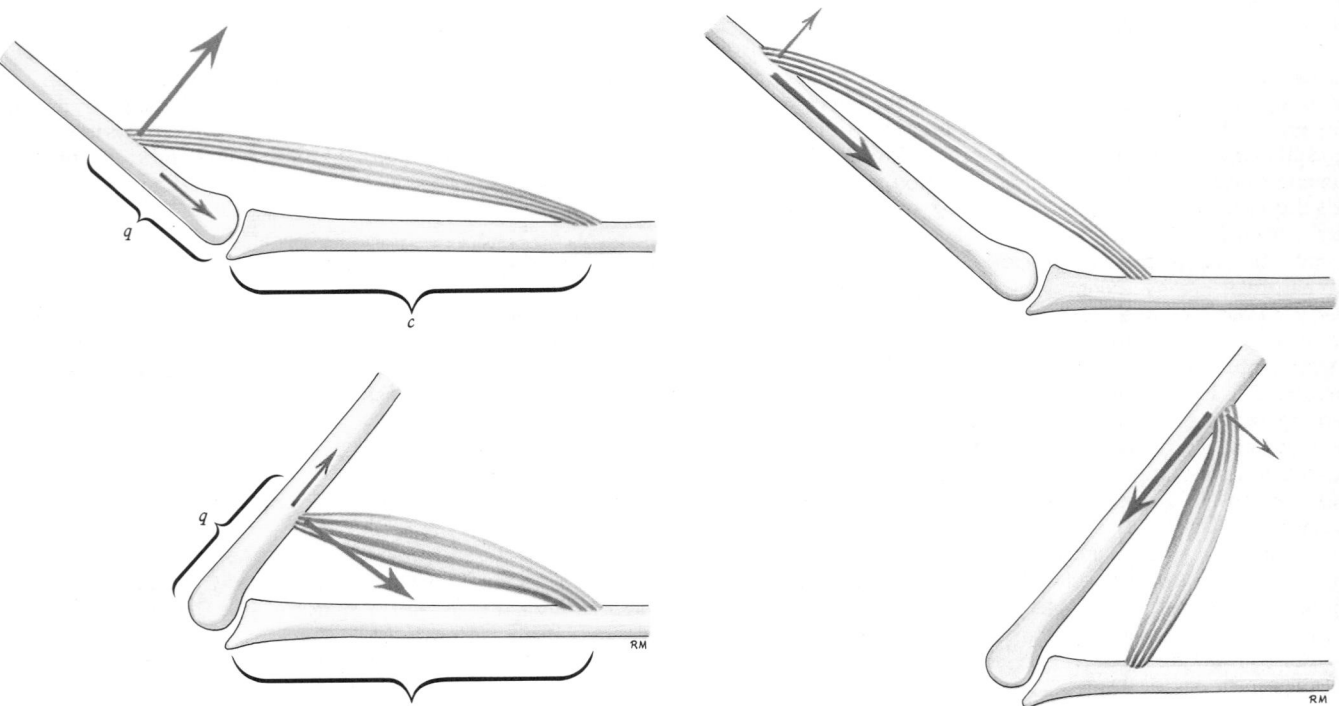

5.19A Some features of a 'spurt' muscle. Note that it has a large 'swing' component of its force which acts to impart an angular acceleration to the mobile bone. Its small 'shunt' component acts towards the joint up to 90° of flexion and thereafter away from the joint. c = cisarticular length; q = transarticular length.

5.19B Some features of a 'shunt' muscle. The large shunt component of its force acts towards the joint in all positions. In Fig. **5.19A** and B no specific human muscle or bone is intended; in each case it is clear that the right bone is regarded as fixed whilst the left bone is mobile. Clearly their mobilities (and designations) could be interchanged.

(**5**.18B,C,D). For example, when a muscle is markedly eccentric in attachment to or even spirals around the mobile bone, it has a spin component; e.g. the transversely wrapped pronator quadratus is an almost pure spin muscle, while the longer, oblique pronator teres has spin, transarticular and swing components. In contrast, femoral flexors exert a powerful swing force (flexion) on knee and a weaker spin (rotation) on the semiflexed leg. Spin components may be used as prime movers, e.g. in pronation/supination of forearm, medial/lateral rotation at hip, etc.; or they may be used synergically to counteract another spin force, ensuring pure swing (*antispin*, p. 480).

Ignoring spin components, relative values of transarticular and swing components vary with distances of fixed and mobile attachments from a joint (**5**.19), termed *cisarticular (c)* and *transarticular (q) lengths* respectively, *c/q* denoting the *partition ratio (p)*. When the fixed attachment is more distant from a joint (**5**.19A), i.e. $p > 1$, there is a powerful swing (spurt) component and a small transarticular (shunt) component; such a muscle is termed a *spurt* muscle (formed by *spurt myonemes*). A typical example is brachialis brachii acting at the elbow joint from a fixed humerus. Such muscles, acting alone, are most effective in initiating and continuing slow movements of swing, whether weak or powerful, and in postural readjustments demanding graded changes of tension and length. The shunt component is often sufficient to maintain articular contact; but when angulation exceeds 90° the component of force along the bone is directed away from the joint.

In contrast, a *shunt* muscle (of *shunt myonemes*) acting from a fixed base near a joint and with a more distant mobile attachment, i.e. $p < 1$, shows a large persistent (shunt) vector along the bone towards the joint in all positions but only a trivial swing component (**5**.19B); an example is brachioradialis acting at the elbow joint from a fixed humerus. Shunt myonemes or muscles are necessary wherever other factors act to separate articular surfaces, which may be due to loading, e.g. of a pendent arm. But it is suggested that during *rapid* swings centrifugal force tending to separate joint surfaces is too great for spurt musculature alone, which may be assisted synergically by shunt muscles exerting a centripetal force towards the articulation. Generation of either force would be independent of direction of swing; shunt activity of brachioradialis has been shown electromyographically in both rapid flexion *and* extension of forearm (vide infra). In this brief analysis one attachment is regarded as fixed (i.e. functional 'origin'), the other ('insertion') attached to a mobile unit. Relative fixity and mobility of attachments are often reversed in natural use and a spurt myoneme would then operate as a shunt. Moreover, both attachments may be in motion. For these and other reasons the terms *origin* and *insertion* have been here abandoned in descriptions of individual muscles, the general term *attachment* being preferred. (For other vocational purposes, however, see the comments on colour coding for muscle attachments in Osteology, p. 325.)

Since its first proposal, refinement and mathematical development (MacConaill & Basmajian 1977, MacConaill 1978a,b,c), the anatomicokinetic classification of myonemes and whole muscles into spurt and shunt types has prompted much discussion and some adverse criticism (Stern 1971, Rozendal & Molen 1972, Joseph 1973, Jackson et al 1977, Stanier 1977). However, MacConaill has strongly argued the validity of his *classification* but has withdrawn some *hypotheses* based upon it. In an attempt to provide a 'generalized mechanics of articular swing' the interesting proposal has been made, that centripetal force acting along a

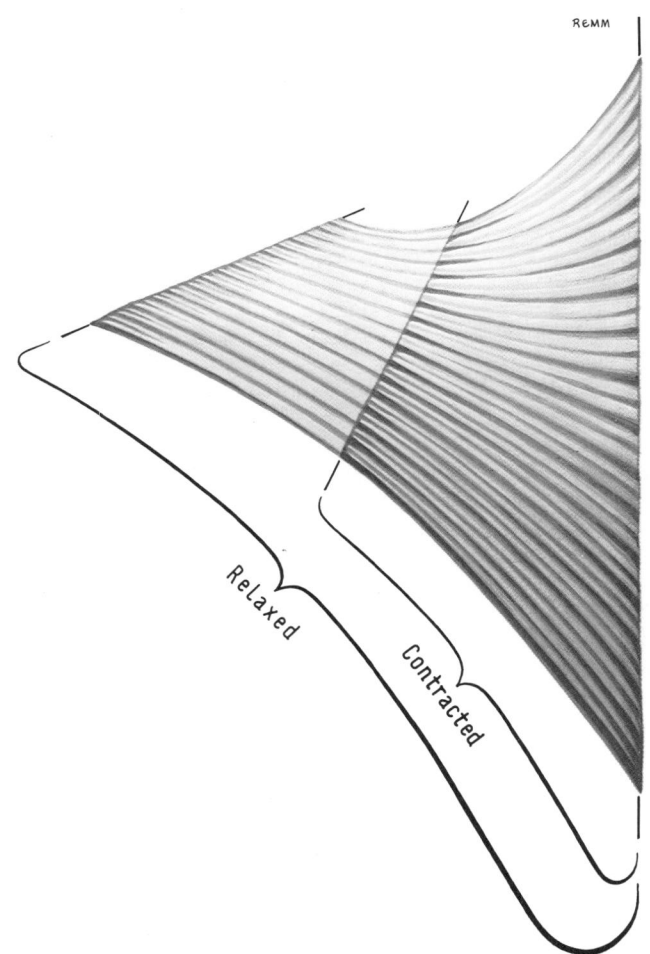

REMM

Relaxed

Contracted

5.20 The 'detorsion' or untwisting which follows the contraction of a spirally disposed muscle.

bone moving at *any* speed is a *consequence* of its curvilinear motion, not a *cause* of it.

Another absorbing problem currently concerning thoretical and experimental kinesiologists is analysis of *patterns* of sequential muscle contractions occurring when two or more muscles can produce the same movement (Joseph 1973, Jackson et al 1977). The latter contend that a 'primary' muscle always initiates movement and that 'secondary' muscles are recruited only when a critical level of activity is reached in the primary, as applied to both isotonic and isometric contraction. Variations in force and velocity as determinants of sequential patterning are emphasized. However, values of parameters which constitute in combination a critical level of 'activity' have not yet been adequately determined. Few observers have discussed the phenomenon of *increasing length* accompanied by *increasing tension* in a myoneme or whole muscle, a combination commonplace in habitual movements and clearly not 'relaxation'.

These recent considerations of activities of myonemes, muscles and groups of muscles render much customary myological nomenclature redundant, often inaccurate and misleading.

FASCIAE AND MUSCLES OF THE HEAD

Muscles of the head are customarily divided into *facial muscles*, related to orbital margins and eyelids, external nose, lips, cheeks and mouth, pinna, scalp and cervical skin (all often grouped as 'muscles of facial expression') and *masticatory muscles*, primarily concerned with temporomandibular movements. This division reflects different embryonic origins and innervations; but in most activities, such as mastication, deglutition, vocalization, communicative and emotional expression, respiration, ocular, aural and nasal actions the two groups are interdependent.

The other muscle groups of the head are described elsewhere: these include *ocular* and *extraocular muscles* (pp. 1193, 1188, 1207), *auricular* and *tympanic muscles* (pp. 1220, 1227), *lingual, palatal* and *upper pharyngeal muscles* (pp. 1320, 1289, 1326).

Craniofacial Muscles

These muscles all receive their innervation from branches of the facial nerve. Those described here are grouped (topographically and functionally) as epicranial, circumorbital and palpebral, nasal and buccolabial. With the same innervation but described elsewhere are platysma (p. 583), posterior belly of digastric and stylohyoid (p. 584) and stapedius (p. 1227). Most muscles clearly belong to one of the major groups named above; others, however, cross the boundaries between groups and functionally may affect both; their allocation to one of these groups alone is arbitrary. Variation between individuals is common and some pronounced racial differences are recognized. Furthermore, investigators specializing in craniofacial myology often use different criteria for categorizing parts of a muscle (often multiple 'heads' of attachment) or of groups of muscles; in parallel, over the decades, the associated nomenclature has changed. The latter, sometimes imprecisely defined, compounded in some locations by insufficient data, adds to the difficulty of making comparisons between different accounts. What follows is a brief account of widely accepted features; for further details of this highly complex structural and functional region consult the references with bibliographies quoted below; see also 'anatomy of speech' p. 1257. For a general survey see 5.21.

(A) EPICRANIAL MUSCULATURE

Epicranius consists of M. occipitofrontalis and M. temporoparietalis. The **superficial fascia** in the scalp is firm, dense, fibro-adipose and closely adherent both to skin and the underlying epicranius including its *epicranial aponeurosis* (galea aponeurotica); posteriorly it is continuous with the superficial fascia of the neck. Laterally, looser in texture, it is prolonged into the temporal regions.

Occipitofrontalis, a broad, musculofibrous layer, covers the cranial vault from the highest nuchal lines to the eyebrows. It has

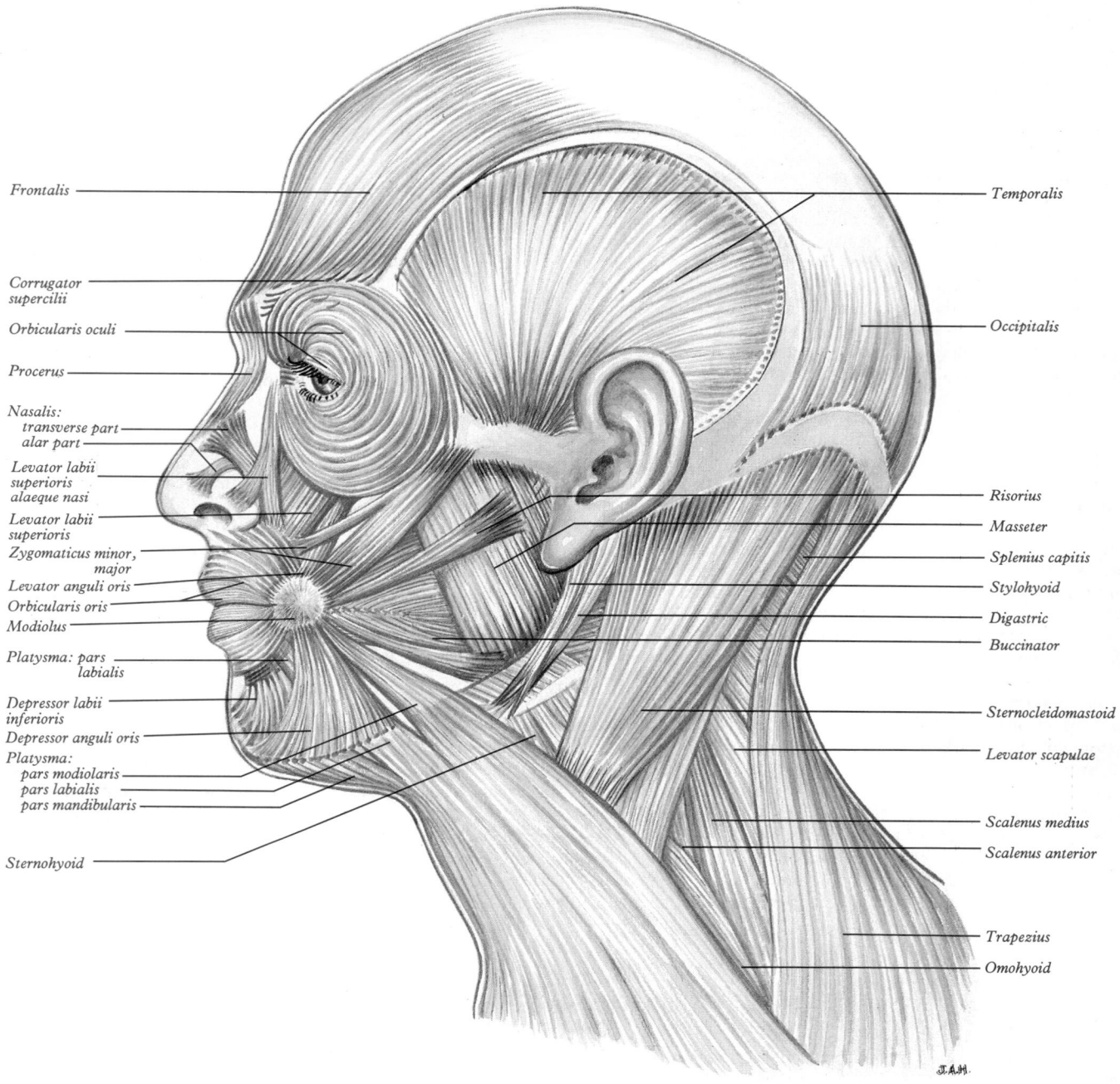

Frontalis
Corrugator supercilii
Orbicularis oculi
Procerus
Nasalis:
    transverse part
    alar part
Levator labii superioris alaeque nasi
Levator labii superioris
Zygomaticus minor, major
Levator anguli oris
Orbicularis oris
Modiolus
Platysma: pars labialis
Depressor labii inferioris
Depressor anguli oris
Platysma:
    pars modiolaris
    pars labialis
    pars mandibularis
Sternohyoid

Temporalis
Occipitalis
Risorius
Masseter
Splenius capitis
Stylohyoid
Digastric
Buccinator
Sternocleidomastoid
Levator scapulae
Scalenus medius
Scalenus anterior
Trapezius
Omohyoid

5.21 Muscles of the head and neck (superficial lateral view) including these groups: circumorbital, buccolabial, nasal, epicranial, masticatory and cervical. The auricular muscles are omitted. Risorius, a variable muscle, here has two fasciculi; the lower is unlabelled. The nature of the modiolus, the modiolar muscles and their co-operation in labial control is detailed in the text. The direct labial tractors to both upper and lower lips have been transected, showing their coronal laminae and revealing subjacent orbicularis oris.

two occipital and two frontal muscular parts, connected by a fibrous aponeurosis. Each *occipitalis*, thin and quadrilateral, is attached by tendinous fibres to the lateral two-thirds of the highest nuchal line and mastoid temporal bone. It ends in the aponeurosis. Each *frontalis* (**5**.21), also thin and quadrilateral, is adherent to overlying superficial fascia, particularly of the eyebrows, but it is broader and its fibres longer and paler than occipitalis. It has no direct bony attachment but its medial fibres are continuous with procerus, the intermediate blending with corrugator supercilii and orbicularis oculi, the lateral also blending with orbicularis over the zygomatic process of the frontal bone. All fibres ascend to join the epicranial aponeurosis anterior to the coronal suture. The medial margins of these frontal bellies blend above the nasal root, the occipital bellies being separated by a variable interval occupied by the aponeurosis.

The temporoparietalis is a variably developed muscular sheet between frontalis and the anterior and superior auricular muscles.

The epicranial aponeurosis covers the cranial vault, forming with the epicranial muscle a continuous fibromuscular sheet from occiput to eyebrows. Posteriorly, between the occipital parts of occipitofrontalis, it is attached to the external occipital protuberance and highest nuchal lines; anteriorly it encloses the frontales, with a narrow prolongation between them. On each side the anterior and superior auricular muscles are attached to it; here it is thinner and continued over temporal fascia to the zygomatic arches. Over the cranial vault, the aponeurosis is intimately united to the skin by fibrous superficial fascia but it is loosely connected to the pericranium by areolar tissue and hence moves with the skin of the scalp (**5**.22). Subdivision of this subaponeurotic tissue into three layers has been described (Chayen & Nathan 1974), the intermediate being dense and like the aponeurosis in its attachments.

Nerve supply. Occipitals is supplied by the posterior auricular branch, the frontalis by temporal branches of the facial nerve.

Actions. The occipitales retract the scalp; during contraction the frontales raise the eyebrows and nasal skin; simultaneously they protract the scalp, thus progressively creasing the forehead (at five to ten major transverse 'lines' and variable intermediate minor ones). Although loosely termed 'transverse' these lines form a series of roughly concentric arches above the orbits and parallel to the superior orbital margins, i.e. convex upwards. The medial ends of many of the arches are continuous across a median strip about 2 cm wide but here the curvature is reversed, i.e. convex downwards. The two parts of the muscle, acting alternately, move the entire scalp back and forwards. The frontales elevate the eyebrows commonly in glancing upwards, also in expressions of surprise, horror, fright, etc.

A thin slip, *transversus nuchae*, occurs in about 25% of people, extending from the external occipital protuberance or superior nuchal line, passing superficial or deep to trapezius, usually to fuse with auricularis posterior but it is sometimes blended with sternocleidomastoid.

Applied Anatomy. The scalp has five layers: skin, subcutaneous tissue, epicranius and its aponeurosis, subaponeurotic areolar tissue and pericranium (**5**.22). The first three form a unit; when torn off in accidents, they remain firmly connected. Due to the density of subcutaneous tissue, inflammatory swelling is slight; wounds which do not involve epicranius or its aponeurosis do not gape. Contraction and retraction of arteries is impeded by this dense tissue and haemorrhage from scalp wounds is often copious. Subaponeurotic areolar tissue is surgically important; it is through this that avulsion occurs and surgical exposures are made. Vessels remain in the partly avulsed tissue and anastomose freely; hence necrosis is unusual.

(B) CIRCUMORBITAL AND PALPEBRAL MUSCULATURE

Of the muscles in this group, orbicularis oculi, corrugator supercilii and levator palpebrae superioris, the last is described with the ocular adnexa (p. 1207).

Orbicularis oculi (**5**.21,23) is a broad, flat, elliptical muscle invading the eyelids, surrounding the orbital opening and spread-

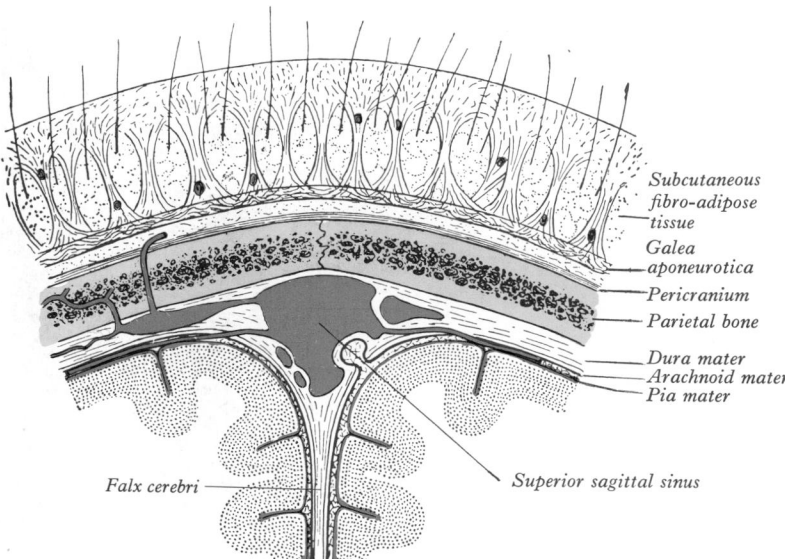

5.22 Coronal section through the scalp, skull and brain. Note: loculated fat between fibrous septa blending with dermis and galea; loose subaponeurotic areolar tissue; emissary, diploic, dural and neuropial veins. The superior sagittal sinus and lateral lacunae are more complex than depicted, see p. 800.

ing into the anterior temporal region, the infraorbital cheek and the superciliary region. It has orbital, palpebral and lacrimal parts.

The *orbital* part, reddish and thicker than the palpebral fasciculi, is attached to the nasal part of the frontal bone, the frontal process of the maxilla (**5**.24) and to the medial palpebral ligament between them. Its fibres form ellipses, complete laterally, where there is no bony attachment, their upper parts blending with frontalis and corrugator supercilii and many reaching skin and subcutaneous superciliary tissue, forming a **depressor supercilii**. Inferiorly, the ellipses overlap or partially blend with the attachments of levator labii superioris alequae nasi, levator labii superioris and zygomaticus minor; medially some ellipses may reach the upper lateral fibres of procerus. Most laterally and peripherally sectors of complete (but sometimes parts of incomplete) ellipses have a loose areolar connection with the temporal extension of the epicranial aponeurosis. (The inferomedial margins of the orbital part may be continuous with the variable musculus malaris, vide infra.)

The *palpebral* part, thin and pale, starts from the medial palpebral ligament, chiefly its superficial but also its deep surface, though not its lower margin; it is attached to bone above and below the ligament and its fibres sweep across the eyelids anterior to the orbital septum (p. 1215) interlacing at the lateral commissure as a *lateral palpebral raphe*. A slip of fine fibres, near the margin of each eyelid behind the lashes, is the *ciliary bundle*.

The *lacrimal* part is behind the lacrimal sac, separated from it by lacrimal fascia; it is attached to this fascia, to the upper part of the lacrimal crest and adjacent lateral surface of the lacrimal bone (**5**.24). Passing laterally behind the sac it divides into upper and lower slips, whose fibres are partly attached to the tarsi near the lacrimal canaliculi, most fibres continuing anterior to them to the lateral palpebral raphe.

The *medial palpebral ligament*, about 4 mm long and 2 mm broad, is attached to the frontal process of the maxilla, anterior to the nasolacrimal groove. Crossing the lacrimal sac, separated by lacrimal fascia, it divides to be attached to the medial ends of both tarsi. The *lateral palpebral raphe*, a weaker structure, consists of the interlacing lateral palpebral fibres of orbicularis oculi, strengthened by the orbital septum. A few lobules of the lacrimal gland, but more often of fat, may separate it posteriorly from the lateral palpebral ligament.

Nerve supply: temporal and zygomatic branches of the facial nerve.

Right frontal sinus

Lacrimal sac

Lacrimal part of orbicularis oculi

Probe in nasolacrimal duct

Middle concha

Inferior concha

Probe passing into nasolacrimal duct

Nasal septum, curved posterior free margin

Orbital part ⎱ Orbicularis
Palpebral part ⎰ oculi

Superior tarsus

Inferior tarsus

Zygomatic bone

Infra-orbital nerve

Maxillary sinus

5.23 Dissection to expose the right orbicularis oculi from behind.

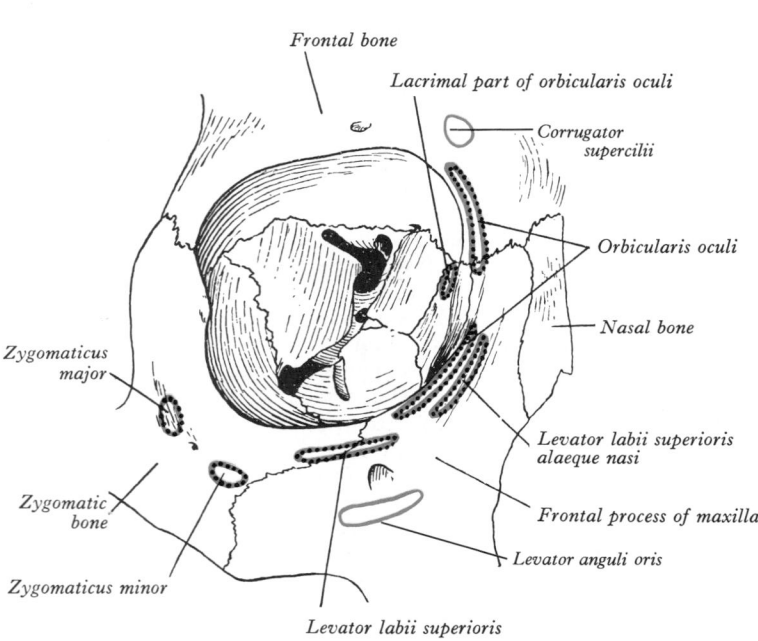

Frontal bone

Lacrimal part of orbicularis oculi

Corrugator supercilii

Orbicularis oculi

Nasal bone

Zygomaticus major

Zygomatic bone

Zygomaticus minor

Levator labii superioris alaeque nasi

Frontal process of maxilla

Levator anguli oris

Levator labii superioris

5.24 Attachments of muscles around the right orbital opening.

Actions. Orbicularis oculi is the palpebral sphincter. The palpebral part, voluntarily or by reflex, closes the lids gently in sleep or rapidly in blinking; the orbital part is usually voluntary in action. Although the major factor in closure is lowering the upper lid, there is some elevation of the lower. The palpebral muscle has upper *depressor* and lower *elevator* fascicles. When the entire muscle contracts, the frontal, temporal and malar skin is drawn towards the *medial* orbital angle; the lids are firmly closed and displaced medially in toto; in this activity skin folds are formed, radiating from the lateral palpebral angle; in the middle decades these wrinkles often become permanent. Levator palpebrae superioris opposes the superior palpebral depressor fibres. The lacrimal part also draws the lids and lacrimal papillae medially, also pulling on the lacrimal fascia and perhaps dilating the sac (p. 1219) to aid drainage of tears. It is said to influence pressures in the lacrimal gland, ducts and orifices and may affect the sinuous

flow of tears across the cornea; it may also direct the puncta lacrimalia into the lacus lacrimalis and express secretions of ciliary and tarsal glands. The orbicularis oculi is also involved in facial expression and various ocular reflexes. Narrowing the palpebral fissure with protrusion of the eyebrows reduces the entry of light; the upper orbital fibres may thus cause largely vertical furrows above the nose (vide infra), a feature markedly developed in people much exposed to strong sunlight.

The corrugator supercilii, a small paired pyramidal muscle at the medial end of each eyebrow, is deep to frontalis and orbicularis oculi. Attached to bone at the medial end of the superciliary arch it ascends laterally, interlacing and blending with the foregoing muscles, thus exerting traction on the skin above the midpart of the supra-orbital margin.

Nerve supply: temporal branches of the facial nerve.

Actions. The muscle draws the eyebrow medially and down, producing, with orbicularis oculi, mainly *vertical* wrinkles on the supranasal strip of the forehead. (Commonly most prominent are a bilateral pair of major vertical wrinkles of variable length, 2–4 cm, above eyebrow level and about 1 cm apart; however, their lower ends curve laterally around or through the medial end of the eyebrows. Lesser vertical wrinkles occur in variable numbers, mainly lateral to the major ones.) The corrugator assists in shielding the eye in bright sunlight. It is also involved in frowning.

(C) THE NASAL MUSCULATURE

This group comprises Mm. procerus, nasalis and depressor septi (**5.21**).

Procerus is a small pyramidal slip near and often partially blended with the medial side of frontalis. From a fascial aponeurosis covering the lower part of its nasal bone and adjoining lateral nasal cartilage it passes to the skin of the lower forehead between the eyebrows. Normally its lower aponeurosis blends with that of nasalis transversus; occasionally, however, a few muscular fascicles of procerus continue to the nasal ala, some even reaching the upper lip.

Actions. It depresses the medial end of the eyebrow, producing transverse wrinkles over the nasal bridge and root. Active in frowning and 'concentration', it aids in reducing the glare of bright sunlight.

Nasalis has transverse and alar parts, sometimes continuous. The *transverse part, compressor naris*, is attached to the maxilla lateral to its nasal notch, and proceeds superomedially to a thin, expanded aponeurosis, continuous with its fellow across the dorsum nasi and also blends with the aponeurosis of procerus. The *alar part, dilatator naris*, attached to the maxilla, inferomedial to and partly blended with the compressor, passes to the cartilaginous ala nasi. (However, see also depressor septi, below.)

Actions. The transverse part compresses the nasal aperture at the junction of the vestibule and nasal cavity; the alar part depresses the ala laterally, widening the anterior nasal aperture, visibly in deep inspiration, and thus is associated both with exertion and emotional states.

Depressor septi (often described as part of dilatator naris) is attached to the maxilla above the central incisor; it ascends to the mobile part of the nasal septum, deep to the superior labial mucous membrane.

Actions. It assists the alar part of nasalis in widening the nasal aperture.

Nerve supply. All muscles of this group are supplied by superior buccal branches of the facial nerve.

(D) THE BUCCOLABIAL MUSCULATURE

The buccal orifice and labial postures are controlled by an intricate three-dimensional complex of muscular slips (5.21), including: the *elevators, retractors* and *evertors* of the upper lip and buccal angle (levator labii superioris alaeque nasi, levator labii superioris, zygomaticus major and minor, levator anguli oris and risorius); the *depressors, retractors* and *evertors* of the lower lip and buccal angle (depressor labii inferioris, depressor anguli oris and mentalis); antagonists of the foregoing, a complicated *sphincter* (orbicularis oris with its subregions, incisivus superior and inferior) and buccinator. Some labial fascicles are clearly intrinsic, others extrinsic; a large proportion are, however, so closely interwoven, or even continuous, that their satisfactory investigation and terminological allocation have proved difficult, sometimes controversial and, in some locations, as yet impossible.

Levator labii superioris alaeque nasi is attached to the upper part of the frontal process of the maxilla, then descends inferolaterally, dividing into a medial slip attached to the greater alar cartilage and the skin over it and a lateral slip prolonged inferolaterally to blend with levator labii superioris and orbicularis oris. The superficial fibres of the lateral slip curve laterally across the ventral aspect of levator labii superioris and attach successively to the dermal floor of the upper part of the nasolabial furrow and ridge.

Actions. The lateral slip raises and everts the upper lip and raises, deepens and increases the curvature of the nasolabial furrow's superior part; the medial slip dilates the nostril and displaces laterally and modifies the curvature of the inferolaterally convex *circumalar furrow*.

Levator labii superioris descends from the inferior orbital margin, being attached to the maxilla and zygomatic bone above the infra-orbital foramen, and converges into the upper lip between the lateral slip of levator labii superioris alaeque nasi and zygomaticus minor with, more deeply, levator anguli oris.

Actions. It elevates and everts the upper lip. Its contraction indirectly modifies the variably prominent, but permanent *nasolabial furrow* (*groove* or *sulcus*), often deepened in sad or serious expressions. The furrow descends laterally from the lateral nasal margin at the upper limit of the ala, to become indistinct, in an adult, about 1.25 cm lateral to the labial commissure (buccal angle), over the *modiolus* (vide infra). In some faces the furrow is a highly characteristic feature. The furrow's prominent superolateral rim, the *nasolabial ridge*, delimits the *infra-orbital cheek*; its flatter, more deeply placed inferomedial rim bounds *the superior labial area*. The disposition of the furrow varies with the position of the modiolus and posture of the lips and nasal ala. These indirect influences are supplemented directly by the state of contraction of levator labii superioris alaeque nasi (lateral slip), zygomaticus minor, levator anguli oris and, when present, malaris. In the nasolabial ridge the buccal panniculus adiposus is maximally thick; in the labial area the panniculus is extremely thin, often virtually absent.

Zygomaticus minor, attached to the zygomatic bone behind the zygomaticomaxillary suture, descends medially into the upper lip; it is separated superiorly from levator labii superioris by a narrow triangular interval (5.21) but inferiorly blends with this muscle.

Actions. It elevates the upper lip exposing maxillary teeth and, as mentioned, assists in deepening and elevating the nasolabial furrow. With the other main elevators it curls the upper lip in smiling and in smugness, contempt or disdain.

(Some earlier investigators regarded the foregoing three muscles as individual 'heads', named respectively angularis, infra-orbitalis and zygomaticus, of one compound *musculus quadratus labii superioris*.) For further discussion of corporate activities see below.

Levator anguli oris (*caninus*) is attached to the canine fossa below the infra-orbital foramen, whence it converges and mingles near the buccal angle (at the modiolus, vide infra) with zygomaticus major, depressor anguli oris and other muscular bands including orbicularis oris. Between it and levator labii superioris are the infra-orbital vessels and nervous plexus (5.26). Some of the levator's superficial fibres have curved directly anteriorly to gain a dermal attachment to the floor of the lower part of the nasolabial furrow.

Actions. It raises the modiolus and buccal angle, incidentally displaying the teeth in smiling, and contributes to the depth and contour of the nasolabial furrow. It assists in momentarily fixing a translocated modiolus.

Zygomaticus major extends from the zygomatic bone, in front of the zygomaticotemporal suture, to the modiolus near the buccal angle, blending here with levator anguli oris and orbicularis oris and also, more deeply, with other modiolar muscles.

Actions. Zygomaticus major retracts and elevates the modiolus and buccal angle, as in laughing. It is also a fixator of the modiolus. **Malaris** is subject to much individual and racial variation; sometimes absent, when present it may be represented by a few superficial fascicles, a partial or complete thin sheet, or occasionally a substantial complete sheet. It is continuous with the inferior limit of orbicularis oculi from the temporal fascia to its medial palpebral attachment; it is possibly derived from orbicularis oculi (Lightoller 1925). Its fibres incline medially and downwards, covering zygomaticus major and minor and levator labii superioris, its deep fascicles blending and distributed with these muscles. Of its superficial fascicles, a proportion have a dermal attachment to the nasolabial ridge and sulcus; some pass directly to the modiolus; the remainder enter the outer third of the upper lip and intersect with bundles of orbicularis oris.

Nerve supply. All six preceding muscles are supplied by buccal branches of the facial nerve.

The inferior labial area is bound by the free red-lip margin above, by the centrally transverse *mentolabial sulcus*, separating it from mental skin and dense fasciae below, while laterally are the short curved *inferior buccolabial sulci*; the latter descend from the buccal angle and sweep medially to meet the down-curved ends of the mentolabial sulcus. The inferior labial area may be arbitrarily considered as completed by a projection from the centre of the inferior buccolabial sulcus passing a short distance upwards and laterally to the modiolus.

Mentalis, a conical fasciculus lying lateral to the inferior labial frenulum, is attached in the mandible's incisive fossa and descends to the mental skin.

Actions. It raises the mental tissues, mentolabial sulcus, (wrinkling the mental skin) and base of the lower lip, aiding its protrusion/ eversion, as in drinking, speech and also expressing doubt or disdain. Electromyography shows prolonged activity in mentalis, even in sleep, an unexplained finding.

Depressor labii inferioris is quadrilateral and attached to the mandible's oblique line, between symphysis menti and the mental foramen, ascending medially into the skin and mucosa of the lower lip (vide infra), blending and intersecting with its fellow and with orbicularis oris. It is continuous below and laterally with platysma (pars labialis); its superficial part contains some admixed fat but its overlying panniculus adiposus is very thin.

Actions. It depresses the lower lip laterally in mastication, may assist its eversion and contributes to expressing irony, sorrow, melancholy, doubt etc.

Depressor anguli oris ascends from the mandible's mental tubercle and its continuation, the oblique line, inferolateral to depressor labii inferioris, then converges into a narrow fasciculus blending with other muscles at the modiolus near the buccal angle. It is continuous below with platysma and cervical fasciae; at the modiolus it is continuous with orbicularis oris and risorius; some fibres continue into levator anguli oris. (Some earlier investigators regarded many, or all, of these continuations and blendings as accessory parts or 'heads' of one large compound *musculus triangularis.*) In association with but below the fibromuscular bundles attached to the mental tubercle, some fascicles may bypass the tubercle inferiorly then, crossing the midline, interlace with their contralateral fellows as the *musculus transversus menti* (or 'mental sling').

Actions. It depresses the modiolus and buccal angle laterally in opening the mouth and in expressing sadness. During opening of the mouth the buccolabial sulci are stretched, flattened and become indistinct; the mentolabial sulcus becomes more horizontal and its central part deepened.

Nerve supply. Mentalis, depressores labii inferioris and anguli oris are supplied by the mandibular marginal branch of the facial nerve.

The buccinator (5.25), a thin quadrilateral muscle between maxilla and mandible, is attached linearly to the external surfaces of the alveolar processes of maxilla and mandible, opposite the molar teeth; curving posteromedially around the sites of the third molar teeth, the *upper* part crosses the maxillary tuberosity then a tendinous band to the pterygoid hamulus, the *lower* crossing the junction of the ramus and body of the mandible to the posterior end of the mylohyoid line, and thence from the anterior border of the pterygomandibular raphe, which is interposed between it and the superior pharyngeal constrictor. Between the maxillary tuberosity and the upper end of the raphe a few fibres spring from the tendinous band between the maxilla and pterygoid hamulus; this encloses the tendon of tensor veli palatini that pierces the pharyngeal wall in a small gap behind this band (5.25). Thus, the posterior part of buccinator is deeply placed and internal to the mandibular ramus and its attachments and in the plane of the medial pterygoid plate; its anterior part *curves out* behind the third molar teeth to assume a submucosal position in cheek and lips. Buccinator fibres converge towards the modiolus near the buccal angle, the central fibres decussating, so that lower and upper ones continue into upper and lower parts of orbicularis oris respectively, while the highest and lowest fibres of buccinator enter their corresponding lips without decussation.

Relations. Buccinator *posteriorly* is in the plane of the superior constrictor, and here is covered by buccopharyngeal fascia. *Superficially* over its posterior part fat separates it from the mandibular ramus, masseter and part of temporalis. (The fat is not clearly linked with suckling.) *Anteriorly* the superficial surface of buccinator is related to zygomaticus major, risorius, levator and depressor anguli oris and the parotid duct, which traverses it near the third upper molar; the facial artery and vein and branches of both facial and buccal nerves cross it. Its *deep surface* is related to the buccal glands and mucous membrane.

Nerve supply: lower buccal branches of the facial nerve.

Actions. The buccinators compress the cheeks against the teeth, passing food between them in mastication, or expelling air when the cheeks are distended (*buccinator* = a trumpeter!). Its labial extensions are mentioned below.

The pterygomandibular raphe is a strand of tendinous fibres passing from the pterygoid hamulus to the posterior end of the mylohyoid line. *Medially* it is covered by buccal mucous membrane, where it is easily palpable; it is separated *laterally* from the mandibular ramus by adipose tissue. *Posteriorly* the superior pharyngeal constrictor is attached to it; *anteriorly* the central part of buccinator arises from it (5.25).

ORAL SPHINCTERS, TRACTORS AND MODULATORS

The almost endless variety of neuromuscular controls of the lips and oral fissure during speech and non-verbal expressive communication are commonly integrated with fluctuating patterns in the masticatory, linguopharyngeal, laryngeal, circumnasal, circumorbital and palpebral, and also intraocular and extraocular muscle groups. Yet to be detailed are: the compound oral sphincter, *orbicularis oris*; its closely associated bilaterally paired *incisivus superior and inferior*; the patterned labial attachments of upper lip elevators and lower lip depressors; also the mode of force-transmission to the lips from those muscles that converge towards dynamic palpable loci (the *modioli*), their centres about 1.25 cm lateral to the buccal angle. Before providing these descriptions it is convenient, despite some repetition, to summarize, but with more information, the limits of the labial area and its superior and inferior parts, the general architecture of a lip including the oral fissure and the free 'red-lip' margin and the concept of bilateral modioli (and their distinction from 'the buccal angles'). See 5.21,26A–E.

The labial area. When the face is in repose, the lips in gentle contact and the teeth maintaining a narrow interocclusal clearance, the area is approximately a hexagon (borders: superior, inferior, paired superolateral and inferolateral). The *superior*

Maxillary artery
Lateral pterygoid plate, partly excised
Maxilla

Tensor veli palatini

Mandibular nerve

Middle meningeal artery

Spine of sphenoid

Levator veli palatini

Pterygoid hamulus

Superior pharyngeal constrictor

Stylopharyngeus

Glossopharyngeal nerve

Styloglossus

Middle pharyngeal constrictor

Stylohyoid ligament

Greater cornu of hyoid bone

Lateral thyrohyoid ligament

Inferior pharyngeal constrictor

Tendinous arch

Recurrent laryngeal nerve

Oesophagus

Tuberosity of maxilla

Buccinator

Parotid duct

Pterygomandibular raphe

Hygoglossus

Mylohyoid

Geniohyoid

Lesser cornu of hyoid bone

Thyrohyoid membrane

Internal laryngeal nerve

Superior laryngeal vessels

Cricothyroid ligament

Cricothyroid

5.25 Buccinator and the muscles of the pharynx. The zygomatic arch, the masseter, the ramus of the mandible, the temporalis and a large part of the lateral pterygoid plate and the pterygoids have all been removed. In addition, the upper parts of the stylopharyngeus and styloglossus have been excised, together with the posteroinferior part of the hyoglossus and all the infrahyoid muscles.

border is between the attached margin of the lower external nose and the upper lips and includes the bilateral curved circumalar sulci, the ridge forming the posterior rim of each nostril; it then continues beneath the junction of the anterior nasal spine of the maxillae and the mobile part of the nasal septum. (Although not so-named this superior border is strictly nasolabial.) The *superolateral boundaries*, described above, incline downwards and laterally from the upper end of the circumalar sulcus to the modiolus and correspond to the so-called nasolabial sulci (superior buccolabial would be more accurate, but transgresses long-established custom). The *inferolateral boundaries* extend downwards and medially from the lateral angles of the hexagon over the modioli to the down-curved lateral ends of the centrally transverse mentolabial sulcus, the latter forming the *inferior boundary*. A transverse line between the external angle (i.e. between the equilibrium 'resting' positions of the modioli) separates the larger (upper two-thirds) superior labial area from the inferior area, and crosses the level of the usually slightly undulant line of contact between the apposed free 'red-lip' surfaces at the closed oral fissure. The considerable variation (between individuals, sexes and races) in the dimensions and curvatures of the exposed red-lip surfaces is commonplace. The junction between external hair-bearing skin and red hairless surface in the upper lip almost invariably takes the form of a double-curved Cupid's bow; centrally convex downwards, then bilaterally rapidly rising to an apex corresponding to the lower end of each ridge of the philtrum, it descends thereafter, laterally retaining a gentle superolateral convexity, to its termination which is often horizontal but sometimes slightly curved upwards, infrequently downwards. The line of contact between the red-lip surfaces is often almost horizontal but quite frequently takes the form of a much less undulant cupid's bow. In the lower lip the skin/red-lip junction varies greatly between individuals in its central vertical depth; in all individuals, however, the lateral extremities descend medially for a few millimetres. In 'shallow' lower red-lips the junctional line curves to form an upward convexity almost parallel to the line of the closed oral fissure. With the deepest lower red-lips, the junctional line follows a continuous downward convexity; many intermediate states exist, the deep downward convexity having a flat (i.e. horizontal) section centrally; the latter may have a varying degree of reversed curvature with an upward convexity.

It may be noted that the highest parts of the superior labial area are paired bilateral, curved, pointed cornua that embrace the external attached margins of the nasal alae. The principal factors affecting the disposition of the lips, oral fissure and circumlabial hexagon are mentioned further below.

Labial architecture. In both upper and lower lips, about the central three-fifths consists mainly of a thick epithelialized fold of tissue, which has free external (cutaneous) and internal (mucous) surfaces and is continuous through a specialized red-lip transitional zone bounding the oral fissure. Above the reflexion of labial mucosa into gingival mucosa the upper lip's perimeter has a narrow band of non-cavitated tissue related to the subnasal maxillae. Similarly, the reflexion in the lower lip approximately coincides with the mentolabial sulcus and here labral and mental tissues are continuous. Radiating from the lateral ends of the oral fissure and buccal angles the labial epithelia and internal tissues cross the external angles and adjacent boundaries of the labial hexagon to become continuous with those of the cheek.

Some aspects of general lip structure, either upper or lower, may be examined in simple vertical parasagittal section; however, this presents some difficulties because, as indicated, the red-lip surface is not simple, plane and horizontal; many prominent muscle bundles run transversely *parallel* to the red surface, whereas, in some lip regions, the principal labial tractors run at right angles to the surface and transverse muscle bundles. Elsewhere, the tractors enter the lip with varying degrees of obliquity. For these reasons histological geometry is most effective when full thickness sections of the lip are made at right angles to the red-lip surface and in the long axis of the tractors, e.g. the segment extending 1 cm lateral to the philtral ridge. The sectional profiles of the lips differ, neither being a simple fold of uniform thickness throughout, although their upper and lower halves extending from their attached bases approach this. The rimal half of

the upper lip has a bulbous sectional profile that is asymmetrical, the adjoining skin and red lip being slightly convex ventrally, while the adjoining red lip and mucosa have a pronounced internal convexity. (This mucosal ridge or 'shelf' can be wrapped around the incisal edges of the parted teeth.) The lower lip is on a more posterior plane than the upper and, in the position of neutral lip contact, its ventral surface curves forwards (with a ventral concavity) to the prominent skin/red-lip junction; posteriorly, a mucosal elevation is minimal or absent.

A thick, approximately central stratum in each lip consists of striated muscle, either parallel bundles of contractile fibres or their dermal, mucosal or intermuscular connective tissue terminal attachments. The muscles are the elements grouped collectively as the *orbicularis oris*, with *incisivus superior and inferior* (all having multiple putative attachments or continuations, vide infra), also the patterned attachments of the direct labial tractors (described below). The skin covering the external lip surfaces has a thin epidermis but, with all the layers that characterize fully keratinized compound epithelia, the dermis is well vascularized and accommodates numerous hair follicles (many large in the male), sebaceous glands and sweat glands. Subcutaneous panniculus adiposus is scant. The mucous membrane lining the labial aspect of the oral vestibule is clothed with a thick non-keratinizing stratified squamous epithelium. Its flattened dying or dead surface cells contain some eleidin and keratohyalin, retain their nuclei and are continuously shed into the saliva, to be replaced by mitosis in the deep germinative layer. The substantial submucosa is well vascularized and lodges numerous labial mucous glands, up to a few millimetres in diameter, the largest being palpable with the lingual tip. Because of the thickness of its semi-opaque epithelium the everted lip's mucosa appears moist, glistening and pink. Between the skin and mucosa, the free, red-lip margin is covered with a specialized stratified squamous epithelium which is thin near the skin, increasing in thickness slightly as the mucosa is approached, then with an abrupt increase in thickness when true mucosa is reached. The epithelium is covered with dead squames that are filled with eleidin and transparent; its deep surface is deeply pitted and grooved receiving the abundant long dermal papillae. The latter carry a rich capillary plexus accounting for the dusky red colour of these labial surfaces; their rich innervation accords with their high discriminative sensitivity. These surfaces are, of course, hairless and the dermis carries no sebaceous, sweat or mucous glands; they are intermittently moistened with saliva by the lingual tip.

Orbicularis oris, so named because for long it was assumed that the oral fissure was surrounded by a series of complete ellipses of striated muscle and that by their concerted action, a simple sphincteric compression of the lip margins, labial mucosa against teeth and indrawing of the buccal angles followed. This was, perhaps, surprising because simple functional inspection during different activities, the effects of neurological deficits, and improving stimulating and recording techniques clearly demonstrate that the muscle consists of independent quadrants (upper, lower, left and right) that co-operate in many contrasting ways in the great variety of labial actions. There is some interlacing and transmedian overlap of corresponding right and left quadrants, while there is, at least, close juxtaposition of upper and lower quadrants at their lateral (modiolar) attachments. Each quadrant consists of a larger *pars peripheralis* (common to many mammals) and a *pars marginalis* (absent in non-primate mammals, slightly marginally different in non-human higher primates, uniquely developed in vocalizing mankind). Formally, then, orbicularis oris as a whole is composed of eight segments each systematically named according its location; as is the case with most modiolar muscles (vide infra), each segment resembles a fan with its stem at the modiolus. Each fan is open in peripheral segments and almost closed in marginal segments. The region of apposition of marginal and peripheral parts is indicated by the lines of junction between the red-lip area and the skin ventrally and the mucosa posteriorly.

Pars peripheralis in each quadrant has a lateral stem attached to the full thickness (apex to base with its related upper or lower cornu) of the labial aspect of the modiolus. The majority of these stem fibres are thought to have a terminal attachment within the tissues of the modiolus, although the fine structural details and

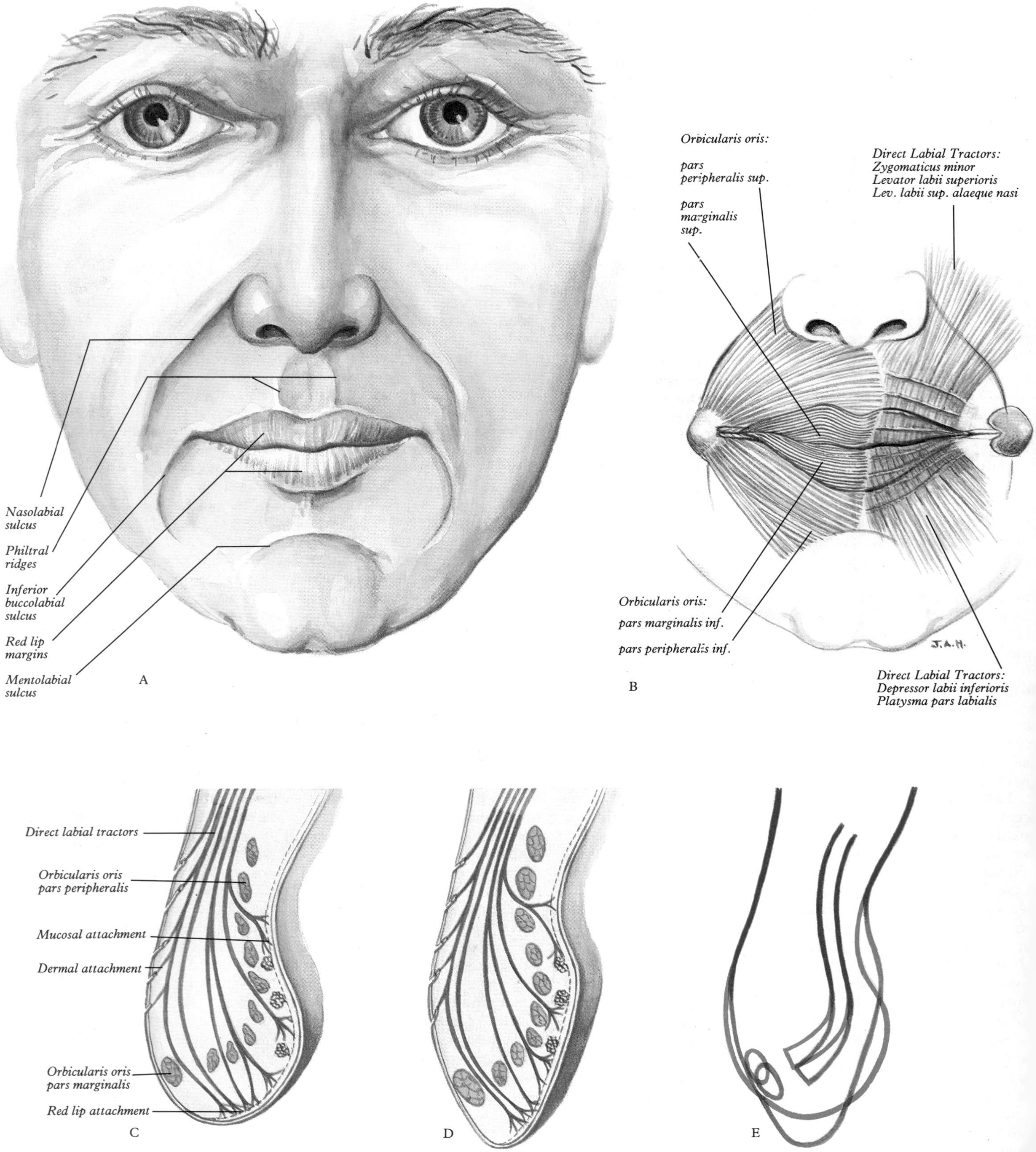

Orbicularis oris:

pars peripheralis sup.

pars marginalis sup.

Direct Labial Tractors:
Zygomaticus minor
Levator labii superioris
Lev. labii sup. alaeque nasi

Nasolabial sulcus

Philtral ridges

Inferior buccolabial sulcus

Red lip margins

Mentolabial sulcus

A

Orbicularis oris:
pars marginalis inf.
pars peripheralis inf.

Direct Labial Tractors:
Depressor labii inferioris
Platysma pars labialis

B

J.A.H.

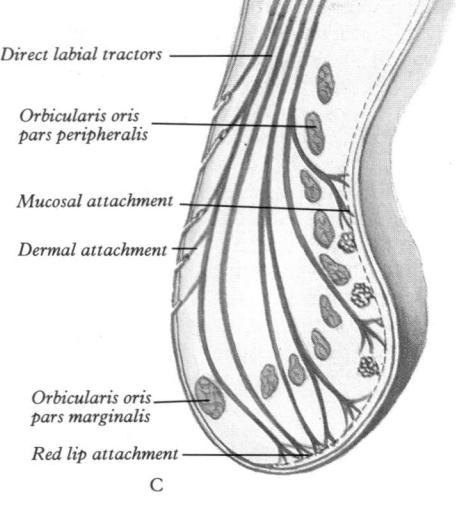

Direct labial tractors

Orbicularis oris pars peripheralis

Mucosal attachment

Dermal attachment

Orbicularis oris pars marginalis

Red lip attachment

C

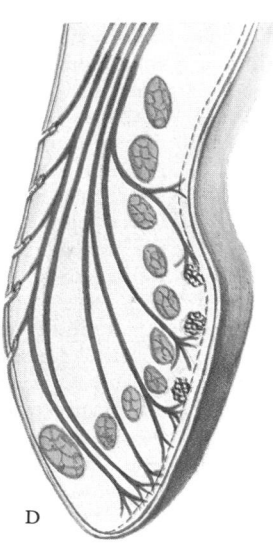

D

E

5.26 A. The principal sulci, creases and ridges of the face which are noted at various points throughout the text. Note particularly those surrounding the 'labial hexagon', the superior and inferior labial areas and one form of the red-lip margins. B. The disposition of the modiolus and orbicularis oris pars peripheralis and pars marginalis (on the left); the successively transected laminae of the direct labial tractors of both upper and lower lips (on the right). C–E are parasagittal sections of the upper lip. In each, on the left is thin skin with oblique hair follicles; on the right thick mucosa with mucous glands and mucosal shelf; between them the red-lip margin. C repose, D slightly contracted forming a narrowed profile of a labial contour. E. Superimposed outlines of C (magenta) and D (blue).

geometry of these have not, as yet, been satisfactorily demonstrated. Neither has the derivation of a limited number of orbicular fibres from direct continuations of one or more (perhaps all) the other modiolar muscles (vide infra). It is commonly stated but clearly an oversimplification that orbicularis has direct reinforcements in the upper lip from buccinator ('upper' or maxillary fibres, decussating 'lower' central or pterygomandibular fibres) and depressor anguli oris; in the lower lip from buccinator ('lower' or mandibular fibres, decussating 'upper' central or pterygomandibular fibres), levator anguli oris and the superficial modiolar attachment of zygomaticus major. (Upper and lower incisivus muscles have a medial osseous attachment and are mentioned below).

The lateral stems (upper and lower) of orbicularis oris enter the lateral angles of their respective superior and inferior labial areas and diverge into bundles arranged as muscular sheets that occupy the central tissue stratum of their triangular hemilabia. These triangular sheets of muscle extend vertically from the level of the red lip/cutaneous junction where the sheet is thickest, then progressively to the thinner region until their respective upper or lower labial borders (vide supra) are reached. The greater part of each sheet enters the free lip where its fibres aggregate into a series of cylindrical bundles, each approximately concentric and parallel to the oral fissure's red-lip margin. Adjacent bundles are separated by extensions of the labial tractors passing to their submucosal attachments; similar extensions separate the rimal bundle of pars peripheralis from the basal bundle of pars marginalis. In the upper lip, the highest fibres course near the nasolabial sulcus, a few fibres attaching to the sulcus; while the others have occasional attachments to the nasal ala and septum. In the lower lip the lowest fibres reach and attach to the mentolabial sulcus. While crossing its quadrant of free lip, the main body of fibres is often held to have a small population that ends in the labial connective tissue, dermis or submucosa. This has not been demonstrated satisfactorily. Most fibres continue medially and as the median plane is approached they start for some 5 mm into the opposite half-lip, until the fibres gain their mainly dermal but partly submucosal attachments. Interlacement occurs beneath the hollow of the upper lip's philtrum and the less marked corresponding depression in the loci in the lower lip. During oral fissure compression a median philtral vertical sulcus becomes more prominent, as do a pair of paramedian vertical wrinkles in the lower lip. (For studies on fetal lips see Latham & Deaton 1976.)

Pars marginalis of the orbicularis oris, as mentioned, is uniquely developed in human lips and closely associated with speech and the production of some varieties of musical tone. In each quadrant the pars marginalis consists of a single (or in some individuals and races double) band of narrow diameter muscle fibres lodged within the tissues of each red-lip margin. Medially, the marginal fibres meet and interlace with their contralateral fellows, then gain attachment to the red-lip dermis a few millimetres beyond the median plane (cf. pars peripheralis). Laterally, the fibres converge and continue to an attachment at a horizontal strip level with the buccal angle, across the deepest (submucosal) part of the modiolar base.

The relations between pars marginalis and peripheralis are not simple. In a full thickness section of an upper lip at right angles to the red-lip margin, the cylindrical bundles of peripheralis fibres are not all in one plane. Their profiles, considered together, are slightly convex forwards above but reversing to a more pronounced forward concavity in their rimal half; i.e. they are classically likened to the shank and initial curved part of a hook. Beyond the rimal bundle of peripheralis, the hook is completed by the blunted triangular (elliptical or oval) profile of marginalis; its base is near the lower border of peripheralis, the remainder extending upwards and forwards occupying the core of the red-lip, its apex approaching the red-lip/cutaneous junction. In the lower lip, peripheralis bundles form, in section, a continuous curve concave forwards; this is surmounted by the flattened, curved, triangular marginalis that descends anteriorly, it apex again nearing the red-lip/cutaneous junction. In both lips, therefore, throughout their rimal red zones, marginalis lies substantially anterior to the rimal bundles of peripheralis. Passing laterally, however, beyond the red-lip and crossing the buccal angle, this relationship alters.

Marginalis is spirally wrapped around the rimal edge of peripheralis to reach its deep (submucosal) surface and this deep position of marginalis (with the vertical positioning of its fibres now reversed) is maintained as it continues to its attachment at the modiolar base. The functional roles implicit in these structural arrangements and other co-operate labial controls are mentioned below (see Lightoller 1925, Burkitt & Lightoller 1926, Duckwork 1947).

Incisivus labii superioris, an accessory muscle of the oral orbicular complex, has an osseous attachment to the floor of the maxilla's incisive fossa superior to the eminence of the lateral incisor tooth and at first lies deep to the upper part of pars peripheralis of the orbicularis. Arching laterally, its fibre bundles become intercalated between and parallel to the orbicular bundles. Approaching the modiolus it segregates into superficial and deep parts; the former partly blends with levator anguli oris and attaches to the intermediate (body) and apical modiolar zones; the latter attaches to the superior cornu and basis moduli.

Incisivus labii inferioris, also an accessory orbicular muscle, has many features in common with superioris. Its osseous attachment is to the floor of the mandible's incisive fossa, lateral to mentalis and below the eminence of the lateral incisor tooth. Curving laterally and upwards, it partially blends with pars peripheralis inferior of orbicularis oris, then has similar modiolar attachments; superficial bundles reach the apex and body; deep bundles reach the base and inferior cornu.

Before a more detailed consideration of the attachments of the direct labial tractors and a brief account of the modiolus are attempted, mention of three further, but variable, muscle aggregates must be made.

Platysma: a general account of this muscle is furnished on p. 583. At its superior end mention is made of the submental transmedian interlacement with its contralateral fellow; thereafter its *pars mandibularis* attaches to the lower border of the body of the mandible. Posterior to this a substantial flattened bundle diverges superomedially to the lateral border of depressor anguli oris, where a few fibres join this muscle while the remainder continue deep to the muscle to reappear at the depressor anguli's superomedial border. Here they continue within the tissue of the lateral half of the lower lip as its direct tractor *platysma pars labialis.* The latter occupies the interval between depressor anguli oris and depressor labii inferioris; it is in the same plane as the labial depressor and their contiguous margins blend. They have similar labial attachments. *Platysma pars modiolaris* constitutes all the remaining bundles posterior to pars labialis except for a few fine fascicles that pass directly to end in the buccal dermis or submucosa. The modiolar bundles curve superomedially posterolateral to the depressor anguli oris, pass deep to risorius and attach to the apex and subapical part of the modiolus.

Risorius: an ineptly named muscle subject to great variability ranging from one or more slender fascicles to a wide, thin superficial fan. Its peripheral attachment may include some or all of the following: the zyomatic arch, parotid fascia, masseteric fascia anterior to the parotid, fascia enclosing the modiolar part of platysma and the mastoid fascia. Its fibres converge to an apical and subapical attachment at the modiolus. It is not particularly associated with laughter any more than many other modiolar muscles; equally it is active in numerous facial activities in addition to laughter. (Some authorities, as mentioned, recognize a large compound *musculus triangularis* which, when fully developed, has the following 'heads' or *capita* of attachment peripherally: *c.menti*—cf. transversus menti; *c.longum*—from the mental tubercle; *c.latum*—from fascia and the oblique line, cf. depressor anguli oris; *c.buccale*—widely called risorius.)

Direct labial tractors pass, as their name suggests, directly into the tissues of the lips without intervention of the modioli (although, of course, activity of the modiolar muscles profoundly affects, and is integrated with, activity of the tractors). Broadly, the contraction of tractors exerts a force at, or approximately at, a right angle to the oral fissure serving to elevate and/or evert the whole or part of the upper lip, also to depress and/or evert the whole or part of the lower lip. In the upper lip the tractors are, from medial to lateral; the labial part of levator labii superioris alaeque nasi, the levator labii superioris and zygomaticus minor. These blend into a continuous sheet that, as it enters the free lip, divides into a series of superimposed coronal sheets anterior to the muscle bundles of pars peripheralis orbicularis oris. In the lower lip the tractors are, medially, depressor labii inferioris and, laterally, platysma pars labialis. Again, these form a continuous sheet that divides into coronal sheets ventral to pars peripheralis orbicularis oris. In both upper and lower lips the sheets may be considered as three groups increasingly deep from the surface skin and each with a distinct zone of attachment. The superficial group of sheets gives a succession of fine fibre bundles that curve ventrally a short distance before gaining attachments as a series of horizontal rows to the dermis between the hair follicles, sebaceous glands and sweat glands. The intermediate sheets are attached to the dermis of the red-lip rimal modified skin which they reach by two routes. The more anterior sheets continue past the skin/red-lip junction, then curve posteriorly over the anterior aspect of pars marginalis orbicularis oris to their punctate attachments on the ventral half of the red-lip dermis. The deeper intermediate sheets pass, first posteriorly between the rimal bundle of pars peripheralis and the base of pars marginalis, then curve anteriorly to their, also punctate, attachments to the dorsal half of the red-lip dermis. The deepest sheets of the tractors are closely applied to the ventral aspect of pars peripheralis orbicularis oris, between the parallel bundles of which fine tractor terminals curve posteriorly to their submucosal and periglandular connective tissue attachments. these various attachments may act in concert or selectively and their effects may involve a complete labial quadrant or be restricted to a short labial segment. Thus bilateral activity involving upper and/or lower lips, balanced against varying activity in their antagonists, the compound orbicularis oris and modified by the modiolar muscles, may range from delicate adjustments of the tension and profile of the lip margins to large increases in the oral fissure with lip eversion. Lip movements are modified to accommodate separation of the teeth by mandibular depression at the temporomandibular joints. Beyond a certain range of mouth opening, labial movements are almost completely dominated by mandibular movements.

The modiolus. As mentioned above, an early view, long subscribed to, regarded the oral sphincter, orbicularis oris, as consisting of complete circumoral ellipses supplemented by bundles from the major extrinsic muscles converging on the buccal angles. Ultimately it became clear that complete ellipses were rare, or indeed absent, and the bundles of roughly parallel transverse labial fibres were *wholly* allocated as *direct* extensions from the extrinsic muscles. Some decussated 'at the buccal angle'; others passed without decussation into their corresponding lips. This view is often perpetuated in elementary accounts. However, meticulous dissection of the buccolabial musculature, with particular reference to the tissues embracing the lateral extremities of the oral fissure and radiating about 2 cm superiorly, inferiorly and laterally from the buccal angle, despite a minority of exceptions, in the main failed to support the earlier propositions (Lightoller 1925, Burkitt & Lightoller 1926, 1927). The dense, compact, mobile, fibromuscular mass, the *modiolus*, consists of the interlacing arrays of fibre terminals of muscles converging towards or diverging from it. Bidigital palpation is most effective with the opposed thumb and index finger compressing the mucosa and skin simultaneously. The name modiolus was adopted because, in surface view, the mass was fancifully thought to resemble the hub of a wheel and the radiating stems of its component muscles, the spokes thereof. Unlike spokes, however, the muscles do not all occupy the same plane and their modiolar stems are often spiralized and most divide into at least two (superficial and deep) bundles, (some into three or four), each with its distinctive interlacements and zones of attachment. Since nine muscles (or more depending on the classification employed) are attached to each modiolus, and they undergo the transformations mentioned, each mass has a high degree of three-dimensional structural complexity and has proved correspondingly difficult to analyse. Sound quantitative stereology has not yet been achieved, while the results of careful microdissection, although often leading to similar broad conclusions, has also led to controversy about some details. Such minutiae will be disregarded here.

The shape and dimensions of the modiolus are necessarily approximations because there are individual, sexual, racial and, of course, age variations; also there are no precise distinct histological boundaries but rather a highly irregular zone where dense, compact interlacing tissue grades into the stems of individually recognizable muscles. The modiolus has some features resembling a blunt 'cone' but with modifications and additions. The base of the cone, lateral to a vertical line through the buccal angle, is adjacent and adherent to the mucosa; this *basis moduli* is roughly elliptical in outline and, as intimated, extends vertically about 2 cm above, and 2 cm below, a horizontal through the buccal angle; it also extends laterally a similar distance from the angle. The blunt *apex* of the cone, about 4 mm across, is centred about 1 cm lateral to the buccal angle, deep and adherent to the panniculus fibrosus; panniculus adiposus is absent here. From mucosa to dermis, therefore, this mass presents axially a base, a central part or body and an apicosubapical part, each about 3.5 mm thick, i.e. the whole is approximately 1 cm thick. The body has an oblique fibrous cleft or channel that transmits the facial artery and possibly limits its compression during buccolabial musculature interplay. The cone shape is further modified by the presence of two blunt, round-edged, flanges that extend into the lateral free lip tissues above and below the lateral extremity of the oral fissure. These superomedially and inferomedially directed flanges (or *cornua*) are extensions of the body but more prominently the base of the modiolus. The tip of the superior cornu extends 0.5–1.5 cm medial to the buccal angle, whereas the tip of the inferior cornu only 0.3–0.5 cm. Thus, the modiolar base including the cornua, with its profile viewed from the mucosal aspect, is sometimes dubbed crescentic but is more reniform, with the buccal angle projecting towards the hilum. Another useful analogy is the resemblance of the whole modiolus and its attached muscular stems to a short conifer and its main branches. The muscles traced outwards from their pericentric modiolar stems have been picturesquely described as an array of fans, some almost closed and strap-like, others widely open and varying from sagittal, through degrees of obliquity, to almost coronal. In some, the 'rays' of the fan do not remain in a single plane but curve in conformity with buccal and labial tissue planes. On occasion muscles may be grouped and considered as forming a large compound fan. Thus, when well developed, zygomaticus major, risorius, platysma pars modiolaris and depressor anguli oris together have been likened to a wide hemicircular fan centred on the modiolus and extending peripherally from the zygomatic bone to the mental tubercle. Spiralization of muscle stems as they converge towards the modiolus is usual, as noted, but particularly clear examples are provided by depressor anguli oris and the quadrants of orbicularis oris. The peripheral attachment of the depressor, long and linear, lies in the plane of the body of the mandible whereas its bundles of modiolar attachment are to parts of an apicobasal column. Similarly the orbicular bundles in the free lip are arranged as a vertically curved but roughly coronal fan; its spiralized stem, however, thickens dorsoventrally to reach a series of modiolar attachments arranged from apex to base. Both muscles have undergone an overall spiral of at least 90° between their attachments. Detailed analysis shows that this general pattern is further complicated by the mutual spiralization of adjacent individual muscle bundles.

Modiolar muscles: classification. Some investigators have found it helpful to group the muscles in terms of their general geometric relation to the modiolus (with the face in repose, the teeth just short of full occlusion, the lips in light contact and the modiolus in its 'resting' or neutral position).

Cruciate modiolar muscles whose profiles, when viewed from the apical aspect, resemble a compound X, are: zygomaticus major, levator anguli oris, depressor anguli oris, platysma pars modiolaris.

Transverse modiolar muscles include buccinator, risorius, the various parts of orbicularis oris, incisivus superior and inferior.

Of the foregoing, some writers have segregated a few of these muscles as a separate group of *accessory modiolar muscles*; any sound logical basis for this is obscure.

Modiolar muscles: attachments. As intimated, a satisfactory complete three-dimensional analysis of modiolar attachments is not yet available. Heavy reliance has been placed on careful dissection unsupported by microreconstruction, and results are often presented as long, detailed verbal descriptions in which the terms used to denote relative positioning are sometimes open to misinterpretation. Some notes concerning the main conclusions will be given or reiterated here. A minority of muscles (risorius, platysma pars modiolaris, occasionally malaris and pars marginalis orbicularis oris) may be regarded as having a single zone of attachment each; the remainder have superficial and deep (or multiple) zones. Attachment involves interlacement of neighbouring bundles but most profusely with their direct physiological antagonists; thereafter the majority of fibres end within the modiolus. Occasional bundles may be traced from one muscle to another, although few in number, theoretically such modiolar strands could form tenuous links between any pair of muscles. They are not regarded as vectorially active generators of modiolar forces but merely as one structural feature resisting modiolar disruption during activity of its dominant, interlacing, terminating bundles.

Zygomaticus major inclines to the modiolus and divides into superficial and deep parts, the two being separated by the converging fibres of levator anguli oris, many of which interlace with ascending bundles of depressor anguli oris (at the 'angular complex'; i.e. the caninotriangular complex of earlier anatomists). The superficial fibres of zygomaticus major spread into the apex moduli where they interlace with the attachments of risorius, modiolar parts of platysma and malaris and with the underlying angular complex. Any or all of these muscles may interchange scattered fascicles apically and subapically. The deep fibres of zygomaticus cornu attach to the centre of the modiolar body and to the superior cornu of its base. The central fibres interlace with corresponding ones of buccinator and with the overlying angular fibres; interchange may occur with depressor anguli oris. Both the superficial and deep parts of zygomaticus major may send extensions into pars peripheralis orbicularis oris.

The modiolar attachments of orbicularis oris (marginal and peripheral parts), incisivus superior and inferior and the muscles forming the *apical rosette* have been mentioned above and in previous paragraphs. Levator anguli oris, as indicated, enters the modiolus between the superficial and deep parts of zygomaticus major where it is closely adherent to the superficial part. Thereafter its fibres radiate in many planes and intermesh with adjacent bundles. Many pass superficially and blend with depressor anguli oris. Deep bundles attach to the modiolar base inferior to the buccal angle. Some fascicles are continuous with orbicularis oris pars peripheralis inferior, others with buccinator. Most of its fibres end blindly in the modiolus. Depressor anguli oris also has superficial and deep bundles: the superficial interlace with all the muscles of the apical rosette; both superficial and deeper bundles interweave with levator anguli oris; basal fibres attach near the deep fibres of zygomaticus major. Some continuous fascicles pass into the territories of buccinator and orbicularis oris peripheralis, both superioris and inferioris. Buccinator blends with the whole lateral aspect of the basis moduli and continues to interlace with many of the muscles attached to the base. However, there is little convincing evidence of a significant number of buccinator fibres passing directly into the tissues of the lips. In contrast, as buccinator courses ventrally through the cheek and modiolar base there is sound evidence that substantial numbers of its fibres successively curve internally to punctuate attachments on the buccal and modiolar submucosa.

The panniculus fibrosus is adherent to the apex moduli and extends posteromedially as a thin sloping sheet to the *buccal angle* where its free border forms a crescentic, narrow, flexible, subcutaneous fibro-elastic cord that accommodates the varying postures of the modioli, lips and oral fissure and jaws.

Modiolar Movements and Shape Changes

The complex, controlled, three-dimensional mobility of the modioli enables them to provide functional loci for integrated activities of the cheeks, lips and oral fissure, the oral vestibule and the jaws bearing teeth. Such activities include: biting, chewing, drinking, sucking, swallowing, changes in vestibular contents and pressure, the innumerable subtle variations involved in speech, the modulation (and occasional generation) of musical tones, harsher sounds in shouting and screaming, crying, and the equally endless permutations of changes in facial expression ranging

from the merest 'hint' to gross distortion, either symmetrical or asymmetrical. Adequate consideration of many of these activities would merit individual monographs and cannot be attempted here. However, a few general points, sometimes ignored, will be indicated. During a complex action there is required simultaneous recording of length, direction and tensional force of numerous muscles (or parts of muscles) over an adequate time interval. Such data are not yet available. Theoretical predictions from an estimated 'average line of pull' *if* the muscle acted in isolation are of little value. The resultant of vectorial combinations of two or more such lines affords limited assistance in the absence of experimental verification. With major modiolar movements it appears likely that many if not all its associated muscles are involved to some extent. While the most obvious determinant of modiolar position and mobility is the state of balanced contraction of its directly attached muscles, another factor having a pronounced influence is the degree of separation or 'gape' between the upper and lower teeth. Starting from the occlusal position and maintaining the lips in contact, the teeth may be separated, near the midline by about 1.25 cm, and the mentolabial sulcus descends a similar distance. The lips now part and gape continues to its maximum: interlabial and interdental distances now approach 4.0 cm and the mentolabial sulcus has descended a further 2.0 cm. In this posture the modiolus has *descended* about 1.0 cm to lie over the interdental space and into which its basal and surrounding buccal mucosa projects a few millimetres. Thus positioned the modiolus is *immobile* and its cornua (and their attached stems of orbicularis oris peripheralis, superior and inferior) diverge into their respective lips at an obtuse angle to each other; the dispositions of the other modiolar muscles change accordingly. During mouth closure teeth glide over the buccolabial mucosa which is progressively retracted from their occlusal surfaces by buccinator and the other submucosal attachments mentioned above. The general shape of the hexagonal labial area is accordingly modified with progressive opening of the mouth and jaws. In maximal opening, its central vertical height (between superior and inferior boundaries) has increased by 3.0–3.5 cm; the transverse distance between its lateral angles is reduced by about 1.0 cm and the angles are obtuse; the nasolabial sulci are longer, straighter and more vertical; and the inferior buccolabial sulci are less deep and curved. These soft tissue changes radiate from the bilateral modiolar loci.

With the lips in contact and the teeth maintained in tight occlusion, modiolar mobility, although moderately free, is not maximal but restricted by a few millimetres in all directions. Maximal mobility is present when there is an interdental clearance of 2–3 mm. Movement may occur from the neutral 'resting' position to any location within the boundaries of its *envelope of movement* which approximates to half a wide, flattened ovoid (or spheroid). The apex of the modiolus (and its apicobasal axis) may move vertically upwards about 1.0 cm, downwards 0.5 cm, posterolaterally 1.0 cm and anteromedially 1.0 cm, these movements occurring in the curved planes of the cheek and lips. Specific movements may be of lesser extent or may occur along any oblique path between the orthogonal ones just mentioned. As indicated above, with wide mouth opening there occurs a mandatory increased lowering and interdental sinking, with fixation of the modiolus. From the neutral position the modiolus may be displaced superficially along its apicobasal axis for distances up to 0.5–1.0 cm by the presence of liquid or a solid bolus in the vestibule, or by increased air pressure in part or the whole vestibule 'ballooning' the cheek and lips.

In all movements (elevation, depression, retraction, protraction or any oblique intermediate combination) the modiolus with its cornua, the buccal angle and lip margins bordering the lateral segments of the oral fissure, maintain the same fundamental relationships. Lateral projection of a line that bisects the buccal angle (and therefore the intercornual angle) passes through the centre of the modiolus, intersecting its apicobasal axis at a right angle. Thus during elevation of the modiolus (and angle) the line of intersection inclines upwards and laterally and is accompanied by medial rotation of the modiolus around its apicobasal axis. Lowering of the modiolus is combined, similarly but to lesser degree, with lateral rotation.

Many activities involve: first, the dominance of a particular modiolar muscle group over its antagonists with rapid relocation of the modiolus; secondly the transient fixation of the modiolus in this new site by simultaneous contraction of principally the cruciate muscles; thirdly, acting from this fixed base the main physiological effectors, buccinator and orbicularis oris engage in their specific actions. Buccinator, having fixed posterior and anterior attachments and by variations in length and tension of its contractile bundles, can control pressure and volume changes in gaseous, liquid or solid contents in the oral vestibule; it can direct materials between grinding molars or compress the cheek against the teeth and gums.

Like buccinator, orbicularis oris may first initiate or permit movement of the modiolus to an advantageous location, where secondary fixation occurs, and from this momentarily static base orbicularis operates. These activities are commonly integrated with partial separation or closure of the jaws and with variable activity in the direct labial tractors. All these factors operate in determining the momentary dispositions of the lips and oral fissure. Modiolar movements and repositioning may be bilaterally symmetrical, unilateral (the other remaining neutral or static) or bilateral but in contrasting directions. Thus the modioli (and buccal angles) may be at their (simultaneous) posterolateral limits and the oral fissure at its greatest horizontal length; then activity in orbicularis oris causes compression between 'thin', stretched, smooth lips. In contrast, with modioli and the angles at their anteromedial limits, fissural length is minimal and orbicularis compresses 'thick' pursed i.e. wrinkled and protruded lips. Fine close wrinkles cross the ventral half of the (mainly lower) red-lip margin and prominent cutaneous wrinkles, more widely spaced, are vertical centrally but elsewhere decline laterally. Lip protrusion is passive in its initial stages and may be suppressed by powerful contraction of the whole of orbicularis; conversely protrusion may be increased by combined selective activity in parts of the direct labial tractors. The functional posterolateral or anteromedial extremes of the modioli and buccal angles mentioned in bilaterally symmetrical movements are not their absolute extremes; the latter are achieved in horizontal asymmetrical movements, e.g. both modioli (and angles) may move either in succession or simultaneously to the right. The ranges of anteromedial movement of the left modiolus and posterolateral movement of the right modiolus are now maximal. The oral fissure is markedly eccentric, the rimal ends of the philtrum with its ridges and the paired vertical sulci (originally paramedian) of the lower lip curve sharply to the right of the midline; and the effectiveness of lip compression is highest in the (original) left quadrants and lower in the right quadrants.

One modiolus may remain static or slightly lowered while the other is raised; other possible combinations and quantitative variations are endless. It is to be appreciated that in many expressive modifications the *movement* (e.g. modiolar translocation) is often more functionally significant than the *position* reached.

During the last 2.5–3.0 cm interincisal distances of wide jaw separation, strong contraction of orbicularis oris cannot effect lip contact but causes full-thickness inflection of upper and lower lips including their red-lip rims, around the incisal edges, canine cusps and premolar occlusal surfaces towards the oral cavity.

The multitude of temporospatial labial controls involved in speech, intonation and less patterned sounds are mentioned elsewhere (p. 1257) and cannot be pursued here. Nevertheless, views concerning some aspects of the actions involving orbicularis oris pars marginalis deserve comment. The disposition of pars marginalis, a sharply recurved hook in the upper lip and a progressive curved 'half' hook in the lower lip, the spiral relation of marginalis to the rimal bundle of peripheralis and their linear, equatorial, deep submucosal attachment to the basis modioli has been surveyed. Contraction of marginalis is considered to alter the sectional profile of the red-lip rim; both the gentle bulbous profile of the rimal part of the upper lip and the smooth posterosuperior convexity of the lower lip transform to a narrower blunt, truncated isosceles triangular one. The narrow, blunt 'apices' (i.e. rims) of variable, but delicately controlled length and tension have been evocatively named *labial cords*. Their formation in the production of some (labial) consonantal sounds has been

confirmed. A labial cord may also have a function as the so-called 'lip' of a closed organ pipe or as a frequency variable vibrating 'reed' in whistling or playing a wind instrument such as the trumpet.

Localized triangular elevation of a segment of the upper lip exemplifies partial contraction of the superior labial tractors (levators) resulting in a postural expression reminiscent of the 'canine snarl' of a number of mammals.

Masticatory Muscles

Directly concerned with mandibular movements in mastication (and speech) are the masseteric, temporal and pterygoid muscles; all are innervated by branches of the mandibular division of the trigeminal nerve. (Sharing this innervation, and in the first two instances less direct functional associations, are the mylohyoid and anterior belly of digastric, also the tensor veli palatini and the tensor tympani.) A layer of deep cervical fascia, the **parotid fascia**, covers masseter, which is firmly connected to it and then attached to the lower border of the zygomatic arch; it invests the parotid gland (p. 1291).

Masseter (5.34), a quadrilateral muscle, has three layers blended anteriorly. The *superficial layer* is largest and attached by a thick aponeurosis to the maxillary process of the zygomatic bone and anterior two-thirds of the inferior border of the zygomatic arch; it descends backwards to the mandible's angle and lower posterior half of the lateral surface of its ramus; intramuscular tendinous septa in this layer mark the ramus. The *middle layer* extends from the medial aspect of the anterior two-thirds of the zygomatic arch and lower border of its posterior third to the central part of the mandibular ramus. The *deep layer* extends from the deep surface of the zygomatic arch to the upper part of the mandibular ramus and its coronoid process. Middle and deep layers together constitute the *deep part* in Nomina Anatomica (MacDougall 1955); they form a *cruciate* muscle (p. 565). The muscle is easily palpable. MacConaill (1975) described superficial fibres as continuous at the lower mandibular border, with medial pterygoid fibres. For more complex accounts of lamination consult Schumacher (1961) and Yoshikawa & Suzuki (1969).

Relations: *superficial* are skin, platysma, risorius, zygomaticus major and parotid gland; the parotid duct, facial nerve and transverse facial vessels cross the muscle; *medial* are temporalis and the mandibular ramus. Fat separates it anteriorly from buccinator and the buccal nerve. The masseteric nerve and artery reach its deep surface via the mandibular incisure. The *posterior margin* is overlapped by the parotid gland; the *anterior margin* projects over buccinator and is crossed below by the facial vein.

Nerve supply: a branch of the anterior trunk of the mandibular nerve.

Actions: occlusion of teeth in mastication. Electrical activity with the mandible at rest is minimal. Masseter has a small effect in side-to-side movements, protraction and retraction. An analysis of its fibre numbers, diameters and distribution has been made by Rinqvist (1974).

The temporal fascia, covering temporalis, is a strong aponeurosis, overlapped by auriculares anterior and superior epicranial aponeurosis and part of orbicularis oculi. Superficial temporal vessels and the auriculotemporal nerve ascend over it. Above, it is single and attached to the whole superior temporal line; below, it has two layers attached to the lateral and medial margins of the upper border of the zygomatic arch. Some fat, the zygomatic branch of the superficial temporal artery and zygomaticotemporal branch of the maxillary nerve are between them. To its deep surface superficial fibres of temporalis are attached.

The temporalis (5.27) extends like a fan from most of the temporal fossa (except its zygomatic part), and from the deep surface of the temporal fascia. Its fibres converge and descend into a tendon that passes between the zygomatic arch and cranial wall and is then attached to the medial surface, apex, anterior and posterior borders of the coronoid process, and anterior border of the mandibular ramus almost to the third molar tooth. The anterior fibres of temporalis descend vertically; traced posteriorly th

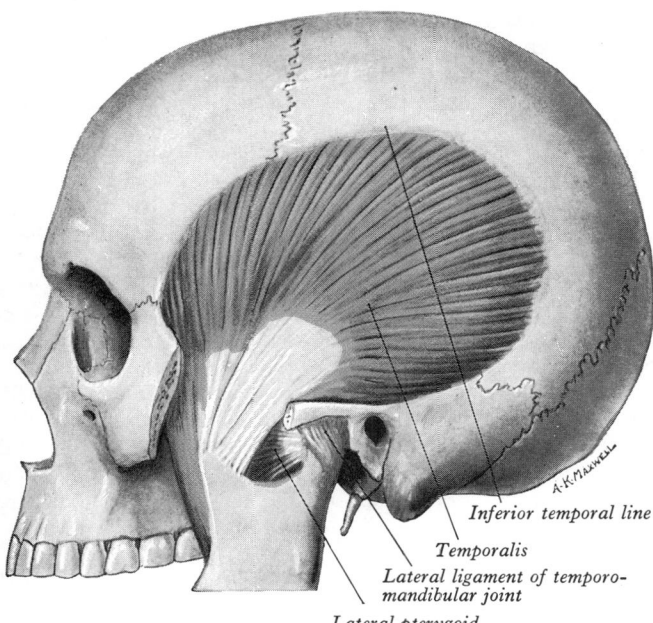

5.27 Left temporalis: the zygomatic arch and masseter have been removed. Note the changing obliquity of muscle fibres vertical anteriorly, horizontal posteriorly.

fibres are increasingly oblique, while the most posterior fibres are almost horizontal.

Relations: *superficial* are skin, auriculares anterior and superior, temporal fascia, superficial temporal vessels, auriculo-temporal nerve, temporal branches of facial nerve, zygomaticotemporal nerve, epicranial aponeurosis, zygomatic arch, and masseter; *medial* are the temporal fossa, lateral pterygoid, superficial head of medial pterygoid, a small part of buccinator, the maxillary artery and its deep temporal branches, deep temporal nerves and buccal nerve and vessels. *Behind* the tendon masseteric vessels and nerve traverse the mandibular incisure. Fat separates its *anterior border* from the zygomatic bone.

Nerve supply: deep temporal branches of the anterior trunk of the mandibular nerve.

Actions: elevation of mandible to close the mouth and approximate the teeth. This requires elevation by anterior fibres and retraction by posterior fibres, since the mandibular condylar head is on the articular eminence when the mouth is open. Posterior fibres retract the mandible from protrusion. The muscle is concerned in lateral grinding movements. Despite much electromyography, little can be added in further analysis; much unresolved disagreement exists between observers. Vitti & Basmajian (1977) suggest that temporalis is active in forcible elevation but not in slow elevation without occlusion: otherwise they confirm the above description. The contracting muscle is easily palpable despite the temporal fascia.

The lateral pterygoid (5.28) is a short, thick muscle with an upper *head* from the infratemporal surface and crest of the greater wing of the sphenoid, and a lower *head* from the lateral surface of the lateral pterygoid plate. It passes posterolaterally to a depression on the anterior aspect of the mandibular neck and to the articular capsule and disc of the temporomandibular joint.

Early in the third month in utero the muscle is attached to mesenchyme condensing round the developing mandibular condyle; part of its tendon sweeps back above the condyle to the part of Meckel's cartilage forming the head of the malleus (Harpman & Woollard 1938). This part of the tendon is absorbed into the articular disc; attachment to the malleus is lost (Rees 1954).

Relations: *superficial* are the mandibular ramus, the maxillary artery crossing deep or superficial to the muscle, the tendon of temporalis and masseter; *medial* are the upper part of the medial pterygoid, sphenomandibular ligament, middle meningeal artery and mandibular nerve. The *upper border* is related to the temporal and masseteric branches of the mandibular nerve; the *lower border* to the lingual and inferior alveolar nerves. The buccal nerve and

maxillary artery are between the muscle's two parts (5.29).

Nerve supply is from the mandibular nerve's anterior trunk.

Action. The muscle aids opening the mouth by protracting the mandibular condyle and articular disc while the mandibular head rotates on the disc (Posselt 1952). In closure backward gliding of the disc and condyle is controlled by slow elongation of the lateral

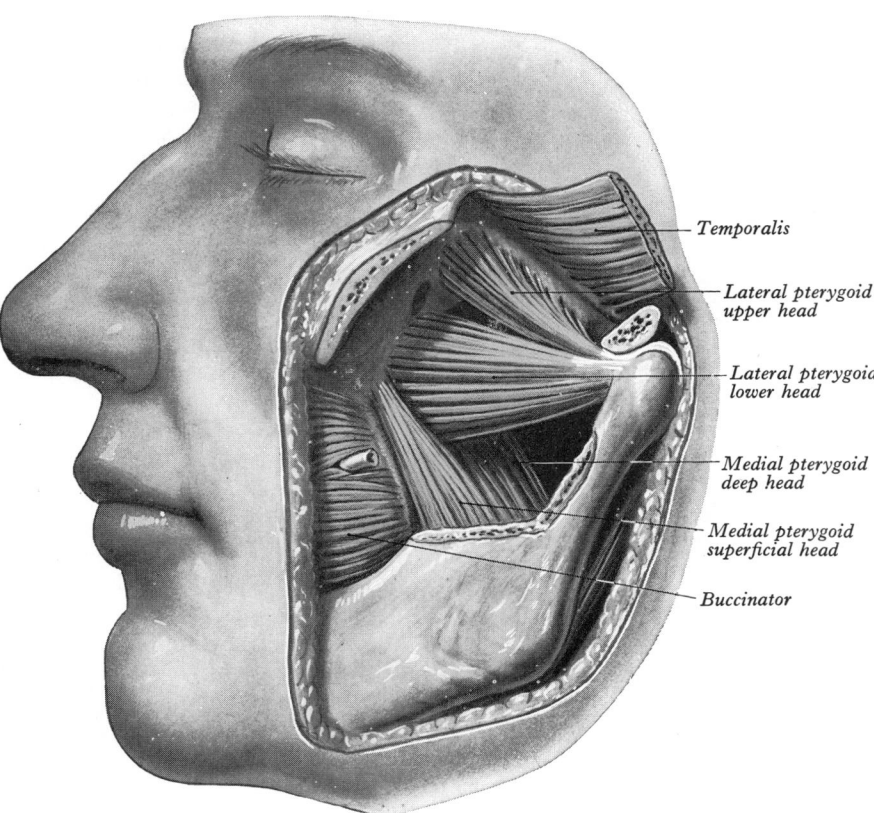

5.28 Left pterygoid muscles: the zygomatic arch and part of the ramus of the mandible have been removed.

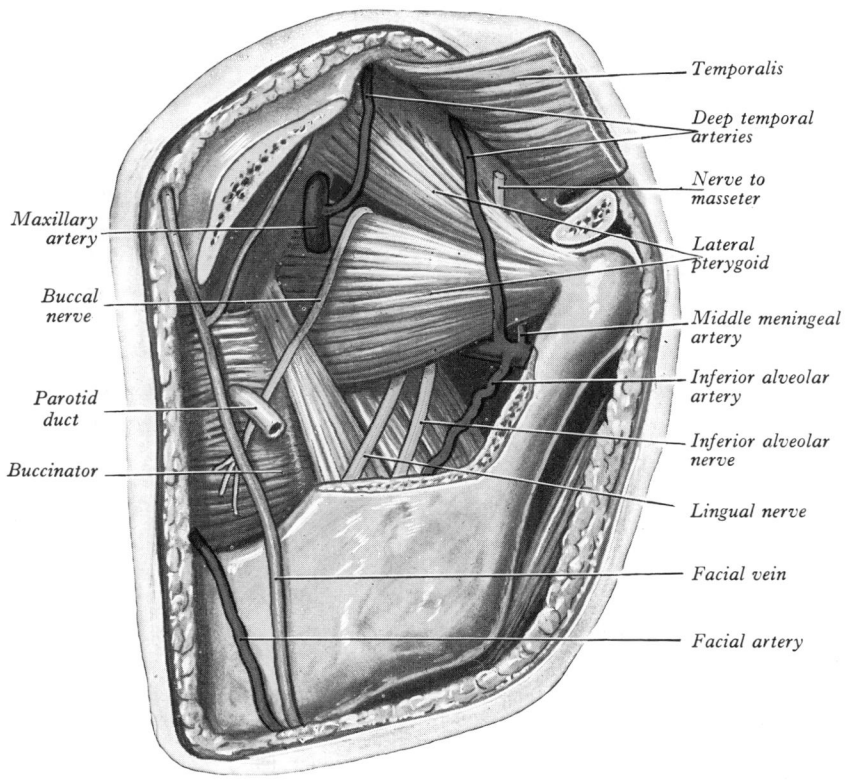

5.29 Structures related to the left pterygoid muscles.

pterygoid and by masseter and temporalis, restoring the teeth to occlusion. Acting with the ipsilateral medial pterygoid the lateral advances its condyle, the mandible rotating on a vertical axis through the opposite condyle. Both medial and lateral pterygoids together protrude the mandible. Different actions have been ascribed to the two parts of the lateral pterygoid, the superior head being involved in chewing, the inferior in protrusion. Electromyographic records in rhesus monkeys (Macnamara 1972) support this. For a review consult Grant (1973).

The medial pterygoid (5.28), a thick quadrilateral muscle, is attached *deeply* to the medial aspect of the lateral pterygoid plate and grooved surface of the palatine's pyramidal process and, more *superficially*, to the lateral surfaces of the maxillary pyramidal process and the maxillary tuberosity. It descends posterolaterally to a strong tendinous lamina attached postero-inferiorly to the medial surfaces of the mandibular ramus and angle, as far as the mandibular foramen and almost to the mylohyoid groove (3.81). This area is rugged, due to tendinous fasciculi in the attachment.

Relations. *Lateral* is the mandibular ramus, from which the muscle is separated by the lateral pterygoid, sphenomandibular

ligament, maxillary artery, inferior alveolar vessels and nerve, lingual nerve and a process of the parotid gland. The *medial surface* is related to tensor veli palatini, and separated from the superior constrictor by styloglossus, stylopharyngeus and areolar tissue.

Nerve supply is from the mandibular nerve. Schumacher et al (1976) have investigated nerves of masticatory muscles in detail. They observed a similar mode of branching in all of them.

Actions. The medial pterygoids assist in elevating the mandible; with the lateral pterygoids they protrude it. The pterygoid muscles of one side together rotate the mandible forwards and to the opposite side, with the opposite mandibular head as a vertical axis (p. 487); alternation of both sets of muscles, in side-to-side movements, are used to triturate food. For further analysis see pp. 487–488.

The pterygospinous ligament, sometimes partially muscular, stretches between the sphenoidal spine and posterior border of the lateral pterygoid plate near its superior end. Sometimes ossified, it then completes a foramen for branches of the mandibular nerve to the temporalis, masseter and lateral pterygoid.

ANTEROLATERAL MUSCLES AND FASCIAE OF THE NECK

Anterolateral muscles of the neck may be grouped as follows:
1. Superficial and lateral cervical. 3. Anterior vertebral.
2. Suprahyoid and infrahyoid.   4. Lateral vertebral.

Superficial cervical fascia, usually a thin lamina covering platysma and scarcely a separate layer, may contain considerable adipose tissue, usually more so in females. Like all superficial fascia it is not a separate stratum but a zone of loose connective tissue between the dermis and deep fascia, continuous with both.

Superficial lamina of deep cervical fascia

Sternohyoid — Sternothyroid — Thyroid gland

Thyroid cartilage
Arytenoid cartilage
Pretracheal lamina
Inferior pharyngeal constrictor
Prevertebral fascia
Carotid sheath
Longus colli
Fifth cervical vertebra
Articular process

Multifidus
Semispinalis cervicis

Sternocleidomastoid
External jugular vein
Common carotid artery
Internal jugular vein
Vagus nerve
Sympathetic trunk
Scalenus anterior
Fifth cervical spinal nerve
Vertebral vessels
Scalenus medius
Iliocostalis cervicis
Levator scapulae
Splenius cervicis
Longissimus capitis and cervicis
Accessory nerve

Trapezius

Semispinalis capitis   Splenius capitis   Deep cervical vessels

5.30 Transverse section through the left half of the neck to show the arrangement of the deep cervical fascia. (Specimen provided by Professor R E M Bowden, Royal Free Hospital School of Medicine.)

Deep cervical fascia (5.30), internal to platysma and investing the nuchal muscles, is fibro-areolar tissue between muscles, viscera, vessels, etc. It forms certain well-defined fibrous sheets; elsewhere it is loosely arranged but condensed around blood vessels as fibrous sheaths which here, as elsewhere, enclose arteries and accompanying veins. Its *superficial (investing) lamina* is continuous with the ligamentum nuchae and periosteum of the seventh cervical spine. It thinly covers trapezius and continues from its anterior border as a loose areolar layer over the posterior triangle to the posterior border of sternocleidomastoid, where it becomes denser. It encloses this muscle and at its anterior margin unites as a single lamina over the anterior triangle to the midline and so to the opposite side; it adheres to the symphysis menti and the body of the hyoid bone.

Deep fascia is fused with periosteum along the superior nuchal line, the mastoid process and the whole mandibular base. Between the mandibular angle and anterior edge of the sterno-cleidomastoid it is very dense. Between the mandible and mastoid process it ensheathes the parotid gland; superficial to the gland it ascends as *parotid fascia* to the zygomatic arch; from the deep parotid layer a strong *stylomandibular ligament* (p. 486) ascends to the styloid process. Inferiorly, deep fascia is attached to the acromion, clavicle and manubrium sterni, fusing with their periostea. A little above the manubrium it splits into superficial and deep layers attached to the anterior and posterior manubrial borders and the interclavicular ligament. The slit-like *suprasternal 'space'* between the layers contain areolar tissue, lower parts of anterior jugular veins and the jugular venous arch, the sternal heads of the sternocleidomastoid muscles and sometimes a lymph node. Over the posterior triangle, inferiorly between trapezius and sternocleidomastoid, the deep fascia also has superficial and deep layers, the former attached to the superior clavicular border, the deep layer surrounding the inferior belly of omohyoid and, deep to the sternocleidomastoid, its intermediate tendon, finally blending with fascia around subclavius and the periosteum on the posterior aspect of both the clavicle and anterior end of the first rib.

The *carotid sheath*, a condensation of cervical fascia, encloses the common and internal carotid arteries, internal jugular vein, vagus nerve and constituents of the ansa cervicalis. Thicker around the arteries, it is peripherally continuous with adjacent loose areolar tissue (5.31).

The *prevertebral lamina* covers the anterior vertebral muscles, extending laterally on scaleni anterior and medius and levator scapulae as a fascial floor of the posterior triangle. The subclavian artery and brachial nerves, emerging from *behind* scalenus anterior, carry prevertebral fascia inferolaterally behind the clavicle as the *axillary sheath*. Laterally, the lamina is thin and areolar, ceasing to be a definite layer under trapezius. Superiorly it is attached to the cranial base and inferiorly descends in front of longus colli into the superior mediastinum, blending with the anterior longitudinal ligament. Anteriorly the prevertebral lamina is separated from the pharynx and buccopharyngeal fascia by an areolar zone, the so-called *retropharyngeal space*; laterally this tissue connects the lamina to the carotid sheath and fascia on the deep aspect of sternocleidomastoid. Ventral rami of cervical spinal nerves are at first behind the prevertebral lamina; nerves to the rhomboids and serratus anterior and phrenic nerve retain this position in the neck but the accessory nerve is superficial (p. 1119).

The *pretracheal lamina* is very thin but ensheathes the thyroid gland, is attached to the arch of the cricoid cartilage and continues into the superior mediastinum with the inferior thyroid veins.

Applied Anatomy. The superficial lamina of deep cervical fascia opposes the superficial extension of abscesses, and pus tends to extend laterally. In the anterior triangle it may reach the mediastinum, anterior to the pretracheal lamina; but due to the thinness of the fascia here it more often 'points' above the sternum. Pus behind the prevertebral lamina may extend laterally into the posterior triangle or may perforate the lamina and buccopharyngeal fascia, bulging into the pharynx as a retro-pharyngeal abscess.

1. SUPERFICIAL AND LATERAL CERVICAL MUSCLES

These include platysma, trapezius and sternocleidomastoid.

Platysma (5.21) is a broad sheet spreading from its fascial attachments over the upper parts of pectoralis major and deltoid and ascending medially across the clavicle to the side of the neck. Anterior fibres interlace across the midline inferoposterior to symphysis menti. Intermediate fibres are attached to the lower border of the mandibular body or pass upwards and medially deep to depressor anguli oris to their attachments in the lateral half of the lower lip (*pars labialis*, see p. 577). The posterior fibres cross the mandible and anterolateral part of the masseter to the skin and subcutaneous tissue in the lower face, blending with muscles at the modiolus near the buccal angle. Deep to the platysma the external jugular vein descends from the mandibular angle to the mid-clavicular point. The muscle varies much in extent and may even be absent. Slips from the mastoid process, occipital bone or fascia over the upper trapezius may join it (occipitalis minor).

Nerve supply is from the cervical branch of the facial nerve.

Actions. Platysma wrinkles the nuchal skin obliquely, diminishing the concavity between jaw and neck. Its anterior, thickest part may assist mandibular depressors. By drawing down the lower lip and buccal angle it helps to express horror and surprise. It is electromyographically active in sudden deep inspiration (de Sousa 1964) and often very much contracted in sudden, violent efforts.

Trapezius is described on p. 608 with scapular musculature.

The sternocleidomastoid (5.32) descends obliquely across the neck (5.33), and is a prominent landmark when contracted. It is thick and narrow centrally and broader and thinner at each end; inferiorly it has two 'heads'. The *medial (sternal) head*, a rounded tendinous fasciculus, is attached to the upper anterior surface of manubrium sterni and ascends posterolaterally. The *lateral (clavicular) head*, variable in width and composed of muscular and fibrous fasciculi, ascends almost vertically from the medial third of the clavicle's superior surface. The two parts are separated near their attachment by a triangular interval which corresponds to a surface depression, the *lesser supraclavicular fossa*. As they ascend the clavicular part spirals behind the sternal, soon blending with its deep surface to form a thick, rounded mass. Superiorly this is attached by a strong tendon to the mastoid process's lateral aspect, from its apex to its superior border, and by a thin aponeurosis to the lateral half of the superior nuchal line. Clavicular fibres mainly reach the mastoid process; sternal fibres are more oblique, superficial and reach the occiput. Traction by the two heads therefore differs; the muscle is 'cruciate' and slightly 'spiralized' (p. 566).

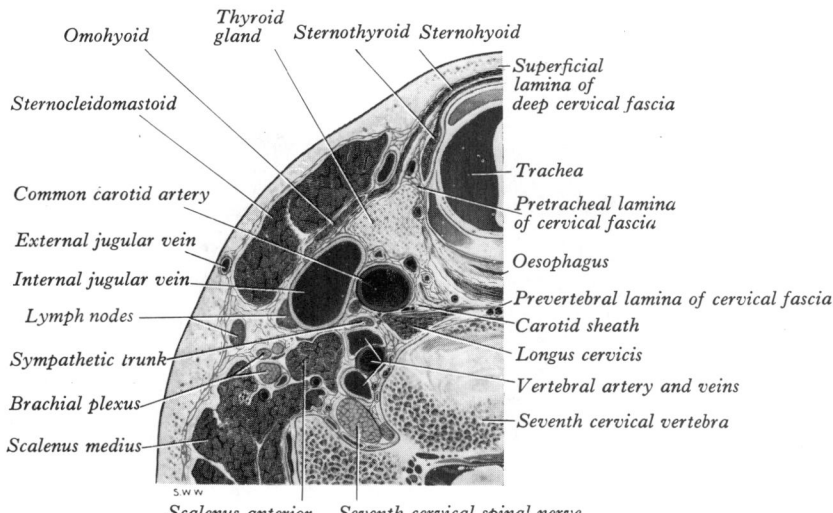

5.31 Part of a transverse section through the lower part of the neck at the level of the seventh cervical vertebra to show the arrangement of the deep cervical fascia. (Specimen provided by R E M Bowden.)

This muscle divides the anterolateral aspect of the neck into principal *anterior* and *posterior triangles*. Boundaries of the *anterior cervical triangle* are the median ventral line, the base of the mandible and a line from its angle to the mastoid process and the anterior sternocleidomastoid border; its apex is at the superior sternal border. Boundaries of the *posterior cervical triangle* are the posterior border of sternocleidomastoid, the middle third of the clavicle's superior surface and the anterior margin of trapezius; its apex is where sternocleidomastoid and trapezius approach or meet at the occiput. The subdivisions and contents of these major triangles are described on pp. 741–743.

Relations. *Superficial* are skin and platysma and deep to these the external jugular vein, great auricular and transverse cervical nerves and superficial lamina of the deep cervical fascia. The muscle is overlapped by a small part of the parotid gland. *Deep* are the sternoclavicular joint, sternohyoid, sternothyroid, omohyoid and anterior jugular vein crossing superficial to the infrahyoid

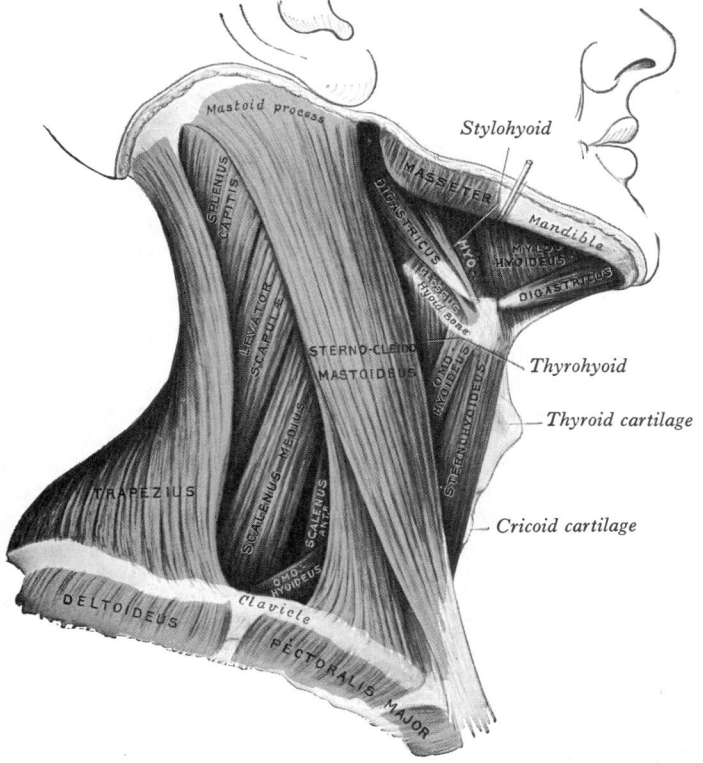

5.32 Muscles of the neck: right lateral aspect.

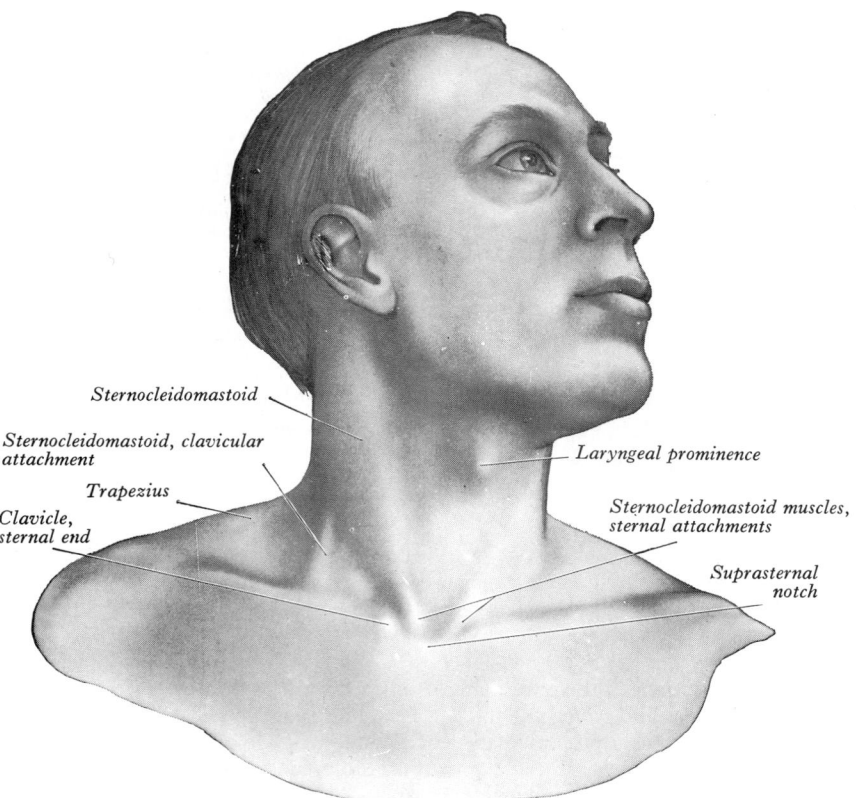

Sternocleidomastoid

Sternocleidomastoid, clavicular attachment

Trapezius

Clavicle, sternal end

Laryngeal prominence

Sternocleidomastoid muscles, sternal attachments

Suprasternal notch

5.33 Surface landmarks of the neck.

muscles and with the carotid sheath and subclavian artery deep to them. Between the omohyoid and the posterior belly of the digastric sternocleidomastoid is superfical to the common, internal and external carotid arteries, the internal jugular, facial and lingual veins, the deep cervical lymph nodes, vagus nerve and rami of ansa cervicalis. The sternocleidomastoid branch of the superior thyroid artery crosses deep to the muscle at the upper border of omohyoid. The posterior part of the sternocleidomastoid is related internally to splenius, levator scapulae and the scaleni, the cervical plexus, upper part of brachial plexus, phrenic nerve, and transverse cervical and suprascapular arteries. The occipital artery crosses deep to the muscle near the lower border of the digastric, where the accessory nerve, piercing the muscle, descends laterally deep to it. Superiorly, the muscle is superficial to the mastoid process, splenius, longissimus capitis and the posterior belly of digastric.

Nerve supply is the accessory nerve, which usually traverses the muscle and branches of ventral rami of the second, third and sometimes fourth cervical spinal nerves; although these cervical rami were long considered solely proprioceptive, clinical evidence suggests that some of these cervical fibres are motor (see Fitzgerald et al 1982).

Actions. Acting alone, the muscle tilts the head towards the ipsilateral shoulder, turning the face to the opposite side, as in an upward, sideways glance. But level rotation, a much commoner visual movement, is probably the most frequent use of the sternocleidomastoids. Together they also project the head forwards, assisting the longi colli in cervical flexion, a common movement in feeding. They also raise the head from the supine position. With the head fixed, they aid thoracic elevation in inspiration. Electromyography (e.g. de Sousa 1973) suggests that sternal fibres are more active in contralateral rotation but that both parts are involved in all the above movements; the muscle may also, apparently, be involved in cervical *extension*.

Applied Anatomy. Torticollis is due to permanent contracture of sternocleidomastoid. Spasmodic torticollis, in adults, begins with tonic or clonic spasm of one muscle, followed by spasm of trapezius. Such abnormal conditions illustrate the isolated muscle's action, but this is a caricature of its ordinary activities; these are invariably modified by synergists and antagonists, such

584

as splenius capitis (p. 587) and others.

2 (A). THE SUPRAHYOID MUSCLES

The *suprahyoid* group include digastric, stylohyoid, mylohyoid and geniohyoid muscles.

Digastric (5.32,34,35) has two bellies and an intermediate tendon. Inferior to the mandible, it angles from the mastoid process to the chin. Its longer, *posterior belly* is attached in the temporal mastoid notch and slopes down and forwards; the *anterior belly* is attached to the mandible's digastric fossa near the midline and inclines backwards. The two parts meet in a rounded tendon, perforating the stylohyoid and held to the hyoid's body and greater cornu by a fibrous loop, sometimes lined by a synovial sheath. An aponeurosis extends from the tendon to the hyoid body and greater cornu. The tendon may be absent and the muscle is then attached midway along the mandibular body. Its posterior belly may be augmented by a slip from the styloid process or be wholly attached to it. The anterior belly may partly cross the midline and is sometimes partially fused with mylohyoid.

Relations. *Superficial* yoidare platysma, sternocleidomastoid, part of splenius, longissimus capitis, the mastoid process, stylohyoid, retromandibular vein and parotid and submandibular salivary glands. *Medial* to the anterior belly is mylohyoid; medial to the posterior belly are superior oblique, rectus capitis lateralis, the transverse process of atlas, accessory nerve, internal jugular vein, occipital artery, hypoglossal nerve, internal and external carotid, facial and lingual arteries and hyoglossus (5.34).

Nerve supply: the anterior belly by the mylohyoid branch of the inferior alveolar nerve; posterior belly by the facial nerve. These supplies accord with derivations from the mesenchyme of the first and second brachial arch (p. 172). Lennartsson (1979) has described proprioceptors.

Actions: mandibular depression and hyoid elevation. Electromyography indicates that the muscles always act together and are secondary to the lateral pterygoids in mandibular depression, assisting maximal depression (Moyers 1950); the posterior bellies are active in swallowing and chewing.

Stylohyoid (5.34,35) is attached by a tendon to the posterior aspect of the styloid process near its base, descending forwards to the hyoid body at its junction with the greater cornu, above the omohyoid, where it is perforated by the digastric tendon. Occasionally the muscle is absent or double; it is medial to the external carotid artery; it may end in digastric or in supra- or infrahyoid muscles.

Nerve supply: a branch of the facial nerve.

Actions. Stylohyoid elevates and retracts the hyoid bone, elongating the buccal floor. With other hyoid muscles it fixes the hyoid bone, the basis for lingual muscles attached to it. Its roles in masticatory and vocal movements have not been adequately analysed (p. 1273).

The stylohyoid ligament, a fibrous cord, extends from the tip of the styloid process to the lesser hyoid cornu. The highest fibres of the middle pharyngeal constrictor are attached to it. It is intimately related to oropharynx (5.25). Below, it is overlapped by hyoglossus. Derived from the cartilage of the second branchial arch (p. 171), it may be partially ossified; in many mammals it is a distinct *epihyal* bone.

The mylohyoid (5.32,35), superior to the anterior digastric belly, forms with its fellow a muscular floor for the oral cavity. It is flat, triangular and attached along the mandible's whole mylohyoid line. Posterior fibres pass medially and slightly down to the hyoid body's front, near its lower border. Middle and anterior fibres from each side decussate in a median fibrous raphe stretching from symphysis menti to the hyoid bone; but this may be absent, the muscles then being a continuous sheet. Sometimes mylohyoid is fused with the anterior belly of digastric.

Relations. The *inferior (superficial) surface* is related to the platysma, anterior digastric belly, superficial part of the submandibular gland, facial and submental vessels, mylohyoid vessels and nerve. The *superior (internal) surface* is related to the geniohyoid, part of hyoglossus, styloglossus, hypoglossal and lingual nerves, the submandibular ganglion, sublingual gland, the deep part of the submandibular gland and duct, lingual an

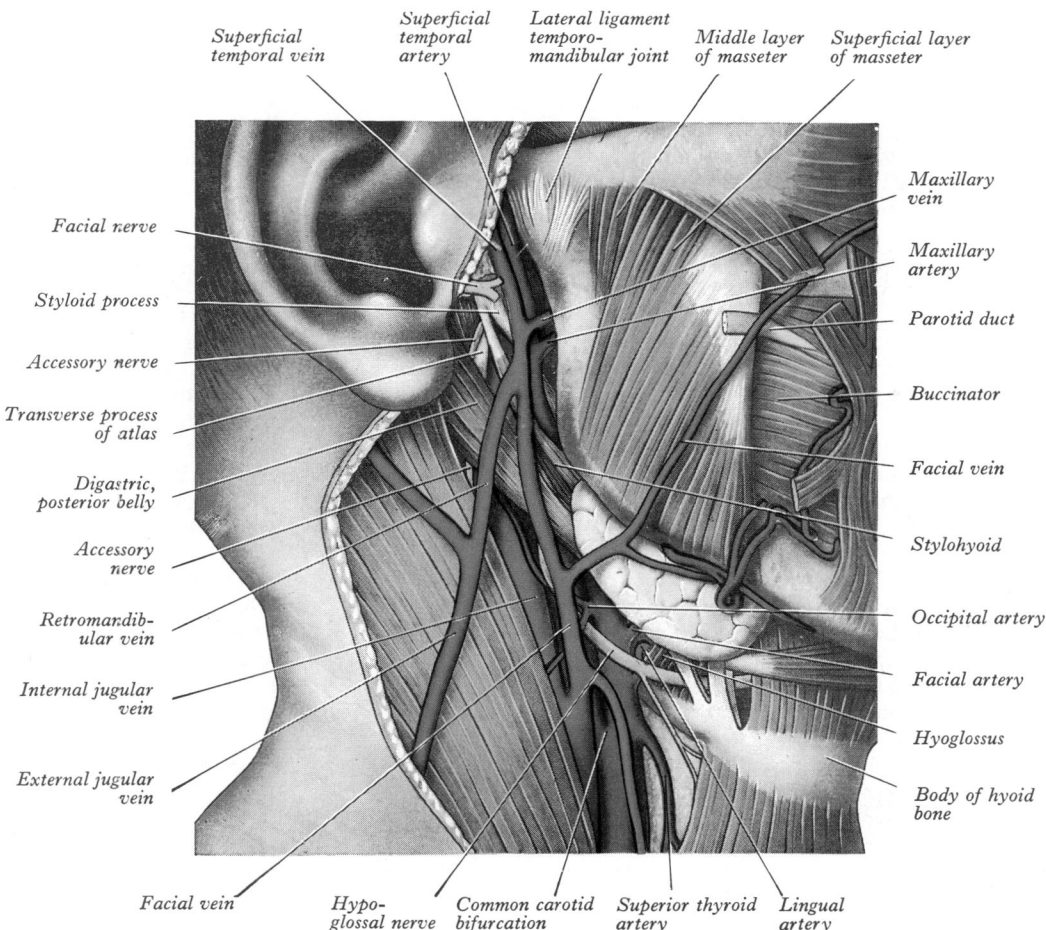

Superficial temporal vein — Superficial temporal artery — Lateral ligament temporo-mandibular joint — Middle layer of masseter — Superficial layer of masseter

Facial nerve

Styloid process

Accessory nerve

Transverse process of atlas

Digastric, posterior belly

Accessory nerve

Retromandibular vein

Internal jugular vein

External jugular vein

Maxillary vein

Maxillary artery

Parotid duct

Buccinator

Facial vein

Stylohyoid

Occipital artery

Facial artery

Hyoglossus

Body of hyoid bone

Facial vein — Hypoglossal nerve — Common carotid bifurcation — Superior thyroid artery — Lingual artery

5.34 Relations of many faciocervical structures, exposed by the removal of only skin, fasciae, parotid glands and cutaneous branches of the cervical plexus. Focal regions to be noted are the carotid triangle and some of its principal contents and the styloid process in the floor of the mandibulo-mastoid gap. The labial modiolus is omitted and details of labial musculature simplified.

sublingual vessels and, posteriorly, the oral mucous membrane. A hiatus in the muscle, through which a process of sublingual gland protrudes, was noted in one-third of cases by Gaughran (1963).

Nerve supply: mylohyoid branch of the inferior alveolar nerve.

Actions: elevation of the oral floor in the first stage of deglutition (Whillis 1946). It may also elevate the hyoid bone or depress the mandible.

The geniohyoid (5.35), a narrow muscle above the medial part of mylohyoid, extends from the inferior mental spine behind symphysis menti to the anterior aspect of the hyoid body; it adjoins its fellow and may fuse with it or genioglossus.

Nerve supply: the first cervical spinal nerve, through the hypoglossal nerve (p. 1119).

Actions. Geniohyoid elevates the hyoid bone and draws it forwards as a partial antagonist to stylohyoid. When the hyoid is fixed, it depresses the mandible.

2 (B). THE INFRAHYOID MUSCLES

The *infrahyoid* group includes sternohyoid, sternothyroid, thyrohyoid and omohyoid muscles. As antagonists to the suprahyoid group they depress the hyoid bone but may act as fixators or co-operate in cyclic hyoid movements.

Sternohyoid (5.32,35), a thin, narrow strap, extends from the posterior aspect of the clavicle's medial end, the posterior sternoclavicular ligament and the upper posterior aspect of manubrium sterni, ascending medially to the inferior border of the hyoid body. Below, it is separated from its fellow but the two usually meet at midpoint and are contiguous above this. The muscle may be absent, double, augmented by a clavicular slip (cleidohyoid) or interrupted by a tendon.

Nerve supply: ansa cervicalis (c.1,2,3).

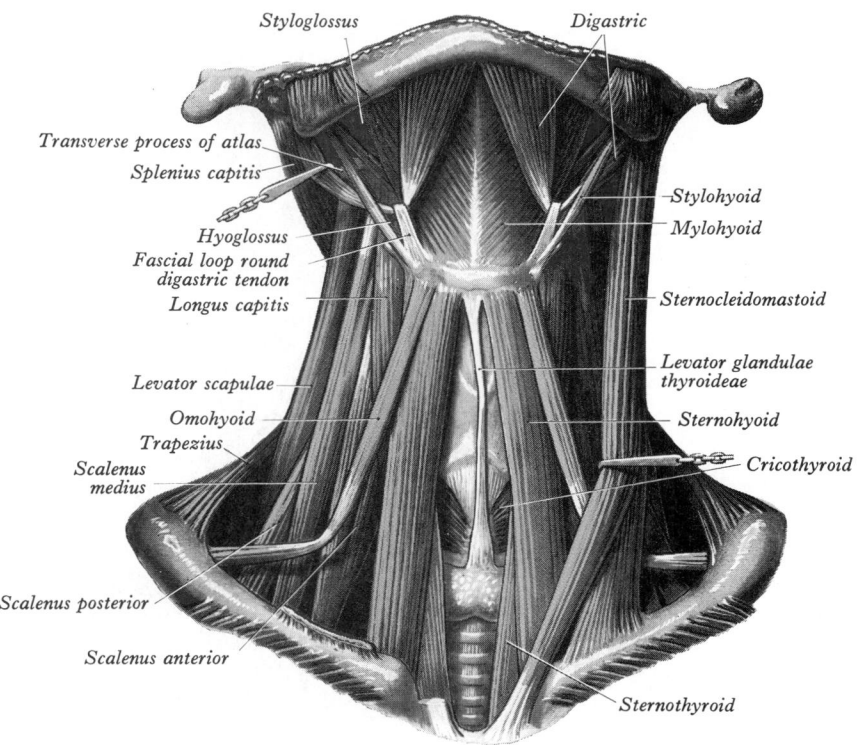

Styloglossus — Digastric

Transverse process of atlas — Splenius capitis — Hyoglossus — Fascial loop round digastric tendon — Longus capitis

Levator scapulae — Omohyoid — Trapezius — Scalenus medius

Scalenus posterior

Scalenus anterior

Stylohyoid — Mylohyoid — Sternocleidomastoid — Levator glandulae thyroideae — Sternohyoid — Cricothyroid — Sternothyroid

5.35 Muscle of the front of the neck. On the right side the sternocleidomastoid has been removed. In this subject, the origin of the scalenus medius extended up to the transverse process of the atlas.

Actions: hyoid depression after elevation in deglutition, speech and mastication.

Sternothyroid (5.32,35), shorter and wider than sternohyoid, is deep and partly medial to it. It is attached below to the posterior manubrial surface inferior to the sternohyoid and to the posterior edge of the first costal cartilage; and above to the oblique line on the thyroid lamina. Inferiorly it is in contact with its fellow but diverges as it ascends. It is anterolateral to the thyroid gland.

Nerve supply: from ansa cervicalis (c.1,2,3).

Actions: depression of the elevated larynx after swallowing or in vocal movements.

Thyrohyoid, a small quadrilateral muscle, appears as an upward continuation of the sternothyroid. From the thyroid lamina's oblique line it ascends to lower border of the greater cornu and adjacent hyoid body.

Nerve supply: a branch of the hypoglossal nerve. Like the branch to geniohyoid this contains fibres from the first cervical spinal nerve, which supply the muscle.

Actions: hyoid depression or elevation of the larynx, or varying mixtures of both.

Omohyoid (5.32,35) has two bellies united at an angle by an intermediate tendon. Its inferior belly, a flat narrow band, extends from the upper scapular border, near the scapular notch, and occasionally the superior transverse scapular ligament (p. 501), inclining forwards and slightly up across the lower neck to pass deep to the sternocleidomastoid as the intermediate tendon. It divides the posterior triangle into upper and lower parts (p. 742). The *superior belly* ascends almost vertically from this tendon, near the lateral border of sternohyoid to the lower border of the hyoid body, lateral to sternohyoid's attachment. The variable intermediate tendon is usually near the internal jugular vein, opposite the cricoid arch. It is ensheathed by a band of deep cervical fascia that descends to the clavicle and first rib, maintains the angulation between its bellies and may contain some striated muscle. Either belly may be absent or double; the inferior may be attached to the clavicle; the superior is sometimes fused with the

sternohyoid (Buntine 1970).

Nerve supply. Superior and inferior bellies are supplied respectively by branches from the superior ramus of the ansa cervicalis (c.1) and the ansa itself (c.2,3).

Actions: hyoid depression in prolonged inspiratory efforts; omohyoid may also tense the lower deep cervical fascia, lessening the inward suction of soft parts. Electromyographic data for hyoid musculature are sparse.

3. THE ANTERIOR VERTEBRAL MUSCLES

This group includes longi colli and capitis and recti capitis anterior and lateralis, all partly flexors of head and neck (5.36).

M. longus colli, placed anterior to the vertebral column, extends between the atlas and the third thoracic vertebra. Divisible into inferior oblique, superior oblique and vertical parts, all are attached by tendinous slips. The *inferior oblique part* is smallest and ascends laterally from the fronts of the first two or three thoracic vertebral bodies to the anterior tubercles of the fifth and sixth cervical transverse processes. The *superior oblique part* ascends medially from the anterior tubercles of third to fifth cervical transverse processes to a narrow tendon attached to the anterolateral surface of the atlantal anterior tubercle. The *vertical*, intermediate part ascends from anterior aspects of the upper three thoracic and lower three cervical vertebral bodies to those of the second to fourth cervical.

Nerve supply: ventral rami of second to sixth cervical vertebrae spinal nerves.

Actions: forward flexion of the neck acting as a whole; additionally the oblique parts flex it laterally and the inferior oblique rotates it to the opposite side. Electromyography (Fountain et al 1966) largely confirms the above, except for lateral flexion. The main antagonist is longissimus cervicis.

M. longus capitis is broad and thick above, narrow below, where tendinous slips are attached to the *transverse processes* of the third to sixth cervical vertebrae at their anterior tubercles, extending above to the inferior surface of the basilar occipital bone.

Nerve supply: ventral rami of the first to third cervical spinal nerves.

Action: flexion of the head.

Anterior vertebral muscles vary chiefly in numbers of vertebral slips.

M. rectus capitis anterior, a short, flat muscle behind the upper part of longus capitis, ascends from the anterior surface of the lateral atlantal mass and root of its transverse process almost vertically to the inferior surface of the basilar occipital bone immediately anterior to its occipital condyle.

Nerve supply: the loop between the ventral rami of the first and second cervical spinal nerves.

Action: flexion of the head at the atlanto-occipital joints.

M. rectus capitis lateralis is a short, flat muscle extending from the upper surface of the atlantal transverse process to the inferior surface of the occipital jugular process. From these attachments and its relation to the ventral ramus of the first spinal nerve, the muscle is homologized with posterior intertransverse muscles.

Nerve supply: the loop between the ventral rami of the first and second cervical spinal nerves.

Action: lateral flexion of the head.

The actions of these deep-seated muscles can only be deduced; their precise roles in natural activities await adequate technical demonstration.

4. THE LATERAL VERTEBRAL MUSCLES

The scaleni, anterior, medius and posterior, extend obliquely like *scaling* ladders between the upper two ribs and cervical vertebral transverse processes (5.36).

Scalenus anterior is situated deep (posteromedial) to the sternocleidomastoid. From the third to sixth cervical anterior tubercles musculotendinous fascicles converge, blend and then descend almost vertically to a narrow, flat tendon attached to the scalene tubercle on the first rib's inner border and a ridge anterior to the subclavian arterial groove.

Rectus capitis anterior

Longus colli, upper oblique part

Longus colli, vertical part

Longus colli, lower oblique part

Scalenus medius

Scalenus posterior

Serratus anterior

Rectus capitis lateralis
Transverse process of atlas
Longus capitis
Splenius capitis
Levator scapulae
Scalenus medius
Scalenus anterior
Scalenus posterior

1st rib

5.36 The anterior and lateral vertebral muscles. On the right side the scalenus anterior and the longus capitis have been removed.

Relations. *Anterior* are the clavicle, subclavius, sterno-cleidomastoid, omohyoid, the lateral part of the carotid sheath, transverse cervical, suprascapular and ascending cervical arteries, subclavian vein, prevertebral fascia and phrenic nerve. *Posterior* are the suprapleural membrane (p. 1269), pleura, roots of the brachial plexus, and subclavian artery; the latter two separate it from scalenus medius. *Below* its attachment to the sixth cervical vertebra, the medial border of the muscle is separated from longus colli by an *angular interval* (5.36) in which vertebral artery and vein ascend to the sixth foramen transversarium. The inferior thyroid artery crosses the interval near its apex and the sympathetic trunk and its cervicothoracic ganglion are here close to the posteromedial side of the vertebral artery (6.63). The thoracic duct crosses the left triangular interval level with the seventh cervical vertebra and usually contacts the medial edge of scalenus anterior. The musculotendinous attachments of scalenus anterior to anterior tubercles are separated from those of longus capitis by the ascending cervical branch of the inferior thyroid artery.

Nerve supply: ventral rami of the fourth to sixth cervical spinal nerves.

Actions. Acting from below, scalenus anterior bends the neck anterolaterally and rotates it contralaterally; from above it helps to elevate the first rib.

Scalenus medius, largest and longest of the scaleni, runs from the transverse process of the axis (and sometimes atlas) and lower five cervical posterior tubercles (5.35) to the first rib's upper sur-face, between its tubercle and the groove for the subclavian artery.

Relations. Its *anterolateral surface* is related to sterno-cleidomastoid; it is crossed by the clavicle and omohyoid; *anteriorly*, it is separated from scalenus anterior by the subclavian artery and ventral rami of the cervical spinal nerves. Levator scapulae and scalenus posterior are posterolateral. The upper two roots of the nerve to serratus anterior and the dorsal scapular nerve (to the rhomboids) pierce the muscle and appear on its lateral surface.

Nerve supply: third to eighth cervical spinal ventral rami.

Actions. From below, scalenus medius bends the neck ipsilaterally; from above it helps to raise the first rib. Scalene muscles, particularly the medius, are active during inspiration, even quiet breathing in the erect attitude (Campbell 1955).

Scalenus posterior, the smallest and most deeply situated, extends from the fourth, fifth and sixth cervical posterior tubercles to a thin tendon attached to the second rib's external border and its surface behind serratus anterior. It is occasionally blended with scalenus medius.

Scalene muscles vary in the number of their attachments, degree of separation and segmental innervation.

Nerve supply: lower three cervical spinal ventral rami.

Actions: ipsilateral flexion of the lower cervical vertebral column with the second rib fixed; from above, elevation of the second rib.

The scalenus minimus pleuralis is associated with the suprapleural membrane and cervical pleura (see p. 1269).

FASCIAE AND MUSCLES OF THE TRUNK

Muscles of the trunk may be grouped as follows:

1. Deep muscles of the back.
2. Suboccipital muscles.
3. Muscles of the thorax.
4. Muscles of the abdomen.
5. Muscles of the pelvis.
6. Muscles of the perineum.

1. Deep Muscles of the Back

The deep (intrinsic) dorsal muscles are a complex extending from pelvis to skull, including extensors and rotators of the head and neck (splenius capitis and cervicis), short segmental muscles (interspinales and intertransversarii), and spinal extensors and rotators (erector spinae and transversospinalis, the latter comprising semispinales, rotatores, and multifidus). Collectively these muscles control the vertebral column (5.37), acting in concert with numerous other muscles.

Superficial and *deep fasciae* of the neck are described on p. 608.

The thoracolumbar (lumbar) fascia covers these deep muscles. *Above*, it is anterior to serratus posterior superior and continuous with the superficial lamina of deep posterior cervical fascia. At *thoracic levels* it is a thin covering for vertebral extensors, separating them from the muscles connecting the vertebral column to the upper extremity. It is attached *medially* to the thoracic spines, *laterally* to the costal angles. At *lumbar levels* it is trilaminar (5.38); its posterior layer is attached to the lumbar and sacral spines and supraspinous ligaments, the middle *medially* to tips of lumbar transverse processes and intertransverse ligaments, *below* to the iliac crest and *above* to the twelfth rib's lower border and lumbocostal ligament (p. 497). The anterior layer covers quadratus lumborum and is attached *medially* to the anterior aspects of the lumbar transverse processes behind psoas major, *below* to the iliolumbar ligament and adjoining iliac crest; *above* it forms the lateral arcuate ligament (p. 592). Posterior and middle layers unite at the lateral margin of erector spinae and at the lateral margin those of quadratus lumborum are joined by the anterior layer, together forming a posterior aponeurosis for transversus abdominis. Bogdule & Macintosh (1984) describe the posterior lamina as being itself bilaminar; they have developed interesting concepts of the fascia's functions.

Splenius capitis (5.58) is attached below to the dorsal edge of the lower half of ligamentum nuchae, the spines of the seventh cervical and upper three or four thoracic vertebrae and their supraspinous ligaments (deep to rhomboids and trapezius). The muscle ascends laterally deep to sternocleidomastoid, reaching the mastoid process and rough surface below the lateral third of the superior nuchal line. It partly floors the posterior triangle, above and behind levator scapulae.

Nerve supply: dorsal rami of the middle cervical spinal nerves.

Splenius cervicis ascends from the third to the sixth thoracic spines to the upper two or three cervical posterior transverse tubercles, anterior to the attachments of levator scapulae. It may be absent, variable in attachments or possess accessory slips.

Nerve supply: dorsal rami of the lower cervical spinal nerves.

Actions. Together the splenii retract the head; separately, they draw it to one side, with slight rotation to turn the face to the same side, in synergy with the contralateral sternocleidomastoid.

THE ERECTOR SPINAE (SACROSPINALIS)

This muscle and its prolongations to thoracic and cervical levels are in a groove flanking the vertebral column (5.37), covered in the lumbar and thoracic regions by thoracolumbar fascia, serratus posterior inferior below and rhomboid and splenial muscles above. It is a large musculotendinous mass, varying in size and composition at different levels. In the sacral region it is relatively narrow and U-shaped but its attachments are chiefly tendinous and of great strength. In the lumbar region it expands into a thick *fleshy* mass, readily palpable, its lateral border flanked by a visible surface groove (5.39). As it ascends across the thorax, the groove follows the costal angles, first inclining laterally, then vertically, finally inclining medially until it is obscured by the scapula. (Parts of this muscular mass have been subjected to an interesting investigation of patterns of types of muscle fibres by Surca & Kostere 1985.)

Erector spinae springs from the anterior surface of a broad, thick tendon attached (medial limb of the U) to the median sacral crest, lumbar and eleventh and twelfth thoracic spines, their supraspinous ligaments and (lateral limb of the U) to the medial aspect of the dorsal part of iliac (p. 424) and lateral sacral crests (p. 325), where it blends with the sacrotuberous and dorsal

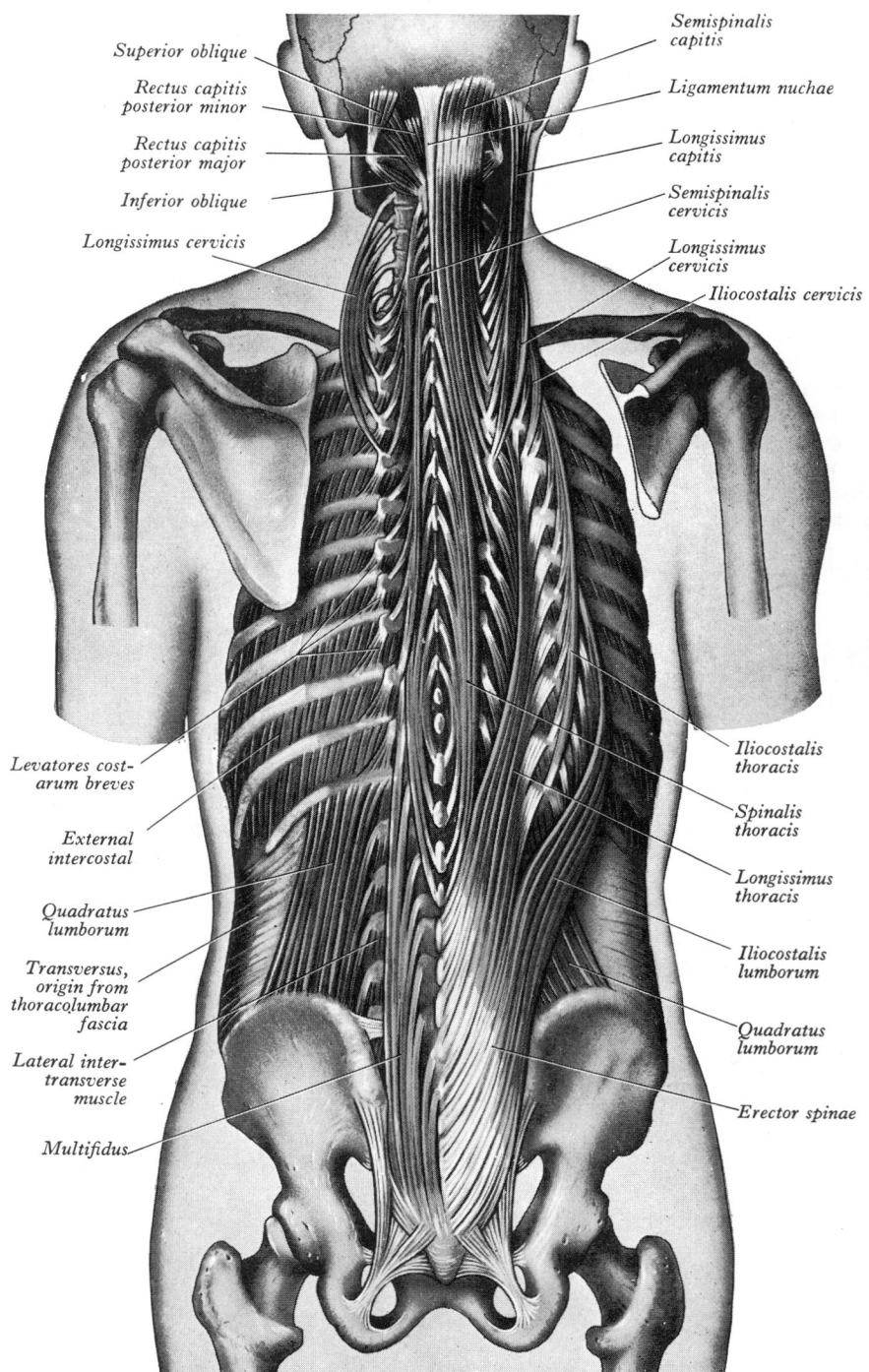

Superior oblique

Rectus capitis posterior minor

Rectus capitis posterior major

Inferior oblique

Longissimus cervicis

Semispinalis capitis

Ligamentum nuchae

Longissimus capitis

Semispinalis cervicis

Longissimus cervicis

Iliocostalis cervicis

Levatores costarum breves

External intercostal

Quadratus lumborum

Transversus, origin from thoracolumbar fascia

Lateral intertransverse muscle

Multifidus

Iliocostalis thoracis

Spinalis thoracis

Longissimus thoracis

Iliocostalis lumborum

Quadratus lumborum

Erector spinae

5.37 The deep muscles of the back. On the left side the erector spinae and its upward continuations (with the exception of the longissimus cervicis, which has been displaced laterally) and the semispinalis capitis have been removed.

sacro-iliac ligaments. Some fibres are continuous with gluteus maximus and many, centrally placed, with multifidus. From all aspects of the tendon muscle fibres arise and form a large mass which divides in the upper lumbar region into three columnar muscles: *iliocostocervicalis* (lateral), *longissimus* (intermediate) and *spinalis* (medial); each is regionally subdivided as follows:

Iliocostocervicalis

Iliocostalis lumborum has flat tendons which reach the inferior borders of the lower six or seven costal angles.

Iliocostalis thoracis starts from the upper borders of the lower costal angles, *medial* to the above; it ascends to the superior borders of the upper costal angles and the back of the seventh cervical transverse process.

Iliocostalis cervicis ascends from the third to sixth costal angles, *medial* to the tendons of iliocostalis thoracis and to the

posterior tubercles of the fourth to sixth cervical transverse processes.

Nerve supply: dorsal rami of the cervical, thoracic and upper lumbar spinal nerves.

Actions: extension and lateral flexion of the vertebral column (p. 590).

Longissimus

Longissimus thoracis is the largest part of erector spinae. In the lumbar region, already separate from iliocostalis lumborum, it is partly attached to the entire posterior surfaces of the lumbar transverse and accessory processes and to the middle layer of thoracolumbar fascia. Its rounded tendons reach the tips of all thoracic transverse processes and the lower nine or ten ribs between their tubercles and angles.

Longissimus cervicis, medial to l.thoracis, extends from the long tendons attached to the upper four or five *thoracic* transverse processes to again form tendons attached to the second to sixth *cervical* transverse process posterior tubercles.

Longissimus capitis, between l.cervicis and semispinalis capitis, is attached by tendons to the upper four or five thoracic transverse and lower three or four cervical articular processes; it reaches the posterior margin of the mastoid process, deep to splenius capitis and sternocleidomastoid. It usually has a tendinous intersection near its upper end.

Nerve supply: dorsal rami of the lower cervical, thoracic and lumbar spinal nerves.

Actions. The longissimi thoracis and cervicis bend the vertebral column back and laterally; longissimus capitis extends the head, turning the face to the same side.

Spinalis

Spinalis thoracis, the medial part of erector spinae, is scarcely a separate muscle, lying medial to longissimus thoracis and blended with it. It springs by three or four tendons from the eleventh thoracic to the second lumbar spines which unite to form a small muscle attached by separate tendons to the upper thoracic spines, varying from four to eight. It blends with semispinalis thoracis which lies anterior to it.

Spinalis cervicis, an inconstant muscle, ascends from the lower ligamentum nuchae, the seventh cervical spine (and sometimes first and second thoracic) to reach the spine of the axis (occasionally the two below it). It is often absent.

Spinalis capitis is variably blended with semispinalis capitis.

Nerve supply: dorsal rami of the lower cervical and the thoracic spinal nerves.

Actions. The spinales are extensors of the vertebral column.

TRANSVERSOSPINALIS

This muscular group consists of the following:

Semispinalis thoracis.   Multifidus.   Rotatores thoracis.
Semispinalis cervicis.   Rotatores cervicis.
Semispinalis capitis.   Rotatores lumborum.

These muscles ascend obliquely and medially from the transverse processes to adjacent but sometimes more distant vertebral spines.

Semispinalis thoracis consists of thin fasciculi between long tendons extending from the lower attachments to the sixth to tenth thoracic transverse processes, thence to the upper four thoracic and lower two cervical spines.

Semispinalis cervicis, a thicker muscle, is attached inferiorly by tendinous and fleshy fibres to the upper five or six thoracic transverse processes, extending superiorly to the fifth to second (axis) cervical spines. The axial fasciculus is the largest and mainly muscular.

Semispinalis capitis (5.37) is deep to splenius and medial to longissimi cervicis and capitis. It springs inferiorly by tendons attached to tips of the upper six or seven thoracic and seventh cervical transverse processes, the fourth to sixth cervical articular processes and, occasionally, the seventh cervical or first thoracic spine. Tendons are succeeded by a broad muscle, ascending to the medial area between the superior and inferior nuchal lines. Its medial part, more or less distinct, is the *spinalis capitis*, sometimes called the *biventer cervicis*, having an incomplete intermediate tendon or intersection.

Nerve supply: dorsal rami of the cervical and thoracic spinal nerves.

Actions. Semispinales thoracis and cervicis extend the thoracic and cervical vertebral regions, rotating them contralaterally; semispinalis capitis extends the head, also minimally turning the face likewise.

Multifidus, a mixture of muscular and tendinous fasciculi, is deep to the foregoing muscles, in the groove lateral to the vertebral spines from sacrum to axis. Its caudal fasciculi are attached to the sacral dorsum as low as its fourth foramen, to the aponeurosis of erector spinae, the posterior superior iliac spine and dorsal sacro-iliac ligaments and to all lumbar mamillary processes, all thoracic transverse processes and the lower four cervical articular processes. Each fasciculus ascends medially to the whole length of

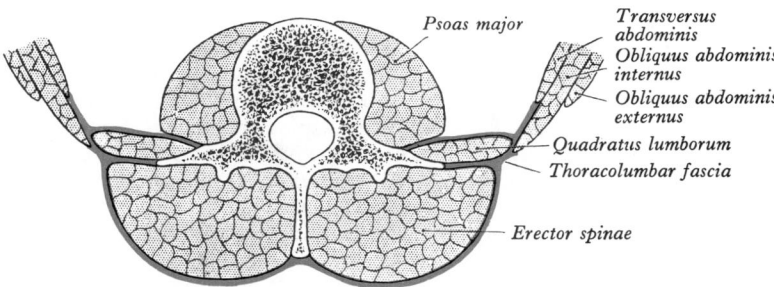

5.38 Transverse section through the posterior abdominal wall, showing disposition of the thoracolumbar fascia. Note that all other connective tissue strata have been omitted.

the spine of a vertebra above; fasciculi, however, vary in length, the most superficial passing from one vertebra to the third or fourth above, deeper ones from a vertebra to the second or third above, the deepest connecting adjacent vertebrae.

Nerve supply: dorsal rami of the spinal nerves.

Rotatores thoracis are deep to multifidus and are the only fully developed rotatores; there are eleven pairs. Small and quadrilateral, they connect the upper posterior part of a transverse process to the lower border and lateral surface of the lamina immediately above. The first is between the first and second thoracic vertebrae, the last between the eleventh and twelfth. One or more may be absent from the upper or lower ends of the series.

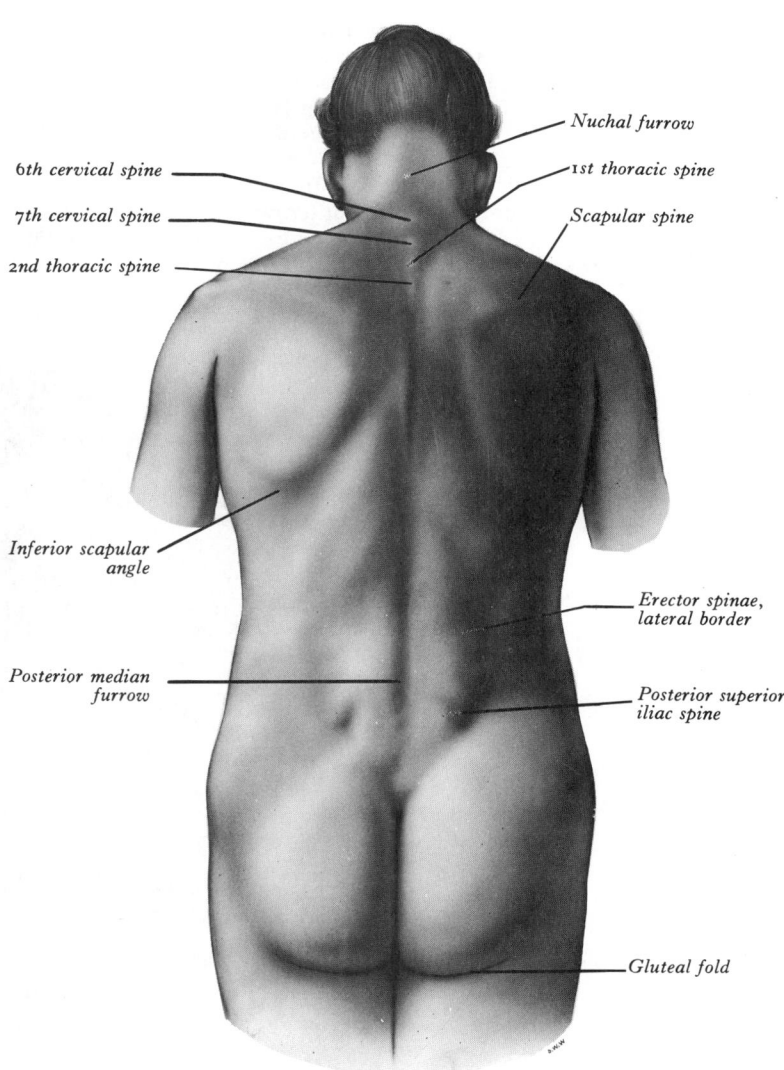

5.39 Dorsal aspect of the trunk, showing principal surface landmarks.

Rotatores cervicis et lumborum are irregular, variable muscles with similar attachments.

Nerve supply: dorsal rami of the spinal nerves.

Interspinales are short paired fasciculi between adjacent vertebral spines, flanking the interspinous ligament. Six *cervical* pairs occur between the axis and the first thoracic vertebra. They are small bundles, attached between apices of spines. *Thoracic pairs* occur between the first and second spines, sometimes the second and third and the eleventh and twelfth. Four *lumbar* pairs exist between the five lumbar spines. Occasionally pairs occur at thoracico-lumbar and lumbosacral junctions. Sometimes cervical interspinales span more than two vertebrae.

Nerve supply: dorsal rami of the spinal nerves.

Intertransversarii, small muscles between transverse processes, are best developed at *cervical level*, consisting of anterior and posterior slips, separated by ventral rami of the spinal nerves. *Posterior intertransverse muscles* are divisible into medial and lateral slips, respectively supplied by dorsal and ventral rami. Each *medial*, 'proper' intertransverse muscle is often divided into medial and lateral parts by the passage of a dorsal ramus. The *anterior intertransverse muscles* and *lateral* parts of the posterior connect the costal processes of contiguous vertebrae; *medial* parts of the posterior muscles connect true transverse processes. Seven pairs of intertransversarii exist, from the atlas to the first thoracic vertebra; but the pair between atlas and axis is often absent. At *thoracic levels* they are single muscles between the last three thoracic and first lumbar transverse processes. At *lumbar level* there are *intertransversarii mediales*, connecting accessory processes of one vertebra to the mamillary processes of the next, and *intertransversarii laterales*, divisible into ventral and dorsal parts (Cave 1937, Morison 1954); ventral parts connect lumbar transverse processes (costal elements); dorsal parts connect accessory processes to the transverse processes of succeeding vertebrae. Thoracic intertransverse muscles and ligaments are homologous with the medial slips of cervical 'proper' posterior intertransverse muscles; *levatores costarum* are homologous with their lateral moieties. The lateral branch of a dorsal spinal ramus separates thoracic intertransverse from levator costae muscles. The lumbar levatores are medial intertransverse muscles, lateral intertransverse being homologous with intercostal muscles. For other views on such homologies consult Sato (1973).

Nerve supply: lumbar intertransversarii mediales, thoracic intertransversarii and medial parts of cervical posterior intertransversarii are supplied by the dorsal rami, the others by ventral rami of the spinal nerves.

Actions. Shorter dorsal muscles are probably postural. The vertebral column is, in effect, a series of short jointed levers, unstable to compression and tending to buckle unless movements at individual joints are controlled. The short muscles may steady adjoining vertebrae and control them during motion of the whole vertebral column, ensuring effective action in the long muscles. Theoretically, short muscles may produce extension (multifidus, spinales), lateral flexion (multifidus, intertransversarii) and rotation (multifidus and rotatores) but details are obscure. Deep muscles are certainly involved in the control of posture, contracting intermittently during swaying movements from an upright position. The erectores spinae extend the trunk, control of extension depending largely on the recti abdominis. Conversely, flexion is initiated by the recti; but as the relevant centre of gravity moves forwards control is transferred to the erectores spinae. Lateral flexion is controlled by the contralateral erector spinae. However, in full flexion the erectores spinae are electromyographically quiet; flexion may be then limited by the spinal ligaments and compression of the intervertebral discs (Floyd & Silver 1955, Joseph 1960). Electromyographic activity in dorsal muscles is greater during work carried out on low surfaces from standing positions (Jonsson 1974), as would be expected.

2. Suboccipital Muscles

These small muscles (5.40) extend the head at atlanto-occipital joints (recti capitis posteriores major and minor) and rotate it with the atlas on axis (obliqui capitis superior and inferior).

Rectus capitis posterior major starts as a pointed tendon from the axial spine and broadens as it ascends laterally to the inferior nuchal line and occipital bone below it. The two diverging muscles leave a triangular space, revealing parts of recti capitis posteriores minores.

Actions: extension and ipsilateral rotation of the head.

Rectus capitis posterior minor, attached by a pointed tendon to the posterior atlantal tubercle, widens as it ascends to the

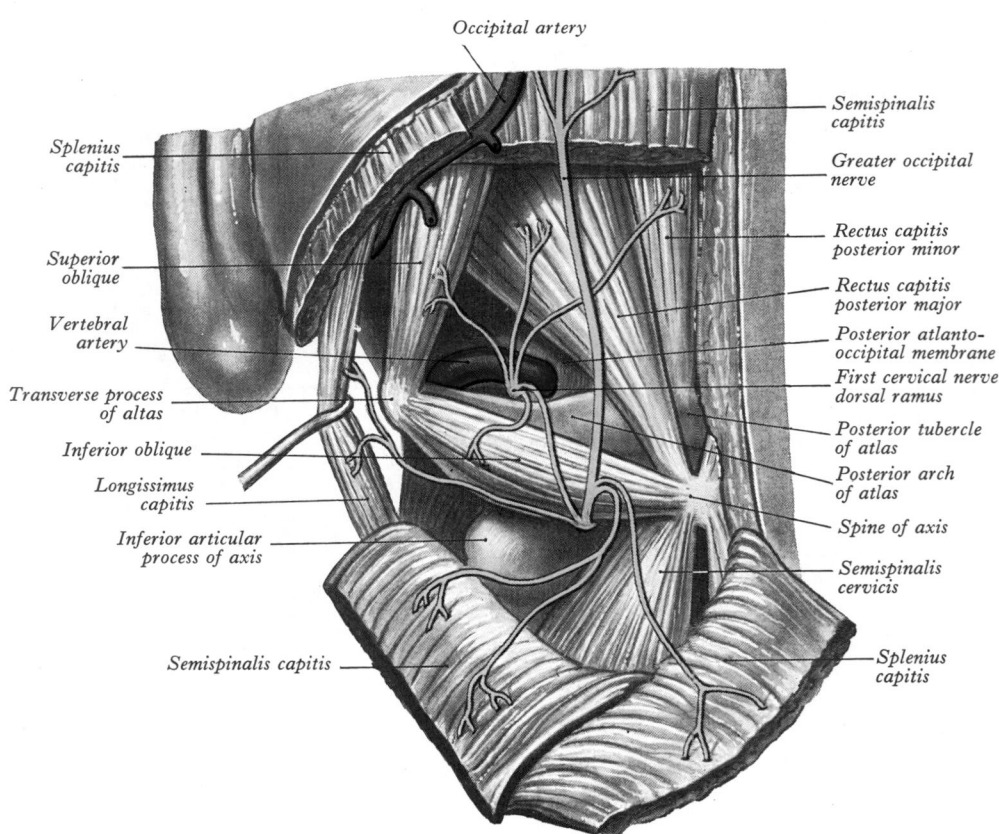

Occipital artery

Splenius capitis

Superior oblique

Vertebral artery

Transverse process of altas

Inferior oblique

Longissimus capitis

Inferior articular process of axis

Semispinalis capitis

Semispinalis capitis

Greater occipital nerve

Rectus capitis posterior minor

Rectus capitis posterior major

Posterior atlanto-occipital membrane

First cervical nerve dorsal ramus

Posterior tubercle of atlas

Posterior arch of atlas

Spine of axis

Semispinalis cervicis

Splenius capitis

5.40 The left suboccipital triangle and its contents.

medial part of the inferior nuchal line and occipital bone between this and the foramen magnum (p. 373). It may be doubled.

Action: extension of the head.

Obliquus capitis inferior, the larger oblique, ascends laterally from the lateral aspect of the spine and adjacent lamina of the axis to the inferoposterior aspect of the atlantal transverse process.

Action: ipsilateral rotation of the head, occurring at a mechanical advantage due to the width of the atlas.

Obliquus capitis superior, attached by tendinous fibres to the upper surface of the atlantal transverse process, ascends dorsally, expanding to the occipital bone between the superior and inferior nuchal lines, lateral to semispinalis capitis and overlapping rectus capitis posterior major.

Actions: posterolateral flexion of the head. The muscle and its neighbours are probably postural rather than prime movers but observation is difficult.

Nerve supply of all suboccipital muscles is from the dorsal ramus of the first cervical spinal nerve.

The Suboccipital Triangle

This region is bounded superomedially by rectus capitis posterior major, superolaterally by obliquus capitis superior, inferolaterally by obliquus capitis inferior. Medially it is covered by dense adipose tissue, deep to semispinalis capitis, laterally by longissimus capitis and sometimes splenius capitis, overlapping obliquus capitis superior. The 'floor' is the posterior atlantooccipital membrane and the posterior atlantal arch; the vertebral artery and dorsal ramus of the first cervical spinal nerve (5.40) are in a groove on the arch's upper aspect.

3. Muscles of the Thorax

These comprise (5.41,42,43) muscles connecting adjacent ribs (intercostales), spanning several ribs (subcostales), connecting the ribs to the sternum (transversus thoracis) or to vertebrae (levatores costarum, serratus posterior superior et inferior) and the diaphragm. All are concerned in costal movements and hence, potentially, respiration.

Intercostales

The intercostales (5.42) are thin musculotendinous layers occupying intercostal spaces, named from their spatial relations: intercostales externi, interni and intimi.

Intercostales externi, in eleven pairs, extend from the costal tubercles, where they blend with the posterior fibres of the superior costotransverse ligaments almost to the costal cartilages, where each continues as an aponeurotic *external intercostal membrane* which reaches the sternum. Each muscle connects the adjacent borders of two ribs. In the lower two spaces they reach the free ends of costal cartilages but in the upper two or three do not reach the ends of the ribs. They are thicker than intercostales interni; their fibres ascend medially at the back of the thorax, laterally at the front.

Intercostales interni, in eleven pairs, have attachments commencing at the sternum between the cartilages of true ribs and at anterior extremities of cartilages of 'false' ribs, extending back to the posterior costal angles, where each becomes an aponeurotic *internal intercostal membrane*, continuous posteriorly with the anterior fibres of a superior costotransverse ligament, anteriorly with fascia between the internal and external muscles. Each muscle descends from a costal groove and adjacent costal cartilage to the upper border of the rib below. Their fibres are also oblique and nearly at right angles to those in the external intercostal muscles.

Actions. Opinions are not unanimous; the following views have been proposed: that both groups act as costal elevators, that external intercostals are elevators and internal intercostals depressors, that intercartilaginous parts of internal intercostals act with external intercostals in inspiration, and that all intercostals are elastic supports preventing the contents of intercostal spaces from bulging in or out during respiration. Experiment shows impulses traversing the nerves to the external intercostals

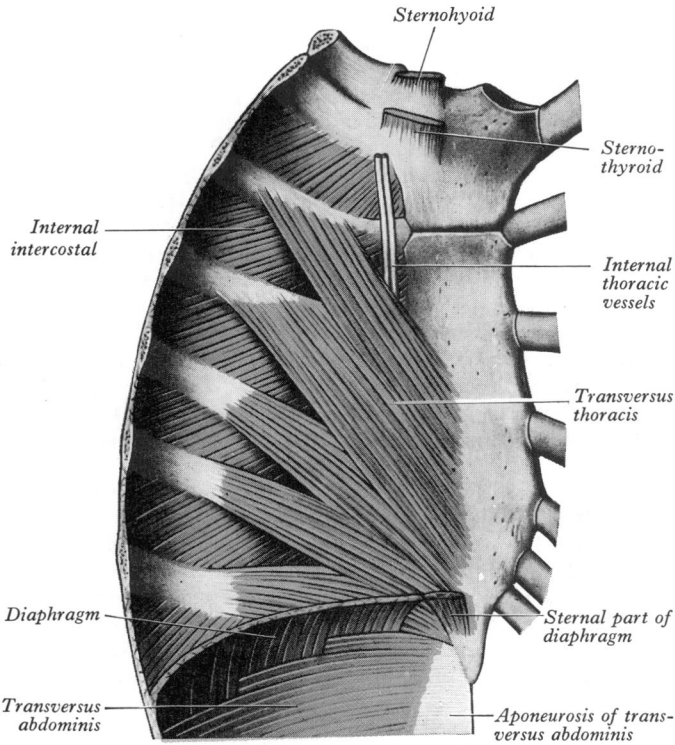

5.41 The left transversus thoracis, exposed and viewed from its posterior aspect. Note that, in the interval between the sternal and costal origins of the diaphragm, the lower border of the transversus thoracis is in contact with the upper border of the transversus abdominis. Note and contrast the relatively 'high' anterior diaphragmatic attachments with the 'low' lumbar level of its posterior attachments (see 5.43).

of lower spaces (5–9) in normal inspiration and continuing into early expiration (Bronk & Ferguson 1935, Draper et al 1960). In forced inspiration contraction is obviously stronger and more sustained. Electromyography yields divergent results. Some claim that intercostals of the sixth to eighth spaces contract in the inspiratory phase of quiet breathing and relax in expiration (Campbell 1955) and that in quiet breathing intercostals of the upper four spaces show sustained, not rhythmic activity; in forced inspiration this is increased and continues at a diminished level in

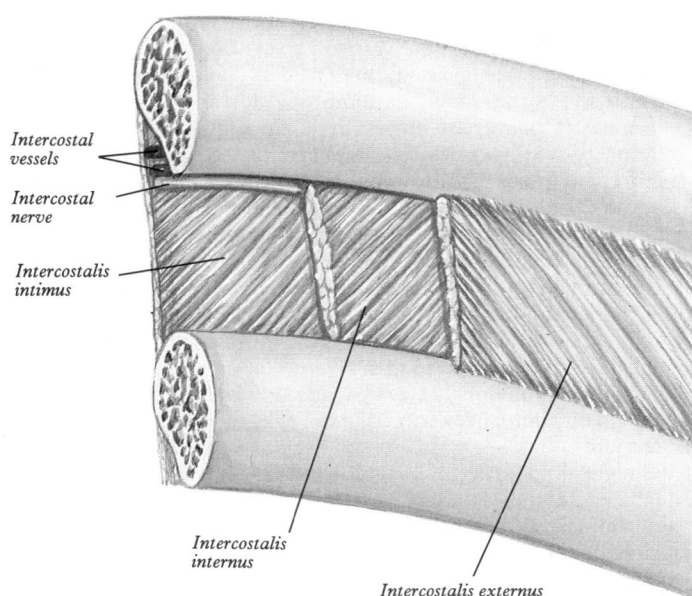

5.42 Dissection of a part of the thoracic wall, showing the position of the intercostal vessels and nerve relative to the intercostal muscles.

forced expiration (Jones et al 1953). It is possible that distribution of activity in intercostals varies with the depth of respiration (p. 594). These controversies have not been resolved (Hoskiko 1962, Basmajian 1967). Intercostals are always active in speech.

Intercostales intimi were once described as the internal laminae of internal intercostal muscles, having fibres coinciding in direction. Each is attached to the internal aspects of adjoining ribs. They are insignificant, sometimes absent, at higher thoracic levels but increase below, extending through the middle two quarters of the lower spaces. Walmsley (1915) considered them in the plane of transversus abdominis and suggested the term musculi intracostales, supported by Davies et al (1931). Posteriorly, in spaces where they appear, they may blend with corresponding subcostales, which Davies et al regarded as a fourth layer; general opinion favours Walmsley's three-layered arrangement. The intercostales intimi are related to endothoracic fascia and the parietal pleura, and externally to the intercostal nerves and vessels.

Actions. Without real information these muscles are presumed to act with intercostales interni (vide supra).

Subcostales are musculo-aponeurotic fasciculi, usually well developed only in the lower thorax; each descends, parallel with intercostales interni, from the internal surface of one rib, near its angle, to that of the second or third below it. Like intercostales intimi they separate intercostal vessels and nerves from pleura. Their incidence and distribution have been recorded by Satoh (1974).

Actions: probably costal depression.

Transversus thoracis (sternocostalis), a bilateral muscle, is spread internally on the anterior thoracic wall (5.41). It extends from the lower third of the posterior sternal surface, the xiphoid process and costal cartilages of the lower three or four sternal ribs near the sternum. Its fibres diverge and ascend laterally as slips attached to the lower borders and inner surfaces of the second to sixth costal cartilages. The lowest fibres are horizontal and adjoin the highest ones of transversus abdominis; the intermediate fibres are oblique, the highest almost vertical. Attachments vary even on opposite sides in one individual. Like intercostales intimi and subcostales, the muscle separates intercostal nerves from pleura.

Actions: depression of the costal cartilages.

Nerve supply: adjacent intercostal nerves.

Levatores costarum (5.37) are 12 paired, strong bundles attached to the tips of the seventh cervical to the eleventh thoracic transverse processes, descending laterally, parallel with the posterior borders of the external intercostals, to reach the upper edge and external surface of the rib below, between its tubercle and angle (*levatores costarum breves*). The four lower muscles divide into two fasciculi, one attached as described, the other descending to the *second* rib below (*levatores costarum longi*).

Nerve supply: lateral branches of the dorsal rami of corresponding thoracic spinal nerves (Morison 1954).

Actions: costal elevation, but a role in respiration is disputed (Primrose 1952); possibly vertebral rotation and lateral flexion.

Serratus posterior superior, a thin, quadrilateral muscle external to the upper posterior aspect of the thorax, is attached by a thin aponeurosis to the lower part of the ligamentum nuchae, the seventh cervical and upper two or three thoracic spines and their supraspinous ligaments. It descends laterally, ending as four digitations attached to the upper borders and external surfaces of the second to fifth ribs, lateral to their angles. It is superficial to the thoracolumbar fascia and deep to the rhomboid muscles. Digitations vary from three to six; the muscle may be absent.

Nerve supply: second to fifth intercostal nerves.

Actions. It obviously elevates the ribs but experiments in dogs show no respiratory function (Ogawa et al 1960); its human role is uncertain.

Serratus posterior inferior (5.58) is a thin, irregularly quadrilateral muscle attached by a thin aponeurosis to the lower two thoracic and upper two or three lumbar spines and their supraspinous ligaments; the aponeurosis also blends with the thoracolumbar fascia. Ascending laterally, the muscle's four digitations reach the inferior borders and outer surfaces of the lower four ribs, anterolateral to their angles; their number may be reduced; the whole muscle may be absent.

Nerve supply: ventral rami of the ninth to twelfth thoracic spinal nerves.

Actions. The muscle draws the lower ribs down and backwards, but perhaps not in respiration.

THE DIAPHRAGM

The diaphragm (5.43) is a modified half-dome of musculofibrous tissue between thorax and abdomen, its mainly convex postero-superior aspect facing the former and concave antero-inferior surface the latter. Its muscular periphery is attached to (and posteriorly beyond) the highly oblique circumference of the thoracic outlet, whence it converges to a central tendon. Its muscular fibres form sternal, costal and lumbar groups, which are, however, particular dispositions in an otherwise continuous sheet.

Diaphragmatic shape. It is common to designate this partition as a dome; this is misleading, particularly when viscera, serous membranes and apertures are considered in terms of relations, vertebral levels and surface anatomy. The term, modified *half-dome*, is a useful simplification for the overall shape, but needs some amplification. Considering the skeletal thoracic outlet (p. 326) and upper three lumbar vertebrae, its attachments include, serially: the xiphoid process and symphysis, ventral rib ends and costal cartilages of the (vertebrosternal) seventh, (vertebrochondral) eighth to tenth and (vertebral, 'free' or floating) eleventh and twelfth ribs; the relative positions of the twelfth rib, first lumbar transverse process, the upper lumbar vertebral bodies and associated symphyses. From these the levels of lumbar, costal and sternal attachments can be assessed and the relative lengths, curvatures and planes of its muscular sectors are more easily understood; also the comparative vertebral levels of the aortic, oesophageal and vena caval orifices (vide infra). Thus the attachments are low (lumbar *levels*) posteriorly and laterally but anteriorly they rapidly ascend medially to the xiphoid levels. The superior surface is centrally almost flat where it is related to the pericardium covering most of the diaphragmatic surface of the heart. On each side of this *cardiac plateau* is raised a smooth dome or *cupola* (higher on the right), each reflecting the shape of superjacent lung and pleura and subjacent viscera (listed below).

Diaphragmatic attachments. The *sternal part* is attached by two slips to the back of the xiphoid process; the *costal part* to the internal surfaces of the lower six costal cartilages and their adjoining ribs, interdigitating with transversi abdominis (5.41); the *lumbar part* attaching to the aponeurotic medial and lateral arcuate ligaments (lumbocostal arches) and to some lumbar vertebrae by *crura*. The xiphoid slips are sometimes absent.

The *lateral arcuate ligament*, a thickened band in the fascia on quadratus lumborum, arches across this muscle and attaches medially to the front of the first lumbar transverse process, laterally to the inferior margin of the twelfth rib near its midpoint.

The *medial arcuate ligament* is a tendinous arch in the fascia covering psoas major; medially it blends with the lateral tendinous margin of the corresponding crus and is thus attached to the side of the first or second lumbar vertebral body; laterally it is attached to the front of the first lumbar transverse process.

The *crura* are tendinous at their attachments, blending with the anterior longitudinal vertebral ligament. The *right crus*, broader and longer, springs from anterolateral aspects of the bodies and intervertebral discs of the upper *three* lumbar vertebrae, the *left crus* from corresponding parts of the upper *two*. Their medial tendinous margins converge in a median arch across the aorta at the level of the thoracolumbar disc; this *median arcuate ligament* is often poorly defined.

From this circumferential attachment fibres converge to a central tendon; those from the xiphoid process are short, with only a slight inclination, and sometimes aponeurotic; those from arcuate ligaments, and especially from ribs and cartilages, are longer, initially almost vertical, then progressively curved as they ascend to their central attachment. Fibres from crura diverge, the most lateral continue ascending laterally to reach the central tendon. Medial right crural fibres ascend *left* of the oesophageal opening; sometimes a muscular fasciculus from the medial side of the left crus crosses the aorta and runs obliquely through the fibres of the right crus towards the vena caval opening but does

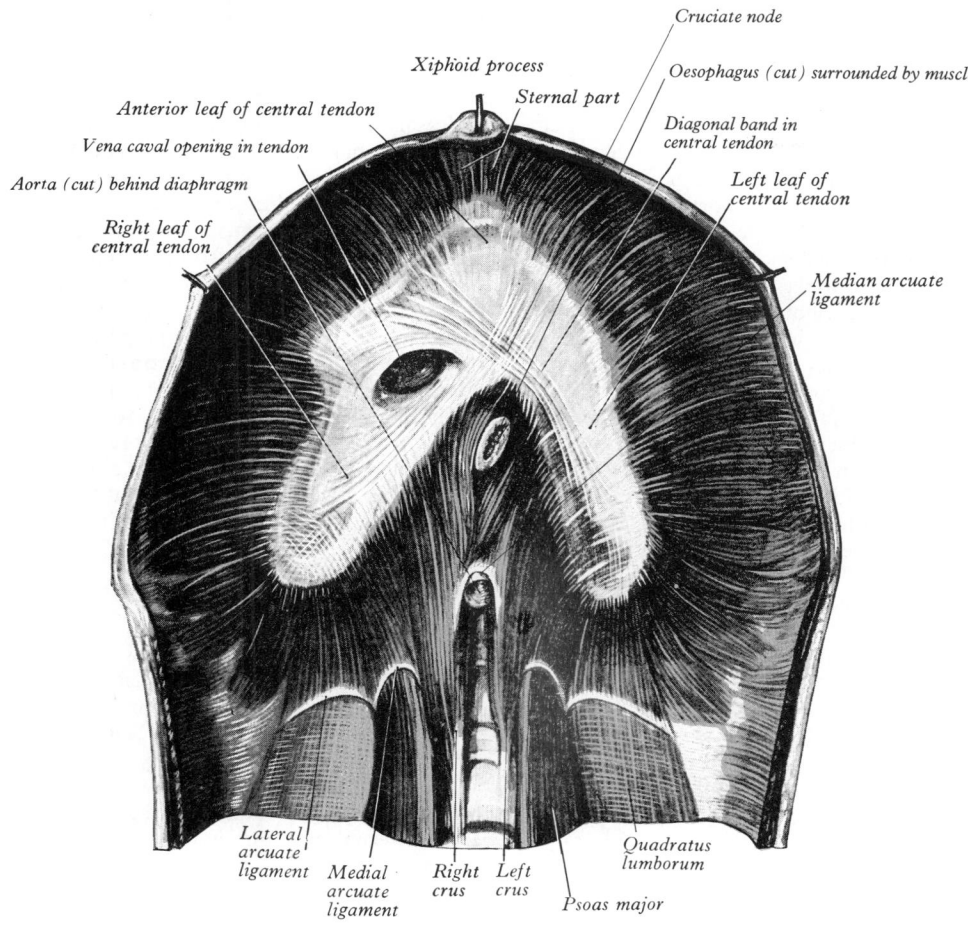

Cruciate node

Xiphoid process

Oesophagus (cut) surrounded by muscle

Sternal part

Anterior leaf of central tendon

Diagonal band in central tendon

Vena caval opening in tendon

Left leaf of central tendon

Aorta (cut) behind diaphragm

Right leaf of central tendon

Median arcuate ligament

Lateral arcuate ligament

Medial arcuate ligament

Right crus

Left crus

Quadratus lumborum

Psoas major

5.43 Abdominal aspect of the diaphragm. Contrast the relatively 'high' anterior sternocostal attachment descending steep and obliquely to its complex 'low' lumbar posterior attachment (see **5.41**).

not approach the oesophageal opening (Low 1907). The latter's right margin is covered by deeper medial *right* crural fibres.

The *central tendon* is a thin, strong aponeurosis of interwoven collagen fibres, with part of its margin nearer the front of the diaphragm, so that the posterior muscular fibres are longer. Its central area is beneath the pericardium and partly blended with it. The whole is trifoliate; its three leaves are partly separated by slight indentations. The middle leaf is an equilateral triangle, its apex towards the xiphoid process. Right and left leaves are long and linguiform, curving posterolaterally, the left being narrower. The central area shows four *diagonal bands* expanding from a thick *node* where decussation of compressed tendinous strands occurs in front of the oesophageal aperture and left of the vena caval opening.

The Diaphragmatic Apertures

The diaphragm is traversed by many structures; three large openings, aortic, oesophageal and vena caval (**5.43**) and a number of smaller avenues exist.

The *aortic aperture*, lowest and most posterior, is level with the caudal border of the twelfth thoracic vertebra and the thoracolumbar intervertebral disc, slightly left of midline. It is actually a vertical osseo-aponeurotic and symphysio-aponeurotic channel, between the crura laterally, the vertebral column posteriorly and the diaphragm anteriorly. It is thus *behind* the diaphragm or its median arcuate ligament when present. Occasional tendinous crural fibres pass *behind* the aorta, forming a fibrous ring. The opening transmits the aorta, thoracic duct and sometimes azygos and hemiazygos veins (p. 808); lymphatic trunks also *descend* through it from the lower posterior thoracic wall.

The *oesophageal aperture* is surrounded by diaphragmatic muscle and is sited at an intermediate level, opposite the tenth thoracic vertebra; it is elliptical (long axis slightly oblique, ascending to the left of the midline) and between the medial fibres derived from the right crus (Low 1907), which have now crossed the midline. It is above, in front and slightly left of the aortic opening. (Strictly, because of the obliquity of this elliptical aperture, its lower end is almost directly anterosuperior to the aortic orifice, whereas its upper end is further anterosuperior and a few centimetres to the left.) It transmits oesophagus, gastric nerves, oesophageal branches of left gastric vessels and lymphatic vessels. Muscle of the oesophageal wall and that of the opening remain separate. Inferior diaphragmatic fascia, continuous with transversalis fascia (p. 603) and rich in elastic fibres, ascends the opening like a flattened cone to blend with the oesophageal wall about 2 cm above the gastro-oesophageal junction; some elastic fibres penetrate to the submucosa. This fascial expansion, the *phreno-oesophageal ligament*, flexibly connects oesophagus and diaphragm, permitting some freedom during swallowing and respiration (Allison 1951, Hayward 1961).

The *vena caval aperture*, highest of the three and about level with the disc between the eighth and ninth thoracic vertebrae, is quadrilateral and sited *within* the central tendon, at the junction of its right leaf with the central area, so that its margin is aponeurotic. The inferior vena cava is adherent to the margin as it traverses the opening and is accompanied by branches of the right phrenic nerve.

Two lesser apertures in each crus transmit the greater and lesser splanchnic nerves. The ganglionated sympathetic trunks are usually behind the medial arcuate ligaments. Openings for small veins are frequent in the central tendon.

In two small bilateral areas muscular tissue is replaced by areolar tissue. One, between sternal and costal parts, contains the superior epigastric artery and lymph vessels from the abdominal wall and liver; the other, between the costal part and fibres from the lateral arcuate ligament, is less constant; when it exists, the kidney (posterosuperior aspect) is separated from the pleura only by areolar tissue.

Relations. *Superiorly* are three serous membranes: on each side pleura separates it from a pulmonary base, between them pericardium over the middle folium (and apically some diaphragmatic muscle) separates it from heart. The central area (*cardiac plateau*) is slightly *lower*, flatter and more horizontal than the smooth convex summits (*cupolae*) of the lateral regions. Thus in anteroposterior view the superior profile of the diaphragm consists of a high smooth convex right cupola, a flat low cardiac plateau that extends more to the left than right and also inclines gently ventrally and towards its left limit, finally a less high, slightly narrower, but smooth convex left cupola. (For variations in levels see below.) The inferior surface is mostly covered by peritoneum, that of the *right cupola* being in contact with the convex surface of the right hepatic lobe, right kidney and suprarenal gland; the peritoneum of the *left cupola* is in contact with the left hepatic lobe, gastric fundus, spleen, left kidney and suprarenal gland. (As mentioned, the right cupola is higher and less mobile than the left; also their abdominothoracic visceral relations differ. Clinically, therefore, the profile of a cupola should be specified, right or left. Unqualified terms, e.g. *the dome*, or cupola *of the diaphragm*, should be abandoned.)

Nerve supply. Motor fibres travel in the phrenic nerves. The lower six or seven intercostal nerves distribute sensory fibres to the peripheral diaphragm. A controversy concerning the crural motor supply suggested an intercostal innervation but Collis et al (1954) indicated the phrenic nerves as sources; this was confirmed by Shehata (1966). Both state that all crural fibres (irrespective of origin) right or left *of the oesophageal opening* are supplied by the ipsilateral phrenic nerve.

Actions. The diaphragm is the essential respiratory muscle. Its general concavity towards the abdomen is flatter and centrally attached above; its periphery is muscular. During inspiration the lowest ribs are at first fixed; from these and the crura, muscular fibres draw down the central tendon with its attached pericardium. The curvature is scarcely altered, the cupolae also moving down and a little forwards almost parallel to their original positions, displacing the abdominal viscera, the descent of which is permitted by stretching of the abdominal wall; the limit of this is soon reached, the central tendon becoming fixed for diaphragmatic contraction, which elevates the lower ribs and through them pushes forwards the sternum and upper ribs. The right cupola, above the liver, has greater resistance to overcome but right crus and cupola are more substantial than the left. The balance between descent of the diaphragm and protrusion of the abdominal wall ('abdominal' breathing) and elevation of the ribs ('thoracic' breathing) varies in individuals and with the depth of respiration. Thoracic breathing is usually more marked in females and, of course, in deep inspiration.

In all expulsive acts the diaphragm adds power to such effort. Thus, before sneezing, coughing, laughing, crying, vomiting, urination, defaecation, and during expulsion at birth, a deep inspiration helps. Deep inspiration, followed by closure of the glottis, is also a preliminary to powerful action of the trunk muscles, as in lifting heavy weights.

The right crus has been said to act as a sphincter of the human oesophagus. Expiration immediately after swallowing relaxes these fibres, allowing the oesophageal contents to enter the stomach (Whillis 1931, p.1318). Less emphasis is now placed on this supposed action of the right crus; the intrinsic muscle of the lower oesophagus, below the attachment of the phreno-oesophageal ligament, is considered an effective sphincter (p. 1333).

The level of the diaphragm varies continuously during respiration; it is also affected by distension of the stomach and intestines and the size of the liver. After forced expiration the right cupola is level anteriorly with the fourth costal cartilage, laterally with the fifth to seventh ribs, posteriorly with the eighth; the left cupola is a little lower. Diaphragmatic excursion is about 1.5 cm in quiet breathing; in deep respiration maximal ranges are 6–10 cm (Campbell 1958).

Radiographs show that diaphragmatic level also varies with posture, its profile being *highest* in the supine body, in which respiratory excursions are greatest in normal breathing. With the body erect, the diaphragm is lower, its respiratory movements smaller. The diaphragmatic profile is even lower in sitting and respiratory excursions smallest. With the body horizontal and on one side, the halves of the diaphragm do not behave similarly and they achieve *extreme positions and ranges*; the uppermost sinks lower even than in sitting and moves little with respiration; the lower half rises higher even than in the supine position and respiratory excursions are much increased.

Position of the diaphragm in the thorax apparently depends on: (1) elastic retraction of lung tissue, tending to pull it up; (2) visceral pressure from below (negative in sitting or standing, positive in recumbency), (3) intra-abdominal pressure due to abdominal muscles (contracting in standing but not in sitting), the diaphragm being pushed up higher in the former.

Applied Anatomy. Abdominal organs, usually the stomach, may herniate into the thorax, either alongside the oesophagus or between the lateral arcuate and costal fibres.

The Movements of Respiration

Rhythmic respiratory movements alter thoracic capacity, causing inflow and outflow of air to ventilate pulmonary alveoli, where gaseous exchanges between blood and alveolar air occur. Capacity can be increased in transverse, antero-posterior and vertical dimensions, together or separately.

Transverse and anteroposterior dimensions are increased by costal movement. When elevated, the anterior ends of the vertebrosternal ribs thrust forwards and upwards and this is transmitted to the sternal body, with slight movement at the manubriosternal symphysis (joint, angle) increasing anteroposterior diameter. The same ribs, rotating at vertebral ends (p. 498), are everted, increasing transverse diameter. Elevation of vertebrochondral ribs (4.35) splays them out and backwards, increasing transverse diameter with some widening of the infrasternal angle and enlargement of the upper abdominal cavity. Vertical dimension is increased by diaphragmatic contraction pressing on abdominal viscera, which are permitted to descend by relaxation of the abdominal wall during inspiration; at the limit of this descent abdominal viscera halt the central tendon, enabling the diaphragm to elevate the lower ribs. (Movements of the first rib and manubrium, i.e. the whole sternum, only become appreciable in deep and forced respiration; vide infra and p. 498.)

A respiratory cycle has inspiratory and expiratory phases. At rest the former lasts about one, the latter about three seconds. Increase of thoracic capacity reduces intrapleural pressure and thereby expands the lungs, decreases intrapulmonary pressure and draws air into the respiratory passages. The expiratory phase is largely passive, recoil of the thoracic wall and lungs raising intrathoracic pressure to expel air.

In *quiet respiration* about 500–750 ml of *tidal air* are inspired and expired in each cycle, about one-third occupying 'anatomical dead space', extending from the nostrils to the terminal bronchioles, where epithelium is not respiratory and little gaseous exchange occurs. The remaining inspired air ventilates the alveoli. Inspired air is warmed, humidified and filtered in the 'dead space', especially in the nose.

In *quiet inspiration* the main and often sole active muscle is the diaphragm. In average adults the right cupola descends to the level of the disc between the tenth and eleventh thoracic vertebrae, and the left cupola is level with the disc between the eleventh and twelfth, the central tendon moving about 1.5 cm. This is responsible for almost all the tidal volume of about 500 ml. There is little or no movement of osseous boundaries of the thoracic inlet. Electromyography, however, shows a rhythmic activity of the scalene muscles (anterior and medius), which fix the upper ribs; intercostal muscles in the upper six spaces may also show activity, preventing displacement of soft tissues in these spaces by changes of intrapleural pressure. Intercostals of lower spaces may also show rhythmic contraction, increasing the transverse diameter at this level and probably also resisting pressure changes.

In *inspiration* intrapleural and intrapulmonary pressures fall below atmospheric; at its end intrapulmonary equals atmospheric. Air entering the lung is not uniformly distributed, regional differences occurring in ventilation and blood flow. Ventilation is greater in lower than in upper lobes, due to greater movement in the lower thorax. Regional differences may also be related to the varying elasticity and dimensions of air passage. Alveoli which expand earliest in inspiration may receive more air from the dead space.

In *quiet expiration* elastic recoil of the lungs and thoracic wall is balanced by a slow relaxation of inspiratory muscles; electromyography shows that the diaphragm and intercostals continue to relax into expiration. The thin film of material between alveolar air and the surface of alveolar epithelium (p. 1281) may also be a factor in recoil. With the ventilation rate below 40 litres/min intercostal and abdominal muscles are not active in expiration.

In *deep inspiration* these movements are augmented and additional muscles participate, all contracting with greater force; active motor units increase, discharges along their nerves are for longer periods and motor units are more synchronized. Intercostal muscles become active in progressively higher spaces to move the upper ribs and sternum. The first rib is elevated by the scaleni, anterior and medius, and by sternocleidomastoid indirectly through the clavicle and costoclavicular ligament and directly through its manubrial attachment. Each 1 cm increase in the circumference of the thorax increases the capacity by about 200 ml. The twelfth rib, fixed by quadratus lumborum, enables the diaphragm to press more powerfully on abdominal viscera (Boyd et al 1965); this may also provide a basis for controlled relaxation in precise adjustments of expiration for speech (Taylor 1960). The spinal erectors also diminish thoracic vertebral curvature, slightly increasing the width of the intercostal spaces and allowing the ribs a greater range of movement. Intrapulmonary pressure at the start of *deep expiration* may rise to 30 mm above atmospheric.

In *forced respiration*, breathing against resistance, further muscles act. In forced *inspiration* the diaphragm contracts maximally, the right cupola descends level with the eleventh thoracic vertebra, the left cupola reaching that of the twelfth. This corresponds to a change in level of 9–10 cm. Action of the spinal erectors is augmented and muscles connecting the upper limb to the trunk may participate. Forced *expiration* is augmented by strong contraction of the abdominal muscles, particularly the oblique and transverse, and latissimi dorsi, all contracting forcefully in such efforts as coughing and sneezing. The abdominal muscles raise intra-abdominal pressure, forcing the relaxing diaphragm up and drawing lower ribs down and medially.

The range and balance of thoracic movements exhibit striking individual variations, perhaps dependent on the form of the thoracic skeleton, habit and other factors; such variation must be allowed for when movements are analysed in any individual.

In all phases of respiration changes in nostrils, larynx, trachea and bronchial tree occur. Dilatation of nostrils is evident in forced respiration. During quiet respiration the rima glottidis (p. 1254) is widened by lateral rotation of the vocal processes, abducting the vocal folds; it is widened further in forced inspiration by increasing abduction of the folds and the arytenoid cartilages. Tracheal, bronchial and bronchiolar calibre can be changed by distension or actively by alteration of the tone of their mural muscle, mediated by para-sympathetic (vagal) and sympathetic nerve supplies, so that they dilate in inspiration and contract in expiration.

Applied Anatomy. Changes in the level of the diaphragm during alteration in posture explain why patients with severe dyspnoea are most comfortable when sitting up. In unilateral disease of pleura or lungs interference with the position or movement of the diaphragm is usually observable by X-ray examination.

4. Muscles of the Abdomen

Abdominal muscles are usually considered in anterolateral and posterior groups.

ANTEROLATERAL ABDOMINAL MUSCLES

This is a group of four large flat muscular sheets forming the abdominal wall (obliqui externus and internus, transversus and rectus abdominis) and smaller elements, cremaster and pyramidalis.

Superficial fascia of the superior and central abdominal wall is a single layer containing variable fat; but inferiorly, particularly in the obese, it differentiates usually into two layers, the extent and significance of which is disputed; between them are superficial vessels, nerves and lymph nodes.

The superficial layer (or layers) is thick, areolar and contains variable fat. It passes over the inguinal ligaments into corresponding superficial fascia of the thighs; in males, however, this layer continues over the penis, spermatic cords and scrotum, where it is thin, free of fat, pale red, containing non-striated muscular fibres, the *dartos* muscle supplied by the genitofemoral nerves. Thence it continues into the remaining perineum, including in females the labia majora.

The deep layer (or layers) is more membranous, contains elastic fibres and is loosely connected by areolar tissue to the external oblique aponeuroses; but in the midline it is adherent to the linea alba and symphysis pubis and thickened, and prolonged on the dorsum penis, contributing to its *fundiform ligament* (5.45). It is continuous with superficial thoracic fascia and, over the inguinal ligaments, with a membranous superficial layer in the proximal thigh and it fuses with the subjacent fascia lata in the depths of the (ventral) *inguinal flexures* (5.45); medially, it is continued over penis and spermatic cord to the scrotum and thence to form the remainder of the membranous layer of superficial perineal fascia (p. 606), in females into the labia majora and remaining perineum.

In children the testis can often be retracted into loose areolar tissue between the external oblique and the deep layer of superficial fascia over (ventral to) the inguinal canal, the *superficial inguinal pouch* (Browne 1938).

Obliquus externus abdominis (5.44A,B,53A–E), curved anterolaterally round the abdomen, is the largest and most superficial anterolateral muscle and has eight digitations attached to the external surfaces and inferior borders of the lower eight ribs, alternating with digitations of serratus anterior and latissimus dorsi along an oblique line extending down and backwards; upper digitations are attached close to costal cartilages, the lowest to the apex of the twelfth cartilage, the middle ones to ribs some distance from their cartilages. Fibres diverge inferomedially, the posterior fibres being almost vertical and attached to the outer lip of the anterior half of the ventral segment of the iliac crest (p. 424); the middle and upper fibres are progressively less oblique and end in an aponeurosis, along a line from the ninth costal cartilage to a little below the umbilicus and then turning laterally towards the anterior superior iliac spine. Fibres rarely descend beyond a line from the anterior superior iliac spine to the umbilicus. The muscle has a free posterior border (5.58).

The muscle's *aponeurosis* was for long accepted as a single dense tendinous sheet, its fibres descending medially to end in the median *linea alba* (5.44), a raphe stretching from xiphoid process to symphysis pubis, in which right and left aponeuroses fuse and cover the abdomen. (However, this main part of the aponeurosis, together with those of obliquus internus and transversus, the sheath of rectus abdominis and the linea alba have been reinvestigated and new observations and functional implications recorded, vide infra, 5.44B, 53C–E.) Inferomedially each aponeurosis is attached to the pubic symphysis (upper border), pubic crest and tubercle. Between the anterior superior iliac spine and pubic tubercle it is a thick band, infolded to form a groove; this is the *inguinal ligament*. Expansions from its medial end are attached to the pecten pubis as the *lacunar ligament* and fibres also pass superomedially to the rectus sheath and linea alba, as a *reflected part* of the inguinal ligament (5.50).

Rizk (1980) made extensive investigations into ventral abdominal musculature in 41 human specimens (male and female ranging from birth to 70 years), and in 75 specimens from many other mammalian groups. The main part of the human obliquus externus aponeurosis from costoxiphoid margin to pubis was *bilaminar*, i.e. deep and superficial layers, their fibres

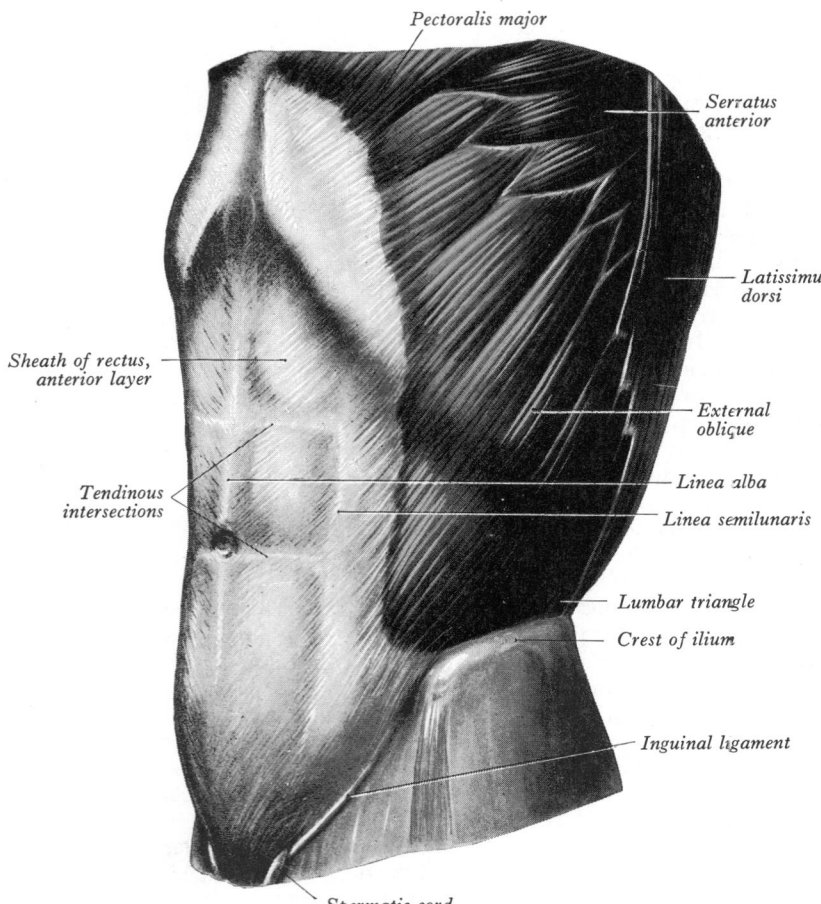

5.44A The left obliquus externus abdominis.

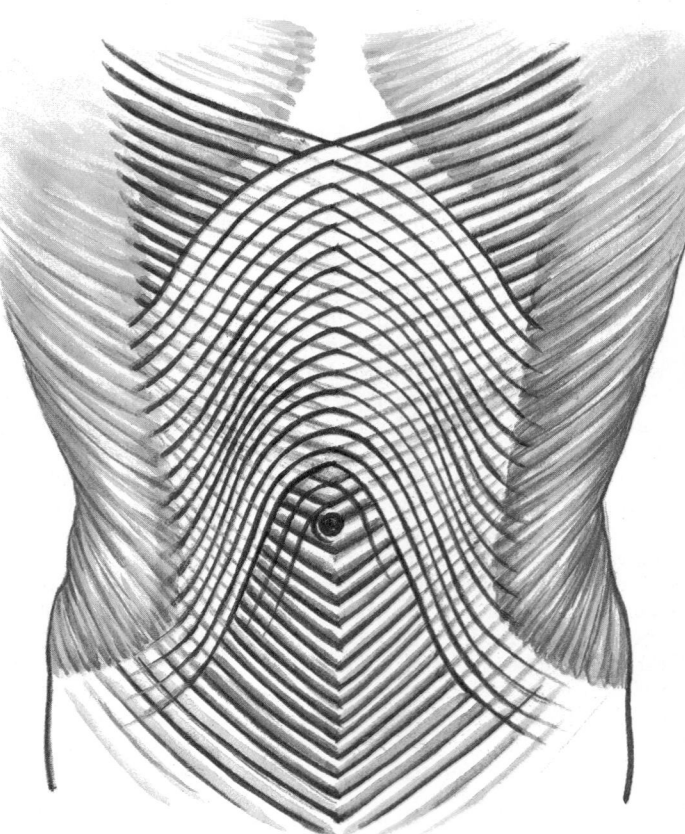

5.44B The concept of bilaminar aponeuroses of the external oblique muscles. Note that the fibres of the superficial and deep layers are approximately at right angles; decussations occur as part of the linea alba (modified from Rizk 1980; see text for further details and reference).

approximately at right angles to each other and neither ending at the linea alba but both crossing the midline to enter the contralateral half of the abdominal wall at linear decussations. The *deep layer* fibres are direct continuations of the ipsilateral muscle's fasciculi; they pass inferomedially and at the midline, which they cross, they separate into two laminae. A considerable proportion are *directly continuous* with the aponeurotic fibres of the anterior layer of the contralateral obliquus *internus* (forming, in effect, a *bilateral digastric muscle*); the remainder form the external's superficial layer of its contralateral aponeurosis. Below the umbilicus most crossed fibres remain deep but some recurve to form the superficial intercrural fibres.

The *superficial layer* derived, as noted, from the contralateral deep fibres, largely forms a series of parallel, wide S-shaped curves, broadly downwards and laterally. Decussation of external oblique aponeurotic fibres was single and midline (30%), or triple (70%) with one midline and two near the medial margins of recti abdominis. The superficial fibres, after decussation, as indicated, passed principally *inferolaterally* (crossing the inferomedially inclined deep fibres); designated S-shaped, but rather gently sinuous, the medial three-quarters, or more, had a superolateral convexity, the lateral extremity a superolateral concavity. The investigator quoted divided the anterolateral abdominal wall into principal levels: costal margin and xiphoid down to umbilicus, from the latter to the iliac crest, from the iliac crest to the inguinal ligaments and pubes. Unfortunately, which part of the iliac crest is not specified. Nevertheless, using this terminology the superficial aponeurotic fibres of obliquus abdominis externus are as described down to umbilical level and below the iliac crest they become the intercrural fibres associated with the superficial inguinal ring. Between umbilicus and iliac crest, however, the anterior laminae of obliquus abdominis *internus* become superficial to externus, vertical lines of decussation between internus and externus occurring along the ventral centre-line of each rectus abdominis. For this limited extent, therefore, the most ventral

(superficial) stratum in the medial half of the ventral wall of the rectus sheath is provided by obliquus internus.

Through the adherence of the superficial stratum to the (contralateral muscle's) deep stratum the two external obliques together form another bilateral digastric muscle.

The external oblique is invested by external and internal fasciae, the former better developed. First and last digitations may be absent; digitations or the whole muscle may be reduplicated. Digitations may also be continuous with pectoralis major or serratus anterior.

Nerve supply: ventral rami of the lower six thoracic spinal nerves.

The inguinal ligament (5.46), the thick, inrolled inferior border of the external oblique aponeurosis, has a grooved abdominal surface ('floor' of the inguinal canal) and stretches from the anterior superior iliac spine to the pubic tubercle. (Synonyms are: Poupart's ligament, crural or superficial crural arch.) Its length is convex towards the thigh and here it is continuous with the fascia lata. In adults it is 12–14 cm in length and inclined 40–35° to the horizonal; its lateral half is rounded and more oblique, its medial half gradually widening towards its pubic attachment, becoming more horizontal where it supports the spermatic cord. Deep fibres of the aponeurosis are *not* initially parallel to the ligament's axis, approaching at an angle of 10–20°, then turning *medially*, the majority then running along the ligament to reach the pubic tubercle. Even deeper fibres splay out posteromedially to the pecten pubis (Lytle 1974).

The lacunar ligament (*pectineal part of inguinal ligament*) (4.58) has been described as composite (Lytle 1974). Its abdominal, 'deep' part (i.e. classic lacunar ligament) extends posterolaterally from the medial part of the inguinal ligament to the medial end of pecten pubis. It is triangular, almost horizontal, a little larger in males and about 2 cm from base to apex. Its thin base, concave laterally, is the femoral ring's medial rim; its apex is attached to the pubic tubercle. Its posterior margin, attached to

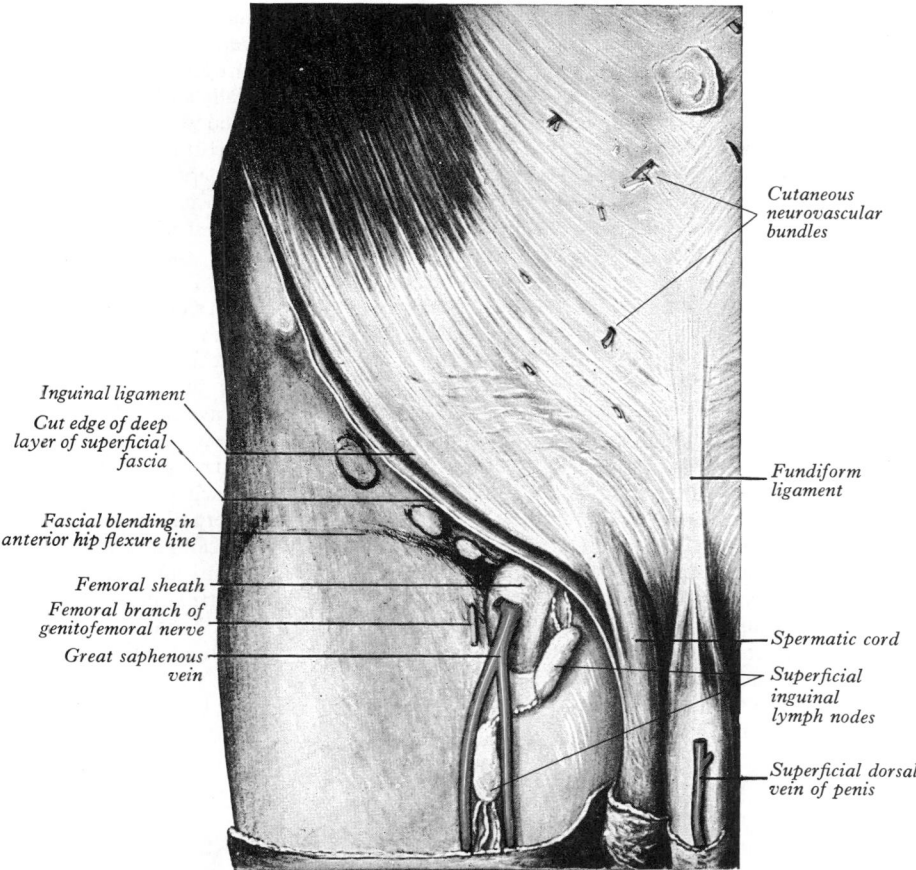

5.45 Superficial structures of the inguinal region and the lower part of the anterior abdominal wall (right side).

Labels on figure 5.45:
- Cutaneous neurovascular bundles
- Inguinal ligament
- Cut edge of deep layer of superficial fascia
- Fascial blending in anterior hip flexure line
- Femoral sheath
- Femoral branch of genitofemoral nerve
- Great saphenous vein
- Fundiform ligament
- Spermatic cord
- Superficial inguinal lymph nodes
- Superficial dorsal vein of penis

pecten pubis, is continuous with the *pectineal fascia*; its anterior margin continues into the main inguinal ligament. It has *superior* and *inferior* surfaces. A strong *pectineal ligament* (of *Astley Cooper*) extends laterally from its base (4.58) along the pecten pubis, augmented by pectineal fascia and by a lateral expansion from the linea alba (*adminiculum lineae albae*, p. 603). The lacunar formation as here described is most obvious, at operation or dissection, from its abdominal aspect. In the femoral approach, in the living, a second lacunar fibrous sheet is visible, which is an inflexion of *fascia lata* which joins the inguinal ligament's posterior border and, reinforced by transversalis fascia, fuses with pectineal fascia for about 1 cm before ascending to fuse with periosteum on the pecten pubis. This *fascial lacunar ligament* also has a thick curved lateral border fitting the medial wall of the femoral sheath; because of its pectineal attachment it is about 1 cm infero-anterior to pecten pubis and 3 cm lateral to the pubic tubercle. Current descriptions of boundaries of femoral hernia may hence need revision, if these findings are confirmed (McVay & Anson 1940, Lytle 1957, 1974).

The reflected inguinal ligament (5.46,50) is an expansion from the lateral crus of the superficial inguinal ring, ascending medially behind the superficial inguinal ring, behind external oblique and in front of the falx inguinalis; fibres of the right and left ligaments decussate in the linea alba.

The superficial inguinal ring (5.45,46), an aponeurotic hiatus superolateral to the pubic crest, is somewhat triangular, with its long axis parallel to the deep aponeurotic fibres. Its size varies; usually it does not extend beyond the inguinal ligament's medial third. It is smaller in females, spermatic cords being absent. Its base is the pubic crest, its sides the aponeurotic *crura*. The lateral crus is stronger and its fibres are attached to the pubic tubercle; it is curved into a groove for the spermatic cord. The medial crus is thin, flat and attached to the front of symphysis pubis, interlacing with its fellow. In or seen through the layer of investing fascia superficial to the external oblique, are variable curved fibrous bands at right angles to those in the aponeurosis. Some may arch above the superficial inguinal ring as *intercrural*

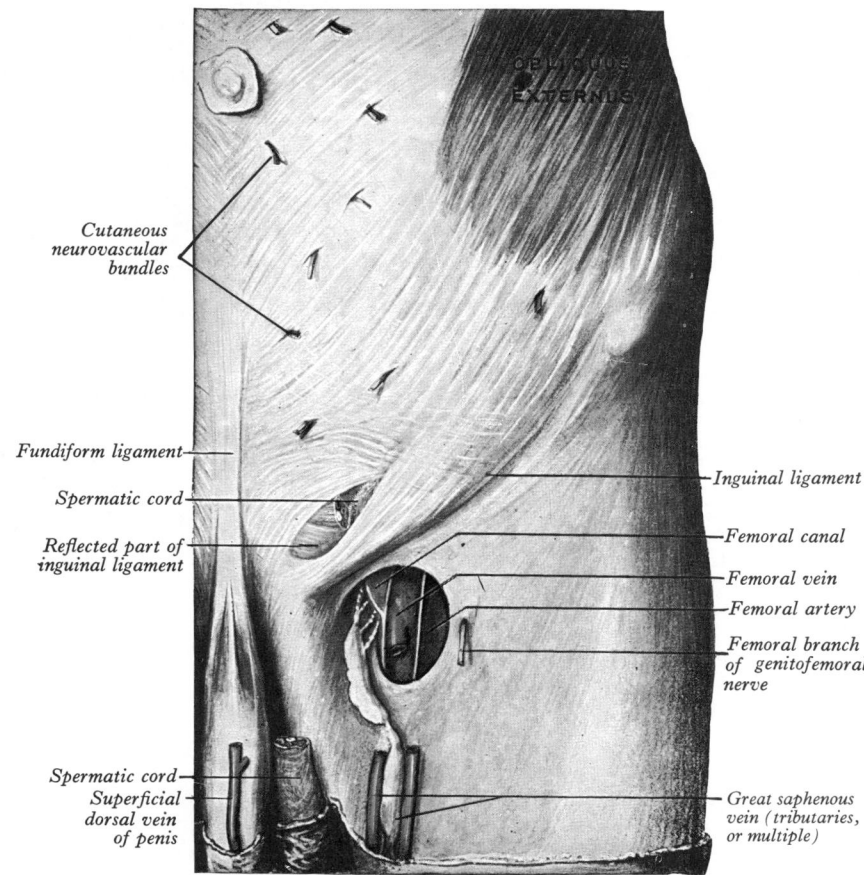

Labels on figure 5.46:
- OBLIQUE EXTERNUS
- Cutaneous neurovascular bundles
- Fundiform ligament
- Spermatic cord
- Reflected part of inguinal ligament
- Spermatic cord
- Superficial dorsal vein of penis
- Inguinal ligament
- Femoral canal
- Femoral vein
- Femoral artery
- Femoral branch of genitofemoral nerve
- Great saphenous vein (tributaries, or multiple)

5.46 Superficial structures of the inguinal region and the lower part of the anterior abdominal wall (left side). Superficial aponeurotic layer (see text) was removed.

fibres. In the account given above, these have been allocated as the lowest fibre bundles of the superficial aponeurotic layer (derived from the contralateral obliquus abdominis externus).

The superficial inguinal ring contains the spermatic cord in males, uterine round ligament in females and ilio-inguinal nerve in both. By invaginating scrotal skin the cord can be followed into the ring, whose crura can be felt and its size assessed. From its margins the external oblique aponeurosis and overlying fascia continue as a tenuous investment around the spermatic cord and testis, the *external spermatic fascia*; the ring is visible only when the continuity of fascia and aponeurosis is severed (Anson et al 1960).

Obliquus internus abdominis (5.47, 53C–E), largely (but not wholly) internal to the external oblique and thinner, has a muscular attachment: to the lateral two-thirds of the inguinal ligament's grooved upper aspect; to the anterior two-thirds of the intermediate line of the ventral segment of the iliac crest; and to the thoracolumbar fascia (5.38). It has been described as attached to the iliac fascia and not directly to the inguinal ligament (McVay & Anson 1940); but fascia and ligament are here adherent (p. 601). Posterior (iliac) fibres ascend laterally to the inferior borders and tips of the lower three or four ribs and their cartilages and are continuous there with the internal intercostals. The lowest fibres, i.e. those from the inguinal ligament, are paler in colour and arch *inferomedially* across the spermatic cord or uterine round ligament, become tendinous and are attached with part of the aponeurosis of transversus abdominis to the pubic crest and medial part of pecten pubis, forming the *falx inguinalis (conjoint*

tendon). Intermediate fibres diverge to end in a bilaminar aponeurosis broadening upwards, the lower fibres being horizontal, while progressively higher fibres are increasingly steep inclining medially and upwards. The uppermost part of the aponeurosis is attached to the outer surfaces and lower borders of the seventh to ninth costal cartilages. Between the costoxiphoid margin and pubis the aponeurosis consists of anterior and posterior layers and their fate is, again, considered at three levels: down to umbilical level; umbilical to iliac crest levels; and between the latter and the pubis. Differing from traditional accounts, some features recorded by Rizk (1980) are given here (in outline only; consult reference for details). The *anterior layer*, down to umbilical level, ascends medially ventral to rectus abdominis, but posterior to the two layers of the external oblique's aponeurosis (providing a *trilaminar* anterior wall to the rectus sheath), then traversing the linea alba; crossing the midline the fibres become directly continuous with the deep aponeurotic fibres of the contralateral external oblique (completing the bilateral digastric muscle mentioned above). Between umbilicus and iliac crest the internal's anterior layer ascends ventral to the rectus to its mid-vertical line where it enters a linear decussation with externus to continue medially as the most superficial (ventral) layer of the trilaminar rectus sheath (i.e. covering the externus aponeurosis). From iliac crest level to pubis the anterior layer blends as one constituent of the falx inguinalis.

The *posterior layer*, down to umbilical level passes medially posterior to rectus, but anterior to the bilaminar aponeurosis of transversus abdominis (providing a *trilaminar* posterior wall to

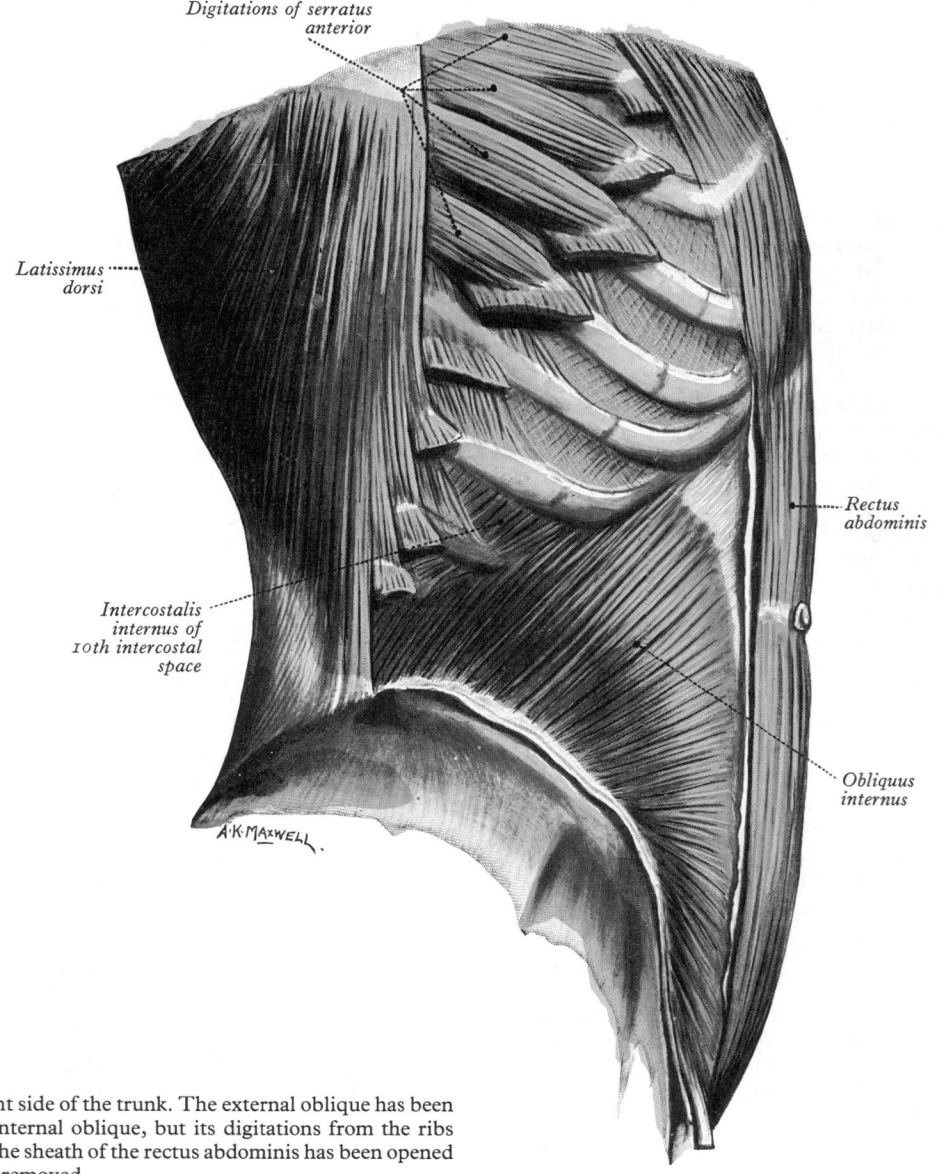

5.47 Muscles of the right side of the trunk. The external oblique has been removed to show the internal oblique, but its digitations from the ribs have been preserved. The sheath of the rectus abdominis has been opened and its anterior lamina removed.

the rectus sheath); crossing the midline in the linea alba the posterior layer fibres of internus become directly continuous with the 'downturned' aponeurotic fibres of the contralateral transversus abdominis (another bilateral digastric muscle). From umbilical to iliac crest levels, however, the posterior layer (aponeurotic fibres of internus) undergoes a vertical linear decussation with the bilaminar aponeurosis of transversus to reach its deep (abdominal) surface. The decussation occurs along the line of the lateral border of rectus, thus between these levels; the most posterior (deepest) stratum of the posterior wall of the rectus sheath is derived from obliquus internus. Below the level of the iliac crest, fibres of the posterior aponeurotic layer of obliquus are gradually and progressively transferred to the anterior layer and thence to the linea alba and (lower fibres) blend with the falx inguinalis (cf. transversus abdominis, vide infra).

Nerve supply: ventral rami of the lower six thoracic and first lumbar spinal nerves.

Cremaster (5.49) consists of loose muscle fasciculi lying along the spermatic cord, united by areolar tissue to form the *cremasteric fascia* around the cord and testis within the external spermatic fascia. Its lateral part, attached to the inguinal ligament, has been variously described as: continuous with the medial edge of the internal oblique; or deep to the internal oblique, extending to the anterior superior iliac spine; or as continuous with either internal oblique or transversus; or as a pointed tendon from the middle of the inguinal ligament, piercing the internal oblique near its medial margin. Sixty subjects all exhibited a tendinous attachment *and* cremasteric fibres derived from obliquus internus and transversus (Blunt 1951). The fibres traverse the lateral aspect of the spermatic cord through the superficial inguinal ring to spread

out into loops of increasing length along its anterolateral aspect. The shortest and highest fasciculi turn inwards in front of the cord to join the medial part; longer fasciculi blend with the fascia over the cord and upper tunica vaginalis. The *medial part* of the muscle is variable and may be absent. It is attached to the pubic tubercle and possibly pubic crest, falx inguinalis and lower border of transversus. Its fasciculi loop posteromedial to the cord, interlacing with those of the lateral part. The whole muscle forms continuous loops from the mid-inguinal region to tunica vaginalis, returning to the pubic tubercle. In females a few fibres descending on the round ligament represent the lateral part.

Nerve supply: genital branch of the genitofemoral nerve, from the first and second lumbar spinal nerves.

Action: elevation of the testis towards the superficial inguinal ring. Although its fibres are striated, cremasteric actions are not usually voluntary. Stimulation of medial femoral skin evokes reflex contraction; this *cremasteric reflex* is more active in children. Doubtless this may be protective, but more significantly the cremaster appears to be essential to testicular thermoregulation. Shafik (1977) considers that the position of the testis is adjusted by the cremaster (which he divides into *cremaster internus* and *externus*, separated by internal spermatic fascia) and dartos, attached to the testis by a '*scrotal ligament*' (p. 1431).

Transversus abdominis (5.48, 53C–E), the innermost of the anterolateral abdominal muscles, is attached to the lateral third of the inguinal ligament, the anterior two-thirds of the inner lip of the ventral segment of the iliac crest, the thoracolumbar fascia between the iliac crest and the twelfth rib, and internal aspects of the lower six costal cartilages, where it interdigitates with the diaphragm (5.41). Whether the muscle's inguinal fibres are attached

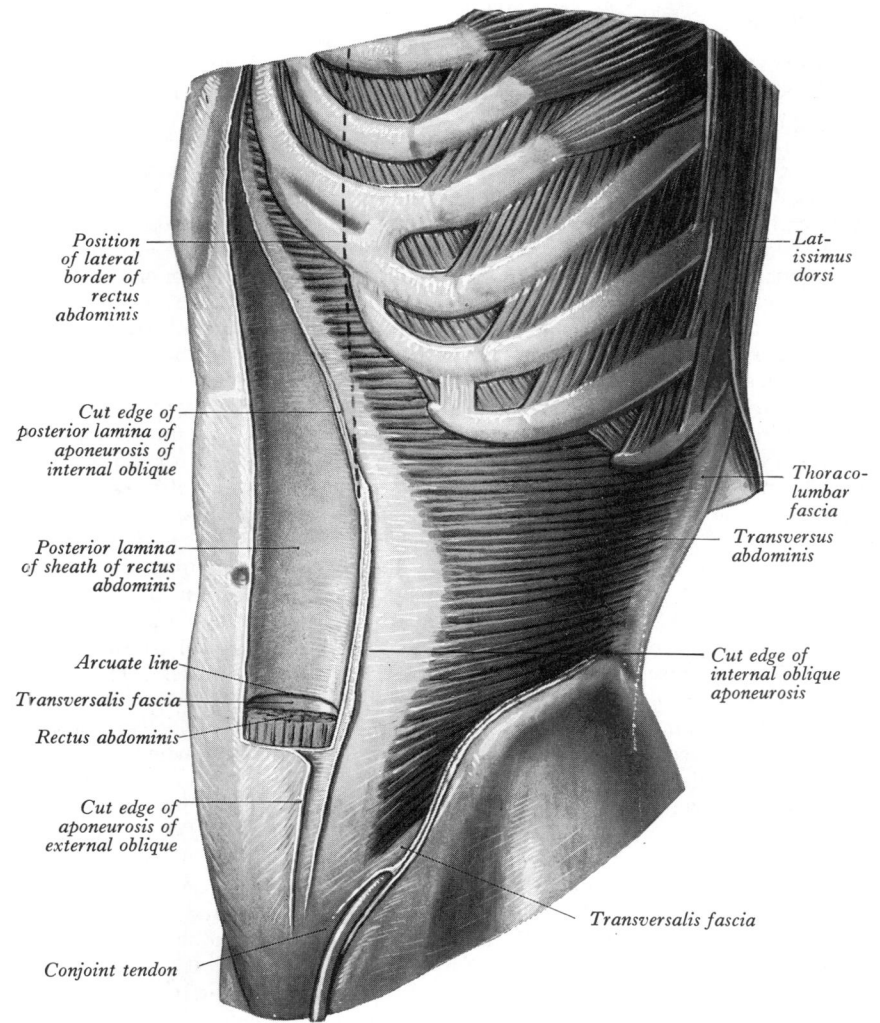

Position of lateral border of rectus abdominis

Cut edge of posterior lamina of aponeurosis of internal oblique

Posterior lamina of sheath of rectus abdominis

Arcuate line

Transversalis fascia

Rectus abdominis

Cut edge of aponeurosis of external oblique

Conjoint tendon

Latissimus dorsi

Thoraco-lumbar fascia

Transversus abdominis

Cut edge of internal oblique aponeurosis

Transversalis fascia

5.48 The left transversus abdominis. This diagram incorporates the traditional features ascribed to an aponeurosis of transversus (see text and 5.53C–E for alternative analysis).

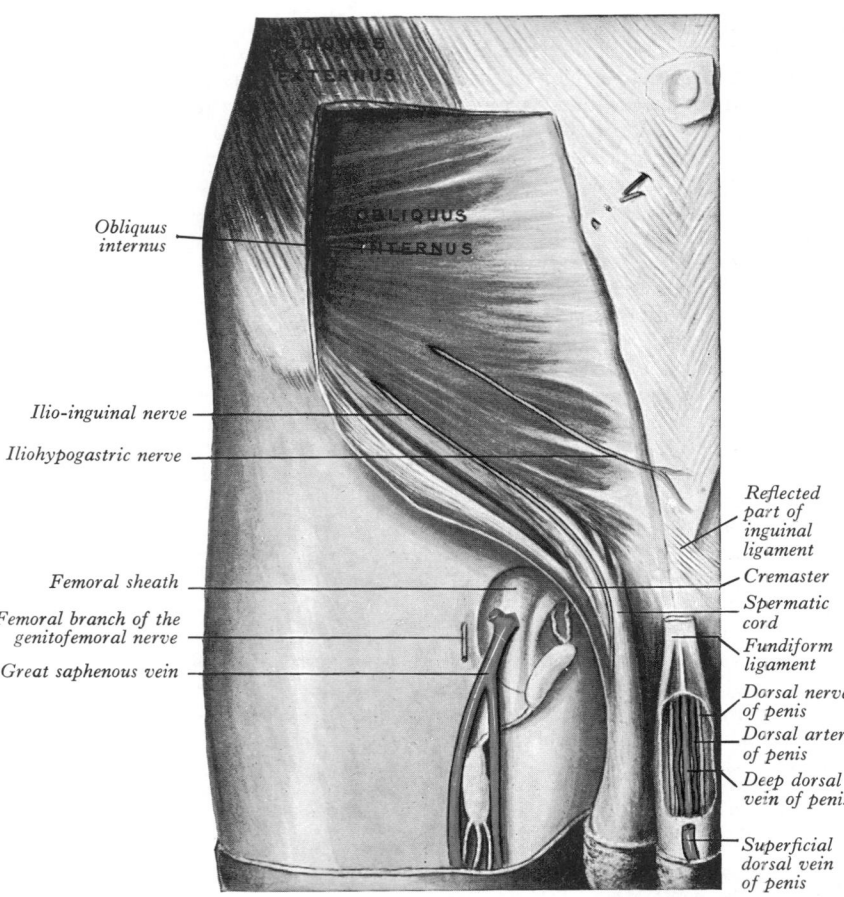

Obliquus
internus

Ilio-inguinal nerve

Iliohypogastric nerve

Reflected
part of
inguinal
ligament

Cremaster

Femoral sheath

Spermatic
cord

Femoral branch of the
genitofemoral nerve

Fundiform
ligament

Great saphenous vein

Dorsal nerve
of penis

Dorsal artery
of penis

Deep dorsal
vein of penis

Superficial
dorsal vein
of penis

5.49 Dissection of the regions shown in 5.45, with part of the obliquus externus removed. (See text for additional recent observations on abdominal aponeuroses.)

directly to the ligament or via the iliac fascia, is disputed (McVay & Anson 1940). The muscle ends in a complex and variable bilaminar aponeurosis whose *lower fibres* curve inferomedially blending with those of the internal oblique to reach the pubic crest and pecten, as the *falx inguinalis*. The rest pass medially, variously related to rectus abdominis, then decussate at and blend with the linea alba. The length of transversus muscle fibres varies considerably in accord with the disposition of its posterior skeletal and fascial attachments and the curvilinear line of aponeurosis formation. Upper costal and anterior iliac fibres are short, lower costal and posterior iliac fibres are longer and thoracolumbar fibres are longest. Aponeurosis formation is only 2–3 cm from the linea alba above (i.e. transversus *muscle* extends behind rectus, and decreasingly so down to the level of the tip of the eleventh rib); thus the lateral limit of the aponeurosis, from above, curves first downwards and laterally, is widest (5–6 cm lateral to rectus) at umbilical levels, then recurves downwards and medially towards, but not reaching, the central position of the superior crus of the superficial inguinal ring. The aponeurosis of transversus abdominis is bilaminar, having *anterior* and *posterior* layers but their dispositions vary above and below umbilical levels (Rizk 1980). *Above the umbilicus* fibre bundles segregate into laminae either with their fibres inclined upwards and medially ('up-turned') or downwards and medially ('downturned'). In 70% of specimens examined the anterior layer was 'downturned' and the posterior layer 'upturned'; in the remaining 30% the reverse obtained. The layers approached the midline (after forming the two posterior laminae of the trilaminar rectus sheath) and decussated to cross the midline in the linea alba. The decussation was linear, single and median in 30%, or triple with one median and two bilateral lines along the medial borders of recti abdominis. After crossing the midline, the downturned fibres of one transversus are continuous with the upturned fibres of the contralateral transversus and with the posterior layer of the contralateral obliquus internus (vide supra).

Below the umbilicus both layers of the transversus are inclined downwards and medially and in the upper part of the region the anterior and posterior layers are related to rectus abdominis, as their names indicate, i.e. directly in contact with the anterior and posterior surfaces of the rectus. Between the umbilicus and iliac crest the relative positions of the laminae enclosing the transversus strata are reversed (as described with obliquus internus). Proceeding from the umbilicus towards the pubis the fibres of the posterior layer of the transversus aponeurosis are slowly and progressively transferred to the anterior layer; the posterior layer diminishes in thickness and becomes a fine attenuated layer before fading a short distance above the pubic level. The change in character of the posterior wall of the rectus sheath does not occur at an abrupt arcuate line, as in traditional accounts, and its progressive attenuation is compensated for by an increased thickness of fascia transversalis. The lowest aponeurotic fibres of transversus form a principal component of the falx inguinalis. Above this the anterior and posterior fibres of transversus cross the midline and are continuous with the anterior and posterior aponeurotic fibres of the contralateral obliquus internus respectively.

Fusiform defects filled with fascia occur in lower parts of the internal oblique and transversus abdominis. The muscles are sometimes fused; transversus may be absent.

The falx inguinalis (5.50,51), the conjoint tendon of internal oblique and transversus, mainly from the latter's compound aponeurosis, is attached to the pubic crest and pecten. It descends behind the superficial inguinal ring, supporting an otherwise weak area. Attachment to the pecten pubis is often absent. Medially it fuses with the *anterior* wall of the rectus sheath. Laterally, it may blend with an inconstant *interfoveolar ligament* (5.51), which sometimes connects the lower margin of transversus to the superior pubic ramus and occasionally contains a few muscular fibres. Muscular fasciculi, attached to the pecten pubis *behind* the falx inguinalis, may reach the transversalis fascia, the muscle's aponeurosis or even the lateral end of the arcuate line.

Nerve supply: ventral rami of the lower six thoracic and first lumbar spinal nerves.

Rectus abdominis (5.52), a long muscular strap, broader but thinner above, descends throughout the abdominal wall, from ventral lower thorax to pubis, and is separated from its fellow by the linea alba. Narrower and thicker below, it springs by two tendons; the larger lateral one is attached to the pubic crest and tubercle and sometimes extends beyond the tubercle to the pecten pubis; the medial tendon, interlacing with its fellow, blends with ligaments anterior to the symphysis pubis and pubic bodies. Some fibres may spring from the linea alba's lower part. The muscle is attached above by three unequal slips to the fifth to seventh costal cartilages; most lateral fibres are usually attached to the ventral end of the fifth rib; this slip may be lacking but conversely the muscle may also reach even the fourth and third ribs; the most medial fibres are occasionally attached to costoxiphoid ligaments and their side of the xiphoid process.

The rectus is interrupted by three *tendinous intersections*: one usually near umbilical level, another near the xiphoid apex, a third about midway between these. Transverse or obliquely zigzag, viewed from the ventral aspect, they rarely extend through the full thickness of the muscle, often only halfway through; they are closely adherent to the anterior compound wall of the rectus sheath. Sometimes one or two incomplete intersections appear below the umbilicus; whether they are myosepta delineating myotomes remains controversial. The muscle is enclosed by aponeuroses of the oblique and transverse muscles, forming a so-called *rectus sheath* (5.44,48,53A–E). **The traditional account** (5.53A,B) of the composition of the rectus sheath is first given here; in part because of its frequent inclusion in introductory courses; also to allow a direct comparison with the modification, revision and strong functional implications that stem from extensive recent investigations; these are summarized below. Traditionally each anterolateral muscle was regarded as initially forming a *single unilaminar* aponeurosis, that was confined to its ipsilateral side and *ended* in the median linea alba. At the lateral margin of the rectus abdominis the internal oblique aponeurosis divides: one lamina passing anterior to the rectus and blended with the external oblique's aponeurosis; one behind it blending with that of transversus. These laminae blend at the medial border, helping to form the linea alba. Some authors differed in their description, e.g. Walmsley (1937), McVay & Anson (1940), but this arrangement was widely agreed to exist from the costal margin to a variable level, usually midway between the umbilicus and symphysis pubis, where the sheath's posterior wall was described as ending in an *arcuate line*, concave downwards. The upper muscular fibres of transversus abdominis (p. 600) continue behind the rectus abdominis nearly to the linea alba (**5**.48,53A), the posterior layer of the sheath being here variably muscular. Below the arcuate line all three aponeuroses are anterior to the rectus, those of transversus and internal oblique intimately fused, that of external oblique bound to them by loose tissue except near the midline; this part of the rectus is separated from the peritoneum only by transversalis fascia (**5**.53B). Since the aponeuroses of the internal oblique and transversus reach only to the costal margin, above that level the rectus rests directly on the costal cartilages, overlapped here merely by the aponeurosis of external oblique.

More recent observations (Rizk 1980) on the aponeuroses of the ventrolateral abdominal muscles, rectus sheath, linea alba, linea semilunaris and the site of the so-called arcuate line have been mentioned in the descriptions of individual muscles. With the exception of the arcuate line the general extent of the sheath differs little from the traditional account. Major differences are as follows:

1. The obliquus externus, internus and transversus each form a *bilaminar aponeurosis*.

2. Of the six laminae three pass anterior to rectus and three posterior; i.e. the walls of the rectus sheath are *trilaminar* down to the umbilicus; below this, layers are changed in their superficial-deep sequence and near the pubis they are modified by the formation of the falx inguinalis.

3. The aponeurotic laminae undergo extensive or limited vertical linear *decussations* with either their antimere or another muscle, some at the linea semilunaris, some at the ventral vertical midline of the rectus, but most at the linea alba, in the midline and/or bilaterally.

4. Each aponeurotic lamina crosses the midline and becomes structurally and functionally continuous either with its antimere or another contralateral muscle.

5. The four principal (perhaps unfortunately named, but informative) *bilateral digastric* muscles are: two external obliques together, two transversus muscles together, one internal oblique (anterior layer) with contralateral external oblique (posterior or deep layer) and finally, one internal oblique (posterior layer) with contralateral transversus abdominis (anterior layer). (See **5**.53C–E.)

From the costoxiphoid margin to the umbilical level the anterior wall of the rectus sheath is, as stated, trilaminar and consists (from the external surface, deeply) of: (a) an external oblique aponeurosis superficial layer (fibres oblique, disposed upwards and medially), i.e. downwards and laterally when described as continuations from the contralateral external oblique; (b) an external oblique aponeurosis deep layer (fibres oblique downwards and medially); (c) an internal oblique anterior layer (fibres oblique upwards and medially). Thus the layers (a) and (c) have parallel fibres (upwards and medially), and layer (b), sandwiched between them, has inferomedially directed fibres at right angles to them both. This construction has been likened to the cross-grain of plywood and it is also characteristic of the posterior wall of the sheath in at least 70% of individuals. Proceeding from the muscle towards the extraperitoneal fascia, it has: (x) an internal oblique posterior layer (upwards and medially); (y) a transversus abdominis anterior layer (downwards and medially); (z) a transversus abdominis posterior layer (upwards and medially).

As indicated with the individual muscles, the sequence of layers is changed between the umbilicus and the iliac crest levels and the patterns of fibre disposition change accordingly. Also noted previously was the absence of a sudden posterior transition at an arcuate line; instead there is a slow attenuation with transference of fibres to the anterior layer and to the falx inguinalis and a posterior compensatory thickening of the fascia transversalis.

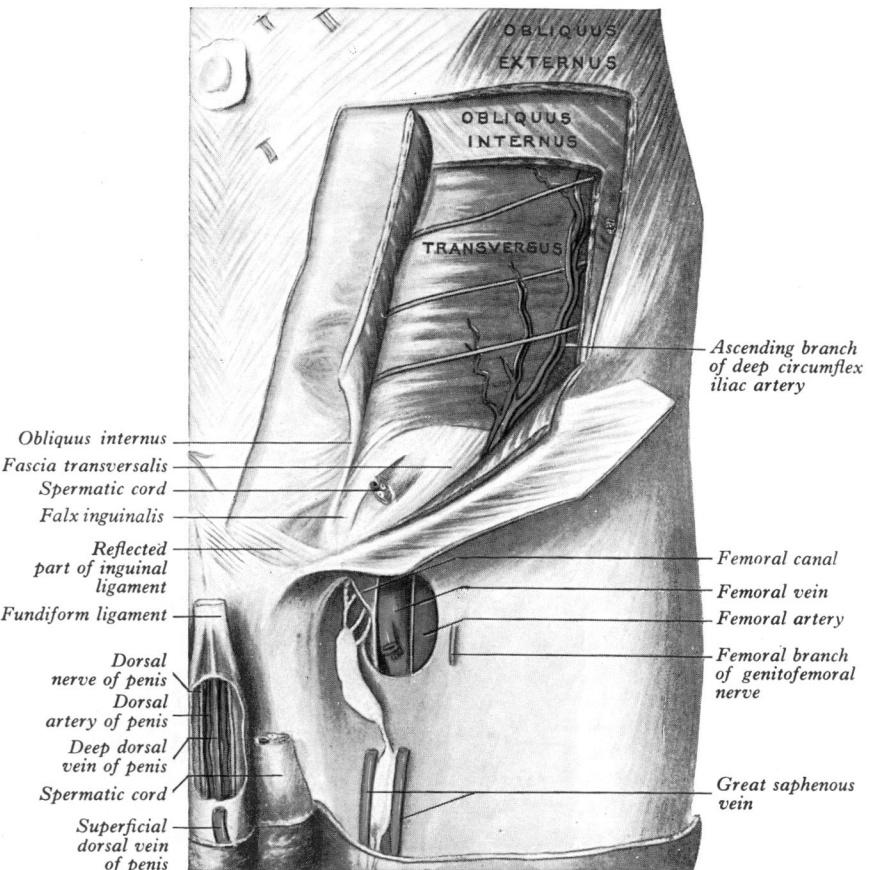

Obliquus internus
Fascia transversalis
Spermatic cord
Falx inguinalis
Reflected part of inguinal ligament
Fundiform ligament
Dorsal nerve of penis
Dorsal artery of penis
Deep dorsal vein of penis
Spermatic cord
Superficial dorsal vein of penis

Ascending branch of deep circumflex iliac artery
Femoral canal
Femoral vein
Femoral artery
Femoral branch of genitofemoral nerve
Great saphenous vein

5.50 Dissection of the regions shown in **5**.46, with parts of the external and internal oblique muscles removed. (See text for additional recent observations on abdominal aponeuroses.)

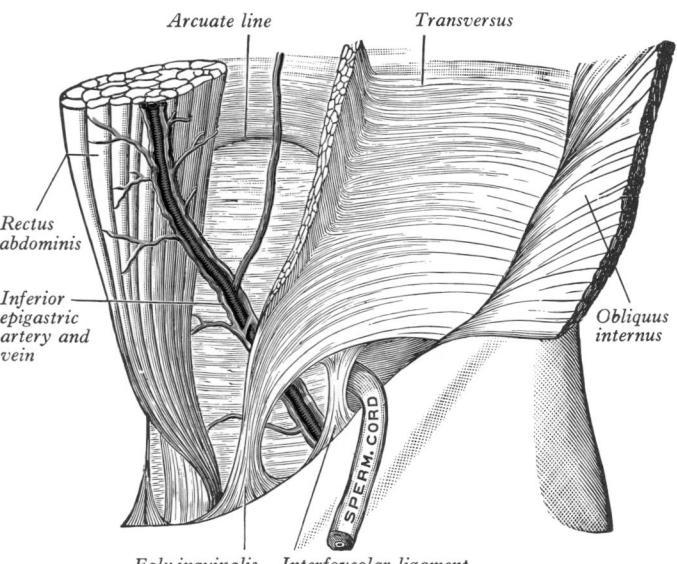

Arcuate line   Transversus
Rectus abdominis
Inferior epigastric artery and vein
Obliquus internus
SPERM. CORD
Falx inguinalis   Interfoveolar ligament

5.51 The lower part of the anterior abdominal wall (left side) showing the relations of the spermatic cord at the deep inguinal ring. (Modified from Braune.) Some authorities do not recognize the sharp arcuate line as the abrupt terminus of the posterior wall of the rectus sheath (see text).

The rectus muscle's medial border adjoins the linea alba: its lateral border may show as a superficial, curved *linea semilunaris* from the tip of the ninth costal cartilage to the pubic tubercle and is visible in muscular subjects even when the muscle is relaxed, but usually obscured in the corpulent.

Nerve supply: ventral rami of the lower six or seven thoracic spinal nerves.

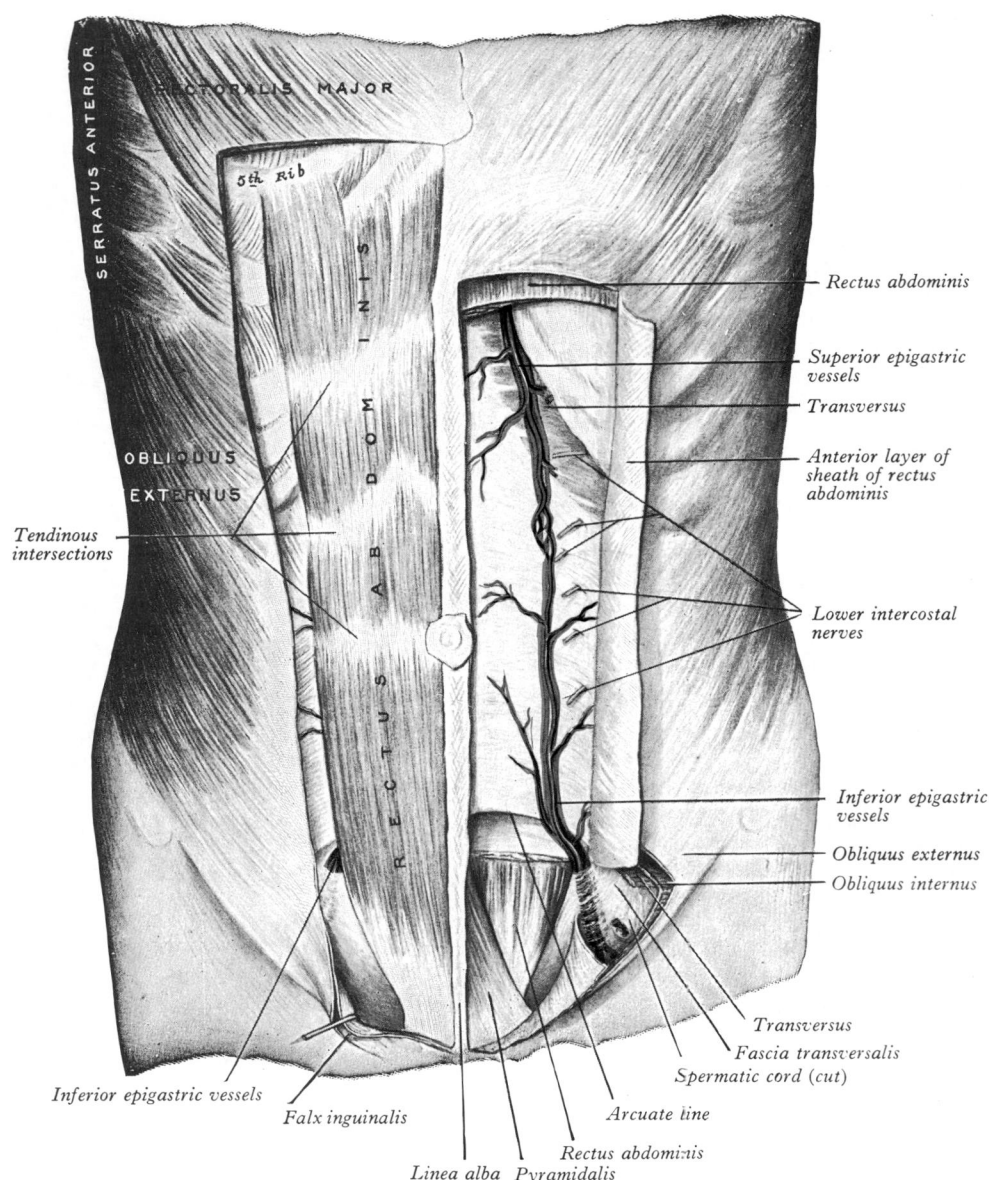

5.52 The right rectus abdominis and the left pyramidalis. The greater part
of the left rectus abdominis has been removed to show the superior and
inferior epigastric vessels. (See text for additional recent observations on
the abdominal aponeuroses, rectus sheath and controversial 'arcuate
line'.)

Pyramidalis (5.52) is a triangular muscle anterior to the lower
part of rectus abdominis and in its sheath; it is attached by tendinous fibres to the front of the pubic body and the anterior
ligaments of its symphysis; it ascends, diminishing in size, with its
lateral border inclining medially and its pointed end embedded in
the linea alba midway between the umbilicus and pubis,
sometimes higher. It varies much and may be larger on one side,
absent, or sometimes double. Anson et al (1938) found it absent in
76 out of 430 cadavers (17.7%).

Besides rectus abdominis and pyramidalis the rectus sheath
contains superior and inferior epigastric vessels and terminal
parts of the lower intercostal nerves.

Nerve supply: subcostal nerve (i.e. the ventral ramus of the
twelfth thoracic spinal nerve).

Actions. The anterolateral muscles are a firm but elastic wall
retaining abdominal viscera and opposing actions of gravity on
them in both erect and sitting postures. However, their most
frequent phasic involvement is in respiration.

When the thorax and pelvis are relatively fixed in position (or
during minimal strongly controlled movement), contraction of
the abdominal muscles 'compresses' abdominal viscera and assists in expelling air during expiration (also faeces, urine, gastric

contents, or a fetus), mainly due to the oblique and transverse
muscles, which tense the rectus sheaths and lineae alba and
semilunaris, the rectus abdominis itself playing a minor role. If
the pelvis and vertebral column are fixed, the external oblique
muscles further aid expiration by depressing and compressing the
lower thorax. When the pelvis is fixed the recti, aided by the
obliqui, *flex* the lumbar vertebral column; with the thorax fixed
they draw the pelvis up and forwards, also flexing the vertebral
column (Floyd & Silver 1950). If the abdominal muscle's action is
predominantly unilateral, the trunk is *laterally flexed* to that side.
The obliques also co-operate as bilateral digastric muscles in effecting *vertebral rotation*; e.g., as noted, the deep fibres of external
oblique decussate at the midline and are continuous with the
anterior aponeurotic layer of the internal oblique of the
contralateral side. Their actions are complementary; the contraction of one external oblique turns the abdomen and rotates the
vertebral column to the contralateral side, while the *opposite* internal oblique turns and rotates them to *its* ipsilateral side, i.e. in the
same direction. Electromyography suggests that in most trunk
movements, sitting or standing, abdominal musculature is little
involved, unless resistance is applied. Extension raises their activity, flexion does not. All activity ceases in the supine position bu

602

the recti at once contract even when the head alone is raised. Further flexion brings the obliques into action, but less forcibly. The obliques appear largely concerned in compressive and twisting movements, the recti in flexion against resistance. In standing there is little activity in abdominal musculature, except for the internal oblique, accessible to electromyography in its lower part (p. 598). Transversus was long assumed to be only compressive and not acting on the vertebral column. However, the more recent analyses of the dispositions of its aponeurotic layers, and particularly the completion of a digastric bilateral muscle between the 'downturned' fibres of one transversus with the posterior layer of the contralateral transversus, makes rotational activity a distinct possibility. Further researches are necessary; unfortunately effective electromyography is difficult. (The *oblique orthogonal* aponeurotic layers cast doubt on the suitability of the name 'transversus'.) Pyramidalis is a tensor of the linea alba.

The linea alba (5.44,53), a complex tendinous raphe extending from the xiphoid process to the symphysis pubis and pubic crest between the recti, is an intermixture of aponeurotic fibres of the oblique and transverse muscles. It was long regarded as a mere moulded feltwork of poorly vascularized collagen that provided a strong structural bond between the rectus sheaths and terminal attachments ('insertions') for essentially unilateral ventrolateral muscles. Its structural status has now been profoundly revised. The three large ventrolateral muscles on both sides each give rise, as noted, to a bilaminar aponeurosis. Having variously contributed to the walls of the rectus sheath, medial to the rectus, six laminae on each side border the interrectal gap. This is occupied by numerous *linear decussations* and together these constitute the linea alba. Each lamina and its antimere may have one or three vertical lines of decussation and this varies between individuals and laminae. When single, the decussation naturally is midline; when triple, one is midline, the others near the medial border of each rectus. The decussations between particular layers are densely packed but occur successively at increasing depths from the surface. After decussation, each layer is continuous with its antimere or with an aponeurotic lamina from another contralateral muscle. Individual examples are mentioned above in the descriptions of particular muscles; note also less prominent sites of decussation, a limited subumbilical length deep to the linea semilunaris and at a similar level ventral to the midline of rectus abdominis. (The positions of the linear decussations and transmedian continuations of aponeuroses are shown in 5.53C–E.) Just below the midpoint of the linea alba is a cicatrix, covered by adherent skin, the *umbilicus*, below which the linea alba is narrow, following the interval between the recti and visible only in the lean and muscular as a slight groove. In its supra-umbilical part, where recti diverge, it is broader and thinner and superficially there is a shallow groove. Inferiorly its superficial fibres pass *in front* of the medial heads of the recti to reach the symphysis pubis; its deeper fibres form a triangular lamella, attached *behind* the recti to the pubic crest on each side as the *adminiculum lineae albae*. Small vessels traverse the linea. The umbilicus transmits umbilical vessels, urachus and, until the third month, the vitelline stalk; it closes a few days after birth, but vestiges of vessels and urachus are attached to its deep aspect (i.e. *hepatic ligamentum teres* and *median* and *medial umbilical ligaments*).

Transversalis fascia is a thin areolar stratum between the transversus and extraperitoneal fat, part of the general layer between the peritoneum and the abdominal walls and continuous with the iliac and pelvic fasciae. In the inguinal region it is thick, dense and augmented by the attenuated aponeurosis of transversus; but it thins as it ascends to blend with the inferior diaphragmatic fascia. Behind, it fuses with the anterior lamina of the thoracolumbar fascia. Below it is attached along the whole iliac crest between the transversus and iliacus and to the posterior margin of the inguinal ligament from the anterior superior iliac spine to the femoral vessels, where it is continuous with the iliac fascia. Medial to these vessels it is thin and fused to the pecten pubis behind the falx inguinalis, with which it blends. It descends anterior to the vessels as the anterior part of the femoral sheath (p. 81); transversalis fascia is here strengthened by transverse arched fibres spreading laterally to the anterior superior iliac spine and diverging medially behind the rectus abdominis, while some

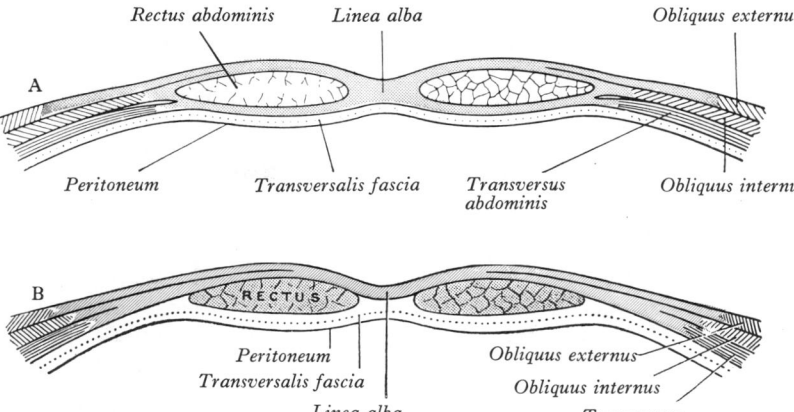

5.53A,B Transverse sections through the anterior abdominal wall, traditional view: (A) immediately above the umbilicus; (B) below the arcuate line. Note the extent to which the external oblique aponeurosis remains as a separate entity, passing medially, ventral to the rectus, before blending with the other aponeuroses; these have already fused lateral to rectus. The traditional view is retained here to maintain historical perspective and for readers whose objectives it remains adequate. (For further analysis see text, 5.53C–E and Rizk 1980.)

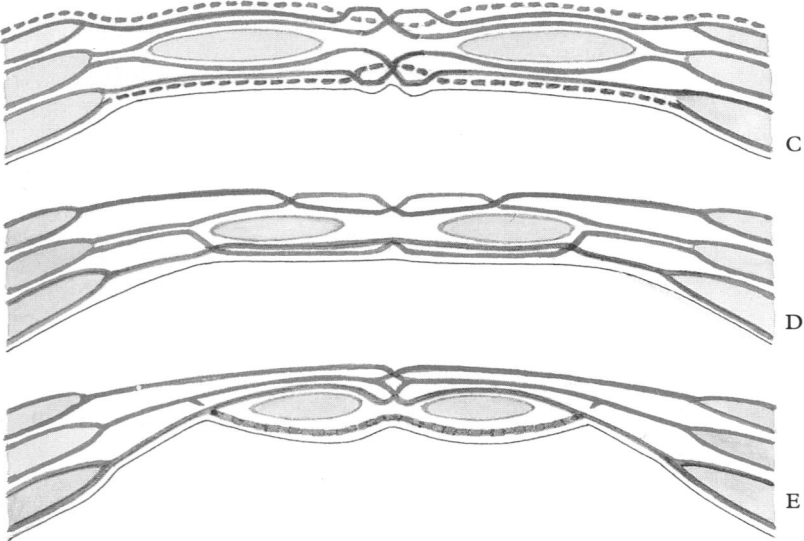

5.53C–E Schematic transverse sections through the ventral abdominal wall, showing bilaminar aponeuroses, external oblique (blue), internal oblique (magenta), transversus abdominis (green), and sites of linear decussation (arrows) which compacted form the linea alba. Tenuous layers are dotted or omitted. (See text for details and reference to Rizk 1980.)

descend to the pecten pubis behind the falx inguinalis. These fibres constitute the *deep crural arch*. The spermatic cord or round ligament passes through the transversalis fascia at the *deep inguinal ring*. The *internal spermatic fascia*, derived from the transversalis, is prolonged along these structures to blend with areolar tissue in the parietal layer of the tunica vaginalis (p. 1424); it sometimes contains muscle fibres (Barrett 1951). The deep crural arch thickens the inferomedial part of the deep inguinal ring.

The deep inguinal ring in the transversalis fascia is midway between the anterior superior iliac spine and the symphysis pubis; it is about 1.25 cm above the inguinal ligament. Oval, its long axis vertical, it varies in size, being larger in males. It is related above to the lower margin of transversus abdominis, medially to the inferior epigastric vessels and interfoveolar ligament, when present. Traction on the ring by the internal oblique is perhaps a valvular safety mechanism when intra-abdominal pressure is raised (Lytle 1970).

The inguinal canal contains the spermatic cord or uterine round ligament and ilio-inguinal nerve. *Oblique*, about 4 cm long, it slants *inferomedially, parallel with* and a little above the inguinal ligament, from the deep to the superficial inguinal rings. *Anterior* to the canal are the skin, superficial fascia, external oblique aponeurosis and, in its lateral third, muscular fibres of the internal oblique. *Posterior* are the reflected inguinal ligament, falx inguinalis and transversalis fascia, separating it from extraperitoneal connective tissue and the peritoneum. *Superior* are the arched fibres of the internal oblique and transversus abdominis. *Inferior* is the union of the transversalis fascia and the inguinal ligament. At its medial end, inferiorly, is the lacunar ligament.

The canal would appear to weaken the lower anterior abdominal wall but this is countered by its obliquity and mural structure. Thus its rings do not coincide; increases in intra-abdominal pressure affect not only the deep ring but also the posterior wall, pressing it firmly towards the anterior. The posterior wall is strengthened by the falx inguinalis and the reflected inguinal ligament that lie directly behind the superficial ring; and, reciprocally, the internal oblique overlaps the deep ring (p. 598). The parts of the internal oblique and the transversus attached to the inguinal ligament (i.e. fibres which 'arch' over the canal) are constantly active in standing; any increase in intra-abdominal pressure (as in coughing or straining) augments contraction of the internal oblique (and probably the transversus).

Extraperitoneal connective tissue is an areolar stratum between the peritoneum and the general fascial lining of the abdominal and pelvic cavities. It varies in distribution, being posteriorly abundant, particularly around the kidneys, where it contains much fat. Anterolaterally it is scanty except in the pubic region, above the iliac crest and in the pelvis. Such fasciae are usually no more than the general connective tissue between differentiated structures. In some regions, however, there occur aligned bundles of collagen, sometimes some elastic fibres and, particularly in the pelvis, bundles or sheets of non-striated muscle. Extraperitoneal tissue is continuous with the epimysium of abdominal muscles and hence with their internal connective tissue.

POSTERIOR ABDOMINAL MUSCLES

Psoas major and minor and iliacus, with their fasciae, though part of the abdominal parietes, are in part muscles of the lower limb and are described with them. Only quadratus lumborum will be considered here.

Enclosing quadratus lumborum are the *anterior* and *middle* layers of the *thoracolumbar fascia* (p. 587). The anterior layer is attached medially to the anterior surfaces of the lumbar transverse processes, below to the iliolumbar ligament and above to the apex and lower border of the twelfth rib. Its upper margin, from the first lumbar transverse process to the lower border of this rib, is the *lateral arcuate ligament* (p. 592). Laterally, this anterior layer blends with the fused posterior and middle strata of the thoracolumbar fascia (5.38).

Quadratus lumborum (5.37,43), irregularly quadrilateral and inferiorly broader, is attached by aponeurotic fibres to the iliolumbar ligament and adjacent iliac crest for about 5 cm; it ascends to the medial half of the lower border of the twelfth rib; it is also attached by four small tendons to the apices of the upper four lumbar transverse processes and sometimes also to the twelfth thoracic. A second layer of muscle may exist anterior to this, passing from the upper borders of the lower three or four lumbar transverse processes to the lower margin and anterior surface of the last rib.

Anterior to quadratus and its fascia are the colon, kidney, psoas major and minor and the diaphragm; the subcostal, iliohypogastric and ilio-inguinal nerves are also anterior to the fascia but bound down by a medial continuation of the transversalis fascia.

Nerve supply: the ventral rami of the twelfth thoracic and upper three or four lumbar spinal nerves.

Actions: fixation of the twelfth rib (a diaphragmatic attachment). With pelvis fixed, there is lateral flexion of the vertebral column to the same side; acting together both muscles may assist lumbar vertebral extension.

5. Pelvic Muscles and Fasciae

Pelvic muscles form two groups: (1) piriformis and obturator internus, described with muscles of the leg (pp. 642, 643); (2) levator ani and coccygeus which, with their antimeres, form the **pelvic diaphragm**. Their fasciae form a connective tissue continuum (5.54), joined to fascial coverings of pelvic viscera above and perineal fasciae below. (For muscles and fasciae of the **urogenital diaphragm** see pp. 606–608.)

Pelvic fascia may be resolved for description into: (1) *parietal pelvic fasica*, sheaths of pelvic muscles; (2) *visceral pelvic fascia*, sheaths of pelvic viscera, their blood vessels and nerves (see individual organs).

Parietal pelvic fascia on the pelvic surface of obturator internus is well differentiated as *obturator fascia*, connected above to the posterior part of the innominate bone's arcuate line and continuous with the iliac fascia. Anterior to this, as it follows the muscle's attachment, it gradually separates from the iliac fascia, continuity being retained only by the periosteum. It arches below the obturator vessels and nerve, completing their canal, and is attached anteriorly to the back of the pubis. Above the *tendinous arch of levator ani* (vide infra) the obturator fascia is markedly aponeurotic and part of the attachment of levator ani (p. 605); below this it is thin and part of the lateral wall of the ischiorectal fossa (p. 606), where it blends with a fascial sheath around the internal pudendal vessels. It is indirectly continuous behind with fascia on the piriformis.

The *piriform fascia* is very thin, fusing with the anterior sacral periosteum at the margins of the anterior sacral foramina, where it *ensheathes* the emerging sacral anterior primary rami, often described as *behind* the fascia. Internal iliac vessels are, however, enclosed in extraperitoneal tissue *anterior* to the fascia and their branches take sheaths of it into the gluteal region, above and below the piriformis.

Fascia of the pelvic diaphragm covers both surfaces of levator ani; below it is the thin, so-called *inferior fascia of the pelvic diaphragm* (anal fascia), which some dispute; it covers the ischiorectal fossa's medial wall and is continuous above with the pudendal sheath and obturator fascia along the attachment of levator ani; it is also continuous with fasciae on the sphincteres urethrae and ani externus. Fascia on the upper surface of the levator ani (*superior fascia of the pelvic diaphragm*) laterally follows the muscle's attachment; anteriorly it is attached to the back of the pubic body, about 2 cm above its lower border, and extends laterally across the superior pubic ramus to blend with the obturator fascia, continuing along an irregular line to the ischial spine. In lower mammals the levator ani springs dorsally (posteriorly) from the pelvic brim but its human attachment is lower (more caudal, or inferior), leaving its aponeurosis as a thickened upper part of the obturator fascia. Tendinous fibres of its attachment may reach the pelvic brim. Medially, the superior pelvic diaphragmatic fascia blends with the visceral pelvic fascia. The fibrous sheets, or the compound 'fascia' on obturator internus above the attachment of levator ani, is therefore composed of (a) obturator fascia, (b) fasciae of levator ani, and between these (c) degenerated levator aponeurosis. The thickened fusion of these structures is the *tendinous arch of the levator ani*. This is to be distinguished from lower *tendinous arch of the pelvic fascia*, a thick white band within the superior pelvic fascia extending from the lower symphysis pubis to the inferior margin of the ischial spine. This is the attachment of the lateral, 'true' vesical ligament. Anteriorly the same fascia forms two thick bands, the paired *puboprostatic (pubovesical) ligaments*.

The levator ani (5.55), a broad, thin sheet attached to the internal surface of the lesser (true) pelvis, unites with its fellow to form most of the pelvic floor. *Anteriorly* it is attached to the pelvic surface of the pubic body lateral to the symphysis, *posteriorly* to the medial surface of the ischial spine and, between these, to the obturator fascia. Its attachment to the fascia corresponds posteriorly to the tendinous arch of levator ani; but in front the muscle is attached to fascia to a varying extent above the arch, sometimes almost as high as the obturator canal. The muscle's fibres pass medially with varying obliquity; the most anterior sweep back across the side of the prostate to the perineal body

(p. 607), forming a *levator prostatae*; but in females they cross lateral to the vagina, forming an additional sphincter, the *pubovaginalis*. Intermediate fibres slope postero-inferiorly lateral to the prostate and turn medially to meet those from the other side, forming a muscular sling at the anorectal flexure; this thick part is the *puborectalis*. Posterior fibres mingle with those of the deep external anal sphincter. Some intermediate fibres blend with the longitudinal rectal muscle, descending within the puborectal sling as the anal canal's conjoined longitudinal layer (p. 1373). The most posterior fibres are attached to the last two coccygeal segments and anococcygeal ligament (p. 1369).

Morphologically levator ani is divisible into *pubococcygeus* and *iliococcygeus*, a useful approximation for description; but their mammalian homologies are controversial (Wendell-Smith 1967).

Pubococcygeus is attached to the back of the pubic body and anterior part of obturator fascia and passes back almost horizontally lateral to the anal canal to reach the front of the coccyx, where it is attached by a tendinous plate continuous with the ventral sacrococcygeal ligament. *Iliococcygeus* springs from the ischial spine and posterior part of levator ani's tendinous arch; its fibres reach the sides of the coccyx and blend with those of the other side in a median raphe inferior to the tendinous plate of pubococcygeus, contributing to the anococcygeal ligament. Iliococcygeus is usually thin and may be aponeurotic; an accessory slip of its posterior part is the *iliosacralis*. In tailed mammals both muscles are attached only to caudal vertebrae; iliococcygeus effects lateral movements and pubococcygeus draws the tail between the hind limbs. Gradual loss of the tail frees these muscles to make a more complete pelvic floor, particularly necessary in bipeds. Levator and depressor caudae of tailed mammals are said to be represented by the rudimentary human *ventral* and *dorsal* *sacrococcygeal* muscles.

Relations. The *superior (pelvic) surface* of levator ani is separated only by fascia (superior of the pelvic diaphragm) from

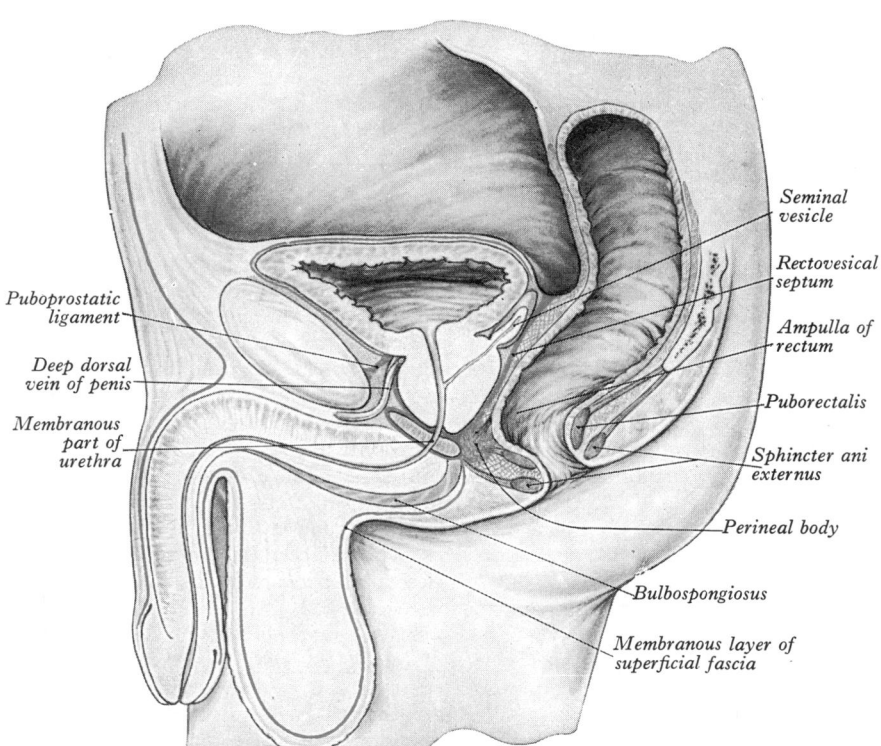

5.54 Median sagittal section through the pelvis, showing the arrangement of the fasciae (blue).

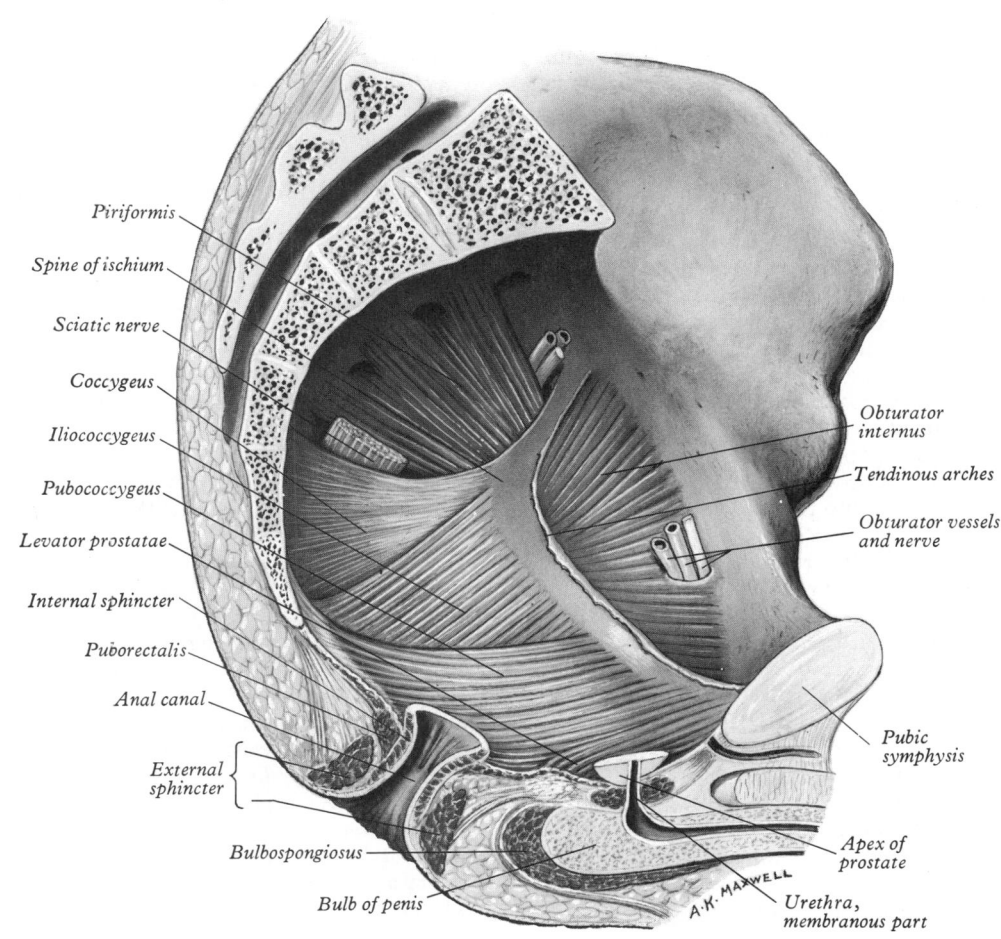

5.55 Pelvic aspect of the left levator ani and coccygeus. The superior gluteal vessels and nerve have been cut close to the upper border of the piriformis; the anal canal has been divided below the anorectal flexure and the greater part of the prostate has been removed. The constituent parts of the levator ani are shown. Note: coccygeus is fused and coextensive with the sacrospinous ligament; the latter is viewed from the gluteal aspect.

the urinary bladder, prostate, rectum and peritoneum. Its *inferior (perineal) surface* is the ischiorectal fossa's medial wall and the superior wall of the fossa's anterior recess (vide infra), both covered by the inferior fascia of the pelvic diaphragm. The muscle's *posterior border* is separated from the coccygeus by areolar tissue. *Medial borders* of both muscles are separated by the *visceral outlet*, through which pass the urethra, vagina and anorectum.

Nerve supply: fourth sacral spinal nerve and from the inferior rectal (p. 1149) or perineal division of the pudendal nerve.

Actions: constriction of the rectum and vagina and probably fixation of the perineal body. With coccygei the levatores form a basin-shaped muscular *pelvic diaphragm* supporting pelvic viscera and opposing sudden increases in, but also an important factor in maintaining and controlling, intra-abdominal pressure. Since they are not easily accessible to direct or electromyographic examination, dogmatic views are to be avoided. Modern scanning techniques, however, support the above views.

Coccygeus (5.55) is posterosuperior to, and *in the morphological plane* of, levator ani. A triangular musculotendinous sheet, its apex is attached to the pelvic surface and tip of the ischial spine and is fused with the sacrospinous ligament; its base reaches the lateral margins of the coccyx and fifth sacral segment. It is occasionally absent, varying in proportions of muscular and fibrous tissues. It *is* the pelvic aspect of the sacrospinous ligament, which is commonly regarded as a degenerate part of it, or its aponeurosis. (In topographical studies it is helpful to note that muscle and ligament are *coextensive*, the former viewed from the pelvic aspect, the latter from the gluteal aspect.) The muscle is well developed and the ligament often absent in mammals with a mobile tail. Morphologically coccygeus (*ischio-coccygeus*) is probably part of the iliococcygeal moiety of levator ani.

Nerve supply: fourth and fifth sacral spinal nerves.

Actions. The coccygei probably pull forward and support the coccyx after defaecation or parturition. With levatores ani and piriformes they compress the posterior part of the pelvic cavity and outlet (also the 'birth canal').

Applied Anatomy. Injuries to muscles of the pelvic floor often occur in parturition. Rupture of the perineal body (p. 607), which includes fibres of pubovaginalis, permits divarication of the levatores and may contribute to uterovaginal prolapse.

6. Perineal Muscles and Fasciae

The perineum overlies the **inferior pelvic aperture** (pelvic outlet). Its skeletal boundaries are: *anteriorly*, the pubic arch and arcuate pubic ligament; *posteriorly* the tip of the coccyx; and *laterally* (from before-backwards) the conjoined inferior pubic and ischial rami, ischial tuberosities and sacrotuberous ligaments. This osseoligamentous opening is trapezoid but its anterior and posterior parts are not coplanar (vide infra, and p. 518). Superficially the male perineum is limited by the scrotum, buttocks and the medial sides of the thighs. A transverse line in front of the ischial tuberosities divides the region into two triangular parts: a posterior **anal region** (*anal triangle*) contains the anal canal's ending; an anterior, the **urogenital region** (*urogenital triangle*), contains external urogenital structures. The urogenital triangle inclines downwards and backwards, whereas the anal triangle inclines downwards and forwards; these sloping planes meet laterally at the lowest (most inferior) parts of the boundaries of the pelvic outlet, the massive ischial tuberosities. It is instructive to consider the relative positions of the tuberosities (and their overlying tissues) with those of the urogenital and anal triangles when in a sitting posture on a firm surface.

Perineal muscles and fasciae form, in both sexes, anal and urogenital groups; but they meet at the perineal body and are derived from a single morphological unit, the *sphincter cloacae*.

MUSCULATURE AND FASCIAE OF THE ANAL REGION

The superficial fascia is thick, areolar and adipose. Adipose tissue extends deeply into the right and left, vertical, cuneiform

ischiorectal fossae between levatores ani and obturatores interni. **The deep fascia** lines each fossa; it includes inferior fascia of the pelvic diaphragm and that part of the obturator fascia below the attachment of levator ani.

The ischiorectal fossa, wedge-shaped, has a superficial, notional base (perineal surface) and a thin edge at the junction of obturator internus and levator ani, covered by the obturator and inferior pelvic diaphragmatic fascia. *Medially* are the sphincter ani externus, inferior pelvic diaphragmatic fascia, *laterally* the ischial tuberosity and obturator fascia. *Posteriorly* the fossa is partly limited by the lower border of gluteus maximus but extends deep to it to the sacrotuberous ligament. *Anteriorly* the fossa is partly bounded by the posterior aspect of the urogenital diaphragm but is prolonged *above* it as a narrow recess reaching almost to the posterior pubic surface. This *anterior recess* retains the same relations between obturator internus and levator ani, but is, as noted, deep (superior) to the muscles and fascia of the urogenital diaphragm; its anterior extent varies, being limited by fusion between the inferior fascia of the pelvic diaphragm and the superior fascia of the urogenital diaphragm. It may reach as far as the retropubic space.

Internal pudendal vessels and accompanying nerves are in the fossa's lateral wall, enclosed in fascia forming the *pudendal canal*; this sheath fuses with the lower part of obturator fascia, which ascends to blend with the inferior fascia of the pelvic diaphragm and descends to fuse with the falciform process of the sacrotuberous ligament (p. 517). It is sometimes termed the *lunate fascia* (8.141B) and regarded as the fossa's roof and lateral wall. It passes forwards above the urogenital diaphragm, blending with its lateral margin at its attachment to the inner surface of the inferior pubic ramus.

Sphincter ani externus (5.56, 8.141B, p. 137), around the anal canal's lowest part, is adherent to skin; above, it overlaps the sphincter ani internus. Details appear on p. 1370. Briefly, it has deep, superficial and subcutaneous parts. Its deep part exchanges fibres with the puborectal sling above it (p. 607) and, via the perineal body, with the deep perineal muscles anterior to it. Similarly, the superficial part exchanges fibres with the superficial urogenital muscles; this part and, to a lesser extent, the others are attached to both the coccyx and the anococcygeal ligament. Above this attachment a median *retrosphinteric space* connects the ischiorectal fossae (Courtney 1949).

The anococcygeal ligament is a median, stratified, musculotendinous structure between ano-rectum and coccyx. With the overlying presacral fascia it has been termed the *postanal plate* (Wendell-Smith & Wilson 1977). From above it comprises the tendinous plate of pubococcygeus, the muscular raphe of iliococcygeus and the posterior attachments of the external anal sphincter.

Nerve supply: perineal branch of the fourth sacral spinal nerve and rami of the inferior rectal branch of the pudendal nerve (second and third sacral spinal nerves).

MUSCLES AND FASCIAE OF THE MALE UROGENITAL REGION

Urogenital musculature consists in both sexes (5.56, 57) of bulbospongiosus, ischiocavernosus, transversi perinei superficialis and profundus and sphincter urethrae, which are concerned in urination, copulation and general support of the pelvic contents. These may be grouped as (1) *superficial urogenital muscles*: a median bulbospongiosus, right and left ischiocavernosi and transversi perinei superficiales, occupying the *superficial perineal space*; and (2) *deep perineal muscles*: a sphincter urethrae and right and left transversi perinei profundi, forming a **urogenital diaphragm**.

Superficial fascia of this region has a superficial adipose and a deeper membranous layer.

The *adipose layer*, thick, areolar, with variable amounts of fat, is continuous with the thin scrotal dartos muscle, with thicker circum-anal subcutaneous areolar tissue, and laterally with the superficial femoral fascia. In the midline it is adherent to skin and the membranous layer.

The *membranous layer* (5.54) is thin, aponeurotic and strong, perhaps serving to bind down muscles of the penile root. It is also continuous with the dartos muscle, fascia penis and membranous superficial fascial layer on the anterior abdominal wall; laterally it is attached to margins of the ischiopubic rami, lateral to the crura penis and as far back as the ischial tuberosities; posteriorly it curves round the transversi perinei superficiales to join the posterior margin of the fasciae of the urogenital diaphragm and perineal body. Between this layer and the inferior urogenital diaphragmatic fascia is the **superficial perineal space**.

The perineal body (the official term, *centrum tendineum*, is unsuitable; it is not an insubstantial centre, nor is it tendinous) is a median pyramidal fibromuscular node in the angle between the anal canal and urogenital apparatus, with the *rectovesical (rectovaginal) septum* at its apex. Below this, muscles and fasciae converge and interlace (cf. the *facial modiolus* p. 578). From above downwards these are the two levatores prostatae (pubovaginales), the deep and some superficial perineal muscles. The deep comprise the sphincter urethrae, deep transversi perinei and deep part of sphincter ani externus; the superficial are the bulbospongiosus, superficial transversi perinei and superficial part of the sphincter ani externus. Attachments of these to the pubes, ischia, and coccyx determine the position of the perineal body and hence of the visceral canals.

Transversus perinei superficialis, a narrow muscular strip, traverses the superficial perineal space anterior to the anus. Variable, often feeble and sometimes absent, it springs by tendinous fibres from the medial and anterior aspects of ischial tuberosity, runs medially to the perineal body and joins its antimere and superficial part of the sphincter ani externus behind, and bulbospongiosus in front. Sometimes the deeper layer of the sphincter ani externus decussates anterior to the anus, becomes more superficial and continues into the transversus, some fibres of which may join the ipsilateral bulbospongiosus or the sphincter ani externus.

Action. Contraction of both muscles probably helps to fix the perineal body, but details are uncertain.

Bulbospongiosus (bulbocavernosus) is a median muscle anterior to the anus, composed of two symmetrical parts united by a tendinous raphe. Muscle fibres spring from the median raphe and perineal body, diverging like the halves of a feather; the most posterior, a thin layer, disperse on the inferior fascia of the urogenital diaphragm; the middle fibres encircle the penile bulb and adjacent corpus spongiosum and are attached to an aponeurosis on their dorsal surfaces; anterior fibres spread out over the sides of the corpora cavernosa, ending partly in them, anterior to ischiocavernosus, partly in a tendinous expansion over the dorsal penile vessels.

Actions. Bulbospongiosus helps to empty the urethra; during urination it relaxes, usually contracting only at its end but is able to arrest it. The middle fibres assist in erection, probably by compressing bulbar erectile tissue, and the anterior fibres by compressing the deep dorsal vein (p. 1433). Like other perineal muscles it has been little investigated by electromyography but results suggest that, with the sphincter urethrae, it contracts repeatedly in ejaculation, as simple palpation reveals.

Ischiocavernosus covers the crus penis, attached by tendinous and muscular fibres to the medial aspect of the ischial tuberosity behind the crus and to the ischial ramus along the crural sides. Muscular fibres end in an aponeurosis attached to the sides and inferior aspect of the crus penis.

Action: compression of crus penis, maintaining penile erection.

Between these muscles is a triangular space, bounded medially by bulbospongiosus, laterally by ischiocavernosus, posteriorly by transversus perinei superficialis; its floor is the inferior fascia of the urogenital diaphragm. Posterior scrotal vessels and nerves and the perineal branch of the posterior femoral cutaneous nerve travel forwards in it; the transverse perineal artery follows its posterior boundary on the transversus perinei superficialis.

The deep stratum of urogenital muscles and their fasciae, i.e. transversus perinei profundus and sphincter urethrae, constitute the **urogenital diaphragm**. Inferior (superficial) to them is the *inferior fascia of the urogenital diaphragm (perineal membrane)*, stretched almost horizontally across the pubic arch. Its posterior

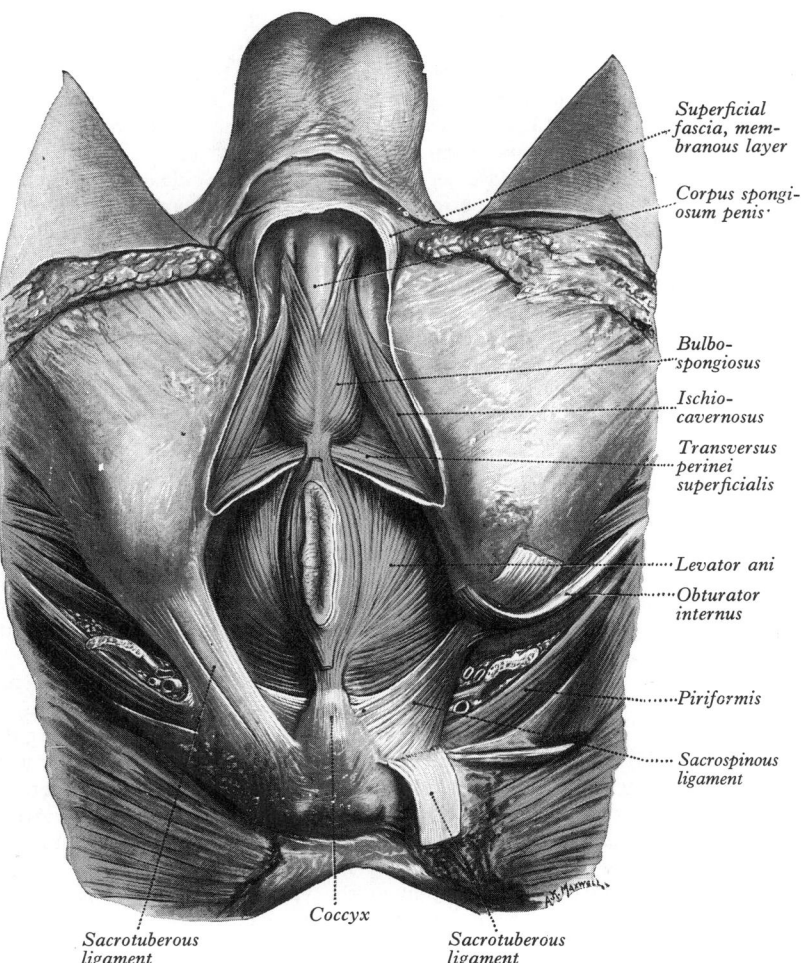

5.56 Muscles of the male perineum. (Modified from Quain's *Anatomy*, 11th edn.)

Labels on figure:
Superficial fascia, membranous layer
Corpus spongiosum penis
Bulbospongiosus
Ischiocavernosus
Transversus perinei superficialis
Levator ani
Obturator internus
Piriformis
Sacrospinous ligament
Sacrotuberous ligament
Coccyx
Sacrotuberous ligament

base is fused with the perineal body and is continuous with the superior fascia of the urogenital diaphragm behind transversi perinei profundi and with the membranous layer of superficial fascia behind the transversi perinei superficiales. Its lateral margins are attached to the ischiopubic rami above the crura penis. Its anterior apex is a thickened *transverse perineal ligament*; between this and the arcuate pubic ligament the deep dorsal vein of the penis or clitoris enters the pelvis and the dorsal nerve leaves it. The diaphragm is perforated as follows: 2–3 cm behind the inferior border of the symphysis by the urethra, its circular aperture about 6 mm in diameter; by arteries and nerves to the bulb and, near the urethra, ducts of the bulbo-urethral glands; by deep penile arteries near the pubic arch and halfway along its margin; by dorsal penile arteries near its apex; in its base also by the posterior scrotal vessels and nerves.

Superior (deep) to the transversi perinei profundi and sphincter urethrae is a less defined *superior fascia of the urogenital diaphragm*, stretching across the pubic arch and continuous laterally with the obturator fascia. Behind, it blends with the inferior diaphragmatic fascia, perineal body and the membranous layer of superficial fascia; it is pierced by urethra and continuous here with the prostatic fascia.

Between the superior and inferior urogenital fasciae is the **deep perineal space**, containing: the membranous part of the urethra, the transversi perinei profundi and sphincter urethrae, bulbo-urethral glands and proximal parts of their ducts, pudendal vessels and dorsal penile nerves (in forward continuations of the pudendal canals), arteries and nerves of the penile bulb and a venous plexus.

Transversus perinei profundus extends from the medial aspect of the ischial ramus to end in the perineal body, joining its antimere, the deep part of the sphincter ani externus behind and

607

the sphincter urethrae in front, usually exchanging fibres with them.

Action: fixation of the perineal body, but clear functional observations are lacking.

The sphincter urethrae surrounds the membranous urethra. Its *inferior fibres* extend from the transverse perineal ligament and neighbouring fasciae, passing back on each side of the urethra to converge in the perineal body. Its *superior fibres*, some from the inner surface of each pubic ramus, pass medially to encircle the membranous urethra.

Actions: compression of the membranous urethra, particularly when the bladder contains fluid (Basmajian & Spring 1955). It relaxes during micturition, like bulbospongiosus, but ejects final drops of urine and is also concerned in ejaculation.

Nerve supply. All urogenital muscles are supplied by the perineal branch of the pudendal nerve (second to fourth sacral spinal nerves).

MUSCLES AND FASCIAE OF THE FEMALE UROGENITAL REGION

This includes the same five paired muscles as in males, with differences in size and disposition due to the vagina and external genitalia (5.57). They are similarly grouped into inferior and superior strata, the latter being the *urogenital diaphragm*.

The transversus perinei superficialis is a narrow muscular slip, differing little from its counterpart in males (vide supra).

Bulbospongiosus surrounds the vaginal orifice, covering the lateral parts of the vestibular bulbs, continuous posteriorly with the perineal body and blending with the sphincter ani externus. Its fibres pass forwards on each side of the vagina to the corpora cavernosa clitoridis; a fasciculus crosses over the dorsum of the body of the clitoris.

Actions: constriction of the vaginal orifice. Anterior fibres probably contribute to erection of the clitoris by compression of its deep dorsal vein.

Ischiocavernosus, smaller than in males, covers the unattached surface of the crus clitoridis. Attached by tendinous and muscular fibres to the inner surface of the ischial tuberosity behind the crus clitoridis, and to strips adjacent to crural margins along the ischial ramus, its fibres end in an aponeurosis attached to the sides and inferior surface of the crus clitoridis.

Actions: compression of crus clitoridis, retarding venous return and thus assisting erection.

The *inferior fascia of the urogenital diaphragm* is less extensive, being traversed by the vagina, with whose external coat it blends, and by the urethra, vessels and nerves as described in males (p. 607). It covers the following: parts of the urethra and vagina,

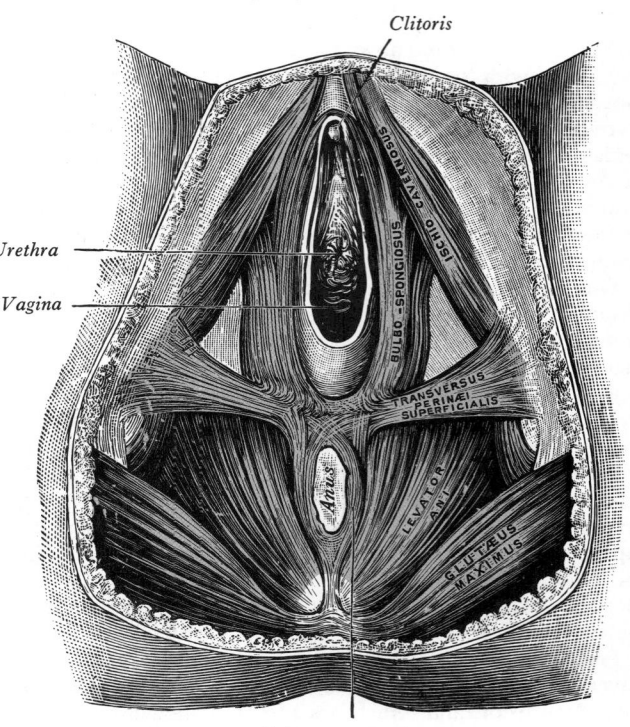

5.57 Muscles of the female perineum.

transversi perinei profundi and sphincter urethrae, internal pudendal vessels, dorsal clitoridic nerves, arteries and nerves of vestibular bulbs and a plexus of veins.

Transversus perinei profundus extends from the internal aspect of each ischial ramus, across behind the vagina, to meet its antimere and other structures in the perineal body. More anterior fibres sink into the vaginal wall.

Action: fixation of the perineal body.

Sphincter urethrae, like the corresponding male muscle, has inferior and superior parts. The *inferior fibres* start from the transverse perineal ligaments and sweep back on each side of the urethra, some interlacing with those of the other side between the urethra and vagina, others passing into the vaginal wall. *Superior fibres* encircle the lower urethra.

Action: urethral compression.

FASCIAE AND MUSCLES OF THE UPPER LIMB

Muscles of the upper limb may be grouped as follows:
1. Muscles connecting the upper limb and vertebral column.
2. Muscles connecting the upper limb and thoracic wall.
3. Muscles of the scapula.
4. Muscles of the upper arm.
5. Muscles of the forearm.
6. Muscles of the hand.

In this section brackets around numerals indicate doubt that the spinal nerves quoted actually provide a substantial *motor* innervation but are principally, or exclusively, sensory and autonomic. Heavy type indicates that the nerve or nerves indicated are predominant sources of motor supply in addition to the afferent modalities (p. 1130).

1. Muscles Connecting the Upper Limb and Vertebral Column

Trapezius, rhomboid major and minor and levator scapulae extend from the cervico-dorsal vertebral column to the pectoral girdle; latissimus dorsi reaches beyond this to the humerus.

The **superficial fascia** of the cervico-dorsal region, thick, strong, and adipose, is continuous with the general superficial fascia. In the upper neck it is tough, with much white connective tissue attaching it to skin.

The **deep fascia** of the region is a dense fibrous layer attached to the superior nuchal lines and ligamentum nuchae, and to all the thoracic vertebral spines below the seventh cervical and the supraspinous ligaments. Elsewhere it is thinner. It is attached laterally to the spine and acromion of the scapula and is continued over the deltoid into the arm; on the thorax it is continuous with the deep fasciae of axilla and chest, on the abdomen with fascia over the abdominal muscles; below, it reaches the iliac crest.

Trapezius (5.58), flat and triangular, extends over the dorsum of the neck and upper thorax; medially it is attached to the medial third of the superior nuchal line, external occipital protuberance, ligamentum nuchae, apices of the seventh cervical and all the thoracic spinous processes and their supraspinous ligaments. Superior fibres descend, inferior ascend and the intermediate fibres approach the horizontal, all converging laterally to the

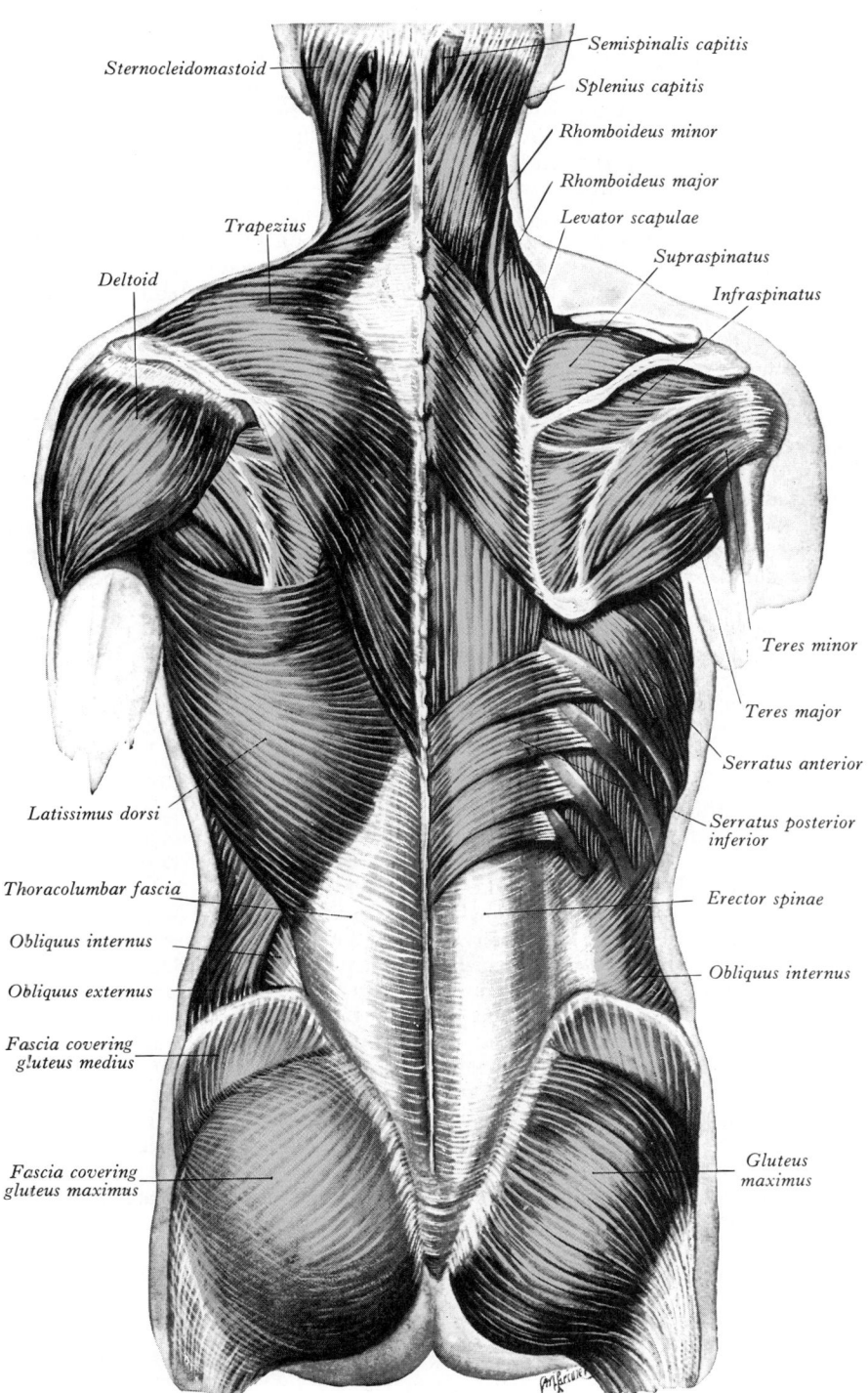

5.58 Superficial muscles of the back of the neck and trunk. On the left only the skin, superficial and deep fasciae (other than gluteofemoral) have been removed; on the right, the sternocleidomastoid, trapezius, latissimus dorsi, deltoid and oblique externus abdominis have been dissected away.

shoulder. Here the superior fibres are attached to the posterior border of the clavicle's lateral third, middle fibres to the medial acromial margin and superior lip of the scapular spine's crest; the inferior fibres are attached to an aponeurosis, gliding over a smooth triangular surface at the medial end of the scapular spine to a tubercle at its lateral apex. Occipital attachment is by a fibrous lamina, which is also adherent to skin; spinal attachment is partly a triangular aponeurosis from the sixth cervical to the third thoracic vertebra and below this contains short tendinous fibres. The two muscles form a trapezoid, its lateral angles at the shoulder tips, the superior 'angle' at the occipital protuberance and superior nuchal lines, the inferior angle at the twelfth thoracic spine.

The clavicular attachment varies, sometimes reaching mid clavicle and occasionally blending with the sternocleidomastoid;

the vertebral attachment sometimes ceases at the eighth thoracic spine, whereas the occipital is sometimes absent. Cervical and dorsal parts are occasionally separated.

Nerve supply. The accessory nerve (spinal part) is the dominant motor supply with sensory (proprioceptive) branches from the ventral rami of third and fourth cervical spinal nerves (Fitzgerald et al 1982.)

Actions. With other scapular muscles trapezius steadies the bone, controlling it during movements of the arm, maintaining the level and poise of the shoulder; but in the 'unloaded' arm electromyography shows little activity. Heavy loads can be suspended with little contribution from its upper part (Bearn 1960). With levator scapulae, its upper fibres elevate the scapula; with serratus anterior it rotates the scapula forward in raising the arm; with the rhomboids it retracts the scapula. With the shoulder fixed, trapezius may bend the head and neck posterolaterally. Trapezius, levator scapulae, rhomboids and serratus anterior combine in producing various scapular rotations (p. 613).

Latissimus dorsi (5.58), large, flat and triangular, extends over the lumbar region and lower thorax, converging to a narrow tendon. It is attached medially by tendinous fibres to the lower six thoracic spines anterior to the trapezius and, through an attachment to the posterior layer of thoracolumbar fascia (p. 587), to the lumbar and sacral spines and supraspinous ligaments and the posterior part of the iliac crest. It also springs by muscular fibres from the posterior part (outer lip) of the iliac crest lateral to the erector spinae and by muscular slips from three or four lower ribs, interdigitating with obliquus abdominis externus (5.44). From this wide attachment fibres converge with varying obliquity (the upper ones horizontal, the middle oblique, and lower almost vertical) to form a thick mass overlapping (and usually receiving some fibres from) the inferior scapular angle. The muscle curves (and spiralizes) round the inferolateral border of teres major to its anterior surface. Here it ends as a quadrilateral tendon, about 7 cm long, in front of the tendon of teres major, attached in the floor of the humeral intertubercular sulcus, with an expansion to the deep fascia; this attachment extends proximal to that of teres major. A bursa sometimes occurs between the muscle and the inferior scapular angle. In curving round teres major, constituent fibres are spiralized: fibres which are *lowest* at their midline attachment are attached *highest* on the humerus, *highest* midline fibres are *lowest* at their humeral attachment. Tendons of latissimus and teres major are united at their lower borders, with a bursa between them near their humeral attachments.

Latissimus dorsi and teres major form the posterior axillary fold. In adduction against resistance, this fold is accentuated; then the whole inferolateral border of latissimus dorsi can be traced to its attachment to the iliac crest.

A muscular *axillary arch*, 7–10 cm in length and 5–15 mm in breadth, may cross from the edge of latissimus dorsi, midway in the posterior fold, over the front of the axillary vessels and nerves to, variously, join the tendon of pectoralis major, coracobrachialis or fascia over the biceps. It occurs in about 7% and may be multiple (Kasai & Chiba 1977). Vertebral and costal attachments of latissimus dorsi may be reduced or, more rarely, increased. A fibrous slip usually connects its tendon near its humeral attachment to the long head of triceps; occasionally muscular, it is the homologue of the *dorso-epitrochlearis brachii* of apes.

Nerve supply: the thoracodorsal nerve from the posterior cord of brachial plexus, C. **6, 7** and 8.

Actions. Latissimus dorsi is active in humeral adduction, extension and especially medial rotation. Humeral adduction and extension are most powerful when the initial position of the arm is one of partial abduction, flexion or a combination of these. With the sternocostal part of pectoralis major and teres major it adducts the arm against resistance. It assists backward swinging of the arm, as in walking or many athletic pursuits. When arms are raised high, as in climbing, it assists in elevating the trunk. It is said to take part in all violent expiratory efforts, such as coughing or sneezing; palpation confirms this. Electromyography suggests a role in deep inspiration. When the muscle is stretched, as in elevation of the arm, it presses on the inferior scapular angle.

The muscle's lower, lateral margin is usually separated from the posterior border of the external oblique by a small *lumbar triangle*, its base the iliac crest and floor the internal oblique (5.58). This is to be distinguished from the *triangle of auscultation* which is medial to the scapula, bounded by the trapezius, latissimus dorsi and medial scapular border; part of rhomboideus major is exposed in it. If the scapulae are drawn forwards by folding the arms across the chest and the trunk is bent, parts of the sixth and seventh ribs and their interspace (overlying the apex of the lower pulmonary lobe) become subcutaneous.

Rhomboideus major (5.58) is attached by tendinous fibres to the second to fifth thoracic spines and their supraspinous ligaments, descending laterally to the medial scapular border between the root of the spine and the inferior angle; most of its fibres usually end in a tendinous band between these two points, joined to the medial border by a thin membrane; occasionally this is incomplete and some muscular fibres are then directly attached.

Bharihoke & Gupta (1986) studied the muscular attachments along the medial border of the scapula in 60 adult Indians; each of the muscles studied (rhomboideus major and minor, levator scapulae, and serratus anterior) had more extensive attachments than usually described, with 'folds' or extensions passing to both *dorsal* and *costal* aspects of the scapula adjoining its medial margin. The aponeurotic attachment of rhomboideus major was largely confirmed but its lower limit ended a short distance above the inferior angle which was enclosed on *all aspects* by the lowest part of serratus anterior. Rhomboideus major also had a muscular *dorsal* attachment immediately above this lowest part of serratus. *Ventrally*, rhomboideus major had an extensive attachment to the fascia covering serratus anterior, fascial fusion occurring from the medial scapular border, 2–3 cm laterally, and over the full vertical length of rhomboideus.

Rhomboideus minor (5.58) is attached to the lower ligamentum nuchae and seventh cervical and first thoracic spines, extending towards the base of the smooth triangular surface at the medial end of the scapular spine. Here, the authors quoted described it as forming a 'double fold' or asymmetrical U-shaped attachment with dorsal and ventral layers that enclose the inferior border of levator scapula (i.e. where the latter is tendinous and extends down to the level of the scapular spine). The dorsal layer of rhomboideus minor is attached to the rim of the triangular surface, dorsolateral and below levator scapulae; the ventral layer is strong and wide extending 2–3 cm medial to and below the levator; here the fasciae of rhomboideus minor and serratus anterior are tightly fused. Rhomboideus minor is usually separate from rhomboideus major but the muscles overlap and are occasionally united. Vertebral and scapular attachments of rhomboids vary. From the upper border of the minor a slender muscle may reach the occipital bone (*rhomboideus occipitalis*).

Nerve supply: the rhomboid branch of the dorsal scapular nerve, C.4, **5**.

Levator scapulae (5.36,58) is attached by tendinous slips to the transverse processes of the atlas and axis and the third and fourth cervical posterior tubercles, descending diagonally to approach the medial scapular border between its superior angle and the triangular smooth surface at the medial end of the scapular spine. It varies much in its vertebral attachments, and separation into slips, with accessory attachments to the mastoid process, occipital bone, first or second rib, scaleni, trapezius, and serrate muscles. Bharihoke & Gupta (1986) described the scapular end of the muscle as consisting of two 'flaps' or 'folds' partly fused just above the superior scapular angle. The posterior fold had an aponeurotic attachment to the dorsal aspect of the medial margin, opposite the supraspinous fossa. The anterior fold had a fascial attachment 2–3 cm wide to the fascial sheath of serratus anterior, below the muscular attachment of the latter, enclosing the superior scapular angle. Below, the anterior fold tapered to a narrow tendon attached to the costal surface of the medial border at the level of the root of the scapular spine.

Nerve supply: direct branches of the third and fourth cervical spinal nerves and from the fifth cervical via the dorsal scapular nerve.

The blood supply is chiefly from the transverse cervical and ascending cervical arteries, its vertebral extremity by rami of the vertebral, an important consideration in transposition of the muscle (Smith et al 1974).

Actions. The rhomboids and levator scapulae assist other scapular muscles in controlling scapular position and movement. With trapezius, rhomboids retract the scapula, with levator scapulae and pectoralis *minor* they rotate it, depressing the shoulder. With the cervical vertebral column fixed, levator scapulae acts with the trapezius to elevate the scapula or to sustain weight on the shoulder; with shoulder fixed the muscle laterally flexes the neck to the same side.

2. Muscles Connecting the Upper Limb and Thoracic Wall

These are serratus anterior and pectoralis minor, connecting the ribs to the scapula; pectoralis major from the ventral thorax (an upper abdomen) to the humerus; and subclavius connecting the first rib to the clavicle.

Superficial fascia of the anterior thoracic region is continuous with that of the neck, upper limb, and abdomen. It contains the mammary gland, sends numerous septa between its lobes and connects it, by fibrous *mammary suspensory ligaments*, to the skin and nipple.

Pectoral fascia is a thin (deep fascial) lamina over pectoralis major, and prolonged between its fasciculi. It is attached medially to the sternum, above to the clavicle and is inferolaterally continuous with the fascia of the shoulder, axilla and thorax. Although thin over pectoralis major, it is thicker between this and latissimus dorsi to which it crosses as the *axillary fascia* and the latter divides at the lateral margin of latissimus dorsi to ensheathe it, attached behind to the thoracic spines. As the fascia leaves the lower edge of pectoralis major to cross the axilla, a layer from it ascends deep to the muscle, splitting around pectoralis minor, at whose upper edge it becomes the *clavipectoral fascia* (vide infra). The axillary hollow is produced mainly by connections from this fascia to the axillary skin, forming a *suspensory ligament of the axilla*. Axillary fascia is pierced by the tail of the mammary gland (p. 1447). In the lower thoracic region the deep fascia is well developed and continuous with the sheath of rectus abdominis.

Pectoralis major (5.59), thick and triangular, is anteriorly attached to the sternal half of the clavicle, to half the anterior sternal surface to the level of the sixth or seventh costal cartilage, to the first to seventh costal cartilages (first and seventh often omitted), the sternal end of the sixth rib and the aponeurosis of obliquus externus abdominis. Clavicular fibres are usually separated from the sternal by a slight cleft. The muscle converges to a flat tendon, about 5 cm broad, attached to the lateral lip of the intertubercular sulcus. It has anterior and posterior laminae usually blended below. The thicker *anterior lamina* is formed centrally by fibres from the sternal manubrium, joined superficially by clavicular fibres and deeply by those from the sternal margin and fifth costal cartilages. (Thus there occurs some 90° of fibre spiralization in the anterior lamina.) Clavicular fibres may be prolonged into the deltoid tendon. The *posterior lamina* receives fibres from the sixth and often seventh costal cartilages and sixth rib, sternum and aponeurosis of the obliquus abdominis externus. Costal fibres join the lamina without twisting; fibres from the sternum and aponeurosis curve at the lower border, turning successively behind those above them, this part of the muscle being so twisted that the fibres that are *lowest* at their medial attachment are *highest* in the tendon (Ashley 1952). The posterior lamina reaches higher on the humerus, with an expansion covering the intertubercular sulcus and blending with the capsular ligament. An expansion from the lamina's deepest part, at its linear attachment, lines the sulcus; from its lower border a third descends into the deep fascia.

Costal attachments vary, as does separation of the clavicular and costal parts. Right and left muscles may decussate across the sternum. Superficially a vertical slip, or slips, may ascend from the lower costal cartilages and rectus sheath to blend with the sternocleidomastoid or attach to the upper sternum or costal cartilages. This is the *sternalis* (*rectus sternalis*). For variations, including agenesis, consult Cihak & Popelka (1961) and Kasai & Chiba (1977).

The muscle's rounded lower border is the *anterior axillary fold*, conspicuous in abduction. When the arm is swung forwards to the horizontal, the clavicular head is visibly contracted, the sternocostal relaxed; when the flexed arm is swung back against resistance, roles are reversed.

Relations. *Anterior* are skin, superficial fascia, platysma, anterior and middle supraclavicular nerves, mammary gland and deep fascia; *posterior* are sternum, ribs and costal cartilages, clavipectoral fascia, subclavius, pectoralis minor, serratus anterior and external intercostal muscles and membranes. It is the superficial stratum of the anterior axillary wall, and hence lies anterior to axillary vessels and nerves and the upper parts of biceps and coracobrachialis. Its *upper border* is separated from the deltoid by the *infraclavicular fossa*, containing the cephalic vein and deltoid branch of the thoraco-acromial artery. Its *lower border* is the anterior axillary fold. On the medial axillary wall, pectoralis major is

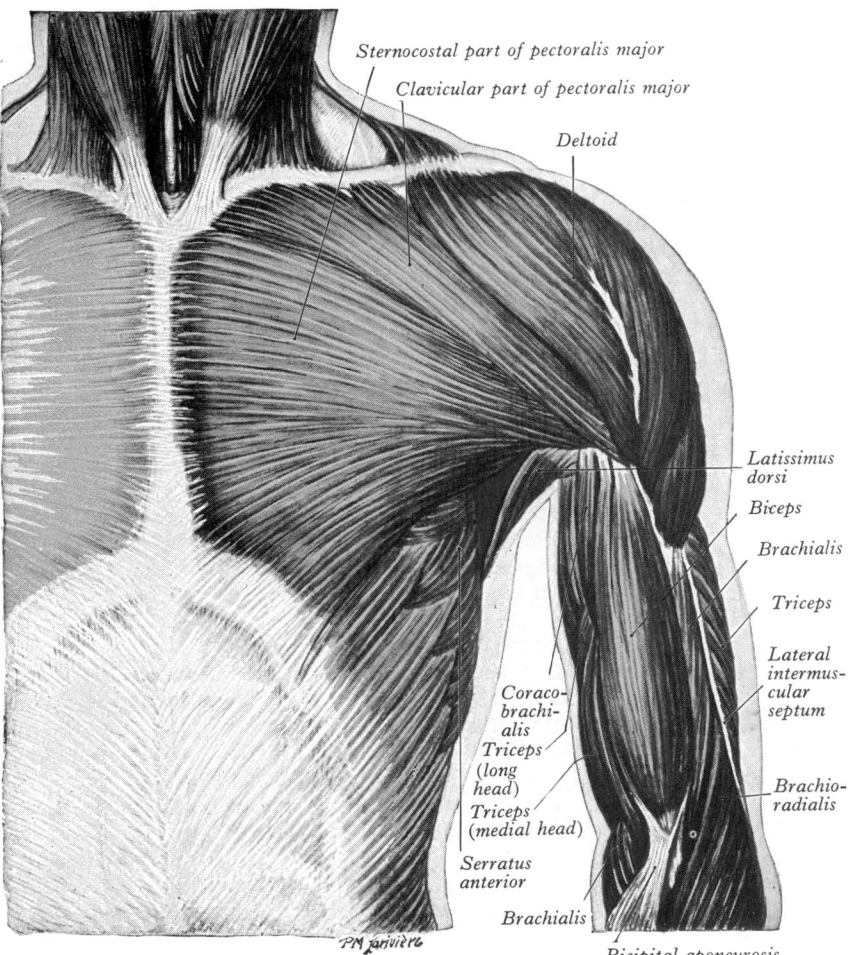

5.59 Superficial muscles of the front of the chest and upper arm (left side).

well separated from latissimus dorsi but these muscles converge towards the axillary lateral wall (i.e. the floor of intertubercular sulcus, between their attachments).

Nerve supply: lateral and medial pectoral nerves; fibres for the clavicular part are from C.5 and **6**, those for the sternocostal from C.**7, 8,** and T.1.

Actions. The muscle's two parts may act separately or together. The whole muscle assists humeral adduction and medial rotation. It swings the extended arm anteromedially, its clavicular part (*portio attollens*) acting with the anterior fibres of the deltoid and coracobrachialis; (in this movement the sternocostal part is relaxed). When the opposite movement, usually assisted by gravity, is resisted the sternocostal part (*portio deprimens*) assists latissimus dorsi, teres major and the posterior fibres of deltoid; with raised arms fixed, e.g. gripping a branch, this muscle combination draws the trunk up and forwards in climbing. Pectoralis major is active in deep inspiration. Electromyography suggests that the clavicular part alone acts in medial rotation.

Clavipectoral fascia is a strong fibrous sheet behind the clavicular part of pectoralis major and fills the gap between pectoralis minor and subclavius, anterior to the axillary vessels and nerves. It splits around subclavius and is attached to the clavicle anterior and posterior to the groove for subclavius; the posterior layer fuses with the deep cervical fascia connecting the omohyoid to the clavicle (p. 586) and with the axillary sheath. Medially it blends with fascia over the first two intercostal spaces and is attached to the first rib medial to the subclavius. Laterally, thick and dense, it is attached to the coracoid process, blending with the coracoclavicular ligament. Between the first rib and coracoid process it is often a thickened *costocoracoid ligament*, below which it is thin, splitting around pectoralis minor and descending to blend with the axillary fascia, spreading laterally to the fascia over the short head of the biceps. The cephalic vein, thoraco-acromial artery and vein and lateral pectoral nerve traverse the fascia.

Pectoralis minor (5.60), thin and triangular, is posterior (deep) to pectoralis major. Attached medially to the upper margins and outer surfaces of the third to fifth ribs (frequently second to fourth), near their cartilages, and from the fascia over adjoining external intercostals, its fibres ascend laterally, converging to a flat tendon attached to the medial border and upper surface of the coracoid process. Part or all of the tendon may cross the process into the coraco-acromial ligament, or may pass through this to the coracohumeral ligament, and is then attached to the humerus. The muscle's slips are sometimes separated and vary in number and level; rarely, one passes from the first rib to the coracoid (*pectoralis minimus*). Anson et al (1953) recorded costal attachments of 1000 muscles as: 2nd to 5th ribs—337; 3rd to 5th—334; 2nd to 4th—193; 3rd to 4th—67; the muscle was never absent.

Relations. Anterior are pectoralis major, lateral pectoral nerve and pectoral branches of the thoraco-acromial artery. *Posterior* are ribs, external intercostals, serratus anterior, axilla, axillary vessels, lymphatics and brachial plexus. Its *upper border* is separated from the clavicle by a triangular gap filled by the clavipectoral fascia, posterior to which are axillary vessels, lymphatics and nerves. The lateral thoracic artery follows the lower border. The medial pectoral nerves pierce and partly supply the muscle.

Nerve supply: both pectoral nerves, C.6, **7** and 8.

Actions. The muscle assists serratus anterior in drawing the scapula forwards round the chest wall. With levator scapulae and rhomboids it rotates the scapula, depressing the shoulder. Both pectoral muscles are electromyographically 'quiet' in inspiration, unless this is very forcible.

Subclavius (5.60) is a small, triangular muscle between clavicle and first rib. It is attached by a thick tendon, prolonged at its inferior margin to the junction of the first rib and its costal cartilage anterior to the costoclavicular ligament (Cave & Brown 1952). It ascends laterally to the groove inferior to the middle third of the clavicle, where it is attached by muscular fibres. It may reach the coracoid process in addition to or instead of the clavicle or scapular upper border.

Relations. Posteriorly it is separated from the first rib by the subclavian vessels and brachial plexus, *anteriorly* from pectoralis major by the anterior lamina of the clavipectoral fascia.

Nerve supply: the subclavian branch of the brachial plexus containing fibres from C.**5**. and 6 (p. 1131).

Action. Subclavius probably pulls the shoulder down and forwards and steadies the clavicle against the sternoclavicular articular disc; but it is inaccessible to palpation and difficult to investigate by electromyography.

Serratus anterior (5.60) is a large muscular sheet curving round the thorax from an extensive multidigital costal attachment to the medial scapular border. Its muscular digitations spring anteriorly from the outer surfaces and superior borders of the upper eight, nine or even ten ribs, and fasciae over the intervening intercostals. (These attachments lie along an extensive, slightly curved line that passes inferolaterally across the thorax.) The first springs from the first and second ribs and intercostal fascia, the others from a single rib; the lower four interdigitate with the upper five slips of obliquus externus abdominis. Applied to the thoracic wall, ventral to the scapula, the muscle reaches the latter's medial border as follows: the first digitation encloses and is attached to a triangular area of both costal and dorsal surfaces of the superior scapular angle; the next two or three form a triangular sheet attached to the costal aspect of almost all the medial border. The lower four or five digitations converge to be attached by musculotendinous fibres to a triangular area on the inferior angle's costal surface; however, they enclose the inferior angle and are also attached to a smaller triangular part of its dorsal surface near its tip (Bharihoke & Gupta 1986). Digitations may be absent, particularly the first and eighth but also sometimes the intermediate part. The muscle may be partly fused with levator scapulae, adjacent external intercostals or obliquus externus abdominis.

Nerve supply: the long thoracic nerve (C. **5**, **6** and **7**), descending on the muscle's external surface.

Actions. With pectoralis minor it protracts the scapula, as prime mover in all reaching and pushing movements. The upper part, with levator scapulae and upper fibres of the trapezius, suspends the scapula but activity is small in the unloaded arm. The lower, major part pulls the inferior scapular angle forwards round the thorax, assisting the trapezius in upward rotation, essential to raising the arm above the head (p. 613). In initial abduction serratus anterior helps other muscles to fix the scapula, so that the deltoid acts effectively on the humerus not the scapula. While the deltoid raises the arm to a right angle with the scapula, serratus anterior and trapezius simultaneously rotate the scapula, so that the arm can be raised to the vertical. To effect upward scapular rotation, forward traction at the inferior angle by the lower digitations of serratus anterior is coupled with superomedial traction on the lateral clavicle and acromion by the trapezius (upper fibres) and inferomedial traction at the base of the scapular spine by its lower fibres. Conversely, slow *downward* scapular rotation assisted by gravity is controlled by lengthening of these muscles; more powerful movement requires balanced contraction of the upper fibres of serratus anterior, levator scapulae, rhomboids, pectoralis minor and the middle part of the trapezius. In the transport of weights serratus anterior prevents backward scapular rotation. Electromyography, it is claimed, has disproved that serratus anterior assists respiration; but Jefferson et al 1960 worked on dogs and a study in human subjects (Catton & Gray 1951) ignored the effects of fixing the scapula by grasping, e.g. a bedrail or railing, as asthmatics and athletes certainly do! The problem is not fully resolved.

Applied Anatomy. When this muscle is paralysed, the scapula's medial border and especially its lower angle stand out prominently. The patient cannot raise the arm fully or push; attempts to do so are followed by further projection ('winging') of the scapula.

3. The Scapular Muscles

The shoulder joint is closely surrounded by six muscles: deltoid, supra- and infra-spinatus, subscapularis and teres major and minor, all extending from the scapula to the humerus. By their proximity to the joint, which they can move in any direction, they maintain contact of articular surfaces in static and dynamic conditions, at a very shallow multiaxial joint (p. 477).

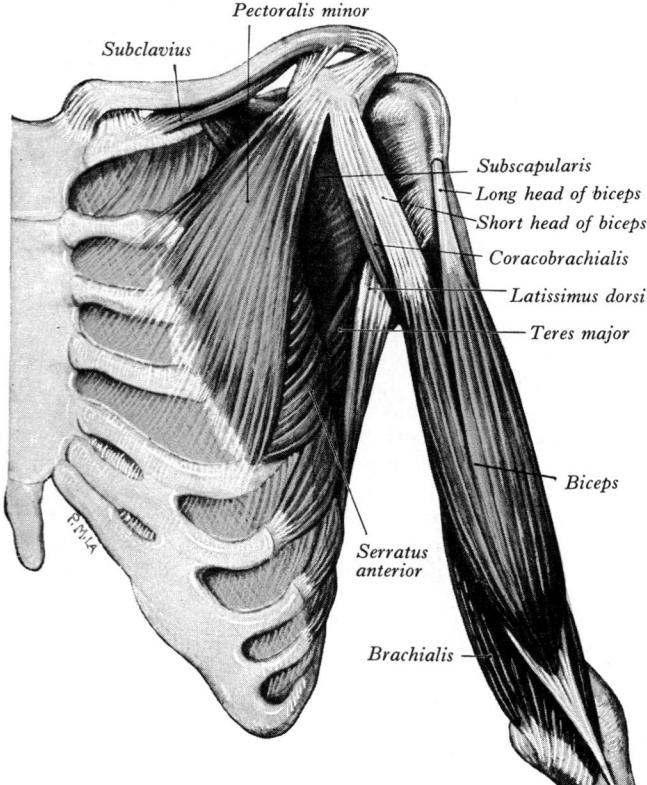

Pectoralis minor
Subclavius
Subscapularis
Long head of biceps
Short head of biceps
Coracobrachialis
Latissimus dorsi
Teres major
Biceps
Serratus anterior
Brachialis

5.60 Deep muscles of the front of the chest and arm (left side).

Deep fascia over the deltoid sends numerous septa between its fasciculi and is continuous with pectoral and infraspinous fasciae; the latter continuation is thick and strong. It is attached to the clavicle, acromion and crest of the scapular spine and is continuous with the brachial fascia.

The deltoid (5.59), a thick, curved triangle of muscle, is attached above to the anterior border and superior surface of the clavicle (lateral third), the acromial lateral margin and superior surface and continues into the lower edge of the crest of the scapular spine (but not its smooth medial triangular surface). The muscle converges inferiorly to a short substantial tendon attached to the deltoid tuberosity, which is lateral on the humeral mid-shaft. Anterior and posterior fibres converge directly to the humeral tendon; the intermediate part is multipennate, four intramuscular septa descending from the acromion to inter-digitate with three ascending from the deltoid tuberosity; short muscle fibres connect these, providing powerful traction. The fasciculi are comparatively large, producing a marked coarse striation. It surrounds the joint on all sides except inferomedially, giving the shoulder its rounded profile. In contraction its borders are easily seen and felt. The tendon has an expansion to the brachial deep fascia which may reach the forearm. Commoner variations are partition into the three parts described, union with pectoralis major and additional slips from the trapezius, infras-pinous fascia or lateral scapular border.

Relations. *Superficial* are: skin, superficial and deep fasciae, platysma, lateral supraclavicular and upper lateral brachial cutaneous nerves. *Deep* are: the coracoid process, coraco-acromial ligament, subacromial bursa, tendons of pectoralis minor, coracobrachialis, both heads of biceps, pectoralis major, sub-scapularis, supraspinatus, infraspinatus, and teres minor, long and lateral heads of triceps, circumflex humeral vessels, axillary nerve and humeral surgical neck and upper shaft, including the tuberosities. Its *anterior border* is separated proximally from pec-toralis major by the infraclavicular fossa, in which are the cephalic vein and deltoid branches of the thoraco-acromial artery; distally the muscles are in contact, their tendons usually united. Its *pos-terior border* overlies the infraspinatus and triceps.

Nerve supply: the axillary nerve, C.**5** and 6.

Actions. Anterior fibres assist pectoralis major in flexing the arm and in medial rotation. Posterior fibres act with latissimus dorsi and teres major in extension and lateral rotation. The multi-pennate, acromial part is a strong abductor; aided by the supras-pinatus it abducts the arm, until the joint's capsule is tense below (in the scapular plane); only thus can scapular rotation have full effect in raising the arm higher. In true abduction (4.5) acromial fibres contract strongly, clavicular and posterior fibres preventing side-sway. While the deltoid effects humeral rotation, scapular rotation is added (p. 500). In early abduction traction by the deltoid is upward, but the humeral head is denied upward transla-tion by the synergic downward pull of the subscapularis, infraspinatus and teres minor. Deltoid and supraspinatus, with subscapularis, infraspinatus and teres minor, constitute a mechanical couple ensuring steady abduction. Electromyography suggests that the deltoid contributes little to rotation but is active in most other shoulder movements. It may aid supraspinatus in resisting the downward traction of a loaded arm (vide infra). A common action of the deltoid is flexion/extension of the arm in walking. Ergonomic study of its clavicular part by Jonsson & Hagberg (1974) shows its value in adjusting the hand to various heights in manual tasks.

Applied Anatomy. The deltoid atrophies after division of the axillary nerve, simulating dislocation of the shoulder by flattening and apparent acromial prominence; the distance between the acromion and the humeral head is increased; the fingertips can be inserted between them.

The **subscapular fascia** is a thin aponeurosis attached to the circumference of the subscapular fossa; to its deep surface the subscapularis itself is partly attached.

Subscapularis (5.60), large and triangular, fills the sub-scapular fossa, attached to the periosteum in its medial two-thirds but also to the tendinous laminae in the muscle, which are attached to ridges in the fossa and to the aponeurosis covering it and separating it from teres major and the long head of triceps. It

converges laterally into a tendon attached to the lesser humeral tubercle and the front of the articular capsule. The tendon is separated from the scapular neck by a large bursa communicating with the shoulder joint, being really a protrusion of its synovial lining. Variation is unusual; a separate slip may pass from the medial scapular border to the glenohumeral capsule or perios-teum medial to the intertubercular sulcus.

Relations. The muscle forms much of the *posterior axillary wall*. Inferomedially its *anterior surface* is related to the serratus anterior, superolaterally to the coracobrachialis and biceps, axil-lary vessels, brachial plexus and subscapular vessels and nerves. Its *posterior surface* is related to the scapula and glenohumeral capsule. Its *lower border* contacts the teres major and latissimus dorsi.

Nerve supply: upper and lower subscapular nerves, C.**5**, **6** and (7).

Supraspinous fascia completes an osseofibrous compart-ment in which the supraspinatus is attached. It is thick medially, thinner under the coraco-acromial ligament. It is attached to the scapula around the limits of attachment of the supraspinatus.

Supraspinatus (5.61) is attached to the medial two-thirds of the supraspinous fossa and supraspinous fascia. It converges, un-der the acromion, into a tendon crossing above the shoulder joint to the highest facet of the greater humeral tubercle; the tendon is adherent to the articular capsule. A slip may pass from it to that of pectoralis major. Fibrocartilage has been described at this muscle's tendinous attachment (Evans & Benjamin 1984), as in other tendons attached to epiphysial bone.

Nerve supply: suprascapular nerve, C.**4**, **5** and 6.

Applied Anatomy. The tendon of supraspinatus is separated from the coraco-acromial ligament, acromion and deltoid by the large subacromial bursa; when this is infected, abduction is pain-ful. The tendon is the most commonly ruptured element of the musculo-tendinous cuff around the shoulder joint.

Infraspinous fascia covers the infraspinatus and is attached to the margins of the infraspinous fossa; the muscle is partly attached to its deep surface. It is continuous with the deltoid fascia along the overlapping posterior border of the deltoid.

Infraspinatus (5.61), a thick triangular muscle, occupies most of its fossa, attached periosteally to its medial two-thirds and by tendinous fibres to ridges on its surface; it is also attached to the infraspinous fascia, which separates it from teres major and minor. Fibres converge to a tendon gliding over the lateral border of the scapular spine and passing across the posterior aspect of the articular capsule to a middle facet on the greater humeral tuber-cle. The tendon is sometimes separated from the capsule by a bursa, which may communicate with the joint; it is sometimes fused with teres minor.

Nerve supply: the suprascapular nerve, C.(4) **5** and 6.

Teres minor (5.61), a narrow, elongate muscle, springs from the upper two-thirds of a flat strip on the scapular dorsum adjoin-ing its lateral border and from aponeurotic laminae separating it from infraspinatus and teres major. Ascending laterally, its upper fibres end in a tendon attached to the lowest facet on the greater humeral tubercle, its lower fibres being attached to bone distally between the facet and the lateral head of triceps. The tendon crosses and is fused to the lower posterior surface of the articular capsule and sometimes with infraspinatus.

Nerve supply: axillary nerve, C. (4), **5** and 6.

Actions. Subscapularis, supraspinatus, infraspinatus and teres minor have an essential steadying effect on the humeral head, maintaining its apposition to the glenoid cavity and checking ex-cessive translation (p. 500). Humeral swings (p. 502) at the glenohumeral joint are accompanied by movements between the articular surfaces which are combinations of rolling, sliding and spinning. Excessive sliding (translation) can easily occur at such a shallow articulation. During initial abduction subscapularis, infraspinatus and teres minor counteract the upward component of pull of the deltoid and thus, aided by supraspinatus, enable the force component *perpendicular* to the humeral shaft to abduct the arm. Infraspinatus and teres minor, with the posterior fibres of the deltoid, are lateral rotators. Subscapularis assists in *medial* rotation. These four muscles form the shoulder joint's '*rotator cuff*'. When the loaded or unloaded upper limb is pendent, a

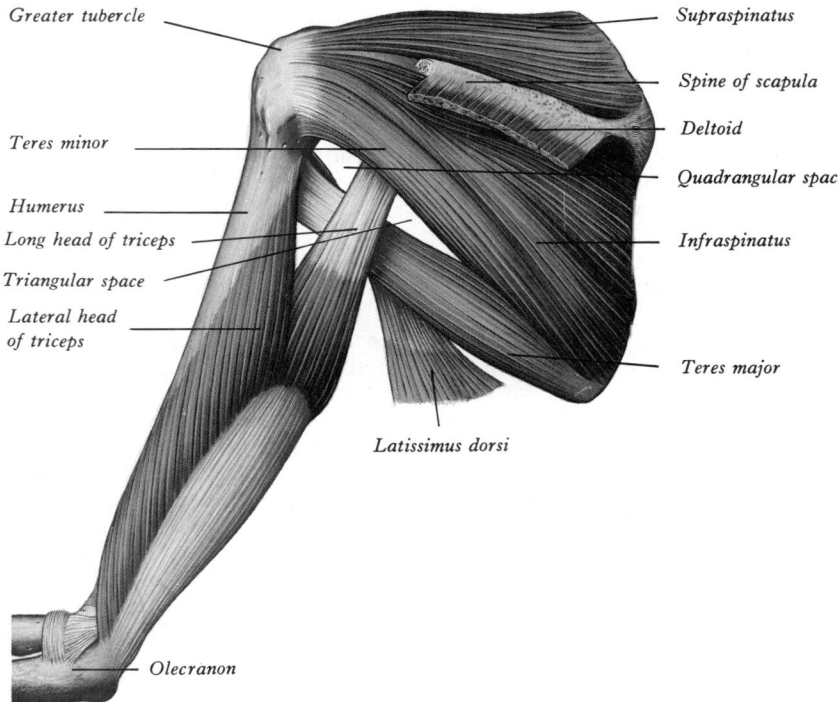

Greater tubercle

Teres minor

Humerus

Long head of triceps

Triangular space

Lateral head
of triceps

Supraspinatus

Spine of scapula

Deltoid

Quadrangular space

Infraspinatus

Teres major

Latissimus dorsi

Olecranon

5.61 The dorsal scapular muscles and triceps (left side). The spine of the scapula has been divided near its lateral end and the acromion has been removed together with a large part of the deltoid. The humerus is laterally rotated and the forearm pronated.

tendency to downward translation of the humeral head in the glenoid cavity is resisted by supraspinatus, rather than the deltoid, according to electromyographic data; the lack of tension in the deltoid can be appreciated by simple palpation. By contrast, anterior and posterior traction on the humerus are resisted by reciprocal parts of the deltoid.

Teres major (5.61) is a thick, flat muscle attached to the dorsal oval area near the inferior scapular angle and to the fibrous septa between it and teres minor and infraspinatus. It ascends laterally to a flat tendon, about 5 cm long, which is attached to the medial lip of the humeral intertubercular sulcus, where it is behind the tendon of latissimus dorsi, separated from it by a bursa; the tendons are united along their lower borders for a short distance. The muscle may be fused with the scapular slip of latissimus dorsi and a slip from it may join the long head of triceps or brachial fascia.

Nerve supply: the lower subscapular nerve, C.**6** and 7.

Actions: posteromedial extension and medial rotation. Electromyographers disagree on the muscle's role even during violent movements but its contribution to the maintenance of static postures and in arm-swinging appears established. Despite a common name, teres major and minor have different nerve supplies and actions.

4. Muscles of the Upper Arm

These include coracobrachialis, acting only on the shoulder joint, biceps and triceps crossing both shoulder and elbow joints, and brachialis acting only at the latter.

Brachial fascia (deep fascia of the upper arm) is continued from the fascia over the deltoid and pectoralis major; it invests muscles of the upper arm, sending septa between them. It is thin over the biceps, thicker over the triceps and humeral epicondyles, strengthened medially by fibrous aponeuroses from the pectoralis major and latissimus dorsi and laterally by the deltoid. Extending from it are strong medial and lateral intermuscular septa, attached to the humeral supracondylar ridges and epicondyles.

The *lateral intermuscular septum* extends distally from the lateral lip of the intertubercular sulcus along the lateral supracondylar ridge to its epicondyle, blending with the tendon of the deltoid. The triceps is attached to it posteriorly, anteriorly the

brachialis, brachioradialis and extensor carpi radialis longus. It is perforated near the junction of the upper and middle thirds by the radial nerve and the radial collateral branch of arteria profunda brachii. The thicker *medial intermuscular septum* extends from the medial lip of the intertubercular sulcus, distal to teres major, and attaches along the medial supracondylar ridge to the medial epicondyle, blended with the tendon of coracobrachialis; it affords attachment to the triceps behind, the brachialis in front and is perforated by the ulnar nerve, superior ulnar collateral artery, and the posterior branch of the inferior ulnar collateral artery. At the elbow the brachial fascia is attached to the humeral epicondyles and olecranon and is continuous with the antebrachial fascia. Medially, just below midarm, the basilic vein and lymphatic vessels traverse it and at various levels branches of the brachial cutaneous nerves.

Coracobrachialis (5.60, 62) springs from the apex of the coracoid process where it is blended with the tendon of the short head of the biceps, and also by muscular fibres from the proximal 10 cm of this tendon; it ends on an impression, 3–5 cm in length, midway along the medial border of the humeral shaft between the attachments of the triceps and brachialis. The muscle appears as a faint rounded ridge in the upper medial side of the arm; pulsation of the brachial artery can be felt and often seen behind it. Accessory slips of it may be attached to the lesser tubercle, medial epicondyle or medial intermuscular septum.

Relations. It is traversed by the musculocutaneous nerve and related, *anteriorly*, to pectoralis major and at its humeral attachment to the brachial vessels and median nerve, which cross it; *posterior* are the tendons of subscapularis, latissimus dorsi, and teres major, the medial head of triceps, humerus and anterior circumflex humeral vessels; *medial* are the axillary artery (third part) and proximal parts of the brachial, median and musculocutaneous nerves; *lateral* are the biceps and brachialis.

Nerve supply: the musculocutaneous nerve, C.5, **6** and 7.

Actions. The coracobrachialis flexes the arm anteromedially especially from a position of brachial extension. In abduction it assists anterior fibres of the deltoid to prevent side-sway.

Biceps brachii (5.60, 62, 63), a large, fusiform muscle in the upper arm's flexor compartment, has proximally two separately attached parts or 'heads'. The *short head* is attached by a thick flat tendon to the coracoid apex together with the coracobrachialis

The *long head* starts in the shoulder joint, as a long narrow tendon, running from the supraglenoid tubercle at the apex of the glenoidal cavity, and continuous here with the glenoidal labrum (p. 501). The tendon, enclosed in a double tubular sheath, an extension of the joint capsule's synovial membrane, curves over the humeral head, emerges behind the transverse humeral ligament and descends in the intertubercular sulcus, retained in this by the ligament and a fibrous expansion from the tendon of pectoralis major. Both bicipital tendons lead into elongated bellies which, though closely applied, can be separated to within 7 cm or so of the elbow joint. Here they end in a flat tendon attached to the rough posterior area of the radial tuberosity, a bursa separating tendon and its smooth anterior area. Approaching the radius the tendon twists (spirals), its anterior surface becoming lateral and applied to the tuberosity. The tendon has a broad medial expansion, the *bicipital aponeurosis*, descending medially across the brachial artery to fuse with deep fascia over the forearm flexors (5.65 and 6.93). The tendon can be split to the tuberosity, its anterior and posterior layers corresponding to the short and long heads.

A third head (10%) may extend from the superomedial part of the brachialis to the bicipital aponeurosis and medial side of the tendon; it is usually behind the brachial artery; it may consist of two slips which descend in front of and behind the artery. Other slips may spring from the lateral humeral aspect or intertubercular sulcus.

Relations. Biceps is overlapped proximally by pectoralis major and the deltoid; distally it is covered only by fasciae and skin and is a conspicuous elevation. Its long head passes *through* the shoulder joint; its short head is *anterior*. Distally it is anterior to the brachialis, musculocutaneous nerve and supinator. Its *medial border* touches the coracobrachialis and overlaps the brachial vessels and median nerve; its *lateral border* is related to the deltoid and brachioradialis.

Nerve supply: the musculocutaneous nerve, C.5 and **6**, separate branches passing to each belly.

Actions. Biceps is a powerful supinator, especially in rapid or resisted movements. It flexes the elbow, most effectively with the forearm supinated and, less effectively, the shoulder joint. Its bicipital aponeurosis, being attached to the posterior border of the ulna, draws the latter's distal end medially in supination (p. 506). The long head helps to check upward translation of the humeral head during deltoid contraction. When the elbow is flexed forcibly, the tendon and bicipital aponeurosis are conspicuous.

Many habitual movements involve *increasing tension* despite *increasing length*, e.g. if the hand is lowered under the influence of gravity, extension at the elbow is controlled by just such activity in *flexors*, such as the biceps. As the hand descends and the elbow extends, the vertical through the gravitational centre of the loaded forearm is carried further from the fulcrum of movement and effective load, (i.e. extending torque) and muscle tension increase.

Applied Anatomy. The long bicipital tendon is sometimes dislocated from its groove; the arm is fixed in abduction but the humeral head can be felt in its proper position. The tendon can usually be replaced by flexing the forearm and rotating the limb.

Brachialis (5.60, 63, 64) is attached proximally to the lower half of the anterior humeral aspect, from the deltoid tuberosity, which it embraces, to within 2.5 cm of the cubital articular surface. It is also attached to the intermuscular septa, more extensively the medial, being separated distally from the lateral by brachioradialis and extensor carpi radialis longus. Its fibres converge to a thick, broad tendon attached to the ulnar tuberosity and a rough impression on the anterior aspect of the coronoid process. It may be divided into two or more parts, fused with brachioradialis, pronator teres or biceps, or send a tendinous slip to the radius or bicipital aponeurosis.

Relations. *Anterior* are the biceps, brachial vessels, musculocutaneous and median nerves; *posterior* are the humerus and capsule of the elbow joint; *medial* are pronator teres and the medial intermuscular septum separating it from the triceps and ulnar nerve; *lateral* are the radial nerve, radial recurrent and radial collateral arteries, brachioradialis and extensor carpi radialis longus.

Nerve supply: the musculocutaneous nerve, C.5 and **6**, and the

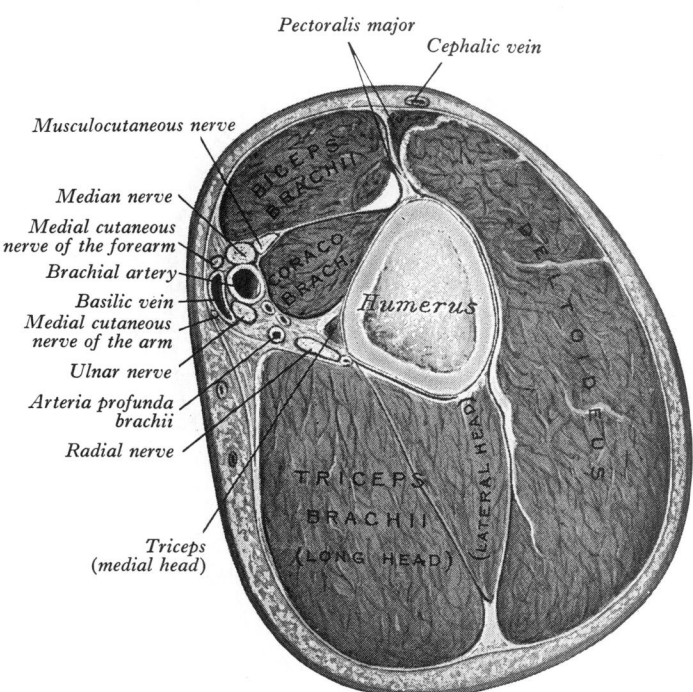

5.62 Transverse section through the right arm at the junction of the proximal and middle thirds of the humerus: superior (proximal) aspect.

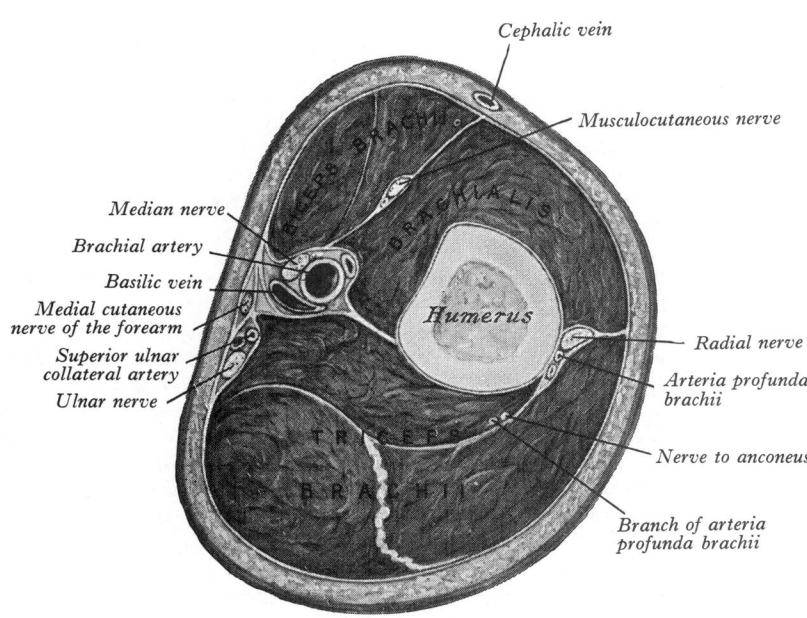

5.63 Transverse section through the right arm, a little below the middle of the shaft of the humerus: superior (proximal) aspect.

radial nerve (C.7) to a small lateral part of the muscle (Ip & Chang 1968).

Action. Flexion of the elbow joint with the forearm prone or supine, with or without resistance.

The **triceps** (5.61, 64), filling most of the upper arm's extensor compartment, has long, lateral and medial heads, hence its name.

The *long head* is attached by a flat tendon to the infraglenoid tubercle, blended here with the glenohumeral capsule; its muscular fibres descend medial to the lateral head, superficial to the medial head, then join them in a common tendon.

The *lateral head* also has a flat tendon attached to a narrow, linear, oblique posterior ridge on the humeral shaft, also to the shaft's lateral border and the lateral intermuscular septum. The attachment ascends with varying obliquity from the lateral border, proximal to the radial groove and behind the deltoid tuberosity, to the surgical neck medial to teres minor. These fibres also converge to the common tendon.

615

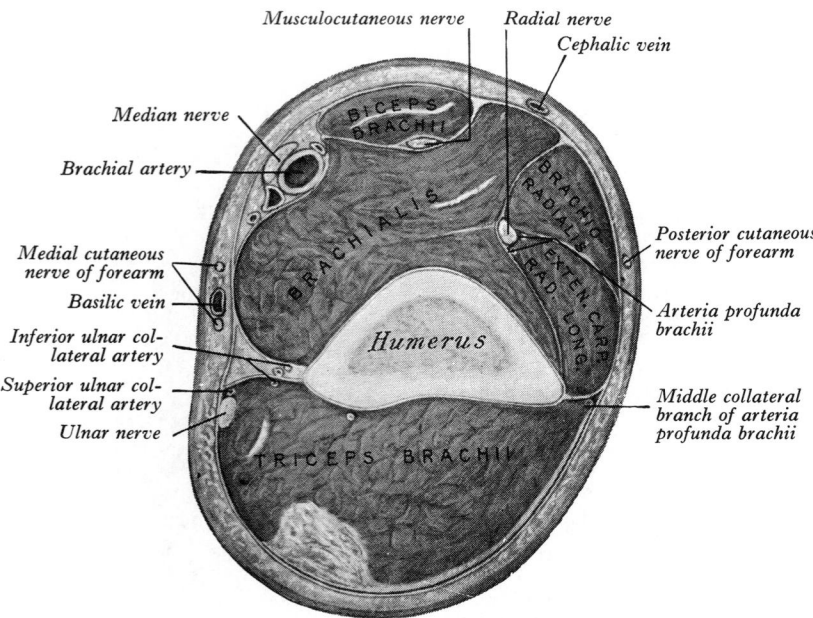

5.64 Transverse section through the right arm, a little proximal to the medial epicondyle of the humerus: superior (proximal) aspect.

The *medial head*, overlapped posteriorly by the lateral and long heads, is attached to a wide posterior surface of the shaft, distal to the radial groove; this attachment extends from that of teres major to within 2.5 cm of the trochlea; the medial head is also attached to the medial humeral border and the back of the distal part of the lateral intermuscular septum. Some muscular fibres reach the olecranon directly, the rest converge to the common tendon.

The tendon of triceps begins near the muscle's midpoint. It has two laminae, one superficial to the muscle's distal half, the other in its substance. Receiving the muscular fibres, the two layers unite above the elbow and are attached, for the most part, to the olecranon's upper surface; a lateral band of fibres continues distally over the anconeus to blend with antebrachial fascia.

The long head descends between teres minor and major, dividing the interval between them and the humerus into triangular and quadrangular parts (5.61). The *triangular space*, containing the circumflex scapular vessels, is bounded above by teres minor, below by teres major, laterally by the long head of triceps. The *quadrangular space* transmits the posterior circumflex humeral vessels and axillary nerve; it is bounded above by subscapularis, teres minor and the articular capsule, below by teres major, medially by the long head of triceps, laterally by the humerus.

The lateral head of triceps, parallel and medial to the posterior border of the deltoid, stands out prominently in extension. The mass medial to it, disappearing under the deltoid, is the long head.

Articularis cubiti (subanconeus) is formed by fibres from the distal deep surface of triceps (medial head) which blend posteriorly with the fibrous cubital capsule.

Nerve supply: radial nerve, C.6, 7 and **8**, with separate branches for each head (p. 1136).

Actions. Triceps is the major forearm extensor. The medial head is active in all forms of extension. Lateral and long heads are minimally active except in resisted extension (Basmajian 1967), as in thrusting or pushing or supporting body weight on the hands with semiflexed elbows. When the flexed arm is extended, the long head may assist in drawing back and adducting the humerus to the thorax. The long head supports the shoulder joint from below, especially when the arm is raised. The articularis cubiti probably retracts the cubital capsule during extension. In forceful supination of a semiflexed forearm, involving contraction of supinator and biceps, the triceps contracts synergically to maintain semiflexion. The human triceps brachii is the subject of an

interesting comparison of geometric and dynamic parameters by Cnockaert & Pertuzon (1974).

5. The Antebrachial Muscles

Antebrachial fascia (deep fascia of the forearm), continuous above with the brachial fascia, is a dense general sheath for muscles collectively and individually; it is attached to the olecranon and posterior ulnar border; from its deep surface septa pass between muscles for their partial attachment, some reaching the bone. Muscles also arise proximally from its internal aspect. Transverse septa, anterior and posterior, separate deep from superficial muscles. Fascia is much thicker posteriorly and in the distal forearm. It is strengthened proximally by fibres from biceps and triceps. Localized thickenings, *flexor* (p. 625) and *extensor* (p. 625) *retinacula* near the wrist, retain the digital tendons in position, increasing efficiency. Fascial apertures exist for vessels and nerves; one, large and anterior to the elbow, transmits a communication between superficial and deep veins.

The antebrachial (forearm) muscles are divided morphologically into flexors and extensors (anterior and posterior) but often they co-operate in more complex activities.

(a). The Anterior Antebrachial Muscles

These are superficial flexors, attached proximally and largely (but not exclusively) to the humerus, and deep flexors attached to the radius and ulna. The superficial 'flexors', however, include the pronator teres, while the deep 'flexors' include pronator quadratus.

SUPERFICIAL ANTEBRACHIAL FLEXORS

Attached to the medial humeral epicondyle by a common tendon are pronator teres, palmaris longus, flexores carpi radialis and ulnaris, and flexor digitorum superficialis (5.65,66). They also have attachments to overlying antebrachial fascia and to septa from this passing between individual muscles.

Pronator teres (5.65, 66) has humeral and ulnar attachments. The *humeral head*, the larger and more superficial, is attached just proximal to the medial epicondyle, to the common flexor tendon, to an intermuscular septum between it and flexor carpi radialis, and to antebrachial fascia. The smaller *ulnar head* springs from the medial side of the ulnar coronoid process distal to the attachment of flexor digitorum superficialis, joining the humeral head at an acute angle. The median nerve (83% of cases) enters the forearm between these heads, separated from the ulnar artery by the ulnar head. The muscle inclines across the forearm to a flat tendon attached to a rough area midway along the lateral surface of the radial shaft (at the 'summit' of its lateral curve). Its lateral border is the medial limit of the triangular *cubital fossa* anterior to the elbow joint. The coronoid attachment may be absent. Accessory slips may arise from a supracondylar process, if present, from biceps, brachialis or medial intermuscular septum.

Nerve supply: median nerve, C.6 and 7.

Actions. Rotation of radius on ulna, turning the palm medially and backwards, i.e. pronation and weak cubital flexion. It acts with pronator quadratus (always active in pronation), assisting only in rapid or forcible pronation, but its activity is said to be always less (Basmajian & Travill 1961). (For a further analysis of pronation/supination, including movements of the ulna, see p. 624.)

Flexor carpi radialis (5.65, 66, 69), medial to pronator teres, springs from the medial epicondyle by the common flexor tendon, from antebrachial fascia and adjacent intermuscular septa. Its belly is fusiform and ends rather more than halfway to the wrist in a long tendon, which traverses a lateral canal roofed by laminae of the flexor retinaculum and floored by a groove on the trapezium within a synovial sheath; it is then attached to the palmar aspect of

616

(p. 624.)

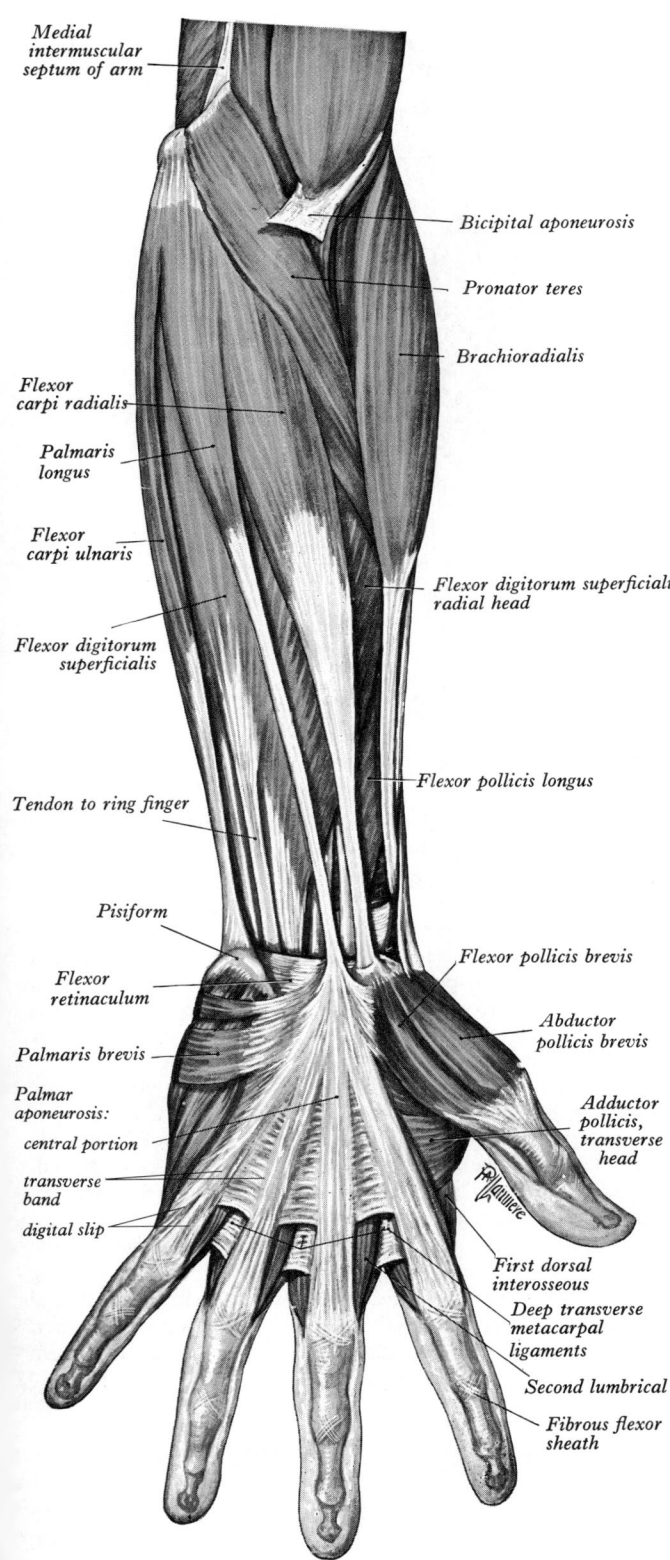

Medial
intermuscular
septum of arm

Bicipital aponeurosis

Pronator teres

Brachioradialis

Flexor
carpi radialis

Palmaris
longus

Flexor
carpi ulnaris

Flexor digitorum superficialis,
radial head

Flexor digitorum
superficialis

Flexor pollicis longus

Tendon to ring finger

Pisiform

Flexor pollicis brevis

Flexor
retinaculum

Abductor
pollicis brevis

Palmaris brevis

Palmar
aponeurosis:

Adductor
pollicis,
transverse
head

central portion

transverse
band

digital slip

First dorsal
interosseous

Deep transverse
metacarpal
ligaments

Second lumbrical

Fibrous flexor
sheath

5.65 Superficial flexor muscles of the left forearm, the palmar aponeurosis and the digital fibrous flexor sheaths. The fibrous flexor sheaths are depicted as in traditional accounts; for modified description consult text and 5.67B.

he second metacarpal's base, sending a slip to the third; these distal attachments are hidden by the oblique head of adductor pollicis (5.80). In the distal forearm the radial artery is between his tendon and that of brachioradialis.

The muscle may be absent or have accessory slips from biceps tendon, bicipital aponeurosis, coronoid process or radius. Distally it may attach also to the flexor retinaculum, trapezium, or ourth metacarpal.

Nerve supply: median nerve, C.6 and 7.

Actions. With flexor carpi ulnaris, and sometimes flexor digitorum superficialis, it flexes the wrist; with radial extensors it abducts the hand.

Palmaris longus (5.65, 66, 79), slender, fusiform and medial to flexor carpi radialis, springs from the medial epicondyle by the common tendon, from adjacent intermuscular septa and antebrachial fascia. Its long slender tendon passes *anterior* to the flexor retinaculum and is attached to the distal half of its anterior surface and centrally to the palmar aponeurosis, often sending a tendinous slip to the thenar muscles. Proximal to the wrist joint the median nerve is deep to the tendon, projecting beyond its lateral edge. Often absent on one or both sides, the muscle is very variable (Machado & DiDio 1967). It may have a proximal tendon or be reduced to a tendinous strand. It may be digastric or reduplicated. It may end in antebrachial fascia, tendon of flexor carpi ulnaris, pisiform, scaphoid, etc. Reimann et al (1944) found the muscle absent in 281 out of 2205 specimens; 15 muscles had accessory slips and four were double.

Nerve supply: median nerve, C.7 and 8.

Actions: carpal flexion; a tensor of palmar fascia.

Flexor carpi ulnaris (5.65,66,69), the most medial superficial flexor, has humeral and ulnar heads connected by a tendinous arch, beneath which pass the ulnar nerve and posterior ulnar recurrent artery. The small *humeral head* arises from the medial epicondyle via the common tendon; the *ulnar head* is an extensive attachment from the medial margin of the olecranon and proximal two-thirds of the posterior ulnar border by a common aponeurosis shared with extensor carpi ulnaris and flexor digitorum profundus, and from the septum between it and flexor digitorum superficialis. A tendon forms along the anterolateral border of the muscle's distal half and is attached to the pisiform bone, thence prolonged to the hamate and fifth metacarpal bones by so-called pisohamate and pisometacarpal ligaments (4.50); a few fibres blend with flexor retinaculum. A coronoid slip may occur; a more substantial attachment to the flexor retinaculum and to the fourth or fifth metacarpal bones may exist. Ulnar vessels and nerve are lateral to the tendon.

Nerve supply: ulnar nerve, C.7 and **8**.

Actions. With flexor carpi radialis, palmaris longus, and flexor digitorum superficialis, it flexes the wrist; with extensor carpi ulnaris it adducts the hand. Flexor and extensor carpi ulnaris are synergists preventing abduction when the pollex is extended at its carpometacarpal joint. It also fixes the pisiform bone during abduction and/or flexion of the minimus. When the wrist flexes against resistance, two tendons stand out: flexor carpi radialis and palmaris longus, approximately central near the wrist. Flexor carpi ulnaris can also be identified by palpation proximal to the pisiform.

The flexor digitorum superficialis (5.65, 66, 69) is deep to the preceding muscles; the largest superficial flexor, it has two heads. The *humero-ulnar head* arises from the medial epicondyle via the common tendon, also from the anterior band of the ulnar collateral ligament, adjacent intermuscular septa and the medial side of the coronoid process proximal to pronator teres. The *radial head*, a thin sheet, arises from the anterior radial border between the radial tuberosity and the attachment of pronator teres. The median nerve and ulnar artery descend between the heads. The muscle usually has two strata; a superficial one, joined laterally by the radial head, divides into tendons for medius and annularis; a deep stratum gives a slip to the superficial fibres for annularis and then ends in tendons for the index and minimus. As the tendons pass behind the flexor retinaculum they are arranged in pairs; the superficial pair are for medius and annularis, the deep for index and minimus. They diverge in the palm, and opposite the bases of proximal phalanges each divides around a tendon of flexor digitorum profundus, the slips twisting to reverse their surfaces and borders before they reunite and partially decussate, thus forming a groove for the profundus tendon. The superficial tendon divides again and ends at the *sides* of the shaft of its middle phalanx. Dimensions of these tendons have been detailed by Shrewsbury & Kuczynski (1974), who have also estimated tendon excursions and bowing ranges during action, correlating these data with an analysis of fibrous architecture, particularly of tendinous *scissural bands* and *chiasmata*.

617

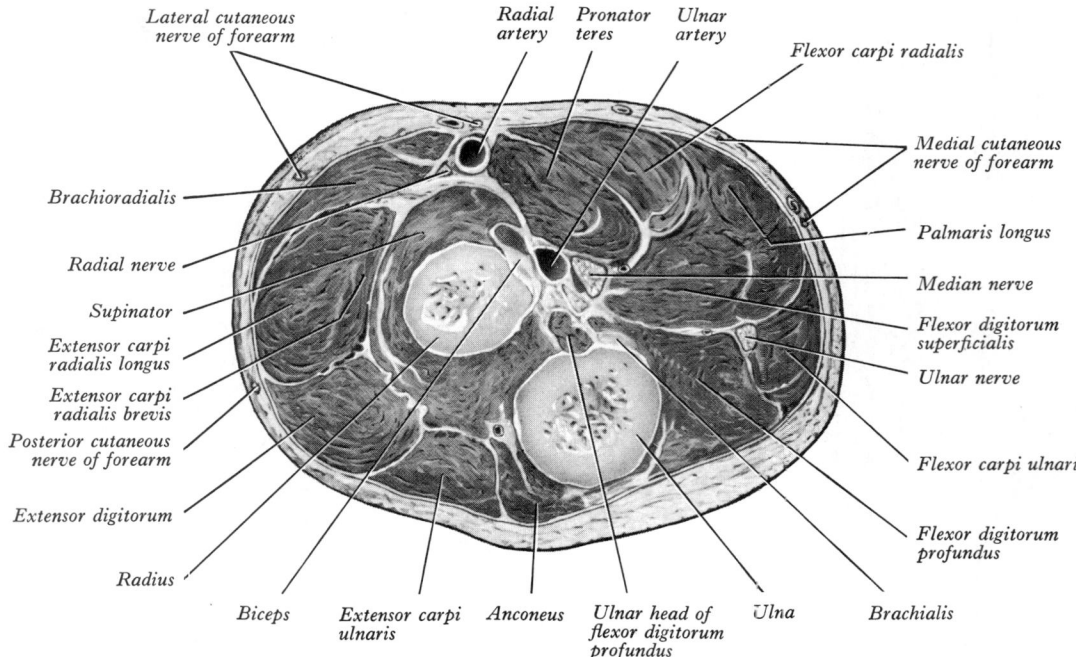

Lateral cutaneous nerve of forearm
Radial artery
Pronator teres
Ulnar artery
Flexor carpi radialis
Brachioradialis
Medial cutaneous nerve of forearm
Radial nerve
Palmaris longus
Supinator
Median nerve
Extensor carpi radialis longus
Flexor digitorum superficialis
Extensor carpi radialis brevis
Ulnar nerve
Posterior cutaneous nerve of forearm
Extensor digitorum
Flexor carpi ulnaris
Radius
Flexor digitorum profundus
Biceps
Extensor carpi ulnaris
Anconeus
Ulnar head of flexor digitorum profundus
Ulna
Brachialis

5.66 Transverse section through the forearm at the level of the radial tuberosity.

The radial head may be absent. A slip from the deep stratum may provide most or all the fibres acting on the index. The part for minimus may be absent and replaced by a separate slip from ulna, flexor retinaculum or palmar fascia. The tendons also vary. Dylevský (1968) has advanced developmental explanations of variations in deep and superficial digital flexors, assigning them to separate initial strata, not a single blastema. However, Chaplin & Greenlee (1975) confirmed the usual description of an initial single blastema for all digital flexor tendons.

Relations. *In the forearm* the muscle is *deep* to palmaris longus, flexor carpi radialis, pronator teres, brachioradialis, radial artery and nerve; and *superficial* to: flexor digitorum profundus, the upper part of the ulnar artery and the median nerve (closely bound to it by areolar tissue) and flexor pollicis longus. *At the wrist* tendons pass *deep* to flexor retinaculum, *superficial* to the tendons of flexor digitorum profundus but in a common synovial sheath (5.76); the flexor pollicis longus tendon and median nerve are lateral. *In the hand* tendons are *deep* to the palmar aponeurosis, the superficial palmar arch and digital branches of the median and ulnar nerves but *superficial* to tendons of flexor digitorum profundus, the lumbrical muscles and the deep palmar arch.

Nerve supply: median nerve, C.7 and **8** and T.1.

Actions: flexion of the middle and then proximal phalanges and carpal flexion, particularly in rapid, forceful flexion of the digits in grasping, but is electromyographically silent in gentle flexion.

DEEP ANTEBRACHIAL FLEXORS

The group comprises flexor digitorum profundus, flexor pollicis longus and pronator quadratus.

Flexor digitorum profundus (5.66, 67, 69) arises deep to the superficial flexors from about the upper three-quarters of anterior and medial surfaces of the ulna, embracing the attachment of brachialis and extending distally almost to pronator quadratus; it also arises from a depression on the medial side of the coronoid process and proximal three-quarters of the posterior ulnar border by a common aponeurosis shared with the flexor and extensor carpi ulnaris; it also springs from the anterior surface of the ulnar half of the interosseous membrane. It ends in four tendons, initially posterior (deep) to those of flexor digitorum superficialis and to the flexor retinaculum. The part acting on the index finger is usually distinct but tendons for other digits are interconnected by

areolar tissue and tendinous slips as far as the palm. Anterior to their proximal phalanges the tendons pass through those of flexor digitorum superficialis (p. 617) to attachments on palmar aspects of distal phalangeal bases. The tendons undergo fascicular rearrangement as they traverse those of superficialis (Wilkinson 1953, Martin 1958). The muscle forms most of the surface elevation medial to the palpable posterior ulnar border.

Accessory slips from radius (acting on index), from flexor superficialis, flexor pollicis longus, the medial epicondyle or coronoid process may join the muscle.

Nerve supply: medial part by the ulnar nerve, the lateral by the anterior interosseous branch of the median nerve, C.**8** and T.1.

Actions. Flexion of distal phalanges, after flexion of the middle by superficialis. Assists carpal flexion. When all fingers are flexed, the digital extensors relax; but when individual digits flex, activity is sustained in the antagonist extensors (Person & Roschina 1958). More complex movements involve the synergic action of the interossei and lumbricals (p. 632). Electromyographically (Long 1968) flexor digitorum profundus is alone active in gentle digital flexion and superficialis added for greater force and/or velocity.

Lumbrical muscles are attached to tendons of flexor digitorum profundus in the palm (p. 630).

FIBROUS SHEATHS OF FLEXOR TENDONS

In the digits both tendons of deep and superficial flexors are in osseo-aponeurotic canals (5.77,79), formed by the phalanges, their joints and by *digital fibrous sheaths* arching anteriorly across the tendons and attached to the periosteal margins of the phalanges and the palmar ligaments of their interphalangeal joints. Similar fibrous dispositions, however, continue proximally into the palm to include the metacarpophalangeal joints. It was early recognized that the sheath had regional variations in thickness and fibre architecture; despite some minor variations, the view widely held for over the first 75 years of this century may be summarized briefly. 'Midway along proximal and intermediate phalanges the bands are strong, their fibres transverse (*annular part*); opposite joints they are thinner and of oblique fibres (*cruciform part*)'; however, vide infra. Each canal, from metacarpal head to terminal phalangeal base, is lined throughout by a double-walled (parieto-visceral) synovial sheath. The parietal layer, complex in form, is closely adherent to the specialized palmar ligaments, periostea, and internal aspects of fibrous flexor sheaths, forming, in effect, the internal free surface of these structures. Thus they exhibit undulations corresponding to constrictions or expansions and the recently observed 'steps' and recesses or pouches between thick and thin zones (Jones & Amis 1987), vide infra. They also noted specific forms of folding and repositioning of zones during flexion and continuing researches are directed, also, to regional structural differences in these parietal strata. The parietal synovium is reflected, at restricted loci, as a 'visceral' layer tightly enclosing and fused to the surfaces of the contained tendons, its shape, therefore, reflecting precisely the exquisite mutual adaptation of the tendons whose profiles change continuously proximodistally as they approach and reach their phalangeal attachments. The dictates of modern hand surgery (Doyle & Blythe 1975, Lundborg & Myrhage 1977, Schneider & Hunter 1982, Lister 1985) demonstrated that the traditional account of the fibrous flexor sheath was inadequate: oversimplified, positionally incorrect and with prominent features lacking. A new terminology has been devised and is currently widely used in surgical literature. Although possibly requiring some future revision, it will be briefly summarized here (5.67A). The fibrous flexor sheath is considered as extending from the metacarpal head to the base of the terminal phalanx and presents a series of thick zones with intervening thin zones. The thick zones, collectively termed pulleys, may be annular or cruciate. The annular pulleys, systematically numbered proximodistally A1–5, alternate with the four cruciate pulleys C0–3, the thin zones occupying the modifiable interpulley gaps. On occasion pulleys C0, C3 and A5 are poorly developed or undetectable and sometimes a cruciate pulley consisted of a single oblique band. In an extended finger the approximate positions of the pulleys in relation to the underlying bones and joints are: A1—ventral to the metacarpophalangeal joint; A2—centred over the junction of the proximal and middle thirds of the proximal phalanx; A3—proximoventral to the proximal interphalangeal joint; A4—ventral to the middle third of the middle phalanx; A5— distoventral aspect of the distal interphalangeal joint; C0 (between A1 and A2)—ventral to the proximal shaft of the proximal phalanx; C1 (between A2 and A3)

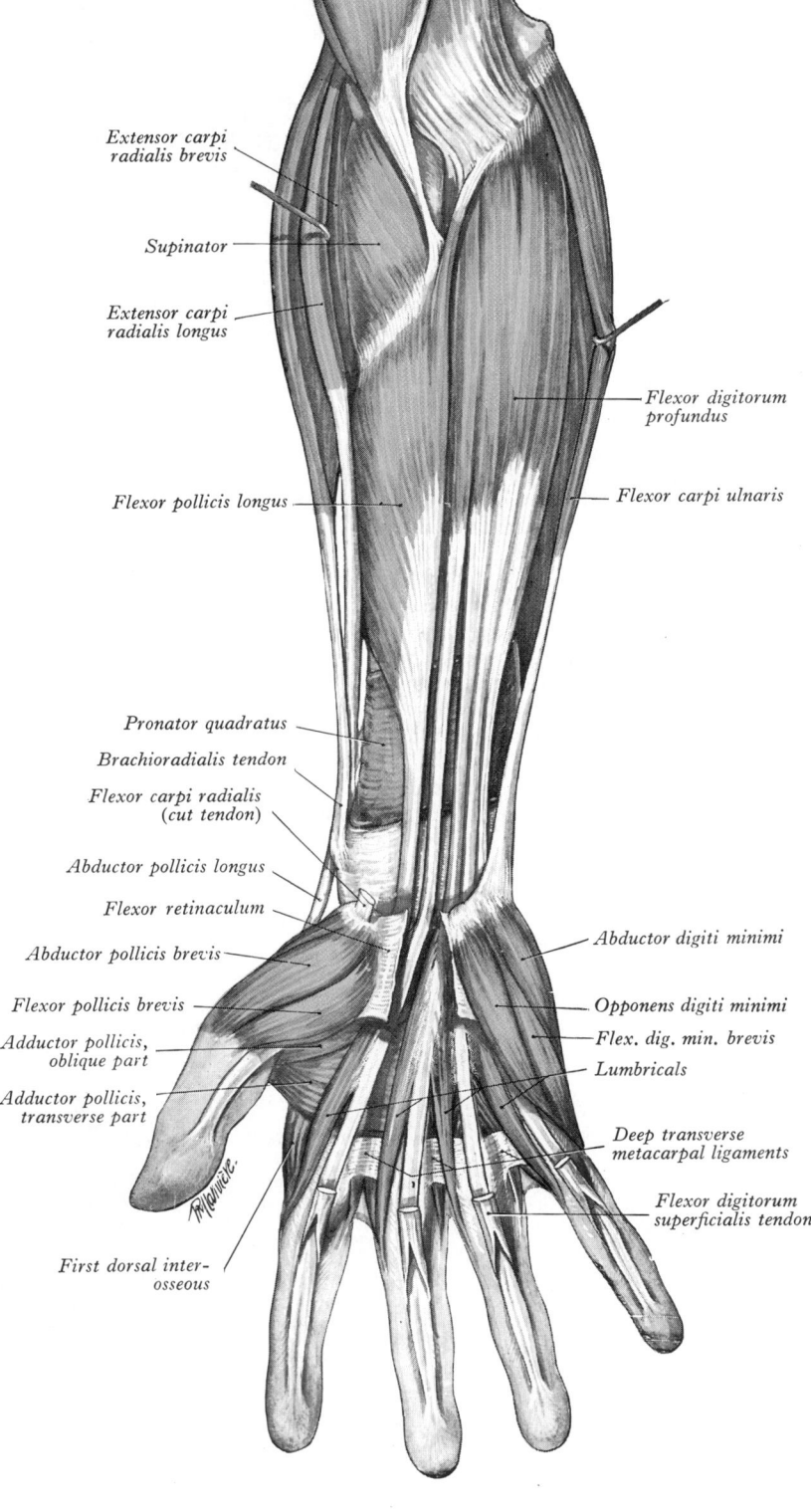

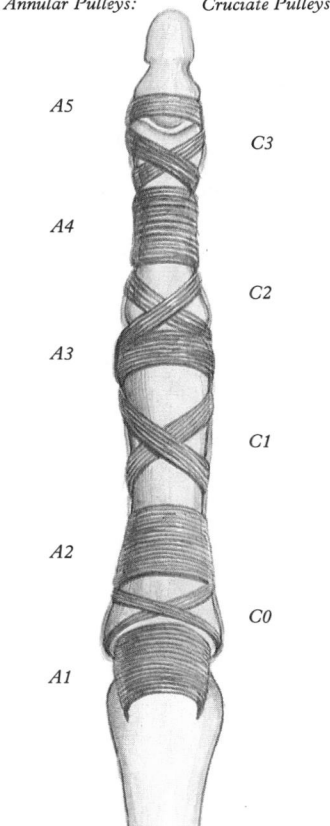

Annular Pulleys:

A5
A4
A3
A2
A1

Cruciate Pulleys:

C3
C2
C1
C0

5.67A The digital fibrous flexor sheath: an increasingly accepted classification of the thickened annular pulleys A1–A5, cruciate pulleys C0–C3 and thin areas (light green) of the sheath.

Extensor carpi radialis brevis

Supinator

Extensor carpi radialis longus

Flexor pollicis longus

Flexor digitorum profundus

Flexor carpi ulnaris

Pronator quadratus

Brachioradialis tendon

Flexor carpi radialis (cut tendon)

Abductor pollicis longus

Flexor retinaculum

Abductor pollicis brevis

Flexor pollicis brevis

Adductor pollicis, oblique part

Adductor pollicis, transverse part

First dorsal interosseous

Abductor digiti minimi

Opponens digiti minimi

Flex. dig. min. brevis

Lumbricals

Deep transverse metacarpal ligaments

Flexor digitorum superficialis tendon

5.67B The deep flexor muscles of the right forearm.

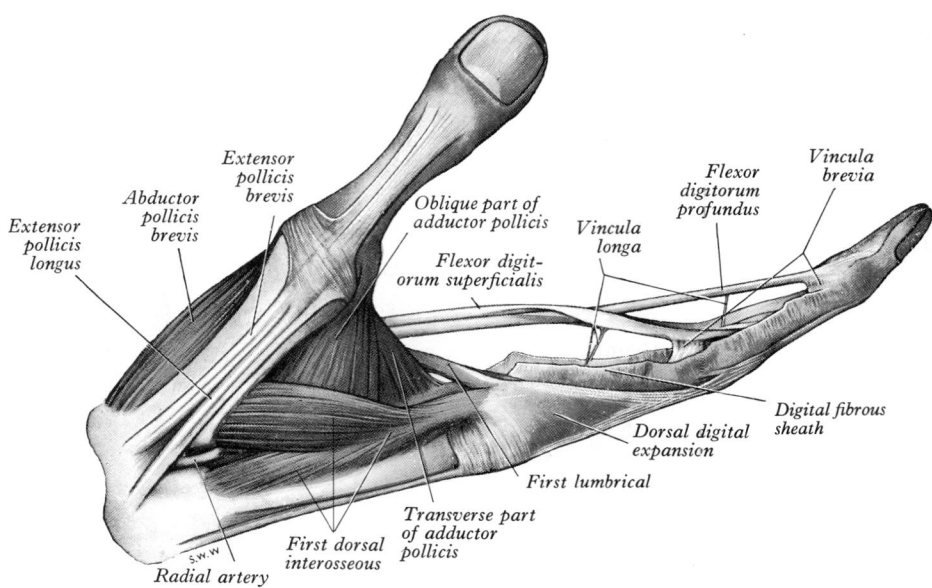

5.68 Lateral part of the right hand showing the tendons and vincula tendinum of the index finger and the muscles in the first intermetacarpal space.

—ventral to the junction of the middle and distal thirds of the proximal phalanx; C2 (between A3 and A4)—ventral to the proximal shaft of the middle phalanx; C3 (between A4 and A5) —ventral to the narrow distal shaft of the middle phalanx. Thus of the five annular pulleys two are related to phalangeal shafts, the remaining three closely related to the three synovial joints; the cruciate pulleys are all related to shafts but not far distant from neighbouring joints. Jones & Amis (1987) in a detailed study of fibrous flexor sheaths confirmed the general nature and distribution of the annular and cruciate pulleys, but extended this to a quantitative (and projected functional) analysis of the 'deep pockets' noted by Lundborg & Myrhage (1977). Best viewed from the internal (parietal synovial) aspect, it was observed that thin zones did not smoothly join edge to edge with their adjacent pulley, but either joined the most superficial stratum of the thick zone or passed for some distance *superficial* to the thick zone before joining its superficial surface. Thus, the thick pulley possesses, internally, margins that are protruding (proximally,

distally or both), either 'step-like' or 'free-edges', external to which a probe may pass on to the thick tissue's superficial surface as far as the attachment of the thin zone. The mean depth of the *overlap (recess or pouch)* was 0.5–1.0 mm, but pooled data for the distal end of A2 in index, medius and annularis was 2.4 mm. During flexion the thin zones fold in a specific pattern *superficially* into potential areolar spaces and *away* from the underlying tendons or a related joint line. Also during continued flexion the margins of the thick pulleys approach each other more and more closely until the tendons are encased ventrolaterally by a virtually continuous curved sheet of tough, deformation-resistant, thick fibrous tissue. (Consult Jones & Amis 1987 for detailed mensuration.) The results of further biomechanical analyses of these regions are awaited with interest.

As the flexor tendons approach their attachments they connect to the dorsal (parietal) aspects of their synovial sheaths and adjoining bone or ligament by triangular or filiform bands of synovial membrane termed *vincula tendinum* (**5.68**), which convey

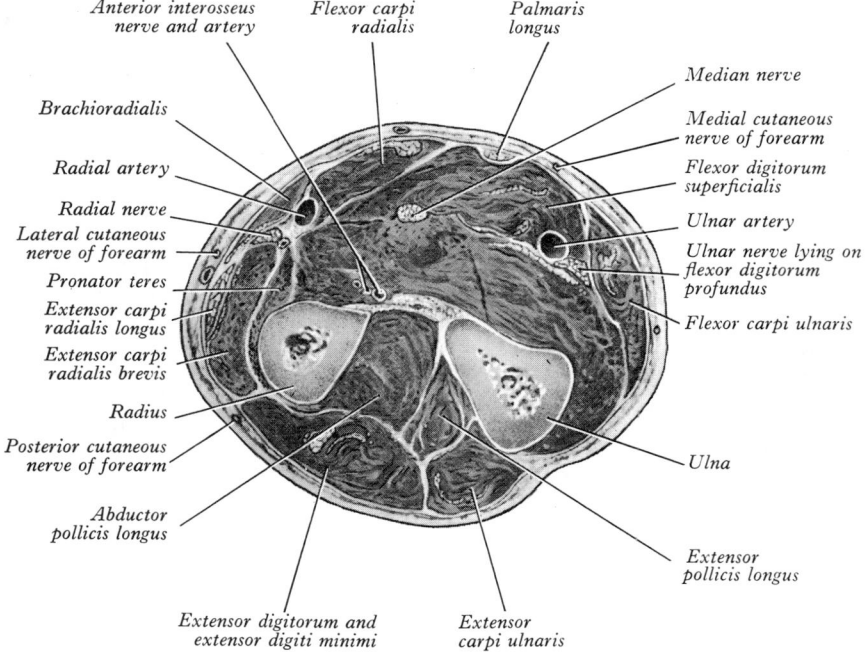

5.69 Transverse section through the middle of the left forearm.

minute vessels to the tendons. They are of two kinds. *Vincula brevia*, two in each finger, are triangular bands attached to deep surfaces of tendons near their attachment; one connects a superficial tendon to its proximal interphalangeal joint capsule and adjacent proximal phalanx, the other connects a deep tendon to its distal interphalangeal joint capsule and adjacent middle phalanx. The *vincula longa* are filiform, usually two attached to each superficial tendon, one to each deep tendon. Those of flexor superficialis are connected to the slips of tendon where they fold over a tendon of flexor digitorum profundus, passing around its sides to attach to the sheath at the lateral margins nearer the proximal end of its proximal phalanx. Vincula longa of flexor digitorum profundus, fixed to its tendons shortly after they pierce the superficial tendons, perforate one of the two slips of the latter tendon or pass between them, thereafter blending with vincula brevia of flexor digitorum superficialis, attached to the dorsal wall of the synovial sheath to the distal ends of proximal phalanges.

Flexor pollicis longus (5.67,76), lateral to flexor digitorum profundus, arises from the grooved anterior surface of the radius between its tuberosity and the upper attachment of pronator quadratus, also from adjacent interosseous membrane, and often by a variable slip from the lateral or, more rarely, medial border of the coronoid process or the medial epicondyle (Martin 1958). The muscle ends as a flat tendon, passing behind the flexor retinaculum, between opponens pollicis and the oblique head of adductor pollicis, to enter an osseo-aponeurotic canal like those in the fingers, to be attached to the palmar aspect of the thumb's distal phalangeal base. The anterior interosseous nerve and vessels descend on the interosseous membrane between flexor digitorum profundus and flexor pollicis longus, which is sometimes connected to either the digital flexor or to pronator teres. The interosseous attachment may be lacking or the whole muscle absent.

Nerve supply: the anterior interosseous branch of the median nerve, C.8 and T.1.

Action: flexion of pollicial phalanges (p. 630).

Pronator quadratus (5.67), flat and quadrilateral, extends anteriorly across the lower parts of the radius and ulna. From the oblique ridge on the anterior surface of the ulna (3.141) and also the medial part of this surface and from an aponeurosis over the muscle's medial third, it passes laterally and a little distally to the distal quarter of the radial anterior border and surface; deeper fibres reach the triangular area proximal to the ulnar notch of the radius.

Nerve supply: the anterior interosseous branch of median nerve, C.8 and T.1.

Action: the principal pronator of the forearm, assisted by pronator teres in rapid and/or forceful pronation. Deeper fibres oppose separation of distal ends of the radius and ulna when thrusts are transmitted through the carpus.

(b) Posterior Antebrachial Muscles

These muscles form superficial and deep groups.

SUPERFICIAL EXTENSOR GROUP

All these muscles (5.72,75), except brachioradialis and anconeus, which act at the elbow alone, extend (or act synergically in other movements) at a series of joints. Extensores carpi radiales longus and brevis and extensor carpi ulnaris act on radiocarpal and carpal joints; extensores digitorum and digiti minimi extend digits and may extend the carpus.

Brachioradialis (5.65,66,69), the most superficial muscle along the radial side of the forearm, forms the lateral border of the cubital fossa. It arises from the proximal two-thirds of the lateral supracondylar ridge and anterior surface of the lateral intermuscular septum. The radial nerve and an anastomosis between arteria profunda brachii and radial recurrent artery are between the septum and brachialis. Its muscle fibres end proximal to mid forearm level in a flat tendon which extends to attach laterally on

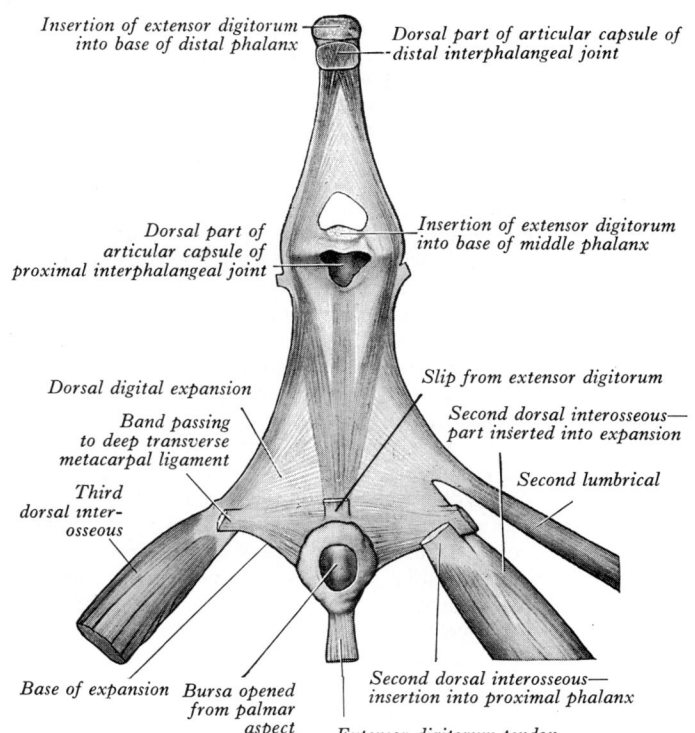

5.70A The dorsal digital expansion of the middle finger, showing some of its principal connections (viewed from the palmar surface, the basal angles having been drawn out and the whole expansion kept taut and flattened). Compare with 5.70B and C. Immediately distal to the bursa which overlies the metacarpophalangeal joint, a small slip from the extensor digitorum tendon is inserted into the base of the proximal phalanx. This slip is not always present.

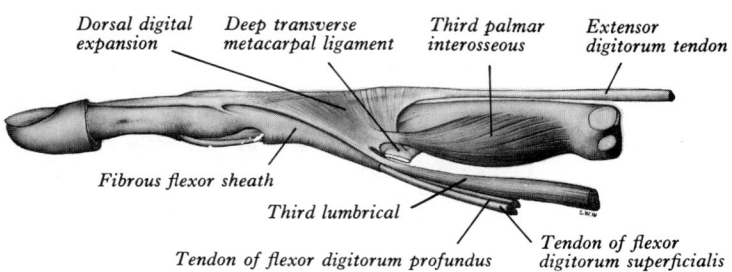

5.70B The ring finger dissected to show the dorsal digital expansion and its principal connections (lateral aspect). Note the position of the base of the expansion when the finger is extended and compare with 5.70C.

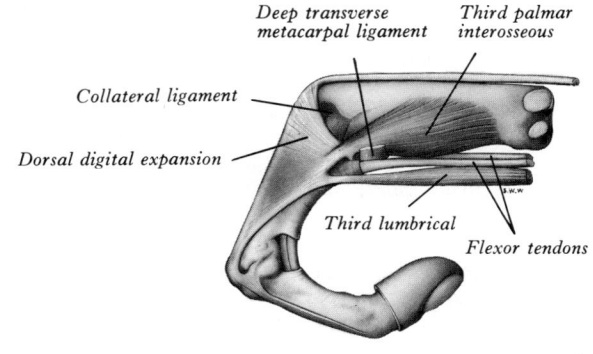

5.70C The dorsal digital expansion of the ring finger in flexion. Compare with 5.70B, and note that the base of the expansion has moved distally. Compare 5.70A, B and C which were prepared by Professor James Whillis. Additional details and nomenclature are shown in 5.71A, B and C.

the distal end of the radius, proximal to its styloid process. The tendon is crossed near its distal termination by those of abductor pollicis longus and extensor pollicis brevis; the radial artery is medial. The muscle is often fused proximally with brachialis. Its tendon may divide into two or three slips. Rarely it is double or absent. Its radial attachment may be much more proximal.

Nerve supply: the radial nerve, C.5, 6 and (7).

Action: *flexion* of elbow (though supplied by an 'extensor' nerve), most effectively in mid-pronation, when it stands out prominently, particularly when carried out against resistance. It is minimally active in slow, easy flexion, or with a supine forearm; but develops a powerful burst of activity in both *rapid flexion and extension*; during these activities it is principally a 'shunt' muscle (p. 569), providing centripetal force towards the elbow to balance the centrifugal force of rapid swings in either direction (Mac-Conaill 1949, Basmajian 1959).

Extensor carpi radialis longus (5.69,72), partly overlapped by brachioradialis, is attached mainly to the distal third of the lateral supracondylar ridge and lateral intermuscular septum and partially to the common extensor tendon. The belly ends at the junction of the proximal and middle thirds of the forearm in a flat tendon along its lateral radial surface, deep to abductor pollicis longus and extensor pollicis brevis; this passes under the extensor retinaculum in a groove on the dorsum of the radius behind its styloid process and is distally attached on the radial side of the dorsal aspect of the second metacarpal base. It may send slips to the first or third metacarpal bones and contributes to intermetacarpal ligaments.

Nerve supply: the radial nerve, C. 6 and 7.

Extensor carpi radialis brevis (5.69,72,76) is shorter than and covered by the preceding muscle. It is attached to the lateral epicondyle by the common extensor tendon (vide infra), to the radial collateral ligament, to a strong aponeurosis covering its surface and to adjacent intermuscular septa. Its belly ends about mid forearm in a flat tendon accompanying that of the long carpal extensor to the wrist; it passes deep to abductor pollicis longus, extensor pollicis brevis and the extensor retinaculum to be attached dorsally to the third metacarpal base on its radial side, distal to its styloid process, and on adjoining parts of the second metacarpal. Under the extensor retinaculum the tendon is in a shallow groove, medial to that of extensor carpi radialis longus, separated by a low variably marked ridge.

Nerve supply: the posterior interosseous nerve, C.7 and 8.

Tendons of the radial carpal extensors pass under the extensor retinaculum in a single, common synovial sheath. They are easily identified as the fist is clenched and relaxed. Both may split into slips, variably attached to the second and third metacarpal bones. The muscles may be united or may exchange muscular slips.

Actions. Both muscles act synergically with the digital *flexors*. With extensor carpi ulnaris they extend the wrist; with flexor carpi radialis they abduct it. They are more often synergic than prime movers, particularly extensor carpi radialis longus, which is less active than the brevis in pure extension (Tournay & Paillard 1953), even when this is rapid. The longus is much more active than the brevis in grasping or clenching.

Extensor digitorum (5.69,72,76) springs from the lateral epicondyle by the common extensor tendon, from the adjacent intermuscular septa and antebrachial fascia. It divides distally into four tendons, which traverse, with the tendon of extensor indicis, a separate tunnel under the extensor retinaculum in a common synovial sheath. These diverge on the dorsum, one to each finger; that to the index is accompanied by extensor indicis, medial to it. On the hand's dorsum adjacent tendons have three variable *intertendinous connections*, inclined distolaterally. The medial connection is strong and pulls the tendon of minimus towards that of the fourth digit; the connection between the middle two tendons is weak and may be absent (Leslie 1954). The function of these bands is not clear; they *may* affect independent extension of digits; a gem of apocryphal anatomy is that their division benefits pianists!

Digital attachments are complicated by a fibrous expansion dorsal to the proximal phalanges in which lumbrical, interosseous and digital extensor tendons all participate.

A dorsal digital expansion is a small aponeurosis covering

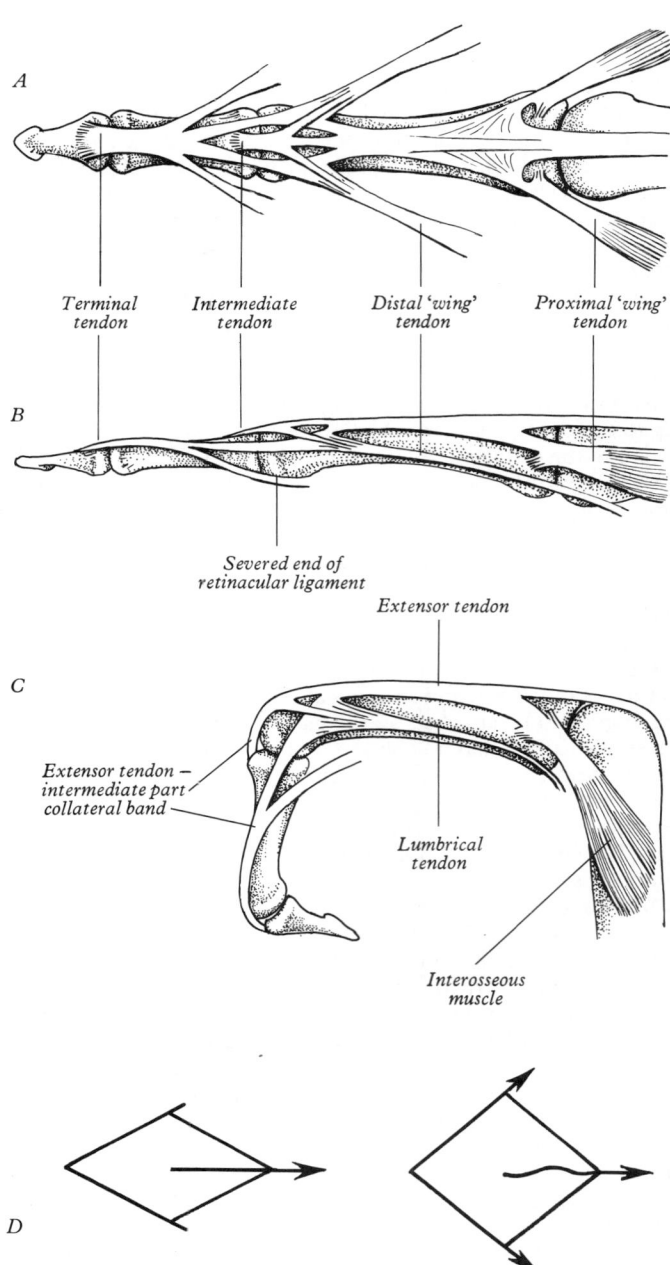

5.71 Detailed diagrams to illustrate the digital extensor mechanisms: A. viewed from the dorsal aspect with digit in full extension; B. viewed from the side, also in full extension; C. viewed from the side, in full flexion. Note particularly the terminal and intermediate parts of the long extensor tendon, the proximal and distal 'wing' tendons, the dorsal extensor expansion, the diamond-shaped complex of tendons dorsal to the proximal interphalangeal joint and the retinacular ligaments. Note also the changing disposition of the tendinous and ligamentous elements in relation to the digital joints, as the digit passes from extension (B) to flexion (C). D is a diagram showing changes in the 'diamond' complex with contraction of the lumbrical musculature. See text for discussion. (Modified from diagrams by Graham Stack 1962, 1963.)

the dorsum of a proximal phalanx and the sides of its base. It is triangular; its proximal base enwraps dorsal and collateral aspects of the metacarpophalangeal joint. A tendon of extensor digitorum blends with the expansion along its axial region, separated from the joint by a small bursa (5.70A). The base connects this tendon on each side with adjoining interosseous muscles and has many transverse fibres. A short band links each basal angle to the deep transverse metacarpal ligaments, separating the *phalangeal* attachment of a dorsal interosseous from the rest of the muscle, and also a palmar interosseous from a lumbrical muscle (5.70A–C).

Margins of these expansions are thickened laterally by tendons of lumbrical and interosseous muscles and medially by an interosseous alone or, in the fifth digit, by abductor digiti minimi. These attachments of intrinsic muscles are termed 'wing tendons'. Proximal and distal 'wings' can usually be identified (5.71) on both sides of each expansion. Attachments of interossei to these has led to a classification as 'distal' and 'proximal' in place of the more usual palmar and dorsal (see also 5.70A–C, 5.71A–D, and p. 630). Between its margins and the extensor tendon the expansion is almost translucent (5.70A). Near proximal interphalangeal joints extensor tendons divide into an axial part and collateral slips. The former, receiving some fibres from lumbrical and interosseous tendons (Landsmeer 1949), passes over the joint to the base of a middle phalanx. Each collateral slip is joined by a corresponding thickened border of the expansion and they unite to be attached dorsally on the base of a distal phalanx. The distal 'wing tendon' is proximal to this attachment.

Each expansion is a movable hood (Bunnell 1949), moving distally when a metacarpophalangeal joint flexes and proximally in extension when it is most closely curved round the joint. Landsmeer (1949) and Haines (1951) have described *retinacular ('link') ligaments* correlating movements at interphalangeal joints (5.71A–C); from proximal attachments to the sides of a proximal phalanx, where the fibrous sheath reaches bone and from the sheath itself they extend distally to blend with the margins of a dorsal expansion (5.71C). They thus reach the bases of terminal phalanges, two in each digit (p. 632).

Tendons of extensor digitorum tendons may be variably deficient, but more often are doubled or tripled in one or more digits, most often the index or medius. A tendon to the thumb occasionally exists.

Nerve supply: the posterior interosseous nerve, C.7 and 8.

Actions: digital extension at metacarpophalangeal and interphalangeal joints, opening the hand to relax the grip or prepare for grasping. The digital extensor is the prime mover with carpal extensors in carpal extension. It tends to spread the index, ring and little fingers but has no such action on the middle. Attachment of the lateral bands of its tendons to deep transverse metacarpal ligaments greatly limits the extensor effect at interphalangeal joints, an action only completed by the attachments of intrinsic muscles, especially lumbricals, to lateral extensor bands *distal* to these joints.

Flexion of metacarpophalangeal combined with *extension* of interphalangeal joints, and its converse, enter into most fine digital movements, especially in the activities of index finger and thumb, as in attempts to thread a needle; but the fine adjustments in approach (flexion at metacarpophalangeal and extension at interphalangeal joints) and withdrawal (extension at metacarpophalangeal and flexion at interphalangeal joints) of thread to needle merely illustrate movements used in a multitude of other common activities. Metacarpophalangeal flexion, due to lumbricals and interossei, allows sufficient relaxation of digital flexors to let extensor digitorum extend interphalangeal joints (Braithwaite et al 1948). However, interossei, working over the distal border of the deep transverse metacarpal ligaments, are advantageously placed to extend interphalangeal joints via dorsal digital expansions and can be trained to do so in radial nerve palsy (Sunderland 1945).

Extensor digiti minimi (5.72), a slender muscle, medial and usually connected to extensor digitorum, arises from the common extensor tendon by a thin tendinous slip and from adjacent intermuscular septa. Its long tendon slides in a separate compartment of the extensor retinaculum just behind the inferior radio-ulnar joint, dividing distally into two, the lateral slip joined by a tendon of extensor digitorum (5.72); all three are attached to the dorsal digital expansion of the fifth digit. The muscle usually arises also from antebrachial fascia. Its tendon may not divide or may send a slip to the fourth digit. It is rarely absent but sometimes fused with extensor digitorum.

Nerve supply: the posterior interosseous nerve, C.7 and 8.

Actions: extension of minimus and carpus, with extensor digitorum.

Extensor carpi ulnaris (5.69, 72) is attached to the lateral epicondyle by the common extensor tendon, also to the posterior ulnar border by the common aponeurosis shared with flexor carpi ulnaris and flexor digitorum profundus and to overlying

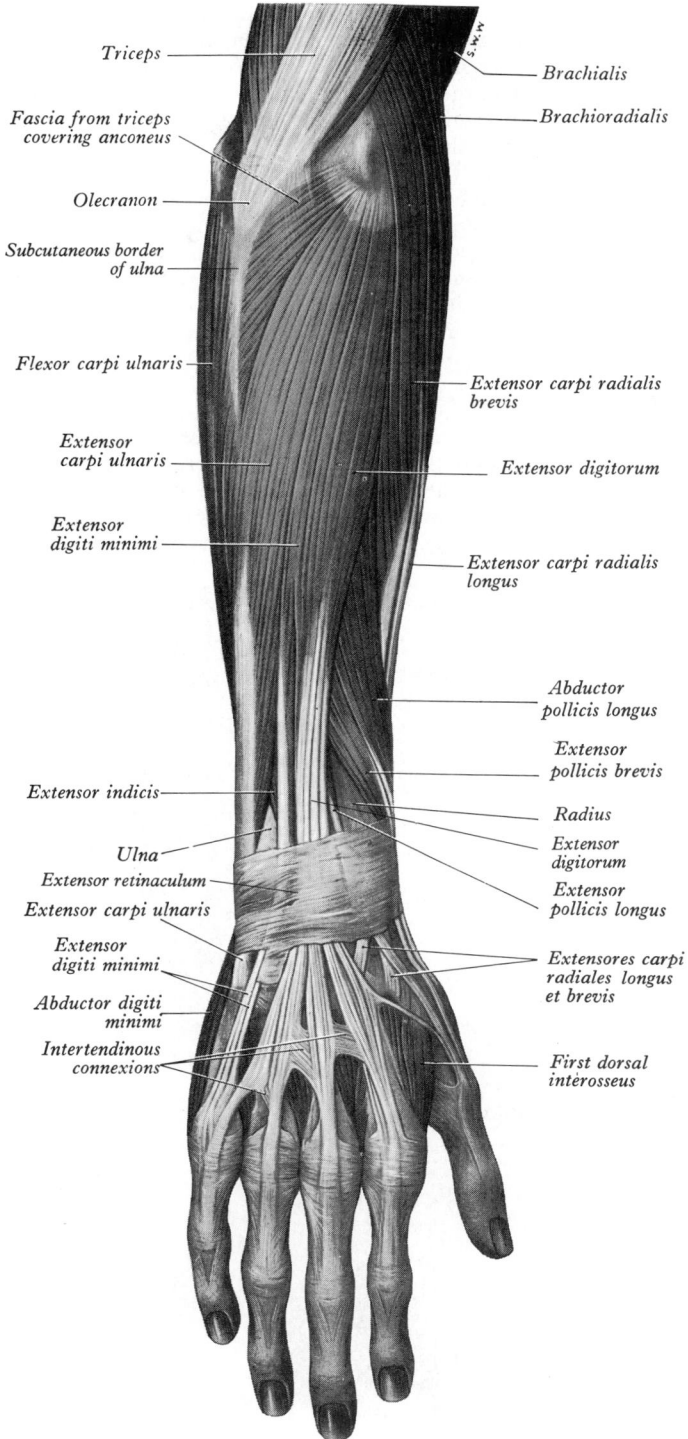

5.72 Muscles of the extensor aspect of the right forearm: superficial layer.

antebrachial fascia. It can be felt lateral to the groove which overlies the ulna's posterior subcutaneous border. It ends in a tendon sliding in a groove between the ulnar head and its styloid process in a separate compartment of extensor retinaculum, and is attached to a tubercle on the medial side of the fifth metacarpal base.

Nerve supply: posterior interosseous nerve, C.7 and **8**.

Actions. With extensores carpi radiales longus and brevis it acts synergically with digital flexors to extend and fix the wrist when objects are gripped or the fist clenched. It is impossible to grip strongly unless the wrist is extended. With flexor carpi ulnaris it adducts the hand.

Anconeus (5.72, 73), small and triangular and behind the cubital joint, is partially blended with triceps, of which it is an integral part in some primates. It has a separate tendon attached posteriorly to the lateral epicondyle; its fibres diverge medially towards the ulna, covering the annular ligament and it is attached to the lateral aspect of the olecranon and proximal quarter of the posterior surface of the shaft of the ulna. The muscle is variably fused with triceps or the extensor carpi ulnaris.

Nerve supply: radial nerve, C.7, 8 and (T.1).

Action. It assists triceps in cubital extension and is often considered responsible for movement of the ulna in pronation (p. 616). Electromyography has not confirmed the latter; adequate investigations are, however, far from complete. Ulnar abduction in pronation, necessary if the forearm is to revolve the hand without translating it medially, can be voluntarily varied in degree. A tool can be revolved 'on the spot' or swept through an arc.

DEEP EXTENSOR GROUP

The deep forearm extensors (5.73,75) comprise three acting on the pollex (abductor pollicis longus and extensores pollicis longus et brevis), together with extensor indicis and supinator. Apart from supinator, all are attached proximally only to forearm bones.

Supinator (5.67,73,74) surrounds the proximal third of the radius and has superficial and deep strata, between which is the posterior interosseous nerve (5.73). They arise by superficial tendinous and deep muscular fibres from the lateral epicondyle, radial collateral ligament, annular ligament, supinator crest of ulna with the posterior part of a triangular depression in front of it and from an aponeurosis over the muscle. Distal attachment is to the lateral surface of the proximal third of the radius, reaching the distal attachment of pronator teres. The radial attachment extends to the anterior and posterior surfaces (p. 411, 3.142), between the anterior oblique line and fainter posterior oblique 'ridge'. The muscle is subject to frequent variation, small parts of it often receiving individual names, e.g. lateral and medial tensors of the annular ligament (Hast & Perkins 1984).

Nerve supply: the posterior interosseous nerve, C.5 and **6**.

Action: radial rotation to bring the palm anterior, acting alone in slow, unopposed supination and in fast or forceful supination joined by biceps brachii. Objects, which may be heavy, are often picked up with the forearm initially pronated. The more powerful supinators lift it against gravity, often combined with increasing flexion to bring the hand's burden towards the eyes.

Abductor pollicis longus (5.69,72,73) is attached, distal to the supinator and close to extensor pollicis brevis, to the posterior ulnar surface distal to the anconeus, to the adjoining interosseous membrane and middle third of the posterior radial surface distal to the supinator. It inclines laterally, becoming superficial in the distal forearm, visible as an oblique elevation (5.72). Its belly ends in a tendon just proximal to the carpus; this is in a groove on the lateral side of the distal end of the radius together with the tendon of extensor pollicis brevis. The tendon usually splits, one slip being attached to the radial side of the first metacarpal base, one to the trapezium. Further fasciculi may continue into opponens pollicis or abductor pollicis brevis. The muscle may be wholly or partially divided.

Nerve supply: the posterior interosseous nerve, C.7 and **8**.

Actions. With abductor pollicis brevis it abducts the pollex and with pollicial extensors extends it at its carpometacarpal joint (p. 629).

Extensor pollicis brevis (5.72,73), medial and connected to abductor pollicis longus, arises from the posterior radial surface

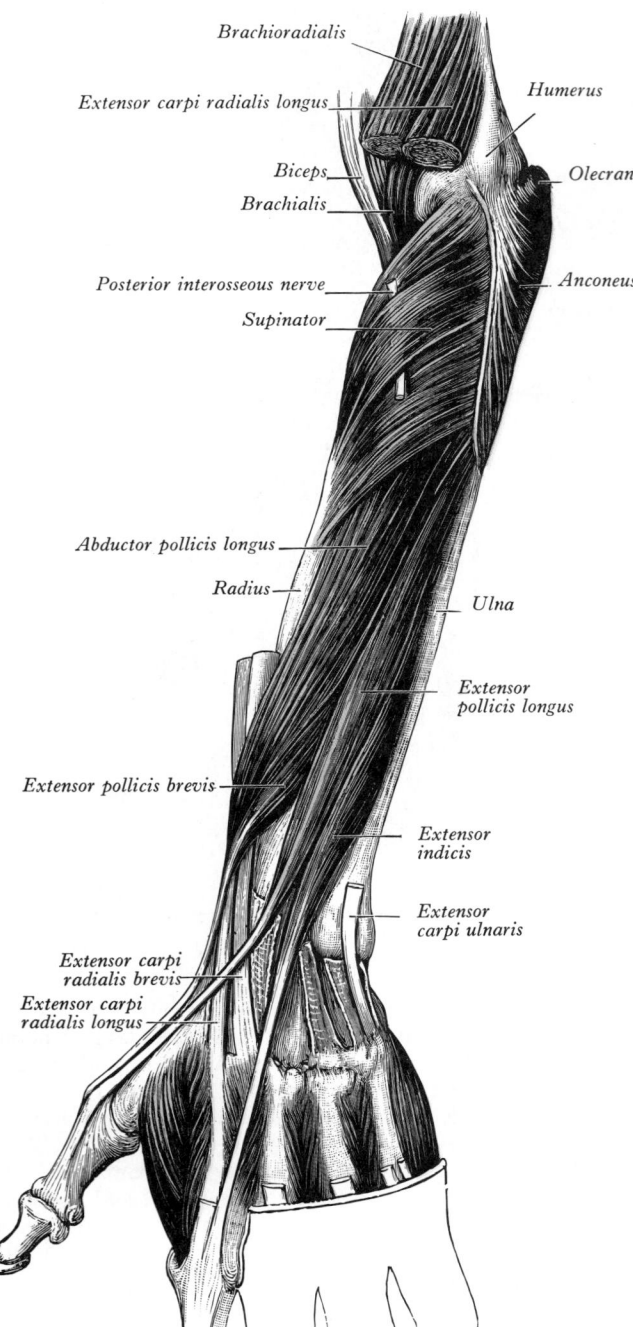

624   5.73 The deep extensor muscles of the left forearm.

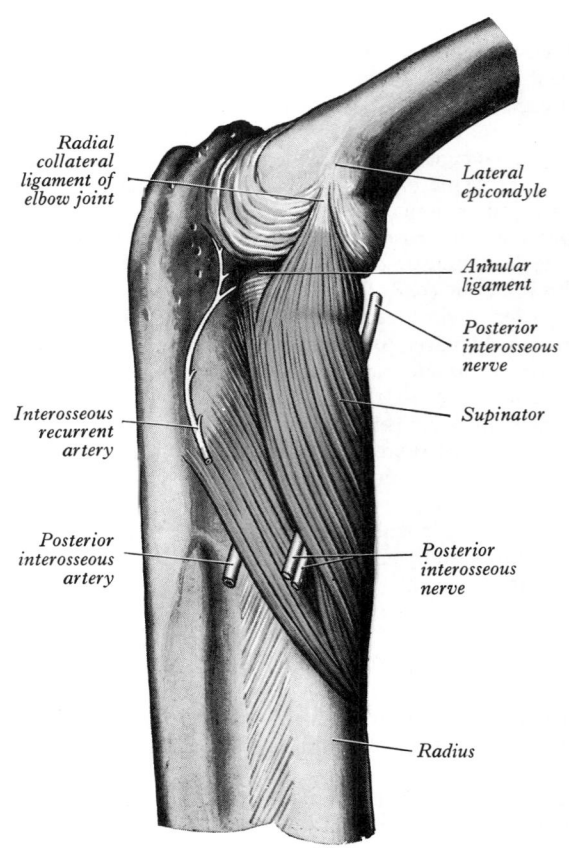

5.74 The right supinator muscle: posterolateral aspect.

distal to it and from the adjacent interosseous membrane. It descends with the abductor, its tendon in the same groove on the lateral side of the distal radius, to be attached to the dorsolateral surface of the base of the proximal pollicial phalanx. Attachment to the base of the *distal* phalanx, usually through a fasciculus joining the tendon of extensor pollicis longus, is common (Muller 1959). The muscle may be absent or fused with abductor longus. Its tendon sometimes unites with the long pollicial extensor.

In the distal forearm abductor pollicis longus and extensor pollicis brevis *emerge* between extensor carpi radialis brevis and extensor digitorum; crossing obliquely the tendons of the radial extensors, they cover the distal part of brachioradialis and traverse the most lateral compartment of the extensor retinaculum in a common single sheath, finally crossing superficial to the radial styloid process and radial artery to reach the base of the proximal pollicial phalanx.

Nerve supply: the posterior interosseous nerve, C.7 and **8**.

Actions: extension of the proximal pollicial phalanx and, in continued action, metacarpal extension.

Extensor pollicis longus (5.72,73) is larger than the brevis, the proximal attachment of which it partly covers. It arises laterally from the middle third of the posterior ulnar surface, distal to abductor pollicis longus, and from the adjacent interosseous membrane. It ends in a tendon traversing a separate compartment of the extensor retinaculum, in a narrow, *oblique groove* on the dorsum of the distal radius. It crosses obliquely over tendons of carpal extensors (5.73) but is separated from extensor pollicis brevis and clearly seen when the pollex is extended by a triangular depression, jocularly termed the 'anatomical snuff-box', in which the radial artery can be felt. (Bony structures felt in the floor of the fossa by deep palpation are proximodistally: the radial styloid, the smooth convex articular surface of the scaphoid, the trapezium and the first metacarpal's base. The latter are more pronounced during metacarpal movement and the scaphoid during adduction – abduction of the hand. The scaphoid and the trapezium are effectively palpated bidigitally between the examining index and thumb, one in the fossa, the other on the palmar tubercle or crest respectively.) The tendon of extensor pollicis longus is attached to the base of the distal pollicial phalanx; dorsal to the proximal phalanx its sides are joined laterally by expansions from tendons of abductor pollicis brevis and medially from the first palmar interosseous and adductor pollicis.

Nerve supply: the posterior interosseous nerve, C.7 and **8**.

Actions: extension of the distal pollicial phalanx; with extensor pollicis brevis and abductor pollicis longus, extension of the proximal phalanx and metacarpal. In continued action, owing to its tendon's obliquity, the muscle adducts the extended thumb and rotates it laterally (p. 513).

Applied Anatomy: avascular necrosis is probably a factor in rupture of the tendon of extensor pollicis longus, which may complicate Colles' fracture. Proximal to the carpus, the blood supply is mainly from the anterior interosseous artery and distally from the radial and pollicial arteries (Davies 1951).

Extensor indicis (5.73), narrow and elongated, is medial and parallel to extensor pollicis longus. It arises from the posterior surface of the ulna distal to extensor pollicis longus and from the adjacent interosseous membrane. Its tendon passes under the extensor retinaculum in a compartment containing tendons of extensor digitorum; near the second metacarpal head it joins the ulnar side of the index tendon of extensor digitorum. It may exchange accessory slips with other extensor tendons and is occasionally interrupted on the dorsum of the hand by an additional belly ('*extensor indicis brevis manus*').

Nerve supply: the posterior interosseous nerve, C.7 and **8**. Distribution of muscle spindles in this muscle has been assessed by Gorp & Kennedy (1974); the greatest density is near the entry of its nerve.

Actions: assists extension of the index and carpus.

The Hand

Human hands, working with eyes in complex synergies mediated by nervous connections of unlimited variation, have been

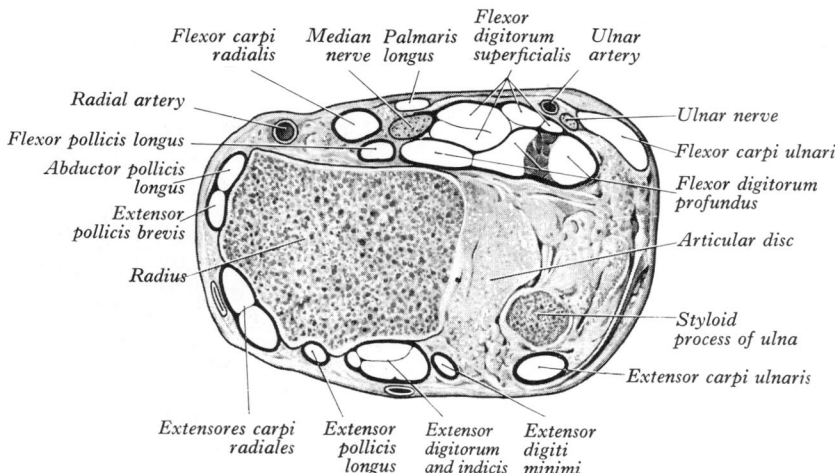

5.75 Transverse section, passing through the distal end of the right radius and the styloid process of the right ulna, made with the hand and forearm in full supination: distal (inferior) aspect.

paramount in the evolution of human skills. Despite an astonishing multitude of tools and techniques produced to extend these skills, hands will continue to refine, develop and invent further tools. Being put to such multiple and frequent uses they are often damaged. For both reasons an immense literature, impossible even to abbreviate here, has accumulated on the structure, actions, injuries and reparative surgery of hands. For guidance the works of Wood Jones (1941), Napier (1966), Kaplan (1965), Stack (1973) and Landsmeer (1976) are valuable, especially since each has contributed notably in this field.

RETINACULA, FASCIAE AND SYNOVIAL SHEATHS OF THE CARPUS AND HAND

Before proceeding to manual muscles, the disposition of local retaining bands, deep fascia and synovial tendon sheaths must be considered.

The flexor retinaculum (5.76,77), a thick strong, fibrous band, converts the anterior concavity of the carpus into a *carpal tunnel* in which are digital flexor tendons and the median nerve (p. 1133). The retinaculum is a mere 2.5–3 cm transversely, with a similar proximodistal length. It is attached medially to pisiform bone and the hook of the hamate; laterally it splits into a superficial lamina attached to the tubercles of the scaphoid and trapezium and a deep lamina attached to the medial lip of a groove on the trapezium (5.76). With this groove the laminae form a tunnel containing the tendon of flexor carpi radialis and its synovial sheath. The retinaculum is continuous, proximally, both with the fascia over flexor digitorum superficialis and with general antebrachial fascia; it is these layers which separate at the trapezium. It is crossed superficially by the ulnar vessels and nerve, and palmar cutaneous branches of the median and ulnar nerves. To its anterior surface tendons of palmaris longus and flexor carpi ulnaris are partly attached; distally some intrinsic muscles of pollex and minimus are attached. It is continuous distally with the palmar fascia, including its central aponeurotic fibres, associated with palmaris longus.

A localized thickening in the antebrachial fascia extends laterally from the pisiform bone as a *superficial part of the flexor retinaculum*, crossing superficial to the ulnar vessels and nerve to blend with the retinaculum lateral to them. (This 'superficial part' is to be carefully distinguished from the laterally placed superficial lamina mentioned above.)

The extensor retinaculum (5.78), a strong, fibrous, oblique band, extends across the radiocarpal dorsum, composed of antebrachial fascia, strengthened by oblique transverse fibres. It is attached laterally to the anterior border of the radius, medially to the triquetral and pisiform bones, and between two ridges on the dorsal aspect of the distal radius.

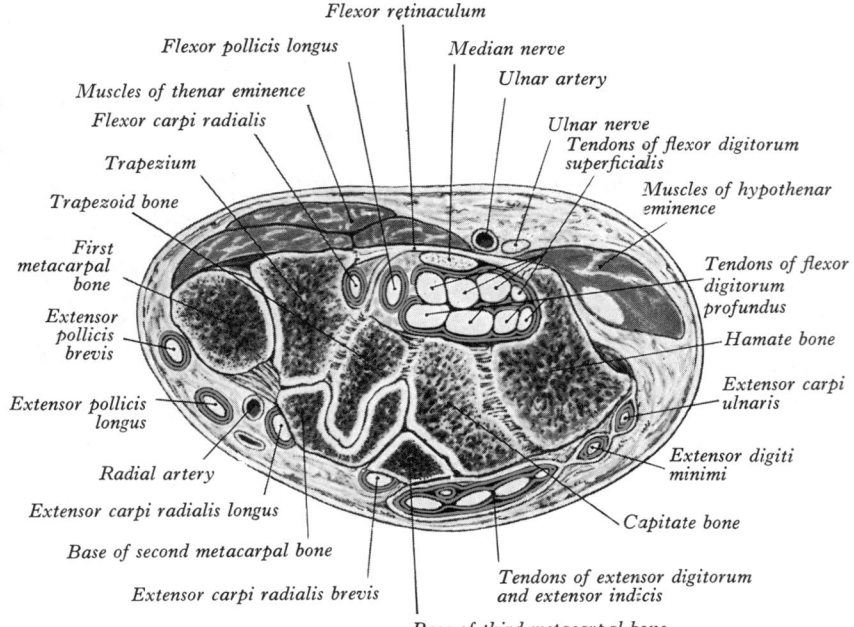

Flexor retinaculum
Flexor pollicis longus
Muscles of thenar eminence
Flexor carpi radialis
Trapezium
Trapezoid bone
First metacarpal bone
Extensor pollicis brevis
Extensor pollicis longus
Radial artery
Extensor carpi radialis longus
Base of second metacarpal bone
Extensor carpi radialis brevis
Median nerve
Ulnar artery
Ulnar nerve
Tendons of flexor digitorum superficialis
Muscles of hypothenar eminence
Tendons of flexor digitorum profundus
Hamate bone
Extensor carpi ulnaris
Extensor digiti minimi
Capitate bone
Tendons of extensor digitorum and extensor indicis
Base of third metacarpal bone

5.76 Transverse section through the left wrist, showing the tendons and their synovial sheaths. The section is slightly oblique and divides the distal row of the carpus and the bases of the first, second and third metacarpal bones. The arrangement of the tendons of the flexors of the fingers shown in the figure is not diagrammatic but represents the actual condition at this level. Observe that the carpometacarpal joint of the thumb is separate from the joint between the trapezium and the base of the second metacarpal bone.

SYNOVIAL SHEATHS OF FLEXOR TENDONS

Two synovial sheaths envelop the digital flexor tendons in the carpal tunnel, one for flexores digitorum superficialis and profundus, the other for flexor pollicis longus (5.76). These extend into the forearm about 2.5 cm proximal to the retinaculum and are occasionally interconnected deep to it. The digital flexor sheath reaches about halfway along the metacarpal bones, ending in blind diverticula around tendons to the index, medius and annularis (5.77), but prolonged around tendons to minimus and usually continuous with the latter's digital sheath. A transverse carpal section (5.76) shows the tendons as if invaginated from the lateral side into the sheath. A parietal layer lines the retinaculum and carpal tunnel's floor and is reflected as a visceral layer over the ventral and dorsal aspects of superficialis and profundus respectively. The visceral layer is again reflected *between* the tendons, passing mediolaterally, as a double fold, on to the dorsal aspect of the tendons of flexor digitorum profundus. Thus from their medial aspect a *recess* of the visceral layer insinuates between the two groups of tendons and passes laterally at a variable distance. The sheath of flexor pollicis longus, usually separate but, as noted, sometimes connected to the common flexor sheath, usually behind the retinaculum, is continued through the pollex to the tendon's end. Fibrous sheaths of the terminal parts of digital flexor tendons are described on p. 618.

SYNOVIAL SHEATHS OF CARPAL EXTENSOR TENDONS

Deep to the extensor retinaculum are six tunnels for extensor tendons, each with a synovial sheath (5.78). They are: (1) lateral to the radial styloid process for tendons of abductor pollicis longus and extensor pollicis brevis; (2) behind the styloid process for tendons of the radial carpal extensors; (3) medial to the dorsal radial tubercle for the tendon of extensor pollicis longus; (4) medial to (3) for tendons of extensor digitorum and extensor indicis; (5) between the radius and ulna for extensor digiti minimi; (6) between the ulnar head and its styloid process for the tendon of extensor carpi ulnaris. The tendon sheaths of abductor pollicis longus, extensores pollicis brevis et longus, extensores carpi radiales and extensor carpi ulnaris cease proximal to the metacar-pal bases; those of extensor digitorum, extensor indicis and extensor digiti minimi are sometimes prolonged somewhat distally along the metacarpus.

FASCIAL ARRANGEMENTS IN THE HAND

No region has provided so fertile an arena for dissection of connective tissue as the hand, with consequent identification of sheets, septa, bands and ligaments of almost tiresome numerosity and often doubtful validity. From a great number of studies, spread over more than two centuries, major features have received general accord, including fibrous flexor sheaths, dorsal digital expansions, tranverse metacarpal ligaments, and palmar fascia or aponeurosis. As Wood Jones emphasized (1941), every structure— muscle, tendon, nerve and vessel—is surrounded by connective tissue separating it from the others to a degree dependent upon local need of mobility. Constant thickenings of such 'sheaths' also serve to restrain or bind together some structures; and in the hand (and foot) subcutaneous tissue is traversed by numerous dense bands or 'skin ligaments', which tie skin to deeper fasciae and hence to bone for obvious functional reasons. These are little developed on the dorsum of the hand but marked in palm and flexor aspects of digits, especially in their pulps. Deeper connective tissue is often compacted and sometimes aponeurotic, blending with periosteum wherever nothing intervenes, and forming septa between some muscles.

Successive studies have added many details. In finer points authorities differ; individual variation doubtless occurs and functional comprehension is not always thus advanced. Such complexities must hence be here reduced to simpler concepts. At simplest it is clear that the deep fasciae of the hand are extensively connected on one side to skin, on the other to bone, including not only carpals, metacarpals and phalanges, but also proximally to the radius and ulna. At well recognized sites, where cutaneous flexion furrows are usually apparent, skin fixation is firmest. Between such sites of maximal anchorage many smaller fixation bands exist, interspersed by cushioning pads of adipose tissue which, though largely restrained by connective tissue bands, permit a small range of tangential movement; and since some displacement is possible, skin and subcutaneous tissue are easil

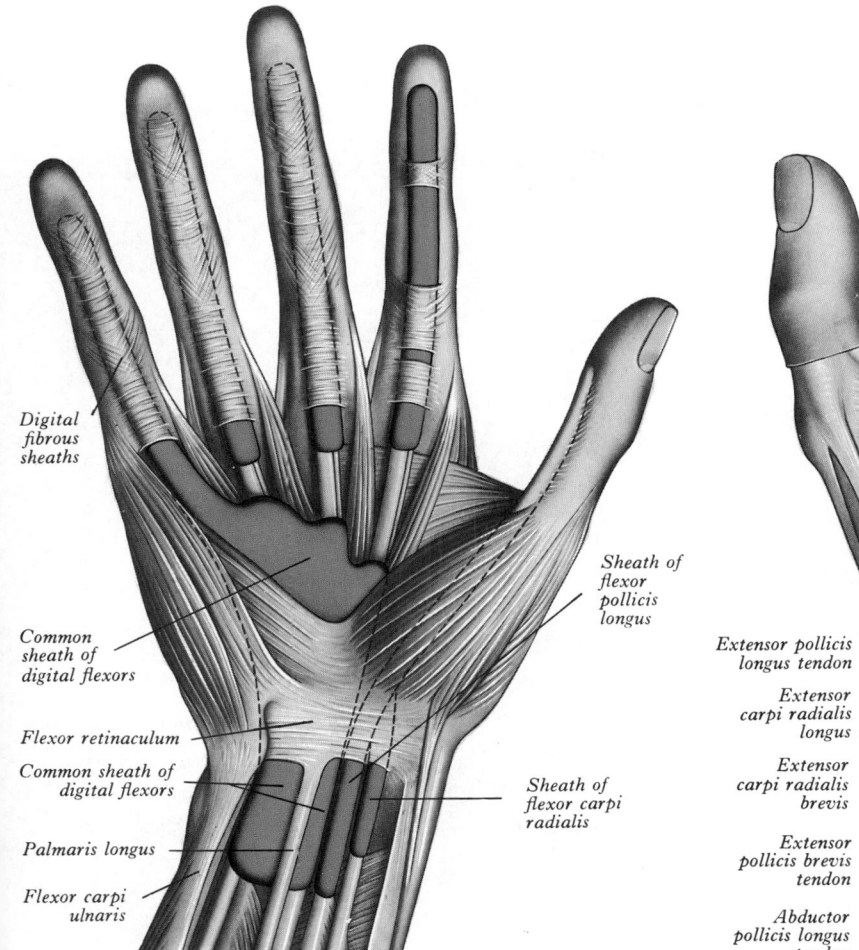

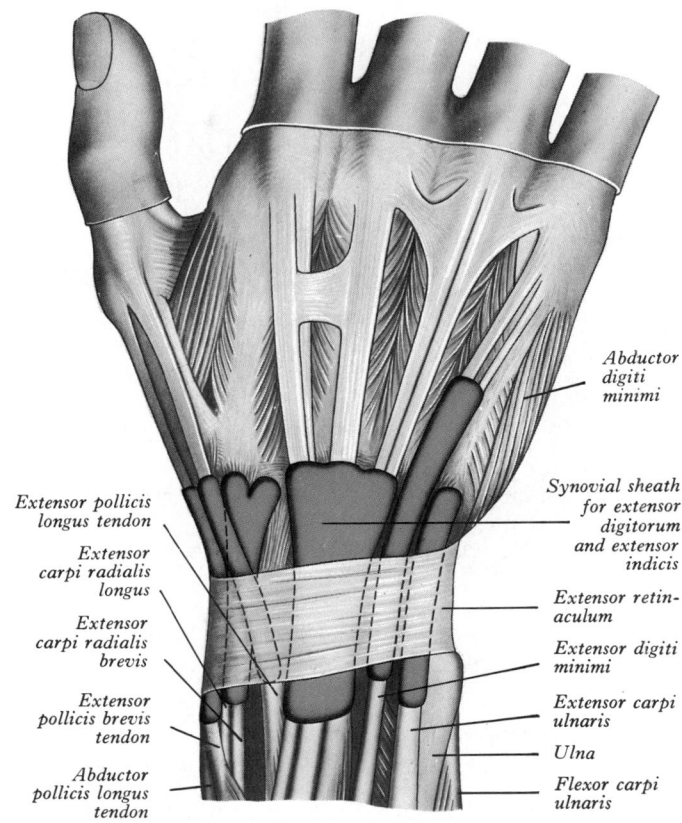

5.77 The synovial sheaths of the tendons on the front of the right wrist and hand (preparation by Professor J C B Grant). Where they are exposed, the synovial sheaths are shown in blue but, where they are hidden by overlying structures, their margins are indicated by interrupted black lines. For simplicity and clarity the digital synovial sheaths are represented as smooth cylinders; in reality they have complex undulant contours (p. 619). The traditional form of the fibrous flexor sheaths shown have been reclassified (see p. 618, **5.**67A).

5.78 Simplified representations of the synovial sheaths of the tendons on the back of the right wrist (preparation by Professor J C B Grant). The synovial sheaths are shown in blue, but they have not been coloured where they lie deep to the extensor retinaculum. In this situation, and where one sheath lies deep to another, the margins of the sheaths are indicated by broken lines.

indented, with resilient recovery almost immediately distorting force ceases. These arrangements have obvious protective value; fat pads in interdigital webs are said to 'protect' digital vessels and nerves from adjacent tendons; thenar and hypothenar pads are clearly useful cushions in applying various tools. Less obvious is pliant conformation of grasping surfaces to the contours of objects grasped, however irregular. The hand can evaluate shape, weight, consistency of an infinity of external objects (p. 634), putting them to limitless uses, often involving large forces, the skilful and safe transmission of which depends as much on skin and fasciae as muscle and bone. Like muscle, skin and fasciae respond to use by hypertrophy, as may the bones while still able to grow. Skin in the hand shows general responsive thickening and also localized callosities (e.g. over metacarpal heads), sometimes characteristic of occupations. Palmar fascia sometimes responds excessively, with resultant disabling contractures. For details consult Kaplan (1965), Stack (1973), Bojsen-Møller & Schmidt (1974) and Landsmeer (1976). Fasciae also help to maintain the palmar concavity to restrain flexor tendons, perhaps to aid venous return.

The palmar aponeurosis (5.79), directly related to palmar muscles and tendons, accordingly consists of central, lateral and medial parts, covering long tendons centrally and thenar and hypothenar muscle masses.

The *central part* is triangular, strong and thick, its apex continuous with the flexor retinaculum; the tendon of palmaris longus expands into it. Its distal base divides into slips, one to each finger, their superficial fibres reaching palmar and digital skin; in the palm they join the dermis at parts of two major transverse

palmar furrows *over* the metacarpophalangeal joints and the multiple transverse folds and furrows at the roots of digits (and a few centimetres *distal* to the metacarpophalangeal joints). Deeper regions of each slip divide into two processes, in part continuous with the fibrous flexor sheaths and also partly attached to the deep transverse metacarpal ligaments, metacarpophalangeal capsules and the sides of proximal phalangeal bases. Short channels are thus formed anterior to the metacarpal heads for flexor tendons. Intervals between the slips are traversed by digital vessels and nerves and lumbrical tendons. Numerous transverse fibres bind slips at their origins. The central palmar aponeurosis is tied to the skin by dense fibrous tissue; to its proximal medial margin palmaris brevis is attached. It covers the superficial palmar arch, the digital flexor tendons, the preterminal median nerve and the superficial terminal branch of the ulnar nerve, as well as deeper structures.

Lateral and *medial parts* of the palmar aponeurosis are thin, covering thenar and hypothenar muscles and continuous with its central part and marginally with the dorsal fascia of the hand. A septum passes dorsally from each border of the central palmar aponeurosis. The *medial palmar septum* is lateral to opponens digiti minimi and reaches the fifth metacarpal bone's palmar surface; it is distally continuous with a slip of the palmar aponeurosis to the lateral side of the flexor sheath of the minimus; proximally the septum reaches the hamate hook and pisohamate ligament and is pierced by deep branches of the ulnar nerve and artery. The *lateral palmar septum* reaches the palmar surface of the first metacarpal bone, passing medial to flexor pollicis brevis and

627

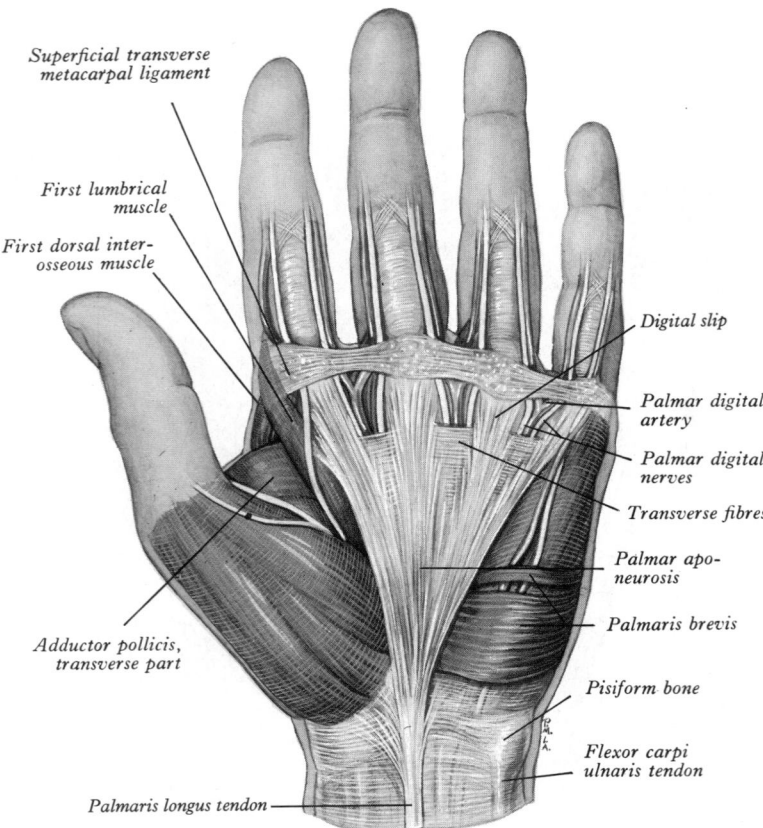

5.79 The left palmar aponeurosis.

Labels for figure 5.79:
- Superficial transverse metacarpal ligament
- First lumbrical muscle
- First dorsal interosseous muscle
- Adductor pollicis, transverse part
- Palmaris longus tendon
- Digital slip
- Palmar digital artery
- Palmar digital nerves
- Transverse fibres
- Palmar aponeurosis
- Palmaris brevis
- Pisiform bone
- Flexor carpi ulnaris tendon

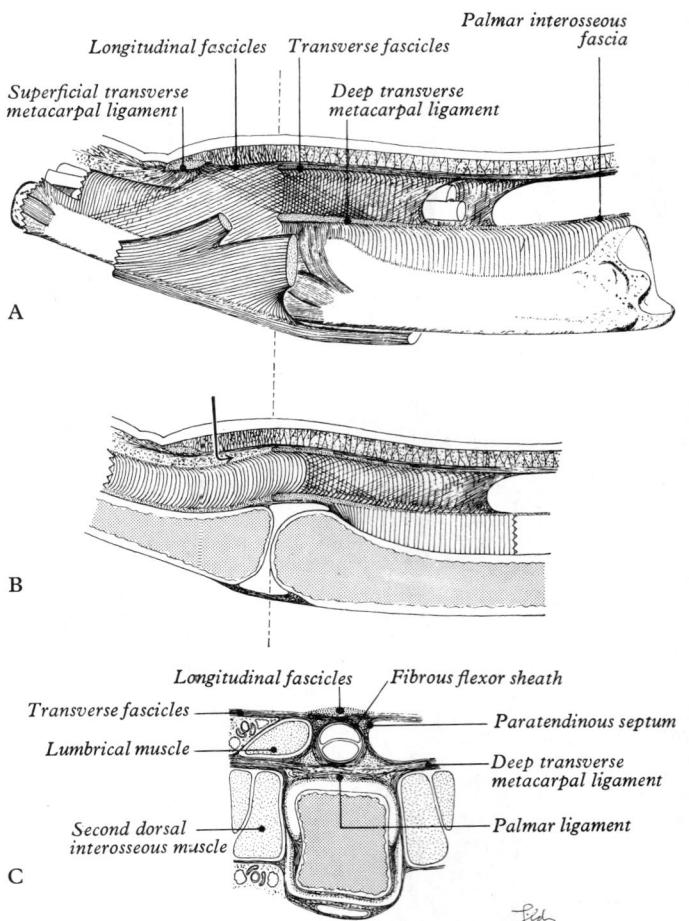

Labels for figure 5.80 A, B, C:
- Longitudinal fascicles
- Transverse fascicles
- Palmar interosseous fascia
- Superficial transverse metacarpal ligament
- Deep transverse metacarpal ligament
- Longitudinal fascicles
- Fibrous flexor sheath
- Transverse fascicles
- Paratendinous septum
- Lumbrical muscle
- Deep transverse metacarpal ligament
- Second dorsal interosseous muscle
- Palmar ligament
- A
- B
- C

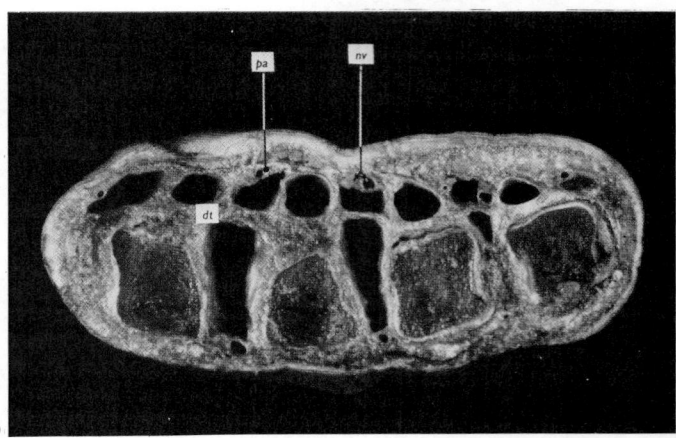

D

opponens pollicis and lateral to the flexor pollicis longus tendon and its sheath. Proximally it reaches the trapezial tubercle and is pierced by a branch of the median nerve to the thenar muscles.

Fascial spaces of the palm. The central palm, deep to the central palmar aponeurosis and between the medial and lateral palmar septa, is divided by a thin intermediate palmar septum into the so-called *middle palmar* and *thenar spaces*. (Perhaps these names are unfortunate and even potentially misleading. Preferable terms would be: the *medial central palmar space* and the *lateral central palmar space*.)

The intermediate palmar septum is between flexor tendons of the index laterally and second lumbrical medially. Distally it blends with a slip of the palmar aponeurosis to the medial side of the indicial flexor sheath and the adjoining deep transverse metacarpal ligament. Dorsally it blends with fasciae over the distal part of the second palmar interosseous and transverse head of adductor pollicis, passing medially on this to the third metacarpal bone. Ventrally it is attached to palmar aponeurosis, but proximally meets the common flexor synovial sheath, blending with connective tissue on its dorsal aspect. Although it is a tenuous structure, pus may gather on either side of it; adjoining areolar tissue may be compressed against it to form an effective septum.

The *middle palmar space* (or *medial central palmar space*), between medial and intermediate septa, is dorsally limited by the third to fifth metacarpal bones, the fascia over the interosseous muscles in the third and fourth spaces and the medial part of the transverse head of adductor pollicis. The central palmar aponeurosis is ventral, the common flexor synovial sheath proximal. The space contains the third to fifth digital flexor tendons and the second to fourth lumbrical muscles, the superficial palmar arch and digital vessels and nerves for the third to fifth digits and ulnar side of the index. Distally it communicates with subcutaneous tissues at the digital webs; proximally, it may extend dorsal to the common flexor sheath. Bojsen-Møller & Schmidt (1974) regard the middle palmar 'space' as divided distally into independent channels for lumbrical muscles and common digital vessels and nerves (5.80D).

The *thenar space* (or *lateral central palmar space*), flanked by lateral and intermediate palmar septa, is ventral to the fascia

5.80 A. Drawing of the course of the fibres of the fasciae around the left third metacarpal bone and proximal phalanx viewed from the radial side. The deep transverse metacarpal ligament connects both with the metacarpal bone and with the longitudinal and transverse fibres of the palmar aponeurosis and, via their retinacula, with the skin. The palmar interosseous fascia is connected by means of a sagittal septum to the shaft of the third metacarpal bone. B. Sagittal section through the centre of the third metacarpal bone. The fibrous sheath of the flexor tendons extends for some distance between the transverse fascicles of the palmar aponeurosis and the deep transverse metacarpal ligament. Between the longitudinal fibres in the palmar aponeurosis and the fibrous sheath, a connective tissue space affords cleavage for demonstration of the paratendinous septa by dissection (arrow). C. Transverse section through the head of the third metacarpal bone showing the course of the fibres between the palmar aponeurosis and the deep transverse metacarpal ligament, and their attachments to the metacarpal bone. D. Section through the hand at the level of the heads of the metacarpal bones (2–5). The subdivision of the distal part of the central compartment into eight narrow compartments is demonstrated. The interdigital nerves and vessels (nv) have been left to show their relation to the lumbrical compartments. The deep transverse metacarpal ligaments (dt) and palmar aponeurosis (pa) are also shown (from Bojsen-Møller & Schmidt 1974, by courtesy of the authors and the Cambridge University Press). Compare A–C with the recent descriptions of the fibrous flexor sheaths mentioned on p. 619 and **5.67A**.

covering the transverse head of adductor pollicis and, beyond this, the first dorsal interosseous muscles, while ventral to it is the palmar aponeurosis. It contains the tendon of flexor pollicis longus and its sheath, indicial flexor tendons, the first lumbrical muscle, palmar digital vessels and the nerves to pollex and the radial side of the index. Distally the space extends to subcutaneous tissues in the web of the pollex and proximally behind the flexor retinaculum.

All these 'spaces' are in fact *regions* in which relatively loose connective tissue allows some mobility but also little opposition to the spread of infection, whereas their boundaries, being denser, may contain an infection.

The superficial transverse metacarpal ligament is a thin strap (5.79) across the digital roots in superficial fascia, attached to the skin in digital clefts and to the fifth metacarpal bone, forming rudimentary webs. Digital vessels and nerves pass dorsal to it. Since it adjoins the bases of proximal phalanges, it is more appropriately named the *palmar digital ligament* (Bojsen-Møller & Schmidt 1974).

Applied Anatomy. The palmar aponeurosis is liable to a progressive disabling contracture (Stack 1973). Owing to constant exposure to injury and contamination, the fingers are liable to infection. In the digital *pulp*, ventral to most of the terminal phalanx, pus may be confined by strong septa radiating between the skin and periosteum, which sometimes entails necrosis of its diaphysis by obstruction of the blood supply from the digital arteries traversing the pulp. The epiphysis may escape, because its vessels leave digital arteries before entering the pulp. Inflammation may spread along flexor sheaths; in synovial sheaths of the pollex or minimus this may be more serious because these may communicate with the principal flexor sheath; infection may hence spread into the palm and behind the flexor retinaculum to the forearm.

Chronic inflammation of the common flexor sheath, a *compound palmar 'ganglion'*, presents a bilobar outline, a swelling proximal to the carpus, another palmar, the constriction corresponding to the flexor retinaculum. Fluid contents can be forced from one to the other behind the retinaculum. Although such pathological significance of fasciae and ligaments in the palm and digits must receive due emphasis, their basic function is to transmit tangential forces from skin to skeleton in the hand. Contraction of palmaris longus and extension at carpal and metacarpophalangeal joints all tense the palmar aponeurosis, with profound and advantageous effects on the mechanics of the normal hand.

6. Muscles of the Hand

The intrinsic manual muscles form three groups: (a) three pollicial muscles in the *thenar eminence*; (b) muscles of the minimus in the *hypothenar eminence*; (c) lumbrical and interosseous muscles of the *central palm* and *intermetacarpal spaces* and finally adductor pollicis with its muscular heads sited with group c, but its main actions associated with group a.

a) THENAR MUSCLES AND ADDUCTOR POLLICIS

The pollex has intrinsic muscles to flex, abduct, oppose and adduct it. The first three form the thenar eminence.

Abductor pollicis brevis, a thin, subcutaneous muscle lying lateral in the thenar eminence, arises mainly from the flexor retinaculum; but a few fibres arise from the tubercles of scaphoid and trapezium and the tendon of abductor pollicis longus (p. 624). Its medial fibres are attached by a thin, flat tendon to the radial side of the proximal phalangeal base in the pollex; its lateral fibres join the pollicial dorsal digital expansion. Accessory slips may spring from the long and short pollicial extensors, opponens pollicis or the radial styloid process.

Nerve supply: the lateral terminal branch of median nerve, C.8 and T.1.

Actions. The muscle draws the pollex ventrally in a plane at right angles to the palm (abduction) also, to some degree, rotating it medially at metacarpophalangeal and carpometacarpal joints. Both actions occur at *both* joints, so that in full abduction the

pollex is *not* in line with its metacarpal bone and different degrees of mutual rotation have occurred. 'Pure' abduction cannot occur because conjunct medial rotation accompanies movement at both joints and the total pollicial movement is considerably more complex.

Opponens pollicis (5.80, 81), deep to the abductor, extends from the trapezial tubercle and flexor retinaculum to the whole lateral border and adjoining lateral half of the palmar surface of the first metacarpal bone.

Nerve supply: the lateral terminal branch of the median nerve, C.8 and T.1 and commonly a ramus of the deep terminal branch of the ulnar (C.8 and/or T.1). Dissections by Day & Napier (1961), Forrest (1967) and electromyographic tests by Harness et al (1974) indicate double innervation in 92 out of 120 hands, to be regarded, therefore, as the usual arrangement.

Actions. The muscle flexes the metacarpal bone medially across the palm, also rotating it medially (p. 632). By this *opposition*, the palmar ('pulpal') aspect of the terminal segment may

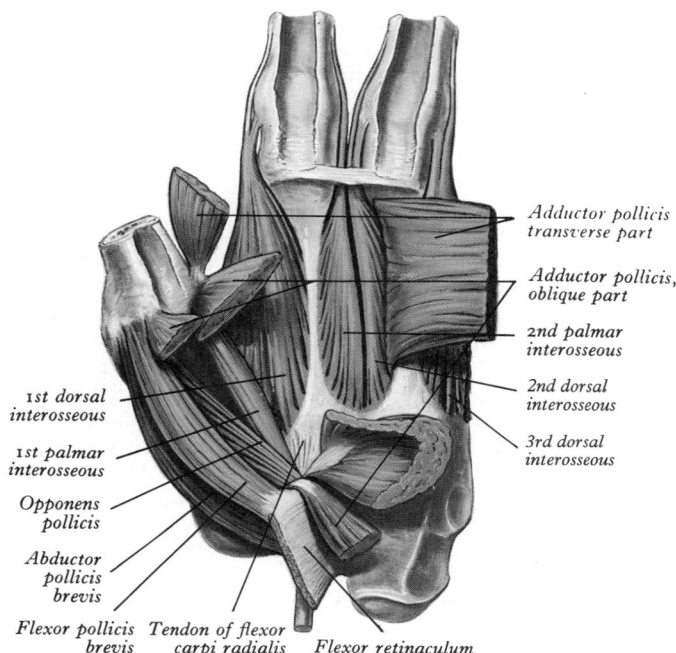

1st dorsal interosseous

1st palmar interosseous

Opponens pollicis

Abductor pollicis brevis

Flexor pollicis brevis   Tendon of flexor carpi radialis   Flexor retinaculum

Adductor pollicis transverse part

Adductor pollicis, oblique part

2nd palmar interosseous

2nd dorsal interosseous

3rd dorsal interosseous

5.81A Dissection of the left thenar eminence and palm over the two lateral intermetacarpal spaces. The parts of the adductor pollicis, including that sometimes designated 'the deep head' of flexor pollicis brevis, have been transected.

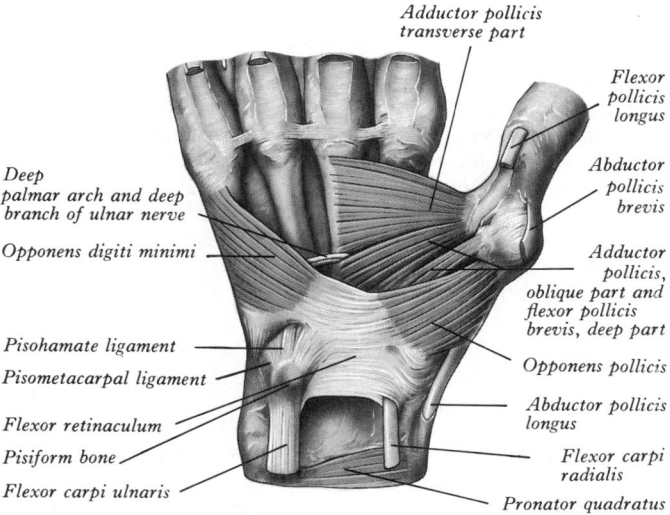

Adductor pollicis transverse part

Deep palmar arch and deep branch of ulnar nerve

Opponens digiti minimi

Pisohamate ligament

Pisometacarpal ligament

Flexor retinaculum

Pisiform bone

Flexor carpi ulnaris

Flexor pollicis longus

Abductor pollicis brevis

Adductor pollicis, oblique part and flexor pollicis brevis, deep part

Opponens pollicis

Abductor pollicis longus

Flexor carpi radialis

Pronator quadratus

5.81B Deep dissection of the right palm, showing the opponens pollicis and opponens digiti minimi and the two parts of the adductor pollicis including the 'deep head' of flexor pollicis brevis. The superficial thenar and hypothenar muscles and palmaris longus have been removed.

contact the flexor aspects of any other digit. Flexion and (varying degrees of) medial rotation are simultaneous and movement also initially involves some abduction. Continuous smooth opposition involves carpometacarpal *and* metacarpophalangeal joints. During opposition, fingers flex at the metacarpophalangeal joints and variably at interphalangeal joints. The pollex may contact any part of a palmar digital aspect from base to tip. Digital flexion also involves some *lateral rotation* to meet the pollicial pulp en face, which is least marked in the index and greatest in minimus, in which displacement also occurs at its carpometacarpal joint (vide infra, opponens digiti minimi).

Flexor pollicis brevis (5.82), medial to abductor pollicis brevis, arises by superficial and deep heads, the superficial from the distal border of the flexor retinaculum and distal part of the trapezial tubercle; it passes along the radial side of the tendon of flexor pollicis longus to be attached by a tendon containing a sesamoid bone to the *radial* side of the pollicial proximal phalangeal base. It is often blended with opponens pollicis. The deep part arises from both trapezoid and capitate bones and the palmar ligaments of the distal carpal row, passing deep to the tendon of flexor pollicis longus to unite with the superficial head at its sesamoid bone. The deep head varies in size and may be absent.

Nerve supply: the superficial head, usually the lateral terminal branch of the median nerve; deep head, deep branch of the ulnar nerve, C.8 and T.1. Variations occur (Day & Napier 1961).

Actions: flexion of the proximal phalanx and, if continued, flexion and medial rotation of the metacarpal in co-operation with opponens pollicis (p. 629).

Adductor pollicis (5.81) arises by the oblique and transverse heads, the *oblique head* being attached to the capitate bone, the second and third metacarpal bases, carpal palmar ligaments and the tendon sheath of flexor carpi radialis. Most fibres converge into a tendon containing a sesamoid bone and joining that of the transverse head to be attached to the *ulnar* side of the proximal phalangeal base. The deepest fibres may reach the medial side of the dorsal digital expansion. Lateral to the oblique head a fasciculus passes deep to the tendon of flexor pollicis longus to join the flexor pollicis brevis as a 'deep head' of flexor pollicis brevis (Day & Napier 1961). The *transverse head* (5.81), deepest of the pollicial muscles, is triangular and arises from the distal two-thirds of the palmar aspect of the third metacarpal, converging with the oblique head and the first palmar interosseous muscle to the base of the proximal phalanx. The two parts vary in relative size.

Nerve supply: the deep branch of ulnar nerve, C.8 and T.1.

Action. The muscle approximates the pollex to the palm, acting at greatest advantage when the abducted, rotated and flexed pollex has been opposed to the fingers in gripping. Thus it is the antagonist of the complex muscular interplay involved in opposition (vide supra).

(b) THE HYPOTHENAR MUSCLES

Apart from palmaris brevis, a superficial muscle, this group flexes, abducts and opposes the fifth digit (5.79,81,82).

Palmaris brevis (5.79) is thin, quadrilateral and subcutaneous on the ulnar side of the palm, attached to the flexor retinaculum and the medial border of the central palmar aponeurosis; it ends in the dermis of the hand's ulnar border, superficial to the ulnar artery and the superficial terminal branch of the ulnar nerve.

Nerve supply: the superficial branch of the ulnar nerve, C.8 and T.1.

Action. It wrinkles the skin on the ulnar side of the palm, deepening a longitudinal hollow or groove along the medial aspect of the hypothenar eminence, and slightly accentuating the eminence. It may contribute to the palmar grip. Clinically, thumbnail pressure on the medial aspect of the pisiform often stimulates its nerve, causing it to contract.

Abductor digiti minimi (5.82) arises from the pisiform bone, the tendon of flexor carpi ulnaris and the pisohamate ligament. Its flat tendon divides into two slips, one attached to the ulnar side of the proximal phalangeal base of minimus, the other to the ulnar border of the dorsal digital expansion of extensor

digiti minimi. The muscle may have two or three slips or be fused with flexor brevis. A slip may also variously arise from the flexor retinaculum, antebrachial fascia, tendons of palmaris longus or flexor carpi ulnaris. It may be partly attached to the fifth metacarpal by a slip from the pisiform.

This muscle has been used to calculate force and work efficiency generated by muscles, using various techniques, by Neto et al (1984).

Action: abduction of minimus from annularis, e.g. in synergic spreading of extended digits. Abduction of minimus may also occur with digits 2–4 tightly adducted and either flexed or extended.

Flexor digiti minimi brevis (5.82), lateral to the abductor, arises from the convex aspect of the hamate's hook and of palmar surface of the flexor retinaculum and is distally attached to the ulnar side of the proximal phalangeal base with abductor digiti minimi. It is separated from the abductor proximally by deep branches of the ulnar artery and nerve. It may be missing or fused with the abductor. A slip may reach the distal end of the fifth metacarpal.

Action: flexion of minimus at its metacarpophalangeal joint, together with some lateral rotation.

Opponens digiti minimi (5.81) is triangular, deep to the flexor and abductor, arising from the convexity of the hamate's hook and the contiguous flexor retinaculum; it ends along the fifth metacarpal's whole ulnar margin and adjacent palmar surface, often divided into two lamellae by deep branches of the ulnar artery and nerve. It is variably blended with its neighbours.

Action. It flexes the fifth metacarpal, rotating it laterally at the carpometacarpal joint, deepening the palmar hollow. With the addition of flexion and lateral rotation at the metacarpophalangeal and interphalangeal joints, it *opposes* the minimus to the pollex.

Nerve supply: all muscles of the minimus are supplied by deep branch of the ulnar nerve, C.8 and T.1.

(c) THE LUMBRICAL AND INTEROSSEOUS MUSCLES

The lumbricals (5.82) are four small fasciculi arising from tendons of flexor digitorum profundus: the first and second from the radial sides and palmar surfaces of tendons to the index and medius; the third from *adjacent* sides of the tendons to medius and annularis; the fourth from *adjoining* sides of the tendons to the annularis and minimus. Each passes to the *radial* side of the corresponding digit and here is attached to the lateral margin of its dorsal digital expansion (5.70). Variations in attachments are common. Any may be unipennate or bipennate, arising from the adjoining tendons of flexor digitorum profundus and from that of flexor pollicis longus, when the first is bipennate (Goldberg 1970). Accessory slips may also be attached to an adjacent tendon of flexor digitorum superficialis.

Nerve supply: first and second, median nerve, C.8 and T.1; third and fourth, deep terminal branch of ulnar, C.8 and T.1. The third is often supplied by the median nerve. This arrangement is remarkably constant but details of muscular rami vary. The first and second are also sometimes supplied by the deep terminal ulnar branch (Mehta & Gardner 1961).

Actions. Lumbrical muscles link *flexor* to *extensor* tendons, a unique arrangement apart from the foot. Actions are hence likely to be complex (p. 633), depending not only on their own length and tension but also on these parameters in their long flexors and extensors. Descriptions of their activities often disregard these facts, being based on incomplete assumptions, the proximal attachments being considered as fixed and their attachment to the dorsal extensor apparatus alone as mobile, and with their line of pull slightly ventral to the transverse axes of the metacarpophalangeal joints. Hence often simply regarded as flexors, they are probably weak in this; since the axes of these joints move ventrally during flexion, lumbrical flexor effect may not be augmented (but, rather, diminished) by increasing flexion. By their attachments to dorsal expansions lumbricals should extend interphalangeal joints. There is direct evidence of this. As long ago as 1954 Backhouse & Catton confirmed this action by inserting electrodes into a number of 2nd lumbrical muscles. Long & Brown (1964) made similar observations. The

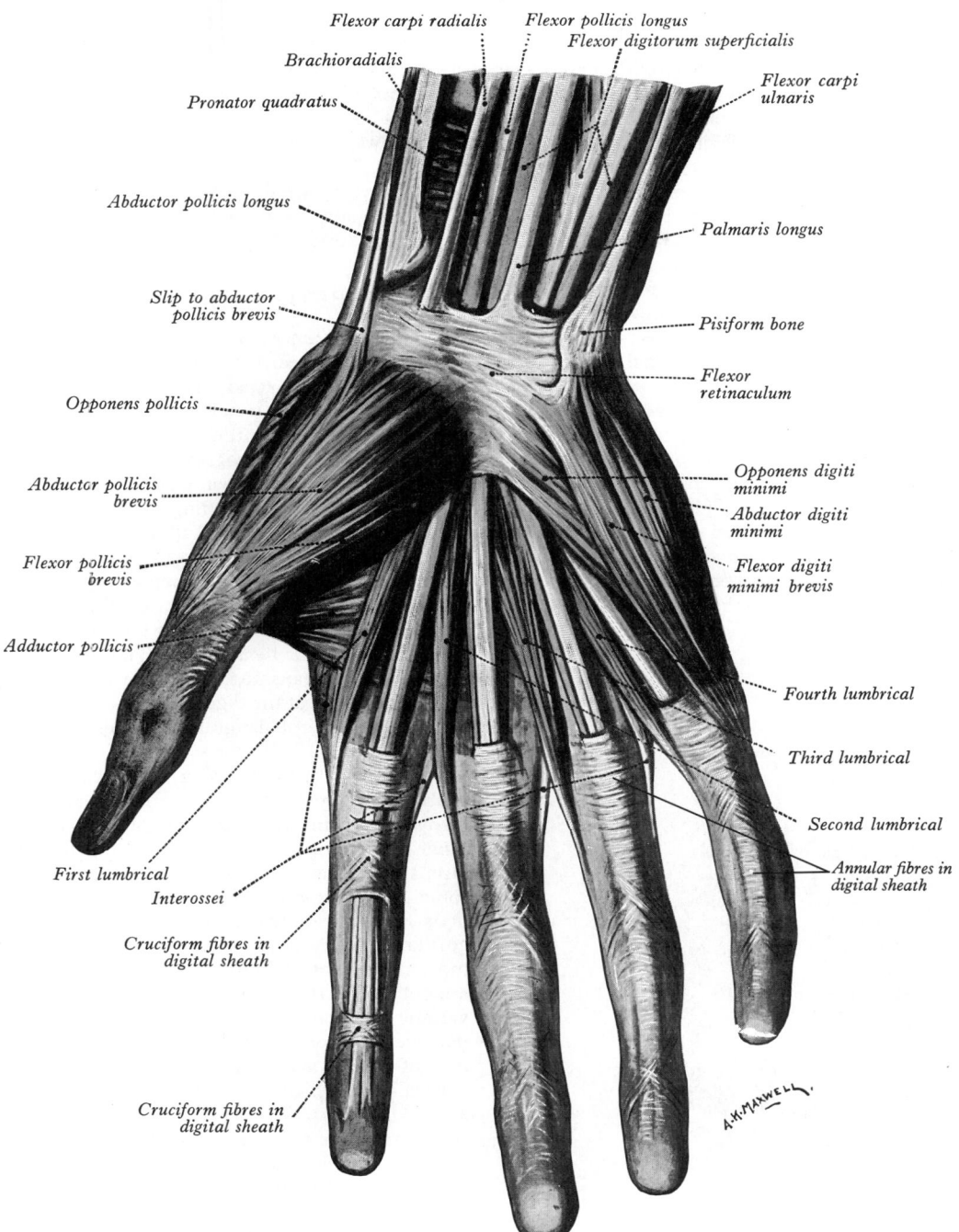

Flexor carpi radialis
Brachioradialis
Pronator quadratus
Abductor pollicis longus
Slip to abductor pollicis brevis
Opponens pollicis
Abductor pollicis brevis
Flexor pollicis brevis
Adductor pollicis
First lumbrical
Interossei
Cruciform fibres in digital sheath
Cruciform fibres in digital sheath

Flexor pollicis longus
Flexor digitorum superficialis
Flexor carpi ulnaris
Palmaris longus
Pisiform bone
Flexor retinaculum
Opponens digiti minimi
Abductor digiti minimi
Flexor digiti minimi brevis
Fourth lumbrical
Third lumbrical
Second lumbrical
Annular fibres in digital sheath

5.82 Superficial dissection of muscles of the palm of the right hand. The annular and cruciform fibres in the digital fibrous flexor sheath are here positioned as in traditional accounts. For a more recent and comprehensive classification see p. 618 and **5.67A**.

well-known 'claw' deformity after ulnar nerve lesions provides further evidence.

When the third, fourth or fifth finger is individually fully flexed at its metacarpophalangeal and proximal interphalangeal joints (with the other digits extended), its distal phalanx can be neither flexed, extended, or 'fixed' voluntarily. This is due to the mode of attachment of extensor tendons; a terminal phalanx can be extended only when the middle one is also extended (p. 633).

The *interossei* occupy the intervals between the metacarpal bones, and are divided into dorsal and palmar sets.

Dorsal interossei (**5.83A**), four bipennate muscles, arise from the adjacent sides of metacarpal bones, more extensively from the metacarpal whose digit they enter. They are attached distally to their proximal phalangeal bases and separately to their dorsal digital expansions (p. 622). Between their proximal attachments are narrow triangular intervals; through the first the radial artery

passes and through the others perforating rami of the deep palmar arch. The *first* and largest muscle (*abductor indicis*) ends at the radial side of the proximal phalanx of the index and capsule of its metacarpophalangeal joint. The *second* and *third* pass to the radial and ulnar sides of the medius; the second usually reaches the digital expansion and proximal phalanx, the third only the digital expansion (**5.71**). The *fourth* may be also wholly attached to the expansion but often also sends a slip to its proximal phalanx.

Palmar interossei (**5.83B**) are smaller and palmar to their metacarpal bones (1,2,4 and 5). Except the first, each arises from the entire length of its metacarpal bone, both arising and passing to the appropriate (adductor) side of its dorsal digital expansion (Salsbury 1937, Landsmeer 1949). (The medius has no palmar interosseous; the remaining digits have palmar interossei on their aspects *facing the medius*.) Thus, the *first* arises from the ulnar side of the palmar surface of the first metacarpal base and passes to a sesamoid bone on the ulnar side of the proximal phalanx, and thence to the phalanx and usually also to the dorsal digital expansion (Lewis 1965). It is ventral to the lateral head of the first dorsal interosseous and is overlapped ventrally by the oblique head of adductor pollicis (5.81A). (Some have preferred to regard this first muscle as a deep stratum of flexor pollicis brevis, a view now largely forgotten.) The *second* arises from the ulnar side of the second metacarpal and is attached to the ulnar edge of the dorsal digital expansion in the index. The *third* arises from the radial side of the fourth metacarpal and is attached in common with the third lumbrical. The *fourth* arises from the radial side of the fifth metacarpal and is attached in common with the fourth lumbrical (vide supra) and additionally to the base of its proximal phalanx. These attachments to the dorsal digital expansions (5.71) stabilize extensor tendons on the convex heads of metacarpal bones during movements at metacarpophalangeal joints.

Interossei do not vary much; they are occasionally reduplicated, perhaps due to the derivation of the palmar interossei from paired short flexors (Lewis 1965).

Stack (1962) and others prefer to classify interossei as 'proximal' and 'distal'. The former, chiefly dorsal interossei, are attached to the proximal parts of extensor expansions ('*hood*' or '*dorsiere*') forming the *proximal wing tendons* and separately to the phalangeal tubercle of bases of proximal phalanges (5.70,71). The palmar or 'distal' interossei, apart from the first, are attached only to extensor expansions, distal to the 'hood' and usually opposite a lumbrical, both attachments forming the *distal wing tendons* (p. 623).

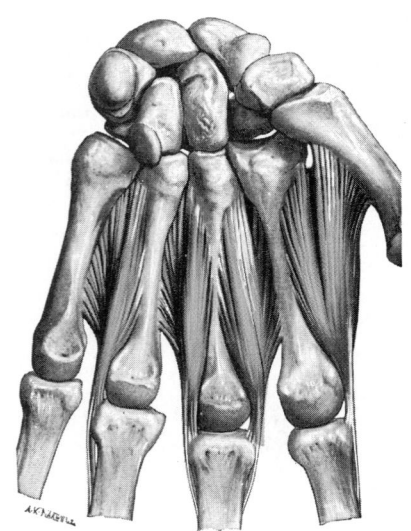

5.83A The dorsal interosseous muscles of the left hand: palmar aspect.

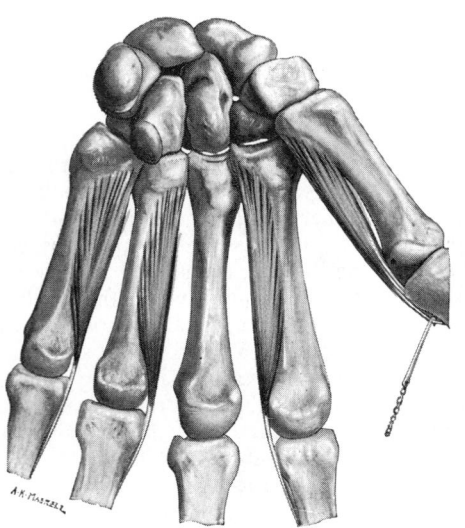

5.83B The palmar interosseous muscles of the left hand: palmar aspect.

Nerve supply: the deep branch of the ulnar nerve, C.8 and T.1. Rarely, the first dorsal interosseous is supplied only by the median nerve (Sunderland 1946).

Actions. Dorsal interossei abduct digits from an arbitrary longitudinal line through the centre of the medius; palmar interossei adduct towards this line (p. 634). With the lumbricals they flex the proximal phalanges; by their attachments to the dorsal digital expansions they are able, in certain conditions (p. 634), to extend the middle and distal phalanges. The first palmar interosseous flexes and adducts the pollicial proximal phalanx.

THE NATURE OF MANUAL DEXTERITY

Human upper limbs, with a mobile, basal shoulder girdle, an extensible, folding arm and forearm and a terminal working tool, the hand, have an extraordinary range and versatility of movement. One has but to watch the limited performance of the most complex mechanical and electronic manipulators elaborated by human ingenuity to recognize the immensely greater skill and repertoire of the human hand. Pre-eminently adapted for reaching, grasping and manipulating, it has also much potentiality for other activities: pushing and adjusting objects without grasping, striking blows and supporting the body in endless positions. When transmitting thrust or supporting weight, the whole upper limb may become a rigid pillar, like the leg, with joints in their close-packed positions; or it may absorb forces by appropriately controlled resilient flexion. The hand may be used as a fist, or forces may be transmitted through fingers extended in close-packed positions at the digital joints and carpus, sometimes with the proximal interphalangeal joints slightly flexed, as in the toes at take-off (p. 658).

But especially in grasping activities has the human hand enabled development of great skills. Despite this, the hand is in many ways primitive and physically capable only of a limited range of movements, though with much skill and versatility. Human hands are not specialized in movement beyond those of some other primates. Some, like the gibbon, have specialized, hook-like hands in which opposition is lost; but the chimpanzee, gorilla and many monkeys have much the same repertoire of hand movements. The difference lies in the range of uses to which they are put (Jones 1941). Mankind incomparably excels in endlessly variable application of manipulative skills, the hands being responsive tools commanded by unrivalled mental development. A note of caution is needed in the use of the terms 'primitive' and 'specialized', which may lead to a semantic morass. Such terms are used by morphologists with little reference to function. In the structural sense the human hand is obviously more generalized than most mammalian extremities, though in many minor details more highly adapted to intricate functions; it is from this point of view extremely 'specialized' (Bishop 1964).

Prehensile activities have been variously analysed, e.g. as cylinder, ball, ring, pliers and pincer grips (Griffiths 1943); but a simpler view is possible, all such grips being variants of *power* and *precision grips* (Napier 1956). Combinations occur in some activities, and a *hook grip* is in some features distinct.

In the **power grip** (5.84) the fingers are flexed round an object, with counter pressure from the pollex, positioned to bring either its pulp or medial border against the object held. The purpose is to hold an object firmly, either as a tool or weapon, or to be worked upon by the other hand. Once grip is made, the hand is held still or moved with the rest of the limb; any skill is due to the limb, including the carpus; relative movements of thumb and fingers are not involved.

In the **precision grip** (5.84), in contrast, not only is grasp more precise but small digital movements are usually the main source of adroit activity. The object is gripped between the tips of the fingers and thumb, sometimes by all, more often by the thumb and index, with the middle finger frequently involved, as in grasping a pen, pencil or the handle of any small tool. The position may be adjusted at the wrist or involve supination, pronation and small adjustments at the elbow and shoulder. In the finest work, however, most skilled manipulations employ the lumbricals and interossei, with metacarpophalangeal flexion and interphalangeal extension, and vice versa. These are characteristically utilized in the motions of advancing and withdrawing any small objects to or from another (e.g. a thread to the eye of a needle), one held still in a power grip, the other manipulated to and fro in a precision grip. Such activities are commonplace in inspecting or manipulating any small object. Flexor-extensor musculature of digits is highly organized for this; individual contributions have been described and investigated in considerable detail (Long 1968, MacConaill & Basmajian 1969, Kapandji 1970).

Relatively small objects may be held steady in a power grip by flexion of the medial three digits against the palm, while the index and thumb carry out manipulations upon them by a precision grip, e.g. a fountain pen may be thus held while its cap is screwed off or on. A single hand may thus be used simultaneously in power and precision grip.

The **hook grip** (5.84) is used to suspend or pull; it may be used like the power grip, to suspend or elevate the body in climbing. Fingers may loop around handles, straps, cords, branches, rocks, etc. and flex towards the palm, to a degree depending upon the dimensions of whatever is grasped. The thumb may or may not be involved. It is a grip for the transmission of forces, not for fine manipulation.

Thus, by flexion and extension of the wrist and digits, adduction and abduction of digits, and the opposition of the pollex, a relatively small repertoire of manual movement is possible. Accessory to this, it is to be noted, are minor degrees of rotation in the fingers and a freer range in the pollex. Though the repertoire is limited, the scope of movements and nicety of control with which they can be varied, especially with practice, are the unequalled prerequisites of mankind. This manual skill, guided by discriminative vision and directed by a highly imaginative and inquisitive mentality into a seemingly endless range of activities, has enabled mankind to master much of the natural environment and to create a culture of art, science and technology without peer. The ever-increasing complexity of the artificial environment thus created has carried us far beyond any other form of life of which we are aware—perhaps in some respects already too far.

Mechanical Analysis of Manual Movements

Mechanics of manual movements involve many factors and not least are the effects of external resistance from objects manipulated. These initiate a continuous feedback, to which the motor apparatus responds in intricate nuances of adaptive adjustment in a complex chain of articulations. Ignoring the carpus, each digit has four joints (three in the thumb), whose individual changes must be concerted, each digit actuated by not less than six muscles (medius and annularis) and as many as nine (minimus), a total of 36. It is scarcely practicable to attempt to deal exhaustively with the myokinetics of manual movements, even if fully supported by experiment or direct observation. In fact, most detailed accounts (Wood Jones 1941, Napier 1966, Kaplan 1965, Stack 1973, Landsmeer 1976) are a mixture of theorization and factual data. Such interpretations of anatomical facts are, however, often plausible and experimental data available are sometimes confirmatory. The following account is necessarily based upon this mixture of real data and derived interpretation.

The hand is a flexor–extensor mechanism, complicated by adduction–abduction and small rotations, the latter usually conjunct. To be added is opposition. The grasping, flexor aspect, the palm, is necessarily hollowed, dynamic palmar arching being carried proximally into a rigid carpal concavity. Analogies between this arched form and that in the foot have been mooted; but the hand is not habitually weight-bearing and such comparison will not be pursued. In a prehensile apparatus flexors would naturally exceed extensors in available power. Electromyography shows that in slow and moderate flexion of digits only flexor digitorum profundus is active, flexor digitorum superficialis assisting in more powerful and rapid gripping. Extension, mainly with digital abduction, is usually preliminary to grasping. But digital manipulation demands more than simple extension and flexion; as already emphasized, it is the dissocation of events at metacarpophalangeal and interphalangeal joints which converts the hand

from a grasping organ to a skilled manipulator. Separate flexors and extensors, operating in opponent pairs at each joint in a digital articular chain, might serve independent flexor–extensor adjustments, as they do in the thumb. But the latter has only two

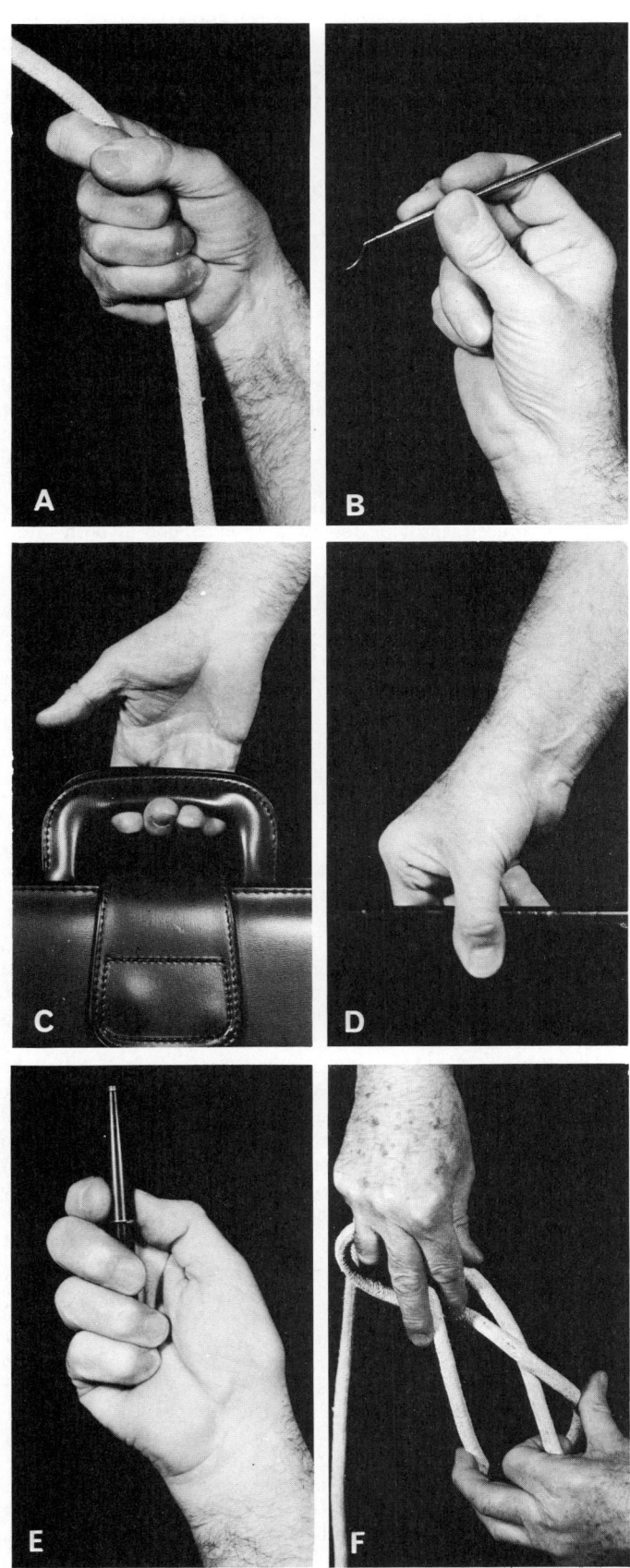

5.84 A few of the many varieties of functional postures which may be adopted by the human hand: A. power grip; B. precision grip; C. hook grip; D. pincers grip; E. simultaneous power and precision grip; F. complex manipulation.

phalanges; the triple phalangeal series of other digits might be difficult, perhaps even impossible, to control completely even with such a multiplicity of long extrinsic tendons. However, only one long extensor is available; and this, through attachment to proximal phalanx and transverse metacarpal ligament (p. 515 and 5.70B), is limited to extension at the metacarpophalangeal joints, at which point retraction of the extensor 'hood' is arrested. This entails that the extensor tendon distal to this must be long enough to allow full interphalangeal flexion, even when the metacarpophalangeal joint is fully extended. In full flexion of an average digit its extensor aspect increases by 2–3 cm. Therefore there must exist some means by which the digital parts of long extensor tendons can be varied in effective length. Mere elasticity might suffice but elongation, to this extent, of largely collagenous structures is impossible. In fact, the extensor apparatus (it is far more than a simple tendon) is arranged to exploit this apparent difficulty (5.70A–C, 5.71A–D). The lateral bands of each tendon, ending at a terminal phalangeal base, can be drawn proximally by the intrinsic muscles attached to them and can also be spread apart by these forces, shortening the distal part of the tendon. The arrangement resembles a diamond (5.71C), with forces applied at its four angles. An increase in width by traction at the lateral angles (by intrinsic muscles) shortens the length of the extensor apparatus. Theoretically this would permit flexion or extension at the interphalangeal joints whatever the state of the metacarpophalangeal; this is precisely what occurs in digital movements. Moreover, proximal *and* distal attachments of the intrinsic muscles provide for separate extension at both interphalangeal joints.

Interosseous and lumbrical muscles thus supply the third and balancing force between extrinsic digital flexors and extensors. Independent control of movements at three digital articulations would otherwise be impossible. This can be stated with confidence; the effects of paralysis of intrinsic muscles are well known, the data of electromyography are also convincing, though limited.

More must be added with regard to the lumbrical muscles. Like interossei, their tractive effect is aligned ventral to the metacarpophalangeal joints; they are therefore somewhat inefficient flexors. Again like interossei, they pull on extensor expansions at their distal wings and exclusively so. Unlike interossei they do not act from a static proximal attachment, but from adjustable bases of deep flexor tendons. It is difficult to envisage precise changes in the length of these muscles which must occur in different states of flexor – extensor balance, nor are direct observations available. However, Stack (1962) has essayed a detailed analysis of such changes in the length of lumbrical (and interosseous) muscles in different digital positions, estimating the proximal positioning of lumbricals from the excursions of deep flexor tendons. Excursions have been assessed at 5–20 mm by Bunnell (1949), Kaplan (1965) and Landsmeer (1955). As will be appreciated from the diagram in 5.71, the lumbrical muscles are shown as most contracted and distal in position in full digital extension; in contrast they appear most elongated and proximal in full flexion. Stack interprets these reconstructions (which cannot be far from reality) as indicating that intrinsic muscles, especially the lumbricals, act as servo-mechanisms, the main power of extension derived from the long extensor tendons, while precise points of application of power at interphalangeal joints is determined and transmitted by the intrinsic muscles. (It is perhaps reasonable, therefore, to enquire whether *extensor* forces cannot equally be exerted by deep digital *flexors* via lumbricals.) Whatever the real activities of lumbrical muscles, theirs is undeniably a key role in digital movements. It has been suggested that their stretch receptors are of special importance, apparently a reasonable assumption, though it is unwise to emphasize any single factor in the complex mechanics of digital movement.

The contribution of the *digital retinacular ligaments* (p. 630) is considered an integration between interphalangeal joints. Flexion at the distal appears to tauten ligaments on each side of the digit and to pull proximal joints into flexion. It is observable that when flexion at a proximal joint is prevented, flexion at the distal is resisted, probably by retinacular ligaments. The two joints are similarly linked in extension. Passive flexion of a distal joint automatically produces a similar movement at the proximal.

Stack (1962) has constructed an interesting model to illustrate these and other digital mechanics.

Nothing has yet been said of slight rotations at digital joints. Interphalangeal joints are often regarded as uni-axial, but slight conjunct rotation occurs (p. 515); more obvious rotation is visible at the metacarpophalangeal joints and the rotational movement of metacarpals at carpometacarpal junctions also contribute to digital movements. When digits are opposed in precise manipulations all except the medius show slight metacarpal rotation; in the fingers these rotations are held to be effected by the appropriate interossei (Braithwaite et al 1948), necessary to approximate the tips of fingers and thumb in various forms of precision grip (vide supra), which can involve more than the pollex and index. These two provide a '*pliers*' or '*pincer*' grip; any or all of the others may be added to produce varieties of '*chuck*' grip (5.84). In these and other flexor actions, the metacarpal movements, though small (except in the pollex), are easily observed. In gripping any object powerfully with the whole hand, the more mobile metacarpals (5th, 4th, 2nd) are flexed to relative extents indicated at the carpometacarpal joints and the fifth at least may also rotate. These movements depend on shape, dimensions and the malleability of the objects grasped, e.g. in grasping a long cylinder across the whole palm metacarpal rotation is unnecessary, unless it is of small diameter or yielding substance, such as a thin rope; in which case flexion and rotation of the fifth metacarpal in particular contribute to the well-known effectiveness of the minimus in gripping small, long objects. In all grasping movements the ultimate events at the articulations involved depend upon the size, contours and resistances presented by an object, as much as upon deployment of muscular forces. For such reciprocally adaptive adjustment the multiplicity of skeletal components, and hence articulations, in the hand appears extremely apt.

The pollex, more accessible to direct observation, presents less difficulty in the description of its activities; but there is more controversy over the definition of its *cardinal* movements (compare Haines 1944, Napier 1955 and Pieron 1973) and especially over how these contribute to the *customary* pollicial activities which are not the selected actions of formal analysis: i.e., flexion, extension, abduction, adduction and circumduction. These activities have, in their turn, dictated the names of the muscles involved, despite the fact that these do not have the 'pure' or cardinal effects indicated by their names. For example, abductor pollicis brevis is not a simple abductor, for it also assists in medial rotation, a combination clearly adapted to opposition; 'pure' abduction is an abstraction, while slightly 'impure' abduction with minimal spin is infrequently required.

It is hence necessary to consider these cardinal movements. Since the pollicial carpometacarpal joint is sellar, planes of its two cardinal 'swings'—flexion – extension and adduction – abduction —must be approximately at right angles; but orientation of these planes relative to the rest of the hand is variably defined in different accounts, perhaps the source of controversies in naming the movements. Commonly the plane of adduction – abduction is regarded as approximately orthogonal to that of the palm, entailing that flexion – extension is parallel to the latter. Inspection clearly shows that the major axes of this sellar joint are not so orientated: the ridge on the trapezial articular surface (along which the metacarpal base slides in adduction – abduction) is inclined at an angle nearer to 60° than 90° relative to the 'palmar' plane. Hence abduction is not only 'away' from the palm but also somewhat lateral. Doubtless this plane varies individually; even to regard it as a plane is probably a simplification. Sellar surfaces at this joint are not perfectly symmetrical; Kuczynski (1974) has examined a short series (15 hands) in great detail, showing that contours of the surfaces are such that pure swings are improbable, being accompanied in fact by slight rotation. He also considers that rotations are even greater in flexion – extension. This is perhaps to be expected in a digit constantly moving from positions of opposition to open grasp with the pollex somewhere between extension and abduction. Both Kapandji (1970) and Kuczynski (1974) emphasize that pollicial movements depend on what is occurring at all joints simultaneously. Medial metacarpal rotation, accompanying flexion and adduction during opposition, is combined with similar rotation at the metacarpophalangeal joint. Medi rotation at the latter is usually compounded with some abduction of the proximal phalanx, as when opposing the tips of pollex and minimus. Movements at both ends of the pollicial metacarpal could be integrated by thenar muscles, all of which, except opponens, act on the carpometacarpal *and* metacarpophalangeal joints. Pollicial muscles, intrinsic (thenar) and extrinsic, have been likened to an array of stays or guy-ropes, alterable in length to adjust either joint to a new position from which the rest of the digit can then be moved. Movements of the terminal phalanx at the interphalangeal joint are controlled by the long flexor and extensor. Having only two phalanges the pollex does not present the same extensor problems as do other digits; no complex extensor expansion is required.

At rest the thumb is in contact with or near to the lateral palmar border and index finger. Other digits are slightly flexed, to a degree increasing from the second to the fifth. Metacarpal heads form an arch, corresponding in palmar concavity to that of the carpus, at which the hand is about half extended. The whole apparatus, though 'open', is semi-flexed or 'cupped', ready for grasping or oppositional effort as required. For larger objects the initial extensor 'spread' may be a preliminary to active progress to some form of power or precision grip. If the latter involves all digits, as in holding a ball between their tips, the variable mobility of metacarpals is particularly noticeable, allowing the fingertips to turn towards the pollex and to space themselves at different angles around such an object. Opponens minimi digiti enables its more mobile metacarpal to bring the fifth digit in complete opposition to the pollex across the diameter of large or small objects. The influence of ligaments in pollicial activity, especially at the carpometacarpal joint, has attracted little interest. The intermetacarpal ligament, between the first and second metacarpal bases (Haines 1944), has been re-examined by Bojsen-Møller (p. 514), who regards it as much involved in such movements.

All movements so far described concern prehension; but the hand can also be used in transmitting force, as in pushing objects or pressing upon surfaces. The resistance to be overcome may be great, amounting to body weight and more. In dynamic situations, where the hand fends off moving bodies or to avoid collision with static obstructions, the forces transmitted through the outstretched, extended hand may be even greater. Most joints in the hand and wrist are then in the close-packed position, and flexor structures, tendons, ligaments and fasciae, are tensed to produce maximal resistance to further extension.

The great variety and skill of manual movements is mediated by a most complex motor equipment and sensory apparatus, cutaneous and neuromuscular. It is not surprising that the hand monopolizes a greater area of motor and sensory cortex (7.1049) than the rest of the arm or the whole of the leg. The thumb's importance is manifest in the allocation to it of about half the cortical 'areas' for the hand.

MUSCLES OF THE LOWER LIMB

Muscles of the lower limb may be grouped as follows:
1. Muscles of the iliac region.
2. Muscles of the femoral and gluteal regions.
3. Muscles of the leg.
4. Muscles of the foot.
For a general survey consult 5.85A.

1. Muscles of the Iliac Region

These three muscles arise from the lumbar vertebral column (psoas major and minor) and iliac bone (iliacus), and two are attached together as femoral flexors. Psoas minor falls short of this and acts on the vertebral spine and sacro-iliac joints.

Iliac fascia covers the psoas and iliacus; thin above, it gradually thickens towards the inguinal ligament. The *part covering psoas* is thickened above as the *medial arcuate ligament* (p. 592). Medially it is attached by a series of fibrous arches with their intervening columns blending with the intervertebral discs, the margins of adjacent vertebral bodies and to the upper sacrum. Laterally, above the iliac crest, it blends with the fascia on quadratus lumborum (p. 604), below the crest, with the fascia covering the iliacus. The *iliac part* is attached to the inner lip of all the ventral segment of the iliac crest to the pelvic brim, blending with the periosteum, and to the iliopectineal eminence, receiving a slip from the tendon of psoas minor. External iliac vessels are anterior to it but branches of the lumbar plexus posterior; it is separated from the peritoneum by loose areolar tissue. Lateral to the femoral vessels the fascia is continuous with the posterior margin of both the inguinal ligament and the transversalis fascia. Medially it passes behind the femoral vessels to become the pectineal fascia, attached to the pecten pubis. Between the lateral and medial regions it is attached to the iliopectineal eminence and capsule of the hip joint, forming a septum between the inguinal ligament and the innominate bone and separating the large osteoligamentous interval into a lateral *lacuna musculorum*, containing the psoas major, iliacus and femoral nerve, from a medial *lacuna vasorum*, transmitting the femoral vessels and lymphatics. Distally the fascia becomes the posterior part of the femoral sheath (p. 638).

The segmental values of the nerves to the muscles in the lower limb are those given by Sharrard (1955).

Psoas major (5.85), a long fusiform muscle lateral to the lumbar vertebral column and pelvic brim, arises: (1) from the anterior surfaces and lower borders of all lumbar transverse processes; (2) by five digitations, each from the localized parts of the bodies of adjoining vertebrae and their intervertebral discs, the highest from the lower margin of the twelfth thoracic body and upper margin of the first lumbar and the interposed thoracolumbar disc, the lowest slip from adjacent margins of the fourth and fifth lumbar bodies and their disc; and (3) from the tendinous arches extending across the concave lumbar bodies between, and blended with, the digitations. Lumbar arteries and veins and rami of the sympathetic trunk pass medial to these arches. The upper four *lumbar intervertebral foramina* bear important relations to these two principal 'lines' or planes of superior attachment of the muscle. The foramina lie *anterior* to the transverse processes and *posterior* to the attachments to vertebral bodies, discs and tendinous arches. Thus, the roots of the lumbar plexus directly *enter* the muscle, the plexus is lodged *within it* and its branches *emerge* from its borders and surfaces. The psoas major muscle descends along the pelvic brim, posterior to the centre of the inguinal ligament, anterior to the capsule of the hip joint (over the acetabulum and femoral head), and converges into a tendon which receives laterally almost all the iliacus and is attached to the lesser trochanter. A large subtendinous iliac bursa, occasionally communicating with the hip joint, separates the tendon from the pubis and articular capsule. These complex attachments display minor variations.

Relations: *above*, the superior tip of the psoas is posterior to the diaphragm (i.e. in the posterior mediastinum) and may contact the pleural sac. In the abdomen its *anterolateral surface* is related to the medial arcuate ligament, which is a superior arched thickening in the general psoas fascia, extraperitoneal tissue and peritoneum, kidney, psoas minor, renal vessels, ureter, testicular (or ovarian) vessels and genitofemoral nerve. The right psoas is overlapped by the inferior vena cava and crossed by the terminal ileum; the left is crossed by the colon. *Posterior* are the transverse processes of the lumbar vertebrae and medial edge of the quadratus lumborum. The lumbar plexus is, as noted, embedded posteriorly between the planes of the muscle (p. 1140). *Medial* are the lumbar vertebral bodies and lumbar vessels. Along its *anteromedial margin* it is in contact with a sympathetic trunk, aortic lymph nodes and, along the pelvic brim, the external iliac artery;

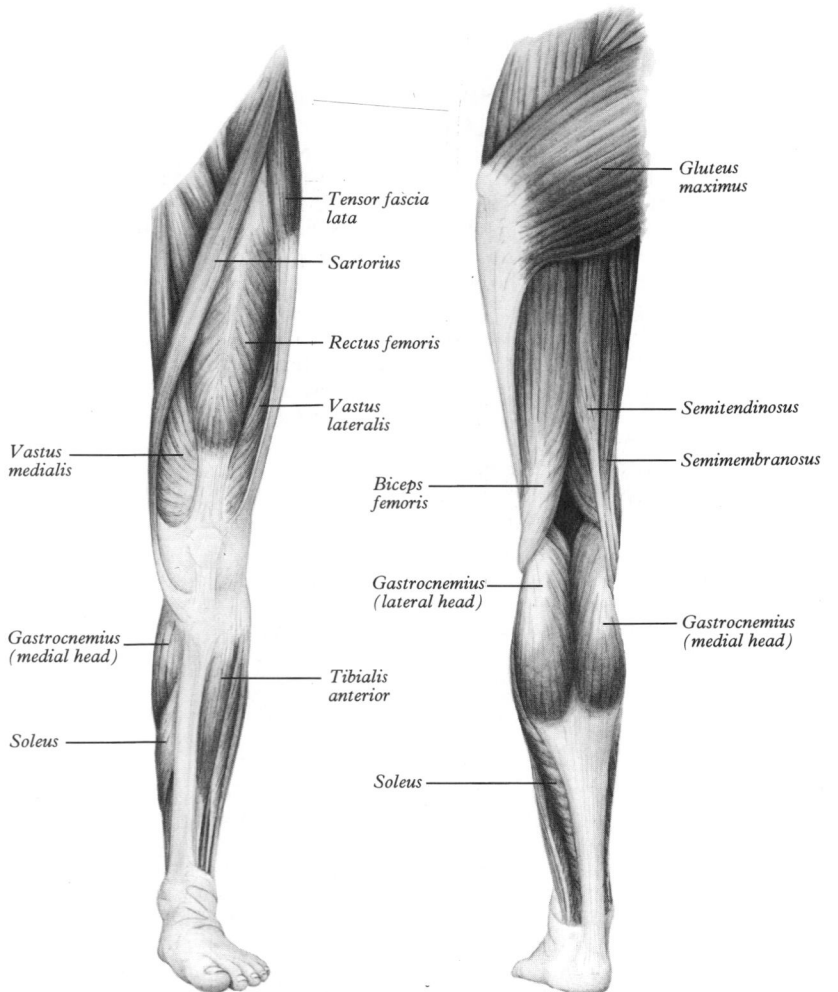

5.85A Muscles of the leg (superficial view): (left) anterior aspect; (right) posterior aspect.

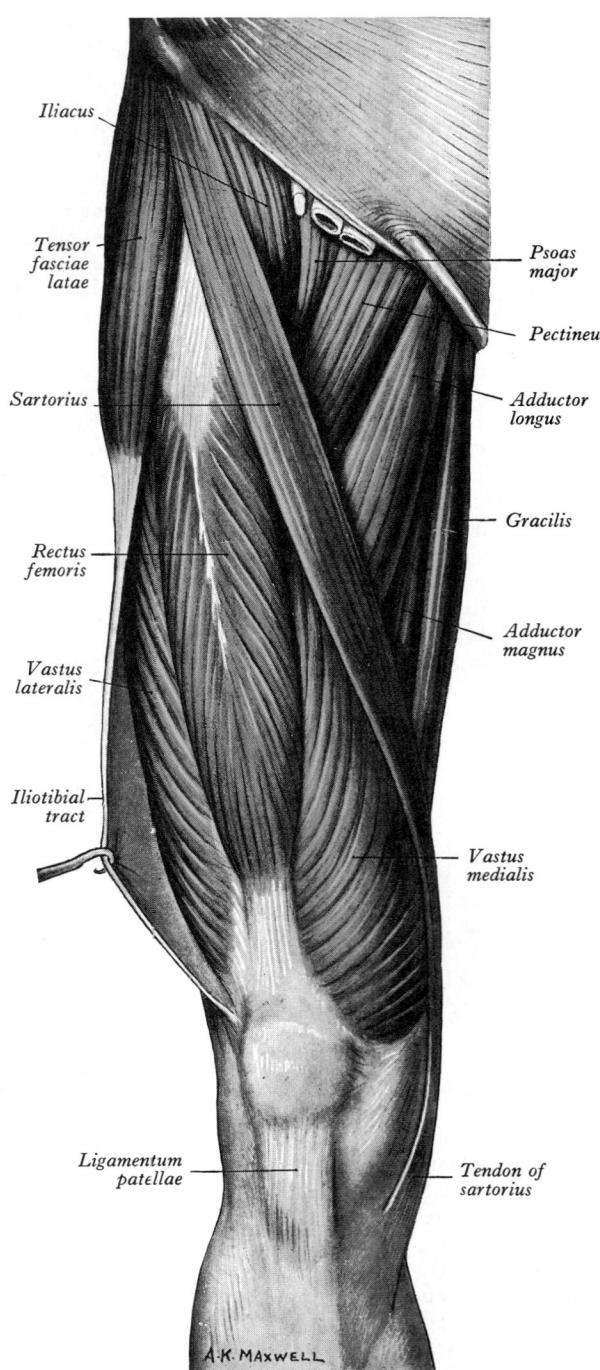

5.85B Superficial muscles of the thigh: extensor aspect (from Quain's Anatomy, 11th edn).

this anteromedial margin is, as noted, covered by the inferior vena cava on the right, whereas on the left it is posterolateral to the abdominal aorta. In the thigh the fascia lata and femoral artery are *in front; behind* is the capsule of the hip joint separated by a bursa; at its *medial border* are the pectineus, medial circumflex femoral artery and the femoral vein, which may overlap it slightly; at its *lateral border* are the femoral nerve and iliacus; the nerve descends at first through its fibres, then between it and the iliacus.

As indicated, branches of the lumbar plexus diverge from the abdominal part of the psoas. From the lateral border of the psoas emerge, from above downwards: the iliohypogastric, ilio-inguinal and lateral femoral cutaneous and femoral nerves; from the anterolateral surface, the genitofemoral nerve; from the medial border, the obturator and accessory obturator nerves and the upper root of the lumbosacral trunk.

Nerve supply: the ventral rami of the lumbar spinal nerves, L.**1, 2** and **3**.

Actions: the psoas major acts largely with the iliacus (vide infra).

The **psoas minor** lies anterior to the psoas major and is confined to the abdomen; it arises from the sides of the twelfth thoracic and first lumbar vertebral bodies and the disc between them. It ends in a long, flat tendon attached to the pecten pubis, iliopectineal eminence and iliac fascia. It is absent in about 40% of subjects.

Nerve supply: a branch from the first lumbar spinal nerve.

Action. It is probably a weak flexor of the trunk.

Iliacus (5.85) is a triangular sheet of muscle attached to the superior two-thirds of the iliac fossa, inner lip of iliac crest, ventral sacro-iliac and iliolumbar ligaments, and upper surface of the sacral ala (3.43); this attachment reaches the anterior superior and inferior iliac spines, receiving a few fibres from the capsule of

the hip joint. Most of the muscle converges into the lateral side of its conjoined strong tendon with psoas major, and together they are attached to the lesser trochanter, but some iliac fibres are attached directly to the femur for about 2.5 cm antero-inferior to this.

Relations. In the abdomen, *anterior* are the iliac fascia separating it from extraperitoneal tissue and the peritoneum and lateral femoral cutaneous nerve; on the right, the caecum, on the left, the iliac part of the descending colon; *posterior* is the iliac fossa; *medial* are the psoas major and femoral nerve. In the thigh its *anterior surface* is in contact with the fascia lata, rectus femoris, sartorius and arteria profunda femoris, its *posterior surface* with the capsule of the hip joint, a bursa common to it and with psoas major interposed.

Nerve supply: branches of the femoral nerve, L.**2** and **3**.

Actions. Acting from above, the psoas major flexes the thigh on the pelvis, assisted by the iliacus. Electromyography does not support the common view that psoas major is a medial rotator the hip joint, but it *is* active in *lateral* rotation, particularly in the young, a view supported by dissections in infants (McKibbin 1968). Conflicting views expressed for many years demand further investigation. When the psoas major and iliacus act from below, they are powerful vertebral flexors against resistance, as in raising the trunk from the recumbent posture. Geometrical reasoning suggests that, with the body erect and lower limb fixed, the psoas major might flex the trunk anterolaterally; but electromyography does not support this, indicating maximum activity when the lumbar curvature is increased. However, direct recording during operations suggests that in addition to being a hip flexor, the muscle is active in balancing the trunk in sitting (Keagy et al 1966).

In symmetrical standing recordings from iliopsoas vary from electrical silence to *slight* intermittent or continuous activity, perhaps reflecting differences between individuals and the techniques of investigation. But most are agreed that the low level of activity correlates with the fact that the line of body weight is behind the transverse coxal axis and that hip joints are near their close-packed position, with spiralization of the ligaments (especially iliofemoral) and marked compression and congruence of articular surfaces, these passive forces largely counterbalancing extending torque of the body weight. However, prolonged symmetrical standing is rare (in non-military personnel).

Applied Anatomy. When the femoral neck is fractured the psoas major acts as a lateral rotator, accounting for the characteristic posture of the lower limb. Occasionally pus from disease of the thoracic and lumbar vertebrae tracks into the thigh within the muscle's fascial sheath, but this is quite uncommon with the use of modern therapeutics.

2. Muscles of Femoral and Gluteal Regions

Muscles of the thigh and hip could be divided into flexors and extensors; but due to rotation of the limb into a new working position in land vertebrates (p. 398), the primitive dorsal, extensor surface of the thigh has become anterior; the original ventral aspect containing the flexor muscles is now posterior. This has not affected the 'girdle' muscles; iliopsoas, a flexor already described, is still ventral and the gluteal muscles, primitive extensors, have remained dorsal. Moreover, development of multiaxial function at the hip has been coupled with the appearance of adductors and abductors, the joint becoming surrounded by muscle forces capable of moving the femur in any direction or maintaining static postures, especially the erect position. These activities are complicated by gravity, aiding some movements, opposing others. Hence muscle groups are unequal in power.

(a) Anterior Femoral Muscles

Included in this group (5.85B) are the tensor fasciae latae, sartorius and rectus femoris, all acting on hip and knee joints, and vasti medialis, lateralis, and intermedius acting only at the knee. The articularis genus, a derivative of vastus intermedius, completes the group. Rectus femoris and the vasti extend the knee by a common tendon and are hence grouped as *quadriceps femoris*.

The **superficial fascia** of the thigh is loose areolar tissue containing variable amounts of fat and in some regions, particularly near the inguinal ligament, it splits into recognizable layers, between which ramify superficial vessels and nerves. In the inguinal region it is thick and between its two layers are superficial inguinal lymph nodes, the great saphenous vein and smaller vessels. The *superficial* layer is continuous with that of the abdominal wall; the *deep* layer, a thin fibro-elastic stratum, is best differentiated medial to the great saphenous vein and inferior to the inguinal ligament, extending between subcutaneous vessels and nerves and the deep fascia, with which it fuses distal to the ligament. This line of fusion lies in the floor of the *ventral flexure line* or *groove* associated with the hip joint.) The membranous fibro-elastic fascia completes the saphenous opening, blending with its spiral circumference and the femoral sheath (p. 781). Over the

opening it is perforated by the great saphenous vein and other vessels, hence the term *cribriform fascia*. Anterior to the patella the superficial fascia contains a subcutaneous prepatellar bursa.

The **fascia lata**, the **deep fascia** of the region, is thicker in proximal and lateral parts of the thigh where tensor fasciae latae and an expansion from gluteus maximus fuse with it. It is thin posteriorly and over adductor muscles but thicker around the knee, strengthened by expansions from the tendon of biceps femoris laterally, the sartorius medially and the quadriceps femoris in front. The fascia lata is attached proximally to the back of the sacrum and coccyx, the iliac crest, inguinal ligament, superior and inferior pubic rami, the ischial ramus and tuberosity and the lower border of the sacrotuberus ligament. From the iliac crest it descends densely over gluteus medius to the upper border of gluteus maximus, splitting into two layers, superficial and deep to this muscle; at its lower border they reunite. Over the lateral femoral aspect fascia lata is compacted into a strong *iliotibial tract*, which splits at its proximal end, where tensor fasciae latae is attached to and enclosed in it, and posteriorly where it receives most of the tendon of gluteus maximus. The superficial layer ascends lateral to tensor fasciae latae to the iliac crest, the deeper ascends medially, deep to the muscle, to blend with the capsule of the hip joint. Distally, the iliotibial tract is attached to a smooth, triangular, anterolateral facet on the lateral tibial condyle, superficial to and blended with an aponeurotic expansion from vastus lateralis. It makes a subcutaneous visible ridge on the anterolateral aspect of the knee, when it is *extended* (Kaplan 1958). Distally the fascia lata is attached to all exposed bony points around the knee, such as the condyles of the femur and tibia and fibular head. These remarks apply particularly to the iliotibial tract (Evans 1977). On each side of the patella, the deep fascia is reinforced by transverse fibres from the vasti, which are thus attached to it; the lateral fibres are continuous with the iliotibial tract. The fascia lata is continuous with two powerful intermuscular septa, attached to the whole linea aspera and its prolongations; the lateral, stronger septum, extending from the attachment of gluteus maximus to the lateral condyle, is between vastus lateralis in front and the short head of biceps femoris behind, providing partial attachment for them; the medial, thinner septum is between the vastus medialis and the adductors and pectineus. Smaller septa pass between individual muscles ensheathing them and sometimes providing attachments.

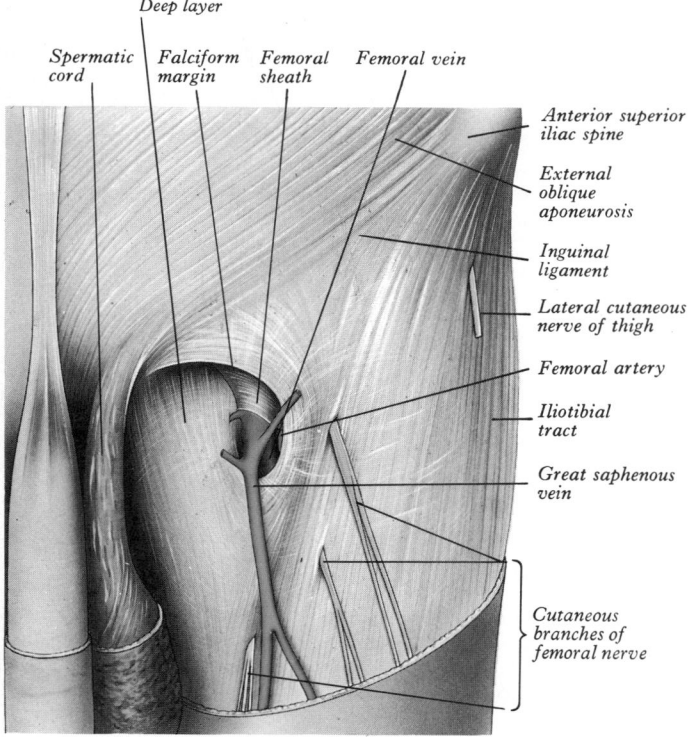

Deep layer

Spermatic cord | *Falciform margin* | *Femoral sheath* | *Femoral vein*

Anterior superior iliac spine

External oblique aponeurosis

Inguinal ligament

Lateral cutaneous nerve of thigh

Femoral artery

Iliotibial tract

Great saphenous vein

Cutaneous branches of femoral nerve

5.86 The left saphenous opening, after the removal of the cribriform fascia. Note the superficial and deep layers of the deep fascia (see text).

Evans (1977) has emphasized the ligamentous nature of the iliotibial tract, regarding it as more substantial than appears necessary for tensor fasciae latae and gluteus maximus. He regards it as a ligament between ilium and tibia, intimately concerned in the standing posture (Evans 1983) (vide infra). This, however, appears to disregard the attachment of these muscles to substantial lengths of the femur through the iliotibial tract and its deep extension, the strong lateral intermuscular septum.

The saphenous opening (5.86) is in the deep fascia lateral and a little distal to the medial part of the inguinal ligament; through it pass the great saphenous vein and other vessels. (Its approximate 'centre' is, in adults, deep to a point 3 cm inferior and lateral to the pubic tubercle.) The cribriform fascia, pierced by these vessels, fills the aperture and must be removed to reveal it. The fascia lata in this region displays the superficial and deep parts (not to be confused with the layers of superficial fascia described above). The *superficial stratum*, lateral to the saphenous opening, is attached to the iliac crest and anterior superior iliac spine, inguinal ligament and pecten pubis (with the lacunar ligament) (p. 596). From the pubic tubercle it turns inferolaterally as an arched *falciform margin*, forming the superior, lateral and inferior boundaries of the saphenous opening (5.86), and is adherent to the anterior aspect of the femoral sheath; the cribriform fascia is attached to it. This margin is described as having *superior* and *inferior cornua*, the latter well defined and continuous behind the great saphenous vein with the *deep stratum* of the fascia lata, which is medial to the saphenous opening and continuous with the superficial stratum at its lower margin. It ascends over the pectineus, adductor longus and gracilis, passes *behind* the femoral sheath and then, united to it, continues to the pecten pubis. The superficial and deep strata are respectively anterior and posterior to the femoral sheath; the saphenous opening, is somewhat spiral and formed by their continuity. The opening varies much, its height from 1.5–8 or 9 cm, its width from 1–3 or 4 cm. Adjacent *subsidiary openings* may occur, transmitting venous tributaries, but are more often in the floor of the fossa.

The **tensor fasciae latae** (5.85) is attached to the anterior 5 cm of outer lip of the iliac crest, to the lateral aspect of the anterior superior iliac spine and part of the notch below it between gluteus medius and sartorius, and to the deep surface of the fascia lata. It descends between, and is attached to, the layers of the iliotibial tract and blends with it near the junction of the middle and upper thirds of the thigh; but the belly is variable in length and some of its fibres may, like the tract, even reach the lateral femoral condyle. Its proximal attachment may spread to the aponeurosis superficial to gluteus medius.

Nerve supply: superior gluteal nerve, L.4 and 5.

Actions. Acting through the iliotibial tract the muscle extends the knee with lateral rotation of the leg; it may assist in abduction and medial rotation of the thigh, but its abductor role has been denied (Kaplan 1958). In erect posture, acting from below, it helps to steady the pelvis on the head of the femur. Through the tract it steadies the femoral on the tibial condyles. When the thigh flexes against gravity, with the knee extended, an angular depression appears distal to the anterior superior iliac spine; its lateral boundary is the tensor fasciae latae, active in flexion. The muscle aids gluteus medius in postural abduction at the hip (p. 642). Kaplan (1958) and Evans (1977) regard the iliotibial tract as more effective in ligamentous steadying of the pelvis than the muscle control mediated (but vide supra, and p. 637).

Sartorius (5.85,87,89), a narrow strap muscle (and incidentally the longest in the body) arises by tendinous fibres from the anterior superior iliac spine and the upper half of the notch below it, crosses the thigh obliquely to its medial side and descends more vertically medial to the knee, where a thin, flat tendon replaces the muscle fibres and curves obliquely forwards, expanding into a broad aponeurosis attached in front of gracilis and semitendinosus to the medial tibial surface (5.98). The proximal part of this aponeurosis curves back over the proximal, curved edge of the preterminal tendon of gracilis to attach behind it. A slip from its upper margin blends with the articular capsule and another, from its distal border, with medial fascia of the leg. The muscle

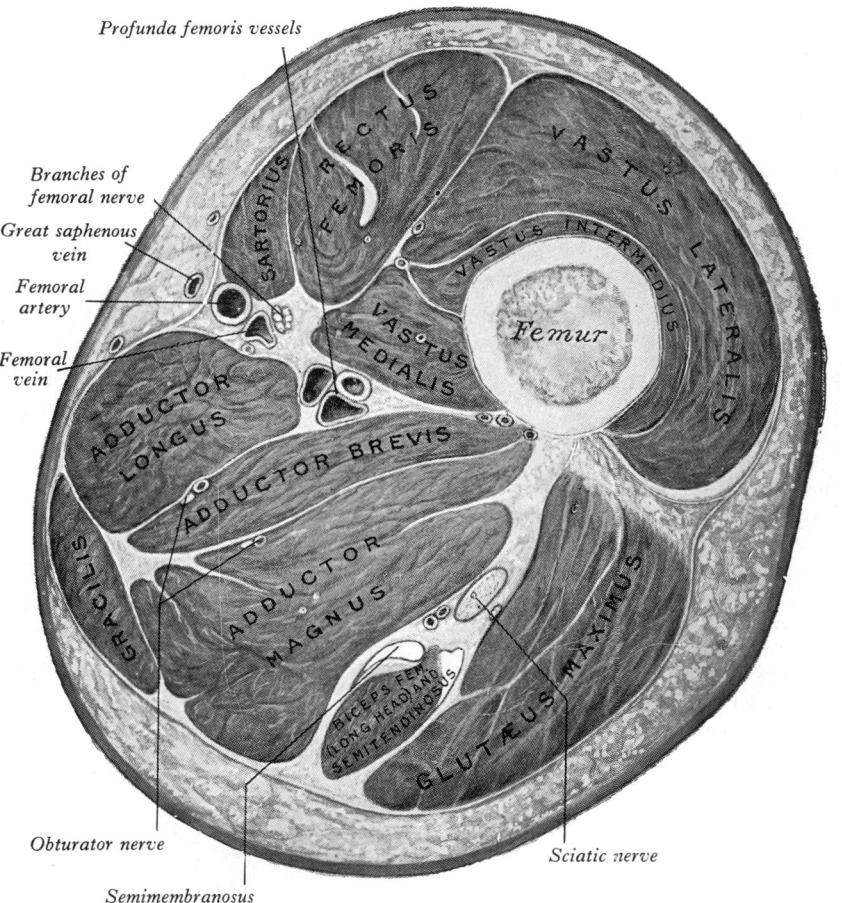

Profunda femoris vessels

Branches of femoral nerve

Great saphenous vein

Femoral artery

Femoral vein

Obturator nerve

Semimembranosus

Sciatic nerve

5.87 Transverse section through the right thigh at the level of the apex of the femoral triangle: proximal (superior) aspect.

varies little but is sometimes bicipital at its proximal end, the extra head being attached to the pectineal line or femoral sheath; it is occasionally absent.

Relations. In the thigh's proximal third the muscle's medial border is the lateral edge of the *femoral triangle*, the medial being the medial border of adductor longus, and its base the inguinal ligament. The femoral artery descends from the centre of the base to the apex of the triangle. In the thigh's middle third the femoral artery is in the adductor (subsartorial) canal, with a stratum of deep fascia and sartorius anterior to it (5.89). This strong fascial layer bridges the interval between adductors and quadriceps; it must be incised to expose the vessels. (For the changing mutual relationships of femoral artery and veins see pp. 782, 814.)

Nerve supply: femoral nerve, L.2 and 3.

Actions. Sartorius assists in flexing the leg at the knee and the thigh on the pelvis, particularly when these movements are combined. It also helps to abduct the thigh and rotate it laterally. These movements, combined with inversion of the foot, bring its sole into direct view. Acting against gravity, as it usually is, the muscle can be seen and felt.

Quadriceps femoris (5.85,87,89), the great extensor of the leg, covers almost all the front and sides of the femur. It can be divided into four parts: one in the middle of the thigh, arising from the ilium, named from its straight course, *rectus femoris*; the other three are attached to the femoral shaft, which they surround, apart from the linea aspera, from trochanters to condyles; lateral is *vastus lateralis*, medial is *vastus medialis* and between them *vastus intermedius*.

Rectus femoris (5.85,87,89) is fusiform, its superficial fibres bipennate, the deep parallel. Its straight tendinous head is attached to the anterior inferior iliac spine and by a reflected tendon to a groove above the acetabulum and the fibrous capsule of the hip joint. These tendons unite at an acute angle and spread as an aponeurosis prolonged distally on the muscle's *anterior* aspect; from this the muscular fibres extend to a broad aponeurosis on the *posterior* aspect of its distal two-thirds, tapering to a flattened tendon attached to the base of the patella as the *superficial central* part of the tendon of quadriceps.

Vastus lateralis (5.85,87,89), the largest component of quadriceps femoris is attached by a broad aponeurosis to the intertrochanteric line, the anterior and inferior borders of the greater trochanter, lateral lip of gluteal tuberosity, and proximal half of the lateral lip of linea aspera (3.185,5.89). This aponeurosis covers the muscle's proximal three-quarters; from its deep aponeurotic surface many additional fibres arise, a few also from both the tendon of gluteus maximus and the lateral intermuscular septum, between vastus lateralis and the short head of biceps femoris. This mass is attached to a strong aponeurosis on the deep aspect of the muscle's distal part; this narrows to a flat tendon, attached to the patella's base and lateral border, blending in the compound tendon of quadriceps (vide infra). It gives to the knee joint's capsule an expansion descending to the lateral tibial condyle and blending with the iliotibial tract.

Vasti medialis and intermedius appear united but when rectus femoris is displaced a cleft appears ascending from the medial patellar border between them, sometimes as far as the lower part of the intertrochanteric line, where they often fuse.

Vastus medialis (5.85,87,89) is attached to the distal part of the intertrochanteric line, spiral line, medial lip of the linea aspera, proximal part of the medial supracondylar line, tendons of adductor longus and magnus and the medial intermuscular septum (p. 637). Its fibres descend forwards mainly to an aponeurosis on the muscle's deep surface which reaches the medial patellar border and tendon of quadriceps (vide infra). An expansion reinforces the knee joint's capsule attached below to the medial tibial condyle. Most distal fibres are almost horizontal, bulging superficially superomedial to the patella.

Vastus intermedius (5.87,89) is attached to the proximal two-thirds of the anterior and lateral surfaces of the femoral shaft and distal part of the lateral intermuscular septum. Its fibres end in an anterior aponeurosis which forms the *deep part* of the tendon of quadriceps, attached to both the lateral patellar border and the lateral tibial condyle.

Tendons of the four components of quadriceps unite distally as

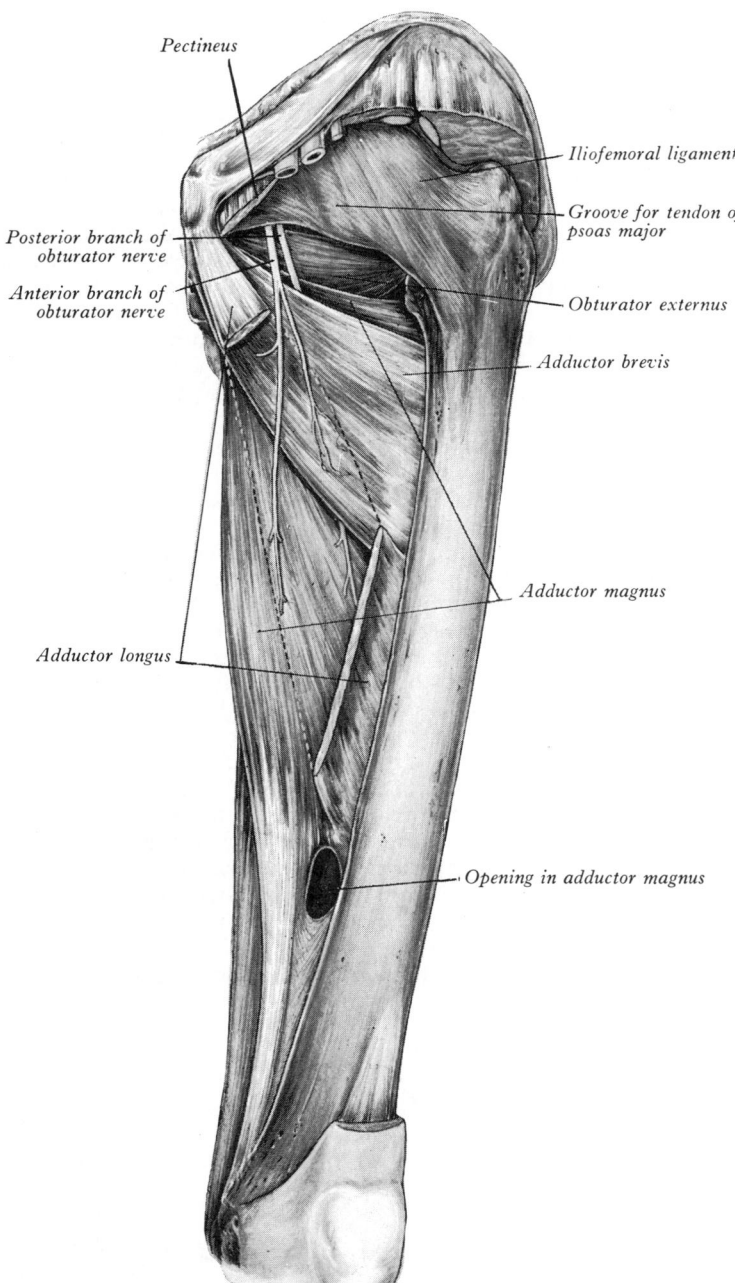

5.88 The adductor muscles of the left thigh: anterior aspect. Most of the adductor longus has been excised; dashed lines indicate its borders.

[Figure labels:] Pectineus; Iliofemoral ligament; Groove for tendon of psoas major; Posterior branch of obturator nerve; Obturator externus; Anterior branch of obturator nerve; Adductor brevis; Adductor magnus; Adductor longus; Opening in adductor magnus

a strong tendon attached to the patellar base, some fibres continuing over this into the ligamentum patellae. The patella is a sesamoid bone in the tendon; the 'ligamentum' patellae, extending from the patellar apex to the tibial tubercle, is the continuation of this tendon, medial and lateral *patellar retinacula* (p. 527) being expansions from its borders. The supra-patellar bursa (a synovial extension of the knee joint) is between the femur and suprapatellar part of the tendon; the deep infrapatellar bursa is between the ligamentum patellae and proximal end of the tibia (4.75). Quadriceps varies little; rectus femoris may arise from the anterior *superior* iliac spine and its reflected head may be absent.

Articularis genus, a small muscle, usually distinct from vastus intermedius but sometimes blended with it, consists of several muscular bundles arising from the distal anterior surface of the femoral shaft and attached to a proximal reflexion of the synovial membrane of the knee joint (DiDio et al 1967).

Nerve supply: femoral nerve, L.2, **3** and **4**.

Actions. The whole muscle extends the knee. Rectus femoris also helps to flex the thigh or, if the thigh is fixed, to flex the pelvis. It is remarkably quiescent in standing (Joseph & Williams 1957).

Rectus femoris may flex the hip and extend the knee simultaneously. The lower fibres of vastus medialis contract late in genual extension to retain the patella in its groove, counteracting its tendency to lateral displacement, due to lateral inclination of the long axis of the thigh relative to that of the leg (Lieb & Perry 1968). The lower fibres of vasti medialis and lateralis may be involved in stabilizing the knee. Electromyography indicates that the vasti are not equally or synchronously active in different phases of extension or rotation, but this awaits confirmation. The muscle's complex architecture offers a wide field for myokinetic analysis; perhaps its complexity explains the paucity of information.

Articularis genus retracts the synovial supra-patellar 'bursa' during extension, presumably preventing the interposition of synovial folds between the patella and femur (cf articularis cubiti, p. 616).

(b) Medial Femoral Muscles

This group is derived, as the nerve supply suggests, from both the flexors and extensors between which it lies. All five muscles, gracilis, pectineus and adductores longus, brevis and magnus, cross the hip joint; only gracilis descends beyond the knee. Often named adductors of the thigh, they are more complex than this.

Gracilis (5.85,87,89), the most superficial, is thin and flat, proximally broad, distally narrow and tapering. Attached proximally by a thin aponeurosis from the medial margins of the lower half of the body of the pubis, the inferior pubic ramus and adjoining part of the ischial ramus (3.171), it descends vertically to a rounded tendon crossing the medial femoral condyle posterior to the tendon of sartorius; it curves round the medial tibial condyle and fans out proximally on the tibial medial metaphysial surface, just distal to its condyle. A few distal fibres continue into the crural deep fascia. The attachment is immediately proximal to that of semitendinosus, its upper edge overlapped by the tendon of sartorius, with which it is partly blended. It is separated from the tibial collateral genual ligament by a tibial intertendinous bursa (p. 530).

Nerve supply: obturator nerve, L.2 and 3.

Actions. Gracilis flexes the leg and rotates it medially; it may also act as a femoral adductor.

Pectineus (5.85), a flat, quadrangular muscle in the femoral triangle, arises from the pecten pubis and the bone in front of it between the iliopectineal eminence and pubic tubercle and from the fascia on its anterior surface; it descends, posterolaterally, to a line from the lesser trochanter to the linea aspera. It may be bilaminar, as in other mammals, layers receiving separate nerve supplies (vide infra). Proximally it may be partially or wholly attached to the coxal capsule.

Relations. *Anterior* is the fascia lata, separating the muscle from the femoral vessels and great saphenous vein; *posterior* are the coxal capsule, adductor brevis, obturator externus and anterior branch of the obturator nerve; *lateral* are the psoas major and medial circumflex femoral vessels; *medial* is the lateral margin of adductor longus.

Nerve supply. Pectineus is supplied by the femoral nerve, L.2 and 3; and the accessory obturator, L.3, when present, occasionally by a branch from the obturator nerve. It may be partly divided into dorsal and ventral strata, supplied respectively by the obturator and femoral (or accessory obturator) nerves. In a total of 800 cases, only 69 displayed a partial supply by an accessory obturator nerve (Woodburne 1960).

Actions: adduction of the thigh; flexion of the pelvis.

Adductor longus (5.88,89), the most anterior adductor, is triangular and in the plane of the pectineus. It has a flat, narrow tendon (but sometimes C-shaped in section) attached to the pubic body, in the angle between the pubic crest and the sheath enclosing rectus abdominis and pyramidalis lateral to the symphysis. This expands to a broad belly, descending posterolaterally to an aponeurosis attached to the linea aspera in the middle femoral third between the vastus medialis and adductores magnus and brevis, usually blending with each. It is occasionally double.

Relations. *Anterior* are the spermatic cord and fascia lata, separating it from the great saphenous vein and distally femoral vessels and sartorius. *Posterior* are the adductores brevis and magnus, the anterior branch of the obturator nerve and distally the profunda femoris vessels. *Lateral* is pectineus, *medial* is gracilis.

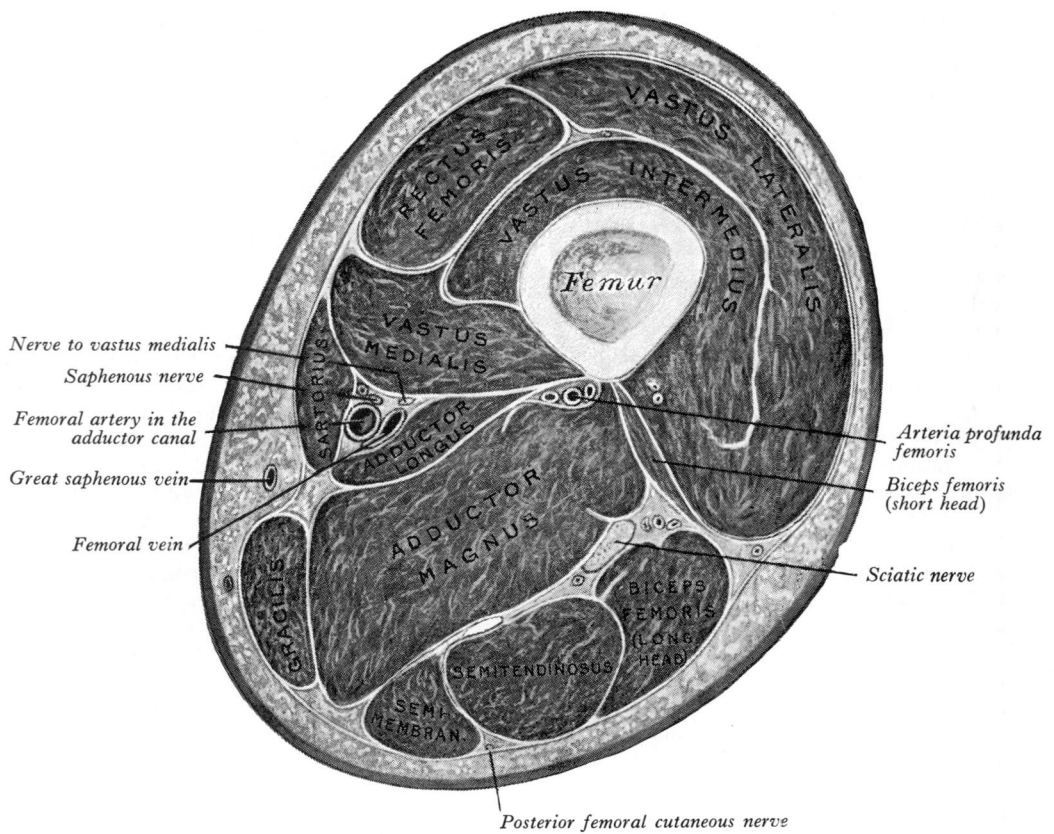

Nerve to vastus medialis

Saphenous nerve

Femoral artery in the adductor canal

Great saphenous vein

Femoral vein

Arteria profunda femoris

Biceps femoris (short head)

Sciatic nerve

Posterior femoral cutaneous nerve

5.89 Transverse section through the middle of the right thigh: proximal (superior) aspect.

Nerve supply: the anterior division of the obturator nerve, L.2, 3 and 4.

Adductor brevis (5.87,88), posterior to pectineus and adductor longus, is also somewhat triangular (but truncated proximally) with a narrow linear (horizontal) attachment to the external aspect of the pubic body and its inferior ramus between gracilis and obturator externus, descending posterolaterally to a vertical aponeurosis attached to the femur along a line from the lesser trochanter to the linea aspera and to the proximal part of the line behind pectineus and adductor longus. It often has two or three parts or may be integrated into the adductor magnus.

Relations. *Anterior* are pectineus, adductor longus, arteria profunda femoris and the anterior branch of the obturator nerve; *posterior* are adductor magnus and the posterior branch of the obturator nerve; at the muscle's *proximal border* are the medial circumflex femoral artery, obturator externus, and the conjoined tendon of psoas major and iliacus; at its *distal border* are gracilis and adductor magnus. Near its femoral attachment the second and sometimes first perforating arteries pierce it.

Adductor magnus (5.87,88,89), a massive triangular muscle, starts from a small part of the inferior pubic ramus, the conjoined ischial ramus, and the inferolateral aspect of the ischial tuberosity. Fibres from the pubic ramus are short, *horizontal* and attached to the medial margin of the gluteal tuberosity, medial to gluteus maximus; this part, anterior to the rest, is sometimes termed the *adductor minimus*. Fibres from the ischial ramus fan out infero-laterally to a broad aponeurosis attached to the linea aspera and proximal part of the medial supracondylar line. The muscle's medial part, mainly fibres from the ischial tuberosity, is a thick mass descending *almost vertically* to a rounded tendon in the distal third of the thigh, which is palpable proximal to its attachment to the adductor tubercle (p. 438) on the medial femoral condyle. This tendon is connected by a fibrous expansion to the medial supracondylar line. This long attachment is interrupted by osseo-aponeurotic openings, bridged by *tendinous arches* attached to the bone. The proximal four are small and transmit perforating branches and the termination of arteria profunda femoris; the distal one is large and admits the femoral vessels to the popliteal fossa. The muscle's vertical, ischiocondylar part is variably separate. The proximal border may be fused with quadratus femoris.

Relations. *Anterior* are pectineus, the adductores brevis and longus, femoral and profunda vessels and the posterior branch of the obturator nerve; a bursa separates the proximal part of the muscle from the lesser trochanter. *Posterior* are the sciatic nerve, gluteus maximus, biceps femoris, semitendinosus and semimembranosus. The *proximal border* is parallel with quadratus femoris, the transverse branch of the medial circumflex femoral artery passing between them. The *medial border* is related to gracilis, sartorius and the fascia lata.

Nerve supply. The adductor magnus is *composite* and doubly innervated by the obturator nerve and by the tibial division of the sciatic, L.2, 3 and 4. The latter supplies the ischiocondylar part. Both nerves are from *anterior* divisions of the lumbosacral plexus, indicating a primitive 'flexor' origin for both parts.

Actions. Extensive or forcible femoral adduction is not a common action and, though adductors can so act when required, they are more commonly synergists in the complex patterns of gait and partly controllers of posture. They are active in flexion and extension of the *knee* (Janda & Stara 1965). Magnus and longus are probably *medial* rotators of the thigh, according to de Sousa & Vitti (1966), who also observed that, while adductors are *inactive* during adduction of the *abducted* thigh in erect posture (where gravity assists), they *are* active in the supine position, or during adduction of the flexed thigh in standing. Adductors are also active at the *hip joint* during flexion (longus) and extension (magnus). In symmetrical easy standing their activity is minimal.

(c) Muscles of the Gluteal Region

Glutei maximus, medius and minimus (5.90,91) are extensors and abductors at the hip joint; a deeper group of smaller muscles, lateral rotators, includes piriformis, obturatores internus and ex-

ternus, gemelli superior and inferior and quadratus femoris. (These actions are, however, enhanced, diminished or altered when the thigh is in different *initial positions*.)

Gluteus maximus (5.90), largest and superficial, is a broad, thick quadrilateral mass, forming with thick superficial adipose fascia the familiar buttock. (However, the borders and extent of the muscle do *not* correspond with those of the adipose tissue; vide infra.) The large size of gluteus maximus is a special feature of the human hip, associated with bringing and keeping the trunk upright. It is coarsely fascicular, its fibres in large bundles separated by fibrous septa. It is attached to the ilium's posterior gluteal line and to the rough bone (including the crest) postero-superior to it, to the aponeurosis of erector spinae, dorsal surface of lower sacrum and side of the coccyx, sacrotuberous ligament, and fascia (gluteal aponeurosis) covering gluteus medius. Its fasciculi *descend laterally*, the upper, larger part and a distal superficial stratum ending in a tendinous lamina lateral to the greater trochanter and continued into the iliotibial tract. Its deeper distal stratum is attached to the femoral gluteal tuberosity between the vastus lateralis and adductor magnus.

A large, usually multilocular *trochanteric bursa of gluteus maximus* separates it from the greater trochanter; a *gluteofemoral bursa* is between its tendon and that of vastus lateralis; an *ischial bursa of gluteus maximus*, often absent, separates it from the ischial tuberosity. Additional slips from the lumbar aponeurosis or ischial tuberosity occur. The muscle may be bilaminar.

Relations. *Superficial*: a thin deep fascia and overlying thick adipose subcutaneous tissue; *deep*: ilium, sacrum, coccyx and sacrotuberous ligament, gluteus medius, piriformis, gemelli, obturator internus, quadratus femoris, ischial tuberosity, greater trochanter, attachments of biceps femoris, semitendinosus, semimembranosus and adductor magnus. The superior gluteal artery (superficial part) passes deep to the gluteus maximus

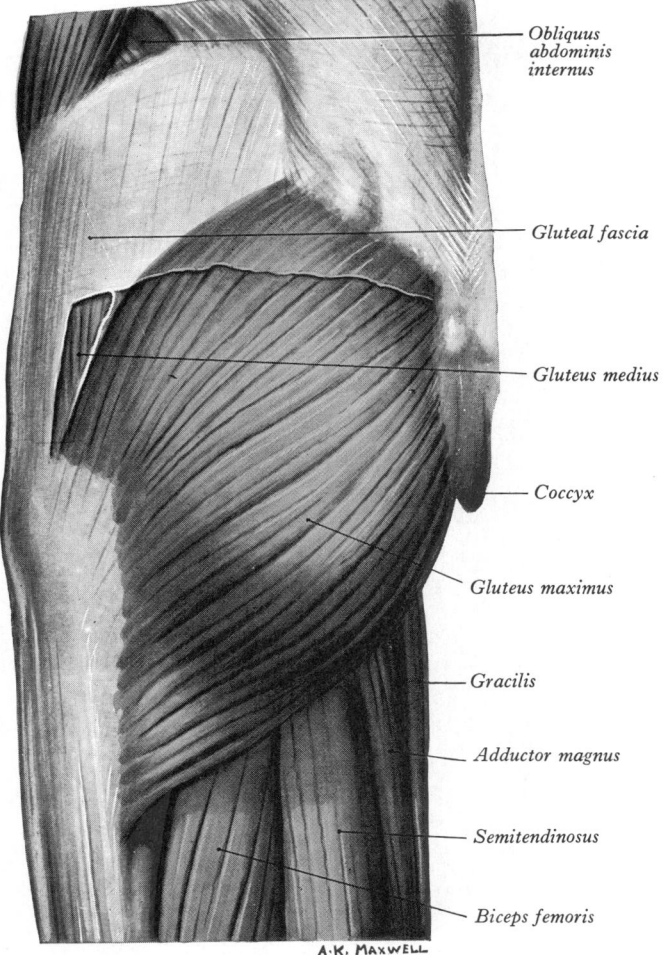

Obliquus abdominis internus

Gluteal fascia

Gluteus medius

Coccyx

Gluteus maximus

Gracilis

Adductor magnus

Semitendinosus

Biceps femoris

A·K· MAXWELL

5.90 The left gluteus maximus. Some of the gluteal fascia has been removed to expose part of the gluteus medius.

between the piriformis and gluteus medius; the inferior gluteal and internal pudendal vessels, sciatic, pudendal and posterior femoral cutaneous nerves and muscular branches from the sacral plexus leave the pelvis below the piriformis. The first perforating artery and terminal branches of the medial circumflex femoral are deep to the lower part of gluteus maximus, whose thin oblique *proximal border* is superficial to gluteus medius. Its prominent *distal border*, sloping down and laterally, is *crossed* by the *horizontal gluteal fold* (the *posterior flexure line* of the hip joint), the latter marking the site of condensed fascial adhesion and indicating the

junction of the thigh and the buttock (5.39). (The *anterior flexure line* corresponds to the fusion between the membranous superficial fascia and the deep fascia distal to the inguinal ligament.)

Nerve supply: inferior gluteal nerve, L.5 and S.1 and **2**. Menning et al (1974) have detailed intramuscular distribution.

Actions. Acting from the pelvis, gluteus maximus extends the flexed thigh in line with the trunk. From its distal attachments it may prevent forward momentum of the trunk flexing at the supporting hip during bipedal gait. It is inactive in standing, forward swaying, and downward bending at the hip. With femoral flexors, however, it raises the trunk from stooping, rotating the pelvis back on the femoral head (Joseph & Williams 1957). It is intermittently active in the walking cycle and in climbing, continuously so in strong lateral rotation. Its upper fibres are involved in powerful abduction; it tenses the fascia lata, stabilizing the femur on the tibia, when the femoral extensors of the knee relax.

Gluteus medius is a broad, thick muscle; its posterior third is deep to gluteus maximus and it is covered in its anterior two-thirds only by deep fascia (5.90). Attached to the external iliac surface between the iliac crest and the posterior gluteal line above, the anterior gluteal line below and to deep fascia superficial to its upper part, it converges to a flat tendon, attached below to a ridge slanting down and forwards on the lateral aspect of the greater trochanter; but it is separated anterosuperiorly from the bone by a *trochanteric bursa of gluteus medius*. A deep slip of the muscle may be attached on the upper border of the trochanter; the muscle is sometimes blended posteriorly with the piriformis.

Gluteus minimus (5.91), smallest of the group and deep to medius, is fan-shaped, attached above to the ilium between the anterior and inferior gluteal lines and behind from the margin of the greater sciatic notch; it converges below to the deep surface of an aponeurosis tapering to a tendon attached to an anterolateral ridge on the greater trochanter, with an expansion to the coxal capsule. A *trochanteric bursa of gluteus minimus* separates the tendon from the medial part of the anterior trochanteric surface. The muscle may show anterior and posterior parts. Separate slips may pass to the piriformis, gemellus superior or vastus lateralis.

Relations. Between the gluteus medius and minimus are deep branches of the superior gluteal vessels and nerve. The reflected tendon of rectus femoris and the articular capsule are deep to the minimus.

Nerve supply. Gluteus medius and minimus are supplied by the superior gluteal nerve, L.5 and S.1.

Actions. Gluteus medius and minimus, acting from the pelvis, abduct the thigh, their anterior fibres rotating it medially. They are essential for keeping the trunk upright when the opposite foot is raised in walking and running, the body weight tending to depress the pelvis on the unsupported side unless counteracted by the gluteus medius and minimus of the supporting limb. They exert such powerful traction that the pelvis actually *rises* on the unsupported side. The muscles are inactive or variably active in varieties of symmetrical standing; in standing 'at ease' with feet separated, abductor muscles are usually electromyographically 'silent'. According to Jonsson & Steen (1962) they *are* active in the position for Romberg's test—feet parallel and close together. Evans (1977) considered the iliotibial tract the major factor in preventing pelvic sag on the unsupported side. (Unmodified, this view must be considered with caution.)

Applied Anatomy. The supportive effect of glutei medius and minimus when the contralateral foot is raised depends on the following: (1) muscles and innervation must be normal, (2) components of hip joint, the fulcrum, must be normal; (3) the femoral neck must be intact and normally angulated. When any one condition is not fulfilled, e.g. (1) paralysis of the glutei, (2) congenital dislocation of the hip, (3) non-united fracture of the femoral neck or coxa vara, the supporting mechanism is deficient, the pelvis sinks in attempts to stand on the affected limb. This constitutes *Trendelenberg's sign.* Paralysis of the gluteus medius and minimus is the most serious coxal muscular disability; sufferers have a peculiar lurching gait. Conversely, when these muscles remain intact, with many others paralysed, walking, even running, is much less affected.

Piriformis (5.91), almost parallel with the posterior margin of gluteus medius, is partly internal to the osteoligamentous pelvis

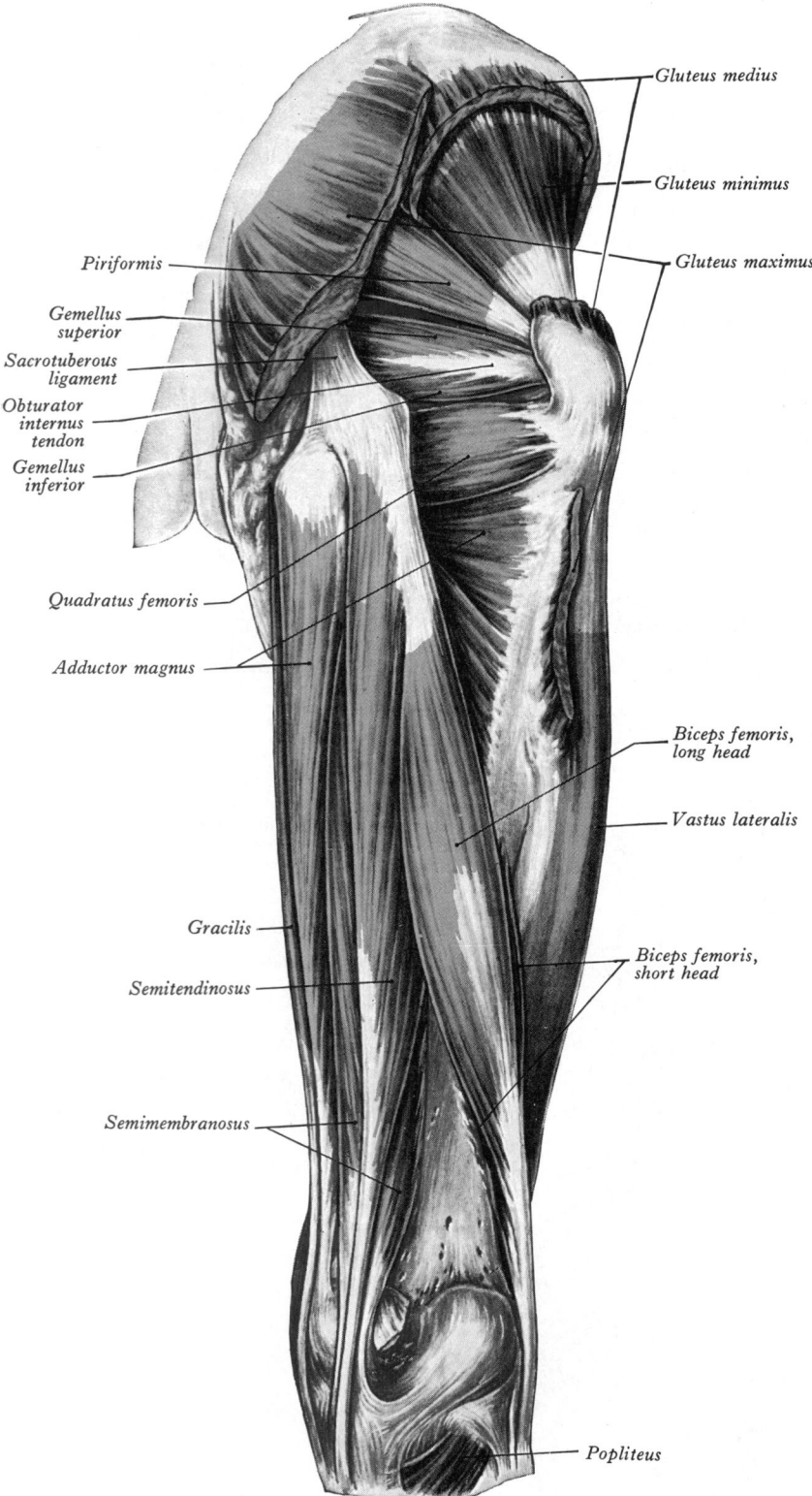

Gluteus medius
Gluteus minimus
Gluteus maximus
Piriformis
Gemellus superior
Sacrotuberous ligament
Obturator internus tendon
Gemellus inferior
Quadratus femoris
Adductor magnus
Biceps femoris, long head
Vastus lateralis
Gracilis
Semitendinosus
Biceps femoris, short head
Semimembranosus
Popliteus

5.91 The muscles of the gluteal region and flexor aspect of the right thigh: posterior aspect (e.g. flexor with respect to the *knee*, they are extensors, abductors and lateral rotators of the *hip*).

and partly external, posterior to the hip joint. Its three digitations are attached anteriorly between the pelvic sacral foramina, to the floor and margins of the grooves leading from them (3.43), from the iliac gluteal surface near the posterior inferior iliac spine, capsule of the sacro-iliac joint, and sometimes the upper pelvic surface of the sacrotuberous ligament. The muscle exits through the greater sciatic foramen, to be attached by a round tendon to the upper border of the greater trochanter, posterosuperior to the trochanteric fossa and often partly blended with the common tendon of obturator internus and gemelli. It may also be blended with gluteus medius.

Relations. *In the pelvis, anterior* are the rectum (especially on the left), the sacral plexus of nerves and branches of the internal iliac vessels; *posterior* is the sacrum. *Outside the pelvis, anterior* are the posterior ischial surface and capsule of the hip joint; *posterior* is gluteus maximus. Its *upper border* is in contact with gluteus medius and the superior gluteal vessels and nerve, its *lower border* with the coccygeus and gemellus superior. The inferior gluteal and internal pudendal vessels, sciatic, posterior femoral cutaneous and pudendal nerves and muscular branches from the sacral plexus appear between the piriformis and gemellus superior. The muscle is frequently traversed by the common peroneal nerve, which may divide it. The sciatic nerve often varies in relation to the piriformis; in 168 dissections Lee & Tsai (1974) observed variation in almost 30%; the nerve emerged *above* the muscle in 2.98%, *through* it in 1.8%, with its divisions above and below in 4.2%, one division between heads of a divided muscle and one division above (1.2%) or below (19.6%), the last being the commonest variation. In a larger series (2250 specimens) reported by Anson (1963), incidence of variation was only 10.7% (242 specimens); in 10% (222) the common peroneal division separately traversed the muscle, the whole nerve in 0.2% (5) and divisions appeared above and below in 0.7% (15); in this series the entire nerve never appeared at the upper border.

Nerve supply: branches from L.5,S.1 and 2.

Actions: lateral rotation of the extended thigh, abduction of the flexed thigh.

The obturator membrane (5.92) is a thin aponeurosis almost closing its foramen. Its fibres interlace, with many oblique bundles but the majority roughly transverse; some form an uppermost bundle between the obturator tubercles, completing the obturator canal for the obturator vessels and nerve. The membrane is attached to the obturator foramen's sharp margin except at its inferolateral angle, where it invades the pelvic surface of the ischial ramus, internal to the foramen. Obturator muscles are attached to its two surfaces; externally some fibres of the pubofemoral ligament are attached to it. It *obturates* its foramen.

Obturator internus (5.93) is partly internal in the osteo-ligamentous pelvis and partly external, behind the hip joint. It is attached to the anterolateral wall of the lesser pelvic cavity, around most of the obturator foramen; to the inferior pubic ramus, ischial ramus and an extensive pelvic periosteal surface postero-inferior to the pelvic brim from the upper greater sciatic foramen posterosuperiorly; to the obturator foramen antero-inferiorly (3.172), and also from the obturator membrane's internal surface, the tendinous arch at its upper border and the obturator fascia. Fibres converge acutely towards the lesser sciatic foramen to end in four or five tendinous bands on the muscle's deep surface, which bend abruptly round the grooved surface between the ischial spine and tuberosity; this is covered by hyaline cartilage and is separated from the tendon by a bursa; it is also ridged in adaptation to grooves in the tendon, whose intervening bands traverse the lesser sciatic foramen to unite into a flat tendon, passing horizontally across the hip joint's articular capsule; fusing with the gemelli it is then attached to an anterior impression on the medial surface of the greater trochanter antero-superior to its fossa. A long bursa usually exists between the tendon and the coxal capsule, sometimes communicating with the bursa between the tendon and the ischium.

Relations. *Within the pelvis, anterolateral* are the obturator membrane and lateral pelvic wall; *posteromedial* are the obturator fascia, levator ani and the sheath of internal pudendal vessels and pudendal nerve (p. 1149), forming the lateral wall of the ischiorectal fossa. *Outside the pelvis* the muscle is covered by gluteus

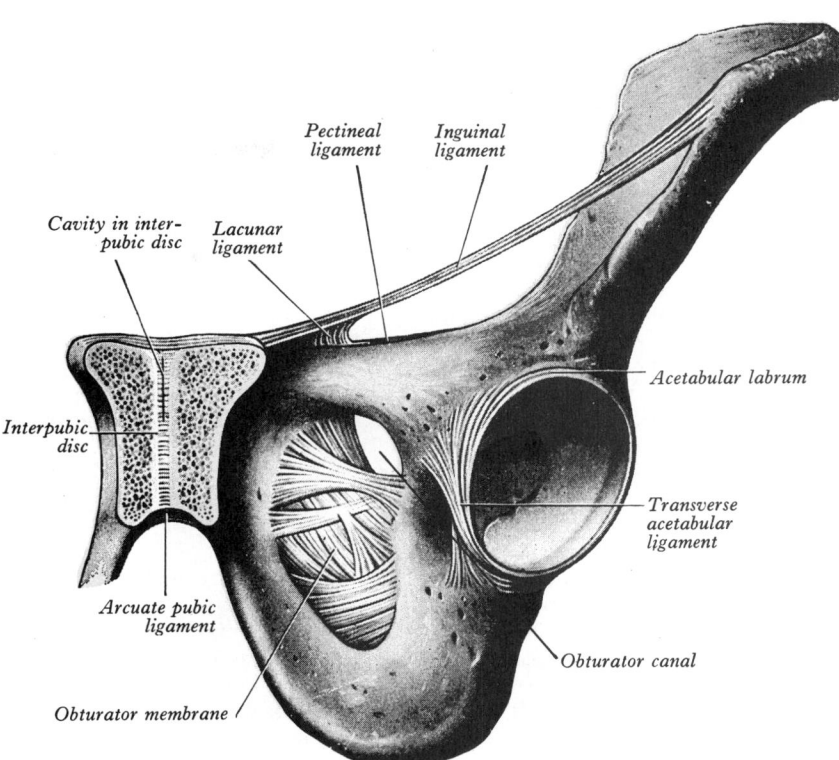

Pectineal ligament *Inguinal ligament*

Cavity in inter-pubic disc *Lacunar ligament*

Interpubic disc

Acetabular labrum

Transverse acetabular ligament

Arcuate pubic ligament

Obturator canal

Obturator membrane

5.92 The left obturator membrane: ventral aspect.

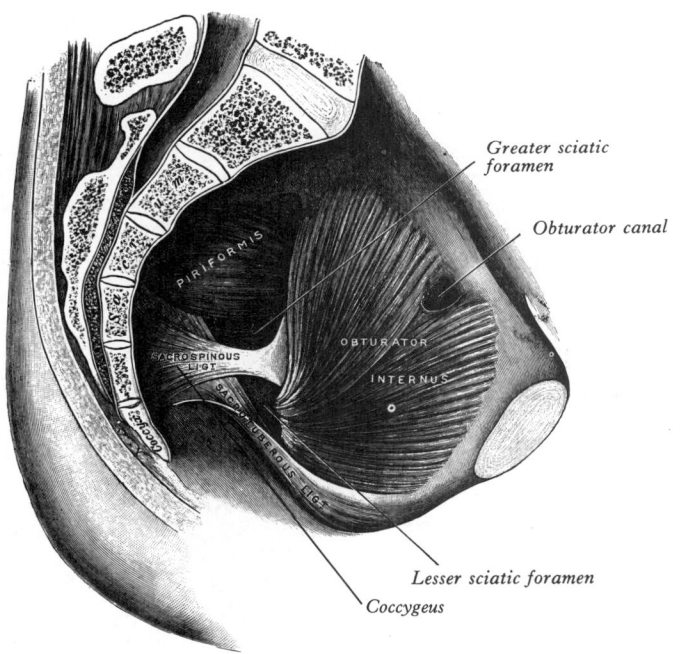

Greater sciatic foramen

Obturator canal

PIRIFORMIS

SACROSPINOUS LIGT

OBTURATOR INTERNUS

Lesser sciatic foramen

Coccygeus

5.93 The left obturator internus: pelvic aspect. Note that the muscle fibres of the coccygeus are blended and coextensive with the sacrospinous ligament (on their gluteal aspect).

maximus, crossed posteriorly by the sciatic nerve, and passes behind the hip joint. As its tendon emerges it is overlapped above and below by the gemelli; near its attachment the gemelli pass anterior to the tendon and form a groove in which it lies.

Nerve supply: nerve to the obturator internus, L.5 and S.1.

Gemellus superior (5.91), the smaller, passes from the dorsal aspect of the ischial spine to blend with the upper part of the obturator tendon, attached with it to the medial aspect of the greater trochanter. It is sometimes absent.

Nerve supply: nerve to the obturator internus, L.5 and S.1.

Gemellus inferior (5.91) is attached to the ischial tuberosity below the groove for the obturator internus tendon, then blends with the tendon's lower part and is attached with it to the medial aspect of the greater trochanter.

Nerve supply: nerve to the quadratus femoris, L.5 and S.1.

Action. Obturator internus and the gemelli rotate the extended thigh laterally but they abduct it when it is flexed.

Quadratus femoris (5.91) is a flat, quadrilateral muscle between the gemellus inferior and the proximal margin of adductor magnus, separated from the latter by the transverse branch of the medial circumflex femoral artery. From the upper external aspect of the ischial tuberosity it passes to the low rounded quadrate tubercle a little above the middle of the trochanteric crest and to the bone a short distance distal to it. It passes transversely, posterior to the hip joint and femoral neck, but separated by the tendon of obturator externus and the ascending branch of the medial circumflex femoral artery. A bursa often exists between the muscle and the lesser trochanter. The muscle may be absent.

Nerve supply: nerve to the quadratus femoris, L.5 and S.1. The arterial supply is chiefly from the medial circumflex and first perforating arteries (Leborgne et al 1973).

Action: lateral rotation of the femur.

Obturator externus (5.94), flat and triangular, covers the external aspect of the anterior pelvic wall. Its fibres converge and ascend posterolaterally to a tendon crossing behind the femoral neck and capsule of the hip joint to end in the trochanteric fossa. (The whole muscle and its tendon *spiral* towards their lateral attachment.) It is attached to the external surface of the obturator membrane (medial two-thirds), the adjacent bone of the pubic and ischial rami and the lip of their *pelvic* aspects between the foraminal margin and the obturator membrane (p. 643). A bursa, communicating with the hip joint, may exist between the tendon, articular capsule and femoral neck. The obturator vessels are between the obturator muscle and the membrane; the obturator nerve's anterior branch enters the thigh in front of the muscle; its posterior branch enters by piercing it (5.94).

Nerve supply.: the posterior branch of the obturator nerve, L.3 and 4.

Action. The muscle is a lateral rotator. The short muscles around the hip joint—pectineus, piriformis, obturators, gemelli and quadratus femoris—are possibly more postural than prime movers. They may act as adjustable ligaments for all positions, but, being largely inaccessible to direct observation, views about their actions are necessarily speculative.

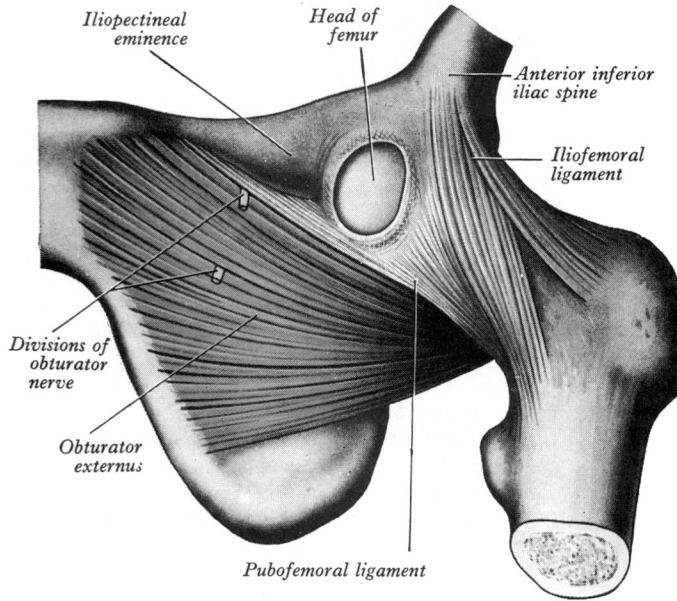

Iliopectineal eminence

Head of femur

Anterior inferior iliac spine

Iliofemoral ligament

Divisions of obturator nerve

Obturator externus

Pubofemoral ligament

5.94 The left obturator externus: antero-inferior aspect. The bursa of psoas major tendon, which in this specimen communicated with the synovial cavity of the hip joint, has been opened to expose the head of the femur.

(d) Posterior Femoral Muscles

This group includes (5.91) biceps femoris, semitendinosus, and semimembranosus, often familiarly termed the 'hamstrings', which span the hip and knee joints, integrating coxal extension with genuflexion.

Biceps femoris (5.87,89,91,95A,B), the posterolateral 'hamstring', has two proximal attachments: a *long head*, attached to an *inferomedial* impression on the proximal area of the ischial tuberosity (3.173) by a tendon common to it and semitendinosus, which blends with the lower part of the sacrotuberous ligament; and a *short head* attached to the lateral lip of the linea aspera between adductor magnus and vastus lateralis and extending proximally almost to gluteus maximus and distally along the lateral supracondylar line to within 5 cm of the lateral femoral condyle and also to the lateral intermuscular septum. Fibres of the long head form a fusiform belly descending laterally across the sciatic nerve to an aponeurosis covering the muscle's posterior surface and receiving fibres of the short head on its anterior aspect; this mass gradually narrows into a tendon (lateral hamstring), most of which splits round the fibular collateral ligament and is attached to the fibular head; the rest splits into three laminae, an intermediate fusing with fibular collateral ligament, the others passing superficial and deep to it to the lateral *tibial* condyle (Sneath 1955). The common peroneal nerve descends along the tendon's medial border, separating it distally from the lateral head of gastrocnemius. The short head may be absent. Additional slips may arise from the ischial tuberosity, linea aspera, or medial supracondylar line.

Nerve supply: sciatic nerve, long head through its tibial part, short head through the common peroneal, L.5, S.1 and 2, which accords with their derivation from both flexor and extensor musculature (cf. the nerve supply of semimembranosus).

Semitendinosus (5.89,91,95A,B), remarkable for its lengthy tendon, is posteromedial and is attached proximally to an inferomedial impression on the upper area of the ischial tuberosity by a tendon shared with the long head of biceps femoris and to an aponeurosis connecting their adjacent surfaces for about 7.5 cm. Its belly is fusiform and ends a little distal to mid-thigh in a long, round tendon, which runs on the surface of semimembranosus and then curves around the medial tibial condyle, skirts the tibial collateral ligament, separated from it by a bursa, to be attached proximally on the medial metaphysial tibial surface behind the attachment of sartorius, distal to that of gracilis. Terminally it is united with the tendon of gracilis, with a prolongation to the adjacent deep fascia. A tendinous interruption usually exists near its midpoint; it may receive a muscular slip from the long head of biceps femoris.

Nerve supply: sciatic nerve through its tibial division, L.5, S.1 and 2 (cf. the nerve supply of semimembranosus).

Semimembranosus (5.89,91,95A,B), so termed from the flat, membranous form of its proximal attachment, is also posteromedial, attached by a tendon to a *superolateral* impression on the ischial tuberosity (3.173). Its fibres are partly interwoven with those of biceps femoris and semitendinosus and its tendon also receives, from the ischial tuberosity and ramus, two fibrous expansions flanking the adductor magnus. This complex proximal tendon expands into an aponeurosis, descending anterior to semitendinosus and the long head of biceps; from this, muscular fibres start and converge to a second aponeurosis on the muscle's distal posterior aspect which tapers into the tendon of the distal attachment. The tendon divides at knee level into five components, the chief being attached to a *tuberculum tendinis* on the posterior aspect of the medial tibial condyle (Cave & Porteous 1958); other slips are attached to the medial tibial margin behind the tibial collateral ligament, to the fascia over the popliteus, as a cord-like tendon to the inferior lip and adjacent groove behind the medial tibial condyle deep to this ligament and a variable fasciculus ascending laterally to the femoral intercondylar line and lateral femoral condyle (oblique popliteal ligament of knee joint). The muscle overlaps the popliteal vessels and is partly overlapped by semitendinosus throughout its extent (5.91). (Distal tendons of semitendinosus and semimembranosus form the medial 'hamstring'.)

Semimembranosus varies much in size and may be absent. It may be double, arising mainly from the sacrotuberous ligament. Slips to the femur or adductor magnus may occur.

Nerve supply: sciatic nerve, tibial division, L.**5**, S.**1** and **2**. Intramuscular distributions of nerves supplying posterior femoral muscles have been detailed by Himstedt et al (**1974**).

Actions. Posterior femoral muscles, acting from above, flex the knee; from below they extend the hip; they pull the trunk upright against gravity, the biceps being the most active. With the knee semi-flexed, biceps femoris is a lateral rotator (of leg on thigh at the knee). Semimembranosus and semitendinosus are medial rotators. The biceps also laterally rotates the *thigh* when the hip and knee are extended, semimembranosus and semitendinosus being then medial rotators. Like quadriceps femoris, the adductors and gluteus maximus, the flexors are quiescent in symmetrical standing; but any action which carries the relevant centre of gravity in front of the transverse coxal axis, e.g. forward reaching, swaying, or bending, evokes an immediate strong contraction in them.

(Gluteus maximus, however, only contracts when powerful extension at the hip joint is required.)

In flexion of the knee against resistance the tendon of biceps is palpable lateral to popliteal fossa, medial to which the tendons of gracilis and semitendinosus are prominent; between them (and also by deep pressure from a 'pincer' grip beyond their margins) semimembranosus can be felt, although less distinctly.

Evidence suggests that semimembranosus, semitendinosus and biceps femoris, though crossing hip *and* knee joints, may produce movement at *one* joint without resistance to antagonists at the other (Markee et al 1955). Each, however, usually contracts as a whole and consequent movement at the hip or knee is determined by fixation by other muscles at one articulation or the other.

Applied Anatomy. Contracture of the flexor tendons is a complication of genual disease, causing flexion and partial posterior dislocation of the tibia, with slight lateral rotation probably due to the biceps femoris. When flexor tendons require surgical manipulation the common peroneal nerve's relation to the medial border of the biceps tendon must be remembered. Relative lengths of flexors vary; sometimes they are so short that they limit flexion at the hips when the knees are extended. Stooping is then largely effected by flexion of the vertebral column or by squatting.

3. Muscles of the Leg

These comprise anterior extensors and posterior flexors, and the lateral peronei, derived from extensors.

(a) Anterior Crural Muscles

This group includes (**5.97**): a common digital extensor, an extensor of hallux and two dorsiflexors of the foot, tibialis anterior and peroneus tertius.

Fascia cruris, the leg's deep fascia, is continuous with the fascia lata and attached proximally to the patella, ligamentum patellae, tibial tubercle and condyles and the fibular head. Posteriorly it becomes the popliteal fascia, strengthened here by transverse fibres and perforated by the small saphenous vein. It receives lateral expansions from the tendon of biceps and multiple medial expansions from tendons of sartorius, gracilis, semitendinosus and semimembranosus. It blends with the periosteum on subcutaneous surfaces of the tibia and fibular head and its malleolus; it is continuous distally with the extensor and flexor retinacula (p. 651). Thick and dense in the proximal and anterior regions, its deep surface is a partial attachment of tibialis anterior and the extensor digitorum longus; it is thinner over gastrocnemius and soleus. Laterally it is continuous with the *anterior* and *posterior crural intermuscular septa*, attached respectively to the anterior and posterior fibular borders. Over flexors and extensors the fascia has several slender extensions enclosing muscles: a *deep transverse fascia of the leg* (**5.99**) passes between

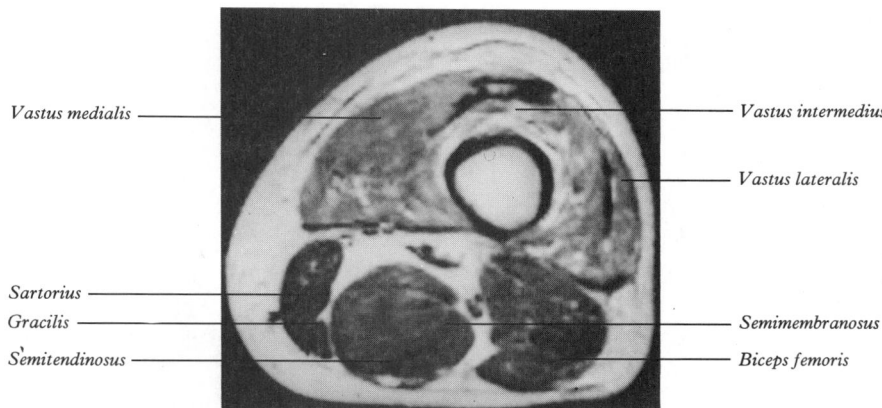

Vastus medialis — Vastus intermedius

— Vastus lateralis

Sartorius —
Gracilis — — Semimembranosus
Semitendinosus — — Biceps femoris

5.95A Transverse magnetic resonance imaging scan of the right thigh, proximal to the section illustrated in **5.95**B: proximal (superior) aspect. (Supplied by Shaun Gallagher, Guy's Hospital; photography by Sarah Smith.)

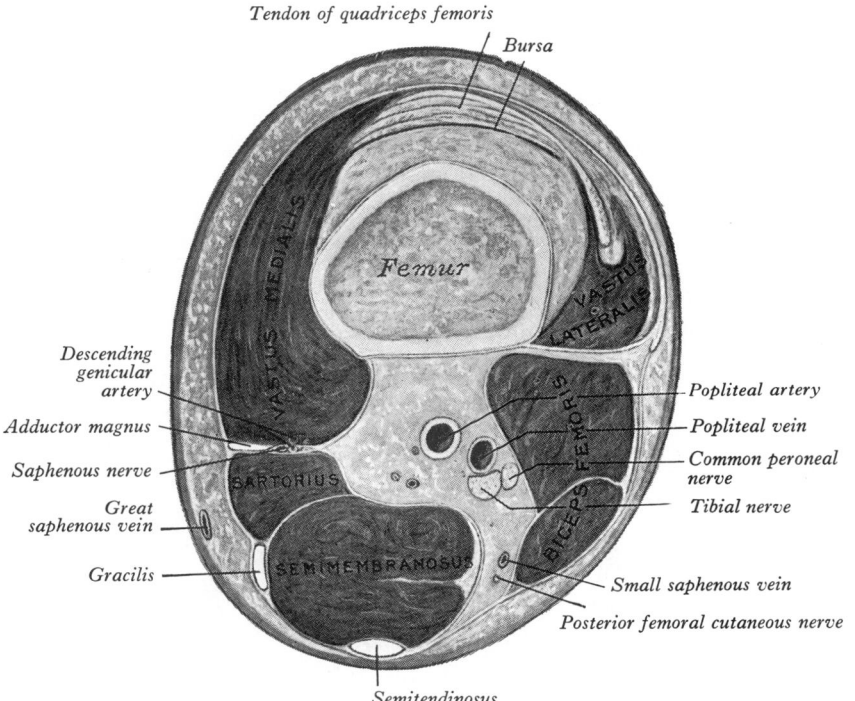

Tendon of quadriceps femoris
Bursa

Descending genicular artery

Adductor magnus
Saphenous nerve

Great saphenous vein

Gracilis

Femur

VASTUS MEDIALIS

VASTUS LATERALIS

BICEPS FEMORIS

SARTORIUS

SEMIMEMBRANOSUS

Popliteal artery
Popliteal vein
Common peroneal nerve
Tibial nerve

Small saphenous vein
Posterior femoral cutaneous nerve

Semitendinosus

5.95B Transverse section through the right thigh, about 4 cm superior to the adductor tubercle of the femur: proximal (superior) aspect.

the superficial and deep muscles in the calf (posterior compartment). All such named sheets are united together and to bone as a connective tissue 'skeleton'.

Tibialis anterior (**5.96**A,B,**97**) arises from the tibial lateral condyle and proximal half or two-thirds of the tibial shaft's lateral surface, adjoining the anterior surface of the interosseous membrane, fascia cruris, and intermuscular septum between the muscle and extensor digitorum longus. It descends vertically to a tendon on its anterior surface in the distal third of the leg; this traverses the medial compartments in the superior and inferior extensor retinacula, inclines medially and is attached to the medial and inferior surfaces of the medial cuneiform and adjacent part of the first metatarsal base. The muscle overlaps the anterior tibial vessels and the deep peroneal nerve proximally in the leg. Further attachments to the talus, the first metatarsal head and the proximal phalangeal base of hallux have been recorded.

Action: dorsiflexor at the talocrural joint and invertor of the foot, most active when both are combined, as in walking. (It is probably operative in the pedal pronation/supination cycle, see p. 657). It elevates the first metatarsal base and medial cuneiform and rotates their dorsal aspects laterally. Lateral to the anterior

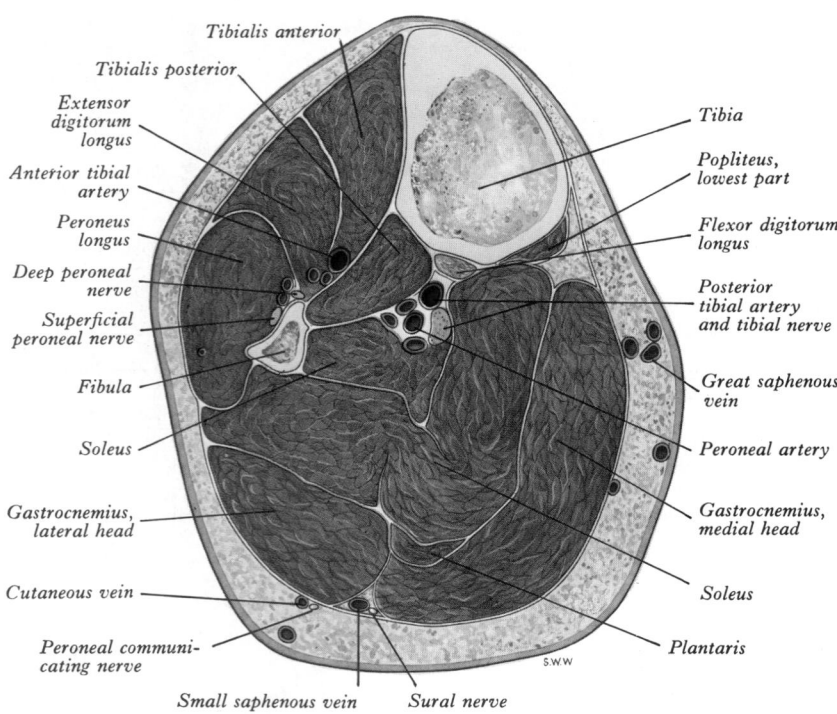

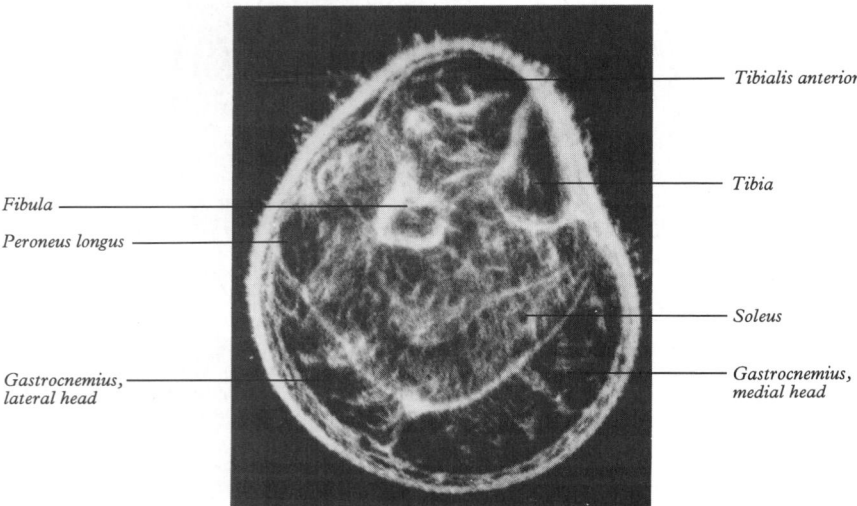

5.96A Transverse section through the right leg, about 10 cm distal to the knee joint: distal aspect.

5.96B Transverse ultrasonic scan of the right leg of an adult man, at a level similar to that of the section illustrated in **5.96A**: distal aspect. (Supplied by H-D Rott and Siemens, Erlangen; photography by Sarah Smith.)

border of the tibia its tendon is visible descending to the foot's medial side. In standing the muscle is usually quiescent, the line of body weight passing anterior to the talocrural joints. It can act from below to assist in balancing the body by flexing the leg forwards at the talocrural joint. It is said to be active in any activity which raises the medial longitudinal arch. Electromyography (under somewhat artificial conditions) suggests that activity in this and other 'arch-supporting' muscles is minimal in standing, but confirms that tibialis anterior contracts powerfully to increase the arch in take-off, 'toeing' in walking, etc.

Extensor hallucis longus (5.97, 99), between and partly deep to tibialis anterior and extensor digitorum longus, is attached to the middle half of the medial fibular surface, medial to the extensor digitorum longus and also to the adjacent interosseous membrane. Anterior tibial vessels and the deep peroneal nerve separate it from tibialis anterior. Its fibres descend into a tendon forming its anterior border and passing deep to the superior and through the inferior extensor retinacula, crosses medial to the anterior

tibial vessels near the talocrural joint and is attached to the dorsal aspect of the distal phalangeal base in the hallux. At the metatarsophalangeal articulation a slip from each side of the tendon covers the joint's dorsal aspect; an expansion from its medial side to the proximal phalangeal base is also usually present. The muscle is sometimes united with extensor digitorum longus and it may send a slip to the second toe. The tendon is identifiable lateral to that of tibialis anterior.

Actions: extension of the hallux, dorsiflexion of the foot. With the hallux actively extended, little force is required to overcome extension of the *distal* phalanx but much is needed to overcome extension of the *proximal* phalanx.

Extensor digitorum longus (5.96A,B,97,99), a pennate muscle, arises from all aspects of the walls of an osteo-aponeurotic tunnel. It is attached to the lateral tibial condyle, the proximal

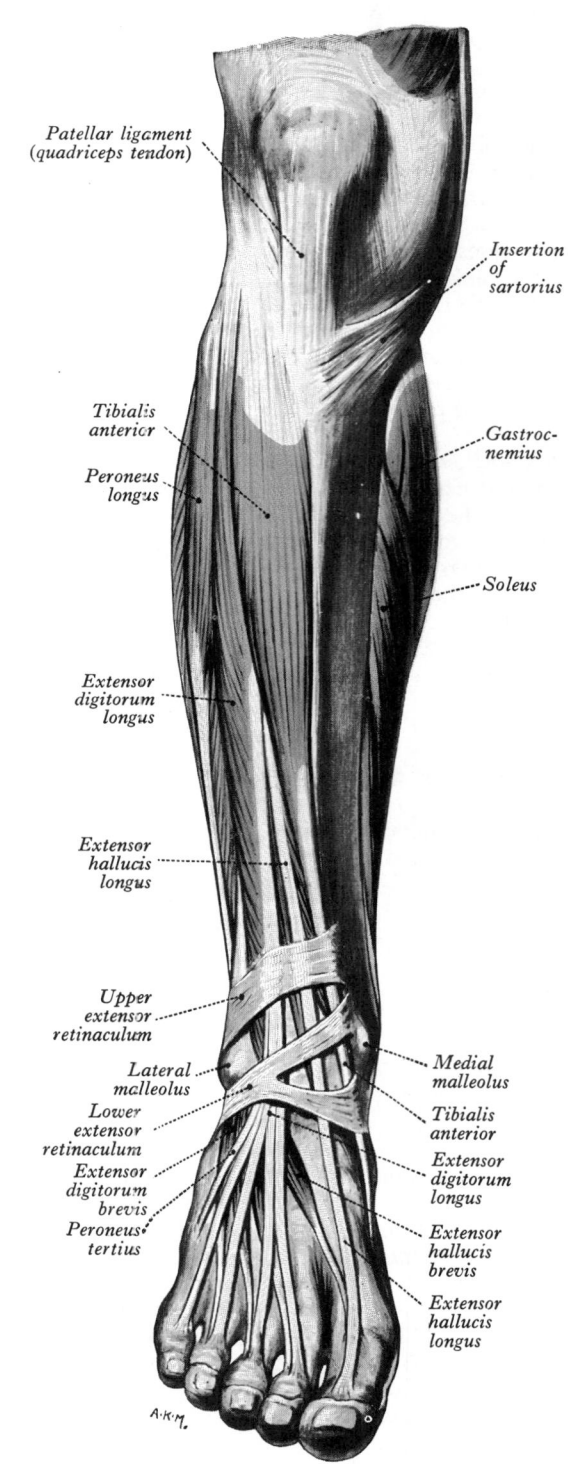

5.97 Muscles on the extensor aspect of the right leg (from Quain's *Anatomy*, 11th edn).

646

three-quarters of the medial fibular surface and the adjacent anterior surface of the crural interosseous membrane, deep surface of fascia cruris, anterior crural intermuscular septum and the septum between it and tibialis anterior. Proximally the anterior tibial vessels and deep peroneal nerve are between the muscle and tibialis anterior, and distally extensor hallucis longus also intervenes. The muscle's tendon passes behind the superior extensor retinaculum and within a loop of the inferior retinaculum (p. 651) with peroneus tertius (5.103). It divides into four slips on the foot's dorsum, attached in a manner that in many respects, in miniature, resembles those of extensor digitorum in the hand (p. 623). Dorsal to the metatarsophalangeal joints the tendons to the second to fourth toes are joined laterally by a tendon of extensor digitorum brevis. The *dorsal digital expansion* thus formed, like that in the fingers, receives contributions from lumbrical and interosseous muscles (pp. 655–66). Narrowing near its proximal interphalangeal joint, the expansion divides into three slips, an intermediate one attached to the base of the middle phalanx, and two collateral slips which reunite dorsal to the middle phalanx and are attached to the base of the distal phalanx. Tendons to the second and fifth toes are sometimes double. Accessory slips may be attached to the metatarsals, including the hallux.

Actions. The muscle extends the toes and dorsiflexes the foot with tibialis anterior and extensor hallucis longus; and with the latter it tautens the plantar aponeurosis (pp. 652, 657).

Peroneus tertius (5.97,104), part of extensor digitorum longus, as if it were its fifth tendon, is attached proximally to the distal third or more of the medial fibular surface, adjoining the interosseous membrane and anterior crural intermuscular septum. Its tendon passes behind the superior and in a loop of the inferior extensor retinaculum with extensor digitorum longus (5.103). It is attached to the dorsomedial surface of the fifth metatarsal base, a thin expansion usually extending along the medial border of its shaft. It is sometimes missing (in 4.4% of dissections, according to Werneck (1957) in a most extensive study of this very variable muscle).

Actions: dorsiflexion of the foot, as part of extensor digitorum longus, and perhaps eversion.

Nerve supply. All anterior crural muscles are supplied by the deep peroneal nerve. Tibialis anterior is innervated by L.4 and 5, the others by L.5 and S.1.

(b) Lateral Crural Muscles

The lateral crural muscles are peroneus longus and brevis, evertors of the foot (5.96–101).

Peroneus longus (5.96,100,101), the more superficial, is attached to the head and adjoining proximal two-thirds of the lateral surface of the fibula, to fascia cruris, anterior and posterior crural intermuscular septa and occasionally by a few fibres to the lateral tibial condyle. Between attachments to the fibular head and shaft is a gap for the common peroneal nerve. The muscle belly ends in a long tendon, descending *behind* the lateral malleolus in a groove shared with the tendon of peroneus brevis, behind which it lies, both covered by the superior peroneal retinaculum in a common synovial sheath (5.104). The tendon runs obliquely forwards lateral to calcaneus, below the peroneal trochlea and tendon of peroneus brevis and beneath the inferior peroneal retinaculum (p. 651); crossing the lateral aspect of the cuboid, it then curves and enters a plantar groove on the cuboid which is closed by the long plantar ligament (4.90). It crosses the sole obliquely to be attached by two slips to the lateral sides of the first metatarsal base and the adjacent medial cuneiform; a third slip may reach the second metatarsal base. The tendon changes direction: (*a*) distal to lateral malleolus, (*b*) on the cuboid bone, and is thickened at both sites with, at the second, a sesamoid fibrocartilage (sometimes ossified) usually developed in it. A second synovial sheath invests the tendon in the sole.

Tendinous slips to the bases of the third to fifth metatarsal bone may occur, or to adductor hallucis. Fusion of peroneus longus and brevis is a rare variation.

Nerve supply: superficial peroneal nerve, L.5, S.1 and 2.

Actions. Doubtless peroneus longus can evert and plantar-flex

the foot, and perhaps act on the leg from the distal attachment. Its oblique course across the sole also favours support of longitudinal and transverse arches. How do these powers contribute to actual movements? With the foot off the ground, overall pedal eversion is obvious. How far this helps to maintain plantigrade contact is less certain; electromyographic evidence indicates little or no activity in standing. But peroneus longus and brevis are strongly active in maintaining a concave foot in take-off and tip-toeing. In lateral swaying peronei contract on that side; but their involvement in the postural activity between the foot and leg remains uncertain. (Pedal movements during maintenance of plantigrade contact, with the leg and trunk in different postures on uneven surfaces and during walking, running, jumping, thrusting etc, involves talocrural activity while inversion and eversion are modified and compounded with pronative or supinative actions, see pp. 656–659.)

Peroneus brevis (5.100,101) is attached to the distal two-thirds of the lateral fibular surface, anterior to peroneus longus, and to the anterior and posterior crural intermuscular septa. It descends vertically to a tendon passing behind the lateral malleolus anterior to that of peroneus longus, both deep to the superior peroneal retinaculum in one synovial sheath (p.651). It passes lateral to the calcaneus, above the peroneal trochlea and the tendon of peroneus longus, to be attached to a lateral tubercle on the fifth metatarsal base.

Lateral to the calcaneus the tendons of peronei longus and brevis occupy *separate* osseo-aponeurotic canals between calcaneus and the inferior peroneal retinaculum, each in a separate prolongation of their proximally common synovial sheath (5.104).

Nerve supply: superficial peroneal nerve, L.5, S.1 and 2.

Action. Peroneus brevis may limit inversion, relieving the ligaments tightened by this movement (the lateral part of interosseous talocalcanean, lateral talocalcanean and calcaneofibular). It aids eversion and may help to steady the leg on the foot (cf. peroneus longus, above).

(c) Posterior Crural Muscles

The muscles in the calf form superficial and deep groups, separated by the *deep transverse fascia* (pp. 645, 649). Superficial is a powerful muscular mass forming most of the calf, flexing the knee and plantar-flexing the foot. Its size is a characteristic feature of human musculature, directly related to upright stance and the mode of progression.

SUPERFICIAL GROUP OF MUSCLES

This group (5.100) comprises gastrocnemius and plantaris, acting on the knee and ankle, the soleus only on the latter.

Gastrocnemius (5.96,98,100), most superficial of the muscles, forms the 'belly' of the calf and is attached by two heads connected to the femoral condyles by strong, flat tendons. The medial and larger head descends from a depression on the upper and posterior part of the medial condyle behind the adductor tubercle and a slightly raised area on the femoral popliteal surface proximal to the condyle. The lateral head is attached to a recognizable area on the lateral surface of the lateral condyle and the adjoining supracondylar line. Both are also attached to subjacent areas of the articular capsule. Each spreads into a tendinous expansion and from their anterior surfaces muscular fibres descend, those of the medial head further distally. The two muscular masses remain separate to their attachment to a broad aponeurosis on the muscle's anterior surface. This gradually narrows and thickens, uniting with the tendon of soleus to form the *tendo calcaneus* (p. 648). The lateral head or whole muscle may be absent; more frequent is a third head from the femoral popliteal surface.

Relations. Fascia cruris separates the muscle's *superficial surface* from the small saphenous vein and the peroneal communicating and sural nerves; the common peroneal nerve crosses the lateral head, partly deep to biceps femoris. The *deep surface* is posterior to the oblique popliteal ligament, popliteus, soleus, plantaris, popliteal vessels and tibial nerve. A bursa, sometimes

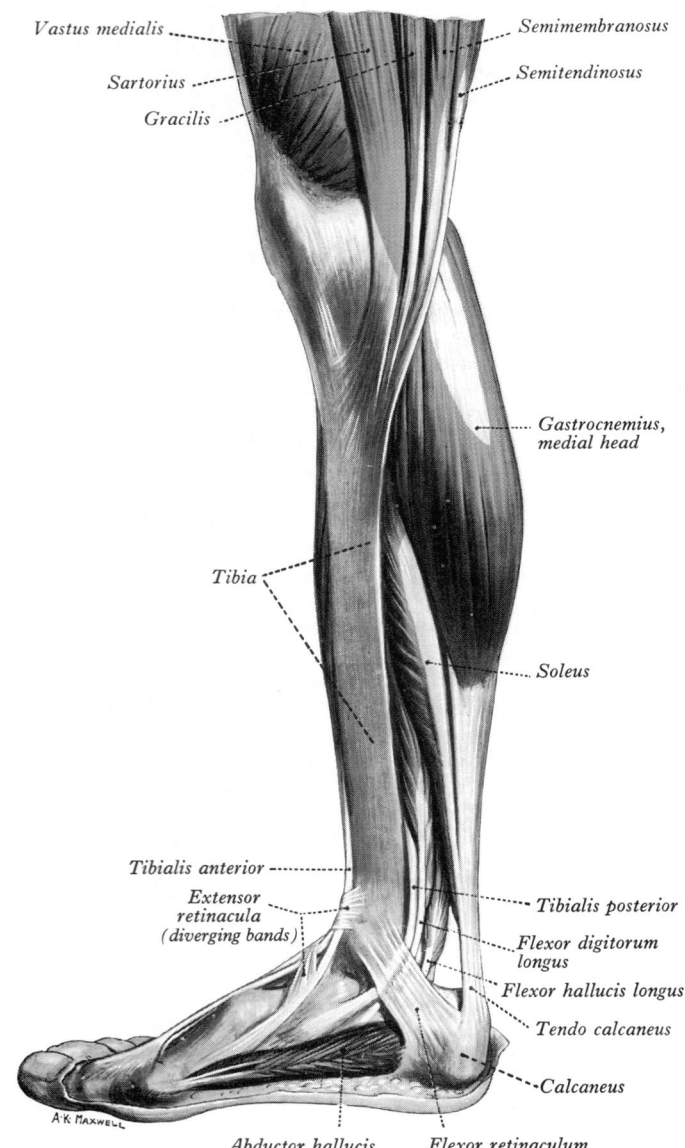

Vastus medialis

Sartorius

Gracilis

Semimembranosus

Semitendinosus

Gastrocnemius, medial head

Tibia

Soleus

Tibialis anterior

Extensor retinacula (diverging bands)

Tibialis posterior

Flexor digitorum longus

Flexor hallucis longus

Tendo calcaneus

Calcaneus

Abductor hallucis

Flexor retinaculum

A.K. MAXWELL

5.98 The muscles of the right leg: medial aspect.

Deep peroneal nerve   *Anterior tibial artery*

Tibialis anterior

Extensor digitorum longus

Extensor hallucis longus

Superficial peroneal nerve

Peroneus tertius

Perforating branch of peroneal artery

Saphenous nerve

Great saphenous vein

Tibia

Fibula

Peroneal artery

Tibialis posterior

Flexor digitorum longus

FLEX. HALL. LONG.

PERON. BREV.

Peroneus longus

Posterior tibial artery

Tibial nerve

Plantaris

Sural nerve

Small saphenous vein

Soleus

Tendo calcaneus

5.99 Transverse section through the right leg, about 6 cm superior to the tip of the medial malleolus: proximal (superior) aspect.

connected to the knee joint, is anterior to its medial tendinous head; the lateral tendon sometimes contains a sesamoid fibrocartilage or bone where it plays over the lateral femoral condyle, and occasionally the medial head has one.

Nerve supply: tibial nerve, S.1 and 2.

Soleus (5.96,98), a broad, flat muscle placed anterior to gastrocnemius, is attached to the posterior aspect of the head and proximal quarter of the shaft of the fibula, the tibia's soleal line and the middle third of its medial border, and to a fibrous arch between the tibia and fibula over the popliteal vessels and tibial nerve. This attachment is a wide aponeurosis; most soleal muscular fibres pass obliquely from the posterior surface of the aponeurosis to the main tendon, which is on the muscle's *posterior* surface; other fibres also start from its *anterior* (deep) aponeurotic surface, being short and bipennate and these reach a narrow, central deep *intramuscular* tendon merging distally with the main one (5.96). The latter, gradually compacted and narrowed, joins the tendon of gastrocnemius to form the *tendo calcaneus*.

Relations. *Superficial* are gastrocnemius and plantaris; *deep* are flexor digitorum longus, flexor hallucis longus, tibialis posterior and the posterior tibial vessels and tibial nerve, all separated from soleus by the deep transverse fascia.

Nerve supply: two branches from the tibial nerve, S.1 and **2**. Intramuscular distributions in gastrocnemius, soleus and plantaris have been detailed by Schumacher et al (1973).

Gastrocnemius and soleus form a tripartite mass sharing the tendo calcaneus, and hence sometimes termed *triceps surae*.

The tendo calcaneus (5.100), the largest and strongest human tendon, is about 15 cm long, beginning near mid-level in the leg, but its anterior surface receives muscle fibres from the soleus almost to its distal end. It gradually becomes rounder to about 4 cm proximal to the calcaneus, expanding then to be attached to the posterior calcaneal surface at its mid-level, a bursa separating it from the bone's proximal area. Tendon fibres spiral through 90° in descending, when the medial fibres become most posterior (White 1943). This permits some elongation and elastic recoil and hence storing of energy, which can be released at an appropriate phase of locomotion. Alexander & Vernon (1975) have studied elastic recoil (in kangaroos), and Alexander has made a mathematical analysis of the contribution of elasticity in tendo calcaneus to bipedal gait.

Actions. Muscles of the calf are major plantar flexors and gastrocnemius also flexes the knee; they possess great power, being usually large.

Gastrocnemius propels in walking, running and leaping. Soleus is said to steady the leg on the foot in standing; this postural function is emphasized but such a rigid distinction between the muscles is probably an over-simplification. In standing, the centre of gravity is on a vertical passing anterior to the talocrural joints (in this erect position they are loose-packed) and a strong force is required behind the joint to maintain posture. Electromyography demonstrates that in symmetrical standing the soleus exhibits continuous activity, whereas the gastrocnemeus is active only intermittently (Joseph et al 1955, Joseph 1960). Phasic activity of the triceps surae in walking has not been satisfactorily analysed into the relative roles of soleus and gastrocnemius.

Plantaris (5.100) is attached proximally on the distal part of the lateral supracondylar line and to the oblique popliteal ligament. Its small fusiform belly, from 7–10 cm long, ends in a long slender tendon, which crosses obliquely between gastrocnemius and soleus to descend along the medial border of tendo calcaneus and gradually fuses with it. The muscle is sometimes double. Occasionally, its tendon merges with flexor retinaculum (p. 645) or the crural fascia. Daseler & Anson (1943) found it absent in 160 out of 1545 legs.

Nerve supply: tibial nerve, S.1 and 2.

Actions. Plantaris is the rudiment of a large muscle, its tendon attached distally to the plantar aponeurosis in some mammals. It presumably acts with gastrocnemeus but details are lacking.

THE DEEP CRURAL GROUP

The deep crural flexors (5.98,101) include one, popliteus, acting solely on the knee joint; the others, flexor hallucis longus, flexor

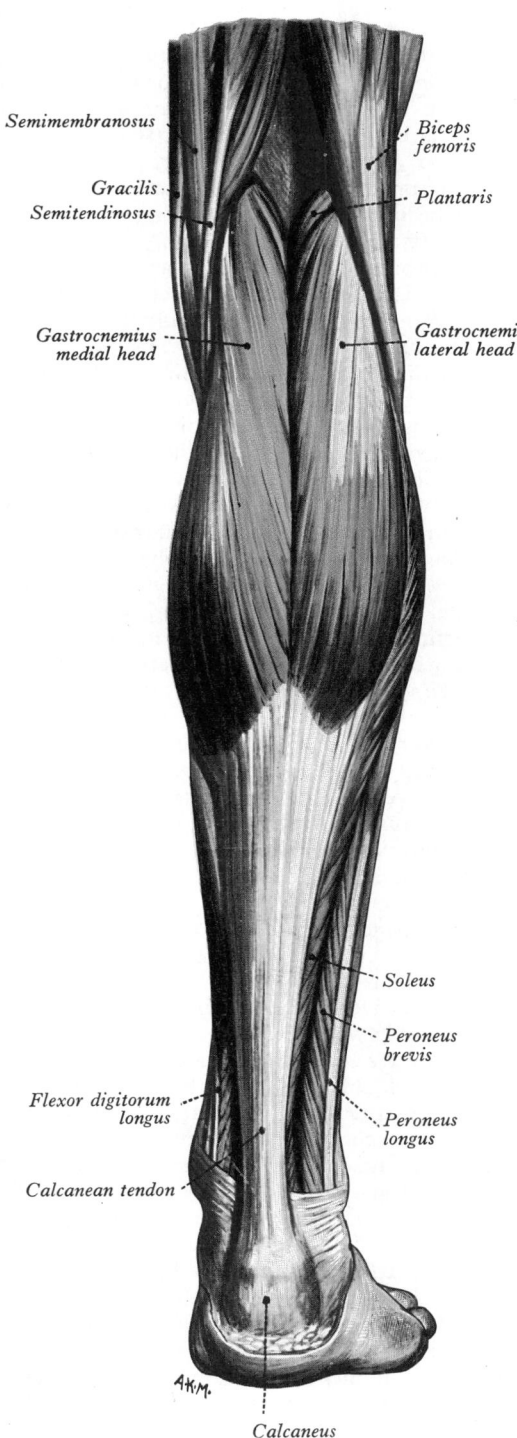

Semimembranosus
Gracilis
Semitendinosus
Biceps femoris
Plantaris
Gastrocnemius medial head
Gastrocnemius lateral head
Soleus
Peroneus brevis
Flexor digitorum longus
Peroneus longus
Calcanean tendon

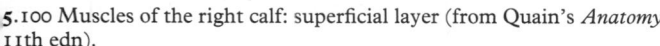

Calcaneus

5.100 Muscles of the right calf: superficial layer (from Quain's *Anatomy*, 11th edn).

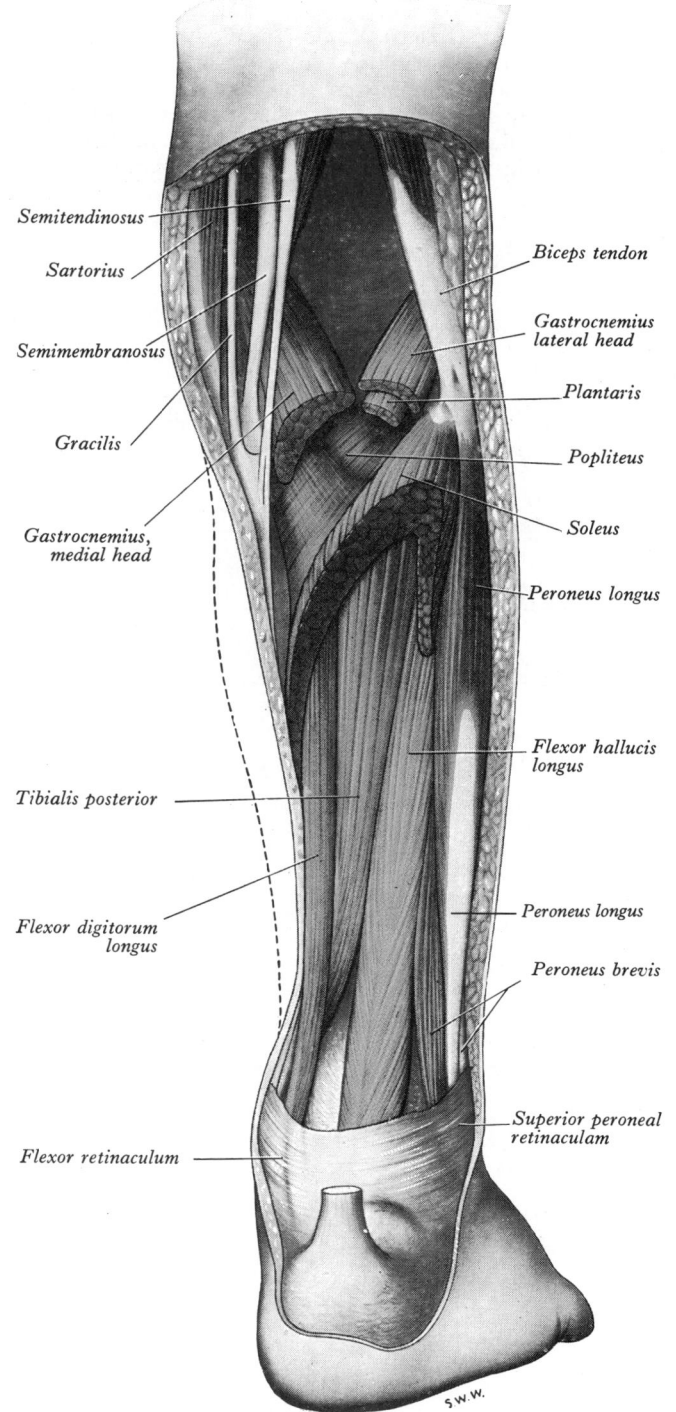

Semitendinosus
Sartorius
Semimembranosus
Gracilis
Gastrocnemius, medial head
Tibialis posterior
Flexor digitorum longus
Flexor retinaculum
Biceps tendon
Gastrocnemius lateral head
Plantaris
Popliteus
Soleus
Peroneus longus
Flexor hallucis longus
Peroneus longus
Peroneus brevis
Superior peroneal retinaculum

5.101 The right posterior crural muscles, deep group, in a child aged eight years.

digitorum longus and tibialis posterior, act on the talocrural and multiple other pedal joints.

Deep transverse fascia, a fibrous stratum between the superficial and deep posterior cural muscles, extends transversely between the medial tibial and posterior fibular borders. Proximally thick and dense, it is attached to the tibial soleal ridge and to the fibula inferomedial to the soleus. Between the bones it is continuous with fascia covering the popliteus and connected to the tendon of semi-membranosus. It is thin at intermediate levels but again is thick distally where it covers the tendons behind the malleoli and is continuous with the flexor and superior peroneal retinacula (p. 651).

Popliteus (5.101), flat and triangular, forms the lower part of the 'floor' of the popliteal fossa. Its lateral, larger part is attached by a strong tendon, about 2.5 cm long, to a depression at the anterior end of a groove on the lateral aspect of the lateral femoral condyle; its medial fibres are attached to the arcuate popliteal ligament (p. 528) where it blends with the fibrous capsule adjacent to the lateral meniscus, to the periphery of which some fibres are also attached. The muscle is attached distally to the medial two-thirds of a triangular area proximal to the soleal line on the posterior tibial surface and to a tendinous expansion covering it. An additional part may start from a sesamoid bone in the lateral head of gastrocnemius.

Relations. The popliteal tendon is intracapsular and overlapped by the fibular collateral ligament and tendon of biceps femoris (4.76). Invested on its deep aspect by synovial membrane, it grooves posteriorly the border of the lateral meniscus and adjoining tibia, emerging inferior to the posterior band of the arcuate ligament (4.72). In the popliteal fossa it is covered by a strong layer of fascia mostly derived as an expansion of the tendon of semimembranosus.

649

Nerve supply: tibial nerve, L.4 and 5 and S.1.

Actions. Popliteus rotates the tibia medially or, with tibia fixed, rotates the femur laterally. It is usually said to 'unlock' the joint when flexion starts from full extension (p. 531); electromyography supports this view. Its connection with the arcuate popliteal ligaments, fibrous capsule and lateral meniscus suggests that it may retract the latter's posterior cornu during

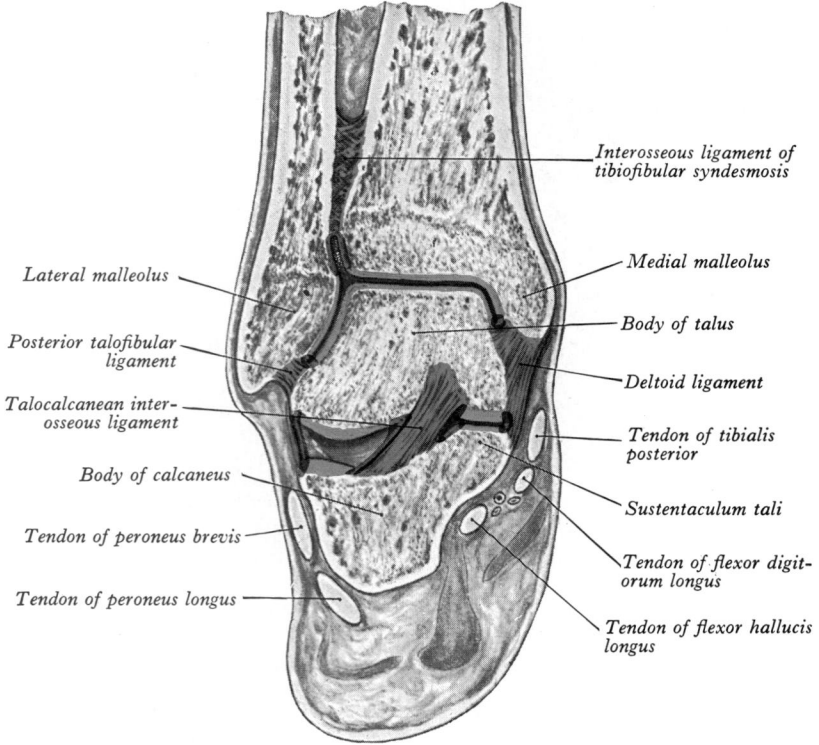

Interosseous ligament of tibiofibular syndesmosis

Medial malleolus

Lateral malleolus

Body of talus

Posterior talofibular ligament

Deltoid ligament

Talocalcanean inter-osseous ligament

Tendon of tibialis posterior

Body of calcaneus

Sustentaculum tali

Tendon of peroneus brevis

Tendon of flexor digit-orum longus

Tendon of peroneus longus

Tendon of flexor hallucis longus

5.102 Coronal section through the left talocrural, talocalcanean and sub-talar joints. Ligaments are shown in green, articular cartilage in blue and synovial membrane in red.

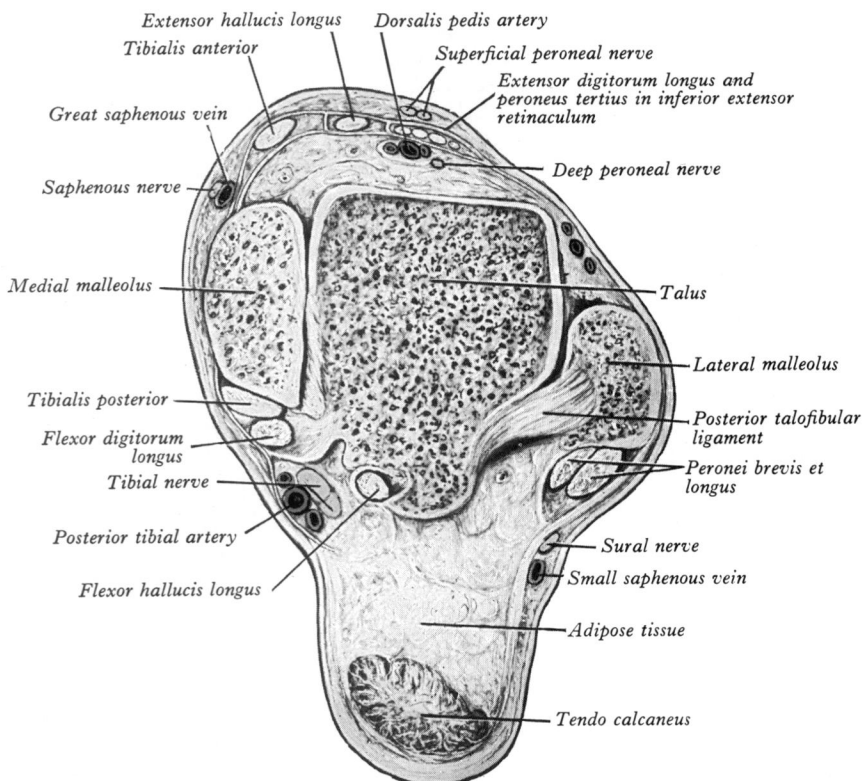

Extensor hallucis longus

Dorsalis pedis artery

Tibialis anterior

Superficial peroneal nerve

Great saphenous vein

Extensor digitorum longus and peroneus tertius in inferior extensor retinaculum

Saphenous nerve

Deep peroneal nerve

Medial malleolus

Talus

Tibialis posterior

Lateral malleolus

Flexor digitorum longus

Posterior talofibular ligament

Tibial nerve

Peronei brevis et longus

Posterior tibial artery

Sural nerve

Flexor hallucis longus

Small saphenous vein

Adipose tissue

Tendo calcaneus

5.103 Horizontal section through the inferior part of the talocrural joint.

lateral rotation of the femur and flexion of the knee to obviate its crushing between the femur and tibia (Last 1950). The muscle is markedly active in crouching, perhaps to aid the posterior cruciate ligament in limiting forward translation of the femur.

Flexor hallucis longus (5.99,101) is attached to the distal two-thirds of the posterior fibular surface, the adjacent interosseous membrane and the posterior crural intermuscular septum and also the intervening fascia covering tibialis posterior, which it overlaps. Its fibres descend obliquely to a tendon along almost all its posterior aspect. This tendon grooves the posterior aspect of the tibia's distal end, then successively the posterior talar surface and the *inferior* surface of sustentaculum tali (5.102,103). Grooves on the talus and calcaneus are converted by fibrous bands into a canal, lined by a synovial sheath. Distal to this, in the sole, it crosses the flexor digitorum longus obliquely, curving *superior* to it; at this crossing the long digital flexor receives a fibrous slip from flexor hallucis longus, whose tendon then continues and crosses the lateral part of flexor hallucis brevis to lie between the sesamoid bones below the first metatarsal head. It runs along the plantar aspect of the hallux in an osseo-aponeurotic tunnel to the plantar aspect of the distal phalangeal base. The tendon is retained over the flexor hallucis brevis by diverging stems of the distal band of the medial intermuscular septum (p. 545). The connecting slip to flexor digitorum longus varies in size, usually continuing into the tendons for the second and third toes, sometimes only the second, occasionally the tendons of the second to fourth toes.

Relations. *Superficial* are soleus and tendo calcaneus, separated by the deep transverse fascia; *deep* are the fibula, tibialis posterior, peroneal vessels, the distal part of the interosseous membrane and talocrural joint; *lateral* are the peronei; *medial* are the tibialis posterior, posterior tibial vessels and tibial nerve.

Nerve supply: tibial nerve, S.2 and 3.

Flexor digitorum longus (5.101), medial to the flexor hallucis longus, thin and pointed proximally, is attached to the posterior surface of the tibia medial to tibialis posterior from just distal to the soleal line to within 7 or 8 cm of the tibia's distal end, and also to the fascia covering tibialis posterior. Its fibres end in a tendon along almost its whole posterior surface, which gradually crosses tibialis posterior and passes behind the medial malleolus in a groove shared with tibialis posterior, but separated from it by a fibrous septum, each tendon in a separate synovial compartment. The tendon then curves anterolaterally, in contact with the *medial side* of the sustentaculum tali (5.102) and under the flexor retinaculum to enter the sole (5.109), where it crosses *superficial* to the tendon of flexor hallucis longus, receiving a slip from it, and crosses the sole to form the whole long flexor tendon of the fifth toe and contributing to those for the second to fourth toes. It may also send a slip to the tendon of flexor hallucis longus. The long flexor tendons of the second to fourth digits receive accessions from the flexor accessorius and variably from the flexor hallucis longus through the slip mentioned. Long flexor tendons of the lateral four toes are attached to the plantar surfaces of their distal phalangeal bases, each traversing a divarication in a corresponding tendon of the flexor digitorum brevis at the level of the proximal phalangeal base (cf. the arrangement in the hand, pp. 618–620).

The lateral head of the flexor accessorius may be attached to the lateral border of the tendon of flexor digitorum longus (p. 654).

Relations. In the leg the muscle's *superficial surface* is related to the deep transverse fascia, separating it from the soleus and distally the posterior tibial vessels and tibial nerve; its *deep surface* is related to the tibia and tibialis posterior. In the foot it is covered by abductor hallucis and flexor digitorum brevis and, as noted, crosses inferior to the flexor hallucis longus.

Nerve supply: tibial nerve, S.2 and 3.

Actions. *With the foot off the ground* both long flexors flex the phalanges, acting primarily on the distal phalanges; they are also plantar flexors. *With the foot on the ground and under the load* they act with the pedal intrinsic muscles and (in the case of flexor digitorum longus especially) with the lumbricals and interossei (p. 655) to maintain the pads of the toes in firm contact with the ground, enlarging the weight-bearing areas and helping to stabilize the metatarsal heads, the fulcrum on which the body is propelled. Both long digital flexors are minimally active in standing, apparently contributing little to the longitudinal arch mechanism. During take-off and tip-toe movements, their activity is marked.

Tibialis posterior (5.96,101) is attached between flexor hallucis longus and flexor digitorum longus, but overlapped by both; it is the most deeply placed flexor. Two pointed heads, separated by an angular interval traversed by anterior tibial vessels, form its proximal end. The medial part of the muscle is attached to the interosseous membrane, except its most distal part, and to a lateral area on the posterior tibial surface between the soleal line and the junction of the shaft's middle and distal thirds. The lateral part comes from a medial strip of the posterior fibular surface in its proximal two-thirds. The muscle is also attached to the deep transverse fascia and adjacent intermuscular septa. In the leg's distal quarter its tendon passes deep to that of the flexor digitorum longus, sharing a groove behind the medial malleolus, but both are enclosed in separate synovial sheaths. It passes deep to the flexor retinaculum (p. 651) but superficial to the deltoid ligament (5.102), to enter the foot; at first inferior to the plantar calcaneonavicular ligament, where it contains a sesamoid fibrocartilage, it then divides. Its superficial, larger division, the direct continuation, is attached to the navicular's tuberosity, from which fibres continue to the inferior surface of the medial cuneiform. A tendinous band also passes laterally and proximally to the tip and distal margin of sustentaculum tali. The deeper, lateral division becomes the posterior tendon of the medial limb of flexor hallucis brevis, continuing between this and the navicular and medial cuneiform bones to end on the intermediate cuneiform and bases of the second to fourth metatarsals. The metatarsal slips are variable, that to the fourth being strongest. Slips to the cuboid and lateral cuneiform are described (Lewis 1964).

Relations. *Superficial* are the soleus, separated by the deep transverse fascia, flexor digitorium longus, flexor hallucis longus, posterior tibial vessels, tibial nerve and peroneal vessels; *deep* are the interosseous membrane, tibia, fibula and talocrural joint.

Nerve supply: tibial nerve, L.4 and 5.

Actions. Tibialis posterior is the main invertor and assists powerful plantar flexion. (It also modifies pronative and supinative activities, p. 656 et seq.) Through attachments to the cuneiform bones and metatarsal bases it may contribute to elevation of the longitudinal arch, but it is quiescent in standing, though phasically active in walking; during this, acting with the intrinsic musculature and peronei, it probably controls pronation and the distribution of weight through the metatarsal heads. It is said that with the body supported on one leg, tibialis posterior, acting from its distal end, assists balance by resisting a tendency to lateral swaying. But any act of balancing demands the cooperation of many muscles, including groups acting on the hip joints and vertebral column.

loop's deep aspect a band passes laterally behind the interosseous talocalcanean and cervical ligaments to be attached to sulcus calcanei (Stamm 1931). From the loop's medial end two diverging bands continue. The *proximal* has two layers, a deep one *under* the tendons of extensor hallucis longus and tibialis anterior, but *superficial* to the anterior tibial vessels and deep peroneal nerve, reaching the tibial malleolus; a superficial layer crosses the tendon of extensor hallucis longus and then adheres to the deep layer, sometimes continuing superficial to the tendon of tibialis anterior to the tibia. The *distal band* extends inferomedially as far as the plantar aponeurosis; it is superficial to the tendons of extensor hallucis longus and tibialis anterior, arteria dorsalis pedis and terminal branches of the deep peroneal nerve.

The flexor retinaculum (5.98) is attached anteriorly to the tip of the medial malleolus and is continuous distally with the deep dorsal fascia of the foot; it continues posteriorly to the medial calcaneal process and plantar aponeurosis. Its proximal border is not clearly demarcated from the adjoining deep fascia, the distal border is continuous with the plantar aponeurosis; fibres of abductor hallucis are attached to it. The retinaculum converts grooves on the tibia and calcaneus into canals for tendons and bridges the posterior tibial vessels and tibial nerve as they enter the sole. Disposed in an oblique curve, medial to lateral, are: the tibialis posterior, tendon flexor digitorum longus, posterior tibial vessels, tibial nerve and flexor hallucis longus (5.103).

Peroneal retinacula hold down the tendons of peroneus longus and brevis as they curve lateral to the ankle. The *superior peroneal retinaculum* (5.101) extends from the back of the lateral malleolus to the deep transverse crural fascia and lateral calcaneal surface. The *inferior peroneal retinaculum*, continuous anteriorly with the inferior extensor retinaculum, is attached posteriorly to the lateral calcaneal surface; some fibres fuse with the periosteum on the peroneal trochlea between the tendons of peroneus longus and brevis.

Synovial sheaths in the talocrural region. Tendons crossing the talocrural joint are all deflected from a straight course, held down by the retinacula and hence enclosed in synovial sheaths.

Anteriorly a sheath for tibialis anterior extends from the proximal margin of the superior extensor retinaculum to an interval between the diverging limbs of the inferior (5.105); a sheath for extensor digitorum longus and peroneus tertius and one for extensor hallucis longus start proximal to the malleolar level, the former a little higher. The sheath for extensor hallucis reaches the first metatarsal base, that for extensor digitorum, the fifth metatarsal base. *Medial* to the ankle (5.105) a sheath for tibialis posterior extends from about 4 cm proximal to the malleolus almost to the tendon's attachment to the navicular tuberosity. A sheath for flexor hallucis longus reaches only malleolar level, that for flexor

Talocrural Fasciae and Retinacula

As the tendons of muscles approach or cross the talocrural joint to the foot, they are bound down by localized retinacula of the deep fascia, comparable in form and function with those at or near the carpus (pp. 625–626). They comprise the superior and inferior extensor, flexor, and superior and inferior peroneal retinacula.

The superior extensor retinaculum (5.97,104) binds down the tendons of tibialis anterior, extensor hallucis longus, extensor digitorum longus and peroneus tertius, immediately proximal to the anterior aspect of the talocrural joint; the anterior tibial vessels and deep peroneal nerve also pass under it. Attached laterally to the distal anterior border of the fibula and medially to that of the tibia, it is proximally continuous with the deep crural fascia; connective tissue also connects its distal border to the inferior extensor retinaculum. Only tibialis anterior has a synovial tendon sheath here (5.104).

The inferior extensor retinaculum (5.97,104) is a Y-shaped band anterior to the talocrural joint; its stem (the lateral end) is attached to the upper (superior) surface of calcaneus, anterior to sulcus calcanei. It loops medially round the tendons of peroneus tertius and extensor digitorum longus (5.103); from the

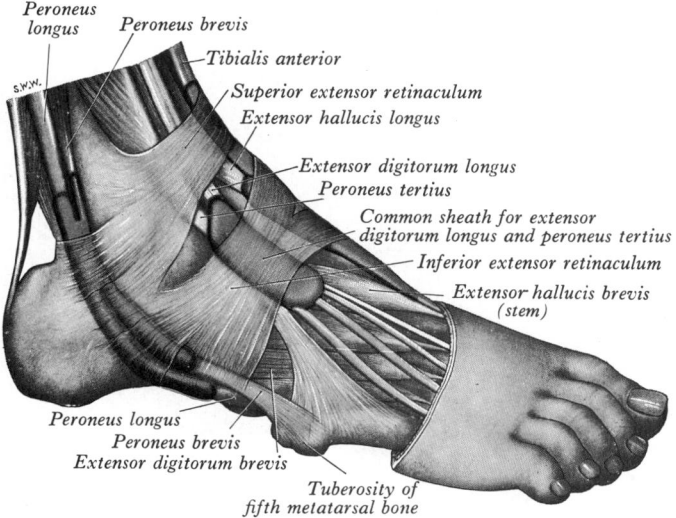

Peroneus longus
Peroneus brevis
Tibialis anterior
Superior extensor retinaculum
Extensor hallucis longus
Extensor digitorum longus
Peroneus tertius
Common sheath for extensor digitorum longus and peroneus tertius
Inferior extensor retinaculum
Extensor hallucis brevis (stem)

Peroneus longus
Peroneus brevis
Extensor digitorum brevis
Tuberosity of fifth metatarsal bone

5.104 The synovial sheath of the tendons at the right ankle: lateral aspect. Note the superior and inferior peroneal retinacula (not labelled).

digitorum longus starting slightly more proximal; the former continues to the first metatarsal base, the latter to the navicular. *Lateral* to the ankle (5.104) a sheath, proximally single but distally double, encloses the peroneus longus and brevis, extending from about 4 cm proximal to the lateral malleolar tip for about the same distance beyond it.

4. Muscles of the Foot

The muscles *intrinsic* or confined to the foot follow a primitive pattern of dorsal extensors and plantar flexors, topographically and functionally associated with the *extrinsic* muscles already considered. Intrinsic extensor musculature is limited but, as in the hand, certain intrinsic flexors are also concerned with extensor activity.

(a) The Dorsal Muscle of the Foot

Deep fascia on the foot's dorsum (*fascia dorsalis pedis*) is thin and continuous proximally with the inferior extensor retinaculum, blending marginally with the plantar aponeurosis; it covers the dorsal tendons.

The **extensor digitorum brevis** (5.97,104), a thin muscle, is attached to the anterior superolateral surface of the calcaneus, anterosuperior to the shallow lateral groove for peroneus brevis, to the interosseous talocalcaneal ligament and the stem of the inferior extensor retinaculum. It slants distomedially across the dorsum to end in four tendons. Its medial part is a variably distinct slip ending in a tendon which crosses the dorsalis pedis artery superficially to reach the dorsal aspect of the proximal phalangeal base of the hallux, hence the term *extensor hallucis brevis*. The other three tendons join the lateral sides of the tendons of extensor digitorum longus for the second to fourth toes. It is subject to much variation by accessory slips (from the talus and navicular), extra tendons, e.g. to the fifth digit, and suppression of tendons. It may be connected to the adjacent dorsal interossei.

Nerve supply: the lateral terminal branch of deep peroneal nerve, S. 1 and 2.

Actions: assists in extension of phalanges in the middle three toes via the tendons of extensor digitorum longus; in the hallux it acts only on the proximal phalanx.

(b) Plantar Fasciae and Muscles

Plantar fasciae, deep and superficial are arranged much as in the hand (5.106A–D), with small differences due to the absence of digital opposition and grip. Fasciae are arranged: to limit the tangential mobility of the skin, to provide 'skin joints', linear flexure lines (p. 80) at which soft tissue angulation is permitted, to 'hold down' muscles and tendons in the sole and digits and to

facilitate their excursions along the most functionally advantageous routes within the constraints of overall skeleto-arthrodial geometry, to reduce compression of plantar and digital arteries and nerves and perhaps to aid venous return. Plantar skin is subject to shearing and impact stresses in locomotion, particularly over the posterior calcaneal tubercles, metatarsal heads and pulps of terminal digital segments. As in the hand, pads of adipose tissue occur in these regions, diffusely pervaded by fine but collectively strong strands, tethering skin and limiting the displacement of fat, thus augmenting a resilient cushioning effect. Direction of the fibrous strands, which extend from the deep fascia through subcutaneous tissues to the dermis, is adapted to the prevailing stresses, as demonstrated in the sub-metatarsal region ('*ball of the foot*') by Bojsen-Møller & Flagstad (1976) (5.106D).

Parts of the deep fascia inferior to the plantar structures are collectively the *plantar aponeurosis* (vide infra); but only the central part is entirely aponeurotic and some authorities reserve the name for this. Medial and lateral parts are proximally aponeurotic for attachment of muscles (vide infra).

The plantar aponeurosis (5.106A) is of densely compacted collagen fibres, chiefly longitudinal but also transverse, divided into medial and lateral parts overlying the intrinsic muscles of hallux and minimus and a larger densest central part over the long and short digital flexors.

The central part is strongest and thickest. Posteriorly narrow, where it is attached to the medial process of the calcaneal tuberosity proximal to the flexor digitorum brevis, it is broader and thinner as it diverges towards the metatarsal heads, proximal to which it divides into five bands, one for each toe. These are united by the transverse and oblique fibres where they diverge (5.106A)

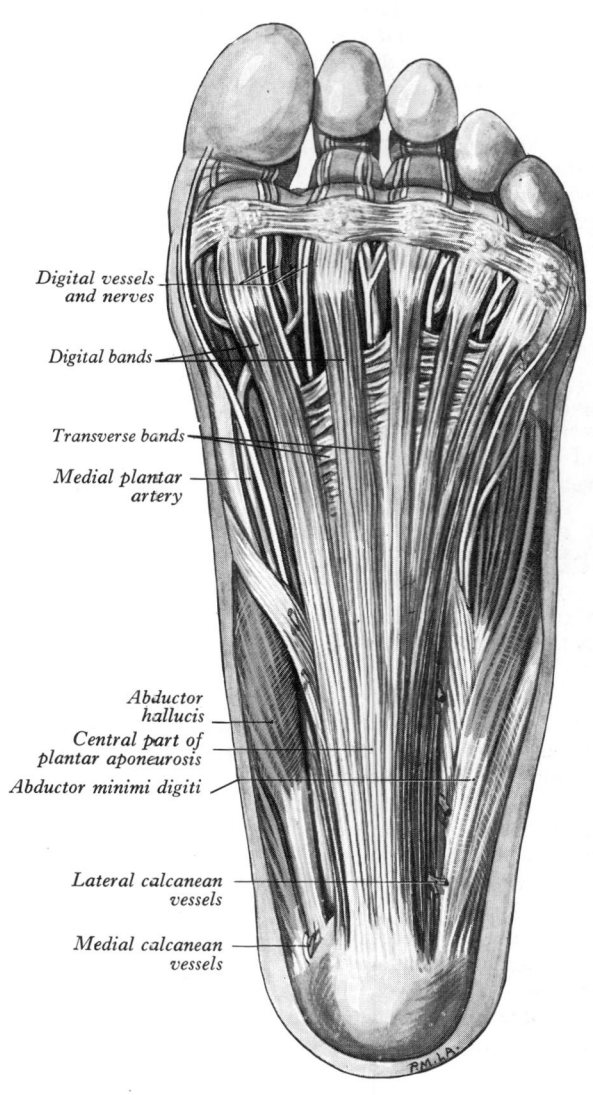

Digital vessels and nerves

Digital bands

Transverse bands

Medial plantar artery

Abductor hallucis

Central part of plantar aponeurosis

Abductor minimi digiti

Lateral calcanean vessels

Medial calcanean vessels

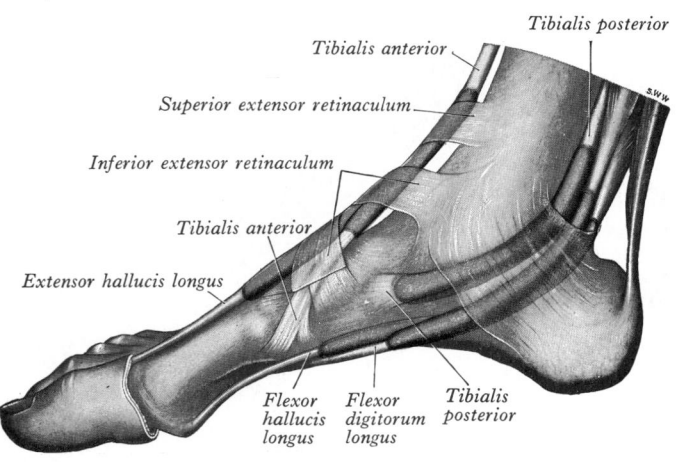

Tibialis posterior

Tibialis anterior

Superior extensor retinaculum

Inferior extensor retinaculum

Tibialis anterior

Extensor hallucis longus

Flexor hallucis longus

Flexor digitorum longus

Tibialis posterior

5.105 The synovial sheaths of the tendons at the right ankle: medial aspect.

5.106A The plantar aponeurosis of the left foot.

below the metatarsal shafts. Proximal, ventral and a little distal to the metatarsal heads and their joints a superficial stratum of each band is connected to the dermis by *skin ligaments (retinacula cutis)*, which reach the skin proximal to, and in the floors of, furrows separating the toes from the sole (5.106D). Proximally the

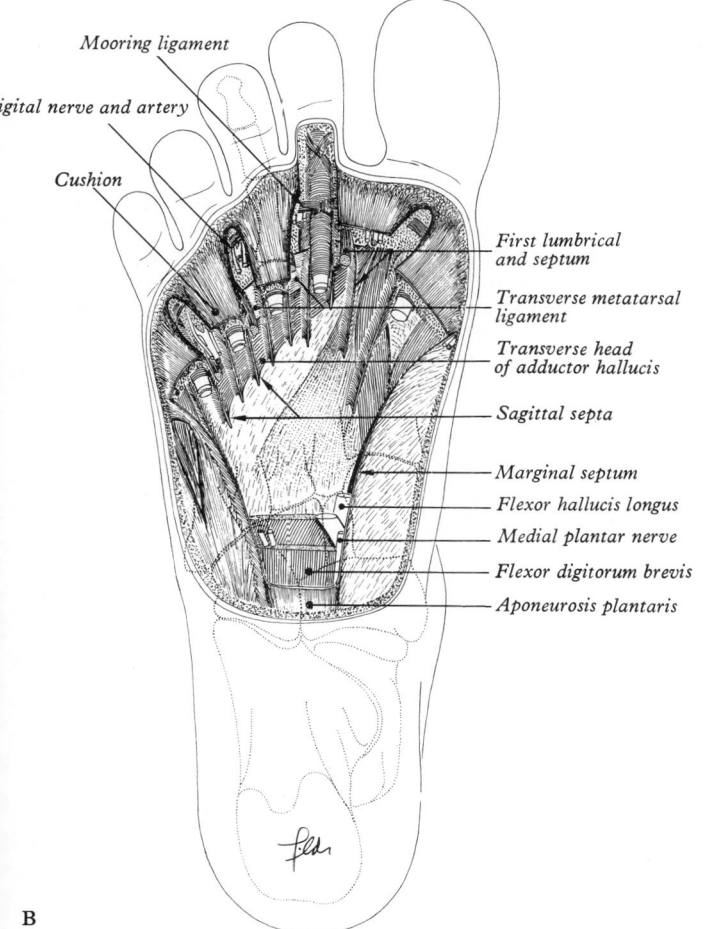

Mooring ligament

Digital nerve and artery

Cushion

First lumbrical and septum

Transverse metatarsal ligament

Transverse head of adductor hallucis

Sagittal septa

Marginal septum

Flexor hallucis longus

Medial plantar nerve

Flexor digitorum brevis

Aponeurosis plantaris

B

5.106B,C,D Details of the tendinous and fibrous architecture of various regions on the plantar aspect of the foot. B. Plantar aspect of the central compartment and the structures forming the ball of the foot. Of the plantar interdigital ligament, only the mooring ligament is shown.

retinacula are condensed to form a sagittal septum but distally diverge into numerous bundles and lamellae; these intersect at right angles bundles of the plantar interdigital ligament (superficial transverse metacarpal ligament, vide infra). A deep stratum of each digital band divides into *two septa* flanking the digital flexor tendons and separating them from the lumbrical muscles and digital vessels and nerves. These septa diverge and fuse with interosseous fascia, deep transverse metatarsal ligaments, plantar ligaments of the metatarsophalangeal joints and the periosteum and fibrous flexor sheaths at each proximal phalangeal base. In webs between the metatarsal heads and bases of the proximal phalanges pads of fat are developed, cushioning the digital nerves and vessels from the tendinous structures and extraneous plantar pressures. These four *fat pads* are tied by *vertical strands* from the digital fibrous flexor sheaths to the superficial stratum of the plantar aponeurosis and thus to the skin. Distal to the metatarsal heads a *plantar interdigital ligament* (superficial transverse metatarsal ligament) blends progressively with the *deep* aspect of the superficial stratum as it enters the toes (5.106C,D). The central plantar aponeurosis is thus an intermediary between the *pedal skin* and the *osteoligamentous* skeleton, joints and tendon sheaths through numerous *cutaneous retinacula* and *deep septa* extending to the metatarsals and phalanges. It is also continuous with its medial and lateral parts, at which junctions *medial* and *lateral intermuscular septa* extend in oblique vertical planes between medial, intermediate and lateral groups of plantar muscles to reach the bone (vide infra). Thinner *horizontal intermuscular septa* pass between muscle layers from these vertical septa.

The *lateral part* covering the abductor digiti minimi is thin distally, thicker proximally where it is a strong band, sometimes containing muscle fibres, between the calcanean tuberosity's

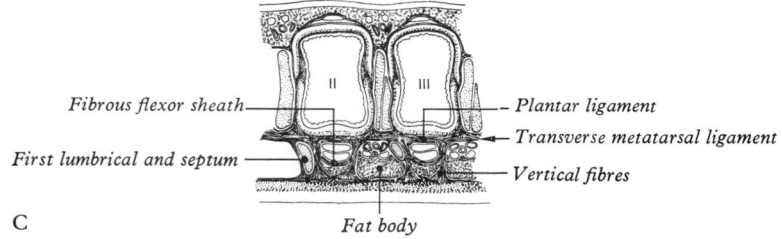

Fibrous flexor sheath

First lumbrical and septum

Plantar ligament

Transverse metatarsal ligament

Vertical fibres

Fat body

C

5.106C Transverse section through the heads of the second and third metatarsal bones showing the course of the collagen fibre bundles in the submetatarsal cushions and around the joints. Fat covers the fibrous flexor sheath inside the cushion; the digital nerves and vessels are lodged between the cushions.

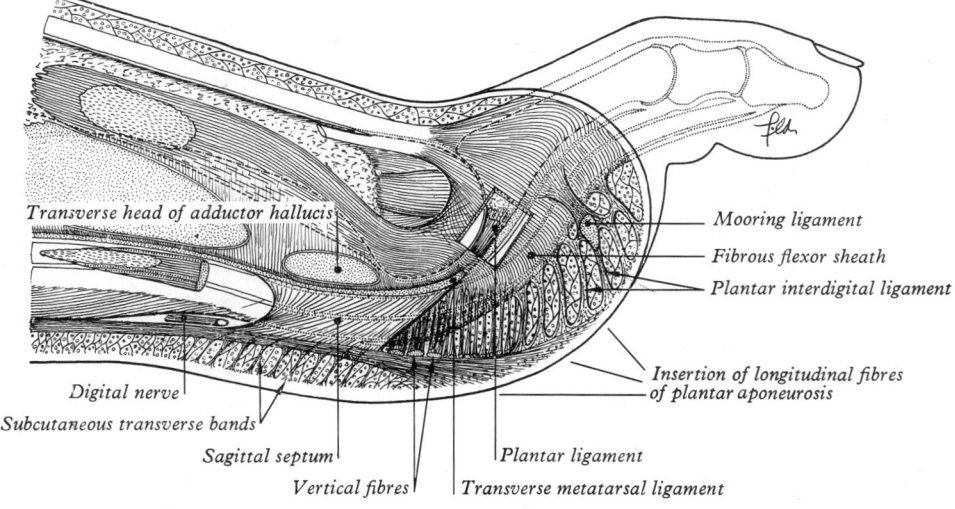

Transverse head of adductor hallucis

Mooring ligament

Fibrous flexor sheath

Plantar interdigital ligament

Insertion of longitudinal fibres of plantar aponeurosis

Digital nerve

Subcutaneous transverse bands

Sagittal septum

Vertical fibres

Plantar ligament

Transverse metatarsal ligament

D

5.106D Sagittal section through the second interosseous cleft showing the internal architecture of the three major areas of the ball of the foot. The sagittal septum is attached to the proximal phalanx through the transverse metatarsal ligament and the plantar ligament of the joint. The vertical

fibres and the lamellae of the plantar interdigital ligament are attached to the proximal phalanx through the fibrous flexor sheath (from Bojsen-Møller & Flagstad 1976, by courtesy of the authors and the Cambridge University Press).

lateral process and the fifth metatarsal base; it is continuous medially with the central aponeurosis, laterally with the dorsal pedal fascia.

The *medial part* is thin, covers the abductor hallucis and is continuous proximally with the flexor retinaculum, medially (then curving dorsolaterally) with the fascia dorsalis pedis, laterally with the central aponeurosis.

The *lateral intermuscular septum* is proximally incomplete; but distally its deep attachments are to the fibrous sheath of peroneus longus and the fifth metatarsal bone. The *medial intermuscular septum*, also discontinuous, divides into bands, proximal, intermediate and distal, each with lateral and medial divisions, as it approaches its deep attachments. The *proximal band*, attached laterally to the cuboid, blends medially with the tendon of tibialis posterior; the *middle band* extends from both cuboid and long plantar ligaments to the medial cuneiform bone and the *distal band* divides to enclose the tendon of flexor hallucis longus and is also attached to the fascia over flexor hallucis brevis.

The plantar muscles form medial, lateral and intermediate groups, as in the hand, the medial and lateral comprising the intrinsic muscles of the hallux and minimus, the intermediate group including lumbricals, interossei and short digital flexors; but it is customary to describe them in four layers, as encountered in dissection. These 'layers' can be overemphasized; functionally the former grouping is more useful. Actions of all are discussed together on p. 656.

FIRST LAYER

This superficial layer (5.107) includes abductores hallucis, digiti minimi and the flexor digitorum brevis. All extend from the cal-

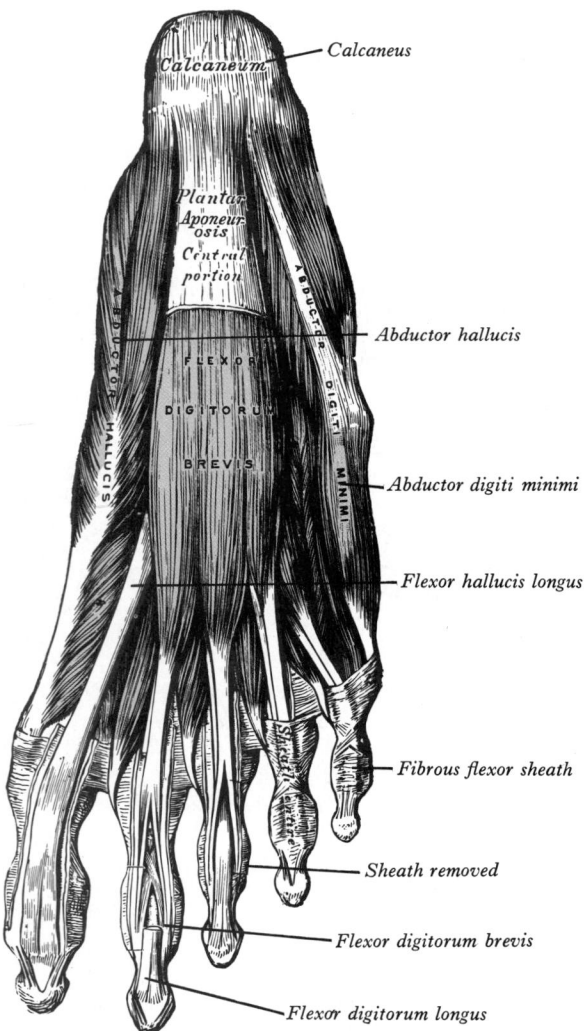

Calcaneus

Plantar Aponeurosis Central portion

Abductor hallucis

Abductor digiti minimi

Flexor hallucis longus

Fibrous flexor sheath

Sheath removed

Flexor digitorum brevis

Flexor digitorum longus

654   5.107 The superficial plantar muscles of the right foot.

canean tuberosity to the toes, being in this case a functional group assisting in the maintenance of a concave sole.

Abductor hallucis (5.107), lying along the foot's medial border, covers the origins of plantar vessels and nerves; it is attached proximally mainly to the flexor retinaculum but also to the medial process of the calcanean tuberosity, plantar aponeurosis and septum between the muscle and flexor digitorum brevis. Muscle fibres end in a tendon attached distally with the medial tendon of flexor hallucis brevis to the medial side of the proximal phalangeal base in the hallux; some fibres are attached proximally to its medial sesamoid; the muscle may derive some fibres from the dermis along the medial pedal border. It flexes and abducts the hallux.

Nerve supply: medial plantar nerve, S.2 and **3**.

Flexor digitorum brevis (5.107) is immediately superior (deep) to the central part of the plantar aponeurosis and separated from the lateral plantar vessels and nerves by a thin fascia. It is attached by a narrow tendon to the medial process of the calcanean tuberosity, and to the central plantar aponeurosis and adjacent intermuscular septa. It divides into tendons for the four lateral toes; at their proximal phalangeal bases, each divides into slips around a tendon of flexor digitorum longus; the two slips, their borders spirally reversed, then reunite, partially decussate, forming a groove (i.e. completing a tunnel) for the long flexor tendon. The short tendon divides again to attach to both sides of the intermediate phalangeal shaft. These details are identical to those in the hand but, of course, in miniature. The tendon for the fifth toe is often absent, sometimes replaced by a muscular slip from the *long* flexor tendon or flexor accessorius. Variations have been surveyed by Nathan & Gloobe (1974), comparing previous findings with observations on 100 feet, 63% of which showed variations, the slip to the fifth toe being most variable (63%), those to the fourth and fifth less so (10%), that to the second almost invariable. Variability from total absence to a supernumerary slip was noted, on occasion, in each toe.

Nerve supply: medial plantar nerve, S.2 and **3**.

Abductor digiti minimi (5.107), near the lateral border, its medial margin being related to the lateral plantar vessels and nerve, is attached to *both* processes of the calcanean tuberosity, the plantar surface between them, the plantar aponeurosis and the septum between the muscle and flexor digitorum brevis. Its tendon glides in a smooth groove plantar over the base of the fifth metatarsal and is attached laterally, with the flexor digiti minimi brevis, to the proximal phalangeal base of the minimus. Some *muscular* fibres from the lateral calcanean process usually reach the tuberosity of the fifth metatarsal (3.203), sometimes forming a separate *abductor ossis metatarsi digiti quinti*. An accessory slip from the fifth metatarsal base also occurs.

Nerve supply: lateral plantar nerve, S.2 and **3**.

Fibrous sheaths of flexor tendons (5.107). The terminations of the long and short flexor tendons are in osseo-aponeurotic canals, as in the fingers, formed by the plantar aspects of phalanges and *digital fibrous sheaths* over the tendons and attached to the phalangeal margins. In traditional accounts proximal and intermediate phalanges are often described as having across their shafts bands that are strong, their fibres transverse (*annular part*); but opposite joints they are thinner and the fibres decussate (*cruciform part*). (However, in recent researches on the fibrous flexor sheaths of *the fingers* Jones & Amis 1987 modified and amplified this account and added hitherto unrecognized features. Their ongoing investigations will necessitate a reappraisal of pedal arrangements. See comments on p. 618.) Each osseo-aponeurotic canal has a synovial lining reflected around its tendon; *vincula tendinum* are said to be arranged as in the fingers (p. 621); here again, the same reservations apply.

SECOND LAYER

This group comprises the flexor digitorum accessorius and four lumbrical muscles (5.108,109). Intimately associated with the intrinsic muscles of the second layer are the preterminal tendons of flexor hallucis longus and flexor digitorum longus.

Flexor digitorum accessorius (5.109) has two heads separated by the long plantar ligament. The medial, muscular and

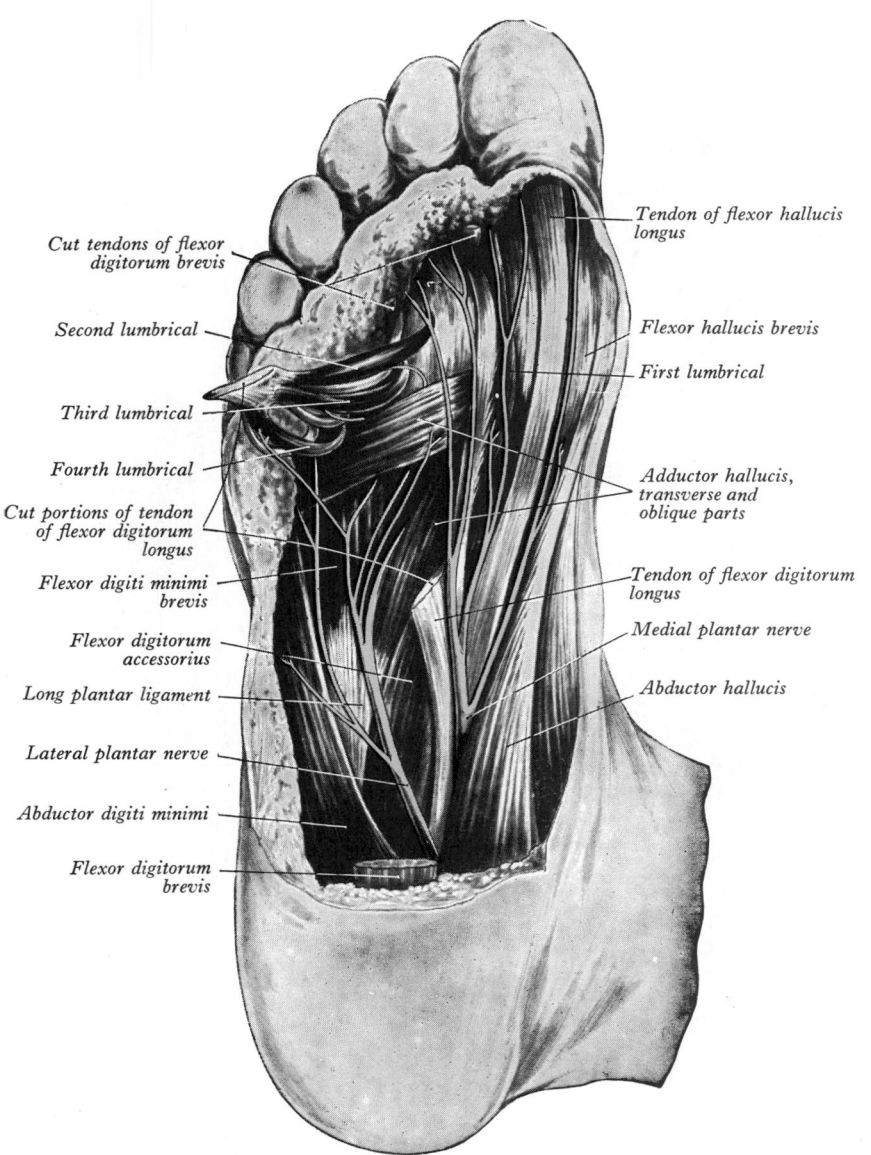

Cut tendons of flexor
digitorum brevis

Second lumbrical

Third lumbrical

Fourth lumbrical

Cut portions of tendon
of flexor digitorum
longus

Flexor digiti minimi
brevis

Flexor digitorum
accessorius

Long plantar ligament

Lateral plantar nerve

Abductor digiti minimi

Flexor digitorum
brevis

Tendon of flexor hallucis
longus

Flexor hallucis brevis

First lumbrical

Adductor hallucis,
transverse and
oblique parts

Tendon of flexor digitorum
longus

Medial plantar nerve

Abductor hallucis

5.108 The plantar muscles and their nerve supply. Most of the flexor digitorum brevis has been removed. The flexor digitorum longus has been divided partially and its distal end has been turned forwards together with the second, third and fourth lumbricals.

larger, is attached to the medial concave surface of the calcaneus below the groove for flexor hallucis longus; the lateral, flat and tendinous, is attached to the calcaneus distal to the lateral process of its tuberosity and to the long plantar ligament. The medial head may end distally in a fibrous lamina, deep to the tendon of flexor digitorum longus, dividing into slips joining the long flexor tendons to the second, third and sometimes fourth digits; but almost as often it joins the lateral border of the long flexor tendon. The lateral head either joins this fibrous *lamina* on its superficial or lateral aspect or may fuse with the lateral border of the long flexor *tendon* (Lewis 1962), perhaps partially 'correcting' its diagonal vector. The muscle is sometimes absent and varies in the digits supplied; the slip to the fourth toe is often absent; a slip to the fifth sometimes occurs.

Nerve supply: lateral plantar nerve, S.2 and **3**.

Lumbrical muscles (**5.**109) are four small muscles accessory to the tendons of flexor digitorum longus, numbered from the medial side. They arise along adjacent sides of the tendons from their angles of separation, except the first which is from the medial side only of the first tendon. The muscles end in tendons which pass distally on the medial sides of the four lesser toes to the dorsal digital expansions on their proximal phalanges.

Nerve supply. The first lumbrical is supplied by the medial plantar nerve, the rest by the deep branch of the lateral plantar nerve, S.2 and **3**.

THIRD LAYER

This comprises the short intrinsic muscles of the hallux and minimus: flexor hallucis brevis, adductor hallucis and flexor digiti minimi brevis (**5.**108,110), the deepest in the sole, except for the interossei which are superior to them.

Flexor hallucis brevis (**5.**110) has a bifurcate tendon, its lateral limb attached medially on the plantar surface of the cuboid proximal to the groove for peroneus longus and to the adjacent part of the lateral cuneiform; the medial limb has a deep attachment continuous with the lateral part of the tendon of tibialis posterior and a more superficial attachment to the middle band of the medial intermuscular septum (Lewis 1964). Its belly divides into medial and lateral parts whose twin tendons reach and attach to the sides of the base of the proximal hallucial phalanx, a sesamoid bone usually embedded near each attachment; the medial part is blended with abductor hallucis, the lateral with adductor hallucis as they reach their attachments. Accessory proximal slips from the calcaneus or long plantar ligament may occur; a tendinous distal slip may reach the second toe's proximal phalanx.

Nerve supply: medial plantar nerve, S.2 and **3**.

Adductor hallucis (**5.**110) has oblique and transverse parts. The *oblique head* springs from the second to fourth metatarsal *bases* and the fibrous sheath of the tendon of peroneus longus, the

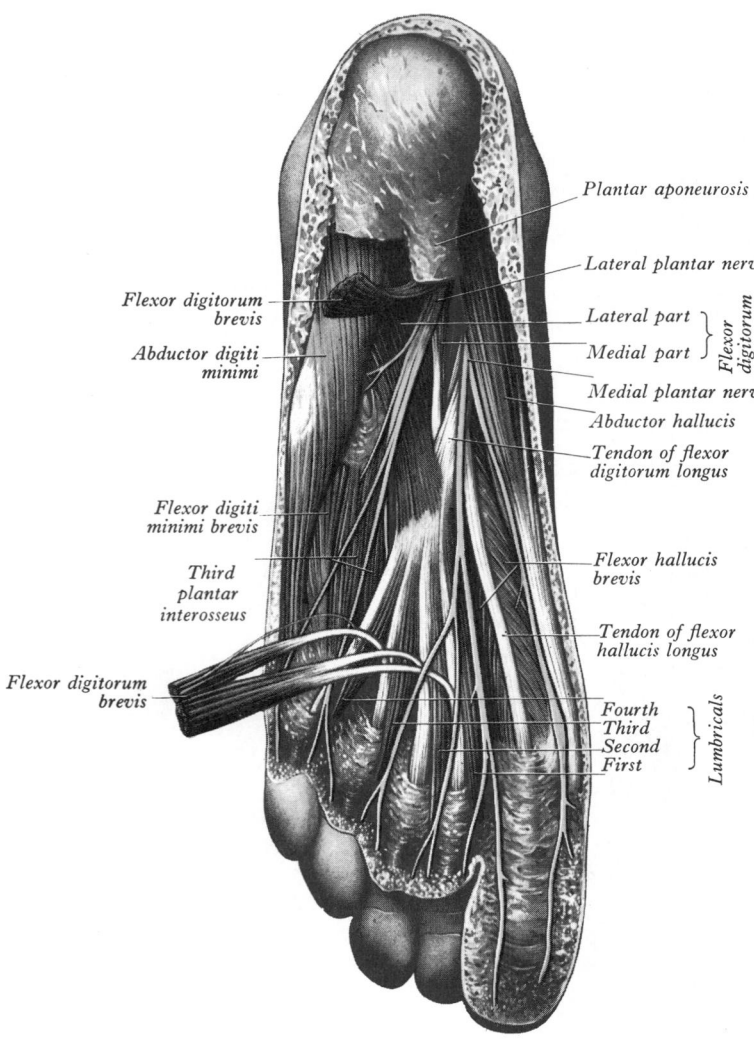

Plantar aponeurosis

Lateral plantar nerve

Flexor digitorum brevis

Abductor digiti minimi

Lateral part

Medial part

} Flexor digitorum accessorius

Medial plantar nerve

Abductor hallucis

Tendon of flexor digitorum longus

Flexor digiti minimi brevis

Third plantar interosseus

Flexor hallucis brevis

Flexor digitorum brevis

Tendon of flexor hallucis longus

Fourth
Third
Second
First

} Lumbricals

5.109 The plantar muscles of the left foot: first and second layers.

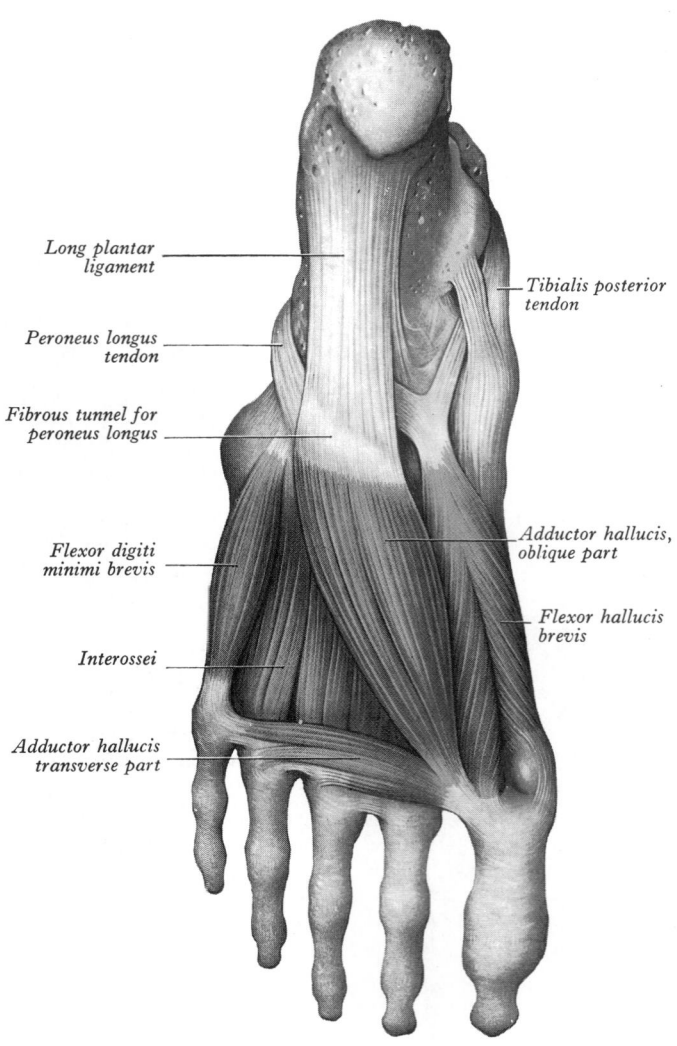

Long plantar ligament

Peroneus longus tendon

Fibrous tunnel for peroneus longus

Flexor digiti minimi brevis

Interossei

Adductor hallucis transverse part

Tibialis posterior tendon

Adductor hallucis, oblique part

Flexor hallucis brevis

5.110 The plantar muscles of the left foot: third layer.

transverse head from the plantar metatarsophalangeal ligaments of the third to fifth toes (sometimes only the third and fourth) and from the deep transverse metatarsal ligaments between them. The oblique head has medial and lateral parts: the medial blending with the lateral part of flexor hallucis brevis and attached to the lateral hallucial sesamoid bone; the lateral part joins the transverse head and is also attached to the lateral sesamoid bone and directly to the base of the first hallucial phalanx. According to Cralley et al (1976) the transverse part is sometimes absent (3 out of 50 feet dissected); they could not identify a phalangeal attachment for the transverse part; fibres failing to reach the lateral sesamoid were attached with the oblique part. Part of the muscle may be attached to the first metatarsal, as an *opponens hallucis*; a slip may also reach the proximal phalanx of the second toe.

Nerve supply: deep branch of lateral plantar nerve, S.2 and **3**.

Flexor digiti minimi brevis (5.110), proximally attached to the medial plantar surface of the fifth metatarsal base and sheath of peroneus longus, has a distal tendon attached laterally on the proximal phalangeal base of the minimus. Some deeper fibres may end on the lateral and distal half of the plantar surface of the fifth metatarsal bone as a distinct *opponens digiti minimi*.

Nerve supply: superficial branch of lateral plantar nerve, S.2 and **3**.

FOURTH LAYER

This comprises plantar and dorsal **interossei**, resembling those in the hand but arranged relative to an axis through the *second* digit, not the third as in the hand, this having the least mobile metatarsal bone.

Dorsal interossei (5.111A), between the metatarsal bones, are four bipennate muscles, each with two heads from the sides of

adjacent metatarsal bones; their tendons are attached to the proximal phalangeal bases and dorsal digital expansions, the first on the medial side of the second toe, the others on the lateral sides of the second to fourth toes. In angular gaps between the proximal heads of each of the lateral three a perforating artery passes dorsally through the same space that in the first muscle transmits the arteria dorsalis pedis to the sole.

Plantar interossei (5.111B) are *inferior* to their metatarsal bones, each of the three being connected to one metatarsal. They pass from the bases and medial sides of the third to fifth, passing to the medial sides of the proximal phalangeal bases of the same toes and to their dorsal digital expansions.

Nerve supply. The dorsal and plantar interossei are supplied by the deep branch of the lateral plantar nerve (S.2 and **3**), except in the fourth interosseous space, where the supply is its superficial branch. The first dorsal interosseous often has an extra ramus from the medial branch of the deep peroneal nerve from the pedal dorsum, and the second a ramus from its lateral branch.

ACTIONS OF INTRINSIC MUSCLES IN THE FOOT

As elsewhere the actions of pedal muscles, or at least their potentialities, can be deduced from the geometry of their fascicular architecture and attachments, as the names applied to some indicate. But purely structural deductions must take into consideration the modifying effects of contact with the ground; they are unlikely to provide more than suggestions of the actual roles of individual muscles in particular activities. Direct observation, by eye or touch, is difficult; even with electromyography only crude data are available, from very few experiments. Despite some attractive speculation (p. 657), much uncertainty persists, as attested by the volume of controversial literature on pedal function.

There are two major conditions in which intrinsic musculature may aid or modify the effects of muscles operating from the leg: these are standing still and maintaining progression. In standing, with feet flat on the ground they are platforms for the distribution of weight, the body's centre of gravity being maintained above them by adjustment of tension and length in the leg and trunk muscles. It might be supposed that the largely longitudinal muscles in the sole would then be sustaining its arches against the flattening effect of body weight; this has long been a canon of anatomical teaching, despite increasing doubts amongst clinical and experimental observers. All evidence from such sources, though limited, indicates that ligamentous ties, including the plantar aponeurosis, are largely sufficient. Similarly, intrinsic muscles, especially abductors and short flexors, have long been believed to exert a 'tonic' effect, steadying the toe pads on the ground. Electromyographic evidence, admittedly scant, contradicts this; simple observation shows that the toes can easily be extended out of contact with little adverse effect upon pedal efficiency in standing. Suzuki (1972), in a study of normal and flat-footed individuals, found the intrinsic muscles relatively inactive in the normal when standing, but active in the flat-footed, as were the long arch-supporting muscles such as peroneus longus.

When the heel is lifted in beginning a stride, in walking or running, all weight and muscular thrust is transferred to the forefoot region of the metatarsal heads and toe pads. This changes the foot from platform to lever, intensifying the forces acting on the forefoot, especially in running and jumping. So much argument has been extended on the nature and behaviour of 'arches' in the foot and on the agents, muscular and ligamentous, which can or do act as 'tie-beams' across them, that the foot's basic role as a lever has been obscured. It appears ill-suited to act as such, being a series of links, though its arched form has good mechanical precedents. As the heel lifts, the sole's concavity is accentuated; electromyographic evidence available shows the intrinsic muscles to be strongly active. This might slacken the plantar aponeurosis, but dorsiflexion of the toes must tighten it. The foot is also inverted and supinated and close-packing of the intertarsal joints is reached as the foot takes the full effects of leverage, losing its pliancy (Inman 1969). The toes are thus held extended at metatarsophalangeal and distal interphalangeal joints and flexed at proximal interphalangeal joints, including that in the hallux. This probably results from contraction of the long and short digital flexors, by drawing terminal phalanges towards the sole and passively buckling the toes into the position just described. Buckling is perhaps controlled by the lumbrical and interosseous muscles, by opposing extension at the metatarsophalangeal and flexion at the proximal interphalangeal joints. Flexor hallucis brevis could prevent excessive extension at the hallucial metatarsophalangeal joint. Abductors of the hallux and minimus act as flexors, with probably little abductor effect, though their geometry varies. Analyses such as this are speculative; in view of the difficulties of electromyography in active feet, present uncertainties may long persist (p. 659). Various imbalances in the actions of toe muscles are held to explain dysfunctions and deformities, such as contracted toes. Actual observation is rare; an evaluation of the weights of lumbrical muscles has shown no correlation with toe contractures (Challey et al 1975).

STANDING, WALKING AND RUNNING

Though birds are in a sense bipeds, as are some reptiles and eutherian mammals, only the human primate has evolved a completely habitual upright posture. Some animals, particularly other primates, may adopt it for considerable periods, but none has reached the degree of specialization as has human adaptation to continuous upright standing and locomotion. Human erect posture is often described as if attained at the cost of such painfully difficult evolution that it is still a precarious habit, prone to all manner of defects and pathology, perhaps due to the influence of medically orientated thinking. A full upright posture existed in most, if not all extinct hominids, foreshadowed in other primates for an even greater period of time. The disadvantages of the quadrupedal habit do not appear to have attracted similar veterinary attention!

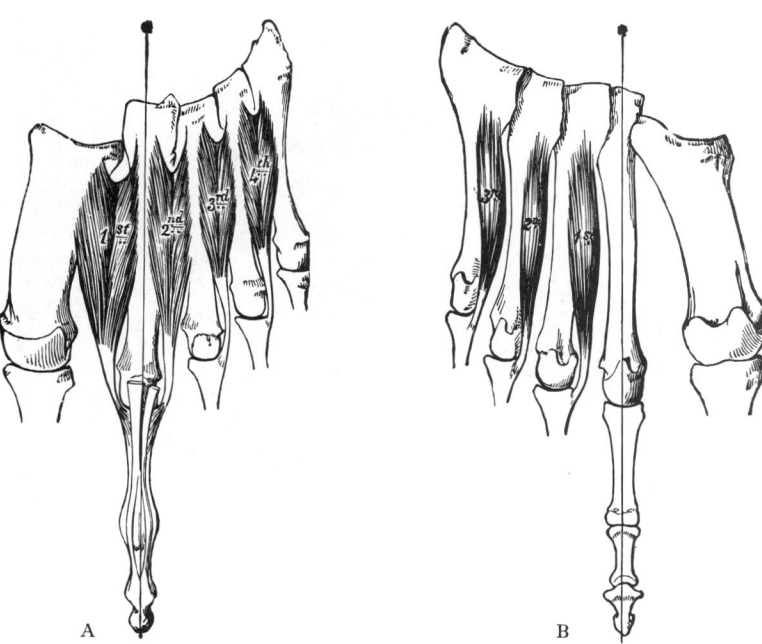

5.111 The interossei of the left foot. A. Dorsal interossei viewed from the dorsal aspect. B. Plantar interossei viewed from the plantar aspect. The axis to which the movements of abduction and adduction are referred is as indicated.

In *posture* the human primate is singular, if not unique, in having straightened the trunk and lower limbs enough to bring the centre of gravity above the joints on which weight impinges; even the head is largely balanced upon the cervical spinal column. A vertical from the centre of gravity (about a centimetre posterior to the sacral promontory) is anterior to the ankle joints. Strong plantar flexion, particularly by the soleus, is hence needed to prevent forward sway or actual falling. But in most individuals standing in a symmetrical manner, the same vertical is so related to the hips and knees that gravity tends to *increase* their extension, partly opposed by ligamentous structures (associated with a nearly close-packed position of the joints, p. 483), rather than by muscle activity. Such a posture, often maintained for long periods, thus requires minimal muscular activity (Akerblom 1948, Joseph 1960), except for brief bursts in appropriate muscle groups when the body sways from it. The standing position is often changed, intermittently relieving stress on ligaments by transferring it to the muscles (Smith 1950), a common habit being to hyper-extend one knee and slightly flex the other; one leg and then the other takes the major stress of body weight, the slightly flexed leg acting temporarily as a prop. Evans (1977) has emphasized the use of the iliotibial tract in achieving and sustaining the 'locked' hyper-extended (close-packed) posture of the weight-bearing leg in asymmetrical standing. The line of gravity is related to spinal curvatures in such a way that they are largely exaggerated by gravity, but controllable by small muscular tensions. However, over details controversy persists. Swaying of the trunk above the pelvis entails not only responsive activity in the leg muscles but also in those of the trunk, such as psoas major, when the body sways back.

Energy-sparing by using gravity to hyperextend (or sometimes flex) joints is applied in some birds, such as the upright penguin, and passerine birds which perch in trees. It is regarded as unique in man among mammals, but many quadrupeds also habitually stand for even more prolonged periods, apparently without great effort (and much less complaint than the human biped).

In *locomotion*, whether walking, running, jumping or climbing, the economical pattern of standing at once ceases. All the muscles of the trunk, arms and legs may be involved, as prime movers, fixators or stabilizers. Arms are swung as balancers and as weighted pendulums to increase the impetus, especially in running and jumping; their use in climbing is obvious.

In studies of locomotion (vide infra) attention has been largely on an analysis of walking movements and the temporal sequences

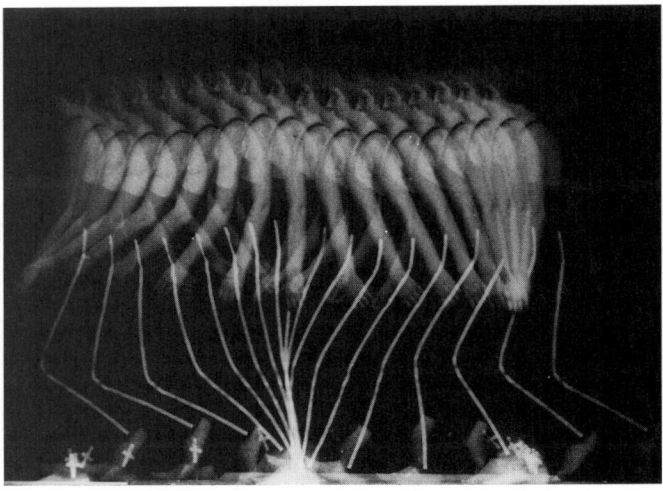

A

B

5.112 Chronocyclograms of two individuals walking relatively fast. In A the support phase, with one foot on the ground, occurs at midpoint of the recording, which occupies one full cycle. In B two such support phases are apparent and the recording occupies somewhat less than two cycles. The bright lines are successive images of a stripe attached to the trousered leg proximal to the camera. Note the concentration of lines at the support phase and also the varying angulation at hip, knee and ankle. (Photographs supplied by D W Grieve, Department of Anatomy, Royal Free Hospital Medical School, London.)

of contraction and relaxation in the muscles involved. As the first step is taken from standing, the trunk leans forwards with the feet slightly inverted and the longitudinal arches accentuated (p. 543). One heel is lifted, transferring the weight progressively along the lateral border of the opposite foot to the metatarsal heads and toes. This is *stance phase*, passing into *swing phase* as soon as the foot leaves the ground, which it then again meets in reverse order, heel first and then a repetition of the roll forwards on to the toes. From one heel's impact to the next is termed a *walking cycle*, which alternates in the legs.

The **stance phase** has been variously analysed (Barnett 1956, Gray & Basmajian 1968, Elftmann 1966); it is essentially a continuous process of weight transference from the heel to the toes, especially the hallux. These pedal movements should not be viewed in isolation; the whole body moves laterally above the leg that is in stance phase, the body's centre of gravity shifting variably (on average about 5 cm). Simultaneously the ipsilateral arm begins to swing forwards with the contralateral leg; the pelvis, at first rotated 'forwards' on the stance side (i.e. relative lateral rotation of the stance femur, its hip being in advance), now advances on the opposite side with an accompanying forward swing of the opposite leg. The foot is fixed on the ground, and hence, as the knee extends and the ankle dorsiflexes, the tibia *rotates medially*, taking the talus with it, and the femur undergoes obligatory conjunct medial rotation at the knee (p. 530) and hence also at the hip joint, medial femoral rotation being accentuated by the pelvic rotation in the opposite direction, as mentioned above. This sequence of congruent medial rotations, combined with dorsiflexion and transient supination of the subtalar skeleton (p. 542), brings all articulations in the leg and foot near to their close-packed positions. The resulting momentary rigidity coincides with maximal thrust by the powerful plantar flexors. Thereafter, as the heel rises again, the foot passes into plantar flexion and pronation, thrust now being transmitted through a loosely packed pliant lever to the head of the first metatarsal and dorsiflexing the hallux, which finally leaves the ground to end the stance phase. (The above should be compared with the analysis of movements of the pedal skeleton, pp. 542–544.)

In the **swing phase** the toes leave the ground as the opposite heel meets it, to transfer support; the leg, relieved of weight bearing, swings forwards with flexion at the hip and knee, dorsiflexion of the foot and lateral rotation of all elements which had undergone congruent medial rotation during their previous stance phase. The swing is pendular, by the hip flexors. As the limit of stride is reached, momentum is overcome by the hip extensors to bring the heel to the ground, to start another cycle.

These events are more rapid than can be described, but this analysis is supported by cinematographic records, from which, of course, it is partly derived. The role of many muscles involved has been investigated at every stage, but details should be sought in original studies (e.g. Alexander 1983,1984). Speed in walking can be increased by augmenting work by the muscles involved and acceleration of their actions. In running, the movements are similar but the heels do not touch the ground, thrust being entirely by the forefoot. Moreover, the body is carried forwards by its momentum *entirely unsupported* between take-off from one foot and descent on the other. A similar unsupported phase characterizes hopping. In jumping, take-off may be from one foot or two, simultaneously or in quick succession; landing is similar.

In all progression, the forces involved are gravitational, inertial and muscular, the latter varying from minimal exertion, sustainable over long periods, to maximal effort for briefer periods; but reliable estimates of the relative expenditures of energy are few.

MECHANICAL ANALYSIS OF LOCOMOTION

The mathematical analysis of the mechanics and energetics of locomotion, studied in many animals including invertebrates, has advanced rapidly. (For introductory literature consult Alexander & Goldspink 1977.) Locomotion in many vertebrates has been analysed, particularly in mammalian quadrupeds. Basic data are usually cinephotographic and force platform recordings (Alexander et al 1977), but X-ray cinephotography and electromyography (Wentink 1976) have also been used. Bipedal locomotion has been studied in amphibia, marsupials and primates including mankind. A most interesting result is confirmation of the storage of recoil energy in muscles *and* tendons (Alexander & Bennet-Clark 1977, Alexander 1984a). Cinematic analysis of human walking and running has already contributed to phasic analysis, as noted above, but mathematical studies go further in accounting for the forces concerned. For example, it is possible to predict a limiting optional velocity for walking, beyond which it becomes more economical, in terms of muscle energetics and oxygen consumption, to introduce a 'floating' phase with a change to running. Elastic recoil in human tendons has been studied by combined light and X-ray cinephotography, but reliable data are lacking. A mathematical interpretation of locomotor mechanics has also renewed interest in bodily proportions, especially those in limbs, which are necessary data for such analyses (McMahon 1977). Not only the dimensions of limb bones relative to body size and weight, but also the strength and proportions of osseous segments and muscle masses have become basic data in comparative mechanical studies of gait (Alexander 1984b). The arrangement of fibres in muscles has acquired renewed significance in quantitative analysis (e.g. Alexander & Vernon 1975). Coordination of such data into even general concepts of quadrupedal or bipedal locomotion is inevitably complex; but progress has been rapid and highly interesting; valuable results are to be expected. For a discussion of the optimal techniques in bipedal walking consult Alexander & Jayes (1978).

The foregoing largely concerns analyses of electromyographic and cinematographic recordings in which data are interpreted qualitatively rather than quantitatively. Some workers (e.g. Grieve & Cavanagh 1974) have made efforts to quantify electromyograms with improved accuracy by computerized techniques, coupled with accurate measurement of cinematic records obtained simultaneously in two or three cardinal planes. Since a temporal parameter is involved, and measurable from recordings, detailed analysis of angular displacements between the segments of limbs in walking (and speeds of movement in each segment) can be undertaken. The data obtained can be integrated by appropriate programmes and may also be graphically illustrated. Such integrated graphs of cycles of movements, in space and time, are 'chronocyclograms'. So-called 'heel-strike' and 'toe-off' points in a cycle are commonly chosen as datum points. Chronocyclography (5.112) has already demonstrated changes in pattern with alternation of speed in walking; comparisons between events in 'normal' records with those in patients afflicted with locomotor disturbances can be made. These developments have facilitated and refined the description of abnormal gaits; it was hoped that they would have applied value but in their original form the techniques are lengthy and tedious. It has been further improved by Grieve and co-workers (1979) by a computer evaluation of cycles of angular change in thigh and foreleg segments, the angles being recorded by impingement of a rotating beam of plane-polarized light upon sensors attached to the limb segments. Since the signals produced by this polarized light goniometer are passed to a suitably programmed computer, a graphic recording is *immediately* available of the subject's angulation cycle in stepping or walking. This renders the technique clinically applicable. A stroboscopic method of recording has also been used; the records shown in 5.112A,B illustrate the cycle in walking.

ANGIOLOGY

6

Introduction

All cells live within the body in a fluid environment upon the stability of which, in terms of physico-chemical characteristics, each cell is dependent for its normal functioning. This fluid environment, whilst being maintained at a steady state within certain limits, also allows for the diffusion of metabolites between cells and between the external and internal environments. In small animals with a relatively simple type of organization, unaided diffusion is rapid enough for nutrients, gases and waste materials to pass to and from the individual cells to satisfy their metabolic demands. In more complex animals diffusion is assisted by the muscular movements of the body and in larger and yet more complex forms, particularly those with a coelom which interrupts the continuity of the tissues, a system of cell-lined tubes separates the body fluids into blood, confined within the tubes and able to circulate by muscular propulsion, and tissue fluid lying in the extracellular spaces outside the tubes. The coelom forms a third compartment which also possesses the function of a circulatory system in some groups of invertebrates. In the vertebrates a fourth division is added in the form of the medullary cavity of the central nervous system which contains the cerebrospinal fluid (p. 1090). A fifth compartment is provided by a system of further vessels which are cell-lined lymphatics, returning some of the tissue fluid to the blood circulation.

In animals with a true blood vascular system (p. 682) a high level of circulatory efficiency is possible. The system of cell-lined tubes also allows for the segregation within the circulation of some elements such as respiratory pigments able to store and carry oxygen and often carried within special cells, as in vertebrates; also present are various types of defensive cell, many migrating into the tissues when required, so achieving rapid access to all parts of the body, and other large molecules such as polysaccharides and proteins, which may gain only limited entry to the tissues. With the development of a pumped circulation the hazard of leakage from damaged vessels is increased; various cellular and chemical *haemostatic* mechanisms ensure that such blood loss is limited.

In mammals most tissues are served by an extensive blood supply circulating by way of the heart through arteries, arterioles, capillaries, venules and veins (p. 684). The capillaries form the main site of exchange by diffusion between blood and tissues. Most cells lie within a few tens of microns of a capillary, active tissues receiving a richer and more intimate supply than less active ones. Some tissues, however, are not penetrated directly by capillaries and diffusion gradients may be several hundreds of microns in length; such tissues are cartilage, some epithelia and the connective tissue of the cornea and dentine.

HAEMAL AND LYMPHOID TISSUE

Before consideration of the detailed organization of the vascular channels, the biology of the circulating cellular components of the blood and lymphatic systems and their development in the bone marrow and lymphoid tissue will be outlined. Collectively, these various cells constitute the **haemolymphoid system**, a complex array of cells and tissues which are closely interdependent in their origins and functions in the body. The **haemal** component, arising in the bone marrow, provides: *red cells (erythrocytes)* which carry oxygen and to a lesser extent carbon dioxide between the lungs and general tissues; a wide variety of defensive cells, *leucocytes*, including the *neutrophil, eosinophil* and *basophil granulocytes*, and *monocytes (macrophages)*, engaged in a plethora of defensive activities, and *platelets* which assist in haemostasis. Of these cells, only two, erythrocytes and platelets, are generally confined to the blood vascular system once they have entered it from haemopoitic tissue. All leucocytes possess the ability to leave the circulation and enter the extravascular tissues, the numbers of cells so doing being greatly increased in local infections and other disease conditions. The **lymphoid** component includes cells which are formed both in the bone marrow and in many sites outside it: the thymus, lymph nodes, spleen and lymphoid tissue associated with the alimentary tract and bronchi (lymphoid nodules). These produce different varieties of *lymphocytes*, which in the blood are included among the leucocytes and like the others of this class are able to migrate into extravascular sites although, unlike haemal cells, lymphocytes may also be found within the channels of the lymphatic system. Lymphocytes are concerned with various types of defence and indeed are the main source of the body's ability to resist infections. Included in lymphoid tissue are various types of phagocytic cell which, with the monocytes in the blood and some other phagocytic cells in other tissues (e.g. macrophages), are often considered as a distinctive component of the body, the *mononuclear phagocyte system* (p. 668).

Tissue Fluid and Lymph

The walls of the vascular system are lined by endothelial cells (p. 683) which allow the free diffusion of water, ions, gases in solution and small organic molecules. Large molecules are mostly retained within the blood, although some leakage does occur. Water and dissolved substances pass into the extravascular space which lies between the tissue cells, being termed *tissue fluid*. Because of the

relatively high pressure at the arteriolar ends of the capillaries tissue fluid is constantly being formed; some of this passes back into the venous end of the capillary and the post-capillary venules, which are at relatively low pressure, but the remainder passes into the neighbouring lymphatic vessels as *lymph*. The lymphatic vessels eventually return the lymph to the blood circulation via the thoracic duct or the right lymphatic duct; en route to these ducts, the lymph is filtered by its passage through successive lymph nodes (p. 822) and whilst in them it picks up a population of lymphocytes and macrophages. Blind-ending lymphatics in the villi of the small intestine contain lymph which picks up great quantities of lipid (chyle) absorbed across the intestinal mucous membrane during alimentation. This is added to the lymph (and eventually to the blood) as minute lipid droplets (*chylomicrons*) each about 0.2–1.0 μm in diameter, which may be seen using light microscopy with dark field illumination. The lymphoid tissue also adds proteins in the form of antibodies (*immunoglobulins*) to the lymph and hence to the bloodstream.

Characteristics of Blood

Blood is an opaque turbid fluid with a viscosity somewhat greater than that of water (mean relative viscosity 4.75 at 18°C), and a specific gravity of about 1.06 at 15°C. When oxygenated, as in the systemic arteries, it is bright scarlet and when deoxygenated, as in systemic veins, it is dark red to purple.

Blood is a heterogeneous fluid consisting of a clear liquid, *plasma*, and formed elements, *corpuscles*; because of this admixture it behaves hydrodynamically in a complex fashion and belongs to that class of fluids termed non-Newtonian. This characteristic has important consequences in the physical study of blood flow in vessels (haemorrheology).

Plasma

This is a clear, slightly yellow fluid which contains many substances in solution or suspension; the *crystalloids* give a mean freezing-point depression of about 0.54°C. Plasma is rich in sodium and chloride ions and also contains potassium, calcium magnesium, phosphate, bicarbonate and many other ions glucose, amino acids, etc. The *colloids* include the high molecula weight plasma proteins, composed chiefly of those associated with clotting, particularly prothrombin, the immunoglobulins an complement proteins involved in immunological defence (p. 672) and glycoproteins, polypeptides and steroids concerned with hormonal activites. Since most of the metabolic activities of the body are reflected in the composition of the plasma, routine chemical analysis of this fluid has become of great diagnostic importance and a considerable body of information on its chemistry is available.

The formation of clots by the precipitation of the protein fibrin from the plasma is initiated partly by the release of specific materials from damaged cells and blood platelets (p. 668) in the presence of calcium ions. If blood or plasma samples are allowed to stand, clot formation occurs to leave a clear yellowish fluid, the *serum*. Removal of the available calcium ions by means of citrate, various organic calcium chelators (EDTA, EGTA) and oxalate prevents clot formation in vitro; heparin is also widely used as an anti-clotting agent, its action interfering with another part of the complex series of chemical interactions leading to fibrin clot-formation.

Blood as a Tissue

Blood has many affinities with connective tissue, as, e.g. in the mesenchymal origin of its cells, the free exchange of leucocytes with the connective tissues and the relatively low cell:matrix ratio. Many of the plasma substances and some of the cells, however, arise from a variety of sources (e.g. many of the proteins associated with clotting are formed in the liver) and so blood is really a composite tissue pool.

The Formed Elements of Blood

Blood contains three groups of formed elements: red and white blood corpuscles and platelets (see Wintrobe 1981, for a detailed description of these cells). Some structural aspects of these elements are visible in fresh blood, but many others are seen only in fixed and stained specimens. The examination of blood cells is of considerable clinical importance since their numbers, proportions of different cell types and structure are valuable indicators of pathological changes in the body. Amongst other techniques, the Romanowsky methods of staining are particularly valuable and are widely used in clinical laboratories. These methods involve staining in aqueous solutions with methylene blue-eosin mixtures which colour both acidic dye binding and basic dye binding structures. The Giemsa and Leishman stains belong to this group. (It should be noted that throughout this section, the figures given for cell dimensions and numbers are approximate ranges only. The data provided by different authorities vary somewhat; further, the dimensions of some cells when measured in the fresh state are substantially smaller than when measured in a dried smear; with erythrocytes the converse applies. In particular the upper limit for the diameter range of the small lymphocyte given here (10 μm) is rather higher than that chosen by some workers, who hold that cells over 8 μm in diameter should not be so classified. The differences stem from methodological and site variations and the somewhat arbitrary nature of the chosen dividing line.)

Red Blood Corpuscles

The red blood corpuscles (*erythrocytes* or *red blood cells*) (**6**.1–3) form the greater proportion of the blood cells (99% of the total number), with a count of 4.1–6.0 × 10^6/μl in adult males and 3.9–5.5 × 10^6/μl in adult females. Each cell is a biconcave disc with a diameter of 6.3–7.9 μm (mean 7.1 μm) and a rim thickness of 1.9 μm; in wet preparations the mean diameter is 8.6 μm. Erythrocytes lack nuclei and are pale red by transmitted light with paler centres because of their biconcavity. They show a tendency to adhere to one another by their rims to form loose piles of cells (*rouleaux*), a character probably determined by the properties of their cell coat. In normal blood a few assume a shrunken star-like, *crenated* form (**6**.1), a shape which can be reproduced by placing normal biconcave erythrocytes in a hypertonic solution, which results in osmotic shrinkage. Such cells are called *echinocytes* (Bessis 1973). In hypotonic solutions erythrocytes take up water and become spherical and may eventually lyse to release their haemoglobin (*haemolysis*); they are then termed *red cell ghosts* (*erythrocytic umbrae*).

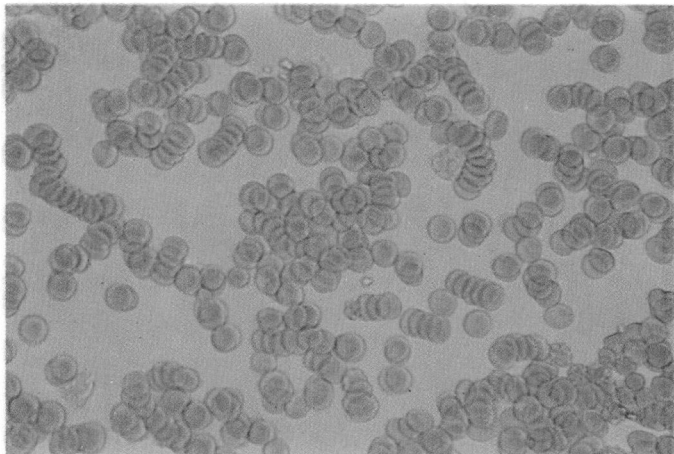

6.1 Fresh preparation of living erythrocytes showing rouleaux formation and red pigmentation. Magnification × 500.

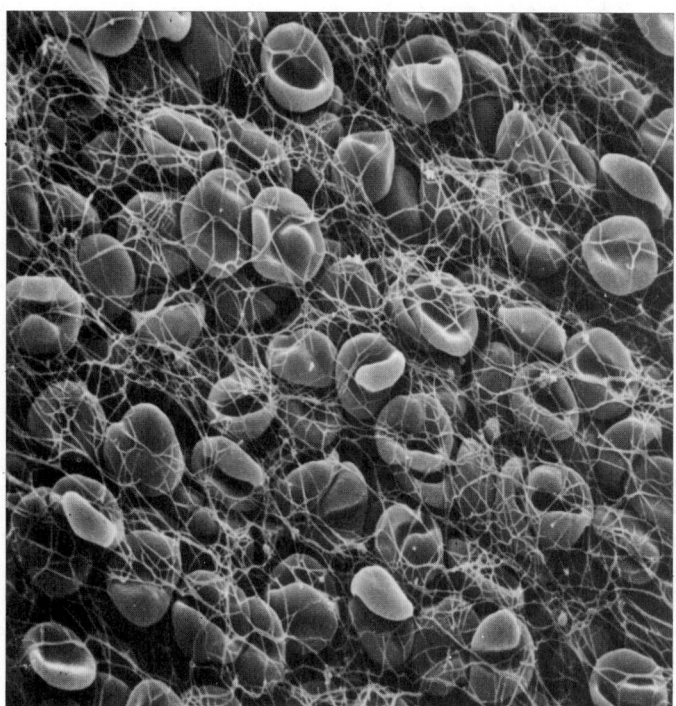

6.2 A scanning electron micrograph of erythrocytes, showing biconcave discoidal and other shapes; the filaments are fibrin resulting from clotting of the plasma after extravasation of blood. Photographed by Michael Crowder, Guy's Hospital, London. Magnification × 1500.

Erythrocytes are bounded by a plasma membrane and consist internally of a single protein, haemoglobin, apart from a few remnants from their initial development. The plasma membrane of erythrocytes has received much attention because of the ease with which it can be obtained for analysis in quantity (Bretscher & Raff 1975). It is about 60% lipid and glycolipid and 40% protein and glycoprotein. More than 15 classes of protein are present, including two major types. Firstly, the glycoproteins *glycophorins A* and *B* (each with an *M*r of about 50 000) span the membrane, and its negatively charged carbohydrate chains project from the outer surface of the cell, conferring most of the fixed charge on the cell exterior by virtue of its sialic acid groups. Secondly, comes the 'Band 2' protein which may bear some antigenic groups; 'Band 3' protein is also a transmembrane macromolecule, composed of four subunits which fit together and leaving a central hole through which ions can pass, thus constituting the important chloride channels in the erythrocyte membrane; these protein assemblies are visible as intramembranous particles in freeze-fractured erythrocyte membranes (p. 1.12); the ABO antigens (p. 681) are

663

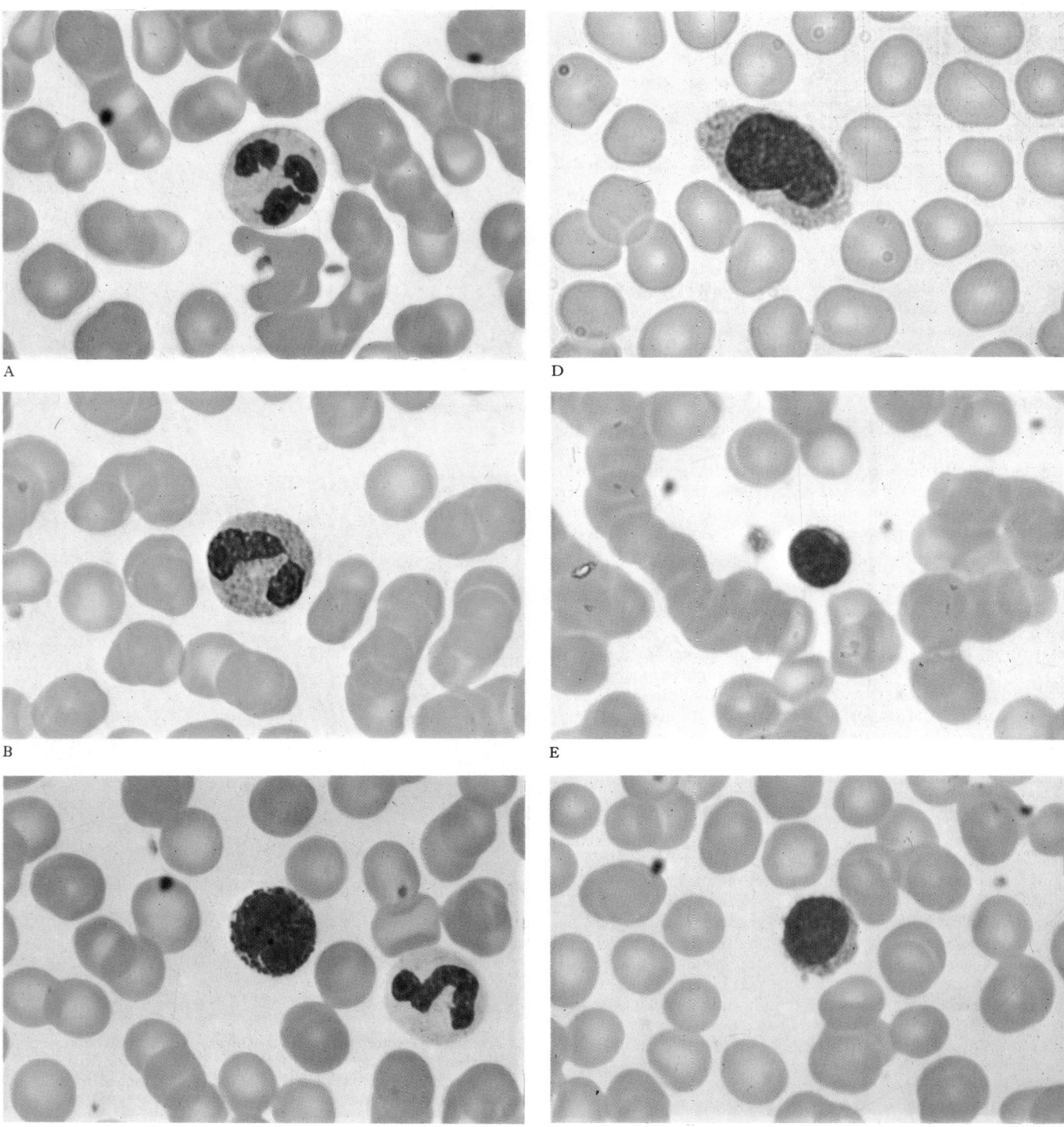

6.3 Blood cell types stained in smeared preparations by the Giemsa method. Erythrocytes are shown in all figures which also demonstrate other cell types: A. neutrophil leucocyte and platelets; B. eosinophil leucocyte; C. basophil leucocyte with prominent densely staining cytoplasmic granules and neutrophil leucocyte; D. monocyte; E. small lymphocyte; F. medium-sized lymphocyte. Material provided by J P Black, Department of Haematology, Guy's Hospital.

all glycolipids (Race & Sanger 1975). Other proteins include several enzyme systems, some concerned with ionic regulation, others with the addition of lipid to the cell membrane from serum lipid. (This is necessary because the cell does not possess its own synthetic apparatus.)

The shape of the eythrocyte is largely determined by the protein *spectrin* (a name which reflects its biochemical preparation from red cell 'ghosts') which is attached to integral membrane proteins (Band III) on the inner surface of the cell membrane via short lengths of actin filaments and other proteins (e.g. 'Band 4.2' protein, and *ankyrin*), forming a stabilizing network. This considerably stiffens the membrane, an effect which is aided by the large amount of cholesterol in the membrane itself. Red cells can thus regain their shape and dimensions after passing through the lumina of the finest ramules of the blood-vascular system; microscopic examination has shown that erythrocytes often pass through capillaries flattened face-first, buckling somewhat to a shield-like shape (Brånemark 1972) rather than rolling up, as might be expected.

Haemoglobin is a globular protein with a molecular weight of 67 000, consisting of *globulin* molecules bound to *haem*, an iron-containing *porphyrin* group (Perutz et al 1960). Each molecule is made up of four subunits, each in turn consisting of a coiled polypeptide chain with a cleft holding a single *haem* group. In normal blood, four types of polypeptide chain can occur, namely α, β, γ and δ. Each haemoglobin molecule contains two α-chains and two others, so that several combinations and hence a number of different types of haemoglobin molecule are possible. Haemoglobin A (HbA), which is the major adult class, contains 2 α- and 2 β-chains. Haemoglobin A₂ (HbA2), a minor component in adults, is composed of 2 α- and 2 δ-chains. Haemoglobin F (HbF), found in fetal and early postnatal life, consists of 2 α- and 2 γ-chains. In the pathological condition *thalassaemia* only one type of chain is synthesized, so that a molecule may contain 4 α-chains (β-thalassaemia) or, more commonly, 4 β-chains (α-thalassaemia) —Haemoglobin H (Wintrobe 1981).

Each polypeptide chain is determined by a separate gene; a number of variant haemoglobins are known in which only one or a few amino acid residues are abnormal, reflecting slight alterations in the corresponding genes. In the Haemoglobin S of sickle-cell disease a single alteration in the β-chains (valine substituted for glutamine) causes a major alteration in the behaviour of the red cell and its oxygen-carrying capacity which, however, may confer some protection against malarial infection, where the disease is endemic. Other common variants include Haemoglobins C and D. The oxygen-binding power of haemoglobin is provided by the iron atoms of the haem groups, and these are always maintained in the ferrous (Fe^{++}) state by the presence of glutathione.

In addition to haemoglobin, erythrocytes possess a number of enzyme systems, notably those concerned with glycolysis and ionic transport, which together maintain low sodium levels within the cell against diffusion gradients, and thus create the appropriate conditions of *p*H and ionic strength for the normal functioning of haemoglobin. As intimated above, glutathione metabolism is also active. Although, of course, in the absence of a nucleus and ribosomes, no protein synthesis takes place in mature erythrocytes, lipid in the plasma membrane can, however, be replaced to some extent from circulating serum lipids, by the activity of membrane enzyme systems.

The iron-containing compound *ferritin* is also often present in newly formed erythrocytes, as are also persisting remnants of the apparatus of protein synthesis (ribosomes and other RNAs) from the stage of differentiation of the cell in the bone marrow. After Romanowsky staining, the residual RNA of young erythrocytes causes a slight bluish tinge; with the supravital stain brilliant cresyl blue it forms a reticulum, giving the name *reticulocyte* to this type of cell. Later in maturation such evidences of basophilia disappear. Other inclusions may be present in red cells, particularly in pathological conditions; amongst these are nuclear remnants (*Howell–Jolly bodies*) and altered haemoglobin (*Heinz–Ehrlich bodies*).

Life Span and Destruction

Erythrocytes which have been labelled radioactively or antigenically and then injected into the circulation, have been shown to last between 100 and 120 days before being destroyed (Berlin et al 1959). As erythrocytes age they become increasingly fragile and their surface charge changes as their content of negatively charged membrane glycoproteins is progressively reduced (Marikovsky & Daron 1969). Perhaps as a result of the changes in charge they are eventually ingested by the macrophages of the spleen and liver sinusoids without previous lysis, and are then hydrolysed. Here, the haemoglobin is broken into its globulin and porphyrin moieties; the globulin is then further degraded into its constituent amino acids which pass to the general amino-acid pool of the body. The iron is removed from the porphyrin and can be used either directly in the synthesis of new haemoglobin in the bone marrow, or stored in the liver as ferritin or haemosiderin; the remaining haem portion is converted in the liver to bilirubin and is then excreted in the bile.

Red cells are produced by the bone marrow (p. 675) and are destroyed at the rate of about 5×10^{11} cells a day.

Fetal erythrocytes up to the fourth month of gestation differ markedly from those of adults, in that they are larger (10 μm), are nucleated and contain a somewhat different type of haemoglobin (Hb-F). From this time they are progressively replaced by the adult type of cell.

The Leucocytes

Leucocytes (*white blood corpuscles* or *cells*) belong to at least five different categories, distinguishable by their size, nuclear shape and cytoplasmic inclusions (**6.3,4,8**). For practical convenience, these types of cell are often divided into two main groups, namely those with prominent stainable cytoplasmic granules, the granulocytes, and those without, the agranulocytes. However, in terms of their biology, this distinction is now known to be quite arbitrary.

The granulocytes or *granular leucocytes* all possess irregular or multilobed nuclei and are often termed **polymorphonuclear leucocytes** for this reason. This group is comprised of three types of cell, the granules of which give different staining reactions with the Romanowsky dyes: they are *acidophil* (or *eosinophil*) leucocytes with granules which bind acidic dyes such as eosin, *basophil* leucocytes the granules of which bind basic dyes (methylene blue) strongly and *neutrophil* leucocytes whose granules stain weakly with both elements by a different type of reaction (see Wintrobe 1981, Zucker-Franklin et al 1981).

NEUTROPHIL LEUCOCYTES

Neutrophil polymorphonuclear leucocytes (*neutrophils, neutrophiles, heterophile leucocytes* or '*polymorphs*') form the largest proportion of the leucocytes (60–70% in adults, with a count of 3000–6000/μl). In dried smears, where the cells have flattened, they have a circular profile with a diameter of 10–15 μm. In the living state the cells may be spherical whilst passively circulating, but can flatten and become actively motile on contact with a suitable surface.

Within the cytoplasm the numerous granules give a variety of colour shades ranging from violet to pink when stained with Romanowsky stains such as Wright's and May-Grünwald-Giemsa, which are commonly employed in haematology. Under the electron microscope, too, the *granules* are heterogeneous in size, shape and content, but all are membrane-bound bodies containing hydrolytic and other enzymes. Two major categories can be distinguished according to their developmental origin and contents (Bainton et al 1971). Firstly, *non-specific* or *primary granules* which are formed early in neutrophil genesis (p. 680); these are relatively large (0.5 μm) spheroidal lysosomes containing myeloperoxidase, acid phosphatase and several other enzymes. With light microscopy they stain strongly with neutral red and azure dyes and are thus said to be azurophilic. Secondly, *specific* or *secondary granules*, formed a little later, assume a wide range of shapes including spheres, ellipsoids and rods. These contain several substances with strong bactericidal actions, including alkaline phosphatase, aminopeptidase, lactoferrin and also collagenase, all of which are lacking in the primary granules. The secondary granules, however, lack peroxidase and acid phosphatase. Some enzymes such as lysozyme are present in both. The presence of these granules correlates well with the phagocytic activity of neutrophils.

In *mature neutrophil* granulocytes the nucleus is characteristically multilobate with up to six segments joined by narrow nuclear strands (the *segmented* stage). Less mature cells have fewer lobes, the earliest to be released under normal conditions being *juveniles (band* or *stab cells)* in which the nucleus is an unsegmented crescentic band. In certain clinical conditions, even earlier stages in neutrophil formation with indented or rounded nuclei (*metamyelocytes* or *myelocytes*, p. 680) may be released from the bone marrow. In mature cells the edges of the nuclear lobes are often irregular; in females about 3% (range 1–17%) of the nuclei of neutrophils show a conspicuous 'drumstick' formation which represents the sex chromatin of the inactive X chromosome (*Barr body*) (see also p. 48).

Mitochondria, a Golgi complex, a sparse endoplasmic reticulum and glycogen are present in the cytoplasm. A conspicuous array of microtubules is often seen between the 'arms' of the nucleus, and it is interesting that locomotion of neutrophils is in the direction in which the free arms point.

Neutrophils form an important element in the defence systems of the body; they can engulf microbes and particles in the

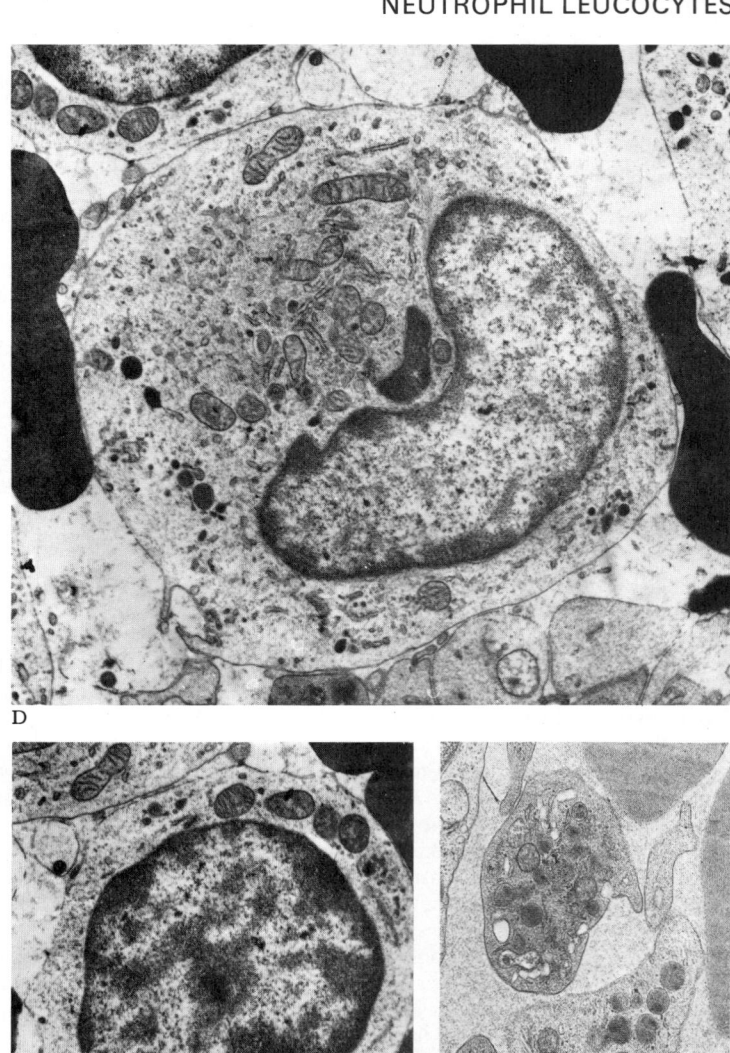

6.4 Electron micrographs: A. an erythrocyte showing a biconcave profile and dense homogeneous contents; B. a neutrophil showing its multilobed nucleus appearing in section as separate profiles. Note the various shapes and densities of the specific granules; C. an eosinophil, with crystalline inclusions in the specific granules; D. a monocyte showing an indented nucleus, endoplasmic reticulum and lysosomes; E. a small lymphocyte; F. a group of platelets. See text for further details.

circulation and, after migrating between the endothelial cells lining capillaries or venules, can perform local phagocytosis in the extravascular tissues, wherever it is needed. The engulfing of foreign objects is followed by digestion through fusion of the phagocytic vacuole, first with the *specific* granules, the pH being reduced to 4.0 by active transport of protons, then with the *non-specific* (azurophilic, primary) granules, which finish the process of bacterial killing and digestion. The chemical reactions occurring in these events have been intensively studied and prove to be quite complex. They involve the oxidation and addition of halide (chloride and iodide) radicals to the engulfed materials by means of enzymic action (myeloperoxidase, etc.) which have the effects of denaturing their proteins and other macromolecules. Lysozyme and lactoferrin are also highly toxic to bacteria. These processes require atmospheric oxygen for successful completion, so that neutrophils active in defence have a high oxygen demand. Phagocytosis, or the release of granules, may be enhanced by the presence of antibodies attached to the surfaces of neutrophils which can bind specifically to target antigens, e.g. in a type of bacterium to which the body has previously been exposed.

Opsonizing antibodies (*opsonins*, p. 672) coating the antigenic target may also promote phagocytosis. The antibodies in both cases are secreted by lymphocytes (p. 672), and the neutrophil granulocyte is just one in a series of defensive cells which form an interrelated system for the elimination of foreign materials from the tissues (**6.6**).

After phagocytosis, the neutrophil's cytoplasmic granules gradually become used up, so that a marked reduction in their number (degranulation) occurs. Granules may also be discharged from the surface of the cell when it is suitably stimulated, to damage or kill neighbouring organisms or cells.

The numbers of circulating neutrophils vary considerably, often rising during episodes of acute bacterial infection. They may circulate freely in the blood (the *circulatory pool*), or they may adhere to the walls of post-capillary venules and other vessels (the *marginal pool*) to re-enter the circulation when suitably recruited e.g. during brief exercise or by exposure to noradrenalin. However, neutrophils are short-lived (half-life 7.5 hours); they may either be destroyed in the bloodstream, or pass through the endothelial walls to the extravascular tissues, engaging in defence alternatively, after entering various secretory ducts such as the bronchi, salivary gland ducts and the urinary tract, they are lost to the body.

EOSINOPHIL LEUCOCYTES

These cells (*acidophil leucocytes*) are similar in size, shape and motile capacity to neutrophils but are present only in small numbers in normal blood (100–400/μl). The granules of the cytoplasm are uniformly large (0.5 μm) and give the living cell a slightly yellowish colour; with Romanowsky stains they are uniformly orange to red. The nucleus has two prominent lobes connected by a thin strand. Ultrastructurally (Zucker-Franklin 1980, 1985), the cytoplasm is seen to be packed with specific granules (**6.4**) which are spherical or slightly ellipsoidal. Each of these bodies is bounded by a membrane and contains an amorphous material (the *matrix* or *externum*) in which is embedded a prominent crystal (the *crystalline core* or *internum*); in man the crystalline core shows a square lattice structure with a 4 nm repeat pattern. The matrix contains several lysosomal enzymes including acid phosphatase, ribonuclease, phospholipase and a myeloperoxidase unique to eosinophils. The crystalline core is composed of the characteristic *major basic protein*, with an Mr of 9200; this protein contains a high ratio of arginine residues and gives the granules their strong eosinophilic staining properties. Apart from these granules, the cytoplasmic organelles are similar to those of neutrophil granulocytes.

Like other leucocytes, eosinophil granulocytes are able to pass into the extravascular tissues from the circulation when suitably stimulated. In small numbers they are typical constituents of the dermis and the connective tissue components of the bronchial tree, alimentary tract, uterus and vagina and thymic medulla. The total life span of these cells is a few days, of which about 30 hours is spent in the circulation and the remainder in the extravascular tissues.

The functions of eosinophils are not yet fully understood. Their ratio to other leucocytes rises greatly (an *eosinophilia*) in certain allergic disorders, and also in worm infestations, and their is now much evidence that they play an important part in the immune system, phagocytosing and inactivating antigen–antibody complexes and also various inflammatory substances, e.g. histamine and leukotrienes I and II; so they may be important in limiting the effects of these substances on the surrounding tissues. It has also been shown in vitro that they can attach themselves to and kill parasitic schistosome larvae which invade the circulation.

BASOPHIL LEUCOCYTES

Like other granulocytes, these are 10–15 μm in diameter (**6.3**), but form only 0.5–2% of the total leucocyte population of normal blood, with a count of 25–200/μl. Their distinguishing feature is the presence of large, conspicuous basophilic granules, varying in number from 10–100 per cell. The nucleus is somewhat irregular but not usually lobated, unlike those of the other granulocytes. Ultrastructurally, the basophilic granules are membrane-bound vesicles with densely stained contents showing a variety of crystalline, lamellar and granular inclusions. Heparin, histamine and several other inflammatory agents are contained in these granules, which closely resemble those of mast cells (p. 60). These substances are apparently associated with polysaccharides, since the granules are also positive to carbohydrate stains, e.g. periodic acid-Schiff (PAS) and Azure A, with which they are *metachromatic*, i.e. they stain a different colour from that of the dye. Their life span in the circulation is long (9–18 months in mice).

Although they resemble mast cells (Selye 1965) and, like these, are formed in the bone marrow, there is much evidence that basophils represent a different, albeit closely related cell line peculiar to the circulation, as shown by their reactions with monoclonal antibodies and differences in cell development (see also p. 681). Their functions in the circulatory system are at present rather obscure.

MONOCYTES

These are the largest of the agranular leucocytes (15–20 μm in diameter in smears) but they form only a small proportion of the leucocytes (2–8% with a count of 100–700/μl of blood). The nucleus, which is euchromatic, is relatively large and has a characteristic indentation on one side. The cytoplasm forms a wide rim around the nucleus, near the indentation of which lies a prominent Golgi complex and vesicles stainable with neutral red. Ultrastructurally (Cawley & Hayhoe 1973, Zucker Franklin et al 1981) many lysosomes are seen to be present, together with some peripheral rough endoplasmic reticulum. Mitochondria are quite abundant, reflecting the highly motile nature of the cell. Monocytes are actively phagocytic.

All of these characteristics are similar to those of macrophages. There is strong evidence that monocytes are mainly or all macrophages in the process of passing from the bone marrow, where they are formed, to the peripheral tissues via the bloodstream, these cells passing into extravascular sites through the walls of capillaries and venules. This topic is discussed further on p. 668–669.

LYMPHOCYTES

These cells (**6.3,4**) are the second most numerous type of leucocyte, forming 20–30% of their total number, i.e. 1500–2700/μl of blood. Like other leucocytes, they are also found in extravascular tissue, but they are remarkable in being formed in large numbers outside the bone marrow as well as within it. They therefore constitute a widely distributed *lymphoid* system (see p. 670 for a fuller description).

Until relatively recently, lymphocytes were regarded as cells of little functional importance, perhaps representing degenerative stages of some other type of blood cell. The work of Miller, Yoffey, Gowans and others in the 1950s showed that these cells play a vital part in the body's defences and since that time they have come to be a focal point of modern immunology (see e.g. Roitt 1984).

In the blood, the term 'lymphocyte' refers to agranular leucocytes 5–15 μm in diameter. This size range represents a heterogeneous collection of various lymphocyte types in different states of maturity and activity, as well as a small percentage of cells ('null cells') which are not apparently true lymphocytes at all but either related cells (e.g. natural killer cells) or some other quite different form, such as circulating haemopoietic stem cells. The various types and conditions of cells can now be at least partially separated by means of immunocytochemical techniques, involving the use of monoclonal antibodies (see p. 675) raised against different components of their cytoplasm. Where these detect specific macromolecules exposed at the surface of lymphocytes, living cells can also be separated into different classes by various techniques.

Small lymphocytes, which form the majority of lymphocytes in circulating blood, each contain a rounded densely staining nucleus, surrounded by a very narrow rim of cytoplasm. The nucleus contains dense, coarse bands of chromatin (a leptochromatic nucleus: *leptos* = a thread), and one or more small nucleoli. The cytoplasm stains slightly blue with Romanowsky dyes. Under the electron microscope (**6.4**), few organelles can be seen in the cytoplasm, apart from a small number of mitochondria, single ribosomes (monosomes), sparse profiles of endoplasmic reticulum and occasional lysosomes; these features indicate a low metabolic rate and such lymphocytes are said to be in the 'resting' phase. However, these cells are freely motile when they settle on to solid surfaces, and can pass between endothelial cells to exit from or enter the vascular system. They may make extensive migrations within the various tissues, including epithelia, even passing into the body's secretions such as saliva. *Larger lymphocytes* in the circulation are a mixture of very immature cells (lymphoblasts) capable of cell division to produce small lymphocytes, and maturing or mature cells which have become functionally active after stimulation of the immune system, e.g. by the presence of antigens. Both lymphoblasts and maturing cells are actively engaged in synthesizing proteins, and thus contain a nucleus which is at least in part euchromatic, and a basophilic cytoplasm with numerous polyribosome clusters. The ultrastructure of these cells varies according to their class and will be described where appropriate (vide infra).

The life span of lymphocytes ranges from a few days to many years and so we can distinguish between *short-lived* and *long-lived* lymphocytes, the latter being of great significance in the mechanisms of *immunological memory* (see p. 674).

As already stated, lymphocytes can be divided into several functional and maturational classes, but two major groups exist in the circulation, designated **B-** and **T-lymphocytes** (or **B** and **T cells**), according to the histories of their cell lineages. These will be considered in detail in another place (p. 671). In normal blood, about 85% of the lymphocytes are T cells.

PLATELETS

Blood platelets, also known as *thrombocytes* (6.3,4), are relatively small (2–4 μm across) irregular or oval discs present in large numbers (250 000–500 000/μl) in blood. In fresh blood samples they readily adhere to each other and to all available surfaces, unless the blood is treated with citrate or other substances which reduce the availability of calcium ions. In Romanowsky-stained preparations, platelets show an outer clear zone (*hyalomere*) and an inner basophilic (or azurophilic) granular region (*granulomere*). Ultrastructurally (Cawley & Hayhoe 1973, Zucker Franklin et al 1981) each platelet is seen to be anucleate, unlike similar cells in submammalian vertebrates (for which the term *thrombocytes* is perhaps best reserved). Each platelet is surrounded by a plasma membrane bearing a thick glycoprotein-rich coat, responsible for the adhesive properties of platelets. Beneath the surface is a band of about 10 microtubules which runs around the perimeter of the cell and probably determines its shape. Associated with these are actin filaments, myosin and other proteins related to cell contraction. Within the cytoplasm are also mitochondria, glycogen, a few profiles of agranular endoplasmic reticulum, including some narrow tubular channels and tubular invaginations of the plasma membrane and various membrane-bounded vesicles. These vesicles include three major types, designated alpha, delta and lambda.

Alpha-granules are the most prominent, with diameters of up to 500 nm; they contain platelet-derived growth factor (PGDF), fibrinogen and other substances. In the smaller (300 nm) *delta*-granules are 5-hydroxytryptamine (serotonin), taken up and concentrated from the blood plasma, calcium ions, ADP, ATP and pyrophosphate. In the smaller (250 nm) lambda set, granules contain several lysosomal enzymes (Bentfield & Bainton 1975).

Platelets are important in haemostasis (the limiting of bleeding after injury to vessels) and a number of more general, systemic functions which at present are rather poorly understood. When a blood vessel is damaged, platelets first aggregate at the site of injury, assisting to staunch the wound with a *platelet plug*. Adhesion between platelets (agglutination) and to other tissues is a function of the thick platelet coat and is promoted strongly by the release of ADP and calcium ions from the platelets as a response to vessel injury. Then, substances released from the alpha-granules, together with factors released from the damaged tissues, initiate a complex sequence of chemical reactions in the blood plasma, leading to the precipitation of insoluble *fibrin* filaments, generating a three dimensional meshwork, the *fibrin clot*, to which more platelets adhere, inserting extensions of their surfaces (filopodia) deep into the spaces between the filaments, to which they adhere strongly. Next, the platelets contract (*clot retraction*) by actin–myosin interactions within their cytoplasm (p. 50), concentrating the fibrin clot and pulling the adhering walls of the blood vessel together, limiting further any leakage of blood. Eventually, on repair of the vessel wall, *clot removal* occurs as a result of complicated enzymic activities, including those of *plasmin* formed by plasminogen activators in the plasma, and probably assisted by lysosomal enzymes derived from the small (lambda) granules of platelets.

Platelets may also be involved in various other biological activities, e.g. they have receptors for class IgE antibody molecules which enable them to adhere to and damage antibody-coated targets by means of their lysosomal secretions. Platelet-derived growth factor is also a potent trophic substance in laboratory cell cultures and may also exert a similar effect elsewhere, e.g. in regenerating tissues after damage (see Longenecker 1985).

Cells of the Mononuclear Phagocyte and Lymphoid Systems

It has already been stated in the description of blood cells that certain leucocytes (monocytes and lymphocytes) have special, interdependent roles in the body's defences against infection and constitute two major systems of cells dispersed throughout the body, present both within and external to the vascular channels.

THE MONONUCLEAR PHAGOCYTE SYSTEM

This term, given by Aschoff in 1924, denotes a diffuse system of phagocytic cells present throughout the body, engaged in a wide range of activities concerned with defence and various 'housekeeping' functions in the tissues where they reside. It is in part equivalent to the *reticulo-endothelial system*, but for various reasons has superceded it in modern literature on this subject (see e.g. Roitt 1984).

Although all cells of the mononuclear phagocyte system are indeed capable of phagocytosis, this is only a part of their overall role in the body, for they also perform many other actions vital to the effectiveness of the immune system and the biology of many tissues of the body. A feature which all mononuclear phagocytes have in common is that their surfaces bear a unique group of Major Histocompatibility Complex molecules (class II MHC antigens, see p. 672) which enable them to interact cooperatively with lymphocytes (vide infra).

As an historical accident, the cells of this sytem were originally given different names according to their location: in connective tissue they include *macrophages* (histiocytes or clasmatocytes); in the liver, *littoral cells of the sinusoids* (*von Kupffer cells*); in the central nervous system, *microglial* cells; and in the meninges, *meningocytes*. They are termed *pleural, peritoneal, alveolar* and *splenic macrophages* in their respective sites and in the synovial joints the *types A* and *B synovial cells* (p. 475). In the blood they are *monocytes*. Cells with similar properties but having a highly branched form (*dendritic* and *interdigitating cells*) are also present in various regions (lymph nodes, lymph nodules, spleen, thymus, epidermis, see e.g. Hoefsmit 1982).

In subserous tissue of the pleura and peritoneum, macrophages often aggregate as *milky spots*, near small lymphatic trunks. In the spleen they also occur in clusters (*ellipsoids*) around the exits of small (penicillar) arterioles (p. 829). Macrophages may fuse to form large syncytia (*foreign body giant cells*) contacting large particles which are too big to be phagocytosed, or when stimulated by the presence of infectious organisms, e.g. tubercle bacilli (*epithelioid cells*). In some situations they appear to be relatively sedentary (*fixed macrophages*), whereas in others they can migrate freely within tissues, or can traverse endothelial linings of blood vessels and lymphatic vessels.

Origins of cell lineages. There is now much evidence that all cells of this system arise from stem cells in the bone marrow related to the production of the monocyte–neutrophil granulocyte line (the CFU-G/M stem cell, see p. 681); indeed it appears that monocytes represent blood-borne cells of these types circulating en route to their final tissue destinations, migrating into them through the endothelial walls of capillaries and venules. Once arrived they may undergo further limited cell division and may also move on to other sites, e.g. lymph nodes (p. 822). Their final form and activities are closely related to the nature of the tissues where they reside but its not yet known whether mononuclear phagocytes undergo this tissue-specific differentiation because of their environment or if it is predetermined in the bone marrow, the cells migrating selectively to their appropriate tissues. Osteoclasts and chondroclasts may also be in some distant way related to these cells, since they have several structural and functional similarities and also arise from stem cells in the bone marrow; however, for various reasons it appears unlikely that they originate directly from the monocyte lineage (see p. 295).

Functions of Mononuclear Phagocytes

Phagocytosis of particulate material and organisms is carried out by macrophages and similar cells in many locations. In general connective tissue they can dispatch invading micro-organisms and remove debris engendered by tissue damage, or in growth and remodelling in undamaged tissue. In the lung, alveolar macrophages constantly patrol the surfaces of alveoli into which they migrate via intercellular junctions of the epithelial cells; they engulf inhaled particles including bacteria and debris and enter the sputum (hence 'dust cells' and, in cardiac disease, 'heart failure cells' full of extravasated erythrocytes). Similar activities occur in the pleural and peritoneal cavities. In lymph nodes they line the walls of sinusoids ('littoral macrophages') and remove particulate matter from the passing lymph as it percolates through these narrow passages.

In the spleen and liver, macrophages carry out similar particle removal but are also involved in the detection and destruction of aged or damaged erythrocytes, whose haemoglobin they begin to degrade preliminary to recycling iron and amino acids (see p. 665).

The part which mononuclear phagocytes play in immune responses is complex and incompletely known. As mentioned elsewhere (p. 673), they bear class II MHC antigens at their surface, which enable them to stimulate different classes of T-lymphocyte through close-range interactions: macrophages phagocytose alien antigens and partially digest them in their lysosomes, passing some of the antigenic remnants to the cell surface where they are bound by MHC molecules. This complex of alien antigen and MHC molecule is then presented to a T-lymphocyte, which, if it possesses the appropriate receptor on its surface, is stimulated in various ways, depending on the type of T cell involved (p. 672). In turn, macrophages are also affected by activated T- and B-lymphocytes: lymphokines (including macrophage-activating factors, Interleukin II, etc) secreted by certain T cells can determine their migration and degree of phagocytic activity (p. 672) and under such influences the macrophages themselves can synthesize and secrete various other bioactive substances, e.g. Interleukin I which stimulates the proliferation and maturation of other lymphocytes, so greatly amplifying the reaction of the immune system to foreign antigens. When activated by Interleukin II and other lymphokines, macrophages synthesize and release many other bioactive molecules, including Tumor Necrosis Factor (TNF) which is able to kill neoplastic cells and appears to be a mechanism for the removal of small tumours which may appear in the body from time to time throughout life; it is interesting that TNF also depresses the anabolic activities of many cells in the body and this may be a major factor in caccexia (wasting) which typically accompanies more advanced cancers. Other macrophage products include plasminogen activator, promoting clot removal (p. 668), and various lysosomal enzymes, several complement and clotting factors and lysozyme (an antibacterial protein). In pathogenesis, such substances may be 'erroneously' released to damage healthy tissues, e.g. in rheumatoid arthritis and various other inflammatory conditions.

Macrophages also have receptors on their surfaces for the Fc ends of antibodies and for the C3 component of complement, and so can bind readily to and avidly phagocytose antibody-coated (opsonized) microbes and other 'alien' material. Such close antibody-mediated binding may also initiate the release of lysosomal enzymes on to the surfaces of the cellular targets to which the macrophages are bound (an example of antibody-dependent cell-mediated cytotoxicity, ADCC, also shown by various other cells including neutrophils), particularly if these are too large to be phagocytosed (e.g. nematode worm parasites such as Wuchereria bancrofti). These reactions are much enhanced when macrophages are stimulated by lymphokines.

Macrophages and related cells also produce several growth and differentiation factors with actions on other tissues. They release factors which stimulate haemopoiesis, including erythropoietin, and have complex metabolic actions through their production of prostaglandins, thromboxanes and other bioactive substances, including the stimulation of bone resorption by osteoclasts.

In summary, these various activities appear to be related to three quite distinct roles: the first associated with defence, the second with repair and regeneration of damaged tissues and the third with the ongoing maintenance of normal proliferation and differentiation of healthy tissues throughout the body. The full extent of macrophage-mediated effects has yet to be determined and some of them are quite unexpected, e.g. high levels of Interleukin I induce sleep, as found in some severe systemic infections. It seems that macrophages may be intricately connected with a vast array of cellular activities in a network of great complexity.

Classes of mononuclear phagocytes. In broad outline, we can group the different mononuclear phagocytes into two major classes: one highly phagocytic, and the other less so but with other important activities. The highly phagocytic cells include all the mononuclear phagocytes named above, apart from the dendritic cells. Their phagocytic function is chiefly to remove from the tissues and circulating fluids any particulate material such as microbes of various kinds, tissue and cellular debris and moribund or dead cells (e.g. aged or damaged erythrocytes by splenic and hepatic macrophages, vide supra). Such phagocytic cells contain numerous lysosomes and can be readily labelled with particulate dyes (e.g. colloidal carbon, trypan blue) and electron-dense tracers (horse-radish peroxidase, colloidal gold, etc,) which they readily take into their cytoplasm. The second major group, the dendritic mononuclear phagocytes, although capable of moderate uptake of particulate material, consists of cells which are specialized to initiate immune reactions in lymphocytes mainly within lymphoid tissue masses. These cells include the follicular dendritic cells and interdigitating cells of lymph nodes and lymphoid nodules, similar cells of the spleen and thymus and the epidermal dendritic (Langerhans) cells of the skin. The microglial cells of the central nervous system may possibly be included in this class, although there is evidence for very active phagocytosis under some conditions (see p. 895).

The main action of dendritic cells appears to be to phagocytose alien antigens and present them in association with class II MHC molecules at their surfaces to the neighbouring lymphocytes. The relative contributions of the dendritic and other mononuclear phagocytes to the general activation of lymphocytes in response to an antigen has yet to be determined. Within the dendritic cell populations, at least two varieties occur, one of which (the epidermal or Langerhans cells and some thymic and non-follicular dendritic cells of lymph nodes) has characteristic 'Birbeck' granules (see p. 77) while the other dendritic cells lack these. It has been suggested that those with the characteristic granules are all epidermal dendritic cells; those present in lymph nodes and the thymus are cells which have migrated in from their peripheral positions, carrying with them antigens which they have picked up in the skin, to trigger appropriate responses in T- and B-lymphocytes. Such stimulated lymphocytes may then either pass to the skin and carry out defensive actions there or populate regional lymph nodes to deal with any infectious agents which might pass into them from the lymphatic drainage of the area.

Because of their branched form and locations, the various, dendritic cells lie in intimate contact with numerous cells in their vicinity and, in view of the known trophic functions of macrophages, it is possible that dendritic cells help to regulate the cells they lie close to. The epidermal dendritic cells may help to control epidermal proliferation (p. 77), those of the bone marrow may assist to control haemopoiesis, etc.

Mononuclear phagocytes vary in structure depending on their location in the body. All are large cells (15–25 μm across) which are moderately basophilic and contain some granular and agranular endoplasmic reticulum, an active Golgi complex, mitochondria, etc. and a large, euchromatic nucleus, signifying an active metabolism and continuing synthesis of lysosomal enzymes, unlike neutrophil granulocytes which virtually cease synthetic activity shortly after leaving the bone marrow. All cells have irregular surfaces with many filopodia and contain varying numbers of endocytic vesicles and lysosomes. Some macrophages are highly motile, whereas others tend to remain attached and sedentary (e.g. the littoral cells of hepatic and lymphoid sinusoids). For further descriptions of these cells, see other sections (pp. 673, 824, 830, 837).

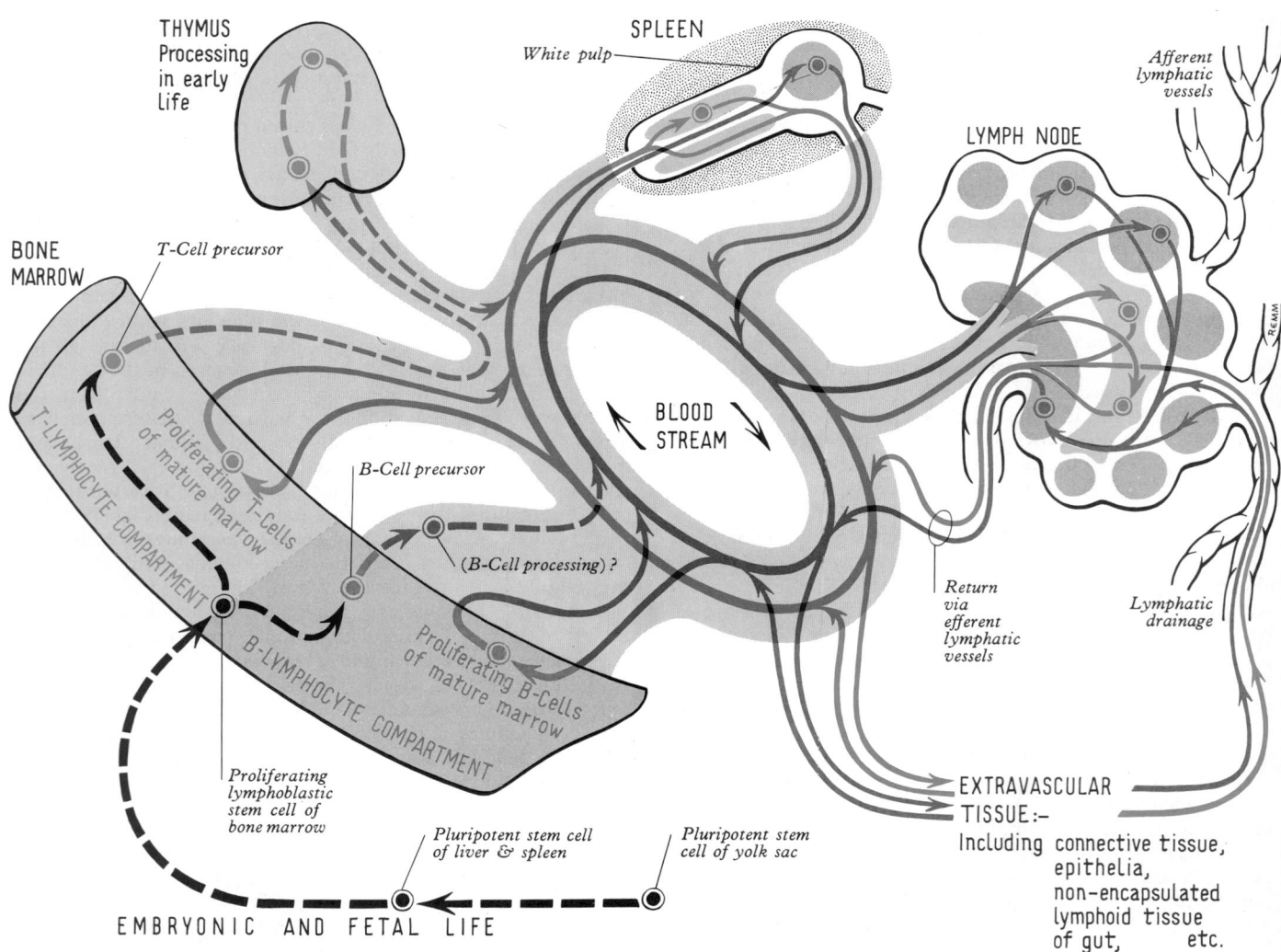

THYMUS
Processing
in early
Life

SPLEEN

White pulp

*Afferent
lymphatic
vessels*

LYMPH NODE

BONE
MARROW

T-Cell precursor

*Proliferating T-Cells
of mature marrow*

T-LYMPHOCYTE COMPARTMENT

B-Cell precursor

(B-Cell processing)?

BLOOD
STREAM

B-LYMPHOCYTE COMPARTMENT

*Proliferating B-Cells
of mature marrow*

*Return
via
efferent
lymphatic
vessels*

*Lymphatic
drainage*

*Proliferating
lymphoblastic
stem cell of
bone marrow*

*Pluripotent stem cell
of liver & spleen*

*Pluripotent stem
cell of yolk sac*

EXTRAVASCULAR
TISSUE:—
Including connective tissue,
epithelia,
non-encapsulated
lymphoid tissue
of gut,   etc.

EMBRYONIC AND FETAL LIFE

6.5 Diagram depicting the current views of the origins and circulation of the two major classes of lymphocyte from the bone marrow to the peripheral lymphoid tissues. B-lymphocytes are shown in red, T-lymphocytes in blue.

The reticulo-endothelial system. The concept of an inter-related system of mononuclear phagocytes has in recent years supplanted that of the *reticulo-endothelial system*, which at its inception was envisaged as a diffuse network of various phagocytic cells, mainly endothelial, lying in or in close proximity to the walls of certain blood vessels and lymphatics. It was found that these cells took up many particulate dyes and other substances infused into blood and lymph and thus seemed well fitted for the removal and destruction of particulate materials from these fluids. However, more detailed study showed that, in general, true endothelial cells are only mildly phagocytic, as indeed are many other tissue cells; the highly phagocytic cells are not endothelial cells at all, but types of macrophages of myeloid origin and, unlike endothelial cells, bear Class II MHC molecules at their surfaces.

THE LYMPHOID SYSTEM

Origins of Lymphocytes

Much of the knowledge of lymphocyte life history (**6.**5,6) has come from experimental situations which include the tracing of radioactively or genetically labelled cells, the latter involving transfusion of lymphocytes with chromosomal abnormalities into normal but inbred strains of laboratory animals. Other experiments are based on the sensitivity of lymphocytes to ionizing radiation, the various components of the lymphocytic system being eliminated by selective irradiation, accompanied by appropriate transfusions or transplants of lymphopoietic tissue from bone marrow, lymph nodes, thymus and so forth. Thirdly, selective surgical removal of various lymphopoietic components and the removal of lymphocytes from the lymphatic channels has

been carried out. In addition to these, a whole battery of immunological techniques has been brought to bear on the problem, including immunoassay, tissue culture methods, techniques for the localization of antibodies by their reactions with antigens or antigen-antibody complexes previously labelled with fluorescent compounds or electron-dense substances (or enzymes) with subsequent examination by fluorescence and electron microscopy respectively.

The picture which has emerged is still far from clear; nevertheless tentative views can be expressed at this stage.

Lymphocytes originate in the embryo from mesenchymal cells of the yolk sac initially, and later in the liver and spleen. These primitive lymphoid *stem cells* subsequently take up residence in the bone marrow, which becomes the only site of stem cell proliferation after birth. Such stem cells divide unequally to throw off lymphoblasts which again divide a number of times and transform into small lymphocytes. Some of these pass in the blood circulation to the thymus where they migrate into the lobules and divide repeatedly (p. 834); the resulting small thymus-processed *T-lymphocytes* then re-enter the bloodstream and can return to the bone marrow or to the *peripheral lymphoid tissues* (tonsils, lymphoid tissue of the alimentary and respiratory tracts, lymph nodes, spleen), which they enter by migrating through the walls of the capillaries or post-capillary venules. Here, the cells are found in the sinusoids of the lymph nodes and in the loose superficial tissue of the lymph nodes (i.e. paracortical areas) and *peri-arteriolar white pulp* of the spleen, i.e. in both cases, tissue neighbouring the *germinal centres* (p. 824). They can enter the general flow of lymph, returning to the bloodstream via the thoracic and right lymphatic ducts and the brachiocephalic veins, and s

670

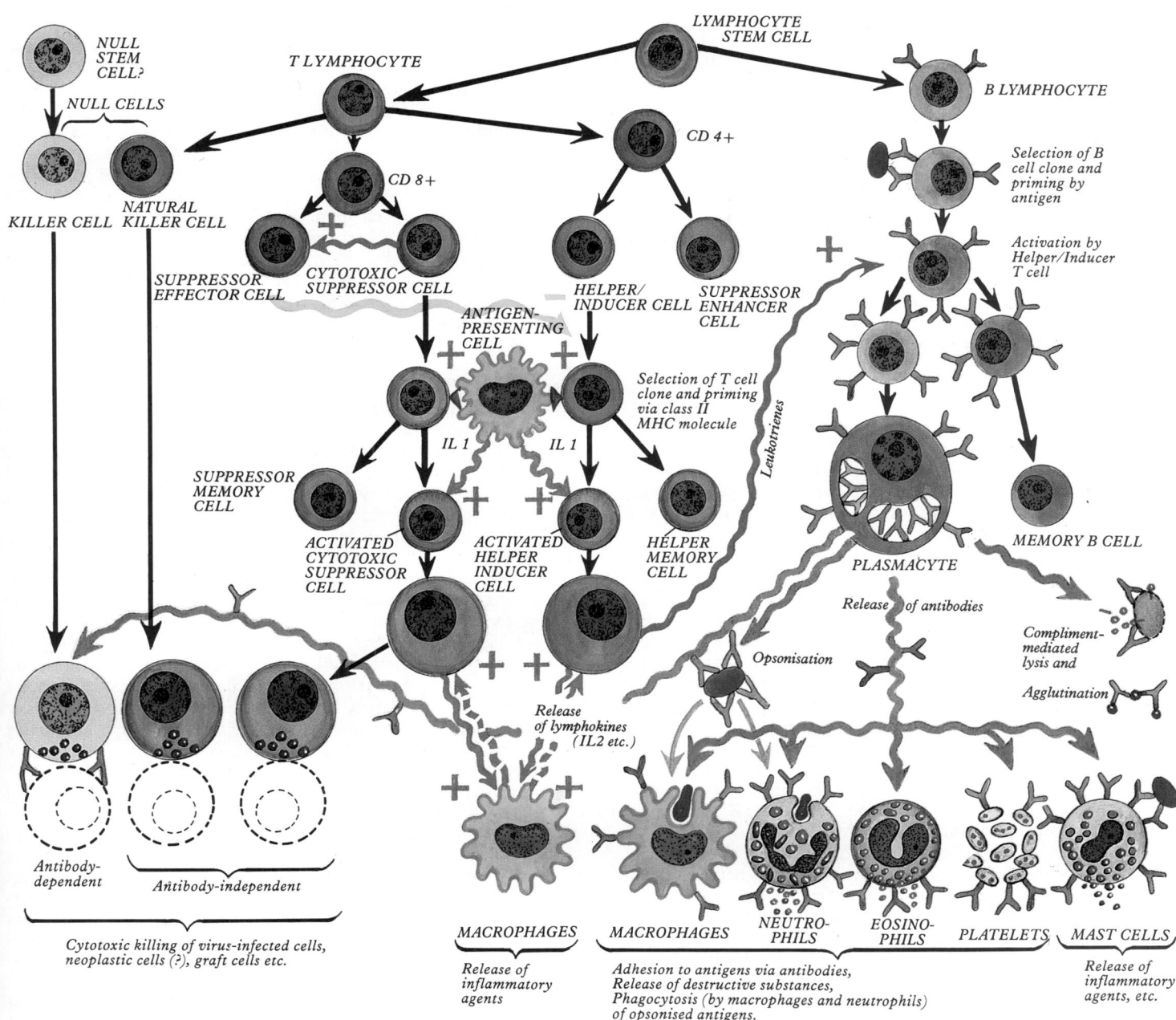

6.6 Schema of the origins and functional interactions by lymphocytes and other cells of the immune system.

eventually back to the lymphoid tissues again. This *circulation of lymphocytes* first established by Gowans, is responsible for the large number of lymphocytes found in the blood (**6.**5). When stimulated antigenically such lymphocytes enlarge and multiply, and their progeny are capable of interacting with tissue cells (vide infra).

The *B-lymphocytes* do not pass through the thymus, but move directly via the circulation to the general lymphoid tissues. where they settle down. En route they may undergo a phase of differentiation similar to that of the T-lymphocytes in the thymus, although it is not clear where this step might take place. Various sites have been suggested, including the intestinal lymphoid tissue; as yet conclusive proof is not available but it appears likely that the bone marrow is the site of B-cell differentiation. In birds, B-lymphocytes are derived from a specialized diverticulum of the cloaca, called the *bursa of Fabricius* (giving this type of lymphocyte its title—the *bursa equivalent*—or *B-lymphocytes*, although subsequently it appears that such a prefix can be even more appropriately applied to denote the bone marrow origin of the B-cell class in mammals). On antigenic stimulation they multiply to form the *germinal centres*; such lymphocytes can, while

still within the lymphoid tissues or after further migration, mature into the large pyroninophilic (i.e. RNA-rich) plasma cell series, which produce antibodies in their extensive rough endoplasmic reticulum.

FUNCTIONING OF THE LYMPHOCYTIC SYSTEM (6.6)

The lymphocytes, together with the phagocytes of the mononuclear phagocyte system, are responsible for the defensive reactions of the body, the former by non-phagocytic interactions of various kinds, the latter mainly by phagocytosis. These defences are directed against foreign invasion by bacteria, viruses, fungi, protozoa, helminth worms, etc. or their metabolites (toxins) and against any unwanted or abnormal materials within the body such as alien or altered proteins and effete, neoplastic or virally transformed cells.

B-lymphocytes

These cells can, when antigenically stimulated, proliferate and their progeny transform to larger cells (*plasmocytes*) which

671

synthesize and secrete *circulating antibodies*; the latter chemically 'recognize' and bind specifically to foreign chemical substances (*antigens*) to inactivate them or cause their destruction. Antibodies may circulate freely in the body fluids (*soluble antibodies*) or may be secondarily attached to a variety of defensive cells (*homocytotropic antibodies*) to enhance their activities or to enable them to carry out a wider range of functions.

Chemically, **antibodies** (or *immunoglobulins*) are proteins with a relative molecular mass (Mr) of 150 000–950 000. Each antibody molecule consists of four polypeptide chains, two of them long (about 15 nm, *heavy* chains) and two shorter (*light* chains) all joined together by sulphydryl links. The whole molecule is Y-shaped, two of the arms of the Y being able to swing around a central hinge. One end of the molecule (the *Fab* fraction) is highly variable in its amino-acid sequences and, since this end is responsible for specific binding to antigens, there is a vast number of possible antibody varieties, each type being able to bind a specific antigen in a manner closely analogous to enzyme-substrate binding. The other end of the molecule, the *Fc* portion, is much less variable and can attach to certain cells if they possess specific *Fc* receptors on their surfaces (or, in the case of IgM molecules, by virtue of hydrophobic *Fc* ends able to insert themselves into membranes), thus conferring antigen-binding properties upon them by virtue of the free *Fab* ends of the bound antibodies.

At present, five classes of antibodies are distinguishable in the blood plasma and elsewhere. These are: Immunoglobulin G (IgG), which forms the bulk of circulating antibodies; Immunoglobulin M (IgM), which is often formed early in the immune response and is usually present as a pentamer, the monomers joined near their *Fc* ends into a star-like aggregate; Immunoglobulin A (IgA), which is present in secretions of the body, particularly saliva and other fluids of the alimentary tract; Immunoglobulin E (IgE), which is a *homocytotropic* antibody found on the surface of mast and perhaps other cells; and finally Immunoglobulin D (IgD) of uncertain significance, although believed to be important in the ontogeny of the immune system. All classes can be soluble antibodies, and IgG and IgE can also be homocytotropic (vide infra).

The defensive functions of antibodies are numerous. They can *agglutinate* antigens by forming cross-links between them so rendering them inactive as infective agents (e.g. viruses). After binding an antigen they may also co-bind *complement*, a protein complex in plasma which punctures cell membranes, so causing the lysis of bacteria and other cells attacked by antibody. If only part of the complement protein complex is bound (e.g. C³), they can form bridges between defensive cells such as macrophages and target cells, causing their destruction. They may also bind to antigens after attachment by their 'safe' *Fc* ends to other defensive cells which are thus armed to detect, adhere to and ingest or enzymatically damage foreign bodies; examples are macrophages and neutrophils both of which bear *Fc* receptors on their surfaces. Such cell-bound antibodies can also activate the complement complex to cause lysis of bacteria, etc. Antibodies may first coat an antigen (*opsonizing antibodies* or *opsonins*) before they attach to a macrophage, thereby stimulating its phagocytic capacity. Homocytotropic antibodies may cause the activation of cells in other ways when they bind antigens. Thus, IgE antibodies trigger the release of histamine and other vasoactive agents and stimulate the synthesis and release of others (e.g. interleukins I and II, prostaglandins, etc.) from mast cells to which they are bound, when activated by antigens; this is seen in certain types of allergy, e.g. to the proteins of pollen but, normally, it plays an important part in defence against microbial invasion.

B-lymphocytes, themselves, can be activated to divide and transform into antibody-secreting plasmocytes by antigenic stimulation of cytophilic IgG bound to their surfaces and T-lymphocytes are also activated in the same manner. In all these cases it should be emphasized, that antibodies are derived from transforming B-lymphocytes (including plasmocytes) and come to be attached to other cells only after passing into the tissue fluids or blood. Some B-lymphocytes may also be very long-lived and so constitute a '*memory*' of previous reactions of the immune system (vide infra).

When circulating antibodies bind to antigens they form *immune complexes* which, if present in abnormal quantities, may cause pathological damage to the vascular system and other tissues either, as in some types of glomerulonephritis, by interfering mechanically with permeability to fluids; or alternatively, by causing local activation of the complement system to attack cell membranes, thus causing vascular disease.

In pregnancy some IgG is transferred across the placenta, conferring a measure of *passive immunity* on the fetus; but in the case of Rhesus factor incompatibility, it brings about the destruction of Rhesus-positive fetal erythrocytes. In some mammals (oxen, sheep), antibodies are transferred in the first-formed milk (colostrum) after birth. In humans, maternal milk contains secretory immunoglobulins (IgA) which help to combat bacterial and viral organisms in the alimentary tract of the baby during the first few months.

T-lymphocytes

These cells include a number of subclasses (see Marrack & Kappler 1986), all derived from stem lymphocytes originating in the bone marrow but later differentiating in the thymus (hence thymus-derived cell) and passing into various other lymphoid organs (**6.6**, **7**). They carry out a wide variety of cell-mediated defensive actions which are not directly dependent on antibody activity and are therefore included under the heading of *cellular immunity*. T-lymphocyte responses can be loosely divided into *effector* and *controlling* actions. Effector responses are direct and indirect attacks against virus-infected tissue cells, fungi, some protozoal infections (e.g. trypanosomes), neoplastic cells (probably) and, unhappily for transplant surgeons, the cells of grafts from other individuals (allografts) when the tissue antigens are not sufficiently similar. Controlling functions are the induction or suppression of immune responses in other lymphocytes, namely B cells and T cells engaged in effector responses and, it appears likely, various non-lymphocytic cells such as those of the bone marrow. Effector cells can be divided into two major classes, those which are responsible for cytotoxic killing of virus-infected cells, etc. and other cells which give rise to a combination of defensive actions by non-lymphocytic cells, seen most clearly though pathologically in delayed hypersensitivity reactions.

Cytotoxic T cells cause cell death in a number of different ways, including the release of toxic lysosomal proteins ('perforins') able to lyse the cell membranes of other cells by forming large pores; such actions occur at close range, the effector cell having contacted the target cell and recognized it as being pathological. This recognition step is of great interest and generally depends on the presence at the surface of the attacked cell of an alien antigen (e.g. of viral origin) in combination with Class I Major Histocompatibility Complex molecules; the latter are expressed at the surfaces of nearly all tissue cells and, although highly variable from person to person, are invariant within the tissues of an individual (p. 682). It is thought that during the development of T-lymphocyte stem cells in the thymus, they in some way 'learn' to recognize the MHC molecules characteristic of the individual to whom they belong. The T cells have specific receptors (also variable, see p. 675) at their surfaces, matched to these MHC molecules and, if at a later time the T cell meets a tissue cell in which the Class I MHC markers are in some way altered, e.g. by binding an alien antigen, or in the case of transplants, by being a different variant, T cells mount an attack on the cell which bears these 'strange' markers. In this way, intracellular pathogens such as viruses beyond the reach of antibodies and perhaps 'transformed' neoplastic cells, can be killed; of course the same applies to the 'alien' cells of an allograft. T-lymphocytes of this kind include some of the larger lymphocytes with cytoplasmic granules containing the cytotoxic substances (large granular lymphocytes), although natural killer cells (vide infra) may have a similar appearance (see Hieberman 1985).

Delayed type hypersensitivity-related T cells also react to the presence of antigens by synthesizing and releasing *lymphokines*, soluble proteins with an Mr of 20 000–80 000 and having a wide variety of actions on other cells. These include chemotactic substances which attract macrophages into the area of release (macrophage chemotactic factor) and then prevent them from

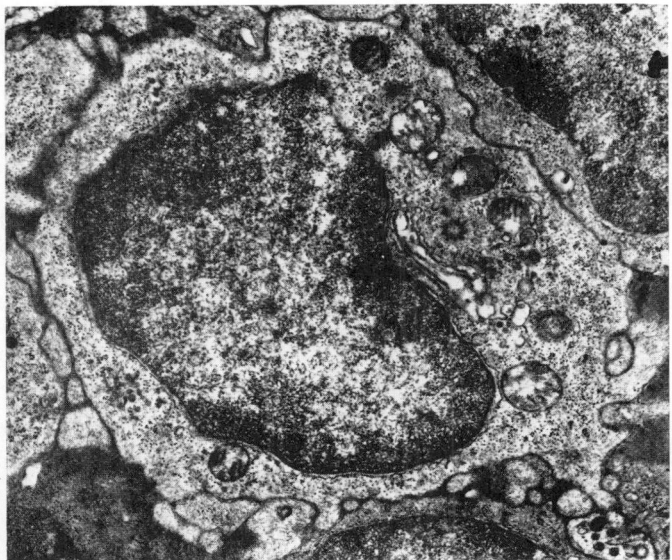

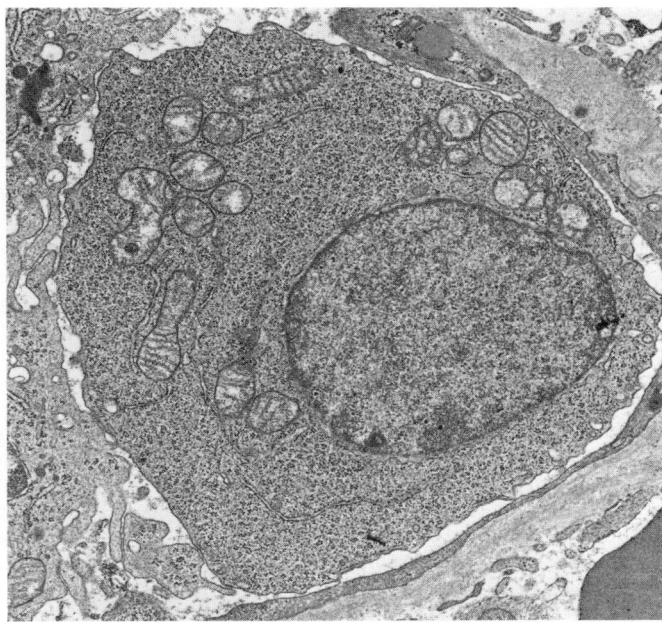

6.7C Transmission electron micrograph of a stimulated T-lymphocyte in the spleen (monkey). Note the numerous free ribosomes and relative paucity of granular endoplasmic reticulum. Magnification × 10 000.

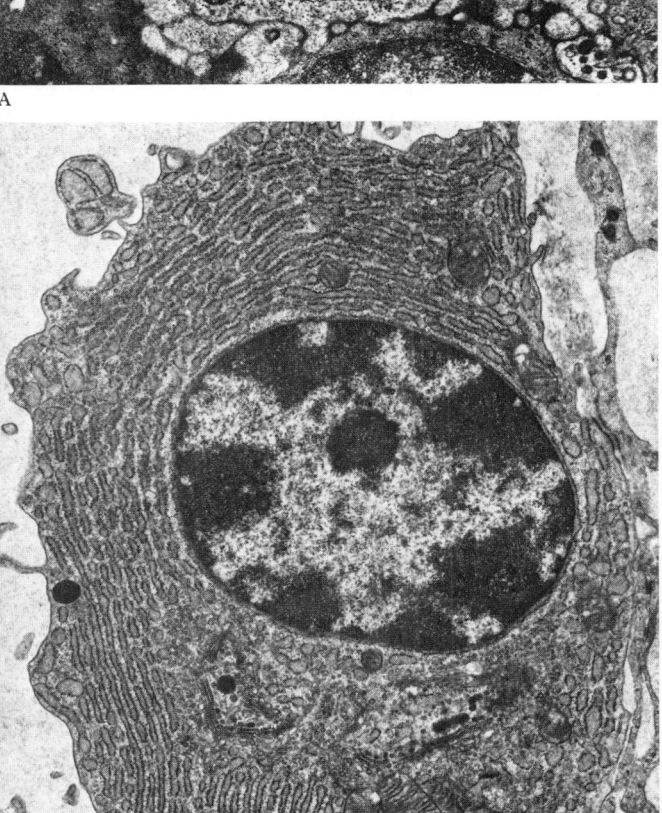

B

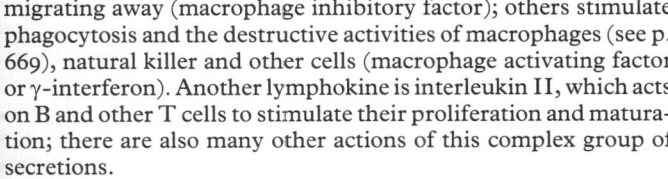

6.7 A, B Electron micrographs of: (A) a small lymphocyte; (B) a plasmacyte in loose connective tissue. A was provided by D R Turner, Guy's Hospital Medical School. Magnification × 6000.

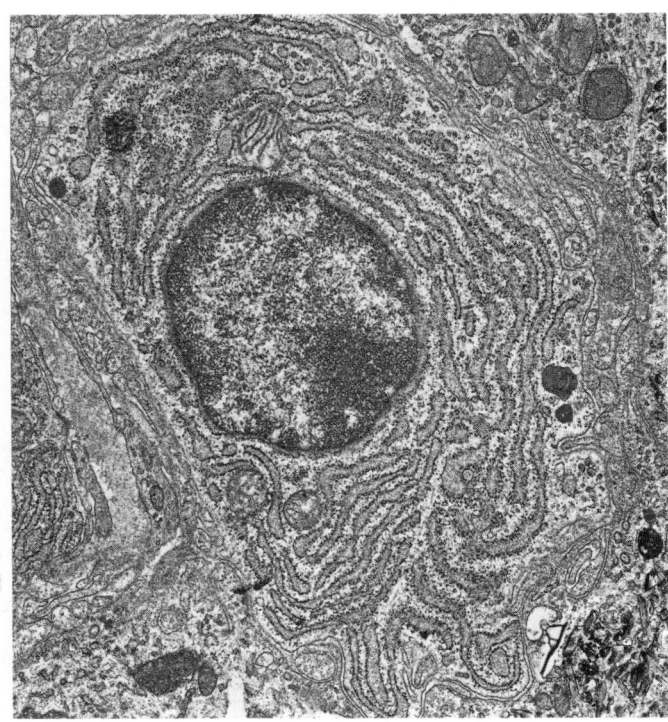

6.7D Transmission electron micrograph of a stimulated B-lymphocyte (plasmocyte) in the spleen (monkey). Note the copious granular endoplasmic reticulum containing newly synthesized antibodies, seen here as finely granular material. Magnification × 10 000.

migrating away (macrophage inhibitory factor); others stimulate phagocytosis and the destructive activities of macrophages (see p. 669), natural killer and other cells (macrophage activating factor or γ-interferon). Another lymphokine is interleukin II, which acts on B and other T cells to stimulate their proliferation and maturation; there are also many other actions of this complex group of secretions.

Helper T cells are vital to cell proliferation and secretion of antibodies by mature B-lymphocytes. These processes are initiated by a foreign antigen being phagocytosed and partially digested by a macrophage, some of the products passing to the macrophage surface where they are bound by a Class II MHC molecule embedded in the plasma membrane (a surface antigen found only on macrophages and a few related cells with similar functions). This combination of antigen and MHC molecule is then presented at the macrophage surface to a helper T cell which bears a (complementary) T-cell receptor. This interaction causes the activation of the T cell and such a cell is able to recognize and activate immature B cells capable of producing antibody to that particular antigen. In this rather indirect way, a clone of B cells producing a single type of antibody matching an antigen can be stimulated to proliferate and secrete their products. A similar activation of cytoxic T cells may occur through a separate group of helper T cells, although delayed-type hypersensitivity-related T cells also have this function, as mentioned above.

If such helper cell activities are suppressed, potentially pathogenic organisms present in the body but normally kept in check by the immune system may become uncontrollable, causing overt pathology and even death. The best known example of this is the Acquired Immune Deficiency Syndrome, AIDS, where a virus (HIV 1) specifically infects and kills helper T cells; similar though less lethal immunosuppression may also be caused by other viral infections and a multitude of other factors, e.g. malnutrition.

Suppressor T cells. These appear to be a separate class of cells which, when antigenically stimulated, suppress the defensive activities of B cells and other T cells. The mechanism of their action is not well understood but may have features in common with cytotoxic effector cells. The existence of both positive and negative controls in the immune system is of great significance, since for normal effective defence against infection, this system must be finely balanced to ensure destruction of alien organisms without the body itself being damaged by the powerful agents released by the various defensive cells. Failure of this delicate control may be seen, at one extreme, in immune incompetence and, at the other, in various pathologies such as autoimmune disease and hypersensitivity reactions (e.g. allergic asthma, allergic dermatitis) etc.

Structurally T-lymphocytes present different appearances depending on their type and state of activity. When 'resting' they are typical small lymphocytes morphologically indistinguishable from small B-lymphocytes; but when stimulated they become large (up to 15 μm), moderately basophilic cells with a partially euchromatic nucleus. In the cytoplasm are numerous free ribosomes, some agranular and granular endoplasmic reticula, a golgi complex and a scattering of mitochondria (**6**.7). Cytotoxic effector T cells also contain dense lysosome-like vacuoles.

Natural killer cells constitute a pool of defensive elements which appear to have similarities to cytotoxic T cells functionally, although they lack some typical lymphocyte features (vide infra). They normally form only a small percentage of all lymphocyte-like cells and are included technically in the 'large granular lymphocyte' category. Natural killer cells, when mature, have a mildly basophilic cytoplasm and a partially euchromatic nucleus. Ultrastructurally, the cytoplasm contains ribosomes, granular endoplasmic reticulum and dense, membrane-bound vesicles 200–500 nm in diameter with crystalline cores (Heiberman 1985). These contain some hydrolases, but the major active component is a protein, cytolysin, capable of inserting holes in other cells, so causing their death. Natural killer cells are activated to attach themselves to and kill target cells of various kinds by a number of factors, including interleukin II from T cells (vide supra). They represent a relatively non-specific means of attacking virus-infected cells, protozoa and other pathogenic cells (see also Henkart & Martz 1984).

Classes of T cells. When monoclonal antibodies were developed in cell cultures against lymphocytes, it was found that different classes of T-lymphocytes could be grouped together according to the characteristic range of monoclonals that they bound. The various species of antibody-binding molecules responsible for these reactions can be divided into a few typical combinations (known as '*differentiation clusters*', CD, see Reinherz 1987). The three major T cell differentiation clusters are termed, respectively: CD3, CD4 and CD8. All 'true' T cells appear to be CD3 positive, but may also be either CD4 or CD8 positive; those which are CD4 positive (termed 'CD4⁺ lymphocytes') include helper-inducer cells important in triggering antibody production from B-lymphocytes, cytotoxic-suppressor T cells and T cells involved in delayed hypersensitivity reactions. Those bearing CD8 ('CD8⁺ lymphocytes') as well as CD3 markers comprise suppressor and cytotoxic-suppressor T-cell lines. Natural killer cells are CD3 positive, but bear neither CD4 nor CD8 markers. This scheme of classification will no doubt give way to a more discriminating taxonomy when more is known

about the functions of T cells and their developmental patterns; the two major groups can already be divided into a number of subsets by their reactions with other monoclonal antibodies.

The CD molecular complexes are believed to act co-operatively with T-cell receptors, CD3 possibly to mediate stimulus transduction, CD4 to initiate 'helper' or other related activities, CD8 for 'suppressor' functions, etc.

Although the plethora of activities carried out by lymphocytes seems highly complex, it is to be expected that the potent and wide-ranging defensive mechanisms of the body should be subject to multiple strict checks, controls and balances. As yet, relatively little is understood about the manner in which the various parts of the whole system of cellular and chemical defences are *integrated*, but it is increasingly clear they they must be viewed as a *single system* of great efficiency and elegance. When, however, such integration breaks down, the effects may be far-reaching as, e.g. in the wide variety of *auto-immune diseases*, and in neoplasia of the immune system, such as myeloma.

Immunological Memory

If after one antigenic response the body is again exposed to the same antigen, the second response is much more rapid and extensive, even after a period of years; this forms the basis of clinical immunization. This phenomenon implies the presence of some type of immunological 'memory'. This appears to be largely retained by long-lived lymphocytes which have been specifically activated by antigens via mononuclear phagocyte contact (p. 669). Such cells can be either T or B cells; when the same antigens are encountered a second time, these memory cells are able to multiply rapidly and initiate a variety of other cell activities, typical of the secondary immune response. Over-reactivity of this system may be associated with various types of hypersensitivity, e.g. anaphylactic shock and delayed-type hypersensitivity.

The Nature of the Antigenic Response

The antibody-antigen reaction is *highly specific*, each antigen requiring its own antibody for binding to occur.

In the life of an individual an enormous range of antigens must impinge on the immunological system, requiring a correspondingly large number of antibodies in B-lymphocytes and an equally large number of receptors for antigens at the surfaces of T-lymphocytes (*T-cell receptors*). How these are coded for by the limited genome of each lymphocyte is a severe intellectual problem. Two possibilities exist: firstly, that the antigen in some way interacts with the cell's DNA or RNA to dictate the type of antibody or T-cell receptor to be formed by the lymphocytes (the *instructive theory*). Alternatively, it is proposed that from a wide range of types of lymphocyte, each capable of synthesizing only one, a limited number of antibodies, only the appropriate lymphocytes are stimulated by a particular antigen to multiply and synthesize antibody (the *clonal selection theory*). The weight of evidence now conclusively favours the second of these two possibilities (p. 675).

Recent studies on the mechanisms of generating a wide enough variety of antibodies capable of recognizing any conceivable macromolecule which might be introduced into the body during infections have indicated that variability results from a number of different processes which operate during the formation and maturation of B-lymphocytes. As already stated, antibody molecules consist of four chains each (two light and two heavy chains), and each of these has an antigen-binding end (Fab) of variable structure, as well as a constant (Fc) region. The constant and variable regions are coded for by separate genes, so that, for IgG antibody, at least eight genes are needed to code for a single molecule, four of which are variable and four constant. Different combinations of only a few types of variable gene would therefore already produce a sizeable number of different antibody structures. In fact, there are large numbers of variable (*V*) genes, since within each of these genes there are many subunits which can be spliced together in different combinations, to give a vast range o possibilities. The possible variations are again increased by somatic mutations in the genes during B-cell development, and probably by other factors such as variations in the translation of DNA into messenger RNA. It has been calculated that according to the known evidence about the system in mice, it would be simple to generate at least 10^9 different antibodies by random combinations of these different factors. Similar considerations apply to T-cell receptor molecules.

The T-cell receptor. This molecule consists of at least five polypeptides, two of which (the Ti α and β chains) are highly variable, and are each encoded by a cluster of variable genes which can be rearranged (probably in the thymus) to generate the wide variety of T-cell receptors encountered. Such receptors are said to be MHC-restricted because they will only recognize an antigen in the presence of an MHC molecule (of Class II, in the case of mononuclear phagocytes, and Class I in other cells). Other types of T-cell receptors have also been described (see Reinherz 1968), and these may play a part in developmental processes.

Clonal selection theory

When the great diversity of antibody specificities became apparent, the mechanism by which the body is able to produce the appropriate antibody to neutralize a particular antigen came under scrutiny and two alternative mechanisms were proposed. The *instructive theory* envisaged some sort of interaction between the antigen and a lymphocyte leading to a change in the antibody it produced, each cell having the capacity to carry out a series of such alterations depending in the prevalent antigen. The alternative *selective theory* proposed that the lymphocyte populations of the body include cells which already have the capacity to make almost any antibody which might be required in the life of the individual; and that an antigen causes only the cell type appropriate to it to proliferate into a series of identical cells (a *clone*) which all produce the same antibody. As an extension of this, it was postulated that clones inappropriate to defence, e.g. those able to produce antibodies against the body's own substances, are in some way suppressed, being either eliminated by induced cell death or kept quiescent.

The discipline of immunology has moved on from that time and it has been amply confirmed that the basic clonal selection hypothesis accurately describes the matching of antibody production to antigen specificity. Of the huge range of possible antibodies a B cell might synthesize, it is found in practice that only one is expressed in a particular B cell (*allelic exclusion*), and that its progeny will express the same antibody, exclusively. When suitably stimulated, such B cells proliferate and mature into plasmacytes synthesizing identical antibodies. Besides its importance in immunology, this phenomenon forms the basis for modern monoclonal antibody technology.

Although the clonal selection concept originally applied to antibody production, T cells being virtually unknown at the time, it has since become clear that the specificity of T-cell responses to antigens are controlled in the same way. As already stated, T helper and cytotoxic T cells, and perhaps other types of lymphocyte, are activated by interaction with antigen-presenting cells (macrophages, etc.), the specificity to a particular antigen being determined by the major T-cell receptor (Ti). Thus, like B cells, a great variety of T cells exists, each with a particular T-cell receptor pre-adapted to recognizing almost any antigen that might enter the body and thus able to launch an immune attack upon it (either by itself, or by helper actions on other defensive cells). When stimulated by the antigen–MHC II complex at the surface of the presenting cell, the various reactions lead to the proliferation of the clone of T cells responsive to that antigen and thus to a greatly multiplied response. Cells not bearing the T-cell receptor corresponding to that antigen are hence unresponsive. The existence of the same phenomenon in two cell types is not accidental, since most B-cell responses are indeed the result of induction by helper T cells responding to the same antigen, recognized by the T cell at the surface of the B-cell, bound to surface-expressed immunoglobulin molecules produced by that lymphocyte. Direct activation of B cells by antigen in the absence of T cells can also occur, but only by a narrow range of rather unusual antigens which can cross-link their surface immuno-

globulins, e.g. pneumococcus polysaccharide; this leads to the production of IgM rather than IgG molecules and must be regarded as of limited importance to the main immune response observed in vivo.

Immunological Tolerance

Since lymphocytes react to foreign antigens and not usually to the proteins and carbohydrates of the body itself, there must be some mechanism which ensures the distinguishing of *self* from *non-self*, that is an *immunological tolerance* to self. This may break down in auto-immune diseases such as Hashimoto's syndrome—self-immune destruction of the thyroid gland—and possibly in disseminated sclerosis which causes demyelination of tracts in the central nervous system. Self-tolerance is achieved in humans during fetal development, involving the action of the thymus (p. 835). Burnet widened the theory of selective antigenic response (vide supra) to attempt an explanation for the mechanism of self-tolerance (the *clonal theory*), suggesting that those cell lines (*clones*) of lymphocytes which produce antibodies (and subsequently, it appears, specific T-cell receptors, in the case of T-lymphocytes) to the body's own tissues are suppressed in fetal life and are then no longer available to multiply at a later stage (see Burnet 1969, Edelman 1974). Self-tolerance may be extended to the antigens of a genetically identical or closely similar individual, so that grafts of tissue between monozygotic twins or inbred strains of animals may be accepted.

All tissues from within a particular species show antigenic classes similar to those seen in the erythrocytes (p. 681) and in man a number of *histocompatibility* tissue antigen groups have been recognized (p. 682). The success or failure of *isologous grafts* (i.e. grafts between genetically dissimilar members of the same species) depends largely on the degree of matching of these antigens between donor and recipient.

Haemopoiesis (6.8)

The earliest sign of blood *vessel* formation in the human embryo occurs when, during the early primitive streak stage of development, angioblastic tissue differentiates almost simultaneously in various extra-embryonic sites, namely, in the mesenchyme of the yolk sac wall and in similar tissue of the connecting stalk and chorion. The earliest formation of blood *cells* (6.8), however, appears to be confined to the wall of the secondary yolk sac, where they differentiate from deeply placed mesenchymal cells which lie next to the yolk sac endoderm. Whilst a mesodermal or endodermal origin for these cells is still debated, they are unquestionably mesenchymal in character and differentiate into the primitive *stem cells* of the haemopoietic line, which give rise directly to fetal blood cells. Beginning in the second month, a number of intra-embryonic sites of haemopoiesis appear and slowly replace the earlier sites. These intra-embryonic sites *succeed but overlap* each other in time, each site gradually increasing in importance and then waning. Initially, the intra-embryonic sites are broadly *intravascular*, but soon *extravascular* loci of haemopoiesis supervene. Rapidly the *liver* becomes the dominant organ of embryonic blood formation, its activities continuing until about the seventh month. Lagging somewhat behind the liver, the *spleen* is then added, its haemopoiesis continuing from the third to sixth months. From late in the third month, an additional source of blood cells emerges, namely the *bone marrow (myeloid tissue)* where *all blood cell types* are formed, and later the peripheral lymphoid tissues, where only lymphocytes are produced (lymphopoiesis). The *thymus* is an active lymphopoietic organ from this stage (whilst initially it also has some general haemopoietic functions). The myeloid and lymphoid tissues become the dominant source of supply by the seventh month and shortly after birth all other sites have regressed completely. Occasionally, clumps of tissue capable of total haemopoiesis are found outside the bone marrow (*extramedullary tissue*) in paravertebral sites; in pathological cases, where there is more demand for haemopoiesis, and especially during early childhood,

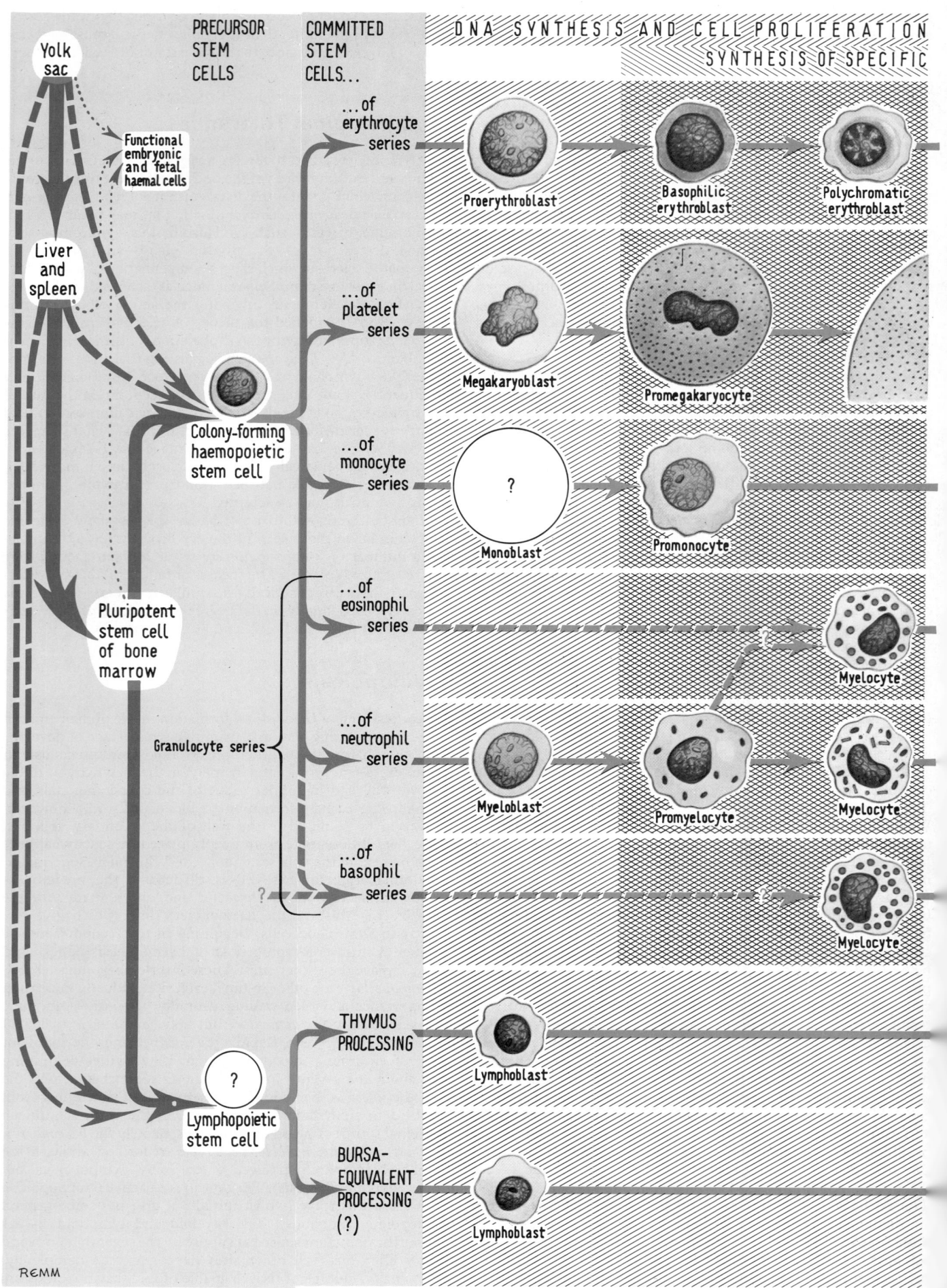

6.8 Haemopoiesis: a schema of the origins, developmental stages and fates of the different classes of haemal cells. Where details of development are uncertain, putative routes are shown in broken lines. Developmental pathways of the prenatal period are indicated (left) in red and differen

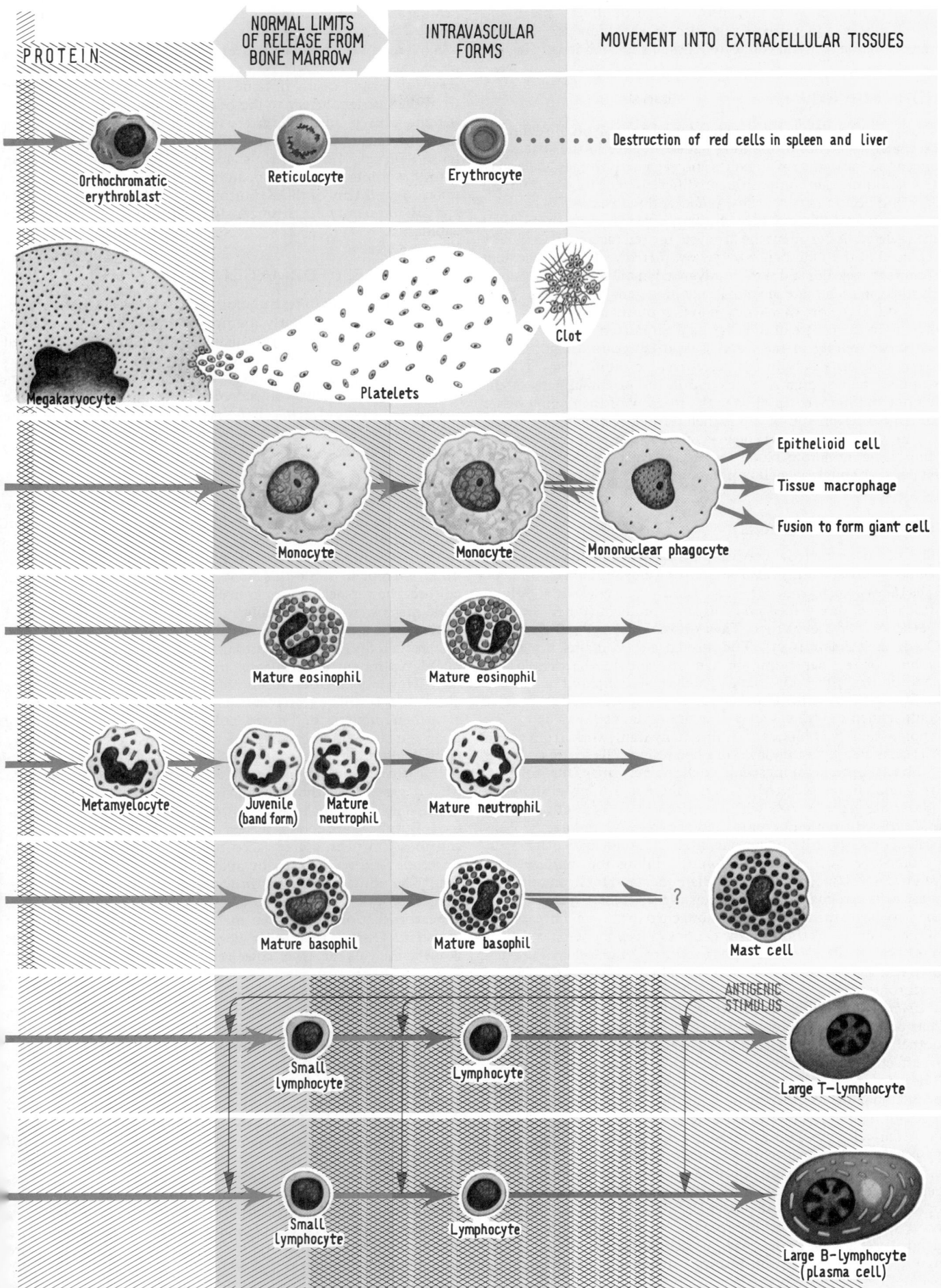

PROTEIN	NORMAL LIMITS OF RELEASE FROM BONE MARROW	INTRAVASCULAR FORMS	MOVEMENT INTO EXTRACELLULAR TISSUES

Orthochromatic erythroblast → Reticulocyte → Erythrocyte .... Destruction of red cells in spleen and liver

Megakaryocyte → Platelets → Clot

Monocyte → Monocyte → Mononuclear phagocyte → Epithelioid cell / Tissue macrophage / Fusion to form giant cell

Mature eosinophil → Mature eosinophil →

Metamyelocyte → Juvenile (band form) / Mature neutrophil → Mature neutrophil →

Mature basophil → Mature basophil → ? → Mast cell

Small lymphocyte → Lymphocyte → ANTIGENIC STIMULUS → Large T–Lymphocyte

Small lymphocyte → Lymphocyte → Large B–lymphocyte (plasma cell)

stages of cellular development are hatched diagonally according to the column headings. It should be noted also that there is evidence for macrophage proliferation in the extravascular tissues, a finding not indicated in this scheme.

complete haemopoiesis may also occur in the liver, spleen and lymph nodes and infrequently also in the kidneys, adrenals, adipose tissue, general connective tissue and even in cartilage.

BONE MARROW

This is a soft pulpy tissue which is found not only in the cylindrical marrow cavities of the long bones but also in the spaces between the trabeculae of all bones and even in the larger Haversian canals. It differs in composition in different bones and at different ages and occurs in two forms, *yellow* and *red marrow*.

During fetal life and at birth there is red marrow throughout the skeleton. After about the fifth year the red marrow is gradually replaced in the long bones by yellow marrow. The replacement commences earlier and is more advanced in the more distal bones. Further, in each bone the replacement, in general, proceeds from the distal to the proximal end, though some maintain that it commences in the centre of the shaft and extends in both directions, but more rapidly in the distal. By 20–25 years of age the red marrow persists only in the vertebrae, sternum, ribs, clavicles, scapulae, pelvis, cranial bones and in the proximal ends of the femora and humeri. In old age the marrow of the cranial bones undergoes degeneration and is then termed *gelatinous marrow*.

The yellow marrow consists of a basis of connective tissue, supporting numerous blood vessels and cells, most of which are fat cells, although a small population of typical red marrow cells persists.

The Red Marrow (6.10,11)

The red bone marrow consists of a network of loosely woven connective tissue, the *stroma*, supporting clusters of haemopoietic cells (*haemopoietic cords* or *islands*) and a rich vascular supply in which large, thin walled *sinusoids* are prominent (for reviews see Quesenberry & Levitt 1979, Tavasolli & Yoffey 1983, Weiss 1984, Golde & Takaku 1985). The stroma also contains a variable amount of fat, depending on age, site and the haematological condition of the body; small patches of lymphoid tissue are additionally present. Thus, the marrow consists of two major compartments, one vascular and the other extravascular. The whole assembly is enclosed within a bony framework, from which it is separated by a thin layer of bone lining cells (p. 295).

The stroma is composed of a delicate network of fine collagen (reticulin) fibres secreted by and adherent to highly branched *adventitial reticular cells* which appear to be a type of fibroblast and derived from embryonic mesenchyme. When haemopoiesis ceases, as in some limb bones in adult life, these adventitial cells become distended with fat globules, filling the marrow with yellow fatty tissue; but if there is a later demand for haemopoiesis, these cells can change back to their earlier stellate form. Also in the stroma are many *macrophages* attached to stromal fibres, some of them being embedded in the centres of haemopoeitic cell clusters. These cells are actively phagocytic of cellular debris created by haemopoietic development, especially the extruded nuclei of erythroblasts and remnants of megakaryocytes (vide infra).

Endothelial cells line the marrow sinusoids, the single layer of cells being supported by reticulin fibres on its basal surface. Endothelial cells are connected by tight junctions, which appear to be effective barriers between vascular and extravascular spaces. The passage of newly-formed haemal cells from the haematopoietic compartment into the blood stream occurs through temporary apertures (large fenestrae) formed in the endothelial cell cytoplasm, the migrating cell fitting tightly as it passes through, the aperture closing immediately behind it.

Haemopoietic tissue. Cords and islands of haematogenous cells consist of clusters of immature haemal cells in various stages of development, typically several different cell lines being represented at each focal group. One or more macrophages of a dendritic shape lie at the cores of each such group of cells, and may contain the iron-bearing molecules ferritin and haemosiderin. Besides the phagocytic functions already mentioned, there is evidence that such macrophages are important in transferring iron to developing erythroblasts for haemoglobin synthesis and may indeed exert control over the rate of cell proliferation and maturation of the neighbouring haemopoietic cells.

No lymph vessels have yet been demonstrated in either yellow or red marrow, as might be expected in a system of spaces enclosed in a rigid casing of bone. The vascular supply is derived from the nutrient artery to the bone and drained by the accompanying vein (p. 299). The artery ramifies in the bone marrow, its branches terminating in thin-walled arteries from which an extensive plexus of irregularly shaped sinusoids arises. These, in turn, drain into disproportionately large veins (Brookes & Harrison 1957). Many of these sinusoids are collapsed at any instant and are frequently but erroneously referred to as intersinusoidal capillaries.

HAEMAL CELL FORMATION

The initial stages of differentiation of haemal cells in mature marrow are far from clear. Early attempts to trace cell lineages in bone marrow preparations solely by histological examination of normal and abnormal tissue, led to much controversy about the origins and relationships of the various types of cell which could be observed. Some investigators considered it likely that all cell lines arose from a single type of stem cell (*haemocytoblast*) in adult tissue (the *monophyletic theory* of Maximow and others), whereas others favoured a multiple origin from many different types of stem cells (the *polyphyletic* theory, propounded by Ferrata and colleagues). Yet others have suggested compromise solutions (see Hardisty & Weatherall 1974). For many years it was also assumed that the lymphocytes arose exclusively in the peripheral lymphoid system and not in the bone marrow.

More recently, many experimental methods have been devised in an attempt to settle this issue; these include cell and organ culture, the replacement or transfusion of radiation-inactivated bone marrow with genetically or radioactively labelled haemal cells and the separation of stem cells from peripheral blood. The picture emerging still presents many uncertainties and applies mainly to rodent experimental models rather than directly to man. However, a brief outline of some recent findings will be attempted. It will be seen that haemopoiesis can best be described as a monophyletic or limited polyphyletic system, depending upon the stage in embryogenesis that is considered the starting point of differentiation.

Haemopoiesis in embryonic life commences at about the third week of gestation in the mesenchymal *blood islands* of the yolk sac, where only giant nucleated erythroid cells, *megaloblasts*, are formed. But these are soon replaced by *nucleated fetal erythroblasts* and these, in turn, are replaced by moderately large *anucleate, biconcave fetal erythrocytes* by about the fourth month of gestation. The adult type of erythrocytes are present by full term. Megakaryocytes, then granulocytes, lymphocytes and monocytes appear in that order between the second and fourth months of gestation. During the earlier developmental stages, stem cells occur in the yolk sac, liver, spleen and early bone marrow (p. 675) and these tissues can give rise to *all haemal cell lines* if they are transplanted to adult spleens (i.e. they are haemally 'totipotent'). Meanwhile, the lymphoid organs start to be colonized by stem cells which under normal circumstances only give rise to lymphocytes and hence appear to be already committed to that particular line of cells. However, the ability of these lymphoid tissues to carry out full haemopoiesis under certain pathological conditions suggests that some uncommitted stem cells have been retained in these sites but are normally suppressed.

In the *bone marrow* it appears likely that two types of *stem cell* have differentiated by early postnatal life; one type forms lymphocytes, whilst the other gives rise to erythrocytes, platelets, monocytes, neutrophils, eosinophils and basophils. The second of these stem cell types readily forms *haemal cell colonies* when transplanted to spleen cultures and each such cell is termed a *spleen colony forming unit* (CFU-S). The pluripotent CFUs are not numerous (about 0.5% of all haemopoietic cells in marrow) but they can circulate freely in the blood, where they appear as small lymphocyte-like cells, 7–10 μm across, with a large pale nucleus containing one or two nucleoli and bearing a narrow rim of cytoplasm.

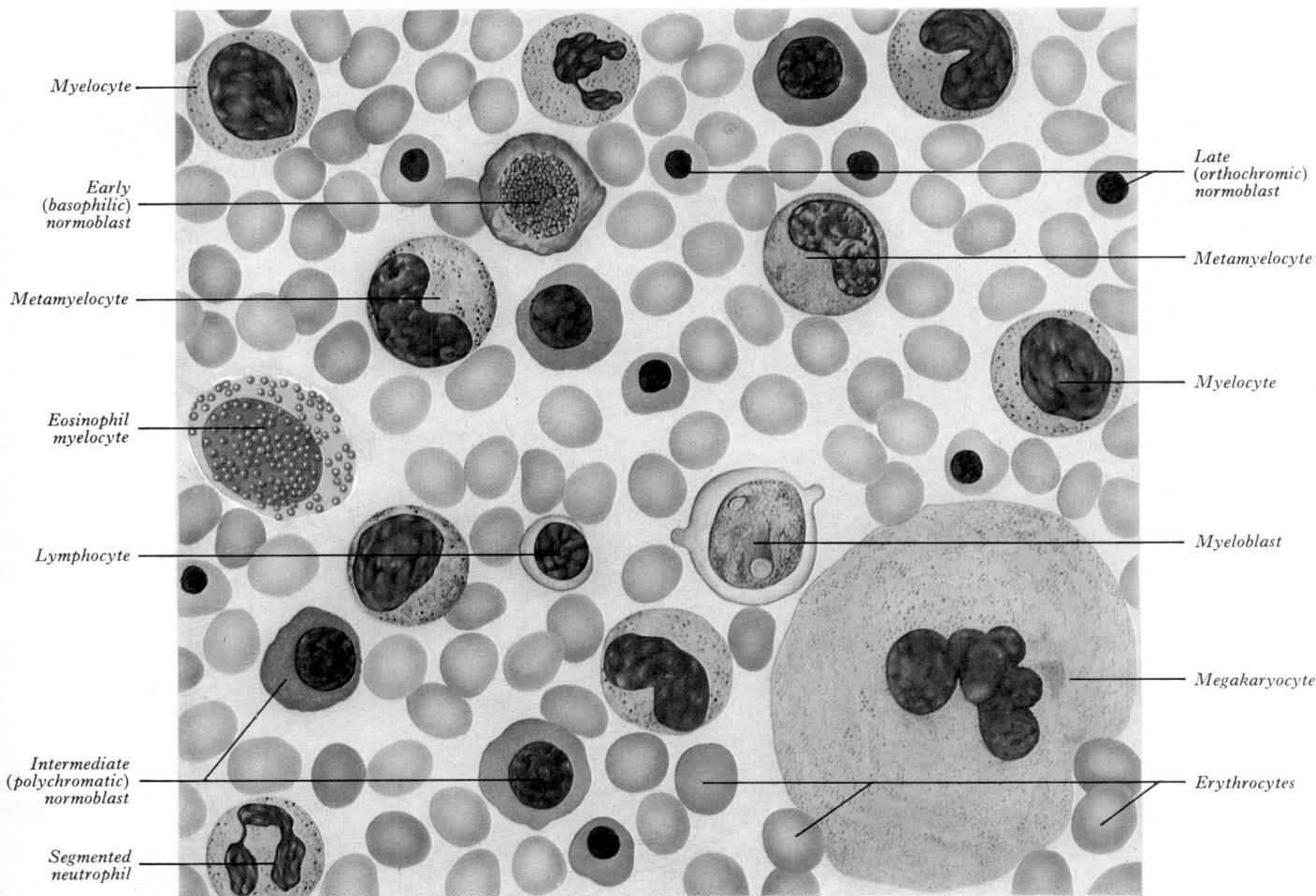

Myelocyte

Early
(basophilic)
normoblast

Metamyelocyte

Eosinophil
myelocyte

Lymphocyte

Intermediate
(polychromatic)
normoblast

Segmented
neutrophil

Late
(orthochromic)
normoblast

Metamyelocyte

Myelocyte

Myeloblast

Megakaryocyte

Erythrocytes

6.9 Smear preparation of normal human red bone marrow, obtained by sternal puncture. This is a composite figure, drawn from a number of normal smears prepared by the late R L Waterfield, Guy's Hospital, and stained with a modification of Leishman's stain.

Next, we turn to the problem of the differentiation of such stem cells into the various cell lines of the blood. The early stages of these lines are difficult to distinguish since usually they cannot be recognized until well after they have begun to differentiate. In the cell line, however, a definite sequence of major stages can be recognized (or ascribed) (see 6.8). These stages are: (1) the *final commitment* of a stem cell to a particular line of differentiation; (2) early cell *proliferation* to form a large pool of dividing cells; (3) *differentiation* as specific proteins characterizing the particular line are synthesized, accompanied by the gradual cessation of cell division; (4) final *maturation*, marked by the gradual closure of protein synthesis; (5) *release* of the cells from the bone marrow parenchyma into the circulation. In the case of some cell lines, cell division and protein synthesis may continue outside the bone marrow (e.g. monocytes or macrophages) and maturation may also be completed after release (e.g. erythrocytes).

The earliest 'committed' stages of each cell line are outwardly similar to the CFU-S, but at the next proliferative stage differences begin to emerge. In all, however, the initial picture is that of a rapidly dividing, relatively undifferentiated cell, in which the nucleus is large and euchromatic, with prominent nucleoli; the moderately basophilic cytoplasm contains unattached polyribosomes and the total cell size is large. As proliferation proceeds, the size of the cell usually diminishes as cell division outstrips cell growth. With continuing differentiation the ribosomes become numerous and the cytoplasmic basophilia increases as the specific RNAs begin to be synthesized, whilst the nucleus gradually becomes more heterochromatic and smaller, as DNA synthesis ceases; ultimately the nucleus becomes multilobed or pyknotic as protein synthesis terminates. The completed cell is then ready to be released, although the precise timing of its passage from the bone marrow varies with metabolic conditions and the 'demand' for more cells, so that relatively immature types of cell may be found in the circulating blood under abnormal conditions. This 'shift to the left' (a graphic convention of haematologists in which the cell is represented as maturing from the left to the right) is a useful concept in diagnostic haematology. These general features of development can be seen in several lines of haemal cells and underlie what appears at first sight to be a highly divergent series of progressions.

Erythropoiesis (6.8,9)

The earliest erythroid progenitor cells have not yet been identified, but the second stage includes cells which after some delay, can multiply very rapidly to form numerous erythroblast cells; they have therefore been named *burst-forming units of the erythroid line* (BFU-E). Third in this lineage is a cell which is sensitive to the hormone erythropoietin (p. 681), which induces it to further differentiation along the erythroid line. This is the *erythropoietin-dependent colony-forming unit* (CFU-E).

The first readily identifiable cell of the erythroid series is the *proerythroblast (pronormoblast)*, a large (14–20 μm) cell with a large euchromatic nucleus and moderately basophilic cytoplasm. The latter already has small amounts of ferritin and bears some of the protein spectrin attached to its plasma membrane (p. 664); both are characteristic of this cell line and can be detected by electron microscopy. Proerythroblasts proliferate, and haemoglobin-RNA synthesis begins, as the smaller (12–17 μm) *basophilic or early erythroblast (basophilic normoblast)* appears, rich in ribosomes; shortly afterwards, haemoglobin synthesis commences so that the cytoplasm becomes partially eosinophilic (the *polychromatophilic or intermediate erythroblast/normoblast*) which has a diameter of 8–12 μm. At this stage, most of the RNA is lost and the nucleus, becoming intensely pyknotic, is finally extruded from the cell, thus forming the anucleate *reticulocyte*. At this point the cell is released into the circulation, losing its residual RNA

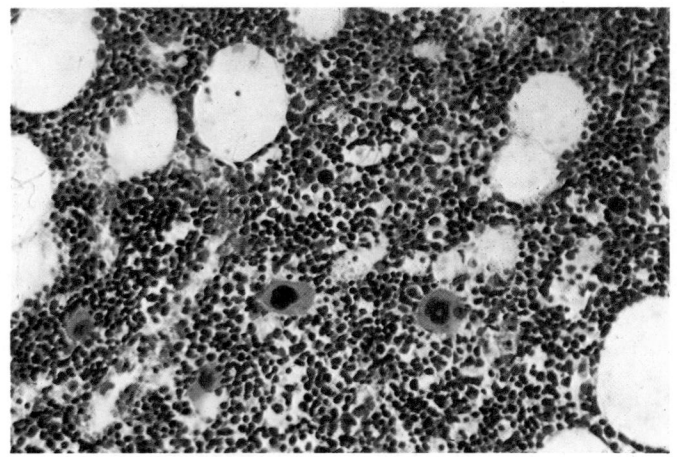

6.10 Photomicrograph of a section of bone marrow from a fetal human long bone. Note the heterogeneous collection of cell types including four large megakaryocytes. Magnification × 150.

in a few days to become a mature *erythrocyte*. The nucleus is removed by a macrophage. The whole process of erythropoiesis takes five to nine days; after release, reticulocytes typically sojourn for up to two days in the marrow sinusoids and then for an equal time in the spleen, perhaps because of their particularly adhesive cell coat. During this period, the remaining ribosomes add a further small amount of haemoglobin to the cell; then all the organelles are finally dismantled by a soluble cytoplasmic enzyme ('ubiquitin').

The cell lineage of normal erythrocytes is often called the *normoblastic series* to distinguish it from abnormal erythroid lines; too few divisions of the early proliferative stages may give rise to abnormally large erythrocytes (*macrocytes*), whereas too many divisions, or insufficient early growth, can lead to *microcyte* formation. Disturbances in haemoglobin synthesis can also give rise to a variety of anaemias and other pathologies.

Granulocytopoesis (6.8,9)

The details of this process are best known for the *neutrophil*, which will be described in some detail. Initially, the putative stem cells transform into the large (10–20 μm) *myeloblasts* which are similar in general size and appearance, though not in internal details, to the proerythroblast (vide supra). These proliferative cells differentiate into the larger *promyelocytes*, in which the first

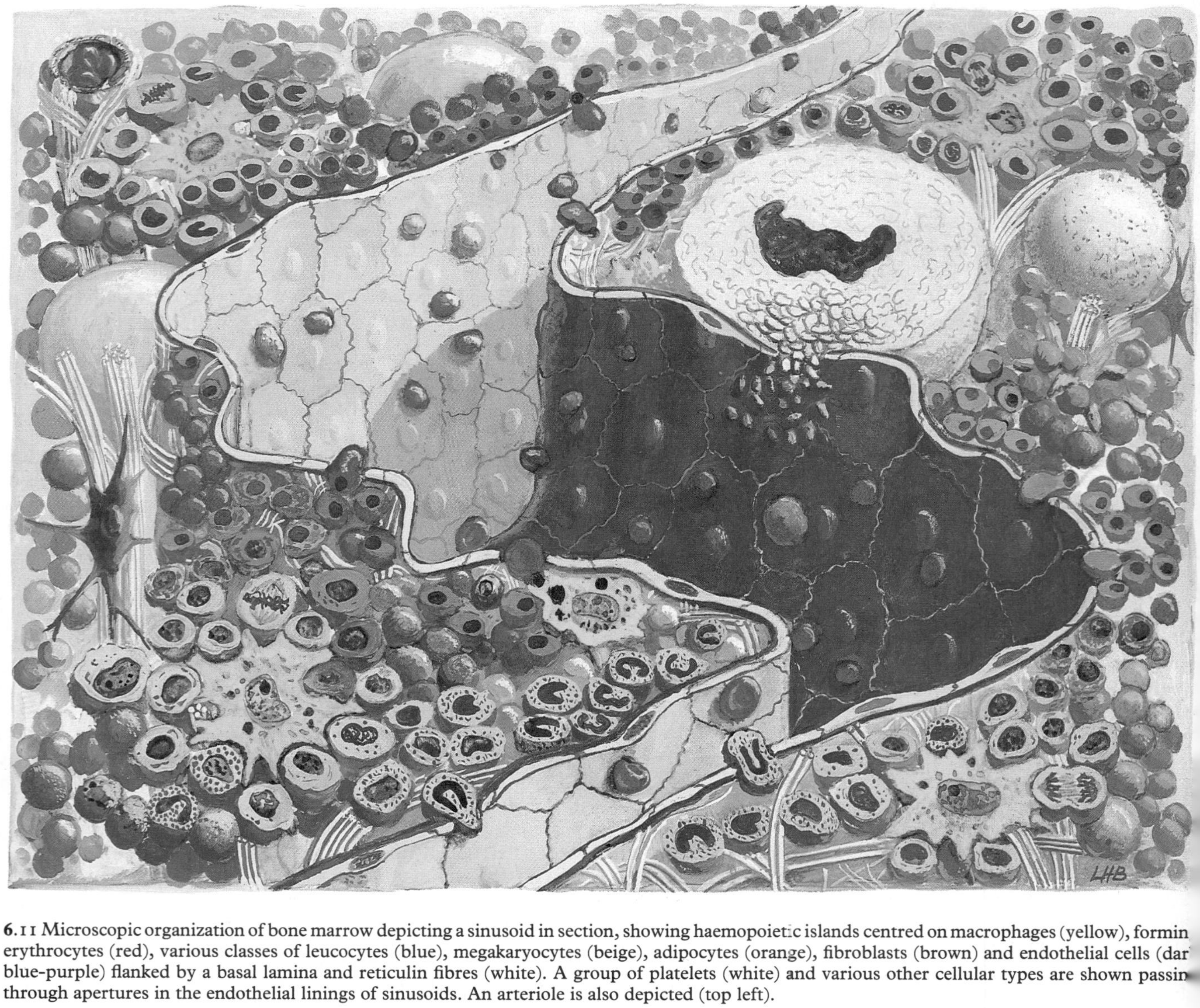

6.11 Microscopic organization of bone marrow depicting a sinusoid in section, showing haemopoietic islands centred on macrophages (yellow), forming erythrocytes (red), various classes of leucocytes (blue), megakaryocytes (beige), adipocytes (orange), fibroblasts (brown) and endothelial cells (dark blue-purple) flanked by a basal lamina and reticulin fibres (white). A group of platelets (white) and various other cellular types are shown passing through apertures in the endothelial linings of sinusoids. An arteriole is also depicted (top left).

group of specific proteins is synthesized in the granular endoplasmic reticulum and Golgi apparatus, both of these organelles being quite prominent. The proteins are stored in large (0.3 μm) *primary ('non-specific') granules*, large lysosomes containing acid phosphatase and having azurophilic staining properties (p. 666). Next, in the smaller *myelocyte* —the last proliferative stage—the smaller *secondary ('specific') granules*, which contain a slightly different enzyme array, are formed in a similar manner, though released into the cytoplasm from the *'cis'* side of the golgi body, whereas the primary granules arise from the other (*'trans'*) side (see p. 28). The nucleus is typically flattened on one side in myelocytes. Subsequently, in the *metamyelocyte* stage the cell size decreases further, the nucleus becomes heterochromatic and horse-shoe shaped and protein synthesis practically ceases. Finally, as the neutrophil is released, the nucleus becomes heavily indented (the *juvenile* or *'stab'* form) and then partially divided into up to six lobes (the *segmented* or *mature neutrophil*). The whole process takes about seven days to complete: the mitotic period about three days and maturation four days. They may then be stored in the medulla for a further four days, depending on demand, before final release into the circulation.

Eosinophils pass through a similar sequence except that their nuclei never become as irregular as that of the neutrophil, and only one set of lysosomal granules is synthesized. it is not certain whether the eosinophil differentiates from the same myeloblast (or promyelocyte) stock as the neutrophil, or whether it is distinct from the colony forming unit (CFU) stage, which at present seems more likely (Zucker-Franklin 1985). In the case of *basophils*, it is not certain that they follow this general sequence at all; they may not even share the CFU as an ancestor.

Monocytes are also formed in the bone marrow, from stem cells of unknown structure; subsequently they pass through a proliferative *monoblast* stage and then form differentiating *promonocytes* in which small lysosomes begin to be formed (these may be demonstrated by neutral red staining). After further divisions, monocytes are released into the general circulation, and at least some are believed to pass to perivascular and extravascular sites, which they then populate as *mononuclear phagocytes* or *macrophages*. Monocytes and neutrophils appear to be closely related cells and arise from the same stem cell, the *colony forming unit for granulocytes and macrophages* (CFU-G/M) (the granulocytes in question being, of course, neutrophils).

Platelets, being fragments of cells, arise in a most unusual manner by the division of the cytoplasm of certain huge cells into many portions (Pennington 1979). The first detectable cell of this line is the highly *basophilic megakaryoblast* (15–50 μm); this is followed by a *promegakaryocyte* stage (20–80 μm) in which synthesis of granules begins; finally, the fully differentiated *megakaryocyte*, a giant cell (35–160 μm) with a large, dense, *polyploid, multilobate* nucleus, emerges. Once differentiation has commenced, mitoses proceed without cytoplasmic division and the chromosomes are retained within a single polyploid nucleus containing 8n, 16n or 32n chromosomes, depending upon how many nuclear mitoses finally occur. Under the electron microscope, the cytoplasm is distinguished by numerous centrioles and spindle microtubules, both of which reflect the repetitive *mitotic* activity. Meanwhile, differentiation proceeds in the cytoplasm with the production of free polysomes, smooth endoplasmic reticulum and fine basophilic granules. Cytomembranes within the cell fuse with one another and with invaginations of the plasma membrane to cut off portions of cytoplasm, which then break away from the parent cell to form platelets. The nucleus of the megakaryocyte eventually disintegrates.

Control of Haemopoiesis

The numbers of cells in the circulation are closely regulated, cell destruction being counterbalanced by cell replacement. How this system operates is known, at least in part, only for erythrocytes. Erythropoiesis is stimulated by a circulating protein *erythropoietin* synthesized by the tissues of the kidney and other parts of the body. The rate of erythropoietin synthesis is inversely proportional to the oxygen content of the tissues; hence low

oxygen tensions, usually consequent upon lowered erythrocyte numbers, stimulate erythropoiesis whereas high oxygen tensions cause the withdrawal of the stimulus. At high altitude the lowering of the partial pressure of oxygen in the atmosphere leads to a raised erythrocyte count.

Many other factors also affect the rate of haemopoiesis, e.g. thyroid hormones, somatotrophic hormone, androgenic steroids and other hormones. In recent years it has also become clear that many other factors affect haemopoiesis, including substances released by T-lymphocytes, macrophages, neutrophils and other cells. The numbers of cells in the blood show a *diurnal rhythm*, probably because of hormonal fluctuations. Infection, haemorrhage and other clinical disturbances also affect the pattern of cell production, as do cytotoxic chemicals and ionizing radiations, to which the dividing cells of the bone marrow are particularly susceptible.

Blood Groups

Early attempts at transfusion of blood led to the discovery that erythrocytes bear antigens on their surface (see Race & Sanger 1975) which can interact with naturally occurring antibodies in the plasma of other individuals, causing agglutination and lysis of the erythrocytes. Such antigens, which are not shared with all members of a particular species, are termed *iso-antigens*; other iso-antigens are found amongst cells of other tissues (vide infra). Erythrocyte antigens are known as *agglutinogens* and the corresponding antibodies as *agglutinins*.

Erythrocytes from an individual can bear several different types of antigen, each type belonging to an antigenic system in which a number of alternative antigens are possible in different persons. So far at least 15 groups have been identified, which vary in their frequency of distribution amongst the various races of mankind, including the ABO, Rhesus, MNS, Lutheran, P, Kell, Lewis, Duffy, Kidd, Diego, Yt, Auberger, Ii, Xg and Dombrock systems. Clinically, only the ABO and Rhesus groups are of major importance. For a full description of blood groups in man, see the comprehensive account by Race & Sanger (1975).

All of these antigens are determined by genes carried by autosomes, except Xg which is borne by the X chromosome. Within each group the antigens are determined by alleles and inheritance is in accordance with simple mendelian principles. Thus, in the ABO system the genome may be homozygous and carry the AA complement, the blood group being A, the BB complement giving blood group B; or it may carry neither (OO), the blood group consequently being O. In the heterozygous condition the following combinations can occur: A B (blood group A B), A O (blood group A) and B O (blood group B). In Caucasians and Negroes group O is the commonest, being present in about 50% of the population, followed in frequency by groups A, B and AB in that order (Mourant 1975). In West Africans the Duffy determinant (Fy) is almost always absent, a lack which confers resistance to *Plasmodium vivax* malaria (Miller & Carter 1976).

The plasma in each case carries naturally occurring antibodies specific to the antigens which are not present on the erythrocytes in the same blood, so that in group A blood, anti-B antibodies are found. Similarly, present in group B blood are anti-A antibodies, in group O blood both anti-A and anti-B antibodies and in group AB blood there is neither type of antibody.

Transfusions succeed only if the recipient's antibodies do not correspond to the donor's antigens and cross-matching of blood antigens is therefore vitally important. Persons with group AB blood, lacking antibodies to both A and B antigens, can be transfused with blood of any group and are termed *universal recipients*; conversely, those with group O, *universal donors*, can give blood to any recipient, the donor's antibodies being diluted to insignificance. Normally, however, blood is only transfused between persons with precisely corresponding groups, since anomalous antibodies of the ABO system are occasionally found in blood and may cause agglutination or lysis.

Within the ABO system several subgroups exist (A_1, A_2, A_{1B}, etc.); the cross-matching of some of these is important in transfusions. The anti-ABO agglutinins, like all others (except those of the Rhesus system), belong to the immunoglobulin M (IgM) class and do not cross the placenta during pregnancy.

The Rhesus antigen system, so-called because of its presence also in the erythrocytes of the Rhesus monkey, is determined by three sets of alleles, namely Cc, Dd and Ee, the most important clinically being Dd. The commonest condition in Britain is CDe and about 83% of the population is Rhesus-positive. Inheritance of the Rh factor obeys simple mendelian laws and it is therefore possible for a Rhesus-negative mother to bear a Rhesus-positive child. Fetal Rh antigens can, under these circumstances, stimulate the production of anti-Rh antibodies by the mother and since these belong to the immunoglobulin G group of antibodies they are able to cross the placental barrier and cause agglutination of fetal erythrocytes. In the first of such pregnancies little damage is usually caused, but in subsequent Rh-positive ones massive destruction of fetal red cells may ensue, causing fetal or neonatal death. Treatment is by exchange transfusion of the neonate infant or, prophylactically, by desensitizing the mother after the first Rh-positive pregnancy with Rh-immune serum, which appears to destroy the fetal Rh antigen in the maternal circulation before the processes of immunological memory can be entrained (see Clarke 1975).

Of the other antigenic systems known, some of which are occasionally of clinical importance, many are restricted to individual genic groups or even families; they can be of great value to anthropologists when tracing demographic relationships, as of course are the major systems described above (Mourant 1983).

Other antigenic systems such as MNS (shared with other tissues in the body) can be used in medico-legal investigations to establish identity of blood, or in parental identification. These antigens can remain intact long after death, and have been detected even in mummified tissues from Egypt over 4000 years old (Harrison et al 1969).

The genetics of blood groups is complicated by gene linkage with other characters which may be of some clinical importance; e.g. duodenal ulcers show a higher incidence in those with group O blood than in the general population.

Leucocytes also bear antigens and about 12 such groups have so far been identified, 10 of them belonging to the same complex system. These are similar to the *histocompatibility antigens* involved in graft rejection (vide infra).

Histocompatibility Antigens

Amongst important components exposed at the surfaces of most cells are the molecules which determine the individuality of tissues from different persons (the histocompatibility locus antigens) and which are intimately related to the functioning of the immune system. They also share the same region of one of the chromosomes, their genes being grouped together as the Major Histocompatibility Complex, expressed in the body as a number of distinctive proteins. This system has come into prominence because of its importance in tissue grafting and transplant surgery and to various other clinical approaches.

In humans, the major histocompatibility complex (MHC) genes lie on the short arm of chromosome 6 and are grouped in a linear sequence close to each other; crossing over during meiosis rarely or never occurs in this sequence, due to either their

proximity to each other or some other intrinsic resistance to this process. The MHC region expresses a number of distinctive classes of molecules in various cells of the body, the genes being, in order of sequence along the chromosome: the Class I, II and III MHC genes. Class I consists of the HLA genes, subdivided into A, B and C subregions, A and B being represented by a number of different alleles; these genes are capable of generating throughout the cells of the body a distinctive set of gene products, expressed at the surfaces of most cells. These appear to be different in all individuals, conferring a unique chemical identity which is the basis of the immune reactions in the body, since lymphocytes recognize and attack cells bearing alien antigens mainly when these are present alongside characteristic 'self'-HLA molecules. This enormous range of possible alternatives is a result of the substructure of the genes coding for them, which are subdivided into many smaller units; these can be spliced together in many different combinations, in a manner resembling the mechanisms for creating diversity amongst antibodies or T-cell receptors (pp. 674, 675), although of course there is a fundamental difference in the outcome, since only one HLA sequence out of a great possible range is selected during development. The HLA sequence also has interesting diagnostic aspects because of chromosomal linkage to sites affecting the frequencies of certain diseases (vide infra).

The genes for Class II MHC molecules include three subdivisions termed DR, DQ and DP. Of their products, the MHC-DR molecules are best known; these occur on the surfaces of antigen presenting cells generally classed as macrophages (p. 673) and including various dendritic cells of lymphoid tissue (e.g. follicular dendritic cells, interdigitating cells) and the epidermis (Langerhans cells). These molecules are anchored into the membrane and like the HLA antigens are highly polymorphic (i.e. have different chemical structure in different individuals). They can bind to alien antigens which the cells have phagocytosed and partially digested, and this combination is presented to helper or cytotoxic T cells where they temporarily bind to the CD 4 molecules (T-cell receptors) present on the lymphocyte surface to activate that cell (p. 675). The genes for Class III MHC products are expressed in various components of the complement system, as well as some other, non-immune related proteins.

The HLA system has attracted much attention lately, because of its importance in transplant surgery and, in a quite different way, because of the association between certain of its subgroups and some types of disease. For example, the subtype HLA-B27 is present in nearly all cases of ankylosing spondylitis; the condition of haemochromatosis, a disturbance of iron metabolism, is strongly associated with HLA-A$_3$ and rheumatoid arthritis with HL$_4$-D$_4$. Other conditions that display a statistical correlation with the HLA system include diabetes in juveniles, Hodgkin's disease and multiple sclerosis. The reasons for these associations are not clear, but are thought to represent some type of loose *genetic linkage* between the HLA determinants and other alleles predisposing the body to a range of diseases. In mice, where this system has been widely investigated, genes that apparently control the responsiveness of the immune system to infection (the Immune related, or Ir genes) are situated between those genes that determine the major histocompatibility (H2) factors.

BLOOD VESSELS

Introduction

For life to be maintained, the composition and properties of the local fluid environment of the body's cells must remain constant within certain limits. This is ensured by the operation of interlocking homeostatic controls which depend for their effect upon adequate circulatory systems. These are composed of great numbers of cell-lined tubes and spaces which surround and permeate the tissues, providing a continuous perfusion of body fluids. The most extensive of these are the *blood-vascular system*, i.e. the heart

and vessels in which blood circulates, and the *lymphatic system* of lymphatic vessels and nodes, which conduct the lymph; this definition also includes the isolated patches of lymphoid tissue, spleen, thymus and various pathways involved in a wider process of the formation and dissemination of lymphocytes (p. 670). Other, more restricted circulations are those of the cerebrospinal fluid, perilymph, various endocochlear fluids, ocular aqueous humour, synovial fluid and the fluids of the coelomic spaces—pericardial, pleural and peritoneal cavities. In this section only the blood-vascular and lymphatic systems are considered.

The rhythmically repeated contractions of the heart, which is essentially a *pair* of valved, muscular pumps arranged in *series* (vide infra), drives blood through the blood vessels to all tissues. Leaving the heart, it is distributed through large tubes, *arteries*, which divide and subdivide until minute vessels, various grades of *arteriole*, are reached; these again divide into dense networks of microscopic vessels, intimately in contact with tissues, which vary in structure and are termed *capillaries* or *sinusoids* depending upon their structure and function. Blood is collected from them via minute *venules* which unite repeatedly to form increasingly large *veins*, returning blood to the heart (**6**.12).

The terms artery, arteriole, capillary, venule and vein are essentially *anatomical* names, based on considerations of size, structure and topographical position. *Functionally* vessels may be classified as *distribution, resistance, exchange* and *capacitance* or *reservoir* types. Varieties of *shunt* vessels provide channels which can bypass capillary beds and sinusoids.

Distributing vessels are the low-volume, high-pressure system of large arteries which leave the heart, and their branches down to arteriolar level. They show gradually changing structural and functional features with increasing division and distance from the heart. As they leave the heart and branch, their walls contain much elastic tissue (*elastic arteries*); their repeated distension and recoil damps the effects of ventricular systole into smoother but still pulsatile flow throughout the arterial tree. Smaller branches, while retaining marked elasticity, display increasing amounts of non-striated muscle in their walls (*muscular arteries*), the tonus of which controls blood flow to large tissue masses in accord with the changing levels of activity. An alternative and perhaps more significant subdivision of arteries can be made into *conducting vessels* for major arteries with more highly elastic walls and *distributing vessels* for more muscular arteries down to the beginning of arterioles, the latter being termed *resistance vessels* (see Simionescu & Simionescu 1984).

Finer control of flow in microcirculatory units of various tissues (vide infra) is exercised by the muscular walls of arterioles and precapillary sphincters which provide the main peripheral resistance to flow. This factor, combined with the prevailing cardiac output, largely determines the general pressure in the arterial tree and, locally, modulates blood flow through the various capillary beds.

Capillaries, sinusoids and postcapillary venules are collectively termed *exchange vessels*: exchange between blood and the tissue fluid around cells, the essential function of circulatory systems, occurs through their walls. This exchange includes oxygen, carbon dioxide, nutrients, water and inorganic ions, vitamins, hormones, metabolic products, antibodies and defensive cells of various kinds.

Larger venules and veins form a co-extensive but variable, large-volume, low-pressure array of *capacitance* or *reservoir vessels*, conveying blood back to the heart.

The heart has four chambers: *right atrium, left atrium, right ventricle* and *left ventricle*. (These somewhat misleading names are discussed on p. 698.) Each atrium leads into a corresponding ventricle, the right and left chambers being separated by *septa*. The right and left sides are thus twin pumps, *topographically combined* in a single organ but interposed *in series* in the vascular system, separating it into *systemic* and *pulmonary* circulations (constituting the so-called *double circulation* typical of birds and mammals, see p. 697). The course of blood from left ventricle through the body at large to the right atrium forms the *systemic circulation*, its passage from the right ventricle via the lungs to the left atrium being the *pulmonary circulation*. The relatively short pulmonary system offers much less peripheral resistance than the systemic circulation, as is reflected in the lower pressures in the pulmonary distribution vessels and in the thinner walls of the right ventricle (p. 698). The average output volume of blood from the right and left sides of the heart must, of course, be the same.

The superior and inferior venae cavae return to the right atrium blood which has become *deoxygenated*, has taken up carbon dioxide and been otherwise modified during circulation through the tissues of the body. This blood then enters the right ventricle, which expels it via the pulmonary trunk to the lungs. In the pulmonary capillaries blood is brought into close proximity to the inspired air, releasing some carbon dioxide and acquiring oxygen. This *oxygenated* blood, returned by the pulmonary veins to the left atrium, enters the left ventricle, which pumps it into the *aorta* for general distribution.

Blood traversing the spleen, pancreas, stomach and intestines is not carried back directly to the heart but passes through the *portal vein* to the liver. This vein divides like an artery, ending in the hepatic sinusoids drained by the *hepatic veins* to the inferior vena cava, whence blood is conveyed to the right atrium. This route is the *portal circulation*; its essential feature is that the blood supplied to the above-named viscera, traverses two sets of capillary vessels before reaching the inferior vena cava; the vessels are: (1) capillaries in the spleen, pancreas, stomach, etc., draining into the portal vein and (2) the hepatic sinusoids, draining into hepatic veins. Passage through these two sets of capillaries enables the blood to transfer the products of digestion directly from the alimentary canal to the cells of the liver. Another venous portal circulation connects the median eminence and infundibulum of the hypothalamus with the pars distalis of the adenohypophysis (p. 1451).

ENDOTHELIUM

The surfaces of all blood vessels, the lymphatic channels and the inner surface of the heart are lined by endothelium, composed of a single layer of squamous, *endotheliocytes* (see Fishman 1982, Simionescu 1983, Simionescu & Simionescu 1984). The properties and activities of these cells are of great importance to the physiology of the vascular system. Although all endotheliocytes resemble each other in their general organization, considerable variations exist in their detailed structure and functions in different parts of the body.

Individual endotheliocytes are flattened cells, polygonal when viewed from the vessel lumen and usually somewhat longer, parallel to the longitudinal axis of the vessel. Typically their cytoplasm is thickened around the nucleus but is for the most part very attenuated, being as little as 0.2 µm in the capillaries. In arterioles and muscular arteries the nucleus and cytoplasm may bulge markedly into the lumen when the vessel is contracted, e.g. at post-mortem (see **6**.17D,E). Within the cytoplasm, organelles are sparse but include small amounts of granular and agranular endoplasmic reticulum, some free ribosomes and mitochondria and a pair of centrioles. Variable bundles of microfilaments and intermediate (vimentin) filaments are also present, presumably helping to determine cell shape (which may change when suitably stimulated in some cases) and reinforcing the cell mechanically. In arterial and arteriolar endothelia there are certain rod-like vesicles about 3×0.1 µm in dimension (Weibel–Palade bodies) which store a clotting factor (the von Willebrand component of factor VIII, Wagner & Marder 1984) which may be released after cell damage as part of haemostasis. A characteristic feature of many endothelial cells is the presence of great numbers of small vesicular invaginations (*plasmalemmal vesicles*) of the plasma membrane along both flat surfaces. These are believed to be important in shuttling materials across the endothelium and in some cases they may fuse to form complete open channels, grading into permanent fenestrae (p. 689). Such structures typify vessels engaged in rapid exchange of fluids or macromolecules between the vascular and extravascular spaces; but their frequency varies directly with vascular permeability, vessels in the brain having very few, and capillaries of the intestine, myocardium and kidneys having many. In some areas there are large gaps between cells, e.g. in linings of splenic sinusoids, which permit free access of plasma (and in the spleen, blood cells) to extravascular tissues.

Endothelial cells are linked to each other laterally by occluding (tight) junctions which may either form continuous belts to create permeability barriers (zonulae occludentes) and hold the cells together or in some cases discontinuous, spot-like contacts (Maculae occludentes), allowing some leakage between cells. Endotheliocytes are also linked to each other and often to other extravascular cells, such as non-striated myocytes, by gap (communicating) junctions, facilitating co-operation between cells connected in this way. They are limited externally by a typical basal lamina, which may also affect the permeability of blood vessels, particularly fenestrated capillaries and sinusoids. In renal glomerula, this basal lamina is particularly important in determining transendothelial filtration (p. 1401).

Endothelial cells lining capillaries and venules are mainly passive, selectively permeable barriers, allowing the passage of water and small ions but retaining larger molecules and red cells within the vascular channels to varying degrees. Various substances in the body also may change vascular permeability, e.g. histamine, which causes a transient separation of intercellular junctions (see Simionescu et al 1982). However, endothelium also

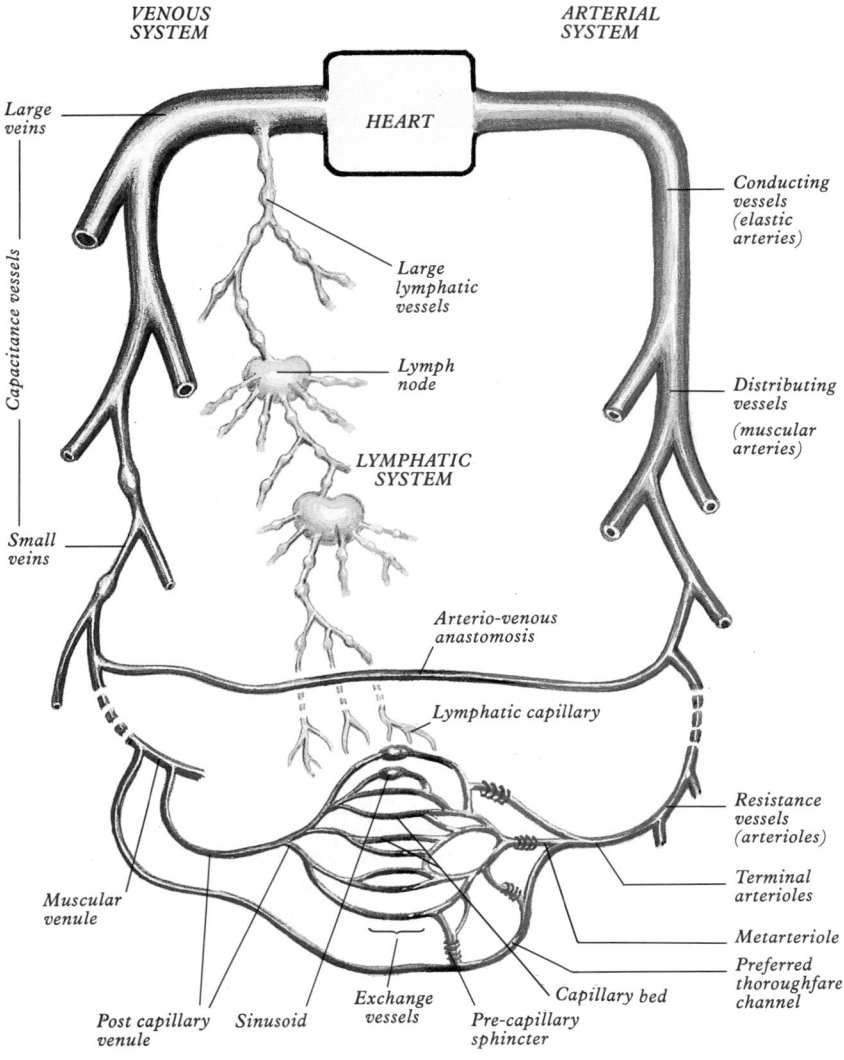

VENOUS SYSTEM

ARTERIAL SYSTEM

HEART

Large veins

Capacitance vessels

Conducting vessels (elastic arteries)

Large lymphatic vessels

Lymph node

LYMPHATIC SYSTEM

Distributing vessels (muscular arteries)

Small veins

Arterio-venous anastomosis

Lymphatic capillary

Resistance vessels (arterioles)

Terminal arterioles

Metarteriole

Muscular venule

Preferred thoroughfare channel

Exchange vessels

Capillary bed

Post capillary venule   *Sinusoid*

Pre-capillary sphincter

6.12 Schema showing the nomenclature of the different vessels in the haemo-lymphoid system.

has many important metabolic functions and a variety of enzyme systems has been located within or at the surfaces of endothelial cells. The capillary bed of the lungs has been most fully investigated and it has been shown that various blood-borne substances are changed during their passage. These include Angiotensin I which is converted to the vasoactive Angiotensin II; many substances are also inactivated, e.g. serotinin, bradykinin, noradrenalin, various prostaglandins (e.g. A, E and F), testosterone, etc. (see Fishman 1982, Stalcup et al 1982). In addition, endothelium strongly inhibits the adhesion of platelets and other thrombus-initiating events; this may be partly due to the negative charge on the endothelial surface which repels other negatively charged cells, but the synthesis of anti-thrombogenic

factors such as prostacyclins and similar prostaglandins appears to be important. If the endothelial surfaces of vessels are damaged, clotting soon occurs because of interactions of plasma-mediated haemostatic mechanisms with the underlying tissues, particularly their collagenous components, and also because of substances released by the damaged endothelial cells themselves. Such disturbances are obviously advantageous in limiting blood loss from breached vessels but also play a key role in thrombosis associated with arterial disease, especially atheroma and hypertensive damage.

Angiogenesis

The microcirculation can readily change its form by adding new vessels and presumeably resorbing old, ill-used pathways. The development of new vascular pathways occurs during the developmental growth of tissues but remodelling continues at a diminishing rate throughout life. This is most clearly seen in wound healing and regeneration, where the repairing tissues are invaded by numerous vascular cords developed both by the migration and proliferation of existing endothelium and, where these are lacking, by the differentiation of connective tissue cells (probably persistent pluripotent mesenchymal cells). Having invaded as solid cords of cells, these canalize (develop a lumen) and join end to end to form vascular loops through which blood flows. Some vessels then acquire fibromuscular coats and innervation to become arterioles, venules and larger vessels. The stimuli for such angiogenetic activity is currently under active investigation; they appear to include various angiogenic factors released from macrophages, activated T cells and other sources (including mast cells), which create chemotactic gradients, stimulate mitotic activity and cause cellular differentiation and maturation (see Folkman & Klagsbrun 1987). Interestingly, some tissues lack an intrinsic network of vessels, e.g. cartilage, stratified epithelia; this may be partly due to active inhibition of vascular growth.

The Structure of Blood Vessels

Blood is ejected from the left ventricle during systole into the aorta, a single tube about 2.5 cm diameter in the adult (Wright 1969), with a wall thickness of about 1.5 mm. Having traversed numerous successive branchings of this trunk, the blood eventually enters a capillary bed in the tissues (**6**.12), consisting of millions of tubes each about 8 μm in diameter, with a wall in places less than 0.2 μm thick, and a total estimated length in adults approaching 60 000 miles (Zweifach 1961). It is often stated that this systemic *arterial tree* progressively divides to create a larger and larger total cross-sectional area; but the evidence in man is not clear, particularly proximal to the arteriolar level. However, there are many progressive changes in the structural and functional properties of arteries with increasing distance from the heart. There is both a geometric reduction of individual lumen cross-sectional areas and an increase in stiffness of the vessel walls. In *arteriolar* and *capillary* branching the total sectional area of vessels increases greatly, with a progressive reduction in the rate of blood flow and of systolic and pulse pressures. However, lining the interior of the whole system from the finest capillaries up to and including the heart, is a smooth, continuous single-layered endothelium. Larger vessels also have surrounding coats composed of non-striated muscle and connective tissue, arranged in characteristic layers. Though laminar organization of these varies greatly with vessel size and type, reflecting their functional roles, three analogous zones (coats or tunics) have classically been described in the walls of all vessels except for capillaries and sinusoids (Cliff 1976, Simionescu & Simionescu 1984). These are from inside outwards the *tunica intima, tunica media* and *tunica adventitia* (**6**.13). These coats confer on the vessels a number of important properties, including an endothelial lining low in friction and connective tissue components able to withstand longitudinal and circumferential stresses due to prevailing blood pressures; they provide elastic recoil for the progressive modification of pulsatile pressure changes or flow variations in arterial trees and the ability to vary vessel calibre by muscular action.

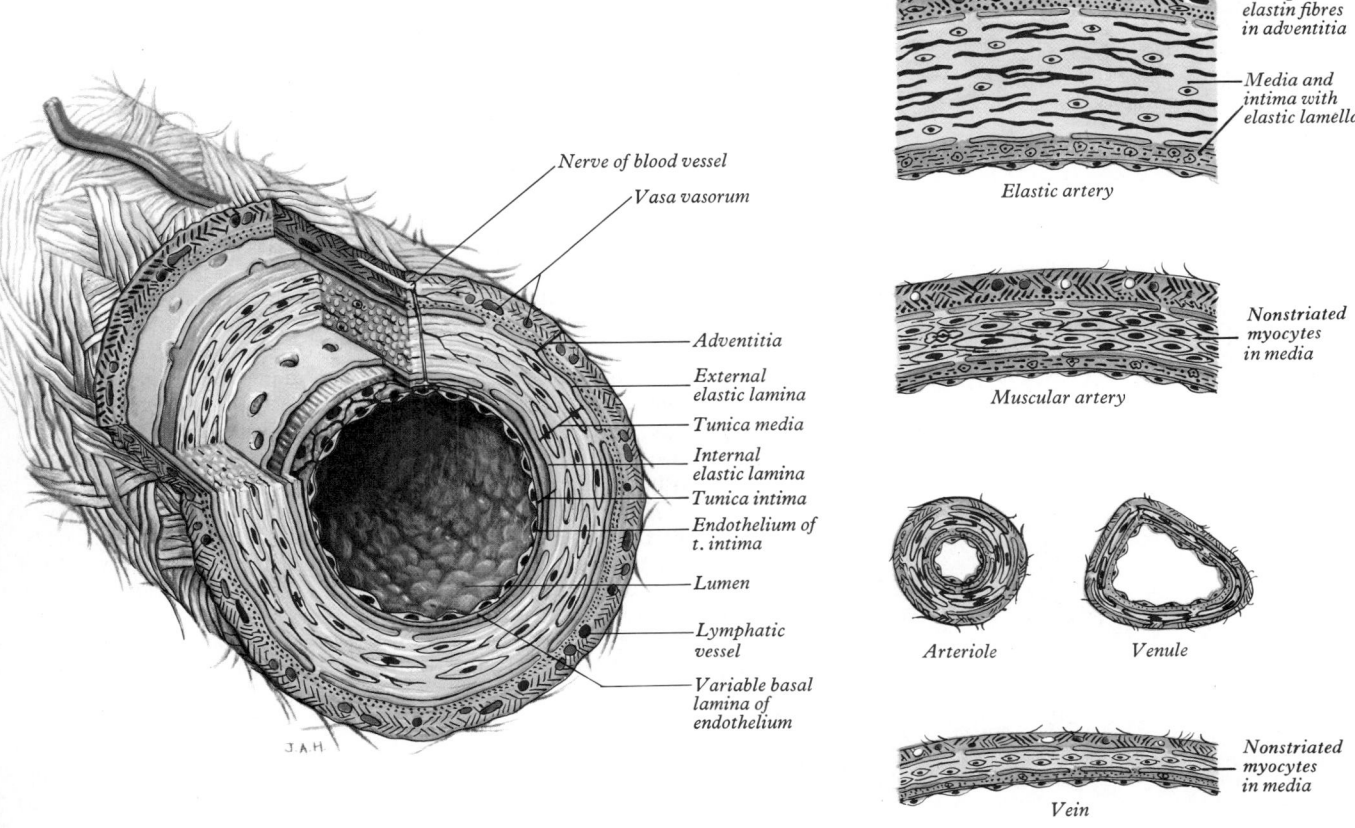

Nerve of blood vessel

Vasa vasorum

Adventitia

External elastic lamina

Tunica media

Internal elastic lamina

Tunica intima

Endothelium of t. intima

Lumen

Lymphatic vessel

Variable basal lamina of endothelium

Collagen and elastin fibres in adventitia

Media and intima with elastic lamellae

Elastic artery

Nonstriated myocytes in media

Muscular artery

Arteriole   *Venule*

Nonstriated myocytes in media

Vein

J.A.H.

6.13 Diagram showing the principal architectural features of the larger blood vessels. On the *left* the major layers and associated structures are depicted; on the *right* the particular features of an elastic artery, muscular artery, arteriole, venule and vein are shown.

The *tunica intima* is generally a single layer of endothelial cells supported externally by fine subendothelial connective tissue of largely longitudinal arrangement. The fibromuscular *tunica media* extends from internal to external elastic lamina (*elastica interna* to *elastica externa*) and is circumferential. The *tunica adventitia*, surrounding the external elastic lamina, is again largely longitudinal or irregular in the orientation of its cells and fibres and blends at its outer limit with the connective tissue of the structures surrounding it.

Nutrient blood vessels, *vasa vasorum*, and lymphatics run in the adventitia and a few in the outer zones of tunica media in larger vessels. Nerve fibres may supply all three tunicae (vide infra). Arteries are enclosed in a thin, delicate *sheath* of loose connective tissue, which usually encloses the accompanying veins, perivascular nerves and sometimes major nerves in the vicinity.

The thickness of vessel walls

The form and composition of blood vessel walls varies largely with blood pressure and, in arteries, with the nature of the pressure wave. Hence arterial walls are proportionately much thicker; and walls in arteries and veins inferior to the heart are in general thicker than those above it, due to the higher hydrostatic pressure in the lower regions when the body is erect. Systemic arteries have thicker walls than pulmonary arteries, reflecting the marked difference in pressure. Arteries within the cranium and vertebral canal (a 'closed box' system) have very thin walls relative to vessel diameter, the external and middle tunics being reduced in relative thickness as is also the condition in the medullary cavities of long bones. Where considerable external pressures or other mechanical forces act upon them, e.g. in coronary and carotid arteries, vessels often have thick muscular walls, mainly due to a wide tunica media.

Elastic arteries usually have walls with a thickness of about one-tenth their calibre; this fraction in muscular arteries is about

one-quarter, in arterioles approximately one-half and in veins the proportion is about one-tenth or less.

ARTERIES

The large elastic arteries (6.14A,B) include the aorta and its main branches, the brachiocephalic, common carotid, subclavian and common iliac arteries, and also the pulmonary arteries and their main branches.

In these vessels, the *tunica intima* is about 100 μm thick in young adults. The endothelium consists of flat, polygonal cells lying on a basal lamina and joined by intercellular junctions, including zonulae occludentes; their general features will be considered with capillary structure. The subendothelial tissue comprises: firstly, a layer of interlacing, mainly longitudinal, collagen and elastin fibres, with a few fibroblasts, non-striated myocytes and macrophages; secondly, a fenestrated elastic membrane, the *internal elastic lamina* of smaller arteries. In large arteries this is not sharply demarcated from the tunica media because both contain a series of similar elastic membranes (Dobrin 1978). Intimal thickness varies much with age and other factors: in children it may be almost insignificant, but in young adults intimal thickening is noticeable, perhaps due partly to migration and proliferation of non-striated myocytes from the tunica media, and otherwise to a variable deposition of lipid material, which often forms irregular fatty streaks. In young adults the tunica intima is about one-sixth of the total mural thickness but, after middle age, lipid deposits may grossly expand it.

The *tunica media* largely consists of circular, concentric, fenestrated elastic lamellae (up to 40 layers in the normal aorta), separated by fibrous tissue with a basophilic proteoglycan matrix permeated by bundles of collagen and elastin fibres, between which are circularly-orientated non-striated myocytes (Somlyo & Somlyo 1968). Most of these myocytes are typical of their cell type

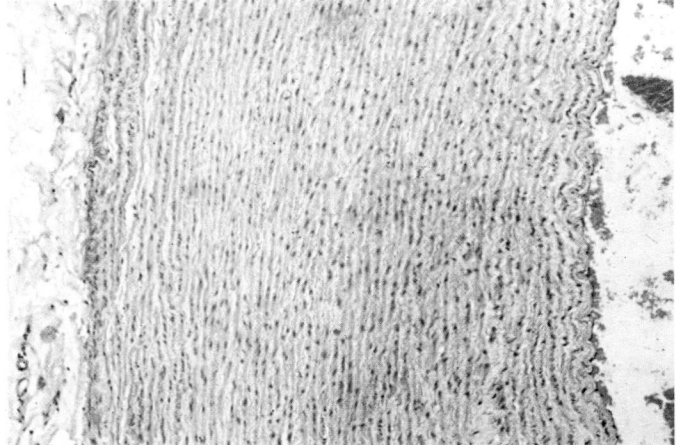

6.14A Part of a transverse section of the aorta of a monkey, stained with haematoxylin and eosin, showing the distribution of cell nuclei. Magnification × 100.

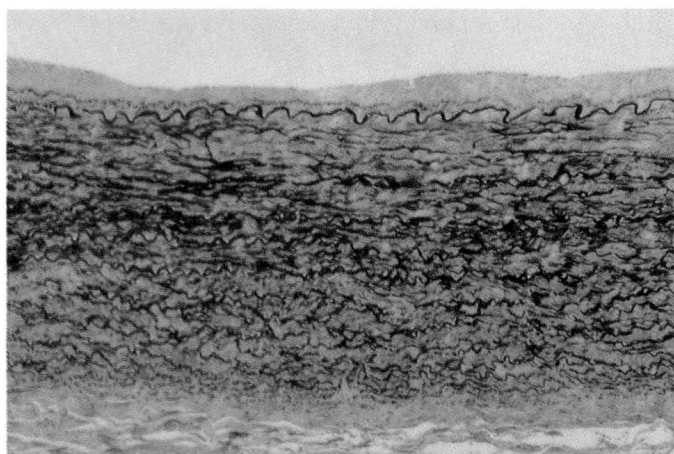

6.14B Part of a transverse section of a young human aorta stained with Verhoeff's stain for elastin. Note the density of the concentric fenestrated elastic laminae. Magnification × 100.

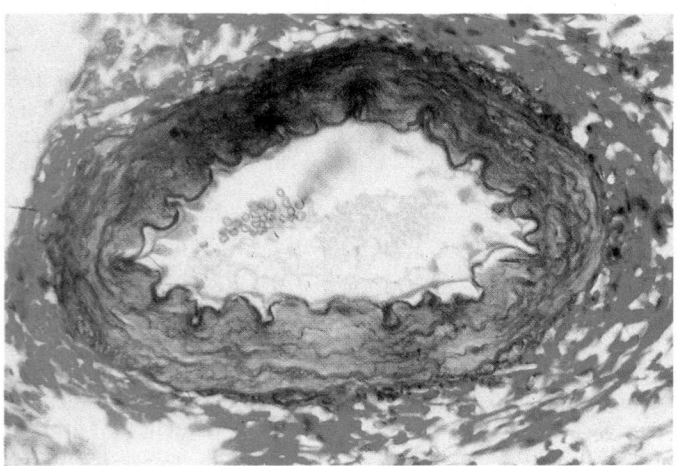

6.14C Transverse section of a small muscular artery, stained with Verhoeff's stain for elastin (purple), and Mallory's stain for collagen (blue). Note the folded, prominent internal elastic lamina. Magnification × 200.

but adjoining intima are often highly branched (Keech 1960), connecting adjacent lamellae. There is a strong correlation between the internal pressure of elastic arteries and the number of their elastic lamellae which, with associated myocytes, form 'lamellar units' (Wolinsky & Glagov 1967a). The myocytes are responsible for the production of lamellae, collagen and proteoglycan components between them (Ross & Klebanoff 1971), an activity which is also shown by non-striated myocytes elsewhere, though to a lesser degree. At birth, the aorta is almost devoid of lamellae and is contractile but later lamellae increase

greatly in number and individual thickness. They also multiply in hypertension due to added mechanical stress (see Cliff 1976).

The arrangement of myocytes varies throughout the vascular system. In the elastic arteries nearest to the heart, they radially or obliquely interconnect the adjacent lamellae ('Spannmuskeln'), whereas more distally they are disposed circumferentially, individual cells lying end to end ('Ringmuskeln') (Benninghof 1927, Wolinsky & Glagov 1964); in both regions they are also slightly helical.

The *tunica adventitia* consists mainly of collagenous fibrous tissue containing a few elastic fibres; it is relatively thin and merges into the surrounding connective tissue.

From middle age onwards structural changes occur in elastic (and other) arteries, in addition to the intimal changes noted above, particularly in the aorta and arteries of the heart and brain. The content of elastin and collagen increases, while the muscle and water content decreases, leading to increased general stiffness. Calcium salts and lipid are also deposited in the intima and media, with splitting of the internal elastic lamina and various other alterations which may become overtly pathological.

Medium and small muscular arteries (6.12,14C,15) have a *tunica intima* which is a single layer of elongate endothelial cells, longitudinally arranged; its external surfaces are covered by a typical basal lamina contiguous with the *internal elastic lamina*, a fenestrated elastic sheet seen in transverse sections as a brightly refractile zone usually undulated by the agonal contraction of the muscular tunica. Since the media contains less elastic tissue than in larger arteries the internal elastic lamina is prominent. Fenestrations in this lamina permit the diffusion of metabolites between the media and the lumen of the vessel.

In the coronary arteries, the internal elastic lamina is bounded by a specialized thick layer of longitudinal muscle and fibrous tissue, presumably enabling them to accommodate to the marked changes in length occurring during the cardiac cycle.

The *tunica media* consists mostly of non-striated myocytes with scattered elastic fibres and a small amount of collagen (mainly type III). The myocytes are generally disposed in tight helices around the circumference of the vessel, although in smaller ones their direction is mainly circular. Adjacent myocytes have communicating junctions which co-ordinate their activities, which are generally under neural and neurohumoral control. Longitudinal myocytes exist in both the intima and media of arteries subject to repeated bending, e.g. the carotid, axillary, uterine, cervicooccipital and palmar arteries. In the tunica media of umbilical arteries, longitudinal muscle occurs in the inner zone, circular in the outer.

The *tunica adventitia* consists of predominantly longitudinal collagenous and elastin fibres. Externally it is rather loosely arranged, merging into the surrounding loose connective tissue and thus allowing considerable movement between the artery and neighbouring structures. Adjacent to the tunica media the adventitia contains much elastic tissue, including the *external elastic lamina*, a feature of variable prominence.

ARTERIOLES, TERMINAL ARTERIOLES, METARTERIOLES AND PRECAPILLARY SPHINCTERS (6.12,16,17)

The smallest arteries terminate in *muscular arterioles* 100–50 μm in diameter, which branch into *terminal arterioles* less than 50 μm in diameter with merely one or two somewhat spiralized layers of non-striated myocytes in their walls (Rhodin 1971). *Metarterioles* (6.17E) are branches of terminal arterioles, 10–15 μm in diameter at their origin and decreasing over 50–100 μm to as little as 5 μm. Where they open into the capillary bed, they are surrounded by a strong circular layer of non-striated myocytes forming *precapillary sphincters*, which effect the final control of blood flow through the capillaries. Precapillary sphincters have been seen to open and close periodically with a cycle of 2–8 seconds.

Both muscular and terminal arterioles have a single layer of elongated endothelial cells, their long axes parallel to the flow of blood. These rest on a basal lamina; they are separated in places from the myocytes of the adjacent tunica media by patches of elastic tissue, homologous with the internal elastic lamina of

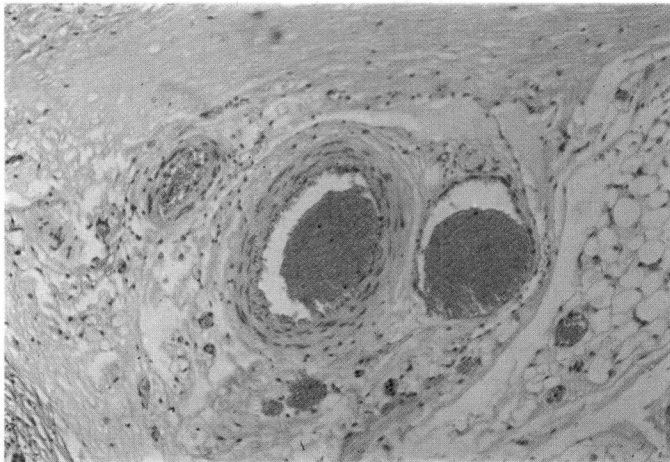

6.15 Transverse section of two small muscular arteries and a small vein, stained with haematoxylin and eosin. Numerous venules and capillaries are included but are indistinct at this magnification. Supplied by D R Turner of the Department of Pathology, Guy's Hospital Medical School.

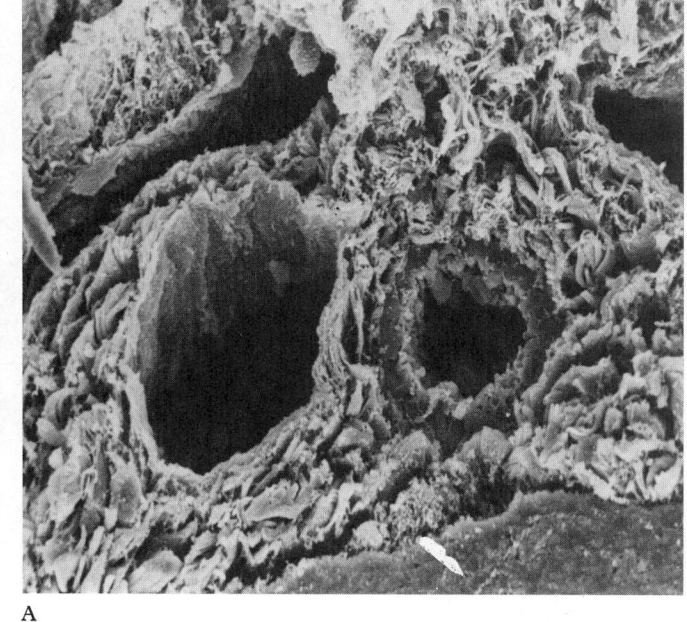

A

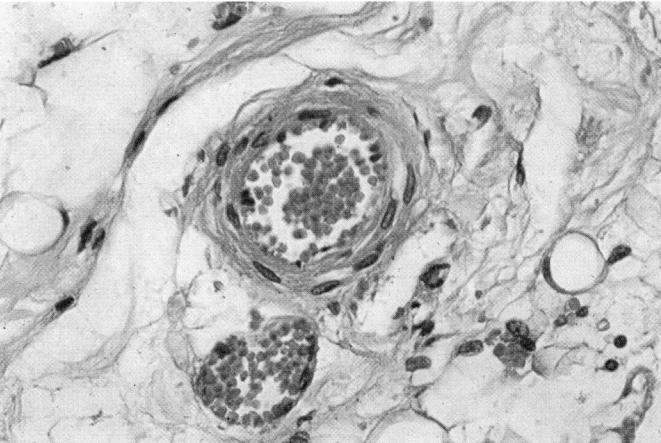

6.16 Transverse section of a large arteriole and venule in loose connective tissue, stained with haematoxylin and eosin. Source as **6.15**.

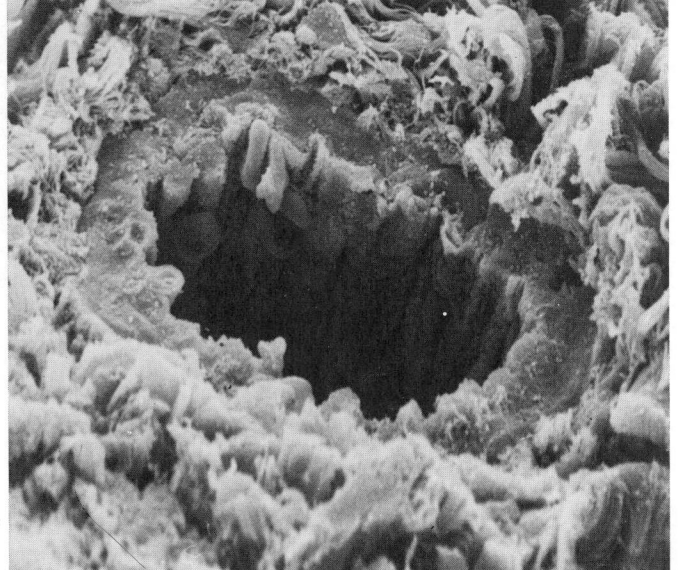

larger vessels. Elastin components decrease and finally disappear in the finest terminal arterioles. The tunica media of larger arterioles has a few circular layers of non-striated myocytes, the most external often being slightly helical. Myocytes form gap (communicating) junctions with overlying endothelium at points where the elastic lamina and basal lamina of the endothelium are lacking.

The adventitia of arterioles consist of a fine reticulum of collagen fibres with a few fibroblasts.

B

Arterial Nutrient Supply

Larger arteries have their own blood vessels, *vasa vasorum* (singular: *vas vasis*), often branches of the artery itself or of a vessel at

6.17 A–C Scanning electron micrographs of small blood vessels in the cut surface of the submucosa of the small intestine (mouse).
A. Low-power view of a group of vessels showing a small artery (mid-right) with an accompanying vein (mid-left) in transverse section. An obliquely sectioned vein is also visible (extreme right) and a lymphatic vessel is present on the upper left; the extravascular tissue consists mainly of irregular dense connective tissue. Magnification × 320.
B. Medium-power view of a transversely sectioned small artery, showing the thick muscular wall and the longitudinally ridged intima with erythrocytes adhering to the endothelial surface. The adventitia blends with the surrounding dense fibrous tissue. Magnification × 1000.
C. High-power micrograph of the cut end of a capillary with three erythrocytes visible. Note the extreme thinness of the endothelial wall (left) and the collagen fibres in the extravascular tissue. Preparations by Michael Crowder, Guy's Hospital, London. Magnification × 6000.

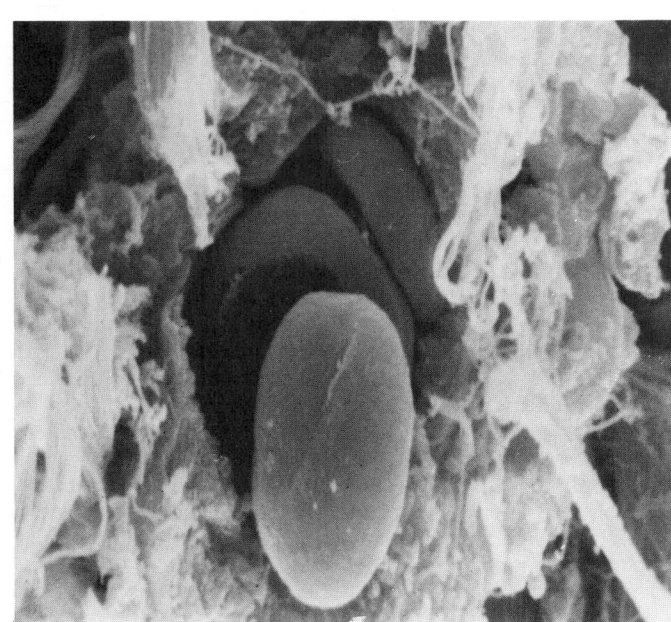

C

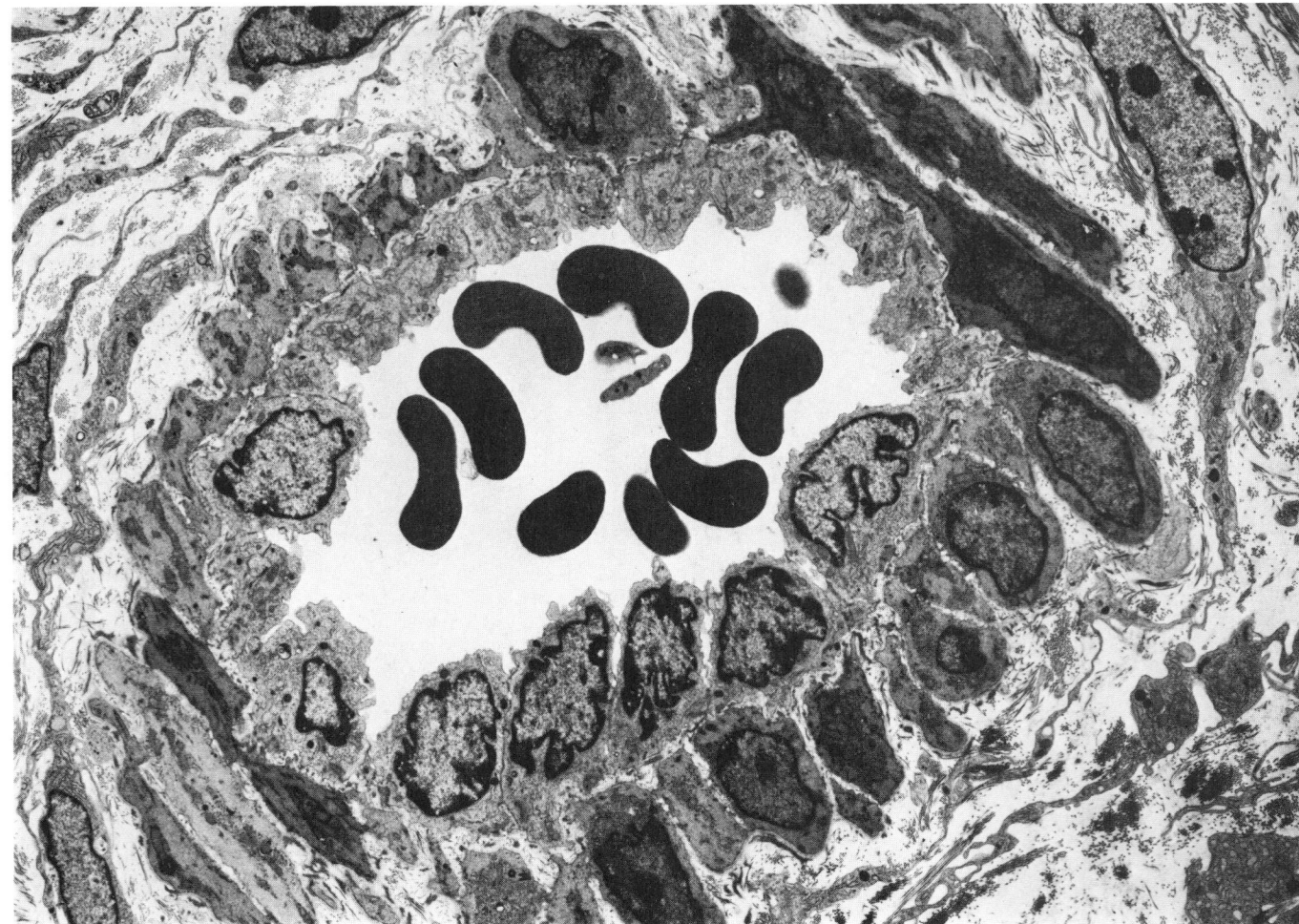

6.17D Transmission electron micrograph of a partially contracted small arteriole in transverse section, showing an outer zone of non-striated myocytes and an inner lining of endothelial cells. Erythrocytes are visible within the lumen. The specimen is from the uterus of a rat. Magnification × 4000. Supplied by Dr Gail ter Haar.

some distance from their distribution; these ramify in the loose connective tissue of the outer sheath and form a dense capillary network in the adventitia and may, in larger vessels, penetrate the outer part of the tunica media. The remainder of the wall is supported by diffusion from blood from these vessels or from the vascular lumen; fenestrations in the elastic laminae are presumably important in allowing this to occur. Minute veins return blood from the adventitia into the artery's companion veins. Lymphatic vessels also occur in the adventitia and drain along parallel channels.

Arterial Nerve Supply

Arteries are innervated mainly by non-myelinated nerve fibres, which are largely efferent and vasomotor to the vessels and conduct a regular flow of impulses maintaining a variable 'vasomotor tone'. Light microscopical evidence of pericellular nerve nets in the tunica media has not been confirmed by electron microscopy, emphasizing the paucity of nerves in the media. Non-myelinated nerve fibres ramify in the adventitia and approach the external myocytes of the media through fenestrations in the external elastic lamina. Their terminals are extensive and contain clusters of mitochondria and numerous vesicles of 30–100 nm diameter, many in the mid-range being dense-cored and typical of noradrenergic synapses. Axons approach to within 60–400 nm of the myocytes (see p. 906) and may relate to several myocytes, each of which is possibly also innervated by more than one terminal (Appenzeller 1964, Lever 1965, 1968, Burnstock et al 1970). The neurotransmitter released from nerve terminals diffuses through the extracellular space and basal lamina to reach receptor sites at myocyte surfaces. A single myocyte may be electrically coupled to

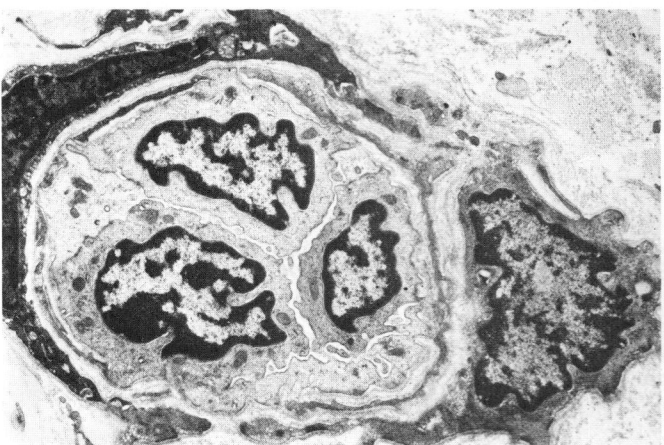

6.17E Transmission electron micrograph of a metarteriole fully contracted. The endothelial cells almost totally occlude the lumen. The dark cells around the periphery are pericytes. A biopsy specimen of human synovial membrane. Magnification × 6000.

many others through gap junctions (p. 24, 562). Myelinated afferent fibres are distributed to the intima and adventitia; they have expanded and varicose endings. The functions of such fibres in the systemic arterial tree is uncertain; some may mediate pain, while in certain sites they are specialized as *baroreceptors* (pressoreceptors), recording general blood pressure (e.g. in the carotid sinus and aortic arch), or *chemoreceptors* responding to increased hydrogen ion and carbon dioxide concentration or decreases in oxygen tension (carotid and aortic bodies). Lamellated (Pacinian) corpuscles, which are mechanoreceptors, occasionally appear in the aortic adventitia.

Most blood vessels have no vasodilator innervation and neurogenic vasodilation follows a reduction of sympathetic vasoconstrictor tone. However, neurally-induced vasodilation occurs: (1) in vessels of skeletal muscle which have a cholinergic sympathetic innervation; (2) in some exocrine glands, following secretomotor activity with secondary release of vasodilator bradykinin; (3) in skin following afferent nerve stimulation, collaterals of which ramify in neighbouring blood vessels as a structural basis for the so-called 'axon reflex'.

The largest arterial trunks receive direct branches from the sympathetic ganglia, but the supply for smaller arteries (brachial, femoral, etc.) is carried in the peripheral nerves and their rami (pp. 1124, 1156). Perivascular autonomic plexuses extend the whole length of the splanchnic arteries.

CAPILLARIES AND SINUSOIDS (6.18C, 19–21)

In many tissues (e.g. muscle, skin, lung, alimentary tract, brain, etc.) branches of terminal arterioles beyond the precapillary sphincters show a transitional zone of 50–100 μm; here a few, spiralling myocytes persist external to the endothelium, the vessels continuing as true *capillaries* of uniform diameter. These ramify and anastomose to create a capillary plexus (6.21), varying in form in different tissues (vide infra). In some sites (e.g. red bone marrow, spleen, liver, suprarenal and parathyroid glands, adenohypophysis, carotid and coccygeal bodies) the smallest vessels are *sinusoids*. These are lined by endothelium, supported externally by only a tenuous coat of connective tissue. They have a wider calibre than capillaries and their lumen may be irregular in cross section. Typically, the endothelium is incomplete or fenestrated, allowing relatively easy diffusion of larger molecules or of cells between the lumen and the adjacent tissues.

Electron microscopy (6.18–20) has revealed marked variation in fine structure among both capillaries and sinusoids (Bennett et al 1957, Rhodin 1962 a, b, Fawcett 1963, Florey 1966, Wisse 1970). Capillaries are tubes of a single layer of polygonal or lanceolate endotheliocytes varying in fine structure and intercellular junctions; external to them is a typical basal lamina, which splits to enclose flat or branching perivascular *pericytes*. An adventitial layer of fine reticular tissue is often associated with the capillary wall, with occasional fibroblasts and mast cells. Capillaries may be broadly classified as *continuous* and *fenestrated* types. Sinusoid endothelial linings are either fenestrated or *incomplete*.

Continuous capillaries (6.18A) occur, e.g. in skin, connective tissues, striated and non-striated muscle, lung and brain. The cytoplasm is more abundant near the oval nucleus. The cell presents a smooth luminal surface with occasional filopodia. Organelles are few but the cytoplasm contains fine actin and vimentin (intermediate) filaments (p. 30) and a little endoplasmic reticulum with attached and free ribosomes; the Golgi apparatus is small. The luminal and basal surfaces of the plasma membrane are often much invaginated by *caveolae* and the adjacent cytoplasm contains many vesicles, 50–70 nm in diameter. These are believed to form on one surface, detach, cross the cell to fuse with the opposite wall and discharge their contents. The cytoplasm around them is rich in phosphatases. At cell junctions plasma membranes may be simple or interdigitated; cells frequently overlap. A gap of about 20 nm, containing electron-dense material, usually separates plasma membranes; but sometimes these approach closely or fuse to form tight junctions (p. 24). True desmosomes are infrequent in human tissue.

Fenestrated capillaries (6.18B) occur in renal glomeruli, intestinal villi, endocrine glands and the pancreas. The cytoplasm on each side of the nuclear region in endotheliocytes is very thin, often approaching 40–60 nm, and apparently perforated at intervals by 'pores' varying greatly in number and size (diameters from 30–100 nm). Pores are often 'closed' by a thin electron-dense diaphragm which appears as a condensation of the basal laminal material. The mechanism of capillary diameter and blood flow is also obscure. Many believe that flow is determined only by the state of contraction of the arterioles and precapillary sphincters, others that endotheliocytes and some types of pericyte possess cytoplasmic contractile elements (Joyce et al 1985).

Sinusoids (6.18C) also vary locally in structure. Their calibre is larger and more irregular, their lining a variety of cell types and they may be *discontinuous, closed* or *fenestrated*. Fenestrated

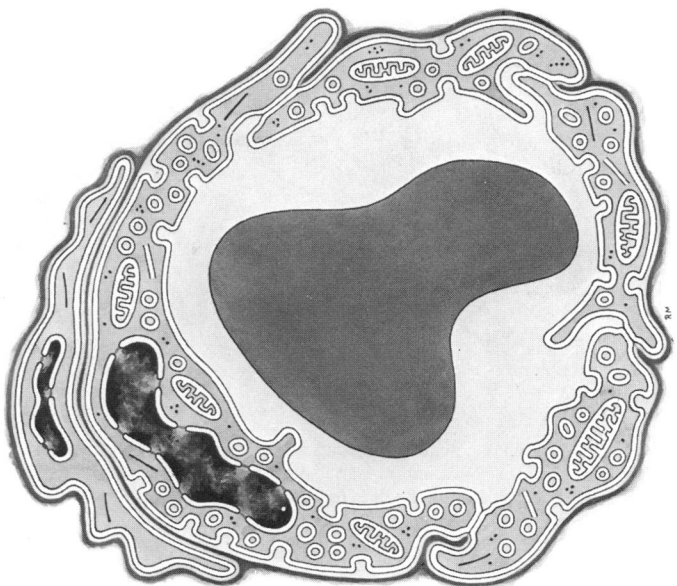

6.18A A 'continuous' capillary in transverse section: a scheme of its ultrastructural appearance.

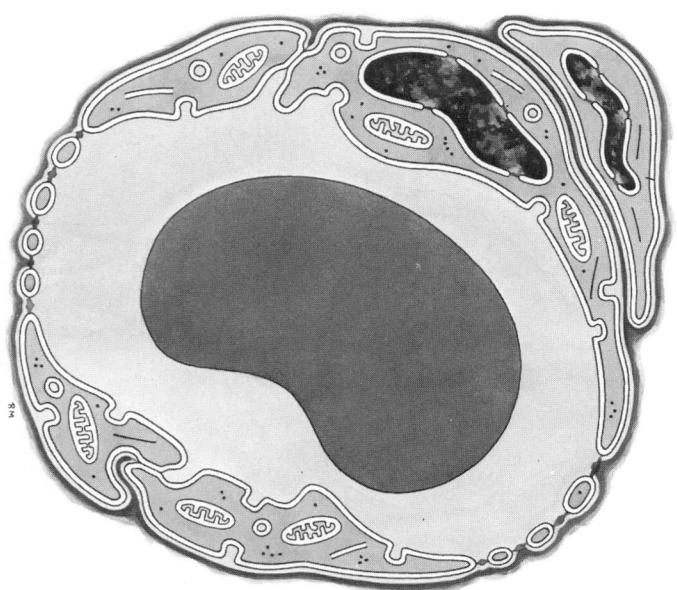

6.18B A 'fenestrated' capillary in transverse section: a scheme of its ultrastructural appearance.

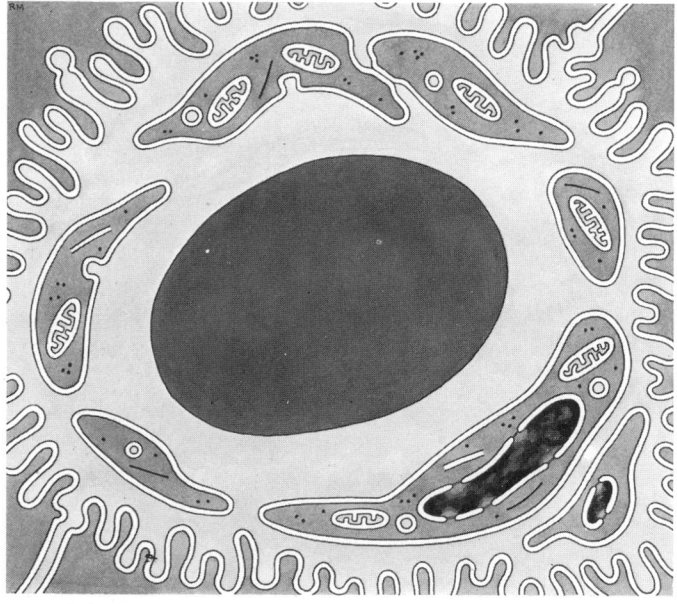

6.18C A 'discontinuous' sinusoid in transverse section: a scheme of its ultrastructural appearance.

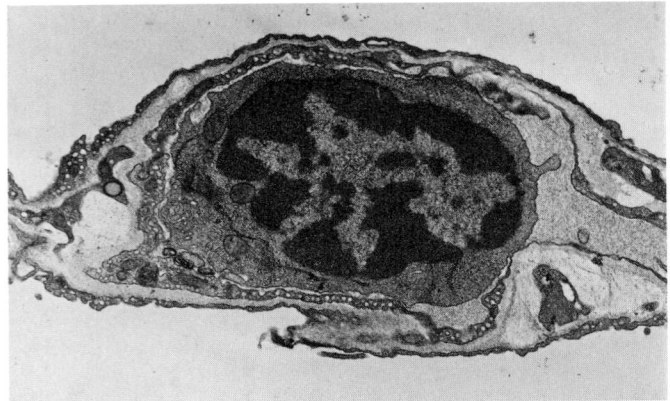

6.19 Transmission electron micrograph of a capillary in the wall of a pulmonary alveolus. Note the presence of numerous pinocytotic vesicles in the attenuated endothelial cell cytoplasm. A small lymphocyte almost fills the lumen. Magnification × 5500.

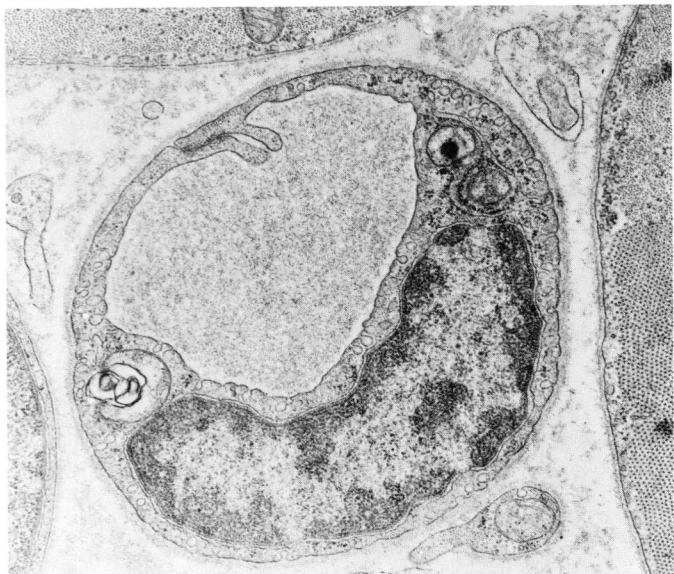

6.20 Transmission electron micrograph of a capillary in transverse section lying in the endomysial space between the skeletal muscle fibres. Note the large endothelial cell nucleus, two intercellular tight junctions and the conspicuous pinocytotic vesicles. A prominent basal lamina surrounds the capillary. Human biopsy specimen. Magnification × 8000.

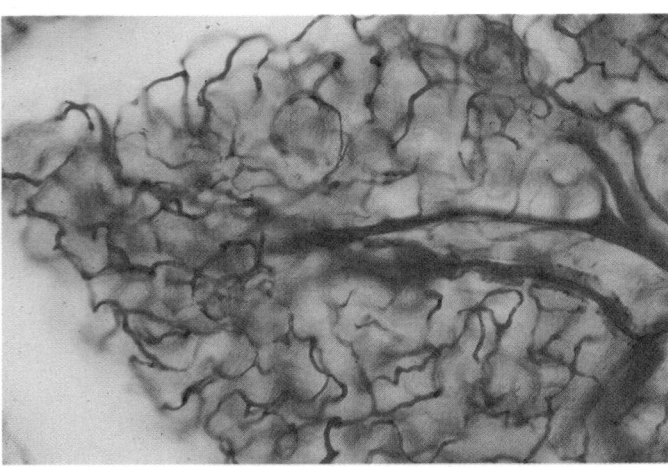

6.21 A capillary bed, injected with red gelatin and cleared to show the vessels. Magnification × 100.

sinusoids are typical of endocrine glands, having much the same ultrastructure as fenestrated capillaries, including the presence of a delicate, complete basal lamina. Sinusoids of the liver are lined by flat attenuated endothelial cells possessing numerous large fenestrations grouped in clusters, 'sieve plates' (p. 1392, 8.16OD). The sinusoids in the red pulp of the spleen provide an extreme example of *incomplete endothelium*, long slits between strap-like endothelial cells allowing blood cells free passage between lumen and peri-sinusoidal spaces (p. 830). Further details will be noted under individual organs (e.g. pp. 829, 1392).

VENULES

A variable number of capillaries form a *venule* (**6**.13), 20–30 μm in diameter and essentially a tube of flat, oval or polygonal endotheliocytes supported by basal lamina and a delicate adventitia of longitudinal fibres and fibroblasts (Rhodin 1968). Some observers (e.g. Simionescu & Simionescu 1984) distinguish between initial, *postcapillary venules* surrounded by pericytes (*pericytic venules*) and larger vessels partially enclosed in a muscular tunica media (*muscular venules*). Postcapillary venules are sites of fluid exchange and leucocyte migration; in lymphoid tissue of the gut and bronchi and in lymph nodes and thymus, postcapillary endotheliocytes are much taller and have intercellular junctions through which lymphocytes can readily pass into and from the extravascular tissue (p. 824). In other tissues these vessels are believed to be a major site of migration of neutrophils, macrophages and other leucocytes into extravascular spaces, and also a region of temporary endothelial attachment for neutrophils, forming *marginated pools* of these cells (p. 666).

The intracellular junctions of venules are sensitive to inflammatory agents which increase their permeability to fluids and defensive cells (see e.g. Marchesi 1961, 1962).

PERMEABILITY OF CAPILLARIES AND VENULES

The capillaries and postcapillary venules form a vast area for *exchange* of nutrient substances, gases, metabolites and water between blood and tissues; this is also the chief route of leucocyte migration from blood to extravascular spaces. In general, the densities of capillary networks vary with the metabolic level of the tissues supplied; e.g. in the myocardium there are about 2000 capillaries per cubic millimetre, about 1000 in skeletal muscle, but only 50 in cutaneous connective tissue (Simionescu & Simionescu 1983). The total surface area is relatively large, as much as 2.4 metres² in skeletal muscle, about 4 metres² per 100 cm³ in lung, with values of about 60 metres² for the total of all systemic capillaries and 40 metres² for the pulmonary circulation (Krogh 1959, Intaglietta & Zweifach 1971).

Capillary and venular permeability has been widely studied in various ways, ranging from physiological measurement of fluid composition to structural investigations with tracers, e.g. Evans' blue, horseradish peroxidase and ferritin (Karnovsky 1967, 1968, 1970). These techniques have shown that although vascular endothelium is a semi-permeable dialysing membrane, allowing water, gases and various ions to diffuse through, while being permeable to large molecules, there *is* some movement of colloids and even large water-soluble molecules through endotheliocytes. Measurements of the passage of large molecules, e.g. dextrans, through capillary walls have shown two types of diffusion, corresponding to hypothetical '*large pore*' and '*small pore*' channels (Grotte 1956, Landis & Pappenheimer 1963, see also Granger & Perry 1983). The precise structural basis of these is uncertain. Tracer studies show that large molecules, or molecular aggregates, cross endothelia in endocytic and exocytic vesicles; indeed, vesicles from both aspects may fuse into continuous channels (Bruns & Palade 1968). Such vesicular transport probably corresponds to 'large pore' diffusion, while 'small pore' diffusion is perhaps mediated by leaky intercellular junctions and perhaps endothelial fenestrae, when present, as in renal glomeruli (see Maul 1971).

Capillary permeability varies greatly among tissues and can be correlated partly with the local type of endothelium. In tissues where large molecules pass easily (e.g. alimentary tract, endocrine glands) fenestrated endothelia exist, with numerous vesicles; intercellular junctions are either incomplete or 'leaky' (Simionescu et al 1972, 1973, 1975, 1980). Where barriers to diffusion of large molecules occur (e.g. brain, thymic cortex, testis and exocrine pancreas), endothelia are complete and not fenestrated, with efficient zonulae occludentes between cells; here, endocytic vesicles are few. Other tissues (e.g. skeletal muscle) show an intermediate condition.

VEINS

Venous walls, like those of arteries, have three layers: an endothelium-lined tunica intima, a muscular tunica media and an external tunica adventitia, which is mainly connective tissue (6.13, 17A). A major difference lies in the comparative weakness of the venous tunica media, which has a smaller content of muscle and elastic fibres, a condition related to the lower venous pressure.

In the smallest veins the three *tunics* are not distinct. Endothelium is supported on a membrane separable into two layers: the external one is thicker and a delicate membrane of collagen fibres and fibroblasts; the internal one is composed of interlacing longitudinal elastic fibres. In veins about 0.4 mm in diameter the middle tunic consists of collagen fibres and a few circumferential non-striated myocytes; connective tissue forms a recognizable *tunica externa*. In veins of medium size the endothelium is like that in arteries but its cells are shorter and broader, supported by a delicate network of branched fibroblasts, external to which is a network of elastic fibres in place of the fenestrated laminae of arteries. This constitutes the *tunica intima*. The *tunica media* is a thick layer of connective tissue with elastic fibres and, in some veins, circumferential non-striated myocytes. White fibres predominate; elastic fibres are less abundant than in arteries. The *tunica adventitia*, as in arteries, is loose connective tissue with longitudinal elastic fibres; in the largest veins it is much thicker than the tunica media and contains longitudinal non-striated myocytes, most numerous in the inferior vena cava, especially at its termination, in the largest portal, hepatic and external iliac, renal and azygos veins. In the inferior vena cava and in the renal and portal veins, myocytes appear throughout the adventitia but in the other veins mentioned, connective and elastic tissues form a layer external to the muscle. In the inferior vena cava, collagen fasciculi in the adventitia form an interwoven, diagonal meshwork spiralling around the vessel. This arrangement, together with the pressure of much elastin, allows it to lengthen and shorten readily with the movements of the diaphragm (Franklin 1937). Large veins entering the heart are encroached upon for a short distance by myocardium, and in the coronary sinus this covering is complete (Coakley & King 1959). Muscular tissue is absent in: (1) maternal placental veins, (2) dural venous sinuses and pial veins, (3) retinal veins, (4) veins of trabecular bone, (5) venous spaces of erectile tissue. Such veins consist of endothelium supported by variable amounts of connective tissue.

Most veins have *valves* to prevent reflux of blood (6.22). A valve is composed of an inward projection of the tunica intima, strengthened by collagen and elastic fibres and covered by endothelium differing in orientation on its two surfaces. Surface cells which are juxtamural are transversely arranged whereas on the luminal surface, over which the main stream of blood flows, cells are arranged longitudinally in the direction of the current. Most commonly, two such valves lie opposite one another, especially in smaller veins or in larger ones where smaller tributaries join; occasionally three valves lie in opposition; sometimes only one is present. The valves are semilunar (cusps) and attached by convex edges to the venous wall; their concave margins are directed *with* the current and apposed to the wall as long as flow is towards the heart but when blood flow reverses the valves close. Centripetal to each valvular flap the wall is expanded into a *sinus*, which fills when blood flow is reversed against a closed valve, giving a 'knotted' appearance to the distended veins, if these have many valves. In the limbs, especially the legs where venous return is against gravity, such valves are of great importance to venous flow, as blood is moved towards the heart by the intermittent pressure produced by contractions of the surrounding muscles.

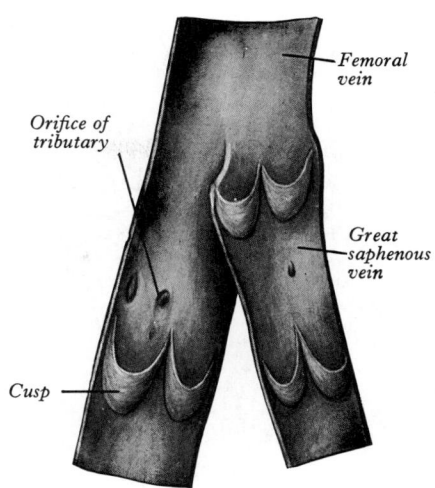

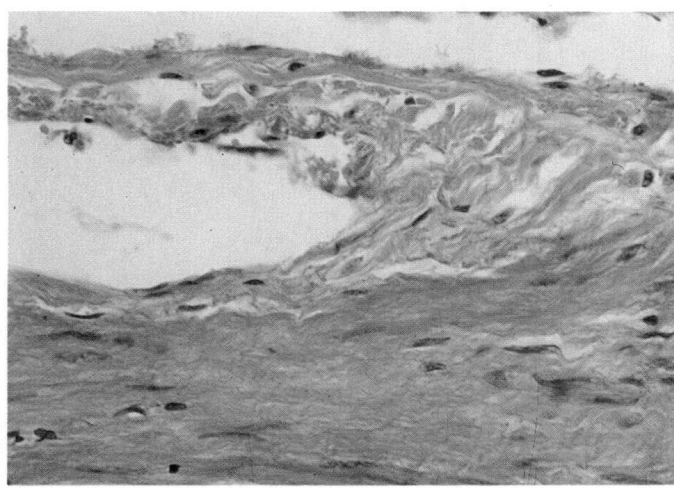

6.22A The upper portions of the femoral and great saphenous veins laid open to show the valves. About two-thirds of the natural size.

6.22B Longitudinal section through a venous valve, showing intimal projections forming the valve leaflet. Magnification × 400.

Valves are absent in very small and in very large veins and in many tissues are rare or absent.

The venous return of blood is influenced by several factors. Smaller veins are filled by the overflow from capillaries. The deep veins in the limbs are massaged by the contractions of surrounding skeletal muscles, and venae comitantes are affected by arterial pulsation in the same way. This compression would drive blood in both directions along the veins, were it not for their valves which only allow flow towards the heart, when they are competent. The force of gravity in the veins of the head and neck and the negative intrathoracic pressure in veins near the heart are also important factors in venous return. Flow in veins is slower than in arteries and veins are larger and more numerous than corresponding arteries, a requirement for an efficient return of blood to the heart.

Walls of the larger veins, like the arteries, are supplied by *vasa vasorum* but these in veins may penetrate the wall deeply, perhaps because of the lower oxygen tension. Postganglionic sympathetic efferent and primary afferent nerves are distributed to the veins, as in arteries, but less abundantly.

Associations between Arteries and Veins

Arteries and veins usually accompany each other in fascial planes, visceral pedicles and neurovascular bundles. The vein may be single, but is often paired; such *venae comitantes* flank an artery, with numerous cross-connections, the whole assembly being enclosed in a single connective tissue sheath.

As noted, the proximity of an artery and vein probably aids venous return. Close apposition may also favour the

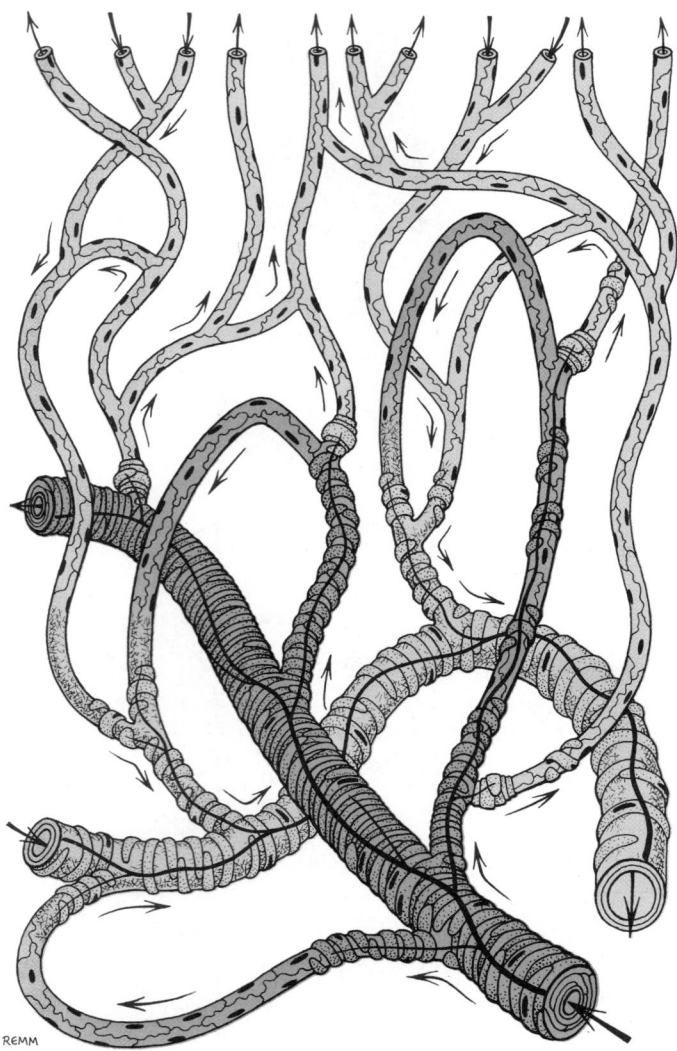

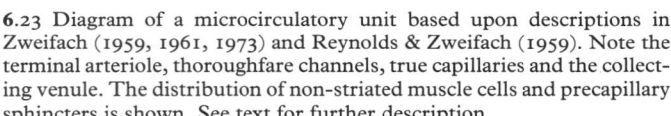

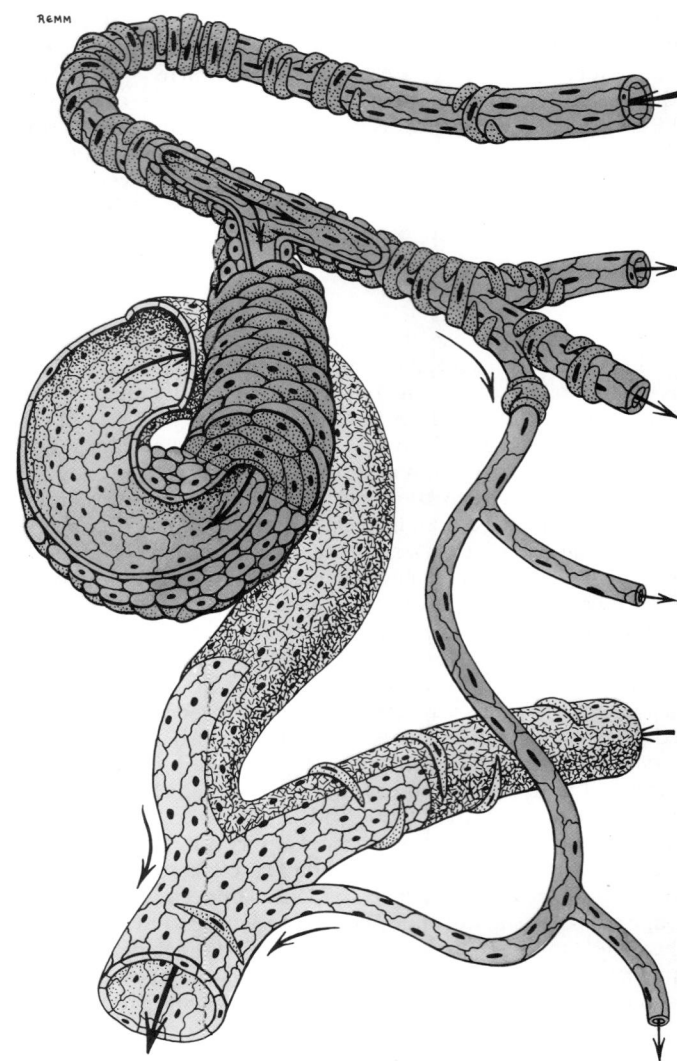

6.23 Diagram of a microcirculatory unit based upon descriptions in Zweifach (1959, 1961, 1973) and Reynolds & Zweifach (1959). Note the terminal arteriole, thoroughfare channels, true capillaries and the collecting venule. The distribution of non-striated muscle cells and precapillary sphincters is shown. See text for further description.

6.24 Diagram of an arteriovenous anastomosis. Note the thick walls of the anastomotic channel composed of many layers of characteristically modified 'epithelioid' non-striated muscle cells. See text for further description.

countercurrent exchange of heat or ions from incoming to outgoing blood, as exemplified by the blood supply of the testis, where tributaries of the testicular vein form a plexus around the artery, promoting heat transfer from arterial to venous blood and thus helping to maintain low testicular temperature (Evans 1949, Grant & Payling Wright 1971, also p. 1429). An analogous arrangement of arterial and venous sinusoids exists in the vasa recta of the renal medulla where countercurrent exchange retains sodium ions at a high concentration in the medulla (p. 1406), efferent venous blood transferring sodium ions to the afferent arterial supply.

Vascular Patterns

The distribution of systemic arteries is like a branching tree, its common trunk, the aorta, emerging from the left ventricle and smaller ramifications extending to viscera and other peripheral regions. The mode of arterial division varies: a short trunk may divide into several branches at one locus, e.g. coeliac and thyrocervical trunks; usually it gives off successive side-branches, continuing as the main trunk, as in arteries of limbs and the head and neck. A branch is smaller than its parent trunk; but where an artery bifurcates, the combined sectional area of the two vessels is almost always somewhat greater than that of the trunk; e.g. the sectional area of the aorta is greatly exceeded by the combined sectional area of all of its branches. These facts are basic to haemodynamics.

ANASTOMOSES

Arteries do not always end in arterioles and capillaries; they may unite, forming *anastomoses* (**6**.12,23,24). Anastomoses between arteries of equal calibre occur in the brain, where two vertebral arteries fuse to form the basilar artery and two anterior cerebral arteries are connected by the anterior communicating artery. A similar fusion is seen where intestinal arteries anastomose by their larger branches. In the limbs, anastomoses are prominent around joints, the branches of arteries proximal to a joint uniting with branches of the vessels below it; lateral anastomoses also form peri-epiphysial circles (p. 300). Arterial anastomoses increase in frequency with distance from the heart and smaller branches of arteries anastomose more frequently than larger ones. Between the smallest twigs, these anastomoses may be so numerous that they form a close network. Anastomoses can equalize pressure differences and provide alternative routes for blood flow, e.g. in peri-articular sites, where circulation may be temporarily impeded by movements of joints. Where arteries join end to end (e.g. palmar and plantar arches, intercostal arteries, gastric and intestinal arteries), section results in bleeding from both ends. In the case of restricted blood flow through an artery, an anastomosis channel may become enlarged (e.g. p. 766), providing a *collateral circulation* e.g. when a vessel is interrupted by accident, disease or ligation. This often develops rapidly. If the increased flow and pressure in the collateral channels are maintained, vessels enlarge, often tortuously, and finally provide a new arterial route, bu

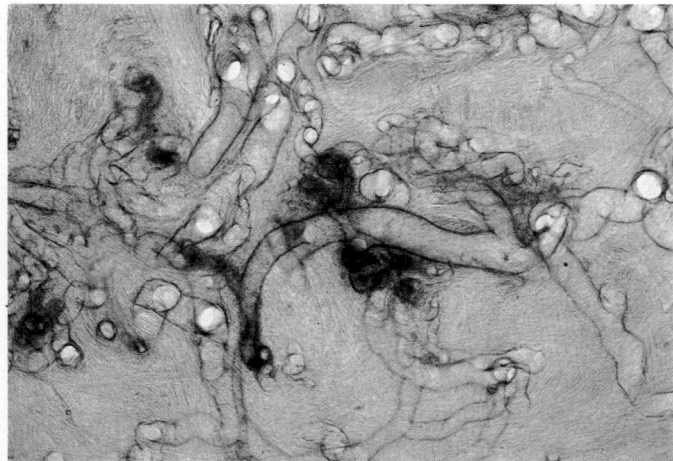
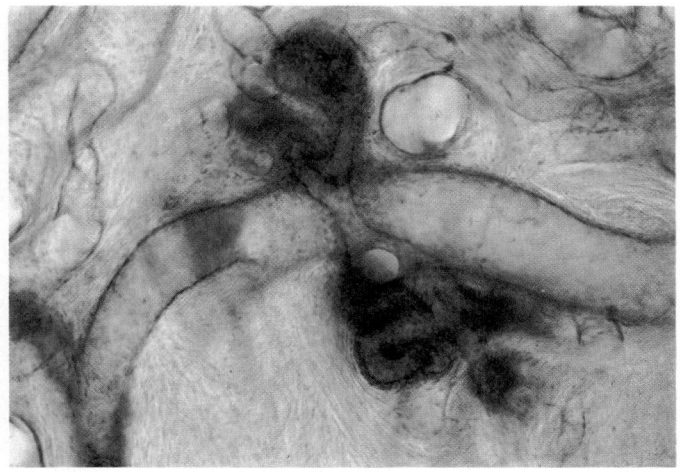

6.25 Human digital arteriovenous anastomoses prepared by intravascular perfusion of haematoxylin and subsequent clearing of a full thickness specimen of skin. The heavily stained, thick-walled, tortuous, anastomotic channels contrast with the central arterial stem and the thin-walled venous outflow channels. See text for further details. The specimens were prepared and provided by R T Grant, Guy's Hospital Medical School.

never in excess of functional needs. In some sites, sudden occlusion causes death of the region supplied, but gradual occlusion may allow the adequate dilatation of anastomosing channels. The collateral circulation develops more rapidly in the young. The mechanisms involved in the dilation of collateral channels are not fully understood; haemodynamic changes consequent on the fall of peripheral resistance, on anoxia, neural factors and the effects of accumulating metabolites have all been implicated in this phenomenon.

END-ARTERIES

In some regions arteries have no anastomoses and are hence *end-arteries*. If an end-artery is occluded, necrosis occurs in the tissues supplied by it, e.g. in the central artery of the retina where arterial occlusion is followed by permanent blindness. Though large arteries *outside* the brain form effective anastomoses, its central arteries in the substance of the brain, communicate only through capillaries, being virtually end-arteries, as are arteries in the spleen, kidneys, lungs and metaphyses of long bones. Branches of coronary arteries do anastomose to some extent, but this is often insufficient to provide an adequate collateral circulation. Occlusive arterial disease in organs with an end-arterial blood supply is hence most serious.

VASCULAR SHUNTS

These are communications between arteries and veins found in many regions of the body where the capillary circulation is bypassed by wider channels. They may be classified according to their dimensions, site and complexity as: (1) preferential thoroughfare channels, (2) 'simple' arteriovenous anastomoses, (3) specialized arteriovenous anastomoses or *glomera*.

Preferential 'thoroughfare' channels. In many tissues true capillaries arise not only as direct side branches of terminal arterioles but also as side branches of a main or 'thoroughfare' channel connecting the terminal arteriole and the venule (Zweifach 1937, 1959, 1961, Reynolds & Zweifach 1959, Maggio 1965, Grant & Payling Wright 1968, 1970). This *thoroughfare channel* has a larger calibre than true capillaries and in ultrastructure resembles typical continuous capillaries, except that widely spaced non-striated myocytes spiral round the endothelium. Each capillary side branch has at its origin a precapillary sphincter. Such a channel and its capillaries form a functional *microcirculatory unit* (**6**.23). When functional demand is low, blood flow is largely limited to the bypass channel, with most precapillary sphincters closed. Periodic opening and closing of different sphincters may irrigate different parts of the capillary net. With increasing demand, blood flow may increase greatly following the opening of many sphincters. The size of the micro-circulatory unit varies greatly, e.g. in skeletal muscle each channel give rise to 20–30 true capillaries, but in some glandular tissues only one or two may be given off. Detailed investigations in the cremaster muscle and biceps femoris tendon of the rat (Grant & Payling Wright 1968, 1970) have shown that, in these sites, bypass channels are confined to perimuscular and peritendinous connective tissues and absent from the muscle itself. The form of the capillary net also varies with the tissue, meshes being either round or elongated. Round or angular meshes are most common and prevail where networks are dense, as in the lungs, mucous membranes and skin. Elongated meshes occur in muscles and nerves, aligned parallel with their fibres. Sometimes capillaries are looped, as in the papillae of the skin and tongue. The number of capillaries and the size of their mesh determine the degree of vascularity; the smallest meshes occur in the lungs and the choroid of the eye.

Arteriovenous anastomoses (**6**.24) are direct connections between smaller arteries and veins (Grant & Bland 1931, Popoff 1934, Clark 1938). Connecting vessels may be straight or coiled, often possessing a thick muscular tunic of peculiar structure and a narrow lumen, about 10–30 μm across. Under sympathetic control through abundant non-myelinated fibres in its wall, the vessel is able to completely close, circulation being then via the capillary bed. When patent, the vessel carries blood from artery to vein, partially or completely excluding the capillary bed from the circulation.

Arteriovenous anastomoses of relatively simple type occur in the nasal, labial and aural skin, nasal and alimentary mucous membranes, coccygeal body, erectile tissue, tongue, thyroid gland, sympathetic ganglia and probably elsewhere.

In the skin of the hands and feet (especially digital pads and nail beds) anastomoses form a large number of small units termed '*glomera*'. They are deep in the corium; each '*glomus*' has one or more afferent arteries, stemming from branches of cutaneous arteries approaching the surface (**6**.24,25). These afferents arise at right angles from their parent vessels, which then continue into the dermal papillary layer, ending in a capillary plexus. A short distance from its origin an afferent artery gives off a number of fine 'periglomeral' branches and then immediately enlarges, makes a sinuous curve and narrows again into a short funnel-shaped vein opening at right angles into a collecting vein. This vein commences on the deep aspect of the glomus, curving round its superficial surface, whence it retraces its course, receiving venules from the dermal papillary layer. Finally, it joins a deeper cutaneous vein.

In the newborn child arteriovenous anastomoses are generally few and poorly differentiated, but they develop rapidly during the early years of life. In old age they atrophy, sclerose and diminish in number.

The vessels concerned in digital arteriovenous anastomoses are unusual (**6.25**). Where these enlarge, the afferent artery has small luminal endothelio-muscular projections but proximal to this structure it is typical. The connecting vessel has an endothelium supported by fine collagenous fibres but no internal elastic lamina. Longitudinal and circular muscle layers are not sharply differentiated but the muscular wall is thick; in sections myocytes appear pale and swollen, with central nuclei, and hence described as 'epithelioid'. The efferent vein has a thin wall lacking muscle but containing many elastic fibrils which pass into the tunica adventitia of the collecting vein.

The mechanisms by which arteriovenous anastomoses regulate local flow are poorly understood. Where they have circular muscle in their walls, epitheliocytes may help to narrow the lumen; where it is absent closure may be due to swelling of these epithelioid cells. Cutaneous arteriovenous anastomoses are essential to the control of general and local body temperature. When a rabbit's ear is raised above 40°C, muscle in the walls of the connecting vessels relax and increased blood flow at body temperature results, with a consequent cooling. When the local temperature is lowered below 15°C, the connecting vessels again relax and increased flow at body temperature then helps to raise local temperature, unless artificial cooling is intensified. When the animal's overall *body* temperature is raised experimentally, a general opening of all subcutaneous arteriovenous anastomoses results, with an increase in heat radiation and consequent drop in body temperature (Grant 1930). The cooling effect of panting in dogs also involves the opening of lingual arteriovenous anastomoses. The paucity and immaturity of arteriovenous anastomoses in the newborn and marked reduction in subcutaneous arteriovenous anastomoses with advancing years may be related to observed less efficient temperature regulation in these two extremes of age.

Arteriovenous anastomoses in alimentary mucous membranes fulfil a different function (Spanner 1931). An arteriole to a human villus has a direct connection with its corresponding venule and, when absorption is in abeyance, the connection is patent and helps to raise portal venous pressure; during alimentary absorption the anastomosis is closed and consequently blood traverses the capillary plexus.

Other suggested functions of arteriovenous anastomoses include regulation of blood pressure, secretion by epithelioid cells and pressor reception.

THE THORACIC CAVITY AND HEART

The thoracic skeleton is described on pp. 335–337. The total thoracic capacity does not correspond with that of the osseous thorax, because its lower part is encroached upon by the diaphragm and upper abdominal viscera. The capacity also varies with respiration, which also affects the positions and relations of the thoracic viscera. Its arbitrary upper limit is usually taken as the plane of the thoracic inlet but the pulmonary apices extend above this into the neck.

THE THORACIC INLET (**6**. 85,86,146)

Osseous boundaries are described on p. 336. Structures traversing the superior thoracic opening form two groups: those in or near the median plane, and those on each side closely related to the cervical parts of pleurae and lungs.

Near the midline, behind the manubrium, the lowest parts of the sternohyoid muscles enter the thorax, behind which are the sternothyroid muscles, thymic vestiges and inferior thyroid veins passing down to the brachiocephalic veins. In children particularly the left brachiocephalic vein itself may be *in* the thoracic inlet. Posteriorly, trachea and oesophagus, with the recurrent laryngeal nerves, occupy the median region; behind the left oesophageal margin the thoracic duct enters the neck. Anterior to the vertebral column are the prevertebral longus colli muscles and anterior longitudinal ligament.

On each side the upper part of the pleura and pulmonary apex occupy the inlet. Between the pleura and neck of the first rib, mediolaterally, are the sympathetic trunk, superior intercostal artery and ventral ramus of the first thoracic nerve passing superolaterally to the brachial plexus. Anteriorly, between pleura and first costal cartilage, the internal thoracic artery enters the thorax; medial to the artery its vein leaves the thorax.

On the right (**6.**85), the brachiocephalic artery leaves the thorax between the trachea and pleura. The vagus nerve, having passed between subclavian artery and vein, is between the pleura and the brachiocephalic artery at the inlet. The right brachiocephalic vein enters the thorax anterolateral to its artery; the right phrenic nerve crosses the internal thoracic artery and is lateral to the brachiocephalic vein behind the first costal cartilage.

On the left (**6.**86), the left common carotid and subclavian arteries leave the thorax between the pleura and trachea, the left vagus nerve descending lateral to the interval between them. Anterolateral to this is the left brachiocephalic vein. The left phrenic nerve crosses the internal thoracic artery at a higher level than on the right and, at the inlet, it is between the left brachiocephalic vein anterolaterally and subclavian and common carotid arteries posteromedially.

THE THORACIC OUTLET

This is wider transversely and slopes obliquely down and backwards, so that the vertical extent of the cavity is posteriorly much longer. The diaphragm (p. 592) closes the opening and forms a convex floor for the cavity; it is flatter centrally than at the periphery and higher on the right; in cadavers this side of the floor reaches the level of the fifth costal cartilage's upper border, on the left only to the sixth. (See p. 594 for further information on diaphragmatic shape and levels.) From the summit of each side the diaphragm slopes abruptly down to the costal and vertebral attachments; this slope is posteriorly much more marked and longer and the space between the diaphragm and the posterior thoracic wall narrows rapidly inferiorly.

DIVISIONS OF THE THORACIC CAVITY

The thoracic cavity is divided by the *mediastinum*, the mass of structures between the lungs extending from the sternum to the vertebral column and from the thoracic inlet to the diaphragm. The heart is in the mediastinum, enclosed by the *pericardium*; the lungs occupy the right and left regions, each covered by a serous membrane, the *pleura*, which also lines the corresponding half of the thorax and additionally forms the lateral mediastinal boundary (**6.**27,29A).

For description the mediastinum is arbitrarily divided. The *superior part* extends from the thoracic inlet to an oblique plane passing through the lower edge of the manubrium sterni and lower border of the fourth thoracic vertebra. The *inferior part,* below this plane, is subdivided into an *anterior part* in front of the pericardium, a *posterior* behind this and the diaphragm, and a *middle,* containing the pericardium, heart and large vessels entering or leaving it. Detailed mediastinal contents are included with descriptions of respiratory organs (pp. 1248–1256), the heart (pp 696–734) and the oesophagus (p. 1331).

The Pericardium

The pericardium (**6.**26,27,29A) contains the heart and juxta cardiac parts of its great vessels. It is posterior to the corpus sterni and second to sixth costal cartilages and anterior to the fifth t eighth thoracic vertebrae. It consists essentially of two oppose surfaces of serous membrane, separated by a film of fluid, providing a complete cleavage between the heart and its surroundings, which allows freedom of movement within the pericardium. This has an outer sac, *fibrous pericardium*, and an inner, double-layered sac, the *serous pericardium*, which lines the fibrous sac and covers the heart. The heart invaginates the serous sac from above and behind and fills its cavity, leaving only a potential circumferential space.

The *fibrous pericardium* is roughly conical, its apex continuous with the exterior of the great vessels and its base attached to the diaphragm's central tendon and a small muscular area of its left half. In some lower mammals the base is completely separate from the diaphragm or joined only by areolar tissue; its human diaphragmatic attachment is also of loose fibro-areolar tissue, easily cleft; but a small area of the central tendon and pericardium are fused. Above, the fibrous pericardium not only blends externally with the great vessels, but is continuous with the pretracheal fascia (p. 583). It is also attached to the posterior sternal surface by *superior* and *inferior sternopericardial ligaments* passing to the corresponding ends of corpus sterni. (In the experience of thoracic surgeons these 'ligaments' are extremely variable; the superior is often indetectible.) By these connections the pericardium is securely anchored and maintains the general thoracic position of the heart. Some say that it prevents 'over-distension' but this view is entirely speculative.

Anteriorly, the fibrous pericardium is separated from the thoracic wall by the lungs and pleurae; but in a small area of the lower left half of the corpus sterni and sternal ends of left fourth and fifth costal cartilages the pericardium is in contact with the thoracic wall. Until it retrogresses, the lower end of the thymus is anterior to the upper pericardium. *Posteriorly* are the principal bronchi, oesophagus, oesophageal plexus, descending thoracic aorta and posterior parts of the mediastinal surfaces of both lungs. *Laterally* are the pleurae and mediastinal pulmonary surfaces; the phrenic nerve with accompanying vessels descends between the fibrous pericardium and mediastinal pleura on each side. *Inferiorly* the pericardium is separated from the liver and gastric fundus by the diaphragm.

Vessels receiving prolongations of fibrous pericardium are the aorta, superior vena cava, right and left pulmonary arteries and four pulmonary veins. The inferior vena cava, traversing the central tendon, has no such covering.

Serous pericardium is a closed lining to the fibrous pericardium invaginated by the heart, with *visceral* and *parietal* layers. The

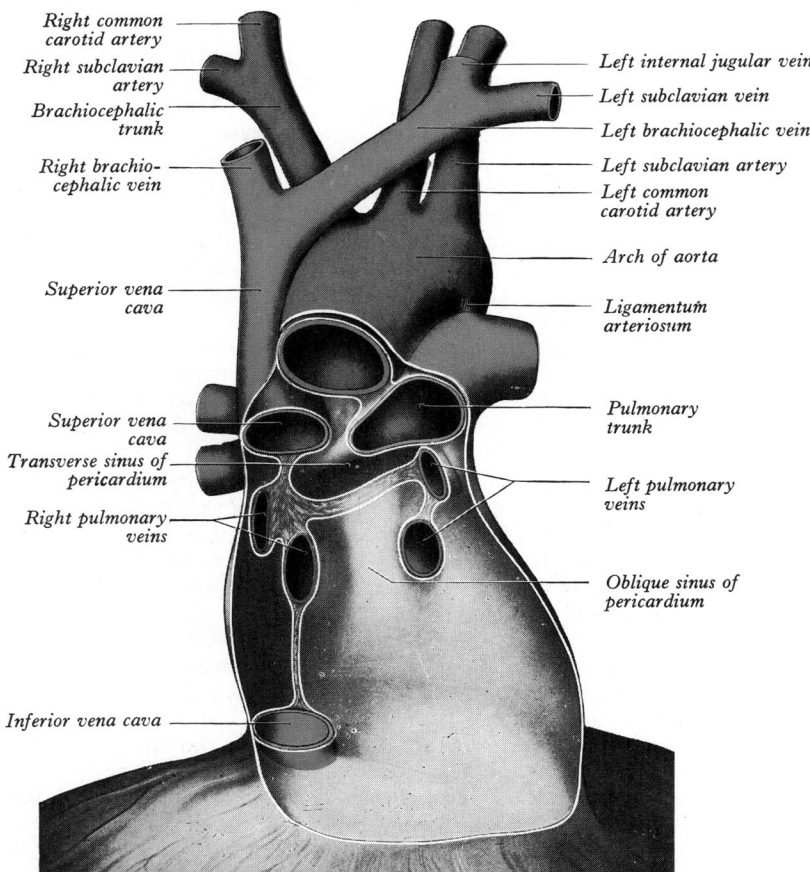

6.26 Interior of the serous pericardial sac after removal of the heart (ventral aspect). See text for additional named recesses of the general serous pericardial cavity and of its transverse sinus.

visceral, or *epicardium*, covers the heart and great vessels and is reflected into the parietal layer, lining the fibrous pericardium. Early in development this reflexion occurs along a double line from the venous to the arterial end of the primitive pericardium, the attachment of the transient dorsal mesocardium (p. 208). With

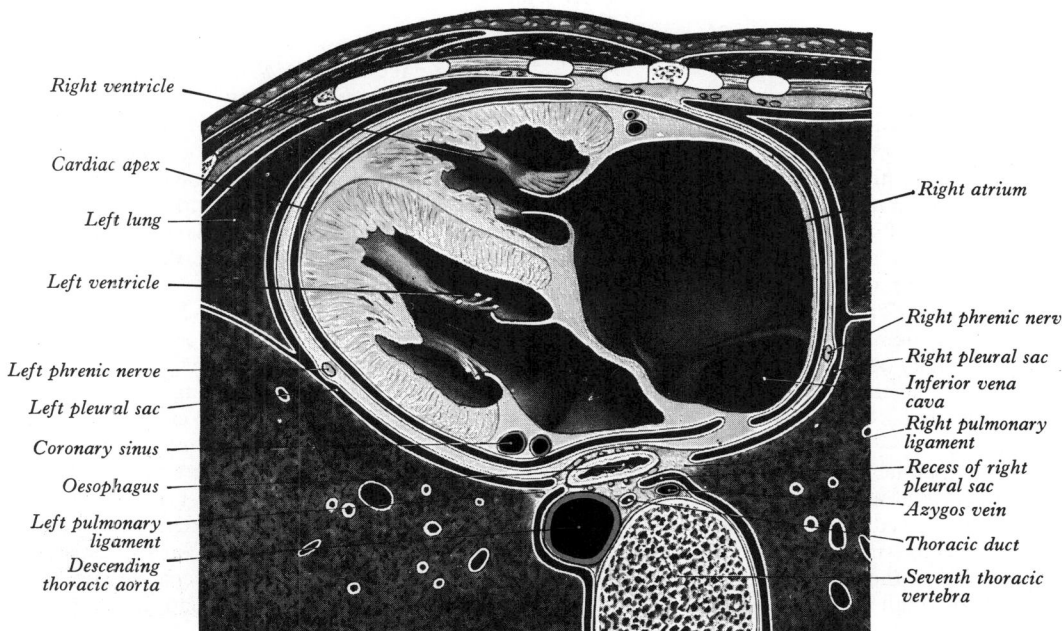

6.27 Transverse section through the mediastinum at the level of the body of the seventh thoracic vertebra, viewed from above. Note the general disposition of cardiac cavities, their intervening septa (about 45° to sagittal and coronal planes) and, orthogonal to this, the plane of the atrioventricular valves. The oesophageal plexus of nerves is clear but not labelled.

the disappearance of the latter the parietovisceral lines of continuity are now restricted to two initially simple, perivascular tubes of inflexion, one periarterial and the other perivenous. This initial simplicity is progressively modified by folding and differential growth of the cardiac tube and great vessels. The serous reflexion is thus now arranged as two increasingly complex 'tubes', the aorta and pulmonary trunk enclosed in one and the superior and inferior venae cavae and pulmonary veins in the other; the latter continuation is J-shaped (**6.26,31**) and the cul-de-sac within its curve is behind the left atrium and termed the *oblique sinus*. The passage between the two so-called pericardial 'tubes' is the *transverse sinus*, which has the aorta and pulmonary trunk in front, the atria and great veins behind (**6.36,59**). In the foregoing simplified account of the serous pericardium, it is clear that, when the complex surface topography of the heart and its main vascular trunks is considered, the developmental analogy of two tubes is insufficient. The oblique and transverse sinuses and the main 'general' cavity are all affected by the development of complex three-dimensional *pericardial recesses* between adjacent structures. (For details, illustrations and bibliography see Vesely & Cahill 1986.) A list adopting their proposed terminology is given, although other apposite terminologies are easily envisaged. Recesses are grouped according to the siting of their orifices or 'mouths'. From the general serous pericardial cavity:

The postcaval recess projects towards the left behind the superior vena cava (atrial end), limited above by the right pulmonary artery and below by the upper right pulmonary vein. Its mouth opens superolaterally to the right. *The right and left pulmonary venous recesses* each project medially and upwards on the back of the left atrium between the superior and inferior pulmonary veins of their side; i.e. they indent the side walls of the oblique sinus; the mouths open downwards and laterally.

From the transverse sinus:

The superior aortic recess, its mouth inferior, ascends posterior to, then right of the ascending aorta to end at the level of the sternal angle.

The inferior aortic recess, its mouth superior, is a diverticulum descending between the lower ascending aorta and right atrium.

The left pulmonic recess, mouth under the fold of the left vena cava, passes to the left between the inferior aspect of the left pulmonary artery and upper border of the superior left pulmonary vein.

The right pulmonic recess lies between the lower surface of the proximal part of the right pulmonary artery and upper border of the left atrium.

A triangular fold of serous pericardium is reflected from the left pulmonary artery to the subjacent upper left pulmonary vein as the *fold of the left vena cava*; it contains a fibrous *ligament of the left vena cava*, a remnant of the obliterated *left common cardinal vein* (left duct of Cuvier, p. 208), which descends anterior to the left pulmonary hilum from the upper part of the left superior intercostal vein to the back of the left atrium, where it is continuous with the *oblique vein of the left atrium* (p. 221). The fold is often the anterior wall of a small recess, its mouth directed left. The *left common cardinal vein* may persist as a left superior vena cava, descending anterior to the left hilum then across the back of the left atrium to enter the right atrium, replacing both the oblique vein of the left atrium and the coronary sinus. Sometimes both common cardinal veins persist as right and left superior venae cavae; the transverse anastomosis between them, normally forming the left brachiocephalic vein, may be small or absent. When the left common cardinal vein persists it is joined by the left superior intercostal vein.

Vessels and nerves. Arteries of the pericardium are from the internal thoracic and musculophrenic arteries and the descending thoracic aorta; *veins* are tributaries of the azygos system. The nerve supply is from the vagus and phrenic nerves and the sympathetic trunks.

Structure. The fibrous pericardium is compacted collagenous fibrous tissue. The serous pericardium is a single layer of flat cells on a subserous areolar layer which blends with the fibrous pericardium and interstitial myocardial tissue, where it contains fat in greatest amounts along the ventricular side of the coronary sulcus, the inferior cardiac border and interventricular grooves. The main coronary vessels and larger branches are embedded in this fat, its amount being related to general body fat and gradually increasing with age.

Applied Anatomy. Pericardial puncture may be performed in the fifth or sixth left intercostal space near the sternum to avoid the internal thoracic artery, or at the left costoxiphoid angle, passing up and backwards into the pericardial sac. The serous pericardium extends on the pulmonary trunk, anterior to the transverse sinus, as far as the ligamentum arteriosum (p. 725).

THE HEART

General Introduction

All triploblastic organisms, including chordates, overcome the limitations of diffusion over long distances by circulating a fluid from regions of high oxygen tension and concentrations of nutrient substances to mesodermal and other cells remote from the external environment. The fluid and its mode of circulation vary amongst invertebrate phyla, but the majority are coelomates with closed vascular systems and some localized means of propelling 'blood' in a true circulation. (There are exceptions: some small annelid worms lack such a system and in some acoelomates, e.g. nematode worms, a body cavity (a *pseudocoel*) does occur but has no endothelium and merely permits tidal movement of contained fluid due to bodily movements. One nematode, *Ascaris*, has a haemoglobin in this fluid.)

In some coelomates, e.g. annelid worms, pulsatile sections of vascular tubes, usually multiple, occur in many regions; the larger of such 'hearts' may be valved and respond to neurogenic 'pacemakers'. In arthropods the single heart is dorsal and circulation 'open', i.e. vessels from the heart open into haemocoels around most organs. (Some small arthropods, e.g. mites, have no circulatory organs.) Aquatic arthropods, such as the *Crustacea*, respire through gill-like structures, the blood acting as the intermediary; but in the *Insecta*, where most of the species has emancipated from water, respiration is via intercellular tubes separated

from the circulation. *Mollusca, Brachiopoda* and *Echinodermata* display an even wider variety of circulatory patterns. These facts are mentioned briefly as a contrast to the phylum *Chordata*, which displays a much more uniform circulatory pattern. Even in the most primitive, *Hemichordata* (acorn worms) and *Urochordata* (tunicates), the heart is single and in *Cephalochordata* (including the familiar *Amphioxus*) there is a *ventral* tubular 'heart' and a basically vertebrate circulation. The 'heart' is assisted by other pulsatile vessels.

The *vertebrate* heart is ventral to the gut—to the pharynx in fishes and embryos of land tetrapods, to the oesophagus in adults of the latter. Being evolved from a tube, it is a single pump in earlier forms. From the beginning (as in simpler extant forms) it is a succession of three or four enlarged segments: a collecting *sinus venosus* for principal veins, a pulsatile but thin-walled *atrium*, a thick-walled, muscular *ventricle* and a *bulbus cordis*, the latter being muscular or elastic and sometimes absent. From the ventricle's conical end (*conus arteriosus*), or bulbus cordis when present, major arteries start. Valves, usually of several flaps (sometimes serially repeated), separate successive cavities. The heart has its own *coronary* circulation (except in *Agnatha*) and is contained in a pericardial coelom, a separated part of the general body cavity. The pulsatile rhythm of vertebrate hearts is basically *myogenic*, but is coordinated with systemic demands by a nerve supply. With increasing specialization from fishes to birds and mammals, 'nodes' and tracts of cardiac muscle differentiate as foci initiating contraction and rapid conductors for the dissemination of cyclic stimuli. (At particular loci the conduction is much slower introducing physiologically imperative delays.) In most vertebrates the cardiac tube outgrows the length of its pericardial sac, developing a sinuous bend; the 'venous end' (sinus venosus and atrium) becomes dorsal to the 'arterial end' (ventricle and conus); moreover, the heart becomes asymmetrical in position with the change from symmetrical cardinal to asymmetrical caval veins. The pericardial cavity is semi-rigid in some fish (e.g. elasmobranchs), being dorsal to the pectoral girdle; the resultant constancy of volume aids atrial filling by suction as the ventricle empties; in terrestrial and aerial vertebrates this effect is largely lost because of caudal 'migration' of the heart and less massive girdles.

In tracing cardiac phylogeny, especially in deriving mammalian from earlier forms of heart, no direct palaeontological data are available and comparison of existing arrangements in extant vertebrate groups is the only source of information. Such comparison can mislead unless allowance is made for several complicating factors. Firstly, hearts must function early in embryonic life, throughout development and into a greatly changed postnatal environment. Environmental demands in adults vary and, if they diminish for a more recent (presumably 'more highly evolved') group, actual cardiac retrogression may occur. For example, in *Dipnoi*, the lungfish (and in crossopterygians, if the extant coelacanth, *Latimeria*, is representative) the atrium is almost completely divided and the ventricle partly so, whereas in *Amphibia* the ventricle is single and the atrium fully divided only in anuran forms. Incomplete atrial division in urodele amphibians is considered structurally retrogressive, but must be a physiological advantage in this more active of these two subclasses. It is orthodox, indeed inescapable, to derive mammals from reptiles, but from no extant reptile can mammalian hearts be directly derived. Though in some reptiles (e.g. *Crocodilia*) hearts are fully divided, the relations of the pulmonary and systemic arteries to ventricles precludes such derivation. Reptiles ancestral to mammals must have had different arrangements. Avian hearts and their outflow vessels are much closer to those of extant reptiles.

A complete phylogenetic history is not available for vertebrate hearts; hence an elaborate comparison will not be attempted, but some evolutionary trends must be mentioned.

The sinus venosus, typically symmetrical in *Pisces*, is asymmetrical in *Cyclostomata*; its right duct only persists in lampreys, the left in hagfish. In *Dipnoi* the sinus opens into the *right* of a partly divided atrium, a condition persisting in subsequent vertebrate classes; in *Anura*, many *Reptilia* and all *Mammalia* this asymmetry is coupled with some retrogression of the sinus and absorption of its vestiges into the right atrium. With complete separation of atria a right-hand systemic venous return is thus established.

The 'arterial end', a *bulbus cordis* in vertebrate embryos (and some adults) or a contractile *conus arteriosus* in adult elasmobranchs, commonly has valve flaps, often in serial rows. Probably derived from these are the spiral valves in *Dipnoi*, which are more elaborate and efficient in *Amphibia*. This change is closely linked with the greatest era of transformation in cardiac evolution, the long series of adaptations which allowed vertebrates to spread from aquatic to terrestrial habitat. This is accepted as occurring in the Devonian Period of the Palaeozoic Era, a period of 50–60 million years, about 300 million years ago. At first were fish, which were chiefly dependent on branchial respiration but able to use, for aerial breathing, a pharyngeal diverticulum, the so-called 'swim bladder' (p. 1248); finally came amphibians, mainly dependent on pulmonary respiration, with lungs developed from some form of pharyngeal air sac, but also respiring through a wet skin. These respiratory changes, inseparable from cardiac and circulatory modifications, also demanded locomotor changes: the evolution of limbs, adapted from aquatic fins to terrestrial progression. For such reasons *Amphibia* are regarded as developed from crossopterygians related to extant *Dipnoi*.

During this transformation, occupying millions of years, an inherent duality already existed in circulations, both dipnoans and amphibians having a *systemic* 'portal', arterial circulation (through gills in one and subcutaneous vessels in the other) and also a parallel *pulmonary* 'portal' circulation through the walls of an air sac or, eventually, true lungs. These separate circuits already return blood to separate atria in dipnoans; and although the ventricle is only partly divided, mixing of atrial inputs is probably diminished by the highly trabecular internal ventricular surface. Of course, dipnoan *ventricular* output, although also largely separated into two streams by spiral valves in a large conus arteriosus, is all conducted to the gills; but blood is returned to the *right* atrium by a dorsal division of the ventral aorta to the more caudal gill clefts and hence largely to the air sacs or 'lungs'.

In amphibian circulation, with loss of the gills, the capillary beds supplying them disappear and respiration is mainly pulmonary. But an undivided ventricle (and only partial division of the atrium in some forms) may permit some mixture of systemic and pulmonary flows; presumably this is an advantage on the return to water for breeding, cutaneous respiration becoming predominant. Thus no sudden change, in structure or function, occurs in the slow cardiac adaptation from aquatic to terrestrial life, but rather a change of balance between two co-existent modes of respiration. Separation of cardiac flows is carried further in *Reptilia*. Though some forms have returned to an aquatic existence, respiration is entirely pulmonary, linked with a complete interatrial septum and at least a partial ventricular division, actually complete in *Crocodilia*; they are unusual in their triple arterial cardiac outflow which is pulmonary from the right ventricle and right *and* left aortic (systemic) arches, each from a contralateral ventricle. Aortic arteries hence cross and communicate at this point. In the cayman this allows an almost complete shunt into the left aorta, foreshadowing the mammalian condition.

In mammals, including mankind, cardiac division is complete and need not be described here; but in the embryo, stages occur (p. 206) very like the final arrangements in earlier vertebrate classes; arrested development at various stages leads to congenital defects resembling conditions in earlier vertebrate forms. However, the resemblance is modified, for the heart must function effectively at all but its initial stages of development. The foramen ovale, a feature of mammalian prenatal development, is a necessary shunt rather than an atavistic indication of earlier incomplete atrial septation. Hence, a persistent patent foramen ovale is due to disturbed *mammalian* rather than *vertebrate* development. Other cardiac abnormalities may be tentatively 'explained' as divergencies from ontogenetic development rather than by reference to phylogeny.

That every vertebrate heart must function in its own circulation from a very early stage, long before either has reached its final arrangement, accords with a continuous cardiac phylogeny, seen here in retrospect. A heart of single serial chambers has become a fully septated double heart, of two muscular pumps, topographically 'in parallel' though pumping in circulatory function 'in series', without departure from smooth evolution and achieved despite great changes from aquatic respiration through gills to wholly pulmonary breathing. This smooth changeover may be ascribed to the existence of *both* modes of respiration in a series of dipnoan and amphibian species over a long period of time. The double circulation, systemic through the gills and pulmonary through the lungs, has gradually evolved from a *parallel* supply by incompletely septated hearts to a *serial* supply with complete septation. Evolution in the vessels served by the heart has also occurred in a harmonious succession of adaptations. For further details consult *Embryology* in this volume, and for extensive bibliographies see Bolk et al (1931–9), Grasse (1954), Foxon (1955), Robb (1965), Romer (1970), Foxon & Bannister (1974).

There is no entirely logical progression in describing the heart; whatever standpoint is used this presumes subsequent details; the sequence adopted here is perhaps a good compromise. General organization precedes external features, surface anatomy and radiology, followed by internal structure, including flow tracts, valves, myocardium, fibrous 'skeleton', specialized conducting tissues and the cardiac cycle.

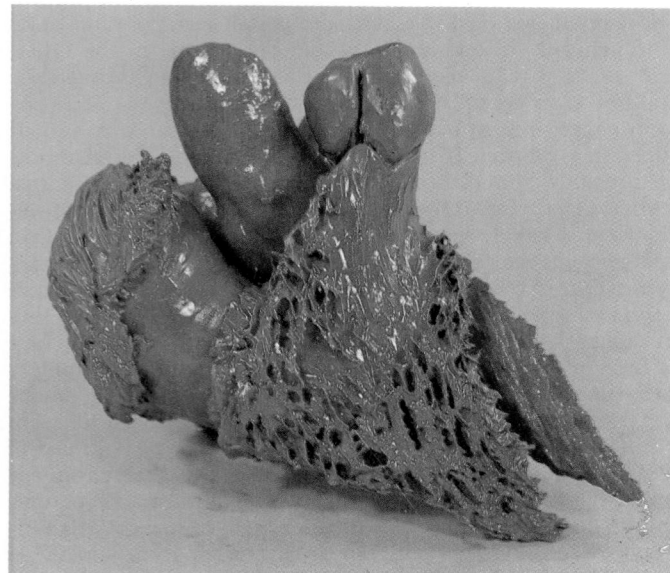

A

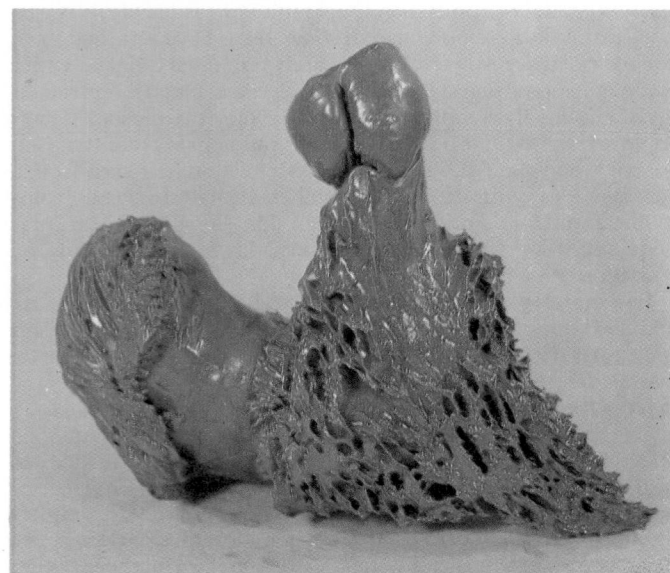

B

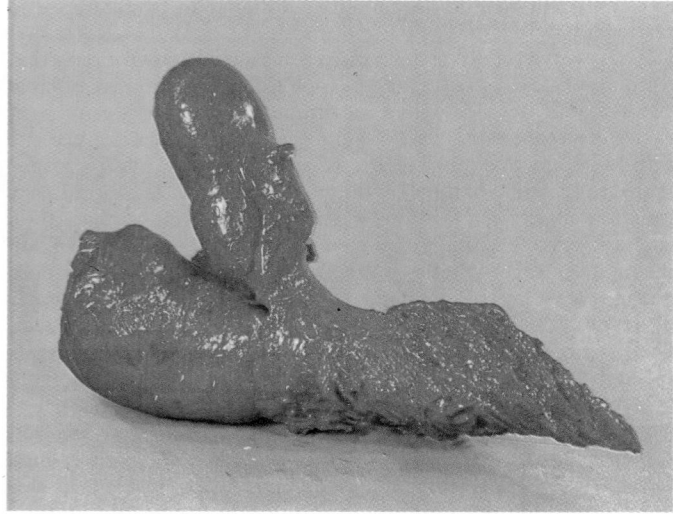

C

6.28 Coloured polyester resin luminal casts of a human heart. Blue = right atrium, ventricle, infundibulum, pulmonary root and valvular sinuses; red = left atrium, ventricle, aortic vestibule, root and sinuses. A. Sternocostal surface with right and left hearts in situ; B. isolated cast of right heart; C. isolated cast of left heart. Note the impressions made by the musculi pectinati, trabeculae carneae (spiralized in the left ventricle), the smooth areas related to the atrioventricular valve leaflets and the dilated sinuses in the walls of the aortic and pulmonary roots. Casts prepared by M C E Hutchinson, photography by Kevin Fitzpatrick, Department of Anatomy, Guy's Hospital Medical School.

General Cardiac Organization

The human heart is a pair of valved muscular pumps combined in a single organ; but, while their fibromuscular framework and conducting tissues are structurally interwoven, each pump (so-called 'right' and 'left' heart) is physiologically *separate* and interposed *in series* at different points in the double circulation. Despite this *functional* serial disposition (vide infra), the two pumps are often described (vide supra) as *topographically* 'in parallel'; even this is an oversimplification, since they have contrasting haemodynamic roles and their outflow channels exhibit a *mutual spiral* of almost 180° as they discharge.

In the four cardiac chambers, two atria receive venous blood as weakly contractile reservoirs for final filling of the ventricles, and the two ventricles provide the powerful expulsive contraction forcing blood into the main arterial trunks.

The 'right heart' of clinical parlance consists of the *right atrium* receiving superior and inferior caval and main coronary inflow from the systemic circulation: this blood traverses the right *atrioventricular orifice*, with its *tricuspid valve*, to the *right ventricular inflow tract*, contraction of which finally closes the tricuspid valve and, with increasing pressure, reaches a level when ejection of blood through the *right ventricular outflow tract (infundibulum* or *conus arteriosus* and *pulmonary valve*) is achieved. (Pressure changes, time relations and valvular events are described below.) Note that the right ventricular outflow tract discharges into the pulmonary trunk and hence to the *relatively* low-resistance pulmonary vascular bed; many structural features of the 'right heart', including overall geometry, vectors of inflow and outflow, myocardial architecture (in particular thickness of wall) and construction and relative strengths of tricuspid and mitral valves all accord with this, being associated with comparatively low changes of pressure.

The 'left heart', in contrast, consisting of the *left atrium* receiving all pulmonary and some coronary venous inflows, contracts to complete filling of the *left ventricle* through the *left atrioventricular orifice* with its *mitral valve*. The valve is the entry to the left ventricular *inflow tract*, contraction of which rapidly raises its pressure, closing the valve (vide infra); this is followed at once by opening of the aortic valve and ejection via the left ventricular *outflow tract* via the aortic orifice into the aortic sinuses, ascending aorta and thence to the whole systemic arterial tree, including the coronary vessels. This vast vascular bed presents high peripheral resistance which, with large metabolic demands, especially the sustained requirements of the cerebral tissues, explains the structural organization of the 'left heart'—its geometry, flow vectors, parameters of mitral and aortic valves and the thickness of the left ventricular myocardium. The ejectional phase of the left ventricle is shorter but its pressure fluctuations are very much greater (**6.55**).

Because of its immense phylogenetic history, complex ontogeny (p. 206) and contrasting functional demands, the human heart is far from a simple pair of parallel pumps, structurally combined, even though the right and left ventricles must deliver the same volume with each contraction. The heart has a complicated, spiralized, three-dimensional organization, not obviously related to cardinal planes; hence terms such as 'left' and 'right', 'apex' and 'base', 'surface', 'aspect' and 'border' do not assist cardiac anatomy. Another source of confusion is the usual study of isolated whole or dissected hearts, with the subsequent difficulty in relating details to the heart in situ. The following preliminary description emphasizes such difficulties to circumvent certain misconceptions before proceeding to more detailed structure.

The principal features of cardiac anatomy can be illuminated by corrosion casts of normal hearts in which both sides have been filled with resins of contrasting colours (**6.28**A,B,C), and by horizontal mediastinal sections or scans at, or near, the seventh thoracic vertebral level (**6.27,29**A,B,C).

In illustrations **6.28**A,B,C the left (red) and right (blue) hearts are viewed from the *ventral aspect* in their normal relation (**6.28**A) followed by right (**6.28**B) and left (**6.28**C) hearts in isolation. Obviously their general forms contrast and unmodified terms are singularly inappropriate. The right heart, while forming a right

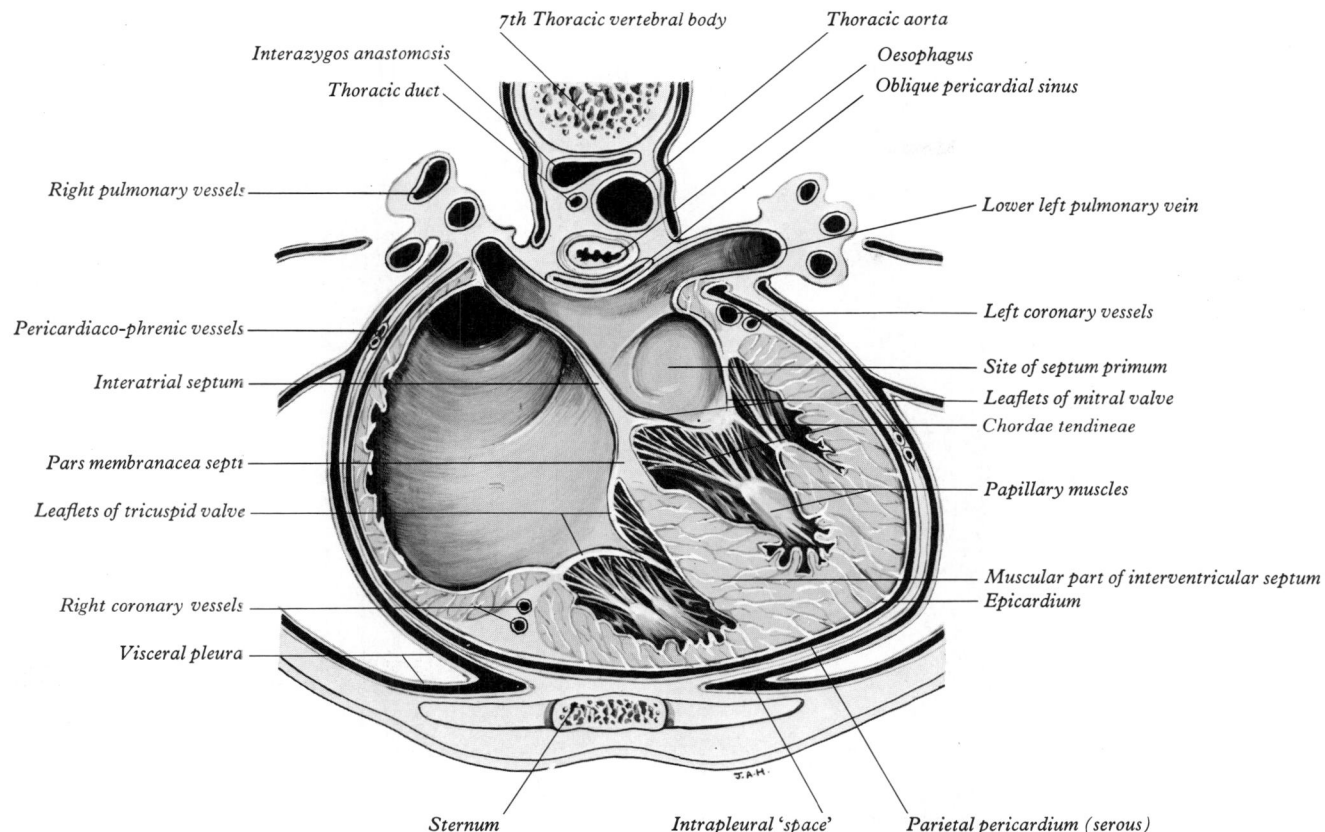

7th Thoracic vertebral body

Interazygos anastomosis

Thoracic duct

Right pulmonary vessels

Thoracic aorta

Oesophagus

Oblique pericardial sinus

Lower left pulmonary vein

Pericardiaco-phrenic vessels

Interatrial septum

Pars membranacea septi

Leaflets of tricuspid valve

Right coronary vessels

Visceral pleura

Left coronary vessels

Site of septum primum

Leaflets of mitral valve

Chordae tendineae

Papillary muscles

Muscular part of interventricular septum

Epicardium

Sternum

Intrapleural 'space'

Parietal pericardium (serous)

6.29A Transverse section through the heart, pericardium and related structures at the level of the lower border of the body of the seventh thoracic vertebra to show the principal dispositions of the cardiac chambers and septa and surrounding structures (viewed from below).

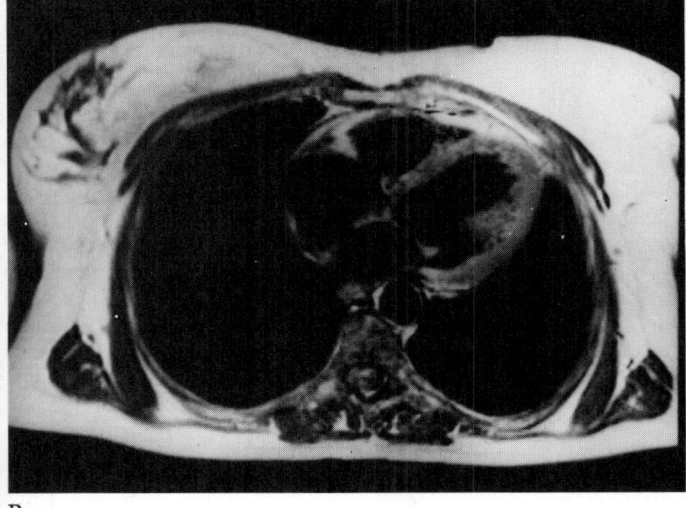

B

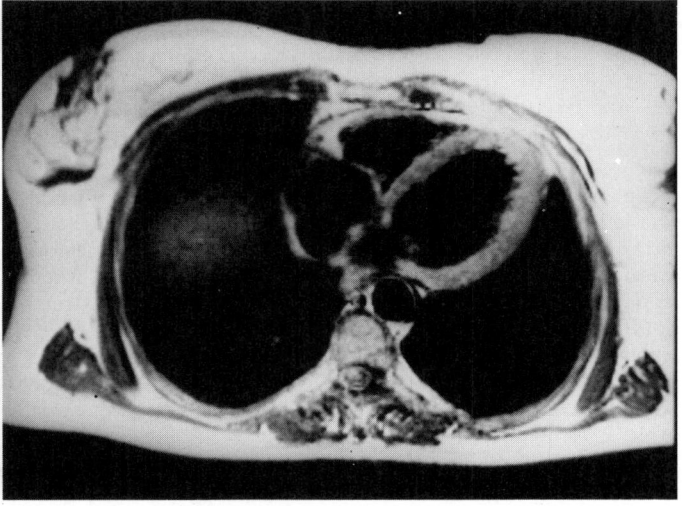

C

6.29B,C Computed tomograms of the thorax: (B) through the body of the seventh thoracic vertebra; (C) through the intervertebral disc between the seventh and eighth vertebrae. Note the overall disposition of the heart, its apex, base, oblique interatrial and interventricular septa and, orthogonal to this, of the atrioventricular valves. Note also the atrioventricular septum, papillary muscles, trabeculae carneae, descending thoracic aorta and contrasting areas of right and left lungs and pleurae. Provided by Shaun Gallagher, Guy's Hospital; photography by Sarah Smith.

aspect or border, follows a gentle curve (almost gastric in form) and covers most of the *anterior aspect* of the left heart (except for a left-sided strip including the apex). Thus, much of the right 'heart' in fact forms a majority of the *ventral surface*, its outflow tract ascending to the left until it terminates on the *left side* of the left ventricular outflow tract. Impressions of tricuspid and pulmonary valves are *widely separated* in different planes and the right ventricle's flat cavity (crescentic in section, vide infra) is splayed out between them. Conversely, the 'left' heart (except the left-sided strip and apex mentioned above) is largely *dorsal* in position and (as seen in **6.28A**) in ventral aspect is obscured by the right chambers. Note that the left ventricular inflow orifice (mitral valve) is *very close* to its outflow (aortic valve), the two being embraced by the wide divarication of the 'right' inflow and outflow orifices. Planes of the left ventricular orifices, though relatively inclined, are more nearly coplanar than those of the right. The left ventricular cavity is narrow, conical, with a spiral trabeculated interior. The conical left ventricular tip occupies the cardiac *apex*, most of the *anatomical base* being the left atrium.

The heart is placed *obliquely* in the thorax, as is clear in horizontal section (**6.27, 29A**) and scans (**6.29B,C**). The interatrial and interventricular septa are virtually in line but inclined *forwards and to the left* at about 45° to a sagittal plane; planes of mitral and tricuspid valves, though vertical and not precisely coplanar,

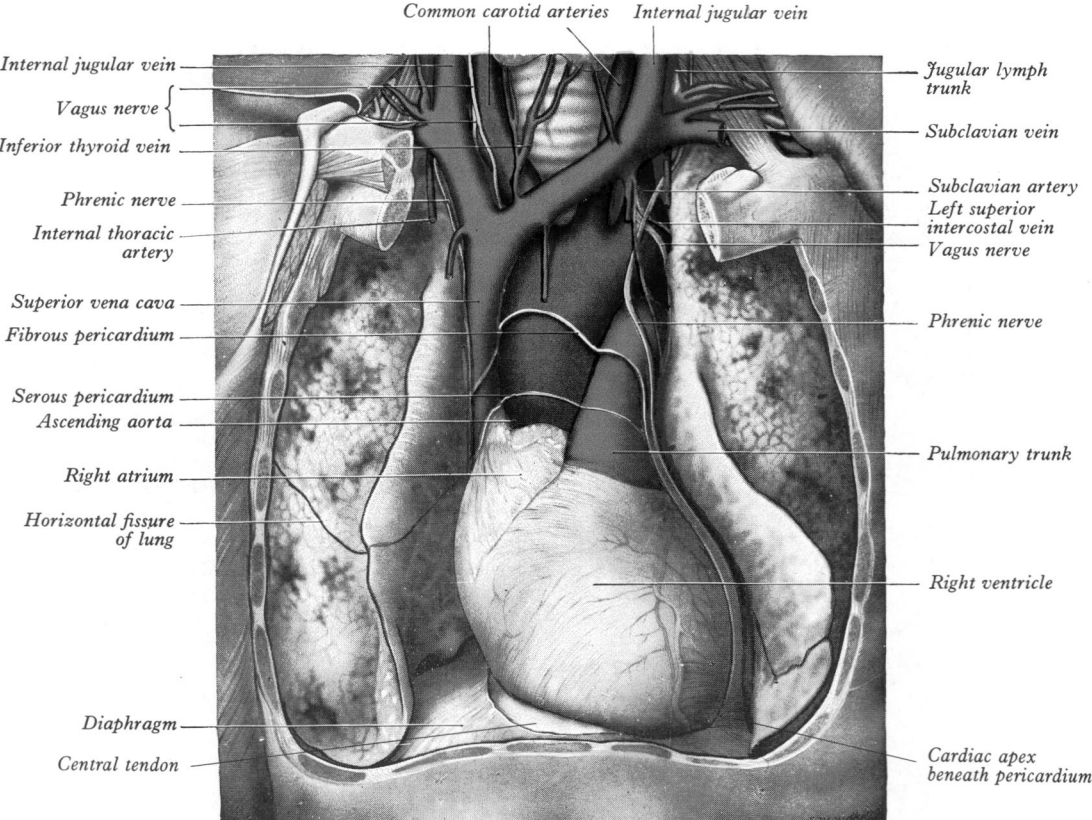

Common carotid arteries   Internal jugular vein

Internal jugular vein

Vagus nerve

Inferior thyroid vein

Phrenic nerve

Internal thoracic artery

Superior vena cava

Fibrous pericardium

Serous pericardium

Ascending aorta

Right atrium

Horizontal fissure of lung

Diaphragm

Central tendon

Jugular lymph trunk

Subclavian vein

Subclavian artery

Left superior intercostal vein

Vagus nerve

Phrenic nerve

Pulmonary trunk

Right ventricle

Cardiac apex beneath pericardium

6.30A Dissection which displays the heart, the great vessels and the lungs in situ. The sternum and the sternal ends of the costal cartilages, together with the parietal pleura on each side, have been excised and the mediastinal pleura and parietal layer of the pericardium over the sternocostal surface of the heart have been removed. The lungs have been displaced to expose the heart and the epicardium dissected off the heart and roots of the great vessels.

On the right side, the inferior cardiac branch of the vagus nerve descends between the brachiocephalic artery and the right brachiocephalic vein. On the left side, a communication descends from the left superior intercostal vein and crosses the aortic arch and the left pulmonary artery to become continuous with the oblique vein of the left atrium.

are broadly at right angles to the septal plane. In brief, therefore, the right atrium is not only to the *right* but *anterior* and *inferior* to the left atrium (and also partly anterior to the left ventricle, an *atrioventricular septum* intervening). The right ventricle forms most of the *anterior aspect* of the ventricular part, only its inferior end being *right* of the left ventricle, its upper left extremity (outflow orifice) being *left and superior* to the aortic valve. The left atrium forms most of the *posterior aspect* supplemented by the right atrium, while the left ventricle is only prominent *inferiorly*, along the left margin and at the apex. The atria are essentially right of and posterior to their ventricles. These general dispositions are of the greatest importance in planning or interpreting radiographs, scans, angiocardiograms and echocardiograms.

Cardiac Size, Shape and External Features

The heart is a hollow, fibromuscular organ of a somewhat conical or pyramidal form, with a base, apex and a series of surfaces and 'borders' (**6**.28,29,30,31); it occupies the middle mediastinum between the lungs and the pleurae (**6**.27,29A,B,C), enclosed in the pericardium (p. 694). It is placed *obliquely* behind the corpus sterni and adjoining left costal cartilages and ribs; approximately one-third lies right of the midline.

An average adult heart is about 12 cm from base to apex, 8–9 cm transversely at its broadest and 6 cm anteroposteriorly. Its weight, in males, varies from 280–340 g (average 300 g), in females, from 230–280 g (average 250 g). Cardiac weight is said to be about 0.45% of body weight in males and 0.40% in females (Hudson 1965); adult weight is achieved between 17–20 years.

The heart's oblique position may be re-emphasized by the analogy of a rather deformed pyramid, with a base facing dorsally and to the

right, the apex anteriorly and to the left. Thus a line from the apex to the approximate basal centre, projected antero-laterally, passes near the left mid clavicular line; projected posterolaterally it emerges near the right mid-scapular line. Some aspects of the cardiac 'pyramid' are flat, others more or less convex, these aspects merging along rather ill-defined 'borders'. Hence precise definition of surfaces and intervening 'borders' is difficult: in the following account official nomenclature (*Nomina Anatomica* 1983) *and* terms from clinical practice will be given as alternatives. The heart is described as having an (anatomical) base and apex, its *surfaces* designated sternocostal (anterior), diaphragmatic (inferior), right and left (pulmonary); its *borders* are termed upper, inferior ('acute' margin or border) and left ('obtuse' margin or border). Some persist in naming the right surface a 'border', despite its extent; the right surface is also as 'pulmonary' as the left. A further avoidable source of confusion is 'posterior' for such unambiguous terms as 'diaphragmatic'; if posterior is to be used for a cardiac surface it should be reserved for its base. Finally, as will be discussed more fully, there are a number of different usages of the term 'cardiac base'.

CARDIAC SURFACE GROOVES AND SULCI

Cardiac divisions into four chambers show externally as grooves or sulci, some deep and obvious, containing prominent structures; others are less distinct, even barely perceptible, sometimes completely obscured, in part, by the major structures crossing them.

Between the atria and ventricles is the *coronary* (atrioventricular) *sulcus* (**6**.30A,B, 31, 37A, 60, 61) containing the main coronary trunks; the sulcus is oblique and its position and plane are basic to cardiac anatomy. On the sternocostal surface it descends to the right, separating the right atrium (and it

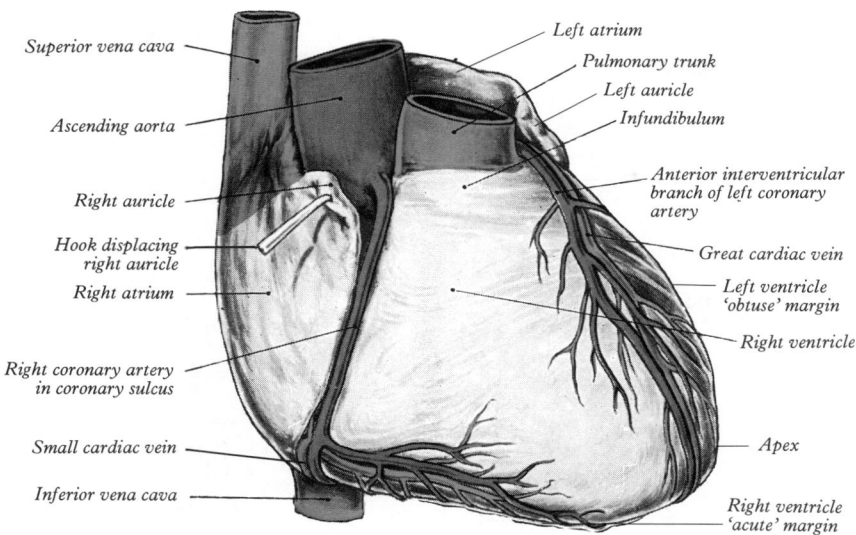

Superior vena cava

Ascending aorta

Right auricle

Hook displacing
right auricle

Right atrium

Right coronary artery
in coronary sulcus

Small cardiac vein

Inferior vena cava

Left atrium

Pulmonary trunk

Left auricle

Infundibulum

Anterior interventricular
branch of left coronary
artery

Great cardiac vein

Left ventricle
'obtuse' margin

Right ventricle

Apex

Right ventricle
'acute' margin

6.30B The anterior or sternocostal surface of the heart.

auricular appendage) from the oblique right margin of the right ventricle and its infundibulum; its upper left part is, however, obliterated where the root of the pulmonary trunk and, behind this, the aorta cross it. Continuing to the left, it curves around the heart's 'obtuse margin' and descends to the right, separating the atrial base from the diaphragmatic (ventricular) surface. Posterior and sternocostal parts of the coronary sulcus curve around the 'acute margin' at its lower right end to become confluent. Thus, both these parts of the sulcus pass from high on the left to the right but the posterior part is a little to the left of the sternocostal. Hence, a section which includes the coronary sulcus is at about 45° to the sagittal plane and at a greater but variable angle to the transverse and coronal planes; it approximately traverses the lines of attachment of the atrioventricular valves and (even less precisely) those of the aortic and pulmonary (**6.**41,42); a line at right angles to the centre of this plane will descend forwards and left to the cardiac apex.

Internally the ventricles are separated by the interventricular septum (p. 769), whose mural margins correspond to the *anterior* and *inferior (diaphragmatic) interventricular grooves*. The former, on the sternocostal cardiac surface, is near and almost parallel to the left ventricular obtuse margin but the latter, which is on the diaphragmatic surface, is nearer the acute border. Interventricular grooves extend from the coronary sulcus (i.e. ventricular 'base') to the *apical incisure* on the acute margin, a little to the right of the true cardiac apex.

CARDIAC BASE, APEX, SURFACES, BORDERS

The *posterior aspect* or *base* (**6.**31) is somewhat quadrilateral, with curved lateral extensions; it faces back and to the right, separated from the thoracic vertebrae (fifth to eighth in the recumbent, sixth to ninth in the erect posture) by the pericardium, right pulmonary veins, oesophagus and aorta. It is mainly formed by the left atrium and only partly by the posterior part of the right. It ascends to the pulmonary arterial bifurcation and descends to the posterior part of the coronary sulcus containing the coronary sinus and rami of the coronary arteries (p. 727). To the right and left it is limited by the rounded surfaces of the corresponding atria; these are separated by a shallow vertical *interatrial groove*. Where the coronary sulcus, interatrial groove and inferior interventricular groove meet is often picturesquely termed *the crux of the heart* (**6.**31). Two pulmonary veins on each side open into the left atrial part of the base: into the upper part of the right atrial basal region open the superior vena cava, and the inferior vena cava into its lower part. The area of the left atrium between the openings of right and left pulmonary veins forms the anterior wall of the oblique pericardial sinus (p. 696). This description applies to the *anatomical base* with the heart in situ. Some confusion is entailed by other current usages of the term: some apply it to the ventricular aspect seen after section through the coronary sulcus

(**6.**47); this should be termed the *base of the ventricles* . In clinical practice auscultation in or near the parasternal parts of the second intercostal spaces is often regarded as applied to the *clinical 'base'* (in contrast to the *clinical 'apex'*). These applications, while unfortunate, seem likely to persist.

The cardiac apex is really the apex of the conical left ventricle, directed down, forwards and left; the left lung and pleura overlap it. It is located most commonly behind the fifth left intercostal space, near or a little medial to the mid-clavicular line.

The anterior, sternocostal surface (**6.**30A,B) faces forwards and upwards, with both right and (more pronounced) left convexities; it consists of an atrial area above and to the right and a ventricular part below and left of the coronary sulcus. The atrial area is almost entirely the right atrium; the left atrium is largely hidden by the ascending aorta and pulmonary trunk (**6.**30B,60); a small part of its auricle, however, projects forwards, left of the pulmonary trunk. Of the ventricular region about one-third is the left and two-thirds the right ventricle; their septum is indicated by the anterior interventricular groove. The sternocostal surface is separated by the pericardium from the corpus sterni, sternocostales and third to sixth costal cartilages; owing to cardiac bulging to the left, more of this surface is behind the left costal cartilages than behind the right ones. It is also covered by pleurae and thin, anterior edges of the lungs, except for a triangular area due to cardiac incisure of the left lung; both lungs and pleurae are variable in their degree of overlap.

The inferior, diaphragmatic surface (**6.**31), is largely horizontal but slopes down and forwards a little towards the apex. It is formed by the ventricles (chiefly the left), resting mainly upon the central tendon but also, apically, on a small area of the left muscular part of the diaphragm. It is separated from the base by the coronary sulcus carrying the coronary sinus and is traversed obliquely by the inferior interventricular groove.

The left surface, facing up, back and left, consists almost entirely of left ventricle, but with a small part of the left atrium and its auricle contributing above. Convex and widest above and crossed here by the coronary sulcus, it narrows to the cardiac apex. It is separated by serous pericardium from the left phrenic nerve and accompanying vessels and by the left pleura from the left lung's deep concavity, antero-inferior to its hilum.

The rounded right surface is formed by the right atrial wall, separated from the mediastinal aspect of the right lung by the serous and fibrous pericardium and the mediastinal and visceral pleurae. Its convexity merges below into the short intrathoracic part of the inferior vena cava and above into the superior vena cava. The sulcus terminalis curves approximately along the junction of the sternocostal and right surfaces of the right atrium.

The cardiac upper border is atrial (principally the left atrium); anterior to it are the ascending aorta and pulmonary trunk. The superior vena cava enters the right atrium at its right extremity.

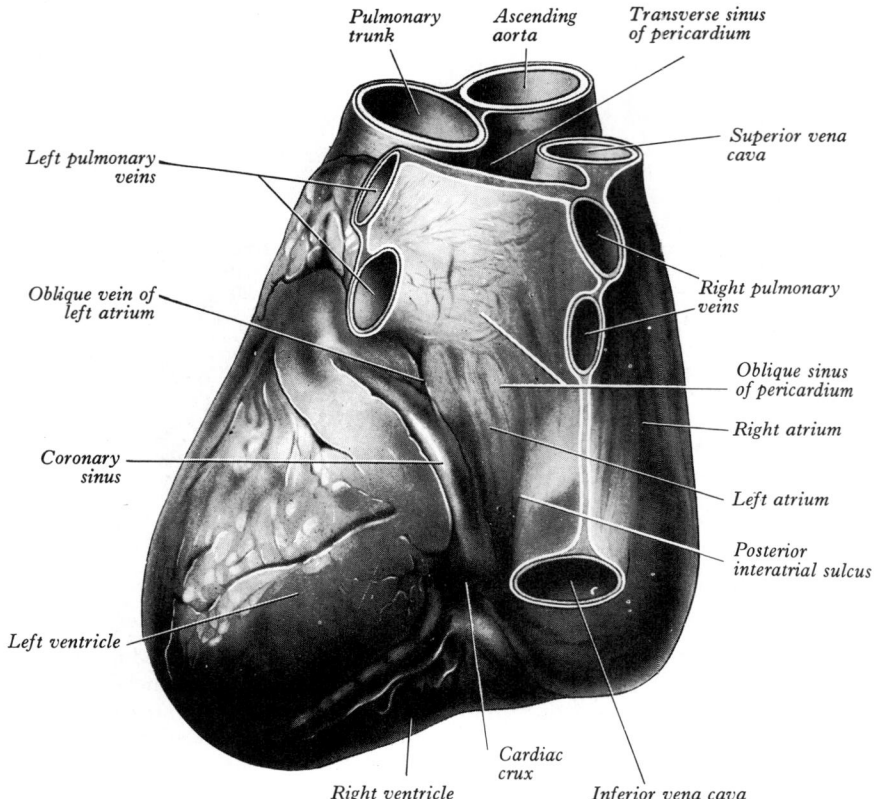

Pulmonary trunk — *Ascending aorta* — *Transverse sinus of pericardium*

Left pulmonary veins

Superior vena cava

Oblique vein of left atrium

Right pulmonary veins

Oblique sinus of pericardium

Right atrium

Coronary sinus

Left atrium

Posterior interatrial sulcus

Left ventricle

Right ventricle — *Cardiac crux* — *Inferior vena cava*

6.31 The base and the diaphragmatic surface of the heart. The serous pericardium is in situ and its cut edge is seen around the great vessels; its disposition is highly schematic (recesses omitted). See text for additional details.

The rounded so-called *right 'border'*, the right atrium's convex profile, approaches the vertical but is obvious on radiographs.

The *inferior border* or *'acute margin of the heart'* is sharp and thin and nearly horizontal; it extends from the lower limit of the right border to the apex. It is formed mainly by the right ventricle but by the left ventricle near the apex.

The *left border* or *'obtuse margin of the heart'* separates the sternocostal and left surfaces; it is round and mainly formed of left ventricle but, to a slight extent above, is formed by the left auricle. It descends obliquely from the left auricle, convex to the left, to the cardiac apex.

The borders, as indicated above, apart from the inferior ('acute') one, are rounded and cannot be precisely defined, being 'aspects' rather than borders. The right and left surfaces are referred to as pulmonary but much of the sternocostal is also related to the lungs. Right and left profiles of the cardiac *shadow* in a postero-anterior radiograph are the radiologist's right and left borders.

Cardiac Chambers and Internal Features

The right and left cardiac chambers will be described seriatim in terms of general form, walls and internal features. They have much in common, such as the structure of valvular leaflets, chordae tendineae, papillary muscles of atrioventricular (inflow tract) valves and the cuspal architecture of pulmonary and aortic (outflow tract) valves. Repetition will be minimized.

THE RIGHT ATRIUM (**6**.30A,B,31,32,36,54)

General and External Features

The *interatrial (atrial) septum* is oblique, so the *right atrium*, a roughly quadrangular chamber, is anterior as well as right of the left atrium, also extending inferior to it. Its walls form the right upper sternocostal surface, the convex right (pulmonary) surface and a little of the right side of the base. The superior vena cava opens into its upper posterior aspect and the inferior vena cava

into its lower posterior part. A small, muscular pouch, *the auricle*, projects to the left from its upper anterior part, overlapping the right side of the ascending aorta; the auricular margins are notched, its interior encroached on by an irregular muscular reticulum. In well-fixed hearts the lateral wall is marked externally by a shallow *sulcus terminalis*, extending between the right sides of the two vena caval orifices; posteriorly a vertical interatrial groove descends to the crux.

Anteriorly, the right atrium is related to the anterior part of the mediastinal surface of the right lung, separated from it by pleura and pericardium, and *laterally* also to the mediastinal surface of the right lung but anterior to its hilum and separated by the pleura, right phrenic nerve and pericardiacophrenic vessels and pericardium. *Posteriorly* and to the left (**6**.13B,15), the right atrium is separated from the left atrium by the septum (its mural attachment indicated by the posterior interatrial groove); posteriorly and to the right are the right pulmonary veins. *Medially* are the ascending aorta and to a lesser extent the root of the pulmonary trunk.

The right atrial interior (**6**.32,34,54) presents two main aspects: a posterior and an anterior.

The posterior part, where the great veins end, is derived from the absorbed right cornu of the sinus venosus (p. 209); it has smooth walls and is the *sinus venarum*. Anterior to this is a ridged region derived from the embryonic *atrium* proper and is continuous anteriorly with the *auricle*.

The *sinus venarum* includes the posterior and lateral wall forwards as far as the crista terminalis (p. 703). Opening into it are the following vessels:

(1) The *superior vena cava* (**6**.30A,B,31,54), returning blood from the body's upper region, opens into it superoposteriorly. Its orifice faces infero-anteriorly and has no valve.

(2) The *inferior vena cava* (**6**.31,32,34,54), larger than the superior, returns blood from the lower body into the lowest part of the atrium near its septum. Anterior to its orifice is a rudimentary semilunar *valve of the inferior vena cava*, its convex margin attached to the orifice's anterior margin. Its concave edge ends in two *cornua*, the left continuous with the anterior edge of the limbus fossae ovalis; the right cornu is lost on the lateral atrial wall

becoming continuous with the lower end of the crista terminalis. The valve is a fold of endocardium enclosing a few muscular fibres. It is prenatally large, directing flow from the *inferior vena cava* into the *left atrium* through the septum's *foramen ovale*. The valve varies in size, being sometimes cribriform or filamentous and occasionally absent.

(3) The *coronary sinus* (**6**.31,32,34,54). This returns the majority of blood from the heart itself, opening between the inferior caval orifice and the atrioventricular opening, protected by a thin, semicircular *valve of the coronary sinus* (**6**.32,34,54) covering the lower part of the orifice. It prevents regurgitation into the sinus during atrial contraction. It may be double or cribriform.

(4) The *foramina venarum minimarum* are the orifices of *venae cordis minimae* which return a small fraction of blood from the heart (p. 792). They are most numerous on the septal wall. The anterior cardiac veins and sometimes the right marginal may enter the atrium (p. 793).

The *intervenous tubercle*, small and variable, projects on the posterior wall below the superior caval orifice; it is distinct in quadrupeds but in man often scarcely visible. In utero it is more prominent and may direct blood from the superior vena cava towards the atrioventricular opening.

The *atrium proper* and the *auricle* are separated from the sinus venarum by the *crista terminalis*, a smooth, muscular ridge, mainly on the lateral wall; it begins on the upper septum and, passing anterior to the superior caval orifice, skirts its right margin to reach the right side of the inferior caval orifice, where it joins the right end of the inferior caval valve. It is the site of the embryo's *right venous valve (sinuatrial valvula*, p. 209), corresponding externally to the sulcus terminalis (p. 702). It marks the junction of the cardiac parts derived from the right cornu of the sinus venosus and the original atrium. Superiorly the crista terminalis accommodates the sinuatrial node (p. 720); its remainder is regarded by some as the route of the posterior internodal tract (p. 722).

The *musculi pectinati*, almost parallel muscular ridges extending anterolaterally from the crista terminalis in the right atrium, incline towards the atrioventricular orifice. In the auricle they interconnect to form a network.

The septal wall presents the *fossa ovalis* (**6**.32,54), an oval depression low on the wall, above and to the left of the inferior caval orifice. Its floor is originally the fetal septum primum (p. 209). The *limbus fossae ovalis* is its prominent margin, representing the edge of the septum secundum (p. 210). It is most distinct above, and deficient inferiorly; its anterior edge is continuous with the left cornu of the inferior caval valve. A small, valvular slit sometimes exists at the fossa's upper margin, ascending beneath the limbus to the left atrium; it is a remnant of the foramen ovale.

Antero-inferior in the right atrium is the large, oval *atrioventricular orifice*, its margin approximately the line of attachment of the *tricuspid valve*.

A triangular zone (*triangle of Koch* **6**.34,54) exists between the base of the tricuspid valve's septal leaflet, the anteromedial margin of the coronary sinus orifice, and the round, collagenous, palpable, subendocardial *tendon of Todaro*, which curves from the *right fibrous trigone' (central fibrous body*, p. 657) to the medial cornu of the inferior caval valve; it is a landmark of surgical import, indicating the septal site of the atrioventricular node and associated juxtanodal bundles (p. 657). Dorsal to the septal leaflet of the tricuspid valve, the septal wall is an extension of the *pars membranacea septi*, which is here *atrioventricular*, intervening between the right atrium and left ventricle.

The septal wall bulges in its anterosuperior area into the right atrial cavity as the *torus aorticus*, due to the proximity of the right posterior (non-coronary) aortic sinus and its cusp (p. 711).

THE RIGHT VENTRICLE

The right ventricle extends from the right atrioventricular (tricuspid) orifice nearly to the cardiac apex; it then ascends to the left as the *conus arteriosus (infundibulum)*, reaching the pulmonary orifice. Thus it has an inflow orifice and tract leading at an obtuse angle to an outflow tract and orifice. Some prefer to include an intermediate 'body' but this view will not be followed here. External features will be followed by considerations of in-

flow and outflow tracts, the tricuspid valvar complex, pulmonary valve and brief functional allusions.

External features. The convex *anterosuperior surface* is a large part of the sternocostal aspect of the heart, separated from the thoracic wall only by pericardium; the left pleura and, to a lesser extent, the anterior margin of the left lung are interposed above and to the left (**6**.30A,41). The *inferior surface* is flat and mainly related to the central tendon and also a small adjoining muscular part (left cupola) of the diaphragm, separated by pericardium. The *left and posterior* wall is the ventricular septum which bulges into the ventricle, so that in sections across the cardiac axis the right ventricular outline is crescentic (**6**.35). A collagenous band (**6**.47,48), the *tendon of the infundibulum (conus ligament)*, connects the pulmonary annulus and muscular infundibulum posteriorly to the root of the aorta, where it blends with the wall and margin of its right posterior (non-coronary) sinus (p. 712); inferiorly this tendon blends with the membranous ventricular septum (p. 709). The ventricle's wall varies much in thickness (vide infra) but is thinner (3–5 mm on average) than the left, in a ratio of about 1 to 3; it is thickest at its atrial end and gradually thins towards its apex.

Internal features. The internal surfaces of inflow and outflow tracts are in complete contrast, the transition between the two being superiorly at the supraventricular crest and antero-inferiorly near the ventricular apex at the septomarginal trabecula (vide infra). The *supraventricular crest*, a massive muscular arch between the right atrioventricular and pulmonary orifices (**6**.33), is oblique, curving forwards and right from the interventricular septum ('*septal limb*') to the right anterolateral ventricular wall ('*mural or parietal limb*'). Its concavity is between these limbs in the line of the crest; at right angles to this, i.e. between the valvar orifices, it is convex. To it posteriorly the anterior tricuspid leaflet is partly attached. Its septal limb often fuses with the septal end of the septomarginal trabecula (vide infra).

The inflow tract has walls roughened by the *trabeculae carneae*, irregular muscular ridges, columns, bands or protrusions covered with endocardium which project into the cavity; these and intervening depressions greatly vary myocardial thickness. Trabeculae are (1) mere ridges, (2) fixed at the ends but free between, or (3) *papillary muscles*, the latter continuous basally with the ventricular wall and at their apices with the *chordae tendineae*, glistening collagenous cords which, with the muscles, are part of the tricuspid valvar complex. The *septomarginal trabecula*, a curved muscular band from the lower part of the interventricular septum to the base of the anterior papillary muscle, forms the lower limit of the inflow tract and contains a continuation of the right bundle-branch of conducting tissue (p. 722). Its alternative name, 'moderator band', records an old idea that it prevents ventricular 'over distention'.

The outflow tract, conus arteriosus *(infundibulum)*, with smooth walls, ascends to the left and posteriorly to the *pulmonary orifice*. In cadavers its lower limit cannot be identified externally but it is the free rounded border of the supraventricular crest (**6**.33) and upper border of the septomarginal trabecula near the ventricular apex. The infundibulum is a persistent part of the bulbus cordis incorporated in the right ventricle; and its persistence as the right outflow channel is attributable, in part, to the support it provides for the pulmonary valve during diastole. It has been shown (in dogs) that, when subjected to increased backward pressure, the pulmonary valve becomes incompetent more readily than the aortic, suggesting that the muscular infundibulum retains contractile tonus throughout ventricular diastole, thus supporting the valvar cusps (Brock 1955).

Trabeculated walls of the inflow tract may assist in slowing entry of blood during early diastole; their contraction and interdigitation may increase the *volumetric efficiency* of ventricular emptying. Smooth walls in the outflow tract are presumed to increase the *velocity* of ejection. Similar considerations apply to the left ventricle.

THE TRICUSPID VALVAR COMPLEX

It is undesirable to describe tricuspid and mitral valves as isolated structures, since they are parts of functionally integrated

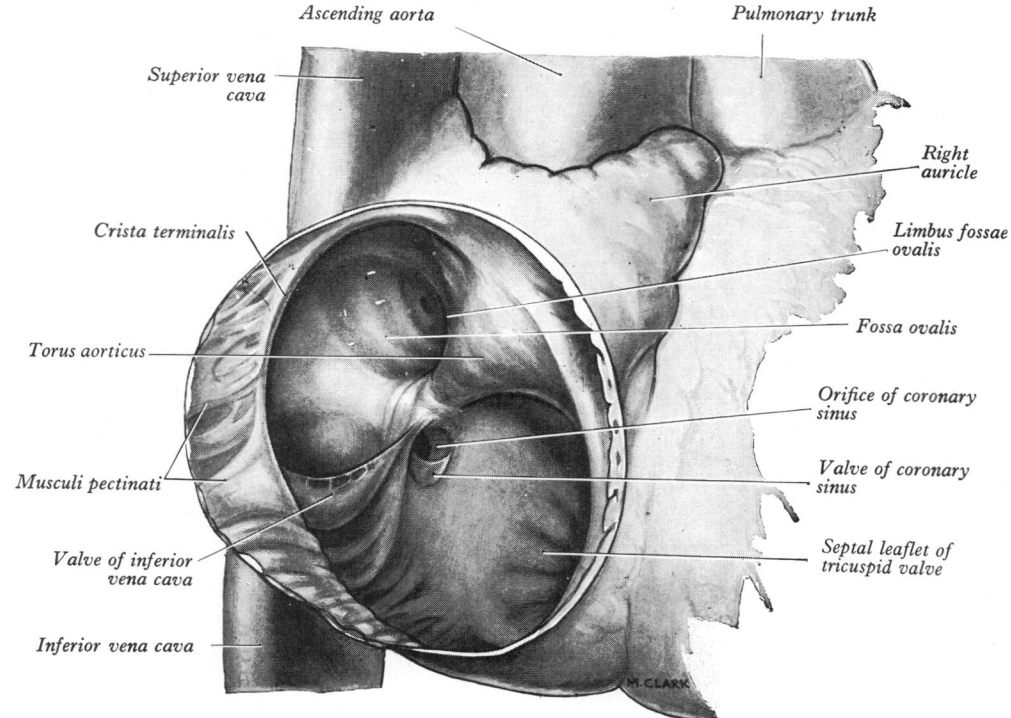

6.32 The interior of the right atrium, viewed from the front.

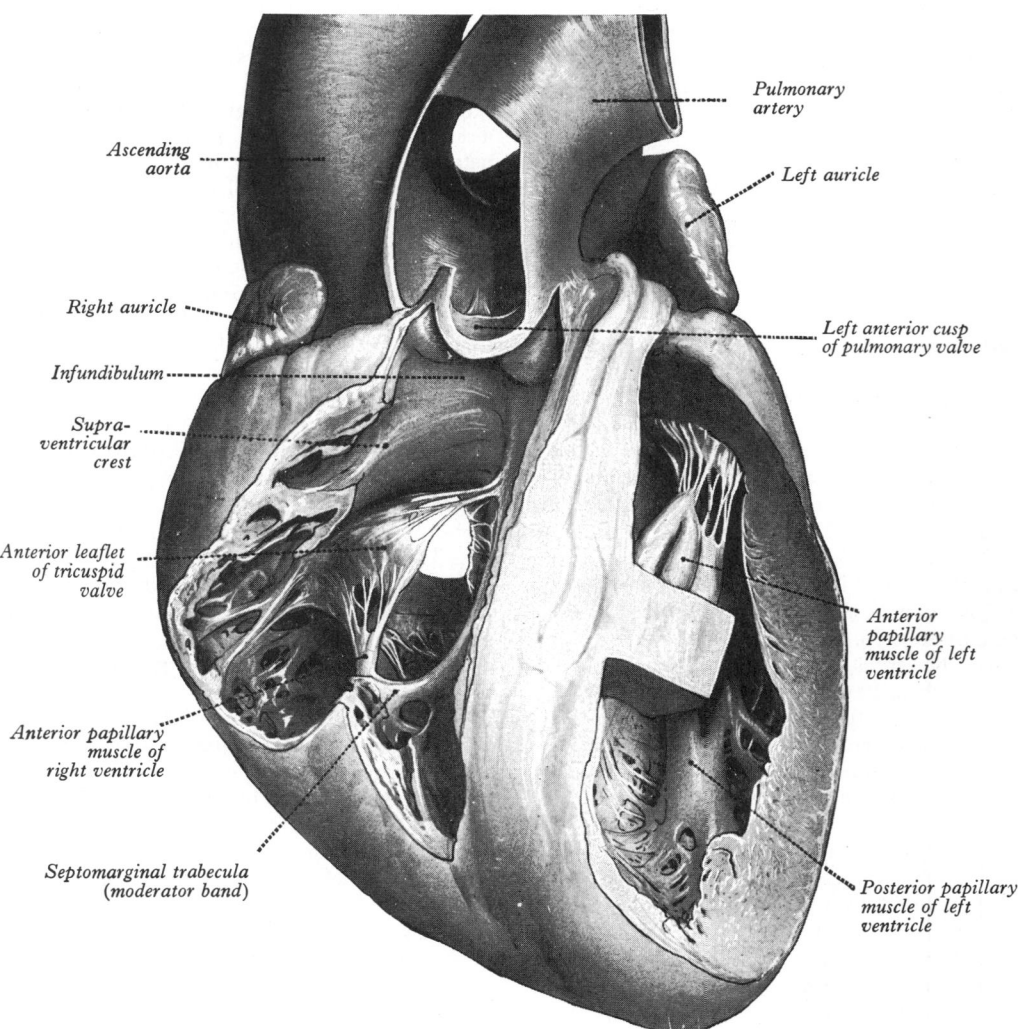

6.33 A dissection opening the ventricles, viewed from the front.

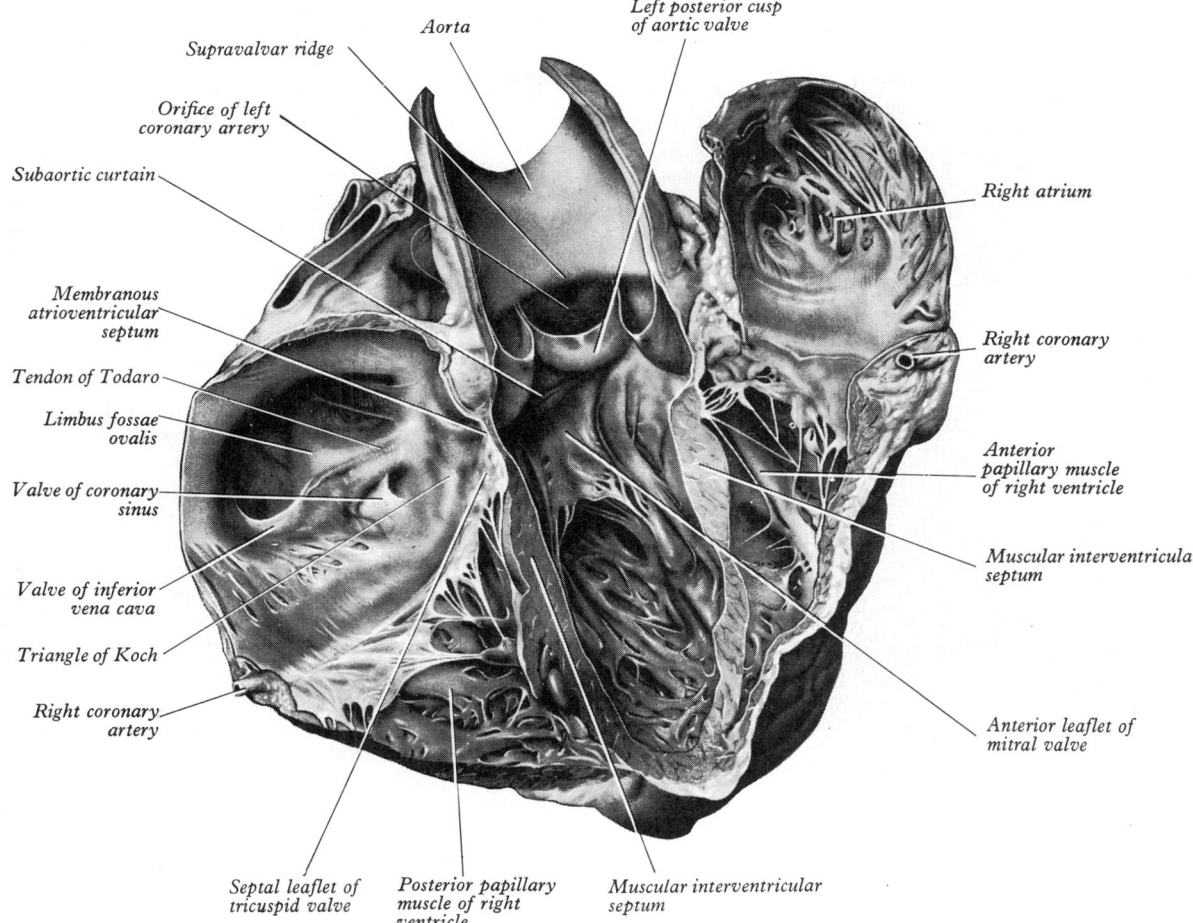

6.34 The interior of the heart revealed by incising it along its right and lower surfaces and excising the pulmonary trunk and infundibulum. The rest of the front of the heart has been turned over to the left.

complexes. The *tricuspid valvar complex* comprises: (1) the so-called *tricuspid atrioventricular 'orifice'* and its associated *annulus*, (2) *valvar leaflets* or *cusps*, (3) *chordae tendineae* of various types, and (4) *papillary muscles*. Harmonious interplay of all these together with the atrial, ventricular and septal myocardium (p. 717), depends on the conducting tissues (p. 720) and mechanical cohesion provided by the cardiac fibro-elastic 'skeleton'. All parts change substantially in position, shape, angulation and dimensions during a cardiac cycle, the in vivo condition being the result, at all times, of the state of myocardial activity balanced against internal changes in pressure and blood flow. Qualitative or

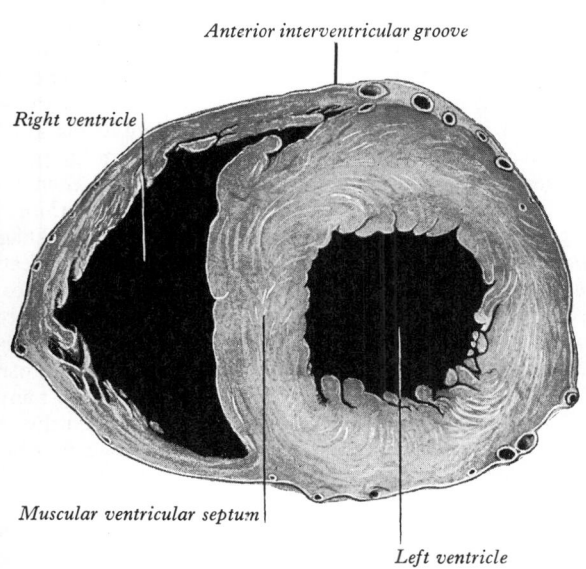

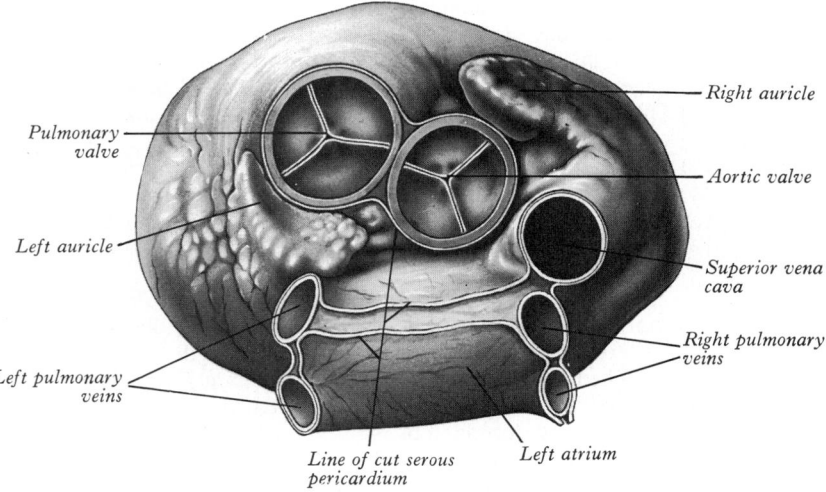

6.36 The heart viewed from above. The two continuous white lines which enclose the pulmonary trunk and aorta on the one hand and the pulmonary veins and the superior vena cava on the other, indicate the continuation of the parietal layer of the serous pericardium with the serous epicardium. The floor of the transverse sinus is seen from above, with the left coronary artery running in it. This diagram, from an earlier edition, has been retained for its pericardial details. However, in some respects it is misleading: the aortic and pulmonary valves are *not*, as shown, coplanar; the pulmonary valve is distinctly higher than the aortic valve; furthermore the planes of the valves 'face' approximately at right angles to each other.

35 Transverse section through the ventricles of the isolated heart, ewed from below. Note that in this illustration the heart is *not* positioned it would be in situ: in the latter position the crescentic 'right' ventricle erlaps most of the *anterior* surface of the 'left' ventricle.

705

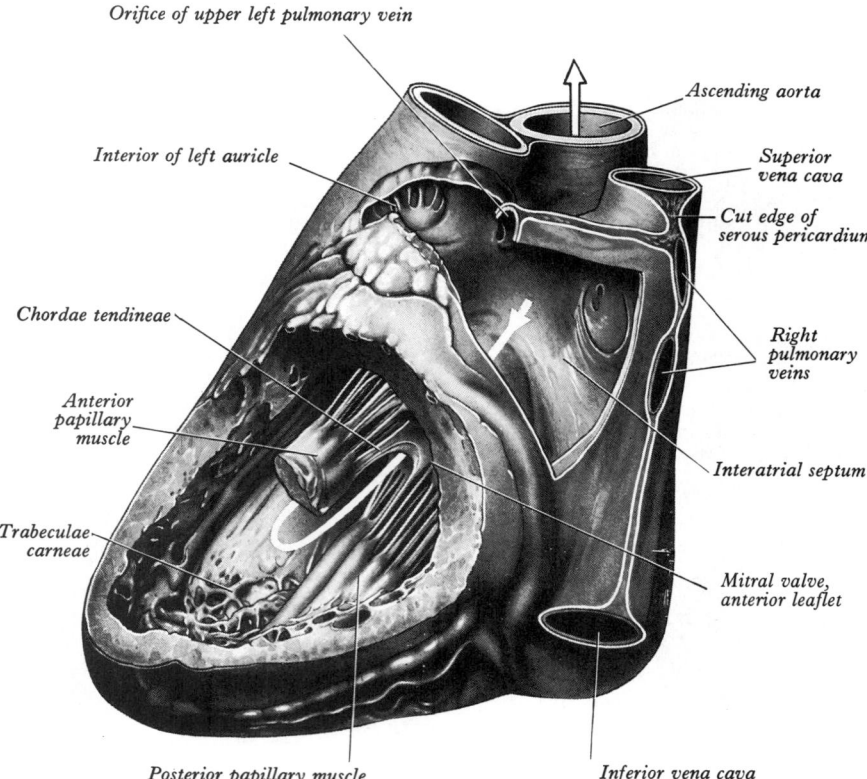

Orifice of upper left pulmonary vein

Ascending aorta

Interior of left auricle

Superior vena cava

Cut edge of serous pericardium

Chordae tendineae

Right pulmonary veins

Anterior papillary muscle

Interatrial septum

Trabeculae carneae

Mitral valve, anterior leaflet

Posterior papillary muscle

Inferior vena cava

6.37A Dissection showing the interior of the left side of the heart. The white arrow indicates the course of blood flow from the left atrium through the left ventricle to the aorta.

quantitative studies of dead hearts, fixed or unfixed, removed or in situ, may be most misleading; similar strictures apply *equally* to observations during cardiac surgery where a diseased heart has been subjected to severe haemodynamic interference.

The *tricuspid valve 'orifice'*, the largest valve orifice, (circumference 11.4 ± 1.1 cm in males and 10.8 ± 1.3 cm in females according to Silver et al 1971), is best seen from its atrial aspect (**6.**47); it has a *clear* line of transition from the atrial wall or septum to the bases of the valvar leaflets (valvulae). It is not a true orifice, becoming so only after leaflet removal; deep to its endocardium are various components of the valvar annulus. This artificial orifice has been described as almost circular, oval or roughly triangular with rounded angles, doubtless reflecting differences in age, normality, preparation and inspection. As detailed below, its margins are not precisely in a single plane; but for present purposes, a near approximation is almost vertical but at about 45° along the sagittal plane and slightly inclined to the vertical, so that it 'faces' (on its ventricular aspect) anterolaterally to the left and somewhat inferiorly. When roughly triangular, its margins are described as anterior, posterior and septal, corresponding to the bases of the valvar leaflets. (The term septal is unambiguous; *anterior* and *posterior*, though simple, are misleading, because of the near-vertical plane of the orifice; right superolateral and right inferolateral would be more accurate although cumbersome.)

The *tricuspid valve annulus* is an ill-defined term used without uniformity. Elementary accounts often describe all four valvar orifices as surrounded by *uniform rings* of collagenous tissue, the rings interconnected by dense masses of collagen; and as providing topographical lines of attachment for the bases of valvar flaps, fusing with their lamina fibrosa and, in mitral and tricuspid valves, situated precisely at the atrioventricular junctions (i.e. separating the myocardia of the cavities). But only some of these assumptions have proved true; in atrioventricular valves, though connective tissue *does* separate atrial and ventricular myocardium completely round the associated 'orifice', it varies in density and disposition around the circumference and with sex and age (Walmsley & Watson 1978). The topographical 'attachment' of the *free* valvar leaflet (valvula) does not *wholly* correspond to the

internal level of attachment of the valve's lamina fibrosa to the atrioventricular connective tissue (though it does in much of its extent). Some limit the term *annulus* to the line of free leaflet attachment; but the collagenous central stratum of the leaflet may pass basally and subendocardially for some distance before merging with this atrioventricular connective tissue, which will be considered to be the valvar annulus in this account. In the tricuspid valve this includes: (*a*) an angled line across the membranous septum, (*b*) the tricuspid aspect of the right fibrous trigone, (*c*) the tapering, dense, fibro-elastic 'anterior' and 'posterior' *fila coronaria* from the trigone, and (*d*) the tenuous sulcal connective tissue completing the annular circumference between the tips of the fila, which will be described below and are illustrated in **6.**48.

Tricuspid valve leaflets, usually three and hence the name, are *septal*, *anterior* and *posterior*, corresponding to the marginal sectors of the atrioventricular orifice so named (vide supra); each is a reduplication of endocardium enclosing a collagenous *lamina fibrosa*, continuous marginally and on its ventricular aspect with diverging fascicles of chordae tendineae (vide infra) and basally confluent with the annular connective tissue. Earlier descriptions of small 'accessory' valvulae, or 'islands of leaflet tissue' between the major leaflets have been modified (Silver et al 1971). Valvular tissue of the tricuspid (and mitral, p. 649) valves is now regarded as a *continuous curtain* arising from or near the annulus, descending into the ventricle but with three principal marginal grades of indentation, dividing it into definable parts. The deepest indentations are bays towards the basal attachment of the main, smoothly arched, *valve commissures*, separating the main leaflets and therefore termed anteroseptal, posteroseptal and anteroposterior. Thus a region may be designated, e.g. 'the anteroseptal half of the septal leaflet'. Commissures are usually well defined (but not as clearly as in the mitral valve) by their attached chordae tendineae and other adjacent features (Silver et al 1971). Intermediate indentations or '*clefts*', with smaller chordae, divide the posterior leaflet into three 'scallops'; one small '*notch*' usually exists on the margins of both septal and anterior leaflets. Thus, from either the atrial or the ventricular aspect, the true valvar orifice is highly irregular.

Palpation, transillumination and histology reveal that (except for parts extending on to the membranous septum) all tricuspid leaflets possess, passing from the free margin, *rough*, *clear* and *basal zones*. The *rough zone* is relatively thick, opaque and uneven on both aspects, particularly the ventricular, where most chordae tendineae are attached, its atrial, 'inflowing' aspect making contact with another leaflet during full valve closure. The *clear zone* is smooth and translucent, receives few chordae and has a thinner lamina fibrosa. The *basal zone*, extending about 2–3 mm from its peripheral attachment, is again thicker, with more connective tissue, is vascularized and frequently contains atrial myocardium.

The *anterior leaflet* (*valvula*, flap or lacinia) is the largest and is attached chiefly to the atrioventricular junction on the posterolateral aspect of the supraventricular crest but extends along its septal limb to the membranous septum, ending at the anteroseptal commissure; one or more notches indent its margin. The *septal leaflet* is smallest; its attachment passes from the posterior ventricular wall across the muscular septum and then angled across the membranous septum to the anteroseptal commissure. The *posterior leaflet* is wholly mural in attachment. Diagram **6.**38A, based on Silver et al (1971), summarizes the main features of the tricuspid leaflets, attachments and chordae; the heart incised along its acute margin, opened and the annulus severed at the anteroposterior commissure and laid flat, with the base–apex ventricular axis vertical. The posterior leaflet and posteroseptal half of the septal leaflet are horizontal, whereas the latter's anteroseptal half and the anterior leaflet slope up to meet on the membranous septum at the anteroseptal commissure. The distribution of commissures, clefts, notches, zones and the pleated form of the septal leaflet are drawn to scale (**6.**38A).

Chordae tendineae are white fibrous collagenous cords; as in the left ventricle most are *true chordae* of the atrioventricular valve complex. *False chordae*, merely connecting papillary muscles to each other or to the ventricular wall including the septum, or passing directly between points on the wall (and/or septum), are irregular in numbers and dimensions and of doubtful significance.

they will not be further discussed here. (However, see brief comments on false chordae associated with the *left* ventricle, also structures designated *left ventricular bands* and a bibliographical source, p. 709. Comparable *right ventricular bands* also occur.) *True chordae* usually arise from small projections on the tips or margins of the apical thirds of papillary muscles but sometimes from papillary bases or general ventricular (including septal) walls; they are attached to some aspect of the leaflets, their clefts, commissures or bases. They were classified by Tandler (1913) into first, second and third order chordae, according to the distance of attachment from margins: subsequent authors have usually concurred, although the schema has little functional or morphological merit. In an extensive study of the mitral valve, Lam et al (1970) proposed a new and more useful classification, applied with additions by Silver et al (1971) to the tricuspid, in which five classes of chordae are described: *fan-shaped, rough zone, free edge, deep* and *basal*.

Fan-shaped chordae have a short stem from which branches radiate to attach to margins (or the ventricular aspect) of main commissures (*commissural chordae*) and to the ends of adjacent leaflets, smaller ones having a like attachment to the clefts between the 'scallops' of the posterior leaflet (*cleft chordae*). *Rough zone chordae* arise from a single stem which splits into three: one cord is attached to the free margin, another on the ventricular aspect of the rough zone at its outer limit (the line of valvular contact in full closure); the third to some intermediate point. *Free edge chordae* are single, threadlike, and often long, passing from either the apex or the base of a papillary muscle into a marginal attachment, usually near the mid-point of a major leaflet or one of its scallops. *Deep chordae*, also long, pass beyond the margin and, sometimes bi- or tri-furcating, reach the more peripheral rough zone or even the clear zone. *Basal chordae*, the most variable, being round cords, flat ribbons, long and slender or short and muscular, arise from the smooth or trabeculated ventricular wall and attach to the basal 2 mm of a leaflet. (The references quoted above give quantitative details and bibliographies.)

Papillary muscles in the right ventricle comprise two principal, *anterior* and *posterior*, and smaller, variable, *septal* muscles. The *anterior papillary muscle* is largest; its base arises from the right anterolateral ventricular wall below the anteroposterior commissure and blends with the right end of the septomarginal trabecula (vide supra); to its apex, occasionally bifid, are attached (apically and subapically) chordae tendineae which extend to the *corresponding zones* of anterior *and* posterior tricuspid leaflets. The *posterior papillary muscle* arises from the ventriculoseptal myocardium below the posteroseptal commissure; it is frequently bifid or trifid, its chordae distributed to the septal and posterior leaflets, their commissure and clefts. 'Accessory' *septal papillary muscles* may be absent or merely irregular fibrous cords; more often a group of small papillary projections arise from the infundibular septal wall below the crista supraventricularis and reach the apical aspect of the anteroseptal commissure, tethering adjacent parts of the anterior and septal leaflets; the highest muscle is often termed the *papillary muscle of the conus*.

In closure the tricuspid valve has much in common with the mitral; since evidence with regard to the latter is clearer, they will be considered together (p. 714).

THE PULMONARY VALVE

The pulmonary valve, the outflow valve of the right ventricle (**6.**33), surmounts the infundibulum, is separated a little from the other three cardiac valves and is anterosuperior to them. Its general plane faces superiorly to the *left* and slightly posteriorly. It has three semilunar *cusps (valvules)*, attached by convex bases to a *trilunate (triple-scalloped, tricrenate, 'coronet-like'* are alternative names) fibrous thickening in the wall of the pulmonary trunk at its junction with the ventricle forming a *valvar annulus* (p. 717 and **6.**48); the cuspal free borders project into the vessel. Two cusps are anterior (right and left), the third posterior. (Cusps of the pulmonary and aortic valves are named here from their approximate in situ positions in adult hearts; this differs from *Nomina Anatomica* where they are named according to the fetal position *before rotation*.) Each cusp is a fold of endocardium, with

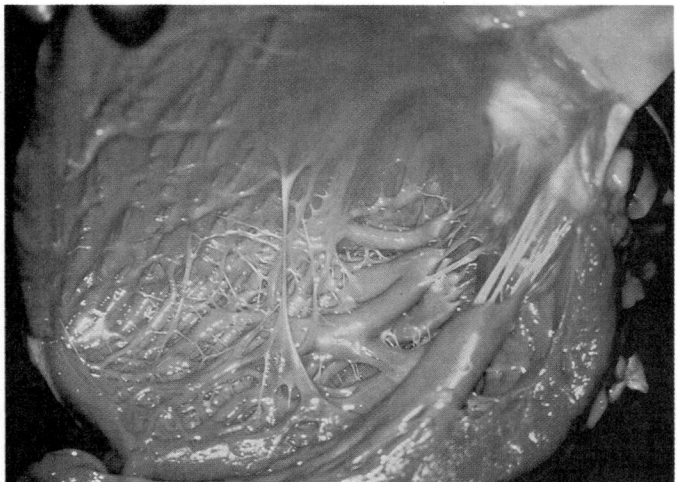

B

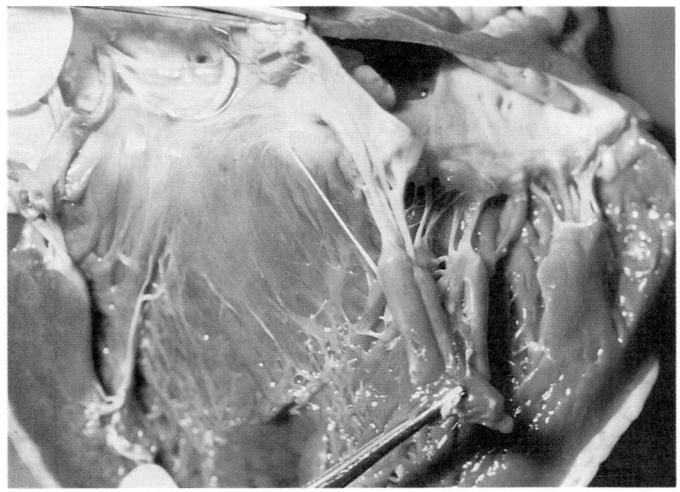

C

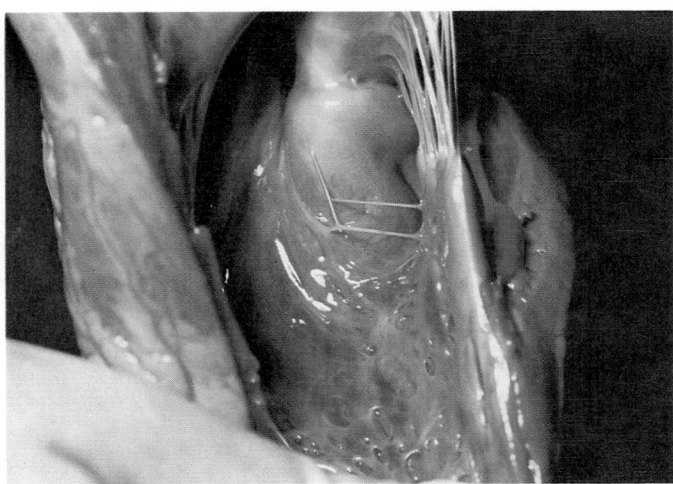

D

6.37B,C,D Photographs of parts of the interior of the left ventricle in freshly dissected human hearts showing: variably prominent trabeculae carneae, papillary muscles with true chordae tendineae springing from their apices and subapical segments and varieties of left ventricular bands (LVBs, one class of 'false' chordae tendineae). The latter in (B) is single, long, slender and between two mural attachments; in (C) between papillary subapex and ventricular wall passing obliquely; in (D) two cross the ventricular cavity horizontally between papillary muscles mid-belly and ventricular septum. Supplied by Blanka Darazs, Faculty of Internal Medicine, University of South Africa. (See also text and Darazs & Taylor 1988.)

an intervening, regionally variably developed *lamina fibrosa*; the lamina is substantial along both free and attached borders and at the latter blends with the fibrous 'annulus'. Central in each cuspal free margin is a localized thickening of collagen, the *nodulus* (Arantii), from which fibres radiate through the lamina to reach the annulus; on each side of a nodule, the collagen is much reduced in the narrow *lunules*. Opposite the semilunar cusps the arterial wall presents three *sinuses* (of Valsalva), whose walls are thinner than elsewhere; although predominantly collagenous near the annulus, their content of elastic tissue increases rapidly in their upper zones. Except for differences in time relation and pressures, opening and closure of the pulmonary valve has much in common with that of the aortic valve (p. 712, **6**.53).

THE LEFT ATRIUM

Though less in volume than the right, the *left atrium* has thicker walls (3 mm on average); its cavity and walls are largely formed by the proximal parts of pulmonary veins incorporated during development (p. 209). (The only clear derivative of the left part of the embryonic atrium is the auricular appendage.) The left atrium is roughly cuboidal and extends *behind* the right atrium, separated by the oblique interatrial septum (p. 702 and **6**.27,29A,B,C). Thus the right atrium is in front, lying anterolateral to the right part of the left atrium; its left part is concealed ventrally by the commencements of the pulmonary trunk and aorta, with part of the transverse pericardial sinus between it and these arteries. Antero-inferiorly and to the left it adjoins the left ventricular base, where the left atrioventricular orifice is sited (vide infra). Its posterior aspect forms most of the cardiac anatomical base and is approximately quadrangular, receiving the terminations of (usu-ally) two pulmonary veins from each lung; it forms the anterior wall of the oblique pericardial sinus (p. 696). This surface ends at the shallow vertical interatrial groove that descends to the cardiac crux.

The left *auricle* is constricted at its atrial junction; it is longer, narrower, and more curved than the right, its margins being more deeply indented. It turns forwards left of the pulmonary trunk and overlapping its commencement. (Relations of the auricle to the left coronary artery are described on p. 730).

The **interior of the left atrium** (**6**.37A,55) presents several features: the four *pulmonary veins* open into its upper pos-terolateral surfaces, two on each side; their orifices are smooth, oval and not valved; the left pair frequently end via a common opening. The *left atrioventricular orifice* is fully described below. *Foramina venarum minimarum* are orifices of venae cordis minimae, returning blood from the myocardium. *Musculi pec-tinati*, fewer and smaller than in the right atrium, are confined to the left auricle. On the atrial septum a lunate impression may appear, bounded below by a crescentic ridge, concave upwards, marking the site of the ostium secundum (p. 209). (Its ridged margin is derived from the septum primum and its floor the sep-tum secundum. Sometimes there are multiple impressions.)

THE LEFT VENTRICLE

General and External Features

In contrast to the right ventricle, construction of the left ventricle accords with its role as a powerful pump to sustain pulsatile flow in high-pressured systemic arteries. Variously described as half-ellipsoid or cone-shaped, it is longer and narrower than the right, extending from its base in the plane of the coronary sulcus to the cardiac apex. Its long axis descends forwards and to the left. In transverse section, at right angles to the axis, its cavity is oval or nearly circular (**6**.35), with walls about three times thicker (8–12 mm) than in the right ventricle. It forms part of the sternocostal, left and inferior (diaphragmatic) cardiac surfaces. Except where obscured by the aorta and pulmonary trunk, the base of the ventricular cone is superficially separated from the left atrium and auricular appendage by part of the coronary sulcus; anterior and

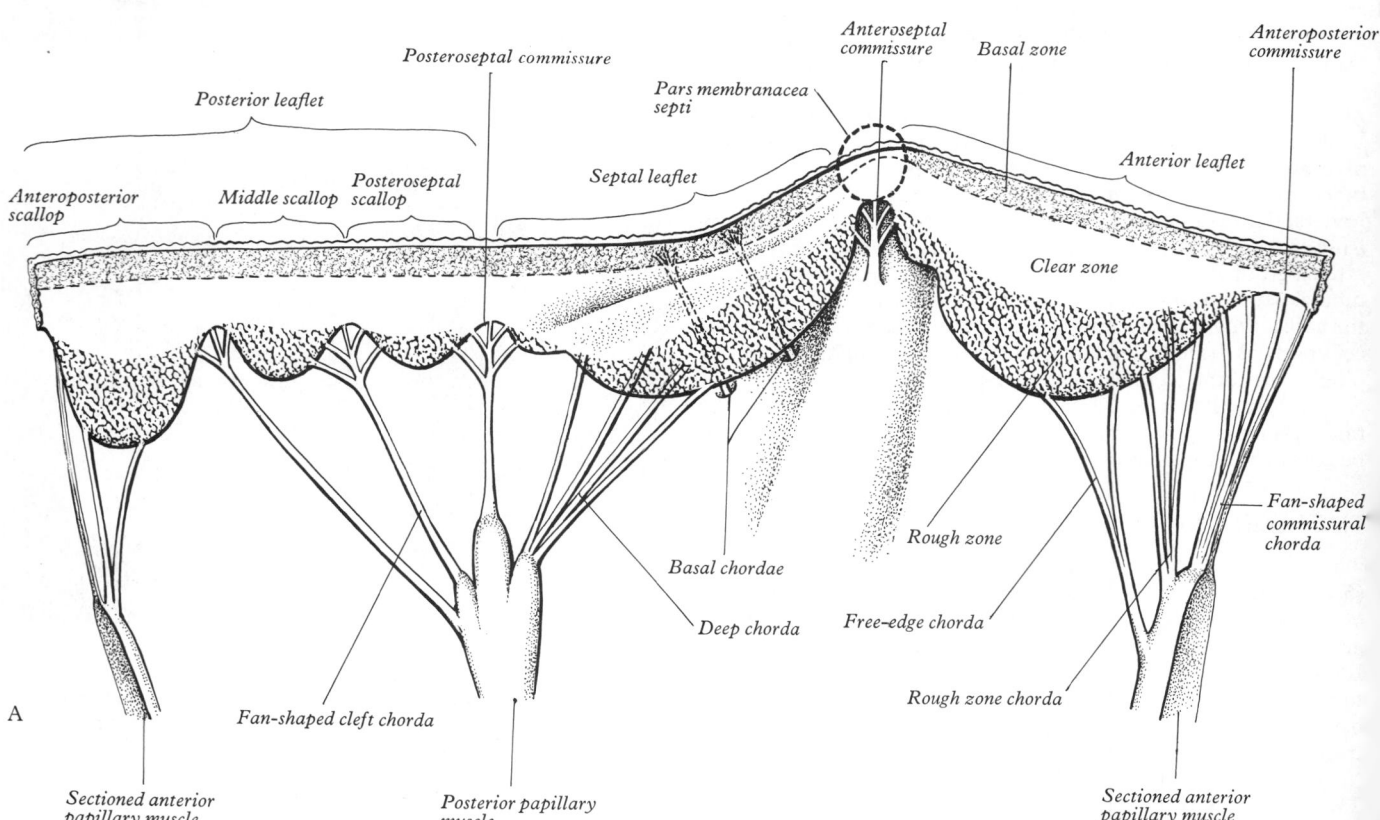

6.38 Diagrams of the tricuspid (A) and mitral (B, facing page) valves, laid flat after division of their fibrous annuli with the atrioventricular axis placed vertically, showing the relative shapes, dimensions and zones of the leaflets and their subdivisions; also the varieties of chordae tendineae. Drawn to scale: A after Silver et al (1971); B after Lam et al (1970). See text for further details.

inferior interventricular grooves indicate the lines of mural attachment of the interventricular septum and the limits of the left and right ventricular territories. The ventricle's sternocostal surface curves bluntly into the left surface at the 'obtuse' margin.

Internal Features (6.33,35,37A–D)

Like the right, the left ventricle has an *inflow orifice* and *tract* continuous with an *outflow tract* and *orifice*, with attendant valvar complexes. The left atrioventricular orifice, with its mitral valve, admits atrial blood during diastole, flow being towards the cardiac apex (*inflow vector*). After mitral closure and throughout the ejection phase of systole (6.55), blood is expelled from the apex through the aortic orifice, defining the *outflow vector*. In contrast to right ventricular orifices, the left inflow and outflow orifices are in *close contact*, in part separated only by the subaortic curtain (intervalvar septum, vide infra) and anterior mitral valve leaflet; thus inflow and outflow vectors are acutely angled.

The anterolateral wall is the interventricular septum (i.e. septal wall as distinct from 'free' ventricular wall) which is mostly (but not exclusively) thick and muscular. The *muscular septum* is concave to the left, its convexity being the posteromedial profile of the right ventricle as seen in section; it thus completes the left ventricle's circular outline (6.35). Towards the aortic orifice the septum becomes the thin, collagenous *pars membranacea septi*, an oval or round area below and partly confluent with the fibrous supports of the anterior (right coronary) and right posterior (non-coronary) aortic cusps (pp. 711, 717 and 6.34,48).

Between the lower limits of the free margins of the mitral valve leaflets and the ventricular apex, including the apical two-thirds of the interventricular septum, the muscular walls are intensely trabeculated. These *trabeculae carneae* are larger and more intricate than in the right ventricle but show the same varieties: mere ridges; terminally attached but centrally free; or papillary muscles. They form a very dense labyrinth towards the apex. Endocardial walls of the outflow tract (as in the right ventricular

infundibulum) are smooth; subendocardial tissue in its final part, the *aortic vestibule*, is increasingly fibrous and less muscular.

THE MITRAL VALVAR COMPLEX

Many *general* comments already made in respect to the tricuspid valve apply equally to the mitral; to avoid repetition, remarks will be confined to features of the following components of the mitral complex: (1) the *mitral atrioventricular 'orifice'* and its *valvar annulus*, (2) *leaflets (valvulae)*, (3) varieties of *chordae tendineae* and (d) *papillary muscles*.

The *mitral atrioventricular orifice*, a well-defined transitional zone between the atrial or aortic wall into the bases of the leaflets, is best viewed from the atrial aspect; it is smaller than the tricuspid (mean circumference: 9.0 cm in males, 7.2 cm in females, according to Ranganathan et al 1970, in a study of 50 Caucasian hearts from 15–85 years). In diastole the orifice, approximately circular, is almost vertical and at 45° to the sagittal plane but with a slight forward tilt; its ventricular aspect faces anterolaterally to the left and a little inferiorly, i.e. towards the left ventricular apex. It is almost coplanar with the tricuspid orifice but postero-superior to it, whereas it is postero-inferior and slightly left of the aortic orifice. All three orifices are intimately connected by the cardiac fibrous skeleton (p. 714).

The *mitral valvar annulus* is again *not* a simple fibrous ring, but comprises elements varying greatly in consistency, with which the valvular laminae fibrosae become continuous. These variations are of the greatest functional significance, allowing major changes in the annular shape and dimensions at different stages of the cardiac cycle, ensuring optimal efficiency in valvular actions of the mitral complex and its equally important role in controlling inflow/outflow patterns through the left ventricle. Such structural variations around the valve's periphery are sometimes accentuated, particularly with age.

The mitral annulus is shown in 6.47,48. It includes internal aspects of parts of the left and right fibrous trigones and,

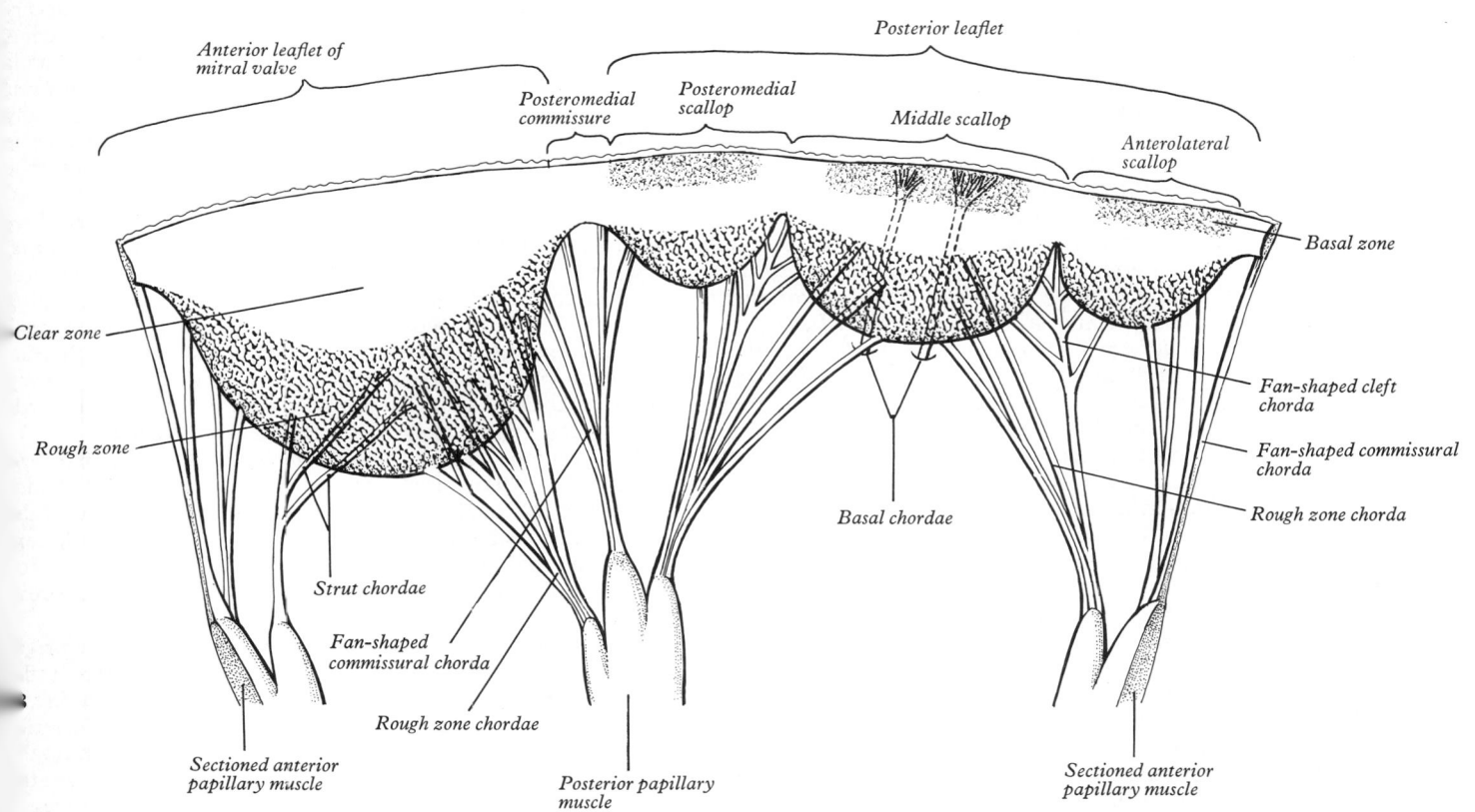

extending from the latter, the anterior and posterior fila coronaria: tapering, fibrous, subendocardial tendons partly encircling the orifice at the atrioventricular junction. Between the filar tips the atrial and ventricular myocardia are separated by a more tenuous sulcal sheet of deformable fibro-elastic connective tissue. Spanning between the trigones, the lamina fibrosa of the central part of the anterior mitral leaflet is a continuation of the fibrous *subaortic curtain (intervalvar septum)*, which descends from the adjacent halves of the left posterior (left coronary) and right posterior (non-coronary) cuspal regions of the aortic annulus; the curtain is specially thick where it passes to the base of the free leaflet. Since the valvar line of connective tissue attachment comprises sequentially a malleable fibrous sheet, bordered by two dense collagenous masses, then two tapering 'tendons' and finally delicate fibro-elastic areolar at different locations, the usual term 'annulus fibrosus' is inappropriate but is deeply engrained and will serve until a more suitable substitute is minted.

Mitral valve leaflets (valvulae), since the earliest descriptions, have been 'two in number'; hence the name *bicuspid valve* is more explicit, though erroneous (the leaflets are *not* cuspid in form) and surely less picturesque than the clinical 'mitral'. However, confusion, controversy and difficulties in quantitation have arisen because one or more small accessory leaflets between the two major ones often occur. Problems have apparently been resolved by the quantitative studies of Lam et al (1970) and Ranganathan et al (1970). The mitral valve is now described as consisting 'of a continuous veil attached around the entire circumference of the mitral orifice' (Harken et al 1952); its free edge bears indentations, two being deep and regularly positioned and receiving unique fan-shaped commissural chordae tendineae (vide infra); the tips of the papillary muscles are guides to them (Rusted et al 1952). These indentations are the *anterolateral* and *posteromedial commissures*; the valvar tissue may therefore be allocated to the two *commissural areas* mentioned and termed *anterior* and *posterior leaflets*. (Names of tricuspid commissures are derived from those of the adjacent cusps, whereas mitral commissures are positionally named. Their official terms, anterior and posterior, though simple, are misleading because of the valve's obliquity; many prefer alternative names (vide infra). General mitral anatomy, with the valve opened and laid flat, is diagrammed in **6.38B** (after Ranganathan et al 1970).

The *anterior leaflet* (aortic, septal, 'greater' or anteromedial, Chiechi et al 1956) is large, semicircular or triangular, with few or no marginal indentations. Its lamina fibrosa is peripherally continuous, beyond the margins of the fibrous subaortic curtain, with the mitral aspects of the right and left fibrous trigones, between these with the fibrous curtain itself and beyond the trigones with the trigonal roots of the fila coronaria (**6.48**). The leaflet has a deep crescentic *rough zone* receiving various types of chordae tendineae (vide infra); the ridge limiting the rough zone's outer margin indicates the maximal extent of surface contact with the posterior leaflet in full closure; from the rough zone to the valvar annulus is a *clear zone*, devoid of chordal attachments, though its lamina fibrosa carries extensions from chordae attached in the rough zone. The anterior leaflet has no basal zone. Hinging on its annular attachment and continuous with the deformable subaortic curtain, it is critically placed *between* inflow and outflow tracts, the whole mechanism having been dubbed the *aortic baffle*. During passive ventricular filling and atrial systole, its smooth atrial surface is important in directing a smooth flow of blood towards the trabeculated ventricular body and apex. After the onset of ventricular systole and mitral closure the ventricular aspect of its clear zone merges into the smooth surface of the subaortic curtain which, with the remaining fibrous walls of the subvalvular aortic vestibule, forms the smooth boundaries of the terminal outflow tract. Dimensions of the anterior leaflet are: mean central vertical height 2.4 cm in males, 2.2 cm in females; mean basal width 3.6 cm in males, 2.9 cm in females; the mean vertical height of the rough zone is 0.9 cm in males and 0.8 cm in females (according to Ranganathan et al 1970, who include further details and analysis of the data of Rusted et al 1952 and Chiechi & Lees 1956).

The *posterior leaflet* (ventricular, mural, 'smaller' or posterolateral) has usually two minor indentations or 'clefts'. Lack of definition of a major intervalvular commissure and of other details, has previously led to disagreement and confusion concerning the territorial extent of the posterior leaflet and the possible existence of an accessory one. Lam et al (1969), however, have provided precise definition of the commissural chordae (vide infra), showing that the posterior leaflet is best regarded as all valvular tissue posterior to the anterolateral and posteromedial commissures. Thus defined, it has a wider attachment to the annulus than the anterior (mean values are 5.4 cm in males, 4.3 cm in females). There are typical cleft chordae tendineae (vide infra) and clefts divide the posterior leaflet into a relatively large *middle 'scallop'* and smaller *anterolateral commissural* and *posteromedial commissural 'scallops'*. Each scallop has a crescentic, opaque rough zone, receiving chordal attachments on its finely serrated margin and ventricular aspect, defining the area of valvular apposition in full closure. From the rough zone to within 2–3 mm of its annular attachment is a membranous *clear zone* devoid of chordal attachment; the basal 2–3 mm is thick, vascular and receives chordae tendineae. The ratio of rough to clear zone in the anterior leaflet is about 0.6, in the middle 'scallop' of the posterior leaflet 1.4; thus, much more posterior leaflet is in apposition with the anterior during mitral closure.

Mitral chordae tendineae resemble those in the right ventricle; false chordae are also irregularly distributed as in the right ventricle. *False chordae*, as noted, may interconnect papillary muscles, extend from the latter to a point on the ventricular wall or interconnect two points on the ventricular walls. They are ignored in many accounts and when mentioned considered of little or no functional significance; further, many names have been applied, confusing rather than clarifying the topic. There has been a recrudescence of interest in structures having these anatomical dispositions and connections; a general name, *left ventricular bands*, has been re-adopted and they are being reinvestigated by Darazs & Taylor (1988), see also Gerlis et al (1984). Their main summary emphasizes: they are defined as discrete fibromuscular bands crossing the left ventricular cavity and having the connections mentioned. They are common, occurring in about 50% of human hearts, and often cross the outflow tract. Many contain Purkinje myocytes. Often left ventricular bands can be identified by two-dimensional echocardiography. Their normal role in cardiac conduction and haemodynamics has still to be determined. A positive association between the presence of left ventricular bands and the occurrence of a vibratory systolic auscultatory murmur has been claimed. The colour photographs **6.37B,C** and D, kindly provided by Dr B Darazs, are of the interior of the left ventricle showing varieties of trabeculae carneae, papillary muscles, true chordae tendineae and left ventricular bands. (As mentioned previously, comparable *right ventricular bands* also occur.) *True chordae* of the mitral complex may first be divided into *interleaflet* or *commissural chordae* and varieties of *leaflet chordae*; of the latter those of the anterior leaflet are *rough zone chordae*, including special *strut chordae*; those of the posterior leaflet include *rough zone chordae*, *cleft chordae* and *basal chordae*. Most true chordae divide into branches from a single stem soon after its origin from the apical third of a papillary muscle, or proceed as single cord dividing into multiple fine cords near their attachment.

Anterolateral and posteromedial *commissural chordae* arise near the tips of papillary muscles by a single stem fanning out at once into radiating strands attached to the smooth free margin of the commissure; some enter the lamina fibrosa of their adjacent leaflets and continue to their annular attachment. Branches of the posteromedial commissural chorda are longer, thicker and spread more.

Rough zone chordae divide into three soon after their origin from the papillary muscle; one cord is attached to the free leaflet margin, one beyond it on the ventricular surface at the line of closure and one between these. Such chordae occur in posterior and anterior leaflets; peculiar to the anterior are two which are the thickest and strongest in the mitral complex; they come from the tips of the anterolateral and posteromedial papillary muscles, attach near the line of valvar closure posteromedially ('4–5 o'clock' position) and anterolaterally ('7–8 o'clock' position). These are *strut chordae*, visible in over 90% of hearts examined by Lam et al (1970); their zone of attachment was earlier termed the *critical point of tendinous insertion* of the anterior leaflet (Brock 195...

Cleft chordae, like miniature fans, have small branches radiating to the free margins of clefts between 'scallops', the deeper branches passing to the ventricular aspect of the adjacent rough zones.

Basal chordae are confined to the posterior leaflet (usually the large middle 'scallop' and one of the two smaller); they arise directly from the mural ventricular myocardium as a single strand which flares into minute branches before attachment to the ventricular aspect of the valvular basal zone. Sometimes basal chordae are partly or wholly muscular.

The *two left ventricular papillary muscles* vary in length and breadth and may be bifid. The *anterior papillary muscle* arises from the sternocostal mural myocardium, the *posterior* from the diaphragmatic region. Chordae tendineae arise mostly from the tip and apical third of each muscle, but sometimes near its base. They diverge and are attached to corresponding points (i.e. points in close apposition in valvar closure) on *both* mitral leaflets.

Opening of the mitral valve at the onset of diastole is passive but rapid, leaflets parting and projecting into the ventricle as left atrial pressure exceeds left ventricular diastolic pressure. Passive ventricular filling proceeds as atrial blood pours to the apex, directed by the pendent anterior mitral flap. Leaflets begin to float passively together, hinging on their annular attachments and partially occluding the ventricular inflow orifice. Atrial systole now occurs, jetting blood apically and causing transient valvar re-opening. As maximal filling is reached, the leaflets again float rapidly together, ventricular systole follows, starting in the papillary muscle and followed rapidly by general mural and septal contraction. Co-ordinated papillary contraction raises tension in the chordae and coaptation of the corresponding points on opposing cusps. With general mural and septal excitation and contraction, left ventricular pressure rapidly rises (**6.55**); the leaflets 'balloon' towards the atrial cavity and the atrial aspects of the rough zones come into maximal contact. Precise papillary contraction and increasing tension in their chordae prevent valvular retropulsion into the atrium, maintaining valvar competence. The *biphasic* nature of mitral closure thus described is vividly apparent in a normal echocardiogram (**6.46**). With the fast drop in ventricular pressure following the ejective phase of systole and the onset of diastole, the leaflets part and the next cycle begins.

It is assumed that, apart from differences in timing and pressures, the general form of activity is similar in the tricuspid valve.

Both atrioventricular orifices and valves undergo large changes in position, form and area during a cardiac cycle. Both valves move anteriorly and to the left during systole, and the reverse in diastole. The mitral valve (for which data are more complete) reduces its orificial (annular) area by as much as 40% in systole; its shape also changes from circular to *crescentic* at the height of systole, the annular attachment of its anterior leaflet being the concavity of the crescent; the attachment of its posterior leaflet, although remaining convex, contracts towards the anterior.

THE AORTIC VALVE (**6.38A,B,40,48**)

The smooth left ventricular outflow tract, or aortic vestibule, ends at the aortic orifice and valve, whose attached margins continue into the aortic root. Although stronger in construction, the aortic valve resembles the pulmonary (p. 707), consisting of a complex fibrous 'annulus', three semilunar cusps (or valvules) attached to it and three dilatations of the aortic wall, the aortic sinuses, one corresponding to each cusp.

Because the annulus is complex, it is difficult to define a valvar 'plane'. Some choose apices of intercuspid commissures as reference points, others select the deepest and most proximal points on the attached cuspal margins, i.e. three nadirs of the fibrous annulus. Approximately the valve faces up, right and a little anteriorly; it is anterosuperior to and slightly right of the mitral orifice; it is like a truncated cone.

The *aortic annulus* consists of three almost semicircular condensations of collagen forming three scallops (or 'crenations'—an indifferent name), the whole encircling the trilunate vestibulo-aortic junction like a three-pronged coronet; the 'prongs' point distally; apically the prongs form the fibrous commissural framework between adjacent cusps (vide infra and **6.48**). The luminal aspect of each scallop fuses with the valvular lamina

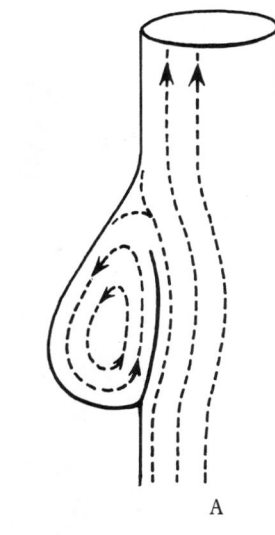

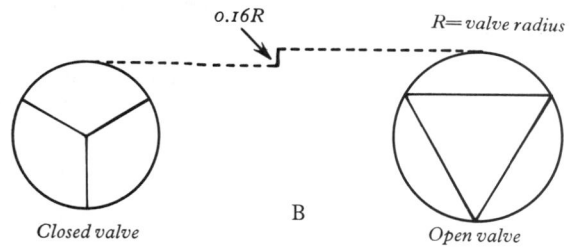

6.39 A. Diagram of the vortices formed within an aortic sinus during the ejection phase of left ventricular systole. Note the position of the valve cusp and sinus wall. B (left). The general position of the apposed valve cusps during diastolic closure of the aortic (and pulmonary) valve. B (right). The open position of the aortic (and pulmonary) valve during the ejection phase of ventricular systole. Note the increase in diameter of the arterial trunk root at the level of the valve commissures and the *triangular* form of the open valve. Consult text for further discussion.

fibrosa at the attached border of its cusp. The proximal (ventricular) aspect of the annulus is thickened at its nadirs (p. 716), while the triangular intervals between the proximal aspects of the scallops' ascending limbs are filled and fuse with extensions of the aortic vestibule's largely fibrous walls, special regions of which include the subaortic curtain (intervalvar septum) and part of the *pars membranacea septi*. The distal, deeply concave aspect of each scallop or crenation is the lower limit of, and continues into, the wall of an aortic sinus. Some maintain that, though largely fibrous, the aortic vestibular walls contain myocardial strands, some crossing the fibrous annulus as fine irregular strands in the walls of the sinuses.

The three aortic *semilunar cusps (valvules)* (**6.40**) are folds of endocardium with a central lamina fibrosa specially thickened in places. With the valve half-open, each cusp equals slightly more than a quarter of a sphere, an approximate hemisphere being completed by the corresponding sinus. Each cusp has a thick basal attached border, deeply concave on its aortic aspect, and a horizontal free margin, which is only slightly thickened except at its mid-point where it has an aggregation of fibrous tissue, the *valvular nodule* (Arantii); fine collagenous bundles radiate from this to the attached border, where the basal fibrous thickening blends with its corresponding annular scallop. At the apices of commissures a few transverse collagen fibres pass between adjacent cusps, but longitudinal fascicles enter the ascending limbs of the annular scallops (p. 716, **6.48**). Flanking each nodule the lamina fibrosa is tenuous, forming the *lunules* of translucent and occasionally fenestrated valvular tissue. The aortic surface of each cusp is rougher than its ventricular (outflow) aspect; endocardial cells are elongated in the direction of blood flow (Clarke & Fink 1974).

Three sets of names for the aortic cusps (valvules) are current. Official terms in *Nomina Anatomica* (1980) refer to fetal positions before full cardiac rotation has occurred; they are *posterior, right*

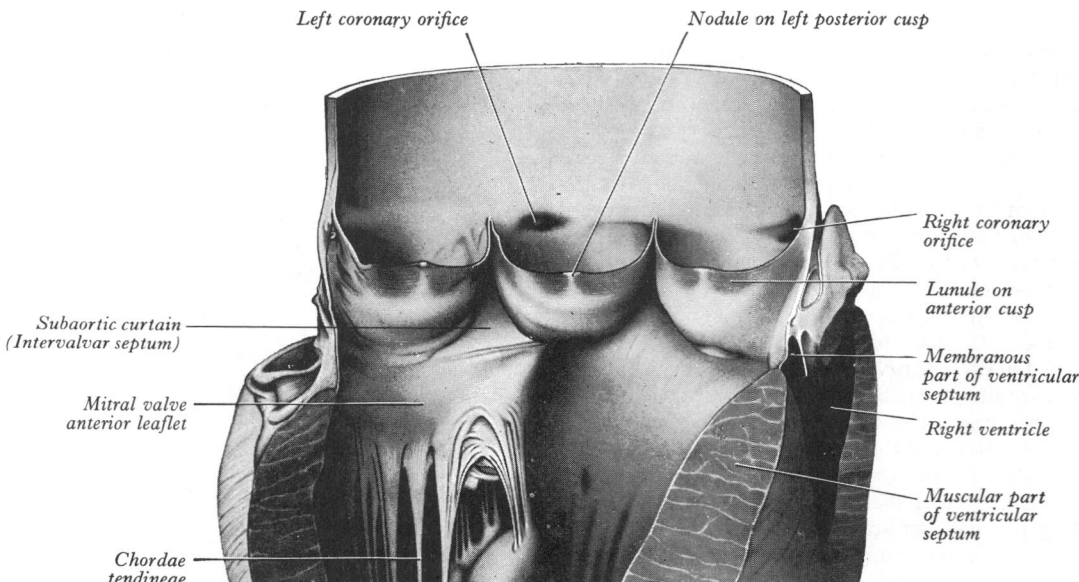

Left coronary orifice

Nodule on left posterior cusp

Right coronary orifice

Lunule on anterior cusp

Membranous part of ventricular septum

Right ventricle

Muscular part of ventricular septum

Subaortic curtain (Intervalvar septum)

Mitral valve anterior leaflet

Chordae tendineae

6.40 The aortic orifice opened from the front to show the cusps of the aortic valve, their nodules, lunules, commissures and the triple-scalloped line of annular attachment. Also shown is the continuity of the subaortic curtain with the mitral anterior leaflet (i.e. 'aortic baffle') and the coronary orifices.

and *left*. Here, terms based on the approximate positions in maturity are retained: *anterior*, *left posterior* and *right posterior*. But a widespread clinical terminology links cusps and sinuses to the origins of the coronary arteries; thus the anterior is *right coronary*, left posterior is *left coronary* and right posterior *non-coronary*(!)

The *aortic sinuses* (of Valsalva) are more prominent than those in the pulmonary trunk; both are dilatations of the arterial wall above the attached base of each cusp. The upper limit of each sinus reaches considerably beyond the level of the free cuspal border and is a well-defined complete circumferential *supravalvar ridge* viewed from the luminal aspect (**6.**34). Coronary arteries usually open near this ridge from the *upper* part of their sinus of origin, the ostium of the left often being a little lower (p. 727). Walls of sinuses are largely collagenous near the annulus, but thence the amount of collagen diminishes and that of lamellated elastic tissue increases; strands of myocardium may enter this fibro-elastic wall. At sinus mid-level its wall is about half the thickness of the supravalvar aortic wall and less than one-quarter of the supravalvar ridge; at this level the mean luminal diameter of the aortic root is almost double that of the ascending aorta's beginning (Reid 1970).

All such details are functionally significant in the valvar mechanism. During diastole the closed aortic valve supports an aortic column of blood at high but slowly diminishing pressure (**6.**53); each sinus and its cusp form a hemispherical chamber; the three nodules are apposed and the margins and lunular parts of adjacent cusps are tightly apposed on their ventricular aspects. From the aortic aspect the closed valve is tri-radiate, three pairs of closely compressed lunular segments radiating to commissures. As ventricular systolic pressure rises it exceeds aortic pressure and the valve is passively opened. The fibrous wall of sinuses nearest the aortic vestibule (i.e. at the nadirs of annular scallops) is almost inextensible; but in the upper parts of sinuses, at the level of the valvular commissures, the wall is fibro-elastic; under left ventricular ejection pressure, the *radius* here increases about 16% in systole. Hence commissures move apart, making the fully open orifice *triangular*, the free margins of cusps becoming almost straight lines between commissures. But they do *not* flatten against the sinus walls, even at maximal systolic pressure, which is probably an important factor in subsequent closure. During ejection most blood enters the ascending aorta, but some enters the sinuses, forming vortices which help to maintain the triangular 'mid-position' of the cusps during ventricular systole and probably initiate their approximation as systole ends; tight, full closure ensues with the rapid drop of ventricular pressure in diastole; commissures narrow, nodules aggregate and the valve reassumes its tri-radiate form. Experiments indicate that about

4% of ejected blood regurgitates through a valve with normal sinuses, while 23% regurgitates through a valve without them (Bellhouse 1968). Normal aortic sinuses may also promote non-turbulent flow into the coronary arteries.

Similar events must occur in the pulmonary valve, though more leisurely, the pressure profiles being less extreme (**6.**53) and valvar structure less substantial.

CARDIAC SURFACE ANATOMY (6.41,42)

The surface projection data set out below apply to an average adult; they are considerably modified by age, sex, stature and proportions, respiration and position. Surface projections of valves are *not* the best sites for *auscultation* (**6.**42).

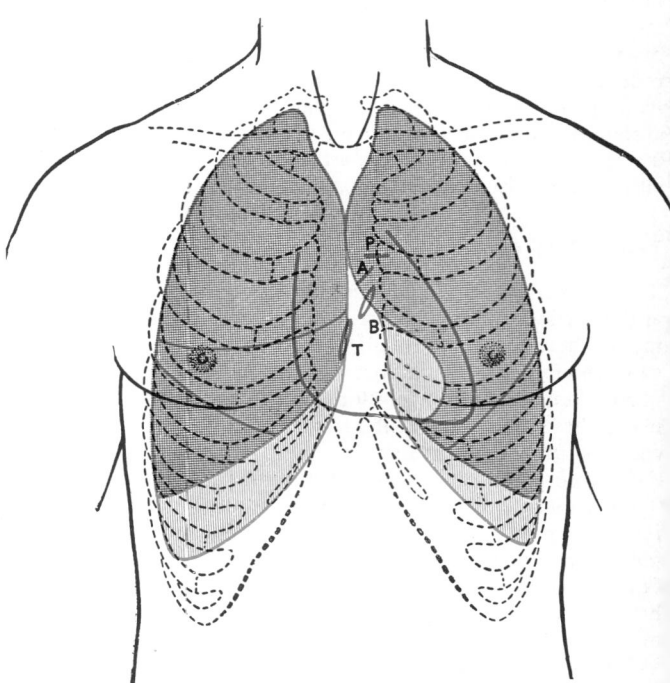

6.41 The front of the thorax, showing the surface relations of the bone lungs (purple), pleurae (blue) and heart (red outline). Compare **6.**42 f further cardiac detail. A = orifice of aorta; B = left atrioventricul (mitral) orifice; P = orifice of pulmonary trunk; T = right atrioventricul (tricuspid) orifice.

ANGIOLOGY 6

The cardiac *apex* almost corresponds to the *apex beat*, which is usually visible and always palpable a little inferomedial to the left male nipple, in the fifth intercostal space, slightly medial to midclavicular line, about 9 cm from midline in average adult males. (The *apex beat* is the most inferolateral point at which a pulsation can be felt, whereas the *true cardiac apex* is a short distance further inferolaterally, does not contact the thoracic wall in systole and its position is seen in profile in appropriate X-rays and scans.)

The cardiac sternocostal surface, projected to the anterior thoracic wall, is an irregular quadrangle. Its *right border* (**6.**41,42) corresponds to a line from the right third costal cartilage's superior border, 1.2 cm from the sternal margin, to the sixth costal cartilage, convex to the right and maximally distant from the midline (about 3–4 cm) in the fourth intercostal space; it represents the right atrium's lateral profile. Upward continuation of this line marks the right border of the superior vena cava and downward that of the inferior vena cava. The *lower border* of the surface projection is a line joining the right border's lower end to the apex beat; it passes over the xiphisternal joint, corresponding largely to the lower (acute) margin of the right ventricle but includes the apical part of the left ventricle. The *left border* is marked by a line from the apex beat to the lower border of the left second costal cartilage 1.2 cm from the sternal margin; it is convex upwards and to the left, corresponding to the obtuse margin of the left ventricle and left auricle above. The surface projection is completed by a sloping line joining the upper ends of the right and left borders, approximating to the upper atrial limits. Left and right borders can be identified by heavy percussion. The surface projection of the anterior part of the *coronary sulcus* is an oblique line joining the sternal ends of the third left and sixth right costal cartilages. This line separates the atrial and ventricular areas; also, although in different planes, the projections of the cardiac valves are sited along this line or are near it. (They are *not* the preferred areas for auscultation.)

The *pulmonary orifice* is partly behind the left third costal cartilage's superior border, partly behind the sternum, represented by a horizontal line, 2.5 cm long; parallel lines from the ends of this, up to the left second costal cartilage, indicate the *pulmonary trunk*.

The *aortic orifice* is below and a little right of the pulmonary, marked by a line 2.5 cm long, from the medial end of the left third intercostal space downward to the right; two parallel lines from the ends of this line, slanting up to the right half of the sternal angle, outline the *ascending aorta*.

The *right atrioventricular orifice* is represented by a line, 4 cm long, commencing medially on the fourth right costal cartilage and passing down and slightly to the right; the centre of this should be midway in the right fourth space. The *left atrioventricular orifice* is behind the left half of the sternum opposite the fourth left costal cartilage, represented by a line, 3 cm long, descending to the right.

The *area of superficial cardiac dullness* is roughly triangular, as mapped out by light percussion, and corresponds to the cardiac area not covered by the lung (**8.**28A).

CARDIAC RADIOLOGICAL APPEARANCES
(**6.**43,44,45)

The heart, being full of blood, casts a shadow, occupying the inferior mediastinum, in sharp contrast to areas occupied by air-filled lungs. Pulsation of this shadow is obvious. In full inspiration the apex is clear of the diaphragm, presenting a blurred outline in radiographs due to movement. The right border's shadow is continuous with those of the venae cavae. Due to the attachment of the pericardium to the diaphragm, the heart elongates during inspiration, shortening during expiration. The cardiac shape also varies with stature and attitude (p. 1335). In lateral radiographs the *retrocardiac space* is a translucent area between the heart and the vertebral column, occupied by the descending aorta and oesophagus (**8.** 105). For detailed study of the cavities and larger vessels angiocardiography is used (Robb & Steinberg 1939, Gardner 1949). A suitable intravenous contrast medium miscible with blood is followed in serial X-ray esposures, usually in anteroposterior or oblique views (**6.**44,45). Computerized radiography and echocardiography are routinely applied to the heart (e.g. **6.**46).

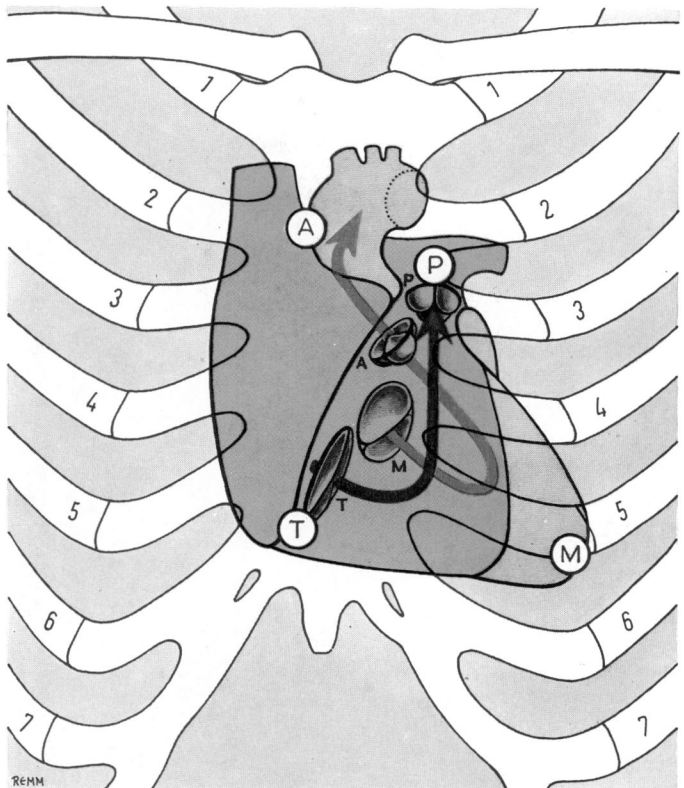

6.42 Diagram illustrating the relation of the sternocostal surface and valves of the heart to the thoracic cage. The right heart is blue, the arrow denoting the inflow and outflow channels of the right ventricle; the left heart is treated similarly in red. The positions, planes and relative sizes of the cardiac valves are shown. The position of the letters, A, P, T and M indicate the aortic, pulmonary, tricuspid and mitral *auscultation* areas of clinical practice.

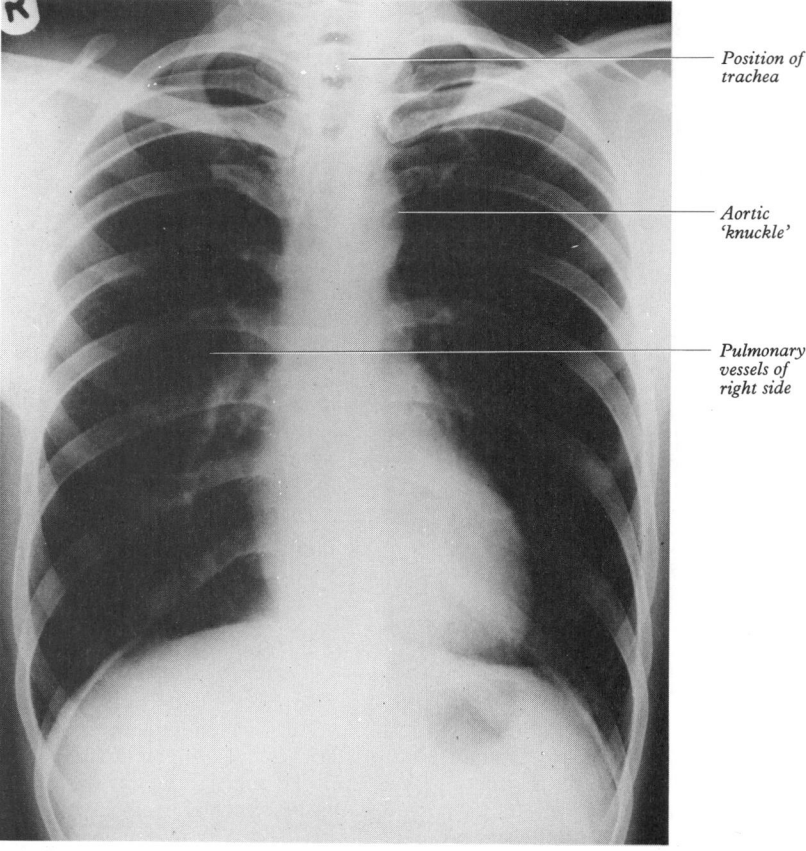

Position of trachea

Aortic 'knuckle'

Pulmonary vessels of right side

6.43 Radiograph of chest, postero-anterior view, of adult male. Note the difference in level of the right and left halves of the diaphragm. Supplied by Shaun Gallagher, Guy's Hospital.

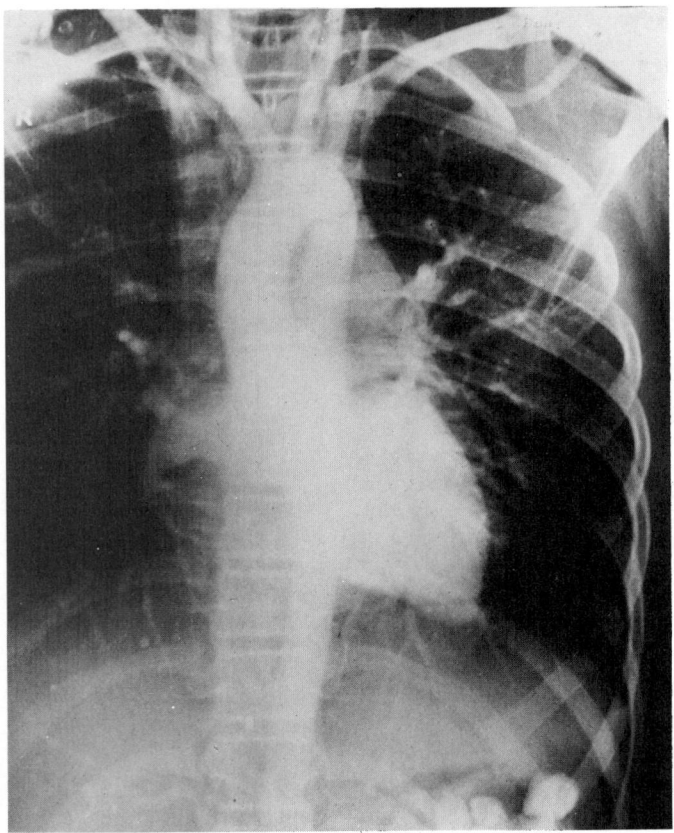

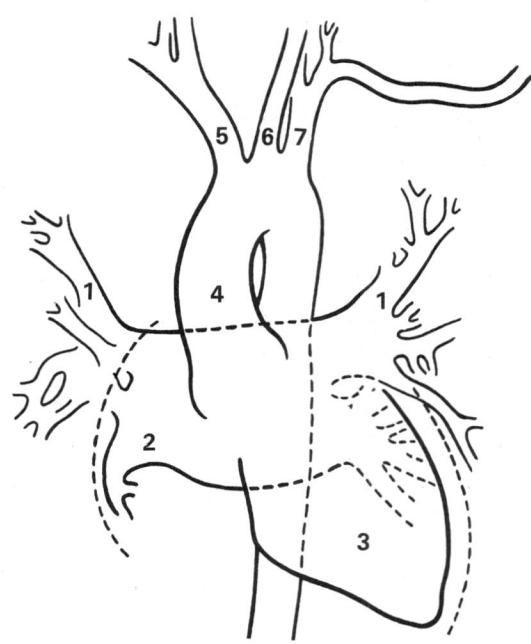

6.44 Angiocardiogram (provided by Frances Gardner), showing the left side of the heart in a child of 11 years: anteroposterior view. 1. Upper pulmonary vein. 2. Left atrium. (Note that owing to the great obliquity of the atrial septum, the left atrium extends to the right behind the right atrium.) 3. Left ventricle. 4. Ascending aorta. 5. Brachiocephalic trunk. 6. Left common carotid artery. 7. Left subclavian artery. (The arms of the patient are raised above the head and as a result the distal end of the artery passes upwards.)

6.45 Angiocardiogram (provided by Frances Gardner), showing the right side of the heart in a child of 12 years; anteroposterior view. 1. Superior vena cava. 2. Right atrium. 3. Right ventricle. 4. Pulmonary trunk. 5. Right pulmonary artery. 6. Left pulmonary artery.

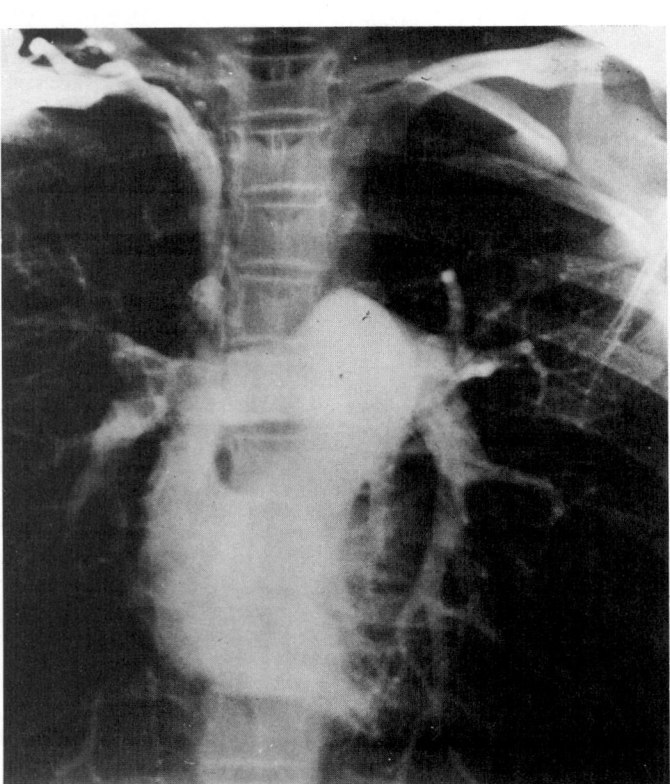

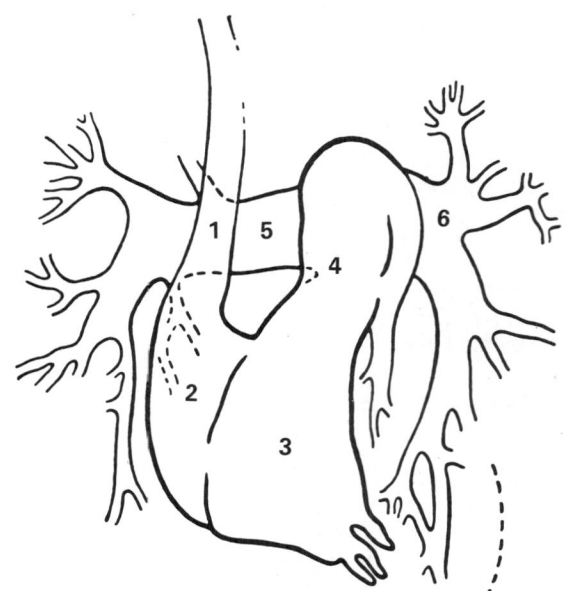

CARDIAC CONNECTIVE TISSUES AND FIBROUS SKELETON (6.47,48)

From epicardium to endocardium and from the orifices of the great veins to the roots of the arterial trunks, intercellular spaces between contractile and conducting elements are everywhere permeated by connective tissue, varying greatly in arrangement and texture in different locations.

Beneath the mesothelium of the serous visceral epicardium over much of the heart is a fine layer of areolar tissue; this accumulates adipose cells, *subepicardial fat*, variable in quantity but increasing

with age, concentrated along the acute margin, coronary sulcus, interventricular grooves and their side channels. The coronary vessels and main branches are embedded in this. Endocardium also lies on a fine areolar tissue rich in elastic fibres. Fibrocellular components of these subepicardial and subendocardial layers blend on their mural aspects with the endomysial and perimysial connective tissue of the myocardium. Each cardiac myocyte is invested by delicate endomysium of fine reticulin, collagen and elastin fibre embedded in ground substance, absent only at desmosomal and gap junctional contacts of intercalated discs (p. 557). Similar arrangements apply to Purkinje myocytes and their extensive

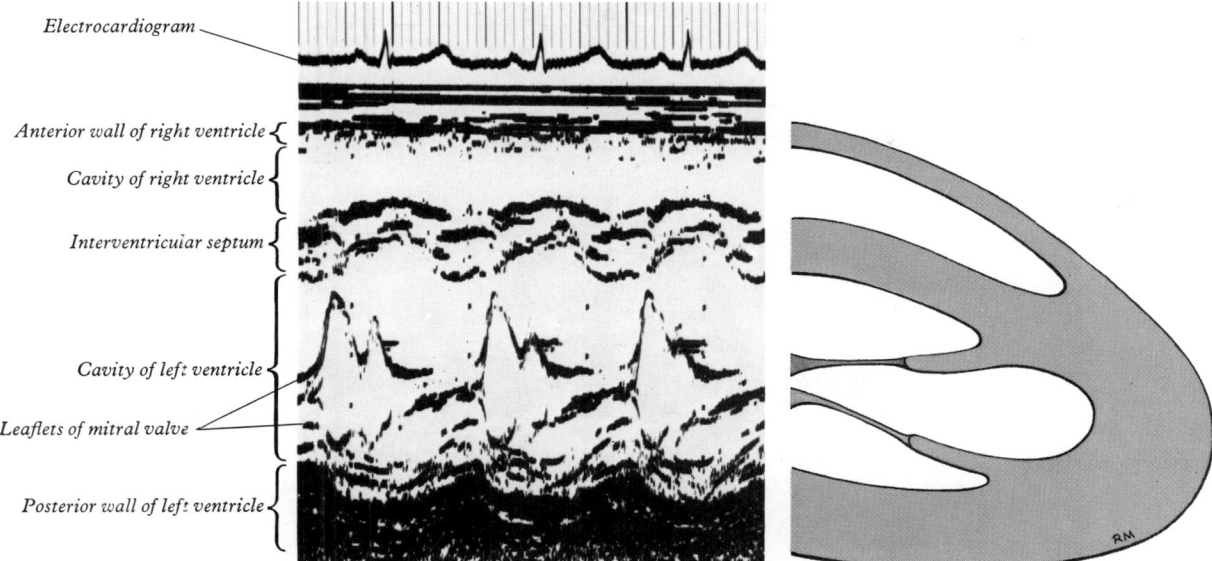

Electrocardiogram

Anterior wall of right ventricle

Cavity of right ventricle

Interventricular septum

Cavity of left ventricle

Leaflets of mitral valve

Posterior wall of left ventricle

6.46 A standard echocardiogram of the heart recorded from the anterior aspect, as increasingly used in clinical practice. The outline diagram shows the corresponding reflecting surfaces. Note the biphasic closure of the mitral valve leaflets.

contacts. It is widely held that cardiac myocytes are structurally and functionally polarized, forming tracts interconnected laterally, and that substantial aggregations of these form bundles, strands or sheets of macroscopic proportions showing complex geometric patterning (p. 720). It has also been claimed that these larger myocardial bundles are 'surrounded' or 'separated' by, and perhaps attached to, stronger perimysial condensations. But more detailed research is needed for a comprehensive view of the three-dimensional array of such tracts and aggregations *and* disposition and properties of associated connective tissues.

Approximately coplanar with the coronary sulcus, i.e. at the 'ventricular base', and intimately related to atrioventricular inflow and arterial outflow orifices, is a complex framework of dense collagen with membranous, tendinous and fibro-areolar extensions, the whole sufficiently distinct to be termed the *cardiac fibrous skeleton*. (For early literature consult Wolff 1781, Henle 1876, Mall 1911, Walmsley 1929, Walmsley & Watson 1978, and detailed analyses by Zimmerman 1959, Zimmerman & Bailey 1962, and Zimmerman 1966.)

As already noted the four cardiac valves depart considerably from a single plane, the fibrous skeleton being a complex, deformable *continuum*. A single section near the ventricular base (**6**.47) provides incomplete data, leading some earlier observers to introduce simple two-dimensional terms such as 'trigones' and 'annuli', which are misleading but in such widespread use that they will be retained here; but wherever possible alternative and additional terms will be noted. Though it is difficult to illustrate this three-dimensional array, Figure **6**.48 may clarify its main features.

Mitral and tricuspid annuli are almost coplanar and inclined to face the cardiac apex. The aortic valve faces up, right and slightly forwards; it is anterosuperior to and right of the mitral orifice. Despite such differences these *three* valvar orifices are *intimately interconnected* through their basal collagenous frameworks. The pulmonary valve is remote from the others, anterosuperior and at right angles to the aortic, its only connection being the long, deformable tendon of the conus arteriosus (infundibulum).

Functions of the fibrous skeleton are: (1) to ensure electrophysiological discontinuity between the atrial and ventricular myocardium except via conducting tissue; (2) as a mechanical attachment for ventricular myocardium and more tenuously for atrial myocardium; (3) to maintain the cardiac position within the pericardium; (4) to establish a stable but deformable base for the attachments of the valvular fibrous cores.

The aortic annulus is central, interconnected with all surrounding fibrous structures; the aortic root always contains a high-pressure column of blood and is exposed to incessant impacts of the full ejection pressure of the left ventricle in each cardiac cycle which, as Mall said, without such strong intermediaries would tear the arterial trunk from the ventricle. The *aortic fibrous annulus* has already been briefly described (p. 711); it is the coaptation of three almost semicircular fibrous scallops (or crenations) like a three-pronged 'coronet'. Illustration **6**.48 shows that each scallop has two lateral 'high points' or zeniths and a central nadir, each with its own specialization. The *distal margin* of each scallop merges with the fibro-elastic wall of an aortic sinus; some myocardial strands may intermix with the fibro-elastic tissue; near the annulus collagen predominates but is distally replaced by increasing quantities of lamellated elastic tissue. The *luminal aspect* of each scallop blends with the lamina fibrosa of a cusp (or valvule). Adjacent *high points* meet and fuse, with some exchange of collagen fasciculi at the three *intercuspid commissures*. The latter do *not* reach the upper limit of the sinuses, which is marked by a circumferential thickening of the aortic wall, the *supravalvar ridge* (**6**.34). Between the commissures the walls of sinuses are quite elastic, allowing the commissures to retreat from each other, with

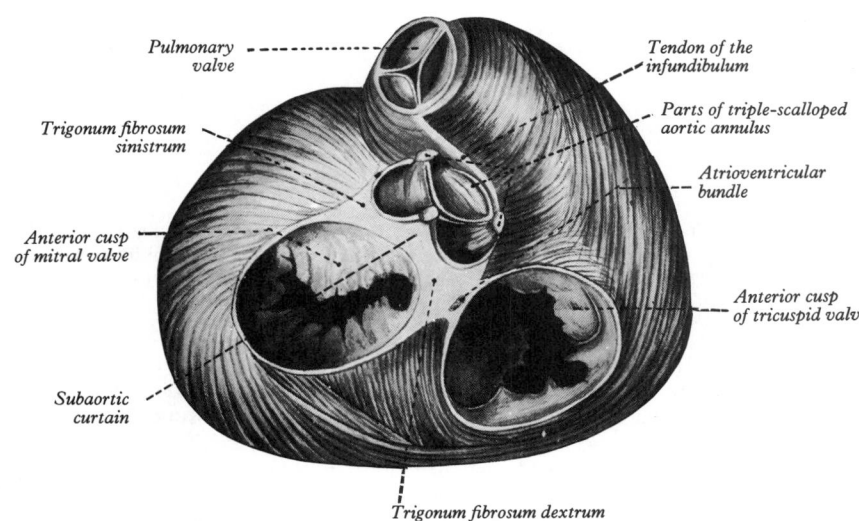

Pulmonary valve

Trigonum fibrosum sinistrum

Anterior cusp of mitral valve

Subaortic curtain

Tendon of the infundibulum

Parts of triple-scalloped aortic annulus

Atrioventricular bundle

Anterior cusp of tricuspid valve

Trigonum fibrosum dextrum

6.47 The base of the ventricles, after removal of the atria and the pericardium. Contrast the planes and positions of aortic and pulmonary valves. (From Walmsley 1929.) Contrast with **6**.48.

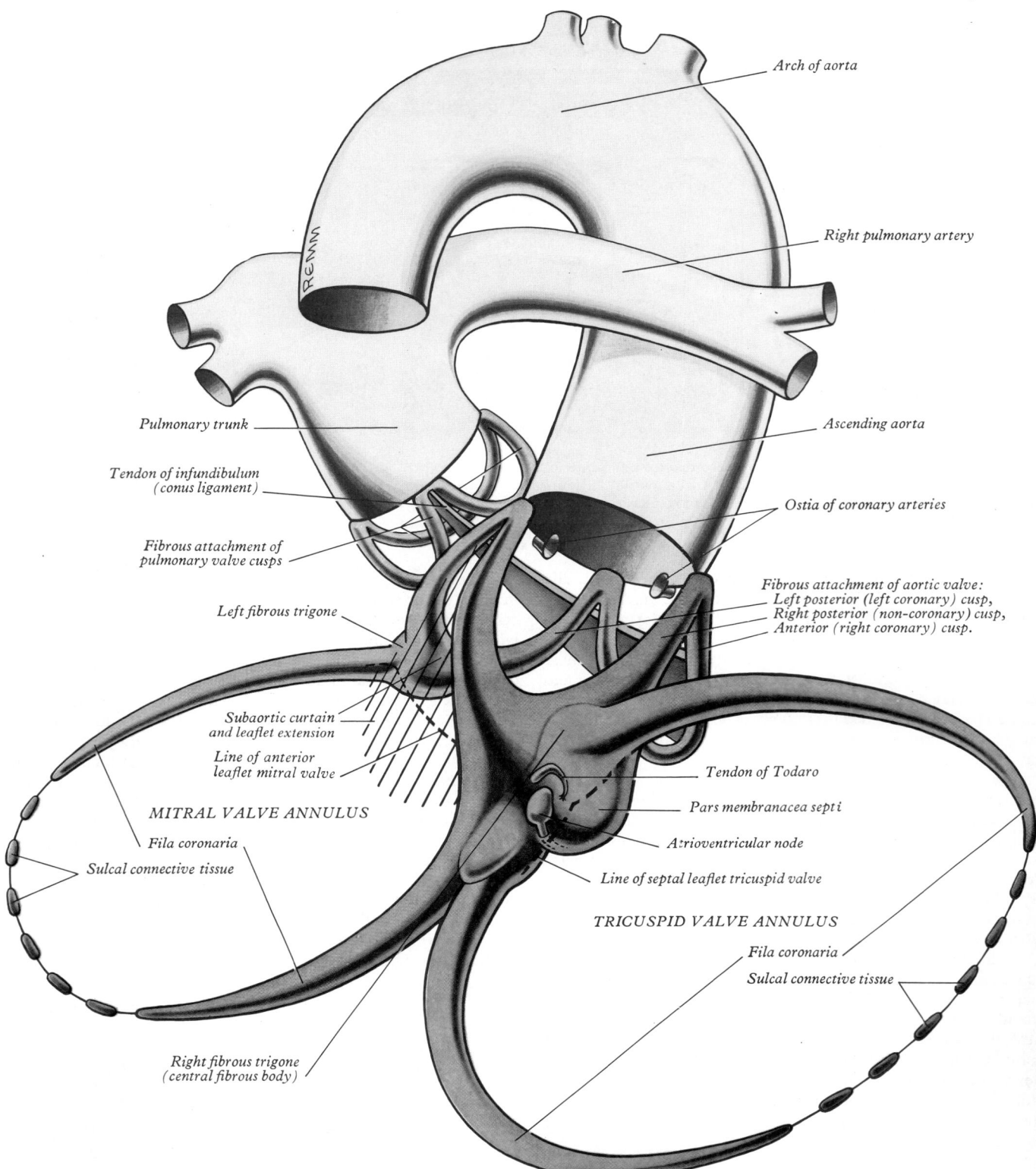

Arch of aorta

Right pulmonary artery

Ascending aorta

Pulmonary trunk

Tendon of infundibulum (conus ligament)

Fibrous attachment of pulmonary valve cusps

Ostia of coronary arteries

Fibrous attachment of aortic valve: Left posterior (left coronary) cusp, Right posterior (non-coronary) cusp, Anterior (right coronary) cusp.

Left fibrous trigone

Subaortic curtain and leaflet extension

Line of anterior leaflet mitral valve

Tendon of Todaro

Pars membranacea septi

MITRAL VALVE ANNULUS

Atrioventricular node

Fila coronaria

Line of septal leaflet tricuspid valve

Sulcal connective tissue

TRICUSPID VALVE ANNULUS

Fila coronaria

Sulcal connective tissue

Right fibrous trigone (central fibrous body)

6.48 Principal elements of the fibrous skeleton of the heart: red = mitral and aortic 'annuli', blue = tricuspid and pulmonary 'annuli', green = tendon of the infundibulum. For clarity the view is from the *right posterosuperior aspect*. Note that due to perspective the pulmonary annulus *appears* smaller than the aortic annulus, whereas in fact the reverse obtains. Based in part on the work of Zimmerman & Bailey (1962) and Zimmerman (1966). Consult text for an extended discussion.

concomitant changes in cuspal shape and disposition in systolic valvar opening, and the converse in diastolic closure. The *nadirs* of the three scallops show collagenous thickenings to very different degrees. Least prominent is that of the anterior (right coronary) cusp, but its margins blend with the aortic end of the infundibular tendon (vide infra). The left posterior (left coronary) cusp's nadir is thickened to an intermediate extent; due its appearance in section in the plane of the atrioventricular valves, it is termed the *left fibrous trigone*. It is an attachment for part of the

anterior mitral leaflet; posterolaterally and left it continues as one of the mitral, tapering, collagenous bundles (*fila coronaria o* Henle) which partially encircle the atrioventricular orifices. The nadir of the right posterior (non-coronary) scallop is the mos massively developed; from its appearance in section earlie anatomists termed it the *right fibrous trigone*. In adult hearts it i a dense irregularly ellipsoidal mass of about 20 × 10 × 5 mm which provides powerful structural and functional links betwee aortic, mitral and tricuspid orifices; in accord with its site an multiple circumvalvar extensions, it is now often termed the heart's *central fibrous body*. It continues into the second mitral filum coronarium, providing a base for both tricuspid fila coronaria. Anteriorly the central body blends with the *pars membranacea septi* (vide infra); posterosuperiorly the *tendon of Todaro*, a palpable subendocardial bundle of collagen about 1 mm in diameter, extends into the right atrial wall, curving across the *torus aorticus* (p. 703) towards the medial end of the inferior caval valve, where its identity is lost. The tendon is one of the boundaries of the *triangle of Koch* (p. 703, **6.34**) and provides an important surgical guide to the atrio-ventricular node.

The *proximal* (ventricular) *aspect* of the triple scalloped fibrous condensation, constituting the aortic annulus, has three almost triangular areas, the *subaortic spans* (Zimmerman 1959, Zimmerman & Bailey 1962). The spans, originally termed *spatia intervalvularia* by Henle (1876), occupy intervals *between* adjacent aortic sinuses; their triangular apices correspond to intercuspid commissures; their walls variously consist of collagen or admixed muscle strands and fibro-elastic tissue; they surround the *subvalvular recesses* of the aortic vestibule. The interval between right (non-coronary) and left (left coronary) posterior sinuses is filled with the deformable *subaortic curtain (intervalvar septum)*, which continues into the base of the central part of the anterior mitral leaflet; down to the latter's base it has a thin covering of left atrial myocardium. Its critical functional role as an *aortic baffle*, between inflow and outflow tracts of the left ventricle, is noted on p. 709. The span between right posterior (non-coronary) and anterior (right coronary) sinuses is continuous with the anterior surface of the right fibrous trigone; it is a variably developed collagenous sheet, the *pars membranacea septi*. This is an oval area of about 1 cm from the aortic annulus to the muscular interventricular septum, about 1.2 cm anteroposteriorly and is normally about 1 mm in thickness. As far posteriorly as the angled attachment of the anteroseptal tricuspid commissure (p. 706), the pars membranacea separates the right and left *ventricles*, whereas posterior to the commissure it separates the left ventricle and right atrium (*atrioventricular septum*). After traversing the right fibrous trigone, the atrioventricular bundle of conducting tissue courses along the postero-inferior margin of the membranous septum to reach the muscular interventricular septum, where it divides into its initial bundles. This cardiac region has a complex developmental history related to the final stages of septation (p. 209) and is a very common site of congenital anomalies (p. 725). Since it lies at the heart's centre, interconnecting the aortic, mitral and tricuspid valves and the right and left ventricles and right atrium, and since it is so intimately related to the atrioventricular bundle and also frequently subject to maldevelopment, the membranous septum is of prime interest in cardiac investigation and surgery.

The third subaortic span, namely that between the anterior (right coronary) and left posterior (left coronary) aortic sinuses, is filled with fibro-elastic tissue blending along its anterior cuspal margin with the *infundibular tendon*; the rest of it is covered by the left ventricular myocardium.

The *pulmonary fibrous 'annulus'*, anterosuperior and almost at right angles to the aortic annulus, is similar in general construction; though of greater diameter, it is considerably less robust. It is triple-scalloped, with commissures at intercuspid apices and barely perceptible thickenings at the nadirs of scallops. The subvalvular spans are filled with fibro-elastic tissue, well-supported by interlacing myocardial extensions from the muscular infundibular wall. The *apex* of the subvalvular span nearest to the aorta becomes confluent with the pulmonary end of the infundibular tendon. Zimmerman & Bailey (1962) suggested that the tendon is a powerful bond between the two arterial trunks and that it 'permits a certain degree of torsional movement between them, while preventing them from being torn asunder by the differently directed ejaculatory forces of the ventricles'.

The *mitral* and *tricuspid annuli* have already been described (pp. 709, 705), but it is re-emphasized here that they are not simple, rigid, collagenous rings but dynamic, deformable lines of valvular attachment that vary greatly at different peripheral points and change considerably with each phase of the cardiac cycle and with increasing age.

MYOCARDIAL ARCHITECTURE

It has long been held that cardiac walls consist of 'fibres' transversely and longitudinally striated (p. 557) and intricately intermingled. They can be classed as (1) atrial, (2) ventricular, and (3) conducting fibres (p. 715). Atrial and ventricular muscle fibres are completely separated by the fibrous annuli, the only connection being via conducting fibres (p. 720). In modern terminology such fibres are presumed to consist of polarized rows, bundles or other aggregations of cardiac myocytes with their connective tissues (p. 557, and below.)

Atrial fibres form two layers: superficial, common to both atria, and deep, proper to each. *Superficial fibres* are most distinct anterior to the atria, crossing their bases transversely as a thin, incomplete layer; some pass into the atrial septum. *Deep fibres* are looped and annular, the former passing over each atrium to its corresponding atrioventricular annulus, in front and behind, while annular fibres surround their auricles and encircle the vena caval openings and fossa ovalis.

Ventricular fibres are complex and few descriptions since those of MacCallum (1897) and Mall (1911, 1912) have been available until recently (vide infra). Superficial and deep layers exist, all but two bands of the latter entering papillary muscles. In the infundibulum superficial fibres are transverse to its long axis and deep fibres are axial; hence, vertical incisions may be preferable. *Superficial ventricular layers* include the following: (1) Fibres from the infundibular tendon (**6.48, 50**) curve across the diaphragmatic surface, sweeping left across the anterior interventricular sulcus to form a vortex at the apex (**6.49**), finally ascending into the left ventricular papillary muscles. From the anterior half of the tendon fibres pass to both anterior and posterior papillary muscles, those from its posterior half only to the anterior muscle. (2) Fibres

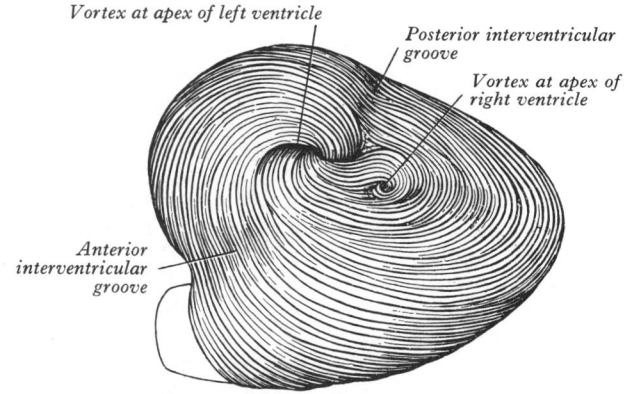

Vortex at apex of left ventricle

Posterior interventricular groove

Vortex at apex of right ventricle

Anterior interventricular groove

6.49 The two vortices in the myocardium at the apex of the heart (after Mall).

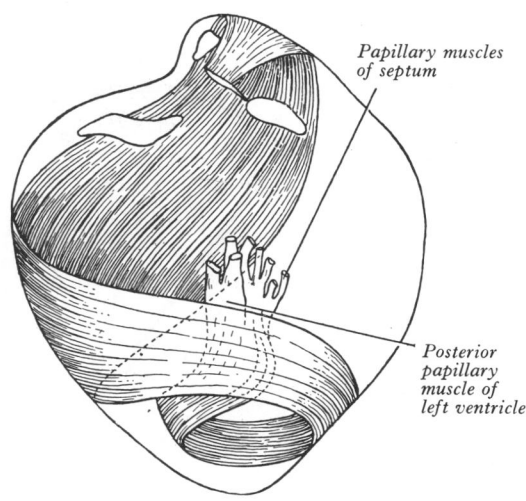

Papillary muscles of septum

Posterior papillary muscle of left ventricle

6.50 The superficial muscular fibres of the ventricles of the heart originating in the tendon of the infundibulum (after MacCallum).

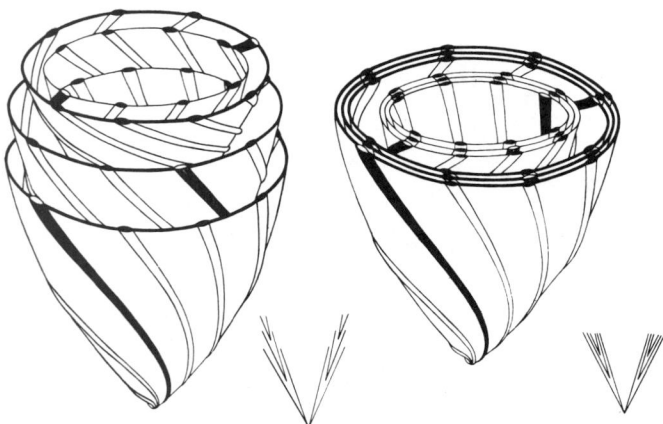

6.51 The arrangement of the deep layers of the muscular fibres of the ventricles, as seen in cross-section (after MacCallum).

from the *right* atrioventricular annulus, which cross the diaphragmatic surface of the right ventricle, reach its sternocostal surface, passing under the fibres just described across the *anterior* interventricular sulcus to wind round the apex, ending in the left ventricular posterior papillary muscle.

(3) Fibres from the *left* atrioventricular annulus cross the *posterior* interventricular sulcus to the right ventricle, ending in its papillary muscles. Three *deep ventricular* layers start in the papillary muscles of one ventricle and curve sinuously into the anterior interventricular groove to end in the papillary muscles of the other (**6.51**), the most superficial in the right ventricle is deepest in the left, and vice versa. The first layer almost encircles the right ventricle, crosses left in the septum, uniting with the superficial fibres from the right annulus to form the posterior papillary muscle; the second layer is less extensive in the right ventricle but greater in the left, joining the superficial fibres from the anterior half of the infundibular tendon to form the septal papillary muscles. The third layer almost encircles the left ventricle, joining the superficial fibres from the posterior half of the tendon to form the anterior papillary muscle. These deep layers are held to synchronize ventricular systole with atrioventricular closure. Two non-papillary bands are also described: one, from the *right* annulus crosses the septum (p. 703), encircles the deep left ventricular layers, to end in the *left* annulus; the second is confined to the left ventricle, passing from the *left* atrioventricular annulus to encircle the aortic orifice (Mall 1911, 1912). Fibres from both left ventricle and atrium also reach the aorta, close to its annulus and cusps. These are said to rotate the aortic base in the general twist of systole.

This account broadly agrees with views of most workers, MacCallum (1900) and Mall (1911) being most quoted. However, concepts of spiralling laminae, overlapping at differing angles in ventricular walls, the existence of apical ventricular vortices through which superficial fascicles spiral inwards—these and other details were all early described by Lower (1669), Borelli (1681), von Haller (1764), Gerdy (1823) and Pettigrew (1864). All these, like their successors, used blunt dissection of hearts which had been boiled (sometimes even roasted!), with varying

concentrations of acetic or nitric acid. It is important to realize that, while such maceration of myocardial tissue is assumed to involve principally a change in endomysial connective tissue, the branching of cardiac myocytes entails that a multitude of minute tears of *muscle* occurs during such blunt dissection. Tears *across* tracts of myocytes may therefore be as likely as *along* them; but from Lower to the most recent exponent, Torrent Guasp (1970), all who have used the technique appear to be confident they could trace predominant orientation of a layer or fascicle throughout its supposed extent. MacCallum's *tour de force* was the 'unrolling' of ventricular muscle of a fetal pig's heart; his short study, accomplished in three weeks, included no human hearts; this aspect was extended by Mall after the former's premature demise. In his extensive investigation Mall (1911) emphasized the helical spread of superficial fibres after plunging deep into ventricular walls. In the porcine heart he described deep fibres which were illustrated like ropes encircling each ventricle; he thought these capable of twisting compression, producing a kind of *wringing* action, as suggested long before by Borelli (1681). Others have used the analogy of ropes in descriptions and particularly Torrent Guasp, whose extensive studies of human, bovine and other vertebrate hearts are somewhat inaccessible, being privately published monographs (1959, 1970). It is at this level of detail that many accounts become difficult to comprehend, as noted by Robb (1934) and Thomas (1957). Robb's brief survey emphasized the difficulty of equating alternative descriptions, except in most general statements. Even in major details there are difficult contradictions. Thomas (1957), who dissected canine and porcine hearts (in one of the few detailed studies of *atrial* musculature), states that all myocardial tracts begin and end by attachment to atrioventricular annuli, except those entering papillary muscles. This is the usual view but denied by Pettigrew (1864), who considered myocardial 'fibres' continuous, a view, with certain differences, also inherent in descriptions by Torrent Guasp (1970) of continuous arrangements of helical bands, which turn like 'figures of eight' in encircling both ventricles, an analogy used by Lower (1669). Pettigrew (1864) proposed a complex seven-layered structure, whereas most are content with a thin superficial layer, an irregular internal one concerned in formation of trabeculae and papillary muscles, and a thick intermediate layer, partly interventricular. Potential artificiality inherent in myocardial unravelling was admitted by Mall (1911), who stated that 'fasciculi and sheets are never fully separated from adjacent fasciculi and sheets'. Torrent Guasp (personal communication, 1978) also admits that the number of dissectable strata is indefinable. However, he has contributed perhaps the most complicated account of myocardial arrangement in mammals, including mankind. Details cannot be included here, but equation of such elaborate architecture with the mechanics of cardiac contraction and conducting pathways is of great interest. Unfortunately, comprehensible harmony between architecture and dynamics of cardiac activity has not been achieved, except in the crudest terms. However, a recent account of myocardial architectonics (Streeter 1979, 1980) confirms the work of Torrent Guasp in major details. Streeter's own approach is a mathematical analysis of the angular orientation between myocardial 'fibres' at different depths in the ventricular wall. He finds his results much in accord with Torrent Guasp's concept of 'nested' spiral laminae (**6.52**), perhaps foreshadowed by Krehl in 1891. The most up-to-date reviews are in a Symposium on the *Structure and Mechanics of the Heart* (1987) edited by Torrent Guasp and containing articles and bibliographies by eight contributors. Although all these efforts have not yet established the *total* arrangement of the myocardium, there is perhaps already enough to satisfy physiological concepts. The accepted existence of 'fibre-pathways' of obliquity, varying from circular to almost longitudinal, provides at least the powers to decrease ventricular cavities in all dimensions; passage of obliquely helical bands into papillary muscles also provides an integration between atrioventricular valves and cyclical changes in longitudinal ventricular dimensions. Some problems require further investigation, such as the 'attachment' of the ventricular myocardium to the misnamed atrioventricular annuli of the cardiac fibrous skeleton (p. 717).

6.52 Diagrammatic representation of one concept of the arrangement of ventricular myocardial 'fibre bundles', from the work of Torrent Guasp (1970).

6.53 Diagram summarizing some of the principal events occurring in the cardiac cycle that are mentioned at various points throughout the text.

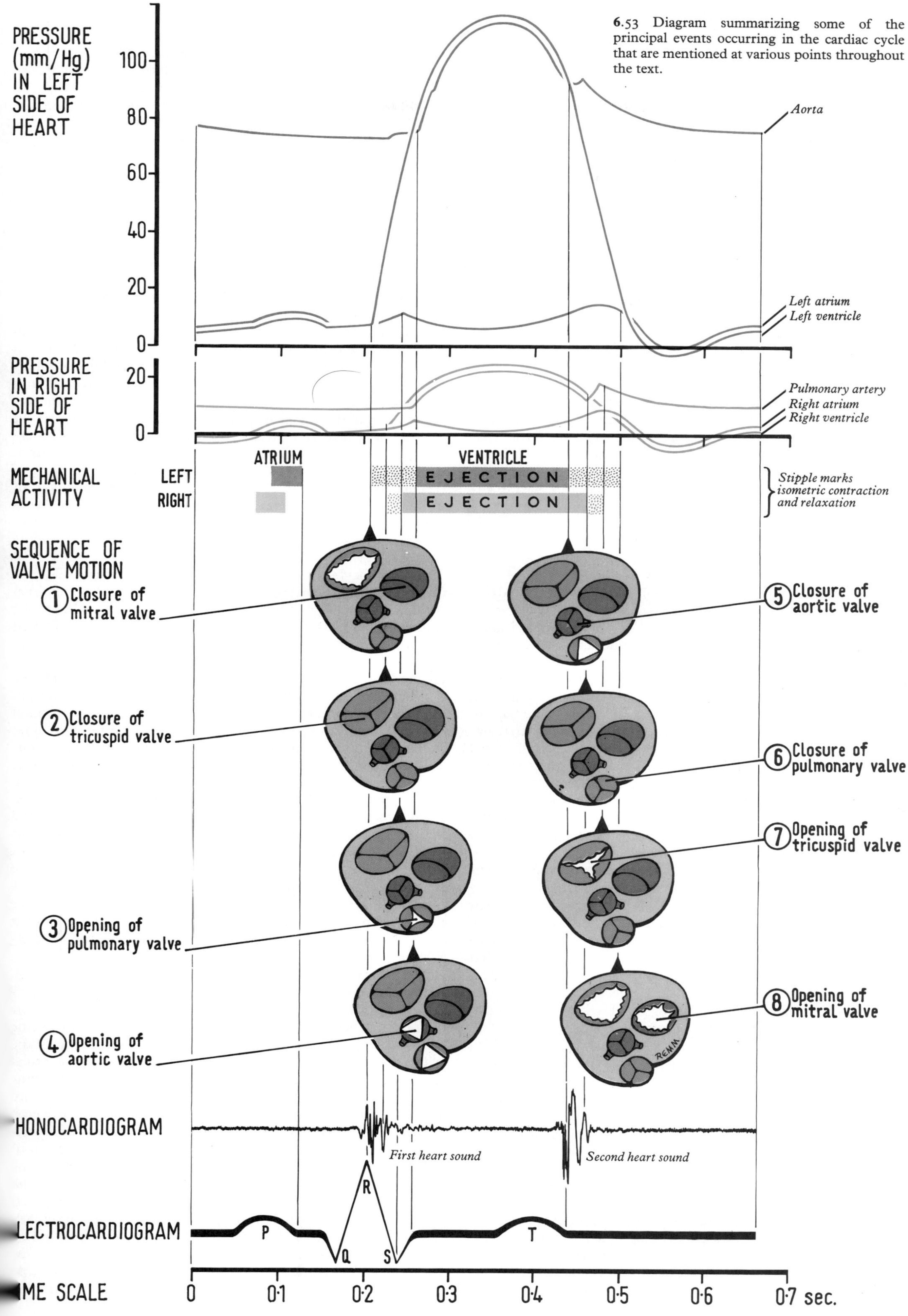

PRESSURE (mm/Hg) IN LEFT SIDE OF HEART

Aorta

Left atrium
Left ventricle

PRESSURE IN RIGHT SIDE OF HEART

Pulmonary artery
Right atrium
Right ventricle

MECHANICAL ACTIVITY

LEFT

RIGHT

ATRIUM

VENTRICLE

EJECTION

EJECTION

Stipple marks isometric contraction and relaxation

SEQUENCE OF VALVE MOTION

①Closure of mitral valve

②Closure of tricuspid valve

③Opening of pulmonary valve

④Opening of aortic valve

⑤Closure of aortic valve

⑥Closure of pulmonary valve

⑦Opening of tricuspid valve

⑧Opening of mitral valve

HONOCARDIOGRAM

First heart sound

Second heart sound

LECTROCARDIOGRAM

P   Q   R   S   T

IME SCALE

0   0·1   0·2   0·3   0·4   0·5   0·6   0·7 sec.

719

COORDINATION OF CARDIAC ACTIVITIES: THE CONDUCTING SYSTEM

The human heart beats ceaselessly at about 60–100 or more cycles every minute for many decades, maintaining perfusion of pulmonary and systemic tissues, the most critical being the cerebral tissues where even transient failure may lead to irreversible changes. The actual rate and 'stroke volume', of course, fluctuate in accord with prevailing physiological demands.

The contrasting roles of 'right' and 'left' hearts, the topography of cardiac chambers, construction and action of valves, the main elements of the fibrous skeleton and of general myocardial architecture have all been examined. The principal events in a cardiac cycle, including the electrocardiogram, mechanical sequences of diastole, atrial systole, isovolumetric contraction, ejection and isovolumetric relaxation in ventricular systole, the main elements of phonocardiogram, pressure profiles of right and left hearts and arterial trunks and the sequences of valvar events are summarized in Figure **6**.53; their study belongs elsewhere but a brief survey of structural elements involved in the cardiac cycle, with some indication of controversial features, must be added. Cardiac *efficiency* basically depends on precise timing of the operation in interdependent structures. Passive diastolic filling of the atria and ventricles is followed by atrial systole to complete ventricular filling. Excitation and contraction of the atria must be synchronous and complete before ventricular contraction; this is effected by an *atrioventricular conduction delay*. Further, ventricular contraction proceeds in a precise manner; structures exist to ensure that closure of atrioventricular valves is followed rapidly by waves of excitation and contraction spreading from the ventricular apices towards the outflow tracts and orifices, rapidly accelerating the blood during ejection.

Vertebrate cardiac contraction is accepted as *myogenic*, originating in the myocytes, the neural influences merely adapting intrinsic rhythm to functional demands. *All* cardiac myocytes (p. 557) are excitable, with autonomous rhythmic depolarization and repolarization of cell membranes, conduction of waves of depolarization via gap junctions to adjacent myocytes and excitation-contraction coupling to their actomyosin complexes (p. 561). But these properties are developed to different degrees in different sites and in different types of myocyte. The *rate* of depolarization and repolarization is slowest in the ventricular myocardium, intermediate in the atrial, fastest in the sinuatrial nodal myocytes derived from the sinus venosus. The latter override those of slower rhythm and, in the normal heart, are the locus for the rhythmic initiation of cardiac cycles. Conversely, conduction *velocity* is slow in nodal and transitional myocytes, intermediate in general 'working' cardiac myocytes, and fastest in Purkinje myocytes.

The nodes, tracts and networks of these cell types (nodal, transitional and Purkinje myocytes) constitute the cardiac *conducting system* or *specialized conducting tissues*. While agreement on their main features is general, their cellular makeup and the existence of other elements remain debatable. Accepted parts are the sinuatrial node, atrioventricular node, common atrioventricular bundle, left and right bundle branches or *crura* and the subendocardial plexus of Purkinje 'fibres'. Advocated by some, rejected by others, are the interatrial and internodal tracts of conducting tissue, atrioventricular 'bypass' fibres and accessory atrioventricular bundles. These constituents are illustrated in Figures **6**.54,55. For reviews of the cardiac cycle consult Rushmer (1976) and Schlant (1978); comprehensive, sometimes controversial accounts of the cardiac conducting system are found in Hudson (1965), Anderson (1974, 1975, 1980), James (1978), James & Scherf (1978).

The sinuatrial node of Keith & Flack (1907), the cardiac 'pacemaker', is generally considered to initiate (in some of its cells) excitation of each cardiac cycle. It is at the junction between parts of the right atrium derived from embryonic sinus venosus and primitive atrium. Central histological features of the node are distinctive but at its periphery are less obvious, doubtless accounting for disagreements on its dimensions, shape and limits. Contrary to earlier views nodal tissue does not fill the full thickness of the right atrial wall from epicardium to endocardium but in most of its extent is about 1 mm from the epicardium and rather more from the

endocardium. It is often covered by a plaque of subepicardial fat. It is elongated, extending on the right from the groove between the right atrial appendage and anterolateral aspect of the terminal superior vena cava, continuing postero-inferiorly into the upper part of the crista terminalis. Variously described as a flattened ellipsoid 'Indian war club', with a 'head, body and tail', it is generally considered to be about 10–20 mm in length and about 1 mm thick and 3 mm wide at its maximal lateral convexity.

In referring to structure of conducting tissues, the classification of cardiac cells proposed by James and co-workers (James 1978) will be followed (p. 557). This was based on abundant macroscopic and microscopic observations, but not all authorities recognize his full range of cardiac myocytes as present in all sites.

A main feature of the sinuatrial node is its central *artery of the sinuatrial node*; it has a surprisingly large calibre and is variable in origin (p. 730). Its tunica adventitia merges into a dense collagenous reticulum, firm to palpation, which permeates the node and surrounds its myocytes. Its small lateral branches supplying nodal tissue are few, the vessel continuing beyond the node to ramify in the atrial myocardium. Its enlarged adventitia serves a sensory function, responding to aortic pressure and pulsation, as part of a feedback stabilizing the appropriate sinus rhythm; pharmacological studies support this view. The nodal myocytes have long been described as 'primitive' cardiac cells, slender and fusiform or branched, typically pale-staining; most earlier observers considered them to make contacts *directly* with the atrial myocytes at nodal margins. James distinguishes the following in the nodal complex and its vicinity: (1) round, polygonal, fusiform or stellate *nodal myocytes* (pale-staining or P cells), (2) a heterogeneous and more complex group of short, slender *transitional myocytes*, (3) *Purkinje myocytes* and (4) atrial contractile *cardiac myocytes*.

Nodal myocytes are confined to the nodal centre, circumferentially arranged around the nodal artery and more irregularly external to this. These cells are now considered 'pacemaker'; they make functional contacts with each other and adjacent transitional myocytes, which are much smaller than the general myocardial cells, their internal organization varying from a simple almost organelle-free type (as in the P cell) to one with a well-developed contractile system and complex sarcoplasmic reticulum; they are slow-conducting and interposed between P cells and either general myocardial or Purkinje myocytes. The latter, shorter and wider than general myocardial myocytes (p. 557), may have a rapid conduction velocity; in human and small mammalian hearts in general they are not markedly wider than general myocytes (contrasting with prominent Purkinje myocytes in larger ungulates). Therefore, some consider Purkinje myocytes absent from the atrial walls, reserving the term for more distal parts of the ventricular conducting system. James, however, considers Purkinje myocytes numerous around all nodal margins, extending also into the interatrial and internodal tracts (vide infra); according to him excitation of atrial myocardium follows this sequence: pacemaker activity of nodal myocytes, slow conduction along transitional myocytes, fast conduction along Purkinje myocytes, spreading excitation-contraction between general atrial myocytes.

The atrioventricular node (Tawara 1906) is oval or elliptical, its long (anteroposterior) axis about 7–8 mm, vertical 3mm, transverse 1 mm. It is under the right atrial septal endocardium in the triangle of Koch (p. 703), dorsal to the basal attachment of the tricuspid septal leaflet, 1 cm or less above the septal margin of the orifice of the coronary sinus and encircled by the subendocardial tendon of Todaro. Its atrial aspect is convex, its left concave, abutting on the base of the mitral annulus. Its dorsal end projects into the septum and narrows into the *nodal crest*; its antero-inferior end close to the central fibrous body, becomes the common atrioventricular bundle. The node is pervaded by an irregular collagenous reticulum enmeshing myocytes, but this is less dense than in the sinuatrial node. Its arterial supply is noted on p. 730. *Nodal myocytes* are relatively few and confined to the deeper parts, particularly near the central fibrous body; the main population is of *transitional myocytes*, irregularly arranged in the dorsal part but increasingly arrayed in longitudinal tracts extending to the atrioventricular bundle. All nodal surfaces are said to be covered by Purkinje myocytes, which (1) receive fine terminals of internodal

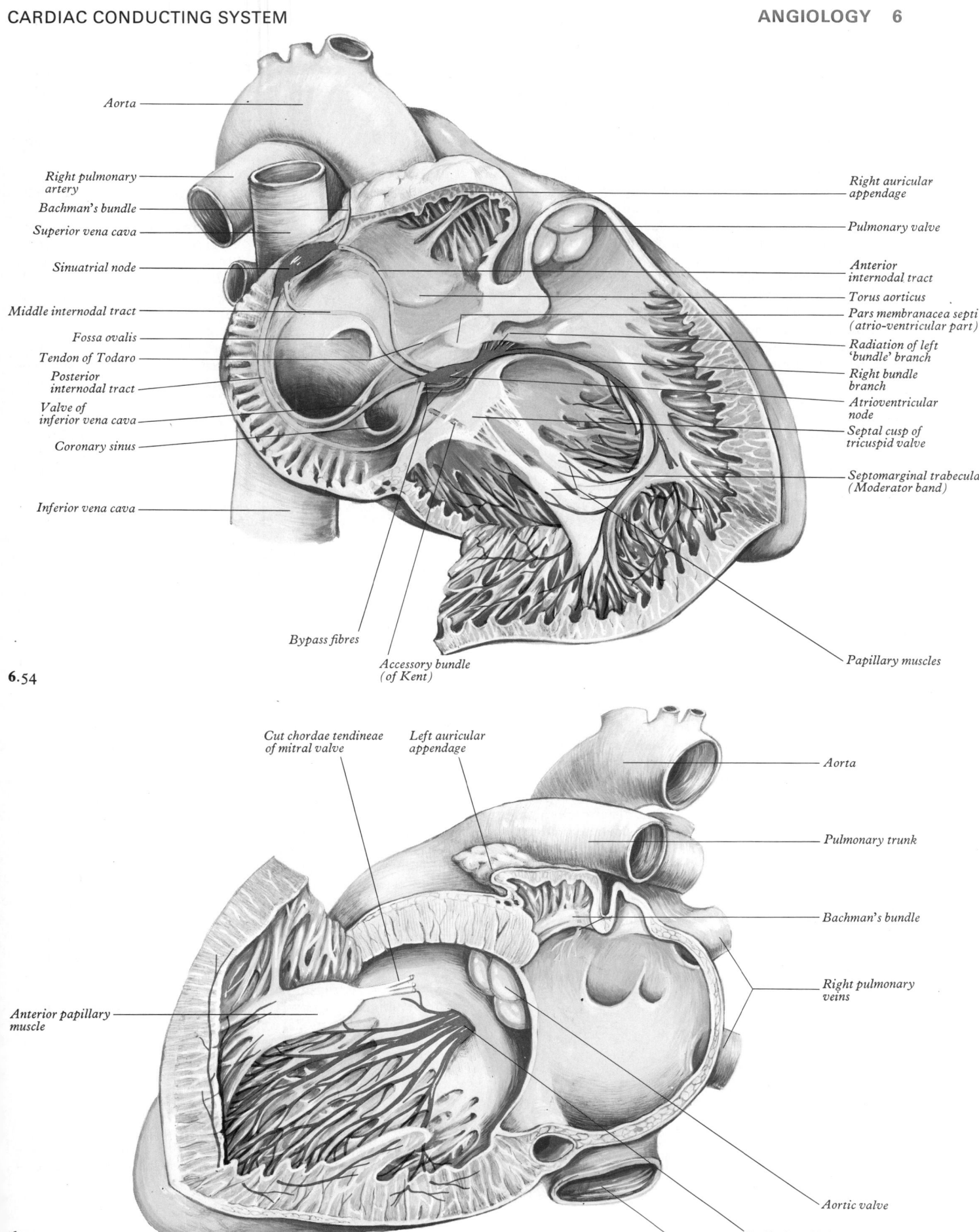

Aorta

Right pulmonary artery

Bachman's bundle

Superior vena cava

Sinuatrial node

Middle internodal tract

Fossa ovalis

Tendon of Todaro

Posterior internodal tract

Valve of inferior vena cava

Coronary sinus

Inferior vena cava

Bypass fibres

Accessory bundle (of Kent)

Right auricular appendage

Pulmonary valve

Anterior internodal tract

Torus aorticus

Pars membranacea septi (atrio-ventricular part)

Radiation of left 'bundle' branch

Right bundle branch

Atrioventricular node

Septal cusp of tricuspid valve

Septomarginal trabecula (Moderator band)

Papillary muscles

6.54

Cut chordae tendineae of mitral valve

Left auricular appendage

Aorta

Pulmonary trunk

Bachman's bundle

Right pulmonary veins

Anterior papillary muscle

Aortic valve

6.55

Inferior vena cava

Radiation of left 'bundle' branch

6.54, 55 Diagram of the conducting tissues of the heart as seen from the right (**6.**54) and left (**6.**55) aspects. Generally accepted elements are shown in red; accessory atrioventricular paths and 'bypass' bundles in blue. The internodal and interatrial pathways shown in yellow are subject to some controversy. Consult text for further discussion. Note the conducting tissue accompanying fine trabeculae carneae and false chordeae; cf. ventricular bands p. 710 and **6.**37B–D.

conduction paths and 'bypass' fibres (vide infra) and (2) continue as one cellular component of the atrioventricular bundle.

The large population of slow-conducting transitional myocytes in the atrioventricular node is probably responsible for the *atrioventricular conduction delay*.

The atrioventricular bundle (His 1893) is formed by tracts of myocytes converging at the antero-inferior pole of the atrioventricular node. Narrowing rapidly, ensheathed in delicate vascular connective tissue, it becomes transversely oval, quadrangular or triangular as it enters a channel in the central fibrous body (right fibrous trigone, p. 717). Traversing this, it reaches the dorsal edge of the pars membranacea septi, coursing along its postero-inferior margin to the muscular interventricular septal crest. It is commonly described as dividing into *right* and *left branches (crura)*, implying simple bifurcation: but this is misleading. The *right branch (crus dextrum)* continues as a narrow, discrete round group of fascicles at first in myocardium and then subendocardially towards the right ventricular apex, entering the septomarginal trabecula to reach the anterior papillary muscle. In its septal course it has few branches to the ventricular walls, but in the papillary muscle it divides profusely into fine subendocardial fascicles, which diverge and first embrace the right ventricular papillary muscles, then return towards the ventricular base to reach all parts of the ventricular parieties. Fascicles of conducting tissue are also carried by trabeculae carneae and *right ventricular bands* (p. 703). The *left branch (crus sinistrum)* is misnamed, because numerous fine fascicles, separately ensheathed, leave the left margin of the parent bundle through much of its course along the membranous septum; these fascicles form a *flattened sheet* arching over the crest of the muscular interventricular septum. The sheet diverges apically and sub-endocardially across the left aspect of the ventricular septum and soon separates into anterior and posterior sheets to the corresponding papillary muscles. Fine branches leave the sheets, forming subendocardial networks, which first encompass the papillary muscles and then curve back subendocardially to all parts of the ventricle. Mention has already been made (p. 710) of structures that have a free course across the ventricular cavity, connecting terminally with papillary muscles and/or points on the ventricular walls. Named *left ventricular bands*, and sometimes confused with false chordae tendineae, they occur in about 50% of human hearts. Many if not all such bands of fibromuscular tissue contain Purkinje myocytes. Their role in the overall pattern of conduction has yet to be determined (see **6.**37B,C,D.)

The atrioventricular bundle, at its start, contains extensions of *transitional myocytes* from the node, but distally and in its branches and terminal parts, cells increase in diameter, assuming the features of polarized tracts of *Purkinje myocytes* (**6.**56), intermingled are tracts of cells with the ultrastructure of general cardiac myocytes, whose conduction properties are unknown.

Principal branches of the bundle are insulated from the

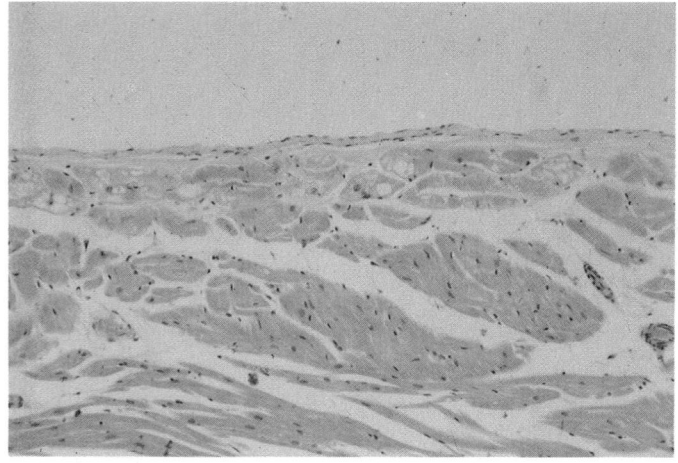

6.56 Section of conducting, Purkinje fibres in the left ventricular wall. Note the paler, enlarged cells among normal myocytes. Haematoxylin and eosin. Magnification × 500.

surrounding myocardium by sheaths of connective tissue, the functional contacts between Purkinje myocytes and ventricular myocardium becoming numerous only in the subendocardial terminal reticula. Hence papillary muscles contract first, followed by a wave of excitation and contraction travelling *from* ventricular *apex* to basal outflow tract and orifice; and, because the Purkinje network is subendocardial, ventricular excitation proceeds from the endocardial to the epicardial aspect. Little is known concerning the deployment of Purkinje myocytes in ventricular walls; even their degree of penetration is uncertain.

Interatrial and internodal conduction paths. It has long been standard teaching in physiological texts that sinuatrial activity *directly* excites general atrial myocytes, with radial spread via the intercellular nexuses to both atria, excitation eventually reaching the atrioventricular node; a further assumption is that the pathways followed merely reflect the complex topography of the atrial walls, in particular the positions of orifices. Indeed, some current texts dismiss any specialized interatrial and internodal conduction pathways (Anderson et al 1974, 1975, Anderson 1975, Rushmer 1976) or consider that such could have no demonstrable physiological role (Marriot & Myerberg 1978).

However, others have postulated such specialized routes (e.g. Wenckeback 1908, Thorel 1909, 1910, Bachmann 1916). Interest re-awakened when it was shown that impulses from the sinuatrial node arrive at the atrioventricular more rapidly than they could have travelled through ordinary myocardium (Hoffman & Cranefield 1960). James (1978), a strong supporter of interatrial and internodal pathways, claims to have shown them by dissection and subsequent microscopy. The following pathways, each an admixture of Purkinje myocytes and ordinary cardiac myocytes, have been described.

The *anterior internodal tract* (including the *interatrial bundle of Bachmann*) leaves the anterior end of the sinu-atrial node, skirts the anterior aspect of the superior caval opening and passes posteromedially to the anterior margin of the interatrial septum. Here it divides, some fibres *penetrating* the septum to diverge in *left* atrial walls (Bachmann's bundle), the others descending across the right side of the septum behind the torus aorticus to divide into terminal fascicles approaching the atrioventricular node (vide infra).

The *middle internodal tract* (of Wenckebach) emerges posterosuperiorly, curving posterior to the superior caval orifice to the interatrial septum and crossing it to splay out near the atrioventricular node.

The *posterior internodal tract* (of Thorel) leaves the postero-inferior part of the node, continues through the crista terminalis and valve of the inferior vena cava and proceeds to the posterior margin of the atrioventricular node; its relation to the ostium of the coronary sinus varies.

Fine terminal branches of these internodal tracts are said to enter the atrioventricular node to make functional contact with transitional myocytes at many levels, including its crest, its convex right atrial surface and even the atrioventricular bundle beyond the node. Such radicles could be 'bypass fibres' circumventing much or all of the node, a possible anatomical basis for a *dual form* of atrioventricular transmission first proposed by Moe et al (1956). It is to be emphasized, however, that mere existence and all details of interatrial and internodal conducting tracts remain controversial.

The existence of **accessory atrioventricular bundles** is often invoked to explain various cardiac arrhythmias. It is a general belief that these may exist near the atrioventricular bundle, or at any point on the circumference of either the mitral or tricuspid 'annuli'. They are generally described as slender, irregular in occurrence and of no functional significance in *normal* cardiac cycles. Their structure is unknown, but current research is directed towards their anatomico-physiological analysis and assessment of applied surgical relevance.

THE CARDIAC NERVE SUPPLY

Initiation of the cardiac cycle may be myogenic or neurogenic in invertebrates; but in vertebrates it is myogenic. In vertebrate hearts, with increasing specialization, initiating foci and conducting tracts have progressively elaborated to integrate the cycle. This *conducting system* has been described (p. 720). The cardiac cycle is harmonized in rate, force and output by an extrinsic nerve supply, operating on cardiac nodal tissue and its prolongations, on coronary vessels and perhaps cardiac muscle. This supply is autonomic, efferent and afferent. Parasympathetic fibres reach the heart through vagal branches (p. 1116), the sympathetic by rami of the sympathetic trunk (p. 1162). Vagal preganglionic fibres proceed from brainstem origins, particularly the medulla, including the nucleus ambiguus (p. 956), the reticular nuclei (p. 989) and possibly the dorsal vagal nucleus (p. 956). Vagal preganglionic axons leave in the cardiac branches to reach the cardiac plexuses without interruption (see, however, p. 1155). Sympathetic preganglionic neurons are in the lateral grey column of the thoracic spinal cord, in its upper five or six segments (Kuntz 1953). Preganglionic fibres from these sources end in cervical and third and fourth thoracic sympathetic ganglia (Mitchell 1953), from all of which bilateral cardiac nerves of postganglionic fibres proceed to the heart (pp. 1116, 1164). (See also general comments and modern reservations concerning the simplification implicit in the terms pre- and postganglionic, p. 1154.)

The central connections of cardiac preganglionic neurons, parasympathetic and sympathetic, are described elsewhere (*Reticular System, Hypothalamus,* and *Cerebral Cortex*). The existence and behaviour of these integrating influences is deducible in terms of function, but precise locations of connecting pathways in the spinal cord, brainstem, and cranial to this, are uncertain, perhaps because they do not form aggregated tracts. Also, many pathways are probably polysynaptic.

Nearing the heart these autonomic nerves form a mixed *cardiac plexus* (p. 1164), usually considered divisible into a ventral (superficial) part, inferior to the aortic arch and ventral to the pulmonary artery, and a dorsal (deep) part between the aortic arch and tracheal bifurcation. These plexuses contain ganglia, which also occur in the heart (**6.57**) along the distribution of branches of the plexus, their neurons considered largely, if not exclusively, postganglionic parasympathetic in nature. With the advent of reliable staining techniques for cholinergic and adrenergic nerve cells, axons and axon terminals, the distribution of cardiac autonomic elements has been subjected to detailed examination, though methylene blue clarified many details prior to this. Even a brief review is impossible here; Kuntz (1953), Mitchell (1956), and Pick (1970) have all summarized the earlier work; more recent observations cited here are merely representative of the large literature.

Cholinergic and adrenergic fibres, arising in or passing through the cardiac plexus, are distributed to nodal tissue (sinuatrial and atrioventricular), to the atrial and ventricular myocardium and to coronary arteries and cardiac veins, in rodents and dogs, according to many observers (Glabella 1976). Whole mounts of the right atrial wall, in pigs, demonstrate a rich supply of cholinergic nerve fibres (Bojsen-Møller & Tranum-Jensen 1971); the same technique shows extensions of the cardiac plexus to nodes and the atrioventricular bundle. These cholinergic fibres presumably originate in small ganglia frequently described in the cardiac plexus and its prolongations. In man a single ganglion in its 'superficial' part has alone been described; but in other mammals numerous small ganglia occur in the atrial walls, especially near both nodes and along the coronary arteries. Adrenergic neurons have not yet been identified in such ganglia but they contain chromaffin cells of the small, intensely-fluorescent type (SIF-cells, see p. 1466), according to Nielsen & Owman (1968). Adrenergic fibres occur in all cardiac chambers; both cholinergic and adrenergic axons appear more abundant in the atria and near nodal tissue than elsewhere; but in cats adrenergic fibres appear more densely distributed in the ventricles. According to Chiba & Yamauchi (1970) cholinergic fibres are more numerous in human atrial myocardium and adrenergic axons in the ventricular wall.

In most reports nerve fibres are described as close to nodal tissue or myocardium; but Thaemart (1969) has reported neuromuscular junctions, with a gap of 78–100 nm, in the atria, ventricles and septa of murine hearts. Such ultrastructural studies have been reviewed by Yamauchi (1973). To summarize, adrenergic and cholinergic terminals occur in nodes, atrial and ventricular myocardium, the atrioventricular bundle and coronary vessels of a wide variety of laboratory mammals. The functions of such supplies are not dif-

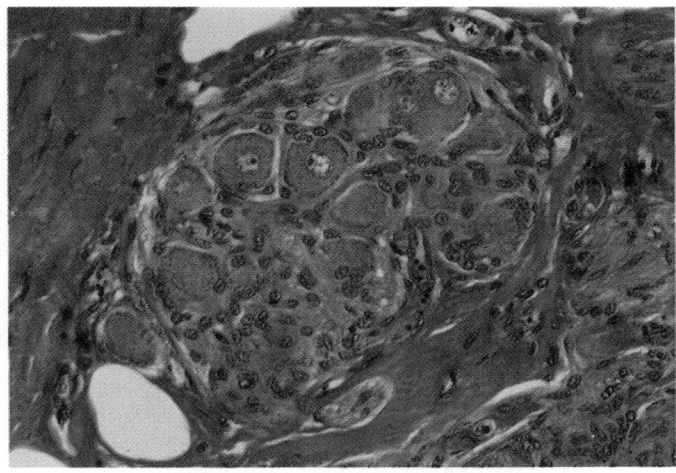

6.57 Postsynaptic parasympathetic (vagal) ganglion in the wall of the left atrium (monkey), showing large ganglion cells surrounded by smaller satellite cells and myocardial tissue. Masson's trichrome stain. Magnification × 600.

ficult to postulate but are not yet confirmed. Confirmation of such innervations in human hearts is uncertain, though it seems probable that similar arrangements exist.

Afferent cardiac nerves have attracted less anatomical interest, though their existence is physiologically certain. Some fibres accompanying coronary arteries have long been considered visceral afferents; afferents from the atria, including sinuatrial and atrioventricular nodes, from the terminations of caval and pulmonary veins and from the root, ascending trunk and arch of the aorta are generally accepted. The precise routes of such visceral afferents are not established (p. 1155) but the cardiac branches of vagi and middle and inferior cervical sympathetic ganglia probably provide major pathways. Most research has concerned the 'terminations' of cardiac visceral afferents. Ultrastructural identification of these largely depends on the mitochondrial accumulations, which are characteristic. By such criteria Kisch (1958) reported subendothelial afferent terminals in the atria and ventricles of mammals, including man; he considered them to mediate the pain phenomena of *angina pectoris*. (*Anginal pain* is almost invariably substernal but in a considerable proportion there is further radiation of pain, most commonly to the *left* side of the neck and medial border of the *left* arm. On occasion *both* arms are involved; exceptionally, the substernal pain is accompanied by radiation exclusively to the *right* arm. This re-emphasizes our poor understanding of the three-dimensional distribution of afferent cardiac innervation.) Similar nerve endings have been demonstrated in the pulmonary valve (Chiba & Yamauchi 1970).

Vessels of the Heart

Arteries of cardiac supply, the aortic coronary branches, are described on pp. 720–731, cardiac veins and the coronary sinus on pp. 792–793 and lymphatic drainage on p. 857.

The Fetal Circulation

Fetal blood reaches the placenta via *two umbilical arteries* and it returns in early fetal life by *two umbilical veins*; later the right one disappears (p. 202). The persisting *left umbilical vein* enters the abdomen at the umbilicus and traverses the falciform ligament's edge to the hepatic visceral surface, where it sends two or three branches into the left lobe, including the quadrate lobe (see discussion on hepatic lobation, p. 1385). At the porta hepatis it then joins the left branch of the portal vein; opposite the junction a large vessel arises and ascends posterior to the liver to join the left hepatic vein near its termination in the inferior vena cava; this is the *ductus venosus*. (For a detailed developmental account, with illustrations, of the circumhepatic veins see p. 221.) The fetal portal vein is small compared with the umbilical and parts of its left branch, proximal

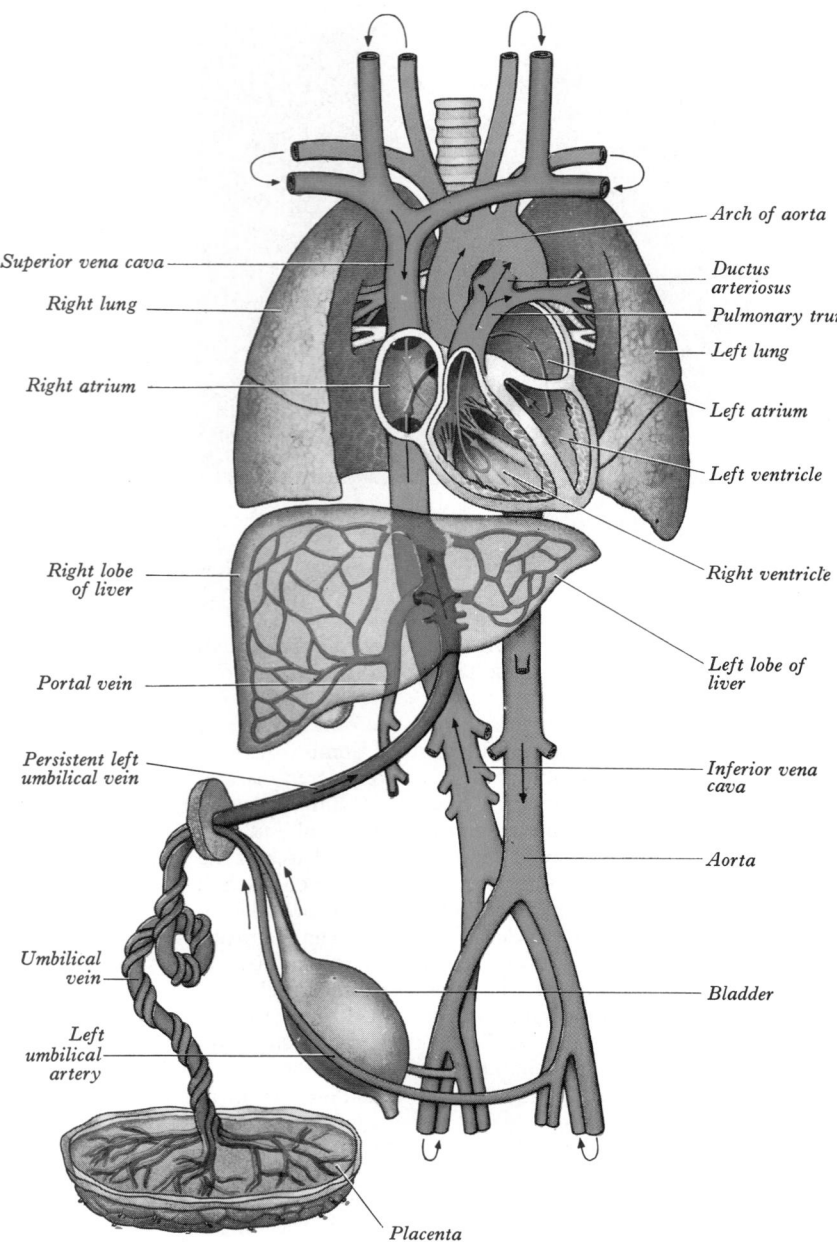

Arch of aorta

Superior vena cava

Right lung

Right atrium

Ductus arteriosus

Pulmonary trunk

Left lung

Left atrium

Left ventricle

Right lobe of liver

Right ventricle

Portal vein

Left lobe of liver

Persistent left umbilical vein

Inferior vena cava

Aorta

Umbilical vein

Bladder

Left umbilical artery

Placenta

6.58 Plan of the fetal circulation. The arrows indicate the direction of blood flow. The placenta is drawn to a greatly reduced scale.

returned via the inferior vena cava (as already mentioned). From the right ventricle this blood enters the pulmonary trunk. Fetal lungs being inactive, only a little traverses the right and left pulmonary arteries and this returns by the pulmonary veins to the left atrium; the greater part of the pulmonary trunk outflow is bypassed via the *ductus arteriosus* to the aorta, where it mixes with the small quantity of blood passed from the left ventricle into this part of the aorta. The mixture descends the aorta and is partly distributed to the lower limbs and viscera of the abdomen and pelvis, but most is returned via the umbilical arteries to the placenta.

This fetal circulation has been confirmed by radiographic observation of radio-opaque substances, and by timing the appearance of tracer material in different parts after injection of isotopically labelled serum albumin into the bloodstream of fetal sheep (Barclay et al 1944, Dawes 1961). The following points, supported by evidence of blood-gas analysis, can be inferred: (1) The placenta serves nutrition and excretion, receiving deoxygenated fetal blood and returning it with oxygen and nutriment. (2) Some blood in the left umbilical vein traverses the liver before entering the inferior vena cava, a fact linked with the relatively large size of the fetal liver, especially in earlier months. (3) Only the pulmonary veins open directly into the left atrium, blood entering it from them being small in volume. Much more enters the right atrium, within which pressure is much higher than in the left atrium; hence the flap-like septum primum (p. 210) is thrust to the left (**2.**83), allowing blood to pass from the right to left atrium. The inferior vena cava's valve is so placed as to direct nearly all blood from the vein to the foramen ovale and left atrium, whereas superior vena caval blood enters the right ventricle directly through the right atrioventricular orifice. (4) Refreshed placental blood, mixed with blood from the portal vein and inferior vena cava, passes almost directly to the aorta for distribution to the head and upper limbs. (5) Blood in the descending aorta, chiefly after circulation through the head and upper limbs, with a small amount from the left ventricle, is distributed to the abdomen, lower limbs but *principally* to the placenta.

Changes in the Vascular System at Birth

At birth, as pulmonary respiration begins, increased amounts of blood from the pulmonary trunk traverse the pulmonary arteries to the lungs and return by the pulmonary veins to the left atrium; consequently its pressure rises. A fall in pressure also occurs in inferior vena cava due to reduction of venous return by occlusion of the umbilical vein. Atrial pressures become equal and the valvular foramen ovale is closed by apposition and later the fusion of septum primum to septum secundum (p. 212). Contraction of the atrial septal muscle, synchronized with that in the superior vena cava, may assist this closure (which occurs after *functional* closure of the ductus arteriosus (vide infra, and Barclay & Franklin 1938). Sometimes fusion is incomplete, a potential atrial communication persisting throughout life; this, unless large, has no functional effect, since the inequality of atrial pressures and the valve-like arrangement of the opening do not favour passage of blood.

When the umbilical cord is ligated, arresting placental circulation, the umbilical vein thromboses (see p. 817), gradually becoming the *ligamentum teres*. Umbilical vessels are muscular but devoid of a nerve supply in their extra-abdominal course. They constrict in response to handling, stretching, cooling and altered oxygen and carbon dioxide tensions. The ductus venosus shuts down by an unknown mechanism; it is absent or already closed in about 30% of newborn infants (Rudolph et al 1961). Its fibrous remnant is the 'ligamentum' venosum. After ligation of the umbilical cord, the umbilical (hypogastric) arteries also thrombose from the origin of their last branches (superior vesical arteries) to the umbilicus, subsequently becoming fibrous cords (medial umbilical ligaments) in the extra-peritoneal fat of the abdominal wall, which raise the peritoneal *medial umbilical folds*.

Obliteration of the ductus arteriosus is essential; it contracts rapidly at first. Blood probably flows intermittently through it for a week or so after birth but flow is reversed, due to a rise in systemic vascular resistance resulting from exclusion of the placental circulation and a fall in pulmonary resistance with the

and distal to their junctions function as branches of the latter, carrying oxygenated blood to the right and left parts of the liver (**6.**58). Hence, blood in the left umbilical vein reaches the inferior vena cava by three routes: some enters the liver directly and reaches the inferior vena cava via the hepatic veins; a considerable quantity circulates through the liver with portal venous blood before also entering the inferior vena cava by the hepatic veins; the remainder is bypassed into the inferior vena cava by the ductus venosus (with its complex developmental history).

In the inferior vena cava, blood from the ductus venosus and hepatic veins mixes with blood from the lower limbs and abdominal wall. It enters the right atrium and, guided by the inferior vena cava's valve, passes mostly through the *foramen ovale* into the left atrium, where it mingles with limited venous return from the pulmonary veins. However, some blood returning via the inferior vena cava, instead of traversing the foramen ovale, joins blood from the superior vena cava passing through the right atrium into the right ventricle. From the left atrium blood enters the left ventricle and thence the aorta, by which it is probably distributed almost entirely to the heart, head (brain and special sense organs) and upper limbs, little reaching the descending aorta. Blood from the head and upper limbs returns via the superior vena cava to the right atrium, all traversing the right atrioventricular orifice, with the small amount expansion of the lungs. When blood flows through a narrowed ductus arteriosus loud cardiac murmurs may be heard on auscultation (Burnard 1959). Closure is due to proliferation of its lining endothelium but takes some months to complete. Initial constriction at birth has been attributed to raised oxygen tension. A nervous factor may also be involved; the muscular wall has afferent and efferent nerve endings and responds to adrenalin and noradrenalin (Barcroft 1941, Franklin 1939, Barclay et al 1942). Before birth the ductus is directly continuous with the pulmonary trunk and of similar calibre. It becomes an impervious *ligamentum arteriosum* connecting the left pulmonary artery (near its origin) with the aortic arch. (For morphological and biomechanical studies of the ligamentum arteriosum consult Dohr et al 1986.) For a general review of perinatal vascular changes see Dawes (1969).

Cardiac Abnormalities

Cardiac malformations are relatively common, amounting to about one-quarter of all developmental abnormalities, with an incidence estimated at 6 per 1000 live births and 2% of stillbirths. Owing to early death, incidence in older children is about a half of this. Some cardiac anomalies can be attributed to genetic or environmental factors; the majority are believed to be multifactorial, genetic factors being effective only in certain environmental conditions (Abbot 1936, Brown 1950, Taussig 1961, Hudson & Wendell-Smith 1966).

Complete absence, *acardia*, is rare; surveys account for 155 cases (Napolitani & Schreiber 1960, Severn & Holyoke 1973), with an incidence of 1 in 34 600 births (Gillian & Hendricks 1957). Afflicted individuals are almost always monochorionic twins, depending for development on the other's heart.

1. Abnormalities of Position

(*a*) The heart may be completely reversed, with an apex directed right instead of left, associated with transposition of great vessels. It may be part of a general transposition of viscera (*situs inversus*) or involve only the heart (*dextrocardia*).

(*b*) In *ectopia cordis* the heart projects on the surface through the lower thoracic wall, associated with breakdown of the thin body wall and anterior part of the pericardium at a very early stage of development.

2. Abnormalities Due to Failure of Normal Development

This group includes most anomalies, the commonest being complete or partial failure of the division of common cardiac chambers. In the complete form there is a common atrium or common ventricle, or both persist (*cor triloculare* and *cor biloculare*). When incomplete, defects involve the interatrial septum, interventricular septum or both, or failure of division of the bulbus cordis and truncus arteriosus.

(*a*) **Atrial septal defects**

(i) *Defects of ostium secundum.* A persistent communication between right and left atria is common, resulting from failure of fusion of the septum primum and secundum. Though rarely of functional importance, the communication may be large if the septum primum cannot occlude the foramen ovale.

(ii) *Persistence of ostium primum.* Normal occlusion of the ostium primum by union of the free edge of the septum primum with the fused atrioventricular cushions fails, often associated with malformation of the tricuspid septal leaflet or mitral posterior leaflet, sometimes with underdevelopment of the septum secundum.

(iii) *Other atrial septal defects.* Rare defects occur dorsal to the fossa ovalis, sometimes associated with drainage of pulmonary veins into the right atrium. They are difficult to explain in terms of development.

(*b*) **Ventricular septal defects**

(i) *Defects of membranous septum.* The commonest defect is in the right wall of the aortic vestibule, below the commissure between the right and anterior aortic cusps. The defect is beneath the tricuspid septal leaflet and below the *supraventricular crest* (p. 703). It results from failed development of the membranous interventricular septum, being usually associated with overriding of the muscular interventricular septum by the aortic orifice, pulmonary stenosis or atresia and hypertrophy of the right ventricle (*Fallot's tetralogy*); rarely the pulmonary trunk is normal or even dilated (*Eisenmenger's complex*). This defect may also be associated with transposition of the great vessels (vide infra), in which the *pulmonary* trunk overrides the muscular interventricular septum (*Tausig–Bing syndrome*). In such septal defects the atrioventricular bundle and its right and left crura are usually along the postero-inferior margin of the defect; occasionally they are on its anterior margin (Campbell 1965).

(ii) *Defects of proximal bulbar septum.* Less commonly part of the interventricular septum formed from the bulbar ridges may be deficient below the commissure between the left and posterior pulmonary cusps and between the anterior and right aortic cusps. It connects the outflow channels of both ventricles.

(iii) *Defects of the muscular septum.* A defect in the muscular part of the interventricular septum is rare (*Maladie de Roget*).

(*c*) **Combined atrial and ventricular septum defects**.

These are usually linked with a persistent common atrioventricular orifice and defects in the adjacent septa. The defect is bounded inferoposteriorly by a free margin of the muscular interventricular septum. Atrioventricular valves are abnormal; frequently a superior and an inferior leaflet common to both sides is draped across the free border of the muscular interventricular septum.

(*d*) **Persistent truncus arteriosus** (Truex & Bishof 1958).

A large undivided channel is above and astride the free margin of the muscular interventricular septum connecting the ventricles. Right and left pulmonary arteries arise independently from it as it continues as the ascending aorta. A persistent truncus usually has a valve with four cusps. The defect is due to a failure of development of the bulbar ridges and the aorticopulmonary septum.

3. Abnormalities Resulting from Defective Progress in some Developmental Process (Truex & Bishof 1958)

(*a*) **Abnormal bulbar and truncal septation**

(i) *Complete transposition*, in which the aorta arises from the right ventricle and the pulmonary trunk from the left. This may be linked with an incomplete interventricular septum or a defective septum of the type described above (*Tausig–Bing syndrome*).

(ii) *Varying degrees of partial transposition of the great arteries*, usually associated with the defective interventricular septum of type described in (2.*b*.i). The pulmonary trunk usually arises from the right ventricle, but aortic position varies from a wholly right ventricular origin to varying degrees of overriding of the ventricular septum.

(*b*) **Anomalous entry of the great veins**

(i) *A persistent left superior vena cava* may drain into the coronary sinus or superior aspect of the left atrium.

(ii) *The inferior vena cava* may be duplicated.

(iii) *The right pulmonary veins*, or less frequently all four, may drain into the right atrium. Any one may drain into some part of the thoracic systemic venous system or, less frequently, into the abdomen.

4. Abnormalities of visceral arches

(*a*) Right aortic arch (p. 214).
(*b*) Patent ductus arteriosus (p. 724).
(*c*) Coarctation of the aorta (p. 766).

Congenital cardiac malformations are often multiple and probably occur more frequently in sibs and children of consanguineous marriages, but there is a low correlation among monozygotic twins. About 20% are ventricular septal defects; persistent ductus arteriosus, atrial septal defects, aortic coarctation, pulmonary stenosis, Fallot's tetralogy and transposition of the great vessels each account for about 10%, aortic stenosis for about 5% of all cases.

THE ARTERIAL SYSTEM

The Pulmonary Trunk

The pulmonary trunk (**6**. 59,60,64) conveys deoxygenated blood from the right ventricle to the lungs (Fishman & Hecht 1969). About 5 cm in length and 3 cm in diameter, it arises from the right ventricle's base (from the complex pulmonary annulus surmounting the conus arteriosus) above and left of the supraventricular crest. It slopes up and back, at first in front of the ascending aorta, then to its left. Below the aortic arch it divides, level with the fifth thoracic vertebra and to the left of the midline, into the right and left pulmonary arteries of almost equal size. Thus the pulmonary trunk bifurcation is *antero-inferior* and *left* of the tracheal bifurcation, with its associated inferior tracheobronchial lymph nodes and deep cardiac plexus of nerves.

and below the tracheal bifurcation (vide supra) and thence in front of the oesophagus and right main bronchus to the right pulmonary hilum where it bifurcates. Its lower, larger branch is distributed to the middle and lower lobes, the upper accompanies the right upper lobar bronchus. (Because the right pulmonary artery commences left of the midline and its relation to the tracheal bifurcation, it is related—at an oblique depth— to the roots of both left and right principal bronchi.)

The left pulmonary artery, shorter and smaller than the right, runs horizontally in front of the descending aorta and the left principal bronchus to the left hilum, dividing into upper and lower lobar branches. It is connected to the concavity of the aortic arch by the ligamentum arteriosum, to the left of which is the left recurrent laryngeal nerve and on the right the superficial cardiac

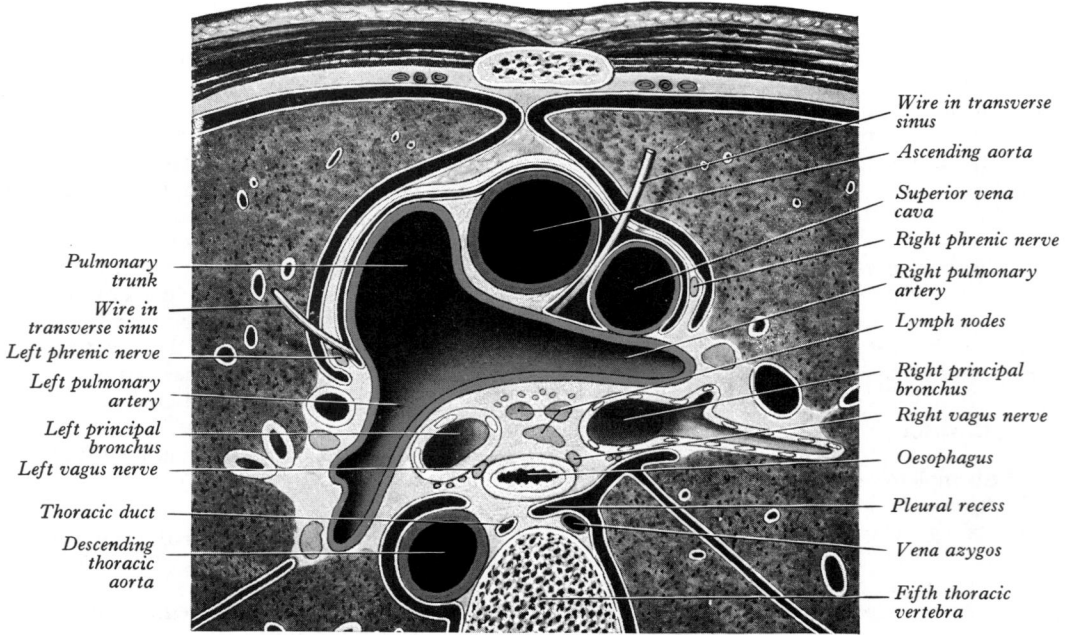

Pulmonary trunk

Wire in transverse sinus

Left phrenic nerve

Left pulmonary artery

Left principal bronchus

Left vagus nerve

Thoracic duct

Descending thoracic aorta

Wire in transverse sinus

Ascending aorta

Superior vena cava

Right phrenic nerve

Right pulmonary artery

Lymph nodes

Right principal bronchus

Right vagus nerve

Oesophagus

Pleural recess

Vena azygos

Fifth thoracic vertebra

6.59 Transverse section through the mediastinum at the level of the upper border of the fifth thoracic vertebra: superior aspect. Note nerve fibres of the deep cardiac and posterior pulmonary plexuses, inferior tracheobronchial and hilar lymph nodes.

Relations. It is entirely within the pericardium, enclosed with the ascending aorta in a common tube of visceral pericardium; fibrous pericardium gradually peters out in the adventitia of the pulmonary arteries. *Anteriorly* it is separated from the sternal end of the left second intercostal space by the pleura, left lung and pericardium. *Posterior* are at first the ascending aorta and left coronary artery, then the left atrium. The ascending aorta is finally on its right. An auricle and coronary artery are on each side of its origin. The superficial cardiac plexus is between the pulmonary bifurcation and the aortic arch; above, behind and right are the tracheal bifurcation, lymph nodes and nerves (vide supra).

The right pulmonary artery, slightly longer and larger, runs horizontally to the right, behind the ascending aorta, superior vena cava and upper right pulmonary vein, then in front

plexus. The pericardial fold of the left vena cava (p. 696) passes from its lower border to the upper left pulmonary vein.

Branching of right and left pulmonary arteries shows little variation (except in dextrocardia, where the pulmonary lobation may also be reversed). Cory & Valentine (1959) found the usual pattern in all 34 right and all 80 left of 114 pulmonary arteries observed at operation (p. 1284). Treatment of congenital pulmonary stenosis, with and without associated 'right to left intracardiac shunt', has advanced rapidly from indirect 'systemic/ pulmonary vessel anastomosis' (Blalock 1947) to a direct blind approach to the narrow pulmonary orifice and right ventricular outflow tract (Brock 1949), and finally via open dry heart techniques employing hypothermia or suitable pump/oxygenator circulatory replacement machines.

THE AORTA

The aorta, the trunk of the arterial tree conveying oxygenated blood to the body, begins at the aortic annulus (pp. 711, 715), part of the base of the left ventricle, where it is about 3 cm in diameter. Passing up and right for about 5 cm, it arches upwards, backwards and to the left over the left pulmonary hilum and then descends in the thorax at first left of the vertebral column, then gradually

inclining towards the midline, to enter the abdomen via the diaphragm's aortic hiatus. Diminished in size to about 1.75 cm, it ends a little left of the midline, level with lower border of the fourth lumbar vertebra, dividing into the right and left common iliac arteries. For convenience it is described as arbitrarily divided into *ascending, arch* and *descending thoracic* and *abdominal* parts.

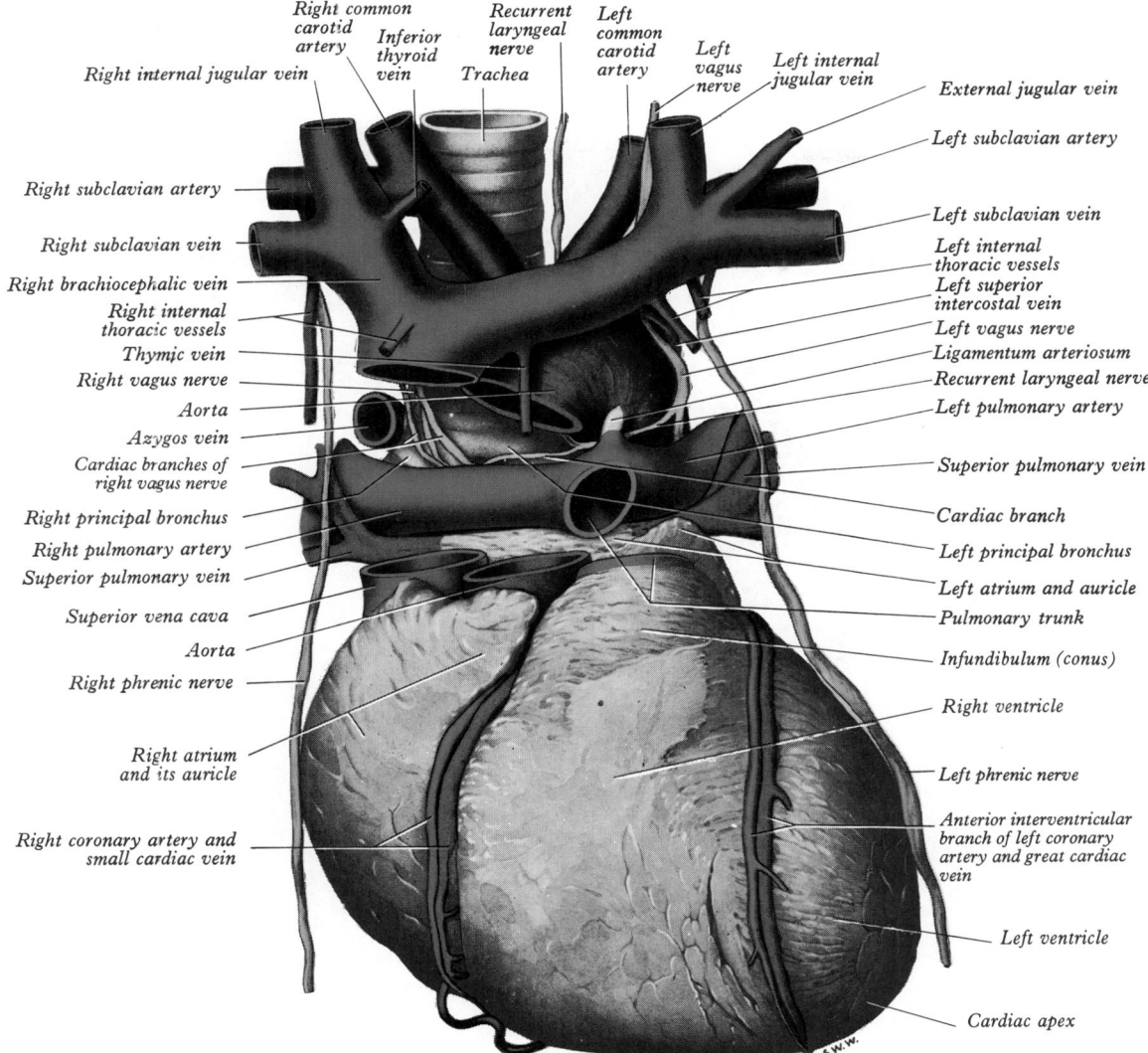

6.60 The relations of the pulmonary arteries and primary bronchi seen from the front. Parts of the ascending aorta, pulmonary trunk and superior vena cava have been removed in the dissection. The right vagal trunk is uncoloured to avoid confusion.

The Ascending Aorta

The ascending aorta (**6**.28,59,60,63), about 5 cm long, begins at the base of the left ventricle, level with the third left costal cartilage's lower border; it ascends obliquely, curving forwards and right, behind the left half of the sternum to the level of the second left costal cartilage's upper border. At its origin, distal to the aortic annulus, are three *aortic sinuses* (p. 711), beyond which the vessel's calibre is slightly increased by a bulging of its right wall; this *aortic bulb* gives the vessel an oval section.

Relations. The ascending aorta is within the fibrous pericardium, enclosed in a tube of serous pericardium with the pulmonary trunk (**6**.26). *Anterior* to its lower part are the infundibulum (p. 703), the commencement of the pulmonary trunk, and right auricle; *superiorly*, it is separated from the sternum by the pericardium, right pleura, the anterior margin of the right lung, loose areolar tissue and thymic remains; *posterior* are the left atrium, right pulmonary artery and principal bronchus; *right lateral* are the superior vena cava and right atrium, the former partly posterior; *left lateral* are the left atrium and, at a higher level, the pulmonary trunk.

At least two structures (reminiscent of the carotid arterial chemoreceptors and baroreceptors, p. 1472) lie between the ascending aorta and the pulmonary trunk. The *inferior aorticopulmonary body* is near the heart and anterior to the aorta; the *middle aorticopulmonary body* is near the right side of the ascending aorta (Boyd 1961).

Branches are the right and left coronary arteries (**6**.61A–E,62), supplying the heart itself.

THE CORONARY ARTERIES (**6**.61A–E,62)

The right and left coronary arteries open from the ascending aorta in its anterior and left posterior sinuses. (Some clinicians call these right and left coronary sinuses, the right posterior being inelegantly termed 'non-coronary'.) Variations are rare but the two may start, separately or in common, from the same sinus; three or even four coronary arteries have been observed; the most common variation concerns a right coronary branch, *arteria coni arteriosi* or 'conus artery', which is usually (64%) its first branch but often arises separately in the anterior sinus (36%), as a third coronary artery. The left coronary opening may be double, leading into major initial branches, usually the circumflex and anterior interventricular; one may lead into a stem common to one such branch and a diagonal ventricular ramus. The levels of coronary orifices are variable; Thebesius (1708) appears to have started a view that aortic cusps obstruct them when fully spread in systole; but the coronary orifices are at a higher level, at or above cuspal margins, though below in about 10% (right coronary) and 15% (left). (Further, as detailed on p. 710 et seq, it is now established that, even at the height of systole, the cusps do not 'flatten' against i.e. co-apt to, the walls of their sinuses.)

The two arteries, as indicated by their name, form an oblique inverted *crown*, with an anastomotic *circle* in the atrioventricular

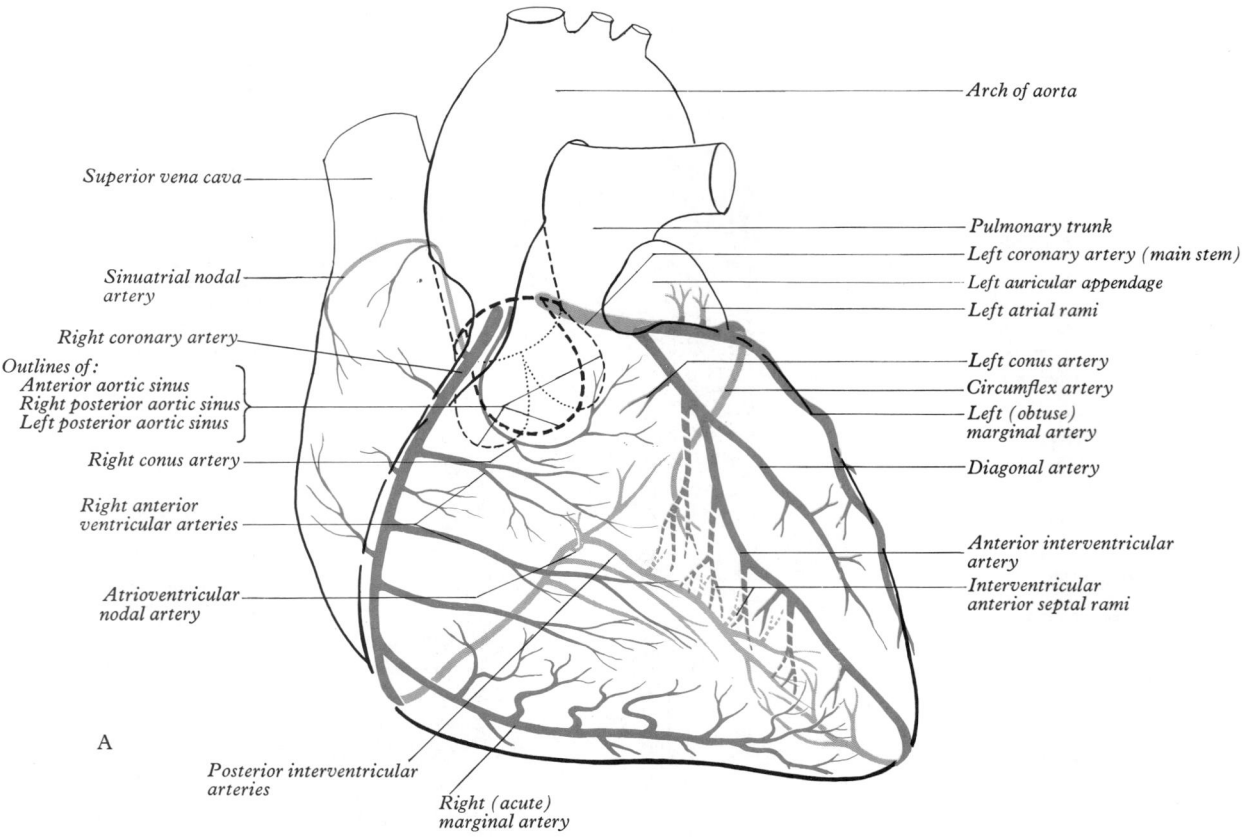

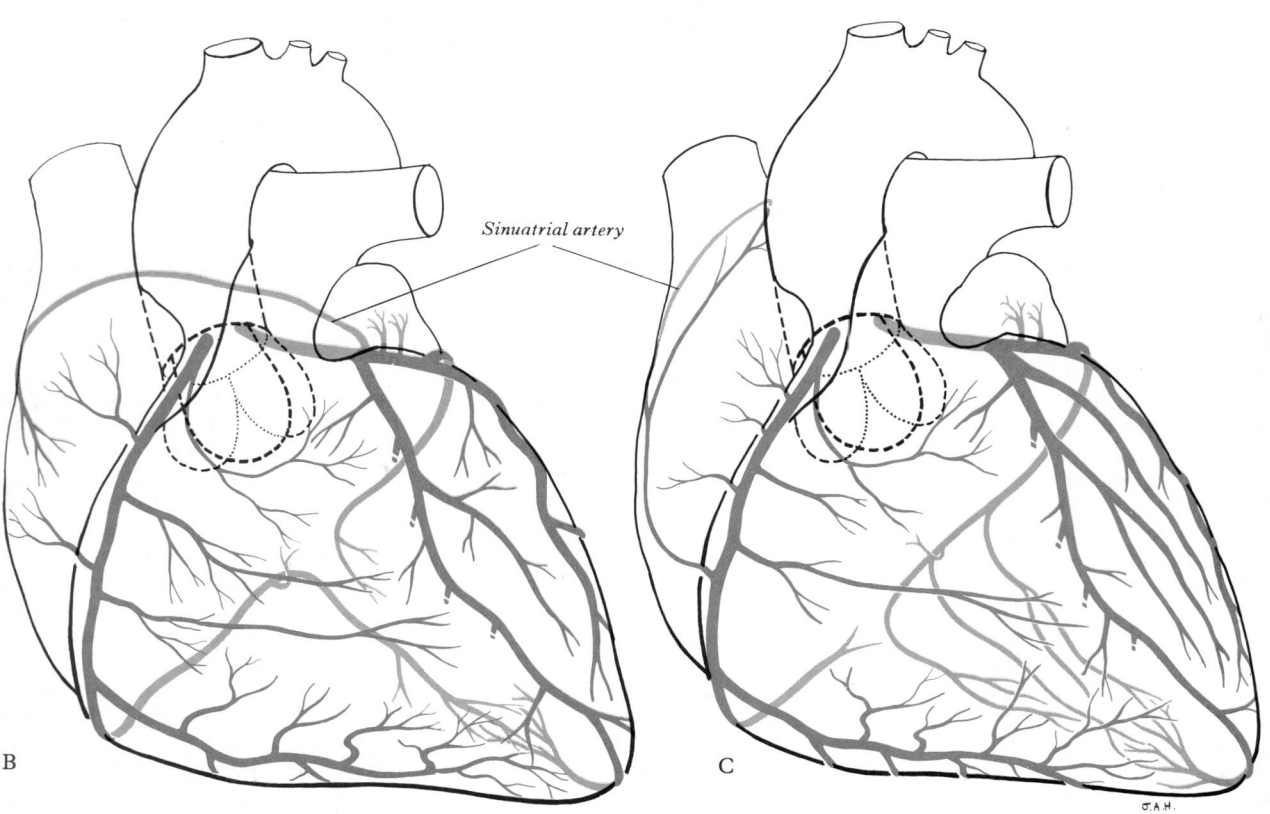

6.61A–C Anterior views of the coronary arterial system, with the principal variations. A. The commonest arrangement. B. A common variation in the origin of the sinuatrial nodal artery. C. An example of *left* 'dominance' by the left coronary artery, showing also an uncommon origin of the sinuatrial artery.

Note that in **6.61**A–E the right coronary arterial tree is shown in magenta, the left in full red. In both cases posterior distribution is shown in a paler shade.

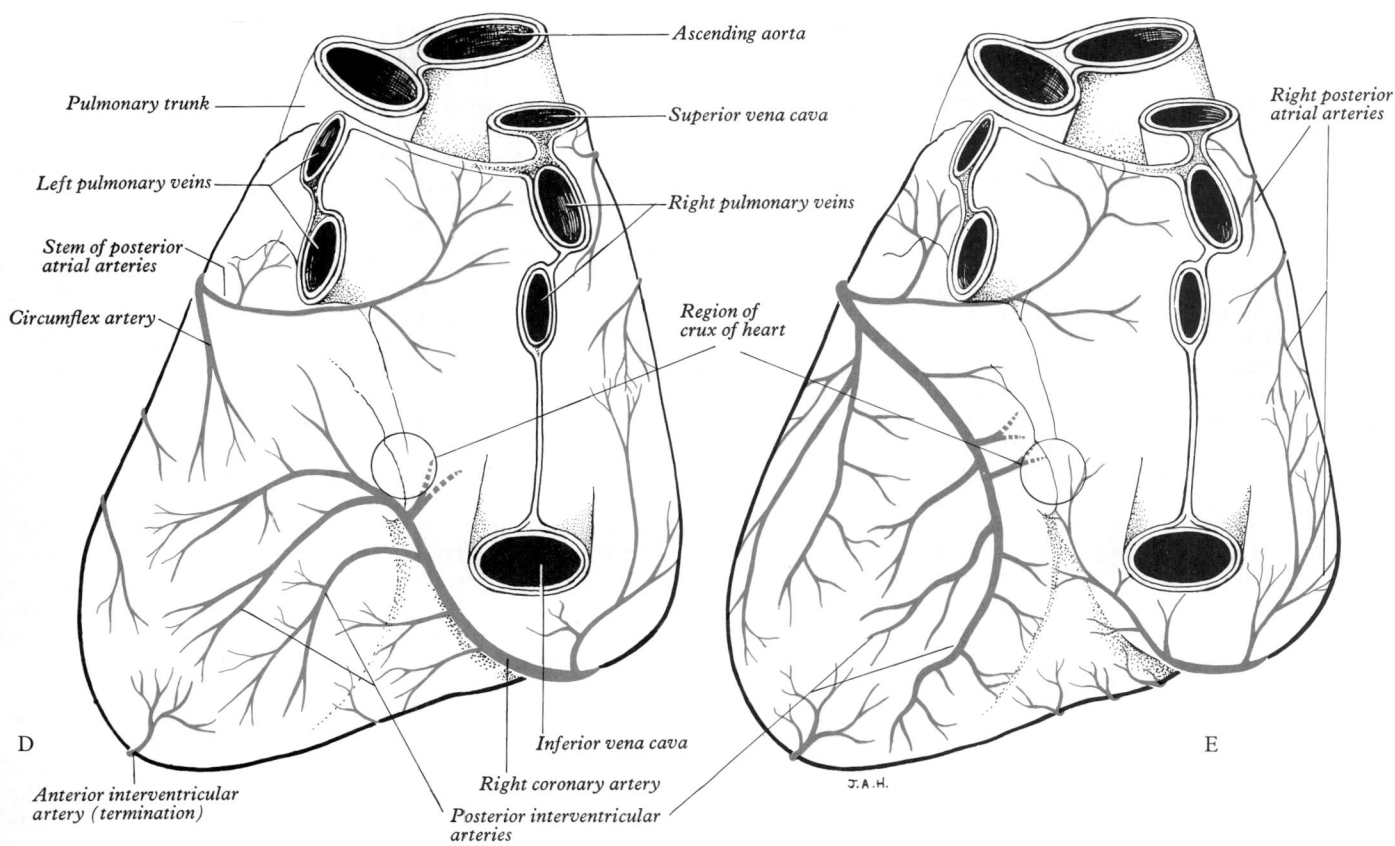

Pulmonary trunk

Ascending aorta

Left pulmonary veins

Superior vena cava

Stem of posterior
atrial arteries

Right pulmonary veins

Circumflex artery

Region of
crux of heart

Right posterior
atrial arteries

D

E

Anterior interventricular
artery (termination)

Inferior vena cava

Right coronary artery

Posterior interventricular
arteries

J.A.H.

6.61D,E Postero-inferior views of the coronary arterial system. D. An example of the more normal distribution in *right* 'dominance'. E. A less common form of *left* 'dominance'.
N.B. In these 'posterior' views the diaphragmatic (inferior) surface of the ventricular part of the heart has been artificially displaced and foreshortening ignored to clarify the details of the so-called posterior (inferior) distribution of the coronary arteries.

sulcus connected by marginal and interventricular *loops* intersecting at the cardiac apex (**6.61**). This is, of course, only an approximation; the degree of anastomosis is most variable and usually insignificant (vide infra). The main arteries and major rami are usually sub-epicardial but those in the atrioventricular and interventricular sulci are often deeply sited, occasionally hidden by overlapping myocardium or embedded in it. Myocardial strands may also cross atrial or ventricular branches; Polacek (1961) found them in more than 80% of ventricles; Bloor & Lowman (1963) have emphasized their importance in interpretation of coronary arteriograms.

The diameters of coronary arteries, both main stems and larger branches, have often been recorded; such figures are of limited value, since technique is not always stated, physiological state often ignored and measurement of external or internal diameters not clearly distinguished. Calibre is usually the basis, most measurements being made on arterial casts or angiograms. The maximum ranges recorded in major studies are 1.5–5.5 mm for coronary arteries at their origins. Baroldi & Scomazzoni (1967) give means of 4.0 and 3.2 mm. The left exceeds the right in about 60% of hearts, the right being larger in 17%, the vessels approximately equal in 23%. Vogelberg (1957) considered that coronary diameters increase up to the thirtieth year.

The Right Coronary Artery

Arising from the anterior ('right coronary') aortic sinus, the artery passes at first anteriorly and slightly to the right between the right auricular appendage and pulmonary trunk, where the sinus usually bulges. Reaching the atrioventricular (coronary) sulcus it descends in this almost vertically to the right (acute) cardiac border, curving around it into the posterior part of the sulcus, where it approaches its junction with both interatrial and interventricular grooves, a region appropriately termed the *crux of the heart*. In about 60% the artery reaches the crux and ends a little left of it by variable anastomosis with circumflex branch of the left coronary. In a minority (*c.* 10%) the right coronary artery ends near the right cardiac border or between this and the crux (*c.* 10%); more often (*c.* 20%) it may reach the left border, replacing part of the circumflex artery.

Branches of the right coronary supply both right atrium and ventricle and, variably, parts of the *left* chambers and atrioventricular septum. The first branch (arising separately from the anterior aortic sinus in 36%) is the *conus artery* (sometimes a 'third coronary'); since a similar vessel comes from the left coronary, this is more correctly named the *right conus artery*. It ramifies anteriorly on the lowest part of the pulmonary conus and upper part of right ventricle; it commonly anastomoses with a similar left coronary branch to form the '*annulus of Vieussens*', a tenuous anastomotic 'circle' around the pulmonary trunk. Descriptions of the conus artery vary (Baroldi & Scomazzoni 1967), some regarding the right conus artery of significance in coronary arterial disease; some consider it to be the right coronary's first ventricular branch, supplying a variable region from the conus to the apex.

Anterior atrial and ventricular rami diverge from the so-called *first segment of the right coronary*, extending from its origin to the right margin of the heart. Both groups diverge widely, approaching a right angle in the case of ventricular arteries, in contrast to the more acute origins of the left coronary ventricular rami. The **right anterior ventricular rami**, usually two or three, ramify towards the cardiac apex, which they rarely reach unless the **right marginal branch** is included, as it is by some; this is then the largest right anterior ventricular ramus, greater in calibre and long enough to reach the apex in most hearts (93%, Baroldi & Scomazzoni 1967). When the right marginal artery is very large, the remaining anterior ventricular rami may be reduced to one, or is completely absent. From the *second segment of the right coronary artery* (between the right border and crux) one to three small **right posterior ventricular rami**, commonly two, supply the diaphragmatic aspect of the right ventricle. Their size is reciprocal to that of the right marginal artery, as in the anterior right ventricular supply, the right marginal usually extending to the cardiac diaphragmatic surface. Posterior right ventricular rami may be absent. 729

As the right coronary approaches the crux, it produces one to three posterior interventricular rami but only one in the interventricular sulcus; this **posterior interventricular artery**, single in about 70%, is otherwise accompanied by parallel *right* coronary branches, to the right or left or on both sides of the sulcus. When these flanking vessels exist, branches of the posterior (descending) interventricular artery are small and sparse; when it exists alone it gives off a few branches, particularly to the right ventricle but also to the left. It is replaced in about 10% by a left coronary branch.

The **atrial rami** of the right coronary artery are sometimes described as anterior, lateral (right or marginal) and posterior *groups* but are most frequently single, small vessels averaging 1 mm in diameter. The right anterior and lateral are occasionally double, very rarely triple and supply chiefly the right atrium. The posterior ramus is usually single, distributed to the right and left atria; but in 40% or more a left posterior atrial branch of the right coronary exists. The **artery of the sinuatrial node** is an atrial branch, distributed largely to the myocardium of both atria, mainly the right. Its origin is variable: from the *left* coronary in about 35% (Hutchinson 1978), arising from its circumflex branch (vide infra); when it is a branch of the right coronary, it usually comes from its anterior stem, less often from its right lateral part, least of all from its posterior atrioventricular part. This 'nodal' artery thus usually passes back in the sulcus between the right auricular appendage and aorta. Whatever its origin the artery usually branches around the superior vena cava's base, commonly as an *arterial loop* from which small rami supply the right atrium. A large '*ramus cristae terminalis*' (Spalteholz 1924) traverses the sinuatrial node (**6.61**), perhaps instead this ramus should be termed 'nodal artery', since most of the currently named vessel actually supplies the atria and is more appropriately named the '*main atrial branch*' (Baroldi & Scomazzoni 1967).

Right coronary **septal rami** are relatively short, leaving its posterior interventricular ramus to supply the posterior interventricular septum. They are numerous but do not usually reach the apical septal parts (supplied by terminal septal branches of the anterior interventricular). The largest posterior septal artery, usually the first, is commonly from the *inverted loop* said to characterize the right coronary artery at the crux, where its pos-

terior interventricular branch arises; this **large posterior septal artery** usually supplies the **atrioventricular node**—in 80% of hearts, according to Hutchinson (1978).

DiDio (1967) described the *atrioventricular rami* of the right coronary artery as consisting of small recurrent branches from each ventricular artery crossing the atrioventricular sulcus to supply the adjoining atrial myocardium, or ventricular twigs from the atrial arteries.

The Left Coronary Artery

The left coronary artery is the larger in calibre, supplying a greater *volume* of myocardium, including almost all the left ventricle and atrium, except in so-called 'right dominance' where the right coronary partly supplies a posterior region of the left ventricle. The left coronary usually supplies most of the interventricular septum. Its *initial stem*, between its ostium in the left posterior ('left coronary') aortic sinus and its first branches, varies in length from a few millimetres to a few centimetres. It lies between the pulmonary trunk and the left auricular appendage, emerging into the atrioventricular sulcus, in which it turns left; this part is loosely embedded in subepicardial fat and usually has no branches; but a small atrial ramus may occur and, rarely, the *sinuatrial nodal artery* may arise from the left coronary artery (James 1961); but when it is a ramus of the left coronary it almost always comes from the circumflex branch. Reaching the atrioventricular or coronary sulcus, the left coronary divides into two or three main rami, its **anterior interventricular (descending) ramus** being commonly described as its continuation; this descends obliquely forward and left in the interventricular sulcus, sometimes deeply embedded or crossed by bridges of myocardial tissue and by the great cardiac vein and its tributaries. It reaches the apex almost always, terminating there in one-third of specimens, but more often turning round the apex into the posterior interventricular sulcus, in which it traverses a third to a half of its length, to meet the terminal twigs of the corresponding right coronary ramus.

The **anterior interventricular artery** produces right and left anterior ventricular, anterior septal and variable, corresponding posterior rami. *Right anterior ventricular rami* are small and rarely number more than one or two, the right ventricle being supplied almost wholly by the right coronary artery.

From two to nine large **left anterior ventricular arteries** branch at acute angles from the anterior interventricular to cross *diagonally* the left ventricle's anterior aspect, larger terminals reaching the rounded (obtuse) left border. One is often large and may arise separately from the left coronary *trunk* (which then ends by trifurcation); this **left diagonal artery**, reported in 33–50% or more cases, is occasionally duplicated (20%). A small *left conus artery* frequently leaves the anterior interventricular near its start, anastomosing on the conus with that of the right coronary and with the vasa vasorum of the pulmonary artery and aorta. The **anterior septal rami** leave the anterior interventricular almost perpendicularly, passing back and down in the septum, of which they usually supply about the ventral two-thirds. Small *posterior septal rami* from the same source supply the posterior septal third for a variable distance from the cardiac apex.

The **circumflex artery**, in calibre comparable to the anterior interventricular, curves left in the atrioventricular sulcus, continuing round the left cardiac border into the posterior part of the sulcus and ending left of the crux in most hearts, but sometimes continuing as a posterior interventricular artery. Proximally the left auricular appendage usually overlaps it. In about 90% a large ventricular branch, the **left marginal artery**, arises perpendicularly from it to ramify over the rounded 'obtuse' margin, supplying much of the adjacent left ventricle, usually to the apex. Smaller anterior and posterior rami of the circumflex artery also supply the left ventricle. **Anterior ventricular branches** (1–5; commonly 2 or 3) course parallel to the diagonal artery, when present, replacing it when absent. **Posterior ventricular branches** are smaller and fewer, the left ventricle being partly supplied by the posterior interventricular artery; when this is small or absent, it is accompanied or replaced by an interventricular continuation of the circumflex. Such a **left posterior interventricular artery** is frequently double or triple. **Atria**

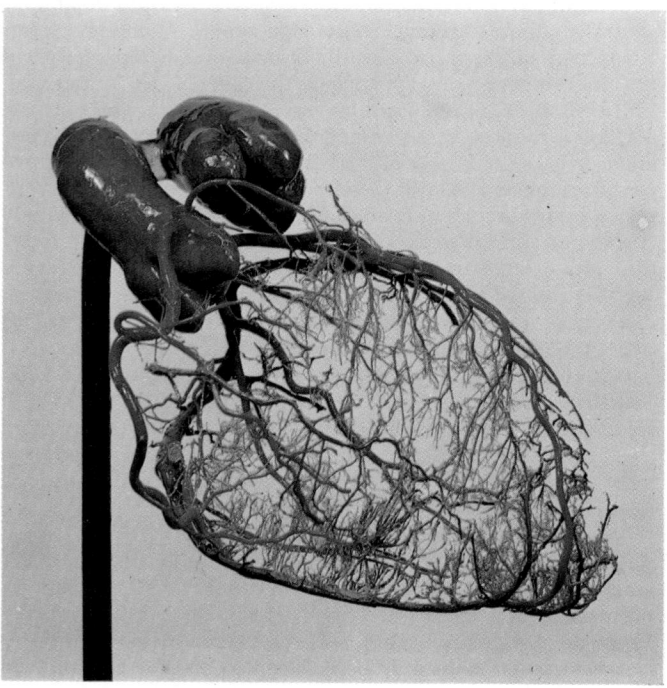

6.62 A composite coloured resin cast: red = the aortic root, aortic sinuses and coronary arteries; blue = the pulmonary trunk, sinuses and the coronary sinus and its tributaries. Note reduplication of the right coronary artery; the smaller of these vessels is known as the artery of the conus. Note particularly the septal arterial and venous rami. See text for detailed account. Photography by Kevin Fitzpatrick, Guy's Hospital Medical School.

rami, anterior, lateral and posterior, from the circumflex, supply the left atrium.

Inconstant branches of the circumflex artery require mention. The **artery to the sinuatrial node** is a branch in about 35% (Hutchinson 1978), usually from the anterior circumflex segment, less often the circummarginal. It passes over and supplies the left atrium, encircling the superior vena cava like a right coronary nodal ramus. It sends a large branch to (and through) the node but is predominantly atrial in distribution. The **artery to the atrioventricular node**, the terminal ramus in 20%, arises near the crux and then the circumflex usually supplies a posterior interventricular ramus, an example of so-called 'left dominance' (vide infra). *Kugel's anastomotic artery*, 'arteria anastomotica auricularis magna' was described by Kugel (1927) as a constant circumflex branch, usually from its anterior part, traversing the interatrial septum (near its ventricular border) to establish direct or indirect anastomosis with the right coronary. This anastomosis is controversial, apparently accepted by James (1974) but denied by Baroldi & Scomazzoni (1967). James considered it an auxiliary supply to the atrioventricular node.

Details of coronary distribution require integration into a concept of total cardiac supply. Most commonly the *right coronary* supplies all the *right* ventricle (except a small region right of the anterior interventricular sulcus), a variable part of the *left* ventricular diaphragmatic aspect, the postero-inferior third of the intraventricular septum, the right atrium and part of the left, and the conducting system as far as the proximal parts of the right and left crura. *Left coronary* distribution is, of course, reciprocal, including most of the *left* ventricle (vide supra), a narrow strip of *right* ventricle (vide supra), the anterior two-thirds of the interventricular septum and most of the left atrium. As noted, variations (**6.61A–E**) chiefly affect the diaphragmatic aspect of ventricles residing in relative '*dominance*' of supply by the left or right coronary artery. The term is misleading, since the left artery almost always supplies a greater *volume* of tissue. In '*right dominance*' the posterior interventricular artery is from the right coronary, in '*left dominance*' from the left. In the so-called '*balanced*' pattern, branches of both run in or near the sulcus. Less is known of variation in atrial supply; the small vessels involved are not easily preserved in corrosion casts. From Hutchinson's results (1978) it is apparent that in over 50% the right atrium is supplied only by the right coronary, the rest receiving a dual supply. More than 62% of left atria are largely supplied by the left and about 27% by the right coronary; but in each group a small accessory supply from the other coronary exists, 11% being supplied almost equally by both. Sinuatrial and atrioventricular supplies also vary. According to James (1961) right and left coronary arteries supply the sinuatrial node respectively in 55% and 45%, corresponding values from Baroldi & Scomazzoni's study (1967) being 51% and 41% (8% receiving bilateral supply), and from Hutchinson (1978) 65% and 35%. For the atrioventricular node James's values are 90% (right coronary) and 10% (left coronary), Hutchinson's 80% and 20% respectively; Baroldi & Scomazzoni merely note that right coronary supply is common and left supply rare.

Coronary Anastomosis

Anastomoses between branches of coronary arteries, subepicardial or myocardial, and between these arteries and extracardiac vessels are of prime medical import. Clinical experience suggests that anastomoses cannot rapidly provide collateral routes sufficient to circumvent sudden coronary obstruction. It is hence traditional to regard coronary circulation as end-arterial. Nevertheless, anastomosis has long been established particularly between the finer subepicardial rami. According to Gross (1921) such anastomoses may improve during individual life. Those who have *investigated* coronary arteries by radio-opaque perfusants (Vastesueger et al 1957 in post-mortem hearts, Laurie & Woods 1958 by in vivo coronary radiography), by perfusion with calibrated spherules (Prinzmetal et al 1947) or by subsequent corrosion casts of plastic resins (Baroldi et al 1956, James 1961) have almost all described anastomoses, and in vessels up to 100–200 μm in calibre. Baroldi & Scomazzoni (1967) have tabulated all results reported since 1880; no study denying

anastomoses has been recorded since 1957. Some describe anastomoses only between branches of individual coronary arteries but the majority record *intercoronary* anastomoses. James (1974) considers evidence conclusive for anastomoses at all levels: subepicardial, myocardial, and subendocardial; the most frequent sites of extramural anastomoses are the apex, the anterior aspect of the right ventricle, posterior aspect of the left ventricle, crux, interatrial and inter-ventricular sulci and between the sinuatrial nodal and other atrial vessels. The functional value of such anastomoses must vary but they appear to become more effective in *slowly* progressive pathological conditions. Their structure is uncertain; most observations depend on corrosion casts, which suggest that anastomotic vessels are relatively straight in normal hearts, but much coiled in hearts subject to coronary occlusion. Little has been recorded of their microscopic structure; they appear little more than endothelial tubes, without muscle or elastic tissue.

Extracardiac anastomoses may connect various coronary branches with other thoracic vessels via the pericardial arteries and arterial vasa vasorum of vessels linking the heart with systemic and pulmonary circulations. The classic study of Hudson et al (1932) showed that coronary injections of India ink could reach the diaphragm through the aortic vasa vasorum. Similar connections along pulmonary trunks reach the mediastinal and bronchial arteries and also exist along pulmonary veins and venae cavae. These results have been confirmed (Baroldi & Scomazzoni 1967) but the effectiveness of such connections as collateral routes in coronary occlusion is unpredictable.

Coronary arteriovenous anastomoses were reported by Nussbaum (1912). His evidence was indirect but Hirsch (1960) described glomeral structures with typical sphincteric appearances (p. 693) in cardiac sulci; these must be regarded as at present unproven. Various other forms of 'arteriovenous' connections have been described; Wearn et al (1933) recorded numerous connections through the very thin-walled 'arterial' vessels between the coronary circulation and cardiac cavities, naming them '*myocardial sinusoids*' and '*arterio-luminal*' vessels. They have been confirmed (Watanabe 1960) and indirect evidence of them, from perfusion experiments, dates back to Vieussens (1706). Their value in coronary disease is uncertain.

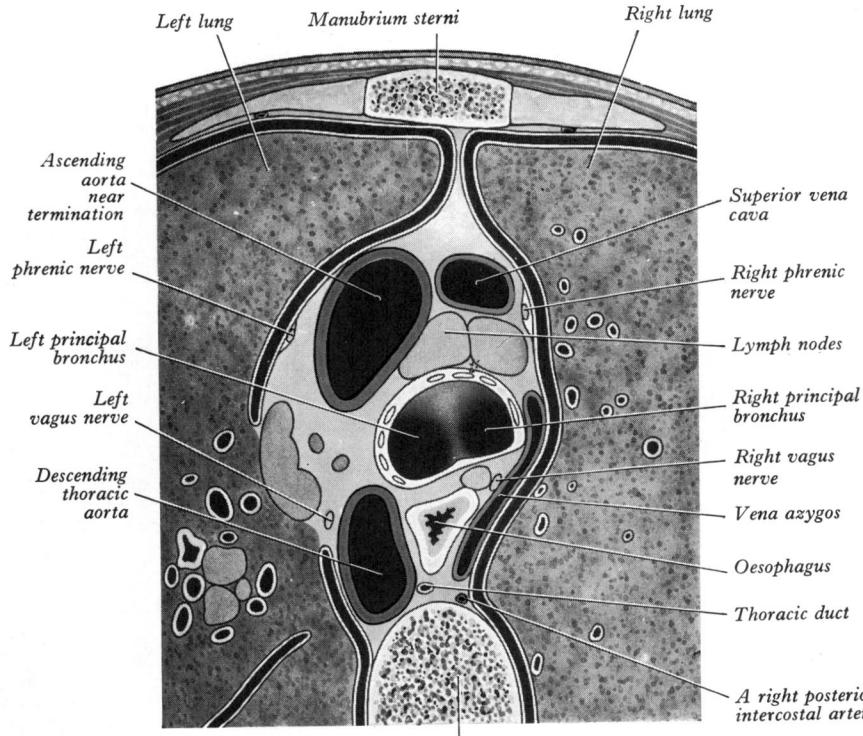

6.63 Transverse section through the mediastinum at the level of the lower part of the body of the fourth thoracic vertebra, viewed from above. The deep cardiac plexus of nerves is omitted.

Left lung *Manubrium sterni* *Right lung*

Ascending aorta near termination

Left phrenic nerve

Left principal bronchus

Left vagus nerve

Descending thoracic aorta

Superior vena cava

Right phrenic nerve

Lymph nodes

Right principal bronchus

Right vagus nerve

Vena azygos

Oesophagus

Thoracic duct

A right posterior intercostal artery

Fourth thoracic vertebra

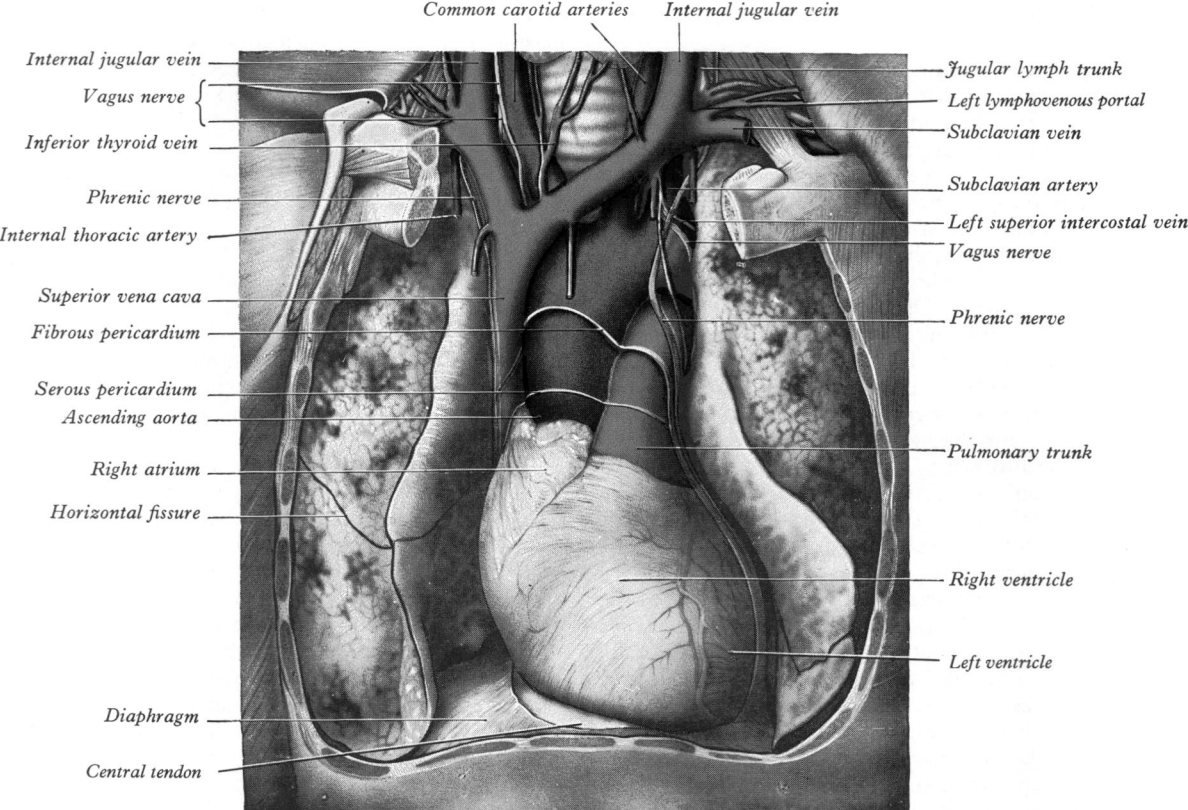

Common carotid arteries *Internal jugular vein*

Internal jugular vein —

Vagus nerve {

Inferior thyroid vein —

Phrenic nerve —

Internal thoracic artery —

Superior vena cava —

Fibrous pericardium —

Serous pericardium —

Ascending aorta —

Right atrium —

Horizontal fissure —

Diaphragm —

Central tendon —

— *Jugular lymph trunk*

— *Left lymphovenous portal*

— *Subclavian vein*

— *Subclavian artery*

— *Left superior intercostal vein*

— *Vagus nerve*

— *Phrenic nerve*

— *Pulmonary trunk*

— *Right ventricle*

— *Left ventricle*

6.64 Dissection to display the heart, great vessels and lungs in situ. The sternum and the sternal ends of the costal cartilages, together with the parietal pleura on each side, have been excised and the mediastinal pleura and parietal layer of the pericardium over the sternocostal surface of the heart have been removed. The lungs have been displaced to expose the heart and the epicardium dissected off the heart and the great vessels.

The Arch of the Aorta

The aortic arch (**6.**59–65) begins behind approximately the *right* half of the manubrium sterni, level with the upper border of the second *right* sternocostal articulation. It first ascends diagonally back and to the left over the anterior surface of the trachea, then back across its left side and finally descends left of the fourth thoracic vertebral body, continuing at its lower border as the descending thoracic aorta. Its end corresponds, in level, with the sternal end of the second, *left* costal cartilage (**6.**42). Thus, the aortic arch lies wholly in the superior mediastinum. It has two curvatures: one convex upwards, the other convex forwards and to the left. Its upper limit is usually at about *mid-level* of the manubrium sterni.

Relations. *Anteriorly and to the left* is the left mediastinal pleura, deep to which it is crossed by four nerves: the left phrenic, left lower cervical vagal cardiac branch, left superior cervical sympathetic cardiac branch and left vagus, in anteroposterior

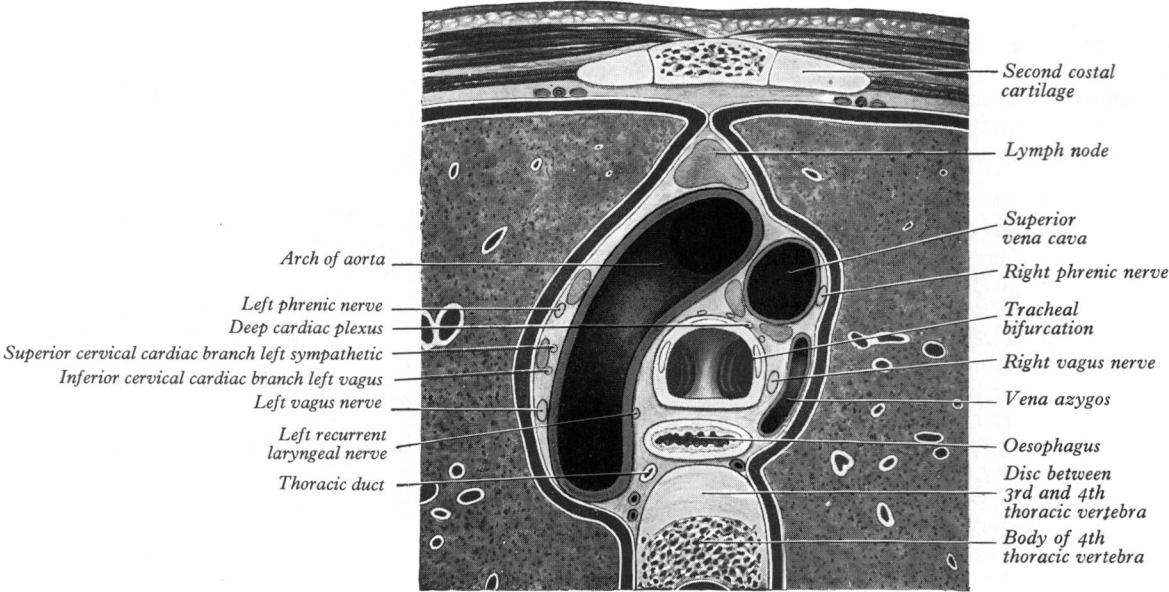

Arch of aorta —

Left phrenic nerve —

Deep cardiac plexus —

Superior cervical cardiac branch left sympathetic —

Inferior cervical cardiac branch left vagus —

Left vagus nerve —

Left recurrent laryngeal nerve —

Thoracic duct —

— *Second costal cartilage*

— *Lymph node*

— *Superior vena cava*

— *Right phrenic nerve*

— *Tracheal bifurcation*

— *Right vagus nerve*

— *Vena azygos*

— *Oesophagus*

— *Disc between 3rd and 4th thoracic vertebra*

— *Body of 4th thoracic vertebra*

6.65 Transverse section through the mediastinum at the level of the upper part of the body of the fourth thoracic vertebra, viewed from above.

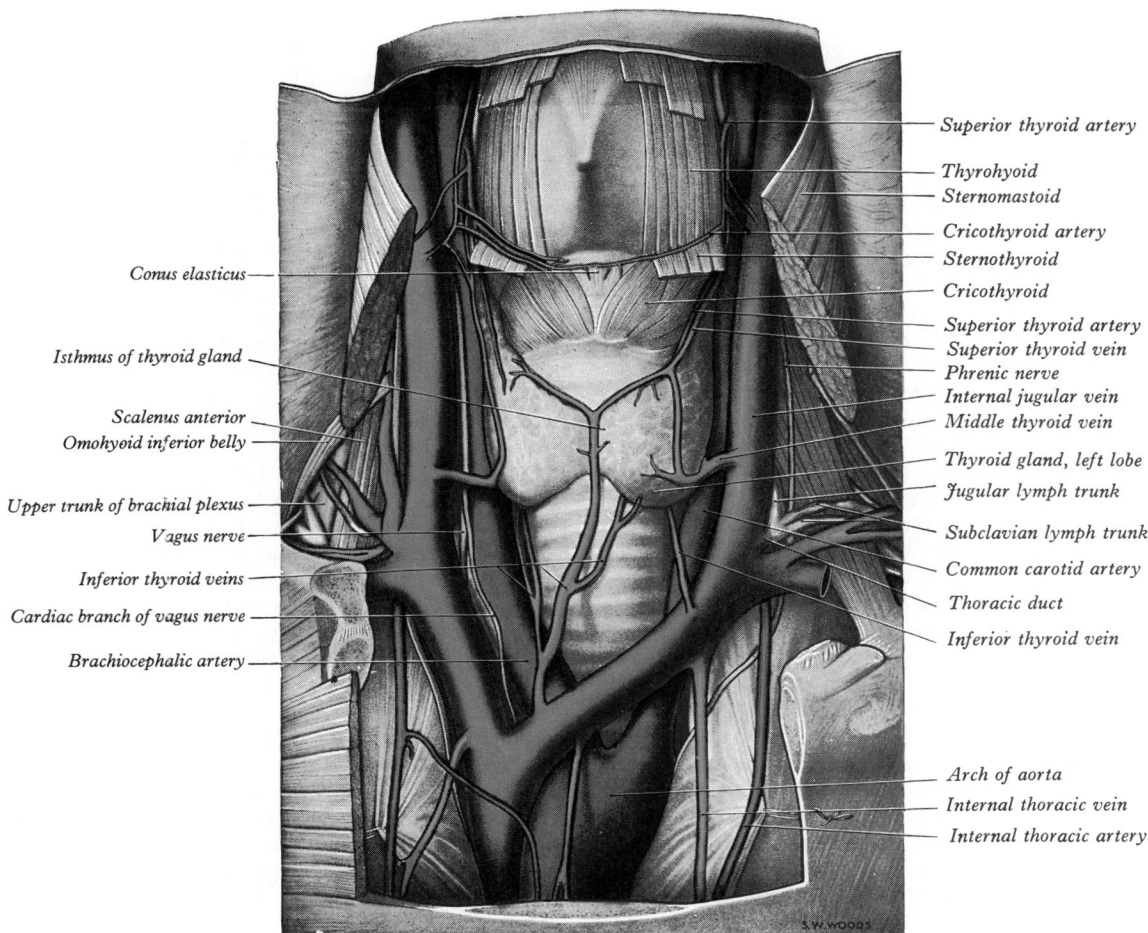

Superior thyroid artery
Thyrohyoid
Sternomastoid
Cricothyroid artery
Sternothyroid
Cricothyroid
Superior thyroid artery
Superior thyroid vein
Phrenic nerve
Internal jugular vein
Middle thyroid vein
Thyroid gland, left lobe
Jugular lymph trunk
Subclavian lymph trunks
Common carotid artery
Thoracic duct
Inferior thyroid vein

Arch of aorta
Internal thoracic vein
Internal thoracic artery

Conus elasticus

Isthmus of thyroid gland

Scalenus anterior
Omohyoid inferior belly

Upper trunk of brachial plexus
Vagus nerve

Inferior thyroid veins
Cardiac branch of vagus nerve

Brachiocephalic artery

S.W.WOODS

6.66 Dissection of the lower part of the front of the neck and of the superior mediastinum. The manubrium sterni and the sternal ends of the clavicles and the first costal cartilages have been removed and the pleural sac and lung have been retracted on each side. In this specimen each superior thyroid artery arose from the common carotid artery.

order. As the left vagus crosses the arch its recurrent laryngeal branch hooks below the vessel left and behind (developmentally caudal to) the ligamentum arteriosum and then ascends on the arch's right. The left superior intercostal vein ascends obliquely forwards on the arch, superficial to left vagus, deep to the left phrenic nerve (**6.64**). Left lung and pleura separate all these from the thoracic wall. *Posteriorly to the right*, are the trachea and deep cardiac plexus, the left recurrent laryngeal nerve, oesophagus, thoracic duct and vertebral column. *Above*, the brachiocephalic, left common carotid and left subclavian arteries arise from its convexity, crossed anteriorly near their origins by the left brachiocephalic vein. *Below* are the pulmonary bifurcation, left principal bronchus, ligamentum arteriosum (p. 725), superficial cardiac plexus and left recurrent laryngeal nerve. (Best viewed from the left, the concavity of the aortic arch is the upper curved limit through which structures gain access or exit through the hilum of the left lung.)

The fetal aortic lumen narrows between the origin of the left subclavian artery and the attachment of the ductus arteriosus, as the *aortic isthmus*; beyond the ductus arteriosus the vessel presents a fusiform *aortic spindle*, the junction of the two parts being marked inferiorly by an indentation; these features persist variably in adults.

Variations. The summit of the arch is usually about 2.5 cm below the superior sternal border but may diverge from this. Sometimes the aorta curves over the *right* pulmonary hilum (as a right aortic arch) descending right of the vertebral column, a condition normal in birds; there is usually transposition of thoracic and abdominal viscera. Less often, after arching over the right hilum, it passes behind the oesophagus to its usual position; this is not accompanied by visceral transposition. The aorta may divide, as in some quadrupeds, into ascending and descending

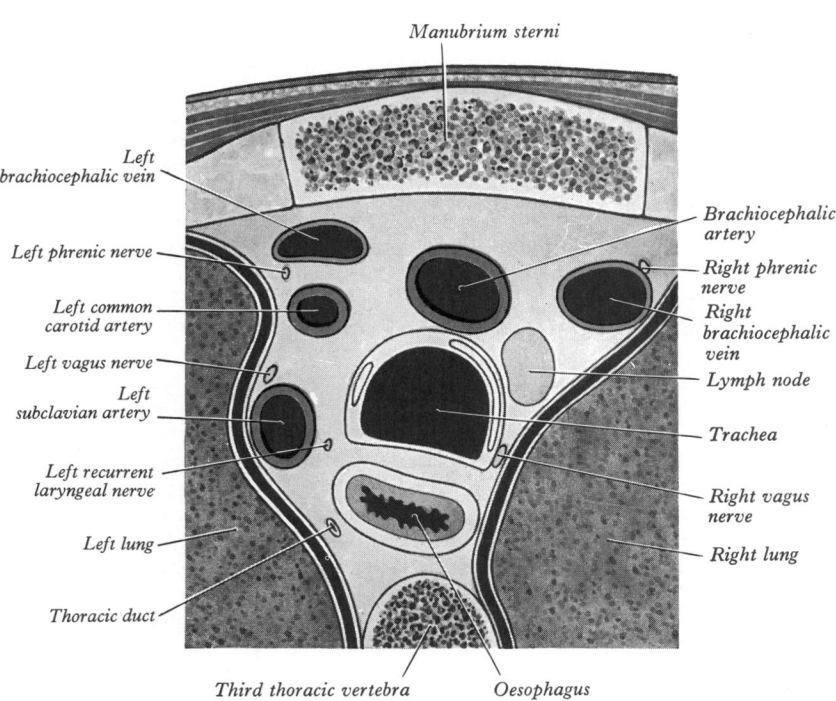

Manubrium sterni

Left brachiocephalic vein

Left phrenic nerve

Left common carotid artery

Left vagus nerve

Left subclavian artery

Left recurrent laryngeal nerve

Left lung

Thoracic duct

Brachiocephalic artery
Right phrenic nerve
Right brachiocephalic vein
Lymph node
Trachea
Right vagus nerve
Right lung

Third thoracic vertebra   Oesophagus

6.67 Transverse section through the superior mediastinum at the level of the body of the third thoracic vertebra, viewed from above.

trunks, the former dividing into three branches to supply the head and upper limbs. Sometimes it divides near its origin, the two branches soon reuniting; the oesophagus and trachea usually pass through the interval between them; this is the normal condition in reptilia and is due to the persistence of a part of the right dorsal aorta which usually disappears (p. 214).

Radiological appearances. The shadow of the arch is easily identified in anteroposterior radiographs (**6**.45) and its left profile is sometimes called the 'aortic knuckle'. The arch may also be visible in left anterior oblique views enclosing a pale space, 'the aortic window', in which shadows of the pulmonary trunk and its left branch may be visible.

Branches (**6**.64,67). Three branches spring from the vessel's convex aspect: the brachiocephalic trunk, left common carotid and left subclavian arteries. They may branch from the beginning of the arch or the upper part of the ascending aorta; the distance between these origins varies, the most frequent being approximation of the left common carotid artery to the brachiocephalic trunk (Wright 1969). Primary branches may be reduced to one, more commonly two, the left common carotid arising from the brachiocephalic trunk (7%), or (more rarely) the left common carotid and subclavian arteries arising from a left brachiocephalic or right common carotid and subclavian arising separately, in which case the latter more often branches from the left end of the arch and passes behind the oesophagus (p. 214). The left vertebral artery may arise between the left common carotid and the subclavian. Very rarely, external and internal carotid arteries arise separately, the common carotid being absent on one or both sides; or both carotids and one or both vertebrals may be separate branches. When a 'right aorta' occurs, the arrangement of its three branches is reversed. The common carotids may have a single trunk, the subclavians separate, the right arising from the left end of the arch. Other arteries may branch from it, most commonly one or both bronchial arteries and the arteria thyroidea ima.

An analysis of variation in branches from 1000 aortic arches (Anson 1963) showed in 65% the usual pattern; in 27% a left common carotid shared the brachiocephalic trunk (contrast percentage quoted above); in 2.5% the four large arteries branched separately. The remaining 5% showed a great variety of patterns, the commonest (1.2%) being symmetrical right and left brachiocephalic trunks.

The Brachiocephalic Artery

The brachiocephalic (innominate) artery, the largest branch of the aortic arch, is from 4–5 cm in length (**6**.60,66,67), arising from the arch's convexity posterior to the centre of the manubrium sterni; it ascends posterolaterally to the right, at first anterior to trachea, then on its right. Level with the right sternoclavicular joint's upper border it forks into the right common carotid and subclavian arteries.

Relations. *Anterior* are sternohyoid and sternothyroid, the remains of the thymus, left brachiocephalic and right inferior thyroid veins, crossing its root and sometimes the right vagal cardiac branches, all separating it from the manubrium. *Posterior* are the trachea below, right pleura above, where the right vagus is posterolateral before passing lateral to the trachea; *right lateral* are the right brachiocephalic vein, the upper part of the superior vena cava and pleura; *left lateral* are the thymic remains, the origin of the left common carotid artery, the inferior thyroid veins and the trachea at a higher level.

Branches. The brachiocephalic artery usually has only terminal branches but occasionally an *arteria thyroidea ima* arises from it, sometimes a *thymic* or *bronchial branch*.

Arteria thyroidea ima, small and inconstant, ascends on the trachea to the thyroid isthmus, in which it ends. It may arise from the aorta, right common carotid, subclavian or internal thoracic arteries.

THE CAROTID SYSTEM OF ARTERIES

The common carotid arteries are the main supply to the head and neck; they ascend to the level of the thyroid cartilage's upper border, where each divides into: an external carotid, supplying the exterior of the head, face and most of the neck; and an internal carotid, supplying the cranial and orbital contents.

The common and internal carotid arteries, with veins and nerves accompanying them, lie in a cleft bounded posteriorly by cervical transverse processes and attached muscles, medially by the trachea, oesophagus, thyroid gland, larynx and pharyngeal constrictors, anterolaterally by the sternocleidomastoid with, at different levels, omohyoid, sternohyoid, sternothyroid, digastric and stylohyoid muscles. (Other styloid structures have a varying relationship depending on the length of the styloid process.)

Common Carotid Arteries

These differ in length and origin, the *right*, exclusively cervical, begining at the brachiocephalic bifurcation behind the right sterno-clavicular joint, the *left* at the highest part of the aortic arch immediately posterolateral to the brachiocephalic trunk and therefore having both thoracic and cervical parts.

The **thoracic part of the left common carotid artery** (**6**.66,67) ascends until level with the left sternoclavicular joint, where it enters the neck. It lies at first in front of the trachea, then later inclines to its left side.

Relations. *Anterior* are the sternohyoid and sternothyroid, the anterior parts of the left pleura and lung, the left brachiocephalic vein and the thymic remnants, separating it from the manubrium; *posterior* are the trachea, left subclavian artery, left edge of the oesophagus, left recurrent laryngeal nerve and thoracic duct. *Right lateral* are (below) the brachiocephalic trunk and (above)

the trachea, inferior thyroid veins and thymic remains; *left lateral* are the left vagus and phrenic nerves, left pleura and lung.

Cervical parts of both common carotid arteries have similar courses (**6**.66,67,69). Each ascends, diverging laterally from behind the sternoclavicular joint to the thyroid cartilage's upper border, where it divides into external and internal carotid arteries (**6**.68, 70). At its division the vessel has a dilatation, the *carotid sinus*, usually involving or restricted to the beginning of the *internal* carotid; the tunica media is thinner here and the tunica adventitia, relatively thick, contains many receptor endings of the glossopharyngeal nerve (p. 1113); these show some degeneration and regeneration, suggesting continuous turnover, which may accelerate in the elderly (Rusager & Weddell 1962). The sinus is responsive to changes in arterial blood pressure, leading to reflex haemodynamic modification. Its situation on the brain's main artery is appropriate to its role as a baroreceptor in control of intracranial pressure. The *carotid body*, behind the common carotid bifurcation, a small, reddish-brown structure, is a 'chemoreceptor'. (See Adams 1958 for a comparative account and p. 1473 for modern views on its ultrastructure and modus operandi.)

In the lower neck the common carotids are separated by a narrow gap into which projects the trachea; above this the thyroid gland, larynx and pharynx project between them. Each is contained in a carotid sheath (p. 582), continuous with the deep cervical fascia and of loose texture, though that actually around the artery is denser. This sheath encloses also the internal jugular vein and vagus nerve, the vein lateral to the artery, the nerve between them and posterior to both. The superior root of the ansa cervicalis is embedded in its anterior wall.

Relations. The artery is crossed *anterolaterally*, level with the cricoid cartilage, by the intermediate tendon (sometimes the

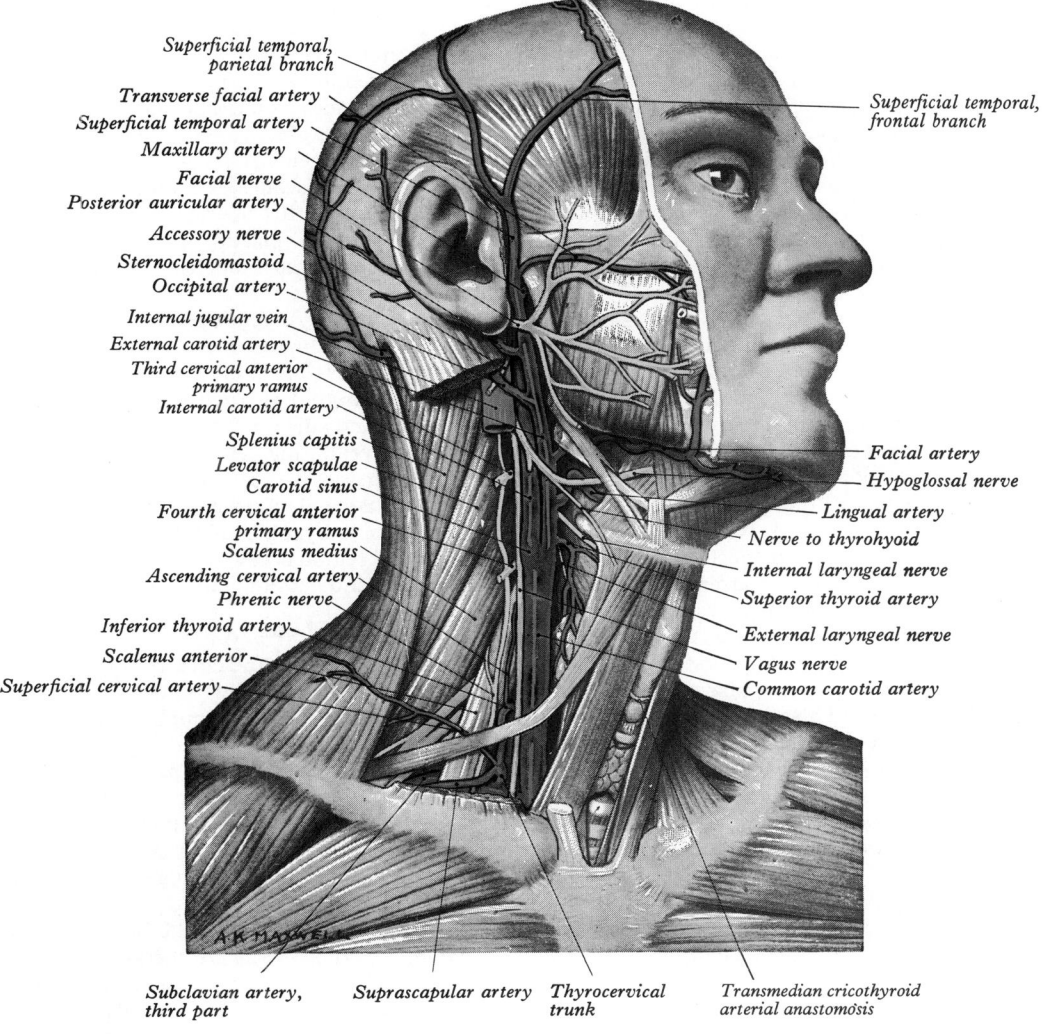

Superficial temporal, parietal branch
Transverse facial artery
Superficial temporal artery
Maxillary artery
Facial nerve
Posterior auricular artery
Accessory nerve
Sternocleidomastoid
Occipital artery
Internal jugular vein
External carotid artery
Third cervical anterior primary ramus
Internal carotid artery
Splenius capitis
Levator scapulae
Carotid sinus
Fourth cervical anterior primary ramus
Scalenus medius
Ascending cervical artery
Phrenic nerve
Inferior thyroid artery
Scalenus anterior
Superficial cervical artery

Superficial temporal, frontal branch

Facial artery
Hypoglossal nerve
Lingual artery
Nerve to thyrohyoid
Internal laryngeal nerve
Superior thyroid artery
External laryngeal nerve
Vagus nerve
Common carotid artery

Subclavian artery, third part
Suprascapular artery
Thyrocervical trunk
Transmedian cricothyroid arterial anastomosis

6.68 Dissection of the right side of the neck, showing the carotid and subclavian arteries and their branches. The parotid and submandibular glands have been removed together with the lower part of the internal jugular vein, most of the sternocleidomastoid and the upper parts of the stylohyoid and posterior belly of the digastric.

superior belly) of omohyoid. Below this muscle it is sited deeply, covered by skin, superficial fascia, platysma, deep cervical fascia, the sternocleidomastoid, sternohyoid and sternothyroid. Above the omohyoid it is more superficial, covered merely by skin, superficial fascia, platysma, deep cervical fascia and the medial margin of sternocleidomastoid and is crossed obliquely from its medial to lateral side by the sternocleidomastoid branch of the superior thyroid artery. In front of, or embedded in, the carotid sheath is the superior root of the ansa cervicalis, joined by its inferior root from the second and third cervical spinal nerves and crossing the vessel obliquely. The superior thyroid vein usually crosses near the artery's end, the middle thyroid vein a little below cricoid level; the anterior jugular vein crosses it above the clavicle, separated by sternohyoid and sternothyroid. *Posterior* are the fourth to sixth cervical transverse processes, and attached to them the longus colli and longus capitis and tendinous slips of scalenus anterior; the sympathetic trunk and ascending cervical artery are between the common carotid artery and the muscles. Below the level of the sixth cervical vertebra the artery is in an angle between the scalenus anterior and longus colli, anterior to the vertebral vessels, inferior thyroid and subclavian arteries, sympathetic trunk and, on the left, thoracic duct. *Medial* are the oesophagus, trachea, inferior thyroid artery and recurrent laryngeal nerve and, at a higher level, the larynx and pharynx; the thyroid gland overlaps it anteromedially. *Lateral* is the internal jugular vein; *posterolaterally* in the angle between artery and vein is the vagus nerve.

On the right, low in the neck, the recurrent laryngeal nerve crosses obliquely behind the artery; the right internal jugular vein

diverges from it below but the left vein approaches and often overlaps its artery.

Variations. In about 12% the *right* common carotid artery arises above the level of the sternoclavicular joint or it may be a separate branch from the aorta; again it may arise with its fellow. The *left* common carotid artery varies in origin more than the right; it may arise with the brachiocephalic (see also p. 734). Division of the common carotid may occur higher, near the level of the hyoid bone, more rarely at a lower level alongside the larynx. Very rarely it ascends without division, either the external or internal carotid being absent. Rarely, also, it is replaced by separate external and internal carotid arteries arising directly from the aorta, on one side or bilaterally.

The common carotid artery usually has no branches but the vertebral, superior thyroid (**6.**66) or its laryngeal branch, ascending pharyngeal, inferior thyroid or occipital may be branches of it.

The External Carotid Artery

This artery (**6.**45,53) begins lateral to the thyroid cartilage's upper border, level with the disc between the third and fourth cervical vertebrae. A little curved, and with a gentle spiral, it first ascends slightly forwards and then inclines backwards and a little laterally, to pass midway between the mastoid tip and mandibular angle where, in the substance of the parotid gland behind the mandible's neck, it divides into the superficial temporal and maxillary arteries. It diminishes rapidly in calibre due to its many large branches. In children it is smaller than the internal carotid but in adults the two are of almost equal size. At its origin, it is in the carotid triangle and lies anteromedial to the internal carotid but becomes anterior, and then lateral to this as it ascends. At mandibular levels the styloid process and its attached structures intervene between the vessels, the internal carotid being deep and the external carotid superficial to the styloid. The carotid triangle may be *seen* as a triangular depression in the living neck; a finger tip placed here perceives *powerful arterial pulsation.* Beneath the finger lie the termination of the common carotid, the origins of external and internal carotids and the stems of the external carotid's initial branches.

Relations. Superficial to the artery *in the carotid triangle* are: the skin, superficial fascia, the loop between the facial nerve's cervical branch and the transverse cutaneous nerve of the neck, deep fascia and the anterior margin of sternocleidomastoid; it is *crossed* by the hypoglossal nerve and its vena comitans and by the lingual (common), facial and sometimes the superior thyroid veins. Leaving the triangle it is crossed by the posterior belly of the digastric and stylohyoid and ascends between this muscle and the posteromedial surface of the parotid gland, which it enters, lying medial to the facial nerve and the junction of the superficial temporal and maxillary veins. *Medial* to the artery are at first the pharyngeal wall, superior laryngeal nerve and ascending pharyngeal artery; at a higher level the internal carotid artery is separated from the external by the styloid process, styloglossus and stylopharyngeus, glossopharyngeal nerve, pharyngeal branch of vagus nerve and part of the parotid gland (**6**.69). The relation of the artery to the parotid gland is controversial, many clinicians asserting that it is often *medial* to it rather than *in* it. A recent study (Guffarth & Graumann 1975), a short series of dissections, indicated that both relations occur at about equal frequency.

Branches of the external carotid artery (**6**.68,72) are:

1. Superior thyroid	5. Occipital
2. Ascending pharyngeal	6. Posterior auricular
3. Lingual	7. Superficial temporal
4. Facial	8. Maxillary

1. The Superior Thyroid Artery (6.68)

This arises from the front of the external carotid artery just below the level of the greater cornu of the hyoid, dividing into terminal branches at the apex of the thyroid lobe, but it may issue from the common carotid (**6**.66).

Relations. From an origin under the sternocleidomastoid it descends forwards in the carotid triangle along the lateral border of the thyrohyoid, covered by skin, platysma and fasciae and then deep to the omohyoid, sternohyoid and sternothyroid. *Medial* are the constrictor pharyngis inferior and external laryngeal nerve; the nerve is often posteromedial.

Branches. It supplies the adjacent muscles and thyroid gland; it anastomoses with its fellow and the inferior thyroid arteries. Glandular rami are *anterior,* along the medial side of the upper pole of the lateral lobe, supplying mainly the anterior surface, a branch crossing above the isthmus to anastomose with its fellow; and *posterior* descending on the posterior border, supplying the medial and lateral surfaces and anastomosing with the inferior thyroid artery. Sometimes a lateral branch supplies the lateral surface. The artery also has named branches: infrahyoid, superior laryngeal, sternocleidomastoid and cricothyroid.

The infrahyoid artery is small, runs along the lower border of the hyoid deep to thyrohyoid and anastomoses with its fellow.

The sternocleidomastoid artery, frequently arising from the *external* carotid, descends laterally across the carotid sheath.

The superior laryngeal artery accompanies the internal laryngeal nerve deep to the thyrohyoid, pierces the lower part of the thyrohyoid membrane, supplies the larynx and anastomoses with its fellow and the inferior laryngeal branch of the inferior thyroid.

The cricothyroid artery is small and crosses high on the cricothyroid ligament, communicating with its fellow.

2. The Ascending Pharyngeal Artery (6.79)

This, the smallest branch of the external carotid, is a long, slender vessel, arising near the external carotid's origin and ascending

between the internal carotid artery and pharynx to the cranial base; it is crossed by the styloglossus and stylopharyngeus with longus capitis posterior to it; it anastomoses with the facial artery's ascending palatine branch. Its named branches are: pharyngeal, inferior tympanic and meningeal.

Three or four **pharyngeal arteries** supply the constrictors and stylopharyngeus. A variable ramus supplies the palate and may replace the facial's ascending palatine branch; it descends forwards between the superior border of the superior constrictor and the levator veli palatini, accompanying the latter to the soft palate; it gives minute branches to the tonsil and one to the auditory tube.

The inferior tympanic artery, a small branch, traverses the temporal canaliculus for the tympanic branch of the glossopharyngeal nerve to supply the tympanic cavity's medial wall.

The meningeal branches are small vessels to the nerves, dura mater and adjacent bone, entering the cranium through the foramen lacerum, jugular foramen and hypoglossal canal. They supply the nerves in these passages and their surrounding tissues.

Numerous small rami supply the longi capitis et colli, the sympathetic trunk, hypoglossal, glossopharyngeal and vagus nerves and cervical lymph nodes, anastomosing with rami of the ascending cervical and vertebral arteries.

3. The Lingual Artery (6.68)

This vessel, bringing the chief supply to the tongue and buccal floor, arises anteromedially from the external carotid opposite the tip of the hyoid's greater cornu, between the superior thyroid and facial arteries. Ascending medially at first, it loops down and forwards to the greater cornu, passes medial to the posterior border of the hyoglossus and horizontally forwards deep to it and, ascending again almost vertically, courses sinuously forwards on the tongue's inferior surface as far as its tip (**6**.71). Its relation to the hyoglossus naturally divides the vessel into descriptive 'thirds'.

Relations. In its *first part* the lingual artery is in the carotid triangle; superficial to it are the skin, fascia and platysma; the middle pharyngeal constrictor is medial. It ascends a little medially, then descends to the level of the hyoid bone, its loop crossed externally by the hypoglossal nerve. Its *second part* passes along the hyoid's upper border, deep to the hyoglossus, the tendon of digastric, stylohyoid, the lower part of the submandibular gland and posterior part of the mylohyoid; the hyoglossus separates it from the hypoglossal nerve and its vena comitans; here its medial aspect adjoins the middle constrictor and crosses the stylohyoid ligament; it is accompanied by lingual veins (p. 797). The *third part* is the *arteria profunda linguae,* which turns upward near the anterior border of the hyoglossus, passing forwards close to the inferior lingual surface near the frenulum, accompanied by the lingual nerve. Medial to it is the genioglossus, lateral to it the longitudinalis linguae inferior, below it the lingual mucous membrane. Near the lingual tip it anastomoses with its fellow.

Branches of the lingual artery:

The suprahyoid artery is very small, it runs along the hyoid's upper border to anastomose with the contralateral artery.

The dorsal lingual arteries, usually two or three small vessels, arise medial to the hyoglossus, ascending to the posterior part of the lingual dorsum to supply its mucous membrane, palatoglossal arch, tonsil, soft palate and epiglottis; they anastomose with the opposite vessels.

The sublingual artery arises at the anterior margin of hyoglossus and goes forward between the genioglossus and mylohyoid to the sublingual gland, supplying this, the mylohyoid and the buccal and gingival mucous membranes. One ramus courses through the mandibular gingiva to anastomose with its fellow; another pierces the mylohyoid and joins the submental branches of the facial artery.

The lingual artery often arises with the facial or, less often, with the superior thyroid artery. It may be replaced by a ramus of the maxillary artery.

4. The Facial Artery (6.72,79)

This arises anteriorly from the external carotid in the carotid triangle above the lingual artery and immediately above the greater cornu of the hyoid bone. Medial to the mandibular ramus it arches upwards and grooves the submandibular gland's posterior aspect; it then turns down again between the gland and the medial pterygoid. Reaching the surface of the mandible it curves round its inferior border anterior to the masseter to enter the face. Here it ascends forwards across the mandible and buccinator to traverse a cleft in the modiolus (p. 574) near the buccal angle. It then ascends the side of the nose and ends at the medial palpebral commissure, supplying the lacrimal sac and joining the dorsal nasal branch of the ophthalmic artery. The artery is very sinuous throughout: in the neck perhaps to adapt to the movements of the pharynx during deglutition and on the face to movements of the mandible, lips and cheeks. Distal to its superior branch (6.72) it is termed the *angular artery*. Facial artery pulsation is most palpable where it crosses the mandibular base and between thumb and finger, near the buccal angle.

Relations. In the neck, at its origin, the artery is superficial, covered by the skin, platysma and fasciae and often crossed by the hypoglossal nerve. It runs up and forwards, deep to the digastric and the stylohyoid and posterior part of the submandibular gland. At first on the middle pharyngeal constrictor, it may reach the lateral surface of the styloglossus, separated there from the tonsil only by this muscle and the lingual fibres of the superior constrictor. Thence it descends to the lower border of the mandible in a lateral groove on the submandibular gland. *In the face*, where, as noted, its pulse can be felt as it crosses the mandible, it is superficial and at first just beneath the platysma. It is covered by skin, the fat of the cheek and near the buccal angle by superficial modiolar muscles (p. 574). Deep to it are the buccinator and levator anguli oris; it may pass over or through the levator labii superioris. Terminally it is embedded in the levator labii superioris alaeque nasi. The facial vein is posterior, in a more direct course across the face; at the anterior border of the masseter the two are in contact; in the neck the vein is superficial. Branches of the facial nerve cross forwards over the artery, which supplies the muscles and tissues of the face, submandibular gland, tonsil and soft palate. Its branches are cervical and facial.

Cervical Branches of the Facial Artery

The ascending palatine artery (6.79), starting near the facial's origin, ascends between the styloglossus and stylo-pharyngeus to the side of the pharynx, along which it ascends between the superior constrictor and the medial pterygoid towards the cranial base. Near the levator veli palatini it bifurcates: one branch follows this muscle, winds over the upper border of the superior constrictor, supplies the soft palate and anastomoses with its fellow and the greater palatine branch of the maxillary artery; the other branch pierces the superior constrictor to supply the tonsil and pharyngotympanic tube, joining with tonsillar and ascending pharyngeal arteries.

The tonsillar artery, the main supply to the tonsil, sometimes arises from the ascending palatine, though usually separate; it ascends between the medial pterygoid and styloglossus and at the latter's upper border it perforates the superior constrictor and ramifies in the tonsil and posterior lingual musculature.

The glandular branches, three or four large vessels, supply the submandibular salivary gland and lymph nodes, adjacent muscles and skin.

The submental artery, the largest cervical branch, arises as the facial quits the submandibular gland, turning forwards on the mylohyoid (6.72) below the mandible. It supplies the surrounding muscles and anastomoses with a sublingual branch of the lingual and mylohyoid branch of the inferior alveolar arteries; at the chin it ascends the mandible, dividing into superficial and deep branches which anastomose with the inferior labial and mental arteries, supplying the chin and lower lip.

Facial Branches of the Facial Artery

The inferior labial artery (6.72), arising near the buccal angle, passes up and forwards under the depressor anguli oris, penetrates the orbicularis oris and runs sinuously near the lower lip's margin between the muscle and the mucous membrane. It supplies the inferior labial glands, mucous membrane and

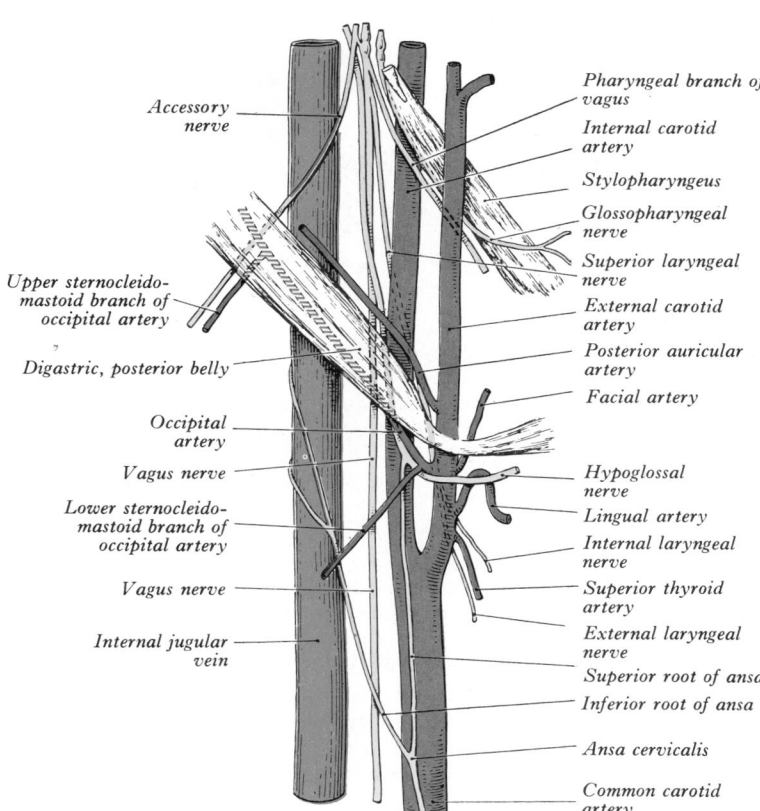

6.69 The structures crossing the internal jugular vein and carotid arteries and those intervening between the external and internal carotid arteries.

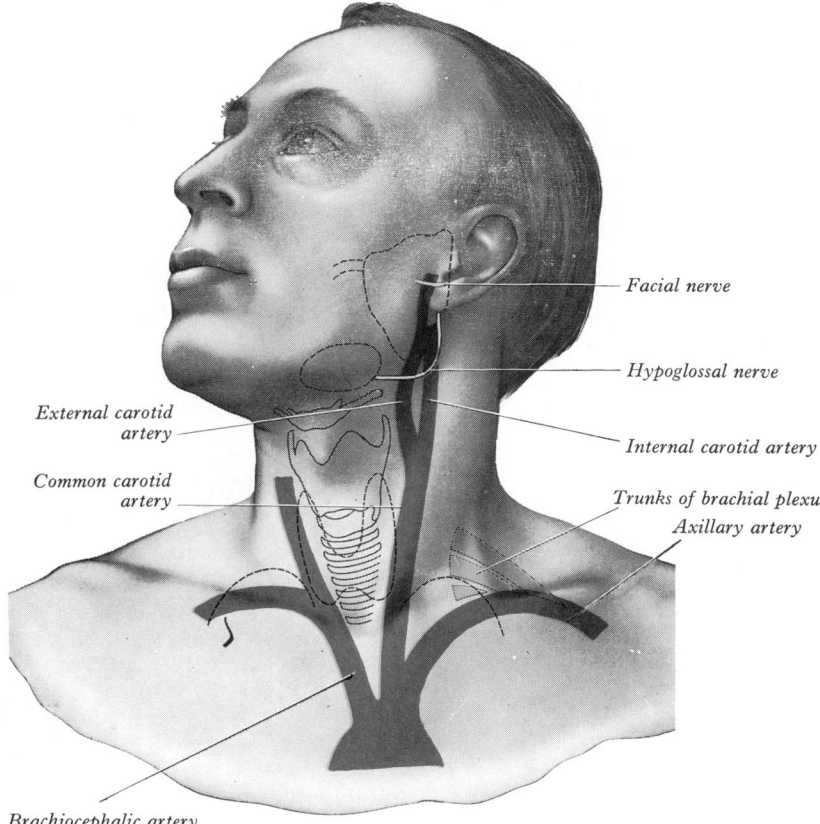

6.70 The surface projection of some of the larger structures in the face and neck. Note that the parotid gland and duct, the submandibular and thyroid glands and the apices of the lungs are shown as interrupted outlines; the hyoid bone and the thyroid, cricoid and tracheal cartilages are indicated by continuous outlines.

737

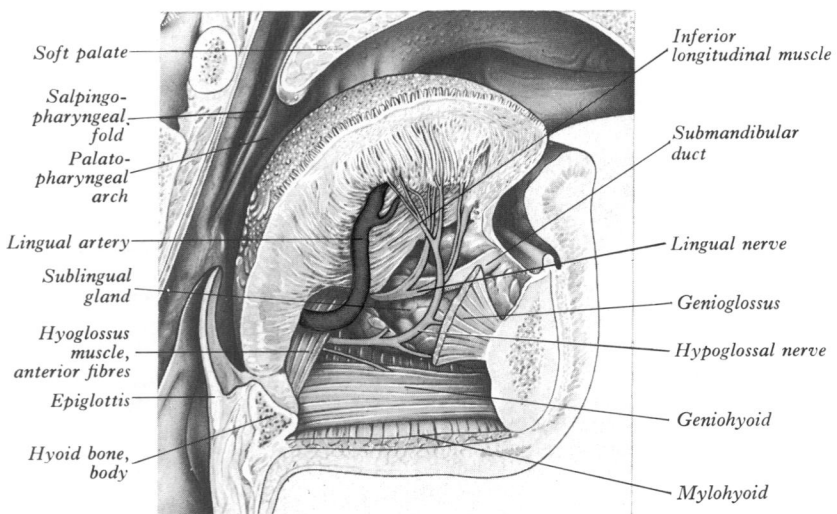

6.71 Dissection of the left half of the tongue from the medial side, exposing the end of the second part and the beginning of the third part of the left lingual artery and adjoining structures, in an edentulous subject.

muscles, anastomosing with its fellow and the mental branch of the inferior alveolar artery.

The superior labial artery (6.72), larger and more tortuous than the inferior, has a similar course along the superior labial margin between the mucous membrane and the orbicularis oris; it anastomoses with its fellow, supplying the upper lip, a *septal branch*, which ramifies antero-inferiorly in the nasal septum, and an *alar branch*.

The lateral nasal artery (6.72), branching from the facial as it ascends the side of the nose, supplies the nasal ala and dorsum, anastomosing with its fellow, the septal and alar branches of the superior labial, dorsal nasal ramus of the ophthalmic and infraorbital branch of the maxillary artery. It may be replaced by several small rami or arise from the superior labial, diverging from its septal branch (as in 6.72).

Facial anastomoses are numerous not only with corresponding contralateral rami but also: *in the neck*, with the sublingual branch of the lingual, ascending pharyngeal and palatine branch of the maxillary; *on the face*, with the mental branch of the inferior alveolar, transverse facial branch of the superficial temporal, infraorbital branch of the maxillary and dorsal nasal branch of the ophthalmic. The anastomoses in the lips are by main trunks, an important fact in labial injuries.

Variations. The facial artery may arise with the lingual, as a *linguo-facial trunk*. It varies in size and supply to the face: it may end as the submental artery and often extends only to the buccal angle. The deficiency is then filled by rami of neighbouring arteries. In 110 human fetuses a common linguo-facial trunk occurred in 43%; in 42% the facial did not reach the medial orbital angle, ending as a superior (20%) or inferior (22%) labial artery (Kozielec & Jozwa 1977).

5. The Occipital Artery (6.73)

This artery arises from the back of the external carotid, opposite the facial; at first medial to the posterior belly of the digastric, it ends posteriorly in the scalp.

Course and relations. At its origin, the artery is crossed superficially by the hypoglossal nerve, winding round it from behind. It goes back, up and deep to the posterior digastric belly, crossing the internal carotid, internal jugular vein, hypoglossal, vagal and accessory nerves (6.73). Between the transverse process of the atlas and temporal mastoid process it reaches the lateral border of

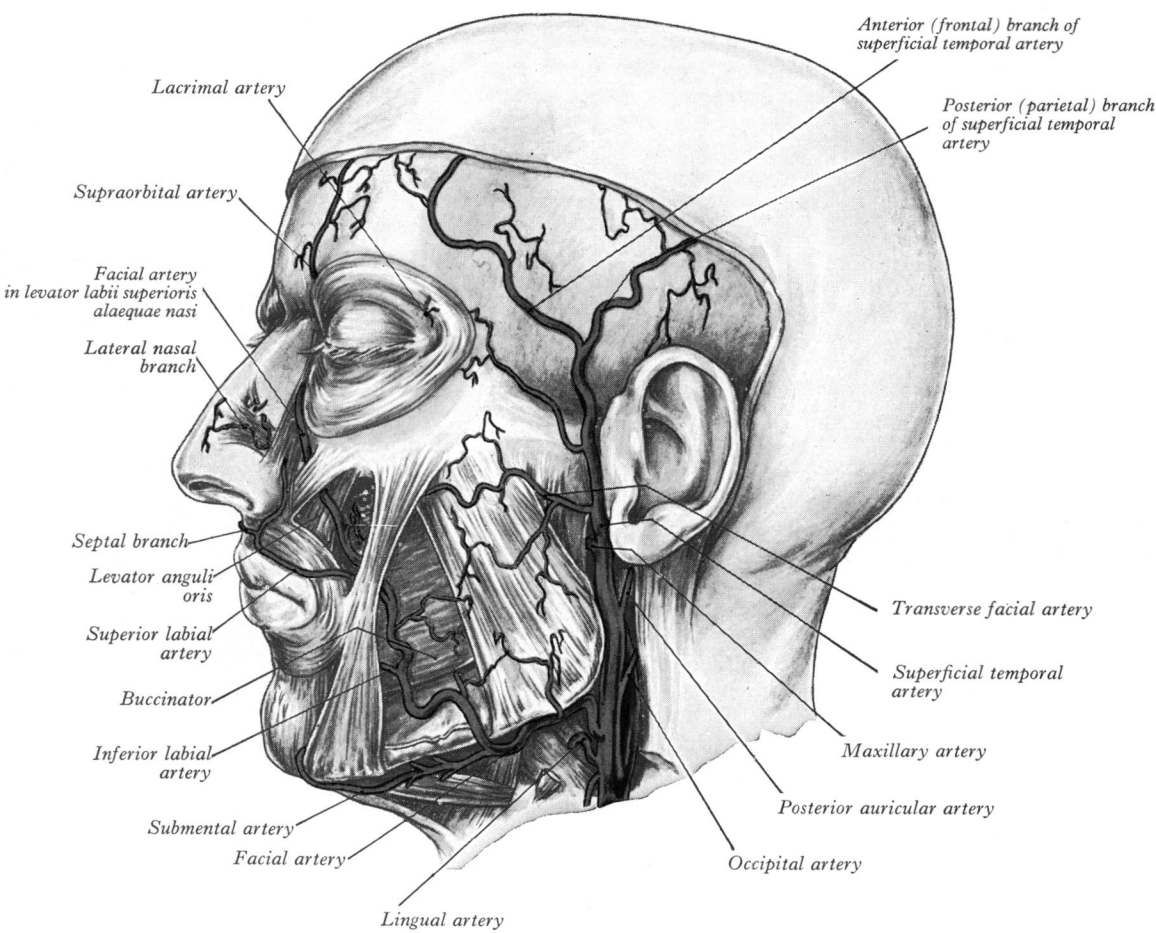

6.72 The arteries of the left side of the face and their main branches. Many of the post-modiolar muscles and part of the modiolus (through which the facial artery passes) have been resected. Note the less usual origin of the lateral nasal branch in this specimen.

the rectus capitis lateralis. It then runs in the temporal bone's occipital groove, medial to the mastoid process and attachments of the sternocleidomastoid, splenius capitis, longissimus capitis and digastric, lying successively on the rectus capitis lateralis, obliquus superior and semispinalis capitis. Finally, it turns up to pierce the fascia connecting the cranial attachments of the trapezius and sternocleidomastoid, ascends tortuously in the dense superficial fascia of the scalp and divides into many branches. Its terminal part is accompanied by the greater occipital nerve and its branches are as follows.

Two sternocleidomastoid branches are usual, *the lower* arising near the origin of the occipital but sometimes directly from the external carotid. It descends backwards over the hypoglossal nerve and internal jugular vein, enters the sternocleidomastoid, and anastomoses with the sternocleidomastoid branch of the superior thyroid. The *upper branch* arises as the occipital crosses the accessory nerve, running down and backwards superficial to the internal jugular vein. It enters the deep surface of the sternocleidomastoid with the accessory nerve.

The mastoid artery, small in size and sometimes absent, enters the cranial cavity via the mastoid foramen, supplying the mastoid air cells and dura mater.

The stylomastoid artery branches from the occipital in two-thirds of subjects (p. 740).

An auricular branch supplies the medial aspect of the auricle, anastomosing with the posterior auricular artery.

Muscular branches supply the digastric, stylohyoid, splenius, longissimus capitis and neighbouring muscles.

A descending branch (6.73) arises where the occipital adjoins the obliquus superior, dividing into superficial and deep rami. The *superficial ramus* passes deep to the splenius, anastomosing with the superficial branch of the transverse cervical artery; the *deep ramus* descends between the semispinales capitis et cervicis, anastomosing with both the vertebral and the deep cervical artery (from the costocervical trunk) (6.79).

Meningeal branches enter the cranium via the jugular foramen and condylar canal to supply the dura mater and bone of the posterior cranial fossa and the caudal four cranial nerves.

Occipital branches, tortuous terminal rami distributed to the scalp as far as the vertex, run between the skin and the occipital belly of the occipitofrontalis, anastomosing with the opposite occipital, posterior auricular and temporal arteries and supplying the occipital belly of the occipitofrontalis, skin and pericranium. One may have a meningeal ramus, traversing the parietal foramen.

6. The Posterior Auricular Artery (6.72)

This small vessel branches posteriorly from the external carotid just above the digastric and stylohoid. It ascends between the parotid gland and the styloid process to the groove between the auricular cartilage and the mastoid process, dividing into auricular and occipital branches. As well as supplying the digastric, stylohyoid, sternocleidomastoid, and parotid gland, the posterior auricular artery has three named branches:

The stylomastoid artery, an indirect branch of the posterior auricular in about a third of subjects (Blunt 1954), enters the stylomastoid foramen to supply the facial nerve, tympanic cavity, mastoid antrum and air cells, and semicircular canals. In the young its posterior tympanic ramus forms a circular anastomosis with the anterior tympanic artery (p. 740).

The auricular branch, ascending deep to auricularis posterior, ramifies on the cranial aspect of the auricle; some branches pierce this, others curve round it to supply its lateral aspect.

The occipital branch passes laterally across the mastoid process, turning back over the sternocleidomastoid to supply the occipital belly of the occipitofrontalis and scalp above and behind the ear; it anastomoses with the occipital artery.

7. The Superficial Temporal Artery (6.68)

This, the smaller terminal branch of the external carotid, begins in the parotid gland behind the mandible's neck, crosses the posterior root of the zygomatic process of the temporal bone and about 5 cm above this divides into anterior and posterior branches.

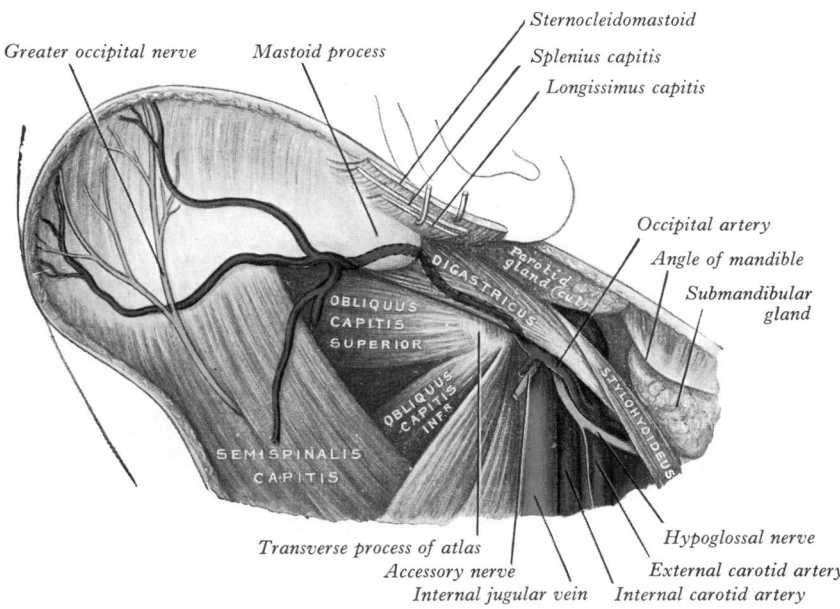

6.73 Dissection to show the course of the occipital artery. The upper and lower sternocleidomastoid branches of the artery have been transected and are not labelled.

Relations. As it crosses the zygoma it is covered by the auricularis anterior; in the parotid gland temporal and zygomatic branches of the facial nerve cross it; in the scalp it is accompanied by corresponding veins, and just posterior to it lies the auriculotemporal nerve.

Branches. The superficial temporal supplies the parotid gland, temporomandibular joint and masseter and it also has several named branches:

The transverse facial artery (6.72) arises before the superficial temporal emerges from the parotid gland; it traverses the gland, crosses the masseter between the parotid duct and the zygomatic arch, accompanied by one or two facial nerve branches, and divides into numerous rami supplying the parotid gland and duct, masseter and skin, anastomosing with the facial, masseteric, buccal, lacrimal and infraorbital arteries.

The anterior auricular branches are distributed to the lobule and anterior part of the auricle and the external acoustic meatus.

The zygomatico-orbital artery, sometimes from the middle temporal, skirts the upper border of the zygomatic arch between two layers of temporal fascia to the lateral orbital angle; supplying the orbicularis oculi, it anastomoses with the lacrimal and palpebral rami of the ophthalmic artery.

The middle temporal artery branches just above the zygomatic arch, perforates the temporal fascia, supplies the temporalis and anastomoses with the deep temporal branches of the maxillary.

The frontal (anterior) branch meanders towards the frontal tuberosity, supplying muscles, skin and pericranium in this region; it anastomoses with its fellow and the supraorbital and supratrochlear arteries.

The parietal (posterior) branch, larger than the frontal, curves up and back superficial to the temporal fascia, anastomosing with its fellow and the posterior auricular and occipital arteries.

Variation in the superficial temporal artery is largely in the relative sizes of the frontal, parietal and transverse facial branches; the first two may be absent, the transverse facial may replace a shortened facial artery. Variations in fetal material have been described by Kozielec & Jozwa (1976).

Applied Anatomy. Crossing the zygomatic process the artery is palpable through skin and fascia and is easily compressed here to control temporal haemorrhage. This vessel and other arteries supplying the scalp from below are well protected by dense tissue. Rarely are all implicated in a scalping injury and its branches

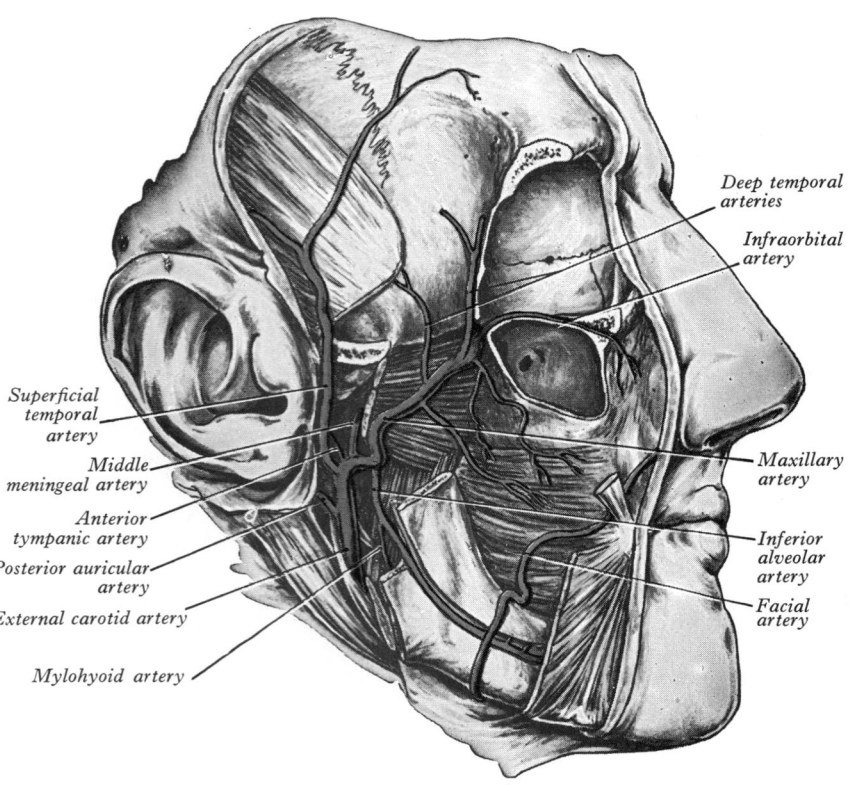

6.74 The right maxillary artery. An extensive dissection has been carried out, involving the removal of the parotid gland, the zygomatic arch, part of the ramus of the mandible, the lateral walls of the orbit and maxillary sinus and the orbital contents.

Labels on figure:
- Deep temporal arteries
- Infraorbital artery
- Superficial temporal artery
- Middle meningeal artery
- Anterior tympanic artery
- Posterior auricular artery
- External carotid artery
- Mylohyoid artery
- Maxillary artery
- Inferior alveolar artery
- Facial artery

gland behind the temporomandibular joint, pierces the cartilaginous or osseous wall of the external acoustic meatus and supplies its cuticular lining, the exterior of the tympanic membrane and the joint.

The anterior tympanic artery ascends behind the temporomandibular joint, enters the tympanic cavity through the petrotympanic fissure and ramifies on the interior of the tympanic membrane, forming a vascular circle around it with the posterior tympanic branch of the stylomastoid; it anastomoses with twigs of the artery of the pterygoid canal and caroticotympanic branches of the internal carotid artery in the mucosa of the tympanic cavity.

The middle meningeal artery, largest of the meningeal arteries, ascends between the sphenomandibular ligament and lateral pterygoid, passes between the roots of the auriculotemporal nerve and may lie lateral to the tensor veli palatini before entering the cranial cavity through the foramen spinosum. It then runs in an anterolateral groove on the squamous part of the temporal bone, dividing into frontal and parietal branches. The *frontal (anterior) branch*, the larger, crosses the greater wing of the sphenoid, reaches a groove or canal in the parietal's sphenoidal angle and divides into branches between the dura mater and cranium, some ascending to the vertex, others to the occipital region. One ascending branch grooves the parietal bone about 1.5 cm behind the coronal suture, corresponding approximately to the precentral sulcus. The *parietal (posterior) branch* curves back on the squamous temporal bone, reaching the lower border of the parietal anterior to its mastoid angle and dividing to supply the posterior parts of the dura mater and cranium. These branches anastomose with their fellows and with the anterior and posterior meningeal arteries.

In the cranial cavity the artery has the following branches: (1) Numerous *ganglionic branches* supply the trigeminal ganglion and roots. (2) A *petrosal branch* enters the hiatus for the greater petrosal nerve and supplies the facial nerve, ganglion and tympanic cavity, anastomosing with the stylomastoid artery (pp. 739, 740). (3) A *superior tympanic artery* runs in the canal for the tensor tympani, supplying both muscle and the canal's lining membrane. (4) *Temporal branches* traverse minute foramina in the sphenoid's greater wing and anastomose with deep temporal arteries. (5) An *anastomotic branch* (p. 746) enters the orbit lateral in the superior orbital fissure, anastomosing with a recurrent branch of the lacrimal artery; enlargement of this anastomosis explains an occasional origin of the lacrimal from the middle meningeal artery. Apart from these and a supply to the dura mater, the middle meningeal artery is predominantly periosteal, supplying bone and red bone marrow.

Surface Anatomy (**6.**119). The *middle meningeal artery* enters the skull medial to the zygoma's midpoint (**6.**119), dividing 2 cm above this. From here the frontal branch runs first up and forwards to the pterion and then up and back towards a point midway between the inion and nasion. The parietal branch runs up and back towards the lambda.

Applied Anatomy. The middle meningeal artery may be torn in temporal fractures or by injuries separating the dura mater from the bone, followed by haemorrhage between them. Trephining may be necessary to reduce cerebral compression.

The accessory meningeal artery may arise from the maxillary or the middle meningeal. It enters the cranial cavity through the foramen ovale, supplying the trigeminal ganglion, dura mater and bone, but its main distribution is *extracranial* (Baumel & Beard 1961), principally the medial pterygoid, lateral pterygoid (upper head), tensor veli palatini, sphenoid bone (greater wing and pterygoid processes), mandibular nerve and otic ganglion. It is sometimes replaced by separate small arteries.

The inferior alveolar (dental) artery descends posterior to the inferior alveolar nerve, to the mandibular foramen. Here it is between bone laterally and the sphenomandibular ligament medially. Before entering the foramen it has a *mylohyoid branch*, which pierces the sphenomandibular ligament to descend with the mylohyoid nerve in its groove on the mandibular ramus; it ramifies superficially on the muscle and anastomoses with the facial's submental branch. The inferior alveolar artery then traverses the mandibular canal with the inferior alveolar nerve and divides into the incisor and mental branches near the firs

anastomose so freely that a partially detached scalp may be replaced with reasonable hope of success as long as one vessel is intact. In craniotomy, incisions should be convex upwards to include the superficial temporal artery in the flap. In carotid angiograms branches of the superficial temporal and middle meningeal arteries are superimposed, but are distinguishable by the straighter course, lack of anastomoses and narrower calibre in the meningeal rami (Domnić-Stošić & Jeličić 1974).

8. The Maxillary (Internal Maxillary) Artery (6.74)

This, the larger terminal branch of the external carotid, arises behind the mandibular neck, at first embedded in the parotid gland; it then passes medial to the mandibular neck and superficial or deep to the lower head of the lateral pterygoid to reach the pterygopalatine fossa, usually passing between the two heads of the lateral pterygoid. It has mandibular, pterygoid and pterygopalatine segments, related sequentially to bone, muscle and bone, a useful indication of its branches.

The *first, mandibular part* is horizontal and passes between the mandible's neck and the sphenomandibular ligament, parallel with and slightly below the auriculotemporal nerve; it crosses the inferior alveolar nerve and skirts the lower border of the lateral pterygoid.

The *second, pterygoid part* ascends obliquely forwards medial to the temporalis and superficial to the lower head of the lateral pterygoid; it is often deep to the latter, lying between it and branches of the mandibular nerve and it may then project as a lateral loop between the two parts of the lateral pterygoid.

The *third, pterygopalatine part* passes between the heads of the lateral pterygoid and through the pterygomaxillary fissure into the pterygopalatine fossa, where it is situated anterior to the pterygopalatine ganglion.

Branches. The artery is distributed to the mandible, maxilla, teeth, muscles of mastication, palate, nose and cranial dura mater. Its branches form three groups, corresponding with its parts.

Branches of the First Part (**6.**74). **The deep auricular artery,** often arising with the anterior tympanic, ascends in the parotid premolar. The *incisor* branch continues below the incisor teeth to the midline, where it anastomoses with its fellow. In the canal the arteries supply the mandible, tooth sockets and teeth by rami entering minute holes at apices of their roots to supply the pulp. The *mental branch* leaves the mental foramen, supplies the chin and anastomoses with the submental and inferior labial arteries. Near its origin the inferior alveolar artery has a *lingual* branch, which descends with the lingual nerve to supply the buccal mucous membrane.

Branches of the Second Part (6.74). **Deep temporal branches**, anterior and posterior, ascend between the temporalis and bone, supplying mainly the former. They anastomose with the middle temporal artery. The anterior connects with the lacrimal by small rami perforating the zygomatic bone and greater wing of the sphenoid.

Pterygoid branches, irregular in number and origin, supply the pterygoid muscles.

The masseteric artery is small and with the masseteric nerve passes behind the tendon of temporalis through the mandibular incisure (notch) to the deep surface of masseter, in which it anastomoses with the masseteric branches of the facial and transverse facial arteries.

The buccal artery runs obliquely forwards with the buccal nerve between the medial pterygoid and the attachment of the temporalis to supply the external surface of the buccinator (and through it the mucosa), anastomosing with rami of the facial and infraorbital arteries.

Branches of the Third Part. **The posterior superior alveolar (dental) artery** leaves the maxillary artery as it enters the pterygopalatine fossa. Descending on the maxilla's infratemporal surface, it divides, some branches entering the alveolar canals to supply molar and premolar teeth and the maxillary sinus, others continuing over the alveolar process to supply the gingivae.

The infraorbital artery often arises with the posterior superior alveolar, entering the orbit posteriorly through the inferior orbital fissure, to run in the infraorbital groove and canal with the infraorbital nerve, both emerging on the face via the infraorbital foramen, deep to the levator labii superioris. In the canal it has (1) *orbital branches*, which supply the rectus inferior, obliquus inferior and lacrimal sac, and (2) *anterior superior alveolar (dental) branches*, which descend via the anterior alveolar canals to supply the upper incisor and canine teeth and mucous membrane in the maxillary sinus. On the face some branches ascend to the medial canthus and lacrimal sac, anastomosing with the terminal rami of the facial; others anastomose with a dorsal nasal branch of the ophthalmic artery and some descend between the levator labii superioris and levator anguli oris, anastomosing with the facial, transverse facial and buccal arteries.

The remaining branches arise in the pterygopalatine fossa. **The greater palatine artery** and nerve descend in their palatine canal; the artery gives off two or three *lesser palatine arteries*, transmitted through lesser palatine canals to supply the soft palate and tonsil, anastomosing with the ascending palatine. The main vessel emerges on the palate's oral surface by the greater palatine foramen and runs in a curved groove near the alveolar border of the hard palate to the incisive canal; it ascends this canal and anastomoses with a branch of the sphenopalatine artery. It distributes rami to the gingivae, palatine glands and mucous membrane.

The pharyngeal artery, very small, runs back through the pharyngeal (palatovaginal) canal with the pharyngeal branch of the pterygopalatine ganglion; it supplies mucosa of the nasal roof, the nasopharynx, sphenoidal air sinus and auditory tube.

The artery of the pterygoid canal, frequently from the greater palatine, passes back in the pterygoid canal with the corresponding nerve, supplying its walls and contents and the mucous membrane of the upper pharynx, pharyngotympanic tube and tympanic cavity.

The pharyngeal artery is medial, that of the pterygoid canal lateral and the trunk of the maxillary artery passes anterior to the pterygopalatine ganglion.

The sphenopalatine artery, the termination of the maxillary, traverses the sphenopalatine foramen into the walls of the nasal cavity posterior in the superior meatus. Here its *posterior*

lateral nasal branches ramify over the conchae and meatuses, anastomosing with the ethmoidal arteries and nasal branches of the greater palatine, supplying the frontal, maxillary, ethmoidal and sphenoidal sinuses. Crossing anteriorly on the inferior sphenoid surface, the artery ends on the nasal septum as the *posterior septal branches*, which anastomose with the ethmoidal arteries; one ramus descends on the vomer to the incisive canal to join the end of the greater palatine artery and septal branch of the superior labial.

Collateral circulation, after interruption of one common carotid, is often established by the connections across the midline between the carotids, intra-and extracranial, and by enlargement of the subclavian branches. Chief extracranial connections are between superior and inferior thyroid arteries, the deep cervical and the descending branch of the occipital; the vertebral artery substitutes for the internal carotid in the cranium. Nevertheless symptoms of cerebral disturbance supervene in about 25%.

After interruption of the *external* carotid, circulation is maintained by anastomoses between most of its large branches (facial, lingual, superior thyroid, occipital) and their fellows, by their anastomosis with branches of the *internal* carotid and of the occipital with branches of the subclavian, etc.

The Triangles of the Neck

Anterolaterally the neck (**6.75**) presents a somewhat quadrilateral area, limited *above* by the base of the mandible and a line continued from its angle to the mastoid process, *below* by the clavicle's upper border, *in front* by the anterior median line, *behind* by the anterior margin of trapezius. This region is divided by the sternocleidomastoid, ascending obliquely from the sternum and clavicle to the mastoid process and occipital bone. The area anterior to this is the *anterior triangle* and that behind it the *posterior triangle*. While these and their subdivisions are emphasized by some as being purely arbitrary because many major structures (arteries, veins, lymphatics, nerves, some viscera) transgress their boundaries without interruption, nevertheless they have a topographical value in description. However, two further points should be made. Some of their subdivisions *are* easily identified by inspection and palpation and provide invaluable assistance in surface anatomical and clinical examination (vide infra). As the neck has a roughly cylindrical form, crossed obliquely by the sternocleidomastoid, the names anterior and

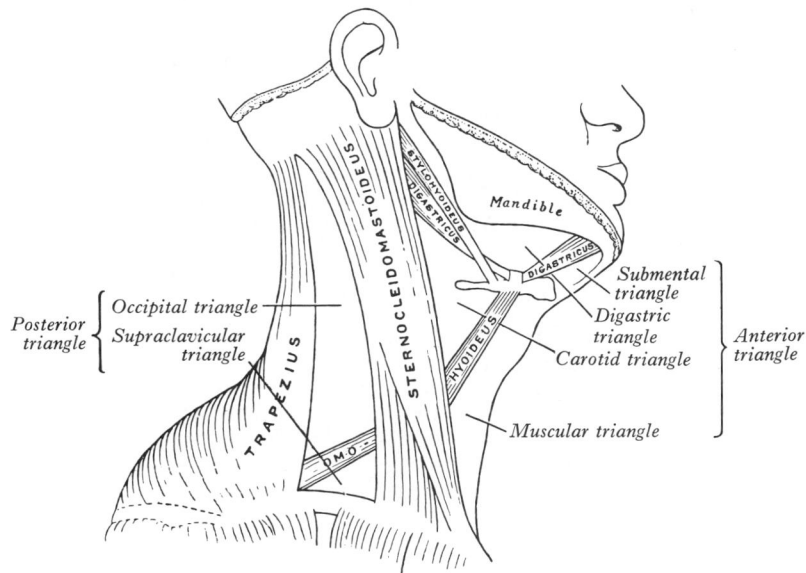

6.75 The triangles of the right side of the neck: a highly schematic two dimensional representation of what in reality are non-planar trigones distributed over a waisted column. Submandibular would be an alternative name for digastric.

posterior are not particularly apt; the triangles are not plane (and coplanar as represented in a two-dimensional diagram such as **6**.75) but both are spiralized regions (trigones) on the surface of the column.

THE ANTERIOR CERVICAL TRIANGLE

This is bounded anteriorly by the median line and posteriorly by the anterior margin of the sternocleidomastoid, its base being the inferior mandibular border and its mastoid extension noted above; its apex is at the manubrium. It may be subdivided into muscular, carotid, digastric and submental triangles.

The muscular triangle is bounded by the median line from the hyoid bone to the sternum, inferoposteriorly by the anterior margin of the sternocleidomastoid and posterosuperiorly by the superior belly of the omohyoid.

The carotid triangle is limited posteriorly by the sternocleidomastoid, antero-inferiorly by the superior belly of the omohyoid and superiorly by the stylohyoid and posterior belly of the digastric. In the living, except the obese, the triangle is usually a small visible triangular depression, sometimes best seen with the head and cervical vertebral column slightly extended and the head contralaterally rotated. Often the latter position is quite unnecessary; judicious oblique lighting (window or lamp) throws the hollow into relief.

It is covered by the skin, superficial fascia, platysma and deep fascia containing branches of facial and cutaneous cervical nerves. The *hyoid bone* forms its *anterior angle* and adjacent floor; its position can be located immediately on simple inspection, verified by palpation. Parts of thyrohyoid, hyoglossus and inferior and

middle pharyngeal constrictors form its floor. It contains the upper part of the common carotid and its division into external and internal carotid arteries, overlapped by the anterior margin of the sternocleidomastoid; the external carotid is first anteromedial, then anterior to the internal. Branches of the external carotid are also encountered: the superior thyroid runs antero-inferiorly, the lingual anteriorly with its upward loop, the facial anterosuperiorly, the occipital posterosuperiorly and the ascending pharyngeal medial to the internal carotid. Massive arterial pulsation greets the examining finger. The veins correspond to the branches of the external carotid artery: superior thyroid, lingual, facial, ascending pharyngeal and sometimes the occipital, all ending in the internal jugular vein. The hypoglossal nerve crosses both carotid arteries, curving round the origin of the lower sternocleidomastoid branch of the occipital, where the superior root of the ansa cervicalis leaves it, descending anteriorly in the carotid sheath. Medial to the external carotid, below the hyoid bone, is the internal laryngeal nerve and, below this, the external laryngeal. Many structures in this region, such as all or part of the internal jugular vein, associated deep cervical lymph nodes, vagus nerve, etc., may be variably obscured by the sternocleidomastoid and, pedantically, are thus 'outside the triangle'; much more importantly, their location is obvious during clinical examination.

The digastric triangle is bordered above by the base of the mandible (and its projection to the mastoid process), postero-inferiorly by the posterior belly of the digastric and stylohyoid and antero-inferiorly by the anterior belly of digastric. It is covered by the skin, superficial fascia, platysma and deep fascia, in which are branches of facial and transverse cutaneous cervical nerves. Its floor is formed by the mylohyoid and hyoglossus. Its anterior region contains the submandibular gland, superficial to which is the facial vein and deep to it the facial artery, crossing the lower border of the mandible at the anterior edge of the masseter; on the mylohyoid are the submental artery and mylohyoid artery and nerve. Variably related to the submandibular gland are the submandibular lymph nodes (p. 844). Its posterior region contains the lower part of the parotid gland; the external carotid, passing deep to the stylohyoid, curves above the muscle and overlaps its superficial surface where it ascends deep to the parotid gland to enter it. The external carotid, which is superficial to the internal carotid, crosses it posterolaterally; deeper and separated from the external carotid by styloglossus, stylopharyngeus and the glossopharyngeal nerve, are the internal carotid artery, internal jugular vein and vagus nerve.

The submental triangle, unpaired, is demarcated by both digastric muscles (anterior bellies); its apex is at the chin, its base the body of the hyoid and its floor the mylohyoid muscles. It contains lymph nodes and small veins uniting to form the anterior jugular.

THE POSTERIOR CERVICAL TRIANGLE

The posterior triangle is delimited anteriorly by the sternocleidomastoid, posteriorly by the anterior edge of trapezius, inferiorly by the middle third of the clavicle; its apex is between the attachments of the sternocleidomastoid and trapezius to the occiput and is often blunted, the 'triangle' becoming quadrilateral. It is crossed, about 2.5 cm above the clavicle, by the inferior belly of the omohyoid, which divides it into occipital and supraclavicular triangles.

The occipital triangle, the upper, larger part of the posterior triangle, has the same borders, except below where its limit is the omohyoid. Its floor is, from above down: splenius capitis, levator scapulae, and scaleni medius and posterior. (Sometimes semispinalis capitis appears at the apex.) It is covered by the skin, superficial and deep fasciae and below by the platysma. The accessory nerve pierces the sternocleidomastoid and crosses on the levator scapulae obliquely down and back to the deep surface of the trapezius; cutaneous and muscular branches of the cervical plexus emerge at the posterior border of the sternocleidomastoid; below, supraclavicular nerves, transverse cervical vessels and the uppermost part of the *brachial plexus* cross the triangle. Lymph nodes are arranged along the posterior border of the cleidomastoid from the mastoid process to the root of the neck.

Platysma, deep surface   Facial artery   Facial vein   Branches of facial nerve   Stylohyoid and digastric   Parotid gland

Lingual artery

Facial vein

External jugular vein

Superior root of ansa cervicalis

Great auricular nerve

Platysma, lower portion

Transverse cervical nerve

Digastric, anterior belly

Submandibular gland

Mylohyoid

Lingual veins, accompanying hypoglossal nerve

Hyoglossus

Internal laryngeal nerves

Thyrohyoid

External laryngeal nerve

Omohyoid, superior belly

Superior thyroid vessels

Sternohyoid

6.76 Dissection of the left anterior triangle. The platysma has been divided transversely; its upper part has been turned upwards on to the face, its lower part turned backwards, exposing the lower part of the sternocleidomastoid. Dominating the centre of the illustration is the carotid triangle, with many of its contents and surrounding structures.

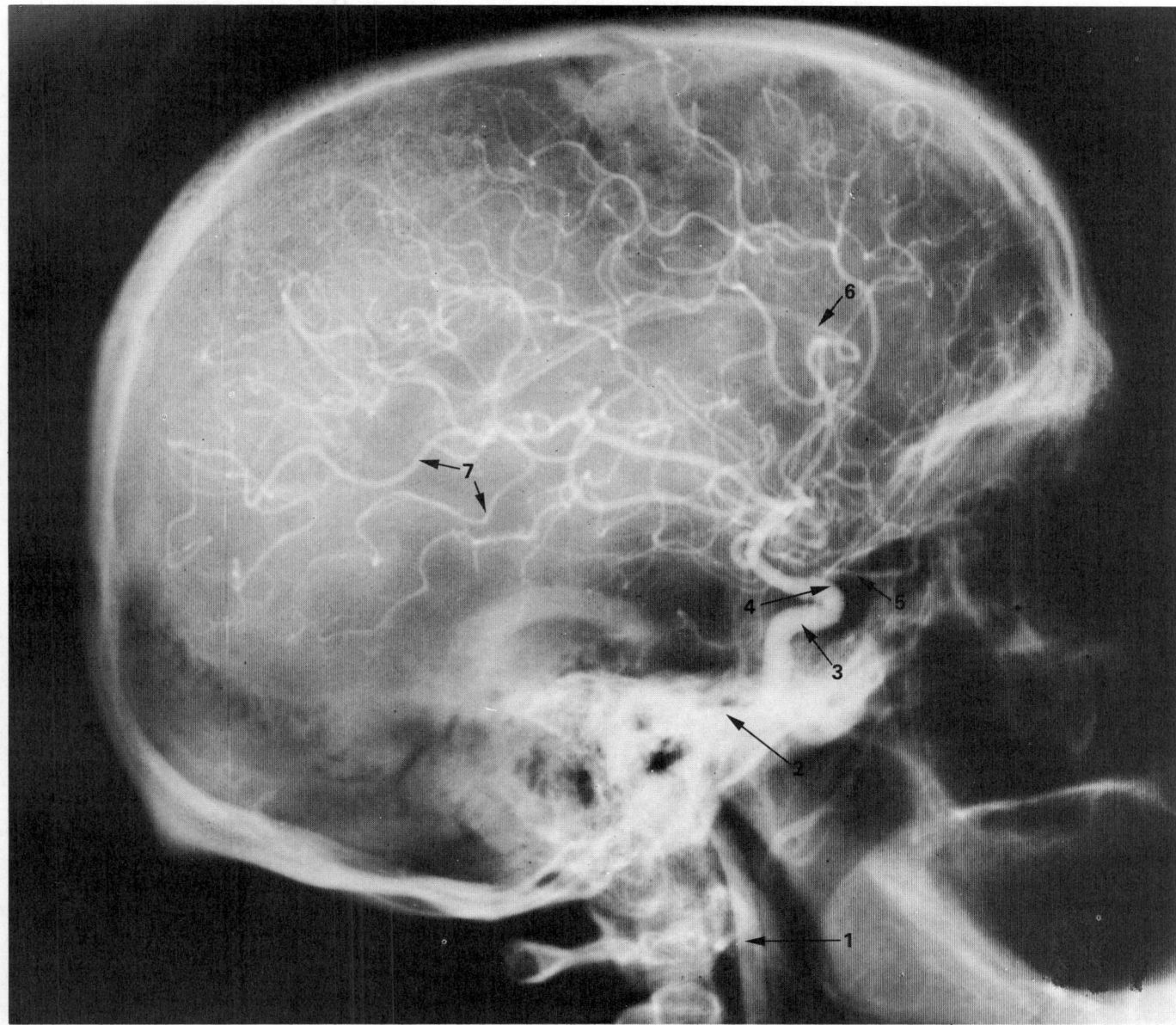

6.77 Internal carotid arteriogram (right): lateral view, in adult male of 33 years. The following can be identified: parts of the internal carotid artery (and individual vessels): 1. cervical. 2. intrapetrous. 3. cavernous. 4. terminal. 5. ophthalmic artery. 6. anterior cerebral artery. 7. branches of middle cerebral artery. Note the absence of radio-opaque injectant from the cerebellar vessels.

The supraclavicular triangle, the lower, smaller division, is bounded like the posterior triangle, except above where the omohyoid limits it. It corresponds in the living neck with the lower part of the deep, prominent hollow, the *greater supraclavicular fossa* (colloquially 'the salt cellar'). Its floor contains the first rib, scalenus medius and the first slip of serratus anterior. Its size varies with the extent of the clavicular attachments of the sternocleidomastoid and trapezius and also the level of the inferior belly of the omohyoid. The triangle is covered by the skin, superficial and deep fasciae and platysma and crossed by the supraclavicular nerves. Just above clavicular level the third part of the subclavian artery curves inferolaterally from the lateral margin of the scalenus anterior, across the first rib to the axilla. The subclavian vein is behind the clavicle and is not usually in the triangle; but it may rise as high as the artery and even accompany it behind scalenus anterior. The *brachial plexus* is partly above, partly behind the artery and closely related to it. The *trunks* of the brachial may easily be palpated here, the neck being contralaterally flexed and the examining finger drawn across the trunks at right angles to their length. With the musculature relaxed, pulsation of the subclavian artery may be felt or the arterial flow controlled by retroclavicular compression against the first rib. The suprascapular vessels pass transversely behind the clavicle; at a higher level are the transverse cervical artery and vein. The

external jugular vein descends behind the posterior border of the sternocleidomastoid to end in the subclavian vein; it receives the transverse cervical and suprascapular veins, which form a plexus in front of the third part of the subclavian artery; occasionally it is joined by a small vein crossing the clavicle anteriorly from the cephalic vein. The nerve to the subclavius also crosses this triangle and some lymph nodes are contained in it.

The Internal Carotid Artery

The internal carotid artery (**6.**77–84) supplies most of the ipsilateral cerebral hemisphere, eye and accessory organs, forehead and, in part, the nose. From the carotid bifurcation, where it usually has a carotid sinus (p. 734), it ascends to the cranial base, enters the cranial cavity by the carotid canal and turns anteriorly through the cavernous sinus in the carotid groove on the side of the sphenoid body, ending below the anterior perforated substance by division into the anterior and middle cerebral arteries.

It may be divided conveniently into cervical, petrous, cavernous and cerebral parts. In the broadest outline its course is: (1) *vertically upwards* in the neck; (2) curving *horizontally forwards and medially* in the petrous carotid canal; (3) *upwards* in the upper foramen lacerum; (4) *horizontally forwards* in the cavernous sinus;

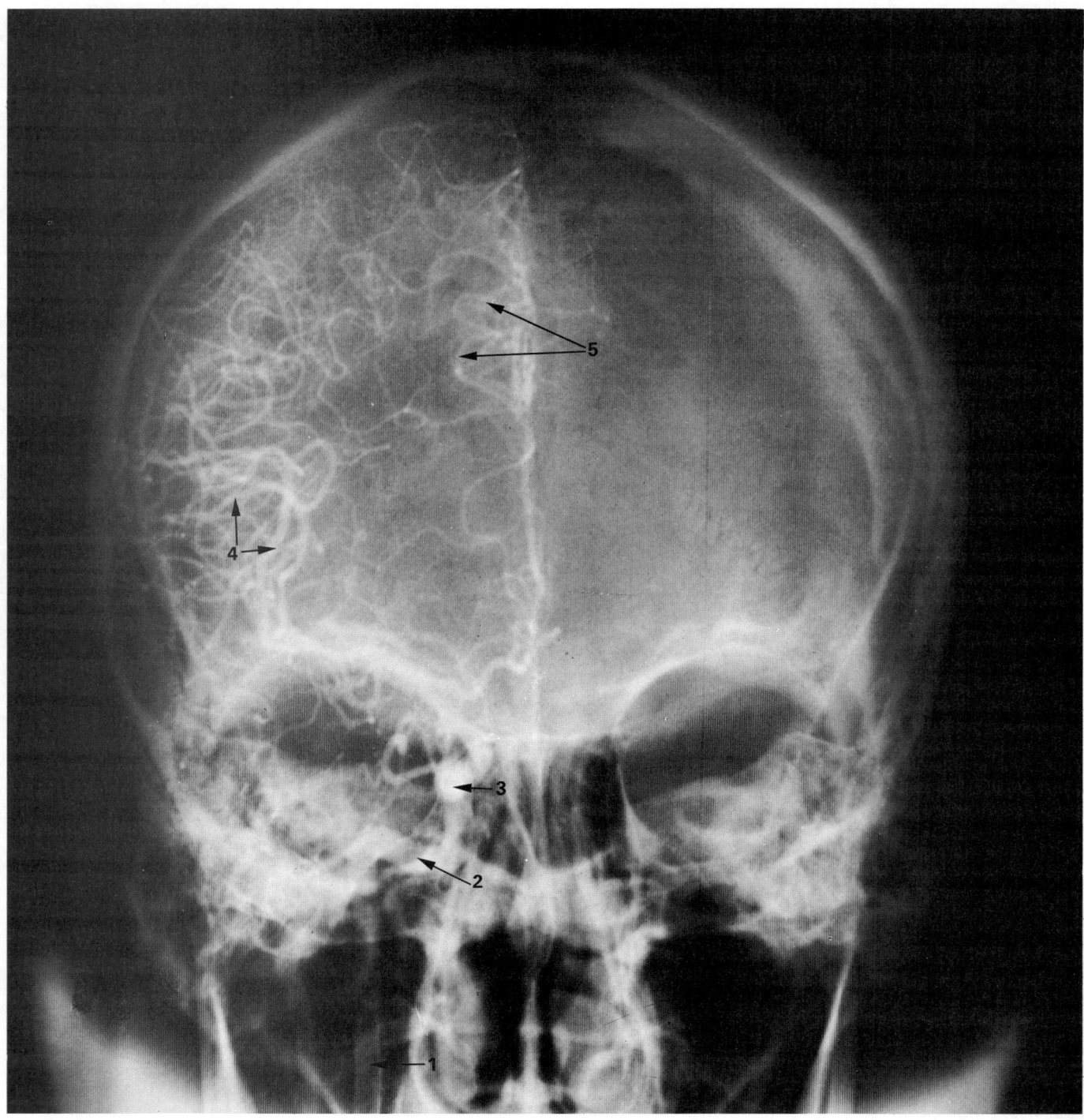

6.78 Internal carotid arteriogram (right): anteroposterior view of same subject as **6.**77. Parts of the internal carotid artery: 1. cervical. 2. intrapetrous. 3. cavernous. 4. branches of middle cerebral artery. 5. branches of anterior cerebral artery. Note the lack of contrast medium on the left side.

(5) *vertically upwards* medial to the anterior clinoid *process*; (6) looping a short distance *backwards and upwards* to its terminal division.

The Cervical Part

This section begins at the carotid bifurcation and ascends in front of the upper three cervical transverse processes to the inferior aperture of the carotid canal in the petrous temporal bone. It is superficial at first in the carotid triangle, then passes deeper, medial to the posterior belly of the digastric. Except near the skull, the internal jugular vein and vagus nerve are lateral; the external carotid is first anteromedial but then curves back to become superficial. The artery has many other relations. *Posteriorly* it adjoins the longus capitis, with the superior cervical

sympathetic ganglion between them and the superior laryngeal nerve crossing obliquely behind it. *Medial* is the pharyngeal wall separated by fat and pharyngeal veins from the ascending pharyngeal artery and superior laryngeal nerve. *Anterolaterally* the artery is covered by the sternocleidomastoid; *below the digastric* the hypoglossal nerve and superior root of the ansa cervicalis and the lingual and facial veins are superficial. *At the level of the digastric* it is crossed by the stylohyoid muscle and the occipital and posterior auricular arteries. *Above the digastric* it is separated from the external carotid by the styloid process, styloglossus and stylopharyngeus, glossopharyngeal nerve, vagal pharyngeal branch and the deeper part of the parotid gland. At the base of the skull the glossopharyngeal, vagus, accessory and hypoglossal nerves are between the internal carotid artery and the internal jugular vein, which here has become posterior.

The Petrous Part

The artery at first ascends in the carotid canal, curves anteromedially and then superomedially above the cartilage filling the foramen lacerum, to enter the cranial cavity, passing between the lingula and petrosal process. It is at first anterior to the cochlea and tympanic cavity, separated from the latter and the pharyngotympanic tube by a thin, bony lamella, cribriform in the young, partly absorbed in old age; anterior to this it is separated from the trigeminal ganglion by the thin roof of the carotid canal, often deficient. The artery is surrounded by a venous plexus and the carotid autonomic plexus derived from the internal carotid branch of the superior cervical ganglion.

The Cavernous Part

In the cavernous sinus the artery is covered by lining endothelium. It ascends to the *posterior* clinoid process, turns anteriorly on the side of the sphenoid and again curves up medial to the *anterior* clinoid process, emerging through the dural roof of the sinus; occasionally the two clinoid processes form a ring round the artery, which is also surrounded by a sympathetic plexus; the oculomotor, trochlear, ophthalmic and abducent nerves are lateral to it.

The Cerebral Part

Having traversed the dura mater the artery turns back below the optic nerve, passing between this and the oculomotor nerve to the anterior perforated substance at the medial end of the lateral cerebral sulcus, where it divides into anterior and middle cerebral arteries.

Variations. The length of the artery varies with the length of the neck and the point of carotid bifurcation. It may arise from the aortic arch and then be medial to the external carotid as far as the larynx but there crossing behind it. Its cervical part is normally straight but on occasion may be very tortuous, being nearer to the pharynx than usual and very near the tonsil. Its absence has also been recorded.

Surface Anatomy. The internal carotid corresponds in position to a broad line from the termination of the common carotid to the back of the mandibular neck (**6.**70).

Branches. The cervical part has no branches. Those from the other parts are:

From the petrous part	1. Caroticotympanic
	2. Pterygoid
From the cavernous part	3. Cavernous
	4. Hypophysial
	5. Meningeal
From the cerebral part	6. Ophthalmic
	7. Anterior cerebral
	8. Middle cerebral
	9. Posterior communicating
	10. Anterior choroid

1. The caroticotympanic artery is small, occasionally double; it enters the tympanic cavity by a foramen in the carotid canal, anastomosing with the anterior tympanic branch of the maxillary artery and the stylomastoid artery.

2. The pterygoid artery, inconstant, enters the pterygoid canal with the nerve of the same name, anastomosing with a (recurrent) branch of the greater palatine artery.

3. The cavernous rami, numerous small vessels, supply the trigeminal ganglion, walls of the cavernous and inferior petrosal sinuses and contained nerves. Some anastomose with middle meningeal rami.

4. The hypophysial rami are small but numerous and important vessels (for details see p. 1455).

5. The meningeal branch is minute and passes over the lesser sphenoid wing to supply the dura mater and bone in the anterior cranial fossa; it anastomoses with a meningeal branch of the posterior ethmoidal artery.

6. The ophthalmic artery (**6.**80) leaves the internal carotid as it quits the cavernous sinus medial to the anterior clinoid process; it enters the orbit by the optic canal, inferolateral to the optic nerve; for a short distance it is then lateral to the nerve, medial to the oculomotor and abducent nerves, ciliary ganglion and rectus lateralis. It crosses between the optic nerve and rectus superior to

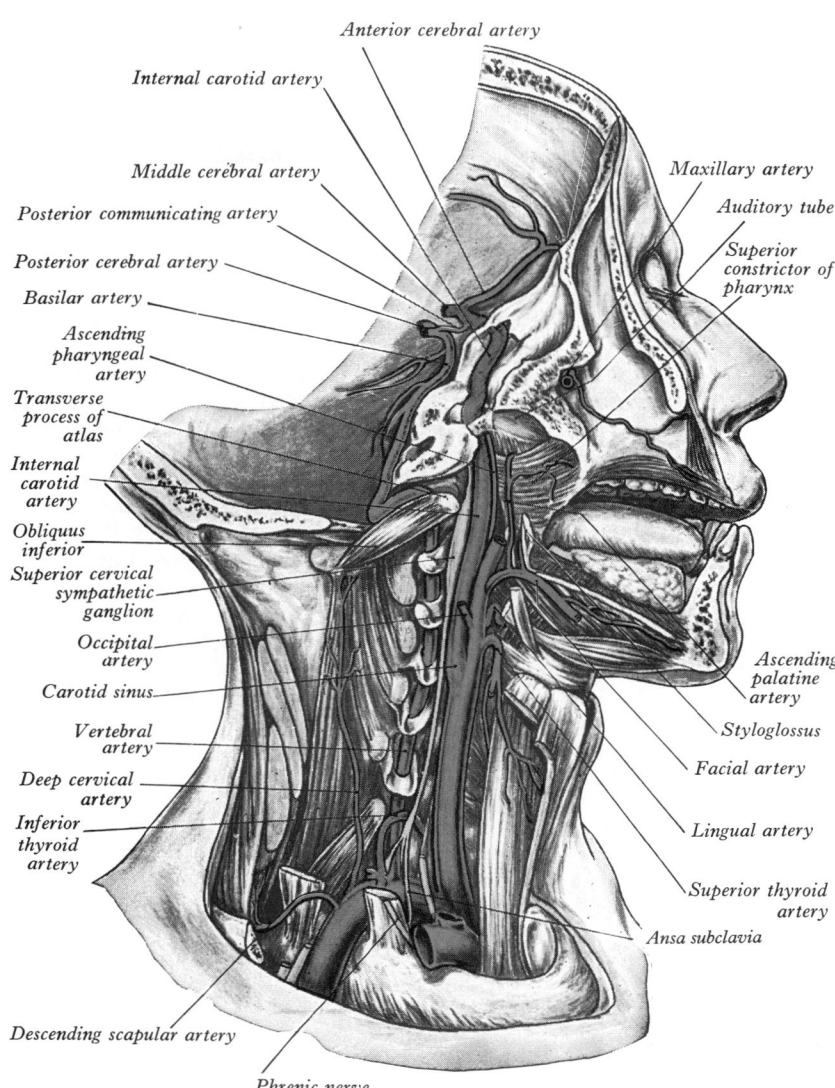

6.79 Dissection to show the course of the right vertebral and internal carotid arteries and some of their branches.

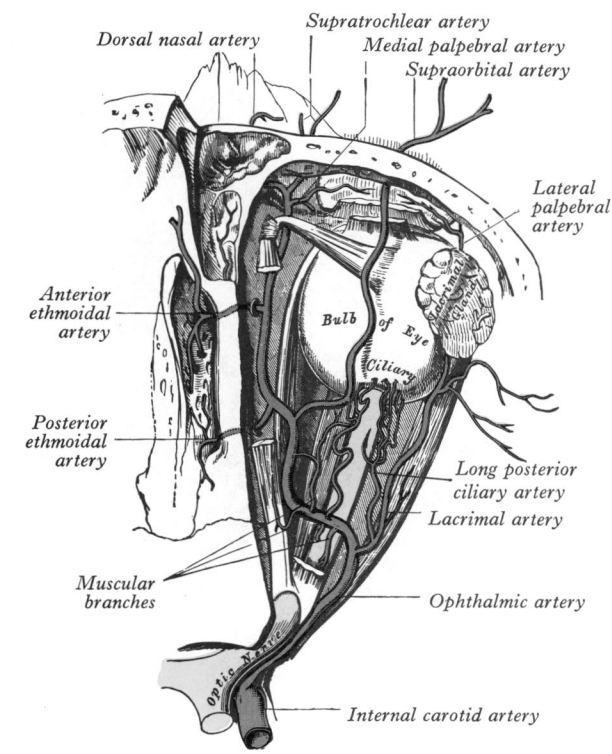

6.80 The ophthalmic artery and its branches in the right orbit, as seen from above.

the medial orbital wall, runs between the obliquus superior and the rectus medialis to the medial end of the upper eyelid, dividing into *supratrochlear* and *dorsal nasal* rami. As it crosses the optic nerve with the nasociliary nerve it is separated from the frontal nerve by the rectus superior and levator palpebrae superioris; its terminal ramus accompanies the infratrochlear nerve. In about 15% the ophthalmic artery is below the optic nerve. Its branches are as follows:

The *central artery of the retina*, a first and small branch, begins below the optic nerve. For a short distance it is in the nerve's dural sheath; about 1.25 cm behind the eye it enters the nerve's inferomedial surface and runs to the retina along its axis. Its distribution is described on p. 1204.

The *lacrimal artery* leaves the ophthalmic near its exit from the optic canal and is a large branch, sometimes arising before the ophthalmic enters the orbit; a branch of the middle meningeal artery (p. 689) may replace it. It accompanies the lacrimal nerve along the upper border of the rectus lateralis, supplying the lacrimal gland, after traversing which it ends in the eyelids and conjunctiva as *lateral palpebral arteries* running medially in the upper and lower lids to anastomose with the medial palpebral arteries. The lacrimal artery gives off one or two *zygomatic branches*: one reaches the temporal fossa via the zygomaticotemporal foramen, anastomosing with the deep temporal arteries; another reaches the cheek by the zygomaticofacial foramen, anastomosing with transverse facial and zygomatico-orbital arteries. A *recurrent meningeal ramus* passes back via the lateral part of the superior orbital fissure to anastomose with a middle meningeal branch; enlargement of this anastomosis may provide an alternative lacrimal artery.

Muscular branches frequently spring from a common trunk but form superior and inferior groups, most accompanying branches of the oculomotor nerve. The inferior, more constant, contains most of the anterior ciliary arteries. Other muscular rami branch from the lacrimal and supraorbital or the trunk of the ophthalmic artery.

Ciliary arteries are divisible into three groups: long and short posterior and anterior. *Long posterior ciliary arteries*, usually two, pierce the sclera near the optic nerve (p. 1194). About seven *short posterior ciliary arteries* pass around the optic nerve to the eyeball; dividing into 15–20 branches, they pierce the sclera around the optic nerve to supply the choroid and the ciliary processes. At the optic disc they anastomose with twigs of the central retinal artery and at the ora serrata with the long posterior and anterior ciliary arteries. *Anterior ciliary arteries* are rami of muscular branches of the ophthalmic; reaching the eyeball on tendons of the recti to form a circumcorneal subconjunctival vascular zone, they pierce the sclera near the sclerocorneal junction and end in the greater arterial circle of the iris (p. 1194).

The *supraorbital artery* leaves the ophthalmic where it crosses the optic nerve, ascends medial to the rectus superior and levator palpebrae superioris, meets the supraorbital nerve and runs with it between the periosteum and levator palpebrae superioris to the supraorbital foramen or notch; traversing this it divides into superficial and deep branches, supplying the skin, muscles and frontal periosteum, anastomosing with the supratrochlear artery, frontal branch of the superficial temporal and its fellow. It supplies the rectus superior and levator palpebrae and sends a branch across the trochlea to the medial canthus. At the supraorbital margin it often sends a branch to the diploë of the frontal bone and may also supply the mucoperiosteum in the frontal sinus.

The *posterior ethmoidal artery* enters the posterior ethmoidal canal, supplies the posterior ethmoidal air sinuses, enters the cranium, sends a meningeal branch to the dura mater and nasal branches descending into the nasal cavity via the cribriform plate, to anastomose with the sphenopalatine branches supplying bone.

The *anterior ethmoidal artery* and nerve traverse their canal, the artery supplying anterior and middle ethmoidal, and frontal air sinuses and, entering the cranium, giving a meningeal branch to the dura mater and nasal branches descending into the nasal cavity with the anterior ethmoidal nerve; they run in a groove on the deep

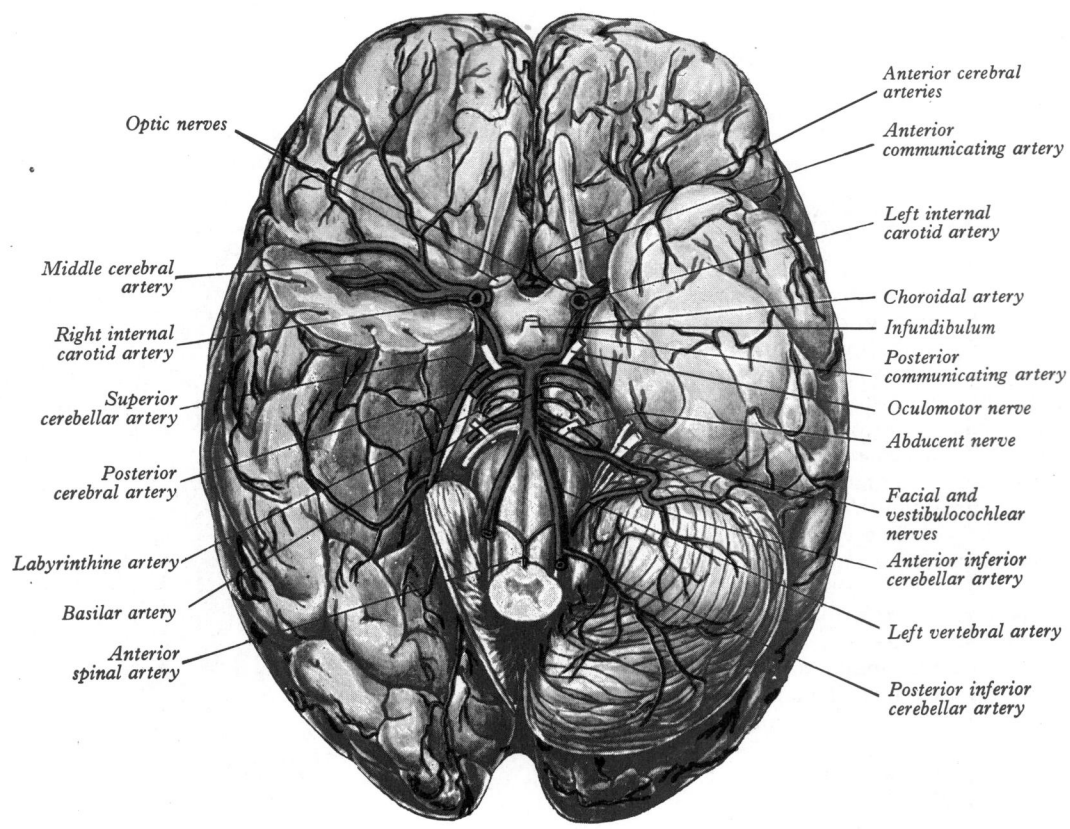

Optic nerves

Anterior cerebral arteries

Anterior communicating artery

Left internal carotid artery

Middle cerebral artery

Right internal carotid artery

Superior cerebellar artery

Posterior cerebral artery

Choroidal artery

Infundibulum

Posterior communicating artery

Oculomotor nerve

Abducent nerve

Facial and vestibulocochlear nerves

Labyrinthine artery

Anterior inferior cerebellar artery

Basilar artery

Left vertebral artery

Anterior spinal artery

Posterior inferior cerebellar artery

6.81A The arteries at the base of the brain. The right temporal pole and most of the right hemisphere of the cerebellum have been removed. Variations in the pattern of these vessels is common.

surface of the nasal bone to supply the lateral nasal wall and septum; a terminal branch appears on the nose between the nasal bone and the upper nasal cartilage. Angiographic studies (Kuru 1967) show the meningeal branch extending to the falx; Muller (1977, 1978) has confirmed this in fetal and adult material; he also derives such *'falciate arteries'* from the recurrent meningeal branch of the lacrimal artery.

The *meningeal branch*, a small artery passing back by the superior orbital fissure to the middle cranial fossa, anastomoses with the middle and accessory meningeal arteries, supplying bone.

Medial palpebral arteries, superior and inferior, leave the ophthalmic below the trochlea, descending behind the lacrimal sac to enter the lids, where each divides into two rami coursing laterally along the tarsal edges, to form the superior and inferior arches, completed by anastomoses with rami of the supraorbital and zygomatico-orbital (superior arch) and palpebral branches of the lacrimal (both arches); the inferior arch also links with the facial artery, thus supplying the mucosa of the naso-lacrimal duct.

The *supratrochlear artery*, a terminal branch of the ophthalmic, leaves the orbit superomedially with the supratrochlear nerve, ascending on the forehead to supply the skin, muscles and pericranium, anastomosing with the supraorbital artery and with its fellow.

The *dorsal nasal artery*, the other terminal branch, emerges from the orbit between the trochlea and medial palpebral ligament, gives a ramus to the upper lacrimal sac and divides; one branch joins the terminal part of the facial artery, the other along the dorsum of the nose supplying its exterior and joining its fellow and lateral nasal branch of the facial.

7. The anterior cerebral artery (6.81A,B,82,83,84A,B), the smaller terminal branch of the internal carotid, starts at the medial end of the stem of the lateral cerebral sulcus and passes anteromedially above the optic nerve to the longitudinal fissure, where it connects with its fellow by a short transverse anterior communicating artery. The two arteries thence travel together in the fissure, curving round the genu of the corpus callosum and back along its upper surface to its posterior end, where they anastomose with posterior cerebral arteries. Occasionally they are a single vessel. There are central and cortical branches.

The *anterior communicating artery*, with an average length of 4 mm, connects anterior cerebral arteries across the anterior end of the longitudinal fissure; it may be double. It has a few anteromedial central branches. According to Crowell & Morawetz (1977) its branches, from three to 13, supply the optic chiasma, lamina terminalis, hypothalamus, parolfactory areas, fornix (anterior columns) and cingulate gyrus.

Central branches arise from the commencement of the anterior cerebral, entering the anterior perforated substance and lamina terminalis to supply the rostrum of the corpus callosum, septum pellucidum, the anterior part of the putamen of the lentiform nucleus and the head of the caudate nucleus. *Cortical branches* are named by distribution: two or three *orbital branches* ramify on the frontal lobe's orbital surface, supplying the olfactory lobe, gyrus rectus and medial orbital gyrus; *frontal branches* supply the corpus callosum, cingulate gyrus, medial frontal gyrus and paracentral lobule, sending twigs over the hemisphere's superomedial border to the superior and middle frontal gyri and upper part of the precentral gyrus (including the 'leg area' of the motor cortex, p. 1049); *parietal branches* supply the precuneus and the adjacent lateral surface.

8. The middle cerebral artery (6.81A,B,82,83 84A,B), the larger terminal branch of the internal carotid, runs first in the lateral cerebral sulcus, then posterosuperiorly on the insula, dividing into branches distributed to this and the adjacent lateral cerebral surface (6.82). Its *central branches* are small and from its commencement they enter the anterior perforated substance, arranged in two sets: *medial striate arteries* ascend through the lentiform nucleus to supply it, the caudate nucleus and the internal capsule; *lateral striate arteries* ascend over the lower lateral aspect of the lentiform nucleus (in the external capsule) and turn medially to traverse it and the internal capsule to supply the

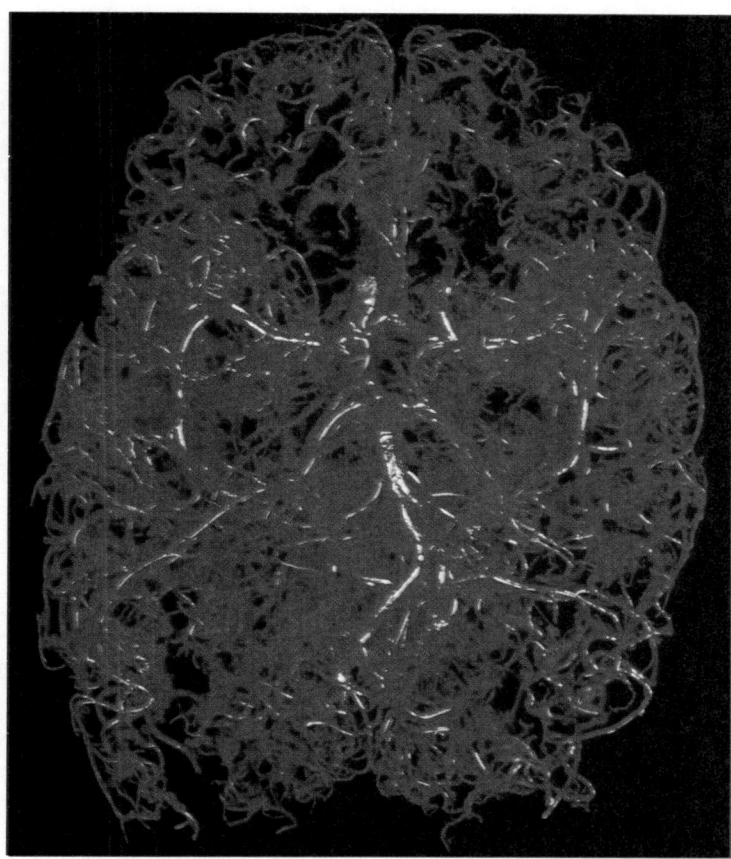

6.81B Resin corrosion cast of the arteries of the whole brain. Cast prepared by M C E Hutchinson and photographed by Kevin Fitzpatrick, Department of Anatomy, UMDS, Guy's Campus, London.

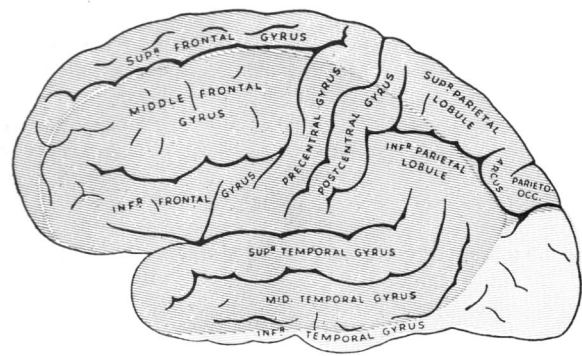

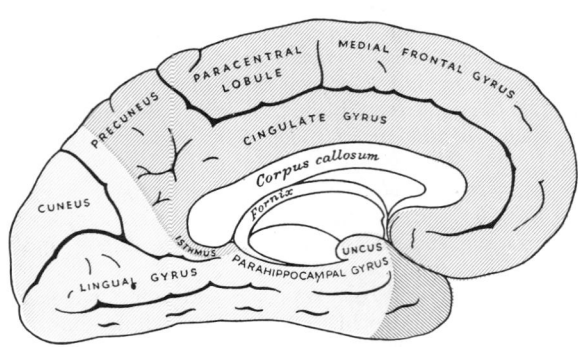

6.82 The lateral surface of the left cerebral hemisphere, showing the areas supplied by the cerebral arteries. In this and the next figure the area supplied by the anterior cerebral artery is coloured blue, that by the middle cerebral artery pink and that by the posterior cerebral artery is yellow.

6.83 The medial surface of the left cerebral hemisphere, showing the areas supplied by the cerebral arteries (see description of 6.82).

caudate nucleus. One ramus, usually the largest, was termed by Charcot the 'artery of cerebral haemorrhage'. *Cortical branches* supply *orbital rami* to the inferior frontal gyrus and the lateral orbital surface of the frontal lobe; *frontal rami* supply the precentral, middle and inferior frontal gyri; two *parietal rami* are distributed to the postcentral gyrus, the lower part of the superior parietal lobule and the whole inferior parietal lobule. Two or three *temporal rami* supply the lateral surface of the temporal lobe. Cortical branches of the middle cerebral thus supply all the motor area (excluding the leg) (p. 1051), the corresponding somaesthetic area (p. 1056) and the auditory area (p. 1060).

9. The posterior communicating artery (6.81A,B, 84A,B) runs back from the internal carotid above the oculomotor nerve, anastomosing with the posterior cerebral, a basilar branch. It is usually small but sometimes so large that the posterior cerebral appears to come from the internal carotid rather than basilar artery. It is often larger on one side. From its posterior half several small *central branches* pierce the posterior perforated substance

with others from the posterior cerebral to supply the medial thalamic surface and walls of the third ventricle (p. 1083).

10. The anterior choroidal artery, small but constant, leaves the internal carotid near its posterior communicating branch (Abbie 1933, 1934), passing back above the medial part of the uncus to cross inferior to the optic tract and reach and supply the crus cerebri. Turning laterally, it recrosses the optic tract, arrives lateral to the lateral geniculate body and supplies it with several branches. Finally it enters the inferior cornu of the lateral ventricle via the choroidal fissure to end in the choroidal plexus. It supplies: the globus pallidus, caudate nucleus and amygdaloid body, hypothalamus, tuber cinereum, red nucleus, substantia nigra, posterior limb of the internal capsule, the optic radiation, optic tract, hippocampus and the fimbria of the fornix.

THE CIRCULUS ARTERIOSUS (OF WILLIS)

Much of the brain is supplied by the two internal carotid arteries (p. 750) and a central anastomosis, the *circulus arteriosus*, exists

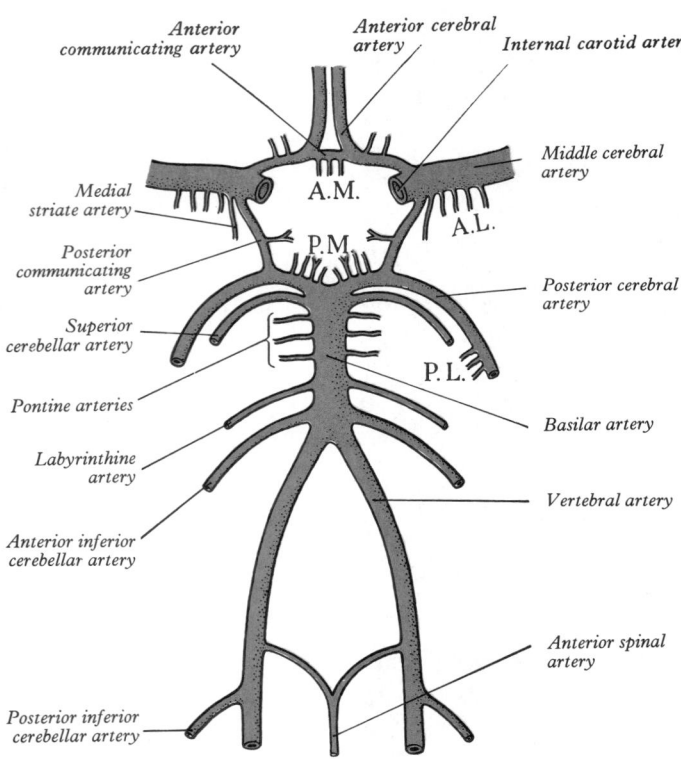

6.84A Diagram of the arteries at the base of the brain, showing the constitution of the arterial circle. A.L. = anterolateral central branches, A.M. = anteromedial central branches, P.L. = posterolateral central branches, P.M. = posteromedial central branches. The arteries constituting this so-called arterial 'circle' are commonly asymmetrical and sometimes a constituent vessel is missing.

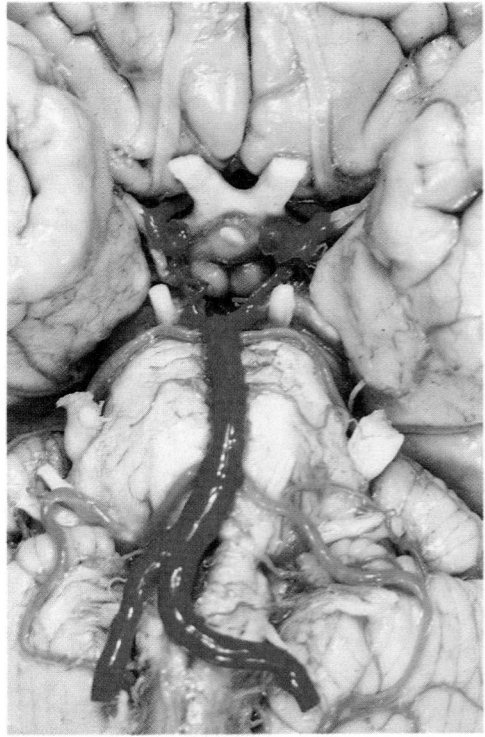

6.84B Resin cast of the arteries at the base of the brain, showing the components of the arterial circle. Cast prepared by M C E Hutchinson and photographed by Kevin Fitzpatrick, Department of Anatomy, UMDS, Guy's Campus, London.

between these and the two vertebral arteries (p. 750) that supplies the remainder. This 'circle', more polygonal than circular, is in the cisterna interpeduncularis, surrounding the optic chiasma, the neural infundibular stem of the hypophysis cerebri and other related neural structures in the interpendicular fossa (7.81). *Anteriorly* the anterior cerebral arteries (from the carotids) are joined by the anterior communicating artery; *posteriorly* the basilar artery (p. 753) divides into two posterior cerebral arteries, each joined to the ipsilateral internal carotid by a posterior communicating artery (**6**.84A,B).

Vessels of this 'circle' vary in calibre, being often maldeveloped, sometimes even absent. About 60% of circles display 'anomalies'; the above description applies to a minority. Variations have been much studied, from Windle's account in 1888 of 200 specimens to that of Puchades-Ort et al (1976) on 62 dissections, the largest series being the 700 dissections of Fawcett & Blachford (1906) and Riggs & Rupp's (1963) 994 dissections. Fields et al (1965) have summarized such studies. Cerebral and communicating arteries, anterior and posterior, may all be absent, variably hypoplastic, double or even triple. In about 90% there is, nevertheless, a complete 'circular' channel but in most one vessel is sufficiently narrowed to impair its role as a a collateral route. The haemodynamic 'balance' is usually disturbed by variation in the calibre of communicating arteries, often with variation in the

segments of anterior and posterior cerebral arteries extending from their origins to their junctions with the corresponding communicating arteries. This segment is particularly 'labile' in the posterior cerebral, being often reduced or absent; where this is coupled with an enlarged ipsilateral posterior communicating artery the posterior cerebral artery is filled from the internal carotid. Abbie (1933), Williams (1936) and Kaplan (1956) have shown that this anomaly accords with the ontogenetic and phylogenetic association of posterior cerebral and internal carotid arteries. Anterior in the arterial circle, agenesis or hypoplasia of the initial anterior cerebral segment is more frequent than anomalies in the anterior communicating, and hence a commoner cause of defective circulation. Angiographic evidence indicates such defective or absent circulation in about a third of individuals (Sedzmir 1959); existence of an effective arterial circle can never be assumed and surgical procedures involving its 'feeders' must be preceded by angiography. Radio-opaque substances may be injected into the internal carotid or vertebral arteries in the neck for radiography of the condition of their intracranial branches (**6**.77,78,88,89).

Further details of the distribution of cerebral arteries and veins appear on pp. 1083–1088 and of intracranial venous sinuses on pp. 799–805.

THE SUBCLAVIAN SYSTEM OF ARTERIES

The stem artery of the upper limb is single as far as the elbow, but its name changes in the regions traversed. From its origin to the outer border of the first rib it is *subclavian*; thence to the tendon of teres major it is *axillary*; and from this to its division at the elbow it is *brachial*.

The Subclavian Arteries

The right subclavian arises from the brachiocephalic trunk, the left from the aortic arch. For description, each is divided into a *first part*, from its origin to the medial border of the scalenus anterior, a *second part* behind this muscle and a *third part* from the muscle's lateral margin to the first rib's outer border, where the artery becomes *axillary*. Each subclavian artery arches over the cervical pleura and pulmonary apex. Their first parts differ, the second and third parts are almost identical.

First Part of the Right Subclavian Artery (**6**.60,85)

The right subclavian, branching from the brachiocephalic trunk behind the upper border of the right sternoclavicular joint, passes superolaterally to the medial margin of the scalenus anterior. It ascends about 2 cm above the clavicle but this varies.

Relations. The artery is deep to the skin, superficial fascia, platysma, anterior supraclavicular nerves, deep fascia, clavicular attachment of the sternocleidomastoid, sternohyoid and sternothyroid. It is at first *behind* the right common carotid's origin; more laterally it is crossed by the vagus nerve, the cardiac branches of the vagus and the sympathetic and by internal jugular and vertebral veins; the subclavian sympathetic loop encircles it. The anterior jugular vein diverges *laterally* in front of it, separated by the sternohyoid and sternothyroid. *Below* and *behind* the artery are the pleura and pulmonary apex but separated by the suprapleural membrane (p. 1269), the ansa subclavia, an accessory vertebral vein (p. 797) and the right recurrent laryngeal nerve (curving round inferoposterior to the vessel).

First Part of the Left Subclavian Artery (**6**.41,44,63,118)

This springs from the aortic arch, behind the left common carotid, level with the disc between the third and fourth thoracic vertebrae; it ascends into the neck, then arches laterally to the medial border of the scalenus anterior.

Relations. (Its branches are considered separately below.) In *the thorax* it is related, *anteriorly* to the left common carotid artery and left brachiocephalic vein, separated by the left vagus, cardiac

(p. 1118) and phrenic nerves. Superficial to these the anterior pulmonary margin, pleura, sternothyroid and sternohyoid are between the vessel and the upper left area of the manubrium sterni. *Posterior* are the left side of the oesophagus, the thoracic

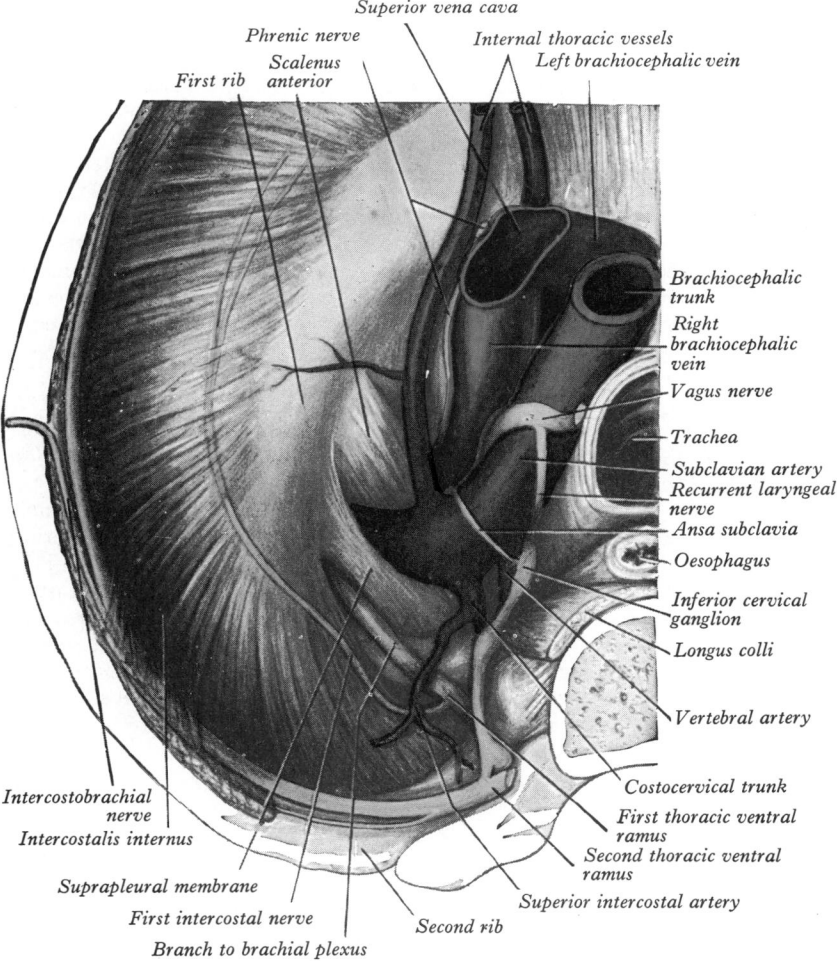

6.85 Structures related to the right cervical pleura, as seen from below.

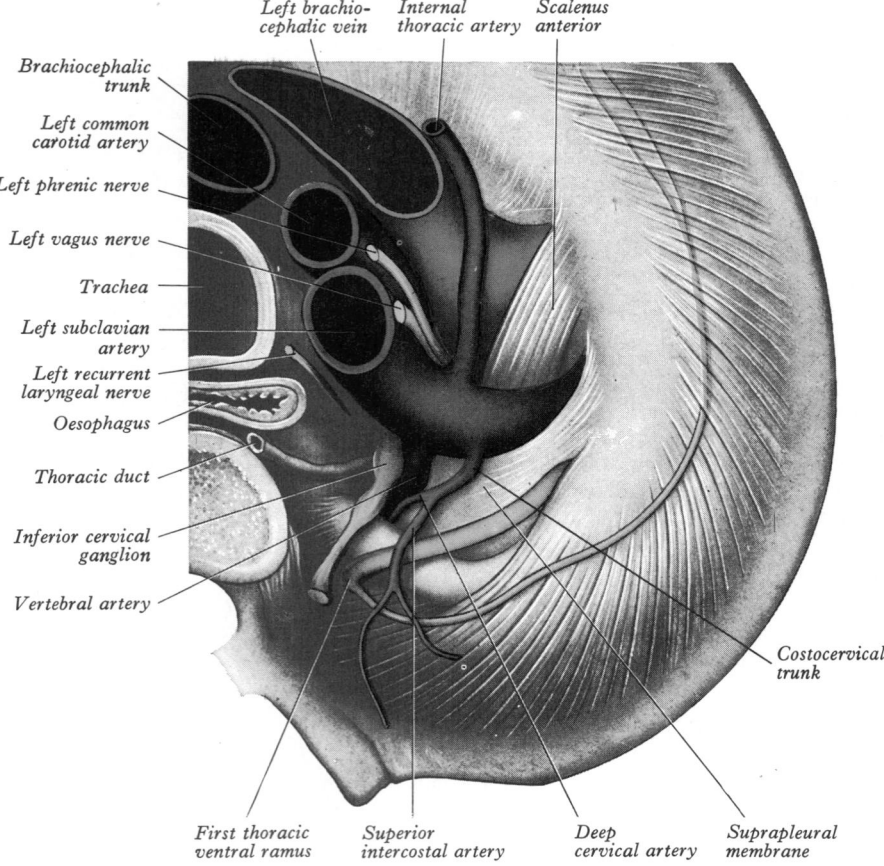

Left brachio-cephalic vein   *Internal thoracic artery*   *Scalenus anterior*

Brachiocephalic trunk

Left common carotid artery

Left phrenic nerve

Left vagus nerve

Trachea

Left subclavian artery

Left recurrent laryngeal nerve

Oesophagus

Thoracic duct

Inferior cervical ganglion

Vertebral artery

Costocervical trunk

First thoracic ventral ramus   *Superior intercostal artery*   *Deep cervical artery*   *Suprapleural membrane*

6.86 Structures related to the left cervical pleura, as seen from below.

duct and longus colli; it is in contact posterolaterally with the left lung and pleura. *Medial* are the trachea, the left recurrent laryngeal nerve, oesophagus and thoracic duct. *Laterally* the artery grooves the mediastinal surface of the left lung and pleura which also encroach on its anterior and posterior aspects.

In the neck, near the medial border of the scalenus anterior, the artery is crossed *anteriorly* by the left phrenic nerve and the termination of the thoracic duct. Otherwise anterior relations are as previously described for the first part of the right subclavian artery. *Posteriorly* and *inferiorly,* the relations of both vessels are identical but the left recurrent laryngeal nerve, medial to the left subclavian artery in the thorax, is not directly related to its cervical part.

Second Part of the Subclavian Arteries (6.68,87)

This is behind the scalenus anterior; it is short and the highest part of the vessel.

Relations. Anterior are the skin, superficial fascia, platysma, deep cervical fascia, sternocleidomastoid and scalenus anterior; the right phrenic nerve is often described as separated from the second part by the scalenus anterior, but crossing the *first* part on the left; Qvist (1977) stated that *both* nerves are anterior to the muscle. *Postero-inferior* are the suprapleural membrane, pleura and lung and the lower trunk of the brachial plexus; *superior* are the upper and middle trunks of the plexus; the subclavian vein is antero-inferior, separated by the scalenus anterior (6.87).

Third Part of the Subclavian Arteries

This descends laterally from the lateral margin of the scalenus anterior to the outer border of the first rib, where it becomes axillary; it is the most superficial part of the artery and lies partly in the supraclavicular triangle (p. 743), where its pulsations may be felt and it may be compressed.

Relations. Anterior are the skin, superficial fascia, platysma, supraclavicular nerves and deep cervical fascia. The external

jugular vein crosses its medial end and here receives the suprascapular, transverse cervical and anterior jugular veins, together often forming a venous plexus. The nerve to the subclavius descends between the veins and the artery; the latter is terminally behind the clavicle and subclavius, where it is crossed by the suprascapular vessels. The subclavian vein is antero-inferior and the lower trunk of the brachial plexus is postero-inferior, between the artery and the scalenus medius (and on the first rib). *Superolateral* are the upper and middle trunks of the brachial plexus (palpable here) and the inferior belly of the omohyoid. *Inferior* is the first rib.

Surface Anatomy. The subclavian artery corresponds to a broad line, convex superiorly, curving from the sternoclavicular joint to the midpoint of the clavicle (6.70).

Variations. The right subclavian may arise above or below sternoclavicular level; it may be a separate aortic branch and be the first *or* last branch of the arch; when first, it is in the position of a brachiocephalic trunk and when last it arises from the arch's left end, ascending obliquely to the right *behind* the trachea, oesophagus and right common carotid to the first rib. In this case the proximal part of the artery represents a persistent part of the right dorsal aorta, the right fourth aortic arch taking no part in its formation (p. 733); hence the right recurrent laryngeal nerve hooks round the common carotid, derived from the third arch. Sometimes, when the right subclavian is the last aortic branch, it passes *between* the trachea and oesophagus. It may perforate the scalenus anterior; very rarely it passes anterior to it. Sometimes the subclavian vein accompanies the artery behind the scalenus anterior. The artery may ascend as high as 4 cm above the clavicle or it may reach only its upper border. The left subclavian is occasionally combined at its origin with the left common carotid.

Applied Anatomy. The subclavian artery's third part is most accessible. Since the line of the posterior border of the sternocleidomastoid approximates to the (deeper) lateral border of the scalenus anterior, the artery is just lateral to the former and, as noted, can be felt in the antero-inferior angle of the posterior triangle. It can only be effectively *compressed against the first rib*; with shoulder depressed pressure is exerted *down, back* and *medially* in the angle between the sternocleidomastoid and clavicle. The palpable trunks of the brachial plexus may be infiltrated with local anaesthetic allowing major surgical procedures.

BRANCHES OF THE SUBCLAVIAN ARTERY

These are (1) vertebral, (2) internal thoracic, (3) thyrocervical, (4) costocervical and (5) dorsal scapular.

On the left all branches except the dorsal scapular arise from the first part; on the right the costocervical trunk usually springs from the second part. The origins of branches proceeding into the posterior triangle are variable but distributions are relatively constant.

1. The Vertebral Artery (6.79,81A,84A,85,86,88,89)

This artery arises from the superoposterior aspect of the subclavian's first part, ascending through the foramina in all cervical transverse processes except the seventh, to curve medially behind the lateral mass of the atlas and then ascending to enter the cranium via the foramen magnum. At the lower pontine border it joins its fellow to form the basilar artery. Occasionally it may enter the bone at the fifth, fourth or seventh cervical transverse foramen (p. 733).

Relations. The vertebral artery has a *first part* ascending back between the longus colli and the scalenus anterior, related anteriorly to the common carotid artery and vertebral vein, and crossed by the inferior thyroid artery, and on the left also by the thoracic duct. *Posterior* are the seventh cervical transverse process, the cervicothoracic ganglion (6.85,86) and ventral rami of the seventh and eighth cervical spinal nerves. The *second part* ascends through the transverse foramina, with a large branch from the cervicothoracic sympathetic (stellate) ganglion and a plexus of veins which form the vertebral vein low in the neck. It is anterior to the ventral rami of the cervical spinal nerves

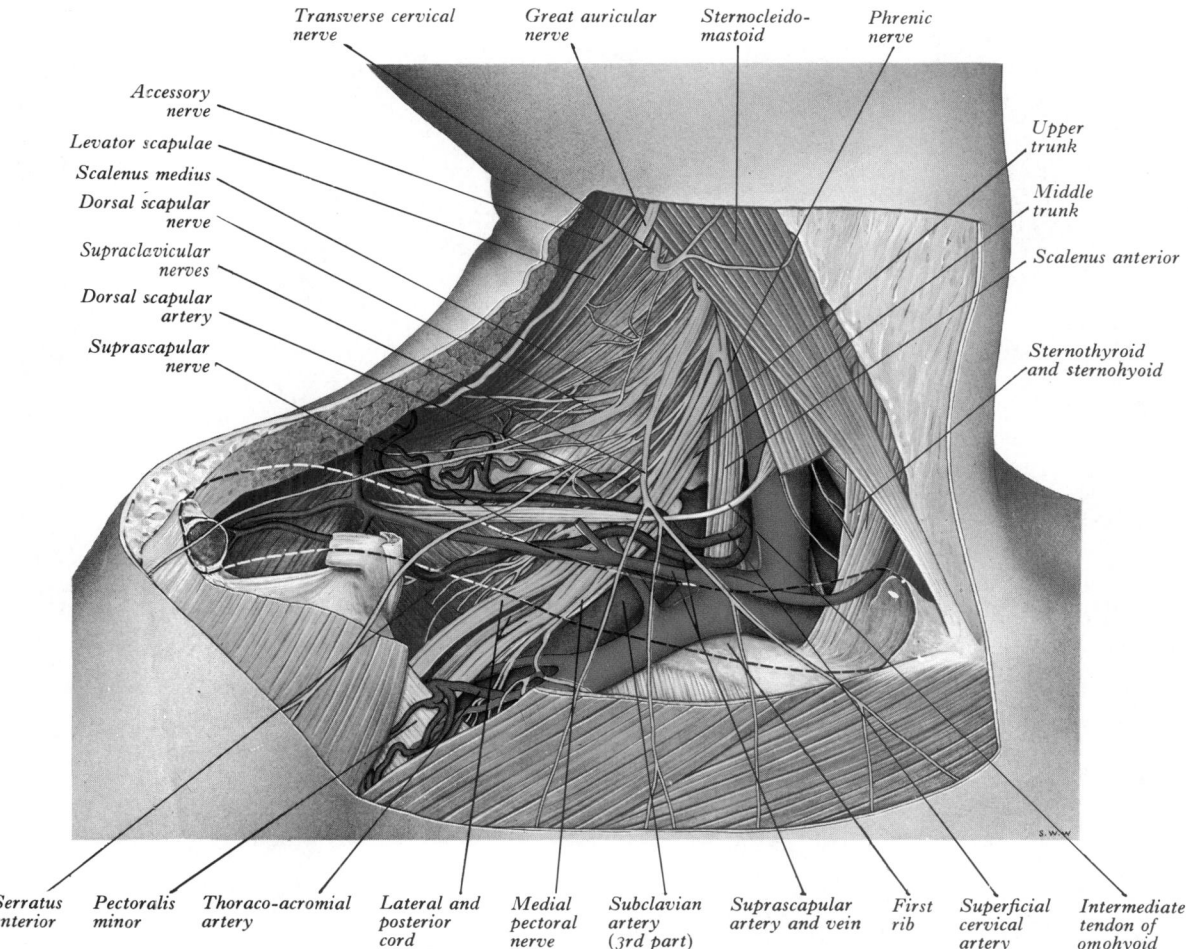

Transverse cervical nerve *Great auricular nerve* *Sternocleido-mastoid* *Phrenic nerve*

Accessory nerve

Levator scapulae

Scalenus medius

Dorsal scapular nerve

Supraclavicular nerves

Dorsal scapular artery

Suprascapular nerve

Upper trunk

Middle trunk

Scalenus anterior

Sternothyroid and sternohyoid

Serratus anterior *Pectoralis minor* *Thoraco-acromial artery* *Lateral and posterior cord* *Medial pectoral nerve* *Subclavian artery (3rd part)* *Suprascapular artery and vein* *First rib* *Superficial cervical artery* *Intermediate tendon of omohyoid*

6.87 The lower part of the posterior triangle showing the relations of the third part of the right subclavian artery. Note the clavicle has been removed but is shown as a dotted outline; the middle trunk of the brachial plexus gives an unusual contribution to the medial cord.

(C.2–C.6), ascending almost vertically to the transverse process of the axis, through which it turns laterally to the atlantal transverse foramen; from here the *third part* issues medial to the rectus capitis lateralis, curving back and medially behind the lateral mass, the first cervical ventral spinal ramus being medial. It is then in a groove on the upper surface of the atlantal posterior arch, entering the vertebral canal below the inferior border of the posterior atlanto-occipital membrane. This part, covered by the semispinalis capitis, is in the suboccipital triangle. The first cervical dorsal spinal ramus is between the artery and the posterior arch. The *fourth part* pierces the dura and arachnoid mater, ascends anterior to the hypoglossal roots (p. 1120) inclining anterior to the medulla oblongata where, at the lower pontine border, it unites with its fellow to form the midline basilar artery (**6**.81,84).

Winckler (1972) has described variation in elastic and muscular tissue in the vertebral artery; in its first and third parts it appears adapted by increased elasticity to the greater mobility and lack of support in these regions.

Cervical Branches of the Vertebral Artery

Spinal branches enter the vertebral canal by the intervertebral foramina, each dividing into a branch passing along the spinal roots to supply the spinal cord and membranes (p. 948); they anastomose with other spinal arteries, which fork into ascending and descending rami, to unite with those above and below, forming two lateral anastomotic chains on the posterior surfaces of the vertebral bodies near the attachment of their pedicles. From these chains branches supply the periosteum and vertebral bodies,

others communicating with similar branches across the midline; from these connections small rami join similar ones above and below, forming a median anastomotic chain on the posterior surfaces of the vertebral bodies.

Muscular branches arising from the vertebral artery as it curves round the lateral mass of the atlas, supply the deep muscles of this region and anastomose with the occipital, ascending and deep cervical arteries.

Cranial Branches of the Vertebral Artery

One or two **meningeal branches** from the vertebral artery near the foramen magnum ramify between the bone and dura mater in the cerebellar fossa, supplying bone, diploë and the falx cerebelli.

The posterior spinal artery may arise from the vertebral near the medulla oblongata but most frequently from its posterior inferior cerebellar branch (vide infra). It passes posteriorly, descending as two branches, anterior and posterior to the dorsal roots of the spinal nerves; these are reinforced by spinal twigs from the vertebral, ascending cervical, posterior intercostal and first lumbar arteries, which reach the vertebral canal by the intervertebral foramina, sustaining the posterior spinal arteries to the lower spinal levels (p. 948, 7.80).

The anterior spinal artery, a small branch arising near the vertebral's end, descends anterior to the medulla oblongata and unites with its fellow at mid-medullary level. The single trunk then descends on the ventral midline of the spinal cord, reinforced by a succession of small spinal rami entering the vertebral canal through the intervertebral foramina from the vertebral, ascending cervical, posterior intercostal and first lumbar arteries. They

751

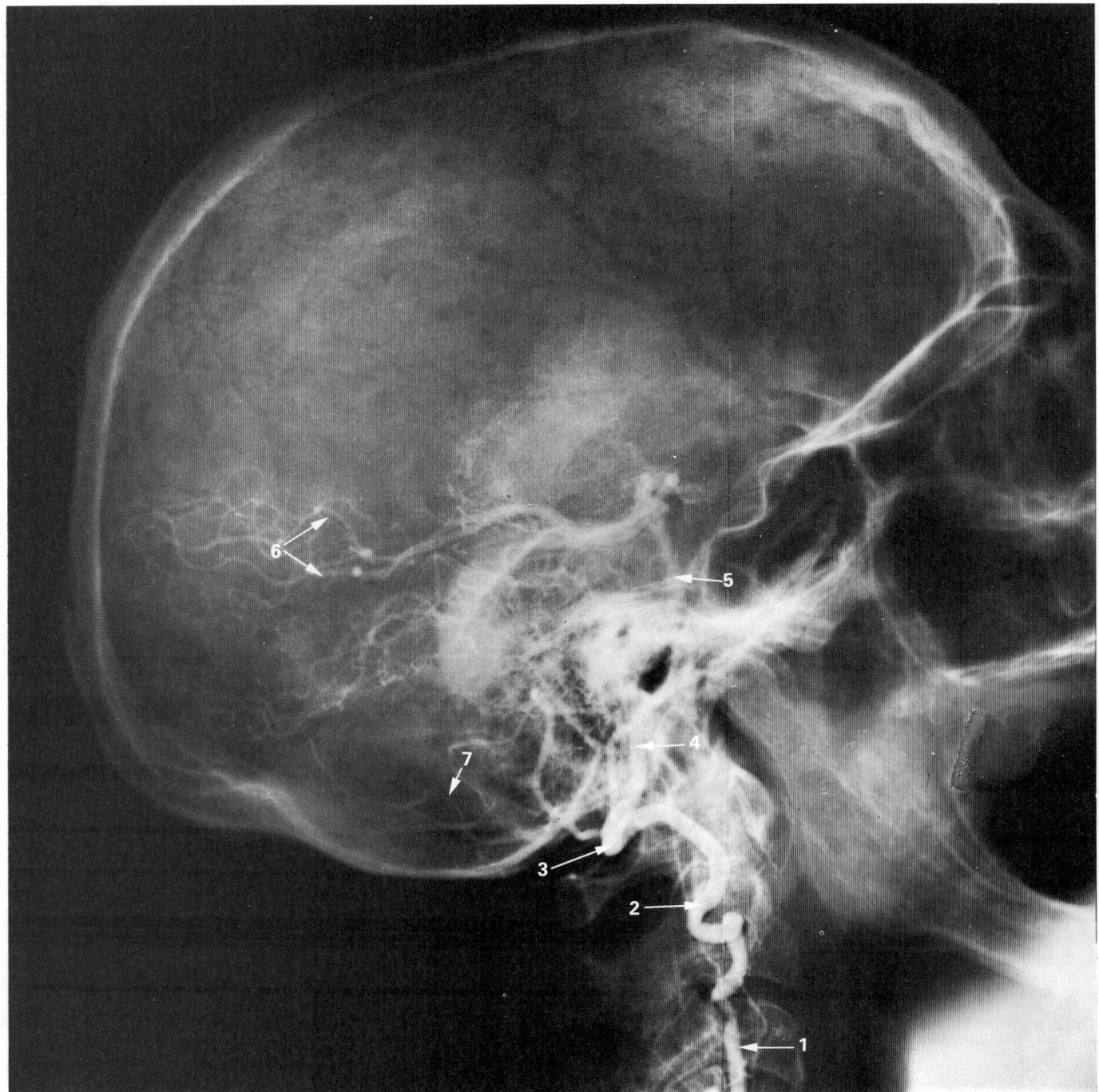

6.88 Vertebral arteriogram (left): lateral view. 1. Vertebral artery, ascending part. 2. Loop between transverse foramina of axis and atlas. 3. Suboccipital part. 4. Intracranial part. 5. Basilar artery. 6. Posterior cerebral branches. 7. Inferior cerebellar branches.

unite by ascending and descending branches as a single anterior median artery, which reaches the lower spinal cord and filum terminale. This median artery is encased in the pia mater along the anterior median fissure; it supplies the spinal cord and inferiorly the cauda equina. Branches from the anterior spinal arteries and beginning of their common trunk pass into the medulla oblongata, with a central distribution sharply limited dorsally by the trigonum hypoglossi (p. 1084, **7**.80).

The posterior inferior cerebellar artery (**6**.81), the largest branch, is sometimes absent. Arising near the lower end of the olive, it curves back around it to ascend behind glossopharyngeal and vagal roots to the inferior border of the pons, where it curves and descends along the inferolateral border of the fourth ventricle. Finally it turns laterally into the cerebellar vallecula, dividing into medial and lateral rami. The *medial branch* runs back between the cerebellar hemisphere and inferior vermis, supplying both; the *lateral branch* supplies the inferior cerebellar

surface as far as its lateral border, anastomosing with the anterior inferior and superior cerebellar arteries (from the basilar artery). Its trunk supplies the medulla oblongata and choroid plexus of the fourth ventricle, sending up a branch lateral to the (cerebellar) tonsil to supply the dentate nucleus. The medullary area supplied is dorsal to the olivary nucleus and lateral to the hypoglossal nucleus and its emerging fila.

Applied Anatomy. The *posterior* inferior cerebellar artery supplies the *lateral* medulla. Thrombosis therefore causes loss of function ('*lateral medullary syndrome*') in ambigual, solitary, vestibular and cochlear nuclei, spinocerebellar tracts, the lateral spinothalamic tract, trigeminal spinal nucleus and tract. The *anterior* spinal artery supplies the *medial* medulla; thrombosis of it ('*medial medullary syndrome*') affects the hypoglossal nucleus and nerve, medial lemniscus and corticospinal tract.

The medullary arteries are minute vessels from the vertebral and its branches, distributed to the medulla oblongata.

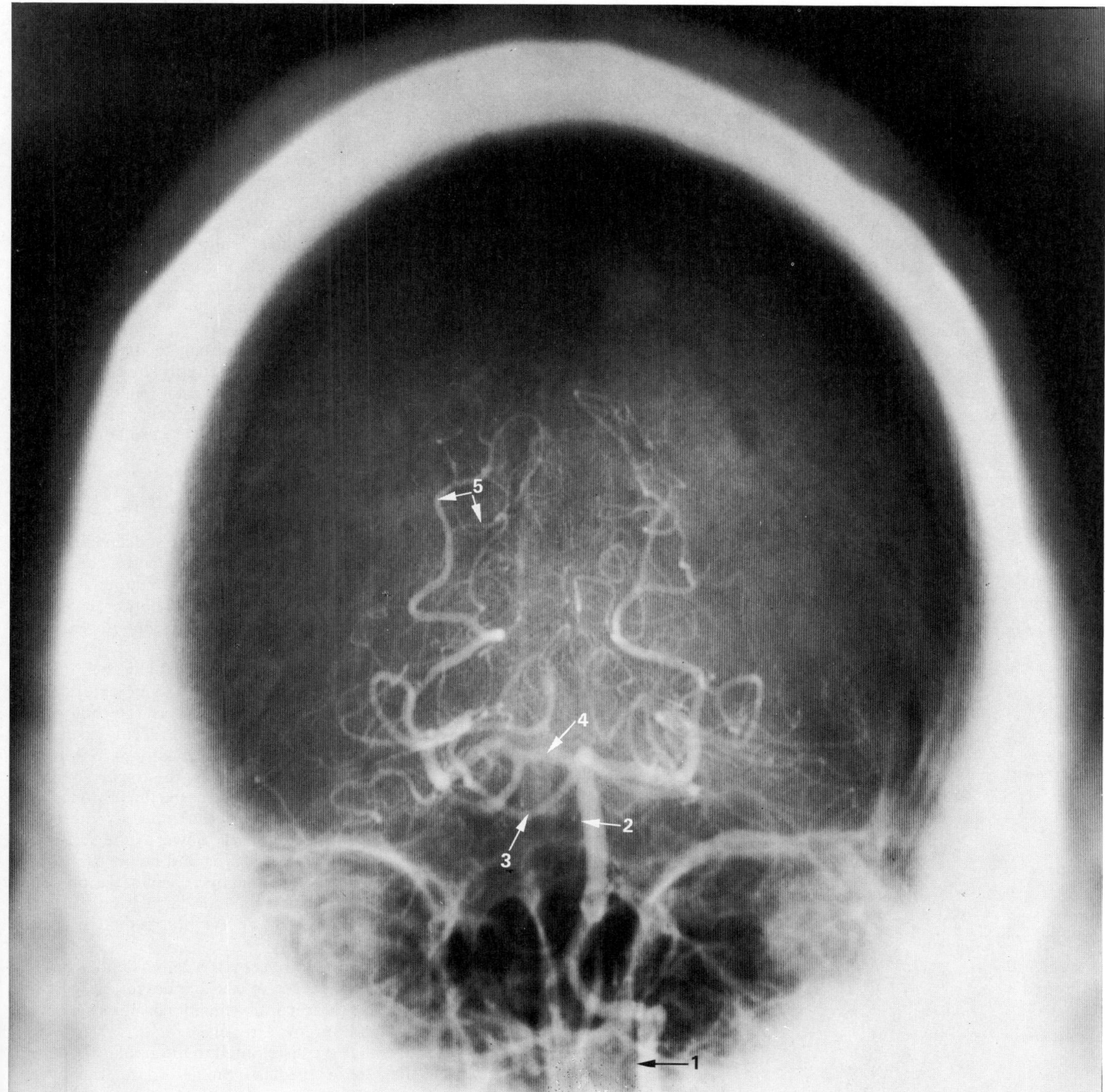

6.89 Vertebral arteriogram (left): anteroposterior view. The circular shape of the field is due to use of a cone to improve resolution. 1. Left vertebral artery. 2. Basilar artery. 3. Right superior cerebellar artery. 4. Right posterior cerebral artery. 5. Branches of right posterior cerebral artery.

The Basilar Artery

This median vessel, the junction of the vertebral arteries, extends from the lower to the upper pontine borders in the cisterna pontis. It adjoins a shallow, median groove on the ventral pontine surface, between the abducent nerves at the lower pontine border and the oculomotor at the upper pontine border, where it divides into two posterior cerebral arteries.

The pontine branches are numerous and leave the front and sides of the basilar to supply the pons and adjacent parts.

The labyrinthine (internal auditory) artery, long and slender, may branch from the lower part of the basilar but more often from the anterior inferior cerebellar artery (vide infra); it accompanies facial and vestibulocochlear nerves into the internal acoustic meatus and is distributed to the internal ear (p. 1228). In a radiographic study of this artery Wende et al (1975) identified the origins of 238. Only 38 (16%) were from the basilar; 108 (45.4%) were from the anterior inferior cerebellar. The superior cerebellar artery accounted for 58 (24.4%), the posterior inferior cerebellar for 13 (5.4%). The remaining 21 (8.8%) were reduplicated and were branches of the basilar and one or other of the cerebellar arteries. They found a unilateral artery in 24 of 316 patients. Others have recorded different incidences. Cavatori (1908) observed a basilar origin for the labyrinthine artery in about 70% in Italians; Stopford (1916) and Adachi & Hasche (1928) recorded it as most often from a trunk common to it and one of the cerebellar arteries. Gillilan (1972) has indicated racial variation.

The anterior inferior cerebellar artery (6.81) branches from the lower part of the basilar and runs posterolaterally usually ventral to the abducent, facial and vestibulocochlear nerves, commonly forming a variable loop into the internal acoustic meatus below the nerves (Sunderland 1945), from which the laryrinthine artery often arises. Emerging from the meatus the

753

artery supplies the anterolateral region of the inferior cerebellar surface, anastomosing with the posterior inferior cerebellar branch of the vertebral. A few branches supply the inferolateral parts of the pons and sometimes the upper medulla oblongata.

The superior cerebellar artery (6.81), arising near the basilar's end, passes laterally below the oculomotor nerve which separates it from the posterior cerebral artery and curves round the cerebral peduncle below the trochlear nerve; arriving at the superior cerebellar surface, it divides into branches ramifying in the pia mater to supply this cerebellar aspect and anastomose with branches of inferior cerebellar arteries. It also supplies the pons, pineal body, superior medullary velum and tela choroidea of the third ventricle.

The posterior cerebral artery (6.81, 82, 83), frequently double, is larger than the superior cerebellar, from which it is separated near its origin by the oculomotor nerve and, lateral to the midbrain, by the trochlear nerve. Passing laterally, parallel with the superior cerebellar, it receives the posterior communicating artery, winds round the cerebral peduncle and reaches the tentorial cerebral surface, where it supplies the temporal and occipital lobes. Its branches are central and cortical.

Central branches. Several small *posteromedial central branches* (6.84) from the beginning of the posterior cerebral, with similar branches from the posterior communicating artery, pierce the posterior perforated substance to supply the anterior thalamus, lateral wall of the third ventricle and the globus pallidus. *Posterior choroidal rami* vary (Abbie 1933); one or more course over the lateral geniculate body and help to supply it before entering the posterior part of the inferior cornu of the lateral ventricle via the lower part of the choroidal fissure. Others curl round the posterior end of the thalamus and traverse the transverse fissure, some going to the third ventricle's tela choroidea, some to traverse the upper choroidal fissure; they supply the choroid plexuses of the third and lateral ventricles and the fornix (Percheron 1977, p. 1084). Small *posterolateral central branches* arise from the posterior cerebral artery beyond the cerebral peduncle; they supply this and the posterior thalamus, colliculi, pineal gland and medial geniculate body.

Cortical branches. The *temporal branches*, usually two, are distributed to the uncus, parahippocampal, medial and lateral occipitotemporal gyri; *occipital branches* supply the cuneus, lingual gyrus and posterolateral surface of the occipital lobe; *parietooccipital branches* supply the cuneus and precuneus. The posterior cerebral artery supplies the visual area (p. 1084) and other structures in the visual pathway.

2. The Internal Thoracic Mammary Artery (6.90)

This arises inferiorly from the first part of the subclavian artery, about 2 cm above the clavicle's sternal end, opposite the root of the thyrocervical trunk. It descends behind the upper six costal cartilages about one cm from the lateral sternal border; level with the sixth intercostal space it divides into musculophrenic and superior epigastric branches.

Relations. At first it descends anteromedially behind the clavicle's sternal end, the internal jugular and brachiocephalic veins and the first costal cartilage. As it enters the thorax, the phrenic nerve crosses it obliquely from its lateral side, usually in front. The artery then descends almost vertically to its bifurcation; *anterior* to it are the pectoralis major, the upper six costal cartilages and intervening external intercostal membranes and internal intercostals and terminations of the upper six intercostal nerves. It is separated from the pleura, down to the second or third cartilage, by a strong layer of fascia and below this by the transversus thoracis. It is accompanied by a chain of lymph nodes and venae comitantes uniting at about the third costal cartilage into a single vein medial to the artery. Its intermediate branches are as follows:

The pericardiacophrenic artery is a long, slender branch accompanying the phrenic nerve to the diaphragm, descending between the pleura and pericardium, finally anastomosing with the musculophrenic and phrenic arteries.

Mediastinal arteries are distributed to the areolar tissue and lymph nodes in the anterior mediastinum and to the thymic vestiges.

Pericardial branches supply the upper anterior region of the pericardium.

Sternal branches are distributed to the transversus thoracis, the periosteum of the posterior sternal surface and the sternal red bone marrow.

The foregoing three groups, with rami of the pericardiacophrenic, anastomose with branches of the posterior intercostal and bronchial arteries to form a *subpleural mediastinal plexus*.

Anterior intercostal branches are distributed to the upper six intercostal spaces. Two in each space, they pass laterally along the borders of the space to anastomose with the posterior intercostal arteries (and their collateral branches). They lie at first between the pleura and the internal intercostals, then between the intercostales intimi and the internal intercostals. They supply the intercostal muscles and send branches through them to the pectoral muscles, breast and skin.

Perforating branches traverse the upper five or six intercostal spaces with *anterior cutaneous branches* of the corresponding intercostal nerves. They enter the pectoralis major and, curving laterally, supply the muscle and skin. In the female the second to fourth branches supply the breast; during lactation they are enlarged.

6.90 The left internal thoracic artery and vein and their main branches. The lateral end of the resected clavicle has been artificially elevated.

The musculophrenic artery passes inferolaterally behind the seventh to ninth costal cartilages, traverses the diaphragm near the ninth and ends near the last intercostal space. It anastomoses with the inferior phrenic and the lower two posterior intercostal and ascending branch of the deep circumflex iliac arteries. Two anterior intercostal arteries branch from it for each of the seventh to ninth intercostal spaces, distributed like those in other spaces. The musculophrenic also supplies the lower part of the pericardium and the abdominal muscles.

The superior epigastric artery descends between the costal and xiphoid slips of the diaphragm, anterior to the lower fibres of the transversus thoracis and the upper fibres of the transversus abdominis. Entering the rectus sheath, at first behind the muscle and then perforating and supplying it, it anastomoses with the inferior epigastric branch of the external iliac. Branches perforate the sheath to supply the abdominal skin; a branch anterior to the xiphoid process anastomoses with its fellow. The artery supplies the diaphragm; on the right small branches reach the falciform ligament to anastomose with the hepatic artery.

3. The Thyrocervical Trunk (6.68,146)

This short, wide artery, from the front of the subclavian's first part near the medial border of the scalenus anterior, divides almost at once into the inferior thyroid, suprascapular and superficial cervical branches.

The inferior thyroid artery is looped; first it *ascends* anterior to the medial border of the scalenus anterior, turns *medially* just below the sixth cervical transverse process, passing anterior to the vertebral vessels and posterior to the carotid sheath and its contents and usually the sympathetic trunk, whose middle cervical ganglion usually adjoins the vessel. It finally *descends* on the longus colli to the lower border of the thyroid gland. As it approaches this, its relation to the recurrent laryngeal nerve is surgically important (p. 1117). Nearing the gland the artery usually passes behind the nerve but nearer, on the right, the nerve is with equal frequency anterior or posterior to or amongst the branches of the artery; the left nerve is usually posterior. Relations between the terminal branches of artery and nerve are very variable (Bowden 1955). On the *left*, near its origin, the artery is crossed anteriorly by the thoracic duct, curving inferolaterally to its end.

Muscular branches supply the infrahyoid muscles, longus colli, scalenus anterior and the inferior pharyngeal constrictor.

The ascending cervical artery, a small branch, arises as the inferior thyroid turns medially behind the carotid sheath and ascends on the anterior tubercles of the cervical transverse processes between the scalenus anterior and the longus capitis. It supplies the adjacent muscles and has one or two *spinal branches* which enter the vertebral canal through the intervertebral foramina to supply the spinal cord and membranes and vertebral bodies, as do the spinal branches of the vertebral artery. The ascending cervical artery anastomoses with the vertebral, ascending pharyngeal, occipital and deep cervical arteries.

The inferior laryngeal artery ascends on the trachea with the recurrent laryngeal nerve, enters the larynx at the inferior constrictor's lower border and supplies the laryngeal muscles and mucous membrane, anastomosing also with its contralateral fellow, and with the superior laryngeal branch of the superior thyroid artery.

Pharyngeal branches supply the lower pharynx, *tracheal branches* the trachea (anastomosing with the bronchial arteries), and *oesophageal branches* the oesophagus (anastomosing with the oesophageal branches of thoracic aorta). Inferior and ascending *glandular branches* supply the posterior and inferior regions of the thyroid gland, anastomosing with the opposite inferior and ipsilateral superior thyroid arteries. The ascending branch also supplies the parathyroid glands.

The suprascapular artery (6.87,88) first descends laterally across the scalenus anterior and phrenic nerve posterior to the internal jugular vein and sternocleidomastoid; it then crosses anterior to the subclavian artery and brachial plexus, posterior and parallel to the clavicle and subclavius and the inferior belly of the omohyoid, to reach the superior scapular border. Here it passes above (sometimes under) the superior transverse ligament,

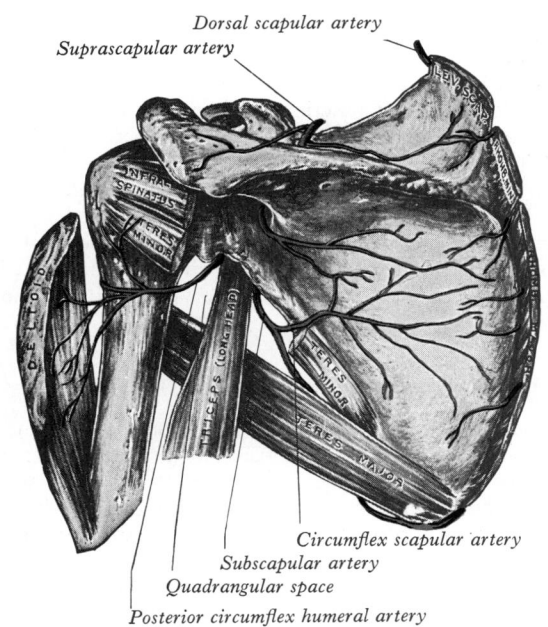

Dorsal scapular artery
Suprascapular artery

Circumflex scapular artery
Subscapular artery
Quadrangular space
Posterior circumflex humeral artery

6.91 Scapular anastomoses of the left side: dorsal aspect.

separating it from the suprascapular nerve, to enter the supraspinous fossa (6.91), where it lies on the bone, supplying the supraspinatus. It descends behind the scapular neck, through the great scapular notch deep to the inferior transverse ligament to the deep surface of the infraspinatus, where it anastomoses with the circumflex scapular and deep branch of the transverse cervical artery. Besides supplying the sternocleidomastoid, subclavius and infraspinatus, it has a *suprasternal branch* which crosses the sternal end of the clavicle to the skin of the upper thorax and an *acromial branch* which pierces the trapezius to supply the skin over the shoulder, anastomosing with the thoraco-acromial and posterior circumflex humeral arteries. As the suprascapular artery passes the superior transverse ligament, a branch of it enters the subscapular fossa beneath the subscapularis; this anastomoses with the subscapular artery and the deep branch of the transverse cervical. It also supplies the acromioclavicular and glenohumeral joints, the clavicle and scapula. It may arise from the third part of the subclavian artery.

The superficial cervical artery (6.87), at a higher level than the suprascapular, crosses anterior to the phrenic nerve, the scalenus anterior and brachial plexus and is covered by the internal jugular vein, sternocleidomastoid and platysma. It crosses the posterior triangle's floor to the anterior margin of the levator scapulae, ascending deep to the anterior part of the trapezius, supplying it, the adjoining muscles and the cervical lymph nodes. It anastomoses with the superficial ramus of the descending branch of the occipital artery. (See also below: variations of superficial cervical and dorsal scapular arteries.)

4. The Costocervical Trunk (6.85)

On the right, this short vessel arises posteriorly from the second part of the subclavian artery, and, on the left, from its first part. It arches back above the cervical pleura to the first rib's neck, dividing here into superior intercostal and deep cervical branches.

The superior intercostal artery descends between the pleura and necks of the first and second ribs to anastomose with the third posterior intercostal artery. Crossing the neck of the first rib it is medial to the ventral ramus of the first thoracic spinal nerve, which it crosses at a lower level (6.85), and lateral to the cervicothoracic ganglion. In the first space it provides the first posterior intercostal artery, similar in distribution to the lower posterior intercostals. It descends to become the second posterior intercostal artery, usually joining a branch from the third; it is not constant, and is more common on the right; when absent, it is replaced by a direct aortic branch.

The deep cervical artery (6.79), usually arising from the costocervical trunk, is analogous in its first segment to a posterior ramus of a posterior intercostal artery: occasionally it is a separate branch of the subclavian. Passing back above the eighth cervical spinal nerve between the seventh cervical transverse process and the neck of the first rib (sometimes between the sixth and seventh cervical transverse processes), it then ascends between the semispinales capitis and cervicis to the second cervical level. (This ascending segment is a longitudinal post-transverse anastomosis, p. 710.) It supplies adjacent muscles and anastomoses with the deep ramus of the descending branch of the occipital artery (p. 739) and branches of the vertebral. A spinal branch enters the vertebral canal between the seventh cervical and first thoracic vertebrae.

5. The Dorsal Scapular Artery (6.87,91)

This arises from the third or less often second part of the subclavian, passing laterally through the brachial plexus in front of the scalenus medius and then deep to the levator scapulae to the superior scapular angle; here it descends with the dorsal scapular nerve under the rhomboids along the medial scapular border to the inferior angle. It supplies the rhomboids, latissimus dorsi and trapezius and anastomoses with the suprascapular, subscapular and posterior branches of some posterior intercostal arteries.

Variations. About a third of the superficial cervical and dorsal scapular arteries arise in common from the thyrocervical trunk as a *transverse cervical artery*, with a *superficial* (superficial cervical artery) and a *deep branch* (dorsal scapular artery); the latter passes laterally anterior to the brachial plexus and then posterior to the levator scapulae.

The Axilla

The axilla is a pyramidal region between the upper thoracic wall and the arm. Its blunt *apex* continues into the root of the neck (*cervico-axillary canal*) between the external border of the first rib, superior scapular border, posterior surface of the clavicle and the medial aspect of the coracoid process; through it pass the axillary vessels and nerves. Its imaginary *base*, facing down, is broad at the chest, narrow at the arm and corresponds to the skin and a thick layer of *axillary fascia*, between the inferior borders of the pectoralis major in front and the latissimus dorsi behind. It is of course convex up, conforming to the armpit's concavity. The *anterior wall* is formed by the pectorales major et minor, the former covering the whole wall, the latter its intermediate part. The interval between the upper border of the pectoralis minor and clavicle is occupied by the clavipectoral fasica. The *posterior wall* is formed by the subscapularis above, teres major and latissimus dorsi below. *Medial* are the first four ribs with their intercostal muscles and the upper part of serratus anterior; this 'wall' is convex laterally. *Laterally* anterior and posterior walls converge, the 'wall' being narrow, consisting of the humeral intertubercular sulcus; the lateral angle lodges the coracobrachialis and biceps.

The axilla contains axillary vessels, the infraclavicular part of the brachial plexus and its branches, lateral branches of some intercostal nerves, many lymph nodes and vessels, loose adipose areolar tissue and in many instances the 'axillary tail' of the breast. The axillary vessels and brachial plexus run from the apex to the base along the lateral wall and nearer to the anterior wall, the axillary vein being anteromedial to the artery. Owing to the obliquity of the upper ribs, the neurovascular bundle, emerging from behind the clavicle, crosses the first intercostal space; its relations are therefore different at upper and lower levels. Thoracic branches of the axillary artery are in contact with the pectoral muscles; along the lateral margin of the pectoralis minor the lateral thoracic artery reaches the thoracic wall. Subscapular vessels descend on the posterior wall at the lower margin of the subscapularis, and subscapular and thoracodorsal nerves cross the anterior surface of the latissimus dorsi at different inclinations; circumflex scapular vessels wind round the lateral scapular border; posterior circumflex humeral vessels and the axillary nerve curve back and laterally around the humeral surgical neck. No large vessel lies on the medial 'wall', which is crossed proximally only by small branches of the superior thoracic artery. The

long thoracic nerve descends on the serratus anterior and the intercostobrachial nerve perforates the upper anterior part of this wall, crossing the axilla to its lateral 'wall'. The position and arrangement of lymph nodes are described on p. 845, nerves on p. 1130 et seq.

Applied Anatomy. When axillary suppuration occurs, fascial arrangement affects the spread of pus. As described on p. 611, the clavipectoral fascia, between the clavicle and superomedial border of the pectoralis minor, splits to enclose the muscle, blending at its lateral border with the axillary fascia in the anterior axillary fold. Suppuration may be superficial or deep to this layer, either between the pectoral muscles or behind the pectoralis minor; in the former an abscess would appear at the edge of the anterior axillary fold or the groove between the deltoid and pectoralis major; in the latter, pus would tend to surround vessels and nerves and ascend into the neck, the direction of least resistance; pus may also track along vessels into the arm. When an axillary abscess is incised a knife should enter the axillary 'base', midway between the anterior and posterior margins and near the thoracic side to avoid the lateral thoracic, subscapular and axillary vessels on the anterior, posterior and lateral walls. Relations of vessels and nerves in the axilla are important when lymph nodes are removed from the axilla in operations for mammary carcinoma; the positions of major structures in the lateral wall must be remembered.

The Axillary Artery

The axillary artery (6.92), a continuation of the subclavian, begins at the first rib's outer border, ending nominally at the inferior border of the teres major where it becomes *brachial*. Its direction varies with the limb's position: thus it is almost straight when the arm is raised at right angles, concave up when the arm is elevated above this and convex up and laterally with the arm pendent. At first deep, it becomes superficial, covered only by the skin and fasciae. The pectoralis minor crosses it and divides it into *three parts*: proximal, posterior and distal to the muscle.

Relations of the first part. Anterior are the skin, superficial fascia, platysma, supraclavicular nerves, deep fascia, clavicular fibres of the pectoralis major and the clavipectoral fascia. This part is crossed anteriorly by the lateral pectoral nerve, the loop of communication between it and the medial pectoral nerve, and by the thoraco-acromial and cephalic veins. *Posterior* are the first intercostal space and external intercostal, the first and second digitations of the serratus anterior, the long thoracic and medial pectoral nerves and the medial cord of the brachial plexus. *Lateral* is the posterior cord of the brachial plexus. *Anteromedial* is the axillary vein. The first part is enclosed with the axillary vein and brachial plexus in a fibrous *axillary sheath*, continuous with the prevertebral layer of the deep cervical fascia.

Relations of the second part. Anterior are the skin, superficial and deep fascia, pectoralis major and minor. *Posterior* are the posterior cord of the brachial plexus and the areolar tissue between it and the subscapularis. Medial is the axillary vein, separated from it by the medial cord of the brachial plexus and medial pectoral nerve. *Lateral* is the lateral cord of the brachial plexus, separating it from the coracobrachialis. The cords of the brachial plexus thus surround the second part on three sides, with the dispositions implied by their names, and separate it from the vein and adjacent muscles.

Relations of the third part. Anterior are pectoralis major, distal to this skin and fasciae. *Posterior* are the lower part of the subscapularis and tendons of the latissimus dorsi and teres major. *Lateral* is the coracobrachialis. *Medial* is the axillary vein. Branches of the brachial plexus are arranged as follows: *laterally* the lateral root and then trunk of the median nerve and, for a short distance, the musculocutaneous nerve; *medially* the medial cutaneous nerve of the forearm between the axillary artery and vein anteriorly, between them posteriorly the ulnar nerve; the medial cutaneous nerve of upper arm is medial to the vein; *anterior* is the medial root of the median nerve and *posterior* are radial and axillary nerves, the latter only to the distal border of the subscapularis.

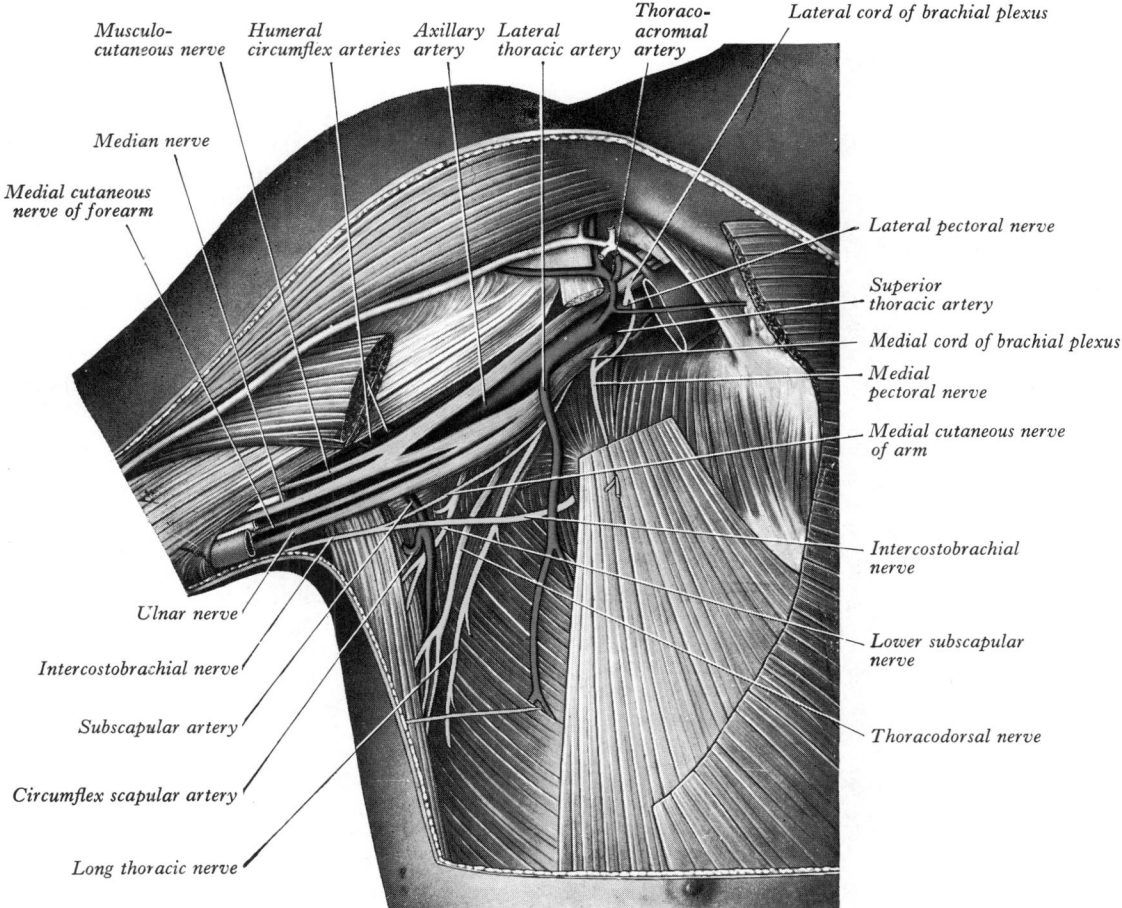

6.92 The right axillary artery and its branches. The pectoralis major and part of the pectoralis minor have been removed. Prominent but unlabelled features are the medial and lateral roots of the median nerve.

The artery's branches are superior thoracic, thoraco-acromial, lateral thoracic, subscapular, anterior and posterior circumflex humeral.

The superior thoracic artery (6.92), a small vessel from the first part of the axillary near the lower border of the subclavius (sometimes from the thoraco-acromial) runs anteromedially above the medial border of the pectoralis minor, then passes between it and the pectoralis major to the thoracic wall. It supplies these muscles and the thoracic wall, anastomosing with the internal thoracic and upper intercostal arteries.

The thoraco-acromial (acromiothoracic) artery (6.87,92), a short branch from the second part, is at first overlapped by the pectoralis minor; skirting its medial border, it pierces the clavipectoral fascia and divides into the pectoral, acromial, clavicular and deltoid branches.

The *pectoral branch* descends between the pectoral muscles, is distributed to them and the breast and anastomoses with the intercostal branches of the internal thoracic and lateral thoracic arteries. The *acromial branch* crosses the coracoid process under the deltoid, which it supplies, pierces the muscle and ends on the acromion, anastomosing with rami of the suprascapular, deltoid branch of the thoraco-acromial and posterior circumflex humeral arteries. The *clavicular branch* ascends medially between the clavicular part of the pectoralis major and clavipectoral fascia, supplying the sternoclavicular joint and subclavius. The *deltoid branch* often arises with the acromial, crossing the pectoralis minor to accompany the cephalic vein between the pectoralis major and deltoid, supplying both.

The lateral thoracic artery (6.92) follows the lateral border of the pectoralis minor to the thoracic wall, supplies the serratus anterior and pectoral muscles, the axillary lymph nodes and subscapularis; it anastomoses with the internal thoracic, subscapular, and intercostal arteries and the pectoral branch of the

thoraco-acromial artery. In females it is large and has *lateral mammary branches* which curve round the lateral border of the pectoralis major to the mammary gland.

The subscapular artery (6.92), largest branch of the axillary, usually arises at the distal (inferior) border of the subscapularis which it follows to the inferior scapular angle, where it anastomoses with the lateral thoracic and intercostal arteries and the deep branch of the transverse cervical. It supplies adjacent muscles and the thoracic wall. It is accompanied distally by the nerve to the latissimus dorsi; about 4 cm from its origin it gives off the *circumflex scapular artery*, usually larger than the continuation of the subscapular. This curves round the lateral scapular border, traversing a *triangular space* between the subscapularis above and the teres major below, and the long head of triceps laterally. It enters the infraspinous fossa under the teres minor and then divides. One branch (*infrascapular*) enters the subscapular fossa deep to the subscapularis, anastomosing with the suprascapular and dorsal scapular arteries (or deep branch of the transverse cervical); the other continues along the lateral scapular border between the teres major and minor and, dorsal to the inferior angle, anastomoses with the deep branch of the transverse cervical artery. Small rami supply the posterior part of the deltoid and the long head of the triceps, anastomosing with an ascending branch of the arteria profunda brachii.

The anterior circumflex humeral artery (6.92), arising from the lateral side of the axillary artery at the distal border of the subscapularis, runs horizontally behind coracobrachialis and the short head of biceps, anterior to the surgical neck of the humerus. Reaching the intertubercular sulcus, it sends an ascending branch to supply the humeral head and shoulder joint. It continues laterally under the long head of biceps and deltoid, anastomosing with the posterior circumflex humeral artery.

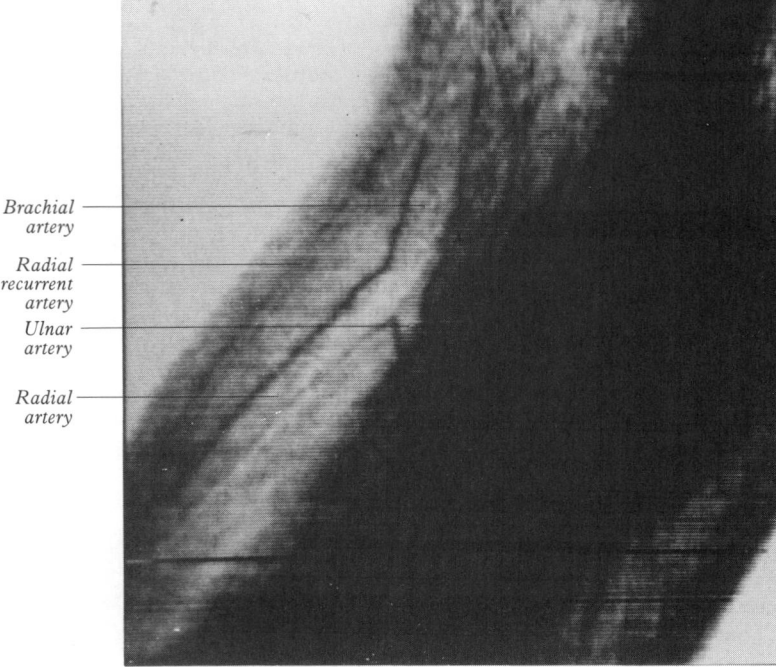

Acromial artery

Deltoid artery

Deltoid

Nutrient artery to humerus

Arteria profunda brachii

Superior ulnar collateral artery

Brachial artery

Biceps

Brachialis

Inferior ulnar collateral artery

Brachioradialis

Tendon of biceps

Radial recurrent artery

Bicipital aponeurosis

Radial artery

Pronator teres

6.93A The right brachial artery and its branches.

Brachial artery

Radial recurrent artery

Ulnar artery

Radial artery

6.93B Ultrasound transmission image of the distal part of the arm and the proximal part of the forearm, showing the radial and ulnar arteries arising from the brachial artery. The radial recurrent artery can also be seen. The image was produced using a 2 MHz ultrasound camera system with a depth of focus of 6 mm (supplied by Peter D Edmonds, SRI International, Menlo Park, California).

The posterior circumflex humeral artery (6.91), larger than the anterior, branches from the third part of the axillary at the distal border of the subscapularis and runs back with the axillary nerve through a *quadrangular space*, bounded by the subscapularis, the capsule of the shoulder joint and the teres minor above, the teres major below, the long head of triceps medially and the surgical neck of the humerus laterally. It curves round the humeral neck and supplies the shoulder joint, deltoid, teres major and minor, and long and lateral heads of triceps, giving off a descending branch to anastomose with the deltoid branch of the arteria profunda brachii and with the anterior circumflex humeral and acromial branches of the suprascapular and thoraco-acromial arteries.

Surface Anatomy. Pulsation of the axillary artery can be felt against the axillary lateral wall. Its upper segment can be mapped out, when the arm is raised, by a line from this to the midpoint of the clavicle.

Variations. Branches vary considerably; an *alar thoracic*, often from the second part, may supply fat and lymph nodes in the axilla. Occasionally the subscapular, circumflex humeral and arteria profunda arise in common and then branches of the brachial plexus surround this instead of the axillary artery. The posterior circumflex humeral artery may be from the arteria profunda brachii, passing back below the teres major instead of traversing the quadrangular space. Sometimes (anomalous 'high division') the axillary divides into radial and ulnar arteries and is occasionally the source of the anterior interosseous artery.

Applied Anatomy. Axillary compression is most effectual against the humerus. Except for the popliteal, the axillary artery is more frequently lacerated by violence than any other, being most susceptible when diseased. It has been ruptured in attempts to reduce old dislocations, especially when the artery is adherent to the articular capsule.

The Brachial Artery

The brachial artery (6.93–96), a continuation of the axillary, begins at the distal (inferior) border of the tendon of teres major and ends about a centimetre distal to the elbow joint (at the level of the neck of the radius) by dividing into radial and ulnar arteries. At first it is medial to the humerus, but gradually spirals anterior to it until it lies midway between the humeral epicondyles. Its pulsation can be felt throughout.

Relations. The artery is wholly superficial, covered *anteriorly* only by skin and superficial and deep fasciae; the bicipital aponeurosis crosses it anteriorly at the elbow, separating it from the median cubital vein; the median nerve crosses it lateromedially near the distal attachment of coracobrachialis. *Posterior* are the long head of triceps, separated by the radial nerve and arteria profunda brachii and then successively by: the medial head of triceps, the attachment of coracobrachialis and the brachialis. *Lateral* are: proximally the median nerve and coracobrachialis and distally the biceps and the muscles overlapping the artery. *Medial* are: proximally the medial cutaneous nerve of forearm and ulnar nerve, distally the median nerve and basilic vein (separated distally by the deep fascia). With the artery are two venae comitantes, connected by transverse and oblique branches.

At the elbow the brachial artery sinks deeply into the triangular intermuscular *cubital fossa*. The fossa's base is an (arbitrary) inter-epicondylar line, the sides being the medial edge of the brachioradialis and the lateral margin of pronator teres; the 'floor' consists of brachialis and supinator. The fossa contains the tendon of the biceps, the terminal part of the brachial artery and accompanying veins, the commencement of the radial and ulnar arteries and parts of the median and radial nerves. The brachial artery is central and it divides near the neck of the radius into its terminal branches, the radial and ulnar arteries. *Anterior* to it are the skin, superficial fascia and median cubital vein, separated by the bicipital aponeurosis. *Posteriorly* the brachialis separates it from the elbow joint. The median nerve is *medial* proximally but is separated from the ulnar artery by the ulnar head of the pronator teres. *Lateral* are the tendon of biceps and the radial nerve, the latter concealed between supinator and brachioradialis.

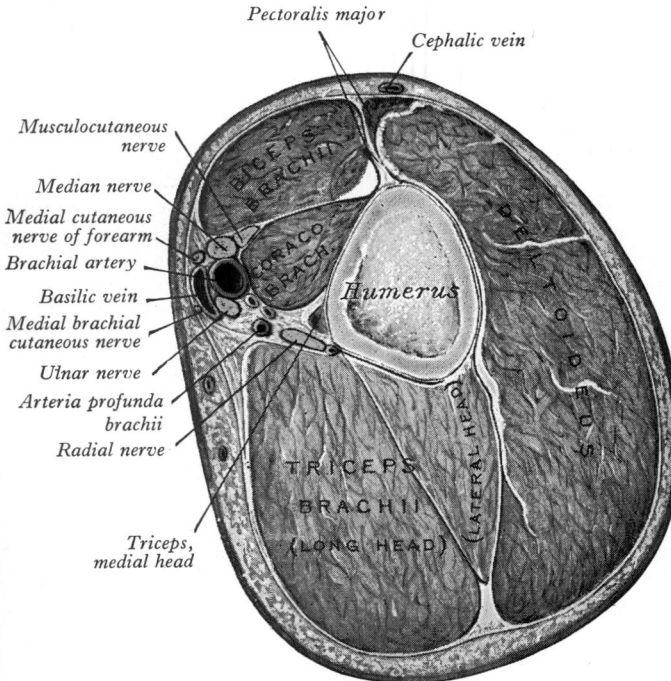

6.94 Transverse section through the right arm at the junction of the proximal and middle thirds of the humerus: proximal aspect.

Variations. The brachial artery, with the median nerve, may diverge from the medial border of the biceps, descending towards the medial humeral epicondyle, usually behind a *supracondylar process* from which a fibrous arch crosses the artery, and which then runs behind or through the pronator teres to the elbow. This resembles the normal arrangement in some carnivores (p. 410). Occasionally the artery divides proximally into two trunks which re-unite. Frequently it divides more proximally than usual into radial, ulnar and common interosseous arteries. Most often the radial branches proximally, leaving a common trunk for the ulnar and common interosseous; sometimes the ulnar arises proximally, the radial and common interosseous forming the other division; the common interosseous may also arise proximally. Sometimes slender *vasa aberrantia* connect the brachial to the axillary artery or to one of the forearm arteries, usually the radial. The brachial artery may be crossed by muscular or tendinous slips from the coracobrachialis, biceps, brachialis or pronator teres.

Branches are: the arteria profunda brachii, nutrient, superior and inferior ulnar collateral, muscular, radial and ulnar arteries.

The arteria profunda brachii (6.93,94,97), a large branch from the posteromedial aspect of the brachial, distal to the teres major, follows the radial nerve closely, at first back between the long and medial heads of the triceps, then in the nerve's groove covered by the lateral head of triceps; here it divides into terminal branches (6.97). Apart from the muscular rami, it supplies the following: the nutrient, deltoid, middle collateral and radial collateral arteries. A *nutrient artery* enters the humerus posterior to the deltoid tuberosity but may be absent. The *deltoid (ascending) branch* ascends between the lateral and long heads of triceps and anastomoses with a descending branch of the posterior humeral circumflex artery. The *middle collateral (posterior descending) branch*, the larger terminal ramus, arises behind the humerus and descends in the medial head of the triceps to the elbow, anastomosing with the interosseous recurrent artery behind the lateral epicondyle; it often has a small branch which accompanies the nerve to the anconeus. The *radial collateral*, the other terminal branch, is the artery's continuation. It accompanies the radial nerve through the lateral intermuscular septum, descending between the brachialis and brachioradialis anterior to the lateral epicondyle, anastomosing with the radial recurrent artery.

The nutrient artery of the humerus arises near the mid-level of the upper arm, entering the nutrient canal near the attachment of coracobrachialis; it is directed distally.

The superior ulnar collateral artery (6.93,95,97) arises a little distal to the upper arm's mid-level, often as a branch from the arteria profunda brachii. It accompanies the ulnar nerve, piercing the medial intermuscular septum to descend between the medial epicondyle and olecranon, ending deep to flexor carpi ulnaris by anastomosing with the posterior ulnar recurrent and inferior collateral arteries; sometimes a branch of it passing anterior to the medial epicondyle anastomoses with the anterior ulnar recurrent artery.

The inferior ulnar collateral (supratrochlear) artery (6.93,97,98) begins about 5 cm proximal to the elbow, passes medially between the median nerve and brachialis and, piercing the medial intermuscular septum, curls round the humerus between the triceps and bone, forming, by its junction with the middle collateral branch of arteria profunda brachii, an arch proximal to the olecranon fossa. As it lies on brachialis it has branches descending anterior to the medial epicondyle to

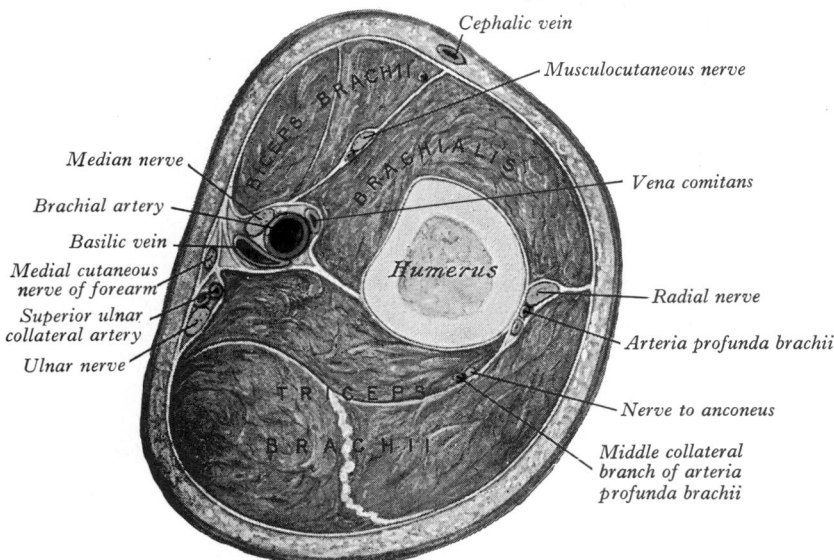

6.95 Transverse section through the right arm, a little below the middle of the shaft of the humerus: proximal aspect.

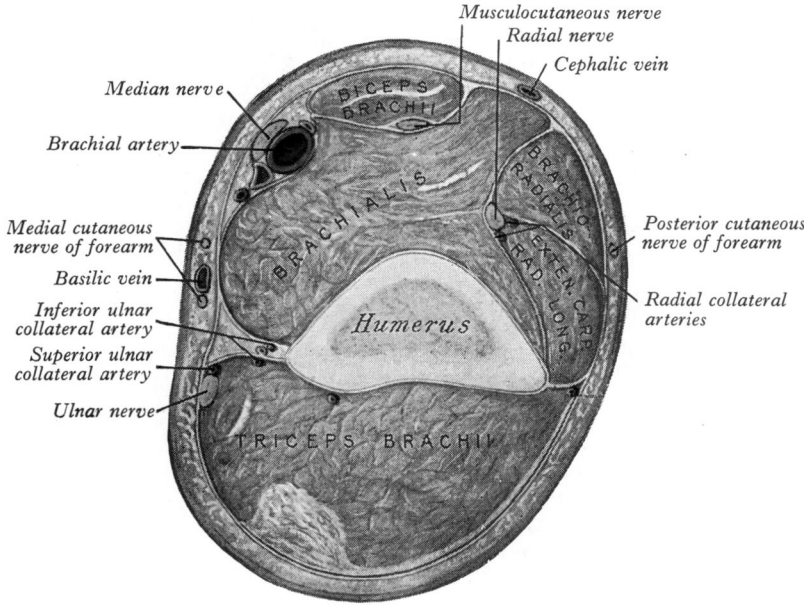

6.96 Transverse section through the right arm, 2 cm above the medial epicondyle of the humerus: proximal aspect.

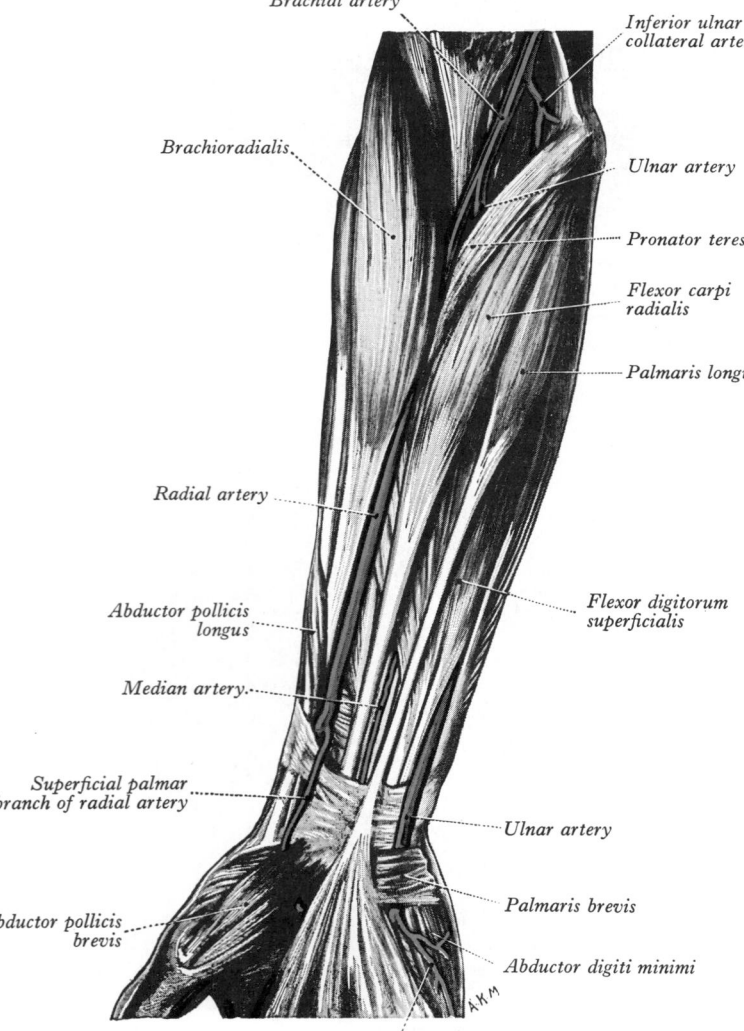

Arteria profunda brachii

Superior ulnar collateral artery

Brachial artery

Radial collateral artery

Middle collateral branch of arteria profunda brachii

Inferior ulnar collateral artery

Anterior ulnar recurrent artery

Radial recurrent artery

Posterior ulnar recurrent artery

Interosseous recurrent artery

Common interosseous artery

Posterior interosseous artery

Radial artery

Ulnar artery

Anterior interosseous artery

6.97 The arterial anastomoses around the (right) elbow joint.

Brachial artery

Inferior ulnar collateral artery

Brachioradialis

Ulnar artery

Pronator teres

Flexor carpi radialis

Palmaris longus

Radial artery

Flexor digitorum superficialis

Abductor pollicis longus

Median artery

Superficial palmar branch of radial artery

Ulnar artery

Palmaris brevis

Abductor pollicis brevis

Abductor digiti minimi

Palmar digital artery

760 6.98 The right radial and ulnar arteries, superficial dissection.

anastomose with the anterior ulnar recurrent artery. Behind the epicondyle a branch anastomoses with the superior ulnar collateral and posterior ulnar recurrent arteries.

Muscular arteries from the brachial are distributed to the coracobrachialis, biceps and brachialis.

Applied Anatomy. Compression of the brachial artery may be effected at almost any level; if proximal, it should be directed laterally, if distal, backwards. The most favourable site is about midway, where the artery is on the tendon of the coracobrachialis and still medial to the humerus; pressure should be exerted slightly posterolaterally.

The Radial Artery

The radial artery (**6**.98,99,100), though smaller than the ulnar, appears a more direct continuation of the brachial. It begins about 1 cm distal to the bend of the elbow, (level of the neck of the radius, **6**.93B,100), then descends along the lateral side of the forearm to the wrist, where it is palpable between the flexor carpi radialis medially and the salient anterior border of the radius. It then curls posterolaterally round the carpus, beneath tendons of the abductor pollicis longus and extensor pollicis brevis and longus, to the proximal end of the first inter-metacarpal space, swerving medially between the heads of the first dorsal interosseous into the palm and then crossing medially to form the deep palmar arch with the deep branch of the ulnar artery. The radial artery is thus divisible into parts: in the forearm, wrist and hand.

In the forearm (**6**.98,99,100) the artery extends from the medial side of the neck of the radius to the front of its styloid process, being medial to the radial shaft proximally, but anterior to it distally. Proximally it is overlapped anteriorly by the belly of brachioradialis, whereas the rest is covered only by the skin, superficial and deep fasciae. Posterior are successively: the tendon of biceps, supinator, the distal attachment of pronator teres, radial head of flexor digitorum superficialis, flexor pollicis longus, pronator quadratus and the lower end of the radius (where its pulsation is most accessible). Proximally pronator teres is medial, brachioradialis lateral; distally the tendon of flexor carpi radialis is medial, that of brachioradialis lateral. The superficial branch of the radial nerve is lateral in the vessel's middle third, and filaments of the lateral cutaneous nerve of the forearm run along its distal part as it curves round the carpus. The vessel is accompanied by paired venae comitantes.

At the wrist (**6**.101,103) the radial artery passes on to the dorsal aspect of the carpus between the lateral carpal ligament and tendons of abductor pollicis longus and extensor pollicis brevis. It crosses the scaphoid bone and trapezium (in the 'anatomical snuff-box'), where again its pulsation is obvious, and as it passes between heads of the first dorsal interosseous it is crossed by the tendon of extensor pollicis longus. Between the pollicial extensors it is crossed by the beginning of the cephalic vein and the (palpable) digital branches of the radial nerve supplying the thumb and index.

In the hand (**6**.104) the radial artery, having traversed the first interosseous space between the heads of the first dorsal interosseous, crosses the palm, at first deep to the oblique head of adductor pollicis and then between its oblique and transverse heads or through the transverse head. At the fifth metacarpal base it anastomoses with the deep branch of the ulnar artery, completing the *deep palmar arch* (**6**.100).

Variations. Sometimes the radial artery arises proximally, usually from the axillary or beginning of the brachial artery. In the forearm it is sometimes superficial to the deep fascia and occasionally superficial to the pollicial extensor tendons (see also under 'Variations of the Brachial Artery', above).

The radial recurrent artery (**6**.97,100) arises just distal to the elbow, passing between superficial and deep branches of the radial nerve to ascend behind the brachioradialis, anterior to the supinator and brachialis; it supplies these muscles and the elbow joint, anastomosing with the radial collateral branch of the arteria profunda brachii.

Muscular branches are distributed to muscles on the radial side of the forearm.

The palmar carpal branch (**6**.100), a small vessel, arises near the distal border of pronator quadratus and crosses the anterior surface of the distal end of the radius, near the palmar carpal surface, passing medially to anastomose behind the long flexor tendons with the palmar carpal branch of the ulnar; this transverse anastomosis is joined by longitudinal branches from the anterior interosseous and recurrent branches from the deep palmar arch, forming a cruciate *palmar carpal arch*, which, by descending branches, supplies the carpal articulations and bones. (Although so named this is usually sited near the wrist joint on the distal forearm bones.)

The superficial palmar branch (**6**.104) arises from the radial artery just before it curves round the carpus, passes through and occasionally over the thenar muscles, which it supplies, sometimes anastomosing with the end of the ulnar artery to complete a *superficial palmar arch.*

The dorsal carpal branch arises deep to the pollicial extensor tendons, runs medially across the dorsal carpal surface under them and anastomoses with the ulnar dorsal carpal branch and also with the anterior and posterior interosseous arteries to form a *dorsal carpal arch.* The carpal arches are both close to bone and supply the distal epiphysial parts of the radius and ulna. From the dorsal arch three *dorsal metacarpal arteries* descend on the second to fourth dorsal interosseous muscles and bifurcate into the *dorsal digital branches* for the adjacent sides of all four fingers; they anastomose with the palmar digital branches from the superficial palmar arch; near their origins they also anastomose with the deep palmar arch by the *proximal perforating arteries* and, near their bifurcation, with the palmar digital rami of the superficial palmar arch by *distal perforating arteries.*

The first dorsal metacarpal artery (**6**.103), a branch of the radial just before it passes between the heads of the first dorsal interosseous, divides almost at once into two branches supplying the adjacent sides of the pollex and index; the radial side of the pollex receives a branch direct from the radial artery itself (vide infra).

The arteria princeps pollicis (**6**.100) arises from the radial as it turns into the palm, and descends on the palmar aspect of the first metacarpal under the oblique head of adductor pollicis lateral to the first palmar interosseous. At the base of the proximal phalanx, deep to the tendon of flexor pollicis longus, it divides into two branches, appearing between the medial and lateral attachments of the oblique head of adductor pollicis to run along both sides of the pollex, forming, on the palmar surface of its distal phalanx, a pollicial arch supplying the skin and subcutaneous tissue. The arteria princeps pollicis is the usual nutrient of supply to the first metacarpal bone.

The arteria radialis indicis (**6**.100,104), often a proximal branch of the arteria princeps pollicis, descends between the first dorsal interosseous and transverse head of adductor pollicis, and along the lateral side of the index finger to its end; it anastomoses with the indicial medial digital artery. At the distal border of the transverse head of the adductor pollicis it anastomoses with the arteria princeps pollicis and links with the superficial palmar arch.

The arteriae princeps pollicis et radialis indicis may be combined as the *first palmar metacarpal artery.*

The Deep Palmar Arch (**6**.100)

This is formed by anastomosis of the end of the radial with the deep palmar branch of the ulnar artery. It crosses the bases of the metacarpal bones and interossei, covered by the oblique head of adductor pollicis, the digital flexor tendons and lumbricals. In its concavity, running laterally, is the deep branch of the ulnar nerve. The arch was incomplete in six of 200 arches (Coleman & Anson 1961). Variation is chiefly in the size of contribution from the ulnar artery.

Surface Anatomy. The deep palmar arch is indicated by a horizontal line about 4 cm long from a point just distal to the hamate's hook (**6**.105). It is about 1 cm proximal to the superficial arch.

Branches of the deep palmar arch are the palmar metacarpal, perforating and recurrent.

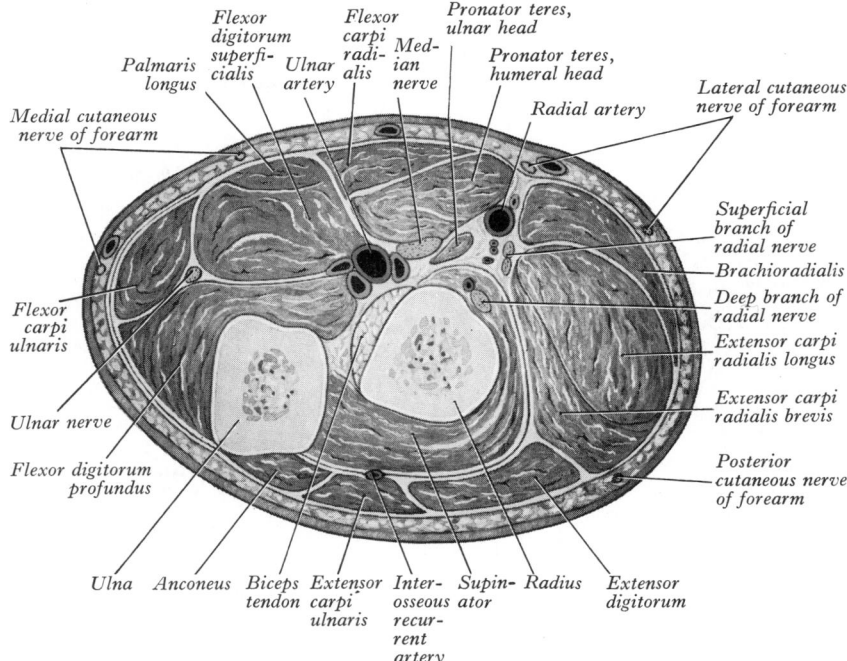

6.99 Transverse section through the forearm at the level of the radial tuberosity.

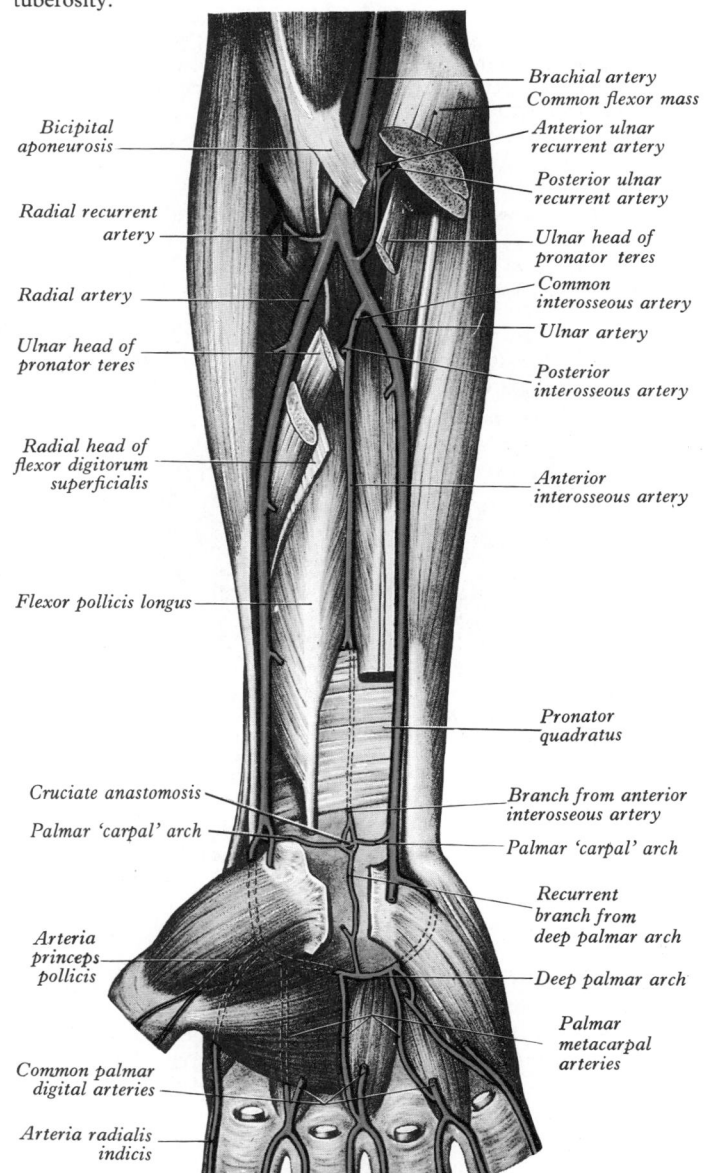

6.100 The arteries of the right forearm and hand: deep dissection. The palmar 'carpal' arch lies across forearm bones.

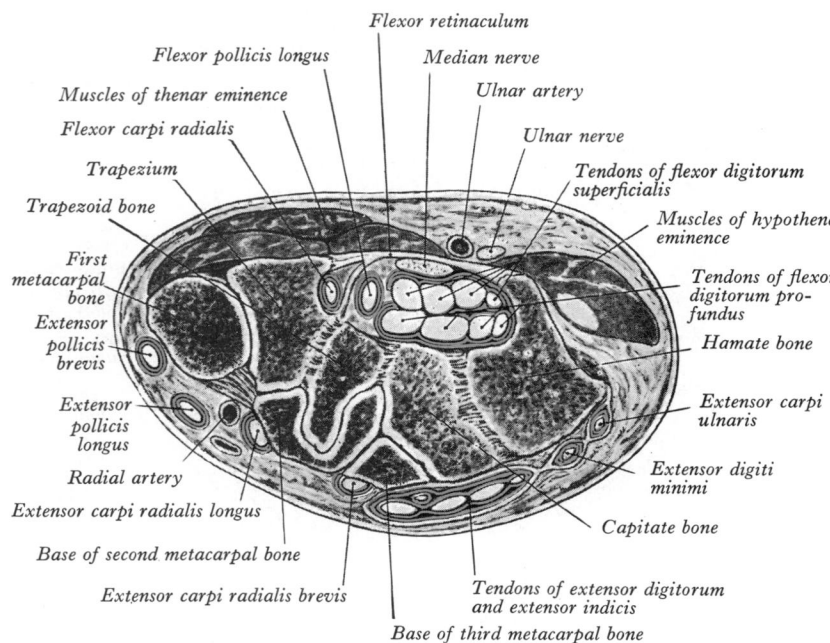

6.101 Transverse section through the left wrist: superior aspect. The section is slightly oblique and divides the distal row of the carpus and the bases of the first, second and third metacarpal bones. The arrangement of the tendons of the flexors of the fingers shown in the figure represents the actual condition in the specimen. Observe that the carpometacarpal joint of the thumb is separate from the joint between the trapezium and the base of the second metacarpal bone.

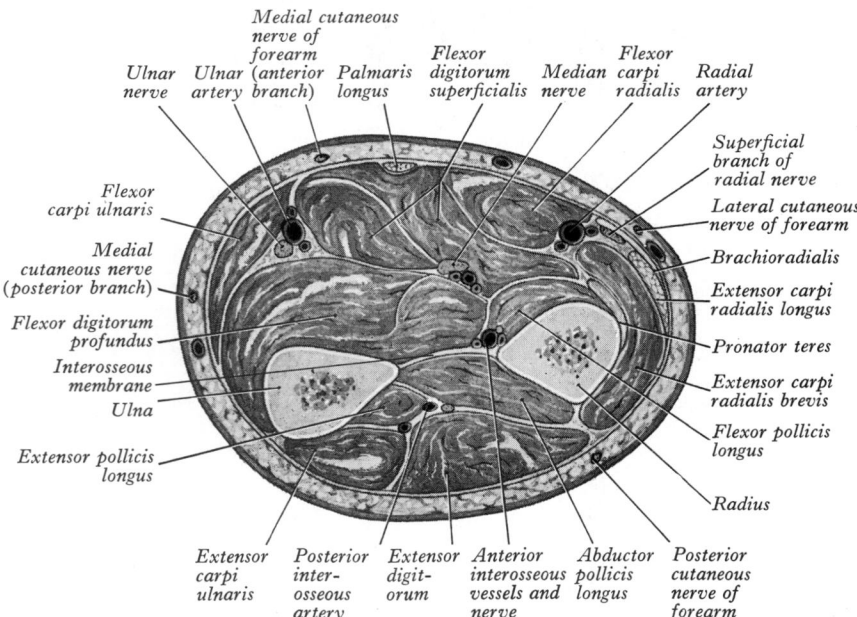

6.102 Transverse section through the middle of the left forearm: proximal aspect.

The three palmar metacarpal arteries (**6.**100), run distally from the convexity of the arch on the interosseous muscles of the second to fourth spaces; at the digital clefts they join the *common digital branches* of the superficial arch. They supply nutrient branches to the medial four metacarpals.

Three perforating branches from the deep palmar arch traverse the second to fourth interosseous spaces between the heads of the corresponding dorsal interossei to anastomose with the dorsal metacarpal arteries.

Recurrent branches (**6.**100) ascend proximally from the deep palmar arch anterior to the carpus to supply the carpal bones

762

and intercarpal articulations, ending in the palmar carpal arch (mentioned above).

The Ulnar Artery

The ulnar artery (**6.**98–104), the larger terminal branch of the brachial, begins just distal to the bend of the elbow. It reaches the medial side of the forearm midway between elbow and wrist, which it passes vertically, crossing the flexor retinaculum lateral to the ulnar nerve and pisiform bone; distal to this it has a deep branch and then continues across the palm as the superficial palmar arch.

Relations. In the forearm the *proximal half* of the artery (**6.**98,99,100,102) passes posterior to the pronator teres, flexor carpi radialis, palmaris longus and flexor digitorum superficialis; medially it is overlapped in its middle third by flexor carpi ulnaris; it lies in front of the brachialis and flexor digitorum profundus. Distal to the elbow the median nerve is medial for about 2.5 cm and then crosses it but is separated by the ulnar head of pronator teres. The artery's *distal half* (**6.**98,100,104) lies on the flexor digitorum profundus, covered by the skin, superficial and deep fasciae, between the flexor carpi ulnaris and flexor digitorum superficialis. It is accompanied by venae comitantes; the ulnar nerve lies medial to its distal two-thirds and its palmar cutaneous branch descends along it to the hand.

At the wrist (**6.**100,101,104) the artery is covered by skin, fasciae and palmaris brevis, and it lies between the superficial and main parts of the flexor retinaculum (p. 625); the ulnar nerve and pisiform bone are medial.

Surface Anatomy. A line from a point in the limb's midline just distal to the elbow's fold descends medially to meet a line stretching from the medial epicondyle to the pisiform bone, from the junction of its upper and middle thirds. Together these represent the artery's upper third and distal two-thirds respectively.

Variations. The ulnar artery may arise proximal to the elbow, the brachial being more often its source than the axillary artery; it is then usually superficial to the forearm flexors, commonly under the deep fascia, and is rarely subcutaneous; the brachial artery then supplies the common interosseous and this the ulnar recurrent arteries.

Branches. The artery supplies medial muscles in the forearm and hand, the common flexor synovial sheath and ulnar nerve (Blunt 1959), including the following named branches:

The anterior ulnar recurrent artery (**6.**97,100) arises just distal to the elbow, ascends between the brachialis and pronator teres, supplies them and anastomoses with the inferior ulnar collateral artery anterior to the medial epicondyle.

The posterior ulnar recurrent artery (**6.**97,100) is larger and arises distal to the anterior recurrent. It passes dorsomedially between the flexores digitorum profundus and superficialis and ascends behind the medial epicondyle; between this and the olecranon it is deep to the flexor carpi ulnaris, ascending between its heads with the ulnar nerve. Supplying adjacent muscles, nerve, bone and elbow joint, it anastomoses with the ulnar collateral and interosseous recurrent arteries (**6.**97).

The common interosseous artery (**6.**100), a short branch of the ulnar, just distal to the radial tuberosity, passes back to the proximal border of the interosseous membrane, dividing into the *anterior* and *posterior interosseous arteries*.

The anterior interosseous artery (**6.**100,103) descends on the anterior aspect of the interosseous membrane with the median nerve's anterior interosseous branch, overlapped by contiguous sides of flexor digitorum profundus and flexor pollicis longus; it has *muscular* branches and *nutrient* rami for the radius and ulna. On the membrane, branches leave to pierce it and supply deep *extensor* muscles. Proximal to pronator quadratus its continuation also traverses the membrane to the back of the forearm where it anastomoses with its own posterior interosseous branch, descending over the carpal dorsum to join the dorsal carpal arch. It is in the extensor retinacular compartment with the tendons of digital extensors. Before it pierces the interosseous membrane, a branch descends behind the pronator quadratus to the anterior 'carpal

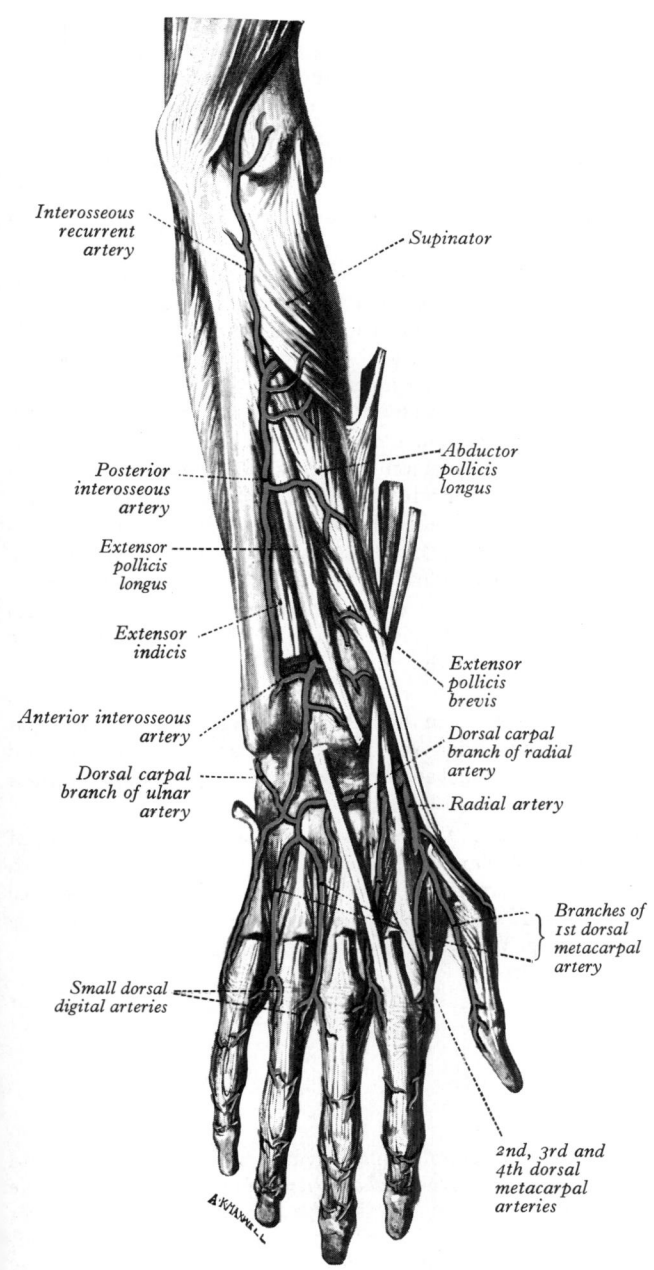

6.103 The arteries of the posterior surface of the right forearm and hand.

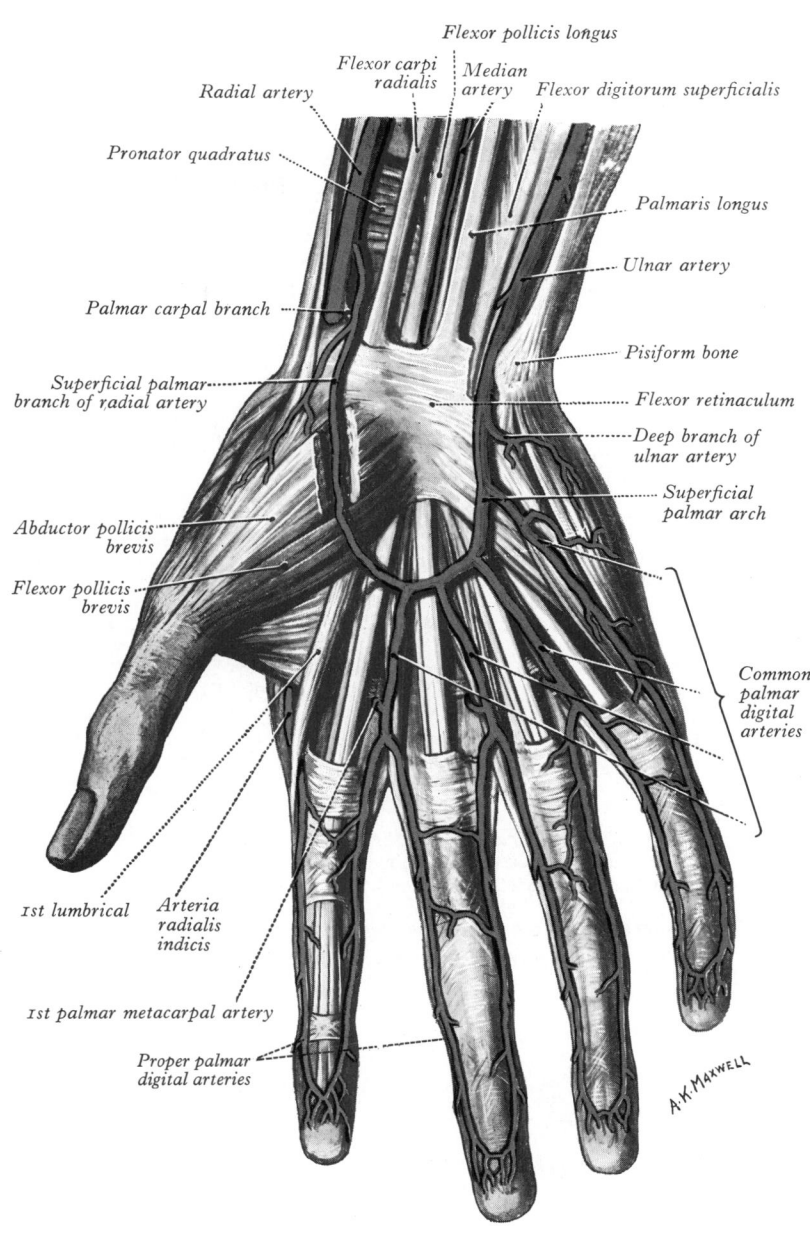

6.104 The superficial palmar arch and its branches. A part of the abductor pollicis brevis has been excised to expose the superficial palmar branch of the radial artery.

arch. (Strictly, as mentioned, the latter is *proximal* to the line of the wrist joint.) The slender *median artery*, from the start of the anterior interosseous, accompanies and supplies the median nerve; it often arises from the common interosseous, sometimes much enlarged, reaching the palm with the nerve (p. 218), where it may join the superficial palmar arch or end as one or two palmar digital arteries.

The posterior interosseous artery (6.100,103), usually smaller than the anterior, passes dorsally between the oblique cord and proximal border of the interosseous membrane and then between supinator and abductor pollicis longus, descending deep to the superficial extensors, which it supplies. On abductor pollicis longus it accompanies the deep branch of the radial nerve. Distally it anastomoses with the end of the anterior interosseous and dorsal carpal arch. Near its origin the *interosseous recurrent artery* leaves it to ascend between the lateral epicondyle and olecranon, either on or through the supinator but deep to anconeus, to anastomose with the middle collateral branch of the arteria profunda brachii, posterior ulnar recurrent and ulnar collateral arteries.

The muscular branches are direct rami of the main vessel, distributed to muscles in the ulnar region.

The palmar carpal branch (6.100), a small vessel, crosses the distal ulna behind the tendons of flexor digitorum profundus; it anastomoses with a palmar carpal branch of the radial to make a so-called palmar carpal arch (p. 761).

The dorsal carpal branch (6.103), arising just proximal to the pisiform bone, curves deep to the tendon of flexor carpi ulnaris to the carpal dorsum to pass laterally across it under the extensor tendons, anastomosing with the radial dorsal carpal branch to complete the dorsal carpal arch (p. 761). Near its origin it sends a small digital branch along the ulnar side of the fifth metacarpal to supply the medial side of the dorsal surface of the fifth finger.

The deep palmar branch (6.100,104), often double, passes between the abductor and flexor digiti minimi, through or deep to the opponens digiti minimi; it anastomoses with the radial, completing the deep palmar arch, accompanied by the deep branch of the ulnar nerve.

The Superficial Palmar Arch (6.104,105)

This anastomosis is fed mainly by the ulnar artery, entering the palm with the ulnar nerve, anterior to the flexor retinaculum and

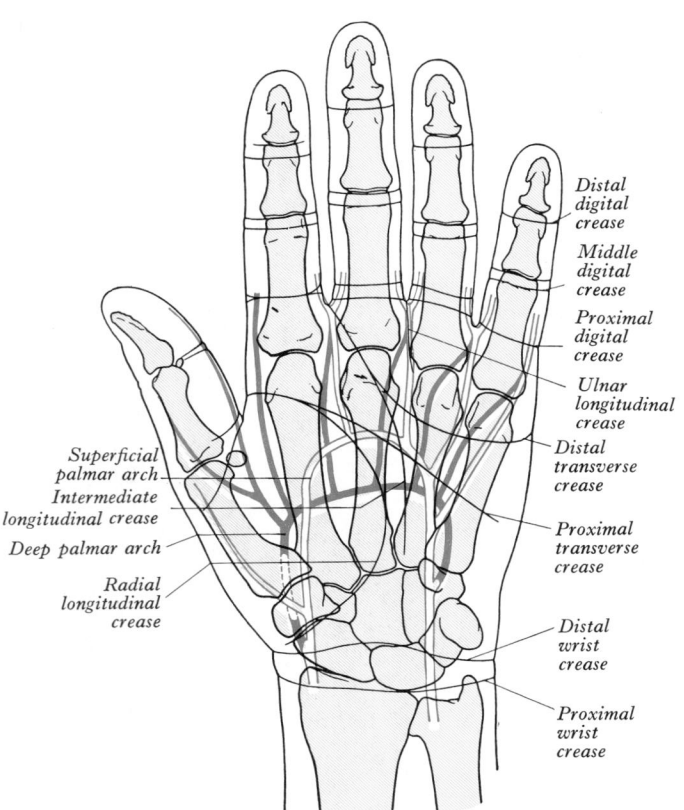

Distal
digital
crease

Middle
digital
crease

Proximal
digital
crease

Ulnar
longitudinal
crease

Distal
transverse
crease

Proximal
transverse
crease

Superficial
palmar arch
Intermediate
longitudinal crease

Deep palmar arch

Radial
longitudinal
crease

Distal
wrist
crease

Proximal
wrist
crease

6.105 The relation of the skin flexure lines and palmar arterial arches to the bones of the left hand (simplified).

lateral to the pisiform, passing medial to the hamate's hook, then curving laterally to form an arch, convex distally and level with a

transverse line through the distal border of the fully extended pollicial base. About a third of the superficial palmar arches are formed by the ulnar alone; a further third are completed by the superficial palmar branch of the radial and a third either by the arteria radialis indicis, a branch of arteria princeps pollicis or by the median artery (Coleman & Anson 1961). It is covered by the palmaris brevis and palmar aponeurosis and it is superficial to the flexor digiti minimi, branches of the median nerve and to the long flexor tendons and lumbrical muscles.

Branches. **Three common palmar digital arteries (6.**104) from the convexity of the superficial palmar arch proceed distally on the second to fourth lumbricals, each joined by a corresponding palmar metacarpal artery from the deep palmar arch and dividing into two *proper palmar digital arteries.* These run along the contiguous sides of all four fingers, dorsal to the digital nerves, anastomosing in the subcutaneous tissue of the finger tips and near the interphalangeal joints. Each digital artery has *two dorsal branches* anastomosing with the *dorsal digital arteries* and supplying the soft parts dorsal to the middle and distal phalanges, including the matrices of the nails. The *palmar digital artery* for the medial side of *minimus* leaves the arch under the palmaris brevis. Palmar digital arteries supply metacarpophalangeal and interphalangeal joints and nutrient rami to the phalanges. They are the main digital supply, the dorsal digital arteries (p. 761) being minute.

Anastomoses between the radial and ulnar arteries occur: (1) at the wrist by the palmar and dorsal carpal arches; (2) in the hand through the superficial and deep palmar arches; (3) between their digital and metacarpal branches.

Applied Anatomy. In wounds of the palmar arches ligature of *one* forearm artery may be ineffective; simultaneous tying of both proximal to the carpus may also fail, because of interosseo-carpal anastomoses. If local pressure fails the brachial artery may be compressed (p. 758) as a temporary expedient.

THE ARTERIES OF THE TRUNK

The Thoracic Aorta

The thoracic aorta (**6.**106) is the segment of *descending aorta* confined to the posterior mediastinum. It begins level with the fourth thoracic vertebra's lower border, continuous with the aortic arch, ending anterior to the twelfth thoracic's lower border in the diaphragmatic aortic aperture. At its origin it is *left* of the vertebral column; as it descends it approaches the midline and at its termination is directly anterior to it.

Relations. Anterior, from above down, are the left pulmonary hilum, the pericardium separating it from the left atrium, oesophagus and diaphragm; *posterior* are the vertebral column and hemiazygos veins; *right lateral* are the azygos vein and thoracic duct and below, the right pleura and lung; *left lateral* are the pleura and lung. The oesophagus, with its plexus of nerves, is *right lateral* above but becomes *anterior* in the lower thorax; close to the diaphragm it is *left anterolateral.* Thus, to a limited degree, the descending aorta and oesophagus are mutually spiralized.

Surface Anatomy. The vessel is projected as a band 2.5 cm broad from the sternal end of the second left costal cartilage to a median position about 2 cm above the transpyloric plane (p. 1334).

Branches. The thoracic aorta provides *visceral branches* to the pericardium, lungs, bronchi, oesophagus and *parietal branches* to the thoracic wall.

Pericardial branches, a few small vessels, are distributed to the posterior pericardial aspect.

The bronchial arteries vary in number, size and origin. Usually one *right bronchial artery,* from the third posterior intercostal or upper left bronchial artery, runs posteriorly on the right bronchus and its branches, supplying them, the pulmonary areolar

tissue and the bronchopulmonary lymph nodes, pericardium and oesophagus. The *left bronchial arteries,* usually two, arise from the thoracic aorta, the upper near the fifth thoracic vertebra, the lower below the left bronchus. They run posteriorly on the left bronchus and are distributed as on the right (p. 1248). Cauldwell et al (1948) found this arrangement in 40% of 150 cadavers; less frequent (at about 20% each) were two left and two right bronchial arteries or one on each side, all direct branches from the descending thoracic aorta, arising near the third and fourth intercostal arteries. In about 10%, *one* left and *two* right bronchial arteries existed. Complex variations consisted chiefly of more numerous aortic branches. Very rarely a bronchial artery arose from the aortic arch.

The oesophageal arteries, four or five, arise anteriorly from the aorta, descending obliquely to the oesophagus, forming a *vascular chain* on it which anastomoses above with the oesophageal branches of the inferior thyroid arteries and below with the ascending branches from the left phrenic and gastric. (Aortic oesophageal arteries are persistent supradiaphragmatic ventral splanchnic foregut arteries.)

The mediastinal branches are numerous small vessels supplying lymph nodes and areolar tissue in the posterior mediastinum. **Phrenic branches** arise from the lower thoracic aorta distributed posteriorly to the superior diaphragmatic surface and anastomosing with the musculophrenic and pericardiacophrenic arteries.

The Posterior Intercostal Arteries

Usually nine pairs of posterior intercostal arteries are derived from the posterior aspect of the descending thoracic aorta. They are distributed to the lower nine intercostal spaces, the first and

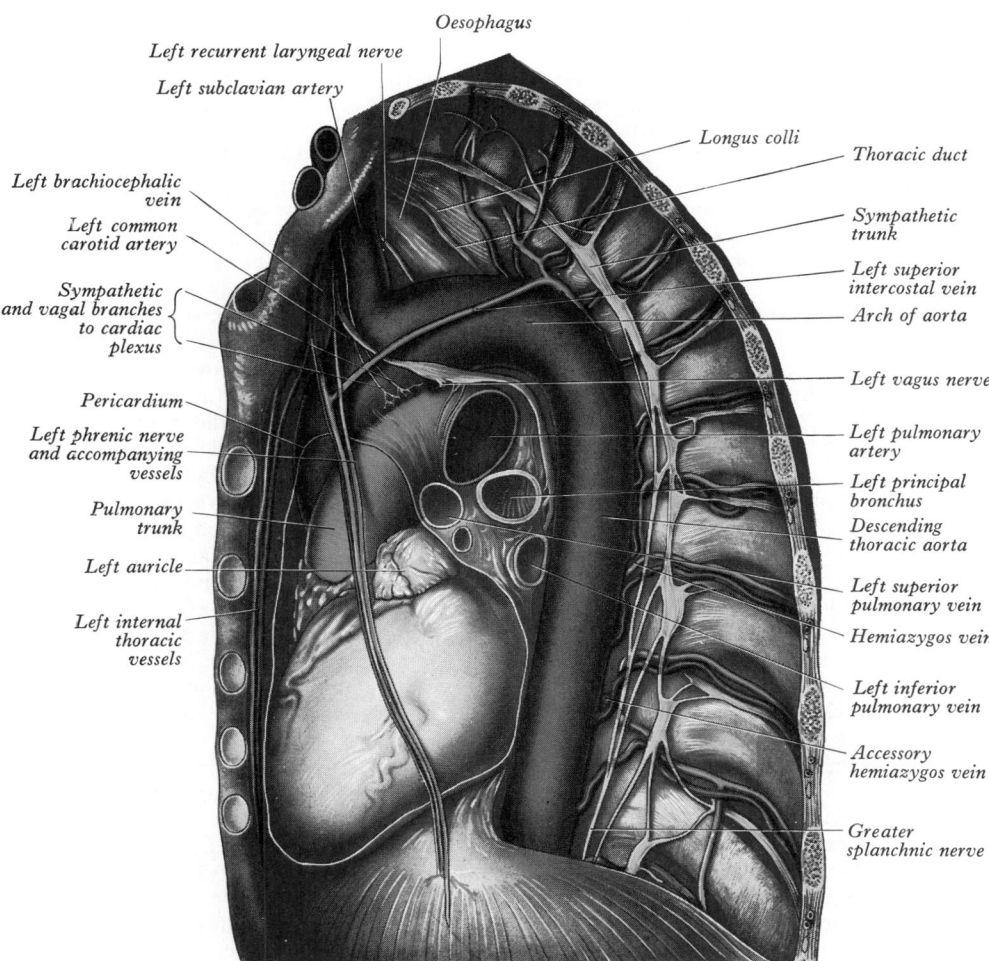

Oesophagus

Left recurrent laryngeal nerve

Left subclavian artery

Left brachiocephalic vein

Left common carotid artery

Sympathetic and vagal branches to cardiac plexus

Pericardium

Left phrenic nerve and accompanying vessels

Pulmonary trunk

Left auricle

Left internal thoracic vessels

Longus colli

Thoracic duct

Sympathetic trunk

Left superior intercostal vein

Arch of aorta

Left vagus nerve

Left pulmonary artery

Left principal bronchus

Descending thoracic aorta

Left superior pulmonary vein

Hemiazygos vein

Left inferior pulmonary vein

Accessory hemiazygos vein

Greater splanchnic nerve

6.106 The left aspect of the mediastinum. The left lung and pleura have been removed and an extensive opening has been made into the pericardial sac to expose the heart. Note the obliquity of the thoracic inlet, and forward inclination of the longus colli, upper oesophagus and thoracic duct.

second being supplied by the superior intercostal artery (p. 755). *Right posterior intercostal arteries* are longer, due to aortic deviation to the left; they cross the vertebral bodies behind the oesophagus, thoracic duct and azygos vein, right lung and pleura. *Left posterior intercostal arteries* turn backwards on the vertebral bodies in contact with the left lung and pleura, the upper two crossed by the left superior intercostal vein, the lower by the hemiazygos and accessory hemiazygos veins. Their further course is the same on both sides. Anterior to the heads of the ribs the sympathetic trunk descends in front of them and additionally the splanchnic nerves in front of the lower arteries.

Each artery crosses its intercostal space obliquely towards the angle of the rib above and continues forward in its costal groove (**6**.106). At first between the pleura and internal (posterior) intercostal membrane as far as the costal angle, it passes between the internal intercostal and intercostalis intimus muscles (p. 592), anastomosing with an anterior intercostal branch from an internal thoracic or musculophrenic artery. It has a vein above and a nerve below, except in the upper spaces where the nerve is at first above the artery. The third anastomoses with the superior intercostal artery and may largely supply the second space. The lower two arteries continue anteriorly into the abdominal wall to anastomose with the subcostal, superior epigastric and lumbar arteries. Each posterior intercostal artery has dorsal, collateral, muscular and cutaneous branches.

The dorsal branch runs dorsally between the necks of adjoining ribs, with a vertebral body and a superior costotransverse ligament lying medial and lateral, respectively. It has a *spinal branch* entering the vertebral canal by the intervertebral foramen to supply vertebrae, spinal cord (p. 948) and meninges; it anastomoses with the spinal arteries above and below and with its fellow. It then crosses a transverse process with the dorsal ramus of a thoracic spinal nerve to supply the dorsal muscles; a cutaneous twig accompanies the cutaneous branch of the spinal

nerve's dorsal ramus.

A collateral intercostal branch leaves its posterior intercostal near the costal angle and descends to the upper border of the subjacent rib, along which it courses to anastomose with an anterior intercostal branch of the internal thoracic or musculophrenic artery.

Muscular branches supply intercostal and pectoral muscles and the serratus anterior, anastomosing with the superior and lateral thoracic branches of the axillary artery. **Lateral cutaneous branches** accompany the same branches of the thoracic spinal nerves. **Mammary branches** from the vessels in the second to fourth spaces supply the pectoral muscles, skin and mammary tissue; they enlarge during lactation.

Un-named branches supply all other tissues constituting the thoracic wall, e.g. costal periosteum, bone and bone marrow of ribs, tissues of synovial and synarthrodial joints and parietal pleura.

The right bronchial artery may arise from the right third posterior intercostal artery (vide supra).

Applied Anatomy. A thoracic puncture needle should not be introduced posteriorly medial to the costal angles, as the intercostal artery (and vein) crosses its space medial to this. Laterally, however, it is in the upper part of its intercostal space; therefore puncture should be through the *lateral* chest wall in the *lower* half of a space.

The subcostal arteries, the last paired branches of the thoracic aorta, in series with the posterior intercostal arteries, are below the twelfth ribs. Each runs laterally anterior to the twelfth thoracic vertebral body and posterior to the splanchnic nerves, sympathetic trunk, pleura and diaphragm; the right is also posterior to the thoracic duct and azygos vein, the left to the accessory hemiazygos vein. Each then enters the abdomen posterior to the lateral arcuate ligament with the twelfth thoracic (subcostal) nerve at the lower border of the twelfth rib, anterior to quadratus lumborum and posterior to the kidney. The right artery courses posterior to the ascending colon, the left to its descending part. Piercing the aponeurosis of the transversus abdominis each proceeds between this and the obliquus internus, anastomosing with the superior epigastric, lower posterior intercostal and lumbar arteries. Each has a dorsal branch, distributed like those of the posterior intercostal arteries.

A small *aberrant artery* sometimes leaves the thoracic aorta on its right near the right bronchial. It ascends to the right *behind* the trachea and oesophagus and may anastomose with the right superior intercostal. It is a vestige of the right dorsal aorta (p. 214); occasionally it is enlarged as the first part of a right subclavian (p. 750).

Variations. The aortic lumen is occasionally partly or completely obliterated, above (preductal or infantile type), opposite or just beyond (postductal or adult type) the entry of the ductus arteriosus. The condition, *coarctation of the aorta*, is congenital; the ductus arteriosus may remain patent, but rarely compensates, systemic blood pressure being usually much higher than pulmonary.

In the *preductal* type, the *coarctation's* length is variable and may involve the left subclavian and even the brachiocephalic artery, with little scope for the development of an effective collateral circulation to regions distal to the stenosis. Many cases are incompatible with survival for more than a few months and surgical problems are great. However, coarctation may be restricted to a short segment between the brachiocephalic and left subclavian arteries, pressures in the left arm being lower than in the right; a collateral circulation may develop through branches of the brachiocephalic.

The *postductal* type of *coarctation* has been attributed to abnormal extension of the ductal tissue into the aortic wall, stenosing both vessels as the duct contracts after birth. This form can permit many years of normal life, allowing the development of an extensive collateral circulation to the aorta distal to the stenosis. High vascularity of the thoracic wall is important and clinically characteristic; many arteries arising indirectly from the aorta, *proximal* to the coarctation segment, anastomose with vessels connected with it *distal* to the block; these become greatly enlarged. In the anterior thoracic wall the thoracoacromial, lateral thoracic and subscapular arteries from the axillary, the suprascapular from the subclavian and the first and second posterior intercostal arteries from the costocervical trunk anastomose with other posterior intercostal arteries; the internal thoracic artery and its terminal branches anastomose with the lower posterior intercostal and inferior epigastric arteries. Posterior intercostal arteries are always involved, and enlargement of their dorsal branches may eventually groove ('notch') the inferior margins of the ribs. The X-ray shadow of the enlarged left subclavian artery is also increased. Enlargement of the scapular vessels and anastomoses may lead to widespread interscapular pulsation (easily appreciated with the palm of the hand, and sometimes heard on auscultation).

The Abdominal Aorta

The abdominal aorta (**6**.107A,B,119) begins at the median, aortic hiatus of the diaphragm, anterior to the twelfth thoracic vertebra's inferior border and the thoracolumbar intervertebral symphysis ('disc'), descending anterior to the vertebrae to end at the fourth lumbar, a little left of the midline, by dividing into two common iliac arteries. It diminishes rapidly in calibre, since its branches are large. Measurements of casts of the abdominal aorta in 100 individuals, from 16–70 years, showed a widening with age. In males superior and inferior ends measured 9.8–14.1 mm and 8.1–14.6 mm; in females luminal diameters were 9.7–15.7 mm and 9.1–14.6 mm (Aleksandrowicz et al 1974). These values conflict with radiological observation of 61 adults (17–41 years) by Leithner et al (1975), who recorded 26 mm and 19 mm (averages) for both ends of the abdominal aorta; they also gave a mean value of 37° for the angle of aortic bifurcation. Dimensions are of interest in attempts to estimate a suspected hydrodynamic ('haemodynamic') factor in the genesis of atherosclerosis (Newman et al 1971, Lallemand & Newman 1973). Theoretically, the pressure pulse wave in arteries

is reflected at any junction, at certain values of combined arterial *luminal areas* of the branch or branches relative to that of the parent vessel; this is the *area ratio* of a junction. At an equal bifurcation, such as the aortic, with an area ratio of 1:15, reflexion of the pressure pulse wave is near to zero; the vessels are said to be 'matched'. Oscillations and possibly turbulence set up by 'mismatching' (at other ratios), perhaps also influenced by asymmetry of bifurcation, may cause intimal damage, predisposing to aortic atheroma. Luminal and other dimensions of the bifurcation may assume special significance, as may changes in these during life. Measurement of aortico-iliac junctions in humans, dogs and domestic fowls (free from vascular disease) has shown area ratios usually close to the theoretical value for 'matching' and independent of age in dog and fowl (Gosling et al 1971). However, the human aortic bifurcation appears to be 'matched' only in infancy; it is 1.11 ± 0.02 at birth, diminishing with advancing age to a value of about 0.7 in the fifth decade, at which theory predicts a 'mismatch' reflecting pulse pressure wave at about one-third of its amplitude. These studies give special interest to a study of the geometry of aortic bifurcation by Shah et al (1978), containing the most extensive data so far recorded, including diameters and angles of deviation, iliac lengths and curvatures and dorsal angulations of these vessels as they enter the pelvis. Unfortunately, diameters were external and only on a small series of cadavers at autopsy, and cannot be compared with those cited above. These interesting observations should be carried further with improved techniques and greater cohesion between different groups involved.

Relations. The abdominal aorta has at first *anterior* to it the coeliac trunk and its branches, with the coeliac plexus and the lesser sac (omental bursa) which intervenes between it and the hepatic papillary process and lesser omentum. Below this the superior mesenteric artery leaves the aorta, crossing anterior to the left renal vein. The body of the pancreas, with splenic vein applied posteriorly, extends obliquely up and left across the abdominal aorta, separated from it by the superior mesenteric artery and left renal vein. Below the pancreas, the proximal parts of its testicular (or ovarian) arteries, and the horizontal part of the duodenum are anterior. In its lowest part it is covered by the posterior parietal peritoneum and crossed by the oblique parietal attachment of the mesentery.

Posterior to the abdominal aorta are the thoracolumbar intervertebral 'disc' (symphysis), the upper four lumbar vertebrae, intervening intervertebral discs and the anterior longitudinal ligament. Lumbar arteries, arising from its dorsal aspect, and the third and fourth (sometimes second) left lumbar veins, crossing behind it to reach the inferior vena cava, separate it from the ligament. It may overlap the anterior border of the left psoas major.

On the right the aorta is related above to the cisterna chyli and thoracic duct, azygos vein and right crus of diaphragm, which overlaps and separates it from the inferior vena cava and right coeliac ganglion. Below the second lumbar vertebra it adjoins the inferior vena cava.

On the left it is related above to the left diaphragmatic crus and left coeliac ganglion. Level with the second lumbar vertebra are the duodenojejunal flexure and sympathetic trunk descending, at its left side, and the ascending duodenum and inferior mesenteric vessels.

Surface Anatomy. The vessel is indicated by a band about 2 cm wide from a median level 2.5 cm above the transpyloric plane to one about 1 cm below and left of the umbilicus. When the abdominal wall is relaxed the aorta may be felt pulsating just above its bifurcation and its pulsation may be visible.

Branches (**6**.107A,B) may be described as ventral, lateral, dorsal and terminal; ventral and lateral are distributed to the viscera, the dorsal branches supplying the body wall, vertebral column, canal and its contents:

Ventral:	Coeliac, superior and inferior mesenteric.
Dorsal:	Lumbar and median sacral.
Lateral:	Inferior phrenic, middle suprarenal, renal, ovarian or testicular.
Terminal:	Common iliac.

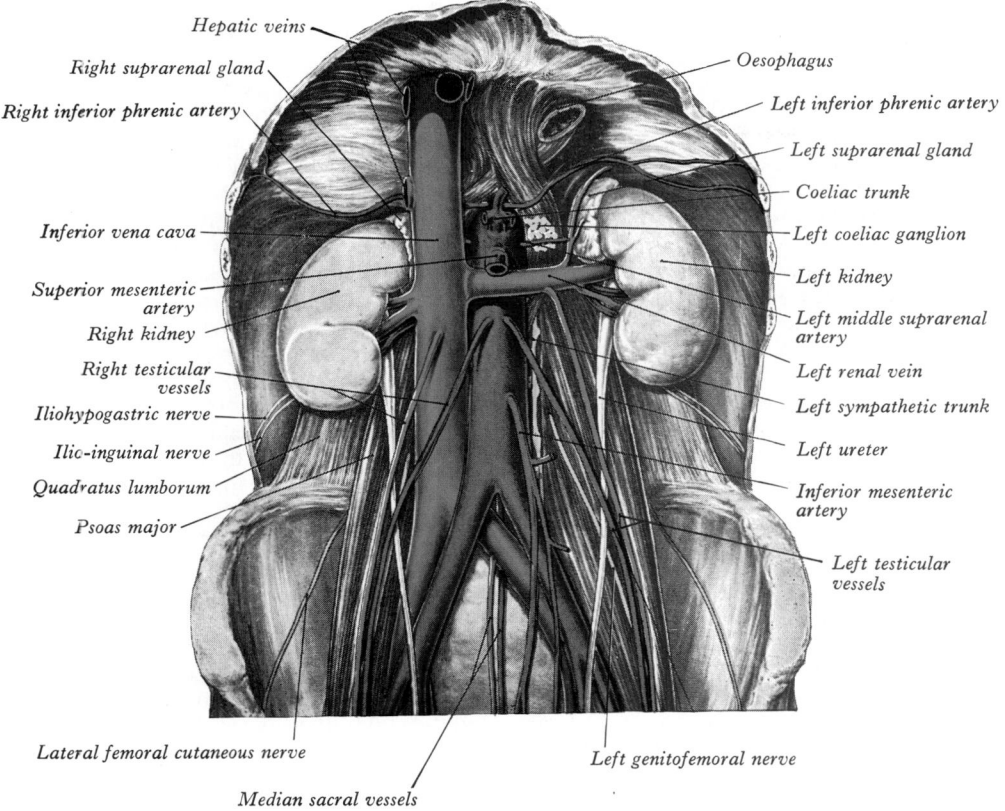

Hepatic veins
Right suprarenal gland
Right inferior phrenic artery
Inferior vena cava
Superior mesenteric artery
Right kidney
Right testicular vessels
Iliohypogastric nerve
Ilio-inguinal nerve
Quadratus lumborum
Psoas major

Oesophagus
Left inferior phrenic artery
Left suprarenal gland
Coeliac trunk
Left coeliac ganglion
Left kidney
Left middle suprarenal artery
Left renal vein
Left sympathetic trunk
Left ureter
Inferior mesenteric artery
Left testicular vessels

Lateral femoral cutaneous nerve
Median sacral vessels
Left genitofemoral nerve

6.107A The abdominal aorta and its branches in the male.

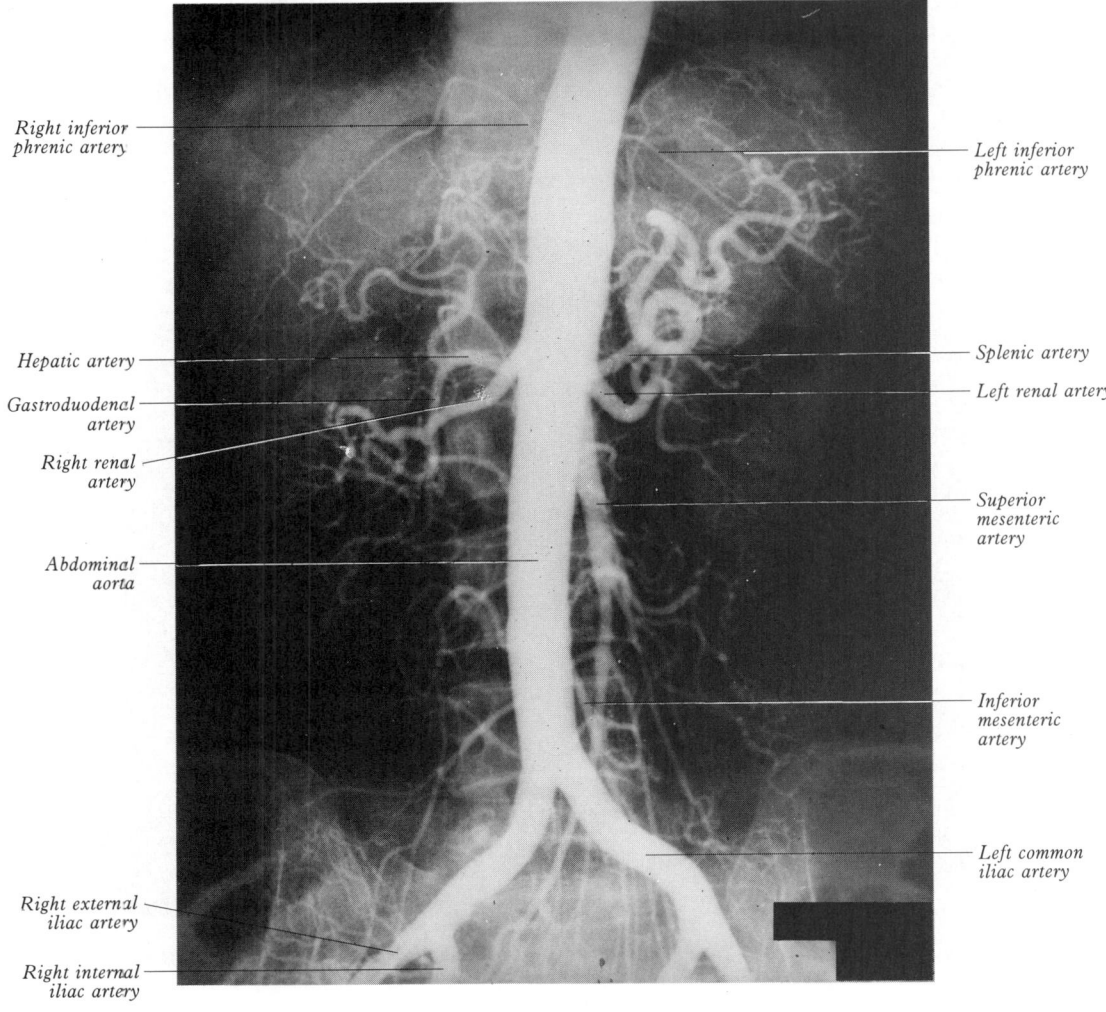

Right inferior phrenic artery
Hepatic artery
Gastroduodenal artery
Right renal artery
Abdominal aorta
Right external iliac artery
Right internal iliac artery

Left inferior phrenic artery
Splenic artery
Left renal artery
Superior mesenteric artery
Inferior mesenteric artery
Left common iliac artery

6.107B Aorto-iliac angiogram. Supplied by Shaun Gallagher, Guy's Hospital; photography by Sarah Smith.

The Coeliac Trunk

The coeliac trunk (**6.**108,109,110), a wide ventral branch, about 1.25 cm long, just below the aortic hiatus, passes almost horizontally forwards and slightly right above the pancreas and splenic vein, dividing into: (1) *left gastric*, (2) *common hepatic* and (3) *splenic arteries*. It may also give off one or both inferior phrenic arteries. The superior mesenteric may arise with the coeliac trunk or the latter's usual branches may be direct independent branches of the aorta.

Relations. Anterior is the omental bursa (lesser sac); the coeliac plexus surrounds the trunk, sending extensions along its branches. *Right lateral* are the right coeliac ganglion, right crus and hepatic caudate process; *left lateral* are the left coeliac ganglion, left crus and cardiac end of the stomach. *Inferior* are the pancreas and splenic vein. The duodenum's suspensory muscle (p. 1357) may encircle the coeliac artery but is usually on its left.

1. The Left Gastric Artery (6.108)

This, the smallest coeliac branch, ascends to the left, posterior to the omental bursa, to the cardiac end of the stomach. It is near the left inferior phrenic artery and medial or anterior to the left suprarenal gland. Near the stomach two or three *oesophageal branches* ascend through the oesophageal opening to anastomose with the aortic oesophageal rami; others supply the cardiac part of the stomach and anastomose with the splenic branches. The artery then turns antero-inferiorly into the left gastropancreatic fold to run (often doubled) curving to the right near the gastric lesser curvature to the pylorus between layers of the lesser omentum; it supplies both gastric surfaces and anastomoses with the right gastric artery. An *accessory left gastric artery* may arise from the left branch of the hepatic, also reaching the lesser curvature through the lesser omentum.

2. The Hepatic Artery (6.108,109,110)

This is intermediate in size between the left gastric and splenic arteries; but in *later* fetal and early postnatal life it is the largest coeliac branch. Accompanied by the hepatic autonomic plexus it first passes forwards and right, below the epiploic foramen to the upper aspect of the superior part of the duodenum (**6.**108). Crossing the portal vein, it ascends between layers of the lesser omentum, anterior to the epiploic foramen, to the porta hepatis, where it divides into right and left branches to the hepatic lobes, accompanying the ramifications of the portal vein and hepatic ducts. In the lesser omentum it is anterior to the portal vein and left of the bile duct, its right branch crossing posterior (occasionally anterior) to the common hepatic duct (**6.**111). The artery may be subdivided into: (*a*) the *common hepatic artery*, from the coeliac trunk to the origin of the gastroduodenal artery and (*b*) the *hepatic artery proper*, from that point to its bifurcation.

In embryonic and *early* fetal life, the hepatic artery arises from the left gastric (in 67% of 56 individuals, Godlewski et al 1975). This condition rarely persists, but the hepatic may arise from the superior mesenteric, or the hepatic's right or left branches may be from other vessels; the former from the superior mesenteric, the latter from the left gastric. For other variations consult Quain (1865, 1899) and Woodburne (1962) (see also pp. 769, 1394). The hepatic artery has right gastric, gastroduodenal and cystic branches, rami to the bile duct from the right hepatic and sometimes the supraduodenal artery (vide infra).

The right gastric artery (**6.**108) arises above the duodenum's superior part, usually before, or sometimes beyond the gastroduodenal, descending in the lesser omentum to the pyloric end of the stomach; it passes left along the lesser gastric curvature, supplying the upper parts of the anterior and posterior gastric surfaces. It ends by anastomosing with the left gastric; the supraduodenal artery may be a branch (vide infra).

The gastroduodenal artery (**6.**108,109,110), arising behind, sometimes above, the superior part of duodenum, is short and wide. It descends between the duodenum and the neck of the pancreas, immediately to the right of the peritoneal reflection from the posterior duodenal surface. It is usually left of the bile duct

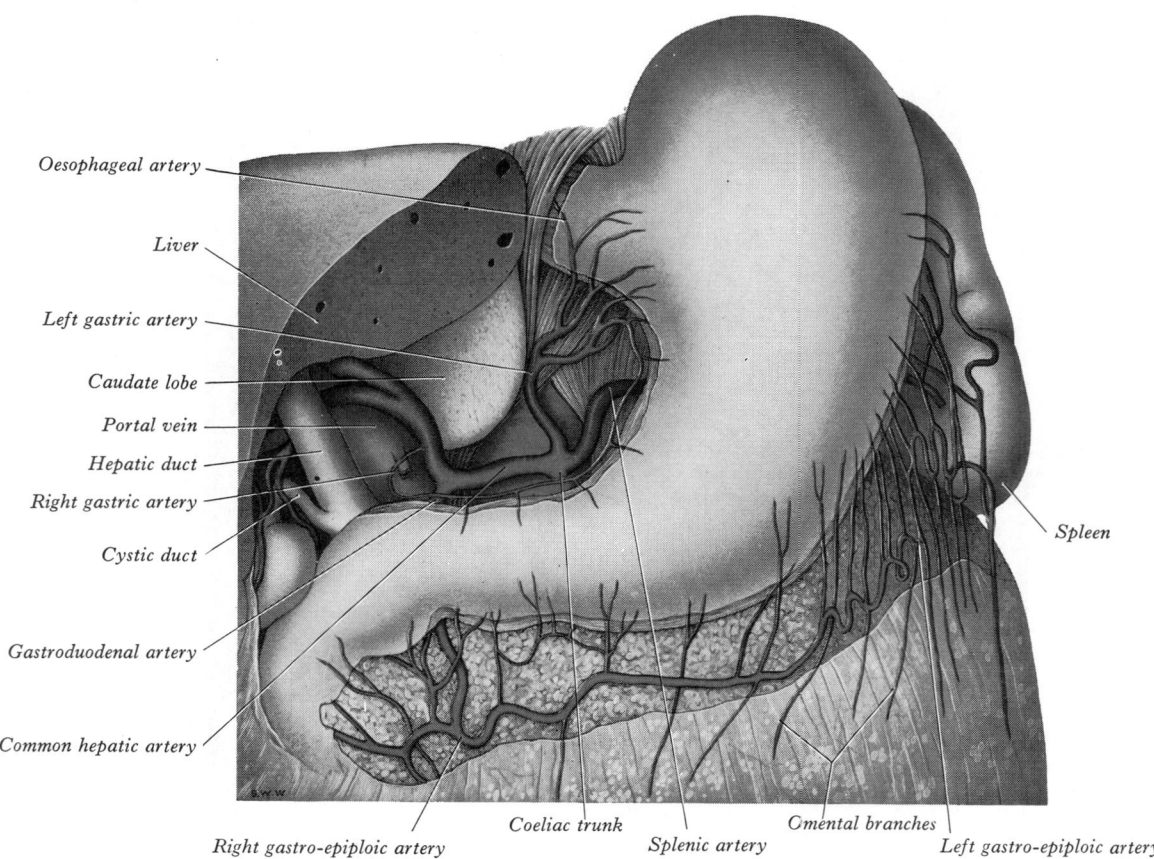

Oesophageal artery

Liver

Left gastric artery

Caudate lobe

Portal vein

Hepatic duct

Right gastric artery

Cystic duct

Gastroduodenal artery

Common hepatic artery

Spleen

Right gastro-epiploic artery   *Coeliac trunk*   *Splenic artery*   *Omental branches*   *Left gastro-epiploic artery*

6.108 The coeliac trunk and its branches. Part of the liver and all the lesser omentum have been removed, as well as the posterior wall of the omental bursa and part of the anterior layer of the greater omentum.

but sometimes anterior. At the lower border of the duodenum's superior part it divides into the *right gastro-epiploic* and *superior pancreaticoduodenal* arteries, after supplying small branches to the pyloric end of the stomach and to the pancreas, retroduodenal branches to the superior part of the duodenum, and sometimes providing the supraduodenal artery (vide infra). The first branch of the common hepatic artery is usually the gastroduodenal artery, but this may come from the superior mesenteric, coeliac trunk or an aberrant right hepatic artery (p. 770); its most invariable feature is its intermediate position between the *neck* of the pancreas and the duodenum, this being clinically important due to its frequent involvement in duodenal ulceration (Bradley 1973).

The *supraduodenal artery*, sometimes double, is variable; it may arise from the gastroduodenal, hepatic (common, proper or the latter's branches) or from the right gastric artery. It supplies the superior half circumference of the proximal half or more of the duodenum's superior part; but the duodenum is often invaded proximally by branches of the right gastric artery (p. 1358).

The right gastro-epiploic artery (6.108,109,111**)**, the larger terminal branch of the gastroduodenal, skirts the right margin of the omental bursa and then turns left along the greater curvature, between the (anterior two) layers of the greater omentum. It ends in direct anastomosis with the left gastro-epiploic branch of the splenic. Except at the pylorus, where it adjoins the stomach, it is about 2 cm from the greater curvature. Of its many branches some ascend to both gastric surfaces, others descend into the greater omentum. It also supplies the inferior aspect of the duodenum's superior part.

The superior pancreaticoduodenal arteries are usually double: the *anterior* descends anteriorly between the duodenum and head of the pancreas. It supplies both and anastomoses with the anterior division of the inferior pancreaticoduodenal branch of the superior mesenteric. The *posterior superior pancreaticoduodenal artery*, which is usually a separate ramus of the

gastroduodenal arising at the upper border of the superior part of the duodenum, descends to the right, anterior to the portal vein and bile duct and then posterior to the head of the pancreas, supplying branches to it and the duodenum; it crosses posterior to the bile duct, piercing the duodenal wall to end by anastomosing with the posterior division of the inferior pancreaticoduodenal artery. (Pancreaticoduodenal arteries form varying vascular arcades before their final distribution, see Michels 1962.) The artery supplies several branches to the lower part of the common bile duct (p. 1394).

The cystic artery (6.111**)**, usually from the right branch of the hepatic proper, passes behind the common hepatic and over the cystic duct to the superior aspect of the gallbladder's neck, on which it descends to divide into *superficial* and *deep* branches. The former ramifies on the inferior, the latter on the superior aspect. The cystic artery may arise from the hepatic artery itself (rarely from the gastroduodenal), crossing anterior or posterior to the bile or common hepatic duct to reach the gallbladder. Variations in the artery's origin are of surgical interest. In 800 specimens Anson (1963) observed the following incidences: from the right hepatic 63.9%, the hepatic trunk 26.9%, left hepatic 5.5%, gastroduodenal 2.6%, superior pancreaticoduodenal 0.3%, right gastric 0.1%, coeliac trunk 0.3%, and superior mesenteric 0.8%. Direct origin from the hepatic artery varied from its beginning to its bifurcation. An *accessory cystic artery* may arise from the common hepatic or one of its branches. The cystic artery supplies the hepatic ducts and upper part of the common bile duct (p. 1394). A comparative study of its *distribution* in various reptilian, avian and mammalian species included 74 injected and cleared human gallbladders (Gordon 1967). Briefly, the human cystic artery reached the gallbladder at its neck; it was never in contact with the cystic duct. Coursing over the neck, one or two pairs of branches encircled it, anastomosing posteriorly. Division into two principal longitudinal branches occurred at any point along the neck; each then gave off four to eight pairs of lateral branches

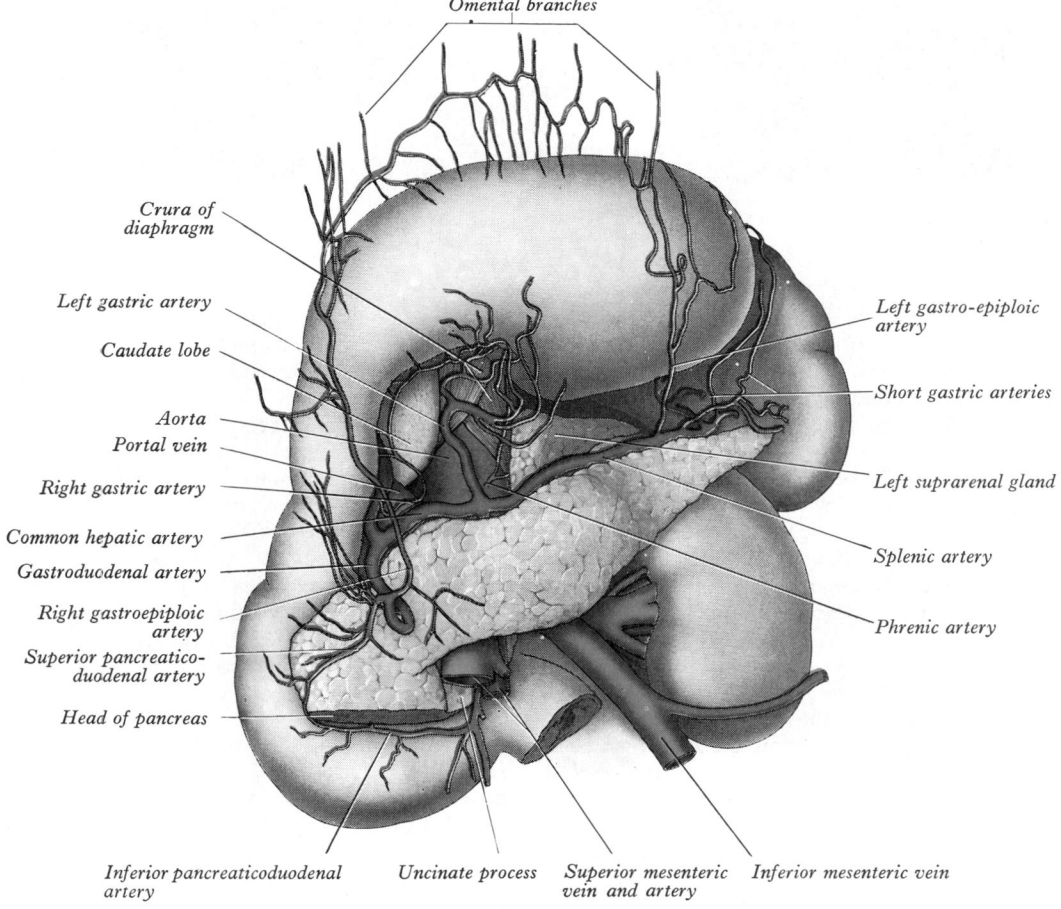

Omental branches

Crura of diaphragm

Left gastric artery

Caudate lobe

Aorta
Portal vein

Right gastric artery

Common hepatic artery

Gastroduodenal artery

Right gastroepiploic artery

Superior pancreatico- duodenal artery

Head of pancreas

Left gastro-epiploic artery

Short gastric arteries

Left suprarenal gland

Splenic artery

Phrenic artery

Inferior pancreaticoduodenal artery   *Uncinate process*   *Superior mesenteric vein and artery*   *Inferior mesenteric vein*

6.109 The coeliac trunk and its branches exposed by turning the stomach upwards and removing the peritoneum on the posterior abdominal wall.

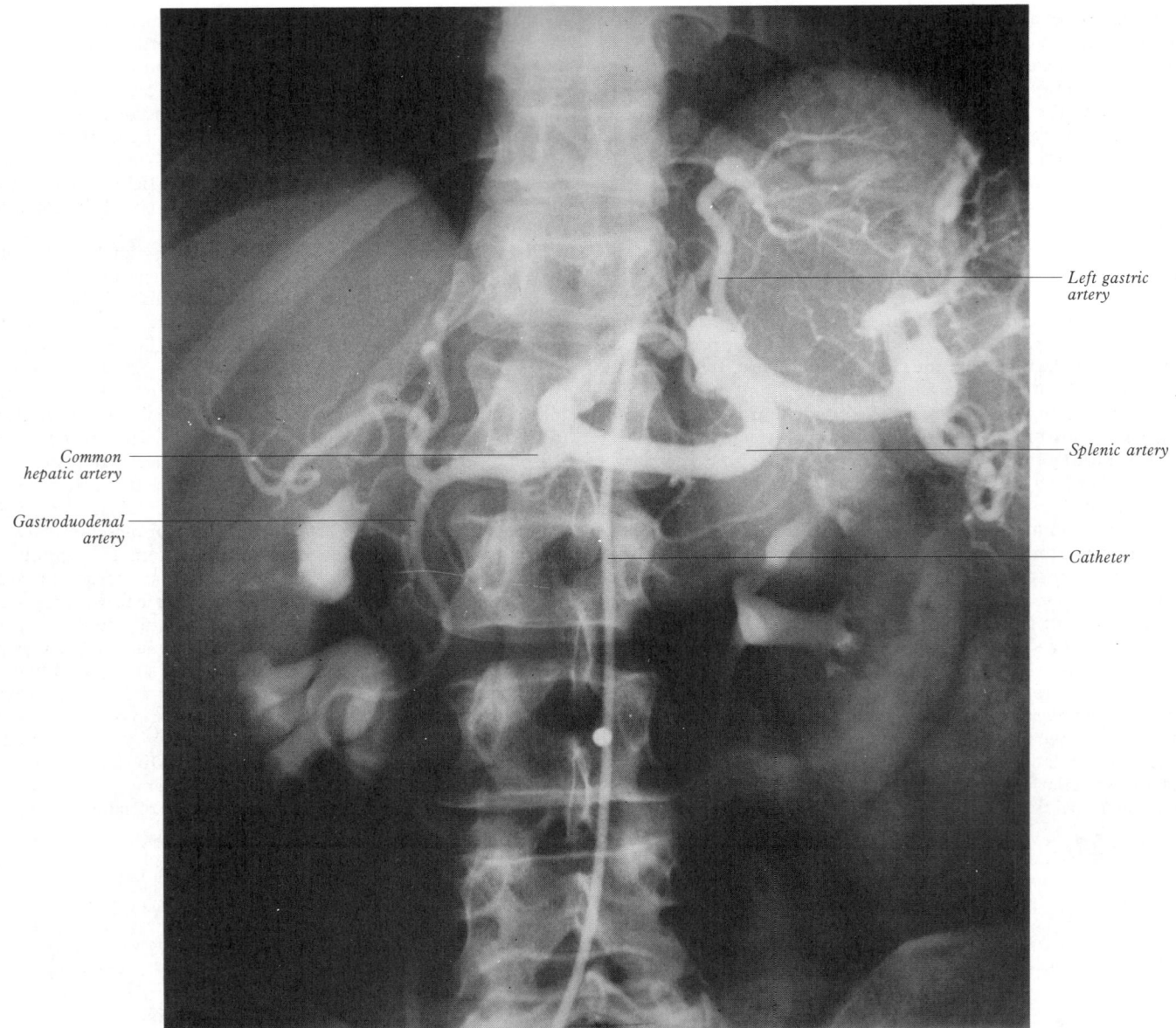

Left gastric artery

Common hepatic artery

Gastroduodenal artery

Splenic artery

Catheter

6.110 Coeliac axis arteriogram. Supplied by Shaun Gallagher, Guy's Hospital; photography by Sarah Smith.

at right angles, anastomoses completing four to eight vascular circles. The principal vessels bifurcated to form a diamond-shaped anastomosis at the junction of the cystic body and fundus, the diamond being traversed centrally by a fine anastomotic net. The overall arterial pattern was described as *bipinnate* (bipennate).

The **terminal, intrahepatic branches** display a pattern of branching relatively constant in its major details (p. 1386), which justifies a **segmental** description of the liver; it is the result, as in other organs, of growth and branching of an epithelial blastema, its pattern accompanied by vascular rami (and innervation). Arterial hepatic segmentation is described on p. 1387. Consult also Woodburne (1962).

3. The Splenic Artery (6.108,109,110)

The largest branch of the coeliac axis, the splenic is remarkably tortuous. Surrounded by a splenic nerve plexus and accompanied by the straight splenic vein, it ascends to the left, behind the stomach and omental bursa, along the superior border of the pancreas; it is anterior to the left suprarenal gland and upper part of the left kidney and enters the lienorenal ligament. Nearing the spleen it divides into five or more *segmental* branches which enter its hilum (p. 832).

Branches of the splenic artery are as follows:

Pancreatic branches (6.109), numerous and small, supply the neck, body and tail of the pancreas, leaving the splenic artery

as it runs along its superior border. A *dorsal branch* (sometimes from the superior mesenteric, middle colic, hepatic or, more rarely, coeliac artery) descends posterior to the pancreas, dividing into right and left rami. The former, usually double, runs between the neck and uncinate process to form a *prepancreatic arterial arch* with a branch from the anterior superior pancreaticoduodenal; the left ramus runs along the inferior border to the pancreatic tail; it anastomoses with branches (*arteria pancreatica magna* and *arteria caudae pancreatis*) from the splenic artery which supply the left part of the body and the tail.

Short gastric arteries (6.109), five to seven, arise terminally from the splenic and its final divisions or from the left gastro-epiploic artery. They pass between layers of the gastrosplenic ligament to supply the gastric fundus, anastomosing with branches of the left gastric and gastro-epiploic arteries.

The posterior gastric artery, from any part of the splenic but most commonly its middle section, has been described by many authorities (e.g. Quain 1844) but many subsequent texts have omitted it. Susuki et al (1978), also surveying reports on it, found it present in 38 (62.3%) of 61 adult cadavers; incidence from 14 reports (1904–1968) varied from 12.7 to 77%, with an average of 58% in a total of 870 cadavers. They described the vessel as ascending behind the peritoneum of the omental bursa towards the gastric fundus to reach the posterior gastric wall in the gastrophrenic fold; it was usually about 2 mm in diameter.

The left gastro-epiploic artery (6.108,109,116), the splenic's largest branch, arises near the splenic hilum and runs antero-inferiorly and right, sending branches through the gastrosplenic ligament to supply the proximal third of the greater curvature; these are necessarily longer than the gastric branches of the right gastro-epiploic artery and may be 8–10 cm long. A large terminal omental branch descends to the right in the greater omentum. The main vessel curves forwards at a higher level to join the right gastro-epiploic. This loop leaves part of the greater curvature devoid of branches. At partial gastrectomy the greater omentum is divided below the right gastro-epiploic artery, cutting all omental branches; the greater omentum survives because its supply from this large omental branch of left gastro-epiploic usually escapes damage (Horton 1952). Vessels supplying the greater omentum are epiploic (omental) branches of the right and left gastro-epiploic arteries. The right, middle and left colic arteries do *not* supply the greater omentum; the transverse mesocolon, though usually adherent to the greater omentum, is separable from it (p. 1344).

Terminal splenic branches enter the organ's hilum in the lienorenal ligament. Their distribution is described with the spleen (p. 832).

Variations in arrangement of the hepatic artery and its branches are common and surgically important. They include: (1) the origin of the common hepatic from the superior mesenteric or, less often, from the aorta; it usually passes behind the portal vein to enter the lesser omentum; (2) an *accessory left hepatic artery* most often from the left gastric artery passing right in the lesser omentum to the porta hepatis; (3) an *accessory right hepatic artery* most often from the superior mesenteric, usually running behind the portal vein and bile duct in the lesser omentum to the porta hepatis. Accessory left or right hepatic arteries may also arise from the gastroduodenal or aorta. They may be combined with

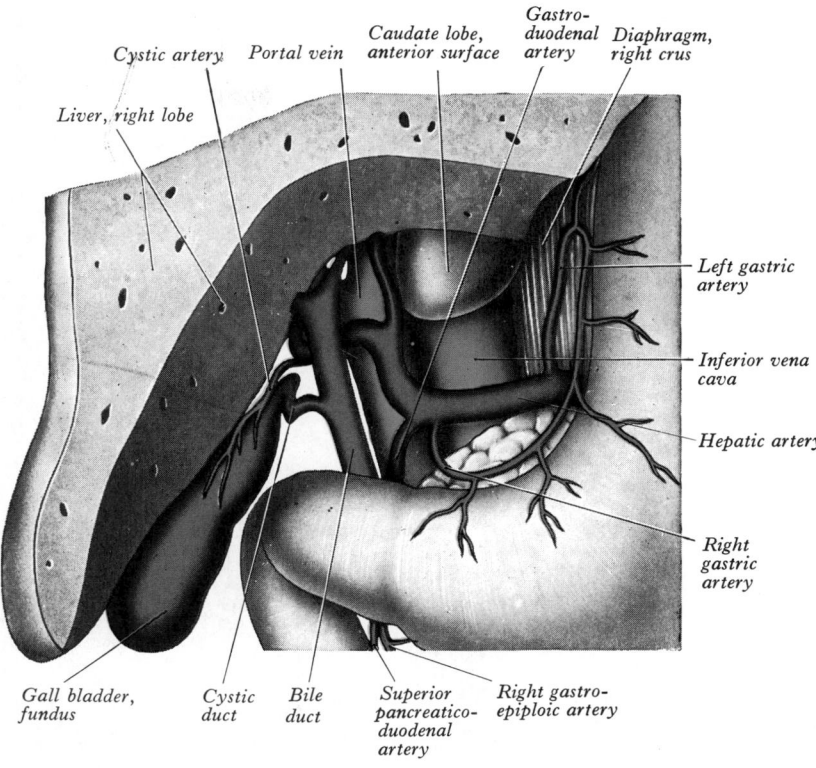

6.111 The relations of the hepatic artery, bile duct and portal vein exposed by removal of the lesser omentum and the peritoneum on the posterior abdominal wall.

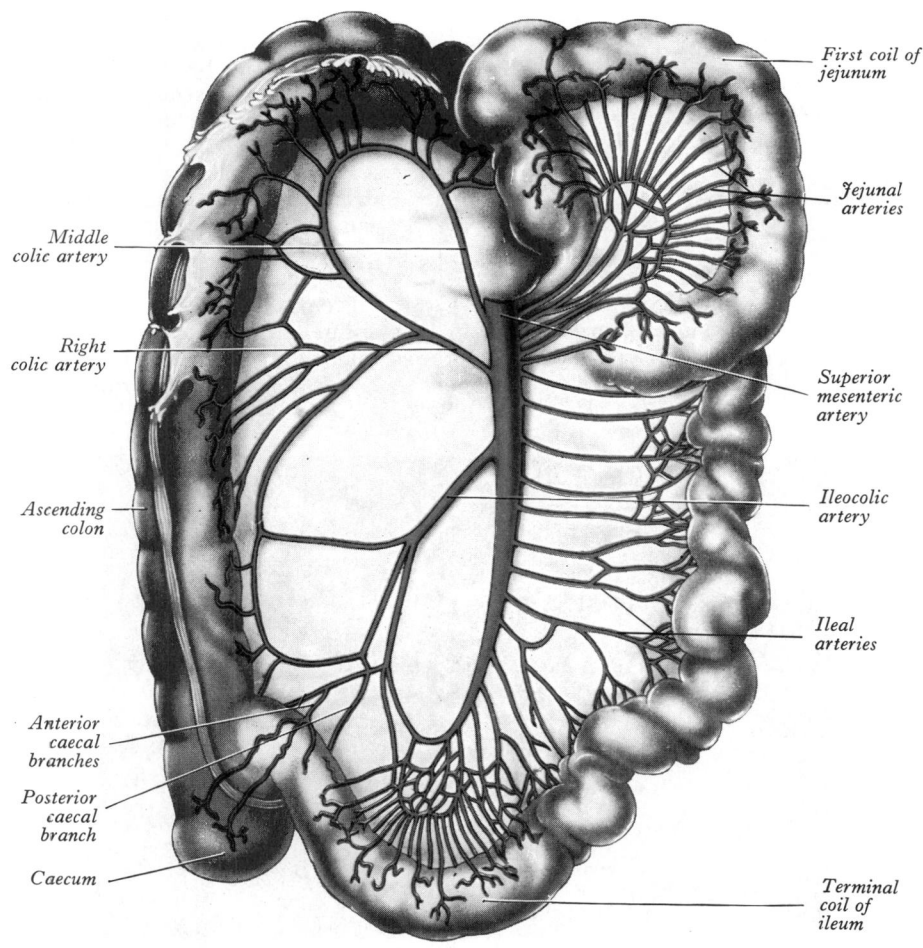

6.112 The superior mesenteric artery and its branches. The first coil of the jejunum and the terminal coil of the ileum have been spread out to show the arrangement of their arteries.

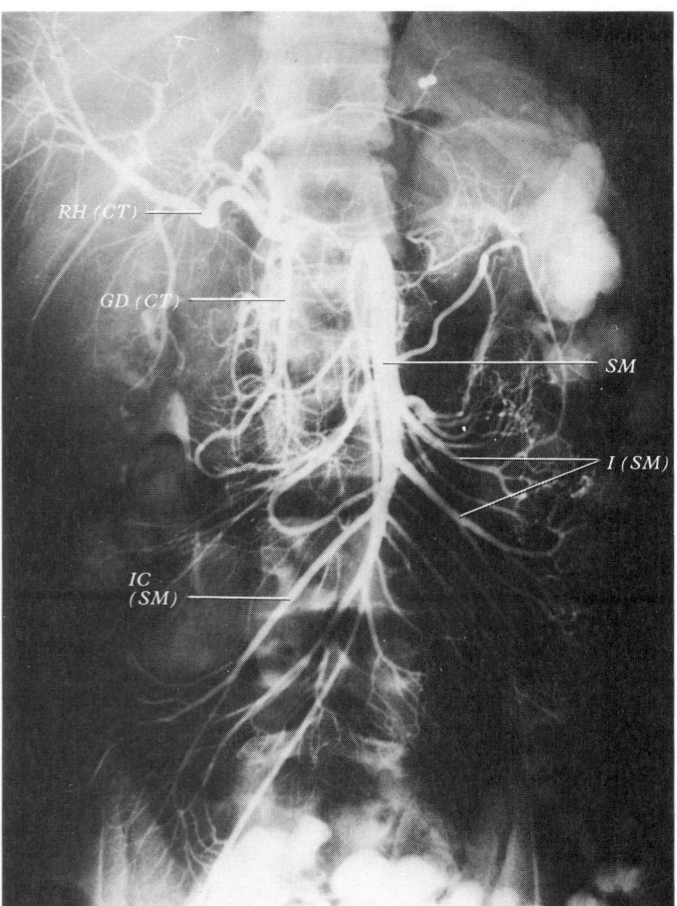

6.113 Superior mesenteric arteriogram with filling of the right hepatic artery. GD(CT) = gastroduodenal artery (coeliac trunk); I(SM) = ileal arteries (superior mesenteric); IC(SM) = ileocolic artery (superior mesenteric); RH(CT) = right hepatic artery (coeliac trunk). Supplied by Shaun Gallagher, Guy's Hospital; photography by Sarah Smith.

'normal' branches of the hepatic artery or *replace* them as the sole supply to parts of the liver, being called *'aberrant replacing arteries'*.

Collateral circulation after hepatic ligature or obstruction. Any blockage may lead to hepatic necrosis, the effects depending on the site of the block. Occlusion of the common hepatic artery, proximal to the origin of the right gastric, allows collateral circulation to the liver through the left and right gastric, left and right gastro-epiploic, pancreaticoduodenal and gastroduodenal arteries, and so necrosis is unlikely. If, however, an obstruction of the hepatic artery proper occurs beyond the origin of the gastroduodenal artery, any collateral circulation is limited to the small inferior phrenic arteries (p. 776).

The Superior Mesenteric Artery

This artery (6.112,113,114) supplies all the small intestine, except the superior part of the duodenum, the caecum, the ascending and most of the transverse colon. It leaves the front of the aorta about 1 cm below the coeliac trunk and is crossed anteriorly at its origin by the splenic vein and body of the pancreas; it is separated posteriorly from the aorta by the left renal vein. It next proceeds down and forwards, anterior to the pancreatic uncinate process and horizontal part of the duodenum, then descends obliquely in the mesentery near its root to the right iliac fossa. Here, much diminished, it joins its own ileocolic branch. As it descends it crosses anterior to the inferior vena cava, the right ureter and psoas major, forming an arch, its convexity forwards, down and to the left. Accompanied by the superior mesenteric vein on its right, it is surrounded by the superior mesenteric plexus of nerves. A fibrous strand may pass to the umbilicus from its terminal part or from a lower ileal branch, a vestige of the embryonic artery along the vitelline duct to the yolk sac.

The inferior pancreaticoduodenal artery (6.109) leaves the superior mesenteric, or its first jejunal branch, near the superior border of the horizontal part of the duodenum, usually dividing at once into anterior and posterior rami. The *anterior branch* passes to the right, anterior to the head of the pancreas, and ascends to anastomose with the anterior superior pancreaticoduodenal artery. The *posterior branch* ascends to the right, posterior to the head of the pancreas, which it sometimes traverses, and then anastomoses with the posterior superior pancreaticoduodenal artery. Both branches supply the pancreatic head, its uncinate process and the adjoining duodenum.

Jejunal and ileal branches (6.112,113) arise from the left side of the superior mesenteric; usually 12–15 are distributed to the jejunum and ileum, except in the latter's terminal part, which is supplied by the ileocolic artery. They run almost parallel in the mesentery, each dividing to unite with adjacent branches in a series of arches (6.112). Rami from these unite to form a second series and this may be repeated three or four times. In the short, upper part of the mesentery one set of arches exists but, as it increases in depth, a second, third, fourth and even fifth series appear. From terminal arches numerous *straight* vessels supply the intestine, distributed alternately to opposite aspects. It has been claimed, but not substantiated, that adjacent branches do not anastomose (Cokkinis 1930). Jejunal arteries are usually

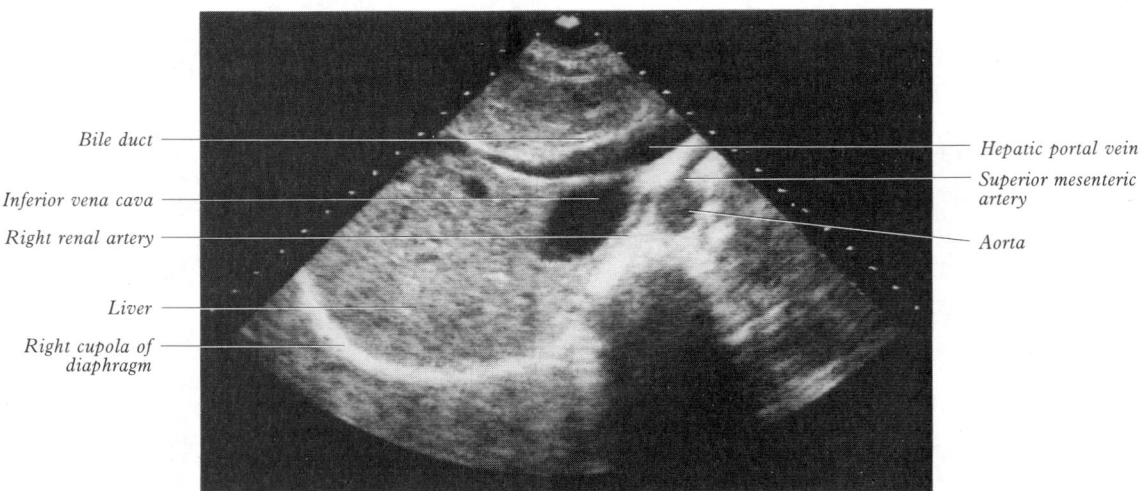

6.114 Ultrasonogram through the origin of the superior mesenteric artery. Provided by Shaun Gallagher, Guy's Hospital; photography by Sarah Smith.

longer and less numerous. Small branches supply regional lymph nodes and other structures in the mesentery.

The ileocolic artery (**6**.112), the last branch from the *right* side of the superior mesenteric, descends to the right under the parietal peritoneum to the right iliac fossa, where it divides; its *superior branch* anastomoses with the right colic, the *inferior* with the end of the superior mesenteric. The ileocolic artery crosses anterior to the right ureter, testicular or ovarian vessels and psoas major. Its *inferior branch* approaches the superior border of the ileocolic junction and branches as follows (**6**.115): (1) *ascending (colic)* passing up on the ascending colon, (2) *anterior* and *posterior caecal*, (3) an *appendicular artery*, descending behind the terminal ileum to enter the mesoappendix; after giving off a recurrent branch anastomosing with one from the posterior caecal artery, it runs close to and then in the edge of the mesoappendix, its terminal part being in actual contact with the appendix; (4) an *ileal* branch ascending to the left on the lower ileum, supplying it and anastomosing with a terminal twig of the superior mesenteric artery.

The right colic artery (**6**.112) arises near the middle of the superior mesenteric, or in common with the ileocolic. It passes to the right behind the parietal peritoneum and anterior to the right ovarian or testicular artery and vein, right ureter and psoas major, towards the ascending colon. Sometimes it is higher and crosses the descending duodenum and right inferior renal pole. Near the colon it divides into a descending branch, which anastomoses with the ileocolic, and an ascending branch anastomosing with the middle colic. These form arches, from which vessels are distributed to the ascending colon, supplying its upper two-thirds and the right colic flexure.

The middle colic artery (**6**.112) leaves the superior mesenteric just inferior to the pancreas; descending in the transverse mesocolon it divides into a right and left branch; the former anastomoses with the right colic artery, the latter with the left, a branch of *inferior* mesenteric. Arches thus formed are 3 or 4 cm from the transverse colon, which they supply.

The superior mesenteric artery may be the source of the common hepatic, gastroduodenal, accessory right hepatic, accessory pancreatic or splenic arteries. It may arise from a common coeliaco-mesenteric trunk (Mangoushi 1975).

The Inferior Mesenteric Artery

The inferior mesenteric artery (**6**.116,117,118) supplies the left third of the transverse colon, all the descending colon, sigmoid colon and most of the rectum. It is smaller than the superior mesenteric, arising 3 or 4 cm above the aortic bifurcation, posterior to the horizontal part of the duodenum. It descends behind the peritoneum, at first anterior to the aorta, then on its left, crosses the left common iliac artery medial to the left ureter and then enters and continues in the sigmoid mesocolon into the lesser pelvis as the *superior rectal artery*. Distally the inferior mesenteric vein is lateral. The artery has left colic, sigmoid and superior rectal branches.

The left colic artery (**6**.116–118) ascends subperitoneally to the left, anterior to the psoas major, and divides into ascending and descending branches. The trunk and its branches cross the left ureter and ovarian or testicular vessels. The ascending branch passes anterior to the left kidney into the transverse mesocolon, where it anastomoses with the middle colic artery; the descending branch anastomoses with the highest sigmoid artery. From arches thus formed, branches supply the left half of the transverse and the descending colon. Territories of supply by middle and left colic arteries show reciprocal variation; the left branch of the middle colic may take over the supply of the splenic flexure (in 19 of 100 cadavers, according to Sierocinski 1975).

The sigmoid (inferior left colic) arteries (**6**.116–118), two or three in number, descend obliquely to the left under the peritoneum anterior to the left psoas major, ureter and testicular or ovarian vessels. Branches supply the lower descending colon and sigmoid colon, anastomosing above with the the left colic artery, below with the superior rectal artery. Anastomoses between branches of the left colic and sigmoid arteries form a con-

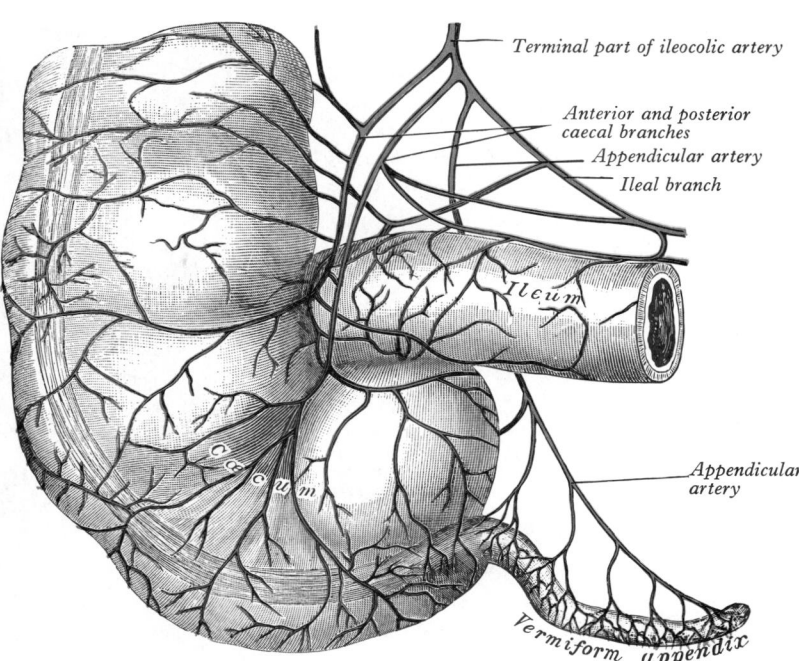

Terminal part of ileocolic artery

Anterior and posterior caecal branches

Appendicular artery

Ileal branch

Ileum

Appendicular artery

Caecum

Vermiform appendix

6.115 The arteries of the caecum and vermiform appendix.

tinuous 'marginal artery' near the colon (Drummond 1914). Continuity of this is *rarely* interrupted, even near the sigmoido-rectal junction, anastomosis at which is sometimes by rather tenuous vessels easily overlooked in dissection; high aortic injections, however, fill the superior rectal vessels; a series of 100 dissections by Sierocinski (1976) indicated *constant* anastomosis between the last sigmoid and superior rectal arteries. Sudeck (1907) introduced uncertainty in this anastomosis as a '*critical point*' in colico-rectal surgery; many authorities have repeated this erroneous view (see Hollinshead 1971 for a review).

The superior rectal artery (**6**.116–118), a continuation of the inferior mesenteric, descends into the pelvis in the sigmoid mesocolon, crossing the left common iliac vessels. It divides, near the third sacral vertebra, into two branches descending one on each side of the rectum; about halfway they divide into smaller branches, which pierce the muscular rectal wall to descend vertically, at submucous level, to the sphincter ani internus; here, by mutual anastomoses, they form loops around the lower rectum, communicating with the middle rectal artery, a branch of the internal iliac, and with the inferior rectal from the internal pudendal (p. 778).

Anterolateral Visceral Arteries

The Middle Suprarenal Arteries

These two small vessels arise laterally from each side of the aorta, level with the superior mesenteric, ascending slightly over the crura of the diaphragm to the suprarenal glands, where each anastomoses with the suprarenal branches of the phrenic and renal arteries. The right passes behind the inferior vena cava and near the right coeliac ganglion; the left is related to the left coeliac ganglion, splenic artery and superior border of the pancreas.

The Renal Arteries (**6**.107A,B,119)

These two large vessels branch laterally from the aorta just below the inferior mesenteric; both cross the corresponding crus at right angles to the aorta. The *right* is longer and often, higher, passing posterior to the inferior vena cava, right renal vein, head of the pancreas and descending part of the duodenum. The *left* is a little lower; it passes behind the left renal vein, the body of the pancreas and splenic vein and may be crossed anteriorly by the inferior mesenteric vein. Nearing its renal hilum, each divides into four or five branches, most between the renal vein and

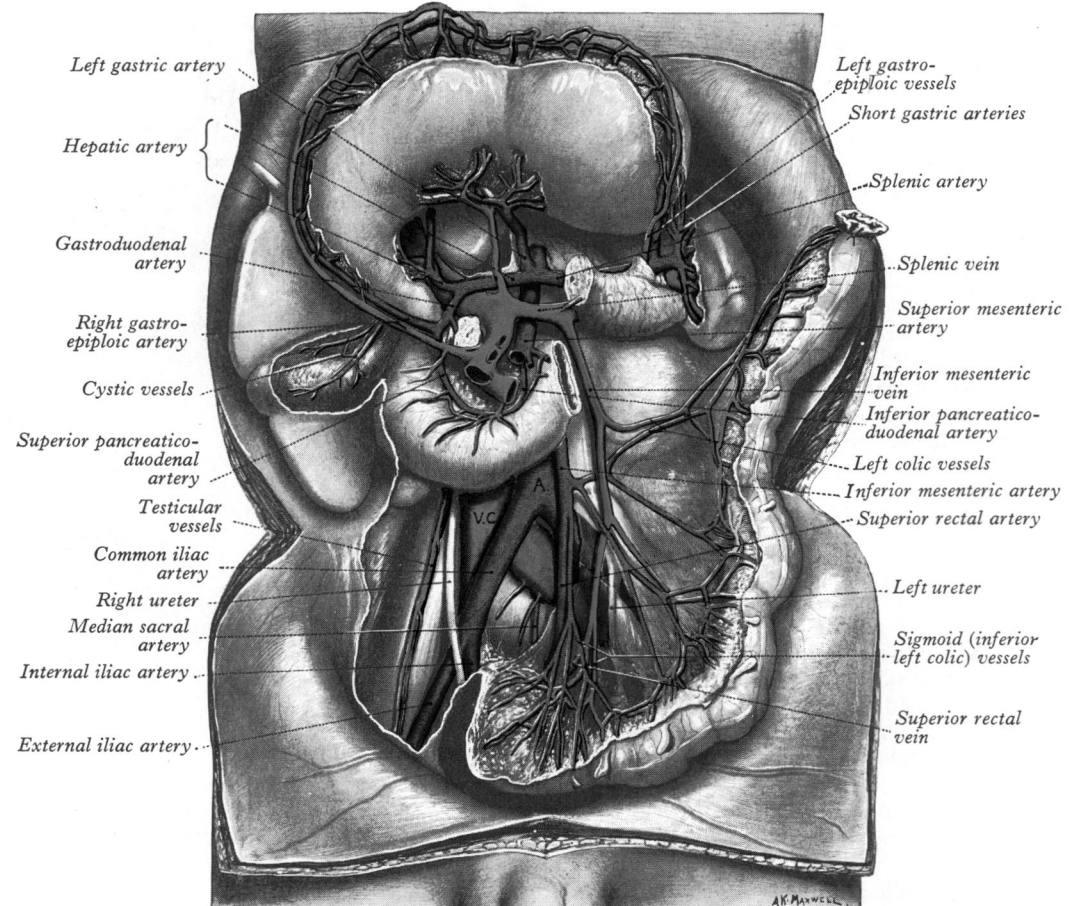

Left gastric artery

Hepatic artery

Gastroduodenal artery

Right gastro-epiploic artery

Cystic vessels

Superior pancreatico-duodenal artery

Testicular vessels

Common iliac artery

Right ureter

Median sacral artery

Internal iliac artery

External iliac artery

Left gastro-epiploic vessels

Short gastric arteries

Splenic artery

Splenic vein

Superior mesenteric artery

Inferior mesenteric vein

Inferior pancreatico-duodenal artery

Left colic vessels

Inferior mesenteric artery

Superior rectal artery

Left ureter

Sigmoid (inferior left colic) vessels

Superior rectal vein

6.116 The inferior mesenteric vessels and their branches (male subject). Note the stomach has been turned upwards and the whole of the jejunum and ileum, the caecum, ascending colon and transverse colon have been removed, together with part of the pancreas.

colic Branches

left colic A.

Superior rectal artery

Marginal artery

Middle rectal artery

6.117 The sigmoid colon and rectum, showing the distribution of the branches of the inferior mesenteric artery and their anastomoses. From a preparation by Hamilton Drummond.

ureteric pelvis, the vein being anterior, the pelvis posterior, but one or more usually behind the pelvis. Each renal artery supplies small *inferior suprarenal branches* (p. 251) and also the ureter, surrounding cellular tissue and muscles. The distribution of the renal arteries is described on p. 1407.

Surface Anatomy. The renal arteries can be projected as broad lines running laterally for 4 cm from the aorta (**6.110**) just inferior to the transpyloric plane; the left inclines across the plane.

One or two *accessory renal arteries* frequently occur, especially on the left, usually from the aorta above or below the main artery, the former slightly more often. They usually enter above or below the renal hilum; if below, the vessel crosses anterior to the ureter and, on the right, usually also anterior to the inferior vena cava.

The Testicular Arteries (6.107A,B,119)

These two long, slender vessels arise anteriorly from the aorta a little inferior to the renal arteries. Each passes inferolaterally under the parietal peritoneum on the psoas major; the right lies anterior to the inferior vena cava and posterior to the horizontal part of the duodenum, right colic and ileocolic arteries, root of the mesentery and terminal ileum; the left testicular artery lies posterior to the inferior mesenteric vein, left colic artery and lower part of the descending colon. Each crosses anterior to the genitofemoral nerve, ureter and the lower part of the external iliac artery, passing to the deep inguinal ring to enter the spermatic cord with other constituents, via which the vessel traverses the inguinal canal to the scrotum. At the posterosuperior aspect of the testis it divides into two branches on its medial and lateral surfaces, which pass through its tunica albuginea to ramify in the tunica vasculosa. Terminal branches enter the testis over its surface. Some pass into the mediastinum testis and loop back before reaching their distribution (Harrison & Barclay 1948). In the abdomen the testicular artery supplies perirenal fat, ureter and iliac lymph nodes; in the inguinal canal it supplies the cremaster.

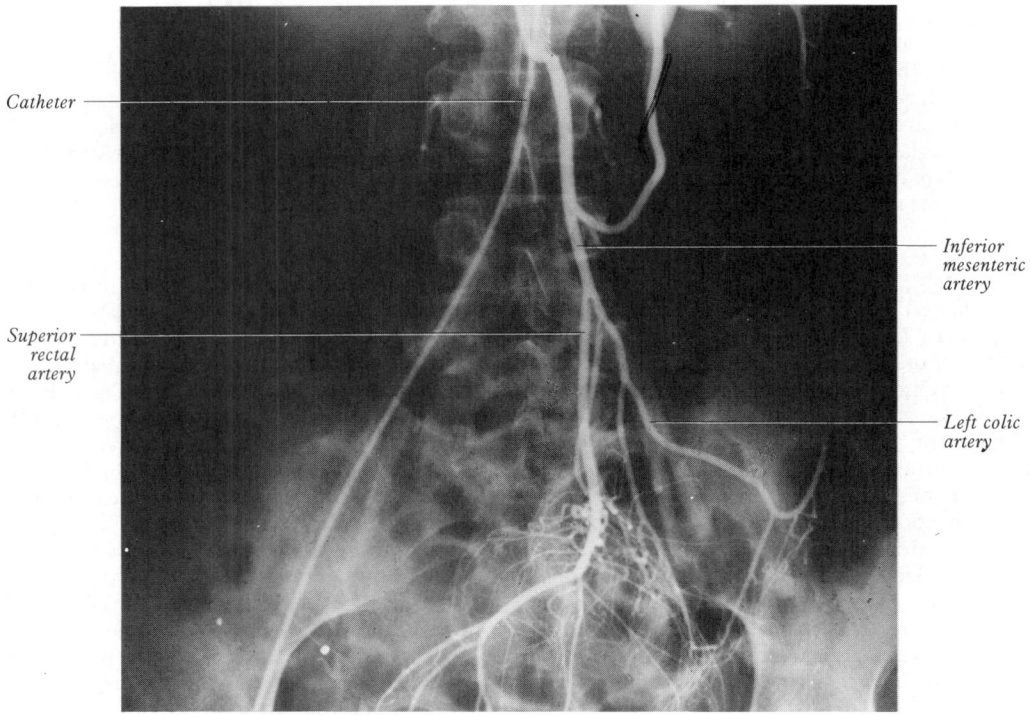

Catheter

Superior
rectal
artery

Inferior
mesenteric
artery

Left colic
artery

6.118 Inferior mesenteric arteriogram. Provided by Shaun Gallagher, Guy's Hospital; photography by Sarah Smith.

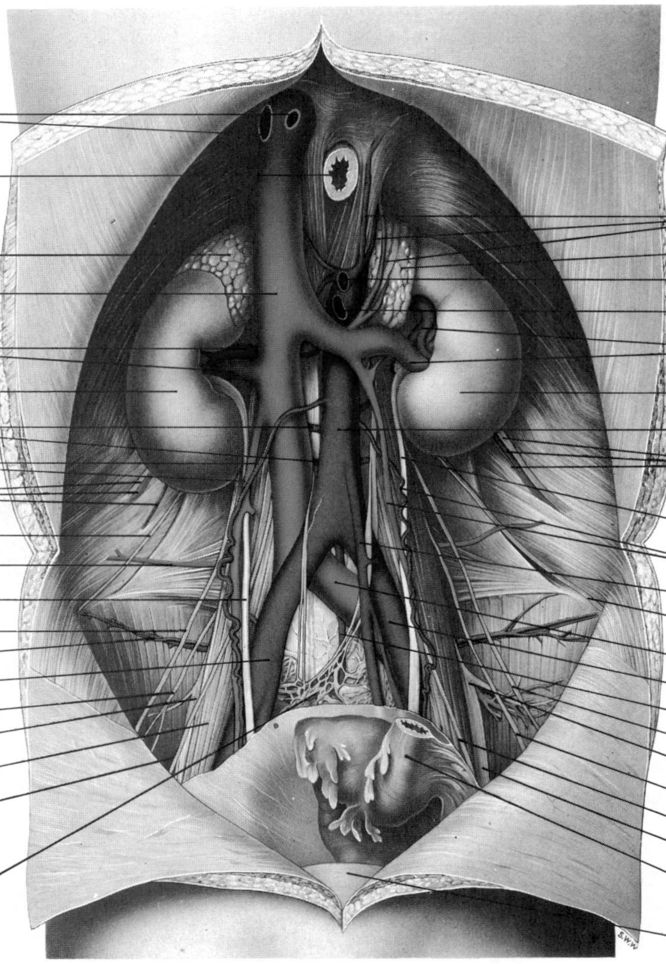

Hepatic veins (cut)

Oesophagus

R. suprarenal gland

Inferior vena cava

R. renal vessels

R. kidney

Lumbar arteries

R. ovarian vessels

Subcostal nerve and vessels

R. iliohypogastric nerve

Lumbar vein

R. ureter

R. ilio-inguinal nerve

Iliac artery and vein

R. common iliac artery

Lateral cutaneous nerve of the thigh

Femoral nerve

Psoas major

R. genitofemoral nerve

Superior hypogastric plexus

L. inferior phrenic vessels

L. suprarenal gland

Coeliac trunk

Superior mesenteric artery

L. renal vessels

L. kidney

Abdominal aorta

Subcostal nerve and vessels

L. ureter

L. ovarian vessels

Inferior mesenteric artery

L. iliohypogastric nerve

L. common iliac vein

L. common iliac artery

Median sacral artery

L. ilio-inguinal nerve

Nerve to iliacus

Lateral cutaneous nerve of the thigh

L. genitofemoral nerve

Femoral nerve

Sigmoid colon

Urinary bladder

6.119 Dissection to show the relations of structures on the posterior abdominal wall (female subject).

775

Sometimes the right testicular artery passes *posterior* to the inferior vena cava. Both arteries represent persistent lateral splanchnic aortic branches (p. 218) which enter the mesonephros and cross ventral to the supracardinal but dorsal to the subcardinal vein. Normally the lateral splanchnic artery which persists as the *right* testicular passes *caudal* to the suprasubcardinal anastomosis forming part of the inferior vena cava (p. 218). When it passes *cranial* to this, the right testicular artery is behind the inferior vena cava.

The Ovarian Arteries

These correspond to the testicular arteries but enter the pelvis to supply the ovaries (6.122). Initially they resemble the testicular arteries but at the brim of the lesser pelvis each crosses the lower external iliac artery and vein to enter the true pelvic cavity, turning medially in the ovarian suspensory ligament to continue into the uterine broad ligament, below the uterine tube. At ovarian level it passes back in the mesovarium and divides into branches to the ovary. Small branches supply the ureter and uterine tube and one passes to the side of the uterus to unite with the uterine artery. Others accompany the round ligament through the inguinal canal to the skin of the labium majus and the inguinal region.

Early in intrauterine life, when testes or ovaries flank the vertebral column inferior to the kidneys, the testicular and ovarian arteries are relatively short; but with descent of the gonads into the pelvis and beyond, they gradually lengthen.

The (Inferior) Phrenic Arteries (6.107A,B, 119)

These two small vessels help to supply the diaphragm. They may arise separately from the aorta, just above its coeliac trunk, by a common aortic stem or from the coeliac trunk; sometimes one is from the aorta, the other from a renal artery. Each artery ascends laterally anterior to a crus of the diaphragm, near the medial border of the suprarenal gland. The left passes behind the oesophagus and forwards on the left side of its diaphragmatic opening. The right phrenic passes posterior to the inferior vena cava then along the right of its opening. Near the posterior border of the diaphragm's central tendon each divides into medial and lateral branches. The medial curves forwards to anastomose with its fellow in front of the central tendon and with the musculophrenic and pericardiacophrenic arteries; the lateral approaches the thoracic wall, anastomosing with the lower posterior intercostal and musculophrenic arteries. The lateral branch of the right artery supplies the inferior vena cava while the left sends ascending branches to the oesophagus. Each has two or three small *superior suprarenal branches*. The liver (p. 773) and spleen also receive small rami from the phrenic arteries.

The Lumbar Arteries (6.119)

These are in series with the posterior intercostal arteries. Usually four on each side, they arise posterolaterally from the aorta, opposite the lumbar vertebrae. A fifth, smaller pair occasionally arise from the median sacral artery but lumbar branches of the iliolumbar arteries usually take their place. The lumbar arteries run posterolaterally on the four upper lumbar vertebral bodies, behind the sympathetic trunks, to intervals between the lumbar transverse processes and continue into the abdominal wall. The right arteries pass posterior to the inferior vena cava; the upper two right and first left are also posterior to the corresponding crus. Arteries of both sides pass under tendinous arches (which span the lateral concavities of the vertebral bodies, p. 635) for attachment of psoas major, proceeding posterior to the muscle and the lumbar plexus. They then cross the quadratus lumborum, the upper three posterior, the last usually anterior to it. At its lateral border they pierce the posterior aponeurosis of the transversus abdominis, advancing between it and the internal oblique. They anastomose with one another and the lower posterior intercostal, subcostal, iliolumbar, deep circumflex iliac and inferior epigastric arteries.

Branches. Each lumbar artery has a *dorsal ramus* passing back between the adjacent transverse processes to supply the dorsal muscles, joints and skin; this also has a *spinal branch* entering the vertebral canal to supply its contents and adjacent vertebra, anastomosing with the arteries above and below it and across the midline. The spinal branch of the first lumbar supplies the terminal spinal cord itself; the remainder supply the cauda equina, meninges and vertebral canal. Branches of the lumbar arteries and their dorsal rami supply the adjacent muscles, fasciae, bones, red marrow, ligaments and joints (symphyses, syndesmoses and synovial joints).

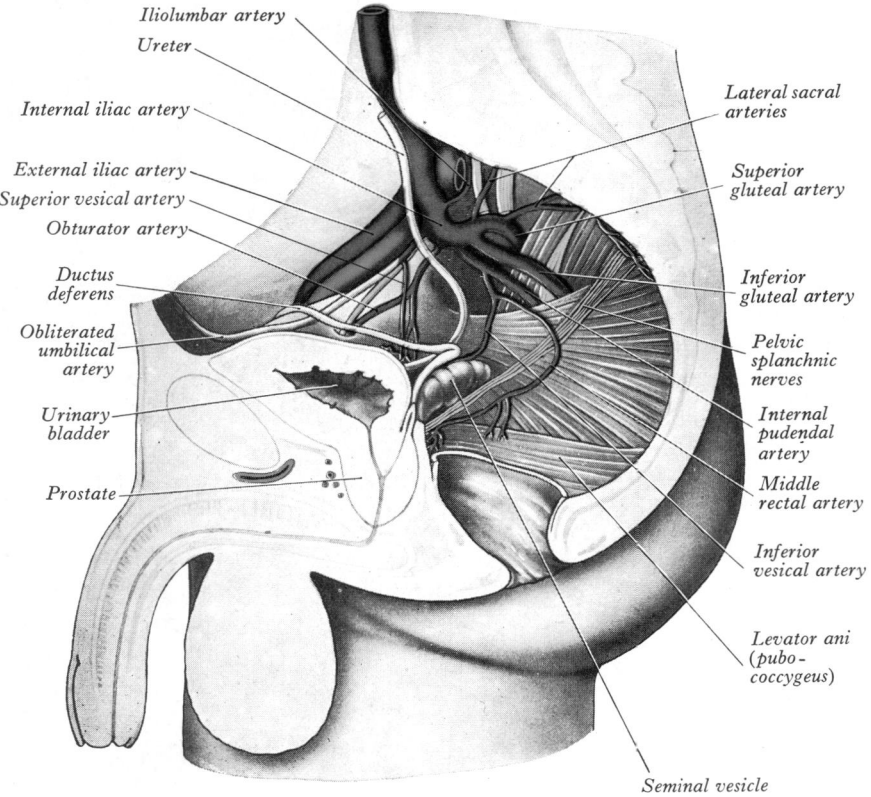

Iliolumbar artery
Ureter
Internal iliac artery
External iliac artery
Superior vesical artery
Obturator artery
Ductus deferens
Obliterated umbilical artery
Urinary bladder
Prostate

Lateral sacral arteries
Superior gluteal artery
Inferior gluteal artery
Pelvic splanchnic nerves
Internal pudendal artery
Middle rectal artery
Inferior vesical artery
Levator ani (pubo-coccygeus)
Seminal vesicle

6.120 The arteries of the male pelvis (right side). The internal iliac vein and its tributaries have been removed; the rectum has been divided just above the anal canal and its upper part has been taken away.

The Median Sacral Artery (6.107A,B,119)

This small posterior branch leaves the aorta a little above its bifurcation. It descends in the midline, anterior to the fourth and fifth lumbar vertebrae, sacrum and coccyx, ending in the coccygeal body. At fifth lumbar level it is crossed by the left common iliac vein and often gives off a small lumbar artery (*arteria lumbalis ima*), minute branches of which reach the rectum. Anterior to the last lumbar vertebra the median sacral anastomoses with a lumbar branch of the iliolumbar; anterior to the sacrum it anastomoses with the lateral sacral arteries and sends rami into the anterior sacral foramina.

The Common Iliac Arteries

The abdominal aorta bifurcates, anterolateral to the *left* side of the *fourth* lumbar vertebral body, into the right and left common iliac arteries (**6.**107A,B,119). These diverge as they descend to divide near the *level* of the lumbosacral intervertebral disc (between the last lumbar and first sacral vertebrae) into *external* and *internal iliac arteries*; the former supplies most of the lower limb, the latter the pelvic viscera and walls, perineum and gluteal region. The *division* of the common iliac is anterior to its *sacro-iliac joint*.

The Right Common Iliac Artery (6.107A,B,119)

This is about 5 cm long and passes obliquely across part of the fourth and the fifth lumbar vertebral body. *Anteriorly*, it is crossed by the sympathetic rami to the superior hypogastric plexus and, at its division, by the ureter; it is covered by the parietal peritoneum, which separates it from the coils of the small intestine. *Posteriorly*, it is separated from the fourth and fifth lumbar vertebral bodies and their intervening disc by the sympathetic trunk, the terminal parts of the common iliac veins and the commencement of the inferior vena cava; the obturator nerve, lumbosacral trunk and iliolumbar artery are also posterior, traversing fatty tissue between the fifth lumbar vertebra and the psoas major. *Lateral* to its upper part are the inferior vena cava and the right common iliac vein; lateral to its lower part is the right psoas major; *medial* to its upper part is the left common iliac vein.

The Left Common Iliac Artery (6.107A,B,119)

The artery is about 4 cm long. *Anterior* are the peritoneum, ileum, the sympathetic rami to the superior hypogastric plexus, the superior rectal artery and, at its terminal bifurcation, the ureter. *Posterior* are the sympathetic trunk, fourth and fifth lumbar vertebral bodies and intervening disc; the obturator nerve, lumbosacral trunk and iliolumbar artery are more posterior (i.e. deeply situated). The left common iliac vein is partly *medial*, partly *posterior* to the artery; *lateral* and closely related is the left psoas major.

Surface Anatomy. The vessel corresponds to the superior third of a broad line from the aortic bifurcation (p. 766) to a point midway between the anterior superior iliac spine and the pubic symphysis. The *external iliac artery* corresponds to the inferior two-thirds of this line, which is laterally slightly convex.

Branches. In addition to the terminal branches, each common iliac artery gives small rami to the peritoneum, psoas major, ureter, adjacent nerves and surrounding areolar tissue; occasionally it has the iliolumbar and accessory renal arteries as branches.

The Internal Iliac Arteries

Each internal iliac artery (**6.**120,121), about 4 cm long, begins at the common iliac bifurcation, level with the lumbosacral intervertebral disc and anterior to the sacro-iliac joint; it descends posteriorly to the superior margin of the greater sciatic foramen, dividing here into: an *anterior trunk*, which continues in the same line towards the ischial spine; and a *posterior trunk*, passing back to the foramen (Braithwaite 1952).

Anterior are the ureter and, in females, the ovary and fimbriated end of the uterine tube; *posterior* are the internal iliac vein, lumbosacral trunk and sacro-iliac joint; *lateral* is the external iliac vein, between the artery and the psoas major and inferior to this

the obturator nerve; *medial* is the parietal peritoneum, separating it from the terminal ileum on the right and the sigmoid colon on the left, and tributaries of the internal iliac vein.

In the fetus the internal iliac artery is twice the size of the external and is the direct continuation of the common iliac. It ascends on the anterior abdominal wall to the umbilicus, converging on its fellow. Having traversed the opening, the two arteries, now *umbilical*, enter the umbilical cord, coil round the umbilical vein and ultimately ramify in the placenta. At birth, when placental circulation ceases, only the pelvic segment remains patent as the internal iliac artery and part of the superior vesical, the remainder becoming a fibrous *medial umbilical ligament* raising the peritoneal *medial umbilical fold* from the pelvis to the umbilicus. In males, the patent part usually gives off an artery to the ductus deferens (vide infra).

BRANCHES FROM THE ANTERIOR TRUNK OF THE INTERNAL ILIAC ARTERY

The superior vesical artery (**6.**120,121) supplies many branches to the vesical fundus (Braithwaite 1951); from one the *artery to the ductus deferens* occasionally starts and accompanies

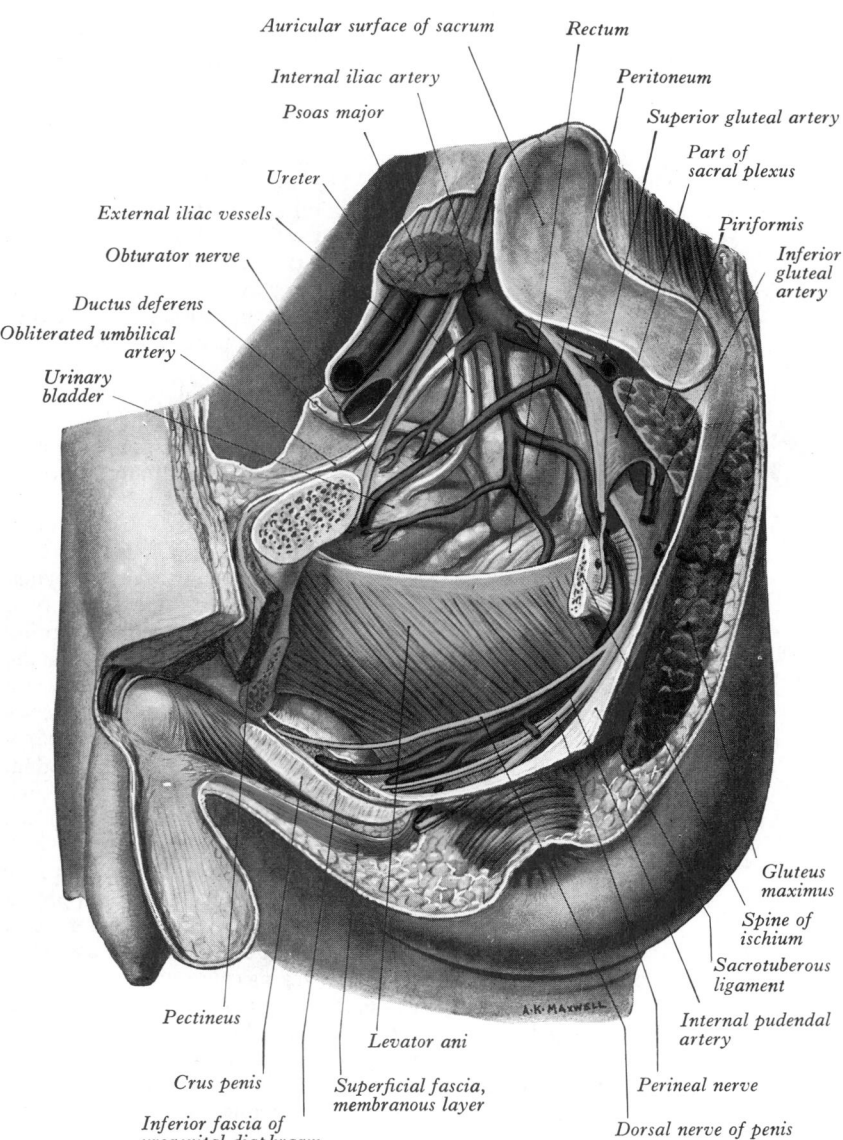

Auricular surface of sacrum
Rectum
Internal iliac artery
Peritoneum
Psoas major
Superior gluteal artery
Part of sacral plexus
Ureter
External iliac vessels
Piriformis
Obturator nerve
Inferior gluteal artery
Ductus deferens
Obliterated umbilical artery
Urinary bladder
Gluteus maximus
Spine of ischium
Sacrotuberous ligament
Internal pudendal artery
Pectineus
Levator ani
Perineal nerve
Crus penis
Superficial fascia, membranous layer
Dorsal nerve of penis
Inferior fascia of urogenital diaphragm

6.121 Structures of the male pelvic contents from the left side. Most of the left innominate bone has been removed together with the obturator internus. The sciatic nerve has been cut away close to its origin from the sacral plexus. All the vessels and nerves exposed are those of the *left* side. Note the superior vesical, obturator, inferior vesical and middle rectal arteries which are, for technical reasons, unlabelled (cf. **6.**120).

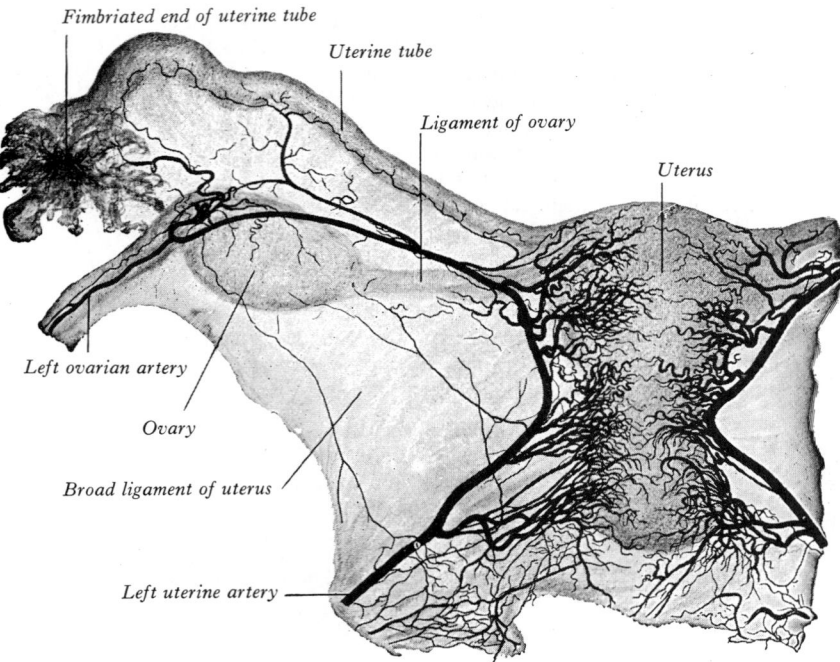

Fimbriated end of uterine tube

Uterine tube

Ligament of ovary

Uterus

Left ovarian artery

Ovary

Broad ligament of uterus

Left uterine artery

6.122 The left uterine and ovarian arteries of a nullipara of 17½ years: posterior aspect. From a preparation by Hamilton Drummond.

the ductus to the testis, anastomosing with the testicular artery. Others supply the ureter. The beginning of the superior vesical artery is the proximal, patent section of the fetal umbilical artery (vide supra).

The inferior vesical artery (**6.**120,121) often arises with the middle rectal, supplying the vesical fundus, prostate, seminal vesicles and lower ureter. Prostatic branches communicate across the midline. The inferior vesical may sometimes provide the artery to the ductus deferens.

The middle rectal artery (**6.**117,120) usually arises with the inferior vesical. It vascularizes muscular tissue in the lower rectum, anastomosing with the superior and inferior rectal arteries. It supplies the seminal vesicles and prostate by branches which join those of the inferior vesical.

The uterine artery (**6.** 122) runs medially on the levator ani to the cervix uteri; about 2 cm from this it crosses above the *ureter*, to which it supplies a small branch, and above the *lateral vaginal fornix*. It ascends tortuously lateral to the uterus in its broad ligament to the junction of the uterine tube and uterus, turning laterally towards the ovarian hilum, and ends by joining the ovarian artery. It supplies the cervix uteri and branches descend on the vagina, anastomosing with branches of the vaginal arteries to form two median longitudinal vessels, the *azygos arteries of the vagina*; one descends anterior, the other posterior to the vagina. The uterine artery supplies the body of the uterus, uterine tube and round ligament of the uterus. Terminal branches in the uterine muscle are tortuous *helicine arteries*.

The vaginal artery, often double or triple, corresponds to the inferior vesical in males; it descends on the vagina, supplying mucous membrane, and sends branches to the vestibular bulb, vesical fundus and the adjacent part of the rectum. It assists in forming the azygos arteries of the vagina (vide supra).

The obturator artery (**6.**120) inclines antero-inferiorly on the lateral pelvic wall to the upper part of the obturator foramen. Leaving the pelvic cavity by the obturator canal, it divides into anterior and posterior branches. In the pelvis it is related laterally to the obturator fascia, separating it from the obturator internus; it is crossed medially by the ureter and the ductus deferens, separating it from the parietal peritoneum. In the nullipara the ovary is medial. The obturator nerve is above, the vein below.

Branches. In the pelvis, the obturator artery provides: (1) *iliac branches* to the iliac fossa, supplying the bone and iliacus and anastomosing with the iliolumbar artery; (2) a *vesical branch* passing medially to the bladder, sometimes replacing the inferior

vesical branch of the internal iliac; (3) a *pubic branch* just before it leaves the pelvis, which ascends over the pubis to anastomose with its fellow and the pubic branch of the inferior epigastric. Outside the pelvis its anterior and posterior terminal branches encircle the foramen between the obturator externus and the obturator membrane. The *anterior branch* curves forwards on the membrane and then down along its anterior margin, supplying branches to the obturator externus, pectineus, femoral adductors and gracilis and anastomosing with the posterior branch and the medial circumflex femoral artery. The *posterior branch* follows the foramen's posterior margin and turns forwards on the ischial ramus to anastomose with the anterior. It supplies the muscles attached to the ischial tuberosity and anastomoses with the inferior gluteal. An *acetabular branch* enters the hip joint at the acetabular notch, ramifies in the fat of the acetabular fossa and sends a ramus along the ligament of the femoral head.

Variations. In 20–30% the obturator artery is replaced by an enlarged pubic branch of the inferior epigastric (p. 781); this descends almost vertically to the obturator foramen. Such an *abnormal obturator artery* is usually near the external iliac vein, lateral to the femoral ring, and is then safe in herniotomy. Sometimes it curves along the edge of the lacunar part of the inguinal ligament, partly encircling the neck of a hernial sac, and may be inadvertently incised at operation. Rarely the artery crosses the femoral ring and may be pushed medially or laterally by a femoral hernia. According to Pick et al (1942), recording 640 dissections of obturator arteries, it is a branch of the internal iliac (24%) or its anterior (21%) or posterior (3%) divisions; incidence of epigastric origin was 27% (vide supra); apart from the superior gluteal (11%) and inferior gluteal artery (9%), abnormal origins include the external iliac, internal pudendal and iliolumbar arteries.

The Internal Pudendal Artery (**6.**120,121,123,124)

This, the smaller of two terminal branches of the anterior division of the internal iliac, supplies the external genitalia. Though similar in course, it is smaller in females and its distribution differs.

In the male the artery descends laterally to the inferior rim of the greater sciatic foramen, where it leaves the pelvis between piriformis and coccygeus and enters the *gluteal region*; then curving around the dorsum of the ischial spine to enter the *perineum* by the lesser sciatic foramen, it traverses the pudendal canal in the lateral wall of the *ischiorectal fossa*, medial to the obturator internus, about 4 cm above the ischial tuberosity's lower limit. Approaching the margin of the ischial ramus, it proceeds above or below the inferior fascia of the urogenital diaphragm along the medial margin of the inferior pubic ramus and ends behind the inferior pubic ligament, dividing into the *deep and dorsal arteries of the penis*. It may descend through the inferior fascia before its division. (The internal pudendal distal to its perineal branch has been named *artery of the penis*, appropriately in view of its distribution, vide infra.)

Relations. In the pelvis the internal pudendal artery crosses anterior to the piriformis, sacral plexus and inferior gluteal artery. Behind the ischial spine it is covered by the gluteus maximus, with the pudendal nerve medial and the nerve to obturator internus lateral. In the pudendal canal (p. 606) it travels at first with companion veins and the pudendal nerve; beyond this the dorsal nerve of the penis is above, the perineal nerve below.

Branches (**6.**123,124) **Muscular branches** leave the artery in the pelvis and gluteal region to supply the adjacent muscles and nerves.

The inferior rectal artery arises above the ischial tuberosity. Escaping from the pudendal canal (p. 606), it divides into two or three rami crossing the ischiorectal fossa medially to supply the anal skin and musculature. Small branches skirt the lower edge of the gluteus maximus to supply the gluteal skin. The inferior rectal anastomoses with its fellow, and with the superior, middle rectal and perineal arteries.

The perineal artery (**6.**123) leaves the internal pudendal near the anterior end of its canal, turns down through the inferior fascia of the urogenital diaphragm (p. 604) and approaches the scrotum in the superficial perineal region, between the bulbospongiosus and ischiocavernosus. Beyond the diaphragm, and near its base, a small *transverse branch* passes medially inferior to the superficial transverse perineal muscle to anastomose with its fellow and the posterior scrotal and inferior rectal arteries, supplying tissues between the anus and the penile bulb. The *posterior scrotal arteries*, distributed to the scrotal skin and dartos muscle, are usually terminals of the perineal but may also arise from its transverse branch; they also supply the perineal muscles.

The artery of the bulb of the penis, short but wide, runs medially through the deep transverse perineal muscle and inferior urogenital fascia to the penile bulb. Penetrating this, it supplies the posterior part of the corpus spongiosum and the bulbo-urethral gland.

The urethral artery traverses the urogenital diaphragm's inferior fascia and enters the corpus spongiosum, reaching the glans penis. It supplies the urethra and erectile tissue around it.

The deep artery of the penis, a terminal branch of the internal pudendal, passes through the inferior fascia of the urogenital diaphragm to enter the crus penis. It traverses the corpus cavernosum and supplies its erectile tissue.

The dorsal artery of the penis, the other terminal branch of the internal pudendal, leaves the inferior aspect of the urogenital diaphragm, *ascends* between the crus penis and pubic symphysis, and traverses the suspensory ligament of the penis to run along its dorsum to the glans, where it forks into branches to the glans and prepuce. In the penis it lies between its dorsal nerve and deep dorsal vein, the latter being most medial. It supplies penile skin and the fibrous sheath of the corpus cavernosum, anastomosing through the sheath with the deep penile artery.

In the female the internal pudendal is naturally smaller but its origin, course and branches are similar, including the *posterior labial branches*, the *artery of the bulb* (distributed to the erectile tissue of the vestibular bulb and vagina), *deep artery of the clitoris*, supplying the corpus cavernosum, and a *dorsal artery* to the glans and prepuce of the clitoris.

Variations. Branches of the internal pudendal are sometimes derived from an *accessory pudendal*, usually a branch of the pudendal before its exit from the pelvis.

The inferior gluteal artery (6.121,125), the larger terminal branch of the anterior internal iliac trunk, chiefly supplies the buttock and thigh. It descends anterior to the sacral plexus and piriformis, posterior to the internal pudendal artery. Passing between the first and second or second and third sacral anterior spinal nerve rami, then between the piriformis and coccygeus, it traverses the lower part of the greater sciatic foramen to reach the gluteal region. Descending between the greater trochanter and ischial tuberosity with the sciatic and posterior femoral cutaneous nerves, deep to the gluteus maximus, it continues down the thigh, supplying the skin and anastomosing with branches of the perforating arteries. The inferior gluteal and internal pudendal arteries are often a common stem from the internal iliac, sometimes including the superior gluteal artery.

Surface Anatomy. The inferior gluteal artery leaves the pelvis near the midpoint of a line joining the posterior superior iliac spine and the ischial tuberosity.

Branches are: (*inside the pelvis*) (1) to the piriformis, coccygeus and levator ani; (2) to the perirectal fat, occasionally replacing the middle rectal artery; (3) to the vesical fundus, seminal vesicles and prostate; (*outside the pelvis*) *muscular branches* supply the gluteus maximus, obturator internus, gemelli, quadratus femoris and the proximal parts of the hamstring muscles, anastomosing with the superior gluteal, internal pudendal, obturator and medial circumflex femoral arteries. *Coccygeal branches* run medially through the sacrotuberous ligament to supply the gluteus maximus and the structures attached to the coccyx. The *artery to the sciatic nerve* runs on the nerve for a short distance, then descends in it to the lower thigh. An *anastomotic branch* descends obliquely across obturator internus, gemelli and quadratus femoris, to join the *cruciate anastomosis* (p. 783) linking with the first perforating and medial and lateral circumflex femoral arteries. An *articular branch*, usually from the anastomotic, is distributed to the hip joint. *Cutaneous branches* supply the buttock and back of the thigh.

BRANCHES FROM THE POSTERIOR TRUNK OF THE INTERNAL ILIAC ARTERY

The iliolumbar artery (6.120) ascends laterally *anterior* to the sacro-iliac joint and lumbosacral trunk, *posterior* to the obturator

nerve and external iliac vessels, to reach the medial border of psoas major, dividing behind it into the lumbar and iliac branches. The *lumbar branch* supplies the psoas major and quadratus lumborum, anastomoses with the fourth lumbar artery and sends a small *spinal branch* through the intervertebral foramen between the fifth lumbar and first sacral vertebrae, which supplies the cauda equina. The *iliac branch* supplies the iliacus; between the muscle and bone it anastomoses with the iliac branches of the obturator. A large nutrient branch enters an oblique canal in the ilium; others skirt the iliac crest, supplying the gluteal and abdominal muscles and anastomosing with the superior gluteal, circumflex iliac and lateral circumflex femoral arteries.

The lateral sacral arteries (6.120) are from the posterior trunk of the internal iliac, usually as a superior and an inferior branch. The *superior* and larger passes medially into the first or second anterior sacral foramen, supplies the sacral vertebrae and

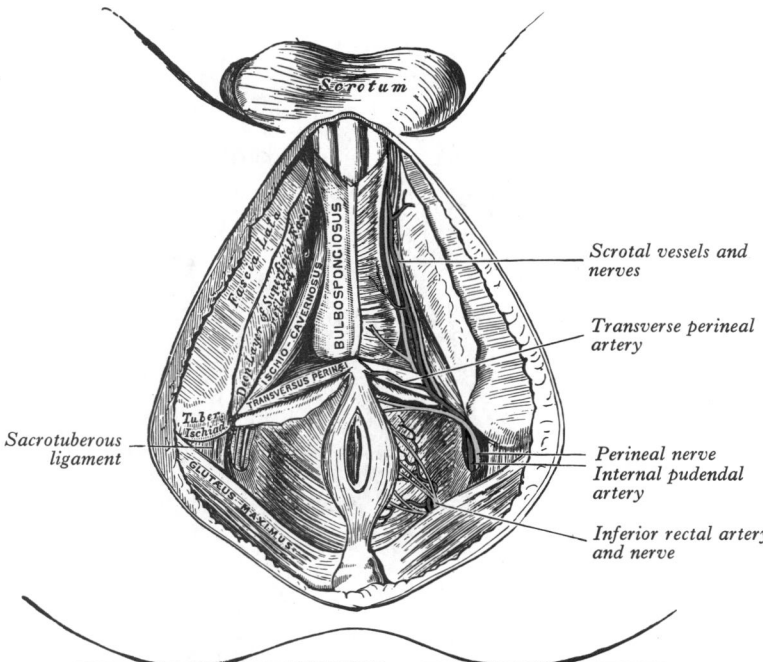

6.123 The superficial branches of the internal pudendal artery, in the male.

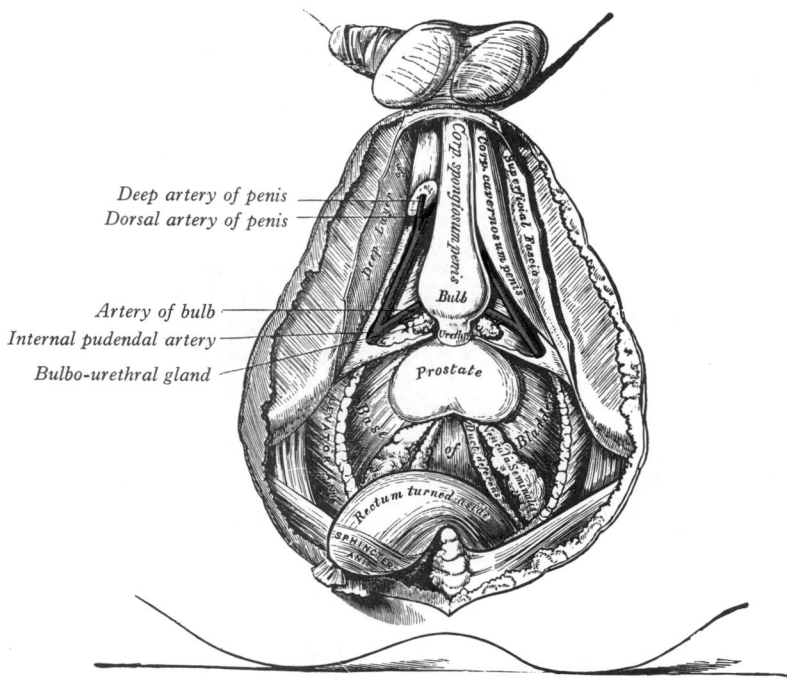

6.124 The deeper branches of the internal pudendal artery, in the male.

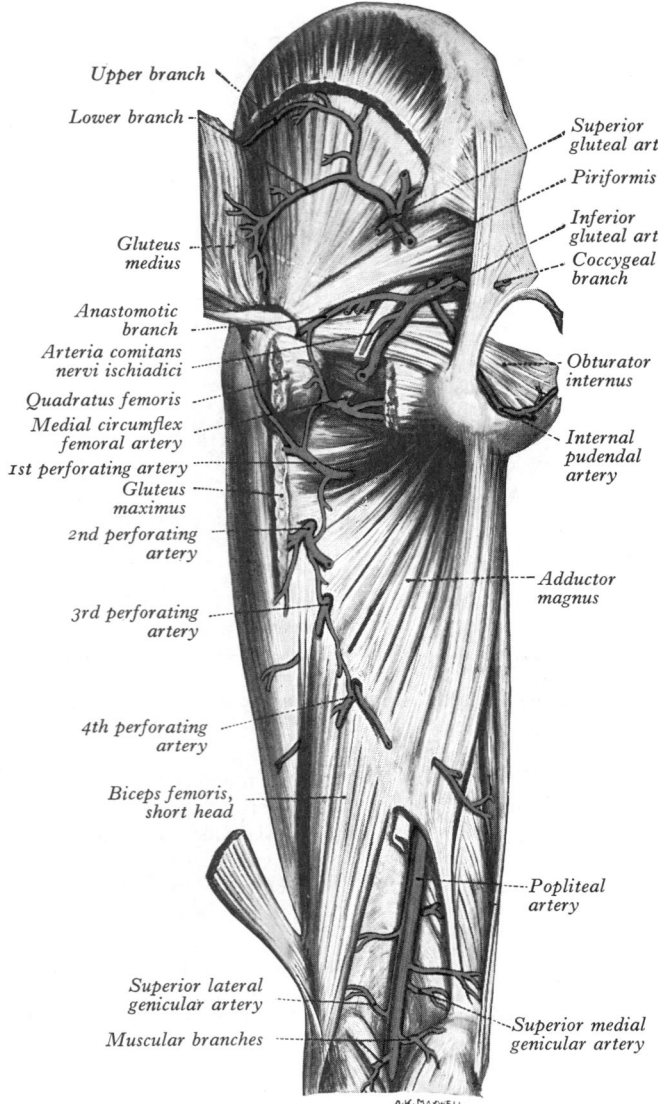

6.125 The arteries of the left gluteal and posterior femoral regions.

Labels on figure 6.125:
Upper branch
Lower branch
Superior gluteal artery
Piriformis
Inferior gluteal artery
Gluteus medius
Coccygeal branch
Anastomotic branch
Arteria comitans nervi ischiadici
Obturator internus
Quadratus femoris
Medial circumflex femoral artery
Internal pudendal artery
1st perforating artery
Gluteus maximus
2nd perforating artery
Adductor magnus
3rd perforating artery
4th perforating artery
Biceps femoris, short head
Popliteal artery
Superior lateral genicular artery
Muscular branches
Superior medial genicular artery
A.K. MAXWELL.

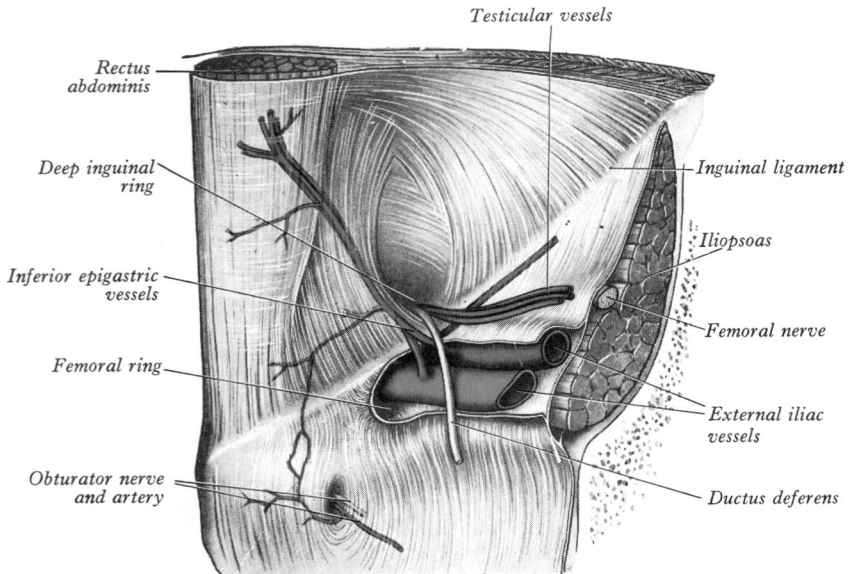

6.126 Dissection of the deep aspect of the lower part of the abdominal wall of the right side with the thinner posterior wall of the rectus sheath. The femoral and deep inguinal rings are displayed together with the vessels and other structures in relation to them and also the opening into the obturator canal.

Labels on figure 6.126:
Rectus abdominis
Testicular vessels
Deep inguinal ring
Inguinal ligament
Inferior epigastric vessels
Iliopsoas
Femoral nerve
Femoral ring
External iliac vessels
Obturator nerve and artery
Ductus deferens

contents of the sacral canal and escapes via the corresponding dorsal foramen to supply the skin and muscles dorsal to the sacrum. The *inferior lateral sacral artery* crosses obliquely anterior to the piriformis and the sacral anterior spinal rami, then descends lateral to the sympathetic trunk to anastomose with its fellow and the median sacral artery anterior to the coccyx. Its branches enter the anterior sacral foramina, distributed like those of the superior artery.

The superior gluteal artery (6.120,125), the largest branch of the internal iliac and the continuation of its posterior trunk, runs back between the lumbosacral spinal trunk and the first sacral ramus or between the first and second rami, leaving the pelvis by the greater sciatic foramen above the piriformis and dividing into *superficial* and *deep branches*. In the pelvis it supplies the piriformis, obturator internus and an innominate nutrient artery. The *superficial branch* enters the deep surface of the gluteus maximus; its numerous rami supply the muscle and anastomose with the inferior gluteal, others perforating its tendinous medial attachment to supply the skin over the sacrum, anastomosing with the posterior branches of the lateral sacral arteries. The *deep branch* is between the gluteus medius and the bone, soon dividing into superior and inferior rami. The *superior* skirts the superior border of the gluteus minimus to the anterior superior iliac spine, anastomosing with the deep circumflex iliac artery and the ascending branch of the lateral circumflex femoral. The *inferior ramus* traverses the gluteus minimus obliquely and supplies it and also the gluteus medius, anastomosing with the lateral circumflex femoral; a branch enters the trochanteric fossa to join the inferior gluteal and ascending branch of the medial circumflex femoral; other branches pierce the gluteus minimus to supply the hip joint.

The superior gluteal artery may arise from the internal iliac with the inferior gluteal and sometimes the internal pudendal.

Surface Anatomy. The artery's pelvic exit corresponds to the junction of the upper and middle thirds of a line joining the posterior superior iliac spine to the apex of the greater trochanter.

The External Iliac Arteries

The external iliac arteries (6.120,121,126) are larger than the internal. Each descends laterally along the medial border of the psoas major from the common iliac bifurcation (anterior to the sacro-iliac joint at lumbosacral disc level) to a point midway between the anterior superior iliac spine and the symphysis pubis, entering the thigh posterior to the inguinal ligament to become the femoral artery.

Anteromedially the artery is related to the parietal peritoneum and extraperitoneal tissue, separating the right from the terminal ileum and often the appendix, the left from the sigmoid colon and coils of the small intestine. At its origin the artery may be crossed by the ureter, in females by ovarian vessels. Testicular vessels are anterior for some distance near its distal end, and it is crossed here by the genital branch of the genitofemoral nerve, the deep circumflex iliac vein and the ductus deferens or round ligament. *Posteriorly* the iliac fascia separates it from the medial border of the psoas major. The external iliac vein is partly posterior to its upper part, medial to it below. *Laterally* it is related to the psoas major, the iliac fascia lying between them. Numerous lymph vessels and nodes lie on its front and sides.

Branches. Besides supplying the psoas major and neighbouring lymph nodes, the artery has inferior epigastric and deep circumflex iliac branches.

The inferior epigastric artery (5.52, 6.126) leaves the external iliac just proximal to the inguinal ligament, curves forwards in extraperitoneal tissue, ascends obliquely along the medial margin of the deep inguinal ring, continues up to pierce the transversalis fascia and the attenuated part of the rectus sheath (p. 600) and ascends between the rectus abdominis and the posterior lamina of its sheath. It divides into numerous branches, which anastomose with those of the superior epigastric and lower posterior intercostal arteries. The artery thus skirts the deep inguinal ring inferomedially, passing posterior to the spermatic cord but separated from it by the transversalis fascia. It raises the parietal peritoneum of the anterior abdominal wall as the *lateral umbilical fold* (p. 1339). The ductus deferens, or round ligament, winds laterally round it. It supplies the following branches:

The *cremasteric artery* accompanies the spermatic cord, supplies the cremaster and other coverings of the cord and anastomoses with the testicular artery. In females it is small and accompanies the round ligament.

A *pubic branch*, near the femoral ring, descends posterior to the pubis and anastomoses with the pubic branch of the obturator. In 20–30% of subjects, the inferior epigastric's pubic branches are large and replace the obturator's pubic ramus (p. 778).

Muscular branches supply the abdominal muscles and peritoneum, anastomosing with the circumflex iliac and lumbar arteries.

Cutaneous branches perforate the aponeurosis of the external oblique, supply the skin and anastomose with branches of the superficial epigastric artery.

Variations. The artery may arise from the femoral and then ascend, anterior to the femoral vein, to the abdomen. It often arises from the external iliac artery in common with an abnormal obturator and, rarely, directly from the obturator artery.

Applied Anatomy. The inferior epigastric artery is a main route, through anastomosis with the internal thoracic, for collateral circulation after ligature of either the common or the external iliac arteries. It is *medial* to the neck of an oblique inguinal hernia but *lateral* to that of a direct inguinal hernia (p. 1377).

The deep circumflex iliac artery branches laterally from the external iliac almost opposite the inferior epigastric. It ascends laterally to the anterior superior iliac spine posterior to the inguinal ligament in a sheath formed by the junction of the transversalis and iliac fasciae. There it anastomoses with the ascending branch of the lateral circumflex femoral artery, pierces the transversalis fascia and skirts the internal lip of the iliac crest; about halfway it perforates the transversus abdominis and runs between this and the internal oblique to anastomose with the iliolumbar and superior gluteal arteries. At the anterior superior iliac spine it has a large *ascending branch*, which runs between the internal oblique and the transversus, supplying them and anastomosing with the lumbar and inferior epigastric arteries.

Collateral Circulation. A collateral circulation may be established, in young adults, after ligature of the common iliac artery; when arterial walls degenerate in older patients it is unlikely to supply the leg adequately.

THE ARTERIES OF THE LOWER LIMBS

The chief artery of the thigh is the continuation of the external iliac, extending from the inguinal ligament to the distal border of the popliteus, where it divides into anterior and posterior tibial arteries. Its proximal section, the femoral artery lies among the genual extensor muscles; its distal continuation, the popliteal artery, is among the genual flexors.

The Femoral Artery

The femoral artery (**6**.128–132), a continuation of the external iliac, begins behind the inguinal ligament, midway between the anterior superior iliac spine and the symphysis pubis, descends the thigh anteromedially and becomes the popliteal as it passes through an opening in the adductor magnus near the junction of the middle and distal thirds of the thigh. Proximally it is in the *femoral triangle*, distally in the *adductor (subsartorial)* canal. Its first 3 or 4 cm are enclosed, with its vein, in the *femoral sheath*.

The Femoral Sheath (6.127)

Distal prolongations, behind the inguinal ligament, of the transversalis fascia *anterior* to the femoral vessels and of the iliac fascia, *posterior*, together form a short funnel, wider proximally, its distal end fusing with the vascular fascia about 3 or 4 cm distal to the ligament. At birth the sheath is shorter, elongating when extension at the hips becomes habitual. Its (longitudinally) vertical lateral wall is perforated by the femoral branch of the genitofemoral nerve; the medial wall slopes laterally and is pierced by the great (long) saphenous vein and lymphatic vessels. Like the carotid sheath the femoral is a mass of connective tissue in which the vessels are embedded. Three compartments are described: a lateral containing the femoral artery; an intermediate for the femoral vein; medial and smallest is the *femoral canal*, containing the lymph vessels and a lymph node embedded in areolar tissue, probably to allow the vein to distend. This canal is conical, about 1.25 cm in length; its proximal end is the *femoral ring*, bounded in front by the inguinal ligament, behind by the pectineus and its fascia, medially by the crescentic edge of the lacunar ligament and laterally by the femoral vein (p. 1379). The spermatic cord, or round ligament, is just above its anterior margin; the inferior epigastric vessels are near its anterolateral rim. It is larger in women than men due partly to the greater breadth of the female pelvis, partly to the smaller size of the femoral vessels. The ring is filled by condensed extraperitoneal tissue, the *femoral septum*, covered by the parietal peritoneum (p. 1377). The femoral septum is traversed by numerous lymph vessels connecting the deep inguinal to the external iliac lymph nodes.

The femoral triangle (p. 638) underlies a depression distal to the inguinal fold. Its apex is distal, its limits being *laterally* the medial margin of sartorius, *medially* the medial margin of the adductor longus; proximally (the base) is the inguinal ligament. It is like a gutter, floored laterally by the iliacus and psoas major, medially by the pectineus and adductor longus. The femoral vessels, passing from midbase to apex, are in its deepest part. Lateral to the artery the femoral nerve divides. The triangle also contains fat and regional lymph nodes.

The adductor (subsartorial) canal (**6**.129) is an aponeurotic tunnel in the thigh's middle third from the apex of the femoral triangle to the opening in adductor mangus, through which femoral vessels reach the popliteal fossa. Triangular in section, it is bounded *anterolaterally* by the vastus medialis, *posteriorly* by the adductor longus, distally the adductor magnus and *anteromedially* by a strong aponeurosis extending between the adductors across the vessels to the vastus medialis. The sartorius is

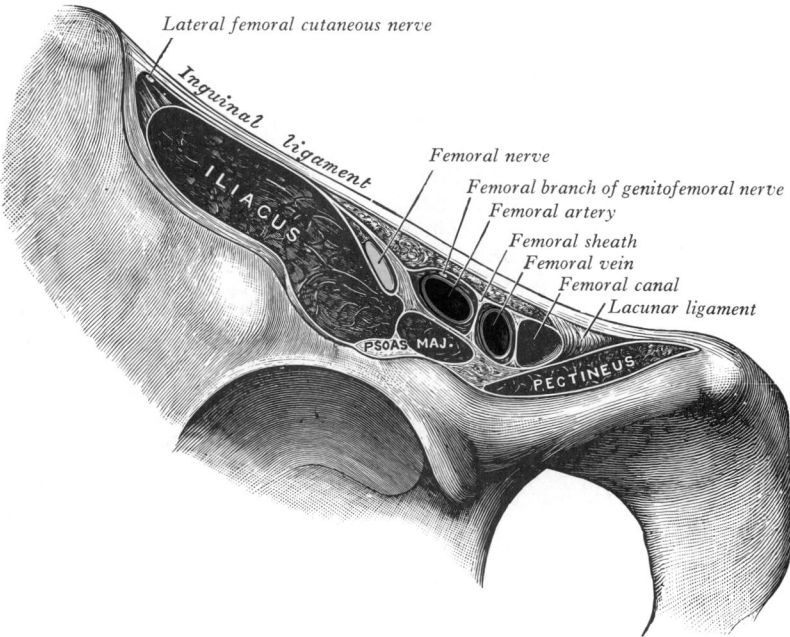

6.127 The structures passing posterior to the right lacunar ligament: inferior (distal) aspect. Note the lacuna musculorum and the lacuna vasculorum.

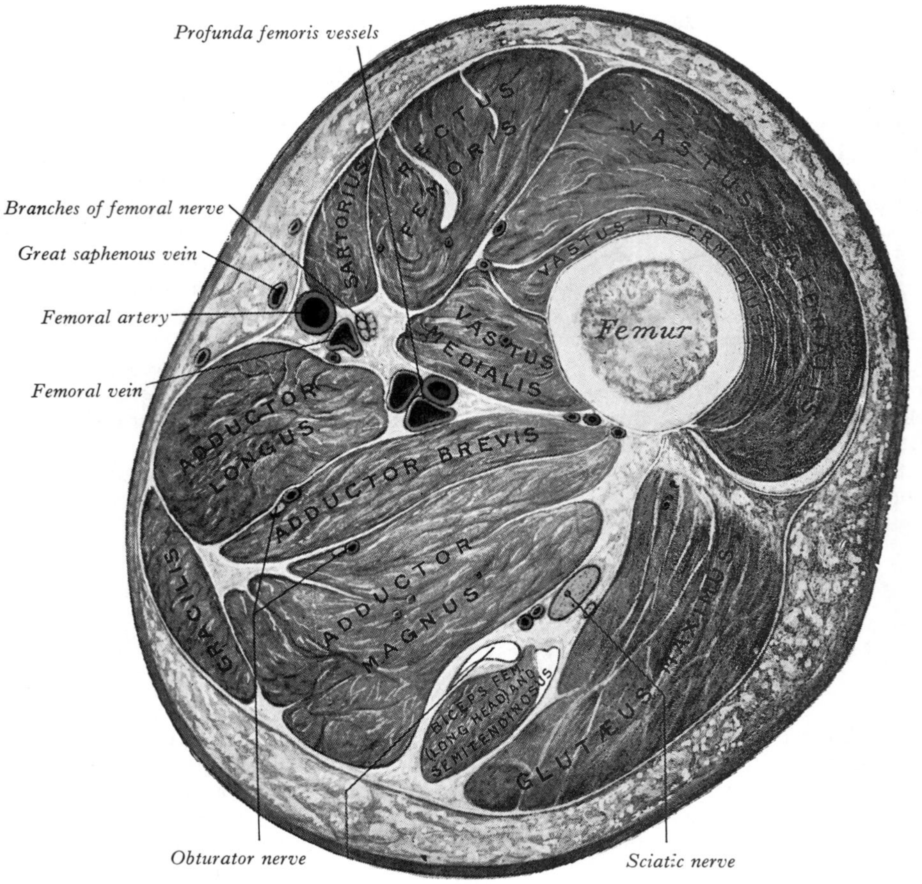

Profunda femoris vessels

Branches of femoral nerve

Great saphenous vein

Femoral artery

Femoral vein

Femur

Obturator nerve

Sciatic nerve

6.128 Transverse section through the right thigh at the level of the apex of the femoral triangle: superior (proximal) aspect. About three-fifths of the natural size. The cutaneous nerves are omitted.

anterior. The canal contains the femoral artery and vein, the saphenous nerve, and the nerve to the vastus medialis until it enters its muscle.

Femoral artery relations:

Relations in the femoral triangle (**6.**130): *anterior* are the skin, superficial fascia, superficial inguinal lymph nodes, fascia lata, femoral sheath, superficial circumflex iliac vein (crossing in the superficial fascia) and the femoral branch of the genitofemoral nerve (at first lateral then anterior). Near the apex the medial femoral cutaneous nerve crosses the artery from the lateral to the medial side. *Posterior* are the femoral sheath, tendon of psoas, pectineus and adductor longus. The artery is separated from the hip joint by the tendon of the psoas major, from the pectineus by the femoral vein and profunda vessels, from the adductor longus by the femoral vein. Proximally, the nerve to the pectineus passes medially behind the artery; *lateral* to it is the femoral nerve. The femoral vein is *medial* in the proximal part of the triangle, becoming *posterior* near its apex, distally.

Relations in the adductor canal (**6.**129,130,131): *anterior* are the skin, superficial and deep fasciae, the sartorius and fibrous roof of the canal. The saphenous nerve is first lateral, then anterior and finally medial. *Posterior* are the adductor longus and adductor magnus; the femoral vein is also posterior proximally, but becoming lateral distally. Anterolateral are the vastus medialis and its nerve.

Surface Anatomy. The artery corresponds to the proximal two-thirds of a line drawn from the midpoint between the anterior superior iliac spine and the pubic symphysis to the adductor tubercle, with the thigh semiflexed, abducted and laterally rotated. Its pulsation is easily palpable in its proximal course.

Variations. Rarely the femoral artery divides, distal to the origin of the arteria profunda femoris, into two trunks reuniting near the adductor opening. It may be replaced by the inferior gluteal artery, accompanying the sciatic nerve to the popliteal fossa and representing a proximal persistence of the original axial artery (p. 218); the external iliac is then small, ending as the arteria profunda femoris.

Applied Anatomy. Compression of the femoral artery is most effective just distal to the inguinal ligament, where it is superficial and separated from the *bone* (iliopubic eminence) only by the psoas tendon.

Branches. **The superficial epigastric artery** (**6.**130) arises anteriorly from the femoral about 1 cm distal to the inguinal ligament, traverses the cribriform fascia to ascend anterior to the ligament and runs in the abdominal superficial fascia almost to the umbilicus. It supplies the superficial inguinal lymph nodes and superficial fascia and skin, anastomosing with branches of the inferior epigastric and its fellow.

The superficial circumflex iliac artery (**6.**130), the smallest superficial branch of the femoral, arises near or with the superficial epigastric. Usually emerging through the fascia lata lateral to the saphenous opening, it turns laterally distal to the inguinal ligament towards the anterior superior iliac spine; it supplies the skin, superficial fascia and superficial inguinal lymph nodes, anastomosing with the deep circumflex iliac, superior gluteal and lateral circumflex femoral arteries.

The superficial external pudendal artery (**6.**130) arises medially from the femoral, close to the preceding branches. Emerging from the cribriform fascia, it passes medially, usually deep to the great saphenous vein, across the spermatic cord (or round ligament) to supply the lower abdominal, penile, scrotal or labial skin, anastomosing with branches of the internal pudendal.

Veins accompanying the superficial epigastric, superficial circumflex iliac and external pudendal arteries *join* the *great saphenous vein* before it enters the saphenous opening.

The deep external pudendal artery (**6.**130) passes medially across the pectineus and anterior or posterior to the adductor longus, covered by fascia lata, piercing it to supply the skin of the perineum and scrotum or labium majus; its branches anastomose with the posterior scrotal or labial branches of the internal pudendal.

Muscular branches supply the sartorius, the vastus medialis and the adductors.

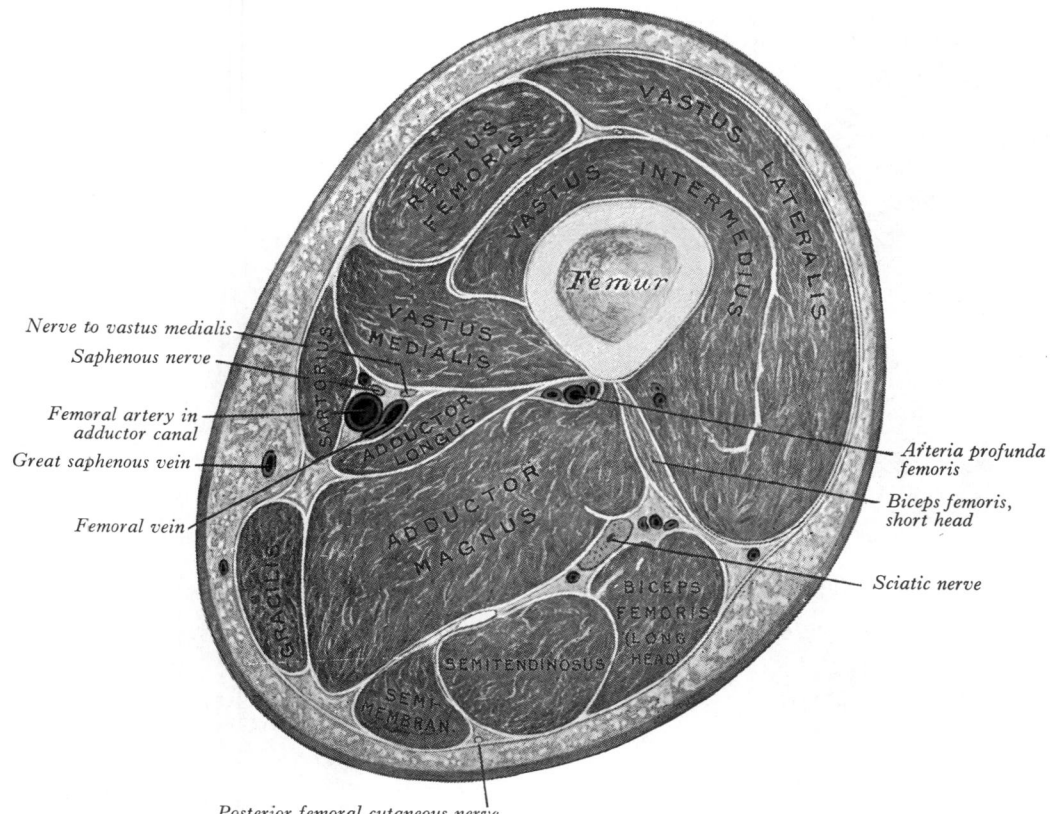

Nerve to vastus medialis
Saphenous nerve
Femoral artery in adductor canal
Great saphenous vein
Femoral vein
Arteria profunda femoris
Biceps femoris, short head
Sciatic nerve
Posterior femoral cutaneous nerve

6.129 Transverse section through the middle of the right thigh: superior (proximal) aspect. About three-fifths of the natural size.

The Arteria Profunda Femoris (6.129,130,131,132)

This large branch arises laterally from the femoral about 3.5 cm distal to the inguinal ligament. At first lateral to the femoral artery it spirals posterior to this and the femoral vein to the medial side of the femur; it passes between the pectineus and adductor longus, then between the latter and the adductor brevis and then descends between the adductor longus and adductor magnus to finally pierce the latter and anastomose with the upper muscular branches of the popliteal. This terminal part is sometimes named the *fourth perforating artery.*

Relations. Posterior, in proximodistal order, are: the iliacus, pectineus, adductor brevis and adductor magnus. *Anterior* are the femoral and profunda veins and distally the adductor longus, separating it from the femoral artery. *Laterally* the vastus medialis separates its proximal part from the femur.

Variations. Its origin is sometimes medial, or rarely posterior on the femoral artery; if the former, it may cross anterior to the femoral vein and then pass backwards around its medial side.

The deep femoral artery is the main supply to the adductor, extensor and flexor muscles; it also anastomoses with the internal and external iliac arteries above and the popliteal artery below. Its *branches* are as follows:

The lateral circumflex femoral artery, a lateral branch near the root of the profunda, inclines laterally between divisions of the femoral nerve, posterior to the sartorius and rectus femoris, dividing into ascending, transverse and descending branches. It may arise from the femoral. The *ascending branch* ascends along the inter-trochanteric line, under the tensor fasciae latae, lateral to the hip joint; it anastomoses with the superior gluteal and deep circumflex iliac arteries, supplying the greater trochanter, and forms an anastomotic ring round the femoral neck with branches of the medial circumflex femoral; from this ring the femoral neck and head are supplied. The *descending branch,* sometimes direct from the profunda or the femoral, descends posterior to the rectus femoris, along the anterior border of the vastus lateralis, which it supplies: a long ramus descends *in* vastus lateralis to the knee, anastomosing with the lateral superior genicular branch of the

popliteal, accompanied by the nerve to the vastus lateralis. The *transverse branch,* the smallest, passes laterally anterior to the vastus intermedius, pierces the vastus lateralis to wind round the femur, just distal to the greater trochanter, anastomosing with the medial circumflex, inferior gluteal and first perforating arteries (*cruciate anastomosis*).

The medial circumflex femoral artery, usually from the posteromedial aspect of the profunda but often the femoral artery, supplies the adductor muscles and curves medially round the femur between the pectineus and psoas major and then the obturator externus and adductor brevis, finally appearing between the quadratus femoris and upper border of the adductor magnus, dividing into transverse and ascending branches. The *transverse branch* takes part in the cruciate anastomosis. The *ascending branch* ascends on the tendon of the obturator externus, anterior to the quadratus femoris, to the trochanteric fossa, where it anastomoses with branches of the gluteal and lateral circumflex femoral arteries. An *acetabular branch* at the proximal edge of the adductor brevis enters the hip joint under the transverse acetabular ligament with one from the obturator artery; it supplies the fat in the fossa, and reaches the femoral head along its ligament. For blood supply of the proximal end of the femur consult Crock (1965).

The perforating arteries (6.125), usually three, perforate the attachment of the adductor magnus to reach the thigh's flexor aspect. They pass close to the linea aspera under small tendinous arches in the muscle, supplying the muscular, cutaneous and anastomotic branches. Diminished, they pass deep to the short head of biceps femoris (the first usually through the attachment of gluteus maximus), traverse the lateral intermuscular septum and enter the vastus lateralis. The first arises proximal to the adductor brevis, the second anterior and the third distal. The *first perforating artery* passes back between the pectineus and adductor brevis (sometimes through the latter), piercing the adductor magnus near the linea aspera to supply the adductor brevis, adductor magnus, biceps femoris and gluteus maximus, anastomosing with the inferior gluteal, medial and lateral circumflex femoral and second perforating arteries. The larger *second perforating artery,* often arising with the first, pierces the attachments of adductor brevis

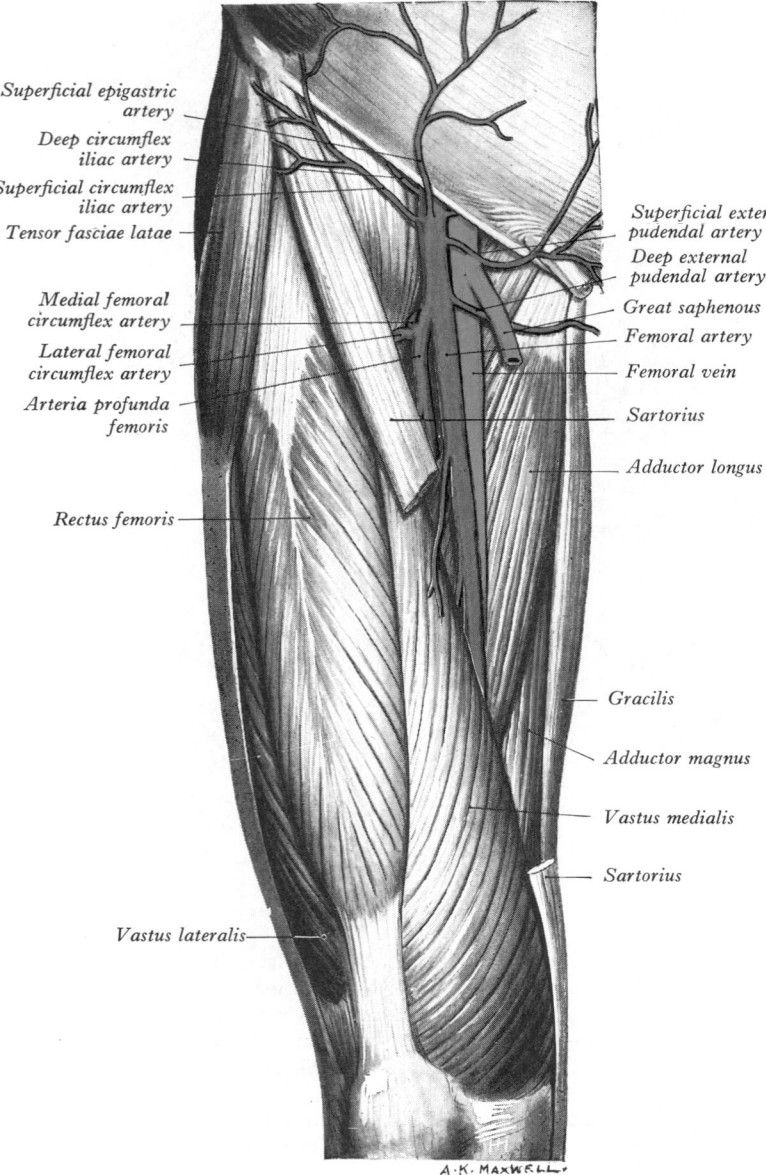

Superficial epigastric artery
Deep circumflex iliac artery
Superficial circumflex iliac artery
Tensor fasciae latae

Medial femoral circumflex artery
Lateral femoral circumflex artery
Arteria profunda femoris

Rectus femoris

Vastus lateralis

Superficial external pudendal artery
Deep external pudendal artery
Great saphenous vein
Femoral artery
Femoral vein
Sartorius

Adductor longus

Gracilis

Adductor magnus

Vastus medialis

Sartorius

A. K. MAXWELL.

6.130 The right femoral vessels and some of their branches.

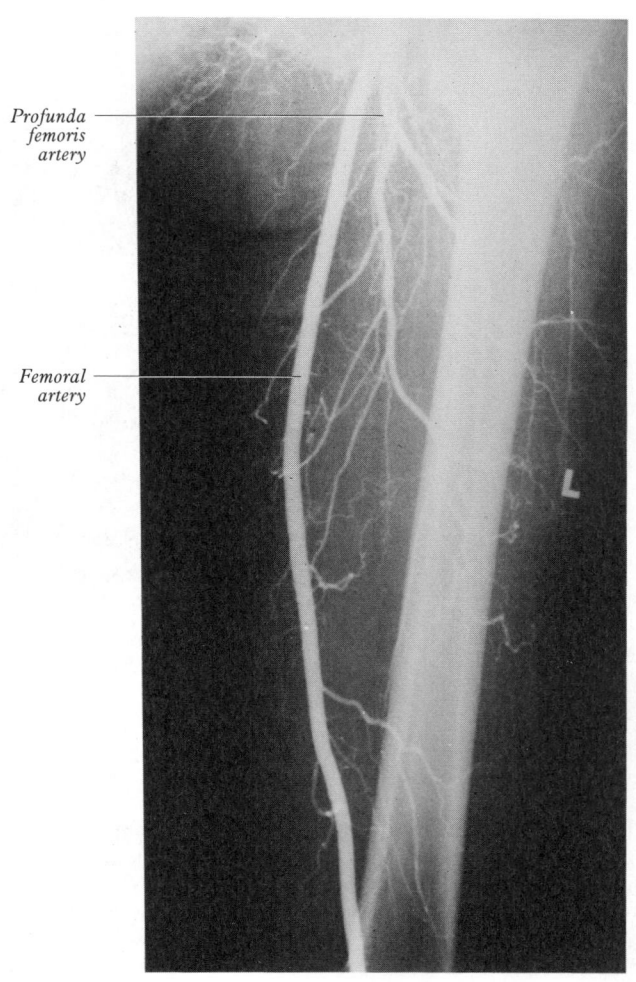

Profunda femoris artery

Femoral artery

6.131 Femoral arteriogram. Supplied by Shaun Gallagher, Guy's Hospital; photography by Sarah Smith.

and magnus, divides into the ascending and descending branches supplying the posterior femoral muscles and anastomoses with the first and third perforating arteries. The femoral *nutrient artery* usually arises from it; when two nutrient arteries exist, they usually come from the first and third. The *third perforating artery* starts distal to adductor brevis, pierces the attachment of adductor magnus and divides into branches to the posterior femoral muscles; it anastomoses proximally with the perforating arteries, distally with the end of the profunda and muscular rami of the popliteal. The femoral nutrient artery may arise from it. Rami of the diaphysial nutrient and other branches of the profunda also provide subsidiary cortical arteries (Crock 1967).

The end of arteria profunda femoris is the *fourth perforating artery*. The perforating arteries form a *double chain of anastomoses*: (1) in adductor muscles and (2) near the linea aspera.

Numerous muscular branches arise from the arteria profunda femoris; some end in adductors, others pierce the adductor magnus, supply the flexors and anastomose with the medial circumflex femoral artery and superior muscular branches of the popliteal. The profunda is thus the *main* supply to the femoral muscles.

Anastomosis on the back of the thigh. This important chain of anastomoses stretches from the gluteal region to the popliteal fossa, formed in proximodistal order by anastomoses between: (1) gluteal arteries and terminals of the medial circumflex femoral; (2) circumflex femoral arteries and the first perforating artery; (3)

perforating arteries and each other; (4) the fourth perforating artery and the superior muscular branches of the popliteal.

The descending genicular artery (6.136) arises from the femoral just proximal to the adductor opening, at once supplying a *saphenous branch* and then descending in the vastus medialis anterior to the tendon of adductor magnus, to the medial side of the knee, anastomosing with the medial superior genicular artery. *Muscular branches* supply the vastus medialis and adductor magnus and have *articular branches*, which anastomose round the knee joint. One articular branch crosses above the femoral patellar surface, forming an arch with the lateral superior genicular artery and supplying the knee joint. The *saphenous branch* emerges distally through the roof of the adductor canal to accompany the saphenous nerve to the medial side of the knee. Passing between the sartorius and gracilis it supplies the skin of the proximomedial area of the leg, anastomosing with the medial inferior genicular artery.

Collateral Circulation. After ligation of the femoral artery proximal to the origin of the arteria profunda femoris, the main anastomotic channels available are: (1) superior and inferior gluteal branches of the internal iliac with the medial and lateral circumflex femoral and the first perforating branch of the arteria profunda femoris; (2) the obturator branch of the internal iliac with the medial circumflex femoral of the arteria profunda femoris; (3) the internal pudendal branch of the internal iliac with superficial and deep external pudendal branches of the femoral; (4) a deep circumflex iliac branch of the external iliac with the lateral circumflex femoral branch of the arteria profunda femoris and the superficial circumflex iliac branch of the femoral; (5) the inferior gluteal branch of the internal iliac with perforating branches of the arteria profunda femoris.

The Popliteal Fossa (5.100, 6.134, 7.259, 7.260)

This is a rhomboidal region posterior to the knee joint, more apparent when disturbed by dissection. *Lateral* are proximally the biceps femoris and distally the plantaris and lateral head of the gastrocnemius; *medial* and proximally are the semitendinosus and semimembranosus, and distally the medial head of the gastrocnemius; *anterior* are the femoral popliteal surface, oblique popliteal ligament, back of the proximal end of the tibia and the fascia covering the popliteus, collectively forming a so-called floor. The fossa is covered *posteriorly* by the popliteal fascia. (Note that 'popliteal fascia' refers to part of the general investing layer of deep fascia that forms a 'roof' for the fossa; to be carefully distinguished from the 'fascia of popliteus' which forms part of the floor.)

Contents (6.134, 7.260). Until disturbed, the popliteal fossa is about 2.5 cm wide and its contents are largely hidden, especially in its distal part, where the heads of gastrocnemius are in contact. When its boundaries are separated its contents are seen to be the popliteal vessels, the tibial and common peroneal nerves, the small saphenous vein, posterior femoral cutaneous nerve, an obturator articular branch, lymph nodes and fat. The tibial nerve descends centrally immediately anterior to the popliteal fascia, crossing the vessels *posteriorly* from lateral to medial. The common peroneal nerve descends laterally near the tendon of biceps femoris. Popliteal vessels are deep on the floor, the vein superficial to the artery, and united by dense areolar tissue. The vein is thick-walled, proximally lateral to the artery, and crossing to its medial side distally; sometimes it is double with the artery between the veins, the latter usually being interconnected. An articular branch from the obturator nerve descends on the artery to the knee. Six or seven popliteal lymph nodes are embedded in the fat, one under the popliteal fascia near the end of the small saphenous vein, one between the popliteal artery and knee joint, others around the popliteal vessels.

The Popliteal Artery

The popliteal artery (6.133,134,135), continuing the femoral, traverses the popliteal fossa; from the opening in adductor magnus it descends laterally to the intercondylar fossa, inclining *obliquely* to the distal border of the popliteus, where it divides into the *anterior* and *posterior tibial arteries* (6.107). This division is at the proximal end of the crural interosseous space (which is asymmetrical) between the wide tibial metaphysis and the slender fibular metaphysis. Thus the popliteal artery extends from the *medial* border of the femur to the laterally placed interosseous space, accounting for its oblique descent (6.135).

Relations. Anterior, proximodistally, are fat covering the femoral popliteal surface, the capsule of the knee joint, and the fascia of popliteus. *Posterior* are, proximally, the semimembranosus and, distally, the gastrocnemius and plantaris. At intermediate level the artery is separated from the skin and fasciae by fat and crossed from lateral to medial by the tibial nerve and popliteal vein, the vein being between the nerve and artery and adherent to the latter. *Lateral* are proximally the biceps femoris, tibial nerve, popliteal vein and lateral femoral condyle and distally the plantaris and lateral head of gastrocnemius. *Medial* are the semimembranosus and medial femoral condyle and distally the tibial nerve, popliteal vein and medial head of gastrocnemius. Relations of the popliteal lymph nodes are described on p. 848.

Variations. The artery may divide into terminal branches *proximal* to the popliteus, the anterior tibial artery then descending *anterior* to the muscle. It sometimes divides into the anterior tibial and peroneal arteries, the posterior tibial being absent or rudimentary; it may divide into the anterior and posterior tibial and peroneal.

Surface Anatomy. The popliteal artery is approximately represented as extending from the junction of the middle and lower thirds of the thigh, 2.5 cm medial to its posterior midline, to

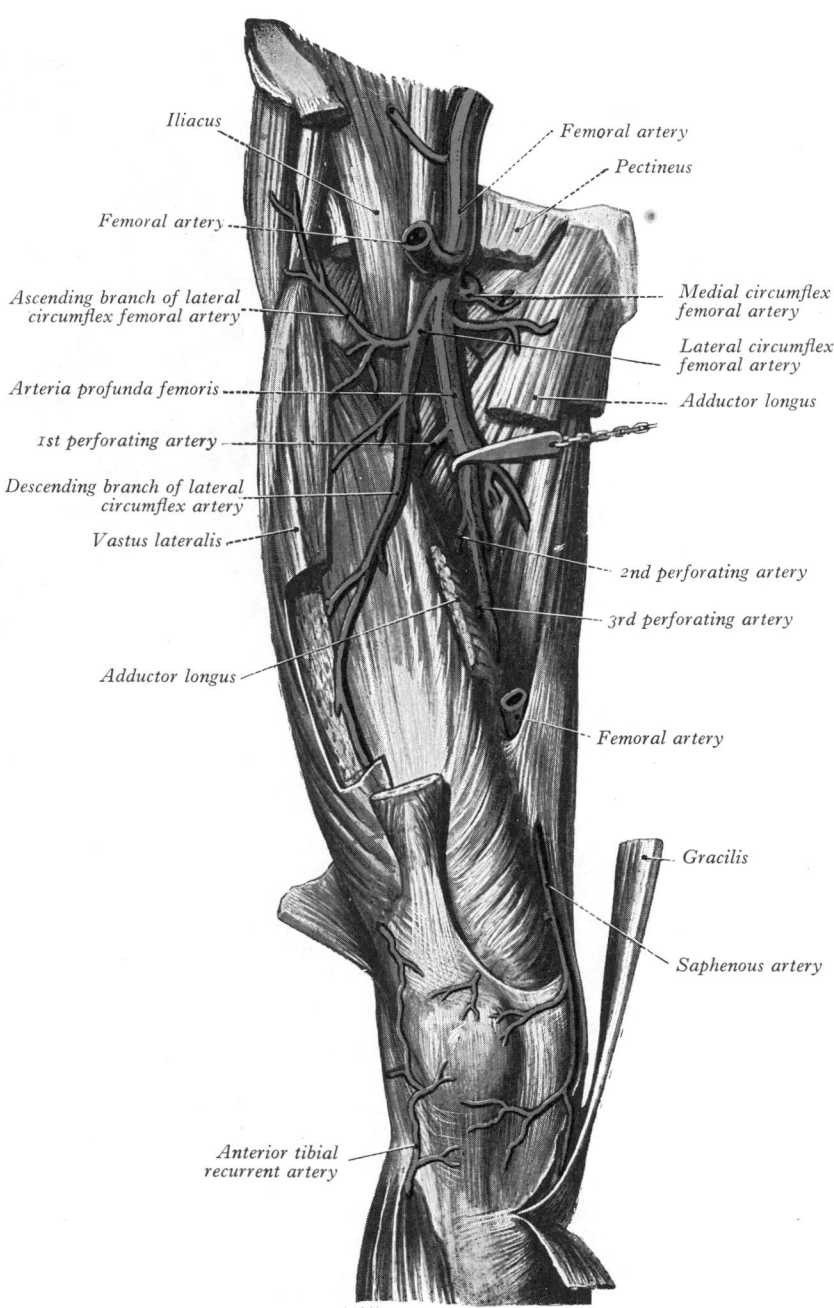

Iliacus

Femoral artery

Pectineus

Femoral artery

Ascending branch of lateral circumflex femoral artery

Medial circumflex femoral artery

Lateral circumflex femoral artery

Arteria profunda femoris

Adductor longus

1st perforating artery

Descending branch of lateral circumflex artery

Vastus lateralis

2nd perforating artery

3rd perforating artery

Adductor longus

Femoral artery

Gracilis

Saphenous artery

Anterior tibial recurrent artery

A.K.MAXWELL.

6.132 The right profunda femoris artery and its branches.

the midpoint between the femoral condyles, continuing inferolaterally to the level of the tibial tuberosity, medial to the fibular neck.

Branches are the cutaneous, muscular and genicular branches which reach the tibiofibular interosseous gap.

Cutaneous branches leave the popliteal or its rami, descend between the heads of gastrocnemius and perforate the deep fascia to supply the skin on the *flexor* aspect of the leg; one usually accompanies the small saphenous vein.

Superior muscular branches (two or three) arise proximally and pass to the adductor magnus and femoral flexors, anastomosing with the termination of the arteria profunda femoris. **Sural arteries** (two in number, and large) arise behind the knee joint to supply gastrocnemius, soleus and plantaris.

The superior genicular arteries (6.134,136) diverge from the popliteal, curving round proximal to both femoral condyles, to the anterior aspect of the knee joint. The *medial superior genicular artery* lies under semimembranosus and semitendinosus, proximal to the medial head of gastrocnemius and deep to the tendon of adductor magnus. It divides into a branch to the vastus medi-alis which anastomoses with the descending genicular and

6.133 Transverse section through the right thigh, 4 cm above the adductor tubercle of the femur: superior (proximal) aspect. About three-fifths of the natural size.

medial inferior genicular arteries, and one ramifying on the femur and anastomosing with the lateral superior genicular artery. Its size varies inversely with that of the descending genicular. The *lateral superior genicular artery* passes under the tendon of biceps femoris, dividing into superficial and deep branches; the superficial supplies the vastus lateralis, anastomosing with the descending branch of the lateral circumflex femoral and lateral inferior genicular; the deep branch anastomoses with the medial superior genicular, forming an anterior arch across the femur with the descending genicular.

The **middle genicular artery** is small and arises from the popliteal near the posterior centre of the knee joint; it pierces the oblique popliteal ligament to supply the cruciate ligaments and synovial membrane.

The **inferior genicular arteries** (**6.**134,136) arise from the popliteal deep to the gastrocnemius. The *medial* is deep to its medial head, descending along the proximal margin of the popliteus, which it supplies, and passing inferior to the medial tibial condyle and under the tibial collateral ligament, at the anterior border of which it ascends anteromedial to the joint; it supplies this and the tibia, anastomosing with the lateral inferior and medial superior genicular arteries and also with the anterior tibial recurrent artery and saphenous branch of the descending genicular. The *lateral inferior genicular artery* runs laterally across the popliteus and forwards over the fibula's head to the front of the knee joint, passing under the lateral head of gastrocnemius, the fibular collateral ligament and tendon of biceps femoris. Its branches anastomose with the medial inferior and lateral superior genicular, anterior and posterior tibial recurrent, and circumflex fibular arteries.

The Genicular Anastomosis

Around the patella and femoral and tibial condyles an intricate anastomosis exists. A *superficial network* spreads between the fascia and skin around the patella and in the fat deep to the ligamentum patellae. A *deep network* lies on the femur and tibia near the adjoining articular surfaces, supplying the bone and marrow, the articular capsule and synovial membrane. The vessels involved are the medial and lateral genicular, descending genicular, the descending branch of the lateral circumflex femoral, circumflex fibular and the anterior and posterior tibial recurrent arteries.

The Anterior Tibial Artery (6.134,135,137,138)

This terminal branch of the popliteal arises at the distal border of the popliteus. At first in the flexor compartment, it passes between the heads of tibialis posterior and through the oval aperture in the proximal part of the interosseous membrane (p. 534) to the extensor region, passing medial to the fibular neck. Descending anteriorly on the membrane it approaches the tibia and, distally, lies anterior to it (**6.**139). At the ankle it is midway between the malleoli, continuing on the dorsum of the foot as the *arteria dorsalis pedis*.

Relations. In its proximal two-thirds the artery descends on the interosseous membrane, in its distal third anterior to the tibia and ankle joint. Proximally it is between the tibialis anterior and extensor digitorum longus, then between the tibialis anterior and extensor hallucis longus. At the ankle it is crossed superficially from the lateral side by the tendon of extensor hallucis longus and is then between this and the first tendon of the extensor digitorum longus. Its proximal two-thirds is covered by adjoining muscles and deep fascia, its distal third by the skin, fasciae and extensor retinacula. Venae comitantes accompany it. The deep peroneal nerve, curling laterally round the fibular neck, reaches the lateral side of the artery where it enters the extensor region but in the middle third of the leg becomes anterior to it and distally again becomes lateral.

Surface Anatomy. Surface projection of the anterior tibial artery begins 2.5 cm distal to the medial side of the fibular head and ends midway between the malleoli. It can be felt pulsating lateral to the tendon of extensor hallucis longus.

Variations. This vessel may be small or even absent, replaced by perforating branches from the posterior tibial or the perforating branch of the peroneal. It occasionally deviates laterally, regaining its usual position at the ankle.

Branches are the anterior and posterior tibial recurrent, anterior medial and lateral malleolar and muscular.

The **posterior tibial recurrent artery**, an inconstant branch, arises before the anterior tibial reaches the extensor compartment, ascending anterior to the popliteus with the muscle's recurrent nerve, anastomosing with the inferior genicular branches of the popliteal. It supplies the superior tibiofibular joint.

The **anterior tibial recurrent artery** (**6.**137) arises near the preceding vessel, ascends in tibialis anterior, ramifies on the front and sides of the knee joint and joins the patellar network anastomosing with the genicular branches of the popliteal and circumflex fibular arteries.

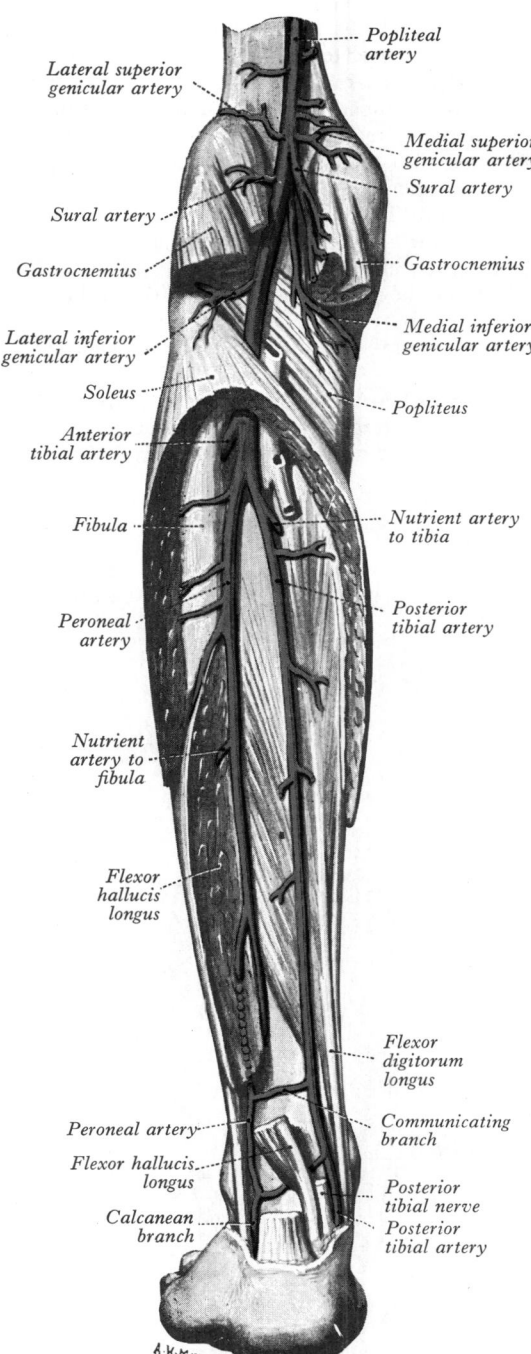

6.134 The left popliteal, posterior tibial and peroneal arteries: dorsal aspect.

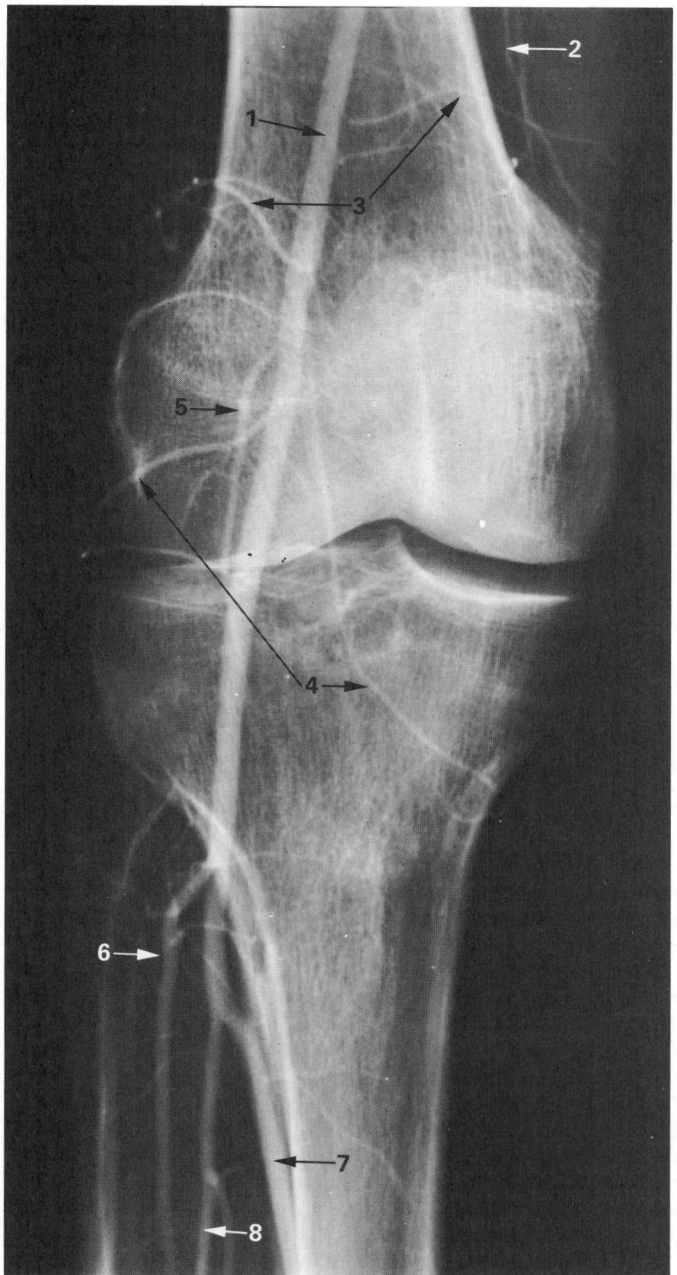

6.135 Popliteal arteriogram: anteroposterior view of adult male of 63 years. The following arteries can be identified: 1. popliteal. 2. descending genicular. 3. superior medial and lateral genicular. 4. inferior medial and lateral genicular. 5. middle genicular. 6. anterior tibial. 7. posterior tibial. 8. peroneal. Note the (normal) obliquity of the popliteal artery.

Numerous muscular branches supply the adjacent muscles; some pierce the deep fascia to supply the skin, others traverse the interosseous membrane to anastomose with branches of the posterior tibial and peroneal arteries.

The anterior medial malleolar artery (6.137,140) arises about 5 cm proximal to the ankle, passing posterior to the tendons of extensor hallucis longus and tibialis anterior medial to the joint, where it joins branches of the posterior tibial and medial plantar arteries.

The anterior lateral malleolar artery (6.137,140) proceeds posterior to the tendons of extensor digitorum longus and peroneus tertius to the lateral side of the ankle, anastomosing with the perforating branch of the peroneal and ascending rami of the lateral tarsal artery.

Anastomosis at the ankle joint (6.140) consists of networks round the malleoli. The *medial malleolar network* is formed by the anterior medial malleolar branch of the anterior tibial, medial tarsal branches of the arteria dorsalis pedis, the malleolar and calcanean branches of the posterior tibial and branches of the medial plantar artery. The *lateral malleolar network* is formed by the anterior lateral malleolar branch of the anterior tibial, lateral tarsal branch of arteria dorsalis pedis, the perforating and calcanean branches of the peroneal and rami of the lateral plantar artery.

The **arteria dorsalis pedis** (*dorsal artery of the foot*) (6.137), the continuation of the anterior tibial distal to the ankle, passes medially along the dorsum to the proximal end of the first intermetatarsal space, where it turns into the sole between the heads of the first dorsal interosseous muscle to complete the plantar arch, where it provides the *first plantar metatarsal artery*.

Relations. The dorsal artery successively crosses the talocrural articular capsule, talus, navicular and intermediate cuneiform, and their ligaments; superficial are the skin, fasciae, inferior extensor retinaculum and, near its termination, extensor hallucis brevis. *Medial* is the tendon of extensor hallucis longus, *lateral* the medial tendon of the extensor digitorum longus and medial terminal branch of the deep peroneal nerve.

787

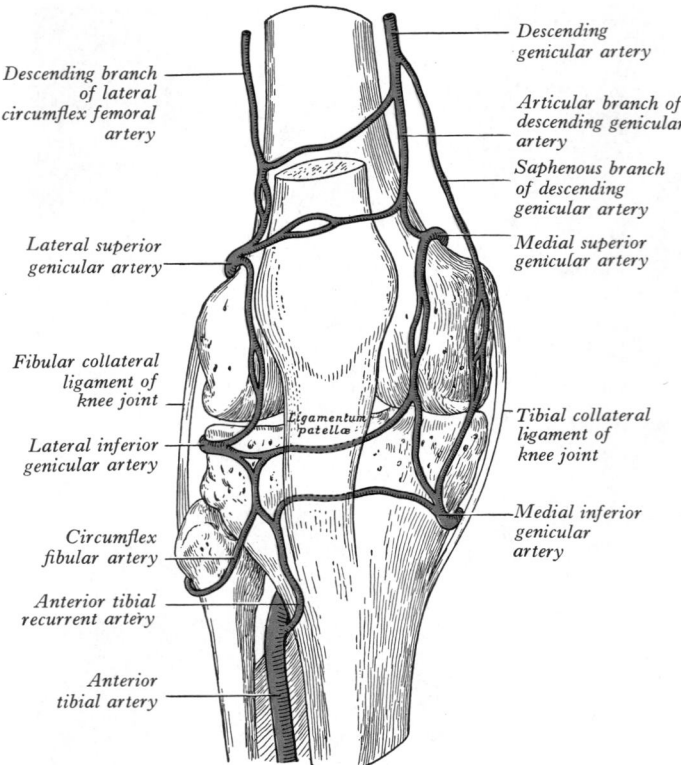

6.136 The arterial anastomosis around the knee joint (schematic).

Surface Anatomy. The vessel's pulsation is palpable from the midpoint between the malleoli to the proximal end of the first intermetatarsal space.

Variations. The artery may be larger to compensate for a small lateral plantar artery or replaced by a large perforating branch of the peroneal. It often diverges laterally from its usual route.

Branches: tarsal, arcuate, and first dorsal metatarsal arteries.

The tarsal arteries, lateral and medial (**6**.137), arise as the arteria dorsalis pedis crosses the navicular; the former runs laterally under the extensor digitorum brevis; it supplies this and the tarsal articulations, anastomosing with branches of the arcuate, anterior lateral malleolar, lateral plantar and the perforating branch of the peroneal. Two or three medial tarsal arteries ramify on the foot's medial border and join the medial malleolar network.

The arcuate artery (**6**.137) arises near the medial cuneiform, passing laterally over the metatarsal bases, deep to the tendons of the digital extensors, and anastomosing with the lateral tarsal and plantar arteries. It supplies the *second* to *fourth dorsal metatarsal arteries*, running distally superficial to the corresponding dorsal interosseous muscles; in the interdigital clefts each divides into two *dorsal digital branches* for the adjoining toes. Proximally, in the interosseous spaces, they receive *proximal perforating branches* from the plantar arch and distally are joined by the *distal perforating branches* from the plantar metatarsal arteries. The fourth dorsal metatarsal sends a branch to the lateral side of the fifth toe.

The first dorsal metatarsal artery arises just before the arteria dorsalis pedis enters the sole; it runs distally on the first dorsal interosseous; at the cleft between the first and second toes it divides, one branch passing under the tendon of extensor hallucis longus and supplying the medial side of the hallux and one bifurcating to supply the adjoining sides of hallux and the second toe.

The Posterior Tibial Artery (6.134,135,138)

This begins at the distal border of the popliteus, between the tibia and fibula, descending medially in the flexor compartment to divide midway between the medial malleolus and the medial tubercle of calcaneus, under abductor hallucis, into the *medial* and *lateral plantar arteries.*

Relations. The artery is successively *posterior* to tibialis posterior, flexor digitorum longus, tibia and ankle joint. Proximally, gastrocnemius, soleus and the deep transverse fascia of the leg are

superficial and distally only the skin and fascia. It is parallel with and about 2.5 cm anterior to the medial border of the tendo calcaneus; terminally it is *deep* to the flexor retinaculum and abductor hallucis. It is accompanied by two veins and the tibial nerve, the latter first medial, but soon crossing posterior, and then largely posterolateral. The arrangement of structures passing from the leg to the sole is described on p. 651.

Surface Anatomy. The posterior tibial artery corresponds to a line joining a point 1–2 cm lateral to the calf's midline at the fibular neck's level, extending downwards and medially to the midpoint between the medial malleolus and the heel (medial calcaneal tubercle).

Branches are: circumflex fibular, peroneal, nutrient, medial and lateral plantar.

The circumflex fibular artery (sometimes from the *anterior* tibial artery) passes laterally round the fibula's neck through the soleus to anastomose with the lateral inferior genicular, medial

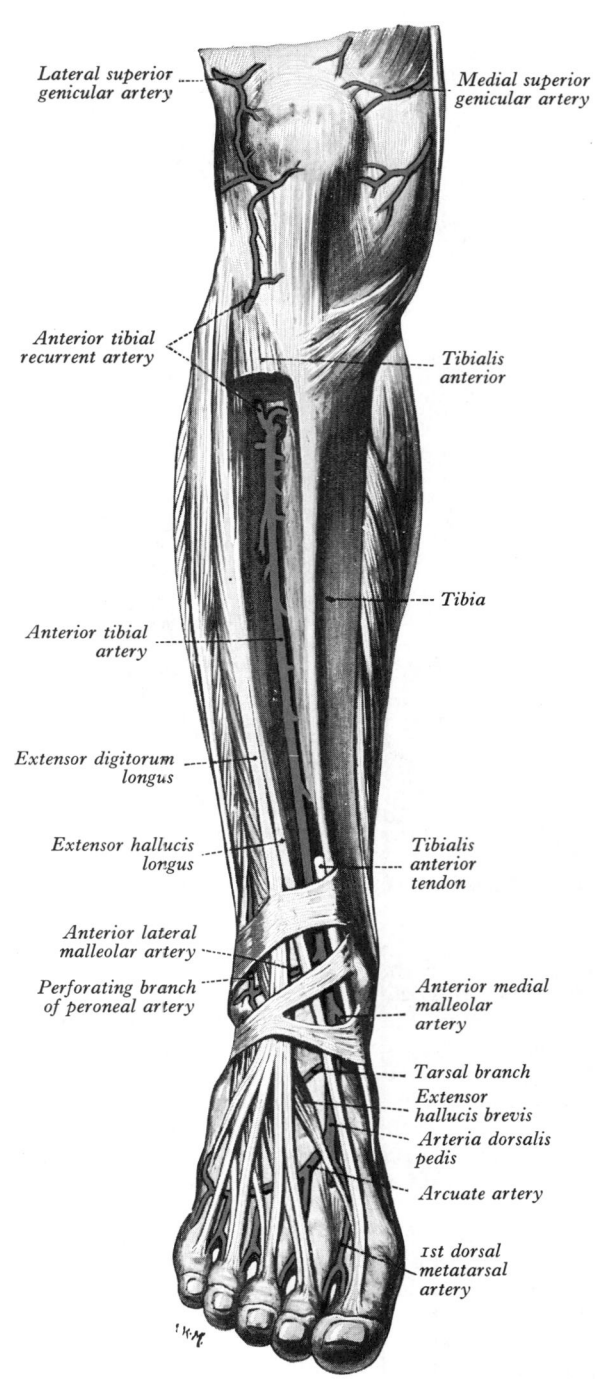

6.137 The right anterior tibial and dorsalis pedis arteries. A large part the tibialis anterior has been excised and the extensor hallucis longus retracted laterally to expose the anterior tibial artery.

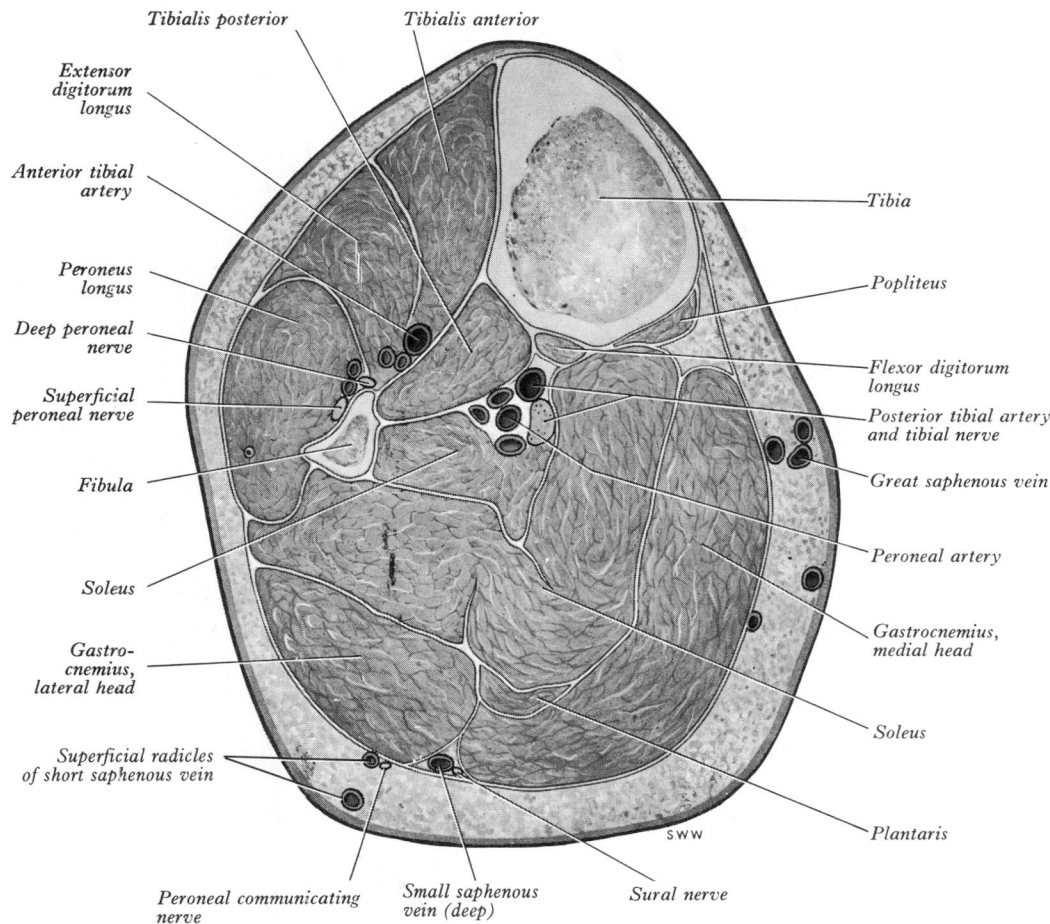

6.138 Transverse section through the right leg, about 10 cm below the knee joint: inferior (distal) aspect. At a slightly lower level the flexor digitorum longus intervenes between the soleus and the fascia on the posterior surface of the tibialis posterior.

genicular and anterior tibial recurrent arteries. It supplies bone and articular structures.

The peroneal artery (6.134), arising about 2.5 cm distal to popliteus, passes obliquely to the fibula, descending along its medial crest in a fibrous canal between tibialis posterior and flexor hallucis longus or in the latter. Reaching the inferior tibiofibular syndesmosis it divides into the calcanean branches, ramifying on the lateral and posterior surfaces of the calcaneus. *Proximally* it is covered by the soleus and deep transverse fascia, between this and the deep muscles; *distally* it is overlapped by flexor hallucis longus.

Variations. The artery may spring earlier from the posterior tibial, or even the popliteal, sometimes 7 or 8 cm *distal* to popliteus. It is more often enlarged and either joins and reinforces the posterior tibial artery or replaces it in the distal leg and foot.

Muscular branches supply soleus, tibialis posterior, flexor hallucis longus and peronei. A *nutrient artery* runs proximally into the fibula. A *perforating branch* traverses the interosseous membrane about 5 cm proximal to the lateral malleolus to enter the extensor compartment, where it anastomoses with the anterior lateral malleolar artery; descending anterior to the inferior tibiofibular syndesmosis, it supplies the tarsus, anastomosing with the lateral tarsal artery. This branch is sometimes enlarged and may replace the arteria dorsalis pedis. A *communicating branch* connects it about 5 cm proximal to the ankle to a *communicating branch* of the posterior tibial. The *calcanean* or terminal branches anastomose with the anterior lateral malleolar and calcanean branches of the posterior tibial artery.

The nutrient artery of the tibia arises from the posterior tibial near its origin; supplying a few muscular rami it descends into bone immediately distal to the soleal line. It is one of the largest of the nutrient arteries. Muscular branches are distributed to the soleus and deep flexors of the leg. The

communicating branch of the posterior tibial runs posteriorly across the tibia about 5 cm above its distal end, deep to flexor hallucis longus, to join a communicating branch of the peroneal. Medial malleolar branches pass round the tibial malleolus to the medial malleolar network. Calcanean branches arise just

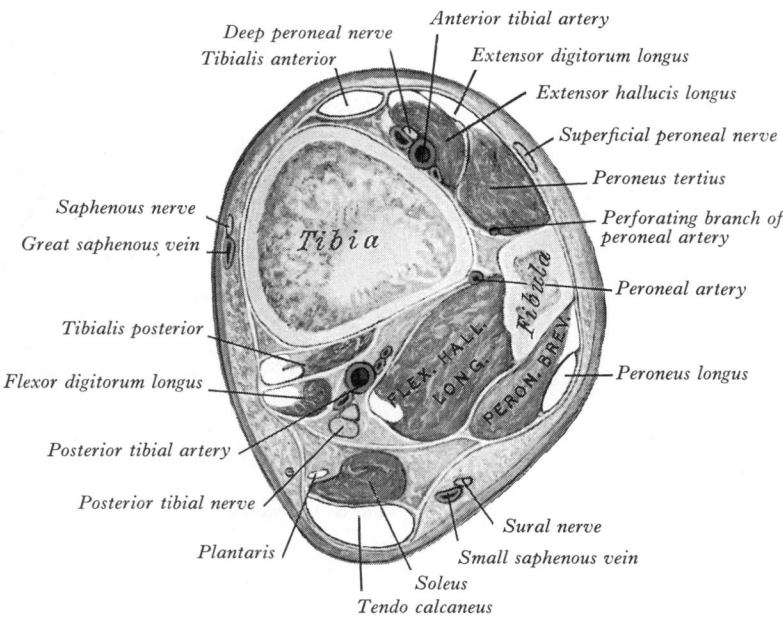

6.139 Transverse section through the right leg, about 6 cm above the tip of the medial malleolus: superior (proximal) aspect.

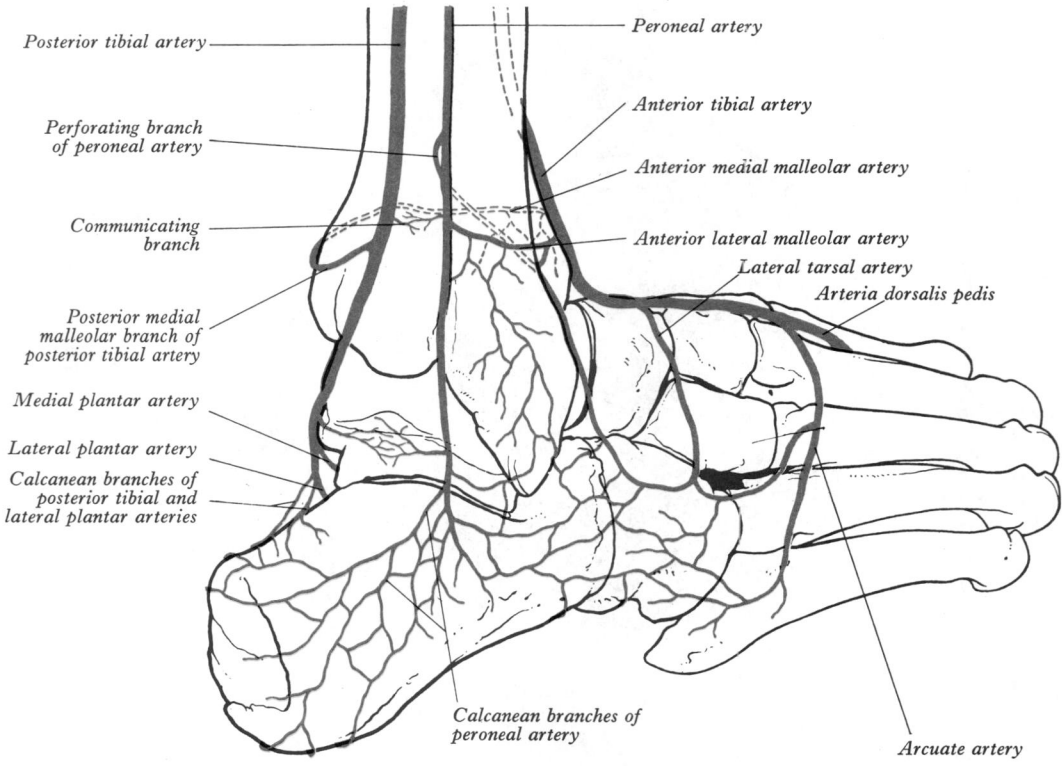

Posterior tibial artery

*Perforating branch
of peroneal artery*

*Communicating
branch*

*Posterior medial
malleolar branch of
posterior tibial artery*

Medial plantar artery

Lateral plantar artery

*Calcanean branches of
posterior tibial and
lateral plantar arteries*

Peroneal artery

Anterior tibial artery

Anterior medial malleolar artery

Anterior lateral malleolar artery

Lateral tarsal artery

Arteria dorsalis pedis

*Calcanean branches of
peroneal artery*

Arcuate artery

6.140 The arterial anastomoses of the ankle, tarsus and metatarsus.

proximal to the terminal division of the posterior tibial; they pierce the flexor retinaculum to supply the fat and skin behind the tendo calcaneus and in the heel and muscles on the tibial side of the sole; they anastomose with medial malleolar arteries and calcanean branches of the peroneal.

The medial plantar artery (6.141,142), the smaller terminal branch of the posterior tibial, passes distally along the medial side of the foot with the medial plantar nerve lateral to it. At first deep to abductor hallucis, it runs distally between this and flexor digitorum brevis, supplying both. Near the first metatarsal base,

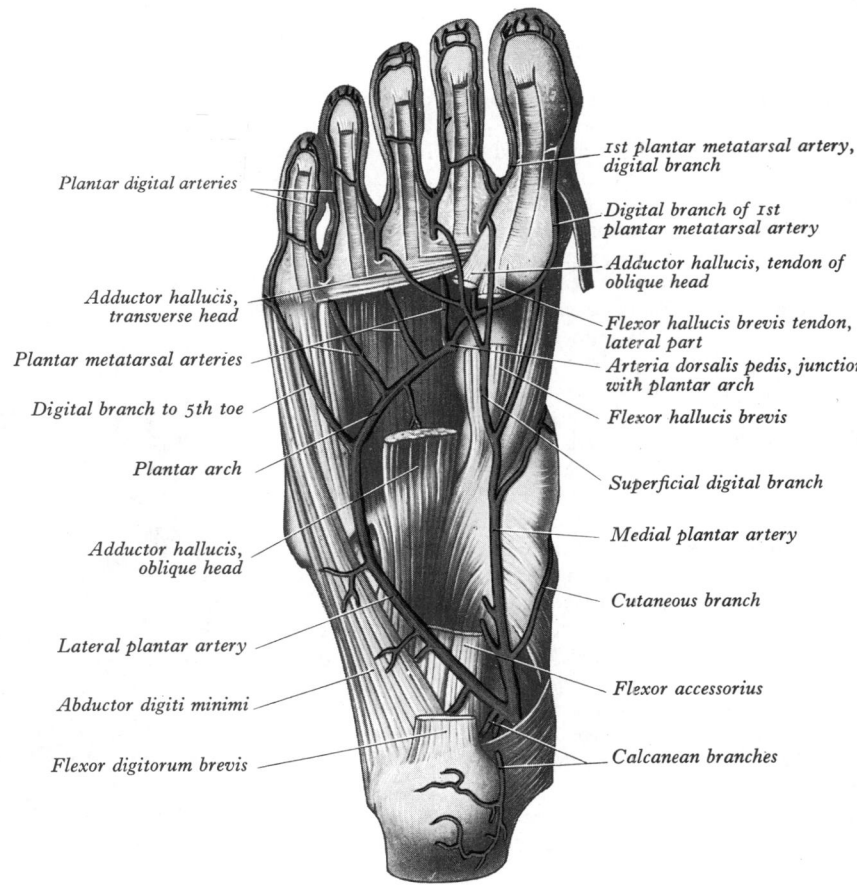

Plantar digital arteries

*Adductor hallucis,
transverse head*

Plantar metatarsal arteries

Digital branch to 5th toe

Plantar arch

*Adductor hallucis,
oblique head*

Lateral plantar artery

Abductor digiti minimi

Flexor digitorum brevis

*1st plantar metatarsal artery,
digital branch*

*Digital branch of 1st
plantar metatarsal artery*

*Adductor hallucis, tendon of
oblique head*

*Flexor hallucis brevis tendon,
lateral part*

*Arteria dorsalis pedis, junction
with plantar arch*

Flexor hallucis brevis

Superficial digital branch

Medial plantar artery

Cutaneous branch

Flexor accessorius

Calcanean branches

6.141 The plantar arteries of the right foot.

its calibre, diminished by muscular branches, is further diminished by a *superficial stem* and it then passes to reach the medial border of the hallux where it anastomoses with a branch of the first plantar metatarsal artery (vide infra). Its superficial stem trifurcates and supplies three *superficial digital branches* accompanying the digital branches of the medial plantar nerve to join the first to third plantar metatarsal arteries.

Surface Anatomy. The trunk of the artery begins midway between the medial malleolus and heel (medial calcaneal tubercle) extending towards the first interdigital cleft as far as the navicular bone.

The lateral plantar artery (6.141), the larger terminal branch of the posterior tibial, passes distally and laterally to the *fifth* metatarsal base, the lateral plantar nerve medial to it. (Note that the plantar nerves lie *between* the plantar arteries.) Turning medially, with the nerve's deep branch, to the interval between the first and second metatarsal bases, it unites with the arteria dorsalis pedis to complete the *plantar arch*. As it passes laterally, it is first between the calcaneus and abductor hallucis, then between flexor digitorum brevis and flexor accessorius; running distally to the fifth metatarsal base it passes between flexor digitorum brevis and abductor digiti minimi and is covered by the plantar aponeurosis, superficial fascia and skin.

Branches. *Muscular branches* supply the adjoining muscles; *superficial branches* emerge along the lateral intermuscular septum to supply the skin and subcutaneous tissue lateral in the sole; *anastomotic branches* run to the lateral border, joining branches of the lateral tarsal and arcuate arteries. Sometimes a *calcanean* branch pierces abductor hallucis to supply the skin of the heel.

The Plantar Arch

This is deeply situated, extending from the fifth metatarsal base to the proximal end of the first interosseous space. Convex distally, it is plantar to the bases of the second to fourth metatarsal bones and corresponding interossei but dorsal to the oblique part of adductor hallucis.

Branches. Three perforating and four plantar metatarsal branches, and numerous rami supply the skin, fasciae and muscles in the sole. *Three perforating branches* ascend through the proximal ends of the second to fourth intermetatarsal spaces, between the heads of the dorsal interosseous muscles, anastomosing with the dorsal metatarsal arteries. *Four plantar metatarsal arteries* (6.141) extend distally between the metatarsal bones in contact with the interossei. Each divides into two *plantar digital arteries*, supply-

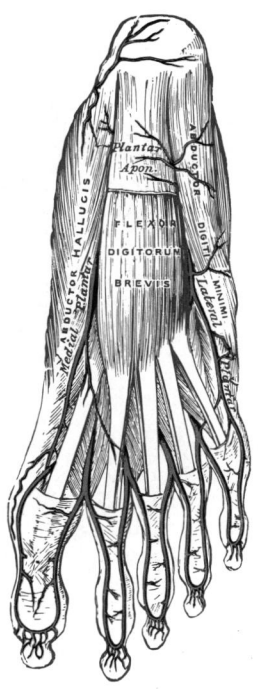

6.142 The plantar arteries: superficial dissection.

ing the adjacent digital aspects. Near its division each plantar metatarsal sends dorsally a *distal perforating branch* to join a dorsal metatarsal artery. The *first plantar metatarsal* artery springs from the junction between the lateral plantar and dorsalis pedis arteries, sending a digital branch to the medial side of the hallux. The lateral digital branch for the fifth toe arises directly from the lateral plantar artery near the fifth metatarsal base.

Surface Anatomy. Beginning between the heel and medial malleolus, the lateral plantar artery crosses obliquely to a point 2.5 cm medial to the fifth metatarsal's tuberosity and with a slight distal convexity reaches the proximal end of the first inter-metatarsal space.

Applied Anatomy. Haemorrhage from the plantar arch is difficult to stem, due to the depth of the vessel and its important close relations. It must be treated like the palmar arches (p. 764).

THE VENOUS SYSTEM

Veins are divisible into three main systems: pulmonary, systemic and portal.

Pulmonary veins contain oxygenated blood, returning from the lungs to the heart. **Systemic veins** return venous blood to the heart from much of the rest of the body. *Superficial veins* are located in superficial fascia and are variable in disposition. *Deep veins* are beneath the deep fascia and are usually enclosed in connective tissue sheaths with accompanying arteries, the latter assisting venous return (p. 691). Smaller arteries are accompanied by *venae comitantes*, paired veins flanking them; larger arteries usually accompany single veins. Some however, run separately. Systemic veins are usually more variable than arteries; anastomoses are less frequent between larger veins. In many regions, such as the pelvis and vertebral column, veins form extensive plexuses devoid of valves. These plexuses are the basis of anastomosis between the veins of the trunk; they may also act as reservoirs of variable capacity. At many points, such as the junctional regions between the trunk and limbs and near joints, valved *connecting veins* join superficial and deep systemic veins.

The portal vein receives tributaries draining venous blood from almost the whole subdiaphragmatic intestinal tract and associated extrinsic glands and spleen: the whole of this blood passes through the hepatic circulation before returning to general systemic venous channels.

Pulmonary Veins

Pulmonary veins return oxygenated blood to the left atrium. Usually four, two from each lung and destitute of valves, they commence in capillary networks in the alveolar walls. By repeated junctions tributary veins finally form a single trunk in each lobe, i.e. three in the right lung, and two in the left. The right middle and superior lobar veins usually join so that two veins, superior and inferior, leave each lung; they perforate the fibrous pericardium and open separately and posterosuperiorly into the left atrium (6.31,35). Occasionally the three right lobar veins remain separate. Sometimes the two left pulmonary veins form a single trunk. Occasionally the two left pulmonary veins, each draining a lobe, may be augmented by an *accessory lobar vein* from each lobe and these may unite to form a *third* left pulmonary vein (Corry & Valentine 1959).

In the pulmonary hilum (pp. 1275, 1284) the superior pulmonary vein is antero-inferior to the pulmonary artery, the inferior being the most inferior hilar structure and also slightly posterior. The principal bronchus is posterior to the pulmonary artery. On the right the superior pulmonary vein passes posterior to the superior vena cava, the inferior behind the right atrium. On the left both pass anterior to the descending thoracic aorta. In the pericardium, they are partly covered by serous pericardium. Between the terminations of the right and left veins is, centrally, the oblique pericardial sinus and, laterally, directed medially and upwards,

smaller, variable pulmonary venous pericardial recesses (p. 696).

The Systemic Veins

Systemic veins form three groups: (1) *cardiac veins*, (2) veins of the upper limbs, head, neck and thorax, all draining to the *superior vena cava*, (3) veins of the lower limbs, abdomen and pelvis, all draining to the *inferior vena cava*.

CARDIAC VEINS

Veins draining the heart may be grouped as: (1) the *coronary sinus* and tributaries, returning blood to the right atrium from the whole heart (including its septa) except the anterior region of the right ventricle and small, variable parts of both atria and left ventricle; (2) the *anterior cardiac veins* draining an anterior part of the right ventricle and a region around the right cardiac border when the right marginal vein joins this group, ending principally in the right atrium; (3) the *venae cordis minimae* (Thebesian veins), opening into the right atrium and ventricle and, to a lesser extent, the left atrium and sometimes left ventricle.

CORONARY SINUS

Most cardiac veins drain to the wide coronary sinus, about 2 or 3 cm long, lying posterior in the coronary sulcus (atrioventricular groove) between the left atrium and ventricle (**6.**31,143A,B). It ends in the right atrium between the inferior vena caval opening and the right atrioventricular orifice, its opening displaying a semilunar *valve of the coronary sinus* (**6.**32,37). Its tributaries are the great, small and middle cardiac veins, the posterior vein of the left ventricle and the oblique vein of the left atrium, all except the last having valves at their orifices.

The great cardiac vein (**6.**143A,B) begins at the cardiac apex, ascends in the anterior interventricular sulcus to the coronary

sulcus and follows this to the left and round posterior to the heart to enter the beginning of the coronary sinus. It receives tributaries from left atrium and both ventricles, including the large *left marginal vein* ascending the left aspect ('obtuse border') of the heart.

The small cardiac vein (**6.**143A,B) is posterior in the coronary sulcus between the right atrium and ventricle and opens into the coronary sinus near its atrial end. It receives blood from the back of the right atrium and ventricle; the *right marginal vein* passes right, along the inferior cardiac margin ('acute border'), and may join the small cardiac vein in the coronary sulcus but more often opens directly into the right atrium.

The middle cardiac vein (**6.**115) begins at the cardiac apex, runs back in the posterior interventricular groove to end in the coronary sinus near its atrial end.

The posterior vein of the left ventricle (**6.**143A,B) is on the diaphragmatic surface of the left ventricle a little left of the middle cardiac vein; it usually opens into the centre of the coronary sinus but sometimes into the great cardiac vein.

The oblique vein of the left atrium, a small vessel, descends obliquely on the back of the left atrium to join the coronary sinus near its left extremity; it is continuous above with the *ligament of the left vena cava* (p. 696); the two structures are remnants of the left common cardinal vein.

ANTERIOR CARDIAC VEINS

The *anterior cardiac veins* drain the anterior part of the right ventricle. Usually two or three, sometimes even five (Baroldi & Scomazzoni 1967), they ascend in subepicardial tissue to cross the right part of the atrioventricular sulcus, passing deep or superficial to the right coronary artery. They end in the right atrium, near the sulcus, separately or in variable combinations. A

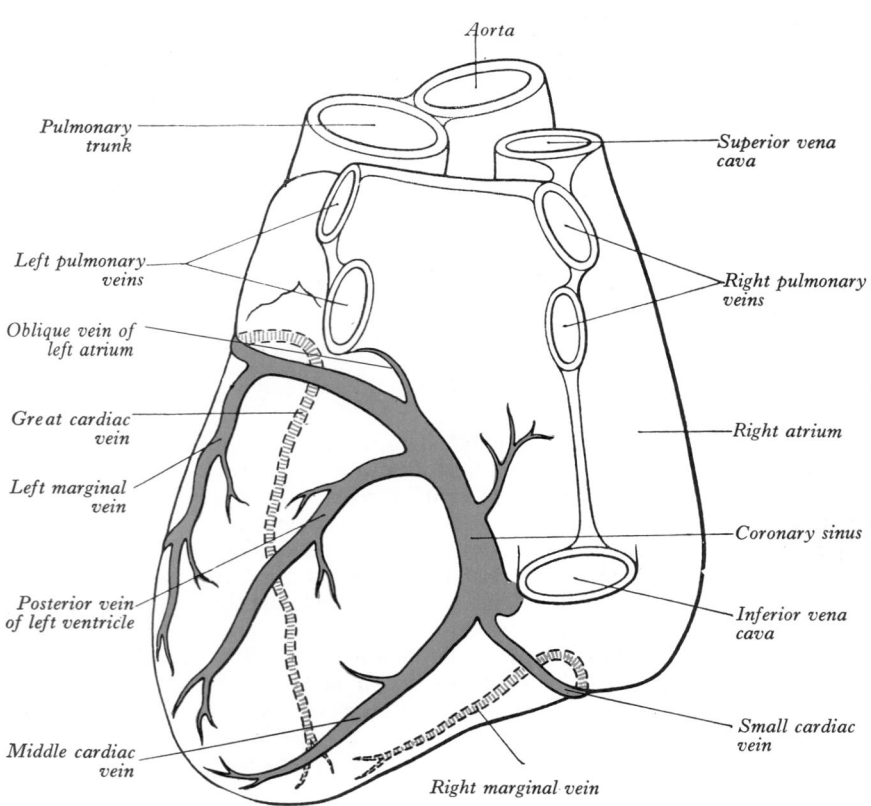

Aorta

Pulmonary trunk

Superior vena cava

Left pulmonary veins

Right pulmonary veins

Oblique vein of left atrium

Great cardiac vein

Right atrium

Left marginal vein

Posterior vein of left ventricle

Coronary sinus

Inferior vena cava

Middle cardiac vein

Small cardiac vein

Right marginal vein

792   **6.**143A The principal veins of the heart.

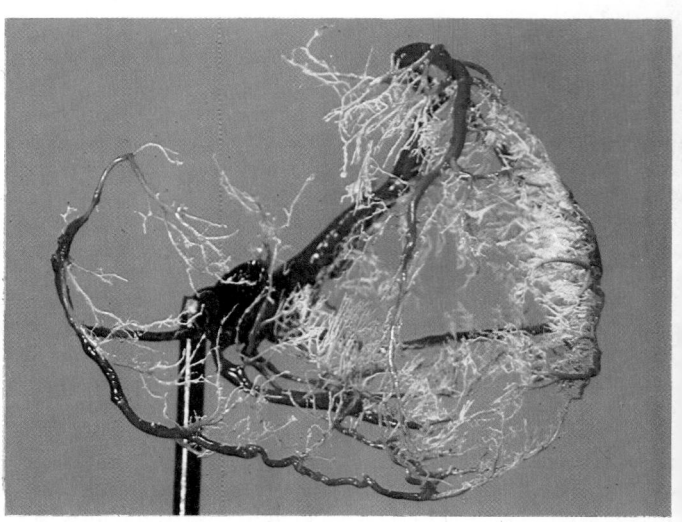

6.143B A coloured resin cast of the coronary sinus and its tributaries. Of course, the venae cordis minimae and anterior cardiac veins are not filled. Compare with **6.**143A. Specimen by D H Tompsett, Department of Anatomy, Royal College of Surgeons of England.

subendocardial collecting channel, into which all may open, has been described (James 1961). **The right marginal vein** courses along the inferior ('acute') cardiac margin, draining adjacent parts of the right ventricle, and usually opens separately into the right atrium but may join the anterior cardiac veins or, less often, the coronary sinus. Because it is commonly independent it is often grouped with the venae cordis minimae but, since it is larger in calibre, it is comparable with the anterior cardiac veins or even wider. It is perhaps better considered one of the latter, which also sometimes drain with it into the coronary sinus. Mechanik (1934) described all cardiac veins as draining into the coronary sinus in the early fetal period.

VENAE CORDIS MINIMAE

The existence of *venae cordis minimae*, opening into all cardiac cavities, has been confirmed by many subsequent to their first recording by Thebesius (1708); they are more difficult to demonstrate than larger cardiac vessels. Their numbers and size are highly variable. Aho (1950) demonstrated 'minimal' veins of up to 2 mm in diameter opening into the right atrium and about 0.5 mm into the right ventricle. He found venae minimae numerous in the right atrium and ventricle, occasional in and often absent from the left atrium, and rare in the left ventricle. Grant & Regnier (1926) considered venae minimae as derived from the intertrabecular spaces of developing mammalian hearts.

Cardiac Venous Anastomoses

Most investigators accept widespread anastomosis at all levels of cardiac venous circulation, on a scale exceeding that of the arteries and amounting to a veritable venous plexus, according to some (Baroldi & Scomazzoni 1967). Not only are adjacent veins often connected but connections also exist between tributaries of the coronary sinus and those of the anterior cardiac veins (Mierzwa & Kozielec 1975). Regions of abundant anastomoses are the apex and its anterior and posterior aspects. Like coronary arteries (p. 731) cardiac veins connect with extracardiac vessels, chiefly the vasa vasorum of the large vessels continuous with the heart.

Variation in Cardiac Veins

Attempts to categorize variations in cardiac venous circulation (Aho 1950) into 'types' have not produced any accepted pattern. Major variations concern the general directions of drainage. The coronary sinus may receive *all* cardiac veins (except the venae minimae), including the anterior cardiac veins (33%), which may be reduced by diversion of some into the small cardiac vein and then to the coronary sinus (28%); the remainder (39%) represent the 'normal' pattern, as described above. Baroldi & Scomazzoni (1967) distinguished two major variants: a majority (70%) in which the small cardiac vein is independent, small or absent and a less frequent pattern (30%) in which this vein, though variable in size, connects with both coronary and anterior cardiac 'systems'.

VEINS OF THE HEAD AND NECK

Veins of the head and neck may be grouped as: (1) veins of the exterior of head and face, (2) cervical veins, (3) diploic, meningeal, intracranial veins and dural venous sinuses.

This classification is particularly significant at cranial level, where veins, like arteries, are arranged as a three-layered system: (*a*) vessels of the scalp, (*b*) dural vessels and (*c*) cerebral and cerebellar vessels. But the arteries and veins differ, largely in interconnections between the three layers. Arteries of the scalp and dura, both from the *external* carotid, intercommunicate to a limited extent, as do corresponding veins; but the latter, though quite variable, usually communicate much more extensively (emissary veins, p. 804). Dural or meningeal arteries, on the other hand, are independent of cerebral and cerebellar arteries, the latter being derived from the *internal* carotid, whereas dural venous sinuses share a drainage to the internal jugular vein which is also common to veins of the cerebrum and cerebellum. It is possible to hypothecate a *fourth* venous tier, the diploic veins; since these, however, drain into dural veins, they are here grouped with them, following Browder & Kaplan (1976). It is to be noted that intracranial veins communicate at many points with extracranial vessels by the emissary and other veins (p. 804).

In accounts of venous sinuses they are usually described as simple arrangements with a single lumen. Developmentally they emerge as venous *plexuses*; and it is clear, from angiographic studies and corrosion casts, that *most sinuses* preserve a *plexiform* arrangement to a variable degree. Browder & Kaplan (1976), examining human venous sinuses in hundreds of corrosion casts, have observed vascular plexuses adjoining, in particular, the superior and inferior sagittal and straight sinuses and, with a lesser incidence, the transverse sinuses. Details show much individual variation; departures from 'average' patterns are frequent in earlier years, e.g. the falx cerebelli may in infancy contain large plexiform channels and venous lacunae, augmenting the occipital sinus. Such variations cannot be detailed in a general text; in any case they must be established for the individual by angiography when clinical necessity arises; but the wide variation possible in the structure of cranial venous sinuses, with their plexiform nature and wide connections with cerebral and cerebellar veins, must be emphasized. Another kind of connection may be noted here; experiment shows (Rowbotham & Little 1962, Browder & Kaplan 1976) that parts of sinuses (and even diploic veins) can be filled by

forcible internal carotid injection, suggesting the existence of arteriovenous shunts. Browder & Kaplan, by injection of the middle meningeal arteries, established a connection between these and the superior sagittal sinus at sites still unknown.

External Veins of the Head and Face

As with most superficial veins these are subject to variations, far too numerous to illustrate. Some major features are, however, relatively constant; a common pattern is shown in **6**.144.

The supratrochlear vein starts on the forehead from a venous network connected to the frontal tributaries of the superficial temporal vein. Veins from this form a single trunk, descending near the midline parallel with its fellow to the radix nasi, across which they are joined by a *nasal arch* draining the dorsum nasi. The veins then diverge, each joining a *supraorbital vein* to form the *facial vein* near the *medial canthus*. Supratrochlear veins may join, dividing again on the radix nasi to form the two facial veins.

The supraorbital vein begins near the zygomatic process of the frontal bone, connecting with radicles of the superficial and middle temporal veins. Passing *medially* above the orbital opening under orbicularis oculi, it pierces this to form the *facial vein* by joining the supratrochlear near the medial canthus. A branch through the supraorbital notch joins the superior ophthalmic vein, receiving in the notch veins from the frontal sinus and frontal diploë.

The facial vein, receiving the supratrochlear and supraorbital veins, descends obliquely near the side of the radix nasi, receding from the ala, and then turns posterolaterally below the orbital opening, passing downwards and backwards behind the facial artery, being less tortuous. It passes under zygomaticus major, risorius and platysma and then descends on to the anterior border and then the surface of masseter, crosses the body of the mandible and runs obliquely back under the platysma but superficial to the submandibular gland, digastric and stylohyoid. A little anteroinferior to the mandibular angle it is joined by the anterior division of the retromandibular vein; descending superficial to the lingual artery's loop, the hypoglossal nerve and external and internal carotid arteries, it enters the internal jugular near the greater cornu of the hyoid bone (i.e. in the upper angle of the

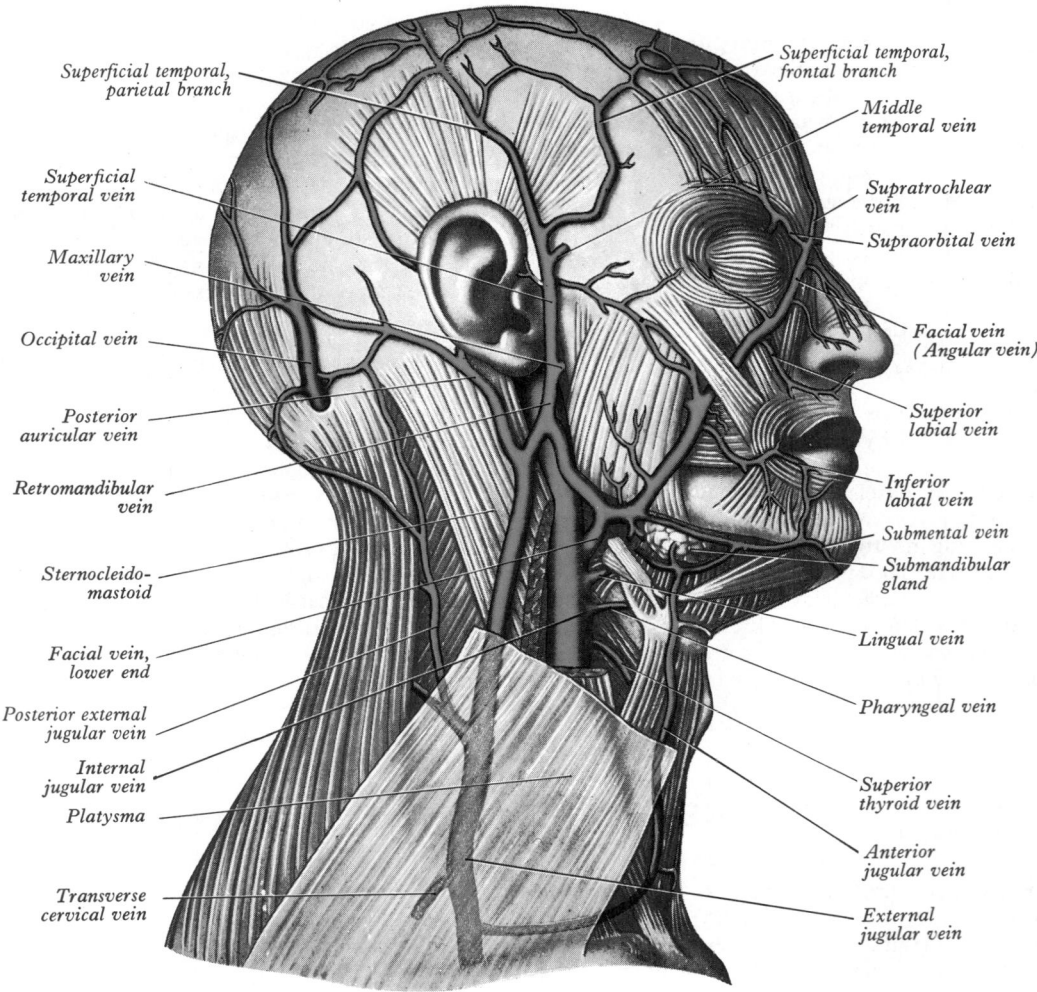

Superficial temporal, parietal branch

Superficial temporal vein

Maxillary vein

Occipital vein

Posterior auricular vein

Retromandibular vein

Sternocleido-mastoid

Facial vein, lower end

Posterior external jugular vein

Internal jugular vein

Platysma

Transverse cervical vein

Superficial temporal, frontal branch

Middle temporal vein

Supratrochlear vein

Supraorbital vein

Facial vein (Angular vein)

Superior labial vein

Inferior labial vein

Submental vein

Submandibular gland

Lingual vein

Pharyngeal vein

Superior thyroid vein

Anterior jugular vein

External jugular vein

6.144 The veins of the right side of the head and neck. Parts of the right sternocleidomastoid and platysma have been excised to expose the trunk of the internal jugular vein. The external jugular vein is visible through the lower part of the platysma.

carotid triangle). Near its end a large branch often descends along the anterior border of sternocleidomastoid to the anterior jugular vein. Its uppermost segment, above its junction with the superior labial vein (vide infra), is often termed the *angular vein*.

Tributaries. Near its beginning the facial vein connects with the superior ophthalmic directly and via the supraorbital; it is thus connected to the cavernous sinus. It receives veins of the ala nasi and, lower, a large *deep facial vein* from the pterygoid venous plexus and also the inferior palpebral, superior and inferior labial, buccinator, parotid and masseteric veins. Below the mandible, submental, tonsillar, external palatine (paratonsillar) and submandibular veins join it and sometimes the vena comitans of the hypoglossal nerve, and the pharyngeal and superior thyroid veins.

Applied Anatomy. The facial vein has no valves. It connects, as noted, with the cavernous sinus by two routes: through the ophthalmic vein or its supraorbital tributary, or by the deep facial vein to the pterygoid plexus and hence the cavernous sinus. Infection may thus spread from the face to the intracranial venous sinuses.

The superficial temporal vein (6.144) begins in a widespread network joined across the scalp to the contralateral vein and to the ipsilateral supratrochlear, supraorbital, posterior auricular and occipital veins, all draining the same network. Anterior and posterior tributaries unite above the zygoma to form the superficial temporal, joined here by the *middle temporal vein*. It crosses the posterior root of the zygoma and enters the parotid gland to join the maxillary vein, to form the *retromandibular vein*.

Tributaries are the parotid veins, rami from the temporomandibular joint, anterior auricular veins, and *transverse facial vein*.

The middle temporal vein, after receiving the *orbital vein* which is formed by the lateral palpebral veins, passes back between layers of temporal fascia, piercing this to join the superficial temporal vein.

The pterygoid venous plexus is partly between temporalis and the lateral pterygoid, and partly between the pterygoids. Sphenopalatine, deep temporal, pterygoid, masseteric, buccal, dental, greater palatine and middle meningeal veins and a branch or branches from the inferior ophthalmic are all tributaries. The plexus connects by the *deep facial vein* with the facial and with the cavernous sinus through the sphenoidal emissary foramen, foramen ovale and foramen lacerum. Its deep temporal tributaries often connect with tributaries of the anterior diploic (p. 798) and thus with the middle meningeal veins.

The maxillary vein, a short trunk, accompanies the first part of the maxillary artery; it is the confluence of veins from the pterygoid plexus, passing back between the sphenomandibular ligament and mandibular neck, uniting with the superficial temporal to form the retromandibular vein.

The retromandibular vein descends in the parotid gland between the external carotid artery and superficially the facial nerve. It divides into an anterior branch going forwards to join the facial and a posterior branch joining the posterior auricular to form the *external jugular vein*. Occasionally it is not connected to the external jugular, which is then small, the anterior jugular often being enlarged.

The posterior auricular vein (6.144) begins in a parieto-occipital network also draining into tributaries of the occipital and superficial temporal veins. It descends behind the auricle to join

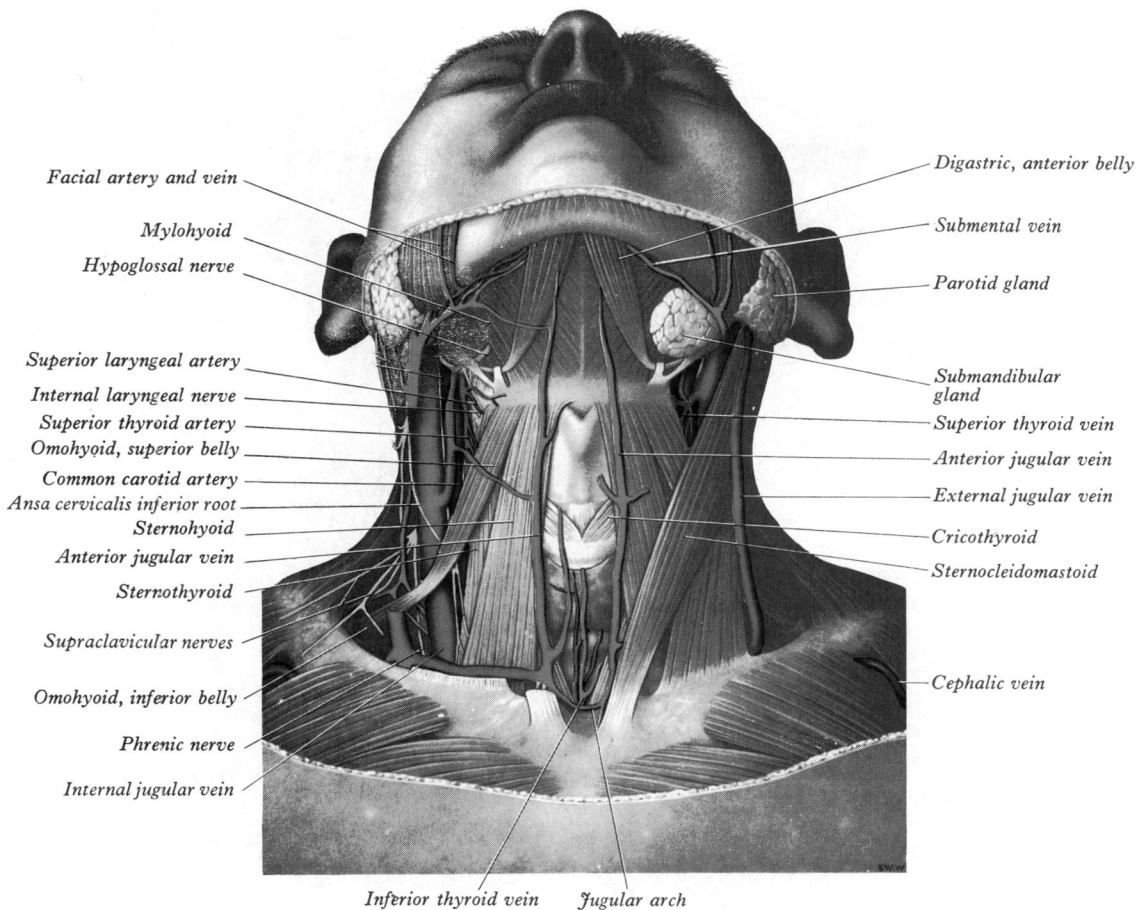

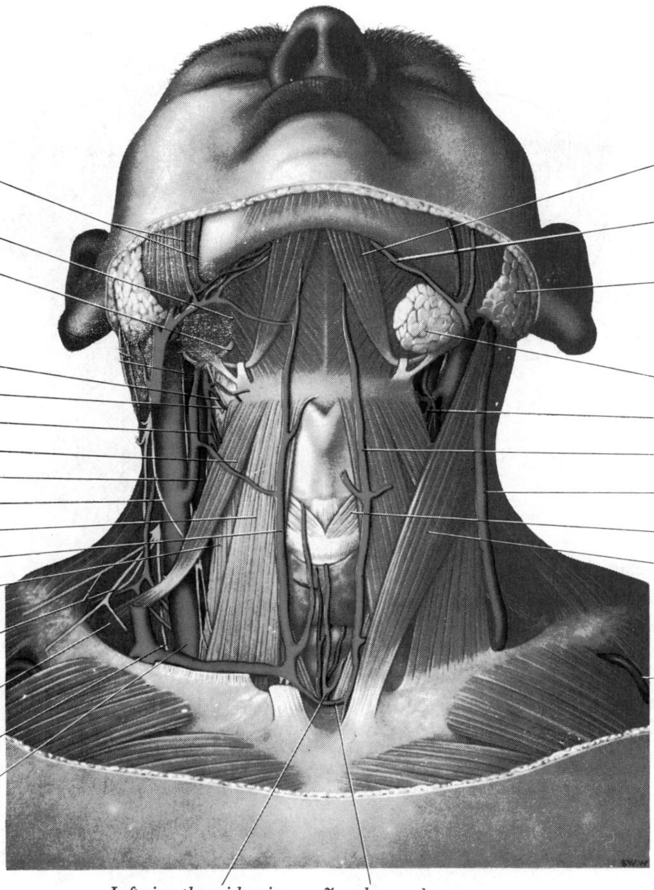
Facial artery and vein
Mylohyoid
Hypoglossal nerve
Superior laryngeal artery
Internal laryngeal nerve
Superior thyroid artery
Omohyoid, superior belly
Common carotid artery
Ansa cervicalis inferior root
Sternohyoid
Anterior jugular vein
Sternothyroid
Supraclavicular nerves
Omohyoid, inferior belly
Phrenic nerve
Internal jugular vein
Digastric, anterior belly
Submental vein
Parotid gland
Submandibular gland
Superior thyroid vein
Anterior jugular vein
External jugular vein
Cricothyroid
Sternocleidomastoid
Cephalic vein
Inferior thyroid vein
Jugular arch

6.145 Anterior view of the veins of the neck.

the posterior division of the retromandibular vein in or just below the parotid gland, to form the external jugular. It receives a stylomastoid vein and tributaries from the cranial surface of the auricle.

The occipital vein (**6**.144) begins in a posterior network in the scalp, pierces the cranial attachment of trapezius, turns into the suboccipital triangle and joins the deep cervical and vertebral veins. It may follow the occipital artery to end in the internal jugular; sometimes it joins the posterior auricular and hence the external jugular vein. Parietal and mastoid emissary veins link it with the superior sagittal and transverse sinuses. The occipital diploic vein sometimes joins it (p. 794).

Veins of the Neck (6.144,145,161)

These are divisible into those either *superficial* or *deep* to the deep fascia but these are not entirely separate. Superficial veins, tributaries (some with specific names, given below) of the external jugular, drain a much smaller volume of tissue than the deep veins, which drain all but the subcutaneous structures, mostly into the internal jugular vein (but some into the vertebral veins).

The external jugular vein (**6**.144) largely drains the scalp and face but also some deeper parts. The union of the posterior division of the retromandibular and posterior auricular veins begins near the mandibular angle just below or in the parotid gland, descending from the angle to the mid-clavicle. It crosses obliquely, superficial to sternocleidomastoid, to the subclavian triangle, where it traverses the deep fascia to end in the subclavian vein, lateral or anterior to scalenus anterior. Its wall is adherent to the rim of the fascial opening. It is covered by platysma, superficial fascia and skin, separated from sternocleidomastoid by deep cervical fascia; it crosses the transverse cervical nerve and is parallel with the great auricular nerve, ascending posterior to its upper half. In size the vein is inversely proportional to other veins in the neck; it is occasionally double. It has valves at its entrance into the subclavian vein and about 4 cm above the clavicle, between which it is often dilated, as a so-called *sinus*. The valves do not prevent regurgitation.

Tributaries. In addition to formative tributaries, the external jugular receives the posterior external jugular and, near its end, transverse cervical, suprascapular and anterior jugular veins; in the parotid gland it is often joined by a branch from the internal jugular. The occipital vein occasionally joins it.

The posterior external jugular vein begins in the occipital scalp and drains the skin and the superficial muscles posterosuperior in the neck. It usually joins the middle part of the external jugular.

The anterior jugular vein (**6**.144,145) starts near the hyoid bone by the confluence of the superficial submandibular veins. It descends between the midline and the anterior border of sternocleidomastoid; turning laterally, low in the neck, posterior (deep) to the muscle but superficial to the hyoid depressors, it joins the end of the external jugular vein or the subclavian vein directly. In size it is usually inverse to the external jugular. It communicates with the internal jugular, receiving the laryngeal veins and sometimes a small thyroid vein. There are usually two anterior jugular veins, united just above the manubrium by a large transverse *jugular arch*, receiving the inferior thyroid tributaries. They have no valves and may be replaced by a midline trunk.

Surface Anatomy. Usually the *external jugular vein* is visible as it crosses the sternocleidomastoid; to assist visibility it can be distended by expiring against resistance or by gentle supraclavicular digital pressure. Similarly the *anterior jugular vein* can often be made visible in the upper two-thirds of the neck. The end of the facial vein runs from a point, where the anterior border of the masseter meets the inferior mandibular border, to the greater hyoid cornu.

The internal jugular vein (**6**.144,145) collects blood from the skull, brain, superficial parts of face and much of the neck. It begins at the cranial base in the posterior compartment of the jugular foramen, continuous with the sigmoid sinus. At its origin

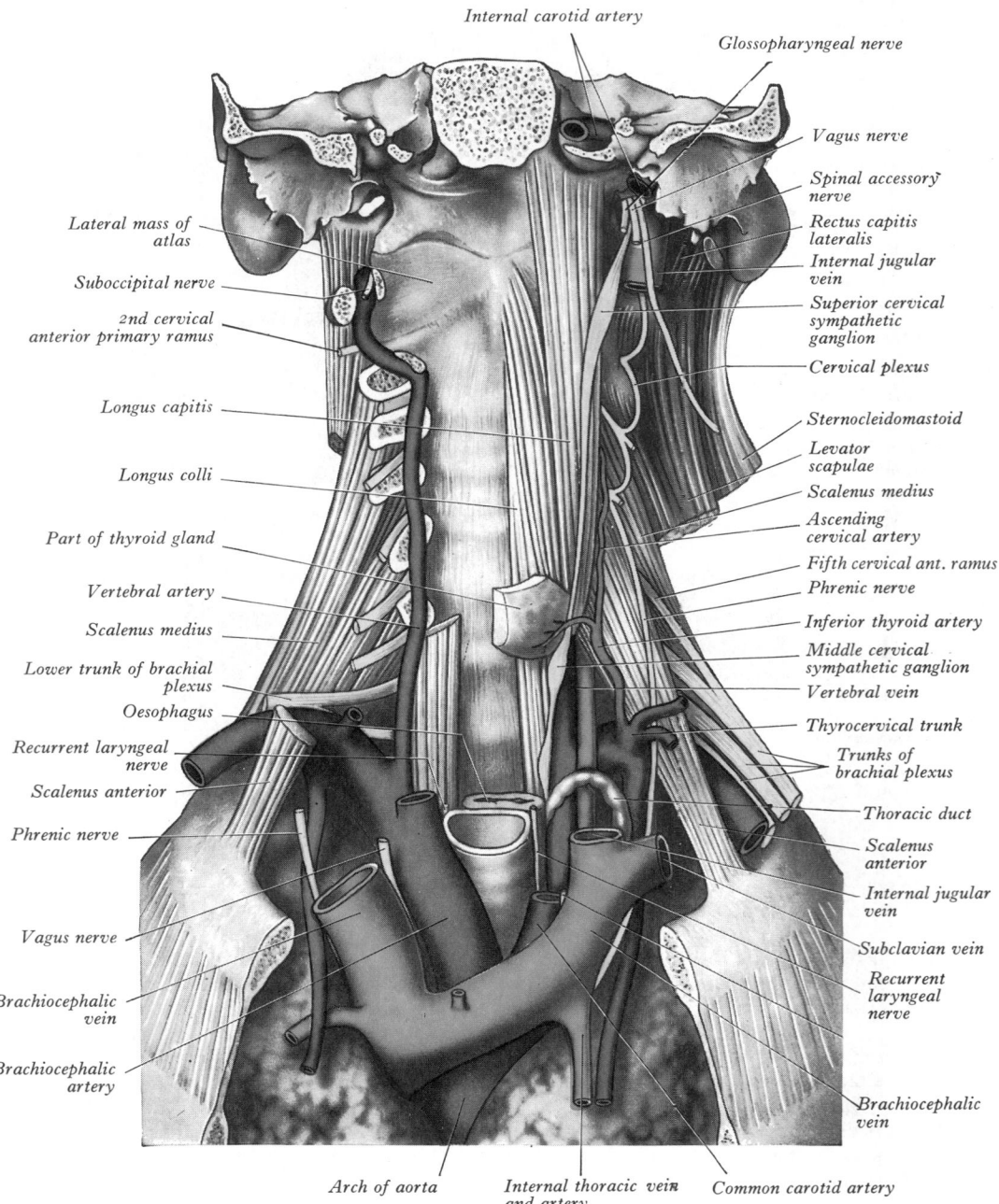

Internal carotid artery

Glossopharyngeal nerve

Vagus nerve

Spinal accessory nerve

Rectus capitis lateralis

Internal jugular vein

Superior cervical sympathetic ganglion

Cervical plexus

Sternocleidomastoid

Levator scapulae

Scalenus medius

Ascending cervical artery

Fifth cervical ant. ramus

Phrenic nerve

Inferior thyroid artery

Middle cervical sympathetic ganglion

Vertebral vein

Thyrocervical trunk

Trunks of brachial plexus

Thoracic duct

Scalenus anterior

Internal jugular vein

Subclavian vein

Recurrent laryngeal nerve

Brachiocephalic vein

Lateral mass of atlas

Suboccipital nerve

2nd cervical anterior primary ramus

Longus capitis

Longus colli

Part of thyroid gland

Vertebral artery

Scalenus medius

Lower trunk of brachial plexus

Oesophagus

Recurrent laryngeal nerve

Scalenus anterior

Phrenic nerve

Vagus nerve

Brachiocephalic vein

Brachiocephalic artery

Arch of aorta

Internal thoracic vein and artery

Common carotid artery

6.146 A dissection to show the prevertebral region and the superior mediastinum. On the right the costal elements of the upper six cervical vertebrae have been removed to expose the cervical part of the vertebral artery. On the left most of the deep relations of the common carotid artery and the internal jugular vein are exposed. Details of the terminal parts of the left lymphatic trunks have been omitted.

is its *superior bulb*, which is below the posterior part of the tympanic floor. The vein descends in the carotid sheath (p. 582), uniting with the subclavian, posterior to the sternal end of the clavicle, to form the brachiocephalic vein. It is also dilated near its end as its *inferior bulb*, above which it contains a pair of valves. *Posterior* to the vein, from above, are: the rectus capitis lateralis, transverse process of atlas, levator scapulae, scalenus medius and cervical plexus, scalenus anterior, phrenic nerve, thyrocervical trunk, vertebral vein and first part of subclavian artery; on the left it also crosses anterior to the thoracic duct (**6.146**). *Medial* to the vein are the internal and common carotid arteries and the vagus nerve between vein and arteries but posterior to them. *Superficially* the vein is overlapped above, then covered below by sternocleidomastoid and crossed by the posterior belly of the digastric and the superior belly of omohyoid. Superior to the digastric, the parotid gland and styloid process are superficial, the accessory nerve, posterior auricular and occipital arteries crossing the vein. Between the digastric and the omohyoid, the sternocleidomastoid arteries and inferior root of the ansa cervicalis cross it, but the nerve often passes between the vein and the common carotid.

Below the omohyoid, it is covered by the infrahyoid muscles and the sternocleidomastoid and it is crossed by the anterior jugular vein. Deep cervical lymph nodes lie along the vein, mainly on its superficial aspect. At the root of the neck the right internal jugular is separated from the common carotid, but the left usually overlaps its artery. At the base of the skull the internal carotid artery is *anterior*, separated from the vein by the *ninth to twelfth cranial nerves*.

Surface Anatomy. The vein is represented in surface projection by a broad band from the ear's lobule to the medial end of the clavicle; its inferior bulb is in the interval (depression) between the sternal and clavicular heads of the sternocleidomastoid (i.e. the bulb bulges into the *lesser supraclavicular fossa*; a needle may be inserted here with precision).

Tributaries are: the inferior petrosal sinus, facial, lingual, pharyngeal, superior and middle thyroid veins, sometimes the occipital. It may communicate with the external jugular vein. The thoracic duct opens near the union of the left subclavian and internal jugular veins; the right lymphatic duct is at the same site on the right.

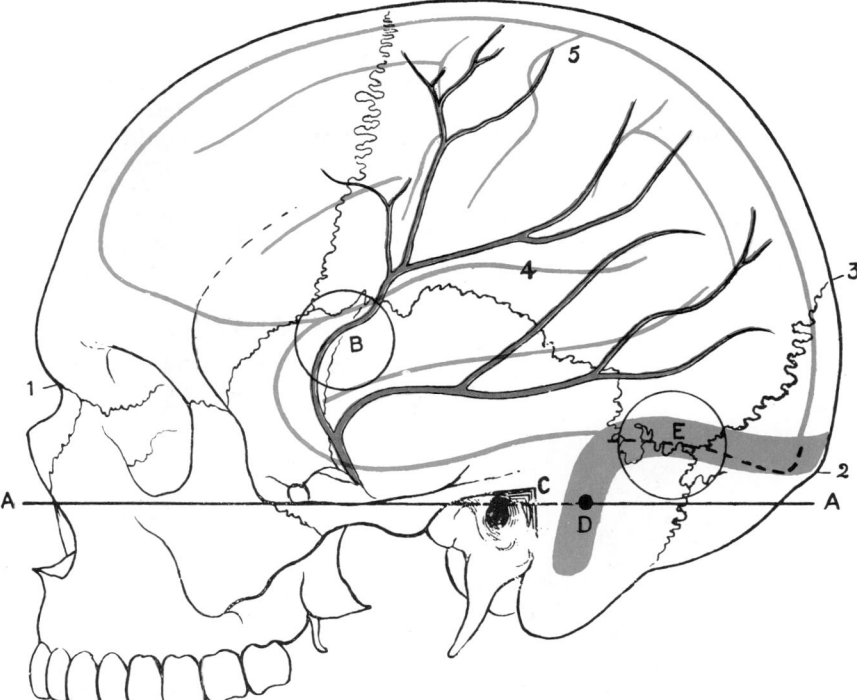

6.147 The relations of the brain, the middle meningeal artery and the transverse and sigmoid sinuses to the surface of the skull. 1. Nasion. 2. Inion. 3. Lambda. 4. Lateral cerebral sulcus. 5. Central sulcus. AA = Frankfurt plane, which traverses the lower margin of the orbital opening and the upper margin of the external acoustic meatus. B = area (including the pterion) for trephining over the frontal branch of the middle

meningeal artery and the cerebral Sylvian point. C = Suprameatal tri-angle, D = sigmoid sinus, E = area for trephining over the transverse sinus, exposing the dura mater of both cerebrum and cerebellum. The outline of the cerebral hemisphere and its major sulci are indicated in blue; the course of the middle meningeal artery is in red.

The inferior petrosal sinus leaves through the anterior part of the jugular foramen, crosses lateral or medial to the ninth to eleventh cranial nerves and joins the superior jugular bulb.

The lingual veins follow two routes: (1) *Dorsal lingual veins* drain the dorsum and sides of the tongue and join the *lingual veins* accompanying the lingual artery between the hyoglossus and genioglossus. Near the greater cornu of the hyoid bone they join the internal jugular. (2) The *deep lingual vein* begins near the tip and runs back near the mucous membrane on the tongue's inferior surface. Near the anterior border of hyoglossus it joins a *sublin-gual vein*, from the salivary gland, to form the *vena comitans nervi hypoglossi* which runs back between the mylohyoid and hyog-lossus with the hypoglossal nerve to join the facial, internal jugular or lingual vein.

The pharyngeal veins begin in a *pharyngeal plexus* external to the pharynx. After receiving meningeal veins and a vein from the pterygoid canal, they end in the internal jugular but sometimes in the facial, lingual or superior thyroid vein.

The superior thyroid vein (**6.**144,145), formed by deep and superficial tributaries corresponding to the arterial branches, ac-companies the artery and receives the superior laryngeal and crico-thyroid veins, ending in the internal jugular or facial vein.

The middle thyroid vein (**6.**145) drains the lower part of the gland and also receives veins from the larynx and trachea. It crosses anterior to the common carotid artery to join the internal jugular vein behind the superior belly of omohyoid.

Facial and occipital veins are described on pp. 793, 794, inferior thyroid veins on p. 808.

Applied Anatomy. When the superior jugular bulb thromboses (e.g. in otitis media), the glossopharyngeal, vagus and accessory nerves may be involved. The internal jugular vein may be endan-gered during removal of tuberculous or neoplastic lymph nodes.

Venous pulsation may be visible in the external jugular at the root of the neck. There are no valves in the brachiocephalic veins or the superior vena cava; hence the right atrial systole causes a wave of distension up these vessels, which may appear as a feeble

flicker over the external jugular vein. This atrial systolic impulse is much increased when the right atrium is abnormally distended or hypertrophied, as in disease of the mitral valve.

The vertebral vein is formed in the suboccipital triangle by many small tributaries from internal vertebral plexuses which leaves the vertebral canal above the posterior atlantal arch. They join small veins from local deep muscles to form a vessel which enters the foramen in the atlantal transverse process to descend, as a plexus around the vertebral artery, through successive trans-verse foramina. This ends as the vertebral vein, emerging from the sixth cervical transverse foramen, whence it descends, at first anterior then anterolateral to the vertebral artery, to open superoposteriorly into the brachiocephalic vein; the opening has a paired valve. The vertebral vein descends behind the internal jugular, passing in front of the first part of the subclavian artery (**6.**146). A small *accessory vertebral vein* usually descends from the vertebral plexus, traverses the seventh cervical transverse foramen and turns forwards between the subclavian artery and the cervical pleura to join the brachiocephalic vein.

Tributaries. The vein connects with the sigmoid sinus by a vessel in the posterior condylar canal, when this exists. It also receives branches from the occipital vein, prevertebral muscles, internal and external vertebral plexuses. It is joined by anterior vertebral and deep cervical veins (vide infra) and sometimes near its end by the first intercostal vein.

The anterior vertebral vein starts in a plexus around the upper cervical transverse processes, descends near the ascending cervical artery between attachments of scalenus anterior and lon-gus capitis and opens into the end of the vertebral vein.

The deep cervical vein accompanies its artery between the semispinales capitis et cervicis. It begins in the suboccipital region from communicating branches of the occipital and veins from suboccipital muscles and also from plexuses around the cervical spines. It passes forwards between the seventh cervical transverse process and the neck of the first rib to end in the lower part of the vertebral vein.

797

Cranial and Intracranial Veins

DIPLOIC VEINS

These veins occupy channels in the diploë of some cranial bones (6.148) and are devoid of valves. They are large, with dilatations at irregular intervals; their thin walls are merely endothelium supported by elastic tissue. Radiographically they may appear as relatively transparent bands 3 or 4 mm wide. Absent at birth, they begin to develop with the diploë at about two years. They communicate with meningeal veins, dural sinuses and pericranial veins. Recognizably regular channels are: (1) *a frontal diploic vein*, emerging from bone in the supraorbital foramen to join the supraorbital vein; (2) *an anterior temporal (parietal) diploic vein*, confined chiefly to the *frontal* bone, which pierces the greater wing of the sphenoid to end in the sphenoparietal sinus or anterior deep temporal vein; (3) *a posterior temporal (parietal) diploic vein*, in the parietal bones, descending to the parietal mastoid angle to join the transverse sinus through a foramen at the angle or mastoid foramen; (4) *an occipital diploic vein*, the largest, confined to the occipital bone, opening into occipital veins or the transverse sinus near the confluence of sinuses or into an occipital emissary vein. Numerous small diploic veins emerge near the superior sagittal sinus to end in its venous lacunae (p. 800).

MENINGEAL VEINS

Meningeal veins begin from plexiform vessels in the dura mater and drain into efferent vessels in the outer dural layer which connect with lacunae of the superior sagittal sinus and with other cranial sinuses, including those accompanying the middle meningeal arteries (p. 740), and with diploic veins.

CEREBRAL AND CEREBELLAR VEINS

Veins of the brain (p. 1085) have no valves; their thin walls have no muscular tissue. They pierce the arachnoid mater and the inner dural layer to open into the cranial venous sinuses. They comprise cerebral and cerebellar veins and veins of the brain stem.

The cerebral veins (6.149), external and internal, drain the surfaces and interior of the hemispheres. **The external cerebral veins** form *superior*, *middle* and *inferior* groups.

The superior cerebral veins, eight to twelve to each hemisphere, drain their superolateral and medial surfaces and mainly follow the sulci, though some cross the gyri. Ascending to the superomedial border, they receive small veins from the medial surface and open into the superior sagittal sinus; anterior veins open almost at right angles; the larger, posterior veins are directed obliquely forwards against the current in the sinus. This may resist the collapse of thin-walled cerebral veins which might result from a rise of intracranial pressure; but another factor is the backward growth of the cerebral hemispheres and the consequent displacement of vessels during development.

The superficial middle cerebral vein begins on the lateral surface, following the posterior ramus and stem of the lateral sulcus to end in the cavernous sinus. A *superior anastomotic vein* runs posterosuperiorly between the middle cerebral vein and the superior sagittal sinus, thus connecting the superior sagittal and cavernous sinuses. An *inferior anastomotic vein* courses over the temporal lobe, connecting the middle cerebral vein to the transverse sinus.

The inferior cerebral veins are small. Those on the frontal orbital surface join the superior cerebral veins and thus drain to the superior sagittal sinus; those on the temporal lobe anastomose with basal and middle cerebral veins, draining to the cavernous, superior petrosal and transverse sinuses.

The basal vein begins at the anterior perforated substance by the union of: (1) a small *anterior cerebral vein*, accompanying the anterior cerebral artery; (2) *a deep middle cerebral vein* receiving tributaries from the insula and neighbouring gyri and running in the lateral cerebral sulcus; and (3) *striate veins* emerging from the anterior perforated substance. The basal vein passes back round the cerebral peduncle to the great cerebral vein (6.149), receiving tributaries from the interpeduncular fossa, inferior cornu of the lateral ventricle, parahippocampal gyrus and midbrain.

The internal cerebral vein drains the deep parts of its hemisphere; it is formed near the interventricular foramen primarily by the *thalamostriate* and *choroid veins*. Numerous smaller veins from surrounding structures also converge here, each runs back parallel to its fellow between the layers of the tela choroidea of the third ventricle and below the splenium, where they join to form the median *great cerebral vein* (6.149).

The thalamostriate vein runs anteriorly between the caudate nucleus and thalamus, receives many veins from both and unites behind the anterior column of the fornix with the choroid vein to form the internal cerebral. The *choroid vein* runs along (curves or 'spirals' along) the whole choroid plexus, receiving veins from the hippocampus, fornix, corpus callosum and adjacent structures.

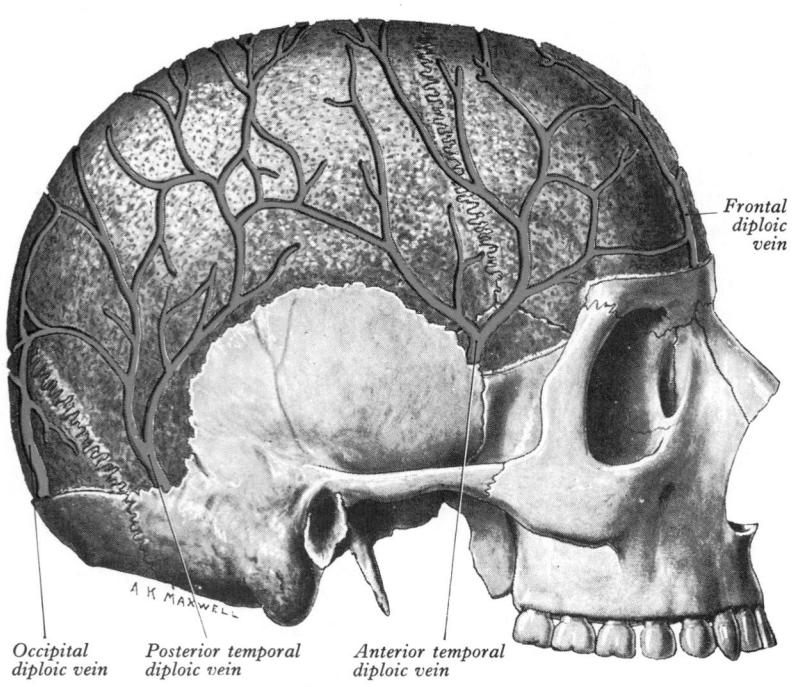

Frontal diploic vein

Occipital diploic vein — *Posterior temporal diploic vein* — *Anterior temporal diploic vein*

6.148 The veins of the diploë, displayed by the removal of the outer table of the skull.

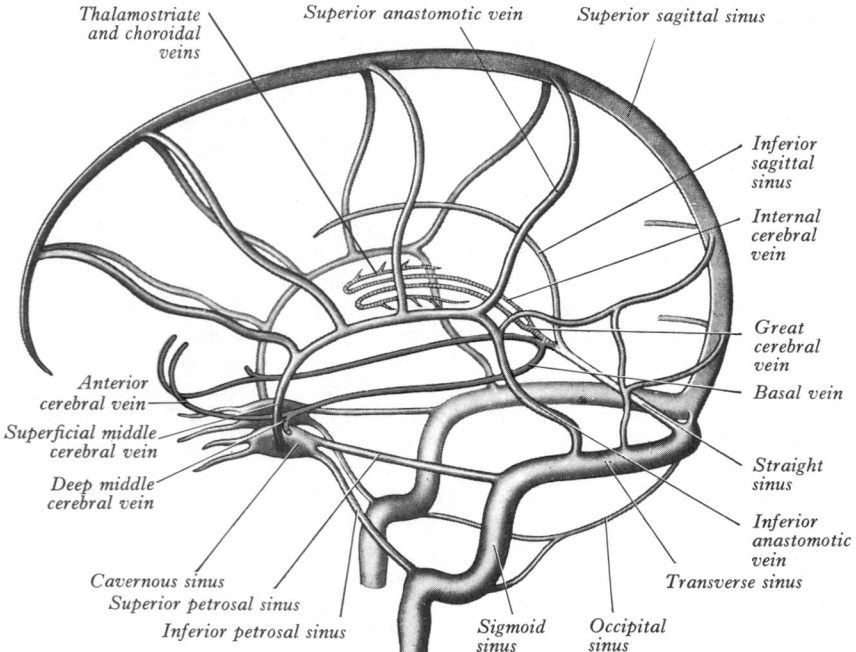

Thalamostriate and choroidal veins — *Superior anastomotic vein* — *Superior sagittal sinus*

Inferior sagittal sinus

Internal cerebral vein

Great cerebral vein

Basal vein

Anterior cerebral vein

Superficial middle cerebral vein

Deep middle cerebral vein

Straight sinus

Inferior anastomotic vein

Transverse sinus

Cavernous sinus
Superior petrosal sinus
Inferior petrosal sinus — *Sigmoid sinus* — *Occipital sinus*

6.149 Schema of the venous sinuses of the dura mater and their connections with the cerebral veins. The more deeply placed cerebral veins are shown in blue and those inside the brain are shown in *interrupted* blue.

The great cerebral vein starts by union of the two internal cerebral veins as a short median vessel *curving sharply up* around the splenium to open into the anterior end of the straight sinus, after receiving the right and left basal veins.

The cerebellar veins course on the cerebellar surface, comprising superior and inferior sets. Some *superior cerebellar veins* run anteromedially across the superior vermis to the straight sinus or great cerebral vein; others run laterally to the transverse and superior petrosal sinuses. *Inferior cerebellar veins* include a small median vessel running backwards on the inferior vermis to enter the straight or (either) sigmoid sinus; laterally coursing vessels join the inferior petrosal and occipital sinuses.

Veins of the brain stem form a superficial venous plexus *deep* to the arteries. Veins of the midbrain may reach the great cerebral or basal vein. Over the pons they tend to form a lateral vein on each side which, with upper medullary veins, may enter the petrosal sinuses, transverse sinus, cerebellar veins or the venous plexus of the (sphenoidal) foramen ovale. A median pontine vein may exist and join one of the basal veins. Veins of the inferior medulla oblongata communicate with spinal veins and drain into the adjacent venous sinuses or along variable radicular veins following the last four cranial nerves to the inferior petrosal or occipital sinuses or the upper part of the internal jugular vein. Anterior and posterior median medullary veins may run along the anterior medial fissure or posterior median sulcus and are then continuous with the spinal veins in corresponding positions (pp. 948, 1085).

CRANIAL DURAL VENOUS SINUSES

Dural sinuses are venous channels, draining blood from the brain and cranial bones, and lying between two layers of dura mater. They are lined by endothelium continuous with that in veins; they have no valves and their walls are devoid of muscular tissue. Although most accounts describe sinuses as largely simple, smooth channels, a complex 'cavernous' or plexiform nature has been emphasized by Browder & Kaplan (1976), at least in some sites (p. 802). They may be divided into: (1) a posterosuperior group and (2) an antero-inferior group on the cranial base.

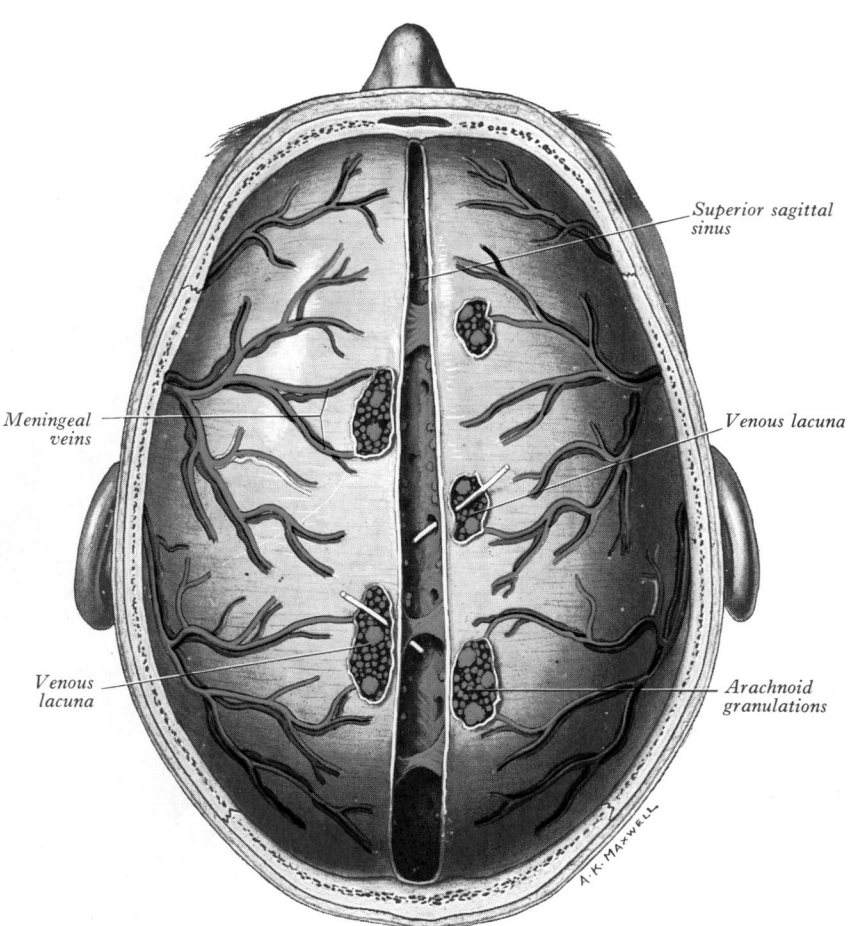

6.150 The superior sagittal sinus laid open after removal of the cranial vault. Some of the fibrous bands which cross the sinus are clearly seen; from two of the venous lacunae, bristles are passed into the sinus.

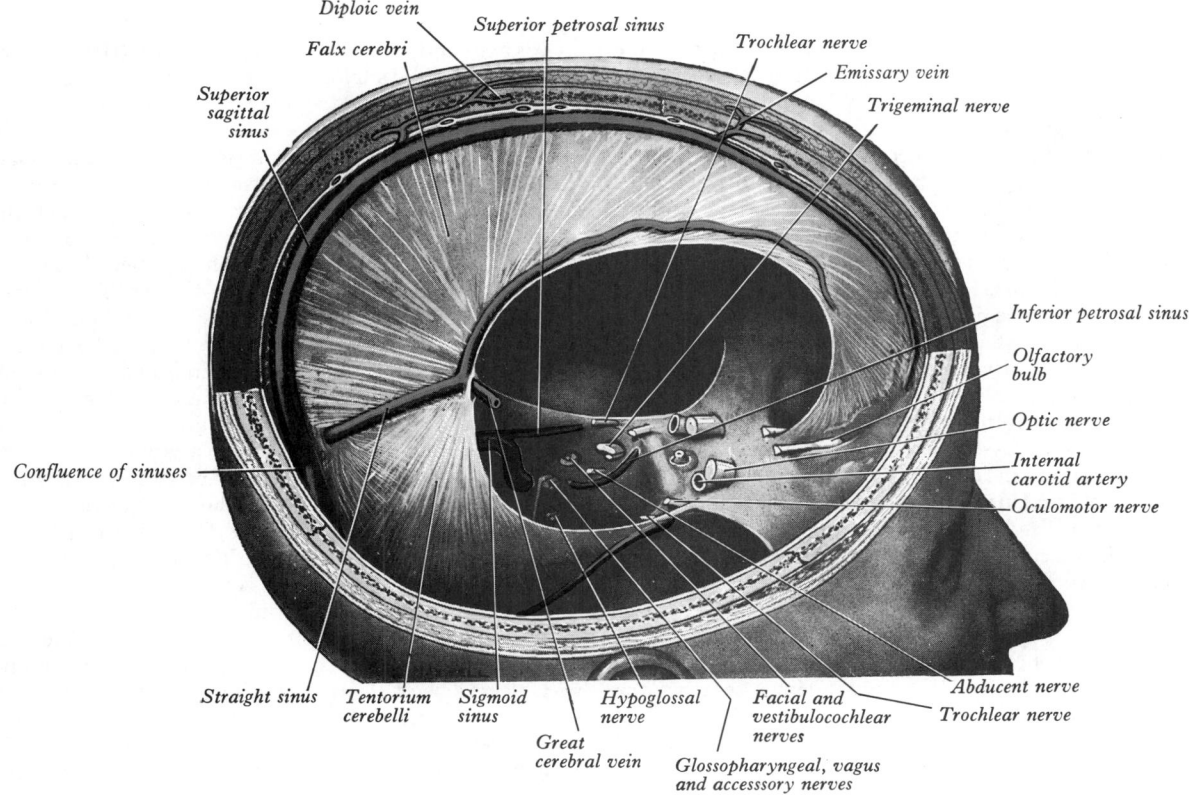

.151 The dura mater, its processes and venous sinuses: right aspect. The avernous and sphenoparietal sinuses are not represented.

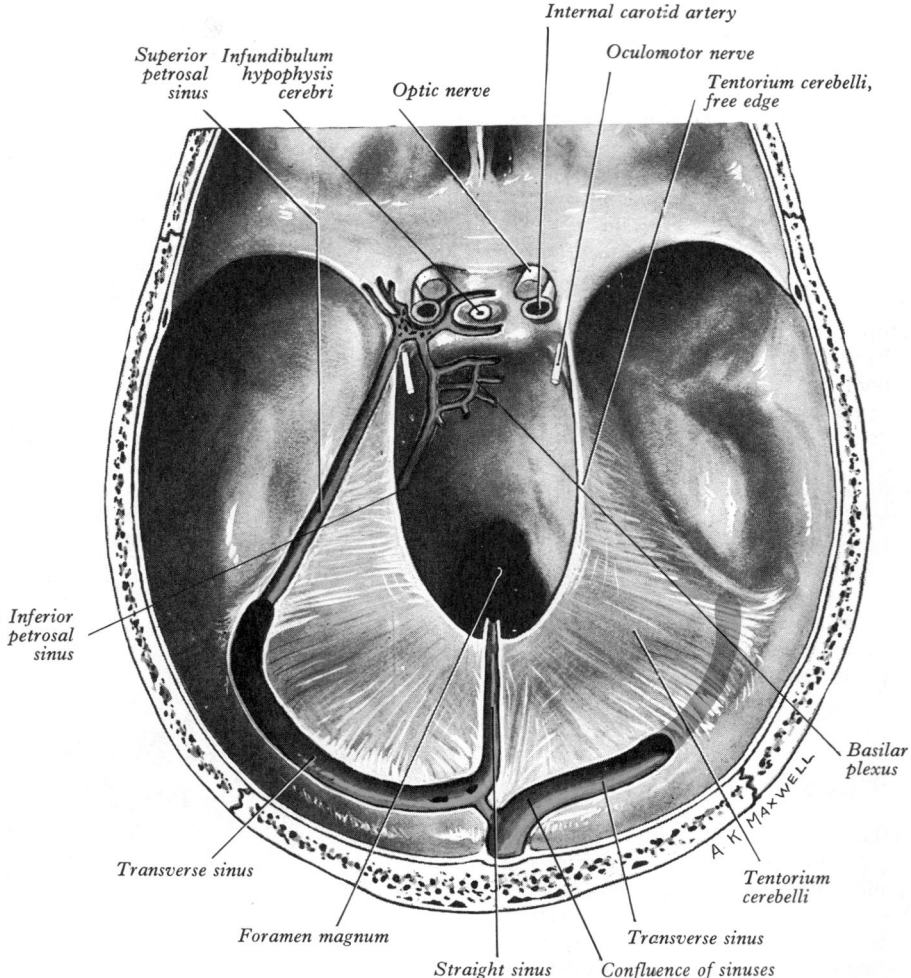

Superior petrosal sinus
Infundibulum hypophysis cerebri
Optic nerve
Internal carotid artery
Oculomotor nerve
Tentorium cerebelli, free edge
Inferior petrosal sinus
Basilar plexus
Transverse sinus
Tentorium cerebelli
Foramen magnum
Straight sinus
Confluence of sinuses
Transverse sinus

6.152 The tentorium cerebelli and venous sinuses: superior aspect. Representation of the cavernous sinuses (or 'plexuses', see text) and their extensions is greatly simplified.

1. *The posterosuperior group of venous sinuses* comprises the superior and inferior sagittal, straight, transverse, petrosquamous, sigmoid and occipital sinuses.

The superior sagittal sinus (6.149,150,151) occupies the attached, convex margin of the falx cerebri. It is said to begin near the crista galli by receiving a vein from the nasal cavity when the foramen caecum is patent; but Kaplan et al (1973) found no such tributary in 201 specimens; in only 9% did the sinus extend as far as the foramen; the first tributaries were cortical veins from adjacent frontal poles, the *ascending frontal veins* of Krayenbuhl (1967). The sinus usually begins a few millimetres posterior to the foramen caecum and runs back, grooving the internal surface of the frontal bone, the adjacent margins of both parietals and the squamous occipital bone. Near the internal occipital protuberance it deviates, usually to the right, continuing as a transverse sinus. Triangular in cross-section, it gradually enlarges backwards. Its interior shows the openings of superior cerebral veins, projecting arachnoid granulations, and many fibrous bands across its inferior angle; it also communicates by small orifices with irregular *venous lacunae*, situated in the dura mater near the sinus, usually three on each side: a small frontal, a large parietal and an occipital intermediate in size. In the elderly, lacunae tend to become confluent as one elongated lacuna on each side. Fine fibrous bands cross them and numerous arachnoid granulations project into them. The superior sagittal sinus receives the superior cerebral veins and, near the posterior end of the sagittal suture, veins from the pericranium passing through the parietal foramina; the lacunae drain the diploic and meningeal veins.

The complexity of these lateral lacunae and of the sinus itself have been obscured by over-simplification in general texts; but this complexity has often been emphasized (LeGros Clark 1920,

Balo 1950) and studies of corrosion casts (Browder & Kaplan 1976) and cerebral angiography have revived earlier descriptions. Lateral lacunae are often so complex as to be almost plexiform and rarely the simple venous spaces usually depicted. All more recent observers have described plexiform arrays of small veins adjoining the sagittal, transverse and straight sinuses. LeGros Clark and Balo regarded these masses as cavernous tissue, which commonly adjoin all sinuses inter-communicating at their confluence. Ridges of such 'spongy' venous tissue often project into the lumina of the superior sagittal and transverse sinuses. Their function can only be conjectured (p. 1090). The superior sagittal sinus is also invaded, in its intermediate third, by variable bands and projections from its dural walls, even extending as horizontal shelves dividing its lumen into superior and inferior channels. Such variable features make it impossible to give a simple description of this or other venous sinuses, whose variations have been detailed by Browder & Kaplan (1974) in a large series of corrosion casts; individual variations can only be shown by angiography.

The confluence of the sinuses (6.152) is a term for the dilated posterior end of the superior sagittal sinus, situated to one side (usually right) of the internal occipital protuberance, where it turns to become a transverse sinus. It also *connects* with the occipital and contralateral transverse sinus. The size and degree of communication of the channels meeting at the confluence are variable. (Consult Browder & Kaplan 1974 for statistics of 21 specimens.) In more than half of the specimens all venous channels converging towards the occiput do interconnect, including straight and occipital sinuses. In many instances, however, communication is nil or tenuous. Any sinus involved may be reduplicated, narrowed or widened near the confluence. Variation is too great for useful description.

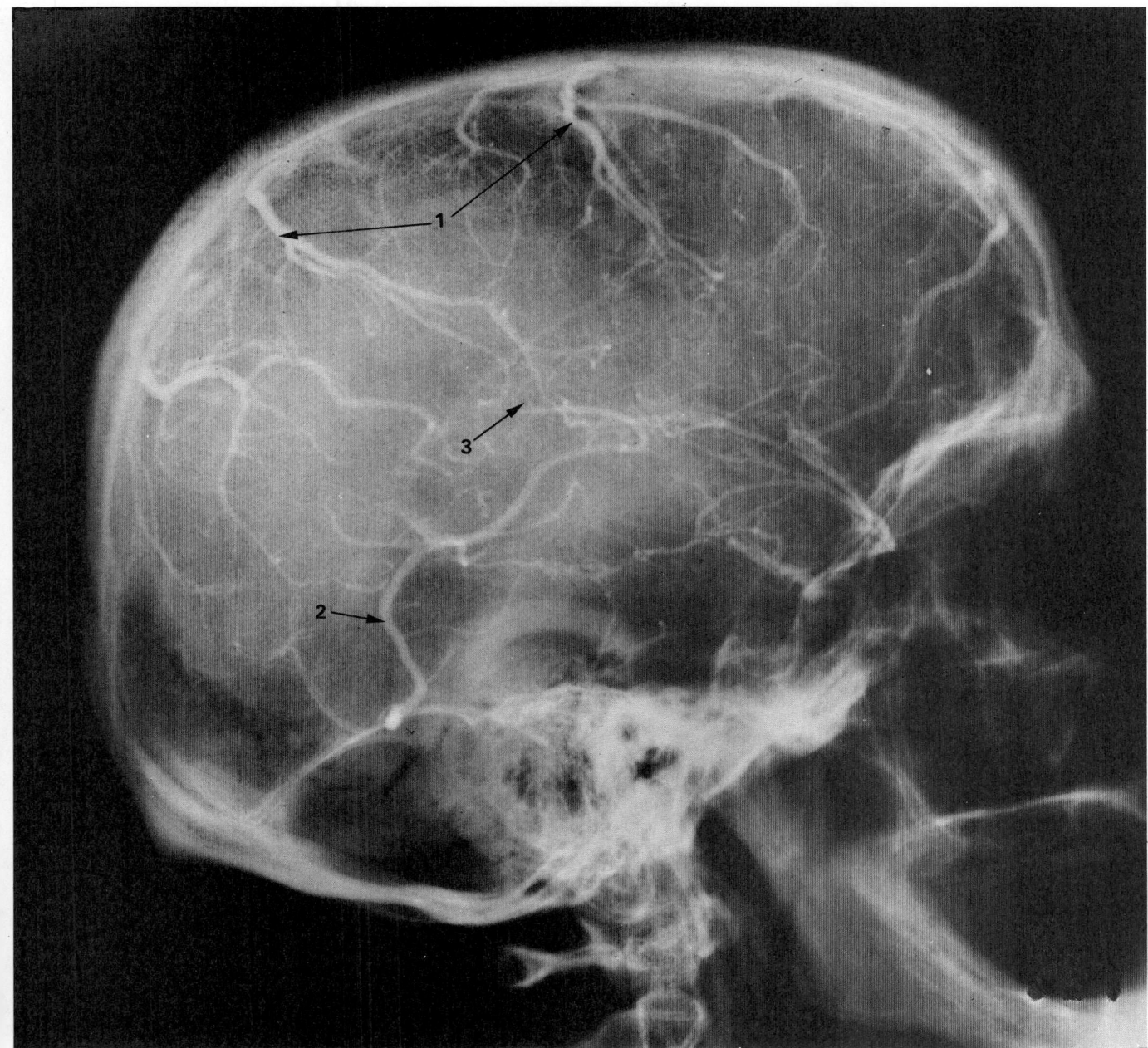

6.153 Internal carotid arteriogram (right), venous phase: lateral view. (Same subject as in **6**.77 and **6**.78, p. 743.) 1. Superior cerebral veins. Note the anterior course at entry into the superior sagittal sinus. 2. An inferior cerebral vein ending in the straight sinus. 3. Region of venous anastomoses.

Applied Anatomy. Connections between the superior sagittal sinus and veins of the nose, scalp and diploë explain the occasional spread of infective thrombosis in these parts.

The inferior sagittal sinus (**6**.151), in the posterior half or two-thirds of the free margin of the falx cerebri, increases in size posteriorly, ending in the straight sinus. It receives veins from the falx and sometimes from the medial cerebral surfaces.

The straight sinus (**6**.151,152) lies in the junction of the falx cerebri with the tentorium cerebelli. Triangular in cross-section, it has a few transverse bands. It runs *postero-inferiorly*, continuing the inferior sagittal sinus into that transverse sinus which is not, or only tenuously, continuous with the superior sagittal sinus. It may communicate terminally, but quite variably, at the confluence. Its tributaries include some superior cerebellar veins and the great cerebral; the site of the latter's opening is marked by a dilatation. A small body projects into the floor of the sinus at its junction with the great cerebral vein. This contains a sinusoidal plexus of vessels; it may become engorged and act as a valve controlling outflow from the great cerebral vein, affecting the secretion of cerebrospinal fluid in the lateral ventricles. As noted, other masses of cavernous tissue are related to other dural sinuses; engorgement possibly influences their blood flow but structural data make this unlikely (vide supra).

The transverse sinuses (**6**.152,154) begin at the internal occipital protuberance, one (right) directly continuous with the superior sagittal sinus, the other with the straight sinus. Each curves anterolaterally to the posterolateral part of the petrous temporal bone, where it turns down as a sigmoid sinus. It is in the attached margin of the tentorium cerebelli, first on the occipital's squama, then on the parietal's mastoid angle. It has a gentle curve, convex upwards, and increases in size as it proceeds forwards. Transverse sinuses are triangular in section and usually unequal in size, the one draining the superior sagittal sinus being the larger. Where they continue as sigmoid sinuses, they are joined by the superior petrosal sinuses; they receive the inferior cerebral, inferior cerebellar, diploic and inferior anastomotic veins (p. 798).

Each petrosquamous sinus runs back in a groove, which sometimes becomes a canal posteriorly, along the junction of the squamous and petrous parts of the temporal bone, opening behind into the transverse sinus. Anteriorly it connects with the

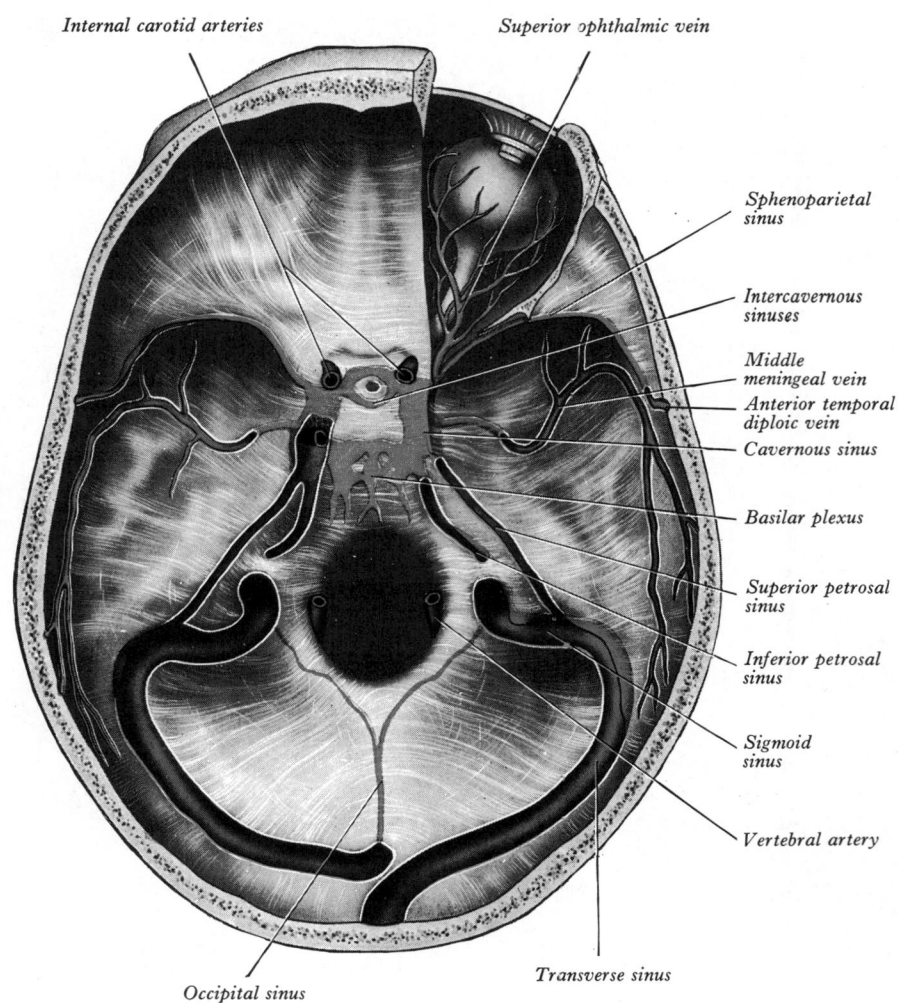

Internal carotid arteries

Superior ophthalmic vein

Sphenoparietal sinus

Intercavernous sinuses

Middle meningeal vein

Anterior temporal diploic vein

Cavernous sinus

Basilar plexus

Superior petrosal sinus

Inferior petrosal sinus

Sigmoid sinus

Vertebral artery

Transverse sinus

Occipital sinus

6.154 The sinuses at the base of the skull. The sinuses coloured dark blue have been opened up. See text and **6.**156, 157 for alternative views on the construction of the cavernous sinuses.

retromandibular vein through a postglenoid or squamosal foramen (pp. 377, 378). The sinus may be absent; it may drain entirely into the retromandibular vein.

The sigmoid sinuses (**6.**154) are continuations of the transverse sinuses, beginning where these leave the tentorium cerebelli. Each sigmoid sinus curves inferomedially in a groove on the mastoid temporal bone, crosses the occipital's jugular process and turns forwards to the superior jugular bulb, lying posterior in the jugular foramen. Anteriorly, a thin plate of bone alone separates its upper part from the mastoid antrum and air cells. It connects with pericranial veins via mastoid and condylar emissary veins.

The occipital sinus (**6.**154), the smallest of the sinuses, lies in the attached margin of the falx cerebelli, occasionally paired. It commences near the foramen magnum in several small channels, one joining the end of the sigmoid sinus; it connects with the internal vertebral plexuses and ends in the confluence of sinuses.

2. *The antero-inferior group of venous sinuses* includes: cavernous, intercavernous, inferior petrosal, sphenoparietal, superior petrosal and basilar sinuses and middle meningeal 'veins'.

The cavernous sinuses (**6.**149,154,155,156,157) lie on the sides of the body of the sphenoid bone; their name indicates their internal structure. It has been asserted that a distended adult sinus contains a few trabeculae, mostly in its periphery near the entry of its tributaries and that these are incorporations of plexiform tributaries during developmental expansion. When the sinus is collapsed, as is usual in cadavers, its cavity is encroached upon by the nerves and arachnoid granulations in its wall, creating a spurious resemblance to cavernous tissue (Butler 1957). From corrosion casts, however, Parkinson (1973) concluded that

the sinus is usually a plexus (as it is during development), a finding in accord with some earlier descriptions; Pernkopf (1963) depicted the 'sinus' as a venous plexus (see **6.**156, 157). Browder & Kaplan (1976), from examination of many casts prepared in cadavers, described the sinus as 'reticulated'. It is not clear whether they meant *plexiform* or *cavernous*. The sinus extends from the superior orbital fissure to the apex of the petrous temporal bone, with an average length of 2 cm and width of 1 cm. The internal carotid artery, with a sympathetic plexus, passes forwards through the sinus, as does the abducent nerve, inferolateral to the artery; the oculomotor and trochlear nerves and ophthalmic and maxillary divisions of the trigeminal (**6.**155) are usually said to be *in the thickness* of its lateral wall; but they are of such diameters that they project into the sinus; while they may be surrounded by dural connective tissue, they are usually covered medially by little more than endothelium (McGrath 1977). The sphenoidal air sinus and hypophysis cerebri are medial; the trigeminal cave is near the inferoposterior part of its lateral wall, extending posteriorly beyond it and enclosing the trigeminal ganglion. The uncus is also lateral.

Tributaries are: the superior ophthalmic vein, a branch from the inferior ophthalmic vein (or the whole vessel), the superficial middle cerebral vein, inferior cerebral veins, and sphenoparietal sinus; the central retinal vein and frontal tributary of the middle meningeal sometimes drain to it. The sinus drains to the transverse sinus via the superior petrosal sinus, to the internal jugular via the inferior petrosal sinus and a plexus of veins on the internal carotid, to the pterygoid plexus by veins traversing the emissary sphenoidal foramen, foramen ovale and foramen lacerum, and to the facial vein via the superior ophthalmic. The two sinuses are

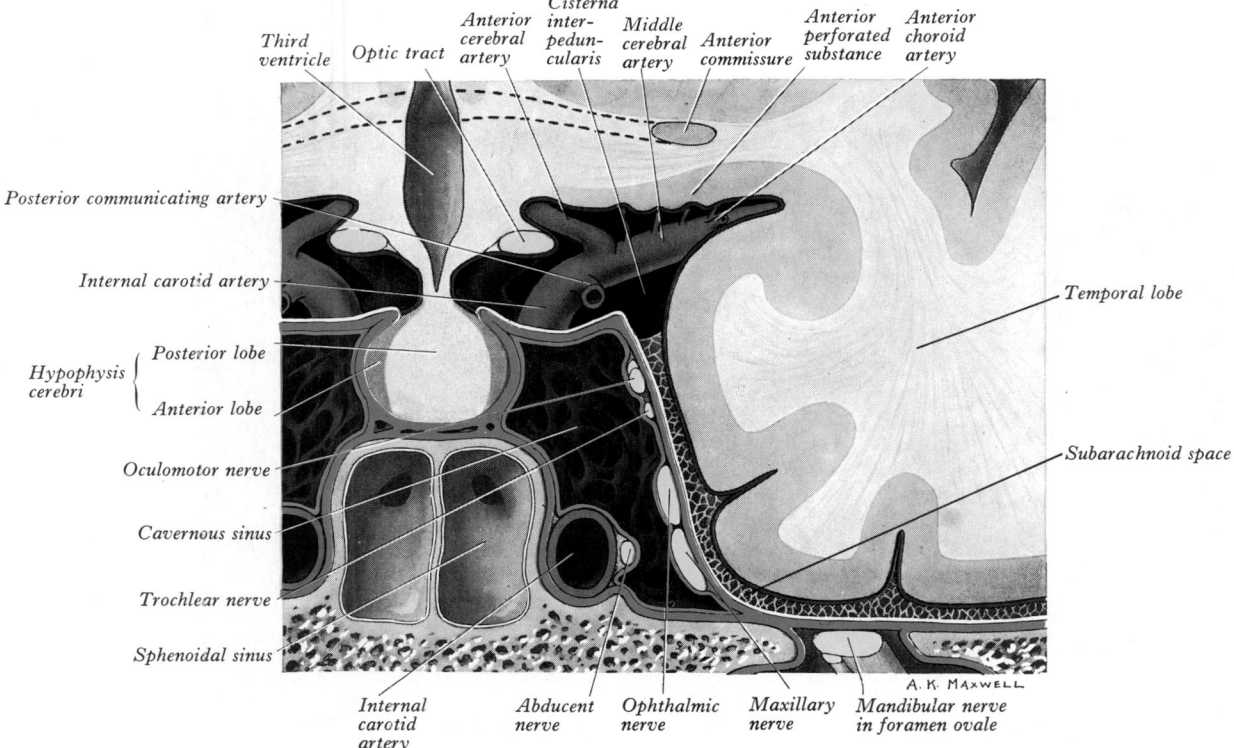

6.155 Coronal, slightly oblique section through the middle cranial fossa, showing the cavernous and cerebral portions of the internal carotid artery and the cavernous sinus: pia mater—mauve; arachnoid mater— white; layers of dura mater—green (the mesothelium of the dura mater is not indicated); endothelium of cavernous sinus—blue.

connected by anterior and posterior intercavernous sinuses and the basilar plexus. All connections are valveless; the direction of flow in them is reversible.

Propulsion of blood in the sinus is partly due to pulsation of the internal carotid artery. It is also influenced by gravity and hence by the position of the head.

Applied Anatomy. An arteriovenous leak may occur between the cavernous sinus and internal carotid artery, causing a pulsating orbital swelling. Ligation of the internal or common carotid artery has sometimes alleviated the condition. Suppuration in the upper nasal cavities and paranasal sinuses or near the medial canthus may lead to septic thrombosis of the cavernous sinuses.

The ophthalmic veins (**6.**154,158), superior and inferior, devoid of valves, link the facial and intracranial veins.

The *superior ophthalmic vein* forms posteromedial to the upper lid from two tributaries connecting anteriorly with the facial and supraorbital veins (p. 793). It runs with the ophthalmic artery, receiving corresponding tributaries, and traverses the superior orbital fissue to end in the cavernous sinus. The *inferior ophthalmic vein* begins in a network near the anterior region of the orbital floor and medial wall, receiving veins from the rectus inferior, obliquus inferior, lacrimal sac and eyelids; it runs back above the rectus inferior and often joins the superior ophthalmic vein but may reach the cavernous sinus. It connects with the pterygoid venous plexus by small rami through the inferior orbital fissure.

The central retinal vein first traverses the optic nerve, leaving it to pursue a long course in the subarachnoid space before entering the cavernous sinus or superior ophthalmic vein. It receives a central vein which drains the nerve while still within it.

The sphenoparietal sinuses (**6.**154) are inferior to the periosteum of the lesser wings of the sphenoid bone, near their posterior edges. Each receives small veins from the adjacent dura mater and sometimes the frontal ramus of the middle meningeal vein; curving medially it opens into the anterior part of the cavernous sinus. It often receives connecting rami, in its middle course, from the superficial middle cerebral vein, sometimes veins from the temporal lobe and the anterior temporal diploic vein. When these connections are well developed it is a large channel.

The intercavernous sinuses, anterior and posterior, interconnect the cavernous sinuses in the anterior and posterior attached borders of the diaphragma sellae; they thus complete a venous *circular sinus* (**6.**154). Small, irregular sinuses inferior to the hypophysis cerebri drain into them. Such *inferior intercavernous sinuses* were studied by Kaplan et al (1976), who emphasized their size and plexiform nature, features important in a surgical trans-nasal approach to the hypophysis.

The superior petrosal sinuses (**6.**154), small and narrow, drain the cavernous to the transverse sinuses. Leaving the posterosuperior part of the cavernous sinus, each runs posterolaterally in an attached margin of the tentorium cerebelli, crossing above the trigeminal nerve to a groove on the superior border of the petrous temporal bone; each ends by joining a transverse sinus where this curves down to become the sigmoid. It receives cerebellar, inferior cerebral and tympanic veins. It connects with the inferior petrosal sinus and basilar plexus.

The inferior petrosal sinuses drain the cavernous sinuses to the internal jugular veins. Each (**6.**154) begins postero-inferiorly at its cavernous sinus and runs back in a groove between the petrous temporal and basilar occipital bones. Traversing the anterior part of the jugular foramen it ends in the superior jugular bulb. It receives *labyrinthine veins* via the cochlear canaliculus and the vestibular aqueduct and tributaries from the medulla oblongata, pons and inferior cerebellar surface. According to Browder & Kaplan (1976) the sinus is more often a plexus and sometimes drains by a vein in the hypoglossal canal to the suboccipital vertebral plexus.

Relations of structures in the *jugular foramen* are as follows: the inferior petrosal sinus is anteromedial with a meningeal branch of the ascending pharyngeal artery, and the sinus descends obliquely backwards; the sigmoid sinus is situated at the lateral and posterior part of the foramen with a meningeal branch of the occipital artery; between the sinuses are in succession, posterolaterally: the glossopharyngeal, vagus and accessory nerves (p. 1118).

The basilar venous plexus (**6.**154) consists of interconnecting channels between layers of dura mater on the clivus; it interconnects the inferior petrosal sinuses and joins with the internal vertebral venous plexus. It also usually connects with the

Hypophysis, Diaphragma sellae

Int. intercavernous sinus

Optic n., Ophthalmic a.

Lesser wing, orbital vein

Oculomotor n.

Trochlear n.

Int. carotid a.

Trigeminal n. (Sensory part)

Maxillary n.

Infundibulum, Post. intercavernous sinus

Ophthalmic n.

Trigeminal n. (Motor part)

Trigeminal ganglion

Post. petroclinoid fold

Middle meningeal a.

Sup. tympanic a. Lesser petrosal n.

Sup. petrosal sinus, Trigeminal fossa

Greater petrosal n.

Abducent n., Int. carotid a. Cavernous sinus

Petrosphenoidal lig., Sup. petrosal sinus

Basilar plexus, Abducent n.

Fac. n. (+ N. intermed.), Vestibulococh. n. and Labyrinthine a.

Glossopharyngeal n.

Basilar sinus

Vagus n.

Hypoglossal n.

Vertebral a.

Accessory n., Inf. petrosal sinus

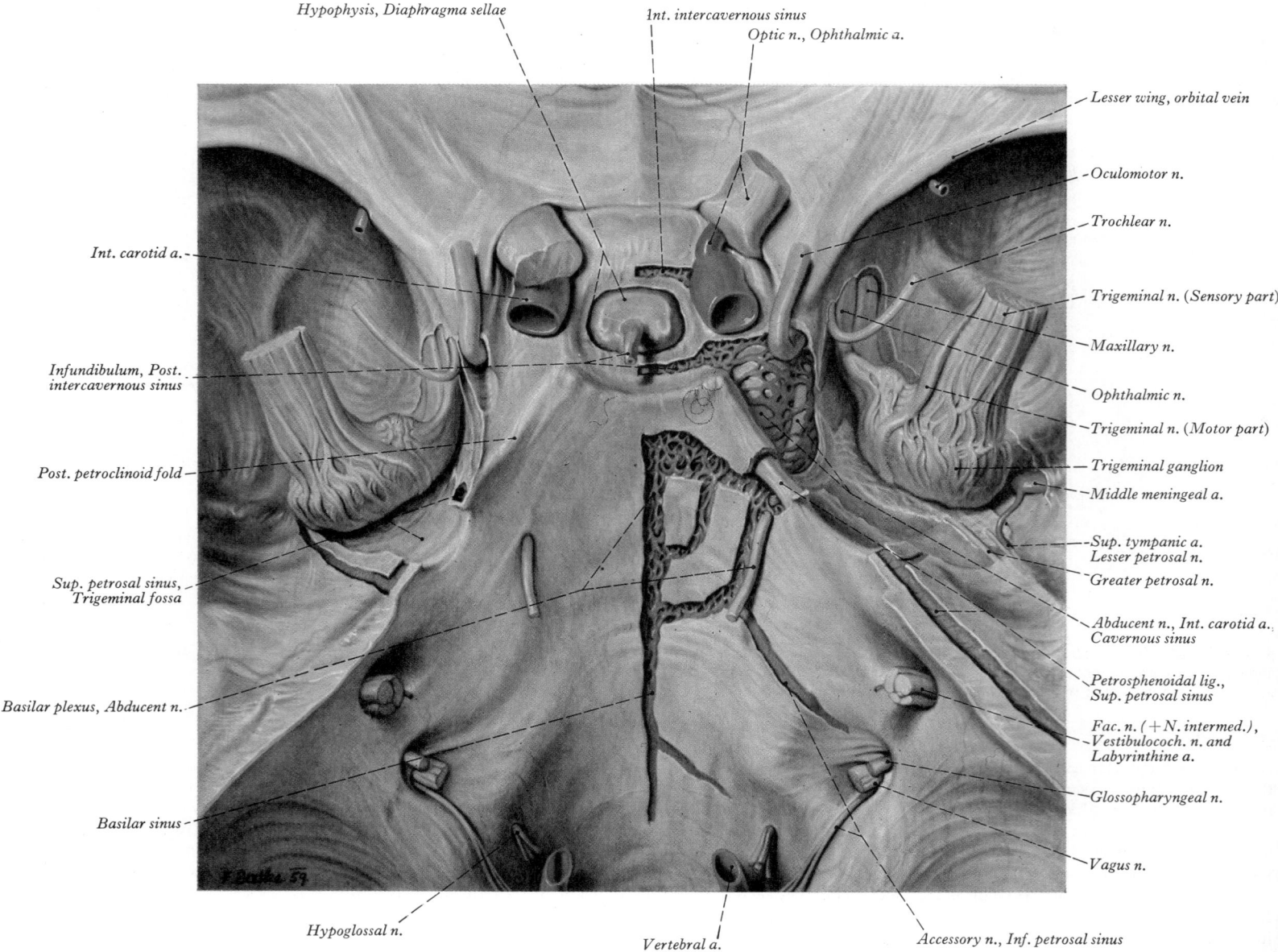

6.156 The middle cranial fossa, viewed from above to show the termination of the internal carotid artery, its branches and the cavernous sinus. Note the plexiform nature of the 'sinus', which communicates with similar venous plexuses in the hypophysial fossa and over the clivus. These have been exposed by partial removal of the dura mater. (See also **6.**157.)

cavernous and superior petrosal sinuses at its anterior end. When marginal sinuses (p. 805) are large they communicate anteriorly with the plexus; an almost complete *circular venous channel* may then surround the foramen magnum, connecting the basilar plexus intracranially to the inferior petrosal, sigmoid and occipital sinuses and to variable extracranial vertebral plexuses in the suboccipital region.

The middle meningeal veins (or sinuses) (**6.**154) communicate above with the superior sagittal sinus through its venous lacunae; *below* they converge and unite as *frontal* and *parietal trunks*, which accompany branches of the middle meningeal arteries in grooves on the internal parietal surfaces; but the veins are closer to bone and sometimes occupy separate grooves. The veins' situation has been said to make them liable to tears in fractures (Wood Jones 1911). Their termination is variable. The parietal trunk may traverse the foramen spinosum to the pterygoid venous plexus; the frontal may also reach the plexus via the foramen ovale or may end in the sphenoparietal or cavernous sinus. Besides meningeal tributaries they receive small inferior cerebral veins and connect with the diploic and superficial middle cerebral veins. Browder & Kaplan (1976) state that middle meningeal 'veins' are histologically *sinuses*, in places almost surrounding the middle meningeal arteries; they also report frequent arachnoid granulations in them.

804

Surface Anatomy. The *superior sagittal sinus* runs from the glabella (p. 342) to the inion (**3.**126). Narrow anteriorly, it widens to about 1 cm. The *transverse sinus* begins at the inion and runs laterally, with slight upward convexity, to the base of the mastoid process, from which the *sigmoid sinus* passes down just anterior to the posterior mastoid border to a point about 1 cm above its tip.

EMISSARY VEINS

Emissary veins traverse cranial apertures and make connections between venous sinuses and extracranial veins. Some are constant, others sometimes absent. (1) A *mastoid emissary vein* in the mastoid foramen unites the sigmoid sinus with the posterior auricular or occipital vein. (2) A *parietal emissary vein* traverses the parietal foramen to connect the superior sagittal sinus with the veins of the scalp. (3) The *venous plexus of the hypoglossal canal*, occasionally a single vein, connects the sigmoid sinus to the internal jugular vein. (4) A *posterior condylar emissary vein* connects the sigmoid sinus with veins in the suboccipital triangle via the condylar canal. (5) A plexus of emissary veins (*venous plexus of foramen ovale*) links the cavernous sinus to the pterygoid plexus via the foramen ovale. (6) Two or three small veins traverse the foramen lacerum connecting the cavernous sinus with the pharyngeal veins and pterygoid plexus. (7) A vein in the emissary sphenoidal foramen (of Vesalius) connects the same vessels

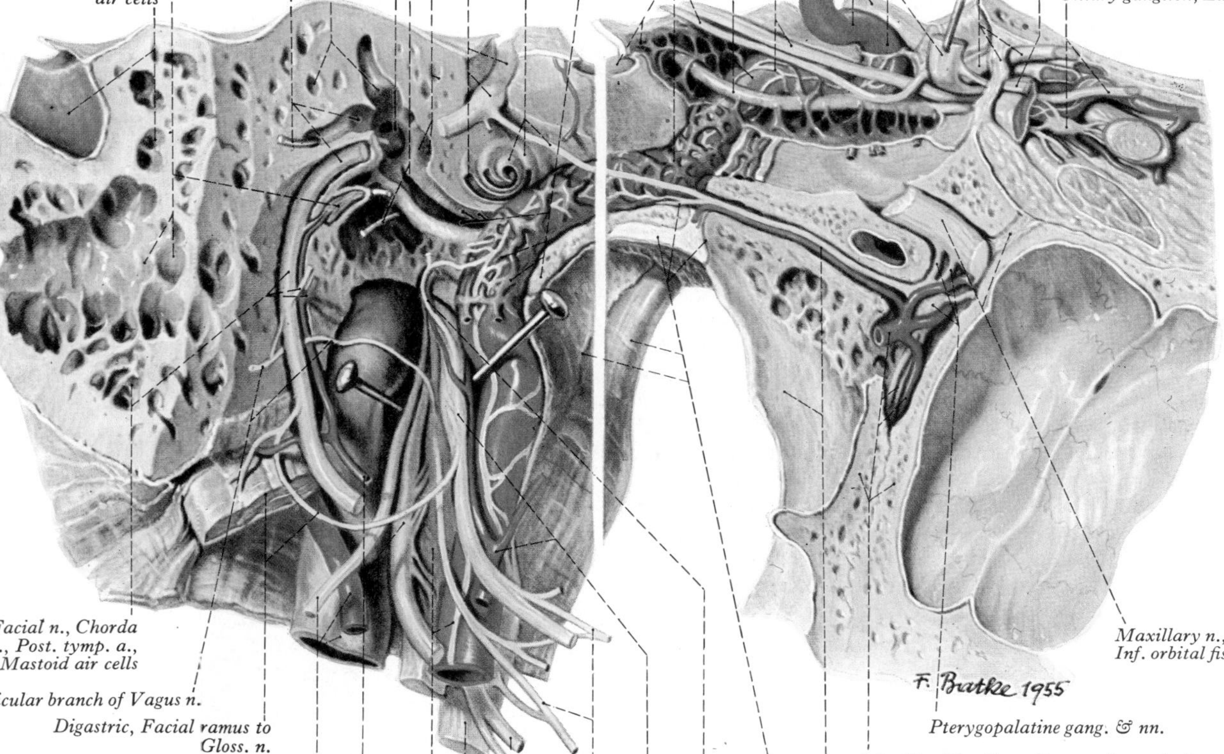

Cochlear canal, Facial n., Genicular ganglion
Spherical recess of saccule, Vestibular crest
Spiral lamina, Fenestra cochlearis
Vestibule
Ant. semicirc. canal & ampulla, Mastoid air cells
Facial n., Lat. semicirc. canal
& ampulla
Stapedial musc. n. & a.
Mastoid air cells
Sigmoid sinus, Mastoid
air cells

Cochlea, Greater petrosal n.
Int. carotid a., nerv. and ven. plexuses,
Cochlea
Sup. petrosal and cavernous sinuses,
Int. carotid a., Trochlear n.
Int. carotid a., Abducent n.
Oculomotor n., Int. carotid plexus
Int. carotid a., Optic n.
Carotid genu, Ophthalmic n.
Common annular tendon, Frontal n.
Lacrimal n.
Ciliary ganglion, Lacrimal n.

F. Batke 1955

Facial n., Chorda
tymp., Post. tymp. a.,
Mastoid air cells
Auricular branch of Vagus n.
Digastric, Facial ramus to
Gloss. n.
Accessory n., Occipital a., Int. jugular v.
Stylomastoid a., Hypoglossal n.
Sup. vagal ganglion, Carotid sinus nerve
Sympathetic trunk, Sup. cerv. symp. ganglion
Sup. laryngeal n.

Maxillary n., For. rotundum,
Inf. orbital fissure
Pterygopalatine gang. & nn.
Maxilla, Maxillary & ascending palatine aa.,
Palatine nn.
A. & N. of pterygoid canal, Medial pterygoid lamina
Basal cartilage, Deep petrosal n., Longus cap.,
Prevertebral fascia
Tympanic n., Inf. tymp. a.
Inf. glossopharyngeal ganglion, Ascending pharyngeal a.
Pharyngeal rami of gloss., vagus, and sympathetic nn.

6.157 An oblique vertical section through the cranial base to display in lateral view the right internal carotid artery and the continuity of the venous plexus around the intra-osseous and cavernous parts of the artery.

(**6.**156 and 157 are from Pernkopf 1963, by kind permission of W B Saunders and Urban & Schwarzenberg.)

(8) The *internal carotid venous plexus*, passing through the carotid canal, connects the cavernous sinus to the internal jugular vein. (9) The petrosquamous sinus (p. 801) connects the transverse sinus with the external jugular vein. (10) A vein may traverse the foramen caecum (patent in about 1% of adult skulls) connecting nasal veins with the superior sagittal sinus. (11) An *occipital emissary vein* usually connects the confluence of sinuses with the occipital vein through the occipital protuberance, receiving also the occipital diploic vein. (12) The occipital sinus connects with variably developed veins around the foramen magnum (so-called *marginal sinuses*) and thus with the vertebral venous plexuses, an alternative venous drainage when the jugular vein is blocked or tied. (13) The ophthalmic veins are potentially emissary, since they connect intracranial to extracranial veins; but parietal emissary veins, included here, are usually minute and do not appear to connect with veins of the scalp in corrosion casts.

These connections are significant in the spread of infection from extracranial foci to venous sinuses. Ligature of the internal jugular vein, to limit the spread of some oral and pharyngeal pathologies, depends on the adequacy of the collateral drainage.

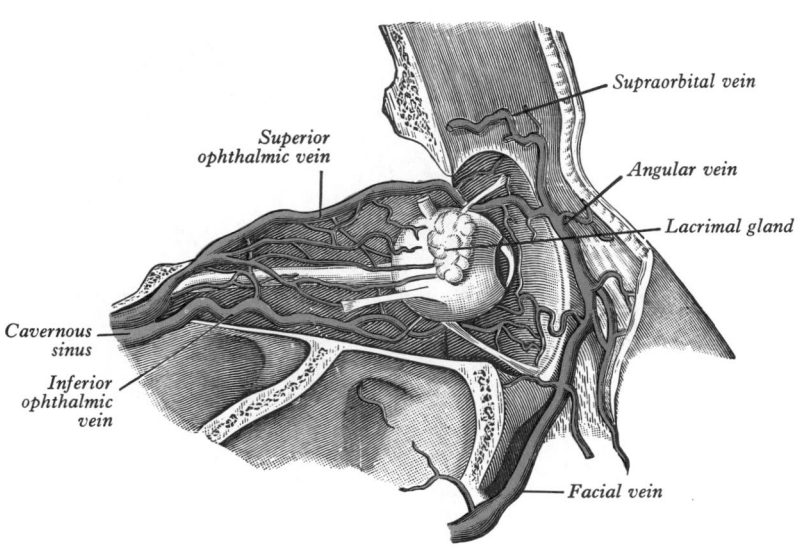

Superior
ophthalmic vein
Supraorbital vein
Angular vein
Lacrimal gland
Cavernous
sinus
Inferior
ophthalmic
vein
Facial vein

6.158 The veins of the right orbit: lateral aspect.

VEINS OF THE UPPER LIMBS

Veins are conveniently grouped as *superficial* and *deep* but these are widely interconnected. The superficial veins are subcutaneous in the superficial fascia; deep veins accompany arteries among the muscles of the limb. Both groups have valves, which are more numerous in deep veins.

SUPERFICIAL VEINS OF THE UPPER LIMB (6.159,160)

Superficial veins include the cephalic, basilic, median cubital and additional antebrachial veins and their tributaries.

Dorsal digital veins pass along the sides of the fingers, joined by oblique branches; they unite from the adjacent sides of digits into three *dorsal metacarpal veins* (6.159), which form a *dorsal venous network* over the metacarpus; this is joined *laterally* by a dorsal digital vein from the radial side of the index finger and both dorsal digital veins of the pollex and is prolonged proximally as the cephalic vein. *Medially* a dorsal digital vein from the ulnar side of minimus joins the network, which drains proximally into the basilic vein. A vein often connects the central parts of the network to the cephalic near mid-forearm. (See also the accessory cephalic vein, and other antebrachial variants below.)

Palmar digital veins connect to the dorsal by oblique *intercapitular veins* passing between metacarpal heads; they also drain to a plexus *superficial* to the palmar aponeurosis, extending over both thenar and hypothenar regions.

The cephalic vein (6.159,160), commonly formed over the 'anatomical snuff box', curves proximally from the radial end of the dorsal plexus round the forearm's radial side to its ventral aspect, receiving veins from both aspects. Distal to the elbow a branch, the *median cubital vein*, joined by a ramus from the deep veins, diverges proximomedially to reach the basilic vein. The cephalic ascends in front of the elbow superficial to a groove between the brachioradialis and biceps, crosses superficial to the

lateral cutaneous nerve of the forearm, ascends lateral to the biceps and between pectoralis major and the deltoid, where it adjoins the deltoid branch of the thoraco-acromial artery. Entering the infraclavicular fossa to pass behind the clavicular head of pectoralis major, it pierces the clavipectoral fascia, crosses the axillary artery and joins the axillary vein just below clavicular level. It may connect with the external jugular by a branch *anterior* to the clavicle. Sometimes the median cubital vein is large, transferring most blood from the cephalic to the basilic vein, the proximal cephalic vein then being absent or much diminished.

An *accessory cephalic vein*, arising in a dorsal forearm plexus or from the ulnar side of the dorsal venous network in the hand, joins the cephalic distal to the elbow. It may spring from the cephalic

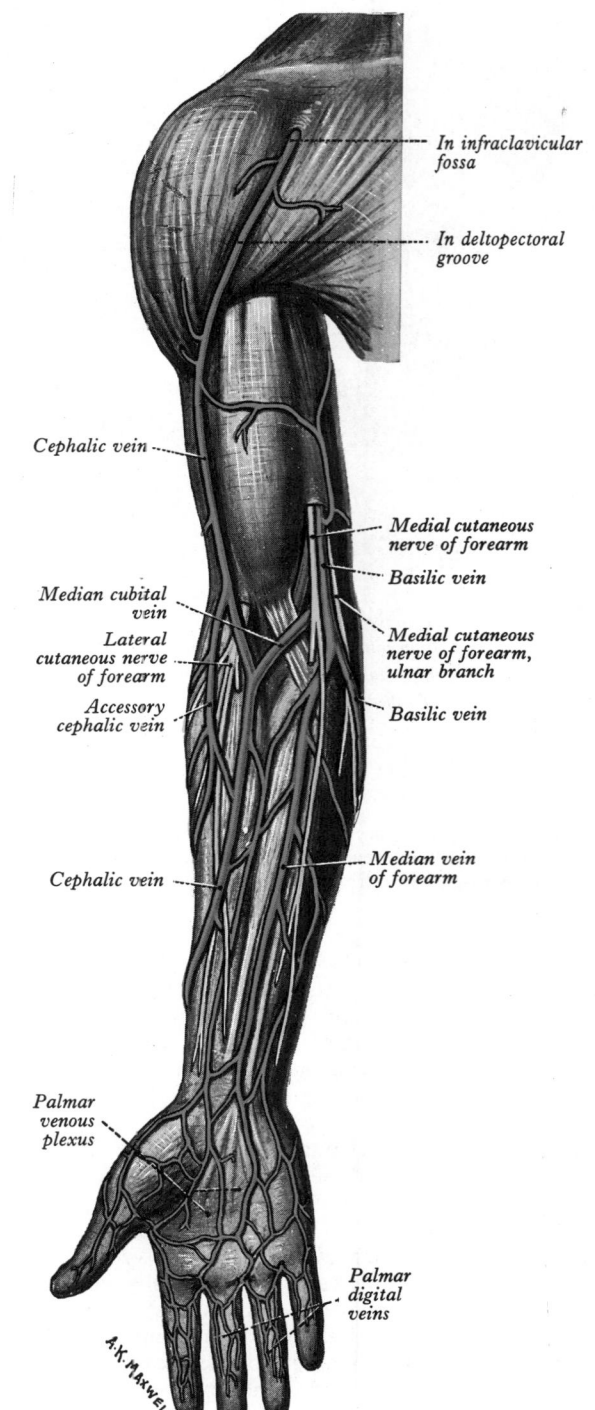

In infraclavicular fossa

In deltopectoral groove

Cephalic vein

Median cubital vein

Lateral cutaneous nerve of forearm

Accessory cephalic vein

Cephalic vein

Medial cutaneous nerve of forearm

Basilic vein

Medial cutaneous nerve of forearm, ulnar branch

Basilic vein

Median vein of forearm

Palmar venous plexus

Palmar digital veins

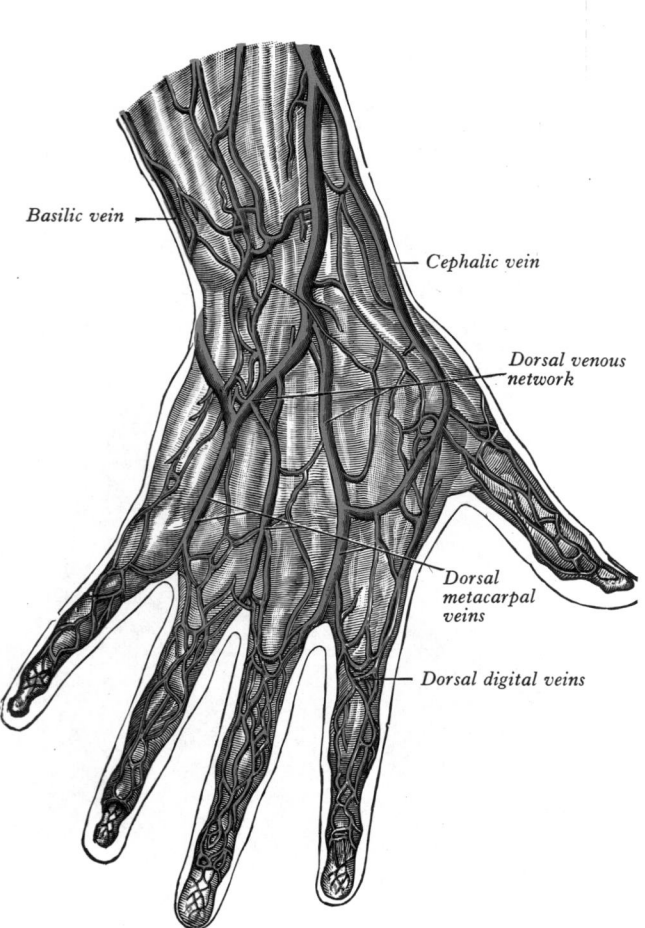

Basilic vein

Cephalic vein

Dorsal venous network

Dorsal metacarpal veins

Dorsal digital veins

6.159 The veins of the dorsum of the hand.

6.160 The superficial veins of the right upper extremity: anterior aspect

proximal to the carpus and rejoin it later. A large oblique vein often connects the basilic and cephalic veins dorsally in the forearm.

The basilic vein (6.160) begins medially in the hand's dorsal venous network, ascending posteromedially in the forearm and inclining forwards to the anterior surface distal to the elbow. Joined by the median cubital vein, it ascends superficial to and between the biceps and pronator teres; filaments of the medial cutaneous nerve of the forearm pass here, in front and behind it. It ascends medial to biceps and perforates the deep fascia about midway in the arm, continuing medial to the brachial artery to the lower border of teres major, there becoming the axillary vein. (Its relation to the brachial veins is variable, vide infra.)

The median vein of the forearm (6.160) drains the superficial palmar venous plexus. It ascends anterior in the forearm to join the basilic or median cubital vein; it may divide distal to the elbow to join both.

Surface Anatomy. Superficial veins are usually visible until they pierce the deep fascia. Larger ones are obvious when the limb is dependent and its muscles contracted, driving blood from the deep to the superficial veins.

Applied Anatomy. Blood sampling, blood transfusion and intravenous injection are commonly done near the elbow or more distally in the forearm; the largest vein is usually the median cubital. The cubital veins are also used for cardiac catheterization for many purposes. Equally useful for such procedures is the cephalic vein where it is superficial to the distal end of the radius in the 'anatomical snuffbox'. The cephalic vein, a little proximal to the snuff box, is the site with many advantages for an indwelling cannula or fine tube when a lengthy period is contemplated; the position of the arm, forearm and hand is optimal for this purpose.

DEEP VEINS OF THE UPPER LIMB

Deep veins (venae comitantes) accompany arteries, usually in pairs, flanking the artery and connected by short transverse links. Since much blood from the upper limb is returned by the superficial veins, the deep ones are relatively small.

Deep veins of the hand. Superficial and deep palmar arterial arches are accompanied by *superficial* and *deep palmar venous arches*, receiving the corresponding branches. Thus *common palmar digital veins* join the superficial arch and *palmar metacarpal veins* join the deep arch. Deep veins accompanying the dorsal metacarpal arteries first receive perforating branches from the palmar metacarpal veins and then end in the radial veins and the dorsal venous network.

Deep veins of the forearm run with the radial and ulnar arteries draining respectively the deep and superficial palmar venous arches; they unite near the elbow as paired *brachial* veins. The radial veins are smaller, receiving the deep dorsal veins of the hand; ulnar veins drain the deep palmar venous arch, connecting with superficial veins near the wrist; near the elbow they receive the anterior and posterior interosseous artery companion veins; a large branch connects them to the *median cubital vein.*

Brachial veins flank the brachial artery, with tributaries similar to the arterial branches; near the lower margin of *subscapularis* they join the axillary vein, the medial one, however, often joining the basilic before it becomes the axillary.

These deep veins have numerous anastomoses with each other *and* with the superficial veins.

The axillary vein, the continuation of the basilic, begins at the lower border of *teres major,* and ascends to the outer border of the first rib, where it becomes the subclavian. Near subscapularis the brachial veins joins it and, near its costal end, the cephalic; other tributaries follow the axillary arterial branches. It is medial to the axillary artery, which it partly overlaps; between them are the medial pectoral nerve, medial cord of the brachial plexus, the ulnar nerve and the medial cutaneous nerve of the forearm. The medial cutaneous nerve of the arm is medial to the vein; the lateral group of axillary lymph nodes is posteromedial. It has a pair of valves near its distal end; valves also occur near the ends of the cephalic and subscapular veins.

The subclavian vein (6.87), continuing the axillary, extends from the outer border of the first rib to the medial border of scalenus anterior, where it joins the internal jugular to form the brachiocephalic vein. *Anterior* are the clavicle and subclavius, *posterosuperior* the subclavian artery, separated by the scalenus anterior and phrenic nerve; *inferior* are the first rib and pleura. The vein usually has a pair of valves about 2 cm from its end. Its tributaries are the external jugular, dorsal scapular and sometimes the anterior jugular; occasionally a small branch ascends in front of the clavicle from the cephalic vein. At its junction with the internal jugular the left subclavian receives the thoracic duct, the right subclavian vein and the right lymphatic duct.

Surface Anatomy. The vein can be projected as a broad band, convex upwards, from just medial to the mid-clavicular point to the medial edge of the clavicular attachment of sternocleidomastoid.

VEINS OF THE THORAX

The brachiocephalic (innominate) veins, two large vessels at the junction of the neck and upper thorax, are the united trunks of the internal jugular and subclavian veins. Both are devoid of valves.

The right brachiocephalic vein (6.161), about 2.5 cm long, begins posterior to the sternal end of the right clavicle, descending almost vertically to join the left brachiocephalic in forming the superior vena cava posterior to the lower border of the first right costal cartilage, near the right sternal border. It is anterolateral to the brachiocephalic artery and right vagus nerve. The right pleura, phrenic nerve and internal thoracic artery are posterior to it above, becoming lateral below. Its tributaries are the right vertebral, internal thoracic, inferior thyroid and sometimes the first right posterior intercostal veins.

The left brachiocephalic vein (6.161), about 6 cm long, begins posterior to the sternal end of the left clavicle, anterior to the cervical pleura. It descends *obliquely to the right*, posterior to the upper half of the manubrium sterni, to the sternal end of the first right costal cartilage, uniting here with the right brachiocephalic to form the superior vena cava. It is separated from the left sternoclavicular joint and manubrium by the sternohyoid and sternothyroid, the thymus or its remains and areolar tissue; terminally it is overlapped by the right pleura. It crosses *anterior* to the left internal thoracic, subclavian and common carotid arteries, left phrenic and vagus nerves, trachea and brachiocephalic artery. The aortic arch is inferior to it. The vein's tributaries are the left vertebral, internal thoracic, inferior thyroid, superior intercostal, sometimes the first left posterior intercostal, thymic and pericardial veins.

Surface Anatomy. The brachiocephalic veins can be projected as broad bands 1.5 cm wide from the sternal ends of the clavicles to the parasternal lower border of the first right costal cartilage.

Variations. The brachiocephalic veins may enter the right atrium separately, the right vein descending like a normal superior vena cava; a *left superior vena cava* may have a slender connection with the right and then cross the left side of the aortic arch to pass anterior to the left pulmonary hilum before turning to enter the right atrium. It replaces the oblique atrial vein and coronary sinus

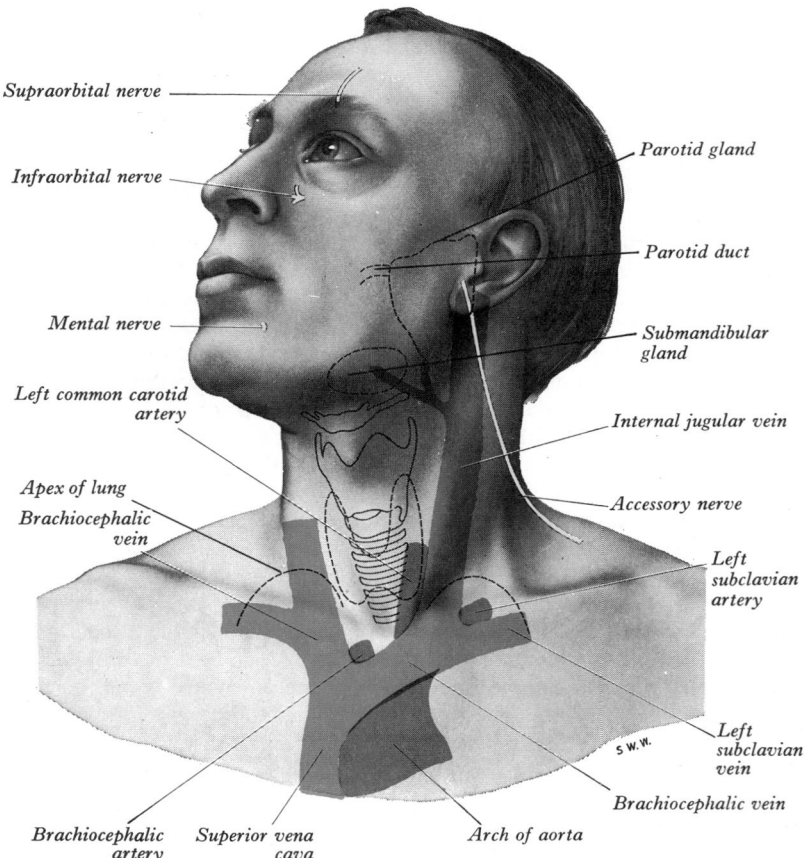

Supraorbital nerve

Infraorbital nerve

Mental nerve

Left common carotid artery

Apex of lung
Brachiocephalic vein

Brachiocephalic artery   *Superior vena cava*

Parotid gland

Parotid duct

Submandibular gland

Internal jugular vein

Accessory nerve

Left subclavian artery

Left subclavian vein

Brachiocephalic vein

Arch of aorta

S.W.W.

6.161 The surface projections of some of the important structures in the face, neck and upper part of the thorax. The apices of the lungs, the thyroid, submandibular and parotid glands and the parotid duct are indicated in interrupted dotted outline; the hyoid bone, the thyroid and cricoid cartilages and the rings of the trachea are shown in continuous outline.

and receives all the latter's tributaries. This abnormality, due to persistence of an early fetal condition, is normal in birds and some mammals. The left brachiocephalic vein sometimes projects above the manubrium (more frequently in childhood), crossing the suprasternal fossa in front of the trachea.

Internal thoracic (mammary) veins (6.66,90) are venae comitantes to the inferior half of the internal thoracic artery; they have several valves. Near the third costal cartilages the veins unite to ascend medial to the artery, ending in their brachiocephalic vein. Tributaries correspond to branches of the artery (p. 754), and include a pericardiacophrenic vein.

Inferior thyroid veins (6.66) arise in a glandular venous plexus, which also connects with the middle and superior thyroid veins. These veins form a *pretracheal plexus* from which the left inferior vein descends to join the left brachiocephalic, the right descending obliquely across the brachiocephalic artery to the right brachiocephalic vein, at its junction with the superior vena cava; the inferior thyroid veins often open in common into the vena cava or left brachiocephalic vein. They drain the oesophageal, tracheal and inferior laryngeal veins and have valves at their terminations.

The left superior intercostal vein (6.162) drains the second and third (sometimes fourth) left posterior intercostal veins, ascending obliquely forwards across the left aspect of the aortic arch, lateral to the left vagus, medial to the left phrenic nerve, to open into the left brachiocephalic vein. It usually receives the left bronchial veins, sometimes the left pericardiacophrenic; it connects inferiorly with the accessory hemiazygos vein.

The superior vena cava (6.60,63,64,65,66) returns blood from the superior half of the body. It is about 7 cm in length, formed by the junction of the brachiocephalic veins, and has no valves. It begins behind the lower border of the first right costal cartilage near the sternum, descends vertically behind the first and second intercostal spaces, ending in the upper right atrium behind the third right costal cartilage; its inferior half is within the fibrous pericardium, which it pierces level with second costal cartilage. Covered anterolaterally by serous pericardium from which projects a *retrocaval recess* (p. 696), it is slightly convex to the right.

Relations. Anterior are the anterior margins of the right lung and pleura, the pericardium intervening below; these separate the vein from the internal thoracic artery and first and second intercostal spaces, and second and third costal cartilages; *posteromedial* are the trachea and right vagus nerves and *posterolateral* the right lung and pleura; *posterior* is the right pulmonary hilum. *Right lateral* are the right phrenic nerve and pleura, *left lateral* the brachiocephalic artery and ascending aorta, the latter overlapping it.

Surface Anatomy. The superior vena cava, 2 cm wide, is partly behind but projects well beyond the right sternal margin, from the lower border of the first to the lower border of the third right costal cartilage. Its lateral border is visible in anteroposterior radiographs.

Tributaries: the azygos vein and small veins from the pericardium and other mediastinal structures.

The azygos vein (6.162A,B,163,164) is inconstant in origin (Gladstone 1929). An origin from the posterior aspect of the inferior vena cava, at or below the level of the renal veins, is to be expected from its development. Such a *lumbar azygos vein* frequently occurs, ascending anterior to the upper lumbar vertebrae. The vein may pass behind the right crus of the diaphragm or pierce it. It may traverse the aortic opening on the right of the cisterna chyli. Anterior to the twelfth thoracic vertebral body it is joined by a large vessel formed by the right ascending

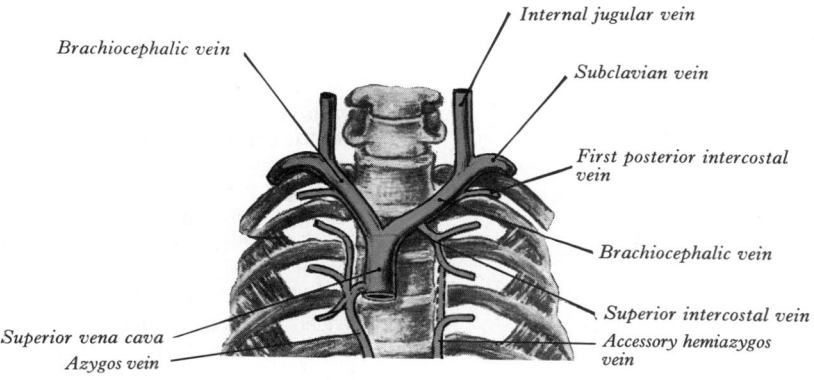

Much variation occurs in the transthoracic parts of the azygos and hemiazygos veins, in terms of numbers of radicles, levels of transmedian crossing, etc. Schemata are usually misleading. That depicted by painting in 6.162B is the most common condition.

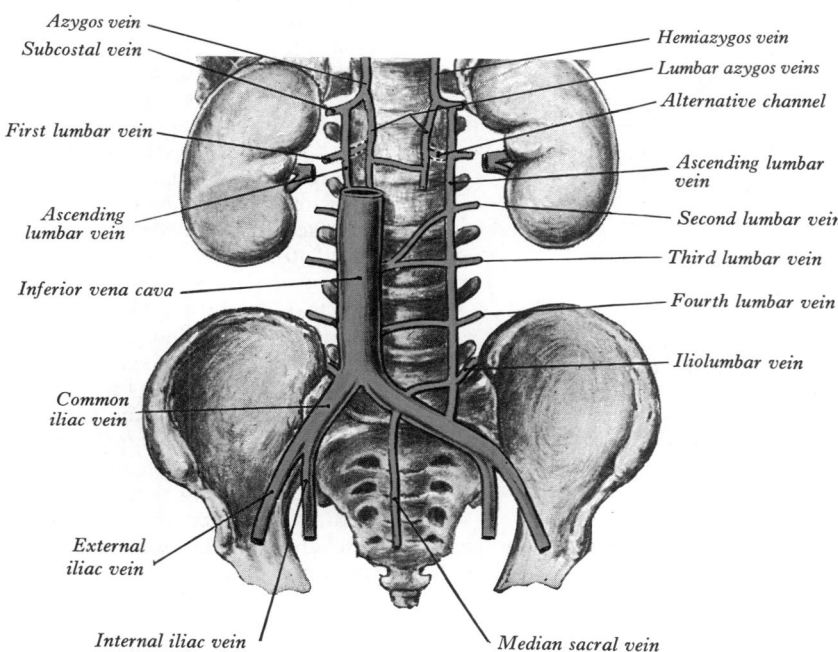

6.162A Schema showing the superior and inferior extremities of the azygos system of veins and their principal associated veins. The intervening parts have been omitted because schemata of this region are often topographically misleading.

lumbar and right subcostal veins, which passes forward and right of the twelfth thoracic vertebra behind the right crus. This common trunk may, in the absence of a lumbar azygos, form the azygos itself. Whatever its origin, the azygos vein ascends in the posterior mediastinum to the fourth thoracic vertebra, arching forward above the right pulmonary hilum to end in the superior vena cava, before the latter pierces (enters) the pericardium. It is *anterior* to the lower eight thoracic vertebral bodies (vide infra), anterior longitudinal ligament and right posterior intercostal arteries. *Right lateral* are the right greater splanchnic nerve, lung and pleura; *left lateral* in most of its course are the thoracic duct and aorta and, where it arches forward, the oesophagus, trachea and right vagus. In the lower thorax it is covered *anteriorly* by a recess of the right pleural sac and oesophagus, emerging from behind the latter to ascend behind the right hilum (**6.164**). In the lower thorax it should be noted that the azygos vein closely follows the right posterolateral aspect of the descending thoracic aorta for a variable distance (Dr M.C.E. Hutchinson, personal communication), thus *often reaching* and even *crossing* left of the midline; aortic pulsations might thus assist venous return in azygos and hemiazygos veins (cf. p. 691).

Tributaries. The azygos vein drains: the right posterior intercostal veins except the first, the veins from the second to fourth intercostal spaces usually via a right superior intercostal vein, the hemiazygos and accessory hemiazygos veins, oesophageal, mediastinal and pericardial veins and, near its end, right bronchial veins. When it begins as a lumbar azygos, the common trunk formed by the right ascending lumbar and subcostal veins is its largest tributary. Imperfect valves occur in the azygos vein, some tributaries having complete valves.

The hemiazygos starts on the left like the azygos; ascending anterior to the vertebral column to the eighth thoracic level, it

809

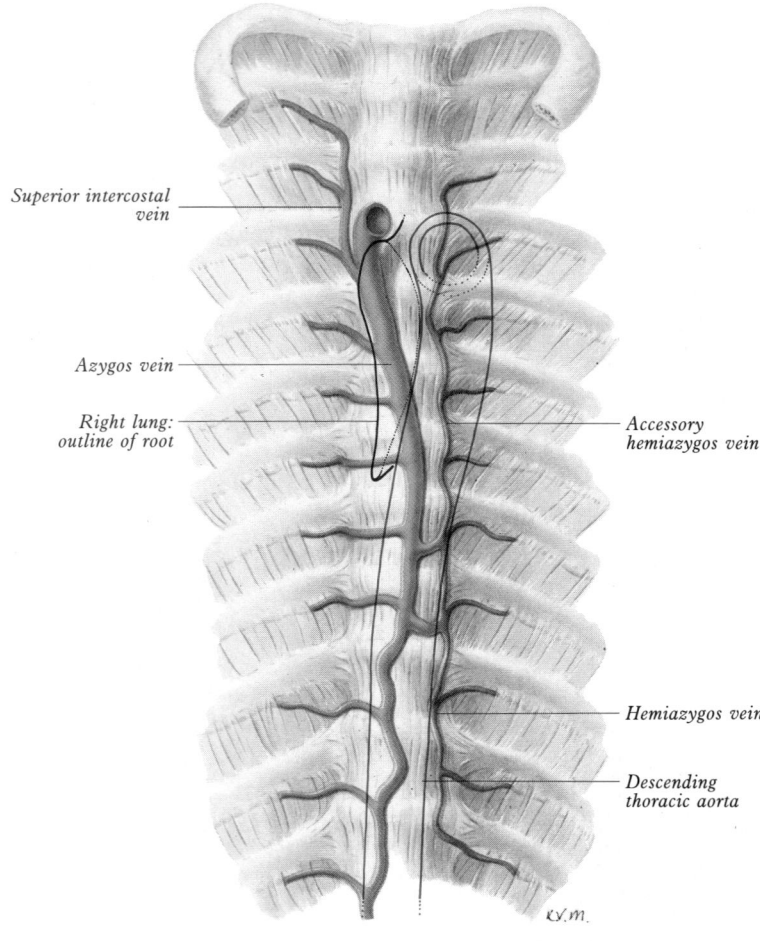

Superior intercostal vein

Azygos vein

Right lung: outline of root

Accessory hemiazygos vein

Hemiazygos vein

Descending thoracic aorta

6.162B A frequent (perhaps the commonest) course followed by the intrathoracic azygos, hemiazygos and accessory hemiazygos veins. Outlines of the root of the right lung and descending thoracic aorta are included. Dissection by M C E Hutchinson, Guy's Hospital Medical School, London.

crosses the column posterior to the aorta, oesophagus and thoracic duct to end in the azygos vein. Its tributaries are the lower three posterior intercostal veins, a common trunk formed by the left ascending lumbar and subcostal veins and oesophageal and

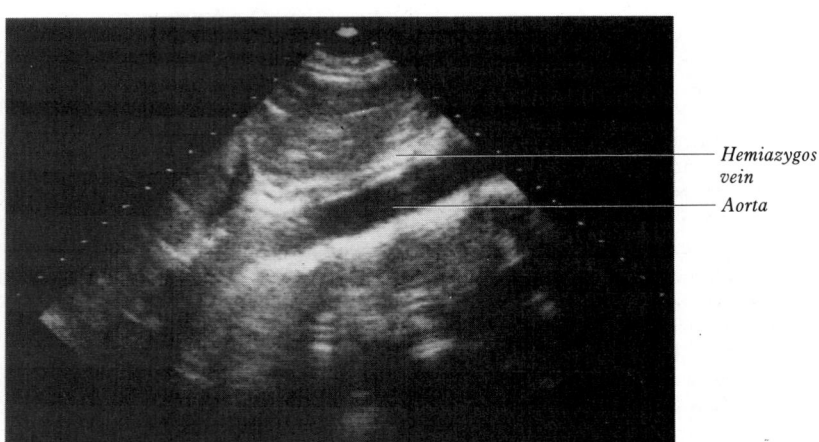

Hemiazygos vein

Aorta

6.163 Ultrasonogram showing the hemiazygos vein accompanying the aorta; both vessels are shown in longitudinal section. Supplied by Shaun Gallagher, Guy's Hospital; photography by Sarah Smith.

mediastinal rami. Its lower end often connects with the left renal vein.

The accessory hemiazygos vein descends to the left of the vertebral column, receiving veins from the fourth (or fifth) to eighth intercostal spaces and sometimes the left bronchial veins. It crosses the seventh thoracic vertebra to join the azygos vein. It sometimes joins the hemiazygos, their common trunk opening into the azygos vein.

The azygos veins vary much in their mode of origin, course, tributaries, anastomoses and termination. For a survey consult Grzybiak et al (1975), who consider the accessory hemiazygos most variable, draining into the left brachiocephalic, azygos or hemiazygos. The arrangement shown in **6.**162B represents a common pattern. In about 1 or 2% according to Anson (1963) there are left and right independent azygos veins (the early embryonic form) and occasionally a single azygos without hemiazygos tributaries, in a midline position. In more than 95% a main 'right-sided' azygos and at least some representative of hemiazygos veins exist. The latter vary, one or other being absent or poorly developed. Retro-aortic trans-vertebral connections from hemiazygos and accessory hemiazygos veins to the azygos are also extremely variable; there may be from one to five, or more; when either hemiazygos is absent, intercostal veins involved cross vertebral bodies to end in the azygos. These trans-vertebral routes are often very short, since the azygos vein is more common-ly *anterior* to the vertebral column (Anson 1963, Dr. M.C.E. Hutchinson, personal communication) and often passes *left* of the midline in part of its course.

Posterior intercostal veins (6.106,164) accompany their arteries in eleven pairs. Approaching the vertebral column each vein receives a posterior tributary returning blood from the dorsal muscles and skin and vertebral venous plexuses. On both sides the first posterior intercostal vein ascends anterior to the first rib's neck, arching forward above the pleural dome to end in the ipsi-lateral brachiocephalic or vertebral vein. On the right the second, third and often fourth form a *right superior intercostal vein* joining the arch of the azygos vein. Veins from the lower spaces drain directly to it. On the left the second and third (sometimes fourth) form a left superior intercostal vein (p. 869). Veins from the fourth (or fifth) to eighth spaces end in the accessory hemiazygos vein, veins from the lower three spaces in the hemiazygos.

Posterior intercostal veins are so called to distinguish them from small *anterior intercostal veins* which are tributaries of the internal thoracic and musculophrenic veins.

Applied Anatomy. In obstruction of the upper inferior vena cava, the azygos and hemiazygos veins and vertebral venous plexuses are the main collateral channels maintaining venous cir-culation, by connecting superior and inferior venae cavae and communicating with the common iliac by ascending lumbar veins and with many tributaries of the inferior vena cava.

Bronchial veins, usually two on each side, drain blood from larger bronchi and from hilar structures. The right bronchial veins join the end of the azygos, the left join the left superior intercostal or hemiazygos vein. Some blood carried to the lungs by bronchial arteries returns via the pulmonary veins (see p. 1285).

VEINS OF THE VERTEBRAL COLUMN (6.165,166)

These form intricate plexuses along the entire column, external and internal to the vertebral canal. Both groups are devoid of valves, anastomose freely with each other and join the inter-vertebral veins. Interconnections are widely established between these plexuses and longitudinal veins early in fetal life (Loginova 1972).

External vertebral venous plexuses, most developed in the cervical region, are anterior and posterior, anastomosing freely. *Anterior external plexuses* are anterior to the vertebral bodies, communicating with basivertebral and intervertebral veins and receiving tributaries from vertebral bodies. *Posterior external plexuses* lie posterior to vertebral laminae and around spines, transverse and articular processes. They anastomose with the internal plexuses and join the vertebral, posterior intercostal and lumbar veins.

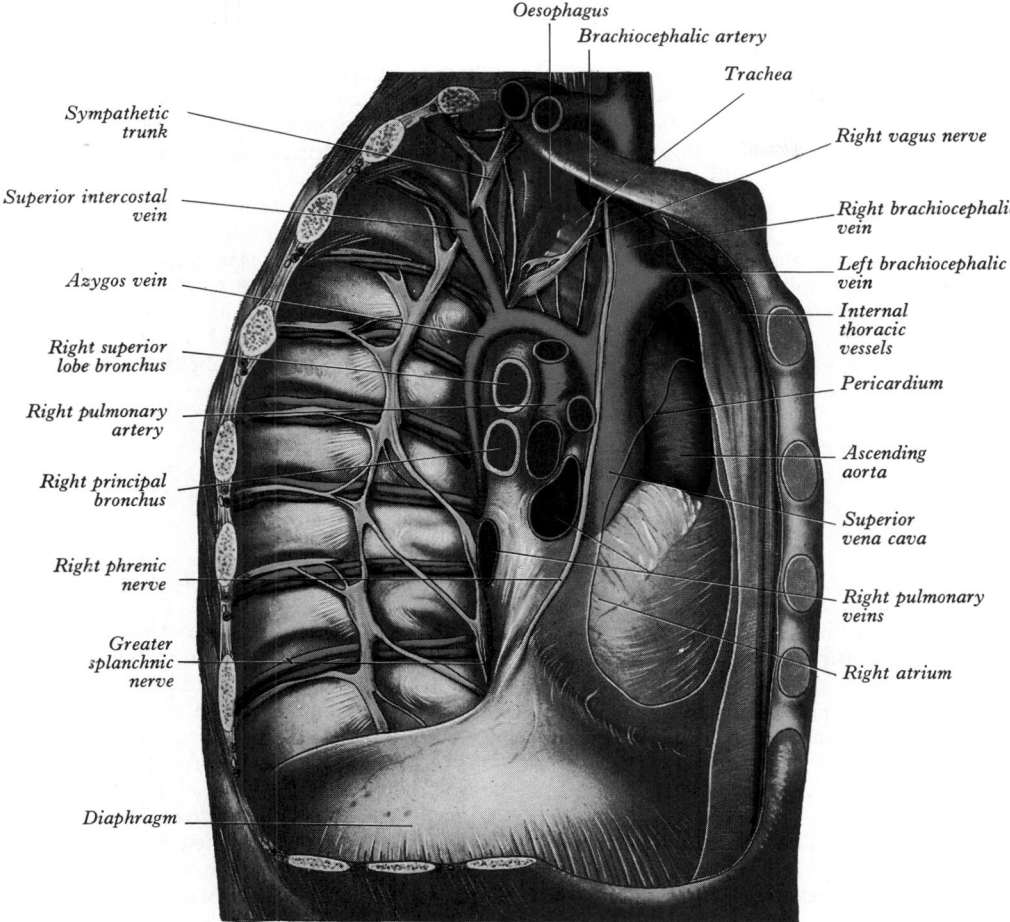

6.164 The right aspect of the mediastinum. The right lung and most of the right pleura have been removed and a large opening made into the pericardial sac to expose the heart. In this specimen the fourth right posterior intercostal vein did not join the superior intercostal vein.

Internal vertebral venous plexuses occur between the dura mater and vertebrae, receiving tributaries from the bones, red bone marrow and spinal cord. They form a denser network than the external plexuses and are arranged vertically as four interconnecting longitudinal vessels, two in front, two behind.

Anterior internal plexuses are large plexiform veins on the posterior surfaces of vertebral bodies and intervertebral discs, flanking the posterior longitudinal ligament; under this they are connected by transverse branches, into which the large *basivertebral veins* open. *Posterior internal plexuses*, on each side in front of the vertebral arches and ligamenta flava, anastomose with the posterior external plexuses by veins passing through and between the ligaments. The internal plexuses interconnect by venous rings near each vertebra. Around the foramen magnum they form a dense network connecting with: vertebral veins, occipital and sigmoid sinuses, basilar plexus, venous plexus of the hypoglossal canal and the condylar emissary veins.

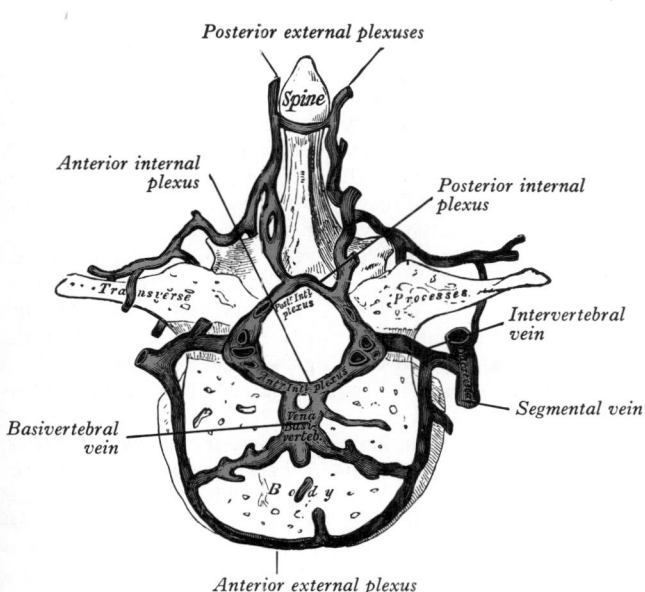

6.165 Transverse section through the body of a thoracic vertebra showing the vertebral venous plexuses and basivertebral veins.

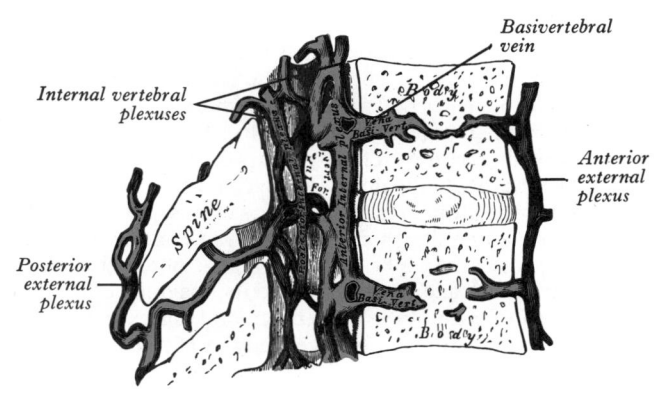

6.166 Median sagittal section through two thoracic vertebrae showing the vertebral venous plexuses and basivertebral veins.

The basivertebral veins emerge from the posterior foramina of vertebral bodies. They are large and tortuous channels in bone, like those in cranial diploë. The trabecular bone in vertebral bodies contains much haemopoietic tissue. The basivertebral veins also drain into the anterior external vertebral plexuses through small openings in the vertebral bodies. Posteriorly they form one or two short trunks opening into the transverse branches uniting anterior internal vertebral plexuses. They enlarge in advanced age.

Intervertebral veins accompany the spinal nerves through intervertebral foramina, draining the spinal cord and internal and external vertebral plexuses, and ending in the vertebral, posterior intercostal, lumbar and lateral sacral veins. Whether the basivertebral or intervertebral veins contain effective valves is uncertain but experiment strongly suggests that their blood flow can be reversed (Batson 1957). This may explain how pelvic neoplasms, in particular, may metastasize in vertebral bodies, the cells spreading into the internal vertebral plexuses by connections with the pelvic veins when blood flow is temporarily reversed by raised intra-abdominal pressure or postural alterations.

Veins of the spinal cord lie in the pia mater, forming a tortuous venous plexus. In this there are: (1) two *median longitudinal veins*, one near the anterior median fissure, the other behind the posterior median septum; (2) two *anterolateral* and two *posterolateral longitudinal veins* respectively behind the ventral and dorsal spinal nerve roots. They drain to internal vertebral plexuses, and thence to intervertebral veins. Near the skull they unite into two or three small veins joined to the vertebral veins and ending in inferior cerebellar veins or the inferior petrosal sinuses.

VEINS OF THE LOWER LIMBS

Veins of the lower limbs can be subdivided, like those of the upper, into *superficial* and *deep* groups, the superficial being subcutaneous in the superficial fascia, the deep veins (beneath the deep fascia) accompanying major arteries. Both have valves, more numerous in deep veins and also more numerous than in the upper limb.

SUPERFICIAL VEINS OF THE LOWER LIMB (6.167,168)

The principal named superficial veins are the great (long) and small (short) saphenous; their numerous tributaries are mostly (but not wholly) unnamed; named vessels will be noted. (For variations consult Kosinski 1926.) As in the upper limb the vessels will be described centripetally from peripheral to major drainage channels.

Dorsal digital veins receive, in the clefts between the toes, rami from the plantar digital veins and then join to form *dorsal metatarsal veins*, which are united across the proximal parts of the metatarsal bones in a *dorsal venous arch*. Proximal to this is an irregular *dorsal venous network* receiving tributaries from deep veins and continuous proximally with a venous network in the leg. At each side of the foot this network connects with *medial* and *lateral marginal veins*, both formed mainly by veins from more superficial parts of the sole. In the sole superficial veins form a *plantar cutaneous arch* across the roots of the toes and also drain into the medial and lateral marginal veins. Proximal to the plantar arch is a *plantar cutaneous venous plexus*, especially dense in the fat of the heel; this connects with the plantar cutaneous venous arch and other deep veins, but drains mainly into the marginal veins.

The great (long) saphenous vein (6.167) starts inferiorly (below) as a continuation of the medial marginal vein and ends in the femoral vein a short distance distal to the inguinal ligament (vide infra), being thus the body's longest vein. It ascends about 2.5–3 cm *anterior* to the tibial malleolus, crosses the distal third of the medial surface of the tibia obliquely to its medial border, then ascends a little behind the border to the knee; proximally it is posteromedial to the medial tibial and femoral condyles, then ascends the medial aspect of the thigh; after traversing the saphenous opening (p. 638) it finally enters the femoral vein. The so-called 'centre' of the opening is often said to be 2.5–3.5 cm inferolateral to the pubic tubercle; and the vein is then held to be represented by a line drawn from this to the femoral adductor tubercle. However, the saphenous opening, as noted elsewhere, varies greatly in size and disposition and its imagined centre has proved a poor indicator of the saphenofemoral junction. Royle et al (1981) made a careful quantitative study in 167 flush ligations in 136 subjects noting, in particular, the relative positions of the pubic tubercle, the venous junction and the inguinal skin crease. They concluded that a correctly placed incision for flush saphenofemoral ligation should be made 1 cm above, and parallel to, the inguinal skin crease, centring the incision at a point 4 cm lateral to and level with the pubic tubercle.

In its course through the thigh the great saphenous vein has branches of the medial femoral cutaneous nerve accompanying it: at the knee the saphenous branch of the descending genicular artery and, in the leg and foot, the saphenous nerve, are anterior to it. The vein is often duplicated, especially distal to the knee. It has from 10 to 20 valves, which are more numerous in the leg than the thigh. One is present just before it pierces the cribriform fascia, another at its junction with the femoral vein. In almost its entire extent the vein lies in superficial fascia, but it has many connections with the deep veins, especially in the leg (vide infra).

Tributaries. At the ankle it drains the sole by medial marginal veins. In the leg it often connects with the small saphenous vein and with deep veins through *perforating veins*. Just distal to the knee it usually has three large tributaries: one from the front of the leg, a second from the tibial malleolar region (connecting with some of the 'perforating' veins) and a third from the calf (communicating with the small saphenous vein). The second of these forms below in a fine network or 'corona' of delicate veins over the medial malleolus and then ascends the medial aspect of the calf as the *'posterior arch vein'* of Dodd & Cockett (1956); the clinical importance of its connections with posterior tibial venae comitantes by a series of perforating (communicating) veins was emphasized by Platz & Adelmann (1976), who proposed the term *'vena arcuata cruris posterior'*. The clinical significance of the latter should be reaffirmed, with relevant points concerning other venous channels in the leg (D Negus, 1988, personal communication). Although Platz & Adelmann (1976) mentioned 3–6 perforating veins, it has been indicated that three are most usual, being equally spaced between the medial malleolus and the mid-calf; more than three was termed 'most uncommon' and an arch vein perforator above mid-calf 'extremely rare'. The posterior crural arch vein was first illustrated by Leonardo da Vinci and his name is often applied to the vein in some surgical circles.

Above the posterior crural arch vein, perforating veins join the great saphenous or one of its main tributaries at two main sites. The first is at a level in the upper calf indicated by its name, the *tibial tubercle perforator (Boyd's perforator)*; the second is in the lower/intermediate third of the thigh where it perforates the deep fascial roof of the subsartorial canal to join the femoral vein (*Hunterian perforator*).

In the thigh the great saphenous vein receives many tributaries; some open independently, whilst others converge to form large named channels that frequently pass towards the basal half of the femoral triangle before joining the great saphenous near its termination. These may be grouped thus: one or more large posteromedial tributaries, one or more large anterolateral tributaries, four or more peri-inguinal veins. *The posteromedial vein of the thigh*, large and sometimes double, drains a large superficial tissue volume in the region indicated by its name; it has (as have the other tributaries) radiological and surgical significance. One of its lower radicles is often continuous with the small saphenous vein. The posteromedial vein is sometimes (perhaps unhelpfully) named the *accessory saphenous vein* with greater emphasis on its variability of form and level of junction with the great saphenous. Some restrict the term *accessory* to a lower (more distal) posteromedial branch when two (or more) are present. Another large vessel, the *anterolateral vein of the thigh (anterior femoral cutaneous vein)*, usually commences from an anterior network of veins in the distal thigh and crosses the apex and distal half of the femoral triangle to reach the great saphenous vein. As the latter traverses its saphenous opening (**6.167**), it is joined by the superficial epigastric, superficial circumflex iliac and superficial external pudendal veins. Their mode of union varies. Superficial epigastric and circumflex iliac veins drain the inferior abdominal wall, the latter also receiving tributaries from the proximolateral region of the thigh; superficial external pudendal veins drain part of the scrotum, one being joined by the superficial dorsal vein of the penis. The deep external pudendal vein joins the great saphenous in its opening.

A *thoraco-epigastric vein* lies along the anterolateral aspect of the trunk and connects the superficial epigastric or femoral vein to the lateral thoracic veins, thus connecting femoral and axillary veins and hence superior and inferior vena caval fields of drainage. It is held to be in line with the primitive mammary ('milk') ridge which extends from the axilla to the pubic region (p. 177).

The small (short) saphenous vein (**6.168**) begins posterior to the lateral malleolus, as a continuation of the lateral marginal vein. In the distal (inferior or lower) third of the calf it ascends lateral to the tendo calcaneus, lying on the deep fascia and covered only by superficial fascia and skin. Inclining medially to the midline of the calf it enters an intra-deep fascial compartment (or tunnel) within which it ascends on the gastrocnemius, only emerging *between* the deep fascia and gastrocnemius gradually at about the junction of the intermediate and proximal thirds of the calf (well *below* the lower limit of the popliteal fossa). Continuing its ascent it passes between the heads of the gastrocnemius, then proceeds to its termination in the popliteal vein, 3–7.5 cm above the knee joint in the popliteal fossa. (These fascial relations are of considerable surgical significance when small saphenous vein ligation is contemplated. Figure **6.168** from an earlier edition is somewhat misleading and should be examined in close conjunction with this textual description.)

The small saphenous vein connects with deep veins on the dorsum of the foot, receives many cutaneous tributaries in the leg and sends several rami proximally and medially to join the great saphenous vein. Sometimes a communicating branch from it ascends medially to the accessory saphenous vein (vide supra); this may be the main continuation of the small saphenous. In the leg the small saphenous lies near the sural nerve. It has 7–13 valves, one near its termination. Its mode of ending is variable; it may join the great saphenous vein in the proximal thigh or it may bifurcate, one branch joining the great saphenous, the other the

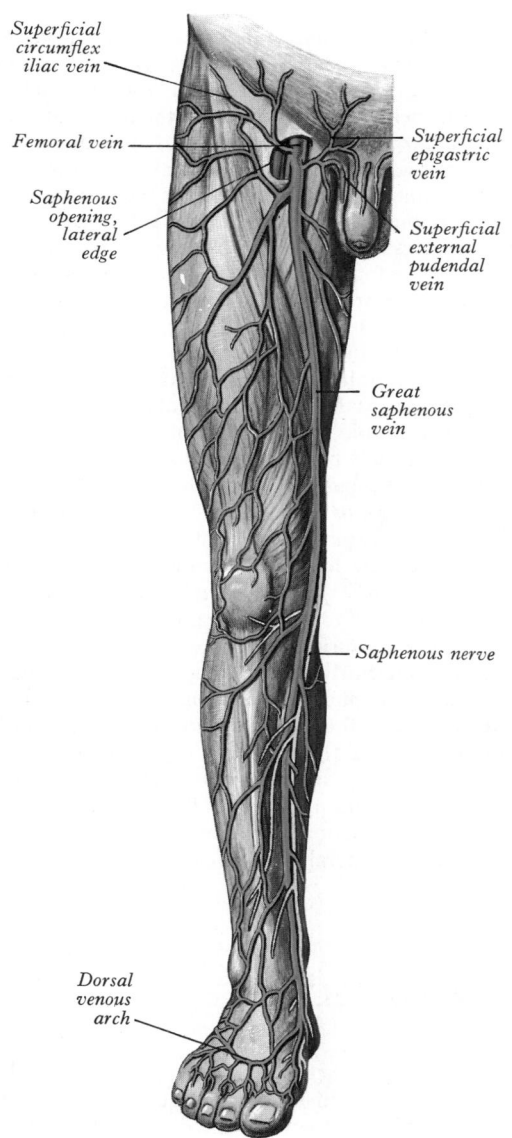

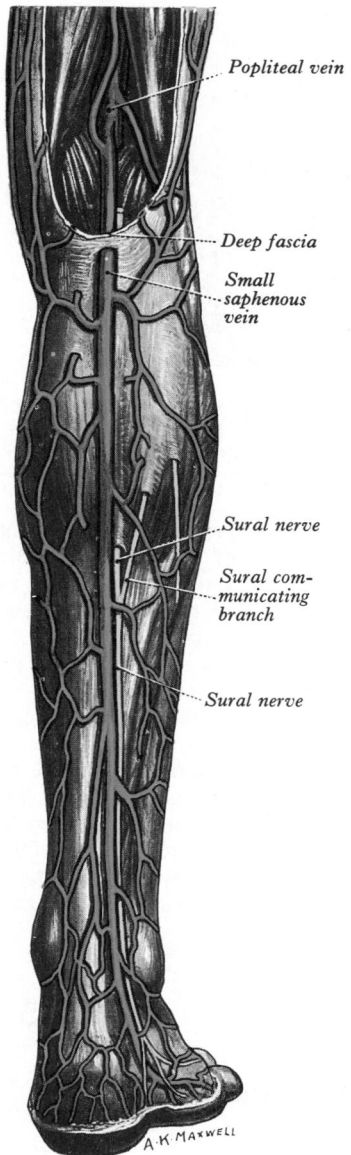

6.167 The great saphenous vein and its tributaries.

6.168 The small saphenous vein and its tributaries.

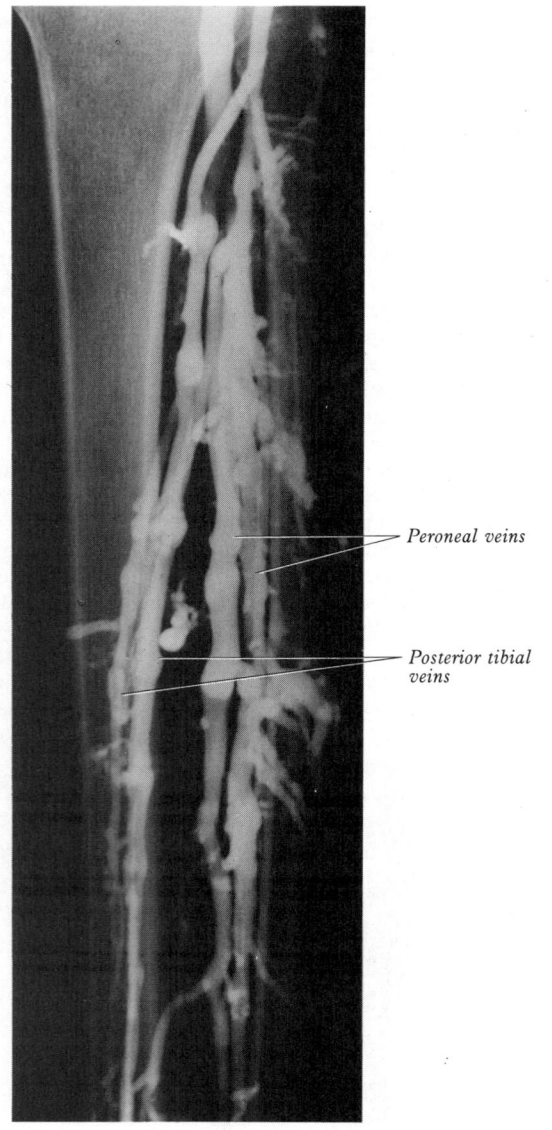

aspirated from superficial into deep veins. If the valves in the perforating veins become incompetent, these veins become 'high pressure leaks' during muscular contraction; this transmission of high pressure in deep veins to superficial veins results in dilatation and venous stagnation in the latter, producing varicosities, devitalization of tissues and ultimately varicose ulceration. In operative treatment of severe varicose veins and ulcers, perforating veins *must* be ligatured. Similar perforating connections occur in the anterolateral region and varicosities may also occur here (Cockett 1956, Green et al 1958). Veins connecting the great saphenous to the femoral vein, in the adductor canal, may also become varicose (Dodd & Cockett 1956, Dodd 1959).

DEEP VEINS OF THE LOWER LIMB

These accompany the arteries and their branches; they have numerous valves (**6.169**). *Plantar digital veins* arise from plexuses in the plantar regions of the toes, connecting with dorsal digital veins and uniting into four *plantar metatarsal veins*; these run in the intermetatarsal spaces and connect by perforating veins with dorsal veins, then continue to form the *deep plantar venous arch*, accompanying the plantar arterial arch. From this *medial* and *lateral plantar veins* run near the corresponding arteries and, after communicating with the great and small saphenous veins, form behind the medial malleolus the posterior tibial veins.

Posterior tibial veins accompany the posterior tibial artery, receiving veins from sural muscles, especially the venous plexus in the soleus, connections from superficial veins and the *peroneal veins*. The latter, running with their artery, receive rami from the soleus and superficial veins.

Anterior tibial veins are continuations of the venous companions of the dorsal pedal artery. They leave the extensor region between the tibia and fibula, through the proximal end of the interosseous membrane, uniting with the posterior tibial veins to form the *popliteal vein* at the distal border of the popliteus.

The popliteal vein ascends through the fossa to an aperture in the adductor magnus, where it becomes the femoral vein. Distally it is medial to the artery; between the heads of gastrocnemius it is superficial (dorsal) to it; proximal to the knee joint it is posterolateral. Its tributaries are: the small saphenous vein, veins corresponding to branches of the popliteal artery and muscular veins, including a large one from each head of gastrocnemius. There are usually four valves in the popliteal vein.

The femoral vein accompanies its artery, beginning at the adductor opening as the continuation of the popliteal, and ending posterior to the inguinal ligament as the external iliac. In the distal adductor canal, it is posterolateral to the femoral artery; more proximally in the canal, and in the distal femoral triangle (i.e. its apex), it is posterior to it; at the triangle's base it is medial (**6.127**, **130**). The vein occupies the middle compartment of the femoral sheath, between the femoral artery and canal, fat in the latter allowing expansion of the vein. It has many muscular tributaries: about 4–12 cm distal to the inguinal ligament the vena profunda femoris joins it posteriorly and then the great saphenous vein, which enters anteriorly. Lateral and medial circumflex femoral veins are usually tributaries. There are usually four or five valves in the femoral vein, the most constant being one just distal to the entry of the profunda femoris and one near the inguinal ligament.

The vena profunda femoris is anterior to its artery, its tributaries corresponding; through these it connects distally with the poplital and proximally inferior gluteal veins. It sometimes drains medial and lateral circumflex femoral veins. It has a valve just before its end.

6.169 Venogram of the leg to show the deep veins; the valves are clearly demonstrated. Supplied by Shaun Gallagher, Guy's Hospital; photography by Sarah Smith.

Peroneal veins

Posterior tibial veins

popliteal or deep posterior femoral veins; sometimes it ends distal to the knee in the great saphenous or deep sural muscular veins.

Applied Anatomy. In a standing position, venous return from the lower limb depends largely on muscular activity (p. 691), especially contraction of the calf muscles, known as the 'calf pump', whose efficiency is aided by the tight sleeve of deep fascia. '*Perforating*' *veins* have been noted that connect the great saphenous with the deep veins, particularly near the ankle, distal calf and perigenual regions. In these channels valves are arranged to prevent flow of blood from deep to superficial veins. At rest, pressure in a superficial vein is equal to the height of the column of blood extending therefrom to the heart. When calf muscles contract, blood is pumped proximally in the deep veins but is normally prevented from flowing into superficial veins by the valves in the perforating veins; during relaxation blood can be

VEINS OF THE ABDOMEN AND PELVIS

The External Iliac Vein

The proximal continuation of the femoral vein is the external iliac: it thus begins posterior to the inguinal ligament, ascends the pelvic brim and ends anterior to the sacro-iliac joint by joining the

internal iliac to form the *common iliac vein*. On the right, it is first medial to the external iliac artery, gradually inclining behind it as it ascends; on the left, it is wholly medial. Medially it is crossed by the ureter and internal iliac artery, and is elsewhere covered by parietal peritoneum. In males it is crossed by the ductus deferens,

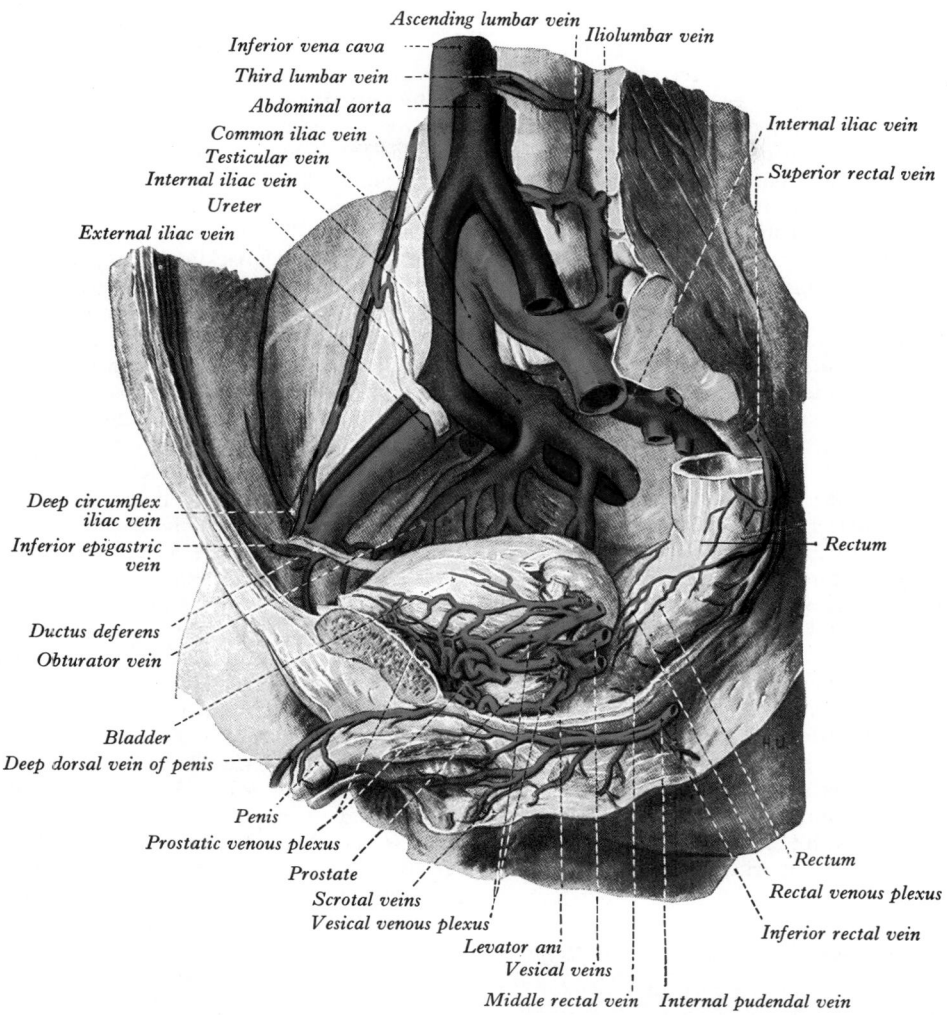

Ascending lumbar vein
Inferior vena cava
Third lumbar vein
Abdominal aorta
Common iliac vein
Testicular vein
Internal iliac vein
Ureter
External iliac vein
Iliolumbar vein
Internal iliac vein
Superior rectal vein
Deep circumflex iliac vein
Inferior epigastric vein
Rectum
Ductus deferens
Obturator vein
Bladder
Deep dorsal vein of penis
Penis
Prostatic venous plexus
Prostate
Scrotal veins
Vesical venous plexus
Levator ani
Vesical veins
Middle rectal vein
Rectum
Rectal venous plexus
Inferior rectal vein
Internal pudendal vein

6.170 The veins of the right half of the male pelvis (after Spalteholz).

in females by the round ligament and ovarian vessels. Lateral is the psoas major, except where the artery intervenes. The vein is usually valveless, but may contain a single value.

Tributaries: inferior epigastric, deep circumflex iliac and pubic veins.

The inferior epigastric vein is the union of the venae comitantes of the inferior epigastric artery, which connect above with the superior epigastric veins; it joins the external iliac about 1 cm proximal to the inguinal ligament. **The deep circumflex iliac vein** is formed from venae comitantes of the corresponding artery; it joins the external iliac about 2 cm proximal to the inguinal ligament after crossing anterior to the external iliac artery. **The pubic vein**, connecting the external iliac with the obturator vein in the obturator foramen, ascends on the pelvic surface of the pubis with the pubic branch of the inferior epigastric artery. It sometimes replaces the normal obturator vein.

The Internal Iliac Vein

Veins converge superiorly in the great sciatic foramen to form the internal iliac vein, which ascends posteromedial to the internal iliac artery to join the external iliac vein, forming the common iliac at the pelvic brim, anterior to the lower part of the sacro-iliac joint. It is covered anteromedially by parietal peritoneum.

Tributaries: (1) gluteal, internal pudendal and obturator veins, with origins outside the pelvis; (2) lateral sacral veins, anterior to the sacrum; (3) middle rectal, vesical, uterine and vaginal veins, originating in the venous plexuses of pelvic viscera.

The superior gluteal veins, venae comitantes of the superior gluteal artery, receive rami corresponding to branches of the ar-

tery; entering the pelvis via the greater sciatic foramen, above piriformis, they join the internal iliac vein, frequently as a single trunk.

The inferior gluteal veins, venae comitantes of the inferior gluteal artery, begin proximally and posterior in the thigh, where they anastomose with the medial circumflex femoral and first perforating veins; they enter the pelvis low in the greater sciatic foramen, joining to form a vessel opening into the distal (lower) part of the internal iliac vein. They connect with the *superficial gluteal veins* by perforating veins (Doyle 1970) similar to those in the calf (p. 812). These *gluteal perforating veins* are, indeed, even more numerous than the sural ones. In addition to a probable venous 'pumping' role, they provide collaterals between the femoral and internal iliac veins.

The internal pudendal veins, venae comitantes of the internal pudendal artery, begin in the prostatic venous plexus (p. 816), accompany the artery and unite as a single vessel ending in the internal iliac vein. They receive veins from the penile bulb and the scrotal (or labial) and inferior rectal veins. The deep dorsal vein of the penis connects with the internal pudendal but ends mainly in the prostatic plexus.

The obturator vein begins in the proximal adductor region, enters the pelvis superiorly in the obturator foramen, and runs back and up on the lateral pelvic wall below the obturator artery covered by peritoneum; it then passes between the ureter and internal iliac artery to end in the internal iliac vein. Sometimes it is replaced by an enlarged pubic vein, which joins the external iliac (p. 814).

Lateral sacral veins accompany the lateral sacral arteries; they are interconnected by a *sacral venous plexus*.

The middle rectal vein varies in size; it begins in the rectal venous plexus, with tributaries from bladder, prostate and seminal vesicle; it runs laterally on the pelvic surface of levator ani to end in the internal iliac vein.

The rectal venous plexus surrounds the rectum, connecting anteriorly with the vesical plexus in males and the uterovaginal plexus in females. It consists of an *internal* part beneath the rectal and anal epithelium and an *external* part outside the muscular stratum. In the anal canal the internal plexus has longitudinal dilatations, connected by transverse branches in circles immediately above the anal valves. The dilatations are most prominent in the left lateral, right anterolateral and right posterolateral sectors. The internal plexus drains mainly to the superior rectal vein but connects widely with the external plexus. The external plexus is drained inferiorly by the inferior rectal vein into the internal pudendal, its middle part by a middle rectal vein into the internal iliac, its superior part by the superior rectal vein, which is the commencement of the inferior mesenteric (a tributary of the portal vein). Communication between portal and systemic venous systems is thus established in the rectal plexus.

Applied Anatomy. Veins of the internal rectal plexus are apt to become varicose, as *internal haemorrhoids*, especially in the sectors indicated above. The vessels are in very loose areolar tissue, less supported by surrounding structures than most veins, and are less able to resist increased blood pressure; the superior rectal and portal veins have no valves; rectal veins pass through muscular tissue and are liable to compression, especially during defecation; they are affected by every form of portal obstruction. The veins in the subcutaneous part of the external plexus may thrombose, the basis of one type of *external haemorrhoid*; rupture of these vessels may lead to acute perianal haematoma.

The prostatic venous plexus is posterior to the arcuate pubic ligament and lower part of symphysis pubis, anterior to the bladder and prostate (**6**.170). Its chief tributary is the deep dorsal vein of the penis; it also receives anterior vesical and prostatic rami, connecting with the vesical plexus and internal pudendal vein. It drains into vesical and internal iliac veins. The plexus is embedded in the lateral fascial prostatic sheath (p. 1434).

The vesical plexus envelops the lower bladder and, in males, the prostatic base, communicating with the prostatic plexus in males and the vaginal plexus in females. It is drained by several vesical veins which usually unite to enter the internal iliac vein.

The dorsal veins of the penis are unpaired, and are superficial and deep: The *superficial dorsal vein* drains the prepuce and penile skin; running back in subcutaneous tissue it inclines right or left, and opens into one of the external pudendal veins. The *deep dorsal vein* is inside the fibrous penile sheath; it receives blood from the glans penis and corpora cavernosa penis, coursing back in the midline between the paired dorsal arteries; near the radix penis it passes deep to the suspensory ligament and through a gap between the arcuate pubic ligament and anterior margin of the perineal membrane (inferior fascia of the urogenital diaphragm), dividing into right and left branches which enter the prostatic plexus after connecting below the symphysis pubis with the internal pudendal veins. The *dorsal vein of the clitoris*, after a similar course, ends in the vesical plexus.

Uterine plexuses extend lateral to the uterus in the broad ligaments, communicating with the ovarian and vaginal plexuses. They are drained by two uterine veins on each side, arising inferiorly in the plexuses, level with the external os and draining to the internal iliac veins.

Vaginal plexuses flank the vagina; they connect with uterine, vesical and rectal plexuses and are drained by vaginal veins, one each side, to the internal iliac veins.

Common iliac veins (**6**.170, 171) result from the union of external and internal iliac veins, anterior to the sacro-iliac joints. They ascend obliquely to end at the right side of the fifth lumbar vertebra, uniting at an acute angle to form the inferior vena cava. The *right common iliac vein*, the shorter, is nearly vertical, ascending posterior, then lateral to its artery. The right obturator nerve passes posterior, descending forward to its foramen. The *left common iliac vein*, longer and more oblique, is first medial to its artery, then posterior. It is crossed anteriorly by the attachment of the sigmoid mesocolon and superior rectal vessels. In the rest of its course it is covered only by peritoneum. Each vein receives iliolumbar and sometimes lateral sacral veins; the left common iliac drains the median sacral vein. There are no valves in these veins.

Median sacral veins accompany the corresponding artery anterior to the sacrum, joining into a single vein usually ending in the left common iliac but sometimes at the common iliac junction.

Variations. The left common iliac vein occasionally ascends *left* of the aorta to the level of the kidney where, receiving the left renal vein, it crosses *anterior* to the aorta to join the inferior vena cava. This vessel represents the persistent caudal half of the left postcardinal or supracardinal vein (p. 223).

The Inferior Vena Cava (6.107A,170,172)

This conveys blood to the right atrium from all structures below the diaphragm. It is formed by the junction of the common iliac veins anterior to the fifth lumbar vertebral body, a little to its right. It ascends anterior to the vertebral column, to the right of the aorta. Reaching the liver, it is contained in a deep groove on its posterior surface or sometimes in a tunnel completed by a band of liver tissue. It perforates the *tendinous* part of the diaphragm between its median and right 'leaves' and inclines slightly anteromedially. Passing through the fibrous pericardium and through a posterior inflexion of the serous pericardium it opens into the inferoposterior part of the right atrium. Anterior and left of its atrial orifice is a semilunar *valve of the inferior vena cava*, relatively less prominent in adults, but large and overtly functional in the fetus (p. 723). The vessel is otherwise devoid of valves.

Relations of the abdominal part. Anteriorly the inferior vena cava is overlapped at its commencement by the right common iliac artery and covered, below the horizontal part of the duodenum, by the posterior parietal peritoneum. It is crossed obliquely by the root of the mesentery and its contained vessels and nerves and by

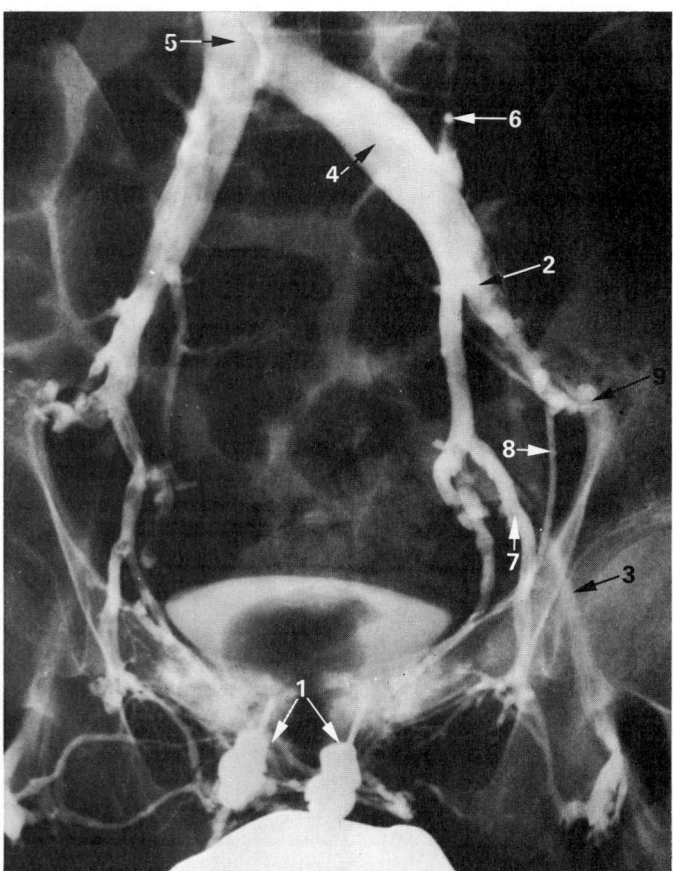

6.171 Venogram showing the veins of the pelvis and groin. The contrast medium has been injected into the bodies of the pubic bones. 1. Injected contrast medium in pubic bones. 2. Internal iliac vein. 3. External iliac vein (faintly outlined). 4. Common iliac vein. 5. Inferior vena cava. 6. Ascending lumbar vein. 7. Obturator vein. 8. Internal pudendal vein. 9. Gluteal vein. Radiograph supplied by M Lea Thomas.

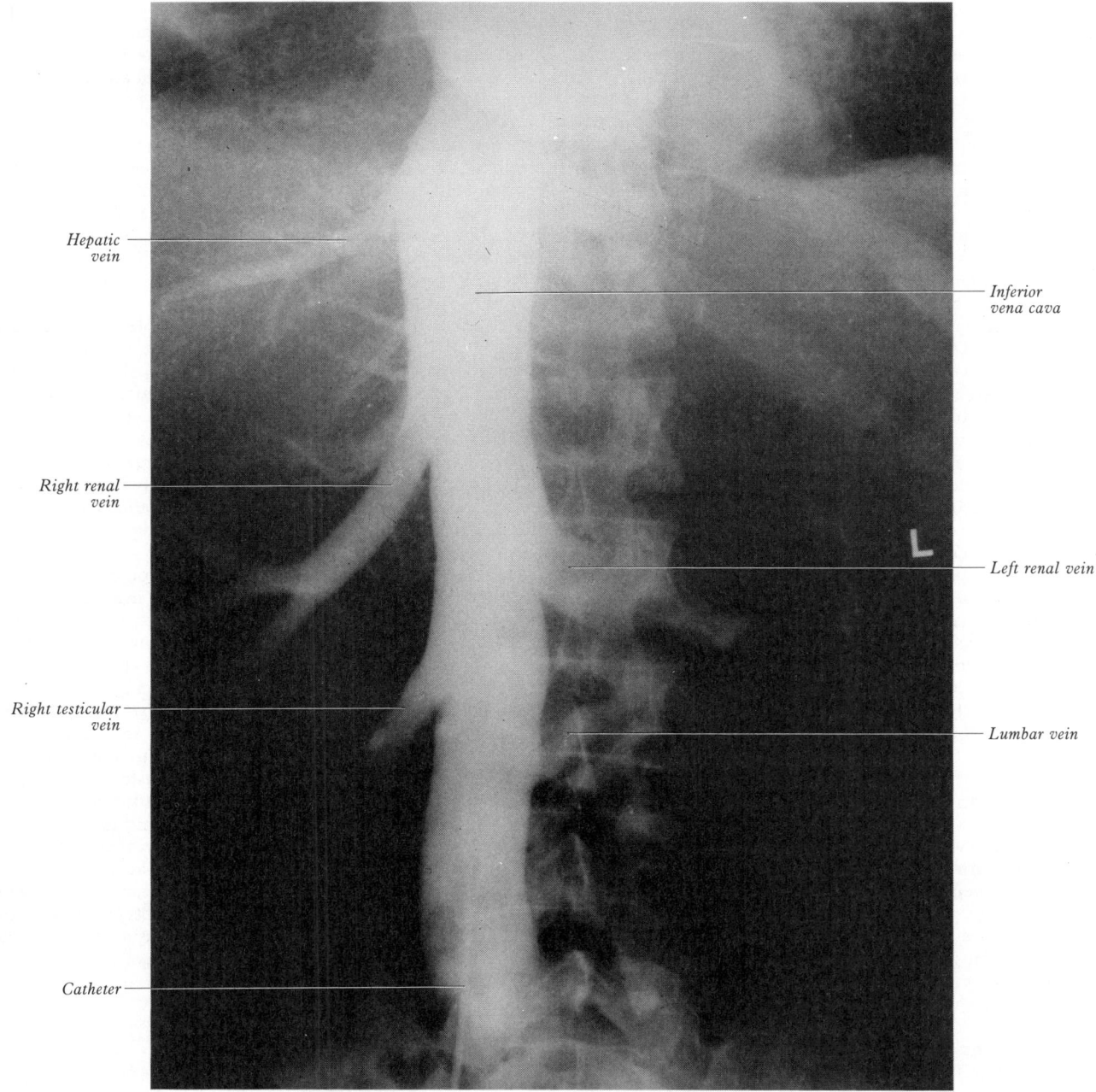

Hepatic vein

Right renal vein

Right testicular vein

Catheter

Inferior vena cava

Left renal vein

Lumbar vein

6.172 Inferior vena cavogram in an adult male while performing the Valsalva manoeuvre. Supplied by Shaun Gallagher, Guy's Hospital; photography by Sarah Smith.

the right testicular or ovarian artery. It ascends behind the head of the pancreas and then the superior part of the duodenum, separated from it by the common bile duct and portal vein. Above the duodenum it is again covered by peritoneum of the posterior wall of the epiploic foramen (8.99), separating it from the right free border of the lesser omentum and its contents. Above this the liver is anterior.

Posterior are the lower three lumbar vertebral bodies, their intervening symphyses ('discs') and the anterior longitudinal ligament, the right psoas major, right sympathetic trunk, and third and fourth right lumbar arteries; superior to these are the right crus (partially separated by the medial part of the right suprarenal gland and the right coeliac ganglion) and the right renal, suprarenal and inferior phrenic arteries.

Right lateral are the right ureter, the descending part of the duodenum, the medial border of the right kidney and right lobe of the liver. *Left lateral* are the aorta and above this the right crus and caudate lobe.

Relations of the thoracic part. This part of the inferior vena cava is very short, partly inside and partly outside the pericardial sac. The *extrapericardial part* is separated from the right pleura and

lung by the right phrenic nerve. The *intrapericardial part* is covered, except posteriorly, by inflected serous pericardium.

Surface Anatomy. The vein begins in, or just below, the transtubercular plane, its centre 2.5 cm right of the midline; about 2.5 cm wide, it ends behind the sternal end of the sixth right costal cartilage. A band from its lower end to a part of the inguinal ligament centred at a point 1 cm medial to the mid-inguinal point indicates the *common* and *external iliac veins* on each side.

Variations. Numerous anomalies occur and are attributable to arrests or errors in its complex formation. It is sometimes replaced, below the level of the renal veins, by two more or less symmetrical vessels, often associated with the failure of interconnection between the common iliac veins, and due to persistence on the left of a longitudinal channel (usually supra- or subcardinal) which normally disappears in early fetal life (p. 223). In complete visceral transposition, the inferior vena cava is left of the aorta.

Applied Anatomy. Thrombosis of the inferior vena cava leads to oedema of the legs and back, without ascites. Collateral venous circulation is soon established by enlargement of either the superficial or deep veins, or both; the epigastric, circumflex iliac, lateral thoracic, thoraco-epigastric (p. 813), internal thoracic, posterior intercostals, external pudendal and lumbovertebral anastomotic veins connect it with the superior vena cava; deep connections are made through the azygos, hemiazygos and lumbar veins. Vertebral venous plexuses may also provide effective collateral circulation between the venae cavae (Batson 1957).

Tributaries: the common iliac, lumbar, right testicular or ovarian, renal, right suprarenal, inferior phrenic and hepatic veins.

Lumbar veins, four pairs, collect blood by dorsal tributaries from lumbar muscles and skin, and by abdominal tributaries from the walls of the abdomen, where they connect with the epigastric veins. Near the vertebral column they drain the vertebral plexuses and are connected by the **ascending lumbar vein**, a longitudinal vessel anterior to the roots of the lumbar transverse processes. The *third* and *fourth lumbar veins* pass forward on the sides of the corresponding vertebral bodies to enter the posterior aspect of the inferior vena cava; the left veins pass behind the abdominal aorta and are therefore longer. *First* and *second lumbar veins* may join the inferior vena cava, ascending lumbar, or lumbar azygos veins; the first does not usually enter the inferior vena cava; it may turn down to join the second and so open into it indirectly, but more often ends in the ascending lumbar vein or passes forward over the first lumbar vertebral body to the lumbar azygos vein (p. 808). The second lumbar vein may join the inferior vena cava at or near the level of the renal veins; sometimes it joins the third lumbar vein or may end in the ascending lumbar. First and second lumbar veins are often connected to each other, to contralateral veins and to right and left lumbar azygos veins by a plexus on the upper lumbar vertebral bodies.

The ascending lumbar vein connects the common iliac, iliolumbar and lumbar veins. It is between the psoas major and roots of the lumbar transverse processes. Superiorly it joins the subcostal vein and the vessel so formed turns forward over the twelfth thoracic vertebral body and, passing deep to the crus, ascends as the azygos vein on the right and as the hemiazygos on the left. There is an angle on the vessel as it turns up; it is usually joined here by a small vessel from the back of the inferior vena cava (or left renal vein on the left). This little vein represents the azygos line (p. 224), already described as the *lumbar azygos vein* (p. 808). Sometimes the ascending lumbar vein ends in the first lumbar, which then skirts the first lumbar vertebra with the first lumbar artery to join the lumbar azygos vein, the subcostal vein then joining the azygos vein on the right and the hemiazygos on the left.

Testicular veins (6.107A) emerge posteriorly from the testis, drain the epididymis and unite to form the *pampiniform plexus*, a chief component of the spermatic cord, ascending anterior to the ductus deferens. Distal to the superficial inguinal ring the plexus is drained by three or four veins traversing the inguinal canal to the abdomen through the deep inguinal ring; they coalesce into two veins, which ascend anterior to the psoas major and ureter, behind the peritoneum, on each side of the testicular artery. These veins join and open into the inferior vena cava on the right at an acute angle just inferior to the level of the renal veins; the left testicular vein opens into the left renal vein at a right angle. The testicular veins contain valves; the left passes behind the lower descending colon and inferior margin of the pancreas and is crossed by the left colic vessels; the right passes behind the terminal ileum and horizontal part of the duodenum and is crossed by the root of the mesentery, ileocolic and right colic vessels.

Applied Anatomy. The testicular veins are frequently varicose; varicocele, which is almost always on the left, is perhaps due to the orthogonal junction of the left testicular and renal veins. It is also overlaid by the descending colon and, when this is distended by constipation, its weight may impede venous return. After removal of a varicocele, venous return is by the small veins of the ductus deferens, cremaster and scrotal tissues.

Ovarian veins each form a plexus in the broad ligament near the ovary and uterine tube, communicating with the uterine plexus. Two veins issue from this and ascend across the external iliac artery with the ovarian artery. Their further course is like that of the testicular veins. Valves may occur in them. Like the uterine veins, they are much enlarged in pregnancy.

Renal veins, of large size and anterior to the renal arteries, open into the inferior vena cava almost at right angles. The *left* is three times the right in length (7.5 cm and 2.5 cm); it crosses the posterior abdominal wall posterior to the splenic vein and body of pancreas and, near its end, is *anterior* to the aorta, just *below* the origin of the superior mesenteric artery. The left testicular or ovarian vein enters it from below and the left suprarenal vein, usually receiving one of the left inferior phrenic veins, enters it above but nearer the midline. The left renal vein enters the inferior vena cava a little superior to the right. The *right renal vein* is behind the descending duodenum and sometimes the lateral part of the head of the pancreas.

Variations. The left renal vein may be double, one vein passing posterior, one anterior to the aorta to join the inferior vena cava, a condition named persistence of the 'renal collar' (p. 225); the anterior may be absent, representing persistence of the posterior limb of the renal collar combined with absence of an intersubcardinal anastomosis.

Suprarenal veins issue from each suprarenal hilum. The right is short, passing directly and horizontally into the posterior aspect of the inferior vena cava; the left descends medially anterior or lateral to the left coeliac ganglion, to pass posterior to the pancreatic body to reach the left renal vein.

Inferior phrenic veins follow the corresponding arteries on the inferior diaphragmatic surface; the right ends in the inferior vena cava; the left is often double, one branch ending in the left renal or suprarenal vein, the other passing anterior to the oesophageal opening to join the inferior vena cava.

Hepatic veins drain the liver, commencing as *intralobular veins*, draining the sinusoids of liver lobules (p. 1381); these lead to *sublobular veins*, which eventually unite into *hepatic veins*, emerging from the posterior hepatic surface to open *at once* into the inferior vena cava in its groove on the posterior hepatic surface. Hepatic veins are arranged in upper and lower groups. The *upper* are usually large veins, right, left and middle, the last from the caudate lobe; the *lower*, varying in number, are small and from the right and caudate lobes. The hepatic veins are contiguous with hepatic tissue and have no valves. Large *'accessory' hepatic veins* of the lower group, draining a variable volume of the right lobe, have been studied in 93 adult livers by corrosion casts; they are usually single, occasionally double, with an incidence of 15% (Sledzinski & Tyszkiewicz 1975).

The Hepatic Portal System

The portal system (6.173,174) includes all the veins draining the abdominal part of the digestive tube (*except* the lower anal canal but *including* the preterminal oesophagus) and spleen, pancreas and gallbladder. From these viscera blood is conveyed by the *portal vein* to the liver, where it ramifies like an artery, ending in the sinusoids, from which blood again converges to reach the inferior vena cava via the hepatic veins. The blood therefore passes through *two* sets of 'exchange' vessels: (1) capillaries of the digestive tube, spleen, pancreas and gallbladder; (2) the hepatic sinusoids. In adults, the portal vein and its tributaries have no valves; in fetal life and for a short postnatal period valves are demonstrable in its tributaries; usually they atrophy but some may persist in degenerate form.

The portal vein (6.111,173,174), about 8 cm long, begins at the second lumbar vertebral level as the junction of the superior mesenteric and splenic veins, *anterior* to the inferior vena cava, *posterior* to the neck of the pancreas. It inclines slightly right as it ascends behind the superior part of the duodenum, bile duct and gastroduodenal artery and is here directly anterior to the inferior vena cava; however, it enters the right border of the lesser omentum, ascending anterior to the epiploic foramen to the right end of the porta hepatis, dividing into right and left stems, which accompany the corresponding branches of the hepatic artery into the liver. In the lesser omentum it is posterior to both bile duct and hepatic artery, the former being to the right; it is surrounded by the hepatic nerve plexus and accompanied by many lymph vessels

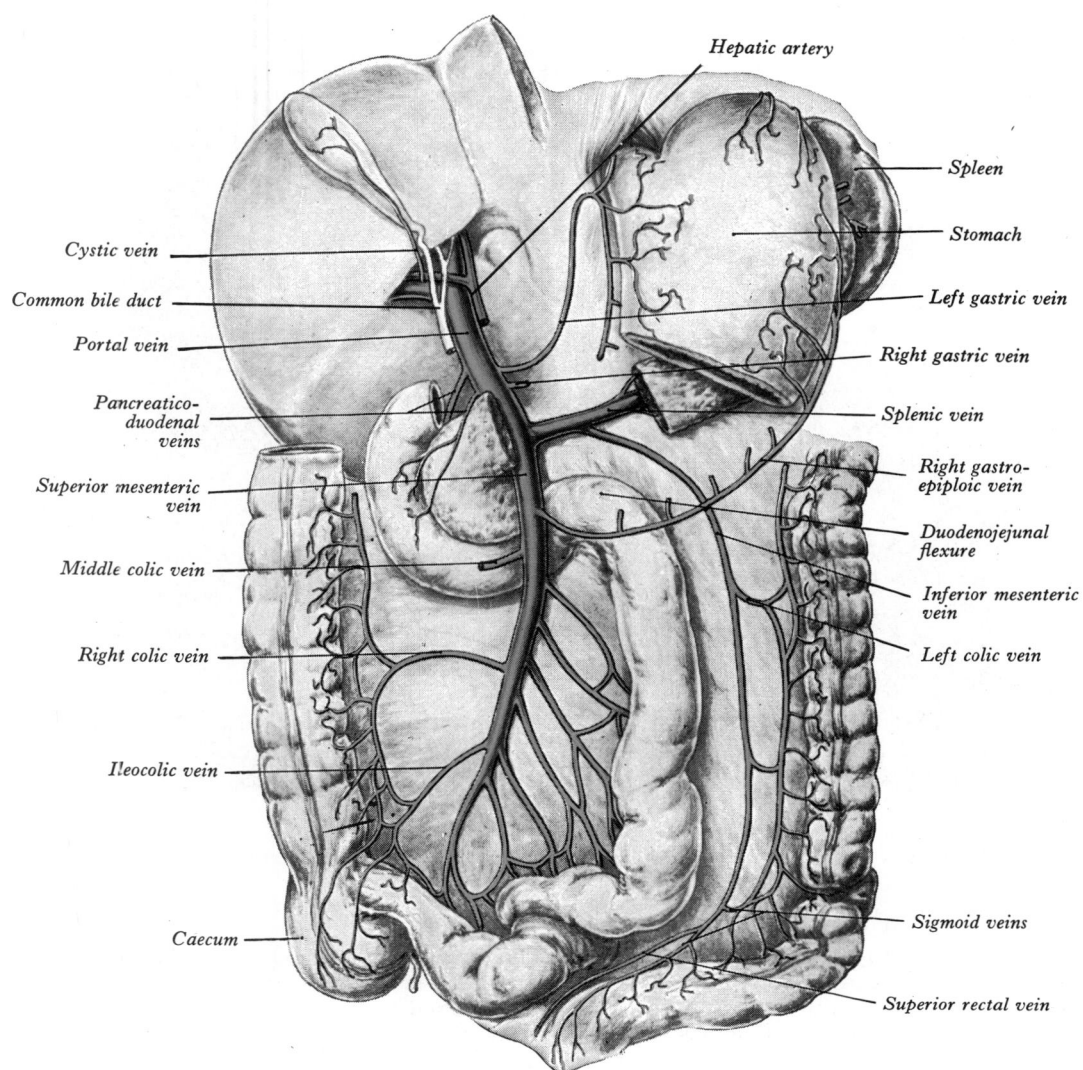

6.173 The portal vein and its tributaries (semi-diagrammatic). Portions of the stomach, pancreas and left lobe of the liver and the transverse colon have been removed.

and some lymph nodes. The *right branch* enters the right hepatic lobe but usually first receives the cystic vein. The *left branch*, longer but of smaller calibre, branches into caudate, quadrate and left lobes. (See critique on hepatic lobation and lobulation pp. 1385, 1387.) As it enters the left lobe it is joined by para-umbilical veins (p. 820) and the *ligamentum teres*, a remnant of the obliterated left umbilical vein. It is connected to the inferior vena cava by the *ligamentum venosum*, a vestige of an obliterated ductus venosus, ascending in a fissure on the liver's posterior aspect (p. 723). The small extrahepatic section of the left branch, from which veins to the quadrate and left lobes arise, is a persistent part of the left umbilical vein. Tributaries of the portal vein are: splenic, superior mesenteric, left gastric, right gastric, para-umbilical and cystic veins.

The splenic vein (**6.**173,174), large and *not* tortuous, is formed by five or six tributaries from the spleen. It traverses the lienorenal ligament with the splenic artery and tail of the pancreas, and descends to the right, across the posterior abdominal wall inferior to its artery and posterior to the body of the pancreas (which it grooves), receiving numerous short rami from the gland. It crosses anterior to the left kidney and its hilar structures (or lower pole of the left suprarenal gland), separated from the left sympathetic trunk and crus by the left renal vessels and from the abdominal aorta by the superior mesenteric artery and left renal vein. It ends behind (lodged in) the neck of pancreas, where it joins the superior mesenteric vein at a right angle to form the portal vein.

Tributaries: short gastric, left gastro-epiploic, pancreatic and inferior mesenteric veins.

Short gastric veins, four or five, drain the gastric fundus and the left part of its greater curvature, traversing the gastrosplenic ligament to reach the splenic vein or one of its large tributaries.

The left gastro-epiploic vein drains both gastric surfaces and the adjacent greater omentum; it runs from right to left along the greater curvature, between the anterior two omental layers, ending in or near the beginning of the splenic vein.

Pancreatic veins drain the body and tail of the pancreas. They may be small and many or large and few. The former empty more or less directly into the splenic vein; in the latter case, superior and inferior arcades receive these larger veins, their ultimate drainage being into the splenic (Sow et al 1976, and p. 1385).

The inferior mesenteric vein (**6.**173) drains the rectum, and sigmoid and descending parts of the colon. It begins as the *superior rectal vein*, from the rectal plexus (p. 816), through which it connects with middle and inferior rectal veins. The superior rectal vein leaves the pelvis and crosses the left common iliac vessels medial to the left ureter with the superior rectal artery, continuing up as the inferior mesenteric vein. This is left of its artery, ascending behind the peritoneum anterior to the left psoas major; it may cross the testicular or ovarian vessels or be medial to them and then passes above, or behind, the duodenojejunal flexure, opening into the splenic vein posterior to the body of the pancreas; sometimes it ends at the union of the splenic and superior mesenteric veins. If a duodenal or paraduodenal fossa exists, the vein is

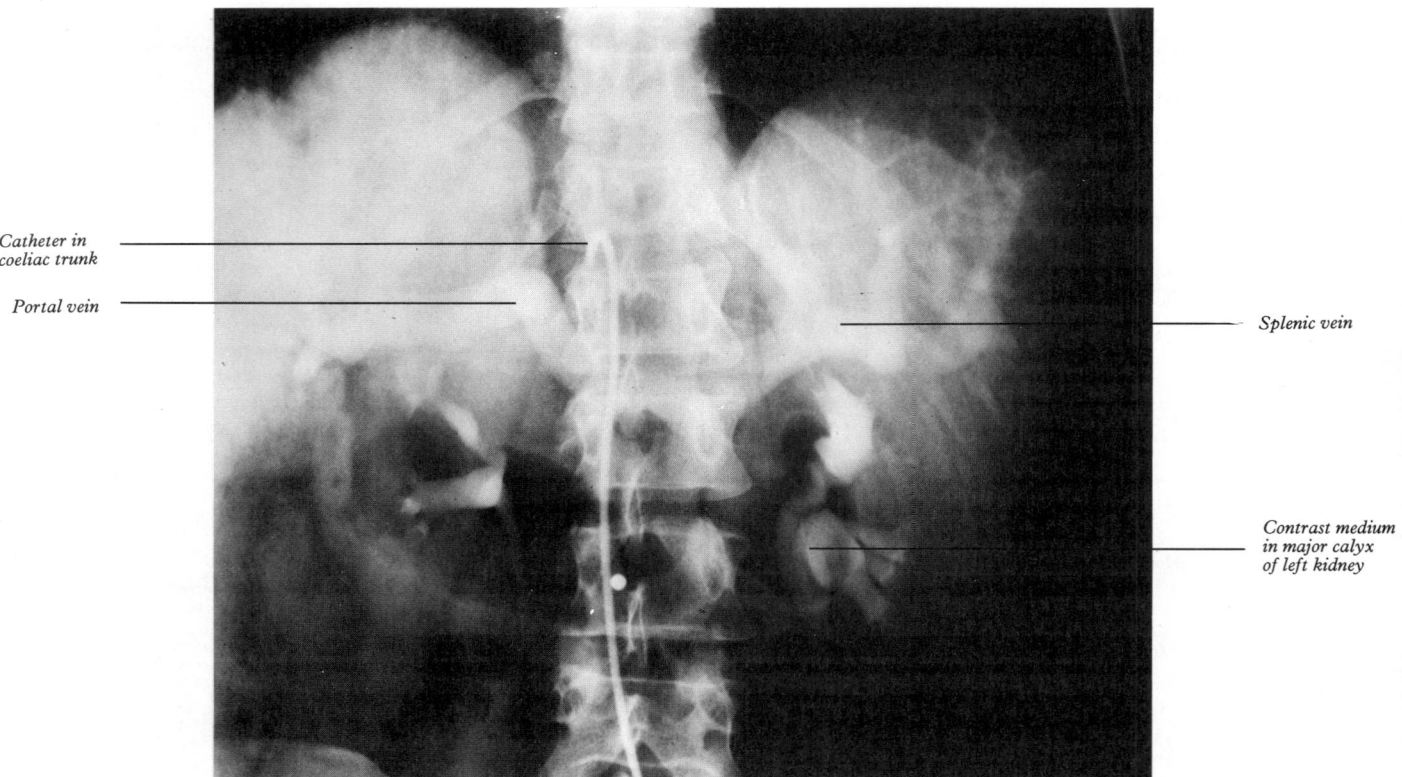

Catheter in coeliac trunk

Portal vein

Splenic vein

Contrast medium in major calyx of left kidney

6.174 Venous phase showing the splenic and portal veins after injection of contrast medium into the coeliac trunk.

usually in its anterior wall (p. 1345). Its *tributaries* are *sigmoid veins* from the sigmoid colon and the *left colic vein* from the descending colon and the left colic flexure.

The superior mesenteric vein (6.173) drains the small intestine, caecum and ascending and transverse parts of the colon. Beginning in the right iliac fossa by the union of tributaries from the terminal ileum, caecum and vermiform appendix, it ascends in the mesentery on the right of the superior mesenteric artery, passing anterior to the right ureter, inferior vena cava, the horizontal part of the duodenum and uncinate process of the pancreas, joining the splenic vein behind its neck to form the portal vein (6.173,174). Its *tributaries* are: jejunal, ileal, ileocolic, right and middle colic, right gastro-epiploic and pancreatico-duodenal veins.

The right gastro-epiploic vein drains the greater omentum and distal part of stomach, passing right on the gastric greater curvature between the anterior layers of the greater omentum to join the superior mesenteric vein below the neck of the pancreas.

The pancreaticoduodenal veins accompany their corresponding arteries: the *inferior* often joins the right gastro-epiploic vein; the *superior* usually ascends to the left behind the bile duct to end in the portal vein. Sow et al (1975) have observed anterior intraglandular and *posterior venous arcades* between the superior and inferior pancreaticoduodenal veins in about 70% of 157 pancreatic corrosion preparations.

The left gastric vein drains both gastric surfaces, ascending the lesser curvature to the left in the lesser omentum, to the oesophageal opening, where it receives the oesophageal veins. It then curves back down and to the right *behind* the omental bursa to end in the portal vein at the upper border of the superior part of the duodenum.

The right gastric vein is small and runs to the right along the pyloric section of the lesser curvature in the lesser omentum, ending in the portal vein. It is joined by a *prepyloric vein* ascending anterior to the pylorus (a surgical guide to the pyloric opening).

Para-umbilical veins connect veins of the anterior abdominal wall and portal vein, extending along the ligamentum teres and median umbilical ligament (p. 1417). Best developed is one begin-

ning at the umbilicus and running in or on (the hepatic) ligamentum teres in the falciform fold to end in the left branch of the portal branch.

Cystic veins draining the gallbladder vary. Those from its superior surface are in areolar tissue between the gallbladder and liver, usually entering the liver through the vesical fossa to join the hepatic veins. The remainder form one or two cystic veins which commonly also enter the liver either directly or after joining the veins draining the hepatic ducts and upper bile duct. Only rarely does a single or double cystic vein drain into the right portal branch.

Applied Anatomy. Portal obstruction may cause ascites, whether obstruction is intra- or extrahepatic. In cirrhosis, radicles of the portal vein are compressed by contraction of the fibrous tissue in their portal canals. In valvular cardiac disease, back-pressure on the hepatic veins, and thus on the whole hepatic circulation, has similar effects. The portal vein may be compressed by hepatic tumours, enlarged lymph nodes in the lesser omentum or carcinoma of the pancreatic head. Portal thrombosis may complicate various conditions. In portal obstruction *anastomoses between portal and systemic circulations*, which may offer effective collateral circulation, are as follows: (1) On the abdominal oesophagus tributaries of the left gastric vein (portal) connect with oesophageal tributaries of the azygos and accessory hemiazygos veins (systemic). Enlargement of these may result in varicosity (oesophageal varices) and even fatal haematemesis. (2) In the anal wall opening up of connections between the inferior and middle rectal (systemic) and superior rectal (portal) veins may result in varicosity. (3) At the umbilicus, veins running on the ligamentum teres to the left portal branch (p. 819) connect with the epigastric veins (systemic); enlargement of these connections may produce varicosities of veins radiating from the umbilicus, the *caput Medusae*. (4) Retroperitoneal veins communicate directly with venous radicles of the colon and bare area of the liver. (5) Very rarely a patent ductus venosus connects the left branch of the portal vein to the inferior vena cava.

Some success has been achieved in treating portal obstruction by anastomosis of the portal vein to the inferior vena cava, or of the splenic to the left renal vein after splenectomy.

THE LYMPHATIC SYSTEM

Introduction

Dispersed widely in the body are the tissues, fluids and cells concerned in a variety of interrelated functions, including the circulation and modification of tissue fluid formed in the capillary beds, the removal by mononuclear phagocytes of cell debris and foreign matter (p. 668) and the long-term specific immune responses of the lymphocytes (p. 667) and other cells. These activities overlap and have a common cellular base with those of the blood vascular system, but it is important to distinguish between *lymphatic vessels* or *'lymphatics'* which are endothelial tubes, lined externally in the case of larger conduits with connective tissue, and *lymphoid tissue*, consisting of aggregates of lymphocytes and associated cells. These cells are in many instances intimately connected with the lymphatic channels, and process or add to their fluid and cellular contents, e.g. in lymph nodes and lymphoid nodules. In other cases, lymphoid *tissue* may be quite separate from lymphatics, e.g. the spleen, which is concerned with modifying the blood, and the bone marrow and thymus, which produce lymphocytes and other cells to populate the lymphoid tissue elsewhere, with immunologically active cells of different classes.

As stated elsewhere, most tissue fluid formed at the arterial ends of capillaries returns to the circulation via their venous ends but 10-20% of such fluid passes into blind ending lymphatic capillaries, then traverses one or more lymph nodes before returning to the venous system and thus the haemal circulation. Before considering the detailed topography of lymphatic channels and lymph nodes, we will consider the general structure of lymphatic vessels, lymph nodes, lymphatic nodules, the spleen and the thymus.

Lymphatic Vessels

Lymphatic capillaries form plexuses in tissue spaces which have much wider meshes than those of the adjacent blood capillaries. They often begin as dilated lymphatics with closed ends; the calibres are larger and cross-sectional appearances are less regular than those of blood capillaries, and they lack a basal lamina, though they have numerous vesicles within their cytoplasm, a typical endothelial feature. (See Leak 1984 for a review of lymphatic microstructive and physiology.) Their endothelium is generally quite permeable to much larger molecules (Allen 1967) and, unlike most haemal capillaries, are readily permeable to colloidal materal and larger particles such as cell debris and micro-organisms from tissue spaces, and to cells. When lymph vessels are obstructed, the surrounding tissues become oedematous, i.e. distended with fluid containing much protein. Experiments suggest that the observed absorption of macromolecules and particles is via gaps between endothelial cells or by micropinocytosis through them. Lymph from most tissues is clear and colourless but from the small intestine it is milky, due to the absorption of fat globules (chylomicrons), and is called *chyle*, the terminal vessels in the small intestine being *lacteals*. Lymphatic capillaries, though present in many tissues, are absent from avascular structures (epidermis, hair, nails, cornea, articular and some other cartilages), and from central nervous tissue and bone marrow. They are very few in the endomysium of skeletal muscle.

Lymphatic capillaries join into larger vessels which pass to local or sometimes more remote lymph nodes. These are arranged largely in *regional groups*, sufficiently regular in position to be named. Each has its region of drainage but a local group is often bypassed. Nodes within a group are often interconnected (Kubik 1974). In general, lymph traverses a series of nodes before reaching the major collecting ducts. There are exceptions to this: lymph vessels of the thyroid gland and oesophagus and of the coronary and triangular ligaments of the liver drain directly to the thoracic duct without passing through lymph nodes (Rusznyak et al 1960). The superficial lymphatics of skin adjoin the deep fascia and accompany superficial veins, but some run independently;

they have few connections with deep lymphatics. Deep lymphatic trunks usually accompany the arteries or veins, almost all reaching either the *thoracic duct* or the *right lymphatic duct* (p. 841), which join the left and right brachiocephalic veins respectively with the veins at the root of the neck. Some observers have also reported additional entry points into the venous system through the inferior vena cava and the renal, suprarenal, azygos and iliac veins. As the lymphatic vessels are closely associated with veins in their development (p. 226), such additional connections would not be surprising, although they are likely to be variable. Most lymphatic vessels anastomose freely and *across the midline*; larger ones have their own plexiform vasa vasorum. If their walls are acutely infected (lymphangitis) this plexus is congested, marking the paths of superficial vessels by red lines, visible through the skin and tender to the touch. Nerve fibres, in and near the walls of larger vessels have been described.

Lymphatic vessels repair easily and new vessels readily form after damage; these are at first solid cellular sprouts from the endothelial cells of persisting vessels, which later canalize.

The microscopic structure of lymphatic vessels. The wall of lymphatic capillaries consists of a single layer of endothelium, as in haemal capillaries. A basal lamina is often lacking and specialized intercellular attachments are few. Fenestrae have been demonstrated in subserous lymphatics, though they are absent in well-fixed subcutaneous lymphatic vessels, except after trauma. Filopodia are frequent on their luminal surfaces and in lacteals similar projections may exist on their external surface. Pericytes are absent (Fraley & Weiss 1961). As in haemal capillaries, there are structural variations between lymphatic capillaries in different tissues (Allen 1967, Leak & Bruke 1968). As they unite into larger vessels, a thin external connective tissue coat supports the endothelium. Larger collecting trunks have three layers, similar to those of small veins. The tunica intima is composed of endothelium with a thin subendothelial layer of fibrous tissue. The tunica media contains some myocytes, mostly circumferential; the tunica adventitia is mainly fibrous tissue with a slight admixture of muscle. Boggon & Palfrey (1973) confirmed this trilaminar structure but did not observe muscle fibres in the adventitia, even with electron microscopy. Elastic fibres are sparse in the tunica intima, but sufficient to form an external elastic lamina in the tunica adventitia. Lymphatics differ from small veins in having many more valves, which are semilunar, generally paired and each composed of an extension of the intima. Their edges point in the direction of the current and the vessel wall downstream of their attached edges is expanded into a sinus, giving the distended vessels a beaded appearance. Valves are important in preventing the backflow of lymph. The thoracic duct is like a medium-sized vein in structure but the non-striated muscle in its tunica media is more prominent and pulsatile movements of its walls have been described (vide infra and pp. 822, 841).

Satiukova & Rassokhina-Volkova (1972) have studied regeneration of lymphatic capillaries in dogs after auto-transplantations of hind legs and lungs; they observed early formation of buds from severed lymphatics in junctional scar tissue, concluding that lymph flow was largely restored.

Movement of lymph. Several factors aid the propulsion of lymph from tissue spaces to lymph nodes and the venous bloodstream: (1) 'Filtration pressure' in tissue spaces, generated by filtration of fluid under pressure from haemal capillaries. (2) Contraction of neighbouring muscles compresses lymph vessels, moving lymph in the directions determined by their valves; extremely little lymph flows in an immobilized limb, whereas flow is increased by either active or passive movements. This fact has been used clinically to diminish dissemination of toxins from infected tissues by immobilization of the relevant regions. Conversely, massage aids the flow of lymph from oedematous regions. (3) The pulsation of neighbouring arteries probably compresses adjacent lymphatic vessels, assisting flow in them. (4) Respiratory movements and the negative blood pressure in the brachiocephalic veins also promote flow of lymph. (5) Non-striated muscle in the walls of the lymphatic trunks is most marked proximal to their

Afferent vessel

Capsule

Trabecula

Reticulum

Valve

Subcapsular lymph space

Lymphatic follicle in cortex

Lymphatic cord in medulla

Artery to node

Efferent vessel

Valve

6.175 Diagram of a lymph node (modified from Maximow & Bloom). In part of the diagram, lymphocytes have been omitted to show the reticulum. This diagram from an earlier source has been retained for historical interest; greater detail and more modern concepts are displayed in 6.176.

valves and stimulation of sympathetic nerves accompanying them results in their contraction. Pulsatile contractions in the thoracic duct also occur (Kinmonth 1982) and, because of the numerous valves along this structure, lymph is forced undirectionally by this muscular action. However, although valves determine the direction of flow, in markedly dilated vessels they may become incompetent, allowing retrograde flow, perhaps explaining the observed retrograde spread of some malignant tumours.

Methods of study. Infective material and neoplastic cells often spread from an affected site along lymphatics, and so the details of their pathways from different regions and organs are clinically important. Dissection is not a suitable method for the tracing of these routes because lymphatic vessels are slender and difficult to see. More reliable information has been obtained as follows: (1) *Experimental injection* of substances into organs or tissues of living or dead animals, including man. These enter the lymphatics draining the site of injection and render them and their related lymph nodes visible. The materials most commonly used for this purpose are suspensions of India ink, Neoprene latex or Prussian blue, the latter employed by Jamieson & Dobson (1907–20) in extensive studies of human pathways. In living animals methylene blue and radio-opaque substances, such as lipiodol, have been injected, the latter requiring radiography. Lymphangiography in human subjects, following the injection of lipiodol into the appropriate peripheral lymphatic channels has much increased our knowledge of their routes and is much used diagnostically (Kinmonth 1964, Kinmonth & Taylor 1964). (2) *Clinical observation* of lymph nodes involved in the spread of known inflammatory or malignant disease. However, it must be cautioned that retrograde spread of tumour cells after blockage of a channel limits the reliability of such observations by altering the normal directions of flow.

Lymph Nodes

Lymph nodes are small, oval or reniform bodies, 0.1–2.5 cm long, lying in the course of the lymphatic vessels. Each usually has a slight indentation on one side, the *hilum*, through which *blood* vessels enter and leave; an *efferent lymphatic* vessel also emerges. Several *afferent lymphatic vessels* enter peripherally. Lymph nodes have a highly cellular *cortex* and a *medulla* containing

numerous, poorly demarcated cavities. The cortex is deficient at the hilum, where the medulla reaches the surface; thus the efferent vessel emerges from the medulla, while afferent vessels empty into the cortex. Lymph nodes are particularly numerous in: the neck, mediastinum, posterior abdominal wall, abdominal mesenteries, pelvis and proximal regions of the limbs. By far the greatest number of these lie close to the viscera, especially in the mesenteries.

THE MICROSCOPIC STRUCTURE OF LYMPH NODES

A lymph node is essentially a continuous framework consisting of the capsule, trabeculae, and the reticulum, with cells enmeshed in it (6.177,179).

The **capsule and trabeculae**. The capsule is composed mainly of collagen fibres, a few fibroblasts and elastin fibres, the latter being more numerous in the deeper layers. In some animals non-striated myocytes occur but these are few in human nodes. From the capsule, trabeculae of dense connective tissue extend radially into the node's interior, continuous with the network of fine collagen (reticulin) fibres, the *reticulum* which supports the lymphoid tissue. At the hilum, dense fibrous tissue may extend into the medulla, with an efferent lymphatic vessel embedded in it.

The reticulum. This network of fine reticulin fibres and associated cells permeates the spaces enclosed in the capsule and trabeculae (6.177) and supports the cell masses within them. Microscopically, the fibres are identifiable with reticulin stains (6.178), which show how their bundles branch and interconnect, forming a very dense network in the cortex, although the germinal centres have fewer fibres. Bundles of these fibres, covered by endotheliocytes, criss-cross the sinuses, providing attachment for various cells, mostly macrophages and lymphocytes. Reticulin fibres with an associated proteoglycan matrix are apparently formed by cells indistinguishable from the fibroblasts, though various names (e.g. 'reticular cells') have been applied in the past.

6.176 (*opposite*) Schema of the general architecture and cellular organization of a lymph node. Particular reference is made to the differential distribution of lymphatic spaces and cell masses. Coloured arrows indicate the circulatory pathways of T- and B-lymphocytes. For details see text.

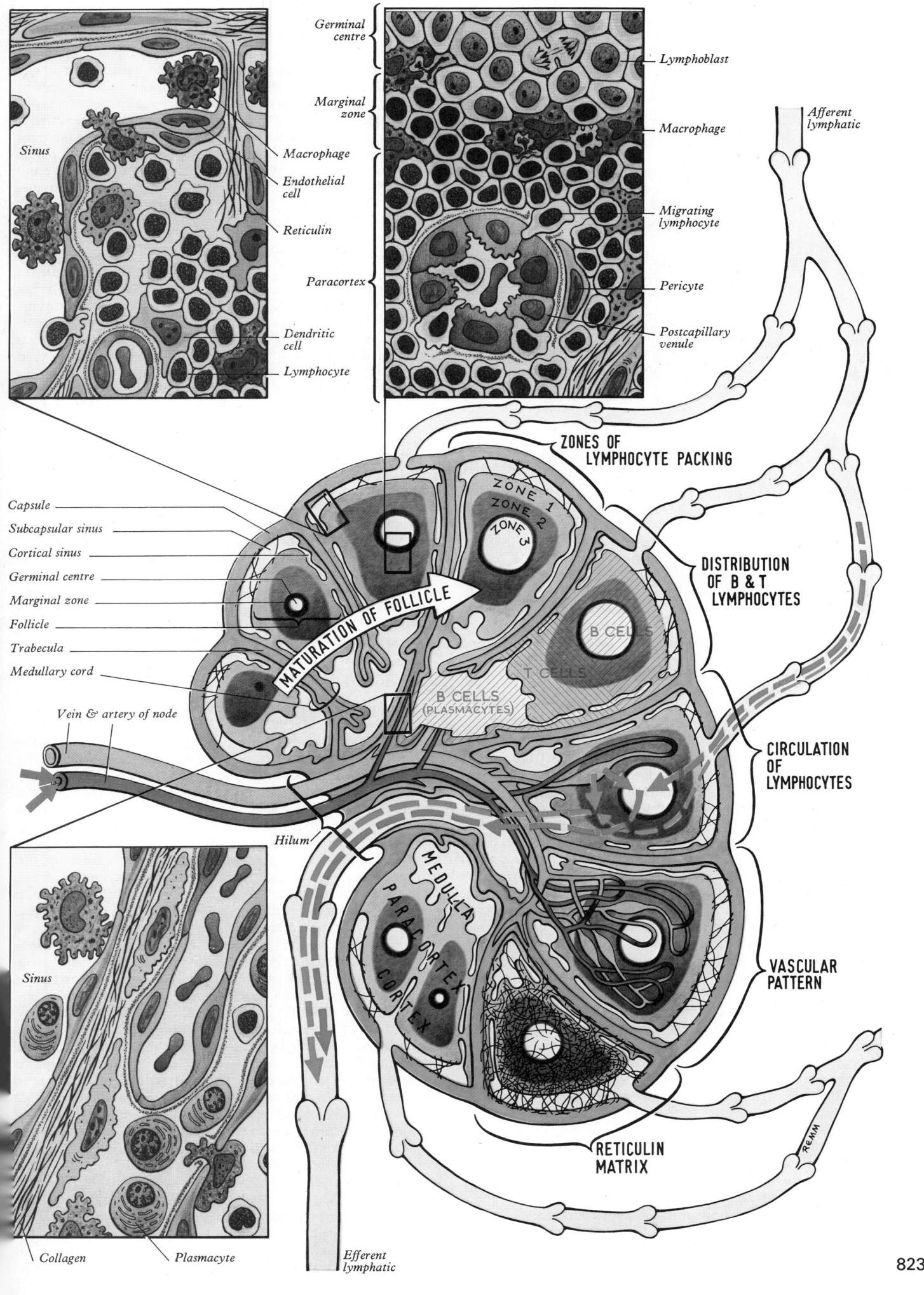

Germinal centre

Lymphoblast

Marginal zone

Macrophage

Macrophage

Endothelial cell

Reticulin

Migrating lymphocyte

Paracortex

Pericyte

Dendritic cell

Postcapillary venule

Lymphocyte

Sinus

Afferent lymphatic

ZONES OF LYMPHOCYTE PACKING

ZONE 1
ZONE 2
ZONE 3

Capsule
Subcapsular sinus
Cortical sinus
Germinal centre
Marginal zone
Follicle
Trabecula
Medullary cord

DISTRIBUTION OF B & T LYMPHOCYTES

B CELLS

T CELLS

MATURATION OF FOLLICLE

B CELLS (PLASMACYTES)

CIRCULATION OF LYMPHOCYTES

Vein & artery of node

Hilum

MEDULLA
PARACORTEX
CORTEX

VASCULAR PATTERN

Sinus

RETICULIN MATRIX

REMM

Collagen   Plasmacyte

Efferent lymphatic

823

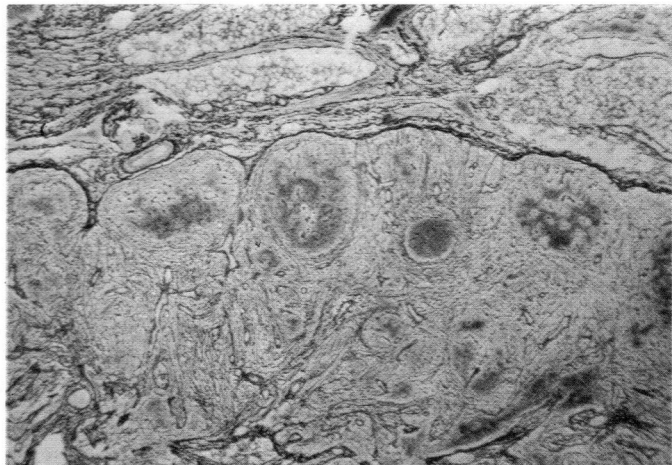

A

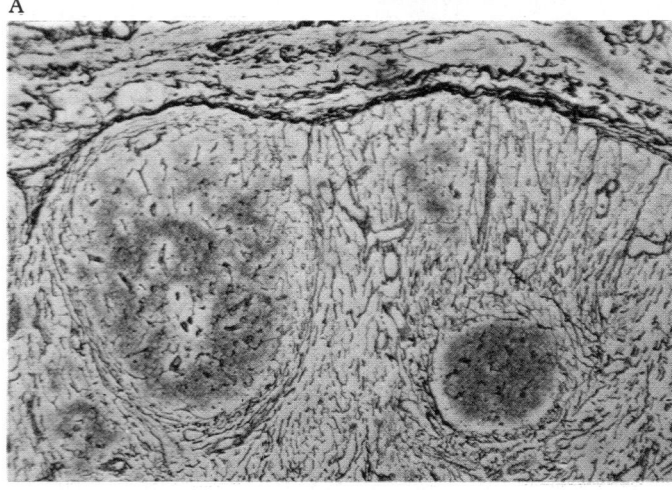

B

6.177 Sections of a lymph node stained by the method of Glees and Marsden for reticulin. Note the heavy concentration of fibres in the capsule and trabeculae. A fine network permeates the rest of the node, with a concentric accumulation surrounding the cortical lymphatic follicles.

Lymphatic channels. Lymph nodes are permeated by channels through which lymph percolates after its entry from the afferent vessels; because macrophages line channels or are entangled amongst the fibres crossing them, lymph is exposed to the action of these phagocytic cells, as well as to the activities of B- and T-lymphocytes adhering to endothelia.

Afferent vessels enter at many points on the periphery, branching to form a dense intracapsular plexus and then opening into the *subcapsular sinus*, a cavity peripheral to the whole cortex except at the hilum (**6.176**). From this sinus numerous radial *cortical sinuses* lead to the medulla, coalescing as larger *medullary sinuses*, which are in turn confluent at the hilum with the efferent vessel draining the node. All these spaces are lined by an endothelium which is continuous despite the constant passage of lymphocytes, macrophages and other cells through the walls of sinuses in both directions. Numerous trabeculae cross them, making their lumina almost labyrinthine and providing large areas for the attachment of various cell types in the spaces (Nopajaroonsri et al 1971, Luk et al 1973).

Lymphatic vasculature. Arteries and veins serving lymph nodes pass through the hilum, giving off straight branches which traverse the medulla and issue minor branches en route. On reaching the cortex, arteries form dense arcades of arterioles and capillaries in numerous anastomosing loops, eventually returning to highly branched venules and veins. Capillaries are specially profuse around the follicles, with fewer vessels within these structures (Herman et al 1973, Blau 1976); postcapillary venules are abundant in the paracortical zones, forming an important site of lymphocytic migration (vide infra). The pattern of vascularization is altered when lymphocytes multiply in response to antigenic stimulation, and then the density of the capillary beds increases greatly (Herman et al 1972).

The structure of these blood vessels is normal except for the *postcapillary venules*, which are lined by tall cuboidal endothelial cells, between which colloids pass readily to perivascular spaces (Mikata & Niki 1970); they also allow extensive movement of lymphocytes from the bloodstream into the paracortex and probably also the reverse (Marchesi & Gowans 1964, Gowans & Knight 1964). The route of migration appears to be through tight junctions which part to allow the passage of cells (Schoefl 1972). Some veins may leave a node through its principal trabeculae and capsule, supplying these and the surrounding connective tissue.

The cells. Most of the cells in the reticulum are *B-* and *T-lymphocytes* but *macrophages* also occur, especially along the walls of sinuses and within germinal centres (vide infra). The distribution of lymphocytes varies in different regions. In the sinuses are some cells which are swept into lymph as it circulates through the node; in the cortex, cells are densely packed and may form isolated *lymphatic follicles* (nodules) (**6.**176–180). The number and isolation of follicles vary according to the prevailing antigenic stimulation. The follicular centre is composed of cells which are larger, less deeply staining and more rapidly dividing than those at its periphery. These areas are *germinal centres*, their cells being mainly *lymphoblasts* which, by prolific mitotic divisions, produce small lymphocytes accumulating at first in the *marginal zones* around germinal centres. The cells pass from follicles into sinuses, which convey them across the medulla to the hilar efferent vessel.

In the medulla, lymphocytes are much less densely packed, forming irregular branching *medullary cords* between which the reticulum is easily seen. Entangled cells include some macrophages, more numerous in the medulla, plasma cells, and a few granulocytes. Under antigenic stimulation, e.g. when the node drains a site of infection or an immunization, the whole node *reacts* by an increase in size and vascularity. The number and size of germinal centres also increases, as lymphocytes and macrophages proliferate, and differentiation of numerous plasma cells occurs in the sinuses. For further details of lymphocyte biology, see pp. 670–675.

CELL ZONES IN LYMPH NODES

As stated above, the cells of lymph nodes are arranged in regions of different packing densities and of distinct cell types. (Nopajaroonsri et al (1971) have suggested the division of the cortex into three zones indicating the packing density of its cells (**6.**176). *Zone 1* is a region of loosely packed cells, predominantly small lymphocytes, macrophages and occasional plasma cells around the extreme periphery of follicles and extending centrally into the medullary cords. *Zone 2* is a denser region internal to zone 1, limited to cortical and paracortical areas and composed mainly of small lymphocytes and macrophages. *Zone 3* comprises the germinal centres of follicles, which are particularly prominent in antigenically stimulated lymph nodes; its cells include large lymphoblasts, some in mitosis, dendritic cells and macrophages. Fibroblasts, reticulin fibres and other cells, mentioned below, appear in all zones. These zones may form a maturational sequence, lymphocytes arising by division in germinal centres (zone 3), passing to the dense zone 2, becoming smaller and finally migrating to zone 1, from which they may traverse the endothelium into the sinuses. But the sequence is complicated by immigration of other cells from lymphatic and haemal sources, and the precise relation between overall cellular architecture and maturation is not clear. In any case, though helpful for descriptive purposes, purely structural schemes also have to note the different functional subclasses of lymphocytes in such zones.

It is also possible to map the distribution of B- and T-lymphocytes within nodes by their immunohistological reactions to monoclonal antibodies (**6.**179). Immunofluorescent staining shows that they occupy distinct territories. Immature B cells occur in the more peripheral parts of follicles, whereas T cells lie mainly between the germinal centres and the medulla, i.e. in the *paracortex* or *thymus-dependent zone*. Mature B cells (plasma cells) exist mainly in medullary cords and sinuses, while some are also peripheral to the cortical follicles. The distribution of T-lymphocytes is clear in animals with a congenital absence of the

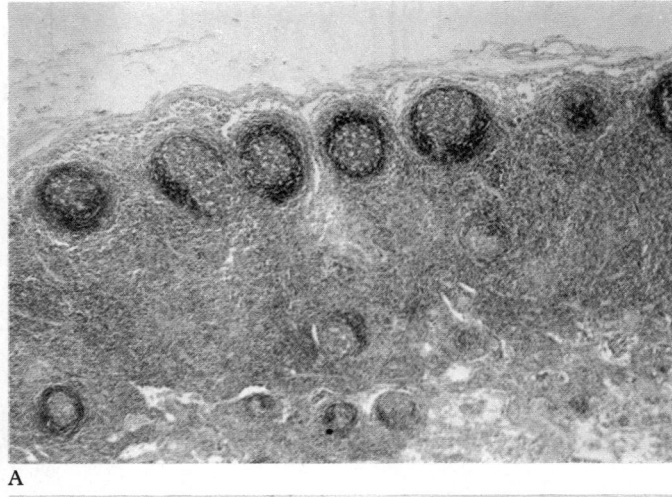

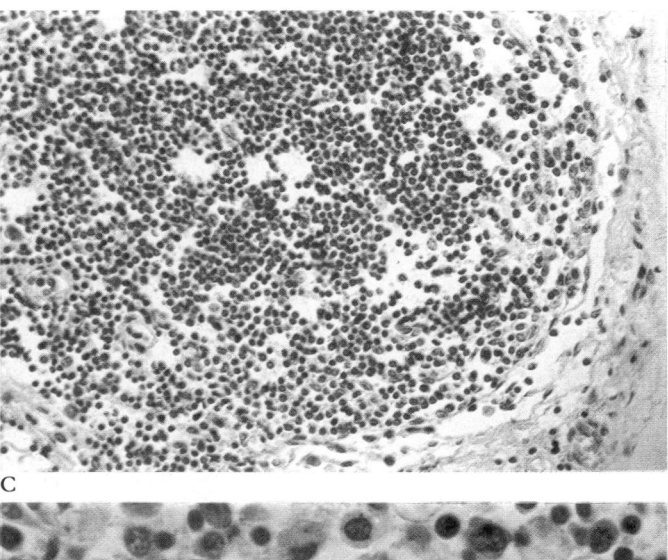

A

C

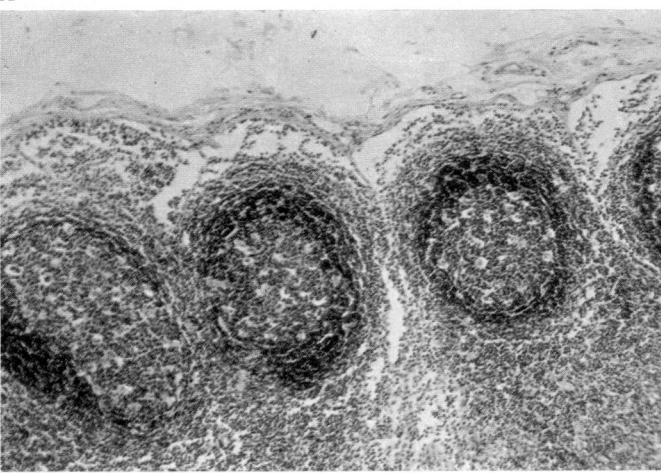

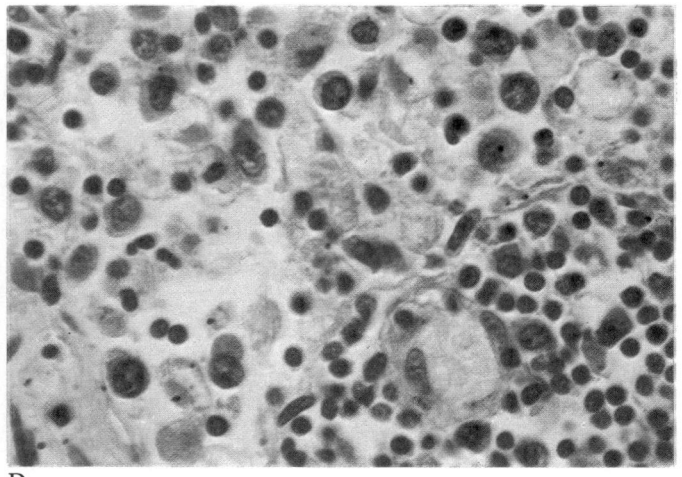

B

D

6.178 Sections of lymph nodes stained with haematoxylin and eosin.

A. Note the round cortical lymphatic follicles with their dense, dark periphery and pale germinal centres and the irregular medullary tissue. Very low power survey micrograph.

B. Low-power view of germinal centres showing the variation in cell density.

C. Higher-power view of the peripheral zone of a follicle showing the densely packed small lymphocytes.

D. Higher-power view of the medulla showing a variety of cell types including small and large lymphocytes and prominent rounded plasmacytes.

thymus (e.g. 'nude' mice), which fail to develop a paracortex. Whether germinal centres contain T or B cells or both is debatable; possibly also cytological markers for their detection are only effective after the lymphocytes have left the germinal centres.

Other Types of Cell in Lymph Nodes

The different types of 'non-lymphocytic' cells have been considerably clarified by recent studies although certain details are as

yet unclear. Following Steinman et al (1974) and other authors, we can distinguish the following: endothelial cells, fibroblasts, typical macrophages, follicular dendritic cells and paracortical dendritic cells ('interdigitating cells of the paracortex'); cells lining blood vessel walls (non-striated myocytes, pericytes) also occur, as do the terminals of non-myelinated nerve fibres. Endothelial cells lining the nodal sinuses appear typical in structure and, contrary to earlier views, do not show the phagocytic ability

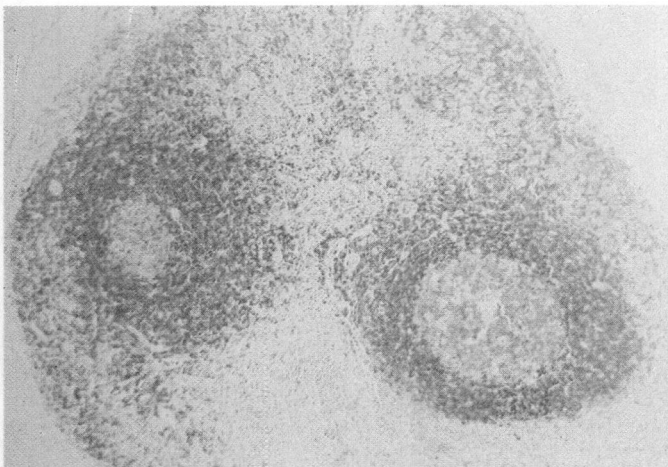

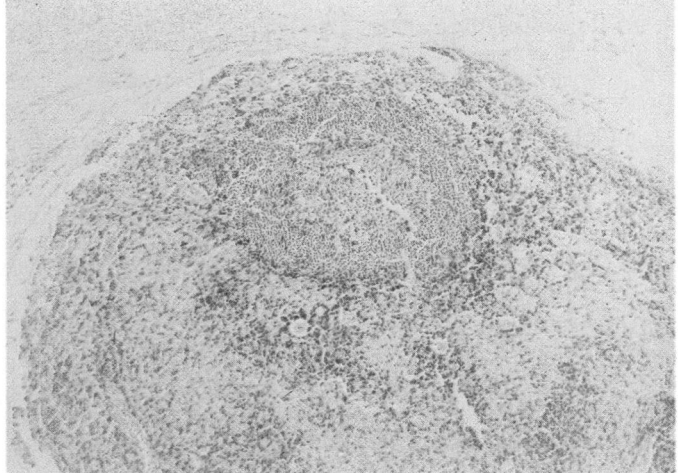

A

B

6.179 Sections through lymph nodes showing different cells stained with the indirect antibody method (second antibody, HRP labelled). In (A), lymphocytes of the B-cell class in two germinal centres are stained brown;

in (B), T cells are demonstrated chiefly in paracortical areas. Provided by R Poston, Department of Histopathology, UMDS, Guy's Campus, London.

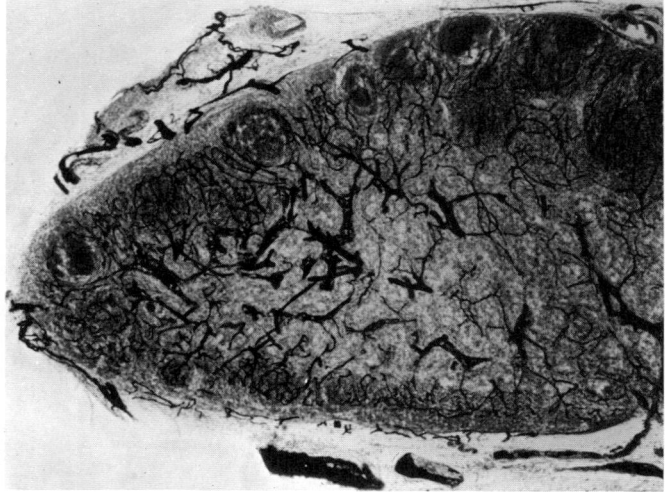

6.180 Lymph node (guinea pig) in section, after blood vessels have been injected with indian ink. Note the large vessels in the medulla, ramifying to form a capillary plexus in the cortex. Provided by N Blau, UMDS, Guy's Campus, London.

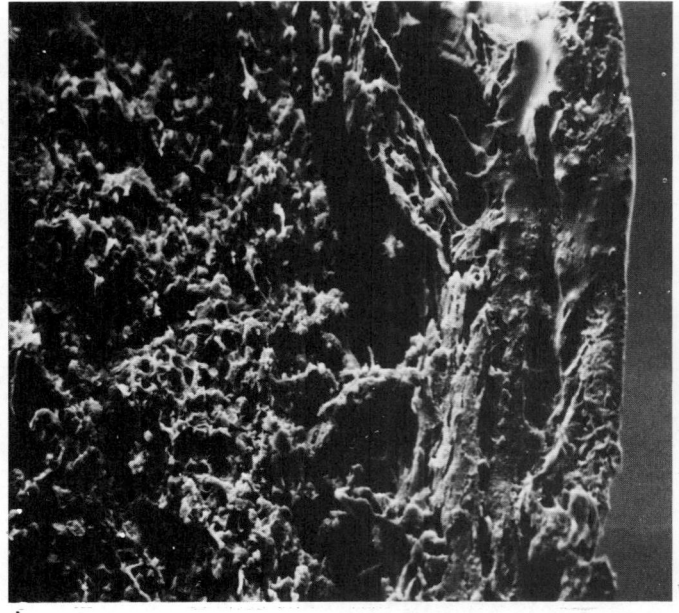

A

which had attracted terms such as 'endothelial macrophages', reticulo-endothelial cells', 'reticular cells' or 'littoral cells' (Nopajaroonsri et al 1971), although true macrophages do occur along the walls of sinuses. The endothelial walls appear to be of the discontinuous type, allowing the free passage of lymphocytes and macrophages. Fibroblasts ('reticular cells' of some authors) produce collagen including reticulin to form the nodal framework, i.e. the capsule, trabeculae and reticulum. Highly phagocytic macrophages (see p. 668) are identifiable in the cortex within and between the follicles and elsewhere.

FUNCTIONS OF LYMPH NODES

Lymphatic capillaries and larger lymphatic vessels returning excess tissue fluid to the venous system carry particulate materials of various kinds to the lymph nodes scattered along their course. The nature of these materials will vary with the region drained; areas rich in microbial flora, e.g. the alimentary tract, are a major potential route of entry of pathogenic organisms and debris into the circulation and of course any area of the body may supply microbes and debris of various kinds, particularly after damage or local infections. In the respiratory tract there is the additional problem of the removal of inhaled particles from the alveoli, which is carried out in part by macrophages re-entering the tissues and passing into the lymphatic pathways. Lymph nodes form a major protection against such materials and organisms, removing them by phagocytic activity and exposing them to a wide variety of powerful defensive actions carried out by lymphocytes resident within them or added to the population of defensive cells circulating in the lymph and blood. Lymph nodes respond dynamically to the presence of such materials and can modulate their activities and structure according to the demands put upon them.

The essential roles of lymph nodes include: (1) The provision of a labyrinth of channels, of large volume and surface area, through which lymph slowly percolates. (2) The exposure of foreign material in the lymph to macrophages in nodal sinuses. (3) The trapping of antigens by different mononuclear phagocytes

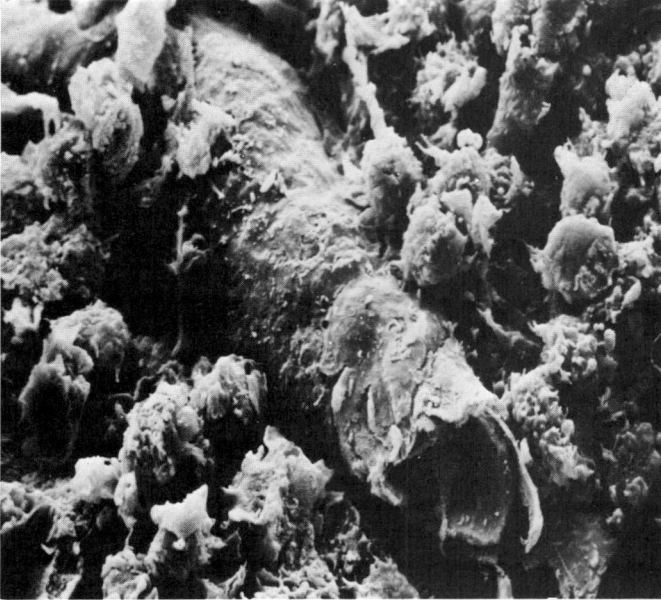

B

6.181 Scanning electron micrographs of the cut surface of a lymph node (guinea pig).
A. Low-power micrograph of the outer cortex showing the capsule (right) together with the subcapsular sinus traversed by reticular fibres. Part of a germinal centre is visible on the left. Magnification × 400.
B. Medium-power micrograph of part of a germinal centre, showing lymphocytes clustered around a capillary. Magnification × 2000.
C. High-power micrograph of the wall of the subcapsular sinus, showing the fine network of reticular fibres with some attached cells. Magnification × 6000.

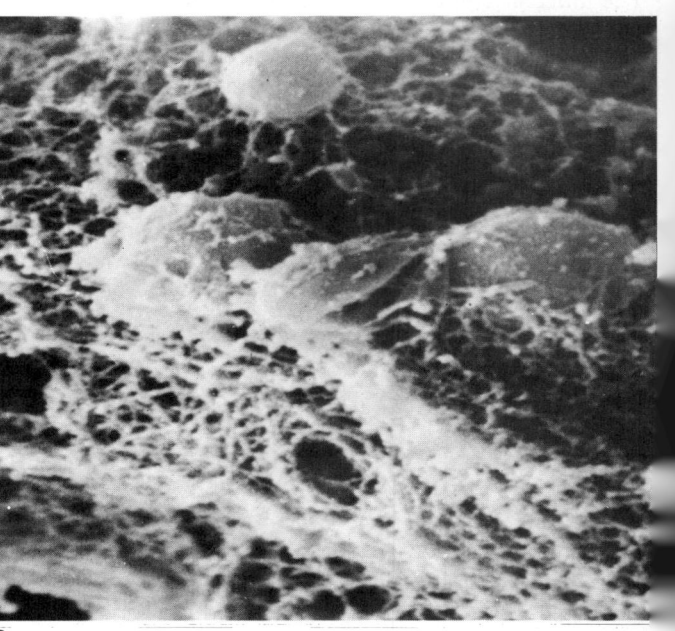

C

including dendritic types. (4) Production of lymphocytes and a pool of stem cells able to become antibody-producing B-lymphocytes and mature T-lymphocytes. (5) Interaction between antigen-bearing mononuclear phagocytes and lymphocytes to produce an immune response, both cell-mediated and humoral. (6) Re-entry of blood-borne lymphocytes into lymphatic channels and thence to the haemal circulation. (7) Humoral antibody production and addition to lymph and, via that route, to blood.

Haemal nodes

In human lymph nodes, erythrocytes may sometimes be found within sinuses. In some animals, small encapsulated lymphoid bodies occur in relation to the abdominal and thoracic viscera, where the sinuses are typically filled with blood, giving the whole structure a red colour; these are termed *haemal nodes* and appear to be more closely related to the blood-vascular than to the lymphatic system, since they lack afferent lymphatics (although they have a single efferent lymphatic vessel). Their structure has been described in detail by Turner (1969), Nopajaroonsri et al (1974) and Hogg et al (1982).

Fast and slow routes of blood circulation have been reported, the former through arterioles, capillaries and venules, the latter through a tortuous sinusoidal system. Specialized post-capillary venules with a high endothelium occur as in ordinary nodes. It is possible that haemal nodes are intermediate between lymph nodes and spleen, and perhaps a basis from which both have evolved. Their human incidence remains uncertain. In some animals, *haemolymph nodes* have been described, with a structure intermediate between that of the lymphatic and haemal nodes and with both lymphatic *and* vascular connections. Some consider them stages in the transformation of lymph into haemal nodes; others deny the existence of such intermediary structures.

Applied Anatomy. Lymphatic vessels and nodes draining infected areas are liable to inflammation, resulting in acute or chronic lymphangitis and lymphadenitis. Chronic lymphangitis, with blocking of vessels by the escaped ova of a minute parasitic worm, *Wuchereria bancrofti*, is the cause of elephantiasis (filariasis), typified by enormous thickening and reduplication of skin, frequently in the lower limbs and scrotum. Blockage of lymphatic vessels may also result from the spread of neoplasms or widespread surgical removal of nodes. Neoplastic cells may spread by minute emboli or may grow along lymphatic vessels in solid masses. Removals of tumours are therefore planned to take away *in one mass* the tumour, the intervening vessels and local nodes. It is important to note that lymphatic vessels from a region may not drain to the local lymph nodes, but to those more remote, often making operative removal difficult if not impractical.

The Spleen

The spleen (**6.**182–188) lies mainly in the left hypochondriac region of the abdomen, its posterior edge extending into the epigastric region. It is situated between the gastric fundus and diaphragm. In shape it varies from a slightly curved wedge (if the colic impression is small) to a tetrahedron (if the impression is large). Its long axis lies in the plane of the tenth rib, its posterior border being about 3.5–4.0 cm from the mid-dorsal line at the level of the tenth thoracic spine and its anterior border reaching the mid-axillary line. It is soft, friable, highly vascular and of a dark purple colour.

Relations. The spleen has diaphragmatic and visceral surfaces, superior and anterior borders and inferior and posterior extremities. The *diaphragmatic surface*, convex and smooth, faces posterosuperiorly and to the left, except at its posterior edge which faces slightly medial. It is related to the abdominal surface of the diaphragm which separates it from the lowest part of the left lung and pleura and the ninth to eleventh left ribs. The pleural costodiaphragmatic recess extends down as far as its inferior border. *The visceral surface* (**6.**182), facing the abdominal cavity, presents gastric, renal, pancreatic and colic impressions. The *gastric impression*, directed anteromedially and upwards, is broad and concave where the spleen abuts on to the posterior aspect of the stomach, from which it is separated by a recess of the greater

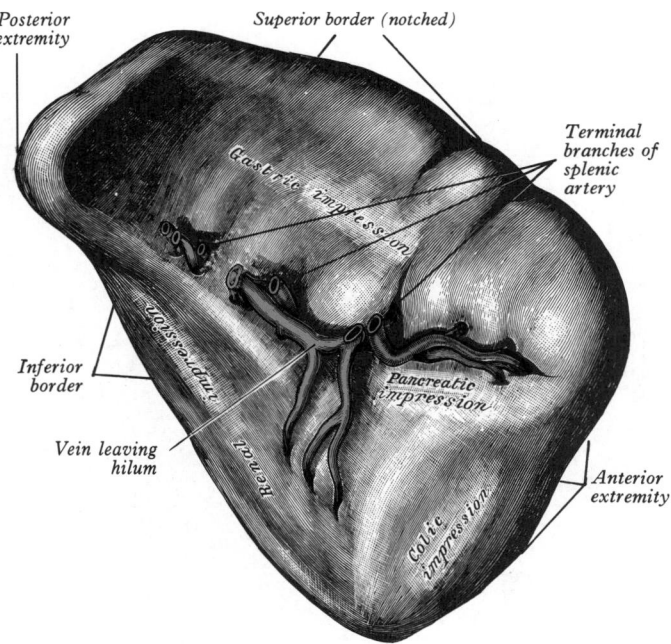

6.182 The visceral surface of the spleen.

sac. Near the inferior limit of the spleen is the *hilum*, a long fissure pierced by several irregular apertures through which vessels and nerves of the spleen pass. The *renal impression*, slightly concave, is located on the lowest part of the visceral surface and is separated from the gastric impression above by a raised margin. It faces inferomedially and slightly backwards, being related to the upper and lateral area of the anterior surface of the left kidney and sometimes to the superior pole of the left suprarenal gland. The *colic impression*, at the extreme lateral end of the spleen, is usually flat and is related to the left colic flexure and phrenicocolic ligament (**8.**125). The *pancreatic impression*, small when present, is situated between the colic impression and the lateral part of the hilum; it is related to the tail of the pancreas which lies in the lienorenal ligament (**8.**114A).

The *superior border*, separating the diaphragmatic surface from the gastric impression, is usually convex and, near its lateral end, has one or two notches indicating the lobulated form of the spleen in early fetal life (p. 238). The *inferior border* separates the renal impression from the diaphragmatic surface and lies between the

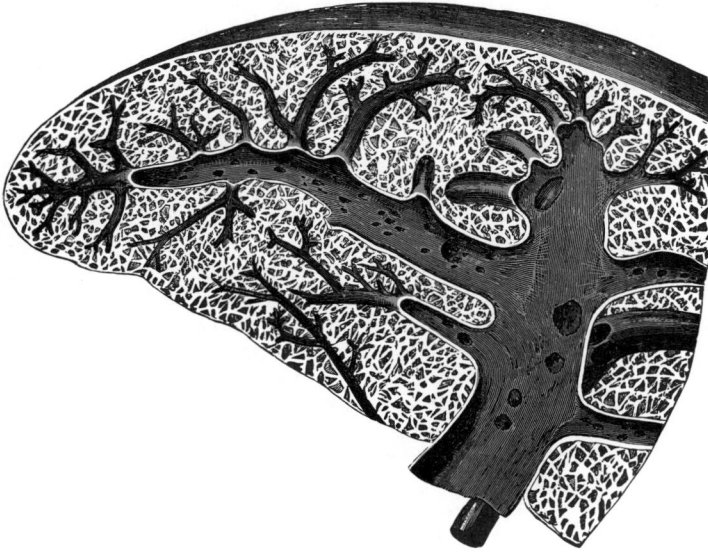

6.183 Transverse section through the spleen, showing the trabecular tissue and the splenic vein and its tributaries. From the first edition of *Gray's Anatomy* (1858).

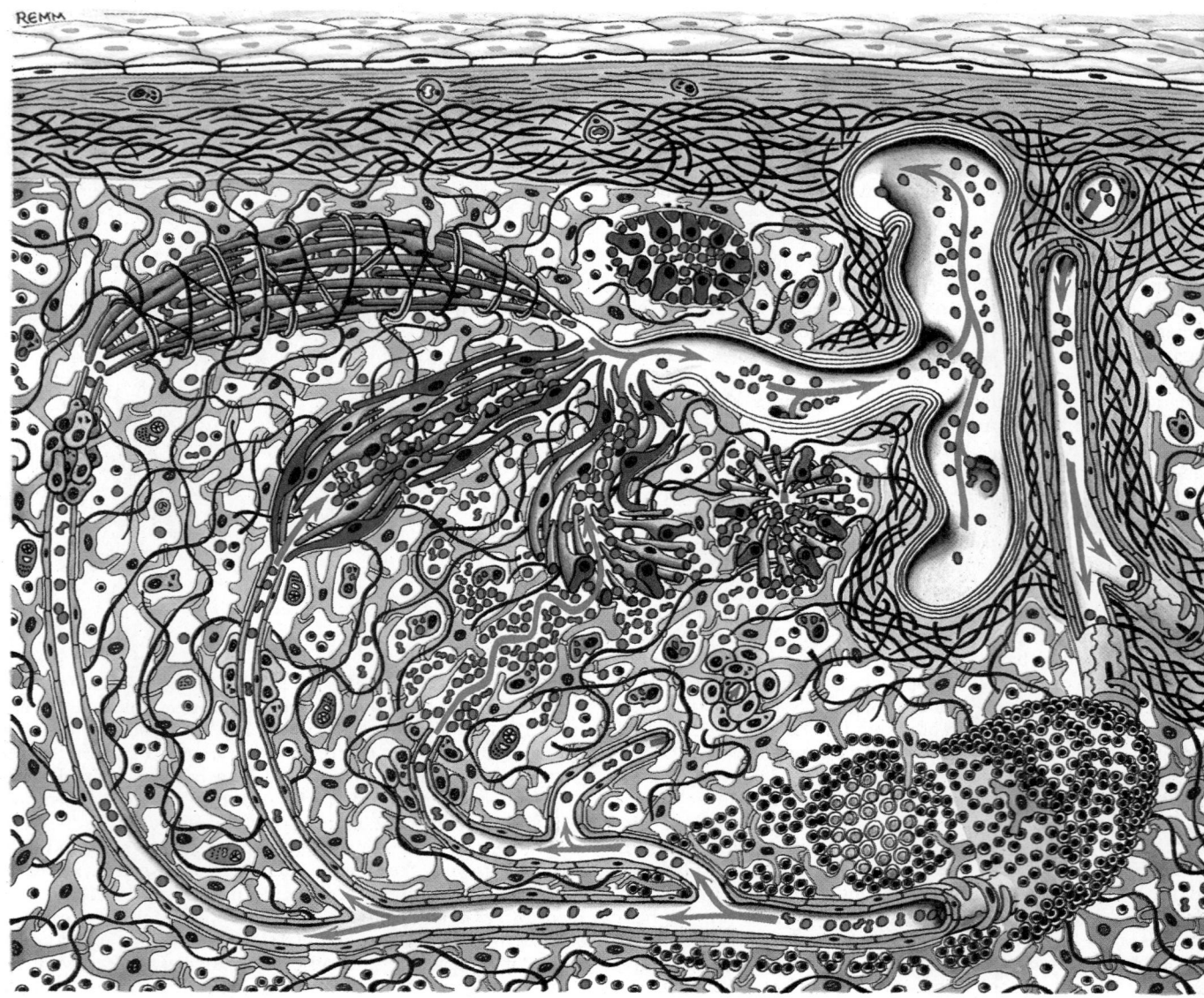

6.184 The main features of splenic structure; the various elements are not drawn to scale, to enable representation on a single diagram. Note the capsule, trabeculae, reticular fibres and cells, the perivascular lymphatic aggregation (white pulp), and the ellipsoids, cell cords and venous sinusoids of the red pulp. The 'open' and 'closed' theories of splenic circulation are illustrated. The venous sinusoids are shown in two states: (1) with their lining of 'stave' cells (bright blue) in close apposition, (2) with intercellular gaps (these have been over-emphasized for clarity). Consult text for further details.

diaphragm and the upper part of the left kidney's lateral border. More blunt and rounded than the superior border, it corresponds in position to the eleventh rib's lower margin.

The *posterior extremity* usually faces the rounded vertebral column. The *anterior extremity* is more expanded and commonly forms a margin connecting the lateral ends of the upper and lower borders. It is related to the left colic flexure and to the phrenicocolic ligament.

The spleen is almost entirely covered by peritoneum, which adheres firmly to its capsule. Recesses of the greater sac separate it from the stomach and left kidney. It develops in the upper dorsal mesogastrium (2.106), remaining connected to the posterior abdominal wall and stomach by two folds of peritoneum, respectively the *lienorenal ligament* and the *gastrosplenic ligament*. *The lienorenal ligament* is derived from peritoneum where the wall of the general peritoneal cavity meets the omental bursa between the left kidney and spleen; the splenic vessels lie between its layers (8.114A). The *gastrosplenic ligament* also has two layers, formed by the meeting of the walls of the greater sac and the omental bursa between spleen and stomach (8.114B); the short gastric and left gastro-epiploic branches of the splenic artery pass between its layers. Most laterally the spleen is in contact with the phrenicocolic ligament.

The size and weight of the spleen vary at different ages, in different individuals and in the same individual under different con-

ditions. *In the adult* it is usually about 12 cm in length, 7 cm in breadth and 3–4 cm in thickness but tends to diminish in size and weight with age. Its average adult weight is about 150 g, ranging between 80–300 g, largely according to its content of blood. It slowly enlarges during alimentary digestion and varies in size with state of nutrition, being large in well-fed animals and small in starved ones.

Near the spleen, especially within the gastrosplenic ligament and greater omentum, small encapsulated nodules of splenic tissue may occur, isolated or connected to the spleen by thin bands of splenic tissue. Such *accessory spleens* or *spleniculi* may be numerous and widely scattered in the abdomen. The spleen may retain its fetal lobulated form or show deep notches on its diaphragmatic surface and inferior border in addition to those usually present on its superior border.

Surface Anatomy. The position of the spleen in the living can be assessed by percussion; the dull area extends over the ninth to eleventh ribs in vertical extent and should not go forward beyond the mid-axillary line. The normal spleen is not palpable.

THE MICROSCOPIC STRUCTURE OF THE SPLEEN
(6.186–187)

The spleen is enclosed in two coats: an outer serous layer (the peritoneum) and an internal fibrous *capsule*. The serous layer

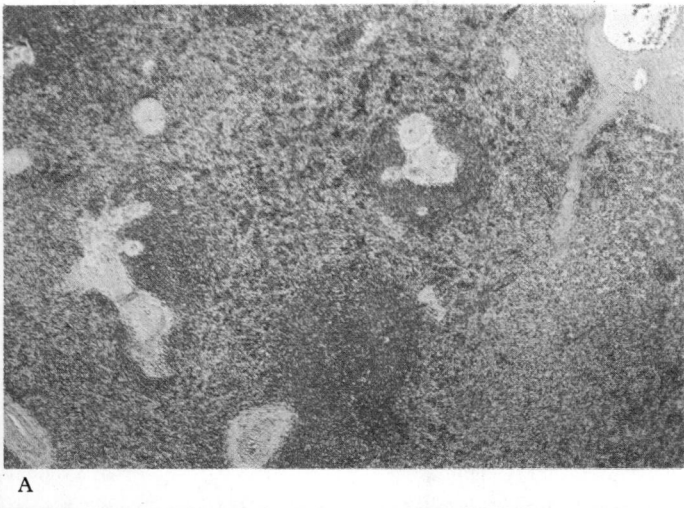

A

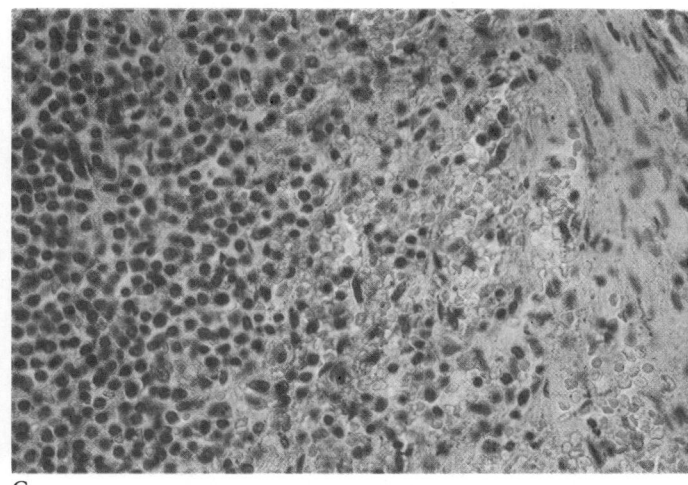

C

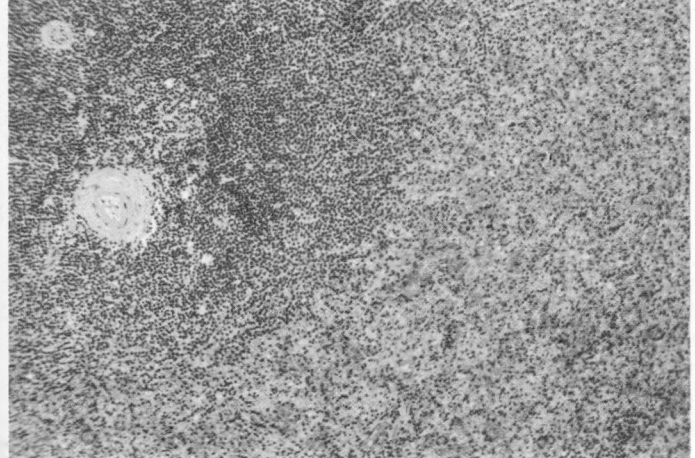

B

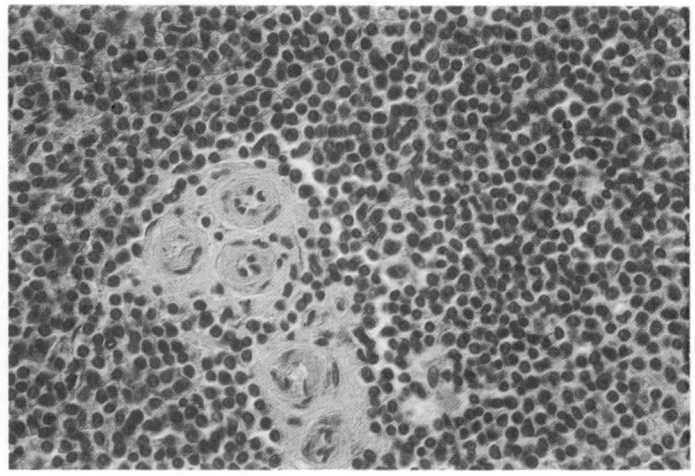

D

6.185 Sections of human spleen stained with haematoxylin and eosin, supplied by D R Turner of the Department of Pathology, Guy's Hospital Medical School.

A, B. Survey photographs at low power showing the general contrast between the white pulp (perivascular lymphatic aggregates, stained blue) and the red pulp (venous sinusoids and intervening cellular cords, stained reddish purple).

c. High-power view of the junctional (marginal) zone between the densely packed lymphocytes of the white pulp and the blood-filled sinusoids and cell cords of the red pulp.

d. A group of small (penicillar) arteries ensheathed by densely packed small lymphocytes.

covers the entire organ except at its hilum and along the lines of reflexion of the lienorenal and gastrosplenic ligaments. The *capsule* or *fibro-elastic layer* sends *trabeculae* into the substance of the spleen, branching within it to form a supportive framework (**6.**184). The largest trabeculae enter at the hilum and ensheathe the splenic vessels, dividing into branches in the splenic pulp (**6.**184–6). The capsule and trabeculae are composed of dense irregular connective fibres, rich in collagen and elastin, the latter more numerous in the trabeculae. In many mammals both capsule and trabeculae contain many non-striated myocytes, the rhythmical contraction of the spleen being attributed to these. In man there are few myocytes, and contraction and distension of the spleen are attributed to the effects of the constriction or relaxation of blood vessels altering the volume of blood in the organ. An increase of intrasplenic blood pressure distends the spleen and stretches the elastic fibres and contraction is due to their recoil.

Within the spleen, branching trabeculae are continuous with a network of fine collagen (reticulin) fibres pervading this organ. Electron microscopy shows this reticulum to be a dense amorphous matrix, probably mainly proteoglycan, with a few fine collagen fibres. Associated with it are fibroblasts ('reticular cells') into which strands of matrix may be invaginated, and macrophages. The reticular spaces are occupied by *red* and *white splenic pulp*. The reticulum is especially dense around the borders of white pulp where they form the framework of the *marginal zone*.

Circulation of blood inside the spleen; white and red pulp (Knisely 1936, MacKenzie et al 1941, Snook 1950, Peck & Hoerr 1951, Lewis 1957, Wennberg & Weiss 1969, & Li-Tsun

Chen & Weiss 1972, 1973) The large, tortuous splenic artery, before reaching the spleen, divides in the lienorenal ligament into five or more rami entering the hilum to ramify in the trabeculae throughout the organ. Likewise, the splenic vein forms in the ligament from five or more tributaries emerging from the hilum. Within the spleen, the finest arteriolar branches pass out of the trabeculae, their adventitia being replaced by a *periarteriolar sheath* of lymphoid tissue accompanying these vessels and their branches (**6.**184). These sheaths constitute the *white pulp* of the spleen, in places being enlarged as *splenic lymphatic follicles* (Malpighian bodies), from 0.25 mm–1 mm across and visible to the naked eye on the freshly cut surface of the spleen as white semiopaque dots, which contrast with the surrounding *red pulp*. Both periarteriolar sheaths and lymphoid follicles are centres of lymphocytic proliferation. When antigenically stimulated, the follicles develop germinal centres as in the lymph nodes, and these become smaller when the infection subsides. Follicles also atrophy with increasing age and may be absent in the very old. Arterioles usually lie to one side of such germinal centres but supply slender branches to follicles. Before arterioles finally lose their sheaths of lymphoid tissue to enter the surrounding red pulp they divide into a series of straight vessels termed *penicilli* (penicillar arterioles). After a course of about 0.5 mm, these pass into or through the *marginal zones* around the white pulp, each penicillus developing slight thickening of its sheath known as an *ellipsoid* (*sheath of Schweigger-Seidel*), formed by an aggregation of macrophages and fibroblasts; here the lumen of the vessel is much narrowed. Ellipsoids are well developed in some mammals, e.g. pig, cat and dog, but not in humans. Beyond the ellipsoids,

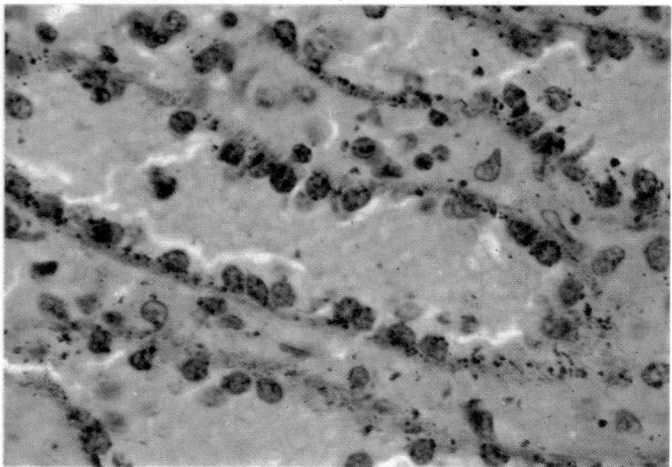

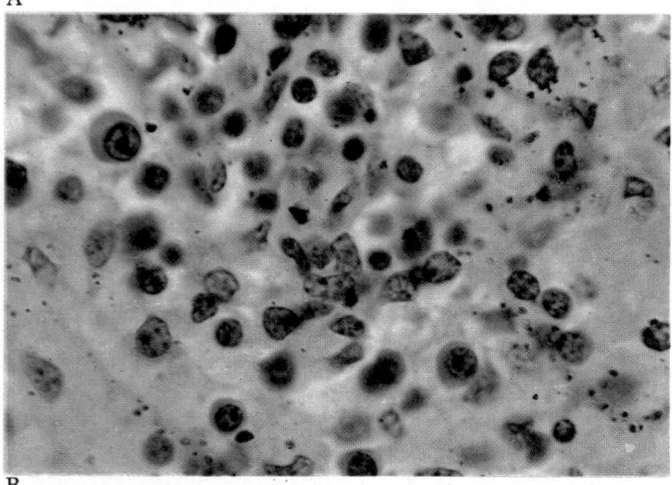

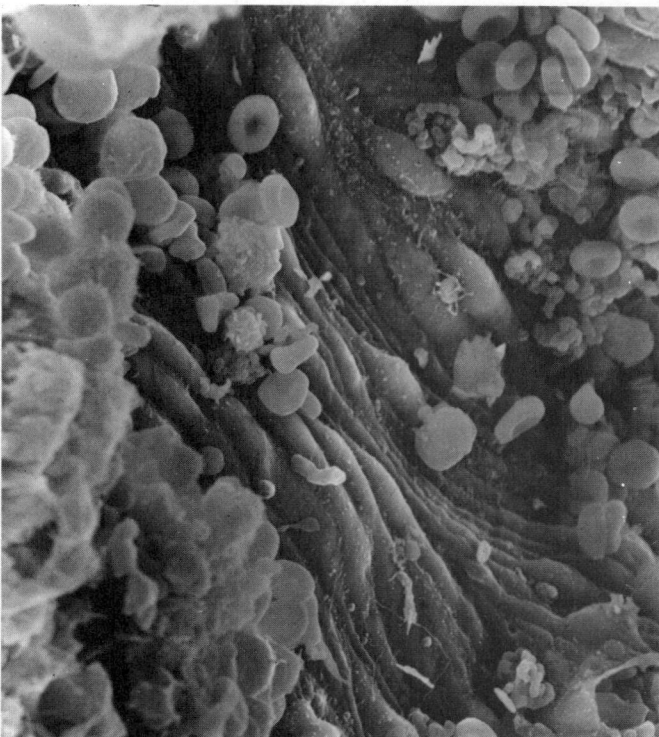

6.187 Scanning electron micrograph of a splenic sinusoid (monkey) to show endothelial cells and associated tissue. Magnification × 1000.

6.186 Section of monkey spleen following intravascular perfusion with a suspension of carbon particles followed later by perfusion fixation, stained by Weigert's haematoxylin and Van Gieson's stain.

A. Showing empty, dilated venous sinusoids and intervening cell cords. The 'stave' cells lining the sinusoids are prominent.

B. High-power view of the cellular region between venous sinusoids. The cell types seen include reticular macrophages with carbon particles in their cytoplasm, small and large lymphocytes and a number of prominent rounded plasma cells.

each vessel continues as a fine arteriole or may divide into two; thereafter the course of the blood is uncertain but, whatever its detailed route, it eventually passes into the red pulp and thence via the sinusoids, venules and small veins (the latter entering the trabeculae) into larger veins which pass out of the spleen at the hilum. Human red pulp consists of numerous *venous sinusoids*, large complex cavities containing blood, embedded in tissue rich in macrophages attached to the splenic reticulum, forming the *splenic cords* (cords of Bilroth).

Splenic venous sinusoids are lined by flat, much elongated fusiform endotheliocytes (*stave cells*, reminiscent of planks in a barrel, 6.187, 188); when sinusoids are inflated by a rise in blood pressure, numerous gaps appear between these cells allowing blood to pass in and out of the tissue of the surrounding splenic cords. The sinusoids are supported externally by circumferential and longitudinal reticulin fibres, which are connected to the fibrous reticulum of the splenic cordal tissue around them.

Ultrastructurally, the endothelial cells of venous sinusoids have some special structural features. Besides possessing numerous endocytic/exocytic vesicles and numerous filopodia on luminal and extraluminal surfaces, they have a complex cytoskeleton which probably determines their elongated shape and may modulate the form of the cell to alter the sizes of the slits in the sinusoid walls, thus regulating the passage of blood through the intercellular gaps (see Chen & Weiss 1972, 1973). The endothelial cells are mildly phagocytic (as are others of this type elsewhere in

the body) but do not appear to contribute strongly to the uptake of particles in the spleen.

Splenic cords consist of all the tissue in the red pulp lying outside the venous sinusoids; the term 'cord' is really a misnomer, as it only describes their appearance in histological sections cut parallel to the long axes of the sinusoids. In three dimensions, this tissue forms a continuous network, penetrated by the vascular channels of the sinusoids and other vessels. Splenic cords contain a network of reticulin fibres and fibroblasts which synthesize them; the elaborate meshes of this network form attachment sites for numerous highly phagocytic *splenic macrophages* and permitting the passage of lymphocytes from the white pulp to the sinusoids and of blood cells as they flow to or from the sinusoid lumen through the gaps in their walls. In some animals sinusoids appear absent and most red pulp is formed by splenic cordal tissue.

The route of the blood from the penicillar arterioles to the sinusoids is controversial, perhaps due to variations in the techniques and species of animals studied. According to the 'open circulation' theory, blood passes directly from arterioles into the splenic cordal tissue, collecting in the venous sinusoids to enter the venous circulation. In the 'closed' theory, blood is envisaged as passing from arterioles directly into venous sinusoids, entering the extravascular cordal tissue only later, in a tidal flow between the sinusoids' endothelial cells. In the spleens of some species of animals, blood appears to flow directly from arteries into sinusoids, escaping through their walls into the surrounding cordal tissue, where it can certainly be demonstrated by electron microscopy. It is probable that blood flow along these routes depends on local control by the intermittent opening and closing of blood vessels at both ends by perivascular myocytic contraction, allowing a tidal flow between the sinusoids and spaces outside them; but this has not yet been demonstrated. However, in animals apparently lacking sinusoids there must be some other type of circulation, presumably of 'open' type.

FUNCTIONS OF THE SPLEEN

The spleen is essentially concerned in phagocytosis, immune responses cytopoiesis and erythrocyte storage. In the fetus it is also an important site of haemopoiesis. However, although of great importance to the defence of the body, it is not absolutely essential, since many of its functions can be assumed by the liver and by other lymphoid tissues if the spleen is removed (see p. 832).

Phagocytosis. Splenic macrophages form a large part of the mononuclear phagocytic system of the body (p. 668). They comprise highly phagocytic cells distributed mainly in the marginal zones of the white pulp and in the ellipsoids and splenic cords. Particulate matter such as carbon is also taken up in small amounts by the fusiform endotheliocytes of venous sinusoids. Beside the removal of microbes, tissue debris and other particulate matter, splenic macrophages remove effete erythrocytes, leucocytes and platelets from circulating blood. The processing of ageing or damaged red cells is especially important. Within splenic macrophages, all stages of erythrophagocytosis, from disintegrating erythrocytes to granules of haemosiderin, can be discerned (Chen & Weiss 1972). Bilirubin, an end-product of haemoglobin catabolism, is conveyed in the bloodstream to the liver for excretion and the iron is largely re-used by bone marrow. Amino acids from the hydrolysis of globin are returned to the amino acid pool of the body. Splenic macrophages are also important in removing microbes and cellular debris from the circulation and their lysosomes possess many powerful enzymes which can hydrolyze these bodies, particularly when activated by lymphokines during immune responses (p. 672). Like other lymphoid organs, the spleen is also involved in phagocytosis of circulating antigens for the initiation of humoral and cellular immune responses (see also p. 673). In addition to such overt phagocytosis, there are numerous dendritic mononuclear phagocytes in the white pulp, particularly in the marginal zones around the perimeters of its lymphoid tissue. Like others of this type, these cells are only moderately phagocytic and are engaged in presenting alien antigens which they have taken up, to lymphocytes, in combination with their Class II MHC surface molecules, to activate both T and B cells (see p. 673).

Immune responses due to lymphocytes. T- and B-lymphocytes occur in different parts of the white pulp, the periarteriolar layers of cells being mainly T-lymphocytes; follicles with germinal centres are largely B-lymphocytes. Cells may migrate from these into splenic cordal tissue and other areas, to perform their specific functions. Some B-lymphocytes mature into plasma cells particularly in the marginal zones, secreting antibodies into the circulating blood when stimulated. T-lymphocytes carry out a wide range of defensive activities, described elsewhere in details (p. 672). Lymphocytes are also added to the general defence of the body by passing into the haemal circulation via venous sinusoids

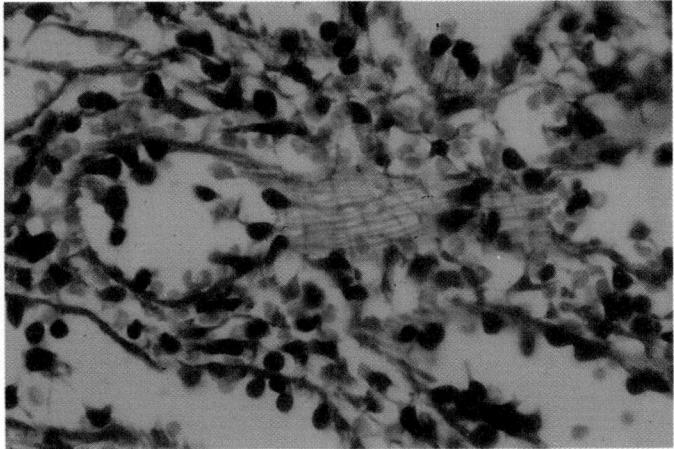

6.188 A splenic sinusoid sectioned tangentially to show parallel strap-like endotheliocytes and surrounding tissue. Gomori's method. Magnification × 600.

and thus the spleen is an important source of these cells. When antigenically stimulated, the white pulp increases in size as lymphocytes proliferate; the (primary) follicles become intensely active in B-cell proliferation, and gain the typical appearance of secondary follicles, as in lymph nodes (p. 822). The presentation of antibody-antigen complexes by dendritic cells of the follicles and marginal zones are involved in these processes and in the generation of 'memory' lymphocytes for future immune responses.

Cytopoiesis. In human fetuses, from the fourth month onwards, the spleen is *haemopoietic*, red pulp containing groups of myelocytes, erythroblasts and megakaryocytes. In some anaemias and myeloid leukaemia, stem cells persisting in red pulp may revert to haemopoiesis. In the mature spleen lymphopoiesis in the white pulp contributes to a circulating reserve of immunologically competent T- and B-lymphocytes and also mononuclear phagocytes to the blood and connective tissues (see p. 670).

Immune responses. With adequate antigenic stimulation, particularly under the action of T-cell derived lymphokines, splenic macrophages proliferate and become highly active; T- and B-cell lymphopoiesis, growth and differentiation increase. In individuals suffering chronic infection of the blood cells, e.g. in

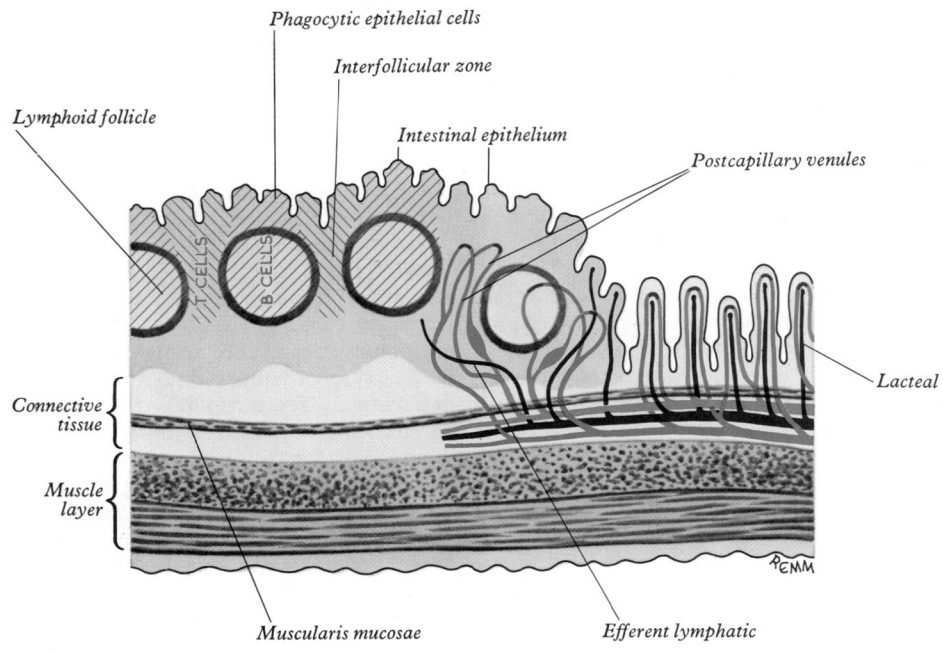

6.189 The organization of an epitheliolymphoid complex in the wall of the small intestine (a Peyer's patch), showing the distribution of B-lymphocytes associated with the follicles and T-lymphocytes in the inter-

follicular zones; the arrangement of the vascular supply of the lymphoid tissue is also shown (right).

malaria, the splenic tissues may be permanently hypertrophied and the spleen greatly enlarged *(splenomegaly)*.

Because of the large volume of blood within the spleen, it can be regarded as a store of this fluid, which can under certain conditions be discharged into the circulation. According to the 'closed' theory of splenic circulation, this occurs in the venous sinusoids, the filling and emptying of which are controlled by sphincters at each end; according to the 'open' theory storage is amongst the reticular meshes of the splenic cords. In states of emergency, especially anoxia, blood is discharged into the circulation. In most mammals, this discharge is due to the contraction of capsular and trabecular muscle, excited by sympathetic innervation and adrenalin; in the human spleen recoil of stretched elastic fibres causes emptying of blood into the circulation, although storage and discharge are not nearly so marked in humans as in other species.

SPLENIC VASCULAR SEGMENTATION

There is evidence (Dreyer & Budtz-Olson 1952) that human spleens, like those in other species, consist of separate 'segments' or 'compartments', each served by a hilar branch of the main splenic artery and splenic vein (Braithwaite & Adams 1957). Adjacent compartments, it is claimed, are connected by intersegmental veins; a congested compartment can thus pass excess blood to those adjacent (vide infra) but, when blood flow is not excessive, splenic segments may act as separate units. (For a review of splenic vascular segmentation since first proposed by Kyber 1870, consult Gupta et al 1976.) Gupta et al studied corrosion casts of 50 adult human spleens. In 42 of these (84%) only two segments existed (superior and inferior); in 8 (16%) three segments (superior, intermediate and inferior) were demonstrated. There was no clear anastomosis between segments. A comparable segmental arrangement of splenic veins was described by Fuld & Irwin (1954). The occurrence of only two or three segments (arterial or venous) was supported by many investigators quoted by Gupta et al; but this does not accord with the usual description of splenic arteries dividing into five or more major branches in the lienorenal ligament before even entering the hilum: no explanation of this discrepancy is advanced by these authors. Braithwaite & Adams described 5–7 vascular splenic segments in rats. Clearly, further investigation is necessary.

Effects of splenectomy. Partial splenectomy is followed by rapid regeneration of lost tissue but even total splenectomy has few obvious effects, its functions being largely assumed by the liver. But especially in the early years of life, splenectomy may entail a general reduction in the rapidity of immune responses and in a consequent increased susceptibility to infection. Splenectomy in later life is followed by leucocytosis with increased lymphocytic, neutrophil, eosinophil and platelet counts in peripheral blood, interpreted as due to removal of humoral factors produced in the spleen which oppose the formation and release of cells from haemopoietic tissues. These effects fade and disappear within a few weeks.

The lymphatic vessels of the spleen, once considered insignificant and merely the drainage of the capsule and largest trabeculae, has been shown (in some species) to occur in *all* parts of the organ. Extensive blind-ended efferent lymphatic vessels occur in the white pulp, running tortuously with arterioles and arteries and emerging at the hilum, the flow of lymph being counter to that of arterial blood. No afferent lymphatic vessels have been identified. This lymphatic pathway drains tissue fluid infiltrating the white pulp and is also a major route for lymphocytes from the spleen into the general circulation, as is the case in lymph nodes (Weiss 1979).

Splenic nerves arise from the coeliac plexus, chiefly nonmyelinated. In humans, their distribution is said to be confined mainly to the branches of splenic arteries; in other mammals they also supply non-striated muscle in the capsule and trabeculae.

Applied Anatomy. Splenomegaly may accompany several conditions, e.g. increased phagocytosis in some generalized infections, bacterial or viral, and in conditions of increased erythrocyte breakdown, such as malaria infections and other haemo-

lytic diseases and in various lipoidoses. Any massive immune response may be accompanied by splenic enlargement, also occurring in many reticuloses. In splenomegaly, the anterior border, anterior diaphragmatic surface and notched superior border become palpable below the left costal margin; the marginal notches are exaggerated and easily palpable. The transverse colon and left colic flexure are displaced downward, no area of colonic resonance remaining over the enlarged spleen, in contrast to a retroperitoneal tumour (e.g. renal), which does not displace the gut and therefore leaves an area of colonic resonance. There is no anastomosis between the smaller splenic arteries, so that their obstruction leads to infarction. During splenectomy the tail of the pancreas is in danger.

Epithelio-lymphoid or 'Mucosa-associated' Lymphoid Tissue

In addition to the encapsulated peripheral lymphoid organs, lymphatic nodes and spleen, large amounts of unencapsulated lymphoid tissue exist as *lymphoid nodules* in the walls of the alimentary respiratory tracts, termed collectively *epithelio-lymphoid* or so-called *mucosa-associated lymphoid tissue (MALT)*. These nodules are located chiefly in the lamina propria subjacent to the epithelium but may when active extend into deeper layers, e.g. the intestinal submucosa; their cells may also migrate widely into neighbouring tissues. Such nodules are often grouped into those associated with the alimentary tract, collectively termed *gut-associated lymphoid tissue (GALT)*: pharyngeal, tubal, palatine and lingual tonsils, oesophageal nodules, aggregated lymphatic nodules in the small intestine (e.g. Peyer's patches) and the vermiform appendix, and smaller nodules in the colon and rectum. Bronchial-associated lymphoid tissue (BALT) is the equivalent lymphoid nodular tissue of the lower respiratory tract.

The form of lymphoid nodules depends on their location but in all sites numerous round follicles similar to those in lymph nodes can be distinguished; they develop germinal centres when antigenically stimulated (Anderson 1977). Between follicles (**6.189**) are cuneiform masses of less closely packed (parafollicular) lymphocytes. The whole mass with numerous macrophages is supported by a fine network of reticulin fibres and associated fibroblasts, with coarser trabeculae in the larger nodules, such as those in the pharyngeal tonsil. Adjacent alimentary epithelium covers the luminal surfaces of nodules, usually producing glandular or other diverticula which penetrate the lymphocytic masses. In some sites (e.g. palatine tonsils, p. 1324) such *crypts* may be loci of infection, leading to general inflammatory response in the whole area. In many sites domes of specialized, thinner epithelium cover the lymphoid nodules (vide infra). Blood vessels branch from the surrounding connective tissue to supply the follicles with a capillary plexus draining to post-capillary venules, which have cuboidal endothelial cells like those of lymph nodes, allowing the migration of lymphocytes between blood and nodule. The lymphatic vessels of nodules are all efferent, draining into the lymphatic channels of the organ in which they are sited.

Immunofluorescence technique has shown that B- and T-lymphocytes of lymphoid nodules are segregated, again resembling lymph nodes, the follicles containing B-lymphocytes and the parafollicular zones, T-lymphocytes. All of these cells can migrate into the lymphatic drainage to rejoin the bloodstream, or into adjacent tissues including the intercellular spaces of neighbouring epithelia. As a rule most of them appear to pass into the circulation before finally migrating into their sites of effective action. In simple epithelia they may eventually enter the alimentary or respiratory lumen. B-lymphocytes which have migrated into the neighbouring lamina propria often mature into plasma cells. In addition, mononuclear phagocytes are present, both as active highly phagocytic macrophages and dendritic, antigen-presenting cells which are Class II MHC positive. These are mainly interdigitating cells similar to those of lymph nodes (p. 825).

FUNCTIONS OF LYMPHOID NODULES

The precise role of lymphoid nodules in lymphocyte biology has evoked much speculation, stemming in part from the avian condition in which B-lymphocytes become committed to their distinctive lineage in lymphoid tissue in a proctodael (hindgut) diverticulum (*bursa of Fabricius*), suggesting that alimentary lymphoid tissues in mammals may have a similar function. However, experiments have failed to support this idea and it seems that lymphoid nodules are regions in which B- and T-lymphocytes can proliferate in association with antigen-presenting mononuclear phagocytes and are thus producers of activated, antigen-specific defensive cells passing to the tissues of the structures they are grouped around.

B-lymphocytes are mainly involved in the synthesis of secretory antibodies of the IgA class occurring in alimentary secretions and in IgE (homocytotropic antibody related to mast cell activities, see p. 672). These are secreted first by plasma cells in the lamina propria and intercellular spaces of simple epithelia and in sub-epithelial glands, the latter being of special significance where epithelium is of the stratified squamous type. Antibodies appear to be absorbed by certain glandular cells (although not goblet cells, at least in the gut) and are modified by the addition of carbohydrate to form secretory IgA (sometimes termed sIgA) which is then passed into the intestinal or bronchial lumen. These antibodies are vital in eliminating pathogenic organisms, although other types of antibody (IgM and IgG), secreted by plasma cells of the lamina propria, are also essential to the destruction of any pathogens which breach the epithelium and infect the adjacent tissues.

If this system is to operate efficiently lymphocytes must be able to detect antigens present on the luminal side of epithelium covering them. Special, probably phagocytic, cells (*follicle associated epithelial cells, FAE, or microfold, M cells*) have been demonstrated in the epithelium covering lymphoid follicles, which may be capable of transferring particulate material to the subjacent lymphoid tissue, thus carrying antigens to the immune system (Bockman & Cooper 1973, Chamberlain et al 1973, Owen & Jones 1974, Smith & Peacock 1980). These cells are columnar in form and bear long, irregular microvilli, numerous apical invaginations and vesicles and an irregular basally-situated nucleus. The lateral borders are deeply indented by small lymphocytes lying within the epithelium.

The Thymus

The thymus (6.190–1956) is one of the two central organs of the lymphoid system, the other being the bone marrow, and is responsible for the provision of thymus-processed lymphocytes (T-lymphocytes) to the whole body. It also has certain humoral secretory functions and so may be regarded as an endocrine gland. It varies considerably in size and activity, depending on age, disease and the physiological state of the body, but remains active even into old age. Its significance was long obscure; but experimentation has revealed the thymus as a primary *central lymphoid organ*. At birth it weighs 10 to 15 g, growing until puberty, when it weighs 30-40 g. Thereafter it gradually diminishes and is progressively infiltrated and replaced by adipose tissue; after middle age it may weigh only 10 g, but in some cases may remain large, weighing 28 and 50 g (Young & Turnbull 1931, Keynes 1954, Kendall et al 1980). Certain diseases greatly accelerate its involution, perhaps accounting in part for this variation.

In early life it is pinkish grey, soft and finely lobulated, consisting of two unequal, pyramidal lobes connected by loose connective tissue. Although it is often described as a single median organ, each lobe develops from the third pharyngeal pouch of the same side (p. 229), so that strictly there are right and left thymic bodies. The thymus lies in the superior and anterior inferior mediastina, extending inferiorly to the fourth costal cartilages, its upper parts tapering into the neck and sometimes reaching the inferior poles of the thyroid gland or even higher. Its shape is largely due to the adjacent structures, to which it is moulded. *Anterior* are the

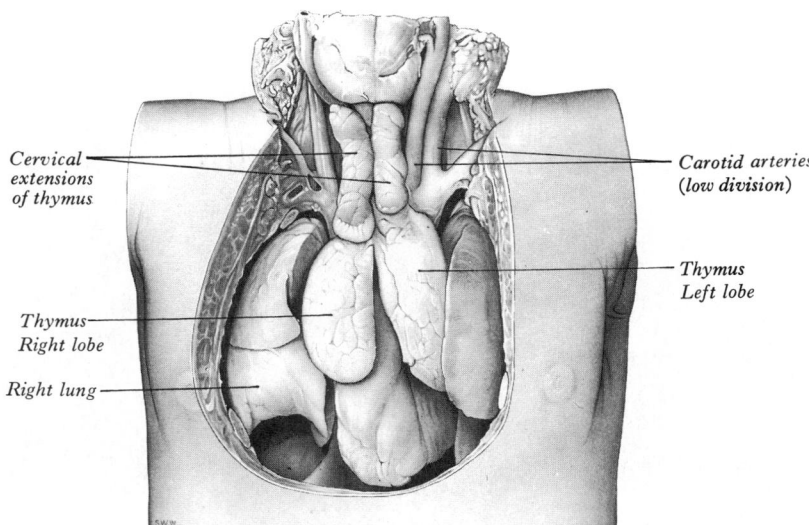

6.190 Dissection to display the neonatal thymus.

sternum, adjacent parts of the upper four costal cartilages and the sternohyoid and sternothyroid muscles. *Posterior* are the pericardium and the aortic arch with its branches, the left brachiocephalic vein and the front and sides of the trachea. After middle age the thymus becomes yellowish due to its gradual replacement by adipose tissue. Small accessory nodules may occur in the neck, representing portions of the thymic diverticula which have become detached during their early descent (p. 229); or the thymus may be found even more superiorly as thin strands along this path, reaching the thyroid cartilage or above. Sometimes its cervical part is represented by strands of connective tissue connecting the thymus to the inferior parathyroid glands. These strands may also approach the superior parathyroids, those related to the superior (parathyroids IV) being developed from the fourth pharyngeal pouches, those to the inferior parathyroids (parathyroids III) from the third pair.

MICROSCOPIC STRUCTURE OF THE THYMUS

Each lobe (6.190) has a loose fibrous capsule, from which septa penetrate to the junction of cortex and medulla, to partially separate the irregular lobules, each 0.5–2.0 mm in diameter. Lobules consist of an external, dark staining, dense, highly cellular *cortex* and an internal, light staining, less dense *medulla*. Lobules are not completely separated and their medullary parts are continuous through a central parenchymatous cord, derived from the embryonic thymic diverticulum. Lobules are connected to this cord by prolongations of the medullary substance; if interlobular connective tissue is removed (in the young), lobules present an irregular 'necklace' appearance. The fibrous capsule is quite thin and contains numerous mast cells.

Both the cortex and medulla have two principal cellular components, differing quantitatively in the two regions. These are: (1) a framework of irregular, interconnected epithelial ('reticular') cells, quite distinct from those in other lymphoid tissues; and (2) lymphoblasts, lymphocytes, macrophages and a few other cells enmeshed in the epithelial framework. Other cells also occur in small numbers including *myoid cells* resembling striated muscle fibres, epithelial cysts and goblet cells. These may represent 'accidents' of embryological development.

The epithelial framework

Unlike other lymphoid structures, where the supportive framework is composed chiefly of collagenous reticular tissue, the thymus is permeated by a network of interconnected epithelial cells (thymic epitheliocytes); there is only a little reticulin and few fibroblasts. Epitheliocytes vary in size and shape in the different regions of the thymus. Typically they have pale, oval nuclei, a rather eosinophilic cytoplasm and desmosomal attachments between cells. Intermediate (keratin) filament bundles lie within the

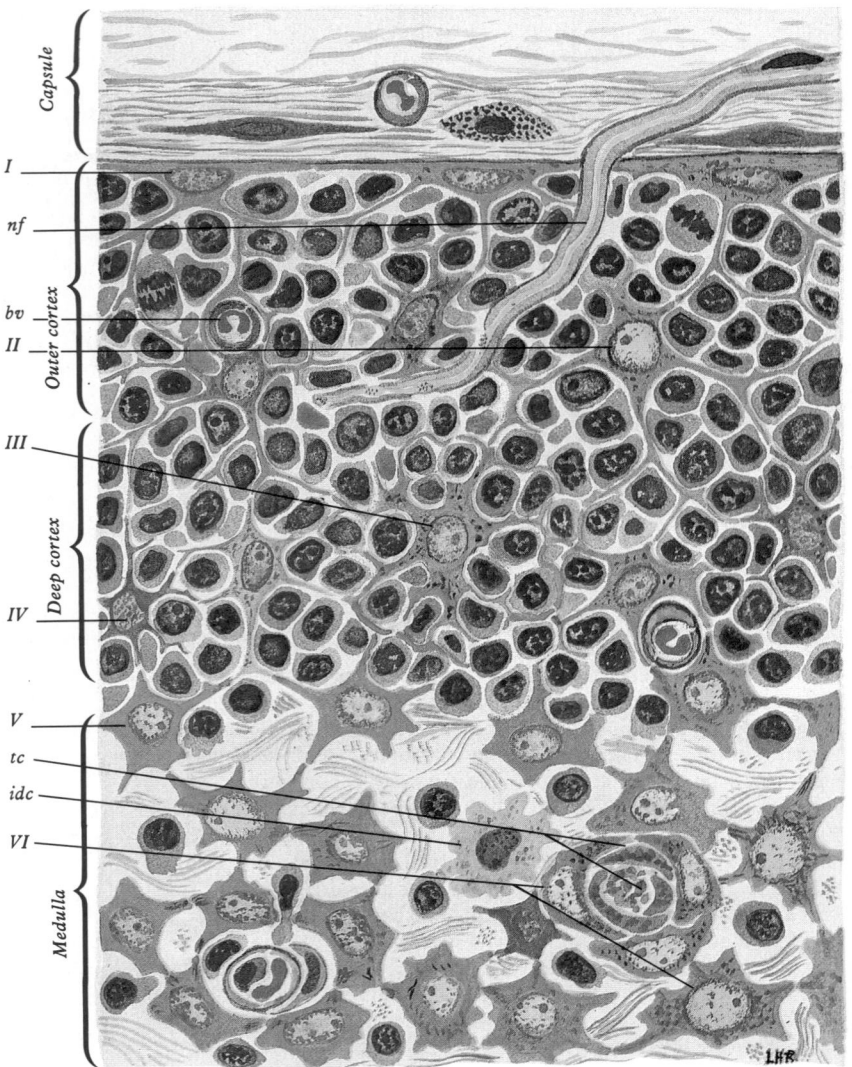

Their cytoplasm shows some granular endoplasmic reticulum, occasional dense bodies and bundles of microtubules. *Type 3 cells*, occurring deeper in the cortex, have distended cisternae of granular endoplasmic reticulum and cytoplasmic vacuoles; while *type 4 cells*, in the deepest cortex and occasionally the medulla, stain densely, have abnormally swollen mitochondria and electron-lucent granular endoplasmic reticulum. It has been proposed that these cells form a graded series whose precise form and function varies according to position in the thymus.

Type 5 cells ('undifferentiated' epitheliocytes) occur in groups in the medulla and show no particularly prominent features; *type 6 (large medullary) epitheliocytes* are usually associated with thymic (Hassal's) corpuscles in the medulla and contain tortuous bundles of microtubules and numerous intermediate (keratin) filaments. The thymic corpuscles themselves are distinctive in their staining and immunochemical properties, although the outermost cells are similar to type 6 cells.

Many investigators have supported the view that the type I cells form a partial haemothymic barrier which limits access of certain extrathymic materials including circulating antigens into the tissue spaces of the thymus. However, recent work has shown that particulate tracers and monoclonal antibodies directed against thymic cells can, after intraperitoneal injection, penetrate the thymus through its capsule (Eggli et al 1986, Stet 1987). Stem cells and prothymocytes also gain access to the thymus through this route, especially in the fetus. (See the further discussion of the thymic barrier, p. 229.)

Embryonic origins of epitheliocytes. Most embryological evidence at present favours the view that the thymic epithelium is derived from both the ectoderm and the endoderm of the third and often the fourth branchial grooves and pharyngeal pouches, with the associated mesenchyme playing an inductive role in the production of the necessary soluble factors (see e.g. Norris 1938, Auerbach 1961, Cordier & Haumont 1980). In the mutant mouse *nude*, where the ectodermal anlage for the thymus is deficient, the thymus is abnormal, cystic and does not support lymphopoiesis, although lymphocytes are still produced in the bone marrow (Pritchard & Micklem 1973). It is not yet known if the epithelial cells are of ectodermal or endodermal origin or both, although ectoderm appears to be necessary for its functional development.

Developing thymocytes

In the *cortex*, massive numbers of densely packed small lymphocytes predominate, occupying the interstices of the epithelial reticulum, which in histological sections they largely obscure, and forming about 90% of the total weight of the thymus. A distinct subcapsular zone is present, housing the thymic stem cells, prothymocytes and lymphoblasts undergoing mitotic division. The first stem cells to enter the thymus in the embryo come from the yolk sac and liver during their haemopoietic phases, possibly, as in birds, being attracted by thymic chemotactic substances. 94During later periods it is probable that all thymic lymphocytes originate in the bone marrow, or at least have sojourned there, before passing in the bloodstream to the thymus.

The cortex has two rather ill-defined zones: an outer cortex with a framework of types 1–3 epitheliocytes and a deep cortex where type 4 cells occur. Thymocytes undergo mitosis in all cortical zones as the clones of differentiating T cells mature, gradually moving deeper in the cortex. In rodents, cell cycling times of eight hours have been recorded in the outer cortex, but no estimates exist for the human thymus. The appropriate conditions for the proliferation and differentiation of thymocytes appear to be produced by their close proximity to neighbouring epitheliocytes (see Janossy et al 1986). Although the nature of these interactions is not clear, it may involve the release from the epitheliocytes of soluble mitogenic and differentiation factors as well as induction of changes through intercellular contact. During this process, not only are thymocytes switched to become T cells but their potential to form different T-cell subclasses is determined. The differentiating thymocytes are pushed towards the medulla by the intense proliferation of cells in the subcapsular cortex, leaving the thymus at the cortico-medullary junction or deeper in the medulla through blood vessels and lymphatics; such cells (*post-thymic thymocytes*) are not yet immunocompetent and in genera[l]

6.191 Cellular organization of the thymus showing thymocytes (blue) and epitheliocyte framework (green) of types I–VI cells. nf = nerve fibre; tc = thymic corpuscle; idc = interdigitating cell; bv = blood vessel.

cytoplasm. These cells form a continuous external lining to the thymus beneath its fibrous capsule, following its lobulated profile and investing the vessels which pass into it. Other cortical cells are branched, with large spaces between them, while the epithelial cells of the medulla tend to form more solid cords as well as the characteristic whorls of (often) partially keratinized stratified epithelium (thymic or Hassal's corpuscles). Lymphocytes lie within the meshes and cords formed by these various cells. There is much evidence that many distinctive functional roles are subserved by the epithelial cells (Ritter et al 1985), some of them related to the differentiation of T-lymphocytes, others with the production of soluble thymic factors or hormones or with barrier and mechanically supportive functions (see Wijngaert et al 1984). At least some possess Class II MHC molecules at the surfaces, indicating an important role in the generation of immune responses.

Type I epitheliocytes (subcapsular-perivascular) create the continuous monolayer around the perimeter of the thymus, extending along the septa to the cortico-medullary boundary; they rest on an (external) basal lamina and send long processes into the cortex to contact other (type 2) epitheliocytes. In rodents these cells form basket-like structures which separate an outer, *subcapsular* cortex from the deeper regions. Pale and darkly-staining varieties of these cells are visible by electron microscopy; in small children they may contain numerous small vesicles. *Type 2 cells*, found in the outer cortex, are stellate and in internal structure are similar to type 1 epitheliocytes; they have a large, pale nucleus.

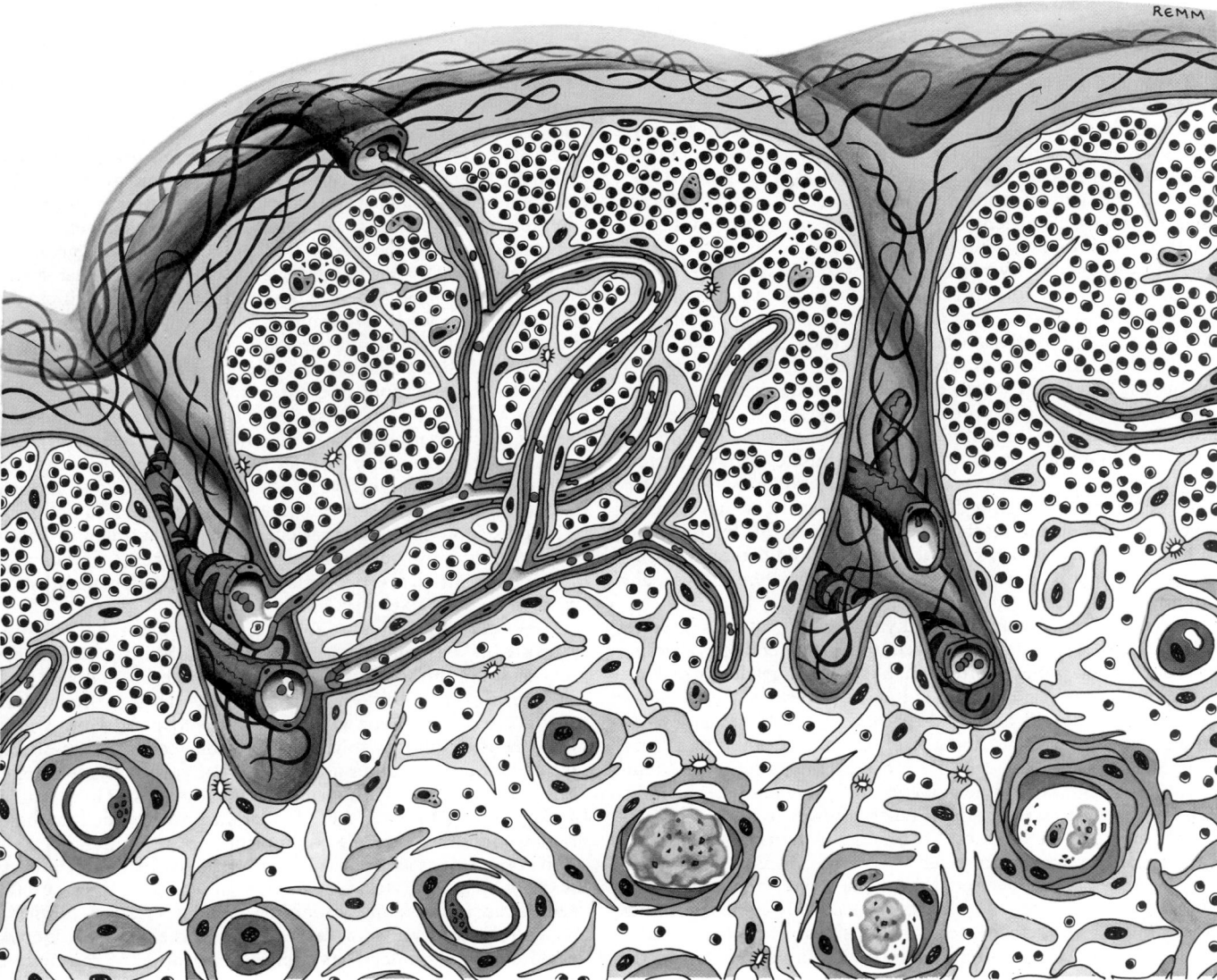

6.192 A scheme of thymic structure; the various elements are not drawn to scale to enable representation in a single diagram. Note the lobular outlines, capsule, delicate interlobular septa, cortical lymphocytes, the epithelial 'reticular' cells and their junctions and the medullary corpuscles of Hassall showing a graded series of increasing maturity; also the transcortical circulation. See text for further discussion.

attain maturity only after leaving the cortex and probably only when they leave the thymus altogether and reach their peripheral destinations.

It has been estimated that 99% of thymocytes never leave the thymus (Scollay et al 1980) although this figure is often regarded as far too high (vide infra). However, during their maturation, thymocytes become able to recognize the HLA (MHC) markers expressed by the unique genome of the individual they belong to, i.e. 'self-antigens', so that when they migrate to the peripheral lymphoid organs and other tissue sites, they and the clone of cells they give rise to can detect alien antigens when these are present in association with 'self'-MHC determinants (see pp. 674, 675) to mount an appropriate immune attack. It has been proposed that the thymocyte's correctness of response to self-MHC antigens is in some way tested by the thymic epitheliocytes before release and that, if they do not interact properly with these cells, the thymocytes do not survive. In this way are selected only T cells which can recognize alien antigens in combination with self-MHC molecules (i.e. they become MHC-restricted, see p. 675). Mechanisms of death of 'inappropriate' thymocytes may involve the loss of control of enzyme systems associated with purine metabolism which, unless corrected by successful interaction with the epitheliocytes, leads to lethal levels of certain metabolites within the cell. However, there is little histological evidence of high levels of cell death and mitotic figures are more common in sections than pyknotic cells; where pyknosis occurs, the cells involved are both epitheliocytes and thymocytes. Nevertheless it is possible that the unwanted thymocytes die rapidly and their remnants are quickly disposed of by phagocytes, limiting the number of observable moribund cells to a minimum.

Mononuclear phagocytes of various types occur in the thymus. Those of the subcapsular cortex may contain phagocytosed nuclei and cell debris and appear to be highly phagocytic macrophages engaged in removing dead thymocytes. Those deeper in the thymus at the cortico-medullary junction and in the medulla are more numerous but only rarely contain phagocytosed particles; some of these are dendritic *interdigitating* cells (p. 825) as they contain characteristic 'Birbeck' granules and bear the Class II MHC antigens typical of such dendritic cells at their surfaces. These latter cells are weakly phagocytic and are presumably concerned with the presentation of antigens to lymphocytes. However, there is much evidence that the thymocytes differentiate in the thymic cortex in an environment free of alien antigens and so it is probable that the antigen-presenting dendritic cells are part of the *medullary compartment* of the thymus into which the more mature thymocytes move after their crucial early development is over. The occurrence of occasional mature B-lymphocytes (plasmocytes) and germinal centres in the medulla (Middleton 1967) also supports the conclusion that at least part of the medulla is more like a secondary ('peripheral') as opposed to a primary ('central') lymphoid tissue, receiving differentiated lymphocytes capable of immune interactions. However, there may be important interactions between the antigen-presenting cells and immature, deep cortical thymocytes, as the thymus

835

may also be involved in the generation of 'self'-tolerance, and is perhaps an observed case of acquired tolerance of persistent 'alien' antigens.

Haemopoietic cells are present in fetal life, when the thymus makes an important contribution to the formation of erythrocytes and leucocytes. Later haemopoietic cells are often present, possibly as a result of the reactivation of persistent fetal stem cells. Normoblasts have been found in the thymuses of many adult open heart surgery patients and immature eosinophils, neutrophils and mast cells have also been observed. Where present in adults, the erythropoietic foci are mainly in the subcapsular and outer cortex.

The concentric thymic corpuscles (of Hassall) first appear in fetal life and then form continuously throughout life. A medullary epitheliocyte enlarges, develops intense eosinophilia and follows a train of degenerative changes. Progressive cytoplasmic vacuolation and nuclear fragmentation are followed by ingestion of nuclear debris by vacuoles which become confluent. More epitheliocytes enlarge, approach the degenerating central cell and form a series of eosinophilic concentric lamellae around the central hyalinizing mass. As the corpuscle grows, further degenerate epitheliocytes join the mass with, it is said, effete macrophages carrying products of lympholysis, which pass between the lamellae to reach the centre and become hyalinized or soluble. Corpuscles vary from 30–100 μm in diameter, increasing in size and number during periods of intense lympholysis and thymic involution.

Thymic involution. The thymus–body ratio is greatest perinatally (**6**.190) but the organ continues to increase in absolute size until about puberty, after which it tends to gradually decline, although a great range of sizes has been recorded throughout post-pubertal life (Kendall et al 1980). Early work from autopsies, often related to patients dying after chronic, debilitating illness, suggested an age-dependent decline in thymic weight, but Steinmann et al (1986) found no evidence of this in autopsies conducted within 24 hours of sudden death. However, such measurements are difficult to assess in functional terms because fatty infiltration occurs progressively from the neonatal period onwards. This process begins in the septa around the lobules and then adipocytes enter the cortex and gradually replace its cells, such infiltration advancing from the capsule towards the medulla. This fatty infiltration often makes the distinction between the thymus and surrounding mediastinal adipose tissue difficult at post-mortem, but careful dissection reveals that encapsulated tissue is always present, albeit fatty.

Because of these changes, the numbers of thymocytes present must be greatly reduced in old age, as was also found in cultured tissue. However, thymocyte production and differentiation persist throughout life, and T cells from this source continue to populate the peripheral lymphoid tissue, blood and lymph (see Steinmann & Muller Heremlink 1984). The activity of the thymus appears to be related to the levels of thymic hormones in the organ; with ageing the functional capacity of the thymus may be reduced, suppressed or not effectively stimulated but is not completely lost.

Vessels and nerves. The arteries are derived mainly from internal thoracic and inferior thyroid artery branches which also supply the surrounding mediastinal connective tissue. There is no main hilum but arterial branches pass either directly through the capsule or, more often, into the depths of the interlobar septa before entering the thymus at the junction of the cortex and medulla. Within the septa the vessels are quite large and are separated from the main substance of the thymus by a continuous sheath of Type 1 thymic epitheliocytes (vide supra) and a perivascular space containing pericytes and other connective tissue elements. After entering the thymus proper, these vessels course along the cortico-medullary junction giving off irregular branches to the medulla and small radially-arranged arterioles and capillary loops to the cortex. Epitheliocyte sheaths and perivascular spaces around arteries and arterioles persist into the medulla and cortex for some distance. The endothelium of vessels in the cortex is generally of the complete type, although medullary and intraseptal vessels may show fenestrations.

Post-capillary venules of the cortico-medullary junction have a thickened endothelium as in lymph nodes (p. 824) permitting the migration of lymphocytes. Venules retrace the arteriolar paths,

converging to principal veins in the interlobular septa; from there the venous return is partly via a capsular plexus and hence circulation from arteriole to radial capillary continues as a *transcortical centrifugal route* through the venules to the small veins, finally joining with the capsular veins (**6**.192). These contrasting microvascular routes may be significant in movements of thymic cells (Blau 1976). *Thymic veins* drain to the left brachiocephalic, internal thoracic and inferior thyroid veins; one or more veins often emerge medially from each lobe of the thymus to form a common trunk opening into the left brachiocephalic vein.

The blood – thymus barrier. When colloidal substances are injected intravascularly they fail to penetrate the extravascular spaces in the thymic cortex, where lymphocytes are proliferating, suggesting that the walls of thymic blood vessels may bar antigens and other substances from passing into the thymic tissue. This tissue would thus be an *immunologically sequestered site* for the differentiation of lymphocytes (vide infra). Raviola & Karnovsky (1972) found that, in rodents, electron-opaque tracers traversed the intercellular junctions of medullary endotheliocytes, particularly in post-capillary venules, diffusing extensively into the medullary tissue. However, the endothelium of cortical capillaries forms a more effective barrier and the small amounts of tracer passed by endothelial vesicular transport were immediately phagocytosed by perivascular macrophages. Presumably macrophagic action also prevents diffusion of particulate substances from medulla to cortex.

Afferent lymphatics are absent from the thymus but **efferent lymphatics** arising from the medulla and cortico-medullary junction drain through the extravascular spaces in company with the arteries and veins entering and leaving the thymus. In rodents, large lymphatic vessels draining to perithymic lymph nodes are often found within the subcapsular cortex but these also receive lymph from other areas of the body; these lymph nodes drain in turn to neighbouring regional nodes (Goldstein & Mackay 1969). Whether there is a similar perithymic lymphatic drainage in humans remains unknown.

Thymic innervation is derived from the sympathetic chain via the cervicothoracic (stellate) ganglion (or ansa subclavia) and the vagus. Branches from the phrenic nerve and descendens cervicalis are distributed mainly to the capsule of the thymus. During development (Hammar 1935), vagal innervation of the thymus commences in the neck before its descent into the thorax. The two lobes are innervated separately through their dorsal, lateral and medial aspects and rich neural plexuses are formed in the medulla. After its descent, the thymus receives the sympathetic nerves along vascular routes, their terminals branching radially and forming with the vagal fibres a plexus at the cortico-medullary junction. Innervation is complete by the onset of thymic function.

While many of the autonomic nerves are doubtless vasomotor, many terminal branches also (at least in rodents) leave their perivascular pathways and pass among the cells of the thymus, particularly the medulla, suggesting that they may have other roles. The medulla also contains a variety of non-lymphoid cells, including cells positive for vasoactive intestinal polypeptide (VIP), acetylcholinesterase (large, non-myoid cells), oxytocin, vasopressin and neurophysin cells and with possible neural crest origin. Clearly, the roles of the nervous system and other neuroendocrine elements in the overall biology of the thymus are far from being understood and suggest many intriguing possibilities.

FUNCTIONS OF THE THYMUS

Thymic functions are not yet completely clarified although in recent years dramatic advances have been made. The evidence comes from a variety of experimental approaches, including studies of the effects of thymectomy or thymic destruction by radiation in neonatal and mature animals and the study of mutant murine strains, e.g. 'nude', in which the thymus is vestigial or absent in homozygous individuals. Attempts have been made experimentally in animals to compensate for thymectomy by the injection of cell suspensions from various lymphoid tissues or the bone marrow, by normal or irradiated thymic tissue, either as free grafts or surrounded by barriers impermeable to cells and by cell-free thymic extracts.

The nature and origins of stem cells capable of re-populating cell-depleted lymphoid tissues or thymic grafts have also been investigated by chromosomal marker or autoradiographic techniques. The restoration of immune competence in animals made immunologically tolerant to certain antigens, by prolonged exposure to small quantities, has been studied in early neonatal animals. It is possible only to summarize the concepts emerging from this busy field of experimentation. These experimental approaches are now being supplemented and greatly clarified by the use of monoclonal antibodies either to characterize different cell types or to demonstrate the presence of particular functional markers in cell interactions.

The major functions of the thymus are to achieve the differentiation of lymphocytes into the different classes of T cells (thymocytes) and to maintain a sufficient supply of them in the general circulation and peripheral lymphoid organs to react to whatever antigenic stimuli may be presented to the body, as described elsewhere (p. 672). During neonatal and early postnatal life, the thymus is essential to the normal development of lymphoid tissues. Thymectomy at this stage leads to a progressively fatal condition, with hypoplasia of the peripheral lymphoid organs, wasting and an inability to mount an effective immune response. By puberty, when the main lymphoid tissues are fully developed, thymectomy is less debilitating, but a reduction in effective responses to novel antigens ultimately ensues.

Secretory activities of the thymus

A second major function of the thymus is the production of various factors and hormones which regulate lymphocyte production, differentiation and activities within the thymus, in peripheral lymphoid tissue and elsewhere. These substances include four major, chemically well-defined polypeptides: thymulin, thymopoietin, thymosin alpha 1 and thymosin beta 4 (Hadden 1987). The first two of these fulfil them as the strict criteria defining them as hormones, whereas the other two are more problematical (vide infra). Thymic humoral factor is also probably of thymic origin and various other substances involved in the later stages of T-cell and mononuclear phagocyte function, also found in other lymphoid tissue, are also produced by the thymus, e.g. interleukins I and II (pp. 669, 673). *Thymulin* appears to be produced exclusively within the thymus; it is a nonapeptide hormone needing the presence of zinc for functional activity. It acts exclusively on T cells, binding to T-cell receptors with high affinity and enhancing the performance of various T-cell subsets, especially (at high doses) T-cell suppressor cells. It can be detected and measured in circulating blood by the appropriate methods of bioassay or radio-immunoassay. *Thymopoietin* is, in vitro, an inducer of the final differentiation of T cells into distinct subclasses and in vivo enhances several T-cell functions, e.g. the generation of cytotoxic T cells. Both thymulin and thymopoietin act systemically to give a finely tuned immuno-regulation of T cells, helping to maintain a balance between the activities of their different subsets.

The nature and biological activities of the other factors produced by the thymus are less clear. *Thymosin alpha 1* stimulates (in mice) lymphocyte proliferation and antibody production, enhances the action of macrophage migration inhibition factor and affects DNA metabolism in thymocytes. It has been localized within epitheliocytes of the thymic medulla but may be synthesized elsewhere, possibly in the central nervous system where it has also been localized in regions related to autonomic and neuro-endocrine control. However, there is some evidence that thymosin alpha 1 may be a degradation product of a precursor present in other tissues, too. *Thymosin beta 4* is a product of mononuclear phagocytes (monocytes) and is therefore not unique to the thymus. *Thymic humoral factor* regulates the proliferation and maturation of T-helper and T-suppressor cells, but its chemical structure has not yet been described; various other factors may be produced by the thymus but their specificity to this organ and their chemical status also have yet to be clarified.

Summary of the origins and fates of lymphocytes in the thymus

The thymus develops before other lymphoid organs as an epithelial structure that begins to receive lymphoid precursor

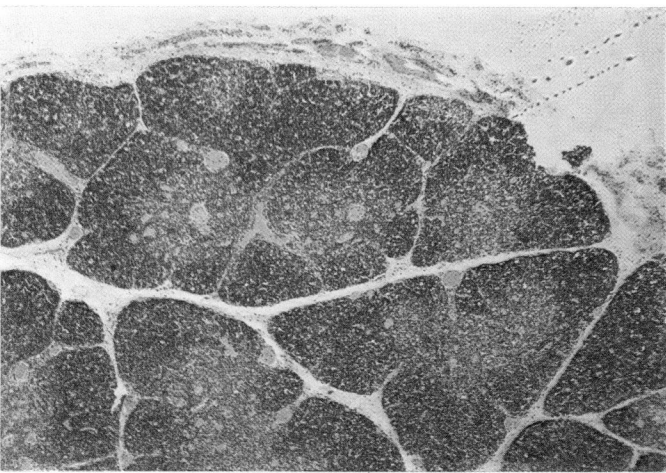

A

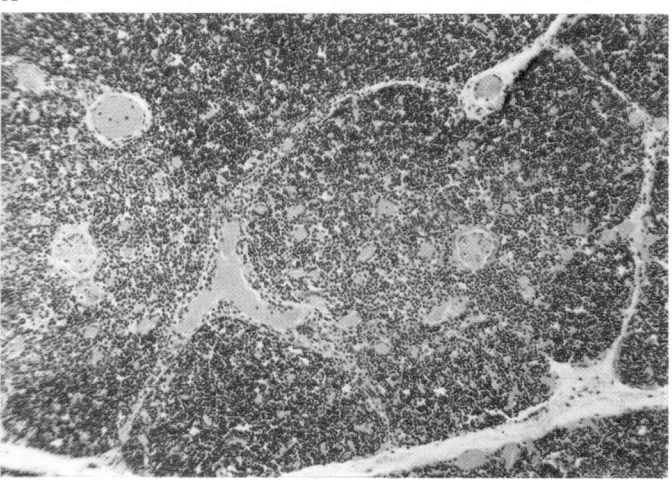

B

6.193 Survey photographs of a neonatal human thymus stained with haematoxylin and eosin. The general lobular architecture is seen; each lobule contains a relatively pale medullary core surrounded by a densely cellular, dark, heavily stained cortex. From a specimen prepared and provided by R O Weller, Department of Pathology, Guy's Hospital Medical School.

cells at 9–11 weeks in utero; by 15 weeks, differentiation into cortex and medulla has occurred and the epitheliocytes now express the typical immunochemical markers of the adult thymus.

Thymocytes arise from stem cells of the bone marrow and, in earlier fetal life, the spleen; they migrate via the bloodstream to the thymus, entering its tissues chiefly through the post-capillary venules. They become pro-thymocytic lymphoblasts in the outer

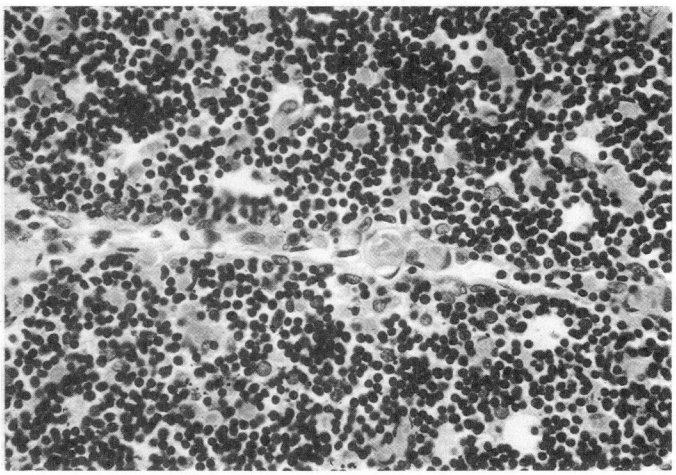

6.194 Neonatal thymus. A cortical field showing a delicate interlobular connective tissue septum, densely packed cortical lymphocytes and scattered profiles of larger, pale, eosinophilic, epithelial, reticular cells.

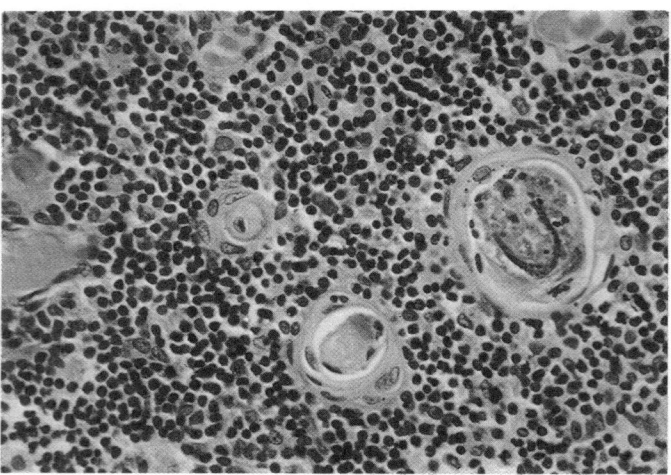

6.195 Neonatal thymus. A medullary field showing three concentric corpuscles of Hassall of varying degrees of maturity, surrounded by many closely packed lymphocytes and a number of reticular cells.

cortex, proliferating extensively and differentiating into different subsets but about 90% of their progeny die in the thymus within a few days; the remainder go on to become fully immunocompetent T-lymphocytes after returning to the circulation to join the reserve pool of T-lymphocytes and to populate the thymus-dependent regions of lymph nodes and other peripheral lymphoid organs where final maturation takes place.

Interactions between the thymus and other endocrine organs

Interestingly, the neonatal thymus appears to have other, non-immunological powers, since neonatal thymectomy (or the con-

genital athymic condition) in rodents is associated with degranulation of cells in the adenohypophysis (probably somatotrophs or prolactin-secreting cells). The levels of circulating luteinizing hormone (LH) and Follicle Stimulating Hormone (FSH) are also reduced but can be corrected by transplantation of a new thymus at birth. Likewise, in girls the development of the ovary is affected in cases of thymic dysfunction, e.g. in ataxia telangiectasia and the Di George and Cri-du-chat syndromes where the thymus fails to develop. Thymectomy in mice later than seven days after birth no longer affects ovarian development; removal of the thymus earlier than this leads to death from wasting disease. There is evidence that thymulin and a thymosin-related factor are involved in these effects.

Interactions with other endocrine glands appear to be widespread (see Michael 1983, Kendall 1988). The activity of the thymus is affected by Adrenocorticotrophic Hormone (ACTH), LH, FSH, prolactin, thyrotrophic hormone (TSH) and somatrophic hormone (STH). ACTH acts indirectly by stimulating the release of glucocorticoid hormones, principally cortisol, from the adrenal cortex; these substances are potent inhibitors of thymic function and cause its involution (human thymocytes bear up to 3000 cortisol receptors per cell and thymic epitheliocytes have 10 times this number). Glucocorticoids induce thymocytes to synthesize endonucleases which damage their own DNA and cause their death. Conversely, adrenalectomy causes thymic enlargement, while neonatal thymectomy (in rodents) stimulates the adrenal glands. The sex hormones also have important effects on the thymus, some of them inhibitory; others have different actions depending on the physiological state, particularly age. In animals, gonadectomy of both sexes leads to an increase in thymic size, while in pregnancy it involutes but is restored during lactation, possibly under the action of prolactin.

STH, which is similar in structure to prolactin and possibly also binds to its receptors on thymic cells, has a similar stimulatory role. Various other hormones (e.g. insulin, insulin-like growth factor, etc.) also affect the production of lymphocytes by the thymus. Like many other aspects of hormonal regulation, the significance of these diverse findings is difficult to assess in the context of the overall strategy of the immune system at different times of life and under various conditions of normal health and disease, but they indicate that the immune and neuro-endocrine system are heavily interdependent and that agents which affect either, including different forms of physiological, neurological and immune stress, are likely to cause complex changes in the operation of both systems.

Applied Anatomy. Castration or adrenalectomy delays thymic involution; hypertrophy of the suprarenal cortex, injection of cortisone or of androgenic hormones all cause atrophy. In young children a large thymus may press on the trachea, causing attacks of respiratory stridor. Thymic tumours may also compress the trachea, oesophagus and large veins in the neck, causing hoarseness, cough, dysphagia and cyanosis. Myasthenia gravis, a chronic auto-immune disease of adults (Castleman 1966), is a diminution in power of certain voluntary muscles for repetitive contraction. Muscles commonly involved are levator palpebrae superioris (leading to ptosis) and extra-ocular muscles (leading to diplopia). Others in the face, jaws, neck and limbs may be involved and in severe cases the respiratory muscles. In many patients thymic abnormalities have been observed, such as hyperplasia or thymoma, a tumour of epithelial cells derived from the reticulum or possibly the lymphocytes. In some cases transient or more lasting improvement may follow thymectomy. Although there may be more than one condition with these signs, myasthenia gravis is now considered an auto-immune disease, in which acetylcholine receptor proteins of neuromuscular junctions are attacked by internal antibodies; the coincidence of the disease with thymic disorders may reflect the role of the thymus in modulating the immune responses of the whole body.

Topography of the Lymph Nodes and Vessels

The detailed architecture of lymph nodes and lymphatic vessels of varying calibre has been given in the preceding pages and brie-

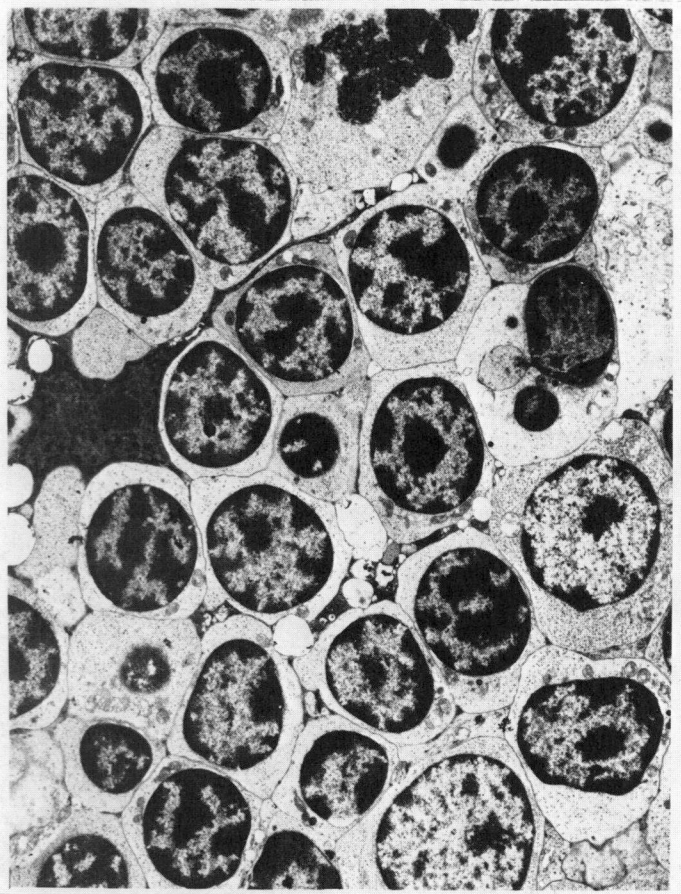

6.196 Transmission electron micrograph of the thymic cortex showing lymphocytes ensheathed in epitheliocytes. Provided by M D Kendall, UMDS, St Thomas' Campus, London. Magnification × 3000.

allusion has also been made to some *major lymph trunks* and the common aggregation of nodes into *regional groups*. The latter have, in most cases, fairly well defined topographical sites, specifically named (but varying criteria have been used), each with its variably well established area of drainage, interconnections with other nodes or groups and the predominant destination of its efferent vessels. A more detailed consideration of these matters is the concern of the remainder of this chapter; this must be prefaced by a summary, or sometimes a reiteration, of a few general principles. The reasons for this cannot be overemphasized. Lymphatic anatomy often appears an almost impossible plethora of topographical names for trunks, groups and subgroups of nodes and their connections. However, an elemental knowledge of general anatomy and a recognition of which of the limited number of general principles apply to the major organ systems make many of the difficulties evaporate. It is particularly useful to appreciate the *overall pattern* present in a particular organ, whole organ system or whole body segment. (Good examples are: the whole subdiaphragmatic alimentary tract, the foregut, the stomach, the tracheobronchopulmonary system, the similarities and contrasting features of the arm and leg and the head and neck as a whole.) These encompass all the main patterns and principles, include all the main *terminal lymph trunks* whereby lymph is returned to the venous system and indicate how helpful reference to main embryological events may be. The *principal groups* of lymph nodes towards which lymph converges from wide tissue areas (often through one or more *subgroups*) and knowledge of which, for many, is mandatory, is clarified. Thereafter, many (but not all) the lesser subgroups, in relation to their formal topographical names, assume a diminished importance in terms of mental retention; nevertheless, their *distribution* can in the majority of instances be *predicted* with confidence. The various criteria for the *topographical naming* of nodes will be mentioned below. Certain *general names* applied to nodes, although not universally used, often prove useful. Lymph circulating in lymphatic capillaries may be returned to the venous system (ignoring intralymphoidal venular events on a microscopic scale, p. 824) almost exclusively at *bilateral sites* at or near the junctions between internal jugular and subclavian veins forming right and left brachiocephalic veins; however their routes vary enormously in *length* and *complexity*. In certain exceptional sites (thyroid, oesophagus, dorsal hepatic 'bare areas', p. 1388) the capillaries drain via a radicle of the thoracic duct with no intervening lymph node (an *anodal* route). In some, a single node provides *uninodal* routes; the majority of routes are *multinodal* with a number of, sometimes many, nodes forming irregular cross-connected *chains*. In such a chain the node or group nearest the tissue drained is variously termed *primary (outlying, or peripheral)* in different circumstances. The last group of the chain, whose efferents form a final uninterrupted principal lymph trunk, is termed *terminal*. Between its primary peripheral and its terminal groups intervening nodes are often segregated into *intermediary* groups; some use the collective term *regional lymph nodes* to include all three groups; others, peripheral and intermediary only. The significance of multinodal pathways is by no means clear; it should be noted that classifications such as the above may impose an artificial simplicity on a potentially highly complex monitoring and reacting system.

Thus briefly a particular node is not one element in a simple chain but, receiving multiple afferent vessels, may be e.g. the primary node for various loci, number three in the chain for other loci, number five for others and so forth. Such considerations, of course, apply to other members of the chain and, summating for the whole chain, there emerges the notion of a system of lymphatic channels and lymphoid stations of great three-dimensional complexity. Differential lymphangiography amply confirms this complexity. This prompts the question: are lymph nodes in general roughly *equivalent* in their ranges of receptivity (monitoring) and reactivity or do they vary? If *variation* exists, to what degree? An extreme (and fanciful) extension of this is the possibility of individual nodes, or even sectors of nodes, being *unique* in their locations and connections. (The writer emphasizes that these ideas are entirely speculative but hopes that they are sufficiently provocative to merit consideration.)

Topographical naming of lymph nodes has not followed a single rigid classification; four main frames of reference have been found convenient, and quite lucid. These are (1) superficial/deep position, (2) related vasculature, (3) related organ's name and architecture, and (4) general topographical location.

Superficial and deep refer, as with a number of other structures, to their location with respect to the deep fascia. As noted below, many superficial nodes are closely applied to prominent superficial veins. An interesting but unexplained fact is that the upper limb has few superficial nodes and its *superficial* lymphatic drainage mostly passes directly to *deep* axillary nodes; in contrast, in the leg the *superficial* lymphatic drainage passes, almost exclusively, to the large *superficial* inguinal nodes before continuing to the external iliac nodes.

Relation to Vasculature. The majority of nodes and node groups are clustered around or abut a prominent blood vessel or one of its branches; from this (with many notable exceptions) the name of the group is derived. The association assists in recalling the location of the group and in many instances is a strong pointer to the main region of lymph drainage. Examples of exceptional superficial nodes associated with veins are: buccal nodes (facial vein), superficial cervical nodes (external jugular vein), anterior cervical nodes (anterior jugular veins), infraclavicular nodes (cephalic vein), supratrochlear nodes (basilic vein), superficial inguinal nodes (great saphenous vein). Deep nodes associated with vessels are so numerous that only a few illustrative examples can be given. The abdominal aorta and common, internal and external iliac arteries are encrusted with nodes. The whole consists of massive chains of nodes, interconnected by lymphatic channels, predominantly vertically but also obliquely and transversely. Thus, the main groups are named with their vessels: *external, internal* and *common iliac* and *circumaortic*. The latter (often grouped with neighbouring nodes particularly scattered over the inferior vena cava as *lumbar nodes*) are divided, on sound developmental and lymphodynamic grounds, into a large median *ventral aortic group*, prominent right and left *lateral aortic groups* and a sparse *retro-aortic group*. Details are furnished in subsequent pages but a *few examples* with comments and one possible synthetic approach to study may be appropriately outlined here. The ventral aortic group aggregates around the three large subdiaphragmatic ventral splanchnic arteries (p. 766) as *coeliac, superior mesenteric* and *inferior mesenteric* groups of nodes, which drain the subdiaphragmatic foregut, midgut and hindgut (and their derivatives), respectively. The foregut provides an excellent framework with respect to its extent and parts (terminal oesophagus, stomach, proximal duodenum); its derivatives (liver, gallbladder and biliary ducts, pancreas and the closely associated spleen); and the mutual disposition of the foregoing and their peritoneal reflexions, omenta and the lesser sac. To this is added the position of the upper abdominal aorta, the coeliac artery, its trifurcation into common hepatic, left gastric and splenic arteries and the courses and main branches of these. Groups of nodes named in relation to these are, e.g. left gastric, right gastro-epiploic, hepatic and pancreaticosplenic; other groups related to these have visceral names, e.g. paracardial and pyloric (stomach), cystic (gallbladder), 'anterior border of epiploic foramen' (bile duct). The general (interconnected) areas of drainage of these groups are evident and their efferents discharge into the coeliac nodes. The latter also receive the efferents of the superior and inferior mesenteric groups, each of which has received the efferents from systematically named groups, aggregated along their branches, or are scattered in the mesenteries; again their areas of drainage are obvious. The *coeliac nodes* are thus the *terminal group* for the whole subdiaphragmatic gut down to mid-rectal level and for most of the liver, the gallbladder and biliary ducts, pancreas and spleen. Their efferents join to form wide right and left *intestinal lymph trunks*; these coalesce and also join the right and left *lumbar lymph trunks* (vide infra) to form the morphologically variable **abdominal confluence of lymph trunks**, the cranial end of which is the entry to the thoracic duct. (Its variability—saccular in a minority of instances only—is mentioned below.) Briefly (details p. 854) the *lateral aortic groups* drain the tissues supplied by the lateral splanchnic and dorsolateral somatic intersegmental aortic branches; caudally they receive the profuse efferents from the common iliac groups which in turn receive the efferents from the internal and external iliac groups, each with their extensive areas of drainage and further associated outlying groups of nodes. These will not be repeated here, as they are given in subsequent pages. The cranial members of the lateral aortic groups are the *terminal nodes* for all these tissues; their efferents converge to form the bilateral lumbar lymph trunks which are the other main avenues forming the abdominal confluence of lymph trunks and thence the initial (caudal) end of the thoracic duct. Mention may be made of the nodes associated with drainage of the leg, often misrepresented. Some drainage first involves a limited outlying group of popliteal nodes (near their vessels), then traverses the superficial or deep inguinal nodes which are *intermediary (not* terminal) groups at the limb's root; thereafter the lymph ascends the chains just described, i.e. via the external and common iliac, the lateral aortic to its upper terminal nodes, then the lumbar lymph trunk, confluence and thoracic duct. (Some deep gluteal lymph follows the internal iliac path to the same destination.) Prominent nodes in the thorax named in relation to vessels are the brachiocephalic group.

Relation to Viscera. These names are self-evident. Examples already mentioned are the paracardial and pyloric, gastric groups; others are the superficial and deep parotid, submandibular and paracolic. The best examples are concerned with, primarily, the drainage of the lower respiratory tract. Passing from the periphery these are named: *pulmonary* (at major bronchial divisions within the lung), *bronchopulmonary* (or simply 'hilar'), *inferior* and *superior tracheobronchial* (p. 857) and *paratracheal*. Their ascending efferents are joined by some from the ipsilateral *parasternal*, *brachiocephalic* and *posterior mediastinal nodes* forming the *right* and *left bronchomediastinal lymph trunks*; these incline over the trachea, then to the ventral aspect of their jugulo-subclavian venous junctions. At or in either great vein, near the junction, the trunks usually open independently, but in about one-fifth of individuals the right trunk may join a **right lymphatic duct**; the left may join the **thoracic duct** or both may occur.

Names related to general topography. The groups of nodes most easily accessible to clinical palpation have widely used general positional names, which vary considerably in their precision. The relation of many (but not all) their subgroups to prominent blood vessels and their branches is close and this provides a more accurate reference system; in some notable sites this is seldom adopted. *Leg*: outlying *popliteal nodes*—here the name is used indiscriminately with respect to the vessels or the fossa; their *palpation* is by finger tips probing the fossa along the line of the popliteal vessels with the passively supported limb gradually moved from extension to semiflexion. *Inguinal nodes*, superficial and deep—here, inguinal simply implies the rather imprecise 'related to the groin'. The deep nodes are few and applied to the medial aspect of the femoral vein; the superficial nodes comprise a lower vertical group clothing the upper great saphenous vein; an upper group parallel to, but *below* the inguinal ligament (related to the superficial circumflex iliac and superficial external pudendal vessels). *Palpation*—with the supported limb slightly flexed, abducted and laterally rotated, along a strip 1 cm below the inguinal ligament and a strip 1 cm medial to the central apicobasal line of the femoral triangle.

Arm: outlying *supratrochlear nodes* (more aptly supra-epicondylar) are adjacent to the basilic vein. *Palpation*—along the line of the vein a few centimetres above the elbow joint; many approaches are satisfactory; facing the subject, an elegant approach is to cup the back of the supracubital arm with the appropriate palm, the semiflexed fingers encircle the medial aspect and their aligned tips effortlessly probe along the vein. The *axillary nodes* have subgroups with alternative names; one system applies to their topographical positioning with respect to the 'walls' of the axilla, the second system applies to their disposition close to the axillary vessels and their branches (especially the *veins*). These are detailed on p. 845 and will not be pursued here. *Palpation*—these must be approached systematically, exploring each wall of the axilla and any attendant vessels and nodes as separate manoeuvres. The supported arm is slightly abducted, the examiner facing the lateral aspect of the shoulder; each fold of the axilla is examined with the appropriate hand, semiflexed finger-

tips invaginating the axillary floor while the thumb grips the fold externally. The fingertips of one or both hands next probe deeply, then down and *laterally* along the axillary vessels. Finally, the fingers of the pronated hand, inserted deeply, are drawn down the medial wall, i.e. the resistant thoracic wall and serratus anterior.

It should be noted that most of the axillary groups are intermediary with their wide areas of drainage and the central group is preterminal; only the *apical group* is terminal. The latter's efferents form the *subclavian lymph trunk* which approaches and, with variable final morphology, opens at or near its jugulo-subclavian junction. The trunk and its opening are on the anterior aspect of the venous walls.

Head and Neck: apart from a few retrovisceral nodes and some deep to the sternocleidomastoid, members of all the nodal groups in the head and neck are clinically palpable when enlarged and all receive regional topographical names. Many of the latter are appropriate and helpful; in the *neck*, however, the major groups have only the most generalized names, despite their principal relationship to large vessels. (Alternative names based on this merit consideration.) The various groups are detailed elsewhere (p. 843); thus a simplified overall plan for the head and neck will be mentioned and some group names added to a suggested approach to their clinical examination. The relationship of craniocervical nodes to the deep fascia (superficial or deep) is discussed subsequently but in some important cases is implied in their names. At and near the junction of the head with the neck an encircling band extends bilaterally from the chin to the external occipital protuberance, the *pericraniocervical ring* (often loosely shortened to 'pericervical ring'). Throughout this encompasses topographically named regional groups with outlying nodes in the face; sequentially the groups are: *submental*, *submandibular* (with outlying *buccal* nodes), *retromandibular* (outlying *parotid*, *retroauricular* (or *mastoid*) and *occipital*. As noted, verbal descriptions of their sites and areas of drainage are given (p. 843) but in general are obvious from their names. *Palpation*—carried out from behind the seated subject, using both hands simultaneously, their fingers semiflexed and adducted and thumbs in partial opposition; the fingers explore systematically: the submental triangle, the submandibular glands and triangles (thumbs probing over buccinators), the retromandibular depressions (thumbs over parotids), the upper attachment of sternocleidomastoid and the occipital attachment of trapezius. Palpation now continues along the approximately *vertical* chains of cervical nodes, superficial and deep. The superficial chains of relatively few, small nodes are associated with the external jugular and anterior jugular veins, the *superficial cervical* and *anterior cervical* groups respectively; both drain finally into deep nodes. (These, and unqualified 'deep cervical' are indifferent, non-specific, unhelpful names; the relative precision of vascular or visceral names are preferable.) The main chain of *deep cervical nodes* is ranged along and embedded in, or in areolar tissue near, the carotid sheath but particularly those aspects surrounding the *internal jugular vein*. Customarily divided into upper and lower groups, they receive, in addition to their direct areas of drainage, all the efferents from the pericraniocervical ring, efferents from the superficial cervical nodes and efferents from other paravisceral deep nodes (e.g. *retropharyngeal*, *infrahyoid*, *prelaryngeal*, *pretracheal*, paratracheal and subclavian). *All* the lymph from the head and neck finally traverses its ipsilateral *lower deep cervical group*, which is the *terminal group*. Efferents from the latter converge, forming (right and left) jugular lymph trunks; each descends on its vein to its termination at or near the jugulosubclavian venous junction.

Lymph Node Numbers: Regional Distribution. Accurate, large statistical surveys are not available; the following are pooled data from many limited sources, nevertheless the overall approximations allow interesting speculation. A normal young adult body contains some 400–450 lymph nodes. Of these the limbs and associated superficial body wall are least well served. The arm and superficial thoraco-abdominal wall (down to the umbilicus) contain about 30 nodes, the leg and superficial buttock, infraumbilical abdominal wall and perineum only about 20 nodes. (Note this does not include the iliac and lateral aortic groups which have numerous additional intra-abdominal afferents.) The head and neck carry some 60–70 nodes. The remainder (about 330) is divided between the thorax (deep walls and contents, some 100 nodes or less), and the abdomen and pelvis (deep walls and contents, some 230 nodes or more). Most richly served by nodes is the gastrointestinal tract; also profusely served is the tracheobronchopulmonary tract.

CERVICAL LYMPHOVENOUS PORTALS

Lymph is returned to the venous blood circulation via the right and left lymphovenous portals which are sited at, or near, the junctions of the large internal jugular and subclavian veins forming the even larger right and left brachiocephalic veins. On the right *three* main lymph trunks converge towards their venous junction; on the left *four* main trunks (three corresponding to the right-sided trunks, but additionally the largest trunk, *the thoracic duct*). The morphology of the venous termination of these trunks is subject to much variation and the account frequently given in textbooks is a fairly uncommon occurrence, hence the introduction of the generalized term lymphovenous portal.

On the right the three trunks converging are: (1) The *right jugular trunk* which extends along the ventrolateral aspect of the internal jugular vein from the terminal lower deep cervical nodes and conveys all the lymph from the right half of the head and neck.

(2) The *right subclavian trunk* from the terminal apical axillary group extending along the axillary and subclavian veins and conveying lymph from the right upper limb and superficial tissues of the right half of the thoraco-abdominal wall down to the umbilicus anteriorly and iliac crest posteriorly (and including much of the mammary gland).

(3) The *right bronchomediastinal trunk* (p. 857), which ascends over the trachea towards the portal and conveys lymph from the deeper thoracic parietes, the right cupola of the diaphragm and subjacent liver, the right lung, bronchi and trachea, the greater part of the 'right heart' (of clinical parlance, *not* the geometric right half, see p. 857) and a proportionately small drainage from the thoracic oesophagus.

The right venous termination of the three lymphatic trunks is subject to great variation. In the great majority of subjects (80%) they open *independently*, their orifices clustered on the ventral aspect of the jugulo-subclavian junction or in the nearby wall of either of the great veins. In a proportion of these one or more of the trunks may bifurcate (or even trifurcate) preterminally and then have multiple orifices. In one-fifth of subjects only, the three trunks fuse to form a short (1 cm) single **right lymphatic duct** that inclines across the medial border of scalenus anterior to the ventral aspect of the venous junction, where its orifice is preceded by a bicuspid semilunar valve. An **incomplete right lymphatic duct** may be present following fusion of, usually, the subclavian and jugular trunks, or any combination of their terminals when divided. In such cases the bronchomediastinal trunk almost invariably opens separately.

Summary. The **right** lymphovenous portal, whatever the final morphology of its trunks, receives lymph from: the *right half* of the head and neck, the thorax and its contents and superficial tissues of the abdomen and trunk down to the umbilicus and iliac crest, *part* of the right cupola of the diaphragm and convex surface (only) of the underlying liver and the *whole* of the right arm. The **left** portal receives much the greater volume of lymph from all the remainder of the body.

On the left the four trunks converging on the lymphovenous portal are:

(1) the *left jugular trunk*, mirroring its right fellow;

(2) the *left subclavian trunk*, also with a disposition corresponding to its contralateral fellow;

(3) the *left bronchomediastinal trunk*, similar to the right trunk, but draining more of the heart (the 'left' and part of the 'right hearts' of clinical parlance, p. 698) and more of the oesophagus;

(4) the *thoracic duct*, which drains all the extensive remaining regions of the body. At its caudal origin as a continuation of the abdominal confluence of lymphatic trunks (vide infra), throughout its course and at its cervical venous termination it is subject to considerable variation.

THE THORACIC DUCT (6.197,198,199)

In adults the thoracic duct *including* the confluence of lymph trunks (or the cisterna chyli in the small proportion in whom the latter is saccular) is 38–45 cm in length, extending from the second lumbar vertebra to the root of the neck. Starting from the superior pole of the confluence near the lower border of the twelfth thoracic vertebra, it traverses the diaphragm's aortic aperture, then ascends the *posterior mediastinum*, right of the midline, between the descending thoracic aorta (on its left) and the azygos vein (on its right). *Posterior* to it is the vertebral column (vertebral bodies, symphyses, anterior longitudinal ligament), the right aortic intercostal arteries and terminal segments of the hemiazygos and accessory hemiazygos veins. *Anterior* to it are the diaphragm and

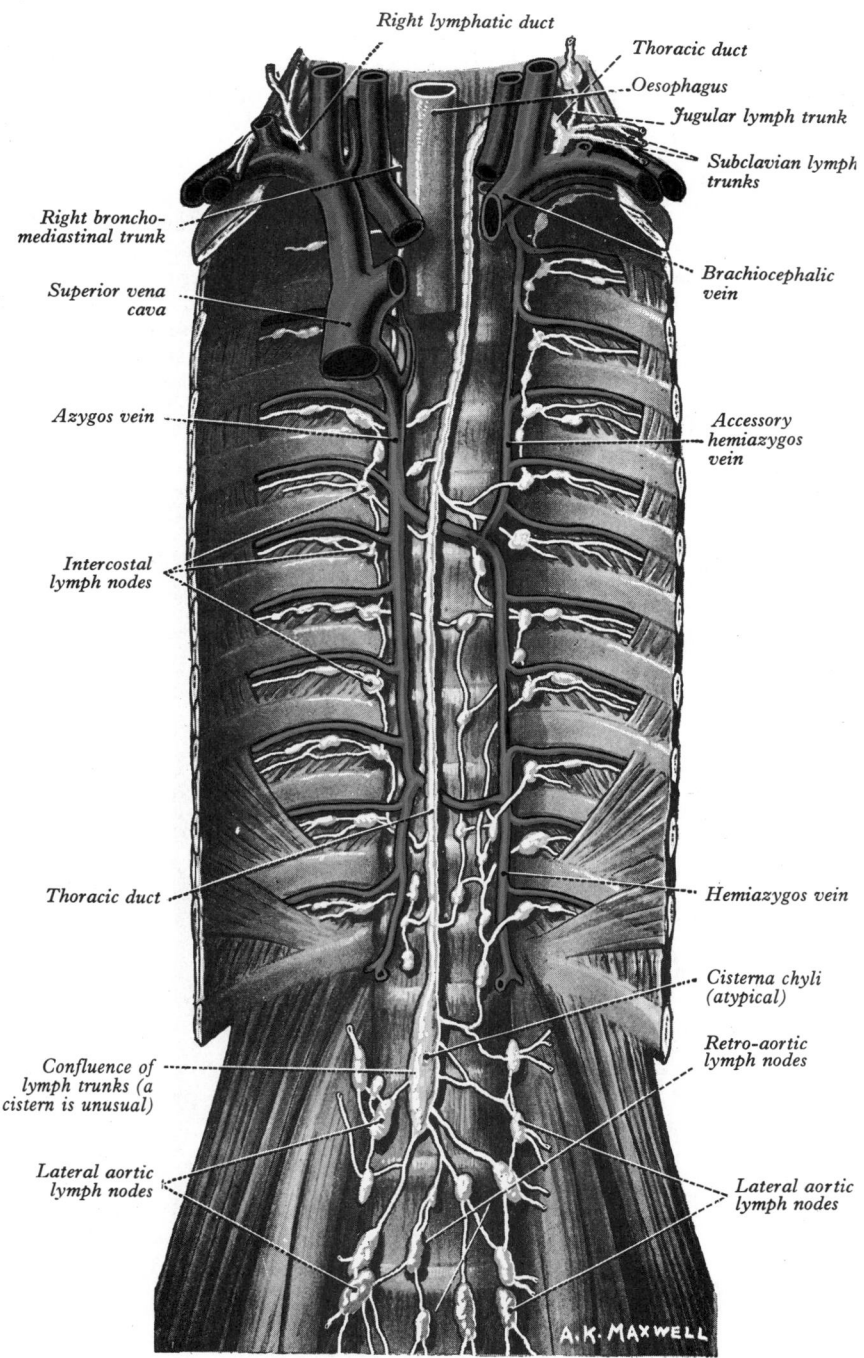

6.197 The thoracic and right lymphatic ducts. The accessory hemiazygos vein is crossing the median plane lower and the hemiazygos higher than usual. Note also the comments concerning the more common course of the azygos vein made in illustration 6.162B and on p. 810. Two features are also uncommon: a single right lymphatic duct (usually two or more trunks open independently); a simple cisterna chyli is infrequent (it is usually a confluence of lymph trunks of varying morphology, p. 842).

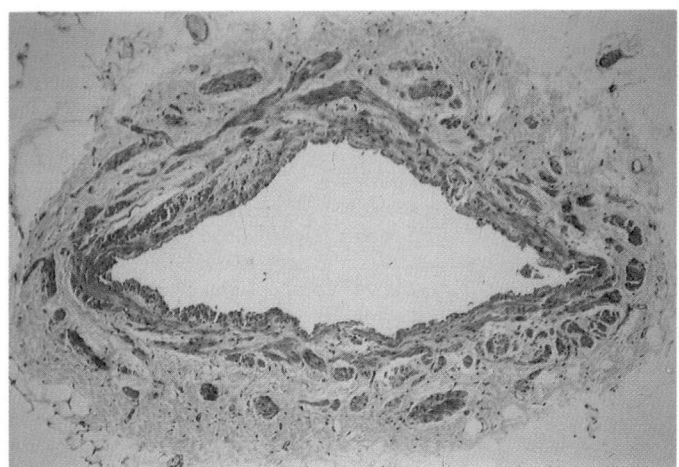

6.198 Transverse section of the thoracic duct showing the fibro-muscular coat (see text). Stained with haematoxylin and eosin. Preparation by Millie Harrison, Department of Anatomy, UMDS, Guy's Campus, London. Magnification × 80.

oesophagus; a recess of the right pleural cavity may separate the duct and oesophagus. Reaching the level of the fifth thoracic vertebral body it gradually inclines to the left, enters the *superior mediastinum* and then ascends to the thoracic inlet along the left border of the oesophagus. In this part of its course the duct is first crossed anteriorly by the aortic arch and it then runs posterior to the left subclavian artery's initial segment, in close contact with

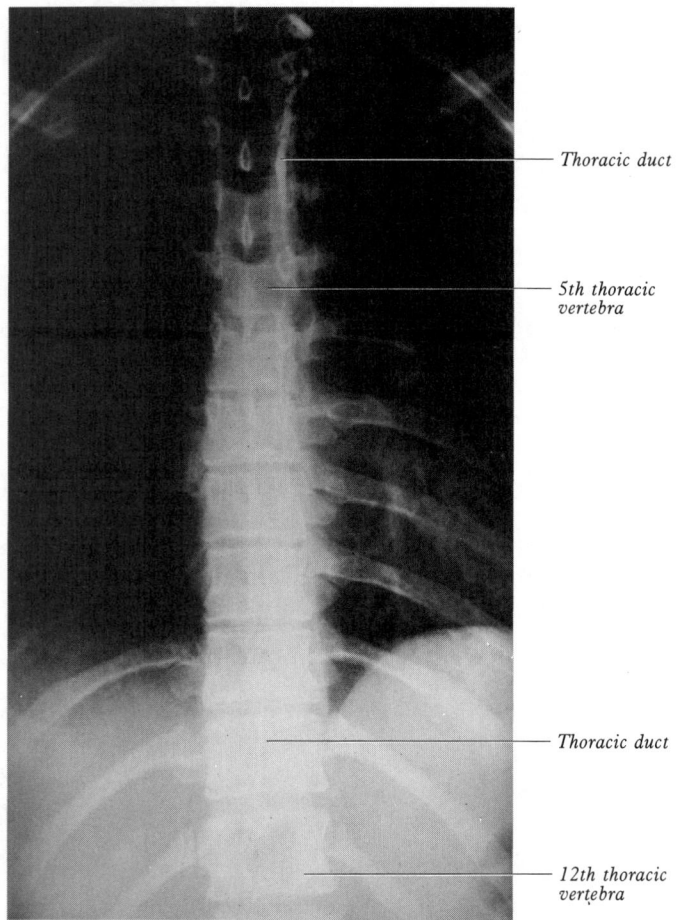

Thoracic duct

5th thoracic vertebra

Thoracic duct

12th thoracic vertebra

6.199 Lymphangiogram showing the entire length of the thoracic duct, approximately 24 hours after injection of lipiodol into a lymphatic vessel on the dorsum of each foot; the cisterna chyli is not evident. Supplied by G I Verney, Addenbrooke's Hospital, Cambridge; photography by Sarah Smith, Department of Anatomy, UMDS, Guy's Campus, London.

the left mediastinal pleura. Passing into the *neck* it arches laterally at the level of the seventh cervical vertebral transverse process. Its *arch* rises 3 or 4 cm above the clavicle and curves *anterior* to the vertebral artery and vein, the left sympathetic trunk, thyrocervical artery or its branches and the left phrenic nerve and medial border of scalenus anterior (but is separated from the nerve and muscle by the prevertebral fascia). The arch passes *posterior* to: the left common carotid artery, vagus nerve and internal jugular vein. Finally, the duct descends anterior to the arched cervical 'first part' of the left subclavian artery and ends by opening into the junction of the left subclavian and internal jugular veins. However, the duct may open into either of the great veins, near the junction, or it may divide into a number of smaller vessels before terminating (vide infra).

At its abdominal commencement the thoracic duct is about 5 mm in diameter but diminishes in calibre at mid-thoracic levels, then in about 50% of subjects is again slightly dilated before its termination. It is slightly sinuous, constricted at intervals and appears varicose. It may divide in its mid-course into two unequal vessels which soon re-unite, or into several small branches which form a plexus before continuing as a single duct. At a higher level it occasionally bifurcates, the left branch ending as usual, the right branch diverging to join one of the right lymph trunks or, when present, a right lymphatic duct; the combined vessel usually opens into the right subclavian vein. The thoracic duct has several valves corresponding to sites exposed to pressure. At its termination a bicuspid valve faces into the vein to prevent or reduce reflux of blood. (After death blood regurgitates freely into the duct, which then looks like a vein.)

Termination. Kinnaert (1973) has collected accounts of 529 dissections (49 his own) of the thoracic duct's termination. In 0–4.5% no thoracic duct appeared on the left. Multiple terminal openings were frequent (10–40%, according to different observers). In Kinnaert's series the preterminal duct was multiple in 66%, but in only 21% were actual terminal openings multiple. Patterns varied greatly in different studies but, in the two largest by Jdanov (1959) and Kinnaert (1973), *sites* of termination were respectively 48% and 36% internal jugular vein, 9% and 17% subclavian vein, 35% and 34% jugulo-subclavian junction. Termination in the left brachiocephalic (innominate) vein occurred in 8% of Jdanov's series, but none of Kinnaert's.

Origin and tributaries. The abdominal origin of the thoracic duct proper is, as stated, situated to the right of the midline at the level of the lower border of the twelfth thoracic vertebral body or the thoracolumbar intervertebral disc. It is the recipient of all the lymph delivered by the four main abdominal lymph trunks which converge to an elongated arrangement of channels of *variable morphology* and hence is given here the generalized name, **abdominal confluence of lymph trunks**. This may be a simple duct-like extension or be duplicated, triplicated or plexiform; when it is wider than the thoracic duct its interior is sometimes irregular and bilocular or trilocular and may surround intercalated lymph nodes. Only in a small proportion of instances is it a simple, fusiform, saccular dilatation, and the widely-used name **cisterna chyli** should be reserved for these. A published thorough statistical study of the origin of the thoracic duct in mankind appears lacking. Anson (1963) depicted variations: in many a cisterna was absent; when present it was usually multilocular or plexiform; no statistics of incidence were given. Kubik (personal communication 1978) observed a 'cisterna' in 14 of 70 dissections. In only six was it single; it was double in five specimens and trilocular in three. In 56 dissections no cisterna was observed; in half of these collecting trunks formed a direct extension of the thoracic duct; in the other half intercalated nodes (also depicted by Anson 1963) simulated the profiles of cisternae.

The abdominal confluence extends from the caudal beginning of the thoracic duct, vertically, for 5–7 cm anterolateral to the right, of the first and second lumbar vertebral bodies (and their intervening disc), immediately to the right of the abdominal aorta. (Thus its site is overlapped by territories containing upper right lateral aortic lymph nodes and right-sided members of the coeliac and superior mesenteric pre-aortic groups.) The upper two right lumbar arteries and the right lumbar azygos vein (p. 808) are between the confluence and the vertebral column. Anterior to it is

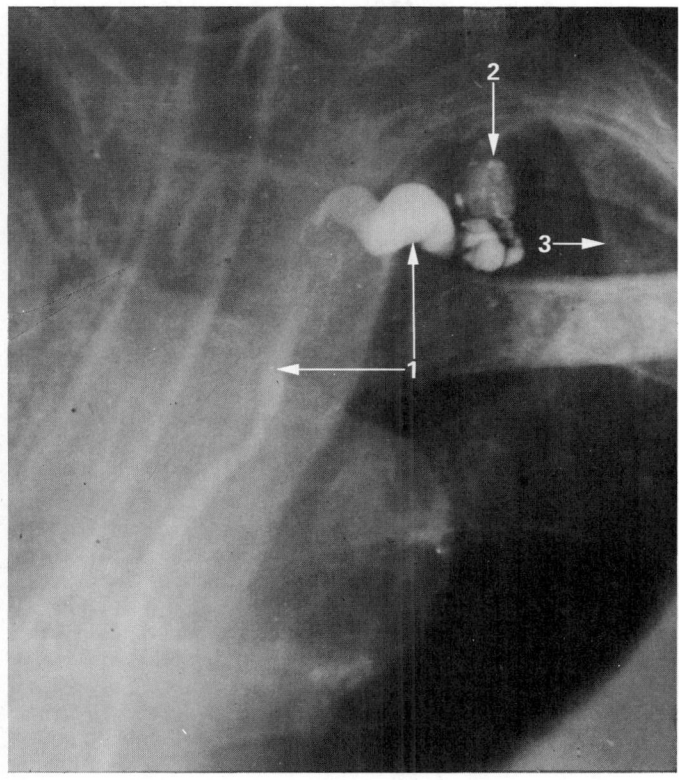

6.200 Lymphangiogram showing the upper part of the thoracic duct. There has been some filling of a cervical lymph node. 1. Thoracic duct. 2. Lower deep cervical lymph node. 3. First rib. Supplied by J B Kinmonth.

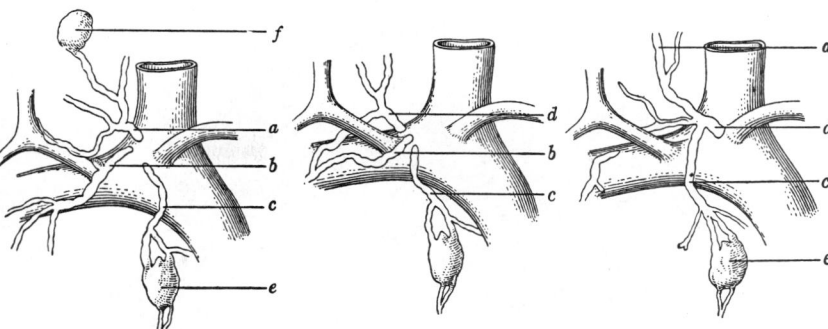

6.201 Variations in the terminal lymph trunks of the right side (after Poirier & Charpy). a. Jugular trunk. b. Subclavian trunk. c. Bronchomediastinal trunk. d. Right lymphatic duct. e. Lymph node of parasternal chain. f. Lymph node of deep cervical chain.

Many of the trunks listed above are described as possessing terminal bicuspid valves which possibly prevent reflux of lymph. However, Sapin & Boryiak (1974) studied the behaviour of radio-opaque masses in thoracic ducts of 180 *cadavers*; they found that reflux into several groups of mediastinal and paravertebral groups was usual (under these conditions!).

Lymphatic Drainage of the Head and Neck

Nodes in the head and neck comprise a terminal (collecting) group and intermediary, outlying groups. The terminal group is related to the carotid sheath and is named *deep cervical*. All lymph vessels of the head and neck drain into this, directly from tissues or indirectly through nodes in outlying groups. Efferents of the deep cervical nodes form the *jugular trunk*, which on the right may end in the jugulo-subclavian junction or right lymphatic duct; on the left it usually enters the thoracic duct but may join the internal jugular or subclavian vein.

Deep Cervical Lymphatic Nodes

These nodes are alongside the carotid sheath; they form superior and inferior groups.

Superior deep cervical nodes (6.202) adjoin the upper internal jugular vein. Most are deep to the sternocleidomastoid; a few extend beyond it. One subgroup, of one large and several small nodes, is in a triangular region bounded by the posterior belly of the digastric and the facial and internal jugular veins; this *jugulodigastric group* is concerned specially with lingual drainage. *Efferents* from the upper deep cervical nodes drain to the lower group or direct to the jugular trunk.

Inferior deep cervical nodes are partly deep to the sterno-cleidomastoid, particularly related to the lower internal jugular vein but some, extending also into the subclavian triangle, are closely related to the brachial plexus and subclavian vessels. One node is on or just above the intermediate tendon of omohyoid, the *jugulo-omohyoid node*, and is concerned especially with the tongue (p. 845). Efferents from this lower group join the jugular lymph trunk.

In lymphatic drainage the tissues of the head and neck, like other regions, can conveniently be considered as (1) superficial and (2) deep. (See also the generalized arrangement of a *pericraniocervical ring* and *vertical cervical chains*, p. 840.)

Lymphatic Drainage of the Superficial Tissues of the Head and Neck (6.202)

Most superficial tissues in the region drain by vessels afferent to local groups of nodes, efferents from these draining to the deep cervical nodes; but some structures drain directly to deep nodes. Groups concerned in superficial drainage are:

(*a*) *In the head*: occipital, retro-auricular (mastoid), parotid, buccal (facial);

(*b*) *in the neck*: submandibular, submental, anterior cervical, superficial cervical.

the medial edge of the right diaphragmatic crus. As mentioned, the confluence (and thence the thoracic duct) receives the right and left lumbar and intestinal lymph trunks. In summary:

(1) The *lumbar trunks* are formed by efferents from lateral aortic lymph nodes. Thus, either directly or after traversing intermediary groups, they carry lymph from: the lower limbs, the full thickness of the pelvic, perineal and infra-umbilical abdominal walls, the deep tissues of most of the supra-umbilical abdominal walls, the pelvic viscera, testes or ovaries, kidneys and suprarenals.

(2) The *intestinal lymph trunks* receive efferents from the coeliac nodes (terminal ventral aortic group) which, after traversing intermediary groups, drain the stomach, intestines (to mid-rectal levels), pancreas, spleen and the (greater) antero-inferior part of the liver.

Tributaries of the thoracic duct proper (in summary):

The *confluence of lymph trunks*, just described, the whole outflow of which enters the origin of the thoracic duct;

The bilateral *descending thoracic lymph trunks* from intercostal lymph nodes of the lower six or seven intercostal spaces of both sides which traverse the aortic orifice and join the lateral aspects of the thoracic duct in the abdomen immediately after its origin;

The bilateral *ascending lumbar lymph trunks* from the upper lateral aortic nodes which ascend and pierce their corresponding diaphragmatic crus, then join the thoracic duct at a variable level within the thorax;

The *upper intercostal trunks* draining the intercostal nodes in the upper five or six *left* intercostal spaces;

The *mediastinal trunks* draining various nodal groups noted below and providing (amongst other tissues) paths to the thoracic duct from the convex diaphragmatic aspect of the liver, the diaphragm, the pericardium, heart and oesophagus;

The *left subclavian trunk* which usually joins the thoracic duct, but may open independently into the left subclavian vein;

The *left jugular trunk* which usually joins the thoracic duct, but may open independently into the left internal jugular vein;

The *left bronchomediastinal trunk* which occasionally joins the thoracic duct, *usually* having an independent venous opening.

Retro-auricular nodes

Occipital nodes

Superficial cervical nodes

Upper deep cervical nodes

Superficial parotid nodes

Buccal node

Submental nodes

Submandibular nodes

Jugulo-omohyoid node

Lower deep cervical nodes

6.202 The superficial lymph nodes and lymph vessels of the head and neck.

Lymphatic drainage of the scalp and ear. Vessels from the frontal region above the root of the nose drain to the submandibular nodes (**6**.202) and are considered with the face. Vessels from the rest of the forehead, temporal region, upper half of the lateral auricular aspect and anterior wall of the external acoustic meatus drain to the *superficial parotid nodes*, just anterior to the tragus, on or deep to the parotid fascia. These also drain lateral vessels from the eyelids and skin of the zygomatic region. Their efferent vessels pass to the upper deep cervical nodes. A strip of scalp above the auricle, the upper half of the auricle's cranial aspect and margin and the posterior wall of the external acoustic meatus all drain to the upper deep cervical and retro-auricular nodes.

The *retro-auricular nodes* (**6**.202), superficial to the mastoid attachment of sternocleidomastoid and deep to auricularis posterior, drain to the upper deep cervical nodes. The auricular lobule, floor of the meatus and skin over the mandibular angle and lower parotid region are drained to the superficial cervical or upper deep cervical nodes. *Superficial cervical nodes* spread along the external jugular vein superficial to sternocleidomastoid, some efferents passing round the anterior border of sternocleidomastoid to the upper deep cervical nodes; others follow the external jugular vein to the lower deep cervical nodes in the subclavian triangle.

The occipital scalp is drained partly to the occipital nodes, partly by a vessel along the posterior border of sternocleidomastoid to the lower deep cervical nodes. *Occipital nodes* are occasionally in the superior angle of the posterior triangle but commonly superficial to the upper attachment of trapezius.

Lymphatic drainage of the face. Lymph vessels draining the eyelids and conjunctiva commence in a subcutaneous plexus and a deep plexus around the tarsal plates; these communicate and

medial and lateral vessels drain from them. Lateral vessels drain the whole thickness of both lids, except their medial parts and *all* the conjunctiva. They pass from the lateral commissure to the superficial parotid nodes and *deep nodes* embedded in the parotid gland, also receiving lymph from the middle ear (vide infra). The medial palpebral vessels drain the whole thickness of the medial parts of the lids and caruncula lacrimalis. Following the facial vein, they end in submandibular nodes.

Submandibular nodes (**6**.202, 203), internal to the deep cervical fascia in the submandibular triangle, are usually three: one at the anterior pole of the submandibular gland, two flanking the facial artery as it reaches the mandible. Other nodes are often embedded in the gland or deep to it. Submandibular nodes drain a wide area, including vessels from the submental, buccal and lingual groups of nodes; their efferents pass to the upper and lower deep cervical nodes. The external nose, cheeks, upper lip and lateral parts of the lower lip drain directly to the submandibular nodes; the afferent vessels may have a few *buccal nodes* along their course and near the facial vein. The mucous membrane of lips and cheeks also drains to the submandibular nodes. The lateral part of the cheek drains to the parotid nodes, the skin over the nasal radix and central forehead drains partly to the parotid nodes, partly to the submandibular.

The central part of the lower lip, buccal floor and lingual apex drain to the *submental nodes*, which are on the mylohyoid between the anterior bellies of the digastric muscles (**6**.203). They receive afferents from *both* sides, some decussating across the chin; their efferents pass to the submandibular and jugulo-omohyoid nodes.

Lymphatic drainage of the neck. Many vessels draining the superficial cervical tissues skirt the borders of sternocleidomastoid to the superior or inferior deep cervical nodes; but some pass over sternocleidomastoid and the posterior triangle to the superficial cervical and occipital nodes. Lymph from the superior region of the anterior triangle drains to the submandibular and submental nodes; vessels from the anterior cervical skin inferior to the hyoid bone pass to the *anterior cervical lymph nodes* near the anterior jugular veins; their efferents go to the deep cervical nodes of both sides, including the *infrahyoid*, *prelaryngeal* and *pretracheal* groups (vide infra). An anterior cervical node often occupies the suprasternal space (p. 582).

Lymphatic Drainage of the Deep Tissues of the Head and Neck

Tissues of the head and neck internal to the deep fascia drain to the deep cervical nodes directly or through outlying groups which include, in addition to those named above: the retropharyngeal, paratracheal, lingual, infrahyoid, prelaryngeal and pretracheal groups.

Retropharyngeal nodes comprise a median and two lateral groups, the latter anterior to the lateral atlantal masses along the lateral borders of the longi capitis. All lie between the pharyngeal and prevertebral fasciae, receiving afferents from the nasopharynx, pharyngotympanic tube and atlanto-occipital and atlanto-axial joints. They drain to the upper deep cervical nodes.

Paratracheal nodes flank both trachea and oesophagus along the recurrent laryngeal nerves. Efferents pass to the corresponding deep cervical nodes.

Infrahyoid, prelaryngeal and pretracheal nodes, beneath the deep cervical fascia, drain afferents from the anterior cervical nodes, their efferents joining the deep cervical nodes. The infrahyoid nodes are anterior to the thyrohyoid membrane, prelaryngeal on the conus elasticus and cricovocal membrane, pretracheal anterior to the trachea near the inferior thyroid veins.

Lingual nodes are small and inconstant, situated on the external surface of hyoglossus and also between the genioglossi. They drain to the upper deep cervical nodes.

Lymphatic drainage of the nasal cavity, nasopharynx and middle ear. Lymphatics of the nasal cavity can be injected from the subarachnoid space, via communications along the olfactory nerves. Lymph vessels from its anterior region pass superficially to join those of the external nasal skin, ending in the submandibular nodes. The rest of the cavity, paranasal sinuses, nasopharynx, and pharyngeal end of the pharyngotympanic tube drain to the upper deep cervical nodes, directly or through the retropharyngeal nodes. The posterior nasal floor probably drains to the parotid nodes.

Lymphatic vessels of the tympanic and antral mucosae drain to the parotid or upper deep cervical lymph nodes; vessels of the tympanic end of the pharyngotympanic tube probably end in the deep cervical nodes; its vessels have been identified in the submucosa by injection and electron microscopy (Pulec et al 1975).

Lymphatic drainage of the larynx, trachea and thyroid gland. Laryngeal lymphatic vessels form superior and inferior groups; on the lateral wall they are distinct, their division being at the level of the vocal fold; the two sets anastomose on the posterior wall. Superior vessels pierce the thyrohyoid membrane to accompany the superior laryngeal vessels, ending in the superior deep cervical nodes; inferior vessels pass between the cricoid cartilage and the first tracheal ring to the inferior deep cervical lymph nodes, or pierce the cricovocal membrane to reach the pretracheal and prelaryngeal nodes.

A dense network of lymph vessels exists in the tracheal wall; its cervical part drains to the pretracheal and paratracheal nodes, or directly to the inferior deep cervical nodes.

Thyroid lymphatic vessels communicate with the tracheal plexus, passing to the prelaryngeal nodes just above the thyroid isthmus and to the pretracheal and paratracheal nodes; some may drain into the brachiocephalic nodes, related to the thymus in the superior mediastinum. Laterally, the gland is drained by vessels along the superior thyroid veins to the deep cervical nodes. Some thyroid lymphatics may drain *directly*, with no intervening node, to the thoracic duct (p. 842).

Lymphatic drainage of the mouth, teeth, tonsil and tongue. *Mouth*. Gingival vessels drain to the submandibular nodes; those of the hard palate continue anteriorly into the superior gingival channels but also run back to pierce the superior constrictor, ending in the superior deep cervical and retropharyngeal nodes; from the soft palate they pass posterolaterally partly to the retropharyngeal, partly to the superior deep cervical nodes. The anterior part of the floor of the mouth drains to the lower nodes of the upper deep cervical group, either directly or via the submental nodes; vessels from the remainder of the floor drain to the submandibular and superior deep cervical nodes. *Teeth*. Dental lymphatics pass to the submandibular and deep cervical nodes.

Tonsil. Vessels from the tonsil pierce the buccopharyngeal fascia and superior constrictor to pass between the stylohyoid muscle and the internal jugular vein to the superior deep cervical nodes. Most end in the jugulodigastric nodes; occasionally one or two vessels run to the small nodes on the lateral aspect of the internal jugular vein, deep or medial to sternocleidomastoid.

Tongue (**6.203**). A lymphatic plexus in the lingual mucosa is continuous with an intramuscular plexus. The anterior lingual region drains into the marginal and central vessels and behind the vallate papillae into the dorsal lymph vessels.

(1) *Marginal vessels* from the lingual apex and frenular region descend under the mucosa to widely distributed nodes: (*a*) Some pierce the mylohyoid in contact with the mandibular periosteum to enter the submental nodes and also pass anterior to the hyoid bone to the jugulo-omohyoid node. Vessels arising in the plexus on one side may cross under the frenulum to end in the contralateral nodes; efferent vessels of submental nodes, which are median, pass to both sides. (*b*) Some vessels pierce the mylohyoid to enter the anterior or middle submandibular node. (*c*) Some pass inferior to the sublingual gland and, accompanying the companion vein of the hypoglossal nerve, end in jugulodigastric nodes; one often descends further, superficial or deep to the intermediate tendon of the digastric, to reach the jugulo-omohyoid node. Some vessels from the lateral lingual margin cross the sublingual gland, pierce the mylohyoid and end in the submandibular nodes; others end in the jugulodigastric or jugulo-omohyoid nodes. Vessels from the posterior part of the lingual margin traverse the pharyngeal wall to the jugulodigastric lymph nodes.

(2) *Central vessels*. The regions of the lingual surface draining into the marginal or central vessels are not distinct. Central vessels descend between the genioglossi, some turning laterally through the muscles; but most pass between them and diverge to the right or left, following the lingual veins to the deep cervical

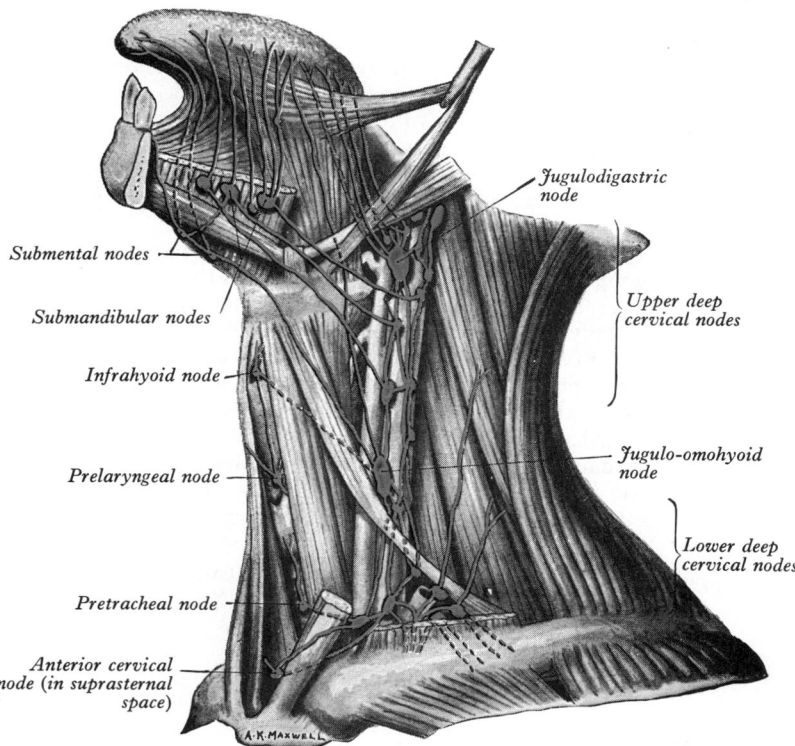

6.203 The lymphatic drainage of the tongue (after Jamieson & Dobson). Removal of the sternocleidomastoid has exposed the whole chain of deep cervical lymph nodes.

nodes, especially the jugulodigastric and jugulo-omohyoid. Some pierce the mylohyoid to enter the submandibular nodes.

(3) *Dorsal vessels*. Vessels draining the region of the vallate papillae and behind them run postero-inferiorly, some near the median plane to both sides. They turn laterally to join the marginal vessels; all pierce the pharyngeal wall, passing around the external carotid arteries to reach the jugulodigastric and jugulo-omohyoid lymph nodes. One may descend posterior to the hyoid bone, perforating the thyrohyoid membrane to end in the jugulo-omohyoid node.

Lymphatic drainage of the pharynx and cervical part of the oesophagus. Collecting vessels from the pharynx and cervical oesophagus pass to the deep cervical nodes, either directly or through the retropharyngeal or paratracheal nodes. From the epiglottic region lymph vessels run to the infrahyoid nodes.

Lymphatic Drainage of the Upper Limbs

All lymphatic vessels from the upper limb (and superficial tissues of a wide area of ipsilateral trunk) drain to the axillary nodes, either directly or (a few) through a more peripheral group. Vessels internal to the deep fascia follow the principal vascular bundles; superficial vessels, except in the hand and dorsum of the forearm, converge towards the superficial veins, which they accompany.

Axillary nodes (**6.205,207**), which drain the whole upper limb and areas of the trunk indicated, are large, varying from 20–30, and may be divided into five not wholly distinct groups. Four of these groups are *intermediary*; only the apical group is *terminal*.

(1) *A lateral group* (**6.204,207**) of four to six nodes is postero-medial to the axillary vein, its afferents draining the whole limb except the vessels accompanying the cephalic vein. Efferent vessels pass partly to the central and apical axillary groups, partly to the inferior deep cervical nodes.

(2) An *anterior* or *pectoral group* of four or five nodes spreads along the inferior border of pectoralis minor near the lateral thoracic vessels. Its afferents drain the skin and muscles of the supra-umbilical anterolateral body wall and mammary gland (centrolateral part, p. 847); efferents pass partly to the central and partly to the apical axillary nodes.

845

(3) A *posterior* or *subscapular group* of six or seven nodes is deployed on the posterior axillary wall's inferior margin, along the subscapular vessels. Afferents drain the skin and *superficial* muscles of the inferior posterior region of the neck and the dorsal aspect of the trunk down to the iliac crest; efferents pass to the apical and central axillary nodes.

(4) A *central group* of three of four large nodes embedded in axillary fat receives afferents from all preceding groups: its efferents drain to the apical nodes.

(5) An *apical group* of six to twelve nodes is partly posterior to the superior part of pectoralis minor and partly above its superior border, extending to the axilla's apex medial to the *axillary vein*. The only *direct* territorial afferents are those with the cephalic vein and some draining the mammary gland (upper peripheral region); but the group drains all other axillary nodes. Its efferents unite as the *subclavian trunk*, draining directly to the jugulo-subclavian venous junction, the subclavian vein, or to the jugular lymphatic trunk or on occasion to a right lymphatic duct; the left trunk usually ends in the thoracic duct. A few efferents from apical nodes usually reach the inferior deep cervical nodes.

Extra-axillary outlying groups in the upper limb are few, comprising: (1) supratrochlear, (2) infraclavicular (both interposed in superficial routes), (3) isolated nodes occasionally appearing along principal blood vessels.

(1) *Supratrochlear nodes*, only one or two, are superficial to the deep fascia proximal to the medial epicondyle and medial to the basilic vein; their efferents accompany the vein to join the deep lymph vessels.

(2) *Infraclavicular nodes* appear beside the cephalic vein, one or two in the groove between the pectoralis major and deltoid, just inferior to the clavicle; efferents pass through the clavipectoral fascia to apical axillary nodes; more rarely some pass *anterior* to the clavicle to reach the inferior deep cervical (supraclavicular) nodes.

(3) Small isolated nodes sometimes occur along the radial, ulnar and interosseous vessels, in the cubital fossa near the bifurcation of the brachial artery, or in the arm medial to the brachial vessels.

Lymphatic drainage of the superficial tissues. Superficial lymphatic vessels in the upper limb begin in the cutaneous plexuses. *In the hand*, the palmar plexus is denser. Digital plexuses are drained along the digital borders to their webs, where they join the distal palmar vessels which pass back to the hand's dorsal aspect (6.204,206). The proximal palm drains towards the carpus, medially by vessels along its ulnar border and laterally to join those of the thumb. Several vessels from the central palmar plexus form a trunk winding round the indicial metacarpal bone to join the dorsal vessels from the index and pollex. *In the forearm and arm*, superficial vessels run with superficial veins. Collecting

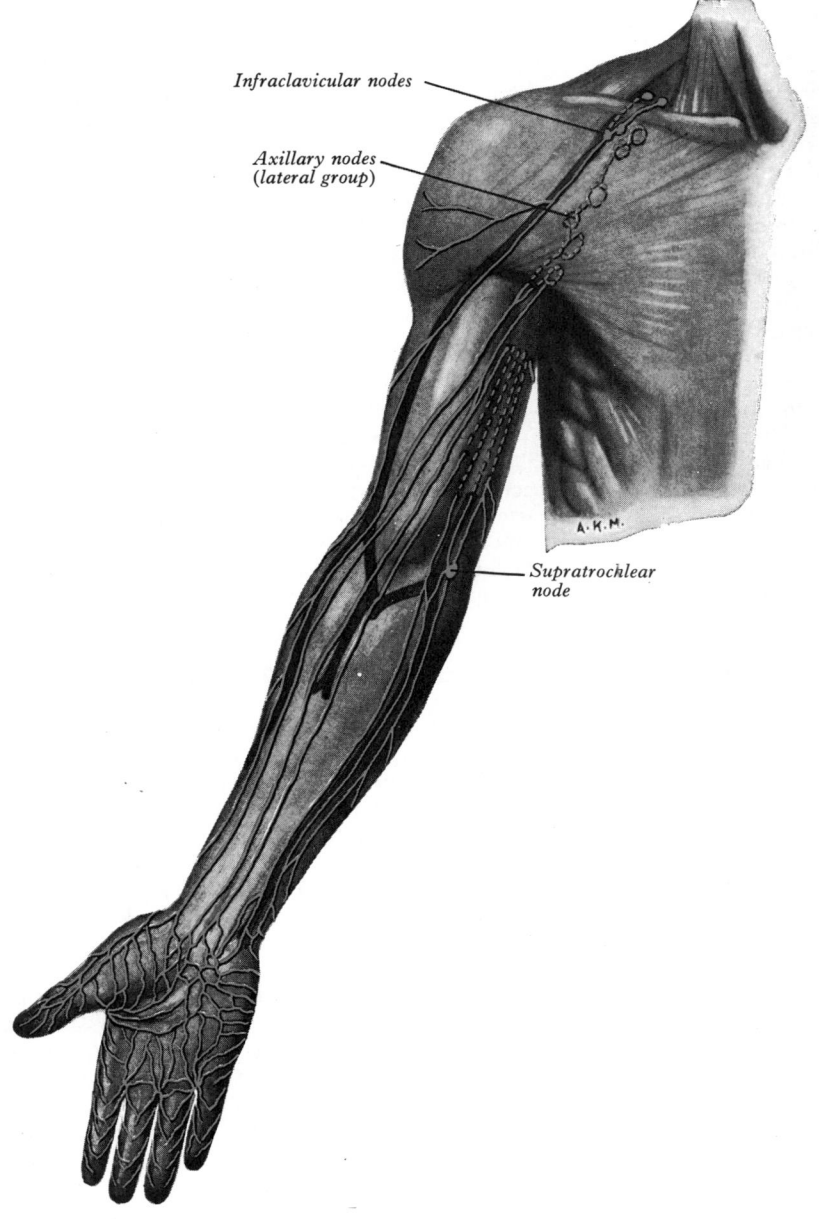

Infraclavicular nodes

Axillary nodes (lateral group)

Supratrochlear node

6.204 The lymphatic drainage of the superficial tissues of the upper limb: anterior aspect (semi-diagrammatic).

vessels from the hand pass into the forearm on all carpal aspects. Dorsal vessels, after running proximally in parallel, curve successively round the borders of the limb to join the ventral vessels (6.206). Anterior carpal vessels traverse the forearm parallel with the median vein of the forearm to the cubital region, proximal to which they follow the medial border of the biceps, then pierce the deep fascia at the anterior axillary fold and end in the lateral axillary lymph nodes.

Vessels which are lateral in the forearm follow the cephalic vein to the level of the tendon of the deltoid, where most incline medially to reach the lateral axillary nodes; a few, however, continue with the vein to the infraclavicular nodes. These lateral vessels receive those curving round the lateral border from the limb's dorsal aspect. Vessels which are medial in the forearm follow the basilic vein. Proximal to the elbow some end in supratrochlear lymph nodes, whose efferents, with the medial vessels which have bypassed them, pierce the deep fascia with the basilic vein to end in the lateral axillary nodes or deep lymphatic vessels. They are joined by vessels curving round the medial border of the limb.

Collecting vessels from the deltoid region pass round the anterior and posterior axillary fold to end in the axillary nodes. The scapular skin drains either to subscapular axillary nodes or by channels following the transverse cervical vessels to the inferior deep cervical nodes.

Lymphatic drainage of the deep tissues of the upper limb. Deep lymph vessels follow the main neurovascular bundles (radial, ulnar, interosseous and brachial) to the lateral axillary nodes. They are less numerous than the superficial vessels, communicating with them at intervals. A *few lymph nodes* occur along them. Scapular muscles drain mainly to the subscapular axillary nodes and pectoral muscles to the pectoral, central and apical nodes.

Mammary lymphatic drainage (6.207). Lymph vessels of the mammary gland start in a plexus in the interlobular connective tissue and walls of the lactiferous ducts, communicating with a cutaneous *subareolar plexus* around the nipple. The gland is also said to connect with a plexus of minute vessels on the subjacent deep fascia; this connection plays little part in normal lymphatic drainage nor in early spread of carcinoma (Turner-Warwick 1959). It offers an alternative route when the usual pathways are obstructed. Efferent vessels directly from the gland pass round the anterior axillary border through the axillary fascia to the pectoral lymph nodes; some may pass directly to the subscapular nodes. From the gland's superior region a few vessels pass to the apical axillary nodes, sometimes interrupted in the infraclavicular nodes or in small, inconstant interpectoral nodes. Axillary nodes receive more than 75% of lymph from the gland, the remainder largely draining to parasternal nodes from the medial and lateral parts of the organ; these vessels accompany perforating branches of the internal thoracic artery. Lymphatic vessels occasionally follow lateral cutaneous branches of the posterior intercostal arteries to the intercostal nodes.

Cutaneous lymphatic drainage is described on p. 856.

Applied Anatomy. Enlargement of the axillary nodes is frequent in malignant disease and infective processes affecting the upper back and shoulder, the front of the chest and mammary gland, upper anterolateral abdominal wall, or upper limb (*palpation*, see p. 840). In operations for mammary carcinoma, pectoralis major, its deep fascia and surrounding muscles were usually removed en bloc because of the wide ramifications of its lymphatics. Axillary nodes, the sternocostal head of pectoralis major and frequently pectoralis minor were also removed, to ensure complete removal of the affected lymphatics and nodes. (Some, particularly with more effective diagnostic techniques, now advocate less radical extirpation.)

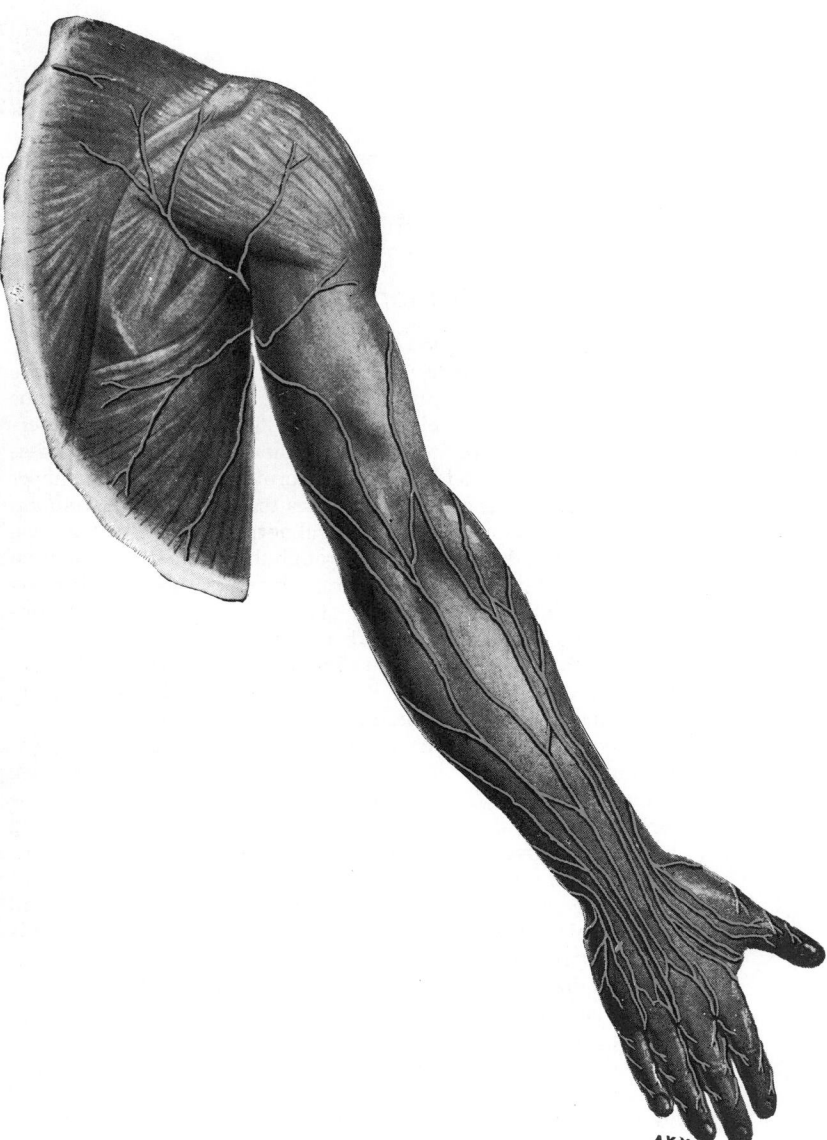

6.205 Normal axillary lymphangiogram, four days after injection of ultrafluid lipiodol into a lymph vessel on the dorsum of the hand. 1. Lateral group of lymph nodes. 2. Pectoral group of lymph nodes. 3. Brachial lymph vessels. Supplied by J B Kinmonth.

6.206 The lymphatic drainage of the superficial tissues of the upper limb: posterior aspect (semi-diagrammatic).

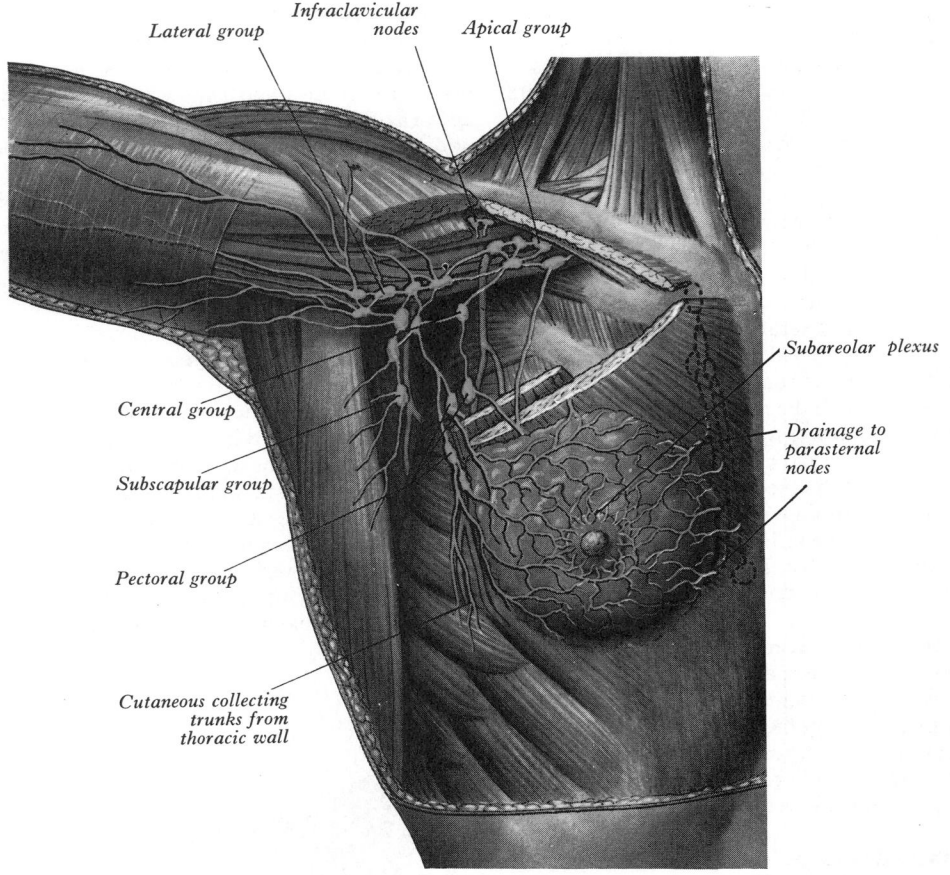

6.207 Lymph vessels of the mammary gland and the axillary lymph nodes.

Lymphatic Drainage of the Lower Limbs

Most lymph from the lower limb traverses a large *intermediary* inguinal group of nodes; some may first traverse a few more peripheral intermediary nodes, however, these are less numerous in the lower limb than elsewhere. The inguinal nodes are superficial and deep to the deep fascia. (Although commonly stated as such, the inguinal nodes are *not* the terminal group for the lower limb; from them the lymph traverses the external and common iliac groups, followed by the lateral aortic group. Deep gluteal lymph reaches the same group through the internal and common iliac chains. The upper lateral aortic nodes *are terminal*, forming bilateral lumbar trunks which discharge lymph into the confluence of trunks, p. 843, and thence the thoracic duct.)

Superficial inguinal nodes (**6**.208,210,211,212) form proximal and distal groups. The *proximal* is usually of five or six nodes just distal to the inguinal ligament. Its lateral members receive afferent vessels from the gluteal region and adjoining infra-umbilical anterior abdominal wall. Medial members receive superficial vessels from: the external genitalia (including the inferior vagina), inferior anal canal and perianal region, adjoining anterior abdominal wall, umbilicus and the uterine vessels accompanying the round ligament. The *distal* group, usually four or five, along the termination of the great saphenous vein, receives all superficial vessels of the lower limb, except those from the calf's posterolateral region. All superficial inguinal nodes drain to the external iliac nodes, some via the femoral canal and others anterior or lateral to the femoral vessels. Numerous vessels interconnect individual nodes.

Deep inguinal nodes vary from one to three, situated medial to the femoral vein. One is just distal to the sapheno-femoral junction, one in the femoral canal; the most proximal one lies laterally in the femoral ring; the middle node is the most inconstant, the proximal node, however, is often absent. All receive deep lymphatics accompanying the femoral vessels, lymph vessels

from the glans penis (or clitoridis) and a few efferents from the superficial inguinal nodes; their own efferents traverse the femoral canal to the external iliac nodes.

Peripheral nodes are few and are all deeply sited. Except for one sometimes proximal on the interosseous membrane near the anterior tibial vessels, they occur only in the popliteal fossa.

Popliteal lymph nodes (**6**.209), usually six of small size, are embedded in popliteal fat. One, near the end of the small saphenous vein, drains the superficial region served by the vein. Another is between the popliteal artery and posterior aspect of the knee joint, receiving direct vessels from the knee joint and those accompanying the genicular arteries. The remainder flank the popliteal vessels, receiving trunks accompanying the anterior and posterior tibial vessels. Popliteal efferents ascend close to the femoral vessels to reach the deep inguinal nodes but some may accompany the great saphenous vein to the superficial inguinal nodes.

Applied Anatomy. Inflammation of the popliteal nodes is often due to lateral lesions of the heel. Superficial inguinal nodes are frequently enlarged in disease or injury in their region of drainage (*palpation*, see p. 840). Thus in malignant or infective disease of the prepuce, penis, labia majora, scrotum, abscess in the perineum, anus and lower vagina or in diseases affecting skin and superficial structures in these regions, or the infra-umbilical part of the abdominal wall, the gluteal region — in all these the proximal inguinal nodes are almost invariably affected, the distal group being implicated only in disease or injury of the limb.

Lymphatic drainage of the superficial tissues in the lower limb. Superficial lymph vessels begin in subcutaneous plexuses. Collecting vessels leave the foot medially, along the great saphenous vein and, laterally, with the small saphenous.

Medial vessels are larger, more numerous and begin on the tibial side of the foot's dorsum, some ascending anterior and others posterior to the medial malleolus; thereafter both converge on the great saphenous vein and accompany it to the distal superficial inguinal nodes. *Lateral vessels* begin on the fibular side, some crossing anteriorly in the leg to join the medial vessels and so to the distal superficial inguinal lymph nodes; others accompany the small saphenous vein to the popliteal nodes. Superficial lymph vessels of the gluteal region circle anteriorly to the proximal superficial inguinal nodes.

Lymphatic drainage of the deeper tissues in the lower limb. Deep vessels accompany the main blood vessels: anterior and posterior tibial, peroneal, popliteal and femoral. Deep vessels

from the foot and leg are interrupted by popliteal nodes; those from the thigh pass to the deep inguinal nodes.

Deep lymph vessels of the gluteal and ischial regions follow their corresponding blood vessels. Those with the former end in a node near the intrapelvic part of the superior gluteal artery, near the superior border of the greater sciatic foramen; those which follow the inferior gluteal vessels traverse one or two of the small nodes below the piriformis and pass to the internal iliac nodes.

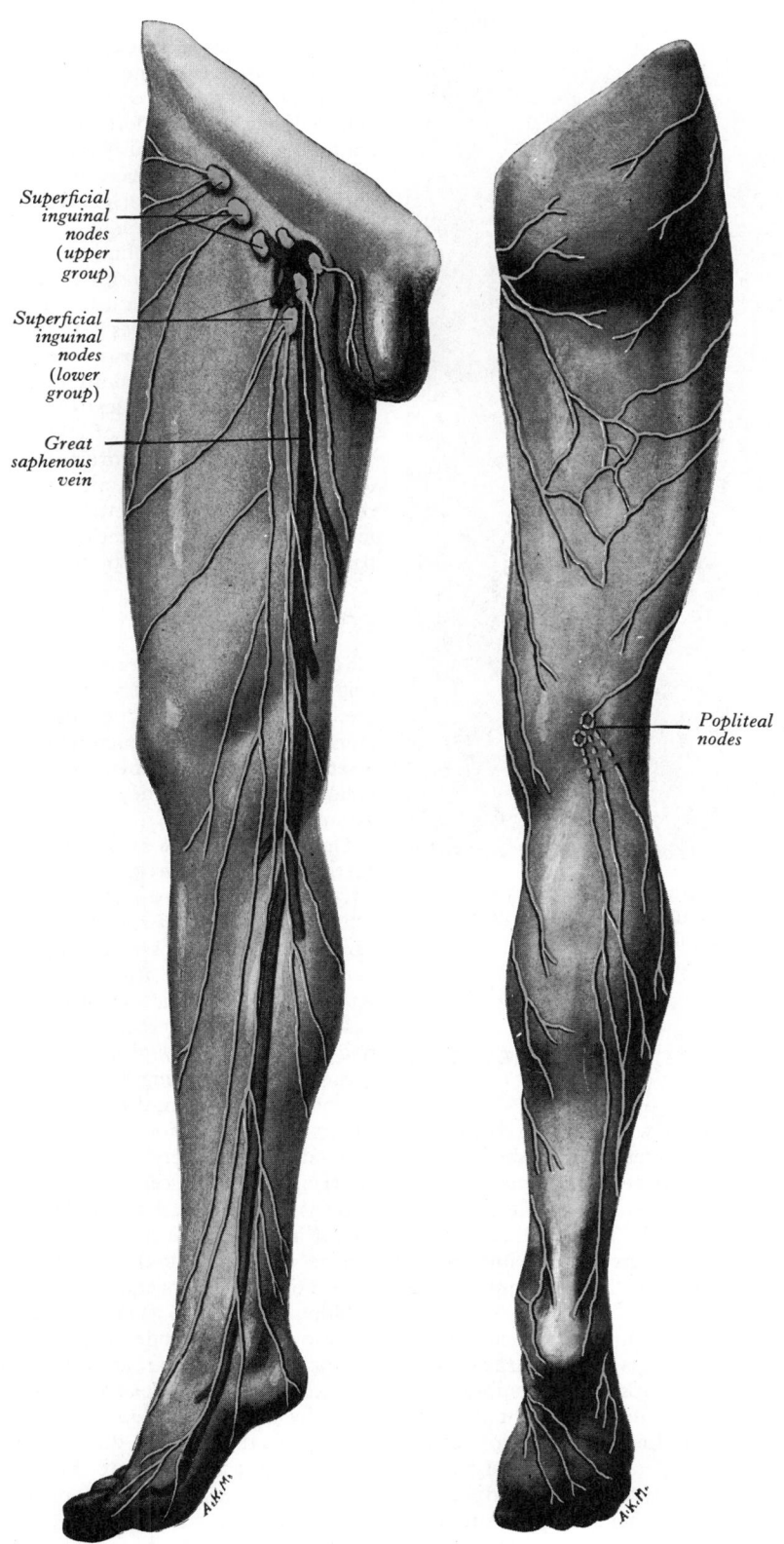

Superficial inguinal nodes (upper group)

Superficial inguinal nodes (lower group)

Great saphenous vein

Popliteal nodes

6.208 The lymphatic drainage of the superficial tissues of the lower limb: anteromedial aspect (semi-diagrammatic).

6.209 The lymphatic drainage of the superficial tissues of the lower limb: posterior aspect (semi-diagrammatic).

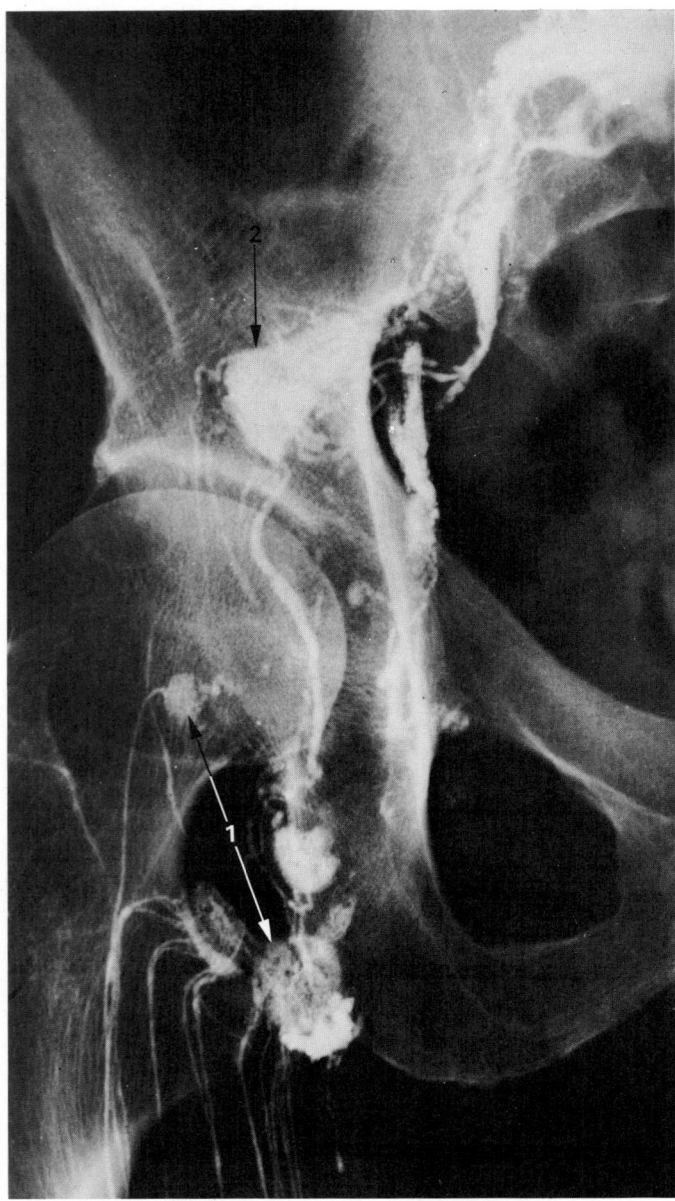

6.210 Lymphangiogram showing the inguinal lymph vessels and nodes taken immediately following injection of ultrafluid lipiodol into a lymph vessel on the dorsum of the foot. 1. Inguinal lymph nodes. 2. External iliac lymph node. Supplied by J B Kinmonth.

Lymphatic Drainage of the Abdomen and Pelvis

Lymph from most of the abdominal wall and all abdominal viscera (except a small hepatic region) is returned via the thoracic duct. Lymphatic vessels run with their corresponding arteries, the lymphatic nodes forming a large number of intermediary groups along the arteries concerned and a few terminal groups near the abdominal aorta. Although referred to as illustrative examples in the introductory paragraphs, they are summarized here with numerous additions.

Lumbar nodes (6.211,212) include three terminal groups, each of which although interconnected has its own large area of drainage, a number of intermediary groups and one 'subsidiary' group. These groups are *pre-aortic*, *lateral aortic* (right and left) and *retro-aortic*. The pre-aortic group drains viscera supplied by the *ventral splanchnic aortic branches*, i.e. the abdominal part of the alimentary canal (down to mid-rectum) and its derivatives. Lateral aortic groups drain viscera and other structures supplied by the *lateral splanchnic* and *dorsolateral somatic* aortic branches, receiving efferents from the large intermediary groups associated with the iliac vesels; their *upper members* are therefore terminal

groups for suprarenal glands, kidneys, ureters, testes, ovaries, pelvic viscera (apart from the gut) and the deeper tissues of the posterior abdominal wall, the full thickness of the subumbilical abdominal, pelvic and perineal walls and the whole of the lower limbs. The retro-aortic group has no special area of drainage; though it may have been primarily associated with drainage of the posterior abdominal wall, it may be regarded as comprising peripheral nodes of the lateral aortic groups and interconnecting surrounding groups.

PRE-AORTIC NODES AND THEIR REGION OF DRAINAGE

Pre-aortic lymph nodes are anterior to the abdominal aorta, receiving lymph from the regional nodes associated with the sub-diaphragmatic part of the alimentary canal, pancreas, liver and spleen. Their cranial efferents form *intestinal trunks* entering the abdominal confluence of lymph trunks (p. 839). They are divisible into coeliac, superior mesenteric and inferior mesenteric groups, being near the origins of these arteries.

In the alimentary canal, lymph vessels begin as minute subepithelial radicles, blind at one end and opening into a *periglandular plexus*. In the small intestine each villus has a *central* vessel, known as a *lacteal* from its milk white appearance. From the periglandular plexuses vessels pierce the muscularis mucosae to join a *submucous plexus*, efferents from which traverse the muscularis, where they connect with or bypass the vessels draining it. The submucous plexus is also joined by vessels from the lymph spaces at the bases of solitary lymphatic follicles. Lymphatics of intestinal muscle drain into a plexus mainly between the longitudinal and circular strata. Collecting vessels leave the gut through the muscle to enter the larger vessels following their mesenteric arteries. Collecting vessels from the alimentary canal pass through local nodes before reaching the pre-aortic group.

COELIAC NODES

Coeliac nodes are anterior to the abdominal aorta around the origin of the coeliac artery. They are a terminal group, their efferents forming right and left intestinal lymph trunks. Their afferents are from the regional nodes along branches of the coeliac artery, forming three main groups: gastric, hepatic and pancreaticosplenic; and they also come from the lower pre-aortic groups.

Gastric nodes (6.213,214) comprise the left gastric, right gastro-epiploic and pyloric groups. *Left gastric nodes*, along the left gastric artery, are divisible into subgroups: *superior* on the artery's stem and *inferior* with descending branches along the cardiac half of the lesser curvature in the lesser omentum and *paracardial*, a chain around the cardiac orifice. They receive lymph both from the stomach and the abdominal part of the oesophagus; their efferents pass to the coeliac group of pre-aortic nodes. *Right gastro-epiploic lymph nodes*, four to seven, lying in the greater omentum along the pyloric half of the greater curvature, receive afferents from the stomach; their efferents mostly pass to the pyloric nodes. Four or five *pyloric lymph nodes* are near the gastroduodenal artery's bifurcation, in the angle between the superior and the descending parts of the duodenum; an outlying node is sometimes sited above the duodenum near the right gastric artery. These nodes drain the pyloric part of the stomach, the first part of the duodenum and finally the right gastro-epiploic nodes; their efferents end in coeliac nodes.

Hepatic nodes (6.213) extend in the lesser omentum along the hepatic arteries and bile duct. They vary in number and site but almost constant are: one at the junction of the cystic and common hepatic ducts, the *cystic node*; and another alongside the upper bile duct, the *node of the anterior border of the epiploic foramen*. Hepatic nodes drain the stomach, duodenum, liver, gallbladder, bile ducts and pancreas; they drain to the coeliac nodes and thence to the intestinal trunks. Enlarged hepatic nodes may press on and obstruct the portal vein.

Pancreaticosplenic nodes (6.214) accompany the splenic artery, near the posterior surface and superior border of the pancreas; one or two are in the gastrosplenic ligament. Their afferents are from the stomach, spleen and pancreas; their afferents join the coeliac nodes.

Lymphatic drainage of the stomach and duodenum. Gastric lymphatics (6.213,214) are continuous at the cardiac orifice with the oesophageal vessels and at the pylorus with the duodenal channels. They largely follow blood vessels and form four groups: vessels of the first group accompany branches of the left gastric artery, receive from a large area on both gastric surfaces and end in the left gastric lymph nodes; a second group drains the gastric fundus and body left of a vertical from the oesophagus, accompanying the short gastric and left gastro-epiploic vessels to end in the pancreaticosplenic nodes; the third group drains the right half of the greater curvature as far as the pylorus, ending in the right gastro-epiploic nodes which drain to pyloric nodes; the fourth group drains the pyloric part of the stomach and drains to the hepatic, pyloric and left gastric nodes.

Although these vessels communicate, their valves direct lymph from the right part of the stomach to the lesser curvature and from the left part to the greater curvature.

Duodenal lymphatics run anteriorly and posteriorly into the small *pyloric lymph nodes*, lying in the anterior and posterior grooves between the pancreatic head and the duodenum. They drain up to the hepatic and down to the pre-aortic nodes around the origin of the superior mesenteric artery.

Lymphatic drainage of the liver. Hepatic collecting vessels are divisible into superficial and deep systems. *Superficial hepatic vessels* run in subserous areolar tissue over the whole surface of the organ, draining in four directions: (1) From the middle part of its posterior surface, the caudate lobe, the posterior part of the convex surfaces of both lobes near the hepatic attachment of the falciform ligament, the posterior part of the inferior surface of the right lobe, vessels accompany the inferior vena cava to nodes

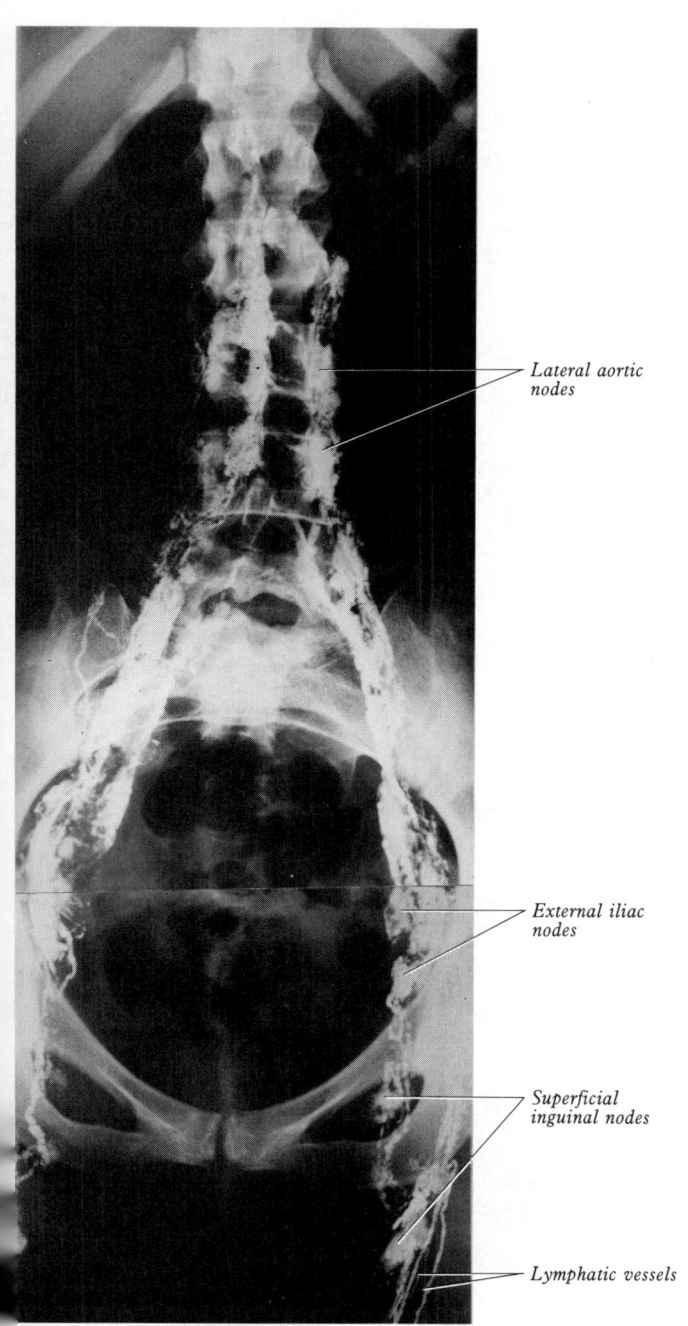

Lateral aortic nodes

External iliac nodes

Superficial inguinal nodes

Lymphatic vessels

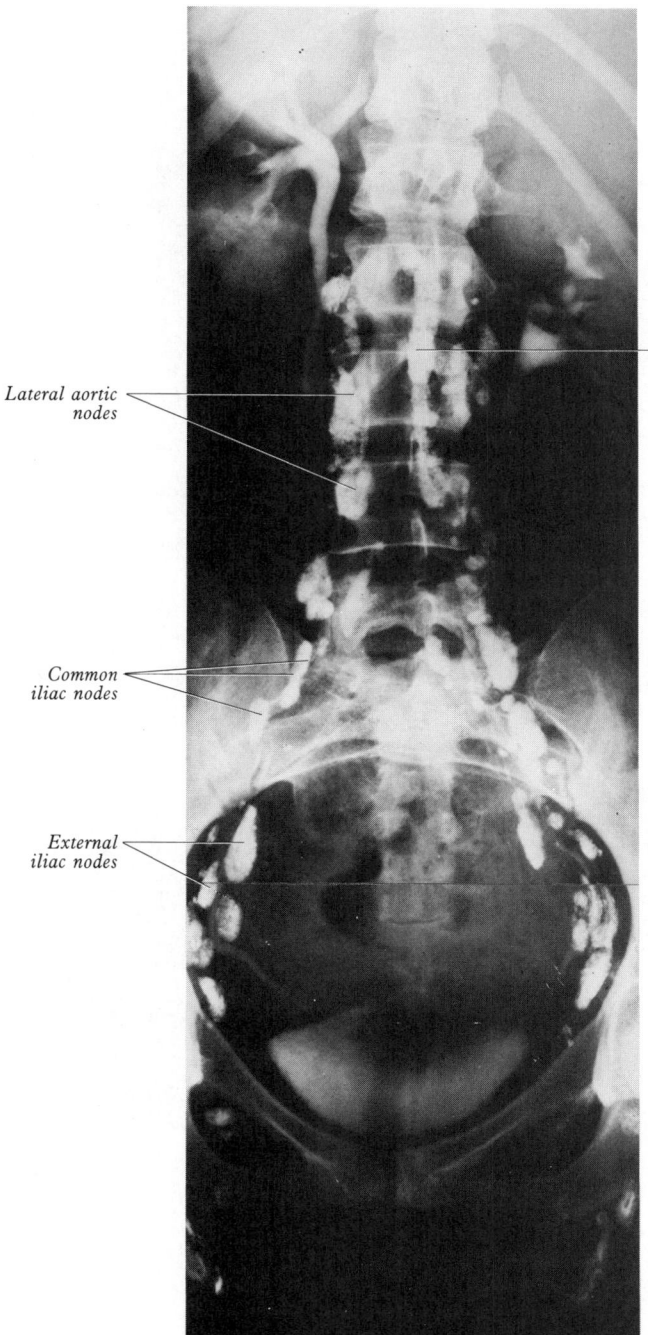

Lateral aortic nodes

Pre-aortic node

Common iliac nodes

External iliac nodes

6.211 Lymphangiogram showing the lymphatic vessels and nodes of the iliac and lateral aortic regions taken approximately 3 hours after the injection of lipiodol into a lymphatic vessel on the dorsum of each foot. Supplied by G I Verney, Addenbrooke's Hospital, Cambridge; photographs prepared by Sarah Smith and K Fitzpatrick, Guy's Hospital.

6.212 Lymphangiogram showing the lateral aortic and proximal iliac lymph nodes, approximately 24 hours after the injection of lipiodol into a lymphatic vessel on the dorsum of each foot. Intravenous contrast was given to show the kidneys and ureters. Supplied and photographed as in **6.211**.

around its terminal part. Vessels in the coronary and right triangular ligaments may directly enter the thoracic duct without any
intervening node. (2) Vessels from the rest of the inferior surface
and anterior part of the convex surfaces of both lobes near the
attachment of the falciform ligament all converge to the porta
hepatis to end in the hepatic nodes. (3) From the posterior region
of the left lobe a few vessels pass towards the oesophageal opening
to end in the paracardial nodes. (4) From the remaining convex
surface of the right lobe one or two trunks accompany the inferior
phrenic artery across the right crus to the coeliac nodes.

Deep hepatic lymphatics form the ascending and descending
trunks; the ascending trunks accompany the hepatic veins and
pass through the vena caval opening to end in the nodes round the
end of the inferior vena cava; the descending trunks emerge from
the porta hepatis to end in the hepatic nodes (p. 1390).

Lymphatic drainage of the gallbladder and bile ducts.
Numerous vessels run from the submucous and subserous
plexuses on all aspects of the gallbladder and cystic duct, those on
the former's hepatic aspect connecting sparsely with the hepatic
vessels. They pass to the hepatic nodes, especially the cystic node
and node of the anterior epiploic border (p. 850). Hepatic nodes
also collect from vessels accompanying the hepatic ducts and the
upper part of the bile duct, those of its lower part draining into the
inferior hepatic and upper pancreaticosplenic nodes.

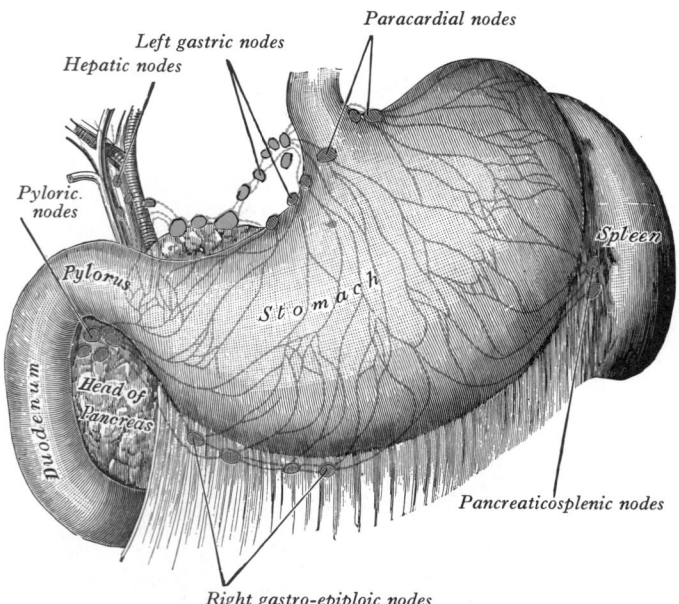

6.213 The lymphatic drainage of the stomach and duodenum (after
Jamieson & Dobson).

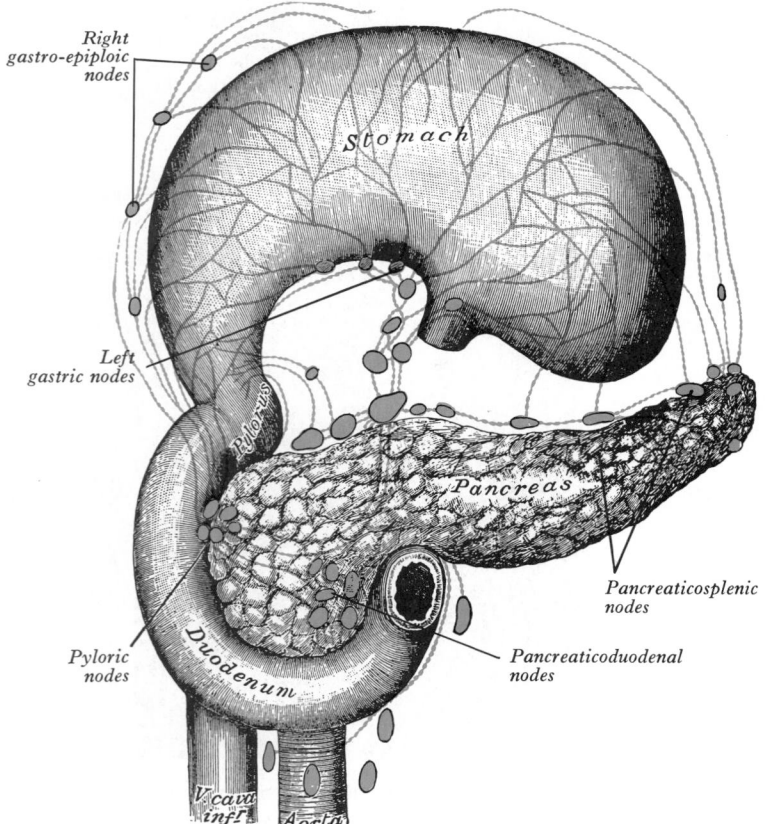

6.214 The lymph vessels and nodes of the stomach, duodenum and pancreas. The stomach has been turned upwards (after Jamieson & Dobson).

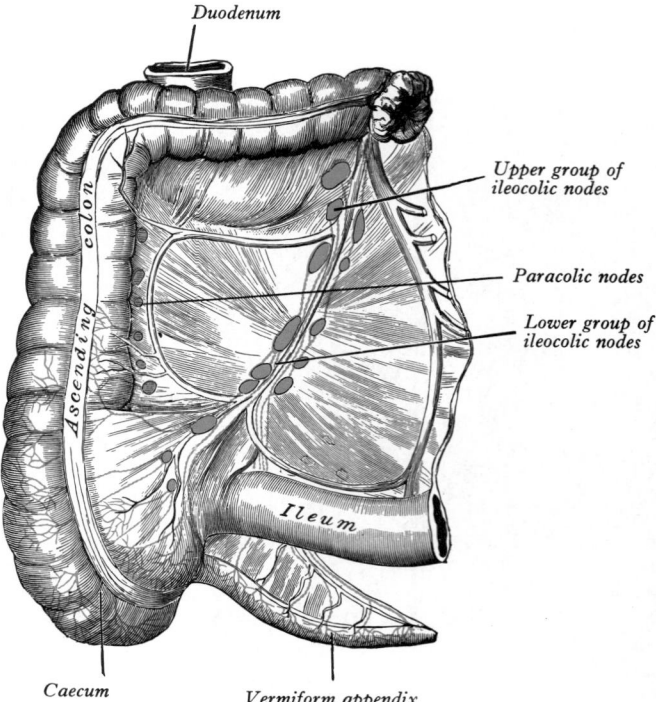

6.215 The lymph vessels and nodes of the caecum and vermiform appendix: anterior aspect (after Jamieson & Dobson).

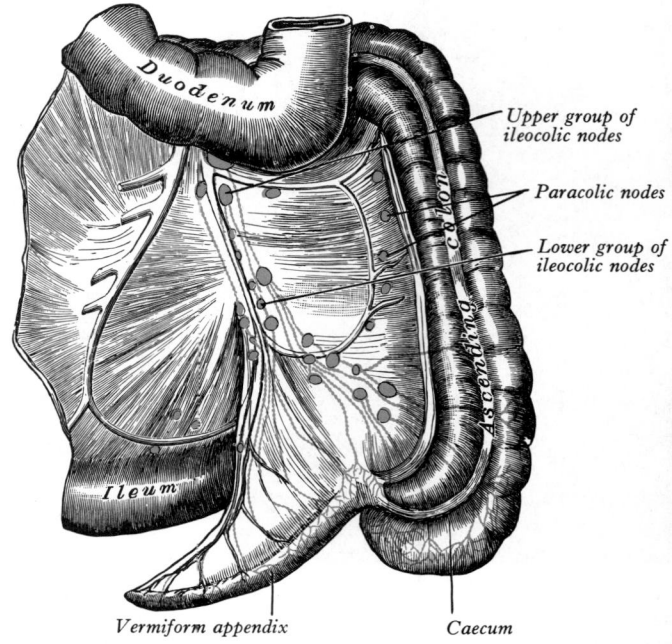

6.216 The lymph vessels and nodes of the caecum and vermiform appendix: posterior aspect (after Jamieson & Dobson).

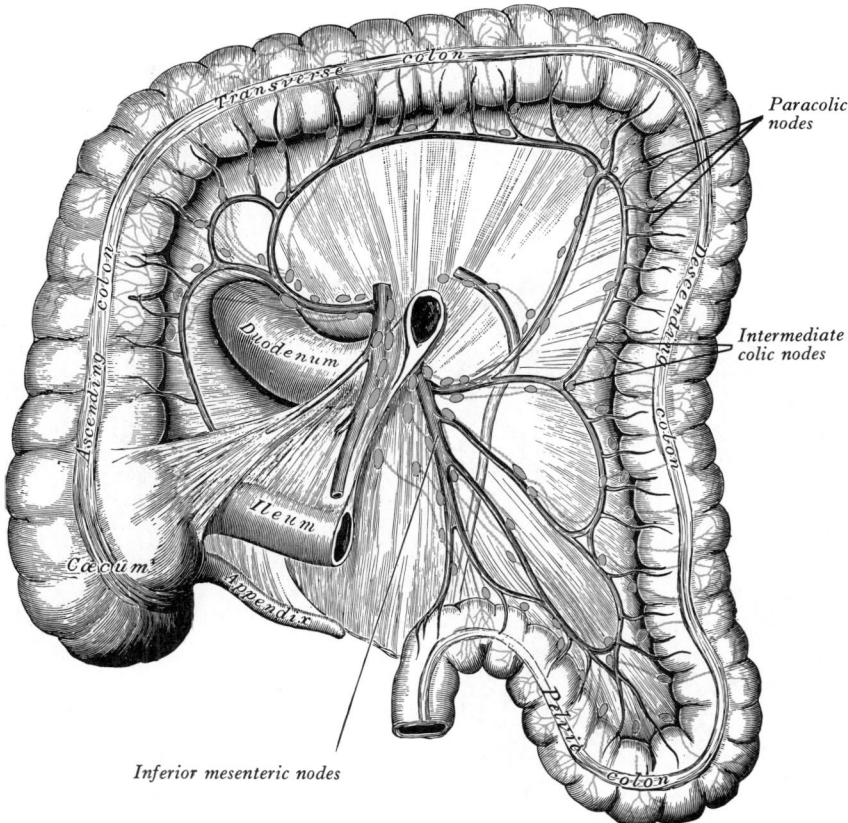

6.217A The lymph vessels and nodes of the colon (after Jamieson & Dobson).

Lymphatic drainage of the pancreas. Lymph capillaries commence around the acini and their continuations, following the blood vessels; there are no lymphatics in the pancreatic islets. Most vessels end in the pancreaticosplenic nodes, some in nodes along the pancreaticoduodenal vessels and others in the superior mesenteric pre-aortic nodes.

Lymphatic drainage of the spleen. Collecting vessels from the capsule end in the pancreaticosplenic lymph nodes.

SUPERIOR AND INFERIOR MESENTERIC NODES

Superior and inferior mesenteric nodes are anterior to the abdominal aorta near the origins of these arteries. They are preterminal groups for the alimentary canal from the duodenojejunal flexure to the upper anal canal and collect from outlying groups, including the mesenteric, ileocolic, colonic and pararectal nodes. They discharge into the coeliac nodes and thence intestinal trunks, confluence and thoracic duct.

Mesenteric nodes, numbering 100–150, comprise three series: one close to the intestinal wall among the terminal rami of the jejunal and ileal arteries (*mural*); a second is among the loops and primary branches of the vessels (*intermediate*); and a third is along the upper trunk of the superior mesenteric artery (*juxta-arterial*). Vessels from the terminal centimetres of the ileum follow the ileal branch of the ileocolic artery to the ileocolic nodes.

Applied Anatomy. Enlargement of the mesenteric nodes occurs in many intestinal diseases, especially typhoid fever, tuberculous ulceration and malignant tumours. Enlarged nodes can often be palpated through the abdominal wall.

Ileocolic nodes (6.215,216) form a *chain* of 10–20 around the ileocolic artery but tend to form two groups: near the duodenum and along the artery's terminal part. The chain divides with the artery, into: (1) *ileal nodes* close to the ileal branch; (2) *anterior ileocolic nodes* (usually 3) in the ileocaecal fold, near the caecal wall; (3) *posterior ileocolic nodes*, mostly in the angle between ileum and colon but partly behind the caecum at its junction with the ascending colon; (4) an appendicular node in the meso-appendix.

Colic nodes form four groups; epicolic, paracolic, intermediate colic and preterminal colic.

Epicolic nodes are merely minute nodules on the colonic wall, sometimes in the appendices epiploicae. *Paracolic nodes* lie along the medial borders of the ascending and descending colon and along the mesenteric borders of the transverse and sigmoid colon. *Intermediate colic nodes* lie along the right, middle and left colic arteries. *Preterminal colic nodes* adjoin the main trunks of the superior and inferior mesenteric arteries, near their corresponding pre-aortic nodes.

Pararectal nodes are in contact with the rectal muscular wall, draining to an intermediate group around the superior rectal artery and thence to nodes near the origin of the inferior mesenteric. Others drain to nodes at the bifurcation of the common iliac artery.

Lymphatic drainage of the jejunum and ileum. Lacteals pass between layers of the mesentery but, before reaching the superior mesenteric nodes, lymph traverses the mesenteric nodes.

Lymphatic drainage of the vermiform appendix and caecum (6.215,216). Lymphatic vessels are numerous, since

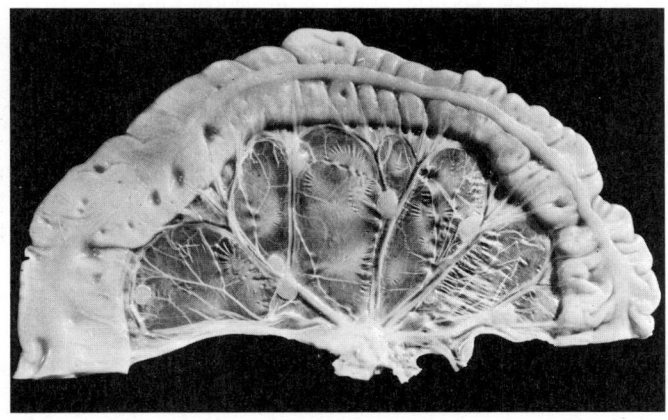

6.217B Preparation of the human colon and mesocolon displaying arterial arcades, neurovascular bundles, lymphatics and paracolic and intermediate lymph nodes. Provided by S Kubic, University of Zürich.

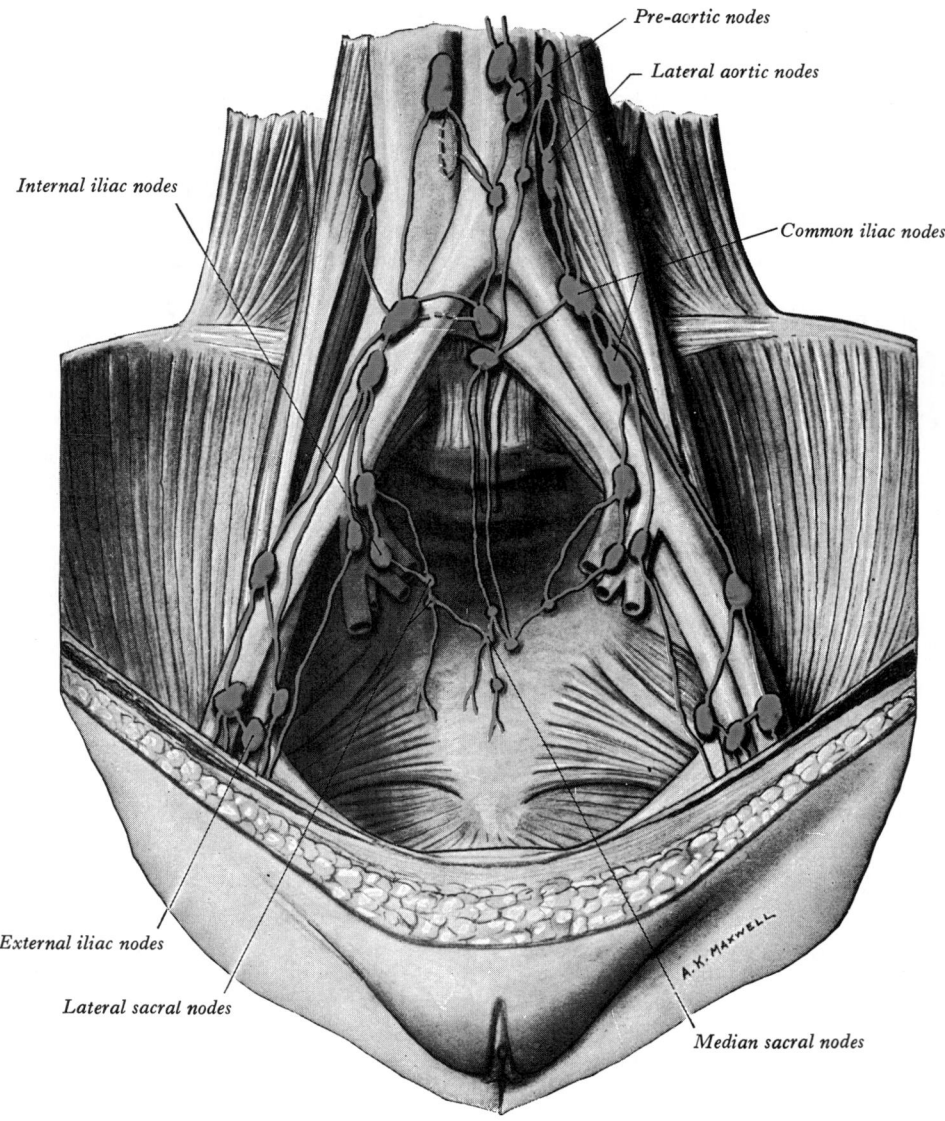

6.218 The lymph vessels and nodes of the pelvis.

Pre-aortic nodes

Lateral aortic nodes

Common iliac nodes

Internal iliac nodes

External iliac nodes

Lateral sacral nodes

Median sacral nodes

lymphoid tissue abounds in their walls. From the body and apex of the vermiform appendix 8–15 vessels ascend in the meso-appendix, a few interrupted by one or more nodes in it. They unite to three or four larger vessels, ending in the inferior and superior nodes of the ileocolic chain. Vessels from the root of the appendix and caecum are anterior and posterior. Anterior vessels pass in front of the caecum to the anterior ileocolic nodes and nodes of the ileocolic chain; posterior vessels ascend behind the caecum to the posterior and inferior ileocolic nodes.

Lymphatic drainage of the colon (6.215,216,217A,B). Lymphatic vessels of ascending and transverse parts of the colon end in the superior mesenteric nodes, after traversing nodes along the right and middle colic arteries and their branches. Those of the descending and sigmoid parts are interrupted by small nodes on branches of the left colic arteries, ending in the pre-aortic nodes around the origin of the inferior mesenteric artery.

Lymphatic drainage of the rectum and anal canal. From the *upper half*, or more, of the rectum the vessels emerge from its wall to *ascend* with the superior rectal vessels through the pararectal nodes to nodes in the lower sigmoid mesocolon and along the inferior mesenteric artery. From the *lower half* of the rectum and the anal canal, *above its mucocutaneous junction*, lymph vessels ascend through the wall to accompany the middle rectal vessels to the internal iliac nodes. Some are said to traverse the levator ani into the ischiorectal fossa, to accompany the inferior rectal and internal pudendal vessels to the internal iliac nodes. Lymphatics of the anal canal *below the mucocutaneous junction* descend to the

anal margin, curving laterally to reach the most medial superficial inguinal nodes.

LATERAL AORTIC NODES (6.211,212)

The lateral aortic nodes flank the abdominal aorta anterior to the medial margins of the psoas major muscles, diaphragmatic crura and sympathetic trunks. On the right some are lateral to the inferior vena cava and anterior to it near the end of the right renal vein. Afferents reach these nodes from structures supplied by the lateral splanchnic and dorsolateral somatic aortic branches and from outlying nodes near the iliac arteries and their branches. Efferents form a *lumbar trunk* on each side, both terminating in the confluence of lymph trunks (occasionally a cisterna chyli, p. 842); a few may pass to the pre-aortic and retro-aortic nodes. Some of the efferents of the right lumbar lymph trunk and its nodes may cross to their left counterparts; or both trunks may divide forming a loose plexus.

Lymphatic vessels from the kidney, suprarenal gland, abdominal ureter, posterior abdominal wall, testis and ovary, uterine tube and upper part of the uterus all pass *directly* to the lateral aortic nodes. Lymphatics from the pelvis, most pelvic viscera and the anterolateral abdominal wall pass first to regional nodes largely related to the internal iliac arteries and their branches. These include the following groups: common, external, internal and circumflex iliac, inferior epigastric and sacral. It must also be emphasized that the external iliac group receives the efferents from the inguinal nodes and the internal iliac group receives deep gluteal lymph; thus the lateral aortic groups ultimately drain the whole of *both lower limbs*.

Common iliac nodes (4–6) are grouped around the artery, one or two inferior to the aortic bifurcation and anterior to the fifth lumbar vertebra or sacral promontory (**6.**218). They drain the external and internal iliac nodes and send efferents to the lateral aortic nodes. They are usually in medial, lateral and intermediate (anterior) chains, the lateral being the main route.

External iliac nodes (**6.**211,212,218) (8–10) are ranged along, usually in three subgroups: lateral, medial and anterior to the external iliac vessels; the anterior is inconstant. The medial nodes are considered the main channel of drainage, collecting from: the *inguinal nodes* (p. 848), the deeper layers of the infra-umbilical abdominal wall, the adductor region of the thigh, the glans penis or clitoridis, the membranous urethra, prostate, vesical fundus, cervix uteri and upper vagina. Their efferents pass to the common iliac nodes. **Inferior epigastric and circumflex iliac nodes** are associated with their vessels and drain the corresponding areas, being outlying members of the external iliac group and inconstant in number.

Internal iliac nodes (**6.**218,219,220) surround the vessels, receiving afferents from all the pelvic viscera, deeper parts of the perineum and gluteal and posterior femoral muscles. Efferents pass to the common iliac nodes. **Sacral nodes** along the median and lateral sacral vessels and an *obturator node*, sometimes occurring in the obturator canal, are outlying members of the internal iliac group.

There is considerable bypassing in the iliac groups of lymph nodes. Lymphangiographic studies have demonstrated the connections between the right and left groups.

Lymphatic drainage of the urinary tract.

(1) *Renal.* Renal lymphatic vessels begin in three plexuses: one around the renal tubules, a second under the renal capsule and a third in the perirenal fat connecting freely with the second plexus. Collecting vessels from the intrarenal plexus form four or five trunks following the renal vein to end in the lateral aortic nodes; as they leave the hilum they are joined by the subcapsular collecting vessels. The perirenal plexus drains directly into the same nodes.

(2) *Ureteric.* Vessels begin in submucous, intra-muscular and adventitial plexuses which intercommunicate. Collecting vessels from the upper ureter may join the renal collecting vessels or pass directly to the lateral aortic nodes near the origin of the gonadal artery; those from its lower abdominal part go to the common iliac nodes; those from its pelvic part end in the common, external or internal iliac nodes.

(3) *Vesical.* Lymphatics (**6.**219) begin in the mucous, intramuscular and extramuscular plexuses. Collecting vessels, nearly all ending in the external iliac nodes, are in three sets: vessels from the trigone emerge on the vesical exterior to run superolaterally; those from the superior surface converge to the posterolateral angle and pass superolaterally across the lateral umbilical ligament to the external iliac nodes (one may go to the internal or common iliac group); those from the inferolateral surface ascend to join those from the superior surface. Minute nodules of lymphoid tissue may occur along the vesical lymph vessels.

(4) *Urethral.* (a) Vessels from the prostatic and membranous urethra in males and the whole female urethra pass mainly to the internal iliac nodes; a few may end in the external iliac nodes. Vessels from the membranous urethra accompany the internal pudendal artery. (b) Vessels of the male spongiose urethra accompany those of the glans penis, ending in the deep inguinal nodes. Some may end in superficial nodes, others may traverse the inguinal canal to the external iliac nodes.

Lymphatic drainage of the male reproductive organs.

(1) The *testis*. Testicular vessels commence in a superficial plexus, under the tunica vaginalis, and a deep plexus in the substance of the testis and the epididymis. Four to eight collecting trunks ascend in the spermatic cord and accompany the testicular vessels on the psoas major, ending in the lateral aortic and pre-aortic nodes.

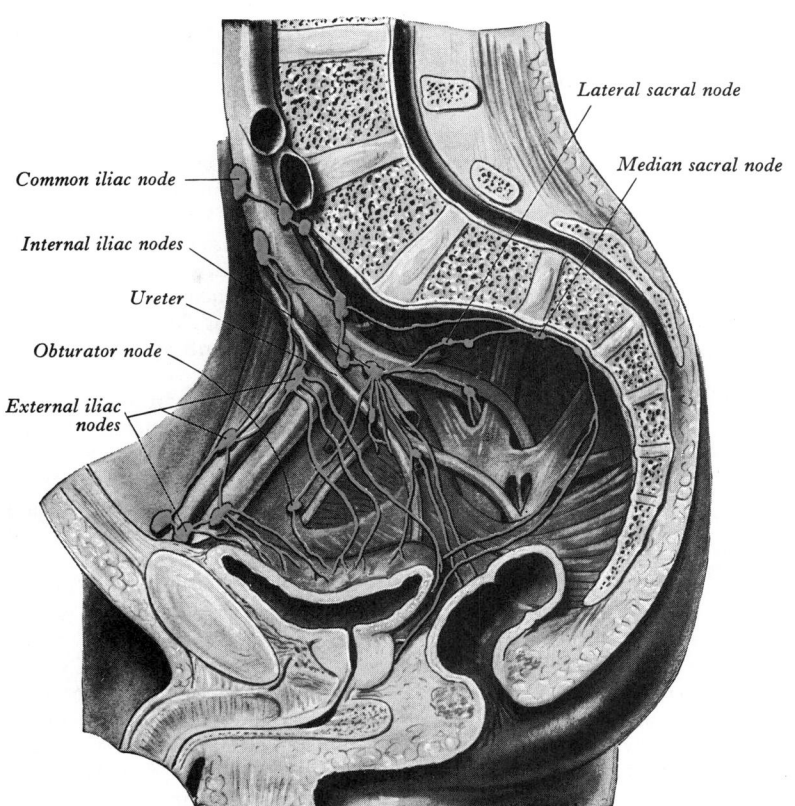

6.219 The lymphatic drainage of the urinary bladder (semi-diagrammatic).

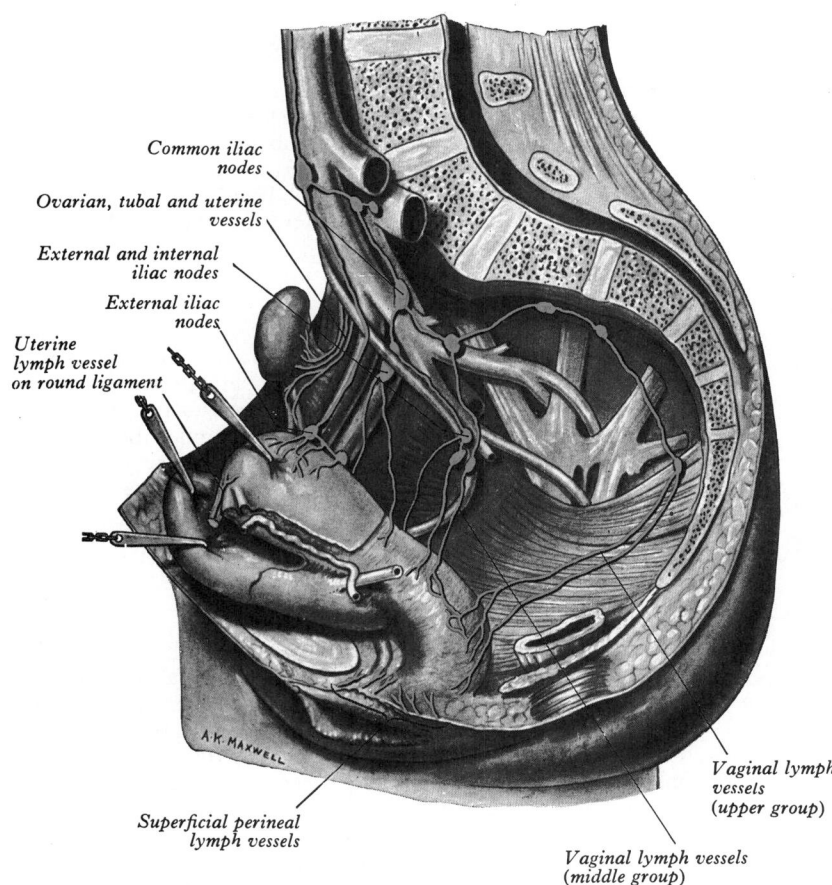

6.220 The lymphatic drainage of the female reproductive organs (semi-diagrammatic, after Cunéo & Marcille).

855

(2) The *ductus deferens, seminal vesicle and prostate gland.* Collecting vessels from the ductus end in the external iliac nodes, while those from the seminal vesicle go to the internal and external iliac nodes. Prostatic vessels end mainly in internal iliac and sacral nodes; a vessel from the posterior surface accompanies the vesical vessels to the external iliac nodes and one from the anterior surface gains the internal iliac group by joining vessels of the membranous urethra.

(3) The *scrotum and penis.* The skin of these parts is drained by vessels which, with those of all perineal skin, accompany the external pudendal blood vessels to the superficial inguinal nodes. Lymph vessels of the glans penis pass to the deep inguinal and external iliac nodes, from the erectile tissue and penile urethra to the internal iliac lymph nodes.

Lymphatic drainage of the female reproductive organs (6.220). (1) The *ovary.* The vessels, like the testicular, ascend along the ovarian artery to the lateral aortic and pre-aortic nodes.

(2) The *uterus and uterine tube.* Uterine lymphatics are superficial (or subperitoneal) and deep in the uterine wall. Collecting vessels from the cervix pass laterally in the parametrium to the external iliac nodes, posterolaterally to the internal iliac nodes and posteriorly in the sacrogenital fold to the rectal and sacral nodes. Some cervical efferents may reach the obturator or gluteal nodes. Vessels from the lower part of the uterine body pass mostly to the external iliac nodes, with those from the cervix. From the upper part of the body, the fundus and the uterine tubes, vessels accompany those of the ovaries to the lateral aortic and pre-aortic nodes, a few passing to the external iliac nodes. The region surrounding the isthmic part of the uterine tube is drained along the round ligament to the superficial inguinal nodes. Uterine lymph vessels enlarge greatly during pregnancy.

(3) The *vagina.* Vaginal lymphatic vessels link with those of the cervix uteri, rectum and vulva. They form three groups but the regions drained are not sharply demarcated. Upper vessels accompany the uterine artery to the internal and external iliac nodes, intermediate vessels accompany the vaginal artery to the internal iliac nodes; vaginal vessels below the hymen, from the vulva and perineal skin, pass to the superficial inguinal nodes but the clitoris and labia minora drain to the deep inguinal nodes and direct clitoridial efferents may pass to the internal iliac nodes (Kubik 1967).

Lymphatic drainage of the abdominal wall. These vessels are either superficial or deep to the deep fascia.

Superficial vessels accompany the subcutaneous blood vessels. Lumbar and gluteal vessels run with the superficial circumflex iliac vessels, those from the infra-umbilical skin with the superficial epigastric vessels. Both drain into the superficial inguinal nodes. The supra-umbilical region is drained by vessels running obliquely up to the pectoral and subscapular axillary nodes, a few to the parasternal nodes.

Deep vessels accompany the deep arteries, the posterior passing without interruption with the lumbar arteries to the lateral aortic and retro-aortic nodes; those from the upper anterior abdominal wall run with the superior epigastric vessels to the parasternal nodes; those of its lower part end in the circumflex iliac, inferior epigastric and external iliac nodes. Vessels of the pelvic wall follow the internal iliac artery and its parietal branches to end in the iliac or lateral aortic nodes.

Lymphatic Drainage of the Thorax

LYMPHATIC DRAINAGE OF THE THORACIC WALLS

Superficial lymphatic vessels of the thoracic wall ramify subcutaneously and converge on the axillary nodes. Those superficial to the trapezius and latissimus dorsi unite to form 10 or 12 trunks ending in the subscapular nodes. Those in the pectoral region, including vessels from the *skin* covering the periphery of the mammary gland and its subareolar plexus run back, collecting those superficial to serratus anterior, to reach the pectoral nodes. Vessels near the lateral sternal margin pass between the costal cartilages to the parasternal nodes but also anastomose across the sternum. A few vessels from the upper pectoral region ascend over the clavicle to the inferior deep cervical nodes.

Lymph vessels from *deeper* tissues of the thoracic walls drain mainly to the parasternal, intercostal and diaphragmatic lymphatic nodes.

(1) **Parasternal (internal thoracic) nodes,** four or five on each side, are at the anterior ends of the intercostal spaces, along each internal thoracic artery. They drain afferents from the mammary gland, deeper structures of the supra-umbilical anterior abdominal wall, the superior hepatic surface through a small group of nodes behind the xiphoid process and deeper parts of the anterior thoracic wall. Their efferents usually unite with those from the tracheobronchial and brachiocephalic nodes to form the *bronchomediastinal trunk*; this may open, on either side, directly into the jugulo-subclavian junction into either great vein near the junction or may join the *right* subclavian trunk, the right lymphatic duct or, on the left, the thoracic duct.

(2) **Intercostal nodes** are posterior in intercostal spaces near the heads and necks of the ribs. They receive deep lymph vessels from the posterolateral aspects of the chest and the mammary gland; some are interrupted by small lateral intercostal nodes. Efferents of nodes in the lower four to seven spaces unite into a trunk *descending* to the abdominal confluence of lymph trunks or to the commencement of the thoracic duct (p. 843). Efferents of nodes in the left upper spaces end in the thoracic duct, those of the right upper spaces end in one of the right lymph trunks.

(3) **Diaphragmatic nodes,** on the thoracic surface of the diaphragm, comprise: anterior, right and left lateral and posterior groups. The *anterior* group consists of two or three small nodes behind the base of the xiphoid process, draining the convex hepatic surface, and one or two nodes on each side near the junction of the seventh rib and cartilage, which receive anterior lymph vessels from the diaphragm. The anterior group drains to the parasternal nodes. The *lateral* groups each contain two or three nodes, close to the entry of the phrenic nerves into the diaphragm. On the right some nodes lie within the fibrous pericardium anterior to the intrathoracic end of the inferior vena cava. Their afferents are from the central diaphragm, the right also draining the convex surface of the liver. Their efferents pass to the posterior mediastinal, parasternal and brachiocephalic nodes. The *posterior* group contains a few nodes on the back of the crura, connected with the lateral aortic and posterior mediastinal nodes.

Lymphatic drainage of the deeper tissues. Collecting vessels of the deeper thoracic tissues include the following:

(1) Lymphatics of muscles attached to the ribs: most end in axillary nodes, some from pectoralis major in the parasternal nodes.

(2) Intercostal vessels draining the intercostal muscles and parietal pleura; those from the anterior thoracic wall and pleura end in the parasternal nodes, the posterior in intercostal nodes. (3) Vessels of the diaphragm form two plexuses, thoracic and abdominal, anastomosing freely and best marked in areas covered respectively by pleurae and peritoneum. The *thoracic plexus* unites with lymph vessels of the costal and mediastinal pleura, its efferents being: *anterior*, passing to the anterior diaphragmatic nodes near the junctions of the seventh ribs and cartilages; *middle*, to nodes on the oesophagus and around the end of the inferior vena cava; *posterior*, to nodes around the aorta where it leaves the thorax. The abdominal *plexus* anastomoses with the hepatic lymphatics and peripherally with those of the subperitoneal tissue. Efferents from its right half end partly in a group of nodes on the inferior phrenic artery, others in the right lateral aortic nodes. Those from the left half of the abdominal diaphragmatic plexus pass to the pre-aortic and lateral aortic nodes and nodes near the terminal oesophagus.

LYMPHATIC DRAINAGE OF THORACIC CONTENTS

Lymph from thoracic viscera traverses one or other of three groups of nodes, brachiocephalic, posterior mediastinal or tracheobronchial, before entering the thoracic duct, the right lymphatic duct or some other lymph trunk entering one of the great veins at the root of the neck.

Brachiocephalic nodes are in the superior mediastinum, anterior to the brachiocephalic veins and large arterial trunks springing from the aortic arch. They drain the thymus and

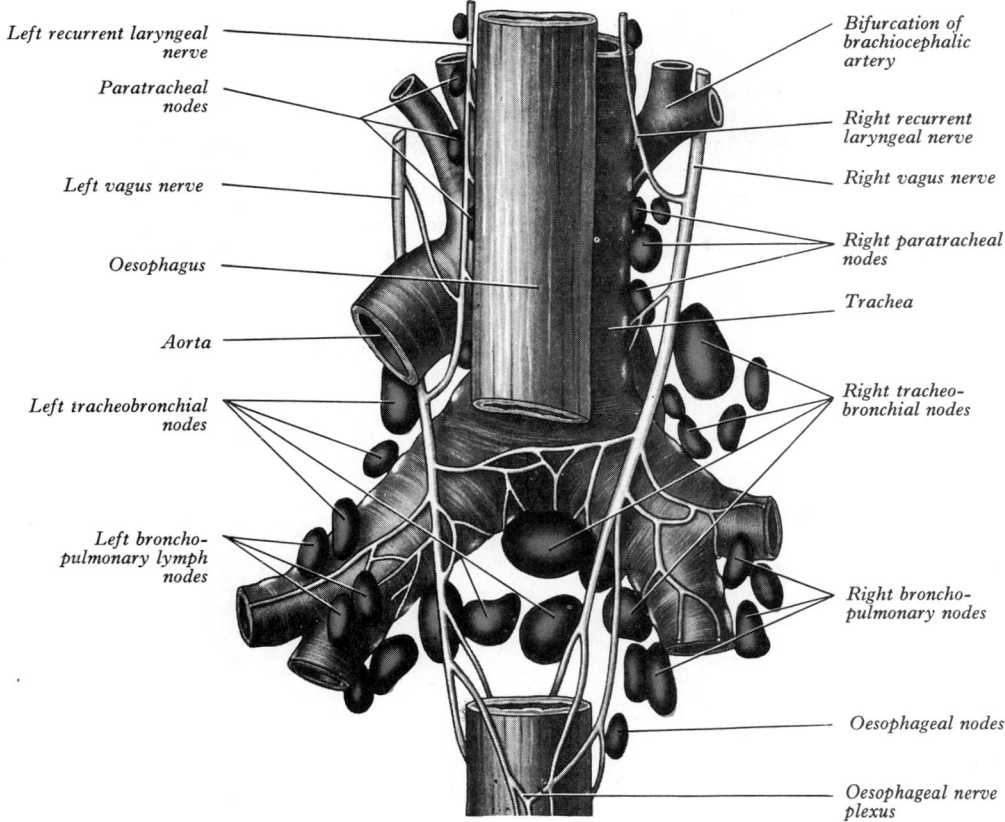

Left recurrent laryngeal nerve

Paratracheal nodes

Left vagus nerve

Oesophagus

Aorta

Left tracheobronchial nodes

Left broncho- pulmonary lymph nodes

Bifurcation of brachiocephalic artery

Right recurrent laryngeal nerve

Right vagus nerve

Right paratracheal nodes

Trachea

Right tracheo- bronchial nodes

Right broncho- pulmonary nodes

Oesophageal nodes

Oesophageal nerve plexus

6.221 The lymph nodes of the trachea, bronchi and lungs. Note the large 'carinate' node lodged between the bifurcation of the principal bronchi.

thyroid glands, pericardium, heart, and lateral diaphragmatic nodes; their efferents unite with those of the tracheobronchial nodes to form the right and left bronchomediastinal trunks.

Posterior mediastinal nodes are behind the pericardium, near the oesophagus and the descending thoracic aorta. Their afferents are from: the oesophagus, posterior pericardium, diaphragm, lateral and posterior diaphragmatic nodes, and sometimes the left lobe of the liver. They drain chiefly to the thoracic duct but some join the tracheobronchial nodes.

Tracheobronchial nodes (**6.221**) are in five main groups, including some of the largest nodes: (1) *paratracheal*, flanking the thoracic trachea but continuous above with the *cervical paratracheal nodes*; (2) *superior tracheobronchial*, in the angles between the trachea and the bronchi; (3) *inferior tracheobronchial* in the angle between the bronchi (commonly termed *carinate nodes* in clinicopathological parlance); (4) *bronchopulmonary*, in the hilum of each lung (*hilar nodes*); (5) *pulmonary*, in the lung substance on larger branches of the principal bronchi. These groups are not sharply demarcated; pulmonary nodes become continuous with the bronchopulmonary and they in turn with the inferior and superior tracheobronchial nodes, continuous with the paratracheal group. Afferents of tracheobronchial nodes drain the lungs and bronchi, thoracic trachea, heart and some efferents of the posterior mediastinal nodes. Their efferent vessels ascend on the trachea to unite with efferents of the parasternal and brachiocephalic nodes as the *right* and *left bronchomediastinal trunks*; the right trunk may occasionally join a right lymphatic duct or another right-sided lymph trunk and on the left the thoracic duct; but more often they open independently in or near the jugulo-subclavian junction on their own side.

Applied Anatomy. In all town dwellers large quantities of dust and carbonaceous pigment may be freely inhaled and are continually swept into these nodes from the bronchi and alveoli.

Lymphatic drainage of the heart. Cardiac lymphatic vessels form subendocardial, myocardial and subepicardial plexuses, the first two draining into the third, efferents of which form the *left* and *right cardiac collecting trunks*. Two or three *left* trunks ascend the anterior interventricular sulcus, receiving vessels from both ventricles; reaching the coronary sulcus, they are joined by a large vessel from the diaphragmatic surface of the left ventricle, which first ascends in the posterior interventricular sulcus and then turns left along the coronary sulcus. The vessel formed by the union of these two ascends between the pulmonary artery and the left atrium, usually ending in an inferior tracheobronchial node. The *right* trunk receives afferents from the right atrium and right border and diaphragmatic surface of the right ventricle. It ascends in the coronary sulcus, near the right coronary artery, and then anterior to the ascending aorta to end in a brachiocephalic node, usually on the left.

Lymphatic drainage of the lungs and pleurae. Pulmonary lymphatic vessels originate in a superficial subpleural plexus and a deep plexus accompanies the branches of pulmonary vessels and bronchi. In larger bronchi the deep plexus has submucous and peribronchial parts; in smaller bronchi a single plexus extends to the bronchioles but not to the alveoli, whose walls have no lymphatic vessels. Superficial efferents turn round borders and the margins of fissures to converge in the bronchopulmonary nodes; deep efferents reach the hilum along the pulmonary vessels and bronchi, ending mainly in the same nodes. There is little anastomosis between the superficial and deep lymphatics, except in the hilar regions. In peripheral parts of the lungs small channels connect superficial and deep lymphatic vessels, capable of dilatation to direct lymph from the deep to the superficial channels when outflow from deep vessels is obstructed by pulmonary disease. Deep in the fissures, lymphatic vessels of adjoining lobes connect; hence, though there is a tendency for vessels from the upper lobes to pass to the superior tracheobronchial nodes and those from lower lobes to the inferior tracheobronchial group, these connections are not exclusive. At the level of pulmonary lobulation the arrangement of lymphatic vessels follows both the central artery of a lobule and its peripheral veins (Kubik 1970), confirming the findings of Celtis & Porter (1952). Policard (1950) has described lymphoid aggregations, non-follicular in appearance, in peribronchial sites and in '*placoid*' formations adjoining pulmonary pleura.

Pleural lymphatic vessels exist in visceral and parietal layers, those of the visceral pleura draining to the superficial pulmonary efferents, forming a plexus beneath the pulmonary pleura (vide supra). Those of the parietal pleura end in three ways: (1) those from the costal region join vessels of the internal intercostal muscles to reach the parasternal nodes; (2) those of the diaphragmatic pleura form a plexus on its thoracic surface (p. 856);

(3) those of the mediastinal pleura end in the posterior mediastinal nodes.

Lymphatic drainage of the thymus. Thymic lymphatic vessels end in the brachiocephalic, tracheobronchial and parasternal nodes.

Lymphatic drainage of the oesophagus. Efferent vessels from the cervical oesophagus drain to the deep cervical nodes, those from its thoracic part to the posterior mediastinal nodes and those from its abdominal part to the left gastric lymph nodes. Some may pass directly to the thoracic duct (p. 842).

NEUROLOGY

7

Introduction

The dependence of organisms upon sources of environmental energy and the essentially dynamic nature of their life processes have been emphasized elsewhere (pp. 3–6). A major aspect of this dynamism is the ability of organisms to interact continually with a fluctuating environment without any loss of structural integrity. Successful adaptations ensuring survival can be related to long or brief scales of time (p. 6): through many generations natural selec-

7.1 The relationsip between a stimulus (S) and a response (R) emphasizing its closed-loop nature.

tion operates on genetic variation engendered by sexual reproduction and occasional gene mutations, resulting in species with greater adaptability to environmental changes. Alternatively, within the lifetime of a complex organism, adaptive responses range from behavioural patterns in mating, rearing, procuring food, escape or combat, etc. to innumerable transient readjustments, e.g. of posture or the maintenance of a constant composition in the body fluids. These are examples of *homeostatic responses*, a central feature of behaviour, simply summarized in illustration 7.1.

An environmental change (S or *stimulus*) to which an organism can react entails an appropriate *response* (R), which is usually to preserve the constancy of the internal state within the limits

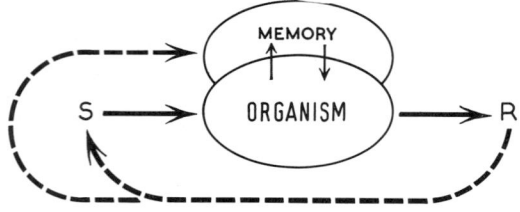

7.2 The stimulus-response loop with the added flexibility of a memory store.

necessary for the continued life of its cells and to preserve the whole organism. It should be noted that the relation of an organism to a stimulus is altered by this response (7.1); the sequence of events is therefore a *loop*, not an *arc*. In the simple case illustrated, the organism's structural organization (the result of its phylogenetic history) limits the range of stimuli to which it can react and its repertoire of responses, whether mechanical, chemical, photic or electrical. Hence environmental disturbances to which it is *unable* to respond, or disturbances too intense for an adequate response, cause the organism to die and dissolve into its environment. In this example, *genetic memory*, selected through countless generations, has determined the range of receptivity and response and hence the fitness to reproduce in a particular environment. Such mechanisms may lead to stereotyped behavioural responses such as those of insects; but in more complex organisms selection has added flexible *memory stores* operating throughout life (7.2). In these, the general sequence of the stimulus–response loop remains similar, but information from a particular event is stored for comparison with previous experience. Thus a choice can be made (from the different alternatives available) of the response likely to be the most effective in the prevailing circumstances. In some way, the effectiveness of a response is assessed and transferred to memory, which is modified accordingly, the probability of this particular response being chosen again in like conditions being raised or lowered. Thus the memory store (or hierarchy of stores) is continually modified and

refined by further experience, i.e. the organism *learns*.

Homeostatic responses are innate in *all* living organisms; but, with increasing size and complexity of structure, the range and flexibility of responses has steadily increased in parallel with the evolution of a *nervous system*.

The human nervous system is the most complex, widely investigated, and yet poorly understood physical system known to mankind. Its structure and activities are inseparable from every aspect of life: physical, cultural and intellectual. Investigators from many disciplines, with many methods, motives and persuasions, converge in its study; there are therefore many more or less appropriate ways of undertaking such a study: developmental, phylogenetic, structural, physiochemical, energetic, cybernetic, behavioural and ethological. In the present account the detailed anatomy of the arbitrary but convenient divisions of the human nervous system is preceded by a consideration of the relevant experimental methods, the biology of its component cells, a brief review of the organization of nervous systems in general and a comparison of nervous systems with information-processing, communication and mechanisms of control of simple homeostatic machines. (For a further discussion consult Young 1964, 1978.)

SIMPLE HOMEOSTATS

In considering mechanisms of homeostasis, it is helpful to draw an analogy with devices such as a simple thermostat. In diagram 7.3 the *energy level* (temperature) of a water bath is continuously *monitored by a sensor* or *receptor* (thermocouple), fluctuations in temperature being converted into variations in the flow of a *pattern of signals* or *coded information* (current flow) along an *afferent communication channel* to a *controller*, which *integrates* the flow of information along its afferent channel, compares it with a reference level of temperature and *decides* between the activities (*responses*) of two *effectors* (steam or cold water). If the bath's temperature falls enough to be detected by the sensor, the change excites signals in the afferent channel resulting, after comparison and computation in the control centre, in an increased flow of steam. It is inherent in the design of *control loop systems* that a return of the temperature to the reference level will not be smooth or immediate; an overshoot occurs, followed by oscillations of decreasing amplitude, gradually approaching reference level. All features in this homeostat have counterparts in a primitive nervous system, which include: various sensors able to monitor the rate and magnitude of some environmental change, *afferent* channels for coded information (nerve fibres bearing temporo-spatial patterns of impulses), computing and decision-making centres (geometric patterns of contact between neurons) and *efferent* channels (nerve fibres) to glands or muscles to stimulate them to ordered responses. All these features will be considered later in further detail.

As depicted in 7.3, such simple control loops, which are present in all organisms (and also widely used in engineering), are frequently *coupled* to others for more complex responses; many features of more complex nervous systems are thus referable to an *integrated hierarchy of control loops*.

Intercellular Communication

Many aspects of intra- and intercellular communication (7.4A,B) are noted elsewhere; these include selective interactions between closely apposed cell surfaces, the phenomena of contact guidance and contact inhibition (p. 115), the structure and informational role of nucleic acids (p. 115) and current theories of modes of action of local cytoplasmic factors or intercellular chemical messengers (embryonic induction factors, hormones, etc.) in changing genotropic control loops, resulting in enhancement or repression of patterns of gene activity (pp. 40, 108). Where cells are particularly close, as at 'gap' junctions (maculae communicantes), some ions and molecules can pass between the cytoplasm of adjacent cells (p. 24). Chemical messengers, such as secretions of endocrine glandular cells, pass into circulation and then diffuse via tissue fluid to remote cells; such mechanisms essentially operate *slowly*

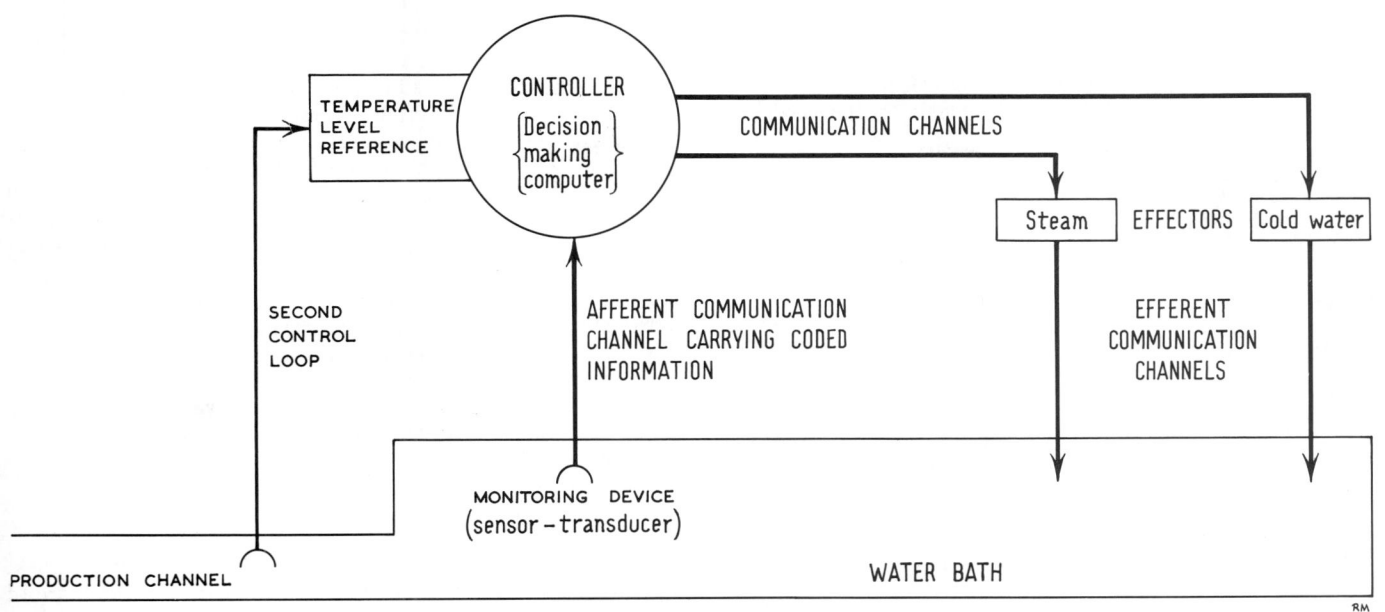

7.3 The essential features of a thermostat coupled to an industrial production line. This is an example of a man-made homeostatic device; see text for further details and discussion. From Young 1964, with permission.

EXCITABLE TISSUES

More rapid mechanisms of intracellular and intercellular communication have evolved in special *excitable tissues* (receptor, neural, muscular and some glandular cells). Some of the general features of excitable cells are shown in 7.4B. All cells possess a plasma membrane separating them from the surrounding tissue fluid (p. 15, 1.5). Due to the membrane's selective permeability the ionic composition of intracellular fluid differs from that outside; hence a large electrical potential exists across it. A suitable *stimulus* applied to its surface leads to a transient, reversible wave of change spreading over the surface from this point. Though details of this change in molecular terms are uncertain, alterations in the permeability of the membrane (conductance) result. Presumably a rapid local redistribution of ions occurs across the membrane as these flow down concentration gradients, ceasing as its permeability returns to its resting state. These ionic fluxes can be recorded as variations in electrical potential and are also accompanied by minor, but detectable thermal exchanges between the cell and its environment. Such properties of excitable cells, discussed in detail elsewhere, entail *trains* or *volleys* of *impulses* caused by prolonged stimulation, the stimulus strength being related to the impulse pattern (number and frequency of impulses). The impulse, spreading over the surface, causes various effects in the cells they innervate, e.g. the release of specific secretory products or the contractions of actin–myosin complexes, depending upon the type of cell. Cells co-operating in nervous function are modified in various ways and exhibit the properties of excitable tissues to a variable extent (7.5A,B,C). Thus, in *protozoa* and some primitive metazoa, *single cells* can give effector responses to stimuli but in most multicellular creatures possessing nervous systems a *division of labour* occurs between the excitable cells in the system. This marks the emergence of *receptor* cells, specialized to react to stimuli: *neurons* for their conduction and integration, and *effectors* (contractile or glandular) for responses. The cytology, evolution and function of these is detailed on p. 904 et seq. but, before their details, the overall pattern of the nervous elements will be considered in general terms.

RECEPTOR CELLS

Within the epithelial layers clothing the body's surfaces and their subepithelial connective tissue, within the walls of blood vessels and of the solid and hollow viscera and in muscles, tendons and ligaments are arrays of *receptor cells*. These have specialized plasma membranes responding selectively to changes in the internal or external environment of the body. Some are responsive to

physical deformation, others to thermal, chemical, electrical or photic variations. Views concerning the *specificity* of individual receptors have been modified much during the last 150 years. Since the law of *specific nerve energies* was proposed by Johannes Muller in 1840, many have held that receptors react preferentially to only one type of energetic change and this determines its normal physiological role. A receptor was held to show a *low threshold* to this form of stimulation, i.e. it reacted to *small* variations in energy level, whereas, to all other changes it showed a *high threshold*, reacting only if stimulation were intense, perhaps

GENERAL TISSUE CELLS

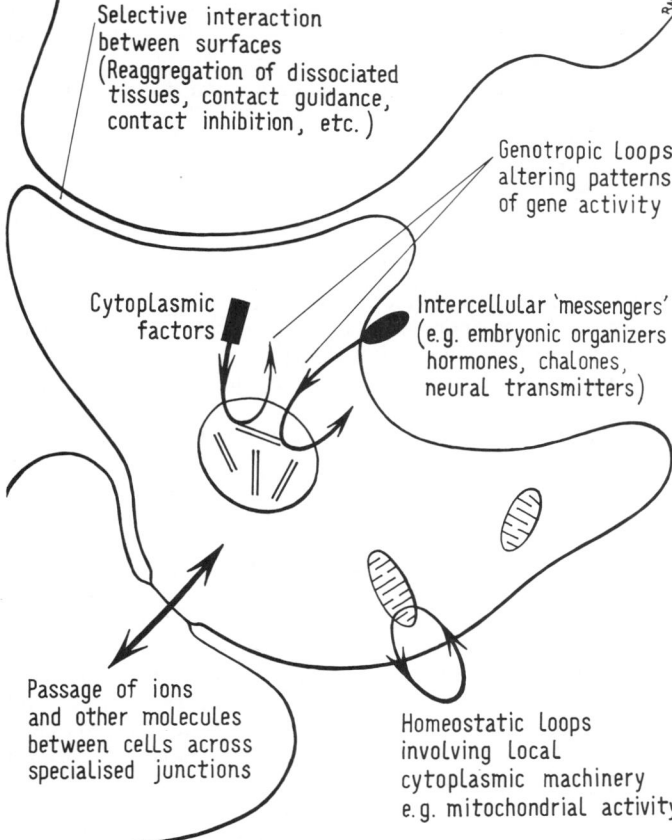

Selective interaction between surfaces (Reaggregation of dissociated tissues, contact guidance, contact inhibition, etc.)

Genotropic loops altering patterns of gene activity

Cytoplasmic factors

Intercellular 'messengers' (e.g. embryonic organizers hormones, chalones, neural transmitters)

Passage of ions and other molecules between cells across specialised junctions

Homeostatic loops involving local cytoplasmic machinery e.g. mitochondrial activity

7.4A Various forms of cellular and intercellular communication systems.

EXCITABLE TISSUE CELL

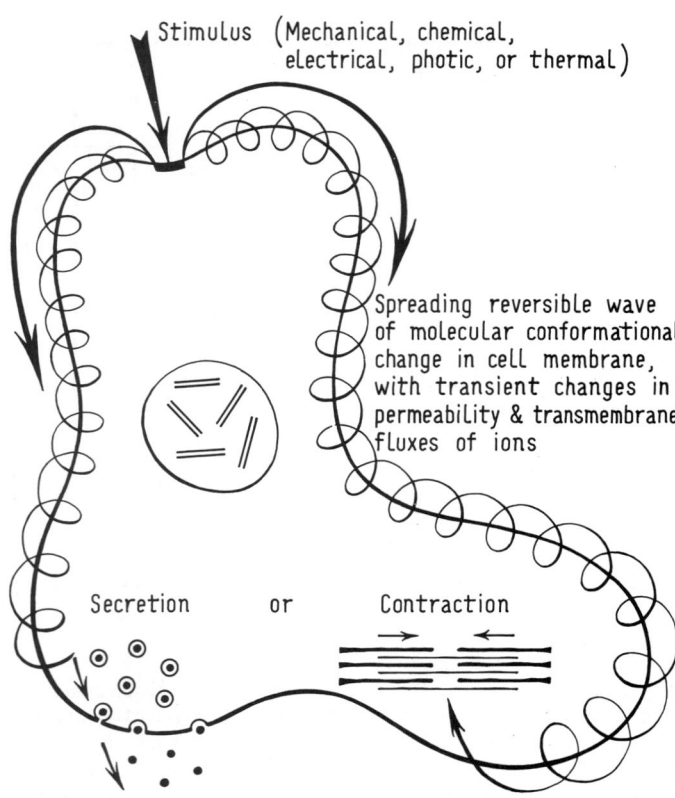

7.4B A summary of the main events which may follow the application of an effective stimulus to the surface of an excitable tissue cell.

abnormally so. Experiments on mammalian cutaneous innervation, however, prompted an opposite conclusion: that cutaneous receptors are *non-specific*, sensory perception depending on *spatio-temporal patterns* of impulses in a *common set* of neural pathways (Weddell 1941). With the subsequent refinement of methods of recording from single fibres and nerve cells there has been a partial return to the specificity theory but with complexities, subtleties and grades of interaction not envisaged in earlier theories (Mountcastle & Powell 1959, Sinclair 1967, Wall 1967). Thus, nerve fibres have been described which serve receptors reacting to more than one kind of change (*polymodal receptors*). Receptors may be fast or slowly adapting; the former respond to a brief stimulus with a sharp volley of nerve impulses ending with the stimulus or, if this is maintained, dropping to a resting frequency. Slowly adapting receptors may generate volleys of impulses for prolonged periods if the stimulus strength persists. Other varieties show a resting level of discharge which may be raised or lowered by changes in environment. These variants will be further considered with individual receptor cells (p. 906).

The molecular mechanisms by which environmental change, mechanical, chemical or thermal, excites a receptor cell remain obscure (excepting photosensitive pigments in retinal receptors, p. 1198). Recordings from many receptors show that exposure to an adequate stimulus leads to a change of ionic permeability in the cell's specialized receptor surface; ions which in the unstimulated state are differentially concentrated in intra- and extracellular compartments flow along concentration gradients, through molecular channels in the receptor membrane; microelectrodes record these ionic fluxes as variations constituting the *receptor potential*. The circuit for ionic flow is completed by an equal and opposite flow across adjacent non-specialized (but excitable) regions of receptor cell membrane, resultant *graded variations* in potential being the *generator potential*. The consequences of fluctuations of the generator potential have been analysed in only a few sites, but they vary with the type and geometry of different receptors. In some sites, e.g. lingual gustatory cells, the generator potential may *directly* affect the levels of activity at the points of

synaptic contact between the receptor cell and a peripheral process of a neuron (vide infra); in others, e.g. Pacinian corpuscles, whose specialized receptor surface is the periphery of a long cytoplasmic conducting process (nerve fibre) proceeding to the central nervous system, other events occur; if the variation in generator potential reaches *threshold* level, a dramatic but rapidly reversible series of changes in ionic permeability ensue in the cell membrane bordering the receptor surface. The consequent ionic fluxes, recordable as an *action potential*, start a self-generating wave of similar changes along a conducting process as a *nervous impulse*. The maintenance of the generator potential at an adequate level above threshold results in a *volley of impulses* passing centrally. The *frequency* of impulses in the volley is related approximately logarithmically to the energy or *intensity* of the stimulus. Receptors thus act as *transducers*, whereby fluctuations in environmental energy levels are transformed to coded volleys of nerve impulses.

SYNAPTIC CONTACTS

Excitable cells (receptors, neurons and effectors) communicate at specialized intercellular contacts termed *synapses*, which are of two main kinds. Most common are *chemical synapses*, where *neurotransmitters* released by one cell stimulate the plasma membrane of another, altering its permeability to certain ions and hence its electrical state. Such junctions occur in all nervous systems. Less frequent is an *electrical synapse*, identical to *communicating junctions* or *nexuses* existing between many types of cell (e.g. cardiac and non-striated myocytes, etc., pp. 557, 562). Here adjacent cells have channels of direct ionic communication between their cytoplasm, one cell exciting the other without release of transmitter.

At most mammalian synapses (as in most other vertebrates and invertebrates) transmission is chemical; cell membranes are

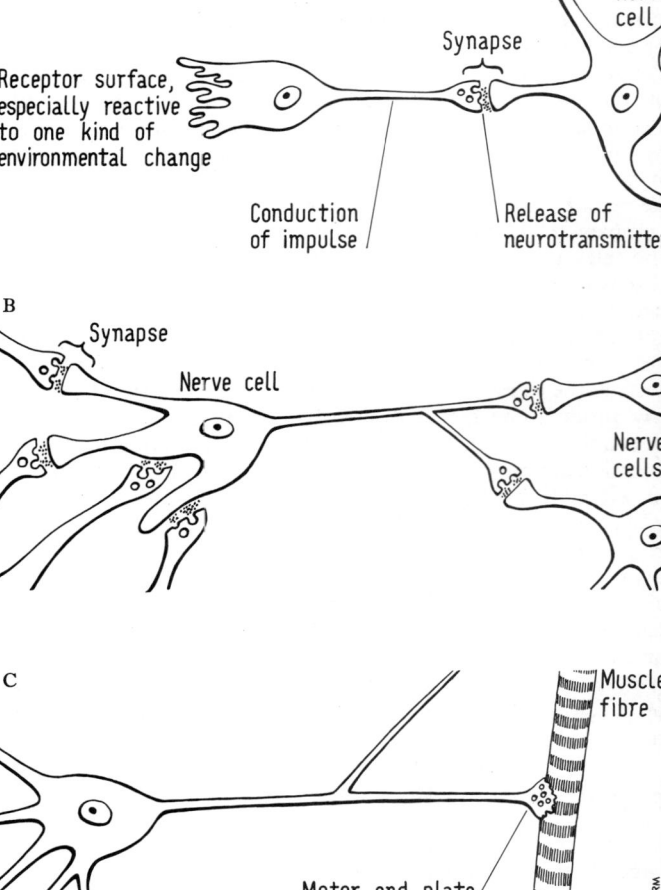

7.5 The three main avenues of differentiation which may be followed by a primitive nerve cell: A. a receptor neuron; B. an interneuron; C. an effector neuron. See text for further discussion.

separated by a *synaptic cleft*. In the great majority of synapses transmission is unidirectional; the synapse is *polarized* and bounded by a *presynaptic* and a *postsynaptic* membrane. As a nerve impulse spreading over the surface of a presynaptic cell (receptor or neuron) approaches the presynaptic membrane, it causes the release of a *neurochemical transmitter* which rapidly diffuses across the synaptic cleft to the postsynaptic membrane, whose ionic permeability it alters. The transmitter is often removed or broken down almost at once by a specific enzyme or is removed by neighbouring cells and inactivated. The type of change in postsynaptic permeability varies with the chemical nature of the transmitter and the detailed composition of the membrane at a synapse. In some cases the change induces a flow of ions to *depolarize* the postsynaptic membrane, i.e. to *reduce* the electric potential difference due to the asymmetrical distribution of ions across a 'resting' cell membrane. Such synapses and transmitters are *excitatory* with respect to postsynaptic cells (vide infra). Other neurotransmitters *hyperpolarize* postsynaptic membranes and are *inhibitory*. These effects, as recorded from postsynaptic cells, are respectively, *excitatory* or *inhibitory postsynaptic potentials* (EPSPs or IPSPs) (pp. 865, 886). In other examples more subtle and complex interactions may occur which may render the postsynaptic membrane less, or more, excitable (or inhibitable) when another neurotransmitter reaches it. These effects are termed *neuromodulation* and not only may occur at synapses in the strict sense but may also be effected at considerable distances from the site of chemical release, which therefore may reach a number of different neurons by diffusion (*paracrine secretion*, see p. 887). To clarify some of the features of these interactions, the general structure of neurons will now be considered.

NEURONS

Neurons are excitable cells, specialized for the *reception, integration, transformation* and *transmission* of coded information. Vertebrate neurons consist of a *cell body* (the nerve cell or *soma*), a mass of specialized cytoplasm with a diploid nucleus and an excitable membrane, from which project one or more *neurites*. These are branching, cytoplasmic processes, enclosed by excitable plasma membrane, which extend for varying distances from the soma. Neurons are classified as *unipolar, bipolar* or *multipolar* according to the numbers of these extensions (7.6A–C). Most neurites conduct towards or directly influence the soma, as *dendrites*; only one (which may branch) conducts away from it, the *axon*.

Neurons vary in shape, size and interconnections, in different species and regions of a nervous system, as will be described later. Here it is to be noted that the dendrites of many bipolar or unipolar neurons either synapse with specialized, peripheral receptor cells or, by means of their dendritic terminals, themselves function as specialized receptors. Almost all neurons of

vertebrate *central* nervous systems (brain and spinal cord) are multipolar (7.7).

The surfaces of dendrites and somata provide large areas for synaptic contacts from other neurons. These may be sparse but usually number many thousands, usually derived from the axonal branches of hundreds of other neurons. Such presynaptic neurons may be widely scattered and highly varied paths may *converge* on a single neuron.

Synapses may thus be *axodendritic* or *axosomatic* in their siting on the postsynaptic membrane; less common are: *axoaxonic* synapses, involving initial axonal segments or presynaptic

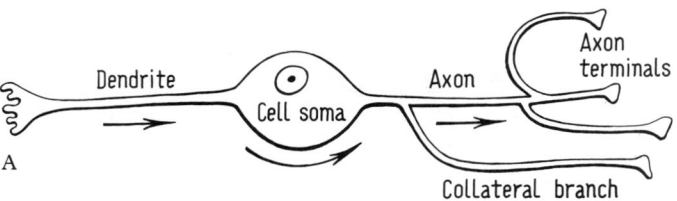

BIPOLAR NEURON

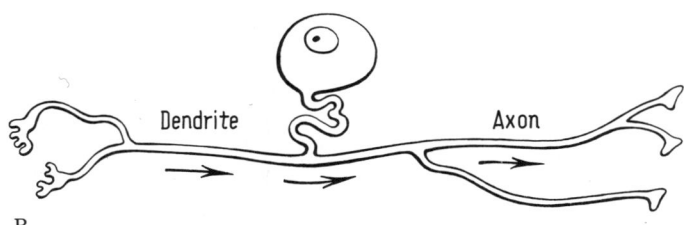

UNIPOLAR NEURON

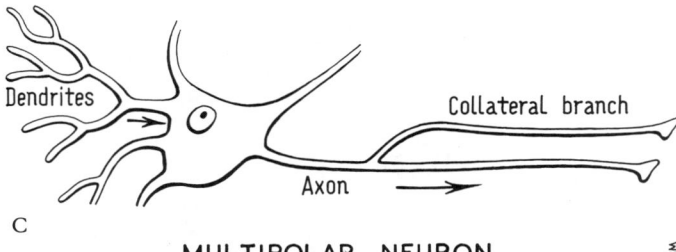

MULTIPOLAR NEURON

7.6 Three general morphological groups of neurons classified according to the number of neurites which arise from the surface of the cell soma: A. bipolar neuron; B. unipolar neuron; C. multipolar neuron. For a more detailed analysis of the branching patterns of neurites, see 7.17.

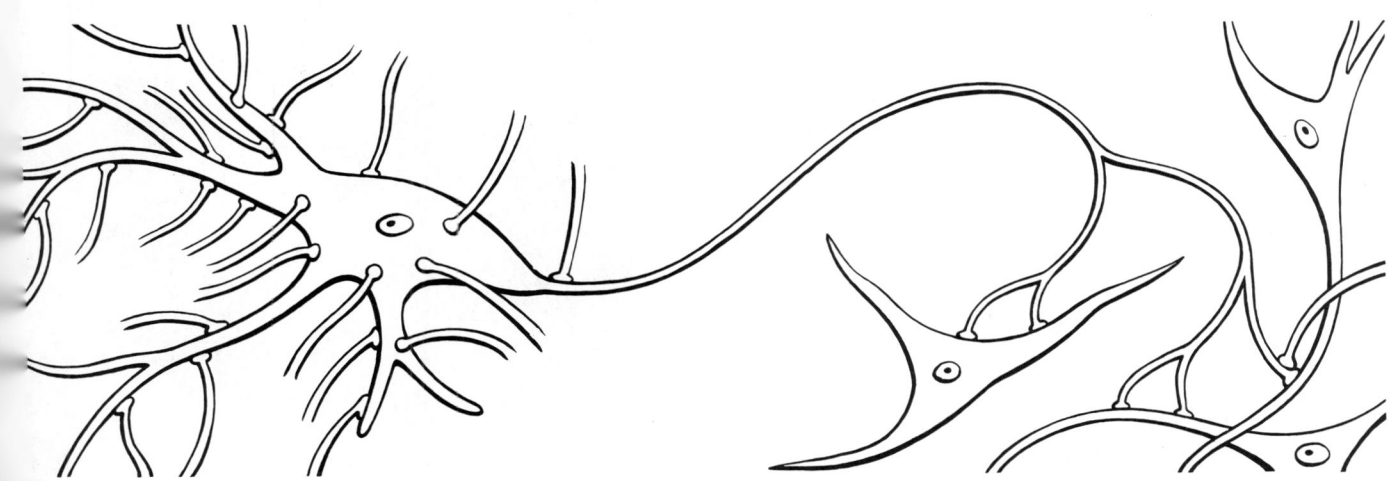

7.7 A generalized multipolar neuron and the parts of its surface upon which synaptic terminals from other neurons may converge. These include the surfaces of the dendrites, the cell soma, the initial segment of the axon and the proximal surface of its own axonal synaptic terminals. For a more detailed representation of the cytology and surface contacts of a multipolar neuron see 7.20.

FEED-FORWARD INHIBITION

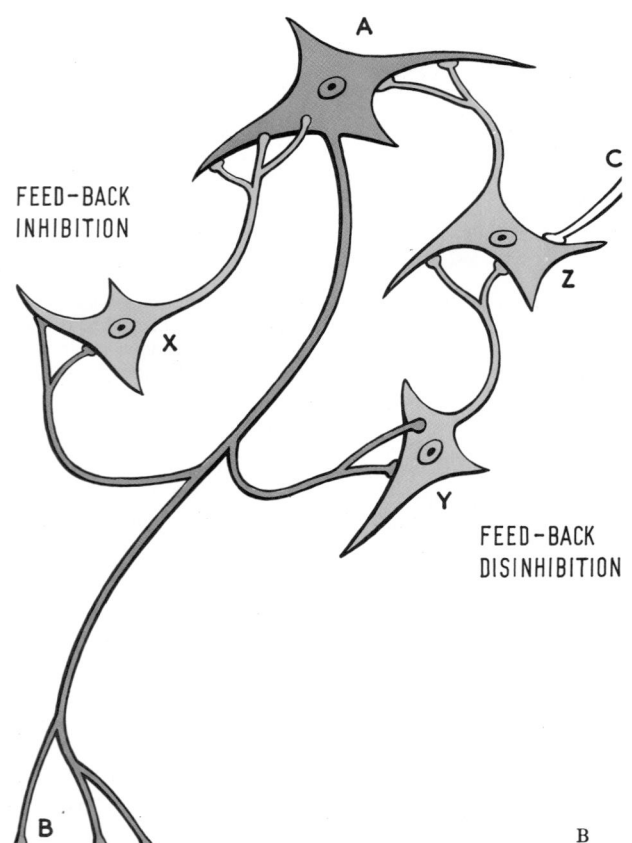

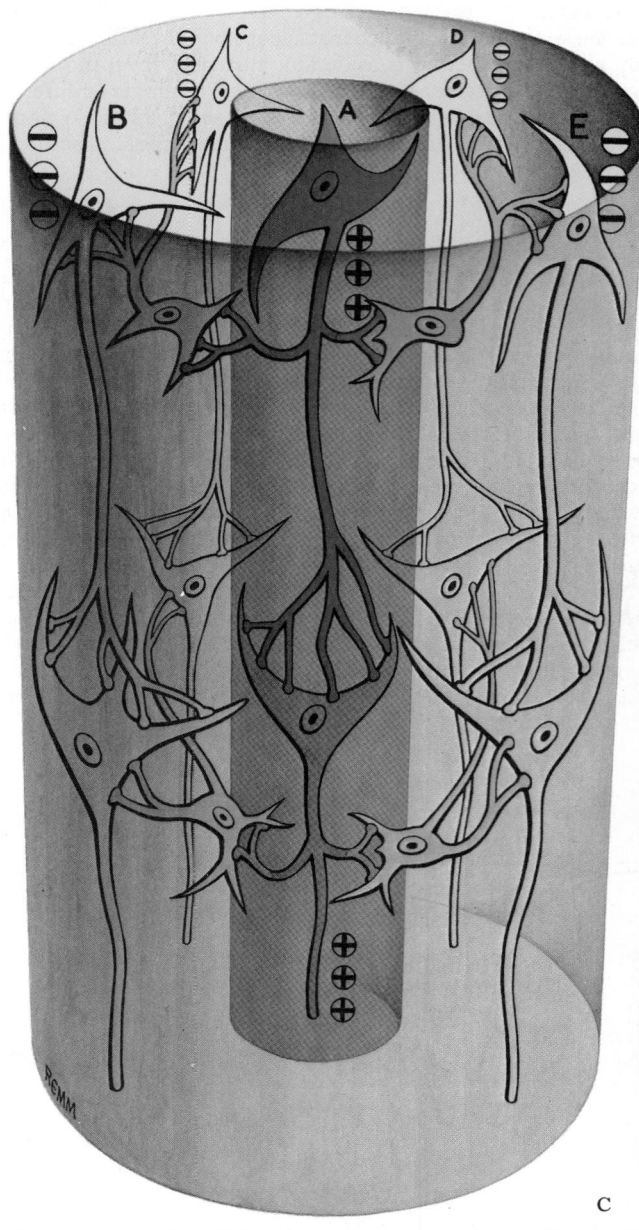

terminals of the neurons concerned, *dendrodendritic*, *reciprocal* and *serial* synapses and synaptic *cartridges* and *glomeruli* around neurites, aggregated in complex geometrical patterns. Glomeruli are *higher integrative units* or *microcircuits*. Cytological details of these and their possible modes of operation are described in pp. 877, 888.

Of the numerous synaptic terminals clustered on dendrites and soma of a multipolar neuron some are *excitatory* while those from other sources are *inhibitory*. Depending on the activity or quiescence of such sources, the ratio of active excitatory and inhibitory synapses continuously varies. Their effects summate; the *excitatory state*, i.e. the levels of depolarization or hyperpolarization of the cell membrane, also fluctuates. When depolarization reaches *threshold level*, a local electrochemical current flows, encircling the dendrites and soma; if it is sufficient to alter the permeability of the *initial axonal segment*, an action potential is generated and spreads along the axon as a *nervous impulse*. If such excitation is maintained, a *volley* of impulses traverses the axon and its branches. Since a neuron's excitatory level *continuously* reflects numerous variations in the patterns of paths converging upon it, it resembles an *analogue computing device*; but

7.8 Examples of elementary circuits (excitatory neurons shown in red, inhibitory neurons in blue): A. a direct excitatory circuit (on the left) is compared with a simple feed-forward inhibitory circuit (on the right); B. feed-back circuits of two orders of complexity with one (on the left) and two (on the right) inhibitory interneurons interposed on the recurrent pathway; C. lateral inhibition. A central excitatory train of neurons i surrounded by a hollow cylindrical zone of inhibition, mediated by shell of inhibitory interneurons which are activated by the central column. Se text for further discussion.

the outflow, the axon, works on an *all or none* basis, being either quiescent or conducting discrete, identical impulses; only their frequency varies. In this the axon acts like a *digital computer*.

Axonal terminals of a particular neuron may be few, concentrating their effects on one or only a small group of postsynaptic cells. Sometimes *collateral branches*, which may be numerous, leave an axon along its course, each arborizing and *diverging* to many destinations. The 'many to one' *convergent* channels integrating many different effects and the 'one to many' *divergent* channels are both basic features of all advanced nervous systems.

INHIBITORY CIRCUITS

Some elementary circuits, proposals from microelectrode studies, are illustrated in 7.8A–C.

In **feed-forward inhibition** (7.8A) an excitatory neuron (X) is linked with two others (Y and Z). However, it is clear that when X is active, the interposition of an inhibitory neuron on the second path reduces the excitatory level of Z, while that of Y rises. In reality other local circuits would also be active but the principle illustrated here probably accounts in part for, e.g. the increase in the force of contraction of one muscle group, while the contraction in an antagonistic group is progressively reduced.

In **feed-back inhibition** (7.8B), impulses from a multipolar neuron (A) traverse its axon to the main synaptic terminals (B). The persistence of the volley is limited by a feed-back loop: an axonal collateral synapses with an inhibitory neuron (X), whose terminals go back to synapse with the dendrites and soma of A, in effect terminating the volley. Such feed-back circuits vary in complexity; in some *two* inhibitory neurons (Y and Z) may form a recurrent loop; the inhibitory effect of neuron Z on A, perhaps from an alternative source (C), may be diminished by the inhibitory neuron Y. Such a release from an inhibitory effect by a second neuron, in series with the first, is called *disinhibition*.

In 7.8C is shown a form of feed-forward inhibition, *lateral inhibition* or an *inhibitory surround*, which is prominent in many sites in the nervous system: parallel paths A–E, carrying functionally similar information, may be imagined to transmit sporadically with a low information content; if the excitatory state of one (e.g. A) significantly exceeds that in the others, activity in B–E is reduced by adjacent 'surrounds' of interposed inhibitory neurons. The central activated path is thus surrounded by a quiescent zone, reducing confusion caused by random activity in the others. The discriminative value of the central channel is thus greatly increased, a phenomenon termed *neural sharpening*, examples of which will be referred to later.

Two further sites of inhibition occur (7.9), though less commonly and with completely contrasting roles. An inhibitory terminal at an initial axonal segment is well placed to *prevent* outflow from a neuron, while *presynaptic inhibitory terminals* may

selectively inhibit at some terminals while others remain unimpeded; examples will also be encountered later.

It should be stated here, however, that our knowledge of the detailed geometry of dendritic trees, cell somata and axonal branching, of three-dimensional interneuronal connections and the distribution of synaptic types, are all very limited, especially in quantitative terms. How individual neurons and co-operative groups transform patterns of information and how they relate to the operation of the whole nervous system and the organism's overall economy remain enigmas. For this reason, much current research is concentrated on the developmental mechanisms of neurogenesis, to determine which neuronal features are specified by the genome and which are modifications occurring during an organism's life.

Some General Features of Neural Organization

An outline of some aspects of the overall patterns of organization in simpler nervous systems is a necessary introduction to complex arrangements in vertebrates, especially humankind. This account, which is too brief to reflect a phylogenetic history, will be limited to the organization of simple neuronal arrays.

In protozoans a single cell acts as receptor, conductor and effector but simple multicellular forms possess *independent effectors*, such as contractile cells around pores in sponges, or the stinging organs of coelenterates. The latter are the simplest metazoans with a nervous system: in various species different orders of complexity can be recognized (7.10). In some sites special receptor cells among other epithelial cells send a conducting process from their bases to contact underlying effectors (myocytes); but many processes of sensory cells discharge into a subepithelial *rete* or *network* of neurons interconnected in a tangential plane; from this neural processes pass to the effectors. Connections or 'synapses' between cells in the network allow stimuli at any point on the surface to activate large areas of the nerve net, to cause widespread motor effects. Superficially the network appears random (7.10C); but some regions have more inputs, some parts of it are denser and in some regions conducting fibres are longer and have a distinct orientation: features which presumably reflect the more complex behaviour in capturing prey or in escape in these simple organisms; these structural patterns foreshadow similar fundamental arrangements which occur in more complex nervous systems, including those of Mammalia.

In simple organisms with *bilateral symmetry* we see the emergence of a *central nervous system* (7.11). Receptors are still largely situated in the surface layers but their basal conducting processes are longer, traversing other tissues and passing towards the median region, where many neuronal somata and their neurites are concentrated as a central nervous system, in which synaptic ends of receptor cell processes terminate. The somata of motor (effector) neurons are located in this concentration, their axons leaving it to reach the effectors. Other neurons connect with receptor terminals and motor neurons. They may have short or long axonal processes and constitute *interneurons* (*intercalated* or *internuncial neurons*). Paths of varying complexity thus connect the receptor cell input to the motor output; some of these pathways are *polysynaptic*, with a chain of interneurons, while in others there is a direct receptor–effector *monosynaptic* path. It is in such a diversity of *connectivity* that the variations in speed and complexity of behavioural response to environmental conditions reside.

In higher invertebrates, with *metameric segmentation*, arrangements are similar. Receptors are still largely epithelial and their *centrifugal* nerve fibres collect into bundles as *sensory nerves* in each segment before entering the central nervous system, which contains masses of interneurons and motor neurons, usually aggregated as a ventral pair of 'ganglia' in each segment. Each of these pairs is connected by axonal processes *across the midline* and also with those of adjacent segments by bilateral groups of *longitudinal* axons, forming a ladder-like structure. A series of *motor nerves* emerges in each segment. The transverse connections are formed by axons of interneurons to mediate bilateral responses.

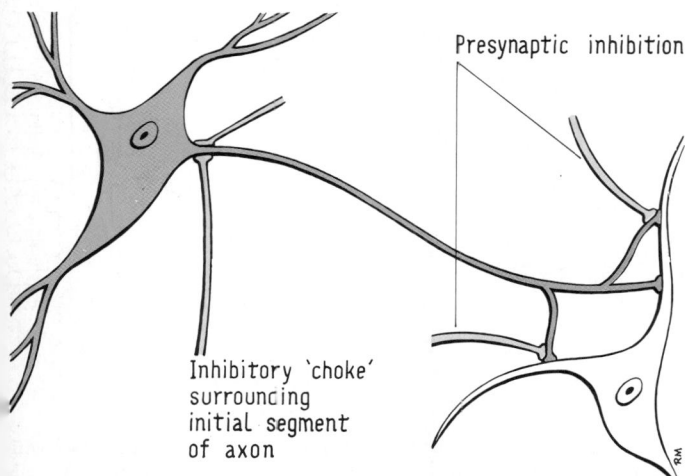

Presynaptic inhibition

Inhibitory 'choke' surrounding initial segment of axon

7.9   The multipolar neuron, (shown in red), has inhibitory synaptic terminals (blue) applied to the initial segment of its axon and others to the surfaces of its own axonal synaptic terminals.

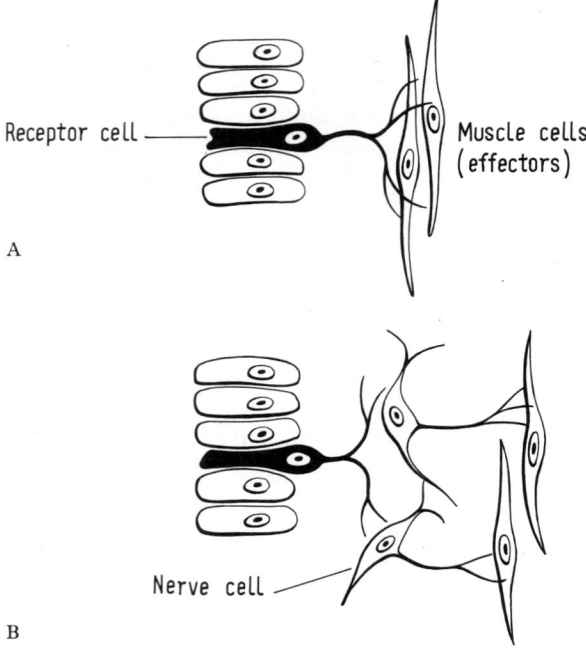

A

Receptor cell

Muscle cells (effectors)

B

Nerve cell

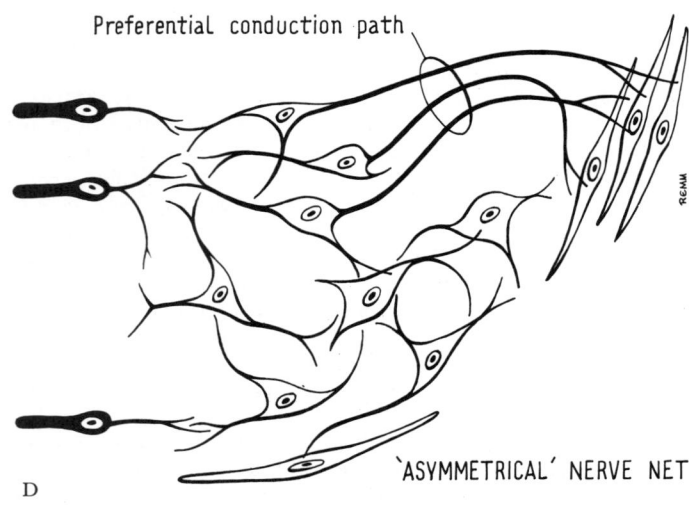

C

Receptor

Receptor

Muscle cell effector

'SYMMETRICAL' NERVE NET

Preferential conduction path

D

'ASYMMETRICAL' NERVE NET

Short axons connect adjacent segments in both directions and longer axons connect more distant segments, including an increasingly important 'head' end whose receptors often dominate behaviour.

During embryogenesis (p. 133) in all classes of *Chordata*, including the mammals and humankind, the central nervous system develops in the dorsal midline as a hollow tube-like structure containing a *central canal*, expanding cranially to form a *brain*, with extensions of the central canal termed *ventricles*. In all vertebrates the brain and spinal cord consist of interneurons and proportionately less numerous motor neurons, with non-excitable glial cells and blood vessels. Two regional varieties of nervous tissue appear, *grey matter* and *white matter*, distinctly recognizable in the freshly cut brain or spinal cord.

Grey matter consists of the somata, dendritic trees and initial axonal segments of neurons and the terminal segments and synaptic endings of axons reaching them from receptors, with the glial cells and blood vessels between these various structures. Some short axons are confined to the grey matter but many leave it in bundles or *tracts*. In the **white matter** there are no neuronal somata or dendrites but numerous nerve fibres, many of them ensheathed by myelin (p. 902). These sheaths are essential for the

← BILATERAL SYMMETRY →

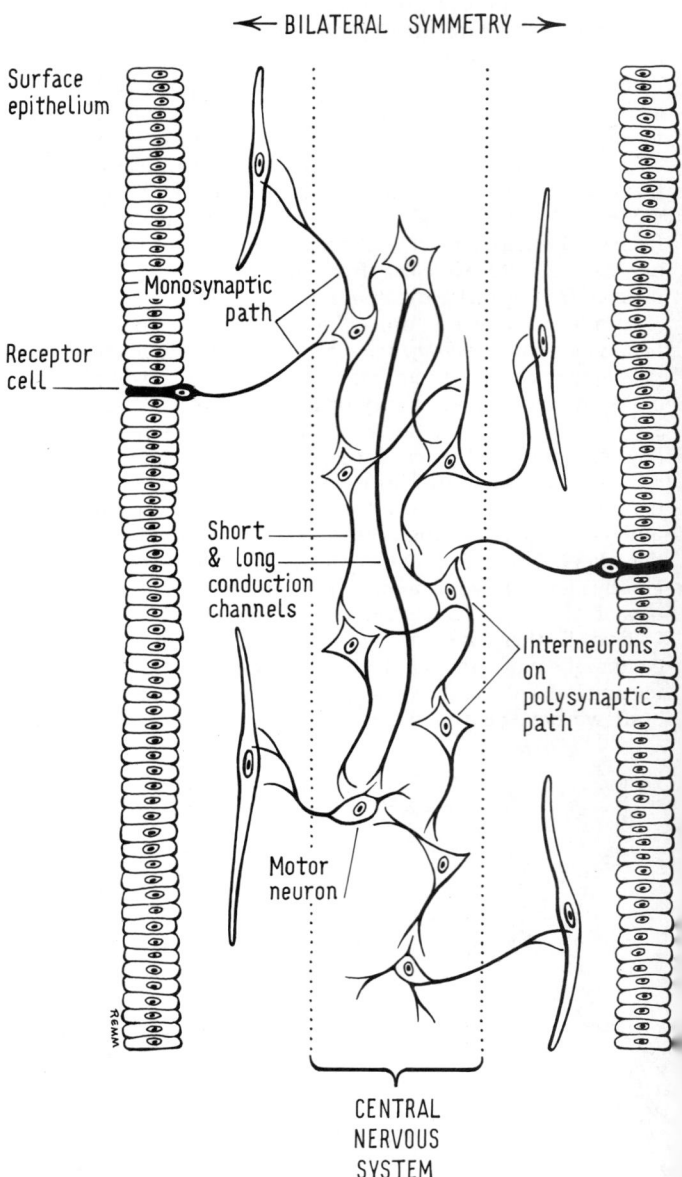

Surface epithelium

Monosynaptic path

Receptor cell

Short & long conduction channels

Interneurons on polysynaptic path

Motor neuron

CENTRAL NERVOUS SYSTEM

7.10 Simple nervous systems: A. direct interaction between a receptor cell and the muscle cell effectors; B. interneurons which make functional contacts with each other are interposed between the receptor cell terminals and the muscle cell effectors; C. receptor cell terminals feed into a fairly symmetrical network of interneurons which in turn makes contact with the effectors; D. the interneurons between the receptors and effectors are organized where 'diffuse' multisynaptic conduction pathways predominate and others where more direct routes are present.

7.11 Some features of the nervous system in a primitive organism with bilateral symmetry. Note the intra-epithelial positions of the receptor cells and the aggregation of interneurons and motor neurons near the plane of symmetry as a central nervous system. Note also the different orders of complexity of the conduction pathways and the possibility of cooperative actions involving both sides of the body.

efficiency of conduction, enhancing the velocity for a given diameter of nerve fibre and reducing its energetic requirements (p. 901). They impart a pinkish-white opalescence to fresh tissue: hence the term white matter.

The central canal in vertebrates is largely lined by ciliated columnar epithelium termed *ependyma*; in early embryogenesis the grey matter develops around it, with white matter more peripherally placed. At first this arrangement is found throughout

the neural tube and in the spinal cord the relationship persists; but with the development of hind-, mid- and forebrain vesicles some masses of grey matter remain in the primitive peri-ependymal position, while others migrate outwards between the developing tracts. Moreover, profuse secondary migrations of developing grey matter occur into the roof of each brain vesicle, where they form a superficial series of laminae, structures of greatest significance in the higher evolution of the nervous system.

TECHNIQUES AND DEVELOPMENT OF NEUROANATOMY

The previous section has been largely an analysis of nervous systems, based on a long accumulation of data, hypotheses and explanations by a succession of observers and experimenters (see e.g. the Anglo-American Symposium 1958, Brazier 1959, Singer & Underwood 1962, Clarke & O'Malley 1968, Meyer 1971, Bellairs & Gray 1974, Neurosciences Study Programs 1967, 1970, 1974). All were limited by the technical means available. It is hence appropriate to note the *methods* by which major advances, leading to our current knowledge, have been achieved, with a brief survey of neuroanatomical and other relevant techniques. The names of innovators should perhaps be kept alive by habitual citation but this is scarcely practicable in the textbooks of today. The ensuing pages may help to correct this deficiency.

Progress in neuroanatomy, always dependent on the contemporary means of observation and experiment, was initially slow: about two millennia separate the earliest recorded observations from the advent of even simple lenses. The great advances in the nineteenth and twentieth centuries coincided with a great flowering of technical innovation. At first physiology lagged centuries behind anatomy, a frustration to all who found function the chief interest of structure; but in the explosion of technical progress functional studies have overtaken the purely structural in many fields, sometimes carrying concepts of function beyond structural knowledge. In neurology this provided a much needed intellectual stimulus to morphologists, as indeed it continues to do.

FROM THE GREEKS TO THE RENAISSANCE

Even allowing for the fact that only simple means were available, the paucity of neurological observation or even (recorded) speculation before ancient Greek times is surprising. Injury, sacrifice and mummification provide obvious opportunities for such studies but the earliest descriptions are limited to optic nerves (Alcmaeon *c.*500 BC) and the recognition in the Hippocratic Collection (*c.*400 BC) of the cleft between the cerebral hemispheres, with a belief that the brain is the seat of the intelligence. Even Aristotle (384–322 BC) confused nerves with tendons (with effects still visible in the term 'aponeurosis') and he set intelligence in the heart, unlike Plato (429–347 BC), who favoured the brain. Biology was already permeated by the Four Elements of Empedocles (*c.*493–433 BC) and the Four Humours of Polybus (*c.*390 BC). The latter were to dominate and obstruct biological thought for almost two thousand years.

Recorded human dissection began in the rising Alexandrian School, led by Herophilus and Erasistratus (*c.*300–250 BC), still regarded (somewhat vaguely) as 'father figures' of anatomy and physiology. How far Egyptian embalming influenced early dissection is unknown (Cave 1958). Herophilus distinguished the cerebrum, cerebellum and fourth ventricle, even its calamus scriptorius, also differentiating motor and sensory nerves. Erasistratus detailed the ventricles, entangling their functions in humoral concepts, his teaching complicated further by the idea of *pneuma* or vital spirit, an impediment lingering long after the Renaissance; but he equated elaborate human gyri with intelligence, a prophetic speculation.

With the Roman invasion (50 BC) the Alexandrian School decayed; but Galen (AD 129–199) studied there and in Rome. His role, as dissector and experimentalist, is well known (Sarton 1954, Singer 1956). He classified cranial nerves, with some errors,

uncorrected in the ensuing 1500 years. He was dogmatic and theoretical but his work on the spinal cord was based on *experiment* (and not improved until Bell and Magendie's observations in the nineteenth century). Section between the first two vertebrae, he concluded, led to instant death and between the third and fourth to respiratory arrest, while at lower levels it caused paralysis of the bladder, intestines and legs. Much of his prolific output was pervaded by humoral and vitalistic ideas, a serious brake on progress for many centuries, due to the veneration accorded him. After Galen, biology and medicine entered a long suspense, the classical legacy being all but extinguished in Europe. Greek texts survived in Arabic translation but with little addition to knowledge. In the general intellectual lethargy, neuroanatomy marked time for a thousand years; indeed, its accuracy decayed. Astrology and the early Christian authority's insistence on the unimportance of man's physical estate displaced interest from his own true fabric. From the eleventh century, Greek medical manuscripts were translated into Latin, at first from Arabic. Though this reawakened European reason, progress was slow, centuries remaining barren of biological advances. Hippocrates, Aristotle and especially Galen were the almost unchallenged sources of medieval anatomy. Even in nascent universities of the twelfth and thirteenth centuries, natural phenomena were neglected; achievements were literary. Dissection reappeared and was practised at first to confirm established dogma. Even the *Anathomia* of Mondino (1316), often styled the first modern text, ascribed the functions dictated by the antique teachings to the brain ventricles, adding nothing to neuroanatomy. Though dissection continued in Bologna, spreading in Italy and to Montpellier, the fourteenth and fifteenth centuries brought little more than direct translations from Greek. The sixteenth century anatomical Humanists were absorbed in the replacement of Arabic anatomical terms by Greek and Latin; but great exceptions from these drab polemics were at work.

FROM THE RENAISSANCE TO THE NINETEENTH CENTURY

The naturalist movement among Renaissance artists in the late fifteenth and early sixteenth centuries culminated in Leonardo da Vinci (1452–1519) (see Belt 1956). Like Dürer, Michelangelo and Raphael, he turned to dissection for direct observation of human structure. Though his contemporary influence is uncertain, he brilliantly symbolizes the revival of enquiry in Europe, rebelling against traditionalism and scholastic orthodoxy. His contribution to neuroanatomy, though limited, was impressive: remarkable figures of the brain depicting wax casts of brain ventricles, a technique invented by him. The movement represented by Leonardo, with the improved techniques of engraving, paved the way for Vesalius.

Vesalius (1514–64) has received almost excessive adulation; his plates, starkly structural compared with Leonardo's functional style, marked a new era of accuracy (see O'Malley 1964). Impatient of Galenic errors and the near sanctity accorded them despite ascertainable facts, he revised human anatomy by dissections. In neuroanatomy he excelled in detail and precision, with somewhat lesser success in describing the peripheral nerves. He sectioned the brain (an innovation), showing the basal nuclei, hippocampus, fornix, internal capsule, pulvinar, colliculi, fourth

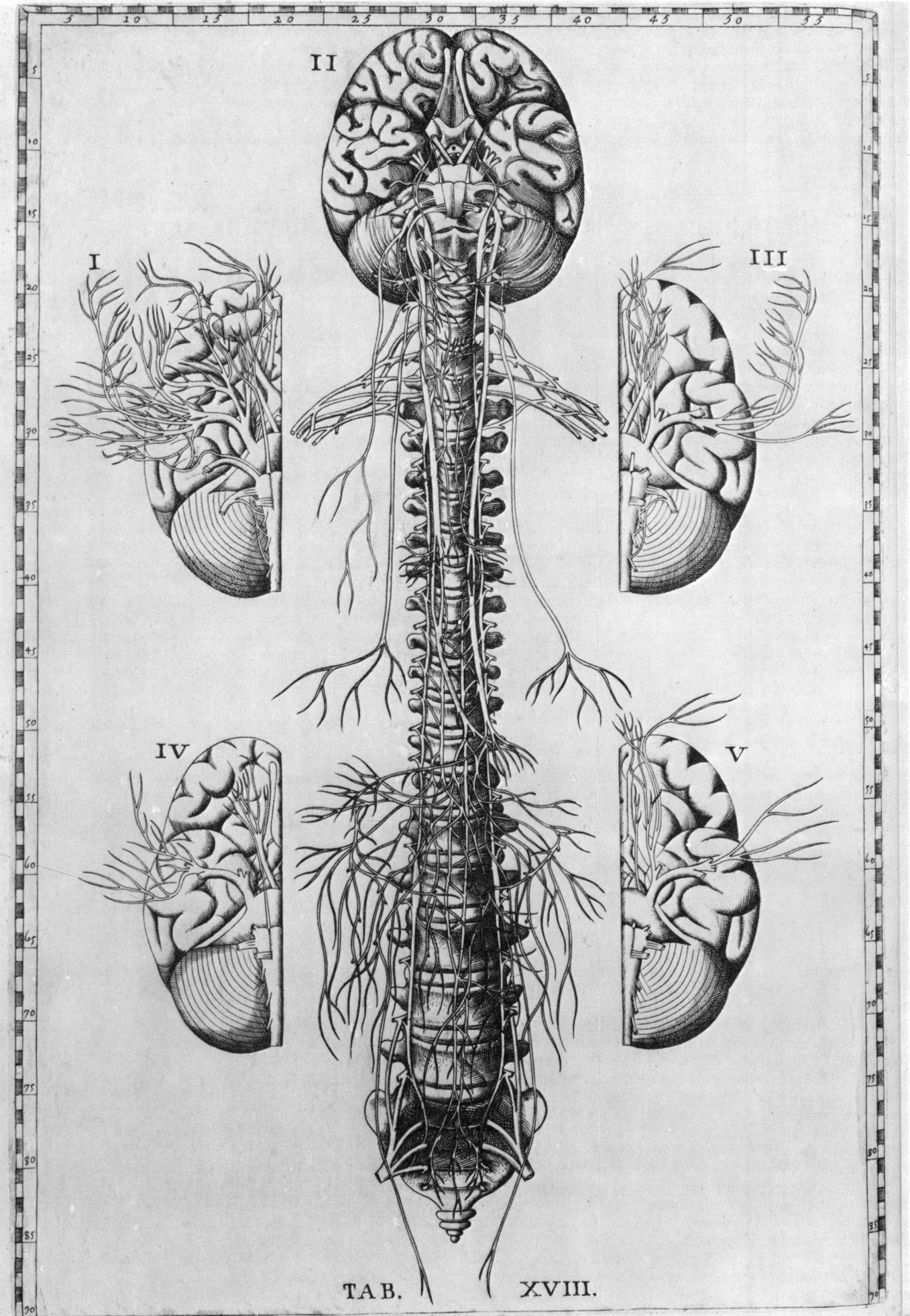

ventricle and many other details, for the first time with almost modern accuracy. Eustachius (1550–74), a contemporary, has received less than just acclaim; his work, like Leonardo's, became generally known only after his death. He depicted the sympathetic nervous system (7.12) and cranial nerves more accurately than Vesalius and the pons (Varolii) more so than Varolius.

Vesalius' less famous successors at Padua nevertheless exerted much influence on visiting scholars. Thus, Coiter of Montpellier (1534–76), an early comparative anatomist, studied under Fallopius, improving on Vesalius and Eustachius by describing the roots of spinal nerves and distinguishing white and grey matter. Fabricius taught Harvey (1578–1647), whose discovery of circulation was oddly combined with Galen's views in other matters, though he did reject his flow of 'spirits' in hollow nerves, a break with dogma worthy of his contemporary, Bacon (1561–1626), whose contempt for scholastic orthodoxy and inductive approach to experiment and factual verification of hypothesis were the first advance beyond Aristotle's logic. This new attitude, though philosophy rather than technique, has perhaps influenced subsequent endeavour as deeply as Harvey's experimental genius. After them the tempo of discovery increased, the focus on 'humours' in blood or nerves, as avenues of communication, slowly transferring to the nervous system.

So far study was only by dissection and the simplest experiments, with the unaided eye; but during the seventeenth and eighteenth centuries discoveries in parallel fields, especially of electrical phenomena and optics, were to become of basic importance in neural investigation. The experiments of Gilbert (1540–1603) on magnetism foreshadowed the electrical stimulation of nerves by Galvani (1737–98) and Volta (1745–1827), heralding neurophysiology. Simple microscopes enabled van Leeuwenhoek (1632–1723) to dispute the hollowness of nerves; compound instruments soon followed, notably exploited by Hooke (1635–1703) but not widely used in neuroanatomy till the early nineteenth century.

Meanwhile topographical neuroanatomy prospered; Willis (1621–75) and Vieussens (1641–1716) hardened the brain by soaking in wine or boiling, allowing its dissection, an advance upon slicing methods. Willis pioneered macroscopic dissection of the brain, the only technique available for almost two centuries. He also came near to formulating modern concepts of reflex activity. Late in the eighteenth century the speculations of Willis, von Haller, Whytt and others regarding reactions to stimuli were being resurrected. Microscopes were improving and in the nineteenth century techniques of fixation, microtomy and staining were evolving. Enquiry was turning from external features to internal structure; some neuroanatomists preferred comparative morphology and development, and workers were becoming more specialized, as also was technical progress.

COMPARATIVE ANATOMY AND EMBRYOLOGY

Though Erasistratus is sometimes dubbed the first comparative neuroanatomist, modern studies stem from Willis (1664) (Cole 1944, Needham 1959). Function rather than evolution became the primary consideration; in this Willis anticipated the nineteenth-century interest in localization, especially in the cerebral cortex. Gall and his pupil, Spurzheim, though achieving in phrenology a notoriety undeserved by Gall, initiated the concept of cortical localization (Gall & Spurzheim 1810–19). Tiedemann (1860) described human fetal brains, Reil (1807) development and comparative anatomy of the cerebellum and insula. Owen (1868) added the concept of cerebral weight and volume to comparative techniques. Others extended the range of species studied, especially primates, and the advent of Darwinism was a spur. Comparative and embryological data were amassed to

bolster up homologies in the now forgotten arguments of the 'localizationists'. All observations were macroscopic, but with Campbell, the Vogts and Brodmann (1909), extending into the twentieth century, microscopy of the cortex burgeoned. Vast cytoarchitectonic investigations, familiar in 'Brodmann maps' were later equalled only by Conel's on the developing cortex (1939–59).

Great advances in neurohistology and neurophysiology in the same period have overshadowed comparative neuroanatomy; but with new histological techniques classic work such as that of C. J. Herrick (1931) and Larsell (1937) on the cerebellum and Polyak (1957) on the visual system continued to appear. The compendia of comparative neuroanatomy by Ariens Kappers must also be noted (Kappers 1920–21, Kappers et al 1936).

Embryology and comparative anatomy were early associated with each other; Arnold (1842) described hippocampal development in primates; Wilhelm His (1874) distinguished 'complete' fissures and secondary sulci. More significant were the embryological studies of growth at the microscopic level; His (1887) demonstrated the axons of neuroblasts, foreshadowing the tissue culture studies by Harrison (1907). Held (1893), confirming neuronal individuality, nevertheless stated that this disappeared in maturation. Flechsig (1876) discovered that myelination proceeds (before and after birth) at different speeds in different tracts, thus introducing a *myelogenetic technique* for the investigation of connections.

DEVELOPMENT OF NEUROHISTOLOGICAL TECHNIQUE AND LIGHT MICROSCOPY
(See Clay & Court 1932, Brachet & Mirsky 1960, Clark 1961.)

Leeuwenhoek (1674) visualized nerve fibres with a simple lens, with which Swammerdam (1675) even dissected the nervous system of mayfly larvae. Jansen (1590) had already constructed a compound microscope, put to non-biological use by Galileo. Kepler is also credited with a similar invention, realized by Scheiner (1611). Huygens (c.1684) introduced a compound eyepiece. Malpighi (1666) is said to have been the first to see nerve cells with a microscope, but reconstruction of his crude microscope has demolished this claim (Clarke & Bearn 1968). Fontana (1781) described nerve fibres but did not distinguish between axons and their sheaths. Further advances depended on improved microscopy. Lister (1837) and Amici (1827) manufactured *achromatic* objectives; but Lister's pioneer efforts were sustained by no British interest, further development passing to Germany, where Abbe, with Zeiss, constructed *apochromatic* objectives in 1886, using the new Jena glasses and also improving the condenser and devising compensating eyepieces. Darkfield condensers (Wenham c.1853) and immersion objectives (Tolles 1874, Abbe 1878) accelerated refinement of late nineteenth-century microscopes. The immediate result was a flood of descriptions of animal tissues. Botanical *cells* had been portrayed long before by Hooke and so named by him (1665). Nerve cells, a discovery often accorded to Schwann in enunciating his cell theory with Schleiden (1839), are now considered first fully defined by Ehrenberg (1833); but undoubtedly Schwann first described the cells which bear his name. Progress now rested on the provision of new optical resources and techniques of preparation and staining.

Preparation of Material for Microscopy

Early microscopists used teased or squashed samples of fresh or, at most, partially preserved tissues to examine the structure of nervous tissue. Hardening being essential for sectioning animal tissues, Reil's introduction of alcohol (1809) for this purpose was an important step; Hannover (1840) used chromic acid as a hardener. Stilling (1842) used both, sometimes with freezing, on the spinal cord (1846) and cut transverse and longitudinal sections, also advocating serial sections, as did Virchow (1846) in studies of neuroglia. Formalin was later introduced by Blum (1893). Stilling & Wallach (1842) devised a simple microtome for cutting sections of tissue and this was improved by Welcher (1856); improvement of design made by Jung, Thoma and others soon followed and Von Gudden & Catsch (1875) were able to cut sections of whole cerebral hemispheres. Rutherford & Cathcart (1873) introduced

7.12 (*opposite*) The brain and autonomic nervous system depicted by Bartolomeo Eustachio (1550–74), a contemporary and rival of Vesalius (1514–64), from *Tabulae Anatomicae*, published posthumously in 1714 by Lancisi. The copperplates of Eustachio (or Eustachius) are not only anatomically more accurate than the woodcuts of Vesalius, but also technically superior. By courtesy of the Trustees of the British Museum.

freezing microtomes. By the 1880s semi-automatic microtomes such as the Cambridge Rocker produced thin serial sections, embedding in paraffin wax having been introduced by Klebs in 1869. Embedding in collodion (Duval 1879) and celloidin (Schiefferdecker 1882) followed and led to the use of such modern materials as soluble waxes, freeze-drying and cryostat techniques.

Tissue Staining Methods in Neuroanatomy

Much was achieved by the early nineteenth-century microscopists without any staining, although these methods were used sporadically from the beginning of microscopy (see the accounts by Baker 1960, Gurr 1965, Pearse 1968). Remak's early description (1836) of axons and their sheaths was made from unstained embryonic material. Using similar methods Purkinje (1837) confirmed this and also described the cerebellar neurons named after him; he used acetic alcohol for clearing and Canada balsam for permanent mounts. Waller's study of neural degeneration (1850) was made from teased nerves, unfixed and unstained, while Kuhne's discovery of motor endings (1862) resulted from the treatment of tissues with weak hydrochloric acid.

The first neurohistological stain was probably the carmine and gold method of Gerlach (1858), which he used to advance the *reticular* or *net* theory of neural organization. The controversy between the protagonists of this and the rival *neuron theory* reverberated down the rest of the nineteenth century into the twentieth. Though Faraday had discovered benzene in 1825 and Perkins the first aniline dye, mauveine, in 1856, the industrial development of dyes was quickly taken up in Germany, largely a by-product being a range of dyes available for histological use, from 1862 onwards (see Conn 1948). Deiters (1865) used carmine to distinguish between dendrites and axons. Ranvier's (1871) classic work on nodes depended on silver impregnation of chromic acid fixed tissue; Golgi (1878) further elaborated silver staining to visualize neurons in their entirety. It is interesting that Golgi's technique, nowadays usually applied only to small samples, was used by him, as by some recent workers, on whole human brains (see Kemali et al 1977). Ramon y Cajal (1908) adopted Golgi's methods, added some refinements and was thus able to accumulate such an overwhelming weight of evidence for the individuality of nerve cells that the neuron theory prevailed, though this proof is often ascribed to Waldeyer. Only with the application of electron microscopy to the nervous system was it possible to adduce more convincing evidence for the correctness of Ramon y Cajal's teaching. Even as late as 1933, a year before the first electron microscope was constructed, Cajal found it necessary to re-emphasize his views. He also used Nissl's methylene-blue method for staining 'chromatin granules' (vide infra); Ehrlich (1886) had used it, as the first vital stain, to delineate peripheral nerve fibres to their terminals; being an axonal stain, it has survived in modification for peripheral non-myelinated autonomic nerve fibres, although it has been largely replaced by more modern methods such as those using fluorescent dyes and horseradish peroxidase (p. 874).

Nissl (1892) discovered the *acute response* of the nerve cell soma to axonal damage, typified by the breakdown of Nissl or *chromatin granules (chromatolysis)* and other manifestations such as swellings of the soma. These retrograde changes are described on pp. 916–919.) This proved most valuable in tracing axons back to their somata. Weigert (1882) had already found a technique for staining *normal* myelin sheaths (consisting of pre-treatment with potassium bichromate followed by acid fuchsin) and was already using collodion for embedding, xylol for cleaning and balsam for mounting sections. Marchi's osmic acid method for staining *degenerating* myelin soon followed, an early histochemical technique (Marchi & Algeri 1885–86). Gudden (1870) had studied the later sequelae of neuronal damage, noting the atrophy of injured neurons, such changes being more rapid in young animals. Both Gudden and Nissl made considerable personal contributions using their own techniques; Gudden's work on thalamocortical connections was carried further by his pupil, von Monakow (1882), who also discovered the optic connections of the lateral geniculate body. It is noteworthy that both problems were re-examined by the same technique half a century later (Le Gros Clark & Penman 1934, Le Gros Clark 1936).

NEURON TRACING TECHNIQUES

The researches of Waller, Gudden, Golgi, Weigert, Marchi and Nissl were all published in the last 30 years of the nineteenth century, apart from the basic observations by Waller in 1850. His demonstration that damage to nerve fibres induces microscopically identifiable changes proved a most fruitful beginning for this brilliant period of neuroanatomy and neuropathology. His immediate successors established the broad basis of neuronal morphology and the major techniques for an unequalled volume of enquiry into the organization of the nervous system. Many modifications and some additions have followed but their methodology remained substantially unchanged until the introduction of the modern alternatives of retrograde or anterograde labelling with axonally transported dyes (vide infra).

The essential features of degeneration in neurites are shown in illustration 7.13. When the soma is ablated or fatally damaged, all of its neurites including the axon and its sheath degenerate and may finally disappear (p. 918); Wallerian changes can be shown by Marchi staining of the degenerating myelin. Terminal branchings of axons and synaptic boutons also degenerate but, being devoid of myelin, these do not stain by Marchi's technique. Glees (1946) and Nauta & Gygax (1951), with modifications by others, developed methods for staining them; and ultrastructural changes have ultimately been utilized.

It is difficult to place small experimental lesions in a mass of nervous tissue; there is no method of ablating *somata alone* within central nervous structures; in all experiments, dendritic trees, synaptic terminals from other neurons, blood vessels and other structures are also affected. Clearer results follow when axons are cut along their course; certain terms are used for various regions of the resultant degeneration, which will be briefly summarized here.

Firstly, effects confined to the parts of a damaged neuron *distal* to the lesion are said to be *anterograde*; those affecting the soma and axon *proximal* to the lesion are *retrograde* changes. Secondly, changes in undamaged neurons connected synaptically with the damaged one are *transneuronal (transynaptic)* changes and may be *primary*, *secondary*, etc, depending on the number of cells interposed; they also may *anterograde* or *retrograde*.

Following axonal damage, anterograde degeneration (of the Wallerian type involving its axon, axon terminals and myelin sheath) occurs distal to the lesion spreading in a wave towards the axon terminals. After suitable intervals, they may be examined by the methods of Marchi, Glees, Nauta, Fink-Heimer and by various histochemical or ultrastructural techniques. Proximal to the site of damage a fibre is often negative to Marchi staining, though Ramon y Cajal described retrograde Wallerian degeneration as far as the first node proximal to the point of injury; this was long disbelieved but it is now accepted that changes do sometimes occur in this direction and may extend to the soma as *retrograde atrophy*, accompanied by the degeneration of *all* neurites (although this late sequel of Nissl's *acute retrograde (chromatolytic) response* is not always followed by chronic atrophy). The degenerative changes in the fibres are thus *retrograde* in the sense that they occur proximal to the lesion but proceed *centrifugally* from the cell soma to the lesion.

The above phenomena are more marked in newborn or young animals, as Gudden noted. Brodal (1940) introduced a so-called *modified Gudden technique*, using young animals in which acute and chronic retrograde changes occur more rapidly. The most suitable methods of approach to any problem and their application require careful consideration and the timing of the interval between placing a lesion and the inspection of its sequelae is critical. Discussions of these problems are given by Glees (1961) and Brodal (1969).

The ultimate fate of a damaged neuron is not always the same and the following remarks apply to the long-term results of natural and experimental injury. When severely damaged, e.g. by focal lesions (vide infra), the whole neuron and all of its processes degenerate and disappear. When the damage is to the axon, an acute retrograde reaction may be followed by recovery over some weeks or alternatively by chronic retrograde atrophy. Such phenomena have been investigated quantitatively by Turner

EXPERIMENTAL NEUROANATOMICAL DEGENERATION TECHNIQUES

7.13 The sequelae of interruption of the fibres of central nervous system neurons. As shown, regeneration or atrophy may follow the response to injury, both in the neuron primarily damaged and in those affected transneuronally. Transneuronal effects may occasionally supervene in neurons more remotely associated with the damaged neuron. See text for further details.

(1943), Bodian & Mellors (1945) and others. If a neuron survives, its pattern of Nissl granules returns and the axon proximal to the injury recovers, its severed part being absorbed, sometimes very slowly; in the spinal cord degenerating tracts have been identified a year or so after injury (Smith 1951).

So far the responses to injury *in a single neuron* have been considered. Early this century, however, it was recognized that degenerative effects are sometimes transmitted from one neuron to another in the same functional pathway, as *transneuronal* or *transsynaptic degeneration*. Anterograde transneuronal degeneration occurs in the visual pathway: retinal lesions not only affect the fibres of the injured ganglion cells as far as the lateral geniculate body, but also the geniculate neurons with which they synapse; the process may even involve neurons in the striate cortex in a chain of *primary* and *secondary* transneuronal degenerations. This kind of response appears only sporadically and may be limited to neurons receiving impulses predominantly, perhaps exclusively, from a single source and are therefore difficult to use for experimental purposes. However, ventral grey column neurons receive afferents from many sources and yet may exhibit transneuronal degeneration when the dorsal spinal roots are cut; such claims

have been severely criticized but, however that may be, the technique offers obvious possibilities in tracing pathways.

Retrograde transneuronal degeneration has received less attention, although its occurrence in Gudden's cortical ablations is now generally beyond doubt; these cause degeneration in anterior thalamic neurons, whose axons reach the cortex, *and* also in other thalamic nuclei known to be in primary and secondary functional relation to the anterior thalamus (Cowan 1970). Gudden used surgical ablation, a method still used where appropriate, although more precise methods of making highly selective lesions have also been devised.

Selective Lesions and Stereotaxis

Tracing techniques require an accurate method of delivering accurate and controlled neural damage. Simple nerve section may suffice in peripheral nerves or superficial central tracts; crushing or repeated injury in the former is sometimes more effective. Superficial targets, such as the cerebral cortex, can be ablated or devascularized by removing the pia mater. Most targets are at some depth and the surgical approach may invalidate some experimental results by unplanned damage to other regions. This

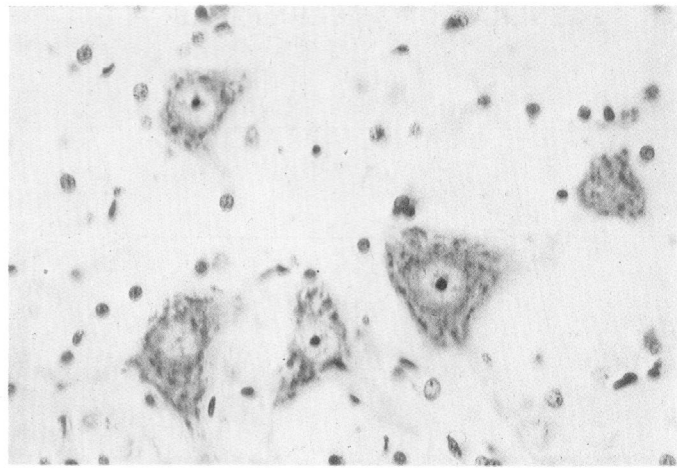

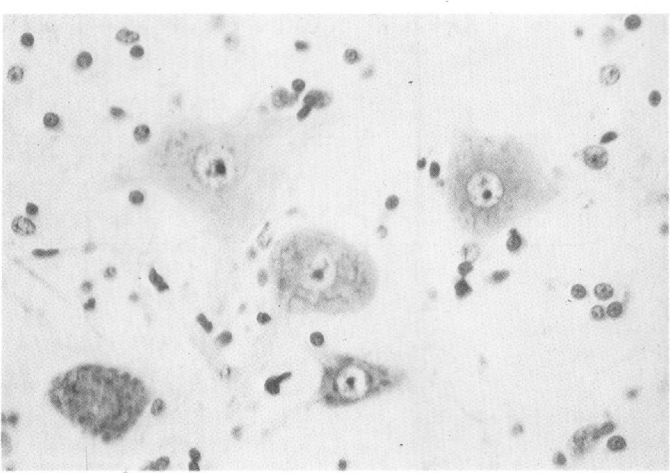

7.14 A.   Large multipolar neuronal perikarya in the magnocellular part of the feline red nucleus, showing prominent Nissl's granules, bases of dendrites and axon hillocks. The nuclei are euchromatic and vesicular, with prominent nucleoli. The small nuclei scattered in the surrounding neuropil are characteristic of the various categories of neuroglial cells.

B.   A field similar to that depicted in (A) but after previous contralateral hemisection of the spinal cord at the level of the fifth cervical segment, thereby severing the rubrospinal tract. The section shows characteristic chromatolytic retrograde changes in the cytoplasm of three large neurons. Two smaller neurons are unaffected. A and B. photography by Kevin Fitzpatrick, Anatomy Dept, Guy's Hospital Medical School, London.

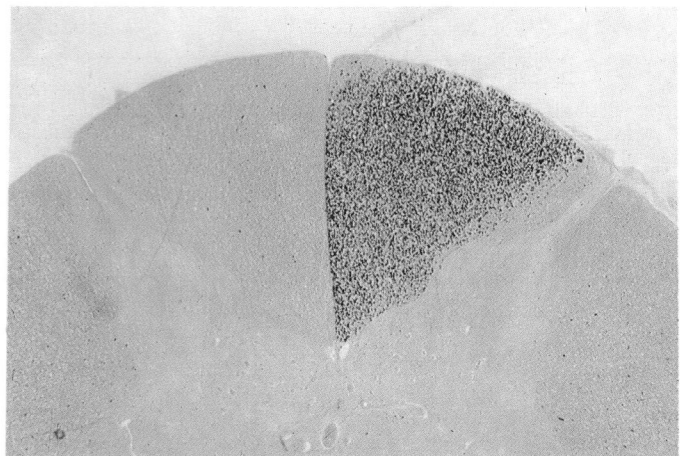

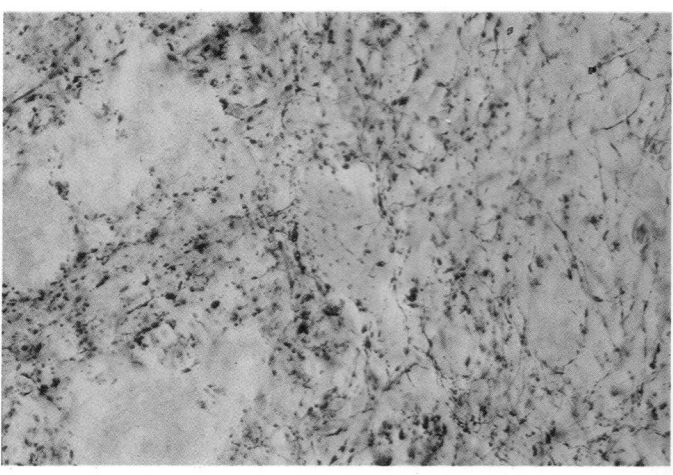

C.   Transverse section through dorsal funiculi of feline cervical spinal cord, after unilateral dorsal column section at a more caudal level. Note anterograde Wallerian degeneration of ascending nerve fibres: degenerating myelin sheaths are stained black by Marchi technique. Provided by E W Baxter, Dept. of Biology, Guy's Hospital Medical School, London.

D.   A high-power light micrograph showing preterminal degeneration of afferent axons ending in relation to neurons of the red nucleus of a rat, after the previous placing of a cerebellar lesion. The preparation was stained by the Nauta-Gygax method. Supplied by K E Webster, King's College, University of London.

difficulty has prompted other techniques, such as the use of fine electrodes, for stimulation *and* electrocoagulation, or of fine probes producing intense cold at their tips. Destructive substances such as alcohol, hydrocyanide, carbon dioxide snow or radioactive yttrium have been implanted. Focused ultrasound (Warwick & Pond 1968) or proton beams (Malis et al 1957) can cause deep damage without any disturbing entry, an advantage when large or irregular structures require repeated overlapping lesions. Antibody-antigen reactions have also been used (to destroy the caudate nucleus).

Most methods of making selective lesions entail the use of *stereotaxis*, i.e. some means of aligning an electrode, beam transducer or the like relative to a target (Carpenter & Whittier 1952). The heads of experimental animals are commonly fixed in a rigid frame in which the electrode, etc. can be adjusted in three planes to co-ordinates established by measurements in serial sections or by an X-ray guidance system, using air-filled ventricles or tracers for reference points. Since co-ordinate data vary a little from brain to brain, the accuracy obtainable is often limited, but is acceptable for all but very small structures. Horsley & Clarke (1908) constructed the prototype stereotactic apparatus; a

succession of improvements have followed and similar apparatus has been used in neurosurgery.

Modern Histochemistry, Electron Microscopy and Neurophysiology

The morphological approach must, of course, be complemented by other methods for an understanding of connectivity and general neural organization, and various combined techniques have been applied with much success in recent years. With improving neurophysiological instrumentation the study of electrical activity has undergone a revolution and the refinement of electrodes allows smaller groups of fibres and single cells to be examined. The techniques of *evoked potential* measurement by controlled stimulation at one point with recording elsewhere are also used extensively and the impact of computer-based analysis has been particularly effective in this area. It is also possible to examine the activity of neuronal populations by relatively non-invasive methods such as positron emission tomography which allows the behaviour of even small areas of the brain to be visualized in conscious human subjects.

These experimental approaches are often departmentalized as either anatomical or physiological but each inescapably contains an element of the other, providing complementary advantages and an important check on each other. With either approach the precise details of routes between points of stimulation or destruction and the sites of distant effects require repeated experiment for their full elucidation. The distance between sites of experimental targets may vary and intermediate blockage is used to elucidate pathways. It is often valuable to use the same electrode to stimulate and then to destroy a localized area, with immediate recording and study of subsequent degeneration.

By combination of all these methods a great volume of data concerning central and peripheral nervous connections has accumulated and continues to grow. More recently the intimate connections of single or small groups of neurons have attracted much attention. The insertion of microelectrodes or micropipettes into single neurons allows the recording of their activity and such cells can be marked with tracers for subsequent structural work by injecting them with dyes or other substances from the recording electrode. By recording simultaneously from several cells linked together in small 'integrative units' or 'microcircuits' and combining the results with those from detailed studies of neuronal connections, a new era of analysis has begun (e.g. Lissak 1967, Shepherd 1974, Rakic 1975, Schmitt et al 1976). As a result of such methods, complex neuronal interactions have been explored, including: excitation, inhibition, facilitation, disinhibition, disfacilitation, and pre- and postsynaptic inhibition, neuromodulation and the like, in a somewhat inelegant neology! The role of synapses in all these events is fundamental; for the analysis of the latter structures other structural methods have allowed more detailed explorations and histochemical (particularly immunocytochemical) methods and electron microscopy have provided techniques of great importance.

PHYSIOLOGICAL TECHNIQUES FOR TRACING NERVE FIBRES

By the 1900s the disciplines of structural analysis of neural tissue by microscopy were well advanced and the general features of nerve cells and fibres had been in large measure established, their types detailed and the neuron theory widely accepted. The overall cellular arrangement of the central and peripheral nervous systems, including their autonomic components, were defined in terms sufficient to be highly useful in clinical diagnosis, though the underlying mechanisms remained largely obscure. Much had been contributed by neurologists, both in the nineteenth century, when combined clinical and laboratory activities were perhaps easier, and in the twentieth, despite the difficulties engendered by specialization and the increasing complexity of experimentation. By this stage, also, neurophysiological studies were proving very successful in localizing function and reflex activity. The spectacular evolution of electronics was beginning to provide accurate instrumentation for examining the functions of nerve fibres, the behaviour of synapses and various aspects of integrated cerebral electrical activity. The study of synapses was even more potently stimulated by the discovery of transmitter substances (Loewi, Dale and others), such as acetylcholine and adrenalin, with immense physiological, biochemical and pharmacological consequences. With these brilliant advances, structural investigation continued with the impetus of new microscopic techniques, including histochemical developments and electron microscopy.

DEVELOPMENTS IN LIGHT MICROSCOPY

The light microscope has become a highly flexible instrument, with refined optics, dark-field illumination, polarization, phase contrast, contrast interference (Nomarski), ultraviolet fluorescence, and other auxiliary techniques including: warmed stages for tissue culture, recording cameras for still, cine and time-lapse photography, computer-linked devices for micromeasurement, microdensitometry and equipment for microdissection and micro-injection. The phase-contrast

microscope (Zernicke 1935) and its congeners, the interference (Francis Smith 1947) and contrast interference (Nomarski) microscopes have initiated a new era in cytology. Phase microscopy is extensively applied to nerve cells and neuroglia (Geiger 1963, Murray 1965). Interference and contrast interference microscopy can assess the refractive index, even in parts of one cell, and concentration, water content and dry mass in living cells (Ross 1967). Incident illumination of living tissue in situ is also a powerful method for analysis, e.g. of myelinated nerves (see, e.g. Williams & Hall 1970, 1971).

Ultraviolet microscopy can also offer increased resolving power and, far more importantly, selective absorption of frequencies by various substances, especially nucleoproteins. Caspersson (1936) evolved a technique applicable to single cells, which was widely applied in quantitative assessments of DNA and RNA in nerve cells by Hyden (1960), although little used in more recent years.

NEUROHISTOCHEMISTRY

Phase, interference and ultraviolet microscopy are adjuncts to histochemistry, especially cytochemistry. Histochemistry perhaps began with Raspail, a botanist, as long ago as 1830 but accelerated greatly with the discovery of methods for the identification of biological compounds (Pearse 1968). Many such tests are available for nervous tissues, both normal and pathological (Adams 1965). Specially valuable are the methods for identification of transmitter substances and associated enzymes in nerve

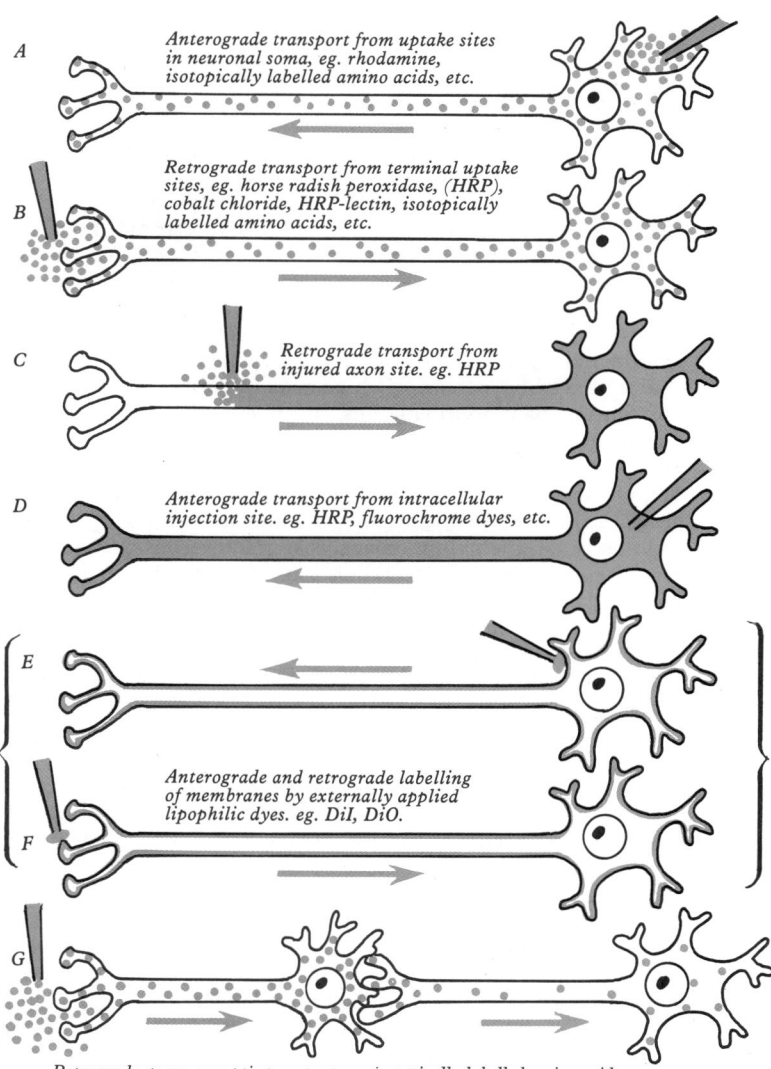

A   *Anterograde transport from uptake sites in neuronal soma, eg. rhodamine, isotopically labelled amino acids, etc.*

B   *Retrograde transport from terminal uptake sites, eg. horse radish peroxidase, (HRP), cobalt chloride, HRP-lectin, isotopically labelled amino acids, etc.*

C   *Retrograde transport from injured axon site. eg. HRP*

D   *Anterograde transport from intracellular injection site. eg. HRP, fluorochrome dyes, etc.*

E

F   *Anterograde and retrograde labelling of membranes by externally applied lipophilic dyes. eg. DiI, DiO.*

G   *Retrograde, trans-synaptic transport. eg. isotopically labelled amino acid.*

7.15   Some widely used methods of tracing the connections of neurons by intracellular labelling.

cells, fibres and synapses, ranging from the original chromaffin reaction of Henle (1865) to modern modifications (Coupland 1965, p. 1465), the Gomori method (1943), fluorescence techniques, and immunocytochemical methods. The latter two techniques in neuroanatomy (Falck & Hillarp 1959, Falck 1962, Fuxe et al 1970, Fuxe & Jonsson 1974, Nieuwenhuys et al 1978) have led to a new classification of nerve cells. Thus, the mammalian brain has been 'mapped' extensively in terms of the distribution of noradrenergic (NA), dopaminergic (DA), serotonergic (ST) and many other classes of transmitter-specific neurons and tracts (p. 889).

In the general field of neurohistology some other innovations must be mentioned. The staining methods such as the silver impregnation techniques of Bodian (1936) for nerve fibres, Glees (1946), Nauta (1950) and Fink-Heimer for terminal axonal degeneration and the dye methods of Einarson (1932) and Kluver (1953) for 'Nissl' granules (Kluver's method also stains axons) have all augmented the armory of experimental techniques.

In a resurgence of the use of Golgi's technique, its various modifications (e.g. Fox et al 1951) have been particularly applied to analyse the interrelations of small neuronal groups, especially interneurons. Golgi staining has an unexplained peculiarity in affecting only a fraction of cells, making the unravelling of intricate connections easier and often allowing the implication of quantitative histology, including microreconstruction and mathematical analysis of dendritic branching (p. 878). Such studies, carried out in a wide variety of vertebrates and invertebrates, have led to a rapprochement between the electrophysiological and anatomical investigation of small networks and individual neurons (e.g. Eccles et al 1967). Another potent technique in drawing together workers in separate disciplines has been electron microscopy.

ELECTRON MICROSCOPY

Though electron microscopes have been in use since the Ruska prototype of 1933, years elapsed before reliably thin sections could be prepared. Pearse & Baker succeeded in doing this in 1948, with an adapted Spencer microtome. Sjostrand (1967) has described these early struggles and the emergence of the modern ultramicrotome. Biological electron microscopy would have been impractable without the development of glass and diamond knives to cut the hard embedding media required: from the 1950s all the basic requirements were met and biological material was submitted to the increasingly high resolutions and magnifications of continuously improving electron microscopes, transforming structural cytology. Nearly four decades of application of supporting techniques has followed: heavy metal shadowing, replication methods, production of homogenate films, heavy metal staining including negative contrast, autoradiography, densitometry, elemental microanalysis, microreconstruction, cell fractionation methods depending upon differential centrifugation (of fundamental importance in neurochemistry), electron diffraction, scanning microscopy (including scanning transmission), freeze-fracture and freeze-etch methods, cytochemical and immunocytochemical staining and other techniques have multiplied the potentialities of electron microscopy. At the same time, there was a rapid advance in the interpretation of greatly enlarged intracellular detail and the growth of an experimental attitude and quantification in electron microscopy; otherwise the combinations of cytology, biochemistry, biophysics and molecular biology which have been so fruitful would not have occurred. In neuroanatomy such bonds have been most effective; the elucidation of the structure of myelin sheaths and of synapses are notable examples of their power. With improved cutting methods and orientation two main defects of electron microscopy have been largely countered: the small scale of sampling and the difficulty in equating the findings of light and electron microscopy. Techniques involving alternate thick and thin sections, for appropriate staining and examination, and even serial ultra-thin sections, have overcome some of these difficulties, and the development of scanning electron microscopy has introduced a new dimension, literally.

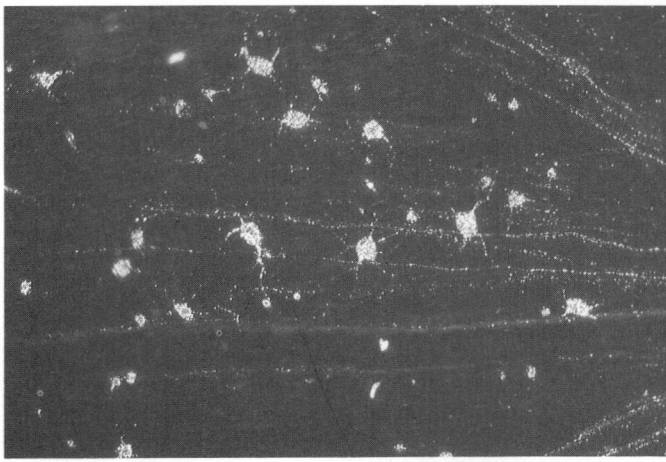

7.16 Part of a whole-mount preparation of a retina (rat) after prior horseradish peroxidase injection into the optic nerve. The tracer has passed in retrograde fashion to fill many ganglion cells and their processes. Polarizing illumination. Provided by E L Rees, Anatomy Department, UMDS, Guy's Campus, London.

OTHER MODERN TECHNIQUES (7.15)

In tracing neuronal circuitry, both central and peripheral, two new techniques have proved particularly valuable: the autoradiographic identification of labelled substances, absorbed by neurons and transported along neurites, and the retrograde transport of identifiable 'marker' substances (e.g. horseradish peroxidase, fluorochrome dyes). Both depend on physiological activities, in contrast to 'degeneration' techniques which use pathological reactions to experimental injury. They are also claimed to be more reliable (Jones & Hartman 1978) but each has peculiar advantages and disadvantages. The autoradiographic technique, introduced by Lasek (1968) and Hendrickson (1969), was popularized by Cowan et al (1972). Radioactively labelled amino acids or sugars are injected among neurons, which absorb and transport them along axons (usually combined in various macromolecules), where they are revealed by autoradiography with light or electron microscopy (Rogers 1973, Hendrickson 1975). The method can be either *anterograde* or *retrograde*, since movements occur in both directions within axons. However, the anterograde labelling technique is usually employed for autoradiography. These methods are now widely applied, confirming many previous findings but also revealing many new ones, e.g. the hypothalamo-spinal autonomic and olivo-cerebellar paths. It has also been shown to produce transneuronal effects (e.g. in the visual pathways).

The horseradish peroxidase, fluorochrome and other dye methods also depend on axonal flow (Matsumoto 1920, Lubinska 1964, and others, p. 881); but not until 1971 did Kristenson & Olsson show that labelled albumin and horseradish peroxidase could be used in tracing axons. The LaVails (1972) quickly confirmed this. Solutions of the enzyme are injected into nervous tissues or the cut surfaces (e.g. a severed peripheral nerve) are immersed in them. Uptake by nerve terminals and axons can be greatly enhanced by using a tracer linked chemically to a lectin such as wheat germ agglutinin, which stimulates endocytosis by the neuron. Iontophoretic injection into neural tissue can be used to reduce local damage. Material absorbed by axons is transported to the somata where it can be identified histologically. Alternatively the tracer can be injected or iontophoresed into neuronal somata through micropipettes and is then transported along axons in the anterograde direction (7.15). The method is applicable to all neurons and can be used with both light and electron microscopy. A large literature on this topic has accumulated (Winer 1977, Jones & Hartman 1978) and a wide variety of tracers is becoming available for this type of study.

Another notable development in research is the widespread applications of immunological techniques to neurobiological problems; e.g. immunohistochemical methods have helped to localize neurotransmitters and their enzymes, pituitary hormones and their releasing factors and other specific proteins, with resolution often superior to customary staining or biochemical assays (Jones & Hartman 1978). These methods involve the production and identification of sites of antibody–antigen interaction in tissue sections. Several factors influence the distribution of such sites, complicating interpretation. These factors are as follows: (1) Production and purification of specific antibodies which bind only to the substance being investigated; however, the use of monoclonal antibodies has largely circumvented this problem, although there are other problems in their use, e.g. cross-reaction with other proteins which may share some amino acid sequences. Recently, molecular genetics has provided a relatively simple means of preparing highly purified antigens which can be used to refine antibody specificity greatly. (2) Adequate preparation of tissues, especially during fixation. Fixation is mostly designed to achieve an optimum compromise between preservation of structure and retention of antigenicity. The commonest fixative in use is formaldehyde; other aldehydes, with stronger cross-linking abilities, have proved less effective, although such combinations as periodate-lysine-paraformaldehyde has been used with advantage (Nakane & Kawaoi 1974). Freeze-drying and freeze-substitution and also cryosectioning can also be employed to minimize exposure to chemicals prior to labelling. (3) Selection of an appropriate visual marker, e.g. fluorescein isothiocyanate (FITC), rhodamine, horseradish peroxidase (HRP) or colloidal gold. FITC emits a high yield of green fluorescence (517 nm) when excited with blue light (peak excitation 490–500 nm), which in combination with modern fluorescence microscopes permits examination of, e.g. fine adrenergic fibres (Hartman 1973); disadvantages are: (a) lack of permanent preparations due to quenching in mountant or photodecomposition with time; and (b) fluorescent markers cannot be used in electron microscopy. HRP is used as a marker in two different ways. As a marker in immunohistochemical studies it can be conjugated directly with antibodies by a number of bifunctional cross-linking reagents, but this results in partial inactivation of peroxidase activity. A more recent development, the peroxidase-antiperoxidase method (PAP), in which antibodies to HRP are bound to peroxidase without inactivating it, had a great impact on immunohistochemistry, especially at ultrastructural level (Sternberger 1974). The large size of the PAP molecule (c.420 000 daltons) may, however, entail problems of penetration (e.g. in fine nerve terminals). A recent extension of the method has involved the use of suitably labelled antigens to

identify different neurons and satellite cells, either in suspension after isolation from whole tissue or growing and differentiating in culture prior to fixation. Thus oligodendrocytes can be labelled in vitro with anti-galactocerebroside, astrocytes with anti-glial fibrillary acidic protein (GFAP), Schwann cells with anti-RAN I and some neurons with antitetanus toxin (Raff et al 1978). Moreover, antisera to certain components of the myelin sheath and to neural growth factor are used in studies of the control of myelinogenesis and of ontogenetic and pathological mechanism in nervous tissues.

Another label which is proving very useful is colloidal gold, to which can be attached antibodies for detecting antigens as described above. The particles can be prepared in a wide range of sizes (5–50 nm) and can be used for both light- and electron-microscopic detection.

Other relevant techniques include: autoradiography (Rogers 1967, 1973), cine and time-lapse photography (Rose 1963), microreconstruction (Gaunt & Gaunt 1978), microphotometry, micromanipulation (Kopac 1960), neuronal counting techniques (Konigsmark 1970, Corsellis et al 1975), morphometric analysis (Weibel & Elias 1969), the construction of models as functional analogues (Weiner & Schade 1963) and various methods of computerized image analysis. Enough has been said to show the dependence of neuroanatomy on technical developments. Despite widespread misapprehension, neuroanatomy (and anatomy in general) has never been an entirely morphological, merely descriptive discipline, though it can descend and frequently has to this sterile level. However, there have always been those, such as Willis, Harvey or Sherrington, who ignore the artificial separation of structure and function in the nervous or any other system. Unfortunately, the contribution of one or the other is often imbalanced. After a considerable and regrettable period early in this century of separation of the disciplines concerned in morphology and function (perhaps chiefly due to exigencies of teaching organization in medical schools), the resulting isolationism now shows signs of dissolution, particularly in studies of the nervous system. Preoccupation with the minutiae of form or with the nervous impulse have borne their separate fruit and we can now pass on to wider prospects. As morphologists become more experimental and functionalists examine the structural arena of their experiments more closely, instrumentation and outlooks are becoming increasingly similar and promise a new cohesion. We may at least be much nearer to a comprehension of the basic units in central nervous function. It is to the individual elements so organized that we must now turn.

CYTOLOGY OF THE NERVOUS SYSTEM

The two previous general sections have concerned the general principles of the organization of the nervous system and methods of approach to it. We will now consider its cellular elements, an understanding of which is essential to any concept of brain function (Schmitt et al 1967, 1970, 1977). Nervous tissue contains two major components: (1) excitable *neurons*; and (2) non-excitable *neuroglia* and *ependyma* in the central nervous system, and Schwann *cells* (lemmocytes) and related cells in its periphery.

The Neuron

Neurons vary considerably in shape and size but they possess many common features. In comparison with other cells, all have a large surface area, specialized by their molecular composition for reception, conduction and transmission of information. The nucleus of each neuron is surrounded by a cytoplasmic mass containing various organelles, together constituting the *soma*, *perikaryon* or *cell body*. The narrow processes extending from the soma, collectively termed *neurites*, are either *dendrites* or *axons*; a neuron's dendrites are usually multiple and conduct excitation towards the soma, while axons are single and conduct away from

the soma. The cytoplasm of the soma contains numerous basophilic *Nissl bodies* or *granules* (7.17), except at a conical projection leading to the axon, the *axon hillock*, its continuation being the *initial segment* of the axon. Sometimes an axon emerges from the base of a dendrite and in a few highly specialized small neurons an identifiable axon is altogether lacking. From other angles of the soma spring *primary dendrites*, branching repeatedly to form a *dendritic tree*. An axon may have *collaterals*, usually of finer calibre than the main stem. Axons usually end in several fine *axon terminals* or *telodendria*, apposed to other neurons at *synapses*, or with effectors such as muscle fibres at *neuromuscular junctions* or with glands; adipose tissue also receives efferent terminals (p. 908). In some cases neurons synthesizing neurohormones have axon terminals which end on blood vessels, into which they release their secretions directly. The axonal plasma membrane is the *axolemma*; its contained cytoplasm is *axoplasm*.

Neurons range in size from microneurons, with somata 7 μm or less across, to some of the largest cells of the body, e.g. spinal ganglia have somata 120 μm in diameter. In some of the larger neurons (e.g. cerebellar Purkinje, cerebral Betz neurons) the soma, which is proportionately large, has a tetraploid chromosomal content (Herman & Lapham 1969). The size of the soma and its nucleus is related to the large volume of cytoplasm contained in its dendrites and axon; the axon may be as much as a metre in length and more than a hundred times the volume of the soma.

HISTOLOGICAL DEMONSTRATION OF NEURONS

Our knowledge of neuronal structure has in large measure resulted from the use of special neurohistological techniques (p. 869) which will now be briefly considered. Though the general methods of light microscopy show neuronal somata, their neurites are not stained sufficiently to trace them with any confidence. Various methods of metallic impregnation, particularly with silver salts, do stain the neurites as well as the somata. The most informative of these techniques, in terms of individual cell structure, are the Golgi methods (p. 870) which entail the precipitation of silver chromate in individual neurons, which can be shown in their entirety in thick sections; since only a small percentage of cells are stained, they may be demonstrated with great clarity, their black spidery profiles contrasting with a clear background (7.17). Although in earlier years the method was attacked as merely demonstrating artefactual deposits of metallic salts, electron microscopy has confirmed that it gives an accurate representation of neuronal structure in finest detail. Other methods of demonstrating entire neurons involve the use of various tracers, either microinjected directly into the somata of living cells or taken up and transported actively by their axons (vide infra).

Various techniques stain *sections* with silver salts, some of them preferentially precipitated by the neurofilaments (*neurofibrillary* stains) (Gray & Guillery 1966), others staining the entire cytoplasm (7.18D); these include the ammoniacal silver nitrate methods of Bielschowsky and subsequent modifications, Bodian's silver proteinate and the silver nitrate techniques by Holmes, Palmgren and others. Ramon y Cajal's block staining method is a similar process except that tissues are bulk stained, then sectioned afterwards. Modified reactions to silver salts have been used to show anterograde axonal and terminal degeneration in sections, the staining of normal axons being suppressed (Nauta and Glees methods) (7.14, see also p. 870). Modifications of such techniques for use in electron microscopy (e.g. Fink-Heimer) have been introduced but have been largely superseded by tracer methods, e.g. with horseradish peroxidase (vide infra).

The sheaths of myelinated fibres (p. 902) require special methods because the myelin, which is mainly lipid, is removed by routine histological processing. Salts of the metal osmium are soluble in lipid and precipitate to give it a black or brown colour, and so can be used in the fixation of neural tissues to preserve myelin. However, myelin sheaths also contain some protein and may be lightly stained with such stains as haematoxylin and Luxol fast blue (7.18C). Staining with chromium salts can demonstrate *degenerating* myelin sheaths (Marchi), now rarely used because of uncertainties of interpretation (p. 870). General cytological stains can be used to demonstrate various organelles, the most useful, Nissl stains, being basic dyes combining strongly with nucleic acids which are abundant in neuronal nuclei and cytoplasm (mainly ribosomal RNA). Cresyl fast violet (7.14) and toluidine blue are also widely used for this purpose. Some stains also demonstrate mitochondria, particularly in axon terminals and modifications of these methods can show *degeneration of axon terminals* due to experimental trauma.

Electron microscopy has revolutionized our concepts of neuronal structure and has provided a wealth of detailed data not obtainable by any other method. However, light microscopy is also an essential tool, particularly when correlated with electron microscopy, since three-dimensional reconstruction from electron micrographs is laborious and limited even when assisted by computer technology.

All the above methods are applicable to dead material; however, much may also be learnt from neurons stained by supra-vital methods. Methylene blue was early found to stain the entire cytoplasm of neurons, showing individual cells as strikingly as Golgi methods; but its use is restricted largely to thin or small whole tissue mounts of e.g. peripheral and embryonic nervous systems, where penetration and staining are rapid.

Tracer methods have been used extensively in recent years to demonstrate the pathways of axons and the positions and structures of their cell bodies. Such techniques are generally applied to living animals; two distinct methods of approach can be used, *anterograde* and *retrograde* (see p. 870). In anterograde methods, a tracer substance is generally injected through a micropipette into individual neuron cell bodies, then time is allowed for axonal and dendrite transport (p. 881) to carry the tracer to its extremities (i.e. in the same direction as axonal conduction). The tissue is then fixed and sectioned and the tracer visualized. In *retrograde* methods, a tracer is placed by injection in the vicinity of the axon terminals or, after slight damage or complete section, close to non-terminal regions of axons. The tracer is taken up by the axon by active processes and transported back against the direction of nerve conduction to the cell body (and dendrites in some cases), where again it can be visualized after fixation. The most widely used tracer is horseradish peroxidase, whose presence can be detected by its action on hydrogen peroxide in the presence of a chemical reactant which produces a coloured precipitate. This can be seen clearly as a red-brown deposit or more conveniently can be visualized by dark-field microscopy.

Other tracers include enzymes or particulate substances conjugated with lectins (e.g. wheat germ agglutinin), which are more readily taken up by axons, and coloured dyes (e.g. procion yellow, true blue). Many of these can fill neurons to give a result strikingly similar to those of the Golgi methods, as well as providing data on the connectivity of neurons. In general, such techniques have in the last 20 years revolutionized the whole subject of detailed neuroanatomy and it has become apparent that many of the 'classic' pathways have to be re-examined by modern methods.

Neuro-immunohistology

Another approach which is bearing much fruit is the application of immunohistochemistry to nervous tissue. These methods use antibodies raised against specific chemicals (e.g. neurofilament proteins, transmitter enzymes), which can detect the presence of these chemicals in tissue sections by immunofluorescence or immuno-electron microscopy. Thus, e.g. it is possible to determine the presence of growing axons in developing or regenerating tissue even when routine neuroanatomical methods fail, and the type of neurotransmitter(s) used by particular neurons can be determined. The use of this technique has also helped to revolutionize our understanding of the nervous system in all its wide variety (see also p. 889).

GENERAL STRUCTURE OF NEURONS

Neurons may be classified according to their pattern of neurites, their physiological actions, transmitter substances and various other features. Several schemes of classification based on cell shape are current, mostly based on the ramification of neurites. A traditional approach (p. 863) distinguishes: (1) *unipolar neurons*, in which a single extension from the soma divides into both dendritic and axonal branches, e.g. mesencephalic trigeminal sensory neurons, dorsal root ganglion neurons (sometimes termed *pseudo-unipolar* because their unipolar form is secondary during development); (2) *bipolar neurons* with an extension at each pole, e.g. retinal bipolar cells, neurons of the cochlear and vestibular ganglia; (3) *multipolar neurons*, with several neurites, e.g. most central neurons. Though this scheme (7.17A) is descriptively useful, it has limited physiological significance. Central neurons may also be grouped as large, Golgi type I neurons with long axons connecting remote regions (Ramon y Cajal 1911) and small Golgi type II neurons (microneurons) with short axons ending near their somata or else entirely absent (7.17). The latter type are often inhibitory, e.g. periglomerular cells in the olfactory bulb (p. 1030). Some microneurons lack an axon, conduction being apparently bi-directional along dendrites, e.g. in *amacrine cells* of the retina (p. 1202) and granule cells of the olfactory bulb. They may be a typical element of all main sensory pathways. Small inhibitory neurons appear to be responsible for an important interaction between neurons in sensory pathways, by which maximally excited neurons inhibit less excited adjacent neurons through inhibitory interneurons, thus 'sharpening' sensory patterns at various levels (7.8 and p.865), a process termed *lateral inhibition*.

Central neurons are also classified by the shape of their dendritic fields (7.17). Those with dendrites filling a spherical volume are termed *stellate*, a type common in the cerebral cortex and some nuclei of the brain stem and spinal cord. The soma of *pyramidal neurons* is appropriately pyramidal or conical and *basal dendrites* emerge from its basal angles to fill a hemispherical volume, an *apical dendrite* often emerging from the soma to give a second dendritic field at some distance from it, e.g. cortical pyramidal neurons. Somata with dendrites emerging at one or both ends are described as *fusiform*.

Many variations exist depending on the pattern of the afferent fibres, the synaptic sites and the mechanical or other effects of neighbouring neurons. A most remarkable dendritic field is that of the cerebellar Purkinje neurons (7.17B), in which a primary apical dendrite branches to form a complex radiating nearly two-dimensional array. *Glomerular neurons* may have a few dendrites with highly recurved short apical branches, where most synaptic contact occurs, e.g. olfactory mitral and tufted neurons (p. 1030), and in the lateral geniculate nucleus, certain thalamic nuclei (p. 1001) and cerebellar granule cells (p. 972).

Attempts to classify dendritic patterns by their geometry, though useful for description, have an uncertain functional significance at present. Perhaps more important is the mathematical analysis of dendritic patterns by their frequency, dimensions, the spatial distribution of the branches and their mode of ramification, their surface and volume, and the frequency, type and spatial location of their synaptic endings, since all these parameters are linked to electrical activity. Such studies are still in their early stages, but computer technology has greatly facilitated this area of research and shows much promise as a powerful approach to understanding the structural basis of electrical activity in neurons.

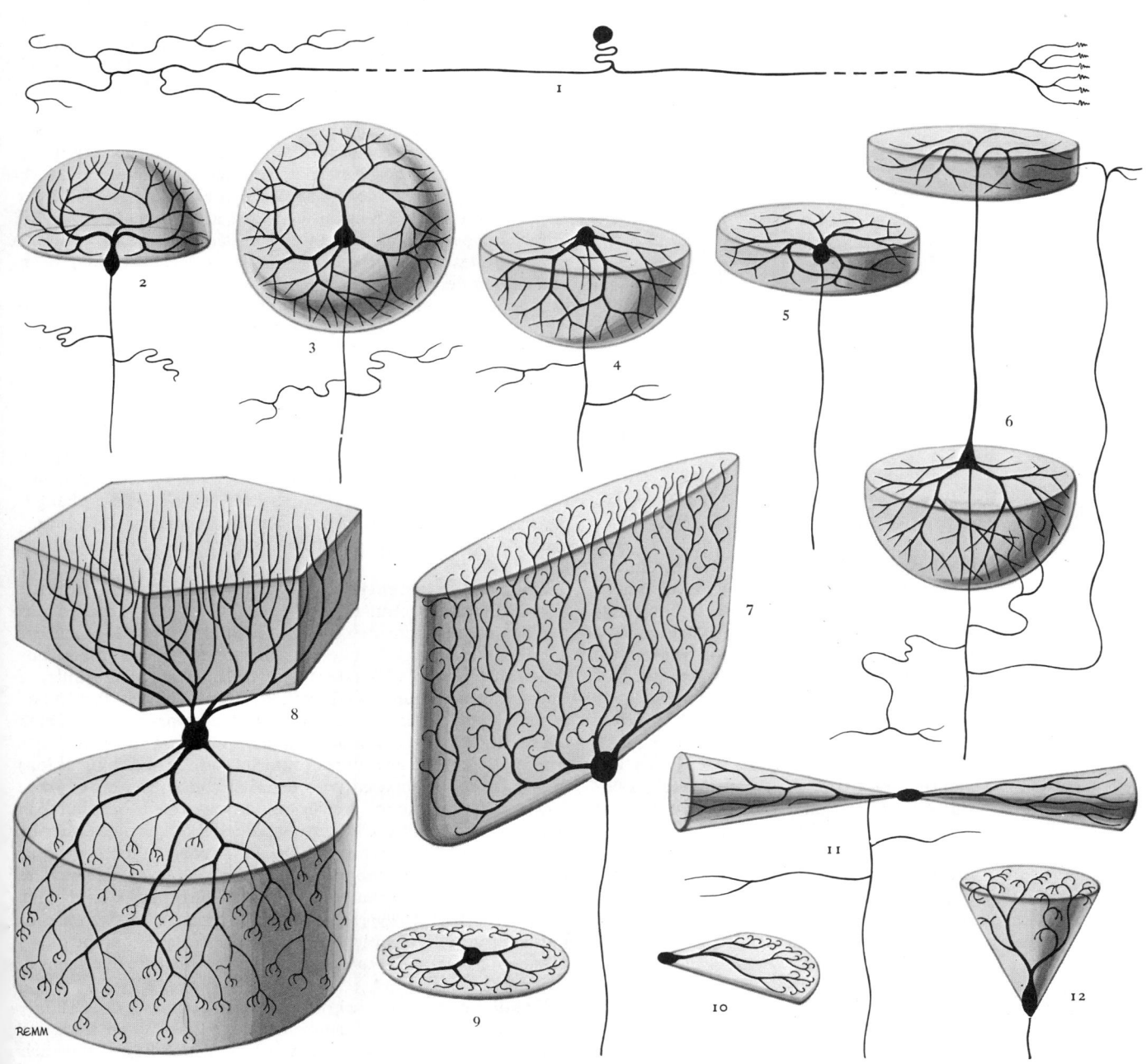

7.17A Scheme showing pattern variations of neuronal geometry: (1) unipolar, sensory ganglionic neuron; (2) bipolar neuron; (3) stellate (isodendritic) neuron, with (4), (5), and (11) which are modifications of this pattern; (6) pyramidal neuron with an apical and a series of basal dendrites and recurrent axon collaterals from the cerebral cortex; (7) Purkinje neuron from the cerebellar cortex; (8) Golgi neuron from the cerebellar cortex; (9) and (10) amacrine cells lacking axons; (12) glomerular neuron (mitral cell) from the olfactory bulb, showing recurved dendritic tips.

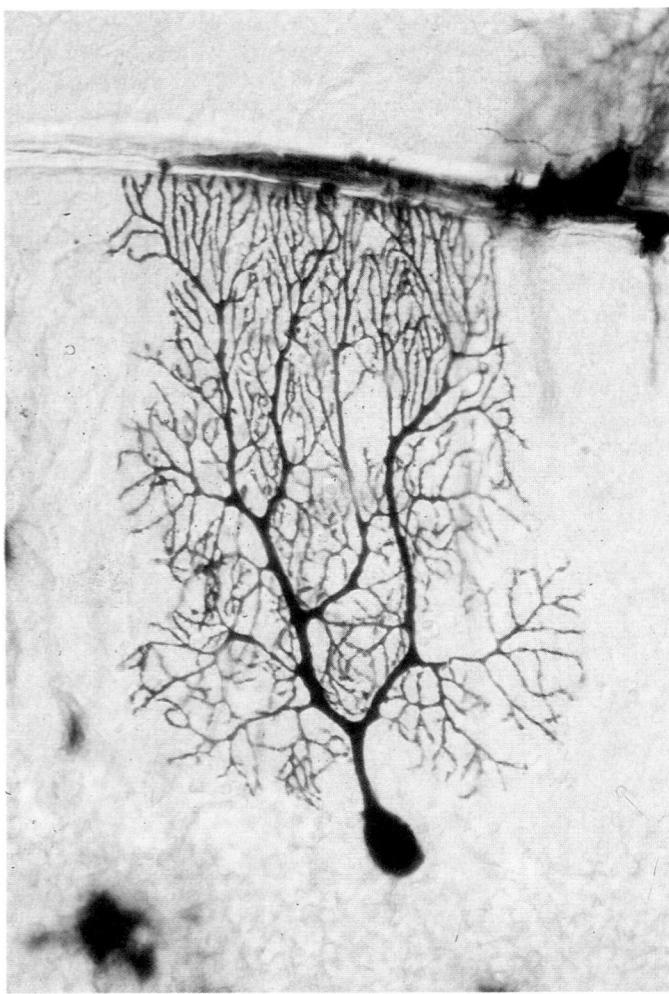

7.17B Purkinje neuron from the cerebellum (rat) stained by the Golgi-Cox method, showing the extensive two-dimensional array of dendrites. Provided by Martin Sadler and M Berry, Anatomy Department, UMDS, Guy's Campus, London.

STRUCTURAL PARAMETERS OF DENDRITES

As already stated, there is considerable variation in the geometry of dendritic trees; however, detailed analysis of their development and adult morphology indicates that there are some common laws which govern their initial generation and final form. These have two aspects: firstly the mechanisms by which dendritic trees are built and maintained; secondly, the functional significance of dendritic patterns in terms of neuronal electrical behaviour.

Techniques of dendritic tree analysis have developed considerably since the pioneering work of Sholl (1953) (see also Aitken & Bridges 1961) who studied the mathematical features of cortical stellate neurons by counting the frequencies of intersections between dendrites and a series of imaginary spheres constructed concentrically around the cell body. This technique proved difficult to apply to less symmetrical neurons and more detailed quantitative studies were made by serial reconstruction (e.g. Mannen 1968) or by examining special types of neurons which lent themselves particularly well to planar measurement. A neuron that has provided a useful model for dendritic analysis has proved to be the cerebellar Purkinje cell, which has an extremely flat dendritic tree accessible to computer-assisted assessment of different parameters (length of dendrites, frequency of branching, types of branching, etc., see the review of this subject by Berry 1980). The mathematical basis of such studies (including the subject of network analysis) is beyond the scope of this account (for details, see Ten Hoopen & Reuver 1970, Hollingsworth & Berry 1975, Berry 1980).

To summarize these results briefly, it has been found that more than 80% of the neuronal surface area (excepting the axon) is situated in the dendritic tree. Most branches are dichotomous in form and occur randomly within a volume determined by various factors including the proximity of other neurons which may compete for space. The initial growth of primary dendrites from the cell body appears to be under epigenetic control (p. 113), i.e. it is determined by interactions between the genome of the cell and its immediate environment, in a manner not well understood; there is much evidence that subsequently the branching pattern reflects the formation of contacts between the dendrite and the growth cones of axons it is destined to form synapses with, dendrites branching upon meeting a growth cone and continuing until the next contact (Berry 1980). The frequency of branching therefore reflects the frequency of synaptic interactions. Once established, dendritic trees may be continually remodelled (Sadler & Berry 1988), as noted in observations on living cells in vivo, labelled with intravital tracers and examined at intervals over a period of months (Purves et al 1986), providing for detailed adjustment of neuronal behaviour.

Counts of *dendritic spines* (an indication of synapses, p. 882) show that in some neurons many thousands are spaced over much of the dendritic tree (Gelfan & Rapisarda 1964, Gelfan et al 1970). Such data indicate that different populations of neurons, with the same general mode, have their own ratios between somatic and dendritic surface volumes, spine counts, etc. (Lindsay & Scheibel 1974). The import of these parameters is in numbers of afferent synapses, their positions and patterns of electrical disturbance at cell membranes following their activation (Rall 1977). A basic feature is the relative distribution of excitatory and inhibitory synapses; this can only be estimated by electron microscopy, with limited statistical validity (p. 886). Microelectrode studies are a possible means of analysis but dendrites are too small for ease of intracellular recording and only elaborate mathematics can extract any meaning from the records. However, it has long been known that dendritic defects can be correlated with behavioural deficits, e.g. neonatal mice, made thyroid hormone-deficient to simulate human cretinism, show behavioural retardation; the dendritic trees of their cerebral cortical neurons are smaller, with fewer branches (Eayrs 1955).

PHYSIOLOGICAL PROPERTIES OF NEURONS

Neurons typically receive, conduct and transmit *information* (p. 860), coded in transient electrochemical changes in their membranes (Katz 1966, Kuffler & Nicholls 1977). These changes consist of rapid fluxes of ions across the plasma membrane, against a background of steady, transmembrane electrical potential difference.

The resting potential (7.19) of a neuron is identical to the transmembrane potential in non-excitable cells. In most neurons it is about 80 mV, inside negative. Such bio-electrical potentials are due to differences in the concentration of anions and cations inside and outside, which depend on the selective permeability of the plasma membrane and the active transport of ions, particularly sodium and potassium, across it. These factors result in a high concentration of potassium and negative, non-diffusible ions (proteins, etc.) and low concentration of sodium and chloride ions *inside* the cell, compared with external concentrations of these ions. Calcium in the cell is also at much lower concentration than in extracellular fluids. The resting potential can change either by *graded potentials* or *action potentials*. Graded potentials occur mainly across the membranes of *dendrites* and *somata*; they are typically transient increases or decreases in resting potential, i.e. the cell is relatively *hyperpolarized* or *depolarized*. Action potentials are transient complete reversals of polarity across the membranes of *axons* (and some large dendrites which are like axons in structure).

Graded potential variations may be *excitatory* or *inhibitory* to the neuron. When excitatory, they accompany an increased permeability to sodium or calcium ions, which flow down their concentration gradient into the cell, progressively *depolarizing* the membrane towards zero potential. This influx results from the transient opening of ionic channels within certain proteins (sodium or calcium gates). The scale and timing of depolarization at a particular locus is related mathematically to the stimulus

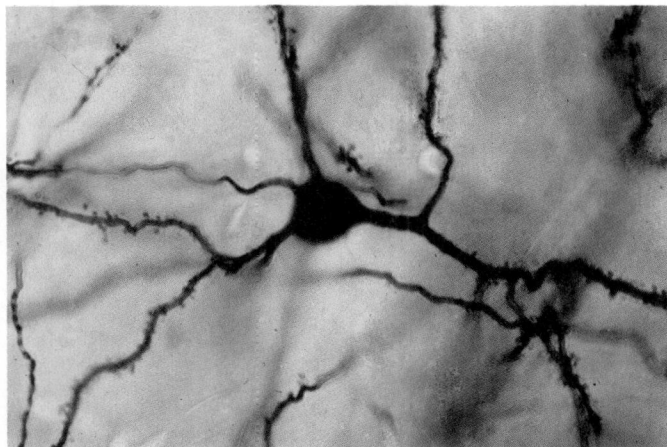

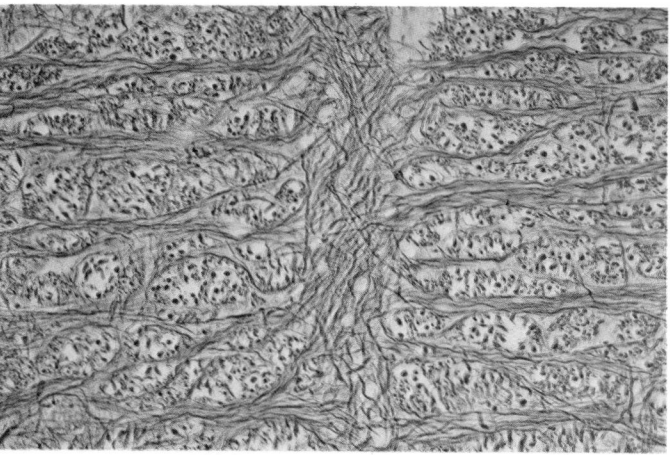

7.18 Different neurohistological methods. A. A high-power light micrograph showing part of a pyramidal neuron from the cerebral cortex of a rat, prepared by the rapid method of Golgi. The bases of several dendrites covered with dendritic spines and a thin axon (left) are visible. Preparation supplied by A. R. Lieberman of University College, London.

B. A low-power light micrograph of groups of interweaving axons from the medulla oblongata of a cat, stained by the Holmes' silver method. Preparations B–D supplied by E L Rees of the Anatomy Dept., UMDS, Guy's Campus, London.

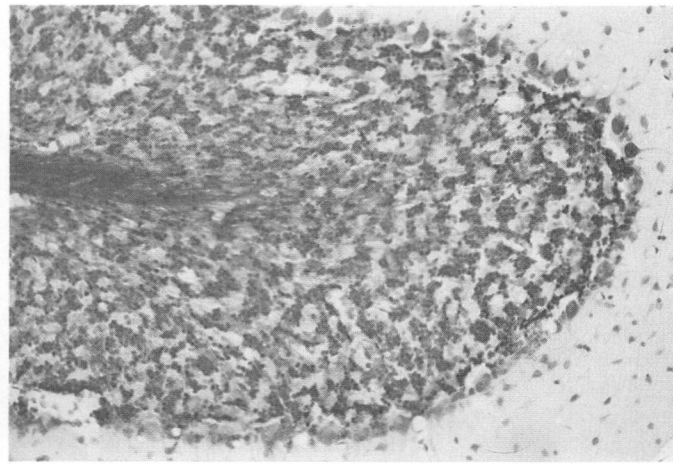

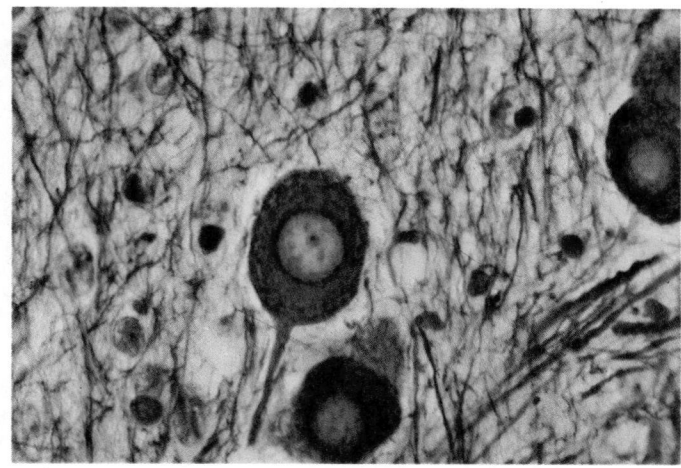

C. A group of neuronal perikarya (purple) amongst bundles of myelinated fibres (blue), from the cerebellum of a rat. Stained by the cresyl fast violet-Luxol fast blue method.

D. A section through the mesencephalic nucleus of the trigeminal nerve (rat), stained with the Holmes' silver nitrate method to show large unipolar neurons from one of which a dendro-axonal process is emerging.

strength. Adjacent regions of membrane are also depolarized, but with decreasing amplitude at increasing distances from the locus. The larger the initial depolarization, the greater is the spread of its effect to the adjacent areas of membrane. This allows a *summation* of the effects of multiple excitatory stimuli applied at different loci of the receiving surface. Inhibitory stimuli, believed to act mainly by an increasing inflow of negatively charged chloride ions, tend to *increase* the membrane potential (*hyperpolarization*), opposing or reducing the *total excitatory state*.

The action potential, nerve impulse or **spike potential**, as seen in peripheral nerves, is in contrast a brief complete, *reversal* of polarity, due to the influx of sodium ions, followed by a rapid return to the resting potential as potassium ions flow out, the whole process being completed in about 5 milliseconds (Hodgkin & Huxley 1952). For a particular neuron, the *amplitude* and *duration* of successive action potentials are identical; however much a stimulus exceeds the *threshold value*, the actional potential is the same size and is thus said to be an all-or-none response. Once initiated, an action potential spreads rapidly; but, unlike graded potentials, its size and timing do not alter; the wave sweeps rapidly over the membrane at a constant rate and does not diminish with distance. It is therefore a *propagated* and *non-decremental* event. After the action potential, however, there is an irreducible *refractory period*, during which action potentials cannot be elicited; this fixes a maximum frequency possible and varies

with the size of the axon diameter. The action potential is the basis of neural conduction and communication within the nervous system which is thus *frequency-coded*. It should be noted that variations may occur in the detail of ionic conductance changes; thus in central axons, the recovery phase of the action potential is not dependent on potassium efflux but on general current leakage.

Graded and action potentials are functionally interrelated at particular regions of neuronal surfaces; e.g. in a motor neuron, the *graded* potentials of the dendrites and soma depolarize the axon hillock to initiate *action* potentials in the initial axonal segment. The greater the graded somatic depolarization becomes, the more frequent are the action potentials; thus, the *state of excitation* of the soma is directly related to the *frequency of the action potentials* in the axon. When an action potential reaches the axonal terminals, it causes a graded depolarization of the presynaptic membrane and as a result discrete amounts of a *neurotransmitter substance* are released to change the degree of excitation of the next neuron, myocyte or glandular cell. Closely grouped action potentials (*trains* or *volleys*) are naturally more effective than single 'spikes' in releasing neurotransmitter; extracellular enzymes often limit the action of neurotransmitters.

Though a change from a graded to an action potential usually occurs in the initial axonal segment, in many peripheral receptors with an axon-like dendrite (e.g. cutaneous sensory neurons) it occurs close to the sensory endings, often at the most peripheral

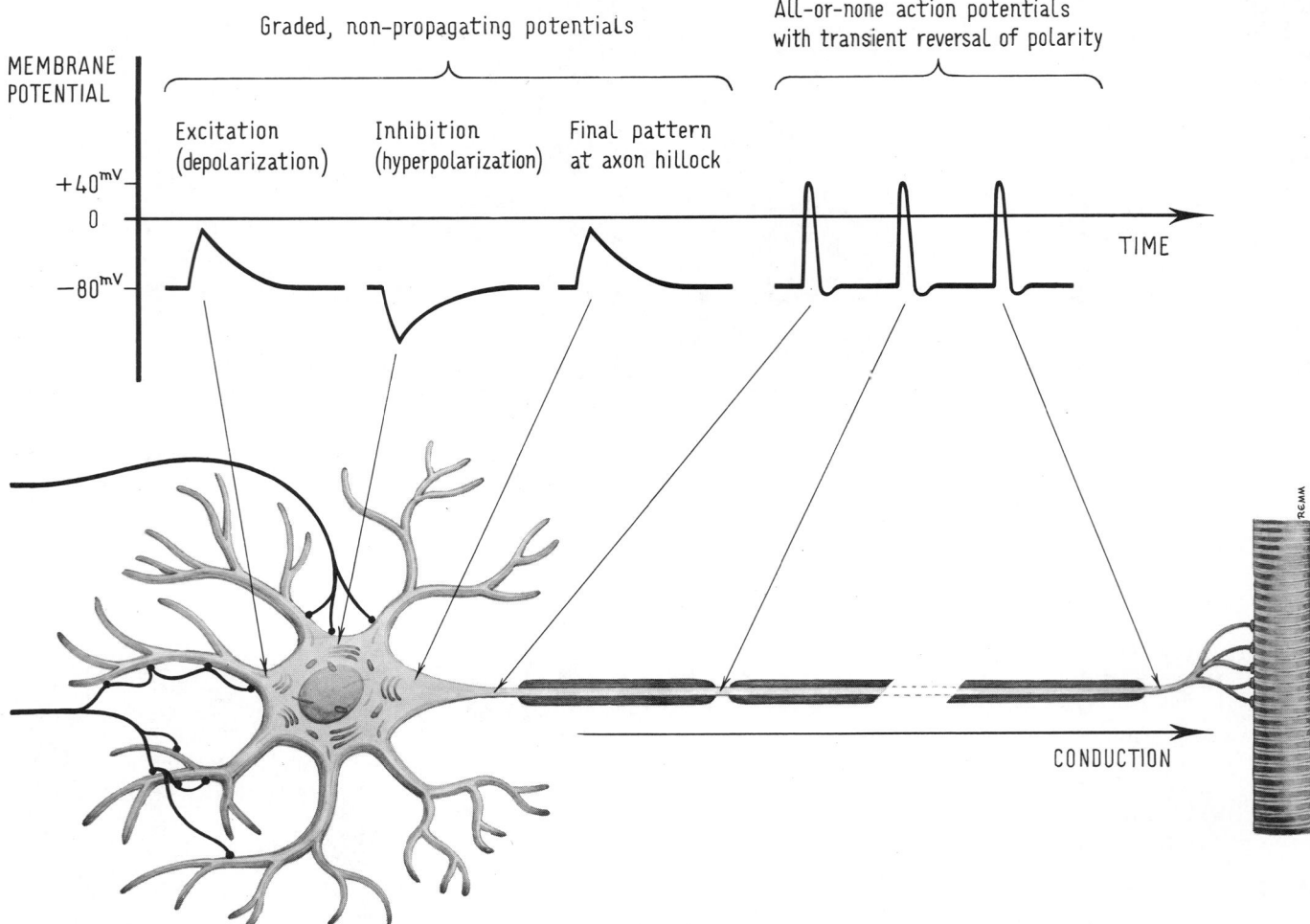

7.19 The types of change in electrical potential which can be recorded across the cell membrane of a motor neuron at the points indicated by the arrows. Excitatory and inhibitory synapses on the surfaces of the dendrites and soma cause local graded changes of potential which summate at the axon hillock and may initiate a series of all-or-none action potentials, which in their turn are conducted along the axon to the effector terminals.

node in myelinated fibres. In such cases, the graded potential itself occurs at the sensory ending (a sensory potential, p. 907) where the sensory stimulus is received.

In amacrine neurons (p. 1202) and other microneurons lacking true axons, *all* electrical activities may be graded; such neurons are able to conduct electrical changes in any direction from the locus of stimulation. In neurons with axons conduction is naturally unidirectional, from dendrites and soma to axon terminals (*orthodromic conduction*); when artificially stimulated in their periphery, axons can conduct centripetally (*antidromic conduction*), such electrical changes invading the soma where they can be detected with a microelectrode.

Due to their mode of synaptic excitation and decremental conduction of dendrites and cell bodies, neuronal shape and the distribution of their excitatory and inhibitory synapses must be related to the neuronal input and output. A full analysis of these parameters may be expected to shed light on the functions of neurons and their interactions, and much attention has been paid to this area of research.

NEURONAL ULTRASTRUCTURE

Since the early studies by Palay & Palade (1955), Robertis & Bennett (1955) and others, our information on neuronal ultrastructure has increased greatly (Peters et al 1976, Palay & Chan-Palay 1977). As in light microscopy it is useful to distinguish between regions consisting mainly of neuronal somata and dendrites ('grey matter') and those dominated by regular arrays of axons constituting tracts ('white matter'). Within the grey matter, the neuronal somata are surrounded by complex regions of axon

terminals, dendrites and neuroglial processes, forming the *neuropil* (e.g. Peters et al 1970).

The cytoplasm of a typical soma is rich in granular and agranular endoplasmic reticulum (7.20,21) and free polyribosome, the latter often congregated in large groups associated with granular endoplasmic reticulum; these aggregates are visible by light microscopy, because of their constituent RNA, as basophilic *Nissl* or *chromatin bodies* or *granules*; these are more obvious in large, highly active cells, such as spinal motor neurons, where *stacks* of granular endoplasmic reticulum occur (7.21). The *nucleus* is characteristically large, round and euchromatic, with one or more prominent *nucleoli*, as in all cells engaged in a marked degree of protein synthesis. Numerous mitochondria act as sources of energy for cellular activities. Lysosomes occur in moderate numbers and Golgi complexes form distinct groups, particularly near the bases of main dendrites and opposite the axon hillock. *Neurofilaments* and *microtubules* are abundant, bundles of the former being the 'neurofibrils' of light microscopy; both occur in the soma and extend along *dendrites* and *axons*, in proportions varying with the type of neuron and cell process. Dendrites are usually richer in microtubules than axons, which may be almost filled by neurofilaments. Though their bundles diverge at neuritic branching, tubules and filaments do not themselves branch. Actin microfilaments have also been described in neurons, as in most cells (Lasek & Hoffman 1976).

Centrioles, formerly believed absent from mature neurons, have been seen in every type examined, perhaps concerned in the generation or maintenance of the microtubular apparatus during development and subsequent life, rather than in cell division. (Neurons cannot undergo mitosis, once they are formed in the prenatal period, see p. 917.) In some neurons centrioles associated with ciliary projections have been reported; though such structures are common on the surfaces of developing neuroblasts, cilia also occur in mature cells; their significance, except at some sensory terminals, is obscure.

Other inclusions are also common in the neuronal cytoplasm; *pigment granules* appear in certain regions (Barden 1969); e.g. the substantia nigra (p. 894) has a *neuromelanin*, probably a waste product of catecholamine synthesis in these neurons. In the locus coeruleus a similar pigment, rich in copper, gives a bluish colour to the neurons. Some neurons are unusually rich in certain metals, e.g. zinc in the hippocampus, iron in the oculomotor nucleus. These metals may form part of special enzyme systems. Ageing neurons accumulate granules of *lipofuscin* (senility pigment), especially in spinal ganglia, a breakdown product of lysosomal action.

Dendrites share some features of the somata, containing: microtubules, neurofilaments, mitochondria, a few lysosomes, ribosomes, agranular and sometimes granular endoplasmic reticulum (7.20, 25). Axons are similar in these respects, except that they contain no ribosomes. The agranular endoplasmic reticulum in neurites often forms perforated transverse septa or longitudinal tubules and spherical vesicles. Various types of vesicle cluster near the points of neurotransmission (vide infra, p. 884).

As stated previously, various features of neurons indicate high levels of protein synthesis. The significance of this is probably manifold. Maintenance and repair of cytoplasm are necessary in all cells and the huge total volume of cytoplasm within the soma and neurites of many neurons must demand a high protein synthesis simply for these routine activities. Other proteins (enzyme systems, etc.) are involved in the elaboration of neurotransmitters and in the reception of stimuli. Various enzymes occur at the surfaces of neurons where they are associated with ionic transport, e.g. sodium and potassium stimulate ATP-ase and the 'second messenger' systems mediated by the adenylcyclase and phosphoinositol pathways. Although the apparatus for protein synthesis (including RNA and ribosomes) occurs throughout the soma and dendrites, it is usually absent from axons, leading to the question of how transmitter substances and other materials are transported to neuronal terminals.

AXOPLASMIC FLOW

Neuronal cytoplasm, as in other cells, is in continual motion; in tissue culture bidirectional streaming of vesicles along axons is clearly visible and there is indisputable evidence that similar movements occur in vivo, resulting in a net transport of materials from the soma to the terminals, with a lesser movement in the opposite direction (Weiss 1970). Experiments with radioactive substances (e.g. from the retina along the visual pathway) and investigations involving the ligature and subsequent examination of central tracts and peripheral nerves (Kapeller & Mayor 1967) show that *two major types of transport* occur: one slow, the other relatively fast. The first is a *bulk flow of axoplasm* which occurs only in the anterograde direction, transporting cytoskeletal proteins and soluble, non-membrane bound proteins. This moves at about 0·1–2 mm a day, the *slow transport*. In contrast, *rapid transport* carries vesicular material at about 100–400 mm a day and in the hypothalamohypophyseal tract (p. 1010) at a maximum of 2800 mm a day; it is bi-directional along the axon. This rapid flow is abolished by local treatment with colchicine, indicating that microtubules are involved in their process (p. 29). Ultrastructurally, (and most clearly demonstrated in lampreys) vesicles with side projections can be seen to line up along the outside of microtubules and may be transported by the shearing forces generated between their side-arms and microtubules. Since ligature experiments show a build-up of vesicles on *both* sides, the mechanism must allow *bi-directional movement* of organelles. Recently, a class of proteins (kinesins) have been shown to move along the external surfaces of microtubules in either direction and appear to correspond to the vesicle–microtubule links seen in electron micrographs. Such proteins may be analogous to the dynein arms of ciliary microtubules (p. 32) and act to cause

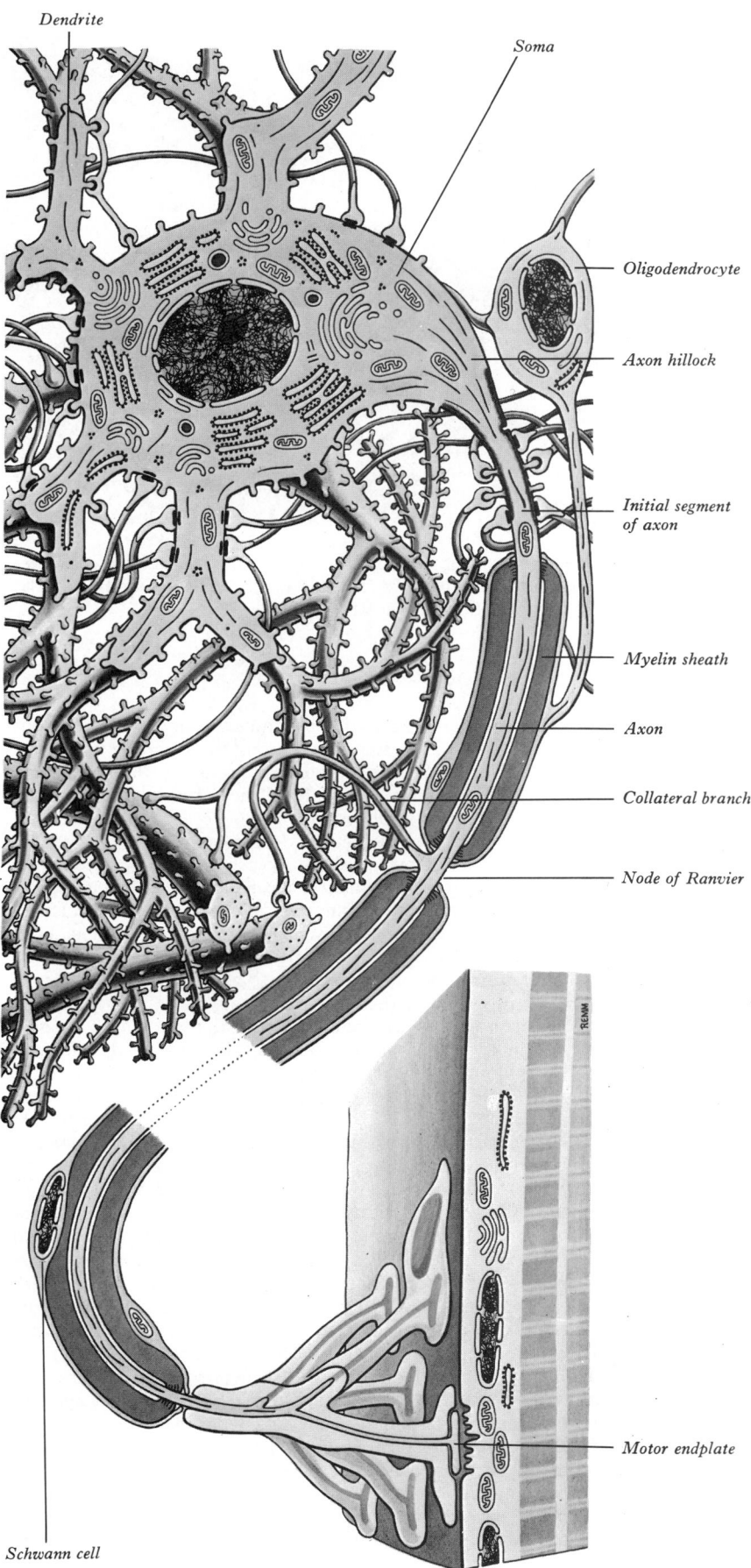

Dendrite

Soma

Oligodendrocyte

Axon hillock

Initial segment of axon

Myelin sheath

Axon

Collateral branch

Node of Ranvier

Motor endplate

Schwann cell

7.20 Schematic drawing of the ultrastructure of a motor neuron, showing part of its dendritic field (above left); the dendrites are studded with spines which are contacted by different types of synaptic terminal. The cytoplasm of the neuronal soma contains stacks of rough endoplasmic reticulum and other organelles. See text for full description.

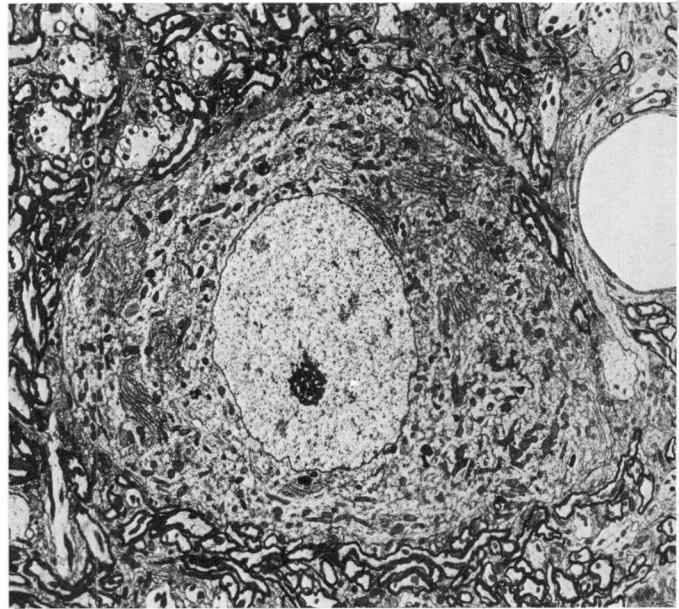

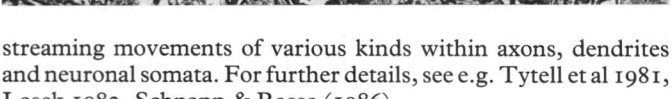

A

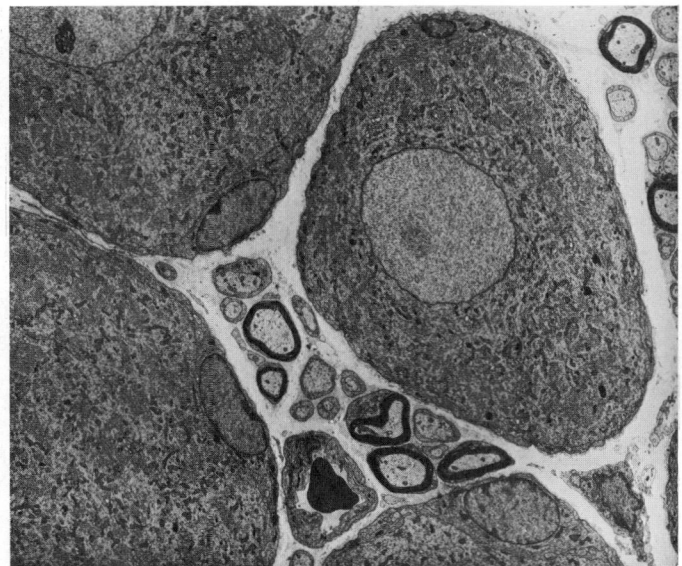

B

streaming movements of various kinds within axons, dendrites and neuronal somata. For further details, see e.g. Tytell et al 1981, Lasek 1982, Schnapp & Reese (1986).

Morphology of Synapses

The concept of an interruption in conduction at specific junctions (*synapses*) arose from the observations that there is always an irreducible delay in reflex responses to sensory stimuli and that conduction normally occurs only in one direction, along a multineuronal pathway (Sherrington 1947, Eccles 1964). With the general acceptance of the neuron doctrine, attention focussed on the localization of synapses. Studies by the Golgi technique (p. 876) demonstrated structural specializations at axonal terminals, while other methods showed similar endings in muscles. More recently, electron microscopy, correlated with physiological, pharmacological and biochemical studies have established the major features of synaptic structure (7.22–27) and molecular organization, although there are still many unsolved questions. Since chemical synapses are unidirectional, there should be some structural asymmetry at such junctions. Synapses can be formed between almost any surface region of two neurons. The most common type (7.22, 23, 25) occurs between an axon and a dendrite or a soma, the axon being expanded as a small bulb or *bouton*, either a terminal of an axonal branch (*bouton terminal*) or one of a row of bead-like endings, the axon making contact at several points, often with more than one neuron (*boutons de passage*). Boutons may synapse with: (1) *dendritic spines* or the *flat surface* of a *dendrite*; (2) *spines* or the *flat surface* of a *soma*; (3) the *axonal initial segment* or (4) *boutons of other axons*. The patterns of axonal termination vary considerably; a single axon may synapse with one neuron, e.g. climbing fibres ending on cerebellar Purkinje neurons (p. 971), or more often, with several, e.g. cerebellar parallel fibres, an extreme instance of this phenomenon (p. 971). An axon may synapse primarily with a dendritic tree or arborize around a soma. Afferent axons of different origins may synapse with specific parts of a neuron, e.g. pyramidal neurons of the visual cortex, where optic afferents synapse chiefly with the basal segments of apical dendrites. (See also afferent terminals in the cornu ammonis p. 1037.) In *synaptic glomeruli* and *synaptic cartridges* groups of axons contact dendrites of one or more neurons in regions encapsulated by neuroglia (7.27); their complex interactions are described below (p. 888).

SYNAPTIC SPINES

The spinous extensions of dendrites and soma (7.18, 20, 22–24), *spines* or *gemmules*, form receptive contacts with many afferent boutons, taking several forms. Spines usually have slender stalks

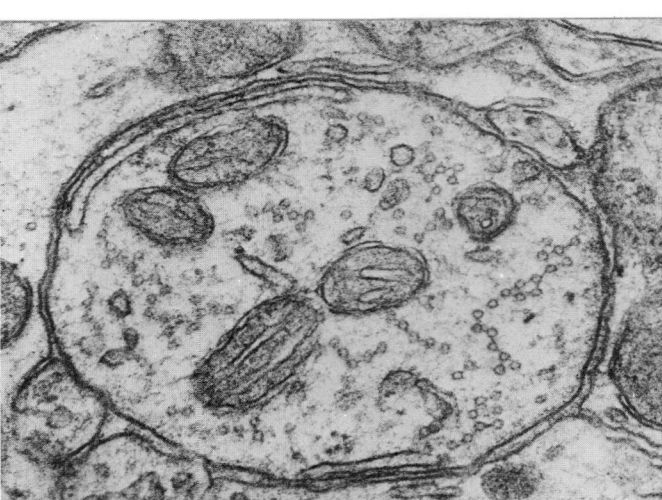

C

7.21 Electron micrographs of neuronal perikarya of the rat. A. Typical ventral grey column multipolar neuron placed amongst numerous profiles of dendrites and axons, including myelinated nerve fibres. Note the prominent nucleolus and in the cytoplasm the clusters of rough endoplasmic reticulum. B. A section through the somata of a number of unipolar spinal ganglionic neurons, associated with which are small flattened capsular (satellite) cells. Myelinated and non-myelinated nerve fibres and a capillary are also present. C. An electron micrograph of a transverse section through the initial segment of an axon, showing the rows of linked microtubules which characterize this region. Specimens 7.21A–C provided by A R Lieberman of University College, London.

and expanded distal ends; most spines are not more than 2 µm long, with one or more terminal expansions; but they can also be short and stubby, branched and bulbous (vide infra). Various experimental and theoretical studies have so far failed to demonstrate unequivocally what role the spines have in dendritic function. It has been proposed that they may limit the impact of individual synapses on the electrical state of the postsynaptic cell so that no single synapse dominates the total input. However, it has been calculated that a change in the width of a spine's stalk could change its effect markedly and it has been suggested that spines may be contractile (Crick 1982), providing a possible means of modulating neuronal excitability, e.g. in memory formation (see Koch & Poggio 1983).

ULTRASTRUCTURE OF SYNAPSES

Ultrastructurally, chemical synapses are specialized appositions between two or more neurons, with some form of asymmetrical

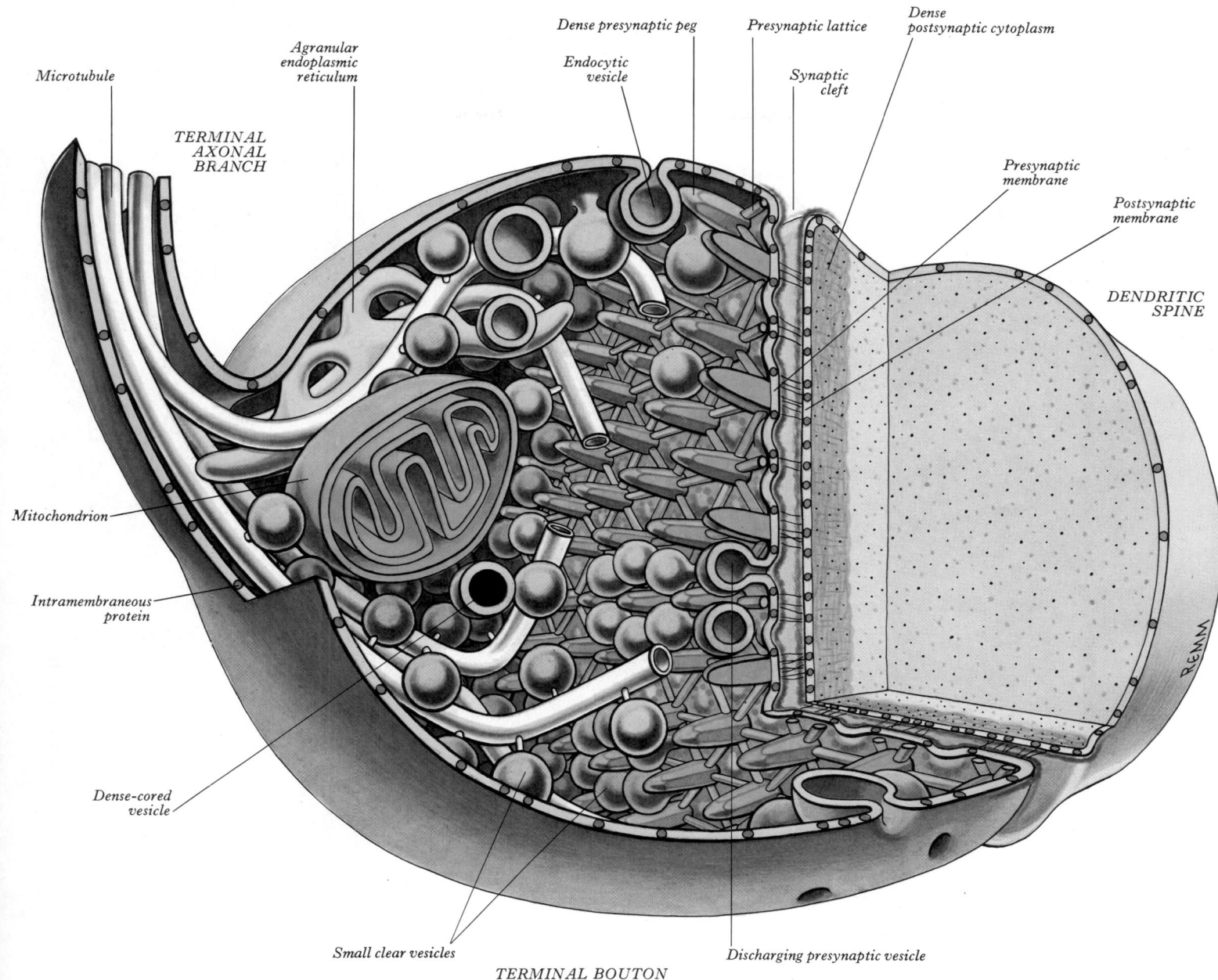

Microtubule

TERMINAL
AXONAL
BRANCH

Agranular
endoplasmic
reticulum

Dense presynaptic peg

Endocytic
vesicle

Presynaptic lattice

Synaptic
cleft

Dense
postsynaptic cytoplasm

Presynaptic
membrane

Postsynaptic
membrane

DENDRITIC
SPINE

Mitochondrion

Intramembraneous
protein

Dense-cored
vesicle

Small clear vesicles

TERMINAL BOUTON

Discharging presynaptic vesicle

7.22 The general organization of an excitatory synapse (asymmetrical, Gray type I) showing a synaptic bouton (left), cut away to show presynaptic structures, and an associated dendritic spine (right).

organization (7.22–25). They can be variously classified, e.g. by the type of neuronal process participating and the direction of transmission; thus synapses may be *axodendritic* (most common), *axosomatic* (also common), or less frequently *axoaxonic*, *dendroaxonic*, *dendrodendritic*, *somatodendritic* or *somatosomatic* (Shepherd 1974, Gray 1974). Axodendritic and axosomatic synapses occur in all regions of the central nervous system and in autonomic ganglia. Axoaxonic synapses occur between boutons of two axonal terminals (p. 888) or terminals and initial segments of other axons (7.23–25). The other types appear restricted to regions of complex interaction between larger sensory neurons and microneurons, e.g. in the thalamus.

For clarity, **axodendritic synapses** will first be detailed (7.22, 23), since they have been subjected to much research, e.g. by Gray (1959, 1961, 1969, 1974), Uchizono (1965), Pappas & Purpura (1972), Shepherd (1974), Rakic (1975) and Jones (1978). Each is an apposition of a *presynaptic bouton* (*synaptic bag*) with a postsynaptic process, separated by a *synaptic cleft*. On both sides are zones of dense cytoplasm, usually broken on the presynaptic side into several groups, and postsynaptically often extended into a filamentous *subsynaptic web*. The cleft shows fine transverse

filaments. The presynaptic bulb has numerous small *synaptic vesicles* clustered against the synaptic cleft, often filling almost the whole bouton; mitochondria, membranous sacs and occasional lysosomes also occur. When fixed in the presence of albumen, boutons show numerous axonal microtubules close to the presynaptic membrane or attached to presynaptic dendrites (Gray 1975), each microtubule surrounded by synaptic vesicles connected to it by short filaments. How these details are revealed by this unusual preparation is uncertain but the complex may be a means of conveying transmitter-containing vesicles close to the presynaptic membrane, to which they may attach and fuse, releasing their contents at specific sites ('*synaptopores*') into the synaptic cleft when the bouton is depolarized. Membrane added by vesicular fusion with the presynaptic surface is re-absorbed for recycling, as endocytic vesicles around the periphery of the bouton (Heuser & Reese 1975).

Cytochemical investigation has also revealed other details of presynaptic organization (Pfenninger & Rees 1976), mainly of the 'type I' class of synapse (vide infra). Staining with ethanolic phosphotungstic acid shows a hexagonal pattern on the cytoplasmic side of the presynaptic membrane, probably consisting of

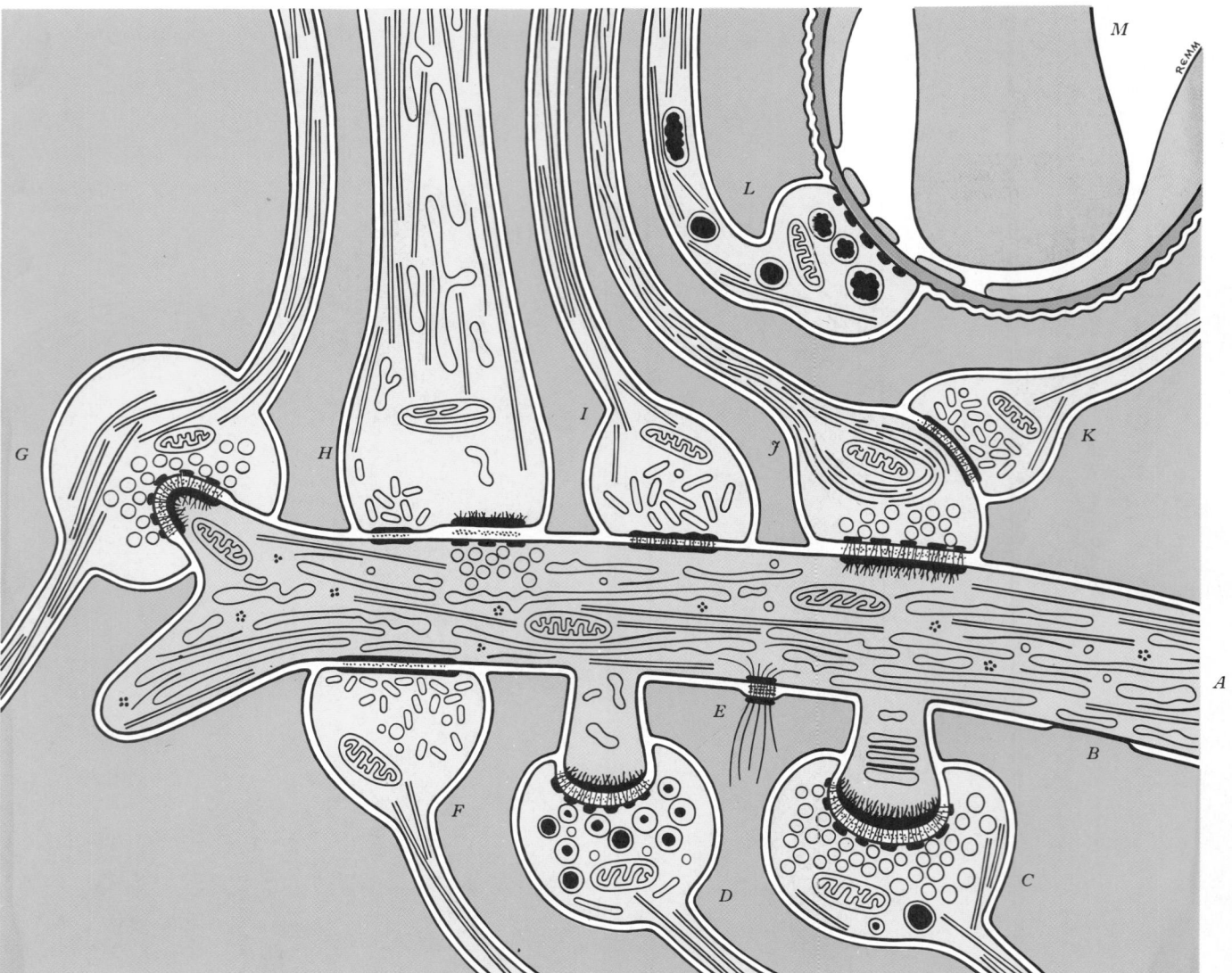

7.23 A scheme of the ultrastructure of chemical synapses, showing various junctional structures, grouped around a dendrite (A). The gap junction (B) and the desmosome (E) are without synaptic significance. Excitatory synaptic boutons are shown (C, G) containing small spherical translucent vesicles. D: a bouton with dense-cored, catecholamine-containing vesicles; F: an inhibitory synapse containing small flattened vesicles; H: a reciprocal synaptic structure between two dendritic profiles, inhibitory towards dendrite A and excitatory in the opposite direction; I: an inhibitory synapse containing large flattened vesicles. J and K: two serial synapses; J is excitatory to the dendrite; K is inhibitory to J; L: a neurosecretory ending adjacent to a vascular channel (M), surrounded by a fenestrated endothelium. All the boutons in this diagram are of the terminal type, except G which is a *bouton de passage*.

proteinaceous pillars guiding vesicles to the membranous attachment (**7.22**). Other methods, such as bismuth iodide staining with enzymic digestion, also demonstrate glycoproteins in the synaptic clefts and postsynaptic proteins in the membranes of adjacent dendrites. Freeze-fracture techniques show rings of particles in presynaptic membranes, at vesicular attachments, and numerous particles in postsynaptic membrane, perhaps representing receptors for neurotransmitter substances (Akert et al 1972) or ionic channels affected by such neurotransmitters.

Another internal constituent recently discovered is the protein synapsin I, a filamentous substance which in its non-phosphorylated state binds strongly to synaptic vesicles and other structures within the bouton. On excitation of the bouton, synapsin I becomes phosphorylated, due to calcium influx, and binds less strongly to the vesicles, perhaps allowing them to reach the presynaptic membrane and fuse with it to release transmitter substances (see Linstedt & Kelly 1987).

In some sites the neurofilaments of an afferent axon enter the bouton to form a variable loop, visible by light microscopy with silver stains. On the postsynaptic side there are often membranous structures in the cytoplasm, such as parallel cisterns forming the *spine apparatus* of some mammalian dendrites. Glial processes commonly surround the synaptic structures but do not normally enter the synaptic clefts. The presynaptic surface is flat where it adjoins the smooth surface of a dendrite or soma but at endings on postsynaptic spines it may be highly curved round one or more of these projections (**7.23, 24, 25**). In some sites the dendritic surface forms a large spike, interdigitating with the presynaptic surface. Conversely, dendritic expansions may be deeply invaginated by axonal terminals.

FUNCTIONAL CLASSIFICATION OF SYNAPSES

Since synapses are inhibitory, excitatory or may affect neurons in other ways and different transmitters are released at them, it is to be expected that various structural varieties exist. Such indeed is the case, although it is as yet far from clear how many or, often, what neurotransmitters they release. However, for practical purposes they may be classed by the shape of their synaptic vesicles and the arrangement of the pre- and postsynaptic cytoplasmic densities into several types. Synapses rich in *catecholamines* (noradrenalin, adrenalin, dopamine) contain large (40–60 nm) *dense-cored vesicles* (**7.23**) like those of adrenal chromaffin cells; at some sites they are inhibitory and at others excitatory, depending on the nature of the receptor molecules in the postsynaptic membrane. *Neurosecretory endings*, like those in posterior pituitary, contain huge (50–200 nm), irregular dense-cored vesicles (**7.25**). Among other central synapses Gray (1961) recognized, in

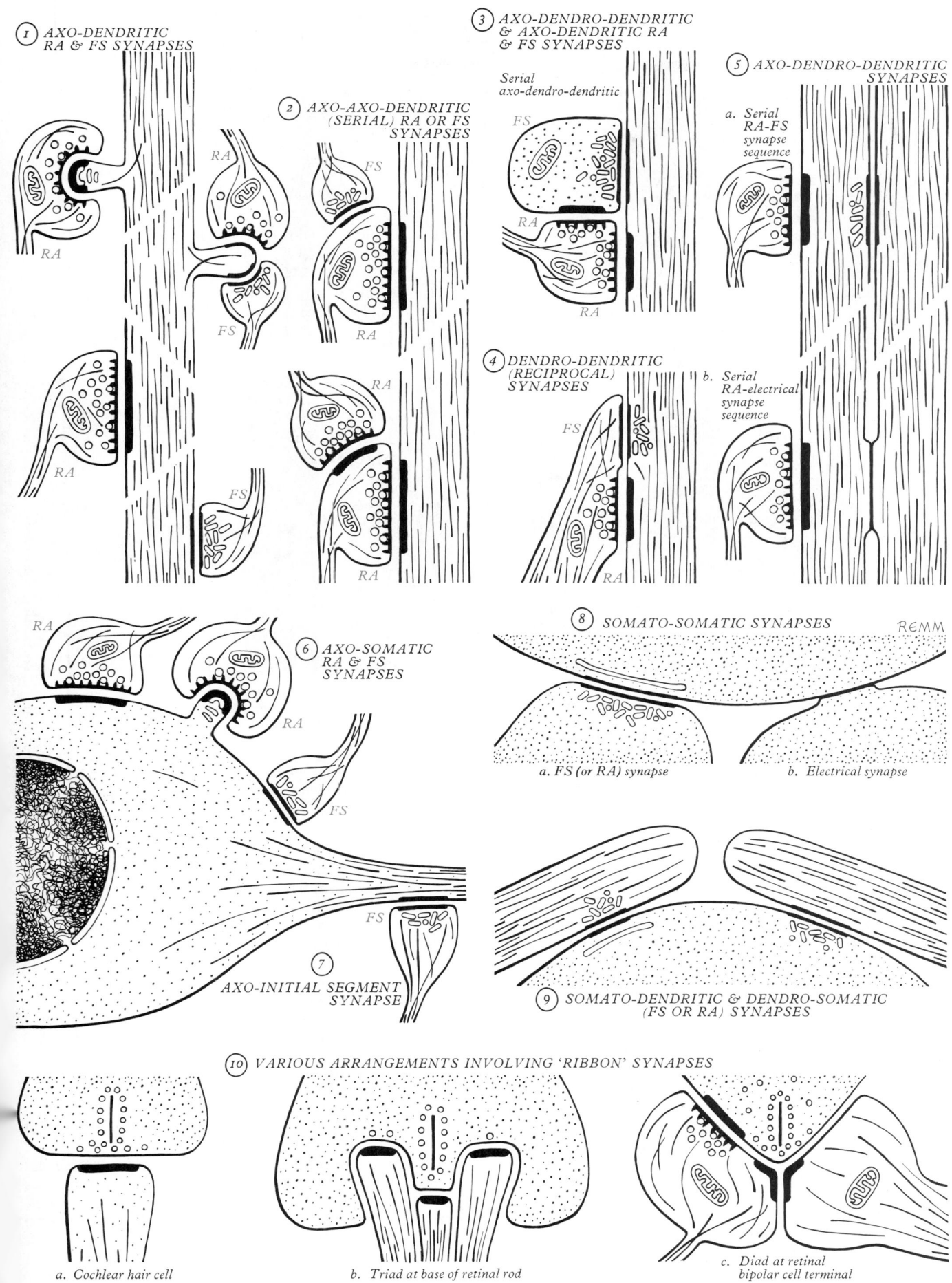

① AXO-DENDRITIC RA & FS SYNAPSES

② AXO-AXO-DENDRITIC (SERIAL) RA OR FS SYNAPSES

③ AXO-DENDRO-DENDRITIC & AXO-DENDRITIC RA & FS SYNAPSES

Serial axo-dendro-dendritic

④ DENDRO-DENDRITIC (RECIPROCAL) SYNAPSES

⑤ AXO-DENDRO-DENDRITIC SYNAPSES

a. Serial RA-FS synapse sequence

b. Serial RA-electrical synapse sequence

⑥ AXO-SOMATIC RA & FS SYNAPSES

⑦ AXO-INITIAL SEGMENT SYNAPSE

⑧ SOMATO-SOMATIC SYNAPSES

a. FS (or RA) synapse   b. Electrical synapse

⑨ SOMATO-DENDRITIC & DENDRO-SOMATIC (FS OR RA) SYNAPSES

⑩ VARIOUS ARRANGEMENTS INVOLVING 'RIBBON' SYNAPSES

a. Cochlear hair cell   b. Triad at base of retinal rod   c. Diad at retinal bipolar cell terminal

7.24 Various types of simple and multiple arrangements of synapse involving 'asymmetrical' synapses with rounded vesicles (RA), 'symmetrical' synapses with flattened vesicles (FS) and electrical synapses. For further details, see text.

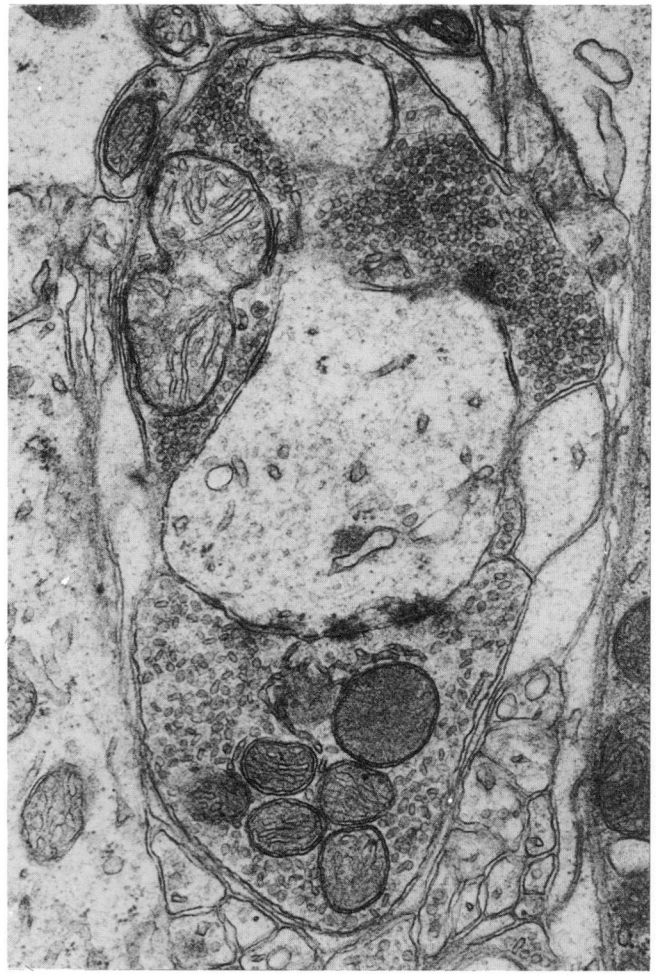

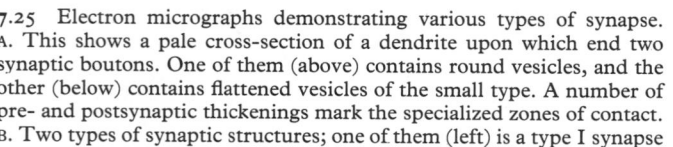

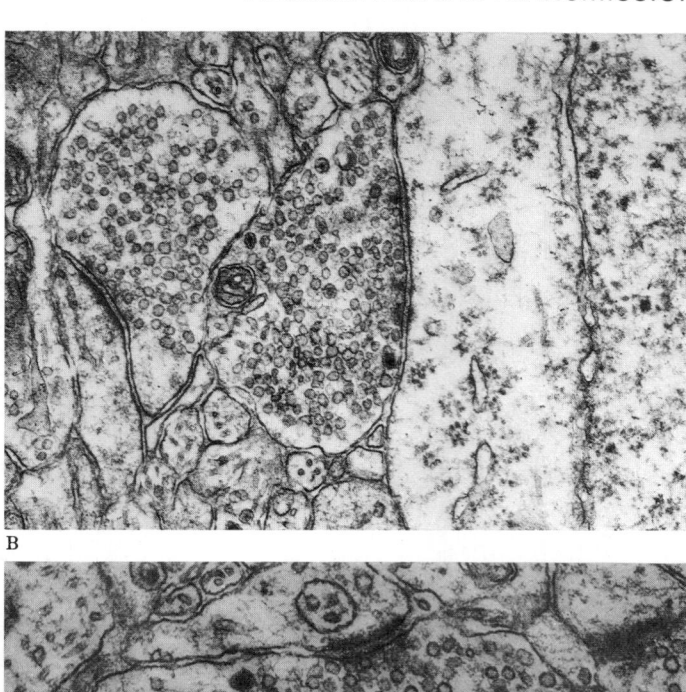

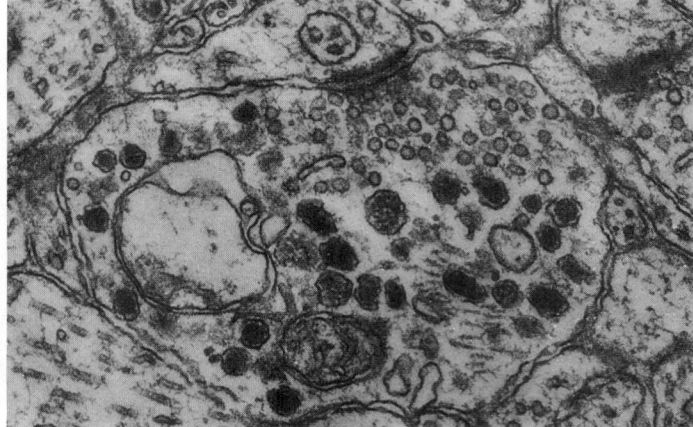

A   C

7.25 Electron micrographs demonstrating various types of synapse. A. This shows a pale cross-section of a dendrite upon which end two synaptic boutons. One of them (above) contains round vesicles, and the other (below) contains flattened vesicles of the small type. A number of pre- and postsynaptic thickenings mark the specialized zones of contact. B. Two types of synaptic structures; one of them (left) is a type I synapse between an axon terminal containing round vesicles and a dendritic spine; the other (right) is a type II synapse between an axon terminal containing pleomorphic vesicles, and the surface of a neuronal soma. C. A type I synapse containing both small, round, clear vesicles and also large dense-cored vesicles of the neurosecretory type. A and B provided by A R Lieberman, Department of Anatomy, University College, London.

osmium-fixed material, two classes: *type I* in which the subsynaptic zone of dense cytoplasm is thicker than on the presynaptic side; and a *type II* in which the two zones are more symmetrical but thinner (7.23, 24). The widths of synaptic clefts also differ, being about 30 nm in type I and 20 nm in type II. With aldehyde perfusion for fixation, these two types appeared to contain distinct kinds of synaptic vesicles, type I having small spherical vesicles about 50 nm in diameter and type II showing a variety of flat forms (7.25A). Where electrophysiological data were available, as in the cerebellum, *type I synapses* could be linked with *excitation* and *type II* with *inhibition* (Uchizono 1965). Other reports indicate a more complex classification; it is possible to distinguish between large (25 × 60 nm) and small (15 × 40 nm) flat vesicles (Pinching & Powell 1971), and also between discoidal, fusiform and irregular types (Dennison 1971). Other studies show that in fresh specimens all vesicles may be spherical, some flattening on exposure to the buffers used in preparative procedures (Bodian 1970). Synaptic endings also show *mixed populations* of flat and spherical vesicles in varying proportions, depending on the osmotic strength of the fixative. Though such appearances are partly artificial, they indicate real chemical differences. Classification of synaptic boutons has been attempted in different regions by criteria such as vesicular shape and size, distribution of pre- and postsynaptic cytoplasmic thickenings, size of bouton, presence of mitochondria, etc. A common scheme (Guillery 1969, Partlow et al 1977) designates the type I synapse of Gray as a round 'vesicle-asymmetrical thickening' (R.A.) and Gray's type II as a flat 'vesicle-symmetrical thickening' (F.S.); various subdivisions of

these two categories have also been suggested (Robson & Hall 1977). While such schemes may have descriptive value, the state of knowledge of transmitters at various synapses and their functional status is generally too slight to justify more than crude classification. Mixed populations of small, clear vesicles and large, dense-cored ones commonly occur, sometimes correlated with peptide transmitter substance P, serotonin, or enkephalins (Beaudet & Descarries 1979, Pickel et al 1979) (vide infra).

Some synapses may lack specialized contact zones, being areas of transmitter release from the varicosities of non-myelinated axons, such effects sometimes being diffuse (paracrine, see p. 887) as in the aminergic pathways of basal nuclei (p. 1076).

The freeze-fracture appearances of type II synapses lack the pre- and postsynaptic intramembrane particles typical of type I, suggesting basic differences in mechanisms of release and action of the transmitter at the postsynaptic membrane (Landis et al 1974).

The junctions between the retinal receptors and bipolar neurons, and between cochlear hair or the vestibular sensory cells and their afferent endings, are both synapse-like structures, differing from the usual pattern in that vesicles are clustered around an internal cytoplasmic rodlet (Flock 1971, Boycott & Kolb 1973).

NEUROCHEMICAL TRANSMISSION

Neurotransmission at all the synapses so far described in this account is chemical, involving the release of neurotransmitters from synaptic vesicles into a synaptic cleft to cause a change in the permeability and therefore the electrical polarization of the postsynaptic membrane. Such alterations may be excitatory (depolarizing) or inhibitory (hyperpolarizing), and are generally rapid and short-lived, as the transmitter is quickly inactivated either by an extracellular enzyme (e.g. acetylcholinesterase) or by uptake into the presynaptic process. The presynaptic ending may also have more complex actions on the target cell, causing much more prolonged or even permanent changes in cell behaviour. Indeed it is becoming increasingly clear that classic, rapid neurotransmission, measured in milliseconds, represents one extreme of a broad spectrum of neural effects which also include slower electrical changes lasting for seconds or minutes, or long-lasting, even permanent cellular changes due to alterations in gene expression, e.g. resulting in the synthesis of more receptor molecules for the postsynaptic surface or altering the numbers of synapses, the extent of the dendritic tree, the growth of axons and the overall metabolism of the cell. Permanent or semi-permanent changes of these types provide a physical basis for memory processes in the nervous system; they are also reflected in the trophic interactions between neurons which occur not only during development (where they are most obvious, see e.g. growth factors, p. 919) but form the constantly fluctuating background to the day-to-day activities of the mature nervous system, seen most dramatically when neurons alter their behaviour and morphology after damage has deprived them of their normal synaptic input. *Neurohormones* must also be included in this range; these are synthesized in neurons and released into the blood circulation by exocytosis at synaptic terminal-like structures and, of course, may act at great distances from their site of secretion. Similarly, neurons may also secrete into the cerebrospinal fluid or into the intercellular spaces of the nervous system (vide infra) to affect many other neurons, diffusely and at a distance. To encompass this wide range of phenomena the general term *neuromediation* has been used, the chemicals involved being *neuromediators*.

Typically, there are two major events during synaptic activity: firstly the release of a neurochemical substance contained within synaptic vesicles by exocytosis (p. 29). This occurs when an action potential (or in some cells, a graded depolarization) reaches the presynaptic terminal, causing an influx of calcium and initiating a series of chemical events which cause synaptic vesicles to fuse with the presynaptic membrane, thus ejecting their contents into the synaptic cleft. Secondly, the chemical diffuses across the narrow gap, binds to receptor molecules embedded in the postsynaptic membrane and alters the permeability of this membrane to various ions (calcium, sodium, potassium or chloride). In some cases (e.g. the 'nicotinic' receptor for acetylcholine), the receptor molecule is also itself a transmembrane ion channel so that, when the neurotransmitter binds to it, there is a rapid, transient opening of the channel, which causes an electrical change lasting only a few milliseconds. In other cases, e.g. in the activation of the 'muscarinic' acetylcholine receptor, the changes take much longer to develop and often last for seconds or even minutes; where they have been examined in detail, all such slow responses operate by means of a receptor molecule in the postsynaptic membrane which activates a 'second messenger' system within the postsynaptic cytoplasm, e.g. involving the activation of the adenyl cyclase-cyclic AMP pathway, leading indirectly to the opening (or closing) of ionic channels at its surface. Therefore the action of a neuromediator depends not only on its chemical structure, but also on the type of receptor system present; different receptors may be linked to particular ion channels, e.g. enkephalin μ receptors to potassium channels and enkephalin κ receptors to calcium channels. A number of classic transmitters have actions both of the fast, direct type and of the slow indirect, e.g. acetylcholine, γ-aminobutyric acid and glutamate.

Second messenger systems are also thought to be responsible for the much wider and long-lasting effects on the postsynaptic cell; after repeated synaptic activity, they may alter the activity of enzymes in the postsynaptic cytoplasm to cause changes in the arrangement of the cytoskeleton in this region and thus the size, shape, stability and electrical properties of the postsynaptic struc-

ture may alter semi-permanently. In some glands, the intercellular 'gap' junctions (p. 24) become non-conducting, so uncoupling adjacent cells from metabolic intercommunication. The numbers of receptor molecules in a postsynaptic membrane may also alter as a delayed response to repetitive activity (or to inactivity), to enhance or diminish the effectiveness of the synapse. Second messengers may reach the nucleus of the cell to activate or suppress DNA transcription and so cause major alterations in the overall metabolism, including the synthesis of structural molecules, neurotransmitters and the enzymes of energy production. All such changes affect the electrical behaviour of individual cells and, if multiplied by similar responses in large neuronal populations, enable neural networks to modulate their behaviour according to the types and magnitudes of the influences reaching them.

Neuromodulators. A further twist in the story is that some neuromediators do not themselves have noticeable effects on the postsynaptic membrane but affect its responses to other neuromediators, either enhancing their activity (increasing the immediate response in size, or causing a prolongation) or perhaps limiting or inhibiting their action. Such substances have been called *neuromodulators*: a single synaptic terminal may contain one or more neuromodulators in company with a neurotransmitter, usually (though not always) in separate vesicles; the mode of action of neuromodulators is still not entirely clear but may involve either a direct binding to, and alteration in the response of, the receptor for the *neurotransmitter* or it may cause a second messenger-mediated change in the postsynaptic membrane which would then alter its response to the neurotransmitter. In some cases there may be a difference in the number of nerve impulses needed for the release, the neuromodulator only being secreted after more sustained stimulation of the ending.

Neuropeptides, which are extremely numerous (see pp. 891–892), are apparently nearly all modulators, at least in some of their actions. They are stored within dense granular synaptic vesicles of various sizes and appearances. Although they are distributed widely through the nervous system, certain areas have high concentrations and large numbers of different types. This finding has been analysed in detail by Nieuwenhuys (1986) in an extensive review and theoretical discussion of neuromediators in relation to classic neuroanatomy (see also p. 992). It is well known that in many parts of the nervous system, both central and peripheral, axon terminals have zones of transmitter release which are not closely apposed to a postsynaptic membrane but separated from it by a variable intercellular gap, perhaps as much as 0·2 μm ('synapses á distance', see e.g. Chan-Palay 1983). Chemicals released from such axons may diffuse quite widely through the intercellular spaces to arrive at receptor sites on the surfaces of several neighbouring cells. Such effects are analogous to those of neuro-endocrine endings, although the chemicals released are of course restricted to a limited volume and are said to be *paracrine*. In a given area of grey matter (or of non-striated muscle), several different neurotransmitters and neuromodulators from different sources may impinge on the mixed population of receptor molecules on a particular target cell, giving a great variety of possible responses, so that the level of excitation and therefore of action potential generation in a neuronal population fluctuates constantly as information from different sources converges on its dendrites and somata. Such areas of multiple interaction are frequently associated with neural activities directed towards homeostasis (e.g. ionic regulation, cardiovascular control), the survival of the individual in more extreme circumstances and the continuance of its genes, i.e. reproduction.

Synaptosomes

The term *synaptosomes* (synaptic 'bags') is the name given to an experimental preparation consisting of axon terminals, separable by fractionation techniques from brain homogenates. Where acetylcholine is the transmitter, enzymes (e.g. choline acetyltransferase) able to synthesize it from choline exist in synaptosomes, the transmitter being stored in or near the synaptic vesicles. Catecholamines and other possible transmitters have also been demonstrated thus in synaptosomes (vide infra).

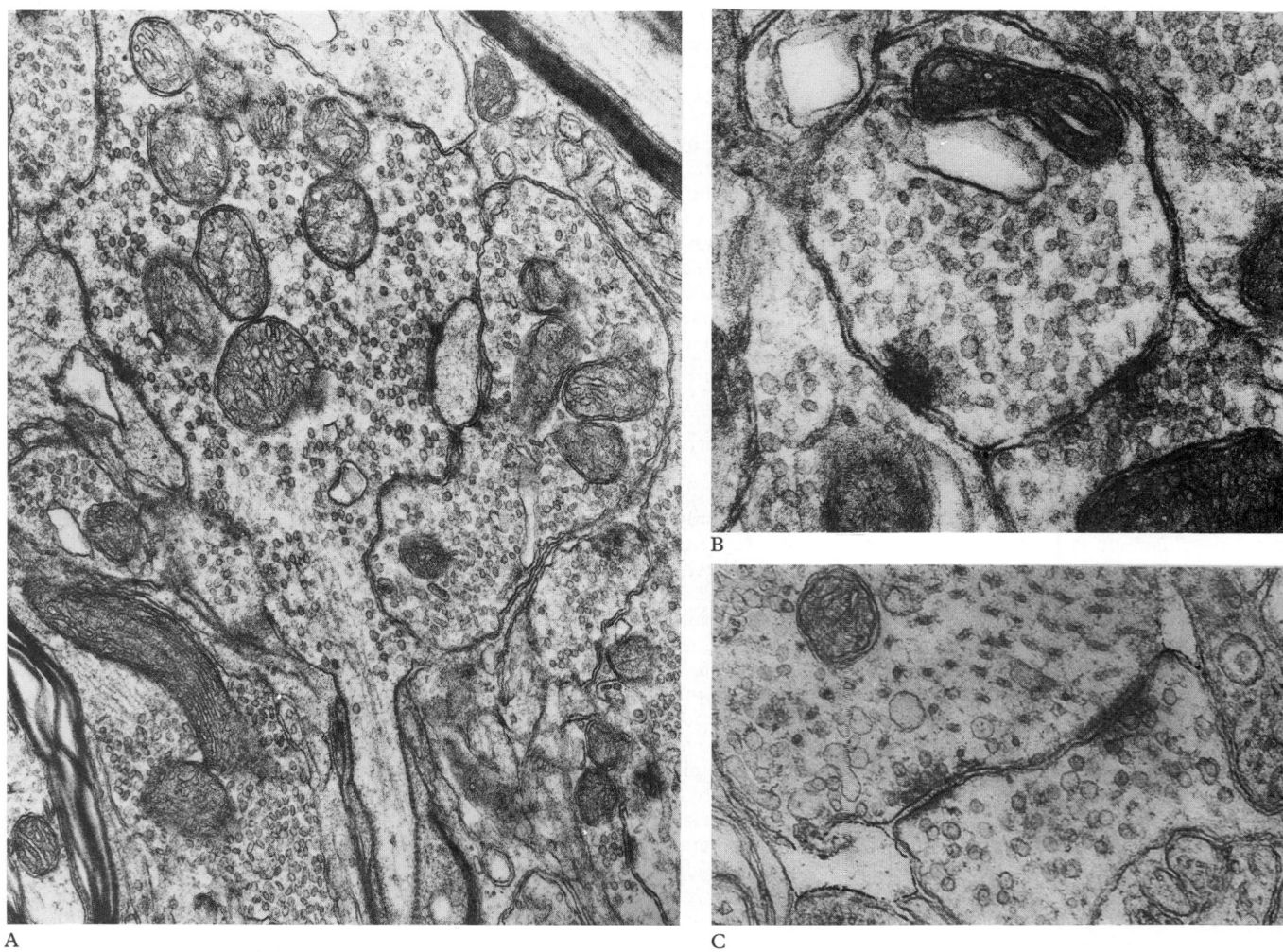

A   C

7.26 Electron micrographs of complex arrangements of synapses. A. This shows a large terminal bouton of an optic nerve afferent fibre, which is making contact with a number of postsynaptic processes, in the dorsal lateral geniculate nucleus of the rat. One of the postsynaptic processes (right) also receives a synaptic contact from a bouton containing flattened vesicles. B. Three neuronal processes in serial contact. On the lower right, a process containing round vesicles synapses with a second process (centre) containing flattened vesicles; in turn the latter makes contact with a third process (lower left): specimen from the dorsal lateral geniculate nucleus of the rat. C. This demonstrates reciprocal synapses between two neuronal processes in the olfactory bulb. A and B provided by A R Lieberman of University College, London.

ELECTRICAL SYNAPSES

In mammals chemical transmission at synapses is by far the most common but in lower vertebrates and some invertebrates electrical synapses are abundant where speed or synchrony of neural response is a requisite. Examples are 'spoon' endings in the chick ciliary ganglion, 'club' endings on giant Mauthner cells in the medulla in many fishes, the electromotor synapses in electrogenic fishes and the giant nerve fibre systems of crayfish and earthworm. Such synapses are like 'gap' or electrical junctions in cardiac (pp. 24, 557) and non-striated muscle (p. 562) and are much faster than chemical synapses. Some invertebrate synapses of this kind are bi-directional. Similar junctions occur in mammals, e.g. in the superior olivary and vestibular nuclei, cerebellar and cerebral cortices; but the corroboration of their transmissive properties is not yet available (Gwyn et al 1977).

ORGANIZATION OF SYNAPTIC GROUPS

The effects of synaptic endings on other neurons partly depend on their arrangement and relation to other synapses. In *serial synapses* (7.23, 24, 26B) some terminal bulbs end on others to modify their response to afferent volleys, a possible basis of *presynaptic inhibition* (e.g. in the spinal cord) and of *presynaptic facilitation*, depending on the type of synapse. *Reciprocal synapses* (7.24, 26C), first found in the olfactory bulb and lateral geniculate body and later in many other sites, transmit both ways by staggered synaptic zones on opposite sides of the synaptic cleft; they are usually excitatory in one direction and inhibitory in the other. They may be the basis of lateral inhibition (Rall et al 1966, Shepherd 1974, 1978). Other varieties of serial synapses between different axonal, dendritic and somal structures have been described, some of them depicted in 7.24 (Gray 1974, Colonnier 1974, Shepherd 1974, 1978).

In **synaptic glomeruli** (7.27A) several boutons synapse with dendrites and sometimes with each other in localized regions of neuropil, usually within layers of gliocytes (Szentágothai 1970). Where microneurons are involved, synaptic patterns may become exceedingly complex, including various excitatory and inhibitory relations between neurites; these may occupy extensive zones as **synaptic clusters**. In **synaptic** or **neuropil cartridges** part of a dendrite is enclosed by a glial sheath to isolate a cylindrical zone of synaptic endings on the spines and dendritic surface between them (7.27B). Astrocytes (p. 893) may also isolate or juxtapose groups of interacting neurons. Though such arrangements of synapses in circumscribed areas is well established, the disposition of different types of synapse on neuronal surfaces is less certain. Inhibitory synapses can reduce the excitability of a neuron by their actions on dendrites and the soma, or may exert a total inhibitory blockade at some strategically-placed sites such as the initial segment of the axon (7.19), or the terminals of basket neuron axons synapsing with the 'pre-axons' of Purkinje neurons (p. 971).

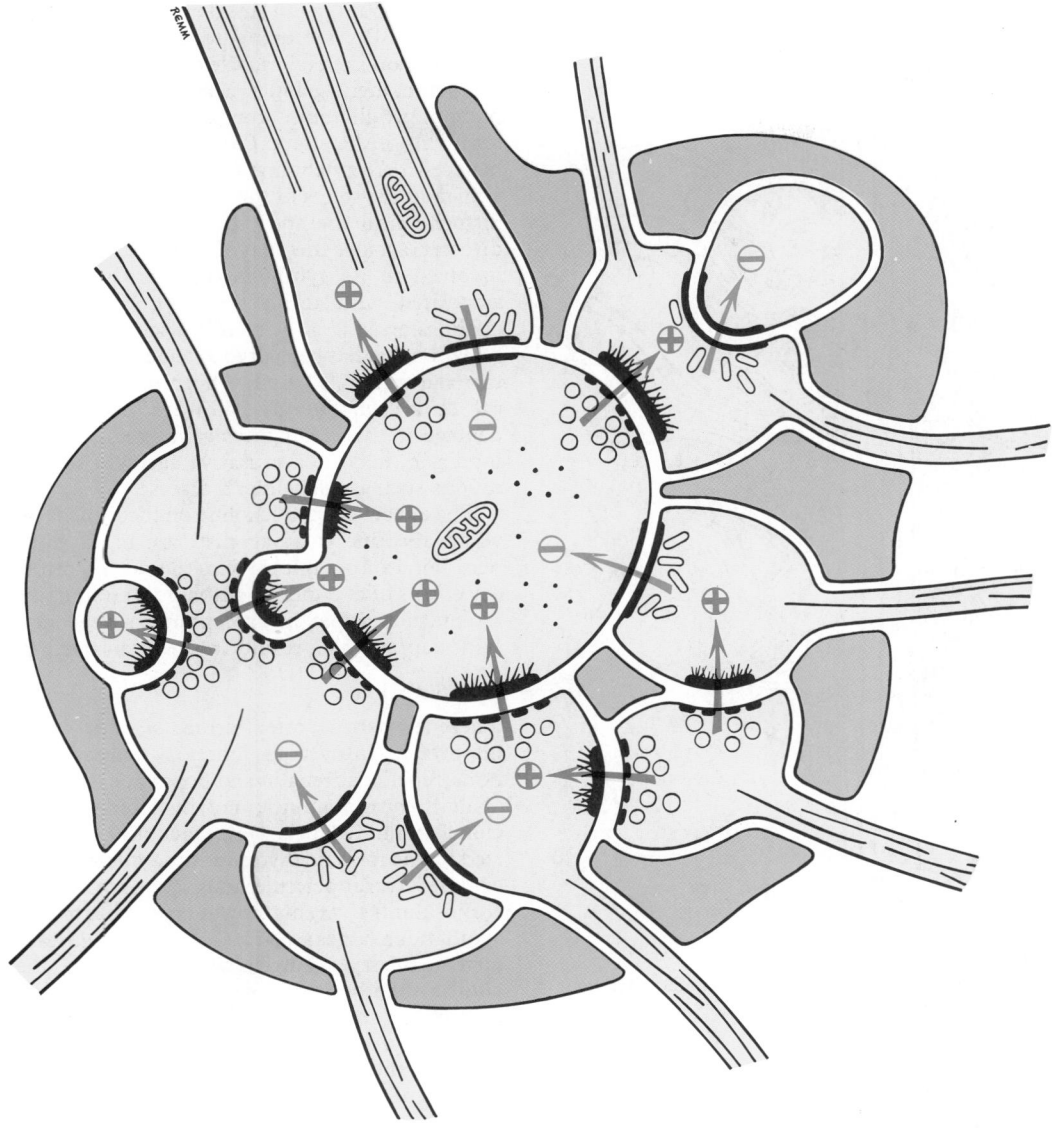

7.27A   A synaptic glomerulus, showing various arrangements of synapses grouped around a centrally placed terminal dendritic expansion, seen in cross-section. Both excitatory (+) and inhibitory (−) synapses are shown; the direction of transmission is indicated by arrows (red for excitation, blue for inhibition). A glial capsule surrounds the whole complex.

DEVELOPMENT AND PLASTICITY OF SYNAPSES

Embryonic synapses first appear as inconspicuous dense zones flanking synaptic clefts (Bodian 1970, Pfenninger & Rees 1976). Immature synapses often appear after birth, suggesting that they may be labile, recruitable for transmission and dispensable when redundant. This is implicit in some theories of memory, which postulate that synapses are modifiable by frequency of use, to establish *preferential conduction pathways*. Recording from hippocampal neurons suggests that in some locations, even a brief synaptic activity (e.g. 1 second) can increase the strength of the synapse for some hours or longer (long-term potentiation). The mechanism of this appears to involve cytoskeletal changes modulated by calcium activation of protein kinases within the postsynaptic process (see e.g. Smith 1987). However, such dramatic changes may be limited to regions of the nervous system involved in memory storage; in other areas smaller adjustments may occur with the use or disuse of synapses, e.g. in the number of postsynaptic receptor sites. During early postnatal life it is well-established that a normal increase in numbers and sizes of synapses and dendritic spines depends on the degree of neural activity, as shown in the visual cortex of young animals temporarily blinded in one eye, where dendritic spines of cortical neurons show a greatly impaired development (Rothblat & Schwarz 1979).

NEUROMEDIATORS

Until recently the classes of chemicals known to be involved in chemical synapses were limited to a fairly small group of 'classic' neurotransmitters: acetylcholine, noradrenalin, adrenalin, dopamine and histamine, all with quite well defined 'fast' effects on other neurons, muscle cells or glands and satisfying the criteria laid down by the classic pharmacologists such as Henry Dale. With increasing research into the chemistry of the brain and more detailed electrophysiological and pharmacological studies, it became increasingly obvious that within the nervous system there are many synaptic interactions that cannot be explained in terms of these neurotransmitters and that other substances, particularly some amino acids such as glutamate, glycine, aspartate, gamma-aminobutyric acid and the monoamine, serotonin, were good transmitter candidates. With the development of monoclonal antibody technology it was found that many substances which hitherto had been observed only in relation to the hormonal secretions of the hypophysis, or of the entero-endocrine system of the alimentary tract, can be detected widely and often systematically throughout the central and peripheral parts of the nervous system (see e.g. Björklund et al 1984, Nieuwenhuys 1986). Many of these are peptides, some of them closely related to each other chemically or even derived from a single large 'parent' molecule by its fragmentation. When isolated and tested by

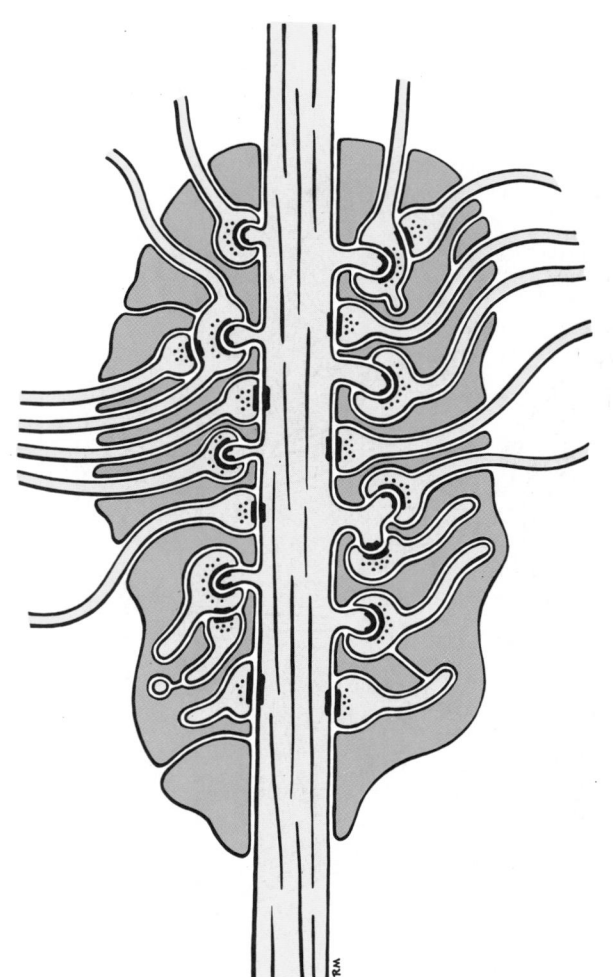

7.27B A synaptic cartridge, with synapses grouped around a segment of a dendrite and enclosed within a glial capsule (green).

neuropharmacological methods, these substances have been shown to have a great range of effects on neurons, non-striated muscle and glands, frequently as neuromodulators (vide supra). There are in excess of 60 known neuropeptides at the time of writing and their numbers are likely to increase as the techniques of molecular genetics are applied to the nervous system. However, it is highly problematical as to which of these substances actually represent functional neuromediators; some of them may be intermediate metabolites, or even the breakdown products of other neuromediators. Because more than one may be present at a single synapse, it has proved difficult to apply the rigorous criteria of classic neuropharmacology and many of these chemicals are therefore often referred to as only *putative* neuromediators. Nevertheless, various of these chemicals have been localized within synaptic vesicles of specific terminals and it has been possible to relate their presence to the physiological properties of known cells. Some of the postsynaptic receptor molecules which bind these substances have also been characterized, giving the corresponding neuromediator a measure of pharmacological respectability.

A comprehensive view of this field, were it indeed possible, is clearly outside the scope of the present account. It must be noted, though, that these substances, their neurobiology and distributions within the nervous system are of major importance in the functional organization of the brain, spinal cord and peripheral nerves and of neural interactions with the rest of the body. Before discussing these topics in the context of neuroanatomical pathways and centres (see e.g. p. 992), we will consider some of the better known neuromediators (see Table, p. 891).

Acetylcholine is perhaps the most extensively studied neurotransmitter of the classic type; its precursor, choline, is synthesized in the neuronal cell body and transported to the axons terminals where it is acetylated and stored in 'clear' spherical vesicles about 50 nm in diameter. This neurotransmitter is synthesized by motor neurons and released at all their motor terminals on skeletal muscle and at synapses in parasympathetic and sympathetic ganglia. Many parasympathetic and some sympathetic ganglionic neurons are also cholinergic, as are some of the terminals of the efferent olivo-cochlear tract ending on cochlear hair cells. Acetylcholine has been demonstrated in many major non-motor systems of the brain and spinal cord, using immuno-histochemical methods, by labelling neurons with antibodies directed against the enzyme choline acetyl transferase which is involved in its synthesis; in some sites acetylcholine is also associated with the degradative extracellular enzyme acetyl cholinesterase (e.g. at neuromuscular junctions, p. 904), an association that has been used extensively to indicate the presence of the transmitter. There has been much recent clinical interest in the central cholinergic pathways, since there is evidence that damage to this neurochemical system, particularly in the basal forebrain, may be a causative agent in some degenerative conditions such as Alzheimer's disease.

The effects of acetylcholine on nicotinic receptors (i.e. those in which nicotine is an agonist) are rapid and excitatory; in the peripheral autonomic system, the slower, more sustained excitatory effects of cholinergic autonomic endings are mediated by different (muscarinic, m_1) receptors via a second messenger system (vide supra) which results in the closure of a potassium channel.

Monoamines

These substances, often termed *biogenic amines* because of their importance to nervous function, consist of: the *catecholamines* noradrenalin, adrenalin and dopamine; the *indoleamine* serotonin (5-hydroxytryptamine); and histamine. Before suitable monoclonal antibody labels were available, catecholamines were localized by the formaldehyde-induced fluorescence (Falck et al 1962) or the glyoxylic acid methods (e.g. Axelsson et al 1973, Lindvall & Björklund 1974) but now it is possible to detect specifically the synthetic enzymes associated with each substance, using immuno-histochemistry. Neurons which synthesize the monoamines include sympathetic ganglia and their homologues, the chromaffin cells of the adrenal medulla and paraganglia; within the central nervous system, their cell bodies lie chiefly in the brain stem, although their axons spread and ramify exceedingly widely into all parts of the central nervous system. Monoamine cells are also present in the retina (p. 1202).

Noradrenalin is the chief transmitter present in sympathetic ganglionic neurons and is released at their endings in various tissues, notably non-striated muscle and glands but also in other sites including adipose and haemolymphopoietic tissue (p. 836) and the corneal epithelium. This neuromediator is also present at synaptic endings widely distributed within the central nervous system; many of them are terminals of neuronal somata situated in the locus coeruleus in the medullary floor. The actions of noradrenalin depend on its site of action, varying with the type of postsynaptic receptor. In some cases, e.g. the neurons of the submucosal plexus of the intestine and of the locus coeruleus, it is strongly inhibitory due to its actions in closing a potassium channel via the α2 adrenergic receptor, whereas the β receptors, e.g. of vascular non-striated muscle, mediate depolarization and therefore vasoconstriction.

Adrenalin is also present in central and peripheral nervous pathways and occurs alongside noradrenalin in the adrenal medulla. Both of these monoamines are found in dense-cored synaptic vesicles about 50 nm across. Dopamine is a neuromediator of considerable neurobiological and clinical importance, present mainly in the central nervous system, where it occurs in neurons with cell bodies in the telencephalon, diencephalon and mesencephalon. A major dopaminergic neuronal population in the midbrain constitutes the *substantia nigra*, so called because its cells contain neuromelanin, a black granular by-product of dopamine synthesis (p. 982). Dopaminergic endings are particularly numerous in the corpus striatum, limbic system and cerebral cortex and pathological reductions of dopaminergic activity have widespread effects on motor control, affective behaviour and other neural activities, a seen in Parkinson's syndrome. Structurally, dopaminergic synapses contain numerous dense-cored vesicles resembling those of noradrenalin.

Serotonin and *histamine* occur in neurons mainly in the central nervous system; serotonin is synthesized chiefly in small median neuronal clusters of the brain stem (mainly in the raphe nuclei), but their axons spread and branch extensively throughout the entire brain and spinal cord. Synaptic terminals contain rounded, clear vesicles about 50 nm across and are of the asymmetrical type. Histaminergic neurons appear to be relatively few and are restricted largely to the hypothalamus.

The amino acids. The best understood of these is γ-amino butyric acid (GABA) which is a major inhibitory transmitter released at the terminals of local circuit neurons within the brain stem and spinal cord (e.g. the recurrent inhibitory Renshaw loop, p. 865), within the cerebellum (as the main transmitter of Purkinje neurons) and elsewhere. It is located in flattened or pleiomorphic vesicles within symmetrical synapses, at which it may be inhibitory to the postsynaptic neuron or may give either presynaptic inhibition or facilitation, depending on the synaptic arrangement (p. 888). A major synthetic enzyme is glutamic acid decarboxylase (GAD) which can be used as an immunochemical marker for its neuroanatomical study.

Glutamate and *aspartate* are considered to be major excitatory transmitters present within widely distributed cell bodies and fibres of the central nervous system, including the major projection pathways from the cortex to the thalamus, tectum, substantia nigra, pontine nuclei, as well as many other parts of the brain and spinal cord. They have been located in the central terminals of the auditory and trigeminal nerves, and glutamate is present in the terminals of parallel fibres ending on Purkinje cells in the cerebellum, amongst other locations. Structurally, they are associated with asymmetrical synapses containing small (30 nm) round, clear synaptic vesicles. *Glycine* is another important, well established inhibitory transmitter of the central nervous system, particularly the lower brain stem and spinal cord, where it is mainly located in local circuit neurons. *Taurine*, another inhibitory neuromediator, is present widely in the central nervous system including the cerebral cortex and cerebellum; there is evidence, however, that it may act as a neuromodulator rather than a transmitter at these sites.

Neuropeptides

As already mentioned, the number of peptide neuromediators, either established or putative, is extremely large (see Table). Many of them co-exist with other neuromediators in the same synaptic terminals, from one to three types often sharing a particular ending with a well-established neurotransmitter; in some cases they have been shown to be present within the same synaptic vesicles. Some of them occur both centrally and peripherally in the nervous system and are particularly represented in the ganglion cells and peripheral terminals of the autonomic system (see Burnstock 1986), where their physiological actions are more accessible to study; others are entirely restricted to the central nervous system. In view of the vast scope of this subject, only a few examples will be considered; for further details the reader is referred to some recent reviews of this subject (e.g. Emson 1983, Martin & Barches 1986, Nieuwenhuys 1986).

For convenience of description, most of the neuropeptides are classified according to the site where they were first discovered. Thus there are the 'gastrointestinal' peptides found initially in the gut wall (p. 1376), then a group first associated with the hypophysis cerebri (including releasing hormones, adenohypophyseal and neurohypophyseal hormones) and finally peptides discovered first in the nervous system. Some of these peptides are closely related to each other in their chemistry because they are derived from the same gene products (e.g. the pro-opiomelanocortin group) which are cleaved in various ways to provide smaller peptides. The significance of the presence of these substances in both neural and non-neural tissue can at present only be guessed at: perhaps it represents a genetic economy in that the same molecule can be used for quite different purposes in different biological situations; alternatively it has been suggested that at least in certain instances the same peptide is present in regions with related biological functions.

A summary of the main classes of neuromediators or putative neuromediators found in the nervous system

Non-peptidergic neuromediators
  Acetylcholine

Monoamines
  Noradrenalin
  Adrenalin
  Dopamine
  Serotonin
  Histamine

Amino acids
  Glutamate
  Aspartate
  Glycine
  Gamma aminobutyric acid (GABA)
  Taurine

Purines
  Adenosine triphosphate (ATP)

Peptides
Peptides first found in the gastrointestinal tract
  Bombesin
  Cholecystokinins (CCK)
  Gastrin
  Glucagon
  Insulin
  Motilin
  Neurotensin (NT)
  Pancreatic polypeptide (PP)
  Substance P (SP)
  Vasoactive intestinal polypeptide (VIP)

Peptides first associated with the hypothamalo-hypophyseal complex
  Hypothalamic releasing hormones:
    Corticotropin-releasing factor (CRF)
    Growth hormone releasing hormone (GHRH)
    Luteinizing hormone releasing hormone (LHRH)
    Somatostatin (SST)
    Thyrotropin-releasing hormone (TRH)
  Neurohypophyseal peptides:
    Vasopressin (arginine vasopressin; AVP)
    Oxytocin (OXT)
  Pro-opiomelanocortin (POMC) derivatives:
    Corticotropin (ACTH)
    Corticotropin-like intermediate lobe peptide (CLIP)
    β-endorphins (β-END)
    β-lipotropin (β-LPH)
    γ-lipotropin (γ-LPH)
    Met-enkephalin (M-ENK)
    Leu-enkephalin (L-ENK)
    α-melanocyte stimulating hormone (α-MSH)
    γ-melanocyte stimulating hormone (γ-MSH)
  Prodynorphin derivatives:
    Neodynorphins α and β (α, β-NE)
    Dynorphins (DYN) A and B

Other peptides
  Angiotensin II (ANG-II)
  Bradykinin
  Calcitonin-related gene peptide (CGRP)
  Carnosine
  Gallinine
  Natriuretic peptide
  Neuropeptide Y
  Sleep peptide(s)

Growth factors
  Nerve growth factor (NGF)
  Platelet derived growth factor (PDGF)

Substance P (SP) was the first of the peptides to be characterized as a 'gastrointestinal' neuromediator. It consists of 11 amino-acid residues and is a major neuromediator in the brain and spinal cord. It occurs in about 20% of spinal and trigeminal ganglion cells, in particular those small neurons giving off narrow unmyelinated (class C) or myelinated (Aδ) axons, nociceptive in function. It is also present in some fibres of the facial, glossopharyngeal and vagal nerves. Within the central nervous system, SP is present in several apparently unrelated major central nervous pathways. It is contained within large granular synaptic vesicles in multi-neuromediator synapses; its known action on neurons is prolonged postsynaptic excitation.

Vasoactive intestinal polypeptide (VIP), another gastrointest-

inal peptide, is also widely present in the central nervous system, where it is probably an excitatory neurotransmitter or neuromodulator. Its distribution includes, among many other areas, distinctive bipolar neurons of the cerebral cortex, small spinal ganglion cells, particularly of the sacral region, and the median eminence of the hypothalamus where it may be involved in endocrine regulation. It is also present in intramural ganglion cells of the gut wall and sympathetic ganglia.

Somatostatin (ST, somatotropin release inhibiting factor) has a broad distribution within the nervous system and may be a central neurotransmitter or neuromodulator. It has also been detected in small spinal ganglion cells.

β-endorphin, leu-enkephalin and met-enkephalin and the dynorphins belong to a group of peptides ('naturally-occurring opiates') which have aroused much interest because of their pharmacological properties in relation to analgesia. They bind to opiate receptors in the brain and can induce analgesia if infused into appropriate areas of the brain stem. Although their distribution in the brain is complex, they coincide frequently with the opiate receptor sites, as determined by biochemical and radioactive labelling methods. In general their action seems to be inhibitory. The enkephalins have been localized in many areas of the brain, particularly the septal nuclei, amygdaloid complex, basal ganglia and hypothalamus; they therefore appear to be important mediators in the limbic system and in the control of endocrine function. They are also strongly implicated in the central control of pain pathways, including the peri-aqueductal grey matter of the midbrain, a number of reticular raphe nuclei (e.g. nucleus raphe magnus, the medial reticular formation of the rhombencephalon) and the spinal nucleus of the trigeminal with its spinal continuation into the substantia gelatinosa. Among other means of modulating the nociceptive channels, the enkephalinergic pathways appear to exert an important presynaptic inhibitory action on the nociceptive afferents in the spinal cord and brain stem (see p. 942). Like many other neuromediators, the enkephalins also occur widely in other parts of the brain, though in lower concentrations, being present throughout the neocortex, the olivo-cochlear bundle and many other sites in the nervous system.

Cholecystokinin is present throughout the brain but is particularly concentrated in the cerebral cortex, amygdaloid complex, hippocampus, peri-aqueductal grey and the dorsal grey columns of the spinal cord. Its precise functions in these sites are as yet uncertain.

Neuroglia

Non-excitable cells constitute a major component of nervous tissue (7.28, 29, 30, 31) with many important functions in the operation of the nervous system as a whole. In the **peripheral nervous system** they include a number of different classes, among which the Schwann cell (lemmocyte) is most prominent (p. 901); but in the peripheral ganglia of spinal and autonomic pathways *capsular gliocytes (satellite cells)* ensheath ganglion cells, terminal *lemmal cells* occur around encapsulated sensory terminals and *teloglial cells* ensheath motor terminals; various types of supporting cells (*sustentaculocytes*) are present in sensory epithelia. These various types of cell will be considered later in this section (p. 901, 911).

In the **central nervous system** the non-excitable cells include various classes of *neuroglial cells* and also cells which line the internal cavities of the brain and spinal cord (*ependymal cells, choroidal cells,* etc.).

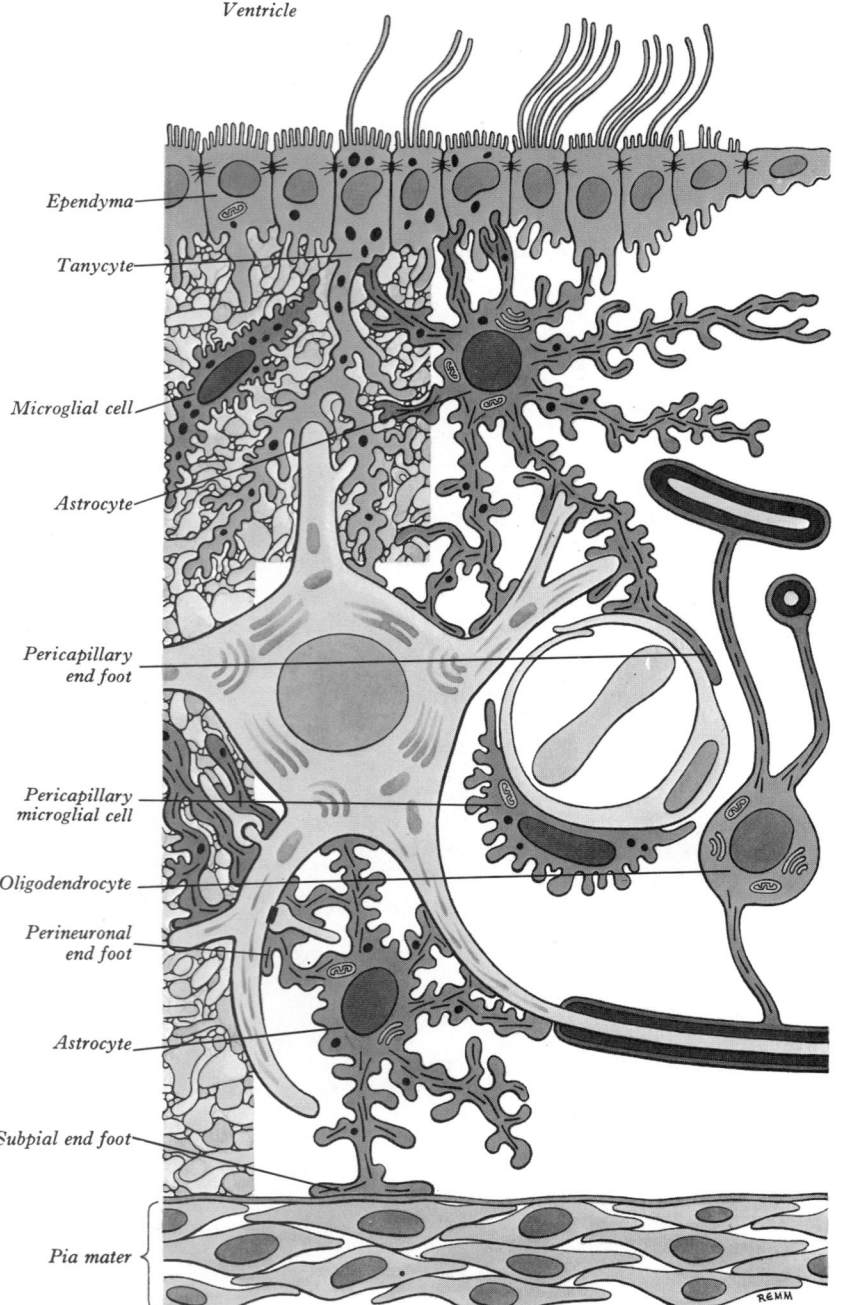

Ventricle

Ependyma

Tanycyte

Microglial cell

Astrocyte

Pericapillary end foot

Pericapillary microglial cell

Oligodendrocyte

Perineuronal end foot

Astrocyte

Subpial end foot

Pia mater

7.28 Schema showing the types of non-neuronal cells in the central nervous system. The ependymal and glial cells are shown in green. The ependyma includes examples of ciliated and non-ciliated cells and one tanycyte, with a centrally directed basal process. Two astrocytes are shown apposed to a neuronal soma and dendrites; one (above) also contacts a capillary, the other (below) expands on the pial surface. An oligodendrocyte (middle right) provides myelin sheaths for two axons. Two flattened microglial cells, one adjacent to a capillary (middle right), and the other within the neuropil at the top left, are also illustrated.

NEUROGLIAL CELL TYPES

Neuroglial cells vary considerably in abundance and types in different regions of the central nervous system. In earlier times, their small size and variable reactions to the special staining methods needed to demonstrate them made them less accessible than neurons to histologists. It was only with the application of electron microscopy, biochemistry and, more recently, immunohistochemistry and cell culture methods that any grea advances on the classic descriptions (e.g. by del Rio Hortega 1924) have been made. Even now there are many uncertainties about the number of classes of neuroglial cells and of their origins and functions in the nervous system. However, they are divisible into two major groups according to their origins: *macroglia* which arise within the neural plate in parallel with neurons and which constitute the great majority of neuroglial cells; and *microglia*, smaller cells derived from mesodermal tissues around the nervous system, which enter relatively late in development (and may continue to do so in postnatal life in small numbers). Microglial cells are generally considered to be a type of mononuclear phagocyte (p. 668).

Macroglia

The cells of this division consist of *astrocytes* and *oligodendrocytes*;

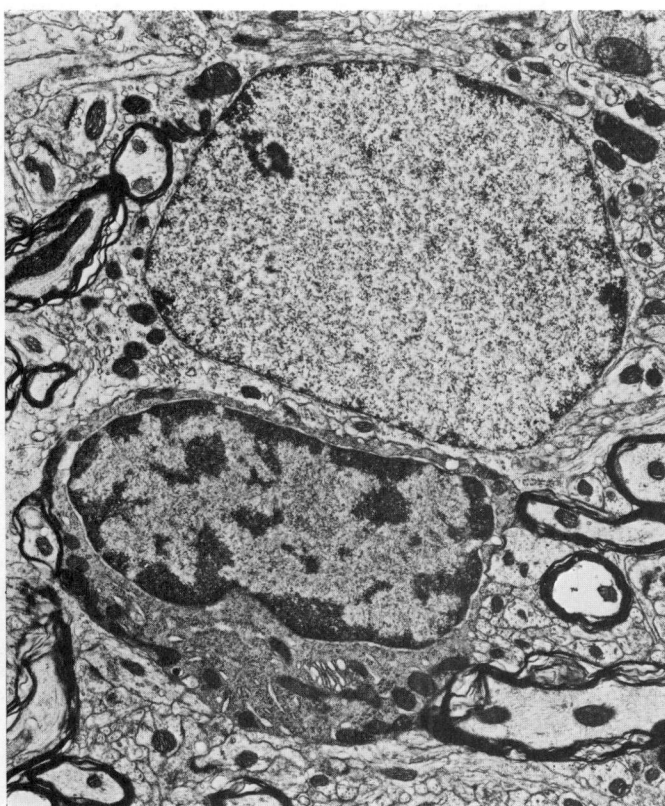

7.29 An electron micrograph of two neuroglial cells situated amongst myelinated and non-myelinated nerve fibres, in the rat thalamus. An oligodendrocyte containing numerous mitochondria, a well-developed endoplasmic reticulum and an indented nucleus is shown below; a larger astrocyte with a vesicular nucleus and scanty cytoplasm is demonstrated above. Provided by A R Lieberman of University College, London.

this term may also be extended to include specialized types such as *retinal glyocytes* (*Müller cells*), *Bergmann glia* of the cerebellum, *pituicytes* of the neurohypophysis, *ependymal cells* and *choroidal cells*. In addition to these there are the stem cells from which macroglial cells originate during development and throughout postnatal life.

Astrocytes (7.28) have small somata (in humans about 8 μm across) with many dendritic extensions. Traditionally they have been subdivided into star-like *protoplasmic astrocytes*, with broad symmetrically spread processes confined to the grey matter and *fibrous astrocytes* located chiefly in the white matter, their processes spread asymmetrically, often in a radial direction among the fasciculi of nerve fibres. A third type, the *perinodal astrocyte* has recently been proposed as occurring in relation to the node of Ranvier in central myelinated fibres, constituting a cell type intermediate between an astrocyte and an oligodendrocyte. Fourthly, immature *stem astrocytes* capable of mitosis are present in certain areas such as the subventricular zones of the nervous

system (see p. 195). Astrocytic processes have fine, foliate extensions which partly surround and separate neurons and neurites, often ending in plate-like expansions on blood vessels, ependyma (vide infra) and the pia mater limiting the surface of the central nervous system (the *glia limitans externa*). Ultrastructurally they are typified by a pale nucleus with a narrow rim of heterochromatin (7.29), pale cytoplasm rich in glycogen, lysosomes (*granules* or *gliosomes* of light microscopy), Golgi complexes and (in fibrous astrocytes) bundles of intermediate (10 nm) filaments throughout their processes (Mori & Leblond 1970, Ling et al 1973). Glial filaments contribute much to the total protein content of brain. Characteristically, these intermediate filaments are composed of *glial fibrillary acidic protein* (GFAP) which can be readily detected by immunocytochemistry (7.30). These filaments are probably skeletal in function (Schacher et al 1977). Desmosomes and gap junctions (pp. 18–19) form contacts between astrocytes and sometimes between astrocytes and neurons. Such cells show several forms; isotopic labelling suggests that astrocytes which can divide in mature animals undergo structural transformations before disintegration. In injured brain they proliferate (gliosis) and can then also act as phagocytes and later form scar tissue.

Oligodendrocytes (7.28, 29), as the name implies, have fewer processes than astrocytes. They occur as *intrafascicular* cells in myelinated tracts and as *perineuronal* oligodendrocytes where their processes adjoin neuronal somata. Their major role is to lay down myelin around central nervous axons and thus they are the central counterpart of peripheral myelinating Schwann cells. However, the composition and detailed form of central and peripheral myelin differ (Gregson 1975) and the relation between the axon and the myelinating cell is also dissimilar: an oligodendrocyte may enclose several axons in separate myelin sheaths, whereas Schwann cells ensheath only one axon. Ultrastructurally oligodendrocytes have a round nucleus and a cytoplasm rich in mitochondria, microtubules and glycogen. Like astrocytes, they vary from cells with large euchromatic nuclei and pale cytoplasm to those with heterochromatic nuclei and dense cytoplasm. Tritium labelling shows that they can proliferate in *mature* animals and thus show stages of maturation and degeneration.

Glioblasts are pre- or postnatal stem cells differentiating into macroglia. They have a pale nucleus and cytoplasm, free ribosomes and are usually rotund. They are numerous, even in adults, in a diffuse subependymal zone. It is not known whether they are multipotent in adults, i.e. able to form astrocytes or oligodendrocytes, or if individual gliocytes for each line may exist. They appear to be the main source of macroglial cells by mitotic division in fetal and postnatal periods (Mori & Leblond 1970, Ling et al 1973, Kaplan & Hinds 1980, Sturrock 1982, Reyners et al 1986, Berry 1986).

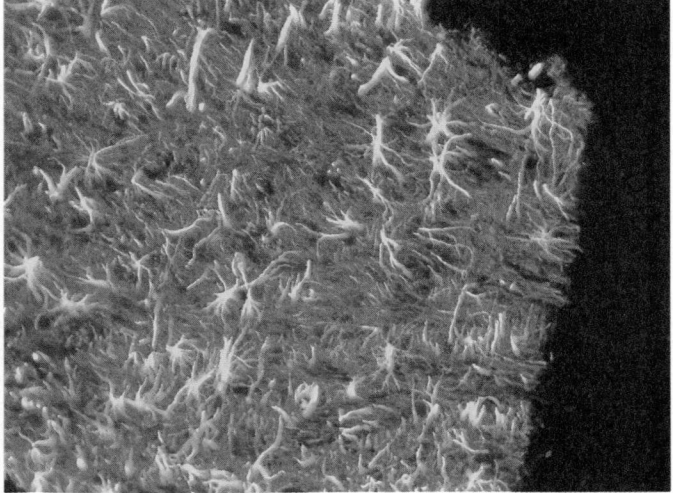

7.30 Section of the brain stem (rat) stained by immunfluorescence to show astrocytes containing glial fibrillary acidic protein (GFAP). Preparation by Martin Sadler, Anatomy Department, UMDS, Guy's Campus, London. Magnification × 150.

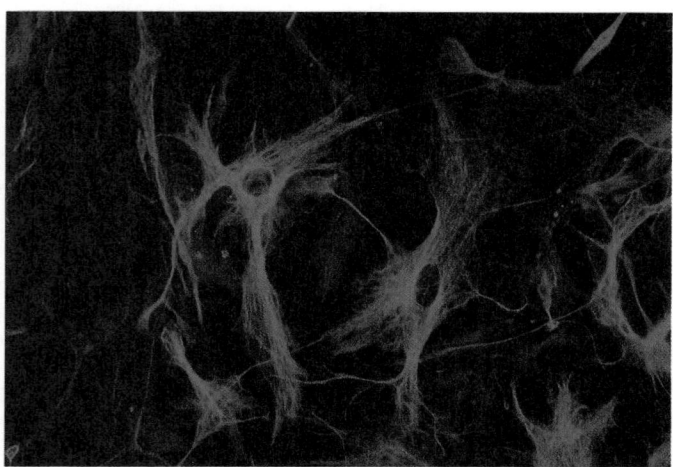

7.31 Astrocyte in tissue culture, demonstrating the presence of glial fibrillary acidic protein (GFAP), shown by rhodamine immuno-fluorescence. Provided by Caroline Wigley, Anatomy Department, UMDS, Guy's Campus, London. Magnification × 600.

Pituicytes of the neurohypophysis resemble astrocytes but their processes end mostly on vascular endothelial cells. **Müller** cells of the retina (p. 1199), or *radial gliocytes*, have much in common with astrocytes and have a similar origin. Cerebellar **Bergmann glial cells** have somata arranged in a row some distance below the pial surface. A single apical process branches to end in pial expansions (end-feet).

Ependymal cells (ependymocytes) form a unicellular epithelium lining the ventricles and central canal (7.28). They vary from squamous to columnar in form according to their locality. At the ventricular surface they are in contact by means of gap junctions and occasional desmosomes; their surfaces have numerous microvilli and cilia, the latter often motile and contributing to the flow of cerebrospinal fluid. Ultrastructurally (Brightman & Palay 1963, Bleier 1977) the nucleus is heterochromatic and indented; their cytoplasm abounds in mitochondria, lysosomes, microtubules and microfilaments. There is much regional variation in the ependymal lining of the ventricles but four major varieties have been distinguished in mammals (see Scott et al 1974, Page et al 1979, Page & Leure-duPre 1983).

These consist of: (1) general ependyma overlying areas of grey matter; (2) general ependyma lining white matter; (3) specialized areas of ependyma in the third and fourth ventricles; (4) choroidal epithelium.

Firstly, the ependyma overlying areas of grey matter are cuboidal, each cell bearing about 20 cilia in its apical centre, surrounded by short microvilli. The cell margins are not extensively folded, and cells are joined by gap junctions and desmosomes; their contents are as stated above. They do not have a basal lamina, but beneath them is a subependymal zone from two to three cells deep, consisting of cells resembling ependymal cells in their organelles and nuclei; the blood vessels beneath them have no fenestrations and few transcytotic vesicles, i.e. they are typical of the central nervous system (p. 691). Where the ependyma lines myelinated tracts, cells are much flattened and even squamous and fewer of them are ciliated; again there are gap junctions and desmosomes but the lateral margins of cells are highly folded and interdigitating. No subependymal zone is present but blood vessels are as described for the previous type. It has been suggested that these differences may be related to a greater role in the exchange of metabolites between grey matter and cerebrospinal fluid than occurs for white matter, the former being metabolically more demanding than the latter.

Specialized areas of ependymal cells are found in four areas around the margins of the third ventricle (the circumventricular organs) including the lining of the median eminence of the hypothalamus and forming the subcommissural organ, the subfornical organ and the organum vasculosum of the lamina

terminalis and other related areas (see p. 1005). At the inferoposterior limit of the fourth ventricle is the area postrema, which has a similar structure. In these sites the ependymal cells are only rarely ciliated and their ventricular surfaces have many microvilli and apical blebs. They have many mitochondria and well-formed golgi complexes lying apical to a rather flattened, basal nucleus. The cells are joined laterally by tight junctions, forming a barrier to the passage of materials across the ependyma, and desmosomes. Many of these cells are *tanycytes*, with basal processes projecting into the perivascular spaces surrounding underlying capillaries, which are fenestrated and therefore do not form a blood–brain barrier. It is thought probable that these areas constitute special zones by which substances can pass from the nervous tissue and vascular supply into the cerebrospinal fluid within the ventricle by active transport through the ependymal cells; many of the neurons in the vicinity of such areas liberate neuropeptides and other biogenic substances, which may therefore pass into the ventricles to gain access to a much wider population of neurons via the permeable ependymal lining of the rest of the ventricle; a reverse process is also likely by this route (see also p. 1453). High concentrations of receptor sites for certain neuropeptides are present at these surfaces.

Finally, the ependyma is highly modified where it lies adjacent to the vascular layer of the *choroid plexuses* (p. 1080). Here the cells resemble those of the circumventricular organs, except that they do not have basal processes but constitute a cuboidal epithelium resting on a basal lamina adjacent to the externally applied pia and capillary layer (with fenestrated endothelium). Cells have numerous long microvilli with a few cilia interspersed; they have many mitochondria and large golgi complexes and their nuclei are basally placed. Tight junctions forming a trans-epithelial barrier and desmosomes occur between cells and the lateral margins of cells are highly folded. All of these structural features accord with the secretory activity of these cells, which are responsible for the formation of most of the cerebrospinal fluid.

Tanycytes (*ependymoglial cells*, *ependymal astrocytes*) have already been mentioned in connection with the possible transport of substances between the vascular system, neurons and the ventricular contents. This form of cell, often with a highly elongated basal process which in some cases reaches the pial surface, is encountered during early development of the central nervous system, where it appears to play a vital part in guiding the migration of neurons (pp. 195, 918). Such cells also occur widely in the nervous systems of lower vertebrates and may be regarded as perhaps the most primitive type of glial cell, phylogenetically. They also persist in mammals in the form of retinal Müller cells and somewhat modified Bergmann glia (vide supra), which may also be involved in guiding neuronal migration; it is interesting that these cells share certain immunochemically demonstrable antigens not present in other astrocytes (see e.g. Schachner 1982).

The functions of glial cells may be numerous, though not yet fully explored (Kuffler & Nicholls 1976): (1) They are a supporting component in nervous tissue; glial filaments, microtubules and surface contacts suit them to this role. (2) They are insulators, separating neurons and their processes, grouping synaptically interacting regions and limiting the ionic spread from neurons. Electrical studies in leeches, where gliocytes are large, confirm this, showing their inability to conduct surface depolarization (Nicholls & Kuffler 1964). (3) They are phagocytes of foreign material or cellular debris and provide limited repair as glial scar tissue or fill gaps left by degenerated neurons. (4) They absorb and store neurotransmitters from nearby synapses and may metabolize them; thus transmitters are removed from the sites of action, as shown by autoradiographic studies of e.g. gamma aminobutyric acid uptake by gliocytes (Schon & Kelly 1974). Transmitters may sometimes be released from them, though this has not yet been demonstrated as a normal activity; changes in external potassium, occurring when nearby axons are active, cause GABA release (Iversen & Kelly 1975), which may modify local synaptic processes (Kuffler & Nicholls 1976). (5) Oligodendrocytes form and maintain myelin sheaths of the larger central axons like Schwann cells in the periphery. (6) Ependymocytes are associated with secretion into and uptake and transport from the cerebrospinal fluid, e.g. in regions associated with hypophysis cerebri, where they may transport hormone-controlling factors in the median eminence and hypophysial stalk (Bleier 1977). (7) Macrogliocytes may help in the control of physiological activity in the neuronal groups which they infiltrate, by regulating their metabolic and ionic environment. Astrocytes (and Schwann cells) have both cation and anion channels at their surfaces and can take up potassium and chloride readily (see Gray & Ritchie 1985); metabolic alterations in glial cells may also affect the ionic environment of neurons and hence their patterns of excitability (Kuffler & Nicholls 1976). Such alterations may occur when anions such as chloride, applied experimentally to the brain surface, cause the *spreading depression of Leao*; various cerebral metabolic changes may also affect electrical behaviour in this way, either normally or pathologically. (8) Golgi first suggested that macrogliocytes, by contacts with blood vessels and ventricular surfaces and neurons, could be an intracellular route for nutrient diffusion from blood and cerebrospinal fluid to neurons, but there is little firm evidence of this; consistent evidence of a neuroglial connection between the nutrient supply and the neuronal surface is lacking and other routes are available, intercellular diffusion and intracellular transport being adequate for this. While glial nutrient routes cannot be excluded, they need not be invoked to explain neuronal nutrition. (9) An important function of glial cells may be to prevent other cell types (e.g. fibroblasts) from growing into the central nervous system; this may be brought about by passive exclusion or, as seems more likely, by active inhibition; this may be related to the failure of axons to regenerate within the central nervous system (see p. 918).

Microglia

Microgliocytes (7.28), the smallest glial cells, have flattened profiles and give off numerous fine, short, dendritic processes insinuated between neurons or applied to capillaries. Ultrastructurally, they show heterochromatic, flat or indented nuclei and many lysosomes, features identical with those of extra-neural macrophages. Experiments with labelled medullary or circulating monocytes show that such blood cells can migrate as macrophages into damaged or degenerating central nervous tissues to phagocytose cellular debris; they are indistinguishable from indigenous microgliocytes activated to perform a similar role. Indeed, macrophages may remain as sedentary microglia, though most apparently migrate into the brain and spinal cord in late fetal and early postnatal life (Imamoto & Leblond 1977, 1978, Ling 1978). See also Hayes et al (1988).

ORIGINS OF NEUROGLIA AND EPENDYMA

As stated elsewhere (pp. 178, 917), ideas about the origins of the different cell types of the nervous system have undergone a number of major changes over the last century and are still not worked out in detail. However, it is now generally agreed that cells of the neural plate begin to differentiate early in the development of the nervous system into neuronal and glial cell lines and that the first recognizable type of the latter is probably the radial glial cell (also termed an ependymal astrocyte or early tanycyte). The perinuclear cytoplasm of this cell forms part of the ventricular lining but it also has a highly attenuated process which stretches away from the ventricle as far as the outer, pial surface of the central nervous system, so bridging the entire wall. Radial glial cells are considered to be of major importance in directing neuroblast migration (see p. 195 for further details); they also appear to give rise to astrocytes which migrate into the nervous tissue and also, by losing their basal processes, they form the ependymal cell layer lining the ventricle. (In some cases the basal process is retained, as in the mature tanycytes of the endocrine hypothalamus, p. 1453.) Immunohistochemical and electron microscopic methods have also shown that the periventricular zone of cell proliferation found in the early neural tube includes cells with glial characteristics (e.g. they are positive for glial fibrillary acidic protein, GFAP), so it is likely that glial cells begin to be formed at the same time as neuroblasts but, unlike neuroblasts, they are able to continue proliferating albeit at a reduced rate in postnatal life. The subependymal layer of the periventricular region is a major source of such cells, new astrocytes migrating out into the neural tissue to their final destinations.

The further differentiation of the classes of macroglial cells is even less clear. Recently, several immunohistochemical markers of the different glial cells have been developed and in particular have been applied to cells in tissue culture as well as in the intact nervous system. Although the most detailed analysis applies only to the cell lineages of the optic nerve, the results may also apply to other parts of the nervous system (Miller & Raff 1984). Astrocytes of the optic nerve are divisible into two classes on the basis of their antigens reactive to monoclonal antibodies. Type I astrocytes probably correspond to the classic protoplasmic and fibrous astrocytes, while Type II is thought to be a perinodal cell. Type I and II cells are formed from separate stem cells but Type II cells share the same stem cell as oligodendrocytes, differentiating along one or other of the latter pathways depending on environmental factors.

It appears that new astrocytes can be formed throughout life, particularly in early postnatal growth and in response to injury; likewise oligodendrocytes have a limited ability to proliferate. Whether mature cells retain this ability, or if persistent stem cells are the source of these new cells is at present unknown. This ability has one unfortunate aspect in that the great majority of brain tumours are of glial origin. Schwann cells and various other accessory cells of the peripheral nervous system (ganglionic capsular cells, etc.) are derived from the neural crest and migrate into position in advance of growing neurites. Schwann cells are also able to multiply throughout life; they do this in early postnatal growth of the body but later only divide as a response to injury.

THE BLOOD–BRAIN BARRIER

In early experiments it was discovered that certain dyes, when injected into the blood, failed to stain the parenchyma of the brain and spinal cord, though they passed easily into non-nervous tissues. Subsequently observations confirmed that access of various substances from the bloodstream to the extracellular space of the central nervous system is restricted as if by a physiological barrier. The structural nature of this '*blood–brain barrier*' has been much discussed in relation to the chemotherapy of nervous disorders and anaesthesia (Davson 1970). However, many materials injected directly into the *ventricles* of the brain do enter brain tissue, indicating that the barrier exists at capillary level. Experiments with peroxidase as a colloidal tracer (Brightman & Reese 1969, Brightman et al 1970) show that diffusion occurs freely *within* the extracellular spaces of the central nervous system and that, when injected into the central cavities, the tracer passes freely through the ependymal intercellular junctions (which are punctate and of the communicating or 'gap' variety), traversing the nervous tissues to reach the pial surface. Such diffusion stops at the capillary endothelium; the capillaries are unusually impermeable to *large molecules*, due to non-leaky tight junctions, a lack of fenestration and only small numbers of membrane vesicles (p. 691). Diffusion from the choroid plexus is also restricted by the combined impermeability to macromolecules of the vascular endothelium *and* the choroidal epithelium; cerebrospinal fluid therefore contains only materials secreted selectively by the choroidal epithelium (Cserr 1971).

EXTRACELLULAR SPACE IN THE BRAIN

The amount of extracellular space in the central nervous system is of much interest because of its relevance to the diffusion of substances through nervous tissue, including the dissemination of neurotransmitters and neuromodulators, the passage of substances to and from the blood and many other factors determining the local environment of neurons and glial cells. Central nervous tissue totally lacks lymphatic drainage but cerebrospinal fluid draining to the dural vessels may be regarded as analogous to lymph in some respects. Electron microscopy has shown that in general there is relatively little extracellular space (apart from that within the vascular system), cells being separated by only a 20 nm gap. This has been confirmed by freeze-fracture studies where rapidly frozen tissue is thought to preserve its structure more faithfully than with chemical fixation. However, it appears that some areas of the brain have larger intercellular spaces.

The lack of large diffusion channels through neural tissue is perhaps not a metabolically limiting factor, given the rich capillary beds which are found in the central nervous system; the restriction of extracellular movements is likely to be important in maintaining a tight control on the movements of chemicals and cells in nervous tissue, providing a highly stable environment and a mechanically coherent framework for neural activity.

Peripheral Nerve Fibres

As in the tracts of the central nervous system, both myelinated and non-myelinated fibres occur peripherally but they are ensheathed by *Schwann cells* (*lemmocytes*, *peripheral glia*) in place of oligodendrocytes. Both central and peripheral fibres present some difficulties to the light microscopist, because of the small size of the non-myelinated fibres and the disruption of the sheaths of myelinated fibres caused by treatment with lipid solvents prior to sectioning. Various methods have been devised to overcome these problems. For myelin sheaths these depend on fixing and staining myelin in teased material for individual fibres or in processed and sectioned blocks. Methods using osmium are valuable, because osmium tetroxide in aqueous solution or chromate and haematoxylin solutions fix and stain simultaneously. Frozen sections may be stained with lipid-soluble dyes. Polarized light, phase or interference microscopy are useful for freshly teased or living nerves. Time-lapse cinemicrography and videotape recording techniques with closed-circuit television have aided the analysis of living nerve fibres (Williams & Hall 1970, 1971a, b). The methods used for demonstrating central neurons (p. 870) have also proved valuable and electron microscopy has added to our knowledge perhaps more than any method.

Structure of the Peripheral Nervous System

The peripheral nervous system includes the craniospinal and autonomic nerves and their associated ganglia, together with their connective tissue sheaths. All lie peripheral to the pial covering of the central nervous system, through which the central and peripheral nerve fibres are continuous, although with contrasting detailed structure and behaviour (7.32–46).

The **sensory ganglia** of dorsal spinal roots (7.36) and the ganglia of the trigeminal, facial, glossopharyngeal and vagal cranial nerves are enclosed in periganglionic connective tissue, resembling perineurium (p. 897). Their neurons are *unipolar*, with spherical or ovoid somata of varying size, aggregated in groups interspersed with fasciculi of myelinated and non-myelinated nerve fibres. Their structure is described elsewhere (p. 1093); a single non-myelinated neurite (a 'dendro-axonal' process) leaves each soma, highly convoluted at its origin before bifurcating at a T-junction into the central and peripheral processes of a sensory nerve fibre. In myelinated fibres the junction occurs at a node of Ranvier (p. 903). The peripheral neurite reaches a sensory ending; since it conducts towards the soma it is functionally an *elongated dendrite* but has the structural and functional properties of a *peripheral axon* and, following common usage, it will be so termed here. Each unipolar soma has a *nucleated capsule* of flat, epithelioid *capsular cells*, variously termed ganglionic gliocytes or *satellite cells*. (The term *satellite* is applied in different ways; some use it for small round extracapsular ganglionic cells, others include ganglionic capsular cells and Schwann cells or include all non-neuronal cells, central and peripheral, which are perineuronal. The name is also applied to cells associated with striated myocytes, p. 562.) The cytoplasm of capsular cells resembles that of Schwann cells (vide infra); their deep surfaces interdigitate with reciprocal irregularities in the

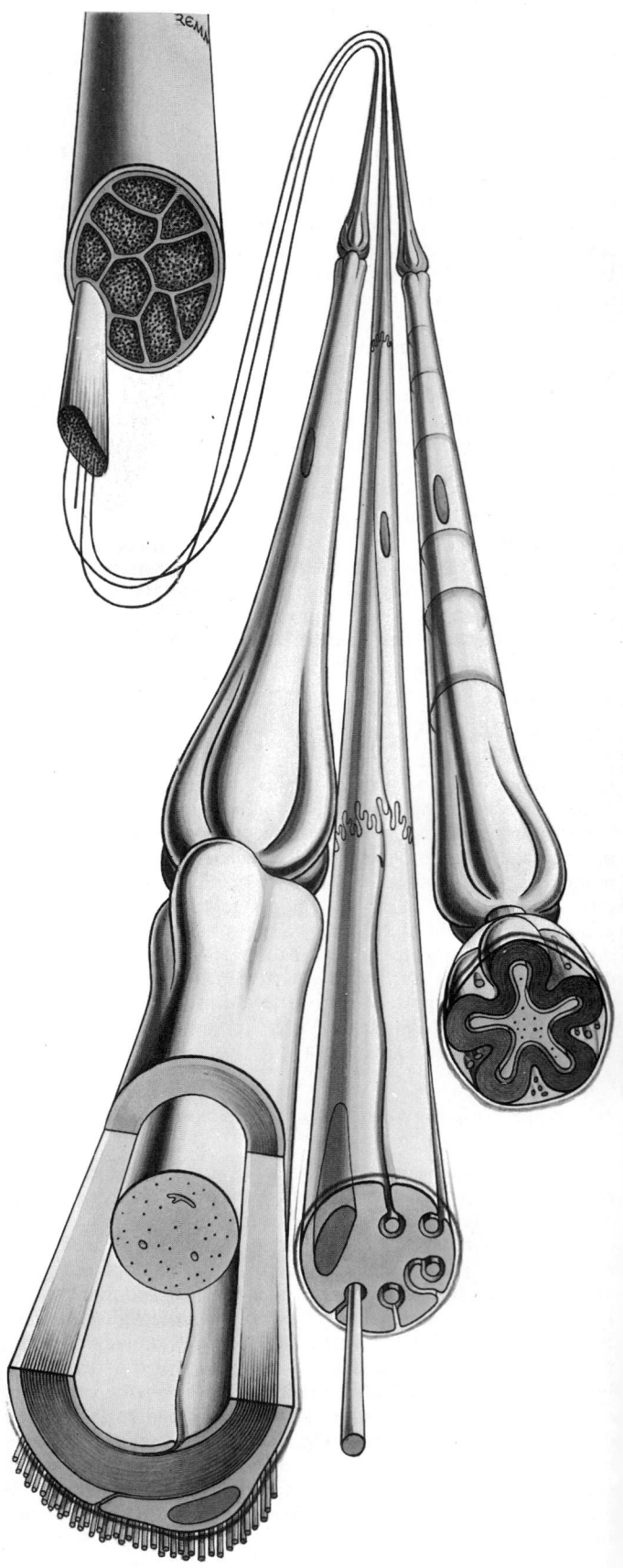

7.32 Some structural features of peripheral nerve fibres. A nerve trunk (top left) is cut away to expose a single fasciculus, from which three fibres are indicated in detail. These include two myelinated axons, one on each side of a group of non-myelinated axons enclosed within a Schwann cell sheath. The myelinated fibre on the left has been cut away at various points to demonstrate the relationship between the axon, the Schwann cell and its sheath of myelin.

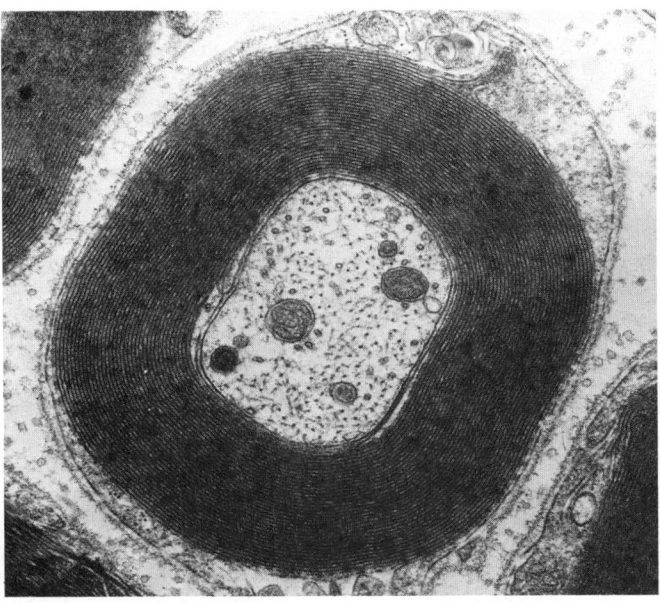

7.34 An electron micrograph of a myelinated peripheral nerve fibre, showing an axon containing neurofilaments, microtubules and mitochondria, surrounded by a myelin sheath, enclosed in turn by Schwann cell cytoplasm and endoneurial spaces. Provided by S Standring, Department of Anatomy, UMDS, Guy's Campus, London. Magnification ×25 000.

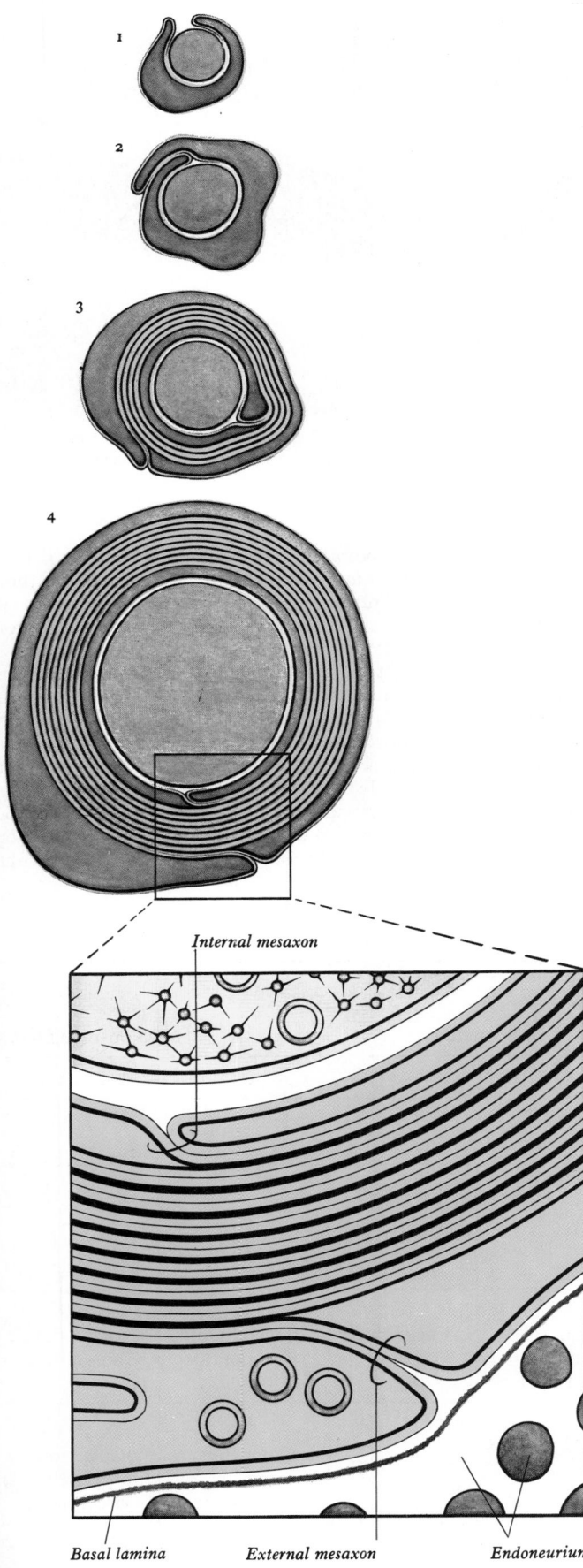

Internal mesaxon

Basal lamina   External mesaxon   Endoneurium

7.33 The development and organization of the myelin sheath of a peripheral nerve fibre. In stages 1–4 myelin formed by a Schwann cell (blue) progressively envelops the growing axon (yellow), to form the final pattern of spirally disposed myelin lamellae: see enlarged detail at the base of the diagram.

exteriors of the subjacent nerve cells. The capsular layer is continuous with similar cells enclosing the convoluted part (*initial glomerulus*) of the dendro-axonal process and then with the Schwann cells of the peripheral and central processes. Outside these layers lies a delicate vascular connective tissue continuous with the endoneurium of the peripheral nerve and nerve root.

Sensory ganglionic neurons are not entirely confined to the discrete craniospinal ganglia; singly or in small groups they often occupy '*heterotopic*' positions distal or proximal to their ganglia.

Autonomic ganglia have a different structure; their neurons are multipolar, with dendritic trees receiving synapses from preganglionic autonomic motor fibres, and are surrounded by a mixed neuropil of afferent and efferent fibres, dendrites, synapses and non-neural cells (p. 1157).

Nerve trunks and their principal branches consist of parallel bundles of efferent and afferent axons and are ensheathed by Schwann cells which in some cases elaborate *myelin sheaths*; these are surrounded by connective tissue sheaths at different levels of organization. Nerve fibres (i.e. axons with their ensheathing Schwann cells) are grouped into *fasciculi* of widely variable numbers. The size, number and pattern of fasciculi vary in different nerves and at different levels along their paths; their number increases and their size decreases some distance proximal to a point of branching. Similarly, where nerves are subjected to pressure, e.g. deep to a retinaculum, fasciculi are increased in number but reduced in size and the associated connective tissue and vascularity also increase; such a nerve shows a pink, fusiform dilatation, sometimes termed a *pseudoganglion* or *gangliform enlargement*.

A dense, irregular, connective tissue sheath, the *epineurium*, surrounds the trunk and a similar though thinner *perineurium* encloses each fasciculus, in which the spaces between nerve fibres contain a loose delicate connective tissue *endoneurium*. These connective tissue planes provide routes for vessels, capillary beds and associated lymphatics, most of which run largely parallel to the nerve fibres, within endoneurial spaces, with cross-connections passing between them.

The epineurium is a collagenous adventitial coat with little regularity of organization (Thomas 1963); the perineurium in contrast has regular laminae of fibroblasts alternating with collagenous sheets. Mast cells and macrophages are relatively frequent in this sheath. Fibroblasts form junctional complexes, which from some tracer experiments appear to impede to some extent the diffusion of large molecules across the sheath (Wag-gener & Beggs 1967); however, more recently this has been denied. The fibroblasts also have numerous pinocytotic vesicles, an indication of active transport across the perineurium.

CLASSIFICATION OF PERIPHERAL NERVE FIBRES

Several schemes of classification of peripheral nerve fibres have been used, based on various parameters including conduction velocities, functional nature, fibre diameters and other attributes. Of two in common use (see accompanying table), the first, devised by Erlanger & Gasser (1937) from conduction speeds in amphibian axons, divides fibres into three major classes, designated A, B and C, corresponding to peaks in the distribution of their conduction velocities. Group A fibres are subdivided into α, β and γ subgroups; B group fibres are preganglionic autonomic efferents, and C fibres non-myelinated. Since diameter and conduction velocity are in most fibres proportional (in myelinated fibres conduction in metres per second is approximately six times the fibre diameter expressed in micrometers), the group Aα fibres are widest and most rapid and C fibres the narrowest and slowest. In mammals Aβ fibres were found to be negligible but a subclass of non-autonomic fibres similar in conduction speeds to B fibres, termed Aδ fibres, was discovered.

In Erlanger & Gasser's scheme *afferent fibres* are confined to Aα, Aδ and C groups, the largest (Aα fibres) including the axons of the encapsulated cutaneous, joint and muscle receptors and some large alimentary enteroceptors. Aδ fibres belong to nociceptors, including those in dental pulp, skin and connective tissue; C fibres have thermoreceptive, nociceptive and interoceptive functions. *Somatic efferent fibres* include Aα, Aβ and Aγ axons. Aα fibres innervate only extrafusal muscle, being maximally 22 μm in diameter and conducting at a maximum of 120 m/s. Fibres to 'fast' twitch muscles are larger than those to 'slow' muscle. Aβ fibres are restricted to collaterals of Aα fibres, forming plaque endings on some intrafusal muscle fibres; Aγ fibres are exclusively fusimotor to plate and trail endings on intrafusal muscle fibres (vide infra).

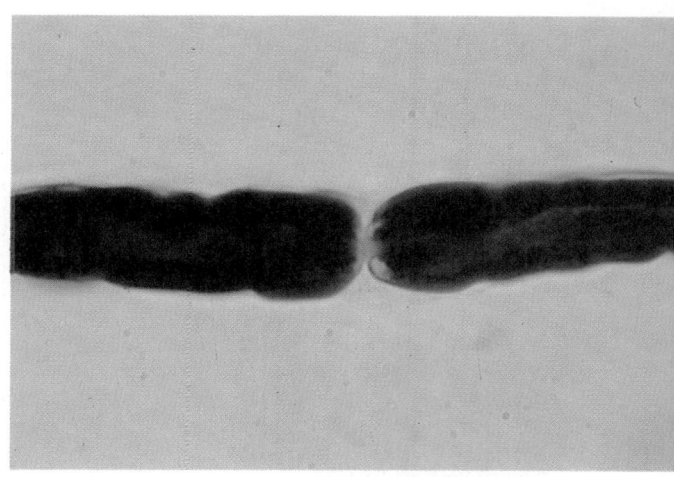

7.35 A myelinated fibre demonstrated by teasing an osmium-fixed peripheral nerve. In this preparation a node of Ranvier is visible. Magnification × 1000.

Autonomic efferents comprise preganglionic B fibres and postganglionic sympathetic and parasympathetic axons of the C (non-myelinated) group. This scheme can be applied to all fibres of spinal and cranial nerves except perhaps those of the olfactory nerve, whose fibres form a uniquely small and slow group.

Another classification, used for afferent fibres of somatic muscles, was introduced by Lloyd in 1943: myelinated fibres are divided into groups I, II and III, non-myelinated fibres forming group IV. *Group I* fibres are large (12–22 μm) and include primary sensory fibres of muscle spindles (Group Ia) and smaller fibres of Golgi tendon organs (Ib). *Group II* comprises fibres of secondary sensory terminals of muscle spindles, with diameters of about 6–12 μm. *Group III* fibres, 1–6 μm in diameter, have free sensory endings in the connective tissue sheaths around and within

CLASSIFICATION OF PERIPHERAL NERVE FIBRES

CHARACTERISTICS OF FIBRE Type of sheath	Myelinated				Non-myelinated
Fibre diameter	22 μm			1·5 μm	2·0–0·1 μm
Conduction speed (Metres/second)	120 60	50	30	4	0·5
CLASSIFICATION 1. Erlanger & Gasser *All fibres*		A		B	C
Subclasses: Efferent	Aα Skeletomotor	Aβ Fusimotor collaterals of A fibres	Aγ Fusimotor	B Preganglionic autonomic	C Postganglionic autonomic
Afferent	Aα and smaller Muscle and tendon; cutaneous			Aδ Cutaneous, muscle visceral, etc.	C Cutaneous, muscle visceral, etc.
2. Lloyd Afferent— skeletal muscle and articular	I *a* primary spindle ending *b* tendon ending	II Secondary spindle ending		III Free ending (nociceptor, etc.) paciniform ending?	IV Free ending (nociceptor, etc.)

It should be noted that the scale for conduction velocities is not arithmetic.

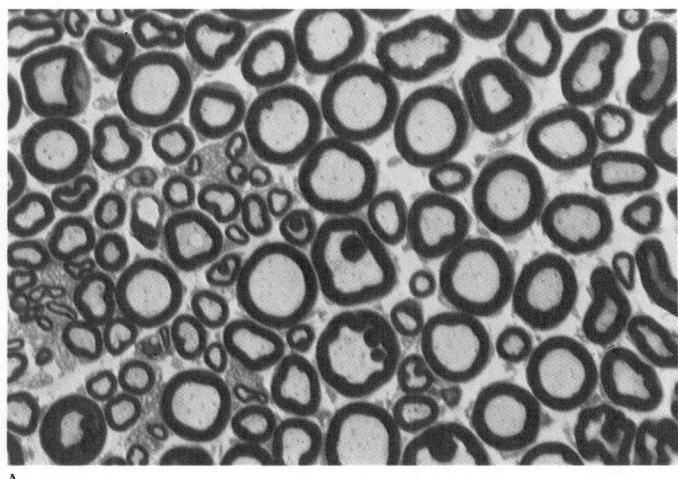

A

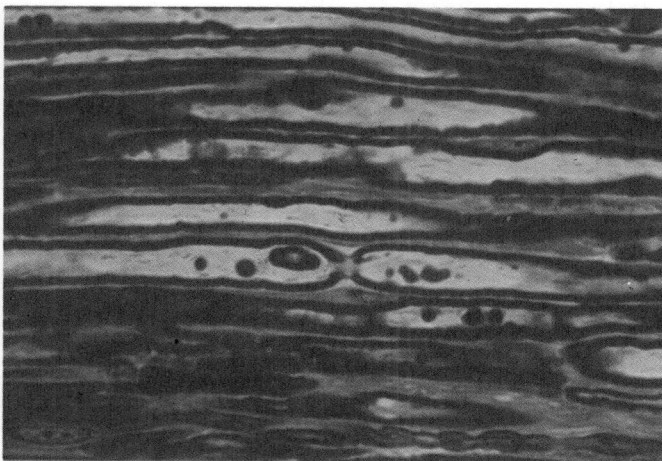

B

7.36 Structure of peripheral nerve fibres.

A. Transverse section of part of a mixed peripheral nerve from a mouse, showing a wide range of diameters of myelinated nerve fibres. Myelin: dark blue; axoplasm: pale blue, embedded in which are slightly darker blue dots, the axonal mitochondria. Note particularly the wide range of *axon* diameters; the smaller axons possess thinner myelin sheaths. The fibre with a particularly thick, double-contoured myelin sheath (bottom left) is sectioned through an incisure of Schmidt-Lanterman. The large fibres with more complex profiles (centre and bottom) are sectioned near the commencement of paranodal bulbs. On the left, between the myelinated fibres, groups of non-myelinated axons, enclosed by Schwann cells (medium blue) are just visible. A 1 μm epoxy resin section, stained with osmium tetroxide solution and toluidine blue, photographed using oil immersion optics.

B. Longitudinal section of material prepared in a manner similar to that in (A). Note particularly the fibre just below the centre of the field which is sectioned through a node of Ranvier. As the internodal myelin sheath approaches the node, its diameter increases to a maximum (the paranodal bulb) and the myelin sheath then curves sharply inwards, to terminate at the limits of the 'nodal gap'. The myelin profiles, apparently lying free within the paranodal bulbs, are demonstrated to be paranodal shelves of myelin, when serial sections are examined (compare with 7.37). Note also the constriction of the nodal axon; long, narrow axonal mitochondria are also visible.

C. Transverse section of an immature peripheral nerve of a rat, taken one week after birth, showing the profiles of numerous axons. Some of the latter are non-myelinated and, either singly or in groups, are invaginated into the surfaces of cytoplasmic processes of adjacent Schwann cells. Other axons are in the process of myelination. Note the Schwann cell nucleus (right) to one side of a myelin sheath and the ultrastructure of the Schwann cell cytoplasm at this stage of development.

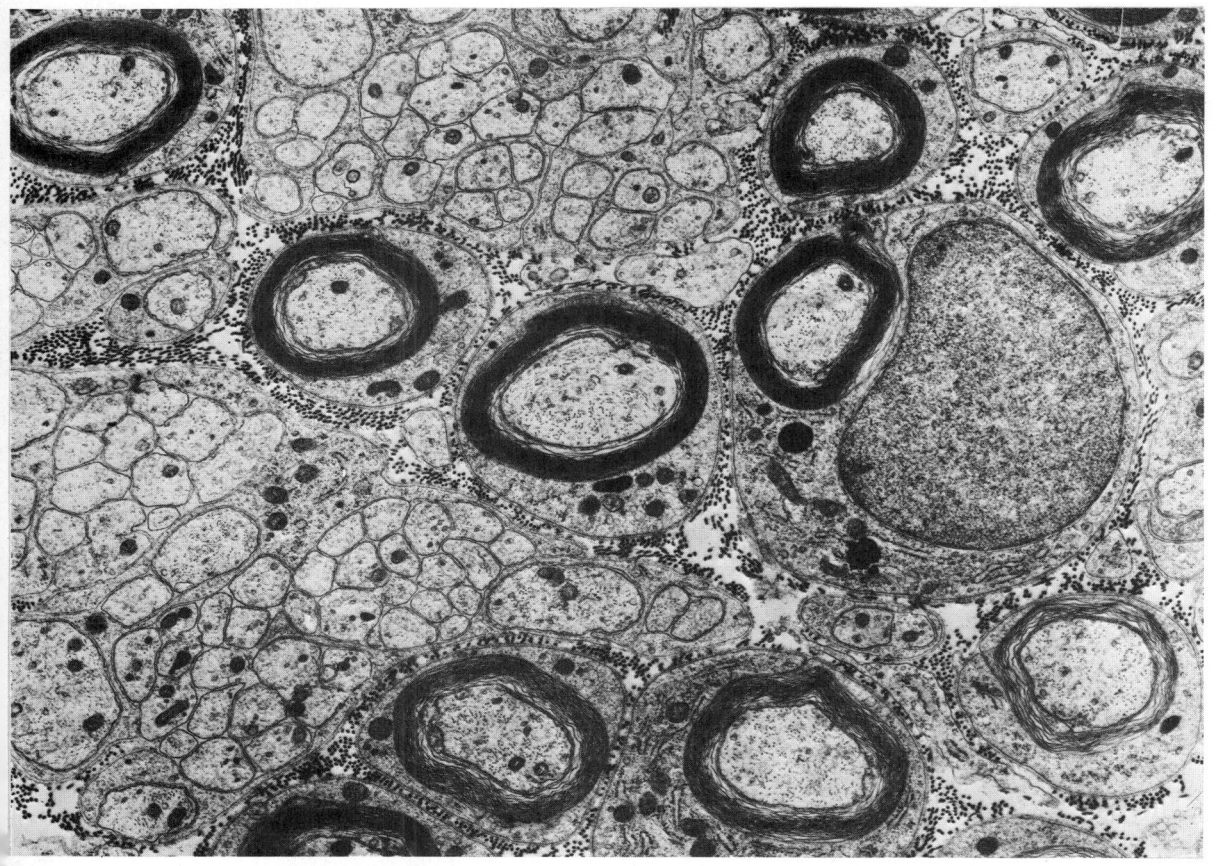

C

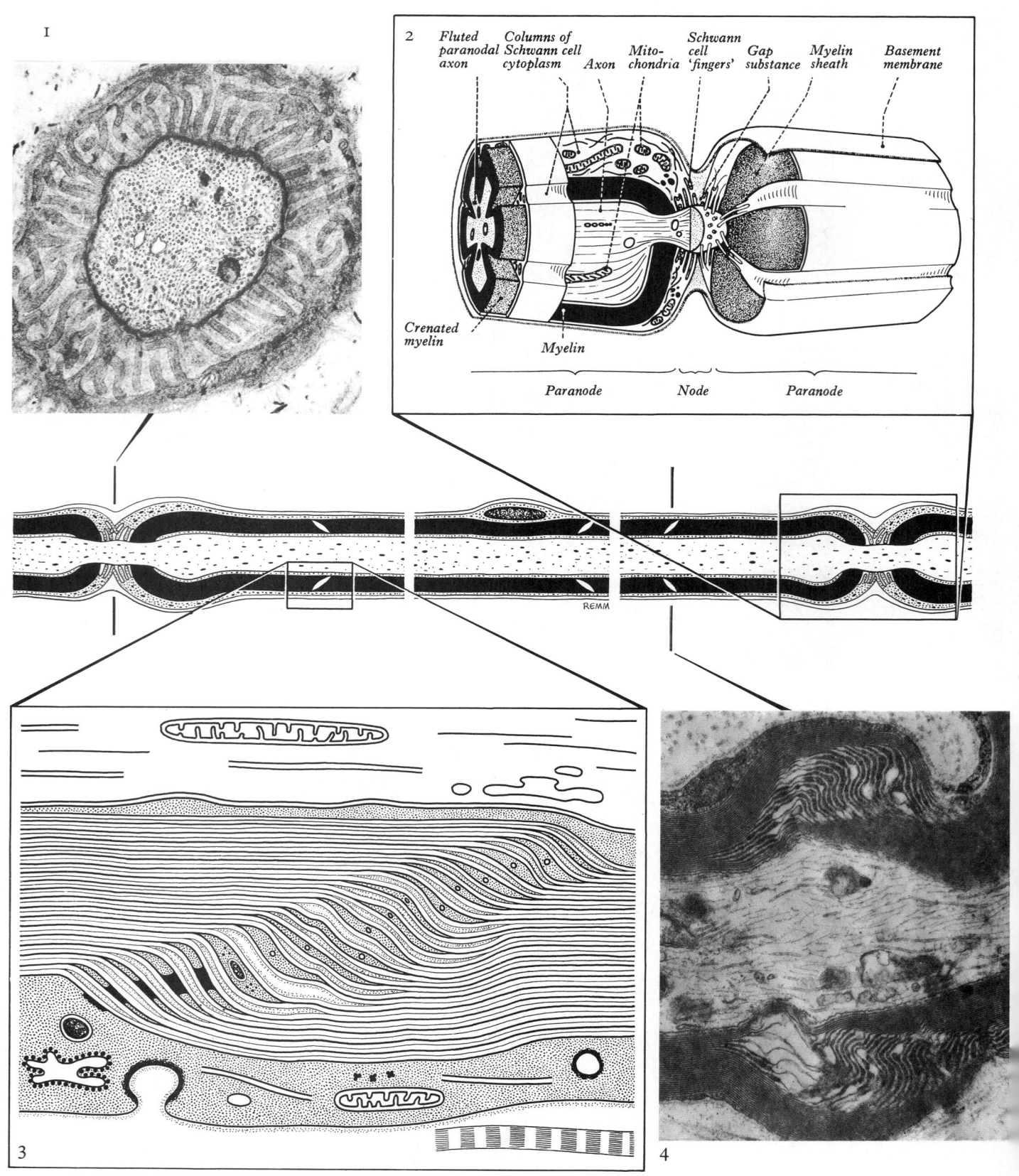

7.37 General plan of a myelinated nerve fibre in longitudinal section including one complete internodal segment and two adjacent paranodal bulbs, used as a key for the more detailed microarchitecture of specific subregions. 1: a transverse section through the centre of a node of Ranvier, with numerous finger-like processes of adjacent Schwann cells converging towards the nodal axolemma. Many microtubules and microfilaments are visible within the axoplasm. 2: a diagram showing the arrangement of the axon, myelin sheath and Schwann cell cytoplasm at the node of Ranvier in the paranodal bulbs. Supplied by P L Williams and D N Landon. 3: detailed substructure of one-half of an incisure of Schmidt-Lanterman consult text for further information. Supplied by Susan M Standring and P L Williams. 4: longitudinal section of part of a myelinated nerve fibre (mouse), including an incisure of Schmidt-Lanterman; this appears as oblique zones in the myelin sheath on both sides of the fibre. Consult text for structural details. Nos. 1 and 4 supplied by Susan M Standring, Dept of Anatomy, Guy's Hospital Medical School, London.

muscles and appear to be nociceptive, related to 'pressure-pain' in externally stimulated muscles. Paciniform (encapsulated) endings of muscle sheaths may also contribute fibres to this class. *Group IV* includes non-myelinated fibres below 1·5 μm with 'free' endings in muscles, being chiefly nociceptive. For a discussion of the sensory nerves of muscle, consult Matthews (1973), Barker (1974) and Hunt (1974).

While these general schemes of classification are valuable in neurophysiology, a single structural parameter such as diameter, measured at one point, cannot adequately categorize nerve fibres throughout the whole course of a nerve. Many values quoted are related only to velocity peaks or sampled diameters. The figures quoted are relative, with some overlap between the functional classes; but these schemes enable us to detect a general pattern of organization in peripheral nerves, which is of considerable interest.

CONDUCTION IN PERIPHERAL NERVES

When a nerve is stimulated, the sum of its action potentials, the *compound action potential*, can be recorded with electrodes placed on the surface. Since action potentials travel at different speeds in fibres of different diameters (large fibres creating greater electrochemical fluxes than small fibres), the compound action potential measured at a distance from the point of stimulation is a complex mixture, with at least four sequential waves of different amplitude and velocity, known as α, β, γ and δ waves. As each nerve has its own spectrum of nerve fibres, so is its wave pattern distinctive. In general, fibre types are classified primarily by their conduction velocities, because fibres of different function may have similar velocities (vide supra).

A number of different factors govern conduction velocities of nerve fibres; in non-myelinated fibres, the action potential sweeps continuously over the axolemma as depolarization of one area of membrane triggers depolarization of adjacent areas. The rate of spread is proportional to the axonal cross-sectional area (Hodgkin 1964); but in myelinated fibres the myelin sheath limits propagative excitation to the nodes of Ranvier, from which ionic currents spread to other nodes in sequence. Since the longitudinal flow of ions in the axoplasm and along the outside of the fibre is more rapid than electronic spread along the axonal membrane, much higher velocities are possible by this method, which is termed *saltatory conduction* (Stampfli 1954).

During nodal excitation the permeability of the exposed nodal axolemma to sodium ions increases rapidly and these ions flow in, along their concentration gradient, generating longitudinal ionic currents inside the *internodal axon* to the next node, the electrical circuit being completed by currents in the reverse direction along the outside, in the endoneurial tissue fluid. As the outflow current density at the *next node* reaches *threshold level*, depolarization causes the voltage-sensitive sodium gates in the nodal axolemma to open, repeating the cycle of excitation. Myelin is an imperfect insulator and some radial leakage occurs between the axoplasm of internodes and surrounding tissue fluid, with some loss of efficiency. For a given diameter, maximal efficiency occurs when the axonal diameter and thickness of the myelin sheath are in a ratio of 0·6:1. Only the largest mammalian fibres reach this ratio. Some other dimensions are equally related to conduction in myelinated nerve fibres, including the *internodal distance* and *area*, the molecular *characteristics* and the immediate *micro-environment* of the nodal axolemma. These parameters naturally vary between species, different categories and along individual fibres and branches (Williams & Wendell-Smith 1971).

SCHWANN CELLS (LEMMOCYTES)

Schwann cells are satellite cells of the peripheral nervous system; all peripheral axons are ensheathed by them and are separated from the endoneurium by the Schwann cell plasma membrane. The large volume and surface area of many neurons and the extensive spread of their neurites has, since the early study of these cells, suggested to investigators that there may be some form of functional dependence on the surrounding components such as Schwann cells and there is some experimental evidence to support

this concept. Schwann cells participate in the supply of metabolites and trophic factors to axons (Varon & Bunge 1978), in the maintenance of the ionic state of the periaxonal space and possibly to the distribution of neurotransmitters (Villegas 1975); also to the siting of sodium channels along the axolemma (Ritchie & Rogart 1977). Conversely, Schwann cells are profoundly dependent on neurons for many aspects of their biology. Myelin thickness and internodal length (i.e. the size of the Schwann cell territories) are related to axonal calibre and may reflect the axolemmal spacing of antigenic determinants. Tissue culture of neurites and Schwann cells, and the allo- and xeno-grafting of nerve fascicles, have demonstrated that (1) axons are mitogenic for Schwann cells, and (2) Schwann cell differentiation depends in part on neuronal signals (Wood & Bunge 1975, Aguayo et al 1978); Schwann cells can even form myelin around central axons under experimental conditions.

Schwann cell development can be divided into perinatal *migratory* and *proliferative* phases, succeeded by a final *axon-associated* state, in which myelin formation and maintenance may occur. Immature Schwann cells are large and rounded, with an oval nucleus and dense cytoplasm. As they migrate from the neural crest they are first fusiform in appearance and then irregular, insinuating their processes between the outgrowing fascicles of neurites. In mature myelinated fibres, Schwann cell cytoplasm containing metabolic organelles is disposed as thin, apparently discontinuous strips lying external to the myelin sheath; this cytoplasm sometimes lies internally, between the myelin sheath and the periaxonal space. Only at paranodes in nodal gap fingers, spiral incisures and the perinuclear region do any substantial volumes of cytoplasm persist in the mature cell.

The close interrelation of axons and Schwann cells is confirmed by their mutual reactions to injury. Any disturbance of either structure results in cellular alterations. Crushing or cutting of the nerve fibre produces Wallerian degeneration; in this reaction, axons distal to the site of injury degenerate and an intense proliferation of Schwann cells follows in which myelin and axonal debris are phagocytosed. Axonal regrowth into any persistent tubes consisting of Schwann cells and their surrounding basal lamina, which may survive in the distal stump, allows the re-association of axon-Schwann cells and subsequent remyelination. Any metabolic or chemical damage which may occur primarily to Schwann cells results in demyelination or hypomyelination but not always to axonal degeneration.

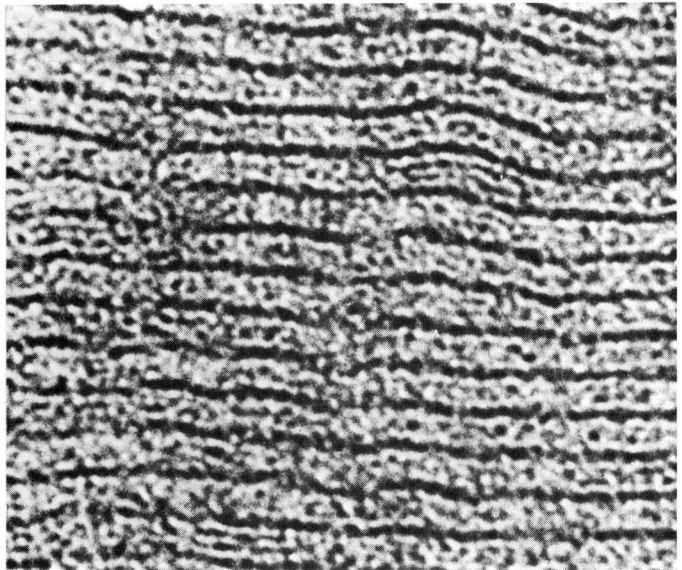

7.38   High-power electron micrograph of part of a myelin sheath in section, showing dense period lines (the apposition of the cytoplasmic aspects of the cell membranes), alternating with less dense intraperiod lines (the apposition of the external aspects of the Schwann cell membranes). The preparative techniques used have caused the intraperiod line to split into two distinct lines in some places.

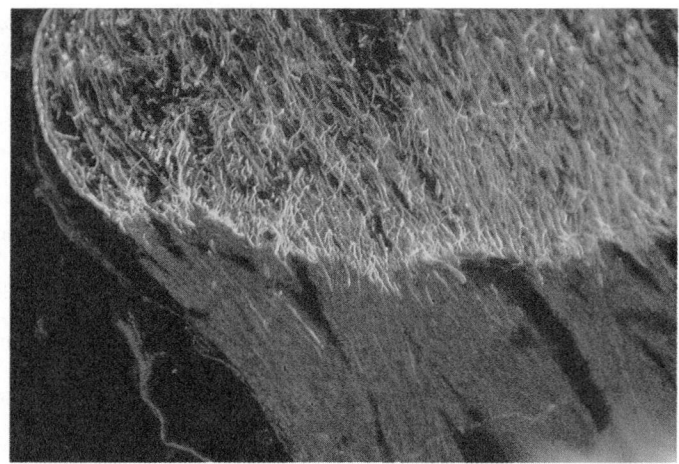

7.39 Section through the junction between the central and peripheral nervous system in the sensory root of the trigeminal nerve (rat) demonstrated by immunofluorescence of glial fibrillary acidic protein (GFAP). This is present in the astrocytes of the central nervous system but not in Schwann cells of the peripheral nerve. Provided by Martin Sadler, Anatomy Department, UMDS, Guy's Campus, London.

All larger mammalian axons are *myelinated* (*medullated*). Myelin is responsible for the glistening whiteness of peripheral nerves and central white matter. Axons smaller than 0·5–1·0 μm in diameter are usually *non-myelinated* (*non-medullated*) fibres. Because of important functional differences between these fibre types they will be described separately.

NON-MYELINATED FIBRES

In cutaneous nerves and dorsal spinal roots, about 75% of mammalian axons are non-myelinated; they form about 50% of the fibres of nerves to muscles and 30% in ventral spinal roots. Autonomic postganglionic axons are almost exclusively non-myelinated and preganglionic nerves contain significant numbers of these fibres. A non-myelinated 'fibre' is really a group of small axons (0·15–2 μm in diameter) within a sequential series of Schwann cells. In mature nerves the mode of enclosure of each group of axons shows variation from site to site, and also differs in different species of animals. Axons are usually separated from each other by tongues of Schwann cell cytoplasm but are sometimes further isolated by separate processes of cytoplasm which converge in the perinuclear region (Gamble & Eames 1964). A simpler organization seen in fetal nerves may also persist in some adult nerves, e.g. the olfactory, where bundles of small axons are collectively invaginated into the Schwann cell cytoplasm (see Landon 1985).

A three-dimensional reconstruction from sections of somatic and autonomic nerves reveals that the spatial relations between axons and satellite cells alter continuously within each cell (Aguayo et al 1973). The transfer of axons between one Schwann cell and the next in line usually occurs at their extremities, where their cytoplasmic processes interdigitate (Gamble et al 1978).

Non-myelinated fibres include the smaller of the central nervous axons, peripheral autonomic postganglionic axons and fine sensory fibres, such as those of nociceptors (pp. 909, 912), olfactory axons and others, invaginated into a longitudinal series of Schwann cells (7.36C). The line of invagination during development is marked by a *mesaxon*, a double layer of Schwann cell plasma membrane, its apposed, intucked surfaces lying parallel to each other and separated by about 15–20 nm. These layers separate to enclose the axon, separated from it by a *periaxonal space* of similar dimensions. At the exterior of the Schwann cell the layers separate and are continuous with the plasma membrane. Because of this arrangement, endoneurial tissue fluid reaches the periaxonal space between the mesaxonal membranes. These intercellular spaces allow the movement of ions when action potentials are conducted along the enclosed axon. In the

absence of a myelin sheath and nodes, saltatory conduction does not occur, and the uninterrupted passage of impulses is relatively slow, with velocities of about 4·0–0·5 m/s. Non-myelinated fibres are classified as types C or IV (p. 898).

MYELINATED NERVE FIBRES

Myelinated fibres (7.32–38) form most of the somatic nervous system, rapid conduction being an obvious evolutionary advantage. The structure of myelin sheaths is best described in terms of their development which begins during prenatal life in many separate sites but is not completed until a considerable time after birth, different pathways maturing at different rates.

Axons of peripheral nerves grow out from neurons in the central nervous system and ganglia in association with Schwann cells which initially multiply as their associated axons lengthen; later they cease dividing and subsequently only increase in size. At first all axons are non-myelinated; later a flap of Schwann cell cytoplasm spirals round the neurite, most of the cytoplasm and the nucleus of the Schwann cell remaining outside (7.33, 36C). As development proceeds, the number of turns increases, transforming the spirals into compacted layers of cell membrane (Robertson 1955). In ultrastructure, mature myelin sheaths are laminated with dense, regular *period lines* (planes of apposition between the *internal* surfaces of plasma membrane), alternating with less dense *intraperiod lines* (appositions between the *external* surfaces of the plasma membrane) (7.33). Treatment with hypotonic solutions causes the sheath to swell, splitting the intraperiod lines and showing that external surfaces are not fused and are apparently able to move on one another (Napolitano & Scallen 1969). On both internal and external aspects of the myelin sheath there are thin layers of Schwann cell cytoplasm which, in many regions along the internodes of mature fibres, are only about 20–30 nm thick. The internal layer is covered by an inner plasma membrane adjoing the 15–20 nm *periaxonal space*, continuous through an *internal mesaxon* with the deepest myelin lamella. Similarly an *external mesaxon* connects the outermost myelin lamella with the superficial Schwann cell plasma membrane. Larger amounts of cytoplasm occur externally at paranodal grooves (p. 903) around the nucleus where it dimples the myelin sheath and also on the internal and external aspects of the incisures of Schmidt–Lanterman (p. 903). These cytoplasmic regions often contain mitochondria, lysosomes, some endoplasmic reticulum and various vesicles and granules. The internal and external cytoplasmic zones are also connected by helical channels containing granular cytoplasm and microtubules at nodes and incisures (vide infra). The external Schwann cell plasma membrane and adjacent basal lamina correspond to the *neurolemma* of light microscopy.

When lipids are removed from the myelin sheath, as in histological processing without prior fixation of lipid components, a network of proteinaceous *neurokeratin* is visible by light microscopy. This represents the denatured remains of the major protein component of myelin (see p. 903), its reticular appearance being an artefact of processing.

THE CHEMICAL COMPOSITION OF MYELIN

Myelin is composed of the compacted plasma membranes of Schwann cells and oligodendrocytes and its composition resembles that of other plasma membranes in containing lipids, proteins and water (which forms at least 20% of the total wet weight) but the relative proportions and dispositions of these are unique to myelin. Myelin can be isolated by techniques involving centrifuging homogenized tissue suspensions in sucrose solutions followed by osmotic shock and subsequent sedimentation or isolation on sucrose or other gradient systems. It can be further analysed by immunochemical, electrophoretic and other chromatographic methods. The proportions of cholesterol, total phospholipid and glycolipid in myelin differ from those in other membranes, and also differ between central and peripheral myelin. On a molar basis, cholesterol is the major lipid (approximately 40 moles per cent of total lipid); although it is not synthesized within the sheaths themselves, inhibition of cholesterol synthesis affects myelin formation as a whole. The most abundant of the glycolipids are galactocerebroside, sulphatide and ganglioside. *Peripheral* myelin has a lower concentration of galactocerebroside but higher ones of choline glycerophosphatide and sphingomyelin compared with central myelin. The choline glycerophosphatides have a higher proportion of unsaturated fatty acids than their central counterparts, although long chain fatty acids (C_{25} and C_{26}) are less than 2% of the total, compared with up to 20% centrally. It is noteworthy that most of the organ-specific complement-fixing or precipitating activity of certain demyelinating antisera appears to be largely antigalactocerebroside. Galactocerebroside has been found to be a cell-surface marker for cultured rat oligodendrocytes (the cell type responsible for central myelinogenesis).

About 30% of adult human central myelin is protein. The myelin proteins have been examined by polyacrylamide gel electrophoresis of isolated myelin, solubilized by phenol or sodium dodecylsulphate. At least three proteins have been identified in central and peripheral myelins and, like myelin lipids, they differ in these two sites. Peripheral myelin contains a low molecular weight glycoprotein, P_0, and two basic proteins, P_1 (*c.* 5%) and P_2 (*c.* 20%). P_1 is considered to be analogous to the encephalitogenic basic protein of central myelin. Analysis of immune responses mounted against the basic proteins has elucidated the position and relations of these proteins within the myelin sheaths of various species. It seems likely that few enzymes accompany myelin; only one, 2:3° cyclic AMP 3'-phosphohydrolase has been examined in any detail and is used as an experimental myelin marker. Its role is unknown.

FORMATION AND GROWTH OF MYELIN

Although the organization of myelin at the molecular level is now understood in some detail, the mechanisms by which it is laid down around nerve fibres is still rather obscure. Early models derived from tissue culture data proposed that Schwann cells first formed a loop of cytoplasm around axons, then migrated around them many times to wrap the axons in a tight spiral of membrane (Geren 1954). In the central nervous system at least this appears unlikely, because a single oligodendrocyte may ensheathe several separate axons (Bunge 1968). Other possibilities for myelin formation include the interstitial deposition of myelin molecular components within the lamellae of the sheath, or an ingrowth of the myelin spiral by a relative sliding of its turns, new myelin being added externally (Webster 1971). The enclosed axon might also rotate to wind the myelin upon itself; however, reports that myelin in adjacent internodes may show a reversed spiralization makes the axonal rotation model improbable. Perhaps excessive ingenuity has been expended on models explaining 'spiralization' with less attention to other dimensional changes. Thus, while the sheath *thickens* from a few lamellae to perhaps 300, the *axon* may also grow from 1 to 15 μm and internodal segments increase from 150 μm to 1500 μm in length (Williams & Wendell-Smith 1971). When myelination begins, multiplication of Schwann cells stops, each elongating in adaptation to the overall growth of the nerve (Vizoso & Young 1948, Williams & Kashef 1966). Clearly, much remains to be learnt about the initiation, mechanisms and cessation of myelination and its continued maintenance in the mature nervous system.

In late fetal and early postnatal development myelination does not occur simultaneously in all parts of the body. Nerves and tracts have their own particular temporal patterns, which can be related to their degree of functional maturity (Ochoa & Mair 1969a, b). Among other factors, which are still largely unknown, the onset of myelination is linked to the growth in diameter of non-myelinated axons; in those animals which have been investigated, the critical point appears to be reached when axons are about 1·0 μm in diameter, although some larger non-myelinated fibres also exist (Matthews 1968) and many myelinated fibres < 1 μm in diameter occur. Once begun, myelination continues in parallel with axonal growth until this stops; the ratio between axonal diameter and thickness of sheath is therefore predictable, its value varying with the function of the fibre and the distance from the parent soma (Williams & Wendell-Smith 1971).

THE NODE OF RANVIER (NEUROFIBRAL NODUS)

In myelinated nerves the axon is exposed to the extracellular fluids at regular intervals or *nodes* (*of Ranvier*) where short gaps exist between adjacent Schwann cells or oligodendrocytes (Hess & Young 1952, Landon 1976, 1985, Ghabriel & Alt 1982), the myelinated segments being *internodes*. The internodal distance varies directly with the diameter of fibre from 150 to 1500 μm (Kashef 1966). When fibres branch, they do so at nodes. In peripheral nerves, the myelin sheaths on both sides of nodes usually expand as *paranodal bulbs* (varicosities) (7.36B, 37), which often show an asymmetry related to growth. The surfaces of these bulbs and the underlying axolemma are fluted as they approach nodes and numerous Schwann cell mitochondria occur in these external cytoplasm-filled grooves (Landon & Williams 1963, Berthold 1968, Landon 1976). Axons also narrow at nodes. Each myelin lamella ends in contact with the paranodal axolemma as an expanded loop containing spiralling microtubules and dense cytoplasm. The external paranodal cytoplasm of the Schwann cells sends a number of digital processes which curve to contact the naked nodal axolemma (7.37); these fingers are numerous, forming a regular hexagonal array in large fibres, but are few and irregular in smaller ones. The depression formed at the node by the axonal narrowing is filled between nodal fingers by an acidic mucosubstance, or *gap substance*, which may be a reservoir of ions or a selective barrier to their flow during conduction (vide supra, p. 902, and Landon & Langley 1971). The narrowing of the axon at the node probably increases the transmembrane current density and hence the efficiency of nodal excitation during saltatory conduction. Sodium channels are concentrated at the nodes, few of them occurring internodally. The structural counterparts of these channels are probably the numerous intramembranous particles shown in the nodal axolemma by freeze-fracturing methods. The dense subaxolemmal coating in this region may represent cytoskeletal structures which stabilize the ion channels and anchor them to the nodal region. Central nodes have been less extensively investigated but they appear to be simpler structures, with few nodal fingers and small or no paranodal bulbs.

MYELIN INCISURES (OF SCHMIDT–LANTERMAN)

Originally described as cytological features, the myelin clefts or incisures of Schmidt–Lanterman were subsequently long regarded as artefacts. Ultrastructural and in vivo observations have reinstated them as characteristic features of central and peripheral myelinated fibres (Hall & Williams 1970, Landon 1976). They are oblique interruptions in the compacted lamellae of myelin (7.37) where funnel-shaped zones of cytoplasm spiral between the internal and external layers of Schwann cell cytoplasm. At an incisure the major dense line (p. 902) separates to enclose a helical band of granular cytoplasm; one or more microtubules follow the spiral with intermediate filaments and occasionally other organelles. The cytoplasm is dense and in places resembles that in a zonula adherens of epithelial and other cells, forming an oblique row or *stack* of *desmosomoid* attachments at one or more points near the incisure's external surface. The intraperiod line also splits at incisures to create an extracellular gap connecting the periaxonal and endoneurial spaces. Incisures possibly conduct metabolites deep into myelin sheath and subjacent axon; they appear to be an early site of cellular change in initial stages of Wallerian degeneration (Williams & Hall 1971) and of focal demyelination.

The *classic view* of Wallerian degeneration was that the first week after crush or section was largely a period of *morphological disruption* of the part distal to the injury, but was *biochemically stable*. The structural changes included the formation of myelin *ovoids* containing 'digestion chambers' with axonal disruption. Only in the second week was there thought to be any biochemical change, principally the degradation of phospholipid in the sheath with the liberation of cholesterol esters. More recently, in vivo examination of degenerating nerves (Williams & Hall 1971) showed that events are more rapid. Initial changes include the retraction of the paranodal myelin sheath, the dilatation of incisures and the collapse and rounding of myelin into ovoids at these points, starting within minutes near the lesion and progressing distally during the next 36–48 hours. The concentration of hydrolytic enzymes increases within 12 hours of injury, associated with the loss of trypsin-digestible *basic protein* from the sheath, demonstrable in histological sections as a loss of *trypanophilia*.

With electron microscopy, Wallerian degeneration shows an accumulation of membrane-bound bodies in the axoplasm proximal and distal to the point of injury. Distal degradation of axoplasmic organelles then ensues and at the dilated incisures the previously compact myelin lamellae collapse, with alterations in the repeat periods in their radial structure. Within four days lipid droplets accumulate round degrading myelin within the Schwann cell cytoplasm and are then extruded into the endoneurial space, to be phagocytosed by invading haematogenous macrophages. The early increases in acid phosphatase activity coincides with the appearance of numerous lysosomes in the cytoplasm of Schwann cells. As degradation of myelin proceeds, Schwann cells begin to proliferate. In nerves containing many large myelinated fibres the total of Schwann cell nuclei visible in transverse sections rises to about 16 times the normal levels by the end of the third week. Proliferating Schwann cells line up in parallel longitudinal chains within the persistent basal lamina, where they take part in any subsequent regenerative phenomena which may ensue (p. 918).

THE CENTRAL-PERIPHERAL TRANSITIONAL ZONE

Crossing the transitional zone between the central (CNS) and peripheral (PNS) nervous system, axons are enclosed in a short glial segment generally close to the surface of the spinal cord or brain stem (7.39). In humans this zone lies more peripheral in sensory than in motor nerves; in both of these the apex of the zone is described as the '*glial dome*', its convex surface being directed distally. Electron microscopy shows the dome's centre to consist of fibres with a typical central organization, surrounded by an outer mantle of astrocytes (corresponding to the external glial limiting membrane). From this mantle numerous processes, the '*glial fringe*', project into endoneurial compartments of the peripheral nerve and interdigitate with the Schwann cells. The astrocytes form a loose reticulum through which axons pass. Peripheral myelinated fibres usually cross the transitional zone at a node of Ranvier, termed a *PNS–CNS compound node* (Carlstedt 1977). At such a node distally there is a corona of Schwann cell microvilli and mitochondria-laden paranodal cytoplasm; centrally are a few astrocyte processes, typically making contacts with the axolemma (Berthold 1978). In the first sacral spinal nerve roots (of cats) an average of four or five nodes are associated with the processes of a single astrocyte and these node–astrocyte relations may be specific to such borderline nodes. In some myelinated fibres, however, the central–peripheral transition occurs at an internode, the central myelin being telescoped within an external peripheral myelin sheath. Considerable rearrangement of axons occurs in the rootlets; moreover, many of the largest non-myelinated axons are invested with a thin myelin sheath as they traverse the transitional region.

Peripheral Endings of Effector Neurons

Of the various types of effector endings the most intensively studied are those which innervate muscle, particularly the skeletal variety. All such *neuromuscular* or *myoneural junctions* are regions of neuronal cytoplasm specialized for the release of neurotransmitter on to surfaces of adjacent muscle fibres, causing a change in their electrical state leading to contraction. Because of their similarity to central nervous synapses and the relative ease with which functional studies can be carried out on them, much of our general knowledge of neurotransmission comes from the ultrastructural, physiological, biochemical and pharmacological analysis of neuromuscular junctions rather than central synapses.

In addition to their innervation of muscle, effector neurons also have terminals in or near other cells capable of various types of activity, e.g. glands (secretomotor endings), myoepithelial cells and adipose tissue. It has been proposed on the basis of

experimental evidence that nerve fibres may also contribute to the regulation of immune responses (p. 836), by their effects on T-lymphocyte maturation in the thymus, and to many other poorly understood activities of tissues not classically considered to be under nervous control. The hormonal contribution of neuro-endocrine cells has, of course, long been known, but more subtle effects elsewhere may play a major role in the maintenance of homeostasis in many other sites in the body.

NEUROMUSCULAR JUNCTIONS OF SKELETAL MUSCLE

The general structure of these has already been outlined on p. 554. The terminals are end branches of somatic motor fibres, each innervating from a few to many hundreds of muscle fibres, a numerical relationship determining the precision of motor control. The precise structure of a motor terminal varies with the type of muscle innervated. Two major endings are recognized: '*en plaque*' terminals typical of extrafusal muscle fibres and *plate endings* of intrafusal fibres, the latter including '*en grappe*' terminals of extraocular muscle fibres and '*trail*' endings on intrafusal fibres (Harker 1972, Barker 1974). In the first type each axonal terminal (*telodendron*) usually ends midway along a muscle fibre in a discoidal *motor end-plate* (7.40–42). In the second type, the axon has numerous subsidiary branches forming a cluster of small expansions extending along the muscle fibre (7.40). *En plaque* endings usually initiate action potentials which are rapidly conducted to all parts of muscle fibres whereas *en grappe* endings, in the absence of propagated muscle excitation, excite the fibre at several points. Both types have a specialized region, the *sole-plate*, in which a number of muscle cell nuclei are grouped within the granular sarcoplasm; ultrastructurally the sole-plate contains numerous mitochondria, endoplasmic reticulum and the folding of the sarcolemma into numerous parallel grooves (7.42). The neuronal terminal expansion is separated by about 30–50 nm from this *subneural apparatus*, the terminal containing mitochondria and many clear spherical vesicles like those in presynaptic boutons and similarly clustered against the membrane over the zone of apposition. Ensheathing the motor terminal are Schwann cells (*teloglia*), with cytoplasmic projections extending into the synaptic cleft. *En plaque* endings of 'fast' and 'slow' twitch muscle fibres differ in details, the sarcolemmal grooves being deeper and the presynaptic vesicles more numerous in the former (Padykula & Gauthier 1970). Freeze-fracture studies show numerous intramembranous particles, which represent the acetylcholine receptors, in the crests of the sarcolemmal folds nearest to presynaptic membrane (Heuser et al 1974). Opposite these the sarcolemmal membrane is slightly invaginated and vesicles cluster along borders of the groove so formed, probably the main site of their release (cf. the '*active zone*' of synapses). Microtubular arrays bearing synaptic vesicles, as in central synapses, have also been described (Gray 1978).

Physiological and pharmacological studies (Katz & Miledi 1965) show that junctions with striated muscle are *cholinergic*, the release of acetylcholine changing the ionic permeability of the muscle fibre. The earliest change occurring after neurotransmitter release is a *graded potential*; if the depolarization of the sarcolemma reaches a particular threshold, it initiates an all-or-none action potential in the sarcolemma, which is then propagated rapidly over the whole cell surface and also within its substance via the invaginations (transverse tubules) of the sarcolemma (see p. 554), causing contraction. As with synapses, the quantity of neurotransmitter released depends on the frequency of the arriving nerve impulses; since the transmitter is inactivated at a constant rate by the enzyme *acetylcholinesterase* present at the sarcolemmal surface of the sole-plate, the amount of neurotransmitter present depends on its *rate* of release, thus reflecting faithfully the *frequency* of action potentials in the nerve fibre. Thus the contraction of a muscle fibre is controlled by the firing frequency of its motor neuron. However, even in quiescent muscle fibres sporadic depolarizations continually occur (*miniature end-plate potentials*), although too small to cause contraction. Their measured amplitudes occur at a series of preferential levels, indicating that multiples of *quanta* of transmitter are released ever

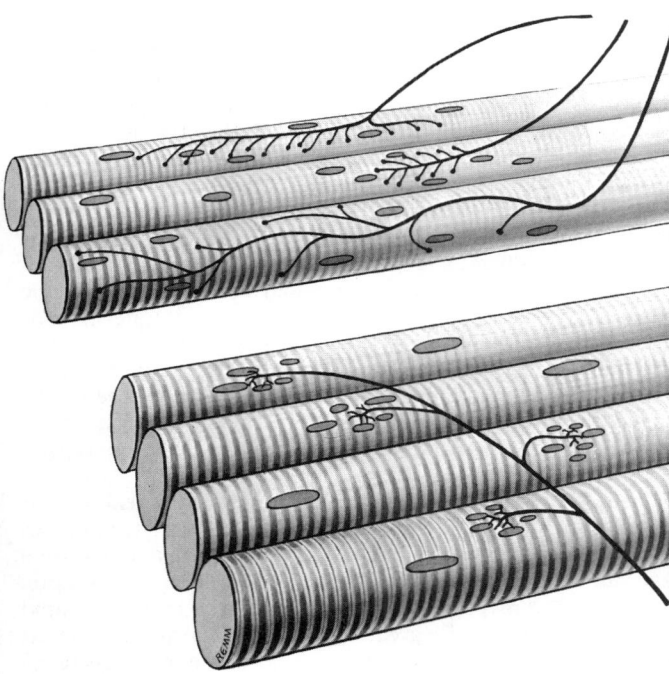

7.40 Diagram showing some types of innervation of striated muscle, including the 'en plaque' terminals of α efferents (below), and the more widely spread 'trail' and 'en grappe' endings of γ efferents (above).

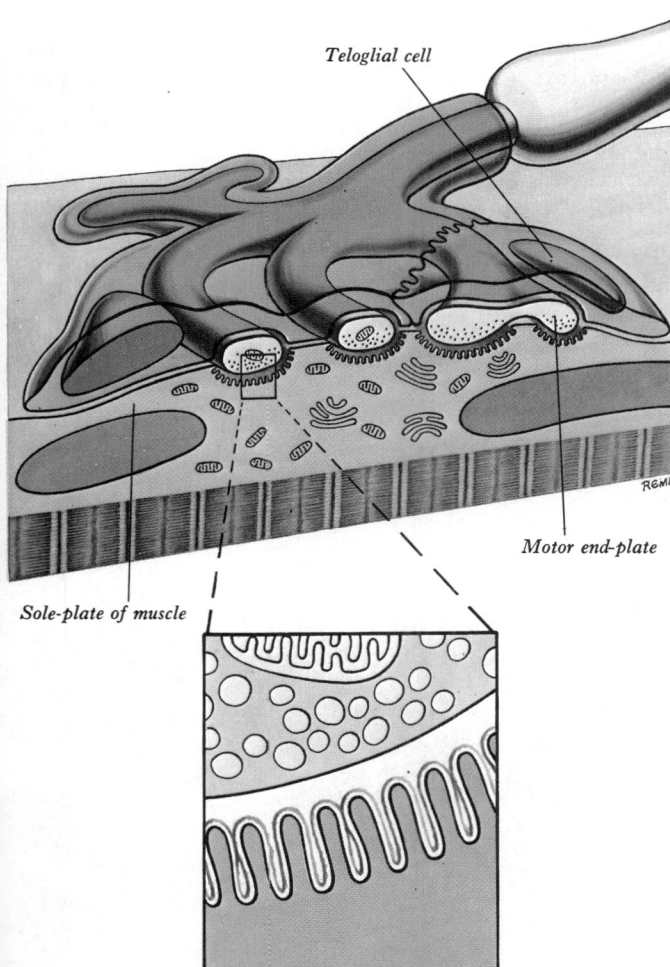

Teloglial cell

Motor end-plate

Sole-plate of muscle

7.41 Diagram of the detailed structure of an 'en plaque' neuromuscular junction. An enlarged portion of the terminal is shown below; note the folding of the sarcolemma to form subsynaptic gutters, the disposition of the basal lamina and the synaptic vesicles within the axon terminal.

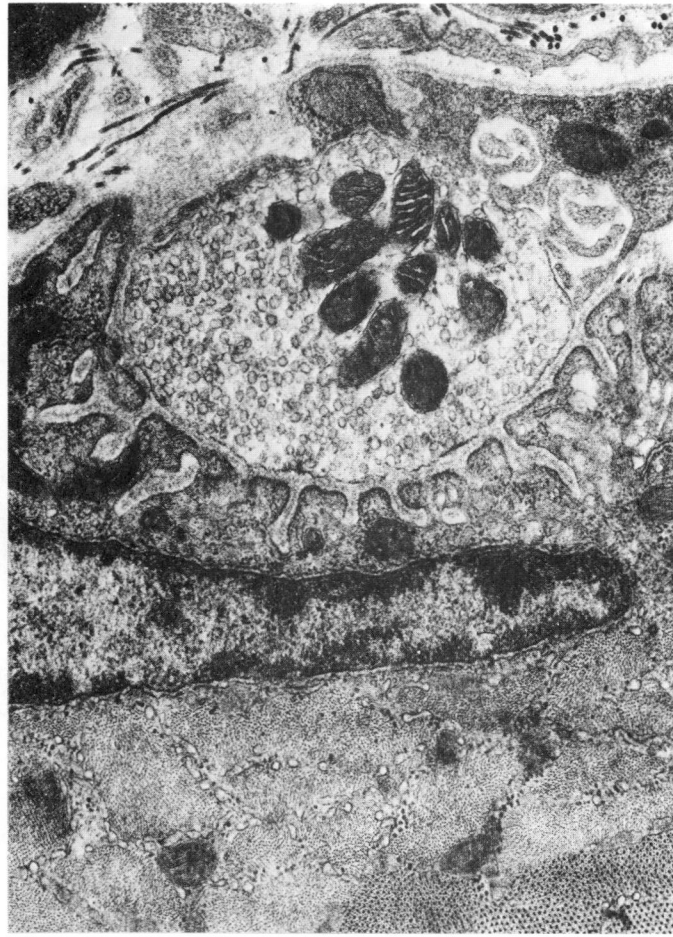

7.42 Electron micrograph of a neuromuscular junction in striated muscle, showing part of a motor end-plate containing synaptic vesicles (above). The latter is situated in a groove in the sarcolemma; this is further convoluted to form the subsynaptic gutters of the sole-plate (below). Provided by D N Landon, Neurobiological Unit, National Hospital for Nervous Diseases, London.

in resting stages, presumably a process of slow leakage. The amount of acetylcholine necessary to cause these unitary changes is calculated to be equivalent to the volume of a single synaptic vesicle (Katz & Miledi 1965).

Acetylcholinesterase can be shown cytochemically to be present in myoneural clefts. Experiments with specialized motor endings in electric organs of the Electric Ray *Torpedo* locate the enzyme to the sarcolemma near the sites of action. The neurotransmitter can be blocked by *curare*, by a sea-snake venom α-bungarotoxin, by black widow spider venom and various other toxins, all combining irreversibly with the acetylcholine receptor molecules to paralyse the neuromuscular junctions (Miledi et al 1971). If acetylcholinesterase is inhibited, e.g. by *eserine* (*physostigmine*), transmitter action is cumulative and tetanus may ensue. Neuromuscular junctions are partially blocked by high concentrations of lactic acid, as in some types of muscle fatigue (Miledi et al 1971). For further review of structure see Duchen & Gale 1985.

AUTONOMIC MOTOR TERMINATIONS

Unlike those in skeletal muscle, autonomic terminals are not usually closely applied to non-striated myocytes but end at variable distances from their surfaces (Richardson 1962). Non-myelinated autonomic axons branch into tapering, varicose collaterals (7.43). At the zone of transmission, clusters of synaptic vesicles exist in expanded regions of axoplasm and the Schwann cell is retracted to leave the axon in a shallow groove, providing a path of diffusion between axon and myocytes. In *sympathetic fibres* the vesicles are usually *dense-cored* (p. 890), and correspond in position to the characteristic fluorescence of

905

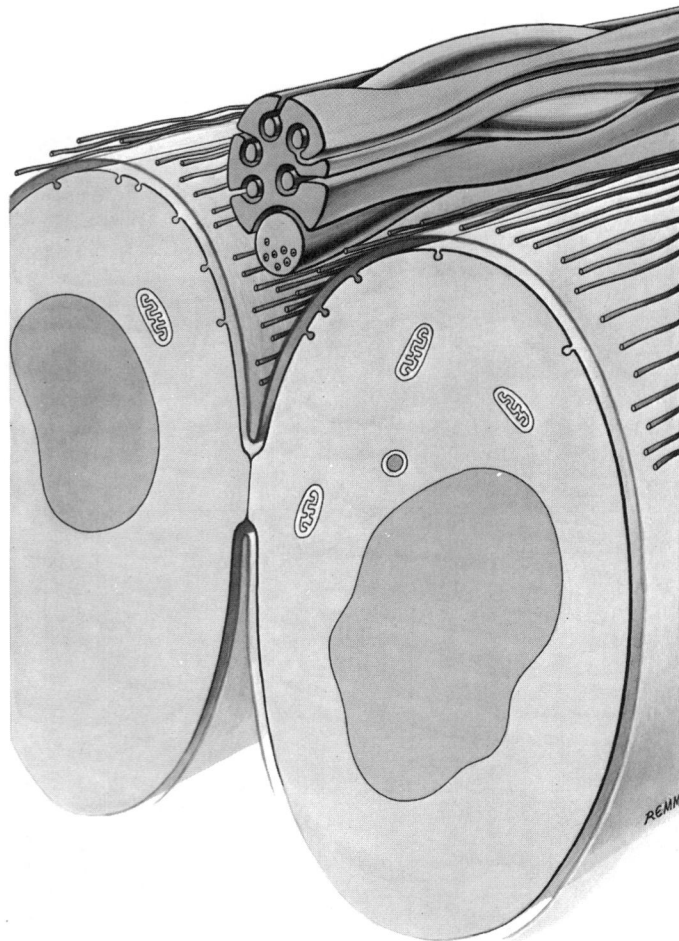

7.43 Diagram of an autonomic neuromuscular junction between a group of non-myelinated axons (above) and smooth muscle cells (below). The Schwann cell (blue) is reflected at intervals to expose enlargements of the axons (yellow) which contain synaptic vesicles.

catecholamines seen when preparations fixed in formalin vapour are viewed with ultraviolet light (Falck & Owman 1965), or to the positive reaction to the glyoxylic acid method. Besides noradrenalin, they contain enzymes involved in catecholamine synthesis and degradation. Other vesicles, 100 μm across, often also occur.

Noradrenalin is the usual transmitter at sympathetic postganglionic endings while *adrenalin* occurs in some of the *chromaffin cells* of the adrenal glands. Synaptic vesicles containing each catecholamine can be distinguished by electron microscopy. Catecholamines can be released experimentally by treatment with the drug, *reserpine*, which is a valuable research tool and also clinically useful. After release some of the transmitter is reabsorbed into the axon by endocytosis and may be re-used, or broken down. Sympathetic endings contain *monoamine oxidases* degrading catecholamines, the control of these enzymes being important in the clinical regulation of sympathetic function both in the peripheral and central nervous systems.

At *parasympathetic* endings are clear spherical vesicles, like those in motor end-plates of striated muscle. Much evidence indicates that they are *cholinergic*; similar terminals also occur in sympathetic endings. In addition to their action on non-nervous tissues, some cholinergic endings may perhaps cause the release of catecholamines from neighbouring sympathetic endings (Burn 1968).

A third category of autonomic neurons differing from noradrenergic or cholinergic cells has been described by Burnstock and colleagues (Burnstock 1975). These can be described as 'non-adrenergic, non-cholinergic endings', and are now recognized to contain a wide variety of chemicals with transmitter properties (see Burnstock 1986), contained within vesicles of different forms. The conjugated purine adenosine triphosphate or ATP (a nucleoside), probably the neurotransmitter at these terminals, is classed as *purinergic*. Typically, their axons contain large, dense vesicles 80–200 nm across ('*large opaque vesicles*'), congregated in varicosities at intervals along nerve fibres. Such fibres occur in many sites in mammals, including the alimentary external muscle layers and sphincters, lungs, vascular walls, urogenital tract and central nervous system. In the intestinal wall their neuronal cell bodies lie in the myenteric plexus, their axons spreading caudally for a few millimetres, chiefly to innervate circular muscle (7.44). Purinergic neurons are under cholinergic control from preganglionic sympathetic neurons. Purinergic endings mainly hyperpolarize myocytes, causing their relaxation, e.g. preceding peristaltic waves, opening sphincters and, probably, causing reflex distension in gastric filling. The release of the transmitter resembles that in other autonomic endings, i.e. by exocytosis from axon varicosities which may be up to 100 nm away from the target cell.

The pattern of terminal branching of autonomic efferents is related to their effects (Burnstock 1970). In non-striated muscles with a slow widespread action, a single neuron may innervate a large number of myocytes, the distance between synaptic varicosities and myocytes being large (50 nm or more). In rapid non-striated muscle, with greater precision of control, e.g. the iris or ductus deferens, innervation is more localized, with less branching and close apposition to myocyte surfaces (15–20 nm), the ending sometimes being invaginated into the sarcolemma (Burnstock 1975). Other tissues innervated by autonomic efferents include glands, myoepitheliocytes, adipose and lymphoid tissue.

Sensory Receptors

As frequently stated, an organism's reaction to changes in its environment requires the presence of suitable receptors to scan the parameters important to survival. Though single-celled organisms such as protozoa have no such specialization, they can detect changes in temperature, osmotic pressure, concentrations, light and other environmental properties by means of the specialized components of their membranes. In more elaborate animals these fluctuations are recorded by cells which have differentiated to respond selectively and with great precision. Such cells are sensory receptors; they take many forms but in mammals, with minor exceptions, they have a common pattern which will be outlined below.

Sense organs have three major forms, according to the relation of the nervous system to the receptor cells. In **neuro-epithelial**

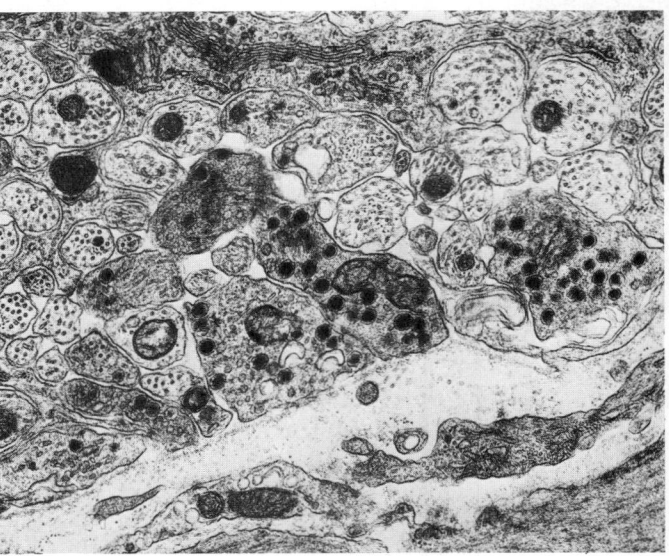

7.44 Electron micrograph of a group of autonomic axons in the myenteric plexus, showing profiles with large dense-cored vesicles typical of the putative purinergic system (see text); smaller, catecholamine-type vesicles are also visible (lower left). Magnification × 40 000.

receptors the receptors are neurons with somata situated peripherally near a sensory surface and axons extending into the central nervous system to connect with second order neurons. The only example of this type in mammals is the sensory cell of the olfactory epithelium but in many invertebrates it is the main type and can thus be regarded as phylogenetically primitive. Secondly, the sensory cell may be an **epithelial receptor**, modified from the cells of a non-nervous sensory epithelium and innervated by a primary sensory neuron, the soma of which lies near the central nervous system. Activity in this type of receptor causes the passage of excitation from the receptor by neurotransmission across a synaptic gap; examples are gustatory and auditory receptors. In gustatory receptors individual cells are constantly being renewed from the surrounding epithelium, as may also occur in other similar sensory systems (p. 1170). Visual receptors are in many ways similar in their form and relations, formed from the ventricular lining in the fetal brain, although no replacement occurs. Thirdly, a **neuronal receptor** is a primary sensory neuron, with a soma in a craniospinal ganglion and a peripheral axon whose end is the sensory terminal. All cutaneous

sensors and proprioceptors are probably of this type, but the sensory terminals may be encapsulated or linked to special mesodermal or ectodermal elements forming a part of the sensory apparatus. These extra-neural cells are not necessarily excitable, but rather create the right environment for the excitation of the neuronal dendrite or modify its excitation in some way.

RECEPTOR RESPONSES

The events which occur between the incidence of a stimulus and the conduction of an electrical signal to the central nervous system have been extensively studied (e.g. Granit 1962), although much remains to be understood. It is convenient to divide these processes into a number of stages: first the preservation of an effective stimulus, then the transduction of the stimulus at the receptor surface into a graded change of electrical potential (*sensory potential*) and next the initiation of an all-or-none action potential passing to the central nervous system. All of these processes may occur in the receptor, where this is a neuron, or some

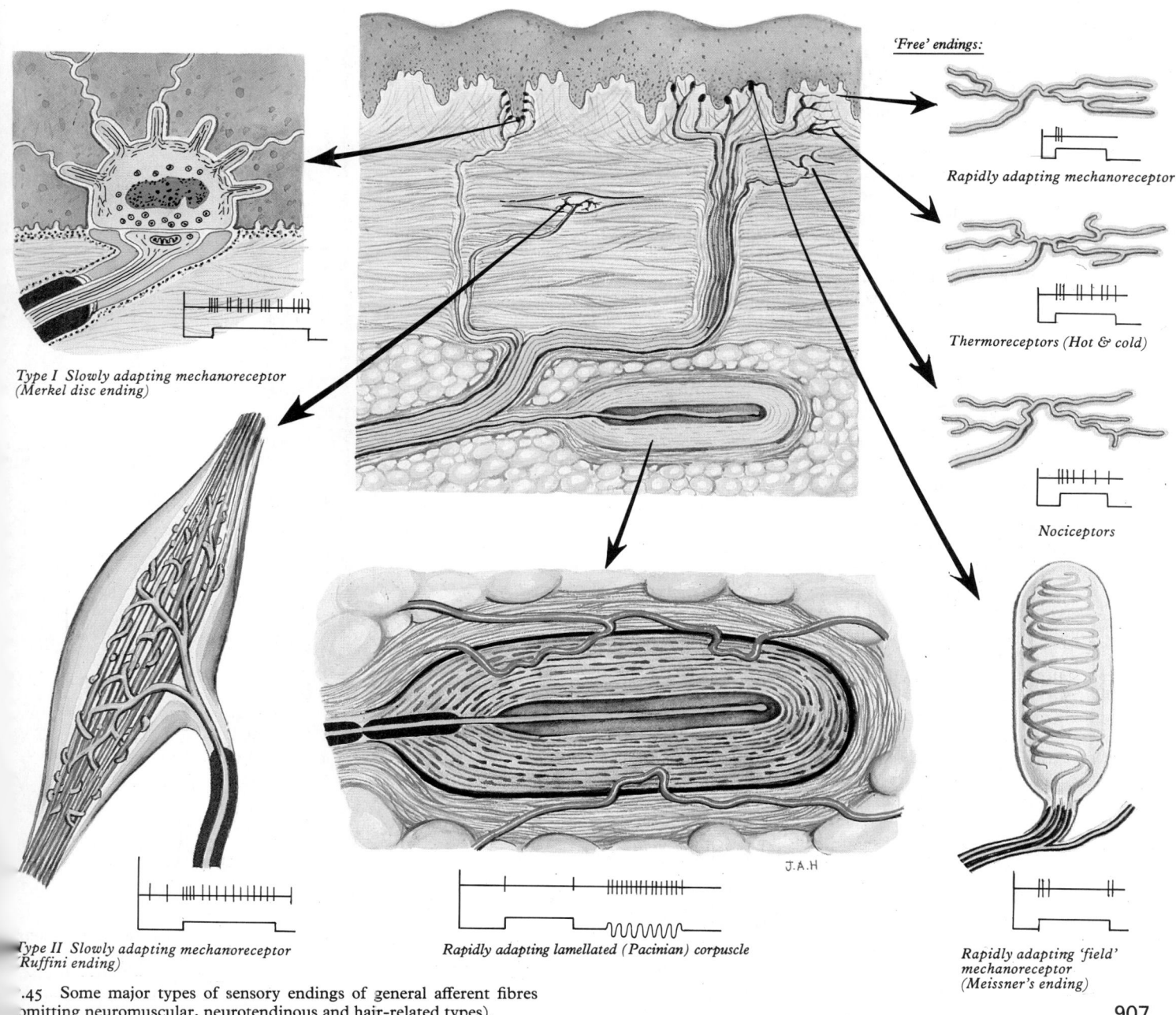

Type I Slowly adapting mechanoreceptor (Merkel disc ending)

'Free' endings:

Rapidly adapting mechanoreceptor

Thermoreceptors (Hot & cold)

Nociceptors

Type II Slowly adapting mechanoreceptor (Ruffini ending)

Rapidly adapting lamellated (Pacinian) corpuscle

Rapidly adapting 'field' mechanoreceptor (Meissner's ending)

.45 Some major types of sensory endings of general afferent fibres omitting neuromuscular, neurotendinous and hair-related types).

may occur partly in the receptor and partly in the neuron innervating it, in the case of epithelial receptors.

Transduction varies with the modality of stimulus, but in all receptors there is a stimulus-induced change in permeability of the receptor membrane to certain ions, usually causing its depolarization (or, in the retina, hyperpolarization). How a stimulus effects this is uncertain. In mechanoreceptors it may involve the deformation of membrane structure perhaps with strain- or voltage-sensitive transducing protein molecules opening ion channels as a result; in chemoreceptors, receptor action may resemble that postulated for acetylcholine at neuromuscular junctions (p. 554), involving receptor proteins altered by binding stimulant molecules, again with the opening of ionic channels. Visual receptors appear to resemble chemoreceptors in many respects; light causes changes in receptor proteins (p. 1198) which may cause the release of *second messengers* to affect membrane permeability.

Osmoreceptors may resemble mechanoreceptors but react instead to mechanical deformation from osmotic inflow or outflow of water. In some fishes a remarkable type of transduction occurs in electroreceptors, capable of sensing changes in electric fields set up by the animal's own bioelectric processes or those of other individuals or prey (Lissmann & Machin 1958). So far, no sensory mechanism of this type has been reported in tetrapods.

The quantitative responses of sensory endings to stimuli vary greatly and give added flexibility to the functional design of sensory systems. Although increase of excitation with the increase of a stimulus level is a common pattern ('*on' response*) some receptors respond to decrease in stimulus (the '*off' response*). Even unstimulated receptors show varying spontaneous activity, against the background of which an increase or decrease in activity occurs with changing levels of stimulus. When stimulation is maintained at a steady level, there is, in all receptors studied, an initial burst (the *dynamic phase*) followed by a gradual *adaptation* to steady level (the *static phase*). Though all receptors show these two phases, one or other may predominate, giving a distinction between *rapidly-adapting* endings accurately recording the *rate* of stimulus onset and *slowly-adapting* endings which signal the constant amplitude of a stimulus, e.g. the position sense. Dynamic and static phases are reflected in the *amplitude* and *duration* of the generator or sensory potential and also in the *frequency* of action potentials in the sensory fibres. The stimulus strength necessary to elicit a response in a receptor (i.e. its *threshold level*) varies greatly between receptors, providing an extra level of information about stimulus strength.

CLASSIFICATION OF RECEPTORS

Receptors are classified in several ways; e.g. by the specific energy forms or '*modalities*' to which they are especially sensitive, such as *mechanoreceptors* responsive to deformation (touch, pressure, sound waves, etc.), *chemoreceptors*, *photoreceptors*, *thermoreceptors* and *osmoreceptors* (reacting to osmotic pressure changes). Some receptors respond selectively to more than one modality (*polymodal receptors*), usually having high thresholds and responding to damaging stimuli, associated with irritation or pain (i.e. *nociceptors*).

Another widely used, though somewhat arbitrary classification divides receptors on the basis of their distribution in the body and their role in its sensory activities into *exteroceptors, proprioceptors* and *interoceptors*. Exteroceptors and proprioceptors are receptors of somatic afferent components of the nervous system; interoceptors are receptors of the visceral afferent pathways.

Exteroceptors respond to external stimuli and are at or close to surfaces (Sinclair 1967). They can be subdivided into the *general* or *cutaneous sense organs* and *special sensory organs*. The general sensory receptors include the 'free' and the encapsulated terminals in skin and near hairs; special sensory organs are the olfactory, visual, acoustic and gustatory receptors.

Proprioceptors respond to stimuli *proper* to deeper tissues, especially of the locomotor system, and are concerned with detecting movement, mechanical stresses and position; they include the neurotendinous organs of Golgi, neuromuscular spindles, Pacinian corpuscles, other endings in joints and vestibular

receptors. Proprioceptors are stimulated by the contraction of muscles, movements of joints and changes in the position of the body or of its parts; they are essential to the co-ordination of muscles, the grading of muscular contraction and maintenance of equilibrium.

Interoceptors include receptors in the walls of the viscera, glands and vessels, where their terminations include 'free' nerve endings, encapsulated terminals and endings associated with specialized epithelial cells. Nerve terminals occur in all the layers of visceral walls and are numerous in the adventitia of blood vessels; the functional nature of many of these endings, and indeed their detailed structure in many cases, is not well established. Encapsulated (lamellated) endings occur in the heart, adventitia, pancreas and mesenteries; 'free' terminal arborizations also occur in the endocardium, loose connective tissue, the endomysium of all muscles and connective tissue generally.

Visceral nerve terminals are not usually responsive to stimuli which act on exteroceptors and do not respond to localized mechanical and thermal stimuli. But tension produced by excessive muscular contraction often causes visceral pain, particularly in pathological states, which is frequently poorly localized and of the deep-seated variety.

Interoceptors include vascular chemoreceptors and baroceptors concerned in the regulation of blood flow and pressure and in the control of respiration.

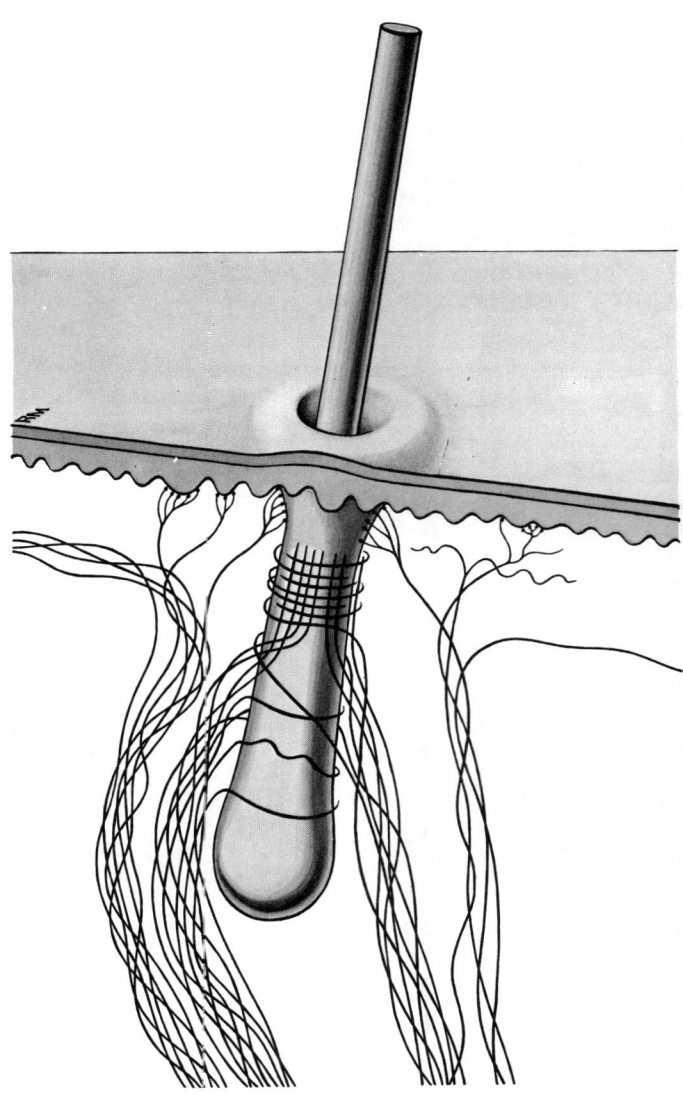

7.46A The innervation of a large hair follicle; fine and coarse axons terminate around the intermediate and superficial regions, some branching in a circular direction and others pursuing a longitudinal course. Nerve terminals associated with Merkel's cells are also shown surrounding the neck of the follicle.

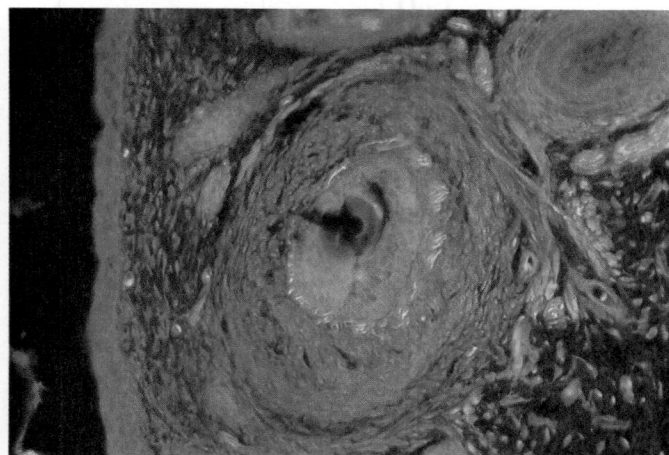

7.46B Sensory endings around a hair follicle, in transverse section. The axon terminals have been demonstrated by immunofluorescence using an antibody directed against neurofilaments. Material provided by Samantha Negus, UMDS, Guy's Campus, London. Magnification ×250.

Irritant receptors responding polymodally to noxious chemicals or damaging mechanical stimuli are widely distributed in the epithelia of the alimentary and respiratory tracts and they may initiate protective reflexes.

This scheme of classification may seem arbitrary, since many *structural* types of end-organ occur in all three classes and their activities may be closely linked in the central nervous system. For convenience, therefore, the structural classification is followed here. But this in turn is arbitrary since in some cases endings with similar structure may be physiologically distinct and a functional classification may be regarded as of deeper significance.

STRUCTURAL CLASSIFICATION OF GENERAL SENSORY ENDINGS

In addition to the division of receptors into functional categories, the receptors (apart from those of the special senses) can be grouped by structural features on morphological grounds (7.45–53). How much functional significance can be attached to these features is debatable. Factors which complicate this issue include variations in receptor structure linked with ageing, regeneration, local topography and with species differences. The terminology is further confused by a lack of agreement on the

structural definition of many sensory endings. Some receptors are limited to one tissue or combination of tissues; others occur in many sites. Generally, one may distinguish between *free nerve endings* which form plexuses or spread without any particular association with other types of cell, *encapsulated endings* where groups of specialized non-nervous cells invest neural process, and *epidermal endings*, where sensory fibres are attached to specific non-nervous cells or tissues (Bannister 1976).

FREE NERVE ENDINGS

Sensory endings, branching to form plexuses, occur in many sites. They are found in all connective tissues including: those of the dermis, fasciae, capsules of organs, ligaments, tendons, the adventitia of blood vessels, meninges, articular capsules, periosteum, perichondrium, Haversian systems, parietal peritoneum, the visceral wall and endomysial spaces of all types of muscle. They also innervate the epithelium of the skin, cornea, buccal cavity, and alimentary and respiratory tracts and their glands, though within epithelia they are devoid of Schwann cells and are enveloped instead by epitheliocytes (and so are perhaps not 'free' endings, see Cauna 1966).

Afferent fibres from free terminals are either myelinated or non-myelinated but are always of small diameter and low conduction velocity, belonging to group III (Aδ) or IV (C) (p. 898). Where afferent axons are myelinated, their terminal arborizations are not and are probably devoid of Schwann cells at their ends. Electrophysiological recordings show that such terminals serve several sensory modalities. In the dermis, some are responsive to moderate cold or heat (*thermoreceptors*), to light mechanical touch (*mechanoreceptors*), to damaging heat, cold or deformation (*unimodal nociceptors*), or to damaging stimuli of several kinds (*polymodal nociceptors*). Similar fibres in deeper tissues may also signal extreme conditions, experienced, as with all nociceptors, as pain. Free endings in the cornea, dentine and periosteum may be exclusively nociceptor in function (see also Kruger 1987).

ENDINGS ASSOCIATED WITH EPIDERMAL STRUCTURES

Terminals associated with hair follicles are those of myelinated fibres in the deep dermal cutaneous plexus; the number, size and form of the endings are related to the size and type of hair follicle innervated (Cauna 1966). In *palisade endings* fibres approach the follicle from different directions just below the sebaceous duct, where they divide to run parallel to the hair in the outer follicular layer (7.46). Some enter it to give branches encircling the hair and terminating as free endings among collagen bundles; others

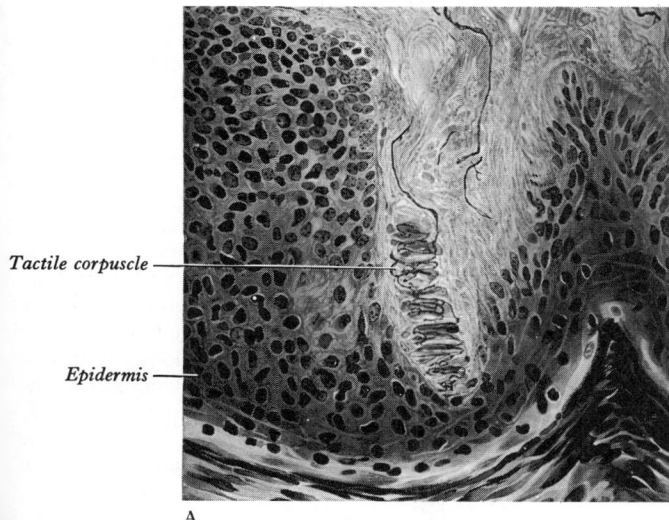

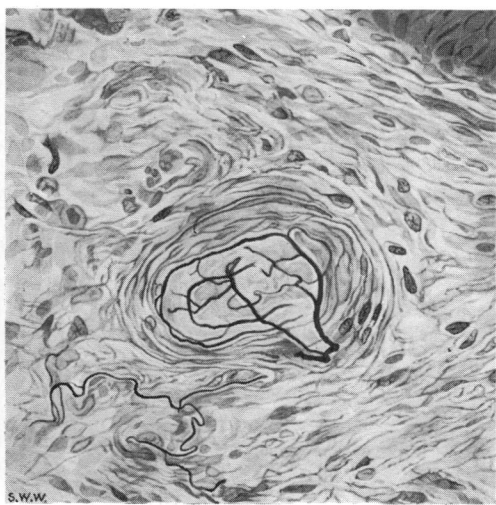

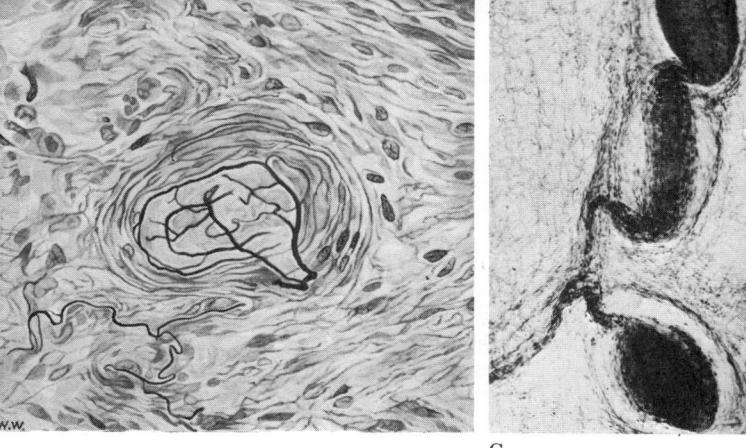

Tactile corpuscle

Epidermis

A

B

C

7.47 Specialized sensory end-organs. A. Tactile corpuscle of Meissner in human skin. Gros-Bielschowsky technique; magnification *c.* ×250. B. Bulbous corpuscle from human anal canal. Gros-Bielschowsky and haematoxylin; magnification *c.* ×480. Material supplied by M J T Fitz-

Gerald, University College, Galway. C. Whole mount of developing Pacinian corpuscles in feline mesentery. Gros-Bielschowsky technique; magnification *c.* ×120. Specimens A and C provided by N Cauna, University of Pittsburgh.

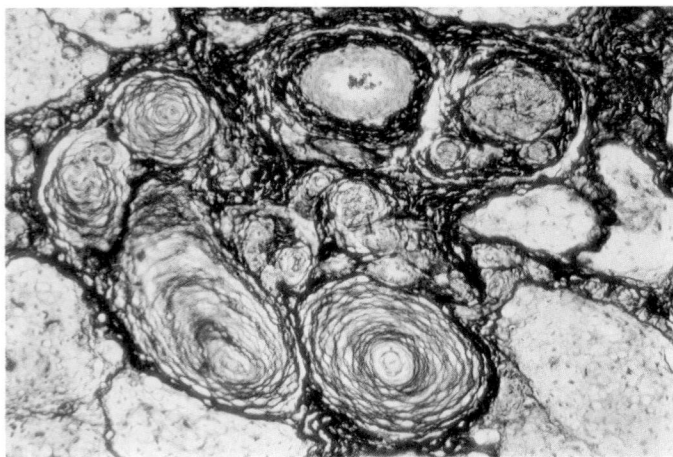

A

7.48 Structure of some peripheral sensory terminals.

A. A cluster of Pacinian corpuscles stained with the Glees and Marsland silver technique to show the capsules and the central axons, sectioned in various planes. (Rhesus monkey finger). Provided by R Bilous, Guy's Hospital Medical School, London. Magnification × 150.

B. Electron micrograph of a Pacinian corpuscle in transverse section, showing the central core region with lamellar cells surrounding the axon. Note the presence of large intercellular spaces between the lamellar cells and the numerous mitochondria in the axon. (Rhesus monkey finger.) Magnification × 5000.

C. Electron micrograph showing a Merkel cell (M) associated with a Type I slowly-adapting cutaneous mechanoreceptor nerve terminal (N) from a rabbit touch-dome. Note the numerous dense vesicles clustered near the sensory nerve ending (arrowheads) and epidermal keratinocytes (K). Material provided by W Hamann, Dept. of Physiology, Guy's Hospital Medical School, London. Magnification × 5000.

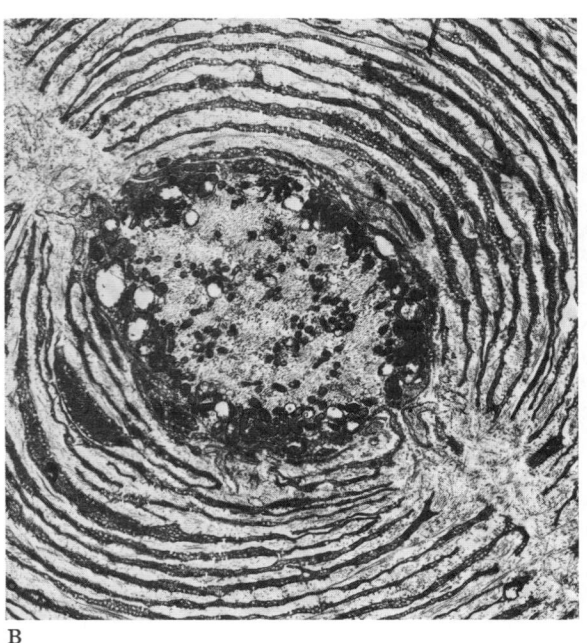

B

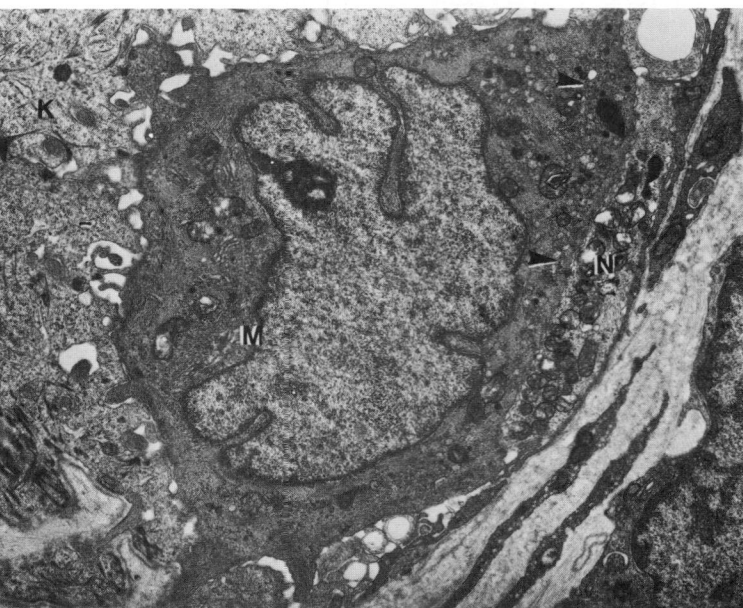

C

enter the hyaline layer and, after losing their myelin sheaths, they divide into fine filaments coursing in parallel branches along the follicular axis, each ensheathed by two Schwann cells or lemmocytes. In many mammals there are fine *down hairs*, coarser *guard hairs* and *whiskers* or *vibrissae* which have erectile vascular tissue around their follicles and are the most complex, with at least three types of ending (see 7.46). Human hair is probably of the first two types. Palisade endings respond mainly to rapid movements when hair is deformed and belong to the rapidly adapting mechanoreceptor group (see also 7.50).

Tactile menisci (Merkel cell (disc) endings) lie just below the epidermis or around the apical ends of some hair follicles (7.48C). Their nerve fibre expands into a disc applied closely to the base of a specialized cell (*Merkel cell*) inserted into the base of the epidermis, bearing spike-like protrusions which interdigitate with the surrounding keratinocytes. The Merkel cells contain many large (50–100 nm) dense-cored vesicles, particularly congregated near the junction with the nerve fibre. In many mammals, groups of such units are assembled at the base of dome-like epidermal discs (*touch domes*) supplied by single, highly branched nerve fibres, also often associated with specialized hairs (*tylotrichs*). These endings are slow-adapting (Type I) mechanoreceptors responsive to vertical pressure and served by large myelinated (Aα) afferents. The ending's structure suggests some form of synaptic transmission but attempts to demonstrate this unequivocally have failed (see Iggo & Muir 1969, English 1978, Diamond et al 1988; see, however, Baumann et al 1988).

ENCAPSULATED NERVE ENDINGS

These are a major group of special end-organs exhibiting considerable variety in their size, shape and distribution but they all have a common feature: the axon terminal is encapsuled by non-excitable cells. Included in this category of ending are lamellated corpuscles of various kinds (e.g. Pacinian, vide infra), neurotendinous endings, neuromuscular spindles and Ruffini endings.

Tactile corpuscles (of Meissner) are found in the dermal papillae of all parts of the hand and foot, the front of the forearm, the lips, palpebral conjunctiva and mucous membrane of the apical part of the tongue, although they occur mainly in thick hairless skin. Mature corpuscles are cylindrical in shape, with their long axes perpendicular to the skin surface, and are about 80 μm long and 30 μm broad. The corpuscle has a connective tissue capsule and central core (7.45, 47), the capsule being only loosely attached to the core and absent at the extremities. Light microscopy indicates an external capsule of fine elastic fibres orientated mainly in the long axis of the corpuscle, interspersed with fibrocytes. The elastic fibres may anchor the corpuscle to the epidermis (Quilliam 1966). Electron microscopy shows the capsule to be continuous with the perineurium of the nerve fibres supplying the corpuscle, consisting of lamellae of greatly flattened cells and their basement membranes. Between successive lamellae there are substantial amounts of collagen, but elastic fibres are absent. Capsular extensions, often one lamella thick, may form complete or incomplete

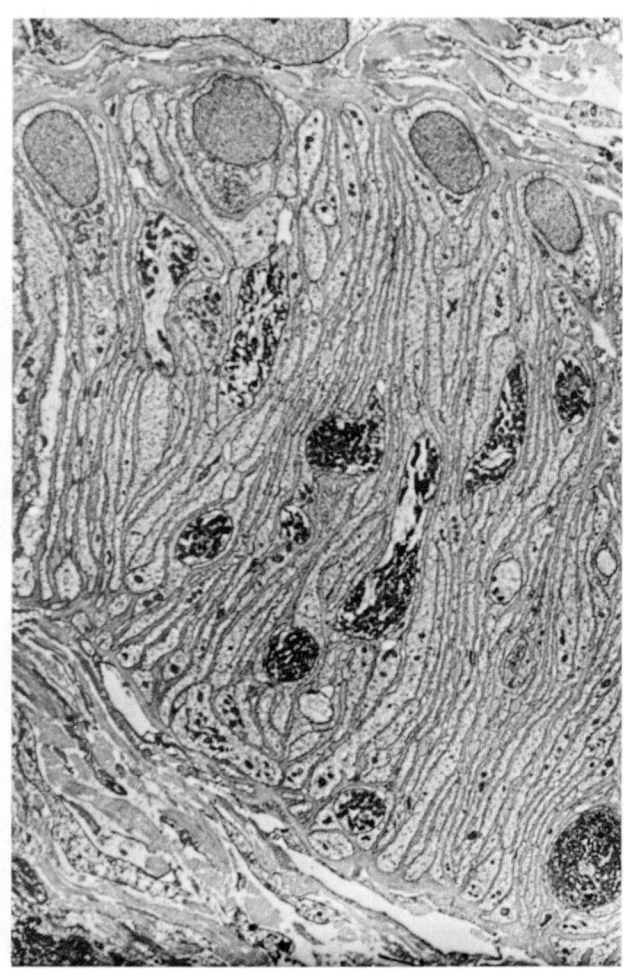

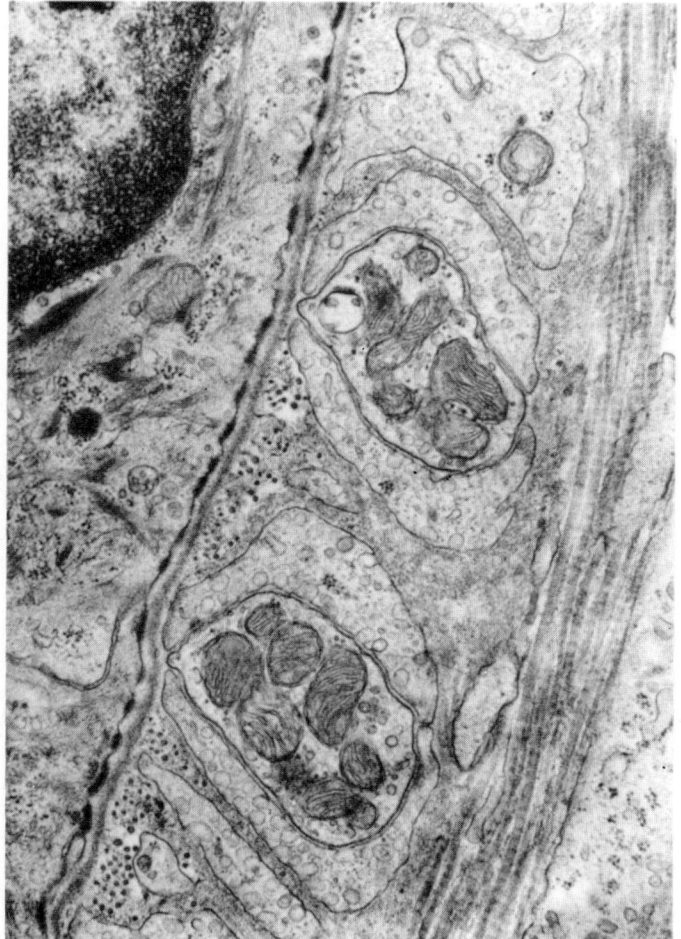

7.49   Ultrastructure, near the apex, of a tactile corpuscle of Meissner in vertical section, showing flattened lemmal cells arranged horizontally, with their nuclei to the right of the field. The sectional profiles of the terminal nerve fibres, appearing between the lemmal cells, contain numerous dense mitochondria. From Cauna 1966 with permission. Magnification c. × 5000.

7.50   Ultrastructure of terminal nerve fibres in close association with a hair follicle in the auricle of a rat. Note (above) part of the nucleus and cytoplasm of a keratinocyte; the plasma membrane adjoins a well-defined basal lamina and presents a series of hemidesmosomes. The two nerve fibres contain numerous mitochondria and are enveloped by Schwann cell processes. Surrounding the latter are basal laminae, reticulin fibres (above) and collagen fibres (below). From Cauna 1969, with permission. Magnification × 35 000.

septa between lobules, particularly at the epidermal end of the capsule. The *core* consists of cells and nerve fibres. At the epidermal end, cells of the core are discoidal, transversely disposed and about 2–4 μm thick and 30–40 μm in diameter; they are claimed to be epidermal. At the other end of the core the cells are irregular with small oval or round nuclei, devoid of nucleoli, with deeply staining chromatin. They are considered to be lemmocytes. Electron microscopy also reveals occasional fibroblasts and collagen fibres, in bundles or singly, mainly parallel to the nerve fibres (7.49). Each tactile corpuscle is supplied by a few thickly myelinated nerve fibres from the deep corial plexus and possibly also by non-myelinated fibres branching from myelinated fibres in the deep plexus of the dermis. Individual sensory fibres may also branch to supply more than one corpuscle. Within the corpuscle the nerve fibres ramify profusely and decrease in size to as little as 0·1 μm in diameter; large numbers of myelinated and non-myelinated axon profiles appear in ultra-thin sections, the former being most numerous at the deep end of the corpuscle, the latter predominating at the epidermal end (7.49) and in smaller lobules. Axons and their terminal branches are all clothed with lemmocytes. Non-myelinated fibres may contain small vesicles. Tactile corpuscles develop around the time of birth, but thereafter about 80% of them eventually disappear before old age is reached; the number of nerves to each corpuscle is also reduced and, with advancing years, nerve endings are confined to the deep end of the corpuscle. Tactile corpuscles are fast-adapting mechanorecep-

tors of the 'field' type, providing information about rapidly fluctuating mechanical forces acting on the ventral surfaces of the fingers, toes and other areas of thick hairless skin (Valbo et al 1979).

The large lamellated corpuscles of Vater–Pacini (Pacinian corpuscles) (7.45, 47, 48) are sited subcutaneously in the ventral aspects of the hand and foot and their digits, the genital organs of both sexes, arm, neck, nipple, periostea, interosseous membranes, near the joints and in the mesentery and pancreas (cat). They are oval, spherical or irregularly coiled and are relatively huge, up to 2 mm in length and 100–500 μm or more across, the larger ones being visible to the naked eye. Each has a capsule, an intermediate growth zone and a central core containing an axon terminal. The capsule is formed by about 30 concentrically arranged lamellae of flat cells about 0·2 μm thick. Adjacent cells overlap at their edges and successive lamellae are separated by an amorphous matrix containing collagen fibres, the latter disposed circularly and closely applied to the surfaces of lamellar cells, especially their external surfaces. The amounts of collagen increase with age. The intermediate zone is cellular; occasional mitoses appear in it and the cells are incorporated into a capsule or core; the intermediate zone is not conspicuous in mature corpuscles. The core consists of about 60 bilateral, compacted lamellae, lying on both sides of the central nerve terminal and separated by two longitudinal clefts. Nucleated cell bodies exist in the outermost core, at the junction with the intermediate zone; from these arise cylindrical cytoplasmic arms passing into the clefts, where they

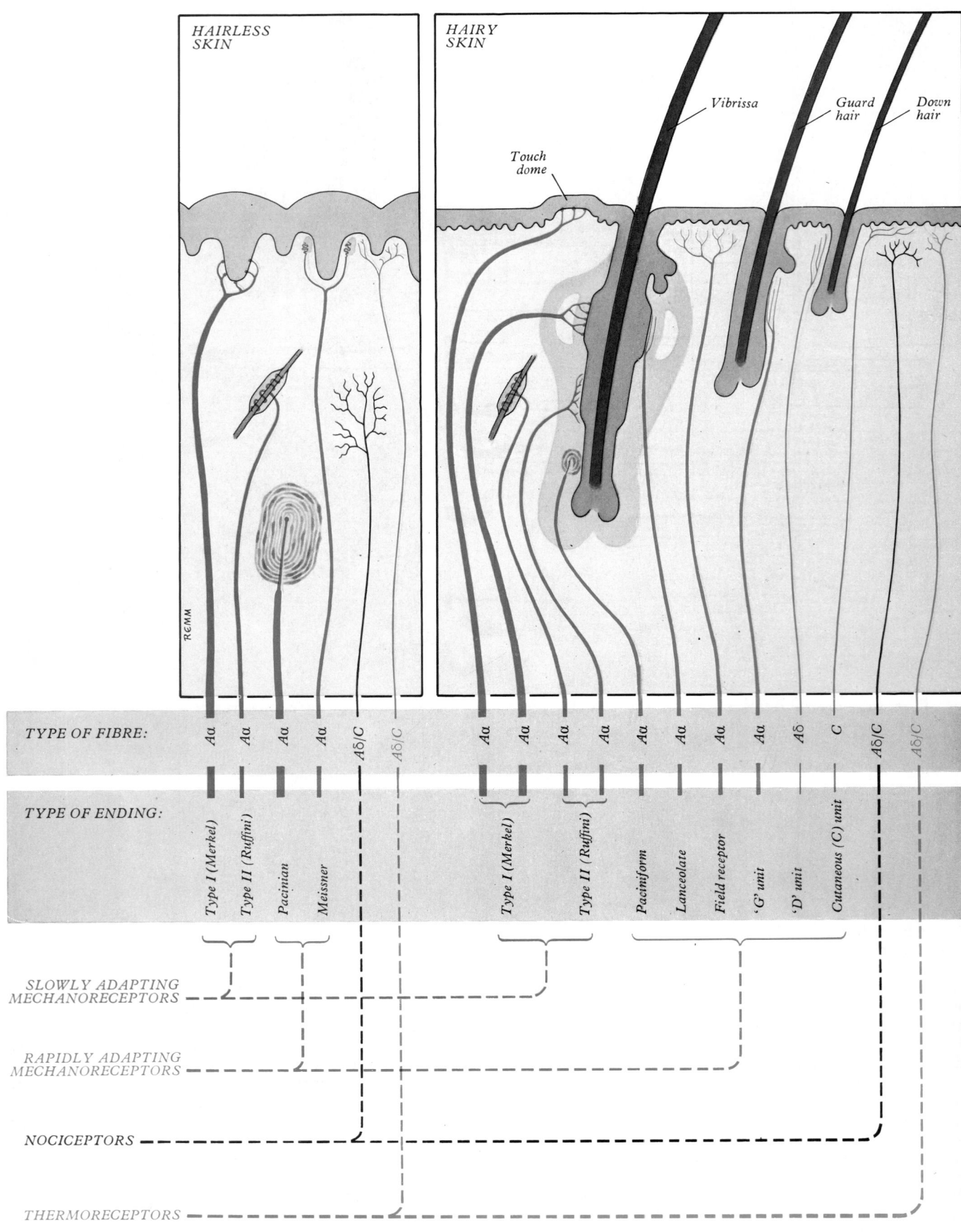

7.51 Schematic survey of the major types of mammalian cutaneous sensory endings and their afferent fibres. The endings are depicted symbolically according to their structural features and their mode of response to specific types of stimulus is indicated by colour coding, i.e. slowly-adapting mechanoreceptors, green; rapidly-adapting mechanoreceptors, red; nociceptors, black; thermoreceptors, blue. The class of afferent fibre is labelled according to the classification of Erlanger & Gasser referring to relative conduction velocities and fibre diameter (see text and table, p. 898). The endings typical of hairless skin are shown on the left and those of hairy skin on the right.

extend as flat processes passing to one or both sides to form core lamellae, interdigitating with the processes from other arms. Adjacent lamellae do not arise from the same arm.

Each corpuscle is supplied by one or, rarely, two thick myelinated fibres (Aα), the unbranched terminals of peripheral nerves. This fibre loses its myelin sheath and at the junction with the core the Schwann cell also ends. The naked axon then runs along the central axis of the core, usually without any branching to end in a slightly expanded bulb. The axon is in contact with the innermost core lamellae and is transversely oval, with the long axis in the plane of the clefts between lamellae. It contains numerous large mitochondria, of which the most superficial are usually arranged radially beneath the axolemma. Minute vesicles of about 5 nm in diameter also occur, aggregated opposite the clefts. Cells of the capsule and core lamellae are actually modified fibroblasts, as distinct from Schwann cells (Pearse & Quilliam 1957). Butyryl cholinesterase has been identified in the core. Lamellated corpuscles may lie close to glomeral arteriovenous anastomoses (p. 693) and are supplied by capillaries accompanying the nerve fibre as it enters the capsule. A condensation of richly elastic fibrous tissue forms an external capsule.

Lamellated corpuscles develop in the third fetal month; the terminal fibre is first surrounded by capsular lamellae which continue to accumulate into adult life, increasing the size of the corpuscle (Cauna & Mannan 1959). The corpuscles are said to exhibit turgidity due to fluid pressure between lamellae. Extensive studies show these endings to behave as very rapidly adapting mechanoreceptors, responding only to sudden disturbances and especially sensitive to vibration (Loewenstein 1971, Gray & Sato 1973). The rapidity may be partly due to the lamellated capsule acting like a 'high pass' frequency filter, damping slow distortions by fluid movements between lamellar cells. If the capsule is removed, the rate of adaptation, though still rapid, is considerably lessened.

Various other *lamellated endings* have been described in various species of mammal and different sites but there is little physiological evidence about their functional nature. Some, which have been described in detail, have a circumferential, concentric, lamellar pattern like the large lamellated (Pacinian) endings but are perhaps less regular and certainly smaller. Examples are found in the endings in the connective tissue sheaths of the vibrissae in cats and in other mammals (Gottschaldt et al 1973) in joint capsules and other such sites (see Polacek & Halata 1970), where they have often been called *paciniform endings*. Endings with similar organization are the *end bulbs of Krause*, *bulbous corpuscles*, *genital corpuscles* and probably the '*innominate corpuscles*' and *Golgi–Mazzoni endings* (see Bannister 1976).

Type II slowly-adapting cutaneous mechanoreceptors (Ruffini endings) occur in the dermis of hairy skin, consisting of highly branched, non-myelinated endings of Aα (or II) myelinated afferents, which invade and ramify among bundles of collagen fibres within a spindle-shaped structure enclosed partly by a fibrocellular sheath derived from the perineurium of the nerve (Chambers et al 1972). Though less organized, these structures resemble the neurotendinous ending of Golgi (vide infra) and appear electrophysiologically similar, being responsive to maintained stresses in dermal collagen. Similar structures appear in joint capsules.

Neurotendinous endings (Golgi tendon organs) are found chiefly near musculotendinous junctions (7.52). More than 50 may occur at any one such site. Each terminal is related closely to a group of muscle fibres (up to 20) which insert into the tendon fasciculus enclosing the ending. Neurotendinous endings are about 500 μm long and 100 μm in diameter, consisting of small bundles of tendon fibres enclosed in a delicate capsule. The collagen bundles (*intrafusal fasciculi*) are less compact than elsewhere in the tendon, the collagen fibres being smaller and the fibroblast larger and more numerous. The capsule consists of concentric cytoplasmic sheets, about 100–300 nm thick, belonging to capsular cells which are closely opposed, the successive layers being separated by intervals of variable width containing basal laminae. Numerous endocytic vesicles occur, suggesting that capsular cells may assist in maintaining the ending's internal environment. Outside is a thin layer of collagen fibres. One or more thickly myelinated group 1β nerve fibres pierce the capsule and divide. Their ramifications, which may lose their Schwann cell sheaths, terminate in clasp- or leaf-like enlargements, rich in vesicles and mitochondria and wrapped round the tendon fasciculi with a basal lamina or Schwann cell cytoplasm intervening (Schoultze & Swett 1972, 1974). The endings are activated by passive stretch of the tendon but are much more sensitive to active contraction of the muscle. Classically, they were considered to initiate myotactic reflexes inhibiting the development of excessive tension during muscle contraction, but it has now been shown that they are important in providing proprioceptive information complementing that from neuromuscular endings (p. 915), in all conditions of muscle activity (see e.g. Prochazka & Wand 1980). Their responses are slowly adapting, signalling maintained tension well (see Matthews 1973). The structural and physiological evidence indicates that the endings may be deformed by surrounding collagen fibres as they straighten under tension.

Special Arrays of Sensory Endings

Now that varieties of individual sensory endings have been considered separately, *arrays* of several types of ending, being functionally significant, must be briefly described. Although our knowledge of such multiple patterns is as yet only partial, there are at least three well studied receptive arrays providing sensory data on the forces acting on a particular locality and analysed, finally, together in the central nervous system. These are the receptors of skin, joints and muscles.

CUTANEOUS ARRAYS

Skin, a major sensory organ, contains the receptors of multitudes of sensory nerve fibres reporting on the mechanical and thermal state of the body surface, including the presence of noxious or damaging stimuli. How this array discriminates between different types of stimulus is much debated. For over a century it has been widely supposed that histologically different types of end-organ had specific modalities, e.g. touch, pressure and cold, though conclusive evidence was lacking. In the 1950s various areas of the body sensitive to different stimuli were seen to lack complex end-organs (e.g. Weddell et al 1955). Each sensory ending was then

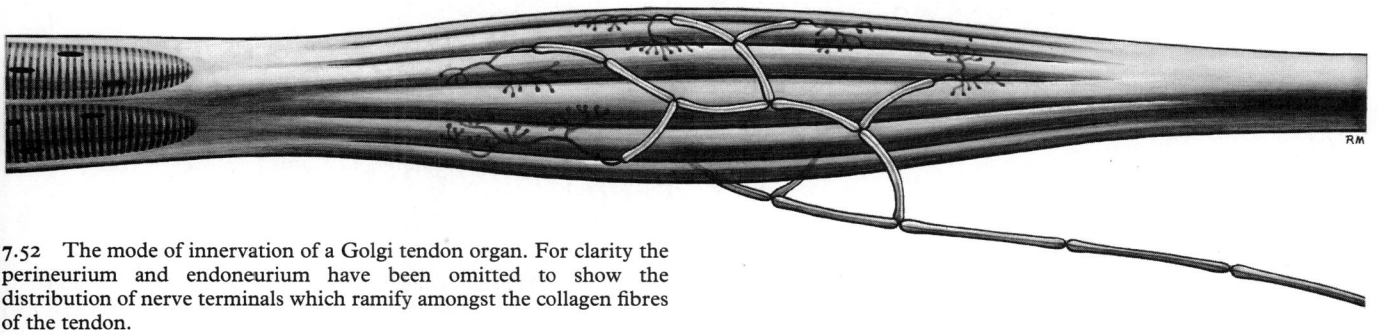

7.52 The mode of innervation of a Golgi tendon organ. For clarity the perineurium and endoneurium have been omitted to show the distribution of nerve terminals which ramify amongst the collagen fibres of the tendon.

considered responsive to a range of stimuli, small differences in optimum sensitivities of a large number of sensory impulses forming a pattern of activity detectable in the central nervous system and each pattern typifying a particular category of stimulus (see Sinclair 1967). Although this idea was attractive, improved electrophysiological and electron microscopic techniques have reinstated the wide range of specific nerve endings, each highly sensitive to a particular stimulus type (Brown 1973, Burgess & Perl 1973, Iggo 1978), although not always structurally distinct. Although all endings fall into three major categories, mechanoreceptors, thermoreceptors and nociceptors, several subdivisions of each of these have been discovered (see 7.51). Mechanoreceptors may be *rapidly adapting endings*, responding only or chiefly to deformational *changes* in mechanical stimulation, or *slowly adapting endings* reacting to steady deformations. Examples of the former are the palisade endings near hair follicles, subcutaneous large lamellated (Pacinian) endings and the tactile (Meissner's) corpuscles adjacent to the epidermis; the second class includes Type I slowly adapting cutaneous mechanoreceptors (Merkel endings), generally associated with a thickened epidermis (e.g. 'touch domes') responding to steady pressures perpendicular to the surface, and Type II slowly adapting cutaneous mechanoreceptor (Ruffini endings) sensitive to steady dermal stresses. Another, distinctly different type has a low threshold but requires prolonged stimulation to be excited; these are 'free' ending class C fibres (7.51).

Thermoreceptors with branched neural endings include those increasing their firing frequency when warmed and those more active when cooled, supplying small restricted areas of skin corresponding to 'warm' and 'cold' spots, first described by Blix in 1882.

Nociceptor fibres have similar free endings, responding to various damaging stimuli, signalled centrally as pain, discomfort or irritation of the skin. Some only respond to extreme cold, others to extreme heat or severe mechanical insult; others can be stimulated by more than one type of noxious stimulus (*polymodal nociceptors*). Nociceptor endings may often be activated by the release of inflammatory products in the surrounding tissues, producing prolonged pain or itching (Iggo 1974).

The precise forms of sensory endings varies with their location and with species. In primates the skin of hairless areas on the palms and soles has its own pattern of endings. The density of endings and therefore the fineness of discrimination vary. But broadly similar arrays of sensory ending occur in each area of skin, ranging from the complex encapsulated or tissue-associated endings of large myelinated nerve fibres, to 'free' endings of fine myelinated or non-myelinated fibres. The entire complex must provide abundant data regarding the state of the body's surfaces, to be further analysed in the central nervous system.

JOINT RECEPTORS

The arrays of receptors placed in and near articular capsules provide information on the position, movements and stresses acting in their vicinity (Wyke 1967). Structural and functional studies have demonstrated at least four types of such receptors, their proportions and distribution varying with site. Three are encapsulated endings, the fourth a free terminal arborization.

Type I endings are encapsulated corpuscles of the type II slowly adapting mechanoreceptor (Ruffini) type (see p. 913), situated in the superficial layers of fibrous capsules in small clusters and supplied by myelinated afferent fibres of the group II class. Being slowly adapting, they provide awareness of joint position and movement, responding, it is thought, to patterns of stress in articular capsules (Skoglund 1973) and particularly common in articulations where static positional sense is necessary for the control of posture (e.g. hip, knee, etc.).

Type II endings are lamellated (Paciniform) receptors, like but smaller than the large (Pacinian) terminals elsewhere in connective tissue (p. 911). They occur in small groups throughout the joint capsules, particularly in the deeper layers and other articular structures (e.g. the fat pad of the temporomandibular joint). They are rapidly adapting, low-threshold mechanoreceptors, sensitive to movement and pressure changes and responding to joint

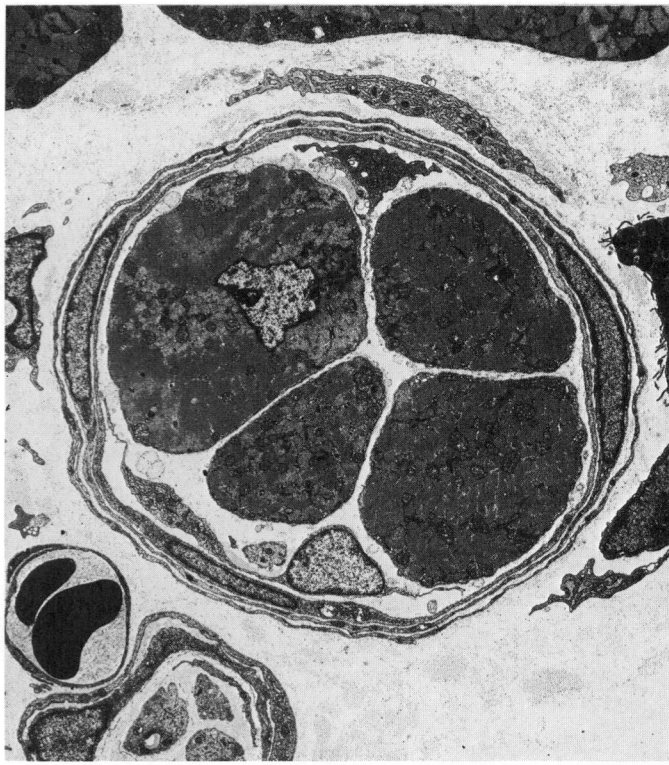

A

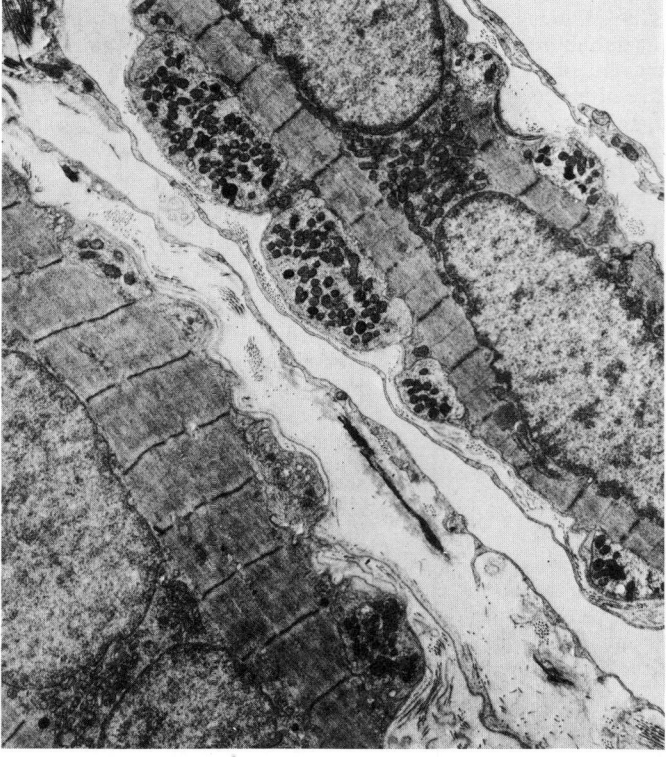

B

7.53 A. An electron micrograph of a neuromuscular spindle of a rat, in transverse section, showing the capsule, capsular space and four intrafusal muscle fibres, one with a centrally positioned nucleus. B. An electron micrograph of a longitudinal section through two intrafusal muscle fibres from a neuromuscular spindle of a rat. Note the primary (annulospiral) afferent nerve fibre endings cut in cross-section as they spiral around the equatorial region of a nuclear chain fibre (top right) and of a nuclear bag fibre (lower left). Note also the large numbers of mitochondria present in the sensory fibres. See text for further description. Provided by D N Landon, National Hospital for Nervous Diseases, London.

movement and transient stresses in the joint capsule. They are supplied by group II or III myelinated afferent fibres, but are probably not concerned in the conscious awareness of joint sensation.

Type III endings are identical with neurotendinous organs (Golgi) in structure and function (p. 913); they occur in articular ligaments, but not their capsules. They are high-threshold, slowly adapting receptors and apparently serve, at least in part, to prevent excessive stresses at joints by reflex inhibition of the adjacent muscles. They receive large myelinated group Ib afferent nerve fibres.

Type IV endings are free terminals of group III myelinated fibres and type IV non-myelinated fibres, ramifying in articular capsules, the adjacent fat pads and around the blood vessels of the synovial layer. These are high-threshold, slowly adapting receptors and are considered to respond to excessive movements, providing a basis for articular pain.

NEUROMUSCULAR SPINDLES

Neuromuscular spindles are essential to the control of muscle contraction, a complex activity reflected in the complicated structure and function of the spindle, which are not entirely understood. Their structure varies much in submammalian species and this account applies only to mammals (Bowden 1966, Landon 1966, Matthews 1971, Barker 1974, Banks et al 1975, Boyd 1985).

Each spindle contains a few, small, specialized *intrafusal muscle fibres*, innervated by both sensory and motor nerve fibres (7.53A,B, 54). This complex is surrounded equatorially by a fusiform *spindle capsule* of connective tissue, arranged as an *external capsule* of flat fibroblasts and collagen, as in the perineurium (p. 897), and an *internal capsule* (axial sheath of Sherrington) forming delicate tubes around individual intrafusal fibres. Between the two sheaths is a fluid rich in glycosaminoglycans of gelatinous consistency.

There are usually 6–14 intrafusal fibres, varying in muscles and species, and two major types of muscle fibre, *nuclear bag* and *nuclear chain fibres*, distinguished by the arrangement of nuclei in their equatorial sarcoplasm. In the former the equatorial cluster of nuclei makes the fibres bulge slightly, whereas in the latter the nuclei form a single axial row. Nuclear bag fibres are greater in diameter than chain fibres and extend beyond the surrounding capsule to the endomysium of nearby extrafusal muscle fibres; nuclear chain fibres are attached at their poles to the capsule or to the sheaths of nuclear bag fibres. Two subtypes of nuclear bag fibres have more recently been demonstrated, the bag_1 (or *dynamic bag_1 fibres*) and bag_2 (static bag_2 fibres), distinguishable by their ultrastructure, histochemistry and physiological properties. Human spindles may contain three or four bag_1 and bag_2 fibres and as many as 10 chain fibres.

In their contractile apparatus the intrafusal fibres resemble the extrafusal, except that the zone of myofibrils is thin where they surround the nuclei. Ultrastructurally, intrafusal dynamic nuclear bag_1 fibres generally lack M lines, possessing little sarcoplasmic reticulum but having an abundance of mitochondria and oxidative enzymes; they have a low alkaline ATP-ase activity and little glycogen. 'Static' bag_2 fibres have distinct M lines and a moderate to high level of alkaline ATP-ase and glycogen. Nuclear chain fibres have marked M lines, sarcoplasmic reticulum and T-tube elements, more glycogen and higher alkaline ATP-ase activity; they also have fewer mitochondria and lower oxidative enzyme levels. There are also differences in the extracellular environments of the fibre types, static bag_2 fibres being surrounded in their polar regions by prominent elastic fibres, whereas dynamic bag_2 fibres are not. These variations in the structural and other properties of intrafusal fibres are reflected in their contractile properties (vide infra).

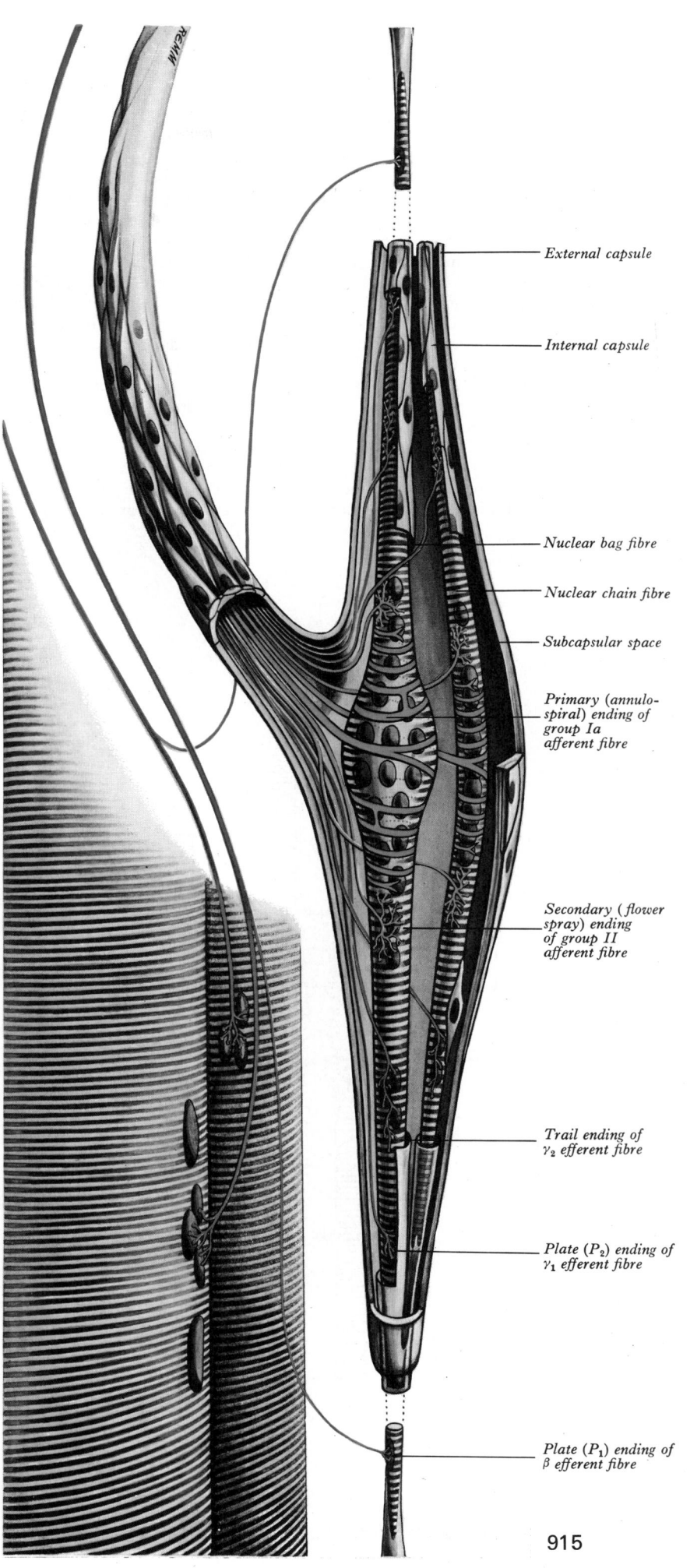

External capsule

Internal capsule

Nuclear bag fibre

Nuclear chain fibre

Subcapsular space

Primary (annulo-spiral) ending of group Ia afferent fibre

Secondary (flower spray) ending of group II afferent fibre

Trail ending of γ_2 efferent fibre

Plate (P_2) ending of γ_1 efferent fibre

Plate (P_1) ending of β efferent fibre

7.54A Schematic three-dimensional reconstruction of a mammalian neuromuscular spindle, showing nuclear bag and nuclear chain fibres; these are innervated by the sensory annulospiral and 'flower-spray' terminals (blue) and by the γ and β fusimotor terminals (red). For further details consult text and 7.54B.

The sensory innervation of muscle spindles is of two types, both of which are the non-myelinated terminations of large myelinated axons. *Primary annulospiral* endings are equatorially placed and form spirals around the nucleated parts of intrafusal fibres; they are the endings of large sensory fibres (group Ia), each of which gives branches to a number of intrafusal muscle fibres. Each terminal lies in a deep sarcolemmal groove beneath the basal lamina and contains mitochondria, vesicles, neurofilaments, microtubules and a pervading flocculent material. *Secondary* (*flower spray* and annulo-spiral) endings, largely confined to nuclear chain fibres, are the branched terminals of somewhat thinner myelinated (group II) afferents; they are beaded and spread in a narrow band on both sides of the primary endings. Ultrastructurally, the varicosities are like the primary endings and lie close to the sarcolemma, though not in grooves. Primary endings are rapidly adapting, while secondary endings have a regular, slowly adapting response to static stretch (Matthews 1972, Hunt 1974, Boyd 1984). However, this statement is an oversimplification of the complex properties of the primary and secondary endings of different types of intrafusal fibres, since these depend on: (1) the mechanical properties of different regions of the bag_1, bag_2 and chain fibres on which they end; (2) the contractile characteristics of these muscle fibres when stimulated by their efferent terminals; and (3) variations in the numbers and patterns of afferent primary and secondary endings and of the efferent terminals. To give an example of this complexity, the primary spiral endings of the dynamic bag_1 fibres are highly sensitive to small stretches but rapidly adapt, probably because of mechanical 'creep' in the bag fibre on either side of it which decreases the tension of the central part of the fibre. In static bag_2 fibres, this tendency is much less marked, presumably due to the presence of elastic fibres surrounding the bag fibre and also because of differences in bag fibre cytology.

Motor endings in muscle spindles include three types, two from fine, but myelinated, fusimotor (γ-) efferents and one from β-efferent (also myelinated) collaterals of extrafusal slow twitch muscle. The first two are situated nearest to the equatorial region, ramifying to form cholinergic motor terminals with no obvious end-plate or sole; the P_2 *endings* placed further away from the equator, have a typical 'en plaque' end-plate and sole. At the extreme ends of nuclear bag fibres are P_1 endings of the β-efferents, forming 'en grappe' end-plates (p. 904). Stimulation of efferent endings causes the contraction of intrafusal fibres, with the activation of their sensory endings. Bag_1, bag_2 and chain fibres contract differently and their motor endings reflect this. The terminals of γ-efferents on dynamic bag_1 and static bag_2 fibres differ in their detailed structure, the bag_1 fibres receiving distinct synaptic end-plates (P_2, or m_b terminals) and bag_2 fibres (usually) long 'trail' (m_a) endings; chain fibres have a third (m_c) type of more complex form or in many cases share branches of m_a terminals. For further details of structure and ultrastructure, see Arbuthnott et al (1982), Boyd (1984).

Muscle spindles signal the length of extrafusal muscle both at rest and throughout activity and relaxation, the velocity of their contraction and changes in velocity. These modalities may be related to the different behaviours of the three major types of intrafusal fibres and their sensory terminals. Dynamic nuclear bag_1 fibres are particularly concerned with signalling rapid changes in length occurring during movement; static bag_2 fibres, in contrast, are less responsive to movement; chain fibres have relatively slowly adapting responses at all times. These elements therefore can detect complex changes in the state of the extrafusal muscle surrounding spindles and can signal fluctuations in length, tension, velocity of length change and acceleration. Moreover, they are under complex central control; recently, sympathetic endings have been described in muscle spindles, adding further possibilities of changes in sensitivity mediated by the central nervous system. *Nuclear bag fibres* are probably concerned with *position*, *velocity* and *acceleration*, giving responses of a rapidly adapting, *dynamic* type, while *nuclear* efferent nerve fibres can adjust the length of the intrafusal fibres and therefore the activity of sensory fibres by causing contraction of the polar regions, thus compensating for the shortening caused by the contraction of extrafusal fibres during normal muscle activity. Such changes in tension can also magnify or reduce the responses of the afferent endings (see Boyd 1984). In summary the organization of spindles is such that they are capable of *actively* monitoring muscle conditions to allow *comparisons* between intended and actual movements and to provide a detailed input to the spinal, cerebellar, extrapyramidal and cortical centres of the nervous system concerning the state of the locomotor apparatus.

The development of neuromuscular spindles has been studied in detail by various investigators (Zelená 1957, Landon 1972, Milburn 1973). Initially, intrafusal fibres form by the fusion of myoblasts, as in extrafusal muscle; the primary sensory endings are established (in rodents) before birth, followed by motor terminals, secondary sensory endings and the opening of the capsular space. Early development is absolutely dependent on the formation of primary sensory endings (Zelená 1957) but later spindles may survive denervation, although they may undergo marked changes even if regenerating crushed nerves reinnervate them (Schiaffino & Pierobon Bormioli 1976).

The Life History of Neurons

The structure of mature neurons, as considered so far, applies only to limited aspects of their life and activities. Their mature

7.54B Schematic diagram of the organization of intrafusal fibres and their innervation within a neuromuscular spindle.

structure must be considered in the context of their development, maturation, reactions to morphogenetic and pathological factors, senescence and death.

ORIGIN AND DEVELOPMENT OF NEURONS

Neurons come from two major embryonic sources: central neurons from the neural plate and tube and ganglionic neurons from the neural crest (p. 178). The neural plate also provides ependymal and macroglial cells, while from the neural crest arise peripheral Schwann cells and chromaffin cells. Olfactory (sensory) neurons and ganglionic neurons of the vestibulocochlear nerve are *placodal* in origin (see p. 138).

The development of cells in the central nervous system has received much attention since the descriptions by His (p. 178). Recently many difficulties in determining the sequence and timing of events in nervous development have been resolved by the use of autoradiography, microinjection cell culture and related techniques for showing the origins and lineages of cell types in the developing nervous system.

The earliest observers described the wall of the neural tube as divided into ependymal, mantle and marginal zones (p. 178), later modified to include a layer of matrix cells (p. 179); more recently (Sidman 1970) the wall of the neural tube has been subdivided into four major strata: the *ventricular, subventricular, intermediate* and *marginal zones* (7.55).

The neural plate (p. 132 et seq.) and early neural tube have a single layer of pluripotent epithelial stem cells which give rise to all cell types in the central nervous system except the microglia. The nuclei of these stem cells lie close to the ventricular surface in the *ventricular zone*, with a nucleus-free cytoplasmic *marginal zone* of the laterally extending 'tails' of the stem cells. As these begin to divide mitotically they change shape, their form becoming rounded as the nucleus moves towards the ventricular surface, then, after division, elongating as the nucleus moves away from the ventricular surface, within the ventricular zone. Since mitoses are not synchronized, nuclei appear at varying levels in this zone. Later, the progeny of some of these divisions abandon mitosis and move away from the ventricle to form an *intermediate* (mantle) *zone* where they differentiate into neuroblasts; others form a *subventricular zone* between the ventricular and intermediate layers, there continuing to multiply to provide further

generations of neuroblasts and later glioblasts. Both of these cell types subsequently migrate into the intermediate and marginal zones but in some regions of the nervous system (e.g. the cerebellar cortex, p. 188) some mitotic subventricular stem cells migrate across the entire neural wall to form a subpial population, thus establishing a new zone of cell division and differentiation. Many cells formed in this site remain subpial in position but others migrate back towards the ventricle through the developing nervous tissue, finishing their migrations in various definitive sites where they finally differentiate into neurons and macroglial cells.

During the genesis of these various cell types, neuroblasts are the first to be formed, followed by glioblasts. The timing of events differs in various parts of the central nervous regions and between species. Most neuroblasts are formed prenatally in mammals but some postnatal neurogenesis does occur, e.g. of the small granular cells of the cerebellum, olfactory bulb and hippocampus. Gliogenesis continues after birth in periventricular and other sites in the nervous system.

Autoradiographic studies show that different classes of neurons develop at specific times. Large neurons tend to differentiate before small ones; but their subsequent migration appears independent of the times of their initial formation. Neurons can migrate extensively through populations of maturing, relatively static cells, to reach their destination e.g. cerebellar granule cells pass through a layer of Purkinje cells en route from the external pial layer to their final, central position. Later, the final form of their neurites, cell volume and indeed their continuing survival depend on the establishment of patterns of functional connection (pp. 860–867).

STRUCTURE OF DEVELOPING NEURONS

Initially neuroblasts are rotund or fusiform, their cytoplasm containing a prominent Golgi apparatus, many lysosomes, glycogen and numerous unattached ribosomes (Tennyson 1969). As maturation proceeds cells send out fine cytoplasmic processes (neurites); these contain neurofilaments, microtubules and other structures, often including centrioles at their bases where microtubules form (p. 29). Internally, endoplasmic reticulum cisternae appear and attached ribosomes and mitochondria proliferate but the glycogen content progressively diminishes.

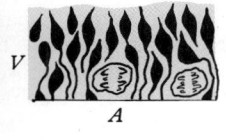
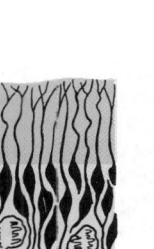
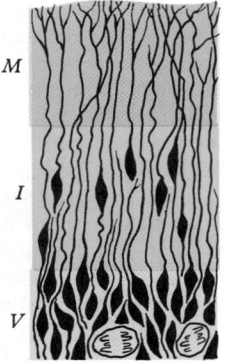
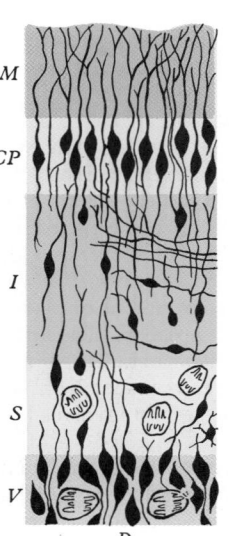
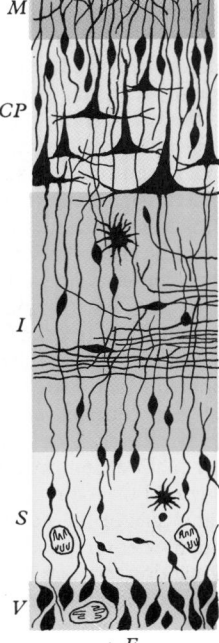

7.55 Schema of five sequential stages (*A* to *E*) in the development of part of the neural tube's wall, in section, to form the cerebral cortex. Modified from Boulder Committee recommendations 1970. Abbreviations: *V* = ventricular zone; *M* = marginal zone; *I* = intermediate zone; *S* = subventricular zone; *CP* = cortical plate. See accompanying text and also p. 179 et seq for details.

One neurite becomes the axon and other processes establish a dendritic tree. Axonal growth, studied in tissue culture, may be as much as 1 mm in a day. At the tips of growing axons and dendrites are bulbous *growth cones*, containing actin microfilaments (p. 184) and numerous membranous vesicles, with small surface filopodia ('microspikes') constantly 'exploring' the surrounding intercellular space (Tennyson 1970, Bunge 1976, Pfenninger & Rees 1976). Like other developments, axonal growth appears dependent on the elaboration of a cytoskeleton.

The direction of axonal growth is the subject of much current research. It appears to be partly governed by extracellular factors such as the architecture and composition of the intercellular matrix, the presence of other axons and Schwann cells along which generations of axons grow to form bundles and hence peripheral nerves (see p. 184). Within the central nervous system there is evidence that axonal growth is at least partly determined by the appearance of various *adhesion molecules* at the surfaces of ependymal astrocytes, there being complementary adhesive receptor molecules on the axonal surfaces. Adhesion molecules may vary in chemical nature in different parts of the nervous system; the nerve cell adhesion molecule (N-CAM) has been characterized in some areas (Edelman 1985). Others include the proteins *contactin* and, peripherally, laminin (see p. 67). The final direction and termination of axons appear to be related to an overall genetically determined plan. Much evidence indicates that growth patterns are determined by embryonic chemical concentration gradients, 'field effects' and chemical recognition of other cells (vide infra). Experiments on various vertebrates, in which grafting or displacement of parts is possible to achieve in the embryonic nervous system, indicates that axons can precisely contact target regions in spite of mechanical interference, as though guided by surrounding structures and stopped by chemical recognition of their target cells. Similar considerations apply to motor innervation. The mechanisms of specific connections are unknown in detail but may involve sensing of metabolites, such as 'nerve growth factor' (NGF) produced by target tissues, and recognition of particular cells by their surfaces (Sidman 1974, Moscona 1976) or impulse firing characteristics. There is much evidence (Hughes 1968) that if an axon fails to make the correct contacts, its parent soma atrophies and dies, probably as a result of toxic materials liberated within it at a programmed time of development. Such mechanisms may explain the numerical correspondence between neurons in a motor pool and the muscle fibres innervated (Tennyson 1969).

The final growth of dendritic trees is also influenced by patterns of afferent connections and their activity; if deprived of afferents experimentally, dendrites fail to develop fully and, after a critical period, may become permanently affected even if functional inputs are restored, e.g. in the visual systems of young animals visually deprived (Blakemore 1974). Metabolic factors also affect the final branching patterns of dendrites; e.g. thyroid deficiency in perinatal rats results in a small size and restricted branching of cortical neurons. This may be analogous to the mental retardation of cretinism (Eayrs 1955).

Once established, dendritic trees appear remarkably stable and partial deafferentation affects only dendritic spines or similar small details. If cells lose all afferent connections or are totally deprived of sensory input (see Guillery 1974), atrophy of much of the dendritic tree and even the whole neuron ensues, though different regions of the nervous system vary quantitatively in their response into such *anterograde transneuronal degeneration* (p. 871). Similar effects occur in retrograde transneuronal degeneration. As development proceeds plasticity is lost and soon after birth a neuron is a stable structure with a reduced rate of growth. Two main alterations may ensue; the first follows trauma to any part of the neuron, the second involves hypothetical permanent memory traces.

Reaction to trauma has been studied most extensively in motor neurons with conveniently accessible peripheral axons, but also centrally where axons form well-defined tracts (p. 872). When an axon is crushed or severed, changes occur proximally and distally to the site of injury (see also p. 901) (Nauta & Ebesson 1970). Distally, axons swell and break up into membrane-bound spheres, this change progressing distally. These *anterograde*

changes, involving also the axon terminals (*terminal degeneration*), continue until the total removal of debris has occurred (p. 876). Proximally, similar changes may occur near the point of injury, followed by sequential (*retrograde*) changes in the soma (Cragg 1970) and firstly causing the removal of much original protein-synthesizing apparatus by autophagic lysosomal action. Cytoplasmic RNA first rises, which is followed by the dispersion of Nissl granules and consequent loss of staining in cytoplasm (*chromatolysis*) due to a reduced amount of RNA (Lieberman 1974). New protein-synthesizing organelles are then formed to produce proteins destined partly for the regrowth of the axon. Nucleoli become peripheral in position and ribosome clusters in the cytoplasm are restored, with the return of staining affinity in the cytoplasm. Meanwhile synapses on the soma may be retracted from the surface membrane and glial processes invade the gap (Matthews & Nelson 1975).

Where axonal regrowth is possible, as in peripheral nerves, an intact endoneurial sheath is essential to the satisfactory regrowth and contact with its end-organ. Degeneration of the myelin sheath distal to the injury has been mentioned (p. 901); this is accompanied by a mitotic proliferation of Schwann cells filling the old endoneurial tube (Webster 1964). Where a gap exists between the severed ends, proliferating Schwann cells emerge from the stumps (mainly distal) to form cords (*bands of Bungner*) across the gap, which may persist for a long time even in the absence of neural regeneration. The proximal axon then swells and many small axonal sprouts grow out from it, the majority ultimately abortive but a successful one entering the endoneurial tube's proximal end to grow down it in contact with its contained Schwann cells. *Contact guidance* (p. 115) involves the growing axonal tip of the axon and Schwann cell surfaces in the tube and Bungner's bands. When a tip reaches and successfully reinnervates an end-organ, the surrounding Schwann cells synthesize new myelin sheaths. If regeneration occurs in mature animals, no increase in the lengths of the internodal segments of myelin occurs; these remain uniformly short (often about 200 µm). Before full *functional regeneration* can occur, a considerable period of growth of both axonal diameter and myelin sheath thickness is necessary. When a high proportion of effective connections has been re-established, the nerve fibre population eventually recovers a normal diameter spectrum but this does not occur without the formation of appropriate peripheral connections, or if regeneration is long delayed.

The regeneration of central axons in mammals does not normally occur. The reasons for this failure are the subject of extensive investigations at present.

A second series of changes occurs throughout the life of the individual. In addition to the continuous turnover of the cytoplasm and surfaces of neurons, constantly changing patterns of afferent nerve fibre activity, circulating hormone levels and numerous other factors affect neuronal behaviour. In every individual, cyclic or intermittent processes, such as memory, feeding, reproduction, aggression and so on, are obviously connected with such varying activity. Possibly changes in the chemical microenvironment can alter a cell's electrical firing pattern probabilistically, changing the responses of whole cell populations. More subtle alterations may occur at the biochemical level to modulate responses (and structure) of neurons with reference to previous events, a probable basis of some acts of memorization, much studied of late. Changes in RNA content and composition are said to accompany learning of conditional responses and experimental inhibition of protein synthesis is also claimed to inhibit certain types of memory (Barondes 1969). But it is uncertain what relevance these data have to the learning processes in intact nervous systems.

During fetal development many neurons degenerate and die, a process which continues postnatally until in senescence it has been estimated that 20% of neurons or more may be lost. In embryos the loss of neurons is a process of 'thinning out' those failing in functional connections (Prestige 1970) or losing in competition with other neurons to innervate structures (Hamburger 1975). Postnatal neuronal loss is less easily understood; it may be a similar process, perhaps related to the stabilization of early behaviour patterns. Certainly, with increasing age, logical efficiency and reaction speeds in general deteriorate, changes perhaps related to this loss of neurons, although whether as a direct result is unknown, and in any case great individual variation exists.

TROPHIC INFLUENCE OF NEURONS

In addition to conduction, neurons have other functions during development and postnatal life in the maintenance or metabolic regulation of some tissues. Nervous tissue is not unique in this respect, since most tissues influence the metabolism of other cells in some way. Neuronal effects are, however, perhaps more marked and far reaching. The most obvious example is the mutual dependence of motor neurons and muscles. If, during development, a nerve fails to connect with its muscle, both degenerate. But if the innervation of slow (red) or fast (white) skeletal muscles, each with peculiar functional properties, is exchanged, the muscles change structure and properties in accordance with innervation, indicating that nerve determines muscle type and not vice versa (Buller 1970, and see p. 553). However, in this case the type of muscle is apparently determined chiefly by the firing pattern of the efferent nerve fibre, rather than by any release of trophic chemical (Lomo & Westgaard 1974). Trophic influences are clearest in lower vertebrates, however, where a whole limb amputated may be regenerated, but only if its nerves are not destroyed (Hamburger 1968). In higher vertebrates the trophic influences of axonal growth on the dendritic trees they innervate are well known (p. 918); the developmental differentiation of

sensory cells, such as seen in the specificity of lingual gustatory sensory cells (p. 1170), appears to be under direct trophic afferent influence (Rosenthal 1977). Conversely, if auditory sensory cells are eliminated, the auditory neurons begin to degenerate and many eventually die, suggesting that in this case the trophic influence is the reverse of the usual situation. In many invertebrates, also, the regeneration of metameric segments is under neurosecretory influence, suggesting that the nervous system may have originated in ancestral metazoans as a system co-ordinating regeneration and development. In this light, the neurosecretory activity of hypothalamohypophyseal neurons may be a primitive survival, other neurons being subsequent specialization for rapid communication.

Role of growth factors. The presence of specific neurotrophic factors in the nervous system has been known for a long period, the best characterized of these being the polypeptide, *Nerve growth factor (NGF)*. This substance is synthesized by various peripheral target organs of the nervous system (e.g. the salivary glands, Levi Montalcini & Angeletti 1968) from which it is taken into the nerve endings and transported back to the neuronal somata. It is necessary for the survival of many types of neurons during early development and for the maintenance of neurotransmitter synthesis during subsequent life. It is becoming increasingly clear that such factors are of great importance for the maintenance of many pathways within the nervous system and their loss may be a precipitating factor in various nervous disorders (see also the review by Korschung 1986).

MAJOR DIVISIONS OF THE NERVOUS SYSTEM

Although essentially continuous, the nervous system is conveniently divided into parts, regions and systems. The *encephalon* or brain (7.56, 57, 58) and *medulla spinalis* or spinal cord form a *central nervous system*. Extending from this in pairs are: 12 cranial and 31 spinal nerves, a *peripheral nervous system*, which includes not only all their ramifications, mediating *somatic sensory* and *motor* functions but also *visceral* or *splanchnic* nerves connected to the central nervous system (which they reach running with the somatic channels) but forming a *peripheral autonomic nervous system*. Division into central and peripheral parts is basically functional; the latter consists of relatively simple conductors connecting peripheral receptor and effector organs to each other through the intermediation of the brain and spinal cord. Though the latter contain proximal extensions of afferent and efferent pathways deployed through the peripheral nerves, the special properties of the central nervous system reside in complex interconnections of neurons, in which arise the appropriate patterns of response to stimuli from external and internal environments.

This same intricate intermediation, between incoming patterns of information and emerging arrays of 'commands' to effectors, is also the domain in which learning, memory and consciousness are intrinsic, each dependent upon the level of development of the central apparatus. Elaboration of these activities, in general increasing along the lines of evolution, especially that leading to the human animal, is clearly related to the increasing populations of central interneurons, rather than to changes in the peripheral afferent and efferent conductors. Nevertheless, the essential continuity and interdependence of all parts of the nervous system is never to be overlooked.

Elements of the nervous system have already been described (p. 875). Whereas peripheral nerves largely contain, except in somatic and autonomic ganglia, roughly parallel, unconnected fibres, central nervous systems contain not only terminations or ('peripheral') neuronal somata of these peripheral axons but also large numbers of complete ('central') neurons, including their dendrites, the ramifications and synapses of which make up most

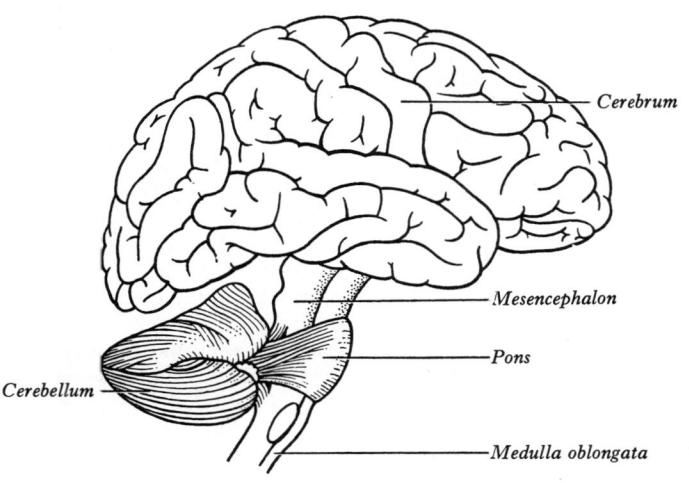

7.56   Semi-diagrammatic scheme of the main divisions of the brain.

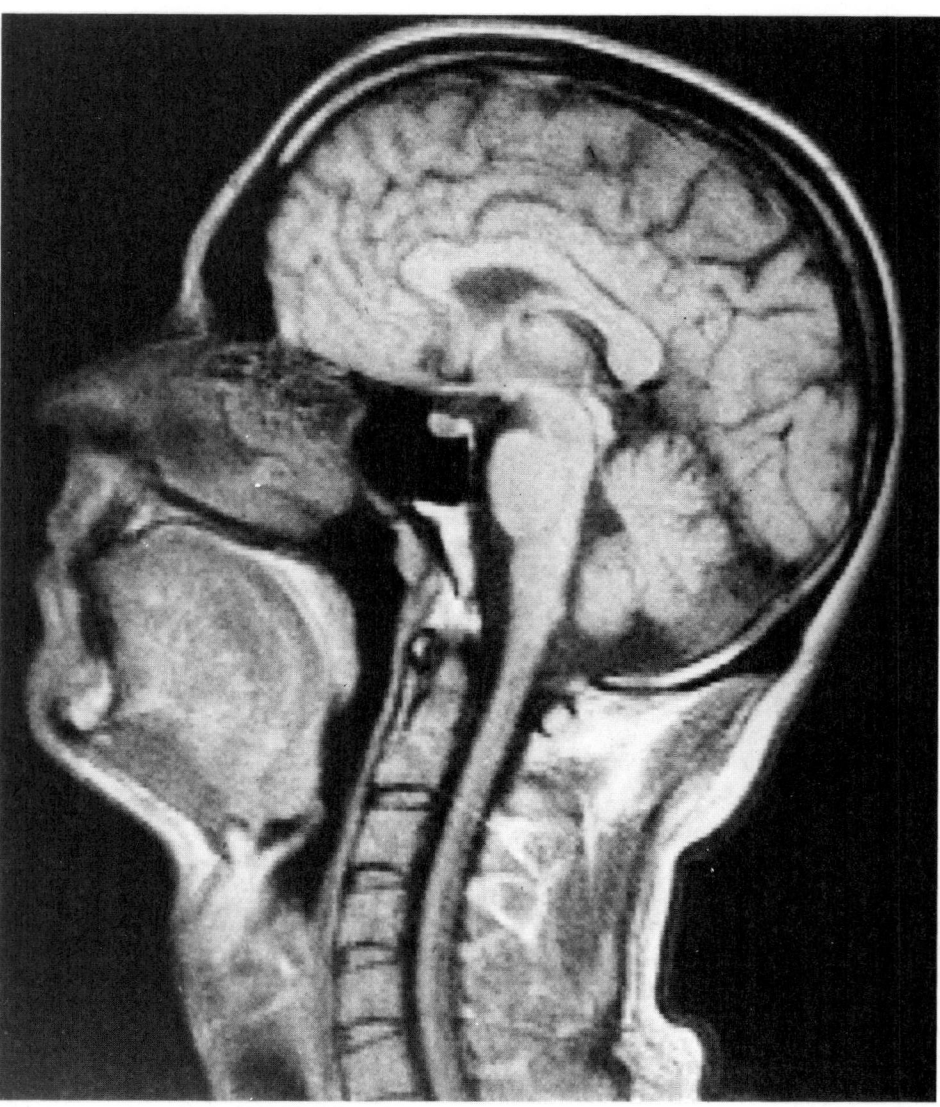

7.57 Midline sagittal magnetic resonance imaging (MRI) scan of the head and neck. Supplied by Siemens, Erlangen, West Germany; photography by Sarah Smith.

of the volume of the brain and spinal cord. Both these parts also contain amounts of connective tissue, neuroglia and numerous blood vessels. Central nerve fibres and somata are so organized that one or the other predominates in particular regions. Somata of neurons are usually (but not always) gathered into masses, termed *nuclei* or *ganglia*; being devoid of myelin they appear darker to the eye than collections of nerve fibres, unless these are non-myelinated. This difference, accentuated by fixation with alcohol, which renders aggregations of nerve cells somewhat grey and myelinated nerve fibres almost white, is the origin of the crude but useful terms 'grey' and 'white matter' (substance), though their colours are buff and cream with the more commonplace formalin fixation. It is not to be supposed, however, that 'grey matter' contains *no* fibres; most central axons are myelinated and every group of nerve cells must contain at least the proximal parts of their own fibres; hence the grey or buff hue is due to a *relatively* low proportion of myelin. Similarly, tracts of myelinated nerve fibres, identified as 'white matter', often contain small numbers of nerve cells; where fibres are largely non-myelinated the colour will be 'grey'. With these provisos the terms are useful in grosser descriptive topography.

The spinal cord occupies about the cranial two-thirds of the vertebral canal and is arbitrarily continuous with the medulla oblongata just below the foramen magnum, the first pair of spinal nerves emerging from the cord immediately caudal to this. Its walls are thick, the central nervous cavity being here an almost microscopic *central canal*, extending nearly throughout the cord.

The encephalon or brain (7.56, 57, 58) wholly within the cranium is divisible, for convenience, into regions on the bases of morphology and function. Ascending from the spinal cord these are: the rhombencephalon or *hindbrain*, **mesencephalon** or *midbrain* and **prosencephalon** or *forebrain*. The rhombencephalon includes the **myelencephalon** or *medulla oblongata*, **metencephalon** or *pons* and the **cerebellum**. The prosencephalon is subdivided into the **diencephalon** ('between brain'), its central connecting part corresponding approximately to the dorsal thalamus and hypothalamus (but also the epithalamus and subthalamus) and the **telencephalon**, mainly comprising the two so-called cerebral 'hemispheres' or *cerebrum* (but with a small transmedian *telencephalon impar*). Midbrain, pons and medulla oblongata form the *brain stem*, connecting the forebrain and spinal cord. The relation of these divisions to each other is appreciated more clearly from their development (p. 184). The pattern of fore-, mid- and hindbrain is not only an expression of ontogenetic growth in individuals; it is also phylogenetic, representing the basic central nervous hierarchy in vertebrates. These levels are sometimes termed 'segments' of the brain; the telencephalon, particularly its cortex and connections, is often described as 'suprasegmental', a term also applied to the cerebellum. Both are phylogenetically later outgrowths of the basically elongated primitive vertebrate brain. (Other authorities use the term suprasegmental in a different sense, i.e. for structures not involved in varieties of segmentation, e.g. metameric, branchiomeric and neuromeric, referred to elsewhere in this volume.)

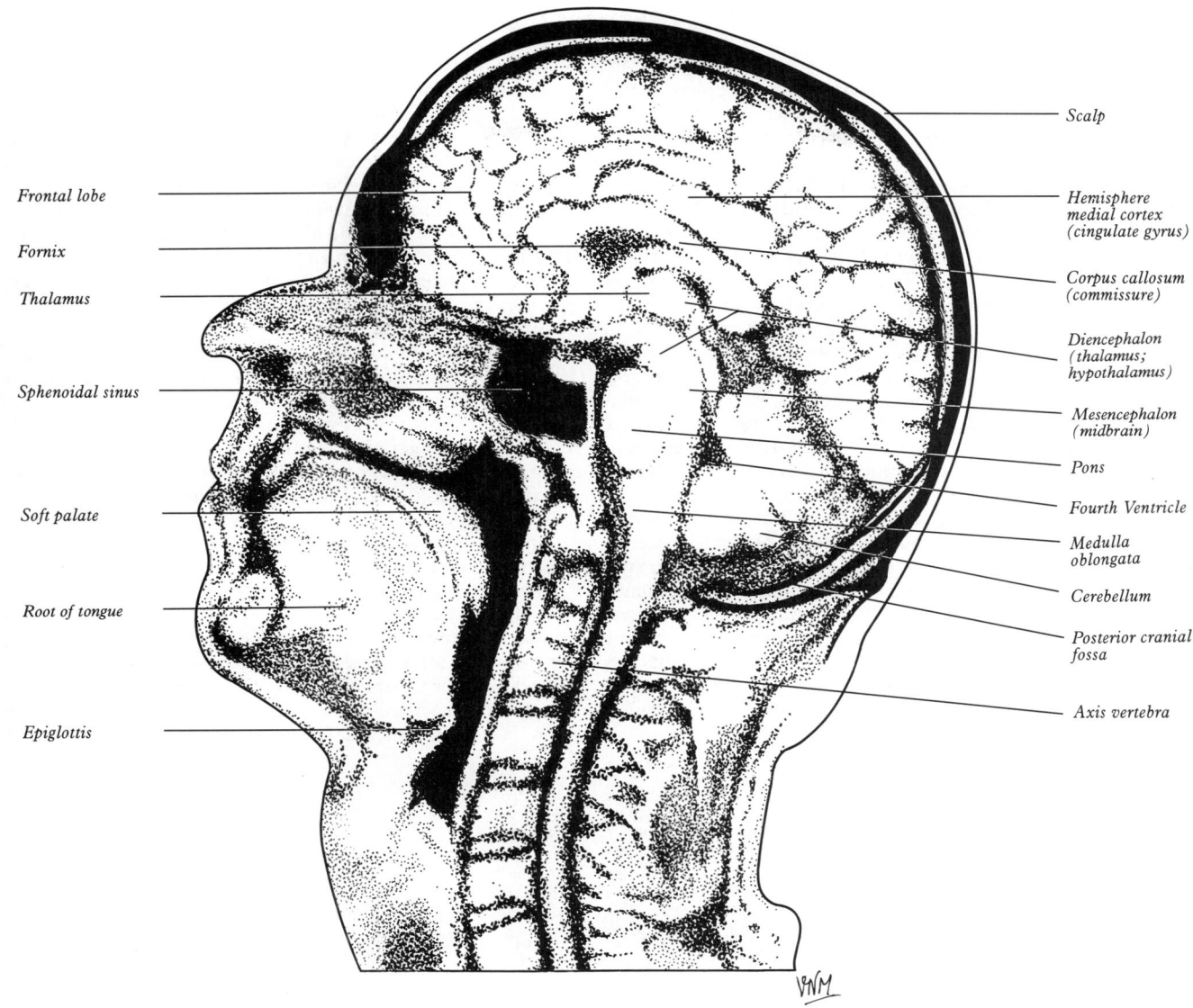

Scalp

Hemisphere
medial cortex
(cingulate gyrus)

Corpus callosum
(commissure)

Diencephalon
(thalamus;
hypothalamus)

Mesencephalon
(midbrain)

Pons

Fourth Ventricle

Medulla
oblongata

Cerebellum

Posterior cranial
fossa

Axis vertebra

Frontal lobe

Fornix

Thalamus

Sphenoidal sinus

Soft palate

Root of tongue

Epiglottis

7.58   Diagram of 7.57. Illustration by V Martin.

The *medulla oblongata* is the most caudal (inferior) part of the brain stem, immediately above the basilar occipital bone and continuous with the pons above and spinal cord below. The *pons* is also related to the basi-occiput, extending to the sphenoidal dorsum sellae; it is greater in transverse and anteroposterior dimensions than the medulla and easily distinguished by the mass of transverse nerve fibres on its ventral aspect. The *cerebellum*, consisting of paired *hemispheres* and united by a median *vermis*, is dorsal to the pons, medulla and caudal midbrain, occupying the posterior cranial fossa; ventrally it is continuous with all three: midbrain, pons and medulla. The cavity of the rhombencephalon is the expanded *fourth ventricle*, which lies dorsal to the pons and upper half of the medulla, and continuous with a canal in the caudal medulla and, through this, with the spinal central canal. The fourth ventricle contracts above to the narrow, mesencephalic and inappropriately termed *cerebral aqueduct*. The *midbrain* is a short segment of brain stem, narrower than the pons but expanding above. The *diencephalon*, almost completely embedded in the cerebrum and largely hidden, contains a median *third ventricle*, communicating caudally with the aqueduct.

The *cerebrum*, the most rostral part, is in mankind a major fraction of the brain's volume, occupying the anterior and middle cranial fossae and directly related to the cranial vault. It consists of two large convoluted *cerebral hemispheres*. (Strictly its halves are *not* hemispherical but together form approximately *one* hemisphere; perhaps the term was originally used in the singular, but the plural however unsuitable is customary.) Each half is

equal to about a quarter of a sphere, containing a crescentic *lateral ventricle*, continuous medially with the third ventricle in the diencephalon. Each has an external grey layer, the *cerebral cortex*, and a central white core of *medullary substance*, in which are several large *basal nuclei* of grey matter (earlier, incorrectly, called 'ganglia').

As noted, the *peripheral nervous system* contains a somatic *cerebrospinal* and a visceral *autonomic system*. In the former, efferent nerve fibres pass from central neurons to effector organs, mostly muscles. Efferent autonomic fibres terminate in the peripheral ganglia, forming synapses with neurons whose axons innervate non-striated muscle and glandular tissue. Autonomic efferent pathways thus have *pre-* and *postganglionic* neurons, though autonomic ganglia also sometimes include a variety of interneurons (p. 865). The arrangement of afferent fibres is similar in cerebrospinal and autonomic systems.

The autonomic nervous system has *sympathetic* and *parasympathetic* divisions, often functionally opposed. Preganglionic sympathetic efferent fibres issue from a spinal region extending from the first thoracic to the second or sometimes third lumbar segments. Preganglionic parasympathetic efferent fibres emerge only in certain cranial nerves (oculomotor, facial, glossopharyngeal, vagus with accessory) and the second to fourth sacral spinal nerves. These groups of autonomic efferents are usually designated *thoracolumbar* (sympathetic) and *craniosacral* (parasympathetic) *outflows*. A detailed description of the autonomic system appears on p. 1121 et seq.

SPINAL MEDULLA OR CORD

The spinal cord (medulla spinalis) is an elongated, approximately cylindrical part of the central nervous system, occupying the superior two-thirds of the vertebral canal (7.60A,B, 61, 62, 63). Its average length in European males is 45 cm, its weight about 30 g. (For dimensional data consult Barson & Sands 1977.) It extends between the levels of the atlantal upper border and the junction between the first and second lumbar vertebrae, the latter level varying, with some correlation with length of trunk, especially in females (Jit & Charnalia 1959). The termination may be as high as the twelfth thoracic vertebra's caudal third or as low as the disc between the second and third lumbar; its position rises slightly in vertebral flexion. The spinal cord is enclosed in the *dura, arachnoid* and *pia maters* and separated by the *subdural* and *subarachnoid spaces*, the former being merely potential, the latter containing cerebrospinal fluid (p. 1050). Continuous cranially with the medulla oblongata, the cord narrows caudally to the *conus medullaris*, from whose apex descends the *filum terminale*, a connective tissue filament, to the dorsum of the first coccygeal segment (7.59). In *transverse width* the spinal cord varies, with a general tapering towards its caudal end, partly obscured by cervical and lumbar enlargements. It is not cylindrical, being greater transversely at all levels and especially in the cervical segments.

The *cervical enlargement* is more pronounced, being the source of large spinal nerves supplying the upper limbs and extending from the third cervical to the second thoracic segments, its maximum circumference (about 38 mm) being in the sixth cervical. (A spinal cord segment provides the attachment of the rootlets of a pair of spinal nerves.)

The *lumbar enlargement* similarly corresponds to the innervation of the lower limbs, extending from the first lumbar to the third sacral segments, the equivalent *vertebral* levels being the ninth to twelfth thoracic. The greatest circumference (about 35 mm) is near the twelfth thoracic vertebral body's lower part, below which it rapidly dwindles into the conus medullaris.

Fissures and *sulci* extend along most of the external surface; an anterior median fissure and a posterior median sulcus and septum almost completely separate the symmetrical right and left halves but these are joined by a *commissural band* of nervous tissue containing a *central canal* (7.62, 63).

The *anterior median fissure* along the whole ventral surface has an average depth of 3 mm, deeper than this at caudal levels. It contains a reticulum of pia mater. Dorsal to it is the *anterior white commissure*. Perforating branches of the spinal vessels pass from fissure to commissure to supply the central spinal region. The *posterior median sulcus* is shallower; from it a *posterior median septum* of neuroglia penetrates more than halfway into the cord, almost to the central canal, varying in anteroposterior extent from 4–6 mm and diminishing caudally as the canal becomes more dorsal and the cord contracts.

A *posterolateral sulcus* exists from 1·5–2·5 mm lateral to each side of the posterior median sulcus; along it dorsal roots (strictly rootlets) of spinal nerves enter the cord. The white substance between the posterior median and each posterolateral sulcus is the paired *posterior funiculus*; in cervical and upper thoracic segments a longitudinal *postero-intermediate sulcus* marks a septum dividing each posterior funiculus into two large tracts: the *fasciculus*

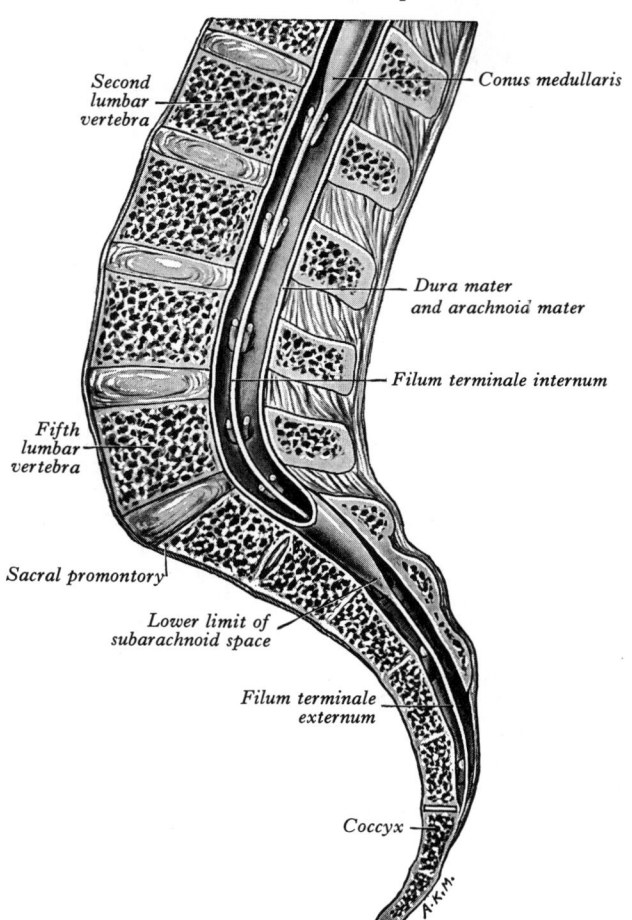

Second lumbar vertebra

Conus medullaris

Dura mater and arachnoid mater

Filum terminale internum

Fifth lumbar vertebra

Sacral promontory

Lower limit of subarachnoid space

Filum terminale externum

Coccyx

7.59 Median sagittal section of the lumbosacral part of the vertebral column to show the conus medullaris and filum terminale. The section has opened up the subarachnoid space as far as the first sacral vertebra. Note the difference in levels between the inferior limits of the spinal cord and its meninges. This illustration retained from an earlier edition has two major inaccuracies. The periosteum lines the vertebral canal and i separated from the dura by internal vertebral venous plexuses etc: a lumbar levels the fibres of interspinous ligaments slope dorsocraniall (p. 491).

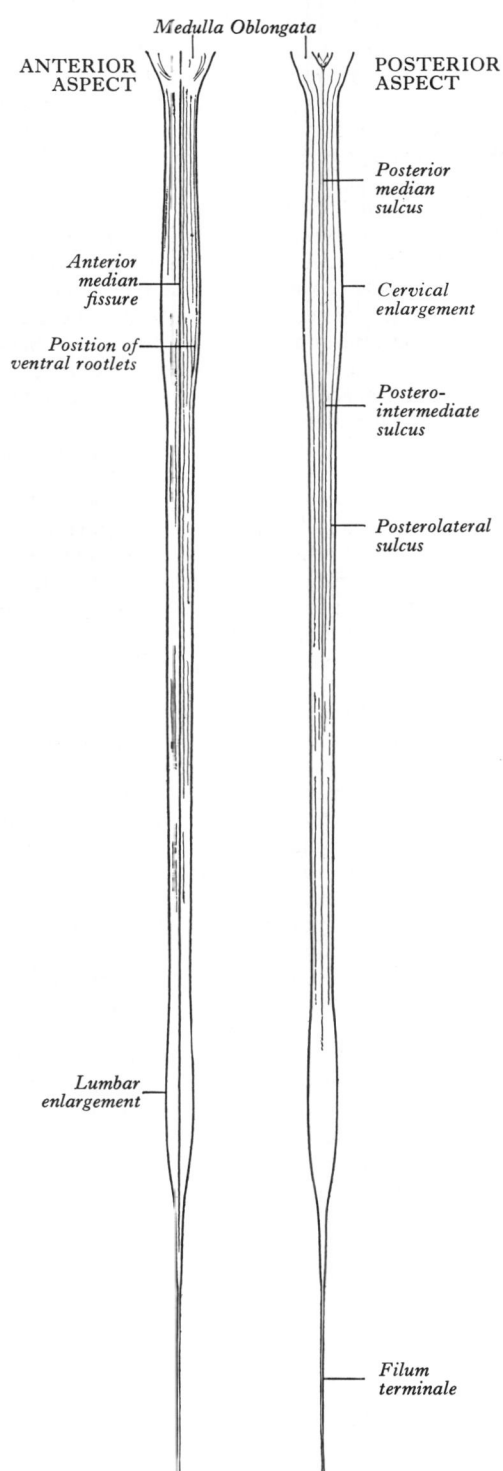

Medulla Oblongata

ANTERIOR
ASPECT

POSTERIOR
ASPECT

Posterior
median
sulcus

Anterior
median
fissure

Cervical
enlargement

Position of
ventral rootlets

Postero-
intermediate
sulcus

Posterolateral
sulcus

Lumbar
enlargement

Filum
terminale

7.60A (*above*) The main features of the spinal cord.

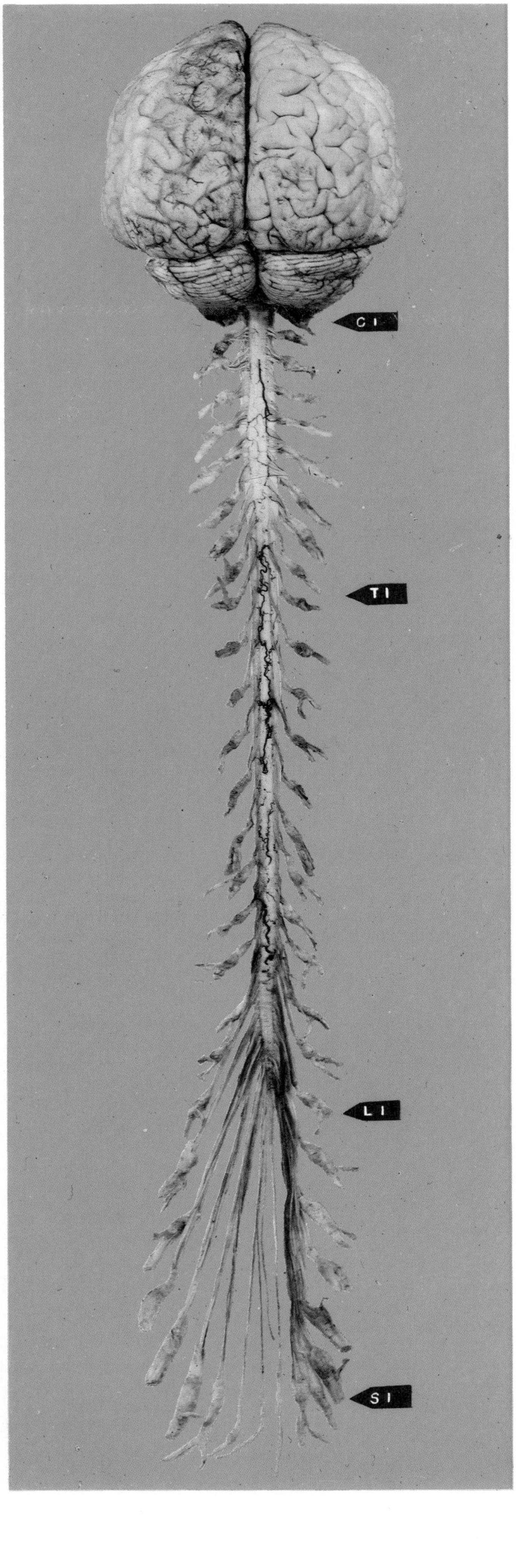

7.60B (*right*) The brain and spinal cord with attached spinal nerve roots and dorsal root ganglia, photographed from the dorsal aspect. Note the relative sizes of the cerebral and cerebellar hemispheres and the fusiform cervical and lumbar enlargements of the spinal cord. The median longitudinal fissure between the hemispheres which receives the falx cerebri and falx cerebelli is visible, together with the horizontal cleft between cerebrum and cerebellum which receives the tentorium cerebelli. Contrast the irregular pattern and dimensions of the cerebral gyri and sulci with the more regular, largely transverse pattern of the smaller cerebellar folia and their intervening fissures. Note also the changing obliquity of the spinal nerve roots in their rostrocaudal progression, the stouter roots attached to the limb enlargements and the formation of the cauda equina and filum terminale. The cauda is undisturbed on the right and has been fanned out on the left to facilitate identification of its individual components. Dissection by M C E Hutchinson, photograph by Kevin Fitzpatrick, both of Dept. of Anatomy, Guy's Hospital Medical School, London.

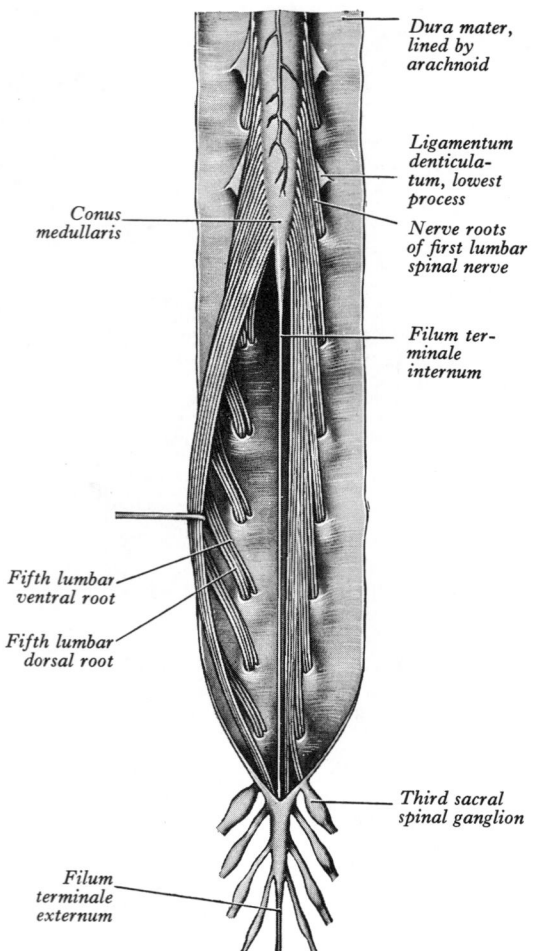

Dura mater, lined by arachnoid

Ligamentum denticula- tum, lowest process

Conus medullaris

Nerve roots of first lumbar spinal nerve

Filum ter- minale internum

Fifth lumbar ventral root

Fifth lumbar dorsal root

Third sacral spinal ganglion

Filum terminale externum

7.61 The lower end of the spinal cord, the filum terminale and the cauda equina exposed from behind. The dura mater and the arachnoid have been opened and spread out.

gracilis (medial) and *fasciculus cuneatus* (lateral). Between the posterolateral sulcus and anterior median fissure is the *anterolateral funiculus*, subdivided into *anterior* and *lateral funiculi* by issuing ventral spinal roots. The anterior funiculus is medial to (and includes) the emerging ventral roots, the lateral funiculus being between the latter and the posterolateral sulcus (7.62, 63). In upper cervical segments nerve radicles emerge *through* each

lateral funiculus to form the spinal accessory nerve, ascending in the vertebral canal lateral to the spinal cord to enter the posterior cranial fossa via the foramen magnum (7.85).

The *filium terminale* (7.59, 61), a filament of connective tissue about 20 cm long, descends from the apex of the conus medullaris. Its upper 15 cm, *filum terminale internum*, is surrounded by extensions of the dural and archnoid meninges and reaches the second sacral vertebra's caudal border. Its final 5 cm, *filum terminale externum*, fuses with the investing dura mater, descending thence to the dorsum of the first coccygeal vertebral segment. The filum is continuous above with the spinal pia mater; adherent to its upper part are a few strands of nerve fibres probably representing roots of rudimentary second and third coccygeal spinal nerves. The central canal is continued into the filum for 5–6 mm. A roomy part of the subarachnoid space surrounds the filum's internal part; it is the site of election for spinal (lumbar) puncture.

Continuous with the cord is a series of paired dorsal and ventral *roots* of spinal nerves (7.63). These cross the subarachnoid space and traverse the dura mater separately, uniting in or close to their intervertebral foramina to form the (mixed) spinal nerves. Since the spinal cord is shorter than the vertebral column, more caudal spinal roots descend for varying distances around and even beyond the cord to reach their corresponding foramina, thus forming, largely distal to the cord, a divergent sheaf of spinal nerve *roots*, the *cauda equina*, which is gathered round the filum terminale in the *spinal theca*. (For developmental changes consult p. 180 and Pearson & Sauter 1971.)

Ventral spinal roots contain efferent somatic and, at some levels, efferent sympathetic nerve fibres emerging from their spinal sources. Each ventral root emerges as a variable number of rootlets which appear over an elongated *vertical elliptical area*. *Dorsal spinal roots* have ovoid swellings, *spinal ganglia*, one on each root proximal to its junction with a ventral root in an intervertebral foramen; each fans out into six to eight rootlets entering the cord in a *vertical row* in the posterolateral sulcus. Dorsal roots are usually said to contain only afferent axons (both somatic and visceral) from unipolar neurons in spinal root ganglia; but a small number (3%) may be efferent (Young & Zuckermann 1936) and autonomic vasodilator fibres may issue in them: views which have received little support. Each ganglionic neuron has a single short 'stem' axon which at once divides into a medial branch entering the spinal cord via a dorsal root and a lateral one passing peripherally to a sensory end-organ. The central branch is an axon, the peripheral is derived from a dendrite (but when traversing a peripheral nerve is, in general structural terms, indistinguishable from an axon). The region of spinal cord associated with the emergence of a pair of nerves is a *spinal segment*, but there is no actual surface indication of segmentation. (Further, the deep neural sources or destinations of radicular fibres may lie far

Posterior funiculus   Posterior median septum

Posterior column

Lateral funiculus

Posterior column

Thoracic nucleus

Central canal

Lateral column

Anterior column

Anterior funiculus   Anterior median fissure

7.62 Typical transverse section of the spinal cord at a mid-thoracic level. Magnification × 8.

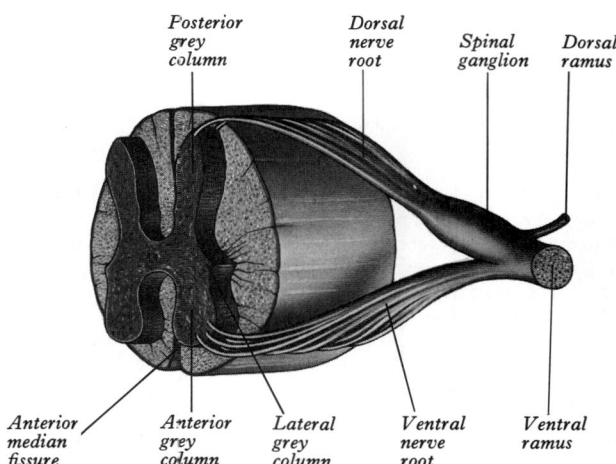

Posterior grey column

Dorsal nerve root

Spinal ganglion

Dorsal ramus

Anterior median fissure

Anterior grey column

Lateral grey column

Ventral nerve root

Ventral ramus

7.63 Diagram of a spinal cord segment showing mode of formation of a typical spinal nerve and the gross relationships of the grey and white matter. Note dorsal nerve rootlets in a single linear row; ventral rootlets in three or more rows.

beyond the confines of the 'segment' so defined.) Recent researches show that ventral spinal nerve roots contain only one neuromediator (acetylcholine), whereas dorsal roots contain at least seven (glutamate/aspartate, substance P, VIP, CCK, somatostatin, dynorphin and angiotensin II; see p. 891).

Internal Structure of the Spinal Cord

Internally the spinal cord may be considered from several complementary points of view. Ignoring supply and support tissues, blood vessels and neuroglia, the arrangement of neurons and their processes can be studied by dissection and macroscopic inspection of sections, by light and electron microscopy, neurohistochemical approaches of great variety, by experimental techniques and combinations of these. Dissection reveals little more than the general layout of fibres and cells; microscopy details cellular types, fibre calibres and their intimate disposition. While light and electron microscopy have greatly clarified connections between neurons, their potential can only be fully exploited with experimental manoeuvres, such as degeneration, microelectrode and chemical tracing methods. Spinal organization is hence considered in steps: (1) the general arrangement of grey and white matter, (2) the distribution of neurons in grey matter, (3) the deployment of tracts of fibres in white matter, and (4) the detailed organization of the spinal neurons.

GENERAL ARRANGEMENT OF GREY AND WHITE MATTER

Spinal grey matter is central (i.e. deep to the white matter) and shaped like a fluted column, except where modified at its continuation into the medulla oblongata and conus medullaris. In transverse sections (7.64) this column has symmetrical right and left comma-shaped masses connected by a narrow transverse *grey commissure*, the whole resembling a letter H. The commissure is traversed by the central canal, which may be just visible to the unaided eye. Each lateral crescentic mass has a lateral concavity and *anterior* and *posterior columns*. At some levels a small, intermediate *lateral column* projects from the concavity. In transverse sections columns appear as projections and are often called 'horns' (*cornua*), a picturesque nomenclature misleading and unnecessary; being elongated longitudinally they are better referred to as *columns*. Central grey matter is surrounded by white matter, the latter consisting largely of nerve fibres, many (but by no means all) being longitudinal and grouped into *funiculi* or *white columns*. The general arrangement of these as dorsal, lateral and ventral funiculi has been noted and further details appear later (p. 930 et seq.).

The anterior or ventral grey column projects ventrolaterally from the grey commissure. It is short, broad and separated from the surface by part of the anterior funiculus (7.64). Its anterior and posterior regions are sometimes named its head and base.

The posterior or dorsal grey column projects dorsolaterally, is transversely narrow and extends almost to the surface near the posterolateral sulcus, separated from it by a thin *dorsolateral tract* (p. 936). It also is considered to have a *base*, continuous with the *intermediate* grey region, a constricted neck expanding into an oval or fusiform *head* and an apex, capped by a crescent of semitranslucent nervous tissue, the *substantia gelatinosa* (*Rolandi*), intimately connected with the afferent nerve fibres (p. 941).

The lateral grey column is a small, angular projection between the second thoracic and first lumbar *spinal cord segments*. (These must be carefully distinguished from vertebral levels.)

The boundary between white and grey matter is usually clear; but at cervical levels strands of grey matter invade the lateral funiculus from the base of the dorsal grey column, separated by interlacing nerve fibres like a net, whence its name, the *reticular formation*. Similar regions appear at lower spinal levels; brainstem reticular formations also exist, where physiological investigations have led to a concept of an extensive reticular 'system' widely deployed throughout the neuraxis (pp. 988–996).

This main pattern of white and grey matter is what might be expected in simple terms of peripherally placed conductors, i.e.

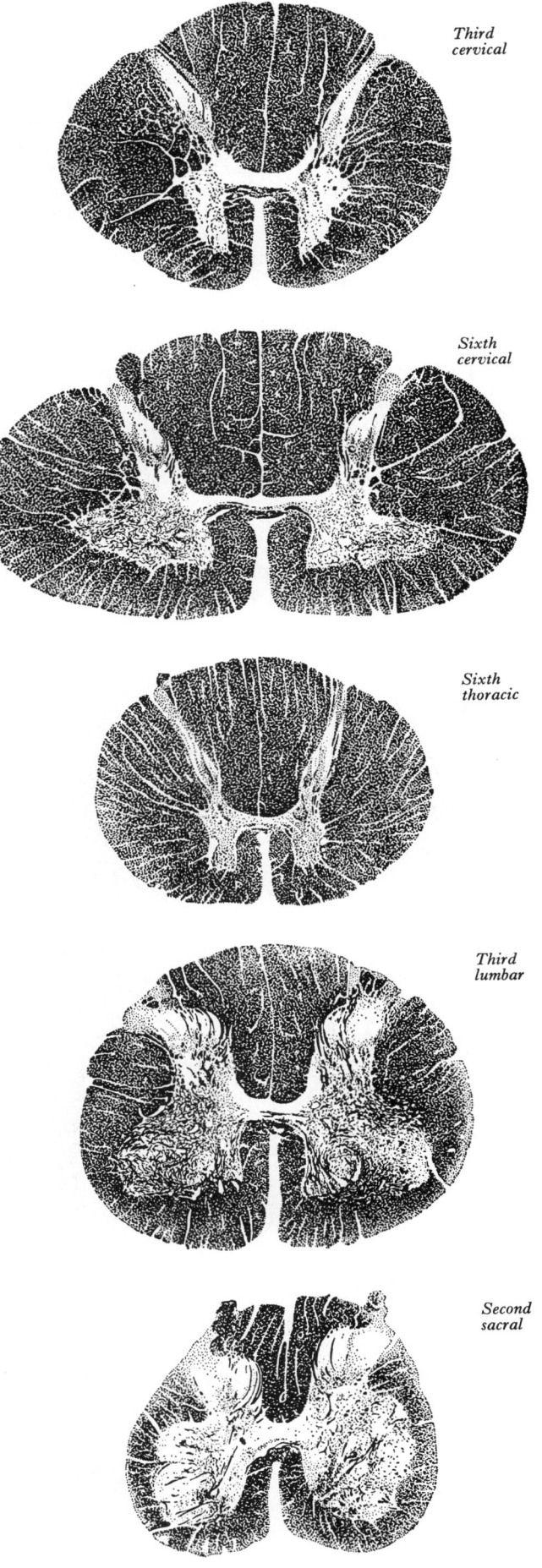

Third cervical

Sixth cervical

Sixth thoracic

Third lumbar

Second sacral

7.64 Transverse sections through the spinal cord at representative levels. Note changes in overall profile and the relative changes in grey and white regions, their shape, size and proportions. Magnification × 5.

longitudinally arranged spinal tracts of fibres, but with their sources and interconnections centrally placed; however, it is also an expression of spinal development (p. 178). Dimensions and relative volumes of peripheral (white) fibres and centrally aggregated (grey) neurons at different levels (7.64) can be explained partly by the amounts of muscle, skin and other tissues innervated by different segments and by the sequential expansion of longitudinal tracts (both ascending and descending) as progression is made in a cranial direction. At thoracic levels the grey matter is absolutely and relatively small in volume, while the white substance shows an ascending increase. In the cervical and lumbar enlargements grey matter, especially of the ventral columns, is augmented by large accumulations of neurons innervating limbs; but while white matter at cervical levels is also pronounced, in the lumbar enlargement and particularly the conus medullaris, white funiculi contain many less fibres passing through; 7.64 shows the details in representative segments; it is obvious that various levels can easily be distinguished. It is, however, more useful to recognize the explanation of these differences.

The central canal traverses the whole spinal cord and caudal half of the medulla oblongata, opening above into the fourth ventricle (p. 979). In the conus medullaris it expands as a fusiform *terminal ventricle*, triangular in section with a ventral base, 8–10 mm in length, obliterating at about 40 years. At cervical and thoracic levels the canal is slightly ventral, central in the lumbar enlargement but more dorsal in the conus medullaris; it extends for 5–6 mm into the filum terminale. It contains cerebrospinal fluid, is lined by columnar, ciliated *ependyma* and encircled by a zone of neuroglia containing a few neurons and a network of fine nerve fibres, the *substantia gelatinosa centralis*, which is traversed by processes spreading from the basal aspects of ependymal cells. Grey matter around it, external to the *substantia*, is the *grey commissure*. Ventrally the commissure is thin. Anterior to it is a thin *ventral white commissure* (p. 875). The grey commissure is traversed by two longitudinal veins (p. 948) and, dorsal to the canal, is continuous with the posterior median septum; it is thinnest at the thoracic level, thickest in the conus medullaris; it is permeated by transverse myelinated nerve fibres, sometimes termed the *dorsal white commissure*.

Internal Structure of the Grey Matter

Spinal grey matter (7.65A,B) is a complex mixture of neurons and neurites, neuroglia and blood vessels. The predominance of neuronal somata is responsible for the so-called *grey* appearance. Neuroglia (p. 892) forms an intricate lattice among the somata and neurites, particularly condensed as gelatinous substance round the central canal. Neurites will be described later, with formation of tracts (p. 891) and the organization of interneurons (p. 938); they also include initial or terminal parts of efferent and afferent peripheral fibres, with collaterals from all these sources, resulting in a complex neuropil of innumerable neurites confined to the grey matter or at least to the spinal cord. Many neurites cross in the commissures; the right and left halves of the cord, including

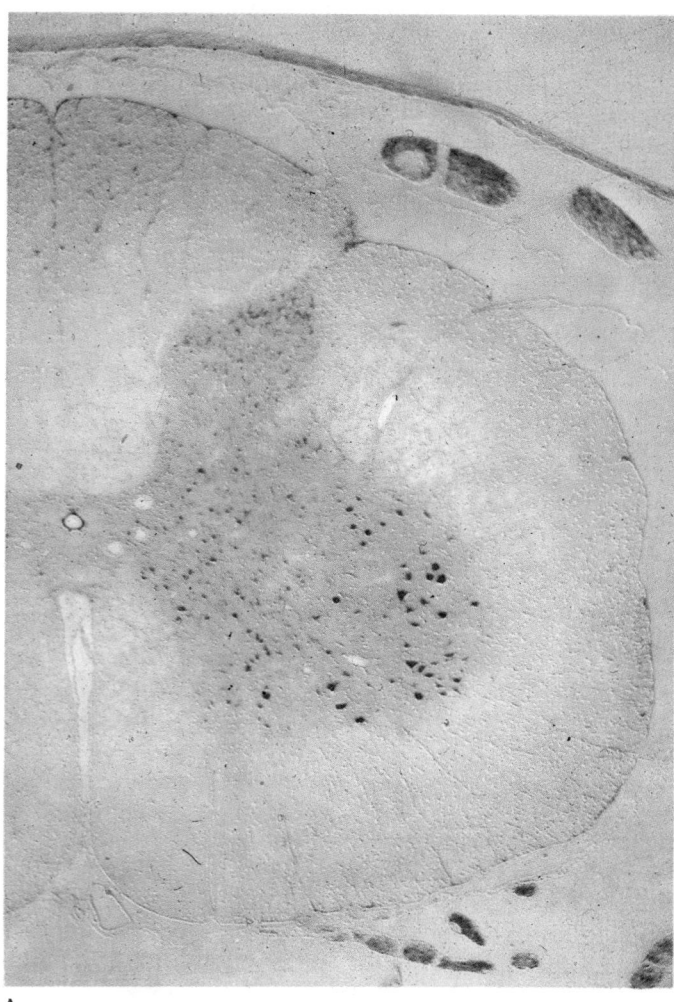

A

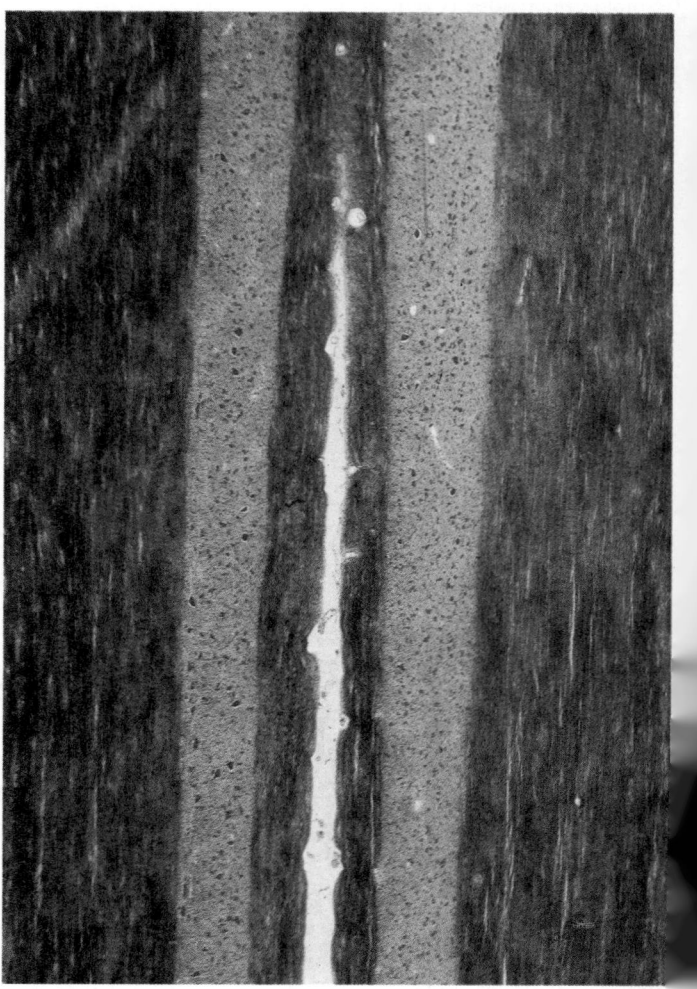

B

7.65 Transverse and longitudinal sections of the spinal cord. A. Transverse section of left half of human spinal cord at a mid-lumbar level. Note dorsal and ventral grey columns and commissural grey mass. The larger motor neurons in the ventral grey column are visibly grouped. For details see text. Stained with cresyl fast violet. B. Longitudinal section of feline spinal cord showing the anterior median fissure and anterior white columns and lateral to these the ventral grey columns, in which motor neurons show some degree of grouping into longitudinal columns. Material prepared by the late L Laruelle and supplied by J André Balisaux, Institut Neurologique Belge, Brussels.

its grey matter, are, therefore, a functional continuum. Neurons in the grey substance are multipolar, varying in size and other features, particularly in length and the arrangement of axons and dendrites. Many are Golgi types I and II neurons (p. 876), axons of the former passing out of the grey matter into ventral spinal roots or spinal tracts. Axons and dendrites of Golgi type II neurons are largely confined to the nearby grey matter. Some neurons are *intrasegmental*, deployed within a single segment; others spread through several, thus being *intersegmental* in distribution (p. 946).

In much of the central nervous system nerve cell somata are grouped, often in large numbers, usually indicating common function. A large group may be divided with a constancy justifying specific names. Such constant patterns of distribution inevitably suggest functional implications, though the influence of developmental processes may sometimes be of greater significance. Neurons of the spinal grey matter are not distributed uniformly, occurring in major and minor aggregations, some with obvious functional significance, which in others is subject to controversy. The following cytoarchitecture of the spinal grey substance is first described in purely topographical terms and, for convenience, separately in ventral, dorsal and lateral grey columns. Functional interpretations of some groups in them will be considered later.

NEURONAL GROUPS OF THE ANTERIOR GREY COLUMNS

Neurons in the ventral columns vary in size; largest are the multipolar cells exceeding 25 μm in average dimensions of soma, with axons emerging in ventral roots to innervate the striated skeletal muscles as α-efferents. Large numbers of smaller neurons of 15–25 μm also occur and some of their axons are γ-efferents innervating the intrafusal fibres of neuromuscular spindles. By the evidence of retrograde degeneration, many smaller neurons in this column do not have such efferents and *most* are interneurons (p. 939).

As in other spinal regions, the ventral column neurons are in elongate groups, a number of separate longitudinal columns extending through several segments, seen most easily in transverse sections (7.65A); longitudinal sections are rarely depicted but these clearly show that the neuronal columns are not uniformly continuous (7.65B), each being a series of small aggregations, too diminutive for segmental significance (Laruelle & Reumont 1933). Basic division of the ventral grey region is into *medial*, *central* and *lateral* columns and all exhibit subdivision at certain levels, usually into dorsal and ventral parts. This promotes a nomenclature; but while agreement on the organization is satisfactory, confusion persists in terminology. The terms employed here are derived from an authority widely consulted (Crosby et al 1962); only recently (1977) have these terms been accepted into *Nomina Anatomica*. As can be seen in Figure 7.66, a **medial group** extends through most segments, perhaps absent from the fifth lumbar and first sacral; in thoracic and the upper three or four lumbar segments it is subdivided into **ventromedial** and **dorsomedial groups**; in segments cranial and caudal to this, the medial group has only a ventromedial moiety, except in the first cervical where only the dorsomedial group exists. The **central group**, the least extensive, is identifiable only in some cervical and lumbosacral segments. In the cervical cord, through the third to seventh segments, is a central columnar **phrenic nucleus**; abundant experimental and clinical evidence shows that its neurons innervate the diaphragm, being probably the least controversial motor pool in the entire cord. For experimental data in some mammals, including primates, see Kohnstamm (1898), Sharrard (1955), Keswani & Hollinshead (1956), Warwick & Mitchell (1956), Ullah (1978), Kuypers 1985. The **lumbosacral nucleus**, in the second lumbar to first sacral segments, is also central; the distribution of its axons is unknown. Neurons whose axons are said to enter the spinal accessory nerve form an irregular **accessory group**, in the upper five or six cervical segments, at the ventral border of the anterior grey column and intermediate in position but lateral to the dorsomedial group in the first cervical segment; the ventral siting of this nucleus may be due to the absence of lateral groups from the first three cervical segments (7.66).

The **lateral group** in the ventral column is subdivided into **ventral, dorsal** and **retrodorsal groups**, all largely confined to the spinal segments innervating the limbs. Their extents are indicated in Figure 7.66 and their significance will be discussed later (p. 829). Onufrowicz (1899) described a ventrolateral group in the first and second sacral segments, considered to innervate the perineal striated muscles. This 'nucleus of Onufrowicz' has been confirmed in mankind by Mannen et al (1977) and by Konishi et al (1978) in cats.

NEURONAL GROUPS OF THE POSTERIOR GREY COLUMNS

Groups in the dorsal regions of spinal grey matter comprise two extending the cord's whole length and two limited to the thoracic and upper lumbar segments. No less than 15 neuromediators have been identified in the spinal dorsal grey column.

Substantia gelatinosa (of Rolando), present at all levels, consists of small Golgi type II neurons with some larger ones. Connections with afferents of the dorsal roots and the spinothalamic tract complex have long been accepted, but a revision of the details has followed experimental work (pp. 941, 942). A second long group of large neurons, ventral to the *substantia*, is the **dorsal funicular group** or *nucleus proprius*. As in the ventral column, a thin lamina of neurons, distinguishable from those of the substantia by their larger size, is the *marginal zone* of some observers, dorsal to the substantia (7.65A, 66).

Nucleus dorsalis or *thoracicus* ('Clarke's column') is basal in the posterior grey column, immediately dorsal to the intermediate zone in laminae VI–VII (p. 944). At most levels it is near the dorsal white funiculus and may project into it. In the human spinal cord it can usually be identified from the eighth cervical to the third or fourth lumbar segments. Similar groups have been described at cervical levels superior to the nucleus dorsalis; also prolongations at caudal levels appear to exist in some long-tailed monkeys (Chang 1951). But these 'cervical' and 'sacral' nuclei consist of different neurons and have been described only in some mammals; hence it is premature to extrapolate such groups to the human spinal cord (p. 944). Neurons of the dorsal nucleus itself vary, most being large, especially in the lower thoracic and lumbar segments; some send axons into the dorsal spinocerebellar tracts (p. 935) while some are interneurons. Petras & Cummings (1977) have described the neurons and connections of this nucleus in neonatal dogs, confirming a role as a 'relay' in the dorsal spinocerebellar paths; they ascribed a similar role to the **nucleus centrobasalis**, which is in the dorsal grey column, like the nucleus thoracicus, but situated in the lower cervical and lumbosacral segments. These groups have not been established in human spinal cord.

Lateral to the nucleus dorsalis and dorsal to the intermediolateral column (7.66), a small region of neurons of medium size extends through the same segments (first thoracic to third lumbar) as the intermediate columns (Takahashi 1913); this group is identifiable in the human cord, but functionally obscure.

NEURONAL GROUPS OF THE INTERMEDIATE GREY MATTER

The intermediate region of spinal grey matter (7.65A, 66), including the lateral grey columns, contains small neurons, many like autonomic preganglionic cells, developing in the embryonic cord at first dorsolateral to the central canal. Many migrate lateral to it, forming an **intermediolateral column**. An **intermediomedial column** is formed from neurons nearer the central canal. The intermediolateral is the projecting lateral grey column proper; many of its neurons send axons into ventral spinal roots and via white rami communicantes to the sympathetic trunk (p. 1156). Preganglionic fibres are similarly derived from some cells of the intermediomedial column (the remainder being interneurons). Both groups extend from the eighth cervical or first thoracic segment to the second or third lumbar, corresponding approximately to the thoracolumbar outflow. In the second to fourth sacral segments a similar group, intermediate in position, is the source of the sacral outflow of

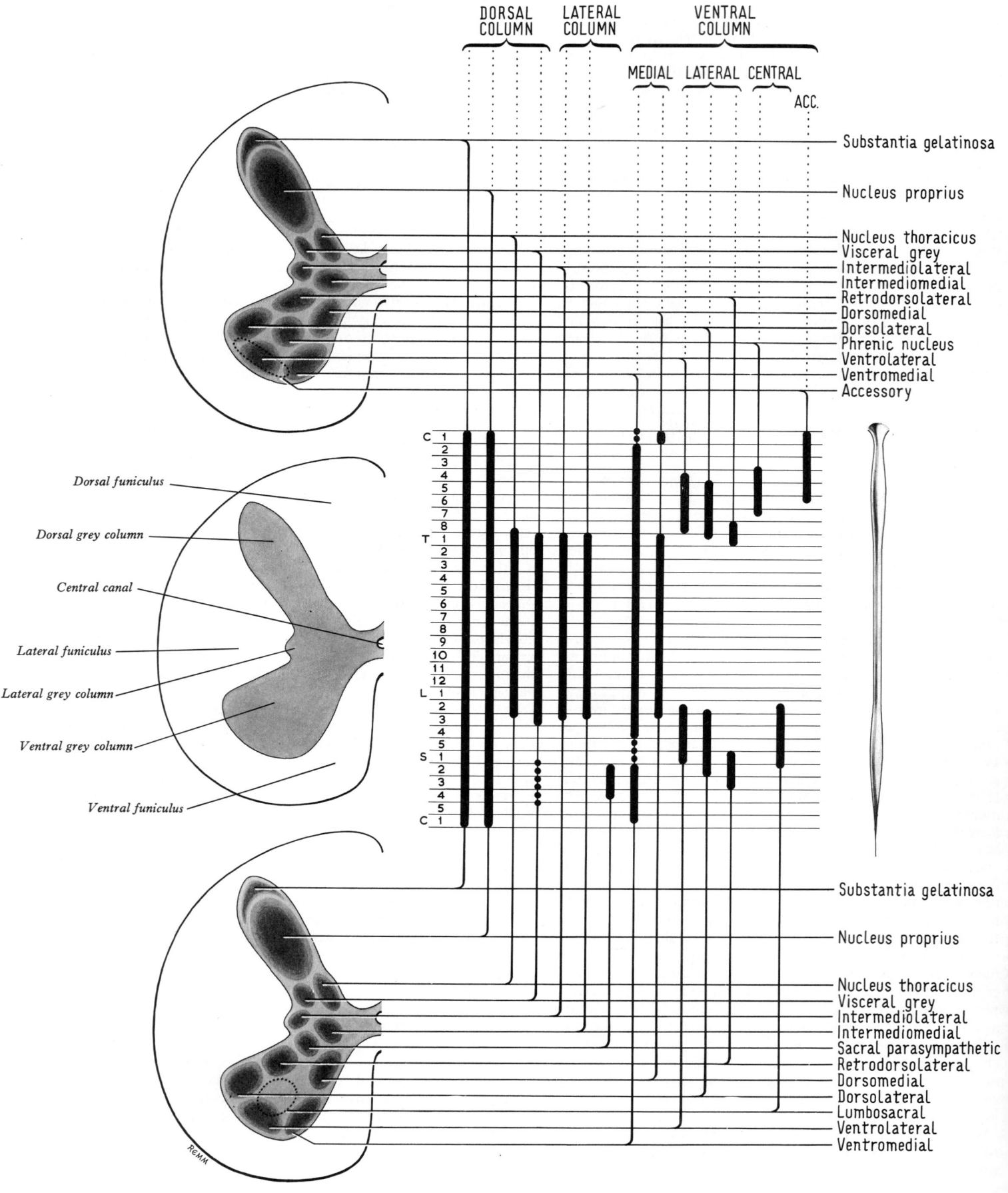

7.66 The groups or nuclei of nerve cells in the grey columns of the human spinal cord as generally accepted. Relative positions of these columnar groups, as well as their extension through varying series of spinal segments, are as indicated. Adapted from data in Crosby et al 1962, with permission of the authors and publishers.

parasympathetic preganglionic nerve fibres (p. 1155). This **sacral parasympathetic grey column** also lies lateral to the central canal and substantia gelatinosa centralis in the zone between the bases of the anterior and posterior grey columns and shows no mediolateral division, nor does it project like the thoracolumbar lateral grey column. The emergence of parasympathetic preganglionic nerve fibres from other segments has been described by Kure et al (1930) and Sheehan (1933), their origins being ascribed to the basal region of the dorsal grey column and perhaps the intermediate grey zone; the fibres were stated to issue in the *dorsal* roots, to be vasodilator and to synapse in the corresponding dorsal spinal root ganglia (Kiss 1932, Kure et al 1934). These interesting views have received neither general acceptance nor confirmation.

This description of spinal cytoarchitecture largely depends on material stained to show the *somata* of neurons rather than their processes; it has been amplified by a *laminar concept* of spinal organization (p. 938). More widely based on interconnections, its structural data are correlated with the results of degeneration experiments and microelectrode studies. The laminar pattern yields a more precise definition of spinal cord activities; but the two modes of description are not exclusive, the older scheme of columnar groups being in most features adaptable into the newer description. The main modifications concern the structural relations between the dorsal grey column neurons and the fibres of dorsal roots and spinal tracts.

FUNCTIONAL IMPLICATIONS OF ANTERIOR GREY COLUMN CELL GROUPS

Even in the earliest accounts of spinal columnar arrays of neurons, mostly based on Nissl-stained transverse sections (from a miscellany of animals, including tadpoles, an ostrich, a gorilla and even man!), a somatotopic interpretation was advanced with confidence (Elliot 1942). Thus, an early tenet was that medial groups innervate the axial musculature (by posterior primary rami), the limbs being innervated from lateral groups. This attractive speculation, based initially on structural data, has been confirmed to some degree by subsequent experiment, as will appear. Few investigators have inspected the grouping of ventral grey column cells in longitudinal sections and errors arising in tracing elongated aggregations through transverse sections may account for some disagreements between earlier topographical results. Obviously an agreed pattern of groups is a necessary prelude to attempts to assess whether a group represents, e.g. innervation of an individual muscle. Considerable agreement is apparent in more recent work, particularly among observers examining the distribution of groups in fetal material, where they are more discrete and identifiable (Romanes 1941, Elliot 1943). Such results proved markedly similar in feline and human material, though only one human fetus was examined. In contrast motor cell grouping in ventral columns is much *less* developed in amphibians and reptiles or even mammals such as bats and moles, all possessing complex forelimb musculature, than in whales, in which the same muscles are much simplified (Romanes 1953). Comparative observations are of limited value, being mostly on a heterogeneous collection of vertebrates, in extremely small series, with little reference to taxonomy. Moreover, little such information has been added in recent decades and certainly no major series.

These difficulties must be mentioned to introduce some caution into the functional interpretations of neuronal groups, not to deny their existence. Pattern is always significant and a topographical pattern certainly exists in the ventral column. The question is: What is its physiological meaning? That it concerns the innervation of muscles seems inescapable. It would appear simple, if tedious, to cut the nerves supplying individual muscles and observe the distribution of the resultant retrograde degeneration. Few such experiments have been carried out on an extensive scale. Since they have involved different species and different spinal levels, they provide only limited checks. In an investigation of lumbosacral cord in cats, various nerves being divided in the hind limb (Romanes 1951), affected groups were all lateral in the anterior columns and cells innervating more distal muscles were

dorsal to the 'motor columns' for proximal muscles. These findings generally corroborated earlier speculations. Limited experimental confirmation, also in cats, has been recorded for anterolateral leg muscles (Balthasar 1952). Romanes concluded that topographical cell groups were not usually *motor pools*, but more often represented groups innervating muscles involved in some *common effect on a joint*.

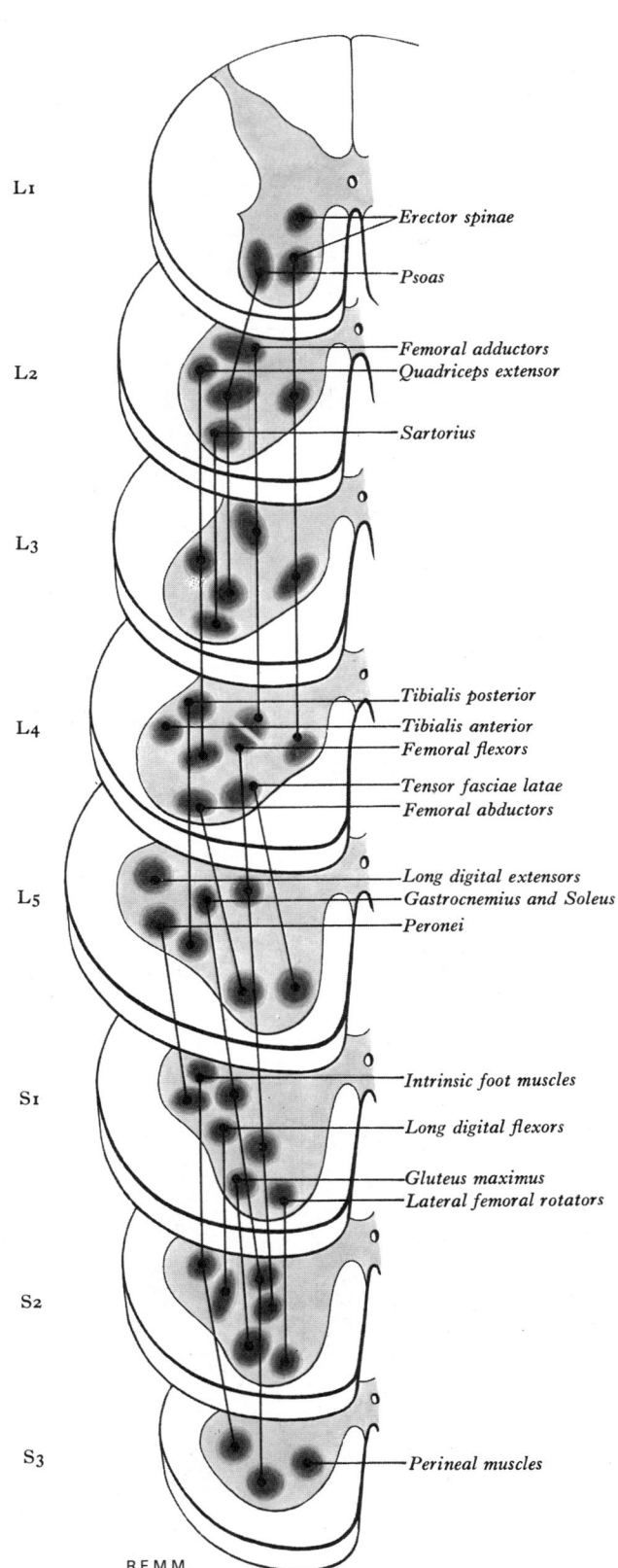

Erector spinae
Psoas
Femoral adductors
Quadriceps extensor
Sartorius
Tibialis posterior
Tibialis anterior
Femoral flexors
Tensor fasciae latae
Femoral abductors
Long digital extensors
Gastrocnemius and Soleus
Peronei
Intrinsic foot muscles
Long digital flexors
Gluteus maximus
Lateral femoral rotators
Perineal muscles

R.E.M.M.

7.67A The approximate location in the transverse plane, and in longitudinal extent, of the nerve cell groups innervating muscles, chiefly in the leg, in the lumbosacral segments of the human spinal cord. Based on clinicopathological studies of poliomyelitis; see Sharrard 1955.

THE INNERVATION OF THE LOWER LIMB MUSCLES

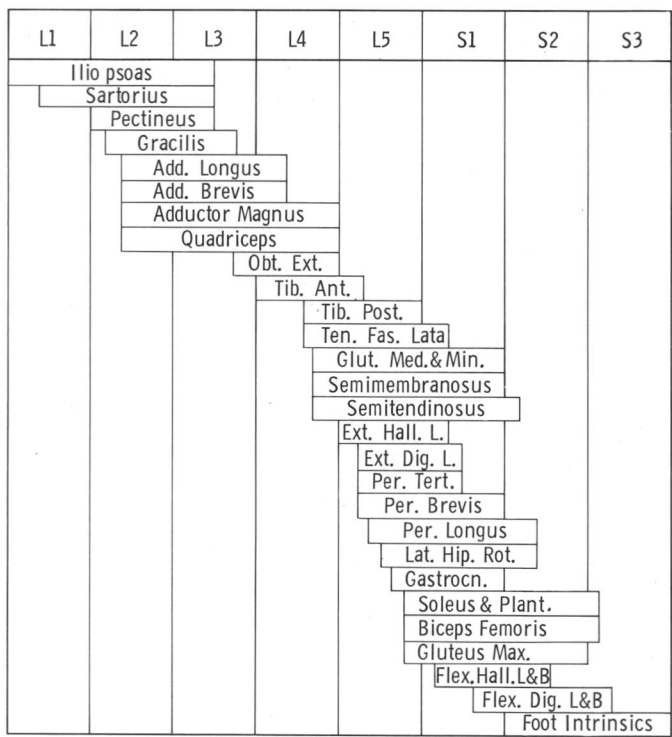

L1	L2	L3	L4	L5	S1	S2	S3

Ilio psoas
Sartorius
Pectineus
Gracilis
Add. Longus
Add. Brevis
Adductor Magnus
Quadriceps
Obt. Ext.
Tib. Ant.
Tib. Post.
Ten. Fas. Lata
Glut. Med.& Min.
Semimembranosus
Semitendinosus
Ext. Hall. L.
Ext. Dig. L.
Per. Tert.
Per. Brevis
Per. Longus
Lat. Hip. Rot.
Gastrocn.
Soleus & Plant.
Biceps Femoris
Gluteus Max.
Flex. Hall. L&B
Flex. Dig. L&B
Foot Intrinsics

7.67B The segmental arrangement of the innervation of the lower limb muscles (see Sharrard 1964).

Another investigator claimed that experiments in monkeys indicated that ventral column groups can be explained not only as above, but equally well in terms of peripheral nerves, limb segments and muscles grouped on a developmental basis (Sprague 1948). The effects of the division of dorsal and ventral primary rami were studied in segments innervating limbs and at thoracic levels, leading to a denial of differentiation into medial and lateral groups in thoracic segments. Even in spinal limb enlargements, where both groups existed, both appeared sources of fibres entering dorsal and ventral rami, refuting the classic assignment of axial and limb musculature to the medial and lateral groups (Kaiser 1891). Sprague later re-emphasized that concepts of columnar organization had become so ingrained as to obfuscate

interpretation. While perhaps an extreme view, this does underline the unsatisfactory state of knowledge and need of a wider re-investigation in primate animals, a formidable undertaking. More precise information might prove of limited interest but would clearly be of some value. The effects of poliomyelitis on the neurons in ventral columns have been correlated with the resultant paralysis by several clinical observers. The most detailed report (Sharrard 1955, 1956) is of special interest, firstly in concerning human conditions and secondly in its partial confirmation of the experimental findings in cats (7.67A,B). Further data of this kind are much needed.

To summarize: despite a copious literature, a plethora of topographical speculations and a much smaller corpus of experimental data, large uncertainties remain. That the topographical arrangement of ventral grey column neurons is columnar is well established, but perhaps columnar organization (to some extent inevitable in an elongated spinal cord) has been over-emphasized. Somatotopic organization related to muscles, individually or in functional groups, appears generally confirmed; but evidence does not favour discrete motor pools for individual muscles, few being satisfactorily identified. Overlapping 'pools' for associated muscles seems a more common pattern of organization. Most workers have apparently concentrated on the *somata* of neurons, ignoring the *dendritic regions* between them, which occupy a much greater volume of spinal grey matter. To this aspect of organization we shall return after consideration of spinal 'white matter'.

Nerve Fibre Tracts of the Spinal Cord

The spinal 'white matter' contains nerve fibres, neuroglia and blood vessels, surrounds the fluted column of grey matter and owes its hue to large numbers of myelinated nerve fibres. Its arrangement into anterior, lateral and posterior funiculi has been noted (7.62). Fibres vary in calibre, many being small and lightly- or non-myelinated. Some tracts are typically of small fibres, e.g. the dorsolateral tract, fasciculus gracilis and central part of the lateral funiculus. Nerve fibres in the fasciculus cuneatus, anterior funiculus and peripheral zone of the lateral funiculus all contain many large fibres. Most 'white' regions show a wide *spectrum of fibre diameters*, from 1 μm or less to about 10 μm. Fibres of 3 μm or less predominate; the larger (few exceeding 10 μm) form a small fraction. Detailed studies of the distribution of fibre diameters in human spinal cord have been few (Haggqvist 1936, Giok 1956) but many tracts are claimed to be identifiable on this basis. The proportions of fibres of particular diameters have been

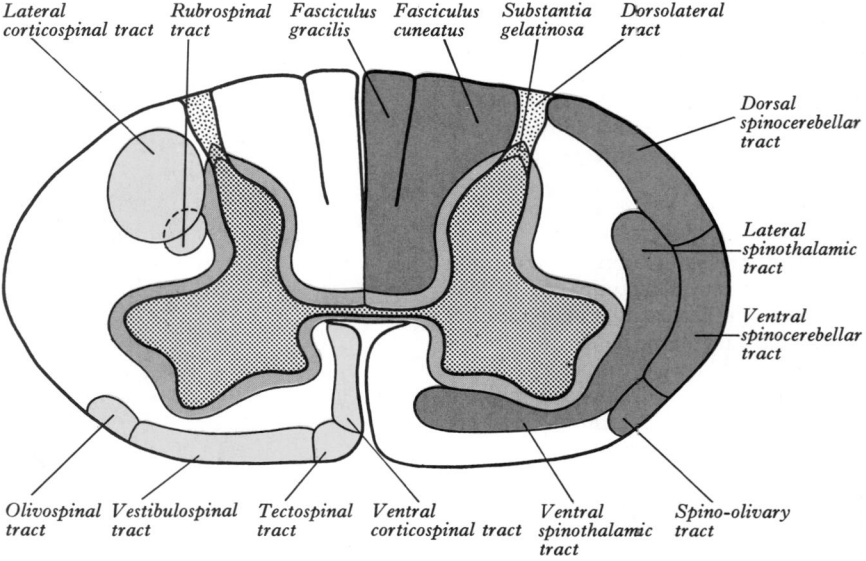

Lateral corticospinal tract | Rubrospinal tract | Fasciculus gracilis | Fasciculus cuneatus | Substantia gelatinosa | Dorsolateral tract

Dorsal spinocerebellar tract

Lateral spinothalamic tract

Ventral spinocerebellar tract

Olivospinal tract | Vestibulospinal tract | Tectospinal tract | Ventral corticospinal tract | Ventral spinothalamic tract | Spino-olivary tract

7.68 Simplified diagram of the main tracts of the spinal cord. The ascending tracts are shown in red on the right side of the figure; the descending tracts are shown in yellow on the left side; the 'intersegmental' tracts are in orange on both sides. Many tracts are omitted (see 7.69A,B).

occasionally estimated (Szentágothai-Schimert 1941) but precise data of this kind are few. Delineation of tracts by such inspection in transverse sections of normal material must be confirmed by experimentation. Most information regarding spinal tracts is derived from selective spinal damage in experimental animals; but the effects of damage to dorsal nerve roots and lesions in the brain stem, cerebellum and cerebrum have also provided information. In such experiments retrograde degeneration indicates the neurons from which particular axons proceed, anterograde terminal degeneration indicating the sites of termination. Suitable staining to reveal Wallerian degeneration at intermediate levels can demonstrate the position of a tract and the degree of compaction, dispersal or overlap of its fibres. While certain tracts are largely discrete and their central regions regularly located in funiculi, reciprocal overlap at their fringes (often much more) is usual (vide infra). This partly accounts for variation in extent in diagrams of transverse spinal sections according to different authorities; in all such diagrams arbitrary boundaries delineate the *supposed* limits of tracts, especially in the human cord, where deliberate experimentation is impossible; the data available result only from disease and injury, neither producing clear-cut lesions. Such investigations, particularly experiments (see p. 870) on other primates, are likely to provide results similar to human arrangements; but it is unwise to assume that they are identical. Evidence from the examination of human spinal cords affected by disease or injury does support the presumption that the general layout of tracts in the human primate is much the same as in others. Information from this source is surprisingly abundant,

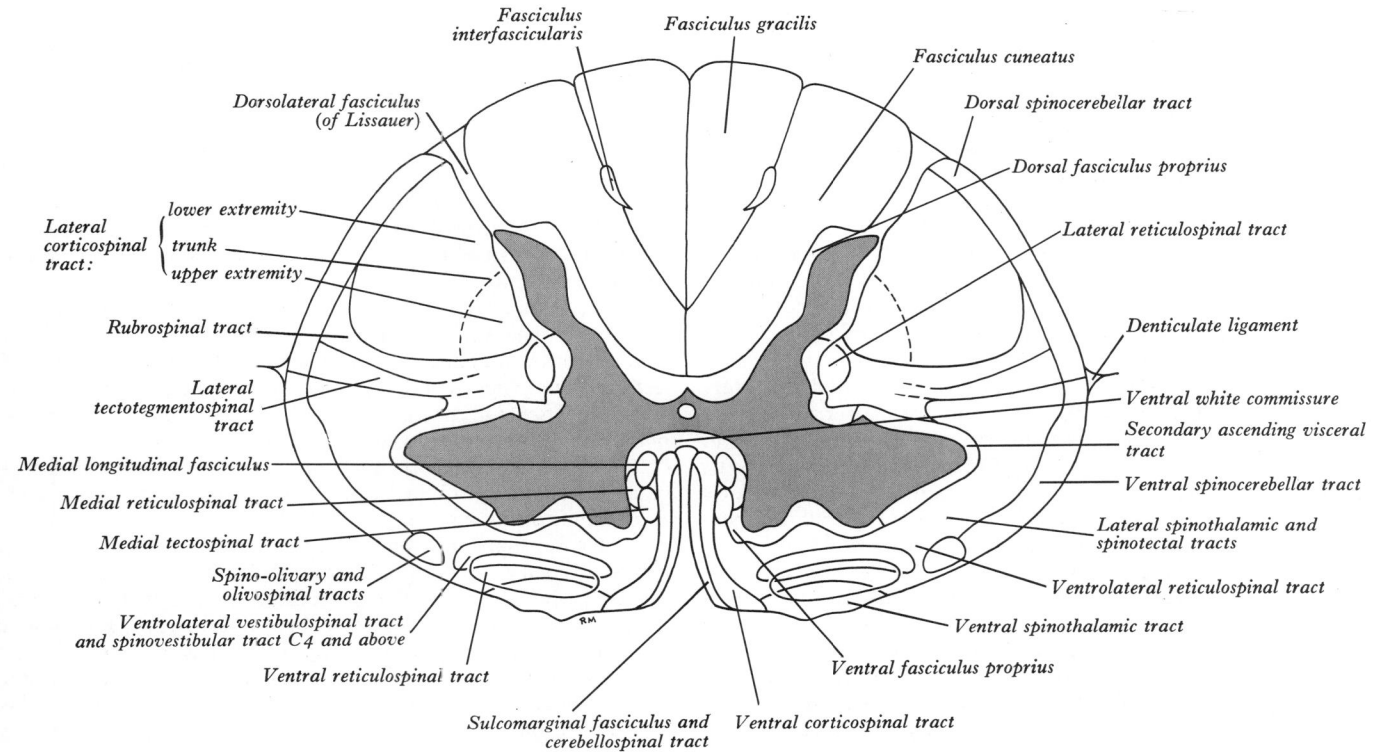

A

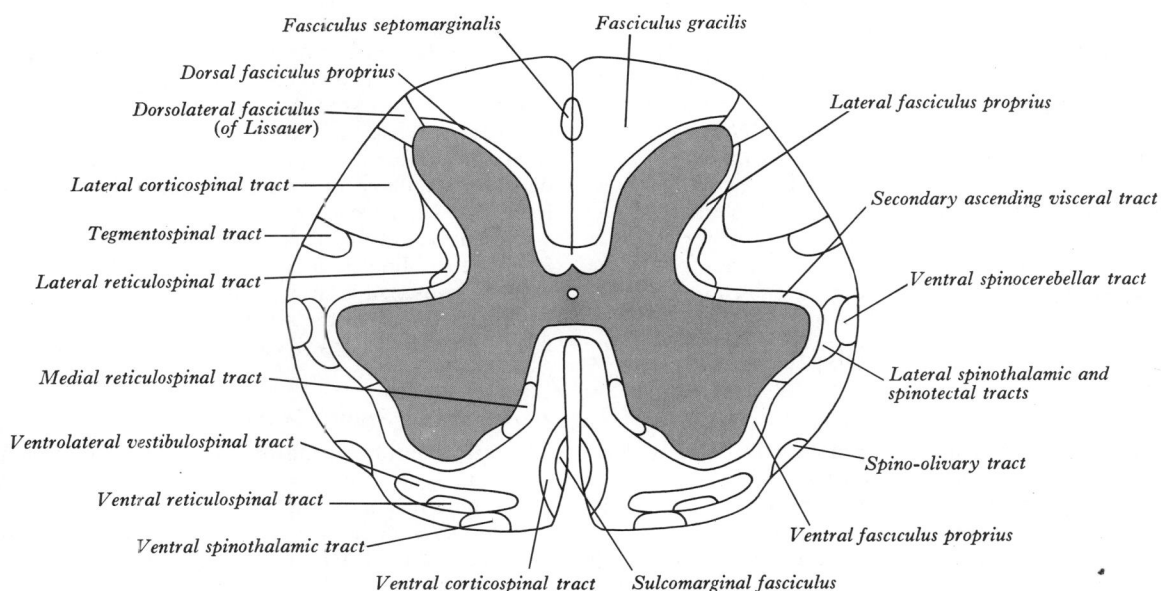

B

7.69 The approximate relative positions of nerve fibre tracts of the human spinal cord at mid-cervical (A) and lumbar (B) levels. Adapted from Crosby et al 1962.

though usually less precise; it was the principal source of early data and a large number of reports is scattered through neurological literature (Nathan & Smith 1955, 1959). The predominance of clinical data has produced a curious vicious circle in associating the observed *syndromes* of sensory and motor disturbance with injury or disease in particular spinal regions but the exact extent of the trauma is often imprecisely assessed. Because some tracts are undeniably concentrated in certain parts of the white funiculi, delineation has become artificially crystallized. The fringe overlap mentioned above may be extensive; some tracts are almost completely mingled, others are a concept rather than a circumscribed reality, their fibres scattered far and wide. Experimental observation, in parallel with clinical deduction, has now outstripped the latter in the recognition of intermingling of tracts, whose supposed separateness is still accepted in much clinical practice: an impediment rather than an aid to accurate diagnosis.

The following account of spinal tracts is concerned with human arrangements; but reference to findings in animal experimentation must often be made where adequate clinicopathological data are unavailable. In a later section (p. 938), dealing with the finer details of neuronal organization in the spinal cord, evidence from animals other than man must predominate, but it is probable that deductions from this source are in many instances applicable to the human spinal cord (but with some notable exceptions).

As a convenient simplification, nerve fibres in spinal white matter are assigned to five groups: (1) afferent fibres from neurons in dorsal root ganglia entering by dorsal roots and extending variable spinal distances; (2) long ascending fibres, derived from spinal neurons conducting afferent impulses to supraspinal levels; (3) long fibres descending from supraspinal sources to synapse with spinal neurons; (4) fibres effecting intrasegmental and intersegmental connections; (5) fibres from motor neurons in ventral and lateral grey columns issuing in ventral nerve roots. Fibres in all except the last category form *longitudinal tracts*. Arrangements are not, however, so simple. Examination in different planes reveals that many fibres proceed in oblique and even horizontal directions, particularly across the midline in grey and white commissures, many being *decussating* fibres. Many others are *commissural* intrasegmental connections linking the neurons in grey columns to contralateral neurons. Some longitudinal tracts are polysynaptic, i.e. composed of trains of neurons. Most, if not all, fibres entering by the dorsal root divide into horizontal ascending and descending rami, all with collateral branches extending into the grey matter, in which the number of neurons sending axons into ventral nerve roots or long projection tracts is a mere fraction of the total, most being local interneurons (p. 939).

In the following account an arbitrary order has been adopted: tracts in the anterior, lateral and posterior funiculi are described in that order, in each of which descending, ascending and intersegmental tracts are successively described. This is primarily to present a topographical picture useful in the context of medical diagnosis, experiment and phylogenetic studies. However, tracts vary in relative positions at different spinal levels (and in different species). For a detailed analysis of tracts in the cat and monkey, see Verhaart (1953, 1954), van Beusekom (1955), Verhaart & van Beusekom (1958). The general positioning of major tracts is illustrated in 7.68 and in greater detail at two levels in 7.69A,B. Some features are further summarized in 7.72A,B, 73.

TRACTS IN THE ANTERIOR FUNICULUS

1. Descending Spinal Tracts

(*a*) **The anterior corticospinal tract,** usually small and inverse in size to the *lateral* corticospinal tract, adjoins the anterior median fissure but is separated from it by the sulcomarginal fasciculus (vide infra). Present in the upper spinal cord, it diminishes as it descends and cannot be traced below middle thoracic levels. Its fibres are considered with those of the lateral corticospinal tract (p. 933). It occurs only in primates, is of uncertain function and very variable (Nyberg-Hansen & Rinvik 1963); it may be absent or, very rarely, contains almost all corticospinal fibres, but usually about 10–30% of them. Its variability may be

linked with its late phylogenetic and ontogenetic development (Humphrey 1960).

(*b*) **The vestibulospinal tract,** principally derived from large and small neurons in the lateral vestibular nucleus (p. 958), descends superficially in the periphery of the anterior funiculus to end around the neurons in the anterior grey column. Being uncrossed, it brings ipsilateral anterior column neurons under control of the vestibular nuclei, as an efferent path for equilibration. The most medial fibres have, however, been ascribed to medial and perhaps lateral and inferior vestibular nuclei (Rasmussen 1932) but only the medial nucleus is confirmed (Pompeiano & Brodal 1957). These medial fibres, descending by a different course in the medial longitudinal fasciculus (p. 985), constitute the *medial vestibulospinal tract*, partly crossed and probably ceasing above the lumbar cord, perhaps concerned only with the upper extremity and neck. In the anterior funiculus it is dorsal to the tectospinal tract and next to the anterior median fissure, sited as a so-called *sulcomarginal fasciculus* (7.69A,B). Most vestibulospinal fibres are uncrossed and descend from the lateral nucleus as a *lateral vestibulospinal tract*, which extends to all levels. Fibres of both tracts end medially in the anterior grey column (laminae VII and VIII, p. 939), influencing α and γ motor neurons through the interneurons (Gernandt 1959); some may end monosynaptically on motor neurons. There is some degree of somatotopic representation in the lateral vestibular nucleus.

(*c*) **The tectospinal tract,** variably sited lateral to the anterior median fissure, arises in the contralateral superior colliculus, ending in synapses with the interneurons (Szentágothai 1948) in cervical segments of the anterior column, with terminations probably in laminae VI–VIII (p. 939). A suggested bilateral origin from superior colliculi may be due to confusion with a *lateral tectotegmentospinal tract* of some authors; when this is described, tectospinal fibres in the anterior funiculus become a *medial tectospinal tract*. Autoradiographic fibre-tracing in primates (Frankfurter et al 1976) suggests that only a minority of tectal fibres reach the spinal cord, most ending in various brain-stem nuclei. For further discussion see pp. 946, 985.

(*d*) **Reticulospinal fibres** scattered in the anterior funiculus, largely in its medial part, arise from ipsilateral pontine reticular nuclei (p. 991), some perhaps decussating shortly before their spinal termination (Torvik & Brodal 1957). Descending the whole cord in the anterior funiculus, they probably synapse with interneurons medially in the ventral grey column (laminae VII and VIII, Nyberg-Hansen 1965, p. 946), perhaps facilitating the motor neurons. Commonly known as the *medial* (pontine) *reticulospinal tract*, its pontine origin corresponds to the 'facilitatory area'. (Contrast the lateral reticulospinal tract, p. 934.) These details are unconfirmed in mankind (Nathan & Smith 1955, Brodal 1957, Webster 1978, and p. 991).

(*e*) **The interstitiospinal tract,** uncrossed, descends from the interstitial nucleus (p. 985) in the medial longitudinal fasciculus and its spinal continuation, the ipsilateral fasciculus proprius (anterior intersegmental tract). In cats the tract is considered traceable to sacral levels (Nyberg-Hansen 1966) but no details are available for the human spinal cord, where nevertheless it has long been recognized (Muskens 1914).

(*f*) **The solitariospinal tract** (Cajal 1909) is a small group of fibres descending in part from the caudal part of the solitary nucleus (p. 954); it is clear in cats (Torvik 1957) but has also been equated with part of the medial reticulospinal tract (Crosby et al 1962). It may be involved in visceral reflexes involving the oesophagus and stomach.

2. Ascending Spinal Tracts

The anterior spinothalamic tract is continuous with the lateral (p. 935) but may be considered the anterior funicular part of the spinothalamic complex, lying medial to fibres of the ventral spinal roots and dorsal to the vestibulospinal tract (7.68). It overlaps with all its other neighbours. Largely on clinical evidence, it is said to be concerned with crude tactile and pressure sensibility but this is not confirmed by physiological investigations (p. 944). According to Applebaum et al (1975) segregation by modality in the anterior/lateral spinothalamic complex does not accord with experimental data and they are to be considered

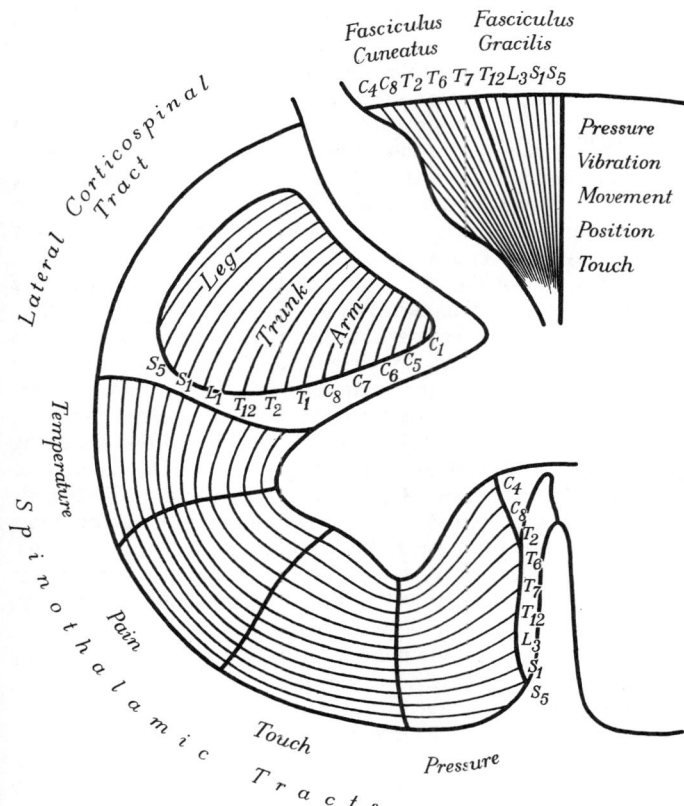

Fasciculus Cuneatus
Fasciculus Gracilis
$C_4 C_8 T_2 T_6 T_7 T_{12} L_3 S_1 S_5$

Pressure
Vibration
Movement
Position
Touch

Lateral Corticospinal Tract

Leg Trunk Arm

$S_5 S_1 L_1 T_{12} T_2 T_1 C_8 C_7 C_6 C_5 C_1$

$C_4 C_8 T_2 T_6 T_7 T_{12} L_3 S_1 S_5$

Temperature

Spinothalamic Tracts

Pain

Touch

Pressure

7.70 The general plan of segmental organization of the fibres in the posterior funiculus, the lateral corticospinal tract and the lateral and anterior spinothalamic tracts. The probable cross-sectional areas of these tracts are enlarged to provide adequate space. This general plan applies to *all* segmentally organized tracts whether ascending, descending, ipsilateral or contralateral. After Foerster 1936, with permission.

as a continuum structurally (as stated) and functionally. However, neurons of spinothalamic axons are diverse in distribution and modalities and simple segregations of sensory mediation in the spinothalamic system (or other ascending tracts) are not likely. But, functionally, neurons in the spinal laminae of spinothalamic origin provide some data; thus neurons in laminae I and V respond to cutaneous 'noxious' mechanical and thermal stimuli and some lamina V neurons to other cutaneous stimuli. Laminae VII and VIII are linked with muscular and articular receptors and some with cutaneous stimulation. The precise neuronal locations of origin of this tract are uncertain, though usually described as 'secondary neurons' in the dorsal grey column, receiving direct synapses from axons entering dorsal roots from primary sensory neurons in dorsal root ganglia. Cajal (1909) showed that axons of neurons in all grey columns (dorsal ventral and intermediate) *decussate* before ascending in the anterolateral funiculus. Investigators using Golgi techniques have claimed that they are concentrated in laminae IV–VII, especially lamina V, with contributions from pericornual neurons, superficial to the substantia gelatinosa (lamina I) and lamina VIII (Webster 1977). some details have been physiologically confirmed in cats and rats (Dilley et al 1968); electrical stimulation near the entry of the medial lemniscus and spinothalamic tracts into the thalamus, combined with intracellular recording of antidromic responses from spinal laminae, showed responding somata in laminae V and VI, providing some confirmation of spinothalamic origins. But the limited sampling inherent in the technique may have missed neurons in other laminae. Further, all spinothalamic fibres are not necessarily axons of 'secondary neurons'; one or more interneurons may be interposed between primary dorsal root afferents and spinothalamic neurons.

Evidence from human cordotomies suggests that most spinothalamic fibres decussate in the anterior white commissure, probably within or near their segment of origin. Though physiological studies (p. 944) suggest some separation of

spinothalamic neurons by sensory *modalities* in the dorsal grey column, the arrangement of ascending fibres in these tracts has long been considered *somatotopic* (Nathan 1963, Morin et al 1951). Fibres crossing at any level join the medial aspect of those already crossed and both tracts are hence segmentally laminated (7.70), fibres from the lower segments, of greater total length, being most superficial. A slight spiral twist is also described, the superficial fibres becoming progressively more dorsal as they ascend. This continues through the medulla and pons to the thalamic nucleus ventralis posterior lateralis (p. 1002). However, their precise arrangement, whether somatotopic, dermatomic or otherwise, is uncertain. Mingled with the anterior spinothalamic are descending *reticulospinal fibres* (already noted) and ascending *spinoreticular fibres*. Spinothalamic fibres may send collaterals to brain-stem reticular nuclei but most of their afferents are now said to form separate spinoreticular paths, some probably polysynaptic. As in other ascending tracts, the spinal origins of spinothalamic tracts, or the associated interneurons, may be selectively inhibited by fibres from neurons in sensory and motor parts of the cerebral cortex and others in the anterior cerebellar lobe and brain-stem reticular formation; but spinal distribution and the termination of inhibitory fibres are uncertain (Carpenter et al 1963, Webster 1977). A small *spinovestibular tract*, mingled with vestibulospinal tracts, conveys exteroceptive and cutaneous information to vestibular nuclei. It may be confined to cervical levels; its degree of decussation is unknown but it may end in the lateral and descending vestibular nuclei. Its origins probably correspond to those of the spinocerebellar tracts.

3. Intersegmental Tracts

The remaining anterior funicular axons form an **anterior intersegmental tract**, partly decussated and probably with intermingled reticulospinal and descending autonomic fibres. The tract is continuous with the medial longitudinal fasciculus. Though all intersegmental or propriospinal fibres in the anterior funiculus are usually thus combined under one term, separate entities elsewhere in the cord have been described by classic neuroanatomists (Nathan & Smith 1959). Details of such fibres are considered with the dorsolateral tract (p. 936).

TRACTS IN THE LATERAL FUNICULUS

1. Descending tracts

(*a*) **The lateral corticospinal tract** descends through almost the whole spinal cord, progressively diminishing to end about the third or fourth sacral spinal segment. It occupies an oval area anterolateral to the posterior grey column and medial to the posterior spinocerebellar tract (7.69A,B); in lumbar and sacral regions, where the latter is absent, the lateral corticospinal reaches the surface. Both corticospinal tracts are motor but physiological evidence suggests other functions (pp. 944, 945). They together contain, on each side, about a million fibres of varying diameter, 70% myelinated; about 90% have a diameter of 1–4 μm, about 9% 5–10 μm, and less than 2% 11–22 μm. However, these figures, particularly those for small fibres, may need substantial revision with systematic quantitative electron microscopy. Many fibres are from neurons in frontal cortical area 4; fibres of 11–22 μm are believed, from cell and fibre counts, to be axons of giant pyramidal neurons of Betz (p. 1044). Some issue from other layers in area 4 but at least two-thirds of corticospinal fibres do not, some being from other frontal and parietal areas, especially 3, 1, 2 and 6. However, the origins of about 50% only can be thus explained (Lassek 1954). All fibres descend via the internal capsule, traverse the cerebral peduncle and pons to enter the medullary pyramid, in the caudal part of which a variable number (usually about two-thirds) decussate and turn caudally in the lateral funiculus as the *crossed lateral corticospinal tract*, the remainder continuing ipsilaterally as the *anterior corticospinal tract*. The lateral tract contains also some ipsilateral *uncrossed lateral corticospinal fibres* (Fulton & Sheenan 1935).

Corticospinal tracts, thus defined, are commonly dubbed the *pyramidal pathway*, a usage likely to persist. They certainly traverse the medullary pyramids (p. 951) but the expression usually includes *corticobulbar* fibres diverging above this level to end in the motor nuclei of cranial nerves (p. 1049). But the chief objection to 'pyramidal' lies elsewhere: corticospinal tracts have a single distinction; they descend without interruption from the cerebral cortex to the spinal grey matter in which they contrast with the rubrospinal, vestibulospinal and other descending tracts, which have intermediate subcortical synapses in the deep cerebrum and brain stem and are collectively often labelled 'extrapyramidal'. The functional distinction between these so-called 'systems' is diminishing; clinical concepts misplaced upon them are dissolving. The subsequent revelation of a so-called reticular system (p. 988) has added to this confusion, since structurally and functionally it overlaps the 'extrapyramidal system', a concept probably obsolescent. Hence the term 'pyramidal' is better avoided.

Throughout the cerebrum and brain stem, except perhaps in the pons, somatopy in these tracts is established, chiefly by clinical observations (7.70). Details are to be found in the appropriate sections. Whether the arrangement is identical in medullary pyramids and the spinal cord is uncertain (Foerster 1936, Barnard & Woolsey 1956). In the cord longer fibres are said to be the most superficial, the shorter ones internal. Most corticospinal fibres, in both anterior and lateral tracts, probably synapse with *interneurons* in the *dorsal* grey column's base, which spread ventrally through the intermediate zone to the ventral grey column, laminae IV–VII (Hoff & Hoff 1934, Chambers & Liu 1957, Nyberg-Hansen 1969 and p. 939). In experiments, mostly in cats, no corticospinal fibres reach the spinal motor neurons directly; but some monosynaptic connections occur in monkeys (Liu & Chambers 1964), suggesting similar human arrangements, perhaps in greater numbers. But it seems certain that most corticospinal fibres, exciting neurons which innervate skeletal muscle, do so through chains of interneurons. Physiologically this influence is facilitatory for flexors and inhibitory for extensors, the reverse of the effects of the lateral vestibulospinal tracts (p. 932). (It is imperative to note that here, and elsewhere, such functional allusions assume, without being stated, that the influences flow from an *increased* flux of potentials in fibres of the tract. With a net reduction, of course, the reverse obtains.) The α and γ motor neurons share these corticospinal influences; fibres from the precentral motor cortex may end ventral to those from the postcentral sensory cortex (Scheibel & Scheibel 1966) and many (perhaps a quarter) terminate ipsilaterally. The effects of corticospinal activity on sensory transmission are discussed elsewhere (pp. 944, 945). Modifications of sensory inputs were discussed in extenso by Schmidt (1973), Gordon (1973), Tone (1973) and McCoskey & Torda (1975). Phillips & Porter (1977) have surveyed the evolutionary status, neuroanatomy, electrophysiology and functional roles of corticospinal neurons.

Ascending spinopontine and *spinocortical fibres* have been described in both anterior and lateral corticospinal tracts in cats (Walberg & Brodal 1953) and in man (Nathan & Smith 1955), derived from all spinal levels but especially the cervical and perhaps concerned in the transmission of cutaneous information (Brodal & Walberg 1952). (These further confuse the traditional concepts of 'pyramidal' and 'extrapyramidal' systems.)

(*b*) **The rubrospinal tract** contains fibres of variable diameter descending from large and small neurons in the red nucleus; caudal to this they decussate and form in the lateral funiculus a compact band ventral to the lateral corticospinal tract (7.68, 69A,B). Fibres may also arise from other tegmental neurons, caudal to the nucleus, forming a *tegmentospinal tract* near the ventral border of the rubrospinal (Woodburne et al 1946). First established in cats, in which it has been chiefly studied, a rubrospinal tract also appears in macaque monkeys, in both species descending to lumbosacral levels. Axons of smaller neurons reach the most caudal segments: this is important to note, since an erroneous view, once current, considered the largest neurons (p. 984) to be the sources of the longest fibres; since large neurons are few in human red nuclei, a persuasion developed that the human tract is negligible, an opinion aided by a dearth of clinical information (Nathan & Smith 1955). Since it is well attested in monkeys and the chimpanzee, it is likely to exist in mankind. The details stated here apply to other mammals; the human condition must await clarification.

At their origin in the red nucleus, rubrospinal fibres show, in cats (Pompeiano & Brodal 1957), a dorsoventral, somatotopic lamination, the shortest fibres descending from the most dorsal cells to cervical levels to influence, presumably, the muscles of the neck and upper limbs, probably through interneurons. Axons of most ventral rubral neurons are considered to be concerned with the lower limbs. Some leave the tract to reach the motor neurons of cranial nerves, probably through interneurons; some such *rubrobulbar* axons may end in the ipsilateral inferior olivary nucleus (Walberg et al 1958).

Distribution of the rubrospinal terminals resembles those of the corticospinal tracts, being largely to laminae V–VII but they do not extend so far ventrally, none reaching the motor neurons directly. This terminal overlap of two tracts suggests close interaction on the motor neurons through a common interneuronal pool. Note that afferents to the red nucleus come not only from the cerebellum but also the cerebral cortex, including area 4 and other corticospinal origins (p. 984) (also many other subcortical regions, p. 984). The corticorubral projection is somatotopic in cats (Mabuchi & Kusama 1966) and primates (Kupyers & Lawrence 1967), like the corticospinal arrangements. Both tracts appear to exercise similar effects on the spinal motor neurons, such as facilitation of those for the flexor musculature; thus it is all the more undesirable to contrast them, physiologically or clinically, as 'pyramidal' or 'extrapyramidal'. However, a revised outlook upon their functions *in man* would be premature, despite suggestive data from other primates.

Neurons of the *tegmentospinal fibres*, noted above, probably synapse with the terminals of *tectotegmental fibres* from both superior colliculi; some direct tectospinal fibres from both may accompany the tegmentospinal fibres. This descending complex (near the rubrospinal tracts) is accordingly termed the *lateral tectotegmentospinal tract*.

(*c*) **The lateral reticulospinal tract**, usually described as medial to both rubro- and corticospinal fibres, is in fact dispersed among the adjacent tracts; in some mammals the reticulospinal fibres descend *lateral* to the rubro- and corticospinal fibres (7.73) and such paths are likely to exist in the human spinal cord, but evidence of this is sparse. The lateral tract differs, anatomically and physiologically, from the medial (p. 932); its axons are largely from large neurons in the medullary reticular formation (*nucleus reticularis gigantocellularis*), some from smaller neurons, explaining the varying diameters of its fibres. Unlike the pontine reticulospinal fibres in the anterior funiculus, the lateral tract is largely crossed but partly ipsilateral. *Each half* of the medullary reticular formation hence exerts bilateral spinal effects. Like the medial tract the lateral's terminals synapse with interneurons in the intermediate and medial zones of the anterior grey columns but are more dorsal (chiefly lamina VII), though some may reach the motor neurons (lamina IX). These details apply to cats (Nyberg-Hansen 1965); reticulospinal fibres were also found to reach all levels, contrary to earlier views (Torvik & Brodal 1957). The medullary origin of the lateral tract approximates to the brain-stem 'inhibitory area', contrasting with the facilitatory effects of the medial tract. In addition to the motor effects of both on α and γ neurons, reticulospinal axons may also modify afferent transmission (7.23 and pp. 988–996).

(*d*) **Descending autonomic fibres**, for ipsi- and contralateral connections between brain-stem visceral nuclei and spinal preganglionic neurons, are amply attested by physiological evidence; but their positions and terminations are uncertain. That they are largely in the lateral funiculus is agreed; but some may descend in the anterior funiculus. Failure to define any spinal grouping of degenerating fibres, as a result of hypothalamic or brain-stem lesions, implies that the descending autonomic connections are diffusely dispersed. Various experimenters have ascribed them to the lateral corticospinal, reticulospinal and intersegmental tracts, some to a superficial position in the lateral funiculus (Enoch & Kerr 1967). They are considered to be mostly fibres of small diameter, probably in polysynaptic chains. Since attempts to interrupt the reticulospinal paths have disturbed visceral functions, such as vasomotor and sudomotor control (Johnson et al 1952), some fibres may be autonomic. Vasomotor effects follow stimulation of the precentral cortex; hence some corticospinal fibres may be autonomic. The evidence, much of it negative, suggests that descending autonomic fibres *are* dispersed in the anterolateral spinal region, in which lesions may disturb visceral control.

(*e*) **The olivospinal tract** is uniformly described as a small triangular area in transverse sections, external to the most lateral fibres of ventral spinal nerve roots, and said to be confined to the upper cervical segments and to contain fibres ending in the ventral grey column, some ascending as a *spino-olivary tract* (Jansen & Brodal 1954, p. 936). The tract was first identified in the human spinal cord, but no experimental evidence connects it with the inferior olivary nucleus (Brodal 1969), suggesting the term *bulbospinal* which merely displaces the uncertainty elsewhere.

2. Ascending Tracts

(*a*) **The posterior spinocerebellar tract**, a flat, posterior band lying peripherally in the lateral funiculus, adjoins medially the lateral corticospinal tract and dorsally the dorsolateral tract. It begins about level with the attachments of the second or third lumbar spinal nerves, enlarges as it ascends and enters the inferior cerebellar peduncle. Its fibres are from the ipsilateral thoracic nucleus, which receives *afferent* impulses from: (i) many long ascending fibres in the posterior funiculus with collaterals synapsing with neurons in this nucleus; and (ii) terminals of intermediate ascending fibres of the posterior column, especially in thoracic segments. Which source preponderates is uncertain.

(*b*) **The anterior spinocerebellar tract**, in transverse section, is a crescentic area peripherally placed in the lateral funiculus, anterior to the posterior tract. Its precise origin is uncertain but it is usually said to be large neurons in the posterior grey column, which are thus secondary in the spinocerebellar path. The primary neurons are probably like those of the posterior tract (vide supra) but ascending axons of these secondary neurons mostly decussate, few remaining ipsilateral (Jansen & Brodal 1954). Experiment indicates that such 'secondary neurons' are dorsolateral in the *ventral* grey column (Ha & Liu 1968). The tract starts in the upper lumbar region and reaches a high pontine level, whence it descends dorsally into the superior cerebellar peduncle.

The spinocerebellar tracts are laminated; fibres carrying impulses from the lower limbs are the most superficial (Yoss 1952/3, Smith 1957). They convey exteroceptive and proprioceptive impulses from dermal and locomotor receptors, essential for synergic control during voluntary movements. The thoracic nucleus diminishes upwards (p. 927) and does not extend above the first thoracic or last cervical segments, suggesting that the posterior spinocerebellar tracts are concerned chiefly with the trunk and lower limbs. Evidence shows that proprioceptive impulses from the upper limbs travel by the posterior external arcuate fibres, originating in the accessory cuneate nucleus in the medulla oblongata (p. 953) as a *cuneocerebellar tract*. In addition to these classic spinocerebellar routes, at least two indirect paths, interrupted in the inferior olivary and lateral medullary reticular nuclei, may project to the cerebellum and receive afferents respectively from the dorsal funiculi and spinothalamic tracts (Brodal 1969).

Spinocerebellar are long established proprioceptive tracts. Physiological evidence shows that the posterior also conveys impulses from touch and pressure endings. Both tracts display a marked somatotopic cerebellar termination in experimental animals, including monkeys, and in man (p. 975). The posterior ends in the cerebellar cortex concerned with the hind limb, its associated cuneocerebellar tract reaching a fore-limb area. Anterior spinocerebellar fibres end only in the hind-limb cortex; but physiological evidence (Oscarsson & Uddenberg 1964, Oscarsson 1965) shows that fibres in the *anterior* tract may serve tendon receptors in *fore-limb* muscles; these form a *rostral spinocerebellar tract*. A spinocerebellar projection for proprioceptors in the neck muscles has been ascribed to the *nucleus cervicalis*; chiefly identified in young mammals, but not in the human spinal cord, it exists at high cervical levels. According to Cummings & Petras (1977) cerebellar ablations in newborn dogs produce a retrograde response in this nucleus, especially the *nucleus cervicalis centralis*, but not in the *nucleus cervicalis lateralis*, as claimed by Rexed & Brodal (1951). Petras investigated the spinal origins of spinocerebellar tracts in monkeys by retrograde degeneration and retrograde transport of horseradish peroxidase, with similar results.

Some neurons contributing to spinocerebellar tracts have been subjected to intracellular recording. Some receive only one type of exteroceptive or proprioceptive afferent, other neurons show the convergence and interaction of different afferents. Parameters of transmission are modifed by activity in the descending brainstem tracts.

(*c*) **The lateral spinothalamic tract**, medial to the anterior spinocerebellar in the lateral funiculus, is continuous with the anterior tract (p. 932). As in the latter (in the anterior white column, 7.69), the simple zonal, 'modality-specific' pattern still widely ascribed to both tracts contrasts with the variation in physiological responses of cells investigated in laminae IV–III, some probably contributing spinothalamic fibres. Variations include marked differences (pp. 944-945) in the receptive fields of different neurons and in the degree of convergence of different primary afferent fibres on one interneuron. At least some spinothalamic fibres may transmit complex patterns; that they are all modality-specific is a simplification. Nevertheless, somatotopic lamination in the anterior is also evident in the lateral tract (an arrangement of surgical importance) and maintained throughout the medulla oblongata and pons. In the midbrain, fibres from the lower limbs spread dorsally and here it is claimed possible to divide pain and temperature fibres from the upper limb and trunk without injury to those from the lower (Earl Walker 1942).

Though the lateral spinothalamic tract is accepted as a major path serving somatic pain and thermal sensibilities, an alternative route is suggested in a series of intersegmental fibres from neurons in the spinal grey matter (p. 936). The spinotectal tract has also been regarded (Earl Walker 1943) as an alternative for pain and thermal sensibility.

Brief reference must be made to the *spinocervical tract* and an associated *spinocervico-thalamic 'system'* of connections. The *lateral cervical nucleus*, small in man, is in the lateral funiculus lateral to the dorsal grey column in the first two cervical segments (Truex et al 1970). The tract is derived from neurons in the column, particularly laminae IV–VI, among those of the spinocerebellar tracts. It ascends from lumbosacral levels. Its projection from the lateral cervical nucleus was first said to be cerebellar but more recent anatomical and physiological data indicate that its axons ascend with the medial lemniscus to the nucleus ventralis posterior lateralis of the contralateral thalamus, mediating light tactile and pressure sensation (perhaps also joint sensibility); hence they are to be associated with the spinothalamic tracts (Morin & Catalano 1955, Ha & Liu 1966). Some of its neurons also respond to noxious stimuli (Brown & Gordon 1977). Gwyn & Waldron (1968) have described a column of similar neurons in rats throughout the spinal cord.

Little is known of routes of impulses arising in painful pathological conditions of viscera. Primary axons travel in splanchnic nerves and almost certainly enter the spinal cord via *white rami communicantes* and *dorsal spinal roots*. Whatever their subsequent path, all are interrupted by the bilateral division of the lateral funiculi in the first thoracic segment (Hyndman & Wolkin 1943), suggesting that they are in the lateral spinothalamic tract.

Since fibres of the lateral spinothalamic tract cross at once, lesions affecting the commissural regions, e.g. *syringomyelia*, produce bilateral loss of pain and thermal sensibilities in areas represented in the diseased segments.

Note that the spinal lemniscus, usually defined as the superior continuation of both spinothalamic tracts, is now more accurately described as continuous only with the lateral tract.

(*d*) **The spinotectal tract** is medial to the anterior spinocerebellar and anterior to the lateral spinothalamic, the three remaining intimately related throughout their ascent in the cord and brain stem. Its fibres start deep in the contralateral grey matter and soon cross to the opposite lateral funiculus. Most easily identified at cervical levels, its fibres ascend to the midbrain to end in the superior colliculus, providing routes for spinovisual reflexes, the superior colliculi being the reflex stations in the visual path but not concerned with transmission to the cerebral cortex. Spinotectal impulses evoke movement of the head and eyes towards the source of stimulation. Some spinotectal fibres reach the ventral thalamus and may mediate cutaneous pain and perhaps other modalities.

(e) **The dorsolateral tract** is a small strand of fine myelinated and non-myelinated fibres between the apex of the posterior grey column and the surface of the spinal cord, close to the dorsal roots; it is formed in part by axons from the lateral bundles of dorsal roots, which bifurcate into ascending and descending branches. The former ascend one or two segments in the tract, giving collaterals to and terminating around neurons in the posterior grey column (pp. 941, 944). Though the tract exists at all levels, no fibres travel more than a few segments; it is hence a *fasciculus*, a term all the more desirable in view of experimental confirmation (Poirier & Bertrand 1955, Earle 1952, Szentágothai 1964) of an old suggestion (Flechsig 1876) that it contains many propriospinal fibres, the contribution from dorsal root fibres being no more than 25% at any level (Earle 1952). (Flechsig's suggestion considerably predated Lissauer's description in 1886.) Many of its fibres are from small neurons of the substantia gelatinosa (laminae II and III, p. 939).

(f) **Spinoreticular fibres**, described in the feline lateral funiculus (Brodal 1949), intermingle with spinothalamic fibres. Few of these ascending fibres cross in the cord but appear to project bilaterally to brain-stem reticular nuclei at all levels, particularly in the pons and medulla. The existence of such connections is claimed in the human brain stem (Mehler 1962).

(g) **The spino-olivary tract**, described as arising from neurons in the deeper laminae of spinal grey matter, decussates, then ascends superficially at the junction of the anterior and lateral funiculi and ends in the 'spinal' regions of dorsal and medial accessory olivary nuclei (Mizuno 1966). The tract probably mediates cutaneous and proprioceptive receptors and those in the muscle and tendon (Grant & Oscarsson 1966). A functionally similar route, *the dorsal spino-olivary*, has been deduced from physiological data as ascending with the dorsal white funiculi, relaying in dorsal column nuclei and projecting to accessory olivary nuclei (Oscarsson 1967). Interest in such paths has been augmented by increasing indications of olivocerebellar climbing fibres with specific localized excitatory effects on the Purkinje cells (pp. 965, 968).

3. Intersegmental Tracts

The lateral intersegmental tract fills the rest of the lateral funiculus, separated ventrally from the *anterior* intersegmental by emerging fibres of ventral spinal nerve roots. It contains intersegmental fibres, some crossed, and possibly reticulospinal and descending autonomic fibres.

TRACTS IN THE POSTERIOR FUNICULUS

1. Ascending Tracts

This funiculus contains the large fasciculi gracilis and cuneatus, separated by the postero-intermediate septum. The two dorsal funiculi connect dorsal root ganglia *and* posterior grey columns to the gracile and cuneate nuclei; but many fibres do not reach the nuclei and many are *propriospinal*. Some axons come by other routes.

The fasciculus gracilis, commencing at the caudal end of the spinal cord, contains mainly long ascending branches of axons of medially placed bundles of fibres of dorsal spinal roots; but a considerable fraction (15% according to Rustioni & Dekker 1974) are axons of *secondary* neurons in the ipsilateral dorsal grey column (Laminae III and IV). All ascend the posterior funiculus, augmented by each dorsal root. Fibres entering in coccygeal and lower sacral regions are pushed medially by successive additions. The fasciculus gracilis, derived from the lower thoracic, lumbar, sacral and coccygeal segments, is medial to the fasciculus cuneatus, in the posterior funiculus of the upper spinal cord (7.70, 71).

The fasciculus cuneatus, commencing at mid-thoracic level and derived from the upper thoracic and cervical dorsal spinal roots, is lateral to the fasciculus gracilis. Some of its axons are secondary, from neurons in laminae III and IV of the ipsilateral dorsal grey column.

Both fasciculi are myelinated, the cuneate fibres being larger probably because proprioceptive fibres (of great diameter) remain in the cuneate but leave the gracile fasciculus at lumbar levels (vide infra). Both contain central processes of neurons in spinal ganglia, i.e. *primary afferent* neurons, some of which ascend *without synapse* or *decussation* to the medulla oblongata, ending in *gracile* and *cuneate nuclei*, in which secondary neurons begin; most of their secondary fibres sweep ventrally round the central grey matter (7.89) as *internal arcuate fibres* and part of the *lemniscal decussation*. Thereafter, as *medial lemnisci*, they ascend on each side to the ventral thalamic nucleus (p. 1002) and are relayed there to the post-central gyrus (areas 3, 1 and 2). Some secondary neurons form *posterior external arcuate fibres* (p. 953), entering the cerebellum. Presynaptic dendrites, making dendro-dendritic connections, have been described in the cuneate nucleus of macaques (Wen et al 1977) but their origin remains obscure.

These two tracts, almost filling the posterior funiculus, mediate proprioception, including vibration, pressure and some exteroceptive *tactile* information, fibres for which all reach the medulla oblongata in their ipsilateral posterior funiculi, with fibres concerned with posture and movement, active and passive. In the cord some fibres ascending in the dorsal white columns send collaterals into the dorsal grey column.

The somatotopic lamination of gracile and cuneate tracts is more intricate than the segmental pattern described above; fibres are also segregated by modality: those from hair receptors being the most superficial, those from tactile and vibrational receptors in the deeper layers (Uddenberg 1968). Experiments show that this pattern reaches the nuclei gracilis and cuneatus; these contain numerous interneurons, unit recording from which indicates a high specificity in fibres of both tracts. Stimuli just outside the

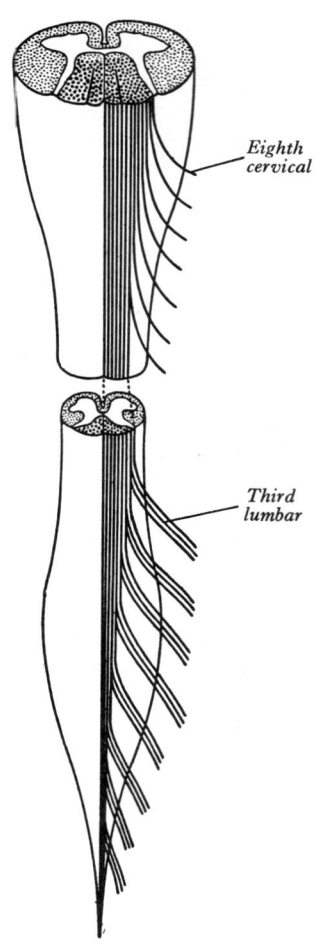

Eighth cervical

Third lumbar

7.71 The lamination of the fibres in the posterior funiculus. The spinal cord is viewed from the dorsal aspect. The drawing shows that the posterior funiculus is formed (in part) by the long ascending fibres of the dorsal roots and that the sacral fibres adjoin the median plane, the lumbar to their lateral side, the thoracic more laterally and the cervical most lateral of all. For dorsal fibres arising elsewhere see p. 935.

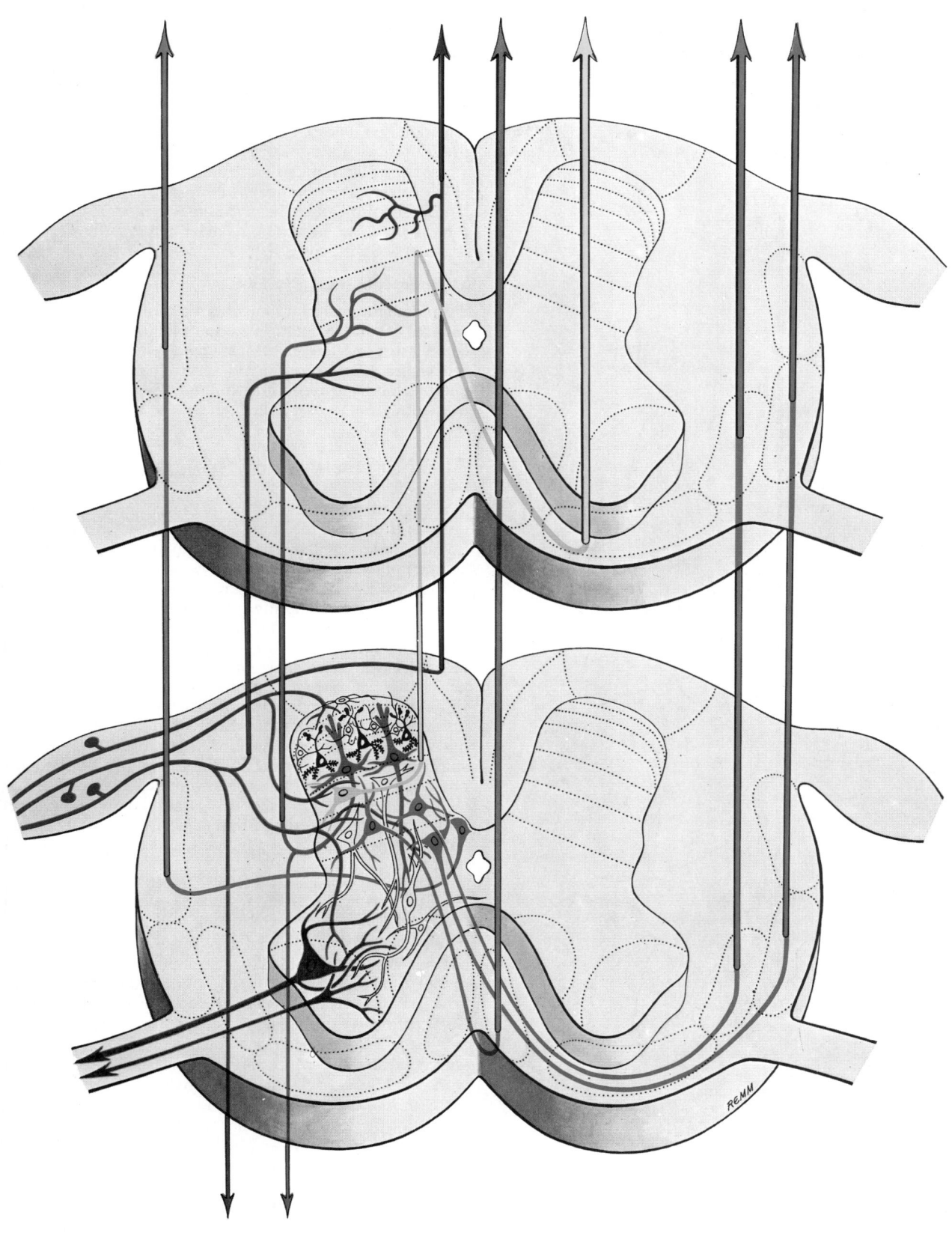

7.72A Simplified scheme of some of the major ascending tract systems of the spinal cord and some features of grey matter organization. Within the grey matter the dotted lines show the laminar pattern; within the white matter they are an approximate guide to the topography of the tracts. Attempts have been made to indicate in a simplified manner the overlapping of dendritic fields described in the text. An alpha and a gamma motor neuron (grey) are included, together with some of the structural features of the substantia gelatinosa which are described and illustrated more fully in 7.75. Some of the small substantia gelatinosa neurons are uncoloured, as are some interneurons in the deeper laminae. The larger substantia gelatinosa neurons are solid black. Large lamina IV cells and associated ascending and descending intersegmental fibres are green. Primary sensory afferent fibres, including a fibre in the fasciculus gracilis, are purple. Spinotectal fibre—orange; anterior spinothalamic fibre—yellow; lateral spinothalamic fibre—magenta; and dorsal and ventral spinocerebellar fibres—blue. Compare with 7.73.

receptive field of a single unit may inhibit. Fibres from pre- and post-central cerebral cortex, descending in corticospinal projections, may facilitate or inhibit gracile and cuneate interneurons. Similar projections originate in brain-stem reticular nuclei (pp. 988–996). Such effects probably contribute to discrimination in the dorsal funicular paths; these tracts, like others, are not simple sensory routes (p. 944).

2. Descending Tracts

Confusingly numerous small descending tracts have been described in the dorsal funiculi, chiefly in the late nineteenth

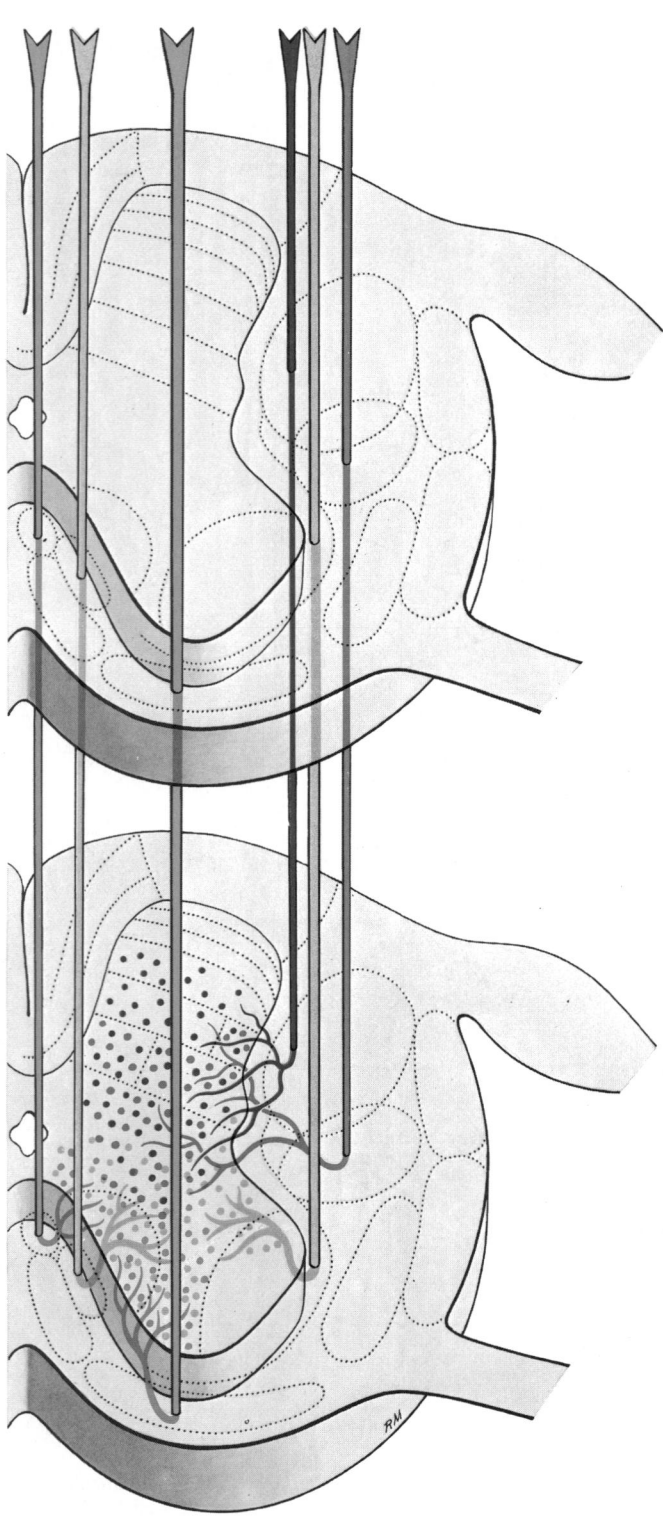

7.72B A simplified scheme of some of the major descending tract systems of the spinal cord including their overlapping zones of termination in the grey matter. The significance of the dotted lines is the same as for 7.72A. Corticospinal tract—mauve; rubrospinal tract—magenta; reticulospinal tracts—yellow; vestibulospinal tracts—blue.

century, in accounts based on pathological appearances in human spinal cords; e.g. in cervical and upper thoracic levels, medial in the cuneate tract, is the *comma tract* (Schultze), *semilunar tract* or, more recently, *interfascicular fasciculus*. In the lower thoracic segments a superficial strand, almost certainly the upper end of the *septomarginal tract*, has been ascribed to a deeper site near the posterior median septum. The so-called '*oval field of Flechsig*' and '*triangular field of Gombault and Philippe*' are also different levels of the septomarginal tract, which alters in shape and position remarkably as it descends from the lower thoracic segments. Continuity of the interfascicular fasciculus and septomarginal tract seems likely, but remains uncertain. The former is said to consist only of descending branches of entering dorsal root fibres; the latter may also contain many intersegmental fibres. Almost all reports on these descending dorsal column fibres are clinical (Nathan & Smith 1959).

3. Intersegmental Tracts

Deepest in the posterior funiculus is a small **posterior intersegmental tract**, crescentic in section and just posterior to the grey commissure (7.68, 69A,B). Most obvious in the lumbar region, it can be traced to thoracic and cervical levels. Its fibres, from neurons in the posterior grey column, divide into ascending and descending branches which re-enter and ramify in grey matter.

Further Aspects of Spinal Organization

The preceding introduction to organization of the spinal cord must now be expanded and partly modified in the light of re-investigation by old and new techniques, both neuroanatomical and neurophysiological (Eccles & Schadé 1964a, b, Lissák 1967, Ralston 1974). These intensive researches include comparisons in many vertebrates, alternative plans of spinal grey matter based on cytoarchitectonics and precise analyses of dendritic and axonal arborizations by modified Golgi techniques. Newer tracing techniques have also augmented our knowledge of the terminations of dorsal root afferents and collaterals, and of descending supraspinal projections. Minute focal lesions have been placed to unravel the intricacies of spinal cord circuitry, while light microscopial techniques have been improved by quantification and detailed ultrastructural studies of synapses. Distribution of putative transmitters has also shown much progress.

This renewed neuroanatomical attack, combined with refined 'unit' and other recording physiological techniques, has proved most stimulating (in both directions, as is apparent in the following brief account).

LAMINAR ARCHITECTURE

The *general outline* of spinal grey regions, as seen in transverse section, has long provided a useful but arbitrary basis for morphology and experiment. Dorsal, lateral and ventral columns and arbitrary subdivisions have been recognized (p. 926) and named. Such entities as the substantia gelatinosa, thoracic nucleus and subgroups of anterior grey column neurons have been named, with a terminology which, if arbitrary, is still used. Much basic work has been carried out within this topographical scheme but increasing analysis, structural and functional, reveals a lack of precision, prompting attempts to re-describe cordal structure (Rexed 1952, 1954, 1964). The currently adopted scheme stems initially from studies in thick and thin sections, stained for neuronal somata in feline spinal cords, neonatal, young and adult. Based on neuronal size, shape, cytological features and density in different regions, nine *laminae* have been distinguished, arrayed more or less parallel with the dorsal and ventral limits of the grey matter, at all levels, with a distinctive pericanalicular region. These laminae, as they appear in the transverse section of a fifth lumbar segment, are shown in 7.74. General confirmation of the laminar pattern in *human* material has been provided by Schoenen (1973). Briefly the structure of laminae is as follows:

Lamina I, a very thin layer with an ill-defined boundary, adjoins the white matter (which outlying neurons invade) and has a reticular appearance due to bundles of coarse and fine nerve fibres intermingling in many directions. It contains small, intermediate and large neuronal somata, many fusiform. Alternative names are *lamina marginalis* or *layer of Waldeyer* (who recognized a similar zone in 1888).

Lamina II contains densely arrayed small neurons crossed, especially medially, by numerous bundles of fibres from the dorsal funiculus. An outer (dorsal) zone, darkly stained due to its high density of the smallest somata, contrasts with an inner (ventral) pale zone.

Lamina III consists of somata mostly larger, more variable and less closely arrayed than those in lamina II.

Lamina IV, a thicker layer, is a loosely packed, heterogeneous zone permeated by fibres. Neuronal somata vary much in size and shape, from small and round, through intermediate triangular, to very large stellar types.

Laminae I–IV correspond to the dorsal column's *head*. Lamina II (including part or all of lamina III) corresponds to the *substantia gelatinosa*; the ill-defined *nucleus proprius* of the dorsal grey column corresponds partly to laminae III and IV.

Lamina V, a thick layer including the dorsal column's *neck*, is divisible into a lateral third and medial two-thirds. Both have a mixed population but the former contains many prominent well-stained somata interlaced by numerous bundles of transverse, dorsoventral and longitudinal fibres: hence the term 'formatio reticularis' for this region, particularly prominent at cervical levels and recognized for over a century (Deiters 1865). (The term is now used more widely, pp. 988–996.)

Lamina VI is most prominent in the limb enlargements, particularly in young animals; it has a medial third of small densely packed neurons and a lateral two-thirds containing larger, more loosely packed, triangular or stellate somata. The medial hence stains more heavily, in contrast to lamina V, where the converse applies. Lamina VI corresponds approximately to the *base* of the dorsal column.

Laminae VII–IX show a variety of complex forms in limb enlargements (7.74A); to assist explanation, the simpler arrangement at thoracic levels is included for comparison (7.74B).

Lamina VII includes much of the intermediate grey column. It contains prominent neurons of the *thoracic nucleus* and *intermediomedial* and *intermediolateral columns* at their spinal levels (p. 927). Remaining areas (between these columns and, in limb enlargements, between lamina VIII and groups of IX) contain a uniform array of medium-sized triangular or stellate somata.

Lamina VIII spans the base of the thoracic ventral column but is restricted in limb enlargements to its medial aspect. It contains a mixture of somata.

Lamina IX is a complex array of columns (p. 927) including very large somata of α motor neurons and numerous smaller ones. Smaller somata include motor neurons with small-diameter efferent fibres (γ efferents) for muscle spindles and numerous interneurons, some perhaps being inhibitory Renshaw cells (vide infra). The location of γ motor neurons was long in doubt but studies of the retrograde changes following peripheral trauma (Nyberg-Hansen 1965) and intracellular recording (Eccles et al 1960) show them to be dispersed among α neurons in the motor columns. The sites and features of Renshaw cells remain uncertain. Intracellular recordings indicate inhibitory interneurons in the ventral extension of lamina VII, insinuated between VIII and IX (Willis & Willis 1966). Golgi staining in this region (Scheibel & Scheibel 1966) has not revealed typical Golgi type II neurons, long assumed to be the basis of 'Renshaw loop' inhibition. Nevertheless, Szentágothai (1967) considers that neurons with longer axons are not incompatible with inhibitory function and that the extension of lamina VII receives most initial collateral branches from axons of α motor neurons (assumed to synapse with Renshaw cells).

Lamina X surrounds the central canal, including the *dorsal and ventral grey commissures* and the *substantia gelatinosa centralis*.

FURTHER ASPECTS OF SPINAL LAMINAR ARCHITECTURE

The preceding description outlines a new scheme of topography; further details will be found in the original papers quoted. (The scheme was originally applied only to feline spinal cords, with limited confirmation in mankind (Schoenen 1973). Similar laminar organization is *expected* to apply to the spinal cords of *all* higher mammals (Rexed 1964). The originator proposed the following tentative functional analysis (though some conclusions have been revised, vide infra).

Laminae I–IV are the main receiving areas for cutaneous primary afferent terminals and collateral branches (vide infra). From this region start many complex polysynaptic reflex paths, ipsilateral and contralateral, intrasegmental and intersegmental; and from it many long ascending tracts to higher levels are considered to arise (but vide infra).

Laminae V and VI receive most of the terminals of proprioceptive primary afferents and profuse corticospinal projections from the motor and sensory cortex and subcortical levels, suggesting intimate involvement in the regulation of movement.

Lamina VII (lateral part) has extensive ascending and descending connections with the midbrain and cerebellum (via the spinocerebellar, spinotectal, spinoreticular, tectospinal, reticulospinal and rubrospinal tracts) and is hence involved in the regulation of posture and movement. Its medial part has numerous propriospinal reflex connections with the adjacent grey matter and segments concerned both with movement and autonomic functions (p. 1011). Its ventral extension may contain inhibitory interneurons (p. 865, Renshaw 1941, 1946).

Lamina VIII, a mass of propriospinal interneurons, receives terminals from the adjacent laminae, many commissural terminals from the contralateral lamina VIII and descending connections from the interstitiospinal, reticulospinal and vestibulospinal tracts and the medial longitudinal fasciculus. Their axons influence motor neurons of both sides, perhaps directly but more probably by excitation of small neurons supplying γ-efferent fibres to muscle spindles.

Lamina IX contains α and γ motor neurons and many interneurons. Large α *motor neurons* supply motor end-plates of extrafusal muscle fibres in the *motor units* of striated muscle (p. 556); they vary in size. Recording techniques have demonstrated *tonic* and *phasic* α neurons (Granit et al 1956). The former, with a lower rate of firing and lower conduction velocity, are assumed to be smaller. Attempts to correlate these varieties with structural and functional types of striated muscle fibre, e.g. *slow* and *fast* (p. 555) have been frustrated by lack of histological differentiation. Some large motor neurons (β *motor neurons*) have been held to supply extrafusal *and* intrafusal fibres.

There are also several functionally distinct types of γ *motor neuron*, fusimotor fibres of which innervate the intrafusal fibres in muscle spindles. The 'static' and 'dynamic' responses of muscle spindles (p. 556) have separate controls mediated by *static* and *dynamic fusimotor fibres* distributed variously to *nuclear chain* and *bag* fibres; but it is impossible histologically to differentiate types of γ motor neuron somata, though *plate* and *trail endings* have been recognized and two varieties of γ-efferent myelinated nerve fibre (γ *1* and γ *2*) in nerves to muscle (Boyd 1962). Correlation of these varieties is disputed (Barker 1974).

GEOMETRY OF SPINAL NEURONS

Since Cajal's classic investigations (1908), foreshadowing so much and still a primary source of information, increasingly precise data on spinal neurons have accumulated: on sizes, shapes, distribution, packing density and the cytological features of neuronal *somata*, as shown by the Nissl technique or its modifications. However, such techniques are limited, requiring complementary techniques permitting analyses, preferably quantitative, of the ramifications, arborizations and terminations of axons, dendrites and collateral branches. Degenerative and other techniques (p. 870) are also necessary for the precise determination of connections.

Quantitative analyses of lumbosacral regions of feline spinal cord were made by Aitken & Bridger (1961) and Schadé (1960), who found that in the *anterior zone* of the ventral grey column of one spinal segment (lumbar 6) the neuronal population totalled about 7000; 700 were considered small neurons of γ-efferent fibres to muscle spindles; 3276 larger ones were propriospinal

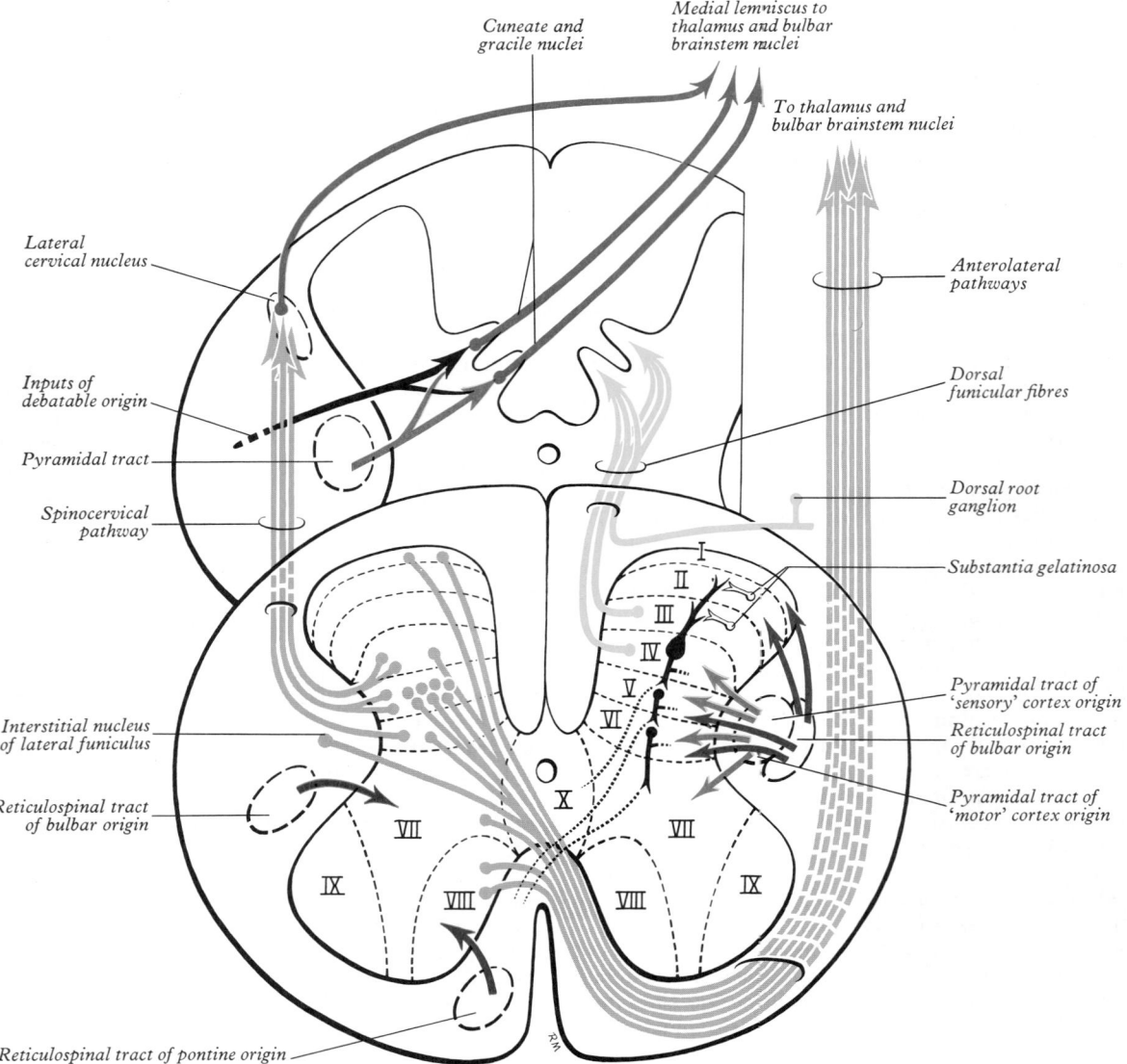

7.73 A more detailed analysis of the principal somaesthetic pathways. Descending corticospinal and reticulospinal tracts involved in sensory modulation are also indicated. Modified from data provided by K E Webster, King's College, University of London.

(interneurons), 126 were 'spinal border' cells and 2898 α motor neurons. Thus, about half were interneurons and half motor neurons; in the whole ventral column, the ratio was about seven interneurons to one motor neuron and, with the intermediate zone and base of the dorsal column included, the ratio rises to 13 to 1. Neuronal density was highest in the intermediate zone (7 cells per 100 μm cube of tissue), lowest in the ventral column (1–2 cells per 100 μm cube). The *total* surface area of cell somata and their dendrites ranged from 11 000 to 97 000 μm², the surface area of *cell somata* alone ranging up to 25 000 μm²; stem dendrites varied to a maximum of 13, *dendritic* surface area ranged up to about 76 000 μm². Dendritic surfaces formed about 80% of the total *receptive* neuronal surface (i.e. excepting axonal surface), dendrites sometimes extending as far as 1000 μm from the parent soma, where terminal dendrites *entered* the adjacent white matter; as much as 50% of the estimated receptor surface was more than 300 μm from the parent soma. The latter figure is especially interesting; many types of neuron have very wide dendritic trees, but neurophysiological evidence (Eccles 1957) has cast doubt on the possible effectiveness of synapses situated more than 300 μm from the soma; their functional role remains to be determined. Schadé (1964) estimated separate percentages of grey matter occupied and the surface areas of somata and dendrites. A mean value of 7500 μm³ was estimated for the volume of international soma and 29 000 μm³ for a motor neuron. The surface area of the 'average' soma ranged from 4.4 to 5.9 × 10³ μm² and that of the 'average' dendritic tree from 59 to 73 × 10³ μm².

Other investigations (Romanes 1964, Sprague & Ha 1964) have also examined the complex dendritic patterns of ventral column motor neurons and interneurons in adjacent cell laminae, emphasizing the wide dendritic trees of motor neurons, better appreciated in transverse *and* particularly longitudinal sections. Not only do dendrites of the adjacent *cells* interweave *horizontally* in complicated patterns, but they also spread even more widely, overlapping the territories of other motor neuronal *columns*; some penetrate adjacent cell laminae (VII and VIII), interweaving with dendrites and somata, others pass deeply into the white matter between its longitudinal myelinated nerve fibres, ending at varying depths (sometimes traversing one-third to one-half of the white matter). Even more marked is the *longitudinal* spread of dendrites from α motor neurons *in* their own column, overlapping the territories of hundreds of adjacent motor neurons. Such data again call in question the functional role of axodendritic synapses remote from a neuron's soma. The causes of circumscribed grouping of motor neuron *somata* contrast with those involved in their *dendritic* interlacing. The latter's significance is obscure but interpretation of the patterns of overlap has been attempted in relation to the arrays of primary afferent terminals from muscle and to interneurons in the adjacent cell laminae, both of which spread longitudinally in three or four segments (Sprague & Ha 1964, Sterling & Kuypers 1967). Scheibel & Scheibel (1968) have analysed the dendritic fields in spinal grey matter, Szentágothai & Albert (1955) and Bohme (1962) have similarly examined Clarke's column, and Szentágothai (1964) and Ralston (1965) have analysed the substantia gelatinosa. Ralston (1974) has reviewed the field.

The general neuronal arrangement in the **substantia gelatinosa** is as follows: neurons of thin lamina I usually have large dendritic trees spreading tangentially across the dorsal aspect of the posterior grey column, mixing with the deepest fibres of the dorsolateral tract and some dorsal nerve root afferents; their axons mainly ramify in the subjacent laminae, but longer ones decussate and ascend to the brain stem and diencephalic levels in the ventrolateral white columns (p. 935). Dendrites of small neurons in lamina II and larger ones in lamina III are predominantly radial, occurring in longitudinal 'sheets' perpendicular to the dorsal column; a few may pass ventrally to the deeper laminae but most remain in the substantia gelatinosa (vide infra). These fine axons ascend or descend a little in lamina II or pass into the dorsolateral tract and return to either lamina II or III before ramifying. Between the radial dendrites of small neurons are two other radial components: the terminal parts of long dendrites of larger, more deeply sited neurons in lamina IV and terminals of many primary cutaneous dorsal root fibres. The fine non-myelinated afferents approach the substantia gelatinosa from its dorsal aspect, their terminal branches travelling ventrally; afferents of larger diameter curve round the substantia to its ventral aspect.

Another detailed study of the substantia gelatinosa (Réthelyi & Szentágothai 1969) was based on Golgi technique and ultrastructure combined with dorsal root transection or isolation of the dorsal grey columns. Neurons in laminae II and III were examined with special reference to the larger pyramidal neurons at the junction between laminae III and IV; these have recurrent axons entering lamina II, where they expand as cores of large *glomerular synaptic complexes*, containing axodendritic synapses between axons of pyramidal neurons and dendrites of small gelatinosal somata, and axoaxonic contacts between 'pyramidal' axons and terminals of dorsal root afferent fibres. Features of these synapses, with details of inputs from primary dorsal root afferents, axonal ramifications and terminals of gelatinosal neurons, main output channels via lamina IV neurons and terminals of fibres from supraspinal sources, are summarized in illustration 7.75. Further researches show that the detailed roles of the substantia gelatinosa in part remain obscure (Nathan 1976, Wall 1978), but also vide infra.

TERMINATIONS OF DORSAL ROOT AFFERENT FIBRES

The formation, topography and division of dorsal spinal roots have been described (p. 927) and confirmed in man (Sindou et al 1974). Terminations of these in spinal grey matter must now be detailed. Approaches to this problem have been anatomical and neurophysiological. In the former terminations at different levels have been followed by such techniques as those of Glees, Nauta and Fink-Heimer after the severance of individual dorsal roots. In the latter, the focal electrical potentials generated in spinal grey matter have been explored by microelectrodes to record the results of stimulation of the muscle, cutaneous nerves and dorsal roots. Most of the informative studies of dorsal root terminals (Sprague 1958, Sprague & Ha 1964) were on feline spinal cords; similar arrays probably exist in other mammals but *details* cannot be transferred without reserve to the human spinal cord. The findings were related to laminar architecture (p. 938), with the proviso that the degenerating terminals seen are only those near cell somata and dendrite trunks, although the branches of many dendrites radiate widely (p. 940). Dendrites are covered by synaptic contacts along all their course (Armstrong et al 1956, Wyckoff & Young 1956, Rasmussen 1957, Young 1958, Illis 1964, Gelfan & Rapisarda 1964), numbering thousands in large neurons, but these are *not* revealed by the degeneration techniques used and such studies must be combined with others, including Golgi techniques.

The fields of termination, after section of cats' sixth right lumbar dorsal roots, are shown at three levels (lumbar 5, 6 and 7) in 7.76A,B,C. All large dorsal funicular fibres (except some placed medially) have collaterals traversing the medial two-thirds of laminae I, II and III, many curving around their medial aspects to form a dense plexus of degenerating fibres of passage and terminals round most neurons in lamina IV. Many fibres of passage recurve into the substantia gelatinosa from its ventral aspect, showing as degenerating terminals between the radial dendrites of small neurons in laminae II and III and terminal segments of long dorsally directed dendrites from lamina IV. Degeneration also occurs in some fine fibres of the dorsolateral tract through three segments in both directions; from these, collaterals pass into laminae I, II and III, appearing in all as degenerating terminals. From lamina IV many larger fibres pass to the medial zones of V and VI (containing commissural interneurons), many others forming a mass of degenerating terminals in the central zones of these laminae. From this concentration 'fingers' of degenerating terminals radiate into laminae VII and VIII; running with them are fibres of passage converging on motor neuron pools and their interneurons in lamina IX. Sprague (1958) demonstrated degenerating synaptic terminals from dorsal root afferents terminating on the somata and dendrites of large multipolar motor neurons, interspersed between numerous terminals from interneurons in other laminae and levels. In cats such monosynaptic terminals (of dorsal root afferents) on motor neurons extended two segments above and below a severed root; terminals on interneurons in other laminae were identified one or two segments beyond this. Similar studies in rhesus monkeys (Shriver et al 1968, Carpenter et al 1968) show similar findings but differ in detail.

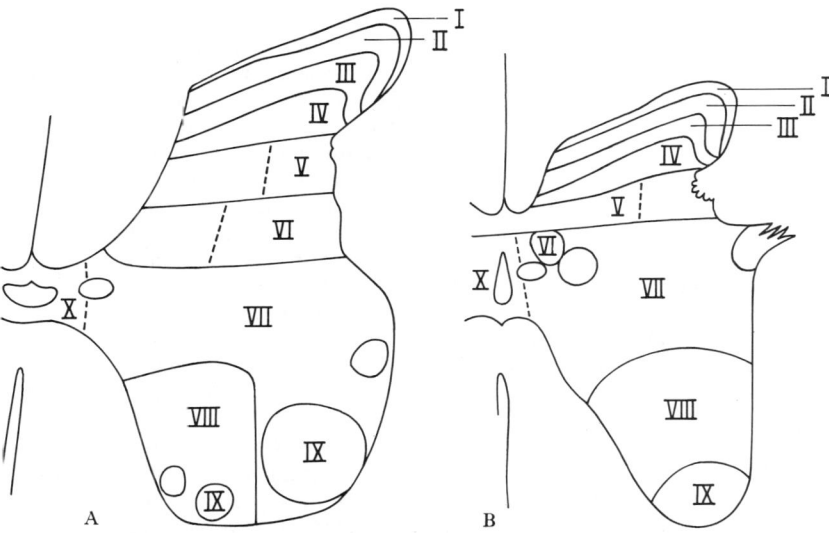

7.74 The pattern of lamination proposed by B Rexed (1964) for the spinal cord grey matter of the cat, viewed in transverse section: A. the fifth lumbar segment, B. the third thoracic segment. Reproduced with the permission of B Rexed and Elsevier.

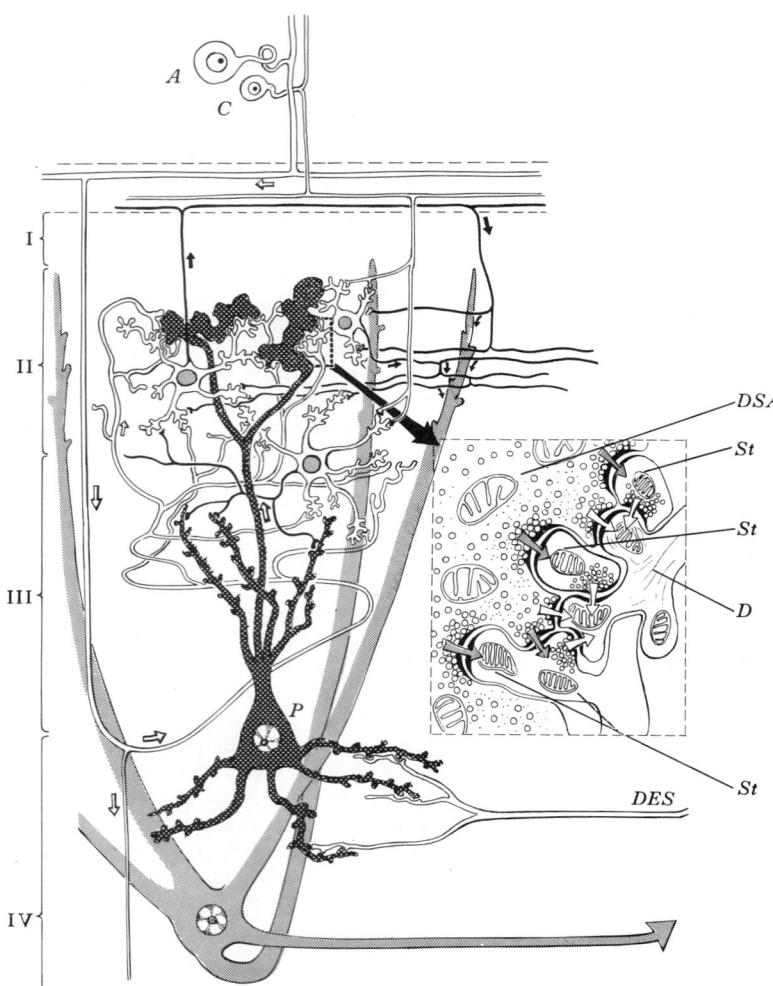

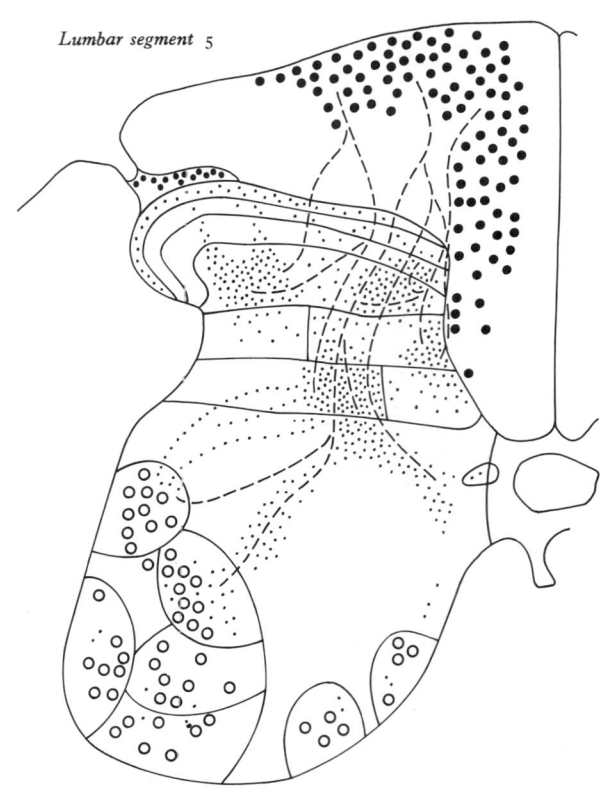

Lumbar segment 5

A

7.75 The arrangement of neurons and their interconnections in a longitudinal section through the dorsal grey column of the spinal cord, which includes the substantia gelatinosa and adjacent neuronal laminae (Roman numerals I–IV). Inset (right) shows synaptic detail of the area indicated. Two primary sensory afferent fibres are shown: a cutaneous afferent of large calibre (A) and a small calibre non-myelinated afferent (C). Small substantia gelatinosa interneurons are in white with black outline; their axons are single black lines. A pyramidal cell with dendrites and spines and a recurrent axon expanding into synaptic complexes is shown in black with white dots. A large multipolar neuron of lamina IV with long radial dendrites and initial axon is in cross-hatch. DES = axon descending from a supraspinal source. Arrows on main diagram show presumed direction of impulse conduction. Note (1) the different sites of synaptic termination of primary afferents A and C, (2) the axonal pattern and terminals of substantia gelatinosa interneurons, (3) the synaptic complexes formed between the recurrent axon expansions of the pyramidal cell, the small gelatinosa interneuron dendrites and the primary sensory axon terminals. See inset for details. DSA = pyramidal cell axon terminal; D = dendrite of gelatinosa interneuron; St = primary afferent fibre axon terminal; white arrows = axo-dendritic synapses; cross-hatched arrows = axoaxonic synapses. Consult text for a discussion of the 'gate' theory and structural details. From Szentágothai 1975.

7.76 The pattern of degeneration of nerve fibres and their terminals, demonstrated with the Nauta-Laidlaw technique in the ipsilateral half of the spinal cord at various segmental levels, five days after surgical division of the *sixth* lumbar dorsal spinal nerve root of the cat. A. The fifth lumbar

In recent years many attempts to correlate architectonics of spinal cord grey matter have been based on classic degeneration techniques and the use of tracers (p. 870), with more critical assessment of the terminations of tracts and complexities of propriospinal neurons, combined with neurophysiological experiments. Findings by Eccles (1957), Eccles & Schadé (1964a, b), Ralston (1974) and others are summarized below.

NEUROPHYSIOLOGICAL CORRELATIONS

In addition to being a prime site for degeneration studies (p. 870), large multipolar neurons of motor columns in lamina IX were the targets for classic studies of reflex responses (Sherrington 1906), from which emerged the first clear evidence of *integration* of contacts between neurons, and concepts of *excitation* and *inhibition*.

Much later their large size made them most suitable for microelectrode recording; also they can be excited *orthodromically* by stimulation of the peripheral nerves or dorsal spinal roots and invaded *antidromically* by stimulation of the ventral roots. Thus detailed analyses are possible of electrical and ionic events at synapses during the generation of *excitatory* and *inhibitory postsynaptic potentials* (EPSPs and IPSPs). Similarly, recognition of adjacent inhibitory interneurons (Renshaw 1941, 1946) has prompted enquiry into *lateral* and *feedback* inhibitory phenomena.

Since 1940 evidence of *presynaptic inhibition* has grown; in this spinal neurons have also been a major experimental field (Eccles 1964). Synaptic terminals (A) impinging on a neuron (B) are themselves subject to synaptic terminals (C) from a nearby interneuron, i.e. these are axoaxonic synapses. When active, inhibitory terminals (C) are said to cause relative depolarization of terminals (A), reducing their power to cause postsynaptic change in the neuron (B). Usually two or more interneurons are thought to exist in a presynaptic inhibitory pathway. Like postsynaptic excitation and inhibition, presynaptic inhibition may be effected by descending fibres from supraspinal sources.

Once regarded as uncommon, *presynaptic inhibition* has been revealed in relation to many, possibly all primary afferent terminals. Thus sensory information entering the central nervous system does not simply reflect environmental changes but is continually modified at the first synapses depending on local conditions in the grey matter. A much investigated site of presumed presynaptic effects is in laminae II and III, the substantia gelatinosa (Wall 1964, Mendell & Wall 1964, Melzack & Wall 1965, Mendell 1966, Heimer & Wall 1968, Ralston 1974, Nathan 1976, Wall 1978, 1984). As noted elsewhere (p. 944) it has been proposed that impulses from cutaneous (and other) afferents are here subjected to *tonic control* involving the relative depolarization and hyperpolarization of primary afferent terminals, mediated by small and pyramidal neurons of the substantia gelatinosa; but precise

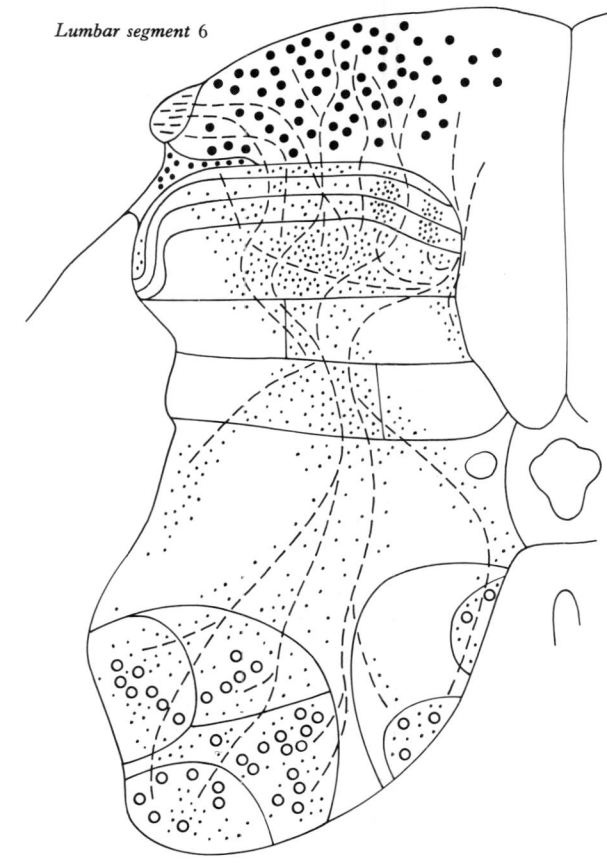

Lumbar segment 6

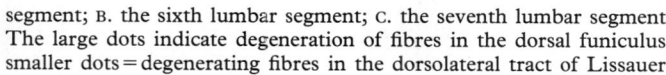

B

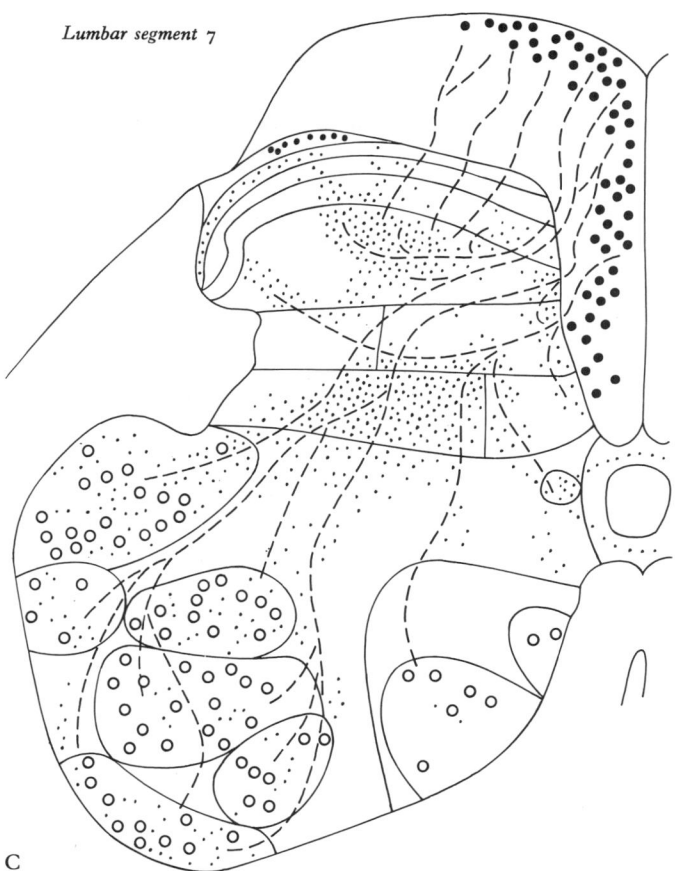

Lumbar segment 7

C

segment; B. the sixth lumbar segment; C. the seventh lumbar segment. The large dots indicate degeneration of fibres in the dorsal funiculus; smaller dots = degenerating fibres in the dorsolateral tract of Lissauer; smallest dots = degenerating nerve terminals; dashed lines = degenerating fibres of passage; circles = neuronal somata of motor neurons. From Sprague & Ha 1964, with the permission of the authors and Elsevier.

synaptic events are still under investigation (Réthelyi & Szentágothai 1969). A *gate control theory*, first proposed by Melzack & Wall (1965) as a possible gelatinosal mechanism providing a 'gate' to inflow along nociceptive and other afferents, has promoted much discussion. The *mechanisms initially* envisaged are summarized in **7.77**.

It *was* proposed that large-diameter afferents (e.g. from hairs and touch corpuscles) are excitatory to gelatinosal interneurons (SG) and larger neurons (T cells) of lamina IV (from which spinothalamic fibres arise), whereas fine non-myelinated afferents are excitatory to T cells but inhibitory to SG cells; axons of the latter were presumed to inhibit presynaptically terminals of all afferents synapsing with T cells. In such a system, low activity in fine afferents inhibits SG cells so that they are prevented from inhibiting T cells; hence the 'gate' to T cells in lamina IV is *open* to transmit intermittent volleys from large fibres. A prolonged high-frequency volley in large-diameter afferents, however, is transmitted to lamina IV T cells initially, but this soon ceases as activity in SG cells *closes* the gate. Conversely, persistent high activity in fine afferents opens the gate to massive bombardment of neurons of lamina IV, which include some neurons of high threshold activated only by such bombardment; it is presumed that onward transmission in the lateral spinothalamic tract will evoke pain at supraspinal levels. Pain would thus result from an *imbalance* between the varieties of afferent impulses when there is abnormally large traffic along the fine afferents, which should perhaps not be regarded as '*specific pain afferents*' but as information mediators of the *state of the tissues* innervated. Thus, in addition to *interaction* and *comparison* of traffic along different fibres, other features of gate control theory involve the *presynaptic inhibition* of primary afferent terminals. Overall 'sensitivity' of the gate may be varied by *descending supraspinal control systems* (vide infra and p. 945).

Since its origin the theory has been progressively refined (Melzack 1973, Wall 1973, 1974, 1976). Nevertheless, Nathan (1976) was prompted to write a lengthy criticism. Subsequently Wall (1978) presented a *re-statement* of the gate control theory of pain mechanisms: 'In 1965, we proposed that transmission of information about injury from the periphery to the first central neurons was under control. The setting of this control or gate was influenced by peripheral afferents, other than those which signalled injury. The gate was also influenced by impulses descending from the brain. Subsequent work has fully supported and enlarged this view. All the cells so far discovered which transmit information from nociceptors are inhibited by low threshold afferents and by descending controls. The mechanism by which this control is achieved remains completely unknown. Pre-synaptic inhibition as a phenomenon isolated from post-synaptic inhibition is in doubt. Whether the inhibitions and facilitations are pre-synaptic or post-synaptic or both is unknown. The role of the substantia gelatinosa in any function is unknown. That a gate control exists is no longer open to doubt but its functional role and its detailed mechanism remain open for speculation and for experiment.'

The Endogenous Analgesia System

Since the above was written many data have accumulated concerning a descending pain control system (for an extensive review see Basbaum & Fields 1979, 1984). Three principal, interconnected regions are involved and each receives a variety of afferents and contains an array of neuromediators.

In the midbrain the *periaqueductal grey matter* (PAG) together with the nucleus raphes dorsalis, and part of the nucleus cuneiformis comprise neuron populations containing serotonin, GABA, substance P, CCK, neurotensin, enkephalin and dynorphin. The PAG receive afferents from the frontal somatosensory and cingulate neocortex, the amygdala, numerous local reticular nuclei and the hypothalamus (afferents from the latter are separate bundles carrying histamine LHRH, vasopressin, oxytocin, ACTH, γ-MSH, endorphin, and angiotensin II). From the PAG fibres descend to rhombencephalic centres (and possibly some direct to the spinal cord).

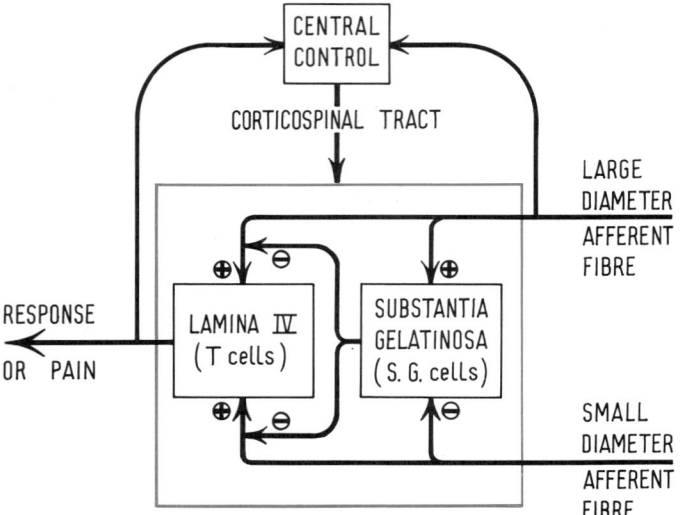

7.77 The sensory 'gate' mechanism as *originally* proposed by R Melzack and P D Wall for the modus operandi of the dorsal laminae of grey matter of the spinal cord. See text for a discussion of the effects of an imbalance in the sensory inflow along the large and small diameter afferent fibres. From an illustration in Melzack & Wall 1965, with permission of the authors and publishers. Consult references for a critical review of the gate theory by Nathan (1976) and a comprehensive reappraisal of the evidence and subsequent re-statement of the theory by Wall (1978).

In the rhombencephalon, nucleus raphes magnus and the medial reticular column is an important multineuromediator centre, containing serotonin, substance P, CCK, TRH, enkephalin and dynorphin. Some neurons contain two or even three neuromediators. Descending bulbospinal fibres pass to the nucleus of the spinal tract of the trigeminal nerve and its continuation, the substantia gelatinosa, throughout the length of the cord. Again there are populations of neurons containing many different neuromediators: GABA, substance P, neurotensin, enkephalin and dynorphin.

There is now abundant physiological and pharmacological evidence that these regions are intimately concerned with nociceptive (and probably other modality) input controls. Their highly complex interactions are, however, under constant reinvestigation and theoretical reappraisal.

Other general electrophysiological investigations have involved the exploration of spinal grey matter by microelectrodes during stimulation of muscle and cutaneous nerves (Eccles et al 1954, Coombs et al 1956) and during 'natural' stimulation of peripheral receptors (Kolmodin 1957, Kolmodin & Skoglund 1960). The focal extracellular potentials thus evoked correlate with the specific laminar regions and terminations of dorsal spinal afferents determined by degeneration techniques.

UNIT RECORDING WITH INTRACELLULAR MICROELECTRODES

Recordings from microelectrodes in neurons deep in the spinal grey matter have provided much information, often unexpected, leading to radical reappraisal of many views. Examples are recorded here of data from cells in laminae IV–VI during natural dermal stimulation, passive limb movements, stimulation of cutaneous, muscular and visceral nerves in various combinations, sometimes with the simultaneous stimulation of tracts from supraspinal sources or after their removal (Wall 1967, Pomeranz et al 1968, Wall 1970, 1978). Earlier experiments confirmed a *functional lamination* corresponding largely to cytoarchitectonic findings (p. 939). Neurons in all three laminae responded to cutaneous stimulation, only lamina VI to movement. Neurons in lamina IV had small receptive fields and responded as if *different* specific cutaneous afferents *converged* upon an *individual neuron* (e.g. to hair movement, touch, skin cooling, etc.). Lamina V neurons responded as if many in lamina IV converged upon each

of them, having larger and more diffuse receptive fields. Many in lamina VI responded to passive movement and cutaneous stimulation. Simultaneous corticospinal stimulation affected the neurons in all three laminae, sometimes facilitating, sometimes inhibiting responses. Brain-stem activity inhibited responses to cutaneous stimulation, but enhanced responses to movement, even *switching the 'modality'* of some lamina VI neurons from cutaneous to proprioceptive. In later investigations the findings were refined. In lamina V *fine myelinated afferents* from *three* sources (cutaneous, muscular, visceral) converged on *individual cells* and interacted. They had large, non-discriminative receptive fields, perhaps in some way signalling information concerning the general 'state' of the tissues. These results were summarized by Wall (1970); cells in laminae IV–VI were compared with those in nuclei gracilis and cuneatus. These laminae are now considered to provide a source of crossed spinothalamic fibres (Dilly et al 1968), a principal sensory path receiving some terminals of fasciculi gracilis and cuneatus, a second major sensory path. Their respective roles have long been debated and comparison emphasizes the important aspects of sensory function.

COMPARISON OF THE DORSAL COLUMN—MEDIAL LEMNISCUS COMPLEX WITH THE SPINOTHALAMIC PATH

It is instructive to compare, in these two 'systems', the size of the receptive fields, somatotopy, specificity of channels or convergence and interaction among them, and forms of control of transmission.

Neurons of dorsal column nuclei (Mountcastle 1968, Norton 1968) receive terminals of long, ipsilateral fibres of fasciculi gracilis and cuneatus (p. 936), the primary afferents for deformation of skin, movement of hairs, joint movement and vibration. In cats fibres from hair receptors are superficial, those for touch and vibration deeper. The localization of fibres in dorsal white columns has been noted (p. 936); the connections of dorsal column nuclei are further detailed with the medulla oblongata (p. 936).

Unit recording in the neurons of dorsal column nuclei shows that tactile *receptive fields* (i.e. excitable skin areas) vary in size, being mostly small and smallest in the digits. Some fields have *excitatory centres* and *inhibitory surrounds*; stimulation just outside its field inhibits a neuron. Neurons in the nuclei are spaced to represent the periphery accurately (in accord with their localization in the dorsal columns). *Specificity* is high; each neuron responds only to *one type* of stimulation; some respond to movements of a hair or joint, never to both; and fibres converging to one neuron are always modally alike.

Conduction from the dorsal columns to the medial lemniscus is controllable (Jabbur & Towe 1961, Anderson et al 1964). The adjacent dorsal column fibres may be presynaptically inhibited by depolarization of the presynaptic terminals. Stimulation of sensorimotor cortex modifies transmission by pre- and postsynaptic inhibition, sometimes by facilitation, mediated by the corticospinal tracts. The reticular formation exerts like effects. Dorsal column nuclei are not merely 'relay nuclei', as long supposed. They have been pictured as a highly reliable telephone system in which afferent information is separated in channels, discrete for spatial origin and stimulus specificity. In this the dorsal columns strongly contrast with the spinothalamic tracts (p. 941).

Neurons in spinal laminae IV–VI have different receptive fields. Those in IV have small fields and somatotopy is marked; those in V have large diffuse fields and respond to variable strength of stimulus. *Specificity* of channels, such as that present in dorsal column nuclei, is absent in the spinal laminae. The *convergence of different functional types* of afferent fibres to common neurons is usual in the cord, its degree varying among laminae. Lamina IV neurons receive large cutaneous afferents, lamina V fine afferents from the skin, muscle and splanchnic nerves, and lamina VI large afferents from muscle and skin. Convergence from different sources results in *interaction* between excitation and inhibition, summating to determine neuronal output. Transmission is variously modified. Firstly, cutaneous afferents may be influenced by the tonic mechanism in the substanti gelatinosa previously discussed. Transmission is also much affected by tracts from the sensorimotor cortex and brain stem (vide supra). Control varies from altered excitability in lamina IV to reversed modality in lamina VI.

The roles of dorsal columns and spinothalamic tracts, have been much debated. A classic view (Mountcastle 1968) regards the dorsal columns as the *essential discriminatory pathway*, without which mechanical stimulus is recognized but assigned no specific location, intensity or pattern, a view strongly challenged (Wall 1970) because complete section of the dorsal column does not abolish discrimination of weight, texture, two-point stimulation, vibration or position. Environmental stimuli may be classified into those 'passively impressed' and those which 'must be actively explored by motor movement or sequential analysis' for successful discrimination.

The former is considered the spinothalamic role, that of the dorsal columns is to *initiate and plan exploration* of the stimulus for the subsequent transmission of information.

SPINAL TERMINATIONS OF TRACTS DESCENDING FROM SUPRASPINAL SOURCES

Degeneration techniques such as Nauta's have much illuminated the spinal terminations of descending tracts (Nyberg-Hansen 1964a, b, 1965, 1966a, b, 1969, Nyberg-Hansen & Brodal 1963, 1964, Nyberg-Hansen & Mascitti 1964, Brodal 1969). Most investigations have been in cats, rarely primates. Results, while probably generally applicable, cannot be transferred in toto to human structures. The principal results are summarized in Figure 7.78, in which the distributions of terminals are related to laminar architecture. The following points, some already briefly mentioned, are noteworthy.

Corticospinal fibres almost all terminate, in cats, on interneurons in laminae IV–VII; but because of the widespread dendrites of multipolar neurons in lamina IX, some penetrating lamina VII, the existence of a few axodendritic contacts with motor neurons cannot be ignored. Corticospinal fibres from the 'sensory' cortex terminate chiefly in laminae IV and V, those from the 'motor' cortex in V–VII, with the densest concentration later-

ally in lamina VI. Thus, despite some overlap, corticospinal fibres from these two regions end differently—findings of interest in view of increasing evidence involving corticospinal tracts in the supraspinal *modulation of sensory inflow*; this may be mediated by presynaptic inhibition of primary afferent terminals or the postsynaptic inhibition or facilitation of subsequent neurons. In contrast there is evidence, anatomical (Hoff & Hoff 1934, Kupyers 1960, Liu & Chambers 1964) and physiological (Bernhard et al 1953, Preston & Whitlock 1961, Landgren et al 1962), that in monkeys some corticospinal fibres end monosynaptically on large α motor neurons. Less is known, quantitatively, about human synaptic terminals; most end on interneurons but some (Kuypers 1958) may end directly on motor neurons. Physiological studies (Corazza et al 1963) indicate that feline corticospinal tracts influence α and γ motor neurons, both via interneurons. The *increased* flux of impulses in corticospinal axons is excitatory to motor neurons of flexors and inhibitory to those supplying extensors. *Converse* effects of *decreased* flux are equally essential in normal function.

Rubrospinal fibres arise from large and small neurons in the contralateral red nucleus, with some degree of somatotopic order. Degeneration after nuclear lesions in cats shows the tract descending to lumbosacral levels, with terminals on interneurons in laminae V–VII. Its terminal zones correspond to those of corticospinal fibres from the 'motor' cortex, with similar effects on α and γ motor neurons. In some animals a somatotopic *corticorubral projection* has been demonstrated from the sensorimotor cortex, suggesting dual routes from cortex to cord, *direct corticospinal* and *indirect corticorubrospinal*, with similar spinal terminations and common functions. But the origin, localization, termination and functions of human rubrospinal connections are not defined, though often stated to be rudimentary.

The vestibulospinal tracts, medial and lateral, much investigated in other animals, are less clarified in man. The lateral, from the ipsilateral *lateral vestibular nucleus*, descends the whole cord, ending at successive levels largely in laminae VII and VIII and less so in lamina IX. The medial tract, mainly from *both* medial vestibular nuclei, descends perhaps only to mid-thoracic levels, terminating less widely in parts of laminae VII and VIII.

7.78 The spinal terminations of various descending tracts of the spinal cord determined experimentally in the cat and referred to the laminar pattern of the grey matter which is described elsewhere in the present section. Redrawn from Brodal 1969, by courtesy of the author and publishers. The original papers and illustrations stemmed from the numerous publications by R Nyberg-Hansen and his collaborators to whom we are grateful. Consult bibliography under 'Spinal Terminations of Descending Tracts from Supraspinal Sources'.

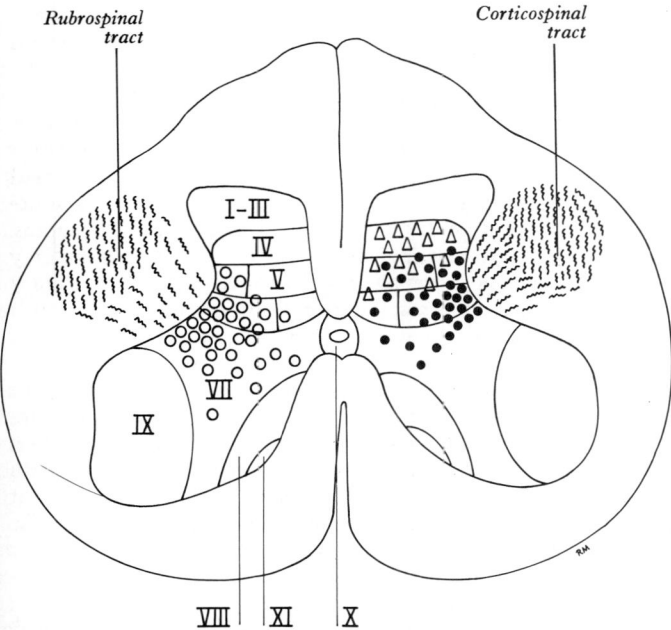

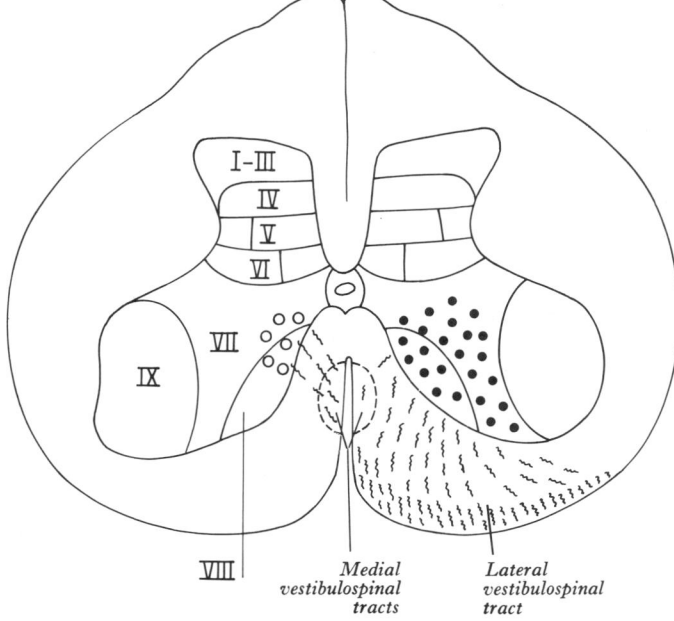

A. Termination of corticospinal fibres from 'motor' areas of the cerebral cortex—black dots; corticospinal fibres from 'sensory' areas of the cerebral cortex—white triangles; rubrospinal fibres—white dots.

B. Terminations of vestibulospinal fibres; those from the lateral vestibulospinal tract—black dots; those from the medial vestibulospinal tract are shown on the opposite side of the cord as white dots.

Activation of the lateral tract excites extensor and inhibits flexor motor neurons. Excitation is monosynaptic, indicating that vestibulospinal terminals synapse with the extensive dendrites of some motor neurons penetrating laminae VII and VIII. Gamma motor neurons are also probably facilitated; the inhibitory effect on flexors is presumably by inhibitory interneurons in their laminae of termination.

The reticulospinal tracts, notably difficult to evaluate in human spinal cords, have been mainly examined in cats, by retrograde degeneration in spinal neurons after spinal trauma and by anterograde effects following brain-stem lesions. Pontine and medullary reticulospinal fibres from one side apparently pass to *both sides* of the spinal cord; pontine fibres are ipsilaterally more concentrated; medullary fibres have also a substantial contralateral component. The zones of termination are summarized in 7.78C. (For non-aminergic, noradrenergic, serotonergic, enkephalinoid and other neuromediator substances associated with reticulospinal fibres, see pp. 992, 995, 996.) Similar experiments indicate that **tectospinal** and **interstitiospinal tracts** both terminate in laminae VI–VIII.

SUMMARY OF SPINAL ORGANIZATION

Szentágothai (1967) has considered spinal grey matter as a *central core* with paired *dorsal* and *ventral appendages*, each with distinctive features (7.79), terms which of course cut across the usual description but are held to be less arbitrary. The core has a diffuse, non-discriminative, reticular organization, with great divergence and convergence of paths; most interneurons connect with hundreds, perhaps thousands of others distributed in a substantial length of cord. In contrast dorsal and ventral appendages are more discriminatively organized in terms of somatotopy and functional localization. The precise limits to core and appendages are indefinable. The concept will undoubtedly need revision and perhaps lose relevance as knowledge increases. The central core includes interneurons of laminae VII and VIII (i.e. the intermediate zone and the areas between the motor neuron columns). But interneurons of dorsal columns and in motor neuron columns may be later included in this postulated reticular core. The ventral appendage corresponds to the neuronal columns of lamina

IX, the dorsal appendage includes laminae I–VI.

The dorsal appendage (p. 941) is a main receptive zone of exteroceptive, proprioceptive and interoceptive dorsal spinal root fibres. Laminae I–IV are the main cutaneous receptive areas; lamina V receives fine afferents from the skin, muscle and viscera; lamina VI receives proprioceptive and some cutaneous afferents. But few appropriate investigations have yet been made; hence the functional boundaries are uncertain and interneuronal complexities are postulated but as yet undemonstrated. Intracellular recording has its difficulties; some neurons may long remain out of technical reach.

Nevertheless the available recordings suggest that neurons in the dorsal appendages abstract data from internal and external environments. Neuronal groups show varying somatotopic array, size and response of receptive fields, specificity, convergence and interaction. Clearly, simple views of spinal neurons as merely relays in invariant, discrete, 'unimodality' channels, transmitting a punctate replica of the environment to the spinal cord or brain, are inadequate. Complex transformations of input occur in the dorsal appendage, whose output spreads to many destinations: directly to the ventral motor neuron columns, or indirectly via laminae including complex interneurons of the reticular core, to both sides of the cord, cranial and caudal to the input. Neurons in the dorsal appendage and reticular core contribute to the ascending tracts, reaching many brain-stem centres.

Transmission to and through the dorsal appendage may be modified by mutual interaction and control. Facilitatory and inhibitory effects, due to simultaneous activity in various afferent fibres converging on a single neuron, have been noted (p. 944), as has tonic modulation by the substantia gelatinosa. Much evidence suggests that transmission in all laminae is influenced by supraspinal sources, with excitation or inhibition provided by a wide array of neuromediators. The functional significance of these mechanisms is obscure; they may eliminate redundancy, reduce confusion or be linked to central 'states of readiness' or temporary 'preoccupation' with immediately significant transformations. Such hypothecation is fortunately complemented by continuing investigation. It is increasingly clear, however, that most descending tracts which influence motor behaviour do so by modifying transmission in the primary afferents, and in interneurons in the spinal laminae, less often than by direct influence on motor neurons.

Interneurons of the reticular core form an intricate net in which each neuron receives from and transmits to large numbers of others (pp. 988-996) and is characterized by one or more neuromediators. The core receives some axons from the dorsal appendage, proprioceptive dorsal root fibres and some descending fibres. This plethora of connections makes investigation difficult but an earlier view of *random nerve networks* is receding with newer methods, which include minute stereotactic lesions, intracellular recording, the inspection of thousands of Golgi-stained sections (Brazier 1969) and the mapping of somata or terminals by their content of established or putative neuromediators (p. 995). An aspect of organization has emerged that concerns the quantitative analysis of connectivity ('transmitting power') of different interneurons and some descending tracts. Some axonal terminals have very long courses in the grey matter and through successive branches give two or three synaptic end bulbs to each of many hundreds of interneurons encountered; others concentrate many end bulbs on one neuron or on a small group; yet others have widespread paracrine effects. Diffuse, *'non-discriminative'* connections hence contrast with *'discriminative'* ones. Many proprioceptive terminals in the core are 'segmentally' localized in *transverse* 'sheets' of grey matter. (Dorsal appendage cutaneous afferents terminate in *longitudinal* sheets.) In such complex interneuronal congeries, sensory input probably interacts with supraspinal influences to set in train the multivarious locomotor response. This analysis is largely speculative but the modern technical armamentarium will doubtless effect some progress.

The ventral appendage (pp. 927, 929, 939), is a columnar array of α and γ motor neurons and interneurons. Evidence points to 'tonic' and 'phasic' types of α neurons innervating striated muscle and to different types of γ neurons for 'static' and

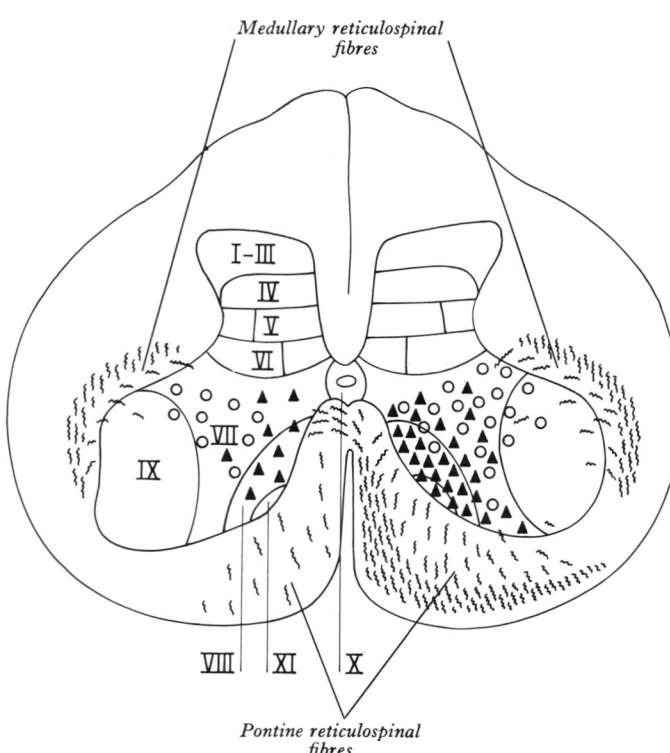

Medullary reticulospinal fibres

I–III
IV
V
VI
VII
IX
VIII XI X

Pontine reticulospinal fibres

7.78C Terminations of reticulospinal fibres; those originating in the medulla oblongata (white dots) are in general more dorsally placed than those originating in the pons (black triangles).

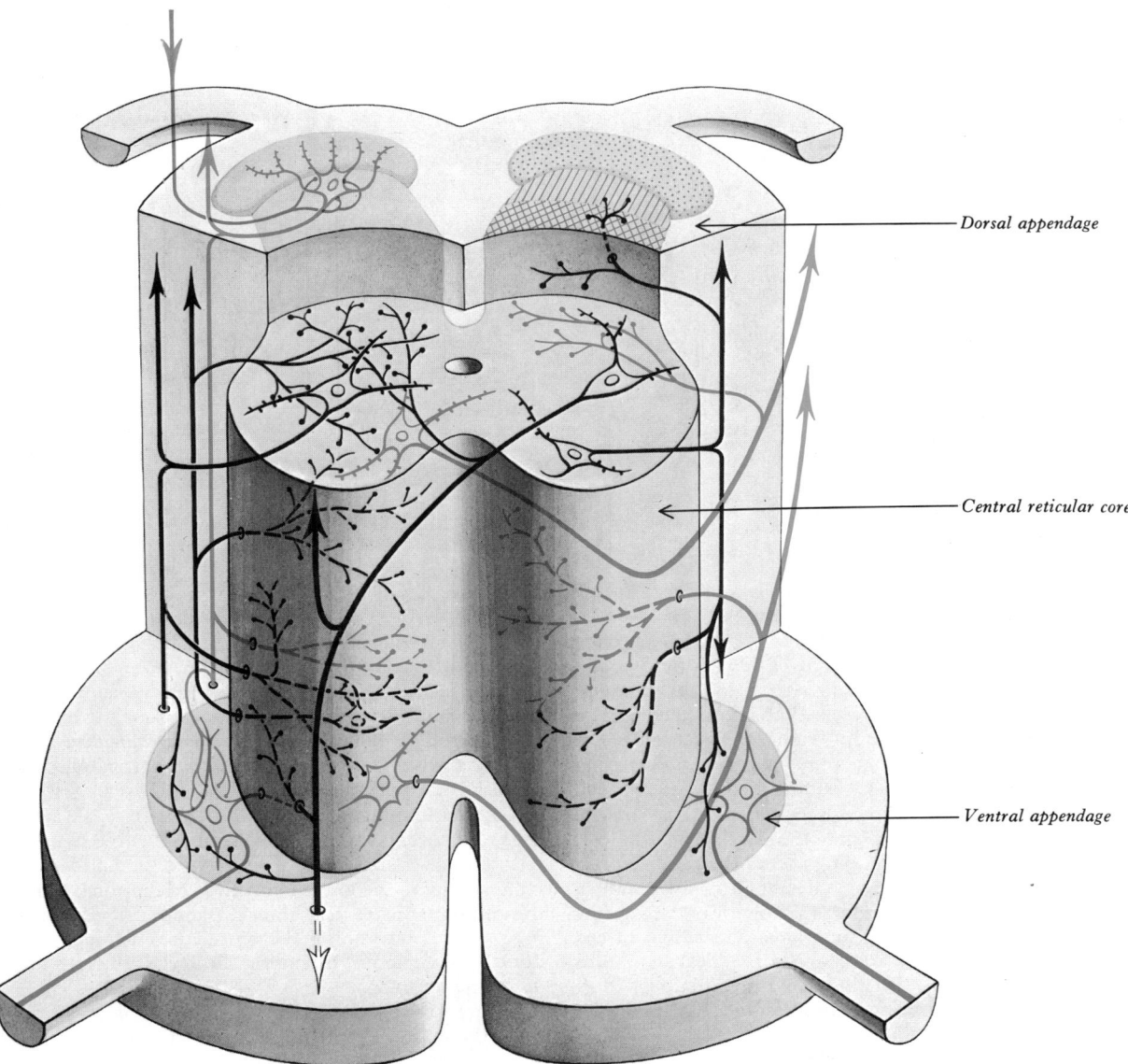

Dorsal appendage

Central reticular core

Ventral appendage

7.79   A highly simplified stereodiagram illustrating the concept of the spinal cord as consisting of a central 'reticular' core of grey matter, with related dorsal and ventral 'appendages' of grey matter. Many structural features are omitted and only a few examples, relevant to the concept, are included. A dorsal column neuron and others, more ventrally placed, which give rise to descending, long ascending and local collateral bran- | ches, are shown in blue. Varieties of interneuron are in black. Two motor neurons are shown in red; also in red is a single example of a fibre descending from a supraspinal source. See text for a more detailed description; see also **7.73** for the origin and termination of some tracts and **7.75** for one view of the fine structure of the substantia gelatinosa. Redrawn from Szentágothai 1967 by courtesy of the author and publishers.

'dynamic' responses in muscle spindles and their independent controls; but their detailed synaptic patterns remain uncertain.

The main connections of the *motor neurons* are: (1) direct monosynaptic terminals of proprioceptive dorsal root afferents in the same or nearby segments, (2) terminals from axonal collaterals from dorsal appendage interneurons, (3) terminals of interneurons of the reticular core: 'discriminative' from the same segment, 'non-discriminative' from adjacent segments, and (4) direct monosynaptic terminals from the vestibulospinal and (in man) corticospinal tracts. The interaction of such converging channels for integrated motor behaviour is obscure. A few generalizations can be made; descending paths can be grouped into those wherein impulses are excitatory to flexors, inhibitory to extensors (cortico-, rubro- and medullary reticulospinal tracts) and those with the opposite effect (vestibulo- and pontine reticulospinal tracts). But this simple dualistic view ignores much investigation reported on the complex modification of reflex activities by descending tracts.

Muscle is made to contract or relax by *two routes*: the α pathway produces a direct, immediate change in the excitatory level of α motor neurons innervating motor units, but probably acts infrequently, in sudden forceful responses; usually a γ *pathway* also operates by sequential activity in local interneurons and activity (or inhibition) of the γ efferents to muscle spindles which, via local

muscle 'servo-loop' mechanism (p. 916), causes the appropriate change in tonic and phasic α motor neurons. How this so-called α–γ *linkage* is maintained or broken in different conditions is under investigation (Granit 1970). Available reports indicate that during voluntary actions initiation of activity in the α system may be more frequent than previously recognized (p. 916).

Applied Anatomy. The segmental level of spinal injury may be determined from clinical data and accurate anatomical knowledge. Complete division above the fourth cervical level causes respiratory failure by loss of activity in the phrenic and intercostal nerves. Lesions between C_5 and T_1 paralyse all four limbs (quadriplegia), effects in the upper limbs varying with the site of injury: at the fifth cervical segment paralysis is complete; at the sixth the arms are between abduction and lateral rotation, with elbows flexed and forearms supinated, due to unopposed activity in the deltoids, spinati, rhomboids, bicipites and brachiales (all supplied by the fifth cervical spinal nerves). In lower cervical lesions upper limb paralysis is less. Lesions of the first thoracic segment paralyse small muscles in the hand and damage sympathetic outflow, resulting in contraction of the pupil, recession of the eyeball, narrowing of the palpebral fissure and loss of sweating in the face and neck (Horner's syndrome); sensation is retained in areas innervated by segments above the lesion; cutaneous sensa-

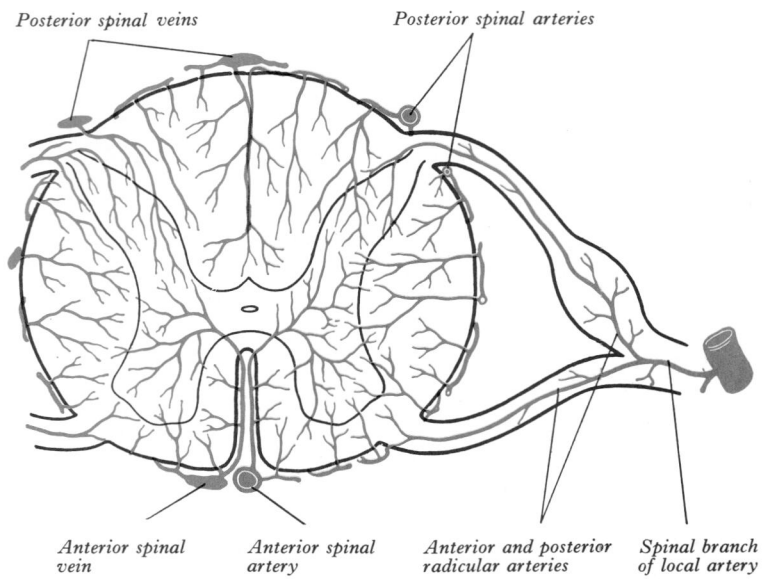

Posterior spinal veins   *Posterior spinal arteries*

Anterior spinal vein   *Anterior spinal artery*   *Anterior and posterior radicular arteries*   *Spinal branch of local artery*

7.80   The intrinsic blood vessels of the spinal cord. The position of the veins is quite variable.

tion is retained in the neck and chest down to the second intercostal space, because this area is innervated by the supraclavicular nerves (C_3 and C_4). At thoracic levels division paralyses the trunk below the lesion's level and both lower limbs (paraplegia). The first sacral *neural* segment is approximately level with the thoracolumbar vertebral junction; injury, commonly occurring here, paralyses the urinary bladder, rectum and muscles supplied by the sacral segments; cutaneous sensibility is lost in the perineum, buttocks, the flexor aspect of the thighs, legs and feet. The roots of lumbar nerves descending to join the cauda equina may be damaged, causing complete paralysis of both lower limbs. Lesions below the first lumbar vertebra may divide or damage the cauda equina; severe damage is uncommon and usually confined to the spinal roots at the level of trauma. Neurological symptoms may be due to interference with the spinal blood supply, particularly in the lower thoracic and upper lumbar segments.

Vertebral Levels of Spinal Cord Segments

The level of spinal segments relative to the vertebrae is clinically important. A useful approximation is: at cervical level a vertebral spine's apex corresponds to the succeeding segment (i.e. the *sixth* cervical spine is opposite the *seventh* spinal segment); at upper thoracic levels a vertebral apex corresponds to the cord two segments lower (i.e. the *fourth* spine is level with the *sixth* segment); in the lower thoracic region there is a difference of three segments (i.e. the *tenth* thoracic spine is level with the *first* lumbar segment). The *eleventh* thoracic spine overlies the *third* lumbar segment, the *twelfth* is opposite the *first* sacral segment. The neonatal spinal cord extends to the upper border of the third lumbar vertebra; Barson (1970), in a series of 258 pre- and postnatal subjects, found the perinatal level at the third lumbar vertebra, rising during the first two postnatal months; variation was marked: first to fourth lumbar at birth, first to third in children (three months to 15 years of age).

Vasculature of the Spinal Cord

Blood reaches the spinal cord along spinal branches of the vertebral, deep cervical, intercostal and lumbar arteries; these, with the anterior and posterior spinal arteries, form longitudinal anastomotic channels along the cord (pp. 751, 765, Gillilan 1972). Spinal arteries send anterior and posterior radicular branches to the spinal cord along ventral and dorsal roots. Most anterior radicular arteries are small, ending in the ventral nerve roots or the cord's pial ganglia. Posterior radicular arteries supply the dorsal root ganglia; according to Somogyi et al (1973) ganglionic ramules enter at both ganglionic poles and are distributed around ganglion cells and nerve fibres; the same authors and Undi et al (1973) have described the arterial supplies of spinal roots. Some radicular arteries (usually four to nine), mainly in the lower cervical, lower thoracic and upper lumbar regions, are large enough to

reach the anterior median sulcus, dividing into slender ascending and large descending rami, which anastomose with the anterior spinal arteries to form a single or partly double, longitudinal vessel of uneven calibre along the anterior median sulcus. The largest anterior radicular artery, *arteria radicularis magna*, varies in level; it arises from an intersegmental aortic branch at the lower thoracic or upper lumbar level; about 65% arise on the left. Reaching the spinal cord, it sends a branch to the anterior spinal artery below and another to anastomose with a ramus of the posterior spinal artery anterior to the dorsal roots. It is sometimes the main supply to the lower two-thirds of the cord. Central rami of the anterior spinal artery enter the anterior median fissure, turning right or left to supply the ventral grey column, the base of the dorsal grey column, including the dorsal nucleus and adjacent white matter (7.80).

Each posterior spinal artery contributes to a pair of longitudinal anastomotic channels, anterior and posterior to the dorsal spinal roots. These are reinforced by posterior radicular arteries, variable in number and size but more numerous than the anterior radicular. The anterior channel is also joined by a ramus from the descending branch of the arteria radicularis magna. Lazorthes et al (1971) have largely confirmed this description but emphasize the uneven calibre and interruptions in longitudinal spinal arteries, which are conjoined at the conus medullaris by anastomotic loops. They also emphasized anastomoses other than those between the pial or peripheral spinal arterial branches, such as a posterior spinal series of anastomoses between rami of the dorsal divisions of segmental arteries near spinous processes.

Central branches of the anterior spinal artery supply about two-thirds of the cross-sectional area of the cord. The rest of the dorsal grey and white columns and peripheral parts of lateral and ventral white columns are supplied by numerous small radial vessels from posterior spinal arteries and the pial plexus (Torr 1957, Gillilan 1958). In a microangiographic study of human spinal cord at cervical levels (Turnbull et al 1966) there were 1–6 anterior and 0–8 posterior radicular spinal arteries. From each centimetre of anterior spinal artery arose 5–8 central branches. No internal anastomoses were seen but overlapping of territories of central arteries was confirmed; similar longitudinal overlap was also emphasized.

Spinal veins drain into six tortuous, often plexiform longitudinal channels, one each in the anterior and posterior median fissures and four others, often incomplete, one pair being posterior, the other anterior to the ventral and dorsal nerve roots. These vessels connect together freely and above with the cerebellar veins and cranial sinuses. Veins with the anterior spinal artery receive large venules from the central grey matter; much blood from the cord's periphery enters the pial veins (p. 812).

THE RHOMBENCEPHALON OR HINDBRAIN

The rhombencephalon includes the medulla oblongata, pons and cerebellum; its cavity is the fourth ventricle. The medulla and pons contain tracts connecting all parts of the central nervous system and also nuclei of several cranial nerves and many other groups of neurons. Emerging superficially from the pons and medulla are: the trigeminal, abducent, facial, vestibulocochlear, glossopharyngeal, vagus, cranial accessory and hypoglossal nerves. Among the nuclei and tracts is the *reticular formation* (p. 988), a mass of neurons and axons continuous with its spinal counterpart; some of its nuclei are concerned in cardiac, respiratory and alimentary control, others in some aspect of all other neural activities (p. 995 et seq.).

THE MEDULLA OBLONGATA

The medulla oblongata is arbitrarily defined as extending from the lower pontine margin to a transverse plane above the first pair of cervical spinal nerves, which is at a level corresponding dorsally with the upper atlantal border and ventrally with the centre of the dens; but the medulla is uninterruptedly continuous with the pons and spinal cord. Its internal structure is similar to the spinal cord below but changes *gradually* with increasing distance from the spinal cord. Its anterior surface is separated from the basilar part of the occipital bone and apex of the dens by the meninges and occipito-axial ligaments. Dorsally it is in the cerebellar notch, between its hemispheres; in its upper part it forms the lower half of the fourth ventricle's floor. It is somewhat piriform (7.81, 82), its broad (superior) end merging into the pons and its narrow lower end continuous with the spinal cord. It is about 3 cm in length, 2 cm transversely at its widest and sagittally some 1·3 cm thick. The spinal central canal is prolonged into its lower half, expanding above as the fourth ventricle. The medulla is hence divided into a *closed part* containing the central canal and an *open part* containing the lower half of the fourth ventricle. Its anterior and posterior surfaces have median fissures.

The anterior median fissure, containing a shallow fold of pia mater, extends throughout but is partly obscured; it is continuous below with the spinal anterior median fissure, ending above at the lower pontine border in a small triangular *foramen caecum*. Inferiorly it is interrupted by fascicles crossing obliquely, the *pyramidal decussation*. Anterior external arcuate fibres emerge from the fissure above this and curve laterally over the medullary surface.

The posterior median sulcus exists only in the closed part, continuous below with the spinal posterior median sulcus; it rapidly shallows above, ending at the mid-level of the medulla, where the central canal expands into the fourth ventricle. (The external posterior median sulcus must be carefully distinguished from the median sulcus on the floor of the fourth ventricle; they are, of course, neither continuous nor homologous.)

Cranial nerves emerging from the medulla appear in line with *both* roots of the spinal nerves. Hypoglossal rootlets are in line with the ventral spinal roots, emerging from an *anterolateral sulcus*. The accessory, vagus and glossopharyngeal nerves are in line with the dorsal spinal roots (p. 924), passing through a *posterolateral sulcus*. These features are used to divide each half medulla into anterior, middle and posterior regions. Though these *appear* to be continuous with the corresponding spinal funiculi, they do not contain precisely the same fibres; some spinal tracts end or begin in the medulla, with others altering course in it.

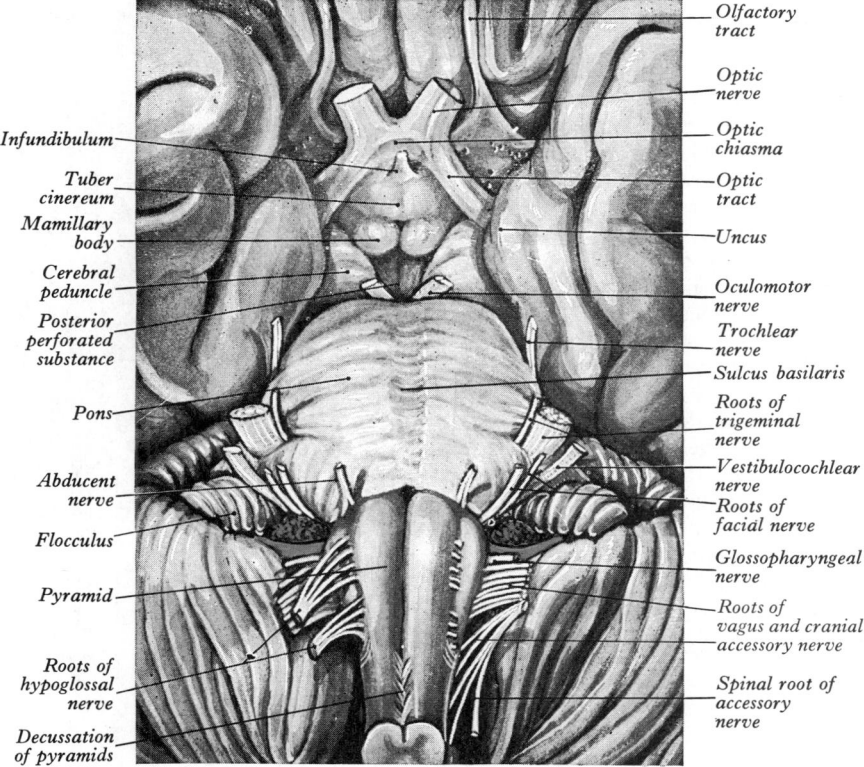

Olfactory tract
Optic nerve
Optic chiasma
Optic tract
Uncus
Oculomotor nerve
Trochlear nerve
Sulcus basilaris
Roots of trigeminal nerve
Vestibulocochlear nerve
Roots of facial nerve
Glossopharyngeal nerve
Roots of vagus and cranial accessory nerve
Spinal root of accessory nerve

Infundibulum
Tuber cinereum
Mamillary body
Cerebral peduncle
Posterior perforated substance
Pons
Abducent nerve
Flocculus
Pyramid
Roots of hypoglossal nerve
Decussation of pyramids

7.81 The ventral aspect of the brain stem and the interpeduncular fossa. The wall of the lateral recess of the fourth ventricle is shown in blue and the choroid plexus, which protrudes through the foramen of the lateral recess into the subarachnoid space, is coloured crimson. Note that the lateral recess covers the medial part of the flocculus and is itself partially obscured by a rootlet of the glossopharyngeal nerve.

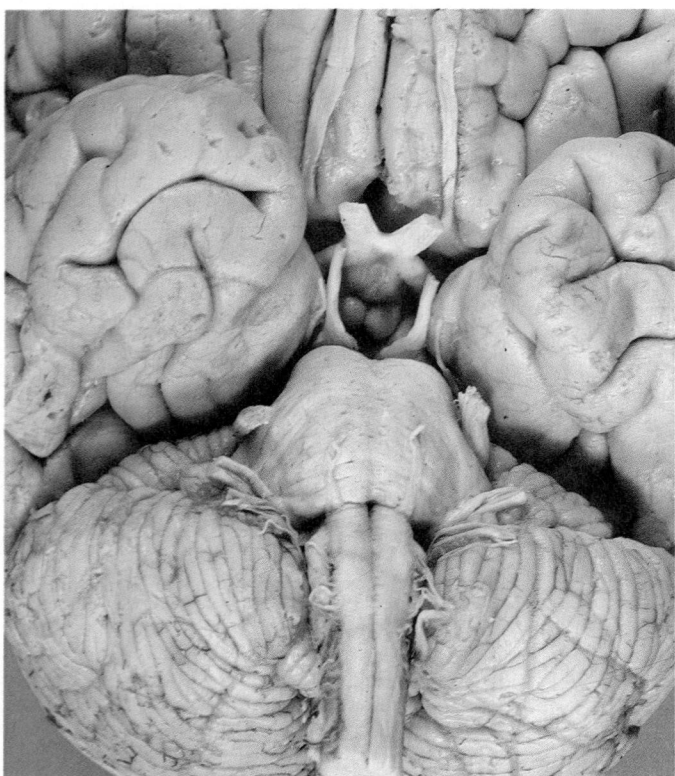

7.82 The ventral aspect of the brain stem, interpeduncular fossa and adjacent parts of the cerebellar and cerebral hemispheres. For identification of the various structures, compare with 7.81. Dissection by E L Rees, photography by Kevin Fitzpatrick, both of the Department of Anatomy, Guy's Hospital Medical School, London.

The anterior region (7.81, 82), between the anterior median fissure and the anterolateral sulcus, shows an elongated surface inappropriately named the *pyramid*. At its slightly narrowed pontine end the abducent nerve emerges; below it tapers into the spinal anterior funiculus, superficially continuous with it. The two pyramids contain corticospinal fibres, approximately 70–90% of which leave the pyramids in successive bundles and which cross in and deep to the anterior median fissure as the *pyramidal decussation*. They then descend dorsally in the spinal lateral funiculus as the crossed lateral corticospinal tract. The remaining fibres, lateral in the pyramids, do not cross; some descend as the anterior corticospinal tract (7.83) into the ipsilateral anterior funiculi, others incline posterolaterally to join the lateral corticospinal tracts as their lesser uncrossed component (p. 832). The corticospinal tracts display somatotopy at almost all levels; in the pyramids the arrangement is like that at higher levels, the most lateral fibres being concerned with the legs, the most medial with the arms and neck. How far this pattern is carried into the anterior corticospinal tracts or decussating fibres is unknown; but similar somatotopy is usually ascribed to the lateral corticospinal tracts as they traverse the spinal cord.

The lateral region (7.84A,B) is between the anterolateral sulcus (with the hypoglossal roots) and the posterolateral sulcus (with roots of the accessory, vagus and glossopharyngeal nerves). Its upper part is largely a prominent, oval *olive*; its lower part matches the spinal lateral funiculus and appears to be its direct continuation, but only part of the latter continues into this region; the lateral corticospinal tract is mainly from the contralateral pyramid and the posterior spinocerebellar tract mostly leaves it to enter the inferior cerebellar penduncle. The lateral intersegmental and anterior spinocerebellar (and other) tracts continue into the lateral region.

The olive, a smooth, oval elevation between the anterolateral and posterolateral sulci, is due to the subjacent *inferior olivary* complex of *nuclei* (p. 956). Lateral to the pyramid, separated by the anterolateral sulcus and hypoglossal fibres, it is about 1·25 cm long; dorsolateral to its upper end is a slight depression at the

lower pontine border; in this, roots of the facial nerve appear. Anterior external arcuate fibres emerge from the anterior median fissure, curving back across the pyramid and olive to the inferior cerebellar peduncle (7.93).

The posterior region (7.84A,B) dorsal to the posterolateral sulcus is, like the lateral, divisible into upper and lower levels. The *lower part*, limited behind the posterior median sulcus, is an upward continuation of the *fasciculi gracilis* and *cuneatus*. The former, with its fellow, flanks the posterior median sulcus, separated from the cuneate fasciculus by a cranial continuation of the posterointermediate sulcus and septum of the cervical spinal cord (p. 922). Both fasciculi are first vertical; but at the lower end of the fourth ventricle they diverge from their fellows, each developing an elongated swelling, the *gracile* and *cuneate tubercles*, produced by the subjacent **nuclei gracilis** and **cuneatus**. Most fibres in both fasciculi synapse with neurons in their respective nuclei. A *tuberculum cinereum* sometimes appears below (7.84A,B), between the fasciculus cuneatus and the roots of the accessory nerve; narrow below, it widens upwards and is produced by a nucleus continuous below with the substantia gelatinosa; the spinal trigeminal tract ends in it, separating it from the surface (pp. 953, 978). The upper posterior medullary region is the *inferior cerebellar peduncle*, a rounded ridge between the fourth ventricle and the glossopharyngeal and vagal roots. The two peduncles diverge and incline from the dorsolateral medulla into the cerebellum. Ascending, they form the inferolateral ventricular boundaries; above they curve dorsomedially into the cerebellar hemispheres, where they are crossed by the *striae medullares*, running to the median ventricular sulcus (7.115, 116). The inferior peduncle is not a continuation of the gracile and cuneate fasciculi, whose fibres end in the gracile, cuneate and accessory cuneate nuclei. The peduncle's composition is described on p. 965.

INTERNAL STRUCTURE OF THE MEDULLA OBLONGATA

Like other neuraxial levels the medulla has been much studied in transverse sections and also by experiment and reconstruction

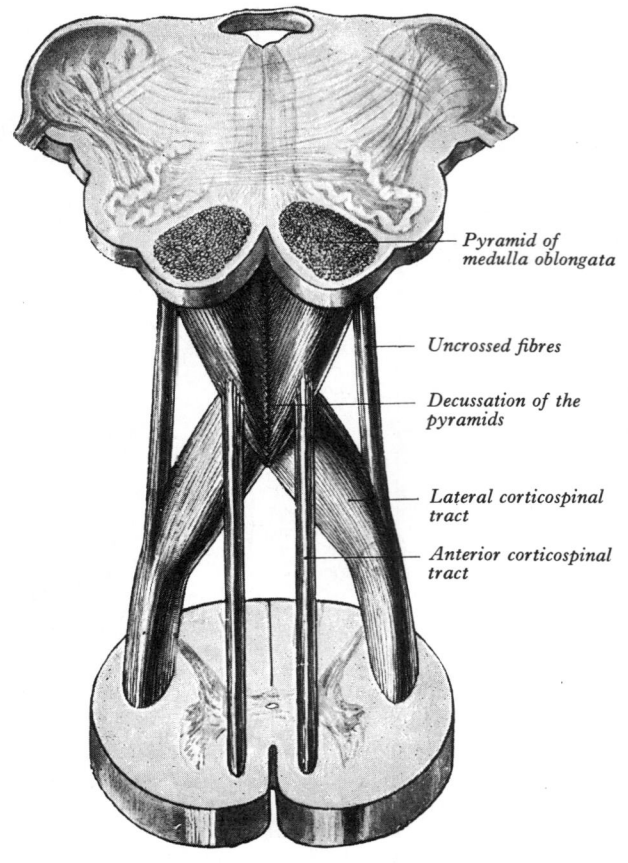

Pyramid of medulla oblongata

Uncrossed fibres

Decussation of the pyramids

Lateral corticospinal tract

Anterior corticospinal tract

7.83 Schematic dissection to show the decussation of the pyramids.

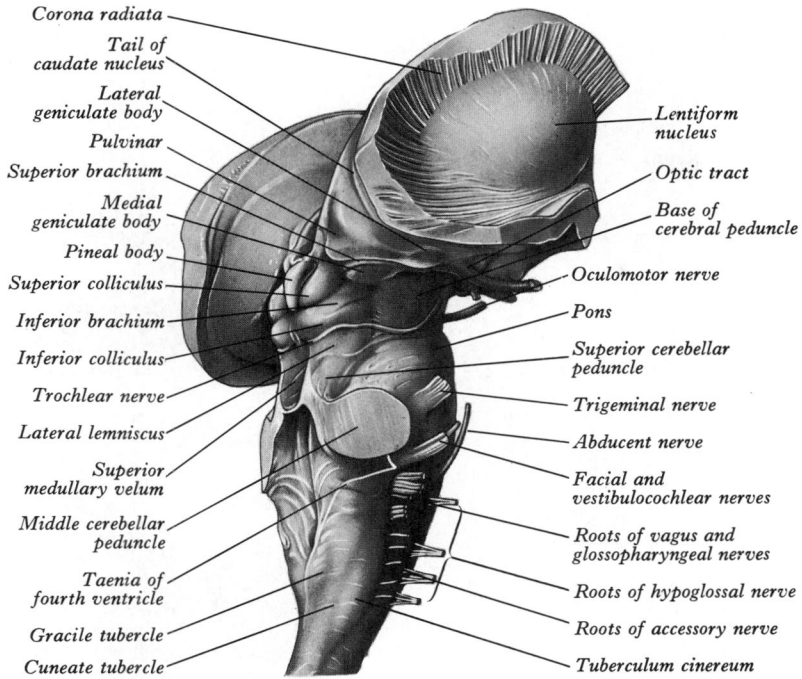

Corona radiata
Tail of caudate nucleus
Lateral geniculate body
Pulvinar
Superior brachium
Medial geniculate body
Pineal body
Superior colliculus
Inferior brachium
Inferior colliculus
Trochlear nerve
Lateral lemniscus
Superior medullary velum
Middle cerebellar peduncle
Taenia of fourth ventricle
Gracile tubercle
Cuneate tubercle

Lentiform nucleus
Optic tract
Base of cerebral peduncle
Oculomotor nerve
Pons
Superior cerebellar peduncle
Trigeminal nerve
Abducent nerve
Facial and vestibulocochlear nerves
Roots of vagus and glossopharyngeal nerves
Roots of hypoglossal nerve
Roots of accessory nerve
Tuberculum cinereum

7.84A The brain stem, posterolateral aspect.

(largely based on transverse sections). It is customary and convenient to describe it in sample transverse sections at several levels. In this account, both of the medulla and other brain-stem levels, this custom is followed, describing four successive medullary levels. The general disposition of some principal brain-stem nuclei is shown in Figures 7.86, 87, 122.

(1) **A transverse section in the lower medulla oblongata** is similar to the adjoining spinal cord, as would be expected (7.88). Dorsal, lateral and ventral funiculi exist and contain the same tracts; however, the grey matter shows two marked changes. Ventral grey columns are separated from the central grey matter by corticospinal fibres decussating dorsolaterally to the opposite lateral funiculus. In the upper medulla these fibres are ventromedial in the pyramids but below most cross, inclining dorsally (7.83) and decussating ventral to the central grey matter. The decussation is orderly: the fibres ending in the cervical segments decussate first. This *pyramidal decussation* is the most striking feature in caudal sections. Usually at least 75% of fibres cross and descend in the lateral funiculus as the crossed lateral corticospinal tract. Of the rest some descend ventromedially in the ipsilateral anterior funiculus as the anterior corticospinal tract; other uncrossed fibres descend with the crossed fibres in their *ipsilateral* lateral funiculus (7.83). The decussation displaces the anterior intersegmental tract towards the central grey matter, which, with its canal, becomes more dorsal. Continuity between the ventral grey column and central grey matter, maintained throughout the spinal cord, is severed; the column rapidly diminishes as it ascends, subdivided into a *supraspinal nucleus*, the efferent source of the first cervical nerve and a *spinal nucleus of the accessory nerve*, which is dorsolateral and, as its name indicates, provides some spinal accessory fibres. The supraspinal nucleus is continuous above with that of the hypoglossal. The spinal nucleus continues into the upper five cervical segments, where it is dorsolateral in the ventral grey column; rostrally it merges with the nucleus ambiguus (p. 956).

The dorsal grey column's outline is still recognizable, but modified. A narrow grey lamina appears in the fasciculis gracilis, continuous ventrally with the base of the posterior column, the lower end of the *nucleus gracilis*, which extends upwards to the caudolateral limit of the fourth ventricle, forming the gracile tubercle (p. 950). A second projection from the dorsal grey column, more laterally placed and beginning and ending at a higher level, invades the fasciculus cuneatus from its ventral aspect, as the *nucleus cuneatus*.

The substantia gelatinosa is prominent as an upward continuation of the apex of the dorsal grey column, becoming continuous with the inferior end of the *spinal trigeminal nucleus* (pp. 961, 1122); fibres of the *trigeminal tract* itself are between the nucleus and the surface (7.88) (cf. the dorsolateral tract at spinal levels).

(2) **A transverse section just above the pyramidal decussation** shows an increase in changes already noted and some new features (7.89).

The nucleus gracilis is broader and the fibres of its fasciculus are groups on its dorsal, medial and lateral surfaces; the *nucleus cuneatus* is also larger. Both, here, retain continuity with the central grey matter but lose it at higher levels. Gracile and cuneate

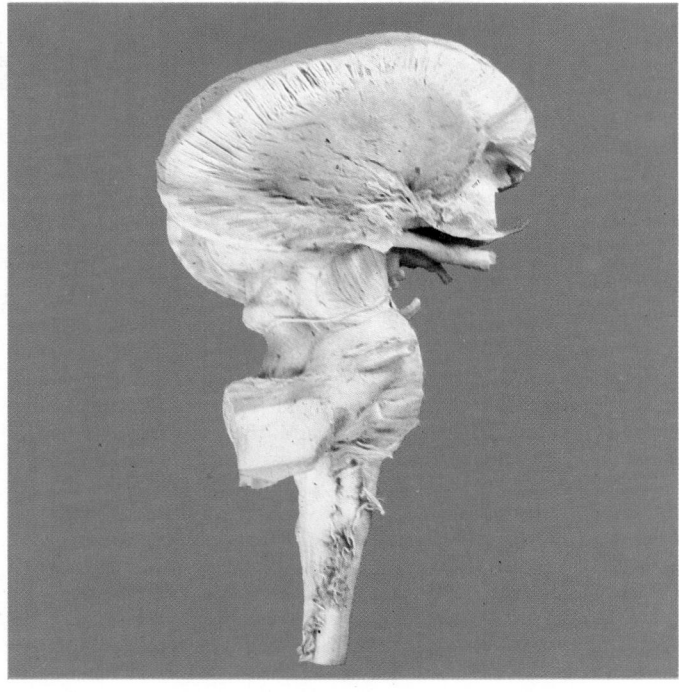

7.84B The right lateral aspect of the brain stem, lentiform nucleus and corona radiata. For identification compare with 7.84A. Dissection by E L Rees, photography by Kevin Fitzpatrick, both of the Department of Anatomy, Guy's Hospital Medical School, London.

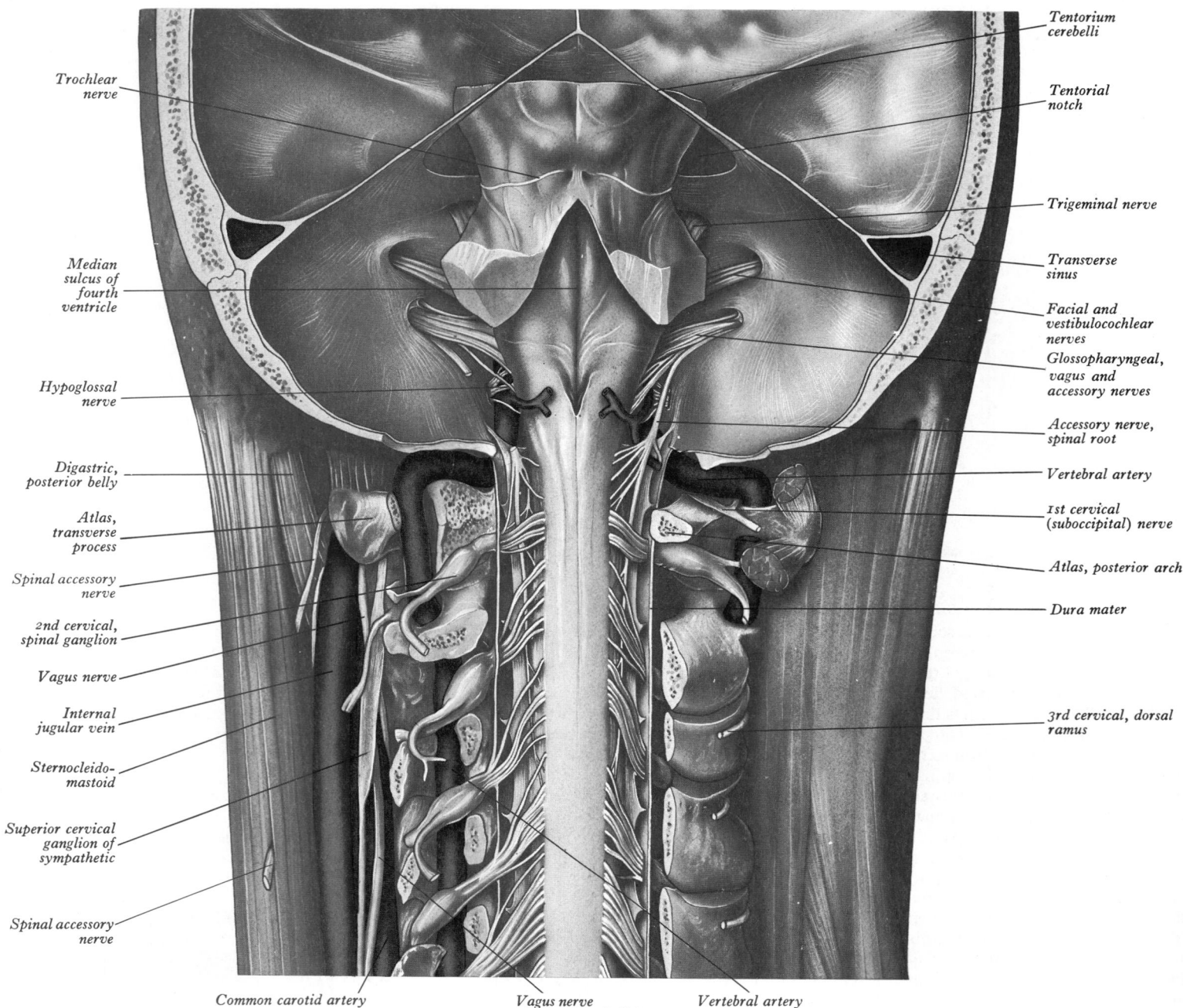

7.85 Dissection exposing the brain stem and upper five cervical spinal segments after removal of large portions of the occipital and parietal bones and the cerebellum together with the roof of the fourth ventricle. On the left the foramina transversaria of the atlas and the third, fourth and fifth cervical vertebrae have been opened to expose the vertebral artery. On the right the posterior arch of the atlas and the laminae of the succeeding cervical vertebrae have been removed.

fascicular fibres (*uncrossed*) synapse in their respective nuclei at different levels. Axons (secondary) *from* the nuclei emerge ventrally *as internal arcuate fibres*, at first curving ventrolaterally around the central grey matter and then ventromedially to *decussate* (7.89), thereafter *ascending* near the median raphe as the *medial lemniscus*. This *lemniscal decussation* is dorsal to the pyramids, ventral to the central grey matter and thus the latter is even more dorsally displaced. Internal arcuate fibres sweep ventrally between the trigeminal spinal tract and the central grey matter. The gracile and cuneate nuclei are not simple sensory relay stations on a path widely considered the major route for the discriminative aspects of tactile and locomotor sensation. Much evidence has changed such concepts. The nuclear neuronal populations have long been considered uniform, despite the early distinction of types among them by Cajal (1900). Many are multipolar but vary in size and distribution. Taber (1961) and Kuypers & Tuerk (1964) have described several types of neurons

in the nuclei and Biedenbach (1972) defined two zones of contrasting cytoarchitecture. In all the species examined, including monkeys, the upper regions of both nuclei are *reticular*, containing small and large multipolar neurons with long dendrites, and large round neurons with short and profusely branching dendrites, the last predominating below in both nuclei in clusters of various form. These zones differ in their connections; both receive terminals from the dorsal spinal roots at all levels but inferiorly they have a denser input. Dorsal funicular fibres from neurons in the spinal grey matter terminate only in the superior, reticular zone. In all connections there is variable ordering of terminals on the basis of spinal root levels, like the dorsal funiculi; but variable overlap has been observed (Millar & Basbaum 1975). In monkeys, the hind leg and tail are represented medially, the trunk ventrally and digits dorsally. There is also modal specificity; lower levels respond to low threshold cutaneous stimuli, reticular (upper) levels to inputs from fibres serving receptors in

Labels on figure:
Trochlear nerve
Median sulcus of fourth ventricle
Hypoglossal nerve
Digastric, posterior belly
Atlas, transverse process
Spinal accessory nerve
2nd cervical, spinal ganglion
Vagus nerve
Internal jugular vein
Sternocleido-mastoid
Superior cervical ganglion of sympathetic
Spinal accessory nerve
Common carotid artery
Vagus nerve (displaced medially)
Vertebral artery
Tentorium cerebelli
Tentorial notch
Trigeminal nerve
Transverse sinus
Facial and vestibulocochlear nerves
Glossopharyngeal, vagus and accessory nerves
Accessory nerve, spinal root
Vertebral artery
1st cervical (suboccipital) nerve
Atlas, posterior arch
Dura mater
3rd cervical, dorsal ramus

the skin, joints and muscles. Both nuclei contain interneurons (Andersen et al 1964, Rustioni & Sotelo 1974), many are inhibitory. But primary spinal afferents synapse with multipolar neurons forming the major nuclear efferent projection, as a relay between the spinal cord and higher levels. Descending afferents from the somatosensory cortex (p. 1003), reaching the nuclei through the corticospinal tracts, appear restricted to the upper, reticular zones; they may inhibit or enhance activity at nuclear level, clearly a region of sensory modulation. The reticular zones also receive connections from the reticular formation proper (Spacek & Lieberman 1974). It is probable that 'feed back' from the gracile and cuneate nuclei to the spinal cord also exists. By antidromic stimulation (Dart 1971), retrograde horse-radish peroxidase (Kuypers & Maisky 1975) and anterograde isotopic transport (Burton & Loewy 1977), it has been shown in cats and monkeys that neurons of these nuclei project to the ipsilateral spinal dorsal appendage, by fibres presumably involved in the observed depression of dorsal column activity (Hillman & Wall 1969).

The *accessory cuneate nucleus*, dorsolateral to the cuneate, contains large neurons like those in the spinal thoracic nucleus; these form the *posterior external arcuate fibres* (p. 966), which enter the cerebellum by its ipsilateral inferior peduncle. The nucleus receives lateral fibres of the fasciculus cuneatus, derived from cervical segments. It provides the *cuneocerebellar tract* for proprioceptive impulses from the upper limb which enter the spinal cord above its thoracic nucleus. A group of neurons, 'nucleus Z', was identified in cats by Brodal & Pompeiano (1957) between the upper pole of the nucleus gracilis and the inferior vestibular nucleus and said to be present in the human medulla (Webster 1977). Its input is probably from the dorsal spinocerebellar tract (Rustioni 1973); it may be a separated part of the gracile reticular zone, which it resembles; it is concerned in proprioception from the ipsilateral hind limb (Landgren & Silfvenius 1971).

The *nucleus of the spinal tract of the trigeminal nerve* (7.86) is separated from the central grey matter by internal arcuate fibres and from the lateral medullary surface only by the trigeminal spinal tract, which ends in it, and by some dorsal spinocerebellar tract fibres; the latter progressively incline dorsally and at a higher level enter the inferior cerebellar peduncle (p. 965).

Two other nuclei occur at this level. One is dorsolateral to the pyramid, the other medial to it and near the median plane; these are parts of the *medial accessory olivary nucleus*, described with the inferior olivary nuclear complex (p. 956).

The central grey substance, at this level near the dorsal medullary surface, contains three bilateral nuclei. The prominent *hypoglossal nucleus*, of large motor neurons interspersed with myelinated fibres, is ventromedial in the central grey matter. It extends into the open part of the medulla, subjacent to the *trigonum hypoglossi* in the fourth ventricle's floor. Near it are several smaller groups, perhaps misnamed 'perihypoglossal complex' or 'perihypoglossal grey', for none is known to be connected with the hypoglossal nerve or nucleus. (However, the location of hypoglossal-related interneurons remains to be clarified.) These small groups include the *nucleus intercalatus, sublingual nucleus, nucleus prepositus hypoglossi* and *nucleus paramedianus dorsalis (reticularis)*, all containing neurons suggestive of reticular connections, definitely ascribed to the paramedian nucleus (Brodal 1957). Gustatory and visceral connections are attributed to the nucleus intercalatus; there is more convincing evidence that the perihypoglossal nuclei project to the cerebellum, at least in cats (Torvik & Brodal 1954) and monkeys (Mehler et al 1960). Representation of lingual musculature has been described in the hypoglossal nucleus (p. 1119).

Dorsolateral to the hypoglossal is the *dorsal vagal nucleus* containing neurons of at least two types, the larger with fine fibres innervating non-striated muscle and smaller fusiform neurons possibly concerned with visceral afferents. But many believe that

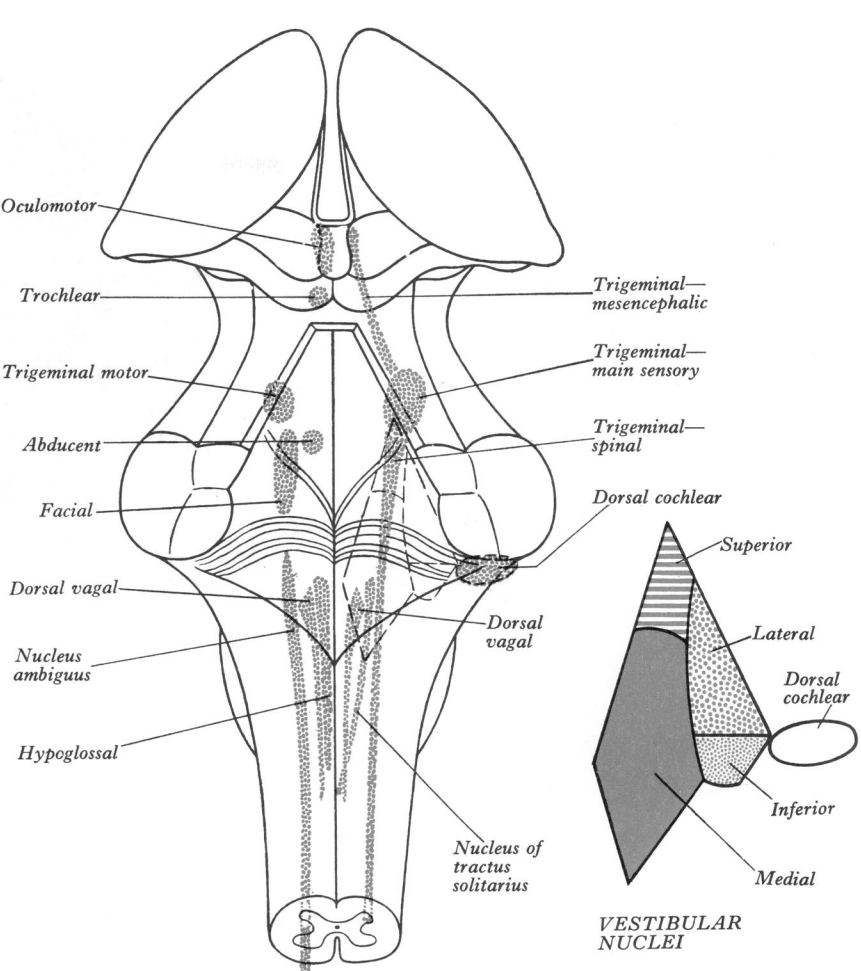

7.86 *Surface projection* of some cranial nerve nuclei on the dorsal aspect of the brain stem. Motor nuclei are in red, sensory in blue. The vestibular nuclei are indicated in the main diagram by interrupted lines and are shown in detail in the small diagram. The olfactory and optic centres are not shown.

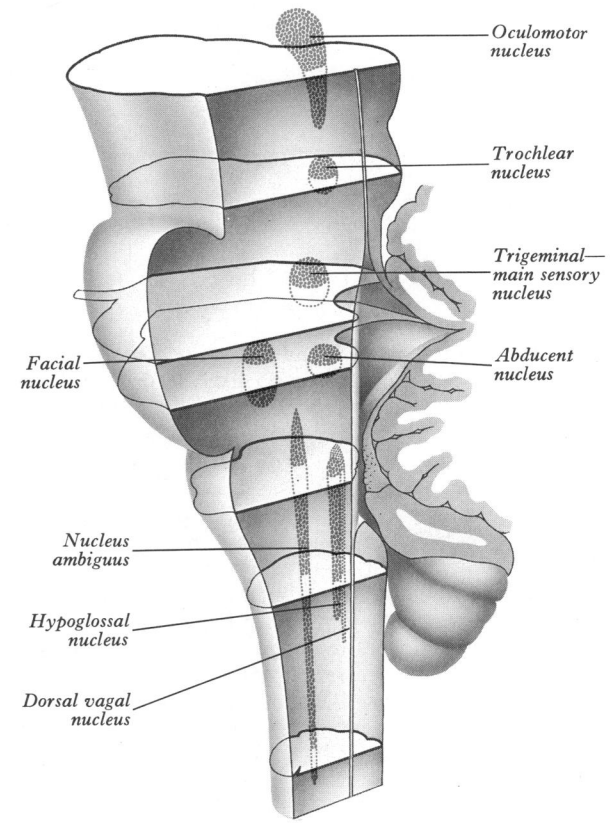

7.87 Motor nuclei of the cranial nerves. The sectional planes shown correspond to those depicted elsewhere in the text.

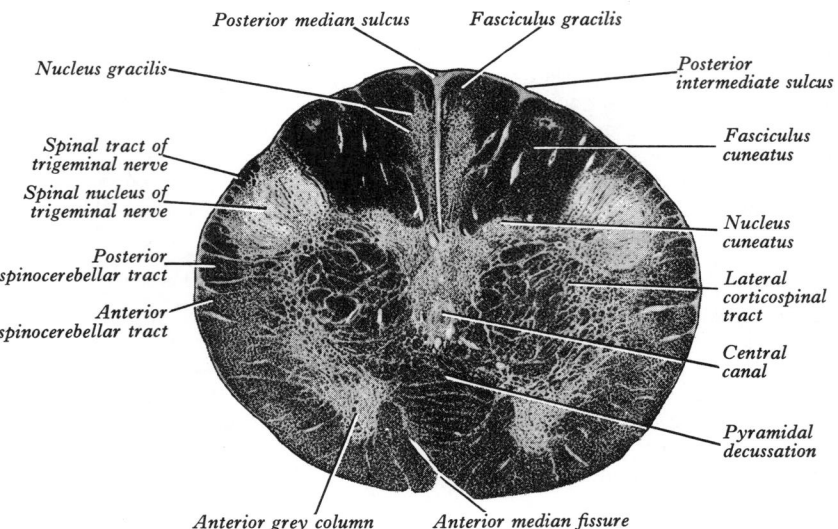

7.88 Transverse section through the medulla oblongata at the level of the pyramidal decussation. Weigert Pal preparation. Magnification × 7.

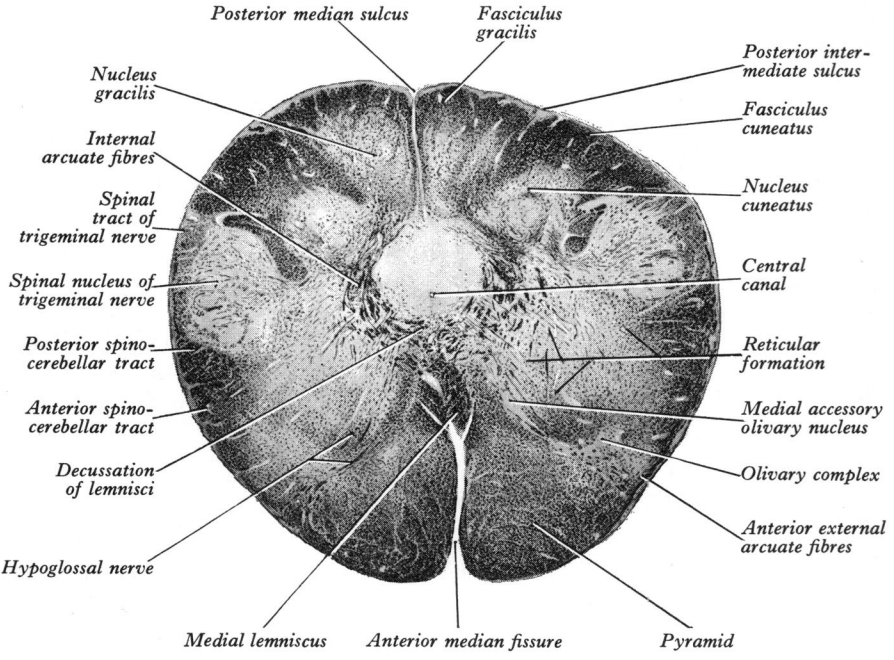

7.89 Transverse section through the medulla oblongata at the level of the decussation of the lemnisci. Weigert Pal preparation. Magnification × 6.

all vagal visceral afferent fibres end in the nucleus solitarius (vide infra). Superiorly the dorsal vagal nucleus is lateral to the hypoglossal, in the fourth ventricle's floor subjacent to the *trigonum vagi*. Dorsolateral to it at this level is the *nucleus solitarius* (strictly: *nucleus of the tractus solitarius*) (7.92), intimately related to the descending *tractus solitarius*. At the medulla's lower end these nuclei fuse dorsal to the central canal. As the nucleus solitarius ascends it lies deeper in the medulla, ventrolateral to the dorsal vagal nucleus and almost co-extensive with it. The tractus solitarius receives afferent fibres from the facial, glossopharyngeal and vagus nerves, entering it in descending order and conveying gustatory information from the lingual and palatal mucosa and, according to many, visceral impulses from the pharynx (glossopharyngeal and vagus) and from the oesophagus and abdominal alimentary canal (vagus). In this vertical representation there is some overlap (Schwartz et al 1951, Kerr 1962). Neurons in the solitary nucleus are smaller than those in the dorsal vagal nucleus; their axons may project to the thalamus and

thence to the cerebral cortex, but attempts to establish this in cats have failed (Morest 1967). The dorsal vagal nucleus and nucleus ambiguus are connected. The solitary nucleus may project to the upper levels of the spinal cord through a *solitariospinal tract* (p. 932) in man (Collier & Buzzard 1903) and cat (Torvik 1957). This nucleus is considered to receive fibres from the spinal cord, cerebral cortex and cerebellum (Angaut & Brodal 1967). A neuronal group ventrolateral to the solitary nucleus has been termed the *nucleus parasolitarius* (Crosby et al 1962).

Numerous islets of grey matter are scattered centrally in the ventrolateral medulla, an area intersected by nerve fibres in all directions and hence termed the *reticular formation*. It exists at all medullary levels, extending to the pontine tegmentum and midbrain. The reticular formation at all brain-stem levels is considered in continuity (p. 988).

The medullary white matter is rearranged above the corticospinal decussation. The *pyramids* contain corticospinal and corticonuclear fibres, the latter distributed to nuclei of cranial nerve and other medullary nuclei; they form two large ventral bundles flanking the anterior median fissure; dorsal are the accessory olivary nuclei and lemniscal decussation.

The *medial lemniscus* (p. 952) ascends from the lemniscal decussation on each side as a flattened tract, near the median raphe. The tracts ascend to the pons, increasing as fibres join from upper levels of the decussation. Ventral are the corticospinal fibres and dorsal the medial longitudinal fasciculus and tectospinal tract. In the decussation fibres are rearranged; those from the gracile are ventral to those from the cuneate nucleus; above this the medial lemniscus is also rearranged (p. 983):

gracile fibres *become lateral*, cuneate fibres *medial*. At this level medial lemniscal fibres in monkeys (Ferraro & Barrera 1936) and chimpanzees (Walker 1937) show a laminar somatotopy on a segmental basis. This is almost certainly present in mankind.

The *medial longitudinal fasciculus*, a small compact tract near the midline and ventral to the hypoglossal nucleus, is continuous with the ventral spinal intersegmented tract; at this medullary level it is displaced dorsally by the pyramidal and lemniscal decussations. It ascends in the pons and midbrain in the *same* relation to the central grey matter and midline and is therefore

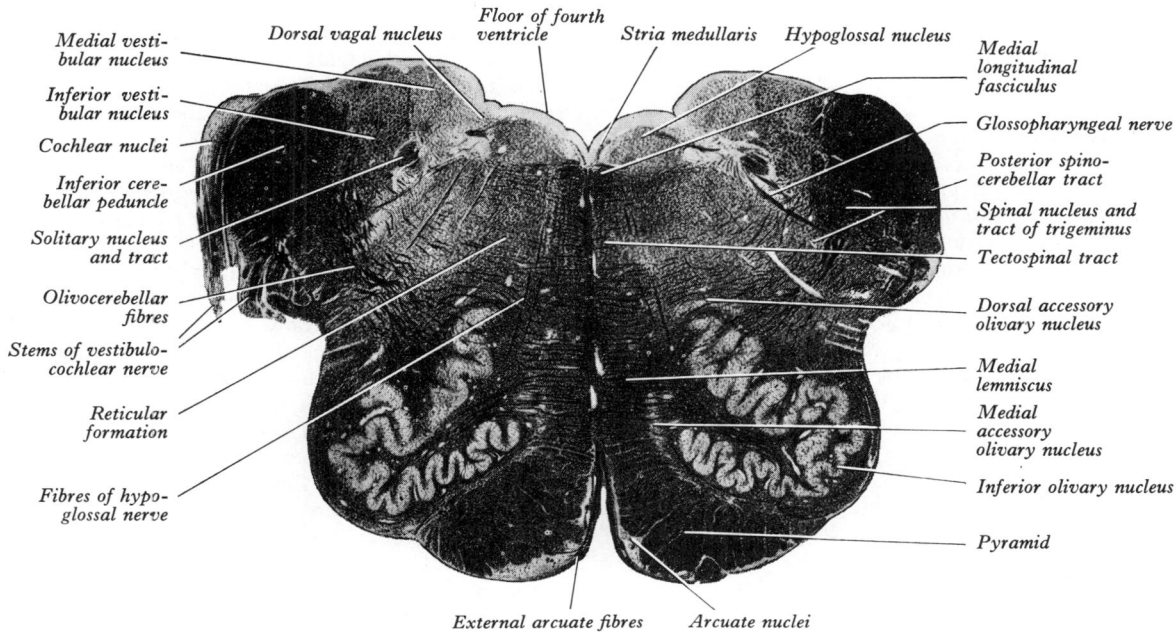

7.90 Transverse section through the medulla oblongata at mid-olivary level. Weigert Pal preparation. Magnification × 4·5.

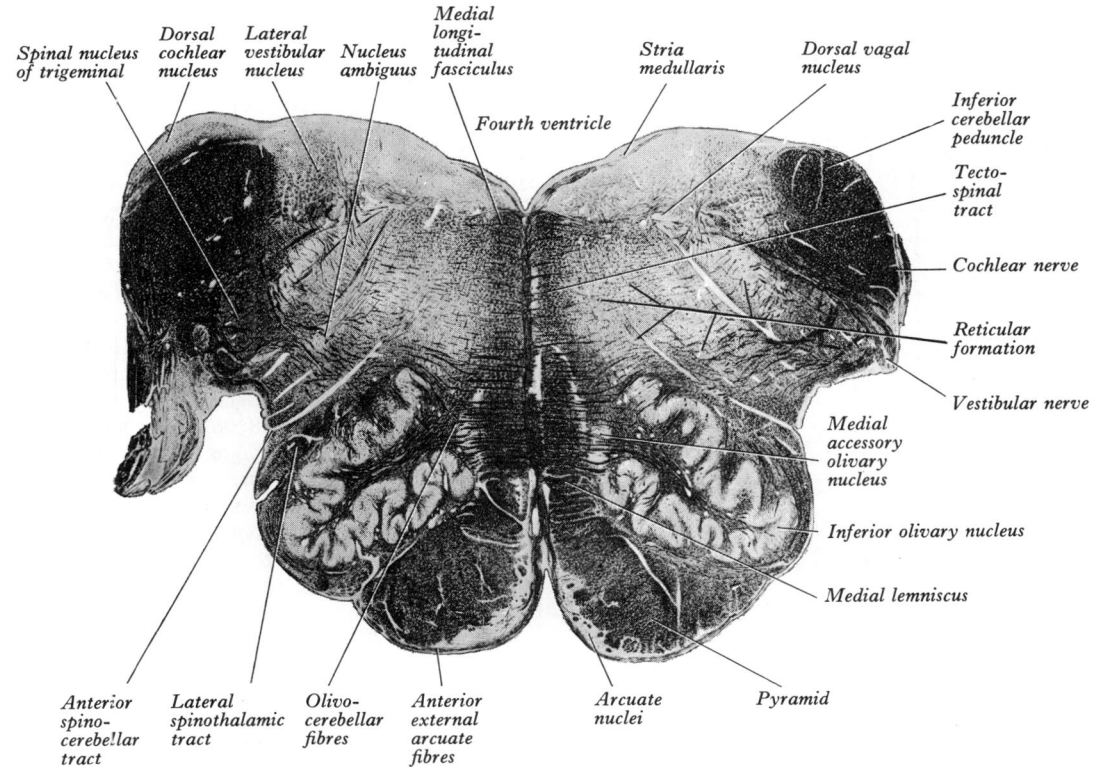

7.91 Transverse section through the superior half of the medulla oblongata. Weigert Pal preparation. Magnification × 4·5.

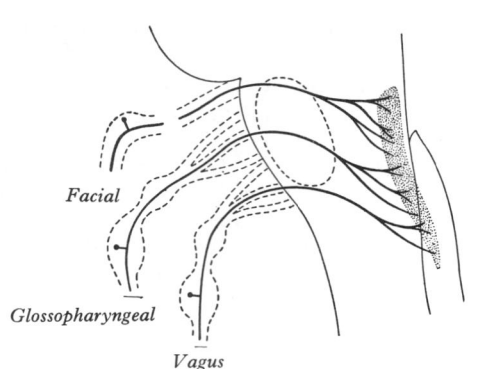

7.92 Afferent fibres of the facial, glossopharyngeal and vagus nerves, conveying gustatory impulses to the nucleus of the tractus solitarius.

near the somatic efferent column. Its fibres have short courses in it, being from a variety of sources, and are detailed on p. 985.

The *spinocerebellar, spinotectal, vestibulospinal, rubrospinal and lateral spinothalamic (spinal lemniscal)* tracts are all in the ventrolateral area, limited dorsally by the spinal trigeminal nucleus, ventrally by the pyramid.

(3) **A transverse section level with the fourth ventricle's lower end** (7.90) shows some new features with most of those already described. The total of grey matter is increased by the large olivary nuclear complex, arcuate nucleus and nuclei of the vestibulocochlear, glossopharyngeal, vagus and accessory nerves.

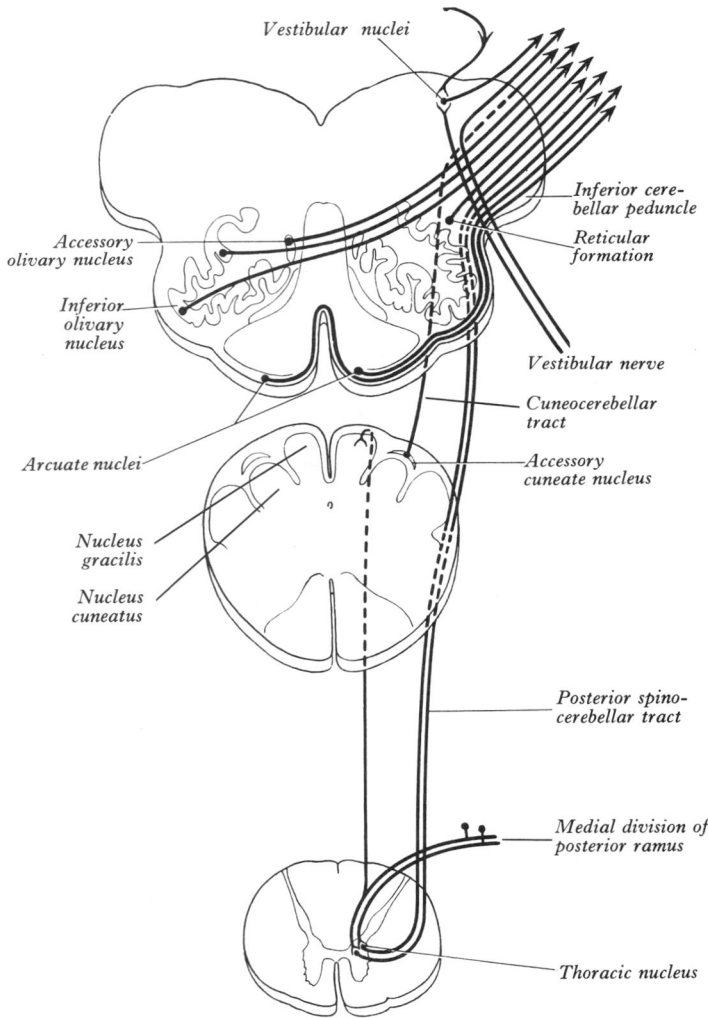

7.93 Some of the afferent components of the inferior cerebellar peduncle; the efferent components have been omitted. See text for further details.

The *inferior olivary nucleus* is hollow, irregularly crenated, with a longitudinal medial hilum and surrounded by myelinated fibres forming the *olivary amiculum.* Dorsolateral to the pyramid, it underlies the olive but ascends to the pons. It contains small neurons, most of which form the *olivocerebellar tract,* emerging from the hilum or through the adjacent wall to run medially, intersecting the medial lemniscus (7.90, 91). Its fibres *cross the midline,* decussate and sweep dorsal to or traverse the opposite olivary nucleus, intersecting the lateral spinothalamic and rubrospinal tracts and spinal trigeminal nucleus to enter the contralateral inferior cerebellar peduncle. The afferent olivary connections are ascending and descending; ascending fibres, mainly crossed, arrive from all spinal levels in the *spino-olivary tracts;* ascending connections also arrive via the dorsal white columns (Hand & Liu 1966). Descending fibres come from the cerebral cortex, thalamus, basal nuclei, red nucleus and central grey of the midbrain (Walberg 1960), some perhaps in the *central tegmental fasciculus* (Jansen & Brodal 1950).

The *medial accessory olivary nucleus* is a curved grey lamina, concave laterally, between the medial lemniscus and pyramid and the ventromedial aspect of the inferior olivary nucleus. The *dorsal accessory olivary nucleus* is a similar lamina dorsomedial to the inferior olivary nucleus. Both nuclei are connected with the cerebellum. The accessory nuclei are phylogenetically older than the inferior, connecting with the paleocerebellum (p. 963); the inferior nucleus occurs only in mammals, enlarging caudally during its evolution. In all connections, cerebral, spinal and cerebellar, olivary nuclei sometimes display very specific somatotopy, particularly in their cerebellar connections, detailed later (pp. 974, 975, 978). The *arcuate nuclei* are curved, interrupted bands, anteromedial to the pyramids; they are said to be displaced nuclei pontis (Rasmussen & Peyton 1946). Anterior external arcuate fibres are derived from them.

The central grey matter, at this level spread over the ventricular floor, contains the *hypoglossal* and *dorsal vagal nuclei* and the *nucleus solitarius* ventrolateral to the vagal; these and the inferior cerebellar peduncle are near the lower ends of the *inferior and medial vestibular nuclei* (p. 958). Between the hypoglossal and dorsal vagal nuclei is the *nucleus intercalatus* (p. 954).

A group of large motor neurons, *the nucleus ambiguus,* is isolated deep in the reticular formation and descends as far as the upper end of the dorsal vagal nucleus. Fibres emerging from it above join the glossopharyngeal nerve, those at a lower level join the vagus and cranial accessory nerves. Below, it continues into the spinal accessory nucleus (p. 951). It has large motor neurons, whose fibres innervate striated muscle of branchial origin (p. 1113). These first pass dorsomedially, then curve laterally to join the emerging fila of the cranial accessory, vagus and glossopharyngeal nerves. The nucleus displays several groups of neurons in man and other mammals; some representation of the muscles innervated has been established (Lawn 1966, and p. 1114). At its upper end, between this and the facial nucleus, is a small *retrofacial nucleus;* though in line with the special visceral efferent nuclei, it is a reputed source of general visceral efferent vagal fibres.

The gracile and cuneate nuclei, diminishing and irregular, fill the dorsolateral region; ventral is the *spinal trigeminal* tract and nucleus. The *cochlear nuclei* are superficial to the inferior cerebellar peduncle (pp. 958, 963). The white matter shows little change except for the development of the inferior cerebellar peduncles lateral to the fourth ventricle. The pyramid, medial lemniscus, tectospinal tract and medial longitudinal fasciculus are unchanged in position. Olivocerebellar tracts sweep across the midline, turning dorsally to join the inferior peduncles (p. 965). *Anterior external arcuate fibres,* bilateral in origin, arising from both arcuate nuclei, emerge from the anterior median fissure to pass dorsolaterally over the surface of the pyramid, olive and spinal trigeminal tract to reach the posterior spinocerebellar tract, then together ascending into the inferior cerebellar peduncle (7.93).

Hypoglossal radicles emerge ventrally from their nucleus, traverse the reticular formation lateral to the medial lemniscus and medial to (sometimes through) the olivary nucleus and curve laterally to emerge superficially through the ventrolateral sulcus.

A small lesion in the ventral medulla at this level may involve the corticospinal tract and hypoglossal nerve, causing characteristic paralyses in ipsilateral lingual muscles and contralateral limbs. The reticular formation is traversed dorsally by vagal radicular fibres and their dorsal nucleus, the nucleus ambiguus and nucleus solitarius, all emerging at the dorsolateral sulcus.

The *spinal lemniscus* (p.935), dorsal to the inferior olivary nucleus, is separated from the surface by the anterior spinocerebellar and spinotectal tracts. Clinical and experimental evidence indicates somatotopy among its fibres: those for the lower limb are superficial, for the upper limb deep and for the trunk intermediate. At it ascends in the medulla, the lemniscus is near the nucleus ambiguus; a ventral lesion in the reticular formation may hence paralyse the ipsilateral vocal fold and soft palate, with contralateral loss of pain and temperature sensibility.

(4) **A transverse section of the medulla at its upper limit** shows little change. The dorsal surface is relatively flat but undulant and may show a few fibres of the stria medullaris (p.960). The inferior olivary nucleus is unchanged, but accessory olivary nuclei are diminishing (7.91). The medial vestibular nucleus is wider and dorsolateral to the dorsal vagal nucleus, separating the latter from the floor of the rhomboid fossa. The inferior vestibular nucleus is between the medial vestibular nucleus and the inferior cerebellar peduncle. At the pontomedullary junction the *lateral vestibular nucleus* (p.958) replaces the inferior vestibular nucleus and the *cochlear nuclei* usually appear. The nucleus solitarius, spinal trigeminal tract and nucleus and the nucleus ambiguus are almost unchanged in position.

The white substance at this level is also little changed: the lateral spinothalamic tract (spinal lemniscus) ascends dorsal to the olivary nucleus, its fibres retaining their somatotopic arrangement (p.935). The *inferior cerebellar peduncle* (p.965) is a large dorsolateral bulge. The *medial lemniscus* widens ventrally and moves between the pyramid and the narrowing olivary nucleus (7.97), its dorsal part receding from the tectospinal tract and medial longitudinal bundle. Entering the pons, it spreads coronally (7.96) in the ventral tegmentum (p.959). It contains fibres for proprioceptive and tactile sensibility. As it ascends in the medulla it is probably joined by the anterior spinothalamic tract (p.932). In the pons, therefore, it contains fibres serving proprioception, tactile and pressure sense from the contralateral lower limb, trunk and upper limb; fibres from the lower limb are lateral, from the upper limb intermediate and from the neck medial in position.

The medullary reticular formation is described on pp.988–996.

THE PONS

The pons lies ventral to the cerebellum, below the midbrain and above the medulla with which it is continuous, but superficially demarcated from it by a transverse ventrolateral sulcus in which the abducent, facial and vestibulo-cochlear nerves appear. Its *ventral surface* (7.81,82) is markedly convex transversely, less so vertically. Its transverse fibres 'bridge' the midline, converging on each side into a large but compact *middle cerebellar peduncle*. It adjoins the sphenoid's dorsum sellae and basilar occipital bone. The surface has a shallow vertical median *sulcus basilaris*, usually containing the basilar artery and bounded bilaterally by prominences due partly to the corticospinal fibres descending in the pons. Lateral to these, near mid-pontine level, the trigeminal nerves emerge, each with a superomedial motor root and a larger, inferolateral sensory root; verticals just lateral to these roots are arbitrary boundaries between the ventral pontine surface and middle cerebellar peduncles. The *dorsal surface*, hidden by the cerebellum, contributes to the rostral half of the rhomboid fossa (p.980). Transverse sections show a dorsal *tegmentum* and a *ventral (basilar) part*, the former a continuation of the medulla, excluding its pyramids. The ventral pons contains bundles of *longitudinal fibres*, some continued into the pyramids, some ending in many pontine or medullary nuclei; also present are many *transverse* fibres and scattered *nuclei pontis*.

('Pons' is commonly used in two senses: to denote the external protuberance of the ventral pons, or to include both this and the tegmentum, i.e. the whole brain stem between the medulla and midbrain.) In mammals, the degrees of development of the cerebral hemispheres, ventral pons and cerebellum (neocerebellum, p.963) are correlated. The ventral pons is absent in non-mammalian vertebrates, present in all marsupials (*Metatheria*) and mammals (*Eutheria*) and possibly present in monotremes (*Prototheria*). It enlarges *pari passu* with the cerebrum and cerebellum, ascending the mammalian evolutionary scale.

INTERNAL STRUCTURE OF THE PONS

The ventral (basilar) pons is similar in structure at all levels. Its longitudinal fascicles (7.97) contain corticopontine, corticonuclear and corticospinal fibres, descending from the midbrain's crus cerebri (p.981), which enter the pons compactly but rapidly disperse into fascicles, separated by *nuclei pontis* and transverse pontine fibres. *Corticospinal fibres* traverse the pons to the medullary pyramids, converging again here into compact tracts (p.950); *corticonuclear fibres* accompany them, some diverging to both contralateral (and some ipsilateral) nuclei of cranial nerves and other nuclei in the pontine tegmentum, the rest reaching the pyramids. Clinical evidence supports the contention that the facial and other nuclei (p.961) also receive ipsilateral corticonuclear fibres. *Corticopontine fibres*, from the frontal, temporal, parietal and occipital cortex, end in the nuclei pontis (7.94); axons from the latter are the *transverse pontine fibres* (pontocerebellar) which after midline decussation continue to form the contralateral middle cerebellar peduncle. Frontopontine axons end in the nuclei pontis above the level of the emerging trigeminal roots and are relayed to the contralateral cerebellum forming the upper transverse pontine fibres. A few pontocerebellar fibres do not cross and end as mossy fibres in the cerebellar cortex (p.971). A degree of somatotopy is maintained in these connections. Axons from the midbrain tectum (*tectopontine fibres*) may also relay in the nuclei pontis (Pearce 1960), perhaps some also from the spinal levels (*spinopontine fibres*).

The *nuclei pontis* include all the neurons scattered in the ventral pons. Varying in size and shape, they are, as indicated, relays on paths from the cerebral cortex to the cerebellum; they migrate ventrally and upwards from the rhombic lip (p.980) but do not all reach the ventral pons; some remain as an oblique dorsolateral ridge across the inferior cerebellar peduncle, forming the *nucleus of the circumolivary bundle* (corpus pontobulbare), whose axons are said to ascend vertically on the surface between the emerging facial nerve and eighth cranial nerve on its lateral side (p.981). Afferents to the nucleus have been claimed to traverse the pons descending longitudinally with the corticospinal fibres, leaving in the medulla to *ascend*, recurring dorsally over the olive to the nucleus, as part of the *fasciculus circumolivaris pyramidis* (7.103).

The tegmental pons varies at different levels, especially in its cytoarchitecture but is adequately described for general purposes by examples of the lower and upper sections.

Transverse section through the pontine lower tegmentum transects the facial colliculus (7.94), containing the motor nucleus of the abducent and the geniculum of the facial nerves. More deeply placed is the facial nucleus, the nearby vestibular and cochlear nuclei and other isolated neuronal groups. The *medial vestibular nucleus* thus continues a little from the medulla into the pontine tegmentum, with the *lateral vestibular nucleus* between it and the inferior cerebellar peduncle.

The vestibular nuclei, subjacent to the roughly rhomboid vestibular area (p.980), are laterally placed in the fourth ventricle's rhomboid fossa. They comprise medial, lateral, superior and inferior vestibular groups receiving fibres from the eighth cranial nerve's vestibular part and sending axons to the cerebellum, medial longitudinal fasciculus, spinal cord and lateral lemniscus. The *medial vestibular nucleus* broadens, then narrows as it ascends from the upper olivary level into the lower pons, separating the dorsal vagal nucleus from the fourth ventricle's floor. It is crossed by the striae medullares nearer the floor. Below, it is continuous with the nucleus intercalatus. The *inferior vestibular nucleus* (the smallest) lies between the medial vestibular nucleus and inferior cerebellar peduncle, from the level of the upper end of the nucleus gracilis to the pontomedullary junction. It is traversed by descending fibres of the vestibular nerve and the vestibulospinal tract. The *lateral vestibular nucleus* is just above the inferior, ascending almost to abducent nuclear level, and is composed of large multipolar neurons, which are the main source of the vestibulospinal tract. The *superior vestibular nucleus* is small and lies above the medial and lateral nuclei.

Vestibular fibres of the vestibulocochlear nerve enter the medulla between the inferior cerebellar peduncle and the trigeminal spinal tract and approach the vestibular area, separating into ascending and descending fascicles. Fibres ending in medial, lateral and inferior vestibular nuclei descend medial to the inferior cerebellar peduncle and mingle with neurons of the inferior nucleus. Ascending fibres enter the superior or medial nuclei; a few enter the cerebellum directly via its inferior peduncle (superficially in the *juxtarestiform body*), to end in the nucleus fastigii, flocculonodular lobe and uvula (p. 965). Vestibular nuclei project extensively to the cerebellum (p. 966) and also *receive* axons from its cortex and fastigial nuclei (p. 975). Their spinal projections via the vestibulospinal tracts have been described (p. 945); vestibular axons also reach the spinal cord in the medial longitudinal fasciculus (p. 939) and some also reach cerebral levels, possibly for bilateral cortical representation. The vestibular nuclear complex also projects to the pontine reticular nuclei (p. 990) and, via the medial longitudinal fasciculus, to motor nuclei of the ocular muscles.

Minor neuronal groups near the vestibular nuclei have been observed; only the *interstitial nucleus* is known to receive primary vestibular axons. The *nucleus parasolitarius* (p. 954) may be linked to the vestibular complex by afferent fastigial connections

(p. 975). Nuclear vestibular neurons much exceed afferent fibres in the nerve, which reach only limited regions of the nuclei; many cells may therefore be interneurons and many also sources of the projections already mentioned. Evidence favours spatial representation of the vestibular apparatus in the nuclei (p. 1111). For appraisal of earlier experimental research on vestibular paths see Rasmussen & Windle (1960) and Brodal et al (1962).

Fibres of the vestibulocochlear nerve's cochlear division partially encircle the inferior cerebellar peduncle laterally and end in *dorsal* and *ventral cochlear nuclei*, projecting slightly on its surface. The *dorsal cochlear nucleus*, or auditory tubercle (p. 959), is posterior on the peduncle and continuous medially with the rhomboid fossa's vestibular area. The *ventral cochlear nucleus* is on its ventrolateral aspect, between the cochlear and vestibular fibres of the eighth cranial nerve, both parts of which, with the cochlear nuclei, are usually at the level of the pontomedullary junction.

The ventral cochlear nucleus, complex in cytoarchitecture, contains many neuronal types (e.g. giant, large and small spherical, multipolar, granular), the variants being segregated in separate nuclear regions (Cajal 1909, Lorente de No 1933 and p. 1112). Marked topographical order has been demonstrated in cochlear nerve terminals within the nucleus, different parts of the spiral organ and differing stimulation frequencies being related to neurons serially arrayed antero-inferiorly in the ventral nucleus (Schuknecht 1960, Whitfield 1960). All cochlear nerve fibres enter the nucleus, bifurcating into ascending branches which end in the ventral nucleus and descending ones traversing it to reach the dorsal nucleus (vide infra). Human cochlear fibres are estimated at about 25 000; since the ventral nucleus alone contains at least three times as many neurons in cats, whose cochlear fibres are half the human complement, cochlear neurons greatly exceed these fibres. Ferraro & Minckler (1977) also found fibre numbers in the lateral lemniscus much more numerous (p. 1112). A minor fraction of the cochlear neurons receive terminals from the nerve, though each fibre may connect with several (Lewy & Kobrak 1936). Terminals are limited to the antero-inferior region of the ventral nucleus, whose elements are probably in large part local interneurons.

The dorsal cochlear nucleus is almost continuous with the ventral, separated only by a thin stratum of nerve fibres. It also

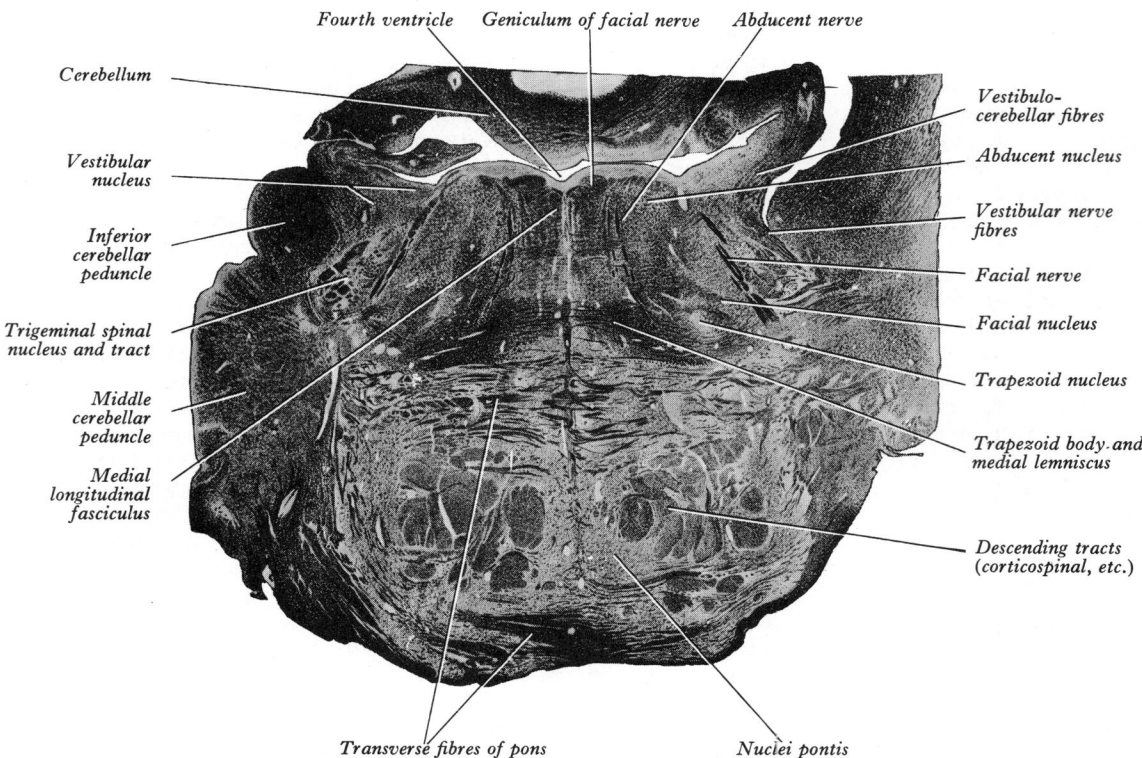

7.94 Transverse section through the pons at the level of the facial colliculus. Weigert Pal preparation. Magnification × 3·5.

Fourth ventricle   *Geniculum of facial nerve*   *Abducent nerve*

Cerebellum

Vestibular nucleus

Inferior cerebellar peduncle

Trigeminal spinal nucleus and tract

Middle cerebellar peduncle

Medial longitudinal fasciculus

Vestibulo-cerebellar fibres

Abducent nucleus

Vestibular nerve fibres

Facial nerve

Facial nucleus

Trapezoid nucleus

Trapezoid body and medial lemniscus

Descending tracts (corticospinal, etc.)

Transverse fibres of pons   *Nuclei pontis*

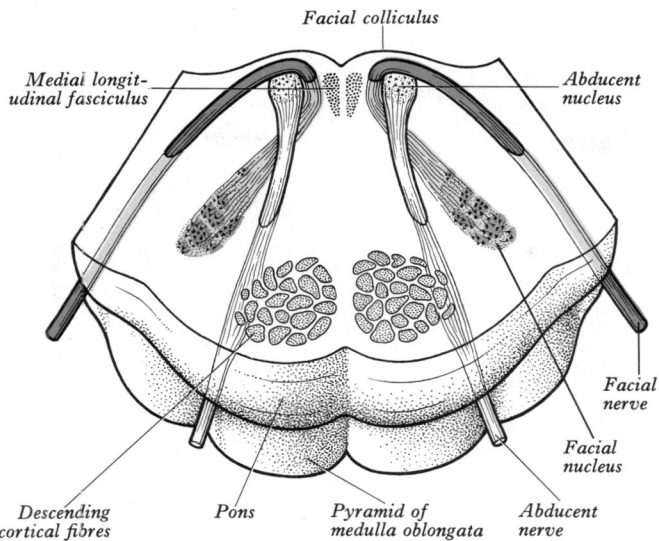

Facial colliculus

*Medial longit-
udinal fasciculus*

*Abducent
nucleus*

*Facial
nerve*

*Facial
nucleus*

*Descending
cortical fibres*

Pons

*Pyramid of
medulla oblongata*

*Abducent
nerve*

7.95 Diagram of the central course of the fibres of the facial nerve,
superior aspect, in a transverse section of the pons.

has a complex laminar pattern, with a range of cell types like those
in the ventral nucleus but also a fusiform or pyramidal type
peculiar to it. Again, the terminals appear limited in distribution,
being, as mentioned, descending branches of cochlear fibres (Olsen 1969).

Though, quantitatively, their cellular origins are not precisely
known, axons of most neuronal types (all having efferent axons,
with few confined to the nucleus) nevertheless leave to end at
pontine levels in the superior olivary, trapezoid and lateral lemniscal nuclei (vide infra). They leave the cochlear nuclei by three
routes. (1) A ventral contingent, the most numerous, decussates
as the *trapezoid body*, level with the pontomedullary junction
(7.96). These axons ascend slightly while decussating; a few, however, do not cross, synapsing in the ipsilateral superior olivary
nuclei. Those decussating relay in the contralateral nuclei; and in
both the next order of axons ascend in the corresponding *lateral*

lemniscus. A few decussating fibres traverse the contralateral
superior olive into the lateral lemniscus to relay in lemniscal
nuclei. (2) Some axons from ventral cochlear neurons pass dorsally, superficial to the descending trigeminal spinal fibres, the
cerebellar fibres in its inferior peduncle and to axons of the *dorsal*
cochlear nucleus (vide infra). These ventral cochlear fibres, smaller than those in the trapezoid decussation, swerve ventromedially
across the midline, ventral to the medial longitudinal fasciculus as
intermediate acoustic striae (Held 1893); their further path is uncertain; they probably ascend in the opposite lateral lemniscus.
(3) The most dorsal fibres, from the dorsal nucleus, curve dorsomedially round the inferior cerebellar peduncle towards the
midline as *dorsal acoustic striae* (Monakow 1905), not to be confused with the striae medullares (p. 960), to which they are ventral.
They incline ventromedially and cross the midline to ascend in the
opposite lateral lemniscus, probably relaying in its nuclei.

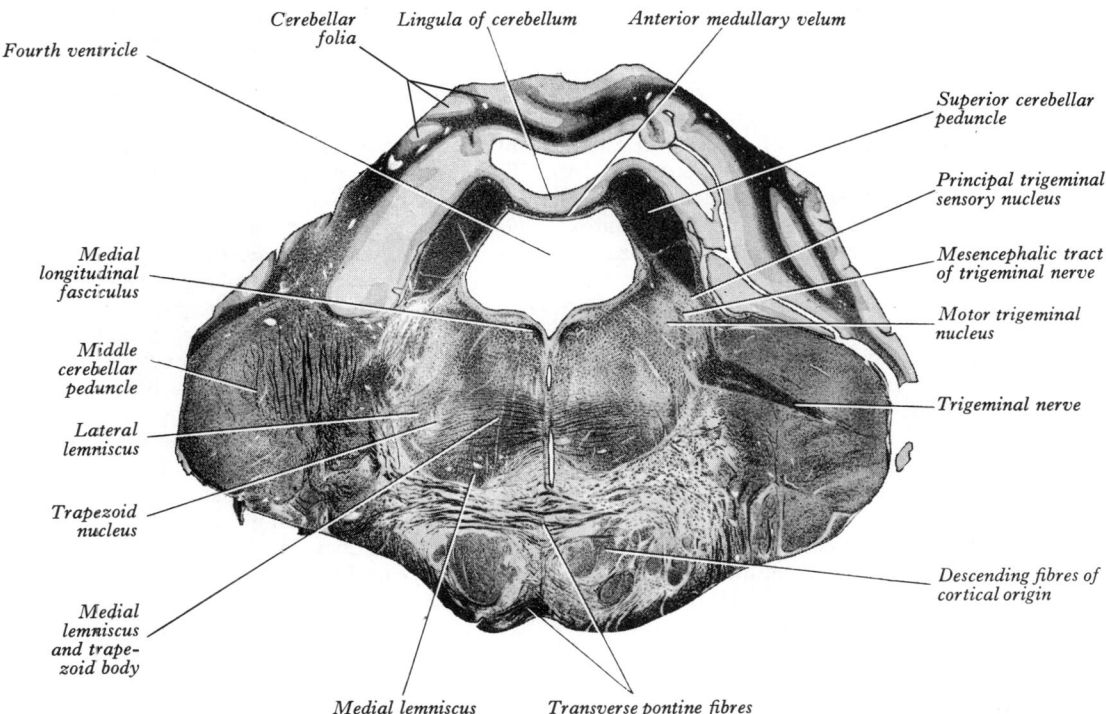

Fourth ventricle

*Cerebellar
folia*

Lingula of cerebellum

Anterior medullary velum

*Superior cerebellar
peduncle*

*Principal trigeminal
sensory nucleus*

*Mesencephalic tract
of trigeminal nerve*

*Motor trigeminal
nucleus*

Trigeminal nerve

*Medial
longitudinal
fasciculus*

*Middle
cerebellar
peduncle*

*Lateral
lemniscus*

*Trapezoid
nucleus*

*Medial
lemniscus
and trape-
zoid body*

*Descending fibres of
cortical origin*

Medial lemniscus

Transverse pontine fibres

7.96 Transverse section of the pons at the level of the trigeminal nerve.
Weigert Pal preparation. Magnification × 2·5.

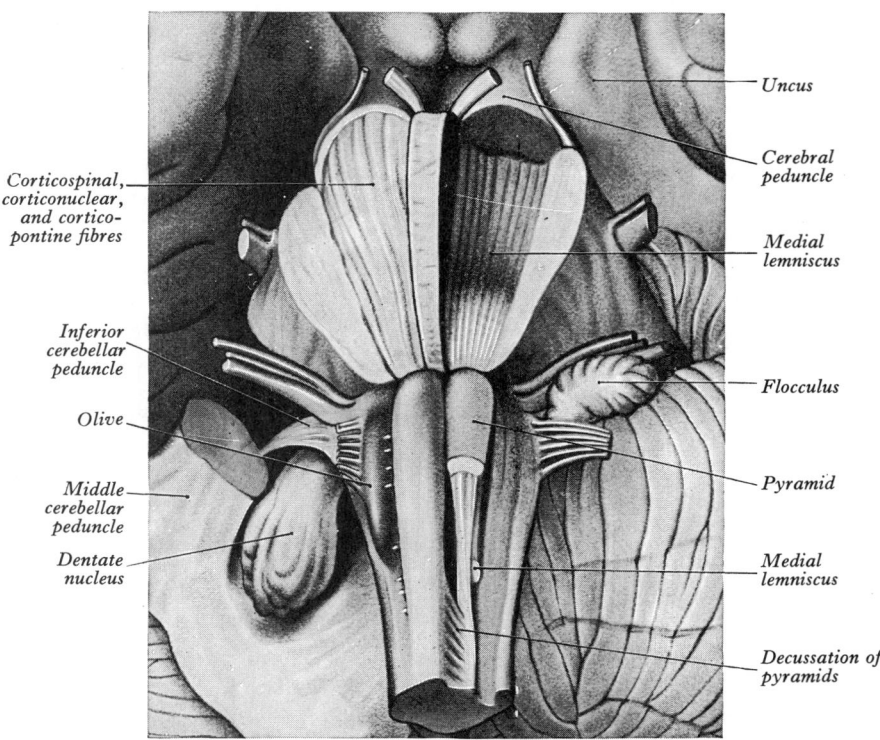

Corticospinal,
corticonuclear,
and cortico-
pontine fibres

Inferior
cerebellar
peduncle

Olive

Middle
cerebellar
peduncle

Dentate
nucleus

Uncus

Cerebral
peduncle

Medial
lemniscus

Flocculus

Pyramid

Medial
lemniscus

Decussation of
pyramids

7.97 Ventral aspect of a dissection of the pons, the medulla oblongata and the right cerebellar hemisphere. In the pons and medulla, the dissection is deeper on the left (right side of figure). Note the spiralling of the medial lemniscus from sagittal (below) to oblique coronal (in pons).

The *superior olivary complex* is lateral in the reticular formation, level with the pontomedullary junction. Medial to it are the trapezoid nucleus and body, inferior to it the much larger inferior olivary group (p. 956). The superior complex includes several named nuclei and nameless smaller groups (Irving & Harrison 1967). The *lateral superior olivary nucleus* ('S-shaped segment' of the feline olivary complex) is small in primates. The *medial (accessory) superior olivary nucleus* (para-olivary nucleus of Minckler) is larger in man. Medial again to this is the trapezoid nucleus (vide infra). Dorsal in the complex is a *retro-olivary group*, the reputed origin of efferent cochlear fibres described below. Some internuclear connections have been described. Unit recording from the feline lateral nucleus (Warr 1966) demonstrated a tonotopical organization, adjoining neurons being connected to the ipsilateral cochlear fibres concerned with different but related frequencies. The medial superior olivary nucleus receives impulses from both spiral organs; evidence suggests its involvement in auditory sound source localization. With the trapezoid nuclei, superior olivary complexes are the main relay stations of the ventral and the largest cochlear projection. These intricate connections are not yet well established in man. Neuronal counts of certain human nuclei have been recorded by Ferraro & Minckler (1977).

The *trapezoid nucleus* has a ventral component of neurons scattered among the trapezoid fascicles and a more compact dorsal nucleus, medial to the superior olivary complex. Though usually regarded as auditory relays, human trapezoid nuclei remain in this an uncertain issue. Some trapezoid axons may enter the medial longitudinal fasciculus, ascending to end in trigeminal, facial, oculomotor, trochlear and abducent nuclei, mediating reflexes involving the stapedius, tensor tympani and extraocular muscles.

The *nucleus of the lateral lemniscus* consists of small groups of neurons among lemniscal fibres. Lateral and medial groups may receive afferent axons from both cochlear nuclei; their efferents enter the midbrain along the lemniscus, with further connections considered later (pp. 986, 1015). Total neuronal counts of 18 000–24 000 have been recorded in human lemniscal nuclei (Ferraro & Minckler 1977).

Efferent cochlear axons travel in the cochlear nerves to the spiral organ (Held 1983, Rasmussen 1967). Though few (about 500 in cats), they may be involved in hearing, perhaps by inhibitory and excitatory reflexes via cochlear nuclei (Allanson & Whitfield 1955). Lateral inhibition has been demonstrated by unit recording at trapezoid level. Efferent cochlear fibres appear to come from retro-olivary neurons in the superior olivary complex, fibres from both sides proceeding to both cochleae. Borg (1973) has confirmed projection to the cochlea, in rabbits, from neurons both dorsal and ventral to the superior olivary nucleus, describing also a descending connection from the inferior colliculus to the olivary complex (confirming Munzer & Wiener 1902). Since a descending projection from the medial geniculate body to the inferior colliculus has been demonstrated, a complete efferent corticocochlear pathway appears established (p. 1015).

Striae medullares of the fourth ventricle are *not* auditory nerve fibres; evidence suggests that they are an *aberrant cerebropontocerebellar* connection, in which arcuate nuclei (p. 956), the pontobulbar body (p. 957) and external arcuate fibres (p. 956) are involved. Disagreement exists regarding the pontobulbar body. This description merely presents a consensus of views, of variable reliability. Embryological evidence (p. 185) suggests that some neurons, migrating ventrally to the pons from the rhombic lip to form the nuclei pontis, remain near the fourth ventricle as the pontobulbar nucleus. Others migrate further, scattered ventrally over much of the surface of the pyramids as arcuate nuclei, just below the pons. Both groups may receive corticobulbar projections, descending to the arcuate nuclei in the pyramids with the corticospinal fibres. Axons from both arcuate groups spread round the medulla, above and below the inferior olive, their fascicles being superficially visible. All these fibres, collectively the external arcuate fibres, enter the inferior cerebellar peduncle. Some so-called '*external*' arcuate fibres pass dorsally from arcuate nuclei *through* the medulla near its midline; near the fourth ventricle's floor they decussate and then turn laterally under the ependyma, entering the cerebellum by its inferior peduncle. They are striae medullares, also known as the *arcuatocerebellar tract*. This may end in the flocculus (Szentágothai 1955); some fibres are said to end in the pontobulbar nucleus but may be confused with its projection fibres; some may travel ventrally in the *circumolivary fasciculus*, which skirts below the inferior olive and are usually visible on the surface. The fasciculus and pontobulbar nucleus have been found absent in pontine aplasia, a confirmation of their pontocerebellar association (Baumgarten 1959). Efferent circumolivary fibres, passing ventrally, join the arcuatocerebellar tract, also reaching the striae medullares to enter the contralateral inferior cerebellar peduncle. However, as noted (p. 960), pontobulbar afferents have also been ascribed to the circumolivary fasciculus; the precise relation of efferent and afferent fibres to the fasciculus remains uncertain.

The abducent nucleus is paramedian in the central grey matter, in line with the trochlear, oculomotor and hypoglossal nuclei, it forms a *somatic motor column* (p. 185). It is near the medial longitudinal fasciculus, which lies ventromedial; via this fasciculus, vestibular and cochlear nuclei and those of other cranial nerves, especially the oculomotor, connect with the abducent, also related to the emerging facial nerve (vide infra). Efferent abducent axons pass ventrally, descending through the reticular formation, trapezoid body and medial lemniscus and traversing the ventral pons to emerge at its inferior border (p. 1107).

The facial nucleus is deeper, lying ventrolateral in the pontine reticular formation, posterior to the dorsal trapezoid nucleus; the trigeminal spinal tract and nucleus are dorsolateral. The facial nucleus receives both crossed corticonuclear fibres and a smaller ipsilateral number (p. 1111) and ipsilateral rubroreticular tract fibres. Axons of its large motor neurons form the facial nerve. At first they incline *dorsomedially* towards the fourth ventricle, below the abducent nucleus (7.95) and ascending medial to it near the medial longitudinal fasciculus, through which the facial may communicate with other cranial nerves. Its fibres now curve anterolaterally round the upper pole of the abducent nucleus, as the *geniculum of the facial nerve*, and descend anterolaterally through the reticular formation. Finally they pass between their own nucleus medially and the spinal trigeminal nucleus. This unusual course provided apparent evidence for neurobiotaxis (p. 185). In 10 mm human embryos the nucleus is in the fourth ventricle's floor, in the branchial (special visceral) efferent column, and is above the abducent nucleus. As growth proceeds, the facial nucleus migrates at first downwards, dorsal to the abducent, then ventrally to its final position. As it migrates its axons elongate, their subsequent course marking the path of migration. ·

The facial nucleus receives corticonuclear fibres for volitional control and (among other sources) afferents from its own sensory root (via the nucleus solitarius) and the spinal trigeminal nucleus. These infracortical afferents complete local reflex loops, as in the spinal segments. It is opined that the facial nucleus migrated to remain near the nucleus solitarius and the spinal trigeminal nucleus.

Facial neurons are grouped; those innervating muscles in the scalp and upper face are dorsal and believed to receive bilateral corticonuclear fibres (Papez 1927, Buskirk 1945). The groups may represent discrete motor pools (p. 1109).

The salivary nucleus is near the upper pole of the dorsal vagal nucleus, just above the pontomedullary junction (Lewis & Shute 1959) and near the inferior pole of the facial nucleus. It is customarily divided into *superior and inferior salivary nuclei*, sending secretomotor fibres to the salivary and, perhaps, lacrimal glands via the facial and glossopharyngeal nerves (pp. 1110, 1113).

The trigeminal spinal nucleus ascends in the lower pons, its tract still applied to its lateral aspect. It is ventral to the lateral vestibular nucleus and traversed by vestibular nerve fibres. The inferior cerebellar peduncle is lateral, but inclines dorsally as it ascends into the cerebellum; the trigeminal spinal tract and nucleus are then lateral to the middle cerebellar peduncle. The nucleus continues up to expand into the superior sensory trigeminal nucleus. The spinal 'tract' consists not of endogenous fibres but of descending axons from the trigeminal ganglion. Their collaterals and terminals enter the spinal nucleus, which is continuous below with the spinal substantia gelatinosa (p. 1099). It is mainly concerned with the mediation of pain and thermal sensibility in the trigeminal area and shows well-established topographical organization (p. 1099).

In addition to the tracts already noted at lower levels, the lower pontine tegmentum contains the trapezoid body, lateral lemniscus and emerging fibres of the abducent and facial nerves. The *medial lemniscus* is ventral; its now *transverse* outline, a flat oval, extends laterally from the median raphe (7.94); its vertical fibres are traversed by horizontal trapezoid axons; it is laterally related to the *lateral spinothalamic tract* and *trigeminal lemniscus*, which are from neurons of the contralateral spinal nucleus, serving pain and thermal sensibility in facial skin and mucosae of the conjunctiva, tongue, mouth, nose, etc. The lemnisci together here form a transverse band composed, in lateral order from the midline, of medial and trigeminal lemnisci, the lateral spinothalamic tract and lateral lemniscus.

The *trapezoid body* contains cochlear fibres, mainly from the ventral cochlear and trapezoid nuclei. These ascend transversely in the ventral tegmentum; traversing or passing ventral to the vertical medial lemniscal fibres, they decussate with the contralateral fibres in the median raphe. Below the emerging facial axons, the trapezoid fibres turn up into the *lateral lemniscus*, as an *ascending* auditory path.

The *medial longitudinal fasciculus* is paramedian, ventral to the fourth venticle, near the abducent nucleus and facial fibres ascending medial to this. Its proximity to many structures suggest its connections (p. 985). In the lower pons it receives fibres from vestibular and perhaps dorsal trapezoid nuclei (p. 960), through its peduncle. Vestibulocochlear contributions form its greater part (p. 960). It is the main 'intersegmental' tract in the brain stem, particularly for interactions between nuclei of extraocular muscles and the vestibular system.

A transverse section at an upper pontine tegmental level contains trigeminal elements (7.96), but otherwise shows no notable alteration.

The trigeminal motor nucleus is in the reticular formation, near the lateral part of the fourth ventricle's floor in line with trigeminal fibres traversing the ventral pons (7.96).

The principal (superior) trigeminal sensory nucleus is lateral to its motor nucleus, between it and the middle cerebellar peduncle, continuous below with the spinal nucleus. Secondary axons from it decussate to ascend in the medial lemniscus to the thalamus (p. 1099).

The small *lateral lemniscal nucleus* is medial to its tract in the upper pons, receiving some lemniscal terminals; some of its efferent fibres enter the medial longitudinal fasciculus, others return to the lemniscus. It is a relay station associated with the trapezoid nucleus.

Tegmental white matter in the upper pons contains the lateral lemnisci, replacing the trapezoid body; its dorsolateral parts are invaded by the superior cerebellar peduncles.

The *medial lemniscus* (7.97), ventral in the pontine tegmentum, is a little lateral to the median raphe and is joined medially by fibres from the principal trigeminal sensory nucleus serving proprioceptive, tactile and pressure receptors in its area. Dorsolateral are the trigeminal lemniscus and lateral spinothalamic tract, serving pain and thermal senses, and the lateral lemniscus and its nucleus. As the lateral lemniscus ascends it is near the brain stem's dorsolateral surface. Above, its fibres enter the inferior colliculus and medial geniculate body. The *medial longitudinal fasciculus* retains its paramedian position.

The *superior cerebellar peduncle*, a mass of fibres, many from the intracerebellar dentate nucleus (p. 975), ascends into the lateral part of the fourth ventricle's roof. It also inclines ventromedially into the dorsolateral tegmentum; the *ventral spinocerebellar tract* is intimately associated with it. In the medulla it is dorsal to the olivary nucleus, separated from the surface only by anterior external arcuate fibres. In the lower pons it inclines dorsally between the trigeminal sensory nucleus and middle cerebellar peduncle to the superior peduncle's lateral aspect. Its fibres then recurve to descend dorsally into the cerebellum.

The *reticular formation* is continuous throughout the pons and is considered on pp. 988–996.

THE CEREBELLUM

The cerebellum, the largest part of the hindbrain, is posterior to the pons and medulla, its median region being separated from them by the fourth ventricle. It occupies the posterior cranial fossa, covered by the tentorium cerebelli (p. 1086). Somewhat ovoid, it is constricted in its median region and flattened, the greatest diameter being transverse. Average weight in males is about 150 grams. In adults the ratio of cerebellum to cerebrum is about 1 to 8, in infants about 1 to 20.

GENERAL FORM OF THE CEREBELLUM

The cerebellum has two so-called *hemispheres* joined by a sagittal, transversely narrow median *vermis*. On its *superior surface* there is no deep grooving; the superior vermian surface, a slight median ridge, is continuous with the hemispheres (7.98). Ventrally the *superior vermis* projects beyond the free edge of the tentorium cerebelli and thence slopes postero-inferiorly below the straight sinus. The superior surface of each hemisphere adjoins the tentorium, sloping down laterally from the superior vermian aspect. Its ventrolateral *margin* corresponds to the tentorial attachment to the superior border of the petrous temporal bone; its curved dorsolateral margin adjoins the transverse sinus in its tentorial attachment.

Inferiorly the cerebellar hemispheres are separated by a deep *vallecula* (7.101). The inferior hemispheric surface is irregularly convex and adjoins the posterior surface of the petrous temporal bone, sigmoid sinus, mastoid temporal bone and occipital squama. The *inferior vermis* projects into the vallecula, limited on each side by a *vallecular sulcus*.

A wide, ventral, *anterior cerebellar notch* is adapted to the pons and upper medulla but separated by the fourth ventricle. In the notch's floor the peduncles enter the white core of the cerebellum.

Dorsally the hemispheres are separated by the *posterior cerebellar notch*, deep, narrow and containing the dural falx cerebelli.

SURFACE CEREBELLAR TOPOGRAPHY

The cerebellar surface is wholly occupied by numerous curved, transverse *fissures*, giving it a laminated appearance and separating its *folia*. Deeper fissures divide it into *lobules*, the most conspicuous being the *horizontal fissure* which extends around the dorsolateral border of each hemisphere from the middle cerebellar peduncle to the posterior cerebellar notch; it separates the superior and inferior surfaces.

Superior surface (7.98, 111). Deepest is the *fissura prima*, somewhat V-shaped, with a dorsal apex cutting into the superior vermian surface at the junction of the anterior two-thirds with the posterior. The fissure curves ventrolaterally around the superior surface to meet the horizontal fissures.

Deep fissures divide the superior vermian surface into *lingula, central lobule, culmen, declive* and *folium vermis* in ventrodorsal order. Each, except the lingula, is continuous bilaterally with an adjoining lobule in each hemisphere (7.100, 111). The fissura prima crosses between the culmen and declive. The lingula is a single lamina of four or five shallow folia, its white core continuous with the superior medullary velum. It is separated from the central lobule by a *postlingual fissure*, the lobule being bilaterally continuous with its *alae* and limited dorsally by the *postcentral fissure*. Between this and the fissura prima is the culmen and laterally the *quadrangular lobules*.

Superior hemispheric surfaces and the vermis *dorsal* to the fissura prima are divided by a curved *postlunate fissure* into a ventral region consisting of the *declive* and its lateral extensions, the *lobuli simplices*, and a dorsal region, the *folium vermis* with adjoining *superior semilunar lobules* limited dorsally by the horizontal fissure.

Inferior surface (7.101, 111). The inferior vermis is divided into the *tuber vermis, pyramid, uvula* and *nodule*, in dorsoventral order. The vermian tuber is continued laterally into *inferior semilunar lobules*, bounded dorsally by the horizontal fissure and ventrally by a *prepyramidal fissure*. The *pyramid* is separated from the *uvula* by a *postpyramidal fissure* (*fissura seconda*) and continuous laterally with a *biventral lobule* visible on each inferior hemispheric surface. Ventral to the uvula, separated by the median part of a *dorsolateral sulcus*, is the *nodule* (7.99, 101).

On each inferior hemispheric aspect, ventral to the biventral lobule, is a deep *retrotonsillar fissure*, passing laterally from the sulcus valleculae opposite the fissura secunda to curve forwards to the ventral surface. With the anterior part of the sulcus valleculae it bounds a circumscribed *tonsil*, connected to the uvula across the sulcal floor by a strip of cortex, the *furrowed band* (7.101). Above, the tonsil is close to the inferior surface of the inferior medullary velum.

The *nodule*, most ventral on the inferior vermian surface, is separated from the uvula by the dorsolateral fissure and connected on each side to the flocculus and its hemispheric core by the inferior medullary velum. Its anterosuperior aspect faces the fourth ventricle. Ventrally, covered by grey matter, it is crossed by two or three shallow fissures, separated from the ventricle by

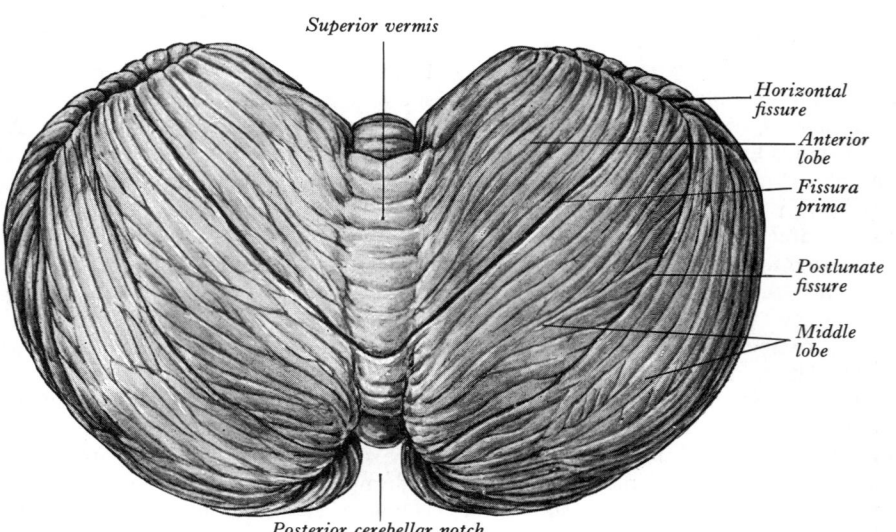

Superior vermis

Horizontal fissure

Anterior lobe

Fissura prima

Postlunate fissure

Middle lobe

Posterior cerebellar notch

7.98 Superior aspect of the cerebellum to show major fissures and lobes. Compare with 7.100.

a double layer of pia mater, its choroid plexus and ventricular ependyma (7.115). Grey matter is dorsally deficient, the white surface being covered only by the neuroglia and ependyma (7.115). The nodule's sides are ventrally free and covered by grey matter; it is continuous dorsally with the inferior medullary velum.

The small *flocculus* is partially detached and just below the vestibulocochlear nerve as it enters the brain stem; it is crossed ventrally by glossopharyngeal and vagal radicles near the jugular foramen. It is oval, with a crenated margin and from its medial end a narrow *peduncle* of fibres emerges; it is covered ventrally by the fourth ventricle's lateral recess and part of the choroid plexus projecting from its aperture (7.81). The peduncle contains afferent and efferent fibres; at the fourth ventricular lateral angle it forms dorsal and ventral parts, the former connecting the flocculus with the nodule and uvula; the ventral part passes medially, turning upwards near the lateral border of the pontine part of the ventricular floor, many of its fibres being afferents from vestibular nuclei and perhaps the medial accessory olivary nucleus; but others are efferent to vestibular nuclei and some may ascend to higher levels.

LOBES OF THE CEREBELLUM

The cerebellum can be primarily divided into a flocculonodular lobe and the corpus cerebelli, the latter having anterior and middle lobes (7.100, 111A). Such division is embryological, morphological and functional. The *flocculonodular lobe* comprises the nodule, both flocculi and their peduncles. The remainder is the *corpus cerebelli*, separated from the flocculonodular lobe by the *dorsolateral fissure*, which is the first to appear in phylogeny and ontogeny. The corpus is subdivided by the *fissura prima* into *anterior* and *middle* lobes. The former is anterior to the fissure and includes the lingula, central lobule and its alae, the culmen and its quadrangular lobules. The rest, the *middle lobe*, includes the median declive, folium vermis, tuber vermis, pyramid, uvula and, serially, their bilateral lobuli simplices, superior and inferior semilunar lobules, biventral lobules and tonsils.

Some sectors are phylogenetically older. The flocculonodular lobe, mostly *vestibular* in connections, and lingula, with *spinocerebellar* and *vestibular* connections, form the oldest, the *archicerebellum*. The anterior lobe, except the lingula but including the pyramid and uvula of the middle lobe, largely *spinocerebellar* in connections, is the phylogenetically intermediate *palaeocerebellum*. At this evolutionary stage, this newer lobe separates the archicerebellum into the lingula (ventral and rostral) and the flocculonodular lobe (dorsal and caudal). With evolution of the neopallium in mammals, further cerebellar expansion with the addition of a larger middle lobe (excluding the

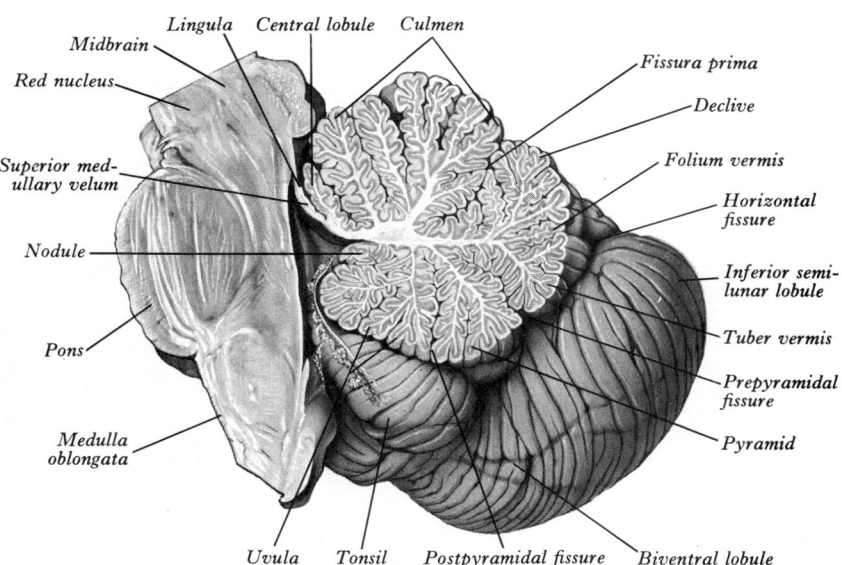

7.99 Median sagittal section of the cerebellum and brain stem.

pyramid and uvula), forms the *neocerebellum*, mainly *corticopontocerebellar* in its afferent connections. Like the palaeocerebellum, the neocerebellum expands between the anterior and flocculonodular lobes (7.100, 111A).

The superior medullary velum, a thin white lamina, stretches between the superior cerebellar peduncles (brachia conjuctiva), forming the upper part of the fourth ventricle's roof; its ventricular surface is covered by ependyma. The velum is narrow above between the inferior colliculi, broader below where it is continuous with the superior vermis. The lingula's folia are prolonged dorsally on its lower half; a median *frenulum veli* descends on it from between the inferior colliculi. The trochlear nerves emerge beside the frenulum (7.113).

The inferior medullary vela are bilateral, crescentic sheets flanking the nodule. Each is a thin layer of nerve fibres and neuroglia, surfaced internally by the ventricular ependyma, externally by the pia mater. The internal surface is the inferior limit of the lateral dorsal recess (p. 979); the external surface adjoins the tonsil's upper aspect. The convex peripheral velar limit is continuous with the cerebellar white core and sides of the pyramid, uvula and nodule; the ventral (sometimes inferior) border is free (7.113) and from it the ventricular ependyma is prolonged in apposition to the pia mater to form a thin part of the ventricular roof, reaching the taeniae. (The taeniae are narrow ridges of white

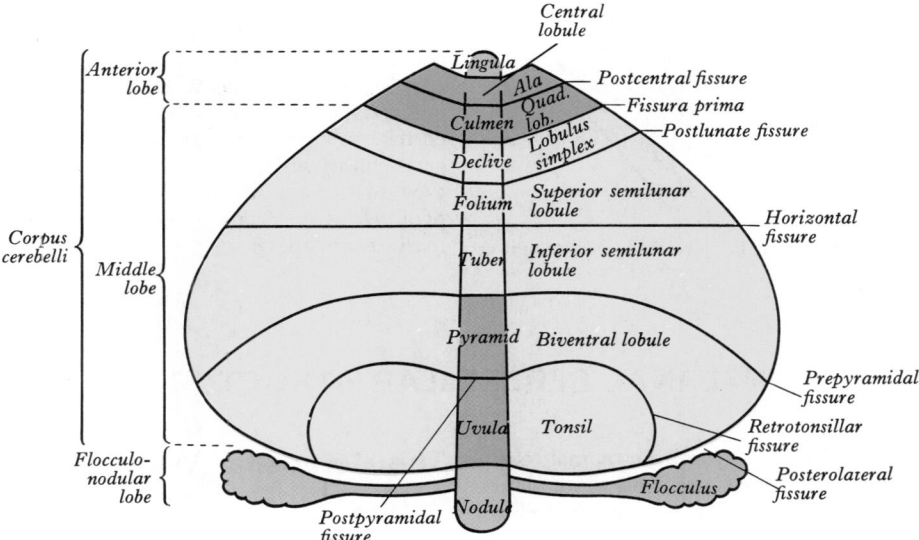

7.100 The fissures, lobes, lobules and main morphological subdivisions of the cerebellum. Blue = archicerebellum. green = paleocerebellum. yellow = neocerebellum. See also illustration 7.111 for further details.

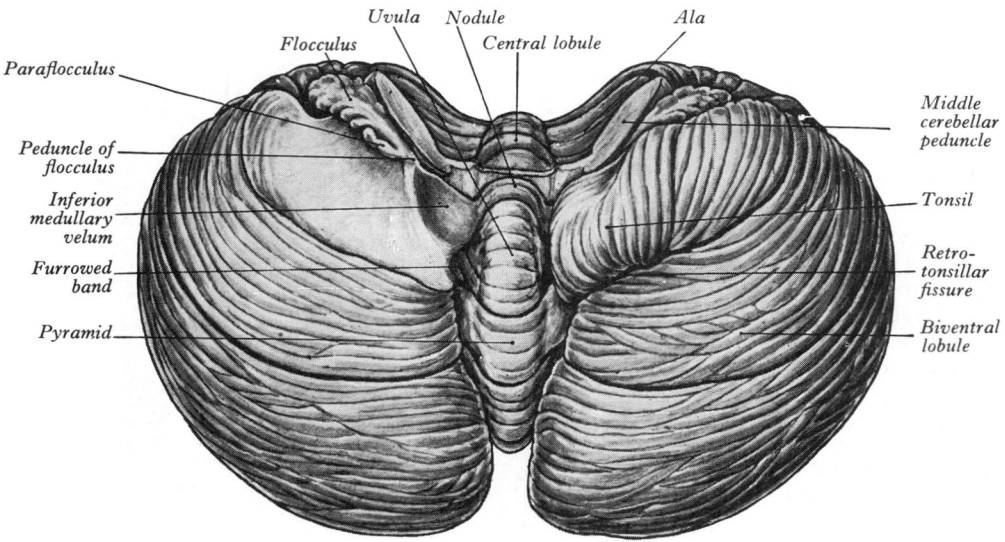

Uvula · Nodule
Flocculus · Central lobule · Ala
Paraflocculus
Middle cerebellar peduncle
Peduncle of flocculus
Inferior medullary velum
Tonsil
Furrowed band
Retro-tonsillar fissure
Pyramid
Biventral lobule

7.101 Inferior aspect of the cerebellum. The tonsil and the adjoining part of the biventral lobule of the right side have been removed.

matter forming the inferolateral limits of the fourth ventricle; here nervous tissue and the tela choroidea of the ventricle's roof commences, p. 979.) At each ventrolateral corner it continues into the dorsal part of the floccular peduncle, from which most, if not all, its nerve fibres issue (7.101).

GENERAL CEREBELLAR DEVELOPMENT

For initial cerebellar development see p. 187. Early in the third month it is a mass across the roof of the hindbrain vesicle, soon

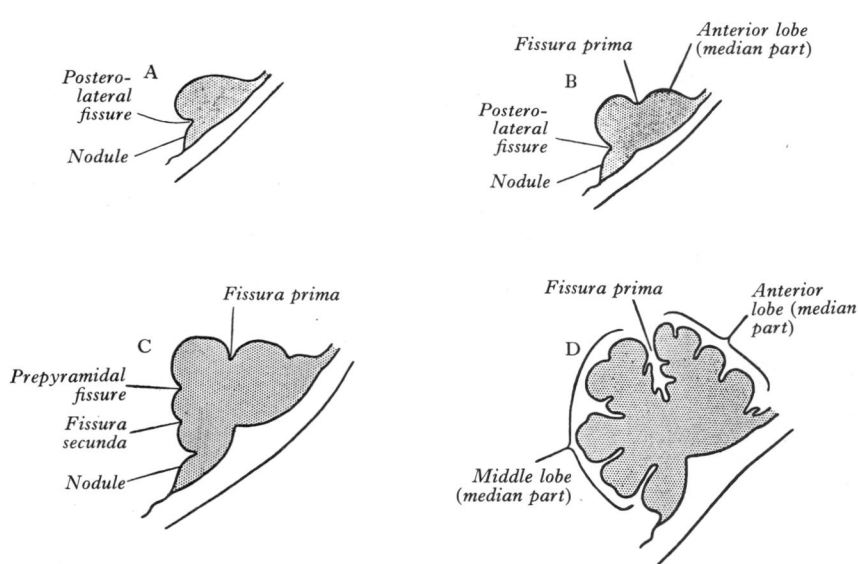

Postero-lateral fissure
Nodule
A

Fissura prima
Anterior lobe (median part)
B
Postero-lateral fissure
Nodule

Fissura prima
C
Prepyramidal fissure
Fissura secunda
Nodule

Fissura prima
Anterior lobe (median part)
D
Middle lobe (median part)

7.102 Median sagittal sections through the developing cerebellum at four successive stages.

becoming transversely bilobar; its narrow median part is the predecessor of the vermis, its expanded bilateral lobes becoming the hemispheres. As growth proceeds transverse dorsal grooves appear on this rudiment, forming superficial fissures (7.101, 102). The lateral parts of the *dorsolateral fissure* appear first, demarcating an inferior (caudal) region from the rest; by this flocculi can be identified. The right and left parts of this fissure then meet in the midline, demarcating the nodule. Flocculonodular lobes can now be recognized as the most inferior (caudal) at this stage; but by adjoining growth (massive radial expansion) they become *antero-inferior* in adults. They form near the attachment of the epithelial roof, i.e. the superior rhombic lip (p. 185). As the third month ends a transverse sulcus on the rudiment superior slope deepens to form the *fissura prima*, invading the vermis and both hemispheres and separating the uppermost (rostral) region as the anterior lobe. About the same time two transverse grooves appear on the inferior vermian surface dorsal to the postnodular fissure. Initially the *fissura secunda* demarcates the uvula, then the *prepyramidal fissure* demarcates the pyramid (7.100–102). The whole cerebellum expands dorsally, rostrally, caudally and laterally; the inferior aspects of the hemispheres enlarging much more than their adjacent vermian surface, which is buried in the *vallecula*. Numerous other transverse grooves develop, most having little taxonomic significance; the most extensive, being an exception, becomes the *horizontal fissure*.

In many mammals part of the hemisphere above the flocculus becomes demarcated, in some most markedly. This is the *paraflocculus*, but the term is topographical; unlike the flocculus, it has mainly afferent connections from the cerebral cortex. A paraflocculus is not identifiable in the human cerebellum; whether it is represented by occasional patches of grey matter on the inferior aspect of the middle cerebellar peduncle is uncertain. (Cerebellar *cellular neurogenesis* is summarized on pp. 187–189.)

INTERNAL CEREBELLAR STRUCTURE

The cerebellum differs profoundly from the spinal cord, medulla and pons, the arrangement of its grey and white matter being reversed. A grey cortex covers the whole surface, following all fissures. Internal masses of grey matter also exist, but the core is predominantly white. In this the cerebellum resembles the cerebrum and this change has led to enormous evolutionary expansion of these parts in the human nervous system.

The Cerebellar White Core

The white core is thicker in the hemispheres than in the vermis, where it is a mere strip between the lateral parts. It is prolonged by many radially divergent primary laminae projecting towards the surface, with secondary laminae branching at very obtuse angles; these may divide again into shorter tertiary laminae, all covered by a grey stratum. Sagittal cerebellar sections cut across primary laminae at right angles; the cut surface has a highly branched pattern, the *arbor vitae* (7.99).

The white core contains: (1) fibrae propriae, (2) projection fibres, (3) myelinated axons of Purkinje neurons, (4) 'climbing' afferent fibres and (5) 'mossy' afferent fibres, requiring separate descriptions.

Fibrae propriae, intrinsic fibres, connect cerebellar regions. *Association fibres* connect the cortical folia, including the vermis; they do not decussate and most are short but some longer; they have been little studied (p. 969). Many *commissural fibres* connect the hemispheres, many being grouped in an *anterosuperior commissure*. A *posteroinferior commissure* also crosses near the fastigial nuclei (vide infra). Decussating spinocerebellar and cerebello-vestibular are intermingled with intrinsic commissural fibres.

Projection fibres, connecting the cerebellum with many other regions, form three large *peduncles* (7.103) on each side, issuing from the ventral cerebellar notch. Superior peduncles connect the cerebellum to the midbrain, middle peduncles to the pons and inferior peduncles to the medulla. These connections are shown highly simplified in 7.104.

THE INFERIOR CEREBELLAR PEDUNCLES

Each inferior cerebellar peduncle is a thick mass of fibres, entering and leaving between many sources and destinations. As noted elsewhere (p. 957) it forms on the dorsolateral aspect of the upper medulla. The pair diverge as they ascend; at the ventral cerebellar notch each turns dorsomedially into its hemisphere, then curves dorsally between the middle peduncle on its lateral aspect and the superior peduncle on its medial aspect. Some investigators divide the inferior peduncle into a small, medial, *juxtarestiform body* and a large, lateral *restiform body*. The relation of the oblique *pontobulbar body* to the peduncle has been mentioned (p. 957).

Tracts enter the cerebellum via its inferior peduncle from many medullary and spinal sources, including the olivo-, parolivo-, vestibulo-, reticulo-, dorsal spino-, cuneo- and trigemino-cerebellar tracts with the ventral external arcuate and other arcuatocerebellar fibres (striae medullares). Tracts *leaving* via the inferior peduncle include the cerebello-olivary, -vestibular and -reticular paths, with minor fascicles of cerebellospinal and cerebellonuclear fibres.

(1) **The posterior (dorsal) spinocerebellar tract** (p. 935) starts from neurons of the ipsilateral thoracic nucleus, shows somatotopic lamination and conveys impulses from cutaneous exteroceptors and proprioceptors of the hind-limb and lower trunk. Some are 'modality-specific', others carry impulses from multiple sources; transmission of both may be modified by

descending brain-stem tracts. Fibres end in 'hind-limb' regions of the cerebellar cortex (vide infra) including the central lobule, culmen, part of the declive and nearby parts of adjacent lobules; some fibres end in the pyramid and uvula.

(2) **The cuneocerebellar tract** (dorsal external arcuate fibres) from the external (accessory) cuneate nucleus reaches the ventral and dorsal regions of the ipsilateral vermis. Tract and nucleus are somatotopically organized and functionally similar to the posterior spinocerebellar tract, but concern the fore-limb and upper trunk. Receiving areas of the cerebellar cortex and localization within them are illustrated in 7.111A,B.

(3) **The olivocerebellar tract** contains axons from the inferior olivary nucleus and (4) **the parolivocerebellar tracts** contain axons from dorsal and medial accessory olivary nuclei. A close correspondence between points in the inferior olive and contralateral cerebellar cortex has long been mooted. Degenerative changes and experimentation have confirmed this, showing a precise point-to-point interconnection between all parts of the olivary complex and opposite hemisphere. Dorsoventral and mediolateral correspondence exists, the inferior olive projecting to larger, more lateral hemispheric areas, accessory olives to vermian and paravermian regions. Illustration 7.111E simplifies largely clinical findings (Jansen & Brodal 1954, Brodal 1969). Most olivocerebellar fibres decussate but a few fascicles are said to terminate ipsilaterally. They end as *climbing fibres*, latterly regarded as functionally highly important (vide infra). Afferents converge on the olivary nuclei from the spinal cord, cerebral cortex and subcortical regions. The spino-olivary tracts (p. 936) convey tactile and proprioceptive afferents to accessory olivary nuclei, which project largely to 'spinal regions' of the contralateral hemisphere, completing a *spino-olivocerebellar* path. Cortico-olivary fibres descend largely from the motor cortex to end in inferior and accessory olivary nuclei (Sousa-Pinto & Brodal 1969) completing a *cortico-olivocerebellar path*. Other subcortical nuclei, including the corpus striatum, red nucleus and

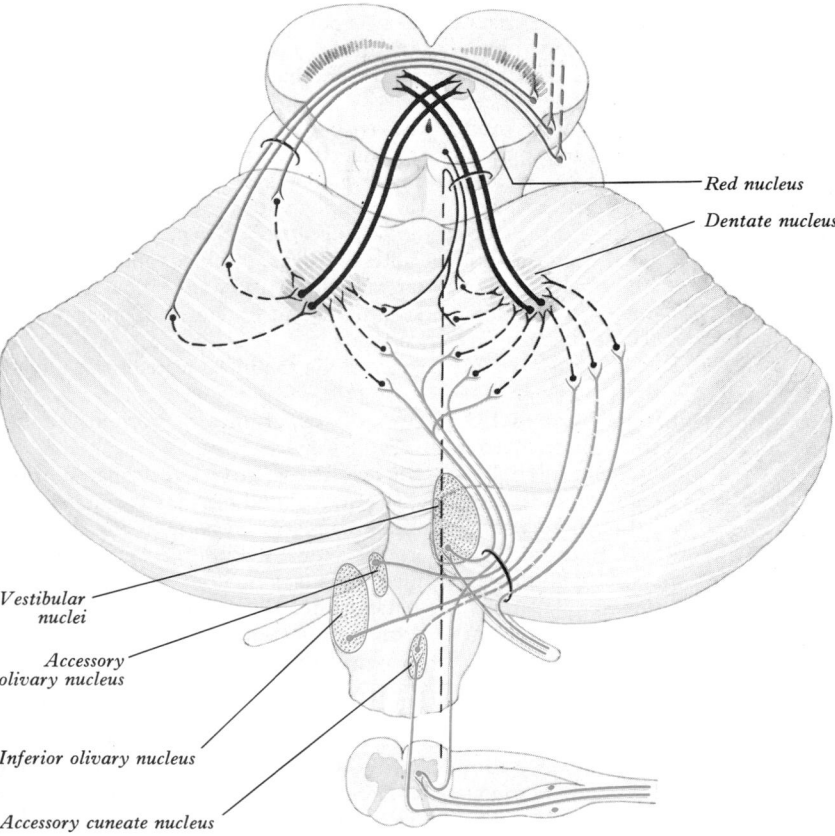

Red nucleus

Dentate nucleus

Vestibular nuclei

Accessory olivary nucleus

Inferior olivary nucleus

Accessory cuneate nucleus

7.104 A simplified diagram of a few of the connections of the cerebellum showing some components of the peduncles: inferior (blue), middle (magenta) and superior (black). The important cerebellothalamic fibres and other efferent systems have been omitted, as have olivary, reticular and tectal connections.

Cerebral peduncle   *Superior cerebellar peduncle*

Trigeminal nerve

Inferior cerebellar peduncle

Middle cerebellar peduncle

Pyramid

Olive

Fasciculus circum-olivaris

Vestibulo-cochlear nerve   *Inferior cerebellar peduncle*

7.103 Dissection of the left cerebellar hemisphere and its peduncles (by the late E B Jamieson, University of Edinburgh).

brain-stem reticular formation, all project to restricted regions of the olivary complex and may modify cerebellar cortical activity.

Not all olivocerebellar fibres end as climbing fibres; some reach contralateral fastigial, emboliform, globose and dentate intracerebellar nuclei and are somatotopically organized. These nuclei are described below.

(5) **The reticulocerebellar tract** is mainly from large and small neurons of the medullary *lateral reticular nucleus (nucleus of lateral funiculus,* p. 990), but other *perihypoglossal* and *paramedian* reticular nuclei also contribute (p. 953). Parts of the latter, in particular neuron group A2 (in part *nucleus commissuralis,* a perihypoglossal component), but also part of group A6 (nucleus coeruleus) which are noradrenergic (p. 993), connect via the inferior peduncle with intracerebellar nuclei and wide areas (perhaps all) of the cerebellar cortex. Other *general* reticulocerebellar fibres end in the vermian 'spinal' regions mentioned above. Indirect connections to the cerebellum thus exist through the medullary reticular nuclei, by convergence on them of spino-, vestibulo- and corticoreticular tracts. (The fibres from groups A2 and A6 are the *noradrenergic reticulocerebellar tract,* in part, the *coeruleocerebellar tract.*)

(6) **The vestibulocerebellar tract** consists mainly of primary afferent fibres from the vestibulocochlear nerve (vestibular part); but some secondary fibres stem from medial and inferior vestibular nuclei. Fibres terminate ipsilaterally in the archicerebellar cortex, principally the flocculus and nodule, but a few may end in the uvula and lingula. Other fibres reach both fastigial nuclei. Stein & Carpenter (1967) held that primary vestibulocerebellar fibres are mainly those with sensory endings in the ampullae of their semicircular ducts; but the nuclei of origin of secondary fibres receive the terminals of the rather small spinovestibular tract (p. 933). Vestibulocerebellar fibres partly occupy the juxtarestiform body (vide supra).

Other paths to the cerebellum via its inferior peduncle are less well investigated.

(7) **Ventral external arcuate fibres** arise from arcuate nuclei of both sides and, with some superficial reticulocerebellar fibres, course over the inferior peduncle's surface into the cerebellum. Arcuate nuclei are possibly homologous with pontine nuclei and hence their fibres may be part of a *cortico-arcuatocerebellar* path, whose cerebellar terminations and functions remain uncertain.

(8) **Striae medullares** (p. 960) may also originate in arcuate nuclei; they partially decussate and reach the flocculus and are sometimes termed *the arcuatofloccular tract* (fibres of Piccolomini), probably to be grouped with ventral external arcuate fibres and those from the pontobulbar body (p. 957).

(9) **Trigeminocerebellar fibres,** crossed and uncrossed, enter the cerebellum from superior sensory and spinal trigeminal nuclei. Their human distribution and the status of other nucleocerebellar tracts are uncertain.

Tracts leaving the cerebellum via the inferior peduncle include the following:

(10) **Cerebello-olivary fibres,** of uncertain origin, leave to end around inferior olivary neurons.

(11) **Cerebellovestibular fibres,** from the ipsilateral flocculus, nodule and fastigial nucleus (and partly its antimer), run with other elements of the juxtarestiform body to end in all the vestibular nuclei. Others are axons of Purkinje neurons in the ventral and dorsal vermian regions which do *not* synapse in the intracerebellar nuclei, passing directly to the lateral vestibular nucleus. The ventral vermian projection is somatotopic, corresponding in this to the origin of the lateral vestibulospinal tract (p. 932). Other vermian projections pass to the fastigial and thence to the lateral vestibular nucleus via a fastigiovestibular tract, all showing somatotopic order (Brodal et al 1962).

(12) **Cerebelloreticular fibres,** from both but mainly from the contralateral fastigial nuclei, partially decussate and proceed as the *hook bundle of Russell,* curving round the cerebellar end of the superior peduncle before joining the inferior peduncle. These fascicles end mainly in the pontine and medullary reticular formation. Different zones of the nucleus fastigii (vide infra) and *fastigiovestibular* and *fastigiobulbar* tracts have hence been recognized. But the two tracts are partially mixed in the direct juxta-

restiform route and less direct hook bundle, some of whose fibres may traverse the medulla to enter the ventral spinal funiculus as a *cerebellospinal tract.* Evidence for this and for direct *cerebellonuclear tracts* to brain-stem motor nuclei remains less than convincing.

THE MIDDLE CEREBELLAR PEDUNCLES

Each middle peduncle (brachium pontis), the most massive of the three, passes lateral to the others from the pons, curving dorsally to radiate into white laminae in the core of its cerebellar hemisphere (7.103); it is composed almost wholly of axons of secondary neurons in the *corticopontocerebellar path* which are distributed as nuclei pontis, scattered throughout the ventral pons. These fibres almost all cross, traversing the opposite middle peduncle to the contralateral cerebellar cortex. Much of the vermis probably receives fibres from both sides but the lingula, adjacent small regions of the anterior lobe and the flocculonodular lobe have no pontocerebellar projection.

The middle peduncle's fibres are arranged in superior, inferior and deep fasciculi. The *superior fasciculus* contains upper transverse pontine fibres; it passes dorsolaterally superficial to the others and is distributed mainly to inferior cerebellar lobules and to superior parts adjoining the dorsolateral margins. The *inferior fasciculus* contains inferior transverse pontine fibres; it passes inferomedial to the superior fasciculus, more or less parallel with it, to folia on the inferior cerebellar surface near the vermis. The *deep fasciculus* comprises most dorsal transverse fibres. At first covered by superior and inferior fasciculi, it crosses obliquely to appear medial to the superior, from which it receives a bundle; its fibres spread out to the superoventral cerebellar folia.

Distribution of pontocerebellar fibres from different pontine regions has been little investigated. In general, lateral and medial pontine nuclei project to the vermis, the intermediate to the contralateral hemisphere.

Pontine nuclei receive corticopontine terminal fibres from many, perhaps all parts of their ipsilateral cerebral cortex and also collaterals from some corticospinal fibres. Though corticopontine projection is extensive, quantification as to areas and possible somatotopy are unknown in primates. Somatotopy among corticopontine sensorimotor projection exists in cats (Brodal 1968a, b). Feline *spinopontine* and *tectopontine* connections are unconfirmed in primates. Another class of afferents in the cerebellum traverse the middle peduncle to reach cerebellar nuclei and wide areas (perhaps all) of the cortex. These aminergic fibres form a *cerebellar serotonergic path* (p. 994) derived from cell groups B5 (part of nucleus raphe pontis) and B6 (part of central pontine reticular formation). Their terminations and functions are undetermined.

SUPERIOR CEREBELLAR PEDUNCLES

Superior cerebellar peduncles (7.103, 116) emerge from the ventral cerebellar notch, overlapped by the anterior lobes; displacement of these reveals the superior medullary velum connecting them and their ascent lateral in the fourth ventricular roof to disappear below the inferior colliculi. Some fibres *enter* the cerebellum via the superior peduncle (vide infra), but most are *leaving* and largely from nucleus dentatus. **Efferent dentate fibres** emerge from the nuclear hilum and, joined by some from emboliform, globose and fastigial nuclei, ascend ventromedially, at first covered by medial fibres of inferior and deep fibres of middle peduncles. Ascending in the ventricular roof, they turn into the tegmental midbrain medial to the lateral lemniscus, sweeping medially to decussate with their contralateral fibres. Some do not decussate, leaving a small ipsilateral component with the crossed majority on each side. Both divide into ascending and descending fascicles, a *crossed ascending bundle* being much more prominent and the descending one often ignored.

The uncrossed component, in part from the fastigial nucleus, mainly reaches the brain-stem reticular formation. *Ascending terminals* end in reticular nuclei of the tegmental midbrain and circumaqueductual grey matter, *descending ones* in reticular nuclei of the pons and medulla.

Crossed descending fibres end mainly in inferior and accessory olivary nuclei and adjacent reticular formation. Some may reach the spinal cord, accompanying the medial longitudinal fasciculus to motor nuclei of extraocular muscles.

Dentatothalamic fibres cross the midline, bypass the red nucleus and end in the thalamic nucleus ventralis anterior (p. 1002), from which relay fibres radiate to cortical 'motor' regions. Thus large areas of the cerebellar cortex may influence the motor cortex via the dentate nuclei.

Crossed ascending fibres in the superior peduncle have been described as terminating in other thalamic nuclei, including the nucleus centromedianus, reticular nucleus, nuclei ventralis posterior and lateralis posterior, and in subthalamic and hypothalamic nuclei (Hassler 1950, Cohen 1958, Nimi et al 1962, Snider 1967).

Afferent fibres in the superior peduncle include the ventral spino-, tecto-, coeruleo- and descending hypothalamocerebellar tracts. **Ventral spinocerebellar tracts** are described on p. 935. In the spinal cord this laminated tract contains some ipsilateral but mostly crossed fibres; it ascends superficially and curves dorsally along the superior peduncle's lateral side to the cerebellum, where many crossed fibres cross back in the inferior cerebellar commissure, ending in 'hind-limb' cortical areas of ventral and dorsal vermian and paravermian regions (7.111A,B). **A superior spinocerebellar tract**, fore-limb counterpart of the ventral spinocerebellar, has been recognized by recording techniques; anatomical verification in primate brains is awaited.

Tectocerebellar tracts, descending from the tectum to the cerebellum, are paramedian in the superior medullary velum and probably arise in all four colliculi, ending in the intermediate vermian and paravermian cortex, including the dorsal region of the ventral lobe, declive and lobulus simplex, folium, tuber and pyramid (7.111D). The exact origins and constitution are unknown. Though they are widely assumed to convey auditory and visual information direct to the cerebellum; alternative routes are the *tectopontocerebellar* and *occipitopontocerebellar* pathways.

The coeruleocerebellar tract contains noradrenergic fibres from aminergic neurons of the brain-stem reticular group A6 (central core of nucleus coeruleus, pp. 980, 993). Traversing the superior peduncle they are spread to the intracerebellar nuclei and cerebellar cortex, details being unknown.

Hypothalamocerebellar tracts, composed in part of cholinergic fibres, were described by Shute & Lewis (1965) as descending from posterior hypothalamic nuclei with the dorsal longitudinal fasciculus to the cerebellum, via its superior peduncle, to uncertain destinations. De Feudi (1974) has reviewed cerebellar cholinergic connections.

Summary of Cerebellar Connections

The description of the contents of cerebellar peduncles is tedious, the plethora of detail confusing. In the following summaries some details are ignored, others emphasized, in an attempt to clarify cerebellar connections.

THE CEREBELLAR INPUT

Input is *direct* or *indirect*. *Fibres entering without intervening synapses*, mostly to the cerebellar cortex, come from neurons in the spinal cord, medulla, pons and midbrain. *All* other connections influence cerebellar activity *indirectly* through one or more of these direct paths.

Direct cerebellar inputs are: (1) from the spinal cord via spinocerebellar tracts (including cuneocerebellar); (2) from olivary, reticular, vestibular and arcuate nuclear complexes of the medulla; (3) from ventral, rapheal and other pontine reticular nuclei; (4) from the nucleus coeruleus; (5) from the mesencephalic tectum.

A useful approximate division into archi-, paleo- and neocerebellum has been noted (7.111A). Apart from minor inconsistencies and overlap, the *archicerebellum* ('vestibulocerebellum') has largely vestibular connections; the *paleocerebellum* ('spinocerebellum') has direct terminals from the spinal cord,

medullary reticular formation and accessory olivary nuclei; the *neocerebellum*, in the middle lobe's large hemispheres, receives massive *pontocerebellar* and inferior olivary connections and, in vermian and paravermian regions, a *tectocerebellar* input. (In accord with the imprecise but useful 'vestibulocerebellum' and 'spinocerebellum', addition of '*pontocerebellum*' and '*tectocerebellum*' may seem appropriate.) This crude parcelling of the cerebellar cortex into receiving areas for main *direct* afferent inputs may clarify complexities introduced by indirect inputs; it has already facilitated clinicopathological description (vide infra).

Indirect routes modifying cerebellar activity converge to the sources of the direct paths summarized above. They include other spinal routes, subcortical nuclei and extensive projections from the cerebral cortex. In addition to direct spinocerebellar tracts, accessory olivary nuclei and medullary reticular formation both project to the spinocerebellum and both receive spinal projections via spino-olivary and spinoreticular tracts. Limited regions of vestibular nuclei and the tectum receive fibres from spinovestibular and spinotectal tracts; but how much activity in vestibulocerebellar and tectocerebellar tracts is modified by spinal influences is uncertain. Nevertheless, spinal activity may reach the cerebellum in different orders of complexity. Direct spinocerebellar routes convey simple information; single fibres may carry patterns from stimulation of a single cutaneous or proprioceptive end organ; others carry patterns formed by the convergence and interaction of connections from two or more receptors, but even here transmission at the first synapse may be modified by descending paths such as the corticospinal. Indirect spinal paths presumably mediate further integration of spinal impulses with those converging by other routes to the medullary reticular formation, accessory olivary nuclei, etc., before entering the cerebellum (i.e. sites of '*cerebellar pre-integration*').

Similar generalizations apply to numerous routes by which the cerebral neocortex may modify cerebellar activity, the most massive being the corticopontocerebellar; with this may be linked to a cortico-arcuatocerebellar path. Other cell stations from which axons reach the cerebellar cortex also receive from the cerebral cortex, via corticotectal, corticoreticular and cortico-olivary tracts; neurons of spinocerebellar tracts are influenced by corticospinal fibres. Hence cerebral cortical activity may influence cerebellar input along most of its afferent paths.

Various brain-stem nuclei projecting to the cerebellum are *not homogeneous* in their connections; e.g. there is much evidence that the olivary complex is not only somatotopically organized in relation to the cerebellar cortex, but varies regionally in terms of afferent connections. Thus, some parts may be largely concerned with *through routes*, e.g. the spino-olivocerebellar path; others may be sites for *pre-integration* of input converging along several afferent channels to the olivary nuclei, including some from the spinal cord and cerebral cortex, others from the corpus striatum, red nucleus and brain-stem reticular formation.

Clearly fuller comprehension of cerebellar activity requires greater knowledge of input and its presentation in the cerebellar cortex. Spinocerebellar and vestibulocerebellar tracts have been much analysed; but regarding the significance of cerebral cortical control and the degree of pre-integration in the brain stem both remain obscure.

THE CEREBELLAR OUTPUT

This starts with axons of Purkinje neurons in the cerebellar cortex and involves intracerebellar nuclei (fastigial, globose, emboliform and dentate) on both sides, which are described later. Most Purkinje axons synapse with neurons in these nuclei, whose axons are the main cerebellar outflow; but some in the flocculonodular cortex and others in the ventral and dorsal vermis have axons which *bypass* intracerebellar nuclei, passing direct to extracerebellar destinations.

The most obvious effects of cerebellar dysfunction are disturbances in muscular integration. It is hence instructive to relate cerebellar outflow to the regions concerned with locomotor control. A few cerebellospinal and cerebellonuclear fascicles have been reported but direct cerebellar connections to motor neuron pools in the spinal cord or brain stem are either *non-existent* or functionally *negligible*. But *all major motor control regions of the brain stem and cerebrum* receive, directly or indirectly, profuse cerebellar connections; these include the vestibular nuclear complex, brain-stem reticular formation, red nucleus, tectum and, via the thalamus, the corpus striatum and cerebral motor cortex. From these arise all the main descending tracts: cortico-, tecto-, vestibulo-, rubro-, reticulospinal, etc. by which (usually through local interneurons) patterns in α and γ motor neurons are varied. It is evident, without further detail, that cerebellar control involves simultaneous activity at all levels. Despite the prominence of *locomotor* defects in cerebellar dysfunction, cerebral and cerebellar cortices may co-operate in other, as yet unrevealed activities. Certainly *autonomic* functions may be affected by cerebellar disease or experimental damage. Increasing evidence points to the cerebellar influence on electrical activity in *sensory*

cortical areas; by connections with reticular formation and the thalamus it may also influence *ascending* sensory paths (Snider 1967).

Grey Matter of the Cerebellum

The cerebellar grey matter is distributed as the cortex and intracerebellar nuclei in the white core.

THE CEREBELLAR CORTEX

The cerebellar cortex (p. 971) is folded by many predominantly transverse curved fissures, closely arranged, approximately parallel and varying in depth; these separate lobes, lobules and folia,

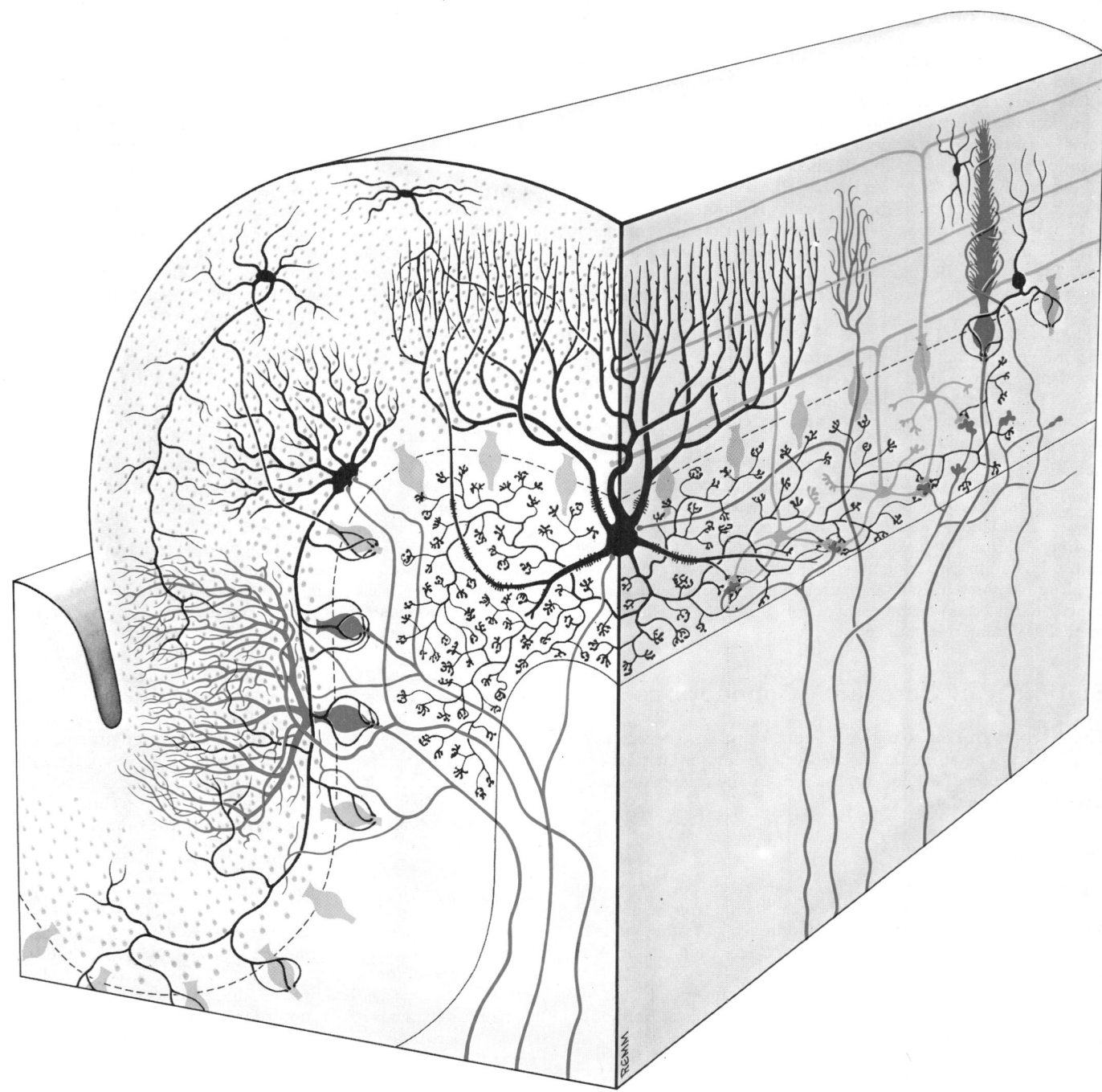

7.105 The general organization of the cerebellar cortex; a single cerebellar folium has been sectioned vertically, both in its longitudinal axis (right part of the diagram) and transversely (on the left). Note: (1) Purkinje cells (red); (2) inhibitory interneurons (black) including outer stellate, basket and Golgi cells; (3) granule cells and their ascending axons which bifurcate into longitudinally disposed horizontal fibres (yellow); (4) climbing fibres and mossy afferents (blue). Note also the synaptic glomeruli formed between the terminals of the mossy afferent fibres, the complex dendrite tips of the granule cells and the ramifications of the Golgi cell axon. Redrawn from Eccles et al 1967 with the permission of the authors and publishers.

correspond to the primary, secondary, tertiary and sometimes quaternary laminae of the central white matter. The smallest laminae and their curved covers of cortex are the *cerebellar folia* (7.99, 110). So complexly pleated is the cortex that its 'unfolded' dorsoventral (caudorostral) extent would exceed a metre, and transversely measure about a seventh of this. The whole cortex is almost uniform in microscopic structure. Local differences, typical in the cerebral cortex, do not occur and sections from different areas are indistinguishable. The same homogeneity obtains throughout the mammalia and, with only minor differences, throughout the vertebrates.

The cerebellar cortex shows much geometric order in its elements, precisely arrayed relative to the tangential, longitudinal and transverse planes in individual folia (7.105). It contains: (1) the terminations of afferent 'climbing' and 'mossy' fibres; (2) five varieties of neuron—granular, outer stellate, basket, Golgi and Purkinje (only the latter's axons leave the cortex); and (3) neuroglia and blood vessels.

The cerebellar cortex has two main strata, the external *molecular* and internal *granular*. Some distinguish an intermediate *Purkinje layer* but here it is considered the most internal part of the molecular layer. Anatomical and physiological research on the cerebellar cortex has accumulated a wealth of data which can only be abbreviated here. For details and extensive bibliographies consult Dow (1942), Jansen & Brodal (1954), Dow & Moruzzi (1958), Eccles et al (1967), Fox & Snider (1967).

Brief introductory summary. Multitudes of granular neurons in the internal stratum are involved in one of the two main cerebellar cortical *inputs*, receiving terminals of mossy afferents (i.e. *all* except the olivocerebellar). The large Purkinje neurons are the cerebellar cortical *output*, their axons mostly converging on intracerebral nuclei, from which tracts leave for the numerous destinations already described. Dendritic trees of Purkinje cells receive terminals from a dense array of *parallel fibres* from axons of granular neurons and from *climbing fibres* of the olivocerebellar tract. The remaining neurons, external stellate, basket and Golgi, are *inhibitory interneurons*, connecting the cortical elements in complex geometrical patterns (vide infra).

The cerebellar cortex thus has *two inputs*: climbing olivocerebellar fibres synapsing with Purkinje neurons and mossy fibres which connect with them via granular neurons and their parallel axons. Both excite the Purkinje cells, but contrast in structure and presumed modes of operation. The *sole output* comprises Purkinje axons, inhibitory to the neurons of intracerebral nuclei, which in turn modify *all* the major motor controls in the brain stem and cerebrum. Purkinje activity is complicated by the complex surrounding 'shells' of inhibitory interneurons.

THE LAYERS OF THE CEREBELLAR CORTEX

The molecular layer, about 300–400 μm thick, has a sparse population of neurons, dendritic arborizations, non-myelinated axons and neuroglia. By ordinary histological technique it is almost featureless, contrasting with the granular layer's immense cellular population. It contains the following features (7.105):

(1) Most internal is a single layer of piriform somata of Purkinje neurons, widest in a transverse folial plane, narrow in sections along a folium.

(2) Complex Purkinje dendritic trees extend towards the surface and spread out in a transverse folial plane.

(3) Recurrent collaterals come from Purkinje axons.

(4) Dendritic trees of Golgi neurons, their somata in the granular layer, also reach towards the surface but span the territories of several Purkinje neurons transversely *and* longitudinally.

(5) Somata, dendrites and axons of external stellate neurons are superficially sited.

(6) Somata, dendrites and axons arise from the deeper basket cells.

(7) Axonal terminals of the deeper granule cells radiate into the molecular layer to bifurcate into branches extending in opposite directions along a folium, forming some of the parallel bundles traversing the dendritic trees (vide supra).

(8) Terminals of olivocerebellar climbing fibres ascend through the granular layer to contact Purkinje dendrites in the molecular layer.

(9) Radiating branches from large neurogliocytes in the granular layer surround all neuronal elements except at the synapses; at the surface their conical expansions join into an external limiting membrane.

The general disposition of the above elements is summarized in Figure 7.105.

The granular layer (7.105), about 100 μm thick in fissures and 400–500 μm on folial summits, contains an enormous number of 'microneurons', *granular neurons*: about 2·4 million per cubic millimetre in the simian cerebellar cortex, 2–7 million in the human cortex (Fox & Barnard 1957, Braitenberg & Atwood 1958). In summary it consists of:

(1) The somata of granular neurons and parts of their axons ascending to the molecular layer to become parallel fibres (7.18).

(2) The dendrites of granular neurons, with their terminal expansions.

(3) Branching terminals of mossy afferent fibres.

(4) Climbing fibres passing to the molecular layer.

(5) Somata, basal dendrites and complex axonal ramifications of Golgi neurons.

(6) Cerebellar *glomeruli*, synaptic complexes with four types of neurite: a mossy fibre terminal synapses with the granular and Golgi dendrites, while granular dendrites also receive synaptic contacts from Golgi terminals (p. 971).

(7) Somata of large neurogliocytes.

THE PURKINJE (PIRIFORM) NEURON (7.17B)

Purkinje (piriform) neurons have a specific geometry and occur only in the vertebrate cerebellar cortex. Their flattened somata are piriform in sections across a folium and narrow at right angles to this. Arranged in a single stratum, deepest in the molecular layer (7.105) and adjacent to the granular layer, individual Purkinje cells are separated by about 50 μm transversely and longitudinally 50–100 μm. Their somata measure vertically about 50–70 μm, transversely 30–35 μm.

One, or sometimes two large primary dendrites arise from the 'neck', the superficial pole, but branching varies with position. In deep interfoliar furrows the dendrite forks at once into two large stems diverging at nearly 180° in transverse foliar plane; from these a most abundant arborization, with several orders of subdivision, extends towards the surface. Near the foliar summits primary dendrites spread superficially before branching at more acute angles. In both sites branches of each neuron are confined to a narrow 'sheet' precisely in the transverse foliar plane. Purkinje neurons are more densely arranged at foliar summits than in fissures; these variations have been interpreted as Cartesian transformations, corresponding to the convexities and concavities due to 'folding' during development. Primary dendrites and first and second order branches have smooth surfaces; third order branches and beyond show a dense array of short *dendritic spines* for synapses with parallel rami of axons of granular neurons (vide infra). These *spiny terminals* (or *branchlets*) carry about 45 spines on each 10 μm of length; thus, on average (Fox & Barnard 1957), each Purkinje neuron has about 180 000 spines on its dendritic tree.

From a Purkinje neuron's base, near the granular layer, the axon traverses this layer into the subjacent white matter. Its initial 30 μm or so is narrow, amyelinate and like the soma in ultrastructure; its surface synapses with basket cell terminals. Hence some term this initial region a *preaxon*; beyond, it suddenly enlarges, acquires a myelin sheath and has collateral branches. The main axon proceeds in the white matter and finally forms a plexus, the terminals of which synapse in one of the intracerebral nuclei; a few bypass these to reach the vestibular nuclei (p. 957). Collateral branches of Purkinje axons interweave as supra- and infraganglionic plexuses external and internal to their somata. Their terminations are uncertain but few, if any, such recurrent collaterals end on Purkinje neurons; they probably synapse with dendrites of the basket and Golgi neurons in the same, adjacent or even quite distant folia, as so-called association fibres, of unknown function.

Synapses with Purkinje cells (7.106) are made by: (1) parallel fascicles of axons from granular neurons, synapsing with Purkinje dendritic spines; (2) climbing olivocerebellar fibres, each largely

7.106 Detailed cytoarchitecture of part of the cerebellar cortex which includes the layer of Purkinje cell somata and the zones immediately superficial and deep to this. The vertical block face to the right is in the longitudinal axis of the cerebellar folium; the upper face is tangential with respect to the convex summit of the folium. The cell details are shown in the following colours: red = the soma of the Purkinje cell, its 'preaxon' passing from the lower pole and its apical stem dendrite with its first-order branches bearing dendritic spines. Blue = climbing fibres. Orange/ yellow = horizontal fibres derived from the bifurcation of the ascending axons of granule cells. Mauve = Golgi cell soma, dendrites with spines and initial segment of its axon. Pale brown = the descending basket cell axons forming complex synapses on the preaxon of the Purkinje cell. Green = areas occupied by glial cell processes. Note the synapses between the horizontal axons and the dendritic spines of the Purkinje cell and Golgi cell. Compounded and redrawn from information and illustrations in Eccles et al 1967.

restricted to one Purkinje neuron, its numerous branches following the ramifications of Purkinje dendrites to make hundreds, perhaps thousands of contacts with smooth interspinal areas; (3) synaptic terminals from external stellate neurons on smooth surfaces of large primary and secondary dendrites; (4) synapses of collaterals of basket neurons, ending among those of stellate neurons; (5) complex terminals of axons of basket neurons surrounding Purkinje preaxons. These contacts are summarized in 7.106.

The ultrastructure of Purkinje neurons has been described by Herndon (1963), Hámori & Szentágothai (1965, 1966a, b), Léránth & Hámori (1970); they contain all the elements common to neurons (p. 880); distinctive arrays of compressed lamellar membranes assist identification. Apart from the synapses, the whole cell surface is covered by processes of so-called Bergmann gliocytes.

THE GRANULAR NEURON

The varying thickness and immense cell population of the granular layer have been noted. Each neuron has a spherical nucleus, 5–8 µm in diameter, with a mere shell of cytoplasm 0·5–1·0 µm thick, containing a few small mitochondria, ribosomes and a diminutive Golgi complex. From its central aspect branch usually 3–5 (occasionally 1–7) dendrites, each about 10–30 µm in length, often single but sometimes branching before their claw-like terminal expansions; these synapse with the terminals of mossy fibres and other components to form *cerebellar glomeruli* in the intervals between granular neurons and visible by light microscopy. Their fine axons pass from their superficial aspect into the molecular layer and branch at a T-junction into two axons (*parallel fibres*) (7.105, 106), passing in opposite directions for a total of 2–3 mm along the folium. Near the initial dichotomy parallel fibres are smooth, but elsewhere have fusiform swellings and, near their termination, short hook-like projections; both are sites of synapses between the parallel fibre and many kinds of dendrite in the molecular layer. The quantitative aspects of these will be noted; most numerous are the synapses with Purkinje dendritic spines, less frequent those with the spines of Golgi dendrites and dendrites of external stellate and basket neurons. It had been estimated that 150 000–300 000 parallel fibres cross and probably synapse with a *single* Purkinje dendritic tree.

THE BASKET NEURON

Basket neurons of *large stellate type* occupy the internal half of the molecular layer. (They are sometimes termed *internal stellate neurons*.) Their dendrites have no specific form but, as in Purkinje neurons, they are flat arrays in a strictly transverse foliar plane. Dendritic surfaces have a few, irregular, long spines synapsing with parallel fibres which intersect their dendritic trees at right angles. The soma is covered by axosomatic synapses, possibly from recurrent collaterals of Purkinje and climbing axons.

Axons of basket neurons are deep in the molecular layer, superficial to the Purkinje neurons, their *stems* being in the *transverse foliar plane*. Continuing for about 1 mm, each covers the territories of 10–12 Purkinje neurons; but from the second neuron onwards descending collaterals pass towards the Purkinje somata, interweaving with collaterals from other basket neurons to form pericellular networks, or 'baskets', hence their vernacular name. Branches from each descending collateral and main axon also weave *longitudinally* in the folium to a further 3–6 rows of Purkinje neurons flanking the axon. Hence 100–200 Purkinje neurons may receive synapses from a single basket neuron. The somatic surfaces of Purkinje neurons are devoid of synapses; synapses of descending basket collaterals cluster around the Purkinje *preaxon* (7.106). Axons of basket neurons have fewer ascending collaterals, which may synapse with the smooth bases of Purkinje dendrites.

Basket neurons provide a unique synaptic system: activity in a fascicle of parallel fibres intersects and activates a longitudinal row of their dendrites. Because their axons are transverse in the folium they act on the origins of Purkinje axons in longitudinal strips on both sides of the fascicle.

THE EXTERNAL STELLATE NEURON

Smaller than the 'basket' cells, *external stellate neurons* are dispersed in the superficial half of the molecular layer. Their dendritic trees, smaller than in basket neurons, are in general similar, with synaptic spines and transverse foliar orientation. They also are traversed at right angles by parallel fibres, with which their spines *and* smooth areas synapse. The origin of apparent sparse axosomatic synapses is uncertain. The smallest stellate neurons have abundant local arborizations, ending on Purkinje dendritic spines; the longer axons of larger neurons pass transversely in a folium to similar sites on remoter Purkinje neurons. The largest *external* stellate neurons have descending collaterals which accompany those of the *internal* stellate neurons ('basket cells' proper).

THE GOLGI NEURON

The Golgi type, which are the largest stellate neurons, occupy the superficial zone of the granular layer, adjoining the Purkinje somata. Large dendrites radiate all over the soma, many curving into the molecular layer to arborize, their courses predominantly at right angles to the surface. Their dendritic trees are *not* compressed, appearing much the same in transverse and longitudinal foliar section; in both planes they overlap the territories of three Purkinje cells. The dendritic tree of a Golgi neuron thus approaches, but does not overlap, those of its fellows; each intermingles, in the molecular layer, with the dendritic fields of about 10 Purkinje neurons. Some Golgi dendrites, however, do not enter the molecular layer, dividing in the granular layer to join the cerebellar glomeruli (vide infra).

A Golgi neuron's axon arises from its soma's central aspect or from the base of a central dendrite, immediately dividing into a profuse arborization deployed through the full thickness of the granular layer and occupying a volume which, nearer the white matter, corresponds to its dendritic tree in the molecular layer, i.e. it does not overlap adjacent Golgi fields. The main synaptic input to Golgi neurons is from parallel fibres in the molecular layer, synapsing with the few spines on its dendrites. Golgi axonal terminals take part in cerebellar glomeruli. As regards Golgi neurons, the cerebellar cortex thus consists of a series of units, crudely hexagonal in surface view, which adjoin but do not overlap. Within each are about 10 Purkinje cell territories and internal to these a great number of granular neurons and synaptic glomeruli.

INPUTS TO THE CEREBELLAR CORTEX

As already mentioned, two inputs serve the cerebellar cortex, climbing fibres (*neurofibrae ascendentes*) and mossy afferents (*neurofibrae muscoideae*) (7.105).

Climbing fibres, illustrated by Cajal but long of uncertain origin, are now established as largely olivocerebellar, with a point-to-point correlation between olivary complex and cerebellar cortex (p. 976). Electronic coupling (through nexuses or gap-junctions) between their inferior olivary origins has been observed (Llinas et al 1974, Sotelo et al 1974). Each fibre traverses the white matter and granular layer without branching, to reach a *single* Purkinje neuron, where it divides repeatedly with its terminal branches applied to the Purkinje dendrites in much of their length, making numerous synapses with the smooth dendritic areas. A few collateral branches leave the climbing fibres, some descending to synapse with Golgi somata and dendrites, others with nearby basket or external stellate neurons. Both mossy and climbing fibres are excitatory; the former excite single Purkinje neurons, with a weaker effect on adjacent interneurons.

Mossy (muscoid) afferent fibres are widely agreed to include *all* afferents except the olivocerebellar. They also are excitatory but otherwise contrast with the localized climbing input. As each mossy fibre traverses the white matter its collateral branches diverge to several adjacent folia; within each the branch traverses its central lamina, giving numerous ramules to the granular layer on both foliar aspects; these divide again into two or three terminals, which expand into grape-like synaptic

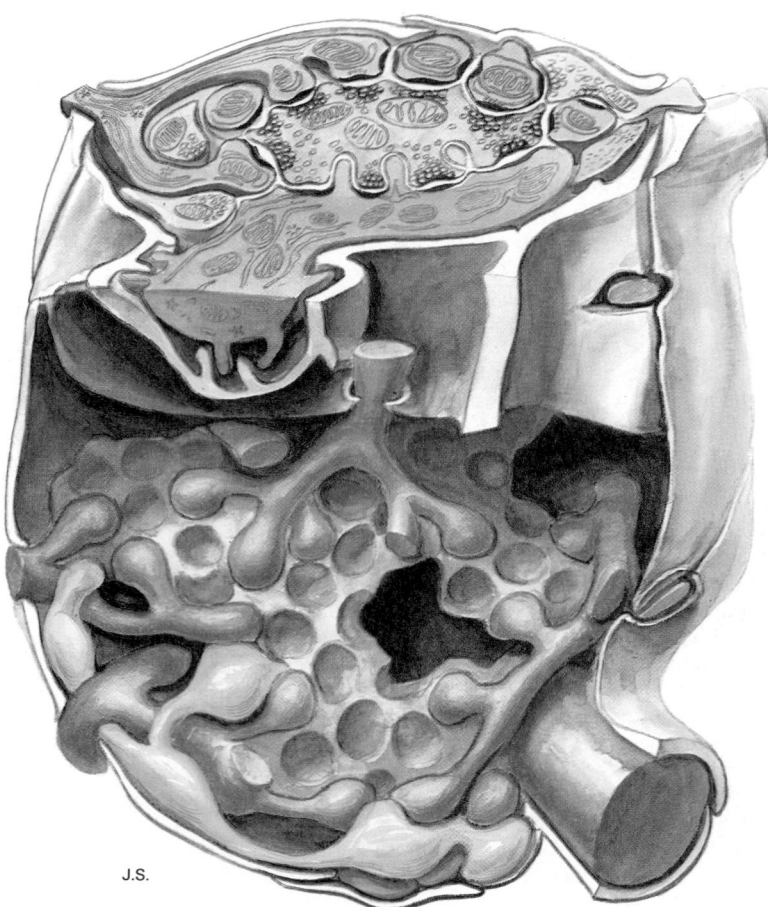

7.107 A stereodiagram illustrating the structure of a cerebellar synaptic glomerulus. Blue = mossy afferent fibre rosette. Red = granule cell dendrites. Yellow = terminals of Golgi cell axon. Green = Golgi cell dendrite. Grey = neurological capsule. Note that the essential synaptic contacts are axodendritic between: mossy afferent fibres and granule cell dendrites; mossy afferents and Golgi cell dendrites; Golgi cell axons and granule cell dendrites. From Eccles et al 1967, by courtesy of the authors and publishers.

terminals (mossy fibre *rosettes*), each in the centre of a cerebellar *glomerulus*.

Cerebellar glomeruli are complex synaptic corpuscles containing a central mossy fibre 'rosette', dendrites of several granular neurons, terminals of Golgi axons and sometimes a Golgi dendrite. Each glomerulus, roughly spherical or ovoid, is only about 10 µm in its greatest diameter; they occur in a ratio to granular neurons of about 1 to 5. All synapses are axodendritic. The central 'rosette' establishes contact with a surrounding spray of up to 20 granular dendrites and, when present, the spine-studded surface of a Golgi dendrite. The Golgi terminals synapse with several dendrites of granular neurons, as shown in diagram 7.107. The mossy ('muscoid') terminals are excitatory to granular neurons and, if included, Golgi dendrites. Conversely, the Golgi neuron, an inhibitory interneuron, can inhibit the granular neuron through its axodendritic contacts.

QUANTITATIVE ASPECTS OF THE CEREBELLAR CORTEX

In its regular, repetitive cytoarchitecture, the cerebellar cortex lends itself to quantitative methods (Eccles et al 1967, Fox & Bernard 1957, Fox et al 1967, Braitenberg 1967, 1977). But throughout it shows extremes in neuronal population, in convergence and divergence or one-to-one transmission and in complex arrays of inhibitory interneuronal circuits.

A vertical column of human cerebellar cortex, 1 mm square in area at a foliar summit, contains about 500 Purkinje, 600 basket, 50 Golgi and perhaps 3 000 000 granular neurons, with about 600 000 glomeruli. It is difficult to estimate the total cerebellar

cortical area; but since its sagittal dimension (p. 969) is said to be a metre, the area must be many thousands of square millimetres. Hence the figures cited must be multiplied by a factor of perhaps 10^5 or more, indicating the vast population of cerebellar components capable of incalculably numerous integrations.

On the *input side* each climbing fibre synapses with *only one* Purkinje neuron (and by collaterals with an undetermined number of interneurons). In contrast a single mossy fibre shows *enormous divergence*, synapsing with 400 or more granular neurons in one folium; if branches to adjacent folia are included, the number is probably several thousand. Conversely, each granular neuron probably synapses with four or five different mossy fibre terminals.

Ascending axons of granular neurons bifurcate into parallel fibres, synapsing with Purkinje, Golgi, basket (internal) and external stellate neurons. A parallel fibre travels about 2–3 mm and synapses with 300–450 Purkinje neurons. Thus divergence from a mossy fibre through the granular neurons is to perhaps hundreds of thousands of Purkinje neurons; there is uncertainty about the amount of overlap between parallel fibre territories, but enormous convergence of paths to individual Purkinje neurons exists, a dendritic tree of one receiving 400 000 synapses from different parallel fibres (Napper & Harvey 1988).

In discussing a 'module concept' of *cerebral* cortical architecture Szentágothai (1975, 1978) has also surveyed the *development* of quantitative knowledge of the cerebellar cortex. Its first definition as a three-dimensional rectangular lattice (Braitenberg

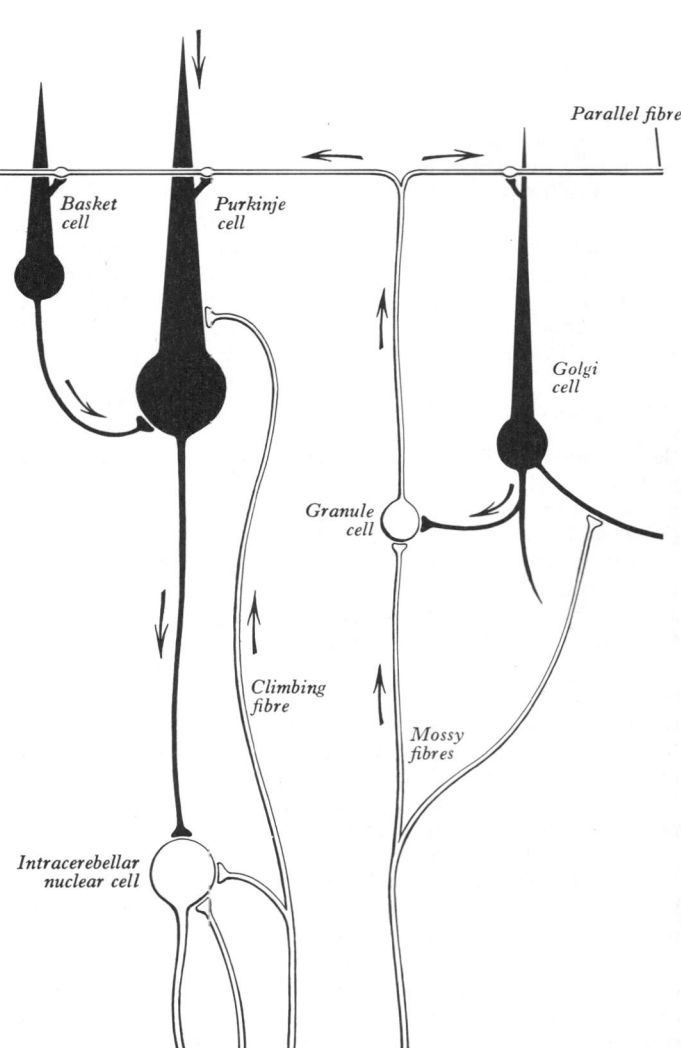

7.108 An analysis of the essential circuitry and synaptic contacts between the climbing and mossy afferent fibres, the main neuronal elements of the cerebellar cortex and the neurons of the intracerebellar nuclei, based upon cytological and microelectrode studies. Excitatory cells, neurites and terminals are white surrounded by a black line; inhibitory elements are solid black. By courtesy of J. C. Eccles.

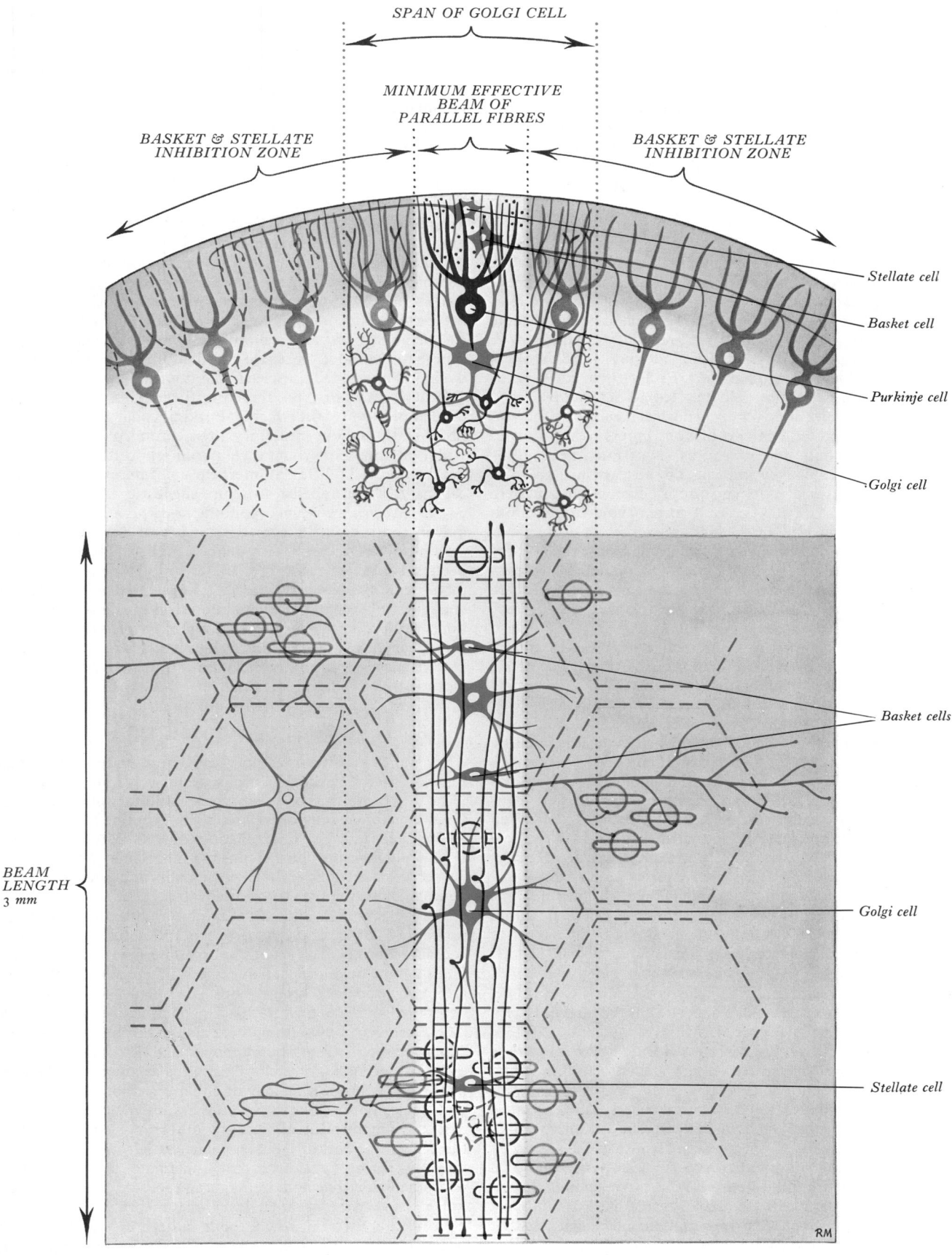

SPAN OF GOLGI CELL

MINIMUM EFFECTIVE
BEAM OF
PARALLEL FIBRES

BASKET & STELLATE
INHIBITION ZONE

BASKET & STELLATE
INHIBITION ZONE

Stellate cell

Basket cell

Purkinje cell

Golgi cell

Basket cells

Golgi cell

Stellate cell

BEAM
LENGTH
3 *mm*

7.109 The concept of the complex neuronal integrative units of the cerebellar cortex. The neurons and their processes are shown in a transverse section of a cerebellar folium (above) and in surface view (below). Neurons in full colour (black, red or blue), or in full coloured outline, are in a state of excitation. Neurons in grey or interrupted outlines are inactive. Full black or grey, black or grey outlines = Purkinje cells, granule cells and their parallel fibre bundles; blue = basket and stellate cells; red = Golgi cells. The white longitudinal strip indicates activation of a locus of granule cells, their bundle of parallel fibres and the associated longitudinal row of Purkinje cells. Simultaneous activation of the basket and stellate inhibitory interneurons by the parallel fibre bundle causes zones of lateral inhibition (grey shading) to flank the activated strip. These zones of inhibited Purkinje cells are in reality about ten rows wide; only four rows are illustrated here (grey profiles). If the active bundle of parallel fibres reaches a critical width, inhibitory Golgi interneurons become active and reduce or stop mossy fibre input in the cerebellar glomeruli. The inhibitory territories of the Golgi cells are indicated by the array of red hexagons. See text for further description. Redrawn from Szentágothai 1967 by courtesy of the author and publishers.

& Atwood 1958) was followed by a hypothetical division into assumed functional spaces (Szentágothai 1963, 1965), later defined physiologically (Eccles et al 1967) and structurally corroborated (Uchizono 1967) in terms of excitatory and inhibitory zones of interaction. Detailed analyses of feline cerebellum (Palkovits et al 1971a,b,c, 1972) helped 'to describe the cerebellar cortical network statistically in numerical and geometric terms, considerably refining the earlier crude qualitative guesses as to spatial and numerical distribution of excitatory and inhibitory interactions'. For reviews, see Ito (1983), Gilman (1985).

MECHANISMS OF THE CEREBELLAR CORTEX

Structural research on the cerebellum, ranging from comparative morphology, developmental analysis and studies of gross connectivity, to quantitative cytology, cell geometry and synaptic relations by ultrastructural techniques, has been paralleled in volume by impressive neurophysiological research (Dow 1942, Dow & Moruzzi 1958, Eccles et al 1967, Ito et al 1964, Ito & Yoshida 1966, Snider 1967, Szentágothai 1967, Eccles 1970, and p. 968). Increasing synthesis in the field has led to several proposed mechanisms, probably interlocking in normal cerebellar function.

The cerebellar cortex has two distinct inputs, climbing and mossy fibres, and only one output, axons of Purkinje neurons (7.108). Both inputs convey information, at least in part, from similar sources: exteroceptors and proprioceptors, brain-stem reticular formation and the cerebral cortex (vide supra). Both inputs are excitatory, but climbing fibres exert a one-to-one, all-or-none excitation on *individual* Purkinje neurons, each mossy afferent having a *diffuse* excitatory effect via hundreds of excitatory granular neurons to *thousands* of Purkinje neurons, which are exclusively inhibitory, their axons exerting varying inhibitory patterns on intracerebellar and vestibular nuclei.

The remaining components, internal (basket) and external stellate and Golgi neurons, each with their particular geometry, input and output, are all inhibitory interneurons. Whatever type of recording or preparation is used, most neurons show a low level of 'background' discharge, even under quiescent conditions. Some use the term 'spontaneous' for this; whether it is intrinsic remains uncertain but upon such a background orderly responses to specific patterns of input must occur. Szentágothai (1967) suggested that climbing fibres and their synapses, largely concentrated on individual Purkinje neurons, do not provide an integrative mechanism, presuming that their informational flow has been *pre-integrated* in the olivary nuclear complexes. He suggested that *integrative units* of cerebellar cortex involve the mossy afferent input, granular neurons, parallel fibres, Purkinje neurons and varieties of interneuron. Their supposed interactions are summarized in 7.108, 109. Mossy afferents may excite a locus of granular neurons causing activity in a narrow fascicle of parallel fibres, about 3 mm in length, which courses along a folium's molecular layer to excite dendritic fields in its path, i.e. Purkinje, Golgi, basket and external stellate dendrites. If the fascicle is very narrow or its activity low, no cells will respond; but as its size increases to the width of the Purkinje dendritic fields, it excites a row of Purkinje and related basket and stellate neurons. Golgi neurons do not respond, since only a small part of their larger dendritic fields receives an input from such parallel fibres. Hence a row of Purkinje neurons, about 3 mm long, in the folium's long axis responds with volleys of impulses. Because of the slow conduction in small parallel fibres (a few decimetres per second) and regular spacing of Purkinje neurons, these are excited *sequentially*. Impulses in parallel fibres arrive at *successive* Purkinje dendrites at intervals of about a tenth of a millisecond, suggesting that the cerebellar cortex is a *timing device*. The transverse disposition of internal and external stellate neurons' axons and terminals ensures *simultaneous* activation by a parallel fascicle, resulting in longitudinal cortical strips containing inhibited Purkinje neurons which flank the active rows on each side. With continual fluctuations of activity in input via mossy fibres, the essential cerebellar response is an ever-changing pattern of innumerable 'unit rows' of excited Purkinje neurons, flanked by inhibitory zones; thus it provides neural 'sharpening', by which the more active parallel bundles are selected from the general background of cortical

activity. Should the active fascicle of parallel fibres become broader than the width of a single row of Purkinje neurons, another mechanism comes into play, mediated by Golgi neurons. The parallel fibres reach a critical width and now excite Golgi dendrites and these reduce the input by the mossy fibres via granular neurons by their inhibitory synaptic terminals on the latters' dendrites in the synaptic glomeruli (p. 972); the principal function of Golgi neurons thus appears to be a 'choke' to prevent the excessive broadening of active fascicles.

Upon this incessantly changing pattern of *sequentially* excited, inhibited or relatively quiescent rows of Purkinje neurons, it is imagined that the sharply localized excitatory effect of climbing fibres is exerted, the resultant at any Purkinje neuron depending on the strength of stimulation by the climbing fibre and the neuron's state. By such means changing inhibitory patterns are transmitted by Purkinje neurons to intracerebellar nuclei and thence to motor control in the cerebrum and brain stem. But how co-operation between localized climbing fibre activity and the time-base provided by active parallel fibres can co-ordinate the dynamics of striated muscles remains an enigma.

Other complexities in intracortical cerebellar circuits include the complication of the effects of Purkinje collaterals and climbing fibres on the various inhibitory interneurons, problems posed by Golgi dendrites in synaptic glomeruli and differential 'on' and 'off' responses in different interneurons. Interesting mathematical analyses of cerebellar structure suggest that it acts as an accurate *biological clock*, incorporating *delay paths*, perhaps important in controlling the correct sequence of events in e.g. rapid 'voluntary' movements (Braitenberg 1967). These *ideas* have been expanded and refined by Braitenberg (1977) and others. The matter, however fascinating, cannot be pursued here, but perhaps enough has been said to excite interest in the detailed investigations, still continuing, into the *higher integrative units* of the cerebellar cortex. We are still far, however, from a comprehensive understanding of the overall relation of the cerebellum to complex motor patterns of the individual.

CEREBELLAR CHEMOARCHITECTURE

The cerebellar cortex is comparatively well understood but gaps remain. Briefly, Purkinje neurons, exclusively inhibitory, contain in various combinations at least three inhibitory neuromediators: GABA (gamma aminobutyric acid), the peptide motilin and the amino acid taurine. Combinations are GABA plus motilin, GABA plus taurine, taurine plus motilin. Some 30% of Purkinje neurons contain none of these; presumably one or more inhibitory neuromediators are yet to be discovered.

The outer stellate and basket inhibitory interneurons are both probably GABA-ergic, but a proportion of outer stellate cells may contain taurine. Golgi inhibitory interneurons are mainly GABA-ergic but a proportion contain enkephalin; whether these are two separate populations of neurons or the two neuromediators co-exist in the same neurons is not established. The cerebellar input through excitatory climbing fibres is considered to be mediated by glutamate and aspartate (but precise technical means for their localization is awaited). Of the mossy fibre input, a proportion may be cholinergic but the status of the vast majority is unknown. The innumerable parallel fibres derived from the bifurcation of granule cell axons are excitatory to the dendrite of Purkinje, outer stellate, basket and Golgi neurons; there is strong evidence that glutamate is the excitatory neuromediator. The noradrenergic and serotonergic fibres form a rich plexus in the granular layer of the cortex, then enter the molecular layer and provide both radial fibres and fibres that bifurcate and run in the long axis of a folium in company with parallel fibres. The aminergic fibres are fine, varicose and form extensive cortical plexuses; their release of noradrenalin and serotonin is assumed to be non-synaptic and their effects are paracrine involving volumes of tissue. The NA coeruleocerebellar projection, when active, inhibits Purkinje cell firing not by direct action but via the enriched generation of cyclic AMP which acts as a 'second messenger'. Thus in terms of neuromediator profiles the cerebellar cortex is relatively simple, compared e.g. with the more complex olfactory bulb, but these

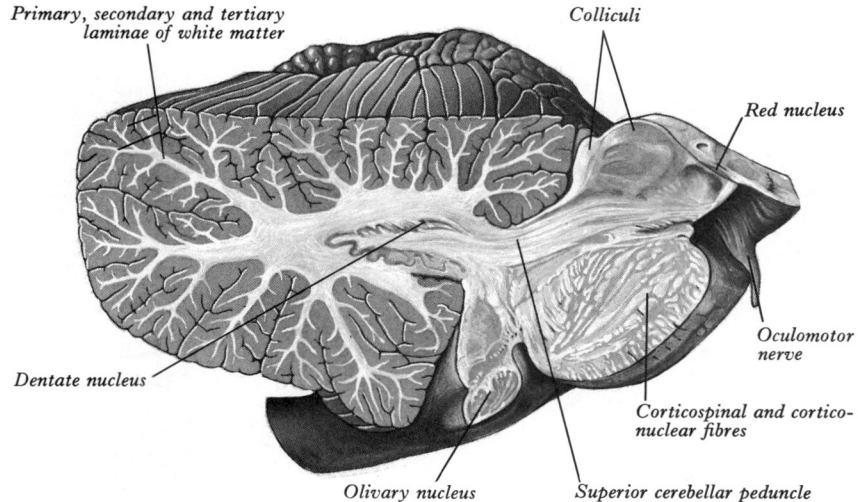

Primary, secondary and tertiary laminae of white matter

Colliculi

Red nucleus

Oculomotor nerve

Corticospinal and cortico- nuclear fibres

Dentate nucleus

Olivary nucleus

Superior cerebellar peduncle

7.110   Oblique vertical section through the right cerebellar hemisphere and right half of the brain stem. Note the arbor vitae.

are dominated by the neuromediator-rich and neuromediator-dense structures grouped by Nieuwenhuys (1986) as a 'paracrine' core and paracores. These include large limbic-associated parts of the basal telencephalic nuclei, septal areas, amygdala and other limbic structures, the medial forebrain bundle and related parts of the hypothalamus, parts and fasciculi of the brain-stem reticular formation, spinal intermediate grey matter (and dorsal grey column).

THE INTRACEREBELLAR NUCLEI

Four grey masses are embedded in the cerebellar white core on each side, the *intracerebellar nuclei*, sometimes termed 'roof' nuclei, though some only adjoin the ventricular roof. Most lateral and largest is the *nucleus dentatus*, medial to which are the smaller *nuclei emboliformis* and *globosus*, the most medial being the *nucleus fastigii*. In most mammals only three appear: a nucleus lateralis corresponding in part to the nucleus dentatus, a nucleus inter-positus corresponding largely to globose and emboliform nuclei and a nucleus medialis corresponding to the nucleus fastigii. Precise homologies of nucleus interpositus are uncertain (Jansen & Brodal 1954). These differences must be noted, because so much information from other mammals is extrapolated to the human cerebellum. The nucleus fastigii is phylogenetically oldest and largely associated with the archicerebellum; the more recent globose and emboliform nuclei are associated with the paleo-cerebellum; the nucleus dentatus, most recent, is associated with the neocerebellum (vide infra).

   The nucleus dentatus (7.110), near the white core of its hemisphere, is an irregularly folded grey lamina containing a mass of white fibres largely derived from neurons in the grey lamina, which is deficient ventromedially; through this so-called 'hilum' fibres stream out to form much of the superior cerebellar pe-duncle (p. 966). **The nucleus emboliformis** partially covers the dentate hilum, **the nucleus globosus** being still more medial and dorsoventrally elongated; **the nucleus fastigii** is near the mid-line and in the ventral (anterior) part of the superior vermis.

   Prominent in these nuclei are large multipolar neurons with simple, stellate but irregular arborizations, the numbers of den-drites increasing from the fastigial to the dentate nucleus; den-dritic trees of adjacent neurons overlap in the fastigial nucleus but are more distinct, with their own territories, in the dentate. Their axons make one or two loops before leaving their nuclei to form **cerebellar outflow tracts** in superior and inferior cerebellar peduncles. In these loops they give off recurrent collaterals end-ing locally in their nucleus. The nuclei also contain numerous small neurons, some probably local Golgi type II interneurons, others contributing to the peduncles (Jansen & Jansen 1955, Flood & Jansen 1966).

   Afferent connections to the intracerebellar nuclei are from intrinsic and extrinsic sources. They make axodendritic and axosomatic synapses with terminals of Purkinje axons, which enter the nuclei and form pericellular nets round several neurons;

each net is derived from more than one Purkinje neuron, i.e. there is divergence *and* convergence in the nucleus. Since Purkinje neurons are all inhibitory, attention is focussed on other afferents, many presumably excitatory. Rubrocerebellar fibres are said to reach these nuclei (Courville & Brodal 1966) and also collateral branches from spino-, ponto-, olivo- and reticulocerebellar tracts (Eccles et al 1967). Noradrenergic and serotonergic fibres from brain-stem reticular nuclei also reach them. Cerebellar outflow is hence perhaps integrated in these nuclei into patterns of excita-tion from such extracerebellar sources, combined with inhibitory patterns from Purkinje neurons, modified by local interneurons.

STRUCTURAL AND FUNCTIONAL CEREBELLAR LOCALIZATION

Since the cerebellar cortex is largely uniform in microstructure and microcircuitry, it may be assumed, as investigations indeed indicate, that its intrinsic mode of operation is also uniform. Functional localization, like that long held to exist in cerebral neocortical areas (but vide infra) does not exist in the cerebellar cortex. Cerebellar localization reflects the manner in which func-tionally dissimilar afferent tracts end in different regions of this *homogeneous* cortex; equally it reflects the differential projection of different cortical cerebellar areas to intracerebellar nuclei and thence to influence diverse brain-stem centres, the diencephalon and cerebral neocortex. (It is becoming increasingly clear, how-ever, that there is also a *fundamental intrinsic uniformity* in the *cerebral neocortex* and that functional localization is conferred on it by extrinsic connections; see pp. 1041, 1052.) Cerebellar localization has been much studied; foundations were laid by comparative, developmental and connectivity studies (Jansen & Brodal 1954, Smith 1903, Riley 1930, Larsell 1937, 1953, Nieuwenhuys 1967) and detailed analysis of patterns of connec-tion (Brodal 1969). Studies of changes in electrical potential evoked by 'natural' or 'artificial' stimulation have included special sense organs, peripheral nerves, the cerebrum, brain stem and spinal cord (Snider 1936, 1940, 1945, Snider & Eldred 1951, 1952, Dow & Moruzzi 1958). These have varied from a crude recording of exposed cerebellum to recording of focal potentials by inserted electrodes and of 'units' by microelectrodes. Cortical stimulation has been used to modify alternatively elicited move-ments. The behavioural effects of selective cerebellar ablation and of human cerebellar disease have provided some information. Examples will be mentioned here; but since experiments have used a range of mammals, with considerable variation in cerebellar morphology, illustrations are simplified and the areas designated vary between species, including mankind. Primary division into *archicerebellum, paleocerebellum* and *neocerebellum* has been noted, and their approximation, with some overlap, to major patterns of afferent cerebellar connections, summarized in the terms 'vestibulocerebellum', 'spinocerebellum' and 'pon-tocerebellum' (including 'tectocerebellum') (7.111A).

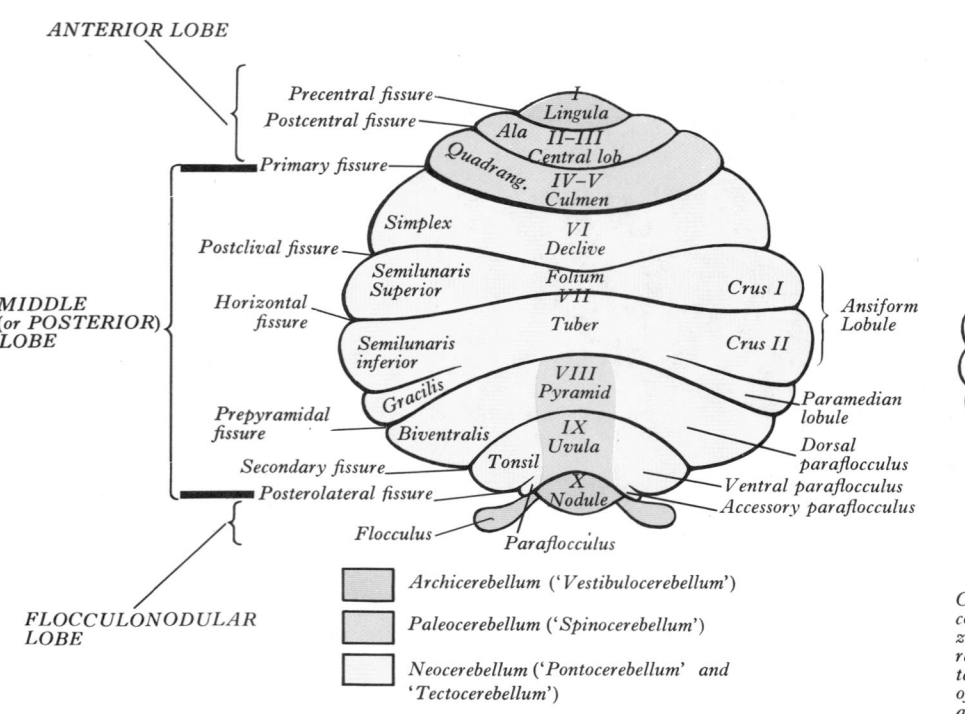

ANTERIOR LOBE

MIDDLE (or POSTERIOR) LOBE

FLOCCULONODULAR LOBE

Precentral fissure
Postcentral fissure
Primary fissure
Postclival fissure
Horizontal fissure
Prepyramidal fissure
Secondary fissure
Posterolateral fissure
Flocculus

I Lingula
Ala
II–III Central lob
Quadrang.
IV–V Culmen
Simplex
VI Declive
Semilunaris Superior
Folium VII
Tuber
Semilunaris inferior
Gracilis
VIII Pyramid
Biventralis
IX Uvula
Tonsil
X Nodule
Paraflocculus

Crus I
Crus II
Ansiform Lobule
Paramedian lobule
Dorsal paraflocculus
Ventral paraflocculus
Accessory paraflocculus

Archicerebellum ('Vestibulocerebellum')
Paleocerebellum ('Spinocerebellum')
Neocerebellum ('Pontocerebellum' and 'Tectocerebellum')

A. Cerebellar terminology. The cerebellar lobes, lobules and fissures and the approximate extent of the archi-, paleo- and neocerebellum are indicated. The subdivisions of the vermis are both named and numbered (after Larsell). On the left are terms widely used in human neuroanatomy; on the right are additional terms often used in general mammalian neuroanatomy.

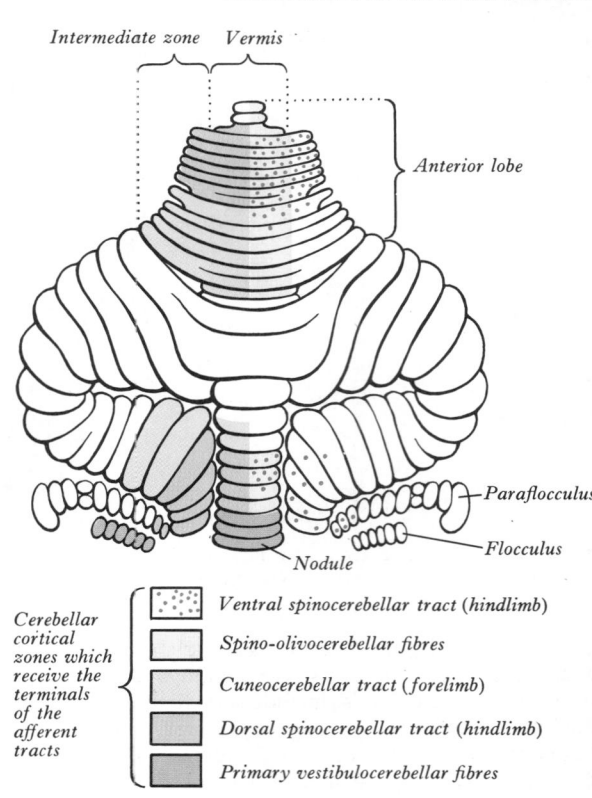

Intermediate zone Vermis
Anterior lobe
Paraflocculus
Flocculus
Nodule

Cerebellar cortical zones which receive the terminals of the afferent tracts

Ventral spinocerebellar tract (hindlimb)
Spino-olivocerebellar fibres
Cuneocerebellar tract (forelimb)
Dorsal spinocerebellar tract (hindlimb)
Primary vestibulocerebellar fibres

B. The cerebellar cortical areas of termination of the afferent tracts indicated, derived from experimental studies.

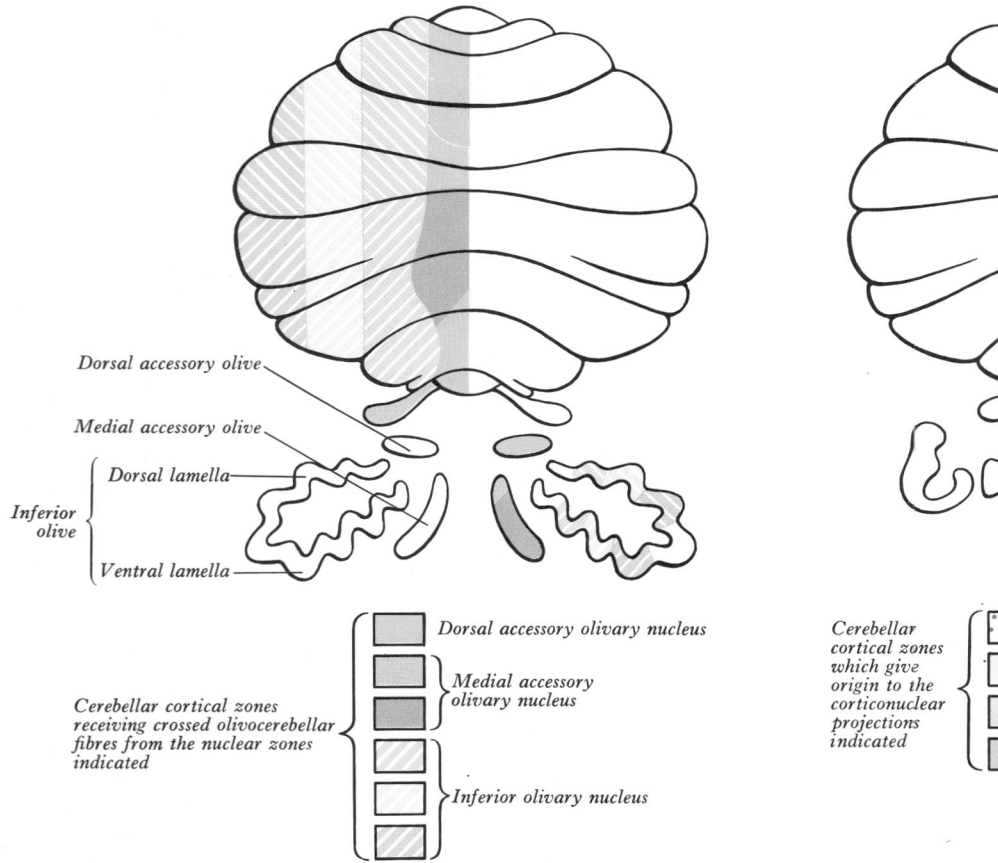

Dorsal accessory olive
Medial accessory olive
Inferior olive { Dorsal lamella
Ventral lamella }

Cerebellar cortical zones receiving crossed olivocerebellar fibres from the nuclear zones indicated

Dorsal accessory olivary nucleus
Medial accessory olivary nucleus
Inferior olivary nucleus

Cerebellar cortical zones which give origin to the corticonuclear projections indicated

Lateral vestibular nucleus
Nucleus fastigii
Nucleus interpositus
Nucleus dentatus

E. An analysis of the topical organization of the cortical zones of the cerebellum which receive crossed olivocerebellar (climbing) fibres from the different parts of the inferior olivary complex of nuclei.

F. An analysis of the topically organized projections of cerebellar cortical Purkinje cell axons on to the lateral vestibular nucleus and the intracerebellar nuclei.

7.111A–H This series of diagrams illustrates certain features of the localization of structure and function in the cerebellum: G is a median sagittal section of the feline cerebellum; in all the remaining diagrams it is

assumed that the cerebellum has been flattened and viewed from the dorsal aspect so that its whole rostro-caudal extent can be seen. A, E and F are approximate outlines of the human cerebellum; B, C, D and H are of the

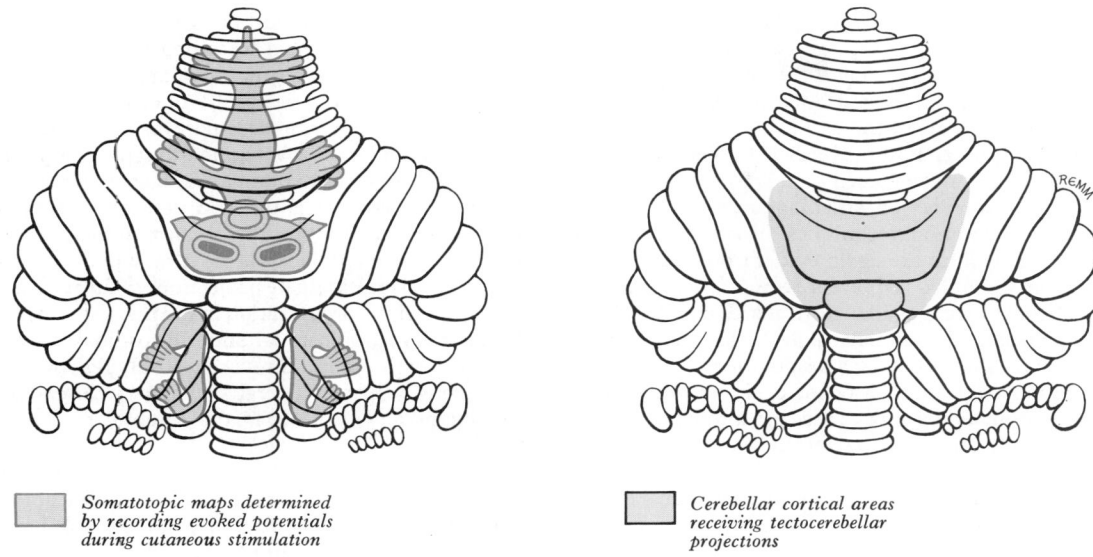

Somatotopic maps determined by recording evoked potentials during cutaneous stimulation

C. The somatotopic arrangements of the evoked potentials recorded from the cerebellar cortex during cutaneous stimulation.

Cerebellar cortical areas receiving tectocerebellar projections

D. The cerebellar cortical areas receiving tectocerebellar projections.

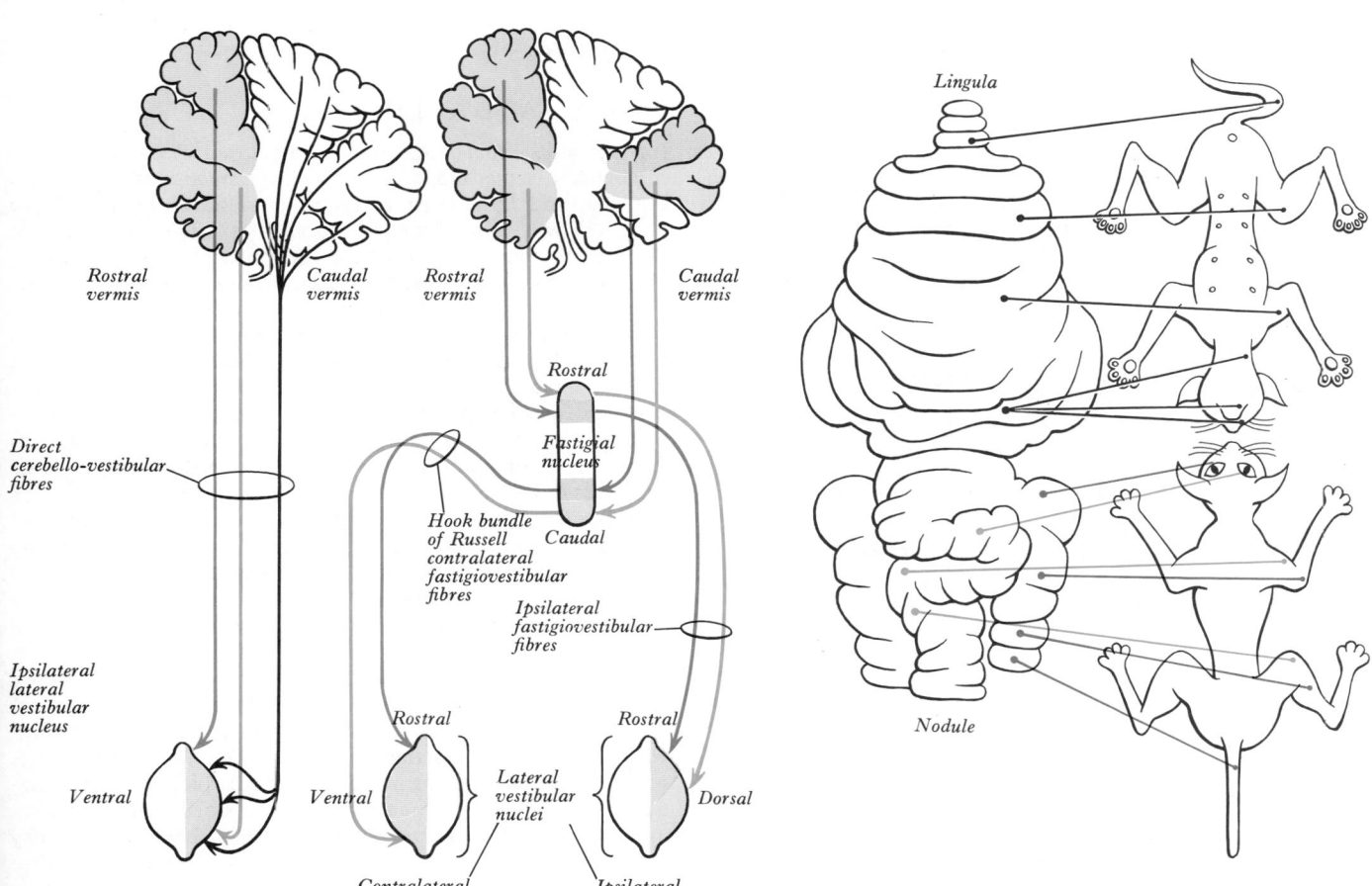

G. The arrangement of direct cerebello-vestibular and indirect cerebello-fastigio-vestibular projections from the cerebellar vermis of the cat.

H. The somatotopic pattern of movements elicited in different parts of the body during stimulation of the cerebellar cortex in the decerebrate cat.

feline cerebellum. Detailed descriptions of the information in these diagrams are in the text and in the references quoted.

Illustrations B, F and G were redrawn and modified from Brodal (1969), C and D were redrawn and modified from Snider (1952) and H from Hampson et al (1952) with permission from the authors and publishers.

The localization, in feline experimentees, of areas of termination of primary vestibulocerebellar fibres and of spino-, cuneo- and spino-olivocerebellar tracts appear in 7.111B. Double distributions of these to separate rostral and caudal, vermian and paravermian zones should be noted; these areas make up the spinocerebellum and within them a general somatotopy of hind- and fore-limbs is shown. Primary vestibular fibres end mainly in the flocculus, nodule and part of the uvula.

Illustration 7.111C is a somatotopic record of potentials evoked during stimulation of cutaneous receptors (Snider 1952, Hampson et al 1952); it corresponds well with diagram 7.111B, based on neuroanatomical degeneration techniques. Rostral and caudal parts of the spinocerebellum are again evident but responses to facial stimulation are included. Different regions are represented in reverse order in rostral and caudal cortical areas; in the former representation is ipsilateral and bilateral in the latter.

Illustration 7.111D shows the area responding to visual and auditory stimulation, and stimulation of occipital areas of the cerebral cortex. Visual and auditory areas are almost coincident and include central regions of the vermis and adjacent paravermian zones, overlapping an area for the face in the ventral spinocerebellum, as determined by tactile stimulation. How far visual and auditory responses are due to tectocerebellar paths or some less direct one is uncertain (p. 987). Punctate correlation between nuclei of the olivary complex and contralateral cerebellar cortex is shown simplified in 7.111E.

These examples are based on the different distributions of afferent tracts to the cerebellar cortex. The main receiving areas are generally arranged in a dorsoventral sequence of *transverse* strips. Evidence is equally clear for the localization of endings of Purkinje axons; but this is *longitudinal* (Jansen & Brodal 1954, Eager 1966, Goodman et al 1963, Voogd 1964, Korneliussen 1968). Most Purkinje axons end in the intracerebellar nuclei but some in the lateral vestibular nucleus. Cerebellar corticonuclear projections are summarized in Figure 7.111F. Direct cerebellovestibular fibres start from the flocculus, nodule and much of the rest of the vermis but particularly its rostral and caudal regions. Projections to the intracerebellar nuclei show an orderly longitudinal arrangement; on each side the cerebellar cortex may be divided into *medial*, *intermediate* and *lateral zones*. Fibres leaving the medial zone (vermis) converge on the nucleus fastigii, their craniocaudal sequence preserved during their convergence and termination. The intermediate zone (a longitudinal paravermian strip) projects to the nuclei globosus and emboliformis, the lateral zone

(cerebellar hemisphere) to the nucleus dentatus. In each zone an orderly craniocaudal sequence of cortical origin of axons is maintained to the nuclear terminations. Almost all these corticonuclear projections are ipsilateral but fastigial nuclei receive bilaterally from the vermian cortex. Recalling the main cerebellar outflow and its nuclei of origin (pp. 967, 968), it is clear that the vermis mainly influences the vestibular nuclei and brain-stem reticular formation. The intermediate paravermian cortex influences the red nucleus and midbrain tegmentum; the cerebellar hemisphere, through the nucleus dentatus and thalamus, mainly affects activity in the corpus striatum and cerebral cortex. In part these outflows are somatotopically organized, including direct and indirect vermian projections to the lateral vestibular nucleus (7.111G) and intermediate projections to the red nucleus, correlating well with the organization in the nuclei themselves and also in the lateral vestibulospinal and rubrospinal tracts (Brodal et al 1962, Courville 1966). Similar localization appears in the thalamic nucleus ventralis lateralis, which receives dentatothalamic projections and projects to the cerebral motor cortex (Walker 1934).

These descriptions, of course, apply to experimental animals but in general will probably be found applicable to the human cerebellum; interesting physiological parallels have already been shown. Cerebellar stimulation often modifies movements generated by simultaneous reflex or cerebral cortical stimulation; cerebellar stimulation in *decerebrate* animals elicits discrete movements (7.111H). Again, rostral and caudal areas corresponding to the spinocerebellum display a somatotopic order (Hampson et al 1952). The rostral area mediates ipsilateral responses, the caudal bilateral responses.

Such investigations, combined with effects of ablation (Chambers & Sprague 1955a,b), suggest that the vermis controls posture, tone, locomotion and equilibrium in the *whole body*, the intermediate zone controlling posture and movements in *ipsilateral limbs*. The lateral zone's role is obscure.

CEREBELLAR DYSFUNCTION

Innumerable reports on the behavioural effects of human cerebellar disease and of selective experimental ablations are available and have been reviewed by Holmes (1939), Wyke (1947), Brown (1949), Dow & Moruzzi (1958), Dow (1969) etc. In essence the cerebellum has an input from cutaneous receptors, proprioceptors, eyes, ears, brain-stem reticular formation and the cerebral cortex; this is integrated and discharged to motor controls in the cerebrum and brain stem. Its normal action is necessary for smooth, co-ordinated, effective movement. The more obvious effects of cerebellar dysfunction include: (1) disturbance in equilibrium; (2) disturbance of muscle 'tone', or resistance to stretch, tendon reflexes and ability to stabilize joints; (3) motor inco-ordination (*ataxia*) due to irregularities in timing, rate and force of the contraction in synergistic muscles.

Disequilibrium shows as a tendency to fall when standing and perhaps unsteady gait, with sensations of spinning, nausea, etc. The affected muscles are soft, tendon reflexes diminished and muscles tire easily (*asthenia*). Lowering of joint control may progress to pendular swinging of a limb segment after displacement or to 'flail' joints. *Muscular inco-ordination* is basic in most cerebellar dysfunction, affecting the regions variably. Hence symptoms vary. *Asynergia* is a diminished capacity for smooth, orderly action between muscle groups. A complex act may become an irregular sequence, *decomposition of movements*. A defect in rapid alternating movements, e.g. supination and pronation, may develop, *dysdiadochokinesis*. Control of range of movement may be lost, *dysmetria*. Locomotor disturbances, a tendency to fall, and (with closed eyes) deviations from the intended course are common. So is an inability to point accurately, *past-pointing*. *Tremor* is usually absent at rest but *intention tremor* may appear, intensified by movement; tremor may also affect the head or trunk. Muscular inco-ordination produces *defects of speech*. *Cerebellar nystagmus* may occur, conjugate drift of the gaze followed by a rapid return. But attempts to correlate clinical phenomena with different cerebellar regions, or with experimental results, have had limited success.

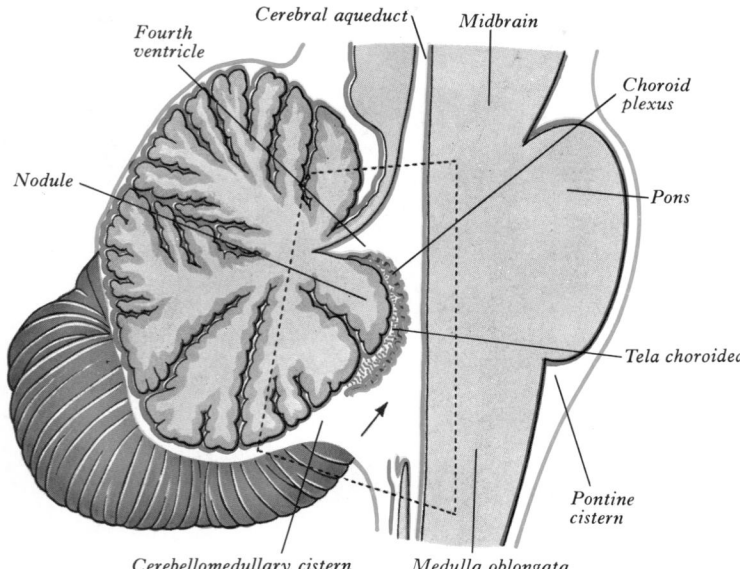

Cerebral aqueduct — *Midbrain*

Fourth ventricle

Choroid plexus

Nodule

Pons

Tela choroidea

Pontine cistern

Cerebellomedullary cistern — *Medulla oblongata*

7.112 Sagittal section through the brain stem and the cerebellum close to the median plane. The black arrow is placed in the median aperture of the fourth ventricle. The area enclosed by the interrupted lines is shown enlarged in 7.114. Blue = arachnoid mater. Red = pia mater. Green = ependyma.

In the *flocculonodular syndrome*, with damage to the nodule, uvula and flocculus in man and other animals, the main feature is imbalance: swaying, staggering and a tendency to fall backwards. Positional nystagmus is often present. These effects are attributable to upset of the integration between vestibular nuclei and the '*vestibulocerebellum*'.

In experimental animals, *ablation* of the vermis in the anterior lobe exaggerates tendon reflexes and rigidity, already present in decerebrate preparations; vermian *stimulation* reduces rigidity. The converse results from ablation or stimulation of paravermian parts of the anterior lobe. These effects show somatotopy but have not been clearly demonstrated in man.

Most human cerebellar disorder is *neocerebellar*, involving one or both hemispheres or their outflows. In unilateral disease, if severe, the hypotonia and inco-ordination noted above appear *on the side* of the lesion, but gross intention tremor and staggering only appear if the dentate nucleus or superior cerebellar peduncle are involved. Small cortical lesions have little effect; even quite extensive disease, though causing transient dysfunction, is followed by rapid improvement in locomotor control. Such *cerebellar compensation* is not yet explained.

Despite the prominence of locomotor effects in cerebellar dysfunction it seems probable that the cerebellum is extensively, but less overtly, involved in other activities such as motor memory, *autonomic homeostasis* and *modulation of sensory transmission*.

The Fourth Ventricle

The fourth ventricle (7.112, 115, 116), a cavity of complex shape, lies ventral to the cerebellum and dorsal to the pons and upper half of the medulla. Developmentally it has three parts: *superior*, part of the isthmus rhombencephali (p. 185); *intermediate*, the metencephalic (pontine); and *inferior*, myelencephalic (medullary). It is lined by ependyma, continuous below with the medulla's central canal and above with the cerebral aqueduct, connecting it to the third ventricle. At its mid-level a narrow, curved *lateral recess* extends on each side between the inferior cerebellar and floccular peduncles then, reaching the flocculus, it is crossed ventrally by the glossopharyngeal and vagal fila. Its lateral end is open, allowing a variable part of the ventricle's choroid plexus to protrude into the subarachnoid space (7.81). In its midline the cavity extends dorsally into the white cerebellar core as a *median dorsal recess* (7.112, 180, 181) above the nodule; on each side a *lateral dorsal recess* extends still further dorsally (7.181), cranial to the inferior medullary velum and caudal to the cerebellar nuclei but separated from them by a thin white lamina. The ventricle has lateral boundaries, a roof and a rhomboidal ventral floor, the *rhomboid fossa*.

Lateral boundaries. Each is formed inferiorly by the gracile and cuneate tubercles, fasciculus cuneatus and inferior cerebellar peduncle, superiorly by the superior cerebellar peduncle.

The roof (7.113) extends dorsally into the median and lateral dorsal recesses. Its upper part is simple, formed by the superior peduncles and superior medullary velum. The peduncles (p. 917), emerging from the cerebellum superoventrally, are at first lateral boundaries but near the inferior colliculi they converge to overlap the ventricle as part of its roof. The *superior medullary velum* (p. 963) fills the angle between them and is continuous dorsally with the cerebellar white core; it is covered dorsally by the lingula of the superior vermis (7.115).

Inferiorly the roof is more complex but mostly a thin sheet, devoid of nervous tissue, formed by ventricular ependyma and the pia mater of the tela choroidea, which covers it dorsally (7.113, 114). The sheet is interrupted by an inferior *median aperture* (7.113, 114), through which the ventricle connects with the subarachnoid space. *The tela choroidea of the fourth ventricle*, a double layer of pia mater, is between the cerebellum and inferior part of the ventricle's roof. Its dorsal layer covers the inferior vermis and, reaching the nodule, is reflected ventro-inferiorly in direct contact with the ependyma. In the tela, vascular fringes form the fourth ventricle's choroid plexus. On each side, the tela choroidea is in contact with the ependyma until it reaches the inferolateral border of the ventricular floor, marked by a narrow,

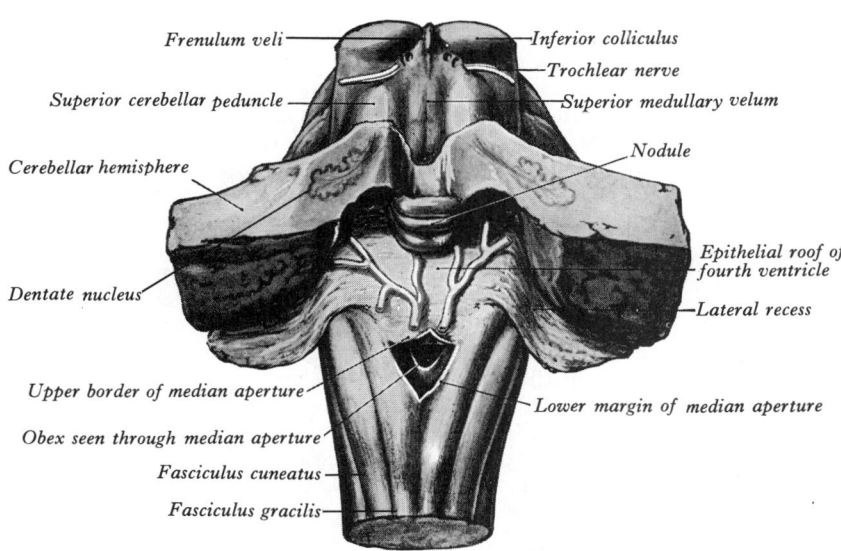

7.113 Dorsal aspect of the roof and the lateral recesses of the fourth ventricle, exposed by removal of parts of the cerebellum.

white ridge, the *taenia*; the paired taeniae are continuous below with a curved margin, the *obex*, which overlaps the ventricle's inferior angle and is covered by ependyma on both its aspects (7.114). Superiorly the taeniae extend laterally along the inferior borders of the lateral recesses.

Apertures in the roof. In the caudal part of the roof are three openings. The *median aperture* is large, inferior to the nodule (7.113, 114) and varies in size, with an irregular upper border drawn dorsally towards the inferior vermian surface to face the cerebellomedullary cistern (7.112). *Lateral apertures*, terminal in the lateral recesses, are partly occupied by the choroid plexus which protrudes into the subarachnoid space (7.81); the ependyma and pia mater are continuous at their margins. These are the only connections between the ventricular system and the subarachnoid space. Occasionally a lateral recess may fail to open but a median aperture is constant.

The choroid plexuses. Bilateral fringes of the vascular tela choroidea contain choroid plexuses; these invaginate the caudal part of the roof and are covered by ependyma which is modified as a secretory epithelium. Each has vertical and horizontal limbs. The two longitudinal vertical fringes flank the midline, fusing at

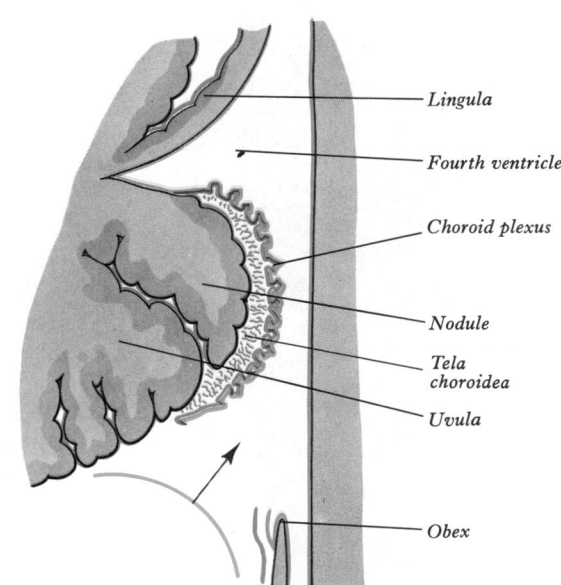

7.114 An enlargement of the part of 7.112 which is enclosed by an interrupted line. Blue = arachnoid mater. Red = pia mater. Green = ependyma. The black arrow traverses the median aperture of the roof of the fourth ventricle.

the cranial margin of the median aperture and often prolonged on the ventral vermian aspect, to which they adhere (Hewitt 1960). The horizontal parts are continuous, project into the ventricle, passing into its lateral recesses, and emerge through the lateral apertures, still covered by ependyma. The entire structure is like a T, with a double vertical limb. This form varies widely (Lang & Schafer 1977). Arterial supply is from the inferior cerebellar arteries (pp. 752, 753). For details of supply and drainage consult Maillot et al (1976).

The rhomboid fossa (7.86, 115, 116), the fourth ventricle's rhombic floor, is formed by dorsal sufaces of the pons and the 'open' cranial half of the medulla, covered by grey matter continuous with that of the walls of the medullary and spinal central canals; superficial to this is a thin lamina of neuroglia covered by ependyma. The floor is divided into superior, intermediate and inferior areas. The *superior* (rostral) area is triangular, limited laterally by superior cerebellar peduncles; its apex is continuous with the cerebral aqueduct's wall, its base an arbitrary line through the upper ends of two small *superior foveae*. The *intermediate* area descends from this to taenial level, where they extend horizontally into the lateral recesses. The *inferior* (caudal) area, also triangular, is continuous below with the wall of the medullary central canal. The fossa is divided vertically by a *median sulcus*, flanking which are paired *medial eminences*, each bounded laterally by a *sulcus limitans*. In the superior area each eminence occupies its half of the floor; but medial to the superior fovea it is expanded into a long *facial colliculus*, superficial to the abducent nucleus but partly produced by radicular facial nerve fibres. In the floor's inferior area each medial eminence is a *hypoglossal triangle (trigonum hypoglossi)*, with medial and lateral parts separated by faint oblique furrows; the medial area corresponds to the upper pole of the hypoglossal nucleus, the lateral to the *nucleus intercalatus* (p. 956).

Each *sulcus limitans* lies lateral to the medial eminence and in its upper part is itself the lateral limit of the floor and presents a bluish-grey *locus coeruleus*, overlying a group of pigmented nerve cells. The *nucleus coeruleus* corresponds partly to the locus but ascends to the lower end of the mesencephalic trigeminal nucleus and may overlap it. Some of its neurons contain neuromelanin, especially in human midbrains (Foley & Baxter 1958); at pontine levels it spreads into the adjacent reticular formation, of which it is usually considered a part (Russell 1955, Webster 1978). All its

central neurons are noradrenergic (Dahlstrom & Fuxe 1964, Fuxe et al 1970), by far the largest group of noradrenergic neurons so far observed in mammalian brains (p. 993). Stereotactic lesions of both nuclei (in rats) show them to be the main and probably sole source of fibres in the *dorsal noradrenergic bundle*, part of the central tegmental fasciculus (Ungerstedt 1971). Ramón-Moliner & Dansereau (1974) demonstrated in the feline nucleus coeruleus high concentrations of acetylcholinesterase, a finding not inconsistent with the above view. Afferent nuclear connections are less certain, but probably include ascending spinal fibres, brain-stem reticular nuclei, trigeminal nuclei, and possibly forebrain centres. Efferent nuclear projections have, in contrast, been much investigated (Maeda et al 1973, Lindvall & Bjorkland 1974, Freedman et al 1975, Kievit & Kuypers 1975, Gatter & Powell 1977, Pickel et al 1977, Bowden et al 1978, Nieuwenhuys 1986) by both immunofluorescence and autoradiography. Briefly, the nucleus has a descending, local brain-stem and profuse ascending projections; descending fibres enter the ventrolateral spinal funiculus and apparently end *bilaterally* in ventral, intermediate and dorsal grey columns *throughout* the cord; laminar destinations are mainly IV–VIII. Local fibres project to many 'specific' and reticular bulbopontine nuclei; others traverse the superior cerebellar peduncle, some ending in intracerebellar nuclei, most in the cerebellar cortex, arborizing around Purkinje somata or continuing into the molecular layer. As indicated, ascending fibres form the *dorsal noradrenergic bundle*; in the upper midbrain the paired

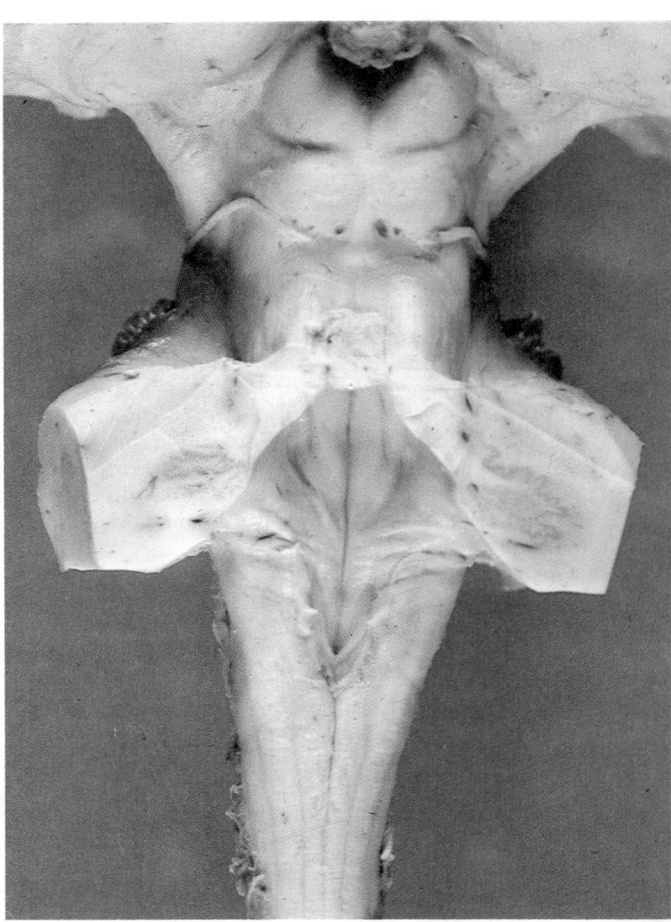

7.116 A dorsal view of the brain stem including the floor of the rhomboid fossa. For identification of structures compare with 7.115. In addition to the structures shown in the latter, note: (1) the crenated outlines of the right and left dentate nuclei in the sectioned surface of the cerebellar white matter opposite the widest part of the rhomboid fossa; (2) the midline pineal gland cranial to the superior colliculi; (3) the rounded pulvinar of the dorsal thalami which encroach on the uppermost part of the photograph; (4) the right and left habenular trigones immediately lateral to the base of the pineal; (5) the medial geniculate bodies, lateral to the superior colliculi. Dissection by E L Rees, photography by Kevin Fitzpatrick, both of the Dept. of Anatomy, Guy's Hospital Medical School, London.

Colliculi

Trochlear nerve

Lingula

Superior cerebellar peduncle

Superior medullary velum

Superior fovea

Middle cerebellar peduncle

Inferior cerebellar peduncle

Striae medullares

Facial colliculus

Vestibular area

Hypoglossal triangle

Inferior fovea

Funiculus separans and area postrema

Vagal triangle

Cuneate tubercle

Obex

Gracile tubercle

7.115 The rhomboid fossa or 'floor' of the fourth ventricle.

bundles exchange some fibres, continuing through the diencephalon in the lateral hypothalamic medial forebrain bundle (p. 1009), its continuation being via the septal region into supracallosal longitudinal stria and cingulum. Numerous complex side branches (p. 1035) pass to a remarkable diversity of destinations, which in the midbrain include the periaqueductal grey matter, reticular nuclei and the colliculi; diencephalic arborizations end in all principal thalamic, geniculate and habenular nuclei and, probably indirectly, hypothalamic nuclei. Telencephalic destinations are major 'limbic' structures, e.g. the septal nuclei, amygdala, hippocampus, subiculum and cingulate, retrosplenial and parahippocampal gyri; all the neocortex is held to receive a uniform, diffuse, bilateral input. The compact, relatively small nucleus coeruleus appears to be concerned with somatic motor and visceral control, the complex hierarchy of sleep 'centres', cognitive mapping, general affective behaviour and in modulating general neocortical and thalamic activity. (For the ventral noradrenergic bundle see p. 993.)

At the level of the facial colliculus the sulcus limitans widens into the *superior fovea* and inferiorly presents an *inferior fovea*. Lateral to the foveae is a rounded *vestibular area*, extending into the lateral recess, where it becomes the *auditory tubercle*, produced by the subjacent dorsal cochlear nucleus and cochlear part of the eighth cranial nerve (p. 958). Winding round the

inferior cerebellar peduncle and across the vestibular area and medial eminence to enter the median sulcus are the *striae medullares* (p. 960). Caudal to the inferior fovea, between the hypoglossal triangle and vestibular area, is the *vagal triangle* (*trigonum vagi*), overlying the dorsal vagal nucleus (p. 1114). The triangle is crossed below by a narrow translucent ridge, the *funiculus separans*; between this and the gracile tubercle is a small *area postrema*. The funiculus is covered by thickened ependyma containing tanycytes; the area postrema has a similar covering over loose, vascular neuroglial tissue containing neurons of moderate size. In these regions the blood–brain barrier may be modified. The specialized ependyma is like areas in the third ventricle and cerebral aqueduct (vide infra, also p. 894). The ependymocytes and tanycytes may be involved in: (1) secretion into cerebrospinal fluid, (2) transport of neurochemicals from subjacent neurons, glia or vessels to cerebrospinal fluid, (3) transport of neurochemicals from cerebrospinal fluid to the same subjacent structures, and (4) chemoreception (area postrema) (Brizzee & Neal 1954, Borison & Borison 1973, Collins & Woollam 1981).

The caudal ventricular floor, towards its inferior apex, resembles a pen nib, hence the term *calamus scriptorius*. (For the phylogenetic and ontogenetic development of the fourth ventricle consult Kier 1977.)

THE MESENCEPHALON OR MIDBRAIN

The midbrain (7.115–118) develops from the intermediate primary cerebral vesicle (p. 189). In human development and phylogenetically it remains simpler than either forebrain or hindbrain. In earlier vertebrates its major feature is the development in its roof plate of higher visual (p. 987) and, later, higher auditory centres. In mammals it is often stated that these become reflex centres, their original function as 'higher' sensory centres being transferred to the cerebral cortex. As this *telencephalization* proceeds, the midbrain is traversed by increasing numbers of neural paths originating or terminating in the cerebral cortex. Midbrain mechanisms also contribute extensively to the reticular system (pp. 932, 988). These statements are merely introductory: some midbrain activities remain reflex, e.g. 'light', 'accommodation', 'visuospinal' and 'auditory' reflexes and many others, but it is misleading to overlook the roles of its intrinsic nuclei, e.g. the colliculi, and reticular elements which mediate, as at other levels, complex interactions between afferent and efferent pathways. It should also be remembered that it is the *only* connection between the forebrain and hindbrain.

EXTERNAL FEATURES

The midbrain traverses the hiatus in the tentorium cerebelli (7.85), connecting the pons and cerebellum with the forebrain. It is the shortest brain-stem segment, not more than 2 cm in length. Lateral to it are the parahippocampal gyri, hiding its sides when the inferior surface of the brain is examined. Its long axis inclines ventrally as it ascends. For description it is divided into right and left halves, *cerebral peduncles*, each demarcated into a ventral *crus cerebri* and a dorsal *tegmental part* by a bilateral pigmented lamina, the *substantia nigra*. The two crura are separate, the tegmental parts united as a single tegmentum and traversed by the *cerebral aqueduct* connecting the third and fourth ventricles. The region dorsal to an oblique coronal plane, including the aqueduct, is the *tectum*, with dorsally four *rounded colliculi* in superior and inferior pairs.

Crura cerebri are superficially corrugated and emerge from the cerebral hemispheres, converging as they descend to meet where they enter the pons, forming here the caudolateral boundaries of the *interpeduncular fossa* (p. 1007). The median caudal part of the fossa is a greyish area, the *posterior perforated substance* (p. 1007), through which pass central branches of the posterior cerebral arteries. The ventral (basilar) crural surfaces are crossed, curving

mediolaterally near the pons, by the superior cerebellar and posterior cerebral arteries; near the crural entry into the hemispheres, the optic tracts wind dorsolaterally around them. Over each crus, near the pons, a thin white *taenia pontis* often appears, extending into the cerebellum between its middle and superior peduncles. Each crus bears a *medial sulcus*, from which roots of the oculomotor nerve emerge (7.81, 82, 118). Its lateral surface adjoins the parahippocampal gyrus and is crossed by the trochlear nerve (7.125); it bears a longitudinal *lateral sulcus*, in which fibres of the lateral lemniscus reach and form a surface elevation which inclines rostrodorsally and in part joins the inferior colliculus; the rest continues into the inferior brachium.

The *colliculi* (*corpora quadrigemina*) (7.115, 116, 125) are four eminences, above the superior medullary velum and inferior to the pineal gland and posterior commissure, the whole sloping ventrally as it ascends. Below the splenium, they are partly overlapped on each side by the pulvinar of the dorsal thalamus. In superior and inferior pairs, they are separated by a cruciform sulcus, whose upper limit expands into a depression for the *pineal gland* (7.125); from its caudal end a median *frenulum veli* is prolonged down over the superior medullary velum; lateral to the frenulum the trochlear nerves emerge; they pass ventrally over the lateral aspects of the cerebral peduncles and traverse the interpeduncular cistern to the petrosal end of the cavernous sinus. The *superior colliculi*, larger and darker, are stations for visual responses (p. 986). The *inferior colliculi*, smaller but more prominent, are associated with auditory paths (p. 986). Difference in colour is due to superficial layers of neurons in the superior colliculi (p. 987).

From the lateral aspect of each colliculus a *brachium ascends ventrolaterally. The brachium of the superior colliculus* passes below the pulvinar, partly overlapping the medial geniculate body, continuing partly into the lateral geniculate body (p. 1016) and partly into the optic tract. It conducts fibres from the retina and optic radiation to the superior colliculus. *The brachium of the inferior colliculus* ascends ventrally, conveying fibres from the lateral lemniscus and inferior colliculus to the medial geniculate body.

INTERNAL STRUCTURE OF THE MIDBRAIN

In transverse section each cerebral peduncle has, as noted above, dorsal and ventral regions separated by the *substantia nigra* (7.117, 118), the dorsal being the *tegmentum* and the ventral the *crura cerebri*. The crura are separated while the tegmental parts are continuous.

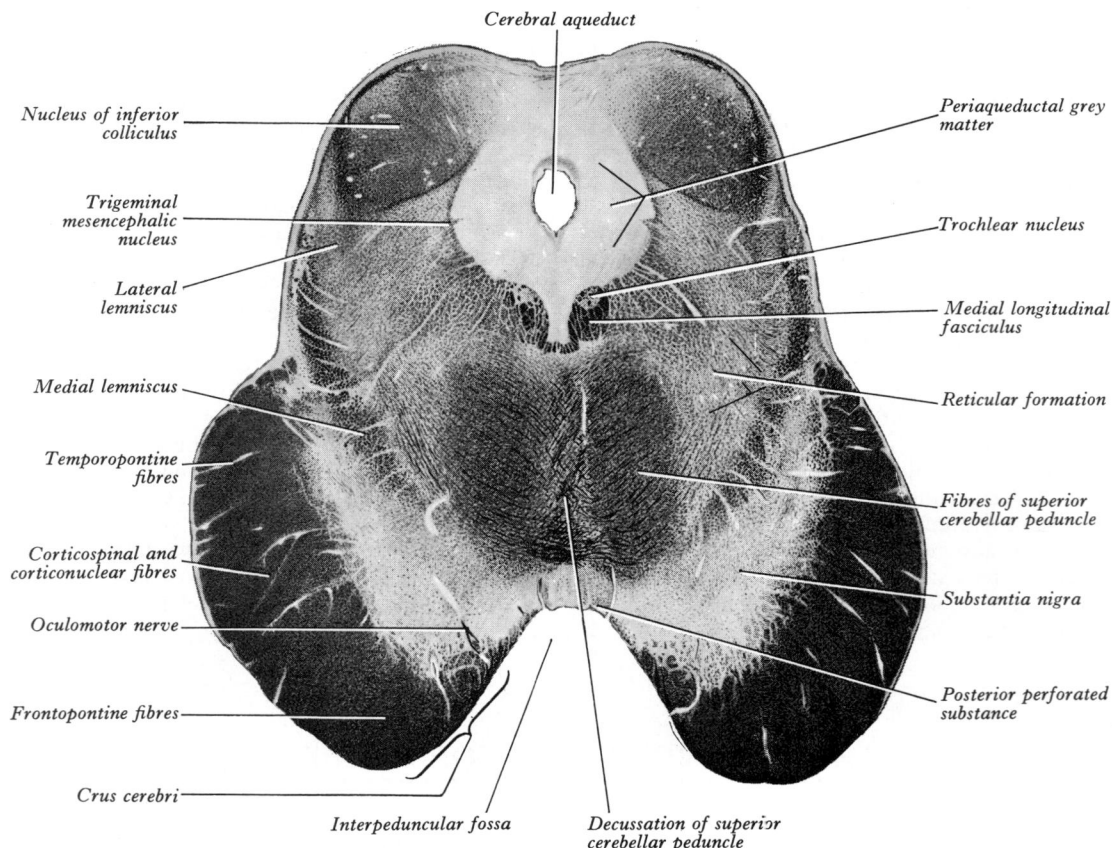

Cerebral aqueduct

Nucleus of inferior colliculus

Trigeminal mesencephalic nucleus

Lateral lemniscus

Medial lemniscus

Temporopontine fibres

Corticospinal and corticonuclear fibres

Oculomotor nerve

Frontopontine fibres

Crus cerebri

Interpeduncular fossa

Periaqueductal grey matter

Trochlear nucleus

Medial longitudinal fasciculus

Reticular formation

Fibres of superior cerebellar peduncle

Substantia nigra

Posterior perforated substance

Decussation of superior cerebellar peduncle

7.117   Transverse section of the midbrain through the inferior colliculi. Weigert Pal preparation. Magnification × 4.

Each **crus cerebri**, semilunar in section, contains corticospinal, corticonuclear and corticopontine fibres (7.117, 118). The first two occupy the middle crural two-thirds, descending via the pons and medulla, where corticonuclear fibres end in nuclei of the cranial nerves and other brain-stem nuclei; corticospinal fibres continue into the medullary pyramid. Corticopontine fibres arise in the cerebral cortex and end in the nuclei pontis. They form two groups: the *frontopontine* from the frontal lobe, principally areas 6 and 4, traversing the internal capsule and then occupying the medial sixth of the crus cerebri; and *temporopontine* fibres from the temporal lobe, also traversing the internal capsule but coming to occupy the lateral sixth of the crus; both end in the pontine nuclei. *Parieto-* and *occipitopontine* fibres are also described in the crus, medial to the temporopontine, which is largely from the posterior region of the temporal lobe; occipitopontine fibres do *not* include a projection from the striate cortex (Verhaart & Mechelse 1954).

A band of fibres, *the tractus peduncularis transversus*, may be visible emerging from the optic tract on the lateral peduncular aspect to pass round its ventral surface midway between the pons and the optic tract and disappearing into the interpeduncular fossa dorsolateral to the corpus mamillare, where it terminates in a small *nucleus of the transverse peduncular tract*, medial to the substantia nigra. The tract, constant in many lower mammals, is identifiable in only 30% of human brains. Since it degenerates after ocular enucleation, it may be part of the visual pathway; it projects to oculomotor nuclei in some mammals (Gillilan 1941, p. 1015).

The substantia nigra, a bilateral lamina of numerous pigmented (neuromelanin-containing) multipolar neurons, extends through the whole midbrain. Its pigmentation is easily visible in transverse or coronal sections (7.117, 118). The pigment is related to the aminergic status of many mesencephalic reticular neurons, increases with age, is more abundant in primates, maximal in man (Marsden 1961) and present even in albinos. The nucleus is semilunar in transverse section, concave dorsally; from its convex ventral surface extensions pass between fibres of the crus cerebri.

Thicker medially, it reaches from the medial to the lateral crural sulcus and from the pons to the subthalamic region; its medial part is traversed by radicular oculomotor fibres streaming ventrally to their oculomotor sulcus. It is divisible into a dorsal *pars compacta* of medium-sized neurons, many pigmented, and a ventral *pars reticularis*, intermingled with fibres of the crus cerebri, containing fewer neurons, some with small amounts of pigment; the ventral part reaches the subthalamic region and is considered continuous with the globus pallidus, which it structurally resembles; both contain iron in unusual amounts. In submammalian vertebrates a well-developed *pars lateralis* is recognizable but is insignificant in man (Huber & Crosby 1933), though it has been shown to project to the tectum in primates (Woodburne et al 1946).

The substantia nigra is connected with the cerebral cortex, spinal cord, hypothalamus and basal nuclei. Corticonigral fibres arise from precentral and probably postcentral gyri, a few terminating on neurons in the *pars reticularis* and many being fibres of passage to the red nucleus and reticular formation. Collaterals from fibres in sensory tracts from the spinal cord are also said to end in the substantia nigra; fibres from the mammillary peduncle and subthalamic nucleus have also been traced to its neurons. Connections with basal nuclei are efferent nigrostriatal, passing to the caudate nucleus, putamen and possibly globus pallidus (the latter often strongly denied); afferent strionigral fibres also exist. Efferent fibres also enter the tegmentum, probably ending in its reticular formation, whence impulses are relayed to spinal ventral column neurons. A major efferent connection in cats is nigrotegmental (Afifi & Kaelber 1965). Lesions medial in the substantia, also in cats (Marcos 1969), indicate that some fibres pass to the ventral thalamus; nigrocortical connections have also been described in cats (Negro 1969). A nigrothalamic projection has been confirmed in monkeys by autoradiographic tracing (Carpenter et al 1976); terminals were identified in the medial part of nucleus ventralis intermedius, the magnocellular part of nucleus ventralis anterior and the juxtalaminar part of nucleus medialis dorsalis (p. 1002). Jayaraman et al (1977), by the same techniques

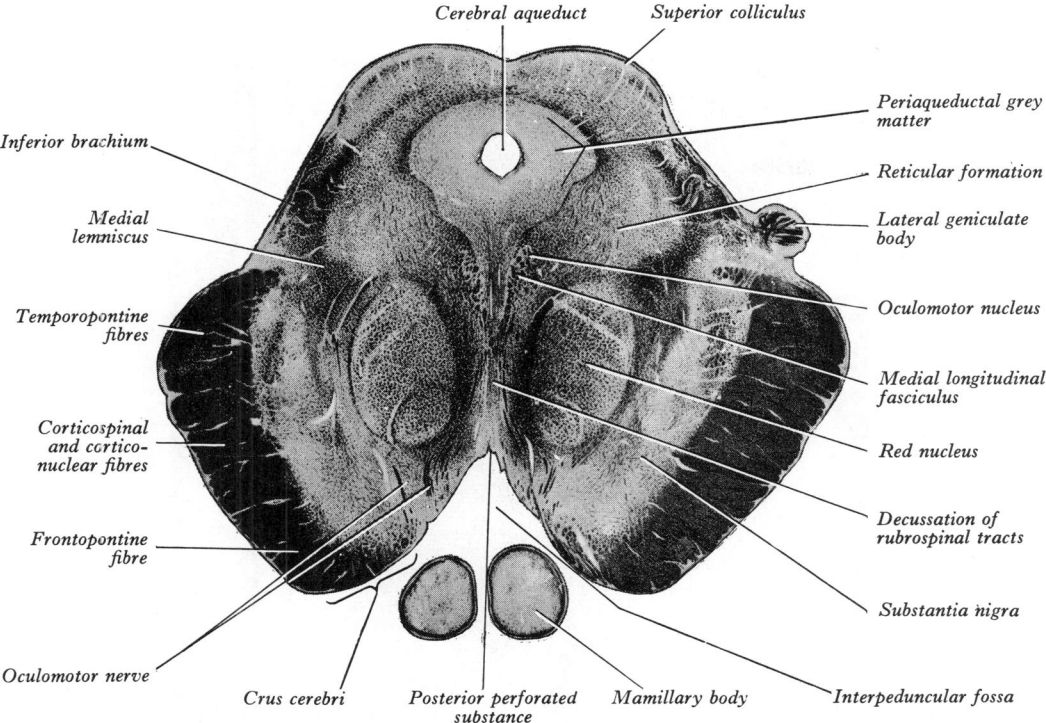

Cerebral aqueduct

Superior colliculus

Inferior brachium

Medial
lemniscus

Temporopontine
fibres

Corticospinal
and cortico-
nuclear fibres

Frontopontine
fibre

Oculomotor nerve

Crus cerebri

Posterior perforated
substance

Mamillary body

Periaqueductal grey
matter

Reticular formation

Lateral geniculate
body

Oculomotor nucleus

Medial longitudinal
fasciculus

Red nucleus

Decussation of
rubrospinal tracts

Substantia nigra

Interpeduncular fossa

7.118 Transverse section of the midbrain through the superior colliculi.
Weigert Pal preparation. Magnification × 3.

and also in monkeys, have described a tectal projection, fibres derived largely from the superolateral region of the substantia nigra ending in intermediate strata of lower levels in the ipsilateral inferior colliculus. The pars lateralis of the substantia nigra includes the dopaminergic cell group A9; the pars compacta constitutes group A10 (p. 993).

The mesencephalic tegmentum differs according to the level of section. It is directly continuous below with the pontine tegmentum and contains the same tracts. At inferior collicular level grey matter is restricted to environs of the cerebral aqueduct, scattered collections in the reticular formation (7.117, 118, 122) and the tectum (p. 986). The *trochlear nucleus* is in the ventral grey matter near the midline, in a position corresponding to abducent and hypoglossal nuclei. With the medial longitudinal fasciculus running immediately ventral, it extends through the lower half of the midbrain, just caudal to the oculomotor nucleus. In some primates these nuclei are merged and distinguished only by the arrangement of their neurons, the trochlear cells being also smaller. Trochlear *efferent fibres* pass *laterodorsally* round the central grey matter, then descend medial to the trigeminal mesencephalic nucleus to reach the upper end of the superior medullary velum, where they decussate and emerge lateral to the frenulum (7.115, 116). A few fibres may not decussate in all vertebrates. The trochlear nerve may originally have supplied muscles of a pineal eye, accounting for its dorsal course. Embryological evidence (Pearson 1943) and phylogenetic data support a more dorsal origin of its nucleus.

The *trigeminal mesencephalic nucleus* is lateral in the central grey matter, ascending from the upper pole of the main trigeminal sensory nucleus in the pons to superior collicular level in the midbrain. It is accompanied by a tract of both peripheral and central branches from its axons. Its large ovoid neurons are unipolar, as in sensory ganglia, in many small groups which extend as curved laminae on each lateral margin of the periaqueductal grey matter. Neurons are most numerous in its lower level. Attempts to categorize them have failed; they have few dendrites and are usually so close that somato-somatic contacts have been suggested (Hinrichsen & Larramendi 1969). For connections see pp. 1099, 1122. Apart from these nuclei the mesencephalic tegmentum contains many scattered neurons; most are included in the reticular formation (p. 988).

The *white matter* contains all the tracts mentioned in the pontine tegmentum; prominent is the large decussation of the superior cerebellar peduncles (p. 966), which enter the tegmentum and pass ventromedially round central grey matter to the median raphe, where most fibres cross in the *decussation of the superior cerebellar peduncles* and then separate into ascending and descending fascicles. Some *ascending* fibres end in or give collaterals to the red nucleus, which they encapsulate and penetrate; many others ascend to the nucleus ventralis lateralis of the thalamus. Some, uncrossed, have branches believed to end in the circumaqueductal grey matter and reticular formation, interstitial nucleus and posterior commissural nucleus (sometimes, perhaps mistakenly, see Ingram & Ranson 1953, called the nucleus of Darkschewitsch); the latter nucleus may send efferent fibres to the medial longitudinal fasciculus and posterior commissure. *Descending* fascicles end in the pontine and medullary reticular formation, the olivary complex and possibly cranial motor nuclei (pp. 966, 967).

The *medial longitudinal fasciculus* adjoins the somatic efferent column, dorsal to the decussating superior cerebellar peduncles. The *medial*, *trigeminal* and *lateral lemnisci* and *lateral spinothalamic tract* (spinal lemniscus) form a curved band dorsolateral to the substantia nigra. Some fibres of the lateral lemniscus end in the inferior collicular nucleus, encapsulating it and synapsing with its neurons. The remaining direct lemniscal and inferior collicular-derived fibres enter the inferior collicular brachium, which commences at this level and conducts them to the medial geniculate body. Some fibres to the inferior colliculus are collaterals of direct lemniscal fibres.

Superiorly, level with the superior colliculus, the tegmentum contains the *red nucleus*, extending into the subthalamic region. *Central grey matter* round the aqueduct contains ventromedially the *oculomotor nucleus*, elongated and related ventrolaterally to the medial longitudinal fasciculus; below, it reaches the trochlear nucleus (7.117). The oculomotor is divisible into neuronal groups partially correlated with the nerve's motor distribution; they include a visceral efferent *accessory oculomotor nucleus* (Edinger-Westphal), dorsal to the main nucleus (p. 1056).

The red nucleus is an ovoid mass with a pink tinge, about 5 mm in diameter, dorsomedial to the substantia nigra (7.118). The tint appears only in fresh material and is due to a ferric

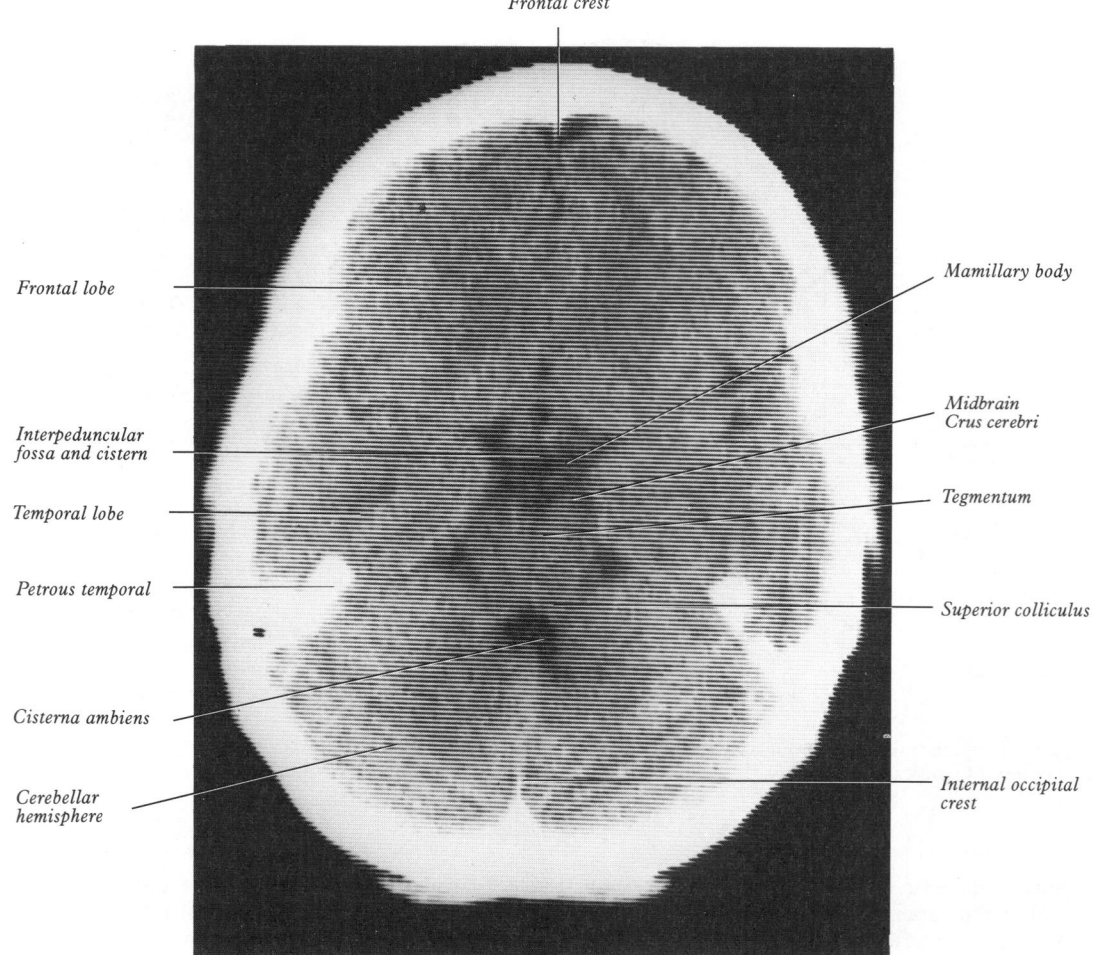

Frontal crest

Frontal lobe

Interpeduncular fossa and cistern

Temporal lobe

Petrous temporal

Cisterna ambiens

Cerebellar hemisphere

Mamillary body

Midbrain Crus cerebri

Tegmentum

Superior colliculus

Internal occipital crest

7.119 Horizontal computed tomogram through the head at the level of the midbrain. Supplied by Shaun Gallagher, Guy's Hospital, London; photography by Sarah Smith.

pigment in many neurons; these are mostly multipolar, of varying size, mostly large or small. Their proportions and arrangements vary in mammalian groups; in primates the *magnocellular* element is decreased, with a reciprocal increase of *parvocellular* complement. Small multipolar neurons occur in all parts, but in human nuclei the larger neurons are restricted to the lower level. Condé & Condé (1973) classified feline rubral cells as multipolar, pyramidal, fusiform and spherical, emphasizing also a heterogeneous nuclear cytoarchitecture; multipolar neurons predominated below, pyramidal and spherical at higher levels, but all occurred at all levels. Superiorly the red nucleus is poorly demarcated, blending into the reticular formation and caudal pole of the interstitial nucleus' lower end (Davenport & Ranson 1930). Inferior *compact* and superior *diffuse* architectonic regions have been described; the whole mass is traversed and surrounded by fascicles of nerve fibres, including many from the oculomotor nucleus (7.118). This makes the red nucleus appear reticular. Its magnocellular element is considered phylogenetically old, which accords with the parvocellular predominance in primates. *Afferent connections* are complex and many; only those well attested are noted here. Some occur in other mammals but are uncertain in man. Uncrossed corticorubral fibres from the sensorimotor cortex appear in cats (and other mammals); an area frontal to the precentral gyrus (p. 1055), termed 'supplementary motor area' in the feline cerebrum, projects bilaterally to the red nuclei; somatotopy has been shown in these fibres and nuclear terminations, conforming with localization in the rubrospinal tract (vide infra). The cortical projection is from primary somatomotor and somatosensory areas. Brown (1974) has described their terminations as axodendritic and largely in peripheral parts of the red nucleus. There are also bilateral, probably reciprocal connections

with the superior colliculi. Feline red nuclei also receive nerve fibres from the contralateral nucleus interpositus, corresponding to human globose and emboliform nuclei and from the contralateral dentate nucleus. All these cerebellar connections and perhaps others (p. 966) traverse the superior cerebellar peduncle and display some somatotopy. Other reputed sources of afferents are the globus pallidus, subthalamic and hypothalamic nuclei, substantia nigra and spinal cord. (Rubral connections are outlined in 7.120.)

The main *outflow* is the **rubrospinal tract** (p. 934), from large and small neurons in other mammals and probably so in mankind. As mentioned (p. 934), it is often regarded as unimportant in man by clinical and academic observers, who usually derive it only from the magnocellular element, with axons which do not travel far in the spinal cord. Evidence for this negative view is exiguous; all studies in other mammals, chiefly cats and monkeys, take it to lumbosacral levels. Tract and nuclear origins show somatotopy, those axons ending at cervical levels being from neurons segregated dorsomedially, those for lumbosacral segments ventrolaterally and those for thoracic levels being intermediate. This somatotopy is confirmed by the recording of antidromic impulses evoked in the nucleus by stimulation of the tract. Some efferent axons form a *rubrobulbar tract* to motor nuclei of the trigeminal and facial nerves and perhaps the inferior olivary and oculomotor, trochlear and abducent nuclei; many 'rubrobulbar' fibres may end throughout the brain-stem reticular formation.

In cats *rubrocerebellar* connections to the dentate nucleus have been described (Courville & Brodal 1966). Efferents may also reach ventrolateral thalamic nuclei from the inferior third of the red nucleus. The rubrospinal tract crosses at once in the *ventral tegmental decussation*, which lies ventral to the *tectospinal*

984

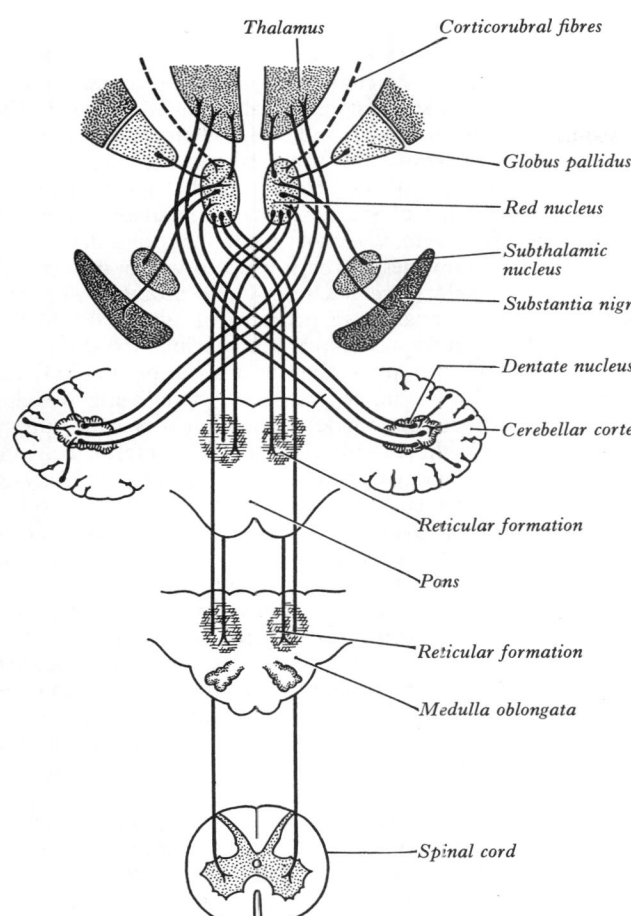

7.120 Simplified diagram of the principal connections of the red nucleus.

pons and upper medulla, but is displaced ventrally by successive decussations of the medial lemnisci and lateral corticospinal tracts, to enter the ventral spinal intersegmental fasciculus.

The proximity of the medial longitudinal fasciculus to oculomotor, trochlear and abducent nuclei, emerging facial fibres, dorsal cochlear fibres and the hypoglossal nucleus, makes it a convenient interconnecting route in the brain stem. Continuity with the ventral spinal intersegmental tract provides for connections between these nuclei and cervical motor neurons, especially those innervating the nuchal musculature. For example, co-ordination between facial and hypoglossal nerves in movements of the lips and tongue in speech is often attributed to connections via the medial longitudinal fasciculus; but this is doubtful. Long established are substantial contributions to it by *vestibular nuclei* and, to a lesser extent, the lateral lemniscal nucleus, its chief function being to co-ordinate movements of the eyes and head in response to vestibulocochlear stimulation (Maciewicz et al 1977, Yamamoto et al 1978, and p. 1112). Fascicular lesions may cause partial ophthalmoplegia. Bilateral vestibular connections ascend or descend the fasciculus or divide into ascending and descending branches, which send collaterals to nuclei of the third to sixth cranial nerves and the spinal nucleus of the eleventh (7.121). All four vestibular nuclei contribute ascending fibres, those from the superior remaining uncrossed (McMasters et al 1966), the others

decussation, then descends ventral to the *decussation of the superior cerebellar peduncles*, through the reticular formation of the pons and medulla to the spinal cord. Since it ends in almost the same laminae as the corticospinal tract, the *cortico-rubro-spinal* projection is one example of an indirect corticospinal path (p. 934).

Associated with rubrospinal tracts are the *tegmentospinal* fibres from tegmental reticular elements lateral to the red nuclei, probably to be grouped with other medullary and pontine reticulospinal tracts (p. 934).

The **tectospinal** and **tectobulbar tracts** also start at this level, from neurons in the superior colliculi, sweeping ventrally round the central grey matter to decussate ventral to the oculomotor nuclei and medial longitudinal fasciculi as part of the *dorsal tegmental decussations*. The tectospinal tract descends ventral to the medial longitudinal fasciculus as far as the medial lemniscal decussation in the medulla, where it diverges ventrolaterally to reach the spinal ventral white column near the ventral lip of the anterior median fissure. The tectobulbar tract, mainly crossed, descends near the tectospinal, ending in the pontine nuclei and motor nuclei of the cranial nerves, particularly those innervating the orbital muscles. It serves reflex ocular movements (see references to lateral tectotegmentospinal tract, p. 932).

The medial longitudinal fasciculus (7.117, 121), a heavily myelinated composite tract, is ventrolateral to the oculomotor nucleus at this level, where its fibres are more dispersed than at lower levels, though its relation to efferent nuclei is retained. The fasciculus ascends to the *interstitial nucleus* (of Cajal) in the third ventricle's lateral wall, just above the cerebral aqueduct. (Closely related to its upper ends and the decussation of the fasciculi are also the *interstitial* and *dorsal* parts of the nuclear groups *of the posterior commissure* which probably exchange fibres with these complex fasciculi and additionally the *nuclei of Darkschewitsch*, whose status remains uncertain.) The fasciculus retains its position relative to the central grey matter through the midbrain,

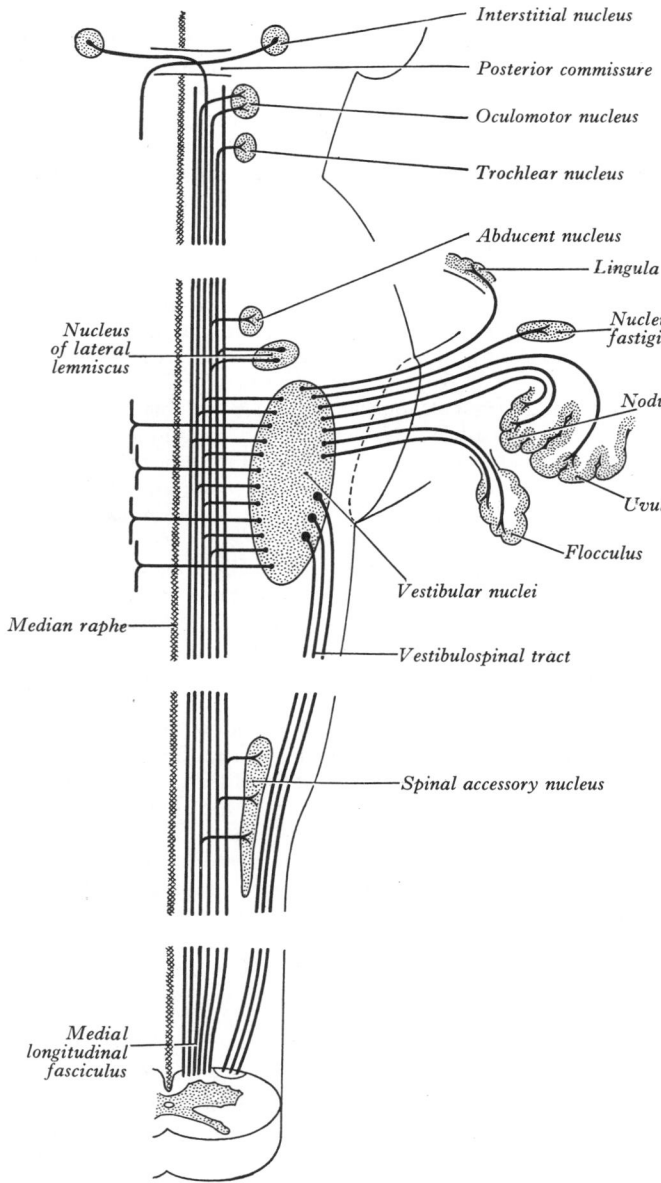

7.121 Simplified diagram of *some* of the components of the *medial longitudinal fasciculus* and of the distribution of its fibres to cranial nerve nuclei.

partly crossed. Some fibres reach the interstitial and posterior commissural nuclei, some decussating to the contralateral nuclei (Mantano 1970, see also p. 1005). Descending axons, from medial vestibular nuclei and perhaps the lateral and inferior, partially decussate and descend in the fasciculus as a *medial vestibulospinal tract* (Brodal et al 1962, and pp. 932, 945). Fibres join from the dorsal trapezoid, lateral lemniscal and posterior commissural nuclei. Therefore the cochlear as well as the vestibular nerve may influence movements of the eyes and head via the fasciculus. Some vestibular fibres may ascend in it up to the thalamus.

Other mesencephalic tracts have been described, some extending in much of the brain stem; only two will be mentioned.

The dorsal longitudinal fasciculus (of Schütz) runs in the central grey matter of the midbrain, pons and medulla, ventrolateral to the aqueduct and fourth ventricle, as a path for descending and ascending connections, largely uncrossed, between hypothalamic and dorsal thalamic and brain-stem nuclei, including the accessory oculomotor, superior collicular, ambigual, salivatory, facial, solitary and hypoglossal (Crosby & Woodburne 1951) and brain-stem reticular nuclei. Some *cholinergic* fibres are described as *descending* from hypothalamic nuclei to enter the cerebellum via its superior peduncle. A *dorsal ascending serotonergic* bundle accompanies the dorsal fasciculus to the diencephalon (p. 994).

The central tegmental fasciculus, prominent in the medulla (p. 956), is more ventral in the midbrain, at first lateral to the medial longitudinal fasciculus and dorsolateral to the red nucleus and decussation of the superior cerebellar peduncles. Descending through the pons into the medulla it swerves laterally to reach and terminate in the inferior olivary complex (Bebin 1956); though few fibres are from the thalamus, it was once called the thalamo-olivary tract. Most descending fibres were said to be from neurons in the central grey zone around the cerebral aqueduct at superior collicular level; but fibres from the basal ganglia and red nucleus also travel in it. Traversing the red nucleus, the ansa lenticularis (p. 1006) and lenticular nucleus, the fasciculus forms a path from the motor cortex to the brain-stem reticular nuclei and inferior olivary nuclear complex. While these components are still largely accepted, additional information has accumulated. Many fibres have proved to be ascending, descending (or both) axons from brain-stem reticular formation, their collaterals and terminals innervating other 'reticular' or adjacent 'specific' nuclei (p. 988). Newer staining techniques have also identified in the fasciculus: *dorsal* and *ventral ascending noradrenergic bundles*, a *ventral ascending serotonergic bundle* and some fibres of *dorsal* and *ventral ascending cholinergic bundles* (pp. 993, 994, 995).

The brachium of the inferior colliculus is a rounded superficial lateral strand in the upper midbrain (7.135A) ascending from the inferior colliculus and lateral lemniscus to the medial geniculate body. It separates dorsolateral fibres of the medial lemniscus from the surface. Quantitative examination by Ferraro & Minckler (1977) in 28 autopsy specimens (from newborn to 97 years) revealed a fibre density of about 68 000/mm², with little variation between left and right and no significant diminution until the ninth decade. The total fibres in an average brachium were 354 314 (right) and 346 810 (left) near the inferior colliculus, rising to 557 122 (right) and 561 359 (left) near the medial geniculate body. These totals must be treated with reserve; the brachium was dissected away from the brain stem. Ramón-Moliner & Dansereau (1974) have examined the 'peribrachial' region at the pontomesencephalic junction and proposed an arbitrary topological division of this part of the tegmental brain stem.

TECTUM OF THE MIDBRAIN

Though the tectum exhibits dorsal swellings from the earliest vertebrate stages, differentiation into superior and inferior colliculi (corpora quadrigemina) appears only in mammals (7.115, 116). In prototherian monotremes, still only a single pair of swellings exists, the *optic lobes* (corpora bigemina). Optic lobes are highly developed in fishes and are larger than the olfactory lobes in Osteichthyes. At this and later levels of vertebrate

evolution the olfactory lobes are, in volume, the major *forebrain* developments; with the emergence of cerebral hemispheres, i.e. further forebrain development with more than olfactory connections, the optic lobes are relatively smaller but still remain substantial in amphibians, reptiles and birds. Their connections and functions show a progressive forward neural migration, passing from reptiles to mammals, a process called *telencephalization*. This initial dominance of the optic lobes (in earlier vertebrates probably concerned with *all* sensory modalities other than olfactory) gradually diminishes, perhaps coupled with a decrease in complexity of their cytoarchitecture. These processes are considered to reach their acme in primates, in whom the highly developed cerebrum has taken over much primal tectal activity. Differentiation of the optic lobes into superior and inferior colliculi appears in eutherian mammals, both being perhaps derivatives of primary optic lobes. Though it is usual to simply link the superior colliculi with visual and the inferior colliculi with auditory behaviour, to accept this as *exclusive* is likely to mislead; superior colliculi in particular are concerned in other ascending paths, from cutaneous receptors and perhaps auditory connections. Inferior colliculi are interconnected with the superior. (See below for many further connections.)

The inferior colliculus has a central ovoid main nucleus, derived from periaqueductal grey matter, with which it preserves some continuity. The nucleus is surrounded by a laminar zone of nerve fibres; many from the lateral lemniscus (p. 959) terminate in it. Small and medium-sized neurons predominate, many stellate or multipolar, and some are scattered in the surrounding nerve bundles, particularly on its lateral and medial aspects; sometimes the latter are regarded as a separate nuclear entity; they almost certainly have a different origin, being continuous with the superficial lamina of the *superior* colliculus. Nuclear and neuronal organization have been detailed in cats (Morest 1966, Geniec & Morest 1971, Rockell & Jones 1973, Meininger & Baudrimont 1977) and to a lesser extent in man. It is accepted that the *central nucleus* has *dorsomedial* and *ventrolateral* zones, chiefly differing in neuronal types. The 'laminar zone', noted above, is lateral and peripheral in cats. Disagreements concern the distribution of different sizes of neuron; Meininger & Baudrimont (1977), by using stereological techniques in a small series of cats, described the dorsomedial nucleus as containing small neurons, those in the ventrolateral being larger. The inferior colliculi are connected by a substantial *commissure* of axons from their own neurons and also crossed lateral lemniscal fibres. The main afferent fibre bundle is the lateral lemniscus (p. 959); efferent fibres largely traverse the inferior brachium to the ipsilateral medial geniculate body; a crossed connection, via the intercollicular commissure, may exist (Ades 1944). Lemniscal fibres relay only in the main collicular nucleus (Goldberg & Moore 1967); some pass without relay to the medial geniculate body; similarly some colliculogeniculate fibres do not relay in the geniculate body but continue with those which do via the auditory radiation (p. 1112) to the auditory cortex, area 42. As in other sensory systems a *descending* projection from this cortical area reaches the inferior colliculus via the medial geniculate body, which some fibres may traverse without relay. This descending path may produce effects at levels from the medial geniculate body downwards; it probably links with efferent cochlear fibres, through the superior olivary and cochlear nuclei (p. 960). Tonotopical projection of lateral lemnisci on the inferior colliculi has been described in cats; some similar orderly arrangement probably pervades *all* auditory connections, ascending and descending (Moore & Goldberg 1966 and p. 1005).

Inferior collicular projections to the brain stem and spinal cord appear to first traverse the superior colliculi, thereby connecting with the origins of tectospinal and tectotegmental tracts. These projections are relatively small; the customary view of the inferior colliculi as reflex centres for auditory responses may require revision. There is, however, much evidence that they are concerned in the ability to localize the source of sounds (Masterton et al 1967).

The superior colliculus (unlike the inferior which is fully developed in mammals) is generally regarded as simplified in higher vertebrates, particularly in primates; but nevertheless in man it still exhibits much of the complex laminar organization of earlier forms. At least six laminae are described in human superior colliculi and even more intricate arrangements in some mammals (Cajal 1909–11, Angaut & Reperant 1976). In this laminar pattern the superior colliculus closely resembles the cerebellar and cerebral cortex which, indeed, it surpasses in complexity in the lower vertebrates. This has prompted the view that the optic lobes, from which superior colliculi are derived, were the first suprasegmental development in vertebrate brains, mediating summation and integration more widely than the term 'optic' indicates. Such considerations are a corrective to the widely current teaching that superior colliculi are exclusively visual in their activities.

Human collicular laminae have been variously termed. From the exterior there are at successive depths the strata zonale, cinereum, opticum and lemnisci. The stratum lemnisci itself is often divided into the strata griseum medium, album medium, griseum profundum and album profundum. These *seven layers* have also been termed: zonal, superficial grey, optic, intermediate grey, deep grey, deep white and periventricular strata (Crosby et al 1962). The two schemes do not completely accord; but as a generalization, layers may be considered alternately composed of neurons or their fibres and dendrites, though some admixture occurs. For example, the *zonal layer* consists chiefly of myelinated and non-myelinated fibres from the occipital cortex (areas 17, 18 and 19, p. 1060), arriving as the *external corticotectal tract*; but among these fibres are a few small neurons, horizontally arrayed. The *superficial grey layer* (stratum cinereum) has many small multipolar interneurons, with which cortical fibres partly synapse, the whole being a cap-shaped lamina, centrally thicker, over the deeper layers. The *optic layer* consists partly of fibres from the optic tract, a much diminished afferent element in higher primates; as they terminate, they invade adjacent layers with numerous collateral branches. The layer also contains some large multipolar neurons; efferent fibres to the retina are said to start in this layer (Wolter & Liss 1956) in man and other mammals. The internal four layers are sometimes considered together as a stratum lemnisci but will be briefly outlined. The next two, *intermediate grey* and *white layers*, are collectively the main reception zone, consisting of neurons of various sizes mixed with axons and dendrites. Its main afferent source is the *medial corticotectal path* from the occipital cortex (area 18) and possibly the preoccipital cortex (area 7), areas concerned with ocular following movements (p. 1060). These layers also receive spinal afferents through spinotectal and spinothalamic routes and probably from the inferior colliculus. The *deep grey* and *deep white* layers, adjacent to the periaqueductual grey substance, again contain a mixture of neurons with dendrites extending peripherally to the optic layer, but their axons form many of the collicular efferents. (For details of the neonatal superior colliculus consult Labrida & Laemle 1977.)

The superior colliculus thus receives *afferents* from a wide area (Wolter & Liss 1956, Meikle & Sprague 1964), including the retina, spinal cord, inferior colliculus and occipital and temporal cortex, the first three of these pathways conveying visual, tactile and probably thermal, pain and auditory impulses, the cortical projection acting as a 'command' and possibly modulating path. Collicular *efferents* pass to the retina and a wide array of brainstem and spinal neurons; *tecto-oculomotor* fibres project to the motor nuclei of ocular muscles. In cats (Edwards & Henkel 1978) neurons in the upper third project to the periaqueductal grey substance dorsal to the upper end of the oculomotor complex, some also synapsing directly with dendrites of oculomotor neurons. Neurons in this central grey region also project to both abducent nuclei. *Tectospinal* fibres (p. 985) descend to cervical segments while *tectotegmental* fibres reach various tegmental reticular nuclei and also the substantia nigra, red nucleus and probably the spinal cord. *Tectopontine* fibres, probably descending with the tectospinal tract, terminate in dorsolateral pontine nuclei, with a relay to the cerebellum (p. 985). *All* descending fibres *cross*, mostly in the *dorsal tegmental decussation*, but to a lesser extent through an intercollicular commissure. Kawamura et al (1974), reviewing these connections, have studied tectoreticular projection in cats by collicular lesions, producing degeneration in ipsilateral mesencephalic and contralateral pontomedullary reticular nuclei (nuclei reticularis gigantocellularis, reticularis pontis caudalis, reticularis pontis oralis and others). Distribution of these projections corresponded to those from cortical, fastigial and vestibular neurons and thus to the main sources of reticulospinal fibres. A tecto-olivary projection, from deeper collicular laminae to the upper third of the medial accessory olivary nucleus, exists in primates (Frankfurter et al 1976); it is crossed and links with the posterior vermis in ocular movements.

The relative paucity of retinal projection to the primate superior colliculus weakens the common view of it as mediating largely visual reflexes, though it so acts in many other mammals. The projection of retinal quadrants on superior colliculi in cats and rodents has been established, by degeneration and stimulation; but results in monkeys are less definite, perhaps due to a greatly reduced retinocollicular projection in primates. Clinical evidence merely suggests a similar limited human pattern of association of the superomedial half of a colliculus with the inferior retinal quadrants and the inferolateral half with superior retinal quadrants. Even the connections to ocular motor nuclei, held to accord with collicular pattern, are uncertain in cats, though perhaps merely interrupted by a pretectal relay. For the present they remain sub judice.

Attention has been much focused on behavioural changes after damage to subcortical parts of the visual pathway (Ingle & Schneider 1969); visual field projection on the optic tectum has also been mapped and compared by recording techniques in vertebrates from fish to mammal. A similar pattern of retinal representation exists even in different classes, the chief difference in mammals being the degree of *ipsilateral* projection. This accords with observations of movement elicited by tectal stimulation; e.g. central collicular stimulation produces 'normal' contralateral head movement in cats (Hess et al 1946). Many similar responses like those of intact animals have been observed in several species, involving the eyes, ears, head, trunk and limbs, thus implicating the superior colliculus in complex integrations between vision and widespread bodily activity (Schaefer & Schneider 1968). In view of the bilateral representation of retinae in the colliculi, results of brain-splitting experiments are of special interest; e.g. tegmental division leads in cats to profound changes in visual behaviour, not so much in reflex responses as in the ability to interpret the environment and adapt to it (Voneida 1965). Response to threatening stimuli and an ability to locate edges and follow them are lost. Similar results appear in monkeys and resemble those of the split-brain syndrome in human patients (p. 1066).

A descending projection from the *auditory cortex* has been described in some mammals, including cats (Diamond et al 1969) and monkeys (Whitlock & Nauta 1956). Paula-Barbosa & Sousa-Pinto (1973) have confirmed this and explored the projection in greater detail in cats, in which a bilateral projection from secondary auditory areas (AII) reaches the superficial laminae in both superior colliculi; no such projection from the primary auditory area (AI) was observed. Though these are from parts of the auditory cortex which, as far as is known, have no visual connections, this corticocollicular pathway may be involved, via known connections between superior and inferior colliculi, in integrations between visual and auditory behaviour.

The pretectal nucleus is a poorly defined mass of neurons at the junction of the mesencephalon and diencephalon, extending from a position dorsolateral to the posterior commissure towards the superior colliculus, with which it is partly continuous. It receives fibres through the superior brachium from the occipital and preoccipital cortex and the lateral root of the optic tract (Kuhlenbeck & Miller 1949, p. 1014). Its efferent fibres reach both accessory oculomotor nuclei (Ranson & Magoun 1933); those which decussate pass ventral to the aqueduct or through the posterior commissure. By this autonomic outflow, with a relay in ciliary ganglia (p. 1097), sphincters of the iris in *both* eyes are made to contract in response to impulses from either (7.139). This bilateral light reflex may be the sole activity of the pretectal nucleus but some of its efferent axons are said to enter the tectobulbar and tectospinal tracts. The significance of the cortical projection is not established.

EXTRAGENICULATE VISUAL PATHWAYS

As a corrective codicil to the above it is apposite to refer briefly to data accumulated in regard to visual paths in mammals (Jones 1974) and particularly birds (Webster 1974). It is almost certain that the vertebrate visual system was primarily mesencephalic in its connections. The concept of 'telencephalization' is part of a greater process of 'encephalization', the evolutionary cranial migration of 'centres' mediating sensorimotor activities, from spinal to *encephalic* levels. Intrusion of visual projections into the *telencephalon* via the thalamus in mammals, with a reciprocal reduction in *mesencephalic* visual connection, remains the major example of *telencephalization*. The change is often regarded as more advanced in mammals, especially primates, than in birds. Compared with mammalian conditions, avian colliculi or optic lobes are relatively larger and more complex in laminar pattern, and the avian corpus striatum is more developed than the superincumbent cortex, proportions apparently reversed in mammalian cerebral hemispheres. The generalizations thus engendered have tended to assign avian and mammalian visual systems to opposite balances in a supposed *duality of pathways*: one integrated at mesencephalic level, vaguely termed 'reflex', the other telencephalic and loosely labelled 'perceptual'. This is an oversimplification. Avian visual structure and function show that both types of path, whether through the midbrain tectum or direct to the dorsolateral thalamus, involve thalamic and cortical projection and striatal connections. In mammals *both* paths exist, in particular a pretectal route (in addition to the superior collicular path) via the pulvinar of the thalamus to the occipital cortex. Functional studies also show that it is erroneous to consider the two visual paths, either in birds or mammals, as being in the one case perceptual and in the other reflexive. In both vertebrate classes the mesencephalic level appears involved in perception and not merely as a level for ocular reflexes; e.g. in avian brains neurons of a tecto-striate projection exhibit responses to movement *and* to colour, though not to shape or orientation. In contrast, a more direct route via the thalamus to a cortical region known as the '*wulst*' in birds (equivalent to the mammalian geniculostriate pathway) has not been shown to evince any special sensitivities, except perhaps to movement. In mammals, including mankind perhaps more than in birds, the complex paths of visual information and multiplicity of collateral and descending interconnections provide a 'system' of intricate projection and 'feed-back' which remains speculative. Despite a mass of data particularly concerning the peculiarities of response in individual neurons at every level, it is difficult to ascribe aspects of visual information to identifiable neuronal groups; perhaps it is misleading to seek to do so. Even in man, clinical evidence from injury to the visual cortex suggests that the customary assignment of 'conscious' vision to cortical levels may impede recognition of the intimate interactions between the occipital cortex, tectum and pretectum. Though no functional synthesis can be discerned in the structural and physiological findings so far amassed, visual function must involve most widespread networks of neurons at all levels.

The cerebral aqueduct, a narrow canal about 15 mm long, connects the third and fourth ventricles. (Being in the midbrain and conducting no fluid to the cerebrum, it is misnamed.) In transverse section it varies, being T-shape below, triangular above and between these oval and slightly dilated (Flyger & Hjelmquist 1957). It is lined by ependyma (with specialized areas, pp. 894, 1005), surrounded by *central grey matter* continuous with the floor of the fourth ventricle and walls of the third. Remains of a *sulcus limitans*, apparent on each lateral wall in the embryonic period, may persist. Dorsally it is separated partly from the collicular grey matter by fibres of the stratum lemnisci; ventral are the medial longitudinal fasciculi and midbrain reticular formation. Scattered through the central grey matter are many and various neurons embedded in fine fibres; it also contains the mesencephalic trigeminal, oculomotor and trochlear nuclei.

THE RETICULAR FORMATION

Long recognized among the more conspicuous tracts and nuclei throughout the brain stem are extensive fields of intermingled neurons and nerve fibres collectively termed the **reticular formation**. Accounts of the central nervous system contain numerous but usually scattered references to this 'system', but it requires a more orderly and overall description. It has attracted a large literature, e.g. Moruzzi & Magoun (1949), Meessen & Olszewski (1949), Olszewski (1954), Olszewski & Baxter (1954), Brodal (1957), Jasper et al (1958), Taber (1961), Valverde (1961a, b, 1962), Ramón-Moliner & Nauta (1966), Horridge (1968), Brazier (1969), Scheibel & Scheibel (1970) and Webster (1978).

GENERAL CONSIDERATIONS

Strict criteria for the designation of nervous regions as 'reticular' are difficult to postulate. Simple inspection of sections allows some measure of identification: many areas in vertebrate nervous systems areas are *predominantly* of grey or white matter (p. 866), their neurons and axons having recognizable, quantitative cytological features, consistently orientated. Thus, many regions of white matter contain fascicles with recognizable spectra of fibre diameters, definable in their directions and conducting between well-defined origins and destinations. Many areas of grey matter contain neuronal populations with somata, axon collaterals, dendritic trees and synapses in distinctive three-dimensional arrays; they also appear to have 'boundaries'. Such are the main nuclei, cortical formations and tracts of the central nervous system, exemplified by well-defined columns and laminae in dorsal and ventral spinal appendages (p. 925), their brain-stem homologues, many brain-stem, diencephalic and telencephalic nuclei, cerebral and cerebellar cortices, tectum and tracts recognized throughout the central nervous system; all are individually described elsewhere. Between these obviously organized *non-reticular* regions are those where grey and white matter intermingle, fascicles of nerve fibres crisscrossing in many directions, dividing neurons in scattered, ill-defined groups; these are the *reticular regions*. Though many agree to the geometric arrangement and non-reticular structure of most main nuclei and tracts, the status of some remains uncertain. Structural extremes of a precise, repeating geometric organization on the one hand and a truly random network on the other are abstractions. Even the highly regular cerebellar cortex shows quantitative variation when examined critically (p. 972) and the most 'reticular' regions show some locally regular structure. A graded range of structural organization exists; inevitably, more diffuse regions are more difficult to investigate. For these reasons, coupled with the sparseness of quantitative data about neuronal forms and connections, views diverge as to which regions should be included in the *reticular formation*. Most agree to the inclusion of parts of the medulla, pons and midbrain (vide infra); some would restrict the term to these. The status of other parts in these regions, such as the olivary complex, parvocellular red nucleus, pars reticularis of the substantia nigra, dorsal commissural nucleus, interstitial nucleus, intracerebellar nuclei and others, is not agreed. Further disagreement concerns the inclusion of central regions of the spinal grey matter, the 'reticular core' of some (p. 946) or inclusion of non-specific thalamic nuclei (p. 1003) and certain hypothalamic nuclei. These debates merely emphasize the undesirability of rigid categorization and need of further research, which may invalidate such controversial parcelling. Precise boundaries between such areas will not be attempted here; but it is emphasized that *all* levels of the neuraxis, spinal cord, brain stem and diencephalon have regions with high levels of ordered structure *interlocked* topographically and functionally with the surrounding reticular regions, where organization is less orderly. Obviously reticular regions are often regarded as phylogenetically ancient, representing a supposed primitive 'random nerve network', upon which background more organized, selective routes have appeared during evolution. But, as noted elsewhere (p. 865), the most primitive nervous systems show *both* diffuse and highly organized regions, which co-operate in response to different demands. Clearly *both* evolved together, each providing *indispensable* and *interdependent* contributions to the total response of the organism.

RETICULAR REGIONS OF THE CENTRAL NERVOUS SYSTEM

The general characteristics of reticular regions are: (1) deeply placed, ill-defined collections of neurons and fibres with diffuse connections; (2) conduction paths which are difficult and often impossible to define, but indicated by physiological evidence as complex and often *polysynaptic*; (3) *ascending* and *descending* components which are (4) partly *crossed* and *uncrossed*, unilateral stimulation often resulting in bilateral responses; (5) possessing *somatic* and *visceral* components.

Earlier investigators reported little structural variation in the different regions, simply dubbing all but well-known nuclei and tracts as 'reticular'. More critical studies have emphasized regional variation in structure (and presumably function), including differences in neuronal size, number, packing density, geometry of neurites, cytochemistry (including putative neurotransmitter production) and in connections. By such criteria (much dependent on the technique employed) many *reticular nuclei* have been named in a variety of animals. However, their limits are often ill-defined, dendritic fields interweaving over wide areas. Interspecific variation exists; varying technique and criteria also yield variable results. The data of electrophysiological or pharmacological exploration, focal stimulation or ablation, must be scrutinized; even greater caution is needed in ascribing functions to reticular zones or nuclei. Complex arrays of reticular neurites has been studied mainly by Golgi techniques which, combined with the cytoarchitectonics of the form and deployment of somata, have defined major zones and nuclei. Smaller groups and variations cannot be pursued here; for these consult the references quoted.

Studies with the Golgi technique (Ramón-Moliner & Nauta 1966, Horridge 1968, Brazier 1969, Scheibel & Scheibel 1967, 1970) show that few brain-stem reticular neurons are classic Golgi type II, with short axons branching locally (vide infra); their dendrites are long and usually spread *across* the brain stem's long axis in transverse sheets. Usually dendrites have a simple pattern likened to 'spokes of a rimless wheel', classified as the *isodendritic configuration* by Ramón-Moliner & Nauta (1966) and described as forming an *isodendritic core*, its territory encompassing many ill-defined, 'diffusely' connected reticular nuclei. Their radiating dendrites may spread into 50% of the cross-sectional area of their half of the brain stem. These dendritic trees are crossed by, and may synapse with, a complex of ascending and descending fibres, the *central tegmental fasciculus*. This has many components, the most prominent being axons of the reticular neurons themselves; their main stems usually ascend or descend or bifurcate to do both, travelling far, perhaps through the whole brain stem and even beyond. As an extreme example, an axon from the magnocellular medullary nucleus (vide infra) may bifurcate, its ascending branch traversing the upper medulla, pons, midbrain tegmentum, subthalamus, hypothalamus or dorsal thalamus, while its other branch descends in the reticular core of the lower medulla and may reach the cervical spinal intermediate grey matter (laminae V and VI). Many reticular neurons have unidirectional, shorter axons, synapsing with the radiating dendrites of innumerable other neurons en route, and also collaterals which synapse with cells in 'specific' brain-stem nuclei or cortical formations, such as the cerebellum. Multitudes of afferent fibres converging on individual neurons and their myriad synapses and destinations provide the structural basis for the polymodal responses elicited by experiment, and also for such terms as 'diffuse' and 'non-specific'. These arrangements perplex investigators; it is misleading to attach single, or even complex, functional labels to individual neurons or to small groups, because the innumerable routes for convergence and divergence in this reticular core make it unwise to assign specific functions to specific structures.

A contrasting dendritic form is an *idiodendritic configuration*, in which dendrites are short, sinuous or curved, branch profusely and pursue re-entrant courses at the perimeter of a nuclear group, defining a boundary between it and its environs. Such neurons are typical of: nuclei of the cranial nerves, basilar pontine nuclei, olivary complexes, nuclei of 'specific' sensory pathways, medullary nucleus funiculi lateralis (the so-called lateral 'reticular' nucleus of the medulla), and mesencephalic tegmental nuclei. Neurons of an intermediate dendritic complexity occur in and near such nuclei and vary in density in much of the remaining reticular formation, being termed *allodendritic*. In different zones the proportion of different sizes of neuronal somata varies, some regions containing only small to intermediate multipolar cells (*parvocellular* regions), while in a few these mingle with large multipolar neurons in 'gigantocellular' or 'magnocellular' nuclei.

In *general terms* the reticular formation is a continuous isodendritic core traversing the whole brain stem, continuous below with the reticular intermediate spinal grey laminae and merging above into the subthalamus and parts of the hypothalamus and dorsal thalamus. Adjacent to or embedded in it are many well-defined nuclei, a few termed reticular (Leontovich & Zhukova 1963, Ramón-Moliner & Nauta 1966). The reticular formation may thus be described as a 'diffuse network' divisible into three bilateral *longitudinal columns*: one median, a second medial (containing mostly large reticular neurons) and a third lateral (containing mostly small to intermediate neurons) columns. Regional localization resides solely in overlapping zones of terminal afferents (vide infra) and sources of efferents. A corollary is that the application of elementary functional localization to individual neurons, small groups named reticular nuclei or groups of nuclei, is absent; the 'system' operates as a whole, enhancing or depressing the 'general background' in the central nervous system. (For a stimulating discussion see Webster 1978.)

In contrast, parcelling reticular formation into named nuclei by cytoarchitectonics has been zealously pursued, the most extreme exponent being Olszewski who described over 40 such nuclei in the mammalian brain stem (Olszewski & Baxter 1954, consult also Brodal 1957). Such entities may be useful for experimenters (if dependable) but a few only have reliably specific patterns of connectivity; hence their individual functional significance is largely obscure.

Details of cytoarchitectonics, neuritic arrays, overall topography, and connectivity in such a plethora of named nuclei would be inappropriate and of dubious value here. A brief summary of principal nuclear groups (identifed by ordinary techniques) with a diagram of their approximate territories (7.122) is sufficient here. Their positions and extents need examination in serial sections, hence their omission in the preceding general description of the brain stem. For details consult Olszewski & Baxter (1954), Brodal (1957), Ungerstedt (1971), Jasper et al (1978), Nieuwenhuys et al (1978).

The median column of reticular nuclei, extended throughout the medulla, pons and midbrain, contains neurons largely aggregated in bilateral, vertical sheets, blended in the midline and occupying the paramedian zones; collectively they are *nuclei of the raphe*. In the upper two-thirds of the medulla and crossing the pontomedullary junction is the *nucleus raphes obscurus* (and associated *nucleus raphes pallidus*); partly overlapping this and ascending into the pons is the *nucleus raphes magnus* and above it the *nucleus raphes pontis*. Also pontine is the *nucleus raphes centralis superior*; finally the *nucleus raphes dorsalis* (*rostralis*) ascends, expanding then narrowing through much of the midbrain. Some neurons in rapheal nuclei are *monoaminergic* (*serotonergic*); some of their connections will be mentioned below.

The medial column of reticular nuclei has a background of reticular neurons of medium size but in some regions they are the largest such neurons; most are isodendritic. In the lower medulla the column is unclear, perhaps represented by the thin lamina

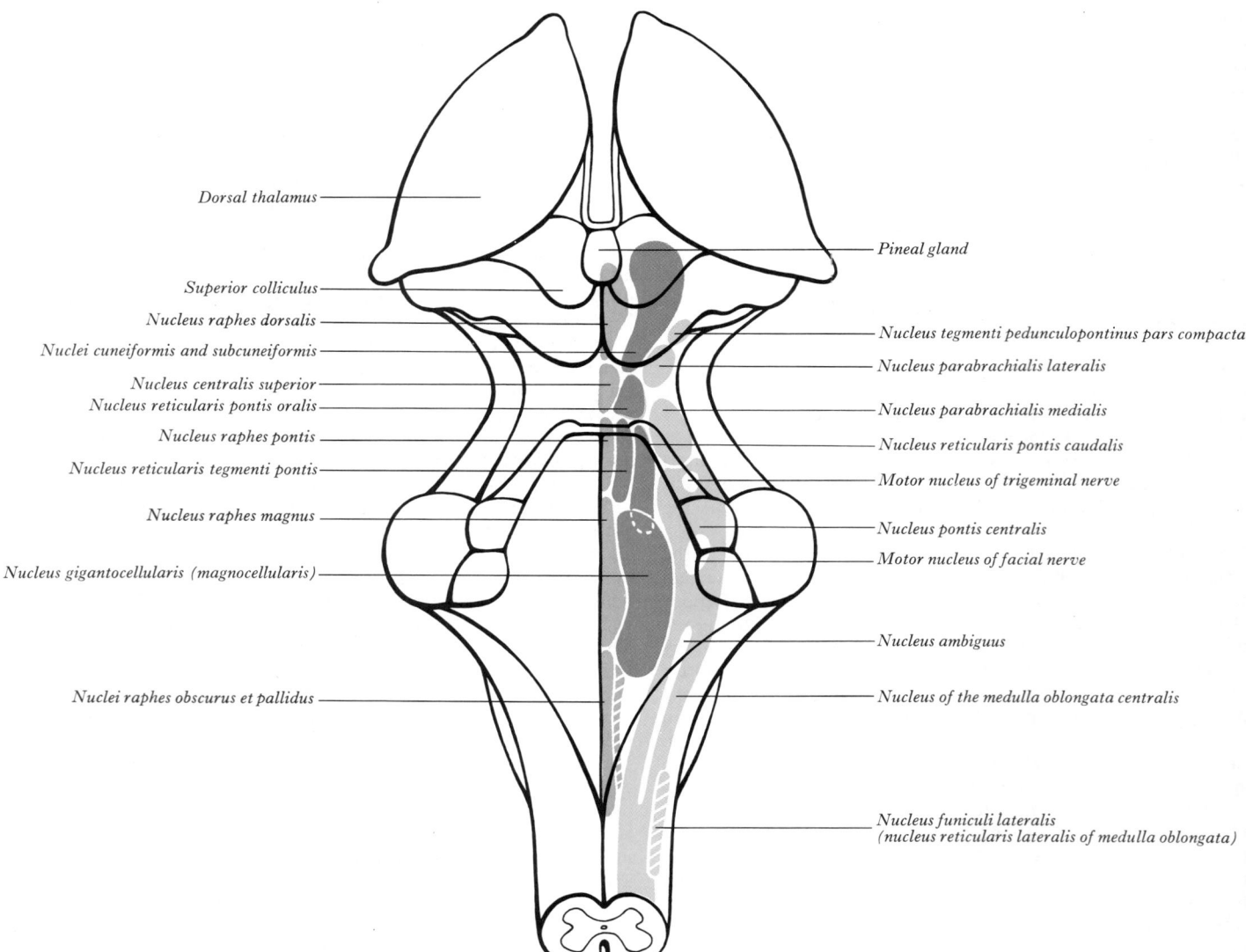

Dorsal thalamus —

Superior colliculus —
Nucleus raphes dorsalis —
Nuclei cuneiformis and subcuneiformis —
Nucleus centralis superior —
Nucleus reticularis pontis oralis —
Nucleus raphes pontis —
Nucleus reticularis tegmenti pontis —
Nucleus raphes magnus —
Nucleus gigantocellularis (magnocellularis) —
Nuclei raphes obscurus et pallidus —

— *Pineal gland*

— *Nucleus tegmenti pedunculopontinus pars compacta*
— *Nucleus parabrachialis lateralis*
— *Nucleus parabrachialis medialis*
— *Nucleus reticularis pontis caudalis*
— *Motor nucleus of trigeminal nerve*
— *Nucleus pontis centralis*
— *Motor nucleus of facial nerve*
— *Nucleus ambiguus*
— *Nucleus of the medulla oblongata centralis*
— *Nucleus funiculi lateralis*
  (nucleus reticularis lateralis of medulla oblongata)

7.122 An outline of the human brain stem (black) extending from the caudal end of the medulla to the dorsal thalami: note the margins of the rhomboid fossa, the lateral angles of which indicate the pontomedullary junction; note also the profiles of the transected surfaces of the cerebellar peduncles, the colliculi and pineal gland. The principal nuclear derivatives of the brain stem reticular formation are indicated in approximate outline, those from the median and paramedian nuclear column are in magenta; medial column derivatives are purple, lateral column derivatives blue. In reality, of course, considerable overlap of the nuclear profiles would be present when the third dimension is considered. A number of 'non-reticular' nuclei are also included.

lateral to the raphael nuclei; in the upper medulla, however, it expands into the *medullary gigantocellular (magnocellular) nucleus*, lateral to the rapheal nuclei, ventrolateral and ventral to the hypoglossal and dorsal vagal nuclei and dorsal to the inferior olivary complex. Ascending further the column continues as the *pontine gigantocellular (magnocellular) nucleus* lying medially in the tegmentum; its neurons suddenly diminish in size forming, in upward order, the almost co-extensive nuclei *reticulares tegmenti pontis* and *pontis caudalis*, then the *nucleus reticularis pontis oralis*, from which the medial column expands into the *nucleus cuneiformis* and *nucleus subcuneiformis*, fading away in the midbrain tegmentum. Other nuclei sometimes classified with the medial column are nuclei coeruleus (p. 980) and subcoeruleus, which cross the pontomesencephalic junction (p. 980); near this, adjoining the mesencephalic dorsal and medial longitudinal fasciculi, are the *dorsal* and *deep tegmental nuclei*, together with numerous groups in the *mesencephalic circumaqueductal grey matter*.

The lateral column of reticular nuclei contains little but scattered *small* neurons, hence often termed the *parvocellular reticular column*, with its contained *parvocellular 'nuclei'*; its neurons are allodendritic or idiodendritic (vide supra). The column extends through the brain stem as a poorly defined 'background' with few 'reticular' nuclei; in or invading it are clearcut 'specific' nuclei, which many exclude from the reticular formation (sometimes on arbitrary criteria). Descending through the lower two-thirds of the lateral pontine tegmentum and upper medulla, the column lies between the gigantocellular nucleus (medial) and sensory trigeminal nuclei (lateral); continuing caudally, it expands to form most of the reticular formation lateral to the rapheal nuclei and continues into the cervical intermediate grey laminae. Its medullary part is sometimes considered as *central and ventral medullary reticular nuclei*, which are confluent and enclose the nucleus ambiguus (pp. 954, 1113). Lateral to them is the *nucleus of the lateral funiculus*, often named *the lateral medullary reticular nucleus*, despite its cytoarchitecture. Similarly, the *nucleus of the ventral funiculus* is often called *the medullary nucleus reticularis paramedianus*. The *pontine parvocellular column* or *nucleus* embraces the motor nuclei of the facial and trigeminal nerves; the column, it is claimed, ascends from the pons into the mesencephalic tegmentum, differentiating here into large, inappropriately named, *lateral and medial parabrachial nuclei*. Above inferior collicular level the lateral column is less marked; but lateral to the decussating superior cerebellar peduncles is the magnocellular *nucleus tegmentalis pedunculopontinus, pars compacta*, above which the parvocellular *pars dissipata* gradually merges into the nuclei cuneiformis and subcuneiformis of the medial column (vide infra). Their parvocellular confluence blends with many subthalamic, hypothalamic and dorsal thalamic nuclei.

CONNECTIONS OF THE RETICULAR FORMATION

Most reticular connections, based on older neuroanatomical techniques, have already been described and here will be merely listed; for a consideration of more recent studies, vide infra.

Afferent projections reach the reticular formation from: (1) the spinal cord via *spinoreticular paths* and *collateral branches of long ascending tracts*, though not as abundantly as once held; (2) collateral branches of some primary and many secondary afferent neurons associated with *cranial nerves*, including the central *vestibular* and *acoustic* pathways; (3) from the *cerebellum* via *cerebelloreticular paths* previously noted (p. 967); (4) indirectly from *visual* and *acoustic* paths and other afferent channels via *tectoreticular fibres*; (5) from *thalamic, subthalamic* and *hypothalamic nuclei* (pp. 1003, 1006, 1010); (6) from the *corpus striatum*, directly and indirectly via its complex outflow; (7) direct *corticoreticular fibres* from the *sensorimotor* cerebral cortex and from other cortical areas; an unknown proportion are collaterals of corticofugal tracts such as the corticospinal; (8) from various parts of the *limbic system*, including the *septal areas, amygdaloid nuclei* and *hippocampus* via various descending routes.

Multiple afferents thus converge on the reticular formation; while there may be *some* zonal localization at their terminations, much evidence exists for overlap and convergence. Unit recording has strongly supported this view; single 'units' have been studied during the multimodal stimulation of cutaneous receptors and proprioceptors in many regions and in many peripheral nerves, special sense organs and diverse central nervous regions, including the spinal cord, cerebral and cerebellar cortices, hippocampus, corpus striatum, etc. For example, some units respond solely to nociceptive stimuli applied in different regions; others respond to a wide range of cutaneous stimuli, from extensive bilateral fields; others respond to the stimulation of many separate regions of the cerebral cortex, and so forth. Some units also respond to the stimulation of several different channels, e.g. visual, acoustic, cortical, cerebellar, etc. Different afferents also interact, some with mutual enhancement and others mutual inhibition. Though different levels of the reticular formation are *not* uniform in input and output, there is always a convergence of afferent channels and a rich interconnection between reticular neurons. How such a structurally complex and functionally fluctuating array is used to direct co-operative responses remains an intractable problem (Klimer et al 1968, Webster 1978).

Efferent connections of the reticular formation are: (1) to autonomic and locomotor control 'centres' and interneuronal pools of the *spinal cord* via *reticulospinal tracts*; (2) by short descending paths to similar centres in the *brain stem*; (3) to the *cerebellum*; (4) to the *red nucleus, substantia nigra* and *tectum*; (5) to numerous nuclei in the *subthalamus, thalamus* and *hypothalamus*; (6) indirectly, through radiations of these diencephalic nuclei, to the corpus striatum and cerebral cortex, including *most areas of the neocortex* and many *limbic structures*.

Further comments on reticular connectivity. As already noted, most neurons of the *medial reticular columns* and many *rapheal nuclear neurons* are isodendritic (vide supra), with radiate dendritic trees crossed at right angles (i.e. longitudinally) by innumerable ascending and descending axons synapsing with successive neurons. The dendrites also have connections with ascending spinoreticular fibres (vide infra), cranial nerve sensory nuclei, cerebellar (particularly fastigial) efferents, descending fibres from the limbic forebrain complex and hypothalamus, neocortex and tectum, and medially directed axons from some lateral column parvocellular nuclei. Ascending, descending, or bifurcating axons and collaterals of these isodendritic neurons may remain within the reticular system to complete local reflex loops or as intermediaries in longitudinal polysynaptic chains; others project to spinal, brain-stem and forebrain neurons, i.e. they are the sources of most longer reticular efferent axons. The allodendritic and idiodendritic neurons of parvocellular lateral column nuclei appear to receive mainly afferents from nuclei of the sensory cranial nerves and collaterals of neocortical corticoreticular fibres. Some axons pass medially, presumably to modify activity in the medial column; but most surround, interweave and synapse

with motor cranial nerve nuclei, completing reflex loops or mediating corticonuclear control.

As examples of zonal or nuclear varieties of reticular activity we may summarize components assumed to be involved in a so-called *ascending activating system*, and some nuclei with well-defined connections. The *ascending activating system* (vide infra) is widely believed to be the substrate for the *arousal reaction* and *desynchronization* (as shown by electroencephalogram), essential to wakening, attention and 'specific' cortical activities (p. 996). It comprises: (1) spinoreticular inputs, (2) collaterals from other long ascending tracts, (3) afferents from sensory cranial nerves, (4) reticulo-diencephalic projections, and (5) hypothalamo-limbic and diffuse 'non-specific' thalamo-cortical radiations; probably the thalamo-striato-pallido-thalamic connection should be included. Most connections are described elsewhere, but a few points need emphasis. Spinoreticular projections reach most parts of the medial reticular column; but at many levels their terminals are sparse and are only abundant in the medullary gigantocellular nucleus, the oral part of the central medullary reticular nucleus and the nucleus reticularis pontis caudalis (inferior). Similar zones receive most terminals from pontomedullary trigeminal sensory nuclei and the nucleus of the tractus solitarius; but vestibulocochlear and retinotectal inputs are concentrated on mesencephalic reticular nuclei. The origins of reticulospinal tracts and some constituent fascicles making reticulo-diencephalic connections are considered on pp. 1003, 1006, 1010; in general they correspond to the inputs just described.

Certain reticular nuclei have clear limits and definite afferent or efferent projections. Thus the lateral reticular nucleus, magnocellular nucleus and nucleus reticularis paramedianus, all medullary groups, and the nucleus reticularis tegmenti pontis all have precise cerebellar connections. The nucleus coeruleus is described on p. 980; the nucleus tegmenti pedunculopontinus, pars compacta, receives a discrete pallidofugal input.

COMPARTMENTALIZATION IN THE RETICULAR FORMATION

Preceding passages have shown that investigators have, with varying techniques, compartmentalized the reticular formation in contrasting ways, each with apparent advantages and limitations. These range as follows:

(1) A view of the formation as a diffuse, 'non-specific', *random nerve network*, with multiple inputs and outputs, polymodal interaction between neurons: perhaps a suitable 'system' for the 'network theory', combined with computer analysis, initiated by Scheibel & Scheibel (1967).

(2) A view of the formation as a central isodendritic core of the neuraxis extending through (and including) the spinal cord to the diencephalon, surrounded by and surrounding 'islands' of allodendritic and idiodendritic neurons and 'specific' nuclei.

(3) A view that the formation, essentially bilateral but cross-connected, has three bilateral, longitudinal columns: 'median', 'medial' and 'lateral', each with its typical neurons.

(4) An extreme parcelling of the foregoing columns into named nuclei by familiar cytoarchitectonic techniques.

(5) Attempts to locate columnar 'zones', individual nuclei or parts thereof as terminations of definable bundles of afferent fibres or as sources of identifiable efferent bundles with distinct destinations, studied by older degeneration techniques (p. 870) and modern fibre tracing methods (p. 874).

(6) An analysis of the results of micro-recording and the structural and behavioural effects of micro-stimulation or ablation of zones or nuclei. Such manoeuvres often result in spectacular focal effects at nodal points often termed 'centres' by neurophysiologists; but much overlap is often present and the relation to named nuclei is vague.

(7) A fashionable view of the formation as 'a collection of more or less specific neuron populations, largely mixed together rather than separated into spatially isolated groups' (Webster 1978): the criteria employed entail the use of techniques for the mapping of the distribution of groups of neurons producing a common neurotransmitter substance (p. 889). Thus, for a while, terms such as 'cholinergic nervous system' and 'aminergic nervous system', etc., were employed with increasing frequency.

(8) The latter terms became redundant with the spectacular recognition of numerous neuroactive substances, in many cases new means for their localization, and the emergence of neural chemoarchitectonics.

CHEMOARCHITECTONICS

The brief references in the 36th edition of this book to the location and projections of cholinergic and monoaminergic neurons, based on early techniques, and the possibility of future 'maps' of the distribution of other neuroactive substances have been in the main confirmed but with greater precision and also massive amplification of the substances studied. The new techniques have permitted, in many but not all cases, the accurate localization of neuronal somata, axons and arborizations associated with one or more neuromediators; the latter term was introduced because of their wide range of activities envisaged. Neuromediators (many, and increasing in number) are grouped as peptidergic or non-peptidergic, some regarded as 'established' others, as yet, 'putative'; the distinctions to be drawn between neurotransmission and neuromodulation, and the great temporo-spatial variation in their production, transport, dissemination, receptor sites, effects on target cells and co-operative interaction are reviewed on pp. 889–891. The Table lists a number of the better investigated neuromediators; although lengthy it is not exhaustive, new substances being evaluated and added quite frequently. Locations in which some mediators are listed are also mentioned in the pages noted. The explosive accumulation of data, particularly in the last decade, has led to the firm establishment of a major subject in its own right, *neural chemoarchitectonics*. In general, the recorded observations concerning the detailed distribution of even a single neuromediator lies beyond the scope of the present volume on general anatomy (and, as yet, even beyond large texts devoted exclusively to neuroanatomy). It is indeed fortunate that Nieuwenhuys (1986) has provided a meticulous monograph *Chemoarchitecture of the Brain*; its 246 pages include a compact, thorough survey of published observations, standardized illustrations of great clarity and over 1200 bibliographical references; this merits consultation when any neural aggregate is being considered. His approach has been to detail the distribution of each neuromediator separately and sequentially, then combining these results to the neuromediators present (numbers and varieties) in different neural 'entities', certain tracts, fasciculi and aggregates of grey matter, making reappraisal of some tenets of classic neuroanatomy mandatory in the future; he concludes with stimulating speculations. If persistently reaffirmed, the main elements of such data will be incorporated at many points throughout the neuroanatomical section in future editions of this volume. At present the review on pp. 889–891 may be supplemented by a few brief paragraphs. Without specifying the particular neuromediators identified, the *numbers* found, as yet, in a limited selection of sites are indicated (brackets). *Nerves, tracts, fasciculi*: spinal nerve, ventral root (1); dorsal root (7); cochlear nerve (3); olivocerebellar tract (1); lateral lemniscus (4); inferior brachium (1); corticospinal tract (1); olfactory tract (6); stria medullaris (7); stria terminalis (13); ventral amygdalofugal tract (12); fornix (9); medial longitudinal fasciculus (1); medial forebrain fasciculus bundle (15); dorsal longitudinal fasciculus (12). Numbers of neuromediators in some major neural entities: olfactory bulb (10); basal telencephalon (23); neostriatum (17); hippocampus (13); neocortex (18); dorsal thalamus (18); preoptica-hypothalamic region (25); superior colliculus (10); cerebellum (6); medial reticular formation (17); spinal dorsal grey column (15). For details of the particular mediators and many other features consult the published tables.

The earlier techniques for localizing cholinergic neurons were unsatisfactory and a brief account is given; the earlier techniques for monoaminergic neurons, however, proved remarkably accurate and have been retained except for a few additional notes or deletions.

CHOLINERGIC NEURONS

α, β and γ motor neurons, autonomic preganglionic (and some postganglionic) neurons are all described elsewhere. The neurons which form the source of the olivocochlear bundle are described on p. 960. In the medial reticular regions of the rhombencephalon and the rostral part of the lateral tegmental rhombencephalon, the nuclei, or parts thereof involved include: the central grey matter, parts of the parabrachial nuclei, nucleus pedunculopontinus, nuclei cuneiformis, subcuneiformis, profundus et dorsalis tegmentalis, raphes dorsalis and the non-dopaminergic part of the substantia nigra. Some lateral parts of the complex are classified as Ch6 and medial parts Ch5, both held to be essential parts of the ascending activating system. These various groups project rostrally as dorsal and ventral fasciculi adjoining the central tegmental fasciculus. The dorsal bundles project to the superior colliculus, pretectal area, medial and lateral geniculate nuclei and various thalamic nuclei (midline, intralamina, medial and anterior). Ventral bundles pass to the ventral tegmental grey matter, interpeduncular nucleus, ventral and anterior thalamic nuclei and most hypothalamic nuclei both medial, lateral, mamillary, supraoptic and pre-optic. The cholinergic medial habenular nucleus is bypassed (vide infra).

The cholinergic basal forebrain comprises groups Ch1–4 and extends from the septal area to the nucleus subthalamicus. Ch1 —about 10% of neurons of the medial septal nucleus; Ch2— about 70% of neurons in the (vertical part) nucleus of the diagonal band of Broca; Ch3—a few neurons in the latter's horizontal part; Ch4—a massive concentration in the human brain, consisting of 90% of neurons in the basal forebrain nucleus of Meynert (embedded in the substantia innominata) and continuous above with scattered large cholinergic neurons in the internal and external medullary laminae of the globus pallidus.

Ch1–3 neurons project via the stria medullaris and some end in the cholinergic neurons of the habenular nucleus medialis; the remainder, with accessions of habenular axons continue via the habenulopeduncular (retroflex) tract to the interpeduncular nucleus and tegmental reticular formation. Ch1 and 2 also project through the fornix to the hippocampus (cornu ammonis and dentate gyrus) and Ch2 to nuclei of the lateral hypothalamic region. The smaller anterolateral part of Ch4 projects to the amygdaloid complex; the remainder (i.e. the bulk of) Ch4 has a topically organized projection to the *whole neocortex*. The basal frontal cholinergic parts of the brain receive numerous afferents and are a nodal locus between reticular, limbic and neocortical areas and naturally there have been speculations concerning a host of functional roles. What is established, however, is a correlation between gross diminution of forebrain basal cholinergic neurons and Alzheimer's disease and also Alzheimer-like dementias.

Short axoned cholinergic interneurons are scattered throughout the neostriatum (putamen, caudate nucleus and nucleus accumbens); also, in some species, the neocortex (uncertain in primates), the hippocampus, subiculum and laminae III–VI of the spinal dorsal grey column. Many or all of these sites show positive neuronal immunoreactivity for VIP. The scarcity (or possible absence) of cholinergic interneurons in the primate neocortex re-emphasizes the importance of the cholinergic axonal radiation from the dense basal aggregate Ch4 to the entire neocortex (cf the wide radiations of noradrenergic and serotonergic axons from brain-stem reticular groups) (vide infra).

MONOAMINERGIC NEURONS

The results stemming from earlier fluorescence techniques have been retained not only for historical perspective (e.g. of paracrine effects); despite the greater ease with which monoamines can be individually recognized, the earlier results have been largely confirmed and still form the basis for current *numerical classification*. Any outstanding advances are inserted but for the torrent of subsequent literature consult Nieuwenhuys (1986).

The Falck–Hillarp technique (Falck & Hillarp 1959, Falck 1962, Falck et al 1962) showed that after treatment with formaldehyde neurons and neurites containing certain biogenic monoamines fluoresce in ultraviolet light. Those containing catecholamines, dopamine, noradrenalin or adrenalin, show greenish fluorescence, those containing the indoleamine, serotonin, (5-hydroxytryptamine) a distinct yellow. At first the catecholamines could not be distinguised (except by slight morphological differences, vide infra); but this became practicable by quantitative photofluorometry combined with pharmacological manipulation (now, of course, completely outdated by specific immunofluorescence). These more reliable techniques have led to a wealth of data of differential distribution of dopaminergic, noradrenergic and serotonergic neurons in the brain stem and diencephalon, usually in rats, cats and monkeys (Dahlstrom & Fuxe 1964, 1965, Pin et al 1968, Ungerstedt 1971, Bjorklund & Nobin 1973, Felton et al 1974, Jacobowitz & Palkovits 1974, Palkovits & Jacobowitz 1974, Lindvall et al 1974, Webster 1978). General confirmation was also effected in adult and fetal human brains (Nobin & Bjorklund 1973, Olson et al 1973a, b). (For application of immunocytochemistry to the human brain stem see Pearson et al 1983.) Some difficulties in interpretation require emphasis: axons of such neurons are extremely fine, often branch much, with minute varicosities from which transmitter is assumed to escape. They form innumerable intermediate synapses and may also perhaps produce miniature 'lakes' of transmitter affecting numerous nearby neurons in a volume of neural tissue, a 'paracrine' effect (but see also p. 1376). Varicosities on noradrenergic fibres are almost spherical, regularly spaced and exhibit intense greenish fluorescence, whereas those of dopaminergic fibres are usually fusiform, smaller, widely and irregularly spaced and their fluorescence is less obvious. (These remarks are not relevant to immunochemistry.) Dendrites of these neurons are also fluorescent and may invade the territories of neighbouring nuclei; and even though they may be involved in, e.g. dendrodendritic synapses, they may also give a spurious appearance of an aminergic innervation in a group which is devoid of this.

In rodent brains catecholaminergic groups occur and are classified as A1–15, of which A1–7 are noradrenergic, A8–15 dopaminergic/adrenergic; nine groups contain the indoleamine serotonin and these serotonergic groups are classified B1–9. Considerable correspondence appears amongst all mammals so far studied; but the aminergic groups thus classified accord in extent only rarely with reticular nuclei recognized by classic neuroarchitectonics: usually they are part of, adjacent to or cross the boundaries of one or more of such nuclei.

NORADRENERGIC NEURONS

A brief account of the general location of the main noradrenergic neuronal groups (A1–7) is followed below by a summary of principal projections; such groups were initially identified in rats but mostly have been identified in primates and confined to the medullary and pontine regions.

Group A1, in the lower medulla, is ventral to the lateral reticular nucleus (nucleus of the lateral funiculus, p. 990) and partly extends around it; A2 is part of the nucleus commissuralis, a parahypoglossal nucleus, dorsolateral to the hypoglossal nucleus and near the fourth ventricle's ependymal floor. More recently it has been described as part of the dorsal vagal and solitary nuclei and intervening regions, partly lying under the area postrema and called the noradrenergic dorsal medullary group. Group A3 is part of, and dorsal to, the dorsal accessory olivary nucleus (inferior olivary nuclear complex); its status in primates is unproved. Group A4, the nucleus pigmentosus tegmentocerebellaris, is subependymal, embracing the ventricular aspect of the superior cerebellar peduncle. Group A5, ventral to the subcoerulean nucleus, partially invades the superior olivary territory and adjoins the facial nucleus. Group A6 is a compact core of the nucleus coeruleus (p. 980). Group A7 is the nucleus subcoeruleus, in the rostral pontine parvocellular lateral reticular column partly confluent with groups 4 and 6, particularly in primates (Hubbard & Di Carlo 1973).

Among these noradrenergic (and some serotonergic) groups most exhibit both complex ascending and descending projections to local destinations. Because their ascending connections differ it is convenient to consider group A6 first and the other groups separately, though the difference may be less than first supposed.

Group A6 is the central noradrenergic core of the nucleus coeruleus and has been described (p. 980). Its descending fibres, joined by some from A7 neurons, provide collaterals to the bulbar

nuclei, including the vagal and inferior olivary, then traverse the spinal ventrolateral white column to provide noradrenergic terminals in ventral, intermediate and basal parts of the dorsal grey columns; this is the ventrolateral coeruleospinal tract. Local brainstem terminals are largely collaterals of ascending fibres; but a substantial bundle, the coeruleocerebellar tract, joins the superior cerebellar peduncle, reaching the cerebellar nuclei and cortex (p. 967). Ascending fibres also form the prominent dorsal tegmental noradrenergic bundle (part of the central tegmental fasciculus) traversing the midbrain ventrolateral to the aqueductal grey matter; it continues through the lateral hypothalamus, largely in the medial forebrain bundle, and thence through the septal region to join the cingulum and supracallosal longitudinal stria; others are transported via the fornix, stria medullaris and stria terminalis. Side branches throughout have, sequentially, terminals in mesencephalic aqueductal grey matter, reticular nuclei and the colliculi; diencephalic terminals include dorsal thalamic nuclei, geniculate and habenular nuclei; telencephalic terminals include all the major limbic structures, septal and amygdaloid nuclei, hippocampus, subiculum and cingulate, retrosplenial and parahippocampal gyri; the neocortex has a uniform bilateral input, fibres of the two dorsal bundles having exchanged fibres in the upper brain stem. It has been claimed that the axon of a single coerulean neuron may provide terminals in the spinal cord, brain stem and throughout the forebrain! (Further comments on p. 980.)

Groups A1, 2, 5 and 7 (and perhaps A3 and 4, status being less certain in primates) have ascending branches which converge, termed by some the ventral tegmental noradrenergic bundle (part of the central tegmental fasciculus). Traversing the ventral midbrain reticular regions it approaches the dorsal bundle in the upper midbrain; both exchange fibres, the ventral then continuing mainly in the medial forebrain bundle through the lateral hypothalamus; en route it provides collaterals ending at mesencephalic levels in the adjacent reticular nuclei (p. 990) and circumaqueductal grey matter; in the diencephalon it diverges to the hypothalamic nuclei, with particularly complex terminals in the dorsomedial, periventricular, tuberal, paraventricular and supra-optic nuclei; it appears to end as terminal sprays in the bed nucleus of the stria terminalis and pre-optic area. The dorsal and ventral tegmental noradrenergic bundles thus mainly differ in diencephalic terminations and the extremely wide continuation of the dorsal bundle into limbic and neocortical areas.

Groups A1 and 2 provide local inputs, especially to dorsal vagal and solitary nuclei.

Groups A1, 2, 5 and 7 send descending fibres to the spinal cord, described as an anterior noradrenergic bulbospinal tract, with terminals in the spinal ventral grey column and a dorsolateral noradrenergic bulbospinal tract, with terminals in the intermediolateral grey column and substantia gelatinosa. Both coeruleospinal paths are bilateral, the decussation occurring at spinal levels.

DOPAMINERGIC NEURONS

No dopaminergic nuclei are found in the hindbrain; of eight recognized (A8–15) A8–10 are mesencephalic, A1–14 are diencephalic and A15 is telencephalic.

The three midbrain groups are interconnected plates of dopamine-positive neurons and difficult to define. Group A8 lies in the mesencephalic reticular formation and is also known as the substantia nigra pars lateralis (p. 982), A9 being the substantia nigra pars compact. Group A10, a median complex, abuts ventrally the nucleus interpeduncularis and is also termed the nucleus linearis, nucleus parabrachialis pigmentosus and interstitial nucleus of the ventral tegmental decussation; how far these are synonymous in primates is obscure.

The diencephalic group A11 is subependymal in the caudal hypothalamus, lying low in the third ventricle's wall and partly surrounding the fasciculus retroflexus (habenulopeduncular tract). Group A12 includes the infundibular (arcuate) nucleus; its upward extensions into the rostral periventricular grey matter are classified as A14. Some caudomedial neurons of the zona incerta (p. 1006) form group A13. The solitary telencephalic group (A15) comprised periglomerular neurons of the olfactory bulb (p. 1031).

Some dopaminergic efferents from groups A8 and 9 (subdivisions of the substantia nigra) ascend through the subthalamus and lateral hypothalamus, then diverge and traverse the internal capsular fibres (as part of the *comb* bundle, p. 1017), where they join the ansal system and continue to radiate; they end *throughout* the neostriatum, i.e. the caudate nucleus and putamen, forming a *nigrostriatal dopaminergic system*. But septal fibres from the mesencephalic group A10 ascend more medially via the medial forebrain bundle to other telencephalic structures, including the nucleus accumbens, nucleus of stria terminalis, olfactory tubercle, granular areas of the frontal neocortex, anterior cingulate cortex and entorhinal cortex, forming a *mesolimbic dopaminergic system*. It may be noted that more recent analyses have regarded the mesencephalic aggregates as one complex group giving a major *mesotelencephalic dopaminergic system* from which three prominent radiations arise, namely: *mesostriatal*, *mesolimbic* and *mesocortical*; all are topically organized in the three cardinal planes.

Groups A11,13 and 14 (not classically 'reticular') have short axons confined to the diencephalon; A11 projects to the medial group of hypothalamic nuclei, to the nuclei of the midline of the dorsal thalamus, the habenular nuclei and pretectal area. Axons of A13 and 14 terminate in the caudal regions of the dorsal thalamus, other parts of the zona incerta and nuclei of the dorsal and oral parts of the hypothalamus, collectively termed the *incertohypothalamic dopaminergic system*.

Neurons of A12, with somata mainly in the infundibular nucleus, send axons to the anteroventral part of the median eminence and upper infundibulum, ending in synaptic relation with or merely near axons or terminals of parvocellular releasing-factor neurons, or in relation to tanycytes of the infundibular recess and the median eminence (pp. 1010, 1451). This *tubero-infundibular dopaminergic system* may influence transport and/or exocytosis of releasing factors and their uptake and the uptake of dopamine itself, by the venous portal system capillaries and the exchange and constitution of cerebrospinal fluid to or from the third ventricle by tanycytes.

Some account of group A15 appears on p. 1030.

Recent evidence indicates the possible occurrence of dopamine in the dorsal vagal nucleus, confirmation of a dopaminergic innervation of the hypothalamic nuclei: supraoptic, paraventricular, and dorsomedial; also the presence of a *dopaminergic hypothalamospinal* projection (as yet shown only in rat and rabbit) passing in lamina I of the spinal dorsal column, the posterolateral fasciculus and subependymal central canal, innervating the substantia gelatinosa, neighbouring laminae and the intermediolateral column.

SEROTONERGIC NEURONS

Reticular neurons synthesizing the transmitter serotonin (5-hydroxytryptamine) occur at medullary, pontine and mesencephalic levels; all are in or near the midline (i.e. parts of the rapheal nuclei, vide supra); groups termed B1–9 have been located. B1 is formed by neurons of the nucleus raphes pallidus but also extend laterally, B2 is the nucleus raphae obscurus; both extend ventrolaterally to include neurons of the nucleus paramedianus. Neurons of nuclei raphae pallidus and magnus ascend and converge to the pontomedullary junction, together, there, constituting group B3. Groups B1–3 are thus serotonergic, median or paramedian and largely medullary. Group B4 is in the pontine periventricular grey matter (deep to the upper part of the fourth ventricle); B5 is a subgroup of the nucleus raphae pontis, near the trigeminal motor nucleus; B6 is in the central pontine tegmental reticular grey matter, extending to the pontine superior central nucleus. B7, the largest serotonergic group, occupies much of the mesencephalic nucleus raphae dorsalis, spreading into the periaqueductual grey matter; B8 is a rostral extension of B6 in the nucleus centralis rostralis; B9 is a lamina around the lateral aspect of the nucleus interpeduncularis near the medial lemniscus. Lateral extensions of B6 and 8 pass lateral to the confines of their neighbouring rapheal nuclei and are intermixed with noradrenergic neurons of the locus coeruleus, subcoerulean area and parabrachial nuclei.

Routes and terminations of fibres from serotonergic cell groups resemble the noradrenergic ones, but with some exceptions. They may be classified as (1) descending, (2) local, (3) cerebellar and (4) widespread dorsal and ventral ascending paths. Single serotonergic axons may spread to many destinations, others appear strictly localized. In short, they innervate almost the *entire central nervous system*.

(1) *Descending bulbospinal serotonergic projections* come from medullary groups B1–3, particularly the nucleus raphes magnus; they form two tracts. One traverses the ventral white funiculus with re-entrant terminals to neurons of the ventral grey columns (laminae VIII and IX); the other descends dorsally in the lateral white funiculus, lateral to the corticospinal tract, innervating the intermediolateral column (preganglionic autonomic) and other laminae of the dorsal column (particularly laminae I, II, and V), including the substantia gelatinosa, the latter held to be much concerned with the modulation of *nociceptive inputs*. With respect to the latter it should be noted that *raphespinal* neurons projecting in the manner indicated contain at least five putative neuromediators that may be involved. (Substance P is always present, as are enkephalin-immunoreactive interneurons in the target tissue.) Presynaptic serotonergic terminals on *primary afferents* have been observed and the modulation of mechanoreceptor and thermoreceptor inputs is also under rapheospinal control.

(2) *Intrinsic brain-stem serotonergic projections* involve B3–8; those from B5–8 project to the nucleus coeruleus, tegmental nuclei, with descending fibres to other (probably aminergic and non-aminergic) reticular pontine and medullary zones. Groups B3–5 also connect with adjacent reticular nuclei and perhaps the inferior olivary complex. The largest group, B7, sends the most massive bundle of serotonergic axons yet identified to the largest group of noradrenergic neurons, in the core of the nucleus coeruleus (A6).

(3) The *cerebellar serotonergic path* consists of axons from B5 and 6; entering the ipsilateral middle cerebellar peduncle, they branch and some end in the intracerebellar nuclei but the bulk then diverge widely to reach the cerebellar cortex (cf. the noradrenergic cerebellar path in the inferior peduncle).

(4) The *dorsal ascending serotonergic bundle*, smaller and more restricted in its destinations than the ventral, arises in groups B3–6; its axons converge up and dorsally into the dorsal longitudinal fasciculus (p. 986); ascending in this they give collaterals to the mesencephalic reticular nuclei and the periaqueductual grey matter; on entering the diencephalon, they diverge to their terminations in the nuclei of the caudal zone of the hypothalamus (p. 1009).

The *ventral ascending serotonergic bundle*, relatively massive, stems from B6–8 (rapheal nuclei of the upper pons and midbrain), curves ventrally and up to traverse the ventral midbrain tegmentum in the central tegmental fasciculus, then enters the diencephalon and traverses the hypothalamus laterally in the medial forebrain bundle. Its multiple collaterals radiate widely to diverse destinations: mesencephalic fibres to the substantia nigra, others to the interpeduncular nucleus, some terminating here, the rest continuing in the interpedunculo-habenular tract, to reach the epithalamus and dorsal thalamus; they end in the habenular nuclei, in the medial and median thalamic nuclear groups and its parafascicular nucleus. In the diencephalon, as part of the medial forebrain bundle, its collaterals reach the mamillary and lateral hypothalamic nuclei; others penetrate the hypothalamus obliquely to reach the caudate nucleus and putamen, some continuing to wide areas of the lateral neocortex, a small fascicle enters the postcommissural (column) of the fornix, recurring through its body, crus and fimbria to reach the dentate gyrus and cornu ammonis. At the anterior end of the hypothalamus the ventral serotonergic fasciculus splits into rami, some passing ventrolaterally in the ansa peduncularis to the nucleus of Broca's diagonal band, prepiriform and entorhinal cortex and parts of the amygdaloid complex; some splay out to nuclei of pre-optic and septal areas, to the organum vasculosum of the lamina terminalis (intercolumnar tubercle), to the olfactory tubercle and adjacent grey matter; others continue to the medial frontal neocortex and along the olfactory tract to the olfactory bulb; another contingent radiates to the nucleus accumbens to supply an input to anterior zones of the caudate nucleus and putamen. The substantial remainder of the fasciculus passes forwards, then dorsally to join the cingulum, arching over the corpus callosum and then curving ventrally to run the length of the temporal lobe. It supplies collaterals to the cingulate, retrosplenial and entorhinal (parahippocampal) cortices and sequentially to the cornu ammonis and subiculum, completing a *dual input* to the hippocampal formation (vide supra).

The serotonergic neurons present in nuclei raphes dorsalis and centralis superior in part, give rise to axons that form a widespread *supra-ependymal serotonergic plexus* which is best developed on the ventricular aspects of the fourth and lateral ventricles and the cerebral aqueduct, patchily developed in the third ventricle.and absent over the hypothalamus. It consists of fine, varicose fibres, which presumably release serotonin into the cerebrospinal fluid and on to the ependyma. With other neuromediators, it may be concerned in controlling the production, constitution, flow and regional neural uptake of the fluid with its fluctuating constituents.

Thus the brain-stem serotonergic 'system' projects to: (1) brain-stem reticular nuclei, (2) hypothalamic nuclei, (3) all major limbic structures, (4) some 'limbic' and 'non-specific' nuclei of the dorsal thalamus and epithalamic nuclei, (5) much spinal grey matter, (6) most, perhaps all, areas of the neocortex, (7) the supra-ependymal plexus. Taken together, these have been termed, 'the most expansive neuronal network yet described', Nieuwenhuys (1986).

FUNCTIONAL ASSOCIATIONS OF THE RETICULAR FORMATION

Information converges to the reticular formation from *all* parts of the central nervous system and is projected back to all regions, modulating their activities through complex interactions involving varieties of complex synaptic neurotransmission and local neuroparacrine and wider neuro-endocrine effects. Earlier views of anatomists and physiologists, dismissing the reticular regions as merely ancient 'leftovers', were a product of ignorance and can now be ignored. However, the present has equal problems; for so little do we yet understand the *modus operandi* of the reticular formation that two contrasting and extreme views of it have been advanced. On one hand some find it sufficient to feed into the formation a channel from another neural complex, e.g. the corpus striatum, and to describe the effects elsewhere as *mediated* by the formation, with no attempt to analyse it structurally, as if it were a kind of 'black box'. Similar are those who invoke 'reticular activity' to explain almost *any* phenomenon which cannot be explained by more orderly 'specific systems'. On the other hand, with the present patchy and limited (but rapidly increasing) knowledge it is similarly facile to attach *simple unitary functional labels* to individual neurons or nuclei, by whatever criteria they are 'classified'. The methods available for structural and functional evaluation of this problem have been outlined elsewhere (pp. 870–875); various cytoarchitectonic, impregnational, histochemical, histofluorescent, immunofluorescent, degenerative, autoradiographic and other tracer techniques are beginning to form a picture of its organization and connectivity, albeit incomplete. This provides (as in other neural problems) the reference framework for functional probing, which includes: microelectrode recording, analysis of behavioural changes or deficits due to focal or zonal microstimulation or ablation, pharmacological enhancement or depression of neuromediator action, studies of neuropathological states, genetic and developmental deviations and studies of psychoses. All such have limitations, particularly when applied to such complexities as the reticular formation, hypothalamus, dorsal thalamus, limbic 'system' and cerebral cortex. In the present context real comprehension of the 'reticular' formation would necessitate intimate data of form and flow of information in multitudinous converging channels, including any previous processing and interaction, transformations in the formation itself and the mode of modulation by outflow channels on their multiple destinations, all modifying complex behaviour. We are far from such comprehension. Some of the active fields currently engaging neurobiologists are noted

below; they may all be allocated (a persistent theme in living creatures) to some aspect of long, intermediate or short-term homeostasis.

(1) **Somatomotor control.** The reticular formation is part of a complex of subcortical regions influencing patterns of activity in skeletal muscles. These include simple reflex loops, large or fine postural re-adjustments, positioning and scanning by distance receptors, manipulative skills, locomotion and complicated patterns of personal and group communciation involving speech, gesture and facial expression. The degree and timing of involvement of a multiplicity of central nervous regions, intimately concerned in influencing alpha and gamma motor neuron pools and adjacent interneurons, varies with the speed, force and complexity of the movement. Involvement of the reticular formation may be direct or indirect through other neural apparatus. *Direct* spinal influences are via reticulospinal tracts from the lateral two-thirds of the pontomedullary reticular formation; from the pons they are largely ipsilateral, in the ventral white funiculus; from the medulla they are bilateral, via a wide area of the lateral white funiculus. Stimulation and ablation led to concepts of an *inhibitory motor 'centre'* or *'zone'* and a *medullary pontine facilitatory 'centre'*. But inhibition by medullary reticulospinal fibres results from *increased* release of neuromediator from terminals influencing spinal interneurons, alpha and gamma efferents, involving various pre- and postsynaptic mechanisms. However, *on balance, reduced* medullary activity would have a facilitatory effect, and in any case during normal activities *continual fluctuations* occur, making this terminology arbitrary. The descending fibres involved stem from multipolar neurons of medullary and pontine gigantocellular nuclei, the central medullary reticular nucleus and nucleus reticularis pontis inferior, including groups A6 and 7, and groups B1–7 in rapheal nuclei. Thus, non-aminergic fibres are mingled with inhibitory noradrenergic and serotonergic fibres; also, the nucleus coeruleus (origin of bulbospinal fibres) receives a serotonergic input from the inferior rapheal nuclei, whose serotonergic bulbospinal fibres may mediate general relaxation of body musculature, typical of rapid eye-movement (REM) sleep. Doubtless other neuromediators are involved. There could be similar direct influences on the motor neuron pools of the cranial nerves and on the groups of reticular neurons which control the diaphragm and intercostal musculature, the latter forming the hierarchy of respiratory 'centres' described by neurophysiologists.

Indirect reticular influences on somatomotor control are numerous, as set out in the relevant sections. They include non-aminergic (presumably many cholinergic) and aminergic fibres (see sections on these 'systems'); but the relative numbers, sites of termination and roles in integration are uncertain. Thus, the *cerebellum* receives reticular inputs through: both inferior and middle peduncles, distributed to its nuclei and cortex; other brain-stem complexes, including the olivary nuclei, colliculi, red nucleus and substantia nigra, corpus striatum, in particular the caudate nucleus and putamen through the comb bundle and central tegmental fasciculus; the nuclei ventralis anterior and lateralis of the dorsal thalamus, zona incerta and other subthalamic nuclei, and from the somatomotor cortex, either directly from the reticular nuclei or indirectly via the many regions just mentioned.

(2) **Somatosensory control.** All lower levels involved in processing somatosensory information are subject to reticular influence. Spinal and brain-stem terminals of primary afferent fibres, nuclear destinations and interneurons, dendrites and somata of neurons projecting to suprasinal centres: all these are subject to reticular influence. Effects may be pre- or postsynaptic, facilitatory or inhibitory. The relevance of such controls to theories of sensory 'gating' mechanisms (p. 944) is of special interest: descending bulbospinal and local bulbar reticular fibres, which are serotonergic, may co-operate with others, which are noradrenergic or contain metenkephalins or substance P, and other neuromodulators (p. 892), in an integration modifying the perception of pain (p. 944) and other modalities. (Compare the diencephalic and limbic distribution of monoaminergic and enkephalin-containing reticular terminal fibres with the distribution of opiate receptors, Guillemin 1978,

Chan-Pham 1978, Nieuwenhuys 1986). 'Higher' centres processing somatosensory information and having a reticular input are sensory relay nuclei of the dorsal thalamus and somatosensory cortices.

Similar considerations apply to special senses; reticular afferents reach cochlear, vestibular, tectal, pretectal and geniculate nuclei, pulvinar and other dorsal thalamic nuclei *and* the visual, auditory and olfactory cortices. Spinoreticular and bulboreticular connections are in themselves held by many to provide bilateral, polysynaptic 'slow pain' pathways, discharging into multiple diencephalic nuclei, for both somatosensory and viscerosensory information.

(3) **Visceromotor control.** Cardiovascular readjustments and the activity of non-striated muscle and glandular cells in thoracico-abdominal viscera are usually controlled by postganglionic autonomic neurons; these, either directly or via local interneurons, are partially controlled by the reticulobulbar and reticulospinal fibres. By focal stimulation and ablation, brainstem transection and pharmacological manipulation, a concept of, e.g. cardiovascular 'centres', has gained physiological acceptance. Though the sites of sympathetic and parasympathetic preganglionic neurons involved are accepted, functionally related reticular nuclei remain speculative.

Forebrain control of autonomic activities involves the orbitofrontal, cingulate and entorhinal cortex, major limbic structures (pp. 1009, 1028), medial and anterior groups of dorsal thalamic nuclei, hypothalamic nuclei and descending fascicles of various tegmental bundles with reticular destinations, and thence via the reticulobulbar and reticulospinal tracts mentioned above.

(4) **Neuro-endocrine transduction.** Via the complex of bundles ascending with central and dorsal tegmental fasciculi, cholinergic, all varieties of monoaminergic and other neuromediator-profiled fibres pass in collateral paths to almost all hypothalamic nuclei, including infundibular nuclei of the median eminence and habenular nuclei. Thus, directly via other hypothalamic nuclei or via dopaminergic infundibular neurons, reticular neurons are one source of modification of synthesis and probably transport and exocytosis of 'releasing' or 'release-inhibiting' factors by parvocellular neurons in the mediobasal hypothalamus (p. 1007), thus controlling the adenohypophysial activity. Similar mechanisms probably operate on the neurohypophysis. Neural input to the pineal gland involves the sympathetic nerves via the upper thoracic spinal outflow of preganglionic fibres, the superior cervical ganglion and nervus conarii. There is also a neural connection via the habenulo-pineal tracts. The role of the suprachiasmatic nucleus has been mentioned elsewhere (pp. 1013, 1459). Both main neuro-endocrine transduction centres are hence partly under reticular influence.

(5) **Biological rhythms.** The notion of '*biological clocks*' has now permeated all branches of biology. A time-base exists in all living systems through the turnover rates of macromolecules, including those related to the genetic apparatus, annual seasonal variations and daily environmental changes. Biological 'rhythms' thus involve reproductive *cycles*, development, cell division, cell death and replacement, etc., the list being endless. Many rhythms depend on an intact hypothalamus, with its multiple connections, including those from the reticular formation which are also involved in neural control of the pineal gland, prompting much interest in its neuro-endocrine role in relation to *circadian*

rhythms, as these repetitive cycles or biorhythms are now called (pp. 1012, 1459).

(6) **Sleep, arousal, states of consciousness, perception.** Much effort (philosophical, psychological, neurophysiological, neuropharmacological and neuroanatomical) has been expended in defining states of consciousness and their neural arenas, ranging from stuporose conditions, varieties of sleep, arousal and awakening, to attention, perception and active responses. Research on all these fronts is endless; a vast literature has accumulated, beyond the scope of this or any volume (Jouvet 1969, 1972, Ingram 1976). It is necessary to distinguish different *behavioural* states of unconsciousness, sleep and awakening from *electroencephalographic* records of these states, where they exist. The terminology used in such recordings is widely employed. Thus, two principal forms of sleep are recognized: *deep* or *slow wave (recuperative) sleep* (SWS), in which individuals are deeply but reversibly unconscious, with random changes of limb position and posture, elevation of sensory thresholds and *synchronized*, high-voltage, slow waves; interspersed are periods of *paradoxical sleep* (PS) or *rapid eye-movement* (REM) sleep, in which dreaming normally occurs, with bodily relaxation but rapid oscillatory ocular movement, accompanied by corresponding bursts of electrical activity, similar waveforms being recorded from the pontine reticular formation, dorsal lateral geniculate nucleus and occipital visual cortex (PGO activity). Arousal and awakening is accompanied in intact animals by a rapid, low amplitude, *desynchronized* waveform. An '*ascending activating reticular system*' has already been mentioned (p. 991); it is polysynaptic, with multiple inputs from sensory pathways and via diencephalic synaptic complexes and widespread bilateral radiation to most regions of the neocortex and limbic structures. Investigations indicate that both cholinergic and monoaminergic reticular groups in the upper pons and midbrain are, when stimulated, the most powerful initiators of cortical desynchronization. It is widely assumed that desynchronization is necessary for wakefulness, attention, effective perception and cognition and preparation for appropriate responses. The varieties of sleep are currently considered the result of activity in a *hierarchy* of hypnogenic neuronal groups balanced against the 'activating system', most being monoaminergic. *Slow-wave sleep* depends on serotonin production by rapheal nuclei, whereas *paradoxical sleep*, after priming by serotonin, follows the intermittent activity of cholinergic *and* noradrenergic neurons in the nucleus coeruleus (rostral part), its medial neurons appearing to be a '*pontine pacemaker*', which with the former generates PGO responses. Descending serotonergic bulbospinal fibres from rapheal nuclei, together with noradrenergic fibres from the inferior part of the nucleus coeruleus, are considered responsible for the relaxation of limbs and trunk in paradoxical sleep.

(7) **Spatio-temporal discrimination, cognitive mapping and exploration, reward, learning and memory, emotional content, long-term homeostasis** all depend on an intact hypothalamo-limbic 'system' and will be discussed later. In addition to numerous other afferents, all hypothalamic nuclei and limbic structures receive widespread inputs via aminergic, non-aminergic and numerous other reticular elements; surgical or pharmacological interference with brain-stem reticular elements can have profound effects.

As stated at the outset, the brain-stem's reticular core has reciprocal connections with *all* major parts of the central nervous system and hence influences *all* behaviour.

THE PROSENCEPHALON OR FOREBRAIN

The prosencephalon develops from the foremost primary cerebral vesicle (p. 987). This *forebrain vesicle* and its cavity, the future *third ventricle*, soon differentiate into a caudal *diencephalon* and rostral *telencephalon*. From the latter's sides, right and left diverticula develop into cerebral hemispheres, each with a contained ventricle. The sites of evagination become the interventricular foramina, by which all three ventricles are continuous.

Survey scans and sections are provided (7.57, 58, 123, 124, 140). The diencephalon corresponds largely to most of the third ventricle and its adjacent structures; the *telencephalon* comprises a median *telencephalon medium* (*impar*), containing a limited forward extension of the third ventricle and massive *bilateral telencephalic hemispheres*, each containing a lateral ventricle. (For development see p. 189.)

THE DIENCEPHALON OR 'INTERBRAIN'

The diencephalic cavity, equal to most of the median slit-like third ventricle, narrows caudally to become the mesencephalic aqueduct and extends rostrally into the median telencephalon (p. 1007). More precise limits of the diencephalon are: caudally, a plane including the posterior commissure and posterior margins of mamillary bodies; rostrally, a plane from the interventricular foramina through the dorsocaudal border of the median part of the optic chiasma, rostral to which are structures of the median telencephalon. While these boundaries have approximate developmental and phylogenetic significance, they are arbitrary descriptive aids, to which function is blind.

The diencephalon, being median, has right and left halves. Across each lateral wall of the third ventricle a *hypothalamic sulcus* extends from the cerebral aqueduct to the interventricular foramen (p. 1019); its line is used to demarcate each lateral half into a **pars dorsalis** and **pars ventralis diencephali**. Each *pars dorsalis* consists of a dorsal thalamus, metathalamus and epithalamus; the *pars ventralis* includes the hypothalamus and ventral thalamus. Dorsal to the hypothalamic sulcus most of each lateral wall is formed by the large ovoid *dorsal thalamus*. The unqualified term *thalamus* is widely used as synonymous with the dorsal thalamus and will frequently be so used here; but the thalamus can be divided into major *parts*, each subdivided into *nuclei*. Postero-inferior to and partly continuous with it are the *medial* and *lateral geniculate bodies*, which contain nuclei also so named, forming the *metathalamus*. The diencephalic roof is largely ependyma, continuous with the third ventricle's lining and apposed in most of its extent to the vascular pia mater, with no nervous tissue between; but caudally in the roof and adjoining the lateral walls are the habendular nuclei and commissure, the epiphysis cerebri or pineal gland and posterior commissure, which together constitute the *epithalamus*.

Some include the whole of the *pars ventralis diencephali* in the hypothalamus, including regions functionally dissimilar; a more restricted definition is preferable. The *hypothalamus* extends axially from the lamina terminalis to a vertical plane caudal to the mamillary bodies and dorsoventrally from the hypothalamic sulcus to include structures in the third venticle's side wall and floor, the latter comprising the mamillary bodies, tuber cinereum and infundibulum, and tissue adjacent to the optic chiasma (7.132, 133). The preoptic region, adjacent to the lamina terminalis and strictly telencephalic, is for functional reasons usually included in the hypothalmus. The remaining zones of the pars ventralis diencephali are partly lateral to the hypothalamus, as a thin sheet ventral to the dorsal thalamus and a thick zone merging caudally with the mesencephalic tegmentum. Collectively these zones are the *ventral thalamus* or *subthalamus*, including upper extensions of the red nucleus and substantia nigra, the subthalamic nucleus, prerubral field, zona incerta and associated complexes of nuclei and tracts.

The (Dorsal) Thalamus

The thalami (7.123–126) are large, grey, bilateral, ovoid masses flanking the third ventricle which also extend posterior to it. Each is about 4 cm long with, descriptively, two ends or poles and four surfaces.

The narrow *anterior pole* is near the midline and forms the posterior boundary of the interventricular foramen. The expanded *caudal (posterior) pole*, the *pulvinar*, directed dorsolaterally, overhangs the superior colliculus and its brachium. Inferolateral to the pulvinar is the lateral geniculate body (p. 1016), a small, oval elevation. Inferiorly the pulvinar is separated from the medial geniculate body (p. 1015) by the brachium of the superior colliculus.

The *dorsal (superior) surface* (7.125), slightly convex and covered by a white *stratum zonale*, is separated laterally from the lateral ventricular aspect of the caudate nucleus by a white *stria terminalis* and the thalamostriate vein (p. 798); but it is separated medially from its medial surface only by the line of reflexion of the

ependyma (*taenia thalami*) forming the third ventricle's roof. The dorsal surface is divided into a lateral region, which is part of the lateral ventricle's floor, covered by its ependymal epithelium and partly hidden by a fringe of choroid plexus (7.126), and a medial region covered by part of the tela choroidea of the third ventricle, separating it from the body of the fornix, the lateral edge of which grooves it. Between the lateral edge of the fornix and the dorsal thalamic surface the lateral edge of the tela choroidea, with its choroid plexus, invaginates into the cavity of the lateral ventricle through the choroidal fissure (7.198–200). Anteriorly, the dorsal surface is separated from the medial surface by a narrow ridge, from which the third ventricular epithelium is reflected to the tela choroidea's ventral surface; the ridge covers a small nervous fascicle, the *stria medullaris thalami* (p. 1004). Thus, anteriorly, the taenia thalami and the stria medullaris thalami are co-incident but progressing posteriorly they diverge, the taenia inclining laterally as the width of the tela expands. The stria medullaris, however, continues subependymally along the dorsomedial rim of the thalamus, then finally turns medially as the anterior limit of the *trigonum habenulae* (7.116, 125), separated from the dorsal thalamic surface by the *sulcus habenulae*.

The *ventral (inferior) surface* is continuous with a tegmental prolongation (*subthalamus*) and anterior to this the ventral surface

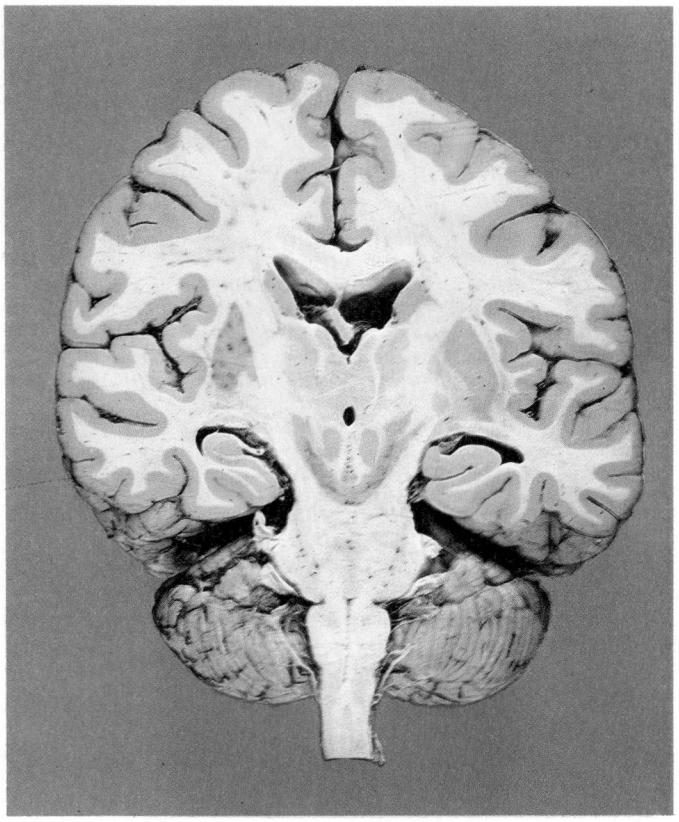

7.123 The dorsal half of a brain sectioned in an oblique coronal plane which passes through the cerebral hemispheres, diencephalon, midbrain, pons and medulla oblongata, to show the general disposition of main structures, some of which are labelled on 7.126. Note: (1) the complex folding of the cerebral cortical gyri and sulci of the frontoparietal, insular and temporal regions; (2) the sectioned surfaces of the corpus callosum, septum pellucidum, body of the fornix, the corona radiata, internal capsule, ventral pons and medulla oblongata; in the latter, part of the decussation of the corticospinal tracts is visible. Note also: (3) the body and inferior horn of the lateral ventricle; (4) the lentiform and caudate nuclei and the dorsal thalami which are fused across the midline. This illustration includes features referred to at many points in the text which are too numerous to include in a caption; these should be studied as appropriate. Dissection by E L Rees, photography by Kevin Fitzpatrick, both of the Dept. of Anatomy, Guy's Hospital Medical School, London.

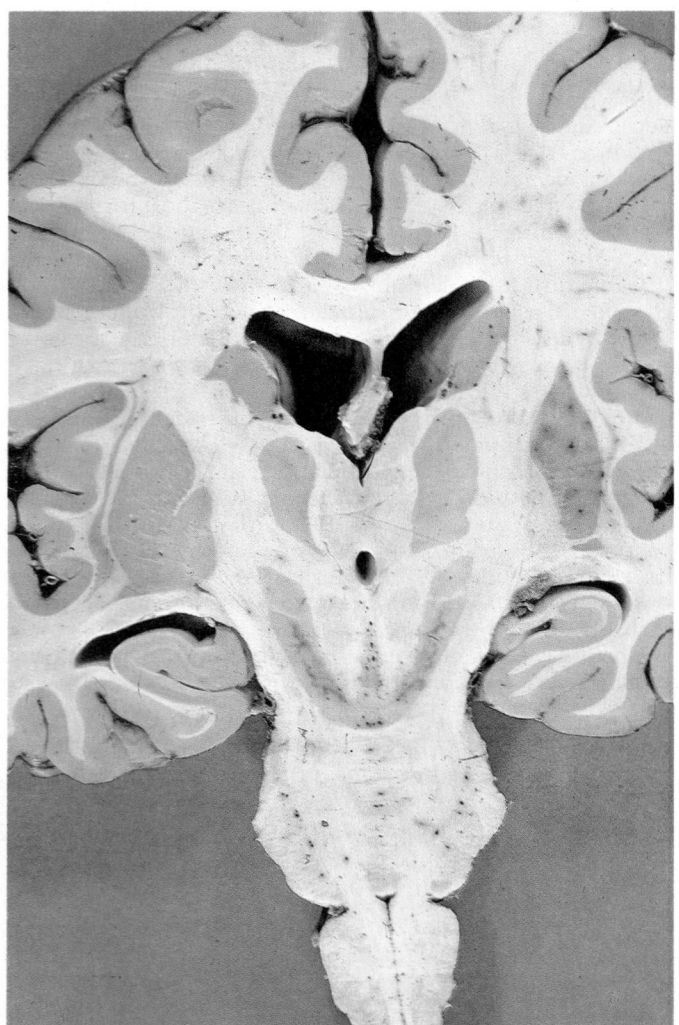

7.124 The central area of the ventral part of the oblique coronal section of the brain shown in 7.123, photographed at higher magnification to show some structural features in greater detail; compare with 7.126 for appropriate labelling. Note in particular: (1) the anterior, medial and lateral parts of the dorsal thalamus, separated by the internal medullary laminae; (2) the relation of the caudate nucleus to the anterior and inferior horns of the lateral ventricle; (3) the lentiform nucleus, divided into an external putamen and an internal globus pallidus, the latter again divided into internal and external parts; (4) the internal capsule, external capsule, claustrum, extreme capsule and insular cortex; (5) the profiles of the sectioned subthalamic and red nuclei and substantia nigra; (6) the hippocampus projecting into the floor of the inferior horn of the lateral ventricle. Other structural features on this section are referred to at many points throughout the text. Dissection by E L Rees, photography by Kevin Fitzpatrick, both of the Dept. of Anatomy, Guy's Hospital Medical School, London.

is dorsal (superior) to part of the hypothalamus in the third ventricle's lateral wall.

The *medial surface* (7.140A,B) is the superior ('dorsal') region of the third ventricle's lateral wall, usually (secondarily) connected to the opposite thalamus by a flat, grey *interthalamic adhesion* (*connexus interthalamicus*), behind the interventricular foramen; its anteroposterior dimension averages about 1 cm, is sometimes multiple, occasionally absent and contains neurons, some of their axons crossing the midline, though many recurve back from this. The medial surface is limited below by an often indistinct *hypothalamic sulcus*, which curves from the upper end of the cerebral aqueduct to the interventricular foramen; it is usually considered a continuation of the sulcus limitans of the spinal cord and brain stem, a view strongly challenged (Christ 1969).

On the *lateral surface* is the thick posterior limb of the internal capsule, separating it from the lentiform nucleus (7.123, 124, 126).

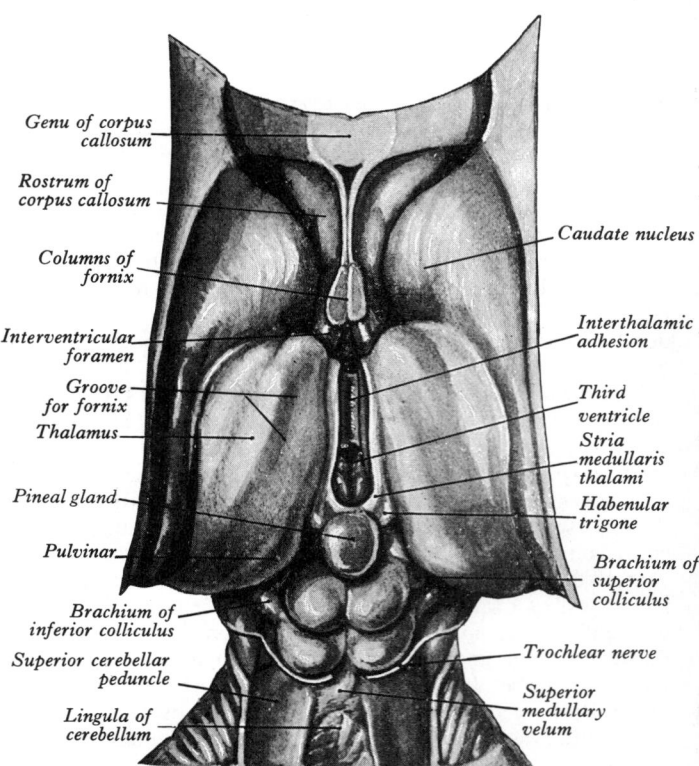

7.125 Dorsal aspect of the caudate nuclei, thalami, pineal gland and tectum, revealed by removal of most of the corpus callosum, the body of the fornix and of the tela choroidea.

The Major Structure of the Thalamus

The thalamus is chiefly grey matter but its superior and lateral aspects are covered respectively by the white *stratum zonale* and *external medullary lamina*. It is incompletely divided into **major parts** by the vertical, white, *internal medullary lamina*, which splits in a Y-shaped manner, dividing the thalamus into *anterior* (rostral), *medial* and *lateral parts*, the lateral being subdivided into dorsolateral and ventromedial regions, all of which contain groups of *thalamic nuclei*, with smaller groups in the lamina and near the medial and lateral thalamic surfaces.

The main nuclear groups are:

(1) *Anterior* (rostral) *group*, forming the anterior part.

(2) *Medial group*, extending medially from the internal medullary lamina almost to (but *not* reaching) the ependymal lining of the third ventricle.

(3) *Lateral group*, dorsolateral moiety of the lateral part, expanding behind as the pulvinar.

(4) *Ventral group*, ventromedial moiety of the lateral part.

(5) *Intralaminar group*, embedded in the internal medullary lamina.

(6) *Nuclei of the midline*, small, interspersed with periventricular fine nerve fibres, part of the grey matter separating the medial thalamic region from the ependyma of the third ventricle and of the variable interthalamic adhesion (p. 1020).

(7) *Reticular thalamic nucleus*, a long, thin, curved lamina, separating the *external* medullary lamina from the posterior limb of the internal capsule.

The thalamic anterior and medial parts, with some smaller nuclear groups, are regarded as phylogenetically older and designated *paleothalamus*, distinguishing it from the lateral *neothalamus*, which reaches full development only in anthropoid apes and mankind; but such evolutionary changes have occurred in the paleothalamus that structural and functional intermingling of various regions of the primate thalamus have left such a division little validity.

Many thalamic regions are connected via the axonal fascicles with the brain stem, spinal cord, cerebellum, hypothalamus and corpus striatum and reciprocally with many cortical areas.

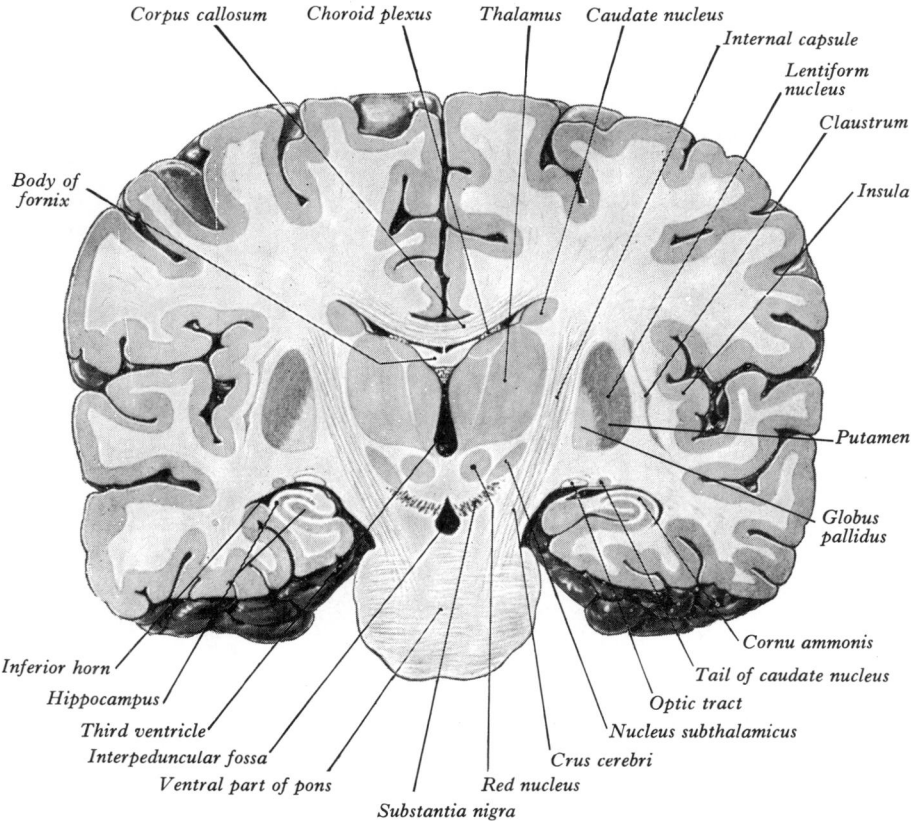

7.126   Coronal section of the brain through the ventral part of the pons.

Classically the reciprocal radiations are described as four **thalamic peduncles**, but they form an almost continuous radiation from ventral, dorsal, posterior and ventrolateral thalamic aspects to most cortical regions, forming much of the internal capsule and corona radiata (7.127, 188, p. 1073). The *frontal thalamic peduncle* interconnects the anterior and medial thalamus with much of the frontal cortex; the *dorsal* (superior) *peduncle* interconnects ventral and lateral thalamic regions with pre- and postcentral gyri and adjacent parts of their lobes; the *posterior* (caudal) *peduncle* interconnects posterior regions of the lateral thalamus, including the pulvinar and lateral geniculate body, with the occipital and posterior parietal cortex. The smaller *ventral thalamic peduncle* interconnects the posterior thalamus and medial geniculate body with the temporal cortex.

Thalamic Nuclei and their Connections

The above thalamic nuclear groups have been much investigated in man and other mammalian and submammalian species, and their cytoarchitectonics, general arrangement, neuronal structure and patterns of connection have been established. The effects of selective stimulation and ablation have also been studied; more recently the detailed ultrastructure of synapses, unit recording and distribution of accepted and supposed neurotransmitters have been observed on an increasing scale. Those researches have revealed immense thalamic complexity, with many often interdependent but functionally distinct zones. Though the broad parcelling of thalamic nuclear groups is agreed, often including individual nuclei, no comprehensive structural and functional design has emerged. Difficulties result from the multiplicity of the nuclei, some being very small and closely sited, their afferents and efferents often traversing or skirting adjacent nuclear territories. Stereotactic lesions hence usually damage more than the selected targets. Retrograde degeneration is difficult to assess in small thalamic neurons and is equally difficult for microelectrode techniques; but copious data of thalamic structure have accumulated, briefly summarized here (7.128A,B, 129) (Le Gros Clark 1932, 1933a, b, 1936, 1937, 1949, Walker 1938, Toncray &

Kreig 1946, Sheps 1945, Dekaban 1953, Kuhlenbeck 1954, Hassler 1959, Schaltenbrand & Bailey 1959, Ajmone-Marsan 1965, Purpura & Yahr 1966, Scheibel & Scheibel 1966, 1967, Van Buren & Borke 1972).

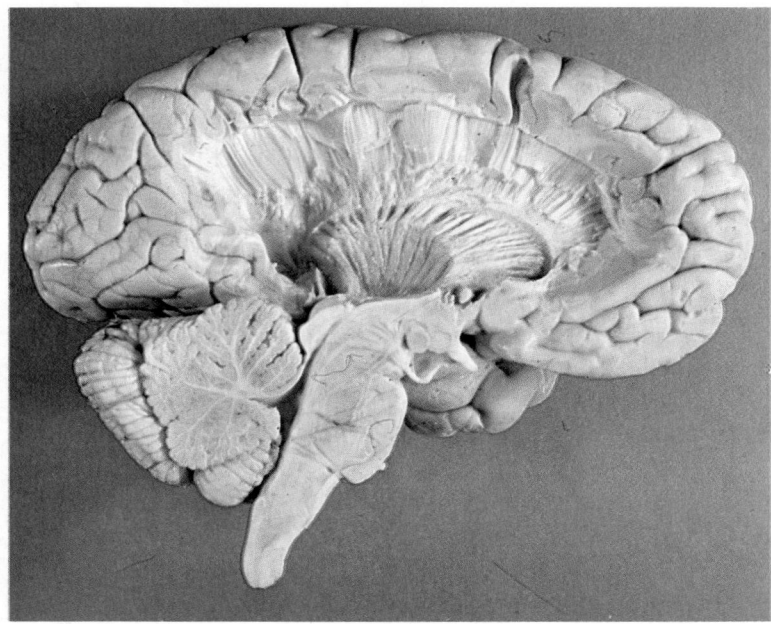

7.127   After median hemisection of the cerebrum, the left cerebral hemisphere has been dissected from its *medial* aspect to display the fibre bundles of the corona radiata and internal capsule. This entailed the removal of the cingulate gyrus and subjacent white matter, much of the paramedian corpus callosum and fornix, the dorsal thalamus and the head and body of the caudate nucleus. The oval depression previously occupied by the dorsal thalamus can clearly be seen within the curved depression left after removal of the caudate nucleus. Dissection by Andrew Seal, photography by Kevin Fitzpatrick, both of the Dept. of Anatomy, Guy's Hospital Medical School, London.

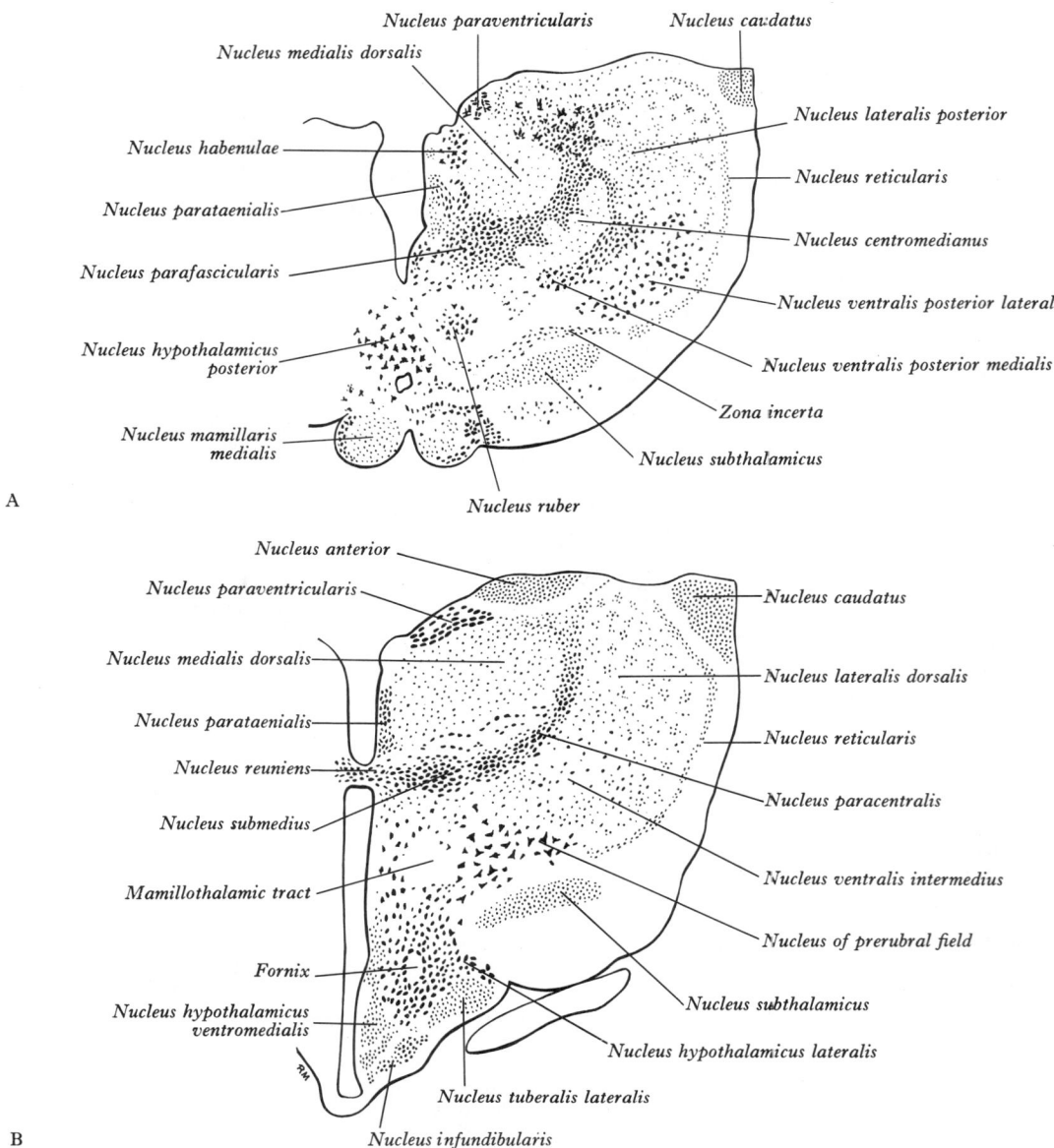

7.128 Drawings of coronal sections through the diencephalon, stained with the method of Nissl to show the main nuclear aggregations of nerve cell somata: A. at the level of the mamillary bodies; B. at the level of the tuber cinereum. Note the variations in cell size, shape and packing density, which characterize the nuclear masses of the dorsal thalamus, subthalamus and hypothalamus at these levels.

A thalamic nucleus may connect with: (1) other thalamic nuclei, (2) adjacent subcortical nuclei, (3) ascending tracts from the brain stem and spinal cord, (4) the cerebral cortex. Intrinsic connections are the most difficult to analyse, and thus are the least clarified. The general thalamic termination of more prominent connections, e.g. long ascending tracts or those from adjacent centres, are well established; but details of their sites and modes of termination and of smaller connections are often uncertain. Many thalamic nuclei degenerate after the removal of certain cortical areas, hence the term *cortically dependent nuclei*. Some nuclei, while locally connected, receive chiefly discrete, often somatotopically ordered inputs with equally ordered cortical projections and are termed *relay nuclei*, in contrast with *association nuclei* which have multiple subcortical connections; but the former are not mere relays as complex transformation is usual in them. By focal thalamic stimulation and cortical recording, thalamic nuclei have been classed as *specific* or *non-specific*. In the former rapid, localized, ipsilateral responses are recorded; in the latter they are widespread, bilateral and with longer latency. Refined methods show that many nuclei, previously considered unconnected with the cerebral cortex, are in fact 'cortically dependent'; also the distinction between specific and non-specific nuclei is fading, as interactions between them become apparent.

SYNAPTIC ORGANIZATION OF THALAMIC 'RELAY' NUCLEI

The ultrastructural organization of main sensory 'relay' nuclei appears complex and analysis of their synaptic patterns is far from complete. A striking feature of all nuclei studied is the *synaptic glomerulus*, an ovoid or irregular unit from 2 μm to 20 μm or more across, which contains a dense mass of diverse axons and dendrites, interconnected by many, usually diverse synapses (Szentágothai 1970, p. 888). Included are the thalamic dendrites and various axonal endings, encapsulated and separated from the surrounding less organized neuropil by extensive astrocytic processes. Specific afferents mainly end in them, e.g. axons of the optic nerve and tract in the lateral geniculate nucleus and medial lemniscal fibres in the nucleus ventralis posterior lateralis. By its synaptic interactions each glomerulus may act as an integrating unit with a corticopetal output, in response to specific afferent impulses, interaction being modified by variable factors such as activity in corticothalamic and other afferent connections and by adjacent specific afferents and glomeruli. Morest (1971) demonstrated *presynaptic dendrites* in thalamic nuclei, containing clusters of synaptic vesicles and synapsing with adjacent dendrites. Some neurons have ordinary and presynaptic dendrites, others resemble amacrine cells of the retina and olfactory bulb, having a

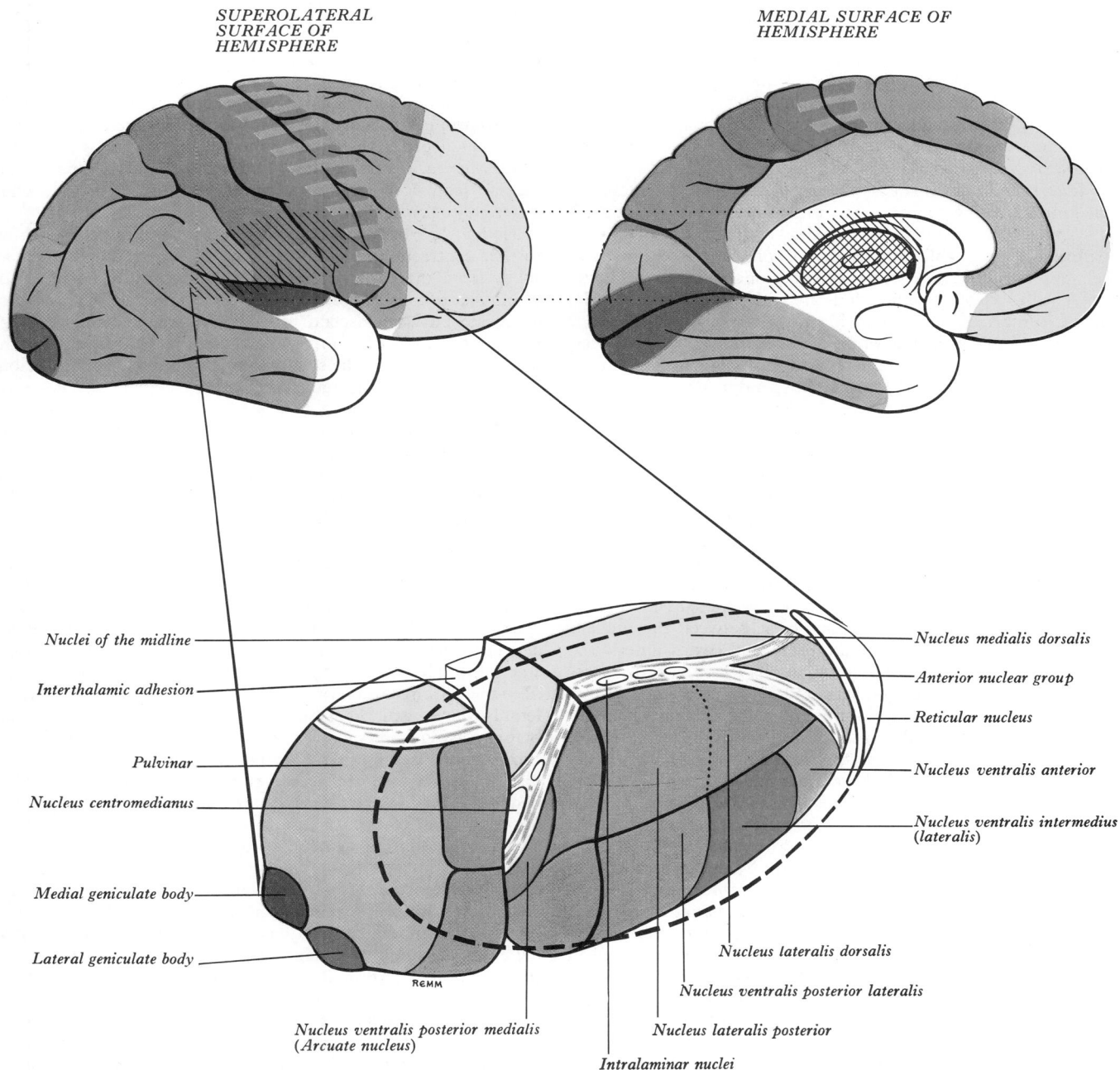

*SUPEROLATERAL
SURFACE OF
HEMISPHERE*

*MEDIAL SURFACE OF
HEMISPHERE*

Nuclei of the midline

Interthalamic adhesion

Pulvinar

Nucleus centromedianus

Medial geniculate body

Lateral geniculate body

REMM

Nucleus medialis dorsalis

Anterior nuclear group

Reticular nucleus

Nucleus ventralis anterior

Nucleus ventralis intermedius
(lateralis)

Nucleus lateralis dorsalis

Nucleus ventralis posterior lateralis

Nucleus lateralis posterior

Nucleus ventralis posterior medialis
(Arcuate nucleus)

Intralaminar nuclei

7.129   The main nuclear masses of the dorsal thalamus (below) have been labelled and colour coded and the same colours have been used to indicate the areas of cerebral neocortex interconnected with these nuclei. The lack of colour in the centromedian, intralaminar and reticular nuclei and in restricted areas of the frontal and temporal lobes are *not* related to the colour code. The boundaries of the coloured cortical zones may well need revision in the future as experimental and pathological data accumulate.

single 'dual-purpose' neurite. Electron microscopy (Lieberman & Webster 1972) shows that presynaptic dendrites have end-bulbs full of distinctive disc-like synaptic vesicles, similar to axonal end-bulbs, forming most of the synapses in thalamic glomeruli. Their significance in processing thalamic activities is discussed by Shepherd (1974, 1978), Schmitt et al (1976) and Rakic (1975).

THE ANTERIOR GROUP OF THALAMIC NUCLEI

The anterior group contains three nuclei named: *anterodorsalis* (AD), *anteromedialis* (AM) and *anteroventralis* (AV); the first is the least prominent. They interconnect with medial and lateral thalamic groups and perhaps with the contralateral anterior group. They receive direct fibres from the postcommissural fornix, the substantial ipsilateral mamillothalamic tract (but including a few contralateral fibres), then also return a few thalamomamillary fibres to the mamillary nuclei. The main anterior radiations are the thalamocortical fibres reaching many parts of the gyrus cinguli (areas 23, 24 and 32), with reciprocal corticothalamic projections from the same areas; some observers add the retrosplenial gyrus (areas 29 and 30) in similar reciprocal connection.

Anterior thalamic nuclei thus link the hippocampus and hypothalamus with other thalamic nuclei and limbic cortical areas (p. 1028). Stimulation or ablation of the mamillothalamic tract affect autonomic control and are presumably concerned in homeostatic cycles involving visceral responses. Normal physiological function in the structural train (hippocampus-fornix-mamillary-body/thalamus/limbic cortex) is considered necessary to *recent memory*. In Korsakow's syndrome, typified by loss of memory for recent events, mamillothalamic lesions are common (Talland 1965, Barbizet 1963); but experiments in rats (Krieckhaus 1967, Krieckhaus & Randall 1968) showed no defect of recent memory after such lesions, but suggested that the 'system' assists in determining the balance between repetitive learned behaviour and new responses.

1001

THE MEDIAL GROUP OF THALAMIC NUCLEI

This includes the large *nucleus medialis dorsalis* (MD) and smaller *nuclei parafascicularis, submedius, paracentralis* and *centralis lateralis*, the connections and significance of which are uncertain in man. (These smaller nuclei are often assigned to the intralaminar group, or to the nuclei of the midline, vide infra.) *Nucleus medialis dorsalis* has an anterior (rostral) *pars magnocellularis* and posterolateral *pars parvocellularis*, considered to make intrathalamic connections with almost all other groups; it has connections from the amygdaloid complex (p. 1034) and piriform cortex, establishing reciprocal connections with parts of the corpus striatum (p. 1076), precise regions in the corpus being uncertain. Its clearest connections are with hypothalamic nuclei and the frontal cerebral cortex (vide infra); reciprocal connections with the former may be part of the *periventricular system* of fibres, fine myelinated or non-myelinated axons in these instances, pursuing vertical or oblique courses subependymally in the third ventricles' lateral wall, intermingled with periventricular grey matter. Some periventricular fascicles may be polysynaptic. The 'system' is often described as principally rostrocaudal (anteroposterior), the fascicles connecting different parts of the medial thalamus with the preoptic, tubero-infundibular and mamillary hypothalamic regions (p. 1007). The most posterior fascicles continue via the *dorsal longitudinal fasciculus* in the periaqueductal grey matter to numerous brain-stem reticular nuclei; but some consider details of periventricular thalamohypothalamic connections less certain (Raisman 1966, Szentágothai et al 1968).

Abundant reciprocal connections exist between the nucleus medialis dorsalis and most of the prefrontal cortex anterior to areas 6 and 32 (p. 1040). These radiations show an orderly arrangement, regions of the nucleus and cortex being specifically linked. In essence, therefore, the nucleus medialis dorsalis is an integral part of the **limbic associated structures**, i.e. those listed under the limbic system, the frontal and cingulate neocortex, the hypothalamus and its basal and brain-stem continuations.

Its complex and copious connections show that the medial thalamus is of great functional significance; but simple functional labels cannot be attached. It probably integrates diverse sensory channels (olfactory, visceral and somatic) by the convergence of paths via the hypothalamus, amygdaloid complex, piriform cortex, corpus striatum and prefrontal cortex but also affecting these reciprocally by divergent paths which reach, additionally, many other regions. Stimulation of the dorsomedial thalamic nucleus in monkeys elicits movements even after ablation of the cortical motor areas; but their patterns disappear with loss of the corpus striatum. Stimulation, disease or removal of the human nucleus results in changes in motivational drive, the ability to solve problems, level of consciousness, general 'personality' and subjective feelings or 'affective tone' of the individual. The effects of ablation partly parallel the results of prefrontal lobotomy.

THE VENTRAL GROUP OF THALAMIC NUCLEI

This group is an anteroposterior series of nuclei *ventralis anterior* (VA), *ventralis intermedius (lateralis)* (VI or VL) and *ventralis posterior* (VP). The last named is subdivided into nuclei *ventralis posterior lateralis* (VPL) and *ventralis posterior medialis* (VPM), with smaller, less well-defined nuclei ventral and anterior in position. Connections of the three main nuclei include intrinsic ones with adjacent thalamic nuclei (centromedianus and the pulvinar, vide infra) and with the subthalamus and corpus striatum; but their major connections are the medial, spinal and trigeminal lemnisci, gustatory paths and two-way links with the cerebral cortex already described. The VPL receives the medial and spinal lemnisci, the VPM the trigeminal lemniscus and gustatory paths. (The station for trigeminal terminals is the *arcuate* or *semilunar nucleus*; gustatory fibres may end in an *accessory arcuate nucleus*.) Structural and functional localization and interactions during transmission in these nuclei have been much investigated.

Somatotopy is general in the ascending tracts themselves, their thalamic nuclei and the latter's cortical connections. Medial lemniscal and spinothalamic fibres from lower body segments (via the nucleus gracilis for the former) end laterally in the VPL, the fibres from successively higher segments ending progressively medially. The more medial VPM receives trigeminothalamic fibres from the face and head. Most medial are the solitariothalamic gustatory fibres from the solitary nucleus. Though such *lateromedial localization* exists, some consider that the terminals of the various tracts are not simply superimposed but end in a rostrocaudal order; the lateral spinothalamic tract may end in the caudal nuclear regions, followed by intermediate terminals of the medial lemniscal and anterior spinothalamic fibres, while trigeminothalamic fibres have a rostromedial termination.

Medial lemniscal and spinothalamic terminations exhibit further contrasts: the lemniscal fibres are crossed and originate solely from the contralateral gracile and cuneate nuclei; their terminals are confined to the VPL. *Most* spinothalamic fibres also decussate but *some* ascend to the ipsilateral thalamus; though many end in the VPL, others go elsewhere, including parts of the medial geniculate nucleus, suprageniculate nucleus and various 'non-specific' thalamic nuclei, the *posterior* (PO) *group* (Poggio & Mountcastle 1960, 1963), which may receive terminals of high threshold spinal neurons transmitting nociceptive impulses (p. 943). The PO nuclei may have reciprocal connections with the secondary sensory area SmII (p. 1056). The structure of terminals in the VPL, viewed by Golgi technique (Scheibel & Scheibel 1966), shows that spinothalamic and reticulothalamic terminals form diffuse networks, that medial lemniscal fibres end as conical bushy terminals in regular laminae, while corticothalamic terminals are confined to disc-shaped 'sheets' around the territories of many lemniscal fibres. Small local interneurons have been identified in VPL (Tombol 1967); synaptic detail of the nucleus has also been explored (Anderson et al 1964a, b, Lieberman & Webster 1972, Shepherd 1974, Rakic 1975). Unit recordings have been made from neurons in the VPL during activity in the medial lemniscal and spinothalamic tracts in monkeys and in man, before placing stereotactic lesions for types of locomotor disease. Simian units responding to lemniscal impulses show features described previously in gracile and cuneate neurons, the VPL neurons being highly specific for the type and source of stimulus. Units respond to contralateral stimulation by distortion of the skin or hairs, joint movement, static joint position and sinusoidal tissue vibration, but each *only to one* variety of stimulus. The receptive fields are small and localized, the smallest being associated with the terminal segments of the limbs. Units responding to spinothalamic activity generally show larger receptive fields; while some are 'modality specific', others show the interaction of impulses from different patterns of stimulus (p. 944). Transmission in the VPL is thus a complex of events. Many units show *lateral inhibition* or 'inhibitory surround' (p. 865), greatly enhancing their discrimination. Transmission may be modified by corticothalamic axons and those converging from other thalamic and extrathalamic sources. Presynaptic and postsynaptic inhibitory and facilitatory effects have also been demonstrated in the VPL.

Recordings in the *human* VPL are inevitably restricted in scope, but substantially support simian somatotopy and modal specificity. Nuclear stimulation in conscious patients evokes localized sensations of tingling, numbness, etc. on the opposite side. The main thalamocortical radiations from the VPL and VPM traverse the internal capsule's posterior limb to the primary somatic sensory cortex (postcentral gyrus, areas 1–3) (p. 1051). Throughout these connections the precise somatotopy in their nuclei of origin is preserved. The same areas project back to the nuclei. Thalamic projections to secondary somatic sensory areas are uncertain; some may stem from the PO group (p. 1004).

Nucleus ventralis intermedius (VI, including the nucleus ventralis lateralis, VL, of many authors) has been less investigated. It connects with the adjacent thalamic nuclei but its main input is from the contralateral nucleus dentatus and from the ipsilateral red nucleus (p. 967, 7.120) and also some fascicles from the globose and emboliform nuclei and the globus pallidus. A nigrothalamic connection exists in monkeys (Carpenter et al 1976). These afferents are all in the *thalamic fasciculus* (p. 1007). The main projection from VI is somatotopic and traverses the internal capsule to the motor and premotor cortex (areas 4 and 6). Its posterior part has often been ablated in cases of severe Parkinsonism.

The *nucleus ventralis anterior* (VA) has become prominent, with VI, through destructive lesions placed in or near them for some locomotor disorders. It has many connections with other thalamic nuclei, particularly the *nucleus centromedianus* (vide infra), other *intralaminar nuclei, midline nuclei* and the *reticular thalamic nucleus*, i.e. all the so-called 'non-specific' thalamic nuclei. In addition VA has an input from the brain-stem reticular formation and wide connections from the globus pallidus via the *thalamic fasciculus* (p. 1007) and uppermost dentatothalamic fibres in the superior cerebellar peduncle. Carpenter et al (1976) have confirmed a nigrothalamic projection to the VA in monkeys; thalamo-cortical fibres radiate from it to the premotor cortex (area 6), less so to the motor cortex (area 4); axons to small regions of the insular cortex have been described. Corticothalamic fibres leave areas 6 and 4, converging on the VA. (For a detailed Golgi study see Scheibel & Scheibel 1966.)

The nucleus ventralis anterior (VA) is thus a focus via which the corpus striatum, ascending reticular formation and non-specific thalamic nuclei and a proportion of the cerebellum may powerfully modify the activities of motor and premotor cortices and probably other cortical regions. How far the impulses converging on the VA are integrated before transmission to the cortex is uncertain. Functional views of the VA are changing: some relation to the motor systems is evident; stimulation increases Parkinsonian rigidity and tremor; ablation (with, doubtless, surrounding tissue) may reduce or abolish tremor. Experimental stimulation of the caudate nucleus reduces activity in the VL and VA and their cortical connections, accompanied by a loss of behavioural responses. Stimulation of the VA also desynchronizes the electroencephalogram; this nucleus is regarded increasingly as a link in the *'ascending activating system'* (p. 996).

THE LATERAL GROUP OF THALAMIC NUCLEI

This group is usually divided anteroposteriorly into *nuclei lateralis dorsalis* (LD), *lateralis posterior* (LP) and most caudally the expanded *pulvinar*, occupying almost all the caudal thalamic quarter. The pulvinar is phylogenetically recent, prominent only in higher mammals and especially primates. Some subdivide it but this and its separation from the LP are imprecise; some prefer to describe an *LP–pulvinar complex*.

These three nuclei are presumed to have connections with other thalamic groups, the details being uncertain. Connections from the lateral geniculate nucleus, possibly the medial geniculate nucleus and amygdaloid complex, have been described, as well as a direct *retinothalamic projection* (p. 1016); these require further evidence. Most certain are the reciprocal connections with specific regions of the cortex, the LD with inferior parietal and posterior cingulate regions, the LP with much of the parietal cortex behind the postcentral gyrus and the pulvinar with wide areas of the parietal, occipital and temporal cortex, each pulvinar subnucleus projecting in orderly array to particular cortical areas, which probably reciprocate to the pulvinar; but connections with *primary* sensory areas are sparse or absent. The pulvinar has abundant connections with Wernicke's speech area; it has been implicated in such diverse activities as the perception of chronic pain, oculomotor control, speech control and as a *'multisensory data processor'* involved in general *'gnosis'* (Ingram 1976).

THE 'NON-SPECIFIC' GROUPS OF THALAMIC NUCLEI

Since Moruzzi & Magoun (1949) observed that stimulation or ablation of regions of the brain-stem reticular formation led to wide changes in electroencephalographic synchronization or desynchronization, possible paths from the reticular formation to the cerebral cortex have been a focus of research. Brain-stem reticular 'centres' were soon suggested to be connected to thalamic nuclei, with a *diffuse thalamocortical radiation* to almost all the cortex. Since stimulation of several thalamic nuclei caused extensive, *bilateral* change in cortical electrical activity, typified by a slow onset and progress to a maximum, followed by fluctuations (recruitment), the responses and nuclei were together dubbed *non-specific*. A wide assumption followed that such effects norm-

ally maintained cortical preparedness for the effective reception of rapid, localized, ipsilateral patterns from larger *specific relay nuclei*, which have somatotopic, thalamocortical and corticothalamic connections. From the large volume of reports some general concepts remain valid; but it is becoming clear that the *reticulocortical paths* involved are *multiple* and more complex than imagined, some involving the ventral diencephalon. Interaction between specific and non-specific thalamic regions is so extensive that, while some terms retain their general usefulness, they do *not* imply a distinction between the two aspects of thalamic function.

Non-specific nuclear groups usually include the intralaminar nuclei, midline nuclei and the thalamic reticular nucleus.

The Reticular Thalamic Nucleus

This thin curved sheet of neurons lies between the external medullary lamina and the posterior limb of the internal capsule. Hence, all corticothalamic and thalamocortical fibres in the capsule are near this nucleus, which was long regarded a final link in the diffuse thalamocortical radiation mentioned above; wide effects on cortical activity follow its stimulation, and nuclear degeneration follows resection of the cerebral cortex. But the Golgi technique (Scheibel & Scheibel 1966) has shown that its axons run *posteriorly*, with collaterals to many thalamic nuclei and also to the midbrain reticular formation. The whole cerebral cortex projects in an orderly manner to the reticular nucleus (Carman et al 1964); degeneration after cortical resection is now confirmed as transneuronal and not retrograde (Rose 1952). Other afferents are from the brain-stem reticular formation and the globus pallidus. Afferents thus converge from many sources; output is mainly to other thalamic nuclei, both specific and non-specific.

Intralaminar nuclei are the small *nuclei paracentralis, centralis lateralis and medialis, parafascicularis*, and larger *nucleus centromedianus* (vide infra).

Nuclei of the midline, complex and well-developed in many mammals, are small in man. Recognizable groups are sometimes seen near the taenia thalami, in the interthalamic adhesion when present, and also scattered in the third ventricular wall. They include *nuclei paraventricularis anterior and posterior*, the *nucleus rhomboidalis* and *nucleus reuniens*. They will not be separately described further here.

Non-specific nuclei are difficult to approach for any investigation. They are held to receive terminals from the brain-stem reticular formation, some connecting with the corpus striatum, cerebellum, spinothalamic tracts, hypothalamus and with other specific and non-specific thalamic nuclei on both sides. The most anterior project to the cerebral cortex, including the phylogenetically old prepiriform and entorhinal areas and parieto-occipital and frontal regions.

The nucleus contromedianus, embedded in the internal medullary lamina, is only prominent in primates and easily recognized in the human thalamus; but disagreement concerning its status and connections persists, though most agree that it is *not* connected with the cortex. Various ascending connections are reported, including a few collaterals or terminals from spinal medial and trigeminal lemnisci, ascending reticulothalamic fibres and some from the superior cerebellar peduncle. The main connections are with parts of the corpus striatum, some being topically organized; abundant connections occur with other non-specific nuclei on both sides and with some 'specific' nuclei, particularly the nucleus ventralis anterior.

The dorsal thalamic connections so far mentioned were established by older techniques, but subsequent data accumulated in a variety of mammals by newer tracing techniques (p. 870) (particulary methods locating somata and neurites of cholinergic and monoaminergic neurons, as detailed in pp. 992, 995) are merely noted here. The *dorsal tegmental ascending cholinergic bundle* provides terminals in the geniculate nuclei, dorsal and anterior nuclear groups of the dorsal thalamus and in midline and intralaminar nuclei, especially the nucleus centromedianus. The *ventral tegmental ascending cholinergic bundle* carries terminals into the ventral and anterior nuclear groups. The *dorsal ascending noradrenergic bundle* is afferent to many, perhaps all, dorsal thalamic nuclei. *Dopaminergic neuron group* A11 sends fine axons among the midline nuclei; groups A13 and 14 similarly reach the caudal thalamic nuclei. The *ventral ascending serotonergic bundle* contains axons reaching the medial nuclear group, midline nuclei and parafascicular nucleus. For brief comments on fibres containing metenkephalins, endorphins, substance P and the distribution of opiate receptor sites, see pp. 892, 995 and vide infra.

SUMMARY

The dorsal thalamus is obviously most complex and must be functionally important; but despite much research, concepts of its activities are rudimentary. A very large number of sensory channels *converge* on it and many *integrate* with each other; the results, of even greater range and complexity, *diverge* to many destinations. It is indeed involved in activities of *all* major regions of the central nervous system; the whole sensory system (except olfactory), most of the cerebral cortex, corpus striatum, cerebellum, hypothalamus, subthalamus and brain-stem reticular formation all *project to* the thalamic and metathalamic nuclei, and most *receive* reciprocal thalamic connections.

The thalamus is not essential to olfactory perception, which involves the primary and secondary olfactory areas (p. 1034); but olfactory information, combined with other modalities, reaches it via the amygdaloid complex, piriform lobe and hippocampus via the mamillary body.

Specific relay nuclei, including the geniculate, receive major sensory tracts and, usually after interaction with other channels, somatotopic thalamocortical radiations pass to sensory cortical areas. Complex transformations include the interaction between parallel and converging channels, pre- and postsynaptic facilitatory and inhibitory effects, phenomena of inhibitory 'surround' and 'neural sharpening' and modulation of transmission by corticothalamic and other axons in converging thalamic connections. Some thalamocortical fibres preserve a high specificity for single modalities and peripheral sites, others transmit more complex orders of information derived by integration. Cutting thalamocortical radiations to the somatosensory cortex lessens the ability to localize tactile stimuli, impairs tactile discrimination and appreciation of texture, weight and shape (*astereognosis*) or of position and movement; but awareness of contact and vaguely localized pain and thermal sensations survive if the thalamus itself remains intact. (Webster 1976 has reviewed somaesthetic pathways.) While some appreciation of pain survives loss of the cortex, its involvement in the intact nervous system has not been clarified (Albe-Fessard & Delacour 1968). Disease in the lateral or central thalamus sometimes causes sudden, 'spontaneous' attacks of ill-defined *thalamic pain*. In contrast, in addition to spinal, medullary or mesencephalic tractotomy for the relief of pain, selective ablation of the nucleus ventralis posterior and surrounding regions, or of intralaminar nuclei including the nucleus centromedianus, have sometimes relieved intractable pain. In summary it appears that involved in sensation of pain are the VPL and VPM, PO group (including parts of the medial geniculate, suprageniculate and intralaminar nuclei), the nucleus centromedianus, probably the nucleus medialis dorsalis and the pulvinar, with many hypothalamic nuclei, the globus pallidus, caudate nucleus and limbic structures, such as certain amygdaloid nuclei and the bed nucleus of the stria terminalis; but how they co-operate remains obscure. Many regions (including: spinal laminae I, II, and V, lateral reticular nuclei, the nucleus ambiguus and solitarius, areas of mesencephalic periaqueductal grey and of the pars compacta of substantia nigra) interchange axons which contain monoamines (particularly serotonin), metenkephalin, endorphin, substance P and other neuromediators; hence the proposal that these co-operate in processing pain perception (Nathan 1977, 1978, Wall 1978, Basbaum et al 1976, Webster 1976, 1978, Snyder in: Reichlin et al 1978). (See also notes and references to the *descending analgesia system* p. 943.)

Copious interconnection of *the nucleus dorsalis medialis* with the frontal cortex, hypothalamus and other specific and non-specific thalamic nuclei has already been noted; it is generally regarded as a main centre for the complex *integration of visceral and somatic functions*. By hypothalamic connections it is involved in autonomic and endocrine activities and also in emotional content, subjective feeling and consciousness of 'self'. Ablation or disease here cause changes in personality, emotional drive and level, intellectual performance, etc. and indifference to pain—like the sequelae of frontal corticectomy.

Anterior thalamic nuclei integrate a complex input along the mamillothalamic tracts from many visceral sources, the hypothalamus and limbic structures (including the hippocampus, amygdala and septal areas), with that from other thalamic nuclei and the cingulate cortex; the anterior group has reciprocal connections with all these. Such arrangements are concerned in complex homeostatic mechanisms, possibly in the efficiency of recent memory and in determining a balance between 'repetitive and stereotyped' or 'novel and exploratory' forms of behaviour.

Non-specific nuclear groups (including the nucleus ventralis anterior) have abundant connections with each other and with the specific thalamic nuclei of both sides, receiving from many sensory paths, the cerebellum and corpus striatum; via diffuse cortical projections they have profound effects, the *arousal reaction*, on the background activity in the cerebral cortex. This reaction involves particularly the parietal, orbital, cingulate and occipital association areas, the most marked experimental responses being in prefrontal areas.

The nuclei ventralis anterior and lateralis are sites of interaction of outflows from the corpus striatum and cerebellum with other thalamic nuclei, projecting to motor and premotor cortices and essential to full locomotor control. The activities of most major regions of the cerebral cortex are under thalamic influence, the cortex equally affecting most thalamic nuclei.

The Epithalamus: Pineal Gland and Habenula

Epithalamic structures occupy the posterior diencephalic roof and adjacent areas of the third ventricular walls. They include right and left *habenular nuclei*, each deep to its *habenular trigone* and receiving a complex *stria medullaris thalami*; also included is the median *pineal gland* (epiphysis cerebri) and *habenular* and *posterior commissures* crossing in anterior and posterior laminae of the pineal peduncle. Pineal development is briefly mentioned on p. 190, its structure, neuro-endocrine roles and innervation on p. 1457 with accompanying illustrations.

The *trigonum habenulae* is a small, bilateral triangular depression, anterior to the superior colliculus, medial to the pulvinar but separated by a *sulcus habenulae*; anteromedial is the posterior end of a ridge occupied by the stria medullaris thalami and by the pineal peduncle.

The habenular nucleus, sometimes described in medial and lateral parts, is, in part, a station on olfactory reflex routes, but probably also concerned with pineal innervation; connections have been described but details are uncertain (see comments on pineal innervation, p. 1457). Many *afferents* to the nucleus run in the **stria medullaris thalami**, formed near the anterior thalamic pole and including connections from the amygdaloid complex via the stria terminalis (p. 1034) and from the hippocampal formation via the fornix (p. 1039); other components are from the olfactory tubercle, anterior perforated substance, preoptic and septal areas and several hypothalamic nuclei. Tectohabenular fibres from the superior colliculi have also been described.

The stria medullaris crosses the superomedial thalamic aspect, skirts medial to the habenular trigone and sends many fibres into the ipsilateral habenular nucleus. The stria has fibres containing seven neuromediators: acetylcholine, noradrenalin, serotonin, GABA, LHRH, somatostatin, vasopressin and oxytocin. Other strial fibres cross in the anterior pineal lamina, decussating as the *habenular commissure* to reach the contralateral habenular nucleus. Some fibres are really commissural and interconnect the amygdaloid complexes and hippocampal cortices; crossed tectohabenular fibres accompany them in the commissure. Additional afferents to the habenular nuclei are fibres containing serotonin (5-hydroxytryptamine) from the *ventral ascending tegmental serotonergic bundle*, which join the habenulopeduncular tract (vide infra) to reach the nuclei; they may exert some control over neurons of the *habenulopineal tract*, thus influencing

innervation of the pinealocytes (p. 1457), as may also the
habenular nuclear afferents from the *dorsal ascending tegmental
noradrenergic bundle* (pp. 986, 993). Though human habenulae
are relatively small, they are a focus of integration of diverse
olfactory, visceral and somatic afferent paths.

The main habenular outflow reaches the interpeduncular
nucleus, thalamic nucleus medialis dorsalis, mesencephalic
tectum and reticular formation, the largest being the
habenulopeduncular tract or *fasciculus retroflexus* (7.131). This
courses ventrally and up, skirts the inferior zone of the thalamic
nucleus medialis dorsalis and traverses the superomedial region
of the red nucleus to the interpeduncular nucleus; this provides
relays to the midbrain reticular formation, from which tectotegmentospinal tracts and dorsal longitudinal fasciculi connect
with autonomic preganglionic neurons controlling salivation,
gastric and intestinal secretory activity and motility; others
pass to motor nuclei for mastication and deglutition. Ablation of habenular complexes causes extensive changes in
metabolism, endocrine and thermal regulation (Szentágothai et
al 1962).

The posterior (dorsal) commissure is a complex
fasciculus decussating in the posterior pineal lamina, relatively
reduced in primates and of unknown constitution in man. It
myelinates early; estimates of fibres in several species have been
made (Tomasch & Malpass 1958). Various nuclei are associated
with it: small groups scattered along it as the *interstitial nuclei
of the posterior commissure*, accumulations in the periventricular
grey matter forming *dorsal nuclei of the posterior commissure*,
nucleus of Darkschewitsch in the periaqueductal grey, and the
interstitial nucleus (of Cajal) near the upper end of the
oculomotor complex, closely linked with the medial longitudinal fasciculus (p. 985). Fibres from all these nuclei and the
fasciculus cross in the posterior commissure. Other contributors
to it include dorsal thalamic nuclei, pretectal nuclei, superior
colliculi and connections between the tectal and habenular
nuclei. The destinations and functions of many of these fibres
are obscure.

Ventral to and below the posterior commissure (i.e. near the
inferior wall of the pineal recess), ependymal cells on the dorsal
aspect of the cerebral aqueduct are tall, columnar and ciliated,
with granular basophilic cytoplasm and specific histochemical
reactions. This patch, possibly secreting into the cerebrospinal
fluid, is the *subcommissural organ* (Keene & Hewer 1935,
Wislocki & Leduc 1953, Mollgard et al 1973). Its cells may be
involved in the transport of materials to the cerebrospinal fluid
from adjoining axonal terminals or capillaries; or substances
may be transported from fluid to neurons, blood vessels or
pinealocytes (p. 1457, and cf the infundibular recess of the third
ventricle, pp. 1019, 1451). Possible neuro-endocrine roles of
these specialized ependymal *tanycytes* have been reviewed by
Knowles (1974), Reichlin et al (1978), Joseph & Knigge (1978),
Collins & Woollam (1981). Other patches of similar ependyma
project into the third ventricle from its secondary roof (body
of fornix) and from its anterior wall between the diverging
fornical columns: the *subfornical organ* and the organum
vasculosum or *intercolumnar tubercle* respectively. These and, in
some vertebrates, other specialized regions of the third
ventricular ependyma are collectively termed **circumventricular organs**. For a review of the disposition,
ultrastructure and possible functional roles of circumventricular
organs see Collins & Woollam (1981); they indicate that circumventricular organs (CVOs) are a mosaic of *three* main cell
varieties: 'basic' ciliated ependymal cells, secretory cells and
tanycytes, the latter, however, varying in their ultrastructure as
each is location specific. In summary, a generalized mammalian
brain displays, in median section, the following specialized
ependymal CVOs: the median eminence (ME), and variable
infundibular recess (IR); the collicular recess organ (CRO); the
aqueductal recess organ (ARO); the subcommissural organ
(SCO); the habenular ependyma (HE); the habenular commissural organ (HCO); the subfornical organ (SFO); the organum vasculosum of the lamina terminalis (OVLT) or intercolumnar tubercle (IT). Recall, also, the area postrema and
funiculus separans in the fourth ventricle (p. 979).

The Ventral Thalamus or Subthalamus

The pars ventralis diencephali, ventral and posterior to the
hypothalamic sulcus, may be divided into the *hypothalamus* and
the *ventral thalamus (subthalamus)*. The hypothalamus, including structures forming the third ventricular floor as far back as,
and including, the mamillary bodies and also structures in its
anteroventral side wall, will be described later. The rest of the
pars ventralis diencephali is the *subthalamus*, merging below
with the midbrain tegmentum; upper extensions of substantia
nigra and the red nucleus project into its lower end and large
tracts pass between the two regions (7.130, 131). Dorsal are the
ventral nuclei of the *dorsal thalamus* and anteromedially the
subdivisions of the *hypothalamus* (7.128A,B). Ventrolaterally the
subthalamus adjoins an expanding, twisting junctional zone
where each cerebral peduncle merges into its internal capsule;
the capsules separate the subthalamus from the medial aspect
of each globus pallidus.

The main neuronal groups are: (1) the upper pole of the
red nucleus, (2) the upper end of the substantia nigra, (3) the
nucleus subthalamicus, (4) the zona incerta, (5) the nucleus of
the prerubral or tegmental field, (6) the entopeduncular nucleus
(nucleus of ansa lenticularis); all, of course, are bilateral.

The main subthalamic tracts are: (1) the upper parts of
the medial, spinal and trigeminal lemnisci and the
solitariothalamic tract, approaching their terminations in the
thalamic nuclei; (2) the dentatothalamic tract from the
contralateral superior cerebellar peduncle accompanied by ipsilateral rubrothalamic fibres; (3) the fasciculus retroflexus;
(4) the fasciculus lenticularis; (5) the fasciculus subthalamicus;
(6) the ansa lenticularis; (7) fascicles from the prerubral field (H
field of Forel); (8) the continuation of the fasciculus lenticularis
(in the H_2 field of Forel); (9) the fasciculus thalamicus (the H_1
field of Forel); all, of course, are bilateral.

Subthalamic topography is complex and best appreciated in
solid models or in coronal and sagittal serial sections. A parallel
study of the corpus striatum (p. 1075), from which prominent
subthalamic tracts are derived, is also helpful. Figure 7.130 may
clarify the region's complicated terminology but three-dimensional topography is difficult to appreciate from flat (two-
dimensional) diagrams.

As the upper ends of the red nucleus and substantia nigra
pass into the subthalamus they diminish in area, ending a little
below the mamillary bodies. The changing relation of the
lemnisci to the red nucleus and substantia nigra as they ascend
in the midbrain has been described (p. 983). Lemniscal fibres
enter the subthalamus largely *lateral* to the red nucleus; as they
ascend they approach the *dorsal* aspect of the nucleus to reach
the *ventral* surface of the thalamic nucleus ventralis posterior,
in which most fibres end. Dentatothalamic and rubrothalamic
run with pallidothalamic fibres in the *thalamic fasciculus*, which
ascends beyond the lemnisci to distribute largely to the thalamic
nuclei ventralis intermedius and anterior (pp. 1002, 1003, and
vide infra).

The nucleus subthalamicus, absent in submammalian
forms, is small in most mammals and is prominent only in
primates. The human nucleus, containing medium to large
multipolar neurons, is biconvex in coronal sections. It is
inferior in the subthalamus, extending below into the midbrain,
where it is dorsolateral to the upper end of the substantia nigra
and lateral to the red nucleus. It is near the medial aspect of
the internal capsule, which separates it from the globus pallidus. Medially, the nucleus adjoins the hypothalamic region
and is dorsally separated from the ventral thalamic nuclei by the
thin *zona incerta*, interposed between continuations of lenticular
and thalamic fasciculi (7.130, and vide infra).

Connections are numerous, some established and many less
certain, the principal being with the corpus striatum; reciprocal
connections by several routes link the globus pallidus and subthalamic nucleus. Connections with the putamen and caudate
nucleus have been described but are less documented. Other
reported connections include the contralateral subthalamic
nucleus and globus pallidus, ipsilateral red nucleus, substantia
nigra, mesencephalic reticular formation, zona incerta and other

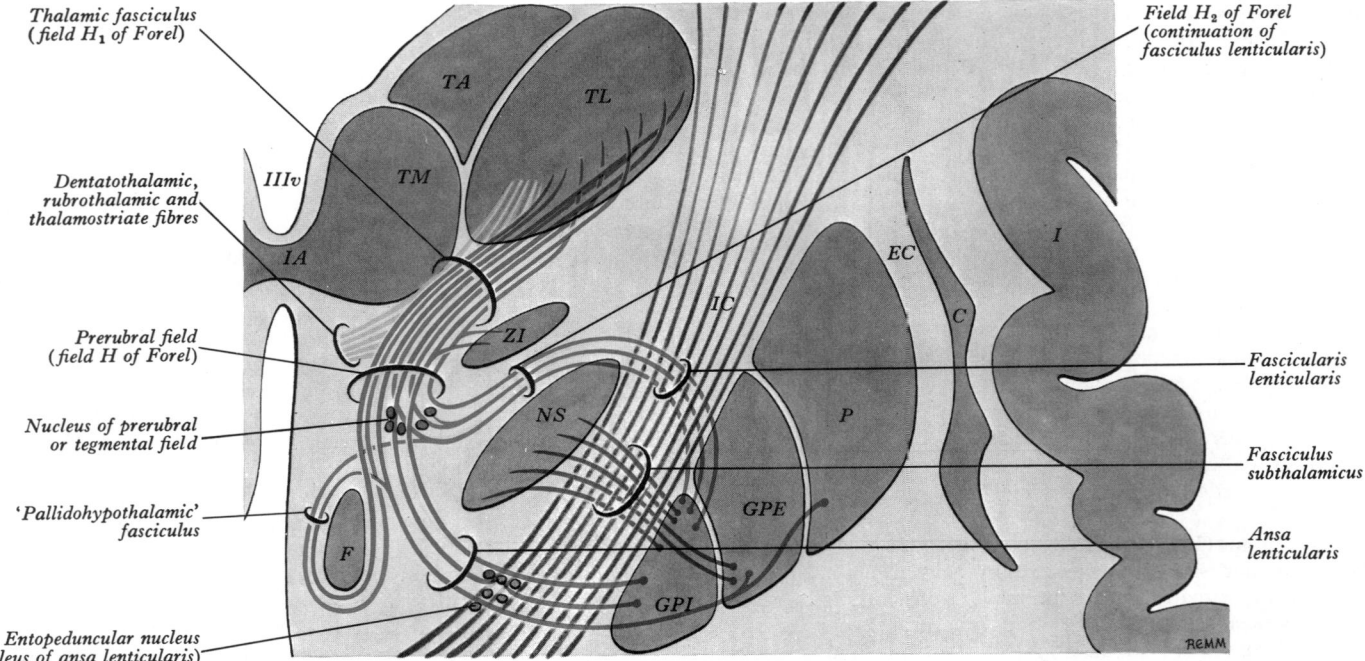

Thalamic fasciculus
(field H₁ of Forel)

Dentatothalamic,
rubrothalamic and
thalamostriate fibres

Prerubral field
(field H of Forel)

Nucleus of prerubral
or tegmental field

'Pallidohypothalamic'
fasciculus

Entopeduncular nucleus
(nucleus of ansa lenticularis)

Field H₂ of Forel
(continuation of
fasciculus lenticularis)

Fascicularis
lenticularis

Fasciculus
subthalamicus

Ansa
lenticularis

7.130 The nuclear masses of grey matter and fibre tract systems associated with, or closely related topographically to, parts of the dorsal thalamus, subthalamus and globus pallidus. The information presented is compounded from a series of closely spaced coronal sections through this region, attempts being made to include what are essentially three-dimensional structures in a two-dimensional diagram; it must be emphasized that all the structures shown would not appear on the same single coronal section. The significance of the letters is as follows: IIIv = third ventricle; M = medial nuclear group of thalamus; TA = anterior nuclear group of thalamus; TL = lateral nuclear group of thalamus; IA = interthalamic adhesion; ZI = zona incerta; F = column of fornix; NS = nucleus subthalamicus; IC = internal capsule; GPI and GPE = internal and external parts of globus pallidus; P = putamen; EC = external capsule; C = claustrum; I = cortex of insula.

small subthalamic nuclei, various hypothalamic and thalamic nuclei and possibly regions of the cerebral cortex. Some afferent fibres are derived from the *ventral ascending tegmental cholinergic pathway* and probably the *incerto-hypothalamic dopaminergic system*; their functional roles remain obscure.

The subthalamic nucleus is clearly a site of integration of motor control centres, particularly by its connections with the corpus striatum and mesencephalic tegmentum. Discrete lesions of one human subthalamic nucleus result in *hemiballismus*, with contralateral, uncontrollable, violent, torsional movements, choreiform in type, often continuing for long periods and usually affecting the proximal musculature of one or both contralateral limbs; facial and trunk muscles are less often involved. Unlike other *dyskinesias*, hemiballismus can be reproduced experimentally; in monkeys it follows the ablation of a quarter or more of the nucleus, provided adjacent fibres of the globus pallidus are intact; it is unaffected by ablation of the rubrospinal, vestibulospinal and reticulospinal tracts or cortical area 6, but *is* abolished by ablation of the globus pallidus or its outflow, or of the thalamic nucleus ventralis anterior, area 4 of the cerebral cortex or the corticospinal tract. This suggests that the nucleus inhibits the globus pallidus and its main outflow via the thalamus to the motor cortex, i.e. the principal origin of the crossed corticospinal tract (Whittier & Mettler 1949, Carpenter 1950, Carpenter et al 1950, 1951, 1958, 1960).

The zona incerta is a thin grey lamina among the fascicles of fine fibres extending through most of the diencephalon a little ventral to the thalamus, separated from it by the thalamic fasciculus (vide infra). On each side it continues laterally into the thalamic reticular nucleus; ventral is the fasciculus lenticularis (vide infra). Functionally associated are neurons grouped along its inferomedial border, collectively the *nucleus of the prerubral* or *tegmental field*, and other groups between fascicles of the ansa lenticularis (vide infra), sometimes regarded as 'detached' parts of the globus pallidus but collectively termed the *entopeduncular nucleus*. These nuclei are mainly relays on discharge pathways from the globus pallidus to the mesencephalic reticular formation; some of their fibres descend in the central tegmental fasciculus to the inferior olivary complex (p. 956), numerous branches leaving to innervate the 'reticular' and 'non-reticular'

brain-stem nuclei. Cortical projections to the zona incerta may exist; various subthalamic groups probably also interconnect with the main nucleus, intralaminar and ventral thalamic nuclei and red nucleus. The zona incerta also has afferents from the *ventral tegmental ascending cholinergic pathway* and is involved in the *incertohypothalamic dopaminergic system* (pp. 992, 994).

In addition to terminal parts of the lemniscal, dentatothalamic and rubrothalamic tracts (p. 1002), the subthalamus is typified by white fascicles, often containing small groups of neurons and of complex topography. Conflicting terminologies have been proposed. The approach adopted here is to consider the fascicles in relation to the main striatal outflows (p. 1078). These are derived partly from the putamen but mostly from the globus pallidus and appear at the latter's surface, fanning out medially; the radiation's dorsal and intermediate fibres intersect those of the internal capsule, while the ventral ones curve round the capsule's posteroventral border. Earlier investigators (von Monakow 1882) termed the whole radiation the *ansa lenticularis*, with dorsal, intermediate and ventral divisions, a term now restricted to the ventral radiation, the intermediate radiation being the fasciculus subthalamicus, the dorsal radiation the fasciculus lenticularis (7.130).

The fasciculus lenticularis is the dorsal division of pallidofugal fibres traversing the internal capsule; it turns medially near the capsule's medial aspect, partly intermingled with the dorsal zone of the subthalamic nucleus and ventral part of the zona incerta, where the fasciculus traverses the H₂ field of Forel. Reaching the medial border of the zona incerta, the fasciculus intermingles with fibres of the ansa lenticularis, scattered elements of the prerubral nucleus and with dentatothalamic and rubrothalamic fibres. This merging of diverse pathways and associated cell groups is variously called the *prerubral*, *tegmental* or *H field of Forel*.

The ansa lenticularis has a complex origin from both parts of the globus pallidus, from the putamen and possibly other adjacent structures; its fibres partly relay in neurons along its course, the *entopeduncular nucleus*. It curves medially round the internal capsule's ventral border, continuing dorsomedially to mingle with other fibres noted above in the prerubral field. Some fibres in both fasciculus lenticularis and ansa lenticularis synapse in the nucleus subthalamicus, prerubral field and zona incerta; the remainder continue laterally with other fascicles into the thalamic nuclei, particularly the nuclei ventralis anterior intermedius and centromedianus.

The thalamic fasciculus is a complex extending from the prerubral field, dorsal to and also partly traversing the zona incerta and related dorsally to the ventral thalamic nuclei. It contains continuations of the fasciculus lenticularis and ansa lenticularis, dentatothalamic and rubrothalamic fibres and thalamostriate fibres. Its territory is sometimes termed the H_1 *field of Forel*.

The 'pallidohypothalamic' fasciculus leaves the main pallidofugal system in the prerubral field and pursues a curious course, curving ventromedially round the column of the fornix towards the hypothalamus (Bard & Rioch 1937, Vidal 1940, Ingram 1940), where it was long assumed to end in the dorsomedial nucleus. But evidence is inconclusive and experiments in monkeys (Nauta & Mehler 1966) show that it recurves laterally below the column, turning dorsally to rejoin the H_1 field of Forel.

The fasciculus subthalamicus is an abundant two-way array of fibres traversing the internal capsule, interweaving with it at right angles, to connect the nucleus subthalamicus with the globus pallidus and, to a lesser extent, the putamen.

The Hypothalamus

The general position and extent of the hypothalamus have been noted (p. 997). It extends from the lamina terminalis to a vertical plane posterior to the mamillary bodies and from the hypothalamic sulcus to the pial surface of the third ventricle's floor. Strictly, this region's anterior part, the preoptic area, belongs to the telencephalon impar but is here included for functional reasons. *Lateral* are the anterior part of the subthalamus, internal capsule and optic tract; *posteriorly* the tegmental part of the subthalamus continues into the mesencephalic tegmentum; *dorsal* are nuclei of the dorsal thalamus; anteriorly the optic chiasma, lamina terminalis and anterior commissure separate the preoptic area from the precommissural septum, i.e. the continuation of the diagonal band into the paraterminal gyrus (p. 1033), anterior to which is the parolfactory gyrus. These 'boundaries' are arbitrary and functionally continuous 'systems' cross many of them.

Structures in the third ventricle's floor reach the pial surface in the **interpeduncular fossa** (7.132); from front to back they are: (1) the optic chiasma, (2) the tuber cinereum, tuberal eminences and the infundibular stalk, (3) the mamillary bodies, (4) the posterior perforated substance, included here for convenience.

The posterior perforated substance, a small grey depression, lies posteriorly in the interval between the diverging crura cerebri, pierced by small apertures for central branches of posterior cerebral arteries. Deep within it is the *interpeduncular nucleus*, small in man and homologous with a more extensive complex in submammalian forms. It receives looped terminals of the fasciculus retroflexus (p. 1005) of both sides and has other connections with the mesencephalic reticular formation and mamillary bodies.

The mamillary bodies are smooth, hemispherical and about the size of two small peas, side by side in the interpeduncular fossa's floor, anterior to the posterior perforated substance, each enclosed in white fascicles largely from the fornix. Internally are a number of nuclei (vide infra).

The tuber cinereum, between the mamillary bodies and the optic chiasma, is a convex mass of grey matter; its ventral aspect is not completely smooth but presents a series of eminences with intervening grooves of varying depth. Behind the optic chiasma a median, conical, hollow **infundibulum** depends ventrally to the solid posterior lobe of the hypophysis cerebri (p. 1451). Around the infundibular base is a **median eminence**, superficially marked by a shallow *tubero-infundibular sulcus* along their junction. The tuber cinereum also has a pair of **lateral eminences** and a median **postinfundibular eminence**; posterolaterally, on each side, is a row of two or three minute projections between the tuber cinereum and mamillary bodies.

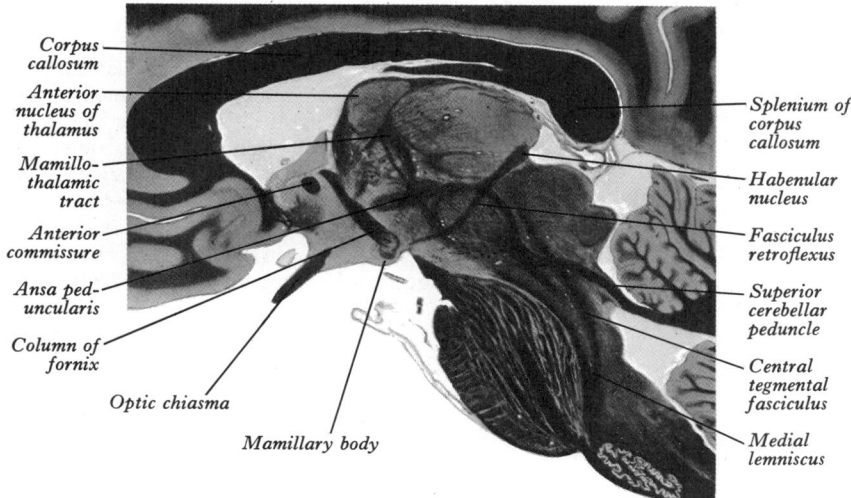

Corpus callosum
Anterior nucleus of thalamus
Mamillo-thalamic tract
Anterior commissure
Ansa peduncularis
Column of fornix
Optic chiasma
Mamillary body
Splenium of corpus callosum
Habenular nucleus
Fasciculus retroflexus
Superior cerebellar peduncle
Central tegmental fasciculus
Medial lemniscus

7.131 Sagittal section of the right cerebral hemisphere through the mamillary body, viewed from the left. Myelin stain. The fasciculus retroflexus is seen crossing the medial side of the red nucleus, which is surrounded by a capsule of white fibres derived chiefly from the superior peduncle. After Foix & Nicolesco 1925.

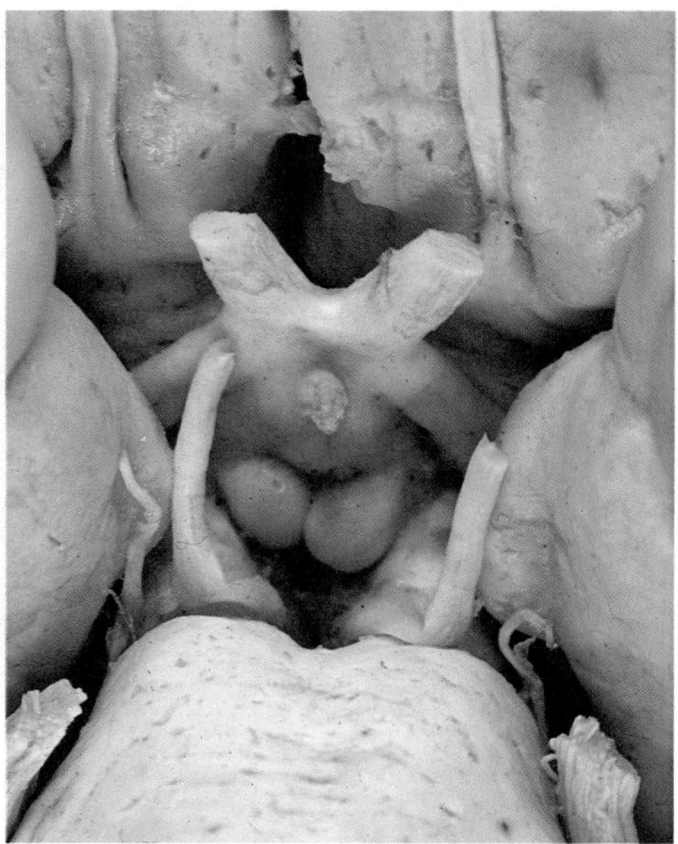

7.132 The interpeduncular fossa and surrounding structures. Note from above downwards: (1) the gyri recti of the frontal lobes and the olfactory tracts, trigones and anterior perforated substances; (2) the optic nerves, chiasma and diverging optic tracts which disappear beneath the medial borders of the unci hippocampi; (3) the tuber cinereum with the attached superior part of the infundibular stem; (4) the mamillary bodies; (5) the deep recess between the mamillary bodies, the diverging crura cerebri and the cranial border of the pons, the floor of which is the posterior perforated substance; (6) on each side the prominent oculomotor nerves, more laterally the slender trochlear nerves and, bordering the pons, the thick trigeminal nerves are visible. Dissection by E L Rees, photography by Kevin Fitzpatrick, both of the Dept. of Anatomy, Guy's Hospital Medical School, London.

HYPOTHALMIC DIVISIONS AND NUCLEI

Most of the hypothalamus contains neuronal groups classified by phylogenetic, developmental, cytoarchitectonic, synaptic and histochemical data into nuclei with specific names, some more clearly defined than others. Several tracts are also visible, the remaining regions being filled by complex connections: afferent, efferent, intrinsic or in transit, some established, others uncertain. This small but complex region is difficult to study, and different classificatory criteria have been applied, with varied terminology. Only major nuclei and connections need be noted. For

details consult Le Gros Clark et al (1938), Crosby & Woodburne (1940), Harris (1955), Haymaker et al (1959), Szentágothai et al (1962) and Reichlin et al (1978).

The hypothalamus may be viewed as a median anteroposterior sequence of *supraoptic, infundibulotuberal* and *mamillary* zones, or as having *lateral* and *medial zones* on each side separated by a paramedian plane, which includes the column of the fornix, the mamillothalamic tract and fasciculus retroflexus. Some prefer to subdivide the medial region into a subependymal *periventricular zone* and a thicker *intermediate* (medial) *zone*. The main nuclear groups are illustrated in 7.133; a more complete list is given

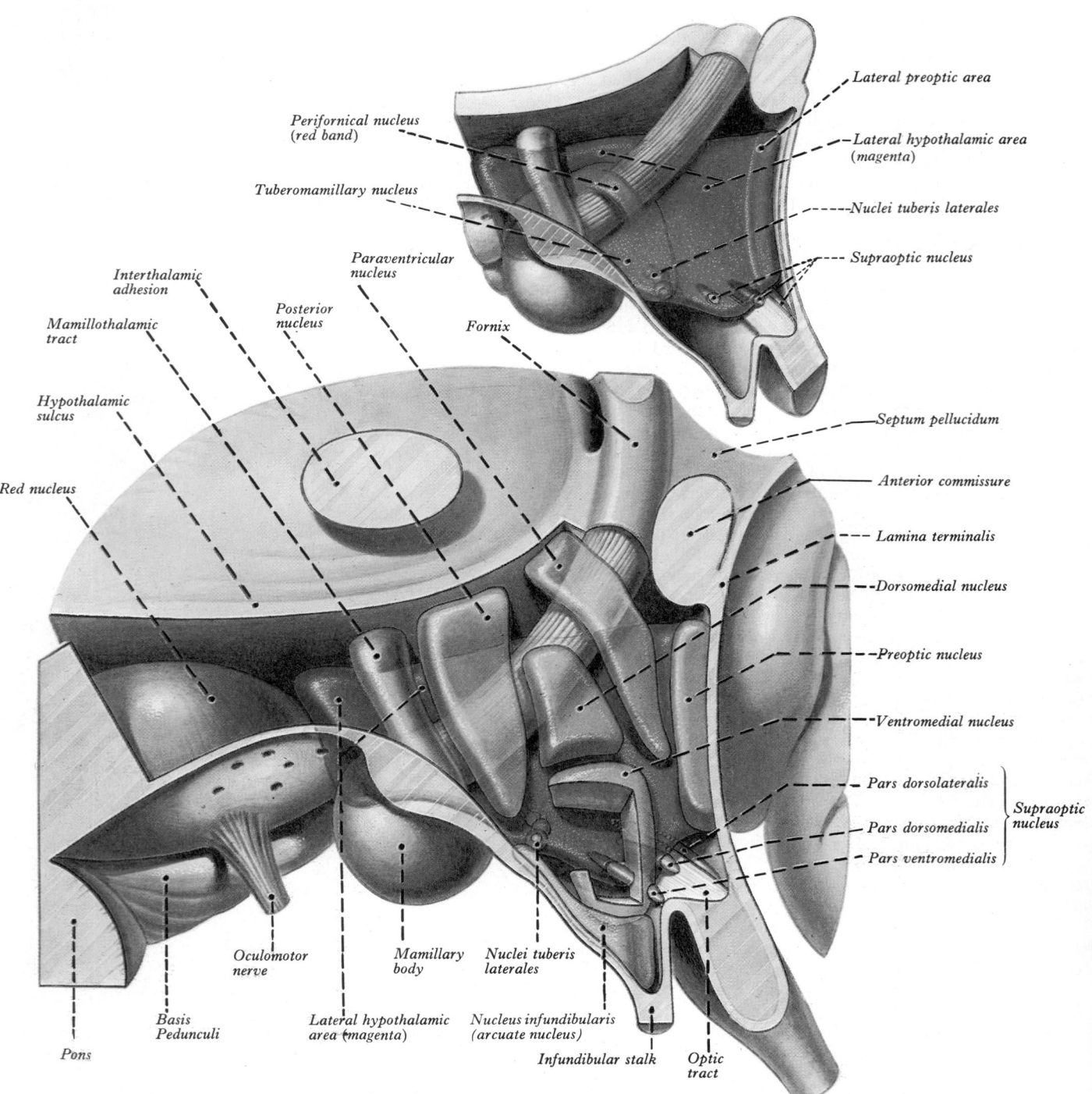

7.133 Schemata of the hypothalamic region of the left cerebral hemisphere from the medial aspect to display the major hypothalamic nuclei. In the upper diagram the medially placed nuclear groups have been removed, whilst in the lower diagram both lateral and medial groups are included. Lateral to the fornix and the mamillothalamic tract is the lateral hypothalamic region (magenta), in which the tuberomamillary nucleus is situated. Situated rostrally in this area is the lateral preoptic nucleus. Surrounding the fornix is the perifornical nucleus (red band),

which joins the lateral hypothalamic area with the posterior hypothalamic nucleus. The medially situated nuclei (yellow) fill in much of the region between the mamillothalamic tract and the lamina terminalis, but also project caudal to the tract. The nuclei tuberis laterales (blue) are situated ventrally, largely in the lateral hypothalamic area. The supraoptic nucleus (green) consists of three parts. See text for further description. From Nauta & Haymaker 1969, by courtesy of the authors and the publishers.

below. Many are small, poorly localized but sometimes subdivided further. Each zone may be considered an anteroposterior nuclear sequence, with some dorsoventral overlap; preoptic grey matter is common to all zones.

The **periventricular zone** consists of: (1) part of the *preoptic* nucleus, (2) a small *suprachiasmatic* nucleus (vide infra), (3) a large *paraventricular* nucleus, (4) the *infundibular* nucleus, (5) the *posterior* nucleus. In general, they are dispersed between the periventricular fibres (pp. 1001, 1010) with which some are connected. Most prominent and functionally best understood is the paraventricular nucleus (vide infra).

The **intermediate zone** (medial zone) consists of: (1) part of the *preoptic* nucleus, (2) the *anterior* nucleus, (3) the *dorsomedial* nucleus, (4) the *ventromedial* nucleus, (5) small *premamillary* nuclei.

The **lateral zone** consists of: (1) part of the *preoptic* nucleus, (2) the *supraoptic* nucleus, (3) the *lateral* nucleus, (4) the *tuberomamillary* nucleus, (5) *lateral tuberal* nuclei.

The mamillary body, with its main *medial* and *lateral mamillary* nuclei and adjacent smaller groups, is between the intermediate and lateral zones, but overlaps both.

These nuclei will not be individually described here; for details see the references quoted. However, some outstanding features must be noted. The preoptic, anterior, dorsomedial and ventromedial nuclei, and part of the posterior, are mainly of small or medium neurons and are ill-defined in man. Larger neurons scattered amongst smaller ones typify the rest of the posterior and much of the lateral nucleus. Lateral tuberal nuclei, though small, are well-defined, encapsulated by fine fibres; each lies in one of the posterolateral projections between the tuber cinereum and mamillary body. The large size of the medial mamillary nucleus is a human characteristic. The supraoptic nucleus, dorsal to the lateral part of the optic chiasma, and the large subependymal paraventricular nucleus are well vascularized and have common distinctive cytological features; they are partly composed of large, bipolar or multipolar neurons, sometimes multinucleate and Nissl-staining more densely than neurons in adjacent nuclei. Their cytoplasm contains granules of neurosecretory material (p. 1010).

Modern tracing techniques, initially employing formaldehyde histofluorescence (p. 992) and now replaced by the extensive use of immunofluorescence (p. 993), have vastly stimulated neuroendocrinology. An immense literature can only be briefly noted. Useful references are included with: neuro- and adenohypophysis (pp. 1451–1454), the latter's circulatory system (p. 1451), the 'diffuse neuro-endocrine system' and the APUD concept (p. 1376), the habenulopineal system and neuro-endocrine transduction (p. 1005). See also Reichlin et al (1978) and Hokfelt and co-workers (1978).

The neuronal somata, axons and the fibre terminals in the preoptico-hypothalamic region contain in toto (and in many subregions) 25 separate neuromediators, each (within technical limits) having been localized ('mapped') and each with a large and increasing literature. These include: angiotensin II, dynorphin, enkephalin, endorphin and MSH, αMSH, ACTH, vasopressin and oxytocin, TRH, somatostatin, LHRH, CRF, neurotensin, CCK, VIP, substance P, glycine, glutamate and aspartate, GABA, histamine, serotonin, adrenalin, noradrenalin, dopamine and acetyl choline. Clearly, it is not possible to describe their locations individually in this volume. (For review see Nieuwenhuys 1986.)

CONNECTIONS OF THE HYPOTHALAMUS

As noted, some hypothalamic connections are discrete fascicles, others diffuse; they are difficult to study and hence uncertain in origin and termination. Broadly, *afferent* hypothalamic axons include the ascending visceral and somatic sensory connections, olfactory paths and numerous tracts from the midbrain, diencephalon, 'limbic' structures and neocortex. *Efferent* paths are reciprocal to many such sources; in particular efferents reach the origins of *peripheral autonomic nerve fibres*. The hypothalamus also controls the secretory cycles of *hypophysis* and *epiphysis cerebri* and, through these, the *endocrine system*. More precisely, different hypothalamic parts connect with: (1) the tegmentum and

periaqueductal grey of the midbrain, (2) the subthalamic nuclei and indirectly the globus pallidus, (3) some thalamic nuclei, (4) the hippocampal formation, (5) the anterior olfactory areas, (6) the amygdaloid nuclear complex, (7) the septal areas, (8) the prefrontal cerebral cortex, (9) the hypophysis cerebri, (10) direct retinal fascicles (vide infra), (11) possibly accessions from the superior cerebellar peduncle and globus pallidus, (12) numerous afferents from pontine and medullary reticular nuclei (pp. 988–996). Some will be briefly outlined.

Afferent Connections

Ascending paths include collaterals of lemniscal somatic afferents but are largely polysynaptic routes for visceral and gustatory impulses from the spinal cord and brain stem, including the **mamillary peduncle**, a convergence of fibres in the mid-brain tegmentum into a discrete fasciculus ending in the mamillary body; afferents also ascend in the **dorsal longitudinal fasciculus** (p. 986 and vide infra) and from the **brain-stem reticular formation**, already described (pp. 991, 996 and below). *Direct* projections reach the hypothalamus from ipsilateral and contralateral subthalamic nuclei and zonae incertae, the latter via supraoptic decussations. Connections pass from the thalamic nucleus medialis dorsalis to many hypothalamic nuclei by **periventricular fibres** previously described (p. 1009), probably also via the medial part of the **inferior thalamic peduncle**.

Certain forebrain structures, many of which were once regarded as mainly olfactory but now recognized as much wider in function, are often grouped as the **limbic system** (p. 1028). These include the hippocampal formation, amygdaloid and septal complexes, piriform lobe and the adjacent regions of the neocortex. Limbic structures have prominent paths to the hypothalamus: the *fornix, stria terminalis, medial forebrain bundle* and *ventral amygdalofugal tracts*.

The **fornix** is complex in structure (p. 1039); it contains diverse commissural and other connections. It is the main outflow from the hippocampal neuronal laminae; as these fibres curve ventrally towards the anterior commissure they are joined by fascicles from the cingulate gyrus and indusium griseum and the septal areas. Continuing ventroposteriorly the fornix divides around the anterior commissure into pre- and postcommissural parts. The *precommissural fornix* distributes in part to preoptic hypothalamic regions; in its descent the *postcommissural (column of the) fornix* connects with dorsal, lateral and periventricular hypothalamic regions before its main termination in the mamillary nuclei. The fornix is further described on p. 1039; its fibres, in diverse bundles, carry at least nine different neuromediators.

Amygdaloid nuclei project to the preoptic area and most hypothalamic nuclei anterior to the mamillary bodies. These **amygdalohypothalamic fibres** have two routes: the complex curved *stria terminalis* and a *ventral amygdalofugal route*; the latter is a diffuse array of fibres, inferior to the lentiform nucleus, which traverse the anterior perforated substance to reach the hypothalamus. The amygdaloid complex and stria terminalis are described below (p. 1034).

The **medial forebrain bundle**, another complex fasciculus, is reduced in human brains, but is still the principal longitudinal hypothalamic fascicle (7.155). With ascending, descending, transverse but mainly longitudinal components, it connects anterior olfactory, hypothalamic and mid-brain tegmental regions and traverses the lateral hypothalamic zone, exchanging small fascicles throughout its course. Descending afferents include numerous *septohypothalamic fibres* from the septal area (p. 1035), *olfactohypothalamic fibres* from the piriform cortex and *corticohypothalamic fibres* from the orbitofrontal cortex. Visceral afferents *ascend* in the posterior part of the medial forebrain bundle, which also contains *descending* fibres which are continuations beyond the main anterior terminations of both the fornix and stria terminalis. Some of these connections are known only in subprimate brains; but their existence in human brains is almost certain. Some of the bundle's connections have been partly clarified by newer fibre-tracing and immunofluorescence techniques (p. 876 and vide supra). Fascicles from the *ventral tegmental ascending cholinergic pathway* are in or adjoin it; most if not all hypothalamic nuclei thus receive a cholinergic input. The *ventral* *tegmental noradrenergic bundle* does the same, with branches to dorsomedial, periventricular, paraventricular, supraoptic and lateral hypothalamic nuclei. A *mesolimbic dopaminergic system* (p. 994) traverses the hypothalamus in the medial forebrain bundle: dopaminergic group A12, in the infundibular (arcuate) nucleus, has terminals in the median eminence and infundibulum, group A11 innervates the medial hypothalamic nuclei, and groups A13 and 14 supply the dorsal and rostral hypothalamic nuclei (p. 994 and **8.224**). The medial forebrain bundle in a recent review (Nieuwenhuys 1986) has been shown to have a fibre bundle containing 15 different neuromediators (also confirming those mentioned above). These are: acetylcholine, dopamine, noradrenalin, adrenalin, serotonin, histamine, substance P, VIP, CCK, neurotensin, CRF, LHRH, somatostatin, ACTH and enkephalin.

Corticohypothalamic fibres reach the hypothalamus from much of the prefrontal cortex by direct and indirect routes, in uncertain relative numbers. *Indirect* cortical fibres converge on the thalamic nucleus medialis dorsalis and relay to the hypothalamus by periventricular routes in some animals; in others *direct* corticohypothalamic fibres may end in lateral, dorsomedial, mamillary and posterior hypothalamic nuclei; some of these connections are questioned by some authors. (For detailed reviews see Nauta & Haymaker 1969, Reichlin et al 1978.)

Efferent Connections

Hypothalamic efferents include: (1) reciprocal paths to the limbic system, (2) descending polysynaptic paths to autonomic and somatic motor neurons, (3) neurovascular links with the hypophysis cerebri. Septal areas and the amygdaloid complex have reciprocal hypothalamic connections along the paths described above. The medial mamillary nucleus produces a large ascending fascicle which diverges into: (a) a *mamillothalamic tract* (7.148) which continues to ascend and ends in all parts of the anterior thalamic nuclei, whence massive projections radiate to the cingulate gyrus (pp. 1003, 1062); and (b) a *mamillotegmental tract*, which curves inferiorly into the midbrain ventral to the medial longitudinal fasciculus and is distributed to tegmental reticular nuclei. Small *mamillosubthalamic fascicles* reach the subthalamic prerubral field, their destinations uncertain. Descending hypothalamic fibres also reach the mesencephalic tegmentum in the *dorsal longitudinal fasciculus* and inferior extension of the *medial forebrain bundle*.

Connections of the hypothalamus with the hypophysis cerebri. In summary these are: (1) *nerve fibres* from hypothalamic neurons ending throughout the median eminence, infundibulum and *posterior lobe* and (2) long and short *portal blood vessels* between sinusoids in the median eminence and infundibulum and plexuses in the hypophysial *anterior lobe* (p. 1451 and **8.224, 229**). Certain points need emphasis: the anterior lobe receives no major nerve supply; almost all its blood comes from hypothalamohypophysial portal vessels; direct observation shows that the predominant direction of flow is from tufts of sinusoids in the median eminence and infundibulum via portal vessels to sinusoids between glandular cells in the anterior lobe. However (p. 1452), *flow reversal* may occur, with widespread neuroendocrine effects.

Neurosecretion is involved in the production and release of posterior lobar *oxytocin* and *vasopressin* and their related proteins: transport *neurophysins* in neurovascular control of the glandular cells of the anterior lobe. Hypothalamic neurons, whose axons end in the median eminence, upper infundibulum or posterior lobe, have their somata in the supraoptic, paraventicular, infundibular nuclei and partly in the ventromedial nuclei.

The supraoptic nucleus, curved over the lateral part of the optic chiasma, is sometimes considered as having three or more subsidiaries; it has a uniform population of large neurons. **The paraventricular nucleus** is a visible bulge into the third ventricle, extending from the hypothalamic sulcus across the medial aspect of the column of the fornix; its ventrolateral angle reaches towards the supraoptic nucleus, whose neurons some of its own resemble, with intermediate and small ones of varying shape. **The infundibular (arcuate) nucleus** is median in the postinfundibular part of the tuber cinereum, extending into the median

eminence and almost encircling the infundibular base; but it is deficient above, where the infundibulum adjoins the median part of the optic chiasma. No glial layer intervenes between the nucleus and the tanycytic lining of the *infundibular recess*. (For the neuroendocrine role of tanycytes consult Knowles 1974, Joseph & Knigge 1978, and pp. 894, 1453.) Its numerous neurons are all small and round in coronal, oval or fusiform in sagittal section.

Thus neurons projecting into or via the infundibulum form two groups: (a) **magnocellular**, including the large and intermediate neurons in the paraventricular nucleus, large cells in the supraoptic, with small groups or rows of internuclear neurons; (b) **parvocellular**, including small infundibular neurons and some in various adjacent nuclei (Nieuwenhuys 1986). Large neurons contain neurosecretory material (vide infra) but both groups are true neurons, each with a soma, dendrites, axon and collateral branches and the usual neuronal organelles. They have axodendritic and axosomatic synapses from unknown sources. Their axons conduct impulses and terminate in the posterior hypophysial lobe and infundibulum at end-bulbs containing mitochondria and vesicles of about 40–60 nm diameter, many being applied to the capillary basement membranes without intervening neuroglia. Neural connections exist to influence the activities of these nuclei, probably from other hypothalamic nuclei and via paths converging on the hypothalamus from other sources, the precise connections remaining uncertain; likewise, the functions of impulses in these neurons and of presynaptic vesicles are not clear: no clear relation to neurosecretion has been shown.

The terminology of neuro-endocrine paths has changed with the increasing knowledge, resulting in an array of imprecise terms. For example *supra-opticohypophysial tract*, often used for *all* fibres entering the infundibulum of whatever origin or destination, is now often restricted to axons from supraoptic and paraventricular nuclei; since these differ in cytology and connections, some authors differentiate *supra-opticohypophysial* and *paraventriculohypophysial tracts*. But many of their fibres reach the infundibulum, not the posterior lobe; *tuberohypophysial* or more precisely **tubero-infundibular tract** denotes fibres of small neurons in infundibular and related nuclei of the ventromedial hypothalamus (vide infra) which end in the infundibulum.

Neuronal somata and axons of **magnocellular neurosecretory** paths from the supraoptic and paraventricular nuclei, internuclear neurons and (some contend) occasional neurons in the suprachiasmatic nucleus, all contain what were first termed *colloid droplets*. These vary in size, sometimes appear between somata and fibres and have an affinity for Gomori's chrome-alum haematoxylin stain, which reveals, in fine axons and branches, rows of stained masses distending them like rosaries. Terminals irregularly distended by masses of neurosecretory material are termed *Herring bodies*. Electron microscopy shows in them membrane-bound aggregations of dense granules from 200–300 nm in diameter. Transection of the infundibulum, hypophysial transplantation, the culture of neurohypophysial explants, autoradiography with labelled cysteine, fractionation and extraction with biological assay or chemical analysis, the effects of dehydration, hydration, drugs and stress, chemical and electrical stimulation and ablation of focal points, all have led to agreement that neurosecretory material is synthesized in the large neuronal somata of supraoptic, paraventricular and adjacent neurons, near their endoplasmic reticulum; also that it is then elaborated in Golgi complexes, transported along axons and branches, released at terminals and absorbed into adjacent capillaries. While *both* supraoptic and paraventricular nuclei are involved in the production of vasopressin *and* oxytocin, vasopressin is *predominantly* a product of the supraoptic oxytocin of the paraventricular nucleus. By immunofluorescence the precise three-dimensional distribution of neurons containing oxytocin and vasopressin has been analysed in several mammals; marked regional and specific differences exist. Many questions that persisted for a time (Donovan 1970, Knowles 1974) are now answered, e.g. the exact chemical constitution of neurosecretory material is now determined; the chemically related hormones, oxytocin and vasopressin (nonapeptides), are transported in an inactive form and conjugated with a specific low molecular weight protein, a *neurophysin.* The rate and mode of transport, presumably by fast axoplasmic flow, however, remain speculative. Details of exocytotic release at axonal terminals are few, but calcium ions and the local synaptic release of acetylcholine may be involved. The vascular relations of supraoptic and paraventricular nuclei are discussed below. (For reviews see Defendini & Zimmerman 1978, Silverman & Zimmerman 1983.)

Neurosecretory paths may also aid the control of anterior hypophysial secretion. Groups of hypothalamic cells produce *peptidergic-releasing hormones,* transported along their axons and discharged into the portal system's upper radicles; the portal vessels carry them to the anterior lobe's dense capillary plexus to influence the appropriate glandular cells. *Releasing hormones* is an inappropriate term; some *inhibit* hormonal release from anterior lobar cells while others stimulate synthesis and promote release. They include *releasing hormones* for corticotropin, luteinizing hormone, follicle-stimulating hormones, thyrotropin and somatotropin and *release-inhibiting hormones* for prolactin, melanocyte-stimulating hormone and somatostatin. The **parvocellular neurosecretory pathway** of the *'tubero-infundibular tract'* is proposed as the main route for the synthesis of releasing hormones and transmission to portal blood. Immunofluorescence has localized some of these peptides (p. 891, table on p. 1451, Renaud 1978). For example: corticotropin-releasing factor is localized to small neurons in paraventricular, anterior and dorsal hypothalamic nuclei, thyrotropin-releasing factor to nuclei dorsomedialis and ventromedialis, luteinizing hormone-releasing factor to nucleus infundibularis (arcuatus), somatotropin-releasing factor to anterior and posterior hypothalamic nuclei and somatostatin to certain periventricular nuclei. Some tubero-infundibular fibres also have high concentrations of acetylcholine, others of catecholamines, in accord with the ultrastructure of their end-bulbs (p. 1451), some containing small clear synaptic vesicles, others larger dense-cored vesicles and some a mixture. There is no evidence of large, electron-dense, membrane-bound granules typical of the magnocellular path. The tubero-infundibular tract may carry releasing hormones in dispersed, active form and also catecholamines or acetylcholine. For such widespread origins *tubero-infundibular* is too narrow a term. Already noted (p. 1010) it is a possible role of tubero-infundibular dopaminergic (and other monoaminergic) paths to regulate synthesis, transport and exocytosis of releasing or release-inhibiting factors and the activities of tancytes (p. 1010 and **8.225**). According to Renaud (1978) tubero-infundibular neurosecretory parvocellular neurons have *afferents* from the amygdala, preoptic nuclei and hippocampus, their *efferent* axons returning collaterals to these structures and to the medial nuclei of the *dorsal thalamus* and other *hypothalamic* and numerous *brain-stem reticular* nuclei, in addition to their better known *infundibular* terminals.

Many axonal terminals of the magnocellular path end in the proximal infundibulum and are perhaps also involved in control of the anterior hypophysial lobe. Thus, magnocellular and parvocellular paths may not be as functionally independent as is sometimes stated.

FUNCTIONS OF THE HYPOTHALAMUS

Hypothalamic lesions have long been linked with widespread and bizarre endocrine syndromes and with metabolic, visceral and motor disturbances; it has been widely assumed that these result from interruption of the paths mediating such 'functions', which, though largely independent, are structurally associated in the hypothalamic region. Though the endocrine role of the hypophysis cerebri was emerging in the late nineteenth and early twentieth centuries (Cushing 1912), the *interrelation* of hypophysis and nervous system has been much more recently established (Harris 1955) with the emergence of *neuroendocrinology* (Scharrer & Scharrer 1963, Heller & Clark 1962, Gabe 1966, Donovan 1970, Knowles 1974, Reichlin et al 1978), which has rapidly progressed, enlarging the accepted role of the hypothalamus as the 'head ganglion' of the autonomic nervous system (Sherrington 1947). Research has confirmed in greater detail the hypothalamic effects on the *endocrine system* and *lower autonomic centres*, also emphasizing the hypothalamic action's dependence on afferent channels, nervous and vascular, interlocking it structurally and functionally with the higher regions, including the complex of structures in the *limbic lobe* (p. 1028) and prefrontal cortex.

One approach to the general study of nervous systems, mentioned earlier (p. 860), depends on their control of *homeostatic cycles* to preserve individual and species in a changeable environment. The mammalian frontal cortex, limbic system, hypothalamus, lower brain stem and spinal cord are regarded as a hierarchy in such cycles, mediated peripherally by the autonomic nervous and endocrine systems through their effector patterns. This concept is germane to widespread use of the terms *lower, intermediate* and *higher* 'centres' for visceral and somatic locomotor controls. 'Centre' is a product of methods of analysing central nervous activities by focal stimulation or destruction of localized topographical targets. Such manoeuvres are then combined with other observations: cytological, biochemical, metabolic, developmental or behavioural, electrophysiological and so forth. Focal stimulation or obliteration causes dramatic changes in certain responses; some regions exhibit apparently spontaneous, sometimes rhythmic electrical changes when isolated; but this does not establish them as autonomous 'centres' controlling particular activities. Most evidence indicates that 'centres' are best regarded as *nodal regions* which, by chemical constitution, structure and intrinsic and extrinsic connections, nervous or vascular, are essential foci in a range of responses following the appropriate stimulus. Hence other regions whose afferents converge on the 'centre', though often less dramatic in their effect on responses when individually disturbed, are collectively as essential as the so-called centre itself. Especially in a confined region such as the hypothalamus, with numerous converging and diverging connections and ill-defined neuronal groups (often with diffuse connections), the putative 'centres' of different investigators may overlap or coincide. Therefore many prefer to discuss the effects of stimulation or ablation of general hypothalamic *zones*, e.g. lateral, preoptic, mamillary, etc., rather than ascribing functions to specific nuclei. Exceptions are the accepted neurosecretory nuclei with their specific products although afferent neurovascular connections are not fully clarified even in these.

Experimental transections generally show that as progressively higher levels are included with the spinal cord below the section, more effective homeostasis is retained (Anand 1970). After transection at upper mesencephalic levels, minor reflex readjustments of cardiovascular, respiratory and alimentary systems survive but are not integrated, nor is normal temperature maintained. After transection above the hypothalamus, separating it from the limbic system but leaving the connections between the hypothalamus, hypophysis, brain stem and spinal cord, a very different picture emerges: effective homeostasis is maintained in a moderate range of conditions; visceral and endocrine control integrates the patterns of responses; 'innate drives' and 'motivated behaviour' are preserved, including feeding, drinking, apparent satiation, and copulatory responses; but homeostasis fails if environmental stresses exceed a certain range e.g. persistently high or low temperatures. Motivated behaviour also becomes abnormal under stress; animals may attack, try to eat, drink or copulate with inappropriate objects. But if the connections between the limbic system (p. 1028) and the hypothalamus survive and the neocortex only is cut off, normal homeostasis is preserved even in a wide range of adverse conditions.

An unwieldy literature has amassed on the functions of the hypothalamus and aspects of neuroendocrinology, unprofitable to detail here; the references quoted here contain extensive bibliographies. To summarize, much experimentation has revealed an intimate involvement of the hypothalamus in the following activities.

(1) **Endocrine control** is effected by releasing or release-inhibiting factors which influence the production of thyrotropin, corticotropin, somatotropin, somatostatin, prolactin, luteinizing hormone, follicle-stimulating hormone and melanocyte-stimulation hormone by cells of the *anterior hypophysis.* Some of these affect target cells in general tissues directly, others through a second endocrine organ, e.g. the thyroid gland, adrenal cortex and gonads, in the sequences summarized in **7.134**. Neural paths

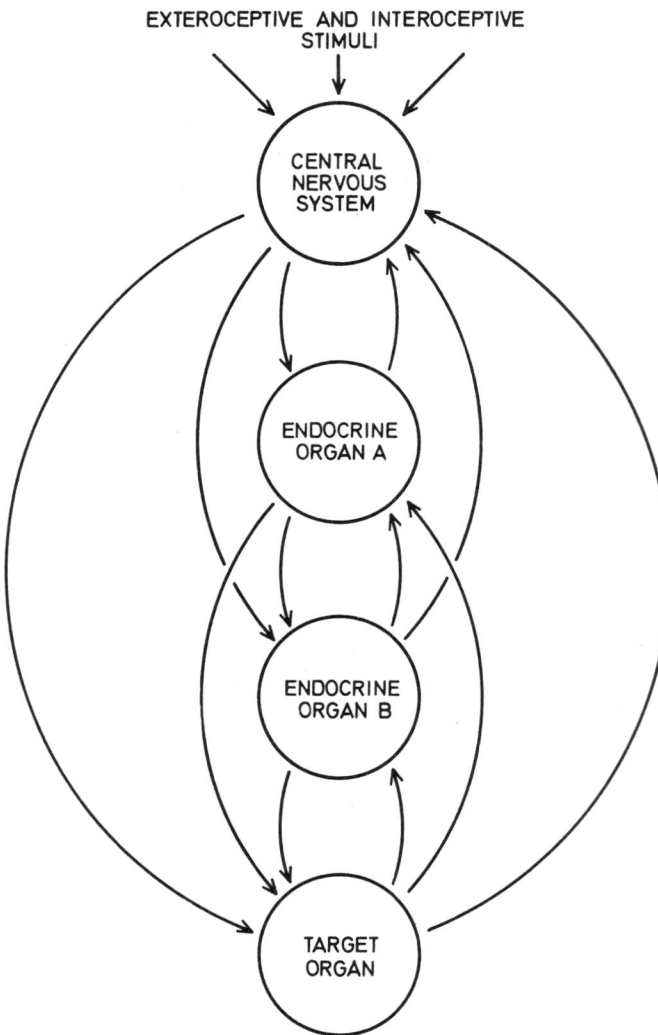

EXTEROCEPTIVE AND INTEROCEPTIVE STIMULI

CENTRAL NERVOUS SYSTEM

ENDOCRINE ORGAN A

ENDOCRINE ORGAN B

TARGET ORGAN

7.134 A flow diagram showing a stimulus pattern impinging on the central nervous system and the possible interactions of the latter with endocrine organs A and B and a target tissue. Note the different orders of feed-forward and feed-back pathways envisaged. From Scharrer & Scharrer 1963, by courtesy of the authors and publishers.

converge to be integrated in the hypothalamus (Nieuwenhuys 1986), modifying the production of all its factors (p. 1011), which in turn influence endocrine glands and, ultimately, target tissues. Each intermediate link may be modified by negative or positive feed-back loops depending on the circulatory levels of secretory products of other links in the chain (Haymaker et al 1969, Donovan 1970, Bargmann & Scharrer 1970, Knowles 1974, Reichlin et al 1978).

(2) **Neurosecretion** of oxytocin and vasopressin (antidiuretic hormone, ADH) is directly into blood in the infundibulum and hypophysial posterior lobe. Vasopressin's effect on vascular non-striated muscle remains uncertain; but it has a powerful effect on renal reabsorption of water. Oxytocin causes contraction of the uterine and mammary muscle, with roles in coitus, parturition and suckling. Stimulation of the nipples and genitalia increases oxytocin release. Vasopressin's rate of release is partly determined by the tonicity and osmotic pressure of the blood traversing *osmoreceptor sites* in the hypothalamus (Verney 1947) and peripheral circulation, probably in the hepatic portal vein (Haberich 1968), from which the impulses generated reach the hypothalamus. Reduced blood volume is believed to stimulate *volume receptors* in the great veins and atrial walls, increasing the release of vasopressin (Gauer 1968); pain, emotional states and stress also modify its rate of release.

(3) **The general autonomic effects** of anterior hypothalamic zones, mediating general parasympathetic activity, and of the posterior zones, mediating sympathetic activity, have been long suspected. While a preponderance of one or other effect in these

zones is undoubted, interaction and overlap occurs; a rigid distinction between parasympathetic and sympathetic 'centres' cannot be maintained. Focal hypothalamic stimulation and ablation intimately affect *cardiovascular*, *respiratory* and *alimentary* control, often causing profound changes in heart rate, cardiac output, vasomotor tone, peripheral resistance, blood pressure, differential flow in organs, the frequency and depth of respiration, motility and secretion in the stomach and intestines.

(4) **Temperature regulation.** In all homeothermic animals production and loss of heat are balanced, the hypothalamus providing a central regulator. Raised temperature is reduced by vasodilatation with increased cutaneous flow, sweating, panting and depressed production of heat; at lower temperatures the reactions are reversed, with shivering and increased thyroid activity. Signals of body temperature reach the hypothalamus from peripheral *thermo-detectors* and also from intrinsic hypothalamic thermo-detector neurons responding to temperature changes in the blood of the hypothalamic circulation. Other hypothalamic neurons react specifically to blood-borne viruses, toxins, pyrogenic drugs, etc. The signals converge to regions of the hypothalamus which evoke the appropriate autonomic, endocrine and muscular responses. An '*anti-rise*' region is hypothecated in the anterior hypothalamus to control vasodilatation, sweating, panting and possibly to depress the production of heat, the posterior hypothalamus containing an '*anti-drop*' region controlling vasoconstriction, shivering and other means of producing heat. The sites and even the distinction of such a *dual hypothalamic thermostat* are obscure. Some investigators (Myers 1969) propose a *dual chemical 'thermostat'* between the inflow of signals and 'antirise' and 'antidrop' regions, consisting of neurons containing 5-hydroxytryptamine or noradrenalin (possibly dopamine). Warmer blood, impulses from heat receptors, antipyretic substances or anaesthetics are all supposed to release noradrenalin (or dopamine) to stimulate 'anti-rise' mechanisms; cooler blood, impulses from cold receptors and pyrogenic substances stimulate the release of 5-hydroxytryptamine, activating 'anti-drop' mechanisms.

(5) **Regulation of the intake of food and water.** Experiments show (Anand & Brobeck 1951, Anand 1961) that regions with opposing actions exist in the medial and lateral hypothalamic zones: ablation of the medial zone promotes over-eating (*hyperphagia*), leading to gross obesity, while ablation of the lateral zone promotes *hypophagia* or complete *aphagia*, with death from starvation. Conversely, stimulation of the medial zone reduces eating, lateral stimulation increases and prolongs it. Hence, a lateral '*hunger*' or '*feeding centre*' balanced against a medial '*satiety centre*' have been proposed. Neurons sensitive to the glucose level in circulating blood exist in the medial zone, the supposed 'satiety centre'.

Water balance and haemal osmolality are partly controlled by an '*osmoreceptor/volume receptor—vasopressin—renal system*' described previously, which determines the rate of water loss in urine. Allied to this is a '*thirst*' or '*drinking centre*' in the lateral zone, regulating water intake; stimulation of this causes copious drinking and, if stimulation persists, gross hyperhydration.

(6) **Sexual behaviour and reproduction.** Through the hypophysial control of gonadotrophin production the hypothalamus controls aspects of reproduction, including gametogenesis, cyclic variations and the development and maintenance of secondary sexual features. Hypothalamic stimulation may induce receptivity in females and simple copulatory movements in males. Some hypothalamic neurons are sensitive to circulating oestrogen or testosterone.

Though elementary *drives* of hunger, thirst and reproduction may stem in part from an intact hypothalamus, full integration into behaviour requires reciprocation between the hypothalamus and the limbic system (p. 1028). Such behaviour includes searching out and procuring food, drink and mate, home-building, rearing of young, etc.

(7) **Biological clocks.** Many tissues and organs show cyclic variation in function, with a period of about 24 hours. Examples of such *circadian rhythms* include fluctuations in temperature, concentrations of several plasma constituents, the eosinophil count, adrenocortical secretory activity and renal secretion. In

some the rhythm is partly intrinsic to the organ or tissue but in many overall control is exerted by an intact hypothalamus. Lesions in other parts of the brain leave this variety of biological timing unaffected; hypothalamic lesions alone cause serious disturbances. The suprachiasmatic nucleus appears in many animals to act as an *endogenous neural pacemaker*, whose rhythmic activity may respond to changes in the level of light (p. 1459). Though it occurs throughout vertebrates (Joseph & Knigge 1978), some doubts have been expressed concerning the prominence (or even presence) of a discrete human suprachiasmatic nucleus (Defendini & Zimmerman 1978). Bilateral experimental destruction permanently interferes with circadian rhythms of, e.g. adrenocorticosterone secretion (Moore & Eichler 1972), drinking and locomotion (Stephan & Zucker 1972), sleeping and wakefulness (Ibuka & Kawamura 1975) and pineal N-acetyltransferase activity (Moore & Klein 1974). Concerning pineal activity Nishino et al (1976) have shown that activity in suprachiasmatic neurons is increased by visual stimulation and that this inhibits the sympathetic neurons supplying pinealocytes, resulting in a decrease in secretion (p. 1459). Though this is still controversial, there is experimental evidence of direct retinosuprachiasmatic connections, perhaps involved in human photic regulation of circadian rhythms. Retinal terminals in the nucleus have been demonstrated by autoradiography after intraocular injection of tritiated amino acids in mammals, from tree-shrew to macaque (Moore 1973), supporting degeneration studies (Moore & Len 1972, Printz & Hall 1974); further evidence is provided by cobalt precipitation techniques (Mason & Lincoln 1976), by electron microscopy (Wenisch 1976) and electrophysiology (Sawaki 1977). *Retinal projections* to the anterior region of the infundibular (arcuate) nucleus have also been shown to exist (Mason & Lincoln 1976). Efferents have been traced from the suprachiasmatic nucleus to ventromedial, dorsomedial and arcuate hypothalamic nuclei, suggesting that their fine control by an endogenous circadian regulator is sensitive.

Alternation of sleep and wakefulness is an outstanding circadian rhythm. Sleep has been much investigated (Jouvet 1962, 1964, 1965, Rossi 1964, Koella 1969); it is no longer regarded as passive 'switching off' or the simple depression of a cerebral activating system. *Two* main types of sleep are recognized. During the waking state, electroencephalograms show a low-amplitude undulatory pattern at about 10 cycles per second, upon which bursts of irregular 'random' activity occur. In sleep the pattern becomes *synchronized*, larger in amplitude and slower. From this *deep* or '*slow-wave sleep*' (SWS) the subject can be aroused easily; it is dreamless. It may be interrupted by periods of *paradoxical sleep* (PS), with a fast, low-amplitude waveform; muscle tone is generally reduced, especially in the neck, and *rapid eye movements* occur (REM sleep); the subject is difficult to rouse and dreams; the electroencephalographic pattern corresponds to ocular oscillation. Three or four spells of REM sleep occur each night. The general view is that the level and type of sleep depend on a *balance* between opposing systems, deployed through much of the brain stem and parts of the thalamus, hypothalamus and limbic system. On one hand is the *reticular activating, arousal system*, with extensions to the cerebral cortex, stimulation of which leads to wakefulness and desynchronization of activity. On the other is a postulated series of *hypnogenic zones* in the medulla, pons, midbrain, thalamus, hypothalamus and basal limbic structures, stimulation of which leads to synchronization of cortical activity and a *variety* of sleep-like states, differing in detailed traits. How these opposing 'hierarchies' of sleep-controlling regions interact is unknown; but the thalamohypothalamic 'centres' seem dominant and those in the medulla, pons and midbrain subordinate, as would be expected. Intact centres in the nucleus coeruleus, rich in noradrenergic and cholinergic neurons, appear important in the generation of paradoxical sleep. Overall the circadian rhythm of sleep and waking is disturbed by lesions in the anterior hypothalamus, with fluctuations in the concentration of serotonin in hypnogenic zones of the brain stem and diencephalon. Whether serotonin functions as a humoral, sleep-inducing factor or a neural transmitter in the 'hypnogenic system' is undetermined. (Further comments appear on p. 996.)

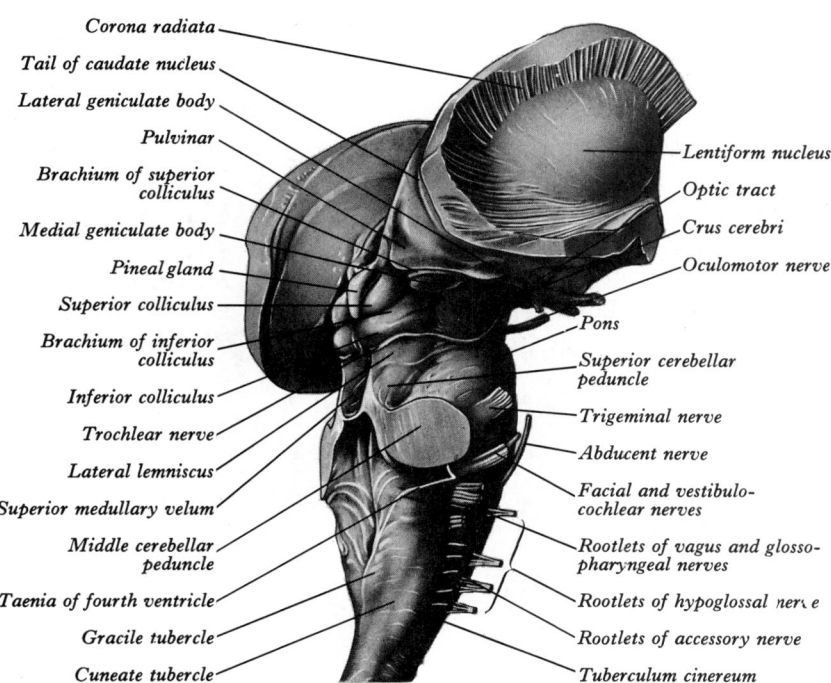

Corona radiata
Tail of caudate nucleus
Lateral geniculate body
Pulvinar
Brachium of superior colliculus
Medial geniculate body
Pineal gland
Superior colliculus
Brachium of inferior colliculus
Inferior colliculus
Trochlear nerve
Lateral lemniscus
Superior medullary velum
Middle cerebellar peduncle
Taenia of fourth ventricle
Gracile tubercle
Cuneate tubercle

Lentiform nucleus
Optic tract
Crus cerebri
Oculomotor nerve
Pons
Superior cerebellar peduncle
Trigeminal nerve
Abducent nerve
Facial and vestibulocochlear nerves
Rootlets of vagus and glossopharyngeal nerves
Rootlets of hypoglossal nerve
Rootlets of accessory nerve
Tuberculum cinereum

7.135A Posterolateral aspect of the hindbrain and midbrain, exposed by removal of the cerebellum and most of the cerebrum.

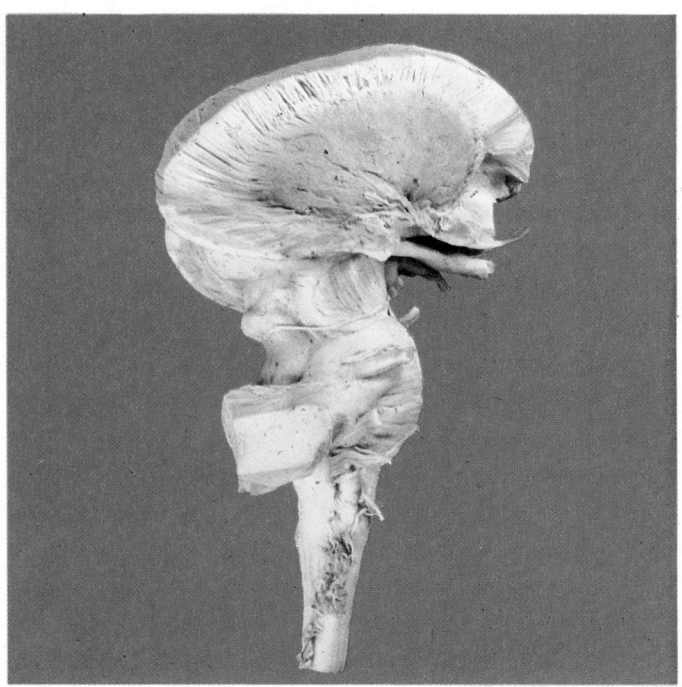

7.135B Right lateral view of the brain stem; compare with 7.135A. Dissection by E. L. Rees, photography by Kevin Fitzpatrick, both of the Dept. of Anatomy, Guy's Hospital Medical School, London.

(8) **Emotions: fear, rage, aversion, pleasure and reward.** Emotional states contain two main elements: *subjective feeling* or *affective tone* and *objective physical accompaniments*, together forming *emotional expression*. For the full integration of both during changes in the internal and external environment and other cerebral activities, essential structures are an intact hypothalamus, limbic system and prefrontal cortex. In such a complex array it is impossible to specify the functional roles to punctate subregions. Some basic information is gained by focal stimulation or ablation of the hypothalamus in intact animals by removal of the 'higher centres' and by observations during human operations. Classic experiments were carried out in decorticate dogs (Goltz 1892) and subsequently analysed in decorticate cats

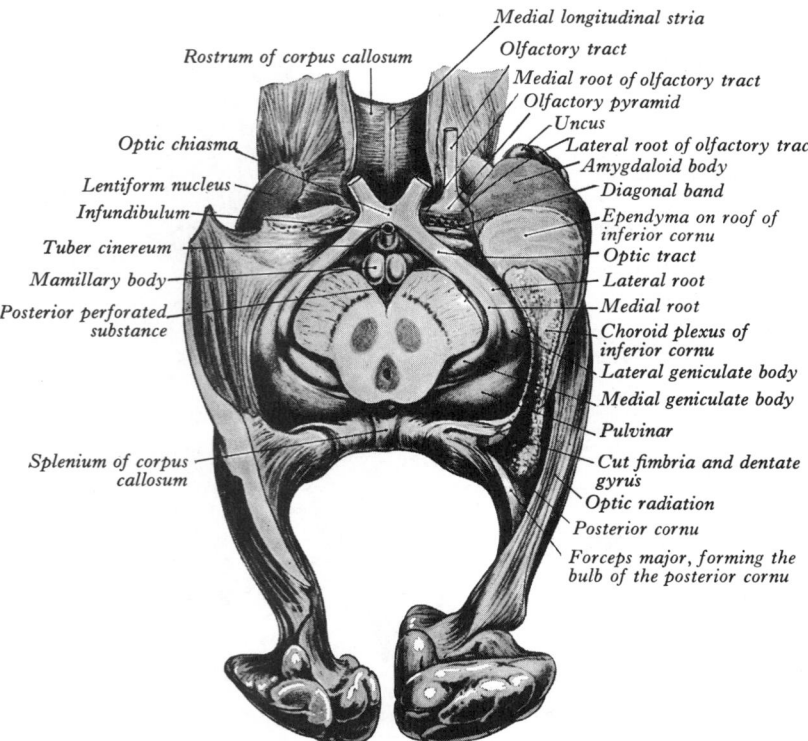

7.136 A ventral dissection of the brain showing the metathalamus and the optic tracts. On the right side of the figure the inferior horn of the ventricle is exposed. The floor has been removed but the choroid plexus is in situ and obscures most of the roof.

Labels (clockwise from top): Medial longitudinal stria · Olfactory tract · Medial root of olfactory tract · Olfactory pyramid · Uncus · Lateral root of olfactory tract · Amygdaloid body · Diagonal band · Ependyma on roof of inferior cornu · Optic tract · Lateral root · Medial root · Choroid plexus of inferior cornu · Lateral geniculate body · Medial geniculate body · Pulvinar · Cut fimbria and dentate gyrus · Optic radiation · Posterior cornu · Forceps major, forming the bulb of the posterior cornu · Splenium of corpus callosum · Posterior perforated substance · Mamillary body · Tuber cinereum · Infundibulum · Lentiform nucleus · Optic chiasma · Rostrum of corpus callosum

(Cannon & Britton 1925), where mild peripheral stimulation evoked *sham rage*, with hissing, growling, baring of claws and fangs, piloerection, arching of the back, dilatation of pupils and lashing of the tail; 'sham' because it was assumed that emotional *feelings* could not occur in decorticate animals. Sham rage was shown later to be abolished by obliteration of the posterior hypothalamus and elicited by stimulation of it in intact animals.

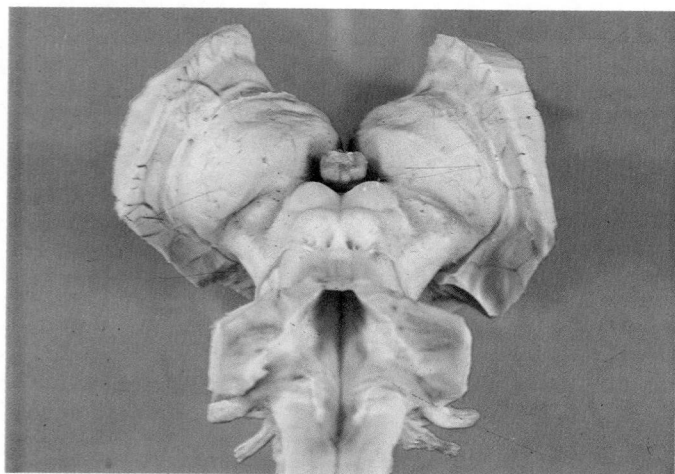

7.137 An oblique view of the dorsal aspect of the brain stem looking cranially. In the foreground is the floor of the rhomboid fossa, bounded laterally by the sectioned white matter of the cerebellum containing the dentate nuclei; the cranial recess of the fourth ventricle passes inferior to the superior medullary velum to continue into the aqueduct of the midbrain. More cranially, the emerging trochlear nerves, the colliculi, superior and inferior brachia, the medial and lateral geniculate bodies and the pineal gland may be identified. Lateral to the pineal gland on each side is the rounded pulvinar of the dorsal thalamus, skirted laterally by the curving body and tail of the caudate nucleus, whilst most laterally is the cut surface of the corona radiata. Dissection by E L Rees, photography by Kevin Fitzpatrick, both of the Dept. of Anatomy, Guy's Hospital Medical School, London.

Subsequent studies showed that stimulation or ablation of the different hypothalamic regions cause changes in emotional expression of the basic drives (thirst, hunger and reproduction), other regions being related to behavioural (and subjective) aspects of rage, fear and pleasure. The existence of broadly opposed 'centres' (*positive* and *negative* 'reward' *centres*) has been popularized. Stimulation of the former ('*pleasure centres*') leads to presumed 'pleasurable sensations' or gratification of a biological drive; in the appropriate experimental situations an animal will *self-stimulate* its hypothalamus until exhausted, ignoring food and drink even after prolonged deprivation. The main regions from which this can be elicited include parts of the catecholaminergic complex of the brain stem, the medial forebrain bundle of the hypothalamus, septal areas, orbitofrontal and parahippocampal (entorhinal) areas of the neocortex. Most effective, in earlier studies, was the medial forebrain bundle (p. 1009), in which ascending and descending fascicles, multiple in source and destination, intermingle with noradrenergic, dopaminergic, serotonergic and non-aminergic paths, which are in close relation. This field has been surveyed, with personal contributions, by Wanquier & Rolls (1976), Olds (1977) and also Routhenberg & Santos-Anderson (1977) who concluded that intact dopaminergic paths ascending from the pars compacta of the substantia nigra were essential to 'normal' patterns of self-stimulation. Stimulation in mankind often causes sensations of well-being, with sometimes a strong erotic content. Such *positive reinforcement* is used to expedite the responses of animals in experiments which demand *learning*. In this regard note the forward projection of *mesolimbic dopaminergic paths* to multiple endings in the limbic system (p. 1028), with a possible relevance to forming memory patterns. Stimulation of negative reward centres presumably causes pain or 'displeasure'; experimental animals make complex efforts to avoid repetitive stimulation.

The Metathalamus and Visual Pathway

The optic chiasma (7.136, 139), a flat, almost quadrilateral mass of decussating nerve fibres at the junction of the anterior wall and floor of the third ventricle, has anterolateral angles continuous with the optic nerves and the posterolateral angles with the optic tracts. The lamina terminalis (p. 1019), continuous with its upper surface, is crossed, dorsal to the chiasma, by the anterior communicating artery. Ventral is the diaphragma sellae, a little behind the 'optic' groove of the sphenoid bone; in about one subject in 10 the chiasma is either in the groove or altogether posterior. It is always near the hypophysis (p. 1451). Posterior to the chiasma are the tuber cinereum and infundibulum, with the third ventricle dorsal to them. Lateral is the end of the internal carotid artery and anterior perforated substance. The third ventricle's optic recess passes over its superior surface to reach the lamina terminalis (8.223A).

Most chiasmatic fibres start in the retina and arrive via the optic nerves; fibres from the nasal half of each retina, including half the macula, cross to the contralateral optic tract, fibres from temporal retinae continuing into the ipsilateral tract. Decussating fibres loop a little, backwards into their ipsilateral optic *tract before* crossing and forwards into the contralateral optic *nerve after* crossing. Macular fibres and those from an adjacent central area form a flat band occupying almost two-thirds of the central chiasma, dorsal to all peripheral decussating fibres, ventral to which are fibres from the extramacular parts of both nasal half retinae; most ventral are the nasal fibres concerned with monocular fringes of the binocular field. Dorsal to and in the chiasma are fascicles not derived from the optic nerve, forming no part of the visual pathways. Though collectively termed *supraoptic commissures*, they are really decussations. One, the *supraoptic commissure* (of Gudden), was incorrectly believed to connect the medial geniculate bodies. Its connections are not close but evidence from mammals suggests that it arises in the brain stem and spinal cord; its existence, in man, is denied by some. It is only one of at least three dorsal connections described in the optic chiasma and reputed to end in various hypothalamic and subthalamic nuclei in some animals, e.g. cats; but it is improbable that they are associated with vision. Yet there *is* convincing evidence of a retinal projection to the suprachiasmatic nucleus, the ventral part of the infundibular (arcuate) nucleus and perhaps the supraoptic nucleus in rats; though not concerned with perceptual vision, these connections mediate, by pineal projection (p. 1459), the influence of light and dark on hypothalamic, pineal and autonomic activity in circadian rhythms. Detached bands of nerve fibres have been noted to leave the optic tract and pass across the cerebral peduncle ventral to the tract; such *transverse peduncular tracts* and other *accessory optic tracts*, present in some lower mammals, have been denied in primates (but see p. 982); in this view *all* myelinated fibres of human optic tracts reach either the lateral geniculate body or the superior colliculus and pretectal nucleus (Polyak 1957). But as noted above and elsewhere (p. 1009), much evidence indicates that the suprachiasmatic and ventral part of infundibular (arcuate) nuclei and a thin lamina in the pulvinar, receive fibres from the optic tract, i.e. as *retinohypothalamic* and *retinothalamic projections*.

The optic chiasma is supplied from a pial plexus receiving branches from the superior hypophysial, internal carotid, posterior communicating, anterior cerebral and anterior communicating arteries; its veins drain into basal and anterior cerebral vessels.

The optic tracts (7.136, 139) continue dorsolaterally from the chiasma, each passing between the anterior perforated substance and tuber cinereum as the ventrolateral boundary of the interpeduncular fossa. Each flattens and curves round its cerebral peduncle, to which it adheres, hidden from view on the ventral cerebral surface by the uncus and parahippocampal gyrus. Reaching the lateral geniculate body it divides into medial and lateral rami ('roots'), the medial believed to contain supraoptic commissural fibres and the lateral consisting of fibres derived from the retina and partially decussating in the chiasma, as described, but also containing a few fine efferent fibres ending in the retinae. Most fibres in the lateral root end in the lateral geniculate body (p. 1016), but some sweep medially ventral to the pulvinar to reach the superior colliculus and pretectal nucleus (p. 987); the termination of some of these in the pulvinar in various mammals, including primates (Campos-Ortega et al 1972) and cats (Berman & Jones 1977), suggests a more direct route to cortical areas outside the striate cortex; the latter workers describe the projection as ending in a thin pulvinar zone lateral to the lateral geniculate nucleus. The arrangement of retinal fibres in the optic chiasma and tract has been minutely detailed in monkeys (Polyak 1957); clinical observations have long suggested similar human organization (Traquair 1948). Figure 7.139 shows these details. A notable feature is that macular nerve fibres are central in the optic nerve and chiasma but assume an eccentric, dorsolateral location in the tract, fibres from both retinae having intermingled. As detailed later, this rearrangement is a prelude to spatial representation in the lateral geniculate body (p. 1017), superior colliculus and visual cortex (p. 1058).

A relay of fibres from neurons in the lateral geniculate nucleus traverses the posterior limb of the internal capsule, emerging as a broad *optic radiation* of axons of secondary visual neurons, curving dorsomedially to the occipital cortex (p. 1058). They are separated from the posterior cornu of the lateral ventricle only by the tapetum of the corpus callosum.

Some fibres in the optic radiation, from neurons in the occipital cortex, descend to the superior colliculus, which therefore receives *cortical and retinal paths*. From this a relay travels by tectobulbar tracts to motor nuclei of the third, fourth, sixth and eleventh cranial nerves and the spinal anterior grey column. Recent studies question such 'oculomotor' connections in primates, including man (p. 987).

Metathalamus: the Geniculate Bodies

The metathalamus (7.136, 137) comprises the *medial* and *lateral geniculate bodies*, bilateral eminences on the posterior parts of the ventral thalamic aspects lateral to each side of the midbrain. Each eminence overlies the subjacent nuclear aggregates. As the collective term suggests, they are late developments in the vertebrate

thalamus. Their basic roles as relay nuclei on the acoustic (medial geniculate) and visual (lateral geniculate) paths are long recognized; subsequent evidence shows that their connections, and hence functions, are much more complex. In both paths, as in most sensory projections, there is a *corticofugal* component modulating *centripetal* inputs and they are no longer regarded as in receipt of simple arrays of impulses from peripheral receptors, proceeding along independent parallel channels. For example, neuronal connections exist, as elsewhere, for collateral inhibition and probably for interaction between different sensory modalities.

THE MEDIAL GENICULATE BODY

The medial geniculate body (7.116, 152) projects ventrally from the thalamic pulvinar (7.116), lies lateral to the superior colliculus and is ovoid in contour, with its long axis ventrolateral. The superior brachium extends from the superior colliculus to the lateral geniculate body and passes between the pulvinar and medial geniculate body. The inferior brachium (p. 981) ascends anterolaterally to reach the medial geniculate body; it contains fibres from the ipsilateral lateral lemniscus (with its complex *bilateral* nuclear origins: cochlear, trapezoid, superior olivary, intrinsic lateral lemniscal and inferior collicular, see pp. 959, 960, 1112).

In **internal structure** the medial geniculate nucleus has no clear subdivisions; the entire mass has a knee-shaped profile in section, hence its name. By size its neurons are largely segregated into a *dorsal pars parvocellularis* or *principal division* and a smaller *ventromedial division*, the *pars magnocellularis*, which also contains some neurons. Wedged between the pulvinar and pretectal region, dorsomedial to the medial geniculate body, is a small *suprageniculate nucleus* (in some species); its connections are uncertain. It may receive fibres from reticular nuclei in the lateral midbrain and pontine tegmentum, concerned in other modalities, including pain (p. 952). Closer study of the medial geniculate neurons, particularly of their dendritic organization and connections, has prompted partition into regions and component nuclei (Morest 1964, 1965). In cats it has been ascribed *dorsal, ventral* and *medial divisions*, with further subdivisons; one part of the ventral division, the '*ventral nucleus*', has a laminar structure, a detail of special interest because this receives most of the inferior collicular projection *in cats* (Moore & Goldberg 1963); moreover, the lateral geniculate nucleus is also laminated, a pattern associated with the spatial organization of sensory terminals. In the feline schema, the medial division appears to be the terminus of direct lateral lemniscal fibres which are probably, in cats and monkeys, spinal afferents concerned in modalities other than acoustic, ending in magnocellular parts of the medial geniculate and suggesting that this, and perhaps the suprageniculate nucleus, may integrate visual, auditory, general somatic and even visceral inputs.

The role of the medial geniculate body in acoustic paths is still partly obscure, apart from being a relay, largely due to its relative inaccessibility to the recording electrode or destructive probe. By analogy, the laminar appearance in its ventral division and a tonotopical organization at cortical level suggest similar arrangements in the medial geniculate body; the evidence is, however, equivocal in experimental animals. It is clearly premature to suggest tonotopical representation in the human structure. The cortical projection is considered to be direct from small-celled regions of the nucleus, terminating in the temporal lobe in an area in monkeys corresponding to human area 41 (p. 1060), in the superior temporal gyrus. Other acoustic areas, as shown by experiment, may also exist in mankind (p. 1060). A primary acoustic area (AI) in cats is directly connected to the medial geniculate body; secondary acoustic areas (AII and others) may receive projections from the parvocellular parts of the geniculate complex (Niimi & Naito 1974). All areas may be concerned in aspects of acoustic function (p. 1061); there is strong evidence of tonotopical representation of the cochlear spiral organ in area AI. The role of the medial geniculate body in the determination of frequencies remains sub judice, evidence being largely negative (Whitfield 1967). Lesions in it do not cause pronounced deafness unless bilateral; *both* cochlear organs are represented in *each* lateral lemniscus (p. 959). As in most receptive areas, the auditory cortex projects to the lower centres. Cajal first suggested a cortico-geniculate projection, demonstrated in monkeys by Whitlock & Nauta (1956). Pontes et al (1975) have detailed projections in cats from A1 and adjoining areas to medial geniculate nuclei (dorsal and magnocellular) and have summarized previous work; they confirm a *descending auditory pathway* (see Rasmussen 1967, and pp. 1005, 1112).

THE LATERAL GENICULATE BODY

The lateral geniculate body is a small nuclear complex visible on the surface as a small ovoid ventral projection from the posterior thalamic region (7.136, 152), its long axis approximately sagittal. Its anterior pole blends with the larger, lateral moiety of the optic tract, the medial ramus (or 'root') of which separates the lateral geniculate elevation from the lateral aspect of the crus cerebri. The medial geniculate is, of course, medial but also dorsal. Superficial aspects of both are overlapped by the temporal lobe, its parahippocampal gyrus being the immediate relation (p. 1029). Proceeding dorsally from the posteromedial part of the lateral geniculate body to reach the pretectal area and lateral side of the superior colliculus, is the *superior brachium* (7.135). This contains uninterrupted retinal nerve fibres mediating optic reflexes through the pretectal nucleus and colliculi and thence the oculomotor nuclei and nerves and the efferent descending tectal tracts (p. 987). The superior brachium passes largely between the medial geniculate body and the pulvinar. From a comparison of the lateral ramus of the optic tract and the superior brachium whose relative sizes represent a balance between forebrain and midbrain optic destinations, it is obvious that the phylogenetically older, mesencephalic termination is reduced in man, as in other primates, but less so in other mammals. The lateral geniculate body emerged during submammalian evolution, when the optic tracts contained only contralateral retinal fibres. This decussation is basic in vertebrates; explanations of it (like other decussations) remain hypothetical (Duke-Elder 1932, Polyak

1957). The advent of uncrossed retinal connections in mammals, especially developed in primates and others with forward-looking eyes and overlapping visual fields, is coupled with increasing specialization in the *dorsal part of the primitive lateral geniculate body*, by volume its major part in higher mammals. In this dorsal region *laminar organization* appears, perhaps associated with combined bilateral termination of retinal projections (vide infra). The *ventral part* is also known as the *pregeniculate nucleus*, represented in man by a small but not exiguous neuronal group (Le Gros Clark 1932), medial to the ventral third of the main, laminar nucleus. Medially, pregeniculate neurons are scattered towards the subthalamic region, marking the evolutionary and embryonic origin of this part of the complex. It has a bilateral retinal projection (possibly by collaterals of fibres passing to the tectum and main nucleus); these are said to be limited in origin to the central retinal region and part of the light reflex route by a *pregeniculo-mesencephalic connection* (Polyak 1957). The arrangement has clinical interest in the Argyll–Robertson phenomenon (p. 1215). However, Pierson & Carpenter (1974) found no evidence of involvement of this nucleus in the light reflex in monkeys subjected to various lesions in ocular, geniculate and pretectal sites. A cortical connection with the pregeniculate nucleus has not been established. Some deny termination of optic nerve fibres in the primate nucleus, which is, however, a definable entity though not as well developed as in ungulates and carnivores (Niimi et al 1963).

The dorsal (main) lateral geniculate nucleus has a well-known laminar structure, with six (and perhaps seven) layers of neurons discernible in some part of the nucleus in primates, such as rhesus macaques (subject to much experiment) and man. It must be noted at once, because of proposals to equate this pattern with function such as trichromatic vision (Le Gros Clark 1940, 1949), that the number and arrangement of laminae varies among mammals, being three in cats (also much experimented upon). Lamination also shows regional variation in the nucleus in most species, including man. Phylogenetically, an original division into *two* laminae has been suggested, serving separate terminations of ipsi- and contralateral retinal fibres (Polyak 1957); but all six

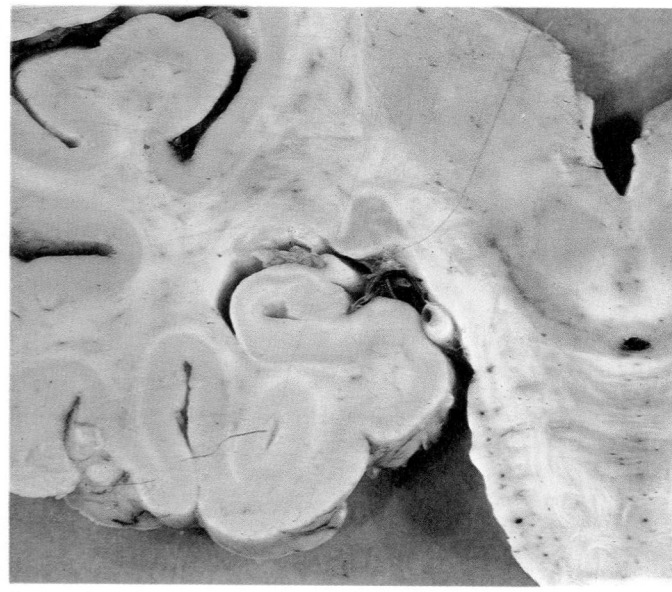

A

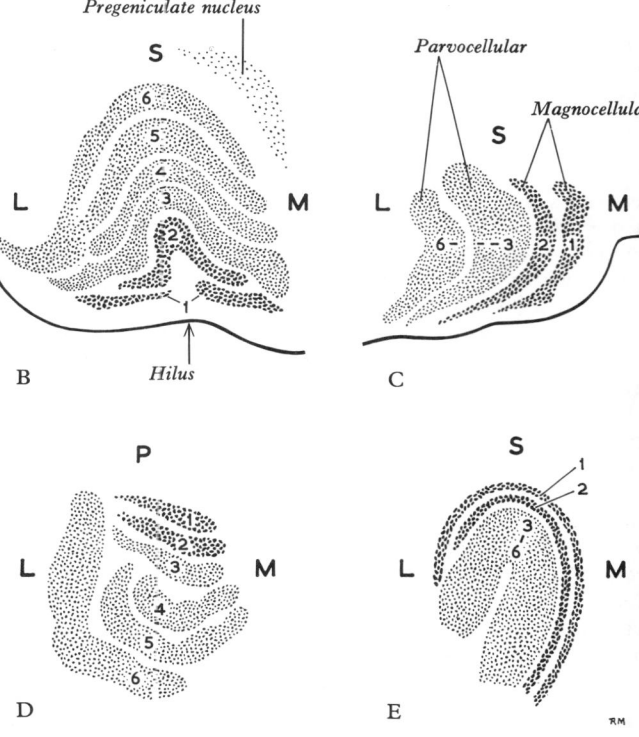

7.138 A. The lateral geniculate body, shown in coronal section in the human brain to display its general position and orientation: it is the cap-shaped mass of grey and white matter near the centre of the field. Note its relationship to the inferior horn of the lateral ventricle, the structures visible on the sectioned surface of the midbrain tegmentum and the dorsal thalamus. Even at this low magnification the lamination of the lateral geniculate body is visible. The four diagrams show the right lateral geniculate nucleus in section: B. a mature human nucleus in coronal section near its central region and (C) near its posterior pole. Note the reduction in the number of discernible laminae in the latter. The lamination is also visible (D) in an approximately horizontal section at the time of birth. In some primates, e.g. *Tarsius* (E), the magnocellular layers (1 and

2) are curved externally around the parvocellular layers (3–6); the higher primate arrangement is the reverse of this when examined in the coronal plane. See text for further details and consult Le Gros Clark (1932), Cooper (1945), Chacko (1955). Dissection and photography as in 7.137.

layers appear simultaneously in human embryos (Cooper 1945). With the development of two classes of neurons, large and small, such primary layers may have split; interlaminar strata of larger cells occur between the three recognized laminae in the feline nucleus (Thuma 1928).

Considering the human lateral geniculate body, particularly as regards its intrinsic and external connections and functional interpretation, it is necessary to emphasize that such details are almost entirely extrapolated from experimental results, observed chiefly in cats and monkeys and linked with a comparative structural study in mammals, including many primates. Commonly used experimental animals, especially monkeys, show such similarity of organization in the lateral geniculate body that data established thus may usually be extrapolated to mankind with some degree of confidence.

The appearance and orientation of the laminae are shown in figure 7.138A–E. Orientation is often confusingly presented in textbooks as stated by Cooper (1945) and Chacko (1955), their relation to other structures being obscured by the omission of reference data. Often depicted in coronal sections (7.138A), they are also shown in sagittal section; in both the lamination is apparent, since the laminae fit like a series of caps set one upon another. The lamination is fully displayed in coronal sections central in the nuclear mass but becomes slanted towards its caudal pole. Rostrally, i.e. near the entry of the optic tract, the layers are blended into a common mass and at the caudal extremity are reduced to four, in a region considered to be monopolized by macular terminations. As illustration 7.138B shows, the laminae are curved, with a ventral concavity as a kind of 'hilum', where some retinal efferents enter from the optic tract. This hilum is more prominent in monkeys than in man. Dorsally, the nucleus is rounded, projecting towards the thalamus; from this dome most fibres of the optic radiation, the *geniculostriate projection*, emerge from the nucleus. Medially and laterally the laminae blend, at the edge of the cap-shaped mass. The lateral border is sharp; blending layers reverse their curvature here, becoming ventrally convex, upturned like the peak of a cap. Concave laminal contours are also reversed in the caudal (macular) part. Hence laminar profiles in the coronal plane are laterally sinuous. (Some of the confusion concerning lateral geniculate lamination is due to reversal of their order in some lower primates, large-celled laminae 1 and 2 being 'external' and *dorsal* to the curvature of the entire structure. In *Tarsius* this is extreme (7.138E), lamination being an *inverted replica* of the human condition. In lemuroids transitional states occur. Much controversy has been entailed thus by the terms inversion and eversion, which need not detain us here.

The most careful estimates (Kupfer et al 1967) suggest a population of about one million neurons in a human lateral genicuate body, with corresponding numbers of fibres in the optic nerve and tract. Estimates in macaque monkeys reach 1·8 million (Le Gros Clark 1941) and 1·7 million (Rakic 1977) but only 1·2 million in the optic nerve (Bruesch & Arey 1942). Compare Chow et al (1950). This 1:1 ratio does not imply that each human optic nerve fibre connects with one geniculate cell. This relationship may obtain for the macula, but details are still uncertain. It is long established (Minkowski 1913, Le Gros Clark 1941, Glees 1941) that each fibre from a retinal ganglion cell ends by dividing into five or six terminals connecting with the same number of geniculate neurons in *one* lamina. (Geniculate cells may also connect with terminals from more than one retinal fibre.) Laminae are usually numbered 1–6, from the concave or hilar aspect. Fibres from the *contralateral* optic nerve end in layers 1, 4 and 6, *ipsilateral* in layers 2, 3 and 5. Early recognition of transneuronal degeneration (p. 871) in the lateral geniculate body (Minkowski 1913), coupled with adequate terminal degeneration technique (Glees 1946), led to detailed plotting of this bilateral retinal representation in each lateral geniculate body. Even with the Marchi technique alone (Brouwer & Zeeman 1926) it was shown in monkeys, by studying results of retinal lesions, that fibres from the retinal quadrants are located correspondingly in the optic nerves (p. 1095 and 7.139), the macular fibres being central. An orderly but somewhat different arrangement was also demonstrated in the optic chiasma and tract (7.139) but, in the tract, fibres from corresponding quadrants in *both* retinae occupy the

same location, a pattern carried into the lateral geniculate body, where macular fibres terminate centrally and caudally in the nucleus, peripheral fibres rostrally, upper retinal quadrants laterally and the lower ones medially (7.139). Transneuronal techniques permitted more precise plotting (Le Gros Clark & Penman 1934) and, since cortical projection fibres of affected geniculate neurons also degenerate, it became clear that similarly precise point-to-point connection continues through the optic radiation to the striate cortex (p. 1058). This precise topical organization in the visual pathway has been repeatedly confirmed in cats and monkeys (Polyak 1957, Meikle & Sprague 1964, Saavedra et al 1969). (Doubts as to factors causing transneuronal degeneration in the lateral geniculate nucleus have been expressed. There is evidence that it is not due solely to deafferentation by e.g. optic tract interruption. Consult Garey et al (1976) for discussion.) As detailed elsewhere (p. 1058), ablation of small parts of the striate cortex leads to highly localized groups of degenerating geniculate cells in *six layers* on the same side. These effects form a *columnar representation* of *corresponding points* in *two retinae*, each in *three* of the six laminae. Results of transneuronal degeneration in the nucleus, following the destruction of retinal ganglion cells or of their axons, have been confirmed by equivalent but less dramatic effects of prolonged visual deprivation in kittens (Wiesel & Hubel 1963, Guillery 1973) and infant monkeys (Von Noorden 1973, Headon & Powell 1973). These experiments (suturing of eyelids) have been performed unilaterally and bilaterally; in the latter (Headon & Powell 1978) an unexplained finding is that the shrinkage of geniculate neurons observed in such experiments was greater in nuclear regions receiving fibres from monocular retinal fields when *both* eyes were subjected to prolonged closure. Autoradiographic tracing of labelled amino acids (subject to orthograde axoplasmic flow and even transneuronal transport) has confirmed these earlier results (see Graybiel 1975, Hubel et al 1976, Rakic 1976), with regard to both laminar terminations and the anatomical orientation of neurons in the nucleus and visual cortex.

Lateral Geniculate Interactions

So far the post-retinal visual pathway has been considered as a series of parallel conductors with no interaction between them, whether from one retina or both. Indeed, it has long been assumed that, apart from the inevitable convergence of far more numerous rods and cones upon the more limited population of retinal ganglion cells (p. 1195) and complex intraretinal interactions (p. 1201 et seq.), the visual pathway consists of functionally isolated conductors with a simple relay in the lateral geniculate body. This has been customary textbook teaching; the assembly of fibres from both retinae in the geniculate body was vaguely dubbed subcortical 'fusion', a naive assumption without data beyond those recounted above. Retinal neuronal organization is well known to be complex; clear evidence, anatomical and physiological (p. 1196), indicates interaction between its neurons. The pattern of impulses in the optic nerve is not the simple imprint of a mosaic of receptor responses, like signals in a cable between a television camera and monitor, even if the effects of corticofugal connections are ignored (p. 1203). Integrative activities in the cat's lateral geniculate body have also been demonstrated (Hubel & Wiesel 1961) between corticofugal fibres and the centripetal sensory projection, as in other sensory paths. While unit recording demonstrates both lateral inhibition and units responsive to specific modes of stimulation, *no clear evidence* indicates that *individual* geniculate cells receive impulses from *both* retinae (Glees 1961).

Increasingly elaborate cytological detail of dendritic fields and synaptic features has been recorded in cats and monkeys, impossible even to survey briefly here; reference to the reports of Peters & Palay (1966), Campos-Ortega et al (1968), Lieberman (1973, 1974) and Lieberman & Webster (1974) will provide an introduction to synaptic organization of the dorsal lateral geniculate nucleus. Lieberman has identified a variety of synapses, pre- and postsynaptic, inhibitory and excitatory, dendrodendritic, and triplet synapses (triads) of glomeruli (vide infra). These and other researches carried out in rodents, cats and monkeys show some species differences but basic resemblances predominate. To

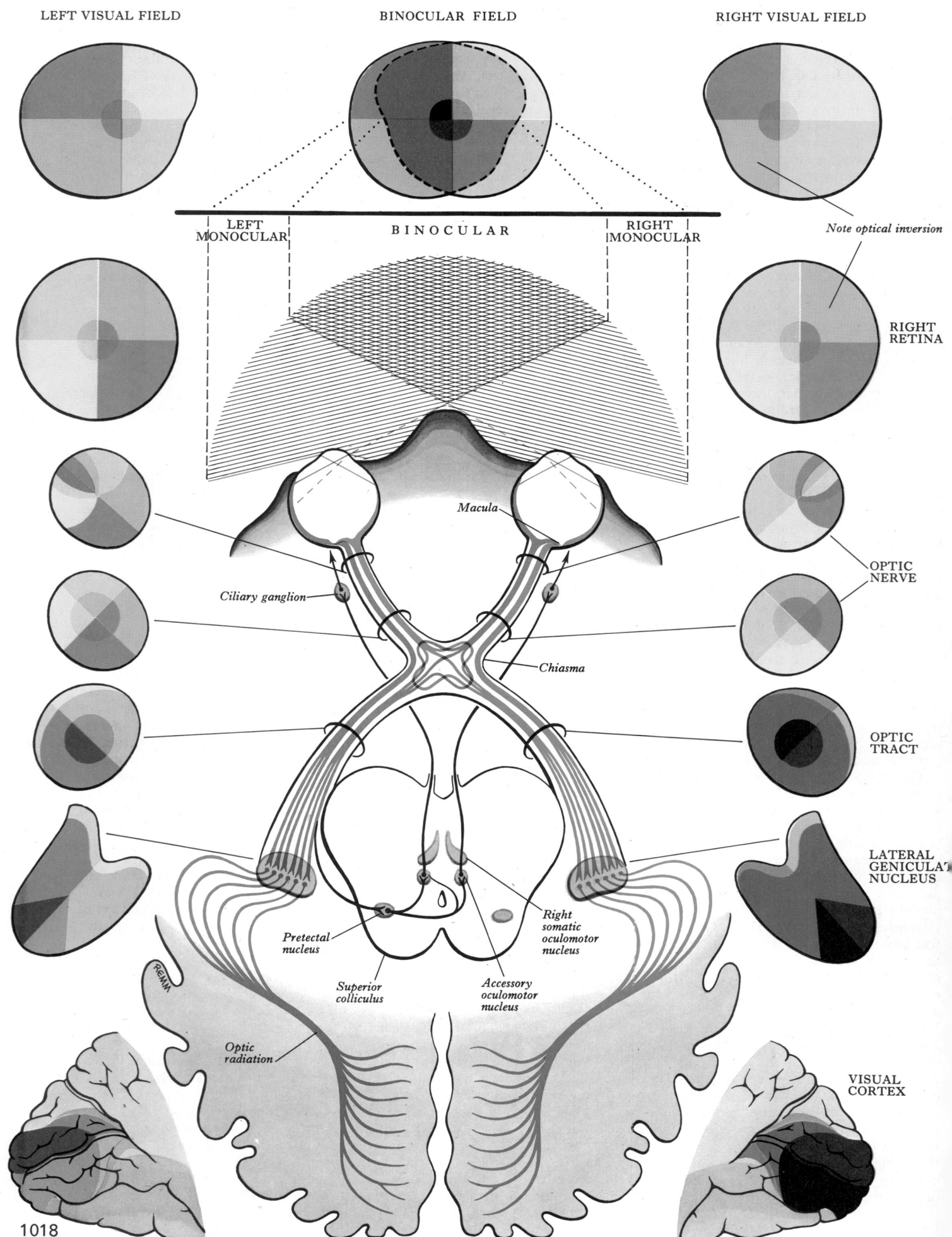

LEFT VISUAL FIELD

BINOCULAR FIELD

RIGHT VISUAL FIELD

LEFT MONOCULAR

BINOCULAR

RIGHT MONOCULAR

Note optical inversion

RIGHT RETINA

Macula

Ciliary ganglion

Chiasma

OPTIC NERVE

OPTIC TRACT

LATERAL GENICULATE NUCLEUS

Pretectal nucleus

Right somatic oculomotor nucleus

Superior colliculus

Accessory oculomotor nucleus

Optic radiation

VISUAL CORTEX

1018

correlate these findings into a model of geniculate activity is not yet possible but the view of this as a simple relay mechanism is completely demolished. Three main elements are involved: (1) *afferents* from the retinae, with some *corticofugal fibres* and perhaps ubiquitous *reticular axons*; (2) the *efferent geniculostriate projection*, (3) an *intrinsic neuronal population*, largely of Golgi type II cells, not all of whose axons are confined to single laminae (and hence some may possibly mediate connections between bilateral retinal afferents). Most accounts say little of the precise arrangement of cells (Golgi type I) whose axons stream through intervening laminae to form the geniculostriate radiation; attention has been largely concentrated, in light and electron microscope studies, on dendritic arrangements and the distribution and attributes of the synapses, and their profusion in the resulting neuropil suggests a highly complex, interneuronal mediation between the afferents and efferents (Szentágothai 1963, Lieberman 1974). A special feature is the occurrence of *glomeruli*, conglomerations of synapses between dendritic and axonal terminals, the latter in part from optic nerve fibres (Peters & Palay 1966, Campos-Ortega et al 1968, Lieberman 1974). Similar to synaptic glomeruli in the cerebellum (p. 971) and in other thalamic 'relay' nuclei (p. 1000), they probably serve similar integrative functions. Glomeruli are complexes of a main and peripheral axons, a main and peripheral dendrites and various other neuronal and glial elements (p. 888). Some axons may be involved in several glomeruli; others, structurally distinguishable, occur in the glomerular periphery, making synaptic contacts both with peripherally placed dendrites *and* with the central axon. The latter type of synapse, like other *axoaxonal* connections, is probably inhibitory (Eccles 1964, Shepherd 1974). Dendrodendritic and axodendritic are, however, the predominant types of synapse (Lieberman 1974). Such observations indicate a structural substrate for inhibitory and excitatory receptive field effects identified by unit recording in the lateral geniculate nucleus (Hubel & Wiesel 1961). Similar glomeruli have been identified in monkeys (Colonnier & Guillery 1964). Brauer & Winkelmann (1976) have surveyed the functions of interneuronal participation in mammalian glomeruli. *Synaptic* glomeruli are not to be confused with groups of nerve cells, also called glomeruli, described much earlier (Taboada 1927); these also exist in all laminae of the simian geniculate nucleus, the numbers increasing from 12–15 in lamina 1 to 60–100 in lamina 6. Peters & Palay (1966) suggested that somata of neurons whose neurites are involved in a *synaptic glomerulus* may be clustered together as a *cellular glomerulus*, a tentative view which equates the latter with the groups noted above. Electron microscopic evidence suggests that synaptic glomeruli may be more complex in outer laminae of the lateral geniculate nucleus.

Findings so far noted largely concern *intra*laminar integration; *inter*laminar interaction is of special interest respecting integration of signals from both eyes. Since each retinal ganglion cell axon terminates, as far as is known, merely in one lamina, any interaction between fibres derived, as is presumed, from ganglion cells near the same locus in one retina but ending separately in three geniculate laminae requires interlaminar connections. Even were these unsupported by anatomical evidence, the results of unit recording at geniculate level indicate their existence. Similar reasoning might apply to the interaction between impulses from corresponding loci in *both* retinae (but vide infra). Again, as far as is known, each lamina projects separately to the striate cortex and not, as some diagrams vaguely depict, by a fictitious form of neuron with a soma in each lamina and a conjoined axon projecting to the cortex! Physiological evidence suggests that impulses from the *two* retinae are integrated *only* at cortical level

(Hubel & Wiesel 1969). However, it has been claimed that unit recording demonstrates binocular interaction at lateral geniculate level; and electron microscopy suggests that in the neuropil between the laminae are synapses between intrinsic lateral geniculate neurons (Guillery & Colonnier 1970, but vide infra). Neurons in laminae, probably secondary interneurons, have also been claimed to mediate interlaminar connections (Guillery 1966, 1971); these *could* channel the effects from both retinae towards single geniculostriate fibres. It is on such problems that most synaptological research on geniculate neurons is currently focused. Peculiarities of the interneurons have attracted special attention (Lieberman & Webster 1974); many, perhaps almost all are amacrine in rodents and some, at least, are anaxonal in monkeys (LeVay 1971). They are integrative, as is clear from their complex presynaptic contacts with retinal and cortical connections in glomeruli. In their most recent summary of activity in visual pathways, Hubel & Wiesel (1977) gauge the evidence strongly *against* the existence of lateral geniculate neurons responding to biretinal impulses and they reject 'fusion' at this level.

The Third Ventricle

The third ventricle (7.125, 140, 191A,B), a derivative of the primitive forebrain vesicle, is a median cleft lying mainly between the two (dorsal) thalami and hypothalami but extending beyond them to the epithalamus and subthalamus and communicating caudally with the fourth ventricle by the cerebral aqueduct and rostrally with the lateral ventricles through the interventricular foramina. A roof, a floor, anterior and posterior boundaries and two lateral walls are ascribed to it.

The *roof* is a layer of ependyma stretched between the upper edges of the lateral walls and is continuous here with the general ependymal ventricular lining. The roof is covered by and adherent to a reduplicated fold of the pia mater, the *tela choroidea*, from the surface of which a pair of parasagittal vascular fringes, *choroid plexuses*, project vertically downwards, invaginating the ependymal roof into the cavity (7.198).

The *floor* (7.133, 140) descends ventrally and is formed mainly by hypothalamic structures; rostrocaudally these are: the optic chiasma, infundibulum and tuber cinereum and the mamillary bodies, caudal to which the floor is formed by the posterior perforated substance and tegmentum of the cerebral peduncles. The ventricle is prolonged into the infundibulum as the *infundibular recess*. The neurohypophysis is continuous with the apex of the infundibulum.

The *rostral boundary* (7.133, 140) is ventrally the *lamina terminalis*, the rostral end of the neural tube. It is a thin layer stretching from the dorsum of the optic chiasma to the rostrum of the corpus callosum. Above this the boundary is formed by the fornical columns, diverging as they descend into the lateral walls of the third ventricle and by the anterior commissure (p. 1034), which crosses the midline rostral to the columns. At the junction of the floor and the rostral wall, just above the optic chiasma, the ventricle has a small angular *optic recess*. At the junction of the roof with the rostrolateral limits of the ventricle are the *interventricular foramina*, by which the third and lateral ventricles communicate. They represent sites of original diverticular outgrowths from the telencephalon to form the cerebral hemispheres, being *relatively* large and circular in a 10 mm human embryo. In the adult, however, they are crescentic slits, between the curving columns of the fornix (anteriorly) and convex anterior thalamic tubercules (posteriorly).

The *posterior boundary* (7.140) consists of the pineal gland, posterior commissure and cerebral aqueduct. A *pineal recess* projects into the pineal stalk, above which is a *suprapineal recess*, a diverticulum of the epithelial ventricular roof.

Each *lateral wall* has an upper part, which is the medial surface of the rostral two-thirds of the (dorsal) thalamus, and a lower formed by the hypothalamus continuous with the grey matter of the floor and part of the subthalamus, the upper and lower areas being separated by the curved *hypothalamic sulcus*, extending from the interventricular foramen to the cerebral aqueduct; the sulcus varies from deep to scarcely perceptible. As noted (p. 997),

7.139 A diagram of the visual pathway to show the spatial arrangement of nerve cells and their fibres in relation to the quadrants of the retinae and visual fields. The proportions at various levels are not exactly to scale and in particular the macula is exaggerated in size in the visual fields and retinae. In each quadrant of the visual field, and the parts of the visual pathway subserving it, two shades of the respective colour are used, the paler for the peripheral fields and a darker shade for the macular part of the quadrant. From the optic tract onwards these two shades are both made more saturated to denote intermixture of neurons from both retinae, the palest shade being reserved for parts of the visual pathway concerned with uniocular vision. The path of the light reflex has also been indicated.

the hypothalamic sulcus divides the diencephalon; above (dorsal to) the sulcus is the *pars dorsalis diencephali* (dorsal thalamus and epithalamus) and below (ventral) is the *pars ventralis diencephali* (hypothalamus and subthalamus). Each lateral wall is limited dorsally by a ridge covering the stria medullaris thalami (7.125).

The columns of the fornix curve ventrally rostral to the interventricular foramina to run in the lateral walls, where at first they are prominent, rapidly sinking into them. The lateral walls are joined across the cavity by the *interthalamic adhesion*, a band of grey matter (p. 998). An extensive account of the phylogeny and

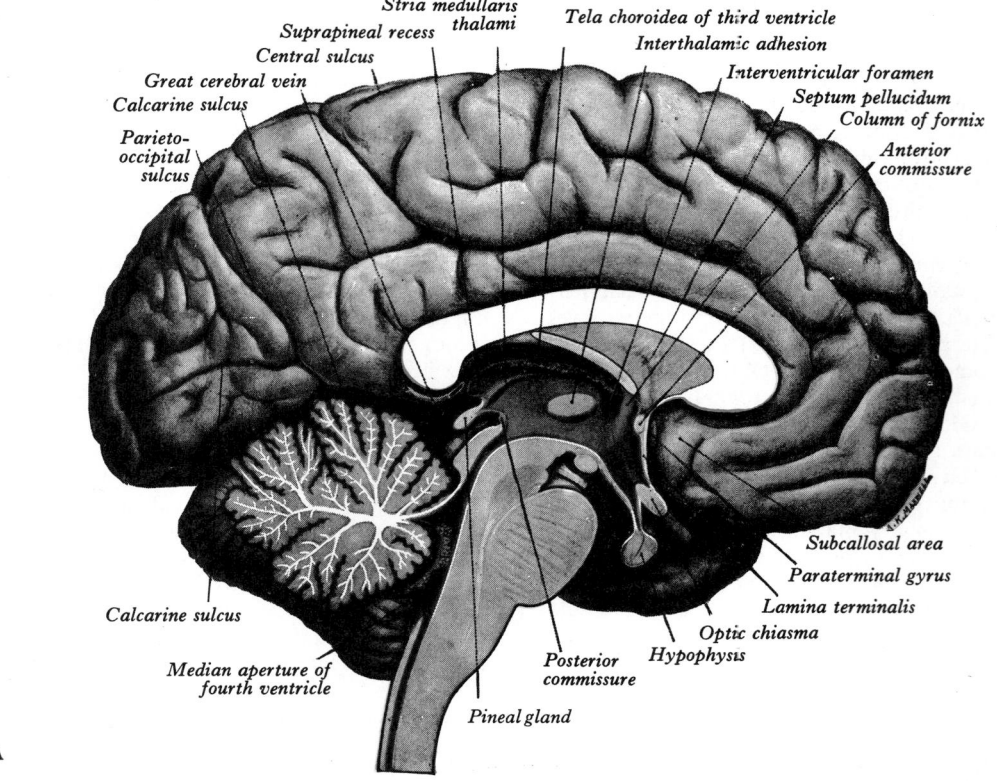

Stria medullaris thalami
Suprapineal recess
Central sulcus
Great cerebral vein
Calcarine sulcus
Parieto-occipital sulcus

Tela choroidea of third ventricle
Interthalamic adhesion
Interventricular foramen
Septum pellucidum
Column of fornix
Anterior commissure

Subcallosal area
Paraterminal gyrus
Lamina terminalis
Optic chiasma
Hypophysis

Calcarine sulcus

Median aperture of fourth ventricle

Posterior commissure

Pineal gland

A

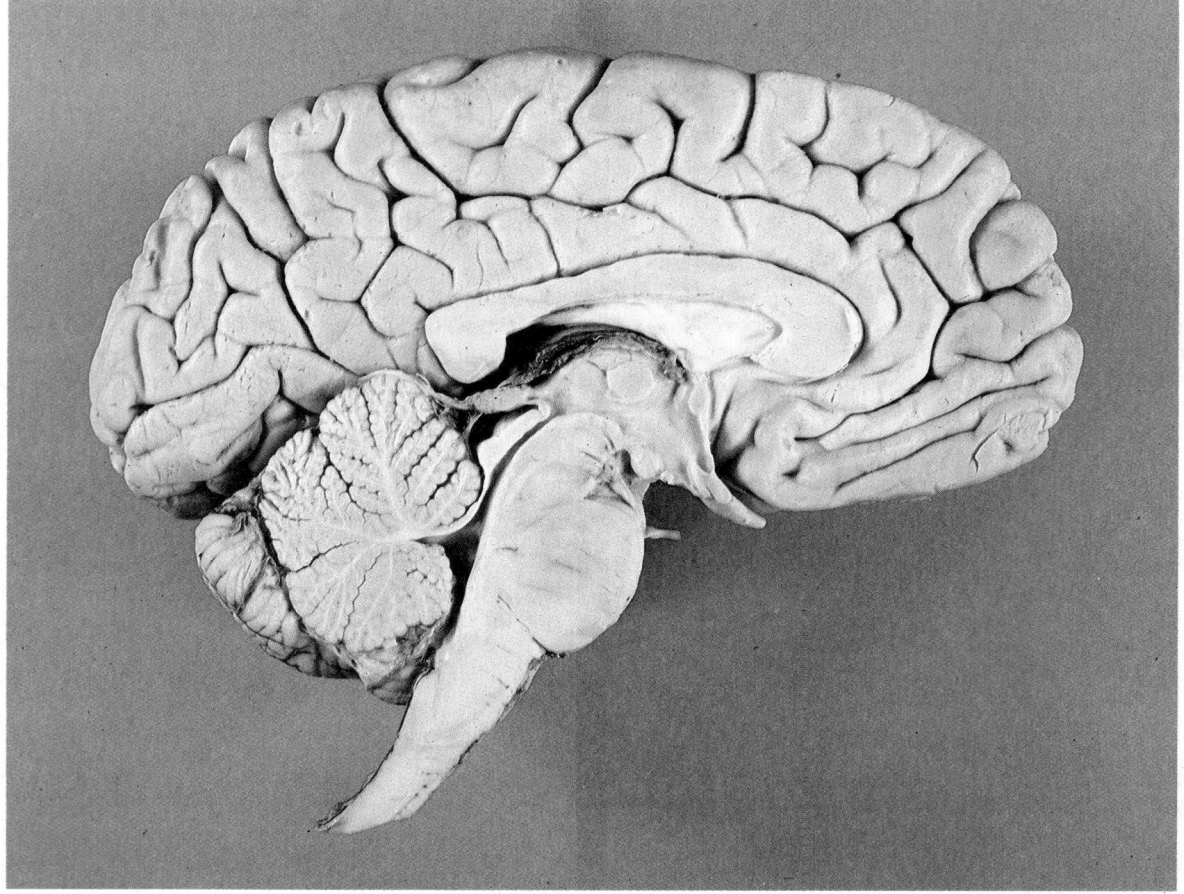

B

7.140 A.   Median hemi-section of the brain to show the third and fourth ventricles. The pia mater is indicated in red, the ependyma in blue. B.   A sagittal hemi-section through the brain. For detailed labelling

of the structures visible on this specimen compare with 7.140A, 7.145A. Dissection by E L Rees, photography by Kevin Fitzpatrick, both of the Dept. of Anatomy, Guy's Hospital Medical School, London.

development of the third ventricle has been contributed by Kier (1977). The specialized ependyma forming *circumventricular organs* is described on pp. 189, 894, 1005.

The interpeduncular fossa (7.81, 132, 146A,B), a trapezoid area, is limited anteriorly by the optic chiasma, posteriorly by the anterosuperior surface of the pons, anterolaterally by the converging optic tracts and posterolaterally by the diverging cerebral peduncles. The structures in it have been described, being the posterior perforated substance, mamillary bodies, tuber cinereum (pp. 1007, 1014), infundibulum and hypophysis cerebri (p. 1009).

THE TELENCEPHALON OR 'ENDBRAIN'

Expansion of the telencephalon and development of the cerebral hemispheres are described in pp. 192-194. In earlier vertebrates each hemisphere is predominantly concerned with olfactory signals, entering it rostrally at the *olfactory lobe*. This lobe is elongated as an *olfactory bulb* connected to the hemisphere by an *olfactory tract*. Basally in each hemisphere masses of grey matter, *basal nuclei*, form early motor centres. The wall of the hemisphere is the *pallium*, where olfactory and other information are presumably integrated. During evolution visual, auditory and other paths have extended through the thalamus to the cerebral pallium, an instance of encephalization. Each hemisphere is thus enlarged by an additional *neopallium*, the largely olfactory pallium being confined to a *piriform lobe*, sited inferolaterally. The hemisphere's medial wall becomes specialized as the *hippocampal formation*, long regarded as primarily olfactory, a view now untenable (see p. 1035). In higher mammals the neopallium is greatly enlarged and the piriform lobe relatively reduced; neopallial motor paths develop from it but the basal nuclei remain essential parts of motor control. This mammalian neopallial expansion is largely due to the growth of *association areas* concerned with interaction between afferent and efferent connections. The hippocampal formation is often termed *archipallium* or *primal* cortex and the piriform lobe *paleopallium* or *ancient* cortex; some, however, group both as *archipallium*.

The telencephalon includes: (1) the cerebral hemispheres, their commissures and their cavities; (2) the anterior part of the third ventricle, including the preoptic regions in the telencephalon impar (p. 1007). Each cerebral hemisphere has an external stratum of neurons, the *cortex*, an internal mass of neuronal processes (*centrum semiovale*), deeply situated *basal nuclei* and a *lateral ventricle*. The cerebral hemispheres are the largest part of the brain and, viewed from above, have an ovoid shape, broader behind, the greatest transverse diameter being between the parietal tuberosities. They are incompletely separated by a deep median cleft, the *longitudinal cerebral fissure*, each hemisphere containing a *lateral ventricle*.

The longitudinal cerebral fissure contains a crescentic fold of the dura mater, the *falx cerebri*, and anterior cerebral vessels; in front and behind, the fissure completely separates the hemispheres; centrally it descends to a large commissure, the *corpus callosum*, connecting the hemispheres.

Surfaces of the Cerebrum

Each cerebral hemisphere presents superolateral, medial and inferior surfaces or aspects.

The *superolateral surface* is adapted to the concavity of its half of the cranial vault. The *medial surface*, flat and vertical, is separated from its fellow by the longitudinal fissure and falx cerebri. The *inferior (basal) surface* is irregular and divided into orbital and tentorial regions; the orbital part of the frontal lobe is concave and lies above the orbital and nasal roofs. The tentorial region is the concavoconvex inferior surface of the temporal and occipital lobes, anteriorly adapted to its half of the middle cranial fossa; posteriorly it is above the tentorium cerebelli which is interposed between it and the superior cerebellar surface.

The surfaces are separated by the following borders: (1) *superomedial*, between the superolateral and medial surfaces; (2) *inferolateral*, between the superolateral and basal surfaces (its anterior part separates the superolateral from the orbital surface in the frontal lobe, as the *superciliary* border); (3) *medial occipital*, between the tentorial region of the inferior surface and the medial; (4) *medial orbital*, separating the orbital region of the inferior surface from the medial. The anterior and posterior hemispheric extremities are the *frontal* and *occipital poles* respectively; the temporal pole is the anterior extremity of the temporal lobe. About 5 cm anterior to the occipital pole on the inferolateral border is the *preoccipital incisure*.

A paramedian line, from a point a little superolateral to the inion to one just superolateral to the nasion, marks the superomedial margin. The superciliary border is at the level of the eyebrows as far as the zygomatic process of the frontal bone, ascending thence to the pterion. The temporal pole corresponds to a line, convex forwards, on the surface of the head, from the pterion to the midpoint of the upper border of the zygomatic arch; it then continues back above the arch to cross the ear a little above the external acoustic meatus; it corresponds to the inferolateral margin, which then curves down to the posterior end of the superomedial margin (**6.**151).

Hemispheric surfaces are moulded into a number of *gyri* or *convolutions*, separated by *sulci* or *fissures*. This familiar appearance is gradually achieved. Until the end of the third fetal month, the surfaces are smooth, as in the brains of reptiles and birds. Thereafter localized depressions appear, deepen and extend over the surfaces as sulci (see p. 192). Each sulcus is an infolding of the cortex, increasing the cortical grey matter about three times. Some sulci develop along the zones separating areas differing in details of structure and functions (Le Gros Clark 1945) and may be termed *limiting sulci*. The central is a limiting sulcus, being between two areas differing in thickness enough to be visible to the eye (7.146A). Some sulci develop in the long axis of a rapidly growing area and are *axial sulci*; the posterior part of the calcarine sulcus is merely a fold in the centre of the striate area. In other sites a sulcus may be between two structurally different areas; but its lip, not its floor, may divide two areas and then a third area may be in the walls of the sulcus without appearing on the surface; such a sulcus is *operculated*, exemplified in the human brain by the lunate sulcus, separating striate from peristriate areas on the surface and containing the submerged parastriate area, which intervenes between them. These varieties include all sulci on the surface, except the lateral sulcus and parieto-occipital sulci; the former results from the slower expansion of the insular cortex with consequent submersion by adjoining areas, which eventually come into contact to delimit the lateral sulcus. The parieto-occipital sulcus is associated with the development of the corpus callosum. The posterior end of this commissure conveys not only fibres from the occipital lobes but also a large number from the temporal lobes. Hence, several smaller axial and limiting sulci are crowded together and some are buried in the walls of the parieto-occipital sulcus. These are really *secondary sulci*, depending on factors other than exuberant growth in the adjoining areas.

Some sulci are deep enough to produce elevations in the ventricular walls; the anterior part of the calcarine sulcus, which thus produces the calcar avis of the posterior cornu, and the collateral fissure, producing the collateral eminence in the inferior cornu, are therefore termed *complete fissures*; but no special morphological or functional significance is attached to the fact that some sulci are complete and others incomplete.

Gyri and their intervening sulci are approximately constant in their general overall arrangement but vary in their dimensions and minor details (or even their occurrence in the smaller cases), not only in different individuals but in the hemispheres of one brain. The gyral pattern is an inevitable result of the greater

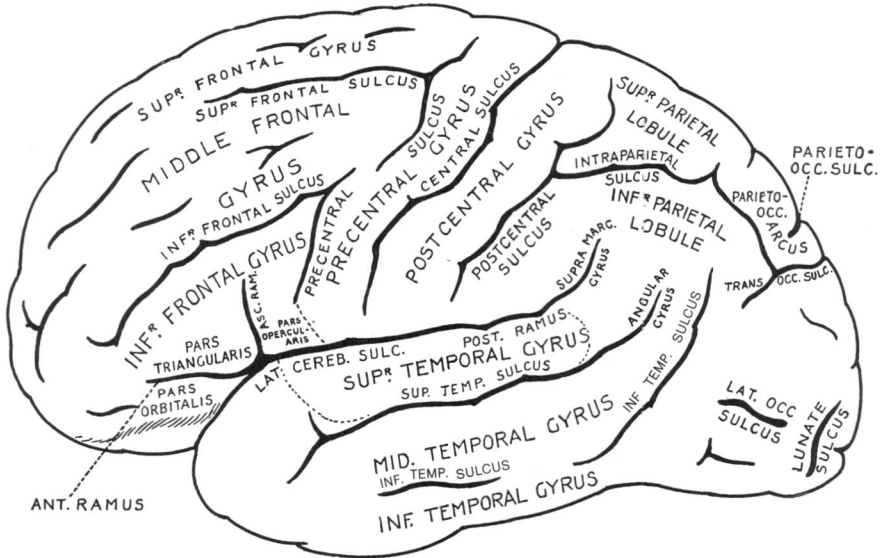

7.141A Lateral aspect of the superolateral surface of the left cerebral hemisphere.

increase in volume of the pallium or mantle of neurons in the cortex compared with the much smaller increase in the subjacent white matter. The area of the human cortex is about 2200 cm², a third of this being visible on the surface while the rest is obscured in the sulci and fissures. By this mode of evolution a large increase in the cortical area entails a lesser change in cranial capacity; but it is misleading to explain the arrangement in this teleological manner. Similarly, a presumed association of high intelligence with great complexity of convolutional pattern is fallacious; the most intricate arrays of sulci and gyri occur in the cerebra of the

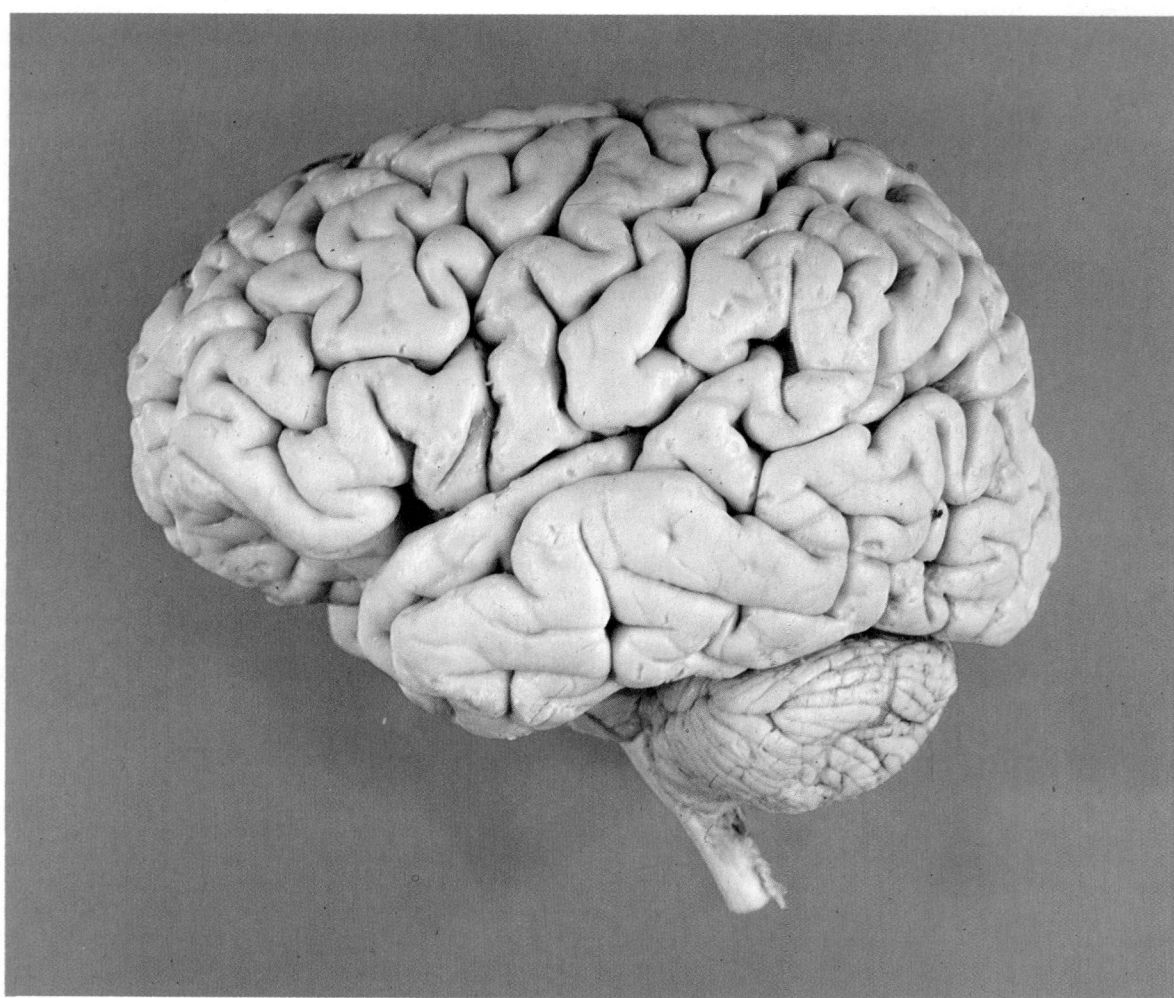

7.141B A left lateral view of the brain to show the pattern of gyri and sulci on the superolateral aspect of the cerebral hemisphere. Compare with 7.141A, which was drawn from a different specimen, for labelling of the many structures visible. Note also the contrasting cortical patterns of the cerebrum and cerebellum. Preparation by E L Rees, photography by Kevin Fitzpatrick, both of the Dept. of Anatomy, Guy's Hospital Medical School, London.

elephant and whale (Hammelbo 1972), both of which also have larger brains than man, though not so in relation to total size. No close relation between gyral complexity and brain size with cerebral abilities has been established in mankind; abundant examples of highly able individuals with relatively small brains and the reverse of this are attested. Attempts to draw deductions from endocranial casts of fossil forms, as an indication of development of certain gyri and hence abilities (e.g. capacity for speech), have likewise proved misleading and are largely abandoned except by anthropologists.

It is convenient to separate the cerebral hemisphere into a number of lobes but this division is purely descriptive; nor do lobes precisely correspond in surface extent to the cranial bones from which they take their names.

THE SUPEROLATERAL CEREBRAL SURFACE

Two sulci, viz. the *lateral* and *central*, are the outstanding features of this surface and the main factors used in limiting its surface divisions (7.141A,B).

The lateral sulcus (7.141, 146), a deep cleft on the inferior and lateral surfaces, has a short stem dividing into three rami. The *stem* commences inferiorly at the anterior perforated substance, extending laterally between the frontal lobe's orbital surface and the anterior pole of the temporal lobe. It is adapted to the posterior border of the sphenoid's lesser wing and contains the sphenoparietal venous sinus. Reaching the lateral surface it divides into anterior horizontal, anterior ascending and posterior rami. The *anterior ramus* runs forwards for 2·5 cm or less into the inferior frontal gyrus while the *ascending ramus* ascends for an equal distance into the same gyrus; the *posterior ramus*, the longest, runs posteriorly and slightly up across the lateral surface for about 7 cm, turning up to end in the parietal lobe. Its floor is the limen insulae and insula; it conducts the middle cerebral vessels from the inferior to the superolateral surface. It is represented by a line sloping back and slightly up for about 7 cm from the pterion, then curving up to the parietal eminence.

The central sulcus (7.141A,B) starts in or near the superomedial border a little behind the mid-point between the frontal and occipital poles (i.e midway between the nasion and inion). It runs sinuously down and forwards for about 8–10 cm to end a little above the posterior ramus of the lateral sulcus, from which it is always separated by an arched gyrus. Its general direction makes an angle of about 70° with the median plane (7.141A,B). It demarcates the primary motor and somatosensory areas of the cortex (pp. 195, 1047). When the sulcus is opened up, its opposed walls are seen to be marked by small gyri which alternate like gears in mesh, hence termed *interlocking gyri*. This provides an additional cortex without the corresponding increase in the surface area. About the middle of the sulcus its walls are usually connected by a transverse gyrus; this is due to the mode of

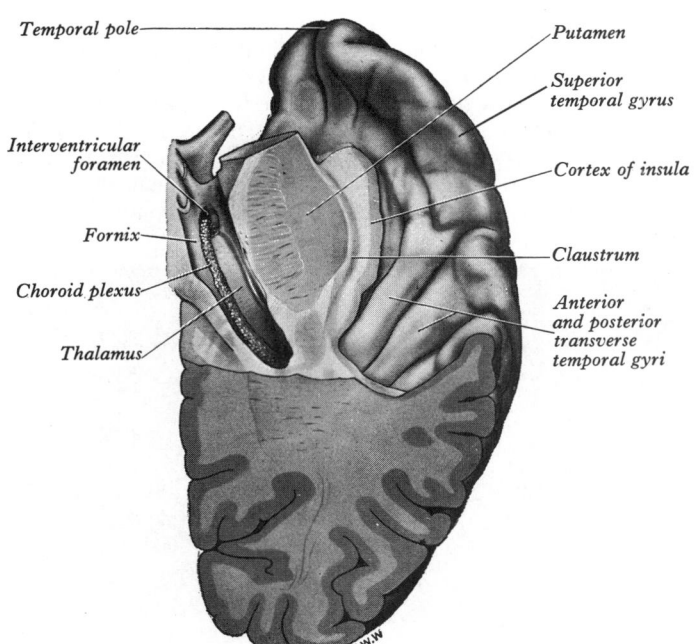

7.142 Horizontal section showing the superior surface of the right temporal lobe.

development of the central sulcus. When it appears in the sixth month, it is in the superior and inferior parts, at first separated by a transverse gyrus connecting the precentral to the postcentral gyrus. The two occasionally remain separate but usually coalesce, the transverse gyrus being buried as a *deep transitional gyrus*.

The frontal lobe, the rostral region of the hemisphere, is limited behind by the central sulcus, above by the superomedial border and below by the superciliary border and stem of the lateral sulcus. Its superolateral surface is traversed by three sulci and four gyri. The *precentral sulcus* is parallel to the central, separated from it by a precentral gyrus and usually divided into upper and lower parts, which may be confluent. The *superior frontal sulcus* curves forwards from about midway in the upper precentral sulcus; the *inferior frontal sulcus* is parallel but at a lower level. The area of frontal lobe anterior to the precentral sulcus is thus divided into the superior, middle and inferior frontal gyri; an incomplete sulcus often divides the middle gyrus (7.141A,B).

The *precentral gyrus*, with the central sulcus posterior and precentral sulcus anterior to it, extends from the superomedial border, where it is continuous with the paracentral lobule on the medial surface, down to the posterior ramus of the lateral sulcus.

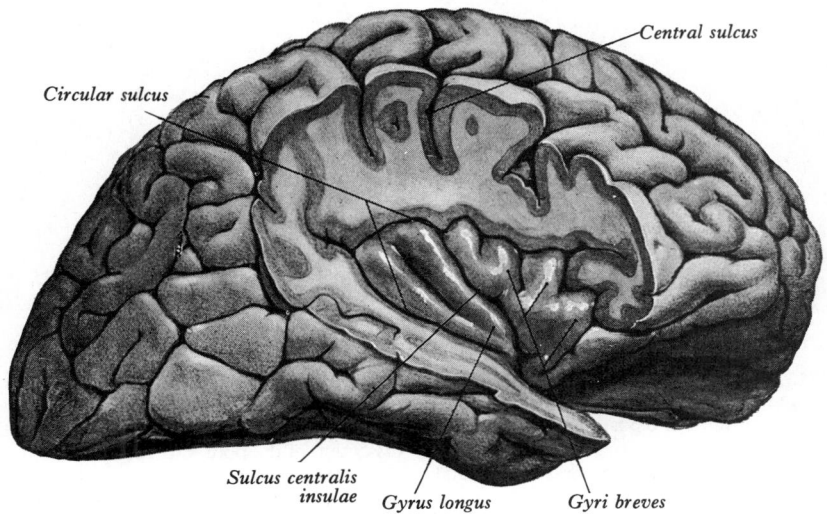

7.143 The right insula, exposed by the removal of its opercula.

Its cortex is the origin of many of the fibres of large corticonuclear and corticospinal tracts, in addition to a multitude of other connections.

The *superior frontal gyrus*, above the superior frontal sulcus, is continuous over the superomedial margin with the medial frontal gyrus. It may be incompletely divided (7.141A,B). The *middle frontal gyrus* is between the superior and inferior frontal sulci. The *inferior frontal gyrus* is below the inferior frontal sulcus and is invaded by the anterior and ascending rami of the lateral sulcus.

The areas around these rami on the left are the *speech area of Broca* (areas 44 and 45), associated with the motor aspects of speech (p. 1055). The region below the anterior ramus is the *pars orbitalis,* curving round the superciliary margin to the orbital surface. The area between the ascending and anterior rami is the *pars triangularis* and that posterior to the ascending ramus is the *pars opercularis (posterior)*, continuous posteriorly with the inferior end of the precentral gyrus.

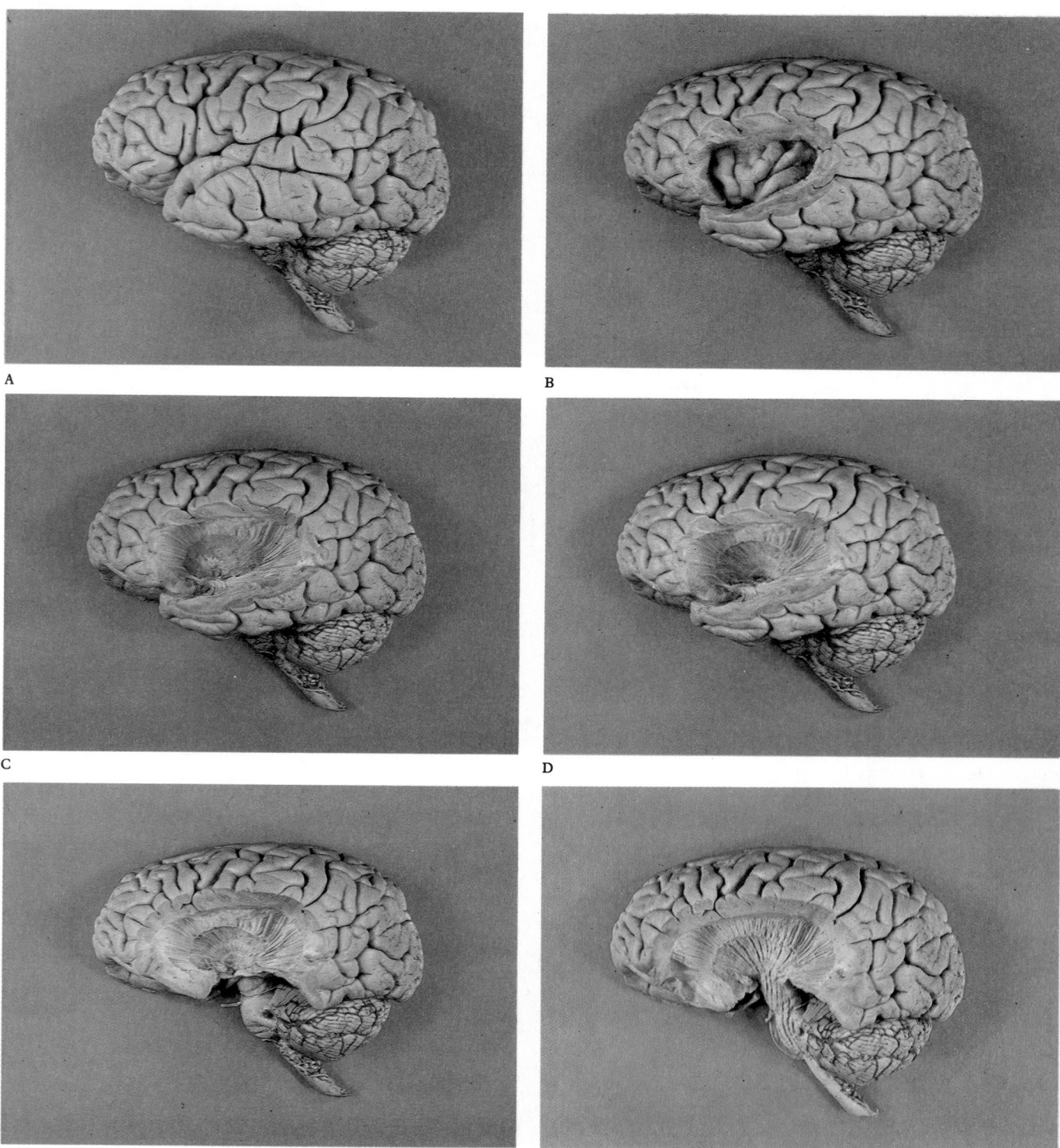

A

B

C

D

E

F

7.144 A series of dissections of the left cerebral hemisphere at progressively deeper levels to demonstrate the insula and subjacent structures: A. the intact brain; note the position of the posterior ramus of the lateral cerebral sulcus on which the dissections are centred; B. the cortical gyri of the insula exposed by removal of the frontal, temporal and parietal opercula; C. the removal of the insular cortex, extreme capsule, claustrum and external capsule has exposed the lateral aspect of the lentiform nucleus (the putamen); D. removal of the lentiform nucleus displays fibres of the internal capsule coursing across its medial aspect; E. removal of part of the temporal lobe shows the internal capsular fibres converging on the crus cerebri of the midbrain; F. removal of the optic tract, and superficial dissection of the pons and upper medulla, emphasizing the continuity of the corona radiata, internal capsule, crus cerebri, longitudinal pontine fibres and the medullary pyramid. Dissection by E L Rees, photography by Kevin Fitzpatrick, both of the Dept. of Anatomy, Guy's Hospital Medical School, London.

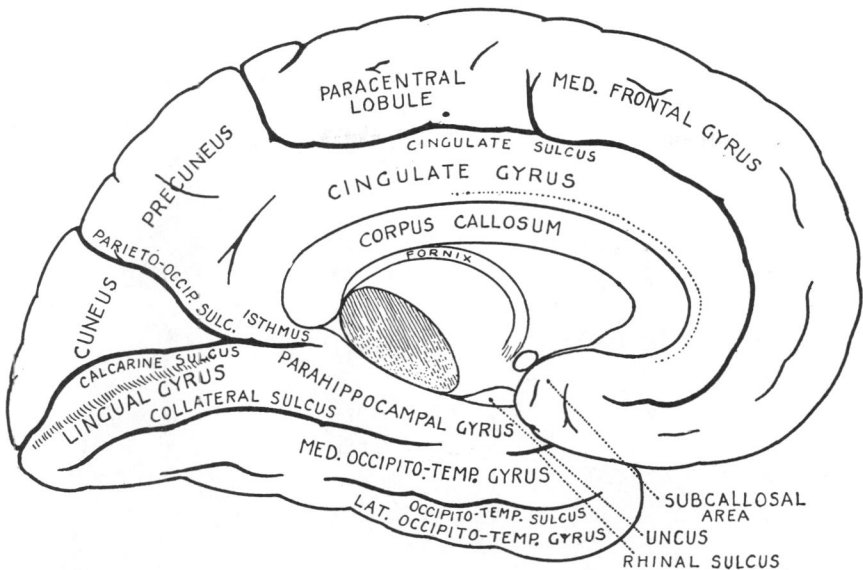

7.145A The medial surface of the left cerebral hemisphere, after sagittal section of the brain and removal of the brain stem.

The temporal lobe is inferior to the lateral sulcus and limited behind by an arbitrary line from the pre-occipital incisure (p. 1021) to the parieto-occipital sulcus, meeting the supero-medial margin about 5 cm from the occipital pole. Its lateral surface is divided into three parallel gyri by two sulci.

The *superior temporal sulcus* begins near the temporal pole and slopes slightly up and backwards parallel to the posterior ramus of the lateral sulcus. Its end curves up into the parietal lobe. The *inferior temporal sulcus* is subjacent and parallel to the superior

and often broken into two or three short sulci; its posterior end also ascends into the parietal lobe, posterior and parallel to the upturned end of the superior sulcus.

The lateral surface is thus divided into three parallel *superior*, *middle* and *inferior temporal gyri*. Along its superior margin the superior temporal gyrus is continuous with gyri in the floor of the posterior ramus of the lateral sulcus; these vary in number, extending obliquely anterolaterally from the *circular sulcus* around the insula as *transverse temporal gyri* (7.142), usually two

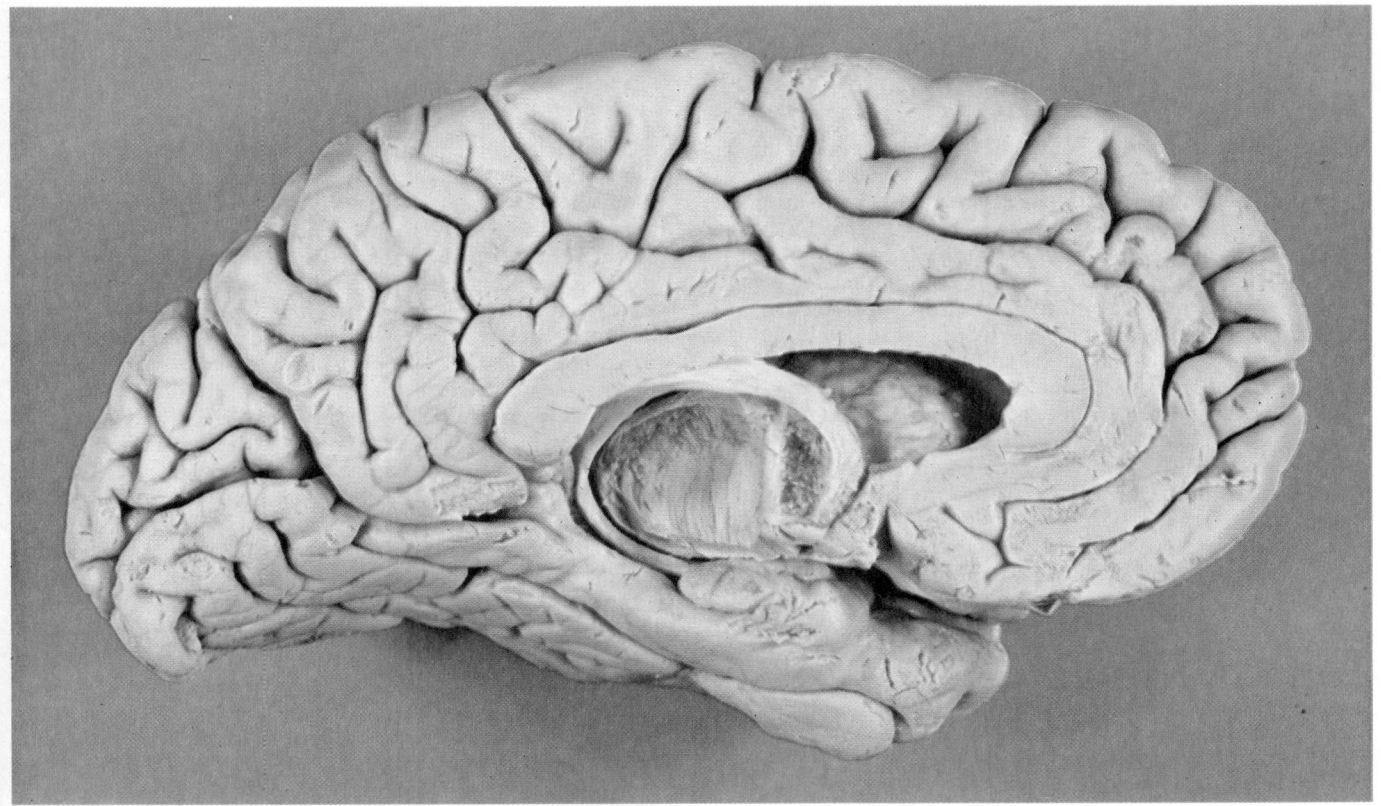

7.145B The medial surface of the left cerebral hemisphere after sagittal section of the brain, followed by removal of the brain stem and septum pellucidum. (For identification of the principal gyri and sulci of the cerebral cortex, compare with 7.145A.) The dissection has been deepened in the region of the dorsal thalamus and hypothalamus to demonstrate the

column, body, crus and fimbria of the fornix and the mamillothalamic fasciculus (compare with 7.148, 149). The head of the caudate nucleus is visible bulging into the floor of the anterior horn of the lateral ventricle. Dissection by E L Rees, photography by Kevin Fitzpatrick, both of the Dept. of Anatomy, Guy's Hospital Medical School, London.

1025

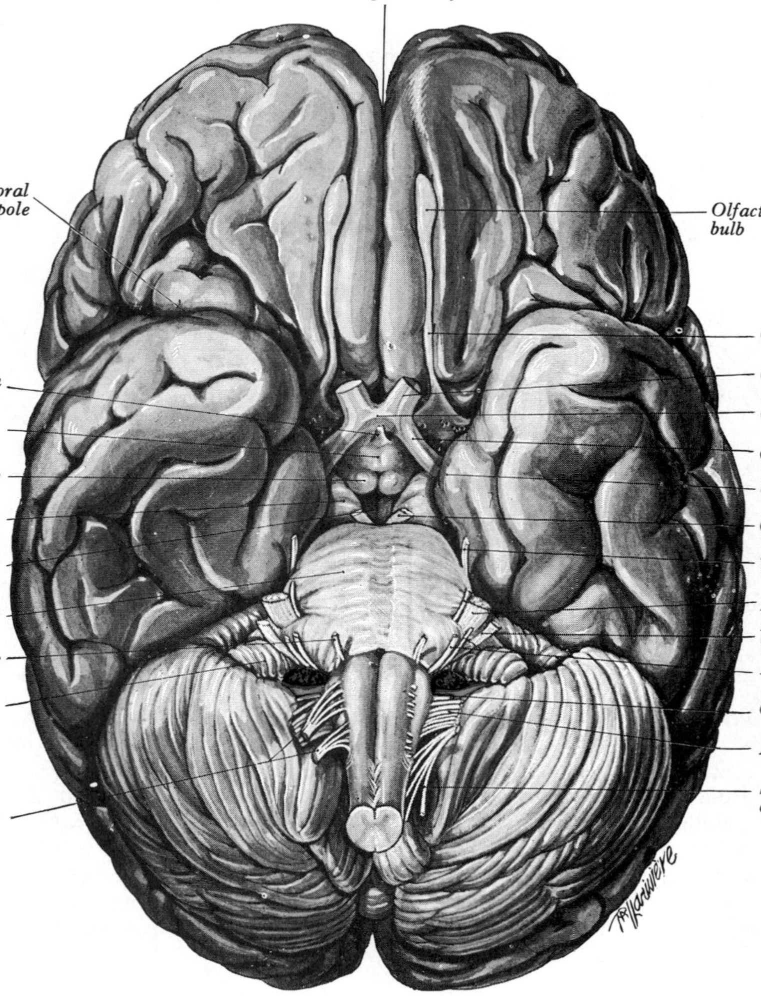

Longitudinal fissure

Temporal pole

Olfactory bulb

Infundibulum

Tuber cinereum

Mamillary body

Midbrain

Posterior perforated substance

Pons

Abducent nerve

Flocculus

Roots of hypoglossal nerve

Olfactory tract

Optic nerve

Optic chiasma

Optic tract

Uncus

Oculomotor nerve

Trochlear nerve

Roots of trigeminal nerve

Vestibulocochlear nerve

Roots of facial nerve

Glossopharyngeal nerve

Roots of vagus nerve

Spinal root of accessory nerve

7.146A Basal aspect of the brain. The anterior perforated substance (unlabelled) is between the diverging lateral and medial roots of the olfactory tract and anterolateral to the optic tract.

(anterior and posterior), but sometimes single on one or both sides (Campain & Minckler 1976). The anterior transverse temporal gyrus and adjoining part of the superior temporal gyrus are auditory in function (p. 1055) and considered to be Brodman's area 42; the anterior gyrus is approximately area 41.

The parietal lobe extends from the central sulcus in front to a lobe joining the pre-occipital incisure to the superomedial margin where it is crossed by the parieto-occipital sulcus. Inferior is the posterior ramus of the lateral sulcus and an imaginary posterior prolongation from its straight part. Parts of its boundaries are thus arbitrary. Its lateral aspect is subdivided into three areas by *postcentral* and *intraparietal sulci*.

The *postcentral sulcus* (7.141A,B), often divided into upper and lower parts, is posterior and parallel to the central sulcus. Inferiorly it ends above the posterior ramus of the lateral sulcus some distance in front of its upturned end, dividing the parietal lobe into the postcentral gyrus and a large posterior area, subdivided by the *intraparietal sulcus*, which usually starts in the postcentral sulcus near its midpoint or at the upper end of its lower part. It extends postero-inferiorly across the parietal lobe, dividing it into superior and inferior parietal lobules. Posteriorly, as its *occipital ramus*, it extends into the occipital lobe, joining the *transverse occipital sulcus* at right angles. The *postcentral gyrus* is between the central and postcentral sulci. Its cortex receives somatic sensory impulses (p. 1002) and has numerous other connections (vide infra).

The *superior parietal lobule*, between the superomedial margin and the intraparietal sulcus, is continuous anteriorly with the postcentral gyrus round the upper end of the postcentral sulcus; posteriorly it often joins the *arcus parieto-occipitalis*, surrounding

the lateral part of the parieto-occipital sulcus (7.141A,B).

The *inferior parietal lobule*, below the intraparietal sulcus and behind the lower part of the postcentral sulcus, is divided into three: the *anterior part* is the *supramarginal gyrus* arching over the upturned end of the lateral sulcus and is continuous anteriorly with the lower part of the postcentral gyrus and postero-inferiorly with the superior temporal gyrus; it may be limited behind by a small *sulcus intermedius primus* descending from the intraparietal sulcus. The *middle part*, the *angular gyrus* (believed to be concerned with the visual element in stereognosis, p. 1017) arches over the end of the superior temporal sulcus and is continuous postero-inferiorly with the middle temporal gyrus; sometimes a small *sulcus intermedius secundus* appear at its posterior end. (Anterior and middle parts of the inferior parietal lobule are subjacent to the parietal tuberosity, p. 381.) The *posterior part* arches over the upturned end of the inferior temporal sulcus on to the occipital lobe, forming an *arcus temporo-occipitalis*.

The occipital lobe is behind the arbitrary line joining the pre-occipital incisure to the parieto-occipital sulcus. The *transverse occipital sulcus* descends from the superomedial margin behind the parieto-occipital sulcus and is joined about its midpoint by the intraparietal sulcus. Its superior part is behind the *arcus parieto-occipitalis*, an arched gyrus surrounding the end of the parieto-occipital sulcus. The lateral occipital sulcus, short and horizontal on the lateral aspect of the occipital lobe, divides it into *superior* and *inferior occipital gyri* (7.141A,B). The *lunate sulcus*, when present, is just in front of the occipital pole, placed vertically and sometimes joined to the calcarine sulcus, but the two are more often separate; its lips, which are opercular, separate *striate* from *peristriate areas* but the *parastriate area* is buried in the sulcus between the other two striate areas. The lunate sulcus is posterior to the *gyrus descendens*, which is behind the superior and inferior occipital gyri. Curved superior and inferior polar sulci often appear near the ends of the lunate sulcus. The *superior polar sulcus* arches up on to the medial occipital surface near the upper limit of the lunate sulcus; the *inferior polar sulcus* arches down and forwards on to the inferior cerebral surface from the lunate's lower limit. These polar sulci enclose semilunar extensions of the striate area (p. 1058), indicating the expansion of the visual cortex associated with the large cortical macular area (Smith 1930).

The insula (7.142-144B), deep in the floor of the lateral sulcus, is almost surrounded by a *circular sulcus* and is overlapped by growth of the adjacent cortical areas and is thus only visible when the lateral sulcus is artificially widely opened; these overlapping areas are therefore termed the *opercula of the insula*, separated by ascending and posterior rami of the lateral sulcus. The *frontal* operculum or *lid* is between the anterior and ascending rami, forming the pars triangularis of the inferior frontal gyrus; it is small when the rami arise by a common stem. The *frontoparietal operculum*, between ascending and posterior rami of the lateral sulcus, is the pars posterior of the inferior frontal gyrus and lower ends of the precentral and postcentral gyri, also the lower end of the anterior part of the inferior parietal lobule. The *temporal operculum*, below the posterior ramus of the lateral sulcus, is formed by superior temporal and transverse temporal gyri. Anteriorly the inferior region of the insula adjoins the pars orbitalis of the inferior frontal gyrus.

When the opercula are removed, the insula appears as a pyramidal area, its apex inferior and near the anterior perforated substance (7.143, 144B), where the circular sulcus is deficient and the medial part of the apex is termed the *limen insulae (gyrus ambiens)*. The *insular surface* is divided into a larger *anterior* and a smaller *posterior* part by the *sulcus centralis insulae*, slanting posterosuperiorly from the apex. The anterior part is divided by shallow sulci into three or four *short gyri*, the posterior being one *long gyrus*, often divided at its upper end. The cortex of the insula is continuous with that of its opercula in the circular sulcus. The insula is approximately co-extensive with the subjacent claustrum and putamen (7.144B,C,D).

The angio-architecture of the primate insula has been studied by Vlahovitch et al (1973). Despite the development of the opercula in man and their exiguous size in, e.g. the baboon or chimpanzee, these observations indicated great similarity of vascular pattern in all primates examined.

THE MEDIAL CEREBRAL SURFACE

The medial surface (7.145A,B) is clearly visible only when the cerebral hemispheres are separated by division of all the commissures and the structures around the third ventricle (7.18). The most conspicuous feature is the great commissure, the *corpus callosum*, a broad arched band in the floor of the central region of the longitudinal fissure (7.140A,B). Its curved anterior part is the *genu*, continuous below with the *rostrum* and narrowing rapidly as it passes back to the upper end of the lamina terminalis. The genu continues above into the *trunk*, the main part of the commissure, which arches up and back to a thick, rounded posterior extremity, the *splenium*. To the concave surfaces of the trunk, genu and rostrum the bilateral vertical laminae of the septum pellucidum are attached, occupying the interval between them and the fornix, a curved, flat band inferior to it. In front of the lamina terminalis, almost co-extensive with it, is a narrow triangle of grey matter, the *paraterminal gyrus* (p. 1033), separated from the rest of the cortex by a shallow *posterior parolfactory sulcus*. A little anterior to this a short, vertical sulcus may occur, the *anterior parolfactory sulcus*; the cortex between these two sulci is the *subcallosal area (parolfactory gyrus)* (7.140, 148, 151). The anterior sloping edge of the paraterminal gyrus is sometimes called the *prehippocampal rudiment* (p. 1035).

The anterior region of the medial surface is divided into outer and inner zones by the curved *cingulate sulcus*, starting below the rostrum and passing first forwards, then up and finally backwards, conforming to the callosal curvature. Its posterior end turns up to the superomedial margin about 4 cm behind its mid-

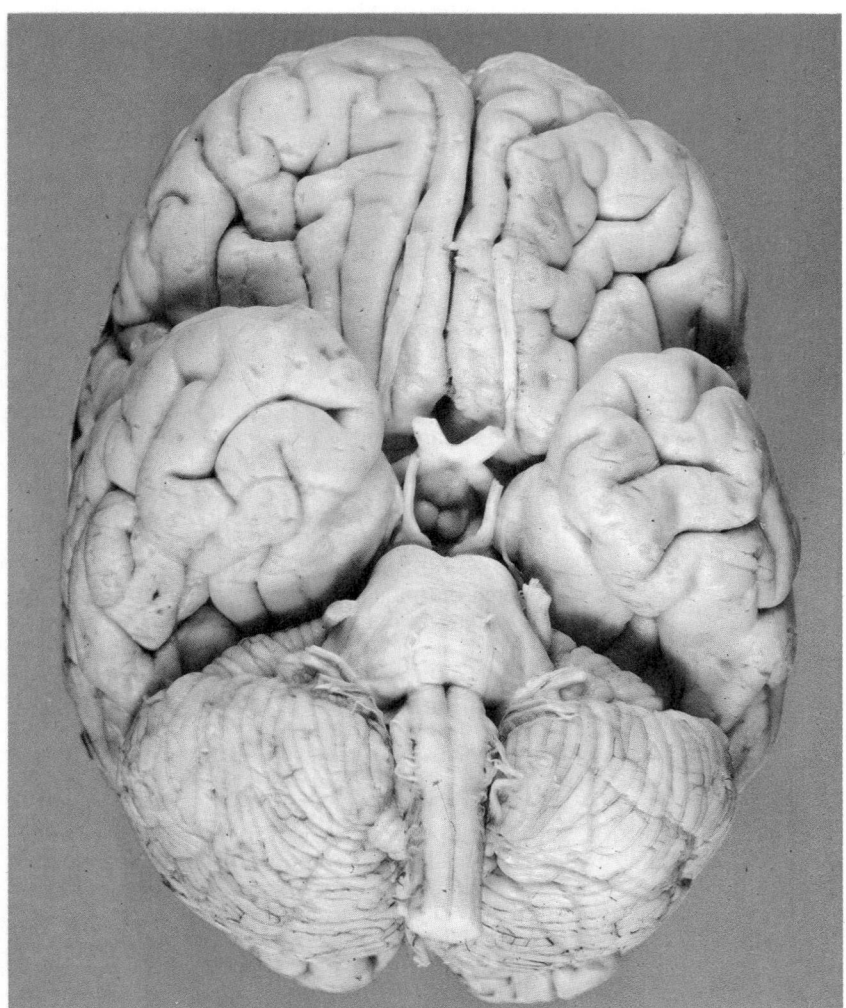

7.146B The base of the brain. For labelling compare with 7.81,146A. Dissection by E L Rees, photography by Kevin Fitzpatrick, both of the Dept. of Anatomy, Guy's Hospital Medical School, London.

point and is posterior to the upper end of the central sulcus (7.140, 145). The outer zone, except for its posterior extremity, is part of the frontal lobe, subdivided into anterior and posterior areas by a short sulcus ascending from the cingulate sulcus above the midpoint of the corpus callosum. The larger, anterior area is the *medial frontal gyrus*, the posterior being the *paracentral lobule*. The superior end of the central sulcus usually invades the paracentral lobule posteriorly and the precentral gyrus is

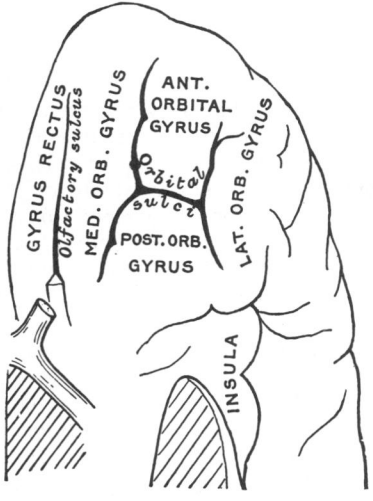

7.147 The orbital surface of the left frontal lobe.

continuous with the lobule. This area is concerned with movements of the contralateral lower limb and perineal region; clinical evidence suggests that it exercises voluntary control over defaecation and micturition (pp. 1375, 1420).

The zone under the cingulate sulcus is the *cingulate gyrus*. Starting below the rostrum this follows the callosal curve, separated by the *callosal sulcus*, and continues round the splenium to the inferior surface, continuing into the parahippocampal gyrus through the narrow *isthmus* (7.148). The cingulate sulcus is interrupted posterior to the paracentral lobule but partially continued by a variable *subparietal (suprasplenial) sulcus*.

The medial surface's posterior region is traversed by two deep sulci converging anteriorly to meet a little posterior to the splenium. These are the parieto-occipital and the calcarine sulci. The *parieto-occipital sulcus* starts on the superomedial margin about 5 cm anterior to the occipital pole, sloping down and slightly forwards to the calcarine sulcus. When opened it is clear that the parieto-occipital and calcarine sulci, though on the surface apparently continuous, are separated by the deeply sited *cuneate gyrus*. The walls of the sulcus also show two or more vertical sulci, originally exposed on the medial surface but included in the parieto-occipital sulcus by growth of the splenium (p. 194). The walls of the parieto-occipital sulcus thus resemble those of the lateral sulcus, though its contained sulci and gyri are fewer and smaller.

The *calcarine sulcus* starts near the occipital pole. Though usually restricted to the medial surface, its posterior end may reach the lateral. Directed anteriorly a little above the inferomedial margin in a slightly curved course with an upward convexity, it joins the parieto-occipital sulcus at an acute angle behind the splenium. Continuing forwards it crosses the inferomedial margin to the interior aspect of the hemisphere, forming the inferolateral boundary of the *isthmus* (which, as noted, connects the cingulate with the parahippocampal gyrus). At its junction with the parieto-occipital sulcus the calcarine sulcus is crossed by a buried *anterior cuneolingual gyrus*. Its posterior part, behind the junction with the parieto-occipital, is an axial sulcus in the long axis of the visual cortex (p. 1058); but the anterior part is a limiting sulcus separating the striate (visual) cortex from that of the isthmus. The anterior part of the calcarine is a complete sulcus, producing the *calcar avis*, an elevation in the posterior cornu of the lateral ventricle.

The quadrilateral area, posterior to the upturned end of the sulcus cinguli, anterior to the parieto-occipital sulcus, inferior to the superomedial margin and superior to the suprasplenial sulcus, is the *precuneus*; with the part of the paracentral lobule behind the central sulcus it forms the medial surface of the parietal lobe. The wedge of cortex bounded in front by the parieto-occipital sulcus, below by the calcarine sulcus and above by the superomedial margin, is the *cuneus* (its surface usually indented by one or two irregular sulci) and is the medial surface of the occipital lobe.

THE INFERIOR CEREBRAL SURFACE

This surface is divided by the stem of the lateral fissure into smaller and larger parts, respectively anterior and posterior to it (7.146A,B, 147). The anterior is the orbital region of the inferior surface, transversely concave and above the cribriform plate of the ethmoid, the orbital plate of the frontal and lesser wing of the sphenoid. A rostrocaudal *olfactory sulcus* traverses it near its medial margin, overlapped by the olfactory bulb and tract. The medial strip thus marked off is the *gyrus rectus*. The rest of this surface bears irregular *orbital sulci*, generally H-shaped, dividing it into *orbital gyri*, usually four: the anterior, medial, posterior and lateral orbital gyri (7.147).

The larger, posterior region of the inferior cerebral surface is partly superior to the tentorium but also to the middle cranial fossa and traversed by the anteroposterior collateral and occipitotemporal sulci (7.145). The *collateral sulcus* starts near the occipital pole, extending anteriorly and parallel to the calcarine sulcus, separated by the *lingual gyrus*. Anteriorly it may continue into the *rhinal sulcus* but they are usually separate. The *rhinal sulcus* runs forwards in the line of the collateral, separating the temporal pole from a somewhat hook-shaped *uncus* posteromedial to it. This sulcus is the lateral limit of the *piriform lobe* (7.151).

The *occipitotemporal sulcus* is parallel to the collateral sulcus and lateral to it. It usually does not reach the occipital pole and is frequently divided.

The *lingual gyrus*, between the calcarine and collateral sulci, passes into the *parahippocampal gyrus*, which commences at the *isthmus* where it is continuous with the cingulate gyrus and passes forwards medial to the collateral and rhinal sulci. Anteriorly the parahippocampal gyrus continues into the uncus, its medial edge lying lateral to the midbrain. The *uncus*, the anterior end of the parahippocampal gyrus, is the posterolateral boundary of the anterior perforated substance. The medial part of the uncus extends laterally above its lateral part and will be described later (7.151); its inferior surface is exposed only when its lateral, more superficial part has been removed (7.152). The uncus is part of the *piriform lobe* of the olfactory system (vide infra), phylogenetically one of the oldest parts of the pallium. (For details of the uncal region and complex terminology, consult illustrations 7.148 and 7.151.)

The *medial occipitotemporal gyrus* extends from the occipital to the temporal poles, limited medially by the collateral and rhinal sulci and laterally by the occipitotemporal. Lateral in this area is the *lateral occipitotemporal gyrus*, which is continuous round the inferolateral margin with the inferior temporal gyrus.

THE LIMBIC LOBE AND OLFACTORY PATHWAYS

During development the superolateral aspects of the diencephalon gradually merge with central areas of the inferomedial surfaces of the hemispheres (p. 192). *Bordering* the whole area of fusion on each side, a series of structures develops in the hemisphere's wall as the **limbic lobe**, a term introduced in 1878 by Broca on a comparative anatomical basis. Many structures of the limbic lobe (literally the *bordering* lobe) are phylogenetically old, with an arched form, topographically interposed between the diencephalon and the cerebral neopallium (7.148, 149). Interest was heightened when Papez (1937) suggested a role in emotional behaviour; subsequent research, most voluminously, has emphasized its profuse connections with the olfactory system, hypothalamus, thalamus, epithalamus and areas of the neocortex. It is intimately involved in the higher integration of visceral, olfactory and somatic information and patterns of complex long and short-term homeostatic responses. These include seeking and capturing prey, courtship, mating, rearing of young, subjective and expressive elements in emotional responses, the balance between aggressive and communal behaviour and (particularly the hippocampus), the formation of memory (Maclean 1958, 1969, Livingston 1970). For reviews, see Isaacson (1974), Isaacson & Pribram (1975), Livingston & Hornykiewicz (1978).

The terminology of these regions is complex; alternative schemes are used and common agreement in terms of a rational basis for scientific definition has not been achieved (White 1965, Brodal 1969, Bargmann & Schade 1963). However, the increasingly widespread popularity of the terms *limbic lobe* and *limbic system* indicates that many investigators currently believe that these collective names and concepts are useful, despite differences in application of the terms by some and strong opposition to their use at all by a minority (Brodal 1969). Most protagonists include a series of subcortical nuclei and their connections, the phylogenetically older areas of the cortex, archipallium and paleopallium and the adjacent neopallium—the cingulate and parahippocampal gyri and associated tracts. Difficulties arise in placing precise boundaries between these and the adjacent

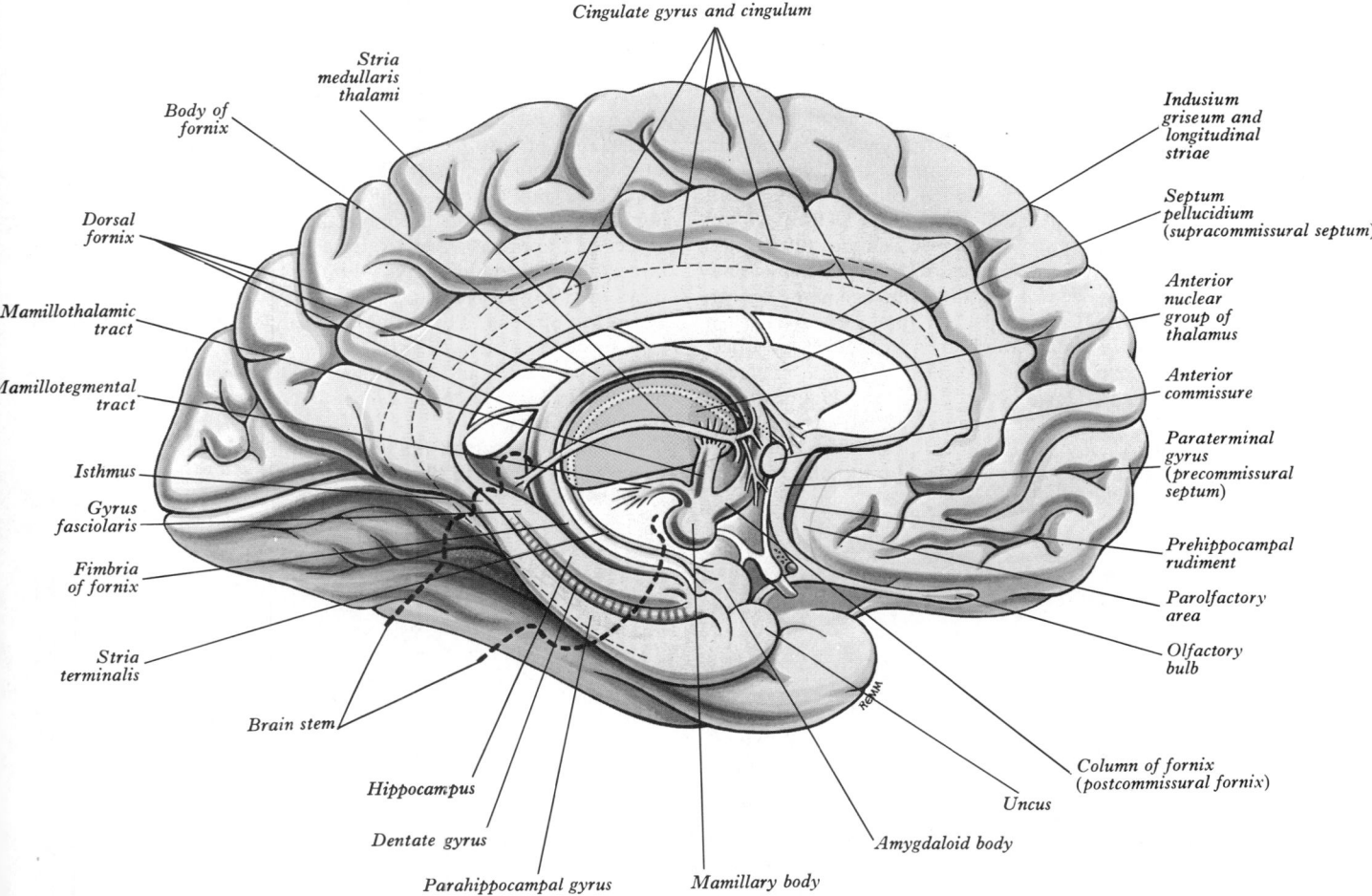

Cingulate gyrus and cingulum

Stria
medullaris
thalami

Body of
fornix

Dorsal
fornix

Mamillothalamic
tract

Mamillotegmental
tract

Isthmus

Gyrus
fasciolaris

Fimbria
of fornix

Stria
terminalis

Brain stem

Hippocampus

Dentate gyrus

Parahippocampal gyrus

Mamillary body

Amygdaloid body

Uncus

Column of fornix
(postcommissural fornix)

Olfactory
bulb

Parolfactory
area

Prehippocampal
rudiment

Paraterminal
gyrus
(precommissural
septum)

Anterior
commissure

Anterior
nuclear
group of
thalamus

Septum
pellucidium
(supracommissural septum)

Indusium
griseum and
longitudinal
striae

7.148 Diagram of a dissection of the medial aspect of a cerebral hemisphere to demonstrate the majority of the structures included under the term limbic system in the present account; these are coloured yellow. The anterior nuclear group of the dorsal thalamus is coloured orange and included with the limbic system; the remainder of the dorsal thalamus is magenta. The approximate position of the brain stem which was removed in the course of dissection is outlined in a heavy interrupted line. This diagram should be compared with the colour photograph of a dissection prepared from the *lateral* aspect of a hemisphere (7.149).

regions, with which some are functionally connected, e.g. opinions differ concerning the inclusion of the frontal cortex and intimately related hypothalamus. There is a need for international nomenclature for as many orders of subdivision of the brain as may be needed, based on the most widely acceptable criteria available, with continuous revision as further knowledge accumulates. In this account 'limbic lobe' and 'system' are used and the structures described under these headings will be listed.

Even greater terminological difficulty attaches to the term **rhinencephalon** (Bargmann & Schade 1963). The name was once used for all brain structures associated with olfaction and it was long assumed that most limbic structures were primarily olfactory paths and centres of integration. But studies in comparative, developmental, physiological, anatomical and behavioural fields have changed the emphasis. Since 'rhinencephalon' has many meanings for different workers, it will not be used here.

The limbic system includes:

(1) Olfactory nerves, bulb and tract (with the nervus terminalis and the transient accessory olfactory bulb and vomeronasal nerve).

(2) The anterior olfactory nucleus.

(3) Medial, intermediate and lateral olfactory striae, with medial and lateral olfactory gyri.

(4) The olfactory trigone, anterior perforated substance and olfactory tubercle and diagonal band of Broca.

(5) The piriform lobe, including: (a) the lateral olfactory gyrus continuing into the gyrus ambiens, together forming the prepiriform cortex; (b) lateral olfactory stria continuing into the gyrus semilunaris (periamygdaloid area); (c) the uncus hippocampi, including the uncinate gyrus, tail of the dentate gyrus

(band of Giacomini) and intralimbic gyrus; (d) the entorhinal area (area 28), cranial part of the parahippocampal gyrus.

(6) The amygdaloid complex of nuclei.

(7) Septal areas, including the septum pellucidum and septum verum (a nuclear complex, much of which is deeply placed but in part superficially corresponding to the paraterminal gyrus).

8) The hippocampal formation, including: (a) the prehippocampal rudiment, indusium griseum and longitudinal striae; (b) the gyrus fasciolaris, Ammon's horn, dentate gyrus, subiculum and related regions.

(9) The fornix and its various ramifications and divisions.

(10) The stria terminalis.

(11) The stria medullaris thalami (stria habenularis).

(12) Cingulate and parahippocampal gyri. The hypothalamus, the medial part of the thalamus, prefrontal cortex and anterior commissure, all intimately related to the foregoing structures, are described elsewhere. The hypothalamus is regarded by some as the *essential centre* of the limbic system (Isaacson 1974).

THE OLFACTORY BULB

The olfactory bulb (7.146A), a flat, oval, reddish-grey mass, is superior to the medial edge of the frontal's orbital plate near the lateral margin of the ethmoid's cribriform plate, inferior to the anterior end of the olfactory sulcus on the orbital surface of the frontal lobe.

Olfactory nerve fibres, central processes of olfactory cells in the nasal mucosa (p. 1094), converge to the inferior aspect of the cribriform plate, collecting into about 20 bundles, which traverse the plate's foramina to enter the inferior surface of the bulb.

The olfactory bulb and tract develop (p. 192) as a hollow diverticulum from the floor of the primitive cerebral hemisphere. The basal part of this elongates to form the olfactory tract and in human embryos the cavity of the olfactory bulb ('olfactory ventricle') and of the elongating tract are gradually obliterated by fusion or their walls. The site of the original cavity is sometimes marked by vestigial groups of modified ependymal cells. Therefore the olfactory bulb has a *radial organization* with a number of superimposed layers. This laminar pattern is well defined in many mammals other than man and in the human fetus, but becomes less distinct as the brain matures. Detailed histology of the olfactory bulb has been known since the turn of the century (Cajal 1890, 1911, 1955, Blanes 1898, Le Gros Clark 1951, 1957, Valverde 1965); the Nissl and Golgi techniques used have been supplemented by experimental studies with degeneration techniques (e.g. Powell & Cowan 1963, Price 1968, Heimer 1968), physiological studies (e.g. Sem-Jacobsen et al 1956, Døving 1967), electron microscopy (Pinching & Powell 1971, Shepherd 1974, 1977), autoradiography and immunofluorescence (Macrides & Davis 1983).

From the surface inwards the olfactory bulb consists of: (1) an olfactory nerve fibre layer; (2) a layer of synaptic glomeruli and interglomerular spaces; (3) a molecular and external granular layer; (4) a layer of mitral cells; (5) an internal granular layer; (6) fibres of the olfactory tract.

Neurons of the olfactory bulb comprise: (1) large *mitral cells* in a layer one or two cells thick; (2) internal, middle and external *tufted cells*, similar to but smaller than mitral cells and diminishing in size throughout the molecular layer from the superficial aspect of the mitral layer to the deep aspect of the glomerular layer; (3) small round or stellate *internal granule cells*; (4) various types of small *periglomerular cell* distributed around and in the synaptic glomeruli. The periglomerular and small external tufted cells form the *external granular layer* of the earlier histologists. Histofluorescence and immunocytochemical techniques show periglomerular cells to be *dopaminergic*, classified as cell group A15 (see p. 993).

Previous accounts of the olfactory bulb, as a simple relay between olfactory nerve fibres and dendrites of mitral cells whose axons enter the olfactory tract, were, as so often happens, too

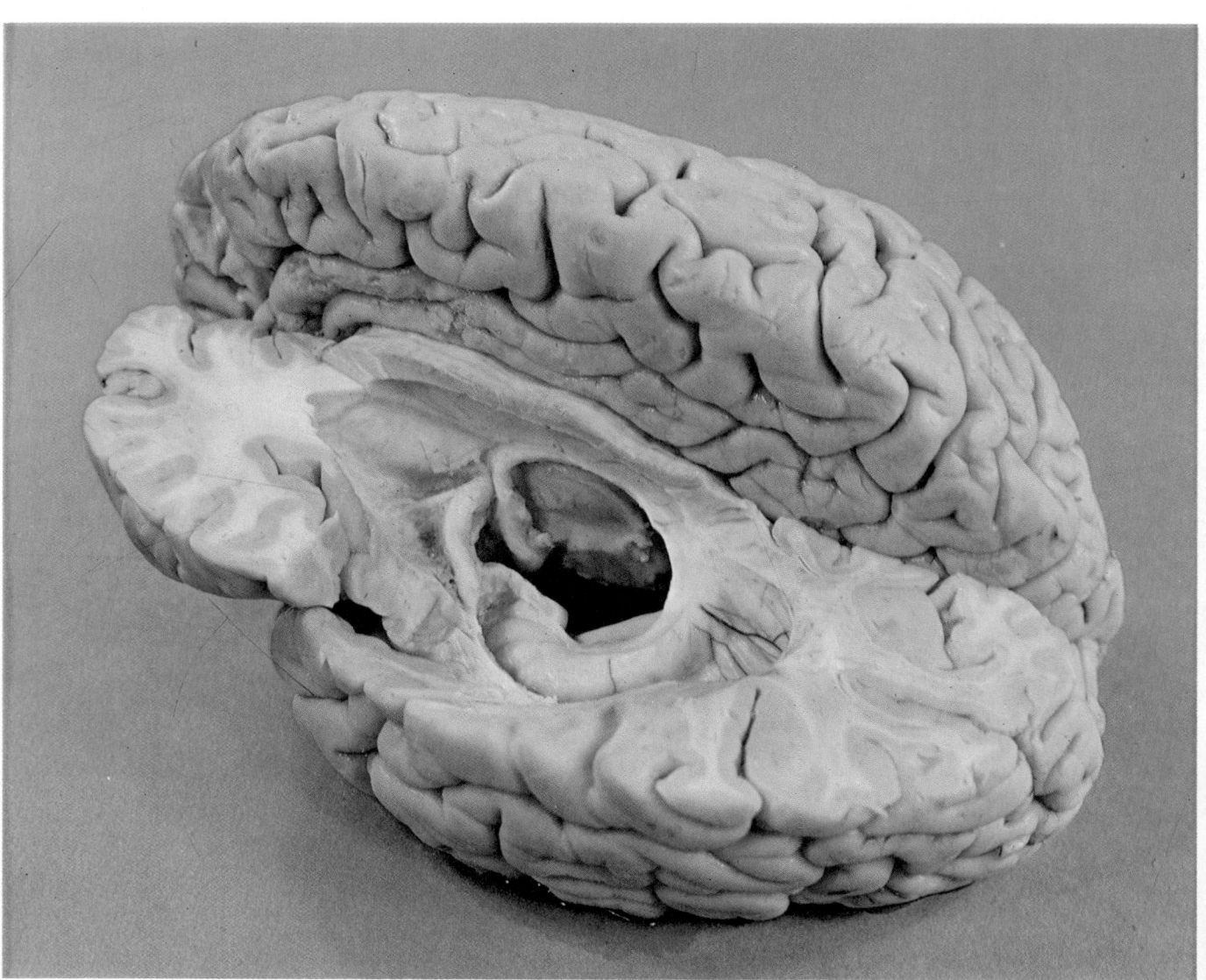

7.149 A dissection of the left cerebral hemisphere from the superolateral aspect to demonstrate various structural features of the limbic system. The corpus callosum is divided sagittally in the region of its body only; the frontal, temporal and occipital lobes have been sectioned horizontally and their superior parts removed. The left lentiform nucleus, much of the caudate nucleus and dorsal thalamus have been removed and the floor of the inferior horn of the lateral ventricle laid open. Note: (1) the horizontally sectioned head of the caudate nucleus; (2) the spiral disposition of the fornix as it curves from the mamillary body through its left column, body, crus and fimbria; (3) the curved elevation of the hippocampus projecting into the floor of the inferior horn of the ventricle and ending anteriorly as the grooved pes hippocampi; (4) the anterior commissure entering the left hemisphere immediately anterior to the column of the fornix and passing laterally, to diverge into small anterior and large posterior components; between the latter the deep aspect of the anterior perforated substance is visible; (5) within the curve of the fornix the medial aspect of the *right* thalamus crossed superiorly by the stria medullaris thalami; (6) coursing above the corpus callosum a longitudinal white stria is visible and above this arches the right gyrus cinguli. Compare with 7.148. Dissection by A M Seal, photography by Kevin Fitzpatrick, both of the Dept. of Anatomy, Guy's Hospital Medical School, London.

THE OLFACTORY BULB   NEUROLOGY   7

simple. The bulb is structurally complicated (see references quoted for details). The following features must be noted: (1) There is great *convergence* of olfactory fibres to dendrites of mitral and tufted cells occupying much of the synaptic glomeruli (vide infra). (2) A number of cells, mitral, tufted and others contribute axons to the tract which proceed to other centres. (3) Several types of interneuron provide for the complex interaction between the main conductors and complicated feedback routes. (4) Transmission in the bulb is modulated by centrifugal axons from the contralateral olfactory bulb, anterior olfactory nucleus and other centres (vide infra). Some of these features are summarized in illustration 7.150. It is convenient to consider first the main path through the bulb, followed by the complications introduced by local interneurons and centrifugal pathways.

In addition to *basal* dendrites, establishing synaptic connections with granule cell processes, the large mitral cells and the more prominent external tufted cells both have large *apical* dendrites passing superficially to the glomerular layer, where they branch to form dense arborizations in several synaptic glomeruli. Incoming olfactory nerve fibres converge on the glomeruli, forming numerous axodendritic synapses with apical dendrites of mitral and tufted cells, axons of the latter passing through the inner granular layer and giving off collateral branches before continuing to form the main constituents of the olfactory tract.

Synaptic glomeruli of the olfactory bulb are large, spherical territories, often exceeding 100 µm in diameter, containing thousands of neurites from various sources, with a wide variety of synaptic connections between them. In addition to the large dendrites of mitral and internal tufted cells just described, a glomerulus also receives dendrites from types of *periglomerular cell* (including external granule cells, external tufted cells and superficial short-axon cells of different authors, see Pinching & Powell 1971). About 100 neurons contribute dendrites to each

glomerulus and (in rabbits) each receives terminals of about 25 000 olfactory nerve axons. Axonal terminals of several periglomerular cells also penetrate the glomerulus. Detailed analyses of synaptic contacts between these elements and possible modes of operation, are still in progress (Price & Powell 1970, Pinching & Powell 1971, Shepherd 1974, 1977, 1978). The following synaptic contacts have been described. Terminals of olfactory axons do not receive axoaxonic (presynaptic) contacts from other sources, but establish axodendritic contacts with mitral and tufted cell dendrites and also dendrites of periglomerular cells. Periglomerular cell dendrites form reciprocal dendrodendritic contacts with mitral and tufted cell dendrites; one or both of the elements in such a double dendrodendritic contact often also receive an axodendritic synapse from an olfactory terminal. Dendrites of periglomerular cells receive several axodendritic synapses from axons of other periglomerular cells.

Interglomerular spaces are partly occupied by cell somata of different types of periglomerular cell and their processes; some have dendrites ramifying in spaces and in glomeruli, as described above. Other interneurons, designated short-axon cells, have wholly extraglomerular dendritic trees. Interglomerular spaces are also penetrated by mitral cell dendrites, many axonal ramifications and terminals of periglomerular cells and terminations of centrifugal axons from the olfactory tract. Again, reciprocal dendrodendritic contacts occur between periglomerular dendrites and those of the mitral and tufted cells, with axodendritic contacts between either the extrinsic or the intrinsic axons and periglomerular cells. Details of these contacts will not be pursued here (see Pinching & Powell 1971, Shepherd 1974, 1978).

The internal granular layer, deep to the layer of mitral cells, contains large numbers of microneurons, *amacrine granule cells* (Rall et al 1966, Price & Powell 1970, Shepherd 1974, 1977, 1978) with no true axons. Their dendrites branch repeatedly; some branch locally, others pass into the molecular layer where, once

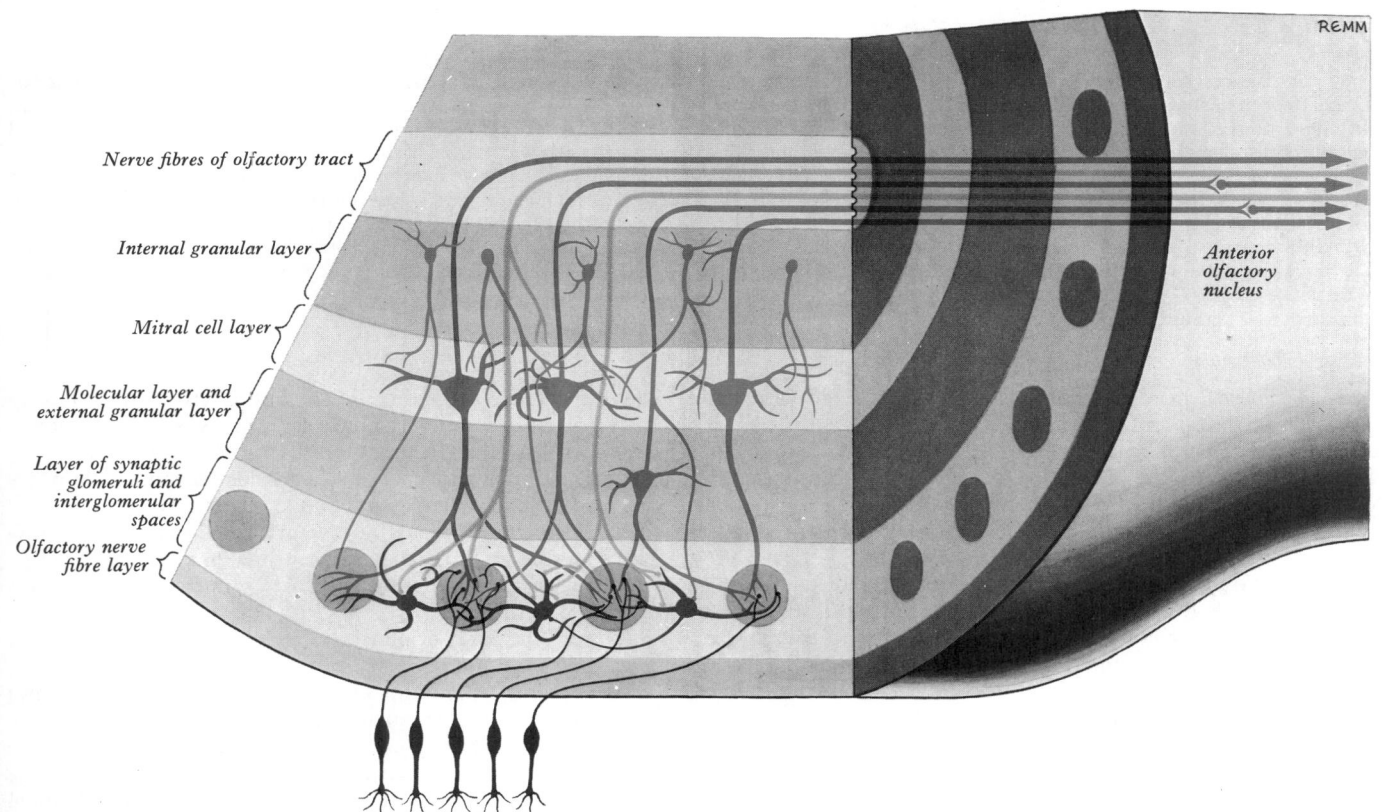

7.150   A scheme of the olfactory bulb based upon neurocytological and experimental studies in a number of mammalian types. The radial organization of the bulb into 'layers', with their principal neuron types, and an approximate indication of their main connectivity patterns is shown. The 'layers' of the bulb which receive the names indicated are, of course, merely a convenient descriptive frame of reference; they should not obscure the functionally essential interconnections which cross the boundaries between these zones. Red = mitral and tufted neurons and their

processes; blue = internal granule neurons; purple = dopaminergic periglomerular neurons; black = olfactory receptor neurons and their processes. Note that the olfactory tract consists of (1) centripetal axons of mitral and tufted cells, some of which synapse with neurons in the anterior olfactory nucleus and (2) centrifugal axons (yellow) which terminate in the different zones indicated. Refer to text for a more detailed description of both the organization of the bulb and the destinations of olfactory tract fibres.

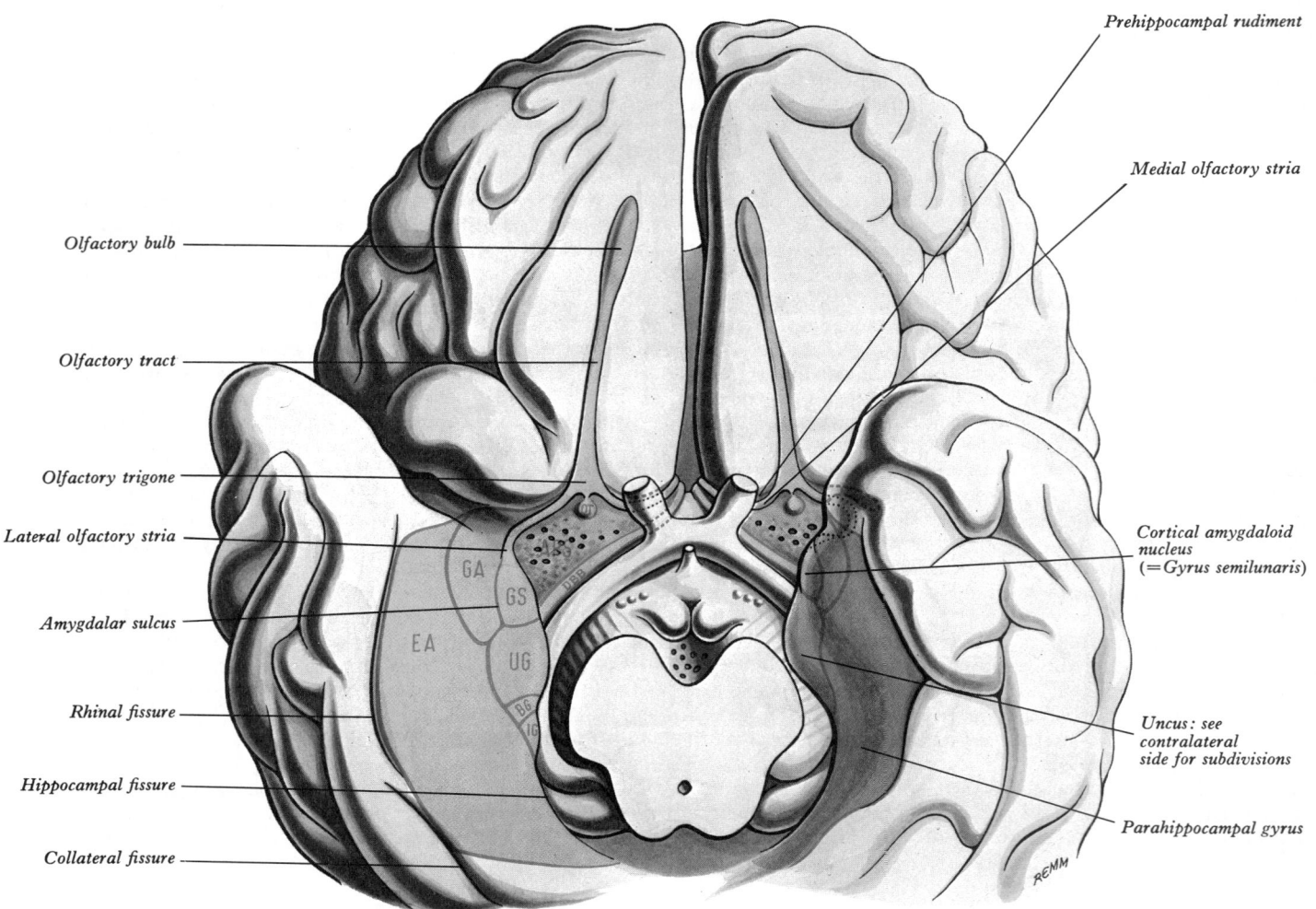

Olfactory bulb

Olfactory tract

Olfactory trigone

Lateral olfactory stria

Amygdalar sulcus

Rhinal fissure

Hippocampal fissure

Collateral fissure

Prehippocampal rudiment

Medial olfactory stria

Cortical amygdaloid nucleus (=Gyrus semilunaris)

Uncus: see contralateral side for subdivisions

Parahippocampal gyrus

7.151 A diagram of structures on the inferior aspect of the human brain in the area immediately surrounding the optic nerves, chiasma, optic tracts and interpeduncular fossa. Many of these structures are intimately related to the olfactory and limbic systems; they are coloured blue. The right temporal pole has been displaced laterally to expose underlying structures. In addition to the features which have been labelled fully, the abbreviations used have the following significance: OT = olfactory tubercle; APS = anterior perforated substance; DBB = diagonal band of Broca. The uncus hippocampi is divided into three areas: IG = the intralimbic gyrus; BG = the band of Giacomini; UG = the uncinate gyrus. The lateral olfactory stria continues into the gyrus semilunaris (GS); this is bordered laterally by the gyrus ambiens (GA); whilst further laterally is the entorhinal area (EA) which is the rostral extension of the parahippocampal gyrus. Note the curved extensions of the prehippocampal rudiments, medial olfactory striae and diagonal bands of Broca on to the medial aspect of the hemisphere. The triangular midline zone between the converging diagonal bands and superior to the optic chiasma is the lamina terminalis. The occasional intermediate olfactory stria which merges with the olfactory tubercle is illustrated but unlabelled. After Kuhlenbeck; redrawn and modified from Nauta & Haymaker 1969, with permission of the authors and publishers.

again, reciprocal dendrodendritic synapses are made with the dendrites of mitral and tufted cells. Granule cell dendrites also receive axodendritic synapses from the recurrent collaterals of mitral and tufted cell axons and from many terminals of centrifugal axons in the olfactory tract.

In short, neurons of the olfactory bulb and their processes present some of the most complex synaptic patterns yet described. Granule and periglomerular cells are considered inhibitory interneurons. By their geometric organization and array of contacts they provide a substrate for interaction between the principal conduction channels through the bulb, mediating feedforward, lateral and feed-back inhibitory processes and inhibitory effects by centrifugal axons entering the bulb from the opposite olfactory tract.

Chemoarchitecture

This is more complex than, e.g. the cerebellar cortex. Olfactory nerve fibres contain the dipeptide carnosine and a specific olfactory marker protein (mol wt 20 000) but their role in neuromediation is conjectural. The mitral neurons and deep tufted neurons and their projection axons have glutamate or aspartate as their neuromediator. Intermediate and external tufted neurons, however, have dopamine or substance P and in some instances contain both. The periglomerular cells and the granule cells may be treated together, the majority contain GABA, the minority dopamine; quite frequently enkephalin co-exists with either in periglomerular cells. In granule cells GABA alone is the outstanding neuromediator; in a small number enkephalin alone has been demonstrated; here they do not co-exist.

The *centrifugal supply* to the olfactory bulb has fibres from different sources containing different neuromediators (see p. 891):

Enkephalin from the anterior olfactory nucleus and precommissural hippocampus.

VIP from the anterior olfactory nucleus, piriform cortex and amygdala.

LHRH from the anterior olfactory nucleus, precommissural hippocampus, medial septal and diagonal band nuclei.

SST from the piriform cortex and diagonal band nucleus.

Substance P from mesencephalic raphe nuclei.

Cholinergic fibres from the ventral nucleus of the diagonal band.

Serotonergic fibres from mesencephalic raphe nuclei.

Noradrenergic fibres from the nucleus coeruleus.

In many non-primate mammals and submammalian forms, a diverticulum with a specialized epithelium, the *vomeronasal organ* (of Jacobson), is sited on each side of the nasal septum. Neural fascicles from this epithelium run with ventral groups of olfactory nerve fibres and then converge to form the *vomeronasal nerve* ending in an *accessory olfactory bulb* on the dorsomedial aspect of the main bulb. Although an accessory bulb and vomeronasal nerve occur on one or both sides in some human embryos, both structures degenerate (McCotter 1915, Hume 1940, Crosby & Humphrey 1941).

THE OLFACTORY TRACT

The olfactory tract (7.146A,B, 151), a narrow white band issuing from the posterior pole of the olfactory bulb, continues on the inferior (orbital) surface of the frontal lobe, along the olfactory sulcus. Triangular in transverse section, its narrow apex recessed into the olfactory sulcus, it consists of centripetal axons of mitral and tufted cells and some centrifugal axons from the opposite bulb and anterior olfactory nucleus which have crossed in the anterior commissure and from neurons in or near the anterior perforated substance and various brain-stem levels (vide supra). The precise contribution of these and other possible centrifugal sources is still under investigation (Cragg 1962, Powell & Cowan 1963, Powell et al 1965, Price 1968, Heimer 1968).

The Anterior Olfactory Nucleus

Posteriorly in the olfactory bulb its characteristic cell layers disappear, but the granule cell layer is represented in the olfactory tract by scattered medium-sized multipolar neurons, the *anterior olfactory nucleus*; these continue into the olfactory striae and trigone (vide infra) to the grey matter of the prepiriform cortex, the anterior perforated substance and precommissural septal areas. Many centripetal axons from mitral and tufted cells relay in, or give collaterals to, the anterior olfactory nucleus, whose axons continue with the remaining direct fibres from the bulb into the olfactory striae.

THE OLFACTORY STRIAE AND TRIGONE

As the olfactory tract approaches the anterior perforated substance it flattens and splays out into the *olfactory trigone* (olfactory pyramid), from whose caudal angles fibres of the tract continue as diverging *medial* and *lateral olfactory striae* bordering the anterior perforated substance (7.151). In some brains a small *intermediate stria* passes from the centre of the trigone to sink into the anterior perforated substance.

The lateral olfactory stria follows the anterolateral margin of the anterior perforated substance as a visible bundle continuing into the limen insulae (p. 1036), where it bends posteromedially to merge with an elevated region, the *gyrus semilunaris*, at the rostral margin of the uncus hippocampi (7.151); the gyrus incorporates some of the *corticomedial* subdivision of the *amygdaloid complex* (vide infra). A tenuous grey layer covering the lateral olfactory stria is the *lateral olfactory gyrus*, merging laterally with the *gyrus ambiens*, part of the limen insulae. The lateral olfactory gyrus and gyrus ambiens form the *prepiriform region* of the cortex, passing caudally into the *entorhinal area* of the parahippocampal gyrus.

The prepiriform and periamygdaloid regions and the entorhinal area (area 28) together comprise **the piriform lobe**, bounded laterally by the *rhinal sulcus* and relatively prominent in macrosmatic mammals and during fetal development. The relative positions of these structures will be clarified by reference to figure 7.151.

The medial olfactory stria, covered thinly by grey matter, the *medial olfactory gyrus*, passes medially along the rostral boundary of the anterior perforated substance towards the medial continuation of the diagonal band of Broca (vide infra). Together they curve up on the medial aspect of the hemisphere, anterior to the attachment of the lamina terminalis (7.151). The diagonal band enters the paraterminal gyrus, the medial stria becoming indistinct as it approaches the boundary zone which includes the paraterminal gyrus, parolfactory gyrus and, between them, the prehippocampal rudiment (7.148).

THE ANTERIOR PERFORATED SUBSTANCE

An important landmark on the cerebral base, this is caudal to the olfactory trigone and diverging olfactory stria, in the angle between the optic chiasma and tract medially and the uncus caudally (7.151). It is continuous medially above the optic tract with the grey matter of the tuber cinereum and, more anteriorly, with the paraterminal gyrus. Laterally, it reaches the limen insulae to continue into the prepiriform cortex; more caudally it merges with the periamygdaloid area (gyrus semilunaris). Superiorly, it is continuous with the grey matter of the corpus striatum and claustrum through aggregations of grey and white matter forming the *substantia innominata*. Part of the latter, with fascicles of the ansa lenticularis and anterior commissure, separate the anterior perforated substance from the globus pallidus. The aggregates of grey matter in the substantia innominata are grouped as the *basal forebrain nucleus* (of Meynert); over 90% of its neurons are strongly cholinergic and these together with similarly reactive large neurons in the medullary laminae of the globus pallidus are classified as cholinergic group Ch4.

The anterior perforated substance is related to the end of the internal carotid artery and origins of the anterior and middle cerebral arteries, from which central arteries pierce the surface to supply deeper structures (p. 747). Caudal to the olfactory trigone the anterior perforated substance displays a small *olfactory tubercule*, variable in prominence, into the base of which the occasional intermediate olfactory stria sinks. The tubercle is large in macrosmatic animals but greatly reduced and sometimes indistinguishable in mankind. The intermediate stria, though sometimes absent, is occasionally two or three fine striae radiating into the perforated substance. The caudal zone of the substance, where it adjoins the optic tract, is the smooth surface of the *diagonal band of Broca*. Caudolaterally the band is continuous with the periamygdaloid area; rostromedially it continues above the optic chiasma into the paraterminal gyrus (precommissural septum).

GENERAL ORGANIZATION OF THE LIMBIC GREY MATTER

Variations in cytoarchitectonic pattern and the numerous and divergent views on connections of the various limbic regions cannot be detailed, but the following salient points may be noted.

The six laminae which, though varying in different regions, are held to characterize the neocortex (p. 1041) do not appear as such in archipallial and paleopallial regions of the limbic system. In some sites a laminar pattern is absent or scarcely distinguishable, in other areas an obvious but different lamination exists, varying from a primitive three-layered to a transitional six-layered variety where the limbic cortex merges with the neopallial regions. Medial and lateral olfactory gyri, the prepiriform cortex and indusium griseum have isolated patches or a film of grey matter and associated fibres, with no laminar pattern, and rostral and caudal zones of the anterior perforated substance, diagonal band of Broca, paraterminal gyrus and periamygdaloid cortex are poorly differentiated without laminar pattern; specific cell aggregations do occur in some of them, including the *nucleus of the lateral olfactory tract* in the periamygdaloid region and the *nucleus of the diagonal band of Broca*. The centre of the anterior perforated substance is more differentiated, with three layers: outer plexiform, intermediate pyramidal and deep polymorphic; this also has characteristic cup-shaped groups of granule cells in the pyramidal layer (sometimes extending into the plexiform layer). These *islands of Calleja* vary considerably, the medial, being largest, is near another prominent group of larger pleomorphic cells, the *nucleus accumbens septi*. The structure of the hippocampal formation is discussed below, but it is notable that the dentate gyrus and Ammon's horn both consist of a primitive trilaminar cortex showing a gradual transition through the subicular region to a modified six-layered variety in the entorhinal area of the parahippocampal gyrus.

TERMINATION OF THE OLFACTORY TRACT

Terminal areas of olfactory fibres have been widely studied in mammals, including monkeys, and, despite minor differences, substantial agreement reigns; what obtains in the human brain is less certain. As we have seen, olfactory fibres are axons of mitral and tufted cells in the olfactory bulb, some synapsing in the anterior olfactory nucleus, axons from this continuing with direct fibres of the tract. Through the *lateral olfactory stria* tract fibres reach and synapse with neurons in the lateral part of the anterior perforated substance, the lateral olfactory gyrus, prepiriform cortex and corticomedial group of amygdaloid nuclei (vide infra), regions often grouped as the **primary olfactory cortex**. In contrast to all other sensory pathways, fibres reach these cortical areas without synapsing in the thalamic nuclei. The entorhinal area (area 28) of the parahippocampal gyrus, in the caudal part of the piriform lobe, receives few or no direct fibres but profuse connections from the primary cortex, hence it is sometimes named the **secondary olfactory cortex**. Primary and secondary olfactory cortices are probably the principal areas for the subjective appreciation of olfactory stimuli.

The variable *intermediate olfactory stria* ends in the anterior perforated substance but the destination of the *medial stria* is less certain: some of its fibres also end in the anterior perforated substance; others may reach the parater minal gyrus and adjacent regions; some of its fibres decussate in the anterior commissure and end in the contralateral anterior olfactory nucleus and olfactory bulb.

The primary olfactory cortex also projects to the basolateral part of the amygdaloid complex, septal areas, nucleus medialis dorsalis of the thalamus and many hypothalamic nuclei.

THE AMYGDALA

The amygdala (amygdaloid body, amygdaloid nuclear complex) is a series of neuronal masses dorsomedial in the temporal pole. It forms the ventral, superior and medial walls of the tip of the inferior cornu of the lateral ventricle. Its relations are complicated. It is partly continuous above with the inferomedial margin of the claustrum; fibres of the external capsule and substriatal grey matter incompletely separate it from the putamen and globus pallidus; it is close to the optic tract. It is partly deep to the gyrus semilunaris, gyrus ambiens and uncinate gyrus (7.151). Transitional zones connect it with the anterior perforated substance, prepiriform cortex and parahippocampal gyrus. Caudally, it is near the ventral part of the hippocampus and it fuses with the apex of the tail of the caudate nucleus; the stria terminalis issues from its caudal aspect. (For a comprehensive review of the amygdala consult Eleftheriou 1972.)

Divisions of the Amygdala

The amygdaloid complex is divided into two main groups, the *corticomedial* and *basilateral* nuclei, with junctional zones where these adjoin or partly fuse with adjacent areas. Details of the subdivisions, connections and homologies are only briefly summarized here (Crosby & Humphrey 1941, Crosby et al 1962, Eleftheriou 1972, Isaacson 1974).

The corticomedial amygdaloid complex has *central*, *medial* and *cortical amygdaloid nuclei, the nucleus of the lateral olfactory stria* and a transitional ill-differentiated *anterior amygdaloid area*. The cortical nucleus is a rudimentary cortex, with irregular groups of pyramidal and granule cells, occupying the surface of the gyrus semilunaris. The corticomedial complex is continuous through the transitional zones with the anterior perforated substance, the diagonal band of Broca, substantia innominata, putamen, caudate nucleus and surrounding cortical areas of the uncus and parahippocampal gyrus. It is small in the human brain.

The basilateral amygdaloid complex, large and well differentiated in the human brain, has *lateral, basal* and *accessory basal amygdaloid nuclei* and is partly continuous with the claustrum and, through a transitional zone, with the cortex of the parahippocampal gyrus.

Connections of the Amygdaloid Complex

These are incompletely established in the human brain; most of the available information is derived from degeneration and electrophysiological studies in mammals, including monkeys (Le Gros Clark & Meyer 1947, Allison 1954, Powell et al 1965, Gloor 1960, Wendt & Albe-Fessard 1962, Eleftheriou 1972).

Afferent connections reach the corticomedial complex via the lateral olfactory stria from the olfactory bulb and anterior olfactory nucleus, with the basilateral complex receiving many afferents from the cortex of the piriform lobe (p. 1033). Other afferents converge from the hypothalamic nuclei, specific and non-specific thalamic nuclei, brain-stem reticular formation and probably areas of the neocortex (vide infra).

The best known **outflow** is via the stria terminalis to the septal areas, preoptic and adjacent regions of the hypothalamus, some continuing into the stria medullaris thalami to reach the habenular nucleus. Other *amygdalofugal fibres* follow a more central direct route (p. 1009) to many hypothalamic nuclei, the cortex of the piriform lobe, nucleus medialis dorsalis of the thalamus and the reticular formation of the midbrain tegmentum. Reciprocal connections probably exist between the complex and the orbitofrontal, cingulate and temporal regions of the neocortex; *interamygdaloid fibres* run through the anterior commissure (vide infra).

THE STRIA TERMINALIS

This small, discrete bundle of fine myelinated nerve fibres is visible to the unaided eye in much of its course. Fibres pass in both directions in it to various destinations. The stria issues posteriorly from the amygdaloid complex, running caudally in the roof of the inferior cornu of the lateral ventricle, medial to the tail of the caudate nucleus, then following its curve to pass ventrally in the floor of the body of the ventricle, in a groove between the caudate nucleus and the thalamus, close to the thalamostriate vein. It passes inferior to the interventricular foramen to the anterior commissure, where it diverges into *supracommissural, commissural* and *subcommissural* components. As it nears the anterior thalamic pole, scattered groups of small neurons occur between the bundles, the *bed nucleus of the stria terminalis*, in which some of its fibres relay. Many fibres in supra- and subcommissural components are amygdalofugal, passing to the septal areas (vide infra), preoptic and anterior hypothalamic nuclei, while others descend to the anterior perforated substance and adjacent regions of the piriform lobe. Some subcommissural fibres recurve to join the column of the fornix and others pass caudally with the stria medullaris to the habenular nucleus. Reciprocal connections probably exist between some of these regions and the amygdaloid nuclei, by returning fibres in the stria. The *interamygdaloid* fibres have an intricate course; leaving one complex they traverse most of the highly curved stria terminalis to the anterior commissure, cross and curve in the reverse direction through the contralateral stria to its amygdaloid complex.

THE ANTERIOR COMMISSURE

This is a compact bundle of myelinated nerve fibres crossing anterior to the columns of the fornix and embedded in the lamina terminalis, where it is part of the anterior third ventricular wall, $1 \cdot 5 - 2 \cdot 0$ cm above the optic chiasma (7.140, 145, 176). In sagittal section it is oval, its long diameter (about $1 \cdot 5$ mm) vertical. Its fibres twist and entwine like the strands of a rope and laterally it separates into two bundles: the smaller *anterior bundle* curves forwards on each side to the anterior perforated substance and olfactory tract; the *posterior bundle* curves posterolaterally on each side for some distance in a deep groove on the antero-inferior aspect of the lentiform nucleus, beyond which it fans out into the anterior part of the temporal lobe, including the parahippocampal gyrus. Commissural fibres have been described in mammals, including primates, as interconnecting the following structures with their fellows: (1) the olfactory bulb and anterior olfactory nucleus, (2) the anterior perforated substance, olfactory tubercle and diagonal band of Broca, (3) the prepiriform cortex, (4) the entorhinal area and adjacent parts of the parahippocampal gyrus, (5) part of the amygdaloid complex, especially the nucleus of the lateral olfactory stria, (6) the bed nucleus of the stria terminalis and nucleus accumbens septi, (7) the middle and inferior temporal gyri in their anterior regions, (8) possibly other neocortical areas, including the frontal lobe.

The inter-temporal neocortical connections are the largest component of the anterior commissure in primates, including mankind. However, many detailed connections in the human commissure remain unknown. Some fibres may not be commissural but decussating pathways between dissimilar structures.

THE SEPTAL AREAS

In subprimate mammals, *the septal areas* are the thick medial walls of the cerebral hemispheres anterior and superior to the lamina terminalis and anterior commissure, consisting of nuclear masses with relatively coarse bundles of fibres, divisible into *pre-* and *supracommissural* parts. In higher primates, particularly in the human brain, septal areas are modified by expansion of the neocortex and corpus callosum. The human supracommissural septum corresponds largely to the bilateral laminae of fibres, sparse grey matter and neuroglia, forming the septum pellucidum. The precommissural septum (*septum verum* of some) consists of well-defined dorsal, ventral, medial and caudal *groups of nuclei*, each divisible into a series of smaller groups, not detailed here. The precise limits of the human septum verum and the terminology applied to its surface topography and nuclear groups are sources of confusion and disagreement. (For detailed reviews consult Stephan & Andy 1962, Andy & Stephan 1968, Nauta & Haymaker 1969.) Most agree that the precommissural septum corresponds partly to the paraterminal gyrus (p. 1027), a narrow vertical strip between the anterior surface of the lamina terminalis and the posterior parolfactory sulcus. Its anterior slope passing into the sulcus is sometimes called the *prehippocampal rudiment*. Inferiorly, the gyrus and rudiment are continuous with the diagonal band of Broca and the medial olfactory stria (7.151); superiorly they narrow, spreading round the rostrum and genu of the corpus callosum to the indusium griseum (p. 1036). But some hold that only the prehippocampal rudiment is continuous with the indusium griseum; the detailed connections of the medial olfactory stria are uncertain in the human brain. Some of the septal nuclei are in the paraterminal gyrus, others are interspersed with fibres of the precommissural fornix (p. 1039). Scattered cell groups interconnect the precommissural septum with the septum pellucidum, substantia innominata and anterior perforated substance.

The main afferent connections to the septal nuclei are from: (1) the amygdaloid complex via the diagonal band and stria terminalis; (2) the anterior perforated substance by fibres probably running with the medial olfactory stria; (3) the hippocampus via the fornix; (4) the midbrain reticular formation and hypothalamic nuclei by ascending fibres in the medial forebrain bundle. (For the monoaminergic status of some of these see pp. 992-994.)

The main outflows from the septal nuclei are: (1) fibres returning to the hippocampal formation via the fornix; (2) descending fibres in the medial forebrain bundle distributed to many hypothalamic nuclei and the midbrain reticular formation; (3) fibres via the stria medullaris thalami to the habenular nuclei. Interconnections are probably established between septal areas and the vestigial indusium griseum and cingulate gyrus, but details remain uncertain. The septal areas are thus focal zones for the interconnection of principal limbic and hypothalamic structures. For long thought to be reduced, atrophic and perhaps functionless in primates, the septal nuclei have been shown to *increase* in prominence throughout primates, reaching their highest degree of development in the human brain (Andy & Stephan 1968).

Hippocampal Formation

The hippocampal formation develops in the medial pallial fringe of the hemisphere, adjacent to the convex border of the choroidal fissure (p. 1081), where it assumes a highly arched form extending from the anterior aspect of the interventricular foramen and lamina terminalis to the ventral extremity of the inferior cornu of the lateral ventricle. (The adjacent 'ends' of the arch are connected by other limbic structures, vide infra.) Modifications in its superior part, accompanying expansion of the neopallium and corpus callosum, have been discussed (p. 193). Essentially, it is a curved band of ancient cortex (*archipallium*), limited on its concave aspect by the choroidal fissure and merging on its convex aspect with the surrounding *neopallium*. Traced radially from the choroidal fissure towards the neopallium, three main zones are distinguished in the archipallium: *the dentate gyrus*, *cornu ammonis* and *subiculum*. The dentate gyrus and cornu ammonis have contrasting features and are regarded as the most primitive *trilaminar* cortex; the subiculum shows a graded variation from four, through five, to a modified six-layered cortex where it merges with the surrounding neocortex. Growth changes occur infolding these archipallial zones towards the adjacent lateral ventricle. These zones and their changes in position are shown in a simplified manner in figure 7.153A–C. The form of infolding is similar throughout Mammalia; but its *degree* and *areas* of hemispheric wall, in which well-differentiated hippocampal formation persists, vary in different groups. These variations reflect differences in the relative expansion of the neopallium,

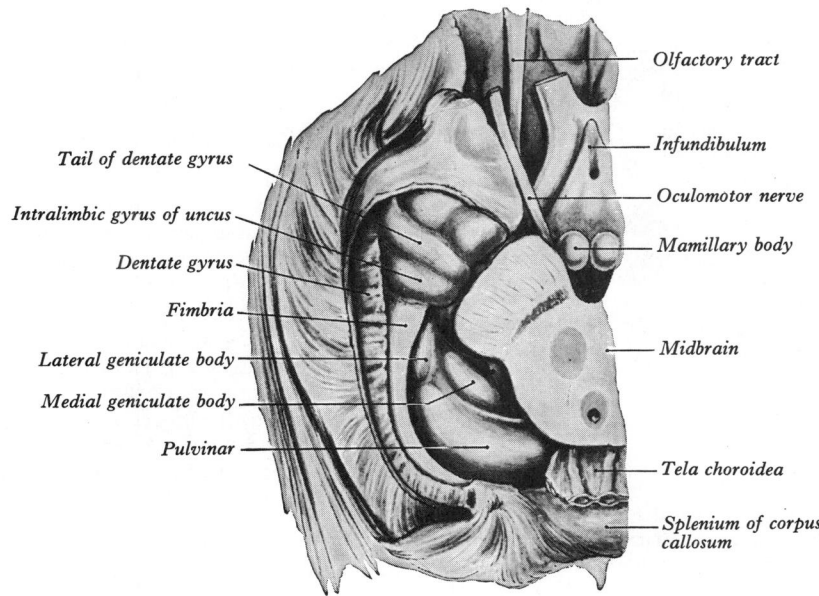

Tail of dentate gyrus

Intralimbic gyrus of uncus

Dentate gyrus

Fimbria

Lateral geniculate body

Medial geniculate body

Pulvinar

Olfactory tract

Infundibulum

Oculomotor nerve

Mamillary body

Midbrain

Tela choroidea

Splenium of corpus callosum

7.152 Basal aspect of part of the brain dissected to display the uncus, dentate gyrus, fimbria, etc.

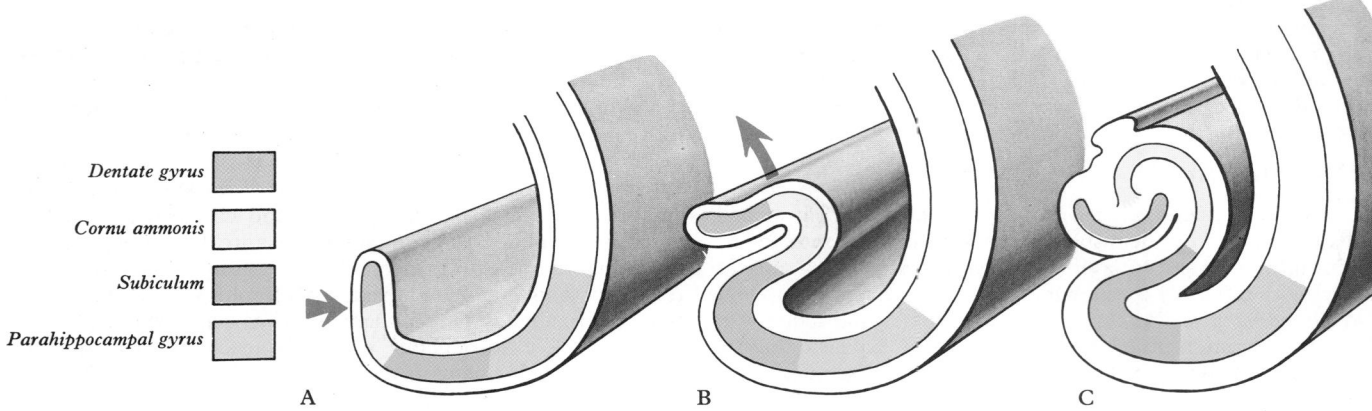

Dentate gyrus

Cornu ammonis

Subiculum

Parahippocampal gyrus

A   B   C

7.153A–C The hippocampus and related structures seen in coronal section: A, B and C are a series of diagrams to assist understanding of the assumption of the definitive positions of the dentate gyrus, cornu ammonis, subiculum and parahippocampal gyrus in the floor of the inferior horn of the lateral ventricle in the human brain. Note that these are *not* tracings from a series of embryonic sections and that they have been somewhat simplified in the interests of clarity. Note also that the amount of curvature and infolding which occurs varies along the length of the hippocampus and in different specimens. It is important to appreciate that, following folding, the original *external* surfaces of the dentate gyrus and part of the subiculum are in contact and that the degree of tissue fusion which occurs along the line of the hippocampal sulcus is quite variable.

corpus callosum and temporal lobe in different groups. In the human brain the part of the hippocampal arch related to the medial wall and roof of the *body* of the lateral ventricle is greatly reduced, due to the relatively enormous size of the corpus callosum, the fully developed hippocampal formation being confined to the floor and medial wall of the inferior cornu. Antero-inferiorly, the ends of the archipallial arch are continuous with the septal area, including the par2terminal gyrus, the anterior perforated substance and part of the piriform lobe (p. 1029). Topographical limits of structures included in the term *hippocampal formation* vary in different accounts. Here, it is considered to include: (1) the indusium griseum, longitudinal striae and their extensions; (2) the gyrus fasciolaris; (3) the dentate gyrus, cornu ammonis and subiculum; (4) parts of the uncus. The term

hippocampus is often restricted to the macroscopic swelling in the floor of the inferior cornu, consisting of the dentate gyrus, cornu ammonis and closely related structures.

THE INDUSIUM GRISEUM

The indusium griseum, the *supracallosal gyrus*, is a poorly differentiated film of grey matter covering the superior aspect of the corpus callosum. Laterally, on each side, it enters the callosal sulcus to continue into the cortex of the cingulate gyrus. Anteriorly, it sweeps around the genu and rostrum to merge with the superior end of each paraterminal gyrus, which is continuous below with the diagonal band of Broca and, through this, with the anterior perforated substance and periamygdaloid area. Posteriorly, the indusium griseum diverges on the splenium to merge with the *gyrus fasciolaris* (*splenial gyrus*), a delicate strip of grey matter curving down, forwards and laterally to the posterior extremity of the *dentate gyrus*. Embedded in the indusium, ridging its free surface, are two narrow bundles of fibres on each side, *the medial* and *lateral longitudinal striae* (7.174). The medial striae are near the midline, the lateral in the callosal sulci. The striae are regarded as the reduced white matter of a vestigial indusium. Anteriorly, they pass towards the paraterminal 'gyri'; posteriorly they continue through the gyrus fasciolaris to the fornical fimbriae (p. 1039). Connections of the striae are uncertain but in their supracallosal course they probably contribute fibres piercing the corpus callosum to form part of the *dorsal fornix* (p. 1039).

THE HIPPOCAMPUS

The hippocampus consists of complexly folded layers of the dentate gyrus and cornu ammonis, the latter being continuous through the subicular region with the parahippocampal cortex. The name stems from its supposed resemblance, in coronal section, to the profile of a sea-horse. The hippocampus is above the subiculum and medial parahippocampal gyrus, forming a curved elevation about 5 cm long along the floor of the inferior cornu of the lateral ventricle. Its anterior end is expanded and here its margin may have two or three shallow grooves giving a paw-like appearance, the *pes hippocampi*. The ventricular aspect is coronally convex and covered by ependyma, beneath which fibres of the *alveus* converge medially on a longitudinal bundle of fibres, the *fimbria of the fornix*. The general order of these cell layers and fasciculi in coronal section are best appreciated in a simplified diagram (7.153D). Note that passing *medially* from the collateral sulcus, the neocortex of the *parahippocampal gyrus* merges with the transitional cortex of the *subiculum*, which curves superomedially to the inferior surface of the *dentate gyrus*, continuing laterally to the laminae of the *cornu ammonis*; this

FIMBRIA

Fimbriodentate sulcus

C O R N U   A M M O N I S

H3   H2 CA2

CA3

H4

H5 CA4

DENTATE GYRUS

CA1

Hippocampal sulcus

H1

SUBICULUM PROSUBICULUM

PRESUBICULUM

PARASUBICULUM

PARAHIPPOCAMPAL GYRUS (ENTORHINAL AREA)

Alveus

Stratum oriens

Stratum pyramidalis

Stratum radiatum

Stratum lacunosum

Stratum moleculare

D

7.153D An analysis of the major topographical zones and the complex terminology applied to the hippocampus and related structures seen in a coronal section of the floor of the inferior horn of the lateral ventricle in a mature human brain. Colour code for tissue zones as in A–C. The approximate limits of the various subdivisions of the subiculum are labelled. CA 1–4 are the cornu ammonis fields of Lorente de Nó (1934) and H 1–5 are the hippocampal fields of Rose (1927).

continues the curvature, first superiorly then medially above the dentate gyrus, and ends pointing towards the centre of the superior surface of the dentate gyrus. The degree of curvature varies somewhat along the length of the hippocampus and in different specimens.

The *dentate gyrus* (7.152,153D) is a crenated strip of cortex related inferiorly to the subiculum, laterally to the cornu ammonis, superiorly to the recurved cornu ammonis, the alveus and, more medially, the fimbria of the fornix (7.154). The form of the fimbria is quite variable, but medially it is separated from the crenated medial margin of the dentate gyrus by the *fimbriodentate sulcus*. The *hippocampal sulcus*, of variable depth, is between the dentate gyrus and the subicular extension of the parahippocampal gyrus. Posteriorly the dentate gyrus is continuous with the gyrus fasciolaris and thus with the indusium griseum. Anteriorly, it is continued into the notch of the uncus, turning medially across its inferior surface, as the *tail of the dentate gyrus* (band of Giacomini), vanishing on the medial aspect of the uncus. The tail separates the inferior surface of the uncus into an anterior *uncinate gyrus* and posterior *intralimbic gyrus* (7.151,152).

THE STRUCTURE OF THE HIPPOCAMPUS

Since the early researches of Cajal (1890, 1911) and Lorente de Nó (1934) a huge literature has accumulated on hippocampal structure and connections. Only the salient points can be mentioned here; for details, monographs and papers must be consulted (Kappers et al 1936, Crosby et al 1962, Brodal 1969). Earlier reviews of complex terminologies for various laminae and subdivisions are to be found in the writings of Lorente de Nó (1934), Rose (1927), and Gastaut & Lammers (1960). Examples of investigations into connectivity patterns are found in papers by Blackstad (1956, 1958), Raisman et al (1965, 1966) and White (1959, 1965), using ultrastructural techniques (Blackstad 1967, Blackstad & Kjaerheim 1967, Blackstad & Flood 1963, Hamlyn 1962) and neurophysiological recording methods (Andersen et al 1964a, b, 1966). More recent comprehensive collections of essays devoted to many aspects of the structure, development, neurophysiology and behavioural aspects, the product of many leading authorities and containing extensive bibliographies are brought together in two volumes entitled *The Hippocampus* and edited by Isaacson & Pribram (1975).

Some aspects of the terminology used and some of the main structural features are shown in outline in figure 7.154.

It should be noted that because of the form of cortical folding which occurs during development, the original *external* surfaces of the dentate gyrus and subiculum (the *stratum moleculare*) are closely applied to each other in the depths of the sulcus hippocampi (or along the obliterated line of the sulcus). Throughout much of its extent, the subicular region merges laterally with the modified six-layered cortex of the *entorhinal* area of the parahippocampal gyrus. (Some authorities prefer to include the subiculum with the parahippocampal gyrus, rather than with the hippocampal formation.) As noted, in its curved course from the entorhinal area to Ammon's horn, the subiculum gradually changes its structure from a modified six-layered to a four-layered cortex, with differences in the patterns of connection. It is hence often divided into zones: *parasubiculum, presubiculum, subiculum* and *prosubiculum*. The prosubiculum merges into the cornu ammonis which, by the same criteria, is a primitive trilaminar cortex with *molecular, pyramidal* and *polymorphic* layers from the original external surface to the ventricular. But, by other methods and criteria, more sublayers are often described in Ammon's horn (vide infra). Though the general cytoarchitectonic design is similar throughout Ammon's horn, there are regional differences in structure and connections. Therefore investigators have proposed varying methods and criteria dividing this region into a series of radial fields (7.153D), including the **cornu-ammonal fields** CA1–CA4 of Lorente de Nó and the **hippocampal fields** H1–H5 of Rose. These do not correspond; the former are limited to Ammon's horn while the H1 field of Rose includes part of the subiculum, prosubiculum and part of Ammon's horn.

The cornu ammonis, following Cajal, is usually considered to have the following strata, starting from its ventricular aspect:

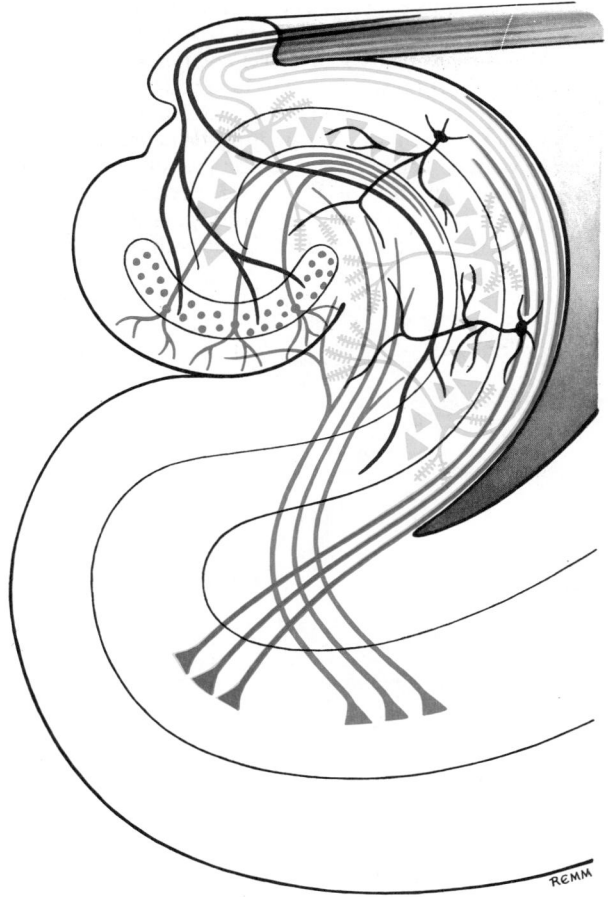

7.154 Some of the main features of the neuronal organization and connectivity patterns of the dentate gyrus, cornu ammonis, subiculum and parahippocampal gyrus. The cell somata, dendrites and axons of the pyramidal neurons of the cornu ammonis are yellow; their axons form the efferent hippocampal fibres of the alveus and fimbria. Afferent fibres to the cornu ammonis from the fimbria are purple, those following the alvear path are green, whilst those following the perforant path are blue. Basket neurons are in black. The neurons of the dentate gyrus and their axons which form the mossy fibres of the hippocampus are in magenta. See text for further details.

(1) the *ependyma*, (2) *alveus*, (3) *stratum oriens*, (4) *stratum pyramidalis*, (5) *stratum radiatum*, (6) *stratum lacunosum*, (7) *stratum moleculare*. Many group the last two as the *stratum lacunosum-moleculare*.

The *alveus* is a subependymal stratum of fibres entering and leaving the hippocampus. *Efferent* fibres are predominantly the axons of large neurons in the stratum pyramidalis but with a smaller number from cells in the stratum oriens and dentate gyrus, whose axons together form the large efferent component of the *fimbria of the fornix* (vide infra). Before joining the fimbria efferent axons have fine collateral branches which re-enter the hippocampus. *Afferent* fibres from other regions, including commissural fibres from the opposite hippocampal formation, approach the cell layers of the hippocampus via the alveus.

The *stratum oriens* is interlaced by axons and collaterals of fibres entering and leaving the hippocampus; it is penetrated by some basal dendrites of the large pyramidal neurons of the adjacent layer, containing cell somata and dendrites of small, irregularly shaped neurons, some termed *basket cells*, and shown to be inhibitory interneurons, receiving axosomatic and axodendritic synapses from some afferent collaterals to the hippocampus and also from efferent fibres. Axons of these interneurons penetrate the radiate and molecular layers, synapsing by axodendritic contacts with the pyramidal cells; but the most distinctive terminals are those of the basket cells which form numerous, crowded, axosomatic synapses on the somata of pyramidal neurons.

The *stratum pyramidalis* is a well-defined *double* layer of large and small pyramidal neurons. Their bases face the stratum oriens and alveus, their apices are directed to the stratum radiatum. The axon of a pyramidal cell issues from its base or a basal dendrite, passing into the alveus, where it has collateral branches, then continuing into the fimbria. Some of these collaterals end on cells in the stratum oriens, but many (*Schaffer collaterals*) are recurrent to the stratum moleculare, terminating on apical dendrites of adjacent pyramidal cells. The dendritic tree of a pyramidal cell has two parts, basal and apical; *basal dendrites* radiate into the nearby pyramidal layer but the majority pass into the stratum oriens and overlying alveus and are beset with dendritic spines.

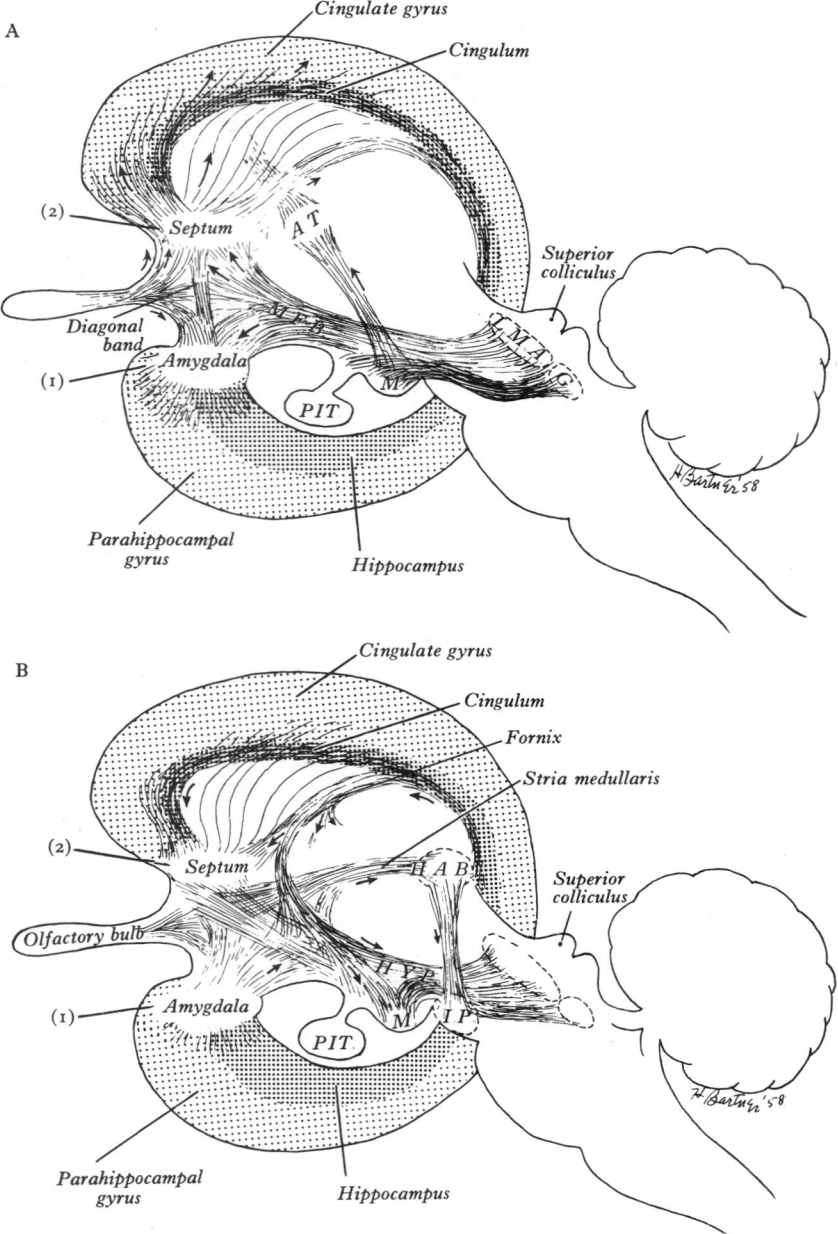

7.155 Schematic drawings of one concept of the 'limbic lobe' (after MacLean) in which emphasis is placed upon the medial forebrain bundle (MFB) as a major line of communication between the limbic cortex, the hypothalamus and the midbrain. Note the relationship between the fornix and the cingulum. The limbic cortex, which is considered as a hierarchical system of concentric strips, is indicated by heavy and light stipple. The neocortex is not included. A. The ascending pathways to the limbic structures, with emphasis on the divergence of fibres from the medial forebrain bundle to the amygdala (1) and to the septal area (2). Note also the input from the olfactory bulb and tract. B. Descending pathways from the limbic system. AT = anterior group of thalamic nuclei; G = tegmental reticular nuclei; HAB = habenular nucleus; HYP = hypothalamus; IP = interpeduncular nucleus; LMA = limbic midbrain area (of Nauta); M = mamillary body; PIT = hypophysis cerebri. From MacLean 1958 by courtesy of the author and publishers.

The *apical dendrites* pass deeply and with associated axons and a few pyramidal cells make up the *stratum radiatum*. Reaching the *stratum lacunosum-moleculare*, apical dendrites branch profusely and these terminal branchlets, with their stem dendrite, are also covered with dendritic spines. Thus, these deepest layers contain the terminal dendrites of pyramidal cells, terminal axons of afferents to the hippocampus, the recurrent collaterals mentioned above and the somata, dendrites and axonal arborizations of scattered, deeply placed interneurons. The connections of the hippocampus are treated later but it is noted here that the zones of pyramidal dendritic trees receive distinct axonal terminals. Thus, fibres from corresponding regions of the opposite hippocampus end on basal dendrites, while those from non-corresponding regions end on apical dendrites in the strata lacunosum and moleculare. Afferents from the entorhinal cortex synapse with the most terminal branches of apical dendrites in the molecular layer, synaptic endings of Schaffer collaterals being in the stratum lacunosum. Mossy fibres from cells of the dentate gyrus form large synaptic terminals enclosing prominent dendritic spines of apical dendrites in the stratum radiatum; but, as noted above, basket cell terminals form a dense array of axosomatic synapses. These zonal variations in axonal terminals have been tentatively correlated with variations in unit recordings from microelectrodes at varying depth in the cornu ammonis (Anderson et al 1964, 1966). Many afferents and pyramidal cells are excitatory; some interneurons, particularly basket cells, are inhibitory.

The dentate gyrus is considered a trilaminar cortical structure. Extending deeply from the hippocampal sulcus it consists of: (1) a superficial *molecular layer*, (2) an intermediate *granular layer*, (3) a deep *polymorphic layer*. Briefly, cell types in the laminae have spine-studded dendritic trees radiating locally or traversing the superficial layer. They receive synaptic terminals from some hippocampal extrinsic sources, also from neighbouring neurons. Some cells are Golgi type II neurons with local axons, others have longer axons which, after collateral branching, join efferent fibres in the fimbria. Most distinctive, however, are the so-called *mossy fibres*, arising from many cells in the granular layer and passing through the polymorphic layer with collateral branches to its neurons. They then proceed along a curved course lying superficial in the stratum radiatum of the cornu ammonis and making a series of 'giant' synapses with spines on the initial segments of pyramidal apical dendrites.

THE CONNECTIONS OF THE HIPPOCAMPUS

Afferent pathways to the hippocampus arise in: (1) parts of the cingulate gyrus, (2) septal nuclei, (3) the entorhinal cortex, (4) indusium griseum, (5) contralateral hippocampal formation, (6) possibly the prepiriform cortex, (7) brain-stem reticular cell aggregates, which provide a dual monoaminergic input (pp. 992–994).

Cingulate fibres reach the hippocampus via the cingulum (p. 1072) and are distributed both directly to the cornu ammonis and indirectly after relays in the subicular region. Afferents from septal nuclei retrace a curved path in the fornix back to the hippocampus. Commissural fibres link both hippocampal formations via the fimbria, crus and commissure of the fornix (vide infra), then recurve through the same structures contralaterally. Some fibres from the indusium griseum run posteriorly in the longitudinal striae, passing through the gyrus fasciolaris to the hippocampus via the fimbria.

Most abundant afferent connections of the dentate gyrus and cornu ammonis are from the entorhinal cortex and subicular regions, long held to have distinct routes. Fibres from the *medial* entorhinal area and parasubiculum follow an *alvear path* through the prosubiculum to the alveus and stratum oriens of the cornu ammonis. In contrast, afferents from the *lateral* entorhinal area follow a *perforant path* through the subiculum, crossing the alvear path to the stratum lacunosum-moleculare of the cornu ammonis and adjacent parts of the dentate gyrus. These routes and those from the fimbria are shown in Figure 7.154; the perforant path predominates and is more certain; some doubt the significance, or even the existence, of an alvear path, at least in some mammals.

THE FORNIX (7.148)

In addition to the afferent hippocampal fibres and commissural fibres described above, the fornix is the sole **efferent system from the hippocampus**, these efferents being mainly continuations of axons of pyramidal cells in the cornu ammonis; but a few come from neurons in the stratum oriens and dentate gyrus. These axons traverse the alveus and converge on the medial border of the ventricular surface of the hippocampus to form the *fimbria*, a flat band of fibres superior to the dentate gyrus and forming the inferior boundary of the choroidal fissure. Its position varies; it may project above the dentate gyrus with a free medial edge, *the taenia fornicis*, and a lateral border merging into the alveus; alternatively, its free border may be twisted over towards the lateral side, uncovering the dentate gyrus (7.152). Anteriorly, it continues into the hook of the uncus (7.145); traced posteriorly on the floor of the inferior cornu, it ascends towards the splenium, most fibres passing forwards *below* the splenium to curve forwards above the thalamus, as the *crus of the fornix*. The two crura are applied to the inferior surface of the corpus callosum and are interconnected by transverse fibres passing between the hippocampal formations of both sides as the *commissure of the fornix* (hippocampal commissure), a thin triangular sheet which, together with converging crura, is sometimes termed the *psalterium* or *lyra* (both varieties of harp!). Between the commissure and corpus callosum a horizontal cleft (so-called *ventricle of the fornix*) sometimes occurs. Anteriorly, the crura meet to form the *body of the fornix*, really a symmetrical bilateral structure above the tela choroidea and ependymal roof of the third ventricle (7.198) and attached to the inferior surface of the corpus callosum and, anteriorly, inferior borders of the laminae of the septum pellucidum. Laterally, the fornical body overlies the medial upper surface of the thalamus; the choroidal fissure is below its free lateral edge. Through this fissure the choroid plexus in the lateral margin of the tela choroidea enters the lateral ventricle (7.198). Above the interventricular foramina the body of the fornix diverges into right and left fascicles curving down towards the anterior commissure anterior to the foramina; approaching the commissure each divides around it into a *precommissural fornix* and a *postcommissural fornix* (column of the fornix). Each column continues to curve inferoposteriorly to sink into the corresponding lateral wall of the third ventricle to the mamillary body. In addition to these bundles, fibres leave the fimbria *near the splenium* to run forwards with the *supracallosal* medial and lateral longitudinal striae, joined by others from the indusium griseum. Some strial fibres descend between the callosal fascicles, joined by others passing into the corpus callosum from the fimbria. These fascicles, intersecting the callosal bundles, are the *dorsal fornix*. Some of the dorsal fornical fibres relay in the scattered grey matter in the septum pellucidum, while others course directly through the septum; direct and indirect fibres largely rejoin the main subcallosal body of the fornix.

Thus, by different fornical fascicles, *hippocampal efferents* pass to: (1) the gyrus fasciolaris, indusium griseum, cingulate gyrus and septum pellucidum via the *dorsal fornix*, (2) the precommissural septum, preoptic and anterior hypothalamic nuclei via the *precommissural fornix*, (3) the anterior thalamic and many hypothalamic nuclei (p. 1010) via direct fibres from the descending *postcommissural column of the fornix*, before proceeding to their main termination in the medial mamillary nucleus, (4) to the habenular nuclei and midbrain reticular formation via other fibres curving caudally from the column of the fornix to join the stria medullaris thalami.

Figures 7.155A,B roughly summarize some main inflow and outflow pathways associated with the limbic system (MacLean 1958, 1969), retained for its historical significance.

Much research has been directed towards the biological roles of the various complex interconnected structures grouped under the term limbic system. Despite this their significance can be described only in the most general terms. Even a brief review is beyond the present scope. Much information and bibliographies exist in the writings of Eleftheriou (1972), Isaacson (1974), Isaacson & Pribram (1975), Livingston & Hornykiewicz (1978). It is also instructive to inspect the data and speculations which stemmed from them of the extensive review of neuromediator distribution by Nieuwenhuys (1985). 25 of the commoner established or putative neuromediators were chosen and the neural sites (e.g. nuclei, certain gyri, 'central grey' tracts, fasciculi) in which they could be located with confidence were listed. Many (but not all) the sites with seven or more neuromediator profiles were noted (19 was the largest figure, present in the midbrain, periaqueductal grey and the nucleus solitarius) and then plotted graphically on a standardized diagram of the brain. The structures so demarcated are termed the *core of the central nervous system* and added tentatively: *the (paracrine?) core*, distinguishing this from the *median paracore* (raphes nuclei) and *bilateral paracores*, a series of grey aggregates usually allocated to the reticular system. In large measure the structures included largely correspond to those described under 'the limbic system' in this volume together with the reticular 'cores' of the brain stem and spinal cord. The principal addition consists of large areas of the neostriatum which it is stressed, however, have profuse amygdalostriate projections, that is the neostriatum can be divided into 'limbic' and 'non-limbic' (somatasensorimotor) parts. (The writer is forced to wonder whether 18 neuromediators per structure are more 'paracrine' beneath a limbic arch than when in mesocortex, neocortex or dorsal thalamus?) Nieuwenhuys' review is a splendid source of limbic and paralimbic data but most especially of provocative thought.

THE CEREBRAL CORTEX

Introduction

The cerebral cortex has been studied over several centuries. It was examined by the first microscopists; as early as 1776 the occipital cortical stria named after him was seen by Gennari and was the first recorded structural detail. Improved microscopes in the 1830s and later made investigation of cortical organization more effective; it has continued uninterruptedly, producing a spate of papers and becoming an unmanageable deluge in this century. No more than the most outstanding contributions can even be mentioned here; our emphasis must be on work in recent decades but it is useful to note the basic discoveries from which the current intricate knowledge has expanded, if only to maintain a perspective amid the flood of detailed description and speculation which makes current activity interesting and yet confusing in its constant sallies and revisions. The accent of endeavour has, naturally, varied with the availability of technique; but major aims remain the elucidation of the *modus operandi* in the cerebral cortex and the localization of different activities in it.

Earlier observations, inevitably microscopic, were stimulated by the technique of Nissl, Golgi, Cajal and Weigert (7.156), amongst others. Even earlier, with inferior methods, Baillarger (1840) and Meynert (1867–8) had ascribed a *laminar pattern* to the cortex, with emphasis on its fibre structure, or *myeloarchitecture*. Later workers using the newer techniques mentioned (Lewis 1878, Campbell 1905, Cajal 1909–11, Brodmann 1909, Vogt & Vogt 1926, von Economo & Koskinas 1929, Lorente de Nó 1949) confirmed this (with notable variations and disagreements, now largely forgotten). The *cortical laminae* described, existed at successive depths and essentially *parallel to* the pial surface. A six-layered schema prevailed (Brodmann), though based only on Nissl-stained material in which all the details of dendrites and axons studied by other workers were invisible; moreover, the data were accumulated from a comparative series which, though

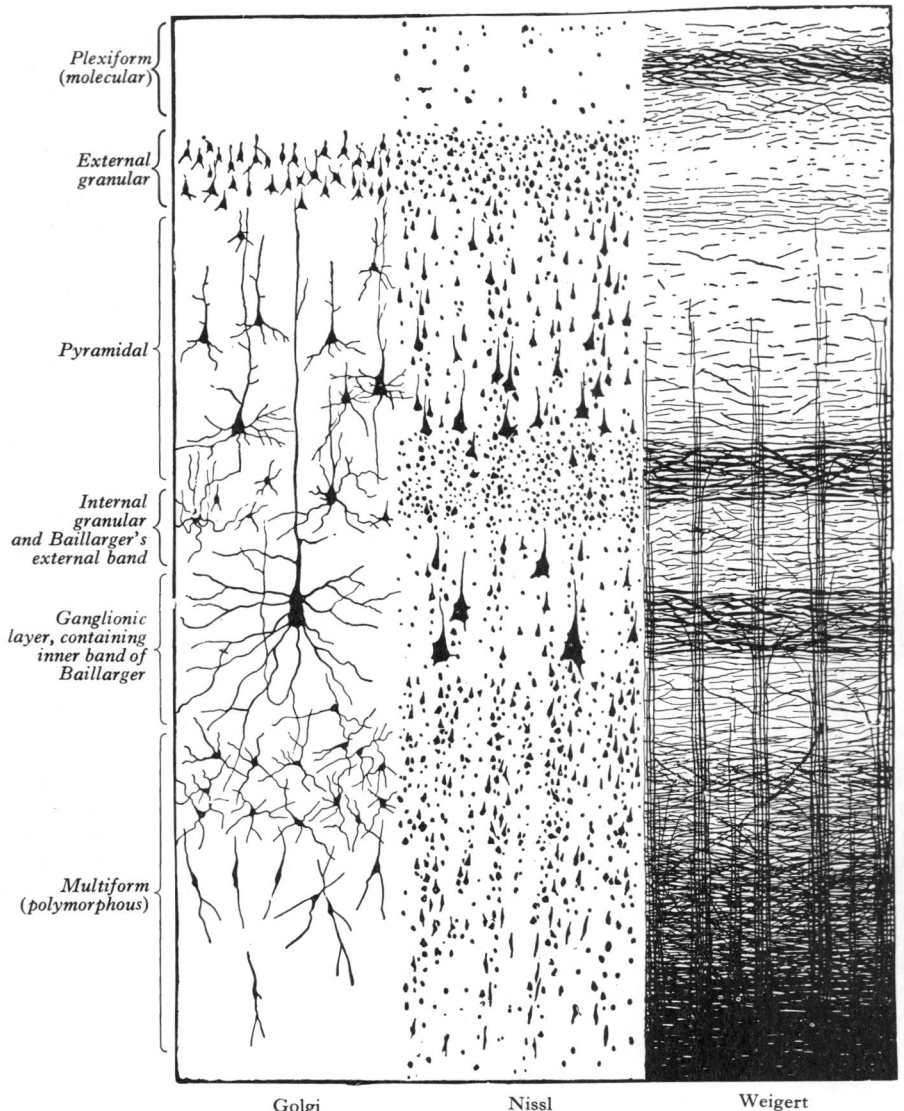

Plexiform (molecular)

External granular

Pyramidal

Internal granular and Baillarger's external band

Ganglionic layer, containing inner band of Baillarger

Multiform (polymorphous)

Golgi   Nissl   Weigert

7.156 Representations of the layers of the human cerebral cortex, as stained by the techniques of Golgi, Nissl and Weigert.

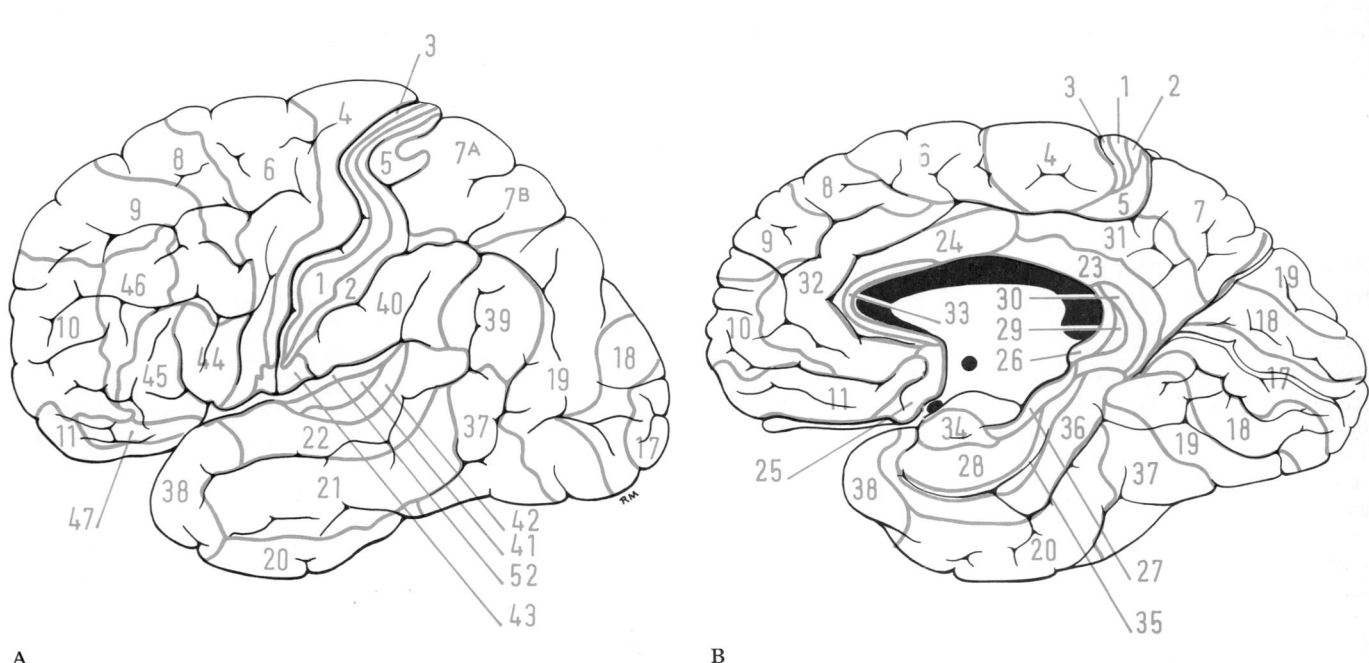

A

B

7.157 The superolateral (A) and medial (B) surfaces of the human cerebral hemisphere demonstrating the cytoarchitectonic areas identified and designated numerically by K Brodmann (1909). See text for further details and references. Compare with 7.158A.

including simian brains, did *not* include the human cortex. These studies, in which sexalaminarity became almost dogma, led to the recognition of great variations in different cortical regions or 'areas', no less than 52 being distinguished. These results, embodied in the familiar 'Brodmann maps', are still widely current and were transferred arbitrarily to man by others, gaining much credence among physiologists, perhaps mostly as a useful numerical guide to the cortex (7.157A,B), as they still are.

Techniques such as Nissl's reveal only the structural variation and distribution of neuronal somata: a *cytoarchitectural* mode of study required additional *myeloarchitectural* methods to display the neighbouring myelinated axons. Only with *both* methods, also coupled with somatic and neurite staining by techniques such as Golgi's, to which must be added the more recent availability of electron microscopy and refined degeneration, histochemical and neurite tracing techniques, could finer details of cortical organization be discerned. Unfortunately, the Nissl-based cytoarchitectonic mapping of the cerebral cortex initially attracted more attention and its continued use, with myelin staining added, has led to an almost excessive parcelling of the cortex into different areas, few of which can be accorded clear physiological significance. One useful outcome was the diversion of interest among younger workers to physiological studies. Rapid advance in this field of cortical investigation gradually led to much discordance between the results of stimulation and architectonic mapping (except in broad details), prompting several notable reviews (e.g. Lashley & Clark 1946, Scholl 1956). These have not entirely deterred the 'architectonicians' and it must be stated that only in the *broader* aspects of their schemata is there some degree of accord between structurally defined areas and the results of experimentation.

The definition of five or six major classes of cortical organization (cf. Campbell 1905, von Economo & Koskinas 1925, 7.158) has proved to have some functional validity, but excessive architectonic geography is reminiscent, except in the admittedly serious structural basis, of the multi-faculty maps of phrenology. Division of the cortex into such a relatively small multiplicity of 'organs' seems basically improbable. As critics have stressed, the variations in a single area in different individuals of one species, the effects of cortical development and folding and other factors, particularly the patterns of subcortical connection, may all contribute to the variable structural appearances in the cortex, variations which themselves are not uniform. The need for quantitative approaches to the problem has also been emphasized but this remains partly an exhortation, though their ultimate necessity is recognized.

That lamination characterizes the cerebral cortex is obvious but dogmatic views on the numbers of layers in this neuronal continuum, or regarding variations in and the equivalence of the customarily numbered six layers in different areas may, if pushed to excess, obstruct rather than enlighten. Nevertheless, the Brodmann schema is still almost universally employed, perhaps as a mere convenience, and will perforce be described below. Of course, an orderly arrangement of neuronal and glial units in the cortex, to some degree stratified, is undeniable and perhaps to be expected. Localization of function to a considerable degree (inasmuch as some areas receive or project large and easily identifiable subcortical connections) also remains undeniable; the polemics which this concept provoked in the late nineteenth century are largely forgotten (Clarke & O'Malley 1968). But the original extent of 'motor' and 'sensory' areas, as defined (and artificially *con*fined) by architectonic studies, have been much modified and augmented by physiological and anatomical experimentation.

Studies of the cortex by Golgi and silver-staining methods, though once somewhat obscured by work with the Nissl technique, have received a renewed stimulus with the development of micro-recording from individual cortical neurons. Prior to this, elaborate structural studies of their neurites, available for half a century, could not be equated with the results of functional experimentation but, as mentioned elsewhere (pp. 872, 873), during recent years the intimate arrangement of neurons and their electrical activities have shown a growing correlation in elaborating concepts of modes of action in small volumes of nervous tissue at many sites in the central nervous system, including the cerebral cortex.

While the classic *cytoarchitectonic* doctrine is now perhaps little more than a useful geographical definition of the cortical areas (though sometimes a source of confusion in the 'translation' of areas from species to species), cyto*architectural* details, as revealed by the methods of Golgi and Cajal, are beginning to fit observed behaviour of 'units' in the living cortex. Classic neurite studies of the cerebral cortex, chiefly in lower mammals such as rodents (e.g. Lorente de Nó 1934), have led to the recognition of many types of neuron. The familiar pyramidal and stellate

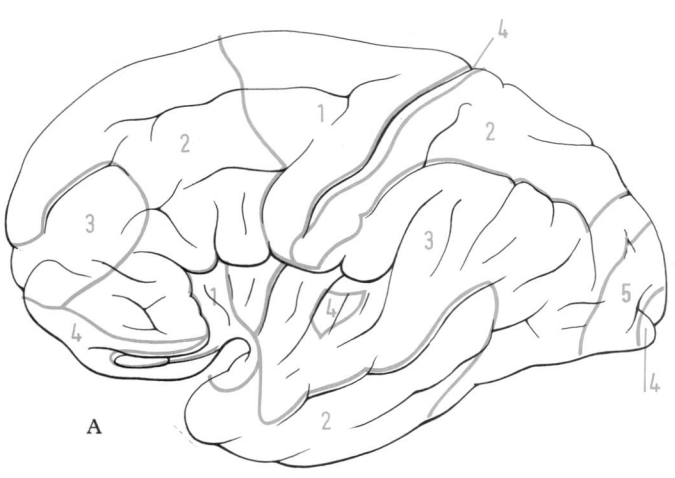

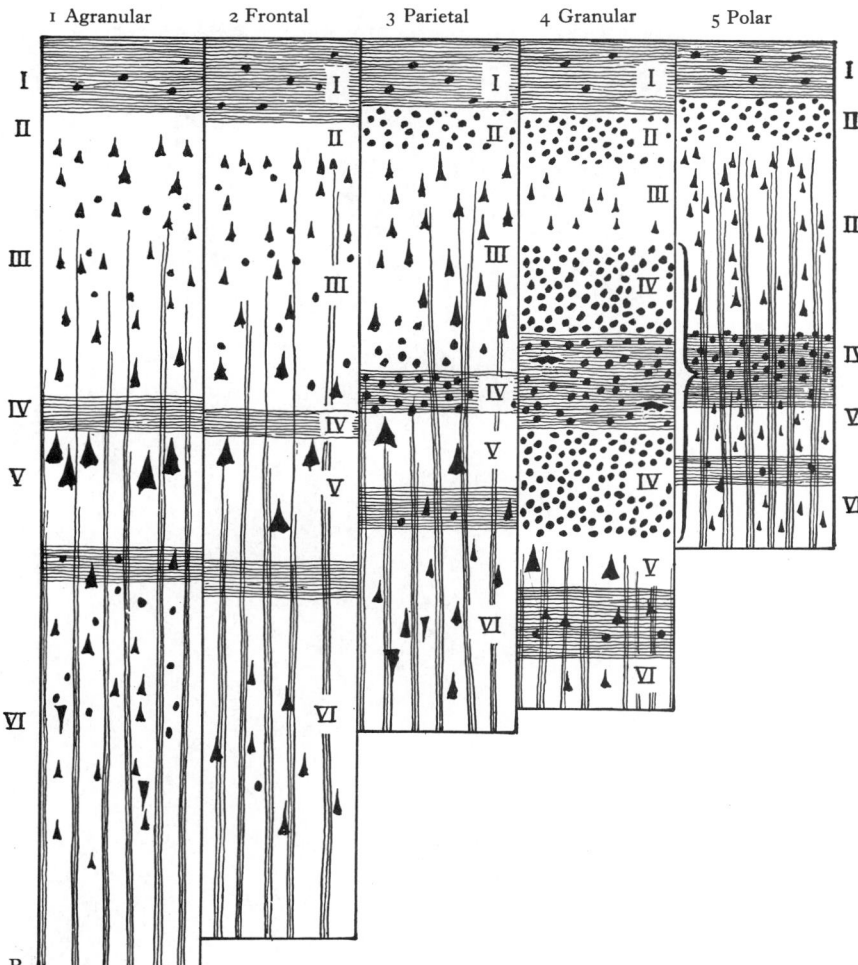

1 Agranular	2 Frontal	3 Parietal	4 Granular	5 Polar

7.158 The distribution of the five major types of cerebral cortex, as projected on to the superolateral surface of the hemisphere, according to von Economo & Koskinas (1925). The numbering of cortical areas (A) corresponds to the cytoarchitectonic types (B). See text for further details and references. Compare with 7.157A,B.

(granule) cells of Nissl-stained material were chiefly sorted by their relative size into unsatisfactory categories (subjectively assessed as large, medium, or small, and so on) without real measurement. Comparatively unilluminating data of this kind are represented in 7.158B, to be compared with 7.159A, B, which illustrates the more usable information provided by techniques revealing the interconnections between cortical neurons. By such criteria a larger range of neuronal specialization can be based; and since it is interneuronal *connections* which must be correlated with living activities this approach is inevitably more appropriate than plain cytoarchitectonics. The results of both methods do not generally conflict; but one is overwhelmingly more informative. Even though light microscopy proved unequal to full visualization of synapses in the cerebral cortex, the general patterns of interconnection of cortical neurons were, in major details, well clarified many years before electron microscopy appeared.

The termination of afferent axons in superficial cortical layers, nearer the surface than the somata of efferent neurons, with the existence of recurrent axon collaterals from the latter which turn back superficially to form connections through interneurons with their own dendrites and the existence of a clear *'vertical'* organization in the cortex, these and other features were all established as one main feature of basic cortical 'circuitry' before microelectrode experimentation began to require a more detailed background to its recordings. As will become apparent, 'chains' of neurons, emphasized by Lorente de Nó, arrayed in innumerable units repeated through the cortex, have fitted aptly into physiological studies of, firstly, the postcentral and striate regions (Mountcastle 1957, Hubel & Weisel 1962, Colonnier 1974, Zeki 1974, and

pp. 1056, 1058), and subsequently many other, perhaps all, cortical regions (vide infra).

As elsewhere in the central nervous system, unit recording and the more precise study of transmission in excitatory and inhibitory synapses have generated a renewed demand for the finest details of interneuronal connections. Various forms of synapse and their ultrastructure are described elsewhere (p. 882); though their identification in the cerebral cortex was comparatively late (Gray 1959), a large literature on cortical synaptology has now accumulated. Axodendritic and axosomatic synapses predominate and for a period were considered the only types present, but dendrodendritic synapses have been identified in some locations. The multitudes of synapses, especially at the dendritic spines of pyramidal cells, provide an obvious structural basis for the most complicated interactions, however far we may yet be from anything more than the simplest interpretations, much less the detailed 'circuitry' of even the most elementary cortical events. Nevertheless, structure and function are rapidly progressing together in current cortical study; even if joint 'models' of cortical activity that are presently evolving evince a tendency to rather rapid change, this is surely a healthier state than the stagnating dogma which has sometimes becalmed neurological research.

In the following simple account of the cerebral cortex and its specific areas, space will not always permit adequate reflection of the intense current research; for this the reader must consult original papers and surveys, e.g. Colonnier (1966, 1968, 1974), Shepherd (1974), Rakic (1975), Szentágothai (1975), Schmitt et al (1976). Two approaches, however, are culminating in such

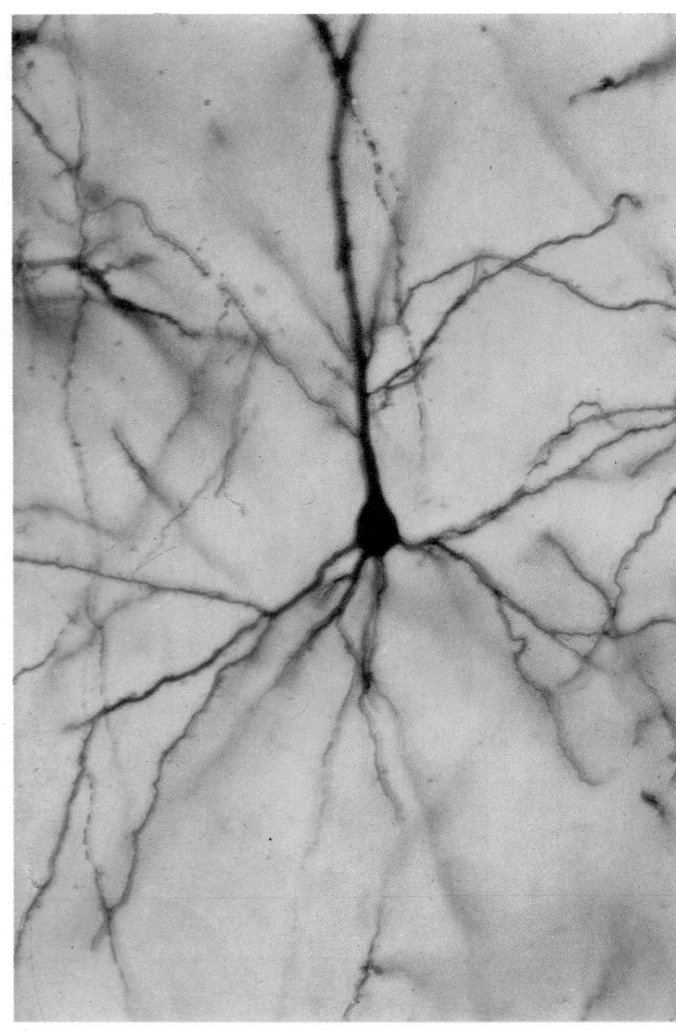

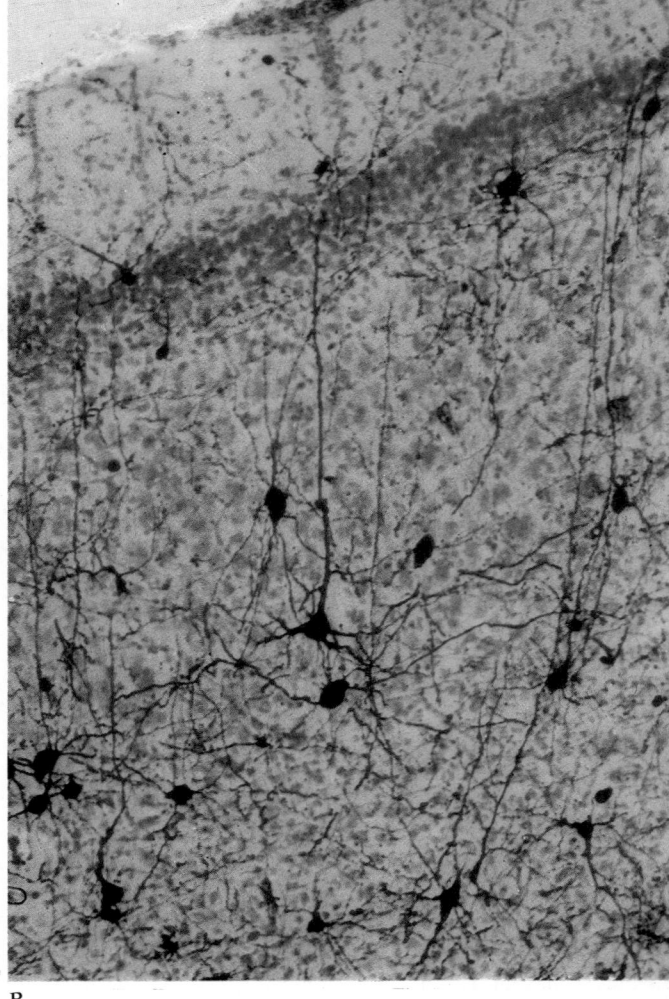

A   B

7.159 Preparations contrasting the Golgi and Nissl methods of staining nerve cells in the cerebral cortex. In (A) a single pyramidal cell stands out amongst many unstained elements. In (B) isolated Golgi-stained neurons are prominent amongst the remaining Nissl-stained cortical elements. Preparations provided by A R Lieberman, University College, London.

powerful influences, particularly in the last decade, that they deserve special emphasis. The overall gross *tangential lamination* mentioned is (except for details) universally accepted; much interest has been diverted to the *vertical organization* (i.e. at right angles to the pial surface). First, tissue units termed *cortical columns* were recognized (initially in certain restricted cortical areas only, pp. 1052, 1054), but a much wider, generalized concept of *cortical modules* (or *higher integrative units*) has been proposed and has stimulated much innovative research. (For reviews see: Hubel & Wiesel 1977, Phillips & Porter, 1977, Szentágothai 1978, Mountcastle 1978, Porter 1980, Phillips 1981, also vide infra). Simultaneously there has occurred an explosive increase in the scope of neurochemistry (see for review the monograph, *Chemoarchitecture of the Brain,* Nieuwenhuys 1985). As mentioned on p. 992 and elsewhere, the number of 'classic' or 'true' neurotransmitters has greatly increased and they may be short or long-acting; some classes of substance may be delivered via intercellular fluid to large numbers of neurons in relatively large volumes of tissue (paracrine effects), many other substances may be grouped as neuromodulators. All such pharmacodynamic materials should be considered in all parts of the nervous system including, of course, the cerebral cortex.

QUANTITATIVE ASPECTS OF CORTICAL STRUCTURE

Quantitative studies of the cerebral cortex were, until recently, comparatively few; a first attempt to ascribe numerical values to some of its grosser features was by von Economo & Koskinas (1925), who recorded data on variation in cortical depth, which are still amongst the most detailed for man. They also computed total surface area as 220 000 mm², a later estimate being 285 000 mm² (Scholl 1956), with a volume of 300 cm³. (In a preliminary account using elementary cartography Walker & James 1977 estimated a much larger cortical surface area, about 800 000 mm². In view of these great differences, it is hoped that these investigations will be expanded and also include alternative techniques.) The total numbers of cortical nerve cells have naturally attracted much computation, and figures of 14 000 million (von Economo & Koskinas 1925), 6900 million (Shariff 1953), 5000 million (Scholl 1956) and 2600 million (Pakkenberg 1966) illustrate a downward trend, perhaps due to changing (improving?) technique. Stereological methods have been employed, e.g. Foh et al (1973) have estimated the total lengths of neuronal and glial processes in unit volumes of the visual cortex, in cats, arriving at 5000 m/mm³ for neurites. Such data are beyond real comprehension; it is easier to grasp the magnitude of such a population by proportions in a smaller sample. For example (Hubel & Wiesel 1977), a column of cells 1 mm square and 2·5 mm deep may contain 100 000 neurons, of which some 66% are *pyramidal neurons* and the remainder *stellate interneurons* (vide infra). In one study of part of the motor cortex (precentral gyrus) each pyramidal neuron was estimated to take part in about 60 000 synapses, connecting with 600 other nerve cells (Cragg 1967, 1975). (However, also vide infra.) Such a small volume, itself containing about 3×10^9 synapses, might be multiplied, perhaps, a quarter of a million times to represent the whole cortex. The wealth of interconnections in such a huge array is clearly very great. In the striate area (Scholl 1955) where about a tenth of the cortical neurons are said to be concentrated, the dendrites of a single neuron may connect with 2000–4000 other cells and an arriving (afferent) projection fibre may ramify through a cortical volume containing 5000 cells. (Even these figures are, of course, dwarfed by cerebellar data, p. 972.)

The density of packing of neurons in different cortical areas and their laminae shows much variation, being densest in the striate area and perhaps least in the precentral gyrus. The ratio by volume of somata of neurons to all other cortical constituents (*grey/cell coefficient*) has been estimated: average ratios of 27:1 (von Economo & Koskinas 1925) and 70:1 (Haug 1956) for man have been cited. The ratio is claimed to show a phylogenetic increase up to man.

Neocortical Homogeneity

The most extensive, including comparative, quantitative studies of cortical and laminar thickness, absolute neuronal numbers in standard full-thickness cortical strips extending from the pia to white matter, and pyramidal/non-pyramidal ratios are by Rockel et al (1980) and Powell (1981). Their data (and reviewed literature) must provoke much radical reappraisal. The cortical areas, (frontal, motor, somatic sensory, parietal, visual (area 17) and temporal) were studied in the mouse, rat, cat, monkey and man; additionally area 17 only received further study in tupaia, galago, marmoset, squirrel monkey, baboon and chimpanzee. Total and differential counts were made in strips 30 μm wide and 25 μm thick cut perpendicular to the pial surface and passing to the subcortical white matter. The outstanding findings were that the *total neuronal populations* of standard strips (and therefore perpendicularly beneath a *unit area* of pial surface), with the exception of primate area 17, are so close as to be regarded as *virtually identical*, whether from different regions in the same brain, the same species or from those as remote as mouse and man. Additionally, *in all cases*, about two-thirds of the neurons are pyramidal, the remainder non-pyramidal (varieties of stellate neurons). Interestingly, the parts of area 17 concerned with primate *binocular* vision have a neuronal population that is about 2·5 times greater than other areas. Later studies (Powell & Hendrickson 1981) showed that *both* monocular and binocular parts of area 17 in the macaque monkey have the same larger population of neurons, whereas area 18 and other visual areas are like all remaining regions of the neocortex. Such findings are in full accord with the neurophysiological results and concepts of functional cortical columns and modules mentioned elsewhere. The *numerical neuronal homogeneity* (both absolute and the ratio of two principal varieties) of almost all the neocortex in a wide range of mammals, in standardized full-thickness strips or columns, despite marked variations in cortical thickness and subjectively described laminar patterns, must profoundly affect the interpretation of the classically described variations in structure ascribed to different regions (vide infra). It also stimulates speculation concerning the phylogenesis, ontogenesis, structural definition and integrative operation of *fundamental cortical units*; the establishment of absolute criteria for their universal definition presents many difficulties.

Various other quantified data regarding mammalian cortex are scattered through the literature, to be consulted for further details. In general, apart from an impression of enormous potentiality for interconnection and interaction, such figures have little immediate application. Knowledge that each afferent or efferent projection fibre may have more than 1000 neurons linked with it, more or less remotely, merely emphasizes the extremely intricate field of interaction between them. The time may be far off when events in such large neuronal assemblies can be defined in precise spatial and temporal terms, if indeed this can ever be expected in more than generalized statistical approximation; and even where accurate multimillion blueprints become available the human mind would inevitably demand combinatory concepts and technologies to aid comprehension.

For the present, limited events in minuscule volumes of cortex (e.g. columns, modules, submodules and their parts and connections) appear to offer a more promising and primarily essential field of investigation. Examples will be mentioned in connection with the particular cortical areas under consideration. Arising from such research, a profusion of evolving concepts of neuronal 'circuitry' and diagrams to illustrate these have appeared in the literature, e.g. Walter (1953), Ashby (1960), Weiner & Schade (1963), Taylor (1964), Young (1964), Arbib (1964), Minsky (1965), Shepherd (1974, 1978), Colonnier (1974), Rakic (1975), Szentágothai (1975, 1978), Hubel & Wiesel (1977), Phillips & Porter (1977), Mountcastle (1978), Porter (1980), Rockel et al (1980), Powell (1981), Phillips (1981), Nieuwenhuys (1985). Naturally, these speculations undergo rapid change and modification as knowledge in the very active fields of neurocytology, synaptology, unit recording

7.160A Typical outlines of characteristic neocortical neurons as seen in sections prepared by the metallic impregnation techniques introduced by Golgi and Cajal. From left to right are shown Martinotti, neurogliaform, basket, horizontal, fusiform, stellate and pyramidal types of neuron. Many other forms and variants have been described. See text for literature.

and cerebral chemoarchitecture advances. Hence, despite their great interest, it is as yet premature to include more than passing reference to these. However, some potentially powerful general hypotheses will be mentioned below.

NEURONS OF THE CEREBRAL CORTEX

To the unaided eye the cerebral cortex forms a complete mantle or *pallium* covering the hemispheres, obviously variable in thickness ($1·5–4·5$ mm) when seen in section and thicker on the summits of gyri than in the depths of sulci (in which most of the cortex is hidden). Such thickness variations might well correspond to microstructural variations in the pallium; and it has been suggested that the positioning of gyri and sulci is conditioned by structural differences (Le Gros Clark 1945), but this cannot be claimed with respect to *functionally* differentiated areas, which often depart in outline from the sulcal pattern. In the freshly cut cerebral cortex some laminar details can be appreciated even without a lens (e.g. the visual stria of Gennari); elsewhere finer horizontal layers of nerve fibres, the inner and outer bands of Baillarger (p. 1046), can usually be discerned. It has even been claimed that more than a score of structurally distinct areas can be identified by simple visual inspection (Smith 1907).

The microscopic structure of the cerebral cortex, like 'grey matter' elsewhere, is an intricate blend of nerve cells and fibres, neuroglia and blood vessels. Neuroglia and vascular arrangements have been dealt with elsewhere (pp. 892, 895). The neuronal features, connections and distributions must now be considered. However, it is to be noted that variation in vascular distribution has also been utilized as a criterion in differentiating cortical areas (Pfeifer 1940) and many other finer features of its microarchitecture show corresponding variations in angioarchitecture (Duvernoy et al 1981).

Neurons of the pallium have been described and categorized in endless detail, but can be assigned to a few classes, most belonging to **two groups:** *pyramidal* cells and *stellate* (granule) cells. Both types may be assorted into variable subdivisions on the basis of size, shape and neuritic array (7.160A, B) and both occur in most cortical levels and areas, though numbers and distribution vary greatly. Other types of stellate cells commonly distinguished are: *basket, fusiform, horizontal, neurogliaform* and the *cells of Martinotti*; considered to be modified pyramidal neurons are the *pleomorphic cells*.

Pyramidal nerve cells, so named from their somatic shapes, vary from small elements about 10 µm across to giant pyramidal cells (of Betz) which reach 70 µm and more. They account for about 66% of the neuronal population in *all* neocortical areas (Rockel et al 1980, Powell 1981). Their apices are usually towards the cortical surface, with a thick *apical dendrite* ascending a variable distance, giving off collateral branches and often ending in a spray of terminal dendritic twigs. From the basal angles of the soma, *basal dendrites* spread laterally into the surrounding neuropil to form dendritic fields of varying shape and extent (Colonnier 1964). Statistical analysis of such branching in pyramidal and stellate cells has been attempted in the visual cortex (Scholl 1953) but such quantitative dendrological study is a neglected and difficult field. On the whole, the vertical extent of dendritic extensions of pyramidal cells is markedly greater than in the horizontal or tangential direction. The dendrites are studded, especially their smaller branches, with myriads of *dendritic spines* (p. 882), now known to be sites of axodendritic synapses (Gray 1959). Since these synapses, even by the crudest estimates, are multitudinous, a physical basis for the most elaborate cortical interneuronal reactions is immediately apparent. Axons of pyramidal cells, invariably smaller in diameter than the trunks of their main dendrites, behave variously. From larger somata, chiefly in lamina V (vide infra), *projection axons* extend centripetally to more or less distant subcortical structures, such as the basal nuclei, thalamus, hypothalamus, reticular formation and other brain-stem nuclei and the spinal grey matter. Some may return into other parts of the *ipsilateral* cortex as *long* or *short association fibres*. Others cross the midline as *commissural fibres*, their destinations being either *homotopic* or *heterotopic* points in the *contralateral hemispheric cortex*. Axons of smaller pyramidal cells usually ramify entirely in the cortex; even those which leave it commonly have a small number of collaterals which remain as intrinsic, *intracortical axons*, which may extend horizontally but more often have oblique recurrent courses to more superficial

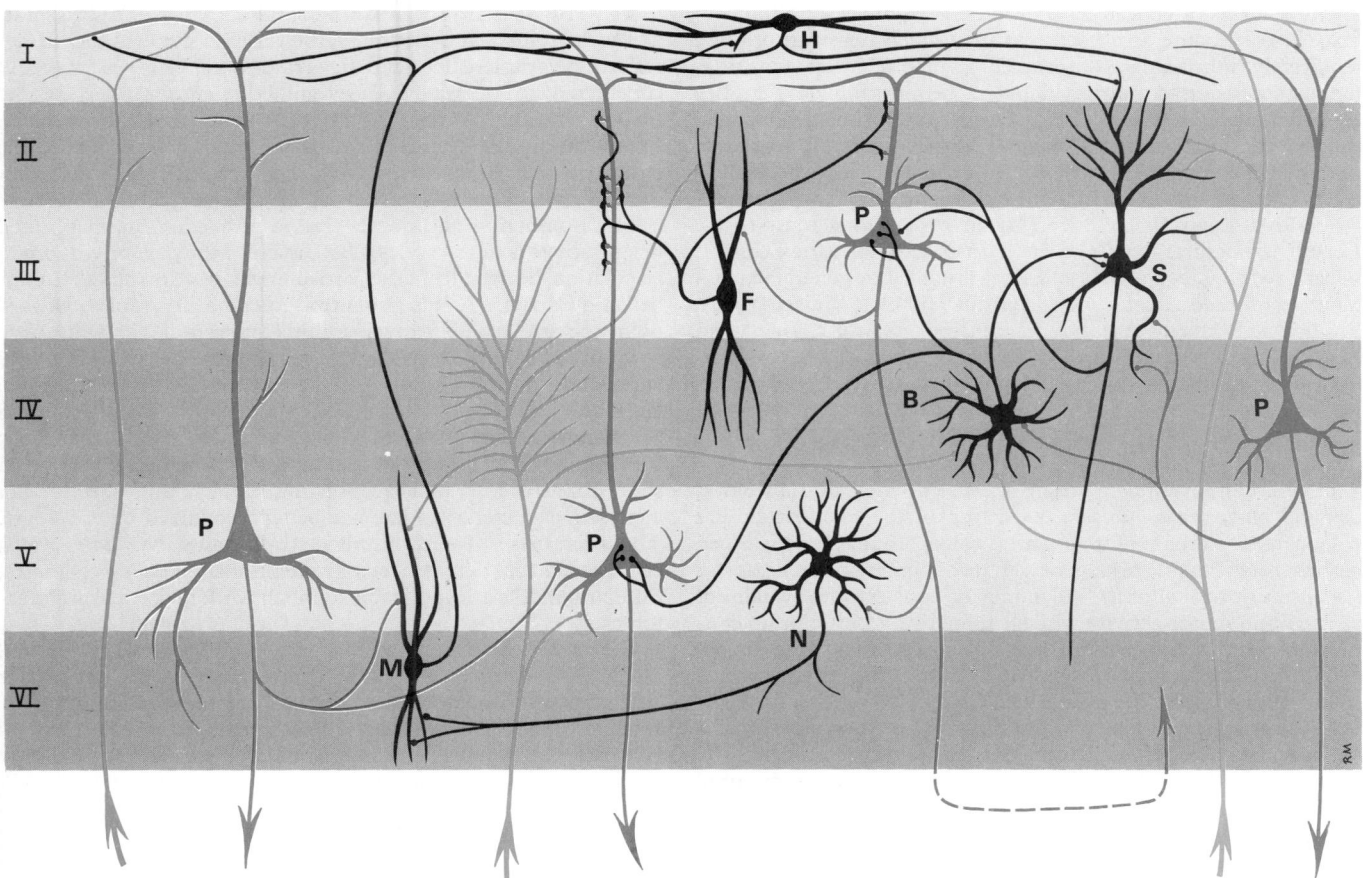

7.160B A diagrammatic representation of the most frequent types of neocortical neuron, showing typical connections with each other and with afferent fibres (blue). Neurons limited to the cortex in their distribution are indicated in black. Efferent neurons are in magenta. The right and left afferent fibres are association or cortico-cortical connections, the central afferent is a specific sensory fibre. Neurons are shown in their characteristic lamina, but many have somata in more than one layer. They are indicated thus: P = pyramidal, M = Martinotti, F = fusiform, H = horizontal, N = neurogliaform, B = basket, S = stellate. See text for details and compare with illustration 7.164 et seq.

cortical laminae. Even this brief description shows that the larger pyramidal cells, whose somata appear in Nissl-stained sections in a *single* lamina, in fact have neurites that extend through most and often *all* cortical levels, their apical dendrites frequently entering the most superficial *plexiform lamina* (vide infra). Thus, each pyramidal neuron forms the core of one fundamental type of *columnar unit* extending right through the cortex, together with its numerous connections to other pyramidal elements, many forms of interneuron and afferent projection fibres; from its base issues its axon (which, as indicated, may be projection, long or short association, intracortical, or commissural; when its collateral branches are considered it may belong to two or more of these categories). Lateral connections are mediated by various intracortical axons and collaterals and horizontal and association neurons.

Stellate nerve cells, often called *granule cells* because of their (usually, but not invariably) small size and granular appearance in Nissl-stained material, appear in variable density in *all* cortical laminae except the most superficial (lamina I), but are usually in greatest abundance in laminae II and IV (vide infra). They account for about 33% of the neuronal population in all neocortical areas (Rockel et al 1980, Powell 1981). Like the pyramidal cells, stellate neurons would probably have been called multipolar if not first described in Nissl-stained sections. They are small, 6–10 μm in diameter, with a round soma drawn out at numerous angles by richly branching dendrites and a single short axon; they are hence Golgi type II neurons. Their dendrites carry many spines, indicating abundant synapses. Varieties of stellate cells have been distinguished, principally on the basis of their axonal disposition, length and arborization. Though confined to the cortex, their axons may travel considerable distances in it, chiefly in the vertical direction but sometimes horizontally. Vertical axons may be centripetal or centrifugal, the latter reaching lamina I. One type, the *basket cell*, is horizontally extended, with a short vertical axon soon dividing into a horizontal family of collaterals, which end in large terminal sprays, synapsing with the somata and proximal dendrites of pyramidal cells. They have a particular interest in their 'horizontality' because of the perhaps excessive attention that was, for a period, concentrated on vertical organization in the cerebral cortex (p. 1059). (Currently the *cortical module* concept is, at its simplest, a variable *four-dimensional unit*; when its submodular regions and their innumerable parameters are contemplated a term such as *multidimensional function space* becomes necessary; cf. p. 1052.) Another type of stellate cell is *fusiform*, due to the emergence of two large dendrites from opposite poles of the soma, dividing at once into elaborate groups of branches extending vertically. The axon, sometimes from a stem dendrite, extends centrifugally to lamina I. The neuron may spread through the whole depth of the cortex, probably synapsing with several pyramidal cells. *Neurogliaform stellate cells* are small, with a dense localized dendritic arborization, within which the short axon also usually ramifies (7.160A).

Horizontal cells, confined to the plexiform lamina I, are small and fusiform; their dendrites spread short distances in two opposite directions in lamina I. Their axons often stem from a dendrite, then divide into two branches which travel away from each other to greater distances in the same layer. **The cells of Martinotti** occur at most cortical levels. They are small, multipolar, with localized dendritic fields and long axons running centrifugally to the plexiform lamina, producing short horizontal collaterals en route. **Pleomorphic cells** are considered to be modified pyramidal cells with axons entering the white matter. Their somata are variously shaped, perhaps in accord with the differences in their dendritic sproutings, which spread widely into the cortex.

Other forms of neuron have been detailed in the cerebral cortex; since even the array of types mentioned here can only be tentatively linked in a co-ordinated scheme of interaction, it is merely confusing to multiply details. The most intensive studies of cortical neurons have been carried out in subprimate mammals; though appearances in the primate cortex (chiefly pursued in macaque monkeys) are similar, investigations of even 'cold' structural arrangements in the human cortex have been few (Economo & Koskinas 1929, Bailey & von Bonin 1951, Conel 1939–1959). For ultrastructural details of the human cerebral cortex consult Braak (1975), Cragg (1976) and Rees (1976). For general experimental and theoretical reviews consult Hubel & Weisel (1977), Phillips & Porter (1977), Szentágothai (1978, 1985, 1987) and Mountcastle (1978).

LAMINAR PATTERN IN THE CEREBRAL CORTEX

The cerebral cortex or pallium is often divided into an older, original part, the *allocortex*, consisting of the archicortex and paleocortex (archipallium and paleopallium), considered elsewhere (p. 1035); and a newer development, the *neocortex* (isocortex or neopallium), which may be equated with systems of sensory and motor activity which originally had little or no connection with the forebrain, but have acquired these by an evolutionary process of prosencephalization. The remarks which follow apply only to the neocortex.

As already stated (p. 1039), the customary description of a sexalaminar cortical structure, successfully promulgated by Brodmann and his followers, has by general acceptance obscured earlier disagreements and dissatisfaction with its arbitrary nature. Similar strictures apply to the excessively detailed architectonic mapping of the cortex derived from the same somewhat dogmatic views; but Brodmann's numbered layers and areas provide a reference grid of practical value and are widely used. Therefore, until more satisfactory systematics emerge, the details in so far as they are useful must be repeated here. The six laminae (7.156, 158) may be described as follows.

I. **The plexiform lamina** (molecular and zonal layer) contains sparsely scattered horizontal cells surrounded by a compacted mass of tangential fibres, derived from pyramidal cells (apical dendrites), stellate cells (vertical axons), cells of Martinotti (centrifugal vertical axons) and cortical afferent fibres, both projectional and associational.

II. **The external granular lamina** contains the somata of stellate and small pyramidal neurons, packed with varying density. Traversing it are vertical dendrites and axons from subjacent layers, intermingling with a dense neuropil of local dendrites and axons. Ascending afferents make extensive multiple synaptic contacts with the apical dendrites of large pyramidal cells (somata in lamina V) both in this and in lamina III.

III. **The pyramidal lamina** contains somata of medium-sized pyramidal neurons, the smaller ones being situated nearer to lamina II. Some stellate cells also occur, including horizontally disposed basket cells and vertically orientated fusiform cells, with dendrites and axons extending far beyond the lamina.

IV. **The internal granular lamina**, usually narrower than other laminae except lamina I, is chiefly populated by the somata of stellate neurons, with occasional small pyramidal cells. The somata are densely aggregated and the lamina is traversed by a concentration of horizontal fibres, long known as the *external band of Baillarger*. As in other laminae, this also contains large numbers of vertical neurites derived from neurons in other cortical areas, in subcortical regions and in adjoining laminae.

V. **The ganglionic lamina** contains the largest pyramidal cell somata but smaller ones also occur, the actual dimensions varying in different cortical areas. In any area, however, the largest pyramidal cells are in lamina V. Small numbers of stellate cells also occur and the lamina is permeated by a dense neuropil of dendrites and axons from its intrinsic elements and cells in other laminae. It is also traversed by the ascending and descending projection and association fibres. A considerable complement of horizontal fibres is apparent, corresponding to the *internal band of Baillarger*.

VI. **The multiform lamina** contains a range of cells, as judged by its somata and processes, the variable shape of the former reflecting variations in dendritic arrays. Most cells are small and considered to be modified pyramidal elements, despite the fusiform, triangular, ovoid and other profiles of their somata. Small, multipolar Martinotti cells are often prominent in this lamina, which is not always well demarcated from a subjacent cortical zone of fibres approaching or departing from the cortex.

The numbering and nomenclature of cortical laminae noted briefly above is in the style of Brodmann. Many synonyms are in circulation, derived from the work of investigators such as Campbell and Cajal; the Vogts also introduced another, more awkward nomenclature, based on myeloarchitectonic studies in which fibre structure is emphasized. Views as to the number of layers have varied and much subdivision has been mooted. For example, lamina VI has been divided into VIa and VIb by the distribution of triangular and fusiform cells and into no less than four sublaminae (VIa^1, VIa^2, VIb^1 and VIb^2) in material stained to show fibre structure. All except lamina II have thus been further analysed, as many as 16 laminae being recognized by the Vogts. The usefulness of such minute stratification has been much criticized; such details are only mentioned here as examples. Similar remarks are applicable to much work in cortical architectonics.

Here again the Brodmann maps have achieved most attention. Their early transference (in a form further elaborated by the Vogts but still not worked out in man) to the human cortex by Foerster has overshadowed the contribution of von Economo & Koskinas. In one respect this is fortunate, because they chose to designate areas by letters, as has Conel (1939); these, and other nomenclatures would have proved less manageable than Brodmann's simple numbers. However, newer views on the extent of major sensory and motor cortical areas have generated a need for more appropriate terminology than Brodmann's (Woolsey 1964). A suggested new series of symbols will be referred to later, for particular areas; many are already in wide use in experimental studies but since a lack of uniformity persists in this new and far from comprehensive terminology, which is only beginning to penetrate clinical neurology, the customary numerical designation of areas must be retained for the present.

REPRESENTATIVE VARIANTS OF CORTICAL STRUCTURE

Note. The following is retained for historical perspective, and because of wide usage. Nevertheless the *numerical neuronal homogeneity* of the neocortex and of its constant ratio of pyramidal/stellate neurons amply demonstrated in many mammals by Rockel et al (1980) and Powell (1981) must be constantly remembered. Individual neurons may vary greatly in size, packing density and in their profusion of efferent and afferent connections with multiple destinations and sources and these underlie the following 'varieties', with their differences in thickness and laminar pattern.

While it is here impossible, and probably unprofitable, to detail and discuss the almost endless nuances of cortical structure based on full-blown architectonics, a smaller number of variant types must be indicated. These were recognized in the classic studies of Campbell (1905) and corroborated by Economo & Koskinas (1929). *Five* fundamental types are described in the neocortex (7.158A,B). While all these are said to develop from the same *sexalaminar pattern*, two are regarded as virtually lacking certain *laminae* when differentiated and are hence *heterotypical*; these are the *granular* and *agranular* types. *Homotypical* variants, in which all six laminae persist, are called *frontal* (premotor), *parietal* (postcentral) and *polar* (visuopsychic), names linking them with specific regions in a misleading manner, as illustration 7.158A shows (e.g. the frontal type occurs in parietal and temporal lobes).

The agranular type is considered to lack, or have diminished granular laminae (II and IV) but always contains scattered stellate somata. The prominent neuronal type is pyramidal and in this form the greatest densities and largest sizes of pyramidal neurons occur. Originally identified in the precentral gyrus (area 4), it also occupies areas 6, 8 and 44 (7.157A,B) and occurs elsewhere.

including parts of the limbic system (p. 1028). It is typified by projection of particularly large numbers of axons from its pyramidal cells; the agranular cortex is thus often equated with 'motor' areas, with the proviso that such areas also receive afferent projections. (Further, *all* neurocortical areas have many varieties of both efferent and afferent projections.)

The **granular type** of cortex (koniocortex) is at the opposite extreme from the main categories of cortical structure. Granular layers are maximally developed and contain densely packed stellate cells, amongst which are dispersed small pyramidal neurons. Laminae III and IV are poorly developed or unidentifiable. This type is particularly associated with afferent projections but, again, there is evidence of less numerous efferent fibres, derived from the scattered pyramidal cells in this otherwise 'granular' cortex.

Despite the relative lack of different laminae, granular and agranular cortices exhibit little qualitative distinction, being rather at opposite extremes of a gradation, in which the pyramidal and stellate neurons are reciprocally prominent. A typical granular cortex appears in the postcentral gyrus, striate area and superior temporal gyrus (acoustic area) and in small parts of the parahippocampal gyrus (p. 1029). Despite its very high density of stellate cells, especially in the striate area, it is almost the thinnest of the five main types. In the striate cortex the external band of Baillarger (lamina IV) is well defined as the *stria of Gennari* (or Vicq d'Azyr).

The remaining three types of cortex are intermediate forms. In the **frontal type** large numbers of small and medium-sized pyramidal neurons appear in laminae III and V, granular layers (II and IV) being less prominent. The relative prominence of these major forms of neuron vary reciprocally wherever this form of cortex exists. It is not confined to the frontal region (7.158A,B). The **parietal type** of cortex contains pyramidal cells which are

mostly smaller in size than in the frontal type; granular laminae are, on the contrary, wider and contain more of the stellate cells. This kind of cortex occupies large areas in the parietal and temporal lobes (7.158A,B). The **polar type** is classically identified with small areas near the frontal and occipital poles, hence its name. It is the thinnest form of cortex (7.158B). All six laminae are represented but the pyramidal layer (III) is reduced in width and not so extensively invaded by stellate cells as in the granular type of cortex. As in the latter, the multiform layer (VI) is more highly organized than in other types.

While subdivision of these five basic types of cortical 'organization' may sometimes be useful, it must be re-emphasized again that in microscopic sections where they are customarily distinguished, whether stained to show somata or processes, the finer and more significant details of true organization are not so apparent in studies of the whole thickness of the cortex. Functional organization is naturally linked to the spatial distribution of cells but it is in their actual patterns of connection that any real enlightenment as to cortical mechanism must be sought. Golgi preparations have yielded an immense amount of information indicating the probable designs of neuronal interaction. Functional hypotheses deduced from such structural data, however intricate, require confirmation in terms of the precise nature of the synapses, both in their distribution *and* mode of action. Such details depend on electron microscopy and unit recording, techniques which deal only with much smaller volumes of cortical tissue, and also on the rapidly expanding techniques of neurohistochemistry. Nevertheless exciting results have accrued in this field in recent years and doubtless more is to come. The cytoarchitectural approach to cortical activity was a necessary prelude and the resulting definitions of forms of organization in structural terms remain the bases of orientation to which ultrastructural and other details must be referred.

THE MAIN CORTICAL AREAS

Before describing the major areas of the human cerebral cortex, customarily distinguished by functional and structural data such as the somatomotor, somatosensory, visual and auditory areas, preliminary general remarks are necessary. As already emphasized, the extreme parcelling deriving from studies by Brodmann has limited its usefulness and can be misleading if taken as much more than a reference grid. Even the simplest division of the cortex into *sensory* areas receiving afferent projections and *motor* areas projecting efferents, the remainder being regarded as 'silent' or *associational*, is no longer appropriate, being an inaccurate over-simplification. Evidence has accumulated to show that areas receiving or originating *specific named groups of* projection fibres are much more extensive than the initial classic studies indicated. Furthermore, division into 'receiving' and 'originating' areas is only relevant in relation to a particular stated tract, fasciculus or radiation, if certain *general structural features* are clearly recognized. *All neocortical areas* are *both* originating and receiving, i.e. they have substantial populations of efferent and afferent fibres. What follows is in no sense comprehensive, but a brief list to provide an ever-present background when a superadded named connection or area is under review; each receives more detailed attention elsewhere. *Efferent fibres* pass *from* all areas *to* the dorsal thalamus (cortico-thalamic), the corpus striatum (cortico-striate), the reticular formation (cortico-reticular), the pons (cortico-pontine), ipsilateral cortex (long and short association fibres), contralateral cortex (homotopic or heterotopic commissural fibres). They are lacking in distal limb areas and narrow strips in visual areas. Other efferents have more restricted origins. *Afferent fibres* pass *to* all areas *from* the dorsal thalamus (thalamo-cortical from both specific and intralaminar nuclei), some elements of the brain-stem reticular formation (reticulo-cortical) including: monoaminergic fibres radiating from the nucleus coeruleus (A6), rapheal nuclei B6–8, medial forebrain bundle and hypothalamus, cortico-cortical afferents,

either ipsilateral (long and short association) or contralateral (commissural).

The named areas in the following account have, for long, been better known because of the obvious and dramatic sequelae when trauma or pathology supervenes and their greater accessibility to investigation. However, intense research effort is currently directed to elucidating the most appropriate arrays of *fundamental cortical units* which are probably repeated countless times over the qualitatively and numerically *uniform neocortex*. On this view, functional localization in the cortex is *not* an intrinsic property but is conferred on it by its mosaic of afferent and efferent channels, their sources, destinations and numbers.

With specific areas a brief historical perspective will be pursued. Thus, the postcentral gyrus is not the only area to which the somatosensory thalamic projection is directed; at least two other areas are similarly involved, as will be detailed later. Similarly, the precentral gyrus is supplemented by a second 'motor' area and it is necessary to qualify the term 'motor', because the distinction of motor and sensory areas which is still customary in most accounts of the cerebral cortex is erroneous and has been known to be so for many years. As long ago as 1933 motor responses to stimulation of 'sensory' areas were demonstrated (Dusser de Barenne 1933); and the projection of efferent ('pyramidal') fibres from this postcentral area was established shortly afterwards (Levin & Bradford 1938). Since these pioneer observations, much confirmatory evidence in various experimental animals has extended such findings to other areas. Moreover, clinical and experimental observations in mankind suggest that similar arrangements obtain. It is hence appropriate to speak of pre- and postcentral areas as *sensorimotor*. Since a mixture of afferent and efferent connections has been shown also in respect of the accoustic and visual 'sensory' areas, they also are sensorimotor.

Recognition that the corticospinal tract is derived from somata in an area far beyond the strict precentral gyrus is paralleled by

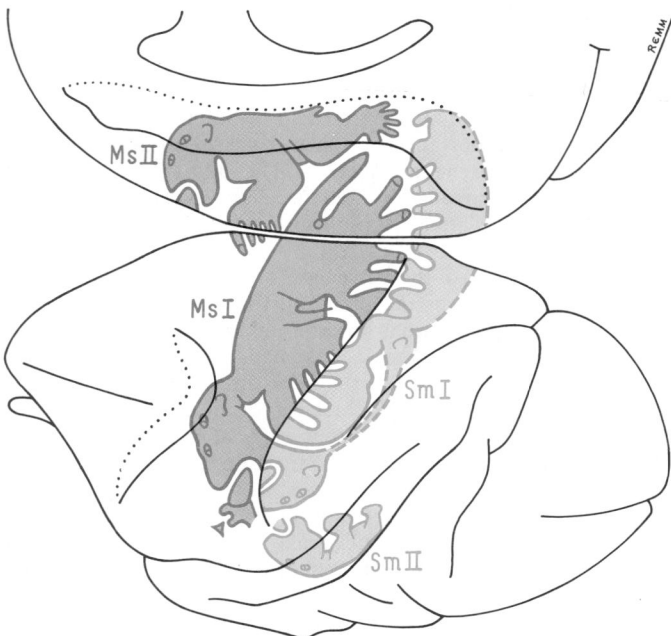

7.161 The main sensorimotor areas projected diagrammatically upon the superolateral surface of the simian cerebral hemisphere. Note the somatotopic arrangement in all four areas. Adapted from Woolsey 1964, see text for details and references.

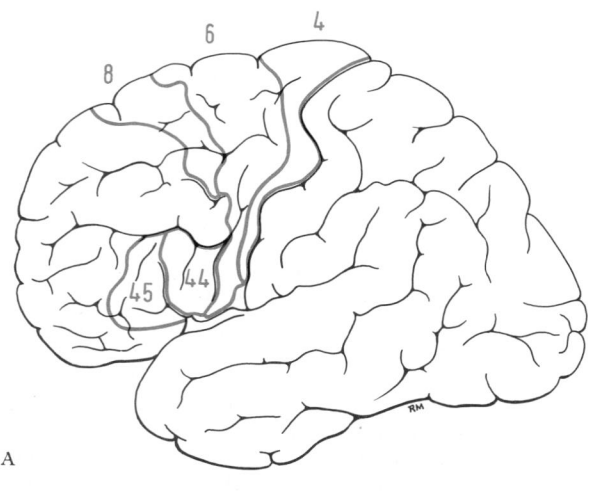

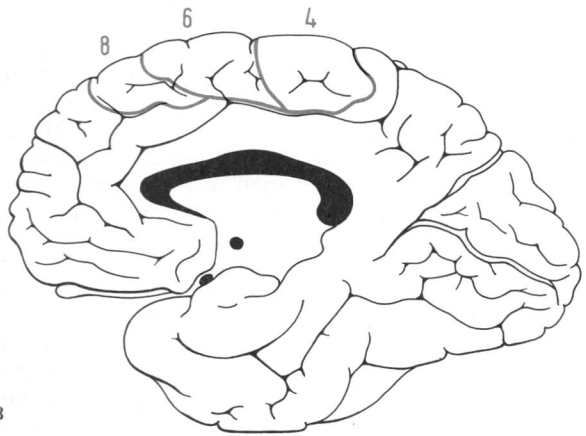

7.162 Superolateral (A) and medial (B) surfaces of the cerebral hemispheres, showing approximate correspondence of the Brodmann areas to the main motor area (4) or MsI, the premotor area (6, 8) and motor speech area of Broca (44, 45). See text for details and compare with 7.157, 169, 170.

findings in various thalamic and geniculate projections on the cortex; these terminate in considerably wider regions than originally described. For example, as intimated, the classic somatosensory area in the postcentral gyrus is supplemented by a second, inferior to it, and even a third, on the medial hemispheric aspect which is also a motor area (vide infra). Similarly lateral and medial geniculate bodies are now known to project to regions of the cortex beyond the visual and auditory areas of older description. Not only the striate cortex (area 17, visuosensory area), but also para- and peristriate areas around it (areas 18 and 19, the 'visuopsychic' cortex) receive projection fibres. In cats the acoustic radiation terminates not only in the *first acoustic area* (41) but also in several others in the temporal cortex.

These modifications of originally simple motor and sensory areas require revised terminology. One commonly used (Woolsey 1964), though not entirely satisfactory nor yet accurately adapted to the human cortex, divides the sensorimotor area into a part in the precentral gyrus termed MsI, because it is the main or *first* predominantly *motor* area but to a lesser extent also sensory (7.161). Conversely, in the postcentral gyrus, SmI is the primary Sensory area, though also partly *motor*. On the medial cerebral surface is another sensorimotor area which, being largely motor, is called the *supplementary motor area*, MsII. Unfortunately, MsII is sometimes also known as the *third* somatomotor area, because there is a second sensorimotor area, SmII (7.161), inferior to SmI. It would have been easier to regard MsII as the second somatomotor area and SmII the third, as some do, but this merely shifts the numerical confusion. Some authors surmount this difficulty by disregarding the abbreviations and speaking of first, second and third motor or sensory areas. This terminology is admittedly perplexing but its shortcomings are merely incidental to the important finding that the frontoparietal sensorimotor region is a complex, each area having its own degree of somatotopic organization. *None* of these 'areas' is exclusively motor or sensory, though one or the other quality always predominates. In the following text ambiguity has been avoided by use of synonyms.

In the occipital lobe striate, parastriate and peristriate regions (areas 17, 18 and 19) are likewise termed *first*, *second* and *third visual areas*, or visual areas I, II and III. More complex terms, applied in particular to the feline cerebrum (Rose & Woolsey 1958), currently designate *four* temporal areas considered to receive auditory projection fibres *in cats*: the *first acoustic area* (AI), the auditory area of usual description, *second acoustic area* (AII) and the *anterior* and *posterior ectosylvian areas* (Ea and Ep). Accessory acoustic areas can be indicated only tentatively in the human cortex (7.173), partly because the acoustic regions are entirely superficial in feline temporal lobes, but partly folded into the lateral sulcus in mankind. All the areas named above not only receive projections *from* their particular thalamic nuclei (though this is not established for the ectosylvian acoustic areas) but also project back to the same nuclei and in some instances to other nuclei at lower levels, connections which in each modality are concerned with inhibitory and sometimes facilitatory effects in their own sensory pathway. It is this afferent–efferent nature of most, probably all, sensorimotor areas which makes the concepts of distinguishable motor and sensory cortical areas anatomically invalid and functionally misleading.

It is clear that much less of the cerebral cortex remains to be dubbed 'associational', in the vague but well-established meaning of the term; but large regions in all cerebral lobes are not accounted for by sensorimotor projections. In the human brain, much clinical evidence suggests that for long the so-called 'silent' areas are indeed concerned in further elaboration and interpretation of sensory information and in the combination of its different modalities, as has long been vaguely implied. Much less is known of the detailed connections of these areas, whether with the brain stem and diencephalon or by association or commissural fibres, although unquestionably *all* are involved. Hence, less can be said of these areas than of the sensorimotor cortical domains. Ablation experiments, coupled with the tracing of degenerating axons and the observation of disturbed behaviour, with similar deductions in human patients affected by damage, disease or surgical intervention, are beginning to provide more coherent concepts of the activities in association areas.

Though it might be functionally appropriate to consider first the predominantly sensory areas and then those mainly motor, widespread involvement of many parts of the cortex in even the simplest activities, and the sensorimotor nature of the areas concerned, render any such considerations of priority almost meaningless. The main areas will therefore be described on the basis of the lobes in which they are located and in an arbitrary order: frontal, parietal, occipital and temporal.

THE FRONTAL LOBE

The frontal cortex may be conveniently divided into *precentral* and the unfortunately named *prefrontal* region. The former is largely sensorimotor, the latter an 'association' cortex.

The precentral area includes the whole precentral gyrus and posterior (caudal) parts of superior, middle and inferior frontal gyri (Brodmann areas 4 and 6). The whole region (7.162A,B) features an almost complete absence of granular layers. Intracortical fibres are numerous and the plexiform layer hence is densely packed. The area is sometimes divided into posterior and anterior parts, *motor* (area 4) and *premotor* (area 6), but the distinction between them differs, according to whether cytoarchitectural or physiological data are applied. As comparison of Figures 7.157A,B and 7.162A,B shows, the boundary between areas 4 and 6 descends on the precentral gyrus, whereas the entire gyrus is functionally the *motor area*. It is less confusing to disregard the Brodmann numeration, or to use it as a merely approximate indication. The premotor area occupies parts of the three frontal gyri, corresponding largely to area 6 but including parts of 8, 44 and 45 (7.162A,B). For the present, it is simpler to designate the whole precentral area as the *first* or leading **somatomotor area** (MsI), remembering that functional variations occur within it which may to some extent coincide with cytoarchitectonic differences, but much more certainly with afferent and efferent connectivity.

A feature of this area (MsI) is the prominence of pyramidal neurons of all sizes. The largest, the *giant pyramidal cells of Betz*, vary in height from 30 to 120 μm and in breadth from 15 to 70 μm, being most numerous in the part of the area extending over the hemisphere's superomedial border into the paracentral lobule (7.162B). They diminish in numbers down the precentral gyrus towards the lateral fissure and also extend less and less forwards in the gyrus, following the boundary of area 4 (to which they are almost limited, being absent from the premotor or anterior part of MsI. In view of the somatotopical arrangement now to be described, it is tempting to suppose that the size of the larger pyramidal cells is associated with the length of their axons. Giant pyramidal cells in the human motor cortex have been estimated at between 25 000 and 30 000 (Campbell 1905, Lassek 1940), a figure conflicting with the million or so fibres in the medullary pyramid (Lassek & Rasmussen 1939, Lassek 1942). The majority of corticospinal and probably corticobulbar fibres are hence derived from the smaller pyramidal neurons. Many such axons do not originate in area 4; it is interesting to note that its ablation in monkeys reduces the fibres in the pyramid by a mere 25% (Lassek 1942). Obviously most 'pyramidal' fibres have other origins, as will be detailed below.

Recognition of simple *contralateral movements*, elicited in different parts of the body by electrical stimulation of loci near the central sulcus, dates from the pioneer study of Fritsch & Hitzig (1870) in dogs. This discovery has been confirmed and elaborated in innumerable subsequent experiments in primates (e.g. Leyton & Sherrington 1917) including man (Ferrier 1874, Penfield & Rasmussen 1950); it has become a tenet of neurology that the area involved is concerned in the mediation of voluntary movements. The order of loci, from the paracentral lobule, is associated with lower limb, trunk (upper part of the precentral gyrus), upper limb, neck and head (7.163A). Much detail of this somatic 'representation' has accumulated, of course, and it is established that the amount of cortex-mediating movement in any region is proportional *not* to the bulk of muscle involved but to the skill with which it is used (7.163A). It is important to emphasize that in the earliest studies of excitability of the 'motor cortex', movements were elicited also from the postcentral gyrus and that the same movement was excited from different loci at different times

in the same brain. More careful control of the parameters of stimulation has later confirmed this apparent 'plasticity' (Liddell & Phillips 1950) which is difficult to represent in textbook illustrations. This and other factors have obscured these early findings to some extent; the polemics over the great volume of reports on cerebral motor function, especially regarding the discrepancies between clinical and experimental observations, were at least useful in so far as they refocused attention on facts rather than theories (Bucy 1944).

It is now well established that while the whole first somatomotor cortex (MsI, areas 4 *and* 6) is a source of corticospinal fibres, it is not the only one. Both in monkeys (Levin &

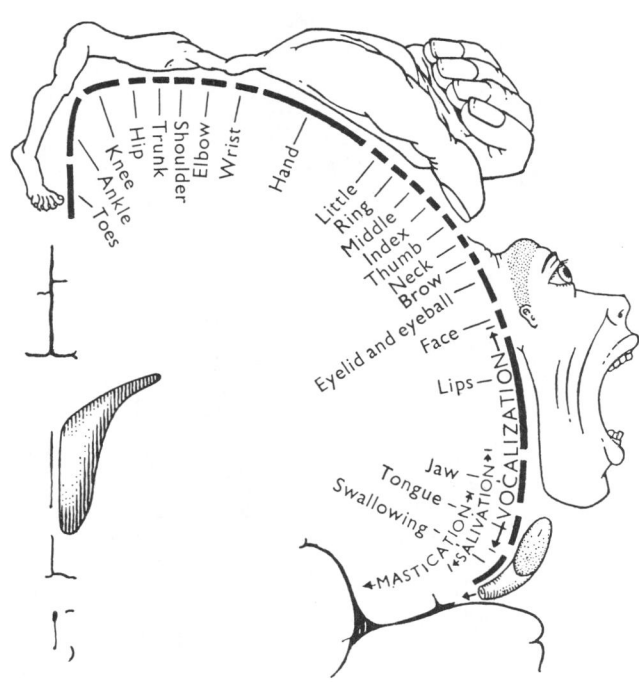

7.163A The *motor homunculus* showing proportional somatotopical representation in the main motor area. After Penfield & Rasmussen 1950.

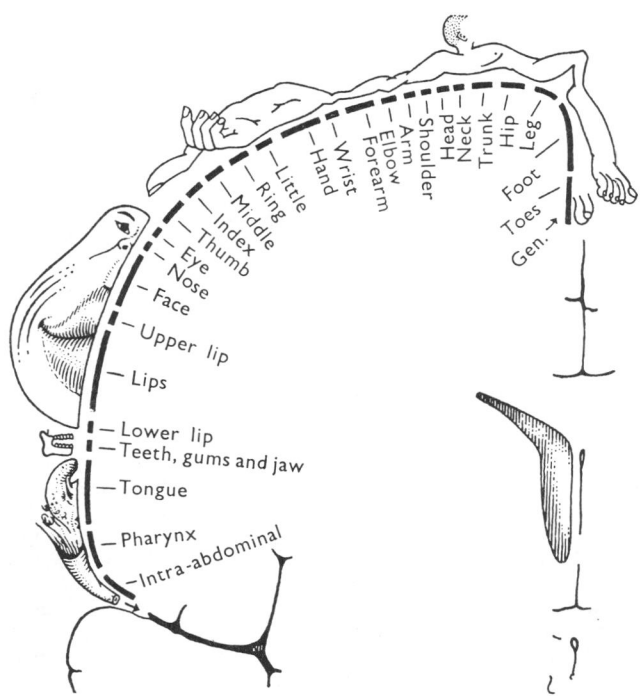

7.163B The *sensory homunculus* showing proportional somatotopical representation in the somaesthetic cortex. After Penfield & Rasmussen 1950.

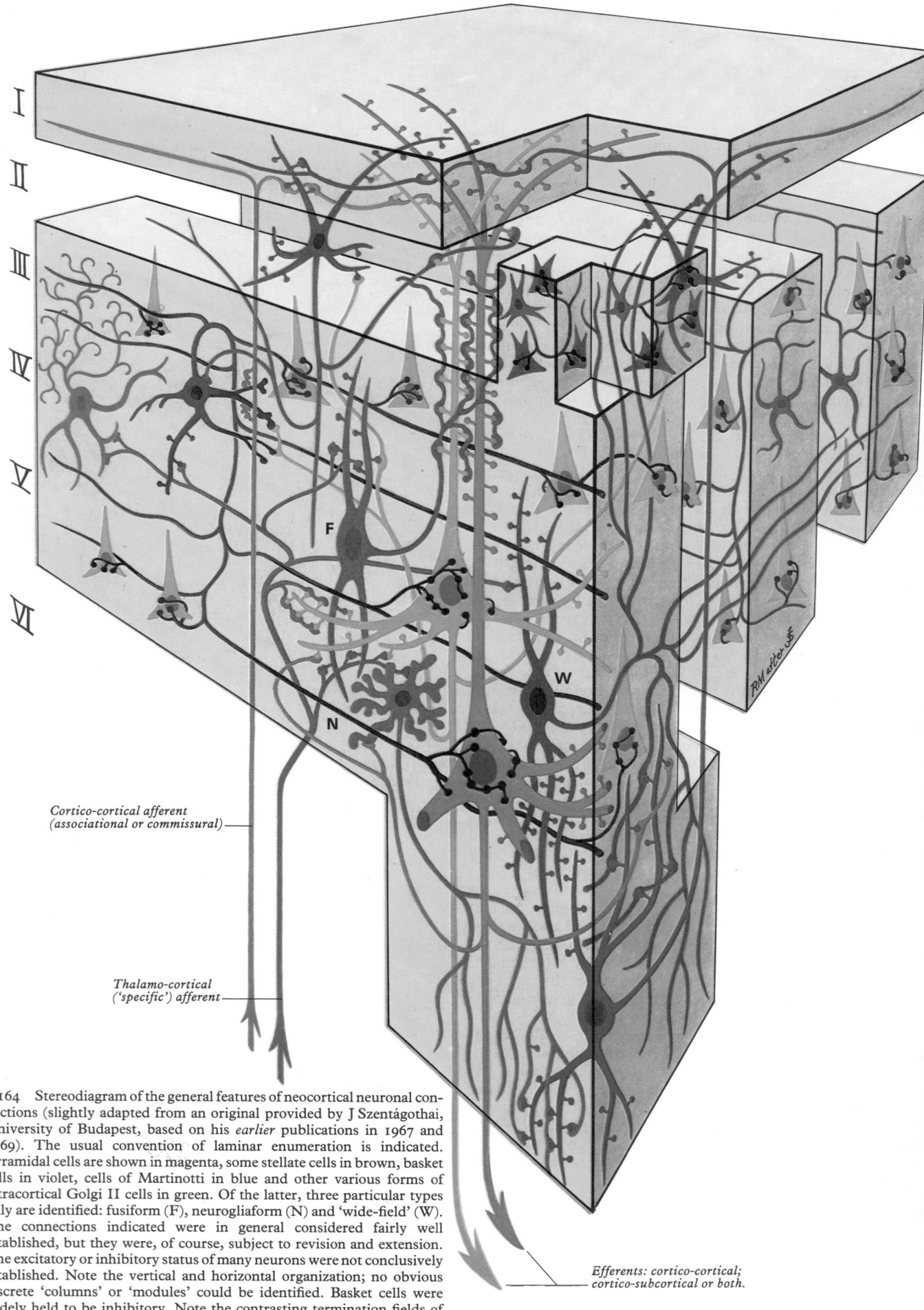

7.164 Stereodiagram of the general features of neocortical neuronal connections (slightly adapted from an original provided by J Szentágothai, University of Budapest, based on his *earlier* publications in 1967 and 1969). The usual convention of laminar enumeration is indicated. Pyramidal cells are shown in magenta, some stellate cells in brown, basket cells in violet, cells of Martinotti in blue and other various forms of intracortical Golgi II cells in green. Of the latter, three particular types only are identified: fusiform (F), neurogliaform (N) and 'wide-field' (W). The connections indicated were in general considered fairly well established, but they were, of course, subject to revision and extension. The excitatory or inhibitory status of many neurons were not conclusively established. Note the vertical and horizontal organization; no obvious discrete 'columns' or 'modules' could be identified. Basket cells were widely held to be inhibitory. Note the contrasting termination fields of 'specific' and cortico-cortical afferents; also of intracortical recurrent collaterals of pyramidal neuron axons.

Cortico-cortical afferent (associational or commissural)

Thalamo-cortical ('specific') afferent

Efferents: cortico-cortical; cortico-subcortical or both.

Bradford 1938) and man (Kuypers 1958) the first somatosensory area (SmI, areas 3, 1 and 2) in the parietal lobe is also a source of corticospinal fibres; possibly the second somatosensory area (SmII, approximately equal to areas 40 and 43, p. 1057), below the first area and above the lateral fissure (posterior limb), also contributes 'pyramidal' fibres. Smaller contributions from occipital and temporal lobes have also been described in cats (Walberg & Brodal 1953). Physiological evidence largely accords with these anatomical findings. Fibres descending from this extensive origin vary much in calibre; in the human pyramid 90% were estimated to be 1–4 μm in diameter, only 2% were 11–22 μm, a fraction corresponding closely with the number of giant pyramidal cells (Lassek 1942). Up to 94% of the fibres are myelinated in man (De Meyer 1959).

The primary, first somatomotor area, or MsI, receives fibres from the cerebellum which relay mainly in the thalamic nucleus ventralis lateralis (p. 1002), distributed particularly to its anterior region (area 6) and adjacent prefrontal cortex (area 8) but some pass to area 4. Fibres from the corpus striatum (pallidofugal fibres) after relay (and pre-integration) in the thalamic nucleus ventralis anterior, reach the same cortical regions. It also receives

afferents concerned in *other sensory* modalities, probably via the thalamus also but in addition from the hypothalamus and also from many other cortical areas. It is also traversed horizontally by monoaminergic fibres from the reticular system (noradrenergic from A6, serotonergic from B6–8) and cholinergic from Ch4, the nucleus basalis (of Meynert). Electrophysiological studies suggest that corticospinal neurons are influenced not only by somatosensory impulses but also by visual, acoustic, 'non-specific', thalamic, reticular and cortico-cortical afferents (p. 1003). Interaction of all (or some) of these on pyramidal neurons in general inevitably demands an intricate deployment of axons, dendrites and synapses. (For a structural/functional review of the primate motor cortex consult Phillips 1981.)

Illustration 7.164 was a tentative epitomization of the kind of structure involved and envisaged as applying to the *neocortex generally*, provided by Professor J. Szentágothai (1969). It is basically *radial* or *columnar* as being demonstrated in other parts of the cerebral cortex (somaesthetic and visual), the pyramidal neurons being extended vertically, with abundant vertical and more restricted horizontal associations with afferents through connections with intrinsic interneurons, some excitatory and

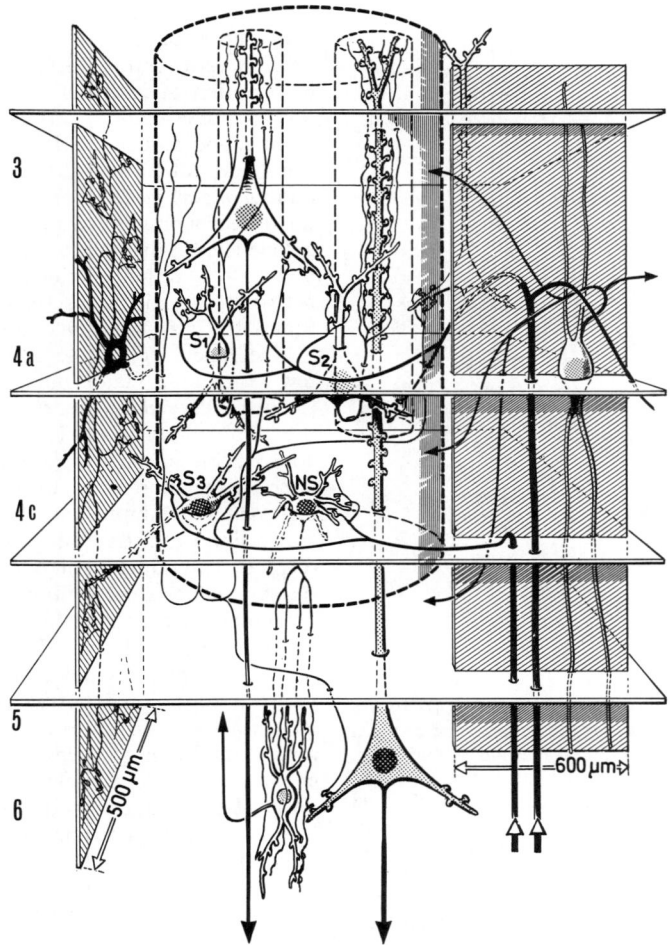

7.165A   A stereoscopic view of the elementary neuron circuit in sensory cortical areas. The horizontal planes are entirely arbitrary and are included to aid stereoscopic visualization; they are unrelated to the cortical lamination (arabic numerals on the left, indicating the 'non-absolute' character of the lamination in such a diagram). Specific sensory afferents (heavy black vertical lines on the right) terminate (separately) on the spiny stellate neurons (S1) and the so-called star pyramids (S2) of upper lamina IV and in lower lamina IV on another type of spiny stellate neuron (S3) as well as on neurogliaform non-spiny stellate neurons (NS). Ascending relay of sensory afferent impulses is twofold: (1) within *narrow* cylindrical spaces (indicated with dashed outlines) by horsetail-shaped axon arborizations of S1 and S2 type neurons and primarily to selected individual

(or small groups of) pyramid cell and (2) within *wider* cylindrical spaces (heavier dashed outline) from ascending S3 type neurons, probably mainly to basal dendrites of a much larger group of pyramidal cells. Descending relay is more widely and more indiscriminately distributed by descending branches of the spiny stellate neurons and star pyramidal neurons and occurs in narrow columns by the vertically orientated axonal lacework of neurogliaform non-spiny stellate neurons (NS). The vertical plane on the left shows the strictly orientated axonal arborization of a large basket cell; the vertical plane on the right, orientated at right angles to the basket cell plane, shows a large stellate neuron with vertically orientated dendrites and part of its extended axonal arborization which is strictly confined to this plane. From Szentágothai 1975 with permission.

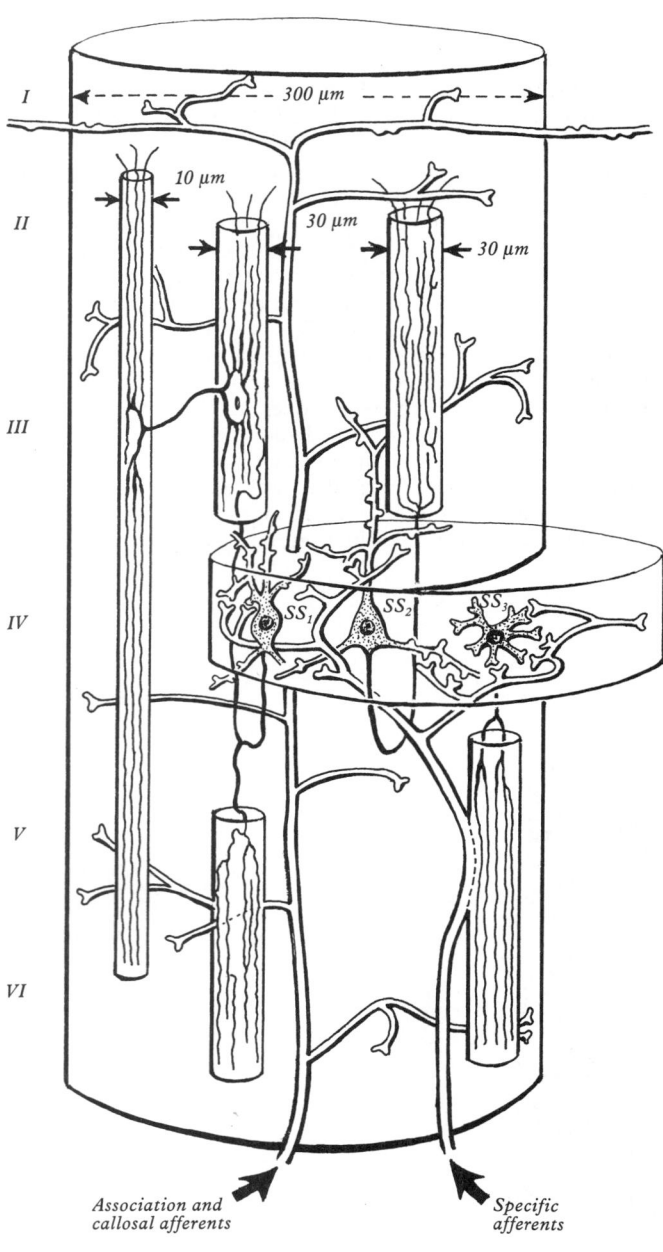

I

— 300 μm —

10 μm

II

30 μm

30 μm

III

IV

SS₁   SS₂   SS

V

VI

*Association and
callosal afferents*

*Specific
afferents*

7.165B Some features of a cortico-cortical module redrawn from a stereodiagram (provided by Szentágothai 1978). All neurons and contacts shown are presumed to be excitatory. Pyramidal neurons, their axonal efferents and inhibitory interneurons are not included for clarity. The module consists of a tall cylinder 300 μm in diameter extending through the full thickness of the cortex; ascending centrally, with lateral and terminal branches is one representative of a cortico-cortical afferent (either association or callosal). The 'flat cylinder' also about 300 μm in diameter is the 'termination space' of a specific thalamocortical afferent; overlap of the latter and the module varies greatly from minimal, through the condition illustrated, to virtual coincidence. Synaptic contacts through one or more excitatory interneurons transforms the flat (almost tangential) termination space of the specific afferent into narrow (10–30 μm), *radial* cylindrical spaces or sub-modules. Stellate interneuron SS2 has a recurrent axon, its arborization ascending to lamina II; the microglioform neuron (right) has a descending arborization; SS1 has both a descending and an ascending arborization, the latter synapsing with a *cellule fusiforme à double bouquet dendritique*, the axon of which again bifurcates forming a long vertical slender (10 μm) arborization and which possibly encloses, and is excitatory to, a single pyramidal neuron. It should be noted, however, that a significant proportion of neurons with all these cytomorphologies have been demonstrated as having their inhibitory counterparts (see 7.167).

1052

some inhibitory. (It is of considerable interest that this early diagram was preceded by an even earlier theoretical attempt to define the co-ordinates needed for the analysis of *higher integrative units* in neural tissues, Szentágothai 1967.) These concepts accord with studies of individual motor neurons by microelectrode techniques in monkeys (Landgren et al 1962, Phillips 1981) and cats (Asanuma & Sakata 1967), which show that some motor neurons at least, though excitable from a larger area, can be most easily stimulated within a cylindrical section of cortex about 1 mm in diameter. At this stage the radial or columnar disposition of the pyramidal neurons dominated the picture and attempts were continued to establish the three-dimensional relationship to them of afferents and stellate interneurons, their synaptic contacts and excitatory or inhibitory status. There was little to indicate that radial arrays formed discrete circumscribed modules or columns and, in Szentágothai's view, if they existed they would 'mutually interpenetrate' into one another's cortical spaces. Electrophysiology of the somaesthetic cortex (Mountcastle 1957 et seq.) and visual cortex (Hubel & Weisel 1962 et seq.) provided, however, powerful evidence for functionally discrete radial columns or modules. The search for their structural counterpart was renewed, but now with particular emphasis on the distribution of *specific thalamocortical afferents*. Illustration 7.165 is a later summary of the '*module-concept*' (*elementary neuron circuits*) in cerebral cortical organization (Szentágothai 1975); many details are mentioned in its lengthy caption. In it: 'An attempt is made to bring earlier circuit models of *primary sensory* cortical areas into better line with recent observations on (1) the distribution of *excitatory feedback connexions* in cortical tissue volume, (2) (the then: Ed.) putative inhibitory interneurons and distribution of *inhibition in well defined space modules*, and (3) direct (monosynaptic) cortical target cells of specific sensory afferents, and modes of relay to secondary neurons. Even though the concept of cortical circuitry on larger "*integrative units*" containing *smaller modules* (or *fields*) of specific (excitatory and inhibitory) neuronal actions, proposed in 1967 and 1969, had gross deficiencies in the light of newly emerging data, the basic idea of how to look at the functional organization of the cortical neuron network may still be useful as a conceptual framework for the functional interpretation of structural data.'

The expectancy in the previous remarks progressed within a few years to a major publication (Goldman & Nauta 1977), further surveyed by Szentágothai (1978) in which it is held that the '*true*' cortical module is not primarily based on thalamocortical afferents but on *cortico-cortical* connections (long and short ipsilateral association fibres, together with homotopic and hetertopic contralateral commissural fibres). These *cortico-cortical modules* are columns, initially regarded as cylinders 200–300 μm in diameter extending through the full thickness of the cortex; each receives *and* originates all the fibre types mentioned and they are presumed to occur in all neocortical areas (and in all mammals yet studied). Some of their features were reviewed in the paper quoted but results of their further investigation were awaited with interest. As pointed out elsewhere (pp. 1071, 1072) details of *human* association fibres are, as yet, only known in relation to gross bundles and quantitative data are lacking; further some cortical areas lack callosal fibres e.g. those representing the 'distal limbs' (hands and feet) and much of the visual areas V1–5, callosal fibres being limited to narrow *strips* between the numbered areas. The following decade involved an explosive expansion of neuroanatomical, neurophysiological and chemoarchitectonic techniques, increasingly, and most powerfully, *in combination* on the same specimen. This has served to confirm and refine the cortico-cortical module as *one* of the dynamic central concepts whose consideration is mandatory when cortical functional microarchitecture is being addressed. Stimulating recent reviews of these topics are provided by Szentágothai (1985, 1987) on whose publications illustrations 7.166 and 7.167 are based. The former shows, in elemental form, the large-scale connectivity involved and, in a single module, one dynamic variation that would accompany one set of inhibitory/excitatory interactions; the latter is an analysis of seven well-established types of inhibitory (GABA-ergic) interneurons variously involved; their lengthy captions should be perused for details. (A complete) acceptance

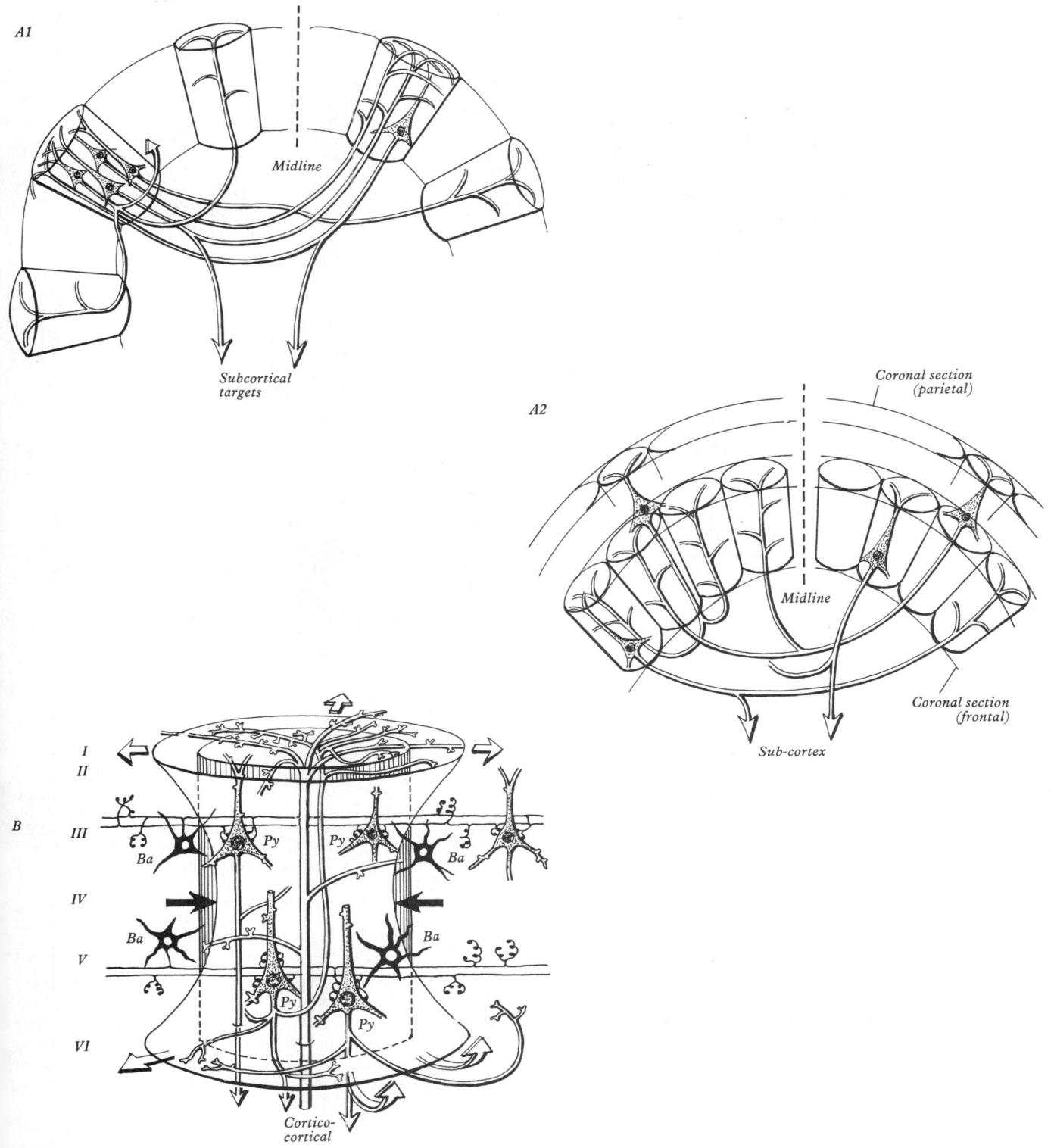

A1

Midline

Subcortical
targets

A2

Coronal section
(parietal)

Midline

Coronal section
(frontal)

Sub-cortex

B

I
II

III

IV

V

VI

Ba   Py   Py   Ba

Ba   Ba

Py   Py

Cortico-
cortical

7.166   Stereoscopic diagram illustrating the principle of (cortico-cortical) modular architecture of the neo-cortex. Part 'A1' shows cortico-cortical connectivity on the large scale: *supragranular pyramidal cells* (in laminae II and III; two cells of the outer row in left middle column) connect preferentially with columnar tissue spaces in the ipsilateral or in the contralateral cortex. *Infragranular pyramidal cells* (lower row of cells in middle column at left and one cell in one column at right) send their axons to subcortical targets, but some of them are addressed to (or give major collaterals to) contralateral cortical columns. Local (intracortical) connectivity extending over approximately 10 neighbouring columns is neglected in this diagram. Part 'B' shows the architecture of a single (cortico-cortical) column (vertically hatched inner cylinder) of 200–300 µm width and extending through the entire depth of the cortex, radically stripped down to a few essentials. A single cortico-cortical afferent arborization is placed into the centre of the drawing (representative for a few of the 10 cortico-cortical afferents coming probably from the same source). Representative pyramidal cells of the supragranular (Py₃)

and the infragranular layers (Py₅) are indicated in outline. Relatively long range inhibitory cells are represented by two large basket cells (Ba₃ and Ba₅) for the upper and the lower row each (see 7.167) in full black. Elements (and direction) of excitatory nature are indicated in outline, inhibitory cells (and directions of connections) are shown in black. Terminal branches of cortico-cortical afferents and pyramidal cell collaterals extend over the limits of the basic cylinder in lamina I and lamina VI; this is emphasized by outline arrows pointing in outward direction in the two layers. Conversely, dark arrows indicate that horizontally oriented basket (and other inhibitory) cells would tend to narrow down activity in the columnar unit. The original cylinder (vertical hatching) would hence become distorted dynamically into the shape of an hourglass (rotation hyperboloid; stippled). Adapted from Szentágothai 1985, 1987, with permission of the author. Part 'A2' is an attempt to illustrate cortico-cortical connections on a large scale, both ipsilateral and contralateral, between two coronal sections at different fronto-occipital positions. Provided by J Szentágothai 1988, personal communication.

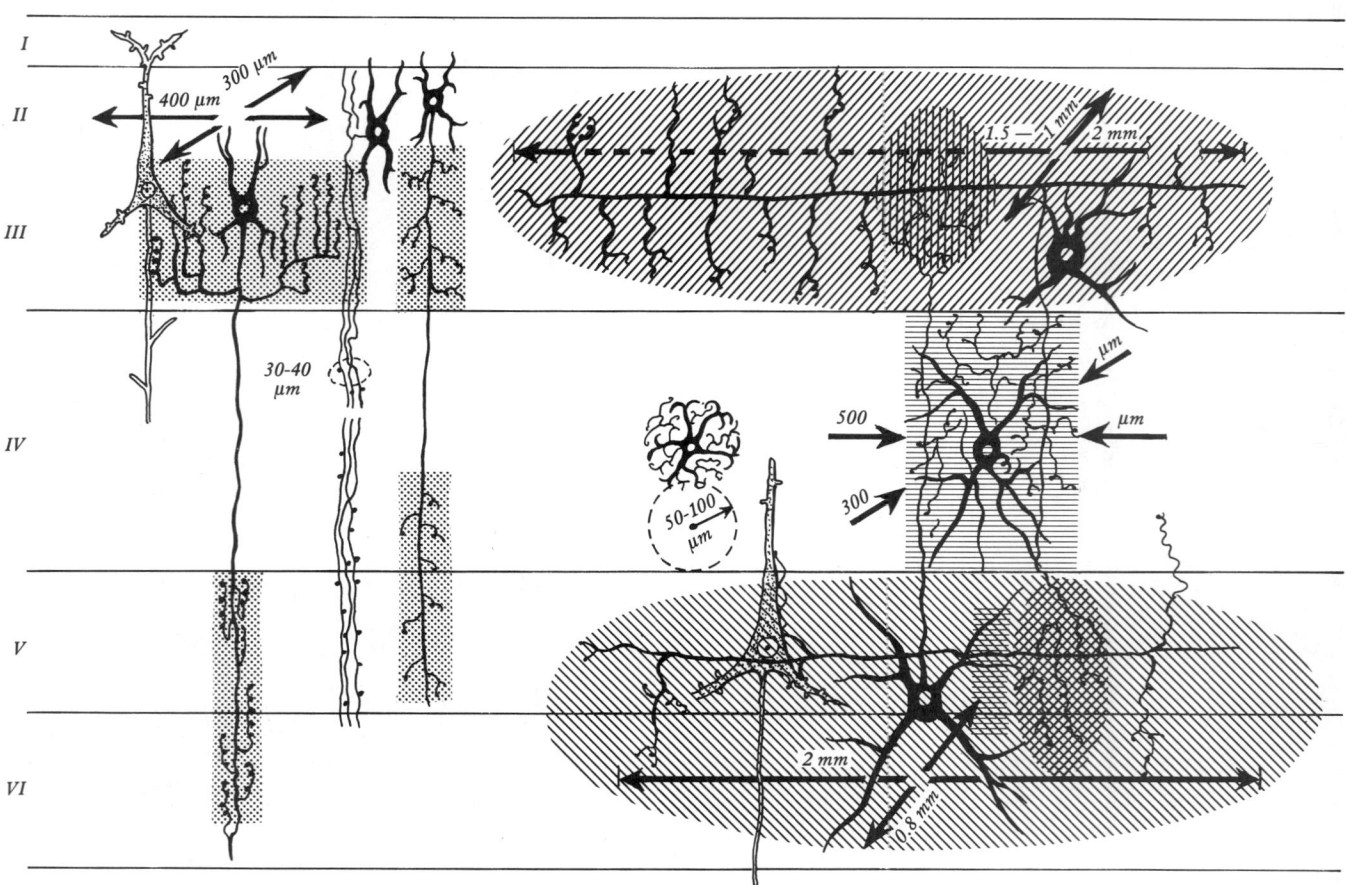

7.167 Diagrammatic illustration of the seven types of well characterized (GABA-ergic) inhibitory interneurons. In order to facilitate recognition of dendrites and axons, the dendritic trees are simplified and drawn with exaggerated thickness. Fields of axonal arborization (where necessary) are indicated by hatching and stippling. Horizontal arrows indicate maximal (observed) extension of the axonal arborizations in the sagittal and oblique arrows in the coronal directions; the extension of the axons in depth can be derived from the cortical layering indicated at left margin. (1) Large basket cell of the upper (supragranular) group (Somogyi et al 1983) (2) large basket cell of the infragranular group (3) 'clutch cell' terminating mainly in lamina IV (Kisvárday et al 1985) (4) columnar basket cell (Szentágothai 1983) (5) microglioform cell (both dendritic and axonal arborization most generally spherical (6) 'cellule a double bouquet' of Ramón y Cajal (7) axo-axonic inhibitory interneuron (Szentágothai 1978). There are more local interneuron types known that are probably also GABA-ergic, but they were not included, because their axonal arborizations have not yet been sufficiently well characterized. The spatial relations between cell types 1–3 have been adapted from a diagram of Somogyi & Soltész (1986). After Szentágothai 1987 with permission.

that a cortico-cortical connection-based unit is *the exclusive universal module* of cortical structure must not render one oblivious of the other structural parameters and connections, e.g. corticostriate, corticoreticular, corticothalamic etc. and sometimes their reciprocals that, if chosen, would generate their own varieties of module, perhaps subordinate to or imposed on the foregoing. Perhaps in cases such as these, the term *apposite cortical module* could be applied, relating it to its particular context: Ed.). The references, in the reviews quoted, to multiple mutual interconnections of (non-discrete) 'hypercolumns' or mutually overlapping 'concentric hypercolumns' and cascade action through an infinity of *re-entrant loops* through corpus striatum, thalamus, hypothalamus, limbic and reticular systems and the cerebellum, each speculatively with their distinctive hierarchies of modules, should perhaps disinhibit innovative thought! Further details of the neuronal population and connectivity of the elementary cortico-cortical module cannot be pursued at length here but a few major points, estimates and relevent references will be noted. The axons of pyramidal neurons more superficially placed ('supragranular') in laminae II, III, and scattered in IV, are exclusively destined for other cortical areas, mainly ipsilateral (association) but some contralateral (callosal). The axons of deeper pyramidal neurons ('infragranular') in laminae V and VI have, in most cases, a target destination in a subcortical centre of grey matter but a substantial proportion have large collateral branches that form callosal fibres destined for the contralateral cortex. Rough estimates of neuronal connections are remarkable (Szentágothai 1987); of the *total cortical neuronal population* over 70% have axons passing some distance from their origin. Of these about 50% of the total establish cortico-cortical connections. The remaining cortical efferents in the *great majority* of cases have destinations in subcortical centres through which, as mentioned, they establish *re-entrant cortical loops* of greater or lesser complexity. The same publication reviews a number of 'general brain theories' based on a host of methodological approaches, but some necessarily incorporating the kind of estimates and concepts just touched on. Particularly relevant to the cortico-cortical module envisaged is the *numerical uniformity* of the mammalian neocortex noted elsewhere (p. 1043) and demonstrated by Rockel et al (1980) and Powell (1981), modular architecture of primate frontal association (prefrontal) cortex by Goldman-Rakic & Schwartz (1982) and Goldman-Rakic (1984) and primate neocortical neurogenesis by Rakic (1984a, 1984b). Slight variations in the basically modular neurogenetic approach could easily explain the regional emergence of the cerebellar cortex and telencephalic limbic (dentate gyrus, cornu ammonis, subiculum) and neocortex.

However the so-called *plasticity* of response of the 'motor' cortex may be explained, anatomical evidence favours a fixed, specific relation between pyramidal cells and their effectors. At all corticospinal tract levels a somatotopical arrangement has been shown. It should be emphasized that experimental stimulation of exposed cortical areas is highly artificial and does not distinguish between direct excitation of pyramidal cells, or even their axons, and of the innumerable intracortical elements connecting with them. However, studies with microelectrode techniques are beginning to point in the same direction.

Before dealing with specific regions of the somatomotor cortex, it is necessary to note that experimental cortical stimuli never produce elaborate co-ordinated movements but only simple activities between synergists and antagonists. The movements elicited are unrelated to particular skills or the motor experience in the individual. This is to be expected; such stimulation leaves out of account other efferent paths and a wide range of afferents from the cerebellum, corpus striatum, thalamus, hypothalamus and other areas mentioned above (see Porter 1985).

Extending into the middle frontal gyrus from a part of the precentral gyrus concerned with facial movement is the **frontal eye field**, occupying much of Brodmann area 8, invading area 6 and probably area 9 (7.168). Stimulation is most effective in area 8 and elicits conjugate ocular movements, especially to the contralateral side but perhaps in other directions. Movements of the head and pupillary dilation may also be elicited from the frontal eye field. For a review of earlier work on these responses see Crosby (1953). Little is certain concerning its efferent projection; degenerating fibres have been traced in monkeys as far as the midbrain but their ultimate end is not established; none have been traced with certainty to motor nuclei innervating extraocular muscles; they may influence these through intermediary cell masses or 'centres' in the brain stem. However, voluntary and reflex eye movements are certainly mediated by the frontal eye field and its projection. Like the rest of the first somatomotor area, this region also receives association fibres, many from the occipital cortex, and also thalamic afferents from the juxtalaminar part of its dorsomedial nucleus (Scollo-Lavizzari & Akert 1963), to which the central zone of the frontal eye field (area 8) reciprocally projects (Rinvik 1968). A **second frontal eye field**, rostral to the above, has been demonstrated in ·monkeys and hypothecated in man (Crosby et al 1952).

The motor speech area (of Broca), an extension of the primary somatomotor cortex into the inferior frontal gyrus, almost coincides with area 44 and part of 45 (7.169). Little reliable knowledge exists of its connections but subjacent to it are large numbers of association fibres for connections between many other cortical areas. Most information as to its function is from results of ablation or stimulation in experimental animals and man (Penfield & Roberts 1959). Ablation may abolish vocalization, usually producing motor *aphasia* or paralysis *of speech* in man, in the left or 'dominant' hemisphere. In some individuals the leading speech area is on the right and in some there is perhaps no dominance (p. 1064).

Damage to the motor speech area does not entail *paralysis of the musculature* involved, all of which is also amenable to stimulation

of the appropriate parts of the primary motor area in the precentral gyrus. Stimulation of the motor speech area leads to various effects, according to the parameters of stimulus; speech in conscious patients may be inhibited or simple acts of vocalization, such as the utterance of a vowel sound, may ensue. Responses in experimental animals are simple movements of the face, lips and larynx and perhaps vocalization. Stimulation in macaques produces contractions in individual laryngeal muscles (cricothyroid and thyro-arytenoid) and combinations of them (Hast et al 1974). It is emphasized that human responses, as in the somatomotor cortex, are always simple. By stimulation of the human cortex similar effects can be elicited from other areas, one in the frontal lobe, the supplementary motor area (MsII), another curving round the caudal end of the lateral sulcus, occupying a large region in the parietal and, more extensively, the temporal lobes. This latter is the **second motor speech area of Wernicke**. All three regions, *anterior, posterior* and *superior motor speech areas*, develop in the dominant hemisphere; while corresponding contralateral parts of the cortex have in some patients apparently taken over the functions of a damaged area in the dominant hemisphere, they are ordinarily uncommitted to speech.

Figure 7.169 shows these areas but other and more complex 'maps' of cortical areas reputed to be concerned in speech have been propounded. Language, in all its permutations (speaking, listening, reading and writing) is a most intricate activity. It would be surprising if this, by far the major form of human communication and perhaps the main distinction of mankind, did not in fact involve widespread volumes of the central nervous system. There is lack of information concerning the connections of various cortical areas believed to mediate these activities; it is likely that other, subcortical mechanisms will prove to be involved, though not yet implicated, despite the intensely active speculation prompted by the problems of aphasia; the pulvinar has been suggested as associated with speech 'centres' (Ojemann et al 1968). The inconclusive nature of evidence in this highly interesting field limits discussion, or should! Consult the excellent reviews available (Brain 1961, Penfield 1966, Millikan & Darley 1967). (See also p. 1064.)

The supplementary motor area (MsII) must now be briefly considered. This is, like MsI, sensorimotor but, being predominantly motor and in the frontal lobe, it is conveniently described here; its sensory activities, about which less is known, will be merely alluded to with the somatosensory cortex. Its situation on the medial hemispheric surface has been noted (p. 1027); it is anterior to, probably confluent with, the medial extension of the first somatomotor area into the paracentral lobule mediating leg and perineal movement. It is thus in the medial frontal gyrus

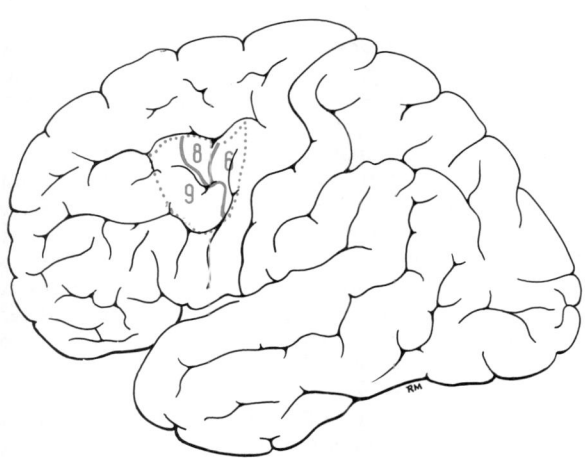

7.168 Superolateral surface of the left cerebral hemisphere showing the frontal motor eye field, corresponding to parts of Brodmann areas 6, 8 and 9. The perimeter of this area is delineated by an interrupted line to indicate uncertainty as to its precise extent.

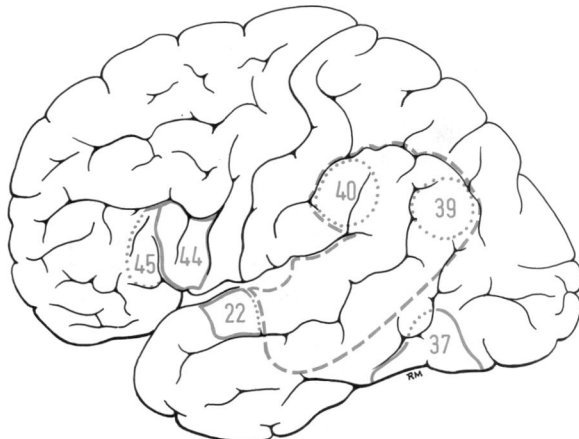

7.169 The superolateral surface of the left hemisphere showing the motor speech areas of Broca (44, 45) and Wernicke. The latter is variously depicted by different authorities and is tentatively indicated by the large parieto-temporal area enclosed in an interrupted outline, which itself includes areas 39 and 40. Areas 22 and 37 are considered by some to be respectively auditory and visuo-auditory areas associated with speech and language.

(area 6 and probably in part area 8). Movements of the contralateral limbs can be elicited from the area in monkeys and mankind (Penfield & Welch 1951) but thresholds are higher than in the primary somatomotor cortex, as in other accessory areas; movements are sometimes ipsilateral or bilateral. A somatotopic pattern (7.161) exists in monkeys (Woolsey et al 1952) but has not been demonstrated in man. Efferent fibres have been traced into the spinal cord in cats (Nyberg-Hansen 1969), most ending contralaterally. There is a *bilateral* projection to the thalamus and to the gracile, cuneate and pontine nuclei, contrasting with similar but unilateral projections from the primary somatomotor cortex. The contributions of these two areas to integration of movement are probably different but details are too ill-defined to warrant discussion (Travis 1955).

The second somatosensory area (SmII), preponderantly sensory, will be considered in detail with the parietal primary sensory cortex. Since, however, it is also *somatomotor*, its less well-established motor features are included here to complete the picture of the 'motor cortex'. Movements of most parts can be elicited from it; the loci of stimulation show somatotopic organization in primates, including man (Adrian 1941). Its projection fibres have been followed to the cervical spinal segments in monkeys (Brodal & Angaut 1967) and to the thalamus, dorsal column and pontine nuclei in cats, in this resembling other somatomotor areas. Little can yet be said regarding its significance in motor function.

The prefrontal area corresponds approximately to all parts of the frontal lobe not so far particularized, hence including much of the frontal gyri, orbital gyri (orbitofrontal area), most of the medial frontal gyrus and the anterior part of the cingulate gyrus, etc. The great development of the frontal lobe is held to be the major distinction of the human cerebrum. So-called *silent areas* have long been regarded a specially notable feature of the human cortex; the prefrontal area is usually assigned to this category. If, however, silent or association areas were to be regarded as cortex with connections predominantly or exclusively by association fibres, the prefrontal region certainly does not qualify. (Neither, in the light of modern knowledge, does *any other area*.) In addition to the elaborate afferent and efferent connections with areas in all cerebral lobes, it has abundant links with the *thalamus*, *corpus striatum*, limbic system, *hypothalamus* and reticular system and projects to the *cerebellum* via the nuclei pontis and perhaps to some *cranial motor nuclei*.

Through *superior* and *inferior fronto-occipital fasciculi* the frontal lobe appears to be connected with the occipital, parietal and temporal lobes. *Commissural* connections link the corresponding parts of the frontal lobes through the corpus callosum and frontal fibres may decussate to other lobes.

Massive *anterior* and *frontal thalamic radiations*, principally from the nucleus medialis dorsalis (MD) of the medial thalamic nuclear group (p. 1002), pass to lateral (areas 9 and 10) and orbital regions of the frontal lobe, both of which reciprocate to the same thalamic nuclei. Anterior thalamic nuclei also establish reciprocal connections with the medial frontal and anterior cingulate cortex. These thalamic nuclei receive many afferents from the mamillary body and thus the hippocampus (p. 1001), which also connects directly with anterior thalamic nuclei via the fornix. Connections of the anterior and medial thalamus with the hypothalamus (p. 1002) therefore bring this also into afferent relation with the prefrontal cortex, *from* which corticohypothalamic projections have also been described. The *prefrontal corticopontine projection*, though less massive than that from frontal motor areas (4 and 6), is equally attested; both relay with specificity in paramedian members of the nuclei pontis (Nyby & Jansen 1951). The orbitofrontal region projects to the paramedian hypothalamic nucleus.

Even this brief survey of prefrontal connections suggests complex interactions in *somatic* and *visceral activities*. The orbital surface has attracted particular attention; simulation in monkeys (Kaada 1960) depresses the respiratory rate, blood pressure and gastric motility; inhibition of induced cortical and reflex movements and emotional reactions have been noted. Ablation of the primate orbitofrontal cortex leads to hyperactivity, restlessness, inattentiveness; somewhat similar results follow ablation of the

anterior cingulate region. The posterior orbitofrontal cortex and anterior cingulate gyrus are often considered 'limbic' (p. 1028). Behavioural studies in monkeys subjected to *anterior* frontal ablations adduce similar but often conflicting findings (Warren & Akert 1964, Sanides 1964); the sum total of fact remains slender.

Functioning of the prefrontal area has been illuminated a little by the results of lobotomy and leucotomy, performed on human patients to ameliorate certain mental disorders now treated pharmacologically. The effects of these operations, usually bilateral, resembled those due to extensive disease of the frontal lobes. Abolition of the distressing aspect of illusions and of severe somatic pain were prominent, results according with the concept that this cortical area is concerned with the 'affective tone' of sensations rather than discriminative or localizing aspects. Some observers reported diminution in intellectual capacity but limitation of the operation to the orbitofrontal area or cingulate gyrus was claimed to affect only the emotional balance of personality, suggesting that the superolateral parts of the prefrontal area are more concerned with 'intellectual' processes (Lewin 1961). Patients with extensive frontal lobe damage, from whatever cause, almost always exhibit permanent and largely undesirable changes. Though more tractable and docile, they also show less initiative, untidiness, disregard for others and/or general tenets of behaviour and a marked lack of concentration, all of which might be summed up as retrogression from the human condition, a vague statement but certainly appropriate to the current dearth of information.

THE PARIETAL LOBE

Posterior to the central sulcus, occupying most of the postcentral gyrus and extending on to the medial surface of the parietal cortex in the paracentral lobule, is the primary or *first somatosensory area* (SmI, areas 3, 1 and 2). Posterior to it is a large 'silent' area of parietal lobe, occupied in part by the second speech area (vide infra). In the lowest part of the postcentral gyrus (extending a little into the precentral) is the *second somatosensory area* (SmII).

Anterior to the medial extension of SmI is the *third somatosensory area*, the supplementary 'motor' area (MsII). Thus, the region of the cortex centrally situated in the hemisphere and the great arrival station for somatic afferents and for the departure of somatic efferents, consists of *four separate*, though *juxtaposed sensorimotor areas*: two, primary and supplementary motor areas, have for long chiefly been regarded as efferent projection cortex, MsI and II; the other two are prominent recipients of afferent projections, primary and secondary somatosensory areas, SmI and II. As 7.161, 170A,B show, all four areas evince a somatotopic pattern in cats and monkeys; evidence that this is largely so in man has been quoted (p. 1041). The motor aspects of all four areas are mentioned above; it remains to consider involvement in sensory functions. (While the foregoing is partly correct, it places excessive emphasis on *one* class of connection and totally ignores the other classes of afferent and efferent common to all neocortical areas.)

The first somatosensory area (SmI) has interesting cytoarchitectonic variations. Its anterior part (area 3), bordering the central sulcus and extending into its depths to meet the *agranular* cortex of area 4, is of *granular* type but also contains scattered medium and small pyramidal cells; the area resembles the striate cortex, e.g. the outer band of Baillarger is broad but not so prominent. However, the posterior part of the postcentral gyrus (areas 1 and 2) differs particularly in its smaller content of less densely packed granular cells (p 1039). (Such subjective comments ignore the relative cortical thickness and surface area involved.) The boundary, if such exists, between pre- and postcentral areas in the central sulcus is disputed but may be important to the interpretation of experimental results (Rinvik 1968).

Projection from the *nuclei ventralis posterior lateralis* and *medialis of the thalamus* (VPL and VPM) to the primary somatosensory area is well attested; it is through this radiation that exteroceptive and priorioceptive modalities of sensation are mediated. Unit recording technique, retrograde degeneration and the newer techniques (p. 1039) have established a specific pattern of localization between the cortex and thalamus, associated with a somatotopic form of organization in the postcentral gyrus like that in the adjoining first somatomotor area, including an allocation of cortex in proportion to skill rather than size in representation of bodily parts. This pattern has been studied in the human cortex by noting the type of sensation and region to which it is referred when loci are stimulated in an exposed cortex at operations (Penfield & Jasper 1954). With carefully gauged stimulation sharp localization can be established. Localization of vesical, rectal and genital sensations has thus been set in the *lowest* part of the *medial* region of the human postcentral area. The sensations aroused are mostly *contralateral* but may be solely *ipsilateral* (oral region) or *bilateral* (larynx, pharynx and perineum).

Experimental studies have largely confirmed the localization demonstrated in the human somatosensory cortex. Natural and artificial stimulation of peripheral receptors have also been shown to evoke potentials in the *precentral* area (Abel-Fessard & Liebeskind 1966). Ablation of the somatosensory cortex revealed a *direct* thalamic radiation to the 'motor' area (p. 1049), an elegant demonstration of the latter's sensorimotor nature. Degeneration studies (e.g. Le Gros Clark & Powell 1953) have provided more exact data on the specificity of the thalamocortical projection; the nucleus ventralis posterior lateralis exhibits a discrete localizational pattern of connection with areas 3, 1 and 2, suggesting a functional significance in the structural differences between them, as mentioned earlier (p. 1051). Unit recording has confirmed this, revealing area 3 as activated only by cutaneous stimuli whereas area 2 is concerned with proprioceptors (Albe-Fessard & Liebeskind 1966). This implies the transverse deployment of modalities *across* the long axis of the precentral gyrus (Mountcastle & Powell 1959), with a segmental or dermatomic projection to such transverse strips of cortex as a further refinement of somatotopic localization (Powell & Mountcastle 1959). Each modality-specific locus appears to be linked with a *vertical column* (module) of cells in the cortex, like those in the visual area (p. 1059). Anatomical evidence in monkeys demonstrated a cortico-cortical connection from the first somatosensory area (Brodmann 3, 1, 2) to area 5 in the parietal lobe (Pandya & Kuypers 1969). This connection, analysed further (Pearson & Powell 1978), is isotopological but reversed in the anteroposterior axis, so that area 3 projects to cells posterior to those receiving fibres from areas 1 and 2. Some anatomical evidence supports modality convergence in area 5, which accords with previous investigations (Mountcastle et al 1975) and provides a basis for 'association'.

The specificity of somatosensory neurons concerned with joint receptors is particularly marked, as in the thalamus. Some cortical units respond to displacement in one direction and continue to discharge even during a static position. Such arrangements, if applicable to man, would provide a neuronal basis for the 'joint sense' hypothecated in much earlier studies (Stopford 1922). Fast adapting units activated by bending hairs and others adapting slowly to deformation of skin have been identified. Similar studies suggest that the afferents from receptors in striated muscle terminate in units sited in the somato*motor* cortex and that impulses may reach them by paths independent of the cerebellum, as part of a basis for awareness of stretch in muscles. These findings imply a difference in sensory properties of pre- and postcentral parts of the main sensorimotor area, which are, however, linked by short association fibres in both directions (Pandya & Kuypers 1969). As in most, if not all, sensorimotor areas, the first somatosensory area has abundant reciprocal thalamic connections (p. 1004), corticothalamic projection comparing in its point-to-point specificity with thalamocortical radiation (Rinvik 1968). In connection with this apparently unvarying specificity at the cerebral level of some modalities, a study of the cortex in mice is most interesting (Woolsey & van der Loos 1970); barrel-shaped groupings of nerve cells were detected in part of the somatosensory area (SmI) related to the head and muzzle; some evidence suggests that these 'barrels' may be equated with individual vibrissae of the animal's muzzle (Waite 1977). This and a study of spatial distribution of synapses in the same cortical area in neonatal dogs (Molliver & van der Loos 1970), though not applicable to the human cortex, illustrate the intricate quantitative kind of analysis under investigation.

Commissural connections with the opposite primary somatosensory area are restricted to representation of the face,

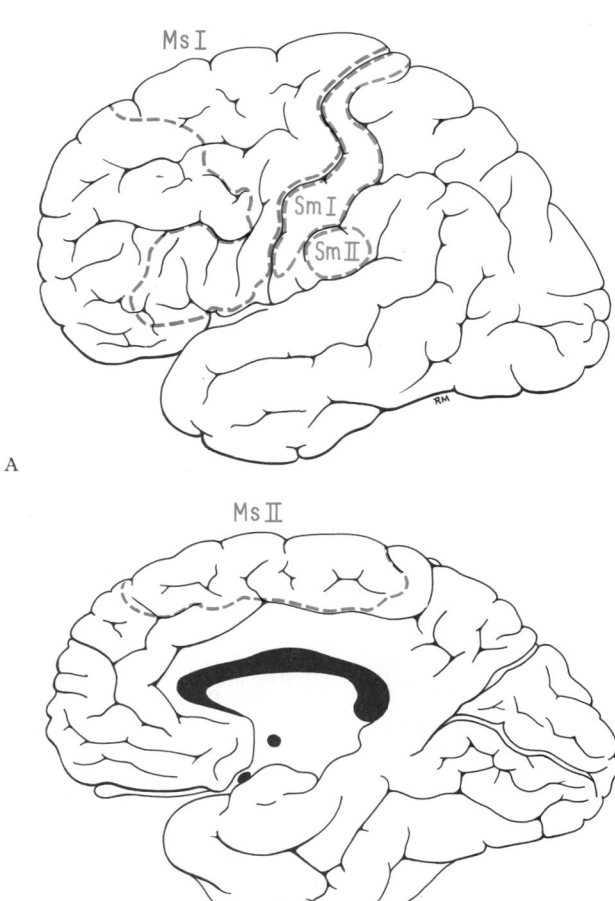

A

B

7.170 Superolateral (A) and medial (B) surfaces of the cerebral hemispheres showing the sensorimotor areas. See text for details and compare with 7.157, 162. These areas can only be applied tentatively to the human cortex, and they are hence delineated with interrupted outlines.

trunk and proximal parts of the limbs (Pandya & Vignolo 1969, Jones & Powell 1969). Commissural projections in monkeys terminate in bands (Jones et al 1975, Shanks et al 1975), as in the visual and motor cortical areas. Shanks et al (1978) have shown that in these bands short intrinsic connections propagate repeated representations of the bodily regions, comparable with retinotopic repetitions in the visual cortex (p. 1058).

Much less is known of the **second** and **third somatosensory areas** (SmII and MsII). The former is in the superior lip of the lateral fissure's posterior ramus, adjoining the insula in monkeys (7.161); it occupies a similar position in man. Evoked potentials indicate somatotopic organization in SmII, the face area being the most anterior, the leg most posterior. Single units associated with tactile and vibrational senses occur in the area; stimulation of Pacinian corpuscles evokes higher potentials than in primary somatosensory area (McIntyre et al 1967). The second somatosensory area projects to the thalamus but its connections have not yet been studied in detail. It also projects to the dorsal column nuclei, like other somatosensory areas. Details of such descending projections in monkeys have been much studied (Weisberg & Rustioni 1977) and are of special interest in the inhibitory modulation of transmission through the dorsal column nuclei.

The third somatosensory area, equivalent of the supplementary motor area, is in sensory functions little understood. It receives thalamocortical fibres and projections leave it, some of which are not motor, but details are uncertain. Stimulation in conscious patients is said to evoke generalized sensations referred to the head and abdomen, but no sensory somatotopic pattern can be ascribed to it.

The second speech area of Wernicke (7.169), being partly in the parietal lobe, must be mentioned here. It occupies only a small parietal area but extends more extensively into the temporal lobe

(p. 1061). The rest of the parietal lobe, between the main somatosensory and the visual areas, is 'silent' and corresponds more or less to the inferior parietal gyrus, areas 39 and 40 (Critchley 1953). Its connections, as far as they are known, are in part association and commissural but concerning these little clear detail is available; however, there is also a reciprocal connection with *all* the parts of the lateral thalamic nuclear group (p. 1003). Corticospinal and corticosegmental projections have been suggested, the latter to brain-stem motor nuclei, especially the oculomotor group, directly or through intermediary nerve cells. These connections are specially apposite to the description of a supplementary motor area in areas 5 and 7 in macaque monkeys (Crosby et al 1959). The strategic position of the parietal association areas close to all cortical regions concerned with general and special modalities of sensation, coupled with multifarious connections, is highly suggestive of complex and integrated activities. Most evidence is derived in man from clinical observations of the behavioural defects consequent on pathological lesions or ablations and extrapolation of similar results in primates deliberately damaged. The reports available are confusing. Muscular debility, slowing of reflex activity, disorders of speech, loss of awareness of bodily parts and other forms of *agnosia* have all been observed. A particular form of agnosia, long associated with parietal lobe damage, is *astereognosis*, failure to interpret the three-dimensional nature of objects when handled without vision. *Interpretation* has been strongly advocated as the prime parietal activity in man (Penfield 1966), in connection with language on the dominant side and in a more generalized sense on the other. The parietal lobe is perhaps a region of high activity in learning processes and certainly has reached a uniquely high development in mankind. Considering the general nature of the evolutionary process it is unwise to attribute a qualitatively unique nature to the operations of human parietal areas but their import in behaviour, especially

discrimination and interpretation, is amply attested by clinical evidence and animal experimentation. (See also p. 1064.)

THE OCCIPITAL LOBE

Almost the whole occipital lobe is occupied by Brodmann areas 17, 18 and 19 (7.171A,B). Though area 17, the **striate** or **visuosensory cortex**, was originally regarded as the sole terminal of the optic radiation, the lateral geniculate body (p. 1016) has been shown to radiate also to areas 18 and 19 in cats (Glickstein et al 1967) and monkeys (Wilson & Cragg 1967); in these, like area 17, spatial arrangement of the radiation is related to the retinae in an orderly manner (Garey & Powell 1967, 1968). All three areas project to the thalamus (lateral geniculate body or pulvinar) or brain-stem motor nuclei or both. Short and long association and commissural fibres link the *three visual areas*, ipsi- and contralaterally, and to other areas in both hemispheres. (For an exhaustive critique consult Polyak 1957.) Occipitopontine fibres are described but their human status is uncertain. Area 17 is co-extensive with the visual stria of Gennari but demarcation between areas 18 and 19 is less easy by the usual cytoarchitectonics; Braak (1977) has introduced a method depending on staining of lipofuscin granules and this sharply distinguishes these two areas and even structural subdivisions in them but their functional significance is not clear.

The striate cortex (area 17), or **first visual area**, occupies the upper and lower lips and depths of the posterior part of the calcarine sulcus (7.171A,B), extending into the cuneus and lingual gyrus. As a cytoarchitectonic area it can easily be defined by the visual stria and its cortical thinness. Posteriorly it is limited by the lunate sulcus (and polar sulci above and below this), extending only to the human occipital pole, though reaching its lateral aspect in other primates. Anterior to the area is the medial parieto-occipital sulcus but below the calcarine sulcus it may extend further. The stria becomes less obvious towards the occipital pole, a change correlated with retinotopical organization (vide infra), its prominence being inversely related to distance from the central retinal area.

Histologically, area 17 is *granular* or *koniocortex* (p. 1046); densely packed stellate cells greatly outnumber its few pyramidal cells. For details consult Polyak (1957), Crosby (1960), Colonnier (1964, 1967, 1968, 1974); only the salient features will be mentioned here. (1) the deeper stratum of the pyramidal layer (III) contains *large stellate cells*, almost replacing the large pyramidal cells of this layer in other types of cortex. (2) The outer band of Baillarger in lamina IV is accentuated as the *visual stria*, containing termination of the optic radiation and probably association fibres. (3) The ganglionic layer (V), commonly consisting of pyramidal cells, here contains a few large, *solitary cells (of Meynert)* of a modified pyramidal shape, about 30 μm in diameter and distributed in a single row. Their dendrites extend widely in the cortex and their axons are the projection of the visual cortex, traversing the optic radiation to the superior colliculus and possibly motor nuclei of the extraocular muscles. A study of Meynert's cells of rhesus monkeys (Chan-Palay et al 1974) showed their potential for widespread, intricate connections; each cell's dendrites were estimated to display 36 000 spines, its apical dendritic field in lamina I being about 400 μm and 800 μm for basal dendrites in lamina V. Distribution density of Meynert cells varies, being greatest (8000/cm²) in the region representing the macula. (For a more extensive reappraisal of the numbers and basal dendritic patterns of Meynert cells in the macaque monkey see Winfield et al 1981, 1983.) (4) External and internal granular layers (II and IV), especially the latter, contain larger numbers of more densely packed small stellate cells than any other cortical area. This laminar pattern is based on Brodmann's; other patterns, particularly Cajal's, have some vogue; a review by Billings-Gagliardi et al (1974) advocates a return to Brodmann.

Although the striate cortex contains only 3% of the cerebral surface, approximately 10% of cortical neurons are said to be in it. In cytoarchitectural details areas 18 and 19 approximate respectively to the polar and parietal types of cortex (p. 1046), the stellate cells being less prominent, the pyramidal cells more so. Myelin and axonal stains have been used to investigate the visual

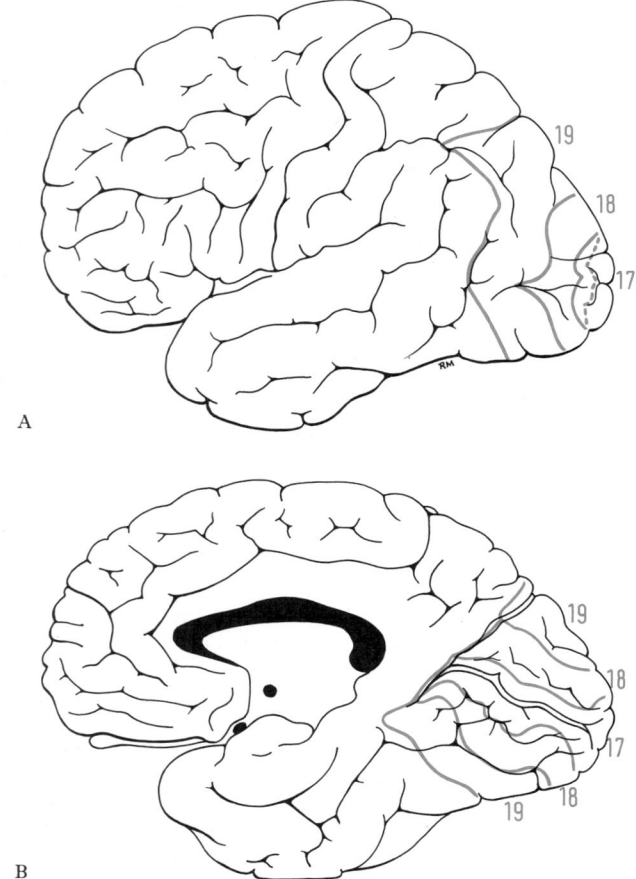

A

B

7.171 Superolateral (A) and medial (B) surfaces of the cerebral hemispheres showing the visual areas in the occipital lobe. The striate (17), parastriate (18) and peristriate (19) areas correspond approximately to the Brodmann areas as indicated and also to visual areas I, II and III. See text for details.

area's fibre structure, especially the striate cortex (Sholl 1955, Colonnier 1964); in general it resembles arrangements elsewhere (p. 1049) but the dendritic fields, in their quantitative aspects, have been studied more intensively in this neocortical area than perhaps any other except the postcentral gyrus, stimulated in part by the demonstration of a columnar organization in the somatosensory area, as already noted, and thereafter in the striate cortex (vide infra). So far physiological techniques are providing more illuminating evidence than structural.

The status of area 17 as the primary visual area is long established, both by electrical stimulation in man and by its connections with the lateral geniculate body, and through this with the retinae (p. 1016). The *retinotopical organization* in the lower parts of this visual pathway has now been shown to obtain to cortical level. Each area receives impulses from two ipsilateral half retinae, representing the *contralateral* half of the binocular visual field. As already described (p. 1016), the patterns of retinal impulses do not undergo simple relay in the geniculate bodies, where some processing occurs, but exclude interaction of a biretinal nature (for details see p. 1017). Such processing has been studied most intensively in the striate cortex, especially by single unit recording technique. The *geniculate radiation* spreads as it swerves through the white core of the occipital lobe, its fibres terminating in strict point-to-point deployment in the striate area: the peripheral parts of retinae activate the most anterior parts in the area, the macular regions activating a relatively large part adjoining its posterior end. Moreover, superior and inferior retinal quadrants are connected with corresponding areas of the striate cortex. In the classic studies of the effects of injuries of the occipital lobe in warfare, similar retinotopic results were obtained (Holmes & Lister 1916, Teuber et al 1960).

Experimental findings in a wide range of mammals, especially primates (see p. 1016 for references), were corroborated by stimulation of the human cortex (Penfield & Jasper 1954); the visual impressions thus elicited were simple, such as flashes of light, but were referred to a specific locus in the visual field according to the location of the cortical stimulus. Eye movements are also produced by such stimulation. When areas 18 and 19 are stimulated more complex visual images are reported, indicating that they are concerned in elaboration of the information reaching area 17.

Most interesting recordings have been obtained from single neurons responding to various forms of retinal stimulus; while earlier studies were chiefly pursued in cats (Hubel & Wiesel 1962, 1963, 1965, 1971, Wiesel & Hubel 1963a,b), there was little reason to doubt that in principle the results were applicable to the human striate cortex, especially since the cytoarchitectonic areas involved in cats can be equated with those in primates (Otsuka & Hassler 1962), in which unit recording has also been studied (Hubel & Wiesel 1977). Each unit appears to correspond to a defined retinal receptive field and can presumably be excited only from it. Units responding to *various modes* of stimulation have been identified; some react to white light, especially in the stripes or edges between light and dark, contrasting with the smaller circular receptive fields demonstrated in the lateral geniculate body. Excitatory units surrounded by inhibitory zones have also been repeatedly described. Many units are sensitive to the orientation of stimulus, responding only to vertical, horizontal or oblique linear stimulation. More complex units occur, some responding to a particular direction of a moving stimulus. These observations represent integration from simpler units. Further studies by Hubel & Wiesel (1968) have revealed striate cortical neurons, in cats and especially monkeys, responding preferentially to stimuli from either the right or the left eye, a phenomenon misleadingly termed *ocular dominance*; vertical (and oblique) exploration by unit-recording has shown that these neurons (with common responses) are deployed in the vertical columns, through *all* layers of the striate cortex, as described below. Certain columns may be binocular in their response in cats but in monkeys this is doubtful, all being right or left orientated. Further evidence of greater development of such 'dominance', perhaps, in the simian cortex resides in the greater facility with which they are found and explored; these workers consider that they are of greater diameter in monkeys. It has proved possible to refer most

neurons involved to particular cortical laminae and to equate their distribution with those of the terminals from the lateral geniculate projection. The significance of these findings is not entirely clear but they confirm segregation of 'information' from each retina at least as far as the striate cortex. The validity of these observations has been confirmed by other techniques, e.g. autoradiographic identification of labelled substances (3H-proline) injected directly into the feline eyeball (LeVay et al 1978). By transneuronal transport in the lateral geniculate body such substances reach the striate cortex, particularly lamina IV, corresponding in distribution most clearly with the columnar arrangement. Studies in cats (LeVay et al 1978) and monkeys (Hubel et al 1977) suggest that these neuronal arrays are to some degree plastic, since they are altered by monocular deprivation in early postnatal life. Later work has confirmed the identification of columnar unitary loci which respond to stimulation from one retina, though some are excited simultaneously from both; a columnar organization coded according to colour responses has also been supposed. Similar results have been obtained in monkeys, increasing the likelihood that they correspond to human visual activities.

These studies, carried out with constant reference to the laminar pattern in the striate cortex, suggest a *columnar* structure in accord with the functional organization in vertical 'units'. The concept of 'chains' of neurons linked vertically in the cortex is, as noted, comparatively old (p. 1041) and more recent work, with the advantages of improved staining and electron microscopy, has confirmed and elaborated this view, and not only in the striate cortex. Synaptological studies and the results of degeneration experiments show that afferents of an area such as the striate cortex feed impulses into a limited columnar group of interneurons, the result of their excitatory and inhibitory interactions, which reach a pyramidal cell for translation to the subcortical neurons or to some other region. Such *columnar units* are not completely independent, being linked to others by recurrent axons, by tangential fibres in the superficial cortical lamina (I) and by spatial overlap. For the development of concepts of neocortical columnar architecture see p. 1051-1054 (and references) and, e.g. compare the publications of Szentágothai in 1967, 1969 and 7.164 with a revised 'module concept' (1975) summarized in 7.165. In the latter, the evolving *cortical module* (with *submodules*) concept was based on the specific thalamocortical afferent and hence seemed particularly appropriate to the striate cortex. However, in a major review published later (Szentágothai 1978) this view was revised and the '*true cortical module*' was held to be centred on the origins and terminations of small bundles of cortico-cortical fibres, i.e. association and commissural fibres (see 7.166, 167). It is appropriate to note that in each cortical unit neurons in cortical laminae II and III project to other cortical areas (such as 18 and 19), those in lamina V to the superior colliculus and in lamina VI back to the lateral geniculate nucleus.

The effects of lateral geniculate lesions in cats provide detailed patterns of the terminations of geniculocortical fibres in all three visual areas (Rossignol & Colonnier 1971); in area 17 degenerating terminals are mainly in lamina IV but a few vertical axons reach lamina I to spread tangentially in it; in area 18 their main termination is also in lamina IV, but in area 19 the afferent terminals are less numerous and spread through layers IV, V and VI. These results confirm the wide region of afferent visual projection in the occipital lobe and accord with the difference of activity in the three visual areas. Such experiments entail the infliction of large, almost total lesions of the lateral geniculate body. Only discrete lesions can demonstrate the specific point-to-point relation between geniculate projection fibres and visual cortical areas. Such specificity can also be shown by observing the results of cortical lesions by retrograde degeneration in the geniculate body, where columnar series of neuronal groups in all six layers are affected (p. 1016). This effect is only achieved when cortical lesions involve the whole thickness of the cortex (Rose & Malis 1965). By a similar combined 'attack', the functional details of visual *areas* II and III are being gradually clarified; in both, details of columnar units are accruing, the units responding to more complex stimulation and being almost all *biretinal* in their receptive fields.

Hubel & Wiesel (1978) have been able to carry the above concepts further, in monkeys, providing at least an initial integration in understanding of the *modus operandi* of the striate cortex. By the transport of radioactively labelled amino acids from one eye via both lateral geniculate bodies to their striate cortices, they have revealed not only neuronal columns correlated with ocular dominance but also a particular arrangement of these throughout the striate cortex (see also LeVay et al 1978). This pattern is based in lamina IV, which receives most of the geniculate projection and contains so-called 'simple' neurons responsive to uni-ocular, circularly symmetrical, retinal stimuli. The pattern comprises curving and branching stripes, about 0·5 mm in width in macaques, separated by unreactive intervals of similar width and similar pattern. The reactive (radioactively labelled) stripes contain cortical 'columns' associated with the injected eye; intervening, nonreactive stripes correspond to the unaffected eye. Therefore interaction between signals from corresponding retinal areas must occur elsewhere in the striate cortex, perhaps in association with 'complex' cells in laminae II, III, V or VI, some of which respond to binocular stimuli. A second type of columnar pattern has been revealed in the striate cortex by combined unit recording and the radioactive tracer technique, showing the sites of high utilization of glucose, the stimulus being orientation of linear objects. Columns associated with such stimuli display a pattern of whorled or 'swirling' 0·5 mm wide stripes like those concerned in uni-ocular reception; but the 'stripes' of the two techniques (radioactively labelled amino acids and unit recording with glucose utilization) form different patterns which intersect at frequent points. Hubel & Wiesel have hence hypothecated a cortical arrangement of repeated unitary columns through the whole thickness of the striate cortex, each unit responding to a stimulus in terms of its position and orientation (its column being concerned, however, purely with uni-ocular input). The theory does not attempt to explain the cortical mechanisms involved in colour, the kinetic or spatial (stereoscopic) aspects of stimuli. These studies have provided the most detailed facts and ideas in manipulation of signals in any part of the cerebral cortex, though they do not, admittedly, approach the integratory processes of perception.

Parts at least of the visual area must be considered sensorimotor, inasmuch as eye movements can be elicited from them. In monkeys this (third) *occipital eye field* (7.171A,B) is confined to visual areas I and II (17 and 18), stimulation producing conjugate deviations in various directions, especially to the opposite side. Head turning, facial movements and sometimes responses in the upper limbs may occur (Rieck 1959, Bender 1964). This occipital motor eye field (III) has reciprocal connections with the *frontal eye fields* I and II which are considered to be involved in the mediation of vision in the voluntary ocular movements held to characterize activity of the frontal eye fields. Motor responses from area 18, including accommodation, are, on the contrary, thought to be reflex in natural activities and involved in following and fixation movements. Area 18 may project *to* the pulvinar, which has a reputed but controversial link with the lateral geniculate nucleus. Projection *from* the pulvinar to areas 18 and 19 is established but with disagreement over details (Locke 1961). Efferent fibres descend from the occipital motor eye field, particularly area 19, to the superior colliculus and nuclei of the ocular motor nerves, to the latter probably through interneurons, such as the para-abducens nucleus (p. 1107), interstitial nucleus, nucleus of Darkschewitsch, and posterior commissural nuclear complex (Carpenter & Peter 1971). These projections must be distinguished from the corticogeniculate projection (pp. 1017, 1203), evidence for which is chiefly physiological (Widen & Marsan 1961). Several workers have detailed corticofugal connections from areas 17, 18 and 19 to the thalamus, tectum and pretectal region, particularly in cats, in which some retinal representation has been demonstrated (Updyke 1977) in the projection from area 19 to the pulvinar.

All three visual areas are intimately linked by short association fibres, which are probably especially specific in a spatial sense between areas 17 and 18; thus preservation of the topographical relation in connections between all three areas has been demonstrated in monkeys by Cragg (1969) and Zeki (1969, 1974); according to Zeki (1977) spatial representation displays a 'mirror-image' reversal between adjoining visual areas, as appears to occur in connections between other adjacent 'sensory' areas

(Somatosensory Area, p. 1056). Zeki emphasized the propagation of 'mosaics' of retinal representation from the striate to 'prestriate' cortex, the latter including not only peri- and parastriate areas (Brodmann 18 and 19) but perhaps more remote regions, in particular the so-called fourth visual area (V4 and V4a), extending between area 19 and the occipital end of the superior temporal gyrus (7.171A). Zeki's concept involves the projection of each unitary retinal area at repeated points in all the visual areas, the response of cortical loci varying in modality with different loci responding to orientation, contour, depth, colours and movement. In this there is a possibility of equating such responsive loci and their well-established columnar architectonics with the results of unit recording. Retinally-linked loci in major visual areas probably serve different visual modalities; e.g. area 17 appears to emphasize orientation, whereas columnar loci in area 18 are more concerned with stereoscopy. Beyond this, in areas 19 and V4, retinal representation extends into the modalities of colour and movement. Picturesquely, one may imagine the total pattern of retinal excitation as propagated from area to area, the cortical units in each adding a particular modality to the total sensory pattern. Association and commissural fibres connect all visual areas, particularly 18 and 19, to many parts in the ipsilateral, and to a much lesser extent the contralateral hemispheres. Particular mention should be made, however, of the commissural connections between the visual cortices. The origins and terminations of *callosal fibres* are restricted to *narrow bands between* areas 17, 18, 19 and the occipital end of the superior temporal gyrus (i.e. bands *between* VI, 2, 3, 4, 4a). These bands process information from the *vertical meridian* of the retina; i.e. the narrow vertical retinal strip whose ganglionic cells have axons that bifurcate and pass into both optic tracts (Dr T P S Powell, personal communication).

Vision and ocular movement may therefore be affected by lesions remote from the visual cortex or from motor fields; e.g. the course of the optic radiation, swerving through the temporal lobe (Meyer's 'loop') entails homonymous field defects in temporal lobe tumours (Falconer & Wilson 1958).

THE TEMPORAL LOBE

This has lateral and inferomedial surfaces, the former with superior, middle and inferior temporal gyri, the latter including medial and lateral occipitotemporal gyri (collectively the fusiform gyrus) separated from the parahippocampal gyrus by the collateral fissure (7.145A, 7.151). The human temporal lobe (apart from limbic lobe structures, p. 1028) is regarded as highly evolved and relatively recent in phylogenetic development. It is much involved in hearing, language and perception.

Part of the superior temporal gyrus is 'rolled' into the lateral fissure's posterior limb during human fetal development, thus hiding two short gyri which ascend obliquely backwards posterior to the insula's limiting sulcus. These anterior and posterior *transverse temporal gyri* almost correspond to areas 52, 41 and 42 (7.142); many other temporal cytoarchitectonic regions have been described, but only those mentioned, together with area 22, adjoining them on the lateral aspect of the superior temporal gyrus, can be reliably equated in man. In Campbell's original description in 1905 the *acoustic area* covered rather more than area 41, to which subsequent observers restricted it when defining a classic acoustic area, a leading feature of the gyrus (7.157A). As will be seen, Campbell's view has been revived.

Area 41 is a variant of the granular cortex, but thicker than the visual and somatosensory cortices (Bailey & von Bonin 1951). Its surroundings, such as area 22 (7.157A), are a parietal type of cortex; beyond this the temporal lobe is largely of frontal type, like the main areas of the parietal and frontal lobes. Both frontal and parietal types of cortex show an admixture of pyramidal cells and (an apparent, subjective) 'diminution' in stellate cells. For further details consult authorities referred to on p. 1039 and Braak (1978).

The acoustic area, classically defined, approximately equals area 41; but for reasons explained below this is now termed the **first acoustic area** (AI). It extends slightly on the superior temporal gyrus but is largely *in* the lateral fissure as the anterior transverse temporal gyrus. Earlier evidence for the acoustic nature of this area was reviewed by Whitfield 1967; its cytoarchitecture has been described in detail by Sousa-Pinto (1973), claiming that area AI so defined corresponds exactly to the projection of the lateral part of the ventral nucleus of the medial geniculate body (in cats). Attempts to define this projection area more accurately have extended it (Rose & Woolsey 1949), and the results of cortical stimulation (evoking head and ear movements) and recording cortical potentials excited by localized stimulation of either the cochlea or the cochlear nerve (Woolsey & Walzl 1942) have confirmed this. Much of this work has been on cats, less often monkeys (Sousa-Pinto 1973); hence details of these accessory areas cannot be confidently translated in their entirety to man (7.173). The **second acoustic area** (AII) is inferior to AI in the equivalent of the superior temporal gyrus, the ectosylvian gyrus in cats. Flanking AII are *anterior* and *posterior ectosylvian areas*, Ea and Ep (7.172); even more areas have been described. There is evidence that all the areas named, except Ea, receive projections from the medial geniculate body. Ablation of AI leads to degeneration only in the rostral part of the medial geniculate body, which receives fibres from the inferior colliculus (p. 986); but other parts in the medial geniculate complex of nuclei project to all areas except perhaps Ea, including part of the insula (Morest 1964). Some fibres in the acoustic radiation may divide to terminate in more than one area and this may in part explain some equivocal results. Short association fibres link AI to AII and Ep, but not Ea. (For detailed development see Krmpotić-Nemanić et al 1979.)

Geniculocortical projection has been 'mapped' in AI in chimpanzees (Walker & Fulton 1938) and its representation in this (Woolsey & Walzl 1942) was earlier considered *tonotopical*, i.e. a localization pattern based on sound frequency; but this was strongly denied in the human first acoustic area (Penfield & Jasper 1954), remaining a major controversy (Evans 1968). Stimulation in conscious patients merely produced ill-localized simple patterns of sound or noise, buzzing, humming or ringing; but in the adjacent accessory areas (perhaps equivalent to those in cats) more complex acoustic phenomena were elicited. The topological relations between acoustic areas and various medial geniculate subnuclei have been investigated by retrograde transport of horseradish peroxidase from small cortical lesions (Winer et al 1977), suggesting that only the laminar part of the ventral medial geniculate nucleus projects to one acoustic area (AI), all other parts of the complex projecting to several acoustic areas. It was also noted that the pulvinar projects to the acoustic areas. Despite much unit recording from AI, this problem has not been settled; but it seems highly likely that there are units responding to specific frequencies, even if not in any particular array. Other units, excited by the *variation* of frequency in time or direction, have been described (Whitfield & Evans 1965), these specific arrays resembling the visual cortex.

Efferent fibres descending from the acoustic areas to the brain stem have been noted elsewhere (p. 1016). These may modulate activities in the medial geniculate body, inferior colliculi and other auditory tract nuclei (pp. 960, 986), but also mediate reflexes by forming connections with the cranial nerve nuclei. Some fibres probably influence the state of contraction of the stapedius and tensor tympani, experiments suggesting that it increases with high-frequency or high intensity sounds (Carmel & Starr 1963). *Association fibres*, in addition to the short ones noted above, link the acoustic areas to other ipsilateral lobes, perhaps particularly the frontal pole, and to the cortex deep in the superior temporal sulcus, which may be loci of convergence of association fibres from all sensory areas. Little is known of the commissural fibres (but the lateral lemniscus is a *bilateral* tract). Lesions in the human medial geniculate body or acoustic cortex do not cause noticeable deafness unless bilateral (an unlikely occurrence). Extensive temporal tumours have minimal effects on hearing but may interfere with acoustic aspects of language interpretation, especially if in the 'dominant' hemisphere. In this connection, note that the posterior motor speech area in the parietal lobe (p. 1057) extends into the temporal lobe posterior to the acoustic region, practically surrounding it according to some (7.169). This proximity doubtless has implications in the cortical organization of speech, hearing and language.

The vestibular area, as regards its site, is a focus of disagreement; but in monkeys it is said to be near the part of the somatosensory area (SmI) concerned with the face, in the postcentral gyrus, area 2 (Jones & Powell 1970). Area 2 is particularly involved in

sensations from deep tissues and joints; the spatial proximity of a vestibular area to this appears interesting; it is said to receive projections from the vestibular nuclei but this is largely without anatomical evidence. Unlike the acoustic pathway, the vestibular projection appears almost entirely crossed. Some degree of somatotopical specificity in the connections of the vestibular cortex and the vestibular nuclear complex have been reported (Massopust & Daigle 1960) but anatomical evidence and physiological data remains sparse and conflicting. Another view must be mentioned; this depends on the recording of equilibratory responses in man and implicates part of the superior temporal gyrus, anterior to the acoustic areas (Penfield 1957).

Apart from the acoustic areas, a possible vestibular area and an extension of the posterior motor speech area, the neocortex of the temporal lobe contains no other functionally identified regions; but the remainder cannot all be simply disregarded as 'silent' cortex; the existence of numerous association fibres connecting parts of the lobe and areas in other lobes is certain, though precise details are few. Connections with visual areas (between areas 18 and 19 and 37) and with the parietal lobe (between areas 7 and 40 and 22) probably mediate complex integration between sensorimotor areas in temporal, occipital and parietal lobes. Thus, the more posterior parts of

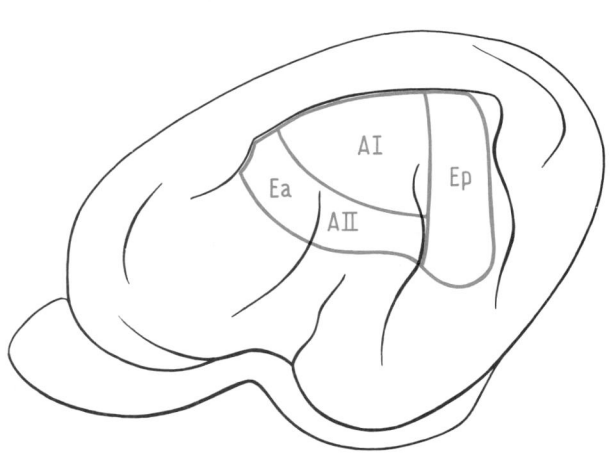

7.172 Lateral aspect of the feline cerebral hemisphere showing the main auditory areas, AI and AII, and accessory areas Ea and Ep. These areas cannot yet be extrapolated with confidence to the human cortex. See text for details. Adapted from Whitfield 1967.

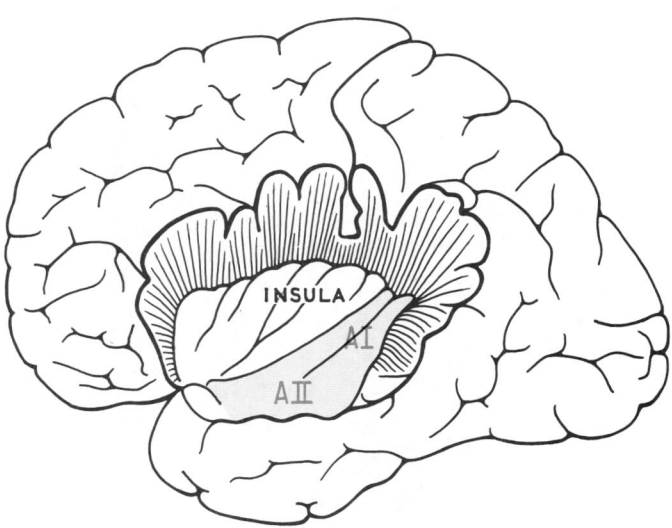

7.173 Superolateral surface of the left cerebral hemisphere from which the opercula have been cut away to expose the insula and the adjoining anterior and posterior transverse temporal gyri and their continuity with the superior temporal gyrus. The area shown in blue contains the classic acoustic area, AI (equal to Brodmann area 41 and parts of areas 42 and 52) and the secondary acoustic area, AII (extending into area 22).

the temporal lobe are perhaps more concerned with somatic activities especially with linguistic communication. The results of human stimulation naturally provide the basis for such views (Blum et al 1950). The anterior region appears even more multiform in its association of somatic and visceral activities; apart from the intrinsic association connections, it is linked to the limbic system (p. 1028) through connections of the parahippocampal gyrus with the hippocampus, with the piriform lobe (p. 1033) and the amygdala (p. 1034); the temporal pole, area 38 (7.125B), is also involved. Effects on blood pressure, respiration and gastric motility, usually depression, have been elicited by stimulation; but the major evidence in mankind depends on defects following partial temporal lobe ablation to relieve epileptiform attacks. Stimulation before lobectomy, which usually leaves the acoustic and motor speech cortex intact, may elicit complex memories of auditory, visual or combined content. After lobectomy a confusing plethora of manifestations have been described in an extensive clinical literature. Effects on speech, auditory memory and interpretation of auditory and visual phenomena have been noted (Milner 1967, Falconer 1967). A form of visual agnosia or 'psychic blindness' in which patients are unable to recognize visual significance, coupled with a change in dietary habits, a tendency to examine objects with the mouth, hypersexual behaviour and loss of emotional responses together constitute a syndrome named after Kluver & Bucy (1937); see also Terzian & Ore (1955); but in these instances lobectomy also involved the amygdaloid body, hippocampus and part of the parahippocampal gyrus; subsequent experiments have shown that these sequelae are only in part due to temporal lobectomy involving only the neocortex.

Of late, the temporal lobe has been loosely associated with memory; but, except for the possible participation of the parahippocampal gyrus in what was first ascribed to the hippocampus and associated structures, i.e. 'short-term' memory, there is little evidence of any entities in the temporal lobe or elsewhere which could be 'sites' of 'long-term' memory.

OTHER CORTICAL AREAS

The insula (7.143, 144A,D, 173) has been described (p. 1027); it adjoins and is overlapped by all lobes save the occipital. Histolog-

ically the posterior part of its cortex resembles the parietal (p. 1047); its anterior region is variously described as agranular or like the piriform cortex (Bailey & von Bonin 1951). Its connections are largely uncertain, but many have been claimed; short association axons between it and all its opercula are described; it is said to be connected to the lateral olfactory gyrus and piriform lobe (p. 1033). Thalamic connections (centromedian nucleus) have been suggested (Le Gros Clark & Russell 1939, Locke 1967). For a review of the numerous efferent and afferent connections with many neocortical areas, limbic areas and dorsal thalamic nuclei see Augustine (1985). Contralateral motor responses in the face and limbs have been elicited in macaques and gibbons, and somatotopy may exist in this possible supplementary field (Showers & Lauer 1961). Stimulation in man excites visceral motor and sensory effects, such as belching, gastric movements, nausea and abdominal 'sensations' (Penfield & Rasmussen 1950). Increased salivation in man and experimental animals has been elicited, with vasomotor effects in the latter. The insula has also been regarded a gustatory area by some. Many other functional associations have been claimed. (For reviews see Kaada 1960, Augustine 1985.)

The 'suppressor areas' reported by earlier experimenters in various cortical sites, particularly a vertical strip anterior to area 4, have been sources of controversy and criticism (Druckman 1952). Suppression of motor activities from such areas is now thought due to a generalized 'spreading depression', not associated with any particular area. This unexplained phenomenon is characterized by a long depression of neuronal activity, perhaps for several minutes, during which all responses are suspended or distorted (Leao 1944). However, the presence throughout the neocortex of fine, sometimes monoaminergic fibres, possibly exerting paracrine effects (p. 993) merits note.

The cingulate gyrus is discussed everywhere with the limbic system (p. 1028) but its anterior part, corresponding to area 24, must be mentioned here. In microscopic structure it has features in common with pre- and postcentral areas. It exchanges projections with the thalamic anterior nuclear complex and may have connections with the corpus striatum, hypothalamus and midbrain tegmentum (reticular formation). Commissural fibres via the corpus callosum have been described and association fibres probably link it with the frontal and parietal cortex anterior and posterior to the main sensorimotor region (Showers 1959).

Stimulation in macaques has been held to evoke contralateral and bilateral movements, with some form of somatotopic pattern (Showers & Crosby 1958). This supplementary motor area may extend into the posterior part of the cingulate gyrus but this requires confirmation. In patients, stimulation of the anterior cingulate area may elicit changes in pulse, respiration and blood pressure but the most reliable information comes from the results of surgical division or removal of the anterior cingulate gyrus, often involved in frontal leucotomy. Cingulectomy alone appears to relieve abnormal aggressiveness and obsession, yielding a milder, more placid personality (Tow & Whitty 1953). Similar effects are recorded in monkeys (Glees et al 1950). Cingulate gyrus ablation has been claimed as a useful manoeuvre for intractable pain (Folty & White 1962). This mixture of somatic motor and sensory, emotional and visceral activities at least accords with the intermediate position of the anterior cingulate gyrus between the limbic system and the neocortex proper.

THE SEPTUM PELLUCIDUM

This is a thin vertical partition (7.148, 176) composed of two laminae, largely separated by a narrow *cavity of the septum pellucidum*, an extrapial space and therefore not communicating with the brain's ventricular system. Seen from the side the septum is a triangle, its base anterior and apex posterior. It is attached: above, to the lower surface of the corpus callosum's trunk; below and behind, to the anterior part of the fornix; below and in front, to the upper surface of the corpus callosum's rostrum; directly in front, to the corpus callosum's genu. Each lamina forms part of the wall of the anterior horn and central part of the lateral ventricle, being therefore lined by ependyma on its ventricular side and pia on its medial aspect. The wall encloses

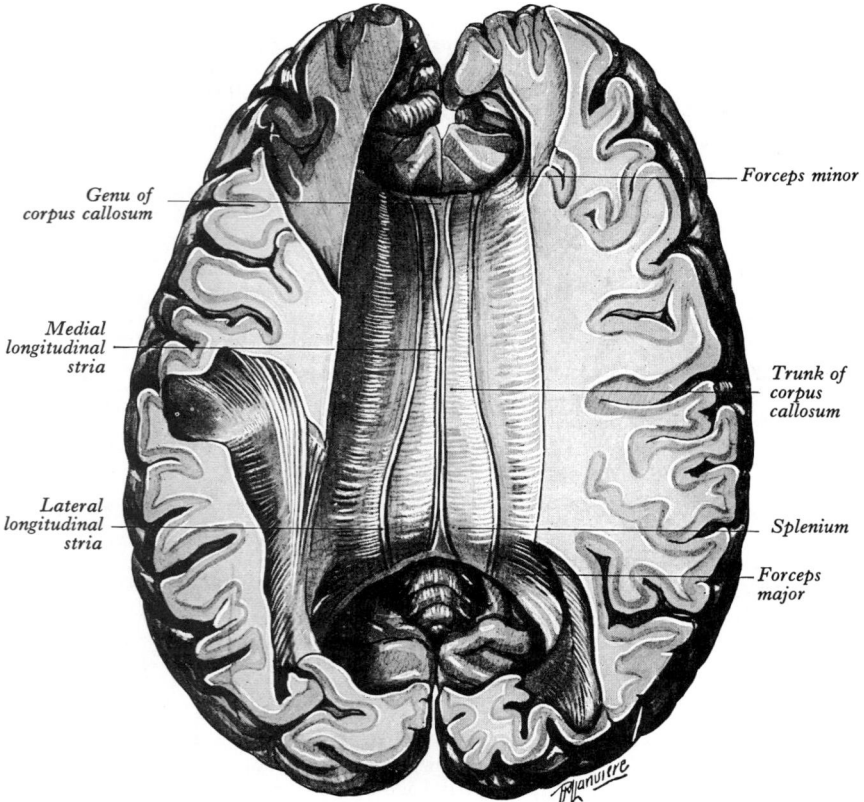

Genu of
corpus callosum

Medial
longitudinal
stria

Lateral
longitudinal
stria

Forceps minor

Trunk of
corpus
callosum

Splenium

Forceps
major

7.174 The corpus callosum, superior aspect, revealed by partial removal and dissection of the cerebral hemispheres.

important groups of *septal nuclei* and their fibre connections (see *septal areas*, p. 1035 and *dorsal fornix*, p. 1039; for development, see p. 194).

THE CORPUS CALLOSUM

The corpus callosum is the major transverse commissure connecting the cerebral hemispheres; it incidentally roofs in the lateral ventricles. Its mammalian development is proportional to the surface area and volume of the neocortex; it is maximal in human brains (Rakic & Yakovlev 1968). Its position and size is well appreciated in median sections (7.57, 140A,B, 145A,B). It forms an arch about 10 cm in length, its anterior end being about 4 cm from the frontal poles and its posterior end about 6 cm from the occipital poles. The *genu*, its anterior end, recurves postero-inferiorly in front of the septum pellucidum, then diminishes rapidly in thickness and is prolonged posteriorly to the upper end of the lamina terminalis as the *rostrum*. The *trunk* arches back and is convex above, ending posteriorly in the expanded *splenium*, its thickest part.

The superior surface of the callosal *trunk* (7.174) is covered by the *indusium griseum*, extending anteriorly around the genu, then on the inferior aspect of the rostrum, to continue into the paraterminal gyrus; in it are narrow longitudinal bundles of fibres, two on each side, the medial and lateral longitudinal striae (p. 1036); posteriorly the indusium griseum is continuous with the dentate gyrus and hippocampus through the gyrus fasciolaris (7.175).

In the median region the trunk is the floor of the great longitudinal fissure, where it is related to anterior cerebral vessels and lower border of the falx cerebri, which may contact it behind. On each side the trunk is overlapped by the gyrus cinguli, separated by the callosal sulcus. Its inferior surface is concave in its long axis but convex transversely. The septum pellucidum is attached to it anteriorly to an extent depending on septal length and disposition of the fornix (7.140). Posteriorly it is fused with the fornix and its commissure. On each side its inferior surface roofs the lateral ventricle (7.192, 198), covered by ependyma.

The *genu* connects the trunk to the rostrum. Its anterior surface, related to the anterior cerebral vessels, is covered by the indusium griseum and longitudinal striae. To its posterior surface the median septum pellucidum is attached; on each side it is the *anterior* wall of the lateral ventricle's anterior cornu.

The *rostrum* connects the genu and lamina terminalis; its superior surface is attached to the septum pellucidum and, on each side, forms part of the narrow floor of the lateral ventricle's anterior cornu (7.145A, B). On its inferior surface the indusium griseum and longitudinal striae pass back to the paraterminal gyrus.

The *splenium* overhangs the posterior ends of the thalami, the pineal gland and tectum but is separated from them by several structures. On each side the crus of the fornix and gyrus fasciolaris (7.175) curve up to the splenium; the crus continues forwards on the *inferior* surface of the callosal trunk but the gyrus fasciolaris skirts *above* the splenium, then rapidly diminishing into the indusium griseum. The tela choroidea of the third ventricle advances below the splenium through the transverse fissure, the internal cerebral veins emerging between its two layers to form the great cerebral vein. Above, the splenium is covered by

the commencing indusium griseum and is related to the falx cerebri with the inferior sagittal sinus and to the cingulate gyrus on each side of it. Posteriorly the splenium is near the tentorium cerebelli, great cerebral vein and the beginning of the straight sinus.

Nerve fibres of the corpus callosum radiate into the white core of each hemisphere, dispersing to the cerebral cortex. Commissural fibres forming the rostrum extend laterally *below* the ventricular anterior cornua, connecting the orbital surfaces of the frontal lobes; fibres in the genu curve *forwards* to connect the lateral and medial surfaces of the frontal lobes, as the *forceps minor*; fibres of the trunk pass laterally across (intersecting with) the projection fibres of the corona radiata (7.187, 188), connecting wide neocortical areas of the hemispheres. Fibres of the trunk and splenium which form the roof and lateral wall of the posterior cornu and the lateral wall of the inferior cornu constitute the *tapetum* (p. 1029). The remaining fibres of the splenium curve back medially into the occipital lobes as the *forceps major*, which bulges into the *medial* wall of the posterior cornu as a curved *bulb of the posterior horn*.

Despite its size and the enormous number of its commissural fibres, limited information is available concerning the precise functional roles of the corpus callosum, apart from the obvious inference that it links the hemispheres, perhaps to ensure that they act as an entity (however, vide infra). The total of its fibres is unknown in man, but in cats there are 700 000 in each square millimetre (Myers 1959). A detailed analysis of callosal connections is lacking for many regions; but initial investigations have been made in the visual area (Zeki 1970, 1974) and postcentral gyrus (Jones & Powell 1969). In somatic sensory areas (SmI and SmII) connections are limited to the same contralateral area, but right and left loci concerned with the hand and foot appear to lack commissural connection. These views have been strongly confirmed and amplified (Dr T P S Powell, Oxford, personal communication 1983): 'there is considerable experimental evidence, and from use of modern techniques, that there are no callosal connections between the cortex containing the representations of the distal limbs. These conclusions are based not only from a correlation of the experimental anatomical work with the maps of the representations from the evoked potential states but also from a correlation of anatomical tracing and microelectrode recording in the same animal. Not only is this so for the somatic sensory cortex in the post-central gyrus but it is also true for the motor cortex and area 4 in the pre-central gyrus.

'It is totally definite that the callosal connections between the visual cortex of the two hemispheres connect only representations of the vertical meridian of the retina (or visual field). The callosal connections are restricted to narrow bands of cortex, each only 1–2 mm wide, precisely at the boundaries between the different visual areas of the occipital lobe. There is such a band at the

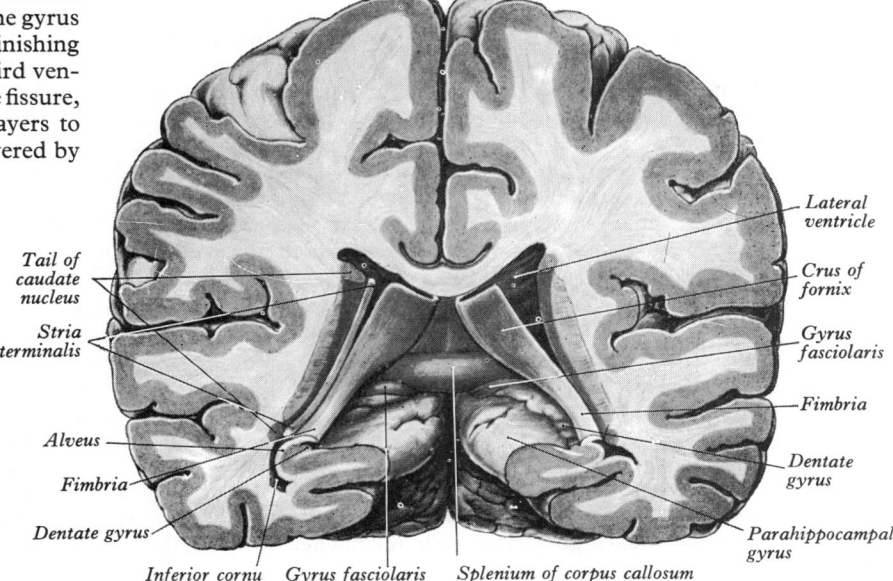

7.175 Anterior aspect of a coronal section of the cerebrum from which the posterior parts of the thalami have been removed to reveal the splenium and parts of the limbic system. (Note that the denate gyrus is not equally exposed on the two sides.)

Tail of caudate nucleus

Stria terminalis

Alveus

Fimbria

Dentate gyrus

Inferior cornu   Gyrus fasciolaris   Splenium of corpus callosum

Lateral ventricle

Crus of fornix

Gyrus fasciolaris

Fimbria

Dentate gyrus

Parahippocampal gyrus

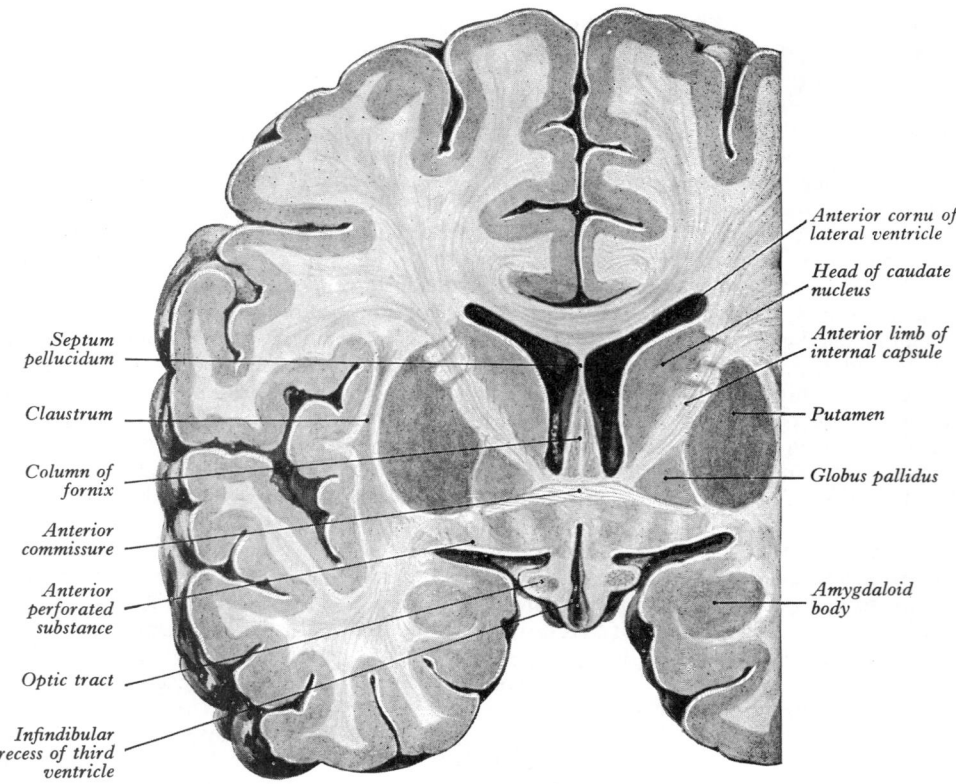

Septum pellucidum

Claustrum

Column of fornix

Anterior commissure

Anterior perforated substance

Optic tract

Infindibular recess of third ventricle

Anterior cornu of lateral ventricle

Head of caudate nucleus

Anterior limb of internal capsule

Putamen

Globus pallidus

Amygdaloid body

7.176 A coronal section of the cerebrum immediately caudal to the optic chiasma and passing through the anterior commissure.

boundary between area 17, the primary visual cortex and the adjoining area 18, and also bands between the other three or four visual areas that have been identified in the pre-striate cortex. It is probably true to say that the regions of cortex connected are related to the very narrow middle region of the retina within which the ganglion cells are now known to send axons which branch and send a branch into the optic tract of both sides.' Commissural linkages are considered highly specific, both associating the 'corresponding' columnar cortical units in bilateral functions, in addition to heterotopic loci.

A congenital absence of corpus callosum is rare, usually found at autopsy, and the clinical history lacks diagnostic features (Unterharnscheidt et al 1968). In recent decades large parts, and in some cases all of the corpus callosum, have been divided with apparently little disturbance of function (Akelaitis 1942). Nevertheless, the accumulated experience of the results of brain damage (Milner 1974), especially to the temporal lobe and corpus callosum, experimental studies of commissurotomized cats (Butler 1966) and chimpanzees (Myers & Henson 1960) and particularly the studies of Sperry (vide infra for references) on the effects of the division of the human corpus callosum—all this evidence has not only illumined the function of this great commissure in the transfer of information (including memorized data) across the midline of the cerebrum but has also confirmed a long-suspected asymmetry of function, leading to a concept of 'dominance' usually by the left hemisphere. This has become a major arena of cerebral research. Though it is impossible even to survey here the profuse and even prolix literature recorded by clinicians, experimenters (anatomical and physiological), psychologists, behaviourists and others, references to major contributions will enable the interested reader to enlarge his knowledge elsewhere.

CEREBRAL ASYMMETRY OR HEMISPHERIC DOMINANCE

A concept of asymmetry of function and structure in the cerebrum is by no means modern, although it has only recently attracted intensive study. It arose in the nineteenth century with developing views on cortical localization of function. Early

protagonists of asymmetry were Bouillard (1825) and Dax (1834), both concerned with speech, and Weber (1834) who believed asymmetry is detectable in sensory functions. But it was Broca's observations (1861), on a case of motor aphasia, which were perhaps not only the first indication of cortical localization but also the first convincing evidence of *unilateral* representation of function in the cortex. The correlation between pathological changes in the area often now known as Broca's area (p. 1055) and an almost complete defect in speech clearly pointed to a close linkage between speech and, in this case, the left cerebral hemisphere. Further patients and similar clinico-pathological deduction by others confirmed this finding. The view was further elaborated into a concept of the left hemisphere as *dominant* and the right as in some manner a *minor* hemisphere, perhaps partly prompted by an exaggerated identification of language with the cognitive processes; perhaps also the high incidence of right-handedness was seen as a corroboration factor, the more skilful hand under command of the more intelligent hemisphere. Others were less dogmatic; Hughlings Jackson (1868), while confirming Broca's clinical findings, was wary of extending the undoubted unilateral motor speech 'centre' in the left hemisphere into a concept of left cerebral dominance or an exclusively 'verbal' hemisphere. He considered that the neglected right or 'minor' hemisphere might be of equal functional import in somewhat vague terms, such as involvement in ideational, non-verbal cognitive activities. As will appear, he was closer to modern concepts than his contemporaries. However, the idea of dominance persisted into the twentieth century, though it did not become part of general neurological teaching until much more recent times. Perhaps this was due to a dearth of data, especially of the experimental type. Further studies of motor aphasia were added in confirmation of Broca's speech 'centre' and defects following damage to other cerebral regions in some instances vaguely suggested asymmetry of cortical function, or sometimes the reverse. Inevitably the corpus callosum demanded attention; a commissure of such size, containing an estimated 200 million nerve fibres in mankind, cannot be ignored. But the observed effects of agenesis or extensive injury proved difficult to assess and many observers accepted that even severe damage produced little change in cerebral performance. Retrospectively it is easy to appreciate that this was due to the

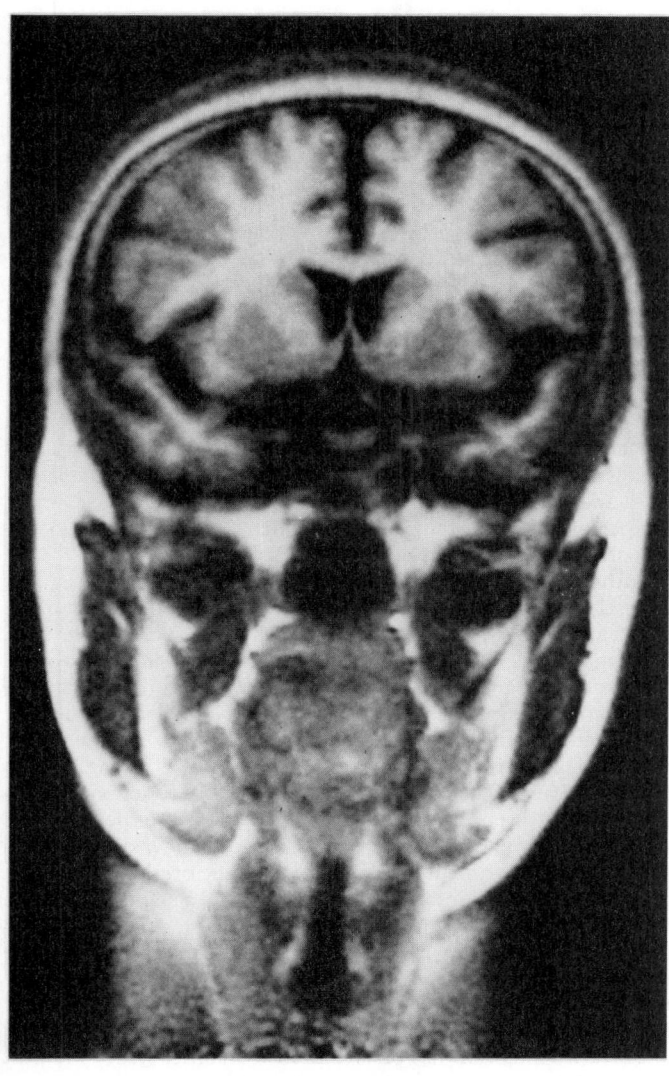

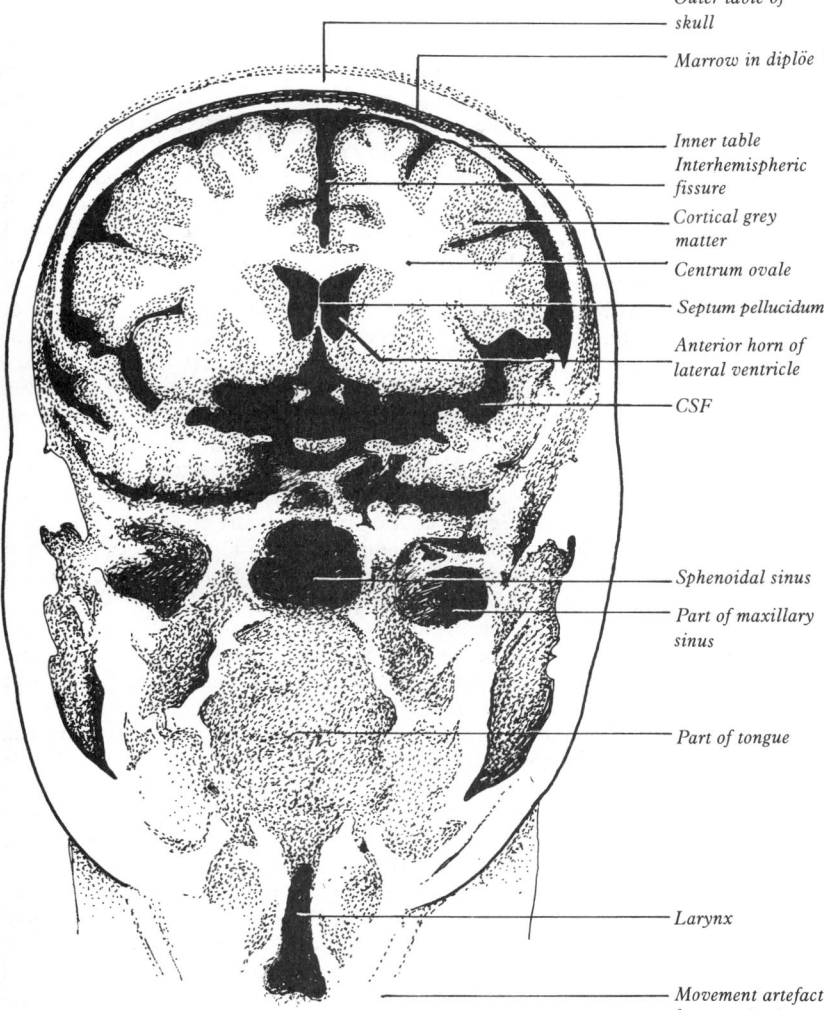

Outer table of skull

Marrow in diplöe

Inner table
Interhemispheric fissure
Cortical grey matter
Centrum ovale
Septum pellucidum
Anterior horn of lateral ventricle
CSF

Sphenoidal sinus
Part of maxillary sinus

Part of tongue

Larynx

Movement artefact from respiration

7.177A Oblique coronal magnetic resonance image of the brain at the level of the anterior cornua of the lateral ventricles. Supplied by Philips Medical Systems; photography by Sarah Smith.

7.177B Diagram of **7.177A** (Illustration: J Halstead).

relatively crude and inappropriate tests available. The introduction of adequate testing transformed the scene; but at this point it is necessary to turn briefly to '*handedness*', the human preference for right or left hand.

The close association of the cortical areas concerned with motor 'control' of speech and the right hand in the same, usually left, hemisphere inevitably evoked much statistical re-examination of the incidence of right and left hand-preference, a fascinating problem in itself. A large literature of data and even more speculation surrounds the subject (Hicks & Kinsbourne 1978). Forelimb preference in mammals other than man is not easy to assess; while such preferences are considered to exist, they are never so clear and universal as they are in human beings; nor is there the same high frequency of a 'dominant' right hand. This is 'preferred' in about 90% in all races and cultures (Chamberlain 1928, Komai & Fukoka 1934, Hécaen & Ajuriaguerra 1964, Annett 1967, Hicks & Kinsbourne 1978). A genetic basis for hand-preference has been confidently postulated for mankind (Nagylaki & Levy 1973) but this is still an issue of some (diminishing) disagreement. Such a genetic lateralization postulate has naturally been extended to the development of a dominant (verbal) cerebral hemisphere and this influence is widely accepted. Single gene (Annett 1964) and double gene (Levy & Nagylaki 1972) models have been proposed to explain the observed data of familial incidence of hemispheric dominance (often coupled with data on hand-preference); but neither model accounts completely for the distribution of variants; such studies have shown that hemispheric asymmetry is far from simple in its incidence. If an individual does not conform

to the large majority of left-sided 'verbal' and right-handed people, he is not automatically the reverse. A few (2% according to Zangwill 1967) who are right-handed nevertheless display evidence of right cerebral dominance and similar discordance between hand-preference and cerebral dominance occurs more often among the left-handed. Occasionally lateralization of function may be difficult to assess, especially in respect of hand-preference. True ambidexterity may occur and a limited degree of bilateral hemispheric function is more often suggested but less easily established. In hand-preference, especially, genetic influence may be complicated by cultural imitation in a largely right-handed population. How far this may, in childhood, produce secondary effects on cerebral dominance is uncertain; but it is certainly unwise to adopt a dogmatic attitude with regard to manual or cerebral asymmetry; the apparent anomalies in their distributions render theorizing on the genesis of asymmetry at present merely speculative.

Studies of cerebral dominance have been greatly improved by the introduction of sodium amytal injections into the cerebral circulation (Wada 1949). Unilateral injection, despite cerebral arterial anastomoses, temporarily suspends, or greatly diminishes, function in the corresponding hemisphere; the 'speech' or dominant hemisphere is thus easily deduced in patients or volunteers, who may then be observed in greater detail by appropriate tests. This method has complemented and sometimes extended the opportunities already afforded by injured patients in the investigation of peculiarities in the two hemispheres. Recorded data have become almost confusingly profuse, especially with the widespread interest of experimental psychologists in this field.

The contribution to the study of cerebral asymmetry by assessment of functional defects following cerebral lesions has been collectively large, though scattered through a large number of necessarily limited observations. Historically this is the oldest mode of study (cf. Broca) and also the newest (cf. Sperry, below), as pointed out by Milner (1974) who has herself made a large contribution, particularly on patients submitted to cerebral lobectomies, frontal and temporal. The association of temporal lobes and hippocampi with the function of memory has been established for many years; a particular interest in the effects of unilateral temporal lobectomies resides in the revelation of asymmetrical memory defects (Milner 1958, Milner et al 1964, Blakemore & Falconer 1967, Milner & Teuber 1968). Left temporal lobectomy impairs learning and memorization of verbal material, visual or auditory, whereas the right lobe appears concerned with the acquisition and retention of visual and auditory experience of a non-verbal nature. Right lobectomy damages the ability to follow mazes, in terms of vision and proprioception. Some lobectomies, of course, involve more than the cortex and may include hippocampal and amygdaloid tissue. Localization of function is therefore sometimes crude and uncertain but there is no doubt of the observed asymmetry.

Though all the avenues of enquiry mentioned above have contributed substantially to unmasking asymmetry of function, by temporary or permanent partial blockage of function in one hemisphere, the most fruitful source of observation has undoubtedly been the *isolation* of otherwise intact hemispheres from each other, though not from the external environment, by therapeutic division of the corpus callosum, and to a lesser extent by the same procedure in experimental animals, including monkeys. Before considering this evidence, however, it is necessary to turn briefly to strictly anatomical contributions.

Though anatomical observation inevitably enters into the interpretation of the effects of disease, injuries and experimental lesions, it might also be supposed that marked functional asymmetries, gradually exposed during a century and more of observation, would be associated with ascertainable structural differences, quantitative and even qualitative. Evidence of such anatomical asymmetry is, however, small and equivocal. Stimulated by the views of Dax (1834) and the discoveries of Broca (1861), some of their contemporaries and also Broca himself (1875) considered the left hemisphere to be heavier; but others recorded the contrary opinion. Not one in the long sequence of structural investigators of the cerebral cortex appears to have noted significant differences between the right and left sides; Braitenberg (1977) has re-emphasized this by his own observations. Because of its proximity to Broca's area (on the left) and its association with hearing, the superior temporal gyrus (and adjoining speech area of Wernicke, p. 1059) has been a focus in the search for asymmetry. Predominance in size of the left superior temporal gyrus has been described (Geschwind & Levitsky 1968, Le May & Culebras 1972, Hyde et al 1973). The first observer cited claimed a left areal superiority in 80% of human brains. (Adequate surveys of transverse temporal gyri are lacking.) A decade earlier two authorities, Von Bonin (1962) and Bodian (1962), had pronounced opposite views on the significance of such structural asymmetries in respect of cerebral dominance, even when convincingly demonstrated, Bodian opining that a search for such differences would be sterile. The general lack of observations in this field, and their uncertainties, corroborates this view so far; but it may be that structural asymmetries are of a more subtle nature. Di Chiro (1962) noted reciprocal differences in calibre between the superior and inferior anastomotic veins of the two hemispheres, which might denote, if confirmed, some divergence of vascular requirement; but this comparatively crude angiographic study has not been amplified. It must be concluded, therefore, that for the present at least a strictly anatomical approach to the problem has proved negative.

Although nineteenth-century interest in asymmetrical control of speech continued into the twentieth (e.g. Penfield & Roberts 1959), the work of Sperry and his collaborators in the sixties and beyond has proved the outstanding contribution and stimulus in

this field (Sperry et al 1969, Gazzaniga 1970, Sperry 1970, 1974, 1977). The fact that he was already investigating 'split-brain' preparations in experimental animals and could collaborate with the pioneers (Bogen & Gazzaniga 1965) in complete division of the corpus callosum to alleviate severe, bilateral epilepsy, proved most productive. Sperry (1961) had elaborated, during work on monkeys, an array of ingenious testing techniques which could be applied, extended and further refined in observation upon human patients, with the great advantage of verbal communication, allowing access to subjective phenomena which could only be hypothecated in monkeys from motor behaviour. The essence of such techniques is the isolated presentation of a variety of sensory cues (visual, auditory, or tactile) to the right or left hemisphere and also isolation from each other of such stimuli or cues presented simultaneously or sequentially in these different modes. The most commonly employed stimuli are visual cues, projected with great care in either the right or left half of the total binocular visual field; but auditory and even olfactory cues have been extensively used and the hands are frequently included (though these are in the performance of motor tasks, in which tactile and proprioceptive stimuli are also highly significant).

Although corpus callosal *commissurotomy* is not a frequent operation, many patients have now been studied postoperatively and a mass of experimental results has been recorded in such patients and also in considerable numbers of patients and volunteers subjected to amytal injection, both by Sperry, his team and other workers. Sperry's findings have been largely corroborated, with some extensions and modifications, often of a semantic nature (Dimond & Beaumont 1973). They are epitomized in 7.178. There has been so much corroboration of functional asymmetry in the cerebrum and such detailed analysis of peculiarities of each hemisphere, that there is a distinct danger of overstatement. As in any fruitful field of discovery, the original impetus is towards *analysis*, to establish differences and details. As will be emphasized below, it is necessary, as a corrective, to preserve also an impulse towards *synthesis*. The two hemispheres are, despite differences, *complementary* and are in normal circumstances undeniably continuous. Phenomena revealed by their artificial separation have now been studied by a host of workers in much detail; only the examples of asymmetric qualities which are best established can be summarily recounted here.

In the surgically 'split' brain, a brief signal presented in the left half of the binocular visual field is received only by the right hemisphere, with no possibility of transfer of information to the left. (The stimulus must be brief enough, usually about 0·1 second in duration, to counter momentary ocular movement from midline fixation.) If the right is the 'minor' hemisphere, as it usually is, no verbal response can be obtained and the patient is unaware (so he says) of the occurrence of the signal when questioned. (He may volunteer statements but their content is unrelated to the experimental situation.) Nevertheless, the patient (or experimentee if sodium amytal is used) must in some manner, apparently outside the consciousness associated with his left hemisphere, recognize signals, especially actual objects, with his right hemisphere. This is shown, e.g. by his ability to *remember* an object displayed to him momentarily and to pick it out subsequently with his *left* hand, unaided by vision, from an array of dissimilar things presented to his searching hand. Whatever comments his isolated left hemisphere likes to make on his performance (of which he has no knowledge in this verbalizing, talkative hemisphere), these phenomena demonstrate that the right hemisphere is able to carry out much that is characteristic of a conscious brain: he 'sees', recognizes, employs memory and directs successfully the corresponding hand. A large series of such experiments, employing a wide variety of visual, auditory and tactile cues and recorded by many observers, demonstrates beyond doubt that each hemisphere can, when separated, act to a marked extent as an independent unit, with its own ability for perception, learning, memorization and motor control; hence, of course, the expression 'split brain' and the assumption that two separate consciousnesses can co-exist in the individual. However, it is only the verbally skilful hemisphere (usually left, sometimes right) which can communicate directly, explicitly and intelligently with other consciousnesses in the medium of language, thus

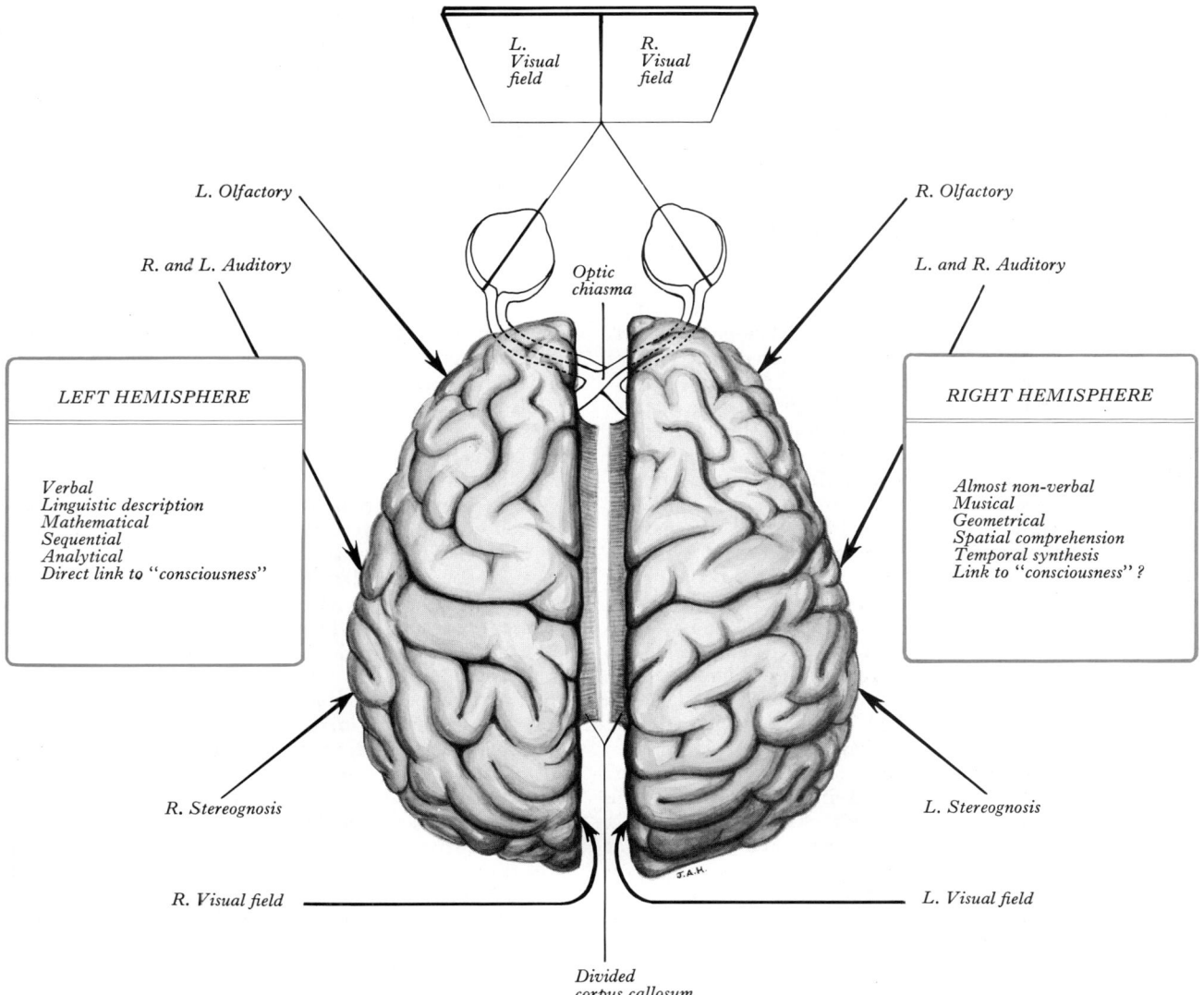

L.
Visual
field

R.
Visual
field

L. Olfactory

R. Olfactory

R. and L. Auditory

L. and R. Auditory

Optic
chiasma

LEFT HEMISPHERE

Verbal
Linguistic description
Mathematical
Sequential
Analytical
Direct link to "consciousness"

RIGHT HEMISPHERE

Almost non-verbal
Musical
Geometrical
Spatial comprehension
Temporal synthesis
Link to "consciousness"?

R. Stereognosis

L. Stereognosis

R. Visual field

L. Visual field

Divided
corpus callosum

7.178 This diagram is an adaptation (with the author's permission) of the original 'split-brain' schema, published by Professor R W Sperry in 1964, of hemispheric attributes as revealed by complete division of the corpus callosum. The right and left halves of the total visual field are projected into the contralateral hemispheres and are completely isolated, the right hand being 'controlled' from the left hemisphere. The olfactory and auditory inputs also project to separated hemispheres, but in the former case ipsilaterally and in the latter bilaterally with a contralateral bias. On each side of the diagram the abilities listed refer to the adjoining hemisphere; they are to be taken as extremely generalized attributes, which are likely to be modified as research in this field progresses. The diagram represents, of course, arrangements in an individual with left cerebral dominance or preference.

uniting its own conscious world with the exterior. The other, the so-called 'minor' hemisphere, is severed or 'split' from such communication and, moreover, from communication with the 'dominant' hemisphere. Nevertheless, this isolated right hemisphere is even able to recognize the meaning of the names of objects, presented either visually or aurally, and to correlate this recognition with memories of similar experiences and to carry out appropriate motor behaviour, though this half-brain, apparently conscious, is out of touch with any observer, who has no access to its activities except through its motor responses (excluding those of speech).

These astonishing findings have persuaded some interpreters to an even firmer belief in the dominance of the verbal hemisphere and to assign to it a major, if not exclusive role in a supposed brain–consciousness linkage. Some workers have even been led to question the need for two such 'brains' (e.g. Teuber 1974). The enthusiasm for analysis has undoubtedly led to an over-emphasis of the peculiarities of each hemisphere, tending to obscure the normal unity of the two halves of the cerebrum. Whatever may be hypothecated with respect to consciousness, it is clear that in some of the attributes of consciousness, such as spatial conception, geometrical perception and perhaps 'musicality', the minor hemisphere is the more able partner and must participate in the conscious activities of the united hemispheres.

It is easy, of course, to overlook the fact that in the diencephalic region and, caudal to this, unity persists even after commissurotomy. As Young (1978) has picturesquely stated, the split personality of commissurotomized patients, dual though it may seem, falls asleep as one, though his two eyes appear to take different views of his environment when he is questioned on awakening. It is not surprising, perhaps, that such patients exhibit so little deficiency or duality in ordinary activities, in which they are also able to explore the visual field freely with both eyes, to use both hands and to rotate the head, thus duplicating in each hemisphere a representative array of signals from the entire sensory field and not merely from one side or the other. It is only by the careful techniques evolved by Sperry and his co-workers that defects are unmasked.

Specialization of hemispheres in an asymmetrical, unilateral manner, so that in the total cerebrum certain abilities are sufficiently correlated with one hemisphere to be assignable to it, is at present beyond explanation, despite interesting theorization. Such arrangements are not, however, wholly unexpected. The existence of two central mechanisms, bilaterally related, for any generalized function is not easy to understand or even accept, without evidence of some linking or co-ordinationg locus. The operation of two such central mechanisms, each dealing with half of what is continuously appreciated as an unbroken visual field has always posed such a problem and appears to evoke some unitary integrating activity, although Sherrington (1906) was prepared to regard the *simultaneous* occurrence of dual sets of neuronal events as sufficient explanation. But evidence for the transference of visual and other information across the midline of the cerebrum is impressive and memory stores of each hemisphere appear to be accessible to the other and to be probably bilateral at least in part.

All such phenomena would in hypothesis demand a large connection between the hemispheres and this of course exists and must provide the anatomical matrix for extensive communication and presumably integration between right and left activities. Though so much research in this field has been focused on functional hemispheric differentiation, with special emphasis at first on the unilateral mediation of speech (thus leading to the concept of dominance), it is nevertheless becoming clearer in later investigation that the laterality of many functions is merely greater on one side and not exclusive to one or other hemisphere. Even in verbalization, some ability has long been assigned to the 'minor' hemisphere (Albert & Obler 1978). Moreover, concepts of what constitutes a discrete 'function' may be in themselves misleading. Though I may *describe* a complex three-dimensional structure to another individual through a series of symbolic sounds called language, chiefly organized into an intelligible sequence by left hemispheric activity, it seems equally probable that I am *conceiving* an image or idea of the object I wish to describe in my right hemisphere and that in the absence of linguistic description the concept may remain largely limited to the 'minor' brain.

The too facile ascription of labelled functions or abilities to one or other side has been criticized by Milner (1974), who has redirected attention to the 'parallelism' of right and left activities. Broadbent (1974) has also emphasized integrative action of the hemispheres (an interesting reminder of the title of Sherrington's classic text!) regarding speech, e.g. as a combined production of both. Such considerations favour a growing tendency to abandon the terms 'dominant' and 'minor'; it is clearly more rational to accept a complex but complementary asymmetry of distribution of ability within the whole cerebrum. Inevitably this will entail a

degree of right and left predominence, though perhaps only in so far as we impress our own semantics upon the complete organization. Eccles (1973 and in Popper & Eccles 1977) has professed a preference for the term 'dominance', from his own interpretation of the dominant, verbalizing hemisphere as the mediator between the external universe and internal consciousness. He has resuscitated an old postulate due to Sherrington (whose student he was) that some part of the cerebrum would prove to be a site of what Eccles calls a *liaison* between brain and consciousness; but he goes further in designating the *dominant* hemisphere as a *self-conscious* cerebral region. Although we are here being returned to the nineteenth-century association of speech and cognitive processes in a common locus, evidence from Sperry's patients and other sources does also support such a view. But the apparent inaccessibility to any verbal communication of the 'minor' hemisphere and its apparent oblivion to its own activities are both revealed by the artificial separation. However useful this isolation may be in the analysis of cerebral abilities, it may also mislead; it must be followed by a re-synthesis to explain the operations of a normally united and singular brain.

These fascinating studies are currently providing a most fertile common ground between neuroanatomists, neurophysiologists, experimental psychologists and even philosophers, in a general endeavour towards understanding the human brain and hence of what we call *our selves*. It is an exciting prospect, particularly because it centres around the great problems of communication between the private world of the individual and his external environment of countless similar individuals, each locked in a private prison except insofar as communication succeeds. Perhaps, as understanding increases, we shall dissolve our differences and disperse our doubts and then perhaps in the happily optimistic words of Young (1968) 'a new form of laughter will echo through the system'. However that may be, it is pertinent to conclude with a statement by Braitenberg (1973), another distinguished student of man's brain: 'The *neuroanatomy* of neuronal networks will be for quite some time, and perhaps forever, the content that relates the *physiology* of single neurons and synapses to the macroscopic events of *psychology*' (Editors' italics). After analysis must come synthesis.

INTERNAL STRUCTURE OF THE CEREBRAL HEMISPHERES

In the interior of the hemispheres are the lateral ventricles, the basal nuclei and many fibre tracts, both projection and intrinsic.

The Lateral Ventricles

The two lateral ventricles (7.179, 180, 181A,B, 182) are irregular cavities lying inferomedially in the cerebral hemispheres, one on

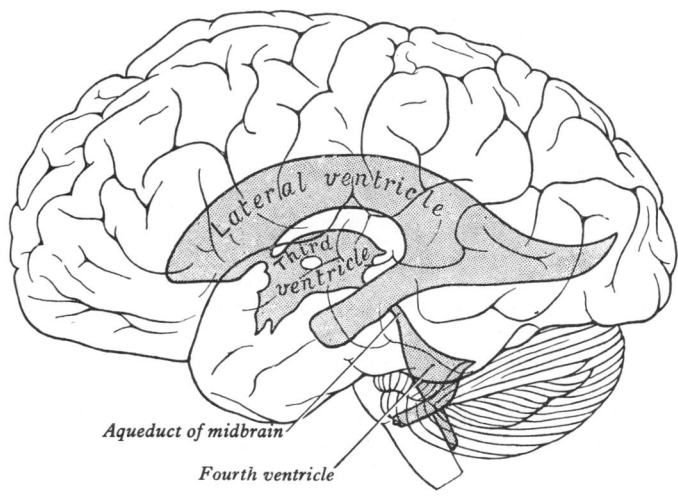

7.180 A drawing of a cast of the ventricular cavities: superior aspect (Retzius). Note that, where the lateral recess joins the fourth ventricle, the lateral dorsal recesses of the roof of the ventricle project dorsally on each side beyond the posterior margin of the median dorsal recess. The superior angle of the fourth ventricle and the aqueduct of the midbrain are hidden by the suprapineal recess.

7.179 Projection of the ventricles on to the left surface of the brain.

each side, and almost completely separated by the *septum pellucidum* but each communicating with the third ventricle and indirectly with the other through its *interventricular foramen* (p. 1019). They are lined by the ependyma and contain cerebrospinal fluid. Each lateral ventricle has a *central part* and three *cornua* or *horns*: anterior, posterior and inferior (7.180, 181A,B, 182, 191, 193).

The central part (7.181B, 182) extends from the interventricular foramen to the splenium. It is a curved cavity with a roof, floor and medial wall; it is triangular in transverse section anteriorly and rectangular posteriorly. The *roof* is the inferior surface of the corpus callosum; the *floor* slopes steeply downwards and medially, presenting two convexities and two grooves; it is formed, in lateromedial order, by the caudate nucleus, the stria terminalis, thalamostriate veins and the lateral part of the superior thalamic surface. The convex body of the caudate nucleus rapidly narrows as it passes back in the floor; its long axis is directed posterolaterally. The stria terminalis, a bundle of white fibres (p. 1034) and the thalamostriate vein occupy a narrow groove following the medial border of the caudate nucleus and separating it from the convex superolateral surface of the thalamus. The latter may be almost hidden by the choroid plexus, which invaginates the ependyma into the cavity through a slit between the fornical edge and a reciprocal groove on the superior thalamic surface, the ependymal invagination occupying this *choroid fissure*. The body of the fornix widens posteriorly; its thin, lateral margin is parallel to the groove for the stria terminalis. The *medial wall* is formed by the posterior part of the septum pellucidum with the body of the fornix in its lower margin. Posteriorly, where the septum ends, roof and floor meet on the medial wall.

The anterior cornu (7.179, 180, 181, 182) passes anterolaterally and slightly down into the frontal lobe and is continuous posteriorly with the ventricle's central part at the interventricular foramen. In coronal section it is a triangular slit below the anterior part of the corpus callosum, bounded anteriorly by the *posterior* aspect of the genu and rostrum. The *roof* is the anterior part of the callosal trunk. Most of the *floor* is the large rounded head of the caudate nucleus, but medially a small part is formed by the upper aspect of the rostrum (7.195). The *medial boundary* is the septum pellucidum containing the column of the fornix in its posterior edge.

The posterior cornu curves posteromedially into the occipital lobe. Its development is variable and often asymmetrical; it may be absent. Its *roof* and *lateral wall* are formed by fibres of the tapetum separating them from the optic radiation (p. 1073). Splenial fibres forming the forceps major pass medially as they sweep back into the occipital lobe, producing a rounded *bulb* of the posterior cornu in the latter's upper *medial wall*, below which is a second elevation, the *calcar avis*, corresponding to the infolded cortex of the anterior calcarine sulcus (7.184). Posteriorly the lateral and medial walls meet. Kier (1977) has shown that the posterior cornu is the most recent extension of the lateral ventricle, present only in the Anthropoidea. It is symmetrical in the fetus but in adults the cornua are usually unequal in size, a perinatal change which Kier ascribes to unequal folding of the calcarine fissures.

The inferior cornu (7.149, 185), largest of the three, traverses the temporal lobe, curving round the posterior aspect of the thalamus. At first it descends posterolaterally, then curves

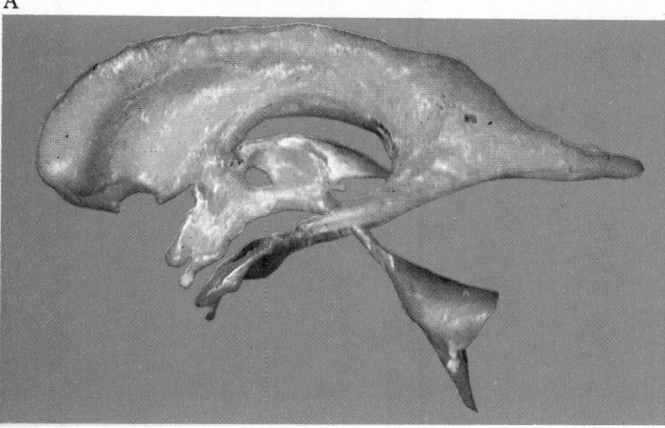

7.181 Resin casts of the ventricular system of the human brain (prepared by D H Tompsett of the Royal College of Surgeons of England): A. ventral view; B. left lateral view. Compare with the superior aspect shown in 7.180.

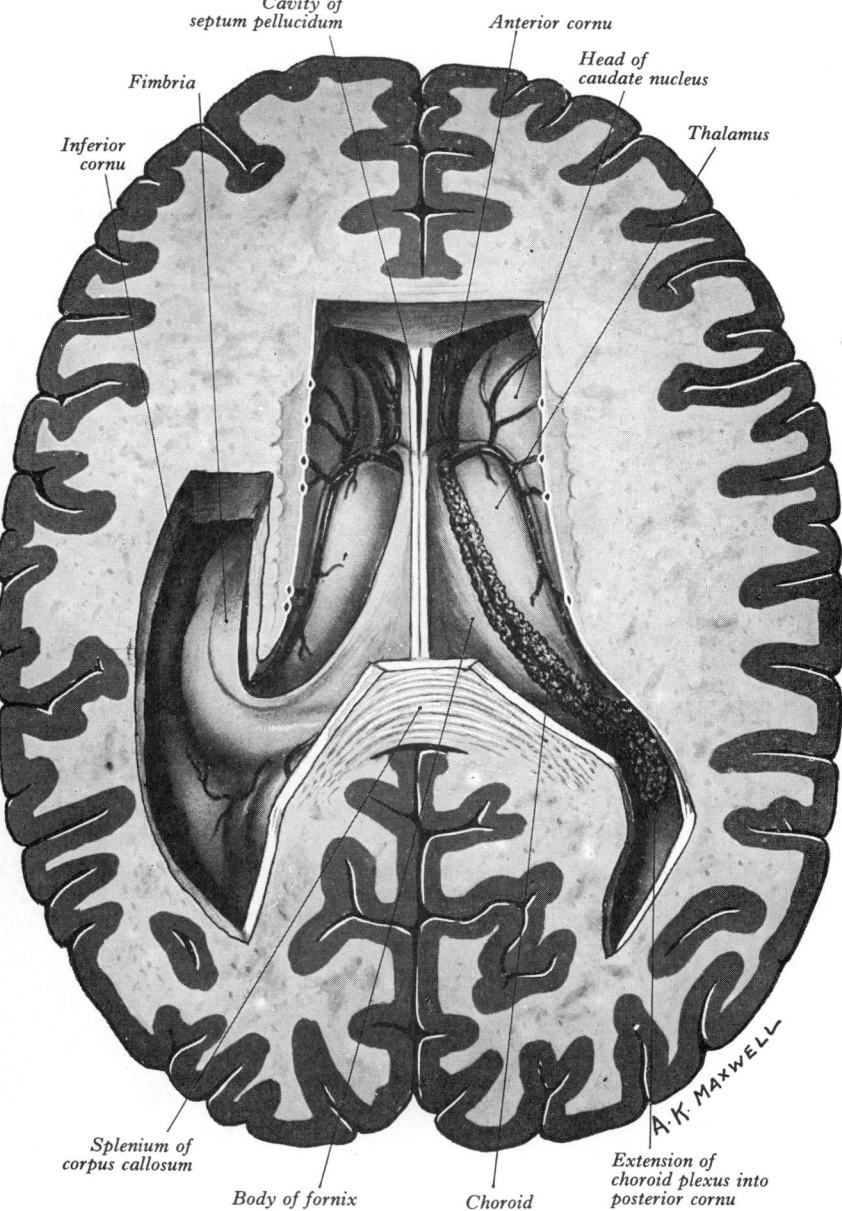

Cavity of
septum pellucidum

Anterior cornu

Head of
caudate nucleus

Fimbria

Thalamus

Inferior
cornu

Splenium of
corpus callosum

Body of fornix

Choroid
plexus

Extension of
choroid plexus into
posterior cornu

A.K. MAXWELL

7.182 Horizontal section of the cerebrum dissected to remove the roofs of the lateral ventricles.

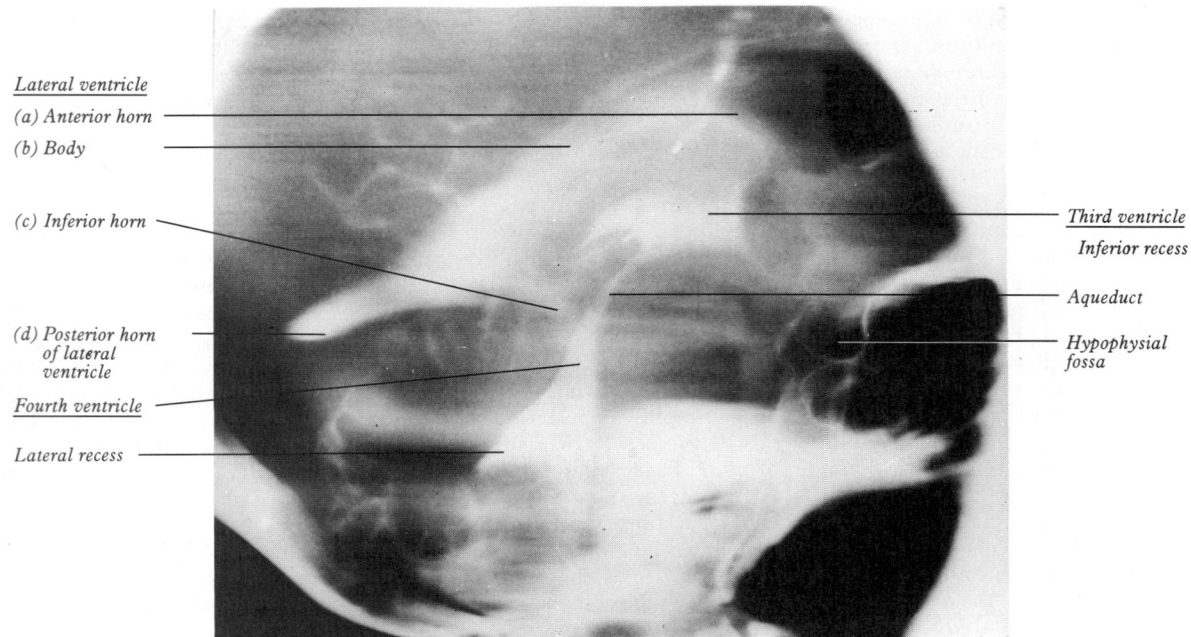

Lateral ventricle
(a) Anterior horn
(b) Body

(c) Inferior horn

(d) Posterior horn
  of lateral
  ventricle

Fourth ventricle

Lateral recess

Third ventricle
Inferior recess

Aqueduct

Hypophysial
fossa

7.183A Positive contrast ventriculogram (tomogram), with some air in the anterior part of the lateral ventricle. Supplied by T D Hawkins, Addenbrooke's Hospital, Cambridge; photography by Sarah Smith.

anteriorly to within 2.5 cm of the temporal pole near the uncus, its position usually corresponding on the surface to the superior temporal sulcus. (A trephine hole, with its centre 3 cm behind and 3 cm above the centre of the external acoustic meatus allows a needle, passed towards the tip of the opposite ear, to enter the inferior cornu at a depth of 5 cm from the surface.

The *roof* is formed chiefly by the tapetum of the corpus callosum but the tail of the caudate nucleus and stria terminalis also

extend forwards in it and they are continuous with the amygdaloid body at its end (tip, or uncal extremity). The *floor* consists of the collateral eminence laterally, the hippocampus medially (the latter covered by the alveus), continuing as the fimbria of the hippocampus, which continues posteriorly with the crus of the fornix (p. 1039). Between the stria terminalis and the fimbria is the inferior part of the choroid fissure, where the lower part of the choroid plexus invaginates the ependyma and closes (i.e.

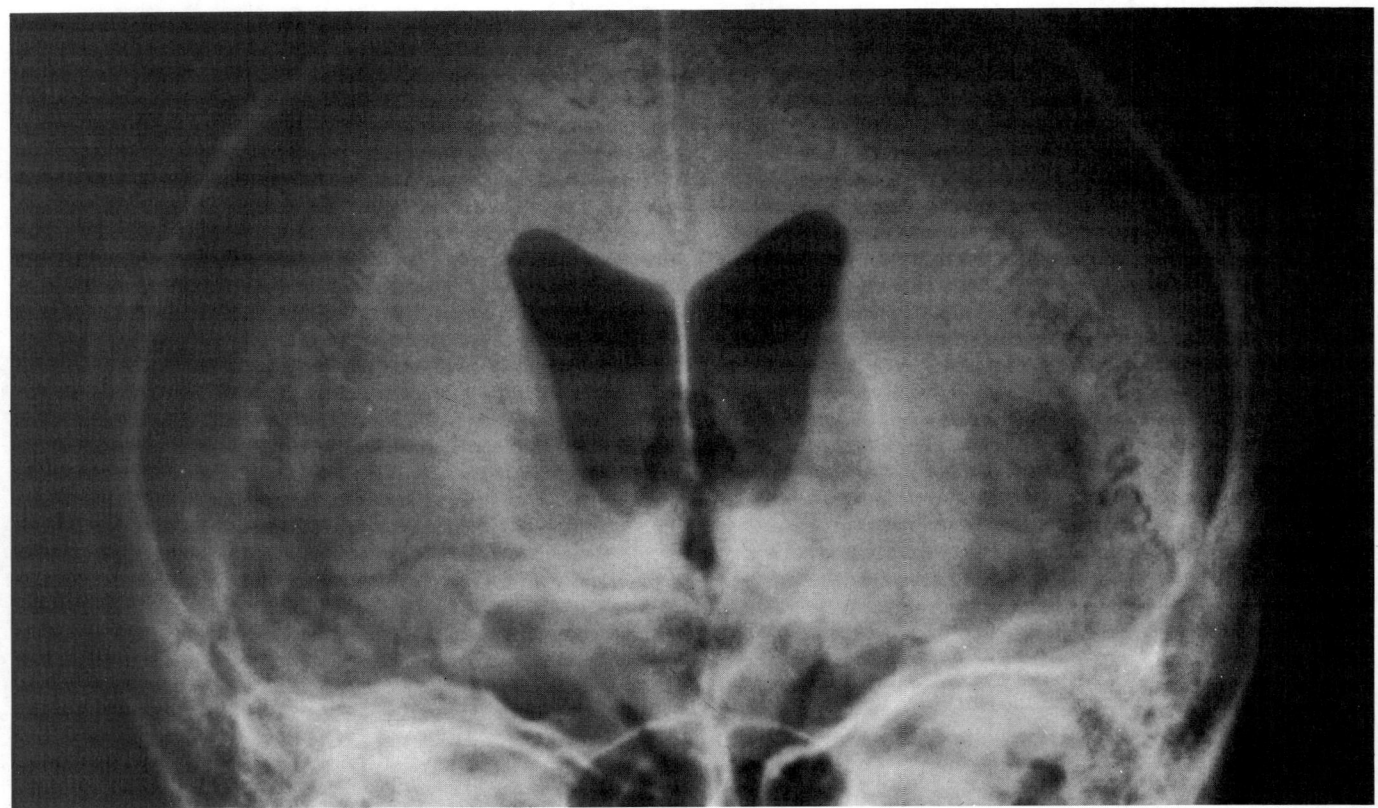

7.183B Anteroposterior radiograph of the head after the introduction of air into the ventricular system of the brain. The outlines of the bodies and anterior horns of both ventricles are separated in the midline by the shadow of the septum pellucidum; directly inferior to the latter, the outline of the third ventricle can be seen. Provided by R D Hoare, Dept. of Diagnostic Radiology, Guy's Hospital, London.

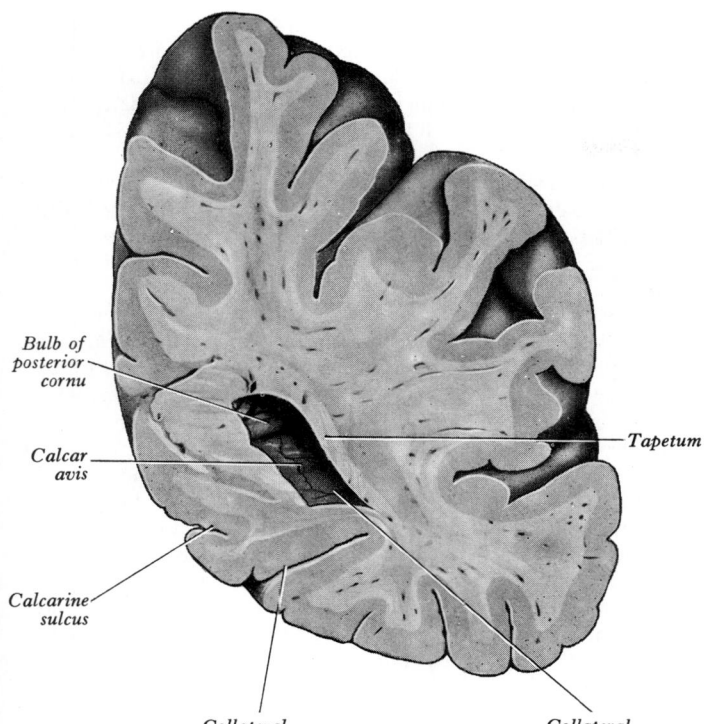

7.184 Anterior aspect of a coronal section through the posterior cornu of the left lateral ventricle.

Nerve Fascicles of the Cerebrum

In a hemisphere sliced horizontally about 1·25 cm above the corpus callosum, its central white substance appears as an oval area surrounded by a narrow, much folded lamina of grey matter and studded by red *puncta vasculosa* due to the escape of blood from divided vessels. Transected at the level of the corpus callosum, it displays the continuity of white matter of the two sides, containing nerve fibres of varying size and supported by neuroglia. These may be described according to their course and connections, as: (1) *commissural fibres*, linking corresponding or *homotopic* loci and *heterotopic* loci in both hemispheres; (2) *association* (arcuate) *fibres* connecting different cortical areas in the same hemisphere; some are collaterals of the projection and commissural fibres but most are main axons; (3) *projection fibres*, connecting the cerebral cortex with the corpus striatum, diencephalon, brain stem and the spinal cord in both directions.

(1) **Commissural fibres** are described on pp. 1005, 1034, 1039, 1062.

(2) **Association (arcuate) fibres** (7.186A, B), confined to one hemisphere and all ipsilateral, are grouped as *short*, connecting adjacent gyri, or *long*, connecting more widely separated gyri. The details of many association pathways appear with their cortical areas.

Short arcuate fibres may be entirely intracortical but many pass subcortically between adjacent gyri, some merely from one wall of a sulcus to the other. They are very numerous and easily, but crudely, displayed by blunt dissection in fixed brain tissue.

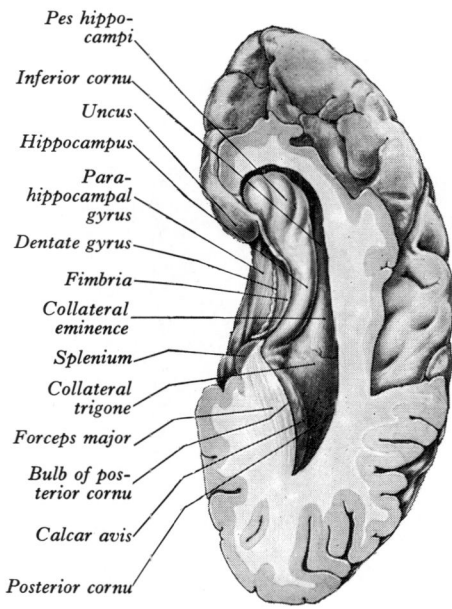

7.185 The posterior and inferior cornua of the right lateral ventricle, exposed from above.

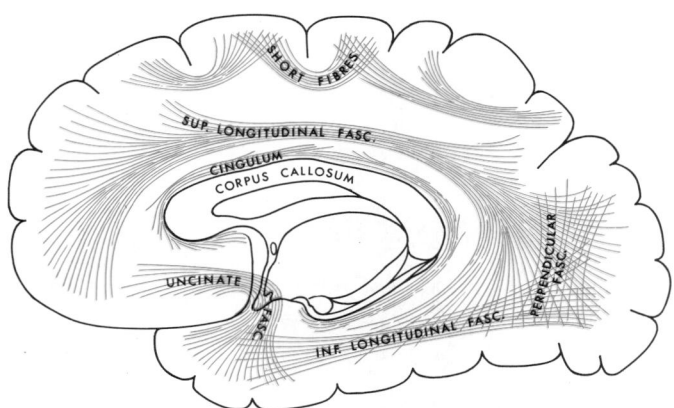

7.186A The principle arcuate (association) fibres in the cerebrum.

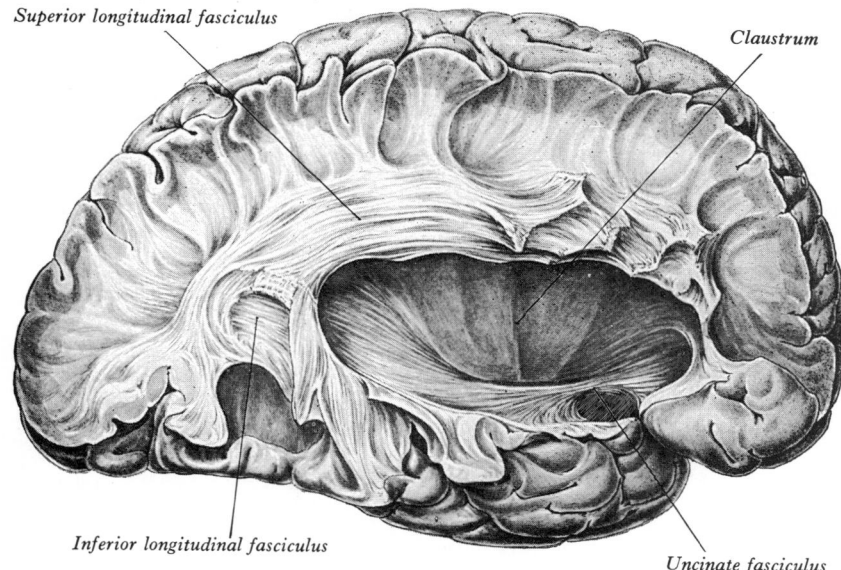

7.186B A dissection showing some of the long arcuate fasciculi of the right cerebral hemisphere.

occupies or 'fills') the fissure and covers the upper surface of the hippocampus.

The hippocampus and fimbria are considered on pp. 1036 and 1038 and the choroid plexus on p. 1080.

The *collateral eminence* (7.185), a long swelling lateral and parallel to the hippocampus, overlies the collateral sulcus, its size reflecting the depth of the sulcus. The eminence continues behind into the flattened triangular *collateral trigone*, flooring the ventricle between its posterior and inferior cornua. The capacity of the lateral ventricles (and other chambers related) can be estimated by anatomical and radiological techniques (Weintraub 1953). Levinger & Kedern (1974) devised a technique for estimating the surface area of cerebral ventricles from casts, but so far only in cats.

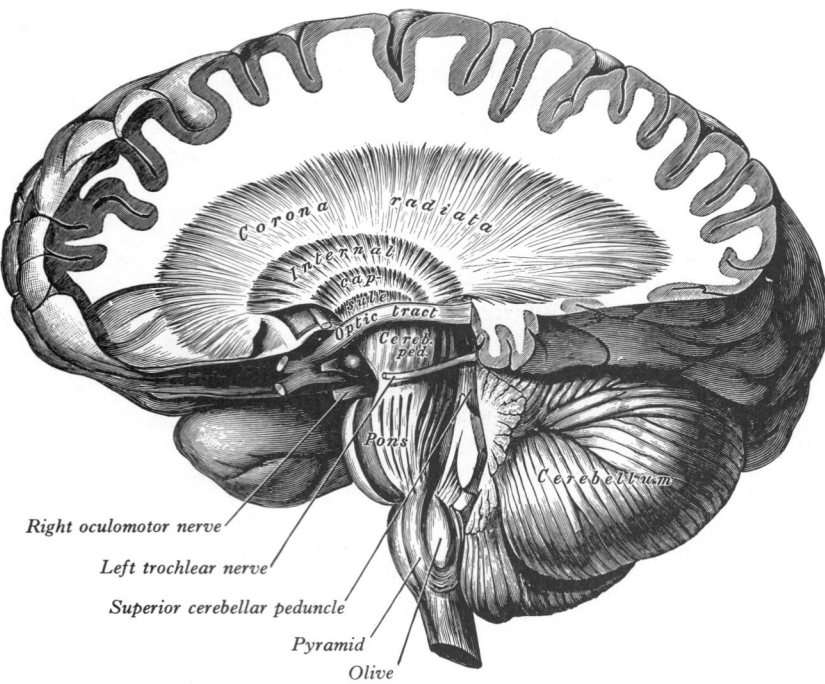

7.187 A dissection showing the convergence of cortical projection fibres through the corona radiata into the cerebral peduncle and pons.

Long arcuate fibres are grouped, somewhat indistinctly, into bundles which can be dissected in formalin-hardened brains after removal of the cortex and subjacent short acuate fibres. Fibres in each fasciculus vary in length, the longest usually being deepest. Concerning the connections of these bundles little information is available; they are difficult to follow by histological technique and any form of dissection reveals only the crudest details. The following large fasciculi can usually be distinguished: (*a*) the uncinate fasciculus, (*b*) the cingulum, (*c*) the superior longitudinal fasciculus, (*d*) the inferior longitudinal fasciculus, (*e*) the fronto-occipital fasciculus.

(*a*) The *uncinate fasciculus* connects the first motor speech area (p. 1055) and orbital gyri of the frontal lobe with the cortex in the temporal pole; the fibres follow a sharply curved course across the stem of the lateral sulcus. They are near the antero-inferior part of the insula (7.186A, B).

(*b*) The *cingulum*, a long, curved fasciculus starting in the medial cortex below the rostrum, then lies in the gyrus cinguli and follows its curve. Inferiorly it enters the parahippocampal gyrus and spreads into the adjoining temporal lobe. From its convexity fibres enter and leave in groups, giving it a spiked irregular appearance when dissected.

(*c*) The *superior longitudinal fasciculus*, largest of the arcuate bundles, commences in the anterior frontal region and arches back above the insular area and lateral to the massive cortical projection fibres of the internal capsule (corona radiata, vide infra). Contributing fibres to the occipital cortex (areas 18 and 19), it curves down and forwards behind the insular area to spread out in the temporal lobe. Like other long fasciculi, fibres leave and join it throughout its extent but it is impossible to determine the precise connections by gross methods (7.186B); even with the modern techniques of fibre tracing little progress has yet been made in this field.

(*d*) The *inferior longitudinal fasciculus* commences near the occipital pole, its fibres derived perhaps mostly from areas 18 and 19. They sweep forwards, separated from the lateral ventricle's posterior cornu by the optic radiation and commissural tapetal fibres and, after being crossed by the superior longitudinal fasciculus, is then distributed throughout the temporal lobe.

(*e*) The *fronto-occipital fasciculus* starts from the frontal pole, passing back deep to the superior longitudinal fasciculus but

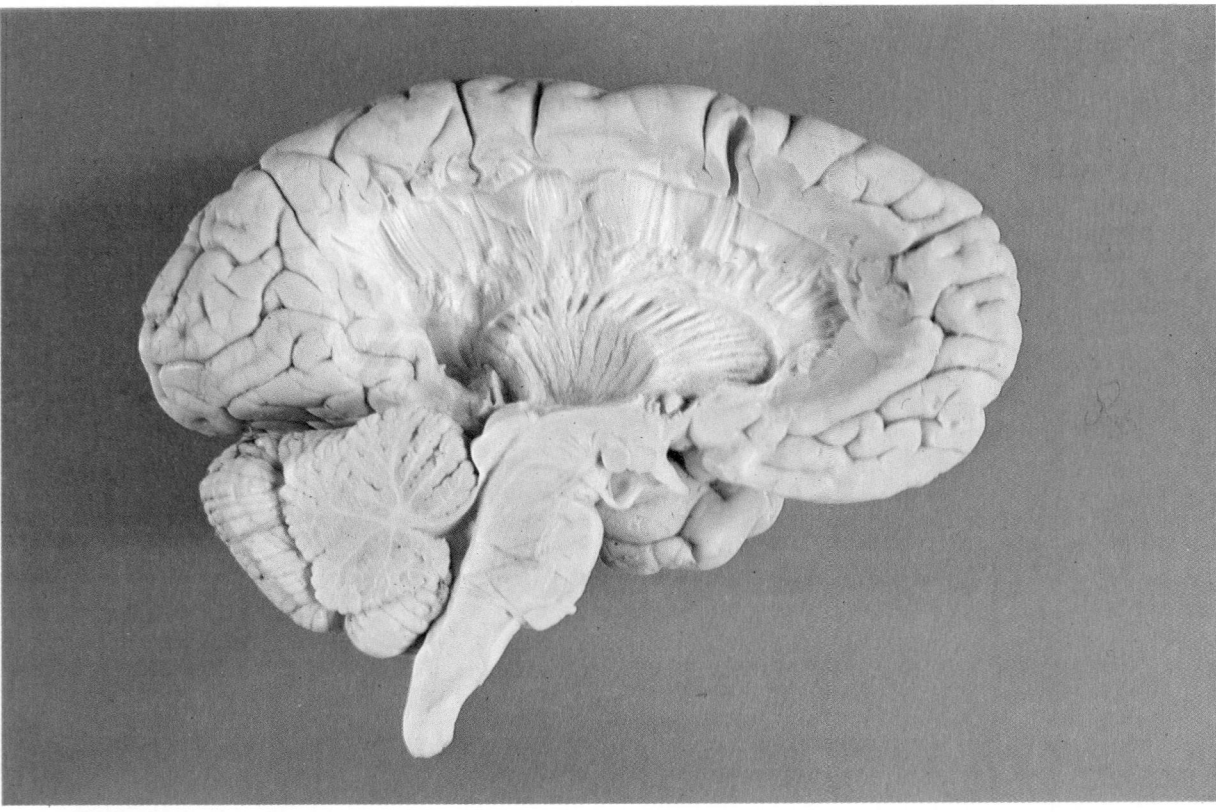

7.188 After median sagittal section of the brain, the left cerebral hemisphere has been dissected from its *medial* aspect to display the fibre bundles of the corona radiata and internal capsule. This entailed the removal of the cingulate gyrus and subjacent white matter, much of the paramedian corpus callosum and fornix, the dorsal thalamus and the head and body of the caudate nucleus. The oval depression previously occupied by the dorsal thalamus can clearly be seen within the curved depression left after removal of the caudate nucleus. Dissection by Andrew Seal, photography by Kevin Fitzpatrick, both of the Dept. of Anatomy, Guy's Hospital Medical School, London.

separated from it by the projection fibres in the corona radiata (vide infra). It is lateral to the caudate nucleus and therefore near the central part of the lateral ventricle. Posteriorly it fans out into the occipital and temporal lobes, lateral to the posterior and inferior ventricular cornua and the criss-crossing fibres of the tapetum.

These details are based on the results of blunt dissection which can only yield crude impressions of the general disposition of fibres en masse. Accurate knowledge of association fibres can only be established by experimental methods not yet applied to most cortical regions. Studies of visual areas (Le Gros Clark 1941, Zeki 1970, 1974) and the main somatic sensory areas (SmI) (see Jones & Powell 1968) suggest a high degree of specificity in such connections. (See also personal communication p. 1063, the comments on modular microarchitecture pp. 1052, 1059 and homogeneity of most cortical areas p. 1043.)

(3) **Projection fibres** connect the cerebral cortex with lower levels in the brain and spinal cord; they are *corticofugal* and *corticopetal*. Projection fibres converge from all directions to the corpus striatum (7.187), mostly (but not exclusively) medial to association fibres, and intersect commissural fibres of the corpus callosum and anterior commissure. At the periphery of the corpus striatum, they form the *corona radiata*; its *medial* aspect is, however, separated from the lateral ventricle by the fronto-occipital fasciculus and its lateral aspect covered by the superior longitudinal fasciculus. It is continuous with the internal capsule, a curved zone including almost all the projection fibres (vide infra).

The Internal Capsule

In horizontal cerebral section the internal capsule is seen as a broad white band, with a lateral concavity adapted to the convex medial aspect of the lentiform nucleus (7.189–194). It has an *anterior limb, genu, posterior limb, retrolentiform* and *sublentiform parts*. Both limbs are, of course, medial to the lentiform nucleus; medial to the anterior limb is the head of the caudate nucleus and medial to the posterior limb the thalamus. Fibres of the capsule continue to converge as they descend, the frontal fibres tending to pass posteromedially, temporal and occipital fibres anterolaterally. At a lower lentiform level, they are crossed by the optic tract and enter the midbrain, many (but *not* all) of the corticofugal fibres lying in the crus cerebri, where the frontal fibres are medially placed while temporal, parietal and occipital fibres are laterally situated.

The *anterior limb* contains *frontopontine fibres*, from the cortex in the frontal lobe, which synapse with cells in the nuclei pontis and the axons of which enter the opposite cerebellar hemisphere. Anterior thalamic radiations interconnect the medial and anterior thalamic nuclei and various hypothalamic nuclei and limbic structures with the frontal cortex.

The *genu* is usually regarded as containing *corticonuclear* fibres mainly from area 4 and terminating in the largely contralateral motor nuclei of cranial nerves. Anterior fibres of the superior thalamic radiation, between the thalamus and cortex, also extend into the genu.

The *posterior limb* includes the *corticospinal tract* in scattered bundles, the fibres concerned with the upper limb being anterior and followed by those for the trunk and lower limbs. This location of corticospinal fibres anterior in the posterior limb has been accepted since the observations of Charcot (1883) and Dejerine (1901); but more evidence from stereotactic lesions in man suggests that these fibres are *posterior* in the posterior limb. The evidence has been reviewed by Smith (1967) and Hanawayu & Young (1977). Other descending fibres include the frontopontine, particularly from areas 4 and 6, *corticorubral* fibres from the frontal lobe to the red nucleus and fibres from the globus pallidus in the subthalamic fasciculus. Most of the posterior limb also contains fibres of the superior thalamic radiation (the somesthetic radiation) ascending to the postcentral gyrus.

In the *retrolentiform part* are the parietopontine, *occipitopontine, occipito-collicular* and *occipito-tectal* and also the posterior thalamic radiation, including the optic radiation, and interconnections between the occipital and parietal lobes and caudal parts of the thalamus, especially the pulvinar.

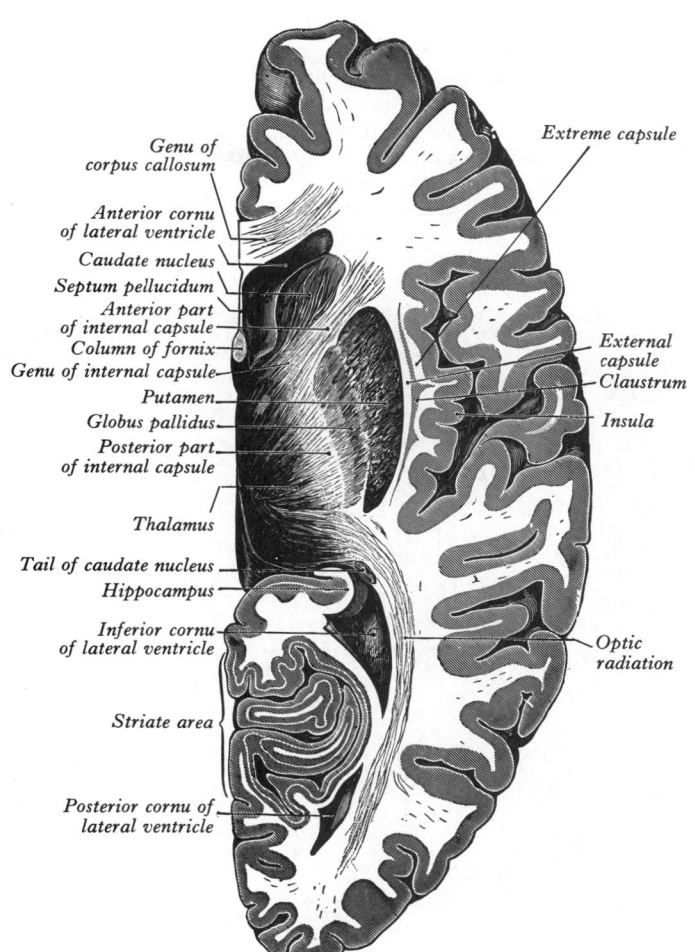

Genu of
corpus callosum
Anterior cornu
of lateral ventricle
Caudate nucleus
Septum pellucidum
Anterior part
of internal capsule
Column of fornix
Genu of internal capsule
Putamen
Globus pallidus
Posterior part
of internal capsule
Thalamus
Tail of caudate nucleus
Hippocampus
Inferior cornu
of lateral ventricle
Striate area
Posterior cornu of
lateral ventricle
Extreme capsule
External
capsule
Claustrum
Insula
Optic
radiation

7.189 Superior aspect of a horizontal section through the right cerebral hemisphere.

The *optic radiation* arises in the lateral geniculate body and sweeps back in the concavity between central and inferior cornual parts of the lateral ventricle, intimately related to the superolateral aspect of the ventricle's inferior cornu and the lateral aspect of the posterior cornu, separated from it by the tapetum.

The *sublentiform part* contains *temporopontine* and some parietopontine fibres, the acoustic radiation from the medial geniculate body to the superior temporal and transverse temporal gyri (areas 41 and 42) and a few fibres connecting the thalamus with the temporal lobe and insula. Fibres of the *acoustic radiation* sweep anterolaterally below and behind the lentiform nucleus to reach the cortex.

Connections between cortex and thalamus are detailed on p. 1000, corticohypothalamic connections on p. 1009 and corticostriate connections on p. 1076. Corticoreticular connections, corticonuclear fibres to all varieties of *sensory* integrating and relay nuclei in the brain stem and spinal laminae are mentioned throughout the text at numerous points. All the foregoing have reciprocal projections but their internal capsular location, in man, is as yet conjectural.

The Basal Nuclei

In each cerebral hemisphere is a group of large subcortical nuclei included in the general term *basal nuclei*. The structures included vary but most commonly are the amygdaloid complex, claustrum, caudate nucleus and lentiform nucleus (7.176, 189, 190, 191, 192). These nuclear masses are heterogeneous in structure, function and phylogenetic history, engendering a confusing terminology among different investigators, the structures being subdivided and regrouped in various ways; commonly used alternatives are mentioned *passim*. (For critical reviews see Mettler 1968, Webster 1978, Marsden 1985.)

The *amygdaloid body* (amygdaloid nuclear complex, amygdala, *archistriatum*) is discussed with the limbic system (p. 1029). The *claustrum*, a tenuous grey sheet surrounded by white laminae separating it from the insular and lentiform nucleus, is described on p. 1080.

The *caudate* and *lentiform nuclei* are commonly grouped as the *corpus striatum*, the lentiform being divisible into an internal *globus pallidus* and an external *putamen*, structurally distinct. The putamen resembles the caudate nucleus in structure; together they are the *neostriatum* or *striatum*. Thus *striato*fugal fibres leave the caudate nucleus or putamen while afferents, with the suffix 'striate' (e.g. thalamostriate), pass to the putamen or caudate nucleus, not to the *corpus* striatum as a whole. The terms will be thus used here but some neuroanatomists refer to the whole corpus when using *striato-*; sometimes their meaning is uncertain. The *globus pallidus* or *paleostriatum* is often named

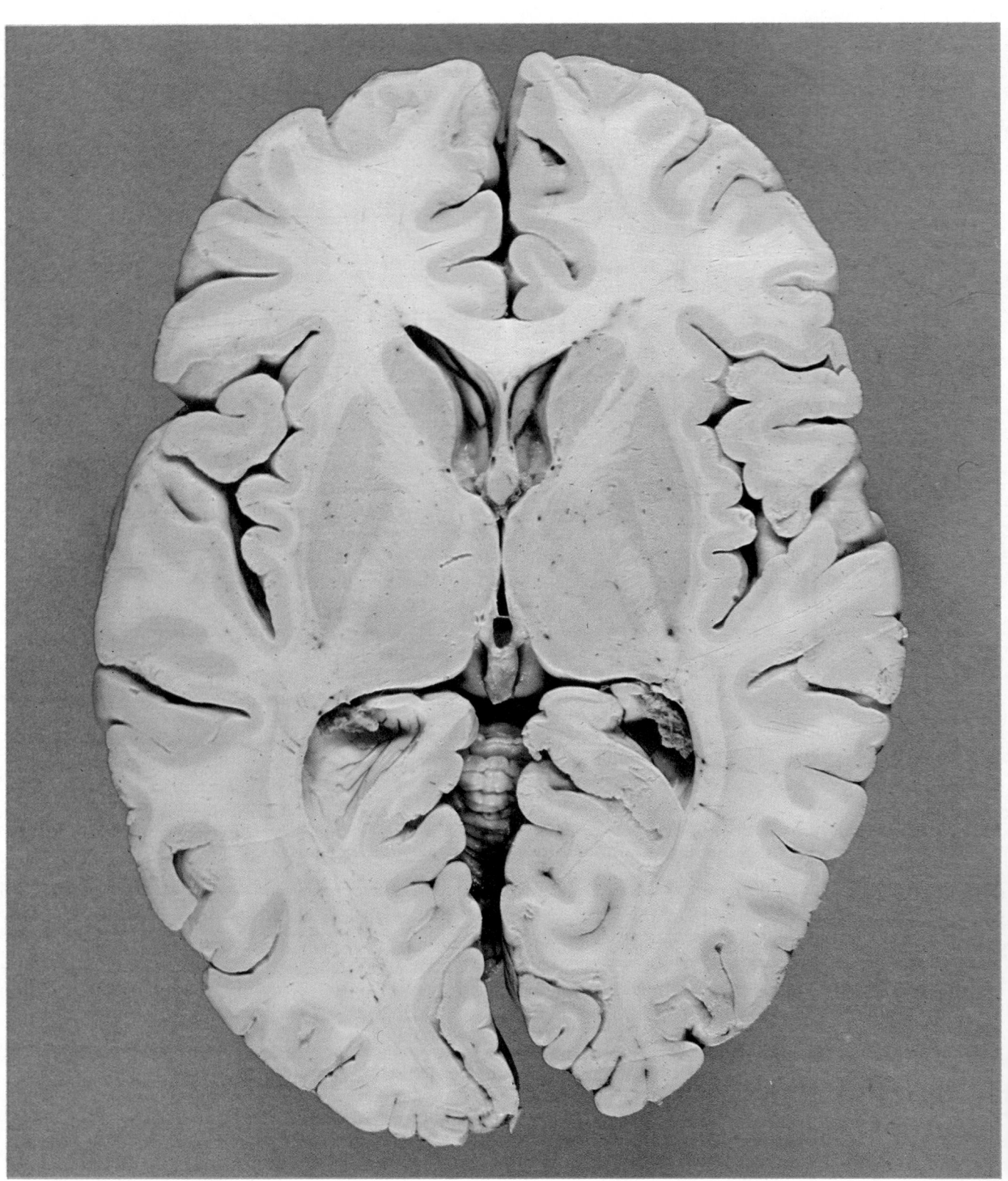

7.190 A horizontal section through the brain including the frontal and occipital poles of the cerebral hemispheres. Features appearing in this section are discussed at many points throughout the text. For appropriate labelling compare with 7.189. Dissection by E L Rees, photography by Kevin Fitzpatrick, both of the Dept. of Anatomy, Guy's Hospital Medical School, London.

the *pallidum*; hence such terms as pallidal afferents or pallidofugal fibres.

TOPOGRAPHY OF THE CORPUS STRIATUM

The caudate nucleus is an arcuate mass in the *floor* of the anterior cornu and central part of the lateral ventricle and the *roof* of its inferior cornu (7.182, 188, 189, 190, 192). Its anterosuperior end is the massive *head*, narrowing at the interventricular foramen into the arched *body* which continues to curve and tapers into the ventrally directed *tail*, merging at its apex with the amygdaloid body. Its whole ventricular aspect is covered by ependyma, while in the ventricle's central part and inferior cornu the stria terminalis lies along the medial border of the caudate nucleus. In the ventricle's central part the stria accompanies the thalamostriate vein in a groove between the caudate nucleus and the thalamus; in the inferior cornu it is the upper border of the choroid fissure. In the lateral ventricle's anterior cornu and central part, the lateral *margin* of the caudate nucleus is related to the corpus callosum, separated by the fronto-occipital arcuate bundle (p. 1072) and subcallosal fasciculus (p. 1076); its lateral *surface* is flat and adjoins the internal capsule. Anteriorly the head of the caudate nucleus is fused with the putamen just above the anterior perforated substance; above the fusion strands of grey matter traverse the anterior limb of the internal capsule between the putamen and caudate head; it is the striped appearance here which engendered the term *striatum*. In the inferior cornu the tail curves forwards behind, below and then lateral to the internal capsular fibres entering the crus cerebri and these separate the caudate tail from the thalamus which lies superomedial. Above, the sublentiform part of the internal capsule and fibres from the *external* capsule separate the tail from the globus pallidus.

The lentiform nucleus is like a biconvex lens, its medial surface having a more pronounced curvature (7.130, 181, 182, 189, 190, 192). In section it has two parts, differing in colour; the larger, lateral, darker part is the *putamen*, the smaller, medial part of a lighter tint is the *globus pallidus*. The nucleus is buried in the hemisphere, covered laterally by a thin layer of white matter, the *external capsule*, lateral to which is the tenuous grey matter of the *claustrum*, between it and the insular subcortical white matter (the *extreme capsule*). Medial to the lentiform nucleus is the internal capsule, separating it from the thalamus behind and the caudate head in front. Anterior, superior and posterior to the nucleus is the corona radiata. Its inferior part is grooved by the anterior commissure passing posterolaterally into the temporal lobe (7.182–192); anteriorly its putamen is continuous with the caudate's head. Anterior to the groove there is irregular fusion of grey matter of the corpus striatum with that associated with the anterior perforated substance; lateral striate arteries, entering the brain here, run laterally and turn up close to the lateral aspect of the lentiform nucleus, before piercing it. The lentiform nucleus is above the lateral ventricle's inferior cornu and separated from it by fibres of the external capsule passing medially to the subthalamic region and by the sublentiform part of the internal capsule (p. 1073), the tail of the caudate nucleus and the stria terminalis. More anteriorly, it is separated from the amygdaloid body by the ansa peduncularis.

STRUCTURE OF THE CORPUS STRIATUM

The caudate nucleus and putamen, similar in structure, are highly cellular, well-vascularized and permeated by small bundles of finely myelinated or non-myelinated, small-diameter fibres; because of this, these regions when freshly sectioned are pinkish-grey, contrasting with the pallor of the globus pallidus, which is encapsulated and traversed by numerous, thick, heavily myelinated fibres.

Neurons of the caudate nucleus and putamen are mainly small multipolar cells with round, triangular or fusiform somata, mixed with a small fraction of large multipolar cells; the ratio is about 20:1. The small neurons are considered to be receptive and associative interneurons, receiving the synaptic terminals of many striatal afferents. The large neurons have large and spherical or ovoid dendritic trees; their axons are one source of striatal

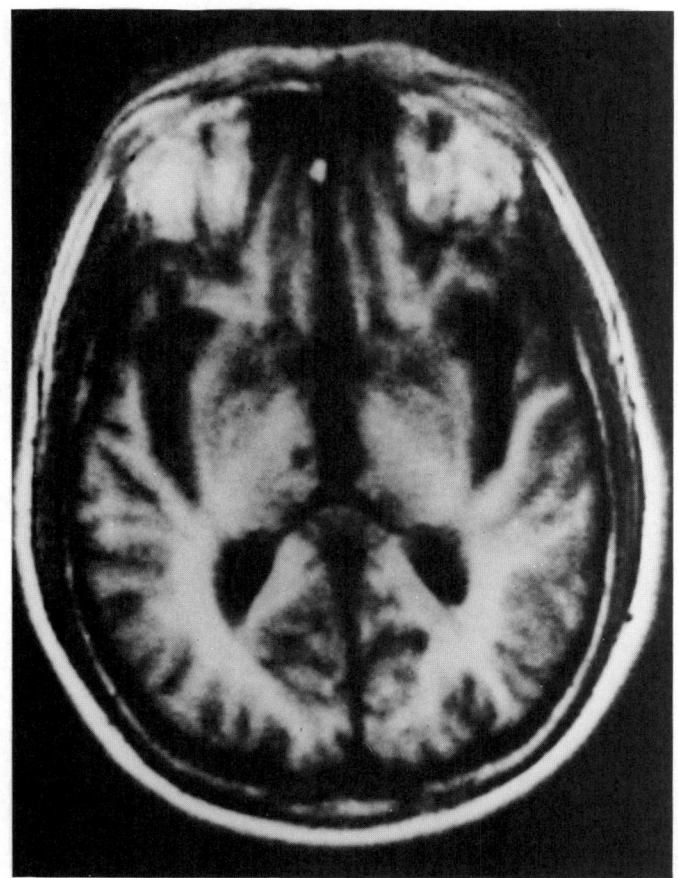

7.191A Horizontal magnetic resonance scan of the head at the level of the third ventricle. Supplied by Philips Medical Systems; photography by Sarah Smith.

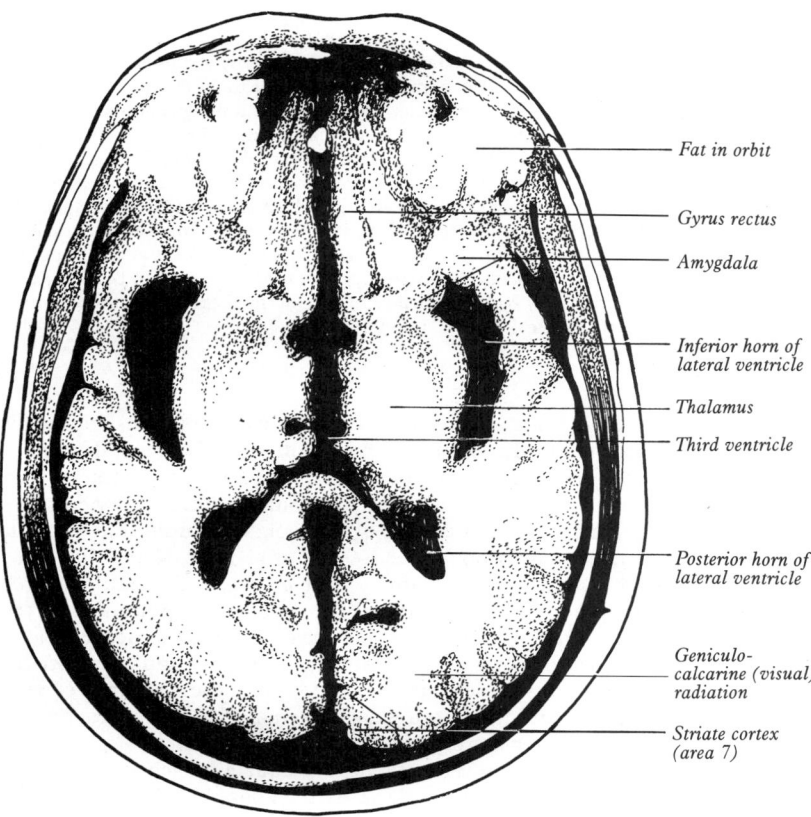

Fat in orbit

Gyrus rectus

Amygdala

Inferior horn of lateral ventricle

Thalamus

Third ventricle

Posterior horn of lateral ventricle

Geniculo-calcarine (visual) radiation

Striate cortex (area 7)

7.191B Diagram of 7.191A. Illustration: J Halstead.

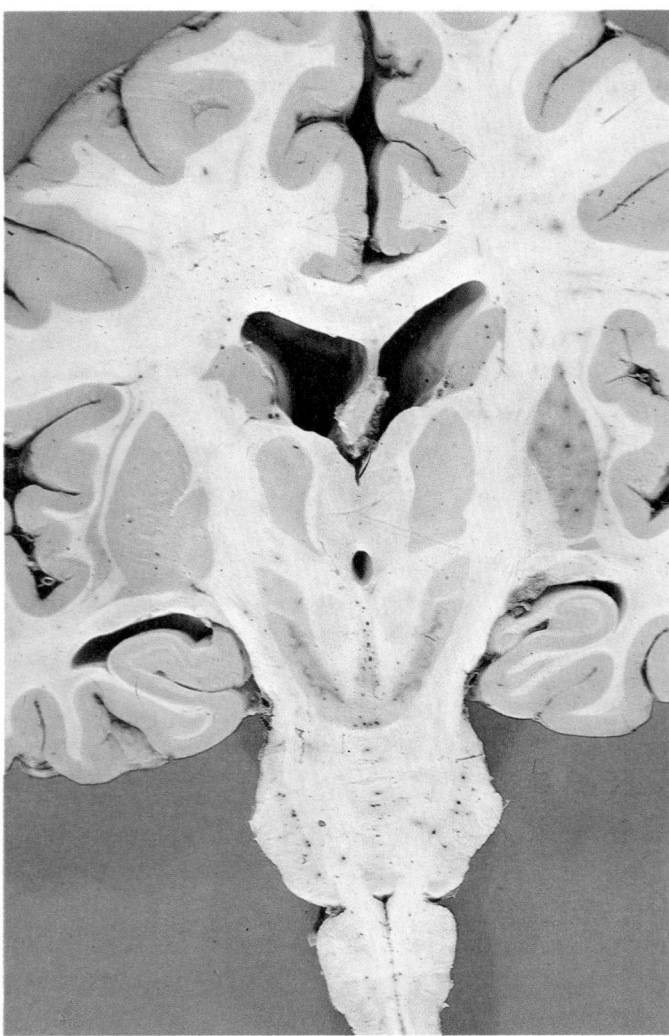

7.192   The central area of the ventral part of the oblique coronal section of the brain shown in 7.122, photographed at higher magnification to show some structural features in greater detail; compare with 7.126, 197 for appropriate labelling. Note in particular: (1) the anterior, medial and lateral parts of the dorsal thalamus, separated by the internal medullary laminae; (2) the relation of the caudate nucleus to the anterior and inferior cornua of the lateral ventricle; (3) the lentiform nucleus, divided into an external putamen and an internal globus pallidus; the latter again divided into internal and external parts; (4) the internal capsule, external capsule, claustrum, extreme capsule and insular cortex; (5) the sectioned profiles of the subthalamic and red nuclei and the substantia nigra; (6) the hippocampus projecting into the floor of the inferior cornu of the lateral ventricle. Other structural features on this section are discussed at many points throughout the text. Compare also with 7.190. Dissection by E L Rees, photography by Kevin Fitzpatrick, both of the Dept. of Anatomy, Guy's Hospital Medical School, London.

efferents (vide infra) but some smaller neurons probably contribute to the striatal outflow. Although the general connections of the striatum with other regions are becoming clearer (vide infra), little is known in detail of its neurons, intrinsic connections and synapses. From the use of the Golgi technique and electron microscopy in various species (Fox et al 1966, Kemp 1968a, b, 1970), some general features have emerged. Striatal dendrites bear spines and axodendritic synapses at the spines and interspinous areas between them, and axosomatic synapses have also been identified; both asymmetrical (Type I) and symmetrical (Type II) synapses occur. Lesions in the cerebral cortex, thalamus or midbrain cause degeneration of asymmetrical synapses, possibly excitatory, on dendrites and (after cortical lesions) also cell somata; but lesions in the caudate nucleus affect type I *and* type II synapses. The latter, possibly inhibitory, are distributed on dendrites, somata and initial axonal segments. Axons of all the striatal neurons studied have collateral branches;

type II synaptic terminals may be derived not only from intrinsic interneurons but also from collaterals of neurons projecting to other regions. Histopharmacological studies (Hokfelt 1968) have revealed that in caudate nuclei of rats many synaptic terminals contain 'granulated' vesicles typical of aminergic neurons, perhaps stores of *dopamine*. Other synapses contain vesicles typical of cholinergic terminals.

The globus pallidus has a scattered population of large multipolar neurons resembling lower motor neurons and the large cells of the substantia nigra. Primary dendrites branch infrequently and have few dendritic spines, at which synapses mainly occur, mostly of type II but with a few of type I; scattered between these dendrites are occasional axosomatic synapses of both types. Lesions in the caudate nucleus produce numerous degenerating symmetrical (type II) terminals at dendrites and somata in the globus pallidus (and substantia nigra). Axons of large pallidal neurons provide a profuse, well-myelinated, pallidofugal system (vide infra). The nucleus is separated by the *external medullary lamina* from the medial aspect of the putamen; an *internal medullary lamina* divides it into smaller medial and lateral parts (*pallidum I and II*).

CONNECTIONS OF THE CORPUS STRIATUM

Summary (see 7.197). The neostriatum (putamen and caudate nucleus) is the main striatal receiving station, which then projects to the globus pallidus; the latter provides the main efferent projection but some efferent paths leave the neostriatum directly (and the globus pallidus also *receives* other afferents).

The afferent connections to the striatum are mainly from the cerebral cortex, thalamus and substantia nigra. Although long in doubt, it is now firmly established that *corticostriate fibres* form a widespread system *topically organized* and converging from almost *all parts of the cerebral cortex* to the caudate nucleus and putamen. Experimental animals have included rats, rabbits, cats and monkeys studied by degeneration techniques of Glees, Nauta and by ultrastructural methods (Glees 1944, Whitlock & Nauta 1956, Webster 1965, Carman et al 1963, Carman et al 1965, Kemp & Powell 1970). Jones et al (1977), by retrograde horseradish peroxidase technique, have shown that corticostriate projection in monkeys is from smaller pyramidal cells of lamina V, scattered sparsely but widely in sensorimotor areas and elsewhere in the frontal and parietal regions, in the case of *ipsilateral* fibres, but only in areas 4, 6, and 8 for contralateral projection. Yeterian & Hoesen (1978), by autoradiographic techniques also in monkeys, obtained different results; following intracortical injection of tritiated leucine or proline, they revealed an interesting phenomenon: that areas with marked strong reciprocal corticocortical connections both projected to common sites in the caudate nucelus. Each part of the cerebral cortex projects to a specific site in the caudate–putamen complex, though some overlap occurs. Most projections are from the ipsilateral cortex alone but some striatal sites receive projections from restricted regions of sensorimotor cortices in *both* hemispheres. Since details in these different mammals vary, they will not be further detailed; obviously they cannot simply be extrapolated to the human brain but similar connections probably exist. The most abundant projections are from the sensorimotor cortex, the least from the visual cortex. Corticofugal fibres reach the striatum through both internal and external capsules and from the temporal lobe by sublenticular routes; some direct and crossed corticocaudate fibres run in the *subcallosal fasciculus* with the fronto-occipital arcuate bundle. It is not clear what proportion of corticostriate fibres has the principal terminations in these nuclei, or whether substantial numbers are collateral branches of other corticofugal projections, e.g. the corticospinal tract. Corticostriate fibres form type I synapses with dendrites and somata of striatal neurons and are probably excitatory.

Thalamostriate fibres form another abundant projection to the striatum, derived from the nucleus centromedianus, other intralaminar and midline nuclei and the nucleus medialis dorsalis, some passing directly to end in the caudate nucleus; others traverse the caudate or skirt it, to the internal capsule, passing between its fibres to the putamen. The details of these connections are not all certain. The nucleus centromedianus is said to

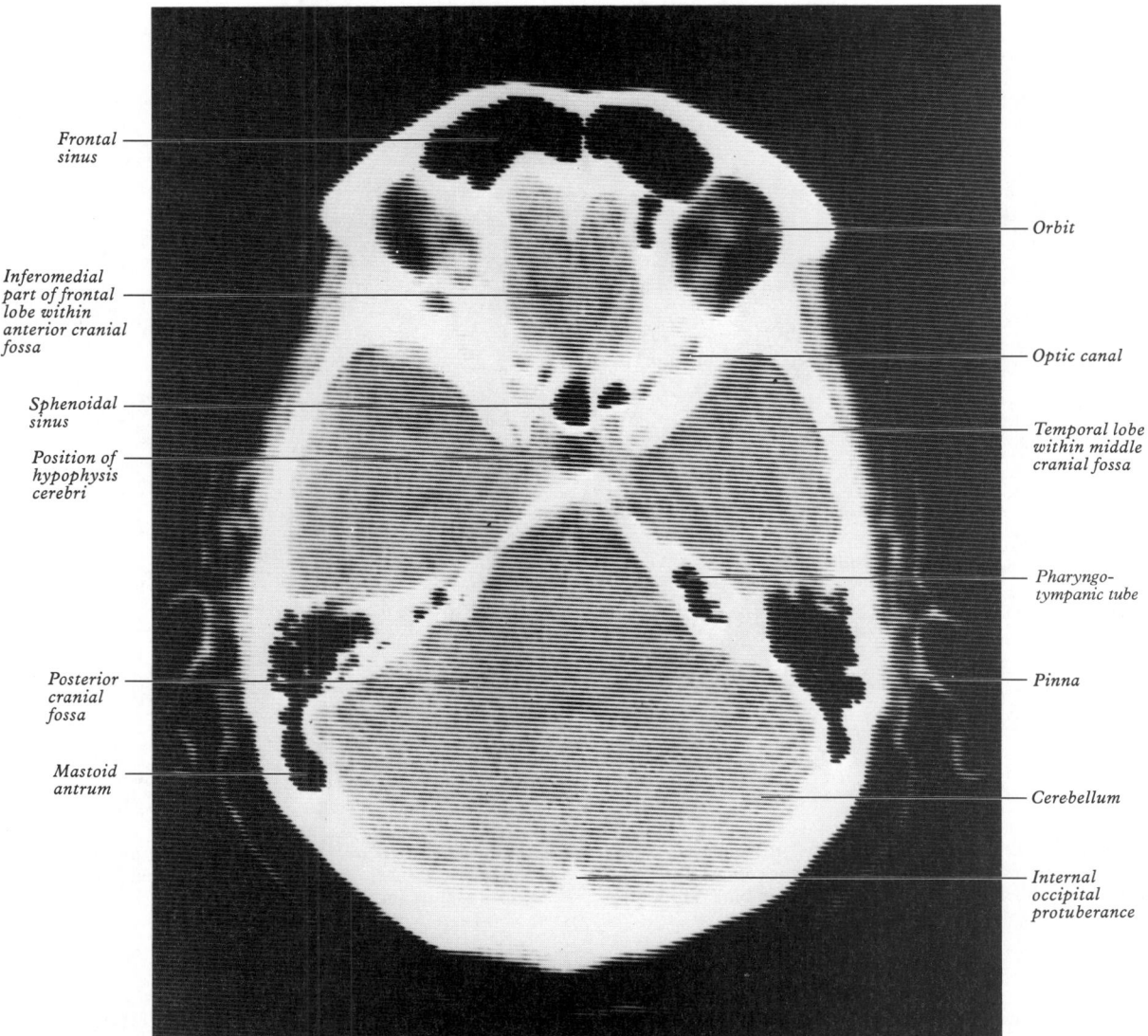

Frontal
sinus

Inferomedial
part of frontal
lobe within
anterior cranial
fossa

Sphenoidal
sinus

Position of
hypophysis
cerebri

Posterior
cranial
fossa

Mastoid
antrum

Orbit

Optic canal

Temporal lobe
within middle
cranial fossa

Pharyngo-
tympanic tube

Pinna

Cerebellum

Internal
occipital
protuberance

7.193   Horizontal computed tomogram through the head at the level of
the orbits and middle ears. (Kindly supplied by Dr Shaun Gallagher,
Guy's Hospital; photography by Miss Sarah Smith.)

project to many parts of the corpus striatum. Studies in monkeys
(Powell & Cowan 1956) showed its projection to be topically organized and confined to the putamen, while others (Mehler 1966)
claimed that its projection was to a band including the whole
mediolateral extent of both putamen and caudate nucleus.
Intralaminar and dorsomedial thalamic nuclei are thought to
project largely to the caudate nucleus.

Nigrostriate fibres, long considered by some to provide an important afferent system to the striatum and globus pallidus, have
become firmly established and prominent in relation to theories of
the genesis of Parkinsonian tremor (Calne 1970). Increasing
evidence, from electrophysiology (Purpura et al 1967, Feltz &
Mackenzie 1969, Connor 1968), neuroanatomy (Adinolfi & Pappas 1968, Nauta & Mehler 1969) and neuropharmacology (Calne
1970) not only confirmed such a pathway but suggested that its
neurons, at least some, utilize *dopamine* as transmitter (via infra).

Nigrostriate fibres from neurons in the partes compacta et
reticularis of the substantia nigra ascend through the caudal subthalamus to the internal capsule, where their bundles interlace
with capsular fibres producing a comb-like appearance, whence
the term *comb bundle*. Some fibres reach the caudate nucleus, the
remainder continuing to the putamen and globus pallidus. Carpenter & Peter (1972) have identified degenerating terminals in
parts of the caudate nucleus and putamen following lesions of the
pars compacta of the substantia nigra in monkeys, describing a
topological nigrocaudate correlation. Further afferents may reach

the striatum from the subthalamic nucleus and nearby neuronal
aggregate.

Amygdalostriate fibres were demonstrated to be rich in the rat
(Kelley et al 1982) and a 'limbic'/'non-limbic' mosaic division of
the neostriatum was mooted (Gerfen 1984). Nieuwenhuys (1986)
pointed out that the 'limbic' part of the neostriatum contained at
least 17 varieties of neuromediator and included it in his paracrine
core of the neuraxis (p. 1039).

Striatofugal connections are mainly to the *globus pallidus*
but *striatonigral* and *striatothalamic* fibres have been described,
retracing the pathways just detailed to reach the substantia nigra
and thalamus. Both caudate nucleus and putamen project in a
topical manner to neurons of the globus pallidus. The lateral
putamen projects only to the external pallidal segment, its more
medial part and the caudate nucleus projecting either to both
pallidal segments or to the internal alone. Other striatofugal connections have been described, including projections to the subthalamic nucleus and restricted parts of the inferior olivary
nucleus; but these are uncertain.

Afferent connections to the globus pallidus are principally
topically organized *striatopallidal fibres* from the putamen and
caudate nucleus, as noted. Other afferents arrive from the subthalamic nucleus, substantia nigra, thalamus and cerebral cortex.
Subthalamic fibres reach mainly pallidum I in the *subthalamic
fasciculus* (p. 1007). *Nigropallidal fibres* have been described as
part of the comb bundle mentioned above; the neurons forming

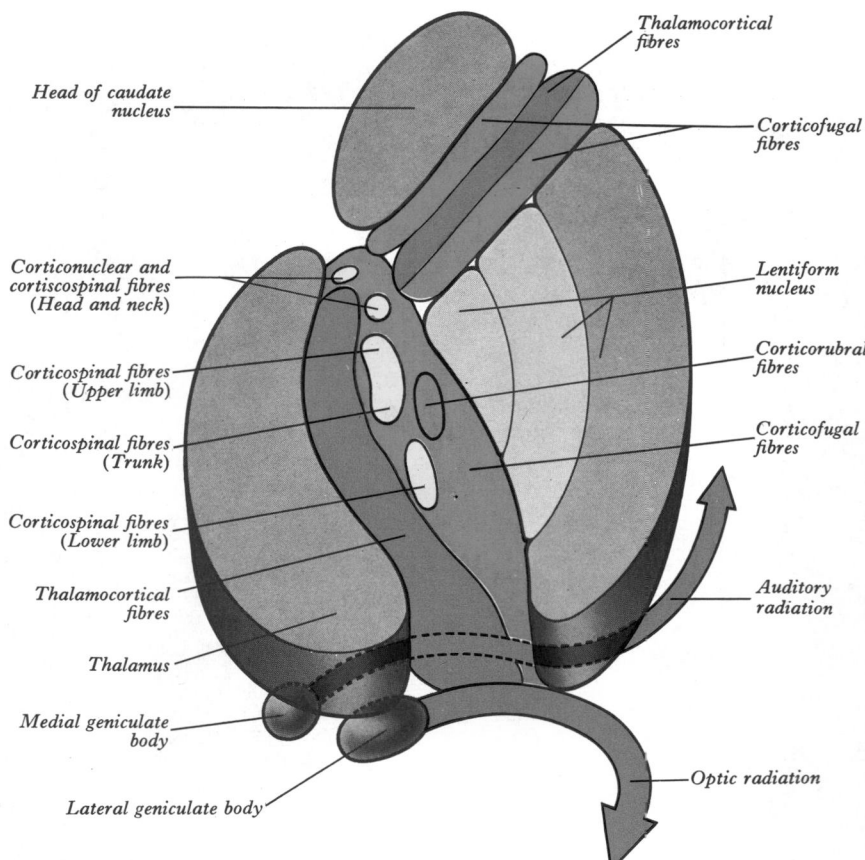

Head of caudate nucleus

Thalamocortical fibres

Corticofugal fibres

Corticonuclear and cortiscospinal fibres (Head and neck)

Corticospinal fibres (Upper limb)

Corticospinal fibres (Trunk)

Corticospinal fibres (Lower limb)

Thalamocortical fibres

Thalamus

Medial geniculate body

Lateral geniculate body

Lentiform nucleus

Corticorubral fibres

Corticofugal fibres

Auditory radiation

Optic radiation

7.194 Diagram of the main components of the internal capsule. Descending motor fibres are shown in yellow, corticofugal fibres to the thalamus and pons, etc. in red and ascending fibres in blue. From Strong & Elwyn's *Human Neuroanatomy* and Kretschmann (1988). However, see also in text references concerning alternative views about the position of the corticospinal and other fibres.

this tract are dopaminergic. Whether nigrostriate or nigropallidal fibres predominate in *human* brains is uncertain. In addition to the extensive corticostriate projection mentioned above, fibres are said to accompany these from many cortical areas, but ending in the globus pallidus. *Thalamopallidal fibres* may also arrive from intralaminar, centromedian and dorsomedial thalamic nuclei. These somewhat variable results have been obtained by examination of several species, sometimes using different methods, and details or even the existence of some connections in human brains requires confirmation.

The pallidofugal system is by far the quantitatively largest outflow; its fibres are thick, well myelinated and form a complex group of paths, the main ones being: (1) *ansa lenticularis*, (2) *fasciculus lenticularis*, (3) *fasciculus thalamicus*, (4) *fasciculus subthalamicus*, and (5) *descending fibres*. These were described and illustrated (7.130) with the subthalamus (p. 1005). Main destinations are: (1) the *thalamus*, mainly the *nucleus ventralis anterior*, with smaller contributions to the nuclei ventralis intermedius and centromedianus; (2) the *subthalamic nucleus* and other subthalamic centres, including the zona incerta, entopeduncular nucleus and nucleus of the prerubral field; (3) the *substantia nigra*; (4) the *red nucleus*; (5) the *midbrain reticular formation*; (6) the *inferior olivary nucleus*. Illustration 7.197 summarizes some of the main connections of the basal nuclei based on an analysis by Webster (1975).

The following points need emphasis. The external pallidal segment projects to the internal, where most pallidofugal fibres start. Afferents to the subthalamic nucleus, however, arise mainly in the external segment. A *pallidohypothalamic tract* has long appeared in accounts of the corpus striatum; studies using newer degeneration techniques yield no evidence of such a connection. The tract so-designated pursues an aberrant course around the column of the fornix, most fibres, however, rejoining the main pallidal outflow.

SUMMARY AND FUNCTIONAL STUDIES

The corpus striatum and associated regions have exceedingly complex interconnections but the details and our understanding of their functions remain elementary. Broadly, information *converges* to the corpus striatum from most of the cerebral cortex, thalamus, subthalamus and some brain-stem centres. 'Processed' information then *diverges* to the thalamus, subthalamus and various brain-stem centres (vide supra).

The main striatal outflow, as part of the 'extrapyramidal motor system', was long considered to *descend* through various polysynaptic paths to the lower motor centres. While reticulospinal and rubrospinal tracts undoubtedly carry information partly from the corpus striatum, it is clear that the major striatal outflow is to the thalamic *nucleus ventralis anterior* (p. 1003) where, after integration with other channels, *ascending* paths radiate to the motor and premotor cortex (areas 4 and 6). Large reciprocal connections with the *subthalamic nucleus* and *substantia nigra* also exist.

Many organic nervous diseases have long been known to affect, variably, different parts of the corpus striatum, their associated nuclei and tracts. These exhibit, in different forms and combinations, the following elements: (1) *disturbances of muscle 'tone'*, i.e. resistance to stretch, which is sometimes reduced; more commonly *rigidity* is present; (2) *loss of automatic associated movements* such as arm-swinging, facial expressions, etc.; (3) *unwanted movements*, uncontrollable and purposeless, which may be *choreiform*, *athetoid* or *ballistic* (p. 1006), or *tremor* due to alternating contractions of opposing muscles; tremor is usually 'static', i.e. when the limb is at rest, but occasionally *intention tremor* occurs, as in some forms of cerebellar dysfunction (p. 978). Attempts to link various combinations with disease in specific locations have had limited success and experimentation has been largely uninformative.

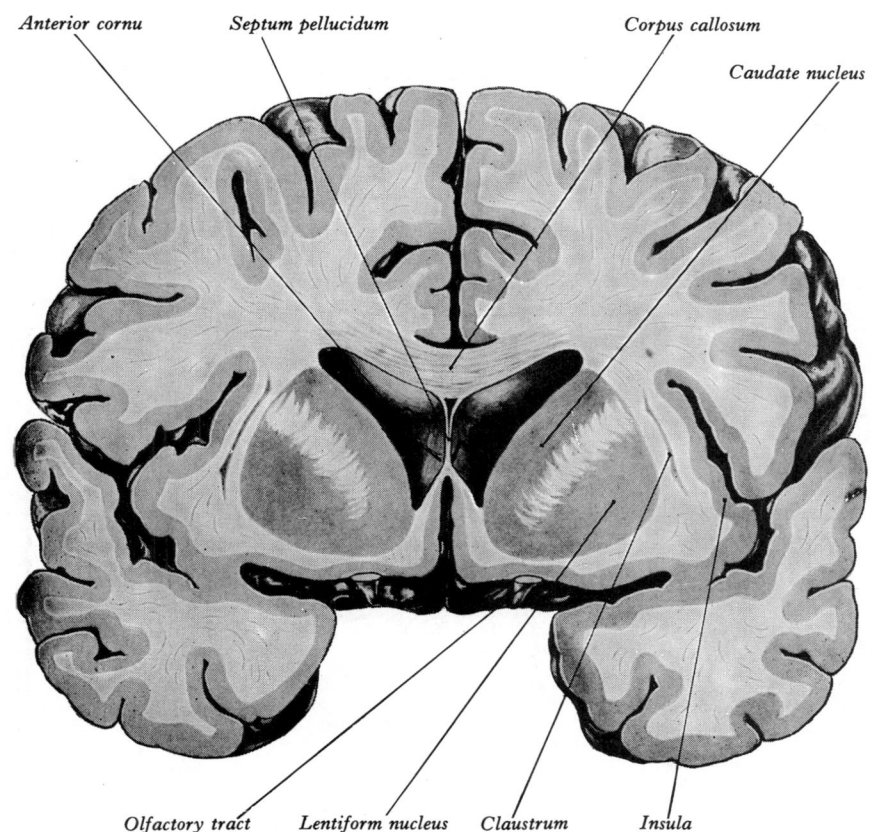

Anterior cornu   Septum pellucidum   Corpus callosum

Caudate nucleus

Olfactory tract   Lentiform nucleus   Claustrum   Insula

7.195 Posterior aspect of a coronal section through the anterior cornua of the lateral ventricles.

Removal of the putamen, caudate nucleus or globus pallidus often causes little obvious motor change, provided the lesions do not affect adjacent structures. Complete bilateral ablation of the globus pallidus in monkeys, however, results in poverty of movement and reduction in manipulative skills.

Rapid stimulation of the caudate nucleus in unanaesthetized animals sometimes elicits movements of the head and limbs, with some indications of a somatotopic nuclear pattern. Low-frequency or chemical stimulation at different sites in the corpus striatum usually inhibits motor responses; it may induce long periods of immobility, inhibition of cortically induced movements and 'arrest reactions' in unanaesthetized animals. Confirmation has been obtained by 'unit' recording from the motor cortex and ventrolateral thalamus: inhibition of unit activity followed caudate stimulation. Inhibition of the responses established by learning processes follows caudate nuclear stimulation.

Surgical ablation of the globus pallidus or ventrolateral thalamus has been performed, often with some success, to diminish contralateral rigidity and tremor. Thalamic ablations seem more effective in the relief of tremor, while pallidal lesions have greater effect in reducing rigidity. In addition to their therapeutic value, such manoeuvres provide material for examination and also allow electro-physiological recording. Most interesting has been the identification of neurons in the ventrolateral thalamus in cases of Parkinsonian tremor, which display rhythmical discharge of impulses, their frequency corresponding to the tremor (Guiot et al 1962, Guiot et al 1964, Jasper 1966, Bates 1969). Similar phenomena have been induced by placing brainstem lesions in monkeys. Intensive study, anatomical, physiological, behavioural and pharmacological, was and is in progress attempting to discover a causal mechanism for this rhythmic neuronal discharge. The demonstration of an *inhibitory dopaminergic nigrostriate pathway*, probably involved in some cases of Parkinsonian tremor (Calne 1970), was highly interesting. Of the various mechanisms initially proposed, it was considered most likely that normal striatal function depends on a balance between the inhibitory afferents just mentioned and cholinergic excitatory aff-

erents. Reduced activity in the inhibitory path leads to excessive excitatory output from the pallidum, leading to oscillating bursts of activity in ventrolateral thalamic neurons which, through their cortical radiations, may generate rhythmical activity in various corticofugal tracts to the lower motor centres. This implied earlier

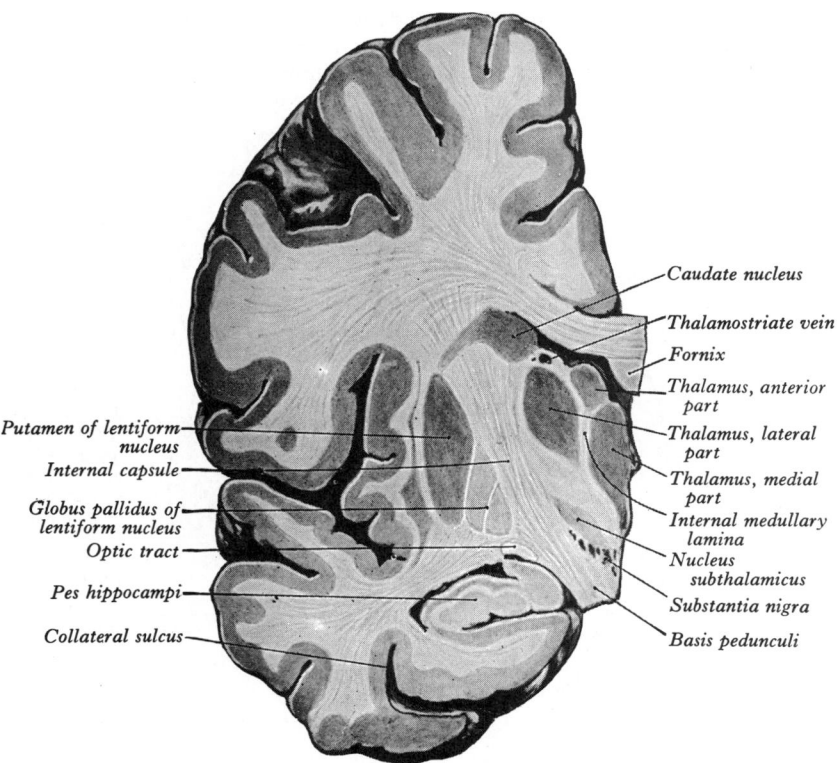

Caudate nucleus

Thalamostriate vein

Fornix

Thalamus, anterior part

Thalamus, lateral part

Thalamus, medial part

Internal medullary lamina

Nucleus subthalamicus

Substantia nigra

Basis pedunculi

Putamen of lentiform nucleus

Internal capsule

Globus pallidus of lentiform nucleus

Optic tract

Pes hippocampi

Collateral sulcus

7.196 Anterior aspect of a coronal section through the right cerebral hemisphere.

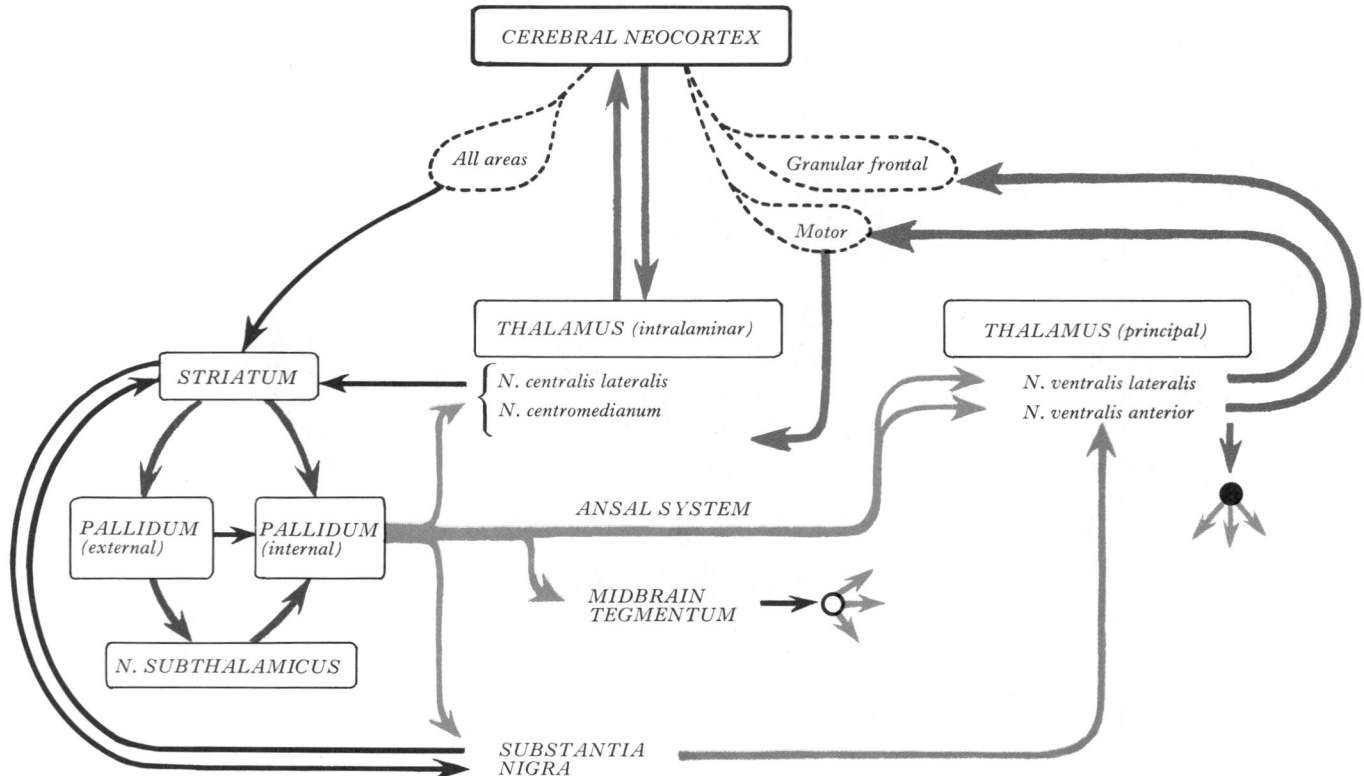

7.197 A scheme of the principal connections of the basal nuclei. At the points indicated the dorsal thalamic nucleus ventralis anterior projects on to several subcortical regions including the thalamic nucleus ventralis lateralis. The midbrain tegmentum probably establishes polysynaptic relationships with the spinal cord grey matter and the thalamic intralaminar nuclei. Colour has been used to assist visualization of the individual pathways. See text for further comment. Based on data summarized by K Webster (1975) of the Dept. of Anatomy, King's College, University of London, by permission of the author.

that dopamine is a normal striatal neurotransmitter. A body of evidence shows that dopamine levels in the striatum are reduced in many cases of Parkinsonism, paralleled by the encouraging clinical responses to replacement therapy in many patients given L-dopa, a precursor of dopamine. The foregoing account is now outdated, but retained for historical reasons. See Nieuwenhuys (1985) for bibliography and review of the complex chemical architecture of the neostriatum (17 neuromediators).

Despite this rapid progress in clinicopharmacological matters, the striatal role in normal behaviour remains conjectural, except

for the established view that the corpus callosum is involved in motor control, with emphasis displaced from a descending path to cortical control by the striatum. (For a review of the structure and function of basal nuclei and an extensive bibliography consult Webster 1975.)

The claustrum is a thin sheet of grey matter co-extensive with the insula and putamen, from which it is separated by the external capsule. Thickest below and in front, it becomes continuous here with the anterior perforated substance, amygdala, and prepiriform cortex. It is regarded by some as belonging to the corpus striatum, by others as a detached part of the insular cortex; but detailed studies suggest that it may have at least two structurally and functionally distinct zones, the *'insular'* claustrum and the *'temporal'* or *'prepiriform'* claustrum. In experimental animals the insular part has been shown to have reciprocal, topically organized *corticoclaustral* and *claustrocortical* connections with many regions of the neocortex. Its connections and functional significance are unknown in the human brain.

The external capsule, a thin layer of white matter interposed between the lentiform nucleus and claustrum, derives its fibres from the insula's frontoparietal operculum; crossing lateral to the lentiform nucleus, these turn medially below it and the ansa lenticularis. Their subthalamic connections are uncertain. Some fibres of the anterior commissure are believed to traverse the external capsule.

The Choroid Plexus of the Lateral Ventricle

Projecting into the lateral ventricle at its medial side is a highly vascularized fringe of pia mater and ependyma (7.198, 199, 200), the choroid plexus of the lateral ventricle, part of a larger structure, the *tela choroidea*, described below. Its pial basis is invaginated during development (p. 189) along a linear region of the medial hemispheric wall where no nervous tissue develops; hence the pia directly contacts the ventricular *ependyma*, the two tissues

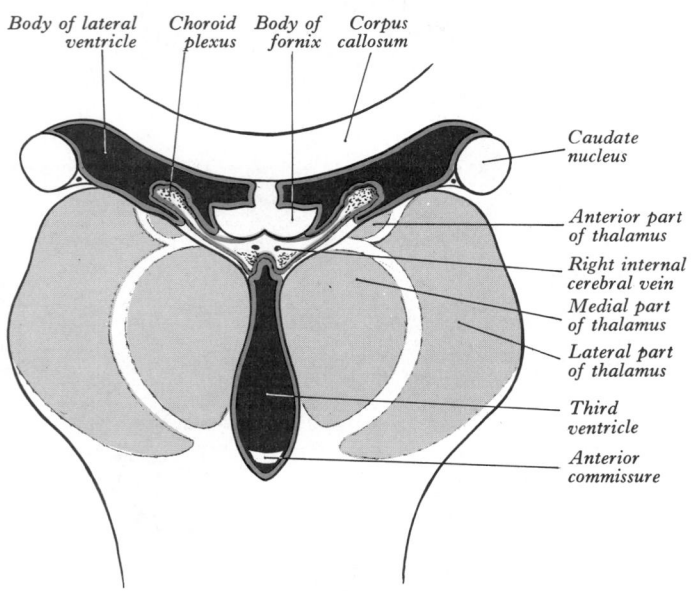

Body of lateral ventricle   Choroid plexus   Body of fornix   Corpus callosum

Caudate nucleus

Anterior part of thalamus

Right internal cerebral vein

Medial part of thalamus

Lateral part of thalamus

Third ventricle

Anterior commissure

7.198 Diagram of a coronal section through the lateral and third ventricles. The pia mater of the tela choroidea is shown in red and the ependyma in blue.

being fused to form the choroid plexus, which otherwise consists chiefly of small blood vessels, capillaries and nerve fibres. It extends anteriorly to the interventricular foramen, where it is continuous across the third ventricle with the plexus of the opposite ventricle. From the foramen it passes posteriorly in contact with the thalamus (7.144) curving round its posterior end into the ventricle's inferior cornu and reaching the pes hippocampi (7.185). When the plexus is torn away from its hemisphere, the line of invagination becomes the *choroid fissure*. Through the central part of the ventricle the fissure is between the fornix superiorly and the thalamus inferiorly (7.198) and in the inferior cornu between the stria terminalis above and the fimbria below (7.199). The fissure is the first groove to appear on the surface of the cerebrum (p. 189); in coronal sections of the brain at eight weeks the choroid fissure is already in contact with the lateral margin of the ependymal roof of the third ventricle and its overlying vascular pia mater. At this stage, before development of the commissures and expansion of the lamina terminalis, one pial layer extends over the third ventricle. The corpus callosum and body of the fornix expand posteriorly above the choroid fissure, carrying a layer of pia on their inferior surface, which overlies the third ventricular pial layer, and fusing with it to form the central part of the tela choroidea; the lateral extensions of this, also double layered, invaginate the choroidal fissures and form choroid plexuses in the lateral ventricles. Posteriorly the two pial layers separate, the inferior (original) layer following the ventricular roof to the pineal gland and tectum, the superior cleaving to the corpus callosum and passing round the splenium to the superior callosal surface.

As described above, two layers of pia mater fuse to form the *tela choroidea of the third ventricle* (as usually termed, though the choroid plexuses of the *lateral* ventricles are extensions of it). From above it is triangular, with a rounded apex between the interventricular foramina, often indented by the anterior columns of the fornix (7.200). Its lateral edges are irregular containing choroid vascular fringes. At the posterior, basal angles of the tela these fringes continue and curve on into the

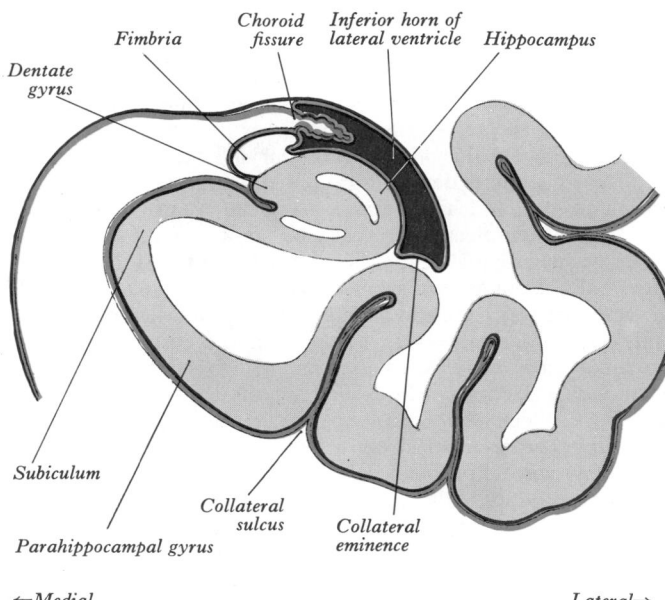

7.199   Diagram of a coronal section through the inferior cornu of the lateral ventricle. The pia mater is shown in red and the ependyma in blue.

inferior cornua, while centrally the pial layers depart from each other here as already detailed. When the tela choroidea is removed a transverse slit is left between the splenium and the junction of the ventricular roof with the tectum, the *transverse fissure* (not a cerebral fissure in the ordinary sense, for cortex is not involved). It marks the posterior limit of the *extracerebral space* enclosed by the posterior extensions of the corpus callosum above the third ventricle; in this, enclosed between the two layers of pia mater, are the roots of the choroid plexuses of the third ventricle (p. 1019) and of the lateral ventricles.

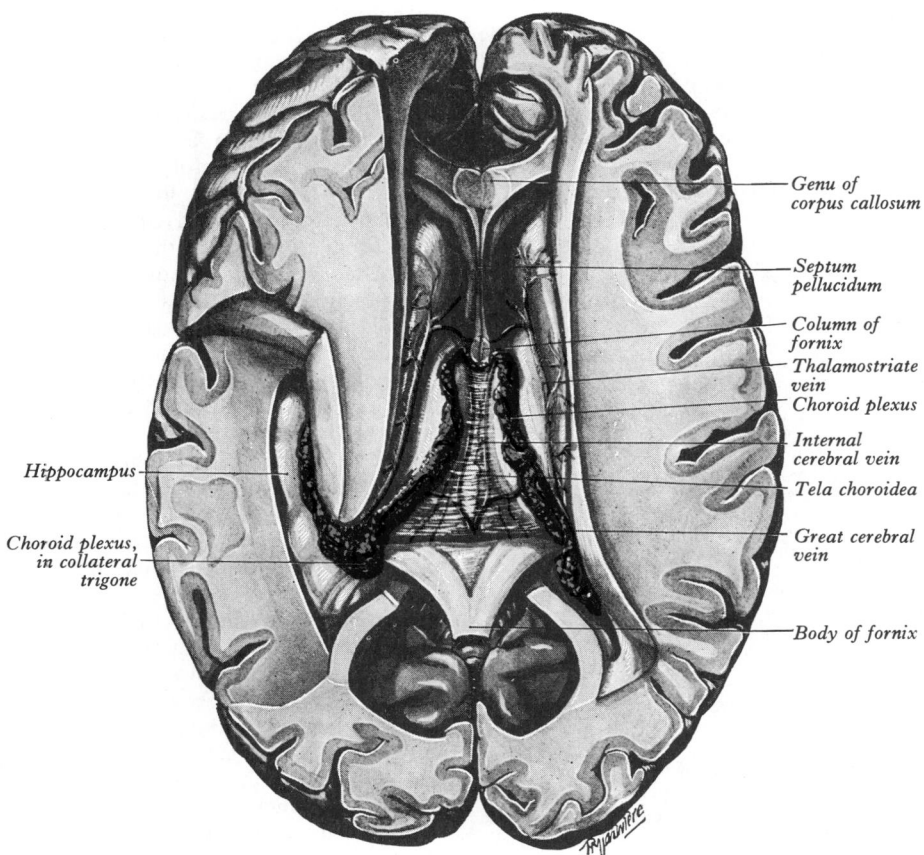

7.200   The tela choroidea of the third ventricle and the choroid plexus of the lateral ventricle.

The microscopic structure of a choroid plexus is essentially as follows. The visible irregular fringes are covered by microscopic *folds* and *villous processes*, each containing afferent and efferent vessels (fine arteries, arterioles, venules and fine veins), and an intervening *capillary plexus*, with supporting connective tissue, and nerve fibres. Villi and folds show varying degrees of complexity (7.201); sections demonstrate the large surface area created, estimated to be at least 200 cm² (Voetmann 1949), ignoring the *microvilli* which electron microscopy shows on the ventricular aspects of ependymal cells (Maxwell & Pease 1956). Ependymal microvilli are not as regular in size and distribution as microvilli in some sites, e.g. the small intestine (p. 30), some having bulbous ends and rather more akin to those in the epithelium of the proximal renal tubules; as in these, the basal aspects of the ependymal cells present invaginations, though to a lesser extent (Pappas & Tennyson 1962). Scott et al 1974 have reported a preliminary scanning electron microscopic study of human fetal and infant choroid plexus, largely confirming the above remarks. These appearances with an absence of secretory granules, unspecialized Golgi apparatus and other features, suggest water transport rather than secretion. However, physiological studies (Davson 1970) show that the cerebrospinal fluid, largely produced by the choroid plexuses, is not a mere filtrate. Ultrastructural studies have shown that some ependymal cells have *cilia*, tight junctions and a distinct *basement membrane* separating them from the adjacent capillaries (Wislocki & Ladman 1958), the latter being sometimes fenestrated. Experiments (Brightman & Reese 1969) with protein peroxidase as a tracer showed that, while capillary endothelium was readily permeable to the tracer, its passage was blocked by tight junctions between the choroidal ependymal cells. All these findings are related to the concept of a *blood–brain barrier* (pp. 895, 1092). (For further comments on the structure and possible functional roles of general ependymal cells and specialized varieties, e.g., tancytes, cells clothing circumventricular organs, choroid plexuses and more recent reviews see p. 894, 1005, 1453.)

The *stroma* of choroidal villi, derived from the pia mater of the tela choroidea, consists of pial fibroblasts but few, if any, collagen fibres. The cells are much flattened and do not form complete sheets between the ependyma and capillary endothelium. Nerve fibres appear to be absent from villi but in their larger stems the connective tissue is greater and myelinated and non-myelinated nerve fibres have been identified (Clark 1934, Cooper 1958); their functional significance is uncertain but the non-myelinated fibres are considered vasomotor postganglionic sympathetic elements; others may be from the vagus and glossopharyngeal nerves. Few studies of the nerves of choroid plexuses have been reported.

However, histochemical studies (Lindvall et al 1977) have revealed cholinergic nerve fibres in association with the vessels and epithelium of all choroid plexuses in various mammals, including baboons.

The *blood supply* of choroid plexuses in the tela choroidea is from the anterior choroidal branch of the internal carotid and choroidal branches of the posterior cerebral artery (pp. 748, 754), the former usually a single vessel, the latter three to five in number (Millen & Woollam 1953). The two sets anastomose to some extent. Capillaries drain into a rich venous plexus, served by a single choroidal vein leaving the tela choroidea, commencing near each basal (posterolateral) angle of the tela. This region corresponds to the frontier between territories of anterior and posterior choroidal arteries and here also (on each side) a solitary so-called 'glomus' (p. 693) is situated, near the free edge of the choroid plexus; many tributaries of the choroidal vein converge towards it; nothing appears to be known regarding its significance in the plexus.

Cerebral Dimensions

The human brain has often been weighed and measured and compared in volume, weight, relation to total body weight and proportionate size of its main divisions with those in brains of many other vertebrates. Even without quantification, comparisons permit some generalizations regarding the relation between brain size and various abilities in mankind and other animals. But even metrical data, where available and reliable, are merely estimates of the bulk of nervous tissue, taking little account of the degrees of organization or even proportions of particular elements, such as nerve cells, nerve fibres, neuroglia and vascular tissue. Such details must obviously be considered if valid comparisons are to be made. Hence, no more than crude deductions are to be expected from data of this kind. The available measurements are usually dependent on small series in any particular species, even sometimes upon a single example. Hence allowance for individual variation can scarcely be made, whether due to sex, age or nutritional state, or indeed any factors which may influence brain size or its relation to body size and weight; and it is particularly this ratio which is the focus of most attention. Even in respect of mankind, for whom much more adequate observations are recorded, the effects of such factors are often overlooked.

An 'average' human male adult's brain is said to weigh about 1450 g and that of his female counterpart about 100 g less; but such figures are not useful without the corresponding ranges: of the order of 1240–1680 in males and 1130–1510 in females. Proportionately males and females differ less in brain weight than in total body weight. Even the ranges quoted would exclude numbers of famous people whose brain weights were outside such extremes, in both directions! (Consult Blinkov & Glezer 1968, for examples and much other data.) Different ranges for races have been stated but the differences between their means are not generally significant, when the much greater variation in body weight is taken into consideration. Chrzanowska et al (1973, 1975), in a study of 1670 human brains (896 male and 774 female, ranging from 20 to 89 years in age), found the average difference between the sexes to be 130–150 g, stating that brain weight is positively correlated with height and recording progressive diminution in brain weight, in both sexes, from 30 to 40 years onwards. They computed a loss in weight of one-ninth in brains of both sexes by the ninetieth year.

Before maturity, human brains vary markedly in weight and ratio to total body weight (Purpura & Schade 1964), showing throughout growth, however, the relatively outstanding development which characterizes adults. Unlike other mammals, such as rodents, in which maximal growth rate may be pre- or postnatal, the cerebral growth spurt in mankind and primates in general is *perinatal*, in late fetal development and the first extra-uterine year, during which the brain at least doubles in weight (Dogson 1962), reaching about 90% of its final weight by the sixth year. Most of this increase is due to myelination rather than any increase in nerve cells; for this reason alone comparison of growing human brains, on the basis of weight, with the brains of adults

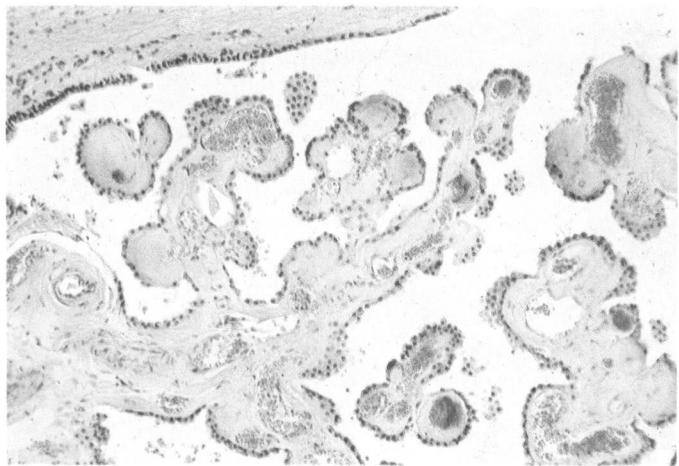

7.201 A section of part of the choroid plexus of the lateral ventricle, stained with haematoxylin and eosin. Note the ependyma lining the ventricular wall above and covering the loose connective tissue cores of the processes of the plexus, which contain numerous small blood vessels, including many capillaries. A calcareous deposit (dark blue) is present in one process. Owing to the complexity of the ramification of the processes, several appear as disconnected islands of tissue.

or of other species, is pointless where correlation with ability is the criterion.

The human brain is obviously large, absolutely and relatively, but surpassed in both respects by those of other mammals. Dolphins, elephants (4000–5000 g) and whales (6800 g is recorded for the blue whale, *Balaenoptera musculus*) have heavier brains but this is offset by brain:body weight ratios of about 1:600 in elephants and about 1:850 in large whales. However, in dolphins the ratio is approximately 1:40, somewhat 'better' than the human average of about 1:50. In some small mammals, such as mice, the ratio may be as low as 1:35, while in smaller primates ratios of 1:12 have been estimated (squirrel monkey). Among primates man occupies an almost average position in brain:body weight ratio, only the larger apes and monkeys falling much below him in this detail. In absolute size of brain, of course, he much surpasses any other primate; a male gorilla (Schultze 1969) may have a cranial capacity of 412–752 cm³, well below the human range (1200–1500 cm³). The largest gorilloid brain recorded weighed 750 g (Holloway 1968), distinctly less than the human range. Extremely small brain weights are recorded in microcephalic idiots; but what is more significant is that brain weights in dwarfs which were *below* the average upper limit for gorillas were nevertheless recorded in humans possessing at least elementary speech; and symbolic abstractions of language are often correlated with a large absolute size of the human cerebrum. This view has been questioned (Lenneberg 1964); while a certain, as yet indefinable, minimal size of brain might be associated with the extraordinary potentialities of mankind in cerebration, it is clear that the data so far noted scarcely illuminate this problem.

As said above, measurements of brain size and brain-to-body weight ratio afford no indication of cerebral organization; nor do they allow for the density of packing or total numbers of nerve cells in any particular brain. While it is said that this density does not vary greatly in mammalian brains when calculated for the whole organ, estimates of this kind were, until quite recently, comparatively few and much subject to error. The figures for samples of the cerebral cortex are subject to similar strictures; it is of interest to note here that a cubic millimetre has been estimated to contain 142 000 cells in mice, 21 500 in macaques and merely 10 500 in mankind. But the area of the human cerebral cortex (p. 1043) is at least twelve times greater than that of macaques, which may nevertheless have a brain-to-body weight ratio similar to man's. However, two separate quantitative investigations are importantly related to these findings; firstly concerning neocortical *surface area* and secondly the distribution of neuronal populations in cortices of different *thickness*. As noted elsewhere (p. 1043) surface area estimates have been surprisingly few, some earlier figures roughly corresponding but a pilot study (employing elementary cartography on published plates in an atlas) resulted in a much higher figure (almost four times greater). Clearly, with these gross differences in such an obvious parameter as surface area, much more extensive investigation with improved techniques is urgently required. The figures for neurons per cubic millimetre of cortex in different species are interesting but become powerfully illuminated when the *total number* of neurons in a column of standard pial surface area and extending through the *full thickness of the* cortex are compared. With the exception of visual area 17, *all other areas*, irrespective of cortical thickness and species of mammal, have virtually identical neuronal numbers (about 66% pyramidal and 33% stellate). Area 17 has a population about 2·5 times greater. In these numerical respects their cortices are uniform or homogeneous, see Rockel et al (1980), Powell (1981), Powell & Hendrickson (1981), and p. 1043.

Another approach to the problems of comparison of mammalian and especially primate brains (Tilney & Riley 1928) is to estimate the weights of the main divisions such as the cerebrum, cerebellum, midbrain, medulla oblongata, olfactory bulb, neocortex, paleocortex, archicortex, corpus striatum and other features, expressing some of these quantities as indices. (As an example of a more recent such study consult Stephan et al 1970.) Large numbers of species of insectivores and primates have been thus compared; though in many species only one sample has been measured, interesting comparisons are possible. It is easy to appreciate, without metrical data, that in the human brain olfactory structures are small, absolutely and relatively, that the cerebrum-pons-cerebellum complex is markedly developed or that the human cerebral cortex is very extensive; but quantitative assessment is much more valuable. In comparing man and gorilla, e.g. ratio by weight of their medullae oblongatae is approximately 9:6:7, though the gorilla weighs at least twice as much as the man. The ratios for the mesencephalon, cerebellum and diencephalon are roughly 2:1, but the human telencephalon is three times as heavy as the gorilloid. The gorilla's olfactory bulb is almost three times the size of man's, while the latter's hippocampus is nearly two and a half times as heavy as the gorilla's. Incidentally the ratio of olfactory bulb to hippocampus in the gorilla is hence about 1:15 and in man 1:90, an interesting commentary on the status of the hippocampus as part of the 'rhinencephalon' (p. 1029). Legait et al (1973) have made some volumetric comparisons of the hypothalamus and hypophysis relative to the total brain volume in a large series of mammals (principally rodents). They observed a constant brain:hypothalamic ratio but some variation between the comparative volumes of hypothalamus and hypophysis, particularly for the anterior lobe of the latter (adenohypophysis).

Space does not permit further exploration of such data and it must be said that, although such considerations may have marked significance in evolutionary comparisons between man and other mammals (particularly sub-human primates) they do not provide any clear definition of what is responsible in man's brain for his enormous complexity of behaviour. Perhaps it is rather in this and his consequently ever more complex culture, that the essence of humanity can best be defined. It is noteworthy that in assessing the human paleontological record the accent has gradually passed from comparison of cranial capacities to that of cultures. The Australopithecinae, with brains little if at all larger than those of gorillas, may leave us in doubt; but in *Homo habilis*, with, it is true, a larger but still very modest cranial capacity, doubts evaporate as to his humanity and perhaps most of all because of his ability, as revealed in surviving artefacts. In more recent ancestors also and again not merely because of increasing *hominoid* brain size (in which *Homo neanderthalensis* even surpassed us, yet is now extinct) but even more because of the increasing evidence of their *human* culture are we convinced of a close relationship to ourselves. Their abilities, which from such simple beginnings have led to the great achievement and communication of practical and abstract creation in our own era, are no more likely than our own to be explicable in terms of gross cerebral mensuration. (For discussions of the significance of brain size consult Tobias 1970, 1971, Jerison 1970, Van Valen 1974.)

Blood Vessels of the Brain

The arterial supply of the brain (pp. 743, 750) is derived from the **internal carotid** and **vertebral arteries** which lie in the subarachnoid space (p. 1089). The vertebral and basilar arteries give branches to the spinal cord, brain stem and cerebellum, the basilar artery ending at the upper border of the pons by dividing into two posterior cerebral arteries. The internal carotid artery divides at its end into **anterior** and **middle cerebral arteries**; the anterior are interconnected by the **anterior communicating artery**. Just before its end the internal carotid artery connects via the **posterior communicating artery** with the **posterior cerebral artery**, completing a vascular circle, the **circulus arteriosus**, around the interpeduncular fossa (**6.84**). The dimensions of vessels forming the circulus arteriosus (circle of Willis) were studied in 100 fixed brains from cadavers by Kamath (1981). The greatest variation in length was found in the anterior communicating artery and in diameter in the posterior communicating artery. Abnormal narrowing of vessels was commoner on the right than on the left, the posterior cerebral artery being particularly affected. In keeping with the dominance of the left hemisphere, it was found that this appeared to have the better blood supply, all arteries except the posterior communicating being of larger mean diameter on the left. It should be appreciated that although the circulus arteriosus offers a potential shunt in abnormal conditions, such as during an occlusion or spasm, in normal circumstances it is not an equalizer and distributor of blood from different sources. For details of collateral circulation following blockage of the main feeders of this circle, see Fields et al (1965) and Gillilan (1974). From the circulus arteriosus, or vessels near it, **central branches** arise to supply the interior of the cerebral hemisphere and the thalamus. These vessels form six principal groups: (1) *the anteromedial group*, from the anterior cerebral and anterior communicating arteries (p. 747); (2) *the posteromedial group*, from the posterior cerebral and posterior communicating arteries; (3 and 4) the right and left *anterolateral groups*, from the middle cerebral arteries; (5 and 6) right and left *posterolateral groups*, from the posterior cerebral arteries. (For details consult Kaplan & Ford 1966.) Lang & Brunher (1978) have described a recurrent central ramus, often double, arising from the anterior cerebral artery beyond the origin of its anteromedial group of central rami and returning to join these; they name this vessel *arteria recurrens anterior*.

The entire blood supply of the **cerebral cortex**, described in detail by Duvernoy et al (1981), comes from **cortical branches** of the anterior, middle and posterior cerebral arteries, which reach the cortex in the pia mater. They divide in its substance, have branches penetrating the cortex perpendicularly and are divisible into long and short rami (Von Bonin 1950). The **long and medullary arteries** traverse the cortex and penetrate the subjacent white matter for 3 or 4 cm (Lewis 1957) without communicating and thus form many small independent systems. **Deep medullary vessels** extending from central branches to the cortex have been described but these are recurrent branches of the long or medullary vessels (Lewis 1957). The **short arteries** are confined to the cortex, forming with the long vessels a compact network in the middle zone of the grey matter, the outer and inner zones being sparingly supplied. Lazorthes et al (1968) and de Reuck (1972) have described the differences in angio-architecture in iso- and allocortical areas, the former being more elaborate with arterioles ending in different strata. Vessels of the cortex are not so strictly 'terminal' as those in the white matter or central system (Sunderland 1938) but, although adjacent vessels anastomose on the surface of the brain, they become end arteries as soon as they enter it. Even superficial anastomoses occur in general only between microscopic branches of the cerebral arteries; there is little evidence that they can provide a vicarious circulation after the occlusion of larger vessels. Owing to the cellularity of the grey matter, its blood supply is more copious than that of white.

The *lateral surface* of the hemisphere is mainly supplied by the *middle cerebral artery*; a strip next to the superomedial border as far back as the parieto-occipital sulcus is supplied by the *anterior cerebral artery*; the occipital lobe and most of the inferior temporal gyrus (excluding the temporal pole) is supplied by the *posterior cerebral artery* (6.82, 83). *Medial and inferior surfaces* are supplied by the *anterior, middle* and *posterior* cerebral arteries, the area supplied by the anterior extending almost to the parieto-occipital sulcus and including the medial part of the orbital surface. (For detailed distribution of the anterior cerebral artery in 53 hemispheres and 300 angiograms consult Farnarier et al 1977.) The rest of this surface and the temporal pole are supplied by the middle cerebral artery. The remaining medial and inferior surface is supplied by the posterior cerebral artery (6.82, 83). The junctional zone near the occipital pole between the territories of the middle and posterior cerebral arteries corresponds to the striate cortex concerned with the macula. The phenomenon known clinically as '*sparing of the macula*' may be due to the collateral circulation of blood from branches of the middle cerebral artery into those of the posterior, when the latter vessel is blocked. The middle

cerebral artery may itself supply the macular area (Smith & Richardson 1966).

Most of the *corpus striatum* and *internal capsule* is supplied by the medial and lateral striate rami of the middle cerebral artery's central branches, the rest being supplied by central branches of the anterior cerebral artery. A ramus of the middle cerebral is Charcot's 'artery of cerebral haemorrhage' (p. 748).

The *choroid plexuses* of the *third* and *lateral ventricles* are supplied by branches of the internal carotid and posterior cerebral arteries.

Finer details of the vessels of some parts of the diencephalic region have been well explored, particularly in relation to the hypophysis cerebri and related hypothalamic nuclei (p. 1008, Haymaker et al 1969). Detailed studies of the vascularization of the lamina terminalis (Duvernoy et al 1969) and of the posterior wall of the third ventricle (Plets 1969) have been reported. The arterial supply to the lamina is from the anterior cerebral arteries and their communicating vessel and is described as supplying a superficial pial capillary plexus which drains into a second, deeper plexus of sinusoidal capillaries, with loops or vortices which drain in turn into the hypothalamic veins. The significance of these arrangements is unknown. The main artery to the posterior parts of the third ventricle is the medial branch of the posterior choroidal artery; this supplies the posterior commissure, habenular region, pineal gland and medial parts of the thalamus, including the pulvinar. The *thalamus* is supplied chiefly by branches of the posterior communicating, posterior cerebral and basilar arteries; their pattern of branches and varying details of angio-architecture in the different thalamic nuclei, have been described in extenso in human brains, with a critique of literature by Plets et al (1970) and Percheron (1977). The latter has denied the often noted thalamic supply by the anterior choroidal artery, deriving almost the entire thalamic supply from branches of the posterior cerebral and basilar arteries (p. 748).

The vessels supplying the *brain stem* have been described in detail by Duvernoy (1978). The *midbrain* is supplied by the posterior cerebral, superior cerebellar and basilar arteries. The crura cerebri are supplied by vessels entering on their medial and lateral sides; the medial vessels enter the medial side of the crus and also supply the superomedial part of the tegmentum, including the oculomotor nucleus, lateral vessels supplying the lateral part of the crus and the tegmentum. The colliculi are supplied by three vessels on each side from the posterior cerebral and superior cerebellar arteries. An additional supply to the crura, the colliculi and their peduncles comes from the posterolateral group of central branches of the posterior cerebral artery.

The *pons* is supplied by the basilar artery and the anterior inferior and superior cerebellar arteries. Direct branches of the basilar enter along the basilar sulcus; branches also enter along the trigeminal, abducent, facial and vestibulocochlear nerves and nervus intermedius. There is also a supply from the pial plexus.

The *medulla oblongata* is supplied by the vertebral, anterior and posterior spinal, posterior inferior cerebellar and basilar arteries. Some branches enter along the anterior median fissure and the posterior median sulcus. Other vessels enter along radicles of the last four cranial nerves to supply the central substance. There is also a supply via a pial plexus from the same main arteries.

The *cerebellum* is supplied by three pairs of cerebellar arteries which, like the cerebral, form superficial anastomoses. Their internal distribution has not been much explored but anastomoses between deeper, medullary branches, as distinct from cortical, have been postulated (Gomes 1969). The anatomy and development of the cerebellar arteries have been reviewed by Gillilan (1974). According to Kielbasinski (1976) the vermis is supplied by one or both inferior cerebellar arteries. The vascularization of the human cerebellar cortex has been investigated in detail by Duvernoy (1983).

The *choroid plexus of the fourth ventricle* is supplied by the posterior inferior cerebellar arteries.

The blood supplies of the optic chiasma, tract and radiation are of marked clinical interest. The chiasma is supplied in part by the anterior cerebral arteries but its median zone depends upon rami from the internal carotids reaching it via the stalk of the hypophysis cerebri. Anterior choroidal and posterior communicating arteries supply the optic tract, while the optic radiation receives blood through deep branches of the middle and posterior cerebral arteries. For further details consult Abbie (1938), Bergland & Ray (1969).

Cerebral Blood Flow and Functional Localization

That the levels of activity, metabolism and blood flow fluctuate in close correspondence has long been known. Though the mechanisms of selective control of blood flow to different regions were then obscure, Schmidt & Hendrix (1937), using thermocouples, demonstrated a marked increase in flow through the visual cortex (in cats), during localized retinal illumination. Subsequent years brought many confirmatory experiments, not only in the visual pathway. In recent years, an increasing range of radio-isotopes and the advent of scanning cameras have stimulated an upsurge of interest in cerebral blood-flow depending upon a new, non-invasive technique, potentially able to localize intracranial pathological states which modify the blood-flow pattern, e.g. neoplasms, post-traumatic loci, intracranial abscess, cerebral infarction and so forth (Maisey 1978). Some observers soon realized the potential of the technique in monitoring fluctuating levels of blood flow (and of functional activity) in different regions of, e.g. the cerebral cortex during a wide range of activities (Lassen et al 1978). An isotope of xenon (xenon 133) dissolved in sterile saline is injected into a carotid artery and the subject's head is scanned (during the arrival and 'washout' of the isotope) by a gamma-ray camera, which carries 254 scintillation detectors each collimated to scan about 1 cm² of the cortical surface. Results from this roster of detectors, processed in a digital computer, are displayed on a cathode colour screen. The colour of each *pixel* or square denotes a range of blood flow rate, shown through the subjacent cortex. The mean flow rate is shown in green, down to 20% below the mean are shades of blue, and up to 20% above the mean are through orange to shades of red. Illustration 7.202A,B shows two examples of the results. Already hundreds of apparently normal hemispheres have been scanned during many activities ranging from uncomplicated voluntary movements or the sensory stimulation of various modalities, to complex activities such as reading aloud and problem-solving. Broadly, the functional areas mapped by traditional means have been confirmed. There will doubtless be refinement of this fascinating technique, with advantage to both clinician and researchers into cerebral activities.

The venous drainage of the brain (pp. 798–805) can be divided into veins receiving from the cerebrum and from the cerebellar-brain stem arena. The veins are thin-walled, devoid of valves and most of them cross the subarachnoid space to join the dural venous sinuses.

The veins of the cerebrum are either external or internal. **External cerebral veins** are grouped as three sets: the *superior* draining forwards into the superior sagittal sinus; the *inferior* draining principally into the transverse and cavernous sinuses; and the *middle* which are subdivided into superficial and deep. The *superficial middle cerebral vein* drains most of the lateral surface of the hemisphere, following the lateral sulcus to end in the cavernous sinus. The *deep middle cerebral vein* drains the insular region and joins the *anterior cerebral* and *striate veins* to form a *basal vein*. Regions drained by the anterior cerebral and striate veins correspond approximately to those supplied by the anterior cerebral artery and the central branches entering the anterior perforated substance. These striate veins have been described in detail by Rosa & Borzone (1973). The basal veins pass back alongside the interpeduncular fossa and midbrain, receive tributaries from this vicinity and join the *great cerebral vein*.

The two **internal cerebral veins** are formed near the interventricular foramina by union of the *thalamostriate* and *choroidal veins* draining the choroid plexuses of the third and lateral ventricles. They travel back parallel to one another between the layers of the tela choroidea and unite to form the great cerebral vein, which enters the straight sinus.

The veins of the *midbrain* join the basal or great cerebral veins. *Pontine* veins drain into the basal vein, cerebellar veins, the petrosal sinuses, transverse sinus or the venous plexus of the foramen ovale. Veins of the *medulla oblongata* drain into the veins of the spinal cord, the adjacent dural venous sinuses or along the last four cranial nerves via radicular veins to the inferior petrosal sinus or superior bulb of the jugular vein. For systematic accounts of the superficial veins of the brain stem consult Tournade et al (1972) and Duvernoy (1975).

The **veins of the cerebellum** drain mainly into sinuses adjacent to them or, from the superior surface, to the great cerebral vein.

Nerves

Though the *innervation* of the intercranial arteries (including those supplying the brain) remains obscure, a considerable supply of postganglionic sympathetic fibres accompany the carotid and

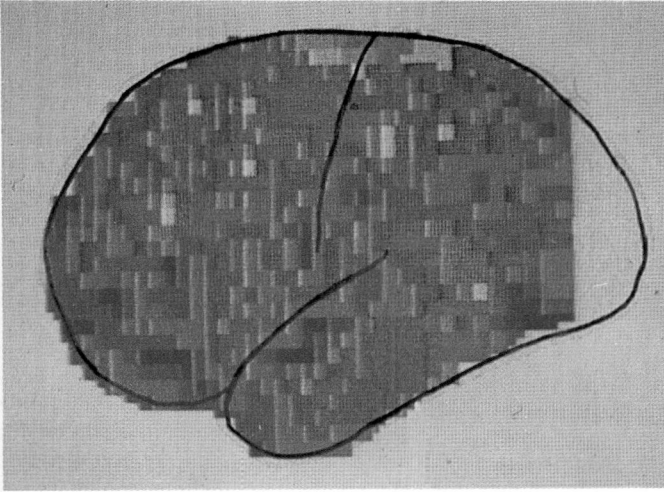
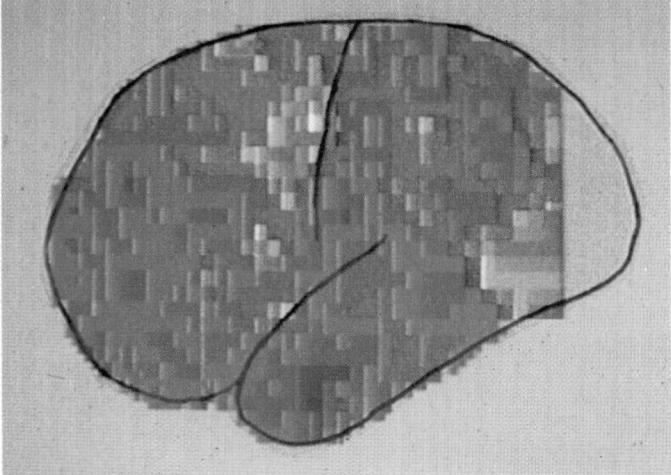

7.202 Patterns of presumed relative levels of neuronal activity in different regions of the cerebral cortex as revealed by measurement of regional blood-flow. A. shows the pattern which accompanies movement of the contralateral foot; B. shows a pattern which accompanies activity of the eyes engaged in a pursuit movement. The colour scales indicate the percentage flow rate above or below the mean (dark green). Radioactive xenon scan pictures supplied by N A Lassen, D H Ingvar and E Skinhoj, Bispebjerg Hospital, Copenhagen (see text).

vertebral arterial trees and some myelinated fibres accompany them. A parasympathetic supply is doubtful (Nelson & Rennels 1970, Purves 1972).

There are *no lymphatics* in the central nervous system. The subarachnoid space is prolonged along the olfactory nerves, providing a route between this space and tissue spaces in the nasal mucoperiosteum. The so-called perivascular spaces around cerebral vessels are controversial (p. 1092). Electron microscopy confirms the extension of pial elements around vessels into the brain in various mammals, e.g. rat (Samarasinghe 1965) and cat (Jones 1970). In the latter study the arachnoid 'space' around small arterioles entering the cortex appeared to contain collagen bundles, pial cells and possibly macrophages, but these do not form a complete sheath; in large areas the basal laminae of the brain and vessel are apposed. The 'space' does not extend around the capillaries, where the two basal laminae fuse. These observations indicate that the perivascular spaces are not continuous with the perineuronal spaces.

THE MENINGES

The brain and the spinal cord are enveloped, from without inwards, by three membranes (meninges): the dura mater, the arachnoid mater and the pia mater.

THE DURA MATER

The dura mater, a thick, dense, inelastic membrane, is the most external of the meninges. *Cerebral dura mater*, enclosing the brain, differs in some features from *spinal dura mater*, surrounding the spinal cord and will be described separately. They are continuous with each other at the foramen magnum.

The cerebral dura mater lines the cranial cavity. It is said to be composed of two layers, an *inner* or *meningeal* and an *outer* or *endosteal*, but these are united, except where they separate to enclose the venous sinuses draining blood from the brain (p. 799). It adheres to the internal surfaces of the cranial bones, sending blood vessels and fibrous processes into them, adhesion being firmest at the sutures, the cranial base and around the foramen magnum. The vessels and fibrous processes are torn across when the dura mater is detached from the bones, its outer surface then becoming rough and fibrillated; the inner surface is smooth. The endosteal layer is continuous through the cranial sutures and foramina with the pericranium and through the superior orbital fissure with the orbital periosteum. The meningeal layer provides tubular sheaths for the cranial nerves as they traverse the cranial foramina; these sheaths fuse with the epineurium as the nerves emerge from the skull. The sheath of the optic nerve is continuous with the ocular sclera (p. 1181).

The meningeal layer is folded inwards as four septa that partially divide the cranial cavity into freely communicating spaces in which the subdivisions of the brain are lodged.

(1) *The falx cerebri* (7.203A) is a strong, crescentic, process of dura mater descending vertically in the longitudinal fissure between the cerebral hemispheres. It is narrow in front, where it is fixed to the crista galli, and broad behind, where it blends in the midline with the tentorium cerebelli; the anterior part is thin and frequently perforated by numerous apertures. Its convex upper margin is attached to the internal cranial surface on each side of the median plane, as far back as the internal occipital protuberance; the superior sagittal sinus (p. 800) runs along this margin in a cranial groove, to both banks of which the falx is attached. Its lower edge is free and concave and contains the inferior sagittal sinus. The straight sinus runs along its attachment to the tentorium cerebelli.

(2) *The tentorium cerebelli* (7.203A,B) is a crescentic, 'tented' lamina of dura mater covering the cerebellum and supporting the

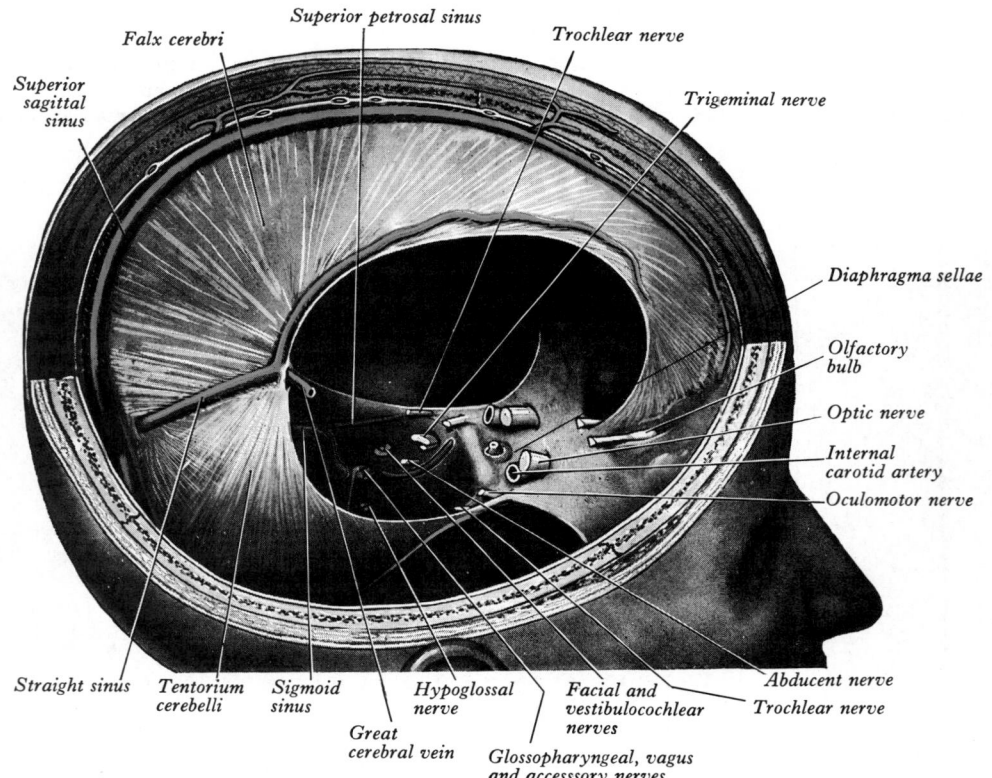

7.203A The cerebral dura mater and its reflexions, exposed by the removal of a part of the right half of the skull and brain.

cerebral occipital lobes. Its concave, anterior edge is free and between it and the dorsum sellae of the sphenoid bone is a large *tentorial incisure*, occupied by the midbrain and the anterior part of the superior aspect of the cerebellar vermis. Its convex outer limit is attached: (*a*) posteriorly, to the lips of the transverse sulci of the occipital bone and the postero-inferior angles of the parietal bones, where it contains the transverse sinuses; (*b*) laterally, to the superior borders of the petrous temporal bones, where it encloses the superior petrosal sinuses. Near the apex of the petrous temporal bone, the lower layer of the tentorium is evaginated antero-laterally under the superior petrosal sinus, forming a recess between the endosteal and meningeal layers in the middle cranial fossa. This recess is the *trigeminal cave* containing the trigeminal roots and ganglion (its posterior part being a little more extensive below the ganglion than above it). The evaginated meningeal layer fuses in front with the anterior part of the ganglion. At the apex of the petrous temporal bone the free border and attached periphery of the tentorium cross each other (7.203B); the anterior ends of the free border are fixed to the anterior clinoid processes and the attached periphery to the posterior clinoid processes. For details of the comparative anatomy and phylogeny of the tentorium cerebelli see Klintworth (1968).

(3) *The falx cerebelli*, a small, crescentic fold of dura mater, is below the tentorium cerebelli, projecting forwards into the posterior cerebellar notch. Its base, directed upwards, is attached to the posterior part of the inferior surface of the tentorium cerebelli in the median plane; its posterior margin, attached to the internal occipital crest, contains the occipital sinus; its apex frequently divides into two small folds, which disappear at the sides of the foramen magnum. Hasan & Das found the falx cerebelli double in 76 of 100 cadavers.

(4) *The diaphragm sellae* (7.203B) is a small, circular, horizontal fold of dura mater, forming a roof to the sella turcica, almost completely covering the hypophysis; a small, central, opening in it transmits the infundibulum.

The arrangement of the dura mater in the central part of the middle cranial fossa is complex (7.203B). The rim of the tentorial incisure converges on the attached periphery, crossing it near the apex of the petrous temporal bone and continuing forwards as a ridge on the dura mater, to attach to the anterior clinoid process. This ridge marks the junction of the roof and lateral wall of the cavernous sinus (7.203A,B). The attached periphery of the tentorium cerebelli follows the superior border of the petrous temporal bone and, crossing under the free border, continues forwards to the posterior clinoid process as a rounded, indefinite ridge on the dura mater. An angular interval thus exists between the anterior parts of the attached periphery and the free border (7.203B); in this interval the dura mater is part of the roof of the cavernous sinus and is pierced in front by the oculomotor and behind by the trochlear nerves, which remain in contact with the dura mater after piercing it and proceed antero-inferiorly into the lateral wall of the cavernous sinus (6.155–157).

In the middle cranial fossa, anteromedially, the dura mater ascends as the lateral wall of the cavernous sinus. Reaching the ridge produced by the anterior continuation of the free border of the tentorium, it turns medially, as the roof of the cavernous sinus, and is here pierced by the internal carotid artery (7.203A,B). Medially, the roof of the sinus is continuous with the upper layer of the diaphragma sellae. At or just below the opening in the diaphragma for the infundibulum, the dura mater, arachnoid and pia mater blend with each other and with the capsule of the hypophysis (p. 1451); within the sella turcica it is impossible to differentiate the meninges, and the subdural and subarachnoid spaces are obliterated (6.155).

Apart from its function as periosteum, the dura mater probably exerts a steadying influence, through the tentorium and falx, upon

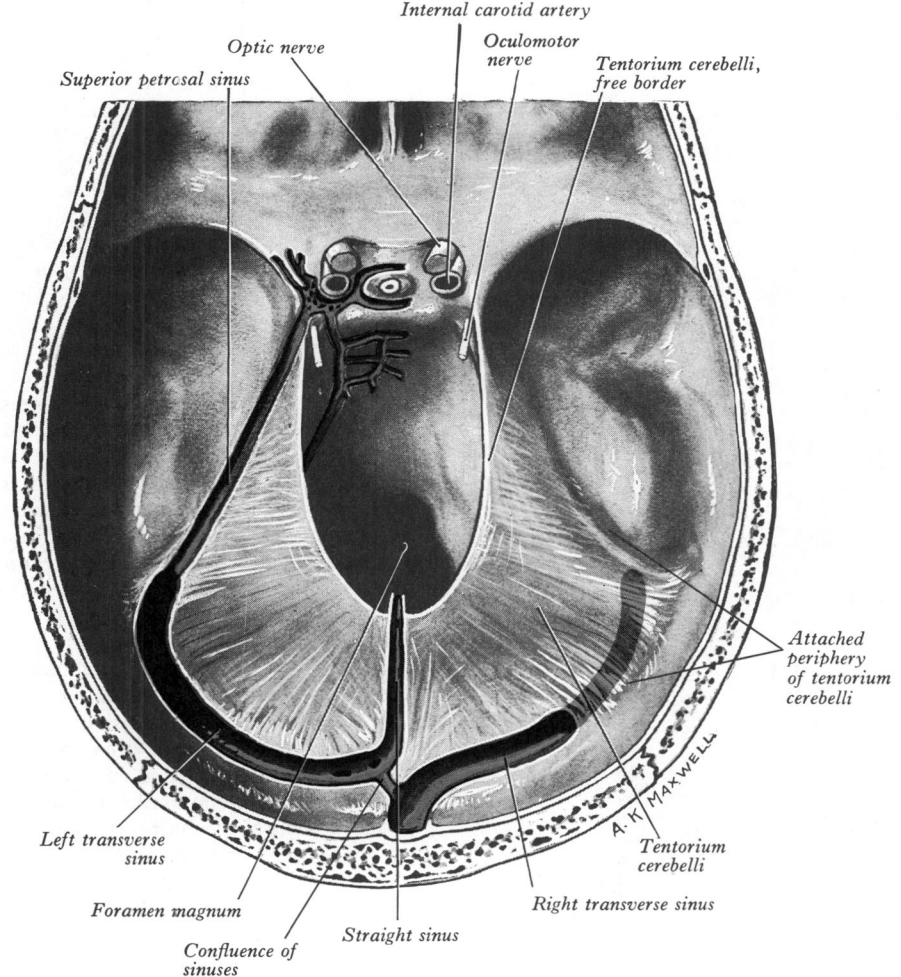

Internal carotid artery

Optic nerve

Oculomotor nerve

Superior petrosal sinus

Tentorium cerebelli, free border

Attached periphery of tentorium cerebelli

Left transverse sinus

Tentorium cerebelli

Foramen magnum

Right transverse sinus

Confluence of sinuses

Straight sinus

7.203B   The superior aspect of the tentorium cerebelli.

the movement of the brain in the cranial cavity; but this is a speculation. As noted, the venous sinuses are enclosed in its thickness and some are in the margins of the dural partitions. The junction of the great cerebral vein and straight sinus is a critical point in the venous drainage of the brain; if the relation between these two channels is much altered for more than short periods of time, as may occur when 'space-occupying' lesions above the tentorium cause the descent of the brain relative to the tentorium, obstruction of the great cerebral vein may ensue, with sequelae of venous back-pressure, oedema of the choroid plexuses and over-production of cerebrospinal fluid.

The structure of the dura mater is basically fibrous, white collagen fibres predominating, with some elastic fibres. The collagen fibres are densely arranged in laminae, in which they often lie parallel in fascicles with wide angles between them in adjacent laminae, producing a latticed appearance easy to see in the tentorium cerebelli. As has been noted, the cerebral dura mater is described as having two layers, *endosteal* and *meningeal*, a description owing more, perhaps, to dural separation where venous sinuses occur and splitting at the foramen magnum and optic canals, than to any marked histological differences. Smaller branches of meningeal vessels are largely in the endosteal layer, since they are (despite their name) primarily periosteal. Fibroblasts occur throughout the dura mater but osteoblasts are confined to the endosteal layer. Elastic fibres separate the laminae of collagen fibres, which are also more extensively separated by lacunar spaces considered by some to be continuous with the subdural space (vide infra); these spaces are mainly confined to the meningeal dura mater. Several features thus distinguish the external and internal layers of the dura but there is no discontinuity indicating a distinct bilaminar nature. At all cranial foramina the endosteal element is continuous with the external periosteum, and with the sutural membranes, until the sutures close; at these locations it is more strongly attached but elsewhere it is more easily detached from the cranial bones. The meningeal element is

continuous with the dural sheaths of the spinal cord and optic nerves. At other foramina it is 'pierced' by the nerve or vessel, but is more accurately described as becoming continuous with the perineurium or adventitial sheaths. Though apposed to the arachnoid mater internally, the dura mater is easily separated from it, as exemplified by the occurrence of subdural haemorrhages between them. Although contrary views have been expressed there is little doubt that a layer of flattened 'mesothelial' cells exists on the internal aspect of the dura and that fluid, though a mere capillary film, usually marks this zone of potential separation. Rascol & Izard (1974) have described a 'neurothelium' in this zone, its cells arranged in several irregular layers, with abundant intercellular spaces devoid of collagen fibres; the cells display tonofilaments and desmosomes. They consider that this fragile epithelium is easily torn apart to produce a 'subdural space'. (The structure of the subjacent arachnoid is described on p. 1089.)

The arteries of the cerebral dura mater are numerous. In the anterior cranial fossa are the anterior meningeal branches of the anterior and posterior ethmoidal and internal carotid arteries and a branch from the middle meningeal. In the middle fossa are: the middle and accessory meningeal branches of the maxillary artery; a branch of the ascending pharyngeal entering by the foramen lacerum; branches of the internal carotid; and a recurrent branch from the lacrimal artery. In the posterior fossa are: the meningeal branches of the occipital artery (one entering the skull by the jugular foramen, another by the mastoid foramen); the posterior meningeal branches of the vertebral artery; and occasional twigs from the ascending pharyngeal which enter by the jugular foramen and hypoglossal canal. The meningeal arteries are chiefly distributed to bone, in contrast to those in the spinal dura mater, and are therefore unsuitably named; only very fine branches are distributed to the cranial dura mater itself.

The veins returning blood from the cranial dura mater are described on pp. 798–805.

The nerves of the cerebral dura mater have been much investigated over a long period (von Luschka 1850, Arnold 1851, Hovelacque 1927, Siwe 1931, Penfield & McNaughton 1940, Kimmel 1961a, b). The most authentic sources of dural innervation are the *trigeminal nerve*, including its ganglion and three divisions, the first three *cervical spinal nerves* and the *cervical sympathetic* trunk. Less well-established meningeal branches have been described from the vagus and hypoglossal nerves and possibly the facial and glossopharyngeal (but vide infra).

In the *anterior cranial fossa* are meningeal branches of the anterior and posterior ethmoidal nerves and anterior filaments of the meningeal rami of the maxillary (*nervus meningeus medius*) and mandibular (*nervus spinosus*) trigeminal divisions. *Nervi meningeus medius et spinosus* are, however, largely distributed to the dura of the *middle cranial fossa*, which also receives filaments from the trigeminal ganglion. A recurrent *tentorial nerve* (a branch of the ophthalmic division) supplies the *tentorium cerebelli*. The dura in the *posterior cranial fossa* is innervated by ascending meningeal branches of the upper cervical nerves which enter it through the anterior part of the foramen magnum (second and third cervical) and through the hypoglossal canal and jugular foramen (first and second cervical nerves). Meningeal branches from the vagus and hypoglossal and possibly other cranial nerves may exist, but strong evidence shows that nerves so described are usually recurrent filaments from the cervical nerves which run in the sheaths of cranial nerves (Kimmel 1961a, b).

All the meningeal nerves contain a postganglionic sympathetic component, either from the superior cervical sympathetic ganglion or by communication with its perivascular intracranial extensions. These components are probably vasomotor; experiments have shown that they exert a vasoconstrictor action on pial vessels (see p. 1085 for cerebral blood flow).

Various dural receptor terminals, including simple end-bulbs and Meissner's and Pacinian corpuscles, have been described in various mammals; as little recent information is available concerning man, the roles of the sensory and autonomic supply to the cerebral dura mater remaining uncertain.

The spinal dura mater (7.204, 208) is a loose sheath, representing only the inner (meningeal) layer of the cranial dura mater; the outer (endosteal) layer is represented by the

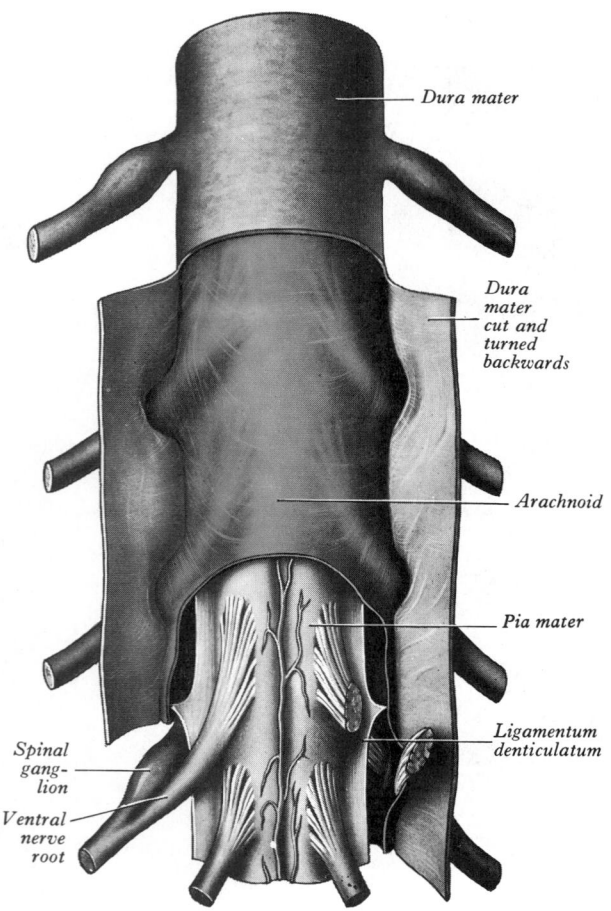

Dura mater

Dura mater cut and turned backwards

Arachnoid

Pia mater

Ligamentum denticulatum

Spinal ganglion

Ventral nerve root

7.204 A part of the spinal cord exposed from the ventral aspect, showing its meningeal coverings.

periosteum of the vertebral canal, which is separated from the spinal dura mater by an *extradural (epidural) space*. Spinal dura mater is attached to the edge of the foramen magnum and to the posterior surfaces of the second and third cervical vertebral bodies and also by fibrous slips to the posterior longitudinal ligament, especially towards the caudal end of the vertebral canal. A midline fold is occasionally present (Parkin & Harrison 1985). The subdural space ends at the lower border of the second sacral vertebra; below this, the dura mater invests the spinal filum terminale, descending to the back of the coccyx to blend with the periosteum. The spinal dura mater has tubular prolongations around the spinal roots and nerves as they traverse the intervertebral foramina (7.208). These prolongations are short in the upper vertebral column but gradually become longer with the increasing obliquity of the spinal roots (7.81). Dural sheaths of spinal nerves fuse with their epineuria in or slightly beyond the intervertebral foramina; this continuity may act as a slight check on movements of the nerves when affected by bodily movement. At the cervical level, where the nerves are short and vertebral movement greatest, their sheaths are bound strongly to the periosteum of the adjacent transverse processes (Sunderland 1974).

The **extradural (epidural) space**, lying between the spinal dura mater and the periosteum and ligaments within the vertebral canal, contains loose connective tissue, fat and a venous plexus. In the lumbar region, the dura mater is apposed to the walls of the vertebral canal, being attached to them by connective tissue in a manner permitting displacement of the dural sac during spinal movement and venous engorgement; fat is deposited posteriorly in recesses between the ligamenta flava (Parkin & Harrison 1985). The connective tissue extends for a short distance through the intervertebral foramina along the spinal nerves. Dyes or other fluids injected into it at sacral level can spread up to the cranial base; local anaesthetics injected near a spinal nerve just outside its intervertebral foramen may spread up or down to affect the adjacent spinal nerves or pass to the opposite side. In each case spread is through the extradural space. (For the nerve supply of the spinal dura mater, see p. 1125.)

The **subdural space** between the dura and arachnoid maters is a potential space containing a film of serous fluid between the surfaces of the apposed membranes. Apparently it does not connect with the subarachnoid space, but continues a short distance along the cranial and spinal nerves, communicating with their lymph spaces. Around the optic nerves it is continued as far as the back of the eyeballs. The functional significance of the subdural space remains controversial; possible connections with venous channels or with hypothetical lymph spaces in the dura mater have been claimed and disclaimed (Hoffmann & Thiel 1956). Electron microscopy shows no cells of an epithelial type in the dura mater, except at its arachnoid surface. The dural lacunae may in fact be artefacts; the evidence refutes any passage of significant amounts of fluid from the dura into the subdural or subarachnoid spaces (Kaplan & Ford 1966), but lymphatic drainage from the spinal extradural adipose tissue, possibly including the spinal dura, has been established (Millen & Woollam 1962).

THE ARACHNOID MATER

The arachnoid, a delicate membrane enveloping the brain and spinal cord, lies between the pial and dural meninges, separated from the dura mater by the *subdural space*, which is traversed by isolated trabeculae, most numerous on the posterior surface of the spinal cord; it is separated from the pia mater by the *subarachnoid space*, which is filled with cerebrospinal fluid. The arachnoid surrounds the beginnings of the cranial and spinal nerves, enclosing them as far as their exits from the skull and vertebral canal.

The *cerebral part of the arachnoid* invests the brain loosely and does not enter the sulci or fissures, with the exception of the longitudinal fissure. On the upper surface of the brain it is thin and virtually transparent, but it is thicker on the basal aspect of the brain and is centrally slightly opaque where it extends between the temporal lobes in front of the pons, leaving a large space between it and the pia mater. It fuses with the other meninges in the hypophysial fossa.

The *spinal part of the arachnoid* (7.204, 208) is a thin, delicate tubular investment for the spinal cord, continuous above with the cerebral arachnoid; below, it widens around the cauda equina and ends at the level of the lower border of the second sacral vertebra. (For structure of the pia-arachnoid see p. 1091.)

The **subarachnoid space,** between the arachnoid and pial meninges, contains cerebrospinal fluid and the larger blood vessels of the central nervous system and is traversed by delicate connective tissue trabeculae, connecting the arachnoid and pia; these two layers come into contact on the summits of the cerebral gyri; but where the arachnoid bridges the sulci, angular spaces are left, filled by subarachnoid trabecular tissue. Wherever the brain and cranium are not closely adapted, the arachnoid is separated from the pia by wide intervals, named *subarachnoid cisterns*, which are continuous with each other through the general subarachnoid space of which they are merely dilatations. Subarachnoid trabeculae are here scant or absent.

The Subarachnoid Cisterns (7.205)

The *cerebello-medullary cistern (cisterna magna)* (7.205) is formed by the arachnoid bridging the interval between the medulla oblongata and the inferior cerebellar surface. It is triangular in sagittal section and is continuous below with the subarachnoid space of the spinal cord. The *pontine cistern* (7.205) is an extensive space ventral to the pons; it contains the basilar artery and below is continuous with the spinal subarachnoid space, behind with the cerebello-medullary cistern and, rostral to the pons, with the interpeduncular cistern. As the arachnoid crosses between the two temporal lobes it is separated from the cerebral peduncles and structures in the interpeduncular fossa by the *interpeduncular cistern*, containing the circulus arteriosus; anteriorly this cistern continues rostral to the optic chiasma and over the surface of the corpus callosum, where the arachnoid stretches between the cerebral hemispheres below the free edge of the falx cerebri, leaving a space containing the anterior cerebral arteries. The *cistern of the lateral fossa* contains the middle cerebral artery and is formed by the arachnoid bridging the lateral sulcus. The *cistern of the great cerebral vein (cisterna ambiens* or *superior cistern)* occupies the interval between the splenium and the superior cerebellar surface; it contains the great cerebral vein and the pineal gland. It is a widely used neurosurgical landmark.

Smaller cisternae have been described, including the *prechiasmatic* and *postchiasmatic cisterns* related to the optic chiasma, the *cistern of the lamina terminalis* and the *supracallosal cistern*, all of which are here included as extensions of the interpeduncular cistern and which contain the anterior cerebral arteries.

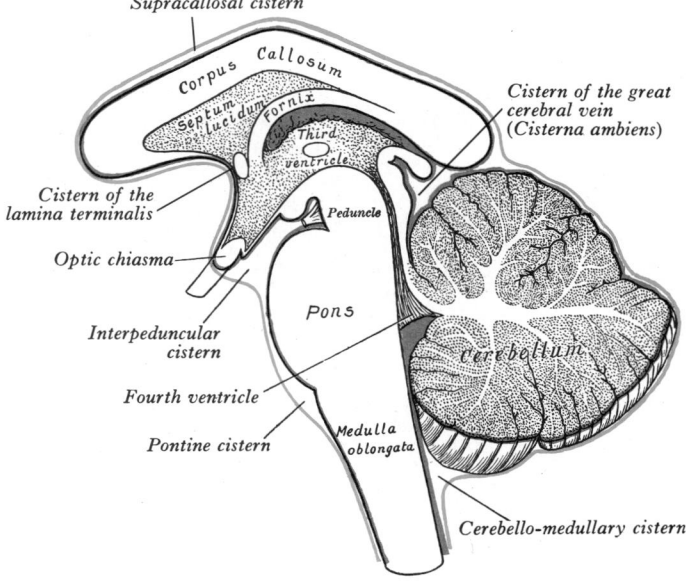

7.205   The positions of the principal subarachnoid cisterns. Red = pia mater, Blue = arachnoid mater.

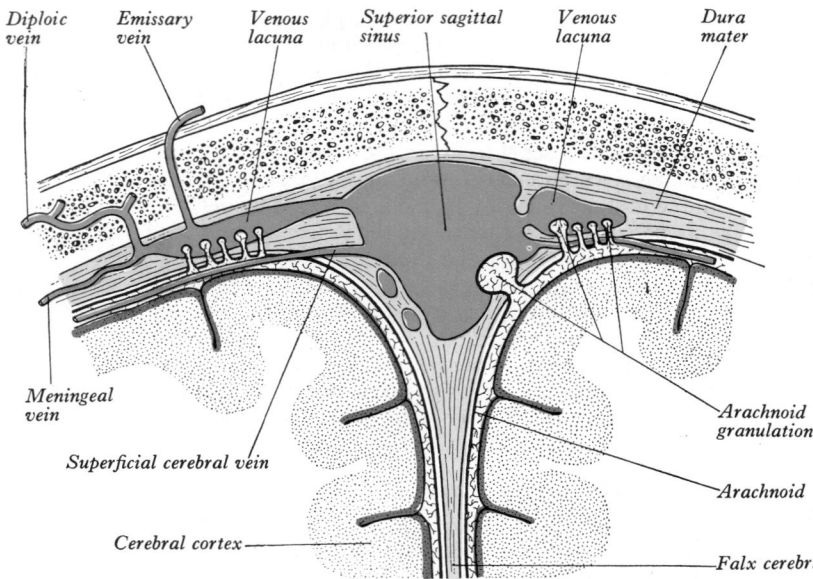

7.206 A coronal section through the vertex of the skull to show the arrangement of the veins and the meninges of the brain and arachnoid granulations.

The subarachnoid space connects with the ventricles of the brain by three openings: the *median aperture* (p. 979) in the median plane in the inferior part of the roof of the fourth ventricle and the two *lateral apertures* at the ends of that ventricle's lateral recesses (p. 979), behind the upper roots of the glossopharyngeal nerves. There is no communication between the subdural and subarachnoid spaces. Communications exist between tissue spaces in the nasal mucous membrane and the subarachnoid space through channels present along the course of the olfactory nerves.

The spinal subarachnoid space is relatively wide, being largest in the lower part of the vertebral canal, where the arachnoid encloses the cauda equina. Continuous above with the cranial subarachnoid space, it ends below at the level of the lower border of the second sacral vertebra. It is partially divided by the subarachnoid septum and the ligamentum denticulatum, described later (p. 1091).

The arachnoid granulations (7.206, 207) are small elevations, usually occurring in clusters near the superior sagittal,

transverse and some other sinuses. They protrude into the sagittal sinus and its venous lacunae (7.207). They are visible at the age of eighteen months and are widely disseminated at three years, increasing in number and size as age advances. They cause pressure atrophy of bone, producing pits on the internal cranial aspect. Arachnoid granulations are macroscopic enlargements, or distensions, of minute projections of the arachnoid mater, termed *arachnoid villi*, normally numerous in the young. The growth and structure of arachnoid villi and granulations have been extensively studied (Le Gros Clark 1920). Each villus is a diverticulum of the subarachnoid space, penetrating into the dura mater, and covered by a layer of flat cells with large nuclei and lightly staining cytoplasm. The subarachnoid space within the granulation has a reticulum of fine fibrous tissue, usually denser peripherally than in the granulation's interior; in advanced age it frequently contains calcareous nodules. At the summit of a villus, mesothelial cells proliferate to form a cap which penetrates the dura mater and fuses with the endothelium of the sinus into which it projects (7.206), pulling out a little stalk of arachnoid with a diverticulum of the subarachnoid space. Except where fused with endothelium, the villus is surrounded by subdural space and dura mater, the latter, covered by a layer of mesothelium, being invaginated into the sinus by the protrusion of the granulation. Fluid injected into the subarachnoid space passes into the granulations and villi and thence into the venous sinuses of the dura by osmosis.

The cerebrospinal fluid (Davson 1970) is a clear, slightly alkaline fluid, with a specific gravity of about 1·007, containing in solution inorganic salts like those in plasma and traces of protein and glucose. It is secreted into the ventricles of the brain by the choroid plexuses and into the subarachnoid space by plexuses in the fourth ventricle's lateral recesses (p. 979). From the ventricles it escapes through the apertures of the fourth ventricle into the cerebello-medullary and pontine cisterns of the subarachnoid space. Intracranially the fluid ascends through the tentorial incisure, spreading over the inferior cerebral surface and ascending over the superolateral aspects of each hemisphere to reach the arachnoid villi of the superior sagittal sinus, there re-entering the bloodstream. There is probably little active flow at spinal level, diffusion and alterations of posture serving to maintain its composition constant throughout the subarachnoid space. Experiments in cats (Howarth & Cooper 1949) suggests that spinal cerebrospinal fluid may return locally to the venous system

7.207A Relations of the pia and arachnoid maters to the dura, brain and vessels, according to Hutchings & Weller (1986) and Alcolado et al (1988). Note the subpial space. Provided by R Weller, Department of Pathology, Southampton University.

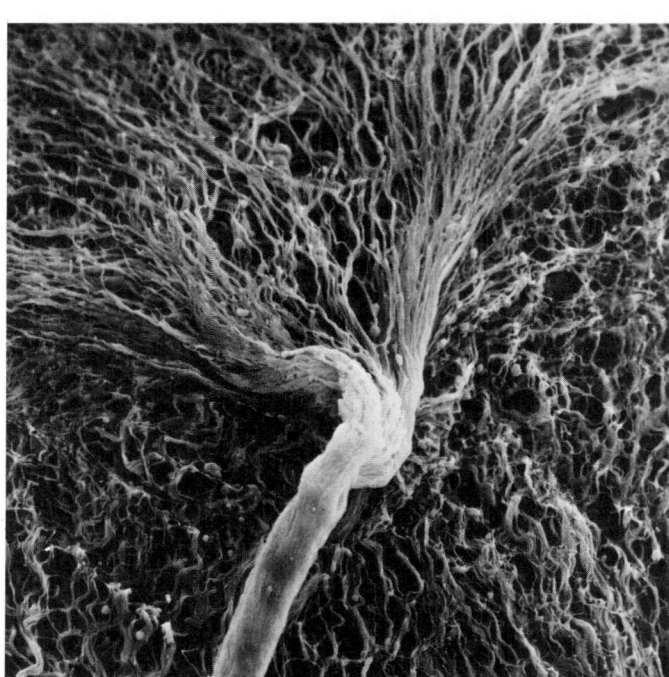

7.207B Scanning electron micrograph (× 240) of an arachnoid. trabecula. Collagen fibres fan out into the internal aspect of the arachnoid. Provided by R Weller, Department of Pathology, Southampton University.

THE PIA MATER

through the vertebral venous plexuses, intervertebral veins and the posterior intercostal and upper lumbar veins into the azygos and hemiazygos veins. The fluid supports the brain and spinal cord and maintains a uniform pressure around them. A brain weighing 1500 g in air weighs 50 g in cerebrospinal fluid, through which the total weight of the system is distributed to the meningeal parieties and their external supports. The circulation of cerebrospinal fluid and the function of the arachnoid villi were classically explored by Weed (1920, 1938). Electron microscopy suggests that there is open communication between the subarachnoid space and the lumen of the superior sagittal sinus by fine endothelial tubules traversing the arachnoid granulations (in sheep). A valvular action has been proposed in these structures but this drainage, even if confirmed, does not exclude filtration (Jayatilaka 1965).

Applied Anatomy. Central nervous diseases often alter the cells which normally occur in cerebrospinal fluid or the concentration of its chemical constitutents. Interference with circulation is indicated by variations in subarachnoid pressure. Specimens of cerebrospinal fluid are obtained by *lumbar puncture*, through the interval between the third and fourth (or fourth and fifth) lumbar laminae or spines. A needle is inserted at the intersection of the intertubercular plane and posterior median line, passed obliquely up and forwards above the upper border of a spine and through or alongside the supraspinous and interspinous ligaments into the vertebral canal. The dura mater and arachnoid are punctured, admitting the needle's tip into the subarachnoid space *below* the end of the spinal cord (7.57). The fluid escapes at the rate of about one drop per second, but when under increased pressure it escapes as a jet. The introduction of a large needle, such as a trocar, between the vertebral laminae is difficult except at lumbar levels. In any case, the introduction of its point accurately enough to withdraw fluid into the narrow subarachnoid space surrounding the spinal cord would be extremely difficult, apart from the danger to the cord itself; but it is primarily the *ease* with which fluid can be tapped from the large subarachnoid space *caudal* to the spinal cord which determines the choice of *lumbar* puncture.

THE PIA MATER

The pia mater closely envelops the brain and spinal cord; it is a vascular membrane, a plexus of minute blood vessels held together by fine loose connective tissue.

The *cerebral pia mater* invests the whole surface of the brain, dipping between cerebral gyri and cerebellar laminae, and is invaginated to form the tela choroidea of the third ventricle and the choroid plexuses of the lateral and third ventricles (pp. 1019, 1080); as it passes over the roof of the fourth ventricle it forms the tela choroidea and choroid plexuses (p. 979). On the cerebral hemispheres it provides a multitude of sheaths around minute vessels which enter the cortex perpendicularly; on the cerebellum it is more delicate, the vessels from its deep surface are shorter and its relations to the cortex less intimate. Recent work by Hutchings & Weller (1986) and Alcolado et al. (1988) defines a subpial space continuous with perivascular spaces around arachnoid vessels (see 7.207A,B).

The *spinal pia mater* (7.204, 208) is firmer and less vascular, but like the cerebral pia it consists of an external 'epi-pia' containing larger vessels and an internal 'pia-glia' or 'pia-intima' in contact with nervous tissue (vide infra). Between the two are some vessels and clefts connecting with the subarachnoid space and a number of blood vessels. It is intimately adherent to the spinal cord. It enters the anterior median fissure, lining its walls with an intervening open meshwork of fine fibrous strands (7.208). A longitudinal fibrous band, the *linea splendens*, extends along the anterior midline. *Ligamenta denticulata* are on each side and a *subarachnoid septum* is posterior (7.208). Beyond the conus medullaris the pia mater continues as the *filum terminale* (p. 922). The pia mater ensheaths the cranial and spinal nerves, blending with their other membranous investments.

The *ligamentum denticulatum* (7.204, 208), a narrow, fibrous sheet situated on each side of the spinal cord between the ventral and dorsal spinal roots, has a medial border continuous with the spinal pia mater, its lateral border having a series of triangular

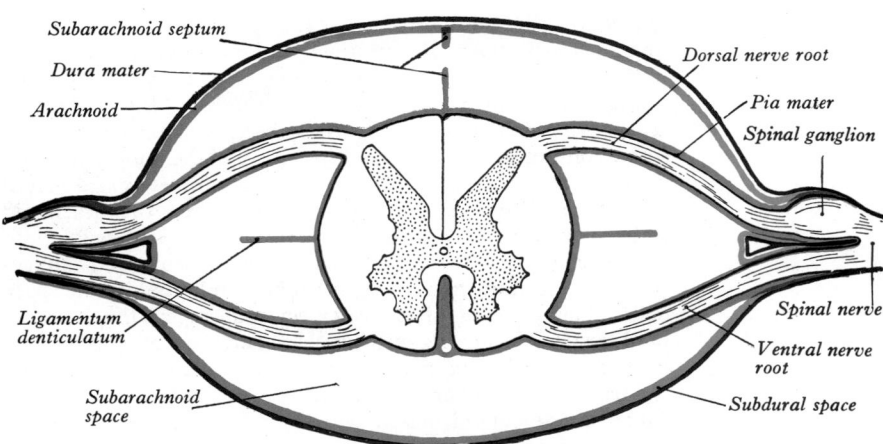

7.208 A transverse section through the spinal cord and its membranes. Black = dura mater, blue = arachnoid mater, red = pia mater.

processes, their apices fixed at intervals to the dura mater. There are usually 21 processes on each side; the first crosses behind the vertebral artery where it pierces the dura mater and is separated by the artery from the first cervical ventral root; it is attached to the dura mater above the rim of the foramen magnum, just behind the hypoglossal nerve; the spinal accessory nerve ascends on its posterior aspect (7.85). The last process is between the exits of the twelfth thoracic and first lumbar spinal nerves and is a narrow oblique band descending laterally from the conus medullaris (7.204). Changes in the form and position of the denticulate ligaments during spinal movements have been demonstrated by cineradiography (Epstein 1966).

The *subarachnoid septum* is a median, interrupted, fibrous sheet connecting the arachnoid to the pia mater opposite the posterior median sulcus (7.208). Incomplete and cribriform in the cervical region, it is a more complete partition at thoracic levels.

The Microscopic Structure of the Leptomeninges

The ultrastructures of the arachnoid and pial meninges are better described together (Winckler 1960); they have a common phylogenetic history, develop embryonically in continuity and preserve this intimate relation in their final differentiation, and indeed, are frequently regarded as a *pia-arachnoid*. They will, however, be described here successively, because they have slight differences. (For review see Hutchins & Weller 1986, Albacado et al 1988.)

The structure of arachnoid mater, like that of the pia, is essentially that of loose, irregular connective tissue. Both *leptomeninges*, like the dura mater (*pachymeninx*), are regarded as generally mesodermal, though the innermost region of pia mater is neurectodermal (Millen & Woollam 1961). In the most primitive vertebrates all three are an undifferentiated *meninx primitiva* but in tetrapods a thick external pachymeninx and more delicate internal leptomeninx are differentiated; in mammals and birds this difference is accentuated. All three meninges develop in mesenchyme around the central nervous system, with the proviso mentioned above.

The *arachnoid mater* consists of collagen, elastin and reticulin fibres, with flat cells usually considered to be mesothelial. On both its dural and internal aspects, cells with long cytoplasmic processes and cytoplasm paler than in the dural fibrocytes, form several layers with a subjacent basement membrane. Tight junctions occur between some of these cells (Nelson et al 1961); collagen fibres are generally interspersed among them and elastic fibres may also appear between the layers. The fibres of the arachnoid are finer than those of the dura mater. Fine *trabeculae* of arachnoid tissue extend across the *subarachnoid space*. These are usually considered to have a covering of 'mesothelial' cells lying over collagen fibres within which are reticulin fibres surrounding numerous small vessels traversing the trabeculae to reach the pia mater and central nervous system. The trabeculae vary much in size and distribution; most are fine strands and are most numerous in the intracranial subarachnoid space but largely absent where this is dilated to form cisternae (p. 1089); they are scarcely evident in the human spinal region. Descriptions of the trabeculae are usually illustrated from other mammals and may convey a misleading impression of their development in humans. Intertrabecular spaces are usually large and are described as coalescing to form a single continuous subarachnoid space; but in the human arrangement large stretches of the space are completely free from trabeculae. Crossing the subarachnoid space the arachnoid trabeculae fuse with the pia mater, which itself is divisible into two layers (vide infra), the external being a vascularized connective tissue stratum covered by mesothelium bounding the space on its inner, cerebral (or spinal) aspect; some regard this as arachnoid tissue, considering only an avascular layer of collagen and other fibres and accompanying cells next to the nervous tissue as the true pia, the subarachnoid space being regarded as existing *within* the arachnoid mater. Others prefer to think of the two leptomeninges as being conjoined as one membrane, the pia-arachnoid.

The arachnoid mater is little vascularized and often described as avascular; but it does support large numbers of small vessels and accompanying nerves which traverse the trabeculae to reach the pia mater. Its mesotheliocytes, and particularly their intercellular tight junctions and the basal lamina, are a barrier to the free diffusion of colloidal substances between the subarachnoid and subdural spaces, the central nervous system and its non-capillary vessels. Return of cerebrospinal fluid is thus largely, if not entirely, limited to the arachnoid villi (p. 1090), where this permeability barrier is modified and possibly circumvented by minute valvular canaliculi. The barrier is also absent where the craniospinal nerves traverse the fused sleeves of the pia, arachnoid and perineurium. At these locations cerebrospinal fluid can diffuse through the arachnoid tissue into the adjacent lymphatics.

The structure of the pia mater is also that of loose connective tissue, containing collagen, elastin and reticulin fibres with flat 'mesotheliocytes' like those of the arachnoid; the cells are considered to be fibrocytic elements by some, especially where they form several layers interleaved with reticulin fibres, next to the central nervous system. Here the innermost cells are apposed to convoluted basal lamina, which lies between them and the subjacent end-feet of astrocytes. This relationship is so close that the internal stratum of the pia, with the basement membrane and glial processes, are sometimes combined as '*pia-glia*' or *pia-intima*, to differentiate it from the rest of the membrane, the '*epi-pia*', which is the pial stratum regarded as arachnoid by some, as noted before. Sleeves of pia and arachnoid, with intervening extensions of subarachnoid space, enter the central nervous system around the blood vessels, forming perivascular 'cuffs', the subject

of so many past polemics.

The pia mater contains large numbers of small blood vessels and their vasomotor nerve supplies. On the spinal surface the more external connective tissue layer of the pia, the epi-pia, is more developed than on the cerebral surface, where it is not easily identified. In this stratum vessels are suspended by trabeculae of pial tissue which blend with arachnoid trabeculae where these occur. Thus many pial vessels project above the surface as they ramify in their pia-arachnoid suspension and are partially surrounded by subarachnoid space. Where they enter the central nervous system, as noted above, prolongations of the space and the two leptomeninges accompany each vessel as far as its capillaries, which are *not* separated from nervous tissue by the 'pia–glial' barrier. Current views deny any significant interchange of fluid between the perivascular extensions of the subarachnoid space and nervous tissue or blood, nor is drainage through them considered likely. It is suggested that these minute fluid-filled spaces merely accommodate to pulsatile variation in the calibre of the vessels which they surround.

The coverings of central nervous capillaries resemble those in the choroid plexuses (p. 1082) in the complete absence of connective tissue elements. The neuronal compartment is separated from blood only by vascular endothelium, the astrocytic end-feet and their joint basal lamina. The individual contribution of these to the *blood–brain barrier* is uncertain in mankind; however, abundant experimental evidence shows this to be the major and (apart from the contributions of the choroid plexuses to the cerebrospinal fluid) probably the exclusive site of entry of fluid and solutes into central nervous tissue. Arachnoid villi thus provide the main exit for cerebrospinal fluid, the drainage of extracellular fluid from the nervous tissue being partly back into the pre-venous ends of capillaries and perhaps partly through the ependyma into the ventricular system. Experiments in mice, using horseradish peroxidase as a tracer, have shed some light on the blood–brain barrier (Brightman et al 1970). In most of the sites examined endotheliocytes are impermeable to large molecules and have very effective occluding (tight) junctions between them, features which prevent passage of tracer. Similar junctions exist between the modified ependymal cells of choroid plexuses. The degree to which these findings apply to other sites has been the subject of much research (see also p. 894).

Apart from their functions in the production, circulation and reabsorption of cerebrospinal fluid, the meninges also collectively provide mechanical support to the central nervous system, especially in the buoyant effect of the contained cerebrospinal fluid. Like connective tissues elsewhere they also provide vital routes of access for vessels and at least a mechanical barrier to infection.

THE PERIPHERAL NERVOUS SYSTEM

The peripheral nervous system comprises the afferent or centripetal nerve fibres connecting receptors to the central nervous system and efferent, centrifugal nerve fibres connecting the central nervous system to the effector apparatus. In mankind these are grouped into 12 pairs of cranial nerves issuing from the brain and 31 pairs of spinal nerves. The sympathetic trunks, ganglia and splanchnic nerves are part of this system but will be described in a separate section (pp. 1154–1169).

In the most primitive vertebrates the spinal cord has a series of ventral motor nerve roots, arising from a ventral (anterior) grey column, and a series of dorsal sensory nerve roots, connected to the dorsal (posterior) grey column. The ventral and dorsal roots do not unite or coincide exactly in position; a ventral root is segmental and distributed to the myotome corresponding to its neuromere of origin while dorsal roots are intersegmental, running in the intersegmental connective tissue to their cutaneous distribution. In most fishes and all higher vertebrates the corresponding ventral and dorsal roots unite to form individual spinal nerves, whose arrangement therefore follows a very primitive pattern, with relatively little modification during evolution.

The arrangement of the cranial nerves, on the other hand, has been profoundly modified. The development and adaptations of the branchial system and suppression of segments during elaborate changes in the head region are largely responsible for this transmutation. In the brain, the corresponding ventral and dorsal nerves *never* fuse, though adjoining ventral *or* dorsal nerves sometimes do. Owing to the disappearance of some myotomes, the corresponding ventral nerves are suppressed; and dorsal nerves, originally sensory to the skin of the head and the mucous membrane of the mouth and pharynx, acquire motor fibres which they distribute to the musculature arising in the branchial region (p. 172). With growth and modification of the brain and elaboration of the head, the cutaneous areas are transferred from one nerve to a neighbour, altering the functions of the individual dorsal nerves.

The incorporation of some precervical segments into the head leads to the fusion of corresponding ventral nerves; the hypoglossal nerve, so formed, is thus added to the cranial series.

The cerebrospinal nerves, as described in detail elsewhere (p. 896), contain numerous nerve fibres collected into bundles,

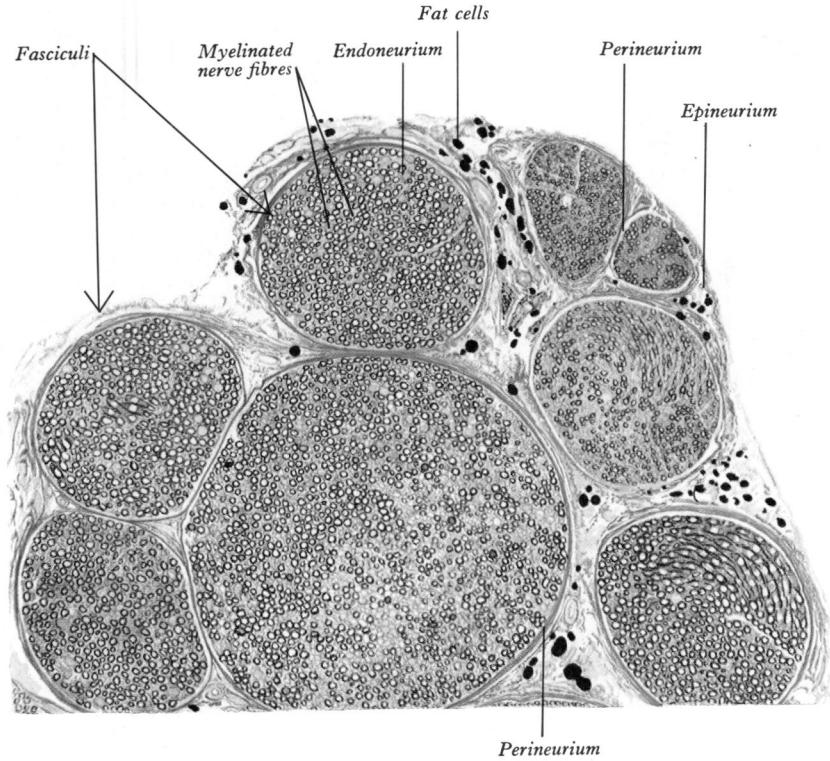

Fat cells
Fasciculi   Myelinated   Endoneurium   Perineurium
nerve fibres
Epineurium

Perineurium

7.209   Transverse section of a peripheral nerve (cat). Stained with osmic acid and van Gieson's technique. Note the nerve fasciculi, the epi-, peri- and endoneurial connective tissue sheaths and the regional variations in the calibre of the myelinated nerve fibres. Magnification *c.* × 55.

and enclosed in connective tissue sheaths (7.209); a small bundle is termed a *fasciculus*. Each fasciculus is surrounded by a connective tissue sheath, the *perineurium*. Individual nerve fibres are held together and supported by a connective tissue *endoneurium*, continuous with septa from the perineurium. The collagen fibres of the endoneurium are longitudinally orientated and bundles are sometimes invaginated into the Schwann cells (lemmocytes) of non-myelinated nerve fibres (Gamble & Eames 1964). A small nerve may be a single fasciculus; but nerves usually contain several fasciculi held together by a connective tissue investment, the *epineurium*. Cerebrospinal nerves contain myelinated and non-myelinated nerve fibres in proportions varying with their functional roles.

The blood vessels supplying a nerve end in a capillary plexus whose members pierce the perineurium and run largely parallel with the fibres, connected by short transverse vessels, to form narrow, oblong meshes, as in muscle. Small non-myelinated vasomotor nerve fibres accompany the larger, and break up into fibrils forming networks around them. Myelinated *nervi nervorum* run in the epineurium, their fibres ending in oval or bulbous corpuscles (p. 913).

Cerebrospinal nerve fibres pursue uninterrupted courses from the centre to the periphery but single fibres or groups may leave one fasciculus to join another, often at acute angles. Nerves divide into branches, frequently communicating with the branches of nearby nerves, forming *plexuses*. Such a plexus is formed by the ventral rami of spinal nerves, e.g. cervical, brachial, lumbar and sacral plexuses; terminal fasciculi may form peripheral plexuses. In forming a plexus, the component nerves divide, join and again subdivide, often in such a complex manner that individual fasciculi are intricately interwoven. Each branch leaving a plexus may hence contain filaments from more than one or even all the 'roots' of the plexus. In smaller plexuses in the periphery there is free interchange of fibres. Through this interchange, every nerve leaving a plexus gains more extensive connections with the spinal cord.

The origin of a nerve is a phrase usually implying its site of exit from or entry into the central nervous system. This is sometimes called its *superficial origin*, its *deep origin* being the assembly of nerve cells from which its peripheral axons come, or

to which they proceed. The superficial origin may be single or by one or several 'roots'. *Efferent nerve fibres* are the peripheral axons of neurons with somata in the central grey matter; *afferent nerve fibres* spring from neurons with somata in special sense organs (e.g. the retina) or in the sensory ganglia of cerebrospinal nerves. Having entered the central nervous system afferent nerve fibres branch and send their terminals to synapse with specific neurons within it.

The peripheral terminations of sensory nerves are described on pp. 906–916 and those of motor nerves on pp. 904–906.

Ganglia are aggregations of neuronal somata associated with some peripheral nerves, occurring in the dorsal roots of spinal

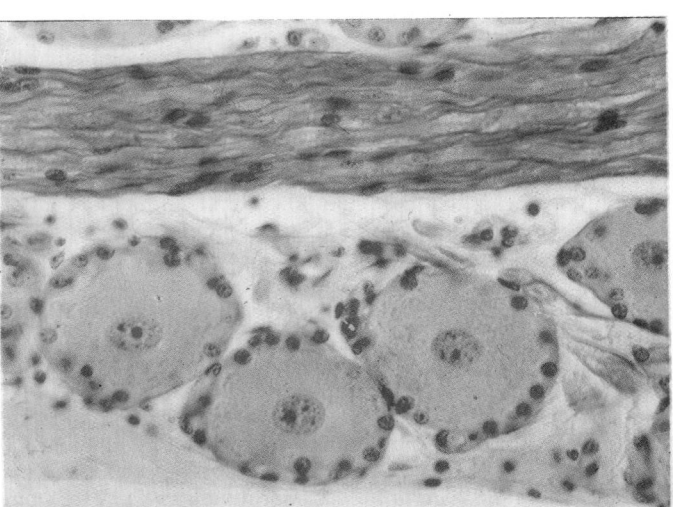

7.210   A typical field in a dorsal spinal nerve root ganglion. Note the characteristic juxtaposition of large ovoid nerve cell somata and the fascicles of myelinated and non-myelinated nerve fibres. Note also the nuclei of the capsular (satellite) cells which surround each nerve cell. Grübler's stain; material provided by Lyn Gregson, Dept. of Anatomy, Guy's Hospital Medical School, London.

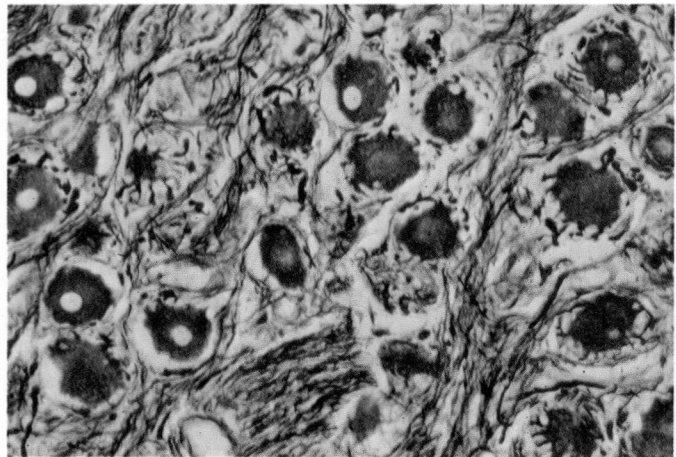

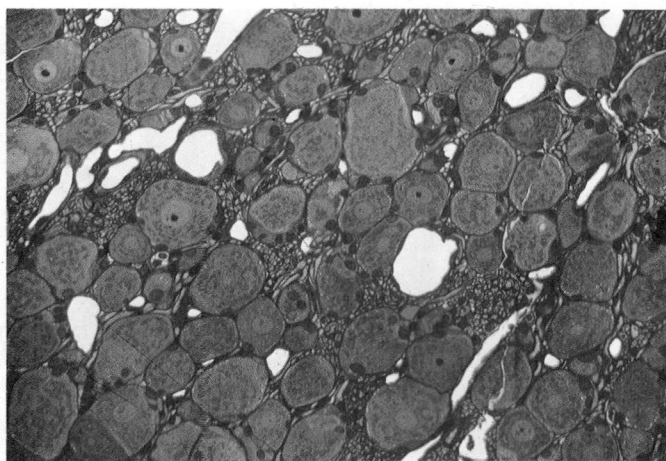

7.211   Typical field in an autonomic ganglion (human ciliary ganglion), showing nerve cells evenly scattered among fascicles of nerve fibres. Stained by Bielschowsky's silver and erythrosin technique in material supplied by N A Locket, Institute of Ophthalmology, London.

7.212   Cells in the superior cervical sympathetic ganglion of rabbit. Semi-thin section of araldite-embedded material stained by toluidine blue. Material supplied by J S Dixon, Dept. of Anatomy, University of Manchester.

nerves and the sensory roots of the trigeminal, facial, glossopharyngeal, vagal and vestibulocochlear cranial nerves. They also occur in autonomic nerves (p. 1157). They vary in form and size, each ganglion being invested by a smooth, firm capsule of fibrous connective tissue continuous with the perineurium and sending numerous fibrous processes into the ganglion.

Each ganglion contains neurons and their fibres and, in some instances, fibres from cells located elsewhere which pass through or terminate in it. Their structure is considered in greater detail on p. 1123. In *spinal ganglia* the neurons are large, unipolar and grouped at the periphery (7.210); in autonomic ganglia they are multipolar and more uniformly scattered (7.211, 212). Each spinal ganglionic neuron has a nucleated capsule (7.236) continuous with the Schwann cells of its fibres. Larger neurons in the spinal ganglia (7.210) are irregularly spherical, with a single dendro-axonal process passing towards the centre of the ganglion and dividing like a T; one neurite (the axon) enters the spinal cord, while the other (the dendrite) passes to the periphery (vide infra).

Near its origin the process is coiled into a form termed a *glomerulus* (p. 888). The smaller neurons, which are more numerous, give rise to fine non-myelinated fibres in peripheral nerves and their spinal roots, each having a thin dendro-axonal process which does not form a glomerulus, instead passing straight to its bifurcation. The peripheral division of the initial process of a unipolar ganglion cell resembles an axon in every respect, but it functions as an elongated dendrite. However, all neurites in peripheral nerves, afferent or efferent, are generally termed axons.

The somata of afferent neurons in primitive invertebrates, such as *Hydra*, lie in the ectoderm, with superficial dendritic processes projecting from them; but in all vertebrates they are close to the central nervous system, their peripheral processes being presumably elongated homologues of the short 'dendrites' of primitive neurons. Primary olfactory neurons retain this primitive position (p. 1173). In the sensory ganglia of the cranial nerves, neurons are unipolar, but in vestibulocochlear ganglia they remain bipolar.

THE CRANIAL NERVES

The craniocaudal sequence of cranial nerves is as follows:

1 Olfactory	5 Trigeminal	9 Glossopharyngeal
2 Optic	6 Abducent	10 Vagal
3 Oculomotor	7 Facial	11 Accessory
4 Trochlear	8 Vestibulocochlear	12 Hypoglossal

These nerves are continuous with the brain, traversing the foramina in the cranial base. The **motor** or **efferent** fibres of cranial nerves arise from groups of neurons in the brain which are their *nuclei of origin*. They are connected with the cerebral cortex by corticonuclear fibres which arise from neurons of cortical motor areas, descending chiefly in the genicular part of the internal capsule to the brain stem, where many but not all decussate and end by arborizing around the neurons of origin of the motor cranial nerves. The **sensory** or **afferent** cranial nerves arise (except for one instance, see p. 1095) from neurons situated outside the brain, grouped to form ganglia or sited in peripheral sensory organs such as the nose and eye. Their processes enter the brain and end around neurons grouped to form their *nuclei of termination*, fibres from the cells of which usually cross before ascending to connect indirectly with the cerebral cortex. Of course, the pathways involved are usually more complex but not different in principle.

Fibres of most cranial nerves begin to myelinate about the fourteenth week in utero, but not until the twenty-second week

i.u. in the sensory part of the trigeminal nerve and cochlear division of the vestibulocochlear, and even later in the optic nerve.

The Olfactory Nerves

The olfactory nerves serving the sense of smell (7.213) have their cells of origin in the olfactory mucosa in the nasal cavity; this olfactory region comprises the mucosa of the superior nasal concha and the opposite part of the nasal septum. The nerve fibres originate as the central or deep processes of the olfactory cells (7.150) and collect into bundles which cross in various directions, forming a plexiform network in the mucosa, finally forming about 20 branches, which traverse the cribriform plate in lateral and medial groups and end in the glomeruli of the olfactory bulb (7.150). Each branch has a sheath consisting of dura mater and pia-arachnoid, the former continuing into the nasal periosteum, the latter into the perineural sheaths of the nerve bundles. Tissue spaces in these sheaths connect with those in the nasal mucous membrane and with the subarachnoid space.

The olfactory nerves are non-myelinated and are bundles of fine axons enfolded within Schwann cells. They are unique in the peripheral origin of their neurons in ectoderm, where they remain throughout life in all vertebrates.

Applied Anatomy. In severe head injuries involving the anterior cranial fossa the olfactory bulb may be separated from the olfactory nerves or the nerves may be torn, producing *anosmia*, loss of olfaction. Fractures may involve the meninges, admitting cerebrospinal fluid into the nose. Such injuries also open up avenues for infection from the nasal cavity. The extensions of the subarachnoid space around bundles of olfactory nerve fibres, sometimes regarded as a lymphatic drainage, may favour the spread of infection into the cranial cavity. Evidence for this route of infection is equivocal.

Associated with the olfactory nerves are two small **nervi terminales** (Pearson 1941). These were discovered in lower vertebrates but they exist in the human embryo and adult, consisting mainly of non-myelinated nerve fibres with small groups of bipolar and multipolar neurons. Each nerve lies medial to an olfactory tract and its branches traverse the cribriform plate to be distributed to the nasal mucosa. The nerve is connected to the brain near the anterior perforated substance and septal areas; in some animals its fibres reach the lamina terminalis, in others the hypothalamic region. Peripherally it passes to the mucosa of the nasal cavity (see p. 1176). Ganglion cells have been associated with the nerve by Pearson (1941) and others and ganglia have been observed in mice by Gruneberg (1973), who did not, however, equate them with the ganglion cells which are scattered along the human terminal nerve. The connections and significance of the latter 'ganglion terminale' are unknown. Bojsen-Møller (1975) considered the nerve to be entirely sensory and recent evidence suggests a role in pheromonal detection in some mammals. The **vomeronasal nerve**, with which the nervus terminalis is frequently confused, is probably absent in adults (p. 1029).

The detailed structure and central connections of the olfactory bulb are described on pp. 1028–1033 and the structure of the olfactory mucosa on p. 1173.

The Optic Nerve

The optic nerve, mediating vision, is distributed to the eyeball. Most of its fibres are afferent, originating in retinal ganglionic neurons (p. 1202), but some may be efferent, their origin uncertain. Developmentally, the optic nerves and retinae are outgrowths of the brain (p. 201), their fibres are covered by oligodendrocytes, not Schwann cells (p. 893). Optic nerve fibres form the internal layer (*stratum opticum*) of the retina, being the axons of cells in its ganglionic layer; they converge on the optic disc, pierce the outer layers of the retina, the choroid coat, and lamina cribrosa near the posterior pole of the eyeball, about 3 or 4 mm nasal to its

centre. As they traverse the lamina cribrosa they develop myelin sheaths and run in fascicles collecting to form the optic nerve. This nerve is about 4 cm long and is directed posteromedially through the back of the orbital cavity, where it runs through the optic canal into the cranial cavity to join the optic chiasma. Its intraorbital part, about 25 mm long, has a slightly sinuous course, the nerve being about 6 mm longer than the distance between the optic canal and the eyeball. Posteriorly it is surrounded by the four recti and separated from them anteriorly by fat, in which the ciliary vessels and nerves are embedded. The ciliary ganglion lies between the optic nerve and the lateral rectus. The nerve is pierced inferomedially about 12 mm behind the eyeball by the central retinal artery and vein, which pass along the centre of the nerve to the optic disc. In the optic canal, which is about 5 mm long, the nerve lies superomedial to the ophthalmic artery and is separated medially from the sphenoidal and posterior ethmoidal sinuses by a thin osseous lamina; anterior to the canal the nasociliary nerve and ophthalmic artery run forward and medially, crossing above the optic nerve, whilst a branch from the inferior division of the oculomotor nerve passes below it to the medial rectus (7.225).

The intracranial part of the optic nerve, which is about 10 mm long, runs posteromedially from the optic canal to the optic chiasma. The posterior parts of the olfactory tract and gyrus rectus and, near the chiasma, the anterior cerebral artery are above the nerve, the internal carotid artery being lateral to it.

The optic nerve is enclosed in three sheaths continuous with the meninges (7.309, 310) and prolonged to the eyeball. The outer *sheath*, from the dura mater, is thick and fibrous, blending with the sclera. The *intermediate sheath* from the arachnoid mater is thin, separated from the outer by the subdural space and from the inner by the subarachnoid space. The inner *sheath* from the pia mater is vascular and closely invests the nerve. From its deep surface septa enter the nerve, dividing and reuniting to enclose, as seen in transverse sections, polygonal areas which are occupied by fascicles of fibres, about 1000 fascicles in all. The inner sheath also invests the central vessels of the retina as far as the optic disc.

The ultrastructure of the optic meninges resembles that of meninges elsewhere (Anderson & Hoyt 1969) but the amount of collagen fibres in the pia and arachnoid is greater than in the intracranial leptomeninges. As elsewhere the subarachnoid space is lined completely by epitheliocytes of the pia-arachnoid, resembling fibroblasts and forming multilaminar membranes of 'mesothelium' or 'meningothelium'.

Near the eye the macular fibres (*papillo-macular bundle*) are lateral in the nerve, but they gradually become medial and just

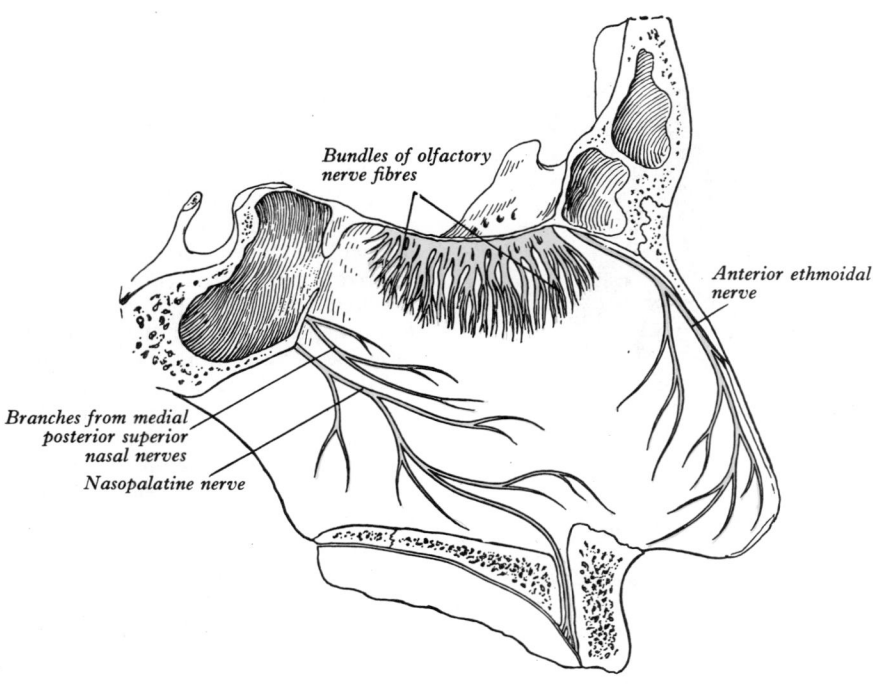

Bundles of olfactory nerve fibres

Anterior ethmoidal nerve

Branches from medial posterior superior nasal nerves

Nasopalatine nerve

7.213   Bundles of olfactory nerve fibres and nerves of the septum of the nose (right side).

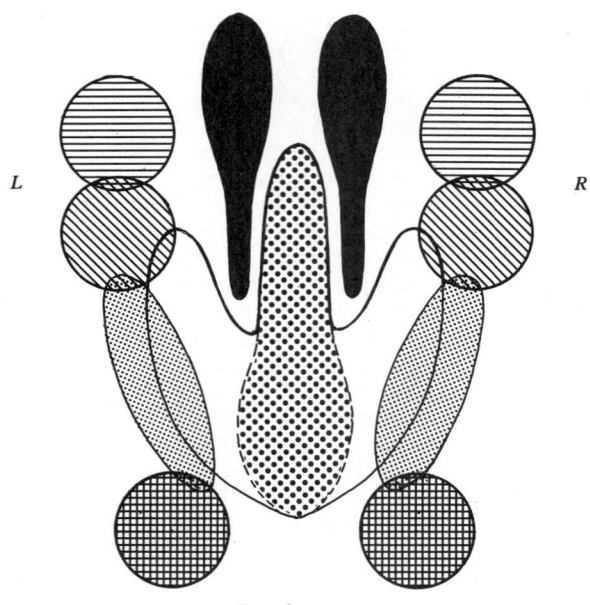

Dorsal aspect

7.214A Representation of the constituent parts of the nucleus of the oculomotor nerve in the monkey. Key as in 7.214B. (Adapted from Brouwer 1918.)

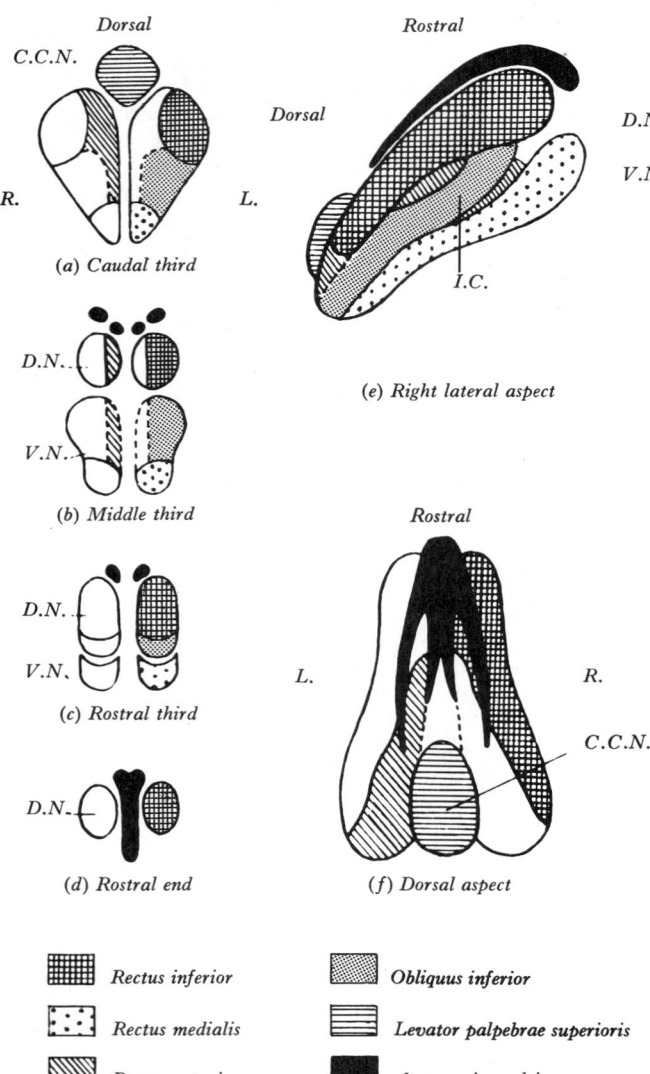

▦ Rectus inferior	▨ Obliquus inferior
⋮ Rectus medialis	▤ Levator palpebrae superioris
▧ Rectus superior	■ Autonomic nuclei

7.214B The constituent parts of the nucleus of the oculomotor nerve in the monkey. From Warwick 1953, by courtesy of the publishers; for significance of abbreviations consult Warwick 1950 and see text.

anterior to the chiasma lie close to the medial margin. Fibres from the upper and lower retinal areas are, respectively, above and below; the fibres from the temporal quadrants are lateral, and those from the nasal quadrants medial.

Counts of optic nerve fibres in man (Oppel 1963, Kupfer et al 1967) show that there are about 1 200 000 myelinated axons, about 92% being about 1 μm in diameter, the rest varying from 2 to 10 μm. About 53% cross in the chiasma. Most end in the lateral geniculate body but some pass to the pretectal nucleus, superior colliculus and hypothalamic nuclei (p. 1009). A small number are efferent but of unknown origin. Such centrifugal fibres reaching the retina have attracted much interest in birds, where they arise from the *isthmo-optic nucleus*, which lies dorsolaterally at the junction of the mid- and hindbrain (Cowan & Clarke 1976).

The optic nerve is supplied with arterial blood by the plexus in its pial sheath and by direct intraneural branches. The pial plexus is supplied by a superior hypophysial and the ophthalmic artery intracranially, by recurrent ophthalmic branches in the optic canal and by posterior ciliary arteries and the extraneural part of the central retinal artery in the orbit. Intraneural branches are from the central artery but their actual contribution to the nerve is probably small (Belmonte 1968, Francois & Neetens 1969). The rich supply of the optic papilla and lamina cribrosa has been emphasized (Henkind & Levitsky 1969). Venous drainage is by the central retinal vein (p. 1204, and Steele & Blunt 1956).

The optic chiasma (p. 1014) and the **optic tract** (p. 1015) have been described elsewhere.

Applied Anatomy. The optic nerve is liable to neuritis or atrophy in some central nervous conditions and its anatomy may partly explain this predisposition. (1) From its development and structure the optic nerve is a prolongation of brain substance, rather than a cranial nerve. (2) It has sheaths from the three cerebral meninges, separated by spaces connecting with the subdural and subarachnoid spaces. The pial sheath sends a process into the nerve and enters intimately into its structure. Thus infection of the meninges or the brain readily extends to the nerve. *Optic neuritis* or *papilloedema*, often a sequela of intracranial neoplasm, is probably caused by increased pressure due to excess fluid in the general subarachnoid space, which extends as far as the lamina cribrosa (p. 1181).

The Oculomotor Nerve

The oculomotor nerve (6.151, 152, 7.139, 152, 216, 226) supplies all the extraocular muscles except the obliquus superior and rectus lateralis; it also supplies, through the ciliary ganglion, the sphincter pupillae and ciliaris. It contains about 24 000 axons. Its fibres arise from a *complex of nuclei* in the grey matter, ventral to the cranial part of the cerebral aqueduct and extending rostrally for a short distance into the floor of the third ventricle. From this region its fibres pass forwards through the tegmentum, red nucleus and the medial part of the substantia nigra, curving with a lateral convexity to emerge from the sulcus on the medial side of the cerebral peduncle (7.86, 87, 118). Peripherally the nerve contains afferent fibres from neurons in the trigeminal ganglion, considered proprioceptive (Bortolami et al 1977, Manni et al 1978). Small neurons, varying in number in different species, occur in the oculomotor nerve; these are usually regarded as ectopic proprioceptive neurons, perhaps displaced from the trigeminal ganglion (Tozer 1912, Bortolami et al 1977).

The nuclear complex from which its efferent fibres arise consists of several paired groups of large multipolar neurons, identifiable in most mammals, with paired masses of smaller multipolar neurons, not so easily identified but well developed in primates and other mammals and the source of the parasympathetic outflow in the oculomotor nerves. On each side the large-celled mass is divided into **dorsolateral**, **intermediate** and **ventromedial nuclei** (7.214), extended in columnar form almost in the long axis of the midbrain (Crosby & Henderson 1948, Warwick 1950); these sub-nuclei are identifiable in early fetal life in man (Pearson 1944). Experiments in monkeys show that they are motor pools of the *rectus inferior*, *obliquus inferior* and *rectus medialis*, in dorsoventral order (7.214). Dorsal to the right and left large-celled masses and level with their caudal extremities, is a median nucleus of similar large neurons, the **caudal central nucleus**, which appears to be the conjoined motor pool of the *levatores* of the upper eyelids (Warwick 1950, 1953). Dorsal to the main oculomotor nuclei are the **accessory** or **autonomic nuclei** (of Edinger and Westphal), composed of smaller multipolar neurons, whose axons travel in the oculomotor nerve to relay in the ciliary ganglion (Warwick 1954); these are largest rostrally, fusing together to arch ventrally over the main oculomotor nuclei (7.214); in their caudal, paired parts these autonomic columns have a tendency to further splitting in man and other primates. Sugimoto et al (1978), using the retrograde horseradish peroxidase technique, suggest that many oculomotor parasympathetic neurons may be sited near but outside these nuclei.

Though a median group of larger motor neurons has been a standard feature of oculomotor topography for decades, this 'central nucleus of Perlia' is most variable in mammals, even in the same species. The function of convergence was early ascribed to it on inadequate and fallacious evidence. Its development in primates is not commensurate with its supposed function, nor is it possible to equate its size with binocular vision (Le Gros Clark 1926). It is often unidentifiable in the human midbrain (Tsuchida 1906, Crosby & Woodburne 1943). There are always a few scattered large motor neurons between the right and left oculomotor masses but these never constitute a clear nucleus, like the caudal central nucleus; in any case, they appear to innervate the superior, not the medial, rectus (Warwick 1955).

Other views of the arrangement of the motor pools of the extraocular muscles have been advanced (Brouwer 1918, Szentágothai 1942, Bender & Weinstein 1943), all suggesting a rostrocaudal pattern rather than the dorsoventral scheme described here; but in both forms of organization, though different in other ways, there is agreement that the issuing oculomotor fibres, somatic and autonomic, are almost completely ipsilateral; at the most, some axons from the median raphe and the caudal central nucleus may cross to the opposite oculomotor nerve to innervate the rectus superior and levator palpebrae superioris; extensive oculomotor decussation, a usual tenet of textbook accounts, is not supported by the experimental studies quoted here, almost all of which were carried out in primates. A more recent study, carried out in cats (Tarlov 1972, 1975), describes longitudinal motor pools for individual muscles but arranged differently from those in monkeys (Warwick 1955). Tarlov has also defined a projection (p. 1112) to the oculomotor from the medial and superior vestibular nuclei, terminals from both providing a partly bilateral and overlapping distribution in some oculomotor 'pools' and suggesting excitatory and inhibitory roles; certain physiological data tend to support this (High et al 1971).

For a discussion of the claims that other nuclei contribute to the oculomotor nerve consult Warwick (1953).

Connections of the oculomotor nuclei include fibres from: (1) *both* corticonuclear tracts, some leaving at oculomotor level, some (aberrant fibres) at a higher level and thereafter descending in the medial lemniscus; they either end on oculomotor neurons or are linked to them via interneurons; (2) the medial longitudinal fasciculus, connecting nuclei of the third, fourth, sixth and eighth cranial nerves (p. 985); (3) the tectobulbar tract, connecting to the visual cortex through the superior colliculus; (4) the bilateral pretectal nuclei for the light reflex. Horseradish peroxidase studies indicate a reciprocal connection between oculomotor and abducent nuclei (Graybiel & Hartwig 1974, Maciewicz et al 1975). Such internuclear connections are expected from the results of experimental stimulation of or damage to the medial longitudinal fasciculus and clinicopathological data of internuclear ophthalmoplegia. These connections are part of the control of horizontal eye movements. (For discussion consult Bender 1964, Highstein 1977. See also p. 1122.) Other experiments using the horseradish peroxidase technique suggest that many neurons in the accessory oculomotor nuclei project to the cerebellum and spinal cord (Sugimoto et al 1978).

Course. As it emerges from the brain the nerve lies in the subarachnoid space covered by the pia mater. It passes between the superior cerebellar and posterior cerebral arteries (6.81A), runs forwards in the interpeduncular cistern lateral to the post-

erior communicating artery, perforates the arachnoid in the triangular interval between the free and attached borders of the tentorium cerebelli and pierces the inner dural layer lateral to the posterior clinoid process to traverse the roof and, anteriorly, descend into the lateral wall of the cavernous sinus, where it lies above the trochlear nerve (6.156). Here it receives one or two filaments from the internal carotid sympathetic plexus and connects with the ophthalmic division of the trigeminal. It divides into superior and inferior rami, which enter the orbit by the superior orbital fissure within the anulus tendineus communis, with the nasociliary nerve between them (7.218, 329).

The smaller *superior ramus* ascends on the lateral side of the optic nerve, supplying the rectus superior and levator palpebrae superioris. The *inferior ramus* divides into three branches: one passing under the optic nerve to the rectus medialis, another to the rectus inferior, the third and longest passing forwards between the rectus inferior and lateralis to the obliquus inferior. The branches enter the muscles on their ocular surfaces, except that to the obliquus inferior, which enters its posterior border. From the nerve to obliquus inferior a short (sometimes double or treble) branch passes to the lower part of the ciliary ganglion as its *motor, parasympathetic root*. It contains finely myelinated fibres from the accessory oculomotor nucleus, which synapse with the ganglionic neurons, whose postganglionic fibres pass in the short ciliary nerves to the sphincter pupillae and ciliaris (pp. 1188, 1194).

The ciliary ganglion (7.215, 218) is a small, flat reddish-grey ganglion about the size of a large pin's head, situated near the orbital apex in loose fat between the optic nerve and the rectus lateralis, usually lateral to the ophthalmic artery. It is a peripheral parasympathetic ganglion. Its neurons, which are multipolar, are larger than in typical autonomic ganglia; a very small number of more typical neurons are also present (Warwick 1954). Its *connections* or *roots* (7.215) enter or leave it posteriorly. Only its parasympathetic function is basic and sympathetic and sensory fibres merely pass through (or are absent in some mammals). The *parasympathetic root*, derived from the nerve to the inferior oblique, consists of preganglionic fibres from the accessory (Edinger–Westphal) nuclei (7.139), which relay in the ganglion, the postganglionic fibres travelling in the short ciliary nerves to the sphincter pupillae and ciliaris. More than 95% of these fibres supply the ciliaris, which is much the larger muscle in volume (Warwick 1954); hence this motor pathway is more concerned with accommodation than with the light reflex. The *sympathetic root* is a branch from the internal carotid plexus, passing direct to the ganglion or joining the sensory root to reach it indirectly; it consists of postganglionic fibres from the superior cervical ganglion, which traverse the ganglion without synapsing, to emerge into the short ciliary nerves. They are distributed to blood vessels of the eyeball but may include axons supplying the dilatator pupillae when these do not follow their usual course in the ophthalmic, nasociliary and long ciliary nerves. The *sensory root* is a *ramus communicans* of the nasociliary nerve, containing sensory fibres from the eyeball which reach the ganglion in short ciliary

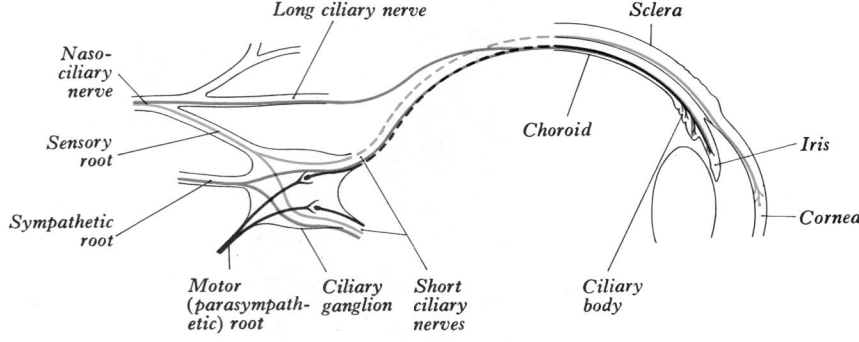

7.215 The ciliary ganglion, with its roots and branches of distribution. Red = sympathetic fibres, heavy black = parasympathetic fibres, blue = sensory (cerebrospinal) fibres. Alternative pathways are given for the sympathetic fibres to the dilatator pupillae. A schematic sagittal section is shown of the upper lateral quadrant of the eyeball but the retina has not been included.

nerves and traverse it without synapsing. The ramus leaves the ganglion posteriorly and runs back to join the nasociliary nerve near its orbital entry.

The *branches* of the ganglion are eight to ten delicate filaments which emerge anteriorly in two bundles, the lower being larger, termed *short ciliary nerves*. The postganglionic parasympathetic fibres in them are myelinated. With the ciliary arteries they run sinuously forwards, above and below the optic nerve, dividing into 15–20 branches which pierce the sclera around the optic nerve and run in small grooves on the internal scleral surface. They contain motor and sensory fibres, the former distributed to sphincter pupillae and ciliaris and to choroidal and iridial blood vessels. The existence of proprioceptive fibres in the oculomotor nerve can no longer be doubted, since stretch endings occur in the extraocular muscles (p. 1122). How far such fibres travel in the nerve, and their central terminations, remain unsettled (p. 1099). The blood supply of the ciliary ganglion, investigated in man by Eliscova (1973), is from small rami of muscular, posterior ciliary and central retinal arteries.

Applied Anatomy. Complete severance of the oculomotor nerve leads to: (1) ptosis, due to paralysis of the levator palpebrae superioris; (2) lateral strabismus, due to the unopposed action of the rectus lateralis and obliquus superior; (3) pupillo-dilatation, because the sphincter pupillae is paralysed; (4) loss of accommodation and of the light reflex, the sphincter pupillae and ciliaris being paralysed; (5) prominence of the eyeball (proptosis) due to muscular relaxation; (6) diplopia, the false image being the higher. Lesions may be partial, with, e.g. a dilated and fixed pupil, with ptosis but no other signs. Irritation of the nerve causes spasm in the muscles supplied by it, e.g. medial strabismus from spasm of the medial rectus, accommodation for near objects only from spasm in the ciliaris or a contracted pupil from spasm in the sphincter pupillae.

The Trochlear Nerve

The trochlear nerve (7.216, 218), the thinnest cranial nerve, supplies the ocular superior oblique muscle. It contains about 3400

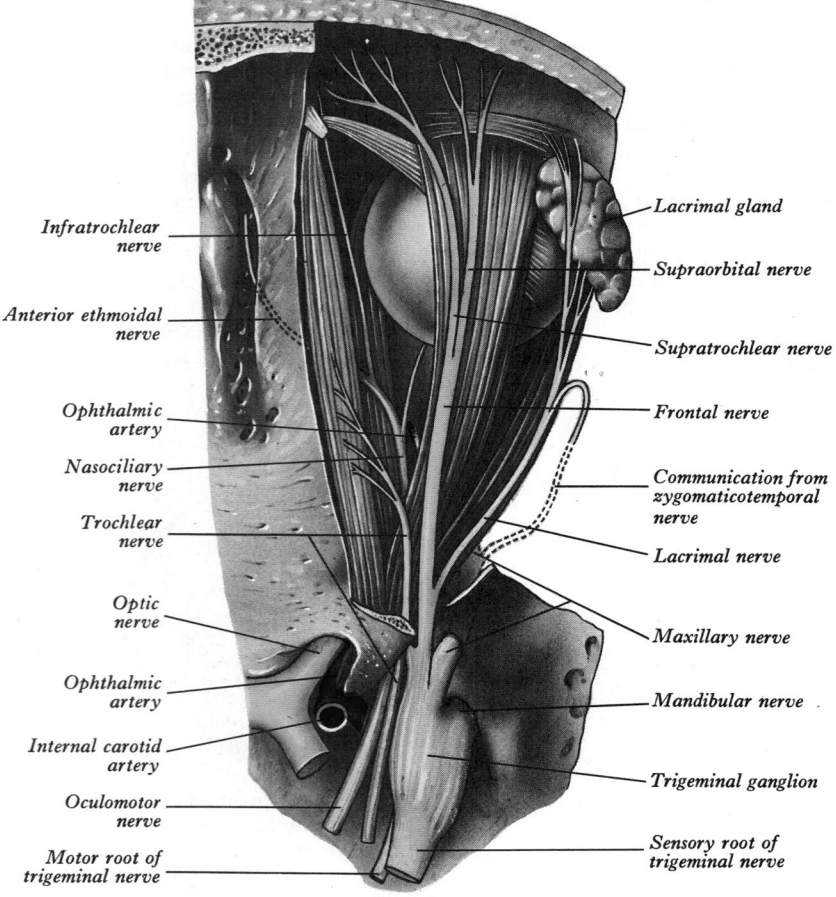

Infratrochlear nerve

Anterior ethmoidal nerve

Ophthalmic artery

Nasociliary nerve

Trochlear nerve

Optic nerve

Ophthalmic artery

Internal carotid artery

Oculomotor nerve

Motor root of trigeminal nerve

Lacrimal gland

Supraorbital nerve

Supratrochlear nerve

Frontal nerve

Communication from zygomaticotemporal nerve

Lacrimal nerve

Maxillary nerve

Mandibular nerve

Trigeminal ganglion

Sensory root of trigeminal nerve

1098   7.216   The nerves of the right orbit: superior aspect.

fibres (Björkman & Wohlfart 1936) but in fetuses the number is greater, as in other cranial nerves. According to Mustafa & Gamble (1979), the adult human nerve contains about 2400 fibres, fetal counts being 4000 (at CR length 9·2 cm), 6000 (at CR length 10 cm) and 3200 (at CR length 24 cm). Many fibres therefore degenerate after an initial increase. The orbital part of the nerve contains more fibres than its stem, a disparity also observed in fetal trochlear nerves.

The trochlear nucleus lies in the floor of the cerebral aqueduct, level with the upper part of the inferior colliculus (7.86, 87, 117), and is in line with the ventromedial part of the oculomotor nucleus, in the position of the somatic efferent column. The medial longitudinal fasciculus is ventral to it. The oculomotor and trochlear nuclei often overlap slightly but can be distinguished by the smaller size of the trochlear neurons.

Connections. The nucleus receives fibres from: (1) *both* corticonuclear tracts, probably in a manner similar to the oculomotor nucleus (p. 1097); (2) the medial longitudinal fasciculus, connecting nuclei of the third, fourth, sixth and eighth cranial nerves (p. 985); (3) the tectobulbar tract, by which impulses arrive from the visual cortex through the superior colliculus (pp. 986, 1122). Tarlov & Tarlov (1975) have described, in cats, GABA-transmitting terminals of axons from the superior vestibular nucleus in the trochlear nucleus, regarding them as inhibitory.

Course. After leaving its nucleus the trochlear nerve pursues an unusual course (p. 983), at first descending laterally through the tegmentum and turning posteriorly round the central grey matter into the anterior medullary velum. Here it decussates with its fellow, crossing the midline to emerge from the velum at the side of the frenulum veli, below the inferior colliculus (7.116). It is the only cranial nerve to emerge dorsally from the brain stem.

The nerve crosses to the lateral side of the superior cerebellar peduncle and winds round the cerebral peduncle just above the pons, between the posterior cerebral and superior cerebellar arteries, appearing between the upper pontine border and the temporal lobe. It pierces the inner dural stratum below the free edge of the tentorium cerebelli, just behind the posterior clinoid process, and then passes forwards in the lateral wall of the cavernous sinus, below the oculomotor nerve and above the ophthalmic division of the trigeminal (6.155). Here it is adherent to the tentorial branch of the ophthalmic nerve, which lies below it. Near the anterior end of the sinus it crosses the oculomotor nerve, entering the orbit by the superior orbital fissure above the annular tendon and medial to the frontal nerve (7.216). In the orbit it inclines medially, above the origin of the levator palpebrae superioris, to enter the orbital surface of the obliquus superior (7.218).

In the lateral wall of the cavernous sinus the nerve communicates with the ophthalmic division of the trigeminal and with the internal carotid sympathetic plexus. In the superior orbital fissure it occasionally sends a branch to the lacrimal nerve. Though an exchange of fibres in the cavernous sinus has been denied (Sunderland & Hughes 1946), subsequent fibre analysis in human material (Zaki 1960) suggests that a substantial component of large nerve fibres, possibly proprioceptive, exists in the trochlear nerve *distal* to the sinus; this part contains 3500 fibres but only 2400 proximal to the sinus. Though it is sometimes assumed, as in the oculomotor nerve, that proprioceptive fibres (from the superior oblique) enter the brain stem in the trochlear nerve, Manni et al (1970) suggest that these leave the nerve peripherally to join the ophthalmic division of the trigeminal.

Applied Anatomy. Interruption of the trochlear nerve paralyses the superior oblique, limiting inferolateral ocular movement; the affected eye rotates medially, producing *diplopia*. Single vision prevails as long as the eyes move above the horizontal; diplopia occurs on looking downwards. To counteract this the sufferer holds his head forwards and inclined to the other side.

The Trigeminal Nerve

The trigeminal, the largest cranial nerve, is the sensory supply to the face, the greater part of the scalp, the teeth, the oral and nasal cavities and the motor supply to the masticatory and some other muscles. It also contains proprioceptive nerve fibres from the masticatory and probably the extraocular muscles. It has three divisions: ophthalmic, maxillary and mandibular. Baumel (1974) has reviewed the functions of the widespread peripheral connections between the trigeminal and facial nerves. Despite the voluminous literature on this topic, uncertainties still exist; however, the preponderant view is that fibres joining the trigeminal from the facial are afferent and arise largely from the facial musculature. The trigeminal nerve emerges from the ventral surface of the pons, near its upper border, as a large sensory and a small motor root, the latter lying anteromedial to the former.

Fibres in the *sensory root* are mainly axons of cells in the **trigeminal** (*semilunar*) **ganglion**, which (7.216–219) occupies a recess (*trigeminal cave*) in the dura mater covering the trigeminal impression near the apex of the petrous temporal bone (p. 397). The ganglion lies at a depth of 4·5–5 cm from the lateral surface of the head at the posterior end of the zygomatic arch; it is crescentic, with its convexity directed anterolateral and on its surface interlacing nerve fascicles are visible. *Medial* to it are the internal carotid artery and the posterior part of the cavernous sinus; *inferior* are the motor root of the nerve, the greater (superficial) petrosal nerve, the apex of the petrous temporal bone and the foramen lacerum. It receives filaments from the internal carotid sympathetic plexus and supplies twigs to the tentorium cerebelli.

The neurites of unipolar cells in the trigeminal ganglion divide into peripheral and central branches, the former being grouped to form the *ophthalmic* and *maxillary* nerves and the sensory part of the *mandibular*. The central branches constitute the fibres of the sensory root, leaving the ganglion's concave margin to run posteromedially under the superior petrosal sinus and tentorium cerebelli to enter the pons. Some fibres from proprioceptor endings in the masticatory muscles traverse the ganglion uninterruptedly to pass to the mesencephalic nucleus of the trigeminal (p. 983). Some sensory fibres from the extraocular muscles may do likewise but electrophysiological observations (Manni et al 1970) indicate that most if not all of such proprioceptive axons have somata in the trigeminal ganglion (p. 1098).

On entering the pons, the fibres of the sensory root run dorsomedially towards the **principal sensory nucleus** situated at this level (7.96); before reaching the nucleus about 50% of the fibres divide into ascending and descending branches, the others ascending or descending without division. The descending fibres, predominantly finely myelinated or non-myelinated, form the **spinal tract of the trigeminal nerve** which reaches the upper cervical spinal cord; as it descends, terminals and collaterals are given off to synapse with neurons in the **spinal trigeminal nucleus** (7.86, 88–90, 91, 94, 217), which contains small and intermediate neurons and is continuous with the substantia gelatinosa of the spinal cord. The tract embraces the nucleus dorsolaterally; in the lower medulla oblongata it is superficial and lies under the tuberculum cinereum. The fibres which synapse in the nucleus are concerned mostly (but not entirely) with pain and thermal sensibility. Experiments show that: (1) the ophthalmic fibres are ventral in the tract and descend to the lower limit of the first cervical spinal segment; (2) the maxillary fibres are central and do not extend below the medulla oblongata; (3) mandibular fibres are dorsal in the tract and do not extend much below the mid-medullary level (7.217). This evidence is confirmed by the clinical results of section of the spinal tract (Smyth 1939, Brodal 1947, Falconer 1949) in cases of severe trigeminal neuralgia; section 4 mm below the obex renders the ophthalmic and maxillary areas analgesic but tactile sensibility, apart from the abolition of 'tickle', is much less affected; to include the mandibular area, section must be made at the level of the obex. By such a section, analgesia is extended to the mucosa of the tonsillar sinus, the posterior lingual third and adjoining parts of the pharyngeal wall (glossopharyngeal nerve) and the cutaneous area supplied by the auricular branch of the vagus. The fibres concerned must, therefore, join the trigeminal spinal tract and end in its nucleus. As a result of such operations the laminar representation in the tract has also been confirmed. A small group of sensory neurons associated with the glossopharyngeal nerve but related to the spinal tract of the trigeminal, has been described (Bossy 1968). Other afferents reach the spinal nucleus from the sensory root of the facial nerve and the dorsal roots of the upper cervical nerves;

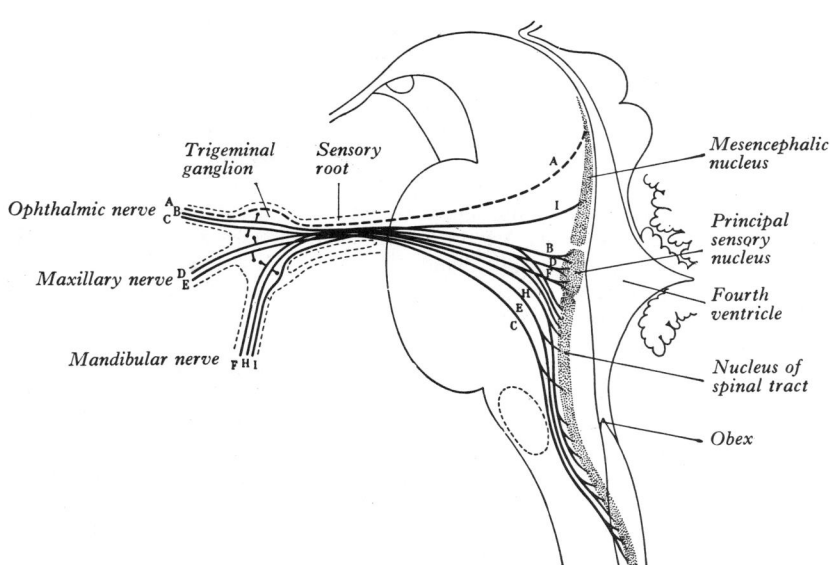

7.217 The nuclei receiving the primary afferent fibres of the trigeminal nerve: A. proprioceptive fibres from ocular muscles; B. tactile and pressure fibres from opthalmic areas; C. pain and temperature fibres from the opthalmic area; D. tactile and pressure fibres from the maxillary area; E. pain and temperature fibres from the maxillary area; F. tactile and pressure fibres from the mandibular area; H. pain and temperature fibres from the mandibular area; I. proprioceptive fibres from the muscles of mastication. Proprioceptive fibres are believed to occur in all three divisions of the trigeminal nerve.

descending fibres from the sensorimotor cortex also pass to it (p. 1122).

It is appropriate to divide the spinal trigeminal nucleus into three levels: *the nucleus oralis* (most rostral, adjoining the principal sensory nucleus), the *nucleus interpolaris* and *nucleus caudalis* (the most caudal part, continuous with the dorsal grey column). They differ in structure and to some extent in their connections. The nucleus caudalis resembles in its neuronal pattern the dorsal grey column, the nuclei interpolaris and oralis being devoid of neurons 'equivalent' to those in the substantia gelatinosa. All levels receive afferents from the whole trigeminal sensory field but all C fibres are believed to end in the caudal nucleus and its neurons alone respond to noxious stimuli; furthermore, its projection is to the contralateral ventroposterior thalamic nucleus, as is that of the nucleus oralis; in contrast, the nucleus interpolaris projects to the anterior lobe of the cerebellum.

Some ascending trigeminal fibres, many of them heavily myelinated, synapse around the small neurons in the principal sensory nucleus, which lies lateral to the motor nucleus and medial to the middle cerebellar peduncle; it is continuous inferiorly with the spinal nucleus (7.86, 217) and is considered to be mainly concerned with tactile stimuli.

Other ascending fibres enter the **mesencephalic nucleus**, a column of *unipolar* cells, whose peripheral branches may convey proprioceptive impulses from the masticatory muscles; it is also stated (Corbin & Harrison 1940, Pearson 1949) that similar impulses reach it from the teeth and from the facial and ocular muscles (7.86, 117, 195, 196). Its neurons are unique in being the only primary sensory neurons with somata in the central nervous system (Johnston 1909). If, however, the primary proprioceptive neurons of extraocular muscles are in fact situated in their motor nerves or in the trigeminal ganglion (pp. 1098, 1122), some mesencephalic trigeminal neurons may be 'secondary' in status. Small multipolar cells, possibly interneurons, occur near the unipolar neurons.

Connections. Most fibres arising in the trigeminal sensory nuclei cross the midline to ascend in the trigeminal lemniscus (p. 983) to the thalamic nucleus ventralis posterior medialis (p. 1002), relaying to the cortical postcentral gyrus (areas 3, 1 and 2, p. 1056). Some, however, ascend to the same nucleus of the ipsilateral thalamus. Collateral branches of primary and secondary afferent trigeminal neurons probably reach many other central regions, such as the other cranial nerve nuclei, the reticular formation, cerebellum, tectum, subthalamus, hypothalamus, etc., but details have not been established in human brains (Humphrey 1969, Webster 1977, Nieuwenhuys et al 1988, p. 1122.)

Nerve fibres ascending to the mesencephalic nucleus may give collaterals to the motor trigeminal nucleus and cerebellum, the 'axons' being, of course, morphologically dendrites. The true axons of mesencephalic neurons possibly descend in part to the principal trigeminal sensory nucleus. Electrophysiological evidence suggests that the mesencephalic nucleus is modulated during masticatory reflexes by connections with the vagus nerve but anatomical confirmation of this is lacking (Manni et al 1977).

The motor nucleus of the trigeminal nerve is ovoid, with characteristic large multipolar cells interspersed with smaller multipolar cells. It lies in the upper pons medial to the principal sensory nucleus, separated from it by fibres of the trigeminal nerve. It occupies the position of the branchial (special visceral) efferent column (7.86, 87, 96) and consists of a number of relatively discrete sub-nuclei whose axons innervate individual muscles (Szentágothai 1949).

Connections. The motor nucleus receives fibres from *both* corticonuclear tracts; these fibres leave the tracts at the nuclear level or higher in the pons (aberrant corticospinal fibres), descending in the medial lemniscus. They may end on motor neurons or interneurons. The motor nucleus receives afferents from the sensory nuclei, including some possibly from the mesencephalic nucleus, forming monosynaptic reflex arcs for proprioceptive control of the masticatory muscles. It also receives afferents from the reticular formation, red nucleus and tectum, the medial longitudinal fasciculus and possibly from the locus ceruleus, by which salivary secretion and mastication may be correlated.

THE OPHTHALMIC NERVE

The ophthalmic nerve (7.216, 219, 224), the superior and smallest trigeminal division, is wholly sensory. It supplies the eyeball, lacrimal gland and conjuctiva, part of the nasal mucosa and the skin of the nose, eyelids, forehead and part of the scalp. It arises from the anteromedial end of the trigeminal ganglion as a flat band, about 2·5 cm long, passing forwards in the cavernous sinus in its lateral wall, below the oculomotor and trochlear nerves (6.155); just before entering the orbit by the superior orbital fissure it divides into *lacrimal*, *frontal* and *nasociliary* branches.

The ophthalmic nerve is joined by filaments from the internal carotid sympathetic plexus and communicates with the oculomotor, trochlear and abducent nerves, thus forming routes by which proprioceptive fibres in these nerves may possibly enter the trigeminal. It has a recurrent meningeal branch (*tentorial nerve*), which crosses below and adheres to the trochlear nerve and is distributed to the tentorium cerebelli (p. 1088).

The lacrimal nerve (7.216), the smallest of the main ophthalmic branches, sometimes receives a filament from the trochlear nerve. The lacrimal nerve enters the orbit through the lateral part of the superior orbital fissure (7.219) and runs along the upper border of the rectus lateralis with the lacrimal artery, receiving a twig from the zygomaticotemporal branch of the maxillary nerve, often said to contain lacrimal secretomotor fibres (p. 1155). Entering the lacrimal gland, it supplies it and the adjoining conjunctiva. It then pierces the orbital septum and ends in the upper eyelid, joining filaments of the facial nerve. It is occasionally absent, being replaced by the zygomaticotemporal nerve; when this is absent it is replaced by a branch of the lacrimal nerve.

The frontal nerve (7.216, 218), the largest branch of the ophthalmic division, enters the orbit by the superior orbital fissure (7.216) above the annular tendon and proceeds between the levator palpebrae superioris and the periosteum, dividing about midway between the apex and base of the orbit into a small supratrochlear and a large supraorbital branch.

The *supratrochlear nerve* runs anteromedially, passing above the trochlea, and supplies a descending filament to the infratrochlear branch of the nasociliary nerve. It then emerges between the trochlea and the supraorbital foramen, curving up on the forehead close to the bone with the supratrochlear artery and supplying the conjunctiva and the skin of the upper eyelid; it then ascends beneath the corrugator and the frontal belly of occipitofrontalis, dividing into branches which pierce these muscles to supply the skin of the lower forehead near the midline.

The *supraorbital nerve* proceeds between the levator palpebrae superioris and the orbital roof and traverses the supraorbital notch or foramen, supplying palpebral filaments to the upper eyelid and conjunctiva. It ascends on the forehead with the supraorbital artery, dividing into a smaller medial and a lateral branch, which supply the skin of the scalp nearly as far back as the lambdoid suture. These branches are at first deep to the frontal belly of the occipitofrontalis; the medial branch perforates it, while the lateral pierces the epicranial aponeurosis. The main nerve and both branches supply small rami to the mucosa of the frontal sinus and to the pericranium; some enter the sinus by foramina in the floor of the supraorbital notch.

The nasociliary nerve (7.216, 218), intermediate in size between frontal and lacrimal, is more deeply placed in the orbit, which it enters through the annular tendon between the two rami of the oculomotor nerve (7.218). It crosses the optic nerve with

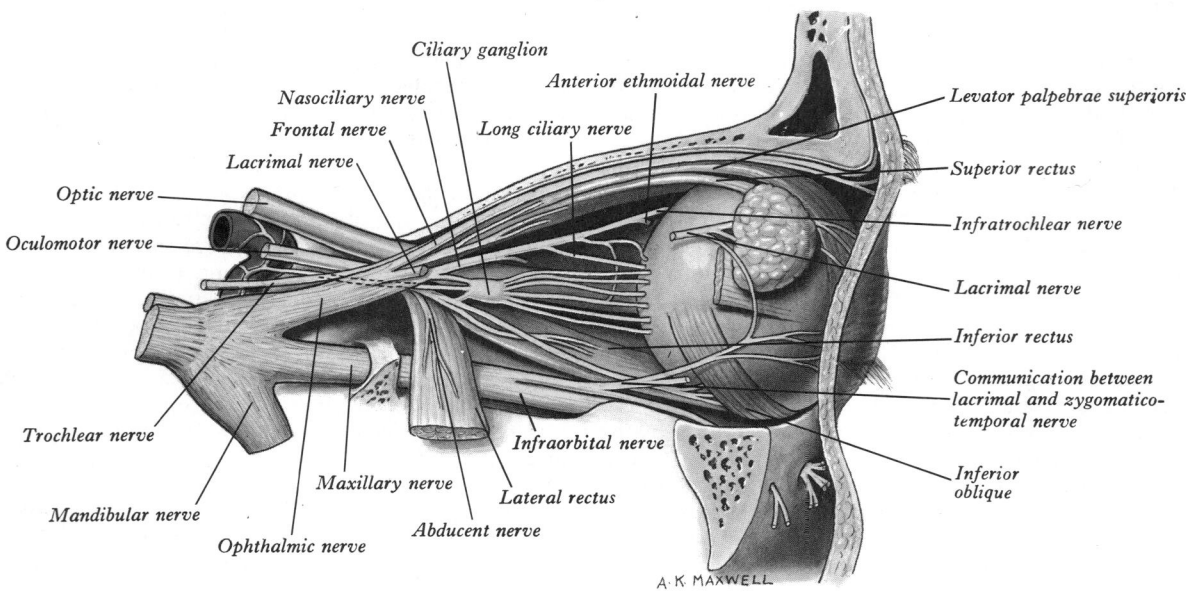

Ciliary ganglion
Nasociliary nerve
Anterior ethmoidal nerve
Frontal nerve
Long ciliary nerve
Lacrimal nerve
Optic nerve
Oculomotor nerve
Levator palpebrae superioris
Superior rectus
Infratrochlear nerve
Lacrimal nerve
Inferior rectus
Communication between lacrimal and zygomatico-temporal nerve
Trochlear nerve
Inferior oblique
Infraorbital nerve
Maxillary nerve
Lateral rectus
Mandibular nerve
Abducent nerve
Ophthalmic nerve

A·K·MAXWELL

7.218 The nerves of the right orbit and the ciliary ganglion: lateral aspect.

the ophthalmic artery and runs obliquely below the rectus superior and obliquus superior to the medial orbital wall. Here, as the *anterior ethmoidal nerve*, it traverses the anterior ethmoidal foramen and canal, enters the cranial cavity and runs forwards in a groove on the upper surface of the cribriform plate beneath the dura mater; it descends through a slit lateral to the crista galli into the nasal cavity, where it occupies a groove on the internal surface of the nasal bone. It supplies two *internal nasal branches*: a medial to the anterior septal mucosa and a lateral to the anterior part of the lateral nasal wall. It emerges, as the *external nasal nerve*, at the lower border of the nasal bone, descending under the transverse part of the nasalis to supply the skin of the nasal ala, apex and vestibule.

The nasociliary nerve connects with the ciliary ganglion and has long ciliary, infratrochlear and posterior ethmoidal branches.

The *ramus communicans* to the ganglion (p. 1097, 7.215) usually joins the nerve as it enters the orbit lateral to the optic nerve, emerging from the ganglion's posterosuperior angle (7.215); it is sometimes joined by a filament from the internal carotid sympathetic plexus or from the superior oculomotor ramus.

Two or three *long ciliary nerves* branch from the nasociliary as it crosses the optic nerve. They accompany the short ciliary nerves to pierce the sclera near the attachment of the optic nerve. Running forwards between sclera and choroid, they supply the ciliary body, iris and cornea and usually contain the sympathetic fibres for the dilatator pupillae (p. 1194), these being postganglionic fibres from neurons in the superior cervical ganglion. In view of the susceptibility of the corneal epithelium to damage, the corneal distribution of the long ciliary nerves is of great importance.

The *infratrochlear nerve* branches from the nasociliary near the anterior ethmoidal foramen; running forwards along the medial orbital wall above the rectus medialis it is joined, near the trochlea, by a small branch from the supratrochlear nerve. It leaves the orbit below the trochlea, supplying the skin of the eyelids and the side of the nose above the medial canthus, the conjunctiva, lacrimal sac and lacrimal caruncle.

The *posterior ethmoidal nerve* leaves the orbit by the posterior ethmoidal foramen and supplies the ethmoidal and sphenoidal sinuses.

THE MAXILLARY NERVE

The maxillary nerve (7.218, 219), the intermediate division of the trigeminal, is wholly sensory; it leaves the trigeminal ganglion between the ophthalmic and mandibular divisions as a flat plexiform band which passes horizontally forwards, low in the lateral wall of the cavernous sinus (**6.155**), to traverse the foramen rotundum, where it becomes more cylindrical and compact. Crossing the upper part of the pterygopalatine fossa, it inclines laterally on the posterior surface of the orbital process of the palatine bone and on the upper part of the posterior surface of the maxilla and enters the orbit through the inferior orbital fissure, as the **infraorbital nerve**. Traversing the infraorbital groove and

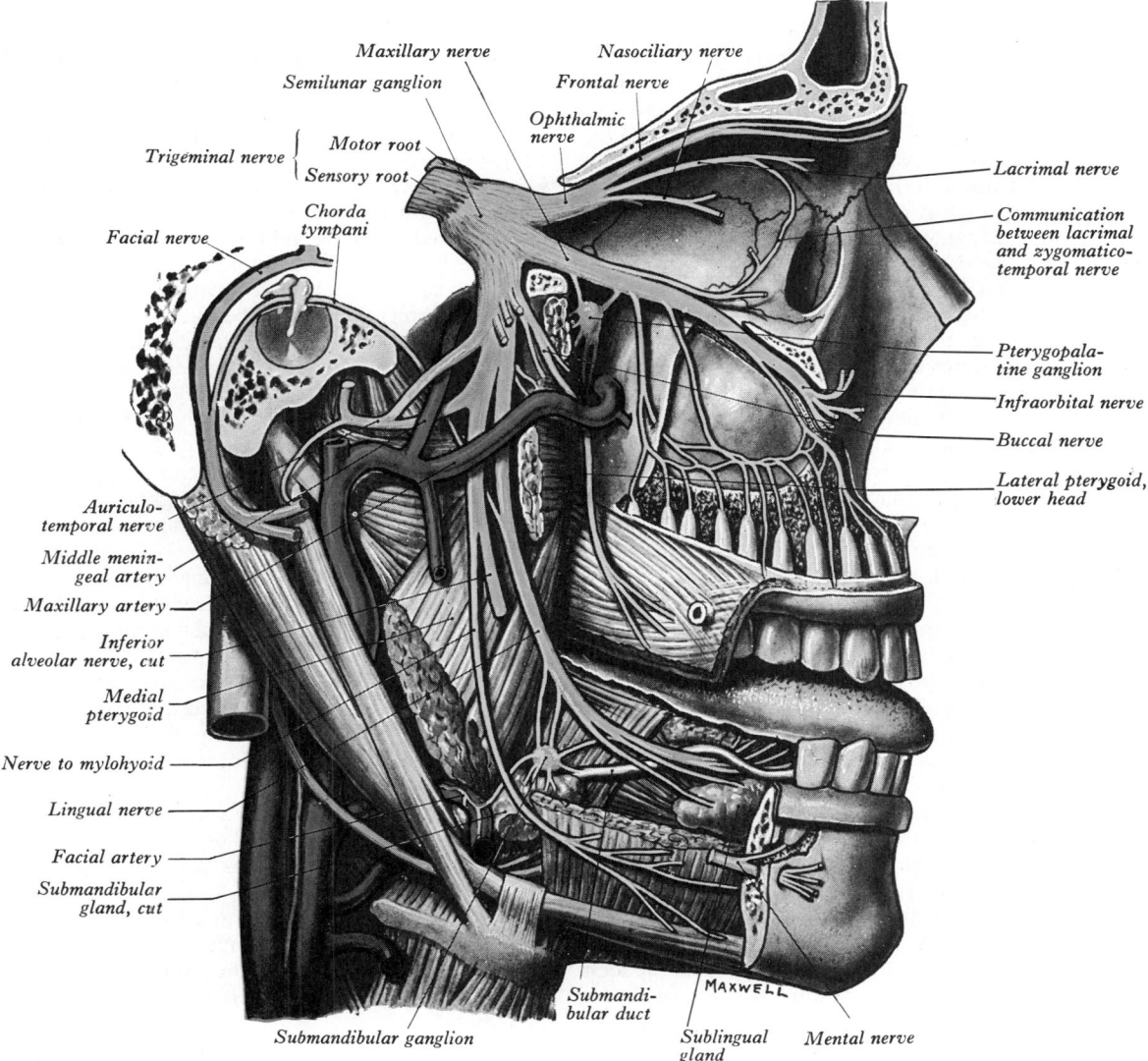

Maxillary nerve
Semilunar ganglion
Trigeminal nerve { *Motor root* *Sensory root* }
Chorda tympani
Facial nerve
Nasociliary nerve
Frontal nerve
Ophthalmic nerve
Lacrimal nerve
Communication between lacrimal and zygomatico-temporal nerve
Pterygopalatine ganglion
Infraorbital nerve
Buccal nerve
Lateral pterygoid, lower head
Auriculo-temporal nerve
Middle meningeal artery
Maxillary artery
Inferior alveolar nerve, cut
Medial pterygoid
Nerve to mylohyoid
Lingual nerve
Facial artery
Submandibular gland, cut
Submandibular ganglion
Submandibular duct
MAXWELL
Sublingual gland
Mental nerve

7.219   The right ophthalmic, maxillary and mandibular nerves and the submandibular and pterygopalatine ganglia. Note that the zygomatic and superior alveolar nerves are not labelled in this diagram.

canal in the orbital floor, it appears on the face through the infra-orbital foramen, under the cover of levator labii superioris, and divides into branches which distribute to the nasal ala and lower eyelid and to the skin and mucous membrane of the cheek and upper lip; these rami communicate with the facial nerve.

Since the mouth is regarded as having evolved from a pair of fused visceral clefts, the maxillary nerve can be described as the pretrematic and the mandibular nerve as the post-trematic branch of the trigeminal nerve. In the early fetus the maxillary nerve primarily supplies structures of the maxillary process but later extends into the adjoining frontonasal process (p. 149 and 7.224).

The branches of the maxillary nerve can be divided into four groups corresponding to their origins, as follows:

In the cranial cavity:	Meningeal
In the pterygopalatine fossa:	Ganglionic, Zygomatic, Posterior superior alveolar
In the infraorbital canal:	Middle superior alveolar, Anterior superior alveolar
On the face:	Palpebral, Nasal, Superior labial

The meningeal nerve or *nervus meningeus medius* (Kimmel 1961a, b) leaves the maxillary nerve near the foramen rotundum; it receives a ramus from the internal carotid sympathetic plexus and accompanies the frontal branch of the middle meningeal artery to supply the dura mater in the middle cranial fossa. Its anterior twigs just reach the anterior fossa.

Two ganglionic branches connect the maxillary nerve to the pterygopalatine (sphenopalatine) ganglion situated just below it in the pterygopalatine fossa (7.219). They contain lacrimal secretomotor fibres (vide infra) and sensory fibres from the orbital periosteum and the mucosae of the nose, palate and pharynx (p. 1328).

The zygomatic nerve (7.219) starts in the pterygopalatine fossa, enters the orbit through the inferior orbital fissure, runs along its lateral wall and divides into two branches, zygomaticotemporal and zygomaticofacial.

The *zygomaticotemporal nerve* passes along the orbit's inferolateral angle, supplies a ramus to the lacrimal nerve (p. 1100), traverses a canal in the zygomatic bone into the temporal fossa, ascends between the bone and temporalis and pierces the temporal fascia about 2 cm above the zygomatic arch, to supply the skin of the temple. It communicates with the facial and auriculotemporal nerves. As it pierces the temporal fascia it sends a slender twig between its two layers towards the lateral angle of the eye. The lacrimal ramus conveys parasympathetic postganglionic fibres from the pterygopalatine ganglion to the lacrimal gland.

The *zygomaticofacial nerve* also traverses the inferolateral angle of the orbit, emerging on the face through a foramen in the zygomatic bone; perforating the orbicularis oculi, it supplies the skin on the prominence of the cheek. It forms a plexus with zygomatic branches of the facial nerve and palpebral branches of the maxillary nerve. Occasionally it is absent. Ruskell (1974) has (in primates) described an *orbitociliary* branch, which leaves the maxillary nerve in the pterygopalatine fossa, to traverse the inferior orbital fissure and reach the eyeball through the ciliary ganglion or retro-orbital plexus (p. 1103); dissections in 25 monkeys, with degeneration experiments, established this course and suggested a predominantly sensory role for its fibres.

The superior alveolar (dental) nerves (7.219) arise from the maxillary nerve in the pterygopalatine fossa or in the infra-orbital groove (canal). They are the posterior, middle and anterior superior alveolar nerves. **The posterior superior alveolar (dental) nerve** leaves the maxillary in the pterygopalatine fossa and runs antero-inferiorly to pierce the infratemporal surface of the maxilla (p. 1305), descending under the mucosa of the maxillary sinus. After supplying the sinus the nerve divides into small branches which link up as the molar part of the *superior dental plexus* supplying twigs to the molar teeth. It also supplies a branch to the upper gum and the adjoining part of the cheek.

The middle superior alveolar (dental) nerve arises from

the infraorbital nerve as it runs in the infraorbital groove, and runs down and forwards in the lateral wall of the maxillary sinus. Like the posterior, it ends in small branches which link up with the superior dental plexus, supplying small rami to the upper premolar teeth. This nerve is variable: it may be duplicated or triplicated or absent (Wood Jones 1939, Fitzgerald 1956, Pacini & Gremigri 1975).

The anterior superior alveolar (dental) nerve (7.219) leaves the lateral side of the infraorbital nerve near the midpoint of its canal and traverses the canalis sinuosus (p. 1304) in the anterior wall of the maxillary sinus. Curving first under the infraorbital foramen, it passes medially towards the nose, turns downwards and divides into branches supplying the incisor and canine teeth. It assists in the formation of the superior dental plexus and gives off a *nasal branch*, which passes through a minute canal in the lateral wall of the inferior meatus to supply the mucous membrane of the anterior area of the lateral wall (as high as the opening of the maxillary sinus) and the floor of the nasal cavity, communicating with the nasal branches of the pterygopalatine ganglion. Finally it emerges near the root of the anterior nasal spine to supply the adjoining part of the nasal septum.

The palpebral branches ascend deep to the orbicularis oculi, piercing the muscle to supply the skin in the lower eyelid and join with the facial and zygomaticofacial nerves near the lateral canthus.

Nasal branches supply the skin of the side of the nose and of the movable part of the nasal septum, joining the external nasal branch of the anterior ethmoidal nerve.

Superior labial branches, large and numerous, descend behind the levator labii superioris, to supply the skin of the anterior part of the cheek, upper lip, oral mucosa and labial glands, and are joined by branches from the facial nerve to form the *infraorbital plexus*.

The pterygopalatine (sphenopalatine) ganglion (7.219, 221), the largest of the peripheral parasympathetic ganglia, is placed deeply in the pterygopalatine fossa, near the sphenopalatine foramen and anterior to the pterygoid canal, It is flattened, reddish-grey and lies just below the maxillary nerve as it crosses the fossa. Though *connected functionally with the facial nerve*, it is so closely related to the maxillary nerve in position that it may conveniently be described here.

Its *motor* or *parasympathetic root* is the *nerve of the pterygoid canal* (p. 1110). Entering the ganglion posteriorly, its fibres probably arise from a special lacrimatory nucleus in the lower pons, emerging in the sensory root of the facial nerve (*nervus intermedius*) to run in its greater petrosal branch (p. 1110), which unites with the deep petrosal nerve (7.221) to form the nerve of the pterygoid canal. These preganglionic fibres relay in the pterygopalatine ganglion, the postganglionic fibres following a complicated course to their destination; leaving the ganglion in one of its branches, they join the maxillary nerve and pass into its zygomatic branch and thence probably into the zygomaticotemporal nerve, leaving it in its communicating ramus to reach the lacrimal nerve (p. 1100). Thus they supply secretomotor fibres to the gland (vide infra). Secretomotor fibres (of uncertain origin) for the palatine, pharyngeal and nasal glands are also believed to follow a similar route to the ganglion, where they are relayed, the postganglionic fibres running in the palatine and nasal branches (7.221).

The *sympathetic root* is also incorporated in the nerve of the pterygoid canal. Its postganglionic fibres arise in the superior cervical ganglion and travel via the internal carotid plexus and deep petrosal nerve.

The *branches* which appear to arise from the pterygopalatine ganglion (7.220) are largely derived from the maxillary nerve and, though intimately related to the ganglion, do *not* synapse in it. They include the orbital, palatine, nasal and the pharyngeal branches.

The *orbital branches* are two or three fine rami which enter the orbit by the inferior orbital fissure and are distributed to the periosteum and orbitalis muscle; some fibres traverse the posterior ethmoidal foramen to supply the sphenoidal and ethmoidal sinuses. The fibres supplying the orbitalis are from the

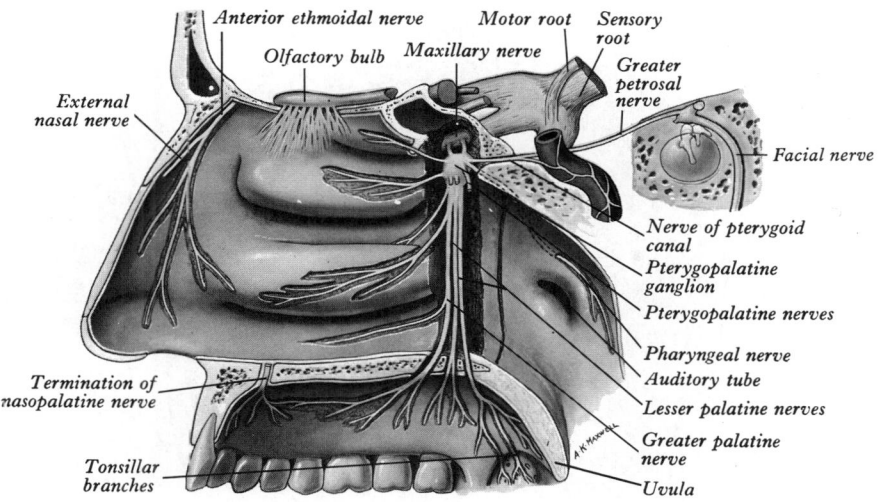

7.220 The right pterygopalatine ganglion and its branches.

sympathetic root. Experiments in monkeys and dissections of human material suggest that the orbital rami form, with branches of the internal carotid sympathetic nerve, a *'retro-orbital' plexus* (Ruskell 1970, 1971) said to supply parasympathetic and sympathetic branches to orbital structures, including the lacrimal gland (p. 1155).

The *palatine nerves* (7.220) are distributed to the roof of the mouth, the soft palate, tonsil and the nasal mucosa. The *greater (anterior) palatine nerve* descends through the greater palatine canal, emerges on the hard palate from the greater palatine foramen, runs forwards in a groove on the inferior surface of the bony palate almost to the incisor teeth and supplies the gums and the mucosa and glands of the osseous palate; it also communicates with the terminal filaments of the nasopalatine nerve. In the greater palatine canal it supplies *posterior inferior nasal branches*, which emerge through the perpendicular plate of the palatine bone and ramify over the inferior nasal concha and the walls of middle and inferior meatuses; at its exit from the canal, the palatine branches are distributed to both surfaces of the soft palate. The *lesser (middle and posterior) palatine nerves* descend through the greater palatine canal, emerge through the lesser palatine foramina and give branches to the uvula, tonsil and soft palate. Fibres conveying taste impulses from the palate probably pass via the palatine nerves to the pterygopalatine ganglion and through it to the nerve of the pterygoid canal and greater petrosal nerve to the facial ganglion, where their somata are situated. The central processes of these neurons traverse the sensory root of the facial nerve (nervus intermedius) to pass to the nucleus solitarius (p. 954).

Nasal branches enter the nasal cavity through the sphenopalatine foramen, forming the lateral and medial groups. (1) About six *lateral posterior superior nasal nerves* innervate the mucosa of the posterior parts of the superior and middle nasal conchae and the lining of the posterior ethmoidal sinuses. (2) Two or three *medial posterior superior nasal nerves* cross the nasal roof below the opening of the sphenoidal sinus to supply the mucosa of the posterior part of the roof and of the nasal septum. The largest of these nerves is the *nasopalatine (long sphenopalatine) nerve*, which runs antero-inferiorly on the nasal septum in a groove on the vomer. It descends to the roof of the mouth through the incisive fossa in the anterior hard palate. When an anterior and a posterior incisive foramen (p. 354) exist in this fossa, the left nasopalatine nerve traverses the anterior and the right nerve the posterior foramen. The nasopalatine nerves supply a few filaments to the nasal septum and end by supplying the mucosa of the anterior part of the hard palate, there communicating with the anterior palatine nerves.

The *pharyngeal nerve* arises from the posterior part of the ganglion, traverses the palatinovaginal canal with a pharyngeal branch of the maxillary artery and supplies the mucosa of the nasopharynx behind the auditory tube.

THE MANDIBULAR NERVE

The mandibular nerve (7.219, 223, 224,) supplies: the teeth and gums of the mandible, the skin in the temporal region, part of the auricle, the lower lip, the lower part of the face and the muscles of mastication; it also supplies the mucosa of the anterior, presulcal part of the tongue and the oral floor. The largest trigeminal division, it has a large, *sensory root* which proceeds from the lateral part of the trigeminal ganglion to emerge almost at once from the foramen ovale and a small *motor root* which passes under the

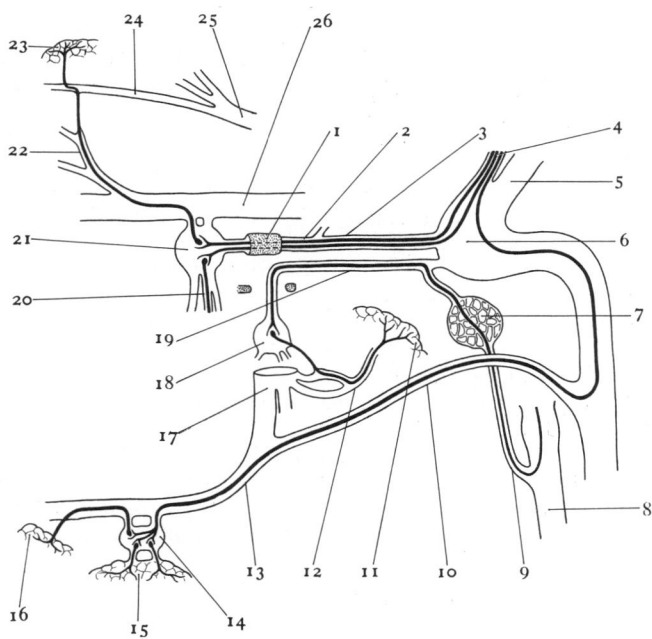

7.221 The parasympathetic connections of the pterygopalatine, otic and submandibular ganglia. The parasympathetic fibres, both pre- and post-ganglionic, are shown as heavy black lines. The parasympathetic fibres in the palatine nerves (20) are secretomotor to the nasal, palatine and pharyngeal glands. Consult text for recent views on the supply to the lacrimal gland.

1. Pterygoid canal.
2. Nerve of pterygoid canal.
3. Greater petrosal nerve.
4. Sensory root of facial nerve.
5. Motor root of facial nerve.
6. Ganglion of facial nerve.
7. Tympanic plexus.
8. Glossopharyngeal nerve.
9. Tympanic nerve.
10. Chorda tympani nerve.
11. Parotid gland.
12. Auriculotemporal nerve.
13. Lingual nerve.
14. Submandibular ganglion.
15. Submandibular salivary gland.
16. Sublingual salivary gland.
17. Mandibular nerve.
18. Otic ganglion.
19. Lesser petrosal nerve.
20. Palatine nerves.
21. Pterygopalatine ganglion.
22. Zygomaticotemporal nerve.
23. Lacrimal gland.
24. Lacrimal nerve.
25. Ophthalmic nerve.
26. Maxillary nerve.

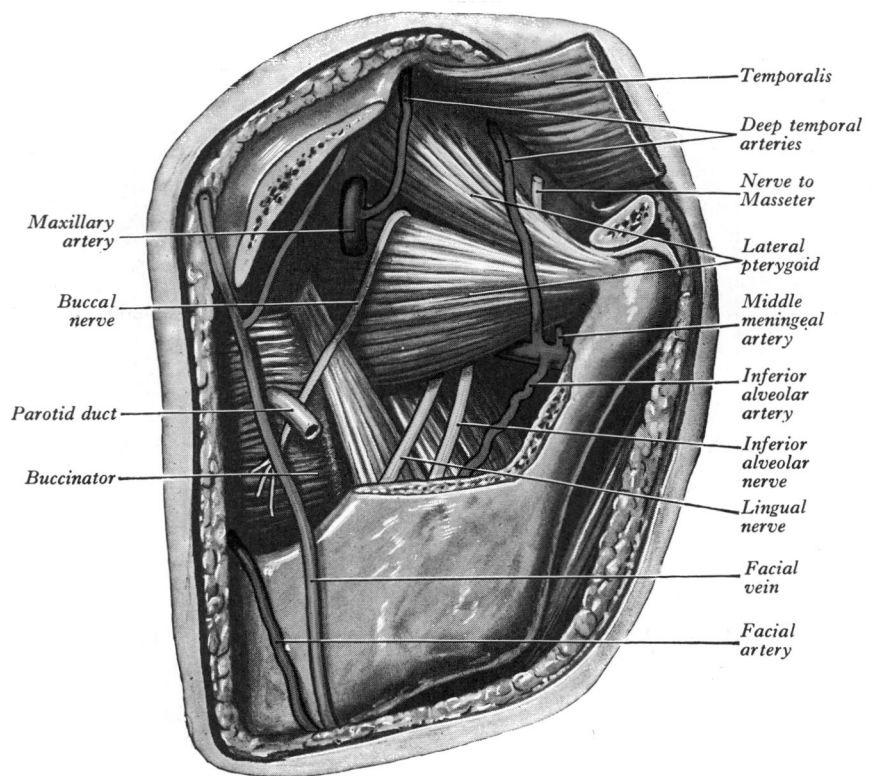

Maxillary
artery

Buccal
nerve

Parotid duct

Buccinator

Temporalis

Deep temporal
arteries

Nerve to
Masseter

Lateral
pterygoid

Middle
meningeal
artery

Inferior
alveolar
artery

Inferior
alveolar
nerve

Lingual
nerve

Facial
vein

Facial
artery

7.222 A dissection of the left pterygoid region, showing some of the branches of the mandibular nerve and the maxillary artery.

ganglion to unite with the sensory root just beyond the foramen ovale, where the nerve passes between the tensor veli palatini (medial) and the lateral pterygoid. Just beyond this junction a meningeal branch and the nerve to the medial pterygoid leave the nerve's medial side; it then divides into a small anterior and large posterior trunk. As it descends from the foramen ovale, the nerve is about 4 cm from the surface and a little anterior to the neck of the mandible.

The meningeal branch (*nervus spinosus*) re-enters the cranium through the foramen spinosum with the middle meningeal artery, dividing into anterior and posterior branches which accompany the main divisions of the artery and supply the dura mater in the middle cranial fossa and to a lesser extent in the anterior fossa and calvarium; the posterior also supplies the mucous lining of the mastoid air cells while the anterior communicates with the meningeal branch of the maxillary nerve. The nervus spinosus also contains sympathetic postganglionic fibres from the middle meningeal plexus.

The nerve to the medial pterygoid is a slender ramus entering the deep aspect of its muscle, supplying one or two filaments which pass *through* the otic ganglion (p. 1113) without interruption to supply the tensor tympani and tensor veli palatini (7.223).

The Anterior Trunk

The anterior trunk of the mandibular nerve gives rise to (1) the sensory buccal nerve and (2) motor branches: masseteric, deep temporal and lateral pterygoid nerves.

The buccal nerve (7.222) proceeds between the two parts of the lateral pterygoid, descending beneath or through the lower part of the temporalis and deep to the mandible and masseter; it emerges from under the cover of the mandibular ramus and masseter anteriorly and unites with the buccal branches of the facial nerve. It supplies the lateral pterygoid while passing through it and may give off the anterior deep temporal nerve. It supplies the skin over the anterior part of the buccinator and the mucous membrane lining its inner surface, together with the posterior part of the buccal gingival surface.

The masseteric nerve (7.222) passes laterally, above the lateral pterygoid, anterior to the temporomandibular joint and posterior to the tendon of the temporalis; it crosses the posterior part of the mandibular incisure with the masseteric artery, ramifies on and enters the masseter's deep surface and also supplies the joint.

The deep temporal nerves, usually an anterior and a posterior, pass above the lateral pterygoid to enter the deep surface of the temporalis. The small *posterior nerve* is posterior in the temporal fossa, sometimes arising in common with the masseteric nerve. The *anterior nerve* is frequently a branch of the buccal nerve; it ascends over the upper head of the lateral pterygoid. A third, middle branch often occurs.

The nerve to the lateral pterygoid enters the deep surface of the muscle. It may arise separately from the anterior division of the mandibular, or with the buccal nerve.

The Posterior Trunk

The posterior and larger, mandibular trunk is mainly sensory but receives a few filaments from the motor root. It divides into auriculotemporal, lingual and inferior alveolar (dental) nerves.

The auriculotemporal nerve usually has two roots, encircling the middle meningeal artery (7.219). It runs back under the lateral pterygoid on the surface of the tensor veli palatini to pass between the sphenomandibular ligament and the neck of the mandible and then laterally behind the temporomandibular joint in relation with the upper part of the parotid gland. Emerging from behind the joint, it ascends posterior to the superficial temporal vessels, over the posterior root of the zygoma, and divides into superficial temporal branches.

It communicates with the facial nerve and otic ganglion. The rami to the facial nerve, usually two, pass anterolaterally behind the neck of the mandible to join the facial nerve at the posterior border of the masseter. Filaments from the otic ganglion join the roots of the auriculotemporal nerve close to their origin (7.226).

The *branches* of the auriculotemporal nerve are: the anterior auricular, branches to the external acoustic meatus, articular, parotid and superficial temporal. Usually two *anterior auricular branches* supply the skin of the tragus (7.224) and sometimes a small part of the adjoining helix. Two *branches to the external acoustic meatus* pass between the osseous and cartilaginous parts of the meatus to supply the skin of the meatus; the upper sends a twig to the tympanic membrane. The *articular branches* are one or

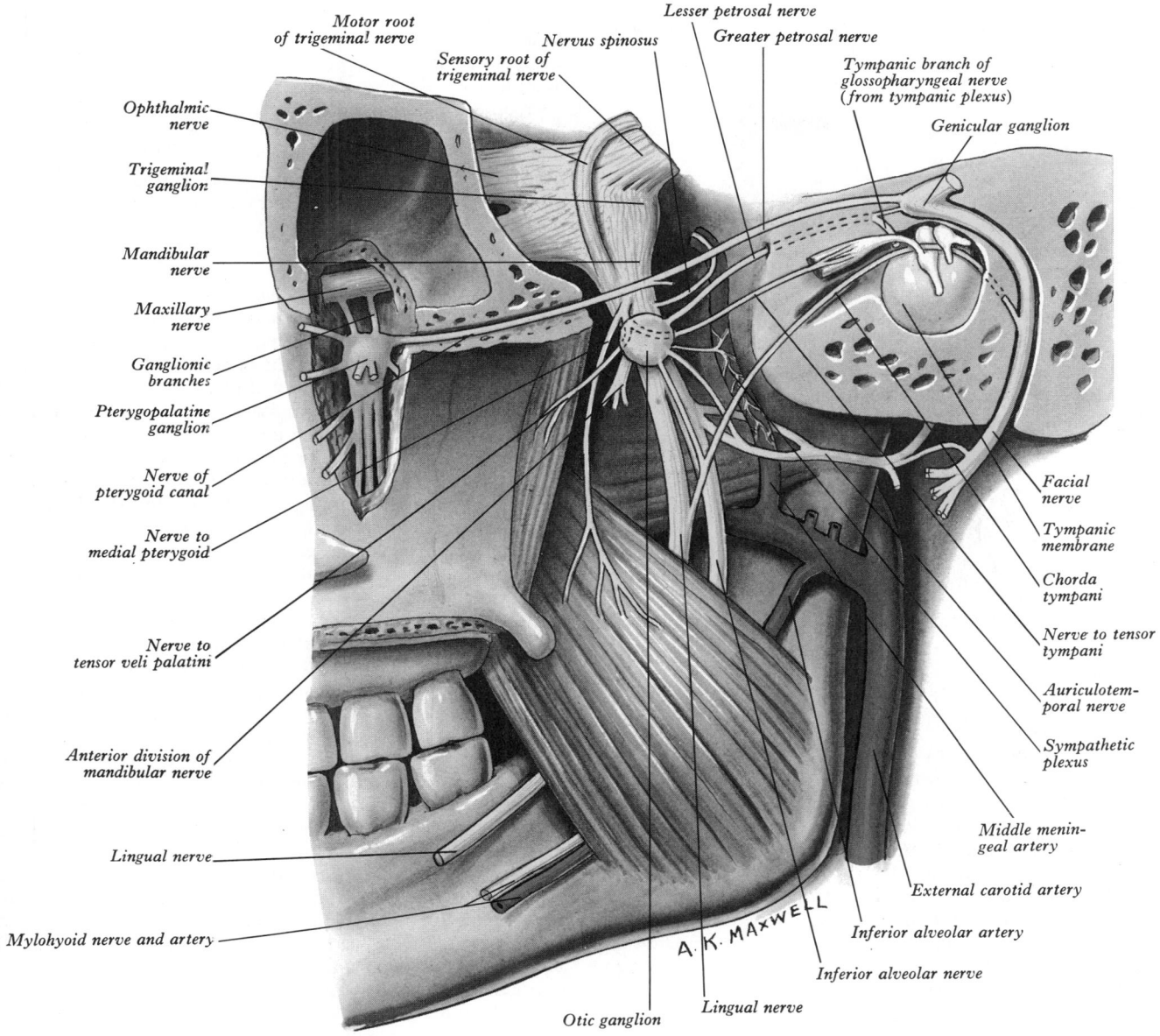

Motor root
of trigeminal nerve

Nervus spinosus

Sensory root of
trigeminal nerve

Lesser petrosal nerve

Greater petrosal nerve

Tympanic branch of
glossopharyngeal nerve
(from tympanic plexus)

Genicular ganglion

Ophthalmic
nerve

Trigeminal
ganglion

Mandibular
nerve

Maxillary
nerve

Ganglionic
branches

Pterygopalatine
ganglion

Nerve of
pterygoid canal

Nerve to
medial pterygoid

Nerve to
tensor veli palatini

Anterior division of
mandibular nerve

Lingual nerve

Mylohyoid nerve and artery

Otic ganglion

Lingual nerve

Inferior alveolar nerve

Inferior alveolar artery

External carotid artery

Middle menin-
geal artery

Sympathetic
plexus

Auriculotem-
poral nerve

Nerve to tensor
tympani

Chorda
tympani

Tympanic
membrane

Facial
nerve

A. K. MAXWELL

7.223 The right otic and pterygopalatine ganglia and their branches
displayed from the medial side (semi-diagrammatic).

two filaments which enter the posterior part of the temporomandibular joint. The *parotid branches* convey secretomotor fibres to
the gland; preganglionic fibres come from the glossopharyngeal
nerve by its tympanic branch, travelling via the lesser petrosal
nerve to the otic ganglion, whence postganglionic fibres pass to
the auriculotemporal nerve to reach the gland (7.221). Vasomotor
fibres to the blood vessels of the parotid gland are from the sympathetic root of the otic ganglion (p. 1113). The *superficial temporal branches* accompany the superficial temporal artery and its
terminal branches, supplying the skin in the temporal region and
connecting with the facial and zygomaticotemporal nerves.

The lingual nerve (7.219) is sensory to the mucosa of the
presulcal part of the tongue, the floor of the mouth and the mandibular gingivae. It arises from the posterior trunk of the mandibular nerve and at first runs between the tensor veli palatini and
the lateral pterygoid, where it is joined by the chorda tympani
branch of the facial nerve and often by a branch of the inferior
alveolar nerve. Emerging from the cover of the lateral pterygoid it
proceeds down and forwards between the mandibular ramus and
the medial pterygoid, lying anterior and slightly deep to the inferior alveolar nerve. It then passes below the mandibular attachment of the superior pharyngeal constrictor and lies against the
medial surface of the mandible near the roots of the third molar
tooth, where it is covered only by the gingival mucosa; here it can
be pressed against the bone by a finger placed inside the mouth. It

leaves the gingiva and passes on to the side of the tongue, where
it crosses the styloglossus and runs on the lateral surface of the
hyoglossus and deep to the mylohyoid, above the deep part of the
submandibular gland and its duct. It proceeds forwards on the the
side of the tongue, lateral to the hyoglossus and genioglossus; and
divides into terminal branches which lie directly under the lingual
mucosa. In the latter part of its course the nerve is near the submandibular duct; it passes downwards and forwards lateral to the
duct, winds below it and then ascends forwards medial to it
(7.219).

In addition to receiving the chorda tympani and a branch from
the inferior alveolar nerve, the lingual nerve is connected to the
submandibular ganglion (p. 1111) by two or three branches
(7.219) and, at the anterior margin of the hyoglossus, it forms
connecting loops with twigs of the hypoglossal nerve.

Branches of the lingual nerve supply the mucosa of the oral
floor, lingual aspects of the gingivae and the mucosa of the presulcal part of the tongue, being overlapped slightly by lingual fibres
of the glossopharyngeal nerve (p. 1113); terminal filaments join at
the lingual apex with twigs of the hypoglossal nerve. It also carries
postganglionic fibres from the submandibular ganglion (p. 1111)
to the sublingual and anterior lingual glands.

The inferior alveolar (dental) nerve descends medial to the
lateral pterygoid and then, at its lower border, passes between the
sphenomandibular ligament and the mandibular ramus to enter

1105

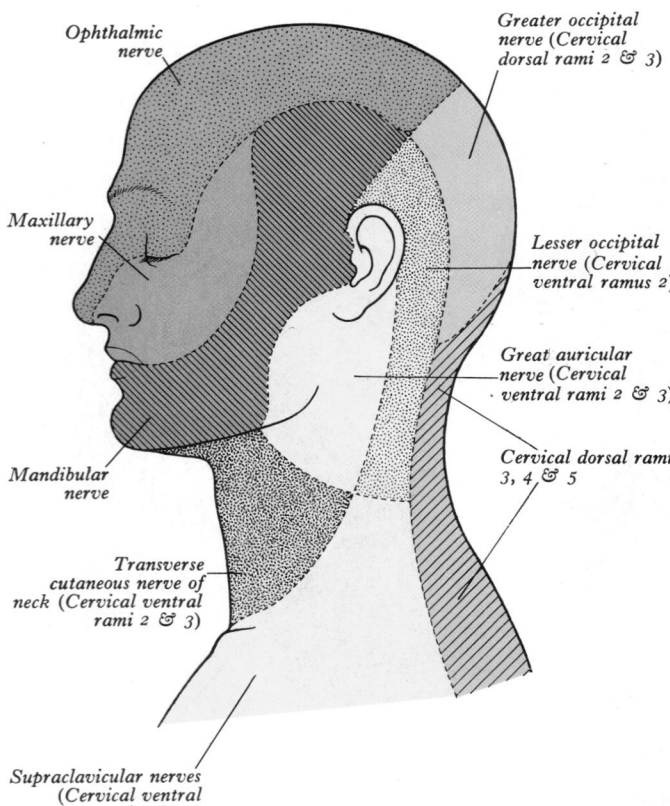

Ophthalmic nerve

Greater occipital nerve (Cervical dorsal rami 2 & 3)

Maxillary nerve

Lesser occipital nerve (Cervical ventral ramus 2)

Great auricular nerve (Cervical ventral rami 2 & 3)

Mandibular nerve

Cervical dorsal rami 3, 4, & 5

Transverse cutaneous nerve of neck (Cervical ventral rami 2 & 3)

Supraclavicular nerves (Cervical ventral rami 3 & 4)

7.224 The cutaneous nerve supply of the face, scalp and neck. Magenta = trigeminal nerve, blue = cervical dorsal rami, yellow = cervical ventral rami.

through the mandibular foramen into the mandibular canal; in this, it runs below the teeth (which it supplies with twigs) to the mental foramen, where it emerges to divide into incisive and mental branches. Below the lateral pterygoid it is accompanied by the inferior alveolar artery. Dissection and radiography show that within most mandibles the inferior dental nerve is *plexiform* and does not occupy a single canal. It is joined directly, or through plexiform branches, by rami entering the bone as parts of neurovascular bundles derived from attached muscles such as the masseter. Such '*accessory' dental nerves* ramify in a plane lateral to

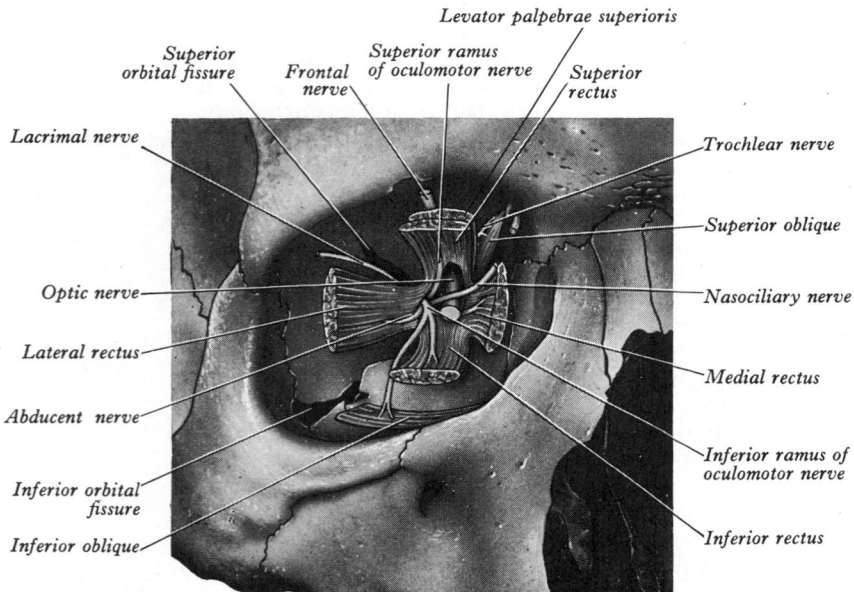

Levator palpebrae superioris

Superior orbital fissure

Frontal nerve

Superior ramus of oculomotor nerve

Superior rectus

Lacrimal nerve

Trochlear nerve

Superior oblique

Optic nerve

Nasociliary nerve

Lateral rectus

Medial rectus

Abducent nerve

Inferior orbital fissure

Inferior oblique

Inferior ramus of oculomotor nerve

Inferior rectus

7.225 A dissection of the right orbit, viewed from in front, to show the origins of the orbital muscles and the relative positions of the nerves of the orbit.

the molar teeth; their common occurrence accounts for incomplete anaesthesia by inferior dental nerve block (Carter & Keen 1971). Variation in the inferior alveolar nerve in man has been described by Khaledpour (1984). The inferior alveolar nerve gives off the mylohyoid nerve, branches to the mandibular molar and premolar teeth and the incisive and mental nerves.

The *mylohyoid nerve* leaves just before the inferior alveolar nerve enters the mandibular foramen. It pierces the sphenomandibular ligament, descends in a groove on the medial surface of the ramus of the mandible and, passing below the mylohyoid line, reaches the inferior surface of the mylohyoid, supplying it and the *anterior belly* of the *digastric*.

Branches to the molar and premolar teeth supply the adjoining gingiva also. Before they enter the dental roots they communicate and form an inferior dental plexus. The *incisive nerve* is often described as continuing on in the bone to supply the canine and incisor teeth. An alternative view is that the nerves supplying the incisor teeth form a plexus external to the mandible after emerging from the mental foramen and before re-entering the bone (Starkie & Stewart 1931). The canine teeth may be supplied from this incisor plexus or from the plexus innervating the premolars.

The *mental nerve*, emerging from its foramen, divides beneath the depressor anguli oris into three: a branch descending to the skin of the chin and two ascending to the skin and mucosa of the lower lip; these communicate freely with the facial nerve (mandibular branch).

Applied Anatomy. A lesion of the whole trigeminal nerve causes anaesthesia of the anterior half of the scalp, of the face (excepting a small area near the angle of the mandible supplied by the great auricular nerve), of the cornea and conjunctiva, the mucosae of the nose, mouth and presulcal part of the tongue. Paralysis and atrophy occur in the muscles supplied by the nerve; when the mouth is opened the mandible is thrust to the paralysed side. Lesions of the divisions of the nerve give a more limited sensory loss and, if the lingual nerve is involved anterior to its junction with the chorda tympani, loss of taste in half of the anterior part of the tongue occurs.

Pains referred to branches of the trigeminal nerve are frequent. Pain is at first confined to one division, though it may radiate over branches of the other divisions. The commonest example is the neuralgia often associated with dental caries. Here, though the tooth may not appear painful, most distressing referred pains may occur and these are at once relieved by treatment directed to the affected tooth.

In the area of the ophthalmic nerve, severe supraorbital pain is often associated with acute glaucoma or with frontal or ethmoidal sinusitis. Neoplasms or empyema of the maxillary sinus often cause maxillary neuralgia. Some of the most striking examples occur in the mandibular zone: earache often occurs where there is no sign of aural disease, the cause being usually a carious mandibular tooth. An ulcer or carcinoma of the tongue often first causes pain radiating to the ear and temporal fossa (the distribution of the auriculotemporal nerve).

The lingual nerve is occasionally divided in order to ease pain due to a lingual carcinoma; this is most easily done where the nerve adjoins the mandible below and behind the last molar tooth (p. 1105).

Intractable trigeminal neuralgia has promoted various surgical manoeuvres. The maxillary and mandibular nerves and trigeminal ganglion have been injected with alcohol and excision of the ganglion has frequently been performed, though this involves serious risks (laceration of the cavernous sinus, etc.) and is now rarely undertaken. The sensory root may be divided behind the ganglion and this is now the usual operation when pain is confined to the maxillary and mandibular areas (7.224). Complete division denervates the cornea, leading to neuropathic keratitis. Consequently, an endeavour is made to preserve the ophthalmic fibres, which are in the superomedial part of the root. The motor root of the nerve is left intact.

When pain is limited to the ophthalmic area or to the ophthalmic and maxillary areas, the preferred operation (Falconer 1949) is division of the spinal tract, where it is most superficial (p. 951) and sometimes forms a recognizable elevation (p. 953) between the lateral margin of the fasciculus cuneatus and the posterior border of the lower olive. Section of the tract 4–5 mm caudal to the obex preserves most mandibular fibres. Pain can be abolished, while retaining tactile sensibility and the corneal reflex.

The Abducent Nerve

The abducent nerve (7.218) supplies the lateral rectus, its fibres arising from a small nucleus in the superior part of the floor of the fourth ventricle, near the midline and beneath the colliculus facialis (7.86, 87, 94, 95, p. 980). They descend ventrally through the pons, emerging in the sulcus between the caudal border of the pons and the superior end of the pyramid of the medullar oblongata (7.81).

The *abducens nucleus* contains large and small multipolar neurons; the former are the sources of the abducent nerve and the latter, known collectively as the *nucleus para-abducens*, is considered to project to the oculomotor nucleus via the medial longitudinal fasciculus. The total number of neurons in the nucleus is about 22 000; only a minority can be radicular, since the nerve contains about 6600 fibres (Konigsmark et al 1969). Claims that all abducent neurons react to division (Warwick 1964 and others) are based on differing considerations of which cells should be included as 'abducent'. A study in kittens using the retrograde horseradish peroxidase technique (Gacek 1974) nevertheless indicated that 50–90% of the cells were radicular. Other studies have demonstrated that some abducent neurons project to the oculomotor and perhaps trochlear nuclei (see Steiger & Büttner-Ennever 1978, for literature). Büttner-Ennever (1977) has identified neurons in the pontine reticular formation, just rostral to the abducent nucleus, which project to regions of the oculomotor nucleus reputedly concerned with innervation of the lateral and inferior recti; and has discussed the considerable evidence for the view that these mediate horizontal and perhaps vertical integration of binocular movement.

Connections. The abducent nucleus receives afferent connections from: (1) the corticonuclear tract, principally contralateral; some of these fibres are aberrant pyramidal fibres which descend from the midbrain to this level in the medial lemniscus (p. 952) and interneurons may link them to the nucleus; (2) the medial longitudinal fasciculus, connecting it to nuclei of the third, fourth and eighth cranial nerves; (3) the tectobulbar tract, linking it with the visual cortex and other centres through the superior colliculus. (See also p. 987.)

Course. After leaving the brain stem, the abducent nerve ascends anterolaterally through the cisterna pontis, usually dorsal to the anterior inferior cerebellar artery. It pierces the dura mater lateral to the dorsum sellae and bends sharply forwards over the superior border of the petrous temporal bone, near its apex and inferior to the petrosphenoidal ligament, a fibrous band connecting the lateral margin of the dorsum sellae with the upper border of the petrous temporal bone near its median end. It next traverses the cavernous sinus, at first lying lateral and then inferolateral to the internal carotid artery (6.155, 156), and enters the orbital cavity via the medial end of the superior orbital fissure within the annular tendon and inferolateral to the oculomotor and nasociliary nerves, finally sinking into the ocular surface of the lateral rectus (7.225, 329).

In the cavernous sinus the nerve is joined by a branch from the end of the internal carotid sympathetic plexus, originally described by Monro (1746) and confirmed by Meckel (1832), but frequently omitted from subsequent textbooks and even denied (Sunderland & Hughes 1946). Its existence has been re-emphasized by Johnston & Parkinson (1974) and Parkinson et al (1978), who have also described a branch of similar calibre, leaving the abducent nerve a few millimetres distal to this sympathetic communication to join the ophthalmic nerve; they believe that sympathetic postganglionic fibres joining the abducent nerve are thus diverted into the ophthalmic division of the trigeminal. This would accord with the passage of sympathetic nerve fibres to the eye via the long ciliary branches of the nasociliary nerve, as suggested by some.

Applied Anatomy. The abducent nerve is sometimes involved in fractures of the cranial base. The resulting paralysis of the

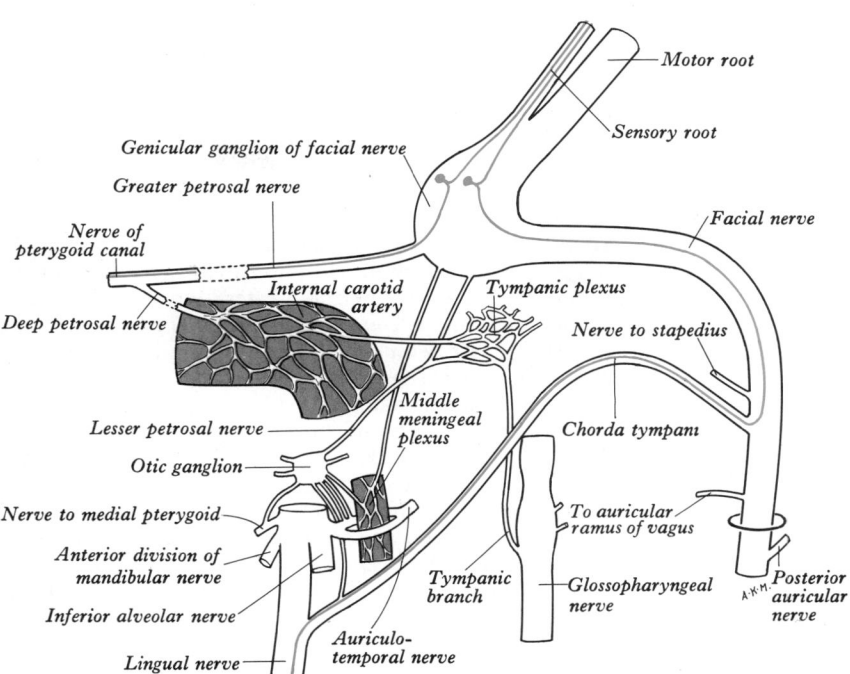

7.226 A plan of the intrapetrous section of the facial nerve, its branches and communications. The course of the taste fibres from the mucous membrane of the palate and from the anterior, oral or presulcal part of the tongue is represented by the blue lines.

lateral rectus leads to convergent squint and diplopia. The long course of the sixth cranial nerve through the cisterna pontis and its sharp bend over the petrous temporal bone make the nerve liable to damage in conditions producing raised intracranial pressure, as a result of which the brain stem may be pushed caudally with consequent stretching of the nerve.

The Facial Nerve

The facial nerve (7.226–229) has a motor and a sensory root, the latter being the *nervus intermedius* (7.227). The two roots emerge at the caudal border of the pons lateral to the recess between the inferior olive and inferior cerebellar peduncle, the motor part being medial; the vestibulocochlear nerve is lateral to the sensory

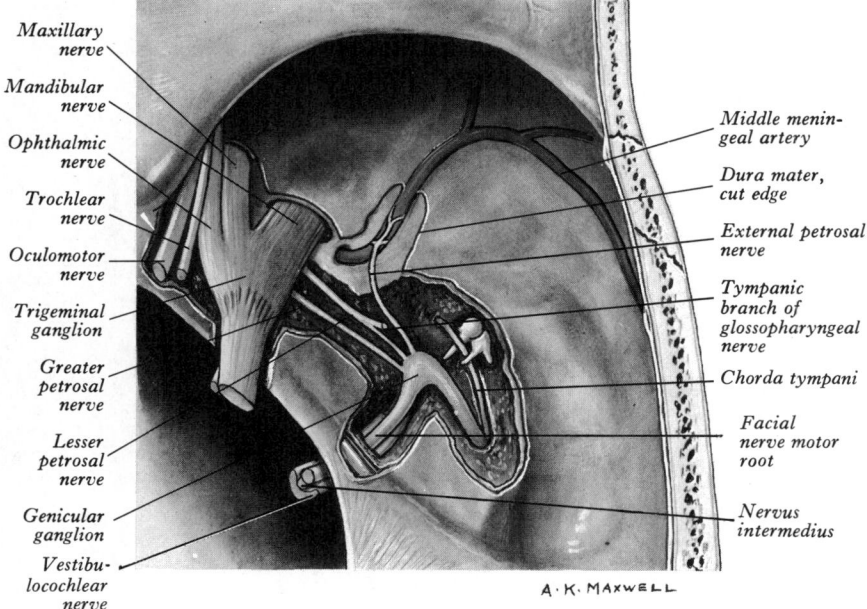

7.227 A dissection of the right middle cranial fossa, showing the course and some of the connections of the facial nerve within the temporal bone.

root. The nervus intermedius usually cleaves at first to the vestibulocochlear rather than the facial nerve, passing to the latter as it approaches the internal acoustic meatus, often as more than one filament. In approximately one-fifth of 73 dissections it was not a separate nerve until the meatus was reached, a point of surgical importance (Rhoton et al 1968).

The *motor root* supplies: the muscles of the face, scalp and auricle, the buccinator, platysma, stapedius, stylohyoid, and the posterior belly of the digastric. The *sensory root* conveys from the chorda tympani gustatory fibres from the presulcal area of the tongue and, from the palatine and greater petrosal nerves, taste fibres from the soft palate; it also carries the preganglionic parasympathetic (secretomotor) innervation of the submandibular and sublingual salivary glands, lacrimal gland and glands of the nasal and palatine mucosae.

The motor nucleus, from which most facial motor fibres are derived, is deep in the reticular formation of the caudal part of the pons (p. 961), posterior to the dorsal trapezoid nucleus (7.86, 87, 94) and ventromedial to the spinal tract nucleus of the trigeminal nerve. It represents the branchial efferent column but lies deeper in the pons than might be expected and its efferent fibres have a most unusual course (7.95). Both these features have been ex-

plained by invoking neurobiotaxis (p. 184). The nucleus receives fibres from both corticonuclear tracts in the lower pons and is reputedly supplied by aberrant pyramidal fibres which descend in the medial lemniscus. Contralateral corticonuclear fibres end in the part of the nucleus innervating the muscles of the lower part of the face (p. 961) while the corticonuclear projection to neurons for muscles around the eyes and forehead is bilateral. Some facial efferent fibres proceed from the *superior salivatory nucleus* (see p. 961), said to be in the reticular formation, dorsolateral to the caudal end of the motor nucleus. Neurons of the facial nucleus have been described as clustered along the intrapontine part of the nerve distal to its loop round the abducens nucleus (Crosby & Dejonge 1963); it belongs to the general visceral efferent column, sending its fibres to the sensory root, by which they ultimately, via the chorda tympani, reach the submandibular and sublingual salivary glands and perhaps also the parotid (p. 1155). Other preganglionic fibres issuing in the sensory root reach the pterygopalatine ganglion in the greater petrosal nerve and the nerve of the pterygoid canal (vide infra). From this double origin the *motor root* passes dorsomedially to the caudal end of the abducent nucleus and ascends superficial to it and deep to the facial colliculus. At the upper limit of the abducent nucleus the nerve

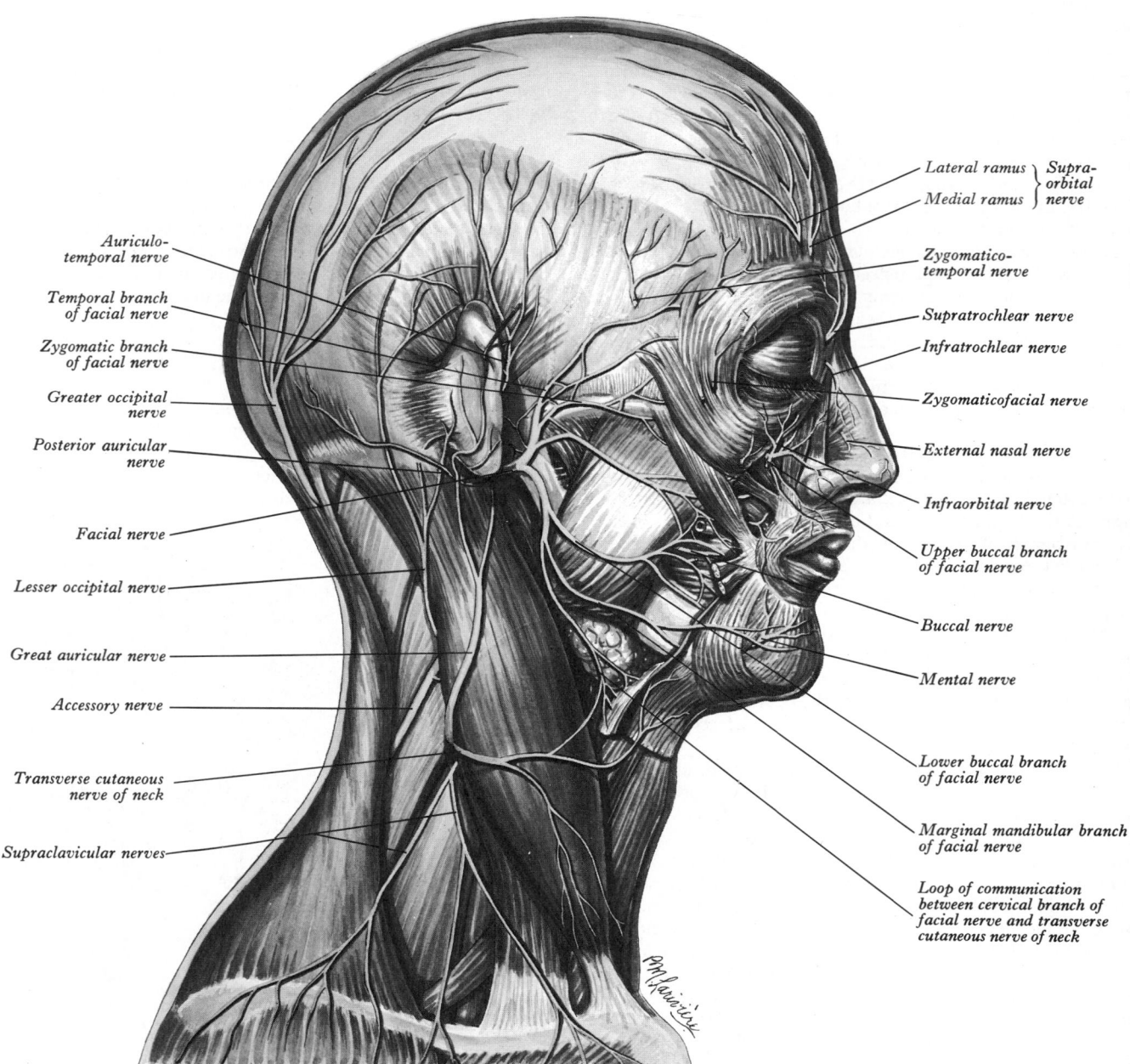

Lateral ramus ⎱ Supra-
Medial ramus ⎰ orbital nerve

Zygomatico-temporal nerve

Supratrochlear nerve

Infratrochlear nerve

Zygomaticofacial nerve

External nasal nerve

Infraorbital nerve

Upper buccal branch of facial nerve

Buccal nerve

Mental nerve

Lower buccal branch of facial nerve

Marginal mandibular branch of facial nerve

Loop of communication between cervical branch of facial nerve and transverse cutaneous nerve of neck

Auriculo-temporal nerve

Temporal branch of facial nerve

Zygomatic branch of facial nerve

Greater occipital nerve

Posterior auricular nerve

Facial nerve

Lesser occipital nerve

Great auricular nerve

Accessory nerve

Transverse cutaneous nerve of neck

Supraclavicular nerves

7.228 The nerves of the right side of the scalp, face and neck. Compare with 7.224.

swerves into a caudoventral course to its emergence between the olive and the inferior cerebellar peduncle (7.81, 95) at the cerebellopontine angle.

The facial motor nucleus is a complex consisting of lateral, intermediate and medial subnuclei (Marinesco 1899, Papex 1927). A further division of the medial nucleus into ventral, dorsal and intermediate groups has been described. (For details consult Vraa-Jensen 1942.) These subsidiary groups of neurons are arranged in columns, like those in the spinal cord or oculomotor nucleus. There is general agreement that they innervate individual muscles or correspond to branches of the nerve, but observers disagree about details. Retrograde changes following division of different branches of the nerve in dogs (Vraa-Jensen 1942, Yagita 1910) and cats (Papex 1927, Courville 1966) have provided most of the evidence; the effects on motor terminals of selective nuclear lesions have also been studied in cats. The lateral subnucleus is said to innervate the buccal musculature, the intermediate sends axons into the temporal, orbital and zygomatic facial branches and the medial group into the posterior auricular and cervical rami and probably the stapedial nerve. Nuclear lesions have produced a roughly similar but more detailed schema (Szentágothai 1948).

The sensory nucleus of the facial nerve is the rostral end of the *nucleus solitarius* in the medulla oblongata (p. 954). It receives gustatory and possibly other afferents from the sensory root and sends fibres to the contralateral ventral lateral thalamic nuclei. As they ascend in the midbrain and subthalamic regions these fibres pass near the midline (Harris 1952). From the thalamus they are relayed to the inferior part of the postcentral gyrus. For other connections, see p.1122.

The sensory root (nervus intermedius) contains the centripetal processes of unipolar neurons in the genicular ganglion, which leave the trunk of the facial nerve in the internal acoustic meatus in one or more slender bundles between the motor root and the vestibulocochlear nerve or adherent to the latter, and enter the brain stem at the caudal border of the pons. The peripheral branches of the ganglion cells are mainly taste fibres travelling in the chorda tympani and greater petrosal nerve and a few somatic afferent fibres from the auricular concha (p. 1219). The sensory root also contains *efferent* preganglionic parasympathetic fibres for the submandibular and sublingual salivary glands, lacrimal gland, and pharyngeal, nasal and palatine glands.

Course. From their emergence from the brain, the two roots pass anterolaterally with the vestibulocochlear nerve to the internal acoustic meatus; here the motor root is in an anterosuperior groove on the vestibulocochlear nerve, with the sensory root between them. At the lateral end of the meatus, the nerve enters the facial canal, proceeding at first laterally above the vestibule and, near the medial wall of the epitympanic recess, bending sharply back above the promontory and arching down in the medial wall of the aditus of the mastoid antrum. It then descends to the stylomastoid foramen. Its sharp bend backwards is the *geniculum*; here the nerve presents a reddish asymmetrical swelling, the *genicular ganglion* (7.226). Emerging from the stylomastoid foramen, the nerve runs forwards in the parotid gland (p. 1290), crosses the styloid process, retromandibular vein and external carotid artery and divides behind the neck of the mandible into branches which pierce the anteromedial surface of the parotid gland and diverge under cover of it. They form a network (*parotid plexus*) which distributes to the facial musculature. At the stylomastoid foramen the nerve is about 2 cm deep to the middle of the anterior border of the mastoid process. Its transparotid course is represented by a horizontal line across the upper part of the auricular lobule (6.70). Angular measurements of the frequent changes in direction of the facial nerve within the facial canal are of surgical importance; Kudo & Nori (1974) and Guerrier (1975) have recorded detailed metrical studies.

Arterial supply and venous drainage. The facial nerve is supplied intracranially by the anterior inferior cerebellar artery and in its canal by the superficial petrosal branch of the middle meningeal artery and the stylomastoid branch of the posterior auricular or occipital arteries; extracranially by branches from the

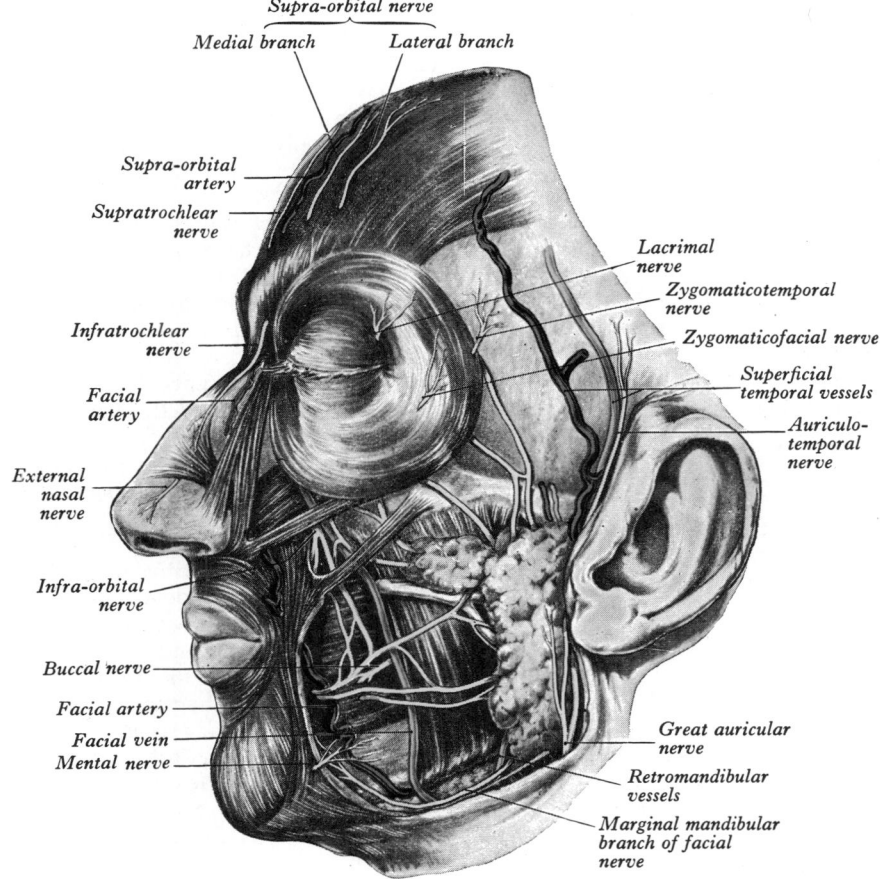

Supra-orbital nerve
Medial branch   Lateral branch

Supra-orbital artery
Supratrochlear nerve
Infratrochlear nerve
Facial artery
External nasal nerve
Infra-orbital nerve
Buccal nerve
Facial artery
Facial vein
Mental nerve

Lacrimal nerve
Zygomaticotemporal nerve
Zygomaticofacial nerve
Superficial temporal vessels
Auriculo-temporal nerve
Great auricular nerve
Retromandibular vessels
Marginal mandibular branch of facial nerve

7.229 The terminal branches of the left trigeminal and facial nerves in the face.

stylomastoid, posterior auricular, occipital, superficial temporal and transverse facial arteries. Venous drainage is into the venae comitantes of the superficial petrosal and stylomastoid arteries (Blunt 1954).

Communicating branches are:

Intracranial:	Vestibulocochlear
Genicular ganglion:	Pterygopalatine ganglion via the greater petrosal nerve
	Otic ganglion via the lesser petrosal nerve
	Middle meningeal sympathetic plexus
Facial canal:	Auricular branch of vagus
At exit from stylo-mastoid foramen:	To glossopharyngeal, vagus, great auricular and auriculotemporal nerves
Post-auricular:	To the lesser occipital nerve
Facial:	With the trigeminal nerve
Cervical:	With the transverse cervical nerve

In the internal acoustic meatus some minute filaments connect the facial nerve with the vestibulocochlear nerve.

The greater petrosal nerve starts from the genicular ganglion and contains mainly taste fibres from the palatal mucosa and also preganglionic parasympathetic fibres travelling to the pterygopalatine ganglion and said to be relayed through the zygomatic and lacrimal nerves to the lacrimal gland (p. 1103) and through the nasal and palatine nerves to the nasal and palatine mucosal glands (7.221). It receives a ramus from the tympanic plexus, traverses the hiatus on the anterior surface of the petrous temporal bone and enters a groove on it, passing under the trigeminal ganglion to the foramen lacerum; here it is joined by the **deep petrosal nerve** (7.226) from the internal carotid sympathetic plexus to form the **nerve of the pterygoid canal**, which traverses the pterygoid canal to end in the pterygopalatine ganglion. Gustatory fibres pass without interruption through or over the pterygopalatine ganglion into its palatine branches.

From the facial nerve near the genicular ganglion, the ganglion itself or the root of the greater petrosal nerve (Vidić & Young 1967), a branch runs to the lesser petrosal nerve (7.226), by which it reaches the otic ganglion. However, fibres in this small communication are *not facial* but come from the vagal auricular branch. In 24 out of 25 human dissections the fibres approached the facial nerve through the stapedius, travelling in its fascial sheath almost to the genicular ganglion, before departing to join the lesser superficial petrosal nerve (Vidić 1968).

The middle meningeal sympathetic plexus is joined to the genicular ganglion by an inconstant *external petrosal nerve*.

Before leaving the stylomastoid foramen the facial nerve receives another branch from the vagal auricular nerve. After leaving the foramen, it receives a twig from the glossopharyngeal, communicating with the great auricular and auriculotemporal nerves in the parotid gland, with the lesser occipital nerve behind the ear, with terminal trigeminal branches on the face and with the transverse cutaneous cervical nerve in the neck.

The branches of distribution (7.226, 228) are:

In the facial canal:	Nerve to stapedius
	Chorda tympani
At exit from stylo-mastoid foramen:	Posterior auricular
	Digastric, posterior belly
	Stylohyoid
On the face:	Temporal
	Zygomatic
	Buccal
	Marginal mandibular
	Cervical

The nerve to the stapedius arises behind the pyramidal eminence of the posterior wall of the tympanic cavity, passing forwards through a small canal to reach the muscle (p. 1227).

The chorda tympani (7.223) leaves the facial nerve about 6 mm above the stylomastoid foramen, ascending forwards in a canal to perforate the posterior wall of the tympanic cavity via its posterior canaliculus near the posterior border of the tympanic membrane's medial aspect, level with the upper end of the handle of the malleus. Passing forwards between the membrane's fibrous and mucous layers, it crosses medial to the handle of the malleus to re-enter the bone via its anterior canaliculus at the medial end of the petrotympanic fissure. It now descends ventrally on the medial surface of the sphenoid spine (which it sometimes grooves) and passes deep to the lateral pterygoid. Here the nerve is posterolateral to the tensor veli palatini and is crossed medially by the middle meningeal artery, the roots of the auriculotemporal nerve and the inferior alveolar nerve. It joins the posterior aspect of the lingual nerve at an acute angle. It contains efferent preganglionic parasympathetic (secretomotor) fibres which enter the submandibular ganglion, from which postganglionic fibres are relayed to the submandibular and sublingual glands; most of its fibres are afferent from the mucosa of the anterior, presulcal part of the tongue, save the vallate papillae, constituting the nerve of taste for this lingual region. Before uniting with the lingual nerve the chorda tympani is joined by a small branch from the otic ganglion.

The posterior auricular nerve arises near the stylomastoid foramen, ascends in front of the mastoid process to be joined by a filament from the auricular branch of the vagus, and communicates with a posterior branch of the great auricular and lesser occipital nerves. As it ascends between the external acoustic meatus and mastoid process it divides into auricular and occipital branches. The *auricular branch* supplies the auricularis posterior and the intrinsic muscles on the auricle's cranial aspect; the larger *occipital branch* passes back along the superior nuchal line to supply the occipital belly of occipitofrontalis.

The digastric branch also starts near the stylomastoid foramen, dividing into several filaments which supply the posterior belly of the digastric, one joining the glossopharyngeal nerve.

The stylohyoid branch, long and slender, frequently arises with the digastric branch; it enters the middle part of the stylohyoid muscle.

The temporal branches cross the zygomatic arch to the temple, supplying intrinsic muscles on the lateral surface of the auricle, the anterior and superior auricular muscles, and join with the zygomaticotemporal branch of the maxillary nerve and the auriculotemporal branch of the mandibular nerve. The more anterior branches supply the frontal belly of the occipitofrontalis, orbicularis oculi and corrugator and join the supraorbital and lacrimal branches of the ophthalmic nerve.

Zygomatic branches cross the zygomatic bone to the lateral canthus, supplying the orbicularis oculi and joining filaments of the lacrimal nerve and zygomaticofacial branch of the maxillary nerve.

Buccal branches pass horizontally to a distribution below the orbit and around the mouth. *Superficial branches* run between the skin and superficial muscles, supplying the latter; some pass to the procerus, joining with the infratrochlear and external nasal nerves. *Upper deep branches* pass under the zygomaticus major and levator labii superioris, supplying them and forming an *infraorbital plexus* with the superior labial branches of the infraorbital nerve; they also supply the levator anguli oris, zygomaticus minor, levator labii superioris alaeque nasi and the small nasal muscles. These branches are sometimes described as lower zygomatic branches. *Lower deep branches* supply the buccinator and orbicularis oris, joining filaments of the buccal branch of the mandibular nerve.

The marginal mandibular branch runs forwards below the angle of the mandible under the platysma, at first superficial to the upper part of the digastric triangle, then turning up and forwards across the body of the mandible to pass under the depressor anguli oris (7.229). It supplies the risorius and the muscles of the lower lip and chin, joining the mental nerve (p. 1106).

The cervical branch issues from the lower part of the parotid gland, runs antero-inferiorly under the platysma to the front of the neck, supplying the platysma and communicating with the transverse cutaneous cervical nerve.

The cutaneous fibres of the facial nerve accompany the auricular branch of the vagus and probably innervate the skin on both auricular aspects, in the conchal depression and over its eminence; however, details of this are uncertain, as is a supply to the external acoustic meatus and tympanic membrane.

The Submandibular Ganglion

This is a small, fusiform body which lies on the upper part of the hyoglossus; there are further ganglion cells in the hilum of the submandibular gland. Like the ciliary, pterygopalatine and otic ganglia, the submandibular is a peripheral parasympathetic ganglion. It is superior to the deep part of the submandibular gland and inferior to the lingual nerve, suspended from the latter by anterior and posterior filaments (7.219). Though related to the lingual nerve, the ganglion is connected functionally with the facial nerve and chorda tympani. Its *motor, parasympathetic root* is the posterior filament connecting it to the lingual nerve; it conveys preganglionic fibres from the superior salivatory nucleus travelling in the facial, chorda tympani and lingual nerves to the ganglion, where they synapse, the postganglionic fibres being secretomotor to the submandibular and sublingual salivary glands. (Some fibres may also reach the parotid gland, see p. 1292.) The *sympathetic root* is derived from the plexus on the facial artery; it consists of postganglionic fibres from the superior cervical ganglion, which traverse the submandibular ganglion without synapsing. They are vasomotor to the blood vessels of the submandibular and sublingual glands. Five or six branches from the ganglion supply the submandibular gland and its duct; other fibres pass through the anterior filament connecting the gland to the lingual nerve and are carried to the sublingual and anterior lingual glands.

Applied Anatomy. Facial paralysis, often unilateral, may be due to: (1) *supranuclear lesions* in the corticonuclear fibres from the frontal lobe, variably combined with numerous other descending fibres converging on the facial nucleus; (2) *nuclear* or *infranuclear lesions* involving lower motor neurons.

Supranuclear facial paralysis involving 'upper motor neuron' pathways is usually part of a hemiplegia. Movements in the *lower* part of the face are usually more severely affected, voluntary movements being weak or absent though emotional expression is little affected. Electrical reactions of affected muscles are unaltered. Occasionally supranuclear lesions may abolish or weaken emotional movements but not voluntary movements. This dissociation shows that the supranuclear control of expressive movements is separate from the corticonuclear path for voluntary movements.

Nuclear or *infranuclear lesions* vary in their effects according to the lesion's site. If the nucleus or facial pontine fibres are involved, neighbouring structures are inevitably also involved. Facial muscles are represented in cell groups within the nucleus, their degree of involvement governing the extent of paralysis, which is ipsilateral; otherwise the effects are identical with those of more peripheral lesions. Lesions due to adjacent damage include paralysis of the lateral rectus because of the involvement of the abducent nucleus (around which the facial nerve loops), paralysis of the masticatory muscles by involvement of the motor trigeminal nucleus, sensory loss on the face from involvement of the principal sensory and spinal trigeminal nuclei or spinothalamic tract and paralysis of the upper or lower limbs due to corticospinal lesions. Due to the proximity of the facial sensory root to the vestibulocochlear nerve, lesions in the posterior cranial fossa or in the internal acoustic meatus may cause loss of taste in the anterior part of the tongue with ipsilateral deafness and facial paralysis. When damage is in the temporal bone, the chorda tympani is usually also involved and in petrous temporal fractures the vestibulocochlear nerve is also usually implicated. The most common cause is inflammation of the facial nerve near the stylomastoid foramen (*Bell's palsy* or *paralysis*), with oedema of the nerve and compression of its fibres in the facial canal or stylomastoid foramen.

If the lesion is complete, the facial muscles are all equally affected; voluntary and associated movements suffer equally. There is facial asymmetry and the affected side is immobile. The eyebrow droops, wrinkles are smoothed out; the palpebral fissure is widened by the unopposed action of the levator superioris palpebrae. Tears fail to enter the lacrimal puncta because they are no longer in contact with the conjunctiva, the conjunctival reflex is absent and efforts to close the eye merely roll the cornea under the upper lid. The ala nasi does not move in respiration. The lips remain in contact and cannot be pursed; in attempting to smile the angle of the mouth is not drawn up on the affected side, the lips remaining nearly closed and the mouth typically triangular. Food accumulates in the cheek, from paralysis of the buccinator, and dribbles or is pushed out between the paralysed lips. The protruded tongue *appears* to turn to the paralysed side though its position relative to the incisor teeth demonstrates that this is not so. Platysma and the auricular muscles are paralysed. The electrical reactions of the affected muscles are typical of degeneration, the degree of which may be a guide to prognosis. Most cases of Bell's palsy recover completely.

For variations in the course of the facial nerve near the middle ear and their surgical importance, consult Durcan et al (1967). Degeneration studies (Naoris 1968) show that the nerve contains no fascicular or other pattern corresponding to its peripheral branches, a finding of significance in suturing the nerve after injury.

The Vestibulocochlear Nerve

The vestibulocochlear nerve (7.86, 90, 91, 94) emerges in the groove between the pons and the medulla oblongata, behind the facial nerve and in front of the inferior cerebellar peduncle (7.81). It has two major sets of fibres, differing in their principal central connections, but both transmitting impulses from the internal ear to the brain. One set forms the vestibular nerve, concerned with equilibration and arising from neurons in the vestibular ganglion in the outer part of the internal acoustic meatus, the other forming the cochlear nerve or auditory nerve, arising from neurons in the spiral ganglion of the cochlea. Both ganglia contain bipolar neurons, each with central and peripheral fibres: the former passes to the brain and the latter to the internal ear. The nerve also contains *efferent fibres* passing from neuronal cell bodies in the brain stem to both parts of the labyrinth. The peripheral arrangement of the vestibulocochlear nerve is described on pp. 1228, 1235–1241.

THE VESTIBULAR NERVE

Vestibular fibres enter the brain superomedial to those of the cochlear nerve, pass back through the pons between the inferior cerebellar peduncle and spinal trigeminal tract to divide into ascending and descending branches ending mainly in the vestibular nuclei, although some go direct to the cerebellum through its inferior peduncle (p. 965).

The vestibular nuclear complex comprises the following: (1) The *medial vestibular nucleus* (p. 958), which lies under the vestibular area of the floor of the fourth ventricle, crossed dorsally by the striae medullares. The largest subdivision of the complex, it extends up from the medulla oblongata into the pons. (2) The *inferior vestibular nucleus* (p. 958), lateral to the medial, reaches to a lower medullary level and lies between the medial nucleus and inferior cerebellar peduncle; descending branches of afferent vestibular fibres end among its cells. (3) The *lateral nucleus* (p. 958) is ventrolateral to the upper part of the medial and it is characterized by its large neurons; its rostral end is continuous with the caudal end of (4) the *superior nucleus*, which extends higher into the pons than other subdivisions, occupying the upper part of the vestibular area. Several minor subdivisions have been described in feline vestibular nuclei (Brodal & Pompeiano 1957) and a tentative somatotopic pattern has been suggested in the lateral nucleus (Løken & Brodal 1970).

Connections. All vestibular nuclei receive *radicular fibres* from the nerve; they may receive *afferent cerebellovestibular fibres* through the inferior cerebellar peduncle, mainly derived from the flocculus and nodule (posterior lobe), together with others perhaps from the uvula, the lingula and fastigial nucleus.

From the nuclei *cerebellar fibres* enter the inferior cerebellar peduncle, most destined for the flocculus and nodule, though some may pass to the uvula, lingula or fastigial nucleus. As noted, some vestibular fibres bypass the nuclei and reach the flocculus and nodule directly via the inferior cerebellar peduncle.

As a whole, the vestibular nuclear complex is a relay station on an afferent cerebellar path and a distributing station for efferent cerebellar fibres. Fibres from vestibular nuclei also enter the medial longitudinal fasciculus (7.121), ascending and descending to motor nuclei of the ocular and nuchal muscles. This 'octavo-oculomotor system' has attracted much attention (see pp. 958, 985) and dissension regarding the vestibular nuclei involved, routes of projections, decussation and destinations. Tarlov (1972, 1975) has described a projection from the medial and superior vestibular nuclei to specific motor pools in the feline oculomotor nucleus (p. 1097) and has also reviewed the literature; it is suggested that excitatory and inhibitory projections exist, mediating complex and subtle integration between vestibular signals and eye movements. From large neurons in the lateral vestibular nucleus efferent fibres descend to form the *vestibulospinal tract* (p. 932), and fibres from the other nuclei join the lateral lemniscus to reach the inferior colliculus and medial geniculate body and, probably, the cerebral cortex. Through its connections the vestibular system influences movements of the eyes and head and muscles of the trunk and limbs to maintain equilibrium.

THE COCHLEAR NERVE

The cochlear nerve consists chiefly of myelinated *afferent* axons of sensory ganglion cells situated in the *spiral ganglion* within the modiolus of the cochlea; peripherally, the bipolar ganglion cells have processes which receive synapses from the auditory sensory cells (hair cells) of the spiral organ. In addition, some *efferent* fibres pass in the cochlear nerve from the brain stem to the spiral organ, constituting the *olivocochlear bundle*.

As the cochlear nerve reaches the brain stem (7.86, 90, 91) it lies lateral to the vestibular nerve but they are soon separated by the inferior cerebellar peduncle. The cochlear nerve curves round the lateral aspect of the peduncle, the vestibular penetrating the brain stem on the medial side of it.

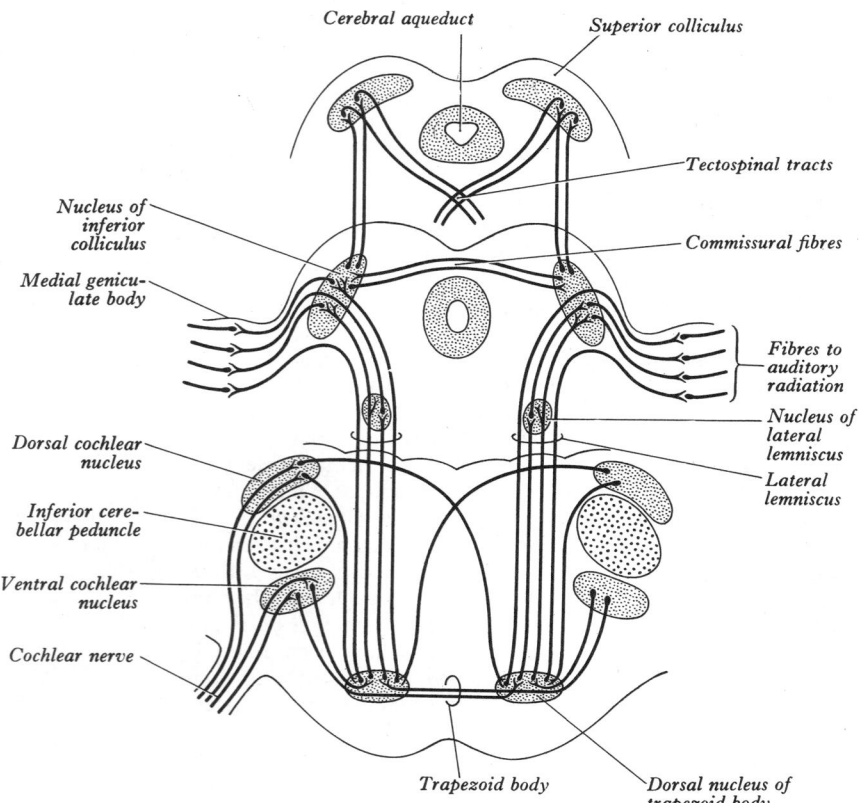

Cerebral aqueduct
Superior colliculus
Tectospinal tracts
Nucleus of inferior colliculus
Commissural fibres
Medial geniculate body
Fibres to auditory radiation
Dorsal cochlear nucleus
Nucleus of lateral lemniscus
Lateral lemniscus
Inferior cerebellar peduncle
Ventral cochlear nucleus
Cochlear nerve
Trapezoid body
Dorsal nucleus of trapezoid body

7.230 A simplified diagram to show some of the central connections of the cochlear nerve and the auditory pathway through the brain stem. Although they are shown in the figure as being of comparable size, the trapezoid body constitutes a much more important and larger commissural bundle than the fibres connecting the nuclei of the two inferior colliculi. *Efferent* fibres in the cochlear nerve and *descending* fibres in the auditory pathway have been omitted.

The cochlear nuclei are two in number and comprise a *ventral cochlear nucleus* on the ventrolateral aspect of the inferior cerebellar peduncle, receiving the larger, ascending branches of the cochlear nerve, and a *dorsal cochlear nucleus* on the dorsal aspect of the peduncle in the lateral part of the vestibular area of the floor of the fourth ventricle, forming the auditory tubercle and receiving the smaller, descending branches of the cochlear nerve. For detailed experimental analysis of the termination and distribution of primary cochlear afferents and a review, consult Osen (1970) and Kane (1974). The former demonstrated a tonotopical organization in the cochlear nuclei and connections (p. 958); both studies concern cats.

Many *ascending fibres* from the *ventral cochlear nucleus* (second neuron fibres in the auditory pathway) end in the dorsal trapezoid nucleus, either ipsilateral or contralateral. They are relayed and the tertiary fibres turn upwards as the *lateral lemniscus* (7.96, 230). Contralateral secondary fibres behave alike; decussations of contralateral fibres of both sides form the *trapezoid body* (7.230). *Efferent fibres* from the *dorsal cochlear nucleus* have similar connections (7.230), tertiary fibres ascending in both lateral lemnisci. Many cochlear efferents, mostly ipsilateral, relay in the superior olivary complex (p. 960); after relay these join both lateral lemnisci.

Each lateral lemniscus hence contains the axons of tertiary neurons from both sides; as they ascend to the midbrain some of these fibres, or their collaterals, synapse in a small group of cells, intimately related to the tract, the *nucleus of the lateral lemniscus*. Reaching the midbrain, some fibres of the lateral lemniscus end in the inferior colliculus, while others bypass it and enter the inferior brachium to reach the medial geniculate body, to be relayed to the auditory cortex (areas 41, 42 and 22). Ferraro & Minckler (1977) have estimated fibres in human lateral lemnisci at about 203 000 (right side) and 185 000 (left side), these being averages from 15 brains (newborn to 90 years).

Commissural fibres link the two auditory pathways at the inferior collicular level; auditory reflexes are mediated via the inferior colliculus to the superior colliculus and thence to the tectospinal and tectobulbar tracts and the medial longitudinal fasciculi (7.121, 230). An *efferent* component in acoustic pathways has also been noted (p. 1238); its peripheral fibres arise from neurons in or near the superior olivary complex (Rasmussen 1942, 1960), forming an *olivocochlear fasciculus* whose fibres terminate on and near hair cells of the cochlear spiral organ. The presence of sympathetic and possibly parasympathetic fibres in the vestibulocochlear nerves has received some support (Ross 1969). For further details of cochlear innervation, see p. 1238.

The vestibulocochlear nerve is soft in texture, its axons proximally ensheathed in *glial* cells. Leaving the medulla oblongata it crosses the posterior border of the middle cerebellar peduncle with the facial nerve, partially separated from the latter by the labyrinthine artery. Both nerves enter the internal acoustic meatus together. At the outer end of the meatus the vestibulocochlear nerve receives one or two filaments from the facial and splits into its *cochlear* and *vestibular* parts, the distribution of which will be described with the internal ear (pp. 1228, 1235–1241).

The Glossopharyngeal Nerve

The glossopharyngeal nerve (7.81, 86, 91, 232, 233) is both motor and sensory, supplying motor fibres to the stylopharyngeus, parasympathetic secretomotor fibres to the parotid gland and sensory fibres to the pharynx, tonsil and posterior (postsulcal) part of the tongue; it is also the gustatory nerve for this part of the tongue. It emerges as three or four rootlets from the rostral part of the medulla oblongata in a groove between the olive and the inferior cerebellar peduncle above the rootlets of the vagus nerve.

The sensory nuclei receive the central processes of unipolar neurons in the nerve's superior and inferior ganglia; fibres concerned with taste end in the *nucleus solitarius* (p. 954) and those concerned with common sensation probably terminate in the *spinal trigeminal nucleus* (Brodal 1947, and see p. 1099).

The motor nucleus, the rostral part of the *nucleus ambiguus* (pp. 956, 1114), is situated deep in the reticular formation of the medulla oblongata, connected with corticonuclear tracts of both sides both directly and through the interneurons; corticonuclear fibres leave their tract at the level of the nucleus ambiguus or in the pons to descend in the medial lemniscus (aberrant pyramidal fibres, p. 955). The glossopharyngeal part of the nucleus ambiguus sends its efferents to the stylopharyngeus.

Parasympathetic fibres join the nerve's motor part from a member of the general visceral efferent column, the *inferior salivatory nucleus* (see p. 961), located in the reticular formation below the superior salivatory nucleus; its fibres travel via the glossopharyngeal tympanic branch and the tympanic plexus (vide infra) to the lesser petrosal nerve and otic ganglion, where they relay; postganglionic fibres join the auriculotemporal nerve to reach the parotid gland (7.221) (cf. pp. 1067, 1292).

From the medulla oblongata the glossopharyngeal nerve passes anterolaterally to the triangular depression for the aqueductus cochleae, on the inferior surface of the petrous temporal bone. At first it lies under the flocculus, resting on the jugular tubercle of the occipital bone, sometimes grooving it. It then turns abruptly down, leaving the skull through the anteromedial part of the jugular foramen, anterior to the vagus and accessory nerves, in a separate dural sheath (7.231); in the foramen it is lodged in a deep groove leading from the cochlear aqueductual depression, separated by the inferior petrosal sinus from the vagus and accessory nerves; the groove is bridged by fibrous tissue, ossified in about 25% of skulls. After exit the nerve passes forwards between the internal jugular vein and internal carotid artery and then descends anterior to the latter, deep to the styloid process and its attached muscles, to reach the posterior border of the stylopharyngeus. It curves forwards on the stylopharyngeus and either pierces the lower fibres of the superior pharyngeal constrictor or passes between it and the middle constrictor (5.25) to be distributed to the tonsil, the mucosae of the pharynx and postsulcal part of the tongue, the vallate papillae, and oral mucous glands.

Two ganglia are situated on the nerve as it traverses the jugular foramen (7.231): superior and inferior.

The superior ganglion is in the upper part of the groove occupied by the nerve in the jugular foramen. It is small, has no branches and is usually regarded as a detached part of the inferior ganglion.

The inferior ganglion is larger and lies in a notch in the lower border of the petrous temporal bone (p. 379). Its cells are typical (pseudo-) unipolar neurons, whose peripheral branches convey gustatory and tactile signals from the mucosa of the tongue (posterior third including the sulcus terminalis and vallate papillae) and general sensory only from the oropharynx, soft palate and fauces.

The glossopharyngeal nerve communicates with the sympathetic trunk, vagus and facial nerves. The inferior ganglion is connected with the superior cervical sympathetic ganglion. Two filaments from the inferior ganglion pass to the vagus, one to its auricular branch and the other to its superior ganglion. A branch to the facial arises from the glossopharyngeal nerve below the inferior ganglion, perforating the posterior belly of the digastric to join the facial nerve near the stylomastoid foramen.

Branches of Distribution

These are: tympanic, carotid, pharyngeal, muscular, tonsillar and lingual.

The tympanic nerve leaves the inferior ganglion, ascends to the tympanic cavity through the inferior tympanic canaliculus (p. 379) and divides in the cavity into branches forming the **tympanic plexus**, contained in grooves on the surface of the promontory. This plexus supplies: (1) a branch to the greater petrosal nerve (p. 1110), (2) branches to the mucosa of the tympanic cavity, auditory tube and mastoid air cells, (3) a branch to the lesser petrosal nerve. For a critique of its final distribution and connections, see Winckler & Cochet (1967).

The lesser petrosal nerve contains parotid secretomotor fibres (vide infra). It enters a canal inferior to that for the tensor tympani, receives a connecting branch from the facial ganglion and reaches the anterior surface of the petrous bone through a small opening lateral to the hiatus for the greater petrosal nerve, passing thence via the foramen ovale or the canaliculus innominatus (p. 374) to join the otic ganglion.

The carotid branch, often double, arises just below the jugular foramen and descends on the internal carotid artery to the wall of the carotid sinus and to the carotid body (p. 341). It may communicate with the vagus (inferior ganglion or one of its branches) and with a sympathetic branch from the superior cervical ganglion. Another branch, either from the preceding or from the main trunk, joins a plexus which also supplies the carotid body; other branches to this plexus come from the sympathetic (superior cervical sympathetic ganglion) and vagus (p. 1116). For distribution of the carotid nerve consult Willis & Tange (1959) and p. 1473.

The pharyngeal branches are three or four filaments uniting, near the middle pharyngeal constrictor, with the pharyngeal branch of the vagus and the laryngopharyngeal branches of the sympathetic trunk to form a *pharyngeal plexus*, through which the glossopharyngeal nerve supplies sensory fibres to the pharyngeal mucosa.

The muscular branch supplies the stylopharyngeus.

Tonsilar branches form a plexus with branches of the middle and posterior palatine nerves around the tonsil; from this, filaments supply the tonsil, soft palate and fauces.

Of the two **lingual branches** one supplies the vallate papillae and mucosa near the sulcus terminalis (p. 1171), the other supplies the mucosa of the postsulcal (posterior) part of the tongue, communicating with the lingual nerve. It is the nerve of special sense (gustation) and general sensibility to the posterior lingual region.

The otic ganglion (7.223, 226, 184), a small, oval, flat reddish-grey ganglion, is situated just below the foramen ovale. It is a peripheral parasympathetic ganglion intimately related topographically to the mandibular nerve, but functionally connected with the glossopharyngeal. *Lateral* to it is the mandibular nerve near its junction with the trigeminal motor root, the ganglion usually surrounding the origin of the nerve to the medial pterygoid; *medial* is the tensor veli palatini, separating the ganglion from the cartilaginous auditory tube; *posterior* is the middle meningeal artery.

The *motor, parasympathetic root* of the otic ganglion is the lesser petrosal nerve, conveying preganglionic fibres from the glossopharyngeal nerve. These fibres originate from neurons in the inferior salivatory nucleus. They relay in the otic ganglion, postganglionic fibres passing by a *communicating branch* to the auriculotemporal nerve and so to the parotid gland (7.221). The *sympathetic root* is from the plexus on the middle meningeal artery; it contains postganglionic fibres, from the superior cervical sympathetic ganglion, which traverse the otic ganglion without relay; emerging with parasympathetic fibres in the connection to the auriculotemporal nerve, they supply blood vessels in the parotid gland.

Branches. A twig connects the ganglion to the chorda tympani, another ascends to join the nerve of the pterygoid canal; these form, so some say, an additional pathway by which gustatory fibres from the presulcal area of the tongue may reach the facial ganglion without traversing the middle ear (p. 1171); they are not relayed in the otic ganglion. Motor branches (mandibular nerve) are supplied to the tensor veli palatini and tensor tympani, derived from the nerve to the medial pterygoid (p. 1104); they have no synapses in the ganglion.

The glossopharyngeal nerve is, approximately, the post-trematic branch of the third branchial arch. The pretrematic branch of the second (hyoid) arch is probably the tympanic branch of the glossopharyngeal. Like the trigeminal and facial nerves, the glossopharyngeal corresponds to a dorsal nerve which has acquired special visceral efferent fibres.

The Vagus Nerve

The vagus nerve (7.81, 86, 87, 90–92, 231, 233) contains motor and sensory fibres and has a more extensive course and distribution than any other cranial nerve, traversing the neck,

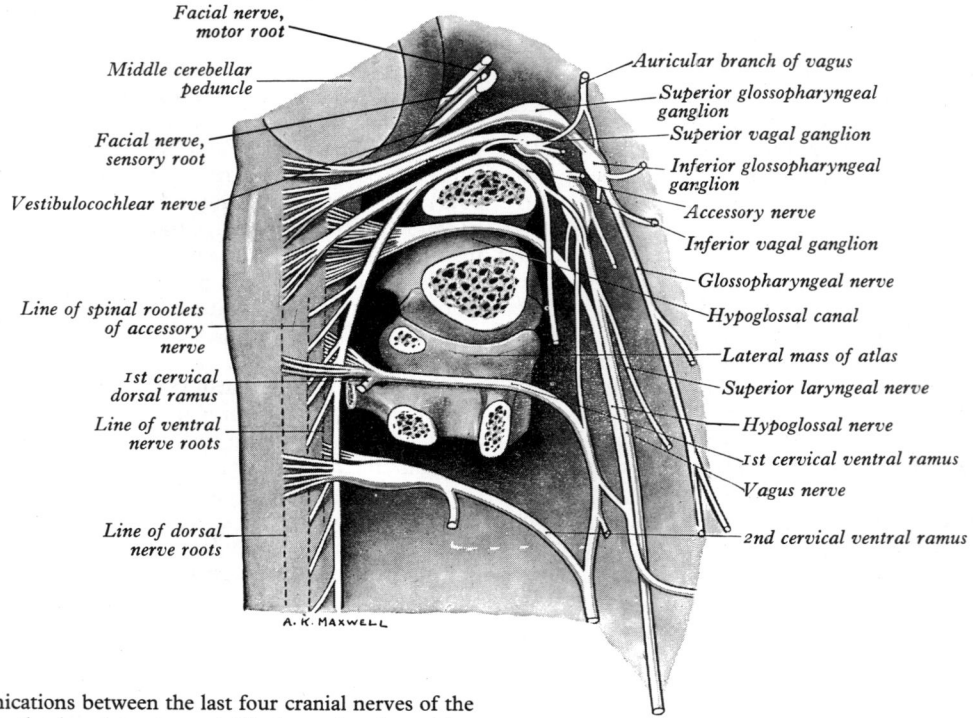

Facial nerve, motor root

Middle cerebellar peduncle

Facial nerve, sensory root

Vestibulocochlear nerve

Line of spinal rootlets of accessory nerve

1st cervical dorsal ramus

Line of ventral nerve roots

Line of dorsal nerve roots

Auricular branch of vagus

Superior glossopharyngeal ganglion

Superior vagal ganglion

Inferior glossopharyngeal ganglion

Accessory nerve

Inferior vagal ganglion

Glossopharyngeal nerve

Hypoglossal canal

Lateral mass of atlas

Superior laryngeal nerve

Hypoglossal nerve

1st cervical ventral ramus

Vagus nerve

2nd cervical ventral ramus

A. K. MAXWELL

7.231 The communications between the last four cranial nerves of the right side viewed from the dorsolateral aspect. The hypoglossal canal has been split in its long axis and the transverse process of the atlas has been divided close to the lateral mass. The descending branch of the hypoglossal nerve is not shown.

thorax and abdomen. It emerges as eight or ten rootlets from the medulla oblongata, below the glossopharyngeal nerve in the groove between the olive and the inferior cerebellar peduncle. It has four nuclei in the medulla oblongata. (1) **The dorsal nucleus** is usually described as a mixed nucleus, representing the fused general visceral efferent and general visceral afferent columns. It is sited in the central grey matter of the lower, closed part of the medulla oblongata and extends up into the open part, under the vagal triangle and separated from the hypoglossal nucleus by the nucleus intercalatus (p. 953). Its *motor fibres* are distributed to the non-striated muscle of the bronchi, heart, oesophagus, stomach, small intestine and part of the colon (p. 1155). Experiments, mainly based on retrograde degeneration, suggest patterns of localization in this visceral motor innervation (Szentágothai 1952, Mitchell & Warwick 1955, Lewis et al 1970, Rao & Sahu 1974) but the results are conflicting; moreover, the origin of vagal visceromotor nerve fibres is uncertain. Some observers (e.g. Kerr 1969) considered that most are derived from the nucleus ambiguus and nucleus retrofacialis (Szentágothai 1952). Sugimoto et al (1979) demonstrated that neurons innervating the heart originate in the nucleus ambiguus (and the reticular formation ventrolateral to it) by injection of horseradish peroxidase into the right cardiac vagal branches in cats. The particular *sensory fibres* which terminate in the nucleus are also uncertain. Though some regard the nucleus solitarius (p. 954) as predominantly a vagal nucleus, there is evidence that afferent fibres from the oesophagus and abdominal alimentary canal terminate in the dorsal vagal nucleus (p. 956). (2) Below the origin of the fibres joining the glossopharyngeal nerve, neurons of the **nucleus ambiguus** (p. 1113) contribute fibres to the vagus for distribution to skeletal muscle: the pharyngeal constrictors and intrinsic laryngeal muscles. It is connected to the corticonuclear tracts of both sides and to many brain stem centres (p. 1122). Detailed studies of the architecture and regional localization in the nucleus ambiguus of man and experimental animals (Szentágothai 1943, Getz & Sirnes 1949, Szabo & Dussardier 1964, Lawn 1966) have demonstrated that the nucleus ambiguus is divisible into subnuclei. Glossopharyngeal fibres arise from a rostral group; individual laryngeal muscles appear to be innervated by relatively discrete groups in more caudal zones; most caudally the ambigual neurons send axons into the cranial accessory nerve (p. 1118). (3) The

caudal part of the **nucleus solitarius** (pp. 954, 1109) receives vagal fibres distributed through the internal laryngeal nerve to the taste buds of the epiglottis and vallecula. Its middle part receives visceral afferent fibres from the tongue, tonsil, palate and pharynx (glossopharyngeal nerve). The rostral part of the nucleus receives gustatory fibres from the presulcal part of the tongue and soft palate (facial nerve). (4) The vagus contains somatic afferent nerve fibres, believed to terminate in the **spinal trigeminal nucleus**.

Vagal rootlets unite to form a flat cord which passes below the flocculus to the jugular foramen, by which it leaves the cranium. As it emerges it accompanies the accessory nerve, sharing an arachnoid and dural sheath, a fibrous septum separating them from the glossopharyngeal nerve which lies anterior (7.231). Here the vagus has a well-marked enlargement, the *superior ganglion*. After exit from the jugular foramen it has a second swelling, the *inferior ganglion*.

The superior ganglion, greyish, spherical and about 4 mm in diameter, is joined by one or two delicate filaments to the cranial root of the accessory nerve, by a twig to the inferior glossopharyngeal ganglion, and to the sympathetic trunk by a filament from the superior cervical ganglion. Its auricular branch gives off an ascending twig to the facial nerve (p. 1110).

The inferior ganglion is cylindrical, reddish, about 2·5 cm long and is connected with the hypoglossal nerve, superior cervical sympathetic ganglion and a loop between the first and second cervical spinal nerves. The cranial root of the accessory nerve passes over it, attached only by fibrous tissue. Beyond the ganglion the cranial accessory blends with the vagus nerve, its fibres distributed mostly in pharyngeal and recurrent laryngeal vagal branches.

Both ganglia contain unipolar sensory neurons (Richardson & Hinsey 1933, Evans & Murray 1954), the only contrary evidence being from old and unconfirmed electrophysiological experiments and unsatisfactory histological observations, both refuted. (For discussion see Mitchell 1956, Lieberman 1968, 1969, 1974). There is no reliable evidence of any significant number of motor neurons in the ganglion, nor have synapses between pre- and postganglionic neurons been demonstrated in it. Preganglionic motor fibres from the dorsal vagal nucleus and the special visceral efferents from the nucleus ambiguus, which descend to the inferior

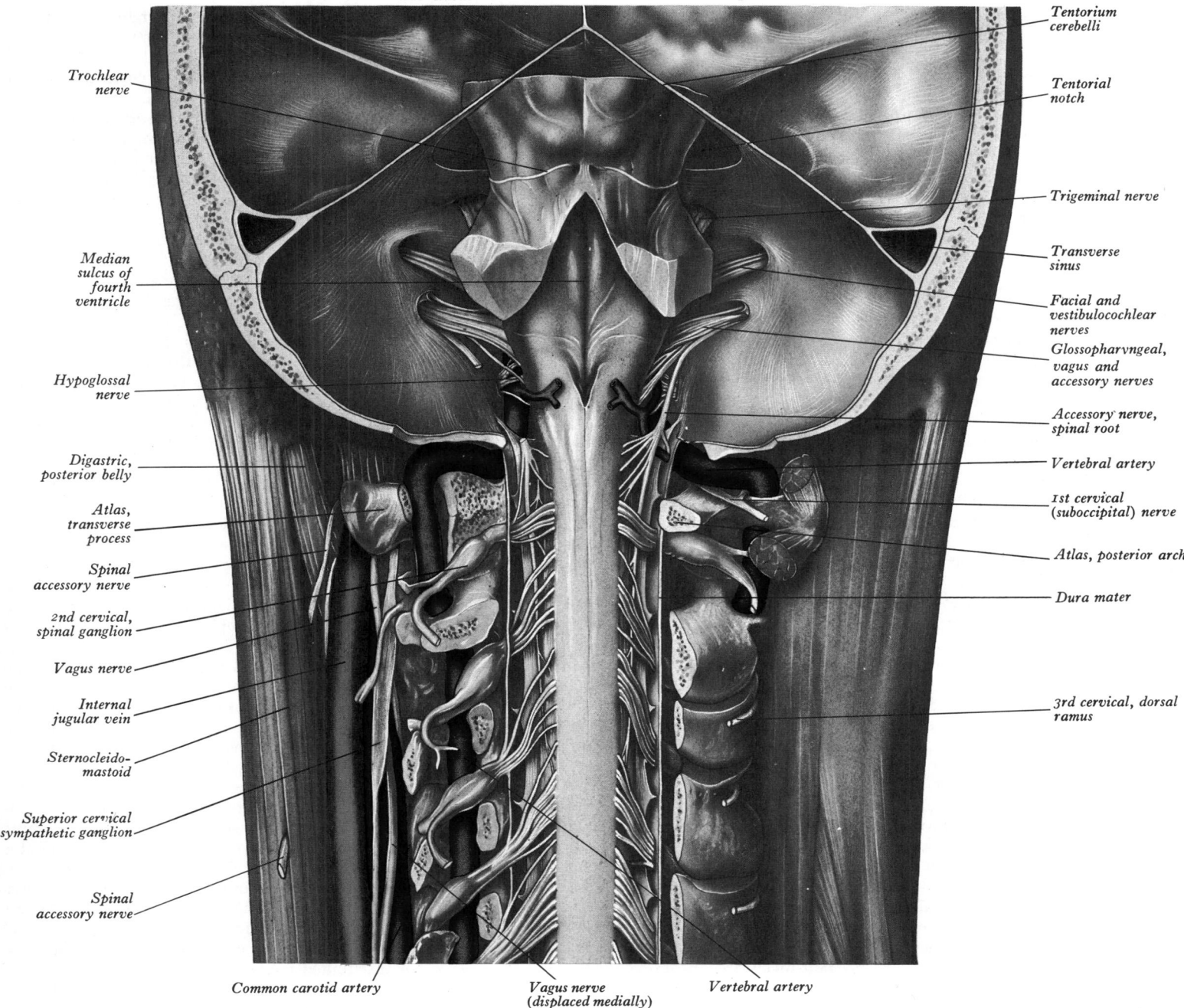

7.232 A dissection exposing the brain stem and the upper part of the spinal cord after removal of large portions of the occipital and parietal bones and the cerebellum, together with the roof of the fourth ventricle. On the left side the foramina transversaria of the atlas and the third, fourth and fifth cervical vertebrae have been opened to expose the vertebral artery. On the right side the posterior arch of the atlas and the laminae of the succeeding cervical vertebrae have been divided and have been removed together with the vertebral spines and the laminae of the opposite side. The tentorium cerebelli and the transverse sinuses have been divided and their posterior portions removed.

vagal ganglion, commonly form a visible band, skirting the ganglion in some mammals (Hoffman & Kuntz 1957, Mei & Dussardier 1966). It is necessary to emphasize these facts, because there is current in some textbooks an unsupported view that the vagal parasympathetic component relays in its inferior ganglion, based with questionable logic on the existence of many thinly myelinated and supposedly non-myelinated fibres in the distal parts of the nerve. Even were it valid to regard autonomic fibres as postganglionic purely on the degree of their myelination, to do so in this instance merely transgresses another tenet of peripheral autonomic morphology, for this view produces two relays: one in the inferior ganglion and one in the wall of the target organ. (See p. 1157 for alternative views.) Retrograde degeneration has been identified in the dorsal vagal nucleus by division of vagal branches as remote as the gastric nerves (Mitchell & Warwick 1955). Reliable analytical counts of nerve fibres in the human vagus nerve are not available; counts in other mammals do not favour the inferior

ganglion as a relay station, a view based almost entirely on an adherence to the supposed axioms of autonomic morphology. The extensive factual information quoted above renders such a view untenable.

Both vagal ganglia are exclusively sensory, containing somatic, special visceral and general visceral afferent neurons. They are, however, traversed by parasympathetic and perhaps some sympathetic fibres. Their neurons are unipolar and bipolar (Cajal 1911, Evans & Murray 1954, Gabella 1976). The occurrence of multipolar neurons in the inferior ganglion has been suggested in the past but was strongly denied by Mitchell (1956) and Lieberman (1968); since they would almost certainly be parasympathetic postganglionic elements, their presence would imply double peripheral relay; further investigation is here required. Synapses between pre- and postganglionic neurons have not been reported. The general consensus is that all efferent fibres, special visceral (branchial) or parasympathetic preganglionic, pass uninterrupted

1115

through both ganglia. (However, see comments on chromaffin cells below.) The disposition of somata and nerve fibres resembles that in spinal posterior root ganglia, the fibres being centrally disposed. The superior ganglion is described as grey, the inferior as pink, a difference which may reflect variation in the distribution of myelinated and non-myelinated fibres or of vessels.

The superior (jugular) ganglion is rounded and about 4 mm across in mankind. In cats (Foley & DuBois 1937) it contains about 8700 unipolar neurons, 73% of which appear to form the auricular nerve, a branch of the ganglion (p. 1117), while 15% contribute to the vagus itself. Counts are apparently not available for the human ganglion; but since the human vagus nerve contains, at mid-cervical level, about 16 500 (Rt) and 20 000 (Lt) myelinated fibres, counted in 17 paired nerves, human ganglionic neurons must be more numerous (Schnitzlein et al 1958). It is connected with the cranial root of the accessory nerve, the inferior glossopharyngeal ganglion and the superior cervical sympathetic ganglion. The significance of these connections is not clear but the first probably contains aberrant motor fibres from the nucleus ambiguus which issue in the accessory nerve, to be distributed to the palatal, pharyngeal, laryngeal and upper oesophageal musculature via the vagus; the sympathetic connection may be like the one between this sympathetic ganglion and the *inferior* vagal ganglion (vide infra). As it leaves the superior ganglion the auricular branch communicates with the facial nerve (p. 1110).

The inferior (nodose) ganglion is larger and elongate, about 25 mm in length and about 5 mm in maximum breadth. In cats (Jones 1937, Foley & DuBois 1937) it is estimated to contain about 30 000 neurons, most are unipolar or pseudo-unipolar but a few fusiform or bipolar; most are in the range of 35–40 µm in perikaryal dimensions but a few measure 20–30 µm (Mohiuddin 1953). Chromaffin cells have been observed (White 1935, Jacobs & Comroe 1951, Grillo et al 1974), which receive agranular synapses but display no efferent synapses. The ganglion is connected with the hypoglossal nerve (p. 1120), the loop between the first and second cervical spinal nerves (p. 1128) and with the superior cervical sympathetic ganglion (p. 1158); the last may be formed by the peripheral axons of sensory neurons resembling nodose ganglion cells, which have been described in this sympathetic ganglion by Terni (1922) as a group of displaced nodose ganglion neurons.

While neurons in the vagal ganglia have received less attention than those in the superior cervical sympathetic ganglion, extensive counts of fibres in the vagus and its periganglionic branches have been made (Gabella 1976). While overall counts show a general agreement, there are discrepancies in the estimated ratios between myelinated and non-myelinated fibres and between fibres of different diameters; both thinly-myelinated and non-myelinated fibres are numerous and while most are preganglionic parasympathetic, a few may be postganglionic sympathetic fibres. Though the feline superior vagal ganglion contains, according to DuBois & Foley (1937), about 8700 nerve cells, about 8800 fibres (chiefly small and non-myelinated) enter the auricular nerve and 1700–3500 continue into the vagus caudal to the superior ganglion. Many discrepancies of this kind await clarification and it would be unprofitable at present to quote further quantitative data. There is wide agreement that the superior ganglion is chiefly somatic, most of its neurons entering the auricular nerve, while neurons in the inferior ganglion are concerned with visceral sensation in the heart, larynx and lungs and the alimentary tract from the pharynx to the transverse colon. In addition, some fibres transmit impulses from taste endings in the vallecula and epiglottis.

Course

The vagus nerve descends vertically in the neck in the carotid sheath, between the internal jugular vein and the *internal* carotid artery to the thyroid cartilage's upper border and then passes between the vein and the *common* carotid artery to the root of the neck. Its further course differs on the two sides.

The *right* vagus nerve descends posterior to the internal jugular vein to cross the first part of the subclavian artery, entering the thorax and descending through the superior mediastinum, at first behind the right brachiocephalic vein, then right of the trachea and posteromedial to the right brachiocephalic vein and superior vena cava. The right pleura and lung are lateral to it above but are separated from it below by the azygos vein, which arches forwards above the right pulmonary hilum (6.164). The nerve then passes behind the right principal bronchus to the posterior aspect of the right hilum and divides into the posterior pulmonary (bronchial) branches, which unite with rami from the second to fifth or sixth thoracic sympathetic ganglia to form the **right posterior pulmonary plexus**. From the caudal part of this plexus two or three branches descend on the dorsal surface of the oesophagus, where, with a left vagal ramus, they form the **posterior oesophageal plexus**, from which a trunk is re-formed and continued posterior to the oesophagus to traverse the diaphragmatic oesophageal opening. This posterior vagal trunk contains fibres from both vagus nerves (p. 1118).

(Details of the **cardiac plexuses** are on pp. 1155, 1158, 1164.)

In the abdomen the **posterior vagal trunk** divides into a small gastric and a larger coeliac branch. The former supplies the postero-inferior gastric surface except the pyloric canal. The coeliac branch ends chiefly in the coeliac plexus but sends twigs to the splenic, hepatic, renal, suprarenal and superior mesenteric plexuses (p. 1165).

The *left* vagus enters the thorax between the left common carotid and subclavan arteries and behind the left brachiocephalic vein. It descends through the superior mediastinum and crosses the left side of the aortic arch to pass behind the left pulmonary hilum. Above the aortic arch it is crossed anterolaterally by the left phrenic nerve and on the arch by the left superior intercostal vein (6.30A). Behind the hilum it divides into the posterior pulmonary (or bronchial) branches, which unite with rami of the second to fourth thoracic sympathetic ganglia; this forms the **left posterior pulmonary plexus**, two branches of which descend anteriorly on the oesophagus forming, with a ramus from the right posterior pulmonary plexus, the **anterior oesophageal plexus**. From this a trunk containing fibres from both vagus nerves continues anterior to the oesophagus, entering the abdomen through the oesophageal diaphragmatic opening (p. 593).

In the abdomen the **anterior vagal trunk** supplies the cardiac antrum of the stomach and then divides into right and left branches. The left group follows the lesser gastric curvature to supply the stomach's anterosuperior surface. The right group has three main branches. The first, sometimes duplicated, proceeds between the layers of lesser omentum towards the porta hepatis, dividing into: (a) upper branches entering the porta, and (b) lower rami supplying chiefly the pyloric canal, pylorus, superior and descending parts of duodendum and the head of the pancreas. The second is distributed to the anterosuperior surface of the body of the stomach; the third branch follows the lesser curvature to the angular notch (see also p. 1355).

Branches of the Vagus Nerve

In the jugular fossa:	Meningeal
	Auricular
In the neck:	Pharyngeal
	Branches to carotid body
	Superior laryngeal
	Recurrent laryngeal (right)
	Cardiac
In the thorax:	Cardiac
	Recurrent laryngeal (left)
	Pulmonary
	Oesophageal
In the abdomen:	Gastric
	Coeliac
	Hepatic
	Renal

The meningeal branch or branches appear to start from the superior vagal ganglion and are distributed to the dura mater in the posterior cranial fossa. However, evidence suggests (Kimmel 1961a, b) that they are in fact recurrent sensory and sympathetic nerves from the upper cervical spinal nerves and superior cervical sympathetic ganglion, which run for a short distance in the sheath of the vagus nerve (pp. 1088, 1125).

The auricular branch also arises from the superior vagal ganglion and is joined soon after by a ramus from the inferior ganglion of the glossopharyngeal nerve. It passes behind the internal jugular vein and enters the mastoid canaliculus on the lateral wall of the jugular fossa. Traversing the temporal bone thus, it crosses the facial canal about 4 mm above the stylomastoid foramen and here supplies an ascending branch to the facial nerve. (Here fibres of the nervus intermedius may pass to the auricular branch, perhaps explaining the cutaneous vesiculation which sometimes accompanies geniculate herpes.) The auricular branch then traverses the tympanomastoid fissure, dividing into two rami: one joining the posterior auricular nerve and the other distributed to the skin of part of the cranial auricular surface and to the posterior wall and floor of the external acoustic meatus and adjoining part of the outer surface of the tympanic membrane. The vagal auricular branch thus contains *somatic afferent* nerve fibres, which probably terminate in the spinal trigeminal nucleus.

The pharyngeal branch, the main motor nerve of the pharynx, emerges from the upper part of the inferior vagal ganglion and consists chiefly of filaments from the cranial accessory nerve. It passes between the external and internal carotid arteries to the upper border of the middle pharyngeal constrictor, dividing into numerous filaments which join rami of the sympathetic trunk and glossopharyngeal and external laryngeal nerves to form a *pharyngeal plexus*, by which vagal fibres are distributed to the pharyngeal and palatal muscles, except the tensor veli palatini. A minute filament, the *ramus lingualis vagi*, joins the hypoglossal nerve as it curves round the occipital artery.

Branches to the carotid body (p. 1473), variable in number, may arise from the inferior ganglion or travel in either the pharyngeal branch or the superior laryngeal nerve, though rarely in the latter. They form a plexus with the glossopharyngeal rami and branches of the cervical sympathetic trunk (Sheehan et al 1941, also pp. 1113, 1158).

The superior laryngeal nerve, larger than the pharyngeal branch, issues from the middle of the inferior vagal ganglion and receives a branch from the superior cervical sympathetic ganglion. It descends alongside the pharynx, first posterior, then medial to the internal carotid artery, and divides into the internal and external laryngeal nerves. The human nerve contains about 15 000 fibres (Ogura & Bello 1952).

The *internal laryngeal nerve* is sensory to the laryngeal mucosa down to the level of the vocal folds. It also carries afferent fibres from the laryngeal neuromuscular spindles and other stretch receptors (Keene 1961, Schever 1964). It descends to the thyrohyoid membrane, pierces it above the superior laryngeal artery and divides into an upper and lower branch; the upper is horizontal and supplies the mucosa of pharynx, the epiglottis, vallecula and laryngeal vestibule; the lower descends in the medial wall of the piriform recess, supplying the aryepiglottic fold, the mucosa on the back of the arytenoid cartilage and one or two branches to the arytenoideus, which unite with twigs from the recurrent laryngeal to supply the same muscle (p. 1258). The internal laryngeal nerve ends by piercing the inferior constrictor muscle to unite with an ascending branch from the recurrent laryngeal. Past observers (see Latarjet 1947, for citations) have described small ganglia on the superior and internal laryngeal nerves. Ramaswamy (1974) has revived this description, describing a ganglion on the internal laryngeal nerve in more than 100 human larynges. Cells in the ganglia were described but without photographic confirmation.

The *external laryngeal nerve*, smaller than the internal, descends posterior to the sternothyroid with the superior thyroid artery but on a deeper plane; it lies at first on the inferior pharyngeal constrictor and then, piercing it, curves round the inferior thyroid tubercle to reach and supply the cricothyroid. It also supplies the pharyngeal plexus and inferior constrictor; behind the common carotid artery it connects with the superior cardiac nerve and superior cervical sympathetic ganglion (Skórnicki et al 1968).

The recurrent laryngeal nerve differs, in origin and course, on the two sides. On the *right* it arises from the vagus anterior to the first part of the subclavian artery, curving backwards below and then behind it to ascend obliquely to the side of the trachea

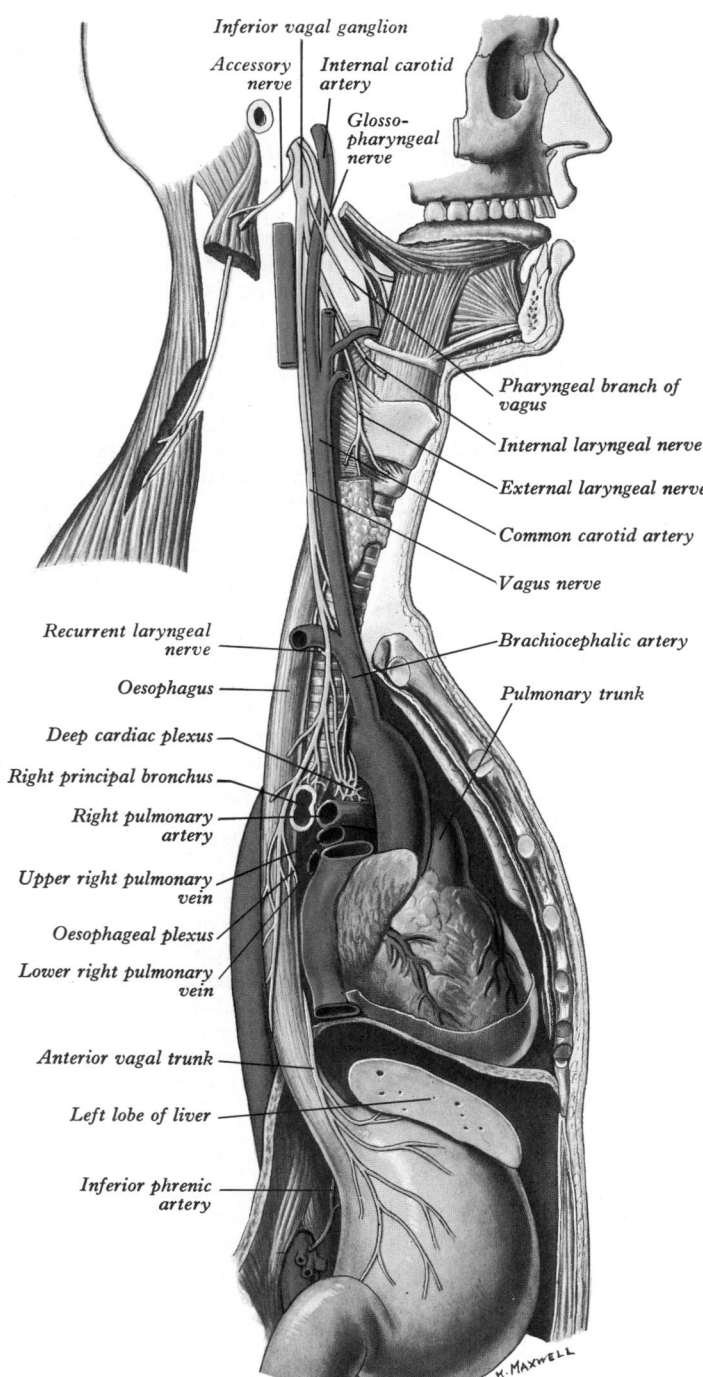

Inferior vagal ganglion

Accessory nerve

Internal carotid artery

Glosso-pharyngeal nerve

Pharyngeal branch of vagus

Internal laryngeal nerve

External laryngeal nerve

Common carotid artery

Vagus nerve

Brachiocephalic artery

Pulmonary trunk

Recurrent laryngeal nerve

Oesophagus

Deep cardiac plexus

Right principal bronchus

Right pulmonary artery

Upper right pulmonary vein

Oesophageal plexus

Lower right pulmonary vein

Anterior vagal trunk

Left lobe of liver

Inferior phrenic artery

A.K. MAXWELL

7.233　The course and distribution of the glossopharyngeal, vagus and accessory nerves.

behind the common carotid artery. Near the lower pole of the thyroid lateral lobe it is near the inferior thyroid artery, crossing *either* in front of or behind it, or between its branches. On the *left*, the nerve arises from the vagus on the left of the aortic arch, curves below it immediately behind the attachment of the ligamentum arteriosum to the arch's concavity and ascends to the side of the trachea. On both sides the recurrent laryngeal nerve ascends in or near a groove between the trachea and oesophagus and is closely related to the medial surface of the thyroid gland before passing under the lower border of the inferior constrictor to enter the larynx behind the articulation of the inferior thyroid cornu with the cricoid cartilage. It supplies all laryngeal muscles, except the cricothyroid; it communicates with the internal laryngeal nerve, supplying sensory filaments to the laryngeal mucosa below the vocal folds. It also carries afferent fibres from laryngeal stretch receptors.

As the recurrent laryngeal nerve curves round the subclavian artery, or the aortic arch, it gives cardiac filaments to the deep cardiac plexus. As it ascends in the neck it supplies branches, more numerous on the left, to the mucosa and tunica muscularis of the oesophagus and trachea and to the inferior constrictor.

The varying relations of the recurrent laryngeal nerves near the larynx are important in thyroid surgery (Bowden 1955, Doyle et al 1967). The nerve does not always lie in a protected position in the tracheo-oesophageal groove but may be slightly anterior to it (more often on the right) and it may be markedly lateral to the trachea at the level of the lower part of the thyroid gland. On the right the nerve is as often anterior to, or posterior to, or inter-mingled with terminal branches of the inferior thyroid artery, while on the left the nerve is usually posterior to the artery, though occasionally anterior to it. The nerve may supply extra-laryngeal branches to the larynx, arising before the nerve passes behind the inferior thyroid cornu. Outside its capsule the thyroid gland has a distinct covering of pretracheal fascia (p. 583), which splits into two layers at the posterior border of the gland. One layer covers the entire medial surface of its lobe and, at or just above the isthmus, has a conspicuous thickening, the *lateral ligament* of the *thyroid gland* (p. 1459) which attaches the gland to the trachea and the lower part of the cricoid cartilage. The other layer is posterior, passing behind the oesophagus and pharynx and attached to the prevertebral fascia. By this splitting of the fascia, a compartment is formed on each side, lateral to the trachea and oesophagus, and it is in the fat that this contains that the recurrent laryngeal nerve and terminal parts of the inferior thyroid artery lie. The nerve may be lateral or medial to the lateral ligament of the thyroid gland, or sometimes may be em-bedded in it.

The cardiac branches, two or three in number, arise from the vagus at superior and inferior cervical levels. The small *superior branches* join sympathetic cardiac branches and reach the deep cardiac plexus. The *inferior branches* arise in the root of the neck, the right passing in front or by the side of the brachiocephalic artery to the deep cardiac plexus, the left descending across the left side of the aortic arch to join the superficial cardiac plexus. Additional cardiac branches arise from the right vagus nerve near the trachea and from both recurrent laryngeal nerves, ending in the deep cardiac plexus. The cardiac plexuses are described on p. 1164.

The anterior pulmonary branches, two or three in number, and small, reach the anterior surface of the hilum of the lung, forming with sympathetic rami the *anterior pulmonary plexus*.

The posterior pulmonary branches, more numerous and larger, pass behind the hilum, forming with sympathetic rami from the second to fifth or sixth thoracic sympathetic ganglia the *posterior pulmonary plexus*, branches from which accompany the bronchial ramifications and supply the constrictor muscles and other tissues of the pulmonary tree (pp. 1165, 1271, 1285).

The oesophageal branches arise above and below the pul-monary, the lower being numerous and larger. They form the *oesophageal plexus*, from which filaments supply the oesophagus and the back of the pericardium (p. 1116).

The gastric branches supply the stomach's anterosuperior surface (mainly from the left vagus) and postero-inferior surface (mainly from the right) (p. 1165). Brizzi et al (1973) described separate 'cardiac' branches to the cardia; they found that both gastric branches supplied the fundus and body, while the pylorus received a complex innervation mainly from the anterior gastric nerve and vagal hepatic branches, less frequently from the pos-terior gastric nerve. Both anterior and posterior gastric nerves lie within the hepatogastric ligament, respectively anterior and pos-terior to the peri-arterial nerves that surround the left gastric artery (Mackay & Andrews 1983).

The coeliac branches of the posterior vagal trunks join the coeliac plexus.

The hepatic branches from both vagi (p. 1165) join the hepatic plexus and supply the liver.

The renal branches, also from both nerves, join the renal plexus (p. 1165).

For details of the abdominal distribution of the vagus nerves, see pp. 1156, 1165, 1166.

Applied Anatomy. The vagus is not often injured but its func-tions may be affected by damage to its nuclei in the medulla oblongata or in its intracranial course. The effects include palpita-tion, tachycardia, vomiting, slowing of respiration and a sensation of suffocation.

Abnormal 'reflexes' in connection with the vagus may occur. The 'ear cough' is perhaps the commonest, where wax in the external acoustic meatus may irritate receptors of the auricular nerve, producing a persistent cough. Syringing the meatus often excites coughing and in children vomiting may occur. Syringing the ear has occasionally caused fatal reflex cardiac inhibition. Another example is persistent cough due to enlarged bronchial lymph nodes in children, irritating the recurrent laryngeal nerve. When peripheral terminations of the superior laryngeal nerve are irritated by a foreign body, reflex spasm of the glottis results. When the nerve is out of action, anaesthesia of the upper laryngeal mucosa ensues, so that foreign bodies can easily enter the larynx; since the nerve supplies the cricothyroid, the vocal folds cannot be tensed; the voice is deep and hoarse. Irritation of the recurrent laryngeal nerves produces spasm of the laryngeal muscles. When both recurrent laryngeal nerves are interrupted, the vocal folds are motionless, and are in the so-called 'cadaveric position', i.e. neither closed as in phonation, nor widely open as in deep in-spiratory efforts. When only one nerve is affected, its vocal fold is motionless, the opposite fold crossing the median plane; hence phonation is possible but the voice is weak. In progressive lesions of the recurrent laryngeal nerve abduction of the vocal folds is abolished before adduction; conversely, during recovery, adduc-tion returns before abduction (*Semon's law*). There is no direct evidence that nerve fibres to the abductor muscles have any special grouping in the recurrent laryngeal nerve.

The Accessory Nerve

The accessory nerve (7.86, 87, 231–233) is formed by the union of its cranial and spinal roots but these are associated for a short distance only. The cranial part joins the vagus and should really be considered part of it; it is a branchial or special visceral efferent nerve. The spinal 'root' may be considered as somatic, special visceral efferent or mixed, depending on the view taken of the embryological origin of the sternocleidomastoid and trapezius, which it supplies. The custom of describing the two parts as a single cranial nerve has been followed here. The spinal part is usually assumed to be purely motor, but evidence for the presence of afferent fibres is provided by the occurrence of ganglia on the nerve in prenatal and early postnatal human material (Pearson et al 1964); the nerve may also communicate with the dorsal roots of the upper cervical spinal nerves, although such observations are unconfirmed in adult material.

The cranial root is the smaller, arising from caudal neurons of the **nucleus ambiguus** (p. 1114) and possibly also of the **dorsal vagal nucleus**. The former is connected with corticonuclear tracts of *both* sides, some fibres from this source descending from the midbrain in the medial lemniscus (aberrant pyramidal fibres, p. 955). The cranial root emerges as four of five fine rootlets from the side of the medulla oblongata below the vagal roots; it runs laterally to the jugular foramen, perhaps interchanging fibres here with the spinal root, with which it unites for a short distance; it also connects with the superior vagal ganglion. It traverses the foramen, separates from the spinal part and continues over the inferior vagal ganglion, to which it adheres. It is distributed main-ly in the pharyngeal and recurrent laryngeal branches of the vagus and is probably the source of the vagal motor fibres which run in the pharyngeal branch to the palatal muscles, except the tensor veli palatini. Some filaments continue into the vagus below the ganglion, to be distributed with the recurrent laryngeal and poss-ibly the cardiac nerves.

The spinal root arises from a column of motor neurons, *the spinal nucleus*, located lateral in the spinal anterior grey column and descending as far as the fifth cervical segment (p. 927). Traversing the lateral white columns of the spinal cord, the fibres emerge midway between the upper cervical ventral and dorsal spinal roots (7.231), forming a trunk ascending between the

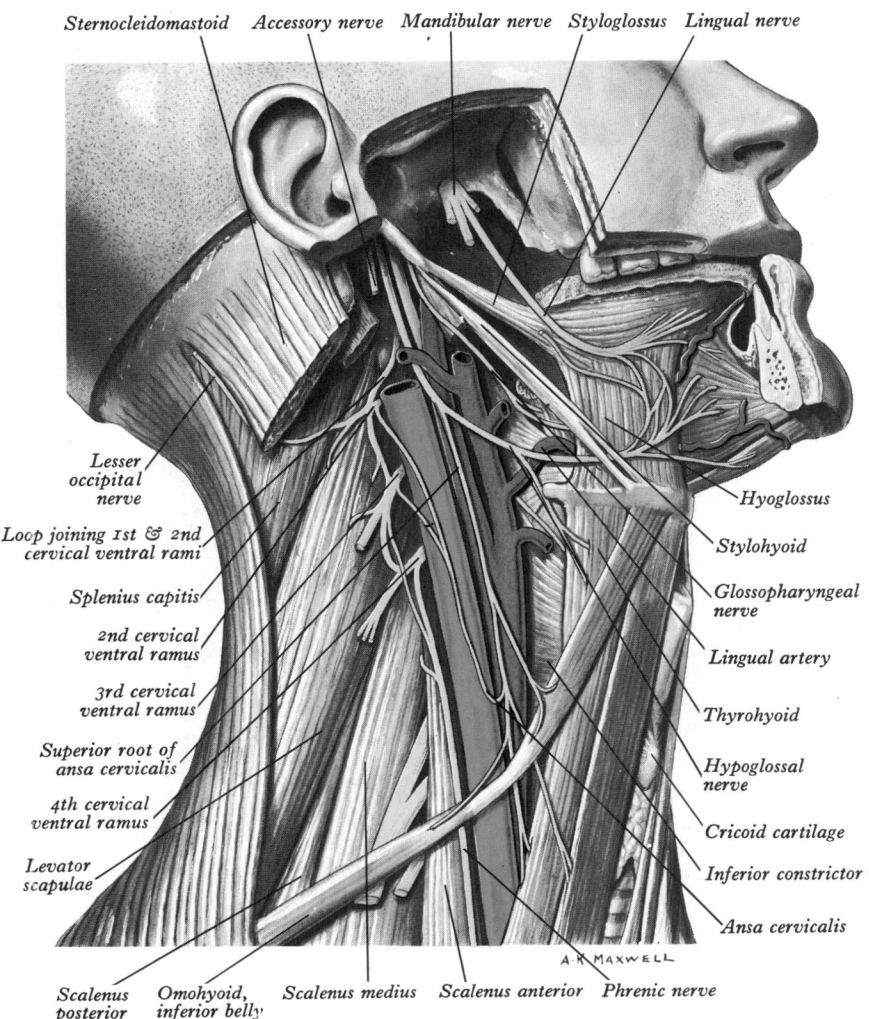

Sternocleidomastoid Accessory nerve Mandibular nerve Styloglossus Lingual nerve

Lesser
occipital
nerve

Loop joining 1st & 2nd
cervical ventral rami

Splenius capitis

2nd cervical
ventral ramus

3rd cervical
ventral ramus

Superior root of
ansa cervicalis

4th cervical
ventral ramus

Levator
scapulae

Hyoglossus

Stylohyoid

Glossopharyngeal
nerve

Lingual artery

Thyrohyoid

Hypoglossal
nerve

Cricoid cartilage

Inferior constrictor

Ansa cervicalis

A.K.MAXWELL

Scalenus   Omohyoid,   Scalenus medius Scalenus anterior Phrenic nerve
posterior   inferior belly

7.234 A dissection to show the general distribution of the right hypoglossal and lingual nerves and the position and constitution of some parts of the cervical plexus of the right side.

ligamentum denticulatum and dorsal spinal roots to enter the skull via the foramen magnum, behind the vertebral artery (7.232). It then turns up and laterally to the jugular foramen, traversing this in a single dural sheath with the vagus but separated from it by a fold of the arachnoid mater. In the foramen it may receive one or two rami from the cranial root or join it for a short distance. At its exit from the foramen it runs posterolaterally, usually posterior to the internal jugular vein but anterior in about one-third of subjects; rarely it passes through the vein. Here the nerve crosses the transverse process of the atlas and is itself crossed by the occipital artery. It then descends obliquely, medial to the styloid process, stylohyoid and posterior belly of the digastric. With the superior sternocleidomastoid branch of the occipital artery it reaches the upper part of the sternocleidomastoid and enters its deep surface, joining branches of the second cervical spinal nerve in supplying it. Emerging a little above the midpoint of the muscle's posterior border it crosses the posterior triangle on levator scapulae (7.228), separated from it by the prevertebral layer of deep cervical fascia and adipose tissue. Here it is comparatively superficial, is related to the superficial cervical lymph nodes, and receives rami of the second and third cervical spinal nerves (Raveau 1968). About 5 cm above the clavicle it passes behind the anterior border of the trapezius, forming a plexus on its deep surface with branches of the third and fourth cervical nerves. From this plexus the trapezius is innervated. The nerve's cervical course follows a line from the lower anterior part of the tragus to the tip of the atlantal transverse process and then across the sternocleidomastoid and the posterior triangle to a point on the anterior border of the trapezius 5 cm above the clavicle. For the significance of the double innervation of the trapezius and sternocleidomastoid see McKenzie (1955) and p. 584.

The cranial and spinal roots, after separating, are also known respectively as the *internal* and *external rami* of the accessory nerve. The spinal root is accepted to be the sole motor supply to the sternocleidomastoid, the second and third cervical nerves conveying proprioceptive fibres from it. The innervation of the trapezius is less certain; some consider the third and fourth cervical spinal supply as purely proprioceptive but others consider that they supply motor fibres to the muscle's lower part.

Applied Anatomy. The functions of the accessory nerve may be affected by central changes or at its cranial exit by fractures across the jugular foramen or by cervical lymphadenitis. Acute torticollis in children is usually due to the latter; central irritation causes clonic spasm of the sternocleidomastoid and trapezius, termed spasmodic torticollis. When all palliative treatment fails, division or partial excision of the nerve may be performed. Where extensive dissection is necessary in the posterior triangle, the nerve should be isolated at the outset from diseased nodes.

The Hypoglossal Nerve

The hypoglossal nerve (7.86, 87, 234, 235) is motor to all the muscles of the tongue, except the palatoglossus (p. 1290); it lies in series with the oculomotor, trochlear and abducent nerves and the ventral nerve roots of the spinal nerves and represents the fused ventral roots of probably four precervical or spino-occipital nerves whose dorsal roots have disappeared.

The hypoglossal nucleus is in line with the spinal anterior grey column. It is about 2 cm long, its rostral part corresponding with the hypoglossal triangle in the floor of the fourth ventricle (p. 980), its caudal part extending into the closed part of the

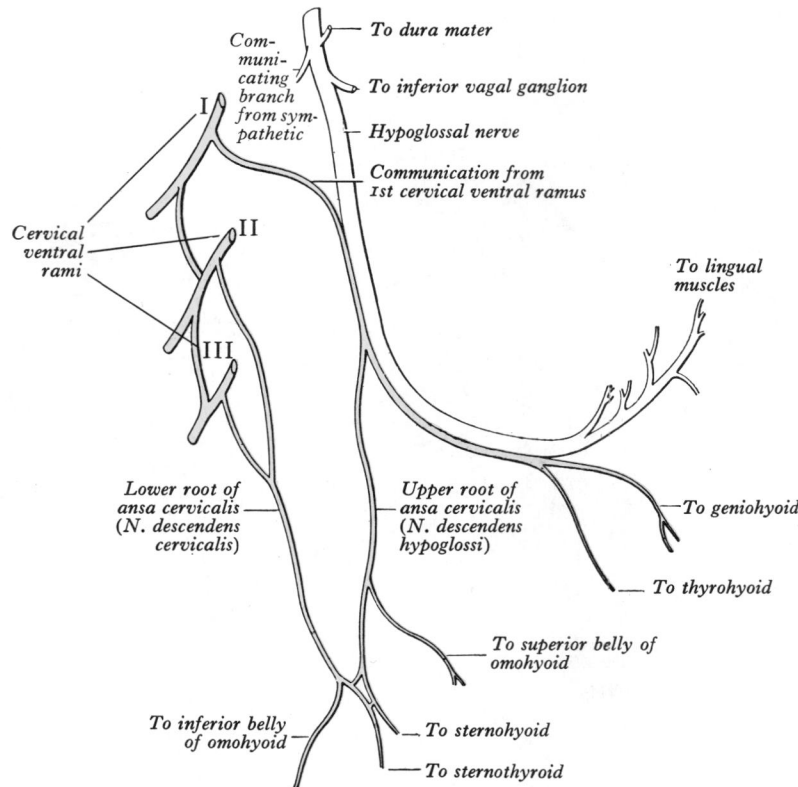

7.235 A plan of the right hypoglossal nerve and ansa cervicalis.

medulla oblongata, where it is ventral and paramedial in the central grey matter (7.89). Its fibres pass ventrally through the medulla to emerge as a linear series of 10–15 rootlets in the anterolateral sulcus between the pyramid and olive (7.81).

The hypoglossal nucleus is divided longitudinally into dorsal and ventral laminae, each subdivisible into a mediolateral sequence of relatively discrete subnuclei, considered to correspond to individual muscles (Kosaka & Yagita 1903, Sturman 1916, Barnard 1940). More recently, Jansen & Korneliussen (1977) have described division into four columns in Cetacea, suggesting that these correspond to four occipitocervical somites and hence, perhaps, to individual lingual muscles.

Connections and Course

The hypoglossal nucleus receives corticonuclear fibres from the precentral gyrus and adjacent areas of mainly the *contralateral* hemisphere, some fibres leaving the tract in the pons to travel in the medial lemniscus; they connect with the nucleus directly or via internuncial neurons. Evidence indicates that the most medial subnuclei receive projections from *both* hemispheres. The nucleus may connect with the cerebellum via adjacent perihypoglossal nuclei (Torvik & Brodal 1954, p. 953) and perhaps also with the medullary reticular formation, the trigeminal sensory nuclei and the solitary nucleus.

The hypoglossal rootlets run laterally behind the vertebral artery, collected into two bundles which perforated the dura mater separately opposite the hypoglossal (anterior condylar) canal in the occipital bone, uniting after traversing it; the canal is sometimes divided by a bony spicule. The separate dural sheaths confirm the nerve's composite character. It emerges from its canal in a plane medial to the internal jugular vein, internal carotid artery, ninth, tenth and eleventh cranial nerves and passes inferolaterally behind the internal carotid artery, glossopharyngeal and vagus nerves to the interval between the artery and the internal jugular vein. Here it makes a half-spiral turn round the inferior vagal ganglion, being united with it by connective tissue. It then descends almost vertically between the vessels and anterior to the vagus to a point level with the mandibular angle, becoming superficial below the posterior belly of the digastric and emerging between the internal jugular vein and internal carotid artery. It

loops round the inferior sternocleidomastoid branch of the occipital artery (p. 736), crosses lateral to both internal and external carotid arteries and the loop of the lingual artery a little above the tip of the greater cornu of the hyoid (7.234) and is itself crossed by the facial vein. It inclines up and forwards on the hyoglossus, passing deep to the digastric tendon, stylohyoid and the posterior border of the mylohyoid. Between the hyoglossus and mylohyoid the nerve is inferior to the deep part of the submandibular gland, submandibular duct and lingual nerve. It then passes on to the lateral aspect of the genioglossus, continuing forwards in its substance as far as the tip of the tongue and distributing fibres to the muscle.

The hypoglossal nerve communicates with the sympathetic trunk, vagus, first and second cervical nerves and lingual nerve. Near the atlas it is joined by branches from the superior cervical sympathetic ganglion and by a filament from the loop between the first and second cervical nerves which leaves the hypoglossal as the *upper root* of the *ansa cervicalis* (7.215). The vagal connections occur close to the skull, numerous filaments passing between the hypoglossal nerve and the inferior vagal ganglion in the connective tissue uniting them. As the hypoglossal nerve curves round the occipital artery it receives from the pharyngeal plexus the *ramus lingualis vagi* (p. 1117). Near the anterior border of hyoglossus it is connected with the lingual nerve by many filaments ascending on the muscle (p. 1105).

Branches

The branches of distribution of the hypoglossal nerve are: meningeal, descending, thyrohyoid and muscular.

The meningeal branch or branches leave the hypoglossal nerve in its canal, returning through it to supply the diploë of the occipital bone, the dural walls of the occipital and inferior petrosal sinuses and much of the floor and anterior wall of the posterior cranial fossa. These meningeal rami (pp. 1088, 1158) may not contain hypoglossal fibres but ascending, mixed sensory and sympathetic fibres from the upper cervical nerves and superior cervical sympathetic ganglion (Kimmel 1961a, b).

The descending branch (descendens hypoglossi or upper root of the ansa cervicalis) leaves the hypoglossal nerve where it curves round the occipital artery and then descends anterior to or in the carotid sheath. It contains only fibres from the first cervical spinal nerve. After giving a branch to the superior belly of the omohyoid, it is joined by the ansa's *lower root* from the second and third cervical spinal nerves. The two roots form the *ansa cervicalis (ansa hypoglossi)*, from which branches supply the sternohyoid, sternothyroid and the inferior belly of the omohyoid; another branch is said to descend anterior to the vessels into the thorax to join the cardiac and phrenic nerves.

The nerves to the thyrohyoid and geniohyoid arise near the posterior border of the hyoglossus and cross obliquely the greater cornu of the hyoid to supply the thyrohyoid and geniohyoid; they contain fibres of the first cervical spinal nerve.

Muscular branches are distributed to the styloglossus, hyoglossus and genioglossus. Numerous slender rami ascend into the tongue to supply its intrinsic muscles; all contain true hypoglossal fibres.

Applied Anatomy. Complete hypoglossal division causes unilateral lingual paralysis and hemiatrophy; the protruded tongue deviates to the paralysed side; on retraction, the wasted and paralysed side also rises higher than the unaffected side. The larynx may deviate towards the active side in swallowing, due to unilateral paralysis of the hyoid depressors. If paralysis is bilateral, the tongue is motionless. Taste and tactile sensibility are unaffected but articulation is slow and swallowing very difficult.

MORPHOLOGY OF THE CRANIAL NERVES

It is now possible to group cranial nerves to conform with their phylogenetic history, components and functions.

Group I includes the oculomotor, trochlear, abducent and hypoglossal nerves. All arise from neurons of the somatic efferent column and are distributed to musculature derived from the *cranial myotomes*. They therefore correspond to *ventral* spinal roots and, with the exception of the trochlear, they emerge from the brain stem in line with them. The identification of the segments with which each nerve is associated presents great problems and is at present beyond experimental or any other type of proof; even the number of cephalic segments is uncertain (p. 139).

Group II includes the trigeminal, facial, glossopharyngeal, vagus and accessory nerves, which innervate the derivatives of the *branchial arches* but differ from the spinal nerves in having motor roots distributed to musculature derived from the *neural crest* (p. 136) and *lateral plate mesoderm* of the branchial region. Some (trigeminal, vagus and accessory) are compound nerves, formed by the fusion of two or more dorsal nerves (p. 1092); in this process cutaneous branches, originally connected with facial, glossopharyngeal and vagus nerves, have been absorbed into the trigeminal; hence these three cranial nerves differ much in humans from their homologues in the lower vertebrates and even more from the dorsal roots of spinal nerves.

Due to this complexity each nerve may possess more than one nucleus of origin and of termination. Neurons of the facial, inferior glossopharyngeal and inferior vagal ganglia, though derived largely from the neural crest, originate partly in *ectodermal epibranchial placodes* developed at the dorsal ends of the first three branchial clefts close to the ganglia (p. 200).

Despite some discrepancies, the homologies of nerves in Groups I and II are generally accepted; but those of the three remaining cranial nerves are entirely uncertain. In its mode of development the optic nerve is unlike other cranial nerves, except for its *special somatic afferent* function; the retinal sources of its fibres are really an outlying part of the brain, although it may be argued that they are derivatives of precursors of neural crest cells. The olfactory and vestibulocochlear nerves may be grouped together or separated, only the vestibulocochlear being considered as the homologue of a dorsal nerve. Both arise, in part at least, from ectodermal cells outside the neural tube and crest; but whereas olfactory neurons remain amongst the epithelial cells of the nasal mucous membrane, the cochlear neurons migrate away from the otic vesicle. Nevertheless, many believe that the contribution made by the neural crest is responsible for both the vestibular and spiral cochlear ganglia and hence consider the vestibulocochlear to be a modified dorsal nerve. In comparing the olfactory and vestibulocochlear nerves it must be remembered that olfactory nerves are restricted to the head, whereas the human vestibulocochlear nerve is the survivor of a series of nerves from lateral line organs, distributed in fish not merely to the head but to the whole length of the body.

For discussion of the phylogeny of vertebrate cranial nerves consult Black (1917), Herrick (1922), Bolk et al (1934), Kappers et al (1936), Kappers (1947), Young (1950), Grassé (1954), Goodrich (1958).

General Considerations of Cranial Nerves

Because of the brain stem's complex structure, changing from level to level, in which many neuronal groups and pathways exist, the difficulties of investigation of this area of the nervous system are great and the finer details of the cranial nerve nuclei are less certain than are the origins of spinal nerves. The descriptions of the site, structure and connections of many cranial nuclei in anatomical texts often convey a misleading simplicity. It may therefore be useful to summarize some of the concepts concerning their organization which are either emerging or are suspected by analogy with spinal cord organization; it must be noted, however, that much has not been established with certainty in primate brains. The olfactory, optic and vestibulocochlear nerves are considered elsewhere.

The motor nuclei of cranial nerves include somatic efferent, special visceral efferent (branchiomotor) and general visceral efferent (autonomic) groups. However, somatic and special visceral efferent nuclei of the oculomotor, trochlear, trigeminal, abducent, facial, glossopharyngeal, vagus, accessory and hypoglossal nerves may be grouped together, since all innervate *skeletal muscle*. Spinal ventral grey matter forms a series of longitudinal columns of neurons, whose significance is uncertain (p. 929), though they exhibit some somatotopic organization. They include α, β, and γ motor neurons (p. 939) innervating extrafusal and intrafusal muscle fibres; the α motor neurons to extrafusal muscle are of 'tonic' and 'phasic' types, 'static' and 'dynamic' types of fusimotor γ neurons being also recognized. Among these motor neurons are many small excitatory and inhibitory interneurons, the latter including the so-called Renshaw cells. Converging on these cell varieties are numerous fibres, completing a wide array of local control loops, both contralateral and ipsilateral, and others descending from supraspinal sources such as the vestibular nuclei, brain-stem reticular formation, red nucleus, midbrain tectum and tegmentum and the cerebral cortex. These regions, which project directly to the lower motor centres, themselves receive essential control fibres from the cerebellum, corpus striatum, thalamus and hypothalamus and many parts of the cerebral neocortex and limbic system. All are detailed elsewhere. Despite uncertainties about the central connections of cranial nerves, it is probable that comparable controls operate in all cases; in view of the precise motor integrations necessary in phonation and ocular activity, their connections may well be of a *higher* order of complexity than those at spinal levels.

The trochlear and abducent nerves each innervate a single small muscle and, reflecting this, their motor nuclei are single small groups of neurons. In contrast, the oculomotor, trigeminal, facial and hypoglossal nerves and the glossopharyngeal-vagus-cranial accessory complex innervate musculature capable of highly precise patterns of movement and in each case their nuclei contain subnuclei perhaps related to major branches or even innervation of single muscles. Some subnuclei show a longitudinal columnar arrangement, the columns varying in their mediolateral and dorsoventral location and in their axial extent. This somatotopic localization is reminiscent of that in spinal ventral grey columns. Though precision of control may be in some way associated with such grouping (p. 929), the ontogenetic, phylogenetic and functional significance of these groups require further exploration. Also, much less is known than in the spinal cord concerning the quantitative degrees of longitudinal and transverse overlap of the *dendritic trees* of motor neurons in adjacent motor nuclei and subnuclei. Though muscle spindles were long considered to be absent or rare in facial, masticatory, lingual, laryngeal and ocular muscles, they have now been confirmed in most of these sites (facial muscles having the least); but the siting and functions of their γ efferent neurons are obscure. Similar uncertainty exists concerning smaller multipolar neurons in or near cranial nerve motor nuclei, many of which are presumably excitatory or inhibitory interneurons, including 'Renshaw' cells.

The afferent connections of cranial nerve motor nuclei include components corresponding to all those converging on spinal grey matter, but some are merely recognized in outline. The best known are: (1) the *medial longitudinal fasciculus*, interconnecting the vestibular nuclei, a longitudinal series of other cranial nerve nuclei and the cervical spinal cord (p. 985); (2) *tectotegmental*, *tectopontine* and *tectobulbar* projections from the superior and inferior colliculi (p. 986); (3) projections from the *red nucleus* (p. 983); (4) interconnections with the *brain-stem reticular formation* (p. 988); (5) *corticonuclear* projections from cerebral sensorimotor areas (p. 1049); (6) projections from *sensory nuclei* of other cranial nerves. However, little is known of how these afferents terminate: e.g. whether directly on α and/or γ motor neurons or through interneurons, and how far they mediate pre- or postsynaptic facilitatory or inhibitory effects. In corticospinal tracts (p. 933) contralaterally-derived fibres predominate, with a variable ispilateral content. Similar considerations apply to cranial nerve motor nuclei. Thus, although mainly *contralateral*, corticonuclear fibres reach most of the hypoglossal nucleus and the part of the facial nucleus innervating the lower face; those to the trochlear nucleus are largely *ipsilateral* and the rest, including the most medial parts of hypoglossal nucleus, receive a *bilateral* projection. The cerebellum, diencephalic centres and corpus striatum are considered to influence the cranial nerve nuclei via indirect paths like those proposed for the control of spinal motor centres.

The general visceral efferent nuclei of the oculomotor, facial, glossopharyngeal and vagus nerves have largely uncertain central connections. (Those of the accessory oculomotor nuclei related to visual reflexes are discussed elsewhere.) It is presumed that the dorsal vagal and salivatory nuclei receive terminals of ascending tracts and those from other cranial nerve nuclei (particularly the solitary nucleus), conveying both somatic and visceral information. They also probably interconnect with the brain-stem reticular formation and receive hypothalamic projections, probably polysynaptic, through which the frontal neocortex, limbic structures and thalamus exert indirect effects. These descending paths have not been satisfactorily analysed. The central connections of **general visceral afferent pathways** (p. 1155) are equally uncertain, though they influence many brain-stem centres, including particularly the brain-stem reticular formation, hypothalamus, limbic lobe and prefrontal neocortex (consult the appropriate sections on these regions). The proportions of primary general visceral afferent fibres terminating in the dorsal vagal nucleus or in the solitary nucleus are undetermined, but the latter are accompanied by **special visceral afferent** (gustatory) fibres of the facial, glossopharyngeal and vagus nerves, which end in a rostrocaudal sequence of overlapping zones in the solitary nucleus. After synaptic relays, the secondary gustatory fibres decussate, ascending to terminate in the thalamic nucleus ventralis posterior medialis and several hypothalamic nuclei. Such *solitariothalamic* and *solitariohypothalamic* routes have not been detailed in human brains,

but the former probably accompanies medial fibres in the *medial lemniscus*, whilst the latter may join the *dorsal longitudinal fasciculus* (p. 986) and *mamillary peduncle* (p. 1009). Collateral branches probably leave these ascending fibres to end in the brain-stem reticular formation and in the nuclei of other cranial nerves. From the caudal levels of the nucleus solitarius fibres descend as a *solitariospinal tract*, but this is not a simple relay on the visceral afferent pathways; other connections converge to the nucleus, including fibres from the spinal cord, vestibulo-cerebellum and cortex, probably interacting in the nucleus, its transmission characteristics being modified by cortical activity.

Somatic afferent nuclei of the brain stem are the spinal, principal sensory and mesencephalic nuclei of the trigeminal nerve. The unique features of the *mesencephalic trigeminal nucleus* (which consists of unipolar *primary* sensory neurons) and its probable proprioceptive role have been described (p. 1099). Electrophysiology supports this role, since rapid responses have been recorded in the nucleus following stretching of the masticatory muscles, passive jaw movement and pressure on the teeth. The peripheral processes of mesencephalic unipolar neurons are hence considered to innervate muscle spindles in the masticatory muscles, temporomandibular joint and periodontal tissues. Possibly some also innervate muscle spindles of the facial, lingual and laryngeal muscles, but this is uncertain, as is the siting of somata of proprioceptive fibres from extraocular muscles (p. 1097). Since the somata of these primary afferents are deep in the midbrain, they present experimental difficulties, the smallest lesions inevitably involving adjacent structures. Hence, the central projections of these neurons are uncertain but some may descend to the principal sensory nucleus of the trigeminal; others may project to the trigeminal motor nucleus, completing a masticatory reflex. Collateral branches are described as a *trigeminocerebellar* tract (p. 966). Other connections are conjectural.

The manner of termination of the primary sensory axons in the three trigeminal divisions, all from neurons in the trigeminal ganglion, have been described. Data from brain-stem disease or neurosurgical manoeuvres suggest that the principal sensory nucleus is a simple relay for tactile information and the spinal nucleus for thermal and nociceptive information; but this is certainly an oversimplification. Bifurcation of many sensory fibres occurs on entry into the pons, one branch reaching the principal and the other the spinal nucleus. The *spinal nucleus*, continuous below with the substantia gelatinosa of the cervical spinal cord, shows regional variation in cytoarchitecture (sometimes divided into three subnuclei termed *oralis*, *interpolaris* and *caudalis*), progressing rostrocaudally. Further, the spinal and principal nuclei not only receive terminals of somatic afferent *trigeminal* fibres but also of *facial*, *glossopharyngeal* and *vagal* axons and others ascending from the cervical spinal cord. Both nuclei interconnect with the brain-stem reticular formation and both receive numerous corticonuclear fibres from the sensorimotor cortex; many afferents thus converge on both and unit recording has confirmed a *wide range* of different cell responses in them, varying from rapidly-adapting neurons with modality-specific, small receptive fields, to those with multimodal responses and larger receptive fields. Neurons responding to light tactile stimuli, others to nociceptive stimuli, occur in *both* nuclei. Some units show an inhibitory 'surround' (p. 865), and both pre- and postsynaptic inhibition and facilitation occur in the nuclei. Their transmission characteristics are modified by stimulation or interruption of the corticonuclear tracts and areas of the brain-stem reticular formation. The clinical conceptions of these nuclei as functionally segregated relay stations are negated by experimental data, which show them to be centres of complex integration. An analogy has been drawn between the principal trigeminal sensory nucleus and the nuclei gracilis and cuneatus, and between the spinal trigeminal nucleus and the substantia gelatinosa, in structural and functional organization. (See pp. 927, 941, 942–943 for details and a theory of action of the substantia gelatinosa in modulating sensory input.)

The ascending efferent connections from the principal and spinal trigeminal nuclei (p. 961) consist of fibres crossing to the thalamus as a 'trigeminal lemniscus'. Experimentation and observations on human material both show that many multiple trigeminothalamic pathways exist. From the principal sensory nucleus a substantial bundle of fibres crosses the midline, a minor part remaining ipsilateral, both ascending through the upper pons, midbrain and subthalamus as the *dorsal trigeminothalamic tract* (dorsal division of the trigeminal lemniscus); in the midbrain this is dorsomedial to the red nucleus and medial lemniscus, ending in the thalamic nucleus ventralis posterior medialis (VPM) (p. 1002). Other fibres from the principal sensory and spinal

nucleus (in its whole length) mainly cross to ascend near the medial lemniscus as the *ventral trigeminothalamic tract* (ventral division of the trigeminal lemniscus); many also end in the VPM but others, mainly from neurons in the caudal part of the spinal nucleus, perhaps end in the medial geniculate body and intralaminar nuclei. Collaterals may leave these ascending tracts to terminate in the brain-stem reticular formation and other cranial nerve nuclei while some probably enter the cerebellum (p. 966).

THE SPINAL NERVES

Spinal nerves are united ventral and dorsal spinal roots, attached in series to the sides of the spinal cord. There are 31 pairs: 8 cervical, 12 thoracic, 5 lumbar, 5 sacral, 1 coccygeal. The abbreviations C, T, L, S and Co, with appropriate numerals, are commonly applied to individual nerves. These emerge through intervertebral foramina; C1 leaves the vertebral canal between the occipital bone and atlas and is hence often termed the *suboccipital nerve*; C8 issues between the seventh cervical and first thoracic vertebrae. Each nerve is continuous with the spinal cord by the ventral and dorsal roots, the latter each bearing a *spinal ganglion*.

Ventral (anterior) roots contain axons of neurons in the anterior and lateral spinal grey columns. Each emerges as a series of *rootlets* in two or three irregular rows in an area about 3 mm in horizontal width. In rats the rootlets of lumbar ventral roots are each ensheathed by a single fenestrated layer of cells. At the spinal cord surface the rootlets are separated by a labyrinth of interradicular spaces lined by this sheath. Between adjacent rootlets the interradicular spaces, which contain small blood vessels, taper distally. The interradicular spaces communicate with the endoneurial spaces of the rootlets and with the subpial space but are isolated from the subarachnoid space by the multilayered, unfenestrated sheath which surrounds the aggregations of rootlets which comprise each ventral root (Kaar & Fraher 1986).

Dorsal (posterior) roots contain centripetal processes of neurons sited in the spinal ganglia. Each consists of medial and lateral fascicles both diverging into *rootlets* entering along the posterolateral sulcus. In many mammals, including humans, the rootlets of adjacent dorsal roots are often connected by oblique filaments, especially in the lower cervical and lumbosacral regions (Pallie & Manuel 1968).

Each root is covered by the pia mater and is loosely invested by the arachnoid mater prolonged to where the roots pierce the dura mater. The dorsal and ventral roots do so separately, each receiving a dural sheath (7.204, 208); where the roots unite to form spinal nerves these sheaths fuse with the epineurium.

Spinal ganglia are large groups of neurons on the dorsal spinal roots. Each is oval and reddish, its size being related to that of its root; it is bifid medially where the dorsal root's two fascicles emerge to enter the cord. Ganglia are usually sited in the intervertebral foramina, immediately lateral to the perforation of the dura mater by the roots (7.61), but the first and second cervical ganglia are on the vertebral arches of the atlas and axis, the sacral inside the vertebral canal and the coccygeal ganglia usually within the dura mater.

The first cervical ganglia may be absent. Small *aberrant ganglia* sometimes occur on the upper cervical dorsal roots between the spinal ganglia and cord. (Heterotopic ganglionic neurons also occur in other sites, see p. 897.)

The internal structures of sensory ganglia, cranial and spinal, is similar (p. 896 and consult Van Gehuchten 1892, Marinesco 1909, Cajal 1911, De Castro 1932, Scharf 1958, Lieberman 1976). Each sensory ganglion has a laminar connective tissue capsule continuous with the epineurium of the spinal root. Endoneurial stroma permeates the ganglion, supporting its neuronal and axonal population; perineurial capsular trabeculae extend between groups of these elements. The stroma is intimately related to the satellite cells (amphicytes); it contains a few mast cells and a dense vascular network, denser around ganglion cells, into which capillary loops may be invaginated (Scharf 1958). These

capillaries are non-fenestrated in rodents (Lieberman 1968) but commonly fenestrated in primates (Olsson 1971). Ganglionic vascular permeability varies among species, amounting to a 'blood–nerve' barrier in some, largely dependent on the junctional complexes between endothelial cells; details are not available for mankind. Ganglionic neurons vary in size (15–110 μm in man, Ohta et al 1974); most are spheroidal but smaller cells appear ellipsoidal or angular in section.

Ganglionic neurons resemble other neurons but vary much in the distribution of cytoplasmic chromatin, from fine disperal to concentration in large masses (Nissl granules); this has been the chief basis of many classifications, which remain conflicting. Melanin and lipofuchsin pigments occur in some but less than in some cranial sensory ganglia. Dense-cored vesicles are common (Lieberman 1968), despite the apparent absence of catecholamines. However, the occurrence of two extreme types is agreed: large 'light' neurons (A cells) and small 'dark' ones (B cells). In 'light' cells the granular endoplasmic reticulum is more dispersed but highly concentrated in 'dark' cells. The latter are sometimes mimicked by artefact, perhaps due to poor fixation. Both types appear in prenatal material. Functional differences remain unclarified but the axons of A cells are said to be myelinated and those of B cells non-myelinated. It has been suggested that B cells are visceral. 'Atypical' neurons have been described in dorsal root and other sensory ganglia, the most interesting being reputed to be multipolar neurons; but since synapses have not been identified in dorsal root ganglia, this is improbable and is unconfirmed. Almost all ganglionic neurons are unipolar (p. 863), many with a coiled 'stem' process close to the parent soma, forming an axonal 'glomerulus' before branching into peripheral and central parts (p. 1094). Spinal ganglionic cells have been submitted to ultrastructural studies (Duce & Keen 1977), assessment of specific intracellular substances (Hokfelt et al 1976), preferential glutamine absorption (Duce & Keen 1978) and specific reactions to neurotoxins (Jancsco et al 1977). Such data, while confirming 'light' and 'dark' cells, collectively suggest a subdivision into further types, particularly of 'dark' cells. Small dark cells may occur in at least two variants, differentiated by production of either substance P or somatostatin. Different cells respond also to different neurotoxins or differ in glutamine metabolism.

Counts of cells in the dorsal root ganglia show *specific* variation, as expected; few estimates exist for human ganglia. Maximal counts are reached about three years after birth; but the subsequent supposed loss has not been confirmed. The distribution of somata and axons is variable, the former often being peripherally concentrated; but grouping of the somata between large fascicles of axons is described as more usual in human ganglia. Such grouping may indicate somatotopic organization, sometimes described in terms of somata and axons. Burton & McFarlane (1973), by injection of labelled amino acids into ganglia, associated local groups of neurons in cats' lumbar ganglia with branches of spinal nerves and also with particular spinal radicles. Horseradish peroxidase injections near peripheral receptor endings have produced more exact results in rodent sensory ganglia; no results are available for primates.

Size and direction of spinal nerve roots vary. The upper four *cervical* roots are small, the lower four large. Cervical dorsal roots have a thickness ratio to the ventral roots of three to one, greater than in other regions. The first dorsal root is an exception, being smaller than the ventral; in about 8% it is absent. The first and second cervical roots are short, running almost horizontally to their exits from the vertebral canal. From the third to the eighth cervical they slope obliquely down, obliquity and length increasing successively; the distance between spinal attachment and vertebral exit never exceeds the height of one vertebra.

Thoracic roots, except the first, are small, the dorsal only slightly exceeding the ventral in thickness. They increase successively in length and in the lower thoracic region descend in contact with the spinal cord for at least two vertebrae before emerging from the vertebral canal (but vide infra).

Kubik & Müntener (1969) consider that the cervicothoracic part of the spinal cord grows more in length than other parts and thus explain their observations which differ from the above remarks. They state that upper cervical roots *descend*, the fifth being horizontal, and that the sixth to eighth actually *ascend*, that the first two thoracic roots are horizontal, the next three ascend, the sixth is horizontal and the rest descend.

Lower *lumbar* and upper *sacral roots* are the largest and their radicles the most numerous, while the *coccygeal* roots are the smallest. Kubik & Müntener (1969) confirm that lumbar, sacral and coccygeal roots descend with increasing obliquity to their exits; since the spinal cord ends near the lower border of the first lumbar vertebra, the lengths of successive roots rapidly increase. The consequent collection of roots is the *cauda equina* (7.60B).

The largest roots and the hence largest spinal nerves are continuous with the spinal cervical and lumbar swellings and innervate the upper and lower limbs.

Immediately distal to the spinal ganglia, ventral and dorsal roots unite to form *spinal nerves*, which emerge through intervertebral foramina and give off recurrent meningeal branches (vide infra), dividing immediately into *dorsal* and *ventral* rami (7.250). (From limited dissections Sato (1974) has described the *trifurcation* of spinal nerves at some cervical and thoracic levels, the third branch being a *ramus intermedius*.) At or distal to its origin each ventral ramus receives a *grey ramus communicans* from the corresponding sympathetic ganglion, while the thoracic and first and second lumbar ventral rami each contribute a *white ramus communicans* to the corresponding ganglia (7.236, 238). The second, third and fourth sacral nerves also supply visceral branches, unconnected with sympathetic ganglia, which carry a parasympathetic outflow direct to the pelvic plexuses (pp. 1156, 1166).

Cervical spinal nerves enlarge from the first to the sixth, the seventh and eighth cervical and first thoracic being like the sixth cervical in size. The remaining thoracic nerves are relatively small. Lumbar nerves are large, increasing in size from the first to the fifth. The first sacral is the largest spinal nerve, thereafter they decrease to the coccygeal, the smallest.

In the intervertebral foramina spinal nerves have clinically important relations. *Anterior* are the intervertebral discs and adjacent vertebral bodies. *Posterior* are the synovial zygapophysial joints. *Superior* and *inferior* are vertebral notches of the pedicles of adjoining vertebrae. Each nerve, accompanied by a spinal artery, a small venous plexus and its own meningeal branch or branches together traverse a foramen.

Applied Anatomy. Spinal roots may be compressed in their course between the spinal cord and their exit from intervertebral foramina. In the cervical region, vertebral disease, the degeneration of an intervertebral disc or disease of the intervertebral joints may affect the roots in their intervertebral foramen, causing pain, diminished cutaneous sensibility and muscular weakness in the field of supply. At the lumbar level, the posterior protrusion of an intervertebral disc or rupture of its annular fibres, with herniation of the nucleus pulposus, is common, particularly in the discs between the fourth and fifth lumbar and first sacral vertebrae. Disc lesions may compress one or both roots in their foramina, resulting in back pain ('lumbago'), with or without radiation of pain to one or both legs ('sciatica'), diminished cutaneous sensibility in the area of supply and weakness of the muscles innervated (e.g. tibialis anterior L.4, extensor hallucis longus L.5, flexor hallucis longus S.1). Less commonly, tumours in the vertebral canal or spina bifida with meningomyelocoele may affect one or more roots in the cauda equina.

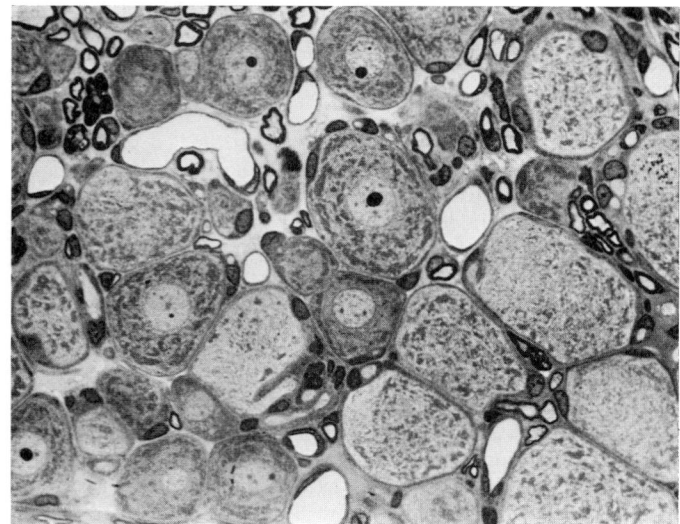

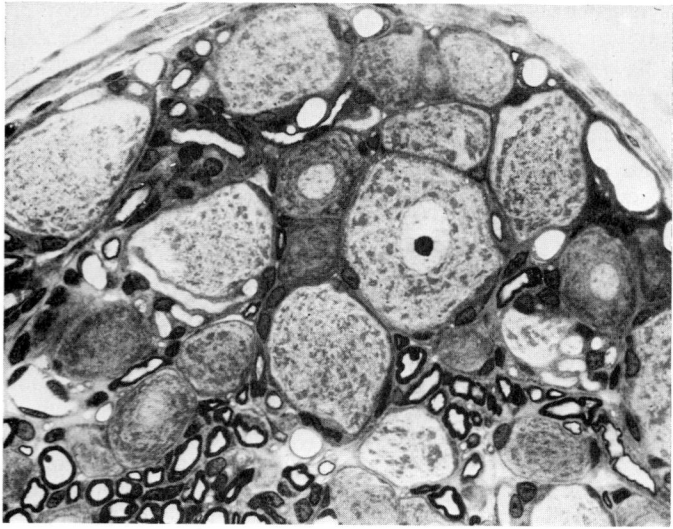

7.236 Two photographs of fields in normal rat cervical dorsal root ganglion to show contrasting features of light and dark neuronal somata. Note the capsules of satellite cells; the darkly stained multiple profiles between many of the nerve cells represent glomeruli in repeated transverse section. Interneuronal capillary profiles are also visible. Cresyl fast violet staining of semi-thin sections of material embedded in araldite. Photographs supplied by J M Jacobs, National Hospital for Nervous Diseases, London.

FUNCTIONAL COMPONENTS OF SPINAL NERVES

Each typical spinal nerve contains somatic and visceral fibres.

The somatic components are efferent and afferent. *Somatic efferent* fibres for the innervation of skeletal muscles are axons of α, β and γ neurons in the spinal anterior grey column. *Somatic afferent* fibres convey impulses into the central nervous system from receptors in the skin, subcutaneous tissue, muscles, fasciae, joints, etc. (pp. 909–915); they are peripheral processes of unipolar neurons in the spinal ganglia.

The visceral components, also afferent and efferent, belong to the autonomic nervous system (p. 1154). They include sympathetic or parasympathetic fibres at different spinal levels. Preganglionic *visceral efferent* sympathetic fibres are axons of neurons in the spinal lateral grey column in the thoracic and upper two or three lumbar segments; they join the sympathetic trunk along corresponding white rami communicantes to synapse with postganglionic neurons distributed to non-striated muscle or glands. The preganglionic *visceral efferent* parasympathetic fibres are axons of cells in the spinal lateral grey column of the second to fourth sacral segments; they leave the ventral rami of corresponding sacral nerves to synapse in pelvic ganglia, the

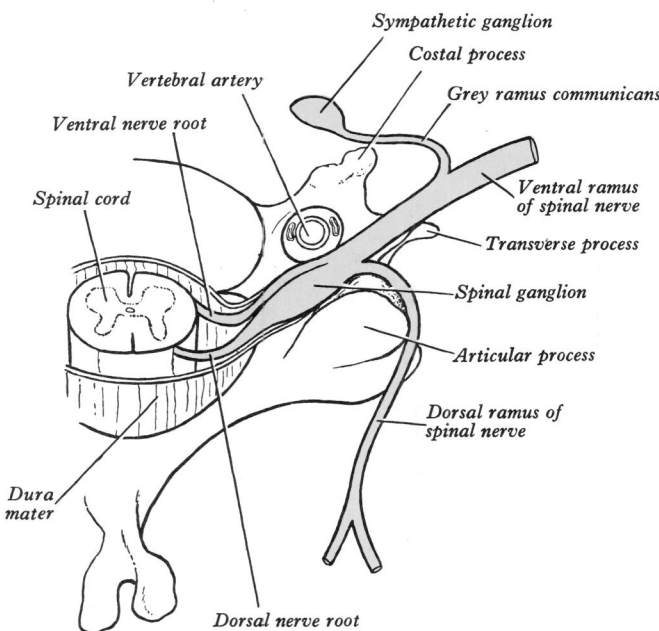

7.237 Scheme showing the relations of a cervical nerve and its ganglion to a cervical vertebra.

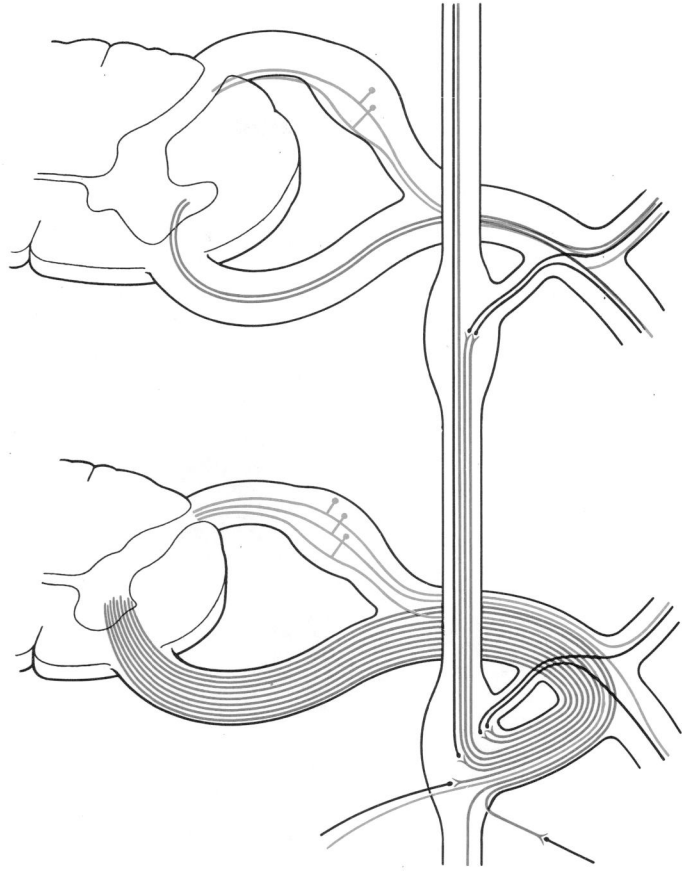

7.238 Scheme showing the constitution of a typical spinal nerve. In the upper part of the diagram the spinal nerve roots show the somatic components; in the lower part of the diagram the spinal roots show the visceral components. Red = somatic efferent and preganglionic sympathetic fibres, blue = somatic afferent and visceral afferent fibres, black = postganglionic sympathetic fibres.

postganglionic axons being distributed mainly to muscle or glands in the walls of the pelvic viscera. *Visceral afferent* fibres are from neurons in the spinal ganglia. Their peripheral processes pass through white rami communicantes and, without synapsing, through one or more sympathetic ganglia to end in viscera. Some visceral afferent fibres may enter the spinal cord in *ventral* roots; Coggeshall et al (1973) claim that almost 30% of fibres in the feline seventh lumbar and first sacral ventral roots are non-myelinated afferents, which may project *into* the cord from neurons in the corresponding dorsal root ganglia.

Central processses of ganglionic unipolar neurons enter the spinal cord by posterior roots and connect with somatic or sympathetic efferent neurons, usually through interneurons, completing reflex paths, or they synapse with other neurons in the spinal or brain-stem grey matter which provide a variety of ascending tracts.

Little detail of the dendrites of spinal autonomic neurons is available. Schramm et al (1976) showed that sympathetic efferent neurons have largely horizontal dendritic arrays in newborn rats, re-orientated into longitudinal arrays in the next five or six post-natal weeks. No such data are available for the human spinal cord.

After emerging from its intervertebral foramen each spinal nerve supplies small *meningeal* branches and splits almost at once into *dorsal* and *ventral rami*, both receiving fibres from both roots.

The meningeal branches (*recurrent meningeal* or *sinu-vertebral nerves*), numbering two to four filaments on each side, occur at all vertebral levels (Kimmel 1961a, b). Each receives one or more rami from a nearby grey ramus communicans or from a thoracic sympathetic ganglion directly; most then pursue a recurrent (often perivascular) course into the spinal canal through the intervertebral foramen ventral to the dorsal root ganglion. Here these mixed sensory and sympathetic nerves divide into transverse, ascending and descending branches distributed to the dura mater, the walls of blood vessels, the periosteum, ligaments and intervertebral discs in the ventrolateral region of the spinal canal. Fine meningeal branches occasionally pass dorsal to the spinal ganglia to the dorsal dura, periosteum and ligaments, others passing ventrally to the posterior longitudinal ligament. Ascending branches of the upper three cervical meningeal nerves are large and distributed to the dura mater in the posterior cranial fossa (p. 1088). Meningeal nerves are important in relation to the referred pain characteristic of many spinal disorders and in occipital headache.

DORSAL RAMI OF THE SPINAL NERVES

Dorsal (posterior primary) rami of spinal nerves, usually smaller than the *ventral* and directed posteriorly, divide (except for the first cervical, fourth and fifth sacral and the coccygeal) into medial and lateral branches to supply the muscles and skin (7.239) of the posterior regions of the neck and trunk.

CERVICAL DORSAL RAMI

Each cervical spinal dorsal ramus, except the first, divides into medial and lateral branches, all innervating muscles; but in general only medial branches of the second to fourth and usually the fifth supply the skin. Except for the first and second, each dorsal ramus passes back medial to a posterior intertransverse muscle, curving round the articular process into the interval between the semispinalis capitis and semispinalis cervicis.

The first cervical dorsal ramus (suboccipital nerve) (5.40), larger than the ventral, emerges superior to the posterior atlantal arch and inferior to the vertebral artery. It enters the suboccipital triangle to supply local muscles: the rectus capitis posterior major and minor, superior and inferior oblique and the semispinalis capitis. A filament from the branch to the inferior oblique joins the second dorsal ramus (5.40). The suboccipital

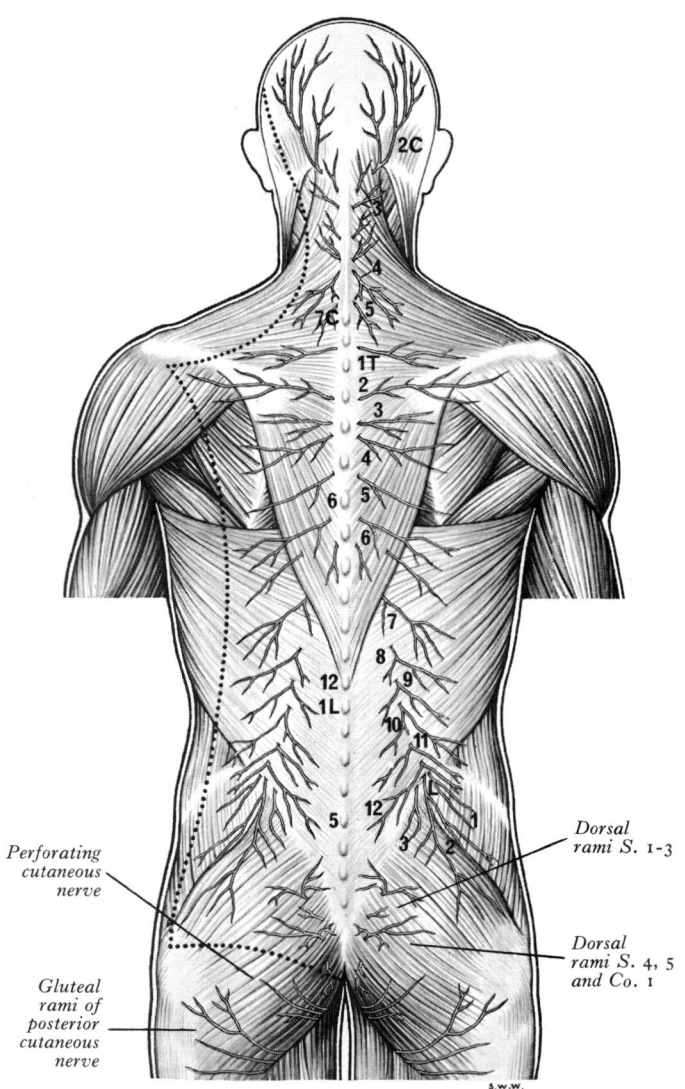

7.239 The cutaneous distribution of the dorsal rami of the spinal nerves. The nerves are shown lying on the superficial muscles; on the left side the limit of the skin area supplied by these nerves is indicated by the dotted line. The nerves are numbered on the right side and the spines of the seventh cervical, sixth and twelfth thoracic and first and fifth lumbar vertebrae are labelled on the left side.

nerve occasionally has a cutaneous branch which accompanies the occipital artery to the scalp, connecting with the greater and lesser occipital nerves.

The second cervical dorsal ramus, slightly larger than the ventral and all other cervical dorsal rami, emerges between the posterior atlantal arch and the lamina of the axis, below the inferior oblique which it supplies, receiving a connection from the first cervical dorsal ramus, and divides into a large medial and smaller lateral branch. Its ganglion is extradural. The *medial* branch, termed the *greater occipital nerve* (7.239, 241), ascends between the inferior oblique and semispinalis capitis, pierces the latter and the trapezius near their occipital attachments and is joined by a filament from the medial branch of the third dorsal ramus; ascending with the occipital artery, it divides into branches which connect with the lesser occipital nerve and supply the skin of the scalp as far forward as the vertex. It supplies the semispinalis capitis and occasionally the back of the auricle. The *lateral* branch supplies the splenius, longissimus capitis and semispinalis capitis and is often joined by the corresponding third cervical branch.

The third cervical dorsal ramus, between the second and fourth in size, courses back round the third cervical articular pillar, medial to the posterior intertransverse muscle, dividing into medial and lateral branches. Its *medial* branch runs between the spinalis capitis and semispinalis cervicis, piercing the splenius and trapezius to end in the skin. Deep to the trapezius it gives rise to a branch, the *third occipital nerve,* which pierces the trapezius to end in the skin of the lower occipital region (7.239), medial to

the greater occipital nerve and connected to it. The *lateral* branch often joins that of the second cervical dorsal ramus.

The dorsal ramus of the suboccipital and medial branches of the dorsal rami of the second and third cervical nerves are sometimes joined by loops to form the *posterior cervical plexus.*

Dorsal rami of the lower five cervical nerves curve back round the vertebral articular pillars and divide into medial and lateral branches. *Medial* branches of the fourth and fifth run between the semispinalis cervicis and semispinalis capitis, reach the vertebral spines and pierce the splenius and trapezius to end in the skin (7.239). The fifth medial branch may not reach the skin. The medial branches of the lowest three cervical nerves are small and end in the semispinalis cervicis, semispinalis capitis, multifidus and interspinales. The *lateral* branches supply the iliocostalis cervicis, longissimus cervicis and longissimus capitis.

THORACIC DORSAL SPINAL RAMI

Thoracic dorsal rami pass backwards close to the vertebral zygapophysial joints to divide into medial and lateral branches. Each medial branch emerges between a joint and medial edges of the superior costotransverse ligament and intertransverse muscle, but each lateral branch runs in the interval between ligament and muscle before inclining posteriorly on the medial side of the levator costae.

Medial branches of the *upper six* thoracic dorsal rami pass between and supply the semispinalis thoracis and multifidus; they pierce the rhomboids and trapezius, reaching the skin near the vertebral spines (7.239). *Medial branches* of the *lower six* thoracic dorsal rami chiefly supply the multifidus and longissimus thoracis and occasionally skin in the median region. *Lateral* branches increase inferiorly in size, running through or deep to the longissimus thoracis to the interval between it and the iliocostalis cervicis, supplying these muscles and the levatores costarum; the lower five or six also have cutaneous branches, piercing the serratus posterior inferior and latissimus dorsi in line with the costal angles (7.239). Some upper thoracic lateral branches also supply the skin. The twelfth thoracic lateral branch sends a filament medially along the iliac crest, then passes down to the anterior gluteal skin. Medial cutaneous branches of the thoracic dorsal rami descend close to the vertebral spines before reaching the skin; lateral branches descend as far as the level of four ribs before becoming superficial, the branch of the twelfth thoracic reaching the skin a little above the iliac crest.

LUMBAR DORSAL SPINAL RAMI

Lumbar dorsal rami pass back medial to the medial intertransverse muscles, dividing into medial and lateral branches.

Medial branches run near the vertebral articular processes to end in the multifidus; they are related to the bone between the accessory and mamillary processes and may groove it, traversing a distinct notch or even a foramen. *Lateral* branches supply the erector spinae (sacrospinalis). In addition, the upper three give rise to cutaneous nerves which pierce the aponeurosis of the latissimus dorsi at the lateral border of the erector spinae and cross the iliac posteriorly to reach the gluteal skin (7.239), some reaching as far as the level of the greater trochanter.

SACRAL DORSAL SPINAL RAMI

Sacral dorsal rami are small, diminishing downwards; excepting the fifth, they emerge through the dorsal sacral foramina. The *upper three* are covered at the exit by the multifidus, dividing into medial and lateral branches. Medial branches are small and end in the multifidus. *Lateral* branches join together and with lateral branches of the last lumbar and fourth sacral dorsal rami form loops dorsal to the sacrum; from these loops branches run dorsal to the sacrotuberous ligament to form a second series of loops under the gluteus maximus; from these two or three *gluteal branches* pierce the gluteus maximus (along a line from the posterior superior iliac spine to the coccygeal apex) to supply the posterior gluteal skin (7.239).

The dorsal rami of the *fourth* and *fifth* sacral nerves are small and lie below the multifidus. They unite with each other and the coccygeal dorsal ramus to form loops dorsal to the sacrum; filaments from these supply the skin over the coccyx. Berthold & Carlstedt (1977) have recorded a detailed study of the feline first sacral nerve at its junction with the spinal cord.

COCCYGEAL DORSAL SPINAL RAMUS

This does not divide into medial and lateral branches. Its connections and distribution are noted above.

VENTRAL RAMI OF THE SPINAL NERVES

The ventral rami of spinal nerves supply the limbs and the anterolateral aspects of the trunk; they are mostly larger than the dorsal rami. The *thoracic* are independent and retain, like all *dorsal* rami, a largely segmental distribution. The *cervical*, *lumbar* and *sacral* connect near their origins to form plexuses. Dorsal rami do not join these plexuses.

CERVICAL VENTRAL RAMI

Cervical ventral rami, except the first, appear between the anterior and posterior intertransverse muscles. The *upper four* form a *cervical plexus*; the *lower four*, with most of the first thoracic ventral ramus, form a *brachial plexus*. Each receives at least one grey ramus communicans, the upper four from the superior cervical sympathetic ganglion, the fifth and sixth from the middle ganglion and the seventh and eighth from the cervicothoracic ganglion (p. 1160).

The *first cervical ventral ramus* (*suboccipital nerve*) emerges above the posterior arch of the atlas, passes forwards lateral to its lateral mass and medial to the vertebral artery. It supplies the rectus lateralis, emerges medial to it, descends anterior to the atlantal transverse process and posterior to the internal jugular vein and joins the ascending branch of the second cervical ventral ramus.

The *second cervical ventral ramus* issues between the vertebral arches of the atlas and axis, ascends between their transverse processes, passes anterior to the first posterior intertransverse muscle and emerges lateral to the vertebral artery generally between the longus capitis and levator scapulae, but when scalenus medius is attached to the atlas it intervenes between the nerve and the levator scapulae. The ramus divides into an ascending branch which joins the first cervical nerve and a descending one which joins the ascending branch of the third cervical ventral ramus.

The *third cervical ventral ramus* appears between the longus capitis and scalenus medius. The remaining ventral rami emerge between the scalenus anterior and scalenus medius.

The Cervical Plexus

The cervical plexus (7.240), formed by the upper four cervical ventral rami, supplies some nuchal muscles, the diaphragm and areas of skin in the head, neck and chest (7.224). Level with the first four vertebrae, it is deep to the internal jugular vein and sternocleidomastoid and anterior to the scalenus medius and levator scapulae. Each ramus, except the first, divides into ascending and descending parts, which unite in communicating loops. From the first loop (C2 and 3) superficial branches supply the head and neck; from the second (C3 and 4) arise cutaneous nerves of the shoulder and chest. Muscular and communicating branches arise from the same nerves. The branches are superficial or deep; the superficial perforate the cervical fascia to supply the skin while the deep branches mostly supply muscles. The superficial form ascending and descending groups and form the deep medial and lateral series.

ASCENDING SUPERFICIAL BRANCHES

These branches of the cervical plexus (7.240–242) include:

Lesser occipital	C2
Greater auricular	C2, 3
Transverse (anterior) cutaneous	C2, 3

The lesser occipital nerve (7.240, 241), from the second cervical ventral ramus and sometimes also from the third, curves

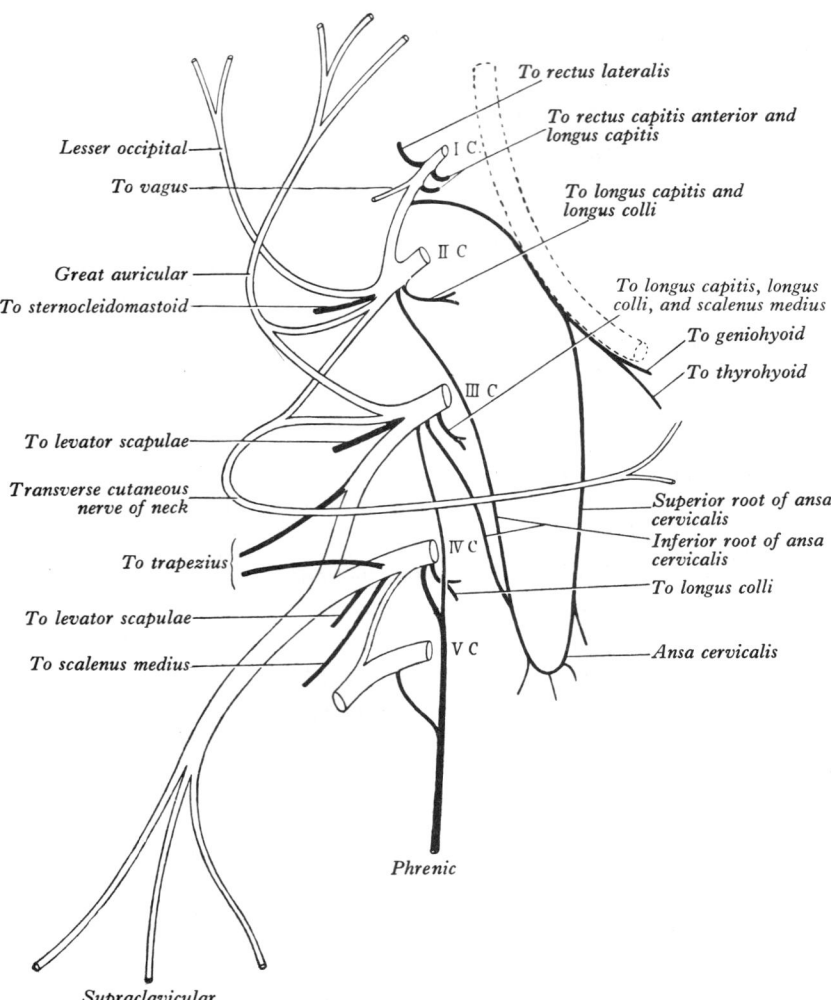

Lesser occipital

To vagus

Great auricular

To sternocleidomastoid

To levator scapulae

Transverse cutaneous
nerve of neck

To trapezius

To levator scapulae

To scalenus medius

I C

II C

III C

IV C

V C

To rectus lateralis

To rectus capitis anterior and
longus capitis

To longus capitis and
longus colli

To longus capitis, longus
colli, and scalenus medius

To geniohyoid

To thyrohyoid

Superior root of ansa
cervicalis

Inferior root of ansa
cervicalis

To longus colli

Ansa cervicalis

Phrenic

Supraclavicular

7.240 A plan of the cervical plexus. The hypoglossal nerve is shown by interrupted lines and the muscular branches by solid black lines. The roman numerals and letters I C to V C indicate the *ventral rami* of these cervical spinal nerves.

around the accessory nerve and ascends along the posterior border of the sternocleidomastoid. Near the cranium it perforates the deep fascia and ascends the scalp behind the auricle, supplying the skin and connecting with the great auricular and greater occipital nerves and the posterior auricular branch of the facial. It varies in size and is sometimes double. Its *auricular branch* supplies the skin on the upper third of the medial auricular aspect and connects with the posterior branch of the great auricular nerve. The auricular branch is occasionally derived from the greater occipital nerve.

The great auricular nerve (7.240, 241), is the largest ascending branch. It arises from the second and third cervical rami, encircles the posterior border of the sternocleidomastoid, perforates the deep fascia and ascends on the muscle beneath the platysma with the external jugular vein. It passes to the parotid gland, dividing into anterior and posterior branches. The *anterior branch* is distributed to the facial skin over the parotid gland, connecting in the gland with the facial nerve. The *posterior branch* supplies the skin over the mastoid process and on the back of the auricle (except its upper part); a filament pierces the auricle to reach its lateral surface where it is distributed to the lobule and concha. The posterior branch communicates with the lesser occipital, the auricular branch of the vagus and the posterior auricular branch of the facial nerve.

The transverse (anterior) cutaneous nerve of the neck (7.240, 241) arises from the second and third cervical rami, curves round the posterior border of the sternocleidomastoid near its midpoint and runs obliquely forwards, deep to the external jugular vein, to the muscle's anterior border. It perforates the

deep cervical fascia, dividing under the platysma into ascending and descending branches distributed to the anterolateral areas of the neck. The *ascending branches* ascend to the submandibular region, forming a plexus with the facial nerve's cervical branch beneath the platysma; others pierce the platysma and are distributed to the skin of the upper anterior areas of the neck. The *descending branches* pierce the platysma and are distributed anterolaterally to the skin of the neck, as low as the sternum.

DESCENDING SUPERFICIAL BRANCHES

These are: supraclavicular (C3, 4): medial, intermediate and lateral.

The supraclavicular nerves (7.240, 241) arise by a common trunk from the third and fourth cervical ventral rami and emerge from the posterior border of the sternocleidomastoid, to descend under the platysma and the deep cervical fascia; they divide into medial, intermediate and lateral (posterior) branches, which diverge to pierce the deep fascia a little above the clavicle.

The *medial supraclavicular nerves* run inferomedially across the external jugular vein and the clavicular and sternal heads of the sternocleidomastoid to supply the skin as far as the midline and as low as the second rib. They supply the sternoclavicular joint. The *intermediate supraclavicular nerves* cross the clavicle to supply the skin over the pectoralis major and deltoid down to the level of the second rib, next to the area of supply of the second thoracic nerve (7.254). Overlap between these nerves is minimal. The *lateral (posterior) supraclavicular nerves* descend superficially across the trapezius and acromion, supplying the skin of the upper and posterior parts of the shoulder.

DEEP BRANCHES—MEDIAL SERIES

These include the following communicating and muscular branches:

Communicating branches with	Hypoglossal	C1, 2
	Vagus	C1, 2
	Sympathetic	C1–4
Muscular branches to	Rectus capitis lateralis	C1
	Rectus capitis anterior	C1, 2
	Longus capitis	C1–3
	Longus colli	C2–4
	Inferior root of ansa cervicalis	C2, 3
	Phrenic nerve	C3–5

Communicating branches pass from the loop between the first and second cervical rami to the vagus, hypoglossal nerve and sympathetic. The hypoglossal branch later leaves it as a series of branches, viz. the meningeal, *superior root of ansa cervicalis*, nerves to the thyrohyoid and probably to the geniohyoid (p. 585). A branch also connects the fourth and fifth cervical rami; the first four cervical ventral rami each receive a grey ramus communicans from the superior cervical sympathetic ganglion.

Muscular branches supply the rectus capitis lateralis, rectus capitis anterior, longus capitis and longus colli.

The inferior root of the ansa cervicalis (nervus descendens cervicalis) (7.240) is formed by the union of a branch from the second with another from the third cervical ramus. It descends on the lateral side of the internal jugular vein, crosses it a little below the middle of the neck, and continues forwards to join the superior root anterior to the common carotid artery, forming the *ansa cervicalis (ansa hypoglossi)* (6.69), from which all infrahyoid muscles are supplied, except the thyrohyoid. In 160 dissections, the inferior root was from the second and third cervical ventral rami in 74%, from the second to fourth in 14%, from the third alone in 5%, from the second alone in 4% and from the first to third in 2% (Poriraer & Chernikov 1965).

The phrenic nerve, sole *motor* supply to the diaphragm, also contains widespread sensory fibres. It arises chiefly from the fourth cervical ramus but also has contributions from the third and fifth (7.240). Formed at the upper part of the lateral border of the scalenus anterior, it descends almost vertically across it *behind the prevertebral fascia* on its anterior surface. It descends posterior

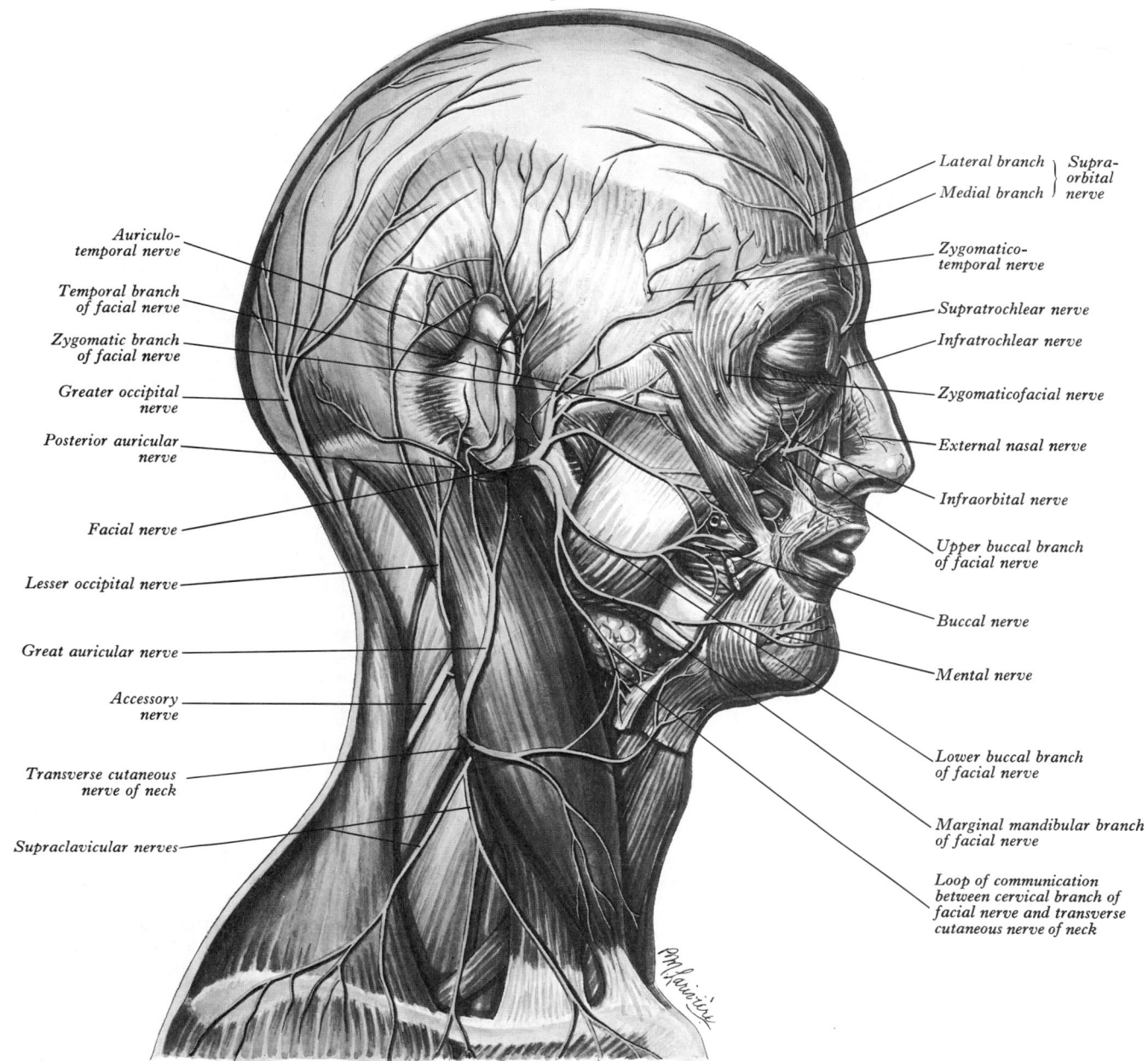

7.241 The nerves of the right side of the scalp, face and side of the neck.

to the sternocleidomastoid, the inferior belly of omohyoid (near its intermediate tendon), the internal jugular vein, transverse cervical and suprascapular arteries (**6.**68) and, on the left, the thoracic duct. It then runs anterior to the subclavian artery, posterior to the subclavian vein and enters the thorax by crossing medially in front of the internal thoracic artery (**6.**90), after which it descends anterior to the pulmonary hilum, between the fibrous pericardium and mediastinal pleura, to the diaphragm, accompanied by the pericardiacophrenic vessels. The right and left phrenic nerves differ in their intrathoracic relations.

The *right phrenic nerve*, shorter and more vertical, is separated at the root of the neck from the second part of the right subclavian artery by the scalenus anterior. It is then lateral to the right brachiocephalic vein, the superior vena cava and the fibrous pericardium covering the right surface of the right atrium and inferior vena cava.

The *left phrenic nerve*, at the root of the neck, is commonly stated to leave the medial edge of the scalenus anterior to pass anterior to the *first* part of the left subclavian artery and behind the thoracic duct. However, Quist (1977) claims that both right and left nerves are *symmetrical* in their cervical course and that at the thoracic inlet the left crosses anterior to the *second* part of the

left subclavian artery, separated by the scalenus anterior. Thereafter the left phrenic crosses anterior to the left internal thoracic artery, descending across the medial aspect of the left lung's apex and its pleura to the first part of the subclavian artery, which it crosses obliquely to reach a groove between the left common carotid and subclavian arteries. It passes anteromedially superficial to the left vagus nerve just above the aortic arch and behind the left brachiocephalic vein. It then passes superficial to the aortic arch and the left superior intercostal vein, anterior to the left pulmonary hilum, to lie between the fibrous pericardium covering left surface of the left ventricle and the mediastinal pleura (p. 750).

In the neck each nerve receives variable filaments from the cervical sympathetic ganglia or their branches and may also connect with internal thoracic sympathetic plexuses (Pearson & Sauter 1971); these connections may represent a devious course of sympathetic fibres to these plexuses (p. 1161). In its thoracic course each nerve supplies sensory branches to the mediastinal pleura, fibrous pericardium and parietal serous pericardium.

Diaphragmatic relations (Merendino et al 1956, Perera & Edwards 1957). The right phrenic nerve traverses the diaphragm's central tendon, either by the caval orifice or just lateral to it. The

1129

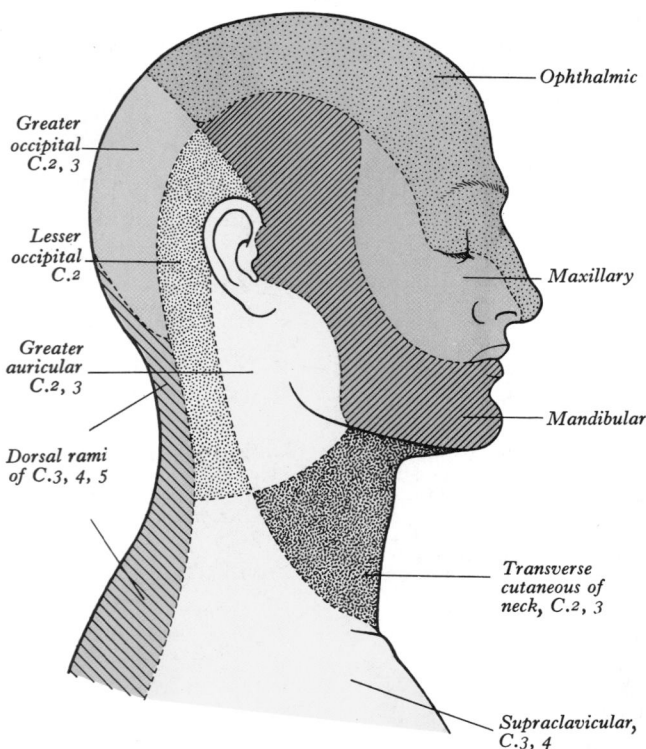

7.242 A diagram showing the cutaneous nerve supply of the face, scalp and neck. Magenta = trigeminal nerve, yellow = cervical ventral rami, blue = cervical dorsal rami.

left phrenic nerve traverses the muscular part of the diaphragm anterior to the central tendon, just lateral to the left cardiac surface and more anterior than the right phrenic. At the diaphragm or slightly above it, each phrenic nerve supplies fine branches to the parietal pleura above, and the parietal peritoneum below, the central diaphragm. The trunk of each nerve then divides as it traverses the diaphragm into commonly three branches arranged as follows, with some variation: (1) An *anterior* (sternal) branch runs anteromedially towards the sternum, connecting with its fellow. (2) An *anterolateral* branch runs laterally anterior to the lateral leaf of the central tendon. (3) A short *posterior* branch divides into a posterolateral ramus coursing behind the lateral leaf and a posterior (crural) ramus supplying the crural fibres; posterolateral and crural branches may arise separately from the phrenic nerve. These main branches are often submerged in diaphragmatic muscle or below it; they supply motor fibres to the diaphragmatic muscle and sensory fibres to the peritoneum and pleura related to the central part of the diaphragm. They also contain proprioceptive fibres from the musculature. Location of the main branches is of importance in avoiding surgical damage. The right crus splits to enclose the oesophagus (p. 592); the right phrenic nerve supplies the part of it to the right of the oesophagus, the left supplying the left crus and the part of the right crus on the left of the oesophagus (Collis et al 1954, Thornton & Schweisthal 1969).

On the diaphragm's inferior surface phrenic rami connect with branches of the coeliac plexuses (p. 1165); on the right, at the junction of the plexuses, is a small *phrenic ganglion*. Rami from the plexuses supply the suprarenal glands and, on the right, the hepatic falciform and coronary ligaments and the inferior vena cava and possibly (via connections with coeliac and hepatic plexuses) the gallbladder (pp. 1165, 1395).

Accessory phrenic nerve. Fibres for the phrenic nerve from the fifth cervical ventral ramus often pass in a branch of the nerve to the subclavius, the *accessory phrenic nerve*. This lies lateral to the main nerve and descends posterior, or sometimes anterior, to the subclavian vein; it usually joins the phrenic near the first rib but may not do so until near the pulmonary hilum or beyond. An accessory phrenic nerve may be derived from the fourth or sixth cervical rami or from the ansa cervicalis (p. 1128).

Applied Anatomy. Division of the phrenic nerve in the *neck* completely paralyses the corresponding half of the diaphragm, which atrophies. If an accessory phrenic nerve exists section or crushing of the main nerve as it lies on the scalenus anterior will *not* produce complete paralysis.

DEEP BRANCHES—LATERAL SERIES

These include:

Communicating—Accessory		C2, 3, 4
	Sternocleidomastoid	C2, 3, 4
Muscular	Trapezius	C2, (3)
branches	Levator scapulae	C3, 4
	Scalenus medius	C3, 4

Communicating branches. Lateral deep branches of the cervical plexus connect with the spinal accessory nerve in the sternocleidomastoid, posterior triangle and under the trapezius.

Muscular branches are distributed to sternocleidomastoid and to the trapezius, levator scapulae and scalenus medius. Trapezial branches cross the posterior triangle obliquely below the spinal accessory nerve.

The Brachial Plexus

The brachial plexus (7.243) is a union of the lower four cervical ventral rami and the greater part of the first thoracic ventral ramus (p. 1127); the fourth ramus usually gives a branch to the fifth and the first thoracic frequently receives one from the second. Contributions to the plexus by C4 and T2 vary; when the branch from C4 is large, that from T2 is frequently absent and the branch from T1 is reduced, forming a *prefixed type* of plexus. If the branch from C4 is small or absent, the contribution of C5 is

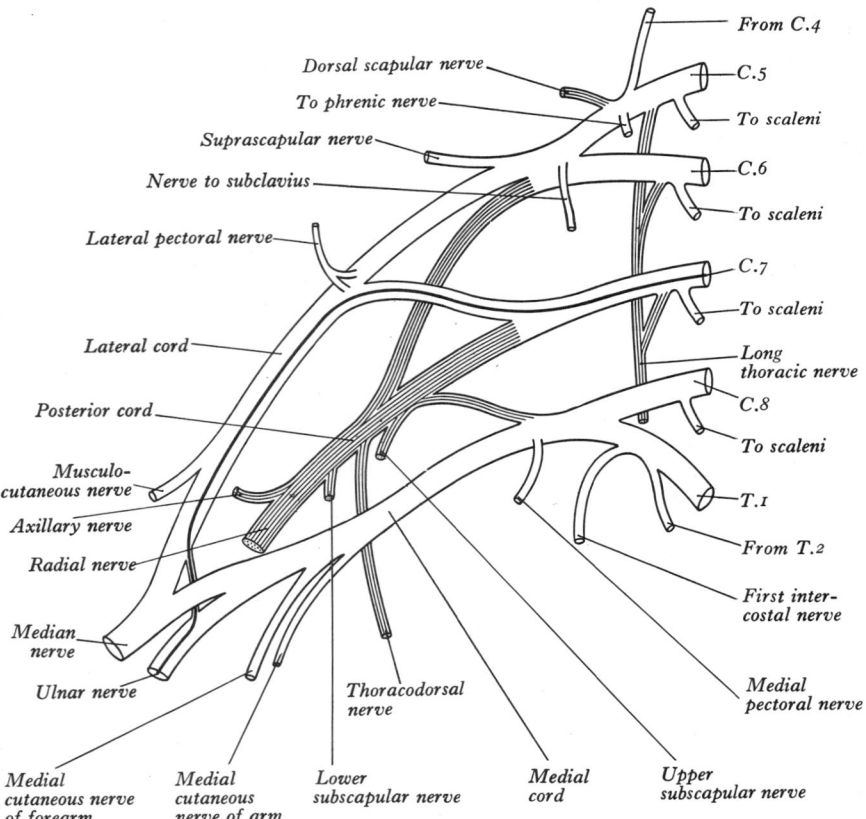

7.243 A plan of the brachial plexus. The posterior division of the trunks and their derivatives are shaded and the fibres from C.7 which enter the ulnar nerve are shown as a heavy black line. Letters and numbers C.4–C.8 and T.1–T.2 indicate the *ventral rami* of these cervical and thoracic spinal nerves.

reduced but that of T1 is larger and one from T2 is always present; this arrangement constitutes a *postfixed type* of plexus. These ventral rami are the *roots* of the plexus, almost equal in size but variable in their mode of junction. The commonest arrangement is as follows: the fifth and sixth rami unite at the lateral border of the scalenus medius as the *upper trunk*; the eighth cervical and first thoracic rami join behind the scalenus anterior as the *lower trunk*; the seventh cervical becomes the *middle trunk*. These three trunks incline laterally; just above or behind the clavicle each bifurcates into *anterior* and *posterior divisions*; the anterior divisions of the upper and middle trunks form a *lateral cord*, lateral to the axillary artery. The anterior division of the lower trunk descends at first behind, then medial to the axillary artery, forming the *medial cord* which often receives a branch from the seventh cervical ramus. Posterior divisions of all three form the *posterior cord*, at first above and then behind the axillary artery. The lower trunk's posterior division is much smaller than the others, containing few, if any, fibres from the first thoracic ramus. It is frequently derived from the eighth cervical before the trunk is formed.

Despite the adaptation to evolutionary changes in upper-limb musculature, the human brachial plexus reflects the original flexor–extensor organization of a primitive fin. The posterior cord is the extensor supply, the medial and lateral cords the flexor supply. Migration of muscle masses has modified this pattern; e.g. brachialis and the anterior part of the deltoid are supplied (the former in part) from 'extensor' nerves. For comparative morphology of the plexus consult Harris (1939).

Relations of the Brachial Plexus

In the neck, the plexus is in the posterior triangle, in the angle between the clavicle and lower posterior border of the sternocleidomastoid, covered by platysma, deep fascia and skin, through which it is palpable. It is crossed by the supraclavicular nerves, the nerve to the subclavius, the inferior belly of the omohyoid, the external jugular vein and the superficial ramus of the transverse cervical artery (**6.87**). It emerges between the scaleni anterior and medius; its proximal part is superior to the third part of the subclavian artery, the lower trunk posterior to it; the plexus passes posterior to the medial two-thirds of the clavicle, the subclavius and the suprascapular vessels and lies on the first digitation of the serratus anterior and the subscapularis. *In the axilla* the lateral and posterior cords are lateral to the first part of the axillary artery, the medial cord being behind it. The cords surround the second part of the artery, related according to their names. In the lower axilla the cords divide into nerves for the upper limb. Except for the median nerve's medial root, these nerves are related to the artery's third part as their cords are to the second, i.e. branches of the lateral cord are lateral, of the medial cord medial and of the posterior cord posterior, to the artery.

Close to their exit from the intervertebral foramina, the fifth and sixth cervical ventral rami receive grey rami communicantes from the middle cervical sympathetic ganglion, the seventh and eighth from the cervicothoracic ganglion (p. 1160). The first thoracic ventral ramus also receives a grey ramus from and contributes a white ramus to the cervicothoracic ganglion.

Branches of the brachial plexus may be described as *supraclavicular* and *infraclavicular*.

SUPRACLAVICULAR BRANCHES

These arise (*a*) from roots or (*b*) from trunks.

From roots:	1. Nerves to scaleni and longus colli	C5, 6, 7, 8
	2. Branch to phrenic nerve	C5
	3. Dorsal scapular nerve	C5
	4. Long thoracic nerve	C5, 6 (7)
From trunks:	1. Nerve to subclavius	C5, 6
	2. Suprascapular nerve	C5, 6

Branches to the scaleni and longus colli arise from the lower cervical ventral rami near their exit from the intervertebral foramina. Anterior to the scalenus anterior the phrenic nerve is joined by a branch from the fifth cervical ramus.

The dorsal scapular nerve, from the fifth cervical ventral ramus, pierces the scalenus medius, passes behind the levator scapulae, which it occasionally supplies, and runs with the deep branch of the dorsal scapular artery to the rhomboids, which it supplies.

The long thoracic nerve (7.247) is usually formed by roots from the fifth to the seventh cervical rami; the last may be absent. (70 dissections demonstrated all three roots in only 42%, Alexandre et al 1968.) The upper two roots pierce the scalenus medius obliquely, uniting in or lateral to it; the nerve descends dorsal to the brachial plexus and the first part of the axillary artery. Crossing the superior border of serratus anterior to reach its lateral surface, it may be joined by the root from C7, which emerges between the scaleni anterior and medius and descends on the latter's lateral surface. The nerve continues downwards to the lower border of the serratus anterior, supplying branches to each of its digitation.

The nerve to the subclavius, small and arising near the junction of the fifth and sixth cervical ventral rami, descends anterior to the plexus and the third part of the subclavian artery and is usually connected to the phrenic nerve. It passes above the subclavian vein to supply the subclavius.

The suprascapular nerve (6.87, 7.248), a large branch of the superior trunk, runs laterally deep to the trapezius and omohyoid, enters the supraspinous fossa through the suprascapular notch inferior to the superior transverse scapular ligament, runs deep to the supraspinatus and curves round the lateral border of the scapular spine with the suprascapular artery to reach the infraspinous fossa, where it gives two branches to the supraspinatus and articular rami to the shoulder and acromioclavicular joints. The suprascapular nerve was found to have a cutaneous branch in six upper limbs out of 61 Japanese cadavers (Horiguchi 1980). When present, it pierced the deltoid muscle close to the tip of the acromion; in the one case where its peripheral distribution was examined, it was found to supply the skin of the proximal third of the arm within the territory of the axillary nerve.

INFRACLAVICULAR BRANCHES

These branch from the cords but their fibres may be traced back to the spinal nerves. They are as follows:

Lateral cord:	Lateral pectoral	C5, 6, 7
	Musculocutaneous	C5, 6, 7
	Lateral root of median	C(5), 6, 7
Medial cord:	Medial pectoral	C8, T1
	Medial cutaneous of forearm	C8, T1
	Medial cutaneous of arm	C8, T1
	Ulnar	C(7), 8, T1
	Medial root of median	C8, T1
Posterior cord:	Upper subscapular	C5, 6
	Thoracodorsal	C6, 7, 8
	Lower subscapular	C5, 6
	Axillary	C5, 6
	Radial	C5, 6, 7, 8, (T1)

The lateral pectoral nerve (7.247), larger than the medial, may arise from the anterior divisions of the upper and middle trunks or by a single root from the lateral cord; its fibres are from the fifth to seventh cervical rami. It crosses anterior to the axillary artery and vein, pierces the clavipectoral fascia and supplies the deep surface of the pectoralis major. It sends a ramus to the medial pectoral nerve forming a loop in front of the first part of the axillary artery (7.247), to supply some fibres to the pectoralis minor.

The medial pectoral nerve, derived from eighth cervical and first thoracic ventral rami, branches from the medial cord while this is posterior to the axillary artery. It curves forwards between the axillary artery and vein; anterior to the artery it joins a ramus of the lateral pectoral nerve, entering the deep surface of the pectoralis minor to supply it. Two or three branches pierce the pectoralis minor and others may pass round its inferior border to end in the pectoralis major.

The superior subscapular nerve, smaller than the inferior, arises from the posterior cord (C5 and 6), enters the subscapularis at a high level and is frequently double.

The inferior subscapular nerve, also from the posterior cord (C5 and 6), supplies the lower part of the subscapularis, ending in the teres major, which is sometimes supplied by a separate branch.

The thoracodorsal nerve, from the posterior cord, derives its fibres from the sixth to eighth cervical ventral rami, arises between the subscapular nerves and accompanies the subscapular artery along the posterior axillary wall to supply the latissimus dorsi, reaching its distal border.

The axillary (circumflex humeral) nerve (7.248) arises from the posterior cord, its fibres being derived from the fifth and sixth cervical ventral rami. It is at first lateral to the radial nerve, posterior to the axillary artery and anterior to the subscapularis, at whose lower border it curves back inferior to the humeroscapular articular capsule and, with the posterior circumflex humeral vessels, traverses a quadrangular space bounded *above* by the subscapularis (anterior) and teres minor (posterior), *below* by teres major, *medially* by the long head of triceps and *laterally* by the humeral surgical neck. The nerve finally divides into anterior and posterior branches. The *anterior branch*, with the posterior circumflex humeral vessels, curves round the humeral neck, deep to the deltoid, to its anterior border, supplying it and giving a few small cutaneous branches which pierce the muscle to ramify in the skin over its lower part. The *posterior branch* supplies the teres minor and the posterior part of the deltoid; on the branch to teres minor an enlargement or pseudoganglion usually exists. The posterior branch pierces the deep fascia low on the posterior border of the deltoid, continuing as the *upper lateral cutaneous nerve of the*

arm and supplying the skin over the lower part of the deltoid and the upper part of the long head of triceps (7.244). The axillary trunk supplies a branch to the shoulder joint below the subscapularis.

The musculocutaneous nerve (7.247), from the lateral cord opposite the lower border of the pectoralis minor and derived from the fifth to the seventh cervical ventral rami, pierces the coracobrachialis and descends laterally between the biceps and brachialis to the lateral side of the arm; just below the elbow it pierces the deep fascia lateral to the tendon of biceps, continuing as the *lateral cutaneous nerve of the forearm*. A line drawn from the lateral side of the third part of the axillary artery across the coracobrachialis and biceps to the lateral side of the biceps tendon is a surface projection for the nerve, but this is varied by its point of entry into the coracobrachialis (Latarjet et al 1967). It supplies the coracobrachialis, both heads of biceps and most of the brachialis. The branch to the coracobrachialis leaves the musculocutaneous before it enters the muscle; its fibres are from the seventh cervical ramus and may branch directly from the lateral cord. Branches to the biceps and brachialis leave after the musculocutaneous pierces the coracobrachialis; the branch to the brachialis supplies the elbow joint. The nerve also supplies a small branch to the humerus, entering with the nutrient artery.

The lateral cutaneous nerve of the forearm (7.244) passes deep to the cephalic vein, descending along the radial border of the forearm to the wrist, supplying the skin of the forearm's anterolateral surface and connecting by branches around its radial border with the posterior cutaneous nerve of the forearm and the terminal branch of the radial nerve. Its trunk gives rise to a slender recurrent branch which extends along the cephalic vein as far as the middle third of the upper arm, distributing filaments to the

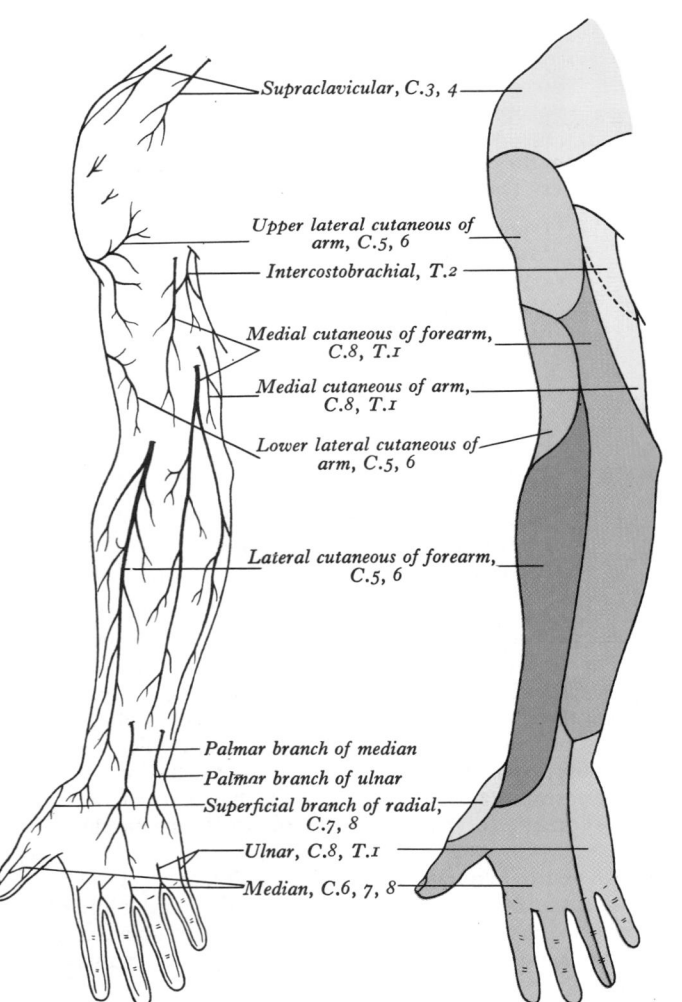

7.244 The cutaneous nerves of the right upper limb, their areas of distribution and segmental origins, viewed from the anterior aspect.

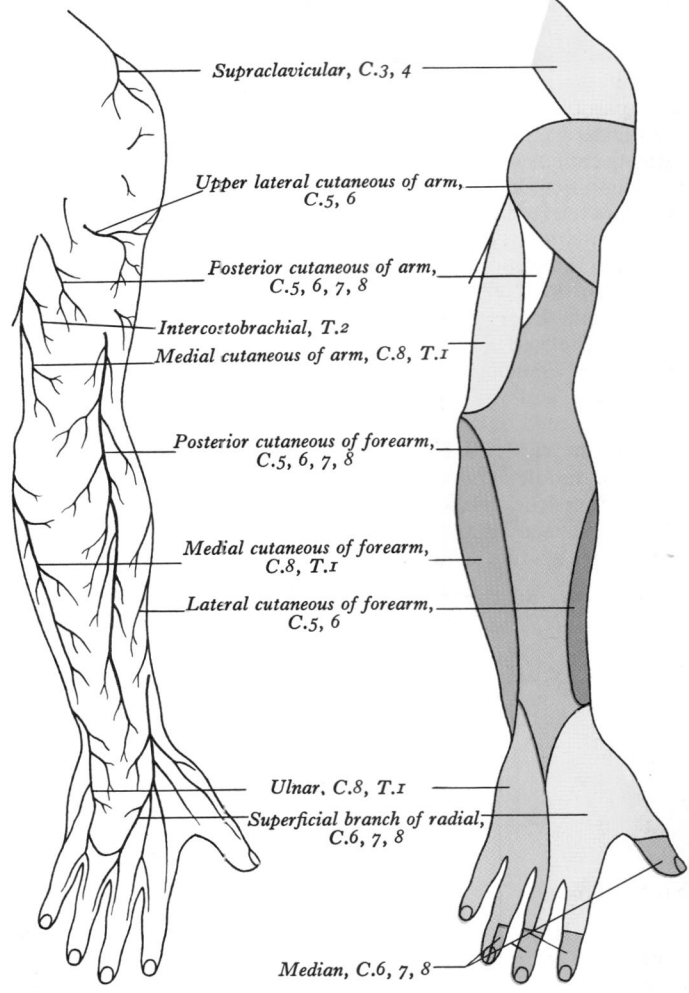

7.245 The cutaneous nerves of the upper right limb, their areas of distribution and segmental origins, viewed from the posterior aspect.

skin over the distal third of the anterolateral surface of the upper arm close to the vein (Horiguchi 1981). Although observed in the nineteenth century (Arnold 1851, Bolk 1898), this recurrent branch has more recently been omitted from most descriptions of the nerve supply of the upper limb. At the wrist joint the lateral cutaneous nerve of the forearm is anterior to the radial artery and some filaments, piercing the deep fascia and accompanying this to the dorsum of the carpus. The nerve then passes to the base of the thenar eminence, ending in cutaneous rami. It connects with the terminal branch of the radial nerve and the palmar cutaneous branch of the median nerve.

The *musculocutaneous* nerve has frequent variations. It may run behind the coracobrachialis or adhere for some distance to the median nerve and pass behind the biceps. Some fibres of the median nerve may run in the musculocutaneous nerve, leaving it to join their proper trunk; less frequently the reverse occurs, the median nerve sending a branch to the musculocutaneous. Occasionally it supplies the pronator teres and may replace radial branches to the dorsal surface of the thumb.

The medial cutaneous nerve of the forearm (7.247), from the medial cord, is derived from the eighth cervical and first thoracic ventral rami. At first it is between the axillary artery and vein and supplies a ramus piercing the deep fascia to supply the skin over the biceps, almost to the elbow. The nerve descends medial to the brachial artery, pierces the deep fascia with the basilic vein midway in the arm and divides into anterior and posterior branches. The larger, *anterior branch* usually passes in front of, occasionally behind, the median cubital vein, descending anteromedial in the forearm to supply the skin as far as the wrist and connecting with the palmar cutaneous branch of the ulnar nerve (7.244). The *posterior branch* descends obliquely medial to the basilic vein, anterior to the medial epicondyle, and curves round to the back of the forearm, descending on its medial border to the wrist, supplying the skin. It connects with the medial cutaneous nerve of the arm, the posterior cutaneous nerve of the forearm and the dorsal branch of the ulnar (7.245).

The medial cutaneous nerve of the arm, supplying the skin in that area (7.244) and the smallest branch of the plexus, arises from the medial cord and contains fibres from the eighth cervical and first thoracic ventral rami. It traverses the axilla, crossing anterior or posterior to the axillary vein, to which it is then medial, and communicating with the intercostobrachial nerve; it descends medial to the brachial artery and basilic vein (7.247) to a point midway in the upper arm, where it pierces the deep fascia to supply a medial area in the arm's distal third, extending on to its anterior and posterior aspects. Some rami reach the skin anterior to the medial epicondyle, others over the olecranon. It connects with the posterior branch of the medial cutaneous nerve of the forearm. Sometimes the medial cutaneous nerve of the arm and the intercostobrachial nerve are connected in a plexiform manner in the axilla. Sometimes the intercostobrachial nerve is large and reinforced by part of the lateral cutaneous branch of the third intercostal nerve, replacing the arm's medial cutaneous nerve and receiving from the brachial plexus a connection representing the latter; occasionally this connection is absent.

THE MEDIAN NERVE

The median nerve (7.247) has two roots from the lateral (C(5),6,7) and medial (C8, T1) cords, which embrace the third part of the axillary artery, uniting anterior or lateral to it. Some fibres from C7 often leave the lateral root in the lower part of the axilla passing distomedially posterior to the medial root, usually anterior to axillary artery, to join the ulnar nerve (7.263); they may branch from the seventh cervical ventral ramus. Clinically they are believed to be mainly motor to the flexor carpi ulnaris. If the lateral root is small, the musculocutaneous nerve (C5,6,7) connects with the median nerve in the arm.

The median nerve enters the arm at first lateral to the brachial artery; near the insertion of the coracobrachialis it crosses in front of (rarely behind) the artery, descending medial to it to the cubital fossa where it is posterior to the bicipital aponeurosis and anterior to the brachialis, separated by the latter from the elbow joint. It

usually enters the forearm between the heads of the pronator teres, crossing to the lateral side of the ulnar artery and separated from it by the deep head of pronator teres. It proceeds behind a tendinous bridge between the humero-ulnar and radial heads of the flexor digitorum superficialis, descending posterior and adherent to the flexor digitorum superficialis and anterior to the flexor digitorum profundus. About 5 cm proximal to the flexor retinaculum it emerges from behind the lateral edge of the flexor digitorum superficialis, becoming superficial just proximal to the wrist between the tendons of the flexors digitorum superficialis and carpi radialis, projecting laterally from behind the tendon of the palmaris longus. It then passes deep to the flexor retinaculum into the palm. In the forearm it is accompanied by the median branch of the anterior interosseous artery. Its course can be marked on the surface by a line from the medial side of the brachial artery in the cubital fossa along the forearm's midline (7.246).

Relations as the median nerve enters the forearm are variable. According to Anson (1963) the route described above occurred in 82·8% of 1000 dissections; in 10·8% the nerve was posterior to the humeral head of the pronator teres, the ulnar slip being absent. In 4·6% the nerve was posterior to both heads, and in 1·8% traversed *through* the humeral head.

Branches of the Median Nerve in the Arm

These are vascular branches to the brachial artery and usually a branch to the pronator teres, a variable distance proximal to the elbow joint.

Branches of the Median Nerve in the Forearm

These are muscular, articular, anterior interosseous, palmar cutaneous and communicating. **Muscular branches**, except one, are given off proximally (near the elbow) to the superficial flexor muscles (except the flexor carpi ulnaris), i.e. to the pronator teres, flexor carpi radialis, palmaris longus and the flexor digitorum superficialis. The branch to the part of the flexor digitorum superficialis for the index finger arises near midforearm but may be from the anterior interosseous nerve. **Articular branches**, arising at or just distal to the elbow joint, supply it and the proximal radio-ulnar joint.

The anterior interosseous nerve branches posteriorly from the median nerve between the two heads of pronator teres, just distal to the origin of branches to the superficial forearm flexors. With the anterior interosseous artery it descends anterior to the interosseous membrane, between and deep to the flexors pollicis longus and digitorum profundus and supplying both; branches to the latter are limited to its lateral part, which sends tendons to the index and middle fingers. Terminally it is posterior to the pronator quadratus, supplying its deep surface, and supplies the distal radio-ulnar, radio-carpal and carpal joints.

The palmar cutaneous branch starts just proximal to the flexor retinaculum, pierces it or the deep fascia and divides into *lateral branches* supplying the thenar skin and connecting with the lateral cutaneous nerve of the forearm; and *medial branches* supplying the central palmar skin and connecting with the palmar cutaneous branch of the ulnar nerve.

A communicating branch, which may be multiple, often arises (sometimes from the anterior interosseous branch) in the proximal forearm and passes medially between flexors digitorum superficialis and profundus and behind the ulnar artery to join the ulnar nerve. It is a factor in explaining anomalous muscular innervation in the hand (p. 1135).

The Median Nerve in the Hand

Proximal to the flexor retinaculum the nerve is lateral to the tendons of flexor digitorum superficialis but further distally it lies between the retinaculum and tendons in the 'carpal tunnel' (see p. 625), i.e. the space between the flexor retinaculum and the anterior carpal surface. The nerve may become compressed here (p. 1137). Distal to the retinaculum the nerve enlarges and flattens, usually dividing into five or six branches but the mode and level of division are variable.

The muscular branch, short and thick, is from the nerve's lateral side; it may be the first palmar or a terminal branch arising level with the digital branches. It runs laterally, just distal to the

7.246 The anterior aspect of the right upper limb, showing the position of the principal nerves and vessels projected on to the surface.

flexor retinaculum, with a slight recurrent curve beneath the part of the palmar aponeurosis covering the thenar muscles. It turns round the distal border of the retinaculum to lie superficial to the flexor pollicis brevis, usually supplying it and either continuing superficial to it or traversing it. It gives a branch to the abductor pollicis brevis, which enters the muscle's medial edge and then passes deep to it to supply the opponens pollicis, entering its medial edge. Its terminal part occasionally gives a branch to the first dorsal interosseous, which may be its sole or partial supply. The muscular branch may arise in the carpal tunnel and pierce the flexor retinaculum, a point of surgical importance (Papathanassion 1968).

The palmar digital branches. (In this account of cutaneous innervation of the digits, the distinction between proximal, penultimate, undivided rami, termed *common palmar digital nerves*, and their ultimate branches to individual digits, the *proper palmar digital nerves*, must be noted. The corresponding terms for the foot are *common plantar digital nerves* and *proper plantar digital nerves*.) The median nerve divides into four or five digital branches, though it often divides first into two: a lateral ramus providing digital branches to the pollex and the radial side of the index finger; and a medial, supplying digital branches to adjacent sides of the index, medius and annularis. Other modes of termination occur (Poisel 1974). Digital branches are commonly arranged as follows. They pass distally, deep to the superficial palmar arch and its digital vessels, at first anterior to the long flexor tendons. Two *proper* palmar digital nerves, sometimes from a common stem, pass to the sides of the pollex, that supplying its lateral side crossing in front of the tendon of flexor pollicis longus. The *proper* palmar digital nerve to the lateral side of the index also supplies the first lumbrical muscle. Two *common* palmar digital nerves pass distally between the long flexor tendons: the lateral divides in the distal palm into two *proper* palmar digital nerves traversing adjacent sides of the index and medius; the medial one divides into two *proper* palmar digital nerves supplying adjacent sides of the medius and annularis. The lateral common digital nerve supplies the second lumbrical muscle; the medial receives a communicating twig from the common palmar digital branch of the *ulnar* nerve and may supply the third lumbrical muscle. In the distal part of the palm, the digital arteries pass deeply between the divisions of the digital nerves; on the sides of the digits the nerves are anterior to the arteries. The median nerve usually supplies palmar cutaneous digital branches to the lateral three and one-half digits (pollex, index, medius and the lateral side of the annularis); sometimes the lateral side of the ring finger is supplied by the ulnar nerve. The proper palmar digital nerves that pass along the medial side of the index, both sides of the medius and the lateral side of the annularis, enter these digits in fat between slips of the central palmar aponeurosis (p. 627). They pass, with the lumbricals and palmar digital arteries, dorsal to the superficial transverse metacarpal ligament (p. 629) and ventral to the deep transverse metacarpal ligament (p. 515). In the digits, the nerves run distally beside the long flexor tendons, outside their fibrous

sheaths, level with the anterior phalangeal surfaces and anterior to the digital arteries. Each nerve gives off several branches to the skin on the front and sides of the digit, many ending in lamellated corpuscles (p. 911), and branches to the metacarpophalangeal and interphalangeal joints. They also supply the fibrous sheaths of the long flexor tendons, digital arteries (vasomotor) and sweat glands (secretomotor). Distal to the base of the distal phalanx the digital nerve gives off a branch passing dorsally to the nail bed, the main nerve dividing to supply the pulp and skin of the terminal part of the digit. Distal to the base of the proximal phalanx, each *proper* digital nerve also gives off a dorsal branch to supply the skin over the back of the middle and distal phalanges (7.245). The *proper* palmar digital nerves to the pollex and the lateral side of the index emerge with the long flexor tendons from under the lateral edge of the central palmar aponeurosis and are arranged in the digits as described above; but in the pollex small distal branches supply the skin on the back of the distal phalanx only.

In addition to the branches of the median nerve described above, variable vasomotor branches supply the radial and ulnar arteries and their branches. Some of the intercarpal, carpometacarpal and intermetacarpal joints are said to be supplied by the median nerve or its anterior interosseous branch, the precise details being uncertain.

THE ULNAR NERVE

The ulnar nerve (7.247) arises from the medial cord (C8, T1) but, as described above (p. 1131), it often receives fibres from the ventral ramus of C7. It runs distally through the axilla medial to the axillary artery and between it and the vein, continuing distally medial to the brachial artery as far as mid-arm; here it pierces the medial intermuscular septum, inclining medially as it descends anterior to the medial head of the triceps to the interval between the medial epicondyle and the olecranon, with the superior ulnar collateral artery. At the elbow it is in a groove on the dorsum of the epicondyle. It enters the forearm between the two heads of the flexor carpi ulnaris superficial to the posterior and oblique parts of the ulnar collateral ligament. It descends the medial side of the forearm on the flexor digitorum profundus, covered proximally by the flexor carpi ulnaris; its lower half, covered by skin and fasciae, is lateral to this muscle. In the upper third of the forearm, it is distant from the ulnar artery but distal to this is close to its medial side (7.247). About 5 cm proximal to the wrist it gives off a dorsal branch which continues distally into the hand, anterior to the flexor retinaculum on the lateral to the pisiform bone and posteromedial to the ulnar artery. It passes behind the superficial part of the retinaculum with the artery and divides into superficial and deep terminal branches. Its relation to the brachial artery and medial epicondyle makes it easy to map out in its proximal course; a line from the medial epicondyle to the lateral edge of pisiform represents its distal course (7.246).

Its *branches* are: articular, muscular, palmar cutaneous, dorsal, superficial terminal and deep terminal.

Articular branches to the elbow joint issue from the nerve between the medial epicondyle and olecranon. Others are described below.

Muscular branches, usually two, begin near the elbow; one supplies the flexor carpi ulnaris (p. 617), the other the medial half of the flexor digitorum profundus.

The palmar cutaneous branch arises about mid-forearm, descends on the ulnar artery (7.247), which it supplies, and perforates the deep fascia to end in the palmar skin, after communicating with the palmar branch of the median nerve. It sometimes supplies the palmaris brevis.

The dorsal branch arises about 5 cm proximal to the wrist, passes distally and backwards, deep to the flexor carpi ulnaris, perforates the deep fascia, descends along the medial side of the back of the wrist and hand and then divides into two, or often three, dorsal digital nerves. One supplies the medial side of the minimus, the second adjacent sides of minimus and annularis, while the third, when present, supplies adjoining sides of the annularis and medius but may be replaced, wholly or partially, by a branch of the radial nerve, always communicating with it on the dorsum of the hand (7.245). In the minimus the dorsal digital nerves extend only to the base of the distal phalanx and in the annularis to the base of the middle phalanx; more distal parts of these digits are supplied by dorsal branches of the proper digital branches of the ulnar and, on the lateral side of the annularis, medial nerves.

The superficial terminal branch supplies the palmaris brevis and the medial palmar skin, dividing into two palmar digital nerves, which can be palpated against the hook of the hamate bone (p. 418); one of these supplies the medial side of minimus, the other (a *common* palmar digital nerve) sends a twig to the median nerve and divides into two *proper* digital nerves for the adjoining sides of minimus and annularis (7.247). The proper digital branches are distributed like those of the median nerve. Murakami (1969) has described articular branchlets from the superficial terminal branch of the ulnar nerve.

The deep terminal branch, with the deep branch of the ulnar artery, passes between the abductor digiti minimi and flexor digiti minimi and then perforates the opponens digiti minimi to follow the deep palmar arch dorsal to the flexor tendons. At its origin it supplies the three short muscles of the minimus. As it crosses the hand, it supplies the interossei and the third and fourth lumbricals; it ends by supplying the adductor pollicis, the first palmar interosseous and usually (p. 630) the flexor pollicis brevis. It sends articular filaments to the wrist joint.

The medial part of the flexor digitorum profundus is supplied by the ulnar nerve, as are the third and fourth lumbricals which are connected with the tendons of this part of the muscle. Similarly, the lateral part of the flexor digitorum profundus and the first and second lumbricals are supplied by the median nerve. The third lumbrical is often supplied by both nerves. The deep terminal branch is said to give branches to some intercarpal, carpometacarpal and intermetacarpal joints, though, as with the median nerve, precise details are uncertain. Vasomotor branches, arising in the forearm and hand, supply the ulnar and palmar arteries.

The Anomalous Nerve Supply of the Hand Muscles

The nerve supply to the short thenar muscles varies. From the results of lesions of the median and ulnar nerves in the forearm (Rowntree 1949, Day & Napier 1961) the following variations were deduced, expressed here in percentages of 226 hands examined and regarding innervation of the thenar muscles by median, ulnar or both nerves: flexor pollicis brevis—median 36, ulnar 48, both 17; abductor pollicis brevis—median 95, ulnar 2·5, both 2; opponens pollicis—median 83, ulnar 9, both 7·5. The usual description is that all three are supplied by the median nerve; the results cited are possibly explicable, in part, by the variable connections between the median and ulnar nerves in the axilla, arm or forearm, by which median fibres are aberrantly conveyed in the ulnar nerve. An arcuate connecting nerve has long been recognized between median and ulnar nerves in the substance of the flexor pollicis brevis (Cannieu 1886), either part of which, superficial or deep, may be supplied by both nerves. In a more recent survey (Harness & Sekeles 1971) of the considerable literature, the results of dissections in 35 hands showed a loop in 77%, suggesting that it should be a feature of 'normal' description. Clinically variations are important; even with a complete lesion of the median nerve some of these muscles may escape paralysis, leading to erroneous conclusions. Clinical evidence reveals that the short thenar muscles receive their segmental supply from the eighth cervical and first thoracic spinal cord segments. Variations in both the motor and sensory supply of the thumb have been examined by Falconer & Spinner (1985).

THE RADIAL NERVE

The radial nerve (7.248) arises from the posterior cord, C5,6,7,8, (T1). The largest branch of the brachial plexus, it descends behind the third part of the axillary artery and the upper part of the brachial, anterior to the subscapularis and the tendons of the latissimus dorsi and teres major. With the arteria profunda brachii and, later, its radial collateral branch, it inclines dorsally between the long and medial heads of the triceps, after which it passes obliquely across the back of the humerus, first between the

lateral and medial heads of the triceps, then in a shallow groove deep to the lateral head. On reaching the lateral side of the humerus it pierces the lateral intermuscular septum to enter the anterior compartment; it then descends deep in a furrow between the brachialis and proximally the brachioradialis, then more distally the extensor carpi radialis longus. Anterior to the lateral epicondyle it divides into *superficial* and *deep* terminal rami.

In the arm the radial nerve is indicated by a posterior line from the start of the brachial artery distally and laterally to the junction of the upper and middle thirds of a line between the lateral epicondyle and the deltoid tuberosity; the line is continued anteriorly as far as the lateral epicondyle, 1 cm or less lateral to the biceps tendon.

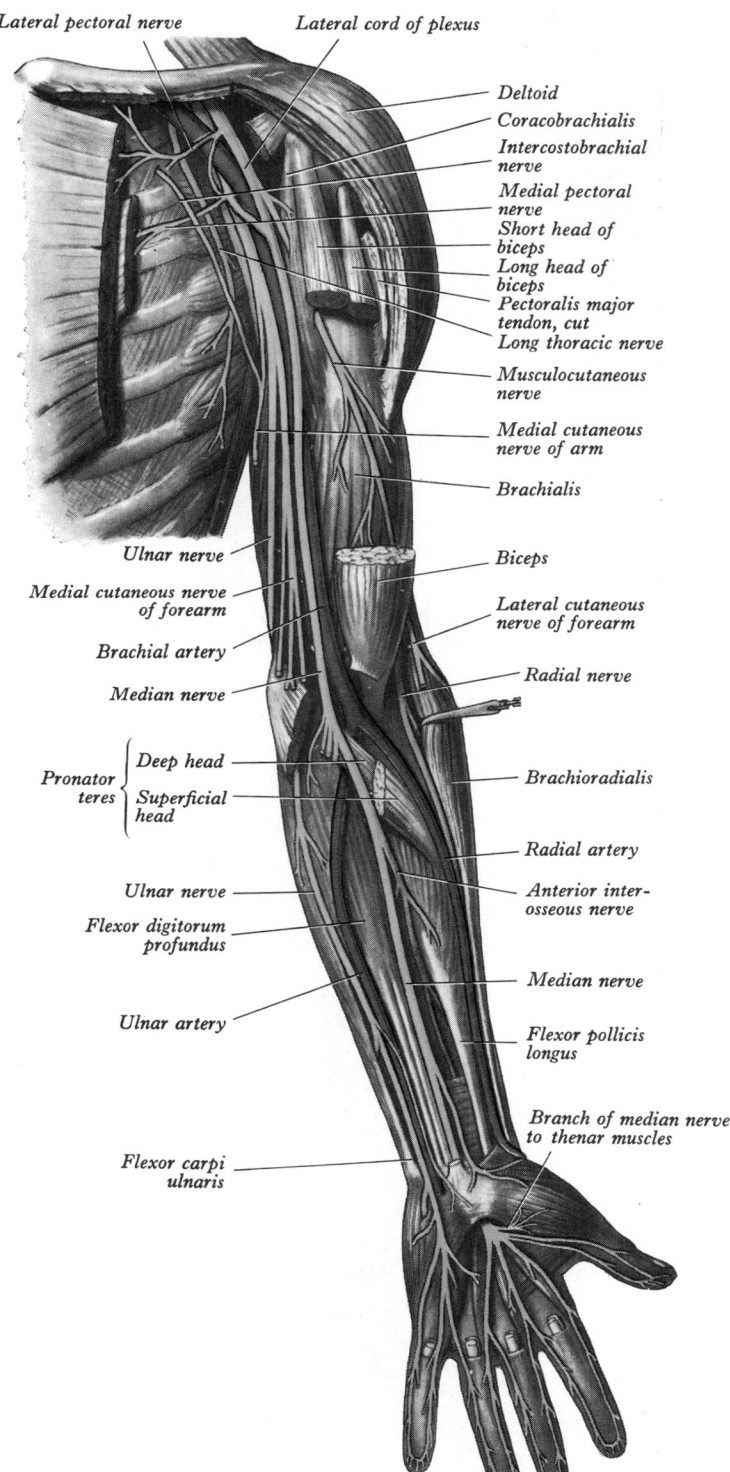

7.247 The nerves of the left upper limb, dissected from the anterior aspect.

The *branches* of the radial nerve are: muscular, cutaneous, articular and superficial and deep terminal.

Muscular branches supply the triceps, anconeus, brachioradialis, extensor carpi radialis longus and brachialis in *medial*, *posterior* and *lateral* groups. *Medial* muscular branches arise from the radial nerve on the medial side of the arm. They supply the medial and long heads of the triceps, the branch to the medial being a long, slender filament which, lying close to the ulnar nerve as far as the distal third of the arm, is often termed the

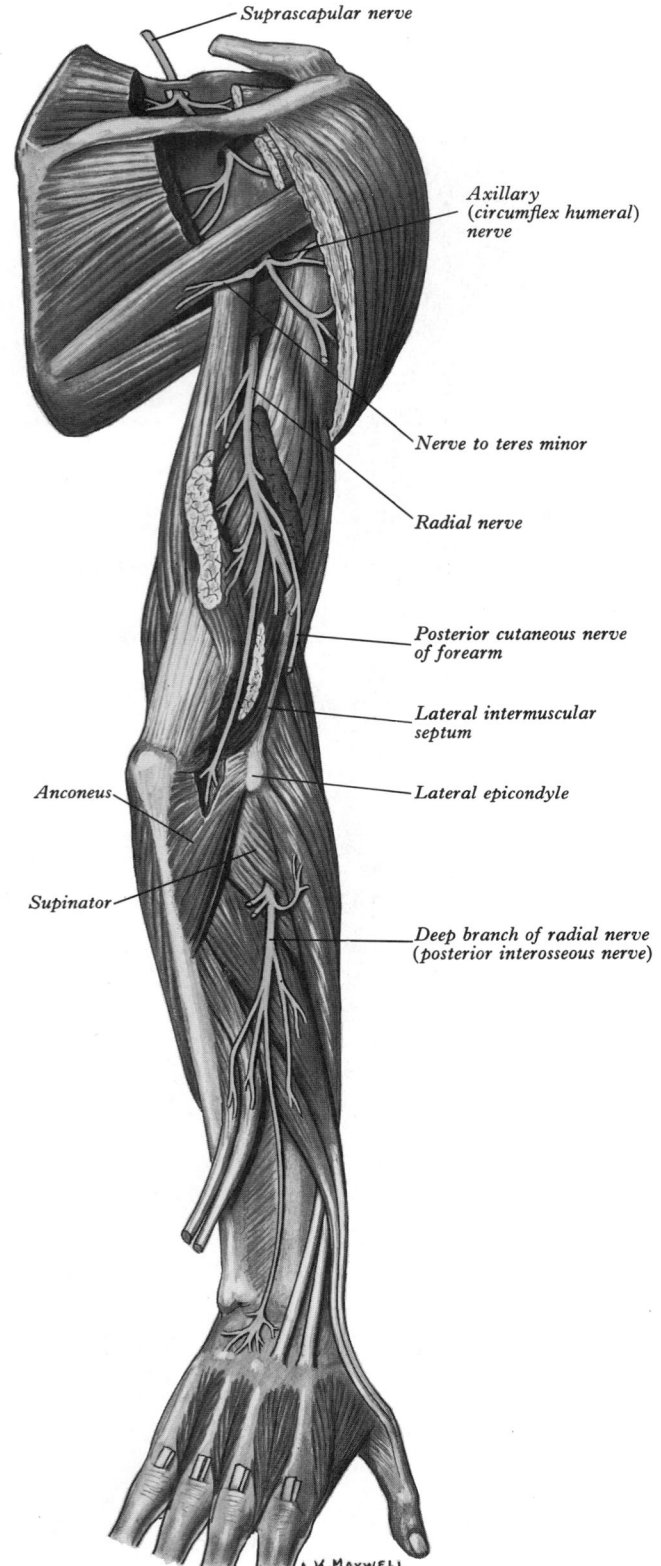

— *Suprascapular nerve*

Axillary (circumflex humeral) nerve

Nerve to teres minor

Radial nerve

Posterior cutaneous nerve of forearm

Lateral intermuscular septum

Anconeus

Lateral epicondyle

Supinator

Deep branch of radial nerve (posterior interosseous nerve)

A·K·MAXWELL

7.248 The suprascapular, axillary and radial nerves of the right upper limb, dissected from the posterior aspect.

ulnar collateral nerve. A large *posterior* muscular branch arises from the nerve as it lies in the humeral groove. It divides to supply the medial and lateral heads of the triceps and the anconeus, that for the latter being a long nerve which descends in the medial head of the triceps and partially supplies it; it is accompanied by the middle collateral branch of the arteria profunda brachii and passes behind the elbow joint to end in the anconeus. *Lateral* muscular branches arise in front of the lateral intermuscular septum; they supply the lateral part of the brachialis, brachioradialis and extensor carpi radialis longus.

The cutaneous branches are the posterior and lower lateral cutaneous nerves of the arm and the posterior cutaneous nerve of the forearm. The small **posterior cutaneous nerve of the arm** arises in the axilla and passes medially to supply the skin on the dorsal surface of the arm nearly as far as the olecranon. It crosses posterior to and communicates with the intercostobrachial nerve. **The lower lateral cutaneous nerve of the arm** perforates the lateral head of the triceps distal to the deltoid tuberosity, passes to the front of the elbow close to the cephalic vein and supplies the skin of the lateral part of the lower half of the arm (7.244). **The posterior cutaneous nerve of the forearm** arises with the preceding nerve. Perforating the lateral head of the triceps, it descends first lateral in the arm, then along the dorsum of the forearm to the wrist, supplying the skin in its course and joining, near its end, with dorsal branches of the lateral cutaneous nerve of the forearm (7.244).

The articular branches are distributed to the elbow joint.

The superficial terminal branch descends from the lateral epicondyle anterolaterally in the proximal two-thirds of the forearm, at first on the supinator, lateral to the radial artery and behind the brachioradialis; in the middle third of the forearm it is behind the brachioradialis, close to the lateral side of the artery, successively anterior to the pronator teres, the radial head of the flexor digitorum superficialis and the flexor pollicis longus. It leaves the artery about 7 cm proximal to the wrist, passes deep to the tendon of the brachioradialis, curves round the lateral side of the radius as it descends, pierces the deep fascia and divides into five, sometimes four, dorsal digital nerves. On the dorsum of the hand it usually communicates with the posterior and lateral cutaneous nerves of the forearm.

The *dorsal digital nerves*, usually four or five, are small. The first supplies the skin of the radial side of the pollex and the adjoining thenar eminence, communicating with branches of the lateral cutaneous nerve of the forearm; the second supplies the medial side of the pollex; the third, the lateral side of the index; the fourth, the adjoining sides of the index and medius; the fifth communicates with a ramus of the dorsal branch of the ulnar nerve and supplies the adjoining sides of the medius and annularis but is frequently replaced by the dorsal branch of the ulnar nerve. The pollicial digital nerves reach only to the root of the nail, those in the index midway along the middle phalanx, those to the medius and the lateral part of the annularis not further than the proximal interphalangeal joints. The remaining distal dorsal areas of the skin in these digits are supplied by palmar digital branches of the median and ulnar nerves (7.245). The superficial terminal branch of the radial nerve may supply the whole dorsum of the hand; for variations consult Sayfi (1967).

The deep terminal branch (posterior interosseous nerve) (7.248) reaches the back of the forearm round the lateral aspect of the radius and between the two planes of the fibres of the supinator. It supplies the extensor carpi radialis brevis and the supinator before entering it; as it traverses the muscle it supplies it with additional branches. The branch to the extensor carpi radialis brevis may arise from the beginning of the superficial branch of the radial nerve. As it emerges from the supinator posteriorly it gives off three short branches, to the extensor digitorum, extensor digiti minimi and extensor carpi ulnaris, and two longer branches, a *medial* to the extensor pollicis longus and extensor indicis and a *lateral*, supplying the abductor pollicis longus and ending in the extensor pollicis brevis. The nerve is at first between the superficial and deep extensor muscles; but at the distal border of the extensor pollicis brevis it passes deep to the extensor pollicis longus and, diminished to a fine thread, descends on the interosseous membrane to the dorsum of the carpus, where it presents a flattened and somewhat expanded termination or 'pseudoganglion', from which filaments supply the carpal ligaments and articulations (7.248).

Articular branches from the deep branch of the radial nerve supply the carpal, distal radio-ulnar and some intercarpal and intermetacarpal joints; digital branches supply the metacarpophalangeal and proximal interphalangeal joints.

Applied Anatomy. The brachial plexus may be injured in falls on the side of the head and shoulder, nerves in the plexus being violently stretched; its upper trunk sustains the greatest injury and paralysis may be confined to muscles supplied by the fifth cervical ramus: the deltoid, biceps, brachialis and brachioradialis, with sometimes also supraspinatus, infraspinatus and supinator. The arm typically hangs in medial rotation, with the forearm extended and pronated. It cannot be abducted; flexion of the elbow and supination of the forearm are lost. This is *Erb's paralysis* and a similar condition occasionally appears in newborn children, from injury to the upper trunk by obstetric forceps or from traction in breech presentations. A second variety is *Klumpke's paralysis*, in which the eighth cervical and first thoracic rami are injured, before or after they join to form the lower trunk. Paralysis mainly affects the intrinsic muscles of the hand and the flexors of the wrist and fingers.

The plexus may also be injured by direct violence or gunshot wounds, by violent traction on the arm or in reducing a glenohumeral dislocation. The extent of paralysis depends on the injury. When the entire plexus is involved, the whole of the upper extremity is paralysed and anaesthetic. Sometimes injury appears to be an avulsion of nerves from the spinal cord; where this involves the first thoracic ramus the ipsilateral pupil may be constricted, due to damage to the preganglionic fibres in it which supply the dilatator pupillae. The plexus is often damaged in the axilla by pressure of a crutch, producing '*crutch paralysis*', in which the radial is most frequently implicated, followed by the ulnar. The median and radial nerves often suffer '*sleep palsies*', paralysis due to pressure during alcoholic or narcotic stupor.

Damage of the long thoracic nerve paralyses the serratus anterior; it may afflict porters carrying heavy weights on the shoulder, the nerve being exposed to injury in the posterior triangle of the neck. The inferior scapular angle is drawn medially by unopposed action of the rhomboids and the levator scapulae and tends to project ('winging' of the scapula) when the horizontal arm is used for forward pushing movements. The arm cannot be raised above the horizontal.

The *axillary (circumflex humeral) nerve*, in its course round the surgical neck of the humerus, is liable to injury in fractures and dislocations; paralysis of the deltoid and anaesthesia of the skin over the lower part of the muscle result. Effective abduction of the arm is impossible. Paralysis of the teres minor occurs but is not easily demonstrated.

The *median nerve* may be involved in forearm injury. When it is divided proximal to its muscular and anterior interosseous branches, flexion of the second phalanges in all digits is lost, and of the terminal phalanges of the index and medius. Terminal phalanges in the annularis and minimus are flexed by the part of the flexor digitorum profundus supplied by the ulnar nerve. Proximal phalanges may be flexed by the interossei. The pollex cannot be opposed or abducted, nor flexed at its interphalangeal

joint (see p. 1134); it remains extended and adducted. Forearm pronation is weakened, for the brachioradialis can effect only mid-pronation. The wrist can be flexed by the flexor carpi ulnaris but this is combined with adduction. Sensation is lost or impaired on the palmar surfaces of all except the medial one and a half digits and on dorsal surfaces of the same over the distal two phalanges, except in the pollex where loss is limited to the dorsum of the distal phalanx. Owing to paralysis of pollicial intrinsic muscles and the unopposed action of the extensor pollicis longus, an '*ape-like*' hand exists. Injury of the median nerve in the mid-forearm may cause only weakness in flexion of the index ('*pointing*' index finger), since the branch to the part of the flexor digitorum superficialis distributed to this finger arises at about that level (p. 1133); more commonly the nerve is injured proximal to the flexor retinaculum, when flexion of the fingers and pronation of the forearm remains intact, unless the flexor tendons are also divided; the main effect of such lesions is usually an inability to oppose the pollex, though the intact abductor pollicis longus and adductor pollicis may combine to imitate this. Deep to the flexor retinaculum the median nerve is in the restricted space between the retinaculum and the carpal bones (carpal tunnel, p. 625). Any condition diminishing this space (e.g. carpal dislocation, arthritis, tenosynovitis, etc.) may cause pressure on the nerve with resultant pain and slight sensory impairment in the digits supplied and sometimes slight wasting of the thenar muscles—the '*carpel tunnel syndrome*'. Commonly no cause is apparent; division of the retinaculum is curative.

The *ulnar nerve* may also be injured in forearm wounds, the commonest site of complete or partial division being behind the medial epicondyle, leading to impaired adduction; in attempts to flex the wrist the hand is abducted by the flexor carpi radialis; owing to paralysis of the dorsal interossei the fingers cannot be spread or flexed at the metacarpophalangeal joints or extended at the interphalangeal joints (especially the annularis and minimus); the hand assumes a 'clawed' shape from the active opposing muscles. Flexion in the fourth and fifth digits is weakened and the pollex cannot adduct. Hypothenar muscles waste. Sensation is lost or impaired, in skin supplied by the nerve.

The *radial nerve* also is often injured but seldom torn in fractures of the humerus, due to its close relationship to this bone. In repair callus formation seldom interferes with the nerve. In fractures of the mid-humerus, the triceps is not paralysed, since its nerve supply arises more proximally from the radial nerve. It may also be contused against bone by kicks or blows or be divided in incised wounds. When paralysed, the hand is flexed at the wrist and flaccid, a condition known as *wrist drop*. The digits are also flexed and in attempted extension only the middle and distal phalanges are extended (by lumbrical and interosseous muscles), the proximal phalanges remaining flexed. Carpal extension is impossible. Supination is completely lost in the extended forearm, but flexion allows the biceps to act as a supinator. Extension of the forearm is lost if injury near the nerve's origin causes the triceps to be paralysed. As the radial nerve has only a small area of exclusive supply, anaesthesia is similarly small and confined to a limited lateral region on the dorsum of the hand.

Dislocations and epiphysial separations at the elbow and supracondylar fractures of the humerus in children often lead to ulnar, median or posterior interosseous injury.

THORACIC VENTRAL RAMI

There are twelve pairs of thoracic ventral rami (7.249, 250); all but the twelfth are between the ribs (*intercostal nerves*), the twelfth being below the last rib (*subcostal nerve*). Each is connected with an adjacent sympathetic ganglion by grey and white rami communicantes: the grey ramus joins the nerve proximal to the exit of the white ramus. Intercostal nerves are distributed chiefly to the thoracic and abdominal walls; the first two supply the upper limb in addition, the next four are limited to the thoracic wall, while the lower five supply the thoracic and abdominal walls (p. 1138). The subcostal nerve supplies the abdominal wall and

gluteal skin. Communicating branches link intercostal nerves posteriorly in the intercostal spaces; the lower five communicate freely in the abdominal wall (Davies et al 1932).

The First to the Sixth Thoracic Ventral Rami

The first thoracic ventral ramus divides unequally; a large branch ascends across the first rib's neck, lateral to the superior intercostal

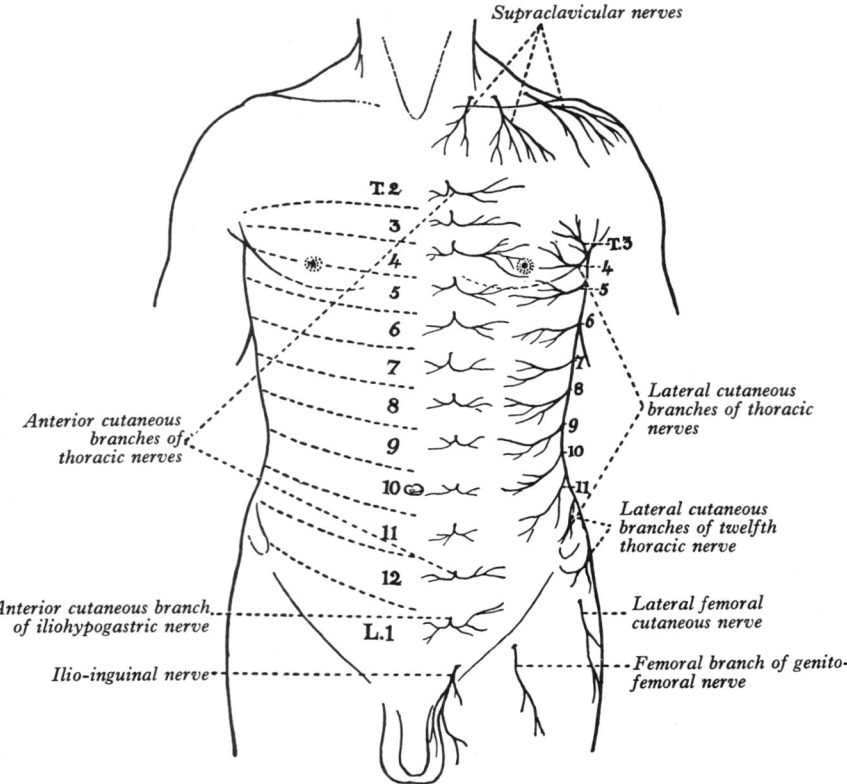

Supraclavicular nerves

T.2

3

4

5

6

7

8

9

10

11

12

L.1

T.3

4

5

6

7

8

9

10

11

Lateral cutaneous branches of thoracic nerves

Anterior cutaneous branches of thoracic nerves

Lateral cutaneous branches of twelfth thoracic nerve

Anterior cutaneous branch of iliohypogastric nerve

Ilio-inguinal nerve

Lateral femoral cutaneous nerve

Femoral branch of genito-femoral nerve

7.249 The approximate segmental distribution of the cutaneous nerves on the front of the trunk. The contribution from the first thoracic spinal nerve is not shown and the considerable overlap which occurs between adjacent segments is not indicated. For the latter see 7.264.

artery, to enter the brachial plexus (p. 1130). The smaller *first intercostal nerve* runs in the first intercostal space and ends as the first anterior cutaneous nerve of the thorax. A lateral cutaneous branch pierces the chest wall in front of the serratus anterior to supply the axillary skin; it may communicate with the intercostobrachial nerve and sometimes joins with the

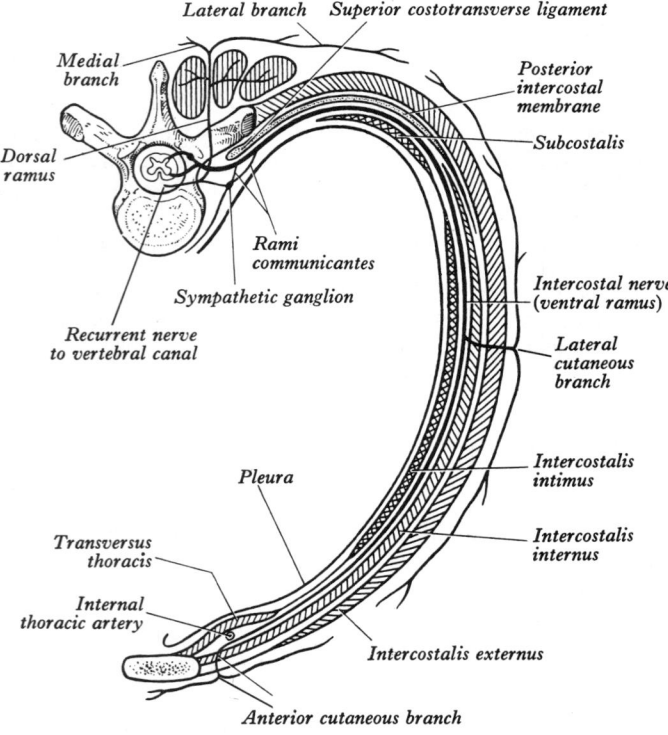

Lateral branch   Superior costotransverse ligament

Medial branch

Posterior intercostal membrane

Dorsal ramus

Subcostalis

Rami communicantes

Sympathetic ganglion

Intercostal nerve (ventral ramus)

Recurrent nerve to vertebral canal

Lateral cutaneous branch

Intercostalis intimus

Pleura

Intercostalis internus

Transversus thoracis

Internal thoracic artery

Intercostalis externus

Anterior cutaneous branch

7.250 The course of a typical intercostal nerve. The muscular and the collateral branches are not shown.

medial cutaneous nerve of the arm (Cave 1929). The first thoracic ramus often connects with the second across the neck of the second rib.

The second to sixth thoracic ventral rami proceed (7.244) in their intercostal spaces below the intercostal vessels. Posteriorly they are between the pleura and posterior intercostal membranes but in most of their course run between the internal intercostals and the subcostales and intercostales intimi (7.249). Near the sternum, they cross anterior to the internal thoracic vessels and the transversus thoracis, pierce the internal intercostals, external intercostal membrane and pectoralis major, ending as the **anterior cutaneous nerves of the thorax**, which supply the skin on the front of the thorax. The second anterior cutaneous nerve may be connected to the medial supraclavicular nerves; twigs from the sixth supply abdominal skin in the upper part of the infrasternal angle.

Branches. Numerous muscular rami supply the intercostals, serratus posterior superior and transversus thoracis. Anteriorly some cross the costal cartilages from one intercostal space to another.

From each intercostal nerve a collateral and a lateral cutaneous branch leave before the main nerve reaches the costal angle. The *collateral branch* follows the inferior border of its space in the same intermuscular plane as the main nerve, which it may rejoin before distribution as an additional anterior cutaneous nerve. The *lateral cutaneous branch* accompanies the main nerve a little way and then pierces the intercostal muscles obliquely; except for the first and second, each divides into anterior and posterior rami which pierce the serratus anterior. *Anterior branches* run forwards over the border of the pectoralis major to supply the overlying skin, the fifth and sixth also supplying twigs to a variable number of upper digitations of the obliquus abdominis externus. *Posterior branches* run back to supply the skin over the scapula and latissimus dorsi.

The second lateral cutaneous branch is the *intercostobrachial nerve* (7.247). It crosses the axilla to the medial side of the arm, joins with a branch of the medial cutaneous nerve of the arm, pierces the deep fascia and supplies the skin of the upper half of the posterior and medial aspects of the arm, connecting with the posterior brachial cutaneous branch of the radial nerve. Its size is in inverse proportion to the size of the medial brachial cutaneous nerve. A second intercostobrachial nerve often branches off from the anterior part of the third lateral cutaneous nerve; it supplies the axilla and the medial side of the arm.

The Seventh to the Twelfth Thoracic Ventral Rami

These lower thoracic ventral rami continue anteriorly from the intercostal spaces into the abdominal wall. Approaching the anterior ends of their spaces, the seventh and eighth curve *supero*medially across the deep costal surface between the digitations of the transverse abdominis to reach the deep aspect of the posterior layer of the internal oblique aponeurosis. Piercing this they are posterior to the rectus abdominis and continue (7.251) for a short distance parallel with the costal margin. Both supply the rectus abdominis and, having traversed it near its lateral edge, pierce its anterior sheath to supply the skin. Both the seventh and eighth nerves cross the costal margin medial to the lateral border of the rectus abdominis and hence enter its sheath from behind.

The ninth to eleventh intercostal nerves pass between digitations of the diaphragm and transversus abdominis to gain the interval between the transversus and the internal oblique. Here the ninth nerve runs forwards almost *horizontally*, while the tenth and eleventh pass *infero*medially. At the lateral edge of the rectus abdominis they pierce the posterior layer of the internal oblique aponeurosis and pass behind the muscle, ending like the seventh and eighth intercostal nerves. The tenth supplies a band of skin which includes the umbilicus (7.249, 264).

These lower intercostal nerves supply intercostal, subcostal and abdominal muscles and the last three supply the serratus posterior inferior. They also provide sensory fibres to the costal

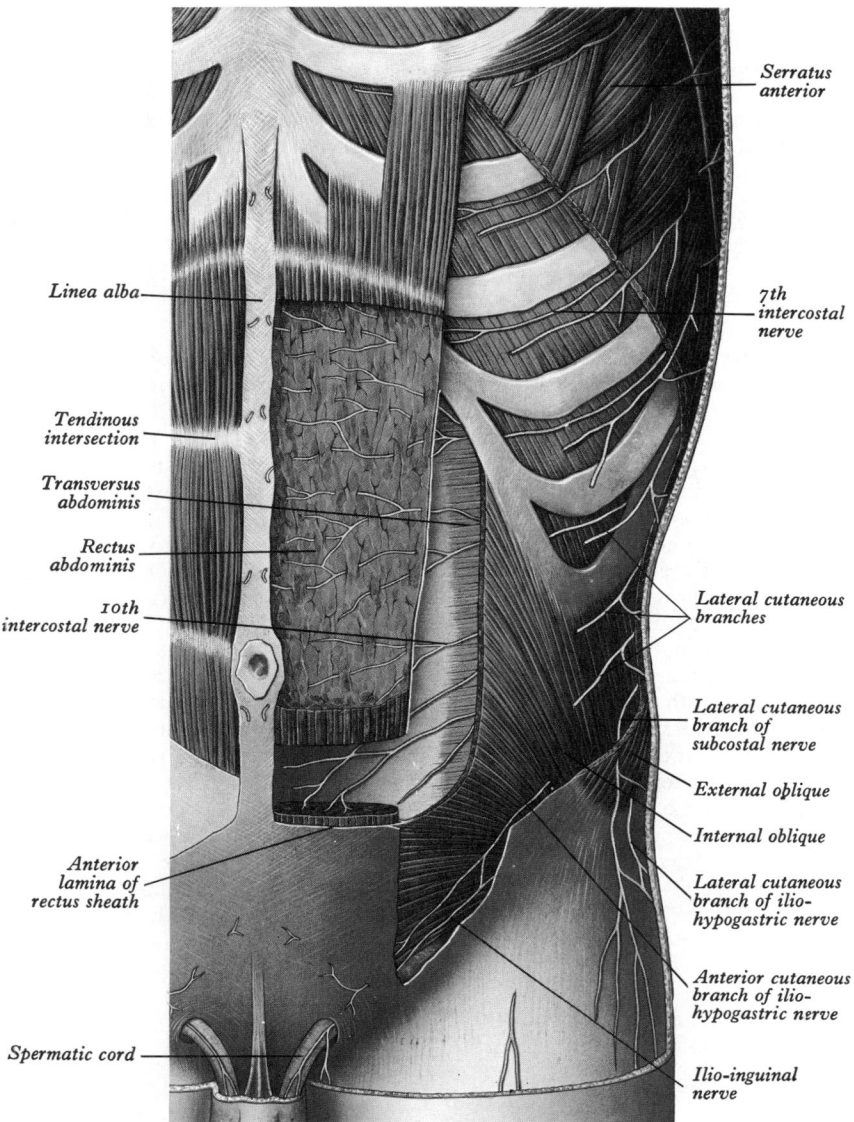

Serratus anterior

Linea alba

7th intercostal nerve

Tendinous intersection

Transversus abdominis

Rectus abdominis

10th intercostal nerve

Lateral cutaneous branches

Lateral cutaneous branch of subcostal nerve

External oblique

Internal oblique

Lateral cutaneous branch of ilio-hypogastric nerve

Anterior lamina of rectus sheath

Anterior cutaneous branch of ilio-hypogastric nerve

Spermatic cord

Ilio-inguinal nerve

7.251 The course of the lower intercostal and the cutaneous branches of some lumbar nerves. Portions of the muscles of the anterior abdominal wall have been removed, including most of the anterior layer of the rectus sheath and parts of the rectus abdominis.

parts of the diaphragm and related parietal pleura and peritoneum. Like the upper intercostal nerves they give off *collateral* and *lateral cutaneous branches* before they reach the costal angles. The collateral may rejoin its nerve but, if it does, it leaves it again near lateral border of the rectus abdominis to run forward below it (7.251), piercing the muscle and its anterior sheath near the linea alba to supply the skin. The lateral cutaneous branches pierce the intercostals and external oblique in the same line as the upper branches, dividing also into anterior and posterior rami, distributed to the skin of the abdomen and back; the anterior also supply digitations of the external oblique, extending antero-inferiorly nearly to the margin of the rectus abdominis; the posterior branches pass back to supply the skin over the latissimus dorsi. Each lateral cutaneous nerve descends as it pierces the external oblique and superficial fascia, reaching the skin on a level with the segment's anterior and posterior cutaneous nerves (p. 1126 and 7.239).

The ventral ramus of the **twelfth thoracic nerve** (subcostal nerve) is larger than the others; it connects with the first lumbar ventral ramus (*dorsolumbar nerve*). Like an intercostal nerve it has a collateral branch. It accompanies the subcostal vessels along the inferior border of the twelfth rib, passing behind the lateral arcuate ligament and kidney (8.168), anterior to the upper part of the quadratus lumborum, perforates the aponeurosis of the origin of the transversus and proceeds

between the transversus and the obliquus internus, to be distributed like the lower intercostal nerves. It connects with the iliohypogastric nerve and supplies the pyramidalis. Its *lateral cutaneous branch* pierces the internal and external oblique muscles, supplies the latter's lowest slip, crosses the iliac crest about 5 cm behind the anterior superior iliac spine (7.258) and supplies the anterior gluteal skin, some filaments reaching the greater trochanter of the femur.

Applied Anatomy. In many conditions affecting spinal rami at or near their origins, pain is referred to their peripheral distributions. Thus, in tuberculosis of the lower thoracic vertebrae, pain is referred to the abdominal wall. When confined to a single pair of nerves the sensation is constrictive, as if a cord were tied round the abdomen; the site of of the fenur this sensation may help to localize vertebral diseases. Where two or more nerves are involved, the pain is more diffused.

Subluxation of the interchondral joints between the lower costal cartilages may trap the intercostal nerves, causing referred abdominal pain. Abrahams (1976) has reviewed the literature on this subluxation or 'clicking rib syndrome'.

Nerves supplying the abdominal skin also supply the planes of underlying muscle, an important fact in protection of the abdominal viscera. A forceful blow will not injure the viscera if the muscles are contracted; but when a blow is unexpected, with the abdominal muscles relaxed, a force insufficient to damage the abdominal wall may rupture some of these viscera. Immediate reflex contraction is obviously important; the origin of the cutaneous and motor fibres from the same spinal segments accelerates the response.

Branches supplying abdominal muscles and skin, derived from from the lower intercostal nerves, are intimately connected with the sympathetic nerves supplying the abdominal viscera through the lower thoracic ganglia, from which the splanchnic nerves are derived. Hence in injuries or acute infection of the abdominal viscera, the muscles of the abdominal wall become firmly contracted, resting and protecting the abdominal contents.

LUMBAR VENTRAL RAMI

Lumbar ventral rami increase in size from first to last and are joined, near their origins, by *grey rami communicantes* from the four lumbar sympathetic ganglia. These rami, long and slender, accompany the lumbar arteries round the sides of the vertebral bodies, behind the psoas major. Their arrangement is irregular: one ganglion may give rami to two lumbar nerves, one lumbar nerve may receive rami from the ganglia; rami often leave the sympathetic trunk between ganglia. The first and second, sometimes the third, lumbar ventral rami are each connected with the lumbar sympathetic trunk by a *white ramus communicans*.

The lumbar ventral rami descend laterally into the psoas major. The first three and most of the fourth form the *lumbar plexus*; the smaller moiety of the fourth joins the fifth as a *lumbosacral trunk*, which joins the *sacral* plexus. The fourth is often termed the *nervus furcalis*, being divided between the two plexuses; but the third is occasionally the nervus furcalis; or both third and fourth may be furcal nerves, when the plexus is termed *prefixed*. More frequently the fifth nerve is furcal, the plexus then being termed *postfixed*. These variations modify the sacral plexus. Piasecka-Kacperska & Gladyskowska-Rzeczycka (1972) have reviewed the variations in primates, including mankind.

The Lumbar Plexus

This plexus (7.252, 253) is in the posterior part of the psoas major, anterior to the lumbar transverse processes and formed by the first three lumbar ventral rami and most of the fourth; the first receives a branch from the last thoracic. The *paravertebral* part of psoas major has a *posterior* mass attached to the transverse processes and an *anterior* mass attached to the lips of the vertebral bodies, intervertebral discs and tendinous arches (p. 635); the lumbar plexus is between these masses and hence in 'line' with the intervertebral foramina.

The plexus varies; in its usual arrangement the first lumbar ramus, joined by a branch from the twelfth thoracic, bifurcates; the upper and larger part divides again into iliohypogastric and ilio-inguinal nerves; the lower unites with a second lumbar branch to form the genitofemoral nerve. The remainder of the second, third and the part of the fourth ramus joining the plexus divide into ventral and dorsal branches. Ventral branches of the second to fourth rami form the obturator nerve. Dorsal branches of the second and third rami each divide into smaller and larger parts; the smaller parts unite as the lateral femoral cutaneous nerve, the larger join with the dorsal branch of the fourth to form the femoral nerve. The accessory obturator, when it exists, arises from the third and fourth ventral branches. For details of the blood supply of the lumbar plexus see Day (1964).

The *branches* of the lumbar plexus may be summarized as follows:

Muscular	T12, L1, 2, 3, 4
Iliohypogastric	L1
Ilio-inguinal	L1
Genitofemoral	L1, 2
	Dorsal divisions:
Lateral femoral cutaneous	L2, 3
Femoral	L2, 3, 4
	Ventral divisions:
Obturator	L2, 3, 4
Accessory obturator	L3, 4

Muscular branches supply the quadratus lumborum (T12, L1–4), psoas minor (L1), psoas major (L2, 3(4)) and iliacus (L2, 3).

The iliohypogastric nerve (L1) (7.247) emerges from the upper lateral border of the psoas major, crosses obliquely behind the lower renal pole and in front of the quadratus lumborum (7.253, 8.168). Above the iliac crest it perforates the posterior part of the transversus abdominis, dividing between this and the internal oblique into lateral and anterior cutaneous branches, also supplying both muscles. The *lateral cutaneous branch* pierces the internal and external oblique muscles above the iliac crest a little behind the iliac branch of the twelfth thoracic nerve; it is distributed to the posterolateral gluteal skin. The *anterior cutaneous branch* (7.249) runs between and supplies the internal oblique and the transversus, pierces the internal oblique about 2 cm medial to the anterior superior iliac spine, and the external oblique aponeurosis about 3 cm above the superficial inguinal ring; it is distributed to the suprapubic skin. The iliohypogastric nerve connects with the subcostal and ilio-inguinal nerves.

The ilio-inguinal nerve, smaller than the iliohypogastric, arises with it from the first lumbar ventral ramus (7.252), emerges from the lateral border of the psoas major, with or just caudal to the iliohypogastric, passes obliquely across the quadratus lumborum and the upper part of the iliacus and perforates the transversus abdominis near the anterior end of the iliac crest, sometimes connecting with the iliohypogastric. It then pierces the internal oblique, supplying it, traverses the inguinal canal below

From 12th thoracic

Iliohypogastric
Ilio-inguinal
Genitofemoral
Lateral cutaneous of thigh
To psoas and iliacus
Femoral
Accessory obturator
Obturator
Lumbosacral trunk

— 1st
— 2nd
— 3rd
— 4th
— 5th

Ventral rami of lumbar spinal nerves

7.252 A plan of the lumbar plexus. The *dorsal divisions* of the second, third and fourth lumbar nerves are shaded.

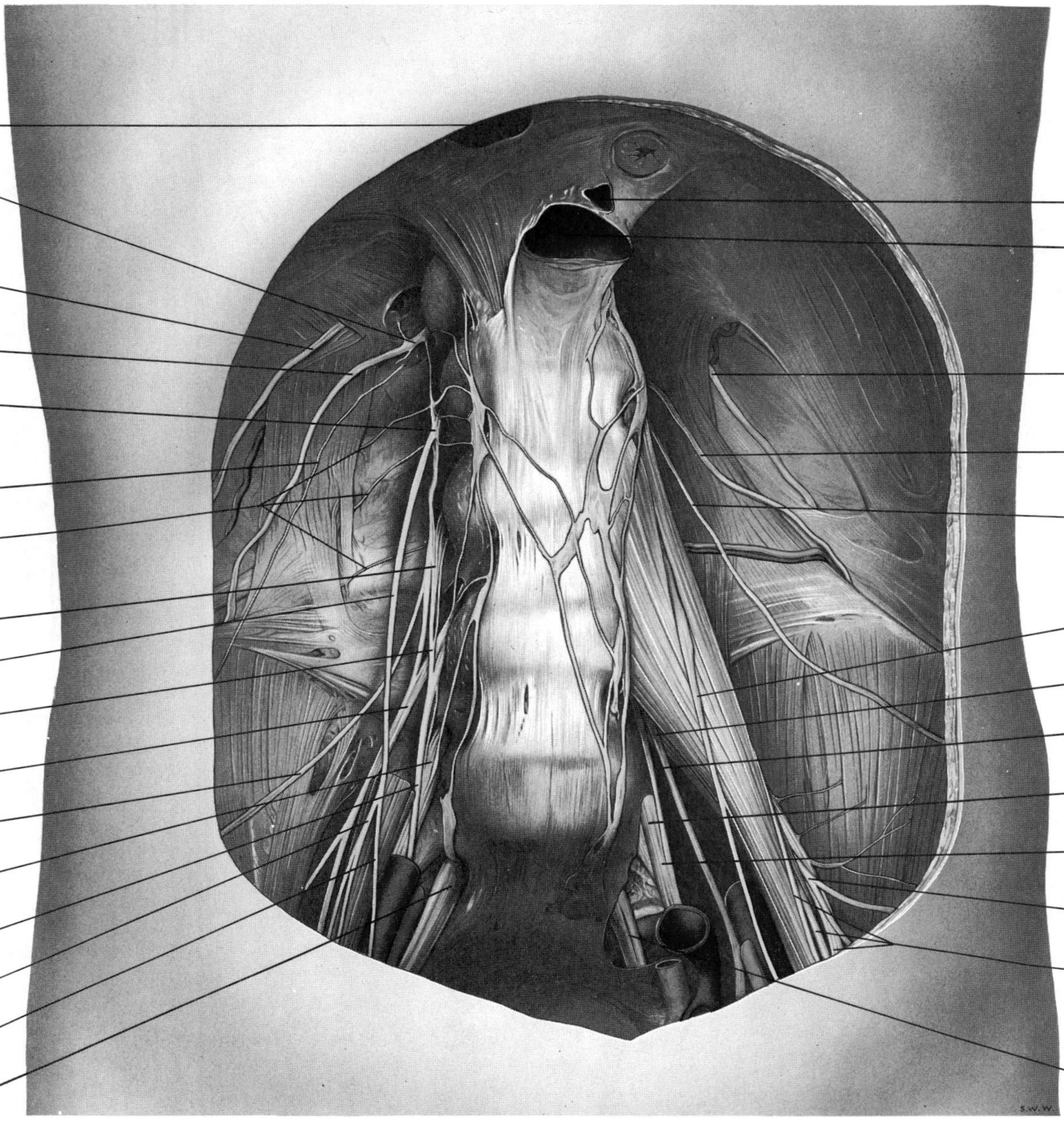
Inferior vena cava
Lumbar 1, anterior primary ramus
Subcostal nerve
Iliohypogastric nerve
Lumbar 2, anterior primary ramus
Ilio-inguinal nerve
Muscular branches
Lumbar 3, anterior primary ramus
Genitofemoral nerve
Lumbar 4, anterior primary ramus
Femoral nerve
Lateral cutaneous nerve of thigh
Lumbar 5, anterior primary ramus
Obturator nerve
Genitofemoral nerve femoral branch
genital branch
Sacral 1, anterior primary ramus
Coeliac artery
Aorta
Subcostal nerve
Iliohypogastric nerve
Ilio-inguinal nerve
Genitofemoral nerve
Obturator nerve
Lumbar 4, anterior primary ramus
Lumbar 5, anterior primary ramus
Lumbosacral trunk
Lateral cutaneous nerve of thigh
Femoral nerve
Sacral 1, anterior primary ramus

7.253 A dissection of the posterior abdominal wall to show the lumbar plexus and sympathetic trunks. The right psoas major has been removed.

the spermatic cord, emerging with it from the superficial inguinal ring to supply the proximomedial skin of the thigh and either that over the penile root and upper part of the scrotum (7.249) or that covering the mons pubis and the adjoining labium majus.

The ilio-inguinal and iliohypogastric nerves are reciprocal in size. The former is occasionally very small and ends by joining the iliohypogastric, a branch of the latter taking its place; or the ilio-inguinal may be absent. By analogy the ilio-inguinal may be regarded as the collateral branch of the first lumbar nerve (Davies 1935) and the iliohypogastric as the main trunk, providing the lateral cutaneous branch.

The genitofemoral nerve (L1, 2) (7.252) descends obliquely forwards through the psoas major, emerging on the abdominal surface near its medial border, opposite the third or fourth lumbar vertebra; it descends subperitoneally on the psoas major, crosses obliquely behind the ureter, dividing variably above the inguinal ligament into genital and femoral branches. It often divides close to its origin, its branches then emerging separately from the psoas major. The *genital branch* crosses the lower part of the external iliac artery, enters the inguinal canal by its deep ring and supplies the cremaster and the scrotal skin. In females it accompanies the round ligament and ends in the skin of the mons pubis and labium majus. The *femoral branch* descends lateral to the external iliac artery, sending a few filaments round it; it then crosses the deep circumflex iliac artery, passes behind the inguinal ligament, enters the femoral sheath lateral to the femoral artery, pierces the anterior layer of the femoral sheath and fascia lata and supplies the skin anterior to the upper part of the femoral triangle (7.254). It connects with the femoral intermediate cutaneous nerve and supplies the femoral artery.

The lateral femoral cutaneous nerve, from the dorsal branches of the second and third lumbar ventral rami (7.252), emerges from the lateral border of psoas major, crossing the iliacus obliquely towards the anterior superior iliac spine. It supplies the parietal peritoneum in the iliac fossa. The right nerve passes posterolateral to the caecum, separated from it by the fascia iliaca and peritoneum; the left passes behind the lower part of the descending colon. Both pass behind or through the inguinal ligament, variably medial to the anterior superior iliac spine (commonly about 1 cm) and anterior to or through the sartorius into the thigh, dividing into anterior and posterior branches (7.254). The *anterior branch* becomes superficial about 10 cm distal to the anterior superior iliac spine, supplying the skin of the anterior and lateral thigh as far as the knee. It connects terminally with the cutaneous branches of the anterior division of the femoral nerve and the infrapatellar branch of the saphenous nerve, forming the *patellar plexus*. The *posterior branch* pierces the fascia lata higher than the anterior, dividing to supply the skin on the lateral surface from the greater trochanter to about mid-thigh. It may also supply the gluteal skin.

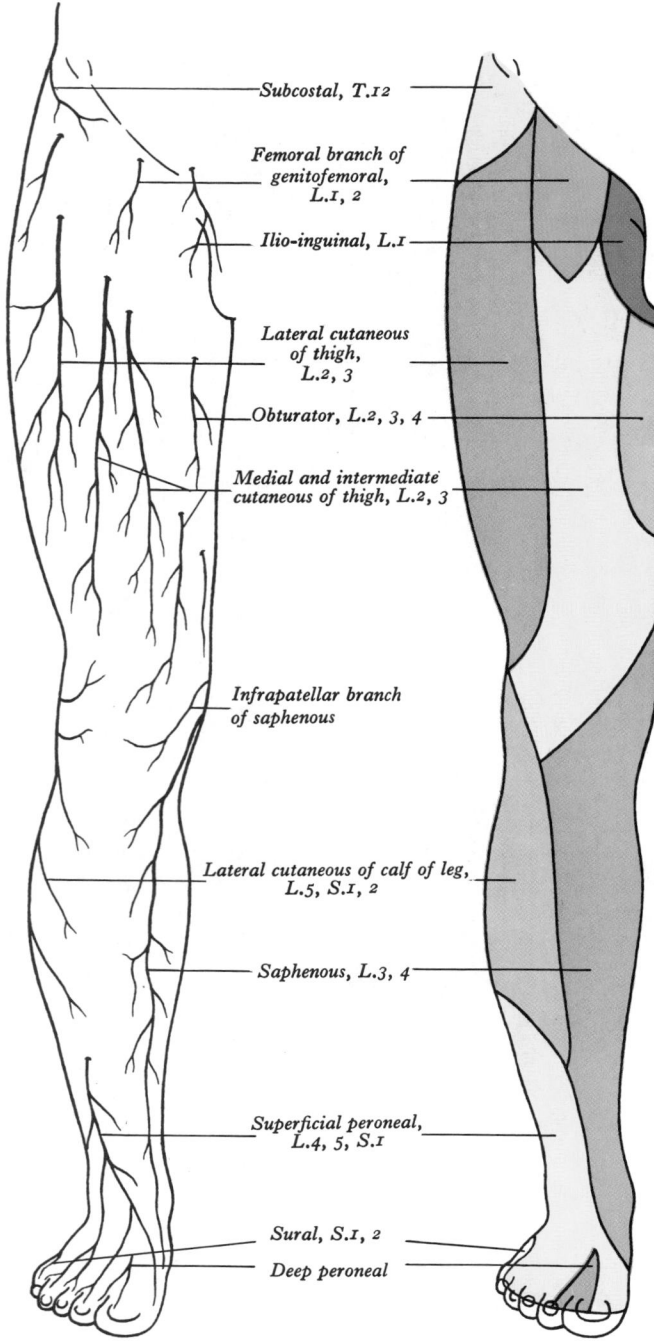

7.254 The cutaneous nerves of the right lower limb, their areas of distribution and segmental origins, viewed from the anterior aspect.

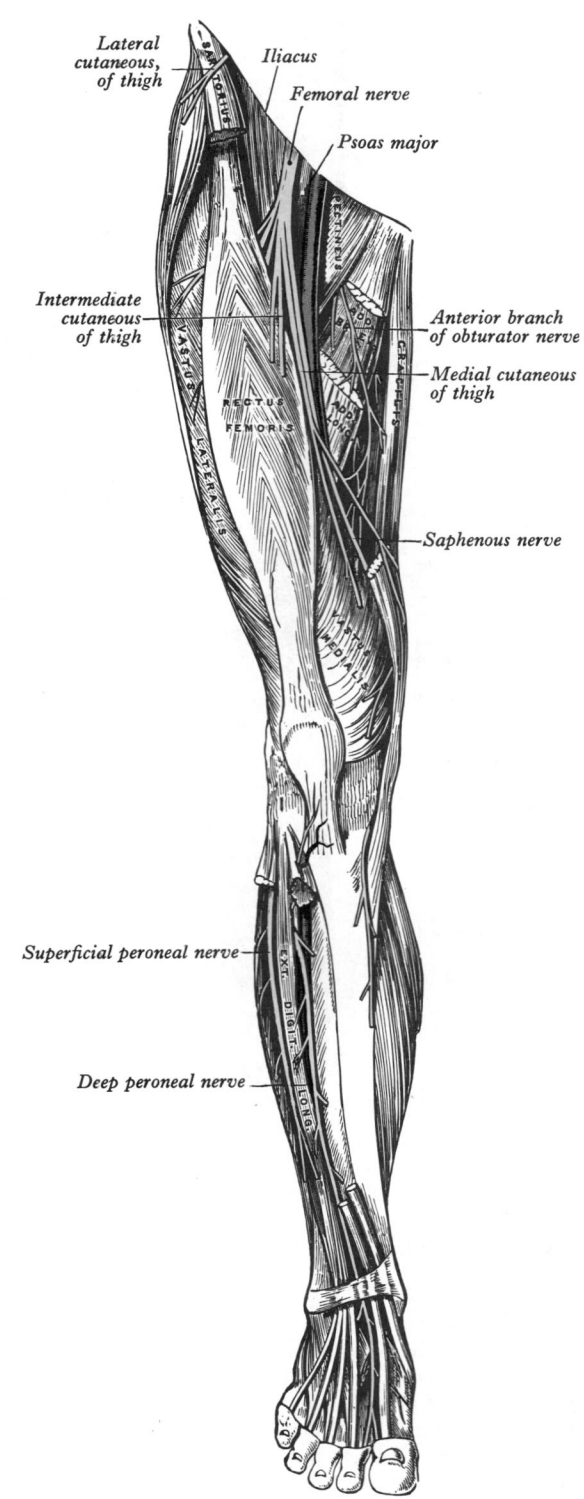

7.255 The nerves of the right lower limb, displayed from the anterior aspect.

THE OBTURATOR NERVE

The obturator nerve arises from the ventral branches of the second to fourth lumbar ventral rami (7.252, 253), that from the third being the largest, the one from the second often very small. It descends in the psoas major, emerging from its medial border at the pelvic brim to pass behind the common iliac and lateral to the internal iliac vessels. It then descends forwards along the lesser pelvic lateral wall on the obturator internus, anterosuperior to the obturator vessels, to the obturator foramen, entering the thigh by its upper part. Near the foramen it divides into anterior and posterior branches, separated at first by part of the obturator externus, lower down by the adductor brevis.

The *anterior branch* (7.255) leaves the pelvis anterior to the obturator externus, descending in front of the adductor brevis, behind the pectineus and adductor longus; at the latter's lower border it communicates with the medial cutaneous and saphenous branches of the femoral nerve, forming a *subsartorial plexus*,

which supplies the skin on the medial side of the thigh (7.254). It descends on the femoral artery, which its termination supplies. Near the obturator foramen the anterior branch supplies the hip joint. Behind the pectineus it innervates the adductor longus, gracilis, usually the adductor brevis and often the pectineus; it connects with the accessory obturator nerve when present. Occasionally the communicating branch to the femoral medial cutaneous and saphenous branches continues as a cutaneous branch to the thigh and leg, emerging from behind the distal border of the adductor longus to descend along the posterior margin of the sartorius to the knee, where it pierces the deep fascia, connects with the saphenous nerve and supplies the skin halfway down the medial side of the leg. The *posterior branch* pierces the obturator externus anteriorly, supplies it and passes behind the adductor brevis to the front of the adductor magnus, dividing into branches to this and the adductor brevis when the latter is not supplied by the anterior division. It usually sends an *articular filament* to the knee joint which perforates the adductor magnus distally or traverses its opening with the femoral artery to enter the popliteal fossa. Here it descends on the popliteal artery to the back of the knee, pierces its oblique posterior ligament and supplies the articular capsule. It gives filaments to the popliteal artery.

The **accessory obturator nerve** (7.252), occasionally present, is small and arises from the ventral branches of the third and fourth lumbar ventral rami. It descends along the medial border of the psoas major, crosses the superior pubic ramus behind the pectineus and divides into branches, one entering the deep surface of the pectineus, another supplying the hip joint and a third connecting with the obturator nerve's anterior branch; sometimes the accessory obturator nerve is very small and supplies only the pectineus. Any branch may be absent and others occur, one sometimes supplying the adductor longus. An accessory obturator nerve appeared in 69 of 800 dissections (p. 640, Woodburne 1960).

THE FEMORAL NERVE

The femoral nerve (7.252, 253), the largest branch of the lumbar plexus, arises from the dorsal branches of the second to fourth lumbar ventral rami. It descends through the psoas major, emerging low on its lateral border, and then passes between the psoas and iliacus, deep to the iliac fascia; passing behind the inguinal ligament into the thigh, it splits into anterior and posterior divisions. Behind the inguinal ligament it is separated from the femoral artery by part of the psoas major. In the abdomen the nerve supplies small branches to the iliacus and pectineus and a branch to the proximal part of the femoral artery; the latter branch may arise in the thigh.

The **nerve to the pectineus** branches from the medial side of the femoral nerve near the inguinal ligament, passes behind the femoral sheath and enters the muscle's anterior aspect.

Anterior division

The *anterior division* of the femoral nerve supplies intermediate and medial cutaneous femoral nerves (7.254, 255) and branches to the sartorius.

The **intermediate femoral cutaneous nerve** pierces the fascia lata about 8 cm below the inguinal ligament, either as two branches or as one trunk which quickly divides into two; these descend on the front of the thigh, supplying the skin as far as the knee and ending in the patellar plexus (p. 1141). The lateral branch of the intermediate cutaneous communicates with the femoral branch of the genitofemoral, frequently piercing the sartorius and sometimes supplying it.

The **medial femoral cutaneous nerve**, at first lateral to the femoral artery, crosses anterior to it at the apex of the femoral triangle, dividing into anterior and posterior branches. Before this it sends a few rami through the fascia lata to supply the skin of the medial side of the thigh, near the long saphenous vein; one

ramus emerges via the saphenous opening, another becomes subcutaneous about mid-thigh. The *anterior branch* descends on the sartorius, perforates the fascia lata beyond mid-thigh and divides into a branch supplying the skin as low as the medial side of the knee and another which crosses to the lateral side of the patella and connects with the saphenous nerve's infrapatellar branch. The *posterior branch* descends along the posterior border of the sartorius to the knee, pierces the fascia lata, connects with the saphenous nerve and supplies several cutaneous rami, some as far as the medial side of the leg. Deep to the fascia lata, at the lower border of the adductor longus, it forms a *subsartorial plexus* with branches of the saphenous and obturator nerves. When the obturator nerve's communicating branch is large and reaches the leg, the posterior branch of the medial cutaneous nerve is small, ending in the plexus and giving rise to a few cutaneous filaments.

The **nerve to the sartorius** arises in common with the intermediate femoral cutaneous nerve.

Posterior division

The *posterior division* of the femoral nerve supplies the saphenous nerve and branches to the quadriceps femoris and the knee joint.

The **saphenous nerve** (7.255), the largest femoral cutaneous branch, descends lateral to the femoral artery into the adductor canal (p. 781), where it crosses anteriorly to become medial to the artery. At the canal's distal end it leaves the artery, emerging through the aponeurotic covering with the saphenous branch of the descending genicular artery. It proceeds vertically along the medial side of the knee behind the sartorius, pierces the fascia lata between the tendons of the sartorius and gracilis and becomes subcutaneous. Thence it descends the leg's medial side with the long saphenous vein along the medial tibial border and divides distally into a branch continuing along the tibia to the ankle and into another passing anterior to the ankle to supply the skin on the medial side of the foot, often as far as the hallucial metatarsophalangeal joint; it connects with the medial branch of the superficial peroneal nerve. Near mid-thigh the saphenous nerve gives a branch to the subsartorial plexus. As it leaves the adductor canal an *infrapatellar branch* (7.254) pierces the sartorius and fascia lata to supply the prepatellar skin; proximal to the knee it connects with medial and intermediate femoral cutaneous nerves while distal to it, it connects with other branches of the saphenous nerve; laterally it connects with the lateral cutaneous femoral nerve, forming a *patellar plexus*.

Muscular branches of the posterior division of the femoral nerve supply the quadriceps femoris. A branch to the rectus femoris enters its proximal posterior surface, also supplying the hip joint. A larger branch to the vastus lateralis forms a neurovascular bundle with the descending branch of the lateral circumflex femoral artery in its distal part and also supplies the knee joint. A branch to the vastus medialis descends through the proximal part of the adductor canal, lateral to the saphenous nerve and femoral vessels; it enters the muscle at about its midpoint, sending a long articular filament distally along the muscle to the knee. Two or three branches to the vastus intermedius enter its anterior surface about mid-thigh; a branchlet from one descends through the muscle to the articularis genu and the knee joint.

Vascular branches of the femoral nerve supply the femoral artery and its branches (p. 1163).

SACRAL AND COCCYGEAL VENTRAL RAMI

The ventral rami of the sacral and coccygeal spinal nerves form the *sacral and coccygeal plexuses*. The upper four sacral ventral rami enter the pelvis by the anterior sacral foramina, the fifth between the sacrum and coccyx, while that of the coccygeal nerve curves forwards below the rudimentary transverse process of the first coccygeal segment. The first and second sacral ventral rami are large, the third to fifth diminish progressively and the coccygeal is the smallest. Each receives a *grey ramus communicans* from a corresponding sympathetic ganglion. *Visceral efferent rami*

leave the second to fourth sacral rami as *pelvic splanchnic nerves* (7.277) containing parasympathetic fibres which reach minute ganglia in the walls of the pelvic viscera (p. 1156).

The Sacral Plexus

The sacral plexus (7.256) is formed by the lumbosacral trunk, the first to third sacral ventral rami and part of the fourth, the remainder of the last joining the coccygeal plexus.

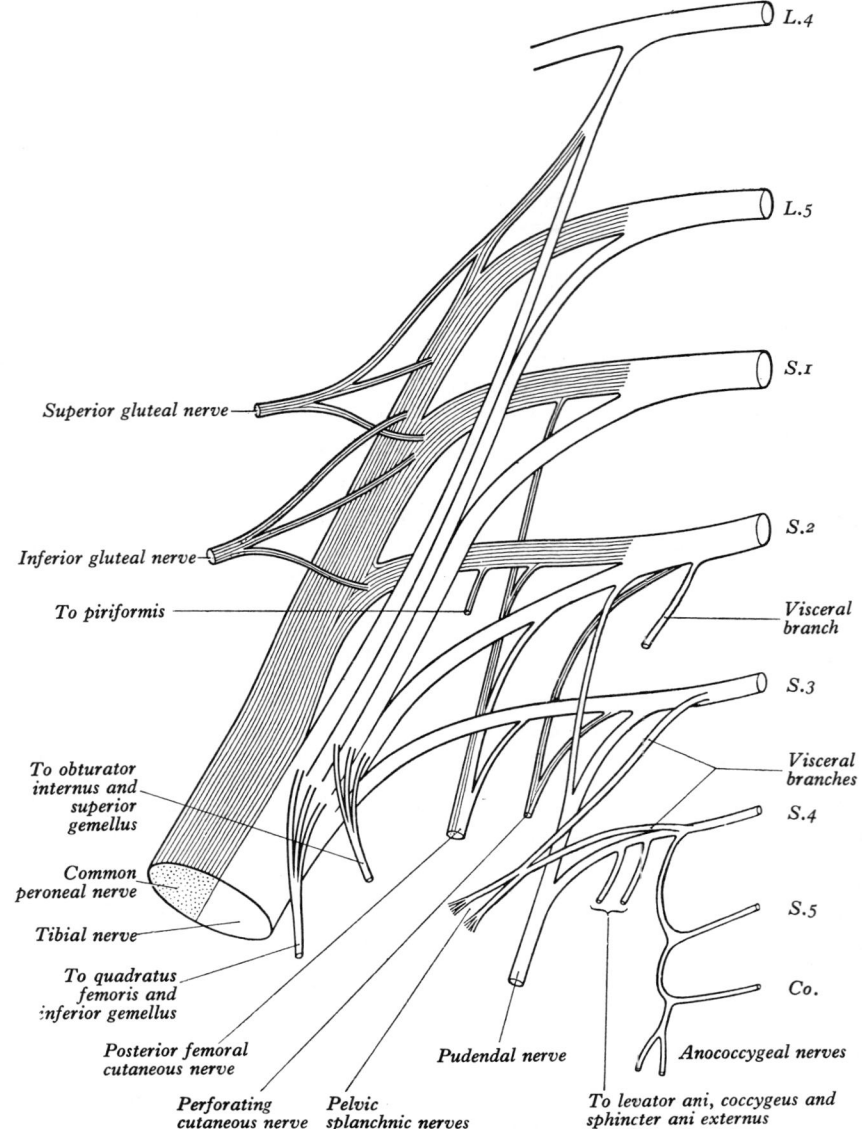

Superior gluteal nerve

Inferior gluteal nerve

To piriformis

To obturator
internus and
superior
gemellus

Common
peroneal nerve

Tibial nerve

To quadratus
femoris and
inferior gemellus

Posterior femoral
cutaneous nerve

Perforating
cutaneous nerve

Pelvic
splanchnic nerves

Pudendal nerve

To levator ani, coccygeus and
sphincter ani externus

Anococcygeal nerves

L.4

L.5

S.1

S.2

Visceral
branch

S.3

Visceral
branches

S.4

S.5

Co.

7.256 A plan of the sacral and coccygeal plexuses. L4–5, S1–5 and Co, indicate the *ventral rami* of these lumbar, sacral and coccygeal spinal nerves. The *ventral divisions* of these rami are unshaded, the *dorsal divisions*, and nerves derived from them, are shaded. The contribution from S2 to the pelvic splanchnic nerves is shown before joining those from 3 and 4.

The lumbosacral trunk comprises part of the fourth and all the fifth lumbar ventral rami; it appears at the medial margin of the psoas major, descending over the pelvic brim anterior to the sacro-iliac joint to join the first sacral ramus. These rami converge to the greater sciatic foramen and unite with little intermingling to form upper and lower bands. The upper, larger one is the union of the lumbosacral trunk with the first, second and greater part of the third sacral rami; it becomes the *sciatic nerve*. The lower band, smaller and more plexiform, is mainly the junction of the smaller part of the third sacral ramus with part of the fourth; it becomes the *pudendal nerve*; it has a small contribution from the second sacral ramus. The sciatic comprises tibial and common peroneal nerves, which usually separate in the thigh but can be pulled apart to their origins, when it can be demonstrated that the tibial is formed by the union of the ventral divisions of the lumbosacral trunk and the first three sacral rami, while the common peroneal is formed by dorsal divisions of the lumbosacral trunk and the first two sacral rami. The sciatic nerve may, however, divide anywhere; when division is at the plexus the common peroneal nerve usually pierces the piriformis in the greater sciatic foramen. (For the blood supply of the sacral plexus see Day 1964.)

Relations of the Sacral Plexus

The sacral plexus adjoins the posterior pelvic wall anterior to the piriformis (7.257), posterior to the internal iliac vessels and ureter and to the sigmoid colon on the left and the terminal ileal coils on the right. The superior gluteal vessels lie between the lumbosacral trunk and first sacral ventral ramus or between the first and second sacral rami, while the inferior gluteal vessels lie between the first and second or second and third sacral rami.

Branches of the Sacral Plexus

These may be summarized as follows:

	Ventral Divisions	Dorsal divisions
To quadratus femoris and gemellus inferior	L4, 5, S1	
To obturator internus and gemellus superior	L5, S1, 2	
To piriformis		S(1), 2
Superior gluteal		L4, 5, S1
Inferior gluteal		L5, S1, 2
Posterior femoral cutaneous	S2, 3	S1, 2
Sciatic—tibial	L4, 5, S1, 2, 3	
—common peroneal		L4, 5, S1, 2
Perforating cutaneous		S2, 3
Pudendal	S2, 3, 4	
To levator ani, coccygeus and sphincter ani externus	S4	
Pelvic splanchnic	S2, 3, (4)	

The nerve to the quadratus femoris and gemellus inferior arises from the ventral branches of the fourth lumbar to the first sacral ventral rami (7.256), leaves the pelvis via the greater sciatic foramen below the piriformis, descends on the ischium deep to the sciatic nerve, gemelli and the tendon of the obturator internus and supplies the gemellus inferior, quadratus femoris and the hip joint.

The nerve to the obturator internus and gemellus superior arises from the ventral branches of the fifth lumbar and first and second sacral ventral rami (7.256), leaves the pelvis similarly to the above, supplies a branch to the upper posterior surface of the gemellus superior, crosses the ischial spine lateral to the internal pudendal vessels, re-enters the pelvis via the lesser sciatic foramen and enters the pelvic surface of the obturator internus.

The nerve to the piriformis usually arises from the dorsal branches of the first and second sacral ventral rami (sometimes only the second) and enters the anterior surface of the piriformis.

The superior gluteal nerve from the dorsal branches of the fourth and fifth lumbar and first sacral ventral rami (7.256) leaves the pelvis via the greater sciatic foramen above the piriformis with the superior gluteal vessels, dividing into superior and inferior branches. The *superior* branch accompanies the upper branch of the deep division of the superior gluteal artery to supply the gluteus medius and occasionally the gluteus minimus. The *inferior* branch runs with the lower ramus of the deep division of the superior gluteal artery across the gluteus minimus, supplying the glutei medius and minimus and ending in the tensor fasciae latae.

The inferior gluteal nerve, from the dorsal branches of the fifth lumbar and first and second sacral ventral rami, leaves the pelvis via the greater sciatic foramen below the piriformis, dividing into branches which enter the deep surface of the gluteus maximus.

The posterior femoral cutaneous nerve, from the dorsal branches of the first and second and the ventral branches of the second and third sacral rami (7.256), issues via the greater sciatic foramen below the piriformis, descends under the gluteus maximus with the inferior gluteal vessels, lying posterior or medial to the sciatic nerve. It descends in the back of the thigh superficial to the long head of the biceps femoris, deep to the fascia lata; behind the knee it pierces the deep fascia and accompanies the short saphenous vein to mid-calf, its terminal twigs connecting with the sural nerve. Its branches are all cutaneous, distributed to the gluteal region, perineum and the flexor aspect of the thigh and leg. Three or four *gluteal branches* curl round the lower border of the gluteus maximus to supply the skin over its inferolateral area. The *perineal branch* supplies the superomedial skin in the thigh, curves forwards across the hamstrings below the ischial tuberosity, pierces the fascia lata and runs in the superficial perineal fascia to the scrotal or labial skin, communicating with the inferior rectal and posterior scrotal branches of the perineal nerve. *Branches to the back of the thigh and leg* are numerous and from both sides of the posterior femoral cutaneous nerve; they supply the skin of the back and medial side of the thigh, the popliteal fossa and the proximal part of the back of the leg (7.258).

The Sciatic Nerve

The sciatic nerve (7.256, 259), 2 cm broad at its origin and the broadest in the body, is the continuation of the upper band of the sacral plexus. It leaves the pelvis via the greater sciatic foramen below the piriformis and descends between the greater trochanter and ischial tuberosity, along the back of the thigh, dividing into the tibial and common peroneal (fibular) nerves, proximal to the knee. Superiorly it is deep to the gluteus maximus, resting first on the posterior ischial surface with the nerve to quadratus femoris between them; it then crosses posterior to the obturator internus and the gemelli, then on to the quadratus femoris, separated by it from obturator externus and the hip joint; it is accompanied medially by the posterior femoral cutaneous nerve and the inferior gluteal artery. More distally it is behind the adductor magnus and is crossed posteriorly by the long head of the

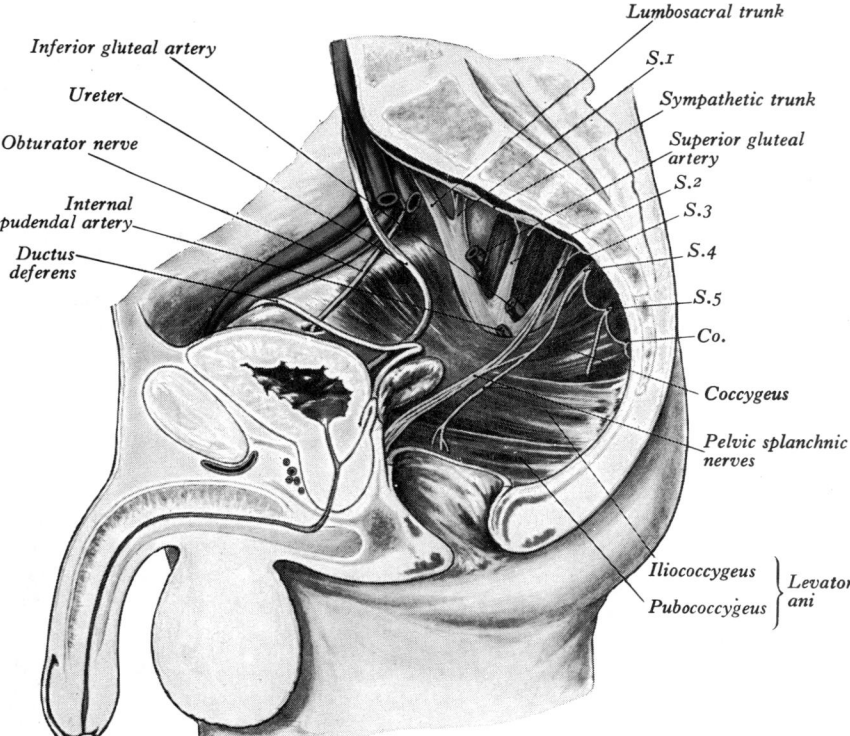

7.257 A dissection of the side wall of the pelvis, showing the sacral and coccygeal plexuses. S.1–5 indicate the ventral rami of the sacral spinal nerves; Co. indicates the ventral ramus of the coccygeal spinal nerve.

biceps femoris. It corresponds to a line from just medial to the midpoint between the ischial tuberosity and greater trochanter to the apex of the popliteal fossa.

Articular branches arise proximally to supply the hip joint through its posterior capsule; these are sometimes derived directly from the sacral plexus. *Muscular branches* are distributed to the biceps femoris, semitendinosus, semimembranosus and the ischial part of the adductor magnus (p. 641), as detailed below.

THE TIBIAL NERVE

The tibial (medial popliteal) nerve (7.259), the larger sciatic division (from the ventral branches of the fourth and fifth lumbar and first to third sacral ventral rami), descends along the back of the thigh and popliteal fossa to the distal border of popliteus, passing anterior to the arch of soleus with the popliteal artery and continuing into the leg. In the thigh it is overlapped proximally by the hamstring muscles, but becomes more superficial in the popliteal fossa, where it is lateral to the popliteal vessels, becoming superficial to them at the knee and crossing to the medial side of the artery (7.260). Distally in the fossa it is overlapped by the junction of the two heads of gastrocnemius. Its surface projection is a line in the midline of the limb, vertical from the apex of the popliteal fossa to the level of the fibular neck and thence to a point midway between the medial malleolus and calcanean tendon.

In the leg the tibial nerve descends with the posterior tibial vessels to lie between the heel and the medial malleolus, ending under the flexor retinaculum by dividing into medial and lateral plantar nerves. Proximally it is deep to the soleus and gastrocnemius, but in its distal third is covered only by skin and fasciae, overlapped sometimes by the flexor hallucis longus. At first medial to the posterior tibial vessels, it crosses behind them, descending lateral to them until it bifurcates. It lies on the tibialis posterior for most of its course, except distally, where it adjoins the posterior surface of the tibia.

Its branches are articular, muscular, sural, medial calcanean and medial and lateral plantar.

Articular branches accompany the superior, inferior medial and middle genicular arteries to the knee joint. The three form a 1145

Labels on figure:
- Lumbosacral trunk
- Inferior gluteal artery
- Ureter
- Obturator nerve
- Internal pudendal artery
- Ductus deferens
- S.1
- Sympathetic trunk
- Superior gluteal artery
- S.2
- S.3
- S.4
- S.5
- Co.
- Coccygeus
- Pelvic splanchnic nerves
- Iliococcygeus
- Pubococcygeus
- Levator ani

plexus with an obturator branch, supplying the oblique posterior ligament; those with the superior and inferior medial genicular arteries supply the medial part of the capsule. Just before the tibial nerve bifurcates it supplies the ankle joint. Champetier & Déscours (1968) have surveyed these articular nerves. Gardner & Lenn (1977) assessed the ratios of myelinated to non-myelinated fibres in genual articular nerves in monkeys as about 4:1; they suggested that most are nociceptive.

Muscular branches, arising between the heads of the gastrocnemius, supply this, the plantaris, soleus and popliteus. The nerve to the soleus enters its superficial aspect; the branch to the popliteus descends obliquely across the popliteal vessels, curling round the muscle's distal border to its anterior surface and also supplying the tibialis posterior, the proximal tibiofibular joint, the tibia and an interosseous branch which descends near the fibula to reach the distal tibiofibular joint.

Muscular branches in the leg, independent or by a common trunk, supply the soleus (on its deep surface), the tibialis posterior, flexor digitorum longus and flexor hallucis longus, the branch to the last-named accompanying the peroneal vessels.

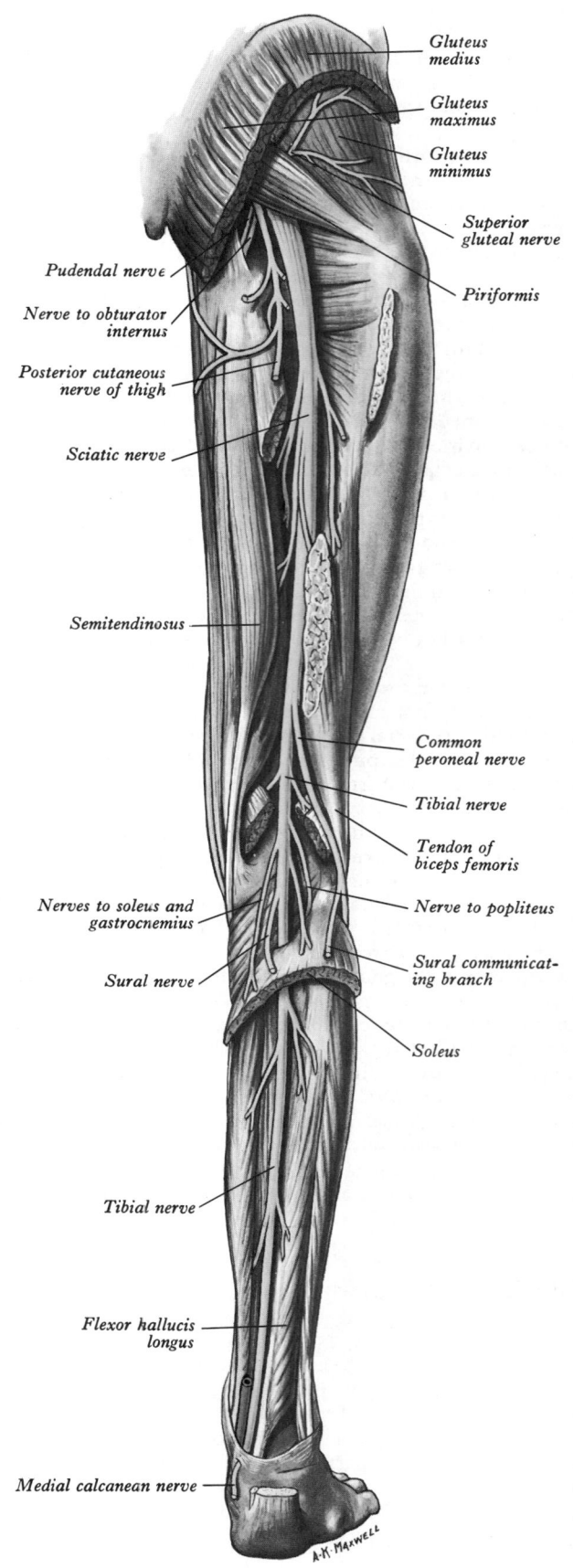

7.258 The cutaneous nerves of the right lower limb and their areas of distribution and segmental origins, viewed from the posterior aspect. The major part of the trunk of the posterior cutaneous nerve of the thigh lies deep to the deep fascia and is therefore shown by an interrupted line.

7.259 The nerves of the right lower limb: posterior aspect. In this figure the gluteus maximus, the gluteus medius and the superficial muscles of the calf of the leg have been removed and the middle part of the long head of the biceps femoris has been excised.

The sural nerve descends between the heads of the gastrocnemius, pierces the deep fascia proximally in the leg and is joined by a sural communicating branch of the common peroneal (7.258). It descends lateral to the tendo calcaneus, near the small saphenous vein, to the region between the lateral malleolus and the calcaneus; it supplies the posterior and lateral skin of the leg's distal third, proceeding distal to the lateral malleolus along the lateral side of the foot and little toe; it connects on the dorsum of the foot with the superficial peroneal nerve, and in the leg with the posterior femoral cutaneous nerve. The fibre spectrum and ultrastructure of fetal sural nerves were described by Ochoa (1971) and in adults by Ochoa & Mair (1969a, b). The three-dimensional distribution of Schwann cells associated with unmyelinated nerve fibres in adult human sural nerves has been analysed by Carlsen & Behse (1980).

Medial calcanean branches perforate the flexor retinaculum to supply the skin of the heel and medial side of the sole.

Vascular branches supply arteries accompanying the tibial nerve and its branches (p. 1163).

The medial plantar nerve (7.261), the larger terminal division of the tibial nerve, lies lateral to the medial plantar artery. From its origin under the flexor retinaculum it passes deep to the abductor hallucis, appears between it and the flexor digitorum brevis, gives off a medial proper digital nerve to the hallux and divides near the metatarsal bases into three common plantar digital nerves.

Cutaneous branches pierce the plantar aponeurosis between the abductor hallucis and the flexor digitorum brevis to supply the skin of the sole of the foot. *Muscular branches* supply the abductor hallucis, flexor digitorum brevis, flexor hallucis brevis and the first lumbrical; the first two arise near the nerve's origin and enter the deep surfaces of the muscles; the branch to the flexor hallucis

brevis is from the hallucial medial digital nerve, and that to the first lumbrical from the first common plantar digital nerve. *Articular branches* supply the joints of the tarsus and metatarsus.

Three common plantar digital nerves pass between the slips of the plantar aponeurosis, each dividing into two proper digital branches, the first supplying adjacent sides of the hallux and secundus, the second adjacent sides of the secundus and tertius, the third adjacent sides of the tertius and quartus and also connecting with the lateral plantar nerve. The first gives a branch to the first lumbrical. Each proper digital nerve has cutaneous and articular branches; near the distal phalanges a dorsal branch supplies structures around the nail; the end of each nerve supplies the ball of the toe. Note that the common digital branches of the medial plantar *are like those of the median nerve;* muscles supplied by both correspond closely. In the hand the median nerve supplies the abductor and flexor pollicis brevis, the opponens pollicis and the first and second lumbricals. The opponens is absent in the foot but the abductor hallucis, flexor hallucis brevis and the first lumbrical are all supplied by the medial plantar.

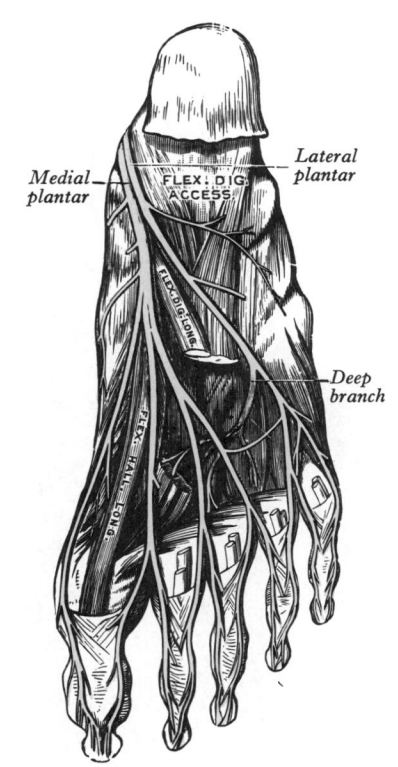

7.261 The plantar nerves of the right foot.

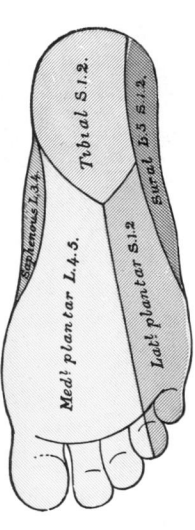

7.262 A diagram showing the distribution of the cutaneous nerves in the sole of the right foot.

Semitendinosus

Popliteal artery

Semimembranosus

Adductor magnus tendon

Superior medial genicular artery

Inferior medial genicular artery

Gastrocnemius, medial head

Popliteus

Soleus

A.K. MAXWELL

Sciatic nerve

Popliteal vein

Superior lateral genicular artery

Common peroneal nerve

Biceps femoris tendon

Sural communicating branch

Inferior lateral genicular artery

Nerve to popliteus

Gastrocnemius, lateral head

Sural nerve

7.260 A dissection of the right popliteal fossa. The two heads of gastrocnemius and the semitendinosus and semimembranosus have been retracted in order to expose the contents of the fossa more fully.

Since the flexor digitorum brevis and flexor digitorum superficialis (median nerve) correspond, only the innervation of the second lumbrical differs.

The lateral plantar nerve (7.261) supplies the skin of the digitus quintus, and of the lateral half of the quartus and most deep muscles, *like the ulnar nerve in the hand*. It passes laterally forwards with the lateral plantar artery lateral to it, towards the fifth metatarsal's tubercle and then between the flexores digitorum brevis and accessorius, to end between the former and the abductor digiti minimi, dividing into superficial and deep branches. Before division it supplies the flexor digitorum accessorius and abductor digiti minimi and gives rise to small branches which pierce the plantar fascia to supply the skin of the lateral part of the sole (7.262). The *superficial branch* splits into two common plantar digital nerves: the *lateral* supplies the lateral side of the digitus quintus, flexor digiti minimi brevis and the two interossei in the fourth intermetatarsal space; the *medial* connects with the third common plantar digital branch of the medial plantar nerve, dividing into two to supply the adjoining sides of the digitus quartus and quintus. The *deep branch* accompanies the lateral plantar artery deep to the flexor tendons and the adductor hallucis, supplying the second to fourth lumbricals, the adductor hallucis and all interossei (except those of the fourth intermetatarsal space); branches to the second and third lumbricals pass

distally deep to the transverse head of the adductor hallucis, curling round its distal border to reach them (7.263).

THE COMMON PERONEAL (FIBULAR) NERVE

The common peroneal (fibular or lateral popliteal) nerve (7.260), about half the size of the tibial, is derived from the dorsal branches of the fourth and fifth lumbar and first and second sacral ventral rami. It descends obliquely along the popliteal fossa's lateral side to the fibular head, medial to the biceps femoris and lying between its tendon and the lateral head of the gastrocnemius. It curves lateral to the fibular neck deep to the peroneus longus and divides into superficial and deep peroneal nerves. Its course can be indicated by a line from the fossa's apex, passing distally medial to the biceps tendon to the back of the fibula's head, where the nerve can be rolled against the bone. Before division it gives off articular and cutaneous branches.

Of the three *articular branches* two accompany the superior and inferior lateral genicular arteries; they may arise in common. The third, termed the recurrent articular nerve, arises near the division of the common peroneal and ascends with the anterior recurrent tibial artery through the tibialis anterior to supply the anterolateral part of the genual capsule and the proximal tibiofibular joint. Two *cutaneous branches* (7.258), often from a common trunk, are the lateral sural and sural communicating nerves.

The *lateral sural nerve* (*lateral cutaneous nerve of the calf*) supplies the skin on the anterior, posterior and lateral surfaces of the proximal leg. The *sural communicating nerve* arises near the fibula's head and crosses the lateral head of the gastrocnemius to join the sural nerve (p. 1147). It may descend separately as far as the heel.

The deep peroneal (anterior tibial) nerve (7.255) begins at the common peroneal bifurcation, between the fibula and the proximal part of the peroneus longus and passes obliquely forwards deep to the extensor digitorum longus to the front of the interosseous membrane, reaching the anterior tibial artery in the proximal third of the leg; it descends with it to the ankle, dividing there into lateral and medial terminal branches. It is first lateral to the artery, then anterior and again lateral at the ankle. It supplies *muscular branches* to the tibialis anterior, extensor hallucis longus, extensor digitorum longus and peroneus tertius and an *articular branch* to the ankle joint.

The *lateral terminal branch* crosses the tarsus deep to the extensor digitorum brevis, enlarges as a pseudoganglion (cf. other sites p. 1132) and supplies the extensor digitorum brevis. From the enlargement three minute *interosseous branches* supply the tarsal and metatarsophalangeal joints of the middle three toes; the first branch supplies the second dorsal interosseous muscle.

The *medial terminal branch* runs distally on the dorsum of the foot lateral to the arteria dorsalis pedis, connecting at the first interosseous space with the medial branch of the superficial peroneal nerve and dividing into two dorsal digital nerves to supply adjacent sides of the hallux and secundus; before dividing it gives off an *interosseous branch* which supplies the hallucial metatarsophalangeal joint and the first dorsal interosseous. The deep peroneal nerve may end as *three* terminal branches (Geller & Barbato 1970).

The superficial peroneal (musculocutaneous) nerve (7.254) begins at the common peroneal bifurcation. It is at first deep to the peroneus longus, passing antero-inferiorly between the peronei and extensor digitorum longus to pierce the deep fascia in the leg's distal third, where it divides into medial and lateral branches. Between the muscles it supplies the peroneus longus, peroneus brevis and the skin of the lower leg. The *medial branch* passes anterior to the ankle, dividing into two dorsal digital nerves, one supplying the medial side of the hallux and the other the adjacent sides of the secundus and tertius. It communicates with the saphenous nerve and deep peroneal nerves (7.254). The smaller *lateral branch* traverses the dorsum of the foot laterally, dividing into dorsal digital branches which supply the contiguous sides of the third to fifth toes, also supplying the skin of the ankle's lateral aspect and connecting with the sural nerve (7.254).

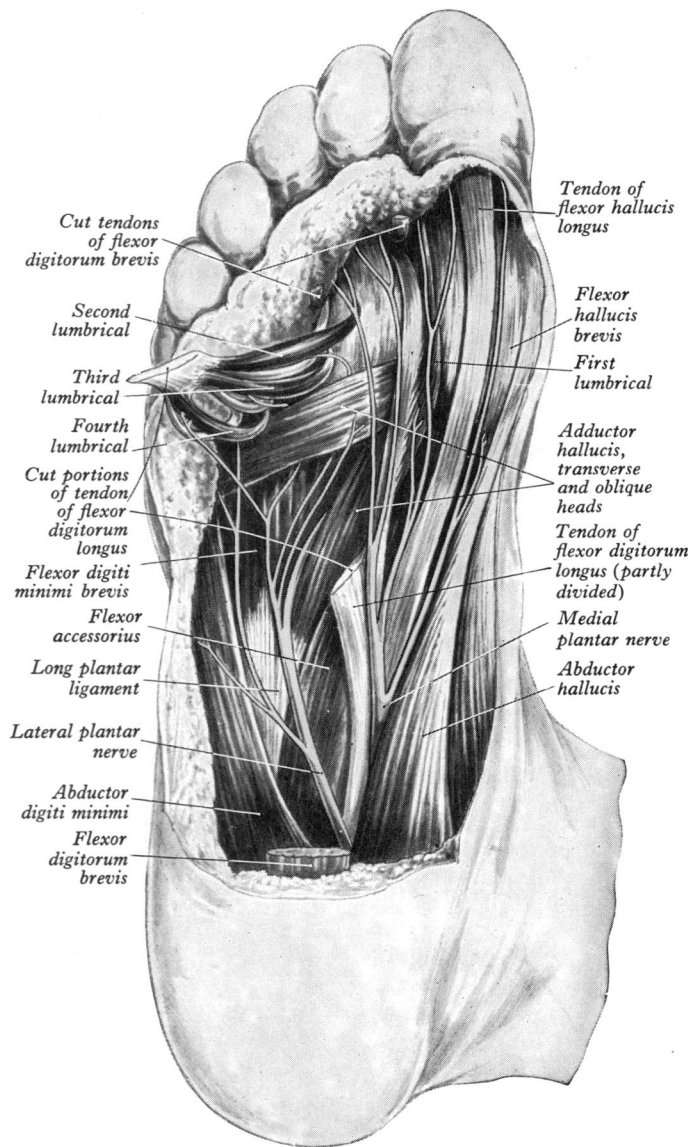

Cut tendons
of flexor
digitorum brevis

Second
lumbrical

Third
lumbrical

Fourth
lumbrical

Cut portions
of tendon
of flexor
digitorum
longus

Flexor digiti
minimi brevis

Flexor
accessorius

Long plantar
ligament

Lateral plantar
nerve

Abductor
digiti minimi

Flexor
digitorum
brevis

Tendon of
flexor hallucis
longus

Flexor
hallucis
brevis

First
lumbrical

Adductor
hallucis,
transverse
and oblique
heads

Tendon of
flexor digitorum
longus (partly
divided)

Medial
plantar nerve

Abductor
hallucis

7.263 A dissection of the lateral and medial plantar nerves of the right foot. Most of the flexor digitorum brevis has been removed. The flexor digitorum longus has been partially divided and its distal end has been displaced together with the second, third and fourth lumbricals.

Superficial peroneal branches supply the dorsal skin of all the toes except that of the lateral side of the fifth and adjoining sides of the hallux and secundus, the former being supplied by the sural, the latter by the medial terminal branch of the deep peroneal. Frequently some of the superficial peroneal lateral rami are absent and replaced by sural branches.

Remaining Branches of the Sacral Plexus

The perforating cutaneous nerve (usually from the posterior aspects of the second and third sacral ventral spinal rami) pierces the sacrotuberous ligament, curves round the inferior border of the gluteus maximus and supplies the skin over the inferomedial aspect of this muscle. The nerve may arise from the pudendal or, if absent, may be replaced by a branch from the posterior femoral cutaneous nerve or from the third and fourth or fourth and fifth sacral ventral rami.

The pudendal nerve (6.121) (derived from the ventral divisions of the second, third and fourth sacral ventral rami) (7.255) leaves the pelvis via the greater sciatic foramen between the piriformis and coccygeus to enter the gluteal region, crossing the sacrospinous ligament close to the attachment to the ischial spine, sited medial to the internal pudendal vessels on the spine. It accompanies the internal pudendal artery through the lesser sciatic foramen into the pudendal canal (p. 606) on the ischiorectal fossa's lateral wall. In the posterior part of the canal it gives off the inferior rectal nerve, which divides into the perineal nerve and the dorsal nerve of the penis (or clitoris).

The *inferior rectal nerve* pierces the medial wall of the pudendal canal, crosses the ischiorectal fossa with the inferior rectal vessels and supplies the sphincter ani externus, the lining of the lower part of the anal canal and the circumanal skin. Its branches connect with the perineal branch of the posterior femoral cutaneous nerve and with the scrotal (or labial) nerves. It sometimes arises directly from the sacral plexus, may perforate the sacrospinous ligament (in 8 out of 40 dissections according to Roberts & Taylor 1973) and may re-connect with the pudendal nerve.

The *perineal nerve*, the inferior and larger terminal pudendal branch, runs forwards below the internal pudendal artery. It accompanies the perineal artery, dividing into posterior scrotal (labial) and muscular branches.

Posterior scrotal or *labial nerves*, medial and lateral, pierce or pass over the inferior fascia of the urogenital diaphragm and run forwards in the lateral part of the urethral triangle with the scrotal (or labial) branches of the perineal artery; they supply scrotal skin or that of the labius majus, connecting with the perineal branch of the posterior femoral cutaneous and inferior rectal nerve.

Muscular branches supply: the transversus perinei superficialis, bulbospongiosus, ischiocavernosus, transversus perinei profundus, sphincter urethrae and the anterior parts of the external sphincter and levator ani. In males a *nerve to the urethral bulb* leaves the nerve to the bulbospongiosus, piercing it to supply the corpus spongiosum penis; it ends in the urethral mucosa.

The *dorsal nerve of the penis* runs forwards in males above the internal pudendal artery along the ischial ramus and the margin of the inferior pubic ramus, deep to the inferior fascia of the urogenital diaphragm. It supplies the corpus cavernosum penis and, at the urogenital diaphragm's apex, is lateral in the hiatus between this diaphragm and the inferior pubic ligament. It accompanies the dorsal penile artery between the layers of the suspensory ligament to the dorsum of the penis, ending in the glans. In females the homologous *dorsal nerve of the clitoris* is very small.

Clinical evidence indicates that the pudendal nerve supplies sensory branches to the lower part of the vagina, probably by fibres in the inferior rectal nerve and posterior labial branches of the perineal nerve. Pudendal nerves can be 'blocked' by infiltration with a local anaesthetic applied via a needle passed through the vaginal wall towards the ischial spine and sacrospinous ligament, both palpable per vaginam (Huntingford 1959, Nakanishi 1967).

Sacral visceral branches arise from the second to fourth sacral ventral rami to innervate the pelvic viscera; they are termed *pelvic splanchnic nerves* (p. 1056).

Sacral muscular branches, from the fourth sacral ventral ramus, supply the levator ani, coccygeus and sphincter ani externus. Those to the levator ani and coccygeus enter the pelvic surfaces of the muscles; the ramus to the sphincter ani externus (*perineal branch of the fourth sacral nerve*) reaches the ischiorectal fossa through the coccygeus or between that muscle and the levator ani, supplying the skin between the anus and coccyx via its cutaneous branches.

The Coccygeal Plexus

The coccygeal plexus is formed by a small descending branch from the fourth sacral ventral ramus and by the fifth sacral and coccygeal ventral rami. The fifth sacral ventral ramus emerges from the sacral hiatus, curving round the lateral margin of the sacrum below its cornu and piercing the coccygeus to reach its pelvic surface, where it is joined by a descending branch of the fourth sacral ventral ramus; the small trunk so formed descends on the pelvic surface of the coccygeus to join the minute coccygeal ventral ramus emerging from the sacral hiatus and curves round the lateral coccygeal margin to pierce the coccygeus and reach the pelvis. This small trunk is the *coccygeal plexus*. Anococcygeal nerves arise from it, a few fine filaments piercing the sacrotuberous ligament to supply the adjacent skin.

APPLIED ANATOMY OF LUMBAR, SACRAL AND COCCYGEAL PLEXUSES

The *iliohypogastric nerve* may be cut (p. 1140) by an incision for appendicectomy; consequent muscular weakness may predispose to the development of a direct inguinal hernia.

The *lateral femoral cutaneous nerve* may be compressed in the inguinal ligament (p. 596) or in the dense fascia lata; the consequent irritation is supposedly a cause of *meralgia paraesthetica*, a rare condition of pain on the lateral side of the thigh.

The *femoral nerve* is occasionally damaged, resulting in paralysis of the quadriceps femoris and sensory loss on the anterior and lateral aspects of the thigh.

The *obturator nerve* is sometimes divided to relieve adductor spasm in spastic paralysis, paraplegia or multiple sclerosis. Because it supplies the hip and knee joints and the medial side of the thigh, hip joint disease may cause referred pain in the thigh (medial side) or knee.

The *sacral plexus* and *lumbosacral trunk* may be compressed by pelvic tumours or the fetal head, causing pain in the lower limbs which, in the former case, may be most severe.

The *sciatic nerve* may be injured in posterior dislocations or fracture dislocations of the hip; if the lesion is complete, which is rare, all the muscles distal to the knee are paralysed and all cutaneous sensibility lost, except in the area supplied by the saphenous nerve. In sciatic nerve lesions in the mid-thigh, flexor muscles often escape because of their more proximal supply.

The *common peroneal* is the most commonly injured nerve in the lower limb because of its exposed position at the fibular neck. Injury here paralyses all the dorsiflexor and evertor muscles of the foot, i.e. the tibialis anterior, extensor hallucis longus, extensor digitorum longus, extensor digitorum brevis, peroneus longus and peroneus brevis, producing a 'drop foot'. There is a variable cutaneous loss on the anterolateral aspect of the leg and dorsum of the foot.

Owing to its deep position the *tibial nerve* is rarely injured although wounds in the popliteal fossa or posterior dislocation of the knee joint may damage it, producing paralysis of the flexors in the leg and of the intrinsic muscles in the sole; loss of sensation in the sole renders it liable to pressure sores.

Loss of sensation in the skin of the gluteal and coccygeal regions through damage to the nerves of supply (vide supra) can also lead to the development of pressure sores.

Morphology of the Spinal Nerves and Limb Plexuses

Spinal nerves conforming to a more primitive arrangement are naturally located in segments largely retaining a *metameric* (segmental) structure, e.g. the thoracic region. Such typical spinal nerves show a common plan. The dorsal, epaxial ramus passes back lateral to the articular processes and divides into medial and lateral branches penetrating the deeper muscles of the back; both innervate the adjacent muscles and supply a band of skin from the posterior median line to the scapular line (vide infra). The ventral, hypaxial ramus is connected to a corresponding sympathetic ganglion by white and grey rami communicantes. It innervates the prevertebral muscles and curves round in the body wall supplying the lateral muscles of the trunk; near the mid-axillary line it gives off a lateral branch which pierces the muscles to divide into anterior and posterior cutaneous branches. The main nerve advances in the body wall, supplying the ventral muscles and terminating in branches to the skin.

The arrangement of ventral rami in segments which have lost their obvious metamerism is greatly modified, especially by the union of adjacent nerves to form cervical, brachial, lumbosacral and coccygeal plexuses.

The cervical plexus. The cutaneous branches are homologous with the anterior terminal and lateral branches of ventral rami of typical spinal nerves. The cervical transverse cutaneous nerve and the medial supraclavicular nerves represent the anterior terminal branches; the lesser occipital and lateral supraclavicular represent lateral branches; the great auricular and intermediate supraclavicular probably represent elements of both.

The brachial plexus. Division of constituent ventral rami into ventral (anterior, flexor) and dorsal (posterior, extensor) branches occurs in both brachial and lumbosacral plexuses. In the brachial plexus division affects all three trunks (p. 1130) and conforms closely to the differentiation of the primitive musculature of the limb into ventral (flexor) and dorsal (extensor) groups. Branches of the ventral divisions largely supply *dorsal* skin in the limb, a discrepancy tentatively explained (Harris 1939) by assuming that each ventral ramus originally divided into ventral and dorsal branches and that, in human evolution, dorsal fibres entered the ventral branches of the *trunks*. As a result, fibres of the median and ulnar nerves have a wide dorsal area of supply in the hand.

The ventrolateral position of the limb buds and the arrangement of the first and second thoracic nerves support the view that the nerves of limb plexuses are only the *lateral* branches of typical ventral rami. The second thoracic ventral ramus gives off a lateral cutaneous branch which goes to the upper limb as the intercostobrachial nerve, its size inversely related to that of the *direct* contribution from the second thoracic to the brachial plexus. The second thoracic is otherwise typical. The first thoracic ventral ramus gives a large contribution to the brachial plexus and this could be homologous with a lateral branch; the remainder, despite its small size, is typical, but its anterior cutaneous branch is reduced and often absent.

The lumbosacral plexus. Division of constituent ventral rami into ventral and dorsal branches is not as clear here as in the brachial plexus, but anatomically the obturator and tibial nerves arise from ventral and the femoral and peroneal nerves from dorsal divisions. Lateral branches of the twelfth thoracic and first lumbar ventral rami are drawn into the gluteal skin but otherwise these nerves are typical. The second lumbar ramus is difficult to interpret; it not only contributes substantially to the lumbar plexus but also has an anterior terminal branch (genital branch of the genitofemoral) *and* a lateral cutaneous branch (lateral femoral cutaneous nerve and the femoral branch of the genitofemoral). Anterior terminal branches of the third to fifth lumbar and first sacral rami are suppressed but the corresponding parts of the second and third sacral rami supply the skin, etc. of the perineum. The posterior femoral cutaneous nerve poses an interesting problem: it arises from both ventral and dorsal divisions and supplies 'flexor' femoral skin *and* 'extensor' gluteal skin.

Segmental Innervation of the Skin

The cutaneous area supplied by *one* spinal nerve, through both rami, is a **dermatome**; typically, dermatomes extend round the body from the posterior to the anterior median line (7.249, 264, 267A,B). Dermatomes of adjacent spinal nerves overlap markedly, particularly in the segments least affected by development of the limbs, i.e. the second thoracic to the first lumbar (7.264, 267A,B).

In some regions, e.g. the upper anterior thoracic wall, cutaneous nerves supplying adjoining areas are *not* from consecutive spinal nerves and the overlap is minimal. When the second thoracic spinal ramus is severed, anaesthesia is sharply demarcated but some overlap for painful and thermal sensibility may exist. Likewise, after section of a peripheral nerve (e.g. the ulnar nerve at the wrist) the area of tactile loss is always greater than that for pain and thermal sensibilities. Hence the area of total anaesthesia and analgesia following section of peripheral nerves is always less than might be anticipated from their anatomical distribution (Foerster 1933, Wollard et al 1940).

Cutaneous Innervation of the Neck and Upper Limbs

The first cervical spinal nerve supplies no skin. The second cervical usually supplies the skin of the head from the vertex to the superior nuchal line, almost all of the lateral aspect of the ear and the skin over the mandibular angle and below the chin (7.224). The third cervical supplies an oblique band from the back of the

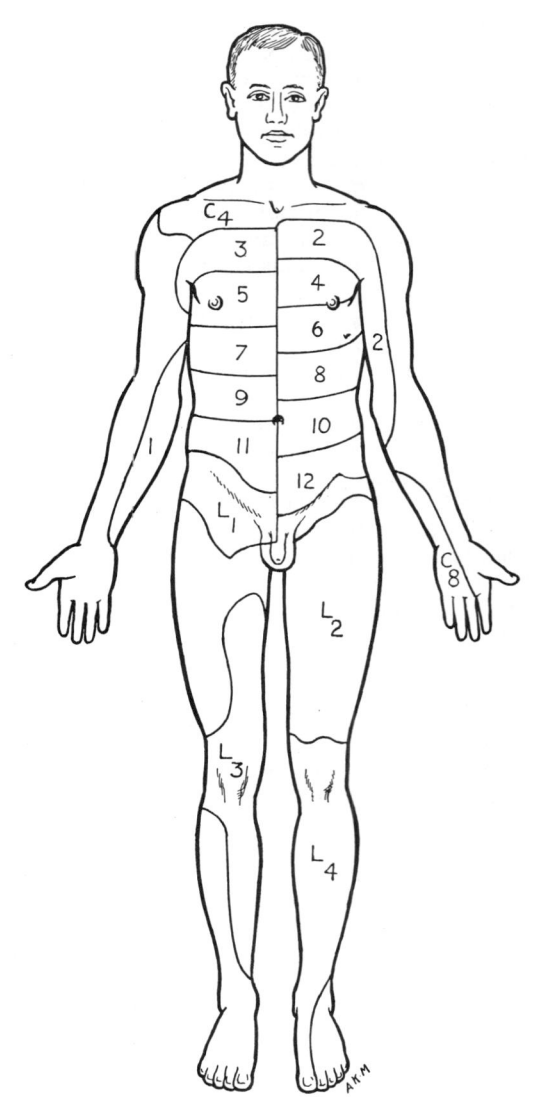

7.264 The cutaneous areas supplied by the ventral rami of the thoracic and upper four lumbar nerves (after Foerster 1933). By comparing both sides the degree of overlapping and the area of exclusive supply of any individual nerve may be estimated. See text for the areas supplied by T1 on the trunk.

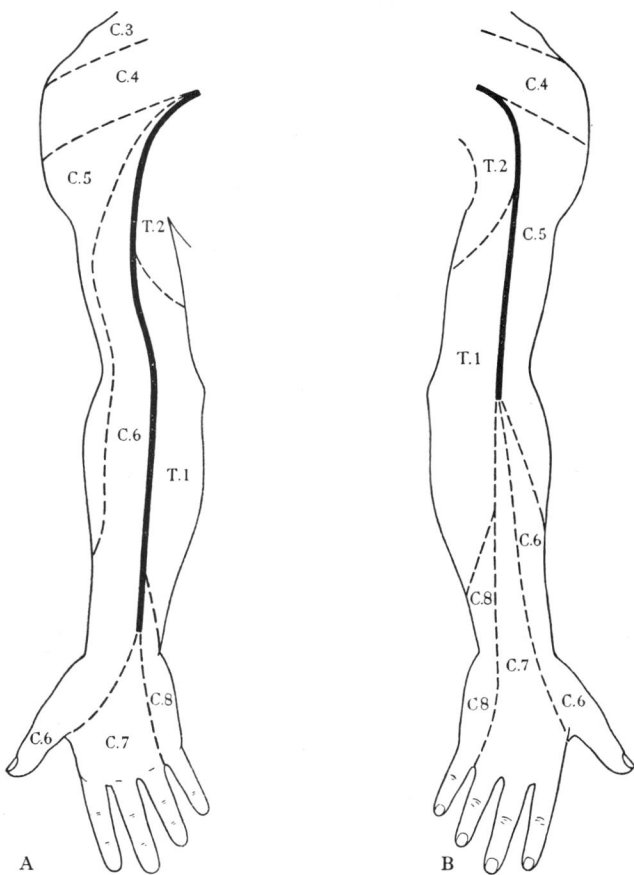

A

B

7.265 A. The arrangement of the dermatomes on the anterior aspect of the upper limb. The heavy black line represents the *ventral axial line* and the overlap across it is *minimal*. Across the interrupted lines, the overlap is considerable. B. The arrangement of the dermatomes on the posterior aspect of the upper limb. The heavy black line represents the *dorsal axial line* and the overlap across it is *minimal*. Across the interrupted lines the overlap may be, and often is, considerable.

distally on the lateral aspect of a supinated limb to the thumb, the *preaxial digit*, while the *postaxial border* runs along the medial aspect to the digitus minimus, the *postaxial digit*. Therefore the cutaneous supply of the lateral aspect of adult limbs is from C4, drawn in at the root of the limb, then from C5 and C6; and its medial aspect is from T2, T1 and C8 (7.265). On the *front of the limb*, areas supplied by C5 and C6 adjoin those supplied by T2, T1 and C8; but at their frontier, the *ventral axial line*, overlap is minimal, for C7 is buried proximally and only reaches the skin a little proximal to the wrist (7.265). On the *back of the limb* the condition is similar; but C7 (in the posterior cutaneous nerve of the forearm) reaches the skin near the elbow, the *dorsal axial line* ending at a more proximal level 7.265). Pronation of the forearm affects these lines (p. 398).

Cutaneous Innervation of the Trunk

The skin of the trunk is supplied by spinal nerves T1–L1, S2–4 and Co1 in consecutive (7.249, 264, 267A,B), curved zones, the upper almost horizontal, the lower oblique. The upper half of

scalp and upper neck, descending across the side of the neck and increasing in width, to the ventral median line from the hyoid bone to the first rib. The fourth cervical supplies the upper half of the back of the neck and an area widening down and forwards round the neck to the anterior thoracic aspect, supplying the skin over the clavicle, first intercostal space, acromion and the upper part of the deltoid. Each of these three areas is overlapped by the succeeding; the overlap is slight but greater for dorsal rami than for ventral.

The cutaneous distribution of those spinal nerves contributing to the brachial plexus is explained by the early development of the upper limbs. In human embryos at the fourth week the upper limb is a small, flattened ventrolateral elevation on the trunk, level with the lower four cervical and first thoracic segments; its ectoderm is continuous with that of the trunk, taking its supply from the nerves of corresponding segments; its mesoderm is also continuous with that in the same segments. Lower limb buds appear slightly later and always lag behind the upper limb buds until after birth.

Limb buds have ventral and dorsal surfaces and *preaxial* (cranial) and *postaxial* (caudal) borders. In the upper limbs the *fifth cervical* ventral ramus supplies a strip of skin on the ventral and dorsal surfaces along the *preaxial* border; the *first thoracic nerve* has a similar distribution along the *postaxial* border. Intervening nerves supply almost parallel strips on the ventral and dorsal surfaces. As the limb elongates, central nerves of the plexus (C6, 7 and 8) become buried proximally, reaching the skin only in distal parts; nerves of segments C4 and T2 and 3 are drawn in to supply proximal skin, i.e. at the root of the limb. During growth the lengthening limb is rotated laterally through about 90° and adducted to the trunk (p. 175). Hence the *preaxial border* runs

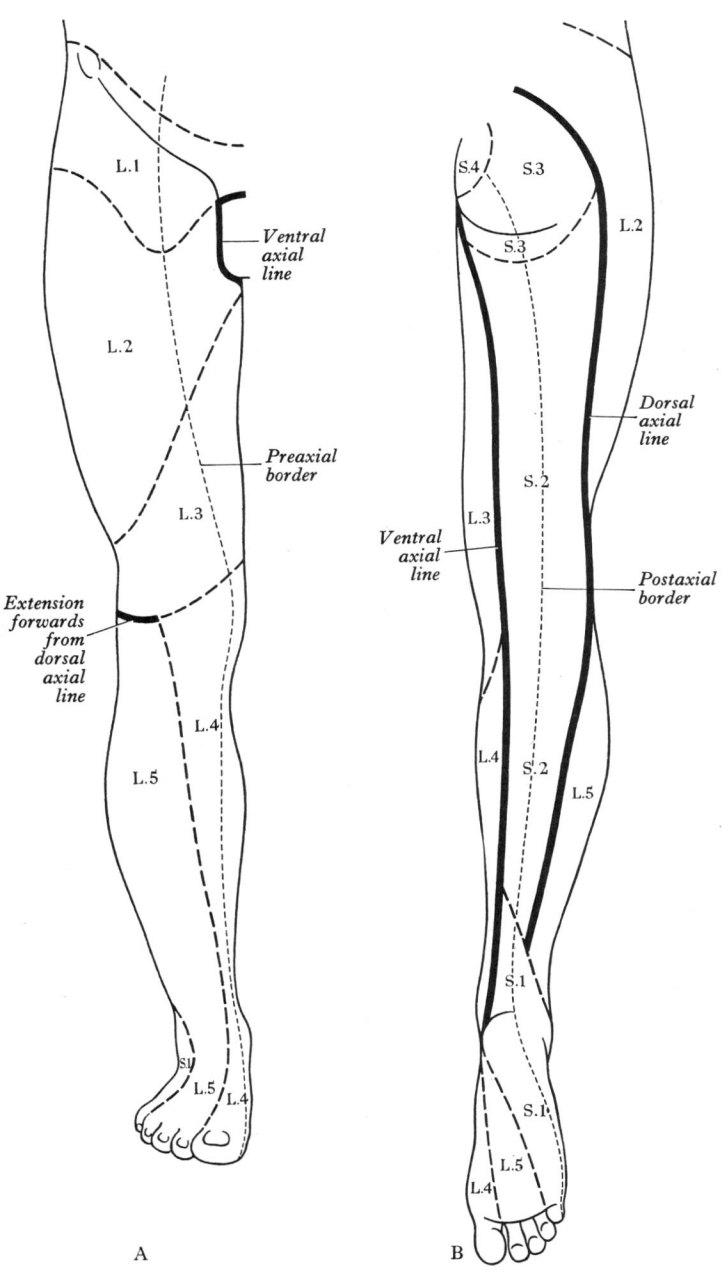

A

B

7.266 A. The segmental distribution of the nerves of the lumbar and sacral plexuses to the skin of the anterior aspect of the lower limb. B. The segmental distribution of the nerves of the lumbar and sacral plexuses to the skin of the posterior aspect of the lower limb. For the significance of the markings see caption to 7.265.

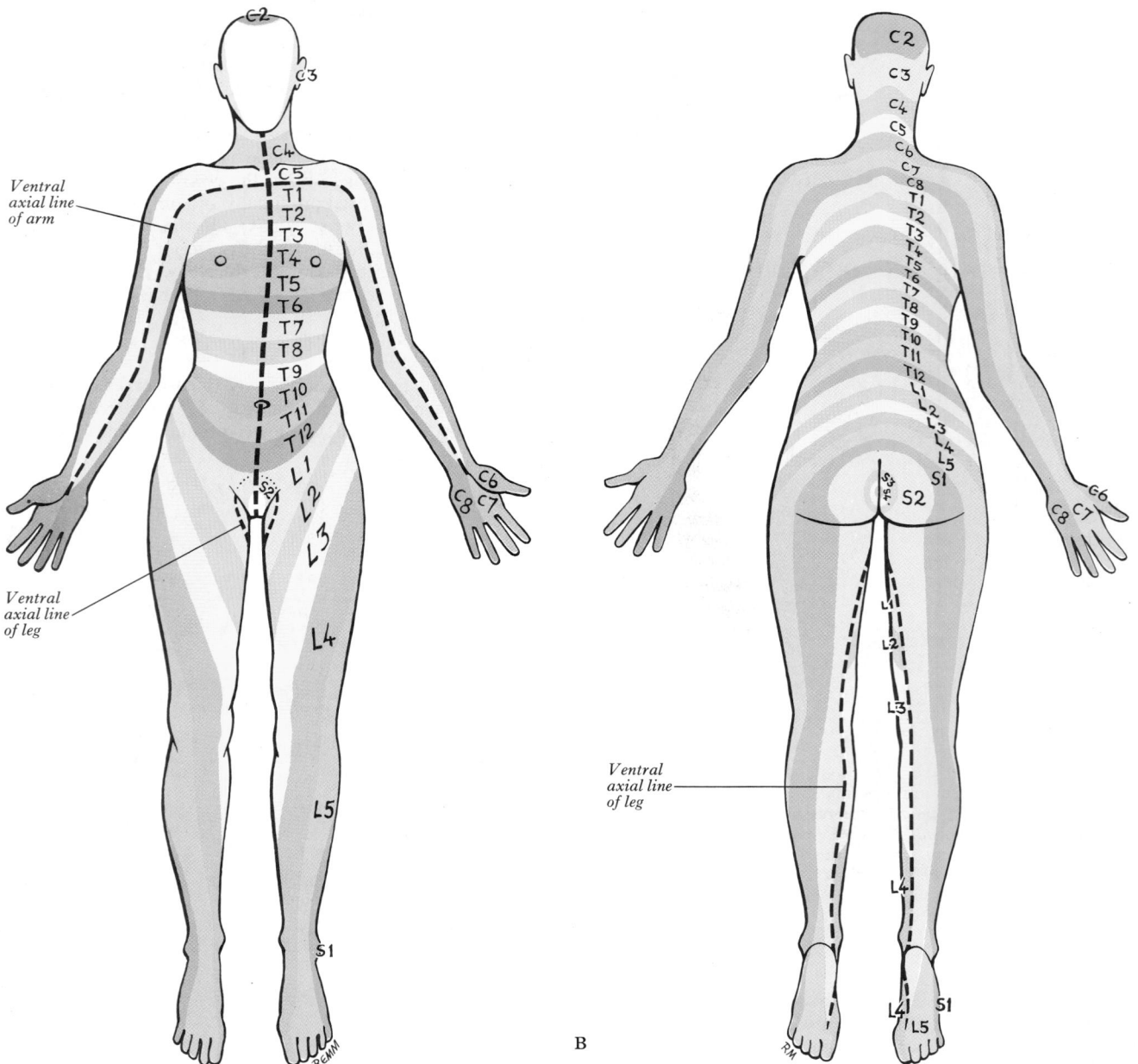

7.267 Ventral (A) and dorsal (B) views of the distributions of dermatomic innervation, modified from the studies of Keegan & Garrett (1948). Diagrams of this kind can only be approximate and cannot clearly indicate the degree of overlap; e.g., the overlap of the supraclavicular nerves (C4 and 5) into the upper pectoral region is not shown and it is possible that the pectoral area assigned to the first thoracic spinal nerve is here exaggerated.

each zone is supplemented by the nerve above, the lower half by the nerve below. No appreciable loss of sensibility follows section of any one spinal nerve. The zone including the subcostal angle is supplied by the seventh thoracic nerve; the umbilicus is in the upper part of a zone supplied by the tenth thoracic.

The area supplied by dorsal rami of these nerves is limited laterally by the dorsolateral line, descending laterally from the occiput to the medial end of the acromion, continued to the posterior aspect of the greater trochanter and curving medially to the coccyx (7.266). Cutaneous strips supplied by dorsal rami do not correspond exactly to those served by ventral rami, differing both in breadth and position.

On the upper ventral thoracic aspect, the third and fourth cervical areas adjoin the first and second thoracic areas (7.249), since the muscles and skin areas supplied by the intervening nerves have grown into the upper limb; a similar, less extensive posterior gap exists.

Similar arrangements exist in the lower trunk, but less obviously, due to approximation of the lower limbs, though apparent in the gluteal region. The first lumbar adjoins the second sacral area

at the root of the penis and scrotum in the male and their homologues in the female (see p. 1446), intervening nerves having been drawn away into the lower limbs during development.

Cutaneous Innervation of the Lower Limb

The skin in the lower limb is innervated by the nerves of its segments of origin, T12–S3 (7.266, 267A,B). The arrangement developmentally begins like that in the upper limbs but its identification in adults is obscured by rotation of the lower limb in early development (p. 175). Initially the *preaxial border* follows the cranial border of the limb bud to the hallux, the *preaxial digit*, the *postaxial border* following its caudal margin to the digitus minimus, the *postaxial digit*. The limb undergoes *medial* rotation, bringing the hallux to lie on the medial side of the foot. The tibia, though homologous with the radius, is also medial. Since rotation occurs at the hip, the gluteal region retains its dorsal (extensor) position.

The *preaxial border* starts near the thigh's mid-point and descends to the knee; it then curves medially, descending to the medial malleolus and the medial side of the foot and hallux. The *postaxial* border commences in the gluteal region and descends to the centre of the popliteal fossa, deviating laterally to the lateral malleolus and the lateral side of the foot. *Ventral* and *dorsal axial lines* exhibit corresponding obliquity, the ventral starting proximally at the medial end of the inguinal ligament, descending the posteromedial aspect of the thigh and leg to end proximal to the heel; the dorsal axial line commences in the lateral gluteal region and descends posterolaterally in the thigh to the knee, inclining medially to end proximal to the ankle (7.266). The segmental cutaneous distribution of the nerves to the lower limb is shown in 7.266.

Knowledge of the extent of individual dermatomes, especially in the limbs, is largely based on clinical evidence, different authorities mapping variable areas for the same dermatomes, partly due to failure to adopt a common method in neurological examination and partly to individual differences in patients with similar lesions. There is more disagreement over dermatomes in the leg, perhaps because injuries to the brachial plexus are more frequent, affording greater opportunities of correlation between cutaneous defects and neural lesions. The figures for limb dermatomes inserted in 7.265, 266 are based on the Medical Research Council's *Report on Peripheral Nerve Lesions* 1942.

In using these figures it must be understood that *broken lines* indicate that the nerves on each side extend considerably beyond them, the overlap being difficult to define. But along *ventral* and *dorsal axial lines*, in *heavy black*, overlap is minimal; the nerves flanking a line are *not* derived from consecutive spinal nerves, the intervening nerve or nerves being buried in the limb and reaching the skin only at more distal points.

Some (Keegan & Garrett 1948) maintain that during the embryonic development of dermatomes in limbs, the sensory nerves grow spirally from the dorsal surface of each limb bud around its preaxial and postaxial borders, to meet anteriorly along the ventral axial line; they deny the existence of a dorsal axial line. By plotting areas of hyposensitivity, particularly hypalgesia following damage to *individual* nerve roots, they have constructed charts of limb dermatomes differing from those in 7.265, 266. Their views (7.267A,B) were based on many observations in hypalgesia due to herniation of intervertebral discs, including 165 cases in the upper limbs (47 verified surgically) and 1264 in the lower limbs (707 verified). They show dermatomes as more continuous strips extending from the trunk along the limbs.

Segmental Innervation of the Muscles

Each spinal nerve originally supplies the musculature derived from its own myotome. Where myotomal derivatives remain entities they retain the original segmental supply; but when derivatives from adjoining myotomes fuse, the resulting muscles do not always retain a supply from each corresponding spinal nerve, although they often do. Since muscles develop in situ in mesodermal cores of developing limbs, it is impossible to identify their original segments merely by a developmental study. The union of spinal nerves and branches in brachial and lumbosacral plexuses renders identification of segmental values impossible. The very concept of immutability of neuromuscular linkage, originally promoted by Furbinger (1873), has been criticized and exceptions recorded (Haines 1935, Minkoff 1974).

Segmental Innervation of the Muscles in Limbs

Most muscles in limbs are innervated from more than one segment of the spinal cord; the segments involved for individual muscles are noted in Myology (p. 569). In the list below the *predominant segmental origin* of the nerve supply is stated; damage to these segments or their motor roots results in maximum paralysis. Data are based chiefly on clinical evidence (Bumke & Foerster 1936, Villiger 1946, Sharrard 1955) but opinion differs for some muscles and not all are included. Though evidence for some muscles is incontrovertible, it is scant and uncertain in many instances. An important example is provided by the intrinsic

muscles of the hand, often regarded clinically as innervated solely by the first thoracic spinal segment, though the eighth cervical is also involved (vide infra).

Upper Limb Muscles

C3, 4 Trapezius, levator scapulae.

C5   Rhomboids, deltoids, supraspinatus, infraspinatus, teres minor, biceps.

C6   Serratus anterior, latissimus dorsi, subscapularis, teres major, pectoralis major (clavicular head), biceps, coracobrachialis, brachialis, brachioradialis, supinator, extensor carpi radialis longus.

C7   Serratus anterior, latissimus dorsi, pectoralis major (sternal head), pectoralis minor, triceps, pronator teres, flexor carpi radialis, flexor digitorum superficialis, extensor carpi radialis longus, extensor carpi radialis brevis, extensor digitorum, extensor digiti minimi.

C8   Pectoralis major (sternal head), pectoralis minor, triceps, flexor digitorum superficialis, flexor digitorum profundus, flexor pollicis longus, pronator quadratus, flexor carpi ulnaris, extensor carpi ulnaris, abductor pollicis longus, extensor pollicis longus, extensor pollicis brevis, extensor indicis, abductor pollicis brevis, flexor pollicis brevis, opponens pollicis.

T1   Flexor digitorum profundus, intrinsic muscles of the hand (except abductor pollicis brevis, flexor pollicis brevis, opponens pollicis).

Lower Limb Muscles

L1   Psoas major, psoas minor.

L2   Psoas major, iliacus, sartorius, gracilis, pectineus, adductor longus, adductor brevis.

L3   Quadriceps, adductors (magnus, longus, brevis).

L4   Quadriceps, tensor fasciae latae, adductor magnus, obturator externus, tibialis anterior, tibialis posterior.

L5   Gluteus medius, gluteus minimus, obturator internus, semimembranosus, semitendinosus, extensor hallucis longus, extensor digitorum longus, peroneous tertius, popliteus.

S1   Gluteus maximus, obturator internus, piriformis, biceps femoris, semitendinosus, popliteus, gastrocnemius, soleus, peronei (longus and brevis), extensor digitorum brevis.

S2   Piriformis, biceps femoris, gastrocnemius, soleus, flexor digitorum longus, flexor hallucis longus, some intrinsic foot muscles.

S3   Some intrinsic foot muscles (except abductor hallucis, flexor hallucis brevis, flexor digitorum brevis, extensor digitorum brevis).

(See also p. 930 for table of lower limb innervation.)

Joint Movements

In terms of movements of joints, segmental innervation of limb muscles may be expressed in general as follows:

Shoulder:	Abductors and lateral rotators.	C5
	Abductors and medial rotators.	C6, 7, 8
Elbow:	Flexors.	C5, 6
	Extensors.	C7, 8
Forearm:	Supinators.	C6
	Pronators.	C7, 8
Wrist:	Flexors and extensors.	C6, 7
Digits:	Long flexors and extensors.	C7, 8
Hand:	Intrinsic muscles.	C8, T1
Hip:	Flexors, adductors, medial rotators.	L1, 2, 3
	Extensors, abductors, lateral rotators.	L5, S1
Knee:	Extensors.	L3, 4
	Flexors.	L5, S1
Ankle:	Dorsiflexors.	L4, 5
	Plantar flexors.	S1, 2
Foot:	Invertors.	L4, 5
	Evertors.	L5, S1
	Intrinsic muscles	S2, 3

THE AUTONOMIC NERVOUS SYSTEM

The autonomic nervous system possesses includes both central and peripheral components, the latter being concerned with the innervation of the viscera, glands, blood vessels and non-striated muscle. It therefore forms the visceral (splanchnic) component of the nervous system. The term 'autonomic' is a convenient rather than appropriate title. The *autonomy* of this part of the nervous system is illusory, since it is intimately responsive to changes in somatic activities; while its connections with somatic elements are not always structurally clear, the functional evidence for visceral reflexes stimulated by somatic events is abundant. (For general information consult Sheehan 1936, White et al 1952, Kuntz 1953, Mitchell 1953, 1956, Pick 1970.)

Visceral efferent paths differ from their somatic equivalents in being interrupted by peripheral synapses, at least two neurons being interposed between the central nervous connections and

visceral effectors (7.268). The somata of the primary neurons are in the visceral efferent nuclei of cranial nerves and in the spinal lateral grey columns; their axons, variably but usually finely myelinated, traverse the cranial and spinal nerves to the peripheral ganglia, where they synapse with the dendrites of somata of secondary neurons. Axons of secondary, effector neurons are usually non-myelinated and supply non-striated muscle or glandular cells. There are therefore in peripheral efferent pathways *preganglionic* and *postganglionic neurons*, the latter being more numerous; one preganglionic neuron may synapse with 15–20 postganglionic neurons, permitting the wide diffusion of many autonomic effects. This disproportion between preganglionic and postganglionic neurons is said to be greater in the sympathetic than in parasympathetic parts of the autonomic nervous system. (In an investigation into human superior cervical ganglia, a ratio of preganglionic

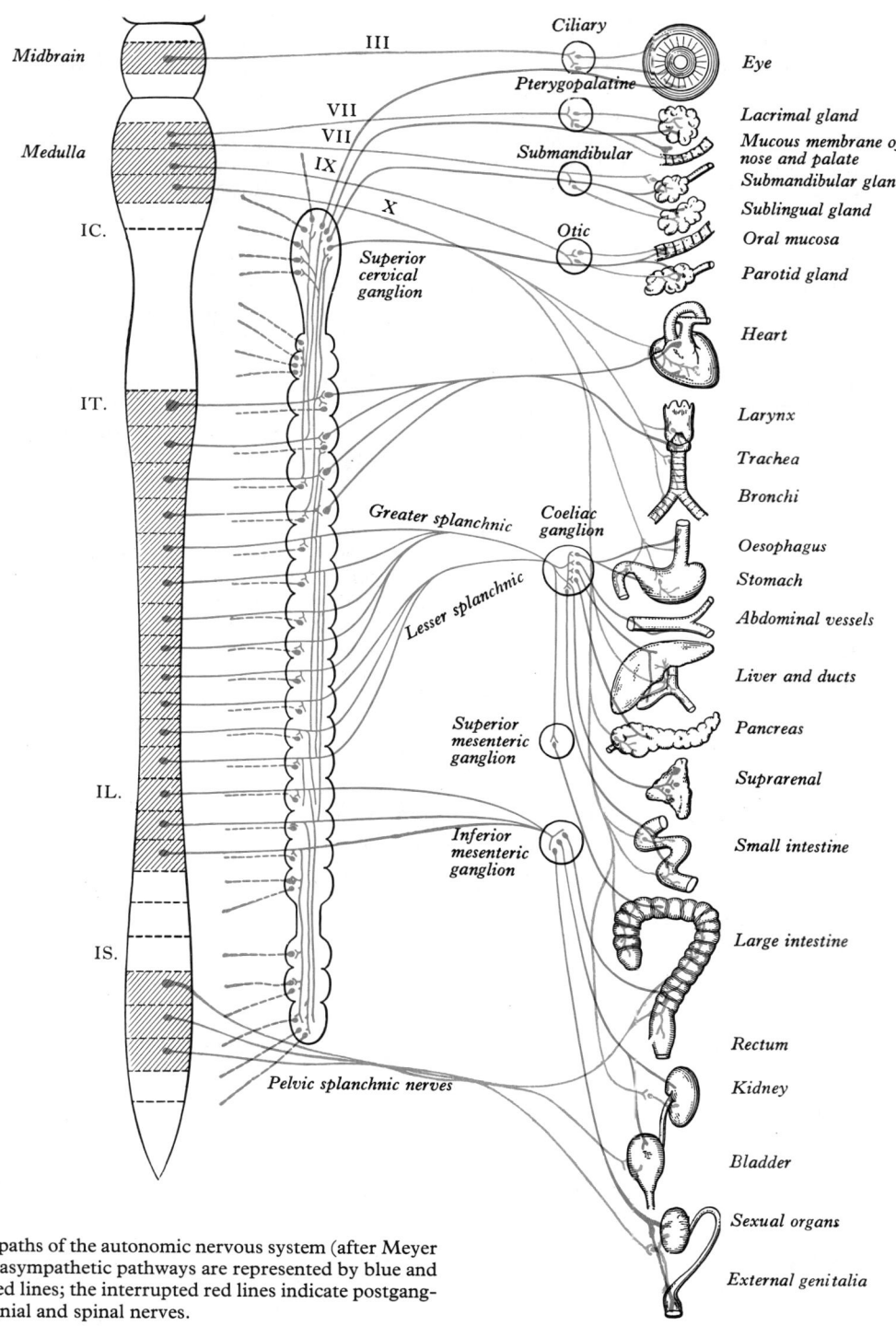

7.268 The efferent paths of the autonomic nervous system (after Meyer & Gottlieb). The parasympathetic pathways are represented by blue and the sympathetic by red lines; the interrupted red lines indicate postganglionic rami to the cranial and spinal nerves.

to postganglionic fibres of 1 to 196 was claimed, Ebbesson 1968.) Terminations of postganglionic neurons are described on p. 905. For the structure of sympathetic ganglia and details of other neuronal types, including interneurons, see p. 1157 et seq.

Visceral afferent paths resemble somatic afferent paths; the cells of origin of their peripheral fibres are unipolar neurons in cranial and dorsal root ganglia. Peripheral processes are distributed through autonomic ganglia or plexuses, or possibly through somatic nerves, without interruption. Their central processes (axons) accompany the somatic afferent fibres through dorsal spinal roots to the central nervous system (p. 1167).

The autonomic nervous system has two complementary parts, *parasympathetic* and *sympathetic*, which differ in structure and function. Parasympathetic preganglionic efferent fibres emerge through certain cranial and sacral spinal nerves as a *craniosacral outflow*, while sympathetic preganglionic efferent fibres emerge through thoracic and upper lumbar spinal nerves as a *thoracolumbar outflow*. The somata of parasympathetic postganglionic neurons are peripheral, sited distant from the central nervous system either in discrete ganglia near to the structures innervated or often dispersed in the walls of viscera. The somata of sympathetic postganglionic neurons are located mostly in ganglia of the sympathetic trunk or ganglia in more peripheral plexuses but they are almost always nearer to the spinal cord than to the effectors innervated.

It has been considered in the past that physiologically the two

systems differed in that parasympathetic reactions are generally localized, whereas sympathetic reactions are mass responses. However, even though widespread activation of the sympathetic nervous system may occur, e.g. in association with fear or rage, it is now recognized that the sympathetic nervous system is also capable of discrete activation and many different patterns of activation of sympathetic nerves throughout the body occur in response to a wide variety of stimuli.

Parasympathetic activity results in cardiac slowing and an increase in intestinal glandular and peristaltic activities, which may be considered to conserve body energies. Sympathetic activities result, e.g. in the general constriction of cutaneous arteries (increasing blood supply to the heart, muscles and brain), cardiac acceleration, increase in blood pressure, contraction of sphincters and depression of peristalsis, all of which mobilize body energies for dealing with increased activity.

Whereas the passage of nervous impulses along *all* preganglionic fibres and also parasympathetic postganglionic and somatic efferent fibres results in the liberation of *acetylcholine* at their terminals, at sympathetic postganglionic terminals *noradrenalin* or *adrenalin* is usually liberated. Hence these nerves are termed *cholinergic* or *adrenergic* respectively. As an exception to the above, the sweat glands are supplied only by postganglionic sympathetic nerves but these are cholinergic. (See also peripheral *purinergic* fibres, p. 906.)

THE PARASYMPATHETIC NERVOUS SYSTEM

Efferent Pathways

Preganglionic parasympathetic axons are myelinated and occur in the oculomotor, facial, glossopharyngeal, vagal and accessory cranial nerves and in the second to fourth sacral spinal nerves. In the cranial part of the parasympathetic system there are four small peripheral ganglia: ciliary (p. 1097), pterygopalatine (p. 1102), submandibular (p. 1111) and otic (p. 1113), all described in this account with their cranial nerves. These are solely efferent parasympathetic ganglia, unlike the trigeminal, facial, glossopharyngeal and vagal ganglia, all of which are concerned exclusively with afferent impulses and contain the somata of sensory neurons only. The cranial parasympathetic ganglia are *traversed* by afferent fibres, postganglionic sympathetic fibres and, in the otic, even by branchial efferent fibres, but none of these are interrupted in the ganglia. Postganglionic parasympathetic fibres are usually non-myelinated and shorter than the sympathetic, since the ganglia in which they synapse are in or near the viscera they supply. Baumann & Gajisin (1975) have emphasized the occurrence of small subsidiary ganglia near those mentioned above, confirming reports by others; they also described minute ganglia at many other sites in fetal material, e.g. along the middle meningeal artery and in some petrosal nerves.

(1) *Oculomotor* preganglionic parasympathetic fibres commence in the midbrain at the accessory oculomotor (Edinger-Westphal) nuclei (p. 1097) and travel in the nerve to leave in its branch to the inferior oblique to reach the *ciliary ganglion*. There they synapse, the postganglionic fibres leaving in the short ciliary nerves which pierce the sclera to run forwards in the perichoroidal space to the ciliary muscle (p. 1188) and the sphincter pupillae (p. 1193). These postganglionic axons are thinly myelinated.

(2) The *facial* nerve contains preganglionic parasympathetic axons of neurons with their somata in the superior salivatory nucleus (p. 1108), emerging from the medulla oblongata in the nervus intermedius. These fibres leave the main facial trunk above the stylomastoid foramen in the *chorda tympani*, which traverses the tympanic cavity to reach the *lingual nerve* (p. 1110). Thus they are conveyed to the *submandibular ganglion*, in which arise postganglionic secretomotor fibres for the submandibular salivary gland. Some preganglionic fibres may synapse around cells in the gland's hilum (pp. 1293, 1297). Postganglionic secretomotor fibres for the sublingual gland continue in the

lingual nerve from the submandibular ganglion (pp. 1111, 1293). Stimulation of chorda tympani dilates the arterioles in both glands in addition to having a direct secretomotor effect. The facial nerve is also usually said to contain efferent parasympathetic lacrimal secretomotor axons, which travel in its greater petrosal ramus and in the nerve of the pterygoid canal, relaying in the pterygopalatine ganglion. Postganglionic axons are said to travel by the zygomatic nerve to the lacrimal gland (p. 1102) and by ganglionic branches to the nasal and palatal glands. Evidence refuting the zygomatic route has been reported by Ruskell (1971), who favours direct *lacrimal rami* from a *retro-orbital plexus* of parasympathetic branches from the pterygopalatine ganglion. Clinical evidence suggests that some facial parasympathetic fibres reach the parotid gland (Diamant & Wiberg 1965 and p. 1108).

(3) The *glossopharyngeal nerve* contains preganglionic parasympathetic secretomotor fibres for the parotid gland. These start in the inferior salivatory nucleus (p. 1113) and travel in the glossopharyngeal nerve and its tympanic branch. They traverse the tympanic plexus and lesser petrosal nerve to reach the otic ganglion where they relay, the postganglionic fibres passing by communicating branches to the auriculotemporal nerve, which conveys them to the parotid gland. Stimulation of the lesser petrosal nerve produces vasodilator and secretomotor effects.

(4) The *vagus nerve* contains preganglionic parasympathetic fibres which arise in its dorsal nucleus (p. 1114) and travel in the nerve and its pulmonary, cardiac, oesophageal, gastric, intestinal and other branches. Some cardiac parasympathetic fibres may originate from neurons in or near the nucleus ambiguus (p. 956). The proportion of efferent parasympathetic fibres in the vagus varies at different levels but is small relative to its sensory content. Efferent fibres relay in minute ganglia in the visceral walls. The disproportion in the numbers of preganglionic to postganglionic fibres is greater in the vagus than in other cranial nerves; this cannot as yet be explained. Cardiac branches slow the cardiac cycle, joining the cardiac plexuses (p. 1164) and relaying in ganglia distributed freely over both atria in the subepicardial tissue (6.57), terminal fibres being distributed to the atria and the atrioventricular bundle and concentrated around the SA and (to a lesser extent) the AV nodes. It has been claimed in the past that only through the latter can the vagi influence ventricular muscle (Cullis & Tribe 1913). The smaller branches of the coronary arteries are innervated mainly via the vagus; larger arteries, with

a dual innervation, are chiefly supplied by sympathetic fibres (Woollard 1926). *Pulmonary branches* are motor to the circular non-striated muscle fibres of the bronchi and bronchioles and are therefore bronchoconstrictor; synaptic relays occur in the ganglia of the pulmonary plexuses. *Gastric branches* are secretomotor and motor to the non-striated muscle of the stomach, with the exception of the pyloric sphincter, which they inhibit. *Intestinal branches* have a corresponding action in the small intestine, caecum, vermiform appendix, ascending colon, right colic flexure and most of the transverse colon; they are secretomotor to the glands, motor to the intestinal muscular coats but inhibitory to the ileo-caecal sphincter. The synaptic relays are situated in the myenteric (Auerbach's) and the submucosal (Meissner's) plexuses (p. 1331).

(5) The anterior rami of the *second*, *third* and often *fourth sacral* spinal nerves issue *pelvic splanchnic nerves* (7.277) to the pelvic viscera. These nerves unite with branches of the sympathetic pelvic plexuses. Minute ganglia occur at the points of union and in the visceral walls. In these ganglia the sacral preganglionic parasympathetic fibres relay synaptically.

The pelvic splanchnic nerves are motor to the muscle of the rectum and bladder wall but inhibitory to the vesical sphincter, supply vasodilator fibres to the penile and clitoridic erectile tissue and are probably vasodilator to the testes and ovaries and vasodilator (and possibly inhibitory) to the uterine tubes and uterus. Filaments from the pelvic splanchnic nerves ascend in the hypogastric plexus to supply the sigmoid and descending colon, the left colic flexure and terminal transverse colon with visceromotor fibres (Telford & Stopford 1934, Mitchell 1935).

THE SYMPATHETIC NERVOUS SYSTEM

The sympathetic system, which is the larger autonomic division, includes the two ganglionated trunks and their branches, plexuses and subsidiary ganglia. It has a much wider distribution than the parasympathetic, for it innervates: all sweat glands, the arrectores pilorum, the muscular walls of many blood vessels, the heart, lungs and respiratory tree, the abdomino-pelvic viscera, the oesophagus, the muscles of the iris in the eye, and non-striated muscle of the urogenital tract, the eyelids and elsewhere.

Efferent Pathways

The *preganglionic fibres* are axons of somata in the lateral grey column of all the thoracic and the upper two or three lumbar

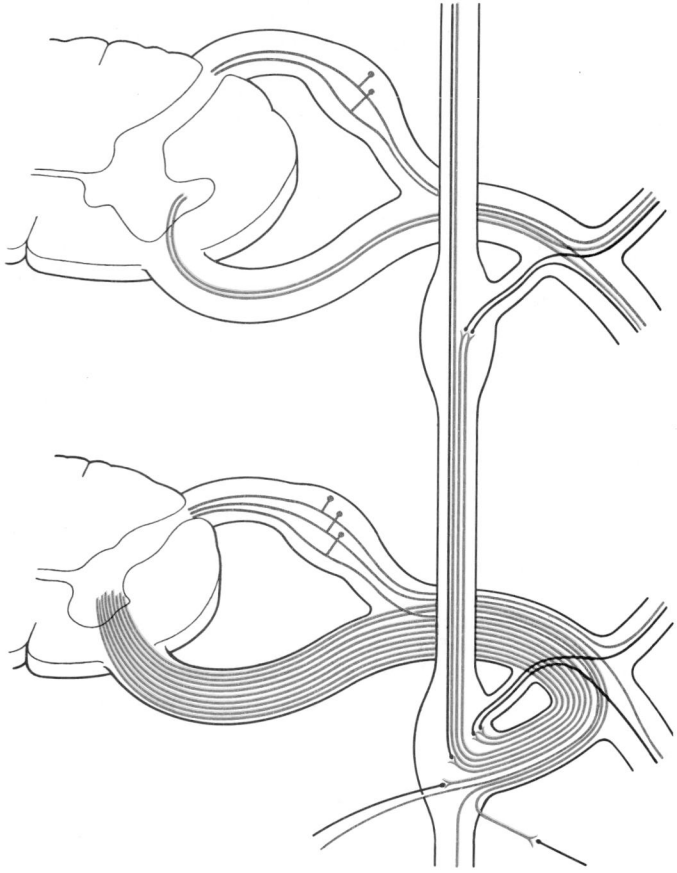

7.269 The constitution of a typical spinal nerve. In the upper part of the diagram the spinal nerve roots show the somatic components; in the lower part the spinal roots show the visceral components. Red = somatic and preganglionic visceral nerve fibres, blue = somatic and visceral afferent fibres, black = postganglionic afferent visceral fibres.

spinal segments, where they form intermediomedial and intermediolateral neuronal groups (p. 927). The axons are myelinated, with diameters of 1·5–4·0 μm, and emerge from the spinal cord through the ventral spinal roots, passing into the spinal nerves at the start of their ventral rami, which they soon leave in *white rami communicantes*, to join either the corresponding ganglia of the sympathetic trunks or their interganglionic segments. This outflow is confined to the thoracolumbar region, the white rami communicantes being restricted to these 14 pairs of spinal nerves, although a limited outflow in other spinal nerves has been suggested. Neurons like those in the lateral grey column exist at other levels of the cord above and below the thoracolumbar outflow (Mitchell 1953) and small numbers of their fibres issue in other ventral roots. Dorsal spinal roots may also contain vasodilator fibres. Reaching the sympathetic trunk, preganglionic fibres may behave in several ways (7.269): (1) They may synapse with neurons in the nearest ganglion. (2) They may traverse this, ascending or descending in the sympathetic chain to end in another ganglion; note however that preganglionic fibres do not *divide* into ascending and descending branches. A single preganglionic fibre may, through collateral and terminal branches, synapse with neurons in several ganglia or terminate in only one ganglion. (3) They may traverse the nearest ganglion, ascend or descend and, without synapsing, emerge in one of the medially-directed branches of the sympathetic trunk to end at synapses in the ganglia of autonomic plexuses (mainly situated in the midline, e.g. around the coeliac and mesenteric arteries, p.1165). Occasionally preganglionic fibres relay in ganglia situated proximal to the sympathetic trunks; these 'intermediate ganglia' are most numerous on grey rami communicantes (vide infra) at cervical and lower lumbar levels (Boyd & Munro 1940); they may be of microscopic size and sometimes occur in spinal ventral roots or trunks. More than one preganglionic fibre may synapse with a single postganglionic neuron (vide infra).

The nervi terminales (p. 1095) may be rostral extensions of the sympathetic system, containing efferent postganglionic fibres distributed to the blood vessels and glands of the nasal cavity, although this view has been challenged (Bojsen-Møller 1975).

The sympathetic ganglia include collections of cells on the sympathetic trunks, in the autonomic plexuses and the 'intermediate' ganglia; some ganglionic cells are dispersed in the plexuses. Originally ganglia on the trunks correspond numerically to the ganglia on the dorsal spinal roots (p. 1125); but adjoining ganglia may fuse and in man there are rarely more than 22 or 23 and sometimes less. Subsidiary ganglia in the major autonomic plexuses (e.g. coeliac, superior mesenteric ganglia, etc.) are derivatives of the ganglia of the trunks. The functional properties of sympathetic ganglia have been investigated extensively over many decades, their peripheral location providing a valuable means of studying interneuronal communication, as well as other aspects of neurobiology (for reviews of the earlier literature, see p. 1154; more recent accounts are given by Gabella (1976) and Eränkö (1978).

The structure of sympathetic ganglia. The classic studies by Langley and his successors led to the view that the autonomic ganglia are relay stations, a concept largely corroborated by anatomical observation, although it was soon recognized that a minor fraction of the fibres traversed one or more ganglia without synapse, some being efferent fibres en route to another ganglion and others afferents from the viscera and glands. This concept remains substantially true but has been modified and extended by electron microscopy, neurohistochemistry and electrophysiology, e.g. a considerable variation in the ratio between pre- and postganglionic fibres has been found. The superior cervical sympathetic ganglion, the most extensively studied, has ratios varying from 1:28 to 1:176 in different mammalian species (Billingsley & Ranson 1918, Samuel 1953, Ebbeson 1968). It has long been accepted that preganglionic axons may synapse with many postganglionic neurons for the wide *dissemination* and perhaps *amplification* of sympathetic activity, a characteristic not shared to the same degree by parasympathetic ganglia. Dissemination may be achieved: (1) by multiple synapses of preganglionic nerve fibres; (2) by the mediation of interneurons; (3) by the diffusion within the ganglion of transmitter substances locally produced (*paracrine effect*) or by a local response to a substance produced elsewhere (*endocrine* effect). There is evidence that all of these mechanisms are involved.

The connective tissue capsule of each ganglion, continuous with the epineurium of its connecting rami, also extends as septa into the ganglion, the surrounding groups of neurons and their fibres. More delicate extensions of this stroma spread amongst the cells, each of which is surrounded by a collagenous intercellular matrix containing a few fibroblasts and many small vessels including capillaries. Satellite cells (amphicytes) encapsulate the somata of ganglionic neurons and their processes. Externally this thin sheath of satellite cells has a continuous basal lamina and the two elements screen neurons from contact with the ganglionic extracellular matrix. Neurons thus have direct access only to the internal faces of satellite cells, the two being separated only by a narrow perineuronal space of 15–20 nm which is, however, linked to the extracapsular spaces by narrow channels between the satellite cells, providing possible routes for the movement of neurotransmitter and hormonal substances between the somata of neurons and the vascular compartment.

Attempts to classify the neurons of the sympathetic (and parasympathetic) ganglia, often on inadequate criteria, have entailed disagreements and confusion. Most are multipolar, with somata ranging from 25–50 μm in mankind; a smaller type, of about 15–20 μm, less angular in shape and present in much smaller numbers, is often clustered in groups (De Castro & Herneros 1945) and probably corresponds to 'small intensely fluorescent' (SIF) cells (vide infra). Multipolar neurons display much more dendritic variation; according to McLachlan (1974) they have (in guinea pigs) a mean of 13 dendrites per cell. The complexity of these dendrites, especially those ramifying in the capsular perikaryal space, is greater in human ganglion cells. Dendritic glomeruli have been observed in many ganglia. In general ultrastructure these glomeruli resemble others (p. 888); clusters of small, granular vesicles, adrenergic in type, are dispersed superficially in the perikaryon and also in the dendrites, probably representing the storage of catecholamines. Ganglionic neurons receive many axodendritic synapses from preganglionic nerve fibres, the axosomatic synapses being less numerous. Each preganglionic fibre forms several synapses with several separate dendrites, providing a mechanism for the dissemination and/or amplification of neural signals. Postganglionic fibres (vide infra) commonly arise from the initial stem of a large dendrite and produce few or no collateral neurites.

The existence of *interneurons* in sympathetic ganglia has been amply confirmed (Williams 1967, Williams & Palay 1969, Libet & Crowman 1974), consisting of the 'small intensely fluorescent' (SIF) cells identified in sympathetic ganglia in many mammals, including mankind (Eränkö & Harkonen 1965, Jacobowitz 1970). Small chromaffin cells also occur in sympathetic ganglia, as recognized by Kohn in 1898. Coupland (1965), amongst

other modern workers, has ascribed them to all ganglia of the sympathetic trunk and to other sites in human neonatal material. The distinction between SIF cells and chromaffin cells appears uncertain in many accounts. The supposed two types have been identified in ganglia by separate techniques (chromaffin reaction and formalin-induced fluorescence) which cannot be applied together to a single cell. In the sympathetic ganglia of rats (Santer et al 1975) SIF cells were found to be more numerous than chromaffin cells and their modes of distribution showed some differences. Both contain catecholamines, some possibly only enough to be revealed by the more sensitive formalin-induced fluorescence technique (Falck-Hillarp), whereas others may have sufficient to produce a positive chromaffin reaction (Gabella 1976). Both types may be interneurons (Santer et al 1975, Gabella 1976). Greengard & Kebabian (1974) have hypothecated that the SIF cells release dopamine, which is then bound by dopamine receptors on ganglionic neurons causing hyperpolarization via a cyclic AMP-dependent 'second messenger' system (p. 887). In the ganglia of some species, two types of SIF cell have been described (Williams 1975): a minority, with long processes end near ganglionic neurons and hence can be regarded as interneurons (Type I), while the more numerous Type II cells have shorter processes ending near blood vessels (Chiba et al 1975). In bovine superior cervical sympathetic ganglia, 24% of SIF cells were described as Type I and 20% in cats. Although the secretory granules in Type I cells may act directly on ganglionic neurons, some SIF cells, presumably Type II, may secrete into local blood vessels (Polonyi et al 1977), exerting more distant effects. The functions of SIF cells in neurotransmission in sympathetic ganglia have been reviewed by Eränkö (1978), and quantification of numbers, dimensions and packing density of ganglionic neurons are reported by Gabella (1976).

The axons of the principal ganglionic cells are narrow, non-myelinated *postganglionic fibres*, distributed to effector organs in various ways: (1) Those from a ganglion of the sympathetic trunk may return to the spinal nerve of preganglionic origin through a *grey ramus communicans*, usually joining the nerve just proximal to the white ramus, to be distributed through ventral and dorsal spinal rami to blood vessels, sweat glands, hairs, etc. in their zone of supply. Segmental areas vary in extent and overlap considerably. The extent of innervation of different effector systems, e.g. vasomotor, sudomotor, etc. by a particular nerve may not be the same. (2) Postganglionic fibres may pass in a medial branch of a ganglion direct to particular viscera. (3) They may innervate adjacent blood vessels or pass along them externally to their peripheral distribution. (4) They may ascend or descend before leaving the sympathetic trunk as (1), (2) or (3). Many fibres are distributed along arteries and ducts as plexuses to distant effectors.

Fusion of grey and white rami may also occur, e.g. in the thoracic region; grey rami may also contain fasciculi of thick myelinated fibres which are somatic efferents using a grey ramus to reach the prevertebral muscles (p. 1158) e.g. in the cervical region. For details of rami communicantes and their variations consult Winckler (1961).

Functional significance. Postganglionic fibres which return to the spinal nerves supply vasoconstrictor fibres to blood vessels, are secretomotor to sweat glands and motor to the arrectores pilorum in their dermatomes. Those which accompany the motor nerves to voluntary muscles are probably only vasodilatory. Most, if not all, peripheral nerves contain postganglionic sympathetic fibres. Those reaching the viscera are concerned with general vasoconstriction, bronchial and bronchiolar dilatation, modification of glandular secretion, pupillary dilatation, inhibition of alimentary muscle contraction, etc. A single preganglionic fibre probably synapses with the postganglionic neurons in only one effector system; hence effects such as sudomotor and vasomotor actions can be separate. This is not necessarily true of visceral *afferent* fibres. Sympathetic and parasympathetic nerves usually exert antagonistic influences on many viscera; but in the urinary bladder normal emptying *and* filling are controlled by parasympathetic innervation, the sympathetic being only vasomotor.

Higher autonomic control. Peripheral autonomic activity is integrated at higher levels in the brain stem and cerebrum, including various nuclei of the brain-stem reticular formation, thalamus and hypothalamus, the limbic lobe and prefrontal neocortex, together with the ascending and descending pathways which interconnect these regions (see details of these given earlier in this section).

The sympathetic trunks are two ganglionated, irregular nerve cords extending from the cranial base to the coccyx. In the neck each lies posterior to the carotid sheath and anterior to the cervical transverse processes; in the thorax it is anterior to the heads of the ribs, in the abdomen anterolateral to the lumbar vertebral bodies and in the pelvis anterior to the sacrum and medial to the anterior sacral foramina. Anterior to the coccyx the two trunks meet in the single, median, terminal *ganglion impar*.

Cervical sympathetic ganglia are usually reduced to three by fusion; from the superior ganglion's cranial pole issues the internal carotid nerve, as a continuation of the sympathetic trunk, accompanying the internal carotid artery through its canal into the cranial cavity. There are from 10–12 (usually 11) thoracic ganglia, four lumbar and four or five in the sacral region.

CRANIAL PART OF THE SYMPATHETIC SYSTEM

This begins on each side as the **internal carotid nerve**, a branch of the superior cervical ganglion containing the postganglionic fibres of its neurons. Ascending behind the internal carotid artery it divides in the carotid canal into branches, one medial and the other lateral to the artery. The larger, *lateral branch* gives filaments to the internal carotid and forms the lateral part of the internal carotid plexus; the *medial branch* also gives filaments to the artery and, continuing on, forms the medial part of the internal carotid plexus.

The internal carotid plexus surrounds its artery and occasionally contains a small, medial *carotid ganglion*; elsewhere it has some scattered postganglionic neurons. Laterally the plexus communicates with the trigeminal and pterygopalatine ganglia, the abducent nerve and tympanic branch of the glossopharyngeal; it also distributes filaments to the wall of the internal carotid artery. One or two filaments join the abducent nerve as it lies on the lateral side of the internal carotid artery. The branch to the pterygopalatine ganglion is the *deep petrosal nerve*, which perforates the cartilage filling the foramen lacerum and forms with the greater petrosal nerve the *nerve of the pterygoid canal*, traversing the canal to the pterygopalatine ganglion. The communication with the tympanic branch of the glossopharyngeal nerve is effected by the *superior* and *inferior caroticotympanic nerves* in the posterior wall of the carotid canal.

The medial part of the internal carotid plexus is inferomedial to the part of the internal carotid artery which indents the cavernous sinus lateral to the sella turcica; it gives branches to the artery and to the oculomotor, trochlear, ophthalmic and abducent nerves and the ciliary ganglion. It also sends vasomotor rami along branches of the internal carotid to the hypophysis cerebri (p. 1455).

The branch to the oculomotor nerve joins the nerve near its point of division and the branch to the trochlear joins it in the lateral wall of the cavernous sinus; filaments also connect with the medial side of the ophthalmic nerve and with the abducent. The filament to the ciliary ganglion, from the anterior part of the plexus, enters the orbit via the superior orbital fissure; this ramus may join the ciliary ganglion directly or unite with the communicating branch from the nasociliary nerve (p. 1097); or it may travel in the ophthalmic nerve and its nasociliary branch. Its fibres traverse the ganglion without synapse and enter the short ciliary nerves to be distributed to the blood vessels of the eyeball. Fibres supplying the dilatator pupillae usually travel via the ophthalmic, nasociliary and long ciliary nerves but occasionally via the short ciliary. Some fibres may also innervate the ciliaris muscle. The preganglionic fibres concerned leave the spinal cord predominantly in T1, pass to and through the cervicothoracic ganglion and ascend in the cervical sympathetic trunk to relay in the superior cervical ganglion.

The internal carotid plexus is prolonged around the anterior and middle cerebral arteries and the ophthalmic arteries, reaching the pia mater along the cerebral vessels; along the ophthalmic artery they pass into the orbit where the plexus accompanies each branch of that vessel. Filaments on the anterior communicating artery connect the sympathetic nerves of the right and left sides and may be associated with a small ganglion. Much of this detail depends on rather old observations; continued disagreement and discrepancy have been reviewed by Mitchell (1953) and Purves (1972). It can, however, be said that the old controversy over the autonomic innervation of the cerebral arterial tree is now settled. Electron microscopy shows that the innervation is like that of other vascular systems and the terminals are confirmed histochemically as adrenergic in various mammals, including man (Iwayama 1970). The sources of these sympathetic vasoconstrictor fibres are the internal carotid and vertebral plexuses but their precise distribution from each is unresolved. (Though cholinergic terminals have been noted in cerebral arterial walls, the existence of a vasodilator mechanism has not been clearly demonstrated.) Falck et al (1965) and Edvinsson et al (1973) have also claimed that some adrenergic axons accompanying pial and intracerebral arteries are *not* derived from known sympathetic sources but from nerve cells at present merely presumed to be intracranial in location.

CERVICAL PART OF THE SYMPATHETIC SYSTEM

The cervical part of each sympathetic trunk contains three interconnected ganglia, the superior, middle and cervicothoracic (7.270), which send grey rami communicantes to all the cervical spinal nerves but receive no white rami communicantes from them; their spinal preganglionic fibres emerge in the white rami communicantes of the upper thoracic spinal nerves which enter the corresponding thoracic sympathetic ganglia, through which they ascend into the neck. In their course, the grey rami communicantes may pierce the longus capitis or the scalenus anterior. For details of the cervical grey rami see Potts (1925), Oxford (1928), Pick & Sheehan (1946), Sunderland & Bedbrook (1949).

The superior cervical ganglion, the largest of the three, adjoins the second and third cervical vertebrae and is probably formed from four fused ganglia corresponding to C1–4. Anterior to it is the internal carotid artery and sheath, while posterior to it is the longus capitis. The internal carotid nerve (vide supra) ascends from it into the cranial cavity; the lower end of the ganglion is united by a connecting trunk to the middle cervical ganglion. Its branches consist of lateral, medial and anterior groups.

The *lateral branches* are the grey rami communicantes to the upper four cervical spinal nerves and to some of the cranial nerves; delicate filaments pass to the inferior vagal ganglion and to the hypoglossal nerve; a branch, *the jugular nerve* ascends to the cranial base and divides into two, one part joining the inferior glossopharyngeal ganglion and the other the superior vagal ganglion; other twigs pass to the superior jugular bulb and associated jugular glomus or glomera and some to the meninges in the posterior cranial fossa.

The *medial branches* of the superior cervical ganglion are the laryngopharyngeal and cardiac. The *laryngopharyngeal branches* supply the carotid body and pass to the side of the pharynx, joining glossopharyngeal and vagal rami to form the *pharyngeal plexus* (p. 1117). A *cardiac branch* arises by two or more filaments from the lower part of the superior cervical ganglion, occasionally receiving a twig from the trunk between the superior and middle cervical ganglia. It is thought to contain only efferent fibres, the preganglionic outflow being from the upper thoracic segments of the spinal cord, and to be devoid of pain fibres from the heart (p. 1164). It descends behind the common carotid artery, in front of the longus colli, crossing anterior to the inferior thyroid artery and recurrent laryngeal nerve. The courses on the two sides then differ. The *right cardiac branch* usually passes behind or sometimes in front of the subclavian artery and posterolateral to the brachiocephalic trunk to the back of the aortic arch where it joins the deep (dorsal) part of the cardiac plexus. It has other sympathetic connections: about mid-neck it receives filaments from the external laryngeal nerve; inferiorly, one or two vagal cardiac branches join it; as it enters the thorax it is joined by a filament from the recurrent laryngeal nerve. Filaments from the

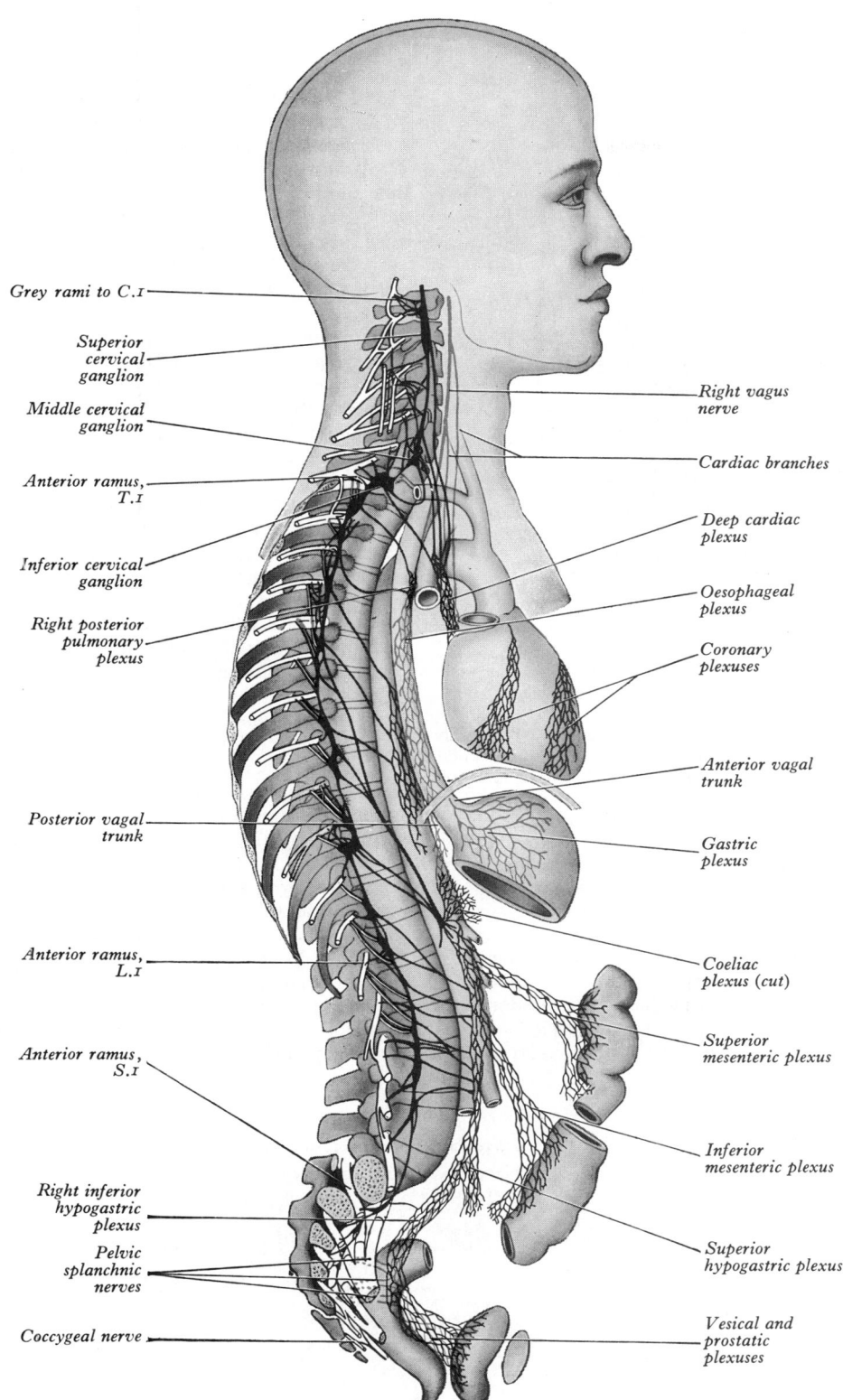

Grey rami to C.I

Superior cervical ganglion

Middle cervical ganglion

Anterior ramus, T.I

Inferior cervical ganglion

Right posterior pulmonary plexus

Posterior vagal trunk

Anterior ramus, L.I

Anterior ramus, S.I

Right inferior hypogastric plexus

Pelvic splanchnic nerves

Coccygeal nerve

Right vagus nerve

Cardiac branches

Deep cardiac plexus

Oesophageal plexus

Coronary plexuses

Anterior vagal trunk

Gastric plexus

Coeliac plexus (cut)

Superior mesenteric plexus

Inferior mesenteric plexus

Superior hypogastric plexus

Vesical and prostatic plexuses

7.270 The right sympathetic trunk and its connections with the thoracic, abdominal and pelvic plexuses. Blue = parasympathetic fibres, black = sympathetic trunk and branches, red = white rami communicantes.

nerve also communicate with the thyroid branches of the middle cervical ganglion. The *left cardiac branch*, in the thorax, is anterior to the left common carotid artery and crosses in front of the left side of the aortic arch to reach the superficial (ventral) part of the cardiac plexus. Sometimes it descends on the right of the aorta to end in the deep (dorsal) part of the cardiac plexus. It communicates with the cardiac branches of the middle cervical and cervicothoracic sympathetic ganglia and sometimes with the inferior cervical cardiac branches of the left vagus; branches from these mixed nerves form a plexus on the ascending aorta.

The *anterior branches* of the superior cervical ganglion ramify on the common and external carotid arteries and the latter's branches, forming a delicate plexus around each in which small ganglia are occasionally found. The plexus around the facial artery supplies a filament to the submandibular ganglion; the plexus on the middle meningeal artery sends one ramus to the otic ganglion and another, *the external petrosal nerve*, to the facial ganglion. Many of the fibres coursing along the external carotid and its branches ultimately leave them to travel to facial sweat glands via trigeminal nerve branches.

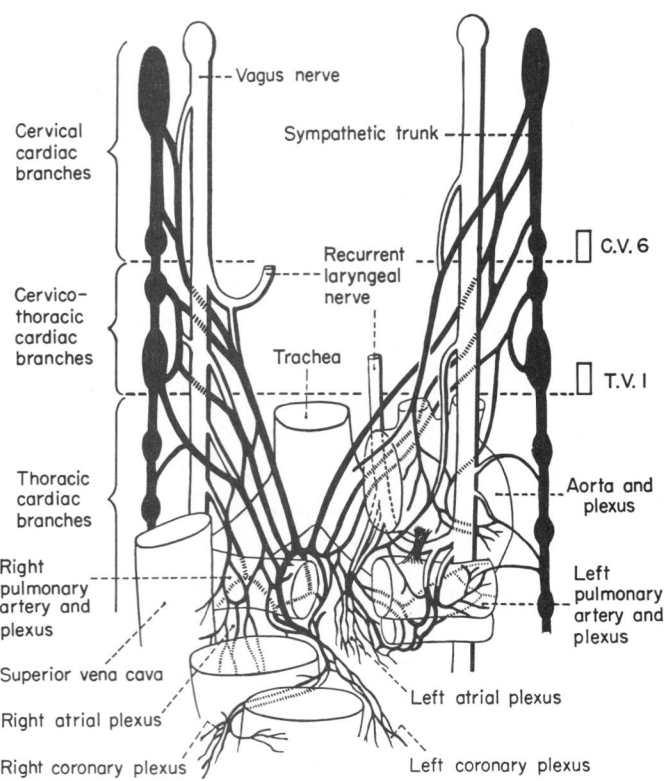

Cervical cardiac branches

Vagus nerve

Sympathetic trunk

Cervico-thoracic cardiac branches

Recurrent laryngeal nerve

C.V. 6

Trachea

T.V. I

Thoracic cardiac branches

Aorta and plexus

Right pulmonary artery and plexus

Left pulmonary artery and plexus

Superior vena cava

Left atrial plexus

Right atrial plexus

Right coronary plexus

Left coronary plexus

7.271 The human cardiac plexus, a semi-diagrammatic representation of its source from the cervical parts of the vagus nerves and sympathetic trunks and of its extensions: the pulmonary, atrial and coronary plexuses. Note the numerous junctions between sympathetic and parasympathetic (vagal) rami which form the plexus. For further information, particularly concerning the frequent variations, consult Mizeres 1963, to the author and publisher of which we are indebted for permission to use this diagram.

The middle cervical ganglion (7.272–274), the smallest of the three, is occasionally absent and may then be replaced by minute ganglia in the sympathetic trunk or may be fused with the superior ganglion. It is usually found at the sixth cervical vertebral level, anterior or just superior to the inferior thyroid artery, or it

may adjoin the cervicothoracic ganglion (vide infra); it is probably a coalescence of the ganglia of the fifth and sixth cervical segments, judging by its postganglionic rami, which join the fifth and sixth cervical spinal nerves but sometimes also the fourth and seventh. The ganglion also has thyroid and cardiac branches. It is connected to the cervicothoracic ganglion by two or more very variable cords: the posterior usually splits to enclose the vertebral artery; the anterior loops down anterior to and then below the first part of the subclavian artery, medial to the origin of its internal thoracic branch, and supplies rami to it. This loop is the *ansa subclavia*; it is frequently multiple, lies closely in contact with the cervical pleura and generally connects with the phrenic nerve. Similar connections with the vagus nerve are of uncertain significance.

Thyroid branches accompany the inferior thyroid artery to the thyroid gland, communicating with the superior cardiac, external laryngeal and recurrent laryngeal nerves, and send branches to the parathyroid glands. Fibres to both glands are in part vasomotor but some reach the secretory cells (Raybuck 1952).

The *cardiac branch*, the largest sympathetic cardiac nerve, either arises from the ganglion itself or more often from the sympathetic trunk cranial or caudal to it. On the *right side* it descends behind the common carotid artery, in front of or behind the subclavian, to the trachea where it receives a few filaments from the recurrent laryngeal nerve before joining the right half of the deep (dorsal) part of the cardiac plexus. In the neck, it connects with the superior cardiac and recurrent laryngeal nerves. On the *left side* the cardiac nerve enters the thorax between the left common carotid and subclavian arteries to join the left half of the deep (dorsal) part of the cardiac plexus. Fine branches from the middle cervical ganglion also pass to the trachea and oesophagus.

The cervicothoracic (stellate) ganglion, irregular in shape and much larger than the middle cervical ganglion, is probably formed by a fusion of the lower two cervical and first thoracic segmental ganglia, sometimes including the second and even third and fourth thoracic ganglia. The first thoracic ganglion may be separate, leaving an *inferior cervical ganglion* above it (7.270, 272). The sympathetic trunk turns backwards at the junction of the neck and thorax and so the long axis of the cervicothoracic ganglion becomes almost anteroposterior. The ganglion lies on or just lateral to the lateral border of the longus colli between the base of the seventh cervical transverse process and the neck of the first rib (which are posterior to it), the vertebral vessels being anterior. Below it is separated from the posterior aspect of the cervical pleura by the suprapleural membrane; the costocervical

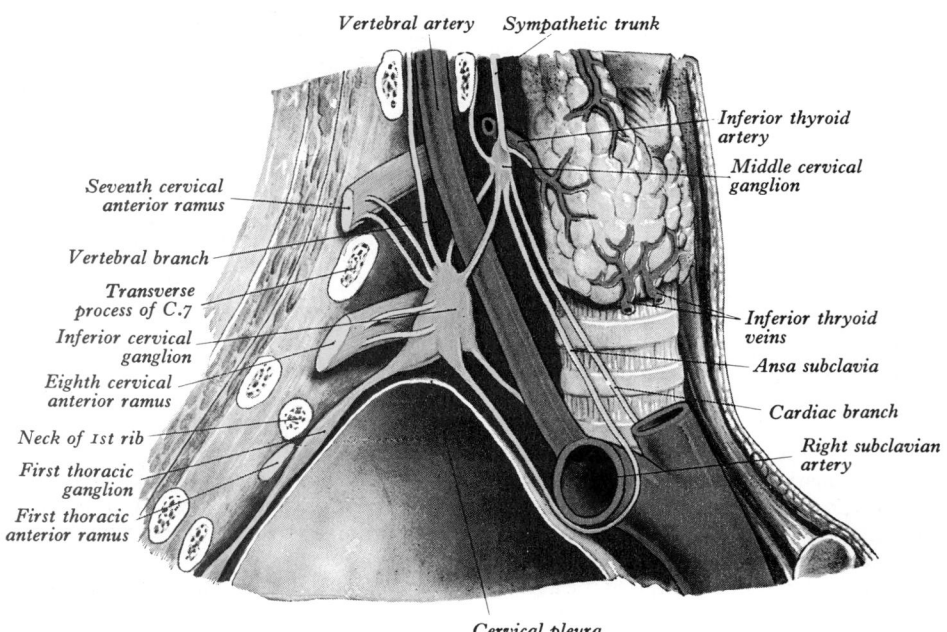

Vertebral artery

Sympathetic trunk

Inferior thyroid artery

Middle cervical ganglion

Seventh cervical anterior ramus

Vertebral branch

Transverse process of C.7

Inferior cervical ganglion

Eighth cervical anterior ramus

Inferior thyroid veins

Neck of 1st rib

Ansa subclavia

First thoracic ganglion

Cardiac branch

First thoracic anterior ramus

Right subclavian artery

Cervical pleura

7.272 The middle and inferior cervical ganglia of the right side, viewed from the right. Note the proximity of the inferior cervical and first thoracic ganglia, usually fused to form a cervicothoracic (stellate) ganglion.

trunk branches near its lower pole. Lateral is the superior intercostal artery.

A small *vertebral ganglion* may be present on the sympathetic trunk anterior or anteromedial to the origin of the vertebral artery and directly above the subclavian. When present, it may provide the ansa subclavia and is also joined to the cervicothoracic ganglion by fibres enclosing the vertebral artery. It is usually regarded as a detached part of the middle cervical or cervicothoracic ganglion. Like the middle cervical ganglion it may supply grey rami communicantes to the fourth and fifth cervical spinal nerves. The cervicothoracic ganglion sends grey rami communicantes to the seventh and eighth cervical and first thoracic spinal nerves and gives off a cardiac branch, branches to nearby vessels and sometimes a branch to the vagus nerve.

The *grey rami communicantes* to the seventh cervical spinal nerve vary from one to five; two, the usual number, are shown in 7.272, 273. A third often ascends medial to the vertebral artery in front of the seventh cervical transverse process, connects with the seventh cervical nerve and sends a filament upwards through the sixth cervical transverse foramen in company with the vertebral vessels to join the sixth cervical spinal nerve as it emerges from the intervertebral foramen. An inconstant ramus may traverse the seventh cervical transverse foramen. Grey rami to the eighth cervical spinal nerve vary from three to six in number.

The *cardiac branch* descends behind the subclavian artery and along the front of the trachea to the deep cardiac plexus. Behind the artery it connects with the recurrent laryngeal nerve and the cardiac branch of the middle cervical ganglion, the latter often being replaced by fine branches of the cervicothoracic ganglion and ansa subclavia.

Branches to blood vessels form plexuses on the subclavian artery and its branches. The subclavian supply is derived from the cervicothoracic ganglion and ansa subclavia, extending to the first part of the axillary artery; a few fibres may extend further. According to Pearson & Sauter (1971) an extension of the subclavian plexus to the internal thoracic artery is joined by a branch of the phrenic nerve (p. 1128). The vertebral plexus is derived mainly from a large branch of the cervicothoracic ganglion which ascends behind the vertebral artery to the sixth transverse foramen, reinforced by branches of the vertebral ganglion or the cervical sympathetic trunk which pass cranially on the ventral aspect of the artery; from this plexus *deep rami communicantes* join the ventral rami of the upper five or six cervical spinal nerves. The plexus contains some neuronal cell bodies and continues into the skull along the vertebral and basilar arteries and their branches as far as the posterior cerebral artery, where it meets a plexus from the internal carotid. Some consider the vertebral plexus to be the main intracranial extension of the sympathetic system (Lazorthes 1949, Mitchell 1952). The plexus on the inferior thyroid artery reaches the thyroid gland, connecting with recurrent and external laryngeal nerves, the cardiac branch of the superior cervical ganglion, and the common carotid plexus.

The preganglionic fibres for the head and neck emerge from the spinal cord in the upper five thoracic spinal nerves (mainly the upper three), ascending in the sympathetic trunk to synapse in the cervical ganglia. The preganglionic fibres supplying the upper limb stem from upper thoracic segments, probably T2–6 (or 7), ascending via the sympathetic trunk to synapse mainly in its cervicothoracic ganglion, whence postganglionic fibres pass to the brachial plexus (mainly its lower trunk). Most of the vasoconstrictor fibres for the upper limb emerge in the second and third thoracic ventral roots; the arteries can thus be denervated by cutting the sympathetic trunk below the third thoracic ganglion, severing the rami communicantes connected with the second and third thoracic ganglia or by cutting the ventral roots of the second and third thoracic spinal nerves (intradurally). The white ramus to the cervicothoracic ganglion is not cut, partly because it does not convey many vasomotor or sudomotor fibres to the upper limb but mainly because it contains most of the preganglionic fibres; these ascend the trunk to the superior cervical ganglion, from which postganglionic branches supply vasoconstrictor and sudomotor nerves to the face and neck, secretory fibres to the salivary glands, dilatator pupillae (and probably ciliaris oculi), non-striated muscle in the eyelids and the orbitalis. Destruction

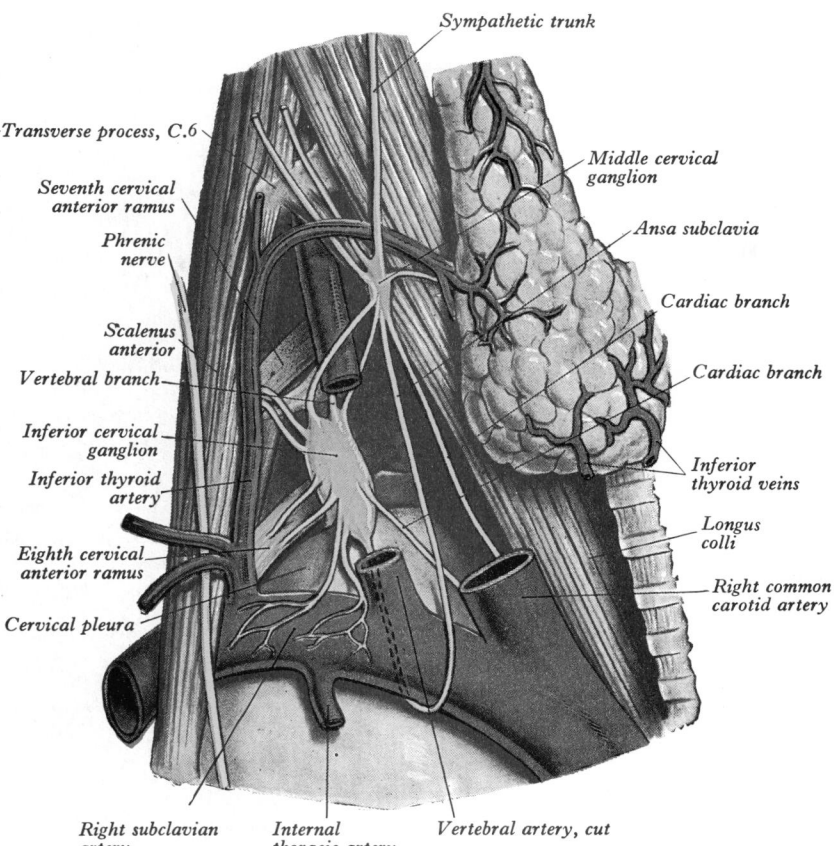

Sympathetic trunk
Transverse process, C.6
Seventh cervical anterior ramus
Phrenic nerve
Scalenus anterior
Vertebral branch
Inferior cervical ganglion
Inferior thyroid artery
Eighth cervical anterior ramus
Cervical pleura
Middle cervical ganglion
Ansa subclavia
Cardiac branch
Cardiac branch
Inferior thyroid veins
Longus colli
Right common carotid artery
Right subclavian artery   *Internal thoracic artery*   *Vertebral artery, cut*

7.273   Anterior view of the same structures illustrated in 7.272. Part of the vertebral artery has been excised to show the inferior cervical ganglion.

of this nerve would thus lead to meiosis, ptosis, enophthalmos and loss of sweating on the face and neck (*Horner's syndrome*) and possibly some disturbance of accommodation. For a review consult Haxton (1954).

Blood vessels of the upper limb beyond the first part of the axillary artery receive their sympathetic supply via branches of the brachial plexus adjacent to them, e.g. the median nerve supplies branches to the brachial artery and palmar arches, the ulnar nerve supplies the ulnar artery and palmar arches and the radial nerve supplies the radial artery.

The first and second (and occasionally the third) intercostal nerves may be interconnected anterior to the necks of the ribs by filaments containing postganglionic fibres from their grey rami; these fibres provide another path by which postganglionic nerves can pass from the upper thoracic ganglia to the brachial plexus.

THORACIC PART OF THE SYMPATHETIC SYSTEM

The thoracic sympathetic trunk (7.270, 274) contains ganglia almost equal in number to those of the thoracic spinal nerves (11 in more than 70%, occasionally 12, rarely 10 or 13). The first thoracic ganglion is usually fused with the inferior cervical, forming the cervicothoracic ganglion; Jit & Mukerjee (1960) found a fused ganglion in 80 out of 100 dissections. Rarely the middle cervical or second thoracic ganglion may be included. The succeeding ganglion is counted as the second in order to make the other ganglia correspond numerically with other segmental structures. Except for the lowest two or three, the thoracic ganglia lie against the costal heads, posterior to the costal pleura; the lowest two or three are lateral to the bodies of the corresponding vertebrae. Caudally, the thoracic sympathetic trunk passes dorsal to the medial arcuate ligament (or through the crus of the diaphragm) to become the lumbar sympathetic trunk. The ganglia are small and interconnected by intervening segments of the trunk. Two or more rami communicantes, white and grey, connect each ganglion with its corresponding spinal nerve, white rami joining the nerve distal to the grey. Sometimes a grey and a white ramus fuse to form a 'mixed' ramus (p. 1156).

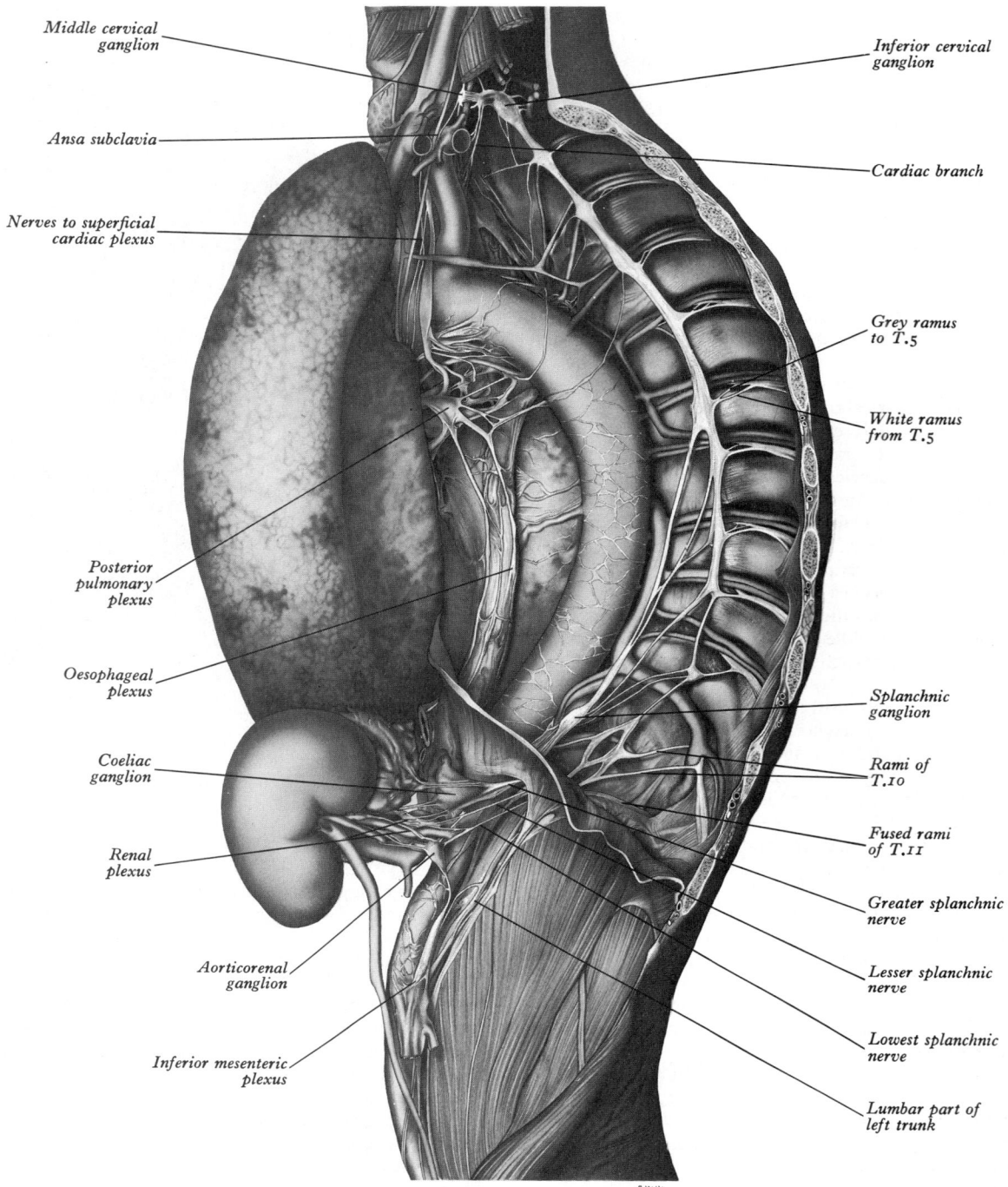

Middle cervical
ganglion

Ansa subclavia

Nerves to superficial
cardiac plexus

Posterior
pulmonary
plexus

Oesophageal
plexus

Coeliac
ganglion

Renal
plexus

Aorticorenal
ganglion

Inferior mesenteric
plexus

Inferior cervical
ganglion

Cardiac branch

Grey ramus
to T.5

White ramus
from T.5

Splanchnic
ganglion

Rami of
T.10

Fused rami
of T.11

Greater splanchnic
nerve

Lesser splanchnic
nerve

Lowest splanchnic
nerve

Lumbar part of
left trunk

S.W.W

7.274 The thoracic part of the sympathetic system of the left side. Drawn from a dissection by the late G D Channell, Guy's Hospital Medical School, London. Note that the diaphragm has been divided to its posterior attachment and the left lung and the left kidney have been drawn forwards and rotated to the right, so as to expose the posterior surface of the left kidney and suprarenal gland.

The *medial branches from the upper five ganglia* are very small, supplying filaments to the thoracic aorta and its branches. On the aorta they form a fine *thoracic aortic plexus* with filaments from the greater splanchnic nerve. Rami of the second to fifth or sixth ganglia enter the posterior pulmonary plexus; others, from the second to fifth ganglia, pass to the deep (dorsal) part of the cardiac plexus. Small branches of these pulmonary and cardiac nerves pass to the oesophagus and trachea. The *medial branches from the lower seven ganglia* are large, supplying the aorta and uniting to form the greater, lesser and lowest splanchnic nerves, the last not always being identifiable.

The *greater splanchnic nerve*, consisting mainly of myelinated preganglionic efferent and visceral afferent fibres, is formed by branches from the fifth to ninth or tenth thoracic ganglia; but fibres in the upper branches may be traced to the first or second thoracic ganglion. Its roots vary from one to eight, four being the most usual number. It descends obliquely on the vertebral bodies, supplies branches to the descending thoracic aorta and perforates the ipsilateral crus of the diaphragm to end mainly in the coeliac ganglion but partly in the aorticorenal ganglion and suprarenal gland. A *splanchnic ganglion* exists on the nerve opposite the eleventh or twelfth thoracic vertebra in 17–68% of *dissections* (Jit & Mukerjee 1960); but Mitchell (1953) reported microscopic evidence that it is always present.

The *lesser splanchnic nerve*, formed by rami of the ninth and tenth (sometimes the tenth and eleventh) thoracic ganglia and the trunk between them, pierces the diaphragm with the greater splanchnic to join the aorticorenal ganglion.

The *lowest (least) splanchnic nerve* (or renal nerve) from the lowest thoracic ganglion enters the abdomen with the sympathetic trunk to end in the renal plexus.

Jit & Mukerjee (1960) described in great detail dissections of the thoracic sympathetic nerves in 50 cadavers and surveyed the previous findings. The incidence of the splanchnic nerves, according to seven observers, is as follows: greater—always present, lesser—94% (86–100%), least—56% (16–98%). A fourth (accessory) splanchnic nerve has been described by de Sousa (1955) but has not been confirmed.

LUMBAR PART OF THE SYMPATHETIC SYSTEM

The lumbar part of each sympathetic trunk (7.274, 275), usually containing four interconnected ganglia, runs in the extraperitoneal connective tissue anterior to the vertebral column and along the medial margin of the psoas major. Superiorly it is continuous with the thoracic trunk posterior to the medial arcuate ligament; inferiorly, passing posterior to the common iliac artery, it becomes the pelvic trunk. On the right side it is overlapped by the inferior vena cava and on the left by the lateral aortic lymph nodes. It is anterior to most of the lumbar vessels but may pass behind some lumbar veins.

The first, second and sometimes third lumbar ventral spinal rami send *white rami communicantes* to the corresponding ganglia. *Grey rami communicantes*, passing from all ganglia to the lumbar spinal nerves, are long and accompany the lumbar arteries round the sides of the vertebral bodies, medial to the fibrous arches to which the psoas major is attached.

Usually four *lumbar splanchnic nerves* pass from the ganglia to join the coeliac, intermesenteric (abdominal aortic) and superior hypogastric plexuses. The first lumbar splanchnic nerve, from the first ganglion, joins the coeliac, renal and intermesenteric plexuses. The second nerve, from the second and sometimes the third ganglion, joins the inferior part of the intermesenteric plexus; the third nerve issues from the third or fourth ganglion, passing anterior to the common iliac vessels to join the superior hypogastric plexus. The fourth lumbar splanchnic, from the lowest ganglion, passes dorsal to the common iliac vessels to join the lower part of the superior hypogastric plexus or the hypogastric 'nerve'.

Vascular branches from all lumbar ganglia join the intermesenteric (aortic) plexus. Fibres of the lower lumbar splanchnic nerves pass to the common iliac arteries, forming a plexus continued along the internal and external iliac arteries as far as the proximal part of the femoral artery. Many postganglionic fibres in the grey rami, connecting the lumbar ganglia to the spinal nerves, travel in the femoral nerve to its muscular, cutaneous and saphenous branches, supplying vasoconstrictor nerves to the femoral artery and its branches in the thigh. Other postganglionic fibres travel via the obturator nerve to the obturator artery. Considerable uncertainties persist regarding sympathetic supplies to the lower limb (Wilde 1951, Wyburn 1956, Pick 1970).

PELVIC PART OF THE SYMPATHETIC SYSTEM

The pelvic sympathetic trunk (7.275) lies in the extraperitoneal tissue anterior to the sacrum, medial or anterior to the anterior sacral foramina, and has four or five interconnected ganglia. Above it continues into the lumbar trunk; below, the two trunks converge to unite in the small *ganglion impar* anterior to the coccyx. *Grey rami communicantes* pass from the ganglia to sacral and coccygeal spinal nerves but white rami communicantes are absent. *Medial branches of distribution* connect across the midline; twigs from the first two ganglia join the inferior hypogastric plexus (pelvic plexus) or the hypogastric 'nerve'; others form a plexus on the median sacral artery. The glomus coccygeum is

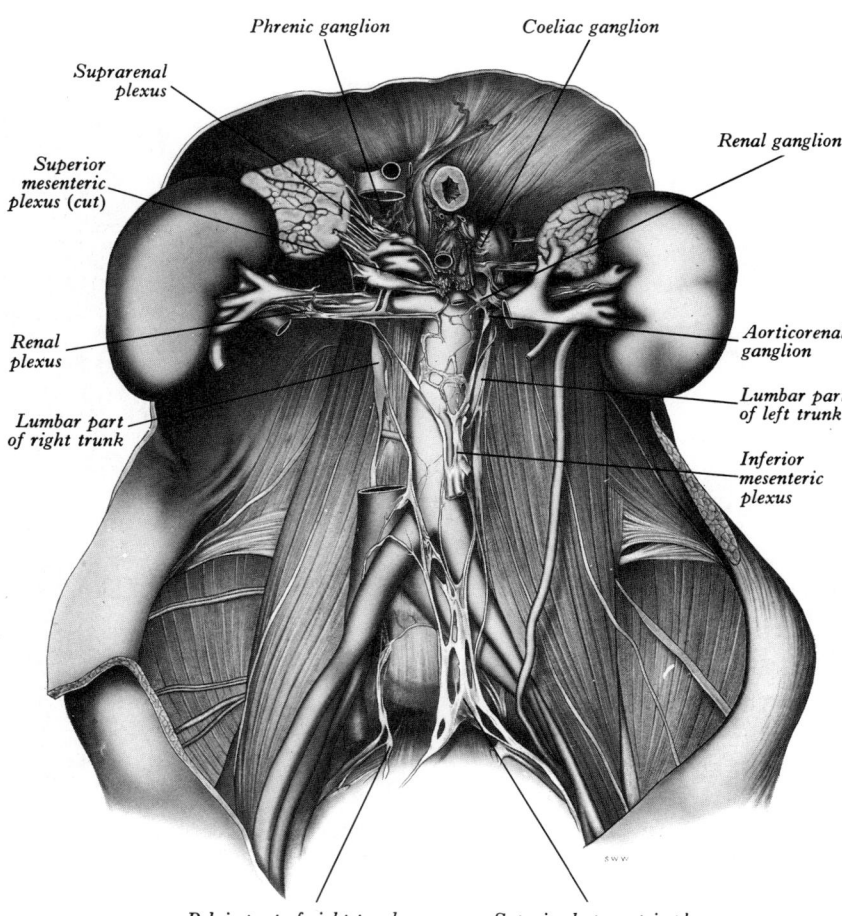

Phrenic ganglion   Coeliac ganglion
Suprarenal plexus
Superior mesenteric plexus (cut)
Renal ganglion
Superior mesenteric plexus (cut)
Renal plexus
Aorticorenal ganglion
Lumbar part of right trunk
Lumbar part of left trunk
Inferior mesenteric plexus

Pelvic part of right trunk   Superior hypogastric plexus

7.275 The abdominal part of the sympathetic system (drawn from a dissection by the late G D Channell, Guy's Hospital Medical School, London).

supplied from the loop between the two trunks. The hypogastric 'nerve', which is usually plexiform, is a redundant term for the right and left connections, between the superior and inferior hypogastric plexuses (p. 1166).

Vascular branches. Through the grey rami many postganglionic fibres pass to the roots of the sacral plexus, especially those forming the tibial nerve, to be conveyed to the popliteal artery and its branches in the leg and foot. Others are carried in the pudendal and superior and inferior gluteal nerves to the accompanying arteries. Branches to lymph nodes are also described (Woźniak 1967).

Preganglionic fibres for the lower limb are derived from the lower three thoracic and upper two or three lumbar spinal segments. They reach the lower thoracic and upper lumbar ganglia through white rami; some descend through the sympathetic trunk to synapse in the lumbar ganglia, whence postganglionic fibres join the femoral nerve to supply the femoral artery and its branches; other fibres descend to synapse in the upper two or three sacral ganglia, from which postganglionic axons join the tibial nerve to supply the popliteal artery and its branches in the leg and foot. Sympathetic denervation of vessels in the lower limb can thus be effected by removing the upper three lumbar ganglia and the intervening parts of the sympathetic trunk, all the preganglionic fibres to the lower limb thus being divided.

PLEXUSES OF THE AUTONOMIC NERVOUS SYSTEM

The larger autonomic plexuses are aggregations of nerves and ganglia situated in the thoracic, abdominal and pelvic cavities. They are the cardiac, pulmonary, coeliac and hypogastric plexuses, supplying the thoracic, abdominal and pelvic viscera. Extensions of these major plexuses pass along most branches of the large vessels which they surround and are usually named after the artery along which they are distributed. This leads to a plethora of named plexuses, often separately described in detail which may overshadow their essential continuity.

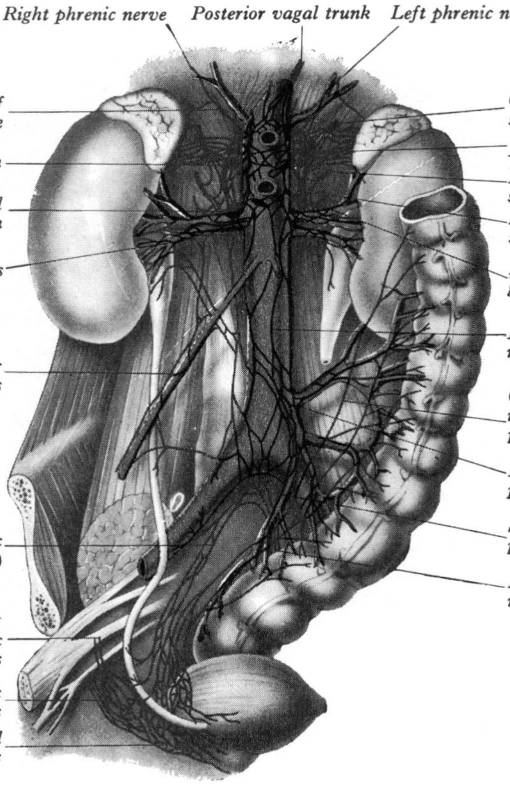

Right phrenic nerve   Posterior vagal trunk   Left phrenic nerve

Suprarenal branch of
phrenic nerve

Right coeliac ganglion

Aorticorenal
ganglion

Renal branches

Superior testicular
nerves

Right hypogastric
nerve (plexus)

Branches from S.1
to inferior hypogastric
plexus

Inferior hypogastric
plexus

Vesical and
prostatic nerves

Greater
splanchnic nerve

Left coeliac ganglion

Lesser
splanchnic nerve

Lowest
splanchnic nerve

Lesser aorticorenal
ganglion

Intermesenteric
nerve

Colic branch from
inferior hypogastric
plexus

Inferior mesenteric
plexus

Superior hypogastric
plexus

Left hypogastric
nerve

7.276 Autonomic nerves and plexuses in the abdomen and pelvis (after Mitchell 1953, by courtesy of the author and the publishers). Note the ascending branches of the inferior hypogastric plexus passing up to supply the descending colon. The sympathetic trunks are not shown on the right side; note the upper, middle and lower ureteric nerves.

CARDIAC PLEXUSES

The cardiac plexus (7.270, 274) at the base of the heart is divided into *superficial* (ventral) and *deep* (dorsal) *parts* which are closely connected. Several small ganglia lie within it, the most constant being the *cardiac ganglion* described below. Mizeres (1963) has emphasized the unity of the cardiac plexus, considering its division into two parts as an artefact of dissection; he was, however, prepared to allow regional names for its coronary, pulmonary, atrial and aortic extensions. Since major concentrations of the plexus are situated as described here, the terms superficial and deep have been retained.

The superficial (ventral) part of the cardiac plexus lies below the aortic arch and anterior to the right pulmonary artery. It is formed by the cardiac branch of the left superior cervical sympathetic ganglion and the lower of the two cervical cardiac branches of the left vagus. A small *cardiac ganglion* is usually present in this plexus immediately below the aortic arch, to the right of the ligamentum arteriosum. This part of the cardiac plexus connects with (1) the deep part, (2) the right coronary plexus, (3) the left anterior pulmonary plexus.

The deep (dorsal) part of the cardiac plexus is anterior to the tracheal bifurcation, above the point of division of the pulmonary trunk and posterior to the aortic arch. It is formed by the cardiac branches of the cervical and upper thoracic sympathetic ganglia and of the vagus and recurrent laryngeal nerves. The only cardiac nerves which do not join it are those joining the superficial part of the plexus.

Branches from the *right half* of the deep part of the cardiac plexus pass in front of and behind the right pulmonary artery; those anterior to it, the more numerous, supply a few filaments to the right anterior pulmonary plexus and continue on to form part of the right coronary plexus; those behind the pulmonary artery supply a few filaments to the right atrium and then continue into the left coronary plexus. The *left half* of the deep part of the cardiac plexus is connected with the superficial, supplying filaments to the left atrium and left anterior pulmonary plexus and then continuing to form much of the left coronary plexus.

The left coronary plexus, larger than the right, is formed chiefly by the prolongation of the left half of the deep part of the cardiac plexus and a few fibres from the right; it accompanies the left coronary artery to supply the left atrium and ventricle.

The right coronary plexus, formed from both superficial and deep parts of the cardiac plexus, accompanies the right coronary artery to supply the right atrium and ventricle.

The atrial plexuses, described by Mizeres (1963), are derivatives of the right and left continuations of the cardiac plexus along the coronary arteries. Their fibres are distributed to the corresponding atria, overlapping those from the coronary plexuses.

All the cardiac branches of the vagus and sympathetic contain both afferent and efferent fibres, except the cardiac branch of the superior cervical sympathetic ganglion, which is purely efferent. The *efferent* preganglionic cardiac *sympathetic* fibres arise in the upper four or five thoracic spinal segments; they pass by white rami communicantes to synapse in the upper thoracic sympathetic ganglia, though many ascend to synapse in the cervical ganglia. Postganglionic fibres from the thoracic and cervical ganglia form the sympathetic cardiac nerves, which accelerate the heart and dilate the coronary arteries. Of the sympathetic fibres from the first four or five spinal segments, the upper pass to the ascending aorta, pulmonary trunk and ventricles, the lower to the atria.

The *efferent* cardiac *parasympathetic* fibres from the dorsal vagal nucleus and neurons near the nucleus ambiguus run in vagal cardiac branches to synapse in the cardiac plexuses and atrial walls. These vagal fibres slow the heart and cause constriction of the coronary arteries (p. 723). In man (like most mammals) intrinsic cardiac neurons are limited to the atria and interatrial septum (Davies et al 1952, King & Coakley 1958); they are most numerous in the subepicardial connective tissue (6.57) near the sinuatrial and atrioventricular nodes. For variations in human sympathetic cardiac innervation consult Ellison & Williams (1969).

PULMONARY PLEXUSES

These are anterior and posterior to the other hilar structures of the lungs, the anterior plexus being much smaller. According to Mizeres (1963) they are extensions from the cardiac plexus along the right and left pulmonary arteries. They are formed by vagal and sympathetic branches. Efferent parasympathetic fibres arise from the dorsal vagal nucleus; efferent sympathetic fibres are postganglionic branches of the second to fifth thoracic sympathetic ganglia.

The *anterior pulmonary plexus* is formed by rami from vagal and cervical sympathetic cardiac nerves as well as direct branches from both sources; the *posterior pulmonary plexus* is formed by the rami of vagal cardiac branches from the second to fifth or sixth thoracic sympathetic ganglia, the left plexus also receiving branches from the left recurrent laryngeal nerve. The two plexuses are interconnected; from them nerves enter the lung as networks along branches of the bronchi and pulmonary and bronchial vessels extending as far as the visceral pleura. Near the hila they contain minute groups of postganglionic neurons, with which efferent vagal preganglionic fibres synapse. In the small intestine *interstitial cells* have been described in terminal autonomic networks, but have not been seen in thoracic organs, apart perhaps from the oesophagus (Dijkstra 1969). Efferent vagal fibres are bronchoconstrictor, secretomotor to bronchial glands and vasodilator. Efferent sympathetic fibres are bronchodilator and vasoconstrictor.

COELIAC PLEXUS

The coeliac (7.270, 275), the largest major autonomic plexus, sited at the level of the last thoracic and first lumbar vertebrae, is a dense network uniting two large *coeliac ganglia*. It surrounds the coeliac artery and the root of the superior mesenteric artery. It is posterior to the stomach and omental bursa, anterior to the crura of the diaphragm and the commencement of the abdominal aorta and between the suprarenal glands. The plexus and ganglia are joined by the greater and lesser splanchnic nerves of both sides and branches from both the vagus and phrenic nerves. They extend as numerous secondary plexuses along adjacent arteries.

The *coeliac ganglia* are irregular masses, one on each side, between the suprarenal gland and the coeliac origin and in front of the crura; the right one is behind the inferior vena cava, the left behind the splenic vessels. The upper part of each is joined by a greater splanchnic nerve; the lower part, more or less detached as the *aorticorenal* ganglion, receives the lesser splanchnic nerve and forms most of the renal plexus; its position is variable but near the origin of the renal artery from the aorta. (For details consult Norvell 1968.) Secondary plexuses from or connected with the coeliac are: the phrenic, splenic, hepatic, left gastric, intermesenteric, suprarenal, renal, testicular or ovarian, superior mesenteric and inferior mesenteric.

The phrenic plexus accompanies the inferior phrenic artery to the diaphragm, with branches to the suprarenal gland. It arises near the upper end of the coeliac ganglion and is larger on the right. It receives one or two phrenic branches. At the junction of the right phrenic plexus with the phrenic nerve is a small *phrenic ganglion*, distributing branches to the inferior vena cava, suprarenal and hepatic plexuses.

The hepatic plexus, the largest coeliac derivative, also receives filaments from the left and right vagi and right phrenic nerve. It accompanies the hepatic artery and portal vein and their branches into the liver, where its fibres are confined to the vicinity of the blood vessels. It follows all branches of the hepatic artery. Branches to the gallbladder form a thin *cystic plexus*; bile ducts are also supplied. Branches accompanying the right gastric artery supply the pylorus. From the gastroduodenal extension of the plexus branches reach the pylorus and the first part of the duodenum. Many follow the right gastro-epiploic artery to supply the right side of the stomach and the greater curvature. The superior pancreaticoduodenal extension of the plexus supplies the duodenum's descending part, the pancreatic head and the lower part of the bile duct. The hepatic plexus contains afferent and efferent sympathetic and parasympathetic fibres; the vagal constituents are said to be motor to the musculature of the gallbladder and bile ducts and inhibitory to the sphincter of the bile duct. Petkov (1968) identified a distinct nerve to the sphincter in 23 out of 25 human dissections.

The left gastric plexus accompanies its artery along the lesser curvature of the stomach, joining with the vagal gastric branches. Gastric sympathetic nerves are motor to the pyloric sphincter but inhibitory to the gastric mural musculature.

The splenic plexus, formed by branches of the coeliac plexus, left coeliac ganglion and right vagus, accompanies its artery to the spleen, giving off subsidiary plexuses along arterial branches. The fibres are mainly, perhaps wholly, sympathetic and terminate in blood vessels and non-striated muscle of the splenic capsule and trabeculae.

The suprarenal plexus is formed by branches from the coeliac ganglion and plexus and greater splanchnic nerve. Relative to its size, the suprarenal gland has a larger autonomic supply than any other organ. Its fibres are commonly described as myelinated and preganglionic. In rats, however, non-myelinated fibres are ten times as numerous and are considered preganglionic; they end in synapses, often deeply invaginated, with large chromaffin cells, the phaeochromocytes, which are thus homologous with the postganglionic sympathetic neurons (p. 1469). A space of 150–200 nm separates the synaptic membranes, which often have electron-dense zones. Small and large vesicles with electron-dense granular contents occur in these endings. Only non-myelinated fibres appear to innervate chromaffin cells, all of which are related to one or more such terminals. Multipolar neurons also occur in the adrenal medulla; some preganglionic non-myelinated fibres form axodendritic synapses with them. The destination of their axons is not known (Coupland 1965a). A preponderance of non-myelinated fibres has also been described in the human suprarenal plexus (Coupland 1965a,b, Grottel 1968).

The renal plexus is dense and formed by rami from the coeliac ganglion and plexus, aorticorenal ganglion, lowest thoracic splanchnic nerve, first lumbar splanchnic nerve and aortic plexus. Small ganglia occur in the renal plexus, the largest usually behind the start of the renal artery. The plexus continues into the kidney around the arterial branches to supply the vessels, renal glomeruli, and tubules, especially the cortical tubules (Norvell 1968). Renal nerves are mostly vasomotor. From the renal plexus branches supply ureteric and testicular (or ovarian) plexuses. The *ureteric plexus* receives, in its upper part, branches from the renal and aortic plexuses, in its intermediate part from the superior hypogastric plexus and hypogastric nerve and in its lower part from the hypogastric nerve and inferior hypogastric plexus. This supply influences the inherent motility of the ureter.

The testicular plexus accompanies the gonadal artery to the testis. Its upper part receives branches from the renal and aortic plexuses. Distally it is reinforced from the superior and inferior hypogastric plexuses. Its rami pass to the epididymis and ductus deferens.

The ovarian plexus accompanies the ovarian artery to the ovary and uterine tube. The upper part is formed by branches from the renal and aortic plexuses; its lower part is reinforced from the superior and inferior hypogastric plexuses.

The nerves in the testicular and ovarian plexuses contain efferent and afferent sympathetic fibres; the efferents are vasomotor and derived from the tenth and eleventh thoracic spinal segments; the parasympathetic fibres, from the inferior hypogastric plexuses, are probably vasodilator.

The superior mesenteric plexus, a downward continuation of the coeliac, receives a branch from the junction of the right vagus and coeliac plexus. It accompanies the superior mesenteric artery into the mesentery, dividing into secondary plexuses distributed to parts supplied by the artery: pancreatic, jejunal and ileal, ileocolic, right colic and middle colic. The *superior mesenteric ganglion* lies superior in the plexus, usually above the superior mesenteric artery's origin. Intestinal sympathetic nerves are motor to the ileocaecal sphincter but inhibitory to the mural musculature; some are vasoconstrictor.

The abdominal aortic plexus (intermesenteric plexus) is formed by branches from the coeliac plexus and ganglia and receives rami from the first and second lumbar splanchnic nerves. It is on the sides and front of the aorta, between the origins of the superior and inferior mesenteric arteries. It consists of four to 12 intermesenteric nerves, connected by oblique branches. It is continuous above with the coeliac plexus and coeliac and aorticorenal ganglia, below with the superior hypogastric plexus. From it parts of testicular, inferior mesenteric, iliac and superior hypogastric plexuses arise; it also supplies the inferior vena cava.

The inferior mesenteric plexus, chiefly from the aortic plexus but also from the second and third lumbar splanchnic nerves, surrounds the inferior mesenteric artery and is distributed along its branches; thus a left colic plexus supplies the transverse colon's left part, the descending and the sigmoid colon; a superior rectal plexus supplies the rectum. Near the origin of the inferior mesenteric artery an *inferior mesenteric ganglion* may occur but more often small ganglia are scattered around the origin of the artery in the proximal part of the plexus. In one study (Southam 1959) an inferior mesenteric ganglion occurred in all of 22 human stillborn infants. The colic sympathetic nerves are inhibitory to mural muscle in the colon and rectum. Branches from parasympathetic pelvic splanchnic nerves ascend occasionally through but usually near the superior hypogastric and inferior mesenteric plexuses to supply the large intestine from the left half of the transverse colon to the rectum (p. 1375 and vide infra); they are motor to the colic musculature. It is to be emphasized that the parasympathetic supply to the distal colon is largely by these direct branches of the pelvic splanchnic nerves, *not* via the hypogastric and inferior mesenteric plexuses (Mitchell 1935, Woodburne 1956).

SUPERIOR HYPOGASTRIC PLEXUS

The superior hypogastric plexus (7.275–277) is anterior to the aortic bifurcation, the left common iliac vein, median sacral vessels, fifth lumbar vertebral body and sacral promontory and between the common iliac arteries. Often termed the *presacral nerve*, it is seldom a single nerve and is prelumbar rather than presacral. It lies in extraperitoneal connective tissue; the parietal peritoneum can easily be stripped off its anterior aspect. It varies in breadth and condensation of its constituent nerves and is often a little to one side of the midline (more often to the left); the

attachment of the sigmoid mesocolon, containing superior rectal vessels, is to the left of the lower part of the plexus. Scattered neurons occur in it. The plexus is formed by branches from the aortic plexus and third and fourth lumbar splanchnic nerves. It divides into right and left hypogastric 'nerves' which descend to the two inferior hypogastric plexuses. The superior plexus supplies branches to the ureteric, testicular, ovarian and common iliac plexuses. In addition to sympathetic fibres, it may also contain parasympathetic fibres (from pelvic splanchnic nerves) which ascend from the inferior hypogastric plexus; but these fibres usually ascend to the left of the superior hypogastric plexus and across the sigmoid branches of left colic vessels. These parasympathetic fibres are distributed partly along the inferior mesenteric arterial branches and also as independent retroperitoneal nerves, to supply the left part of the transverse colon, left colic flexure, descending and sigmoid colon (p. 1156 and 7.275).

INFERIOR HYPOGASTRIC PLEXUSES

The *superior hypogastric plexus* divides caudally into right and left *hypogastic 'nerves'*, each descending in extraperitoneal connective tissue into the pelvis, medial to each internal iliac artery and its branches to become the *inferior hypogastric plexus* (7.277). Each nerve may be single or an elongated plexus of anastomosing filaments. (Hypogastric nerves can scarcely be distinguished from their continuations, the inferior hypogastric plexuses. The latter are joined by pelvic splanchnic nerves, a distinction minimized by the fact that both nerves and plexuses contain sympathetic and parasympatheric fibres. Some authorities prefer to describe a superior hypogastric plexus dividing into two inferior plexuses.) From each hypogastric nerve branches may pass to the testicular, ovarian and ureteric plexus or to the internal iliac plexuses and to the sigmoid colon; each nerve may be joined initially by the lowest lumbar splanchnic nerve from last lumbar sympathetic ganglion.

The inferior hypogastric (pelvic) plexus is in the extraperitoneal connective tissue. In males it is lateral to: the rectum, seminal vesicle, prostate and the posterior part of the urinary bladder; in females each plexus is lateral to: the rectum, uterine cervix, vaginal fornix and the posterior part of the urinary bladder, extending into the broad uterine ligament. Lateral to it are the internal iliac vessels and their branches and tributaries, the levator ani, coccygeus and obturator internus. Posterior are the sacral and coccygeal plexuses and above are the superior vesical and obliterated umbilical arteries. The plexuses contain numerous small ganglia. Each is formed by a hypogastric nerve, conveying most of the sympathetic fibres of the plexus, the remaining few arriving via branches from the ganglia. Parasympathetic fibres are derived from pelvic splanchnic nerves. Preganglionic efferent sympathetic fibres originate in the lower three thoracic and upper two lumbar spinal segments, some relaying in ganglia of the lumbar and sacral parts of the sympathetic trunk, others synapsing in the lower part of the aortic plexus and in the superior and inferior hypogastric plexuses. Preganglionic parasympathetic fibres originate in the second to fourth sacral spinal segments, reach the plexus in the pelvic splanchnic nerves and synapse in it or in walls of viscera supplied by its branches. Numerous branches are distributed to the pelvic and some abdominal viscera, either directly or along their arteries.

Parasympathetic fibres ascend in the hypogastric plexuses or as separate filaments to reach the inferior mesenteric plexus by way of the aortic plexus. By this route the descending and sigmoid parts of the colon receive parasympathetic innervation.

The middle rectal plexus, formed by fibres from the upper part of the inferior hypogastric plexus to the rectum passing directly or along the middle rectal artery, connects above with the superior rectal plexus and extends below to the internal anal sphincter. The rectal and anal nerve supply is from: (1) the superior rectal plexus, (2) the middle rectal plexus and (3) the inferior rectal (haemorrhoidal) nerves, branches of the pudendal nerve. The parasympathetic preganglionic fibres from the rectal plexuses synapse with postganglionic neurons in the well-developed myenteric plexus, while sympathetic afferents pass through it without interruption. Efferent sympathetic fibres in

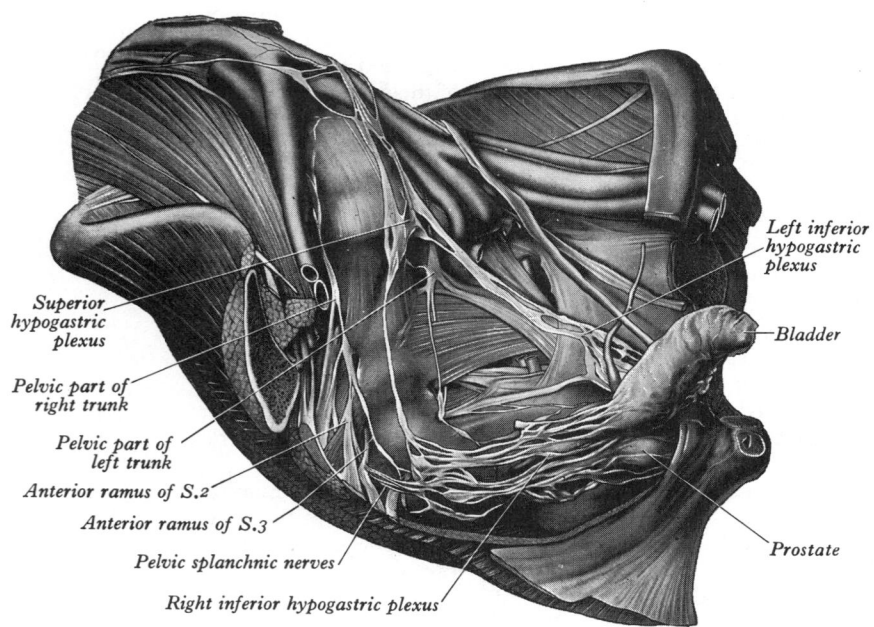

Superior hypogastric plexus

Pelvic part of right trunk

Pelvic part of left trunk

Anterior ramus of S.2

Anterior ramus of S.3

Pelvic splanchnic nerves

Right inferior hypogastric plexus

Left inferior hypogastric plexus

Bladder

Prostate

S.W.W

7.277 The pelvic part of the sympathetic system, viewed from in front and from the right side, a large part of the right hip bone having been removed. The superior hypogastric plexus is seen to divide below into the right and left hypogastric nerves (which are not labelled), which run down to the inferior hypogastric plexuses. Drawn from a dissection by the late G D Channell, Guy's Hospital Medical School, London.

the rectal plexuses inhibit the expulsive musculature and stimulate the sphincter. Pain impulses traverse the sympathetic and parasympathetic fibres but the parasympathetic afferent and efferent fibres are more active in normal defaecation. Inferior rectal nerves supply motor fibres to the striated external anal sphincter and sensory (somatic) fibres to the lower (ectodermal) part of the anal canal (p. 233).

The vesical plexus, from the anterior part of the inferior hypogastric plexus, comprises many filaments which pass along vesical arteries to the bladder. Branches supply the seminal vesicles and deferent ducts. Many small groups of neurons exist among the nerve fibres in the vesical muscular wall. Sympathetic preganglionic fibres in the plexus are from the lower two thoracic and upper two lumbar spinal segments, synapsing with neurons scattered in the superior and inferior hypogastric plexuses and vesical wall. The parasympathetic preganglionic efferent fibres come from the second to fourth sacral spinal segments and synapse near or in the vesical wall with postganglionic neurons which stimulate its detrusor muscle and inhibit its sphincter. Efferent sympathetic nerves are motor to the sphincter and inhibitor to the detrusor muscle; but some maintain that they are mainly vasomotor and that vesical filling and emptying are controlled by parasympathetic nerves.

The prostatic plexus, continued from the lower part of the inferior hypogastric plexus, is composed of large nerves entering the base and sides of the prostate and contains neurons. It supplies: the prostate, seminal vesicles, prostatic urethra, ejaculatory ducts, corpora cavernosa, corpus spongiosum, membranous and penile urethra and bulbo-urethral glands. The nerves to the corpora cavernosa form two sets, the lesser and greater cavernous nerves, arising from the front of the plexus to join branches from the pudendal nerve and then passing forwards below the pubic arch. The precise localization of the autonomic nerves from the pelvic plexus to the corpora cavernosa has been described by Lepor et al (1985) in the adult male pelvis. *Lesser cavernous nerves* pierce the fibrous penile sheath proximally to supply the erectile tissue of the corpus spongiosum and penile urethra. *Greater cavernous nerves* proceed on the dorsum penis, connect with the dorsal nerve and supply the erectile tissue, some filaments reaching the erectile tissue of the corpus spongiosum. Sympathetic supplies to the male genital organs produce vasoconstriction, the parasympathetic being vasodilator. Seminal vesicles are supplied from the vesical and prostatic plexuses and inferior hypogastric nerves; extensions pass to the ejaculatory and deferent ducts. Contraction of the seminal vesicles and ejaculation are considered to be due to the sympathetic supply, which also inhibits the vesical musculature and stimulates the sphincter during ejaculation, preventing reflux into the bladder. Others have suggested that contraction of the seminal vesicles is under parasympathetic control (Matthews & Raisman 1969).

Uterine nerves arise from the inferior hypogastric plexus, mainly the part in the broad ligament, the **uterovaginal plexus**, from which branches descend with the vaginal arteries, while others pass directly to the cervix uteri or ascend with or near uterine arteries in the broad ligament. Nerves to the cervix form a plexus in which are small *paracervical* ganglia, one ganglion sometimes being larger and termed the *uterine cervical ganglion*. Nerves ascending with the uterine arteries supply the uterine body and tube, connecting with *tubal nerves* from the inferior hypogastric plexus and with the ovarian plexus. The uterine nerves ramify in the myometrium and endometrium, generally accompanying the vessels. Efferent preganglionic sympathetic fibres are from the last thoracic and first lumbar spinal segments; the sites of their postganglionic neurons are unknown. Preganglionic parasympathetic fibres arise in the second to fourth sacral spinal segments and relay in the paracervical ganglia. Sympathetic activity may produce uterine contraction and vasoconstriction and parasympathetic activity may produce uterine inhibition and vasodilatation, but these activities are complicated by hormonal control of uterine functions.

Vaginal nerves from the lower parts of the inferior hypogastric and uterovaginal plexuses follow the vaginal arteries to supply the vaginal walls, the erectile tissue of the vestibular bulbs and clitoris (cavernous nerves of the clitoris), the urethra and the greater vestibular glands. The nerves contain many parasympathetic fibres which are vasodilator to the erectile tissue.

Afferent Autonomic Pathways

Efferent autonomic fibres to the viscera and blood vessels are accompanied by their sensory counterparts, the *general visceral afferent* or *autonomic afferent* fibres. These are the peripheral processes of unipolar neurons in some cranial and spinal ganglia. Their central axons enter the central nervous system in the nerves concerned; their peripheral dendrites, myelinated or non-myelinated fibres of varying calibre, are distributed with pre- and postganglionic fibres of both autonomic divisions but do not synapse in their ganglia. Their terminals are described as knobs, loops, rings, tendrils and more elaborate encapsulated endings in the visceral and vascular walls, including the epithelia and serosae. Afferent impulses conducted by these neurons evoke visceral reflexes but usually do not obtrude on consciousness. They also probably mediate organic visceral sensations such as hunger, nausea, sexual excitement, vesical distension, etc. and visceral pain fibres may follow these routes. Although viscera are insensitive to cutting, crushing or burning, excessive tension in non-striated muscle and some pathological conditions produce visceral pain. It is not always easy to distinguish between what is pathological and mere exaggeration of normal activity. 'Abdominal' pain due to strong intestinal contraction is commonplace. In visceral disease vague pain may be felt near the viscus itself (visceral pain) or in a cutaneous area or other tissue whose somatic afferents enter spinal segments receiving afferents from the viscus; this is *referred pain*. If inflammation spreads from a diseased viscus to the adjacent parietal serosa (e.g. the peritoneum), somatic afferents will be stimulated, causing local somatic, commonly spasmodic, pain in this region. Referred pain is often associated with local cutaneous tenderness.

General visceral afferent fibres exist in vagal, glossopharyngeal and possibly other cranial nerves in the second to fourth sacral nerves distributed with pelvic splanchnic nerves and in thoracic and upper lumbar spinal nerves, distributed through rami communicantes and alongside the efferent sympathetic innervation of viscera and blood vessels.

The large vagal general visceral afferent components have their somata in the superior and inferior vagal ganglia, which appear predominantly sensory, despite contrary views regarding the inferior ganglion. Their central processes end in the dorsal vagal nucleus in the medulla oblongata or, according to some, the nucleus solitarius. Their peripheral processes are distributed to terminals in the pharyngeal and oesophageal walls, where with glossopharyngeal visceral afferents in the pharynx, they are concerned in swallowing reflexes. Vagal afferents are also ascribed to the thyroid and parathyroid glands. In the thorax some end in the heart, the walls of the great vessels, the aortic bodies and pressor receptors; in the last they are stimulated by raised intravascular pressure. Some reach the lungs through the pulmonary plexuses, being distributed to: (1) bronchial mucosa, probably involved in cough reflexes, (2) bronchial muscle, where they encircle myocytes and end in tendrils sometimes regarded as 'muscle spindles', believed to be stimulated by change in the length of myocytes, (3) interalveolar connective tissue, where their knob-like endings, together with terminals on myocytes, may evoke Herring–Breuer reflexes, (4) adventitia of pulmonary arteries, where they may be pressor receptors, and the intima of pulmonary veins, where they may be chemoreceptors. Afferent fibres from visceral pleura and bronchi may also travel with the pulmonary sympathetic supply, mediating nociceptive responses.

Vagal visceral afferent fibres also end in the gastric and intestinal walls, in digestive glands, and the kidney. Fibres ending in the gut and its ducts respond to stretch or contraction. Gastric impulses may evoke sensations of hunger and nausea.

Glossopharyngeal general visceral afferents innervate the posterior lingual region, tonsils and pharynx, whose epithelia are endodermal. Innervation of taste buds is not included; these have *special* visceral afferents (p. 1170). Glossopharyngeal afferents also innervate the carotid sinus and body, receptors sensitive to tension and changes in chemical composition of the blood; impulses from these receptors are essential to circulatory and respiratory reflexes. Somata of the glossopharyngeal general visceral afferents are in the nerve's ganglia, their terminations in the medulla oblongata being probably like those of the vagal visceral afferents.

Sensory fibres in pelvic splanchnic nerves innervate pelvic viscera and the distal part of the colon. Vesical receptors are widespread; those in muscle strata are connected with thickly myelinated fibres reaching the spinal cord via pelvic splanchnic nerves and are believed to be stretch receptors, possibly activated by contraction. Pain fibres from the bladder and proximal urethra traverse both pelvic splanchnic nerves and the inferior hypogastric plexus, hypogastric nerves, superior hypogastric plexus and lumbar splanchnic nerves to reach their somata in ganglia on the lower thoracic and upper lumbar dorsal spinal roots. The significance of this dual sensory pathway is uncertain; lesions of the cauda equina abolish pain from vesical overdistension but hypogastric section is ineffective.

Though the fibres of pelvic splanchnic nerves, possibly afferent, are described in the ovary, no supply from this source to the testis is known.

Pain fibres from the uterine body traverse the hypogastric plexus and lumbar splanchnic nerves to reach somata in the lowest thoracic and upper lumbar spinal ganglia; hypogastric division may relieve dysmenorrhoea. Afferents from the cervix, however, traverse the pelvic splanchnic nerves to somata in the upper sacral spinal ganglia. Stretch of the cervix uteri causes pain but cauterization and biopsy excisions do not.

Afferent fibres accompanying pre- and postganglionic sympathetic fibres have a generally segmental arrangement. They end in spinal cord segments from which preganglionic fibres innervate the region or viscus concerned (vide infra). General visceral afferents entering thoracic and upper lumbar spinal segments are largely concerned with pain. Nociceptive impulses from the pharynx, oesophagus, stomach, intestines, kidney, ureter, gallbladder and bile ducts seem to be carried in sympathetic pathways. Cardiac nociceptive impulses enter the spinal cord via the first to fifth thoracic spinal nerves, carried mainly in the middle and inferior cardiac nerves, but a few pass directly to the spinal nerves. There are no general visceral afferents in the superior cardiac nerves. Peripherally the fibres pass through the cardiac plexuses and along the coronary arteries. Myocardial anoxia may evoke symptoms of angina pectoris in which there is typically presternal pain, pain referred to much of the left chest, radiating to the left shoulder and the left arm's medial aspect, along the left side of the neck to the jaw and occiput and down to the epigastrium; pain may be bilateral or even confined to the right. Cardiac afferents are also carried in vagal cardiac branches and are concerned with the reflex depression of cardiac activity. In some animals (e.g. rabbit) a separate depressor cardiac nerve is a branch of the vagus or superior laryngeal nerve. Human depressor fibres, however, run in branches of the superior or internal laryngeal nerves, cardiac branches of the vagus or the sympathetic. Ureteric pain fibres, also running with sympathetic fibres, are presumably concerned in the agonizing renal colic of obstruction by calculi. Afferent fibres from the testis and ovary run through the corresponding plexuses; their somata are in the tenth and eleventh thoracic dorsal spinal root ganglia.

It must be emphasized that autonomic reflexes are not initiated solely by general visceral afferent pathways, nor do impulses in these necessarily activate general visceral efferents. In most situations demanding general sympathetic activity for effort, the afferent element is usually somatic, either from special senses or the skin. Rises in blood pressure and pupillodilatation may result from the stimulation of somatic receptors in the skin and other tissues. Conversely, contraction of the rectus abdominis, a somatic structure, may result from irritation of abdominal viscera. There is also evidence that axon reflexes may be evoked at the terminals of autonomic postganglionic fibres.

Denervation often has no obvious effect on the effector organs, non-striated muscle or glands innervated by autonomic fibres. Contraction may be unaffected by denervation and no structural changes ensue. This is variously attributed to the continued activity of local plexuses or the intrinsic activity of visceral muscle. The severance of preganglionic efferent fibres may result in the hypersensitivity of postganglionic neurons. Denervation does sometimes result in cessation of activity, e.g. in sweat glands, pilomotor muscle, orbital non-striated muscle and the suprarenal medulla (Pick 1970).

DEGENERATION AND REGENERATION IN AUTONOMIC NERVES

Degeneration in the autonomic nervous system resembles that in cerebrospinal nerves. Some evidence suggests that the rate of degeneration differs in different regions or different types of fibre. Regeneration of preganglionic fibres may vary with the site of lesion and, in postganglionic neurons, regeneration may be followed by reinnervation from neighbouring intact nerve fibres. As far as experimental evidence goes, the integrity of Schwann cell sheaths is essential in the regeneration of autonomic nerve fibres (Evans & Murray 1954, Kapeller & Mayor 1967, Williams 1971, King & Thomas 1971, Landon 1976). Some observations suggest that proximity of myelinated fibres is necessary to regeneration in non-myelinated fibres (Evans & Murray 1954, Williams 1971). It is pertinent to mention earlier experiments in which large experimental gaps in the sympathetic trunk, in monkeys and other mammals, have been filled by the growth of fibres, pre- or postganglionic (Tower & Richter 1931, Haxton 1954). Conflicting evidence comes from the human sympathectomies; functional recoveries sometimes observed may be otherwise explained, such as by incomplete interruption of the sympathetic supply or by alternative routes being overlooked. On the whole the experience of surgeons corroborates the experimental findings (Pick 1970).

SEGMENTAL SYMPATHETIC SUPPLIES

Head and neck	T1–5
Upper limb	T2–5
Lower limb	T10–L2
Heart	T1–5
Bronchi and lung	T2–4
Oesophagus (caudal part)	T5–6
Stomach	T6–10
Small intestine	T9–10
Large intestine to splenic flexure	T11–L1
Splenic flexure to rectum	L1–2
Liver and gallbladder	T7–9
Spleen	T6–10
Pancreas	T6–10
Kidney	T10–L1
Ureter	T11–L2
Suprarenal	T8–L1
Testis and ovary	T10–11
Epididymis, ductus deferens, seminal vesicles	T11–12
Urinary bladder	T11–L2
Prostate and prostatic urethra	T11–L1
Uterus	T12–L1
Uterine tube	T10–L1

Applied Anatomy. Various autonomic nervous structures are divided or removed in treating several pathological conditions. In operations on the efferent sympathetic paths, ganglia on the sympathetic trunk are removed or preganglionic fibres cut, rather than postganglionic fibres, since the latter may regenerate. For example, the arteries of limbs may be denervated to alleviate vascular spasm (Raynaud's disease) and the parts removed are as described above (pp. 1160, 1123). In the treatment of hypertension, more extensive sympathectomy has been performed, involving bilateral removal of the sympathetic trunks from the eighth thoracic to the first lumbar ganglia, including the greater and lesser thoracic splanchnic nerves. Sympathectomy is also performed to relieve pain, e.g. in severe angina pectoris (p. 1167). Division of the superior hypogastric plexus (presacral neurectomy) does not relieve all pain in disease of the pelvic organs, because many pain fibres traverse the pelvic splanchnic nerves. But uterine pain fibres pass in sympathetic nerves via the superior hypogastric plexus so that this division does relieve dysmenorrhoea. In males resection of the superior hypogastric plexus leads to loss of ejaculation and sterility, due to interruption of the sympathetic paths to the seminal vesicles, deferent ducts and prostate.

The routes of these nerves between the sympathetic ganglia and the superior hypogastric plexus are uncertain and may vary; but in some individuals an outflow from the first lumbar and possibly the twelfth thoracic ganglia is concerned and in others fibres from the third lumbar ganglion (White et al 1952). For the surgical anatomy of the autonomic nervous system consult Pick (1970).

THE PERIPHERAL APPARATUS OF THE SPECIAL SENSES

In this section the detailed anatomy of the taste buds, the olfactory epithelium and nasal cavity and the eye and ear is considered. The general development of these structures is described in Embryology (pp. 201-206), while their connections with the central nervous system and cranial nerves are detailed on pp. 949-1092 and pp. 1094-1123 respectively.

The Gustatory Apparatus

The peripheral organs of gustation are the *taste buds* (*gustatory caliculi*), composed of modified epithelial cells set in piriform groups (7.278–281) within the epithelia of the tongue, soft palate, palatoglossal arches, posterior epiglottic surface and posterior wall of the oropharynx. They are most numerous in the vallate papillae of the tongue (7.279, 280), although fewer in the walls of the groove surrounding each of these structures. They are also abundant over the foliate papillae and on the posterior third of the tongue, but only sparsely scattered on the lingual fungiform papillae, the soft palate, epiglottis and pharynx. Taste buds occur in greater numbers in infants but gradually atrophy with age; those located most posterior in the tongue and in the epiglottis disappear early in development. There are no taste buds in the central region of the dorsum of the tongue.

MICROSCOPIC STRUCTURE OF TASTE BUDS

Each taste bud (7.280, 281) is separated by a basement membrane from the subjacent lamina propria and opens on to the epithelial surface by a *gustatory pore*. In longitudinal section the cells of taste buds are crescentic, their apices converging to a small cavity connected to the gustatory pore (7.280, 281); some cells have

so-called gustatory hairs, which are groups of microvilli. The base of each bud is penetrated by a small fasciculus of afferent nerve fibres which ramify and spiral around certain of its cells.

Gustatory sensory cells have long been recognized as modified epitheliocytes with synaptic contacts with the terminals of afferent fibres. Reconstructions of taste buds from serial electron micrographs (Murray & Murray 1970) have shown at least five distinct types of epithelial cell, two (or three, see Kinnamon et al 1985) apparently receptors and the others supportive or generative in function (7.281). The receptor cells are typified by the presence of synaptic contacts with the afferent terminals at various points on their surfaces; these show a characteristic asymmetical thickening and presynaptic small (50 nm), clear synaptic vesicles in the receptor cell, although these are less abundant than in many other synapses (p. 884). Also, one epithelial type contains dense-cored vesicles about 70 nm in diameter, identified in other sites as catecholamine vesicles (p. 884); in the other receptor these are absent. These two cell types may be different stages of receptor maturation, as they are continually lost and replaced (vide infra). At their apices few microvilli appear; the 'gustatory hairs' of light microscopy mainly occur on supporting cells enclosing the receptors, except at their apices (Miller & Chaudry 1976). One type of these supportive cells appears to form an encapsulating sheath for the bud, separating it from the adjacent epithelium; another type is located basally and is probably a blastemal cell giving rise to new cells. A third type insulates each receptor from its neighbours and also ensheathes the afferent nerve fibres, which lose their Schwann cells as they enter the taste buds. Ultrastructurally, supporting cells are seen to contain dense secretory bodies, probably the source of polysaccharide material in the bud's apical cavity, into which the sensory terminals project. Gustatory molecules must traverse this to reach the receptor surface; its role in gustation is otherwise unknown. Occluding junctions between the apices of various cells in the bud prevent the access of chemical stimuli to non-apical regions of the receptors (Akisaka & Oda 1978).

7.278 A low-power light micrograph of a sagittal section through the tip and anterior part of a human tongue, showing: muscle fibres orientated in three different directions; a delicate non-keratinized stratified squamous epithelium on the ventral surface; and a partly keratinized stratified epithelium on the dorsum.. The latter is convoluted to produce filiform and fungiform papillae. Dermal papillae project into the deep surface of the epidermal irregularities, Haematoxylin and eosin. Magnification × 10.

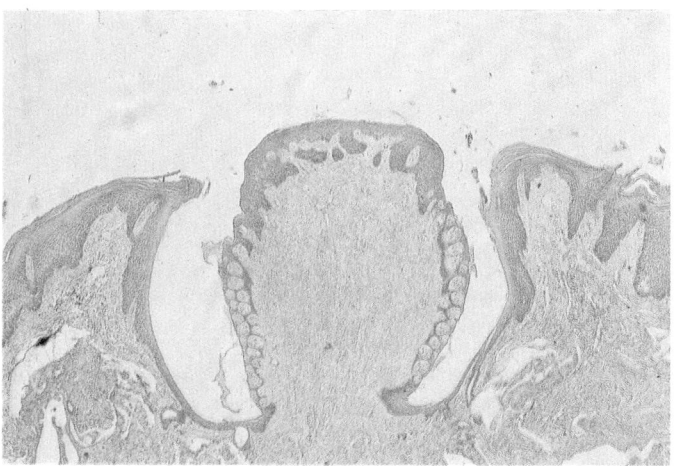

7.279 A light micrograph of a human vallate lingual papilla showing the numerous taste buds clustered along its lateral surfaces. A connective tissue papilla forms the core of the whole structure and serous glands, situated between the muscle blocks deep to the corium, open into the lateral recesses of the papilla. Haematoxylin and eosin. Magnification × 150.

Nerve fibres reaching the taste buds from the subepithelial plexus are complex in their distribution in the tongue, as deduced by recordings from individual nerve fibres in the proximal fasciculi of the chorda tympani and glossopharyngeal nerve. Each fibre may have many terminals, spreading to innervate widely separated taste buds and more than one sensory cell in each bud. Conversely, individual buds may receive the terminals of several nerve fibres. This cross-innervation may be functionally of considerable importance (vide infra and Beidler 1970). No evidence has emerged concerning efferent terminals or inhibitory phenomena in the taste buds.

REPLACEMENT OF CELLS IN TASTE BUDS

The cells of taste buds resemble those in the adjacent epithelium in undergoing continual renewal (Farbman 1980). Isotopic labelling shows that none of the cells of taste buds survive for more than a few days, except the basal, blastemal cells. Newly formed sensory cells must therefore make new synapses with sensory nerve terminals. Experiments indicate that the sensitivity of sensory cells to particular chemicals is determined by the trophic influence of the nerve fibres and not the reverse (p. 919). The continual degeneration and replacement in taste buds implies that their different cells show a scale of structural appearances due to maturation, adding to the difficulties of their classification.

FUNCTION OF GUSTATORY RECEPTORS

In many aquatic vertebrates gustatory receptors are widely distributed over the surface of the body as well as in the alimentrary tract; in terrestrial vertebrates this sense is largely confined to the oral cavity and adjoining pharynx. Human gustation, as commonly understood, is largely mediated by the activity of lingual receptors, though others may participate (vide supra). Four subjective qualities have been classically distinguished, each associated with a region of the tongue, e.g. sweet and salty at the apex, sour (acid) at the sides and bitter at its pharyngeal end. Though these are subjective sensations, it seemed at one time possible that their localization to particular areas might reflect the existence of specific receptors. Electrophysiological evidence suggests that this is, as usual, an over-simplification, since each afferent fibre is connected to widely separated taste buds and may respond to several different chemical stimuli. Some respond to all four classic categories, others to less or only one. Within a particular class of tastes, receptors are also differentially sensitive to a wide range of similar chemicals. It therefore seems that perceived sensations of taste are the results of analysis (presumably central) of a complex pattern of responses from particular areas of the tongue, so that even if there are only a few specific types of taste receptor, much more information about the chemical nature

and intensity of a taste may be gained than if the innervation were simpler (Beidler 1970, Beidler & Smallman 1965, Pfaffman 1970).

GUSTATORY NERVES

Gustatory nerve fibres are the peripheral processes of unipolar neurons in the genicular ganglion of the facial nerve and the inferior ganglia of the glossopharyngeal and vagal nerves. Their central axons form the tractus solitarius (p. 954), their terminals synapsing with neurons in the rostral third of the nucleus of that tract. Secondary gustatory axons cross the midline, many to

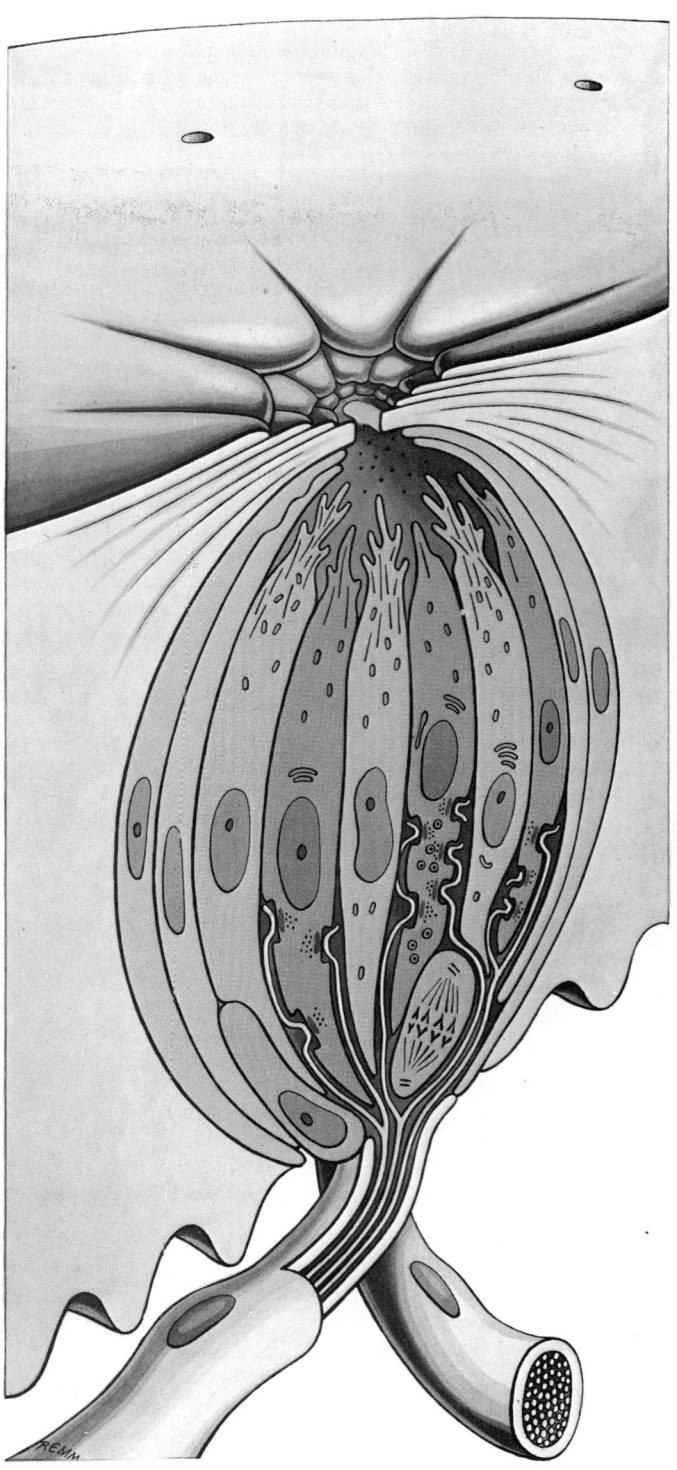

7.281 A schematic reconstruction of the structure of a taste bud, cut away to expose the various cell types. Presumed sensory cells of two types (one with dense-cored vesicles, the other without) are indicated in purple and their innervation in yellow. The supporting cells are indicated in magenta and basal cells in brick red. A dense mucosubstance is present in the apical cavity beneath the apical aperture.

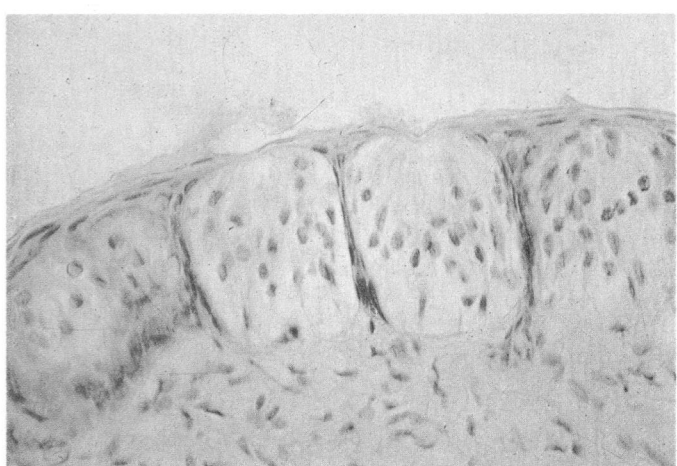

7.280 A light micrograph of a group of taste buds in a vallate lingual papilla, showing the apical cavity and various fusiform cells within the epithelial capsule. Haematoxylin and eosin. Magnification × 400.

ascend the brain stem in the dorsomedial medial lemniscus to synapse with the most medial neurons of the thalamic *nucleus ventralis posterior medialis* (VPM) (in a region sometimes termed the *accessory arcuate nucleus*). From the VPM axons radiate through the internal capsule to the anterio-inferior area of the sensorimotor cortex and the limen insulae. Electrophysiological recording from the VPM and cortex largely confirms these anatomical pathways; some units react to one type of stimulus, others to several different stimuli. Other ascending paths have been described, ending in a number of the hypothalamic nuclei by which gustatory information may reach the limbic system (p. 1028) and allowing appropriate autonomic reactions to be made (p. 1124) (see Stewart & Shepherd 1985).

The gustatory nerve for the anterior part of the tongue, excluding the vallate papillae, is the chorda tympani (via the lingual nerve); in most individuals these fibres traverse the chorda tympani to the facial ganglion but in a few they diverge through connecting branches to the otic ganglion, proceeding thence in the greater petrosal nerve to the facial ganglion (Schwartz & Weddell 1938). The anterior or *oral* part of the tongue is derived from the mandibular branchial arch and is limited posteriorly by the *sulcus terminalis*. Taste buds in the soft palate's inferior surface are also supplied mainly by the facial fibres, through the greater petrosal nerve distributing via the nerve of the pterygoid canal and middle and posterior palatine nerves; the glossopharyngeal also contributes to this region. Taste buds in the vallate papillae and the *pharyngeal* (postsulcal) part of the tongue and in the palatoglossal arches and the oropharynx are innervated by glossopharyngeal fibres, while those in the extreme pharyngeal part of the tongue and epiglottis receive fibres from the internal laryngeal branch of the vagus. (For literature concerning gustatory paths consult Rollin 1979.) These supplies accord with the embryological development of the tongue (p. 228), see also p. 1322).

THE OLFACTORY APPARATUS

The peripheral olfactory organs include the *external nose* and *nasal cavity* which are divided sagitally by a septum into right and left parts. However, these are in reality only accessory organs, the essential olfactory structures being the olfactory epithelium and its nervous connections.

THE EXTERNAL NOSE

Externally the nose is pyramidal, its upper angle or *root* being continuous with the forehead, and its free tip is the *apex*. Inferiorly are two ellipsoidal apertures, the nostrils or external *nares*, separated by the *nasal septum*. The lateral surfaces of the nose are continuous at the median *dorsum nasi*, the shape of which varies greatly between individuals; the upper nose is kept patent by the nasal bones and the frontal processes of the maxillae and below this by cartilages. The lateral surfaces end in rounded *alae nasi*. The supporting framework (7.282) is composed of bone and hyaline cartilages. The *osseus skeleton*, supporting its upper part, consists of the nasal bones, the frontal processes of the maxillae and the nasal part of the frontal bone; the *cartilaginous skeleton* consists of the septal, lateral and major and minor alar cartilages (7.282, 283), all connected to each other and to nearby bones by the continuity of their perichondrium and periosteum. These junctions, though of much interest, have attracted little attention; the cartilages and intervening connective tissue which surround the nares may be regarded as forming a valve, controlling the intake of air (Cottle 1960, Galindo et al 1977).

The *septal cartilage* (7.282B, 283), almost quadrilateral in shape and thicker at its margins than its centre, forms almost the whole of the septum between the anterior parts of the nasal cavity. Its anterosuperior margin is connected above to the posterior border of the internasal suture; the middle part is continuous with the lateral cartilages, and its lowest part is attached to these cartilages by perichondrium. Its antero-inferior border is connected on

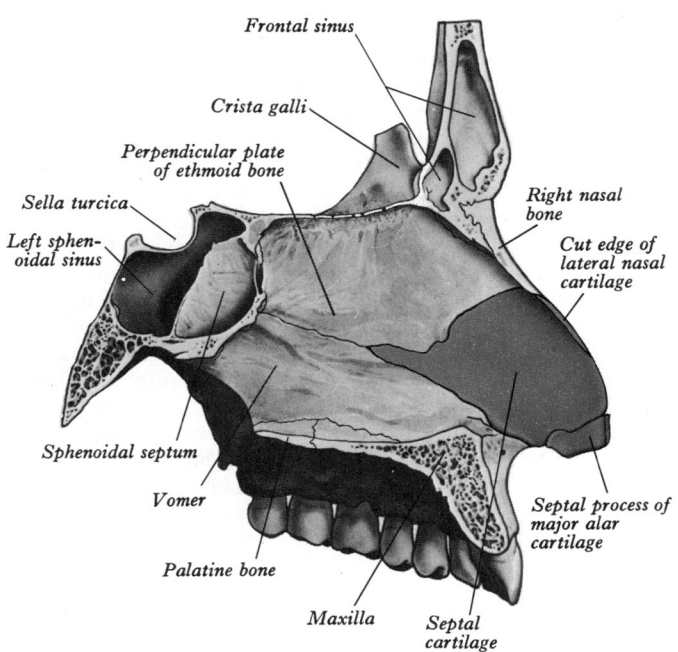

Frontal process of maxilla

Nasal bone

Lateral nasal cartilage

Septal cartilage

Major alar cartilage

Minor alar cartilages

Fibro-fatty tissue

K.V.M.

A

Major alar cartilage

Septal cartilage

Minor cartilages of ala

Alveolar process of maxilla

B

7.282 A. The cartilages of the right side of the nose: lateral aspect. B. The cartilages of the nose: inferior aspect. Note changes in terminology.

Frontal sinus

Crista galli

Perpendicular plate of ethmoid bone

Sella turcica

Left sphenoidal sinus

Sphenoidal septum

Vomer

Palatine bone

Maxilla

Septal cartilage

Right nasal bone

Cut edge of lateral nasal cartilage

Septal process of major alar cartilage

7.283 The right side of the septum of the nose, showing its constituent bones and cartilages. Note the changes in terminology.

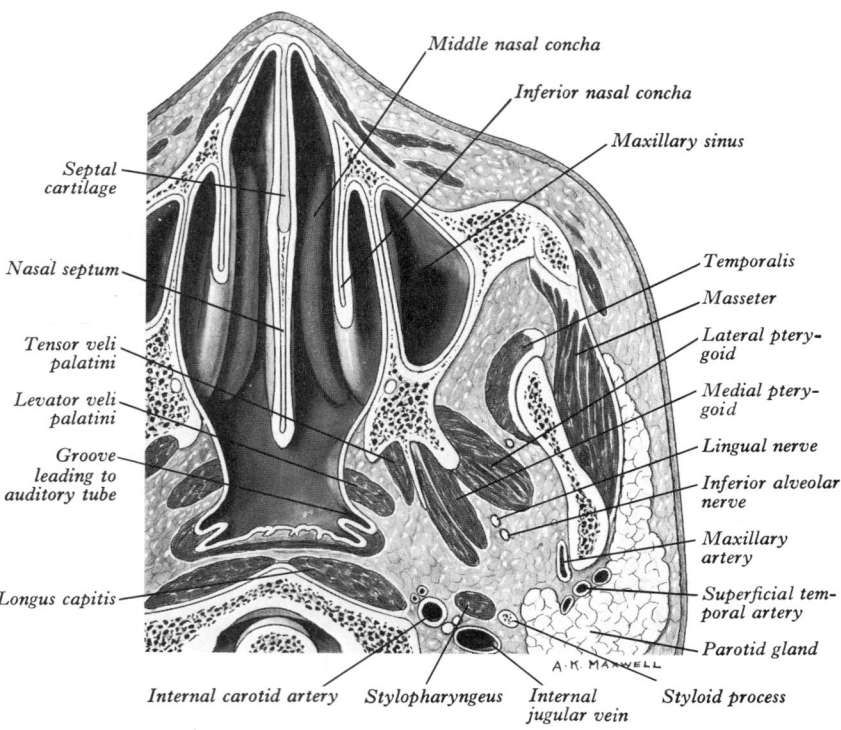

Septal cartilage

Nasal septum

Tensor veli palatini

Levator veli palatini

Groove leading to auditory tube

Longus capitis

Middle nasal concha

Inferior nasal concha

Maxillary sinus

Temporalis

Masseter

Lateral pterygoid

Medial pterygoid

Lingual nerve

Inferior alveolar nerve

Maxillary artery

Superficial temporal artery

Parotid gland

Internal carotid artery   Stylopharyngeus   Internal jugular vein   Styloid process

7.284 A transverse section through the anterior part of the head at a level just inferior to the apex of the odontoid process: inferior aspect.

each side with the septal process of the major alar cartilage. (Galindo et al 1977 suggest that the two lateral and septal cartilages should be regarded as a unit, the *septodorsal cartilage*, an older term used in the Jena Nomina Antomica.) Its posterosuperior border joins the perpendicular plate of the ethmoid; its postero-inferior border is attached to the vomer and to the maxillary nasal crest and anterior nasal spine. The septal cartilage may extend back (especially in children) as a narrow *sphenoidal process* for some distance between the vomer and the perpendicular plate of the ethmoid. The antero-inferior part of the nasal septum, between the nares, is named the *septum mobile nasi* because of its mobility; it is formed by the septal processes of the major alar cartilages and skin and not by the main septal cartilage.

The *lateral* (superior) *nasal cartilage* (7.282A) is triangular, its anterior margin thicker than the posterior and its upper part continuous with the septal cartilage, but separated from it below by a narrow fissure; its superior margin is attached to the nasal bone and frontal process of the maxilla and its inferior margin is connected by fibrous tissue to the major alar cartilage.

The *major alar* (inferior) *cartilage* (7.282A,B), a thin flexible plate lying below the lateral cartilage, is curved acutely around the anterior part of its naris; its medial part, the narrow *medial crus* (septal process), is loosely connected by fibrous tissue to its contralateral fellow and to the antero-inferior part of the septal cartilage, thus forming part of the septum mobile nasi. The *lateral crus*, lateral to the naris, is little more than the inferior border of the major alar cartilage. The upper border of the lateral part of the major alar cartilage is attached by fibrous tissue to the lower border of the lateral nasal cartilage. Its narrow posterior end is connected to the frontal process of the maxilla by a tough membrane containing three or four *minor alar cartilages* (7.282A). According to Drumheller (1969) the junction between the major alar and the lateral cartilages is variable; the two edges may abut or overlap, the lateral cartilage being then the more lateral at the junction. The lower edge (lateral crus) of the major cartilage is shorter than the lateral narial margin, the lower part of the ala nasi being fibro-adipose tissue covered by skin. In front, the major alar cartilages are separated by a notch palpable at the tip of the nose. The *muscles* acting on the external nose are described on p. 572.

Nasal skin is thin and loosely connected to the underlying structures. Over the apex and alae it is thicker and more adherent and bears numerous large sebaceous glands, their orifices usually

being obvious. The *arteries* of the external nose are the alar and septal branches of the facial artery supplying the alae and lower septum; the dorsal nasal branch of the ophthalmic and infra-orbital branch of the maxillary artery supply the lateral aspects and the dorsum. The *veins* end in the facial and ophthalmic veins. The *nerves* for the nasal muscles are from the facial; the skin receives branches from the ophthalmic nerve through its infratrochlear branch and external nasal nerve (p. 1100) and from the infra-orbital branch of the maxillary nerve.

THE NASAL CAVITY

This is divided sagittally by the nasal septum (7.284), its two halves opening on the face through the nares and being continuous posteriorly with the nasopharynx through the posterior nasal apertures. The *nares* are ellipsoid or piriform, narrower in front; each one measures 1·5 cm–2 cm anteroposteriorly and 0·5 cm–1 cm transversely. The *posterior nasal apertures* (choanae) are oval openings, each about 2·5 cm in vertical and 1·25 cm in transverse dimension.

For details of nasal skeletal structures, see p. 365.

Each half of the nasal cavity has a floor, roof and a lateral and medial (septal) wall. It has three regions: vestibular, olfactory and respiratory. The *nasal vestibule* is a slight dilatation just inside the naris (7.285), bounded laterally by the ala and lateral part of lower nasal cartilage and medially by the septal process of this cartilage; it extends as a small recess towards the nasal apex. It is lined with skin bearing coarse hairs (vibrissae) curving towards the naris to arrest the passage of particles in inspired air. In males, after middle age, these hairs increase considerably in size. The vestibule is limited above and behind by a curved elevation, the *limen nasi*, corresponding to the upper margin of the major alar cartilage; along this, vestibular skin is continuous with the nasal mucosa. The *olfactory region* is limited to the superior nasal concha, the opposing septum and the intervening roof. The *respiratory region* is the rest of the cavity.

The lateral wall (7.285, 286) shows three elevations: the *superior, middle* and *inferior nasal conchae*, inferolateral to each being a corresponding *meatus*. Above the superior concha a triangular *spheno-ethmoidal recess* bears the opening of the sphenoidal sinus; sometimes a fourth, *highest nasal concha* appears on the recess's lateral wall (7.285); the *supreme nasal meatus* inferior to it may display an opening of a posterior ethmoidal sinus. The *superior meatus* is a short oblique passage extending about halfway along the upper border of the middle concha; the posterior ethmoidal sinuses open, usually by one aperture, into its anterior part. The *middle meatus*, deeper in front than behind, is inferolateral to the middle concha and continues anteriorly into a shallow fossa above the vestibule, termed the *atrium* of the middle meatus. Above the atrium an ill-defined curved ridge, the *agger nasi* (p. 389), slopes down and forwards from the upper end of the anterior free border of the middle concha; it is better developed in the newborn than in adults. The middle concha must be displaced to display the lateral wall of the middle meatus fully. The main features of this wall are a rounded elevation, the *bulla ethmoidalis*, and below it and extending up in front of it, a curved cleft, the *hiatus semilunaris*. The *bulla ethmoidalis* is formed by the bulging of the middle ethmoidal sinuses, which open on or just above it, its size varying with the contained sinuses. The *hiatus semilunaris*, bounded inferiorly by a sharp concave ridge produced by the ethmoid's uncinate process, leads forwards and up into a curved channel, the *ethmoidal infundibulum*; the anterior ethmoidal sinuses open into it and in 50% or more of subjects it is continuous with the *frontonasal duct* from the frontal sinus; otherwise the infundibulum ends blindly in front in one or more anterior ethmoidal sinuses (*infundibular sinuses*) and the frontonasal duct then opens directly into the anterior end of the middle meatus. The opening of the maxillary sinus lies below the bulla, usually hidden by the flange-like lower edge of the hiatus; this opening is near the roof of the sinus (7.291); an accessory opening frequently exists inferoposterior to the hiatus. The *inferior meatus*, below and lateral to the inferior concha, contains the opening of the nasolacrimal duct under the cover of the anterior end of the inferior concha.

The medial wall or nasal septum (7.283B) is often deflected from the midline, making the nasal chambers unequal in size; ridges or spurs of bone sometimes project from the septum on either side. Above the incisive canal at the lower edge of the septal cartilage there is sometimes a depression which points down and forwards, occupying the position of a canal connecting the nasal and buccal cavities in early fetal life. On each side of the septum near this recess a minute orifice may be seen leading back into a blind tubule, 2–6 mm long, the vestigial *vomeronasal organ*, supported by a *vomeronasal cartilage*; it is lined by a single layer of tall columnar cells and contains many glands. It is prominent in amphibians and reptiles, moderately developed in most mammals including New World monkeys but vestigial in adult primates (Graziadei 1974); in species having a functional vomeronasal organ, it is a chemosensory structure similar to the olfactory epithelium, being connected to the olfactory bulb by the vomeronasal nerve (see p. 1032). **The nasal roof** is narrow transversely, except posteriorly; from behind forwards it can be divided into sphenoidal, ethmoidal and frontonasal regions (pp. 167, 343), according to the bones helping to form it. The ethmoidal part is almost horizontal, the frontonasal part slopes down and forwards and the sphenoidal is inclined downwards and backwards. The cavity is therefore deepest under the cribriform plate of the ethmoid.

The nasal floor is transversely concave, anteroposteriorly flat and almost horizontal, its anterior three-quarters being formed by the palatine process of the maxilla and the rest by the horizontal part of the palatine bone. About 2 cm behind its anterior end a slight depression in the mucosa overlies the incisive canal (p. 389).

The nasal conchae add greatly to the surface area of the nasal cavity. This increases turbulence, perhaps improving olfaction by slowing the passage of air past the olfactoria area. Humidification and warming of the inhaled air are also augmented by the increased mucosal area and turbulence, even in a microsomatic mammal such as man (Cole 1954). Swirling currents also aid the trapping of particles by mucous secretion. In many mammals the conchae are large and highly elaborate structures and the olfactory area spreads over much of their superior and medial surfaces, producing a highly developed olfactory system (Negus 1958; see also p. 1177).

Nasal mucosa lines the nasal cavities, except its vestibules; it is adherent to the periosteum or perichondrium and continuous: with the nasopharyngeal mucosa through posterior nasal apertures, with the conjunctiva through the nasolacrimal duct and lacrimal canaliculi and with the mucosae of the sphenoidal, ethmoidal, frontal and maxillary sinuses through their openings. The mucosa is thickest and most vascular over the conchae, especially at their extremities and also on the nasal septum, but is very thin in the meatuses, on the floor and in the sinuses. Its thickness reduces materially the volume of the nasal cavity and its apertures. The epithelium differs according to local functions; in the *olfactory region*, over the upper 10 mm or so of the septum and over the superior concha and the lateral wall above it, the mucosa is yellowish and the epithelium distinctive in structure.

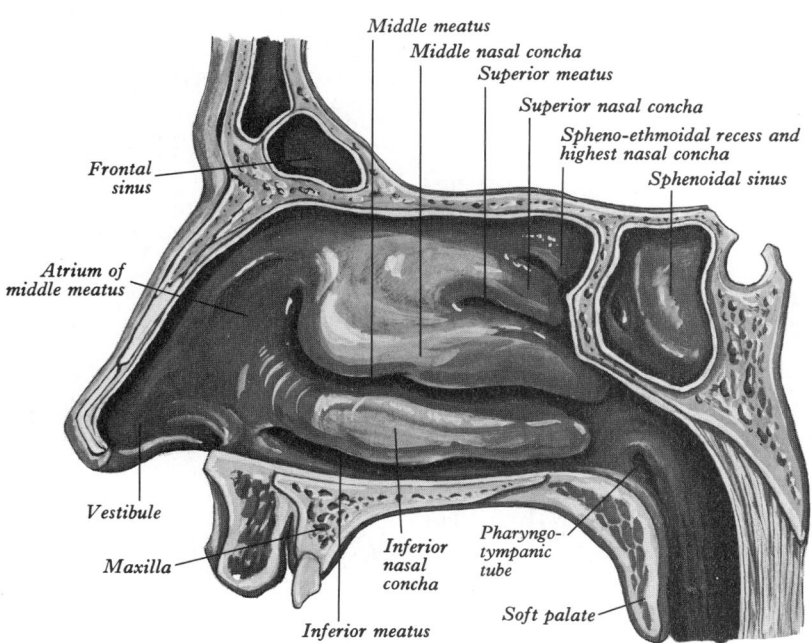

7.285   The lateral wall of the right half of the nasal cavity: internal aspect.

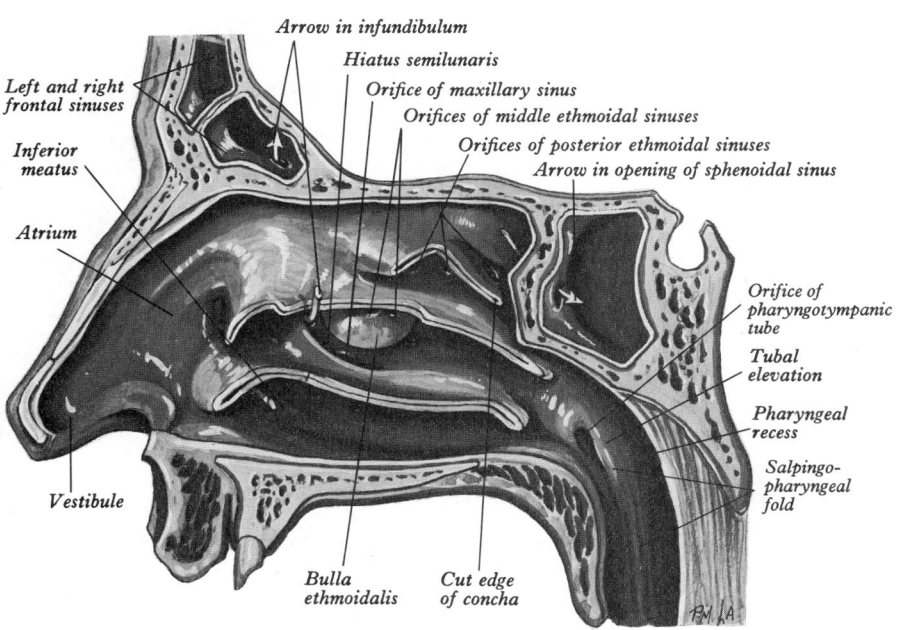

7.286   The lateral wall of the right half of the nasal cavity; the three nasal conchae have been partially removed.

STRUCTURE OF THE OLFACTORY EPITHELIUM

The olfactory epithelium is considerably thicker (up to 100 μm) than the adjacent nasal epithelium and is pseudostratified sensory in type. It contains three principal types of cell (7.287, 288): olfactory receptors, supporting cells (sustentaculocytes) and basal cells (Moulton & Beidler 1967).

Olfactory receptor cells are of much phylogenetic interest, as they are *primary sensory neurons*, with somata situated near the sensory surface, an arrangement common in invertebrates but unique in vertebrates. They are *bipolar* neurons, orientated vertically within the olfactory epithelium, their somata restricted to its basal two-thirds. A basal non-myelinated axon extends from each receptor to run with other axons in small intra-epithelial fascicles among the processes of supporting and basal cells to penetrate the basal lamina, where each fascicle is ensheathed by a Schwann cell (Frisch 1967). Such *fila olfactoria* join to form *olfactory fasciculi* which enter the olfactory bulb to synapse with secondary sensory neurons (mitral cells, basket cells, periglomerular cells, p. 1031). Olfactory axons are amongst the most slender of nerve fibres, about 0·2 μm in diameter. Within the olfactory fila the axons are grouped in bundles of up to 50 or more by invagination into Schwann cells (Gasser 1956), perhaps permitting electrical interaction between adjacent axons.

From the apical pole of each receptor cell (7.290A,B) a single unbranched dendrite extends to the surface, expanding into a bulbous olfactory ending (*rod, knob* or *vesicle*) which projects above it (Frisch 1967, Reese 1965, Graziadei 1971, 1974, Naessen 1971). The endings have numerous *olfactory cilia* projecting into a fluid layer on the epithelium. The cilia have an internal '9 plus 2' pattern of microtubules, as in motile, non-sensory cilia elsewhere (p. 31). Amongst different groups of vertebrates the length and form of olfactory cilia vary. In mammals they have short, thick proximal 'shafts' tapering rapidly to a long, thin, distal tip, with fewer microtubules, thus forming a considerable

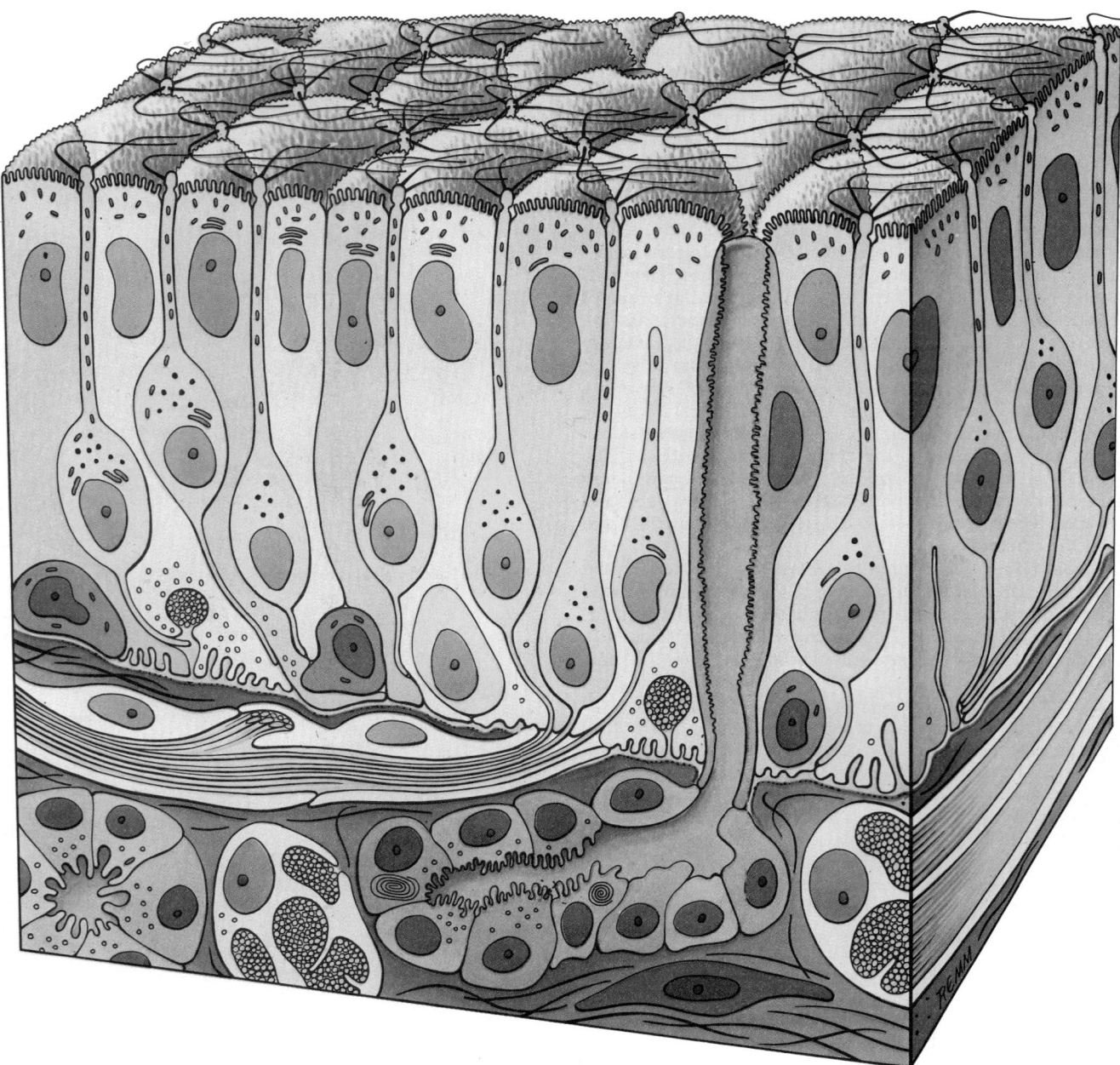

7.287 A schematic reconstruction showing the chief cytological features of the olfactory epithelium. Receptor cells (yellow) are situated among columnar supporting cells. The axons of the receptor cells emerge from the epithelium in groups ensheathed in Schwann cells. Observe the rounded basal epithelial cells and the subepithelial glands of Bowman with their intra-epithelial ducts. One of the basal cells is in process of differentiating into a receptor cell. At the surface are cilia of the receptor cells and microvilli of the supporting cells.

area of receptor surface. The dendrite contains numerous mitochondria, indicative of high energy consumption, and other organelles typical of dendrites: microtubules, smooth and coated vesicles, agranular endoplasmic reticulum and ribosomes; centrioles also often appear, especially in young animals. The *perikaryon* contains abundant granular and agranular endoplasmic reticulum, Golgi bodies and lysosomes, all indicating a high metabolic rate. The nucleus is elliptical and heterochromatic. The exposed regions of olfactory dendrites and cilia are covered by plasma membrane containing high concentrations of intramembranous particles (7.290A) demonstrable by freeze-fracturing (Kerjaschki & Horandner 1976, Menco et al 1975, Menco 1984). These particles are probably the sites of olfactory reception and the ionic movements concerned in sensory activity; their large numbers accord with the extreme sensitivity of olfactory receptors to low concentrations of chemicals (Menco 1976).

Supporting cells are irregular columnar elements separating and partially enclosing the receptors. Their large, elongate, euchromatic nuclei form a layer superficial to the receptor perikarya. Their surfaces are marked by numerous, long, ir-

regular microvilli which, with the olfactory cilia, form a complex meshwork of processes in the mucus at the epithelial surface. Their cytoplasm contains many mitochondria, granular and agranular endoplasmic reticulum and, basally, lamellated dense bodies resembling the lipofuscin granules of neurons (p. 881), representing the remains of autophagic lysosomes and making the whole olfactory area yellowish-brown. A prominent Golgi apparatus and lysosomes also exist in their apical parts. Near the epithelial surface fine microfilaments attached to desmosomes probably give mechanical coherence to the epithelium. Tight junctions also occur between the supporting cells and olfactory receptors at the level of the epithelial surface (Reese & Brightman 1970).

Basal cells are of two types: one probably supportive and the other *blastemal*, able to give rise to new cells by mitotic division (Andres 1966). The first is confined to the vicinity of the epithelial basal lamina, composed of irregular polygonal cells with irregular heterochromatic nuclei and sparse cytoplasm containing numerous tonofibrils attached to desmosomal contacts with adjacent cells. It is likely that they strengthen the epithelial base.

The blastemal cells are confined to a zone at the epithelial base; they are particularly prominent in young animals. Mitotic figures often appear among them and their nuclei can be labelled with tritiated thymidine and examined by autoradiography. Structurally they have many of the features of embryonic neuroblasts, being rounded, with a large euchromatic nucleus and basophilic cytoplasm containing numerous free ribosomes, scattered mitochondria and occasional clusters of centrioles. Some basal cells show an early differentiation into olfactory receptors by developing dendrites and receptor organelles. Various studies have shown that blastemal cells constantly contribute new receptor cells by migrating upwards in the epithelium. During the early development and postnatal period this recruitment is needed for epithelial expansion in both its thickness and area; but a slower, continuous addition to the receptor population appears to occur throughout life, balanced by the degeneration of existing cells, which (in rodents) may only last for two to three weeks (Moulton et al 1970, Graziadei 1974). Degenerating receptors are phagocytosed by the adjacent supporting cells; their remnants, after hydrolysis in lysosomes, contribute to the olfactory pigment (Bannister & Dodson 1975). A continuous turnover of olfactory receptors is required by the exposed position of the olfactory endings and their liability to damage from inhaled noxious agents or viruses; but in the latter case functional regeneration appears to depend on the absence of scar tissue, which may prevent the regrowth of axons (Douek et al 1975). This ability to regenerate primary sensory neurons is unique in mammalian nervous systems, a phenomenon of much neurobiological interest (Graziadei 1978).

The olfactory glands (of Bowman) (7.287, 288) are branched tubular structures beneath the olfactory epithelium, on to whose surface they secrete through narrow, vertical ducts. Their secretory acini have cells with basal nuclei and granular cytoplasm; ultrastructurally two types appear, both containing dense secretory vesicles but one with denser cytoplasm. Their secretions are reported to contain enzymes in some mammals (acid phosphatase, esterase and, in mice, sulphated acid mucosubstances). Since this fluid forms the environment of the receptor endings, it may aid the diffusion of odoriferous substances from the air to the olfactory receptors and regulate the passage of ions to and from sensory cells during electrical activity. it may also be bactericidal.

A variable number of lymphocytes exists, particularly in basal epithelium, as elsewhere in the upper respiratory tract. During life olfactory receptors and their axons gradually diminish in total number, up to 1% being lost each year; the sensory epithelium may be replaced by ciliated columnar epithelium of the respiratory type lining the adjacent nasal regions. These changes parallel a gradual loss of olfactory sense.

DEVELOPMENT OF THE OLFACTORY EPITHELIUM

Olfactory elements develop from the embryonic olfactory placode, resembling the early ontogeny of the neural tube (p. 178). Cuboidal ectodermal cells of the placode gradually lengthen but remain as a single layer throughout development. Mitosis occurs at first only near the placodal surface, the dividing cells rounding up and then re-extending to the epithelial base; but later mitoses appear deeper in the epithelium and are finally confined to the basal layers, where division persists throughout life (Smart 1971, Cuschieri & Bannister 1975a). Electron microscopy shows that receptor cells differentiate early, sprouting apical dendrites and basal axons with large growth cones, growing back to the olfactory bulb to synapse with the secondary neurons (Hinds & Hinds 1972, Cuschieri & Bannister 1975b, Farbman 1975, Constanzo & Graziadei 1986). On reaching the surface, the dendrites expand to form bulbs into which centrioles migrate from the receptor soma, eventually growing cilia. Later the supporting cells, glands and basal cells differentiate, the olfactory apparatus being functional at birth.

In the **vomeronasal organ** (of Jacobson), which is rudimentary in mankind (p. 1173), the cellular components resemble those of the olfactory epithelium except for an absence of basal cells, a larger size of receptor cells and the existence of microvilli instead of cilia (Kolnberger 1971). The absence of cilia is interesting, for

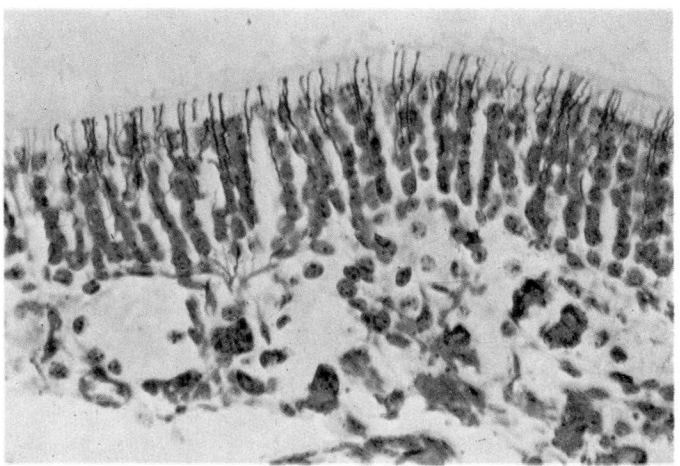

7.288 A vertical section through the olfactory epithelium of the mouse, stained with Holmes' silver method to show the olfactory dendrites and their terminal expansions (above). The nuclei of the receptor neurons are arranged in columns in this preparation; the fila olfactoria can also be seen emerging from the base of the epithelium (below). Provided by A Cuschieri, University of Malta.

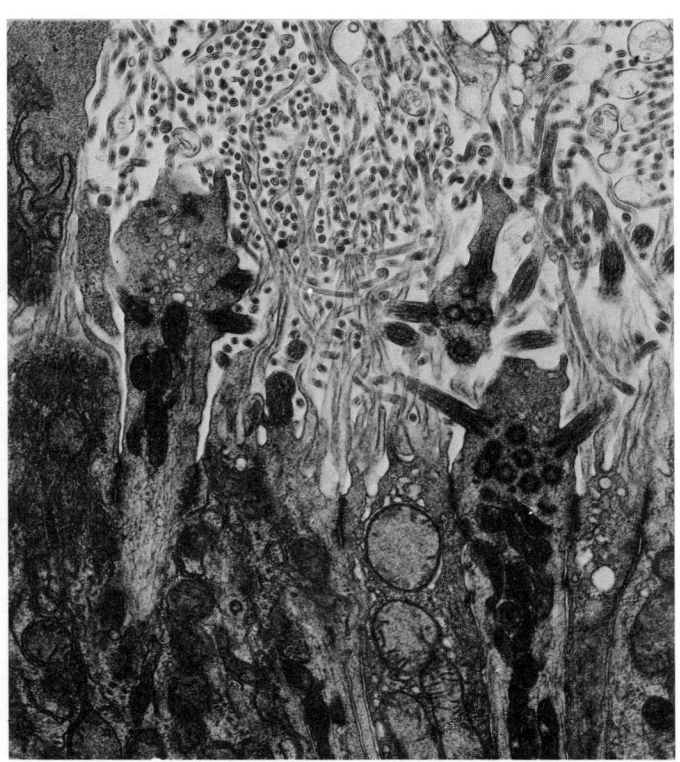

7.289 An electron micrograph showing receptor endings in the olfactory mucosa of a mouse. Note the presence of numerous cilia on the sensory terminals, interspersed by numerous microvilli on the surfaces of the supporting cells. Provided by H C Dodson, Dept. of Anatomy, UMDS, Guy's Campus, London.

they are claimed to be the sites of chemoreception in the olfactory epithelium; however, essentially the exposed plasma membrane is the chemoreceptive area and cilia or microvilli merely increase the receptive surface. Electrical recordings show that vomeronasal receptors respond to odours like the olfactory receptors, but possibly to a narrower range of substances.

Other nasal chemoreceptors include terminals responsive to strong odours and noxious chemicals in aerial solution, e.g. ammonia and sulphur dioxide, even when the olfactory pathway has been destroyed by fracture of the cribriform plate or division of the olfactory tract. The major nerves possibly supplying nasal chemosensory endings include trigeminal fibres from ethmoidal

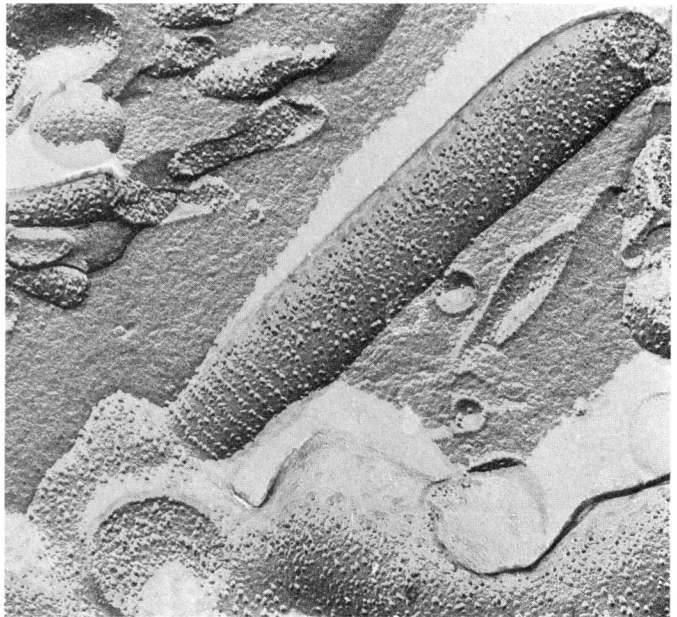

7.290A An electron micrograph of an olfactory cilium (bovine) attached at its basal end to an olfactory dendrite terminal, freeze-fractured to expose the numerous intramembranous particles which are thought to be sites of odour reception. An extensive ciliary necklace is present at the base of the cilium and microvilli of a supporting cell are visible on the left. Supplied by B Menco, University of Utrecht, Netherlands. Magnification × 60 000.

branches of the ophthalmic nerve and various maxillary branches (p. 1177); small bundles of myelinated fibres from these have been detected by Graziadei & Gagne (1973) in the basal olfactory epithelium, losing their sheaths as they penetrate amongst the supporting cells. Another possible chemosensory path is the *nervus terminalis*, whose functions are uncertain but which terminates widely in the nasal cavity in some species (Graziadei 1974 and p. 1095).

FUNCTION OF THE OLFACTORY RECEPTORS

The chemical sense served by the olfactory epithelium differs from the gustatory sense in discriminating between different *airborne* molecules even at very low concentrations. Although the human sense of smell is only moderate compared with that of macrosomatic species in whose behaviour odours play a dominant role, human olfaction may be more significant than imagined. In mammals, including primates, specific scents ('pheromones') affecting social or sexual behaviour are recognized as a common phenomenon, and associations between the central olfactory pathways and cerebral areas concerned with emotional behaviour may be important indications of a more general significance of odours (see the limbic system, p. 1028).

The mechanisms of detection and neural signalling of odours have been studied chiefly by recording electrical changes elicited in or at the surface of the olfactory epithelium by odours (Moulton & Beidler 1967). Two types of electrical change can be recorded; a slow negative wave (*electro-olfactogram*) at the surface and just below it, representing summated ionic fluxes from many receptor dendrites; deeper in the epithelium and in the olfactory nerves *action potentials* can be detected.

Precisely how receptors provide information regarding the quality and quantity of odorous stimuli is little understood. Individual receptors might respond specifically to particular odours

7.290B An electron micrograph of a group of non-sensory cilia from the respiratory epithelium of the nasal cavity (bovine), freeze-fractured to show the distribution of intramembranous particles. Note the presence of ciliary necklaces at the bases of cilia and the relative paucity of particles elsewhere on the cilia (compare with 7.290A). Supplied by B Menco, University of Utrecht, Netherlands. Magnification × 80 000.

but a huge variety of odours can be distinguished and an equally wide range of specific receptors would then be required, each able to detect slight differences in molecular structure, if there is a one-to-one relation between receptor types and odours. It has been shown electrophysiologically that single receptors can respond to several odours and conversely that different receptors exhibit diverse responses to the same odour. It seems likely that olfactory discrimination results from a central analysis of complex patterns of activity in many receptors (Gesteland et al 1965), as in gustatory discrimination.

How odoriferous substances actually stimulate the receptors is also uncertain; many theories have been proposed. The most widely accepted is the stereochemical theory, first elaborated by Amoore (1971); this postulates the presence of various receptor molecules, each responsive to particular chemical groups, on odorous molecules dissolved in the nasal mucus, rather as an enzyme recognizes its specific substrate (a 'lock and key' phenomenon). Thus the shape of some part of an odorous molecule would correspond to a complementary site in a receptor molecule. In early studies of correspondences between molecular shape and subjective olfactory perception, it was postulated that there might be a simple basis for quality coding, with as few as perhaps seven or eight 'primary odours', analogous to primary colours (e.g. floral, fruity, minty, pungent, musky, etc.) and corresponding receptor sites. Later studies have indicated a greater complexity than this. Of particular interest have been investigations into specific anosmias, in which one or more specific classes of odour cannot be sensed, although olfaction is generally unaffected. These suggest that there are perhaps 30 or more 'primary odours' in humans. The receptor sites appear to be all on the surfaces of the olfactory receptor cells, embedded in their exposed plasma membrane (probably as transmembrane proteins). Odorants bind transiently to each site to change the ionic permeability of the membrane to calcium ions. Further studies at a cellular level are necessary for a fuller understanding of the general nature of olfaction, which in the past has been a somewhat neglected sense.

THE RESPIRATORY EPITHELIUM

The non-olfactory regions of the nasal and sinus mucosae have columnar or pseudo-stratified ciliated epithelium containing goblet cells, non-ciliated columnar cells with microvilli and basal cells (Mygind 1975, 1978), collectively termed the *respiratory epithelium* (Negus 1958). In some areas cells may be low columnar or cuboidal. Beneath the basal lamina are groups of serous and mucous glands of variable cytological detail and secretory contents, many of them opening by branched ducts on the epithelial surface. *Cavernous tissue* beneath the respiratory mucosa in such areas is extensive (vide infra) and vascular disturbances, e.g. due to vasomotor autonomic innervation, alter epithelial surfaces visibly by swelling or shrinkage. The endothelium in these cavernous regions is particularly interesting in being fenestrated; the muscular walls of their supplying arterioles may also be under endocrine as well as neural control, providing a basis for cyclical fluctuations in their blood pressure and degree of enlargement (Cauna & Hinderer 1969, Cauna et al 1972). Basal to the epithelium and its basement membranes is a fibrous layer infiltrated with lymphocytes which often form diffuse lymphoid tissue, under which is a nearly continuous layer of mucous and serous glands, their ducts traversing the lymphoid layer to the surface. Abundant mucus from glands and goblet cells makes the surface sticky and dust in the inspired air is deposited on the surface, while the air is humidified and warmed. The mucous film is continually moved by ciliary action backwards into the nasopharynx at a rate of about 6 mm per minute (Proctor et al 1978). Palatal movements transfer the mucus and its entrapped particles to the oropharynx for swallowing, but some also enters the nasal vestibule anteriorly. The secretions of the nasal mucosa also serve many other functions, as they contain antibacterial lysozyme and immunoglobulins (IgA), to inhibit the proliferation of microbes in this very susceptible environment.

In the walls of the sinuses the epithelium is closely bound to the underlying periosteum to form a combined *mucoperiosteum*.

VESSELS AND NERVES OF THE NASAL CAVITY

The *arteries of the nasal cavity* are as follows: the anterior and posterior ethmoidal branches of the ophthalmic artery supplying the ethmoidal and frontal sinuses and nasal roof; the sphenopalatine branch of the maxillary artery, supplying the mucosa covering the conchae, meatuses and septum; the terminal part of the greater palatine artery, ascending through the incisive canal (p. 354); the septal ramus of the superior labial branch of the facial artery, supplying the septum in the region of the vestibule, anastomosing with the sphenopalatine and presenting a common site of epistaxis; the infraorbital and superior (anterior and posterior) alveolar branches of the maxillary, supplying the mucosa of the maxillary sinus; the maxillary artery's pharyngeal branch, distributed to the sphenoidal sinus. The ramifications of these vessels form a dense mucosal and submucosal plexus.

The *veins of the nasal cavity* form a rich submucosal cavernous plexus. Arteriovenous anastomoses also occur (Harper 1947). The plexus is especially dense in the lower part of the septum and in the middle and inferior conchae. Some veins drain to the sphenopalatine; others join the facial vein; some accompany the ethmoidal arteries to end in the ophthalmic veins; a few connect with veins on the orbital surface of the brain's frontal lobes, passing through the foramina in the cribriform plate. When the foramen caecum is patent it transmits a vein from the nasal cavity to the superior sagittal sinus.

The *lymph vessels* are described on p. 844.

The innervation includes branches of the trigeminal and olfactory nerves and autonomic vasomotor and secretomotor fasciculi. The *nerves of ordinary sensation* (7.213, 220) supplying the nasal cavity are as follows: the anterior ethmoidal branch of the nasociliary, supplying the anterior and upper parts of the septum, the anterior part of the roof; the anterior parts of the middle and inferior conchae with the lateral wall anterior to them; the infraorbital nerve, supplying the nasal vestibule; the anterior superior alveolar nerve, supplying part of the septum and the floor near the anterior nasal spine and the anterior part of the lateral wall as high as the opening of the maxillary sinus; the lateral posterior superior and medial posterior superior nasal nerves (including the nasopalatine nerve) which are branches from the pterygopalatine ganglion and the posterior inferior nasal branches of the anterior palatine nerve, supplying the posterior three-quarters of the lateral wall, roof, floor and septum; branches from the nerve of the pterygoid canal to the upper and posterior part of the roof and septum. It is to be noted that all these nerves are derived from the maxillary trigeminal division, with the exception of the nasociliary supply.

Accompanying the sensory fibres in these nerves are postganglionic vasomotor sympathetic fibres to the nasal blood vessels, whilst running with the branches from the pterygopalatine ganglion are postganglionic parasympathetic fibres from this ganglion, providing the secretomotor supply to the nasal glands.

THE PARANASAL SINUSES

The frontal, ethmoidal, sphenoidal and maxillary sinuses (7.284–286, 291–293) vary in size and form in different subjects. Their mucosa is continuous with that of the nasal cavity (a feature unfortunately favouring the spread of infections). The mucosa is similar to that in the respiratory regions of the nasal cavity but is thinner, less vascular and less adherent to bone. Mucus is secreted by glands of the mucosa and is swept into the nose through the apertures of the sinuses by cilia; these are not uniformly distributed but invariably exist near these openings. The functions of the sinuses are uncertain; they lighten the skull and add resonance to the voice; but saving in weight is trivial and the absence of sinuses does not add an equivalent volume of *solid* bone, as the weight of the trabecular bone and enclosed tissue which would occupy such a volume would not be large. It is more probable that sinuses are manifestations of an unusual growth pattern in the bones in which they occur, in part at least. Most sinuses are rudimentary or absent at birth; they enlarge appreciably during the eruption of the permanent teeth and after puberty, markedly altering the size and shape of the face.

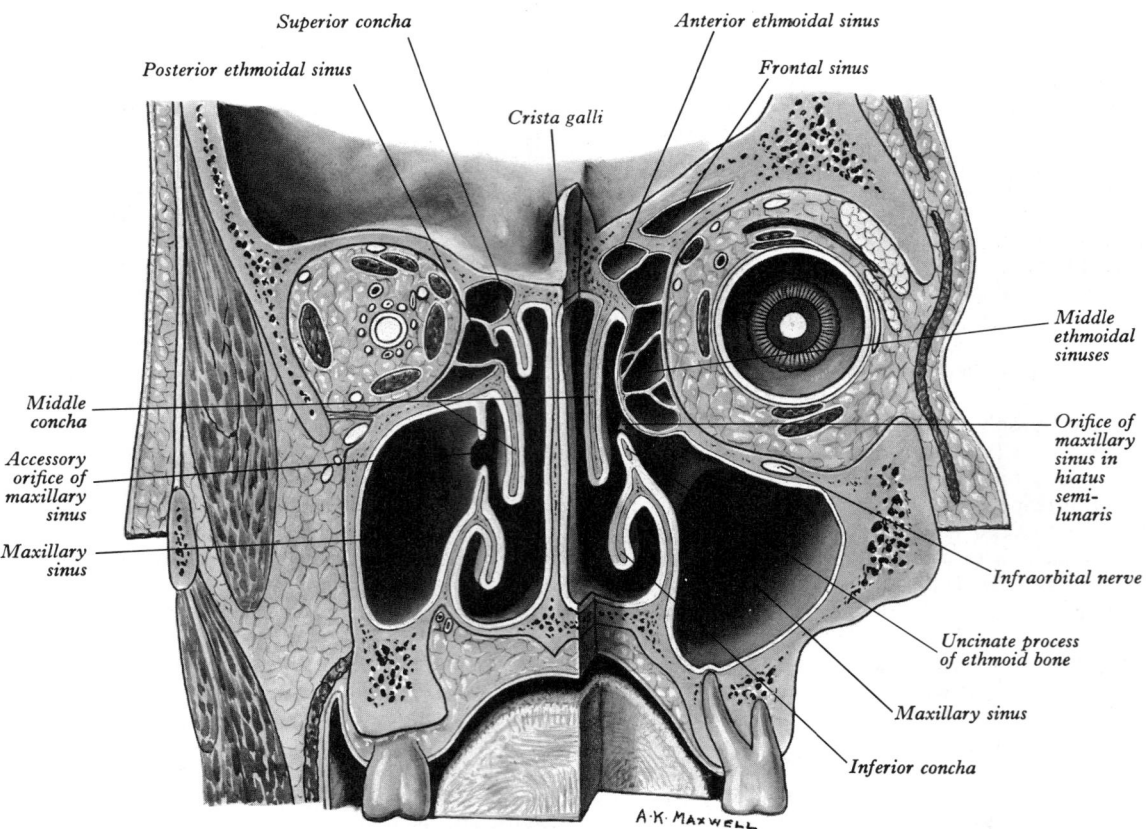

7.291 A coronal section through the nasal cavity, viewed from the posterior aspect. On the right side the plane of the section is more anterior. The normal orifice of the maxillary sinus is shown on the right side and the not uncommon accessory orifice on the left side.

Two frontal sinuses, situated posterior to the superciliary arches, lie between the outer and inner tables of the frontal bone. On average, each underlies a triangular area on the surface, its angles formed by the nasion, a point 3 cm above the nasion and the junction of the medial third with the rest of the supraorbital margin (**3**.166, **7**.292, 293). They are rarely symmetrical, the septum between them usually deviating from the median plane. Their average measurements are: height 3·2 cm; breadth 2·6 cm; depth 1·8 cm. Each extends upwards above the medial part of the eyebrow and back into the medial part of the orbit roof. A sinus is sometimes divided into a number of communicating recesses by incomplete bony septa. Rarely, one or both sinuses may be absent. The prominence of the superciliary arches is no indication of presence or size of the frontal sinuses. The upward extension in the frontal bone may be small and the orbital part large, or vice versa. Sometimes one sinus may overlap in front of the other. Posteriorly a sinus may reach the sphenoid bone (lesser wing) but does not invade it. Each opens into the anterior part of the corresponding nasal middle meatus by the *ethmoidal infundibulum* or *frontonasal duct*, traversing the anterior part of the ethmoid labyrinth. Rudimentary or absent at birth, they are generally well developed between the seventh and eighth years but reach full size only after puberty (p. 383). They are more prominent in males, giving the forehead an obliquity contrasting with the vertical or convex profile usual in children and females. Their blood supply is from supraorbital and anterior ethmoidal arteries and their venous drainage is into the anastomotic vein in the supraorbital notch connecting the supraorbital and superior ophthalmic veins. *Lymphatic drainage* is to the submandibular nodes. The *nerve supply* is from the supraorbital nerve.

The ethmoidal sinuses (see pp. 384, 385) are small, thin-walled cavities in the ethmoidal labyrinth, completed by the frontal, maxillary, lacrimal, sphenoidal and palatine bones. Ranging from three large to 18 small sinuses on each side, their openings into the nasal cavity are also variable in position. They lie between the upper part of the nasal cavity and the orbits, separated from the latter by the paper-thin orbital plates of the ethmoids, a poor barrier to infection, which may spread into the orbit to give orbital cellulitis. They form anterior, middle and posterior groups on each side. (Some anatomists recognize only an anterior and a posterior group, the anterior including those described here as anterior *and* middle.) The groups are not

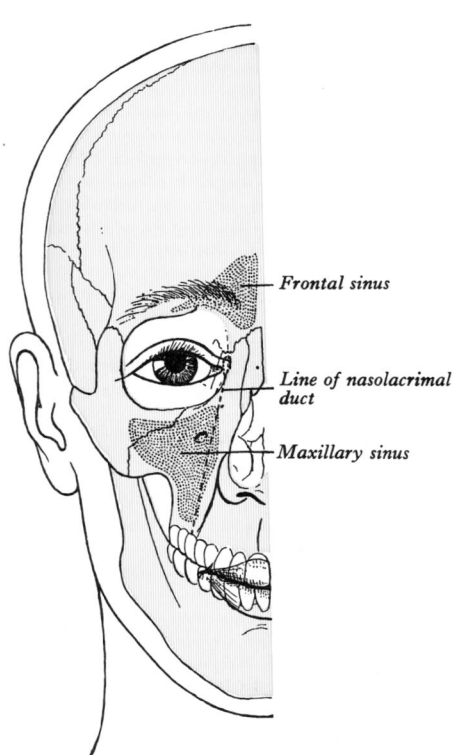

7.292 An outline of the bones of the face, showing the positions of the frontal and maxillary sinuses.

sharply delimited from each other; one may encroach on territory usually occupied by another. They are, however, distinguished by their sites of communication with the nasal cavity. In each group the sinuses are only partially separated by incomplete septa. The *anterior group* (the infundibular sinuses), numbers up to 11 which open into the ethmoidal infundibulum or the frontonasal duct by one or more orifices; one sinus often lies in the agger nasi while the most anterior ethmoidal sinus may encroach on the frontal sinus. The *middle group* (bullar sinuses), usually three in number, opens into the middle meatus by one or more orifices on or above the ethmoidal bulla. The *posterior group* (varying from one to seven), usually opens by a single orifice into the superior meatus inferior to the superior concha though one may open into the highest meatus (when present) and one or more into the sphenoidal sinus. The posterior group is very close to the optic canal and optic nerve. The ethmoidal sinuses are small though of clinical importance at birth and grow rapidly between six and eight years and after puberty. Their blood supply is from the sphenopalatine artery (p. 741) and the anterior and posterior ethmoidal arteries and their drainage is by the corresponding veins. The lymphatics of the anterior and middle groups drain to the submandibular nodes, and those of the posterior group to the retropharyngeal nodes. The anterior and posterior ethmoidal nerves and orbital branches of the pterygopalatine ganglion supply the ethmoidal sinuses.

The two sphenoidal sinuses (p. 374, 3.128, 129) are sited posterior to the upper part of the nasal cavity, within the body of the sphenoid bone. They are related above to the optic chiasma and hypophysis cerebri and on each side to the internal carotid artery and the cavernous sinus. If small, they lie anterior to the hypophysis cerebri. They vary in size and shape and are rarely symmetrical, one often being much larger and extending across the median plane behind the other; occasionally one overlaps above; rarely they intercommunicate. One or both may approach closely or even partly encircle the optic canal. Their average measurements are: vertical height 2 cm; transverse breadth 1·8 cm; anteroposterior depth 2·1 cm. They may extend, if exceptionally large, into the roots of the pterygoid processes, the greater wings of the sphenoid or they may invade the basilar part of the occipital bone. Gaps in their osseous walls may occasionally leave their mucosa in contact with the overlying dura mater. Bony ridges, produced by the internal carotid artery or pterygoid canal, may project into the sinuses from their lateral walls and floor respectively. A posterior ethmoidal sinus may invade the sphenoid to replace largely a sphenoidal sinus. Each sinus connects with a spheno-ethmoidal recess by an aperture high in its anterior wall. They are minute cavities at birth and their main development occurs after puberty. Their *blood supply* is via the posterior ethmoidal vessels; *lymph drainage* is to the retropharyngeal nodes and their *nerve supply* arises from the posterior ethmoidal nerves and orbital branches of the pterygopalatine ganglion.

The two maxillary sinuses, occupying most of the bodies of the maxillae (3.98A, 7.291–293), are the largest accessory air sinuses of the nose. Pyramidal in shape, the base of each is the lateral wall of the nasal cavity, the apex extending into the maxilla's zygomatic process. The roof is the floor of the orbit, frequently ridged by the overlying infraorbital canal; the floor is formed by the maxilla's alveolar process and is usually about 1·25 cm lower than the nasal floor, on a line drawn laterally from the lower border of the ala. Conical elevations corresponding to the roots of the first and second molar teeth project into the floor, which they sometimes perforate. Sometimes the roots of the first and second premolars and third molar, and occasionally the canine root, may also project into the sinus (p. 1318). The size of the sinus varies, even on the two sides of an individual skull; when large, its apex may invade the zygomatic bone. Average measurements are: vertical height opposite the first molar 3·5 cm; transverse breadth 2·5 cm; anteroposterior depth 3·2 cm. The sinus opens into the lowest part of the hiatus semilunaris by an aperture in the anterosuperior part of the sinus' base (7.291); a second orifice is often present in or just below the hiatus. Both are nearer the roof than the floor of the

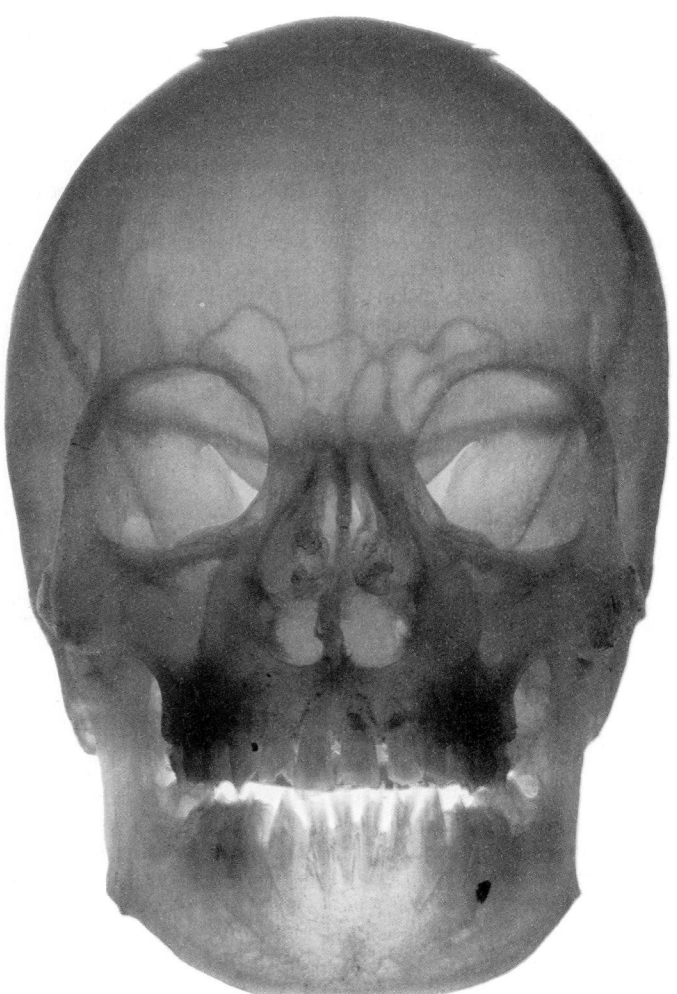

7.293 The skull of an adult woman which has been decalcified and then cleared in methyl salicylate; the specimen was transilluminated and then photographed from the ventral aspect. Note particularly the profiles of the frontal and maxillary paranasal air sinuses, the orbits and superior orbital fissures and the nasal cavities and conchae. The specimen was prepared by D H Tompsett of the Royal College of Surgeons of England.

sinus. During early development, the sinus is a mere groove on the medial maxillary surface in the fourth prenatal month; it reaches full size after the eruption of all the permanent teeth. The *blood supply* is from the facial, infraorbital and greater palatine vessels; *lymph drainage* is to the submandibular nodes and the *nerve supply* is derived from the infraorbital and the anterior, middle and posterior superior alveolar nerves.

For details of the human and comparative anatomy of the paranasal sinuses consult Negus (1958).

RADIOLOGICAL APPEARANCES

Normal sinuses are radiolucent, whereas diseased sinuses show varying degrees of opacity. In anteroposterior views (3.102C, 129) most of the sinuses are visible. The frontal sinuses appear above the nasal cavity and the medial part of the orbits; their asymmetry, vertical extent and the position of their septa can be assessed. The ethmoidal sinuses are superimposed on each other and on the sphenoidal sinuses in this view, lying between the orbits below the cribriform plate. The sphenoidal sinuses are obscured in this view. Each maxillary sinus is a pyramidal translucent area below the orbit and lateral to the lower part of the nasal cavity, extending inferiorly into the maxillary alveolar process. In lateral views, the extent of the frontal sinus both upwards into the frontal bone and back into the orbital roof can be assessed. The ethmoidal sinuses are seen to extend back from the frontal process of the maxilla as far as the sphenoidal sinus, which is clearly visible below and in front of the hypophysial fossa, although the two sphenoidal sinuses appear superimposed; the individual sphenoidal sinuses are seen better from above. The maxillary sinus is clearly seen in a lateral view; it lies below the orbit and its relation to the roots of the teeth is obvious.

Maxillary and frontal sinuses can also be transilluminated. In a dark room, a light is placed in the mouth (for the maxillary sinus) or against the superomedial angle of the orbital opening for the frontal sinus. A red glow normally appears but may be absent when the sinuses are diseased. Both these sinuses show well in cleared preparations of the skull (7.293).

Applied Anatomy. Congenital nasal deformities can occur, e.g. a complete absence of the external nose, with only an aperture existing, or else suppression or malformation on one side. The nasal septum may be displaced by injury or congenital defect;

sometimes deviation may be so great as to bring the septum and one lateral wall into contact, causing complete unilateral nasal obstruction. Suppuration in the paranasal sinuses occurs frequently and pus from the frontal or anterior ethmoidal sinuses may run via the hiatus semilunaris into the maxillary sinus, which thus becomes a secondary reservoir for pus. All paranasal sinuses can be infected from the nasal cavity but a maxillary infection may also spread from the teeth (p. 1318) and this sinus is most often chronically infected, resulting in the loss of mucosal cilia and an impaired flow of mucus. Because its opening is high above its floor, the natural drainage of the maxillary sinus is hindered but may be effected by puncture in the lateral wall of the inferior nasal meatus or through the canine fossa on the anterior surface of the maxilla, which is nearer to its floor.

THE VISUAL APPARATUS

Introduction

The all-pervading nature of sunlight, the environment of almost all animal and plant life, makes a response to this form of electromagnetic radiation inevitable. The photosynthetic processes in plants are paralleled by photochemical receptors occurring almost universally in the animal kingdom. Though the range of frequencies in solar radiation is much wider, the visible spectrum (400–760 nm) is the range within which most animal *photoreceptors* function. Some receptors react outside this range, either to ultraviolet or infra-red frequencies; in vertebrates some vipers and boas (Walls 1963) have facial pit organs responding to infra-red radiation. But in general *visual pigments*, the basis of photochemical response, display absorption maxima within the visible spectrum, e.g. human *rhodopsin* (visual purple), a retinal rod pigment, has a maximum at 497 nm. *Iodopsin*, a cone pigment, occurs in some avian species and several pigments exist in human cones, where differing absorption maxima may account for colour vision, though this is not fully determined (Rushton & Henry 1968). The basis of the sensory response is a light-induced 'bleaching', the pigment changing to another form, with an accompanying electrical flux propagated by chemical transmission from photoreceptors to primary neurons. A rapid restoration of the pigment is obviously essential. Many visual pigments have been identified; further details may be obtained in monographs (e.g. Rushton 1962, Pirenne 1967, Davson 1976).

Next in the elaboration of visual organs is the introduction of a lens to concentrate light energy on photoreceptors and impart spatial discrimination. Many such adaptations have occurred in invertebrates, mostly in two forms: the eyes, as we may now call them, with many separate lens-photoreceptor units, the *compound eyes* familiar in insects and crustaceans; and a *single lens*, focusing light on an array of photoreceptors, as in snails and squids. The latter type is universal in vertebrates.

Such true eyes not only respond to varying *luminance*, a simple function of photoreceptors; by projecting a focused image on an array of receptors, each with neural pathways of some degree of specificity, a new modality appears, sensitivity to *form*. In both modes movement (always of great biological meaning) is detectable but the vertebrate type of focusing eye has potentialities for greater precision.

Primitively, vision is used for reception at a distance, to activate warning systems and to orientate the animal advantageously in light and shade. The paired eyes of most vertebrates, being lateral in position, permit an almost panoramic view. Such *panoramic vision*, with muscles which rotate

the eyes reflexly towards anything of interest, especially movements (of prey, mate, or predator) is typical of mammals, in a few of which, and in some raptorial birds, the orbits have changed position so that two uniocular fields, each subtended by one eye, overlap to a greater or lesser extent. Part of the field in front of the animal is hence binocular. By gradual refinement of the control of ocular muscles, with constant retinal feedback, ocular movements become concerted enough to 'fuse' the two slightly dissimilar retinal images, leading finally to full *binocular vision*. This advance is usual in carnivorous mammals, who may track prey partly by smell but rely on accurate directional vision for the final attack. Primates also have binocular vision; the arboreal factor in their evolution is generally regarded as leading to not merely binocular, but *stereoscopic vision*, allowing a higher motor 'understanding' of three-dimensional space. Olfaction is less useful in trees; acquisition of the skill in not merely climbing but swinging or leaping from branch to branch could only evolve with stereoscopic vision (see also p. 11). An arboreal habitat also favoured the retention and elaboration of pentadactyl, grasping extremities. Though man, and perhaps his immediate ancestors, is not arboreal, the terrestrial specialization of feet has not afflicted his hands. Freed from locomotion by the adoption of a bipedal gait, human hands have formed a most significant partnership with the eyes. This, with the associated development of a large brain, able to process with increasing intricacy highly detailed information from the eyes and other senses and to co-ordinate the eyes and hands in increasingly skilful and subtle tasks, these surely are the leading trio of factors in the extraordinary evolution of human abilities.

The eye, therefore, is not to be viewed in isolation. Its array of modalities—sensitivity to minute changes in luminosity, particularly in dark-adapted, *scotopic vision*, high discrimination of form, movement and colour in light-adapted, *photopic vision* —do not merely provide interesting information. The information is vital; a blind individual cannot long survive outside human society. The eyes continuously guide almost all that we do, especially in manual tasks. Visual communication has proved more useful and lasting than even auditory. The gradual evolution of visual signs, reacting with auditory communication, has led to language in all its permutations; and through language, with its potentialities for the exchange of increasingly precise information and conceptual influences, it becomes possible for generation after generation to profit from recorded knowledge in a unique extension of evolutionary progress. In human culture this now provides the mainstream of evolution. It is against this background that the structure of the visual apparatus should be studied.

THE PERIPHERAL VISUAL APPARATUS

The eyeball, the peripheral organ of vision, is situated in a skeletal cavity, the orbit, the walls of which help to protect it from injury; they also have a more fundamental role in the visual process itself in providing a rigid support and direction to the eye and in forming the sites of attachment to its external muscles. This setting permits the accurate positioning of the visual axis under neuromuscular control and determines the spatial relationship between the two eyes needed for binocular vision and conjugate eye movements. In the following account, the structure of the eyeball itself will first be considered and then certain accessory structures, including extrinsic muscles, fasciae, eyebrows, eyelids, conjunctiva and the lacrimal apparatus, will be described.

The eyeball is embedded in orbital fat, separated from it by a thin *fascial sheath* (capsule of Tenon) (p. 1213). The eyeball can be considered as being composed of the segments of two spheres of different radii. The anterior segment, part of the smaller sphere, is transparent and forms about one-sixth of the whole globe; it is more prominent than the posterior segment, which is part of a larger sphere and opaque, forming the remainder of the globe. The *anterior segment* is bounded by the cornea and the lens and is incompletely subdivided into *anterior* and *posterior chambers* by the iris, being continuous through its pupil. The anterior chamber's periphery is slightly overlapped by the sclera; thus the *angle* between the iris and cornea (p. 1183) forms an annulus of greater diameter than the *limbus*, the junction between the sclera and cornea. The difference between these two varies from 1 to 2 mm, the angle being deeper above and below than at the sides of the eyeball. The posterior chamber, between the posterior surface of the iris and the anterior aspect of the lens and its supporting ligament, the *zonule* (p. 1205), is triangular in section, the apex of the triangle being where the iris touches the lens; its base or zonular region is not the zonule itself, since the posterior chamber extends among the zonule's collagenous bundles and even into a *retrozonular space* (canal of Petit) between the zonule and the vitreous humour in the posterior segment of the eyeball. The *posterior segment* consists of the parts of the eye posterior to the zonule and lens.

The *anterior pole* is the centre of the anterior (corneal) curvature, the *posterior pole* the centre of its posterior (scleral) curvature; a line joining these two points forms the *optic axis*. (By the same convention, the eye has an *equator*, equidistant between the poles; any circumferential line joining the poles is a *meridian*.) The optic axes of the two eyes, in their primary position, are parallel and do *not* correspond with the orbital axes which diverge anterolaterally at a marked angle to each other (vide infra). The optic nerves follow the orbital axes and are therefore not parallel; each enters its eye about 3 mm medial (nasal) to the posterior pole. The ocular vertical diameter (23·5 mm) is rather less than the transverse and anteroposterior diameters (24 mm); the anteroposterior diameter at birth is about 17·5 mm and at puberty 20–21 mm; it may vary considerably in *myopia* (c 29 mm) and in *hypermetropia* (c 20 mm). In females all diameters are on average slightly less than in the male (Stenstrom 1946, Sorsby & Sheridan 1960).

The eye has three *tunics* enclosing its contents. From without they are: (1) a fibrous tunic consisting of the *sclera* behind and the *cornea* in front; (2) a vascular, pigmental tunic comprising from behind forwards the *choroid*, *ciliary body* and *iris*, collectively termed the *uveal tract*, (3) a neural layer, the *retina*.

The Ocular Fibrous Tunic

The fibrous layer of the eyeball (7.294) has an opaque, posterior *tunica sclera* and a transparent, anterior *tunica cornea*.

THE SCLERA (TUNICA SCLERA)

The sclera, so named from its relatively hard consistency, is a dense layer which, when distended by intraocular pressure, maintains the shape of the eyeball. It is thickest (about 1 mm) posteriorly, near the optic nerve's entry point and is thinnest (0·4 mm) about 6 mm behind the corneoscleral junction, near the attachments of the recti (p. 1208). Its *external surface* is white and smooth, except where the tendons of ocular muscles are attached; it is in contact with the inner surface of the fascial sheath of the eyeball (p. 1213). Its anterior part is covered by conjunctival epithelium, reflected on to it from the deep surfaces of the eyelids and continued on to the cornea anteriorly. The scleral *internal surface* is brown and is grooved by the ciliary nerves and vessels. It is separated from the external aspect of the choroid by an extensive *perichoroidal space*, traversed by an exceedingly delicate cellular tissue, the *suprachoroid lamina (lamina fusca sclerae)*. Posteriorly, the sclera is pierced by the optic nerve and is continuous with the nerve's fibrous sheath and hence with the dura mater. Where the nerve pierces the sclera, the latter has the appearance of a cribriform plate, the *lamina cribrosa sclerae* (7.309, 310), the minute orifices in which transmit the optic nerve's fascicles. A larger, central aperture in this structure contains the central retinal artery and vein. The lamina cribrosa is the weakest part of the sclera and bulges outwards in the condition of a 'cupped disc' when intraocular pressure is raised in chronic conditions such as glaucoma. Around the perimeter of the cribriform lamina numerous small apertures transmit the ciliary vessels and nerves; about midway between these and the corneoscleral junction are four or five large apertures for veins (*vernae vorticosae*). The sclera is directly continuous with the cornea at the *corneoscleral junction* (*limbus*). In the sclera near this region is an annular, endothelial canal, the *sinus venosus sclerae*; in section this is an oval cleft, whose outer wall is formed by a groove in the sclera. Posteriorly this cleft extends as far as a rim of scleral tissue, the *scleral spur* which is in section a triangle with its apex directed forwards (7.295). The sinus (the so-called canal of Schlemm) may be double or multiple in parts of its course. Its inner wall, adjoining the aqueous chamber, consists of loose trabecular tissue continuous anteriorly with the posterior limiting lamina of the cornea; among its fibres are spaces through which aqueous humour filters from the anterior chamber into the sinus (7.299), thence draining peripherally into the anterior ciliary veins. Normally the sinus contains no blood; though the channels between the sinus and the veins have no valves, they are oblique and flattened, which may prevent the reflux of blood. However, this valvular mechanism is dubious and pressure gradients are more likely to prevent such a reflux; in venous congestion blood may indeed enter the sinus. To the scleral spur's anterior, external aspect are attached most of the fibres of the trabecular tissue mentioned above and to its posterior, internal aspect, the meridional fibres of the ciliaris. The anterior chamber's *iridocorneal angle* (7.295) lies between the trabecular tissue and the scleral spur anteriorly and externally and the periphery of the iris posteriorly and inwards.

Structurally, the sclera is composed of dense collagenous tissue mixed with fine elastic fibres and interspersed with flat fibroblasts, some of them being pigmented. The fibres are arranged in bundles arranged in patterns characteristic of various scleral regions: they are circumferential around the optic disc, reticular anterior to this and meridional near the attachments of the four rectus muscles (Kokott 1934). Individual *fibrils* vary in diameter from 28 to 280 nm with periodicities of 80 and 21 μm. Collagen forms 75% of the dry scleral weight. The sclera is viscoelastic, a factor important in the regulation of intraocular pressure (Gloster et al 1957, Helen & McEwen 1961). Its *vessels* are few and its capillaries small, uniting at wide intervals. Its innervation is derived from the ciliary nerves.

THE CORNEA

The cornea (7.294) is the anterior, projecting transparent part of the external tunic and the major site of refraction of light entering the eye. Convex anteriorly, it projects from the sclera as a flattened dome. Its curvature varies even in the same eye at different periods of life, being greater in youth than old age. Since it is more curved than the sclera, a slight *sulcus sclerae* marks the

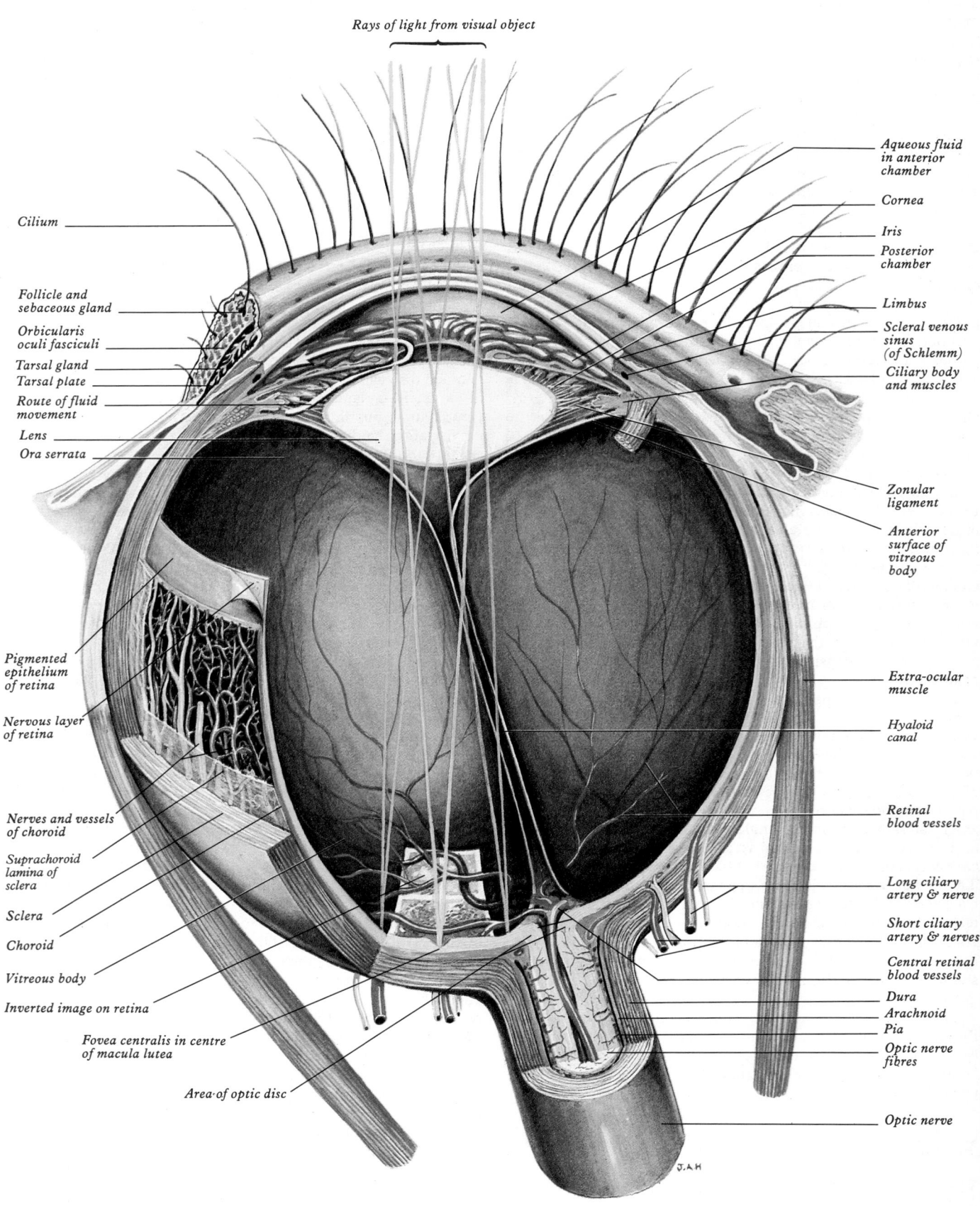

Rays of light from visual object

Cilium

Follicle and
sebaceous gland

Orbicularis
oculi fasciculi

Tarsal gland
Tarsal plate

Route of fluid
movement

Lens

Ora serrata

Pigmented
epithelium
of retina

Nervous layer
of retina

Nerves and vessels
of choroid

Suprachoroid
lamina of
sclera

Sclera

Choroid

Vitreous body

Inverted image on retina

Fovea centralis in centre
of macula lutea

Area·of optic disc

Aqueous fluid
in anterior
chamber

Cornea

Iris

Posterior
chamber

Limbus

Scleral venous
sinus
(of Schlemm)

Ciliary body
and muscles

Zonular
ligament

Anterior
surface of
vitreous
body

Extra-ocular
muscle

Hyaloid
canal

Retinal
blood vessels

Long ciliary
artery & nerve

Short ciliary
artery & nerves

Central retinal
blood vessels

Dura

Arachnoid

Pia

Optic nerve
fibres

Optic nerve

J.A.H

7.294A The organization of the eye, viewed from above. In this illustration the left eye and part of the lower eyelid are depicted in horizontal
section and also cut away to show internal structure (compare 7.294B).

corneoscleral junction. It is dense, about 1·2 mm thick at its periphery and 0·5–0·6 mm at its centre. Its anterior surface is slightly ellipitical, the transverse diameter being a little greater than the vertical; its posterior surface is circular and, because the corneoscleral junction is slightly oblique above and below, it is more extensive than the anterior surface in its vertical axis. The corneal diameter is about 11·7 mm on its posterior aspect; anteriorly it is 11·7 mm horizontally and 10·6 mm vertically. The cornea (7.296–298) consists from in front backwards of five layers: (1) the corneal epithelium, continuous with the conjunctiva; (2) the anterior limiting membrane (of Bowman); (3) the substantia propria; (4) the posterior limiting lamina (of Descemet); (5) the endothelium of the anterior chamber.

The *corneal epithelium* covers the cornea anteriorly and generally has five layers of cells, the deepest being columnar with flat basal surfaces and rounded apices; they have large round or oval nuclei. Cells in the second layer are polyhedral, with oval nuclei; in the more superficial layers the cells become progressively flatter but, unlike those in the epidermis, they contain flat nuclei and are not normally keratinized. Most corneal epitheliocytes are 'prickle' cells, like those in the epidermal stratum spinosum (p. 73). At the corneoscleral junction (limbus) the epithelium thickens to 10 or more layers and is continuous with the scleral conjunctiva. Surface corneal cells display, as in the conjunctiva, microvilli and microplicae (Dohlman 1971), which may help to maintain an unbroken film of secretion (Andrews 1975) at this primary refractive surface of the eye. Scanning electron microscopy (7.297) has confirmed the presence of these features; fine irregularities (microvilli) are mixed with more numerous sinuous, communicating ridges. These two types of surface structure appear to predominate in different cells, some displaying microvilli and some microplicae, while some have both in varying proportions and a few have them only in their central surface.

The *substantia propria* is fibrous, compact and perfectly transparent, composed of about 200–250 superimposed lamellae of Type II collagen fibres (p. 64) and some interstitial proteoglycans, continuous with the matrix of the sclera. Each lamella is about 2 μm thick and of variable width (10–250 μm). Fibres in each lamella are parallel but at large angles to those in the adjacent lamellae. They often pass between the lamellae and vary in size, being larger near the posterior limiting membrane; their diameters vary from 21 to 65 nm. Between lamellae is a little ground substance containing fibroblasts, stellate or dendritic in shape and interconnected. Nomadic macrophages, lymphocytes and polymorphonuclear leucocytes may invade the substantia propria.

Beneath the corneal epithelium is the *anterior limiting lamina* (Bowman's membrane) containing fine interweaving fibrils like those of the substantia propria but lacking fibroblasts. It is about 8 μm in thickness, its collagen fibres being randomly arranged; their width varies from 14 to 35 nm, being much narrower than those of the corneal stroma.

The *posterior limiting lamina*, covering the substantia's posterior surface, is thin, transparent and homogeneous. Microscopy with ultraviolet or polarized light shows it to be stratified and

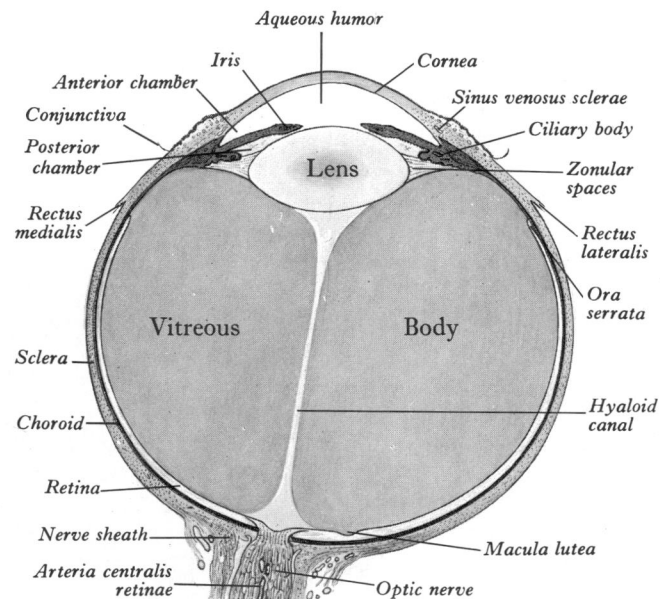

7.294B A horizontal section through a right human eyeball: superior aspect.

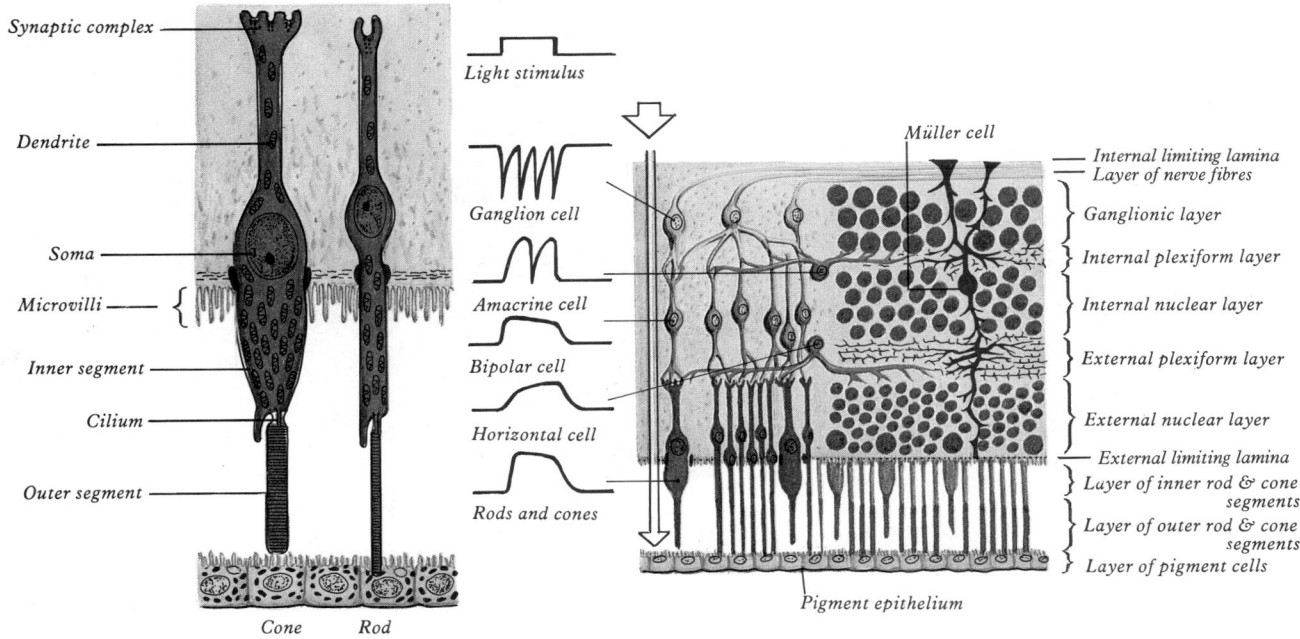

7.294C Retinal structure, showing the organization of the photoreceptors (left) and the layers of the retina (right), including an indication of some of the electrical responses encountered at different depths. Details of neuronal circuitry are not depicted in this diagram; for further information see 7.315 and accompanying text.

fibrillar, as electron microscopy confirms. It is thicker than the subjacent endothelium, of which it is regarded as the basement membrane. It separates easily from both the endothelium and the corneal stroma. At the limbus of the cornea it disperses into the fibres of trabecular tissues adjoining the inner wall of the scleral

venous sinus (7.299); between these trabeculae are the *spaces of the iridocorneal angle*, connecting with the sinus venosus sclerae and with the anterior chamber. Some trabecular fibres continue, internal to the scleral spur, into the iris as the *iridial pectinate ligament*; others connect with the spur's sloping external surface, a few reaching the anterior part of the choroid. The relationships of the trabecular spaces, anterior chamber and scleral venous sinus and their functional interconnections remain the subject of some controversy (see Davson 1976).

The *endothelium of the anterior chamber* covers the posterior limiting lamina, is reflected on to the front of the iris and lines the spaces at the iridocorneal angle. The endothelium is a layer of polygonal, flattened cells; at adjacent borders they have complex, interdigitating profiles when seen by scanning electron microscopy but their surfaces show only a few microvilli.

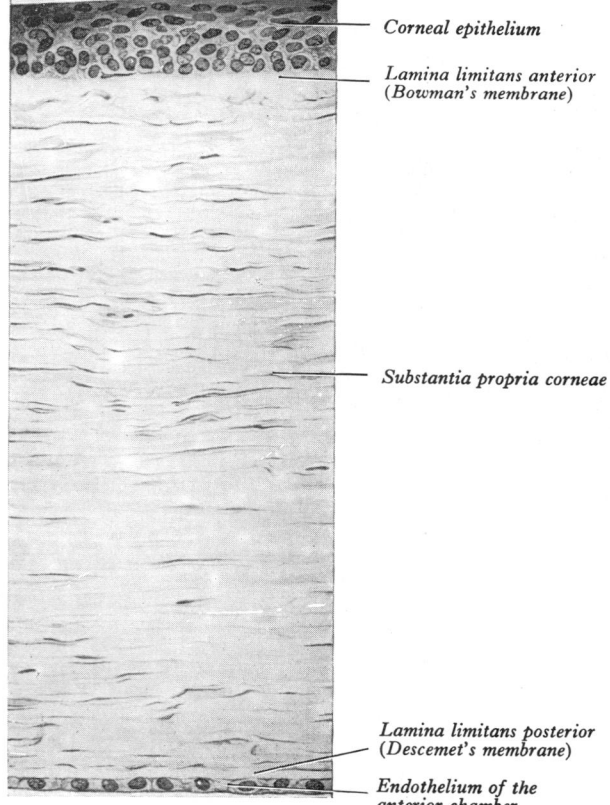

7.295 A general view of a meridional section through the iridocorneal angle.

Iris
Sinus venosus sclerae *Cornea* *Sclera*
Trabecular tissue *Scleral vein*
Radial muscular fibres of iris
Circular fibres of ciliaris *Iridocorneal angle* *Scleral spur* *Meridional fibres of ciliaris*
Oblique fibres of ciliaris

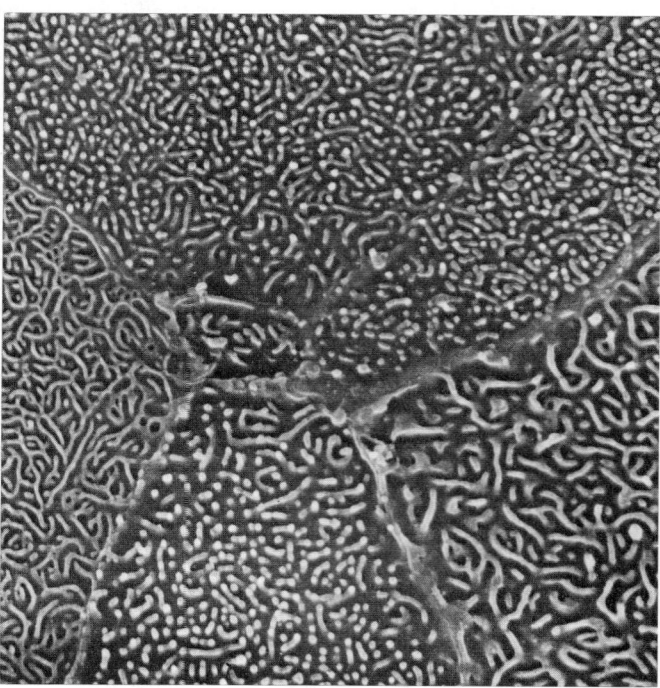

A

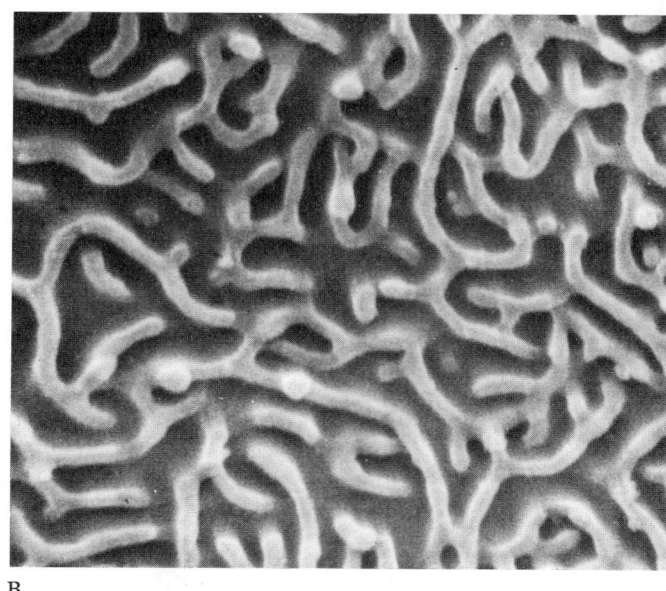

B

7.297 Scanning electrical micrographs of normal human corneal epithelial cells. In A (magnification × 5000) parts of the outlines of several cells are visible; in the upper and lower cells, microvilli predominate but some microplicae are seen. The cells at the right of the field display predominantly microplicae, with only a few microvilli. In B (magnification × 30 000) a small number of microvilli are scattered amongst abundant microplicae. By permission of Pfister & Burstein 1977.

Corneal epithelium
Lamina limitans anterior (Bowman's membrane)
Substantia propria corneae
Lamina limitans posterior (Descemet's membrane)
Endothelium of the anterior chamber

1184   7.296 Radial section through the human cornea. Magnification × 128.

Vessels and Nerves. The cornea is non-vascular, the capillaries of the conjunctiva and sclera ending in loops near its periphery. Lymph vessels are also absent. The *nerves* are numerous and arise as branches of the ophthalmic nerve, particularly the long ciliary nerves. They form an *annular plexus* around the periphery of the cornea, from which radial fibres enter the substantia propria, losing their myelin sheaths and endoneurium (Matsuda 1968) to ramify throughout its matrix in a delicate reticulum. Their terminal filaments form an intricate *subepithelial plexus*, from which fine, varicose fibrils traverse the anterior limiting membrane to form an *intra-epithelial plexus*. There are no specialized end-organs and in the epithelial nerve the fibres are devoid of Schwann cells; they do not arborize. Electron microscopy of simian corneal nerves suggests that stromal nerves do not enter the epithelium, whose nerves arise instead directly from the limbus (Lim & Ruskell 1978).

The fibrous tunic of the eye is often described as being merely protective, but its common, continuous functions are to provide a smooth translatable external surface (vide infra), a suitable attachment for muscles and a resistance to the intraocular pressure. It is thus essential to the maintenance of the shape and dimensions of the eye. For further details of the cornea and sclera see Jakus (1964), Laugham (1969), Davson & Graham (1974) and Davson (1980).

The Vascular Tunic

The vascular tunic, or *uveal tract*, comprises the choroid, ciliary body and iris (7.294), forming a continuous structure. The choroid covers the internal scleral surface, extending forwards to the ora serrata. The ciliary body continues from the anterior edge of the choroid to the circumference of the iris. The iris is a circular diaphragm behind the cornea, presenting an almost central aperture, the pupil.

THE CHOROID

The choroid is a thin, highly vascular, dark brown tissue which lines almost the posterior five-sixths of the eye; it is pierced behind by the optic nerve and is here firmly adherent to the sclera. Posteriorly it is thicker. Its external surface is loosely connected to the sclera by the *suprachoroid lamina* (lamina fusca); internally it is firmly attached to the retinal pigmented layer. At the optic disc it is continuous with the pia-arachnoid tissues around the optic nerve.

Structurally, the choroid consists largely of a dense capillary plexus, with its small arteries and veins of supply. The blood flow through the choroid is high, a feature probably associated with an intra-ocular pressure of 15–20 mmHg, requiring a venous pressure above 20 mmHg to maintain circulation. Metabolic demand is low; the blood loses only 3% of its oxygen (Bill 1970). The warming effect of the choroidal circulation on the eye may be important. Externally is the *suprachoroid lamina*, about 30 μm thick, composed of delicate non-vascular lamellae, each a network of fine collagen and elastic fibres with stellate cells containing dark brown granules. Ganglionic neurons and neural plexuses are enmeshed in the connective tissue. Mesothelium-lined spaces, sometimes described, are unconfirmed by more recent ultrastructural observations and pathological accumulations of fluid may split the suprachoroidal lamellae, exaggerating a potential 'perichoroidal' space. All such potentially weak zones of connective tissue are a great attraction to those in search of ocular lymph spaces.

The choroid proper (7.301) lies internal to the suprachoroid lamina (which is partly scleral tissue) and has a number of layers. Although descriptions of these vary, those generally recognized are: (1) an external *vascular lamina* of small arteries and veins and loose connective tissue, with scattered pigment cells; (2) an intermediate *capillary lamina* (choroidocapillaris); (3) a thin, apparently structureless *basal lamina* (membrane of Bruch). The vascular lamina is sometimes subdivided on the basis of blood vessel calibre, which naturally decreases towards the capillary lamina. But these vessels are relatively small whereas the

capillaries are large and dominant. The vascular lamina contains the terminals of short posterior ciliary arteries (7.306) which extend meridionally from their entry through the sclera near the optic disc. The veins are larger, converging spirally on to four or five principal *vorticose veins*, which pierce the sclera to reach tributaries of the ophthalmic veins. The capillary lamina, separated from the retina only by the thin basal lamina of the choroid, provides a nutritive supply to the layers of the retina, at least in part. It constitutes a close meshwork of vessels, especially in the posterior region of the eye; but its meshes widen towards the ciliary body, where they connect with capillaries in the ciliary processes. The basal lamina is a glassy, homogeneous layer (lamina vitrea) under the light microscope, only 2–4 μm thick. Electron microscopy reveals that it has a more complex substructure. It has a middle stratum of elastic tissue between an internal and an external layer of collagenous tissue, united externally to the choroidocapillary basement membrane and internally to the basement membrane of the retinal pigment cell layer (Lerche 1965, Nakaizumi 1964). Its function is uncertain but is obviously related to the passage of fluid and solutes from the choroidal capillaries to the retina. It is also said to provide a smooth surface for the precise orientation of retinal pigment cells and receptors. It is formed by both the retina and the choroid (Takei & Ozanics 1955), the former contributing the basement membrane of the retinal pigment epithelium. Its external stratum is associated and perhaps identical with the basement membrane of the choroidal capillary endothelium. The functions of choroidal pigment cells are uncertain; they may prevent the passage of light through

7.298 Transmission electron micrograph of the substantia propria of the human cornea; note the geometric precision of the alternation in direction of adjacent layers of collagen fibres. Preparation by John Marshall, see 7.301. Magnification × 48 000.

D

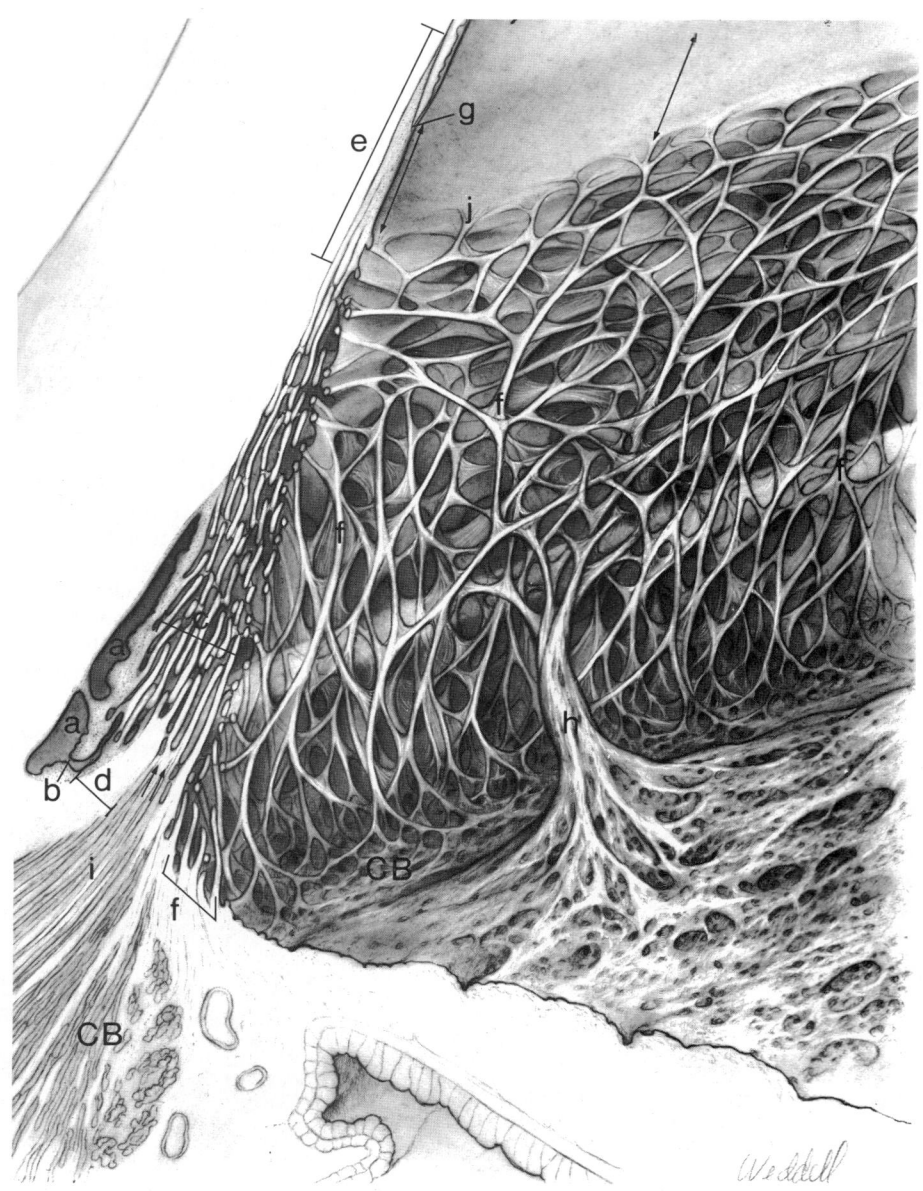

7.299 The iridocorneal angle and adjoining structures, showing the proximity of the scleral venous sinus (aa) to the pectinate ligaments (ff). The trabecular meshwork of the latter is partly uveal, being continuous with the iris (h) and ciliary body (CB) and muscle (i). Anterior to the scleral spur (d), scleral trabecular tissue (c) is even closer to the scleral venous sinus. Aqueous fluid percolates through this trabecular region, reaching the lumen of the sinus through small apertures (b). The pectinate ligament diminishes as it approaches the corneal limbus (e) and in this junctional zone the posterior limiting membrane (of Descemet) also terminates (g). The endothelium of the anterior chamber (posterior corneal epithelium) is continuous with the endothelium of the trabeculae (j) at the limbus. From Hogan et al 1971, by permission of the authors, artist and publishers. 7.300, 303, 304, 305, 321 and 324 are from the same source.

the sclera to the retina but they are more likely to absorb light traversing the retina, preventing internal reflexion within the posterior chamber of the eye. In many mammals, especially those of nocturnal habit, specialized choroidal cells form a reflecting *tapetum*, responsible for the greenish glare in their eyes at night (Walls 1963); its function is uncertain but it may be a mechanism of aggression or a means of augmenting the stimulation of retinal receptors under low light conditions.

Although clinical evidence suggests that the choroid has a nerve supply, the anatomical evidence is scant. Crush lesions of the ophthalmic nerve (Bergmanson 1977) in simians yielded no evidence of any suprachoroidal fibres from ciliary nerves and no degenerating fibres were found terminating within the choroid.

THE CILIARY BODY

The ciliary body is directly continuous with the choroid and extends into the iris; all these regions of the uveal tract have various features in common but also have regional differences related to

variations in their function. The ciliary body is concerned with the suspension of the lens and with accommodation, accounting for the accumulation of muscle fibres which cause it to bulge internally (7.303). It is also a major source of aqueous fluid for the anterior segment of the eye, which its anterior aspect faces. Posteriorly it is contiguous with the vitreous humour and probably secretes some of the glycosaminoglycans of the vitreous body. The anterior and the long and short posterior ciliary arteries meet in the ciliary body (7.305), which is therefore a highly vascular region, involved not only in secretory and muscular activities but also in the blood supply to the iris and limbus. The ciliary body is traversed by the major nerves to all the anterior tissues of the eyeball.

Externally the ciliary body extends from a line about 1·5 mm posterior to the limbus of the cornea (corresponding also to the scleral spur) to a line about 7·5–8 mm posterior to this on the temporal side and 6·5–7 mm on the nasal. The ciliary body is thus slightly eccentric and extends posteriorly from the scleral spur which is its attachment, with a meridional width varying from 5·5

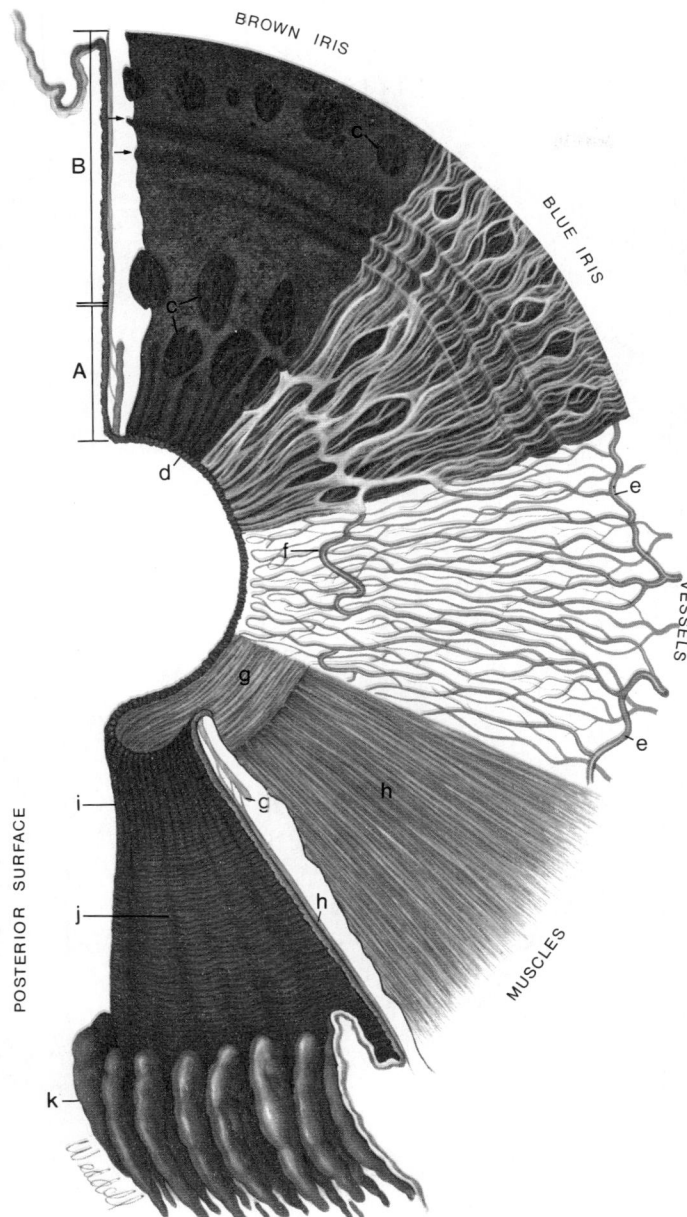

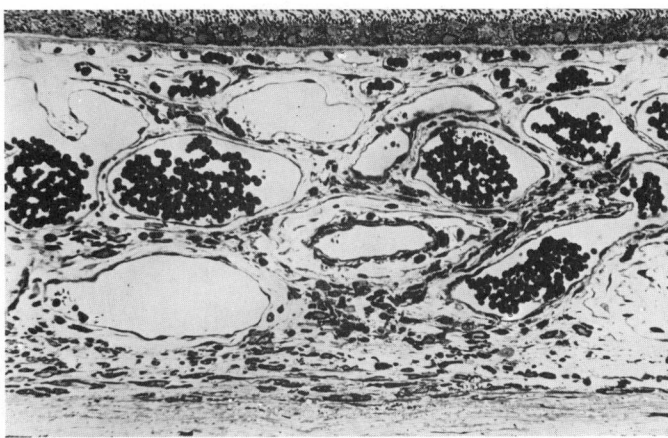

7.301 Light micrograph of a section of human eyeball showing full thickness of the choroid coat and adjacent parts of the retina (above) and sclera (below). Note, from above downwards: rod and cone processes projecting among pigment cells, the basal lamina (membrane of Bruch), layer of capillaries (choriocapillaries), layer of larger vessels and loose connective tissue merging into the sclera. Numerous pigment cells are scattered throughout the choroid. Provided by John Marshall, Institute of Ophthalmology, London. Magnification ×350.

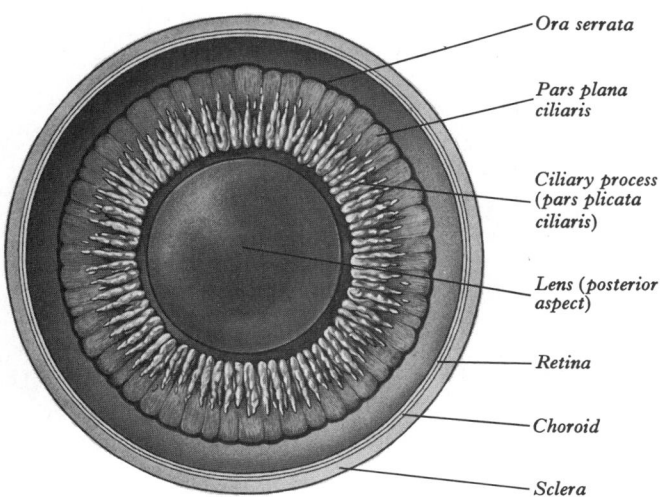

Ora serrata

Pars plana ciliaris

Ciliary process (pars plicata ciliaris)

Lens (posterior aspect)

Retina

Choroid

Sclera

7.302 The interior aspect of the anterior half of the eyeball.

7.300 Composite view of the surfaces and internal strata of the iris. In a clockwise direction from above, the pupillary (A) and ciliary (B) zones are shown in successive segments. The first (brown iris) shows the anterior border layer and the openings of crypts (c). In the second segment (blue iris), the layer is much less prominent and the trabeculae of the stroma are more visible. The third segment shows the iridial vessels, including the major arterial circle (ee) and the incomplete minor arterial circle (f). The fourth segment shows the muscle stratum, including the sphincter (g) and dilatator (h) of the pupil. The everted 'pupillary ruff' of the epithelium on the posterior aspect of the iris (d) appears in all segments. The final segment depicts this aspect of the iris, showing radial folds (i and j) and the adjoining ciliary processes (k). See 7.299 for acknowledgement.

to 6·5 mm. Internally it shows a posteriorly crenated or scalloped periphery where it is continuous with the choroid, termed the *ora serrata*. Anteriorly it is confluent with the periphery of the iris, lateral to which it bounds the anterior chamber's iridocorneal angle. The ciliary body is grey, due to melanin in the deeper layer of its epithelium. It has an anterior plicated part, the *corona ciliaris* (pars plicata), surrounding the base of the iris and posterior to this is a smooth, annular *orbiculus ciliaris* (pars plana, ciliary ring). The orbiculus forms more than half of the meridional width of the ciliary body, being 3·5–4 mm across; its peripheral rim is the ora serrata, at which the *optical* or sensory part of the retina is suddenly reduced to two layers of epithelial cells, extending over the whole ciliary body as the *pars ciliaris retinae* and beyond this on to

the posterior surface of the iris. The corona ciliaris, a smaller annular region within the orbiculus, is ridged meridionally by 70–80 *ciliary processes* radiating from the base of the iris to the orbiculus (7.302). Branching from the sides of these ridges into valleys between them are numerous minor ridges, the *ciliary plicae*, forming a complex pattern of intricate microscopic folds (7.253). Fibres of the *zonule* (the suspensory ligament of the lens) extend into the valleys, passing beyond the ciliary processes to fuse with the basement membrane of the superficial epithelial layer of the orbiculus ciliaris. Their sites of attachment are marked by striae passing back from the valleys of the corona across the orbiculus almost as far as the apices of the dentate processes of the ora serrata (7.303).

The ciliary epithelium is bilaminar, its two layers of simple epithelium being derived from the optic cup's two layers. The *superficial lamina*, made up of columnar cells overlying the orbiculus and cuboidal cells over the ciliary processes, becomes irregular and more flattened between the processes. These cells, containing little or no pigment, are the sole anterior continuation of the *neural* layers of the retina, apart from its pigment epithelium, which is continuous with the *deeper layer* of the ciliary epithelium. The cells of the latter are also approximately cuboidal and are loaded with pigment. The two layers are firmly united, although pathologically fluid may separate them, just as the retina

1187

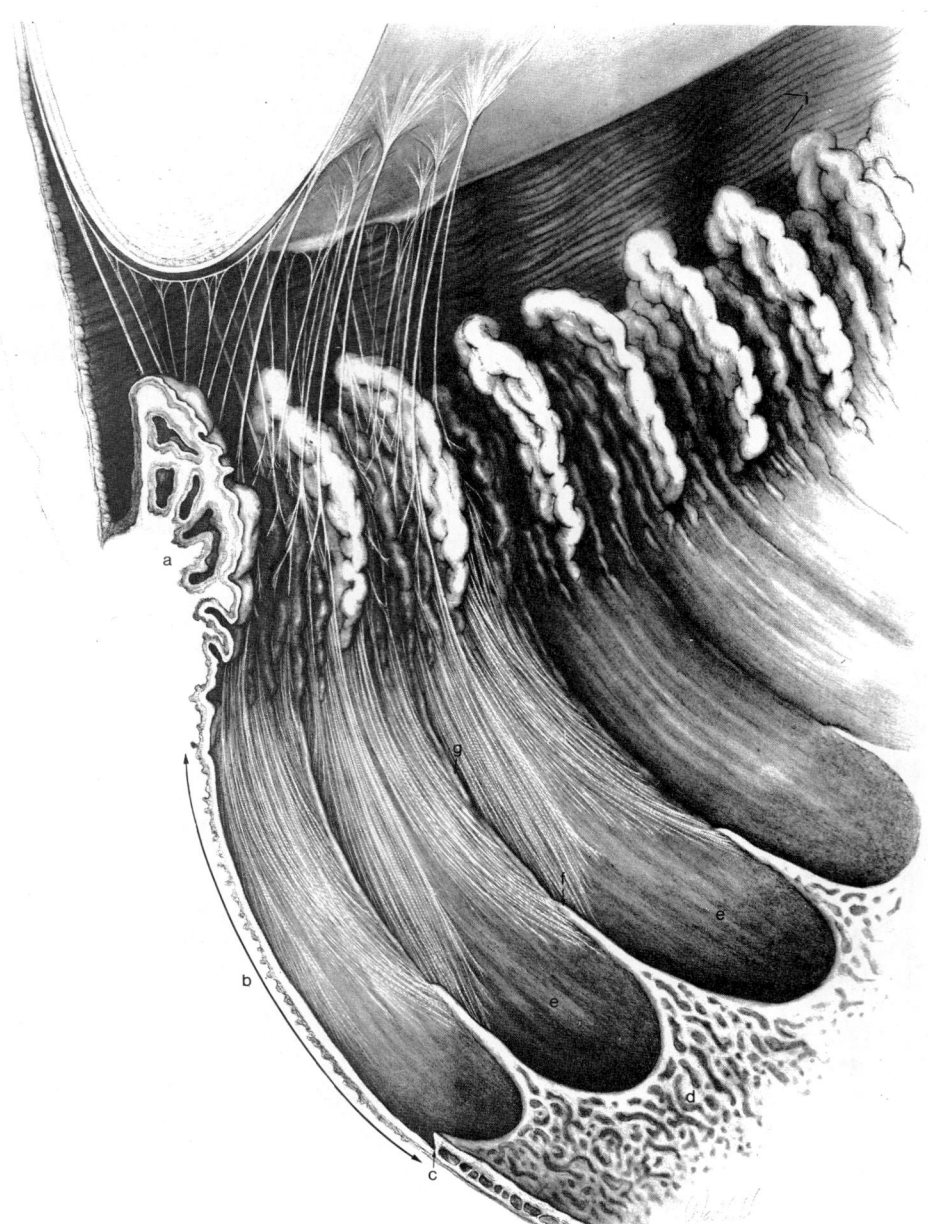

7.303 A magnified view of the ciliary region seen from the ocular interior. Above is the periphery of the lens, attached by the fibres of the *zonule* (suspensory ligament) to the processes of the *corona ciliaris* (pars plicata) of the ciliary body (a). The *orbiculus ciliaris* or pars plana ciliaris (b) has a scalloped boundary, the *ora serrata* (c), which separates it from the retina (d). Flanking the 'bays' (e) of this are the *dentate* processes (f), with which linear ridges or *striae* (g) are continuous. These striae extend forwards between the main ciliary processes, providing an attachment for the longer zonular fibres. The posterior aspect of the iris shows radial (h) and circumferential (i) sulci. See **7.299** for acknowledgement.

detaches from its own pigment epithelium. A basement membrane exists between the two epithelia. Basally the superficial cells are much infolded, like other secretory epithelia. These superficial cells are connected by desmosomes; their cytoplasm contains many mitochondria and a well-developed endoplasmic reticulum, often stacked in perinuclear arrays. Lipid and melanin granules are often present but are not prominent.

The pigment layer is united to the stroma of the ciliary body by its basement membrane, which continues back into the basal lamina of the choroid (p. 1185). The cytoplasm of these cells contains abundant rounded melanin granules 0·6–0·8 μm in diameter. Cells are linked by a few lateral desmosomes, which are more numerous between the two epithelial strata, despite the intervention of a basement membrane. **The ciliary stroma** is composed largely of loose collagen fasciculi, aggregated into a considerable mass between the ciliary muscle and overlying processes, extending into both of them. In this inner stratum of connective tissue are numerous larger branches of the ciliary vessels, with a dense reticulum of large capillaries, most of them adjacent to the epithelium and especially concentrated in the ciliary processes, where they are chiefly of the fenestrated type. Numerous vessels also enter the ciliary muscle but its capillaries show far fewer fenestrations. Anteriorly, near the periphery of the iris is the major arterial circle (7.300, 305), formed chiefly by long posterior ciliary branches of the ophthalmic artery (p. 746) which enter the eye some distance behind the ocular equator, passing between the choroid and sclera to the ciliary body. Ciliary veins, also draining the iris, pass posteriorly to join the vorticose veins of the choroid.

The ciliary muscle has been variously described, the differences being mainly in the subdivision of this small annular mass of non-striated muscle. Most authors recognize three parts: *meridional*, *radial* or oblique and *circular* or sphincteric; but other views have been stated (Calasans 1953, Rohen 1964). Most, perhaps almost all ciliary muscle fibres are attached to the scleral spur (7.304), spreading in several directions, upon which the somewhat arbitrary division of the muscle depends. The outermost fibres are meridional or longitudinal, passing posteriorly into the stroma of the choroid, where many exhibit terminal branchings or *epichoroidal stars*. The innermost fibres swerve acutely from the spur (7.304) to run circumferentially as a sphincteric

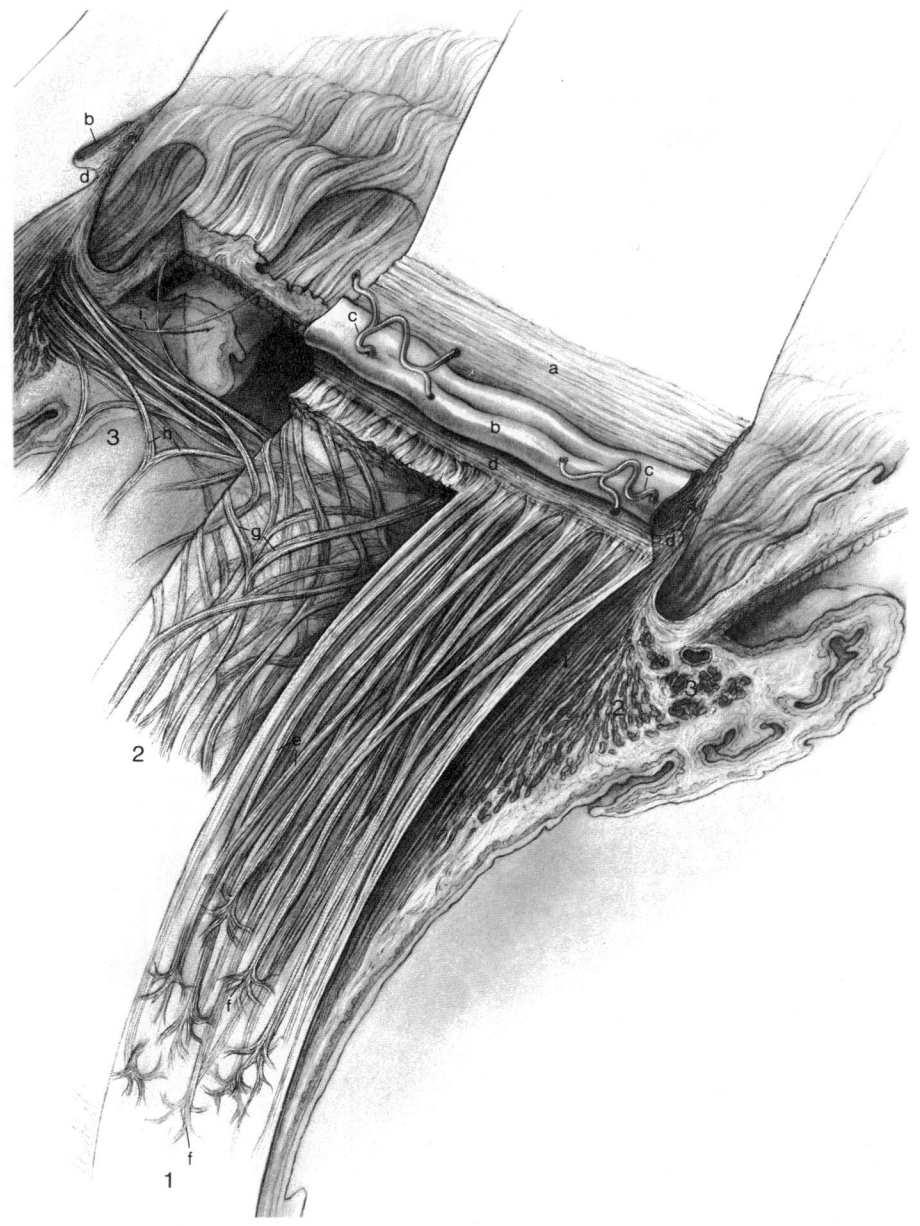

7.304 The ciliary muscle and its components. The meridional or long-itudinal (1), radial or oblique (2) and circular or sphincteric (3) layers of muscle fibres are displayed by successive removal towards the ocular interior. The cornea and sclera have been removed, leaving the pectinate ligament (a), the scleral venous sinus (b), collecting venules (c) and scleral spur (d). The meridional fibres often display acutely angled junctions (e) and terminate in epichoroidal stars (f). The radial fibres meet at obuse angles (g) and similar junctions, at even wider angles (h), occur in the circular stratum of the ciliaris. See 7.299 for acknowledgement.

element near the periphery of the lens. Between these two muscular strata are obliquely interconnecting fibres, frequently forming an interweaving lattice, often referred to as *radial* in direction. In ultrastructure the ciliary muscle differs from other non-striated muscle, the myocytes containing unusually abundant mitochondria and endoplasmic reticulum. A small bundle of fibres is usually surrounded by a common fibroblastic sheath to form units unlike those in any other non-striated muscles. Gap junctions (p. 24) couple the myocytes electrically. Three types of nerve ending have been noted, the most common being an indirect contact of the presynaptic membranes with the interposed basal lamina; contact without the basement membrane also occurs; rarest are larger contacts in depressions in the surfaces of myocytes.

Myelinated and non-myelinated nerve fibres abound in the ciliary muscle and ciliary body, the latter being postganglionic parasympathetic axons from the ciliary ganglion, where they link with the oculomotor parasympathetic outflow; but some fibres appear to be sympathetic, according to much evidence. The parasympathetic supply stimulates the ciliary muscle to contract but the role of the sympathetic innervation is unsettled. Cervical sympathetic stimulation in experimental animals causes the lens to flatten, tantamount to the relaxation of accommodation, but the mechanism of this is uncertain; it may be due to an inhibition of the ciliary muscle or the ciliary body volume may be reduced by vasoconstriction, tensing the zonule and hence the periphery of the lens, the reverse of the slackening effects of the zonule in ciliary contraction (Morgan 1944, Alpern 1969). Electron microscopy of the ciliary autonomic plexus in rhesus monkeys reveals a cycle of degeneration and regeneration which becomes more marked in older animals (Townes Anderson & Raviola 1979).

THE IRIS

The iris is an adjustable diaphragm around a central aperture (slightly medial to true centre), the *pupil*, which controls the amount of light entering the eye. Pupillary diameter varies from 1 to 8 mm at least and has an even wider range under the influence of drugs. This gives an aperture range in excess of $f20-f2\cdot5$, and a ratio of 32:1 in the amount of light permitted to enter the eye.

While this is not enough to save the retina from the effects of intense illumination, it moderates the great range of luminosity encountered in ordinary use, preserving useful vision under highly variable conditions. (The pupillary diameters noted above and the average iridial diameter of about 12 mm are of course estimated through the cornea, introducing a magnification factor of about 12%.) Pupillary *constriction* and *dilatation* are self-explanatory terms, for which meiosis and mydriasis are used clinically, though more properly reserved for the extreme limits of contraction and dilatation. The erudite surveys of the immense literature on the pupil and the iris by Loewenfeld (1958) and Loewenstein & Loewenfeld (1970) should be consulted for further information on this subject.

Though the iris is named after the rainbow, its range of colour extends only from light blue to very dark brown, often varying in the two eyes and even within the same iris. The colour is the combined effect of the iridial connective tissue and the pigment cells in selectively absorbing and reflecting different frequencies of light energy. When pigment is largely absent, as at birth, the colour is light blue; some pigment is necessary to confine light transmission to the pupil and central lens, where optical aberrations are least. The concentration of melanocytes is the main factor determining the hue of the iris, but the distribution of pigment is often irregular, producing a flecked or maculated appearance.

The iris is not a flat diaphragm; the lens causes it to bulge a little, so that it is more accurately a shallow cone, truncated by the pupillary aperture. Sited between the cornea and lens (7.294A) and immersed in *aqueous fluid*, it partially divides the *anterior segment* into an *anterior chamber*, enclosed by the cornea and iris (meeting at the *iridocorneal angle*) and a confusingly termed *posterior chamber*, between the iris and the lens (p. 1181). Into the latter, ciliary processes protrude from the periphery a little between the divisions of the zonular ligament of the lens and here most of the aqueous fluid is produced, traversing the pupil into the anterior chamber to its exit via the scleral venous sinus (p. 1181) at the iridocorneal angle: the 'filtration angle' in clinical parlance.

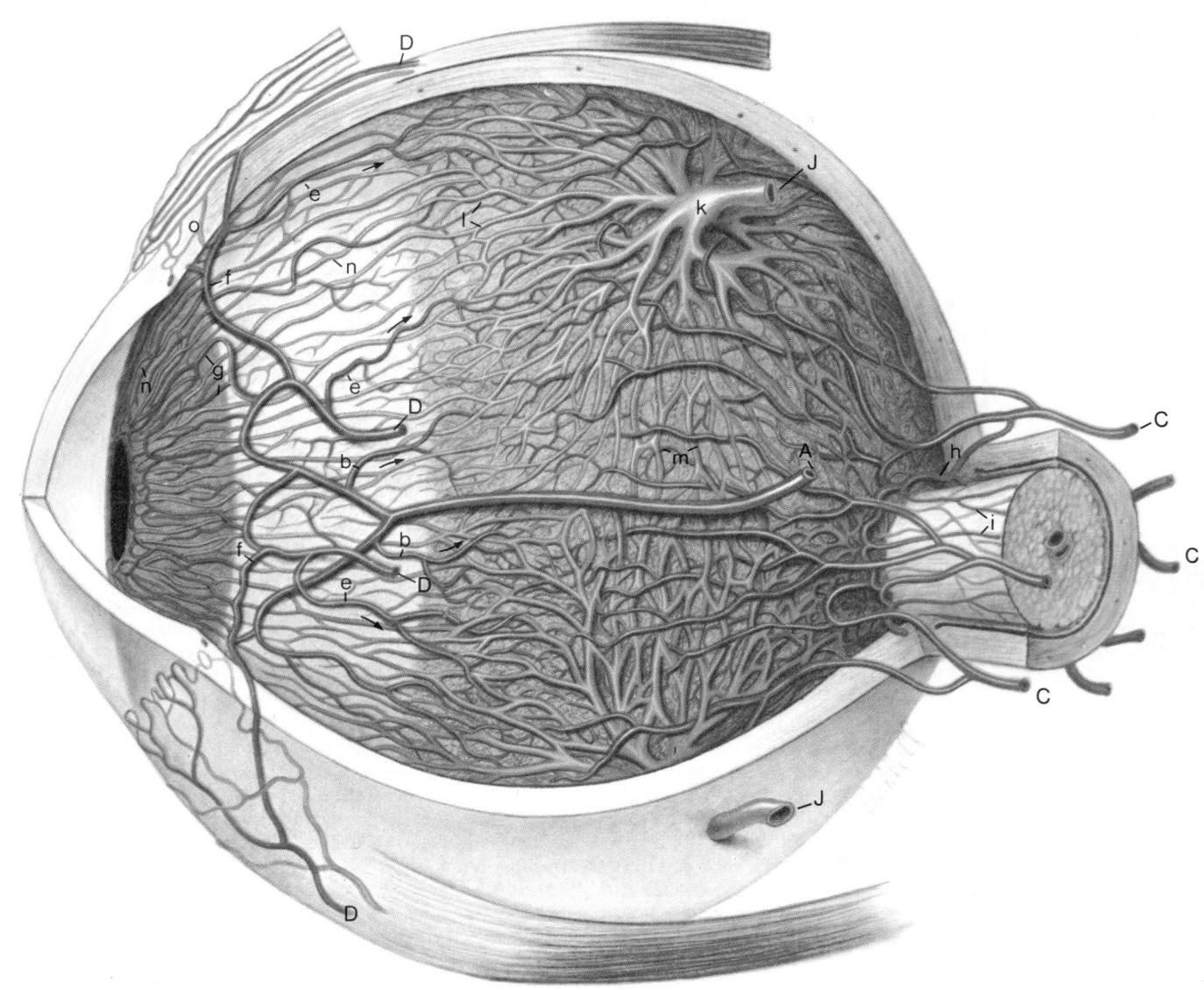

7.305 The vascular arrangements of the uveal tract. The long posterior ciliary arteries, one of which is visible (A), branch at the ora serrata (bb) and feed the capillaries of the anterior part of the choroid. Short posteriorly ciliary arteries (CC) divide rapidly to form the posterior part of the choriocapillaris. Anterior ciliary arteries (DD) send recurrent branches to the choriocapillaris (ee) and anterior rami to the major arterial circle (ff). Branches from the circle extend into the iris (g) and to the limbus.

Branches of the short posterior ciliary arteries (CC) form an anastomotic circle (h) (of Zinn) round the optic disc and twigs (i) from this join an arterial network on the optic nerve. The vorticose veins (JJ) are formed by the junctions (k) of suprachoroidal tributaries (l). Smaller tributaries are also shown (m, n). The veins draining the scleral venous sinus (o) join anterior ciliary veins and vorticose tributaries. See 7.299 for acknowledgement.

The microscopic structure of the iris includes several unusual features (7.300). Its anterior surface, forming the posterior boundary of the anterior chamber, possesses no distinct epithelium, despite statements to the contrary; this surface is merely a modified 'anterior border layer' of the general iridial *stroma*. This stroma contains the regional vessels and nerves and, near the pupillary rim, there is an aggregation of non-striated myocytes forming an annular contractile *sphincter pupillae*. The

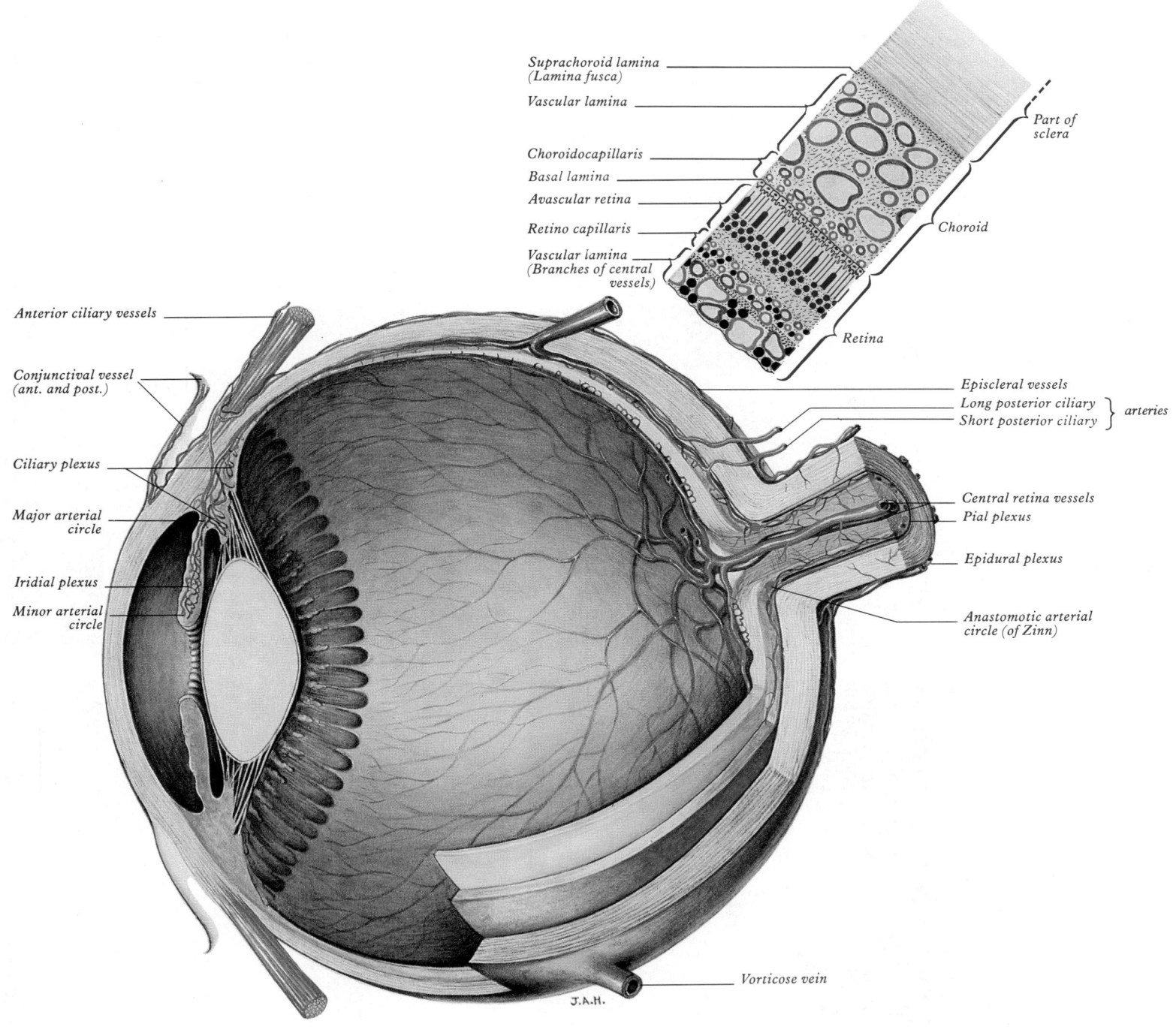

7.306 The main features of the vascularity of the eye. The thickness of the sclera, cornea, choroid and neural retina are exaggerated, although representing relative thicknesses, and the regional variations should be noted. The section of ocular wall at higher magnification is included for particular reference to the avascular zones of the retina (which contain the photoreceptors) and their dual relationship to the vascular laminae and capillary plexuses of the choroid and retina.

epithelial strata of the posterior aspect of the iris is a continuation of the bilaminar epithelium of the ciliary body representing the two layers of the optic cup. The pupil, through which this epithelium curves for a short distance on to the *anterior* iridial surface as the *pigment ruff* or 'border', therefore corresponds to the opening of the optic cup. The deeper and hence on the back of the iris the posterior of these two epithelial layers is confusingly termed the *anterior epithelium*; although it is really *posterior* to its stroma. Its cells are pigmented, as are those of the same layer in the ciliary epithelium; closely associated are the radially arranged non-striated myocytes of the dilatator pupillae, which like the sphincter has a most unusual embryological origin, from the neural ectoderm of the optic cup. Superficial (posterior) to this layer is a stratum of heavily pigmented cells, the so-called *posterior epithelium*, continuous with the internal *non-pigmented*, retinal layer of the ciliary epithelium.

The anterior iridial surface (anterior border layer) has been much studied at low magnification by slit-lamp microscopy; it then appears somewhat fluffy, except when heavily pigmented. *Crypts*, through which vessels may be visible in the stroma, and various radial and circular folds and striae can be observed (Vogt 1942). The constituents of this anterior border lamina are mainly dendritic fibroblasts and melanocytes, with no vestige of the en-

dothelium which covers it at birth but rapidly disappears during the early postnatal years (Vrabec 1952). Electron microscopy confirms this (Tousimis & Fine 1959). Fibroblasts form an almost continuous surface monolayer; their branching processes form no actual junctions (Smelser & Ishikawa 1966). At the iridial periphery they blend with the trabecular connective tissue (pectinate ligament) of the iridocorneal angle. At the pupillary rim they meet the epithelium of the posterior iridial surface. Melanocytes are also intricately branched, again with no special junctions between them. Some capillaries invade the border layer. Naked axons, presumed to be sympathetic, have also been found closely apposed to the melanocytes and fibroblasts, suggesting the release of transmitter substances into the anterior chamber (Ringvold 1975).

The *stroma of the iris* is, like the anterior border layer, derived from the mesoderm between the developing lens and optic cup and is also formed of fibroblasts, melanocytes and a loose collagenous matrix. The intercellular spaces appear to communicate freely with the anterior chamber and interchange of fluid between the two may permit the large changes in volume occurring during contraction and relaxation of the iris. The mesodermal stroma also contains not only abundant blood vessels and nerves but also the ectodermal sphincter and dilatator muscles. There is no elastic tissue. The elastic recoil sometimes

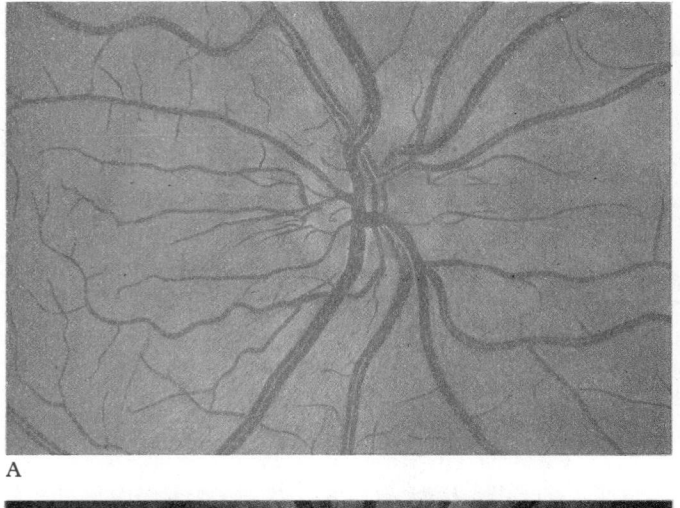

A

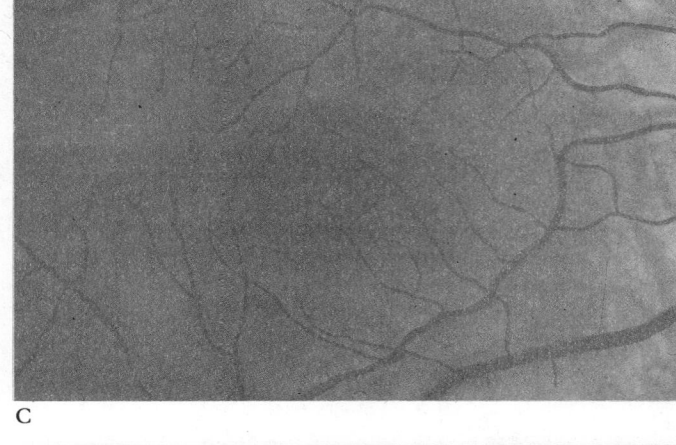

C

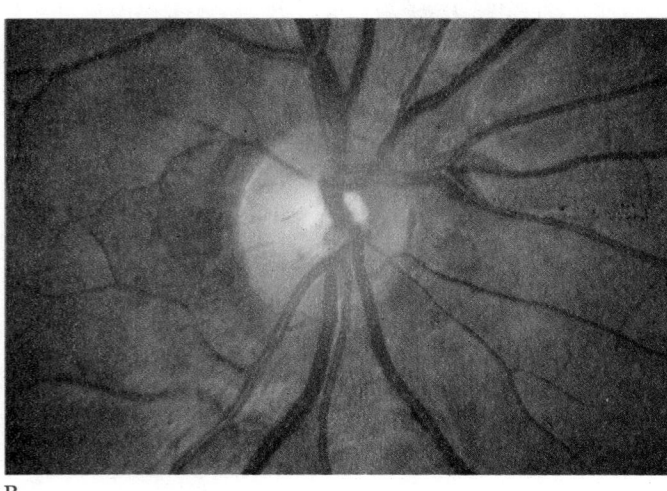

B

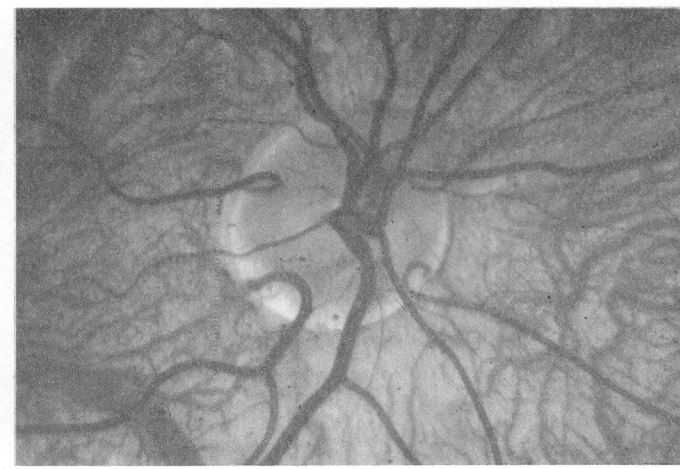

D

7.307 Opthalmoscopic photographs of the right human retina.

A. Note dichotomous branching of vessels, arteries being brighter red and showing a more pronounced 'reflex' to light, as a pale stria along their length. The veins are also larger in calibre; more of them cross arteries superficially than is usual. The optic disc, around the entry of the vessels, is a light pink, with a surrounding zone of heavier pigmentation. Compare with 7.308A from the same Caucasian adult.

B. Appearances in a heavily pigmented individual (Negroid adult), with a paler optic disc than in A. Note accentuation of the edge of the disc by retinal and choroidal pigmentation. The arteries cross the veins superficially in this retina.

C. Normal macula of a young Caucasian subject. Note the fovea, showing as a central, paler, circular area. The macular branches of the central retinal artery are approaching from the right. The macula is largely free of vessels of macroscopic size, but the capillaries here form a particularly close network, except at the fovea.

D. The region of the optic disc in an eye with poorly developed pigmentation. Three cilioretinal arteries are curving round the edge of the disc (two on the left, one on the right). Between the two cilioretinal arteries a single macular artery is apparent. Due to the depressed pigmentation choroidal vessels are also visible, especially veins; and on the left of the photograph two large vorticose venous tributaries can be seen.

attributed to the iris as a dilator force when the sphincter is relaxed must reside in other structures, if it exists at all. Collagen fibrils, about 60 nm in diameter, are loosely arranged, many describing incomplete circumferential loops around the pupil. 'Clump' cells, mast cells, macrophages and lymphocytes have also been described in the stroma.

The sphincter pupillae is a flat annulus of non-striated muscle about 0·75 mm wide and 0·15 mm thick. Its densely packed

fusiform myocytes are often arranged in small bundles, as in the ciliary muscle, and pass circumferentially around the pupil. Collagenous connective tissue lies in front of and behind the muscle fibres and is very dense posteriorly, where it binds the sphincter to the pupillary end of the dilator muscle. Ultramicroscopy shows that muscle fascicles are not pervaded by nerve fibres, usually only one myocyte of each group being innervated, excitation causing contraction presumably spreading to other fibres at

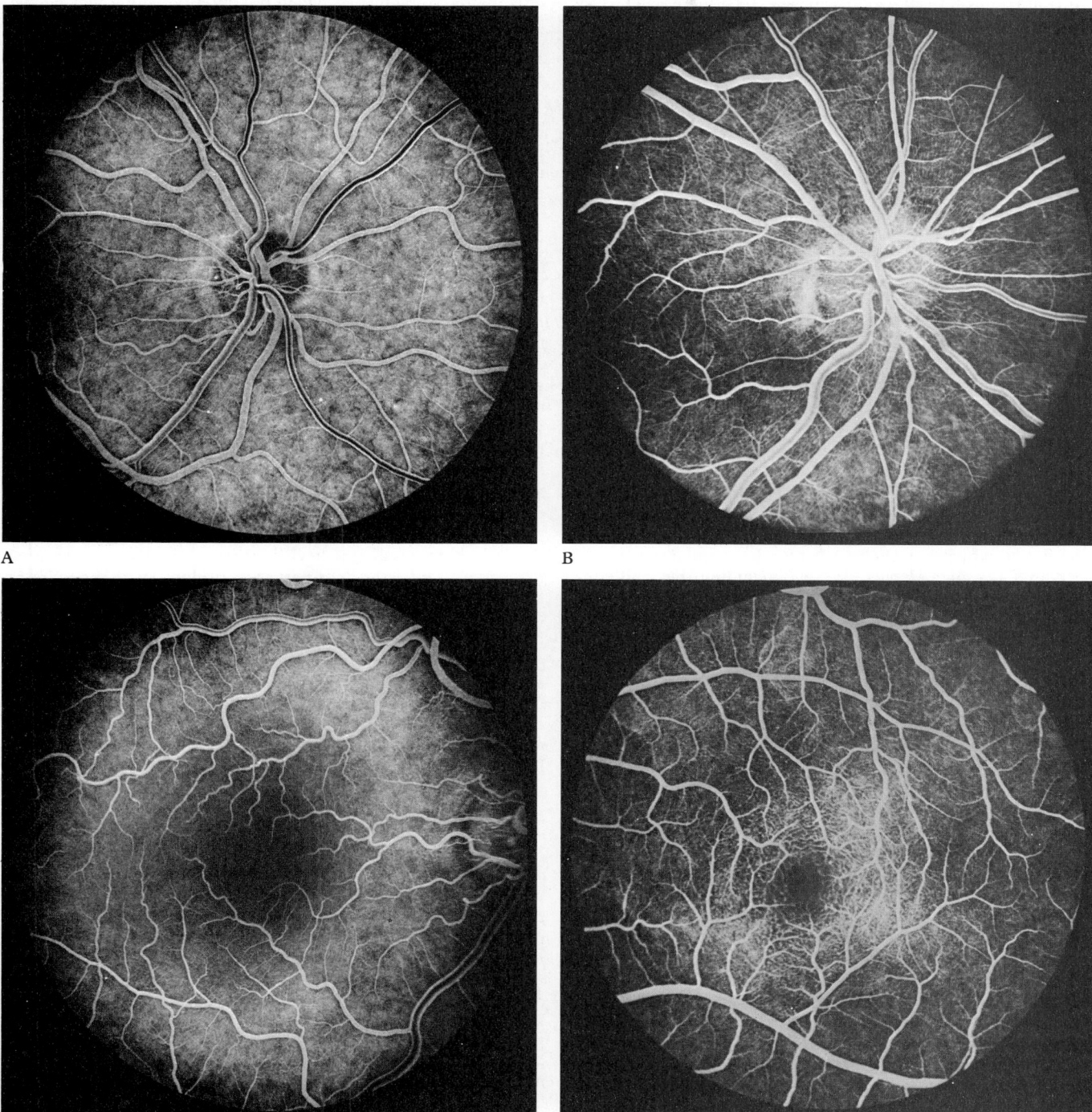

A

B

C

D

7.308 Fluorescence angiograms of the retina. These are produced by photography with a fundus camera at known periods of time following introduction of fluorescein into the circulation. For details of the technique consult Rosen (1969) to whom we are indebted for all the colour photographs and angiograms in this illustration.

A. Angiogram of the same retina as that appearing in 7.307A taken in 'mid-venous' phase. The arteries display an even fluorescence, but the veins appear striped, due to laminar flow. This appearance is the reverse of and not to be compared with the arterial 'reflex' which is seen in 7.307A. The background mottling is due to fluorescence from the choroidal vessels.

B. Angiogram of the left optic disc, showing the major arteries and veins and also their smaller branches. Note particularly the radial pattern in the retinal capillaries. The laminar flow in the veins is less obvious than in 7.308A.

C. Angiogram showing the macular region of a right eye. The main macular vessels are approaching from the right. The subject was an elderly person with considerable macular pigmentation, which masks fluorescence from the choroidal circulation. Compare with 7.308D.

D. Angiogram of the macula of a young subject (left eye) showing the macular capillaries in detail. Note the central avascular fovea. Compare with 7.308C.

gap junctions. Small nerves ramify in the connective tissue between bundles, most of their fibres being non-myelinated and often enclosed in common Schwann cell sheaths. They do not approach the surfaces of myocytes more closely than 0·1 μm.

The dilatator pupillae, its very existence subject to a prolonged, vexed and at times almost ridiculous controversy, is now a well-established entity confirmed by microscopic, physiological and pharmacological observations (Alphen 1963, Loewenfeld 1958). It is a thin stratum lying immediately anterior to the epithelium of the posterior iridial surface. Its fibres are in fact the muscular processes of the anterior layer of this epithelium, whose cells are therefore myoepithelial; their apical processes form the epithelium itself. Myofilaments appear in both parts of these cells but more abundantly in their fusiform basal muscular processes, which are about 4 μm thick, 7 μm wide and 60 μm in length; they form a stratum 3–5 elements thick through most of the iris, from its periphery to the outer perimeter of the sphincter which it slightly overlaps. Here the dilator thins out, sending spurs to blend with the sphincter; unlike the apical parts of the myocytes these have a basal lamina and are joined by gap junctions like those between the sphincteric myocytes. Non-myelinated nerve fibres appear near their muscular processes or 'fibres' and these terminate close (about 20 nm) to their surfaces.

The arteries of the iris (7.300, 305, 306) arise from the long posterior and anterior ciliary arteries and from those in the ciliary processes. Both long ciliary arteries, on reaching the attached margin of the iris, divide into an upper and a lower branch anastomosing with corresponding arteries from the opposite side and with the anterior ciliary arteries, to form a vascular circle, the *circulus arteriosus major*. From this, vessels converge to the free margin of the iris, anastomosing to form a *circulus arteriosus minor*; this is incomplete and some regard its vessels as venous. The smaller arteries and veins are very similar in their mural structure and also share some peculiarities; they are often slightly helical, perhaps to adapt to changes in iridial shape as the pupil varies in size, which may also account for the peculiar structure of their vascular walls. All these vessels, including the capillaries, have a non-fenestrated endothelium and a prominent, often thick basal lamina. In the arteries and veins, there is no elastic lamina and myocytes are few, especially in the veins; connective tissue in the tunica media is loose and external to this a remarkably dense collagenous adventitia appears to form almost a separate tube external to this. The loose stratum of the media has been regarded as a lymph space but this is improbable; it is about 7 μm in width and contains matrix probably derived from the endothelial basement membrane (Hogan et al 1971).

Nerves of the iris, as in the choroid, come chiefly from branches of the long ciliary rami of the nasociliary nerve and from the short ciliary rami of the ciliary ganglion, the latter providing postganglionic non-myelinated axons innervating the sphincter pupillae. The dilatator is supplied by non-myelinated postganglionic fibres from the superior sympathetic ganglion; their routes are not certain and may vary in different species and they may be multiple in man. The internal carotid sympathetic plexus is said to send a branch via the ciliary ganglion, reaching the eye in the short ciliary nerves; but some may travel in the long ciliary nerves. (For other routes in simians, see p. 1102.) The innervation of iridial muscles, as also of the ciliaris, is probably more complex; both the sphincter and the dilatator may have a double autonomic supply. Histochemical stains for acetylcholinesterase and fluorescent techniques have demonstrated cholinergic and adrenergic activity in both muscles (Ehinger & Falck 1966, Lowenstein & Loewenfeld 1970). Though ganglion cells have been noted in the iris, almost all nerve fibres are probably postganglionic and non-myelinated. They form a plexus around the periphery of the iris, from which small nerves and fibres extend to the two muscles, to vessels, the anterior border layer and the anterior epithelium (though not the pigment layer) of the posterior layer of the iris. Some fibres may be afferent and some are vasomotor but little is known of either. (For distribution of non-myelinated autonomic nerve fibres in the choroid consult Ruskell 1971.) It is difficult to identify afferent nerve endings in the iris and ciliary body or to distinguish them from efferent autonomic endings, but such endings have been described in monkeys, using mitochondrial accu-

mulations as a criterion (Bergmanson 1978, Cauna 1966, Macintosh 1974). Trigeminal (ophthalmic division) terminals and also autonomic endings in these sites are said to display this feature.

The pupillary membrane. In fetuses the pupil is closed by a thin, vascular *pupillary membrane* (p. 203). Its vessels are partly from those of the iridial marginal and partly from those of the lens capsule; they end in loops near the membrane's centre, which is avascular. About the sixth month of gestation, the membrane begins absorption from the centre towards the periphery; at birth only scattered fragments remain but exceptionally it may persist and interfere with vision.

The Retina

The retina (7.294) is the neural, sensory stratum of the eyeball. It is thin, being thickest (0·56 mm) near the optic disc, diminishing to 0·1 mm anterior to the equator and continuing at this thickness to the ora serrata (7.302, 303). It is even thinner at the optic disc and fovea of the macula (Spence et al 1969). External to it is the choroid, internal to it the hyaloid membrane of the vitreous body. At the disc it is continuous with the optic nerve. Anteriorly, at the ora serrata (p. 1187), a thin, non-neural prolongation extends forwards over the ciliary processes and iris as the *ciliary* and *iridial parts of the retina*, which consist of retinal pigmented and columnar epithelial layers only (vide supra). From the optic disc to the ora serrata extends the *optic part of the retina*, soft, translucent and purple in the fresh, unbleached state, due to the presence of *rhodopsin (visual purple)* but soon becoming opaque and bleached when exposed to light. (It is hence difficult to demonstrate rhodopsin; in eyes preserved for dissection the *fixed* retina is cloudy white.) Near the centre of the retina is an oval, yellowish area, the *macula lutea* (7.307, 308), which has a central depression, the *fovea centralis*, where visual resolution is highest. At the fovea, the retina is exceedingly thin, some of its layers being absent, and the dark choroid is distinctly visible through it. About 3 mm nasal to the macula lutea the optic nerve is continuous with the retina at the *optic disc*, about 1·5 mm in diameter. The circumference of the disc is slightly raised, while centrally it has a shallow depression, being pierced in this area by the central retinal vessels (7.307, 308). The disc ('blind spot') is insensitive to light. By ophthalmoscopy it is normally pink but it is much paler than the retina and may be grey or almost white. In optic atrophy the capillary vessels disappear and the disc is white. The name 'optic papilla', often applied to the disc, is a misnomer; almost all of the normal disc is level with the retina and centrally *depressed*.

The Structure of the Retina

The retina is derived from the two layers of the invaginated optic vesicle (pp. 189, 201), the outer becoming the lamina of the *pigment cells*; the inner develops into a complex multi-laminar structure containing: the *photoreceptors* (rod and cone cells), their *first order neurons* (bipolar cells) and the somata and axons of the *second order neurons* (ganglion cells) and *interneurons* arranged amongst these centripetal paths (horizontal and amacrine cells). It also contains *neurological elements* and a *vascular system*, chiefly of capillaries. Some authorities separate the region of the retina posterior to the ora serrata (*pars optica retinae*) from its pigment epithelium as the 'retina proper' but, since these two layers are functionally integrated, this arbitrary division into *the stratum pigmentosum* and *stratum nervosum* will be ignored in the present account. Description will, however, be limited to the *sensory* retina, whose ciliary and iridial parts have already been considered (pp. 1186, 1189). The potential space between the pigment cells and photoreceptors can be equated with ventricular cavities since it is developmentally a hollow outgrowth of the brain (p. 201).

Though, following custom, 10 layers in the retina will be described in the present account (7.259–262), this division is merely a convenience in the primary analysis of the highly integrated nervous elements of the retina. For a concise comparative review of structure and function in the vertebrate retina see Boycott (1974).

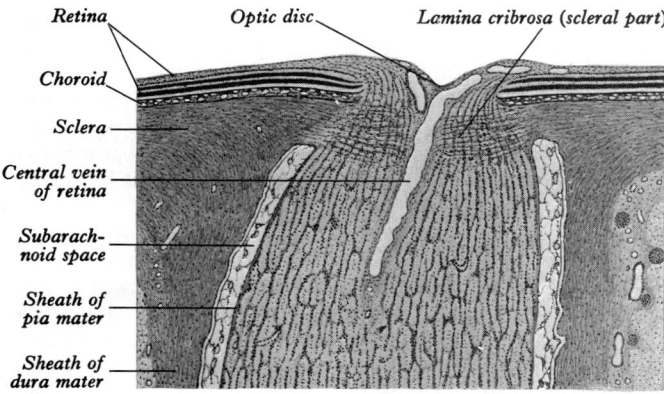

 labels:
Retina
Optic disc
Lamina cribrosa (scleral part)
Choroid
Sclera
Central vein of retina
Subarachnoid space
Sheath of pia mater
Sheath of dura mater

7.309 A horizontal section through the optic nerve at its point of exit from the human eyeball.

THE RETINAL PIGMENT EPITHELIUM

This is a single lamina of regularly arranged cells extending from the periphery of the optic disc to the ora serrata and continuing on into the ciliary epithelium (p. 1187). The cells are flat rectangles in radial sections and hexagonal or pentagonal in tangential aspect (7.313). They number about 4–6 million in the human eye, increasing with age. Near the macula they are about 10 and 14 μm in their radial and tangential dimensions respectively but are much flatter near the ora serrata (Ts'o & Friedman 1967, 1968). Their nuclei are basally placed, near the basal lamina of the choroid (p. 1185) and separated from it by their own basal lamina (Bruch's membrane) which is much infolded into grooves in the basal plasma membrane. Apically they project between the processes of the rods and cones as microvilli 5–7 μm in length; the intermediate cytoplasm contains mitochondria, pigment granules and organelles associated with melanin synthesis, including arrays of granular and agranular endoplasmic reticulum, premelanosomes

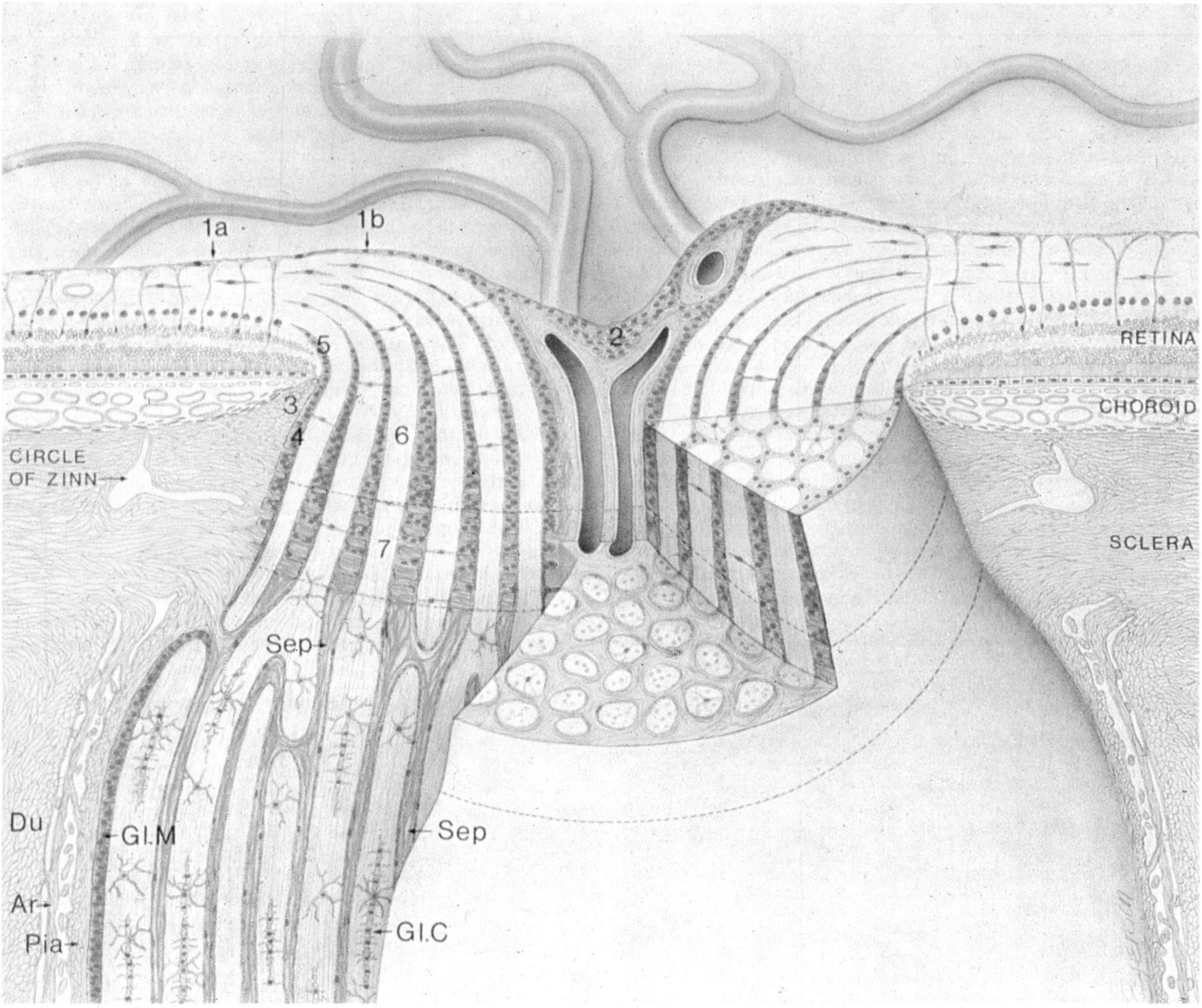

7.310 Schematic representation of the exit of the human optic nerve from the eyeball, showing the distribution of collagenous (blue) and neuroglial (magenta) tissues. Sep = septa of collagenous connective tissue carried into the nerve from the pia mater and dividing the nerve fibres into numerous fascicles. Gl.M = astroglial membrane separating nerve fibres from connective tissue. Gl.C = astrocytes and oligodendrocytes among the fibres in their fascicles. Du, Ar, Pia = dura, arachnoid and pia maters respectively. (1a) is the internal limiting lamina of the retina, which is continuous with an astroglial membrane (of Elschnig) covering the optic disc (1b). An accumulation of astrocytes forms a central meniscus of Kuhnt in the centre of the disc (2). The anterior or so-called 'choroidal part' of the lamina cribrosa (6) is separated from the choroid by a spur of collagenous tissue (3). The 'border tissue of Jacoby' (4), which is largely astroglia, frequently extends beyond the choroid (5) to separate much of the retina from the 'retinal part' of the optic nerve head. The posterior part of the lamina cribrosa (7) contains collagenous tissue derived from the optic nerve septa and fenestrated sheets of collagen fibres continuous with those of the sclera. Reproduced by permission of Anderson & Hoyt 1969.

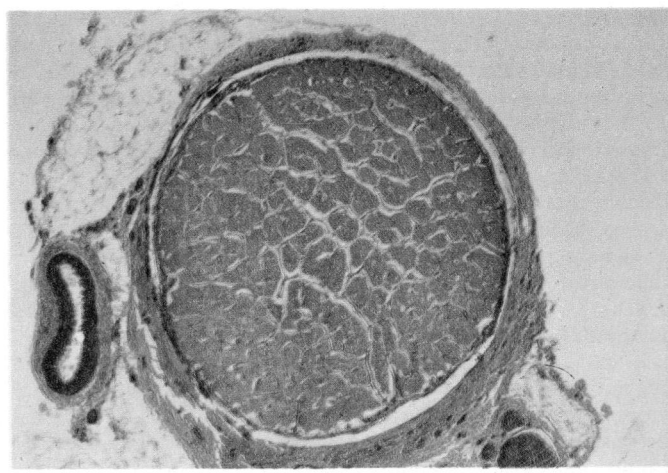

7.311 A transverse section of the optic nerve and its meningeal coverings is visible at the side of the nerve. The dural and pial sheaths are stained green, the subarachnoid pink. Note the fasciculation of the nerve. Material stained by Masson's trichrome technique and provided by N A Locket, Institute of Ophthalmology, London.

and melanosomes (Breathnach & Wylie 1966, Seyi 1967). Lipofuscin also occurs (Feeney et al 1965), probably as an end-product of a phagocytic activity of great importance to retinal function. Autoradiography and ultramicroscopy (Young & Bok 1969, Spitznas & Hogan 1970) have shown that the microvilli of pigment epithelium cells aid the continuous erosion of the external ends of rod and cone processes and that the lamellar inclusions in their cytoplasm are the shed terminal parts of outer segments of these photoreceptors; these phagosomal vacuoles are continually degraded by lysosomal action as they pass deeper into the cells (Marshall & Ansell 1971), the whole process being a vital part of the receptor membrane renewal cycle of photoreceptors (see p. 1198).

The bases of the pigment epithelial cells are much infolded by the basement membrane, as already noted. The latter structure is fibrillar, some fibrils joining the basement membrane of the choroid; this may explain adherence of the pigment epithelium to the choroid rather than to the rest of the retina. The membranes of adjacent pigment cells do not interdigitate much; the variable space between them is sealed off apically from the space between the microvilli and photoreceptors by zonulae occludentes. Other junctions also occur: Miki et al (1975) have described retino-choroidal attachment sites in mammals, including humans; these

include reciprocal interdigitations, as noted above, and desmosomal junctions. A viscous glycosaminoglycan-rich substance fills the space between the pigment epithelium and the neural retina.

Though the functions of pigment epithelium are far from fully understood, it certainly has a phagocytic activity, possibly a nutritive role and some stabilizing action on photoreceptor placement, and the absorption of light, thus preventing back reflection.

THE PROCESSES OF ROD AND CONE CELLS

Retinal photoreceptors (7.312–318) are radially disposed, elongated cells which have a cylindrical photoreceptive portion at one end (nearest the pigment epithelium and choroid), a soma containing the nucleus nearer their other pole and, furthest from the choroid, a terminal synaptic region which is in functional apposition with a bipolar neuron. The photoreceptor part (*outer segment*) is connected to the soma by a region of cytoplasm rich in mitochondria, the *inner segment*. The outer segments are of two types, *rod* and *cone* processes, differentiated respectively by their cylindrical and conical shapes; the cells which bear them are accordingly rod and cone cells or, more simply, rods and cones. They are also differentiated by a constricted *outer fibre* connecting the rod processes to their somata (7.316), a feature absent from cones, though these may show a small 'waist'. At the junctions of rod processes with the outer fibres and of cone processes with their somata is the *external limiting membrane*, through which these photoreceptor processes appear to be thrust (7.316), as if through a sieve; it is described below (see the external limiting lamina).

Rod and cone processes are densely packed and regularly arrayed, but density progressively diminishes towards the ora serrata, where they abruptly cease. In human retinae they are most numerous in and near the *macula*, the region which is best adapted to the discrimination of form and colour, but least adapted to function in low luminosities. They are entirely absent from the optic disc, where centrifugal retinal fibres leave to form the optic nerve. The *central retinal area* is a region 5–6 mm in diameter, containing the *macula lutea*, an area about 2 mm horizontally and 1 mm vertically, its colour being due to the presence of xanthophyll or to a great reduction in the capillary bed or perhaps to other cell inclusions in bipolar and ganglion cells. The macula's *fovea centralis* is a deep conical depression where almost all elements except cones are absent, on its floor at least; its diameter is said to be about 0·4 mm. This *foveola*, as it is often called, is positioned about 4 mm lateral and 1 mm inferior to the centre of the optic disc (the latter corresponding to the 'blind spot' in the uniocular visual field). The minuteness of the foveola accounts for the accuracy with which the visual axes must be directed to achieve the most discriminative vision. The macula has been further divided into peri- and parafoveal areas. The macula

Internal limiting lamina
Lamina of nerve fibres (stratum opticum)
Ganglionic lamina
Internal plexiform lamina
Internal nuclear lamina
External plexiform lamina
External nuclear lamina
External limiting lamina
Lamina of rods and cones
Lamina of pigment cells

7.312 Section through the primate retina, from its vitreous aspect (above) to the choroid tunic (below), showing its layered structure some little distance from the macular region. Diagram to illustrate the customarily recognized strata.

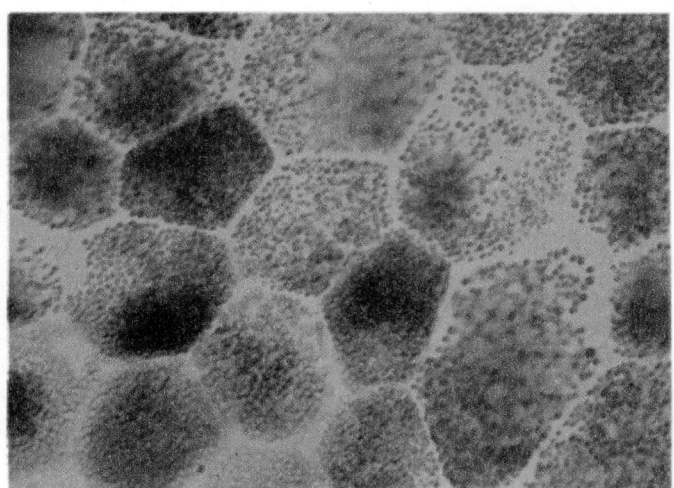

7.313 Colour photography or unstained retinal pigment epithelium seen in surface view (human, aged 40 years). Preparation loaned by John Marshall, Institute of Ophthalmology, London. Magnification ×3500.

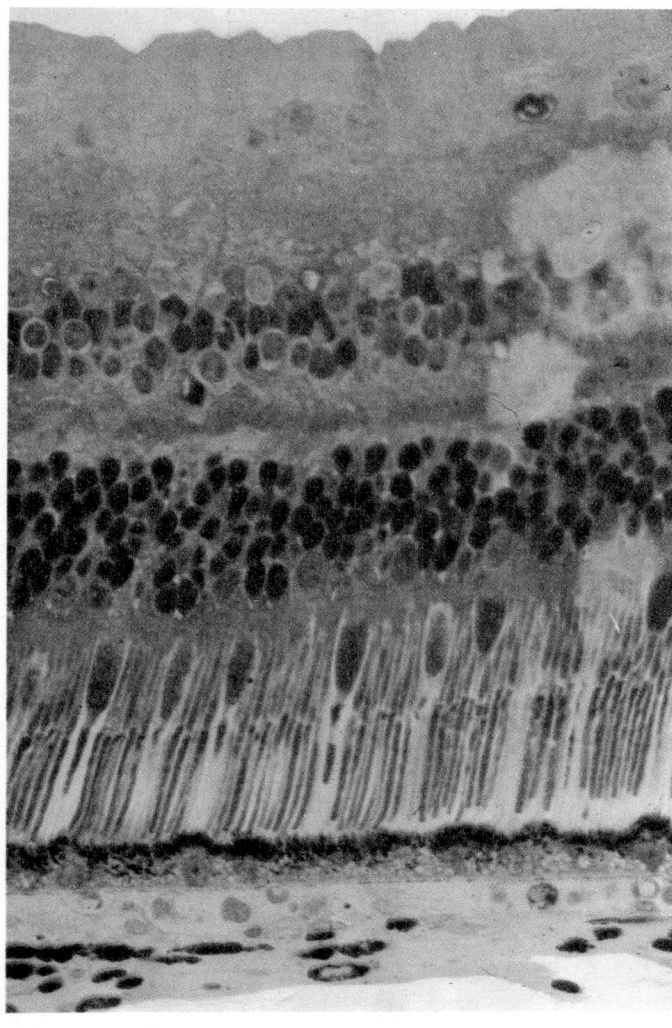

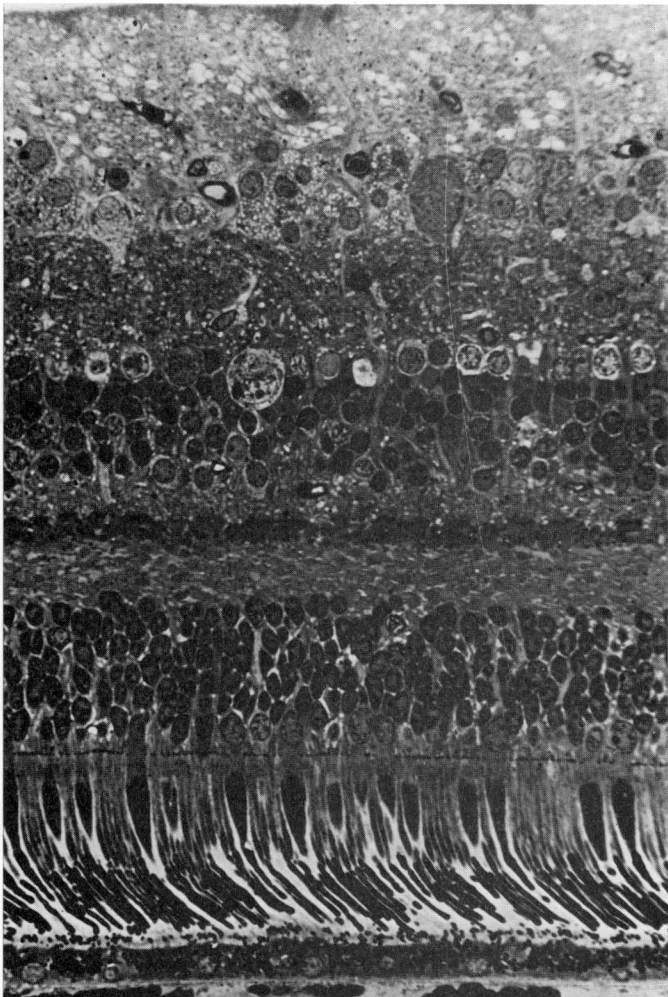

7.314A Thin section of simian retina in araldite-embedded material, stained by toluidine blue. Provided by N A Locket, Institute of Ophthalmology, London. Compare with 7.312.

7.314B Light micrograph of the retina of a 19-year-old male. Toluidine blue stain, resin-embedded. Preparation provided by John Marshall, Institute of Ophthalmology, London. Magnification × 1750.

and fovea centralis will be considered further after the general retinal structure has been described (p. 1203).

The total number of rods in the human retina has been estimated at 110–125 million and of the cones at 6·3–6·8 million (Osterberg 1935). Their distribution differs; the cones are densest at the rod-free foveola (about 147 000 per mm²), but diminish rapidly to a 10° circle round the macula where their density is about 5000 per mm², a frequency which is maintained to the edge of the ora serrata. Rods are almost the reverse of this in their distribution, rising from zero at the foveola to a greater density than the cones at the 10° circle (160 000 per mm²), then diminishing to the peripheral retina, where, however, there are still approximately 30 000 per mm²; the rods are thus six to 30 times more numerous in the peripheral retina outside the 10° circle. This distribution accords well with the phenomena of photopic (cone) and scotopic (rod) vision. The correlation of rod and cone distribution with photopic and scotopic vision is well illustrated in the feline retina; in the area centralis the concentration of cones, though greater than elsewhere, is about one-sixth of the human level but rods, at 275 000 per mm² in the area centralis, reach a maximum of 460 000 per mm² peripheral to this, a level greater than any in the human retinae.

Even under the light microscope, retinal neurons are clearly seen to be less numerous than rod and cone cells; ganglion cells (vide infra, p. 1202), whose axons form the optic nerve, probably total about one million in the human retina. Hence large numbers of photoreceptive cells must activate single axonal paths in the optic nerve and beyond.

Structure of Rod and Cone Processes

These exhibit greater differences by light than is evident by electron microscopy. Even with the former both show an external,

outer segment, which is refractile with the customary stains, positively birefringent and PAS-positive, and an *inner segment* staining deeply and with a fibrillar structure. The combined outer and inner segments of a rod process measure about 100–120 μm long in freshly fixed human retinae, cone processes being about 65–75 μm (Eichner 1957); both diminish in length towards the ora serrata, especially the process of cones, which are also much narrower at the fovea where they closely resemble rods in their dimensions. The outer segments of rods contain *rhodopsins* (visual purple); in cones (Rushton 1962, Pirenne 1967, Dowthall & Knowles 1977) similar photosensitive pigments have been detected but display different absorption qualities, accounting for the distinctive types of behaviour of rods and cones in high and low luminosities. Different absorption maxima have also been observed amongst cones themselves in primate retinae; there is at least some accord between these data and the trichromatic theory of colour vision. (For a survey of vertebrate photoreceptors and pigments consult Crescitelli 1972).

In their *ultrastructure* rods and cones are broadly similar and may be described together; rod processes have been the more extensively studied of the two (Sjostrand 1953, 1961, Villegas 1964, Misotten 1962, Dowling 1965, Cohen 1965, Boycott 1974). The rod *outer segment* appears as a remarkably regular series of discoidal membranous sacs, stacked like thin coins and surrounded by a cell membrane, to the outer end of which are opposed the microvilli of pigmented epithelial cells (vide supra, p. 1195). These *discs*, numbering 600–1000 in rods of various species, are flat vesicles separated from each other by a less electron-dense *intradisc space* (7.317). In some vertebrates the discs are formed as continuous infoldings of the lateral cell membrane but in human rods this continuity is lost. However, in cones, this connection

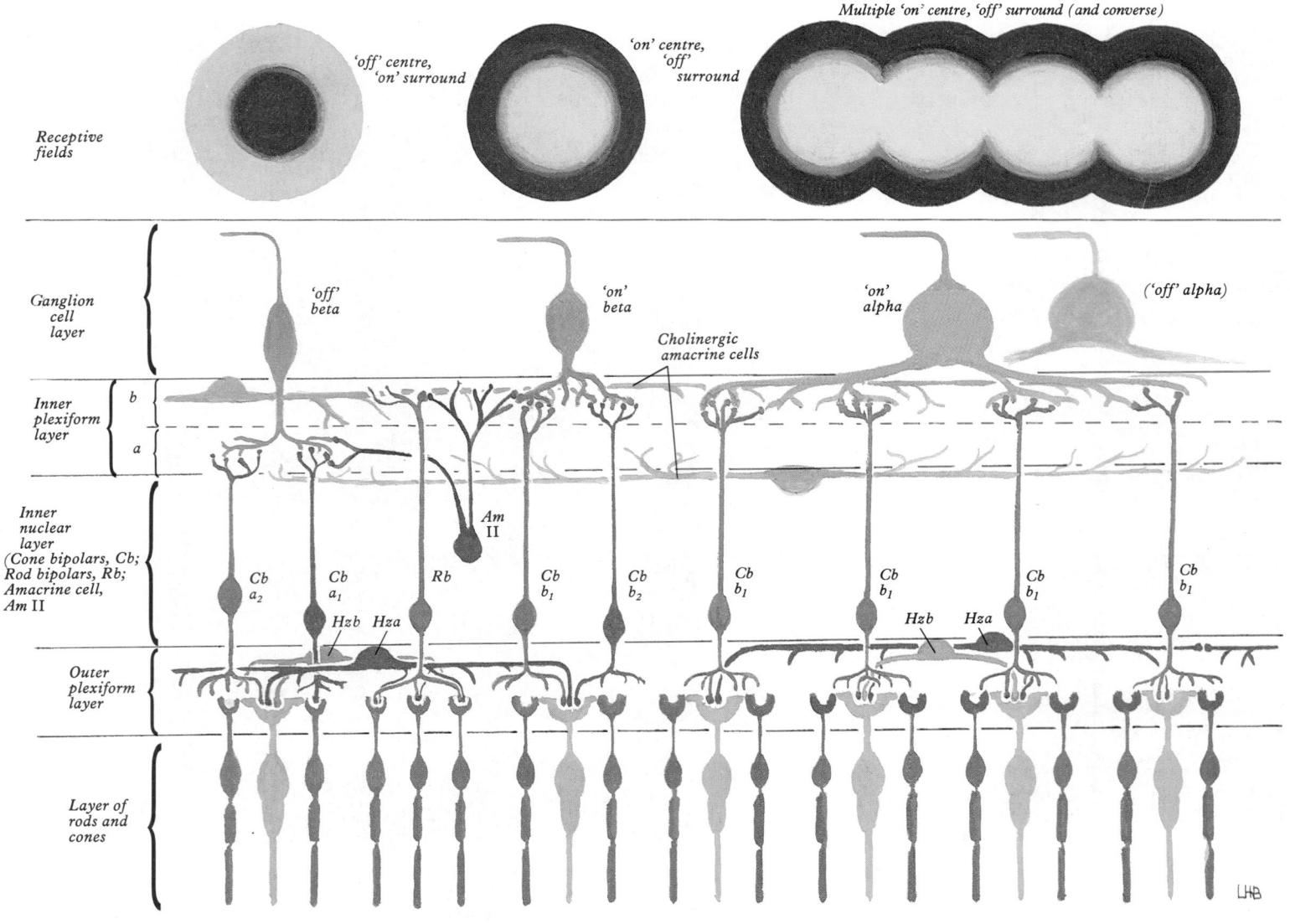

7.315 Simplified diagram of the main cell types and interactions in the mammalian (cat) retina. Cells depicted green are involved in excitatory actions, pink ones are inhibitory. Note also two types (a and b) of horizontal cells (Hza, Hzb). See text for further details.

persists even in mammals (Cohen 1970, Laties & Liebman 1970, Ripps & Weale 1976). In transverse section the discs of cones are circular, whereas those of rods have a scalloped profile (Cohen 1972). Discs of cones are also flatter than those in rods but are separated by a greater interval. Consequently the numbers in unit lengths are similar in both types, at about 30 discs per μm. Both are continually formed at the end closest to the soma and are progressively pushed away from this point towards the pigment epithelium as new discs are added; thus the oldest discs are nearest to the pigment epithelial cells and they eventually break off and are phagocytosed by this epithelial layer (p. 1196). Young (1971) has shown that in rhesus monkeys almost 1000 discs are thus generated proximally and removed peripherally in a rod outer segment in 11 days. (Since each pigment cell may interdigitate with 24–45 rod segments, its voracious 'appetite' may engulf 2000–4000 discs per diem!) This renewal was established by analysing the progression of radioactively labelled disc membranes in rats, mice and monkeys by a technique invented by Droz (1968). Evidence for renewal in the outer processes of cones is less certain, although Hogan (1972) has described similar phagocytosis in cones of the human fovea. Such a rapid turnover of complex intracellular structures is probably related to a limited lifespan in the photoreceptive molecular assemblies which may be degraded gradually by exposure to light. The discal membrane is now considered to consist of a regular array of rhodopsin molecules orientated transversely to the photoreceptor process and embedded in a lipid bilayer (Schmidt 1938, Weale 1970).

Each of these consist of two closely associated components, a transmembrane glycoprotein (Mr about 50 000) and the chromophore, 11-cis-retinal. There are also various other proteins within the outer segment cytosol, concerned with amplifying the light signal greatly. Briefly, when rhodopsin is activated by light, it catalyses the loading of a second protein, *transducin* (G-protein) with guanine nucleotide triphosphate (GTP). This then activates an enzyme, a cyclic guanine monophosphate phosphodiesterase (cGMP-PDE), which in turn depletes cyclic GMP within the cell. The final step is related to the permeability of the membrane to sodium ions; in the resting (dark) state there is a steady inward flow of ions through sodium channels in the membrane but to keep the channel open there must be sufficient cGMP available. When this is reduced by the enzyme activity described above, it causes the sodium gate to close and the electrical state of the cell changes, the outer segment becoming hyperpolarized. This change causes a hyperpolarizing current flow in the synaptic area, inhibiting chemical transmission to the next cell in the series (a bipolar cell). This system, though complex, ensues maximum multiplication of the signal, so that e.g. a rod is able to detect as little as four photons. For further details, see the reviews by Lamb (1986), Baehr & Applebury (1986).

The *inner segments* of rods are longer and broader than the outer segments and even more so in cones. The inner segment is also divisible into an outer *ellipsoid* with acidophilic staining reactions, containing glycogen and many mitochondria, and an inner *myoid*, adjacent to the soma and containing much randomly scattered

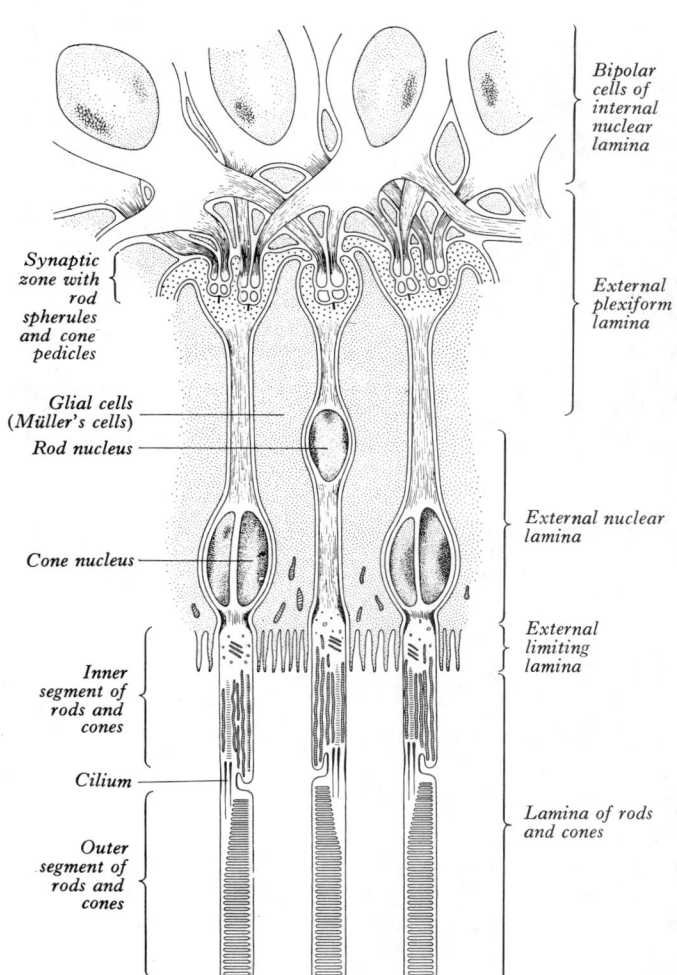

7.316 Schematic representation of the ultrastructure of retinal photoreceptors and of their connections with bipolar nerve cells. Note the stacked discs in the outer segments and refer to 7.319–7.321 and to the text for further details of the synaptic zone. Reproduced from Sjöstrand 1961.

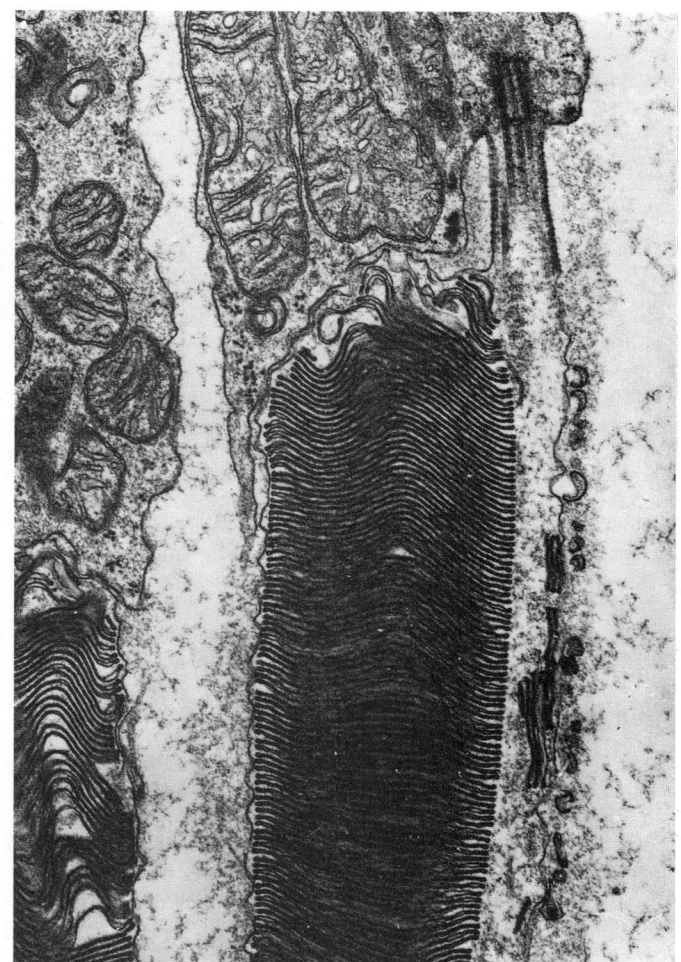

7.317 An electron micrograph of a section of a human retinal rod, showing the junction between the outer and inner segments. The outer segment is made up largely of photoreceptive lamellae (or discs), connected by a short cilium (top right) to the inner segment containing large mitochondria. Provided by N A Locket, Institute of Ophthalmology, London. Magnification × 50 000.

agranular endoplasmic reticulum and free ribosomes. Myoids are basophilic and contain much more glycogen than the ellipsoids. From the inner to the outer segment stretch bands of ellipsoidal cytoplasm covered by plasma membrane, which are closely applied to the outer segment. Another cytoplasmic process, the *cilium*, connects the ellipsoid to the outer segment; it originates in a basal body and has the microtubular structure of a cilium within it, although lacking the central pair of microtubules (De Robertis 1960). The processes of cones are like those of rods, their differences being largely dimensional (Dieterich & Rohen 1970).

Though these processes have been minutely described, their remarkable features have not been entirely correlated with function. The stacked discs have been likened to photomultipliers, intensifying the electrical energy of photochemical processes; but this is merely an analogy. Despite the physiological demonstration of at least three types of human cone based on sensitivity to light, no structural distinctions yet exist. Variations in colour sensitivity in different parts of the retina are not in accord with regional variations in cone shape or dimensions. Known variations in sensitivity (e.g. achromatism at the retinal periphery) may be due to differences in neuronal connections; the duplicity theory of vision and the equation of photopic and scotopic vision with cones and rods may both be explicable in the same terms.

THE EXTERNAL LIMITING LAMINA

At the level of the junctions of the rod and cone processes with the outer fibres of rods and the somata of cones, light microscopy shows a thin densely-staining line which the rods and cones

perforate. It extends throughout the neural retina and was considered for decades to be the fused terminal expansions of the 'fibres of Müller', an elaborate type of neuroglial element, with a soma in the internal nuclear layer, whose long processes stretch radially through the retina from the vitreal surface almost to the pigment epithelium. From these vertical fibres numerous processes spread horizontally in a dendriform manner into the plexiform laminae, embracing the somata of cells in the nuclear and ganglionic laminae. Similar 'fibre baskets' surround the proximal segments of photoreceptors. Their arrangement suggests that these *retinal gliocytes* (Müller cells) provide physical support, like neuroglia elsewhere. Moreover, other processes of these cells form an *internal limiting membrane* on the retina's vitreal aspect. Arey (1932) described the external limiting membrane as serial unions between the cell membranes of rods and cones with the 'fibres of Müller'. Electron microscopy has corroborated this view (see Fine & Zimmerman 1962 for a review of the earlier literature) but there is much variation in the types of junction described. They occur between the glial processes and the rod and cone inner segments and may be zonulae adherentes, which also sometimes unite adjoining Müller cell processes (the classic view) and adjacent photoreceptor processes (Spitznas 1970).

In the plexiform layers the horizontal gliocytic rami intermingle with the dendrites and axons of retinal neurons and may form helical lamellae around their neurites, without, however, producing myelin. Like other neuroglia they also contact blood vessels, especially capillaries; here their basement membranes fuse with those of non-striated myocytes in the media of larger vessels or with endothelial basement membrane in the case of capillaries. These extensive gliocytes form much of the total retinal volume, almost totally filling the extracellular space. All glial processes are cytoplasmic extensions and other functions have been ascribed to them, e.g. the transport of glucose to retinal neurons and the synthesis and storage of glycogen (Cogan & Kuwabara 1967, Radnot & Lovas 1968).

THE EXTERNAL NUCLEAR LAMINA

This layer contains the parts of rod and cone cells that are internal to the external limiting membrane and do not lie in the external plexiform layer; and this refers particularly to their somata. As they traverse the external limiting membrane, photoreceptor processes become the narrower 'outer fibres' of rods and cones, those of rods more slender and elongate (7.315). The outer 'fibres' of cones are more like short 'waists' between the process and soma. The cytoplasm in both types contains long mitochondria, vesicular agranular endoplasmic reticulum and many free ribosomes; as the 'fibres' spread into the soma around the nucleus, aggregations of microtubules are visible. The thickness of the external nuclear lamina and rows of cells in it varies from 27 μm in peripheral retina to 50 μm in the fovea (p. 1203), representing in the former a single row of cones with four of rod nuclei and in the latter about 10 rows of cone nuclei.

THE EXTERNAL PLEXIFORM LAMINA

The 'inner fibres' of the rods and cones pass centrally (i.e. towards the vitreous) forming a most intricate zone of synapses (7.316) with the dendrites and axons of bipolar and horizontal neurons of the internal nuclear lamina. These fibres resemble axons, containing mitochondria and vesicles, free ribosomes, neurofilaments and microtubules. Rod axons are 15–25 nm in diameter and 1 μm or more in length, those of the cones being thicker and with more microtubules. Rod axons end in oval, invaginated *rod spherules*, cone axons as *cone pedicles*; they form complex multiple junctions with bipolar and horizontal neurons, whose neurites approach from the internal nuclear layer. The external plexiform layer is sometimes analysed into *three sublaminae*: an outer one of rod and cone inner fibres or axons, an intermediate one of spherules and pedicles and an internal one of bipolar and horizontal cell neurites. (For reviews consult Dowling & Boycott 1966, Misotten & Van den Dooring 1966, Boycott & Dowling 1969, Boycott 1974).

The rod spherule (7.319) is part of a synaptic complex of three elements, namely, *presynaptic* (the spherule itself), *synaptic* (contacts of the spherular membrane with the dendrites of bipolar cells and neurites of horizontal cells) and *postsynaptic* (neurites of bipolar and horizontal cells). The spherule is invaginated, enclosing in its hollow the terminations of two to seven dendrites or processes; the dendrites are those of *rod bipolar cells* (vide infra) and processes from *horizontal neurons*, the latter conducting in either direction and thus not strictly axons or dendrites. The cytoplasm of rod spherules contains many presynaptic vesicles and a *synaptic ribbon*, a peculiar osmiophilic, lamellar structure (7.319, see also p. 886). The presynaptic (rod) and postsynaptic (bipolar and horizontal cell) membranes are not thickened as in typical synapses. Postsynaptic vesicles may also occur at these sites.

The cone pedicle, also invaginated, is more complex (7.319). Its cytoplasmic organelles are like those of rod spherules but many more neurites are involved. The contacts are of three types: (1) deeply infolded synapses containing three terminal neurites, two more deeply situated than the other, and forming *triads*, of which there are 24 in each pedicle; (2) slightly depressed contacts, on the pedicle's basal aspect, numbering up to 500; (3) interreceptor contacts between the peripheral surfaces of adjacent cone pedicles or rod spherules. Triads contain two terminals from horizontal cells and one dendrite from a *midget bipolar neuron* (vide infra) or from a horizontal process. The contacts of cone pedicles are synapses with the dendrites of 'flat' bipolar cells (vide infra). Interreceptor contacts have no synaptic vesicles and are gap junctions (electrical synapses, see pp. 24, 888) (Raviola &

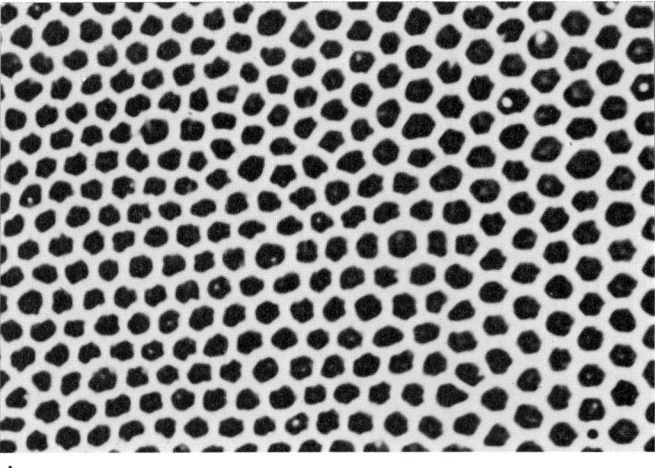

A

B

C

7.318 Tangential sections through the primate retina to show the variations in distribution of rod and cone processes in the foveola (A), fovea (B) and macula (C). Note the *small* cone processes in the foveola, from which rods are absent. The cones, which are elsewhere larger than the rods, predominate in the fovea, rods becoming more numerous towards the periphery. Preparations supplied by John Marshall, Institute of Ophthalmology, London. Magnification × 5000.

Gilula 1973, Witkovsky et al 1974) but do contain microtubules; each pedicle has six to twelve such contacts; they are most frequent among foveal cones and elsewhere are usually between rod and cone cells. In view of the high resolution mediated by the fovea, such contacts have been reluctantly accepted as synapses and perhaps routes of low resistance for electrotonic excitation. If so, a separate mechanism of interaction exists between the photoreceptors, in parallel with the effects of horizontal and amacrine cells (vide infra).

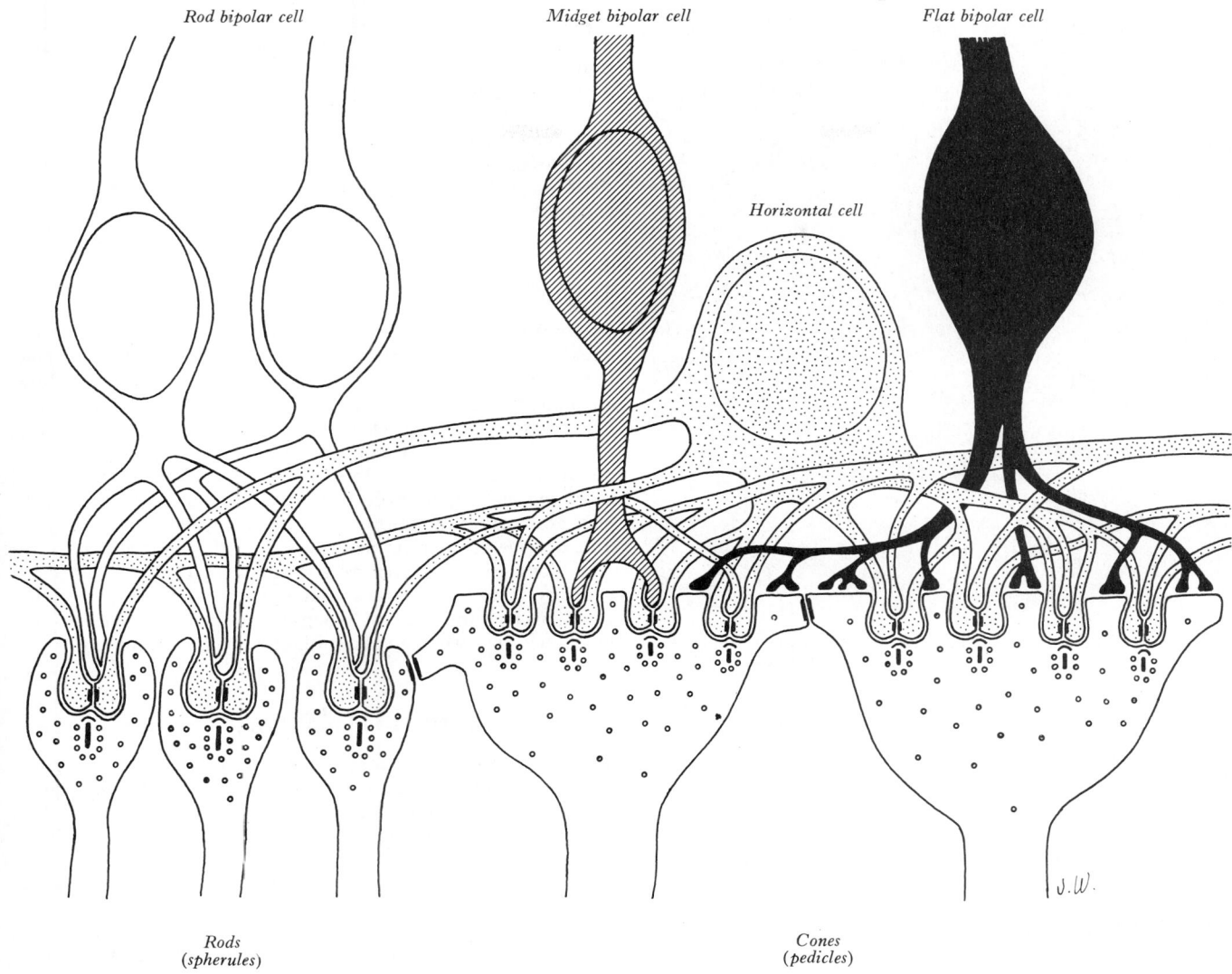

Rod bipolar cell   *Midget bipolar cell*   *Flat bipolar cell*

Horizontal cell

J.W.

Rods
(spherules)

Cones
(pedicles)

7.319 Scheme of the synaptic arrangements involving the rod spherules and cone pedicles of the retina. For details consult text. From Hogan et al 1971, modified from Dowling & Boycott 1966.

To summarize, each rod cell has direct connections with two bipolar and perhaps several cone cells and a horizontal neuron. Cone cells have more complex and numerous contacts, probably between 575 and 600, involving midget and flat bipolar cells, horizontal cells and rod and cone cells. The possible routes for convergence and interaction between photoreceptors are bewilderingly numerous and the precise functional significance of such intricate intercommunications remain uncertain. Inter-photoreceptor contacts are confirmed as gap junctions. These and other details emphasize the central nervous nature of the retina and the complex interactions which are typical of the brain from which it is derived embryonically but do not yet explain the retinal modus operandi.

THE INTERNAL NUCLEAR LAMINA

This layer is perhaps unfortunately named. By light microscopy the retina displays *three* tiers of 'nuclei' or cell somata, the ganglion cells providing the most internal (i.e. nearest the interior of the eye). The 'internal' nuclear layer is therefore *intermediate* between this and the layer of rod and cone 'nuclei' and would be more appropriately termed the *middle* nuclear layer. It contains the somata of the *retinal gliocytes* (Müller cells), and of the bipolar, horizontal and amacrine neurons, arranged in regular strata. The most external in this layer are the somata of the *horizontal neurons*, whose processes extend into the adjacent external plexiform layer

to the rod spherules and cone pedicles (vide supra). Internal to this are the *bipolar neurons*, the primary sensory neurons of the retino-geniculate tract (p. 1019); their dendrites also connect with the rods and cones, their axons passing centrally into the internal plexiform lamina (lamina 7) to synapse with the ganglion cells (lamina 8). Internal again to this are the somata of retinal gliocytes, which have already been considered (p. 1199). Most internal are the somata of the *amacrine neurons*, whose neurites spread into the adjacent internal plexiform layer to contact the dendrites of ganglion cells and the axons of bipolar neurons.

Each of the three categories of neurons in the internal nuclear layer—horizontal, bipolar and amacrine—contains several subtypes, distinguished by their connections and physiology rather than their cytological differences, which are not easily defined by any form of microscopy.

Horizontal neurons are of large inhibitory (*a*) and smaller excitatory (*b*) types; they have multipolar somata, from which extend a single long and several short processes (seven in cone and 10–12 in rod horizontal cells). Their cytoplasm is like that of bipolar cells (vide infra) but contains an organelle rich in ribosomes (Kolmer's crystalloid) which is peculiar to them. The long processes may be up to 1 mm long, a very long neurite by retinal standards, with branches contacting both rod spherules and cone pedicles. The short neurites of *cone* horizontal neurons synapse with seven cone pedicles, taking part in the formation of several triads in each (p. 1200). The short processes of *rod*

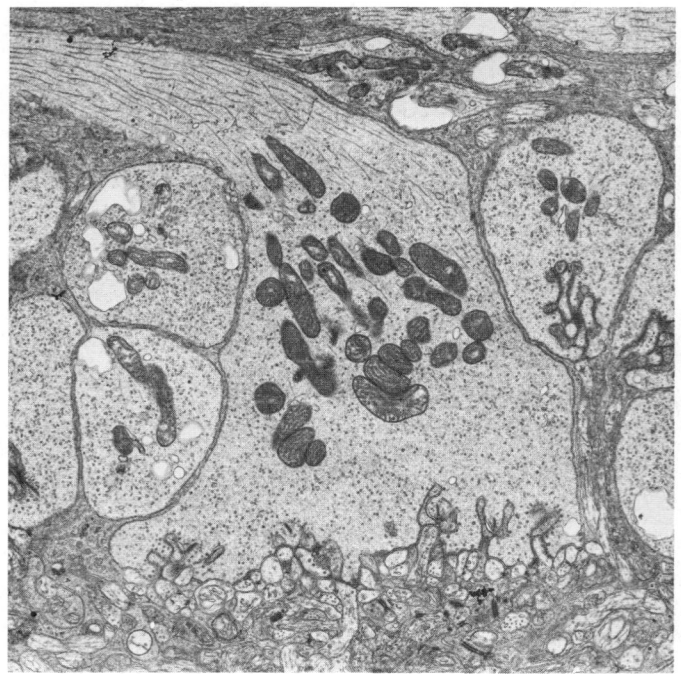

7.320   An electron micrograph of a part of the outer plexiform layer of the retina (monkey), showing synapses at the pedicle base of a cone (centre) and adjacent rod spherules. Note the presence of mitochondria and of synaptic ribbons surrounded by synaptic vesicles in the cone pedicle. The neuropil composed of interweaving dendritic processes of bipolar, horizontal and other cells is shown below. Supplied by N A Locket, Institute of Ophthalmology, London.

horizontal neurons synapse with 10–12 rod spherules. The long and short neurites of horizontal neurons are neither axons nor dendrites and probably conduct in both directions (7.321).

The bipolar neurons are *rod bipolar* cells, synapsing with a number of rod spherules, or *cone bipolar neurons*, involved in forming triad synapses of cone pedicles (7.320). Each rod bipolar axon connects to an amacrine II cell, while the cone bipolars connect with a single midget (X or β) ganglion cell (vide infra). Electrophysiology coupled with dye injection of single cells has shown that there are a number of different types of cone bipolar cells. Cone bipolars a_1 and a_2 connect cones to 'off' β ganglion cells, and cone bipolars b_1 and b_2 connect them with 'on' β ganglion cells and also with the large 'on' α ganglion cells. These pairs operate electrically in opposite directions: when a cone is activated, one cell depolarizes and the other hyperpolarizes. The term 'cone bipolar' is, however, a little misleading from the physiological viewpoint, since rods can communicate with them indirectly via rod gap junctions on cone pedicles. Even by electron microscopy, the somata of these types of bipolar cell are almost indistinguishable; they vary somewhat in size and shape but have similar organelles: abundant mitochondria, free ribosomes, agranular endoplasmic reticulum and microtubules.

Amacrine neurons (7.321), so named because they lack a large axonal neurite, have processes like dendrites, which are said to show the cytoplasmic features of both axons and dendrites; the direction of conduction in any process is determined by polarization of whichever synapses are active. Each neuron has one or two thick processes from which branch dendritic trees of variably complex patterns, on the basis of which at least 12 types of cell have been distinguished, including dopaminergic and cholinergic types (Masland & Tauchi 1986). They contain an indented nucleus, many cisternae of granular endoplasmic reticulum, ribosomes, microtubules and sometimes crystalline bodies and have surface cilia. Superficial cisterns have been described in amacrine, bipolar and ganglionic neurons in simian retinae by Spoerri & Glees (1977). The processes of amacrine neurons contain groups of synaptic vesicles at scattered sites, representing synapses with the axons of bipolar cells and dendrites of ganglion cells. Kolb & West (1977) have shown these to send their

processes into the internal and external plexiform laminae; in the internal plexiform lamina, type II amacrine cells form gap junctions with 'rod' bipolar neurons and chemical synapses with 'off' β ganglion cells. Since they also have presynaptic connections with rod and cone bipolar neurons, they may mediate the feedback between plexiform laminae, although whether excitatory or inhibitory is uncertain. Similar neurons have been described in the retinae of goldfish by Dowling et al (1976), where they are considered to be excitatory.

THE INTERNAL PLEXIFORM LAMINA

Between the internal nuclear and ganglionic laminae is a dense neuropil composed of the neurites of bipolar, amacrine and ganglionic neurons, and their occasionally displaced somata. Outer and inner layers can be distinguished (laminae a and b); the *lamina b* consists of the dendrites of 'on' ganglion cells and synapsing terminals of b_1 and b_2 cone bipolar cells (Cb), plus amacrine dendrites. In *lamina a* are the dendrites of 'off' ganglion cells and incoming cone bipolars (Cb a_1 and a_2). Cone bipolars are in pairs, one inhibitory, the other excitatory (7.315, vide supra) (7.321). The bipolar axons form axodendritic and axosomatic synapses with ganglion cells and axosomatic contacts with amacrine neurons. Amacrine neurites synapse with bipolar cell axons and ganglion cell somata. These synapses are superabundant; near the fovea they are estimated at 2 million per cubic millimetre.

THE GANGLIONIC CELL LAMINA

The ganglion cells are second order neurons of the visual pathway, their dendrites being connected with processes of the bipolar and amacrine neurons in the internal plexiform lamina and their axons forming the layer of nerve fibres (lamina 9) beneath the inner surface of the retina. Here they turn tangentially to the optic disc, by which they leave the eye as fibres of the optic nerve. Ganglion cells form a single stratum in most of the retina and are progressively more numerous from its periphery to the macula, where they are ranked in about 10 rows, diminishing again towards the fovea, from which they are almost absent. They are multipolar neurons, varying from 10–30 µm in diameter, and have a large nucleus. Their dendrites are variable in number and branching patterns, usually emerging opposite the axon. According to their dendritic pattern they have been classified into at least six types (midget, stratified, diffuse, etc., see Cajal 1911, Polyak 1941, Boycott & Dowling 1969, Boycott 1974, Kolb & Nelson 1984, Sharpey & Perry 1986, Dowling 1987).

The midget beta ganglionic neurons are monosynaptic in their connections and possess relatively simple dendrites. They are the commonest type in the central retina, synapsing with the terminal axons of cone bipolar neurons (7.315, 319); since these are usually each connected centrally to a midget ganglion cell and distally to one cone pedicle, especially in the fovea, they provide specific pathways for individual cones. But, as the connections of cones show (p. 1201 and 7.315, 319, 320), this does not entail that a cone discharges exclusively through its midget bipolar and ganglion cell. **Polysynaptic ganglion cells** (rod and flat ganglion cells) include all other ganglionic neurons. They have one or two large dendrites spreading much more widely than those of midget cells; they synapse with the terminal axons of many bipolar neurons, probably with hundreds. The bipolar neurons are obviously more numerous than the ganglion cells (but by less than two orders of magnitude); hence many bipolars must transmit to one ganglion cell; there is also some evidence that the reverse may also be sometimes true. Such intricate connections form the structural bases for summation and lateral inhibition, long established in general terms and more recently explored by techniques of unit recording and accompanying intracellular injection of dyes to relate cell type to physiological response. Their axons are the main component of the nerve fibre lamina, to be considered next.   The arrangement of cone bipolar cells, with their input of both cone and rod activity and their synaptic contacts with midget (β) ganglion cells, or the larger 'giant' (α)

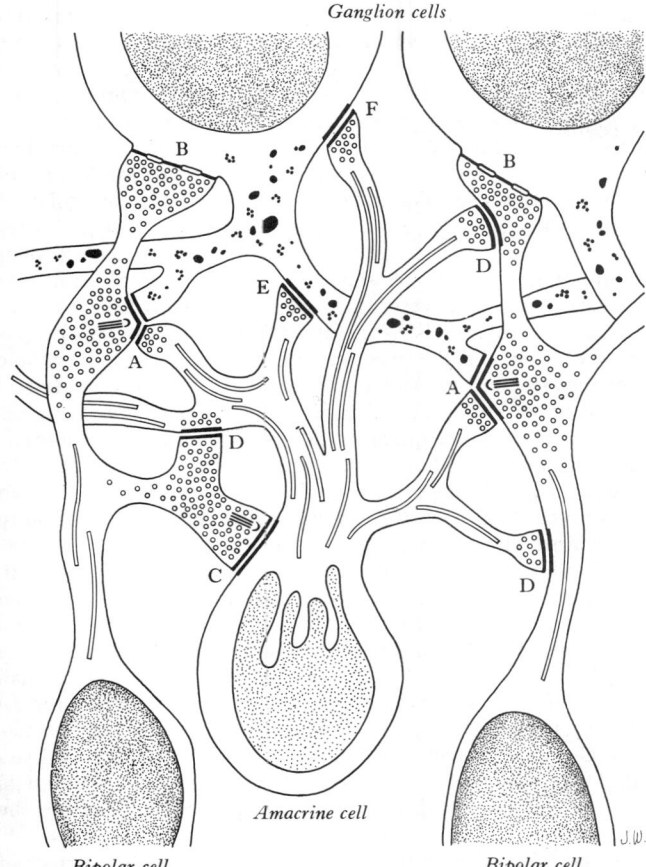

Ganglion cells

Amacrine cell

Bipolar cell   Bipolar cell

7.321 A scheme of the synaptic arrangements in the internal plexiform lamina of the retina. Note that bipolar axonal terminals of three types are shown: *axodendritic* (A) in *dyads* involving neurites of amacrine and ganglion cells, *axosomatic* involving ganglion cells (B) and amacrine cells (C). Similarly the neurites of amacrine cells also make three types of contact: with the axons of bipolar neurons (D) and with the dendrites (E) and somata (F) of ganglion cells. From Hogan et al 1971, modified from Dowling & Boycott 1966.

ganglion cells, permits the sharpening of punctate photic stimuli, with either an 'on' centre and 'off' surround or vice versa (see 7.315). Complex interconnections of amacrine and horizontal cells are involved in these processes.

THE LAMINA OF NERVE FIBRES

The axons of ganglion cells converge on to the optic disc from the whole retina, forming the *stratum opticum*, which is consequently deepest (20–30 µm) at the disc's periphery. They converge in a simple radial pattern from the medial (nasal) half of the retina but the macular area, inferolateral to the optic disc, complicates the course of the lateral (temporal) axons. Those from the macula form a *papillomacular fasciculus* which passes almost straight to the disc; the more temporal fibres, being more peripheral, swerve circumferentially above and below to reach the disc.

Axons of ganglion cells are non-myelinated within the retina, an optical advantage since myelin is refractile. The myelin sheaths commence as the axons enter the optic disc to become the optic nerve. A few small myelinated fibres may appear in human retinae; myelination is usual in parts of the retina in many other mammals. The axons within the retina are surrounded by the processes of retinal gliocytes (Müller cells) and other retinal astrocytes (Wolter 1959). Like their somata, the axons of ganglionic neurons vary in size, ranging from 0·6–2·0 µm in diameter, with a typical axonal ultrastructure.

Centrifugal axons have often been described and almost as often denied (Bowin 1895, Mukai 1970). Their existence in the human retina is still an open issue. Various sources have been suggested

including the lateral geniculate nucleus (p. 1016), superior colliculus, hypothalamus, etc. Such retinal efferents might be vasomotor; but since efferent terminals have been described in relation to amacrine cells by Ramony Cajal and others, it is tempting to assume in the visual pathway an analogue of the cortico-olivo-cochlear connections of the auditory system (p. 1233). The majority of such studies have been in birds (Cajal) or mammals phylogenetically remote from man, a fact rarely made plain in textbooks or even monographs. Nevertheless, there is good evidence for a corticogeniculate pathway, modifying the activity of the afferent visual paths (p. 1019) and it is possible that such a connection reaches the retina.

THE INTERNAL LIMITING LAMINA

Classically, the conical branching terminals of the fibres of retinal gliocytes were held to coalesce at the surface of the vitreous body into a continuous membrane separating the stratum opticum from the vitreous gel and thus 'limiting' the retina. Some early observers denied this, considering the '*membrana hyaloidea*' (the surface layer of the vitreous) to be the sole boundary between the retina and the vitreous. Electron microscopy has led to a combination of both views. Gliocytic processes have at their terminations a basal lamina of about 0·5 µm, sinuously adapted to them externally but smooth on its internal aspect, where they are adjacent to the vitreous; the collagen fibrils of the latter blend with the glial basement membrane, which consequently can be said to be a composite structure. The internal limiting membrane is inevitably involved in fluid exchange between the vitreous and the retina and perhaps through the latter with the choroid.

PECULIARITIES OF THE MACULAR AREA

The general details of the central, macular and foveal retinal areas have already been noted. All the retinal laminae are modified in its central region and to a marked degree in the *fovea*, in the floor of which (the *foveola*, p. 1196) there are no rod cells but about 2500 close-packed elongated cones, greatly resembling rods; the *somata* of even these small cone cells are displaced peripherally to the sloping foveal wall so that only the cone *processes* are present in the foveola. Despite this distorting effect, the foveolar cone processes are strictly vertical and radial in orientation, the only other element present (which light must traverse) being the fibrous cytoplasmic processes of the gliocytes, which even here form internal and external limiting membranes. At the rim of the conical wall of the foveolar pit other modifications appear. The fovea is largely devoid of rod cells or processes, which reach only to its periphery. Its rod-free central part contains approximately 35 000 cones and in the whole fovea (about 1·75 square millimetres) there are about 100 000 of these cells. Hence, in the fovea cones are most slender and most compacted and all other layers absent, a condition which greatly favours photopic vision; here, the cones have their most specific connections (with individual midget bipolar and ganglionic neurons) according with the high degree of spatial discrimination typical of foveal vision.

Because of the general displacement of the outer nuclear lamina to the foveal periphery, the internal processes of the photoreceptors are stretched out tangentially in the external plexiform layer and hence no cone pedicles or rod spherules are present in the central fovea and foveola. The inner nuclear layer is displaced to the edge of the foveal depression and the internal plexiform, ganglionic and nerve fibre laminae are almost absent from the whole fovea. Therefore, even on the foveal wall the retina is thinner and more transparent. Capillaries reach the foveal margin, invading only the ganglionic layer at its circumference. The fovea is thus normally devoid of all blood vessels.

In the *parafoveal region*, extending about 0·5 mm around the fovea (diameter 1·5 mm), the retina is thickest, partly due to the accumulation of displaced bipolar and ganglionic neurons. Outside this a *perifoveal region* is described in which the density of cones begins to diminish rapidly, the incidence of rods increasing.

The fovea is the last part in the primate retina to attain full development, which is not completed until about the fourth post-natal month in mankind. Prenatal development of the primate foveae has been chiefly studied in rhesus monkeys and the details are beyond the present scope of this book; Hollenberg & Spira (1973) and Hendrickson & Kupfer (1976) have reviewed the literature (p. 202).

THE OPTIC DISC AND RETINAL BLOOD VESSELS

The retina is placed between two sets of arteries and veins: the ciliary vessels of the choroid and the branches of the central retinal vessels. It depends on both circulations, neither alone being sufficient to maintain full visual activity in the retina. The choroidal circulation (p. 1185) and the orbital and intraneural parts of the central retinal vessels are described elsewhere (pp. 746, 1095).

The **central retinal artery** enters the optic nerve as a branch of the ophthalmic artery, about 1·2 cm behind the eyeball. It travels in the optic nerve to its 'head', where its fascicles (about 1000 in man) are traversing the lamina cribrosa. This region, where retinal tissues meet the neural elements of the optic nerve (including astrocytes and oligodendrocytes) and also the connective tissues of the sclera and meninges, is a highly complex area. It also provides a point of entry and exit for the retinal circulation and moreover is the only site of anastomoses with other arteries (the posterior ciliary arteries, vide infra). It is visible, by ophthalmoscopy, as the **optic disc** (7.307, 308), a region of much clinical importance since it is here that the central vessels can be inspected directly, the only vessels so accessible in the whole body. Oedema of the disc (papilloedema) may be the first sign of raised intracranial pressure, transmitted into the subarachnoid space around the optic nerve and compressing the central retinal vein where it crosses the space. The optic disc, being superomedial to the posterior pole of the eye, lies away from the visual axis. It is round or oval, usually about 1·6 mm transversely and 1·8 mm in the vertical, and its appearance is very variable; for details see Duke-Elder & Wybar (1961). In light-skinned people the general retinal hue is a bright terracotta-red and the pale pink of the disc contrasts sharply with this; its central part is usually even paler and may be light grey. These differences are due to the degree of vascularization of the two regions, which is much less at the optic disc, and to a total absence of choroidal or retinal pigment cells, the retina being represented by little more than the internal limiting lamina. Even this does not pass far on to the disc, for the retinal gliocytes are replaced by the astrocytes of the optic nerve (Anderson & Hoyt 1969). In individuals with strongly melanized skins both retina and disc are darker (7.307D). The optic disc does not project at all in many eyes and rarely enough to justify the term *papilla*. It is usually a little elevated on its lateral side where the papillomacular nerve fibres turn into the optic nerve; where the retinal vessels traverse its centre there is usually a slight depression.

The central retinal vessels occupy their own apertures in the lamina cribrosa. At this level, which is usually just beyond ophthalmoscopy, the *central retinal artery* divides into two equal branches: first into a superior and an inferior and again, after a few millimetres, into superior and inferior nasal and superior and inferior temporal branches. Each of these four supplies its own 'quadrant' of the retina, their territories being in fact much more than quadrants, since they ramify as far as the ora serrata. Corresponding veins unite to form the central retinal vein but the courses of the venous and arterial vessels do not correspond exactly and arteries often cross veins, usually lying superficial to them. In severe hypertension the arteries may press on the veins and cause visible dilations distal to these crossings. Arterial pulsation is not visible by routine ophthalmoscopy without higher magnification. The branching of the artery is usually dichotomous, equal rami diverging at angles of 45–60°; but smaller branches may leave singly and at right angles. Arteries and veins ramify in the nerve fibre layer, near the internal limiting membrane, accounting for their clarity when seen through an ophthalmoscope (7.307). Arterioles pass deeper into the retina and may penetrate to the internal nuclear lamina, from which venules return to larger superficial veins. A dense capillary bed extends between these vessels and is diffusely organized, showing no laminar pattern. The structure of blood vessels resembles that of vessels elsewhere,

except that the internal elastic lamina is absent from the arteries and myocytes may appear in their adventitia. Capillaries have a non-fenestrated endothelium and numerous mural *pericytes*, extended along the vascular axis outside the endothelium and sharing its same basement membrane. The cytoplasm of both is similar in structure but pericytes do not contain myofilaments; their function remains uncertain. (For the ultrastructure of retinal capillaries see Tominaga & Ikui 1964.) Microcirculatory studies of the human retina in flat preparations, stained after trypsin digestion, reveal many details of capillary arrangement resembling those in the renal glomeruli, a network of capillaries connecting individual arterioles and venules with little connection with nearby vessels. This separation of the capillary beds is not as exclusive as in the larger vessels which feed them; the whole retinal arterial and arteriolar tree is devoid of anastomoses and arteriovenous shunts. For example, the territories of the arteries supplying a particular quadrant do not overlap nor do the branches within a quadrant anastomose with each other. In consequence, a blockage in a retinal artery causes loss of vision in the corresponding part of the visual field; the only exception to this endarterial pattern is in the vicinity of the optic disc; the posterior ciliary arteries enter the eye near the disc (7.305) and, apart from supplying the adjacent choroid, their rami form an anastomotic circle (of Zinn) in the sclera around the head of the optic nerve. Branches from this ring join the pial arteries of the nerve (Hayreh 1969) and from any arteries in this region small *cilio-retinal arteries* may enter the eye to anastomose with a retinal artery (7.307); similarly, small retino-ciliary veins may sometimes also be present. For further details see Singh & Dass (1960), who have surveyed the anastomoses between the central retinal artery and pial branches of the ophthalmic artery in the optic nerve.

Retinal capillaries do not pass towards the external surface of the retina beyond the internal nuclear lamina; they show regional differences in density, being especially numerous in the macula but absent from the central fovea. They become less numerous in the peripheral retina and are altogether absent from a zone about 1·5 mm wide adjoining the ora serrata. The central artery is innervated by sympathetic fibres and probably also has a parasympathetic supply (Ruskell 1970).

The Ocular Refractive Media

The components of the eye which transmit and refract light are the cornea, the aqueous humour, the vitreous body and the lens. Of these, only the refracting power of the lens can be varied.

THE AQUEOUS HUMOUR OR FLUID

The aqueous fluid fills the anterior and posterior chambers. Its total quantity is small; it is formed by active transport and diffusion from capillaries in the ciliary processes, from which it enters the posterior chamber, passes into the anterior through the pupil and escapes at the iridocorneal angle into the anterior ciliary veins via spaces in the angle and the scleral venous sinus. Any interference with its resorption into the sinus increases the intraocular pressure, the condition known as *glaucoma*; if this persists the optic disc becomes 'cupped' and retinal degeneration leads to blindness. Partial iridectomy may re-establish the flow from the posterior to the anterior chamber, in cases where glaucoma is due to obstructive adhesions between the iris and the lens. The aqueous fluid is a nutrient, metabolic avenue for the avascular tissues of the lens and cornea; but it also maintains and regulates the intraocular pressure and hence the constancy of the optical dimensions of the eyeball. It carries glucose, amino acids and respiratory gases. It also has a high content of ascorbic acid.

THE VITREOUS BODY

This fills the vitreous chamber, occupying about four-fifths of the eyeball. It is hollowed in front as a deep concavity, the *hyaloid fossa*, which is adapted to the lens. It is a colourless, apparently structureless, transparent gel, about 99% of water, with some salts, a little glycoprotein and hyaluronate (vide infra). Fibrils of type II collagen (p. 64), about 16 nm diameter and with a 22 nm periodicity, and an interfibrillary substance, the *vitreous humour*, can be distinguished by electron microscopy (Schwarz 1961). Peripherally the gel is condensed into a so-called *vitreous (hyaloid) membrane*. A narrow *hyaloid canal* runs from the optic disc to the central posterior surface of the lens. In fetuses this contains the hyaloid artery (p. 202) which normally disappears about six weeks before birth. The vitreous membrane is attached to the ciliary epithelium and processes and to the optic disc's rim. Anterior to the ora serrata it is thickened by radial fibres as the *ciliary zonule*, where it shows a series of radial furrows in which the ciliary processes adhere; hence some of their pigment remains attached to the zonule when they are removed. The zonule splits into two layers: a thin one lining the vitreous hyaloid fossa; and a thicker one composed of zonular fibres, collectively forming the *suspensory ligament of the lens*, which passes over the ciliary body to be attached to the capsule of the lens a little anterior to its equator. Some of the ligament's fibres are attached behind the equator (7.303) and some delicate ones to the equator itself. The ligament supports the lens and links it to the ciliaris muscle in the activities of accommodation (p. 1214).

Vitreous structure has been studied in much detail, by light and electron microscopy and X-ray diffraction (Suguira 1957, Fine & Tousimis 1961, Brini et al 1968). Its collagen fibrils vary regionally, in particular being condensed cortically. Rounded cells, *hyalocytes*, are revealed by phase contrast microscopy (for details, see Hogan 1963). No blood vessels penetrate the vitreous body and its nutrition is maintained by diffusion from the retinal and ciliary vessels.

THE LENS

The lens (7.322–324), enclosed in a capsule, lies between the iris and the vitreous body. It is encircled by the ciliary processes, which slightly overlap its equator. The *capsule* is a transparent, elastic membrane, thicker anteriorly than posteriorly. It contains no elastic tissue, its elasticity residing perhaps in the disposition of the fine fibrils of which it is formed. It has been described as a very thick basement membrane, being PAS positive. Its fibrils have a periodicity of 60 nm, differing from those of the zonule, with which they are continuous around the equator. Posteriorly the lens abuts the hyaloid fossa of the vitreous body; anteriorly it is in contact with the free border of the iris but recedes from it peripherally, so forming the posterior chamber.

The *lens* is a transparent, biconvex body, its anterior convexity being of greater radius than the posterior. The central points of these surfaces are the *anterior* and *posterior poles*; a line connecting these is the *axis* of the lens and the marginal circumference is its *equator*. Its dioptric power is much less than that of the cornea. All ocular optical media have a refractive index quite close to that of water (1·33) but the corneal surface is in contact with air and most of the 58 dioptres of which an eye is capable are effected here. The value of the lens is its ability to *vary* its dioptric power, a quality partly dependent upon a gradient in its refractive index from 1·386 at its periphery to 1·406 at its core. It contributes about 15 dioptres to the total of the eye. Its *range* in dioptric power, however, does not quite reach this, even at birth. Most young children show minor refractive errors (Sorsby et al 1961) and the available dioptric range decreases with age, being halved at 40 years, and reduced to 1 or 2 dioptres at 60. For further information on physiological optics consult Bennett & Francis (1962) and Duke-Elder (1969).

STRUCTURE OF THE LENS

The lens consists of a soft cortical substance and a firm, central part, the nucleus (7.322). Faint sutural lines (*radii lentis*) radiate from the poles to the equator; in adults there may be six of these lines or more but in fetuses there are only three, diverging at angles of 120° like a Y which is upright on the anterior surface and inverted on the posterior (Mann 1924). These lines, made of an amorphous substance, correspond with the free edges of septa

which dip into the lenticular substance. When hardened the lens appears to be composed of a series of concentric laminae, interrupted at the septa. Each lamina is built of lens fibres (7.323B), like ribbons, with serrated edges, the serrations being reciprocally interlocked and in part connected by desmosomes (Wanko & Gavin 1961); their ends are apposed at the sutures. The contacts between the lenticular fibres have been described as gap junctions in mice (Rafferty & Esson 1974) and in human and bovine lenses (Philipson et al 1975) but Kuwabara (1975) considered that the

7.322 The human lens, hardened and divided (enlarged view).

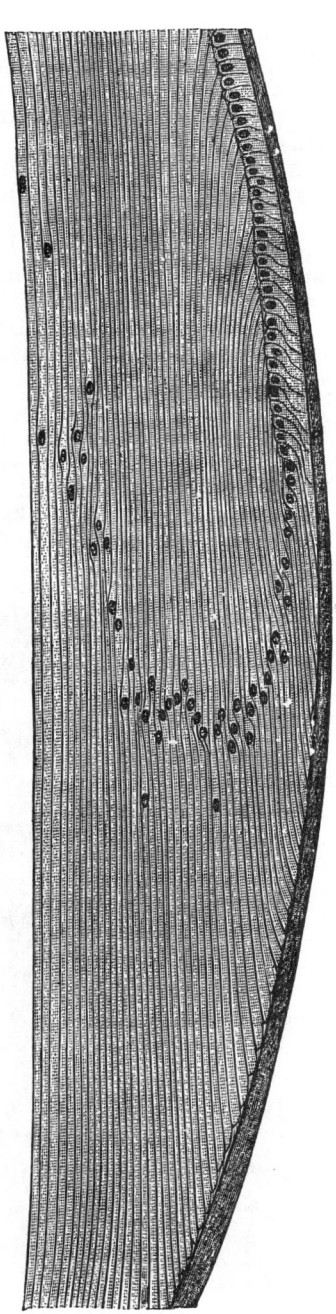

7.323A Section through the margin of the lens, showing the transition of the columnar epithelium into the lens fibres. Note that the lens cells retain their nuclei long after they have assumed the form of a fibre.

cortical cells of the lens adhere by reciprocal 'knob and socket' junctions and the deeper cells by fine ridges. These *lens fibres* curve from the anterior to posterior sutures (7.324). No fibres pass from pole to pole; those beginning near the pole on one surface end peripherally on the other and vice versa. Fibres in the outer laminae are nucleated, forming a nuclear layer, equatorially most distinct. The deeper fibres, which have lost their nuclei, are less flexible and their contours more deeply serrated. Superficial fibres, more recently formed, are more flexible, hexagonal in cross-section, with two opposite sides longer than the rest. Adhesion on these longer sides is considered less firm than elsewhere, perhaps permitting moulding of the lens by its capsule in accommodation. The total number of lens cells or fibres is estimated at about 2000 in adult lenses. They contain semi-stiff transparent proteins (crystallins) which give the lens its transparent, deformable properties.

The anterior surface of the lens is covered by a transparent layer of nucleated cuboidal epithelium. At the equator these cells elongate and their gradual transformation into fibres can be traced (7.323A). In tangential section or surface view these cells are polygonal and about $15\,\mu m$ across. In the anterior central area they may be flattened to a mere $6\,\mu m$. More peripherally they diminish in width but increase greatly in length and mitoses are more frequent. At the equator mitotic activity is maximal. Lens 'fibres' retain their nuclei for a long period; but the oldest and hence the deepest fibres lose them. Despite their very slow growth, the epithelial cells are relatively active; experimental injuries in mice (Rafferty 1976) heal completely within three days.

In fetuses the lens is nearly spherical, with a slight reddish tinge, and is soft, breaking up on the slightest pressure. A hyaloid artery from the central retinal artery traverses the vitreous body to the posterior pole of the capsule of the lens, whence its branches spread as a plexus covering the posterior surface and continuous round the capsular circumference with the vessels of the pupillary membrane and iris. *In adults* the lens is avascular, colourless, transparent and firm in texture. *In old age*, it is less curved on both surfaces, slightly opaque, of amber tint and denser. In *cataract* the lens gradually becomes opaque and blindness ensues.

The dimensions of the lens are optically and clinically important. Its diameter at birth is $6\cdot5\,mm$, increasing to $9\cdot0\,mm$ at 15 years and continuing to grow very slowly throughout life. Its axial

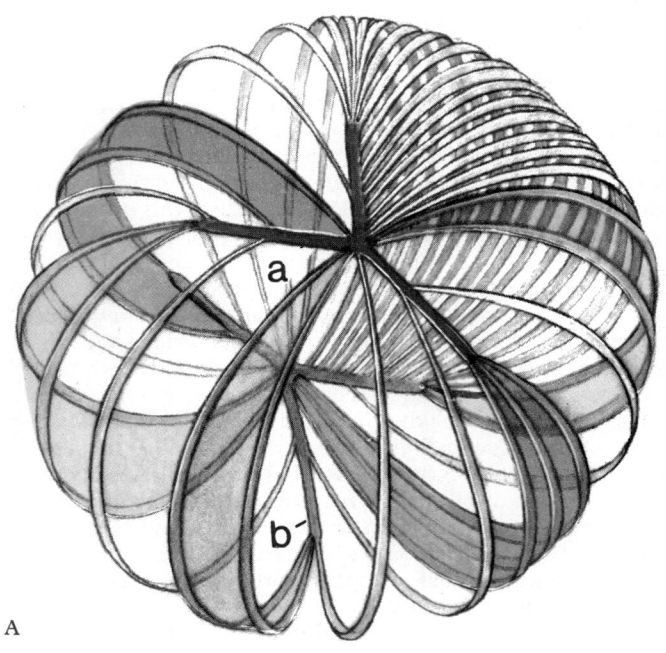

A

7.324   The structure of the fetal (A) and adult (B) human lens, showing the major details of arrangement of the lens cells or fibres. The anterior (a) and posterior (b) triradiate sutures are shown in the fetal lens and it is clear that fibres pass from the apex of an arm of one suture to the angle between two arms at the opposite pole, as shown in the coloured segments. Intermediate fibres show the same reciprocal behaviour, ending nearer to one pole, where they start further from the other, and so on. The suture pattern becomes much more complex as successive strata are added to the exterior of the growing lens, the original arms of each triradiate suture showing secondary and tertiary dichotomous branchings. See 7.299 for acknowledgement.

7.323B   Scanning electron micrograph of a series of lens fibres, depicting the interlocking side projections typical of these structures (rat). Provided by S Gschmeissner, Electron Microscopist, Anatomy Department, Royal College of Surgeons, London. Magnification × 4000.

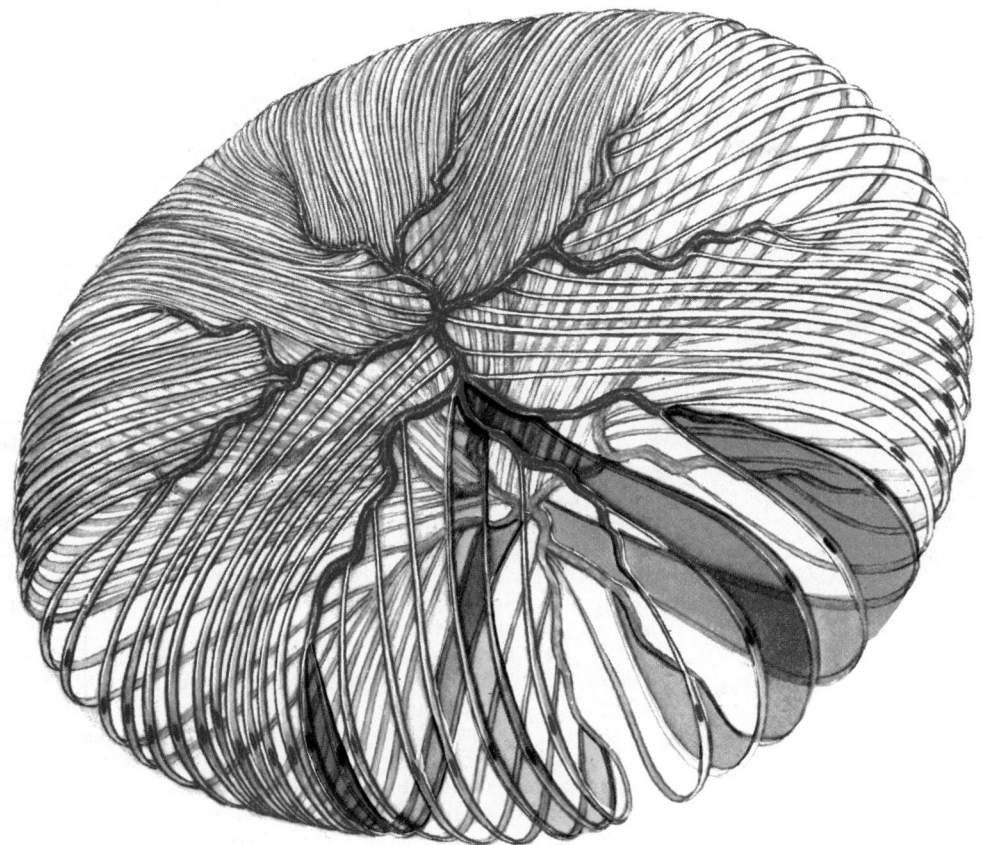

7.324B

dimension increases from 3·5–4·0 mm at birth to 4·75–5·0 mm at 95. Its anterior radius (about 10.0 mm) is greater than its posterior (about 6·0 mm), but both are reduced during accommodation. Continued lenticular growth is due to a continual slow production of new cells in its equatorial epithelium. The ultrastructure of the capsule and epithelium have been studied (Wanko & Gavin 1960, Cohen 1965); the capsule consists largely of fine filaments with a periodicity of 60 nm. The fibre consist essentially of a fine fibrillary substance. In adults there are 2100–2300 fibres and they may be 8–12 mm in length in the cortical zone. Microfilaments of human fibres are 5–9 nm in diameter and, according to Rafferty & Goosens (1978), their arrangements, constitution and sizes in different mammals may be associated with differing accommodative capacities.

THE ACCESSORY VISUAL APPARATUS

This includes the extraocular muscles, fasciae, eyebrows, eyelids, conjunctiva and lacrimal apparatus.

The Extraocular Muscles

Extraocular or extrinsic ocular muscles (7.325–330) include an elevator of the upper eyelid, *levator palpebrae superioris*, and six muscles capable of rotating the eye in almost any direction; these comprise four *recti* (*superior*, *inferior*, *medialis* and *lateralis*) and two *obliqui* (*superior* and *inferior*). This is a phylogenetically ancient pattern of muscles, extending with slight modification through almost the entire vertebrate phylum, apart from the levator palpebrae superioris, a later delamination from the rectus superior to serve the upper eyelid in tetrapods.

In general structure extraocular muscles resemble skeletal muscles elsewhere but differ in some features particularly in their innervation, contraction and pharmacological properties (Bach-y-Rita et al 1971, Hogan et al 1971, Davson 1976). At least two types of fibre occur: a slender (9–11 μm), slow-contracting form and a thicker (11–15 μm), 'twitch' type of fibre. The former have motor endings of the 'en-grappe' category, the latter motor end-plates (Davidowitz et al 1977).

The levator palpebrae superioris (7.325–329), thin and triangular, arises from the inferior aspect of the lesser wing of the sphenoid, anterosuperior to the optic canal and separated from it by the attachment of the rectus superior. At its posterior attach-ment it is narrow and tendinous but anteriorly it becomes broad and fleshy, its medial margin being almost straight and its lateral concave. The muscle ends in a wide anterior aponeurosis splitting into two lamellae; some fibres of the superior lamella are attached to the anterior surface of the superior tarsus (p. 1215) while others

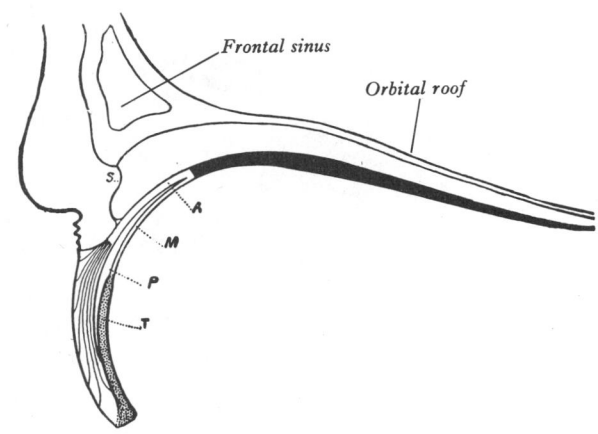

7.325 The levator palpebrae superioris, showing its connections. A = superficial lamella of aponeurosis; M = deep lamella, superior tarsal muscle (of Müller); P = interval between superficial and deep lamella of aponeurosis; T = tarsus; S = orbital septum.

1207

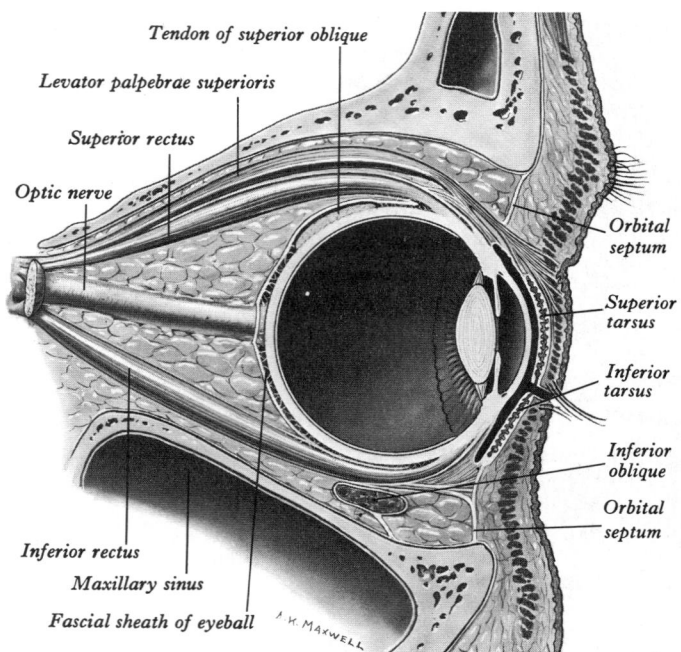

7.326 Sagittal section through the right orbital cavity.

radiate, traversing the overlying orbicularis oculi to the skin in the upper eyelid. The inferior lamella contains non-striated myocytes, the so-called *superior tarsal muscle*; it is attached directly to the upper margin of the superior tarsus and covered inferiorly by conjunctiva. A less prominent stratum of non-striated muscle, the *inferior tarsal muscle*, also exists in the lower lid, uniting the inferior tarsus to the fascial sheath of the rectus inferior and its expansion to the sheath of the inferior oblique. Another thin layer of non-striated muscle, *orbitalis*, bridges the inferior orbital fissure; its function is uncertain. The superior and inferior tarsal muscles presumably assist in the elevation of the upper and depression of the lower eyelid. All three non-striated muscles have a sympathetic innervation.

The connective tissue on the adjoining surfaces of the levator palpebrae superioris and rectus superior is fused; where the two muscles separate to reach their anterior attachments, the fascia between them forms a thick mass attached to the superior conjunctival fornix, which is usually described as an additional attachment of the levator palpebrae superioris. Traced laterally, the aponeurosis of the levator passes between the orbital and

palpebral parts of the lacrimal gland to a tubercle on the zygomatic bone, just within the orbital margin (p. 346); traced medially it loses its tendinous nature as it passes closely over the reflected tendon of the obliquus superior, continuing to the medial palpebral ligament as loose strands of connective tissue. Levator palpebrae superioris raises the upper eyelid but during this process the lateral and medial parts of its aponeurosis are stretched and thus limit its action; the elevation is also said to be checked by the orbital septum (p. 1215). ('Check' mechanisms abound in the orbit but there is little direct evidence that connective tissue structures thus implicated do function in this manner. See also p. 1213.)

The four recti (7.326–330) are attached posteriorly to a *common annular tendon* around the superior, medial and inferior margins of the optic canal (7.329); this fibrous annulus continues laterally across the inferomedial part of the superior orbital fissure and is attached to a tubercle on the margin of the greater wing of the sphenoid. The tendon is closely adherent to the optic nerve's dural sheath and the surrounding periosteum. Within it are: (1) the anterior aperture of the optic canal, transmitting the optic nerve and ophthalmic artery; and (2) the medial part of the superior orbital fissure transmitting the two divisions of the oculomotor and the nasociliary and abducent nerves. The superior ophthalmic vein may pass within or above the annular tendon, the inferior ophthalmic vein within or below it. Two specialized parts of this ring may be discerned: a lower, to which are attached the rectus inferior, part of the rectus medialis and the lower fibres of the rectus lateralis; and an upper, providing attachment to the rectus superior, part of the rectus medialis and the upper fibres of the rectus lateralis; a second small tendinous slip of the rectus lateralis is attached to the orbital surface of the greater wing of the sphenoid, lateral to the annulus. Each muscle passes forwards, in the position implied by its name, to be attached anteriorly by a tendinous expansion into the sclera, posterior to the margin of the cornea, and at average distances from the latter as follows: medialis 5·5 mm; inferior 6·5 mm; lateralis 6·9 mm; superior 7·7 mm.

The obliquus superior (7.327) is fusiform and lies superomedial in the orbit; it arises from the body of the sphenoid superomedial to the optic canal and to the tendinous attachment of the rectus superior. Passing forwards it ends in a round tendon, which plays through a fibrocartilaginous loop, the *trochlea*, attached to the trochlear fossa of the frontal bone. Tendon and trochlea are separated by a delicate synovial sheath. After traversing the trochlea the tendon is deflected posterolaterally and inferior to the rectus superior to be attached to the sclera *behind* the equator in its superolateral posterior quadrant, between the recti superior and lateralis.

The obliquus inferior (7.327), a thin, narrow muscle near the anterior margin of the floor of the orbit, arises from the orbital surface of the maxilla lateral to the nasolacrimal groove. Ascending posterolaterally at first between the rectus inferior and the orbital floor and then between the eyeball and the rectus lateralis, it is attached to the sclera *behind* the equator in its inferolateral posterior quadrant between the rectus inferior and rectus lateralis, near to but slightly posterior to the attachment of the obliquus superior.

Nerve supply. The levator palpebrae superioris, obliquus inferior and recti superior, inferior and medialis are supplied by the oculomotor nerve, the obliquus superior by the trochlear and the rectus lateralis by the abducent nerve.

ACTIONS OF EXTRAOCULAR MUSCLES

Levator palpebrae superioris elevates the upper lid, its antagonist being the orbicularis oculi. The degree of elevation which, apart from blinking, is maintained for long periods during waking hours is a compromise between the adequate exposure of the optical media and control of the entering light. The amount of light, in very bright sunshine, can be reduced by lowering the upper lid, reducing glare. Though much is known of the physiology of blinking (McEwen & Goodner 1962) and its significance in the distribution of lacrimal secretions, the continuous activity of the levator in keeping the lid raised has attracted little interest.

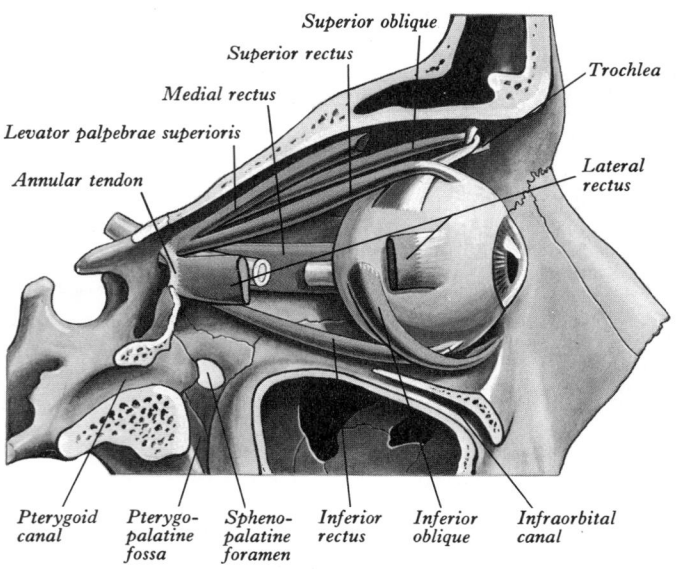

7.327 The muscles of the right orbit: lateral aspect.

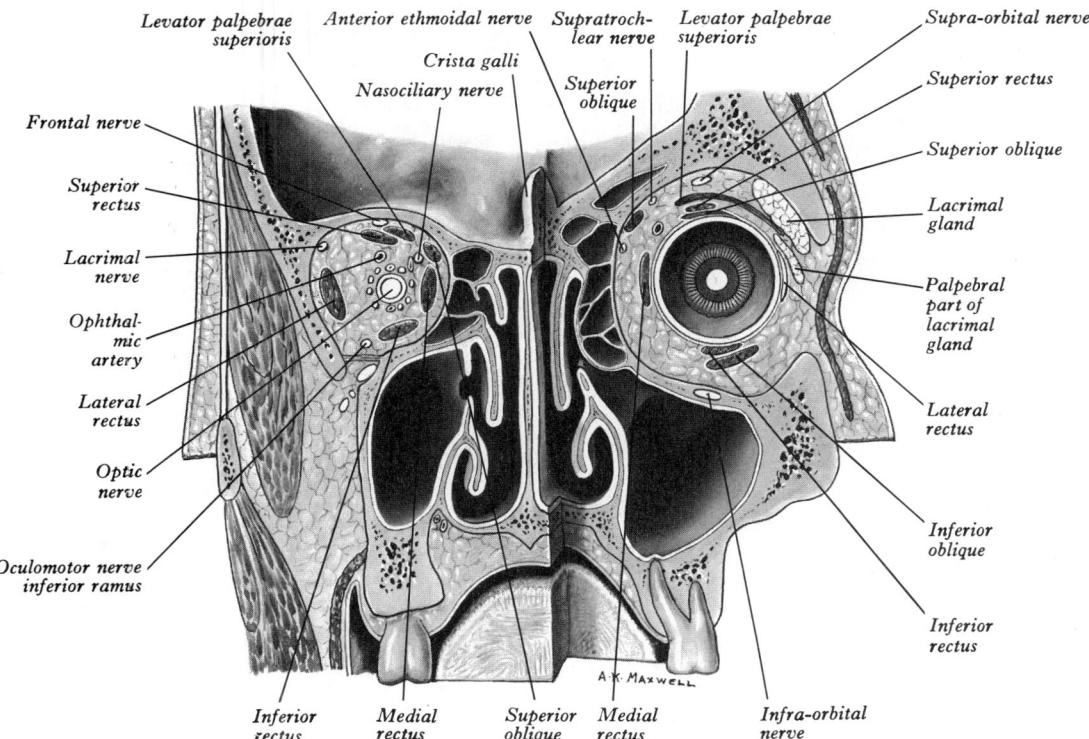

7.328 Coronal sections through the two orbits: posterior aspect. On the left side the plane of the section is more posterior and passes behind the eyeball.

The respective roles of its main, striated, voluntary part and small, inferior non-striated stratum (superior palpebral, tarsal or Müller's muscle) have not been clarified.

The *six extraocular muscles* all *rotate* the eyeball in directions dependent upon the geometrical relation between their osseous and global attachments (7.330), which are, of course, altered by the ocular movements themselves. Before considering their activities it must be recognized that *human* extraocular muscles are not generally accessible to inspection; consequently many opinions regarding them depend on deductions from malfunction due to disturbance of innervation. A complete assessment of the exact extent of neural injury is rarely possible; but this is not to say that, in the vast clinical literature available, there are not at least some valid observations. It is also essential to appreciate that any movement of an eye alters the tension and/or length in *all six* muscles, though direct observation of this has rarely been attempted, even in experimental animals (Sherrington 1905, Szentágothai 1950). It is at least likely that all six muscles are continuously involved and it is therefore merely a preliminary but necessary exercise to consider each in isolation. Because they form more obvious groupings as antagonists or synergists, it is useful to consider the four recti and two obliques as separate groups, remembering always that they act in concert.

Of the four *recti*, the *medial* and *lateral* exert comparatively straightforward forces on the eyeball. Being approximately horizontal when the visual axis is in its primary position, directed to the horizon, they rotate the eye medially (adduction) or laterally (abduction) about an imaginary vertical axis (7.330). They are antagonists; by reciprocal adjustment of their lengths the visual axis can be swept through a horizontal arc. When both eyes are involved, as is usual, the four medial and lateral recti can either adjust both visual axes in a *conjugate* movement from point to point at infinity, their axes remaining parallel, or they can *converge* or *diverge* the axes to or from nearer or more distant objects of attention in the visual field. But since they do not rotate the eye around its transverse axis, the medial and lateral recti cannot effect the extremely frequent act of elevating or depressing the visual axes as gaze is transferred from nearer to more distant objects or the reverse. This is the contribution of the *superior* and *inferior* recti (aided, as will become apparent, by the two oblique

muscles). However, the geometry of these is a little less simple and the key to the rotations which they effect is the obliquity of the orbit (7.330), whose axis does not correspond with the visual axis in its primary position but diverges from it at an angle of approximately 23°, a value varying in different individuals, being dependent upon the angle between the orbital axes and the median plane (7.330A). Hence, the simple rotation caused by an isolated *superior rectus*, analysed with reference to the three hypothetical ocular axes, appears to be complex, being primarily *elevation* (transverse axis), and secondarily a less powerful *medial rotation* (vertical axis) and slight *intorsion* (anteroposterior axis) in which the mid-point of the upper rim of the cornea (often referred to as '12 o'clock') is rotated medially towards the nose. These actions, compounded in fact as a single, simple rotation, are easily

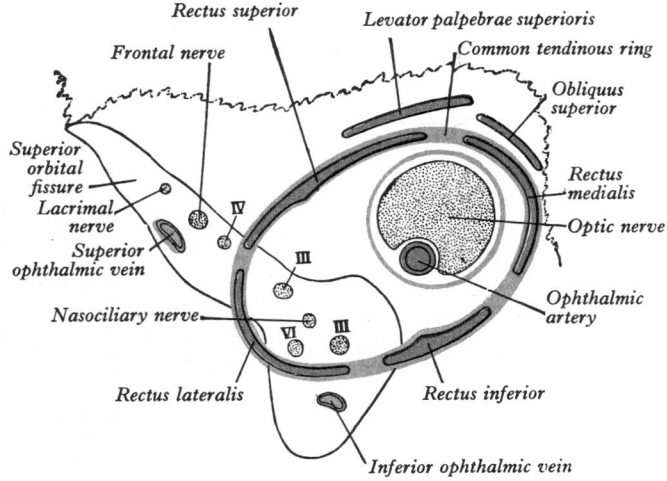

7.329 Scheme to show the common tendinous ring, the origins of the recti and the relative positions of the nerves entering the orbital cavity through the superior orbital fissure. Note that the ophthalmic veins frequently pass through the common tendinous ring. Modified from a figure in Whitnall 1932.

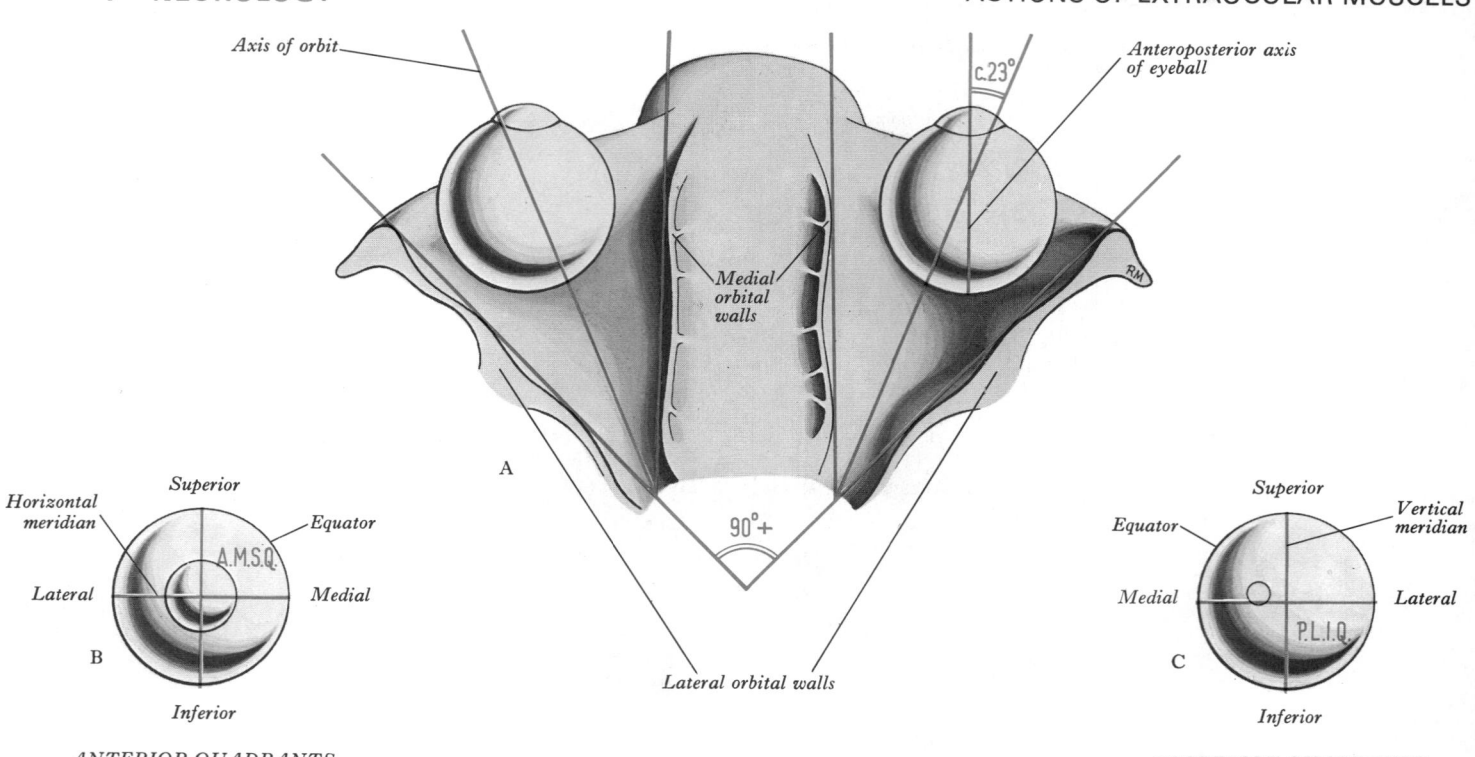

Axis of orbit

Anteroposterior axis
of eyeball

c.23°

Medial
orbital
walls

90°+

Lateral orbital walls

A

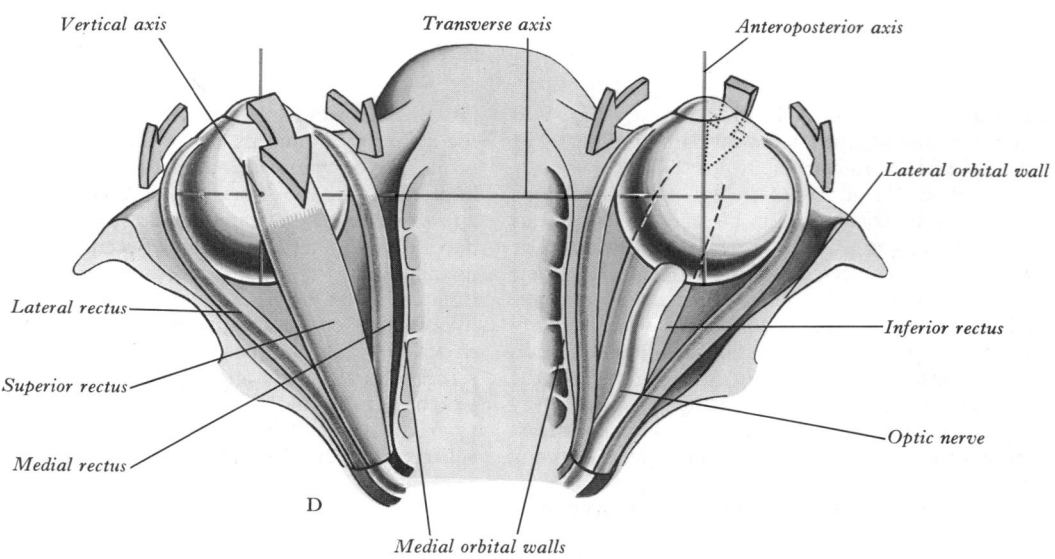

Superior

Horizontal
meridian

Equator

Lateral

Medial

A.M.S.Q.

Inferior

B

ANTERIOR QUADRANTS

Superior

Vertical
meridian

Equator

Medial

Lateral

P.L.I.Q.

Inferior

C

POSTERIOR QUADRANTS

Vertical axis

Transverse axis

Anteroposterior axis

Lateral orbital wall

Lateral rectus

Inferior rectus

Superior rectus

Medial rectus

Optic nerve

D

Medial orbital walls

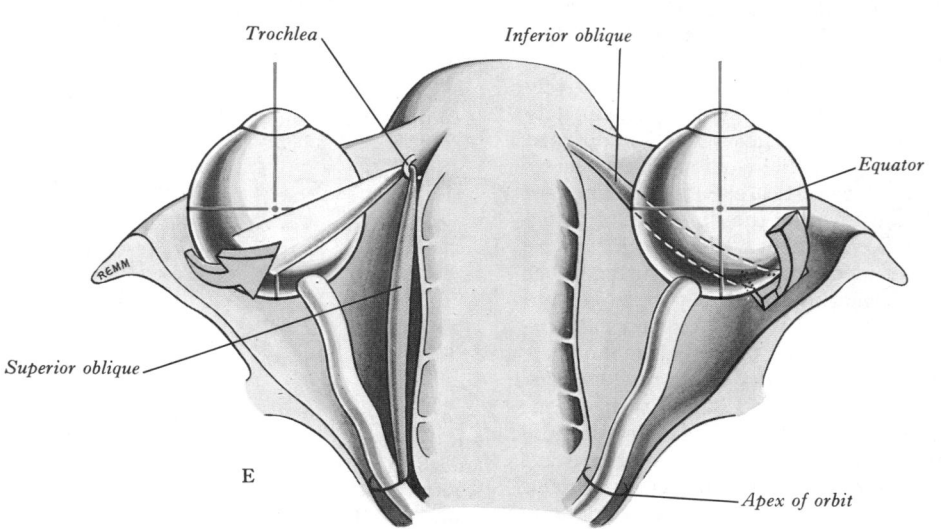

Trochlea

Inferior oblique

Equator

Superior oblique

Apex of orbit

E

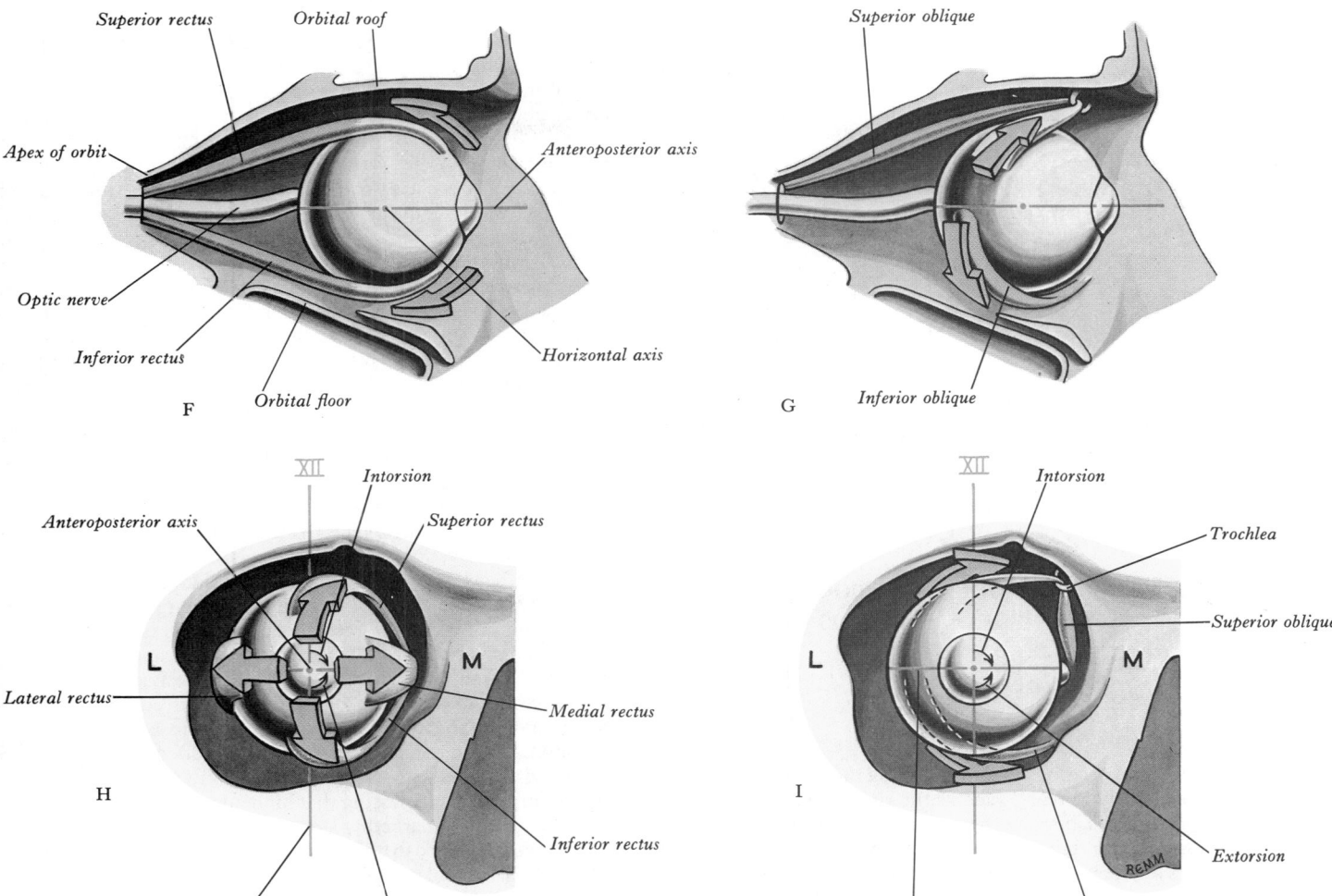

7.330. The geometrical basis of ocular movements.

A. The relationship between the orbital and ocular axes, with the eyes in the primary position of parallel visual axes.

B and C. The ocular globe in anterior and posterior views to show conventional geometry: meridia, equator, etc. A.M.S.Q = anterior medial superior quadrant; P.L.I.Q = posterior lateral inferior quadrant.

D. The orbits from above showing the medial and lateral recti and the superior rectus (left) and inferior rectus (right), indicating turning moments primarily around the vertical axis.

E. Superior (left) and inferior (right) oblique muscles showing turning moments primarily around the vertical and also anteroposterior axes.

F. Lateral view to show the actions of the superior and inferior recti around the transverse axis.

G. Lateral view to show the action of the superior and inferior oblique muscles around the anteroposterior axis.

H. Anterior view to show the medial rotational moment of the superior and inferior recti around the vertical axis. Conventionally the 12 o'clock position indicated is said to be *intorted* (superior rectus) or *extorted* (inferior rectus) as indicated by the small arrows on the cornea.

I. Anterior view to show the torsional effects of the superior oblique (intorsion) and inferior oblique (extorsion) around the anteroposterior axis, as indicated by the small arrows on the cornea.

appreciated when it is seen that the direction of traction of the superior rectus runs *posteromedially* from its attachment *anterior* to the equator and *superior* to the cornea to its osseous attachment near the orbital apex (7.330).

The *inferior rectus* pulls in a similar direction but rotates the visual axis *downwards* about the transverse axis. It is clear, from its comparable geometry, that it also rotates the eye *medially* on a vertical axis but that its action around the anteroposterior axis *extorts* the eye, i.e. rotates it so that the corneal 12 o'clock point turns *laterally*. Superior and inferior recti, therefore, both rotate the eye medially and, since their effects around the transverse and anteroposterior axes are opposed, their combined, equal contractions could only rotate the eye medially. In binocular movements they thus assist the medial recti in converging the visual axes; and by reciprocal adjustment they can elevate or depress the visual axes. It must be added that, as the eye is rotated laterally, the lines of traction of the superior and inferior recti approach the plane of the anteroposterior ocular axis (7.330); hence their rotational effects about this and the vertical ocular axis diminish. In abduction to about 23°, they become almost purely an elevator and depressor of the visual axis. The *superior oblique* acts on the eye from the trochlea and, since the attachment of the *inferior oblique* is for practical purposes vertically below this, both muscles approach

the eye at the same angle, being attached in approximately similar positions in the *superior* and *inferior posterolateral* ocular quadrants (7.330). From this geometry it is easy to understand that the superior oblique elevates the *posterior* aspect of the eye, the inferior depressing it. Hence, the former rotates the visual axis *downwards* and the latter *upwards*, both movements being around the transverse axis. But the obliquity of both is such that their traction, when the eye is in the primary position, is in a direction *posterior* to the vertical axis; both therefore rotate the eye *laterally* around this axis. With regard to the anteroposterior axis it is not difficult to deduce that in isolation the superior oblique *intorts* the eye, the inferior oblique *extorting* it. Like the superior and inferior recti, therefore, the two obliques have a common turning movement around the vertical axis but are opposed forces in respect of the other two. Acting in concert they could therefore assist the lateral rectus in abducting the visual axis, as in *divergence* of the eyes in transferring attention from near to far. Again, like the superior and inferior recti, the directions of traction of the oblique muscles also vary with ocular position; they become more nearly a pure elevator and a depressor as the eye is adducted.

In a short analysis of this kind much must be omitted and nothing can be said of the defects of ocular movement. For further information refer to Whitnall (1932), Cooper et al (1955),

Cogan (1956), Schlossmand & Priestley (1966), Alpern (1969) and Davson (1972). For those requiring the barest data, the actions of extraocular muscles may be summarized as follows:

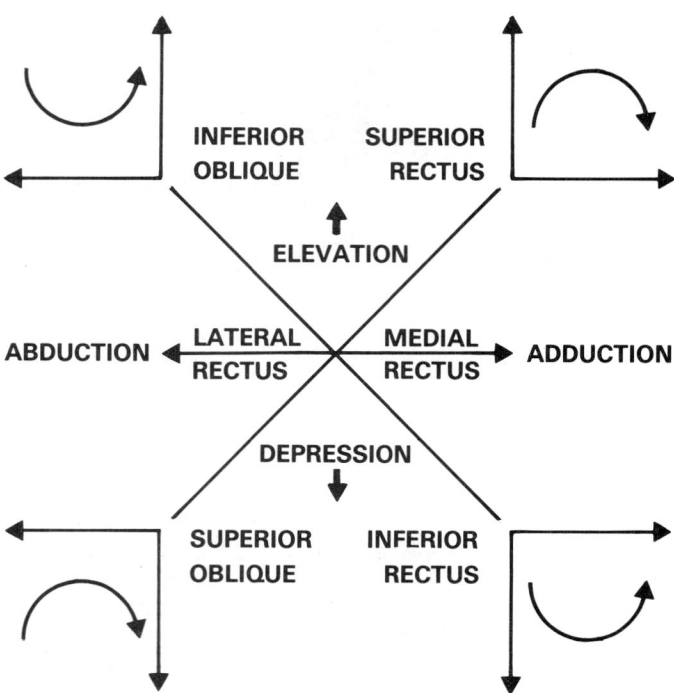

There is a misleading custom of linking the extraocular muscles together, usually in pairs, such as the right superior oblique and left inferior rectus, because they are supposed to deviate the visual axes down and to right, as doubtless they do. Such limited views may have a mnemonic value but they ignore the inescapable fact that in any ocular rotation *all* six muscles must change in length. Since we do not know much of the reciprocal innervational circuits of the extraocular muscles, it is impossible to dogmatize as to whether every muscle is contracted precisely in step with the progressive inhibition of an anatagonist, although the few experiments previously quoted do support such a view. However that may be, it would be more reasonable to link the superior, inferior and medial recti together as adductors or *convergence* muscles and the two obliques with the lateral recti as abductors or *divergence* muscles, a useful concept in view of the endless convergent and divergent movements of the visual axes most commonly required. Even this is an over-simplified view; convergence is usually accompanied by depression of the visual axes towards objects nearby, divergence by elevation. In these much more complex, concerted activities, the torsional effects of the muscles become much more important. Analysis shows that in simple, *cardinal* movements (adduction–abduction, elevation–depression) the torsional effects cancel out and they must also do so in all intermediate positioning of the visual axes to preserve the relation of corresponding retinal loci and hence binocular (single) vision. So far, head movements have been ignored but it is common observation that ocular movements are frequently, perhaps usually, accompanied by movements of the head, which might be likened to the *coarse* adjustment of an optical instrument such as a microscope, the *finer* adjustments being made by the ocular musculature.

It is interesting that while ocular rotations are clearly under voluntary control, *torsional* movements cannot be voluntarily initiated. But when the head is tilted relative to body, reflex torsions occur and are necessary to preserve retinal correspondence. Any small lapse in the concerted adjustment of both retinae entails diplopia. It is indeed surprising that such a complex organization of extraocular, neck and other muscles is learnt so effectively early in life that diplopia is rarely experienced; but the prize, of course, *stereoscopic vision*, is a great one.

All binocular movements are either *conjugate* or *disjunctive*, the visual axes either being trained in parallel or inclined in convergence/divergence activity. While it is a necessary introductory simplification to analyse both forms of action in terms of static

geometry, in reality both are dynamic and hence more complex than simple analysis suggests. In so-called 'fixation' of a focus of attention, whether uni-ocular or binocular, the visual axis is not 'fixed' in a perfectly steady manner but undergoes minute, but observable flicking (of a few minutes or even seconds of arc) across the true line of fixation. Such *micro-saccades* are rapid and surprisingly complex (see Barlow 1952, Riggs et al 1960, etc.). The term *saccade* (a French word of obscure origin meaning a 'jerk on the reins') was introduced by Dodge in 1903 for similar but much larger swings of fixation observed in subjects reading a line of print. Amplitude, velocity and reaction time in saccadic movement were first studied in detail by Westheimer (1954) and Robinson (1964) and investigations carried even further by Childress & Jones (1967), Becker & Fuchs (1969), Hallett & Lightstone (1976) and others. In general, reaction times and movements are measured in microseconds, amplitude varies from seconds to many degrees, with an accuracy of 0·2° or better, and the velocity of a large saccade may reach 500° per second. The speed of the saccades (an obvious survival factor) is assured by an initial contraction of the appropriate muscles which is slightly excessive (the 'pre-emphasis' of Robinson 1964), the necessary deceleration when the target is fixated being apparently due largely to the visco-elasticity of the extraocular muscles and orbital soft tissues, rather than to antagonistic muscular activity.

Clearly, saccadic activity is almost ever-present in human vision, and not merely in reading or the visual target practice of experimental laboratories. Not only are both visual axes endlessly and rapidly transferred to new points of interest in any part of the total visual field, but binocular gaze is very frequently made to travel routes of the most variable complexity in examining objects of some extent in the field; to this must be added the maintenance of *both* visual axes with sufficient accuracy to avoid diplopia. It is interesting to note in passing that binocular movements involving convergence are markedly slower than conjugate movements (Alpern & Wolter 1956), presumably due to the greater complexity of neural control (though the speed of contraction of the ciliaris must be a factor). Conjugate movements, indeed, might almost be 'programmed', in the current electronic jargon; but most visual human activity concerns targets of regard near enough to demand convergence and hence a neuronal intermediation of greater flexibility. Since the prime purpose is clear perception of a 'target', it is not surprising that the visual input is itself utilized in continuous feedback to the correct aiming of visual axes. Disparity in retinal images is demonstrably a strong stimulus to accurate uni-ocular fixation and binocular convergence during saccadic movement; and the sudden withdrawal of experimental stimuli interrupts corrective saccades. An interesting product of such experimentation is the observation that when a target, towards which an eye has commenced a saccade, is rapidly returned to its first position, the eye continues its saccade, as if an appraisal of the target's second position leads to a sequence of neuro-muscular events which is planned and irreversible. This has prompted some observers to liken the control to a 'sampled data system' (Young & Stark 1963) and to various other types of 'computer' apparatus. Whatever else they may achieve, such experiments do demonstrate a discontinuity in visual movement and perhaps therefore in visual perception. We are so accustomed to regarding vision as continuous, that the necessity for some kind of discontinuity, if feedback is to operate discriminantly, is overlooked. Of course, the millisecond scale of events might obscure such discontinuity; the much slower sequence in cinematography easily sustains a similar illusion of continuous vision. Much of the content of consciousness and memory can be dependent upon and evoked by visual stimuli (including the encoding of speech, mathematics and music). Hence the study of saccadic vision has a larger significance than the simpler mechanisms of visual movement. Since it is concerned with elucidating a particularly precise and intricate motor-sensory organization, it may also illumine cerebral activities on a much wider scale.

In addition to these considerations, there is much evidence that the continual movements of the eye are actually essential for vision to occur at all; the retinal and more central neural networks appear to be designed primarily to detect transient events such as movements rather than static, maintained stimuli. Indeed, images which are essentially static, such as those due to retinal blood vessels, are not detectable unless made to move, e.g. by shifting low-angle illumination with an ophthalmoscope. This finding may be related to the complex architecture of the retina and to the presence of circuits which are specifically used to detect movements, e.g. those involving cholinergic amacrine cells (p. 1202).

Since the recti exert a *posterior* traction while the obliques pull the eyeball to some degree *anteriorly*, it is sometimes suggested that they collectively position it in the orbital cavity, preventing anteroposterior movements of the eyeball, assisted perhaps by various 'check ligaments' (vide infra).

The Fascial Sheath of the Eyeball

A thin fascial membrane envelops the eyeball from the optic nerve to the corneoscleral junction, separating it from the orbital fat and forming a socket for its free rotation (7.331). Its ocular aspect is smooth and separated from the sclera by an *episcleral space*, which is traversed by delicate bands of connective tissue extending between the fascia and the sclera. Posteriorly the fascia is traversed by ciliary vessels and nerves and fuses with the sheath of the optic nerve and the sclera around the nerve's entrance. It also blends with the sclera anteriorly, just behind the corneoscleral junction. It is perforated by the tendons of the extraocular muscles and is reflected on to each as a tubular sheath. The sheath of the obliquus superior reaches the fibrous trochlea of that muscle, that of the obliquus inferior the orbital floor, to which it gives off a slip. The sheaths on the recti fade into their perimysium but give off important expansions. That from the rectus superior blends with the tendon of the levator palpebrae superioris, that of the rectus inferior is attached to the inferior tarsus and the sheath of obliquus inferior. The expansions from the sheaths of recti medialis and lateralis are triangular, strong and attached respectively to the lacrimal and zygomatic bones; as they may limit the actions of the two recti, they are the *medial* and *lateral check ligaments*. An inferior thickening of the ocular fascial sheath is the *suspensory ligament of the eye* (Lockwood 1886); it is slung like a hammock below the eye, being expanded centrally and narrowing at its ends; it is formed by the union of the margins of the sheath of rectus inferior with the medial and lateral check ligaments. Anomalies of these fascial arrangements are said to interfere with normal ocular movements, causing various types of squint (Nutt 1955). However, apart from the suspensory ligament, there is no convincing evidence that any of the so-called check ligaments actually limit movement. The link between the fasciae covering the superior rectus and levator palpebrae superioris is of no known significance; it does not impede independent movements of the lid and eye. It is, however, plausible to assume that orbital connective tissue must be arranged to assist in the location of the eye within the orbit, but without obstruction to the activities of the extrinsic muscles. Its connections must also prevent the gross displacement of orbital fat, for this would interfere with the accurate positioning of the two eyes in binocular vision. A reappraisal of the disposition of orbital adipose and connective tissues by Koornneef (1977) emphasizes a series of radial septa which he describes as extending from the ocular fascial sheath to the periorbita. These could provide an excellent suspensory system. His dissections also confirm the presence of separate muscular sheaths, as here described.

The orbital fascia forms the periosteum of the orbit, or periorbita, but is only loosely connected to bone. Behind it is united with the dura mater and the sheath of the optic nerve and is continuous anteriorly with the periosteum at the orbital margin, where it gives off a stratum contributing to the orbital septum. One process of it holds the trochlea in position; another, the *lacrimal fascia*, forms the roof and lateral wall of the sulcus for the lacrimal sac (p. 346). Lang (1975) has detailed the vascularization of the orbital periosteum and fascia.

Visual Reflexes

The nerve impulses concerned with *visual reflexes* effecting movements of the eyes, head and neck in response to visual stimuli

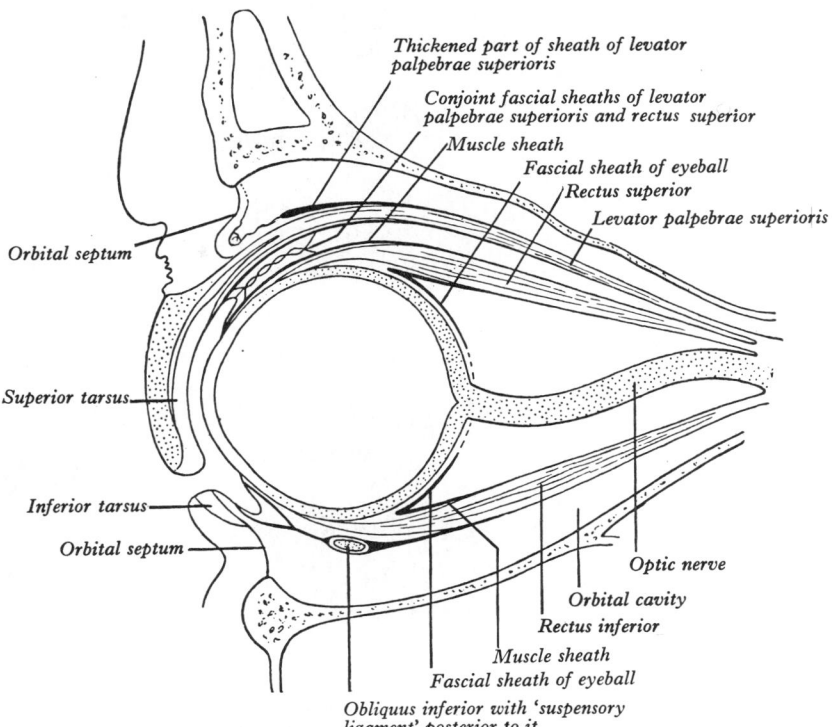

Thickened part of sheath of levator palpebrae superioris
Conjoint fascial sheaths of levator palpebrae superioris and rectus superior
Muscle sheath
Fascial sheath of eyeball
Rectus superior
Levator palpebrae superioris
Orbital septum
Superior tarsus
Inferior tarsus
Orbital septum
Optic nerve
Orbital cavity
Rectus inferior
Muscle sheath
Fascial sheath of eyeball
Obliquus inferior with 'suspensory ligament' posterior to it

7.331A   Scheme of the orbital fascia in sagittal section (after Whitnall 1932).

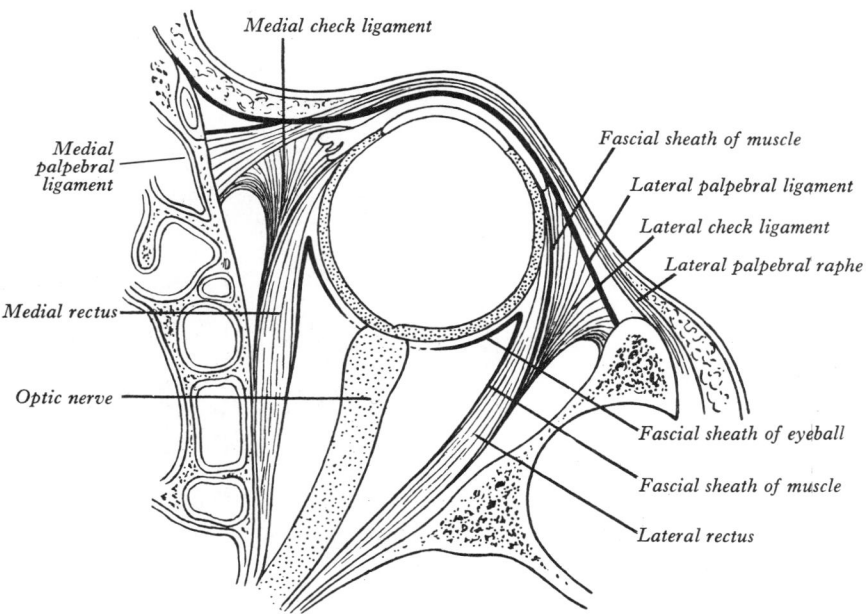

Medial check ligament
Medial palpebral ligament
Medial rectus
Optic nerve
Fascial sheath of muscle
Lateral palpebral ligament
Lateral check ligament
Lateral palpebral raphe
Fascial sheath of eyeball
Fascial sheath of muscle
Lateral rectus

7.331B   Scheme of the orbital fascia in horizontal section (after Whitnall 1932).

follow the optic nerves and tracts to the superior colliculi. Traversing complex connections there, they travel along the tectospinal and tectobulbar tracts to the motor neurons of spinal and cranial nerves (p. 987).

Pupillary Light Reflex

Increased illumination causes reflex meiosis (pupillary contraction), the impulses concerned travelling by the optic nerve and tract to the pretectal nuclei of both sides (p. 987), whence short axons of secondary neurons run close to the central grey matter, conveying impulses to the accessory oculomotor (Edinger–Westphal) nuclei (p. 1097); from there, preganglionic axons reach the ciliary ganglia via the oculomotor nerves and their branches to the inferior oblique muscles. Postganglionic fibres from the ganglia

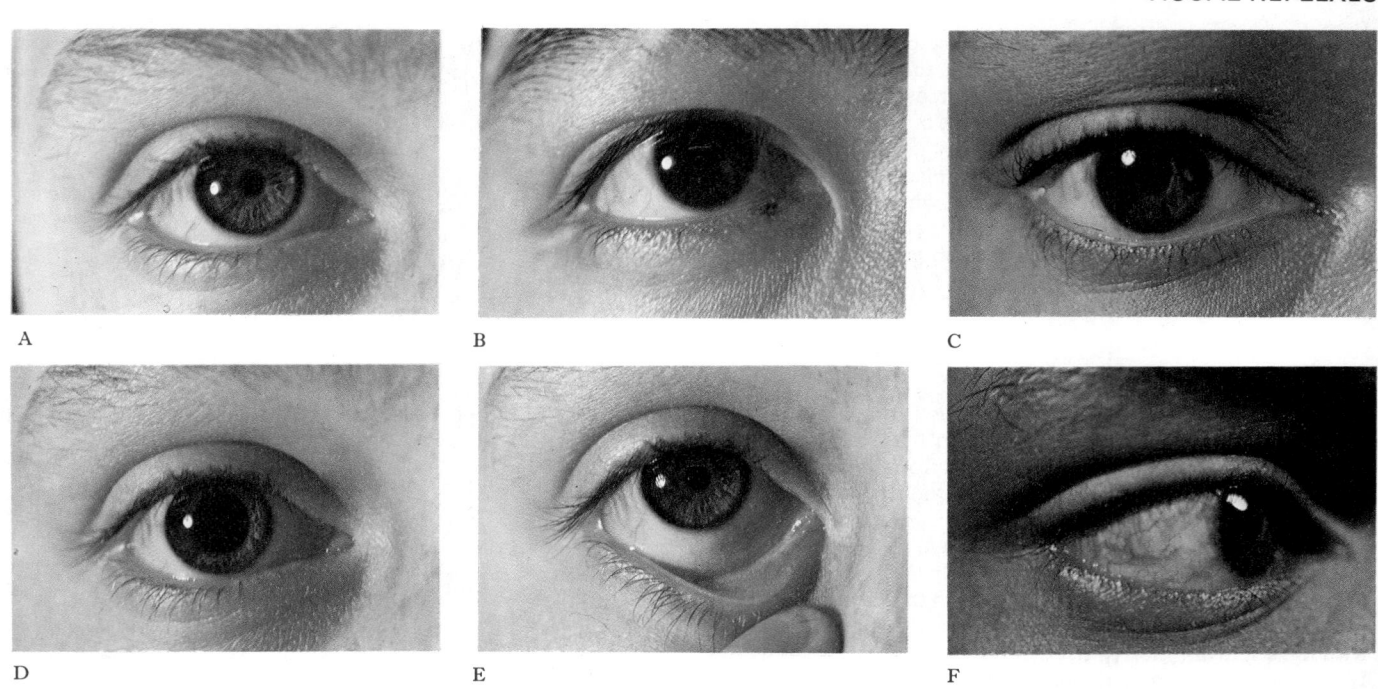

A

B

C

D

E

F

7.332   In the top row of photographs the typical appearances of the living eye are compared in a Caucasian female (A), a Mongoloid male (B) and a Negroid male (C). All subjects are about 20 years of age. Note the pale sclera and grey-blue iris in A, the epicanthus overlapping the medial end of the lower eyelid in B and the dark brown pigmentation of the iris in both B and C, rendering the pupil almost invisible. Compare the size of the pupil in the same eye, photographed under steady bright light in A and by sudden exposure to the same illumination after a period of dark adaptation in D. In E the lower eyelid had been everted somewhat to exhibit the lacrimal punctum and the rich subepithelial network of blood vessels. In F note the circumcorneal pigmentation and the conjunctival blood vessels, deep to which can be seen some details of the episcleral vessels. All photographs are of the right eye. Photography by Kevin Fitzpatrick, Anatomy Department, Guy's Hospital Medical School, London.

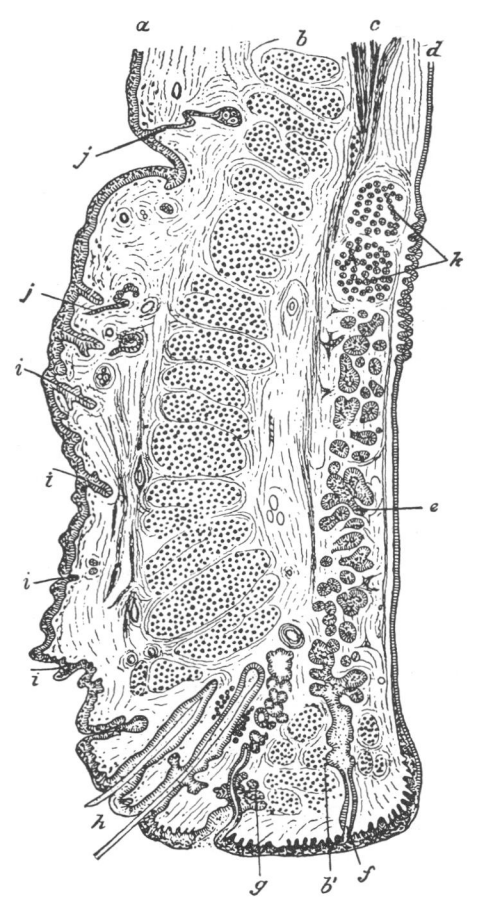

7.333   Sagittal section through the upper eyelid (after Waldeyer). *a*. Skin. *b*. Orbicularis oculi. *b'*. Ciliary fasciculi of the orbicularis oculi. *c*. Levator palpebrae superioris. *d*. Conjunctiva. *e*. Tarsal glands embedded in the tarsal plate. *f*. Opening of a tarsal gland. *g*. Ciliary gland. *h*. Eyelashes. *i*. Small hairs of the skin. *j*. Sweat glands. *k*. Posterior tarsal glands.

traverse the short ciliary nerves to reach the pupillary sphincters. If only one eye is stimulated both pupils nevertheless contract (the *consensual pupillary light reflex*), because fibres from each optic tract pass to *both* pretectal nuclei, decussating in the posterior commissure. The dilatator pupillae is supplied by fibres from the superior sympathetic cervical ganglion. The preganglionic fibres in this pathway arise from neurons in the lateral grey column of the first and second thoracic spinal segments, traversing the upper thoracic spinal nerves and their white rami communicantes to the sympathetic trunk, in which they ascend to the superior cervical ganglion (p. 1158). Since pupillary size results from the *balanced action* of these two innervations, the pupil dilates when the stimulus ceases. The pupil dilates also in response to painful stimulation of almost any part of the body. Presumably fibres of sensory pathways connect with the sympathetic preganglionic neurons described above. Some believe, however, that reflex pupillodilation is largely due to inhibition of the parasympathetic accessory oculomotor nuclei, though the paths involved are uncertain. One manifestation of this reflex is dilatation produced by pinching the skin of the neck, the *pupillary skin reflex*, a reminder that to the above, simplified account must be added other afferent influences, such as reticular connections with the superior collicular nuclei (pp. 987, 991).

Accommodation Reflexes

In accommodating for viewing near objects the eyes converge and at the same time the ciliares contract to modify the shape of the lens , the pupil constricting to increase the depth of focus. Pathways for reflex accommodation comprise the optic nerves and tracts, lateral geniculate bodies, optic radiations and the cortical visual areas, the latter being connected by long association fibres to the frontal cortical eye field, whence fibres descend via the internal capsule to the oculomotor nuclei. From the accessory oculomotor nuclei parasympathetic fibres pass to the ciliaris and sphincter pupillae (relaying in ciliary ganglion). From the ventral part of the oculomotor nucleus (p. 1096) other fibres supply the medial recti for the action of convergence (p. 1209). These routes have not been so clearly defined as those for the pupillary light reflex; it has been suggested (Wilkinson 1927) that contraction of the pupil in the accommodation reflex is secondary to ocula

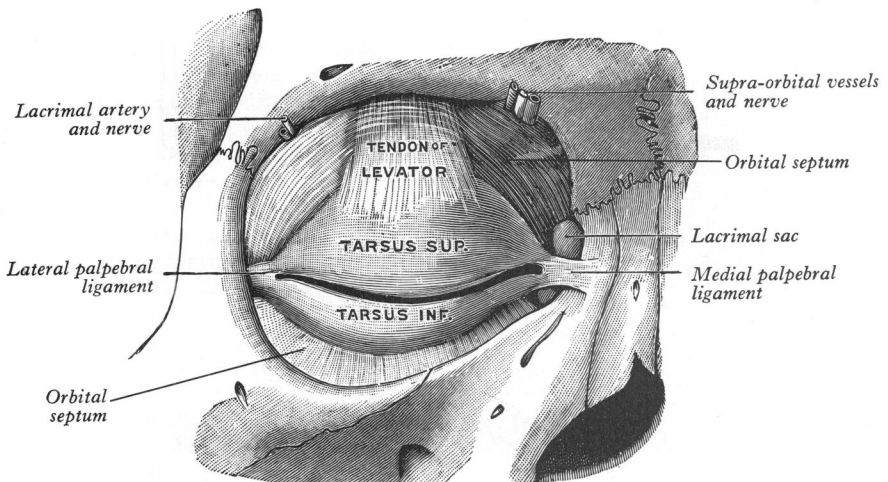

7.334 The tarsi and their ligaments: anterior aspect.

convergence, the afferent impulses arising in proprioceptor endings in the extrinsic muscles and travelling in the oculomotor nerves directly to the accessory nuclei. In certain central nervous diseases (e.g. tabes dorsalis) the pupillary light reflex may be lost but not pupilloconstriction as part of the accommodation reflex (the Argyll Robertson pupil). A lesion producing such an effect would probably be located between the accessory oculomotor nucleus and the lateral geniculate body, where the paths for the two reflexes diverge.

Conjunctival and Corneal Reflexes

Any stimulus to conjunctiva or cornea excites blinking. Afferent impulses travel via the ophthalmic division of the trigeminal nerve and efferent impulses in branches of the facial nerve to orbicularis oculi.

Eyebrows and Eyelids

The eyebrows are arched dermal eminences surmounting the orbits, with numerous short, thick hairs set obliquely in them. Fibres of the orbicularis oculi, corrugator and the frontal part of the occipitofrontalis are attached to the dermis of the eyebrows.

The eyelids or **palpebrae** are thin, movable folds, adapted to the front of the eyes and protecting them from injury, by rapid closure. The upper eyelid is larger and more mobile, being furnished with an elevator muscle, the levator palpebrae superioris (p. 1207). The two eyelids are united at their extremities and when parted an elliptical space, the *palpebral fissure*, appears between their margins. The ends of the fissure are termed the *angles* or *canthi* of the eye.

The *lateral angle* of the eye or *lateral canthus* is more acute than the medial and is closely apposed to the eyeball. The *medial angle* or *medial canthus* is prolonged for a little towards the nose and is about 6 mm from the eyeball; the eyelids are here separated by a triangular space, the *lacus lacrimalis*, in which is a small reddish *caruncula lacrimalis* (7.332). On each palpebral margin, at the basal angles of the lacus, is a small, conical, *lacrimal papilla*, its apex pierced by the orifice of a *lacrimal canaliculus*, an aperture known as the *punctum lacrimale* (7.332).

Eyelashes grow in the palpebral margins, their distribution extending from the lateral canthus to the lacrimal papillae. They are short, thick, curved hairs, arranged in double or triple rows; the upper, which are more numerous and longer, curve upwards while those in the lower lid curve down so that upper and lower lashes do not interlace when the lids are closed. Enlarged and modified sudoriferous, *ciliary glands* are arranged in several rows near the palpebral margins, opening near the lashes (p. 93).

Structure of the Eyelids

From its facial surface inwards each eyelid consists of: skin, subcutaneous loose connective tissue, the fibres of the orbicularis

oculi, the tarsus and orbital septum, tarsal glands and conjunctiva. The upper lid also contains the aponeurosis of the levator palpebrae superioris (7.333). The *skin* is extremely thin and is continuous at the palpebral margins with the conjunctiva. The *subcutaneous connective tissue* is very lax and delicate, seldom containing any adipose tissue.

The *palpebral fibres of the orbicularis oculi* are thin and pale and lie parallel with the palpebral fissure; deep to them is a layer of loose connective tissue, which in the upper lid is continuous with the scalp's subaponeurotic layer (p. 571) so that effusions of blood or pus at this level can pass down from the scalp into the upper eyelid. In this layer are the main nerves; local anaesthetics should hence be injected deep to the orbicularis oculi.

The two *tarsi* (7.334) are thin, elongate plates of dense fibrous tissue about 2·5 cm long, one in each eyelid, contributing to its form and support. The *superior tarsus*, the larger, is semi-oval, centrally about 10 mm in height and narrowing towards its ends. The lowest fibres of the superficial lamella of the aponeurosis of the levator palpebrae superioris are attached to its anterior surface, the deep lamella to its upper margin (7.325). The smaller *inferior tarsus* is narrower, about 5 mm in vertical height. The free, ciliary margins of both tarsi are thick and straight and their attached orbital edges are connected to the orbital margin by the orbital septum. Their lateral ends are attached by a band, the *lateral palpebral ligament*, to a tubercle on the zygomatic bone, just within the orbital margin; this band is separated from the more superficial *lateral palpebral raphe* (p. 571) by a few lobules of the lacrimal gland. The medial ends of the tarsi are attached by a strong tendinous band, the *medial palpebral ligament*, to the upper part of the lacrimal crest and to the adjoining maxillary frontal process in front of it; the lower edge of this ligament is separated from the lacrimal sac by some fibres of the orbicularis oculi which are attached to the ligament.

The *orbital septum* is a weak membrane attached to the orbital rim, where it is continuous with the periosteum. In the upper lid it blends with the superficial lamella of the aponeurosis of the levator palpebrae superioris, in the lower lid with the anterior tarsal surface. It is perforated by vessels and nerves passing from the orbital cavity to the face and scalp, by the aponeurosis of the levator and by the palpebral part of the lacrimal gland.

The *tarsal glands* (7.335), embedded in the tarsi, may be visible through the conjunctiva when the eyelids are everted; they are yellow and are arranged in a single row of about 30 in the upper lid and fewer in the lower. They are embedded in grooves on the posterior tarsal aspect, occupying the full tarsal height and hence being longer in the upper lid, where the tarsus is higher. Their ducts open on the free palpebral margins by minute foramina. They are modified sebaceous glands, each a straight tube with many lateral diverticula. They are supported by a basement membrane and lined at their orifices by stratified epithelium and elsewhere by a layer of polyhedral cells. Their oily secretion

1215

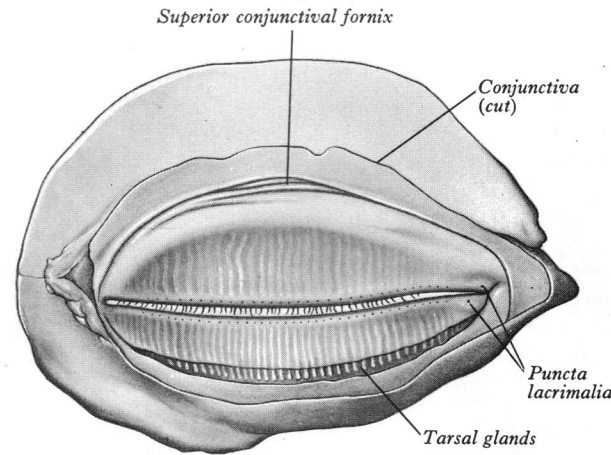

7.335 The posterior surfaces of the upper and lower eyelids of the left side. The orifices of the tarsal glands can be seen on the free margins of the lids.

spreads over the margins of the eyelids and prevents overflow of lacrimal fluid on to the cheek. It is also said to spread over the surface of the tear film to reduce evaporation.

The Conjunctiva

The conjunctiva, a transparent mucous membrane, covers the internal palpebral surfaces and folds on to the anterior sclera and cornea where it is continuous with the corneal epithelium; it is a *conjunction* between these structures.

The *palpebral conjunctiva* is very vascular, has numerous subepithelial connective tissue papillae and contains lymphoid tissue in its deeper parts, especially near the fornices; it is closely adherent to the tarsi. At the free palpebral margins it is continuous with the skin, with the lining epithelium of the ducts of the tarsal glands, the lacrimal canaliculi and lacrimal sac and hence with the nasolacrimal duct and nasal mucosa (an important continuity in the spread of infection). The line of reflection of the conjunctiva from the lids to the eyeball is the *conjunctival fornix*, subdivided into *superior* and *inferior fornices*, the former receiving the ducts of the lacrimal gland. Over the sclera the *ocular conjunctiva* is loosely connected to the eyeball; here it is thin, transparent,

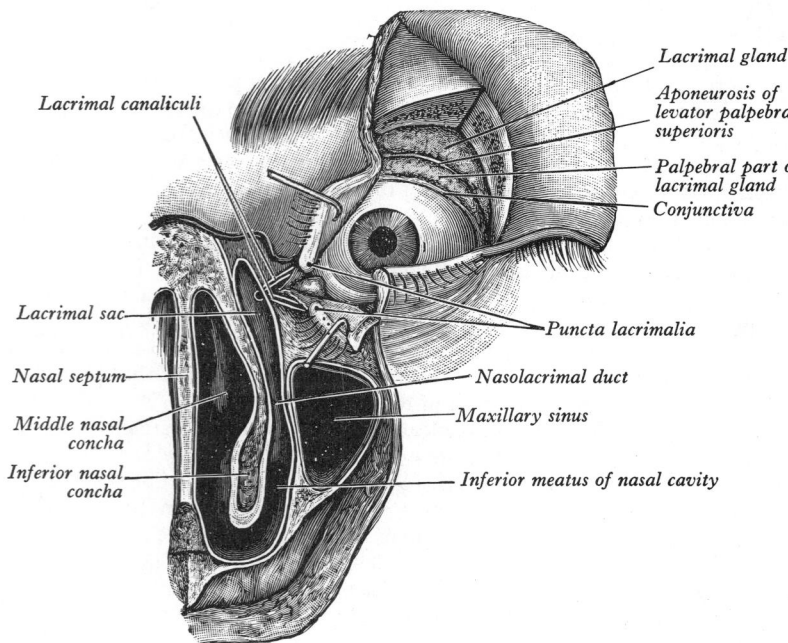

1216 7.336 The left lacrimal apparatus, dissected from the anterior aspect.

without papillae and slightly vascular. Reaching the cornea it continues as the corneal epithelium (p. 1183). The epithelium of the palpebral conjunctiva near the margins of the lids is non-keratinized stratified squamous; about 2 mm from each margin is a groove in which foreign bodies frequently lodge and here the epithelium has two layers, consisting of superficial columnar and deeper flat cells; these persist through most palpebral conjunctiva but near the fornices an intermediate layer of polygonal cells appears and this trilaminar conjunctival epithelium covers the anterior, exposed surface of the sclera. Near the corneoscleral junction the epithelium changes to stratified corneal epithelium (p. 1183). Scattered throughout the conjunctival epithelium are mucus-secreting goblet cells but there are few of these in the palpebral and circumcorneal epithelium.

The *lacrimal caruncle* (7.332) is a small, reddish, conical body in the lacus lacrimalis at the medial canthus; it is a small island of skin, containing sebaceous and sudoriferous glands with a few slender hairs. Laterally and partly obscured by it is a fold of conjunctiva, the *plica semilunaris*, its concave free edge directed laterally towards the cornea. Its epithelium is like that of the scleral conjunctiva but contains numerous goblet cells; subjacent are some fat cells and a little non-striated muscle. The *nictitating membrane* ('third eyelid'), a conjunctival structure in some amphibians, reptiles and mammals, may be homologous with the plica semilunaris.

Vessels and Nerves

The eyelids receive their arteries from the medial palpebral branches of the ophthalmic artery and the lateral palpebral branches of the lacrimal artery (p. 746). The ocular conjunctiva is supplied by the ophthalmic division of the trigeminal nerve; the upper palpebral conjunctiva is supplied by the ophthalmic, the lower by the maxillary division. Many conjunctival nerve fibres end in bulbous corpuscles (p. 913). Lymph vessels of the eyelids and conjunctiva are described on p. 844.

Palpebral Movements

The position of the lids depends on reciprocal tone in the orbicularis oculi and levator palpebrae superioris and on the degree of ocular protrusion. In the usual opened position the margin of the inferior lid crosses the eyeball level with the lower edge of the circumference of the iris, the upper covering about half of the width of the upper iris. The eyes are closed by movements of *both* lids, produced by the contraction of the orbicularis oculi and relaxation of the levator palpebrae superioris. In looking upwards, the levator contracts and the upper lid follows the ocular movement; at the same time, the eyebrows are also usually raised by the frontal parts of the occipitofrontalis to diminish their overhang. The lower lid lags behind ocular movement, so that more sclera is exposed below the cornea and the lid is bulged a little by the lower part of the elevated eye. When the eye is depressed both lids move, the upper retaining its normal relation to the eyeball and still covering about a quarter of the iris. The lower is probably dragged downwards by the pull of the conjunctiva in the inferior fornix. In states of fear the palpebral fissures are widened by contraction of the non-striated superior and inferior tarsal muscles, due to increased sympathetic activity. Lesions of the sympathetic supply result in drooping of the upper eyelid (ptosis), as seen in Horner's syndrome. For further comments on palpebral movement see p. 1207.

The Lacrimal Apparatus

The lacrimal apparatus (7.336, 339) comprises the lacrimal gland, which secretes a complex fluid (tears), and whose excretory ducts convey fluid to the surface of the eye, the lacrimal canaliculi, lacrimal sac and nasolacrimal duct, by which the fluid is collected and conveyed into the nasal cavity.

The lacrimal gland (7.336) is derived phylogenetically from serous secreting glands and those producing oily secretions. In primates the lacrimal (serous) element has migrated from its original position in the lower lid to the upper. The human lacrimal gland is superolateral in the orbit and has a large, upper

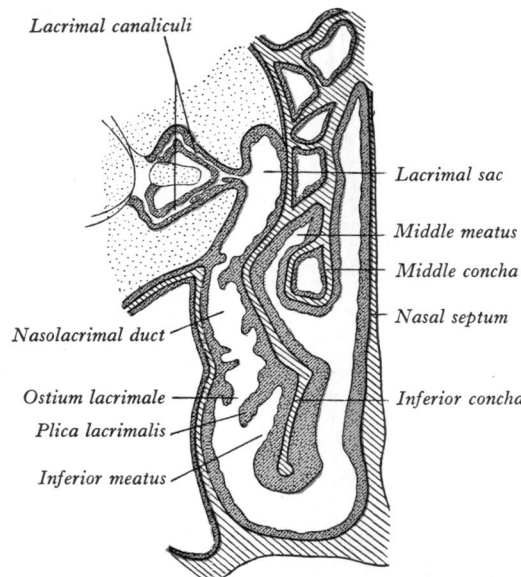

7.337A   Sketch from a coronal section through the right half of the nasal cavity (anterior aspect) to show the relation of the lacrimal passages to the maxillary and ethmoidal sinuses and the inferior nasal concha. The mucous membrane is coloured (after Whitnall 1932).

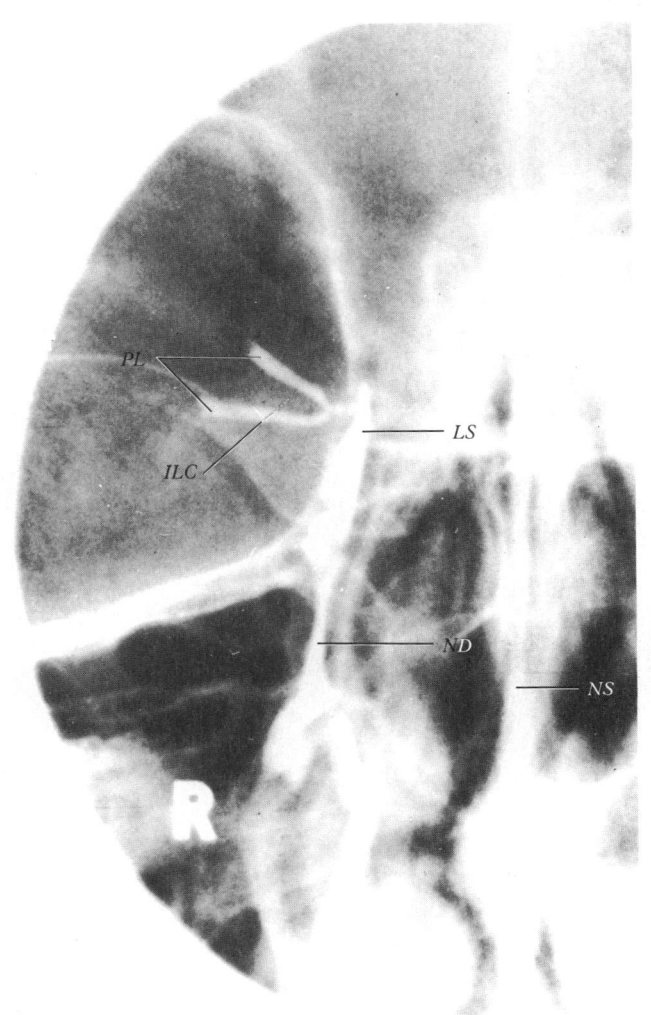

7.337B   Radiograph of the lacrimal drainage pathway, demonstrated by the injection of radio-opaque tracer into the lacrimal duct. Note the puncta lacrimalia (PL), inferior lacrimal canaliculus (ILC), lacrimal sac (LS), nasolacrimal duct (ND) and nasal septum (NS). Provided by T D Hawkins, Addenbrooke's Hospital, Cambridge; photography by Sarah Smith.

orbital and smaller, lower *palpebral part*, the two being continuous posterolaterally around the concave lateral edge of the levator aponeurosis. The orbital part, about the size and shape of an almond, lodges in the lacrimal fossa on the medial aspect of the zygomatic process of the frontal bone, just within the orbital margin. It lies above the levator palpebrae superioris and, laterally, above the lateral rectus; its inferior aspect is connected to the levator's sheath, its upper to the orbital periosteum; its anterior aspect adjoins the orbital septum and its posterior is attached to the orbital fat. The palpebral part, about one-third the size of the orbital, has two or three lobules extending inferior to the levator aponeurosis into the lateral part of the upper lid, where it is attached to the superior conjunctival fornix. It is visible through the conjunctiva when the lid is everted. Its ducts, about 12 in number, open into the superior fornix, those from the orbital part (four or five) passing through the palpebral part to join some of its ducts and the rest (six to eight) opening separately. Thus all ducts traverse the palpebral part so that excision of this part is functionally equivalent to the total removal of the gland.

Many small accessory lacrimal glands occur in or near the fornices; they are more numerous in the upper lid than in the lower. This may explain why the conjunctiva does not dry up after extirpation of the main lacrimal gland.

Structure of Lacrimal Gland (7.338)

The lacrimal gland is lobulated and tubulo-acinar in form (8.1A), its secretory units or 'endpieces' resembling those of salivary glands (p. 1293). The secretion is a watery fluid with an electrolyte content like plasma and containing a bacteriocidal enzyme, *lysozyme*. The primary secretion from the endpieces is modified by the ducts. Though many earlier accounts of lacrimal ultrastructure concerned rodents and lagomorphs (Obayashi 1959, Scott & Pease 1959, Ichikawa & Nakajima 1962), primate lacrimal glands have now received attention (Ruskell 1968, 1969, Egeberg & Jensen 1969, Orzalesi et al 1971, Ruskell 1975, Hirsch-Hoffmann 1976, 1978).

Its glandular cells form two categories in rhesus monkeys (Ruskell 1968, 1969, Hirsch-Hoffmann 1976), and in man two (Ito & Shibaski 1964, Kuhnel 1968), three (Ruskell 1975) or four (Hirsch-Hoffmann 1978), depending on which ultrastructural and histochemical criteria are considered to be significant. In rhesus monkeys, glandular cells with uniformly electron-dense secretory granules are described as *serous* and those with paler heterogeneous granules as *mucous* (Ruskell 1968). Ito & Shibasaki (1964) and Kuhnel (1968) propose two distinct human types of

cell: one, the K cell, contains small, electron-lucent granules and stains like a mucous cell; the other, termed the G cell, contains large, electron-dense granules and stains like a serous cell. According to Allen et al (1972) and Ruskell (1975) most glandular cells are mucous, a surprising report in view of the patently serous nature of tears. Ruskell (1975) distinguishes three categories of human glandular cell, arbitrarily grouping them into 'light', 'medium' and 'dark' according to the electron density of their secretory granules; acini contain either two or all three categories and are closely associated with lymphocytes and myoepitheliocytes. Hirsch-Hoffmann (1978) proposed *four* groups of human glandular cells, distinguishable by the number and electron density of the granules and surrounding cytoplasm; the largest group have pale cytoplasm and numerous granules of varying electron density; a second group, also pale, have fewer granules; a third have darker cytoplasm and numerous granules; the remainder, also dark, have fewer, uniformly electron-dense granules. Hirsch-Hoffmann has suggested that at least some may represent different stages in the secretory activity of one or two distinct types, an interpretation supported by Egebert & Jensen (1969) and Orzalesi et al (1971).

The human lacrimal gland contains interstitial (epilemmal, p. 1297) and parenchymal (hypolemmal, 8.52) nerve terminals (Ruskell 1975). The ultrastructure of most interstitial and all parenchymal terminals accords with cholinergic (parasympathetic) activity. A few interstitial terminals, containing small granular vesicles, may be adrenergic (sympathetic). Only 'dark' or 'serous' glandular cells have a parasympathetic, hypolemmal innervation (Ruskell 1975), as in rhesus monkeys (Ruskell 1969), in which

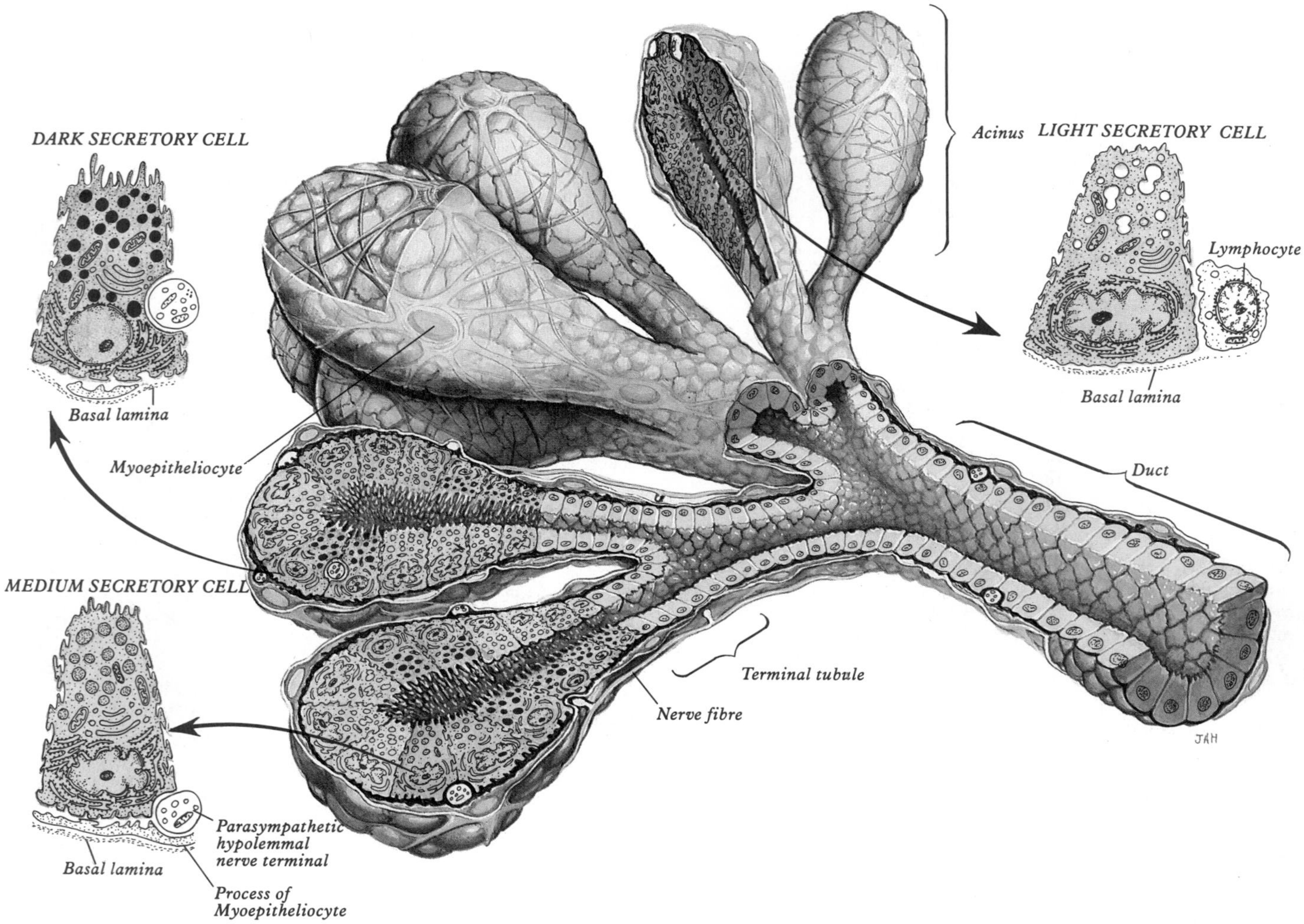

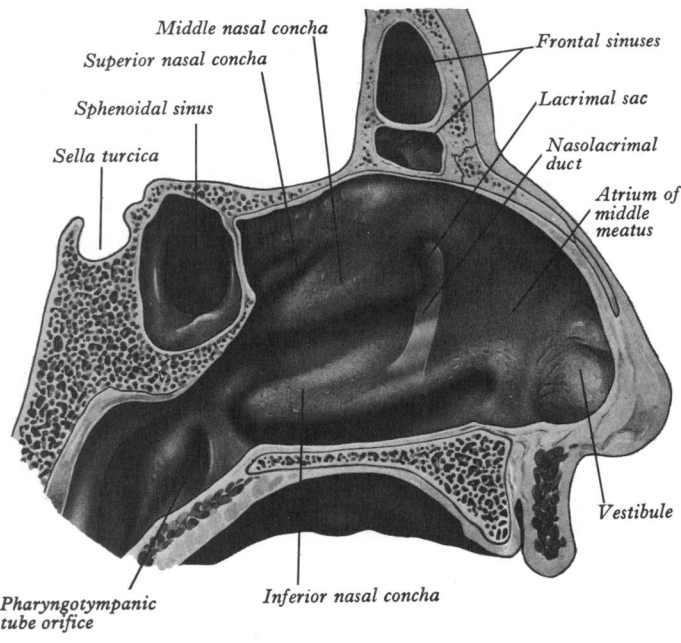

DARK SECRETORY CELL

Basal lamina

Myoepitheliocyte

MEDIUM SECRETORY CELL

Basal lamina

Parasympathetic hypolemmal nerve terminal

Process of Myoepitheliocyte

Acinus LIGHT SECRETORY CELL

Lymphocyte

Basal lamina

Duct

Terminal tubule

Nerve fibre

7.338   The structure of the lacrimal gland, showing the organization of the secretory units.

Middle nasal concha

Superior nasal concha

Sphenoidal sinus

Sella turcica

Frontal sinuses

Lacrimal sac

Nasolacrimal duct

Atrium of middle meatus

Vestibule

Pharyngotympanic tube orifice

Inferior nasal concha

7.339   The left lateral wall of the nasal cavity viewed from the medial side. The lacrimal sac and the nasolacrimal duct of the left side have been projected on to the lateral wall of the nasal cavity to show their positions relative to the middle nasal concha, the middle meatus and the inferior concha.

ultrastructural changes after parasympathectomy suggested that serous cells are under parasympathetic control, mediated by hypolemmal terminals, while mucous cells may function autonomously. Similar parasympathetic terminals occur near the ducts and terminal tubules in the human lacrimal gland; the tubules connect secretory units and ducts, containing in their walls cells intermediate in form between glandular and ductual cells (Ruskell 1975); these cells of the terminal tubules are smaller than glandular cells and contain relatively fewer granules. Most nerve terminals of ducts and tubules are near myoepitheliocytes, perhaps inducing myoepitheliocytic contraction and assisting secretory flow. Myoepitheliocytes in secretory units appear to have no direct, hypolemmal innervation; Ruskell (1975) suggests that nearby interstitial parasympathetic terminals may control them, releasing only sufficient transmitter to induce their contraction in hypersecretion. The role of the interstitial sympathetic terminals in the control of human lacrimal gland is undetermined, as are the ultrastructure and functions of its ducts.

The lacrimal canaliculi, one in each lid, are about 10 mm long; they commence at the *puncta lacrimalia* (7.332, 336, 337). The *superior canaliculus,* smaller and shorter than the inferior, first ascends and then curves acutely inferomedially to reach the lacrimal sac. The *inferior canaliculus* first descends and turns almost horizontally to the sac. At their angles they are dilated into *ampullae.* Their lining mucosa has a non-keratinized stratified squamous epithelium on a basement membrane, beyond which is a lamina propria rich in elastic fibres (the ducts therefore being easily dilated when probed) and a layer of skeletal muscle fibres continuous with the lacrimal part of the orbicularis oculi. At the base of each lacrimal papilla these fibres are sphincteric.

The lacrimal sac (7.334, 336, 337, 339), the closed upper end of the nasolacrimal duct, lies in a fossa formed by the lacrimal bone, the frontal maxillary process and the lacrimal fascia. It is about 12 mm in length, its closed upper end is laterally flattened and its lower part rounded and merging into the duct; the lacrimal canaliculi open into its lateral wall near its upper end.

A layer of *lacrimal fascia*, continuous with the orbital periosteum, passes between the lacrimal crest of the maxilla and the lacrimal bone, forming a roof and lateral wall to the lacrimal fossa; between the fascia and the lacrimal sac is a plexus of minute veins. The fascia separates the sac from the medial palpebral ligament in front and the lacrimal part of orbicularis oculi behind. The lower half of the lacrimal fossa is related medially to the anterior part of the nasal middle meatus; the upper half to the anterior ethmoidal sinuses. (In 100 skulls Whitnall, 1911, observed that in 14 the anterior ethmoidal sinuses were related only to the fossa's posterior wall; in 32 they reached the suture between the lacrimal bone and maxilla; in 54 one irregular sinus extended to the anterior lacrimal crest.)

The lacrimal sac has a fibro-elastic wall, lined internally by mucosa continuous through the lacrimal canaliculi with the contiva and through the nasolacrimal duct with the nasal mucosa.

The nasolacrimal duct (7.337, 339), about 18 mm long, descends from the lacrimal sac to open anteriorly in the inferior meatus at an expanded orifice; a mucosal *lacrimal fold* forms an imperfect valve just above this opening. The duct runs down an osseous canal formed by the maxilla, lacrimal bone and inferior nasal concha; it is narrowest in the middle and directed down, back and a little laterally. The mucosa of the lacrimal sac and the nasolacrimal duct has a bilaminar columnar epithelium, ciliated in places. A surrounding plexus of veins, forming erectile tissue, may, when engorged, obstruct the duct.

Lacrimal fluid enters the conjunctival sac at its superolateral angle and, by capillarity and blinking, is carried across the eye to the lacus lacrimalis, mainly between the lower palpebral margin and the eyeball. From the lacus it enters the lacrimal canaliculi. Contraction of the orbicularis oculi presses the puncta lacrimalia more firmly into the lacus and capillary attraction draws the secretion into the lacrimal sac. Sudden dilation of the sac, produced by the lacrimal part of the orbicularis oculi during blinking (p. 572), probably aids this. Normally the tarsal secretion prevents tear fluid from overflowing and also covers the capillary film of fluid on the cornea and sclera, perhaps delaying evaporation (Mishima & Maurice 1961, Wolff 1976).

THE AUDITORY AND VESTIBULAR APPARATUS

The peripheral auditory apparatus includes the various parts of the ear, but of particular functional importance is the *cochlear part* of its *membranous labyrinth*. Each ear is a *distance receptor* for the collection, conduction, modification, amplification and analysis of complex waves of sound reaching it. These are transduced into coded patterns of impulses in the afferent cochlear fibres of the vestibulocochlear nerve, for transmission and further analysis in the central auditory pathways of the brain (pp. 958, 960, 986, 1015, 1060).

Sound waves consist of the repetitive oscillations of the molecules constituting air; the waves vary in:

(1) the *direction* and *distance* or *location* of their source,

(2) their *intensity* or *energy content*, (3) the mixture of *frequencies* in the train of waves, and (4) the *phase* of their vibrations. The structural and functional design of ears make them, within certain ranges, extremely sensitive to differences in frequency, intensity and phase and, when used binaurally, they are very effective range and direction finders. They are also most responsive to the *rate of change* in all these parameters. Frequencies are expressed as *cycles per second* or *Hertz* (c/s or Hz), which are subjectively appreciated as *pitch*, young adult ears responding to frequencies of about 20–20 000 Hz, although higher and lower values are not uncommon in youth. *Intensity* is expressed as the *quantity of energy* transmitted per *unit time* through a *unit area* perpendicular to the direction of propagation. The subjective appreciation of sound intensity is related to the logarithm of the absolute intensity as just defined but is also dependent on frequency, the human ear being most sensitive to sounds in the range of 1500–3000 Hz; above and below this the threshold rises sharply; e.g. 10 000 times more energy is necessary for an equal perceptive effect at 15 000 Hz than at 2000 Hz. The ear's sensitivity is astounding; sounds may be discerned which involve pressure changes as small as 10^{10} atmospheres, a change equivalent to ascending or descending 1/30 000 of an inch! Because of the great variation in the intensity of sounds commonly experienced, the *decibel* has been introduced for convenience. It is defined as 10 times the logarithm of the ratio of the intensity of the sound, compared to an accepted reference level. In practice this reference level is defined as the Sound Pressure Level (SPL) and decibels are quoted as db SPL. A difference in intensity of about one decibel is usually just perceptible by the human auditory system.

The *quality* of a particular sound depends on the mixture of frequencies of which it is composed. Musical sounds usually consist of one or more fundamental frequencies each with its mathematically related series of *harmonics*. Mixtures of unrelated and *irregular* frequencies are regarded as 'noise'. Human speech has a marked content of noise. Moreover, consonants often involve higher frequencies than do vowels. Hence speech is easily obscured by concomitant noise and, with increasing age and consequent reduction of sensitivity at the high frequency range of the human ear, the elderly find speech less easy to interpret.

The elegant researches into the way in which the ear operates as a peripheral analyser of frequency and intensity and how intensity and phase differences impinging on the two ears are used in detecting the range and directions of the sound source, can merely be mentioned here. For details the reader must consult Rasmussen & Windle (1960), Whitfield (1967), De Reuck & Knight (1968), Wolstenholme & Knight (1970), etc.

The human ear can be divided, according to phylogenetic, developmental, structural and functional criteria, into *external*, *middle* and *internal* parts. Associated topographically with the cochlear, auditory part of the labyrinth, are the sensory receptors of the other major region of the inner ear, consisting of specialized sensory areas of the utricle, saccule and ampullae of the semicircular ducts which, with their contained and surrounding fluids, bony cavities and the vestibular part of the eighth cranial nerve, constitute **the peripheral vestibular apparatus**. The latter provides the central nervous system with information concerning the static position of the head in space and of its linear or angular acceleration or deceleration.

The evolution of the auditory apparatus in land vertebrates has attracted an enormous literature which cannot be even considered here; for a recent survey consult Lombard & Bolt 1977.

The External Ear

The external ear comprises the *auricle*, or *pinna*, and the *external acoustic meatus*. The auricle projects from the side of the head to collect sound waves; the meatus leads inwards from the auricle to conduct vibrations to the tympanic membrane; not that they act solely as an inert ear-trumpet, for they are the first of a series of *stimulus modifiers* in the auditory apparatus, as will become apparent.

THE AURICLE

The lateral surface of the auricle (7.340) is irregularly concave, turns slightly forwards and displays numerous eminences and depressions. Its prominent curved rim, or *helix*, usually bears

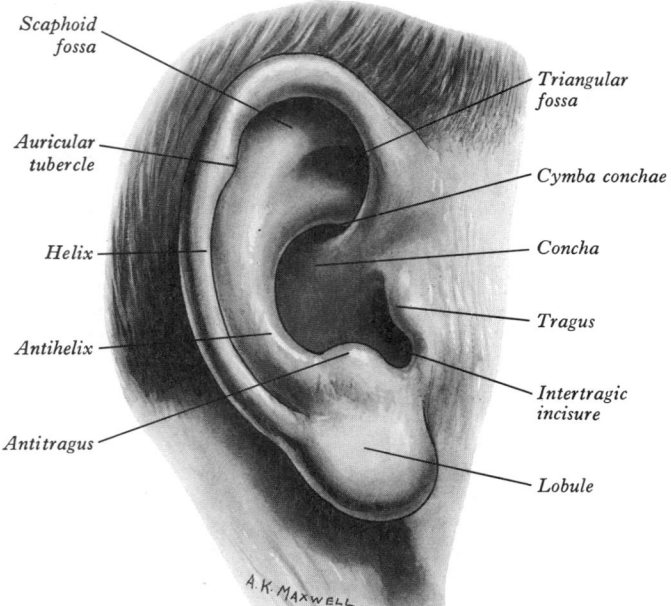

Scaphoid fossa

Auricular tubercle

Helix

Antihelix

Antitragus

Triangular fossa

Cymba conchae

Concha

Tragus

Intertragic incisure

Lobule

A.K. MAXWELL.

7.340 The right auricle: lateral aspect.

postero-superiorly a small *auricular tubercle* (of Darwin), a structure which is quite pronounced around the sixth month of intrauterine life, when the auricle closely resembles that of some adult monkeys. Another curved prominence, parallel and anterior to the posterior part of the helix, is the *antihelix*, dividing above into two *crura* flanking a depressed *triangular fossa*. The curved depression between the helix and antihelix is the *scaphoid fossa*. The antihelix encircles the deep, capacious *concha* of the auricle, which is incompletely divided by the *crus* or anterior end of the helix; the conchal area above this, the *cymba conchae*, overlies the supermeatal triangle of the temporal bone, which can be felt through it (pp. 352, 355), deep to which is the mastoid antrum. Below the crus of the helix and in front of the concha is a small curved flap, the *tragus*, which projects posteriorly, partly overlapping the meatal orifice. Opposite the tragus and separated from it by the *intertragic incisure*, is a small tubercle, the *antitragus*. Below this is the *lobule*; composed of fibrous and adipose tissues, this is soft, unlike the majority which, being supported by elastic cartilage, is firm. The cranial surface of the auricle presents elevations corresponding to the depressions on its lateral surface, after which they are named, e.g. the eminentia conchae, eminentia fossae triangularis, etc.

Though primarily collecting 'trumpets' for sound waves, the auricles, by their asymmetry and variations in thickness, probably introduce *variable delay paths* in sound transmission, which may be important in the efficient binaural (and also the cruder monaural) localization of sources of sound.

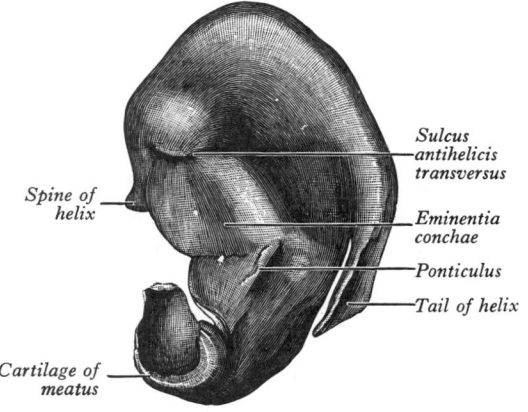

Spine of helix

Cartilage of meatus

Sulcus antihelicis transversus

Eminentia conchae

Ponticulus

Tail of helix

1220   7.341   The medial surface of the right auricular cartilage.

In its **structure** the auricle is a thin plate of *elastic fibrocartilage* covered by skin and connected to the surrounding parts by ligaments and muscles; it is continuous with the cartilage of the external acoustic meatus, which is joined to the margins of the bony meatus by fibrous tissue. Its *skin* is thin, adherent to the cartilage and bears fine hairs furnished with sebaceous glands which are most numerous in the concha and scaphoid fossa. On the tragus, antitragus and intertragic incisure the hairs are strong and numerous, especially in elderly males. The skin is continuous with that of the external acoustic meatus.

The auricular cartilage (7.341) is a single piece of elastic fibrocartilage, its surface moulded by eminences and depressions as described above; it is absent from the lobule and between the tragus and the crus of helix, the gap being filled by dense fibrous tissue. Anteriorly, where the helix curves upwards, is a small cartilaginous projection, the *spine of the helix*, its other extremity being prolonged inferiorly as the *tail of the helix* and separated from the antihelix by the *fissura antitragohelicina*. The cartilage's cranial aspect bears the *eminentia conchae* and *eminentia triangularis*, corresponding to the depressions on the lateral surface. A transverse furrow, the *sulcus antihelicis transversus*, corresponding to the inferior crus of the antihelix on the lateral surface, separates the two eminences. The eminentia conchae is crossed by an oblique ridge, the *ponticulus*, for the attachment of the auricularis posterior muscle. There are two fissures in the auricular cartilage, one behind the crus helicis and another in the tragus.

The ligaments of the auricle form two sets: (1) the extrinsic, connecting it to temporal bone; (2) the intrinsic, interconnecting various parts of its cartilage. There are two *extrinsic ligaments*, anterior and posterior, the *anterior* extending from the tragus and the spine of the helix to the root of the zygomatic process of the temporal bone; and the *posterior ligament* passes from the posterior surface of the concha to the lateral surface of the mastoid process. The chief *intrinsic ligaments are: (a)* a strong fibrous band from the tragus to the helix, completing the meatus anteriorly and forming part of the concha's boundary; (*b*) a band between the antihelix and the tail of the helix. Less prominent bands also exist on the cranial aspect of the auricle.

THE AURICULAR MUSCLES

These are two sets: the *extrinsic*, connecting the auricle to the skull and scalp and moving the auricle as a whole; and the *intrinsic*, connecting the different parts of the auricle.

The extrinsic muscles are the auriculares anterior, superior and posterior. The *auricularis anterior*, the smallest of the three, is a thin fan of pale fibres, arising from the lateral edge of the epicranial aponeurosis, its fibres converging to insert into the spine of the helix. The *auricularis superior*, the largest of the three, is also thin and fan-shaped and converges from the epicranial aponeurosis via a thin, flat tendon to attach to the upper part of the cranial surface of the auricle. The *auricularis posterior* consists of two or three fleshy fasciculi arising by short aponeurotic fibres from the mastoid part of the temporal bone and inserted into the ponticulus on the eminentia conchae.

Nerve supply. Auriculares anterior and superior are supplied by temporal branches of the facial nerve and auricularis posterior by the posterior auricular branch of the same nerve.

Actions. In man these muscles have very little obvious effect: the auricularis anterior draws the auricle forwards and up; the auricularis superior elevates it slightly; and the auricularis posterior draws it back. Despite the paucity of auricular movement, auditory stimuli may evoke patterned responses from these small muscles and electromyography can detect the 'crossed acoustic response' elicited by this means in investigative clinical neurology.

The intrinsic muscles are: helicis major and minor, tragicus, antitragicus, transversus auriculae and obliquus auriculae. They modify auricular shape minimally, if at all, in most human ears. Occasional individuals can, however, modify the shape and position of their external ears.

Helicis major is a narrow vertical band on the anterior margin of the helix, passing from its spine to its anterior border, where the helix is about to curve back. *Helicis minor* is an oblique fasciculus,

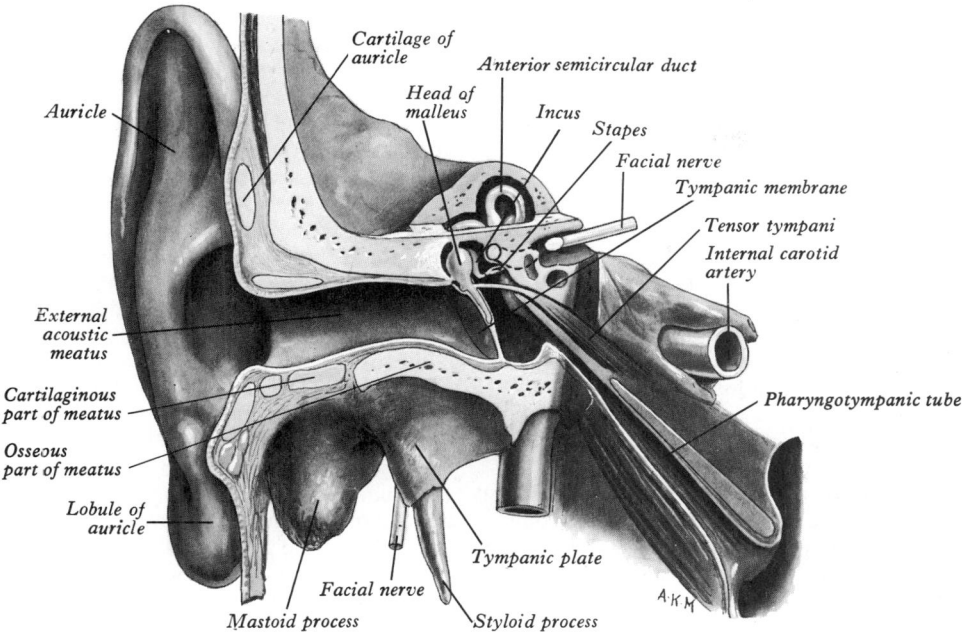

7.342 The external and middle regions of the right ear: anterior aspect.

covering the crus helicis. *Tragicus* is a short, flattened, vertical band on the lateral aspect of the tragus. The *antitragicus* (muscle) passes from the outer part of the antitragus (prominence) to the tail of the helix and the antihelix. The *transversus auriculae*, on the cranial aspect of the auricle, consists of scattered fibres, partly tendinous, partly muscular, extending between the eminentia conchae and the eminentia scaphae. The *obliquus auriculae*, also on the cranial aspect of the auricle, comprises a few fibres extending from the upper and posterior parts of the eminentia conchae to the eminentia triangularis.

The nerve supply to the intrinsic muscles on the lateral aspect is from the temporal branches of the facial and to those on the cranial aspect from the facial's posterior auricular branch.

The arteries of the auricle are: (1) the posterior auricular branch of the external carotid, supplying three or four rami to its cranial surface; twigs from these reach the lateral surface, some through fissures in the cartilage, others round the margin of the helix; (2) the anterior auricular branches of the superficial temporal artery, distributed to the lateral surface; (3) a branch from the occipital artery. **Veins** correspond to the arteries of the auricle. Arteriovenous anastomoses are numerous in the skin of the auricle.

Lymphatics of the auricle drain into: (1) the parotid lymph nodes, especially the node in front of the tragus, (2) the upper deep cervical lymph nodes, (3) the mastoid lymph nodes.

The sensory nerves of the auricle are: (1) the great auricular nerve, supplying most of the cranial surface and the posterior part of the lateral surface (helix, antihelix, lobule); (2) the lesser occipital nerve, supplying the upper part of the cranial surface; (3) the auricular branch of the vagus, supplying the concavity of the concha and posterior part of the eminentia; (4) the auriculotemporal nerve, supplying the tragus, crus of the helix and the adjacent part of the helix; (5) the facial nerve, which with the vagal auricular branch probably supplies small areas on both aspects of the auricle in the depression of the concha and over its eminence. The details of facial cutaneous innervation and whether the facial fibres reach the external acoustic meatus and tympanic membrane require further clarification.

THE EXTERNAL ACOUSTIC MEATUS

This extends from the concha to the tympanic membrane (7.342, 343). Its length, from the floor of the concha, is approximately 2·5 cm and from the tragus about 4 cm. It has two structurally different parts, the lateral third being *cartilaginous* and the medial two-thirds *osseous*. It forms an S-shaped curve, directed at first medially, anteriorly and slightly up (*pars externa*), then posteromedially and up (*pars media*) and lastly again anteromedially and slightly down (*pars interna*). It is oval in section, its greatest diameter being obliquely inclined postero-inferiorly at the external orifice but nearly horizontal at its medial end. There are two constrictions, one near the medial end of the cartilaginous part, the other, the *isthmus*, in the osseous part about 2 cm from the bottom of the concha. The tympanic membrane, which closes

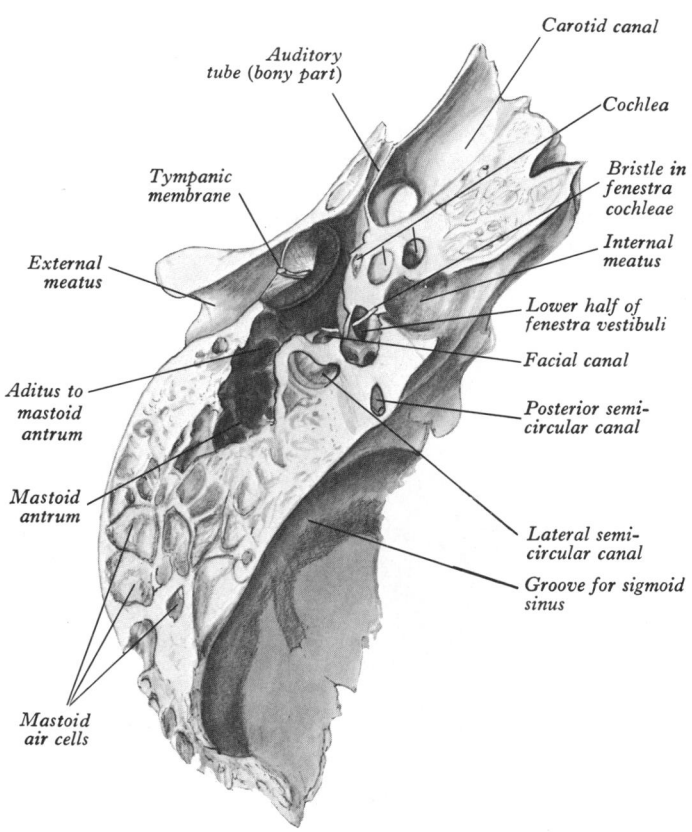

7.343 Oblique section through the left temporal bone viewed from above. From a section prepared by P F Milling. Compare with 7.345 and 7.348.

its medial end, is obliquely set and so the floor and the anterior wall of the meatus are longer than its roof and posterior wall.

The lateral, *cartilaginous part* is about 8 mm long: it is continuous with the auricular cartilage and attached by fibrous tissue to the circumference of the osseous part. This meatal cartilage is deficient postero-superiorly, the gap being occupied by a sheet of collagen; two or three deep fissures exist in its anterior part.

The *osseus part* is about 16 mm long, and is narrower than the cartilaginous part. It is directed anteromedially and slightly downwards, with a slight posterosuperior convexity. Its medial end is smaller than the lateral end and is obliquely placed, the anterior wall projecting medially about 4 mm beyond the posterior and marked, except above, by a narrow *tympanic sulcus*, to which the perimeter of the tympanic membrane is attached. Its lateral end is dilated and mostly rough for the attachment of the meatal cartilage.

The anterior, inferior and most of the posterior parts of the osseous meatus are formed by the tympanic element of the temporal bone, which in the fetus is only a *tympanic ring* (p. 380); the posterosuperior region is formed by temporal squamous bone.

The auricular skin continues into the external acoustic meatus and covers the tympanic membrane's external surface. It is thin, with no dermal papillae, and is closely adherent to the cartilaginous and osseous parts of the tube; inflammation is very painful owing to increased tension in these tissues. In the thick subcutaneous tissue of the cartilaginous part of the meatus numerous *ceruminous glands* secrete ear wax or *cerumen*, their coiled tubular structure resembling that of sweat glands (p. 93). When active the secretory cells are columnar but cuboidal when quiescent; they are covered externally by myoepithelial cells. Ducts open either on to the epithelial surface or into a nearby sebaceous gland of a hair follicle. Cerumen prevents the maceration of meatal skin by trapped water and may discourage invasion by insects. Over-production or accumulation of wax may completely block the meatus or obstruct the vibration of the tympanic membrane. Ceruminous glands and hair follicles are largely limited to the cartilaginous meatus, but a few small glands and fine hairs occur in the roof of the lateral part of the osseous meatus. In addition to the protection given by the cerumen and meatal hairs, the warm humid environment of the relatively enclosed meatal air aids the mechanical responses of the tympanic membrane.

Relations of the Meatus

The mandible's condyloid process is anterior to the meatus and partially separated from its cartilaginous part by a small part of the parotid gland. A blow on the chin may cause the condyle to break into the meatus. Mandibular movements affect the size of the cartilaginous lumen. The middle cranial fossa lies above the

osseous meatus and the mastoid air cells are posterior to it, separated from the meatus by a thin layer of bone. Its deepest part is below the epitympanic recess (p. 1223) and antero-inferior to the mastoid antrum, the lamina of bone separating it from the antrum being only 1–2 mm thick and providing the 'transmeatal approach' of aural surgery.

The meatal arteries are: the posterior auricular branch of the external carotid, the deep auricular branch of the maxillary and the auricular branches of the superficial temporal. The *veins* drain into: the external jugular and maxillary veins and the pterygoid plexus. The *lymphatics* drain with those of the auricle (p. 844). The *meatal nerves* are derived from the auriculotemporal branch of the mandibular, which supplies the anterior and superior walls of the meatus and the auricular branch of the vagus, supplying the posterior and inferior walls.

Clinical Examination of the Meatus

To inspect the meatus and tympanic membrane satisfactorily with light reflected down a speculum, the former's sinuosity can be to some extent straightened by pulling the auricle upwards, back and a little laterally. In this way, most of the meatus and tympanic membrane can be viewed.

At the junction of the osseous and cartilaginous parts the forward bend in the meatus constricts it, a point of importance in attempts to remove a foreign body lodged in the meatus. The shortness of the meatus in children also exposes the tympanic membrane to the danger of damage when an aural speculum is used. Even in adults a speculum should not be inserted beyond the constricted junction of the cartilaginous and osseous parts. Immediately anterolateral to the tympanic membrane is a depression on the meatal floor, bounded laterally by a prominent ridge, where foreign bodies may be impacted. Most of the tympanic membrane is visible through a speculum (7.344). It is pearly-grey, slightly glistening and placed at an acute angle of about 55° with the floor of the meatus and an obtuse angle with the roof. At birth it lies almost in the plane of the cranial base.

A reddish-yellow streak midway between the anterior and posterior margins extends from the centre obliquely upwards and forwards, marking the handle of the malleus, which is attached internally. At the upper part of this streak, near the roof of the meatus, a small, white round prominence marks the position of the lateral process of the malleus, projecting against the membrane. The tympanic membrane is not planar but is drawn in at its centre, or *umbo* (p. 1224), by the handle of the malleus. A bright reflected 'cone of light' appears in the antero-inferior quadrant. Anterior and posterior to the short process of the malleus, the variably prominent *anterior* and *posterior malleolar folds* appear, with the flaccid part of the membrane (p. 1224) between them. Posterior and parallel to the upper part of the handle of the malleus, the long process of the incus is often visible as a whitish streak, sometimes ending below near a round spot, the head of the stapes.

Applied Anatomy. Imperfect development of the external parts, a supernumerary auricle, preauricular cysts, fistulae and sinuses, or the absence of the meatus occasionally occur. In a child up to an age of four or five years a gap exists in the antero-inferior osseous wall of the meatus, the *foramen of Huschke* (pp. 381, 395) which is filled by a membrane; it may persist in the adult. The sources of the meatal innervation, which include the vagus, account for the occurrence of reflex coughing and sneezing when there is irritation of the meatus and the vomiting which may follow syringing the ears in children and the occasional heart failure induced by this process in the elderly. The association of earache with toothache or with lingual carcinoma is due to the involvement of the mandibular branch of the trigeminal, which supplies the teeth and tongue as well as the meatus. The upper half of the tympanic membrane is more vascular: for this reason and to avoid the chorda tympani and ossicles, incisions of this membrane should be postero-inferior.

The Tympanic Cavity (Tympanum)

The tympanic cavity or *middle ear* (7.343, 345, 347, 348) is an irregular, laterally compressed space in the temporal bone, lined

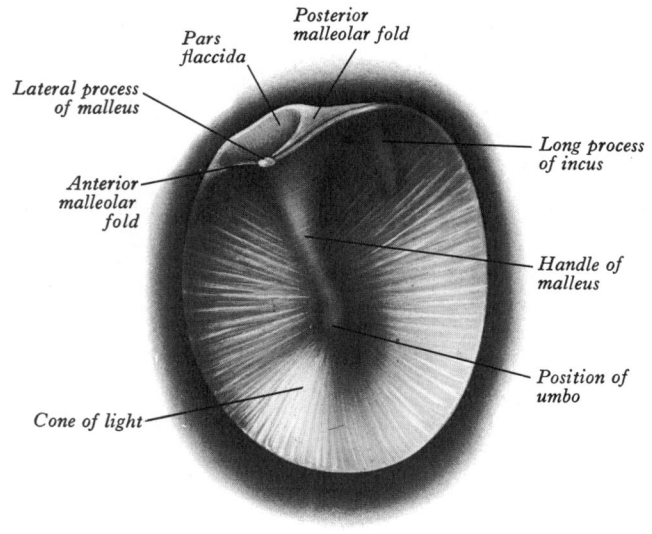

7.344 The left tympanic membrane: external aspect as seen through a speculum.

Posterior malleolar fold

Pars flaccida

Lateral process of malleus

Anterior malleolar fold

Cone of light

Long process of incus

Handle of malleus

Position of umbo

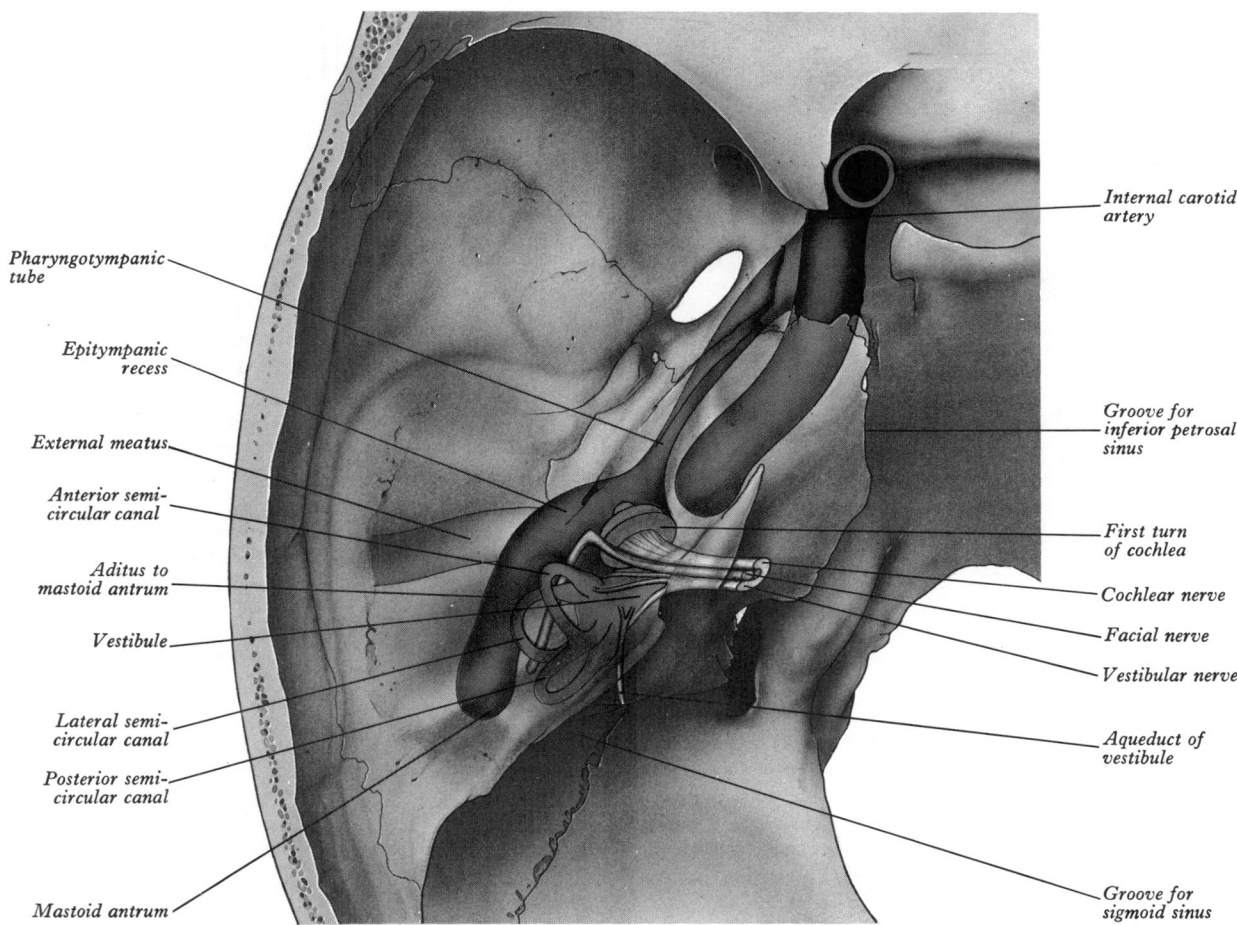

Pharyngotympanic tube

Epitympanic recess

External meatus

Anterior semi-circular canal

Aditus to mastoid antrum

Vestibule

Lateral semi-circular canal

Posterior semi-circular canal

Mastoid antrum

Internal carotid artery

Groove for inferior petrosal sinus

First turn of cochlea

Cochlear nerve

Facial nerve

Vestibular nerve

Aqueduct of vestibule

Groove for sigmoid sinus

7.345 Scheme showing the parts of the left auditory apparatus as if viewed through a semi-transparent temporal bone. Compare with 7.343 and 7.348.

by mucoperiosteum (p. 1226) and containing air derived from the nasopharynx via the auditory tube. It contains a train of movable ossicles, which connect the lateral to the medial wall to transmit vibrations of the tympanic membrane across the cavity of the middle ear to the cochlea; its essential function is to transfer energy efficiently from the relatively weak vibrations in the elastic, compressible *air* in the external acoustic meatus to overcome the inertia of the incompressible *fluids* around the delicate receptors in the cochlea. Mechanical coupling between the two systems must match their resistance to deformation or 'flow', i.e. their *impedance*, as closely as possible. Thus aerial waves of low amplitude and *low force* per unit area arrive at the tympanic membrane, which has 15–20 times the area of the *stapedial footplate* in contact with perilymph. In this manner, the force per unit area generated by the footplate is increased by a similar amount, while the amplitude of vibration is almost unchanged.

Protective mechanisms are also incorporated into the tympanic cavity's design, including the auditory (pharyngotympanic) tube to equalize pressure on both sides of the delicate tympanic membrane, the protective shape of articulations between the ossicles and the reflex contractions of the stapedius and tensor tympani muscles preventing damage due to sudden or excessive excursions of the ossicles.

The cavity has two parts: the *tympanic cavity proper*, opposite the membrane, and an *epitympanic recess*, above its level. The latter contains the upper half of the malleus and most of the incus. Including the recess, the vertical and anteroposterior diameters of the cavity are each about 15 mm; the transverse is about 6 mm above and 4 mm below but opposite the umbo it is only 2 mm. The cavity is bounded *laterally* by the tympanic membrane and *medially* by the lateral wall of the internal ear. It communicates posteriorly with the mastoid antrum and via this with the mastoid air cells and anteriorly with the pharyngotympanic tube (7.345).

BOUNDARIES OF THE TYMPANIC CAVITY

The roof of the tympanic cavity (7.346), is a thin plate of compact bone, the *tegmen tympani*. This separates the cranial and tympanic cavities, forming much of the anterior surface of the petrous temporal bone; it is prolonged posteriorly to roof the mastoid antrum and anteriorly to cover the canal for the tensor tympani. In youth, the unossified petrosquamosal suture (p. 379) may allow the spread of infection from the tympanic cavity to the meninges. In adults, veins from the tympanic cavity traverse this suture to the superior petrosal or petrosquamous sinus (p. 803) and may also transmit infection to these structures.

The floor is a narrow, thin, convex plate of bone separating the cavity from the superior bulb of the internal jugular vein (7.348); bone may be patchily deficient here and the tympanic cavity and vein are then separated only by mucous membrane and fibrous tissue. Near the medial wall is a small aperture for the tympanic branch of the glossopharyngeal nerve. The floor is sometimes thick and may contain some accessory mastoid air cells.

The lateral wall (7.346, 347) consists mainly of the tympanic membrane but is partly also the ring of bone to which the membrane is attached. The ring is deficient or notched above and near this region are the openings of the anterior and posterior canaliculi for the chorda tympani and also the petrotympanic fissure.

The *posterior canaliculus for the chorda tympani nerve* is situated in the angle between the posterior and lateral walls of the tympanic cavity just behind the tympanic membrane and level with the upper end of the handle of the malleus; it leads into a minute canal which descends in front of the facial canal and ends in it about 6 mm above the stylomastoid foramen. Through it the chorda tympani nerve and a branch of the stylomastoid artery enter the tympanic cavity.

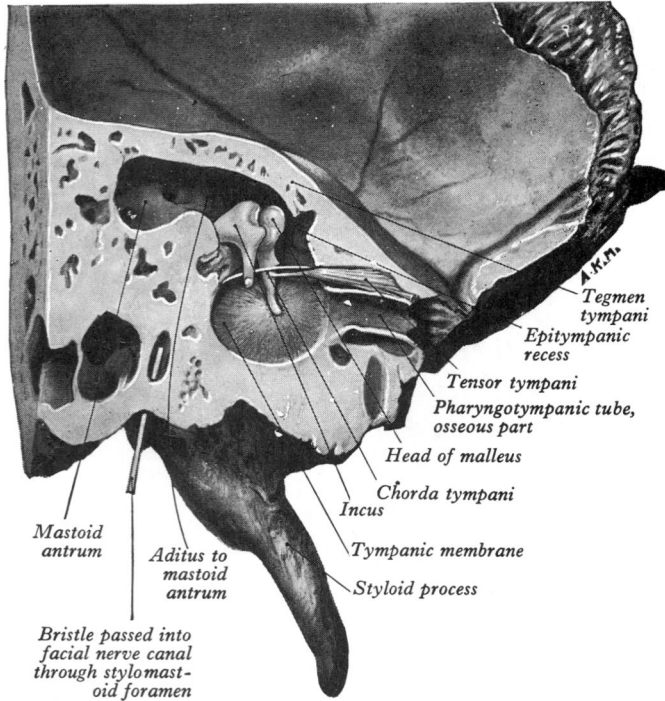

7.346 Oblique vertical section through the left temporal bone, to show the lateral wall of the middle ear and the mastoid antrum.

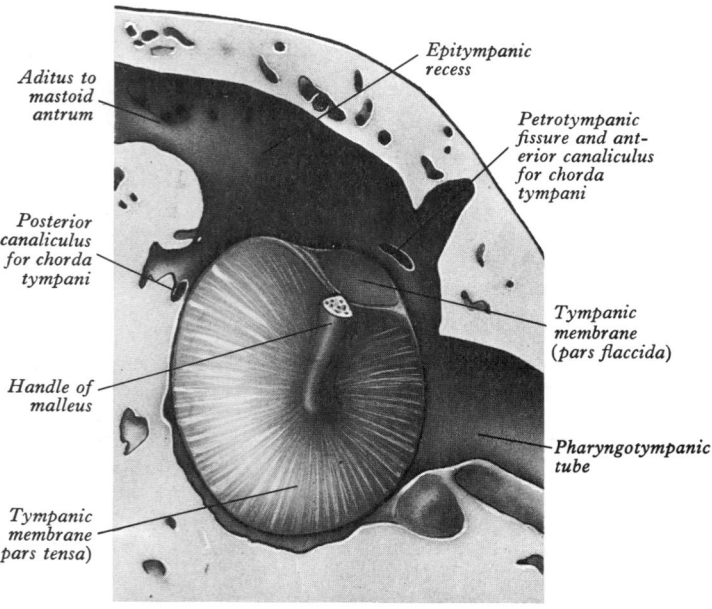

7.347 The lateral wall of the left tympanic cavity.

The *petrotympanic fissure* opens just above and in front of the ring of bone to which the tympanic membrane is attached; it is a mere slit about 2 mm in length, containing the anterior process and anterior ligament of the malleus and it transmits to the tympanic cavity the maxillary artery's anterior tympanic branch.

The *anterior canaliculus for the chorda tympani* opens at the medial end of the petrotympanic fissure; through it the chorda tympani leaves the tympanic cavity.

The tympanic membrane (7.346, 347) separates the tympanic cavity from the external meatus. It is thin and semi-transparent, almost oval, though somewhat broader above than below, and is placed at an angle of about 55° with the meatal floor. Its longest, antero-inferior diameter is from 9–10 mm and its shortest is from 8–9 mm. Most of its circumference is a thickened *fibrocartilaginous ring* attached to the *tympanic sulcus* at the medial end of the meatus; this sulcus is deficient superior-

ly and here, as noted above, the *anterior* and *posterior malleolar folds* pass to the lateral process of the malleus, leaving between them the triangular *pars flaccida*, a thin, lax part of the membrane; a small perforation may occur in this sometimes. The membrane is elsewhere taut, the *pars tensa*. The handle of the malleus is firmly attached to the membrane's internal surface as far as its centre, the *umbo*, which projects towards the tympanic cavity. Though this membrane as a whole is convex medially, its radiating fibres (vide infra) are curved with their concavities directed upwards.

Histologically, the tympanic membrane has three strata: an outer cuticular, an intermediate fibrous and an inner mucous. The *cuticular stratum* is continuous with the thin skin of the meatus and is keratinized, stratified squamous in type, devoid of dermal papillae and hairless. Its subepithelial tissue is vascularized and may develop a few peripheral papillae. The *fibrous stratum* has an external layer of radiating fibres diverging from the handle of the malleus and a deep layer of circular fibres, peripherally plentiful but sparse and scattered centrally in the membrane. Marginally and centrally a fine network of elastic fibres is said to be mixed with the collagen (but vide infra). The *mucous stratum* is a part of the mucosa of the tympanic cavity; it is thickest near the tympanic membrane's upper part and is covered, it has been claimed, by a layer of ciliated columnar epithelial cells. However, cilia occur only in patches or are entirely absent and then replaced by a low cuboidal or simple squamous epithelium.

These appearances, based on light microscopy, have been much amplified by electron microscopy and other techniques: first in guinea-pigs (Johnson et al 1968), in which the *external* epithelium is approximately 10 cells thick and has two zones, a superficial of non-nucleated squames and a deep zone like the epidermal stratum spinosum, with numerous desmosomes between cells, the deepest of which lie on a continuous basal lamina but lack epithelial pegs and hemidesmosomes. The *internal* epithelium is a single layer of very flat cells, with overlapping or interdigitating boundaries carrying desmosomes and tight junctions between cells. Their cytoplasm contains only a few organelles, the luminal surfaces of these apparently metabolically inert cells have a few irregular microvilli and are covered by an amorphous electron-dense material. Ciliated columnar cells are absent. Most interestingly the intermediate stratum contains filaments about 10 µm in diameter, with links between filaments at 25 nm intervals. The filaments are disposed in outer radial and inner non-radial zones, the former more numerous; neither resemble collagen or elastin. In amino acid composition, also, the filaments are distinctive and may consist of a protein peculiar to the tympanic membrane. McMinn & Taylor 1978, reporting on the development of the fibrillar component in guinea-pig and human embryos and fetuses, have confirmed the findings in guinea-pigs mentioned above, but in human fetuses small groups of collagen fibrils were apparent at eleven weeks in utero, with small bundles of elastin microfibrils. Older specimens showed more typically cross-banded collagen fibrils and the development of an amorphous elastin component. Large fibroblasts occur between the external radial fibres and basal lamina of the *external* epithelium, while blood capillaries and their basement membranes lie just deep to the basal lamina of the *internal* epithelium. In the pars flaccida the fibrous stratum is replaced by loose connective tissue.

The *arteries* of the tympanic membrane are from: the maxillary artery's deep auricular branch (to the outer, cuticular stratum) and the stylomastoid branch of the occipital or the posterior auricular artery and the tympanic branch of the maxillary to the internal mucosa. The *superficial veins* drain to the external jugular; those in the *deep surface* drain partly to the transverse sinus and dural veins and partly to the venous plexus of the pharyngotympanic tube. The *nerve supply* is from: the auriculotemporal branch of the mandibular nerve, the auricular branch of the vagus, the tympanic branch of the glossopharyngeal and possibly from the facial nerve.

The medial wall of the tympanic cavity (7.348, 349) is also the lateral boundary of the internal ear. Its features are the promontory, fenestra vestibuli, fenestra cochleae and the facial prominence.

The rounded *promontory*, minutely grooved by the nerves of the tympanic plexus, overlies the lateral projection of the basal turn of the cochlea; a minute spicule of bone frequently connects the promontory to the pyramidal eminence of the posterior wall. Anterior to the promontory the apex of the cochlea is near the medial wall of the tympanum (7.342, 343). Behind the promontory the *sinus tympani* indicates the position of the ampulla of the posterior semicircular canal.

The *fenestra vestibuli* (*f. ovalis*) is a reniform opening posterosuperior to the promontory, connecting the tympanic cavity to the vestibule; its long diameter is horizontal and its convex border directed superiorly. It is occupied by the base of the stapes, the circumference of which is attached to the fenestral margin by an annular ligament.

The *fenestra cochleae* (*f. rotunda*) is postero-inferior to the fenestra vestibuli and separated from it by the posterior part of the promontory. It lies completely under the overhanging edge of the promontory in a deep hollow or niche. It is placed very obliquely; in dried specimens it opens anterosuperiorly from the tympani cavity into the scala tympani of the cochlea. It is closed by the *secondary tympanic membrane*, which is somewhat concave towards the tympanic cavity and convex towards the cochlea, the membrane being bent so that its posterosuperior one-third is at an angle to its antero-inferior two-thirds. The membrane has three layers: an external derived from the tympanic mucosa, an internal from the cochlear lining membrane and a middle, fibrous layer.

The *prominence of the facial nerve canal* indicates the position of the upper part of a bony canal for the facial nerve. The canal, its lateral wall sometimes being partly deficient, traverses the medial tympanic wall from before backwards, just above the fenestra vestibuli, then curves down into the posterior wall of the cavity.

The posterior wall (7.348) is wider above than below, its main features being the aditus to the mastoid antrum, the pyramid and the fossa incudis.

The *aditus to the mastoid antrum*, a large irregular aperture, leads back from the epitympanic recess into the upper part of the *mastoid antrum*, above and behind the prominence of the facial nerve canal, due to the underlying lateral semicircular canal.

The *pyramidal eminence* is just behind the fenestra vestibuli and anterior to the vertical part of the facial nerve canal; it contains the stapedius muscle, its summit projecting towards the fenestra vestibuli; a small apical aperture transmits the muscle's tendon. Its cavity is prolonged down and back in front of the facial nerve canal and communicates with the latter by an aperture through which a small branch of the facial nerve passes to the stapedius (p. 1110).

The *fossa incudis*, a small depression low and posterior in the epitympanic recess, contains the short process of the incus, fixed to the fossa by ligamentous fibres.

The mastoid antrum (7.343, 345–349) is an air sinus in the petrous temporal bone and has relations of considerable surgical importance, being often infected. In its upper *anterior wall* is an opening, the aditus to the mastoid antrum, leading back from the epitympanic recess, with the lateral semicircular canal medial to it. The antrum's *medial wall* is related to the posterior semicircular canal (7.345). *Posterior* is the sigmoid sinus; some mastoid air cells may intervene between this and the antrum. The *roof*, formed by the tegmen tympani, lies below the middle cranial fossa and temporal lobe of the brain. The floor has several openings communicating with the mastoid air cells. *Antero-inferior* is the descending part of the facial nerve canal. The *lateral wall* of the antrum, the usual surgical approach to the cavity, is formed by the postmeatal process of the squamous temporal bone. This is only 2 mm thick at birth but increases at a rate of approximately 1 mm a year to attain a final thickness of 12–15 mm. In adults the lateral wall of the antrum corresponds to the *suprameatal triangle* on the outer surface of the skull (pp. 352, 355), palpable through the cymba conchae (p. 1220). The triangle's superior side, the supramastoid crest, is level with the floor of the middle cranial fossa; the antero-inferior side, forming the posterosuperior margin of the external acoustic meatus, indicates approximately the position of the descending part of the facial nerve canal; and the posterior side, formed by a posterior vertical tangent to the post-

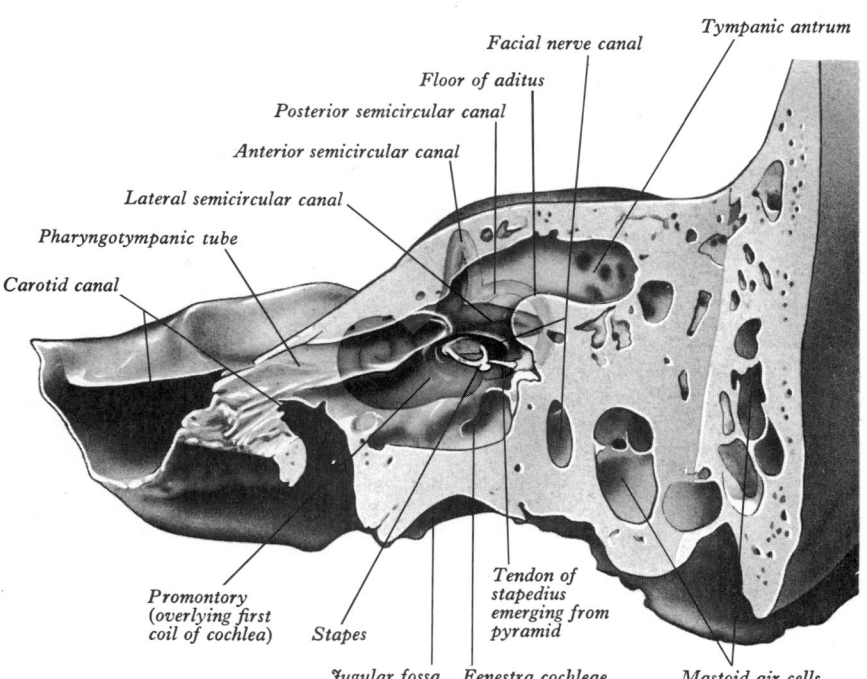

7.348 Oblique section through the left temporal bone, to show the medial wall of the middle ear. The cochlea and the semicircular canals are in blue. Note the relationship of the first coil of the cochlea to the promontory and the closeness of the facial nerve canal and the lateral semicircular canal to the medial wall of the aditus.

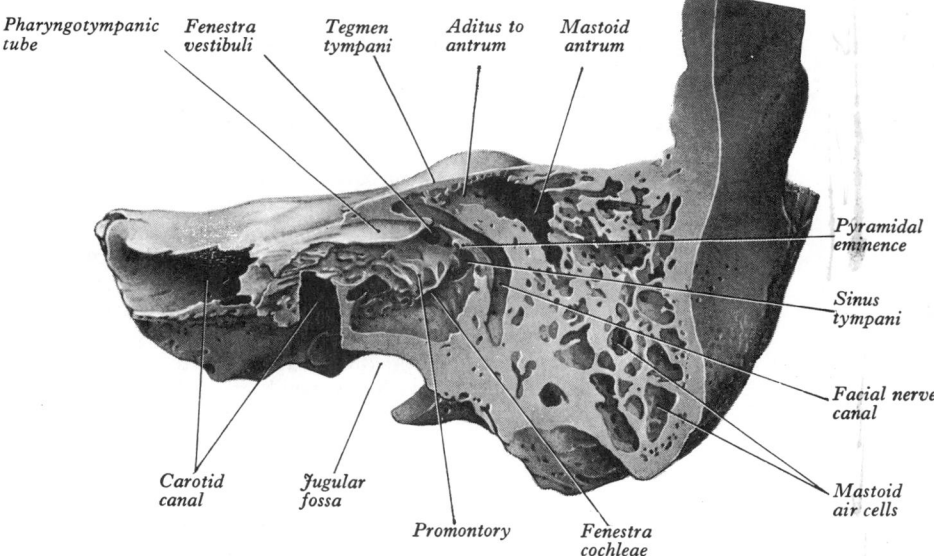

7.349 Oblique section through the left temporal bone, showing the medial wall of the middle ear. Compare with 7.348.

erior margin of the external acoustic meatus, is just anterior to the sigmoid sinus. The adult capacity of the antrum is about 1 ml and its general diameter about 10 mm. Unlike other air sinuses, it is present at birth and is indeed then almost adult in size, although it is at a higher level relative to the external acoustic meatus than in adults. In the very young, owing to the thinness of the lateral antral wall and the absence of the mastoid process, the stylomastoid foramen and emerging facial nerve are very superficially situated.

The mastoid air cells (7.343, 348, 349) vary considerably in number, form and size. Usually they interconnect and are lined by a mucosa with squamous non-ciliated epithelium, continuous with that in the mastoid antrum and tympanic cavity. They may fill the mastoid process, even to its tip, and some may be separated from the sigmoid sinus and posterior cranial fossa only by extremely thin bone, which is occasionally deficient. Some may lie superficial to, or even behind the sigmoid sinus and others may occur in the posterior wall of the descending part of the facial nerve canal. Those in the squamous temporal bone may be separated from deeper cells in the petrous part by a plate of bone in the line of the squamomastoid suture. Sometimes they extend very little into the mastoid, this process consisting largely of dense bone or trabecular bone containing bone marrow. Varieties of the mastoid process occur (3.135) and three types are commonly described: *pneumatic*, with many air cells, *sclerotic*, with few or none, and *mixed*, containing both air cells and bone marrow. Pannoer (1970) proposed a more elaborate scheme, dependent on the degree of pneumatization and the relation of air cells to the lateral sinus; in a series of 100 mastoid processes, the cells most distant from the antrum were largest. Air cells may extend beyond the mastoid process into the squamous temporal bone above the supramastoid crest, into the posterior root of the zygomatic process of the temporal, into the osseous roof of the external acoustic meatus just below the middle cranial fossa and into the floor of the tympanic cavity very close to the superior jugular bulb. Rarely, a few may excavate the jugular process of the occipital bone. An important group may extend medially into the petrous temporal bone, even to its apex, related to the pharyngotympanic tube, carotid canal, labyrinth and abducent nerve; some investigators maintain that these are not continuous with the mastoid cells but grow independently from the tympanic cavity. The extensions of the mastoid air cells described above are pathologically important since infection may spread to the structures around them. Though the antrum is well developed at birth, the mastoid air cells are merely minute antral diverticula at this stage. As the mastoid develops in the second year, the cells gradually extend into it and by the fourth year they are well formed, although their greatest growth occurs at puberty. In about 20% of skulls the mastoid process has no air cells at all.

The anterior tympanic wall is narrowed by the approximation of the medial and lateral walls of the cavity. Its inferior, larger area is a thin lamina forming the posterior wall of the carotid canal, perforated by the superior and inferior caroticotympanic nerves and the tympanic branch or branches of the internal carotid. Opening above on the anterior wall are two parallel canals; superior is the *canal for the tensor tympani* and inferiorly the *osseous part of the pharyngotympanic tube*. These canals incline downwards and anteromedially to open in the angle between the squamous and petrous parts of the temporal bone; they are separated by a thin, osseous septum. The canal for the tensor tympani and the bony septum runs posterolaterally on the medial tympanic wall, ending immediately above the fenestra vestibuli, where the posterior end of the septum is curved laterally to form a pulley, the *processus trochleariformis (cochleariformis)*, over which the tendon of the tensor tympani is turned laterally to its attachment on the upper part of the handle of the malleus.

THE PHARYNGOTYMPANIC (AUDITORY) TUBE

The *pharyngotympanic* or *auditory* tube (7.342, 345, 346), connecting the tympanic cavity to the nasopharynx, allows the passage of air between these spaces to equalize the air pressure on both aspects of the tympanic membrane. It is about 36 mm long and descends anteromedially at an angle of about 45° with the sagittal plane and 30° with the horizontal. It is formed partly by bone and partly by cartilage and fibrous tissue.

The osseous *part*, about 12 mm long, starts from the anterior tympanic wall and gradually narrows to end at the junction of the squamous and petrous parts of the temporal bone, where it has a jagged margin for the attachment of the cartilaginous part; the carotid canal lies medially. It is oblong in transverse section, with its greater dimension horizontal.

The *fibrocartilaginous part*, about 24 mm long, is formed by a triangular plate of cartilage, mostly in the posteromedial wall of the tube. Its apex is attached by fibrous tissue to the circumference of the jagged rim of the osseous tube and its base is directly under the mucosa of the lateral nasopharyngeal wall, forming a *tubal elevation* behind the pharyngeal orifice of the tube. The

cartilage's upper part is bent laterally and downwards, producing a broad *medial* and narrow *lateral lamina*. In transverse section it is hook-like and incomplete below and laterally, where the canal is composed of fibrous tissue. The cartilage is tied to the cranial base in the groove between the petrous temporal bone and the greater wing of the sphenoid, ending near the root of the medial pterygoid plate. The cartilaginous and osseous parts of the tube are not in the same plane, the former descending more steeply than the latter. The diameter is greatest at the pharyngeal orifice, least at the junction of the two parts (the *isthmus*) and increasing again towards the tympanic cavity.

The mucosa of the tube is continuous with the nasopharyngeal and tympanic mucosae; it is lined by a ciliated columnar epithelium and is thin in the osseous part but thickened by mucous glands in the cartilaginous part; near the pharyngeal orifice is a variable, but sometimes considerable lymphoid mass, the *tubal tonsil*.

Relations of the pharyngotympanic tube. *Anterolaterally* the tensor veli palatini separates the tube from the otic ganglion, the mandibular nerve and its branches, the chorda tympani and the middle meningeal artery. Some fibres of the tensor are attached to the lateral cartilaginous lamina and to the fibrous part, forming the *dilatator tubae*. The salpingopharyngeus (p. 1328) is attached to the inferior part of the cartilage near its pharyngeal opening. *Posteromedially* are the petrous temporal bone and the levator veli palatini which arises partly from its medial lamina. The pharyngeal orifice is described with the nasopharynx (p. 1323).

The tube is opened during deglutition but the mechanism is uncertain. The dilatator tubae, aided by the salpinopharyngeus, may be responsible, though some deny the muscle's existence. The levator veli palatini, by elevating the cartilaginous tube, might allow passive opening by releasing tension on the cartilage.

At birth the auditory tube is about half its adult length, its direction being more horizontal and its bony part relatively shorter but much wider. Its pharyngeal orifice is a narrow slit, level with the palate and without a tubal elevation.

Arteries of the auditory tube arise from the ascending pharyngeal branch of the external carotid and from the middle meningeal and the artery of the pterygoid canal (both branches of the maxillary artery). *Veins* drain to the pterygoid venous plexus. *The nerves* are from the tympanic plexus (p. 1228) and from the pharyngeal branch of the pterygopalatine ganglion. The precise contribution from the nerves forming the plexus, i.e. the glossopharyngeal, the cervical sympathetic and possibly the facial, remains uncertain in man.

THE AUDITORY OSSICLES

A chain of mobile ossicles, the *malleus*, *incus* and *stapes*, transfers sound waves across the tympanic cavity from its membrane to the fenestra vestibuli. The malleus is attached to the tympanic membrane and the base of the stapes to the rim of the fenestra vestibuli, while the incus is suspended between them, articulating with both bones.

The malleus (7.350), shaped like a mallet, is 8–9 mm long and is the largest of the ossicles; it has a head, neck, manubrium and anterior and lateral processes. The *head*, its enlarged, ovoid upper end, is situated in the epitympanic recess; it articulates posteriorly with the incus, being covered elsewhere by mucosa. The cartilaginous articular facet for the incus is narrowed near its middle and consists of a larger upper part and a smaller, lower part, orientated almost at right angles to each other. Opposite the constriction the lower margin of the facet projects as the *spur* (*cogtooth*) of the malleus. The neck is the narrowed part below the head and inferior to this is an enlargement from which the processes project. The *manubrium mallei* (*handle of the malleus*) is connected by its lateral margin with the tympanic membrane. It descends posteromedially, diminishing towards its free end which curves slightly forwards and is transversely flat. High on its medial surface a small projection receives the tendon of the tensor tympani. The *anterior process* is a delicate bony spicule, directed forwards from the enlargement below the neck and connected with the petrotympanic fissure by ligamentous fibres (vide infra);

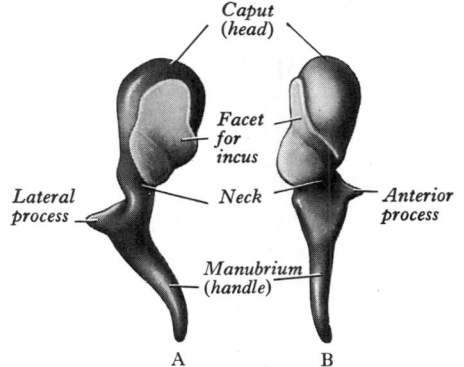

7.350 The left malleus: A. posterior aspect; B. medial aspect.

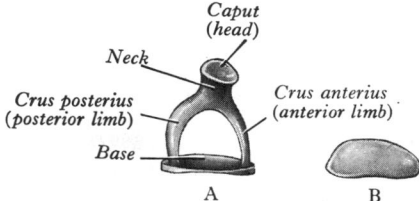

7.352 The left stapes: A. superior aspect; B. basal aspect.

in fetal life this is the longest process of the malleus and is continuous in front with Meckel's cartilage (p. 172). The *lateral process*, a conical projection from the root of the manubrium, is directed laterally and is attached to the upper part of the tympanic membrane and via the anterior and posterior malleolar folds to the sides of the notch in the upper part of the tympanic sulcus.

Ossification. The cartilaginous precursor of the malleus originates as part of the dorsal end of Meckel's cartilage. Excepting its anterior process, it ossifies from a single endochondral centre which appears near the future neck of the bone in the fourth month in utero. The anterior process ossifies separately in dense connective tissue and joins the rest at about the sixth month of fetal life.

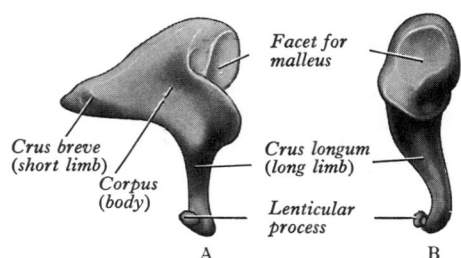

7.351 The left incus: A. medial aspect; B. anterior aspect.

The incus (7.351) is shaped less like an anvil (from which it is named) than a premolar tooth, with its two diverging roots. It has a body and two processes. The *body* is somewhat cubical but laterally compressed; it has an anterior, cartilage-covered, saddle-shaped facet for articulation with the head of the malleus. The *long process*, rather more than half the length of the handle of the malleus, descends almost vertically, behind and parallel to the handle; its lower end bends medially and ends in a rounded *lenticular process*, the medial surface of which is a cartilaginous facet for the head of the stapes. The *short process*, somewhat conical, projects posteriorly and is attached by ligamentous fibres to the fossa incudis in the postero-inferior part of the epitympanic recess.

Ossification. The incus has a cartilaginous precursor continuous with the dorsal extremity of Meckel's cartilage (p. 172). Ossification often spreads from a single centre in the upper part of its long process in the fourth fetal month; the lenticular process, however, may have a separate centre.

The stapes (7.352), or stirrup, has a head, neck two limbs and a base. The *head* (*caput*) is directed laterally and has a small cartilaginous facet for the lenticular process of the incus. To the constricted *neck* is attached posteriorly the tendon of the stapedius. The *limbs* (*crura*) curve away from the neck and merge with a flattened oval or reniform *base* forming the footplate of the stapes and attached to the margin of the fenestra vestibuli by an annular ligament. The anterior limb is shorter and less curved than the posterior.

Ossification. The stapes is preformed in the perforated dorsal moiety of the hyoid arch cartilage of the fetus. Ossification starts

from a single endochondral centre, appearing in the base in the fourth fetal month.

At birth the auditory ossicles have reached an advanced state of maturity.

The Articulations of Auditory Ossicles

These are typical synovial joints. The incudomalleolar joint is saddle-shaped, the incudostapedial is a ball and socket articulation. Their articular surfaces are covered with hyaline cartilage and each joint is enveloped by a capsule containing much elastic tissue and lined by synovial membrane.

The Ligaments of the Auditory Ossicles

The ossicles are connected to the tympanic walls by ligaments: three for the malleus and one each for the incus and stapes. Some of these are mere mucosal folds carrying blood vessels and nerves to and from the ossicles and their articulations; others contain a central, strong band of collagen fibres.

The *anterior ligament of the malleus* stretches from the mallear neck, just above the anterior process, to the anterior wall of the tympanic cavity near the petrotympanic fissure, some of its collagen fibres traversing the fissure to the spine of the sphenoid; others continue into the sphenomandibular ligament which, like the anterior malleolar, is derived from the perichondrial sheath of Meckel's cartilage (p. 172). The ligament may contain muscle fibres (*laxator tympani* or *musculus externus mallei*).

The *lateral ligament of the malleus* is a triangular band stretching from the posterior part of the tympanic incisure's border to the head of the malleus. The *superior ligament of the malleus* connects the head of the malleus to the roof of the epitympanic recess. The *posterior ligament of the incus* connects the end of its short process to the fossa incudis. The *superior ligament of the incus* is little more than a mucosal fold passing from the body of the incus to the roof of the epitympanic recess.

The vestibular surface and the rim of the stapedial base are covered with hyaline cartilage; the rim is attached to the margin of the fenestra vestibuli by an elastic *annular ligament of the base of the stapes*, parts of which are much narrower, acting as a kind of hinge on which the stapedial base moves when the stapedius muscle contracts and during acoustic oscillation (vide infra).

THE MUSCLES OF THE TYMPANIC CAVITY

The tensor tympani (7.342, 346), a long slender muscle, occupies the bony canal above the osseous pharyngotympanic tube, from which it is separated by a thin bony septum. It arises from the cartilaginous part of the pharyngotympanic tube and the adjoining region of the greater wing of the sphenoid, as well as from its own canal. It passes back within its canal, ending in a slim tendon which bends laterally round the pulley-like processus trochleariformis and finally attaches to the handle of the malleus, near its root.

The nerve supply is a branch of the nerve to the medial pterygoid (a ramus of the mandibular nerve), traversing the otic ganglion without interruption (p. 1113). The muscle receives both motor and proprioceptive fibres (Candiollo 1965).

The stapedius extends from the wall of a conical cavity in the pyramidal eminence (and its continuation anterior to the descending part of the facial nerve canal, p. 1110), its minute tendon emerging from the orifice at the pyramid's apex to pass forwards to attach to the posterior surface of the neck of the stapes (7.348). The muscle is of an asymmetrical bipennate form, containing numerous small motor units, each of only six to nine muscle fibres. A few neuromuscular spindles exist near the myotendinous junction.

The nerve supply is a branch of the facial nerve (for details see Blevins 1967).

Actions. The tensor tympani and stapedius usually contract together in reflex response to sounds of fairly high intensity, exerting a 'protective damping effect before vibrations reach the internal ear' (Hallpike 1935). The tensor pulls the tympanic membrane inwards to tense it and also pushes the stapes more tightly into the fenestra vestibuli. The strapedius muscle opposes the tensor in the latter action. Paralysis of the stapedius results in hyperacusis.

MOVEMENTS OF THE AUDITORY OSSICLES

The malleolar manubrium faithfully follows all movements of the tympanic membrane, while the malleus and incus rotate together around an axis running from the short process and posterior ligament of the incus to the anterior ligament of the malleus. When the tympanic membrane and manubrium move medially the long process of the incus moves in the same direction, pushing the stapedial base towards the labyrinth and perilymph (p. 1202). This causes a compensatory outward bulging of the secondary tympanic membrane of the fenestra cochleae. These events are reversed when the membrane moves outwards, but if its movement is large the incus does not follow the full outward excursion of the malleus, merely gliding on it at the incudomalleolar joint and thus preventing a dislocation of the stapedial base from the fenestra vestibuli. When the manubrium moves medially, the spur at the lower margin of the malleolar head locks the incudomalleolar joint, forcing a medial movement of the long process of the incus; the joint is unlocked again when the manubrium moves outwards. The three bones together act as a bent lever so that the stapedial base does not move in the fenestra vestibuli like a piston but rocks on a fulcrum at its antero-inferior border, where the annular ligament is thick. More complex stapedial movements have been described (Békésy 1960). The rocking movement around a vertical axis, which is like a swinging door, is said to occur only at moderate intensities of sound. With loud, low-pitched sounds, the axis becomes horizontal, the upper and lower margins of the stapedial base oscillating in opposite directions around this central axis, thus preventing excessive displacement of the perilymph.

THE TYMPANIC MUCOSA

The mucosa of tympanic cavity is continuous with that of the pharynx, via the pharyngotympanic tube; it covers the ossicles, muscles and nerves in the cavity, forming the inner layer of the tympanic membrane and the outer layer of the secondary tympanic membrane; it also spreads into the mastoid antrum and air cells. It forms several vascular folds extending from the tympanic walls to the ossicles: of these, one descends from the roof of the cavity to the head of the malleus and the upper margin of the body of the incus and a second surrounds the stapedius, others investing the chorda tympani nerve and the tensor tympani muscle. These folds separate off saccular recesses and give the interior of the tympanic cavity a somewhat honeycombed appearance. One such *superior recess of the tympanic membrane* lies between the neck of the malleus and the pars flaccida. Two others, the *anterior* and *posterior recesses of the tympanic membrane*, are formed by the mucosa around the chorda tympani and lie, respectively, anterior and posterior to the manubrium of the malleus. The tympanic mucosa is pale, thin and slightly vascular. It has a ciliated columnar epithelium except over the posterior part of the medial wall, the posterior wall, often parts of tympanic membrane and the auditory ossicles, the cells of these surfaces being flatter and non-ciliated. Near the pharyngotympanic tube goblet cells are numerous; otherwise there are no mucous glands. The mastoid antrum and air cells are lined by flat, non-ciliated epithelium. Undoubtedly there are considerable mucosal variations in the middle ear, which include squamous, cuboidal, columnar and ciliated columnar epithelia. No systematic ultrastructural study has yet been made in this field (Kawabata & Paparella 1969).

1228 It must be re-emphasized that the tympanic cavity and mastoid antrum, auditory ossicles and structures of the internal ear are almost fully developed at birth and subsequently alter little. In fetuses the cavity contains a gelatinous tissue which has practically disappeared by birth, when it is filled by a fluid which is absorbed when air enters via the auditory tube.

THE VASCULATURE OF THE TYMPANIC CAVITY

Six *arteries* supply the walls and contents of the tympanic cavity. Two are larger than the others: the *anterior tympanic* branch of the *maxillary*, supplying the tympanic membrane, and the *stylomastoid* branch of the *occipital* or *posterior auricular* arteries, supplying the posterior tympanic cavity and mastoid air cells. The smaller arteries include: the *petrosal* branch of the *middle meningeal*, entering through the hiatus for the greater petrosal nerve; the *superior tympanic* branch of the *middle meningeal*, traversing the canal for the tensor tympani; a branch from the *ascending pharyngeal* and from the *artery of the pterygoid canal*, accompanying the pharyngotympanic tube; and a *tympanic branch* or branches from the *internal carotid*, arising in the carotid canal and perforating the thin anterior wall of the tympanic cavity. In early fetal life a *stapedial artery* traverses the stapes (p. 217) (a feature which persists in some other mammals). *Veins* terminate in the pterygoid venous plexus and the superior petrosal sinus. From the mucosa of the mastoid antrum a small group of veins runs medially through the arch formed by the anterior semicircular canal, emerging on the posterior surface of the petrous temporal bone at the subarcuate fossa to drain into the superior petrosal sinus; these are the remains of the large *subarcuate veins* of childhood, a route for the spread of infection from the mastoid antrum to the meninges. *Lymph vessels* are described on p. 844.

THE NERVES OF THE TYMPANIC CAVITY

The nerves constitute the *tympanic plexus*, ramifying on the surface of the promontory. They arise from the tympanic branch of the glossopharyngeal nerve and the caroticotympanic nerves of sympathetic origin (Arslan 1960). The former enters the cavity by the tympanic *canaliculus for the tympanic nerve*, ramifying to join the plexus; the *superior* and *inferior caroticotympanic nerves* (from the carotid sympathetic plexus) traverse the carotid canal's wall and also join the plexus. The tympanic plexus supplies: (1) branches to the mucosa of the tympanic cavity, pharyngotympanic tube and mastoid air cells; (2) a branch traversing an opening anterior to the fenestra vestibuli and joining the greater petrosal nerve; (3) the *lesser petrosal nerve*, which may be regarded as the continuation of the tympanic branch of the glossopharyngeal fibres traversing the tympanic plexus. The lesser petrosal nerve occupies a small canal below that for the tensor tympani; it runs past and receives a connecting branch from the facial nerve's genicular ganglion, emerges from the anterior surface of the temporal bone via a small opening lateral to the hiatus for the greater petrosal nerve and then traverses the foramen ovale or the small canaliculus innominatus to join the otic ganglion. Postganglionic secretomotor fibres leave this ganglion in the auriculotemporal nerve to supply the parotid gland.

The *chorda tympani* leaves the facial nerve about 6 mm above the stylomastoid foramen and runs anterosuperiorly in a canal to enter the tympanic cavity via the *posterior canaliculus*. It then curves anteriorly in the substance of the tympanic membrane between its mucous and fibrous layers (p. 1223), crosses medial to the upper part of the manubrium of the malleus to the anterior wall, finally entering the *anterior canaliculus*. (For its further course see p. 1110.) The other nerves which are closely related topographically to the tympanic cavity include: the facial and its genicular ganglion and the stapedial and greater petrosal branches; the auricular branch of the vagus; afferent and efferent terminals of the vestibulocochlear nerve; and the internal carotid sympathetic plexus (all detailed elsewhere in this section). The *meningeal branch* (p. 1106) of the mandibular nerve supplies branches to the mastoid air cells.

Applied Anatomy. Fractures of the middle cranial fossa almost always involve the tympanic roof, accompanied by the rupture of the tympanic membrane or a fractured roof of the osseous external acoustic meatus. Such injuries usually entail prolonged bleeding from the ear, with escape of cerebrospinal fluid if the dura mater has been torn.

The tympanic cavity is a common site of infection, usually spreading to it from the nose and pharynx along the pharyngo-tympanic tube, which usually becomes occluded because of the inflammatory swelling of its mucosa. The products of inflammation, thus confined to the tympanic cavity, may spread into the mastoid antrum, their only escape then being by the rupture of the tympanic membrane, either spontaneous or surgically induced, followed by the free discharge of pus. If the swelling of the pharyngotympanic tube then subsides, normal drainage of the cavity returns and perforations in the tympanic membrane heal; but, if the tube is occluded by enlarged lymphatic tissue, pus may continue to gather and discharge through the perforations as a chronic otorrhoea. Intracranial complications may follow: an abscess may form between the bone and the dura mater, above the tympanic cavity, beneath the dura of the temporal lobe or between the deep aspect of the mastoid process and the sigmoid sinus, possibly extending widely and surrounding the sinus, in which thrombosis may ensue; an infected clot may be detached into the general circulation, causing metastatic abscesses, e.g. in the lungs. Bone disease of the tympanic cavity or mastoid antrum may lead to severe and fatal septic meningitis or to the formation of brain abcesses, particularly in the temporal lobe of the cerebrum and the cerebellar hemisphere. In chronic bone disease in the tympanic cavity, the facial nerve may be exposed in its canal, its inflammation leading to facial paralysis of the infranuclear type (p. 1111).

<h1 style="text-align:center">THE INTERNAL EAR</h1>

The internal ear includes the *osseus labyrinth*, a complex series of cavities in the petrous temporal bone, and the contained *membranous labyrinth*, a corresponding complex set of interconnected membranous sacs and ducts.

0·8 mm in diameter; each has a terminal swelling, an *ampulla* which is almost twice the diameter of the canal. They open into the vestibule by five openings, one of which is shared by two of the canals.

The Osseous Labyrinth

This (7.345, 353, 354) has three regions: the *vestibule, semicircular canals* and *cochlea*. These are all cavities within bone, lined by periosteum and containing a clear fluid, the *perilymph*, within which is the membranous labyrinth. The osseous labyrinth lies in denser, harder bone than that in other parts of the petrous bone; it is hence possible, particularly in young skulls, to dissect out the labyrinth from the petrous temporal.

THE VESTIBULE

The vestibule is the central part of the osseous labyrinth and lies medial to the tympanic cavity, posterior to the cochlea and anterior to the semicircular canals. Somewhat ovoid in shape but transversely narrow, it measures about 5 mm from front to back, the same from above below and about 3 mm across. Its *lateral wall* bears the fenestra vestibuli, closed by the stapedial base and its annular ligament. Anteriorly, on the *medial wall* is a small *spherical recess* containing the *saccule* and perforated by several minute holes, the *macula cribrosa media*; the recess adjoins the inferior vestibular area at the medial end of the internal acoustic meatus (7.354); the foramina transmit twigs of the vestibular nerve to the saccule. Behind the recess is an oblique *vestibular crest*, its anterior end being the *vestibular pyramid*; this ridge divides below to enclose a small depression, the *cochlear recess*, perforated by vestibulocochlear fascicles passing to the cochlear duct's vestibular end. Posterosuperior to the vestibular crest, in the roof and medial wall of the vestibule, is the *elliptical recess* which contains the *utricle*. The pyramid and adjoining part of the elliptical recess are perforated by a number of holes, forming the *macula cribrosa superior*; the holes in the pyramid transmit the nerves to the utricle and those in the recess the nerves to the ampullae of the superior and lateral semicircular ducts. The region of the pyramid and elliptical recess corresponds to the superior vestibular area in the internal acoustic meatus (7.354). The opening of the vestibular *aqueduct* is below the elliptical recess; the aqueduct reaches the posterior surface of the petrous bone, containing one or more small veins and part of the membranous labyrinth, the *endolymphatic duct*. In the posterior part of the vestibule are the five *orifices of the semicircular canals*; in its anterior wall is an elliptical opening leading into the *scala vestibuli* of the cochlea.

THE SEMICIRCULAR CANALS

There are three semicircular canals: anterior (superior), posterior and lateral (horizontal); they lie posterosuperior to the vestibule. They are compressed from side to side and each describes about two-thirds of a circle; they are unequal in length but all are about

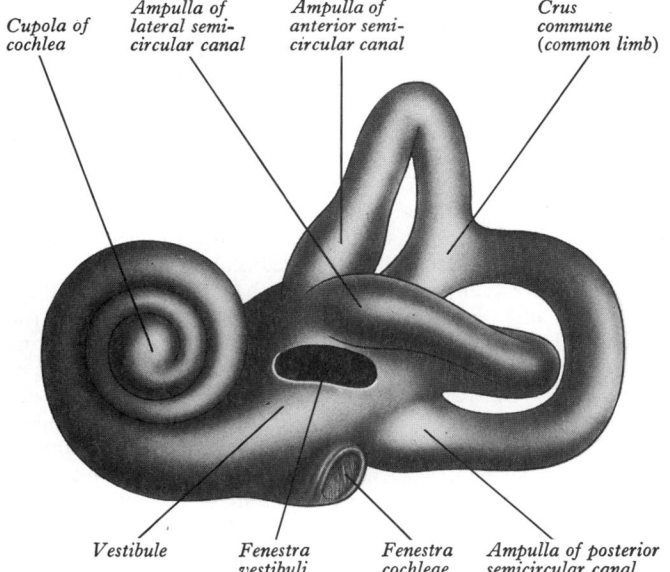

Cupola of cochlea — Ampulla of lateral semicircular canal — Ampulla of anterior semicircular canal — Crus commune (common limb) — Vestibule — Fenestra vestibuli — Fenestra cochleae — Ampulla of posterior semicircular canal

7.353   The left osseous labyrinth: lateral aspect.

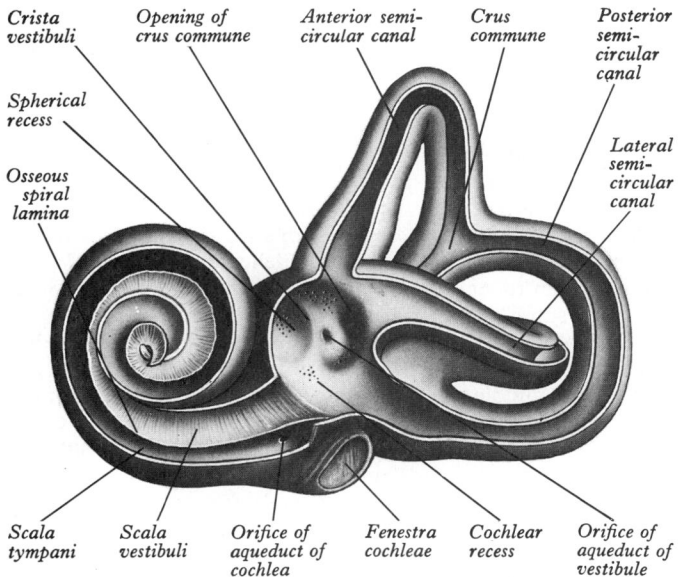

Crista vestibuli — Opening of crus commune — Anterior semicircular canal — Crus commune — Posterior semicircular canal — Spherical recess — Lateral semicircular canal — Osseous spiral lamina — Scala tympani — Scala vestibuli — Orifice of aqueduct of cochlea — Fenestra cochleae — Cochlear recess — Orifice of aqueduct of vestibule

7.354   The interior of the left osseous labyrinth.

The *anterior (superior) semicircular canal*, 15–20 mm in length, is vertical and placed transverse to the long axis of the petrous temporal bone, lying under the anterior surface of its arcuate eminence (p. 379). Some maintain, however, that the eminence does not accurately coincide with this semicircular canal, but is adapted to the occipitotemporal sulcus on the inferior surface of the temporal lobe of the brain. The canal's anterior end is ampullated, opening into the upper and lateral part of the vestibule; its other end unites with the posterior canal's upper end to form the *crus commune*, about 4 mm long, which opens into the medial part of the vestibule.

The *posterior semicircular canal*, also vertical, curves backwards almost parallel with the posterior surface of the petrous bone; it is 18–22 mm long; its ampullated end opens low in the vestibule, where the *macula cribrosa inferior* transmits nerves to the ampulla; this corresponds in position to the *foramen singulare* in the internal acoustic meatus (7.354). The canal's upper end joins the crus commune.

The *lateral (horizontal) canal* is 12–15 mm long, its arch directed horizontally and posterolaterally. Its anterior, ampullated end opens into the upper and lateral angle of the vestibule, above the fenestra vestibuli and just below the superior canal's ampullated end; its posterior end opens below the orifice of the crus commune.

It is often said that the two lateral semicircular canals are in the same plane and that the anterior canal of one side is almost parallel with the opposite posterior canal. Blanks et al (1975) measured the angular relations of the planes of the semicircular osseous canals in 10 human skulls. Although the planes of the three ipsilateral canals were almost perpendicular to each other, some variation was apparent, the angles being as follows: horizontal/anterior $111.76 \pm 7.55°$; anterior/posterior $86.16 \pm 4.72°$; posterior/horizontal $95.75 \pm 4.66°$. The planes of similarly orientated canals of the two sides showed marked departure from being parallel: left anterior/right posterior $24.50 \pm 7.19°$, left posterior/right anterior $23.73 \pm 6.71°$, left horizontal/right horizontal $19.82 \pm 14.93°$. The same observers (Curthoys et al 1977) have also measured the dimensions and radii of the canals; means for the radii of the osseous canals were as follows: horizontal 3.25 mm, anterior 3.74 mm, and posterior 3.79 mm. The functional implications of these data are still speculative. In diameter the osseous canals are about 1 mm (minor axis) and 1.4 mm (major axis). The membranous ducts within them are much smaller but are also elliptical in transverse section, with major and minor axes of 0.23 and 0.46 mm. Representative means for ampullary dimensions are as follows: length 1.94 mm, height 1.55 mm. Curthoys and collaborators have recorded many other labyrinthine dimensions. They attempt to equate the metrical data with theories of labyrinthine mechanics.

THE COCHLEA

The cochlea (7.345, 353, 354, 355), shaped like a conical snail-shell, is the most anterior part of the labyrinth, lying anterior to the vestibule. It is about 5 mm from base to apex and 9 mm across its base. Its apex, or *cupula*, points towards the anterosuperior area of the tympanic cavity's medial wall (7.345); its base faces the bottom of the internal acoustic meatus and is perforated by numerous apertures for the cochlear nerve. The cochlea has a central conical axis, the *modiolus*, with a spiral canal of about two and three-quarter turns around it; a delicate *osseus spiral lamina* projects from the modiolus, partially dividing the canal. Within this bony spiral lies the membranous cochlear duct, attached to the modiolus and to the outer cochlear wall by its other edge. There are thus formed three longitudinal channels within the cochlea, the middle one (the cochlear duct) being blind, ending at the apex of the cochlea and its flanking channels communicating with each other at the modiolar apex at a narrow slit, the *helicotrema* (vide infra). The cochlear duct bears on one of its walls the sensory receptors responsible for audition (p. 1235). The perilymphatic channels on either side of it are the *scala vestibuli* and *scala tympani*. The former is in continuity with the vestibule and the latter is separated from the tympanic cavity by the secondary tympanic membrane at the fenestra cochleae.

The modiolus, the central cochlear pillar, has a broad base near the lateral end of the internal acoustic meatus, where it corresponds to the *tractus spiralis foraminosus*, the multiple foramina of which transmit the fascicles of the cochlear nerve; those for the first turn and a half traverse the small foramina of the tractus spiralis, those for the apical turn traverse the *foramen centrale* in the tract's centre. Canals from the tractus traverse the modiolus and issue centrifugally in sequence into the base of the osseous spiral lamina. Here the small canals enlarge and fuse together to form the *spiral canal of the modiolus* (Rosenthal's canal) which follows the course of the osseous spiral lamina and contains the *spiral ganglion*. The foramen centrale continues as a canal through the modiolar centre to its apex.

The osseous cochlear canal, at its first turn, bulges towards the tympanic cavity where it underlies the promontory (p. 1225). It is about 35 mm long, diminishing in diameter from the base to the summit and ending at the *cupula*, which forms the cochlea's apex; at its start the canal is about 3 mm in diameter. It has three openings at its base: the *fenestra cochleae* (round window) facing the tympanic cavity and closed by the *secondary tympanic membrane*; the *fenestra vestibuli* (oval window) occupied by the base of the stapes (p. 1225); and the *cochlear canaliculus*, a minute funnel-shaped canal opening on the inferior surface of the petrous temporal bone (p. 379), transmitting a small vein to the inferior petrosal sinus and connecting the subarachnoid space to the scala tympani (vide infra).

The osseous spiral lamina is a ledge projecting from the modiolus like the thread of a screw; it reaches about halfway across the cochlear canal, partly dividing it into an upper *scala vestibuli* and a lower *scala tympani*. Its width decreases progressively towards its apex, ending in a hook-shaped *hamulus* partly bounding the *helicotrema*, through which the scalae are continuous. From the spiral modiolar canal many canaliculi radiate through the osseous lamina to its rim, carrying the cochlear nerves. In the lower part of the first turn a *secondary spiral lamina* projects *inwards* from the *outer* cochlear wall but does not reach the osseous spiral lamina, leaving a narrow *vestibular fissure* between them.

The osseous labyrinth was long regarded as lined by a thin fibroserous membrane adherent to the periosteum, its flat epithelium bounding the *perilymphatic space* which is filled with *perilymph* bathing the exterior of the membranous labyrinth. But electron microscopy shows that the perilymphatic space is bounded by fibroblast-like *perilymphatic cells*, with strands of extracellular fibres, the detailed form of the cells varying in different parts of the labyrinth. Where the perilymphatic space is narrow, the cells are essentially *reticular* or *stellate* in form, their sheet-like cytoplasmic extensions crossing and dividing the space into connected intercellular clefts of variable shape and size. This tissue and its spaces occupies the cochlear canaliculus; but where the space is wider, as in the scalae vestibuli and tympani and much of the vestibule, the perilymphatic cells on the periosteum and the external surface of the membranous labyrinth are extremely flat, with a rather featureless cytoplasm; apart from a few projections into the perilymph the arrangement approaches that of a true squamous epithelium. Elsewhere, on parts of the perilymphatic surface of the basilar membrane (p. 1235), the cells are cuboidal. Closely related to the periosteal and labyrinthine aspects of these cells are bundles of collagen fibres, which the cells may form. In some species fibres with a helical substructure, differing from collagen, have been described but their status in mankind is uncertain.

The perilymph resembles the cerebrospinal fluid in its composition but minor differences have been described (Ormerod 1960). Many regard it as an ultrafiltrate of plasma, with perhaps some addition from the cerebrospinal fluid. Its source, rate of production, circulation and absorption are as yet not fully known. The supposed connection between the perilymphatic and subarachnoid spaces via the *cochlear canaliculus* is controversial. The canaliculus was often described as containing a simple, epithelial duct connecting the two spaces, a view rejected by some (Waltner 1948, Mygind 1948, Young 1952, 1953), who surmised that the connective tissue blocked the canaliculus, thus separating the two fluid compartments. However it seems likely that intercellular crevices persist in the canal; electron-dense tracers such as thorotrast, introduced into the subarachnoid space, readily appear in the perilymphatic spaces (Duvall & Quick 1969). Others (Silverstein et al 1969) point out that, in cats, even india ink or avian erythrocytes pass into the perilymphatic spaces via the cochlear canaliculus within 24 hours of their introduction into the subarachnoid space of the posterior cranial fossa. The latter authors regard perilymph as probably derived from: (1) blood vessels surrounding the spaces, (2) fluid spaces surrounding the sheaths of the vestibulocochlear nerve fibres and (3) cerebrospinal fluid arriving along the cochlear canaliculus. The mechanism of removal of perilymph is uncertain.

In summary, the organization of the perilymphatic spaces as follows: the vestibular perilymphatic space connects posteriorly with that around the semicircular canals and opens anteriorly into the cochlear scala vestibuli, which in turn opens into the scala tympani via the helicotrema; the latter scala is separated from the tympanic cavity by the secondary tympanic membrane but is continuous with the subarachnoid space through the cochlear canaliculus (vide supra).

During development, the petrous bone adjoining the labyrinth is developed from the cartilaginous otic capsule by endochondral ossification; it is denser than petrous bone elsewhere, with interglobular spaces containing cartilage cells (7.356). The cochlear modiolus, however, is formed from trabecular dermal bone (Fraser & Dickie 1914).

The Membranous Labyrinth

The membranous labyrinth (7.355), lying within the osseous labyrinth, is much narrower and is filled with *endolymph*, a fluid of unique composition; in its walls the terminal fibres of the vestibulocochlear branches are distributed. It includes: (1) the *utricle* and *saccule*, two small sacs occupying the vestibule, (2) three *semicircular ducts* in the semicircular canals, (3) the *cochlear duct* in the osseous cochlea. These parts form a closed system of channels which communicate freely with one another. The semicircular ducts open into the utricle and this opens into the saccule via the ductus utriculosaccularis, which also joins the ductus endolymphaticus; the saccule opens into the cochlear duct through the ductus reuniens.

The membranous labyrinth is attached to the wall of the osseous labyrinth at certain points, but is separated from the greater part of it by a perilymphatic space. For the ultrastructure of the membranous labyrinth consult Wersall (1956), Engstrom & Wersall (1958), Iurato (1967), Kimura (1969), Babel et al (1970).

THE UTRICLE

The utricle, the larger of the vestibular sacs, is irregularly oblong and occupies the posterosuperior region of the osseous vestibule, in contact with the elliptical recess and the area inferior to it. The part of the utricle in the recess is a pouch or cul-de-sac. The lateral half of its floor and adjoining lateral wall is thickened over an area of about 3 mm by 2 mm as the *utricular macula* (p. 1232), innervated by the utricular fibres of the vestibular nerve. The ampullae of the anterior and lateral semicircular ducts open into its lateral part; the ampulla of the posterior duct, the crus commune and the posterior end of the lateral duct open into its medial part. The lateral duct's posterior end widens into a flattened cone joining the utricle's medial end at a right angle. From the anteromedial part of the saccule the *ductus utriculosaccularis* is given off and gives a side branch opening into the ductus endolymphaticus, the main duct passing on to the saccule.

THE SACCULE

The saccule lies in the spherical recess near the opening of the cochlear scala vestibuli. Viewed from the anterior aspect, it appears almost globular but it is prolonged postero-inferiorly as a cone, the upper surface of which is partly in contact with the under-surface of the utricle, both sharing a common wall. In the saccule's anterior wall is an oval thickening, the *macula of the saccule* (pp. 1232, 1233), in a plane set at right angles to the utricular macula and innervated by the saccular fibres of the vestibulocochlear nerve. Its cavity is connected via a Y-shaped tube with the utricle. The *ductus endolymphaticus* leaves it posteriorly to join the ductus utriculosaccularis, passing inferomedially along the vestibular aqueduct to end in the *saccus endolymphaticus* under the dura mater on the posterior surface of the petrous bone. From the saccule's lower part a short *ductus reuniens* passes inferiorly, widening into the basal end of the cochlear duct (7.355).

THE SEMICIRCULAR DUCTS

The semicircular ducts (7.355, 356) are about one-quarter of the diameter of their osseous canals but are similar in general shape.

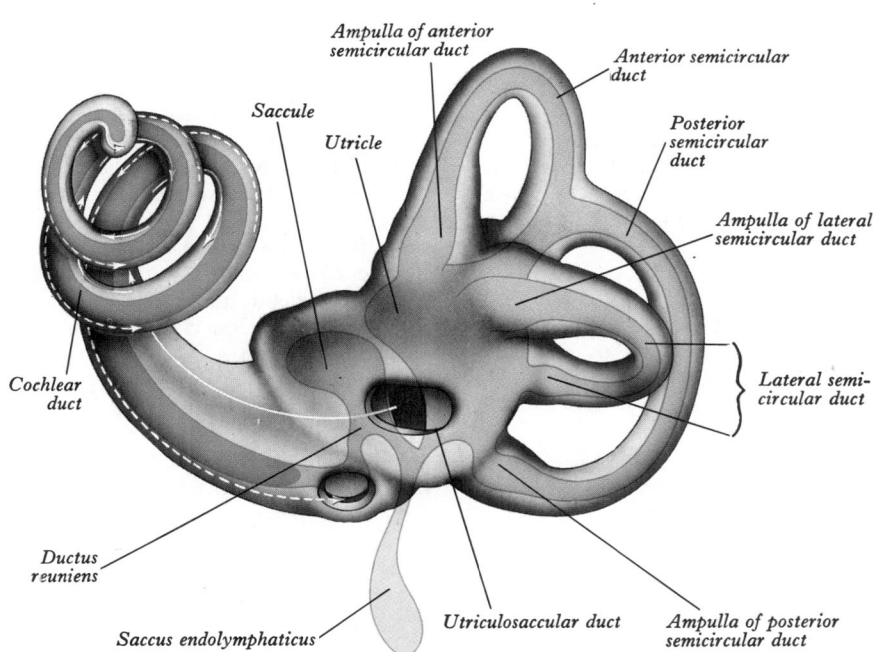

7.355   Scheme of the membranous labyrinth (blue) projected on to the osseous labyrinth. The arrows indicate the direction of sound waves in the cochlea.

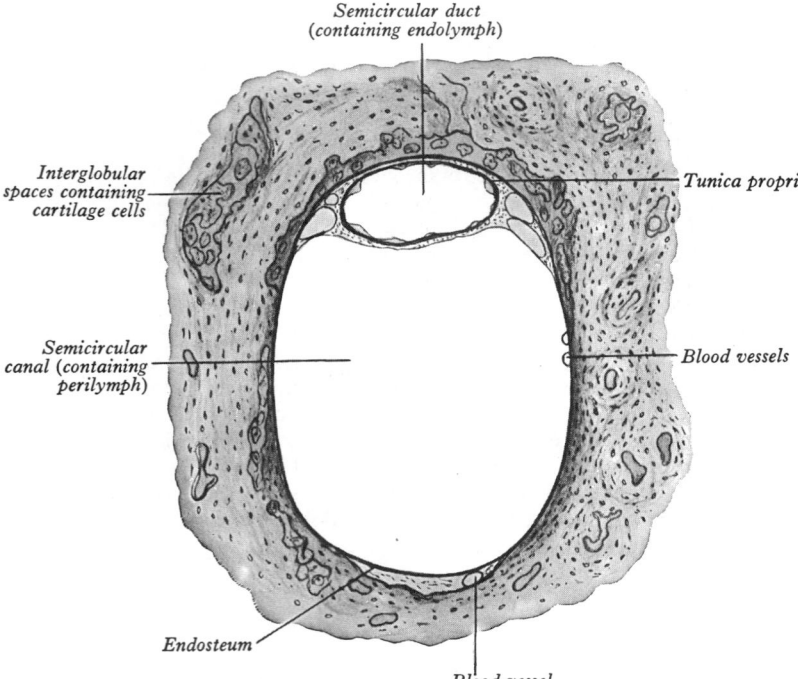

Semicircular duct
(containing endolymph)

Interglobular
spaces containing
cartilage cells

Semicircular
canal (containing
perilymph)

Tunica propria

Blood vessels

Endosteum

Blood vessel

7.356 Transverse section through the left posterior semicircular canal and duct of an adult man (after J K Milne Dickie). Magnification × 50.

Each has an ampulla at one end within the ampulla of its bony canal. The semicircular ducts open into the utricle by five orifices, one being common to the medial end of the anterior and the superior end of the posterior duct. In each ampulla the wall is thickened, projecting inwards as a transverse elevation shaped like an hour glass, the *septum transversum*, its most prominent part being the *ampullary crest*; this projects from the ampullary wall most distant from the concavity of its duct, an arrangement by which 'any movement of endolymph would be caught by the crista to greatest advantage' (Dickie 1920).

The utricle, saccule and semicircular ducts are suspended by fibrocellular bands across the perilymphatic space which stretch from the osseous wall to the membranous labyrinth within; through much of the course of the semicircular ducts, the membranous duct is securely attached by much of its circumference to the osseous wall (7.356).

STRUCTURE OF THE UTRICLE, SACCULE AND SEMICIRCULAR DUCTS

The walls of the utricle, saccule and semicircular ducts are generally described as being trilaminar. The *external* layer is largely fibrous and vascular; its superficial fibres are often covered by flat perilymphatic cells and this outer surface sometimes blends with

the labyrinthine periosteum. The *middle* layer, composed of more delicate vascular connective tissue, has on its internal surface, especially in the semicircular ducts, a number of papilliform projections. The *internal* layer is usually a simple epithelium varying from squamous to cuboidal or polygonal, with a basement membrane. The cells have a *specialized* arrangement in the ampullary crests and maculae (7.357, 358); here the middle layer is also thicker. Flanking each crest, between it and the ampullary wall, is an area of tall epithelium, which appears crescentic in section, known as the *planum semilunatum*.

In the so-called 'non-specialized' areas of the membranous labyrinth, the general epithelium (whose cells vary from squamous to tall columnar) shows ultrastructurally (Kimura 1969) *light* and *dark* cells which are quite dissimilar in their cytology. In many regions the epithelium resembles simple non-secretory epithelia elsewhere. The *light cells* of such areas contain elliptical or crenated heterochromatic nuclei, few mitochondria, occasional ribosomes and micropinocytotic vesicles and show a few microvilli on their luminal aspect. Junctional complexes occur near their luminal borders, with some interdigitation of the cell surfaces. Infolding of the basal plasma membrane is minimal. In contrast, some patches or strips of epithelium in the utricle and the semicircular ducts, including the ampullae, consist of *dark cells*. Their luminal surfaces carry occasional microvilli and show endocytic vesicles; junctional complexes and interdigitations occur; their irregular nuclei are central or near the luminal surface; their dense supranuclear cytoplasm contains: numerous coated vesicles, larger smooth-walled vesicles and mitochondria, free and attached polysomes, lipid droplets, lysosomes, lipofuscin granules, microfilaments and microtubules and an obvious Golgi apparatus. The infranuclear region consists of many long cytoplasmic processes projecting towards the basal lamina. Each process has an elongate, fusiform enlargement, completely occupied by a mitochondrion. The plasma membrane of these processes is covered by basal lamina. Such cells are obviously highly active and their structural similarity to cells in other ion-transporting epithelia (e.g. renal tubules, p. 1401, the ciliary body of the eye, parotid duct, salt-secreting glands of various sub-mammalian forms) suggests an involvement in the control of the ionic composition of endolymph (vide infra).

In *ampullary crests* the epithelium contains sensory *hair cells* and *supporting cells*. **Hair cells** are of two types. *Type I* is piriform with a rounded base and a short neck. Except at its free end it is surrounded by a large goblet-shaped nerve terminal or *calix*, the apposed plasma membranes being separated by a gap of about 20–30 nm but at some points the interval is about 5 nm. The nucleus is basal and surrounded by mitochondria, with a concentration of these organelles near the free surface of the cell. Occasional cisternae of granular endoplasmic reticulum, free polysomes, microfilaments and microtubules, many smooth vesicles of about 20 nm and supranuclear Golgi apparatus all occur. The *type II* hair cell is cylindrical, its nucleus lying at various levels but usually more central than in type I cells. Its cytoplasm contains similar organelles, but the smooth-walled vesicles are more abundant and a supranuclear Golgi apparatus more prominent. The basal part of type II cells is in contact with a number of synaptic end-bulbs, of two varieties, both containing mitochondria and many small vesicles; in *non-granulated terminals* the vesicles are clear, in *granulated endings* they contain electron-dense cores. Non-granular terminals are *afferent* nerve fibres; granular terminals are regarded as rami of *efferent* fibres innervating type II hair cells, probably modifying the effective threshold of the hair cells. Thus, both types of terminal are probably the sites of neurochemical transmission, but are oppositely directed. However, the details remain uncertain. The calix of type I hair cells is regarded as the terminal of an afferent vestibular fibre but whether it is neurochemical or 'electrical' is still to be determined. A number of granular end-bulbs are often applied to the external calicial aspect and these are probably *efferent*, presynaptically modifying transmission by the calix.

Type I hair cells may be regarded as the more 'discriminative' of the two types. Calices are derived from larger and therefore faster vestibular nerve fibres, each innervating a small group of type I cells. But type II cells contact the end-bulbs from several

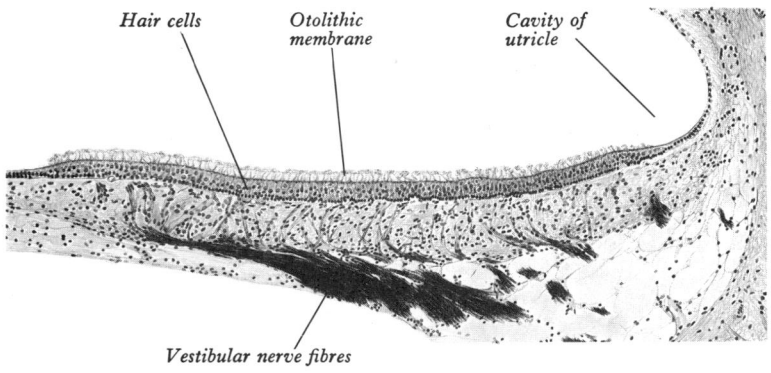

Hair cells

Otolithic membrane

Cavity of utricle

Vestibular nerve fibres

7.357 Section of the macula of the utricle of the cat. Weigert–Pal and iron haematoxylin stain. (For source see 7.358.) Magnification × 112.

thin vestibular nerve fibres, each innervating many type II cells, distributed over a substantial area of membrane.

The apical surfaces of both types are similar, carrying 40–100 'hairs', stereocilia or modified microvilli of varying length and arranged in a regular hexagonal array and polarized with respect to a long *kinocilium* attached to one border of the cell. This kinocilium has a typical basal body within the cell's apical cytoplasm and its shaft contains a ring of nine double microtubules, but the central pair (as in most cilia) is sometimes absent. Despite its name, its motility is uncertain. The stereocilia are non-motile, constricted at their attachment to the hair cell; each contains a longitudinal array of microfilaments. On cell borders opposite the kinocilium, stereocilia are about 1 µm in length but lengthen towards the cilium, reaching about 100 µm close to it. Stereocilia are also interconnected by lateral links which probably ensure that they move together as a unit when deflected during stimulation (Ernstrom 1985).

In each vestibular receptor site the hair cells are arranged in a *precise pattern* of considerable functional significance (vide infra). In the ampullae of the lateral semicircular ducts, the sides of hair cells with a kinocilium are all directed towards the utricular cavity, but in the ampullae of the anterior and posterior ducts they are directed away from it. In the two maculae there is a sinuous *'parting line'* across the central region; structural polarization of the hair cells is reversed on the opposite sides of this line. In the utricle, hair cells are set in curved contours, with the kinocilium of each cell *nearer* to the parting line but in the saccule *remote* from it. For details of the orientation of these arrays consult Iurato (1967), Babel et al (1970), Ades & Engstrom (1974).

The supporting cells are elongate and of variable diameter; each has a basally positioned nucleus and rests on the basal lamina of the epithelium. They have microvilli; their cytoplasm contains large osmiophilic granules, possibly secretory, a well developed Golgi apparatus and vertical microtubules and microfilaments entering a prominent subapical terminal web; mitochondria are abundant. Whether these cells are mainly nutritive or modify the composition of the endolymph remains uncertain. Junctional complexes exist around the subapical parts of the hair and supporting cells; in the ampullae, the processes of both project into a dome-shaped gelatinous protein-polysaccharide mass, the cupula; its precise composition is as yet undetermined. The cupular apex almost reaches the opposite ampullary wall but can swing in response to currents in the endolymph. When the current ceases, its elastic recoil returns the cupula to the intermediate vertical position, but it may overshoot and oscillate before coming to rest.

Epithelial cells of the *planum semilunatum* (Iurato 1967) lie adjacent to the crista on one side and gradually change to ampullary cuboidal epithelium on the other; they have a few microvilli, containing basally abundant parallel, smooth-walled double membranes, arranged perpendicular to the basement membrane. The basal plasma membrane has invaginations ending in vesicular enlargements in the cytoplasm. Between the folds are vesicles and mitochondria in lines. The nucleus is apical with abundant mitochondria, endoplasmic reticulum, well developed Golgi apparatus, polyribosomes and vesicles, some containing granules. In these features the cells of the planum resemble *dark cells* (vide supra) elsewhere in the labyrinth (vide supra). The planum semilunatum may secrete endolymph. *Supporting* and *hair cells* of the *utricular* and *saccular maculae* are like those in the ampullary crests; but the gelatinous mass into which their stereocilia and kinocilia are inserted is the flatter *otolithic membrane* (*membrana statoconiorum*), containing many minute crystalline *otoliths*, *otoconia* or *stataconia*, which consist of calcite (Carlström et al 1953) and associated protein, giving the maculae an opaque white appearance.

The ampullary crests and the maculae are concerned with equilibratory reflexes influencing ocular position relative to *movements* of the head, by vestibular connections via the medial longitudinal fasciculus (p. 985) with the nuclei of the third to sixth cranial nerves and affecting cervical movements through the descending fibres of the same fasciculus (p. 986). They also influence the general body musculature through the vestibulospinal tracts (p. 932) and other central pathways. (For central vestibular

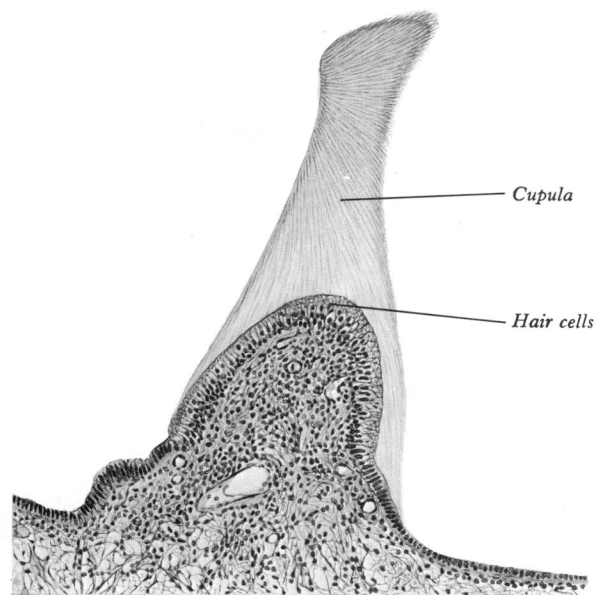

7.358 Section of an ampullary crest of a six-month-old human fetus. Stained with haematoxylin and eosin. From a section lent by E W Walls, Professor Emeritus, Middlesex Hospital Medical School, University of London. Magnification × 75.

and cerebellar connections see pp. 957, 966, 967, 975.) Muscle activity is also influenced by the *position* of the head; macular otoliths, under the influence of gravity, pull on the apical processes of the hair cells in varying positions and are therefore referred to as organs of *static balance* (*statotonic reflexes*); ampullary crests are organs of *kinetic balance*, stimulated by movement or pressure changes in the endolymph due to *angular accelerations* of the head (*statokinetic reflexes*), producing deviation of cupulae. But the maculae may also be involved in *linear acceleration*. The terms static and kinetic are hence not wholly appropriate. The saccular macula, though histologically the same as the utricular, is believed by some to be concerned not in vestibular reflexes but with the cochlea in the reception of low frequency sound.

How deformation of the stereocilia by the movement of the cupulae or otoliths alters the ionic conductance of hair cell membranes is uncertain. Electrophysiological recordings suggest that most vestibular nerve fibres show a continuous basal discharge of afferent impulses when the hair cells are not stimulated. Bending the stereocilia towards the kinocilium raises this frequency and bending in the opposite direction lowers it. Head position, or linear or angular acceleration, is signalled by the balance or imbalance in the patterns of impulses from pairs of receptor sites in the membranous labyrinths of the two sides; e.g. a horizontal swing to the right raises the discharge rate from the ampulla of the right lateral semicircular duct but decreases it in the left. The inertia of the endolymph gives it relative movement to the left in both labyrinths. The positions of the ampullary kinocilia and stereocilia result in compression on the right and decompression on the left ampulla of any bilateral pair of semicircular ducts. (For further discussion see Fischer 1956.)

THE ENDOLYMPHATIC DUCT AND SAC

Throughout the whole endolymphatic duct the surface cells resemble those lining the non-specialized parts of the membranous labyrinth. Where the saccus endolymphaticus expands under the dura mater it is surrounded by a well-vascularized connective tissue, its epithelium changing to tall columnar cells consisting of two main types: one with dense cytoplasm, but otherwise unspecialized; the other less dense, with abundant long microvilli on its luminal surface, its cytoplasm containing numerous mitochondria, pinocytotic invaginations and vesicles and larger smooth-walled vacuoles.

The endolymph filling the membranous labyrinth contrasts in its composition with the perilymph outside it, the latter being

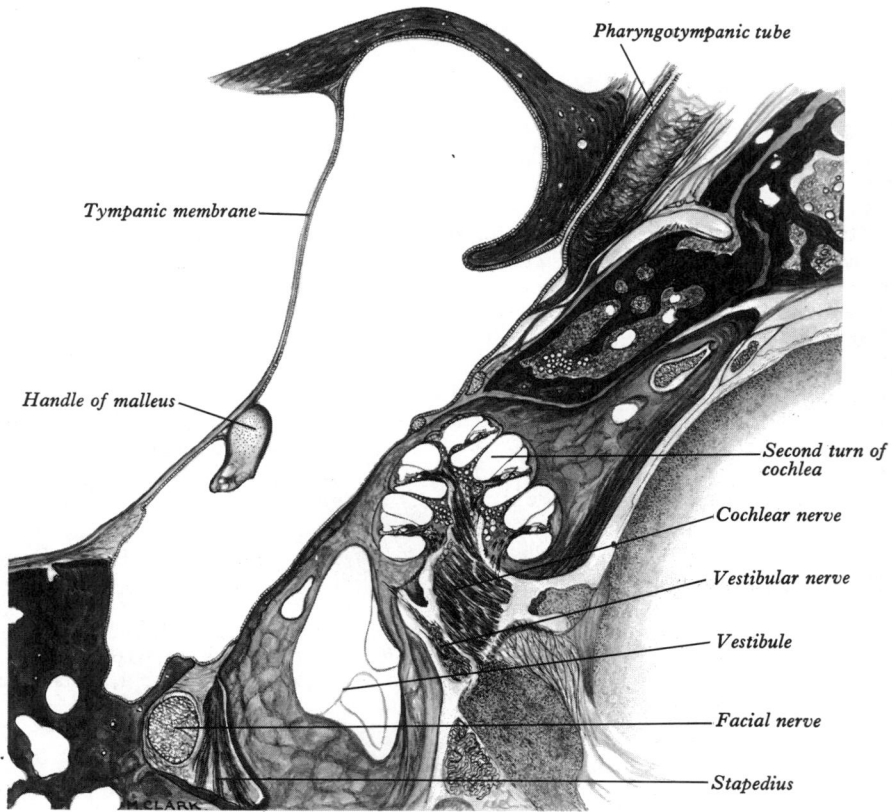

Pharyngotympanic tube

Tympanic membrane

Handle of malleus

Second turn of cochlea

Cochlear nerve

Vestibular nerve

Vestibule

Facial nerve

Stapedius

7.359   Horizontal section through the left temporal bone. Drawn from a section prepared at the Ferens Institute and lent by the late J Kirk.

comparable to extracellular tissue fluid or cerebrospinal fluid, while endolymph resembles intracellular fluid in its ionic composition, being rich in potassium and poor in sodium ions. It is accepted that endolymph is a secretion from a source still unknown. Structures involved in its production may include the *dark cells* of the utricle and semicircular ducts, the columnar cells of the *planum semilunatum* and specialized epithelial cells and related blood vessels of the *stria vascularis* of the cochlear duct (vide infra). Whatever their relative contributions, endolymph probably circulates in the labyrinth, entering the ductus endolymphaticus for removal by the specialized epithelium of its sac into the adjacent vascular plexus. Pinocytotic removal of fluid may also occur in other labyrinthine regions.

A unique positive electrical potential exists in the endolymphatic spaces, varying from + 77 millivolts in the cochlear duct near the stria vascularis to about + 44 millivolts in the utricle, while it is absent or even negative in the ampullae. Thus, a difference of as much as 150 millivolts exists across the membranes of cochlear hair cells, between the cochlear endolymph and the cell interior. This may assist in creating the extreme sensitivity to mechanical deformation found in *auditory* hair cells (vide infra).

THE COCHLEAR DUCT

The cochlear duct (7.359, 360) is a spiral tube running along the outer wall of the osseous cochlea. The osseous spiral lamina (p. 1230) projects for only part of the distance between the modiolus and the outer cochlear wall, the *basilar membrane* completing the roof of the scala tympani. The endosteum of the outer wall is thickened into a *spiral cochlear ligament* projecting inwards and attached to the outer edge of the basilar membrane. A second, thinner *vestibular membrane (Reissner's membrane)* extends from the thickened endosteum on the osseous spiral lamina to the outer wall, attached well above the outer edge of the basilar membrane. The canal thus enclosed, between the scala tympani below and the scala vestibuli above, is the *cochlear duct* (7.360). It is triangular in section, its roof the vestibular membrane, its outer wall the endosteum, its floor the basilar membrane and the outer part of

the osseous spiral lamina. Its closed upper end, the *lagaena*, is attached to the cupola (p. 1230). The duct's lower end turns medially, narrowing into the *ductus reuniens*, to connect with the saccule (7.355). The spiral organ (of Corti), comprising the sensory area of the cochlea, is set on the basilar membrane. The thin vestibular membrane is covered on both surfaces by flat epithelium. The endosteum of the outer wall is greatly thickened to form a *spiral ligament*, which projects inferiorly as a triangular *crista basilaris*, to which the rim of the basilar membrane is attached; immediately above this is a concavity, the *sulcus spiralis externus*, above which the thick, highly vascular periosteum projects as a *spiral prominence*, above which again is a specialized, thick epithelial layer, the *stria vascularis*. (For ultrastructural details see Babel et al 1970.)

THE VESTIBULAR MEMBRANE

The vestibular membrane (of Reissner) has two layers of squamous epithelial cells separated by a basal lamina. The aspect facing the scala vestibuli bears perilymphatic cells, thick in their central perinuclear zone but elsewhere extremely thin; zonulae occludentes occur between them, creating a diffusion barrier. The endolymphatic aspect has typical squamous epithelial cells, also joined by zonulae occludentes and containing mitochondria and many vesicles; their basal surfaces are sometimes smooth but often complexly invaginated; the surface carries numerous, short, irregular microvilli. These cells may be involved in fluid transport.

THE STRIA VASCULARIS

As noted above this is on the cochlear duct's outer wall above the spiral eminence. It has a special stratified epithelium continuing a dense *intra-epithelial capillary plexus* and three cell types: (1) superficial *marginal, dark* or *chromophil cells*, (2) *intermediate, light,* or *chromophobe cells*, (3) *basal cells.*

The endolymphatic surface contains only dark cells; the intermediate and basal cells are cytologically similar, differing in their

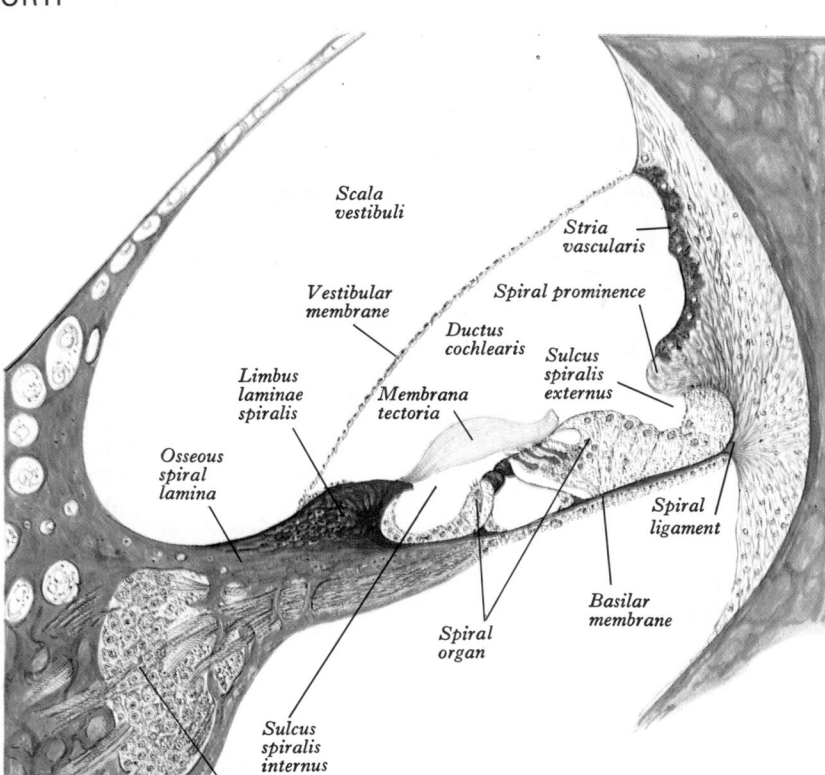

7.360 Section through the second turn of the cochlea indicated in the previous figure. The modiolus is to the left. Mallory's stain.

position, their pale cytoplasm containing scattered mitochondria, many endocytic vesicles and some melanin granules; they send cytoplasmic processes towards the surface, insinuated between the deeper parts of the marginal cells. Marginal dark cells have dense granular cytoplasm with many mitochondria and endocytic vesicles. Their deep parts consist of long cytoplasmic processes separated by deep invaginations of plasmalemma, each containing many mitochondria. Intra-epithelial capillaries are enveloped by the descending processes of dark cells and ascending processes from intermediate and basal cells. The stria vascularis is considered to be ion-transporting, assisting to maintain the unusual ionic composition of endolymph. But other regions of the membranous labyrinth may also be involved in this activity. Exploration of the stria by micro-electrodes shows it to be the source of the large positive endocochlear electrical potential, the maintenance of which is directly dependent upon adequate oxygenation of the epithelial cells, provided by the intra-epithelial capillary plexus.

THE OSSEOUS SPIRAL LAMINA

This consists of two plates of bone, between which are canals for the cochlear nerve filaments. On the upper plate, within the duct of the cochlea, the periosteum is thickened as the *limbus laminae spiralis* (7.360, 362), ending externally in the *sulcus spiralis internus*, which in section is shaped like a C; its upper part, the overhanging limbic edge, is the *vestibular labium* and the lower tapering part is the *tympanic labium*, perforated by foramina for branches of the cochlear nerve. The upper surface of the vestibular labium is crossed at right angles by furrows, separated by numerous elevations, the *auditory teeth* (*dentes acoustici*) (7.362). The limbus is covered by a layer appearing superficially as squamous epithelium, but only the cells over the 'teeth' are flat, those in the furrows (7.363) being columnar (*interdental cells*) and occupying the intervals between the elevations. This epithelium is continuous with that in the sulcus spiralis internus and on the inferior surface of the vestibular membrane. It is considered by some that during development the interdental cells secrete material forming the tectorial membrane (vide infra).

THE BASILAR MEMBRANE

This stretches from the tympanic lip of the osseous spiral lamina to the crista basilaris (7.360, 362). It consists of two zones, a thin *zona arcuata* stretching from the limbus spiralis to the bases of the outer rods and supporting the organ of Corti, and an outer thicker *zona pectinata*, commencing beneath the bases of the outer rods and attached laterally to the crista basilaris. The zona arcuata is composed of compact bundles of small collagenoid filaments 8–10 nm in diameter, mainly radial in orientation. In the zona pectinata the basilar membrane is trilaminar, its upper layer a homogeneous network of transverse fibres, the lower composed of compact bundles of longitudinal fibres with an intermediate structureless layer containing few cells. At its attachment to the crista basilaris the upper and lower layers fuse. The length of the basilar membrane is about 35 mm; its width *increases* from 0·21 mm basally to 0·36 mm at its apex, accompanied by corresponding narrowing of the osseous spiral lamina and a decrease in the thickness of the crista basilaris. Its inferior surface is covered by a layer of vascular connective tissue and elongated perilymphatic cells; one vessel is larger and termed the *vas spirale*; it lies immediately below the tunnel of Corti (*cuniculum internum*).

THE SPIRAL ORGAN OF CORTI

The spiral organ (of Corti) (Smith & Dempsey 1957, Engström & Wersall 1958, Babel et al 1970, Ades & Engström 1974) consists of a series of epithelial structures lying on the zona arcuata of the basilar membrane (7.360–365). The more central of these structures are two rows of cells, the *internal and external rod cells* (*of Corti*) or *pillar cells*. The bases or *foot plates* (*crura*) of the rod cells are expanded, resting contiguously on the basilar membrane but apically widely separated; the two rows incline to each other and come into contact above at the *heads of the pillars*, enclosing between them and the basilar membrane is the *cuniculum internum* (tunnel of Corti) (7.363), which has a triangular cross-section. Internal to the inner rods is a single row of *inner hair cells* and external to the outer rods three or four rows of *outer hair cells,*

7.361 Transverse section of the spiral organ of Corti (cat), stained with the Mallory trichrome method to show the inner and outer hair cells (orange), the basilar membrane (dark blue) and tectorial membrane (light blue) and various supporting cells, including those surrounding the tunnel of Corti. Magnification × 400.

with supporting cells, *phalangeal cells (of Deiters)* and cellulae limitans externae (cells of Hensen). The free ends of the external hair cells and apical processes of phalangeal cells form a regular mosaic termed collectively the *reticular lamina* or *membrane* (Engström et al 1966). The organ is covered by the *tectorial membrane*, a shelf of stiff gelatinous proteinaceous material; a narrow gap separates this from the reticular lamina except where the apical stereocilia of the external hair cells project to make contact with it. In addition to the *inner tunnel* (*cuniculum internum* or tunnel of Corti), other intercommunicating spaces exist around the outer hair cells also connected with the inner tunnel, including an *outer tunnel* (*cuniculum externum*) between the outermost hair cells and inner cells (of Hensen), under the reticular lamina, and also a cuniculum medium (space of Nuel) between the outer pillar (of Corti) and the external hair cells. The latter tunnel is continuous with the extracellular spaces around the apical two-thirds of the external hair cells. This complex of intercommunicating spaces is filled with perilymph which diffuses into it through the matrix of

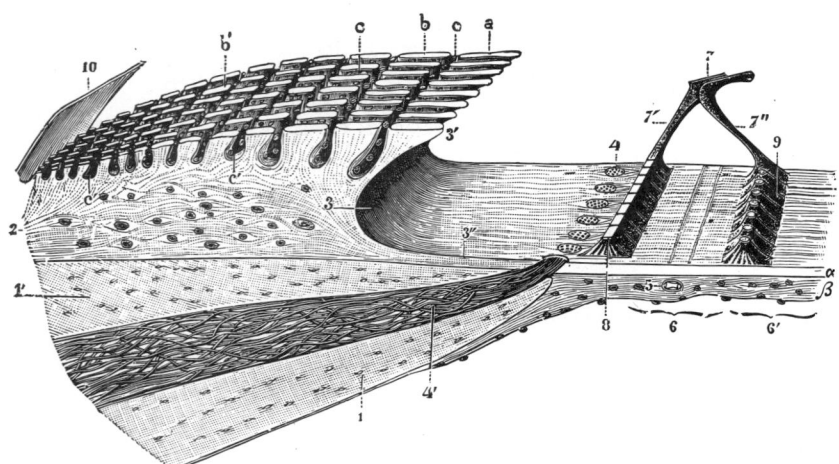

7.362 The limbus laminae spiralis and the basilar membrane (schematic, after Testut). 1, 1'. Lower and upper lamellae of the lamina spiralis ossea. 2. Limbus laminae spiralis, with a, the auditory teeth of the first row; b, b', the teeth of the other rows; c, c', the grooves between the auditory teeth and the cells which are lodged in them. 3. Sulcus spiralis internus, with 3', its labium vestibulare, and 3″, its labium tympanicum. 4. Foramina nervosa, giving passage to the nerves from the spiral ganglion. 5. Vas spirale. 6. Zona arcuata, 6′. zona pectinata of the basilar membrane, with α, its hyaline layer, β, its connective tissue layer. 7. Summit of the tunnel of Corti, with 7′, its inner rod, and 7″, its outer rod. 8. Bases of the inner rods, from which the cells are removed. 9. Bases of the outer rod. 10. Part of the vestibular membrane.

the basilar membrane. The fluid in these spaces is also sometimes called the *cortilymph* and it is possible that minor alterations in perilymphatic composition occur within it, as it is exposed to the activities of synaptic endings and specialized excitable cells.

The pillar cells (rods of Corti) have a base or *crus*, an elongated *scapus* (rod) and an upper end or *caput*; each crus and caput are in contact but the scapi are separated by the inner (Corti's) tunnel. The nucleus is in the triangular crus of each cell. The scapus is finely striated but an oval non-striated part of the caput stains deeply with carmine. Electron microscopy shows many microtubules 30 nm in diameter, arranged in linked parallel bundles of 2000 or more in the scapus and diverging above to terminate in superficial dense granular cytoplasm, the *cuticle*, in the head. Microtubules start in a wide area of limiting membrane in dense cytoplasmic material in the crus. In the body transverse sections show microtubules often arranged in a rectangular lattice with adjacent individuals linked to actin filaments lying parallel to them; many of them curve into the reticular lamina (vide infra) to end near the junctional complexes, with similar expansions from phalangeal cells (supporting cells of Deiters) and the subapical lateral surfaces of outer hair cells. Although the 'rod', as originally interpreted, corresponds only to the cytoskeletal structures and adjacent cytoplasmic densities, this term has now come to include the whole cell which forms these organelles. Their nuclei are situated in the foot-like expansion resting on the basal lamina.

Internal pillar cells (rods) number almost 6000, their bases resting on the basilar membrane near the tympanic lip of the sulcus spiralis internus. Their bodies form an angle of about 60° with the basilar membrane, their heads resembling the ulna's proximal end, with deep concavities for the heads of the outer pillar cells. Their heads overhang those of the latter.

External pillar cells (rods), almost 4000 in number, are longer and more oblique, forming with the basilar membrane an angle of only 40°. Their heads, convex internally, fit into the concavities on the heads of the inner pillar cells and project externally as thin, *phalangeal processes*, which unite with the phalangeal processes of the phalangeal (Deiters') cells to form the *reticular lamina* or *membrane* (vide infra).

The distances between the bases of the internal and external pillar cells increase from the cochlear base to its apex but the angles between them and the basilar membrane diminish.

Hair cells (epitheliocyti pilosi) are columnar or slightly piriform, depending on their position in the cochlea. Their apices are level with the heads of the pillar cells, each surmounted by about 50–100 'hairs' or stereocilia. Their bases reach about halfway down the thickness of the spiral organ and each has a large, basally placed euchromatic nucleus. The basal plasma membrane forms synaptic contacts with the cochlear nerve fibres. They are arranged as inner and outer groups, tilted at an angle to each other and differing in detailed structure and innervation. *Inner* hair cells, almost 3500 in number, form a single row internal to the inner pillar cells; since they exceed the rods in diameter, each inner hair cell contacts more than one. Their free (apical) ends are encircled by a cuticular membrane formed by surrounding supporting (inner phalangeal) cells, fixed to the heads of the inner pillar cells (vide infra). The neighbouring supporting cells continue towards the modiolus as columnar, then cuboidal cells lining the sulcus spiralis internus (*border cells*). The *outer hair cells* number about 12 000 and are nearly twice as tall as the internal. In the basal cochlea they are arranged in three regular rows, to which, in the human cochlea (although not in many other species), is added variably a fourth and sometimes a fifth, less regular in arrangement, near its apex.

Electron microscopy has shown much detail in cochlear hair cells and supporting structures (e.g. Babel et al 1970). **Inner hair cells** resemble type I vestibular hair cells (p. 1232); each has a short piriform body, the basal expansion containing a large euchromatic nucleus (7.363, 364), surmounted by a narrower apex which bears on its surface 50–60 stereocilia but no kinocilium. Cytoplasm contains an abundance of organelles, indicative of the high metabolic rate, including many mitochondria in the apical region, free polyribosome groups, agranular endoplasmic reticulum, various vesicles and lysosomes.

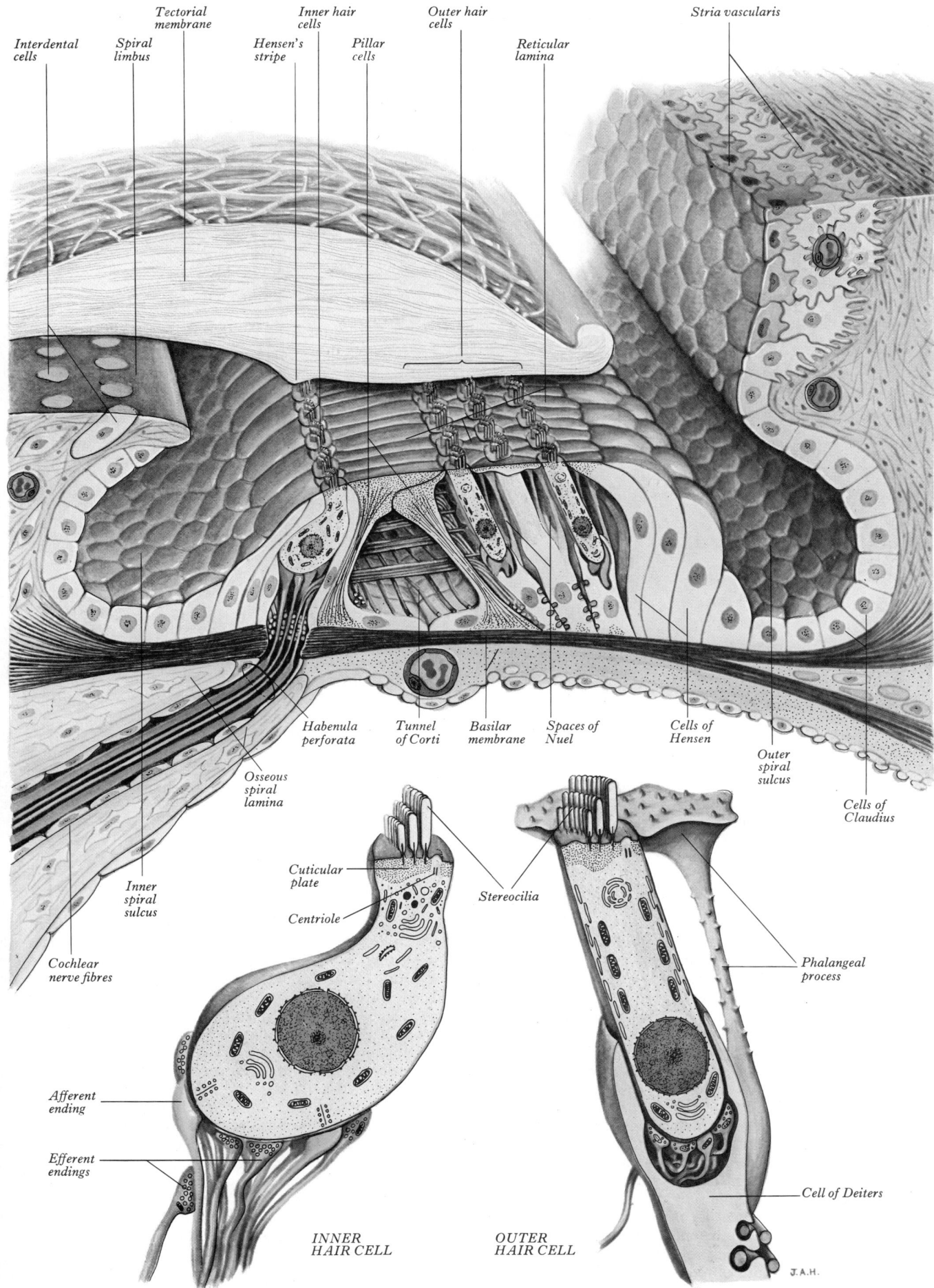

Interdental cells
Spiral limbus
Tectorial membrane
Hensen's stripe
Inner hair cells
Pillar cells
Outer hair cells
Reticular lamina
Stria vascularis

Habenula perforata
Tunnel of Corti
Basilar membrane
Spaces of Nuel
Cells of Hensen
Outer spiral sulcus
Cells of Claudius

Osseous spiral lamina
Inner spiral sulcus
Cochlear nerve fibres

Cuticular plate
Centriole
Stereocilia
Phalangeal process

Afferent ending
Efferent endings

Cell of Deiters

INNER HAIR CELL
OUTER HAIR CELL

J.A.H.

7.363 Three-dimensional schema of the structure of the cochlear spiral organ and stria vascularis, showing the arrangement of the various types of cell and their overall innervation. The organization of the inner and outer hair cells and their synaptic connections are depicted below. Sensory nerve terminals are coloured green and efferent fibres purple. See text for variant terminology.

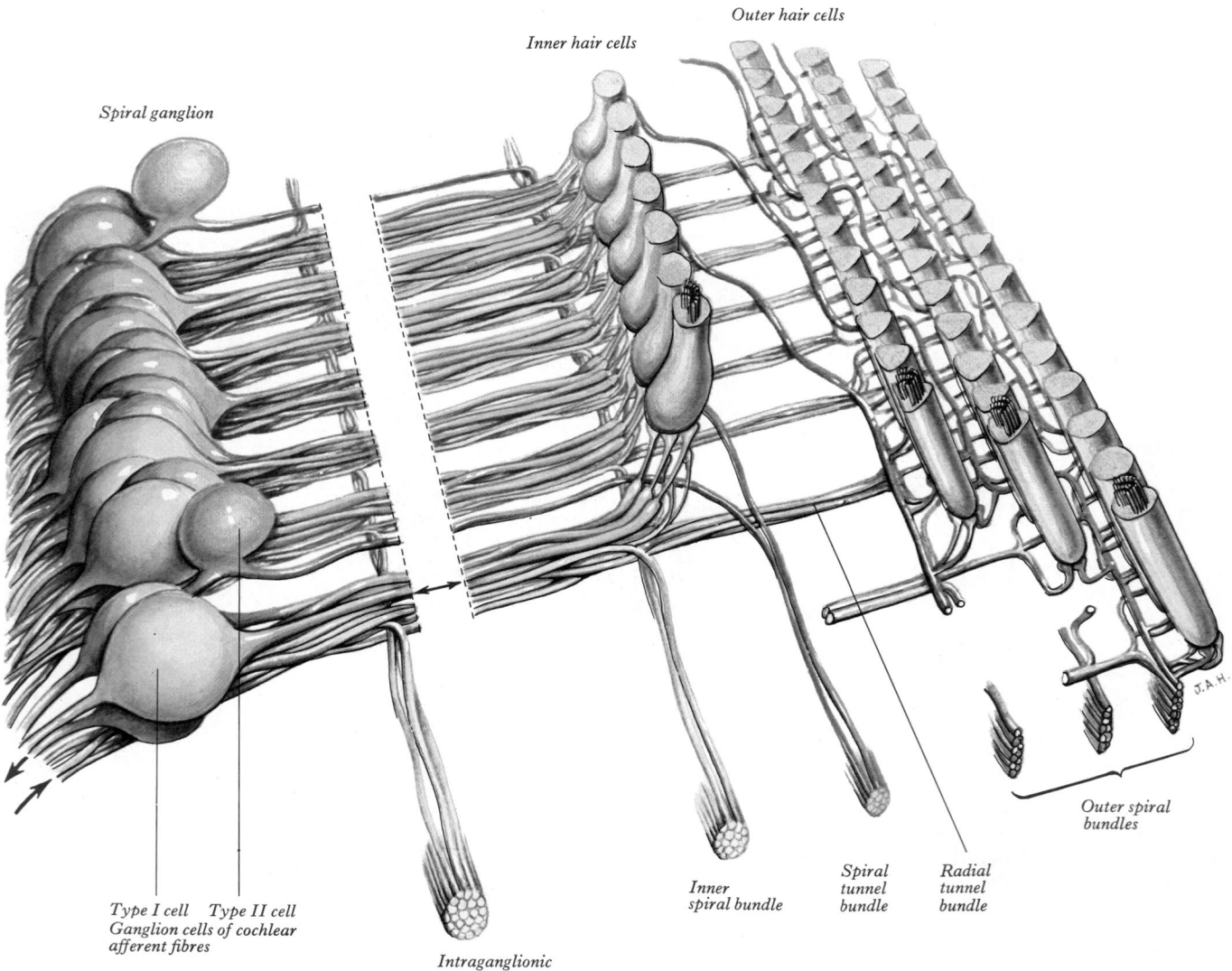

Spiral ganglion

Inner hair cells

Outer hair cells

Type I cell   Type II cell
Ganglion cells of cochlear
afferent fibres

Intraganglionic
spiral bundle

Inner
spiral bundle

Spiral
tunnel
bundle

Radial
tunnel
bundle

Outer spiral
bundles

7.364 The innervation of the spiral organ, showing the distribution of afferent and efferent fibres. The ganglion cells of the sensory nerve fibres include those related to the inner hair cells (dark green) and others innervating the outer hair cells (light green). Efferent fibres are depicted in purple. Note the great contrast between the multiple, convergent afferent innervation of the inner hair cells (about 10 fibres to each cell) and the divergent supply of the outer hair cells (one afferent fibre to about 10 cells).

Microtubules and microfilaments occur, especially in the basal cytoplasm. Under the apical surface a thick layer of filamentous, granular material, the *cuticular plate*, forms a continuous cap except at the apical region nearest the inner tunnel (of Corti) where a centriole is present. The plate is attached around its perimeter by intercellular junctions to the nearby phalangeal and pillar cells.

The sterocilia resemble those of the vestibular hair cells, being very narrow at their base but expanding to a cylindrical process maximally 6 μm long and 0·2 μm across. They are like other microvilli in being enclosed by an apical plasma membrane and they contain numerous longitudinal microfilaments of actin. However, internally they have a highly ordered structure (see e.g. Tilney 1985). Basally an additional dense central axial rootlet penetrates the cuticular plate as an anchorage. Stereocilia are able to yield, at these narrow basal ends, to fluid movements or other mechanical disturbances resulting from auditory stimuli (vide infra). Unlike the vestibular hair cells, a kinocilium is absent from mature cochlear hair cells, though it occurs in early development. In its place is a centriole beneath the apical membrane in the cuticle-free gap, probably a functionless remnant although possibly concerned with microtubule formation.

From the surface, groups of stereocilia appear arranged in a shallow U, its base towards the centriole and thus also the inner tunnel. Each stereocilial cluster has three or four rows, each successively taller towards the tunnel; the tallest may possibly touch the tectorial membrane. The basal plasma membrane synapses with two types of terminal bulbs derived from cochlear nerve fibres. (1) *Afferent* terminals contain few mitochondria or microtubules and clear vesicles of varying size. Both the presynaptic (hair cell) and postsynaptic bulbar membranes are thickened by cytoplasmic densities, the latter more so, and the presynaptic cytoplasm contains clear and dense-cored synaptic vesicles. *Accessory synaptic structures* are electron-dense rods, rings or lamellae around which synaptic vesicles cluster. (2) *Efferent* terminal bulbs contain more mitochondria, microtubules and abundant synaptic vesicles, most being small, spherical and clear but some larger, with dense cores. Although often described as synapsing with cell bases, efferent terminals are never seen to make such contacts in feline and infrequently in rodent and human material. More numerous are the contacts between efferent terminals and the lateral aspect of *afferent* bulbs. The efferent fibre presumably modifies transmission in the afferent fibre terminal. Inner hair cells are, apart from these synapses, wholly surrounded by the cytoplasm of internal phalangeal sustentocytes, whose superficial rims encircle the lateral margins of the apical surfaces of hair cells and form occluding junctions with them. Similar junctions exist between the outer aspect of the rim of phalangeal cells and expansions from the inner pillar cells. These specializations form the inner area of the 'cuticular membrane' of light microscopy.

Outer hair cells are more specialized but have features in common with the inner type. They are longer, more cylindrical

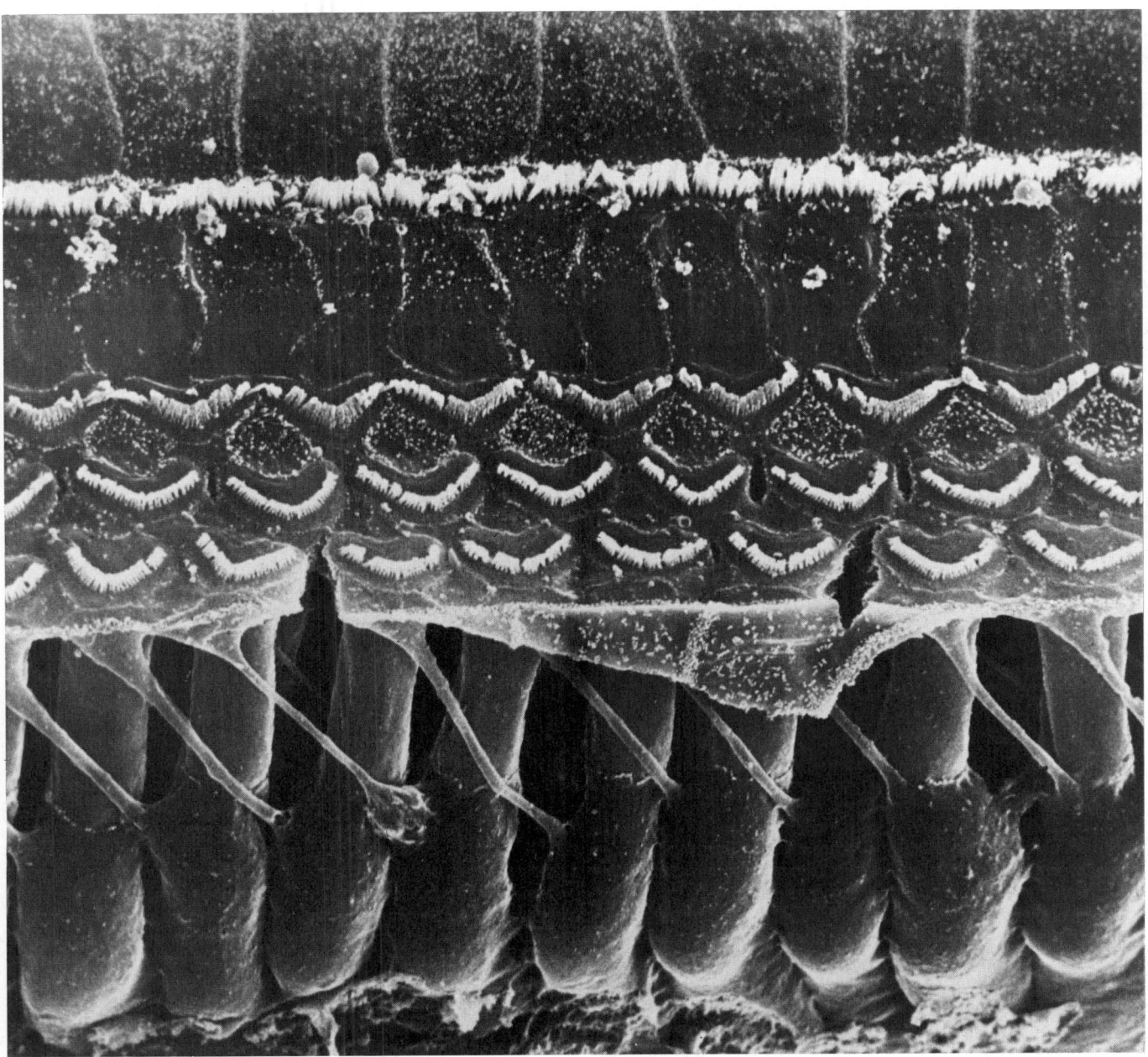

7.365 Scanning electron micrograph of a portion of the spiral organ of Corti (guinea pig) dissected to expose the outer row of outer hair cells and their attendant Deiters' cells with narrow phalangeal processes. The apices of three rows of outer hair cells with their stereocilia are visible in the foreground and behind them are the apices of the rod cells and a row of inner hair cells and their stereocilia. Magnification × 2500.

and have a large basal euchromatic nucleus. Many mitochondria occur near their lateral plasma membranes and in the basal cytoplasm below the nucleus. Otherwise the cell has far fewer organelles than inner hair cells; most notable are the complexes of flat membranous cisternae attached by filaments to the lateral walls or, in the apical cytoplasm, taking the form of concentric whorls (Hensen bodies) and probably precursors of the cisternae. These membranes may help to retain the cell's cylindrical shape but may also be metabolic. Other inclusions are microtubules, microfilaments, polyribosomes, vesicles, lysosomes and glycogen; but most of the cytoplasm is a pale matrix containing fine filaments and granules of unknown composition. A cuticular plate, as in the internal hair cells, caps the cytoplasm and anchors the rootlets of sterocilia, which are arranged in the form of a V or W with its base away from the inner tunnel and towards a centriole in the apical cuticle-free gap. Up to 100 stereocilia have been counted in a single group, arrayed in ranks of graded length (7.365, 366); stereocilia in apical cochlea are taller than those in the basal parts of the cochlear coil but in all groups the longest stereocilia project into corresponding recesses in the tectorial membrane (Hunter-Duvar 1976). It is significant that both inner

and outer hair cells are arranged on the basilar membrane with their centrioles remote from the modiolus (i.e. external), a point of functional interest, indicating a structural polarity which is related to the direction and phase of deformation occurring during auditory stimulation (vide infra). Stereocilia are linked laterally to each other by numerous extracellular filaments; of particular interest are thin strands connecting the apices of short stereocilia to the sides of taller ones ('tip links' Pickles et al 1983, Furness & Hackney 1985) which may be important in sensory transduction.

The base of each external hair cell is cupped in the apex of an external phalangeal cell, except at the synapses. But here afferent *and* efferent terminals, the latter being particularly numerous, are related directly to the plasma membrane of the hair cell (Engström et al 1965).

The external phalangeal cells (of Deiters) (7.363) lie between the rows of outer hair cells, their expanded bases lying on the basilar membrane and their apical ends partially enveloping the bases of hair cells with a finger-like *phalangeal process* extending up diagonally between the hair cells to the reticular membrane, there forming a plate-like expansion completing the gaps

7.366A Scanning electron micrograph of a group of outer hair cells arranged in three rows, showing the arrangement of their stereocilia and related phalangeal processes of the Deiters' cells, collectively forming the reticular lamina (see text). Short microvilli are visible on the surfaces of the non-sensory cells. Magnification × 3000.

between hair cell apices (vide infra). Their cytoplasm contains bundles of microtubules arising from the cell base and continuing up into the phalangeal process; other microtubules reinforce the cup supporting the base of the hair cell. External to them are five or six rows of columnar *supporting cells (of Hensen)* or *external limiting cells* (7.363), beyond which are the *external supporting cells* (of Claudius). Their surfaces bear microvilli. Near the lagaena

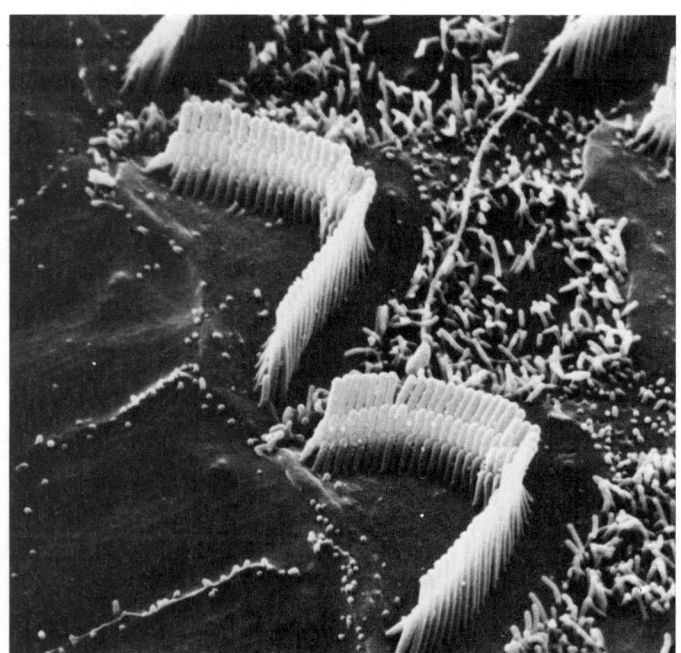

7.366B Scanning electron micrograph of the apices of two outer hair cells showing the different lengths of the stereocilia in the three ranks which constitute each group. Microvilli are also seen on the surface of the Deiters' cells (right). The innermost row of outer hair cells is shown; the tunnel of Corti, roofed by rod cell processes, is on the left. Magnification × 5000. Figs 7.365–369 supplied by Hilary C Dodson and 7.366A, B photographed by Michael Crowder, Guy's Hospital Medical School, London.

they contain fat globules decreasing in number and size down the cochlear duct. Hallpike (1931) suggested that these provide a graduated load tuning the lagaena to low tones.

Near the base of the cochlea, another group of small cells is present among the bases of the phalangeal cells; these are termed Boetschli cells and their functions are unknown.

The reticular lamina (7.363), as seen by light microscopy, is a delicate framework perforated by circular holes occupied by the apices of outer hair cells, extending from the heads of the external rods to the outer row of outer hair cells and completed by several rows of minute cuticular *phalanges*. The innermost row of phalanges are the phalangeal processes of external rods. Ultrastructurally the reticular lamina consists of the horizontal expansions of the outer rods and phalangeal cells carrying bundles of microtubules in an attenuated film of dense cytoplasm. The expansions encircle the apical rims of the hair cells, where occluding zones and desmosomes are formed and the microtubules end. Between adjacent supporting cells are occluding junctions, most apically desmosomes and extensive gap junctions which couple them electrically. Under this delicate support the apical two-thirds of the lateral surfaces of the external hair cells are free and bathed by cortilymph (vide supra). The significance of this arrangement is twofold: the reticular lamina creates a highly im-

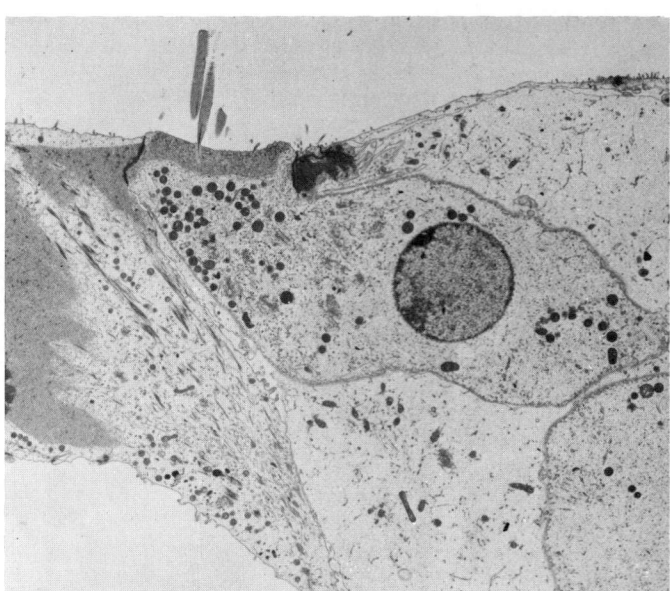

7.367 Electron micrograph of an inner hair cell (guinea pig) showing the apical stereocilia with bases embedded in the dense cuticular plate. Note the centrally placed nucleus and numerous cytoplasmic organelles. The apex of a rod cell is visible on the left and below it the cavity of the tunnel of Corti. Magnification × 3000.

permeable barrier to the passage of ions except through cell membranes during sensory activity; it also forms a rigid support between the apices of the hair cells, coupling them mechanically to the movements of the underlying basilar membrane which causes lateral shearing movements between the cells and the overlying, static tectorial membrane.

If hair cells are lost through trauma by sound or drugs, phalangeal processes rapidly fill the gap, disturbing the regular laminal pattern (*phalangeal scars*) but restoring its function.

The membrana tectoria (7.361, 363), overlying the sulcus spiralis internus and spiral organ, is composed of a stiff, gelatinous plate of proteinaceous composition and fine fibres invisibly embedded in a gelatinous matrix (Iurato 1967, Steel 1978). By electron microscopy filaments of 4 nm in diameter appear, consisting of a collagen-like protein and mucopolysaccharides. In transverse section it has a characteristic shape, the underside being nearly flat and the upper surface convex; it is thin on the modiolar side where it is attached to the vestibular lip of the limbus laminae spiralis, extending centrally as far as the vestibular membrane. Its outer part forms a thickened ridge,

7.368   Scanning electron micrograph of the surface of Reissner's membrane, viewed from the cochlear duct aspect, showing the regular pattern of simple squamous epithelial cells. Magnification × 2500.

overhanging the edge of the reticular lamina. Scanning electron microscopy shows a network of fine ridges on the upper surface, the lower surface being relatively smooth, except where the stereocilia of the outer hair cells leave a pattern of W- or V-shaped impressions. Another feature, in some species, is a longitudinal ridge along its under surface near the inner tunnel, visible by light microscopy as *Hensen's stripe*. The *interdental cells* of the vestibular labium, to which the membrane is attached, have a well developed Golgi apparatus, many mitochondria and free polysomes; they may secrete the membrane. The lips of the stereocilia of the outer hair cells are embedded in the membrane but this attachment is often broken during histological preparation.

The vestibulocochlear nerve divides deep in the internal acoustic meatus into anterior *cochlear* and posterior *vestibular* components. (Their central connections are described on pp. 1111, 1112.)

The vestibular nerve supplies maculae of the utricle and saccule and the ampullae of the semicircular ducts. The *vestibular ganglion* (of Scarpa), from whose bipolar neurons the nerve fibres arise, lies in the trunk of the nerve in the internal meatus. Distal to the ganglion the nerve divides into superior, inferior and posterior vestibular rami. (It sometimes divides on the proximal side of the ganglion, in which case it is divided into three parts, one in each ramus; when this occurs, the posterior one is situated in the foramen singulare.) Filaments of the *superior ramus* traverse the foramina in the superior vestibular area, ending in the utricular macula and ampullary crests of the anterior and lateral semicircular ducts; those of the *inferior ramus* traverse the foramina in the inferior vestibular area, ending in the saccular macula; the *posterior ramus* traverses the foramen singulare, postero-inferior in the meatal fundus, dividing into filaments for the ampullary crest of the posterior semicircular duct.

The cochlear nerve divides into many filaments at the modiolar base; those for the basal and middle coils traverse the foramina in the tractus spiralis foraminosus and those for the apical coil traverse the central canal, bending out to pass between the lamellae of the osseous spiral lamina. The *spiral ganglion* (7.360), containing the cochlear bipolar neurons, occupies the spiral canal of the modiolus. Reaching the rim of the osseous spiral lamina, the nerve fibres lose their myelin sheaths and pass in bundles through minute foramina in the tympanic labium (*habenulae perforatae*); some fibres end around the bases of the

inner hair cells, while others pass between the pillar cells across the inner tunnel to the outer hair cells; they pass between the phalangeal cells, often invaginated in grooves along their sides. The outer hair cells in the basal and middle coils are more richly innervated than those in the apical coil. A vestibular branch of the cochlear nerve supplies the vestibular end of the cochlear duct; filaments of this traverse the foramina in the cochlear recess (p. 1229).

The spiral ganglion neurons are of two types (Spoendlin 1974): type I are large, bipolar cells giving off a large myelinated axon from both poles of the cell, one central and the other peripheral in direction; type II neurons are smaller bipolar cells and have nonmyelinated process, the centrally directed axon being particularly narrow. Experimental studies show that type I cells are exclusively afferent to the inner hair cells, 10 ganglion cells being connected to each sensory cell, whereas type II neurons are afferent only to outer hair cells, each supplying more than 10 of these by extensively branched terminals (Spoendlin 1978). There are also efferent nerve fibres within the cochlear nerve, belonging to the olivocochlear system described by Rasmussen (1946), which contains crossed and uncrossed components, the former from retrolateral olivary neurons and the latter from the S-shaped segment of the ipsilateral olivary complex (Whitfield 1967, Iurato 1974). The pathway is apparently purely inhibitory, probably altering the threshold levels of outer hair cells and modifying transmission via the postsynaptic action on the afferents from the inner hair cells. This system has been studied intensively in recent years, in different species, and it has been found that the inner and outer hair cells receive efferents from two distinctive components of the olivocochlear tract. The lateral olivocochlear tract arises mainly from the ipsilateral superior olivary complex and supplies the inner hair cells, while the medial tract, mainly crossed, innervates the outer hair cells. Different neurotransmitters occur in these two tracts.

The distribution and courses of the nerve fibres to both groups of hair cells is known in some detail and has proved to be quite complex. Most cochlear fibres are distributed to the internal hair cells, each receiving the terminals of several radially disposed fibres, themselves the sites of synapses from collateral branches of radial inhibitory efferent fibres. Outer hair cells receive a minority of cochlear fibres; their efferent fibres are radial, each synapsing with many hair cells. Afferent fibres curve into a regularly organized spiral system, each fibre innervating numerous hair cells. These fibres form distinct groups with specific cochlear orientations and positions, either radial or spiral, i.e. parallel to the organ of Corti (Spoendlin 1974). Among the various cochlear cells is, firstly, the spiral intraganglionic bundle in the modiolus, consisting of non-myelinated fibres considered to be efferent axons; a similar group of efferent internal *spiral fibres* runs under the row of inner hair cells, traversed by groups of axons emerging from the habenulae. More peripherally, *spiral tunnel fibres* run along the inner canal in bundles or scattered. Individual fibres also run along spaces between the external pillar and the phalangeal cells in three rows, flanked and partially invaded by these supporting cells; these ranks are *external spiral fibres*, consisting of afferent and efferent connections of outer hair cells, as do the spiral tunnel fibres already mentioned. Autonomic endings have been observed but appear to be restricted to regions of the spiral organ on the modiolar side of the inner tunnel. Superimposed on this spiral pattern are the radial fibres afferent to the inner hair cells and, crossing in the tunnel, to outer hair cells. A detailed analysis of cochlear innervation is summarized in illustration 7.364.

The arteries of the labyrinth are: (1) the labyrinthine artery (p. 753), arising from the anterior inferior cerebellar artery or more often the basilar artery; (2) the stylomastoid branch of either the occipital artery or the posterior auricular artery (see p. 739). The labyrinthine artery divides at the bottom of the internal meatus into cochlear and vestibular rami. The cochlear ramus divides into 12–14 twigs, traversing the canals in the modiolus and distributed as a capillary plexus to the lamina spiralis and basilar membrane. Vestibular branches are distributed to the utricle, saccule and semicircular ducts.

The veins of the vestibule and semicircular canals accompany

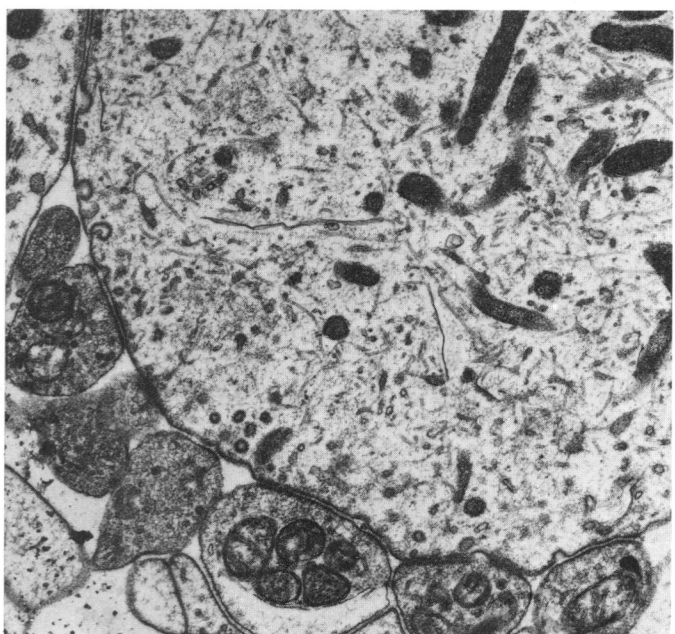

7.369A Electron micrograph of a group of efferent nerve endings at the base of an outer hair cell (guinea pig). Note the numerous mitochondria and vesicles within these endings and the cytoplasmic densities on both the pre- and postsynaptic membranes. Magnification × 10 000.

the arteries; receiving the cochlear veins at the base of the modiolus, they form the labyrinthine vein, ending in the posterior part of the superior petrosal sinus or in the transverse sinus. A small vessel from the basal cochlear turn traverses the cochlear canaliculus to join the internal jugular vein (see p. 795).

THE MECHANISM OF AUDITORY RECEPTION

Much research has been directed to the role of the different auditory components as analysers of sound intensity and frequency patterns and of the sources of the trains of sound waves impinging on them. Sound waves reaching the air column in the external acoustic meatus cause a comparable set of vibrations in the tympanic membrane and auditory ossicles. At the stapedial foot plate the force per unit area of the oscillating surface is increased 20 times. This overcomes the inertia of the perilymph, thus producing in it pressure waves which are conducted almost instantaneously to all parts of the basilar membrane. The latter varies continuously in width, mass and stiffness from the basal to the

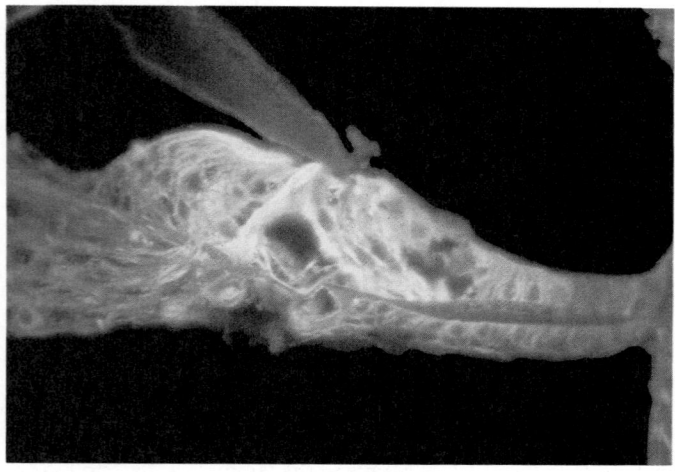

7.369B Section of the spiral organ, labelled by immunofluorescence to show the presence of tubulin, especially in the pillar and Dieters' cells. Provided by Alun Thomas, UMDS, Guy's Campus, London.

apical end of the cochlea; but its fibres are *not* under tension, as was once assumed. The behaviour of such a mechanical system as it reacts to periodic pressure waves in the perilymph varies with the frequency of the vibration. At very low frequencies, say 50 Hz, the whole membrane vibrates in phase at a similar frequency. As the frequency of the driving fluid pressure rises, different parts oscillate less rapidly and increasingly out of phase, from the basal to the apical end. *Travelling waves* hence progress along the membrane from the base to the apex. At intermediate frequencies, their *amplitude* rises slowly as they progress from the basal end to a maximum, after which the amplitude falls rapidly. With increasing frequency, the *locus* of *maximum amplitude* moves progressively from the apical to the basal end of the cochlea. Evidence for such a vibratory distribution in the basilar membrane comes from five main sources; (1) observation of the membrane excited in cadavers through drill holes in the cochlear bone, using stroboscopic illumination; (2) indirect biophysical measurements (e.g. by the Mossbauer effect) of vibrations with a probe placed in the (drained) perilymphatic space; (3) electrophysiological recording of *cochlear microphonic potential*, giving the total electrical activity in short sectors of the basilar membrane; (4) the effects of focal lesions at different loci in the membrane; (5) measurements of evoked otoacoustic emissions (see p. 1243) from the cochlea, combined with (3) and (4).

The pattern of vibrations in the basilar membrane thus varies with the intensity and frequency of the acoustic waves reaching the perilymph. Because of the arrangement of the hair cells on the basilar membrane, such oscillations generate a largely transverse shearing force between the outer hair cells which rest (indirectly) on the basilar membrane and the overlying tectorial membrane, in which the apices of stereocilia are embedded (vide infra). The inner hair cell stereocilia, which probably do not touch the tectorial membrane although they come very close to it, are likely to be stimulated by local movements of the endolymph. At most intermediate frequencies an appreciable sector of the basilar membrane oscillates and hence a specific group of auditory nerve fibres are activated. While the mechanical behaviour of the basilar membrane is mainly responsible for a rather broad discrimination between different frequencies, the very fine tuning of which the cochlea is capable appears to be related to differences in physiology between individual hair cells which make them respond best to particular tones, even when experimentally isolated from the spiral organ. Such individual tuning of cells may be caused by differences in shape, stereocilia length or possibly variations in the molecular composition of their sensory membranes. To these factors may be added the activities of the outer hair cells, which are as yet not fully understood but may play an important part in regulating inner hair cell sensitivity at specific frequencies. This is most dramatically seen in the ability of hair cells to change length when stimulated electrically, at frequencies of many thousands per second. The rapidity of such contractions, and various other features, indicate a novel type of contractile mechanism (see Holley & Ashmore 1988). How such motility is used in hearing is not yet understood, but may be related to the need to adjust the mechanics of the spiral organ constantly, under the influence of the efferent pathway (see Brownell 1983, Kim 1985).

At a particular frequency an increase in the *intensity* of stimulus is signalled by an increase in the *number* of activated cochlear nerve fibres (recruitment) *and* in *their rate of discharge*. For discussion of these hypotheses consult Békésy (1960), Spoendlin (1968), Pickles (1985). Pye (1979) has studied the dimensions of the cochlea in many mammalian orders, including primates, equating the dimensions of the basilar membrane with the recorded auditory range. Her results, and others quoted by her, show no simple relation between the degree of cochlear spiralling and basilar membrane width with the frequency range in any mammalian group.

The roles of the two groups of hair cells have been much debated, particularly since differences in their innervation and physiological behaviour have become obvious. Because of their rich afferent supply internal hair cells have been favoured as the major source of auditory sensation, a view supported by much experimental and clinical evidence; e.g. animals treated with

EXTERNAL EAR
Sound collection and amplification;
source location.

MIDDLE EAR
Amplification of signal (force
per unit area); impedance
matching between air and
water vibrations; neural
reflex and mechanical damping
of excessive vibration; pressure
equalizing through tympanic tube.

INNER EAR
Mechanical and neural filtering and
analysis of signals by spiral organ;
stimulus transduction by sensory
cells; action potential initiation
at synapses between cochlear nerve
fibres and sensory cells; central
control by centrifugal fibres.

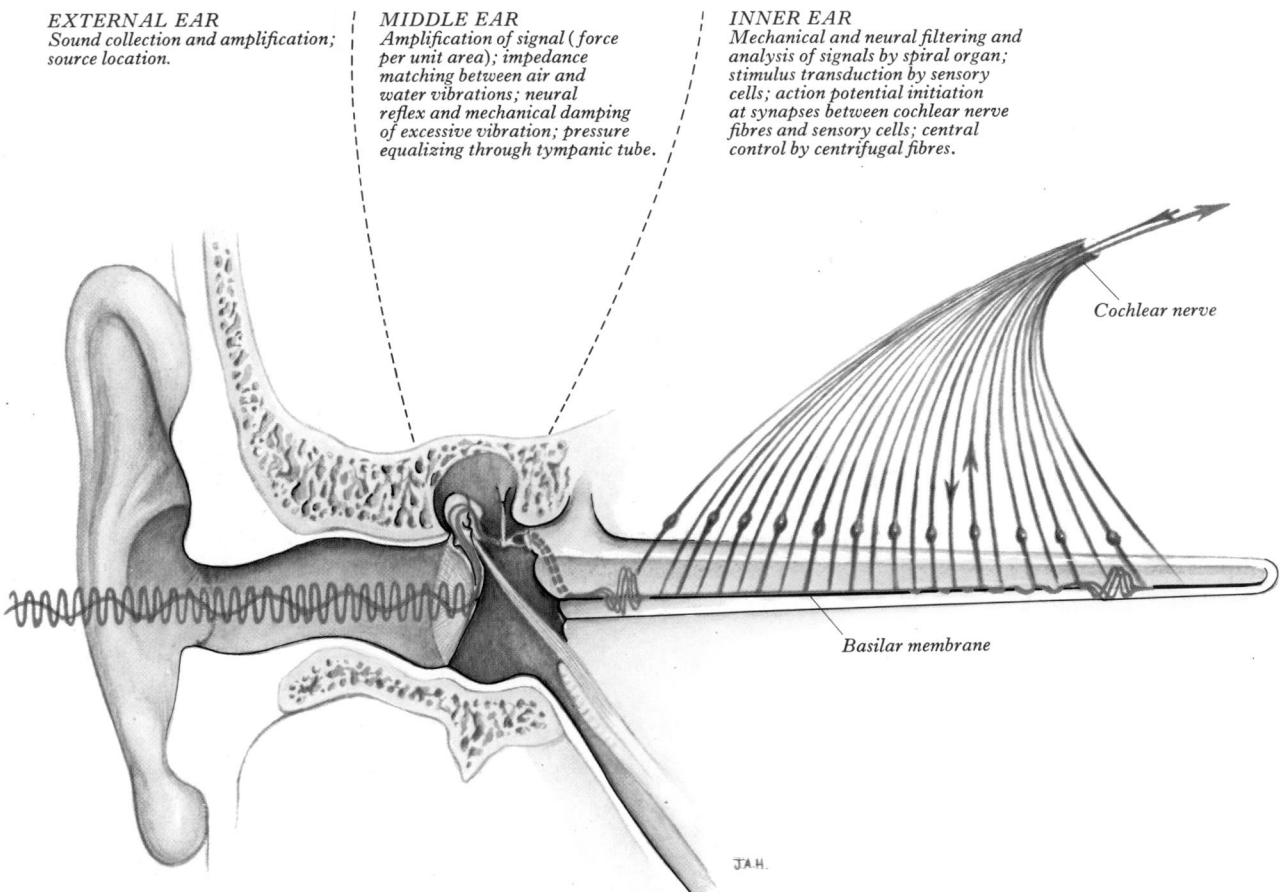

Cochlear nerve

Basilar membrane

JA.H.

7.370 The principal activities of the peripheral auditory apparatus. For clarity the cochlea is depicted as though it had been uncoiled. The points of maximal stimulation of the basilar membrane by high frequency (blue) and low frequency (red) vibrations, together with their transmission pathways through the external and middle ear, are also indicated.

antibiotics specifically toxic to outer hair cells are still able to hear but discrimination is impaired (Harrison & Evans 1979). The possible role of outer hair cells in the fine tuning of auditory responses to particular frequencies may be related to the recent observation that these cells can, when electrically stimulated, change length very rapidly, at frequencies similar to those of sound stimuli, contracting when depolarized and extending when hyperpolarized (Brownell et al 1985, Holley & Ashmore 1987). The cause of this motile response does not appear to be actin-myosin based, as it is too rapid; it may be associated with a network of fine filaments found beneath the cell membrane (Bannister et al 1988, Holley & Ashmore 1988), but further investigations are needed to elucidate this.

Electrical responses of the cochlea are of considerable interest; extracellular potentials can be readily recorded with relatively crude electrode techniques (Evans 1974, Yost & Nielsen 1977). The most significant is the *endolymphatic potential*, a steady potential recordable between the cochlear duct and the scala tympani and caused by the different ionic compositions of their fluids. Since the resting potential of hair cells is about 70 mV (negative inside) and the endolymphatic potential is positive in the cochlear duct, a total transmembrane potential across the apices of hair cells is 150 mV, a higher resting potential than found anywhere else in the body.

Under stimulation by sound, a rapid oscillatory *cochlear microphonic* potential can be recorded, matching the frequency of the stimulus and movements of the basilar membrane precisely. This appears to depend on fluctuations in the conductance of hair cell membranes, probably that of the outer hair cells. At the same time an extracellular *summating potential* develops, a steady direct current shift related to the (intracellular) receptor potentials of hair cells. Cochlear nerve fibres then begin to respond with *action potentials* also recordable from the cochlea. Intracellular recording of auditory responses from inner hair cells (Russell & Sellick 1977) has shown that these cells resemble other receptors, their steady receptor potentials being related in size to the amplitude of the acoustic stimulus. At the same time, afferent nerve fibres are stimulated by synaptic action at the bases of the inner hair cells, firing more rapidly as vibration of the basilar membrane increases in amplitude.

SPLANCHNOLOGY

Introduction

The grouping of the organs of respiration, alimentation, urinary excretion, reproduction and the endocrine glands under the umbrella of 'Splanchnology' is more traditional than rational. There is, perhaps, some reason in dividing bodily fabric and function into somatic or locomotor and visceral or metabolic moieties, but osteology, arthrology, myology, angiology and neurology are not customarily incorporated. In any case there is extensive overlap between all the above and the miscellany of viscera or *splanchnoi* gathered under the label of 'splanchnology'. Again, splanchnology is no sooner formed than dismembered into constituent 'systems', as in this volume. Our only reasons for retaining the term are convenience and, admittedly, tradition.

The viscera are, by definition, partly the structures located within, or closely associated with, the pleural and peritoneal cavities and therefore related to the linings of these cavities. Many of these are tubular or saccular in form (hollow viscera), but some are solid; both are suspended within the coelomic cavities by connective tissue sheets covered on either side by a serous membrane or serosa (composed of a single layer of mesothelium and underlying lamina propria), through which vessels, nerves and ducts gain access to the viscera. Other viscera are located on the body wall external to the serous lining and, in the case of the abdomen, are then *retroperitoneal*; or they are more remote from the coelom-derived cavities and are surrounded by the connective tissue of the head, neck and pelvic wall. The serosa is lubricated by tissue fluid, is smooth and allows free movement of adjacent structures within the pleural and peritoneal cavities. The *hollow viscera* (alimentary, respiratory and urogenital tracts) are also lined internally by a mucous membrane (mucosa) consisting of an epithelial layer with its underlying lamina propria and associated glands and, in some parts of the gut, a thin layer of non-striated muscle. The hollow viscera are enclosed in fibromuscular coats in which non-striated muscle, controlled by the autonomic nervous system, predominates.

Extensions of the epithelial and connective tissue linings of the viscera form various types of exocrine glands which pass their secretions to the mucosal surface. Some of these glands are restricted to the walls of the viscera, but others may form very extensive *solid viscera* themselves, e.g. the pancreas and liver. Endocrine glands secreting into the circulation are in some cases derived from the hollow viscera during development (e.g. the thyroid gland), whereas others are formed quite independently from neural (adrenal medulla, neurohypophysis) or mesodermal (adrenal cortex, Leydig cells) sources.

Glandular tissue can be separated into two major categories: the *exocrine* glands secreting through ducts into the hollow viscera (or body surface), and *endocrine glands* (p. 1450), passing their products into the circulation.

Exocrine glands can be classified according to a number of different criteria (p. 57): (1) their mode of secretion (merocrine, apocrine or holocrine); (2) their secretory chemistry (mucous, serous, zymogen, etc.); (3) their morphology (unicellular, multicellular, tubular, acinar/alveolar, tubulo-alveolar). These different categories are illustrated in **8.1A**.

Mode of secretion. Most glands secreting proteinaceous material release their products by the *merocrine* method, vesicles

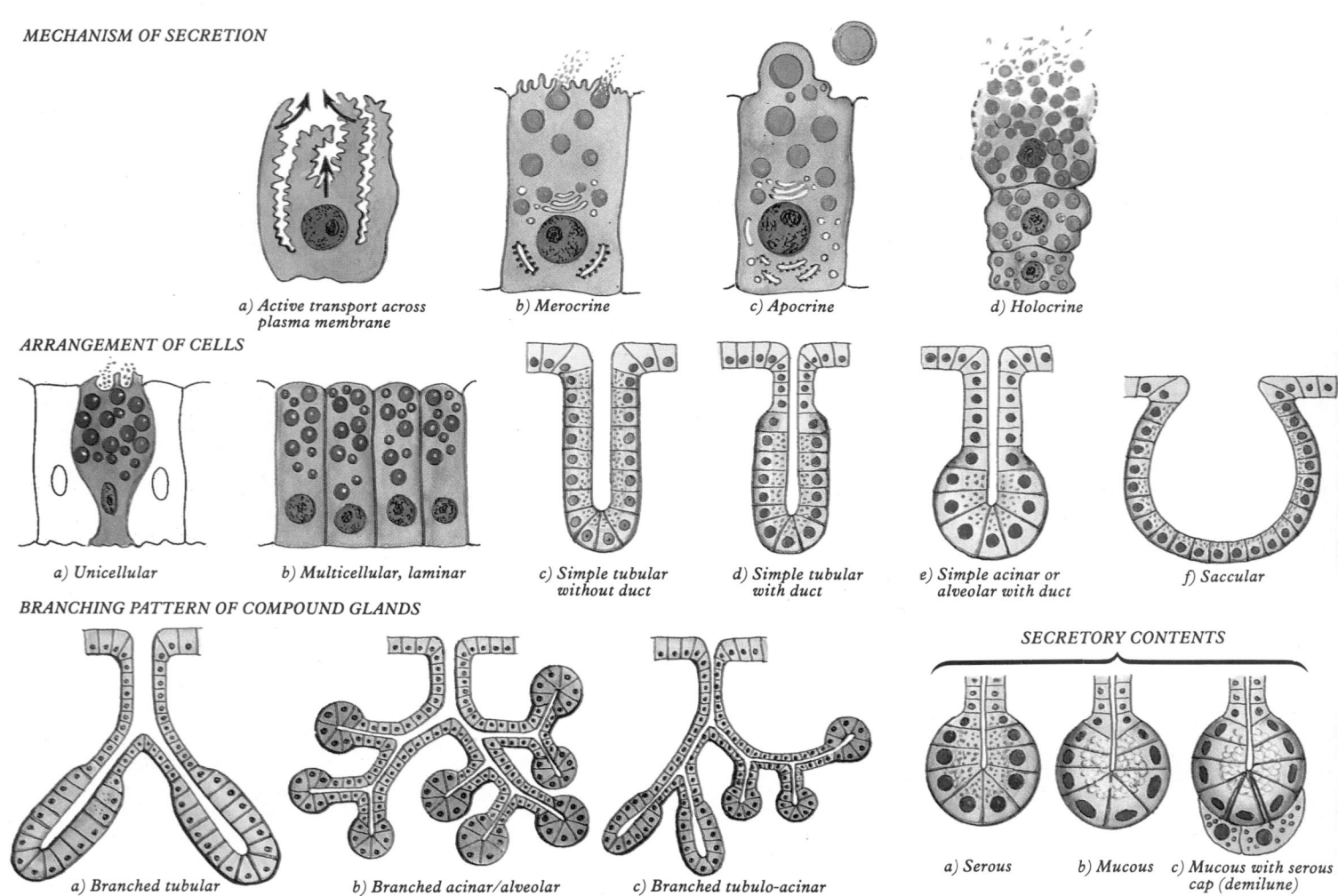

MECHANISM OF SECRETION

a) *Active transport across plasma membrane* b) *Merocrine* c) *Apocrine* d) *Holocrine*

ARRANGEMENT OF CELLS

a) *Unicellular* b) *Multicellular, laminar* c) *Simple tubular without duct* d) *Simple tubular with duct* e) *Simple acinar or alveolar with duct* f) *Saccular*

BRANCHING PATTERN OF COMPOUND GLANDS

a) *Branched tubular* b) *Branched acinar/alveolar* c) *Branched tubulo-acinar*

SECRETORY CONTENTS

a) *Serous* b) *Mucous* c) *Mucous with serous cap (demilune)*

8.1A Schema which shows the different types of glands classified by currently used methods.

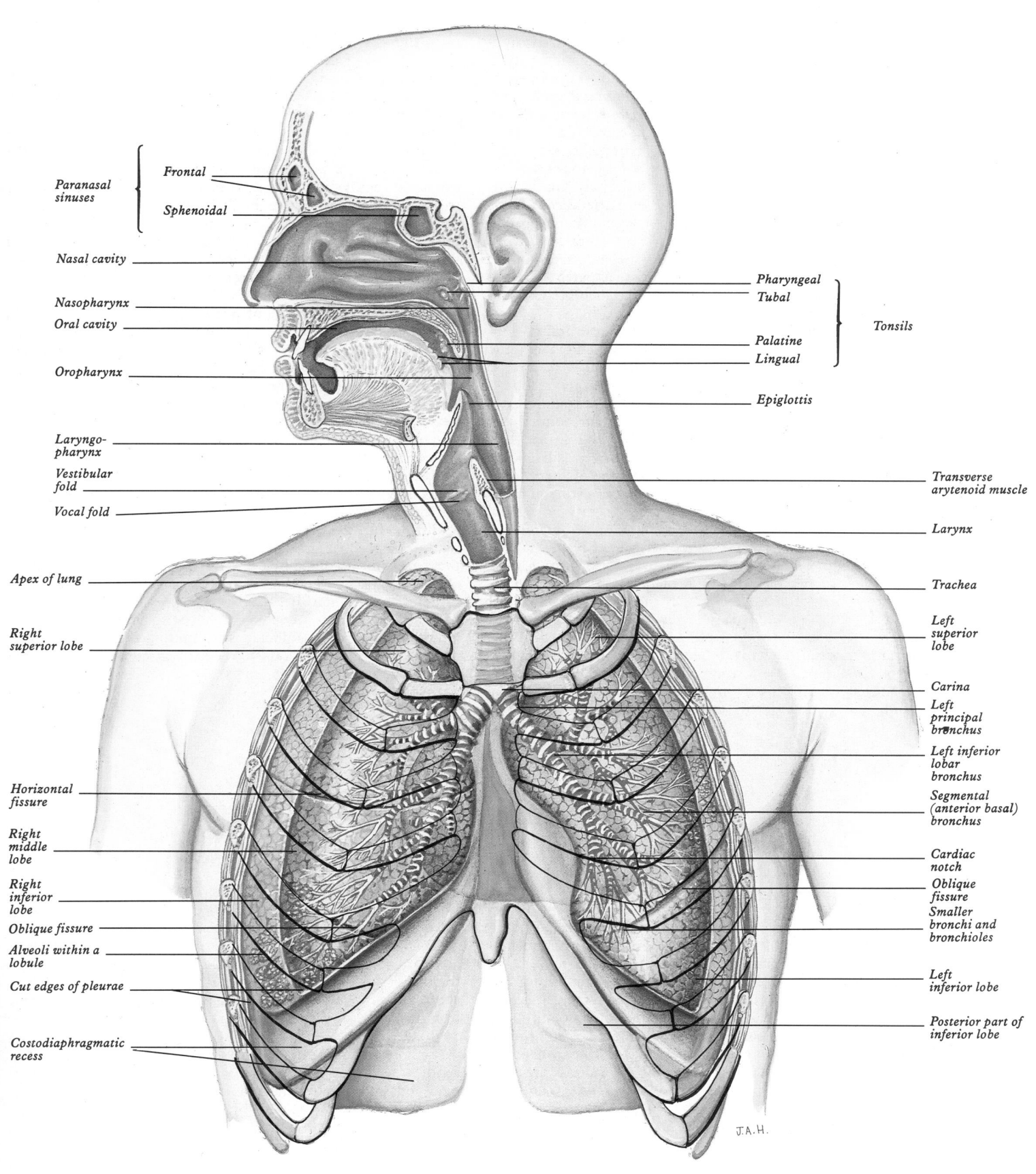

Paranasal sinuses

Frontal

Sphenoidal

Nasal cavity

Nasopharynx

Oral cavity

Oropharynx

Laryngo-pharynx

Vestibular fold

Vocal fold

Apex of lung

Right superior lobe

Horizontal fissure

Right middle lobe

Right inferior lobe

Oblique fissure

Alveoli within a lobule

Cut edges of pleurae

Costodiaphragmatic recess

Pharyngeal

Tubal

Tonsils

Palatine

Lingual

Epiglottis

Transverse arytenoid muscle

Larynx

Trachea

Left superior lobe

Carina

Left principal bronchus

Left inferior lobar bronchus

Segmental (anterior basal) bronchus

Cardiac notch

Oblique fissure

Smaller bronchi and bronchioles

Left inferior lobe

Posterior part of inferior lobe

J.A.H.

8.1B Diagrammatic representation of the respiratory tract. Those parts of the tract in the head and upper neck are shown in sagittal section, in the lower neck turned anteriorly and, in the remainder of the tract, from the anterior aspect.

1247

containing material synthesized in the granular endoplasmic reticulum and Golgi complex being transported to the cell surface where fusion of the vesicle membrane with the plasma membrane occurs, so releasing the secretory product to the outside. All glands of the alimentary and respiratory tracts are merocrine. In *apocrine* secretion, cells pass their secretory vacuoles to the apical region from which portions of cytoplasm detach periodically, to release the secretion when they disintegrate. This mode of secretion is found in only a few regions, including the mammary glands and apocrine sweat glands. In *holocrine* secretion, the cell progressively fills up with secretory droplets and finally disintegrates; this process appears to be restricted to the epidermal sebaceous glands (p. 92). In addition, some cells release their products without visibly detectable exocytosis, particularly when secreting small molecules or ions (e.g. the parietal cells of the gastric mucosa). In endocrine glands polypeptide or protein-secreting cells are merocrine, but steroid-secreting cells, which typically have many of their synthetic enzymes associated with an extensive agranular endoplasmic reticulum, apparently release their products directly through their plasma membranes, since their products are generally lipid soluble.

Secretory products. These vary greatly according to location, physiological state, degree of development and many other factors. Traditionally the glands of the alimentary, respiratory and urogenital tracts are often classified as either serous or mucous. Serous glands typically have eosinophilic granules and a rounded nucleus in the basal half of the cell, surrounded by a basophilic rim of cytoplasm. Mucous cells, in contrast, have nuclei which lie close to the basal margin of the cell (surrounded also by basophilic cytoplasm) and a wide zone of more apical cytoplasm, often occupying most of the cell, consisting of frothy mildly basophilic secretory material rich in mucinogen and therefore staining strongly for carbohydrate. Mixed glands also occur, mucous lobules sometimes being capped by serous cells (serous demilunes). However, more discerning histochemical methods have shown that this simple picture is far from reflecting the great variety of secretions which these cells contain; cells which are classically serous may secrete many different proteins and carbohydrates and mucus can vary widely in chemistry in different regions of the body. The terms mucous and serous are perhaps most applicable to the major salivary glands, but even there they are really only a convenient shorthand for superficial, if rapid, clinical diagnosis (p. 1293).

Glandular morphology. Exocrine glands are either unicellular (e.g. goblet cells, p.53) or multicellular. The latter may be in the form of simple sheets of glandular cells, as at the surface of the stomach, or may be reflected as diverticula of various shapes and extents. Such glands may be single units (simple glands) or branched, compound structures. Simple unbranched tubular glands exist in the walls of many of the hollow viscera, e.g. the small intestine and uterus, while some glands have expanded, flask-like ends (acini or alveoli). Such glands may consist entirely of secretory cells, or may have a blind-ending secretory portion leading through a non-secretory duct to the lumen of the viscus, in which case the ducts may modify the secretions as they pass along them. Glands with ducts may be branched (compound), sometimes forming elaborate ductular trees. Compound glands generally have acinar or alveolar secretory lobules, but the secretory units may alternatively be tubular or mixed tubulo-acinar. While the significance of these different types of geometry is not yet understood, they all constitute ways of creating a larger area for secretion than would be available at a simple secretory surface.

The high metabolic rate of many such glands is reflected in their rich vascular beds which can often be controlled by autonomic vasomotor nerves to modify the activity of the gland. Gland cells are also often under direct secretomotor control by means of synapse-like junctions with efferent nerve endings, which may exert complex actions by a variety of different neuromediator chemicals (see p. 889). Neural control may also be very rapid where contractile myoepithelial cells lie around the secretory units and smaller ducts, since under nervous stimulation the contraction of these can express secretions into the major ducts and so initiate discharge into the visceral lumen, e.g. in salivary glands.

Endocrine glands, secreting directly into the circulation, tend to be grouped around rich beds of capillaries or sinusoids; these are typically lined by fenestrated endothelia (p. 689) which allow the passage of macromolecules through their walls. Endocrine cells may be arranged in clumps around vascular networks or cords between parallel vascular channels or as hollow balls of cells (follicles) surrounding their stored secretions. Single endocrine cells also exist scattered throughout the alimentary and respiratory tracts (the entero-endocrine system, p. 1450).

THE RESPIRATORY SYSTEM

Introduction

Exchange of respiratory gases is essential to life, even in states of suspended animation, hibernation and aestivation, when respiration is greatly diminished. In *Protozoa*, oxygen and carbon dioxide diffuse directly through cell surfaces, establishing concentration gradients effective enough to maintain metabolism. In most *Metazoa* greater mass and hence longer diffusion distances require blood transport systems and specific respiratory areas. Respiratory adaptations assume two roles: the transport of respiratory gases and the development of specialized areas for gaseous exchange. The tracheal system of *Arthropoda* is an example of the former; its chitinous spirals maintain patency in *tracheoles*, and have their equivalents in vertebrates. For *Vertebrata*, and some invertebrate phyla, the two major global environments, *aqueous* and *terrestrial* (or aerial), have posed different problems. In both habitats gaseous exchanges take place between the external environment and circulating blood through the moist epithelium of gills, lungs or skin. Cutaneous respiration is limited to some amphibia, apart from which vertebrate respiratory organs are constantly associated with and largely developed from the branchial arch system and alimentary tract. The first segment of

this beyond the mouth is the pharynx, through which, in fish, water passes to the gills for gaseous exchange; diverticula of the pharynx and oesophagus are common, functioning as hydrostatic organs (swim-bladders) but in a few extant species they are also auxiliary respiratory chambers which can be filled directly from the air. From such '*lungfish*' the earliest terrestrial vertebrates, amphibia, probably evolved, acquisition of air-breathing leading to a host of new forms and adaptations (Young 1962, Romer 1970).

Some members of all terrestrial vertebrate classes (amphibia, reptiles, birds and mammals) have returned to an aquatic existence; but only amphibian larvae and a few adult amphibians, e.g. *Salamander atra*, are able to absorb oxygen directly from water. All others must suspend breathing while immersed, surfacing to empty their respiratory organs and take in fresh air. To keep out water and retain inspired air during submersion, a sphincteric mechanism is needed at the junction of the lungs and pharynx. Even in fish which merely swallow air to fill their swim-bladders and in piscine ancestors of land tetrapods able to 'breathe' from water as well as air, this need existed; and perhaps also there was a need for a dilatator mechanism. This led to the evolution of a *larynx*, a structure as basic as the lung itself. In terrestrial vertebrates the respiratory cavity or *lung* is usually bilateral, though served by a single pharyngeal diverticulum, the *trachea*, and has become highly folded, increasing the epithelial area for gaseous exchange between blood and inspired air. The basic pattern of respiratory organs was thus early established; even in small reptiles, the branching of the trachea, typically dichotomous, into smaller *bronchi* and *bronchioles* and terminal respiratory *alveoli*, is already complex. A tracheobronchial 'tree' of air-transporting tubes and thin-walled alveoli for gaseous exchange is present in all truly terrestrial vertebrates including birds, whose lungs display specializations beyond the scope of this text (Portmann 1950). Though air tubes may be altered in calibre by means of non-striated muscle in their walls, the main respiratory volumetric changes are alveolar. The mechanisms of inflation and deflation of the lungs vary in different vertebrate groups as do the dimensional features of their alveoli and bronchi, bronchial branching and pulmonary lobation. Within groups, however, the structural patterns display considerable constancy; recognition of this in man (p. 1261) has proved important in the diagnosis and treatment of pulmonary disorders.

The *larynges* vary considerably in structure in the different vertebrate groups (Negus 1928, 1949). Although they are primarily sphincteric, they also become *phonatory*, even in the most primitive air-breathers, amphibia. The necessary musculature, skeletal elements and neurovascular supplies have evolved from the pharynx and branchial apparatus (pp. 165, 170, 171, 238). Amphibia still 'swallow' air, drawn in through valvular 'nostrils' in the anterior oral roof, there being no separate nasal cavity. The oral floor is lowered to inspire; when it rises and nostrils close, air is pumped into the pharynx, larynx and lungs. The mouth and pharynx therefore serve both for feeding and breathing. The greater activity of some reptiles demands a more efficient respiration, partly achieved by the development of movable ribs and *inspiration* of air by the reduction of pulmonary pressure. This new mode of breathing is finally improved, in mammals, by the evolution of a muscular diaphragm and a complete musculo-osseous thoracic cavity, whose elaborate musculature varies the thoracic and hence the pulmonary volume and also aids venous return. In mammals the oronasal cavity is divided into respiratory and alimentary regions by a *palate*; but the former, evolving in continuity with primitive nares, is consequently *dorsal* to the alimentary cavity, the reverse of the relation of pharynx and larynx. Hence the mammalian airway and alimentary path cross at mid-pharyngeal level. Many adaptations have occurred in the pharynx, palate and especially the larynx to overcome the difficulties of this curious arrangement, both in structure and in breathing and swallowing reflexes.

The mammalian 'airway' (8.1B) thus includes the nasal cavity, nasal and oral levels of the pharynx, the larynx, trachea, bronchi and their pulmonary ramifications. Breathing via the mouth is, of course, possible; but this negates the olfactory and 'air-conditioning' functions of the nasal cavity. The separation of respiratory and olfactory functions from the mouth allows predatory mammals to breathe while the mouth is obstructed by prey, whose struggles may be prolonged; and their herbivorous prey may also sense warning odours even when feeding. In aquatic vertebrates, such as crocodiles, dolphins and whales, a spout-like elongation of the larynx can project into the nasopharynx, largely separating the airway, so that respiration can continue at the surface with mouth submerged, open and ready for prey; in marine mammals other specializations also occur (Andersen 1969).

In man speech is the outstanding elaboration associated with the respiratory system. The nasal cavity is still respiratory, olfactory and an organ of exchange for heat and water vapour, though also involved in producing the sounds of speech. But the human larynx, though still an essential sphincter, preventing entry of swallowed material and facilitating respiratory blockade to build up pressure for coughing or in extreme muscular efforts, is so frequently phonatory that this is often considered as its prime function. The elaboration of laryngeal musculature and its neural control, integrating laryngeal sound production with movements of respiratory, pharyngeal, palatal, lingual and labial muscles during the complex articulation of speech, provide mankind with a unique ability. Speech, which must of course be heard and

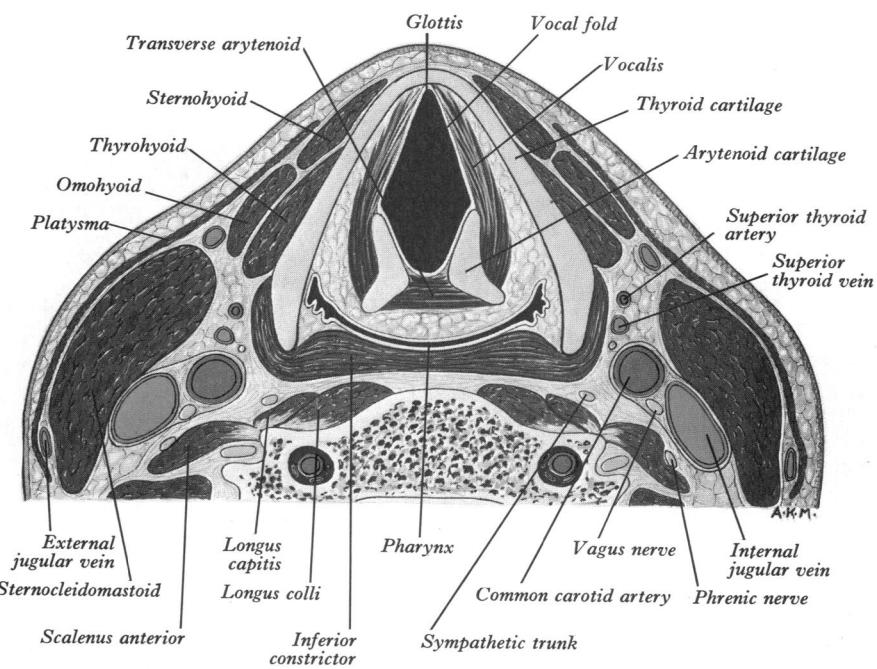

8.2A Transverse section across the ventral region of the neck at the level of the vocal folds, viewed from the superior aspect.

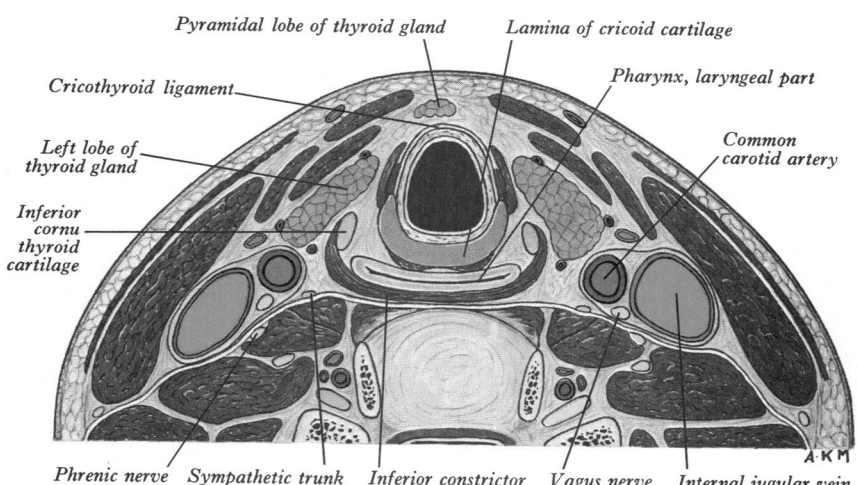

8.2B Transverse section through the ventral region of the neck, between the fifth and sixth cervical vertebrae, viewed from the superior aspect (semi-diagrammatic).

interpreted but may also be expressed in visible, written form, has led to the increasingly complex symbolism of language (with which mathematical and musical notation may be coupled). In this development, unparalleled in any other animal, very high levels of integration between the nervous system and locomotor structures are attained. The vital function of respiration remains; but its primitive phonational potential, leading finally to speech in all its permutations, is fundamental in the evolution of human intelligence. In the other most 'vocal' vertebrate class, birds, the larynx is commonly a sphincteric mechanism alone, sounds in great variety being produced by the *syrinx* at the tracheal bifurcation. Some birds are notable imitators of human speech; but there is no clear evidence of development of true, symbolic language in any vertebrate other than mankind, in whom high manual dexterity, large cerebral development and emancipated vision have together provided the adequate milieu for linguistic evolution.

The Larynx

The larynx, an air passage, a sphincteric device and an organ of phonation, extends from the tongue to the trachea (Terracol et al

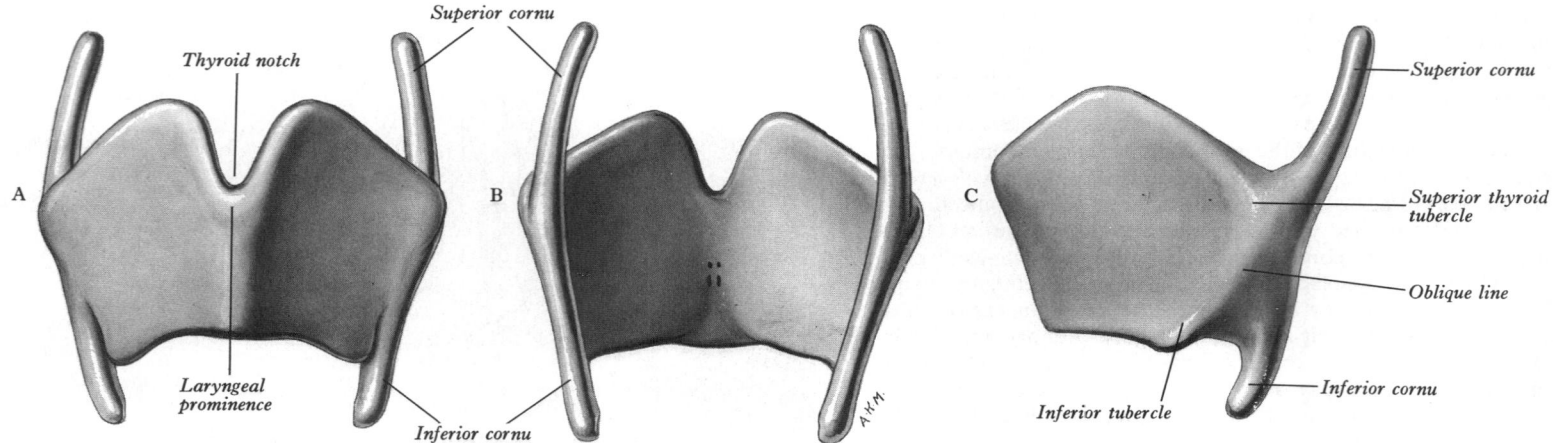

8.3 The thyroid cartilage. The attachments of the vestibular ligaments (above) and the vocal ligaments (below) are shown in B. A. Anterior aspect. B. Posterior aspect. C. Lateral aspect.

1965, Fink & Demarest 1978.) It projects ventrally between the great vessels of the neck and is covered anteriorly by skin, fasciae and the hyoid depressors (**8.2A**). Above, it opens into the laryngopharynx and forms its anterior wall; below, it continues into the trachea. In adult males it lies level with the third to sixth cervical vertebrae, but is somewhat higher in children and adult females. In infants between six and twelve months, the tip of the epiglottis (highest part of the larynx) is a little above the junction of the axial dens and body. Its average measurements in European adults are:

	In males	In females
Length	44 mm	36 mm
Transverse diameter	43 mm	41 mm
Sagittal diameter	36 mm	26 mm

The larger size of the adult masculine larynx is due to greater growth from puberty onwards; all its cartilages enlarge and the thyroid cartilage projects anteriorly in the midline. Its sagittal diameter is nearly doubled during this process.

Cartilages form the skeletal frame of the larynx, connected by ligaments and membranes and moved by muscles. The lining of the larynx is a mucosa continuous posterosuperiorly with that of the pharynx and inferiorly with that of the trachea.

LARYNGEAL CARTILAGES

These comprise the single thyroid, cricoid and epiglottic car-

tilages and the paired arytenoid, cuneiform and corniculate cartilages.

The thyroid cartilage (**8.**1, 2, 3, 13), the largest, consists of two quadrilateral *laminae*, their anterior borders fused inferiorly at a median angle, forming the subcutaneous *laryngeal prominence* ('Adam's apple'). This projection is most distinct at its superior end and is well marked in men but scarcely visible in women. Above, the laminae are separated by a V-shaped superior *thyroid incisure*; posteriorly they diverge, their posterior borders being prolonged as slender horns, the *superior* and *inferior cornua*. On the exterior of each lamina an *oblique line* curves down and forwards, running from a *superior thyroid tubercle* lying anterior to the root of the superior cornu, to an *inferior thyroid tubercle* on the inferior border. To this line, the sternothyroid, thyrohyoid and the thyropharyngeal parts of the inferior pharyngeal constrictor muscles are attached. The smooth *internal surface* is posterosuperiorly slightly concave and covered by mucosa. High in the angle between the laminae the thyro-epiglottic ligament is attached; below this, near the midline, the paired vestibular and vocal ligaments and the thyro-arytenoid, thyro-epiglottic and vocal muscles are attached.

The *superior border* of each lamina is concave behind, convex in front and to it half of the thyrohyoid membrane is attached. The *inferior border* is concave behind, and nearly straight in front, of the inferior thyroid tubercle. At and near the midline it is connected

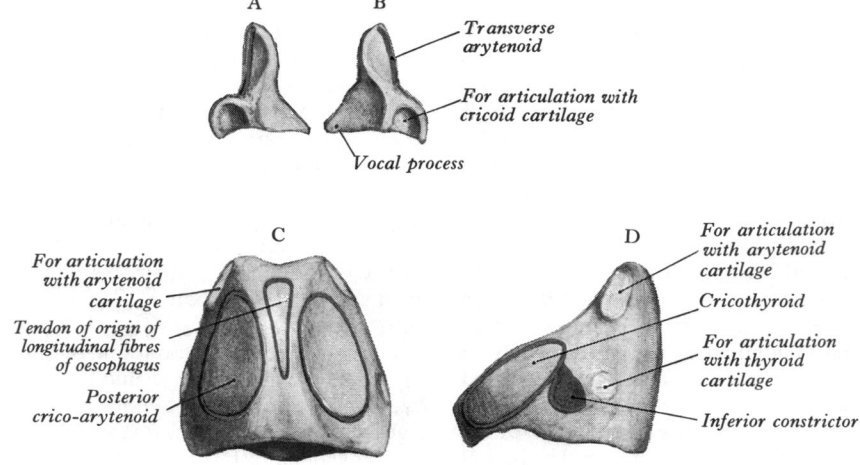

8.4 The arytenoid and cricoid cartilages. A. The left arytenoid cartilage: medial aspect. B. The right arytenoid cartilage: medial aspect. C. The cricoid cartilage: posterior aspect. D. The cricoid cartilage: left lateral aspect. Muscle attachments are indicated in red.

to the cricoid cartilage by the anterior (median) cricothyroid ligament (see p. 1253).

The *anterior* borders fuse at an angle of about 90° in men and about 120° in women. In men the larger laryngeal prominence, the greater length of the vocal folds and the resultant deeper pitch of voice are all associated with a smaller thyroid angle. The thick and rounded *posterior border* receives the fibres of stylopharyngeus and palatopharyngeus. The *superior cornu*, long and narrow, curves up and posteromedially, ending in a conical apex to which the lateral thyrohyoid ligament is attached. The *inferior cornu*, which is short and thick, curves down and slightly anteromedially; on the medial surface of its lower end, a small oval facet articulates with the side of the cricoid cartilage. During infancy a narrow, rhomboidal, flexible strip, the *intrathyroid cartilage*, lies between the two laminae and is joined to them by fibrous tissue.

The cricoid cartilage (8.4, 6, 7, 8), thicker and stronger than the thyroid cartilage and shaped like a signet ring (hence its name), forms the inferior parts of the anterior and lateral walls and most of the posterior wall of the larynx. It has a quadrate posterior *lamina* and a narrow anterior *arch*. Its *lamina* is deep and broad, from 2–3 cm vertically; it has a posterior median vertical ridge, to the upper part of which the two fasciculi of the longitudinal oesophageal fibres are attached by a tendon (p. 1332). Lateral to this are two shallow depressions for the fibres of the posterior crico-arytenoid. The *arch* is narrow in front, from 5–7 mm vertically, but widens towards the lamina. To its front and sides, externally, the cricothyroid is attached and, behind this, part of the inferior pharyngeal constrictor. The arch is palpable below the laryngeal prominence, from which it is separated by a depression containing the resilient conus elasticus. On each side, at the junction of lamina and arch, a prominent circular facet, facing posterolaterally, articulates with the inferior thyroid cornu. The horizontal *inferior border* is joined to the first tracheal cartilage by the cricotracheal ligament, while the *superior border* runs obliquely up and back, giving attachment anteriorly to the thick median part and laterally to the membranous lateral parts of the cricothyroid ligament and the lateral crico-arytenoid muscles. Posteriorly, the lamina has a shallow median notch, on each side of which a smooth, oval, convex facet, directed up and laterally, articulates with the base of an arytenoid cartilage (see, however, Sellars & Keen 1978). The *internal surface* of the cricoid cartilage is smooth and lined by mucosa.

The paired arytenoid cartilages (8.4, 7, 8), sited superolaterally on the cricoid lamina at the back of the larynx, are pyramidal, with three surfaces, two processes, a base and an apex. The *posterior surface*, which is triangular, smooth and concave, is covered by the transverse arytenoid muscle. The *anterolateral surface* is convex and rough; near the apex is an elevation from which a crest curves back, down and then forwards to the vocal process. The lower part of this crest separates two depressions (foveae), the upper triangular, the lower oblong; to the upper is attached the vestibular ligament, to the lower the vocalis and lateral crico-arytenoid muscles. The *medial surface* is narrow, smooth, flat and covered by mucosa; it bounds laterally by its lower edge the intercartilaginous part of the rima glottidis (p. 1254). The *base* is concave, with a smooth surface for articulation with the upper border (lateral part) of the cricoid lamina. Its round, prominent lateral angle, or *muscular process*, projects posterolaterally giving attachment to the posterior crico-arytenoid muscle behind and lateral crico-arytenoid in front. To its pointed anterior angle (the *vocal process*), projecting horizontally forward, is attached the vocal ligament. The *apex* curves posteromedially to articulate with the *corniculate cartilage*.

The corniculate cartilages (8.7, 8, 12) are two conical nodules of elastic fibrocartilage which articulate with the arytenoid apices, prolonging them posteromedially. Lying posteriorly in the aryepiglottic mucosal folds, they sometimes fuse with the arytenoid cartilages.

The cuneiform cartilages (8.8, 12) are two elongated, club-like elastic fibrocartilages, one in each aryepiglottic fold anterosuperior to the corniculate cartilages and visible as whitish elevations through the mucosa.

The epiglottic cartilage (8.5, 7, 8) is a thin foliate lamella of elastic fibrocartilage, projecting obliquely up behind the tongue

and hyoid body, in front of the laryngeal inlet. Its free end, which is broad and round, ascends from a long and narrow stalk, connected by the elastic *thyro-epiglottic ligament* to the back of the laryngeal prominence just below the thyroid notch. Its sides are attached to the arytenoid cartilages by aryepiglottic folds (p. 1253). Its free upper *anterior surface* is covered by mucosa (the epithelium is non-keratinized stratified squamous), reflected on

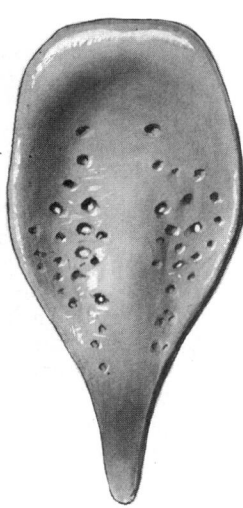

8.5A The epiglottis: posterior aspect. Note its pitted surface.

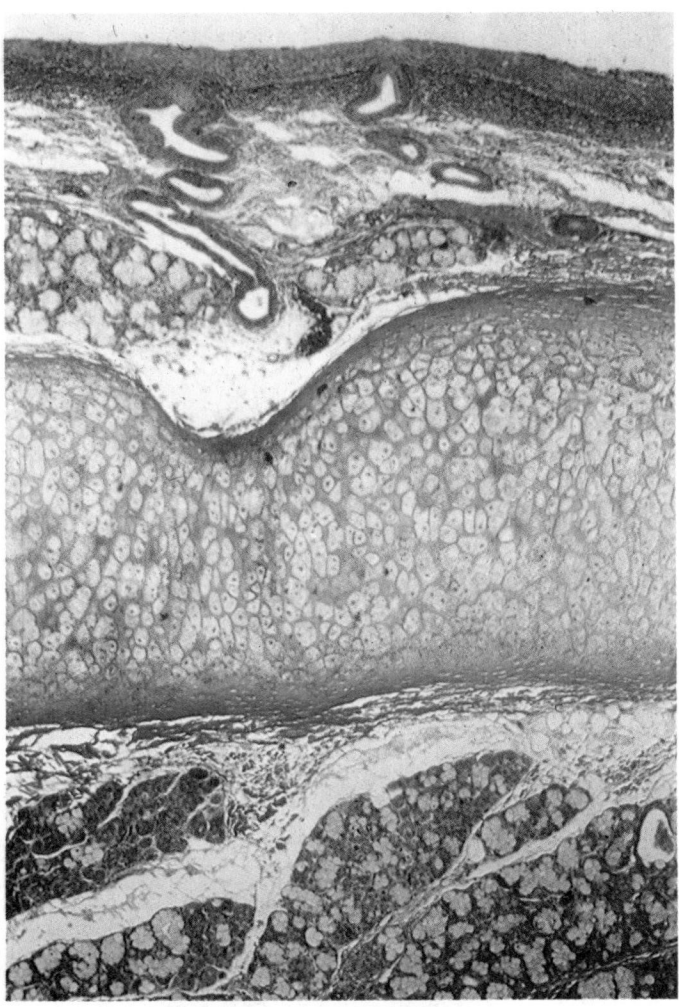

8.5B A low magnification light micrograph of a transverse section of the epiglottis, showing stratified squamous epithelium covering the anterior surface, mucous glands and fibro-elastic cartilage. Mallory's trichrome stain. Prepared by David Ristow; photography by Marina Morris, Department of Anatomy, Guy's Hospital Medical School, London.

to the pharyngeal aspect of the tongue and the lateral pharyngeal walls as a *median glosso-epiglottic* and two *lateral glosso-epiglottic folds*. On each side of the median fold is a depression, the *vallecula*. The lower anterior surface, behind the hyoid bone and thyrohyoid membrane, is connected to the upper border of the former by an elastic *hyo-epiglottic ligament*, and separated from the thyrohyoid membrane by adipose tissue. The smooth *posterior surface* is transversely concave and vertically concavo-convex and covered by respiratory mucosa; its lower projecting part is the *tubercle*. The cartilage is posteriorly pitted by small mucous glands and perforated by branches of the internal laryngeal nerve.

Functions of the Epiglottis

During deglutition (p. 1328) the epiglottis moves upwards and forwards and is squeezed between the base of the tongue and the rest of the larynx; the bolus slips over its anterior surface as it bends back over the laryngeal inlet. In man it is somewhat degenerate in function and is well separated from the soft palate; it is not essential to swallowing, which is normal even if the epiglottis is destroyed by disease; neither is it essential for respiration or phonation. Some mammals, keen-scented even when the mouth is open for feeding, have a large epiglottis which projects into the nasapharynx *above* the soft palate; during feeding, it is drawn down and forward by the hyo-epiglottic muscle (in man a mere *ligament*) against the *upper* surface of the soft palate, keeping the nasal airway clear and the buccal airway closed, thus permitting olfaction even when the mouth is open.

Microstructure of the Laryngeal Cartilages

The corniculate, cuneiform and epiglottic cartilages and the arytenoid apices are composed of *elastic* fibrocartilage with little tendency to ossify or calcify. The thyroid, cricoid and greater part of the arytenoids consist of *hyaline* cartilage and may ossify as age advances, commencing about the twenty-fifth year in the thyroid cartilage and somewhat later in the cricoid and arytenoids; by the sixty-fifth year these cartilages commonly appear, in radiographs, patchily calcified or ossified.

LARYNGEAL ARTICULATIONS

The joints between the inferior thyroid cornua and cricoid cartilage are synovial, each with a capsular ligament strengthened posteriorly by a fibrous band. The cricoid rotates on the inferior cornua around a transverse axis through both joints and, to a limited extent, also glides on them.

A pair of synovial joints exists between the facets on the lateral part of the upper border of the lamina of the cricoid cartilage and the bases of the arytenoids, each joint being enclosed by a capsular ligament; a strong *posterior crico-arytenoid ligament* connects the cricoid to the posteromedial part of the arytenoid base. These joints permit two movements: (1) arytenoid rotation about an oblique axis (dorsomediocranial to ventrolaterocaudal) (Frable 1961), in which each vocal process swings laterally or medially, increasing or decreasing the rima glottidis; (2) a gliding movement, by which the arytenoids approach or recede from one another, the slope of their articular surfaces imposing a forward and downward movement on lateral gliding. The movements of gliding and rotation are associated: medial gliding with medial rotation and lateral gliding with lateral rotation (von Leden & Moore 1961). The posterior crico-arytenoid ligaments limit forward movements of the arytenoid cartilages on the cricoid. (For details consult Sonesson 1959 and Sellars & Keen 1978.) Sometimes the synchondrosis between the arytenoid apex and the corresponding corniculate cartilage is replaced by a synovial joint.

Numerous lamellated (Pacinian) corpuscles, some Ruffini and some free nerve endings occur in the capsules of the laryngeal joints. In cats these respond to mechanical stimuli and are involved in the normal co-ordination of laryngeal muscles during respiration and phonation (Kirchner & Wyke 1965). The human articular supply is chiefly from branches of the recurrent laryngeal nerves, arising independently or from muscular rami (Psenicka 1966).

LARYNGEAL LIGAMENTS AND MEMBRANES

Extrinsic Ligaments

The *thyrohyoid membrane* (**8.6**, 7) is a broad, fibro-elastic layer, attached below to the cranial border of the thyroid cartilage and the

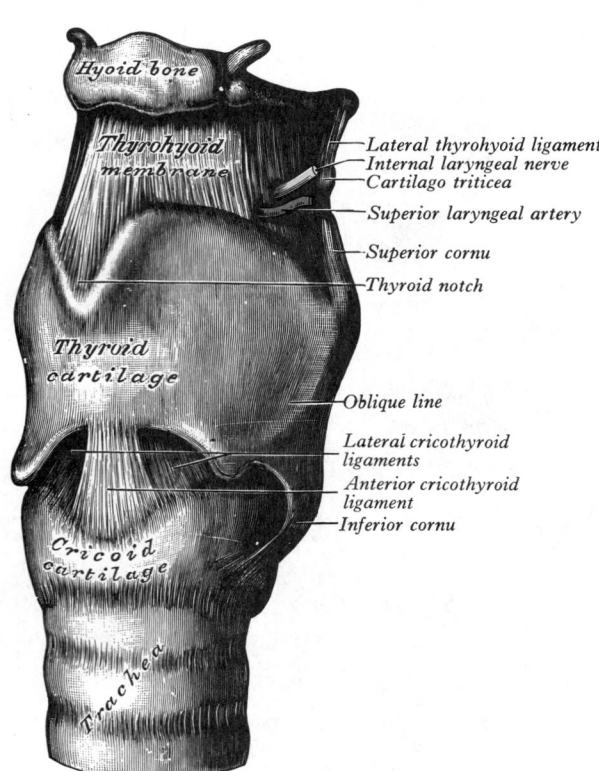

8.6 The ligaments of the larynx: anterolateral aspect.

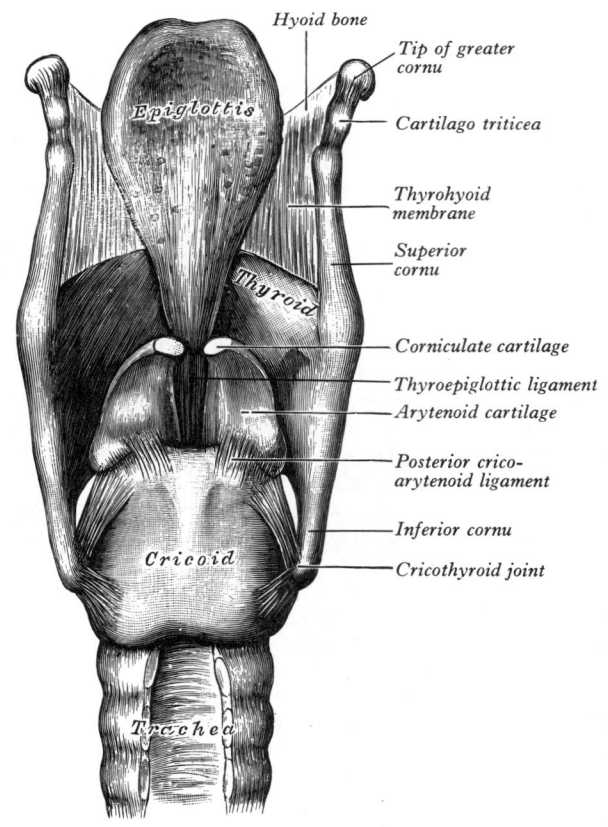

8.7 The ligaments of the larynx: posterior aspect.

front of its superior cornua and above to the *superior* margin of the posterior surface of the hyoid body and its greater cornua. It thus ascends *behind* the concave posterior hyoid surface, separated from the hyoid body by a bursa facilitating laryngeal ascent during deglutition. Its thicker part is the *median thyrohyoid ligament*; the more lateral, thinner parts are pierced by the superior laryngeal vessels and internal laryngeal nerves. Externally it is in contact with the thyrohyoid, sternohyoid and omohyoid muscles and the hyoid body. Its inner surface is related to the epiglottis and pharyngeal piriform fossae. The round, cord-like elastic, *lateral thyrohyoid ligaments* form the posterior borders of the thyrohyoid membrane, connecting the tips of the superior thyroid cornua to the posterior ends of the greater hyoid cornua. A small *cartilago triticea* occurs frequently in each ligament.

The epiglottis is attached to the hyoid bone and thyroid cartilage by the extrinsic *hyo-epiglottic* and intrinsic *thyro-epiglottic ligaments* (p. 1251). The *cricotracheal ligament* unites the lower cricoid border to the first tracheal cartilage, being thus continuous with the tracheal perichondrium.

Intrinsic Ligaments

Beneath the laryngeal mucosa is a fibrous sheet containing many elastic fibres, the *fibro-elastic membrane of the larynx*, interrupted on each side by the interval between the vestibular and vocal ligaments; its upper part, the often ill-defined *quadrangular membrane*, extends between the arytenoids and epiglottis, its lower part forming the well-marked *cricothyroid ligament (cricovocal membrane)*, connecting the thyroid, cricoid and arytenoid cartilages. The articular ligaments are described above.

The cricothyroid ligament (**8.6, 8**), comprising the inferior and larger part of the laryngeal membrane, is mainly elastic tissue, with distinct anterior and lateral parts. The thick *anterior (median) cricothyroid ligament*, broad below, narrower above, connects adjacent margins of the cricoid and thyroid cartilages. An anastomosis between the cricothyroid arteries crosses it and supplies perforating rami to the larynx. The smaller, *lateral*

cricothyroid ligaments* (cricothyroid or cricovocal membranes) are thinner and covered internally by mucosa and externally by the lateral crico-arytenoid and thyro-arytenoid muscles. From the internal rim of the superior cricoid border the ligaments ascend and converge; their superior edges, extending from the dorsal aspect of the thyroid angle (just below its midpoint) to the apices of the arytenoid vocal processes, form the basis of the vocal folds (**8.10**). The thickened edges of the cricothyroid ligamentous complex are the *vocal ligaments* (**8.8**). (The term, *conus elasticus*, is commonly applied to the entire complex but sometimes only its anterior part.)

The laryngeal cavity (**8.9, 10**) extends from the laryngeal inlet, its connection with the pharynx, to the cricoid cartilage's lower border, continuing into the trachea. It is partially divided by paired folds of mucosa projecting medially from its sides. The upper pair are the *vestibular folds*, the fissure between them being the *rima vestibuli*; the lower pair, concerned in vocalization, are the *vocal folds*, the fissure between them being the *rima glottidis*, or *glottis*. The laryngeal *inlet* or 'aditus laryngis' (**8.11**), the aperture between the larynx and pharynx, faces back and somewhat up, the anterior laryngeal wall being much longer and sloping down and forwards in its upper part (**8.9**). The inlet is bounded anteriorly by the upper edge of the epiglottis, posteriorly by the mucosa between the arytenoids and on each side by the edge of a mucous fold between the epiglottis and arytenoid apex. This *aryepiglottic fold* contains ligamentous and muscular fibres; in the posterior part of its margin are oval anterosuperior and postero-inferior swellings, due to the cuneiform and corniculate cartilages respectively, separated by a shallow vertical furrow, continuous below with the opening of the laryngeal sinus (vide infra).

The *laryngeal vestibule* (**8.9, 10**), between inlet and vestibular folds, is wide above, narrow below and longer anteriorly than posteriorly. Its anterior wall is the posterior epiglottic surface, its lower part projecting backwards as its *tubercle* (p. 1252). Its lateral walls, deep in front, shallow behind, are the medial aspects of the aryepiglottic folds; its posterior wall is the inter-arytenoid mucosa, above the vestibular folds.

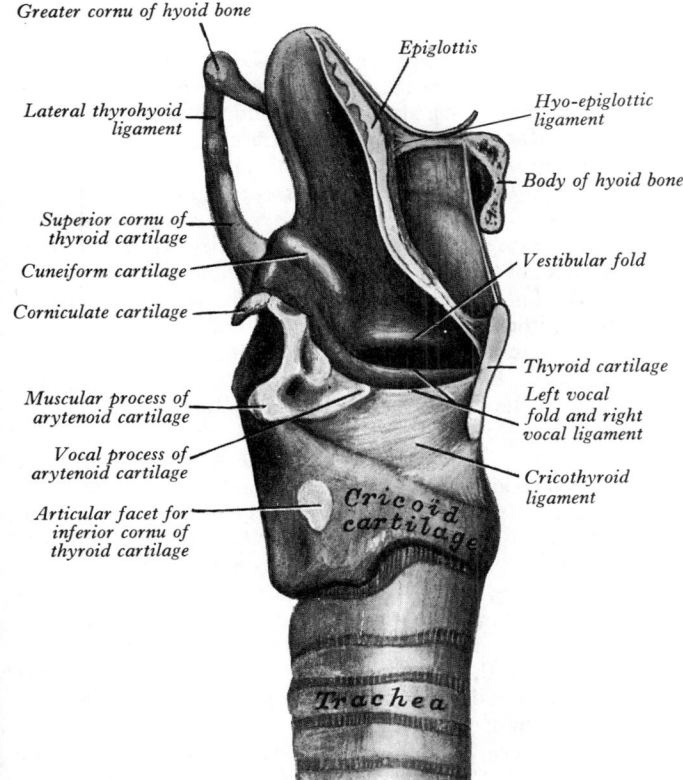

8.8 Dissection to show the right half of the cricothyroid ligament. The right lamina of the thyroid cartilage and the subjacent muscles have been removed.

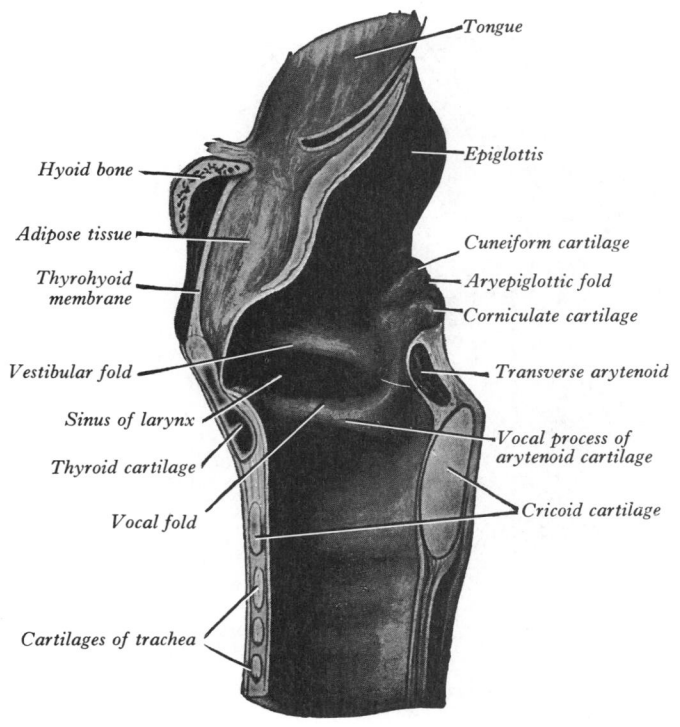

8.9 Sagittal section showing the medial aspect of the right half of the larynx.

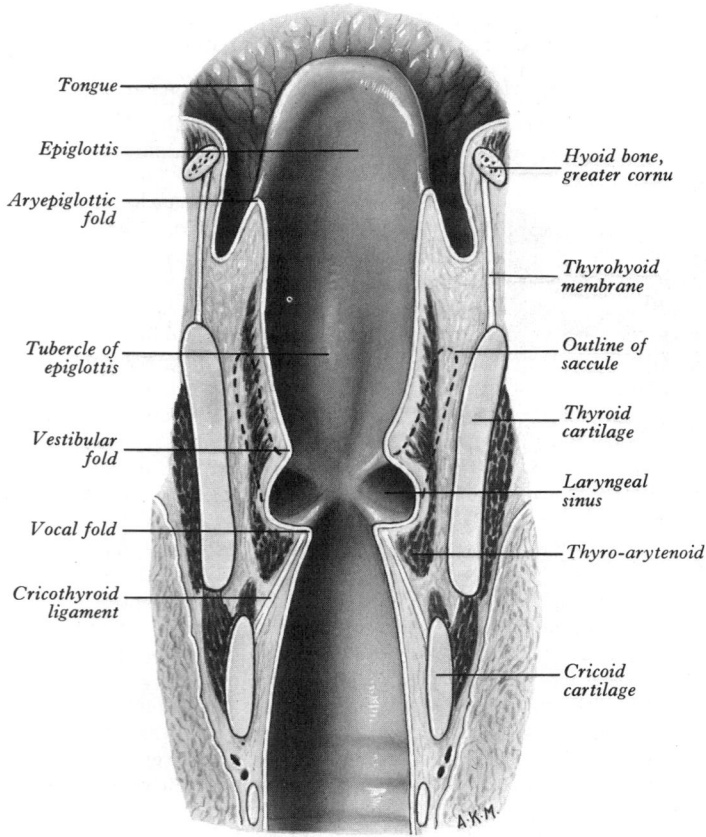

8.10 Coronal section through the larynx and the cranial end of the trachea: posterior aspect.

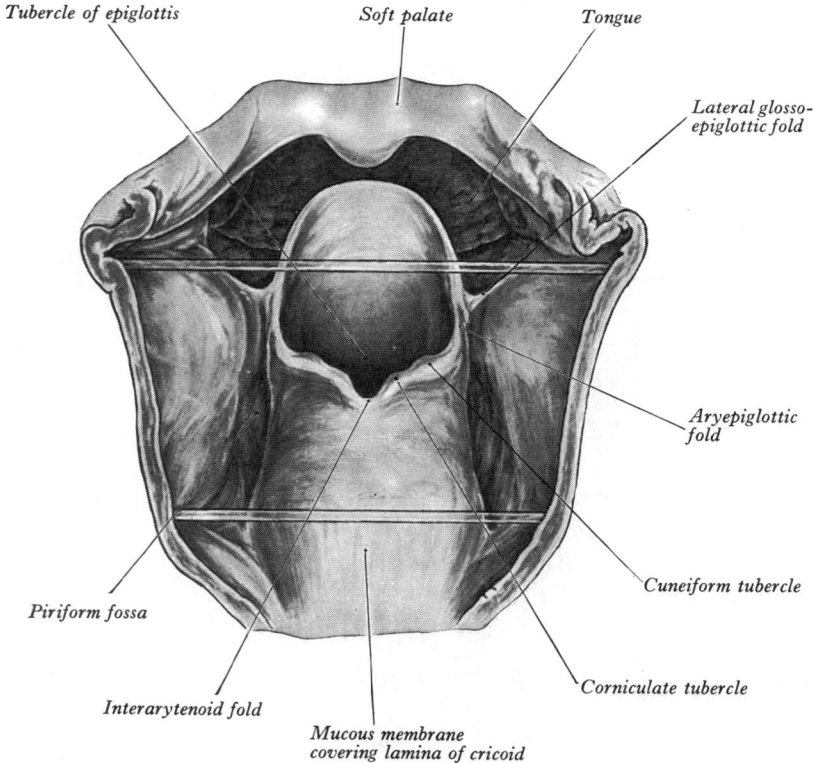

8.11 The inlet of the larynx, viewed from the posterior aspect. The posterior wall of the pharynx has been divided in the median plane and two glass rods have been inserted to keep the cut portions apart.

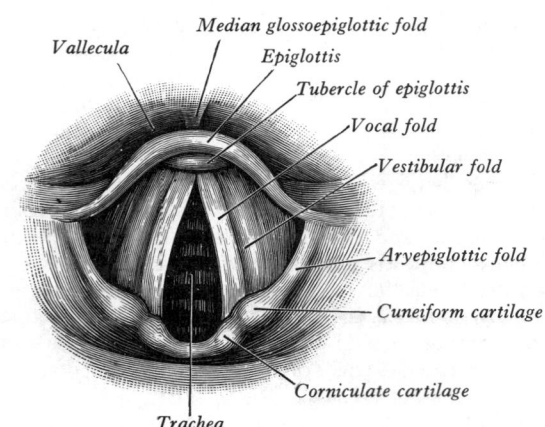

8.12 Laryngoscopic view of the interior of the larynx.

The *middle part* of the laryngeal cavity is the smallest, reaching from the rima vestibuli to the rima glottidis. On each side it opens by a slit between vestibular and vocal folds into the *laryngeal sinus* (**8.**9, 10), a fusiform recess between the folds which ascends lateral to the vestibular fold. It is lined by mucosa and covered externally by the thyro-arytenoid muscle. Anteriorly the sinus opens into the *laryngeal saccule* (**8.**10), a pouch ascending forwards from the sinus between the vestibular fold and thyroid cartilage, occasionally reaching the cartilage's upper border; it is conical, curving slightly backwards. On its luminal surface open 60–70 mucous glands, sited in the submucosa. The saccule has a fibrous capsule, continuous below with the vestibular ligament, and is covered medially by a few muscular fasciculi from the apex of the arytenoid cartilage which pass forwards between the saccule and vestibular mucosa into the aryepiglottic fold. Laterally the saccule is separated from the thyroid cartilage by thyro-epiglottic muscles, which compress it, expressing its secretion on to the vocal folds. In most apes the saccules form air sacs, which may extend into superficial cervical and even axillary tissues; they appear to aid resonance. Human saccules occasionally protrude through the thyrohyoid membrane.

The pink mucosa of each *vestibular fold* (**8.**8, 9, 10) covers a narrow *vestibular ligament*, fixed in front to the thyroid angle below the epiglottic cartilage and behind to the anterolateral arytenoid surface above the vocal process. Like the whole interior of the larynx (except the vocal folds) the epithelium is ciliated pseudostratified, respiratory in type. The white mucosal *vocal folds* (**8.**8, 9, 10) stretch from the mid-level of the thyroid angle to the arytenoid vocal processes. They form the anterolateral edges of the rima glottidis and are concerned in producing sound. They are covered by non-keratinized stratified squamous epithelium which is closely bound to the vocal ligaments; since a submucosa and blood vessels are absent, the folds are pearly white. Each *vocal ligament*, continuous below with the lateral part of the cricothyroid ligament (p. 1253), is a band of yellow elastic tissue related laterally to the vocalis (p. 1255).

The *rima glottidis* or *glottis* (**8.**12), the fissure between the vocal folds anteriorly and the arytenoid cartilages posteriorly, is limited behind by the mucosa passing between the arytenoid cartilages, at the level of the vocal folds. Between these folds is the *intermembranous part* of the glottis, comprising about three-fifths of its length, while between the arytenoids is the *intercartilaginous* part. The average sagittal diameter of the glottis in adult males is 23 mm and in adult females 17 mm. It is the narrowest part of the larynx. Its width and shape vary with the movements of vocal folds and arytenoid cartilages during respiration and phonation (see p. 1257).

The *lower part* of the laryngeal cavity extends from the vocal folds to the lower border of the cricoid. In transverse section it is elliptical above and wider and circular below; it is continuous with the trachea. Lined by mucosa, its walls are supported by the cricothyroid ligament above and the cricoid cartilage below.

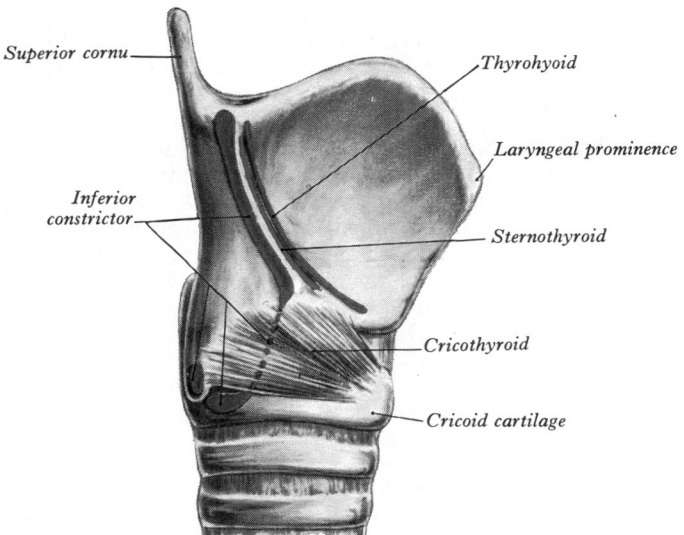

8.13 Lateral view of the larynx, showing the muscular attachments.

LARYNGEAL MUSCULATURE

The muscles of the larynx are divisible into *extrinsic and intrinsic* groups. The extrinsic muscles connect the larynx to neighbouring structures (p. 584), while the intrinsic are confined to laryngeal attachments, being the *cricothyroid, posterior* and *lateral crico-arytenoid, transverse* and *oblique arytenoid, aryepiglotticus, thyro-arytenoid* and its subsidiary part, *vocalis,* and *thyro-epiglotticus.* All but the transverse arytenoid are paired.

The cricothyroid (8.13) is triangular and extends from the anterolateral external aspect of the cricoid cartilage. Its fibres diverge in two groups, a lower '*oblique*' *part* which slants posterolaterally to the anterior border of the inferior cornu of the thyroid and a superior '*straight*' *part* ascending backwards to the posterior part of the lower border of the thyroid lamina. The medial borders of the two muscles are separated by a triangular interval occupied by the conus elasticus.

The posterior crico-arytenoid (8.14) arises from the posterior surface of the cricoid lamina, on the inferomedial area of the depression present to one side of the midline. Ascending laterally its fibres converge to insert on the back of the arytenoid muscular process of the same side. The highest fibres are nearly horizontal, the middle oblique and the lowest almost vertical, some of the last reaching the anterolateral arytenoid surface.

The lateral crico-arytenoid (8.15), smaller than the preceding muscle, ascends obliquely back from the upper border of the cricoid arch to the front of the arytenoid muscular process of the same side.

The transverse arytenoid (8.14) is a single muscle which bridges the gap between the two arytenoid cartilages and fills their posterior concave surfaces. It is attached to the back of the muscular process and adjacent lateral border of both arytenoids.

The oblique arytenoids (8.14) lying superficial to the transverse, are disposed in two crossed fasciculi, each extending from the back of the muscular process of one arytenoid to the apex of the opposite cartilage. Some fibres continue laterally round the arytenoid apex into the aryepiglottic fold, as the *aryepiglottic* muscle.

The thyro-arytenoid (8.15), a broad, thin muscle, lies lateral to the vocal fold, cricothyroid ligament, laryngeal sinus and saccule. It extends from the lower half of the thyroid angle and cricothyroid ligament, ascending posterolaterally to the anterolateral arytenoid surface. Its lower, deeper fibres appear in coronal section as a triangular bundle attached to the lateral surface of the vocal process and to the inferior impression on the anterolateral surface of the arytenoid cartilage. This bundle, the **vocalis muscle,** is parallel and lateral to the vocal ligament; it is said to be thicker behind than in front, because many deeper fibres start from the vocal ligament and so do not reach the thyroid cartilage. Others consider that all its fibres loop and intertwine

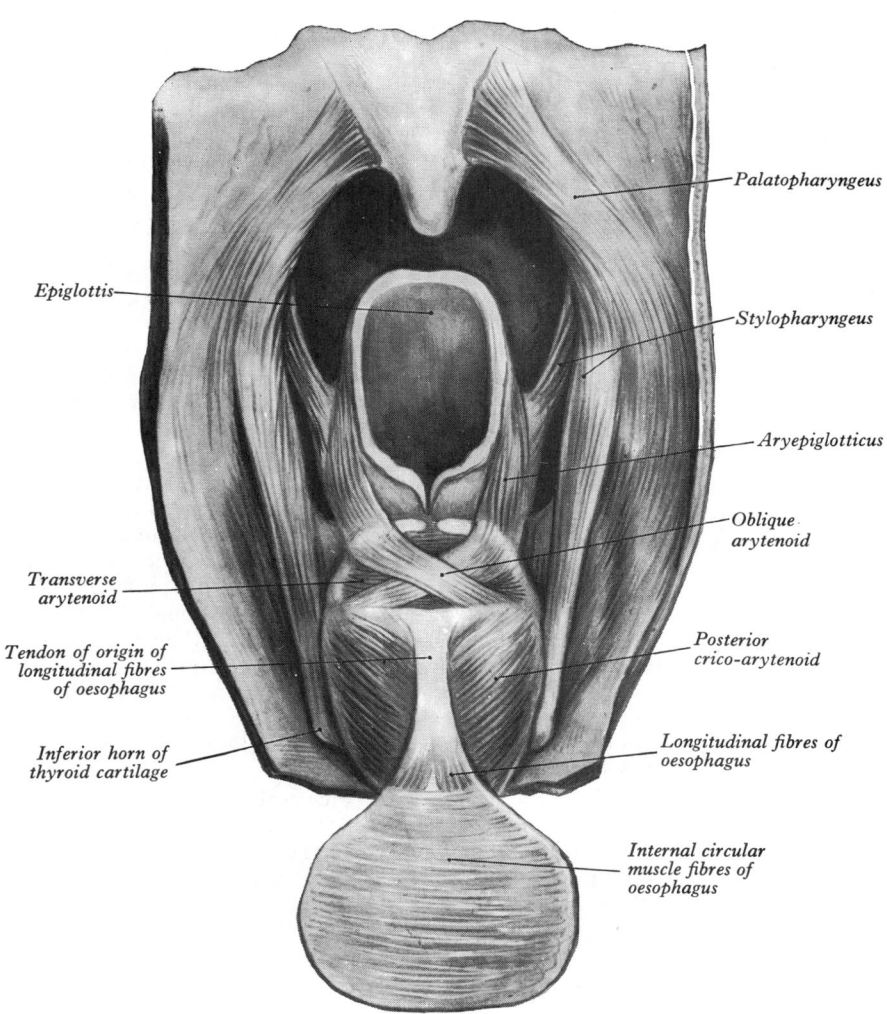

8.14 The muscles of the larynx: posterior aspect.

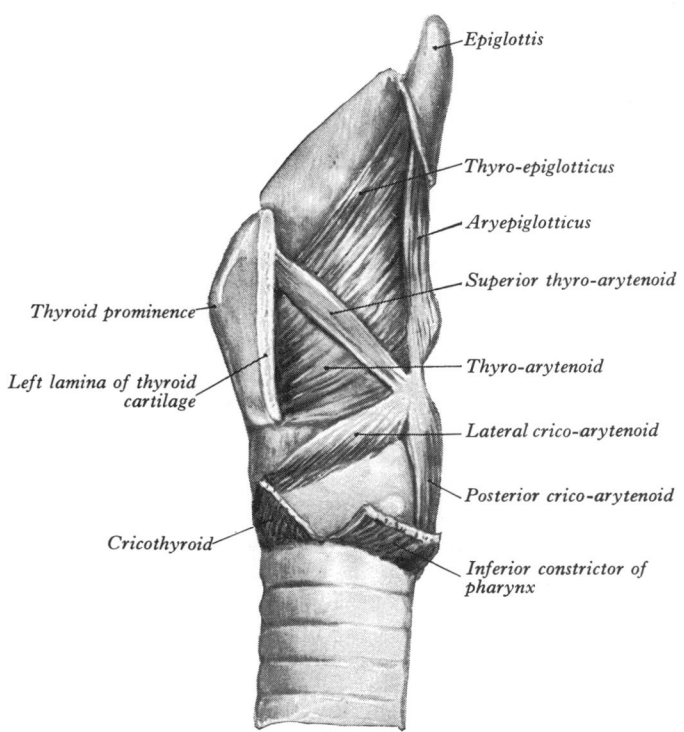

8.15 The muscles of the larynx (most of the left lamina of the thyroid cartilage has been removed): left lateral aspect.

from the thyroid to the arytenoid (Tautz & Rohen 1967). Many of the thyro-arytenoid fibres are prolonged into the aryepiglottic fold, where some cease, others continuing to the epiglottic margin as the **thyro-epiglotticus**. A few fibres extend along the wall of the sinus from the lateral arytenoid margin to the side of the epiglottis. The **superior thyro-arytenoid** (8.15), a small muscle not always present, is on the lateral surface of the thyro-arytenoid, extending obliquely from the thyroid angle to the arytenoid muscular process.

Actions. The laryngeal muscles may be placed in three groups according to their main actions (Negus 1947): (1) those which alter the glottis: *posterior* and *lateral crico-arytenoids* and *oblique* and *transverse arytenoids*; (2) those regulating tension of vocal ligaments: *cricothyroids, posterior crico-arytenoids, thyro-arytenoids* and *vocales*; (3) those which modify the laryngeal inlet: *oblique arytenoids, aryepiglottici* and *thyro-epiglottici* (8.16). Bilateral pairs usually act together.

(1) The *posterior crico-arytenoids* are the only laryngeal muscles which open the glottis, rotating the arytenoid cartilages laterally at the crico-arytenoid joints (p. 1252), thus separating the vocal processes and attached vocal folds. They also retract the arytenoids, assisting the cricothyroids to tense the vocal folds. The most lateral fibres draw the arytenoids laterally, so that the rima glottidis becomes triangular when the posterior crico-arytenoid muscles contract. The *lateral crico-arytenoids* close the glottis by rotating the arytenoids medially, to approximate their vocal processes. The *transverse arytenoid* approximates the arytenoid cartilages, closing the glottis in its posterior part.

(2) The *cricothyroids* stretch the vocal ligaments by raising the cricoid arch and tilting back its lamina, thus increasing the distance between the vocal processes and the thyroid angle, to tense the vocal ligaments, an effect which they also produce by pulling the thyroid cartilage forward; this is probably their principal action, for during phonation the cricoid lamina is held immovably against the vertebral column by the cricopharyngeus (p. 1328). During swallowing, however, the cricopharyngeus relaxes, allowing the cricoid to tilt forwards during laryngeal closure. The *posterior crico-arytenoids* tense the vocal ligaments by pulling the arytenoid cartilages backwards. The *thyro-arytenoids* draw the arytenoids towards the thyroid cartilage, shortening and relaxing the vocal ligaments. At the same time, they rotate the arytenoids medially to approximate the vocal folds. Their deeper fibres, the *vocales*, relax the posterior parts of the vocal ligaments, their anterior parts remaining tense and thus raising the vocal pitch. For details of arytenoid movements consult Sellars (1978).

(3) The *oblique arytenoids* and *aryepiglottici* act as a sphincter of the laryngeal inlet by adducting the aryepiglottic folds and approximating the arytenoid cartilages to the epiglottic tubercle. The *thyro-epiglottic* muscles widen the inlet by their action on the aryepiglottic folds.

Neuromuscular spindles exist in all human laryngeal muscles (Keene 1961), the maximum number (23) being in the transverse arytenoid (Voss 1966).

MOVEMENTS OF THE VOCAL FOLDS

In *quiet respiration*, (8.16) the anterior, intermembranous part of the rima glottidis is triangular when viewed from above, its apex being in front and its base behind, represented by a line (about 8 mm long) connecting the anterior ends of the arytenoid vocal processes; the intercartilaginous part is rectangular, between the medial surfaces of the arytenoids which lie parallel to each other. In *forced inspiration* the vocal folds are fully abducted; the arytenoid cartilages rotate laterally, their vocal processes moving apart. The glottis is then rhomboid, its intermembranous and intercartilaginous parts being both triangular, with

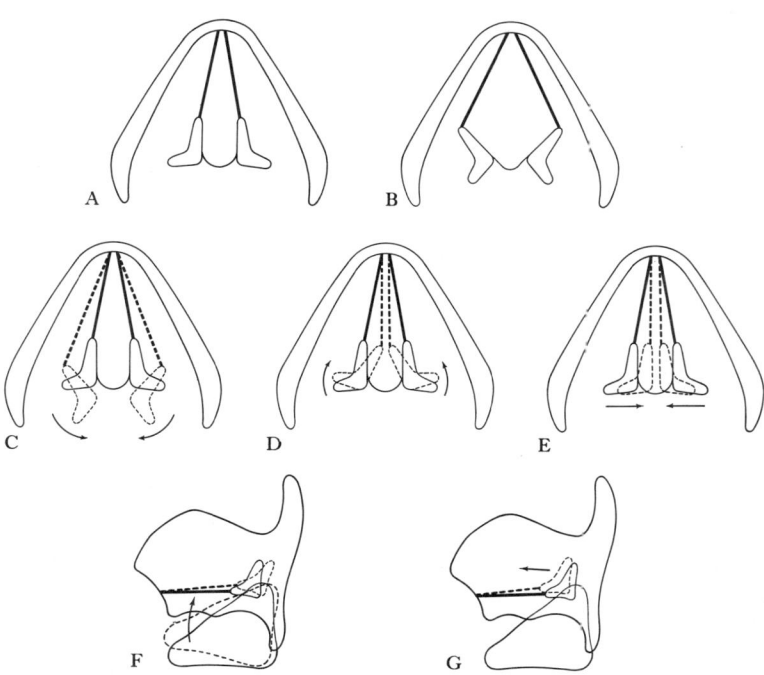

8.16 Different positions of the vocal folds and arytenoid cartilages.

A. Position of rest in quiet respiration. The intermembranous part of the rima glottidis is triangular and the intercartilaginous part is rectangular in shape.

B. Forced inspiration. Both parts of the rima glottidis are triangular in shape.

C. Abduction of the vocal folds. The arrows indicate the lines of pull of the posterior crico-arytenoid muscles. The abducted vocal folds and the abducted, retracted and laterally rotated arytenoid cartilages are shown in dotted outline. Both parts of the rima glottidis are triangular.

D. Adduction of the vocal folds. The arrows indicate the lines of pull of the lateral crico-arytenoid muscles. The adducted vocal folds and the medially rotated arytenoid cartilages are shown in dotted outlines.

E. Closure of the rima glottidis. The arrows indicate the line of pull of the transverse arytenoid muscle. Both the vocal folds and the arytenoid cartilages are adducted, but there is no rotation of the latter.

F. Tension of the vocal folds, produced by the action of the cricothyroid muscles which tilt the anterior part of the cricoid cartilage cranially and so carry the arytenoid cartilages dorsally.

G. Relaxation of the vocal folds, produced by the action of the thyro-arytenoid muscles, which draw the arytenoid cartilages ventrally.

the greatest width being opposite the attachments of the folds to the vocal processes.

Movements of the vocal folds during phonation have been studied by high-speed cinematography (Pressman 1942). Preparatory to phonation, the intermembranous and intercartilaginous parts of the glottis are reduced to a linear chink by adduction of the vocal folds and by adduction and medial rotation of the arytenoids. The folds then tense, the degree of tension determining sound pitch (frequency). To raise the pitch, tension is increased; the folds may lengthen by 50% in the highest tones. Photographs suggest that this lengthening affects both ends of the folds, indicating that the cricothyroid muscles act not only on the cricoid but also tilt the thyroid cartilage down and forwards. In whispering the intermembranous glottis is closed, but the intercartilaginous part remains widely patent, so that air escapes freely.

Fink (1975, 1978) considered that 'laryngeal biomechanics' depend much on the behaviour of various laryngeal *folds* which project into the laryngeal cavity to highly variable degrees dependent upon respiration, physical effort and phonation. In addition to the vocal, vestibular and aryepiglottic folds mentioned above, he identified *median thyrohyoid* and *interarytenoid folds*. The latter consists of the transverse arytenoid muscle and its covering mucous membrane which *folds* into the larynx when the muscle adducts the arytenoids (thus aiding obliteration of the intercartilaginous part of the rima glottidis). The median thyrohyoid fold is more complex: as shown in Fig. **8**.8, the lower part of the epiglottic cartilage is attached to the hyoid bone and thyroid cartilage by elastic ligaments (hyo-epiglottic and thyro-epiglottic p. 1253), separated from the median thyrohyoid ligament by adipose tissue, anterior to which is the elastic anterior part of the thyrohyoid membrane including the median thyrohyoid ligament. During swallowing (p. 1328) the thyroid cartilage and hyoid bone are approximated (in addition to general elevation of the pharynx, larynx and trachea). This causes the structures defined above to bulge posteriorly into the laryngeal inlet as a transverse fold and narrow it during swallowing. The reverse occurs during inspiration, all folds being reduced to a minimum. Intrusion of *all* the folds is important in phonation, since the vocal cavity (vide infra) is thereby altered, modifying its resonant properties. (The *median thyrohyoid fold* contains the so-called 'pre-epiglottic space' of laryngeal surgery, an important region in the spread of supra-epiglottic tumours (see Maguire & Dayal 1974).

The movements of the larynx during deglutition are described on p. 1329.

THE ANATOMY OF SPEECH

Though the respiratory function of the larynx is basic, its adaptation to speech and derivatives of speech is paramount in the emergence of human society and culture; the mute, deaf and illiterate are at great disadvantage. The intelligence, skill and enterprise of the gregarious primates, especially mankind, depends on intercommunication. Language, in all its permutations, firstly in speech for immediate tribal integrations of effort, then in recorded ideas and experience with all consequent accumulation and transmission of knowledge through time and space, is the foundation of human pre-eminence. Acquisition of language is perhaps the most complex sensorimotor achievement in the individual's life. Large cerebral territories are involved in the sensory, perceptual and motor aspects of speech (p. 1055). In addition to the auditory system and the intricate apparatus producing speech, speech involves not only the larynx and other respiratory organs, but also an extensive array of muscles, from the abdominal wall to the lips. These muscles are largely described elsewhere. Here we are concerned with the larynx and associated respiratory spaces.

The larynx is the primary source of the complicated series of sounds which are the basis of speech. In this context, therefore, its structure must now be considered, including also the pharynx, mouth and nasal cavities, which together constitute a 'vocal tract'. The physics of this phonation, the production of sounds and their articulation into *phonemes*, basic units of speech, has attracted less interest than other aspects of linguistics, which have engendered a vast literature, augmented by interdisciplinary approximation between workers in such diverse fields as 'hemispheric' anatomy, neurophysiology and the 'psychological' aspects of speech. Such investigations are outside the compass of this volume; but the acoustics of speech, in terms of the structure in the vocal tract, must be briefly considered.

As in most musical instruments, the mechanism of speech consists of three essentials: a source of energy, structures capable of periodic and aperiodic oscillations and a resonator. Energy is derived from the velocity of the expired air, oscillation primarily from the vocal folds and resonance from the multiform 'column' of air extending from the folds to the lips and nostrils. Variations in all three are intricate, rapid and require complex neural integration but are, nevertheless, subject to the same physical laws as, e.g., an organ pipe. They can be similarly analysed but not with the same simplicity or precision. The vocal folds vibrate with the same interactions of length, mass and tension as a vibrating string. The available increase in length, almost 50%, does not suffice for the range of frequencies characteristic of speech and certainly not for the octaves used in singing. Ranges vary with age and sex, the total for human speech being 60 to 500 cycles per second, with average central frequencies of about 100 for males, 200 for females and 250 for children. An individual's range is achieved by variation in *tension* and *mass* of the vocal folds, as well as length. Muscles such as the cricothyroid and posterior crico-arytenoid (p. 1255) affect tension *and* length. Changes in mass are probably effected by the thyro-arytenoid muscles, whose *pars vocalis* (p. 1255), partly attached to its vocal ligament, may bring about adjustments of length, tension and mass. Other intrinsic muscles also participate; the geometry of their attachments (p. 1255) indicates their effects. Electromyographic and photographic techniques have not yet contributed much to phonational mechanics. The modes of vibration of the vocal folds are complex (Van den Berg 1968, Wyke 1974, Hinchcliffe & Harrison 1976, Fry 1979). Numerous harmonics accompany any fundamental frequency and the orientation of waves in the vocal folds is complex. High-speed cinematography has revealed transverse and longitudinal waves, whose distribution varies even during a single phoneme. Commonly, at the start of phonation the glottis closes to a linear cleft in both intermembranous and intercartilaginous parts (8.16), the latter being narrowed by arytenoid rotation and approximation. The vocal folds come into contact, varying in the precise areas apposed, even during production of a single 'sound'. As air pressure increases below the vocal folds they are suddenly forced apart, recoiling elastically because pressure momentarily drops. This cycle is repeated at a fundamental frequency (and its harmonics) dependent on factors stated above and on the velocity and pressure of the air current. Re-apposition of the folds may be aided by muscular effort and by the 'Bernouilli effect', a kind of suction due to sudden decrease in pressure as the folds open. These events are measured in microseconds and impart pressure waves of like frequencies to the column of air above the vocal folds, which acts as a selective resonator. To term it a 'column' is an over-simplification; not only is it highly variable in shape and dimensions from level to level, and tortuous in its axis, but it is also adjustable within wide limits in respect of these properties *and* in the tension of its walls. The positioning of the tongue, fauces, soft palate and changes in other dimensions, all modify the vocal tract. The sound impressed upon the resonator, in all its dynamic variation of fundamental tone and harmonics, is thus subjected to a second modification, some harmonics being dampened, others enhanced. The analysis of the final 'spectrum' of frequencies in particular sounds may be represented graphically against co-ordinates of frequency and amplitude. Such investigations show that the vocal tract acts as an intricately selective filter and resonator, propagating a remarkably similar and individual pattern or 'envelope' irrespective of the fundamental frequencies produced by the vocal folds. This is essential to maintaining a constant quality or *timbre*, despite a continuously varying tone. The term *formant* is also used, as in musical instruments; e.g. a flute sounds like a flute over a wide register, because the formant is characteristic throughout. Each voice similarly has an identifiable formant or quality; if this constantly varied, especially in the tones given to vowel sounds, intelligibility would be lost.

So far we have considered only production of *tones* but, however complex, these do not constitute speech, in which the phonetic permutation of a small repertoire of vowels is vastly enriched by imposition of consonants, i.e. by *articulation*. Consonants are associated with particular anatomical sites, from which they are designated in the terminology of phonetics, e.g. we speak of *labials (p and b), dentals (t and d), nasals (m and n)* and so on. These sites have two similarities: (1) a partial obstruction at some level of the vocal tract, and (2) production of aperiodic vibration, i.e. noise, superimposed upon or interrupting the flow of laryngeal tones. This complex subject can only be touched upon here. It is easy to ascertain subjectively that in pronouncing dental consonants the tip of the tongue is apposed to the back of the teeth. This momentarily obstructs escaping air, modifies resonant parameters and generates local noise. The vocal tract is considered to consist of a long column from vocal folds to lingual tip combined with a short one from teeth to lips. This 'resonator' may be divided by appositional constrictions into less unequal moieties, e.g. by approximation of the tongue to the palate (hard or soft) and at other levels. Each 'setting' of the anatomical resonator, as with the vocal folds, adds a characteristic to the total acoustic product which must be identifiable as a significant event in the stream of speech if this is to be intelligible.

These anatomical changes in the larynx and resonating vocal tract are effected in small fractions of a second, with a speed, adroitness and subtlety impossible to convey in verbal description. The multiplicity of laryngeal, pharyngeal, hyoidean, palatal, lingual and circumoral muscles, linked in rapidly changing combinations for phonation and articulation, reflects the remarkable complexity of speech. It is not surprising that one philosopher (Chomsky 1965) has suggested that the cerebrum, particularly its 'dominant hemisphere', is genetically programmed for language (p. 1065). Apart from these wider implications, phonetics is a larger study than can be considered here, even superficially in its strictly anatomical connections. Enough has been said, perhaps, to demonstrate the usefulness of the anatomy of speech to phoneticians and others and to stimulate anatomists to a deeper appreciation of the phonetic functions of the larynx and vocal tract.

THE LARYNGEAL MUCOUS MEMBRANE

The laryngeal mucosa, continuous with that of the pharynx, mouth and trachea, is loosely attached to the anterior surface of the epiglottis and to the lining of the valleculae. It covers the aryepiglottic folds, which limit the laryngeal inlet and contain some fibro-areolar tissue. It lines the laryngeal cavity, largely forms the vestibular folds and continues into the laryngeal sinus and saccule. It is firmly attached to the posterior epiglottic surface and to the laryngeal surfaces of the cuneiform and arytenoid cartilages. On the vocal ligaments it is thin and adherent. On the anterior epiglottic surface and its posterior upper half, the upper parts of the aryepiglottic folds and the vocal folds, the epithelium is stratified squamous (8.5B); other such areas occur above the glottis. Elsewhere in the larynx it is ciliated columnar in type. The ultrastructure of the laryngeal epithelium of the rat has been described in detail by Lewis & Prentice (1980).

The laryngeal mucosa has numerous *mucous glands*, especially over the epiglottis, where they pit the cartilage, and along the margins of the aryepiglottic folds anterior to the arytenoid cartilages, where they are termed the *arytenoid glands*. Many large glands in the saccules secrete periodically over the vocal folds during phonation; the edges of these folds are devoid of glands, their stratified epithelium thus being vulnerable to drying and requiring the secretions of neighbouring glands. Hoarseness due to excessive speaking is due to partial temporary failure of this secretion. Scanning electron microscopy demonstrates microvilli and microplicae on the surface epithelial cells of the vocal folds

and elsewhere in the larynx (Andrews 1975, Tillmann et al 1977), features common to other epithelia subjected to drying (e.g. corneal epithelium). Microplicae are regarded as aiding the retention of surface secretions.

Taste buds, like those in the tongue, occur on the posterior epiglottic surface, aryepiglottic folds and less often in other laryngeal regions. Their centripetal pathway is via the vagus nerve.

LARYNGEAL VESSELS AND NERVES

The chief *arteries* of the larynx are rami of the superior and inferior thyroid arteries; their accompanying *veins* join both the superior thyroid vein, opening into the internal jugular, and the inferior thyroid vein, draining into the left brachiocephalic. *Lymph vessels* form groups above and below the vocal folds; the superior accompany the superior laryngeal artery, traverse the thyrohyoid membrane and end in the deep cervical lymph nodes near the bifurcation of the common carotid artery; some of the inferior group of lymphatics pierce the cricothyroid ligament to reach a lymph node in front of the ligament or the upper trachea; others pass below the cricoid cartilage to the deep cervical lymph nodes and to nodes along the inferior thyroid artery. The *nerve supply* is from the internal and external branches of the superior laryngeal and from the recurrent laryngeal and sympathetic nerves. The internal laryngeal ramus is probably entirely sensory and autonomic, although special visceral motor fibres to the transverse arytenoid have been reported (Williams 1951). This nerve enters postero-inferiorly through the thyrohyoid membrane above the superior laryngeal artery; its branches supply both epiglottic surfaces, the aryepiglottic fold and the laryngeal interior as far as the vocal folds. The external laryngeal nerve (p. 1117) supplies the cricothyroid muscle, which it enters via its external surface. Terminally the recurrent laryngeal nerve and the laryngeal branch of the inferior thyroid artery ascend medial to the lower border of the inferior pharyngeal constrictor immediately behind the cricothyroid joint, supplying all the intrinsic laryngeal muscles except the cricothyroid and innervating the mucosa below the vocal folds. Before entering the larynx this nerve usually divides into motor and sensory rami, not 'adductor' and 'abductor' rami as is sometimes asserted (Williams 1954). Motor units in such skilled musculature might be expected to be small: a ratio of 30 muscle fibres to each motor neuron has been estimated (English & Blevins 1969).

Laryngoscopic Examination

The laryngeal inlet, the structures around it and its cavity can be inspected by suitable optics e.g. a laryngoscopic mirror (8.12). The epiglottis is seen foreshortened, but its tubercle is visible. From the epiglottic margins the aryepiglottic folds can be traced posteromedially and the cuneiform and corniculate elevations recognized. The pink vestibular folds and white vocal folds are visible and, when the rima glottidis is wide open, the tracheal mucosa and cartilages may be seen. The piriform fossae can also be inspected.

Radiography

In lateral cervical radiographs (8.20) the epiglottis, aryepiglottic folds, arytenoid, corniculate and sometimes cuneiform cartilages and laryngeal sinus are all visible, as well as ossification in the cartilages.

Surface and Applied Anatomy

Around the midline of the neck the following structures can be identified (5.33). The laryngeal prominence is visible in men, but not always in women. (The vocal folds are nearly level with the midpoint of the prominence.) The superior borders of the thyroid laminae and the thyroid notch are palpable. Above the thyroid, the hyoid bone's body and greater cornua can be felt, the latter most convincingly when the throat at this level is gripped between finger and thumb. The thyrohyoid membrane is in the depression between the thyroid cartilage and the hyoid bone; below the thyroid cartilage, the cricoid can be felt; it is level with the sixth cervical vertebral body. Between the cricoid and thyroid cartilage is the cricothyroid ligament. Below the cricoid is the first tracheal cartilage. Foreign bodies may obstruct the laryngeal inlet or rima glottidis and suffocation may ensue. Small foreign bodies may enter the trachea or bronchi or lodge in the laryngeal sinus and cause reflex glottic spasm with consequent suffocation. In-flammation of the upper larynx may swell the mucosa by effusion of fluid into an abundant, lax submucous tissue (oedema of the glottis). The effusion does not involve or extend below the vocal folds, since the mucosa adheres directly to the vocal ligaments without the intervention of submucous tissue. Laryngotomy *below* the vocal folds through the cricothyroid ligament or tracheotomy may be necessary to restore a free airway. The mucosa of the upper larynx is highly sensitive; contact with foreign bodies excites immediate coughing. Suicidal wounds are usually through the thyrohyoid membrane, damaging the epiglot-tis, superior thyroid vessels, external and internal carotid arteries and internal jugular veins; less frequently they are above the hyoid with damage to the lingual muscles and lingual and facial vessels. (For the results of damage to the laryngeal nerves, see p. 1118.)

The Trachea and Bronchi

THE TRACHEA

The trachea, a tube of cartilage and fibro-muscular membrane, about 10–11 cm long, descends from the larynx (**8.21, 22**), extend-ing from the level of the sixth cervical to the upper border of the fifth thoracic vertebra, where it divides into right and left prin-cipal (pulmonary) bronchi. It lies approximately in the sagittal plane but its point of bifurcation is usually a little to the right. During deep inspiration the bifurcation may descend to the sixth thoracic vertebral level. The trachea is mobile and can rapidly alter in length. It is less than cylindrical, being flattened posterior-ly; its external transverse diameter is about 2 cm in adult males, 1.5 cm in adult females. In children it is smaller, more deeply placed and more mobile. In the living the lumen is smaller than in the dead, its diameter in adults being about 12 mm. In the first year, tracheal diameter does not exceed 3 mm while during child-hood its diameter in millimetres is about equal to age in years. The transverse shape of the lumen is variable, especially in later decades, being round, lunate or flattened (Campbell & Liddelow 1967). Wang & Tai (1965) have studied tracheobronchial dimen-sions extensively; in Chinese subjects (presumably cadavers) the tracheal lumen averaged 16.17 mm (range 9.5–22.0 mm).

Relations of the Trachea

The *cervical part* of the trachea (**6.68**) is covered *anteriorly* by skin, superficial and deep fasciae, crossed by the jugular arch and overlapped by the sternohyoid and sternothyroid muscles. The second to fourth tracheal cartilages are crossed by the isthmus of the thyroid gland, above which an anastomotic artery connects the superior thyroid arteries; below this, in front, are the pretracheal fascia, inferior thyroid veins, thymic remnants and the arteria thyroidea ima (when it exists). In children the brachiocephalic artery crosses obliquely in front of the trachea at or a little above the upper manubrial level; the left brachiocephal-ic vein may also rise a little above the manubrium. *Posterior* is the oesophagus, between the trachea and the vertebral column; the recurrent laryngeal nerves ascend, on each side, in or near the grooves between the sides of the trachea and oesophagus. *Lateral* are the lobes of the thyroid gland, descending to the fifth or sixth tracheal cartilage, and the common carotid and inferior thyroid arteries.

The *thoracic part* of the trachea (**8.17, 18, 19, 23**) descends through the superior mediastinum, related *anteriorly* to: the manubrium sterni, the attachments of the sternohyoid and ster-nothyroid muscles, the thymic remnants, the inferior thyroid and left brachiocephalic veins, the aortic arch, the brachiocephalic

and left common carotid arteries, deep cardiac plexus and some lymph nodes. Diverging as they ascend in the neck the brachiocephalic and left common carotid arteries become respec-tively right and left of the trachea. *Posterior* is the oesophagus, separating the trachea from the vertebral column. On the *right* are: the right lung and pleura, right brachiocephalic vein, superior vena cava, right vagus nerve and the azygos vein; on the *left*: the arch of the aorta, left common carotid and left subclavian arteries. The left recurrent laryngeal nerve is at first between the trachea and aortic arch, then in or just in front of the groove between the trachea and the oesophagus.

THE RIGHT PRINCIPAL BRONCHUS

The right principal bronchus (**8.21, 22, 24**) is wider, shorter and more vertical than the left, being about 2.5 cm long. It gives rise to its first branch, the *superior lobar bronchus* and then enters the right lung opposite the fifth thoracic vertebra. Its greater width and more vertical course explain the greater frequency of foreign bodies entering the right principal bronchus than the left. The azygos vein arches over it and the right pulmonary artery lies at first inferior, then anterior to it. Beyond its first branch, the *superior lobar bronchus*, which arises posterosuperior to the

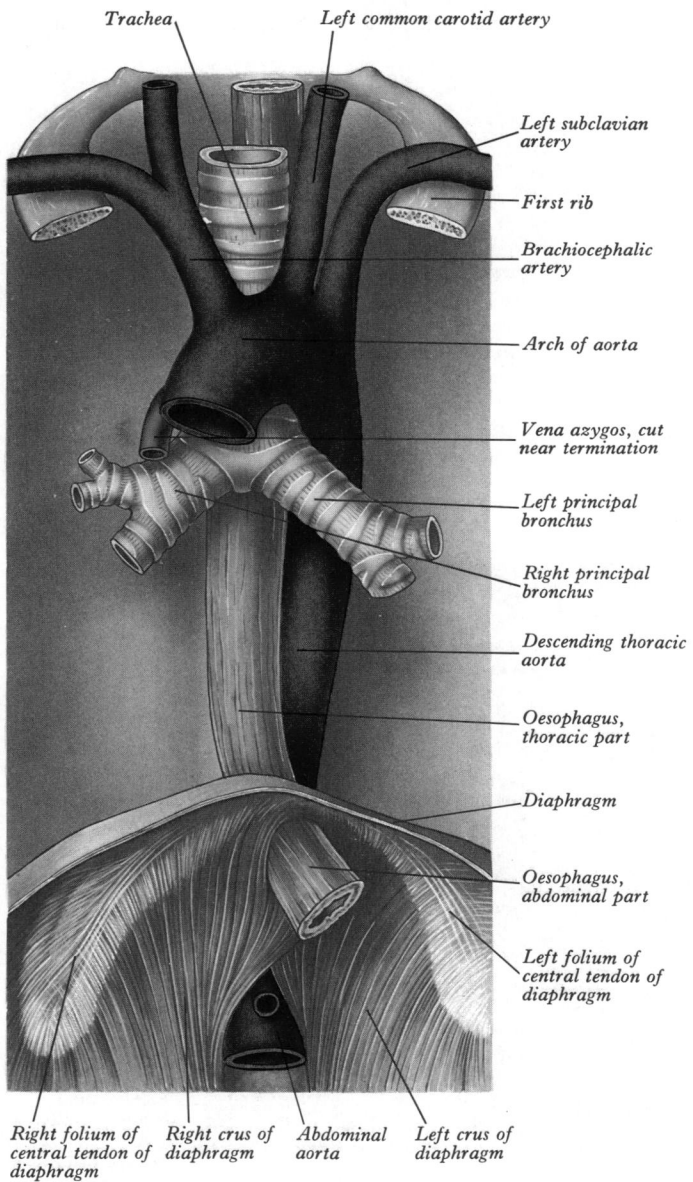

Trachea   *Left common carotid artery*

Left subclavian artery

First rib

Brachiocephalic artery

Arch of aorta

Vena azygos, cut near termination

Left principal bronchus

Right principal bronchus

Descending thoracic aorta

Oesophagus, thoracic part

Diaphragm

Oesophagus, abdominal part

Left folium of central tendon of diaphragm

Right folium of   *Right crus of*   *Abdominal*   *Left crus of*
central tendon of   *diaphragm*   *aorta*   *diaphragm*
diaphragm

8.17 Dissection to show the bifurcation of the trachea and the principal bronchi.

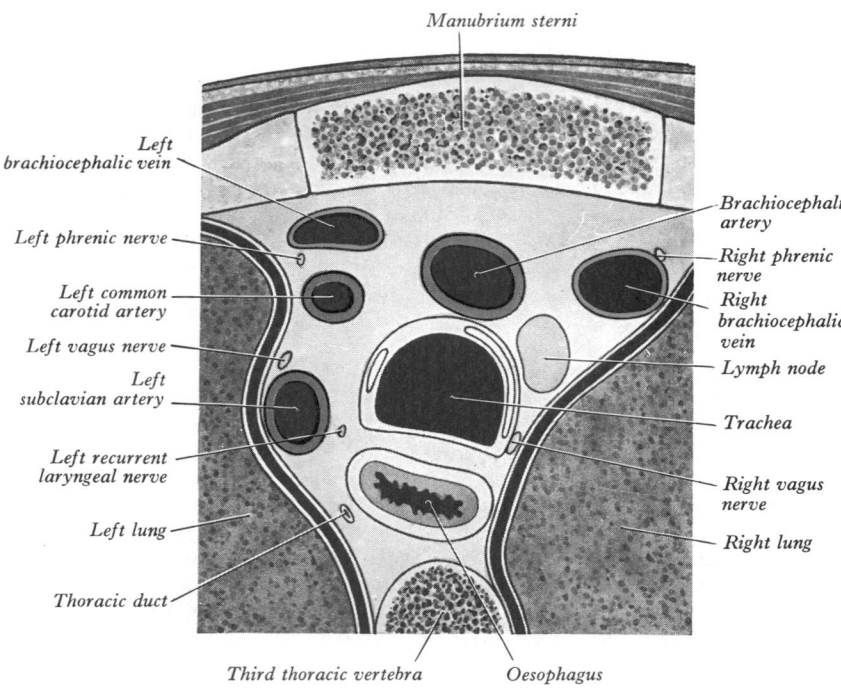

8.18 Transverse section through the mediastinum at the level of the body of the third thoracic vertebra, viewed from the superior aspect.

right pulmonary artery, it crosses the posterior aspect of this artery to enter the pulmonary hilum postero-inferior to the artery and then divides into a *middle* and an *inferior lobar bronchus.*

The right superior lobar bronchus arises from the lateral aspect of the parent bronchus and runs superolaterally to enter the hilum. About 1 cm from its origin it divides into three segmental bronchi: the *apical segmental bronchus* continues superolaterally towards the apex of the lung, which it supplies dividing near its origin into apical and anterior branches; the *posterior segmental bronchus* serves the postero-inferior part of the superior lobe, passing posterolaterally and slightly superior-

ly and soon dividing into a lateral and a posterior branch; the *anterior segmental bronchus* runs antero-inferiorly to supply the rest of the superior lobe; it divides near its origin into a lateral and an anterior branch of equal size.

The middle lobar bronchus starts about 2 cm below the superior, from the front of the parent trunk, descends antero-laterally and soon divides into a *lateral* and a *medial segmental bronchus* to the lateral and medial parts of the middle lobe, respectively.

The right inferior lobar bronchus is the continuation of the principal bronchus beyond the origin of the middle lobar. At or a little below this point, it gives off posteriorly a large *superior* (apical) *segmental bronchus* which runs posteriorly to the upper part of the inferior lobe, subsequently dividing into medial, superior and lateral branches, the first two usually

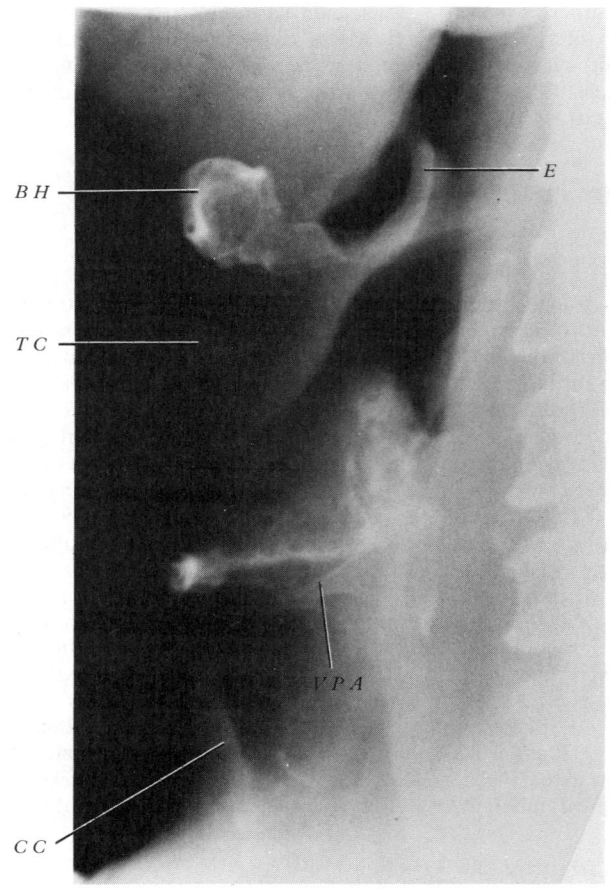

8.20 Lateral soft tissue radiograph of the larynx and adjacent structures in a young adult. BH = body of hyoid bone; CC = cricoid cartilage; E = epiglottis; TC = thyroid cartilage; VPA = vocal process of arytenoid cartilage. Supplied by Shaun Gallagher, Guy's Hospital; photography by Sarah Smith, UMDS, Guy's Campus, London.

8.19 Transverse section through the mediastinum at the level of the upper part of the body of the fourth thoracic vertebra, viewed from the superior aspect.

arising from a common stem. After giving off this branch, the right inferior lobar bronchus descends posterolaterally. The *medial basal segmental bronchus* branches from it anteromedially, running inferomedially to serve a small region below the hilum; the inferior lobar bronchus continues downwards then divides into an *anterior basal segmental bronchus*, which descends anteriorly, and a trunk which soon divides into a *lateral basal segmental bronchus* descending laterally and a *posterior basal segmental bronchus* descending posteriorly. In more than half of all right lungs a *subsuperior* (subapical) *segmental bronchus* arises posteriorly from the right inferior lobar bronchus 1–3 cm below the superior segmental bronchus. This is distributed to the region of the lobe between the superior and posterior basal segments.

THE LEFT PRINCIPAL BRONCHUS

The left principal bronchus (8.21, 24), which is narrower and less vertical than the right and nearly 5 cm long, enters the left hilum level with the sixth thoracic vertebra. Passing left inferior to the aortic arch, it crosses anterior to the oesophagus, thoracic duct and descending aorta; the left pulmonary artery is at first anterior and then superior to it. Having entered the hilum it divides into a superior and an inferior lobar bronchus.

The left superior lobar bronchus, arising from the anterolateral aspect of its parent stem, curves laterally and soon divides into two bronchi. These correspond to the branches of the right principal bronchus supplying the right superior and middle lobes but on the left both are distributed to the left *superior* lobe, there being no separate middle lobe. The *superior division* ascends about 1 cm, gives off an *anterior segmental bronchus*, then continues a further 1 cm as the *apicoposterior segmental bronchus* before dividing into apical and posterior branches. The apical, posterior and anterior segmental bronchi are largely distributed as in the right superior lobe. The *inferior division* descends anterolaterally to the antero-inferior part of the left *superior* lobe, the *lingula*, forming the *lingular bronchus* which divides into *superior* and *inferior lingular segmental bronchi*, unlike the pattern in the right middle lobe.

The left inferior lobar bronchus descends posterolaterally for 1 cm and then the *superior* (apical) *segmental bronchus* arises from it posteriorly and is distributed in essentially the same manner as in the right lung. After a further 1–2 cm it divides into an anteromedial and a posterolateral stem. The former divides into *medial basal* and *anterior basal segmental bronchi*; the latter into *lateral* and *posterior basal segmental bronchi*. The territories supplied resemble those on the right. The medial basal segmental bronchus is an independent branch of the inferior lobar bronchus in about 10% of lungs; this, and the similarity of its territory to that on the right, supports its recognition as a separate segmental bronchus. A *subsuperior* (subapical) *segmental bronchus* arises posteriorly from the left inferior lobar bronchus in 30% of lungs.

BRONCHOPULMONARY SEGMENTATION

Primary branches of the right and left *lobar bronchi* are termed *segmental bronchi* because each ramifies in a structurally separate, functionally independent unit of lung tissue called a *bronchopulmonary segment* (8.24, 25). The main segments are *named* and *numbered* as follows:

Right

Superior lobe:	(1) apical, (2) posterior, (3) anterior.
Middle lobe:	(4) lateral, (5) medial.
Inferior lobe:	(6) superior (apical), (7) medial basal, (8) anterior basal, (9) lateral basal, (10) posterior basal.

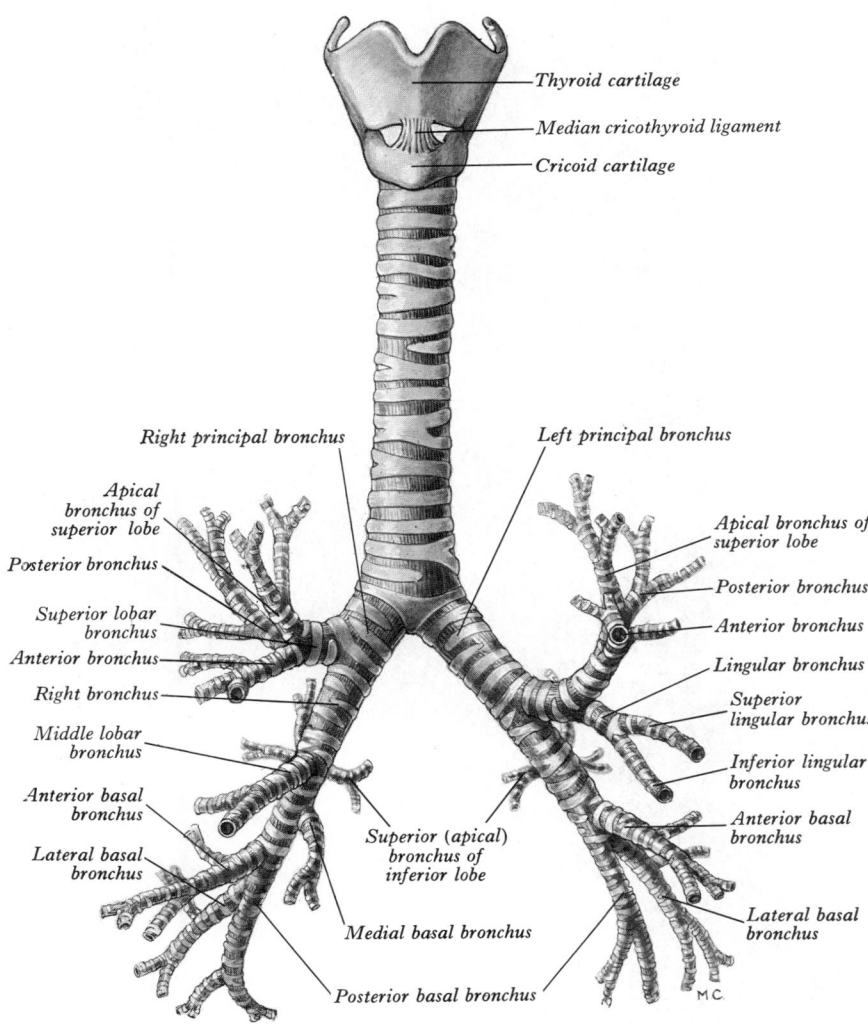

Thyroid cartilage

Median cricothyroid ligament

Cricoid cartilage

Right principal bronchus

Left principal bronchus

Apical bronchus of superior lobe

Posterior bronchus

Superior lobar bronchus

Anterior bronchus

Right bronchus

Middle lobar bronchus

Anterior basal bronchus

Lateral basal bronchus

Superior (apical) bronchus of inferior lobe

Medial basal bronchus

Posterior basal bronchus

Apical bronchus of superior lobe

Posterior bronchus

Anterior bronchus

Lingular bronchus

Superior lingular bronchus

Inferior lingular bronchus

Anterior basal bronchus

Lateral basal bronchus

8.21 The cartilages of the larynx, trachea and bronchi: anterior aspect. Drawn from a metal cast made by Lord Russell Brock, Guy's Hospital, London.

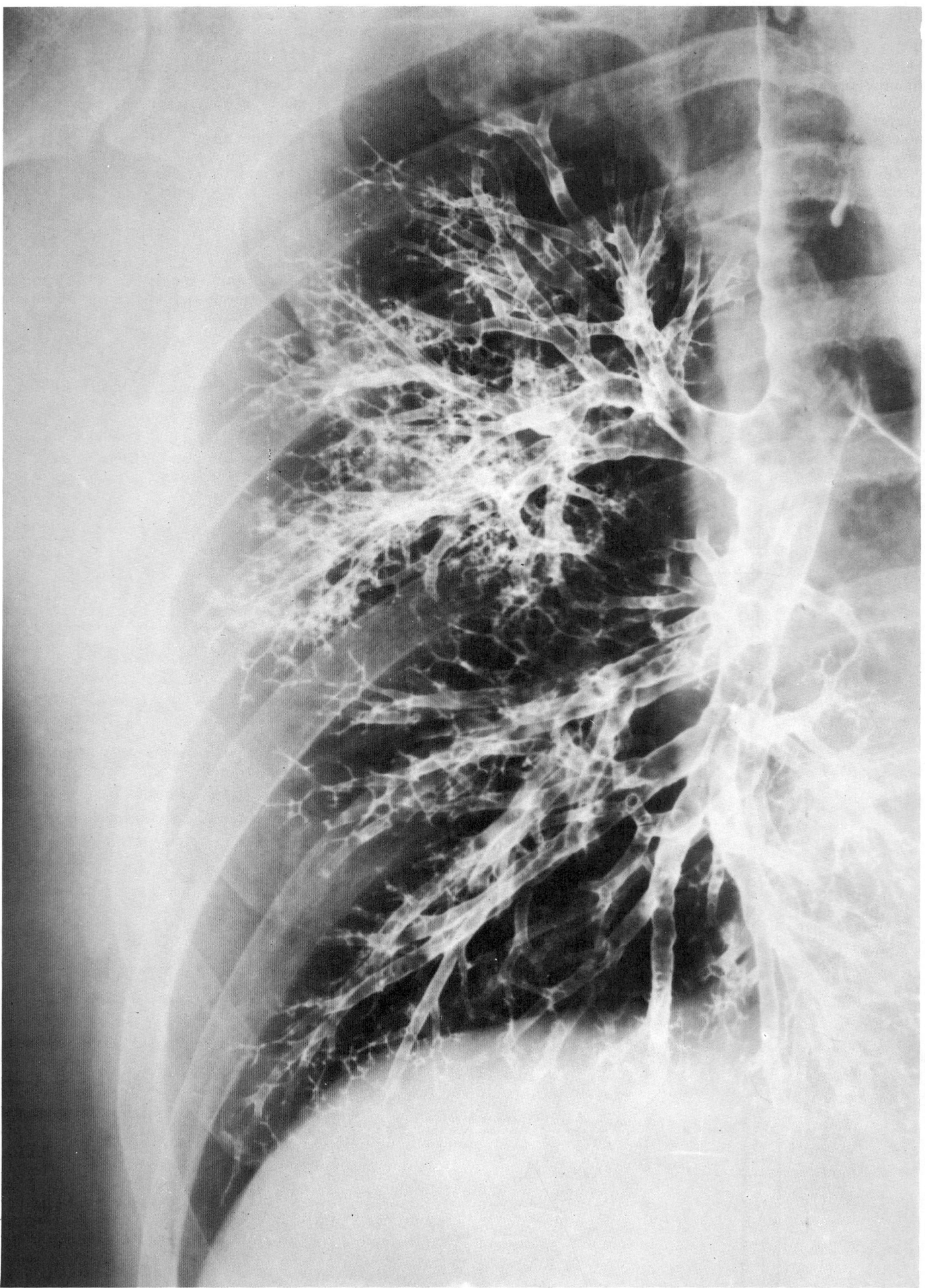

8.22 A slightly oblique, anteroposterior view of the trachea and right bronchial tree, after insufflation of a radio-opaque contrast medium. For identification of the lobar and segmental bronchi in this bronchogram, compare with illustrations of bronchial tree casts (8.21, 24).

Left Lung

Superior lobe: (1) apical, (2) posterior, (3) anterior, (4) superior lingular, (5) inferior lingular.

Inferior lobe: (6) superior (apical), (7) medial basal, (8) anterior basal, (9) lateral basal, (10) posterior basal.

Each segment, surrounded by connective tissue continuous with the visceral pleura, is a separate respiratory unit. The vascular and lymphatic arrangements of the segments are described on pp. 857, 1282. (For further details of segmentation consult Brock 1942, 1943, 1944, 1954, Boyden 1955, Bloomer et al 1960, Volpe et al 1969).

Applied Anatomy. While pathological conditions such as bronchiectasis and some infections may be restricted to one or

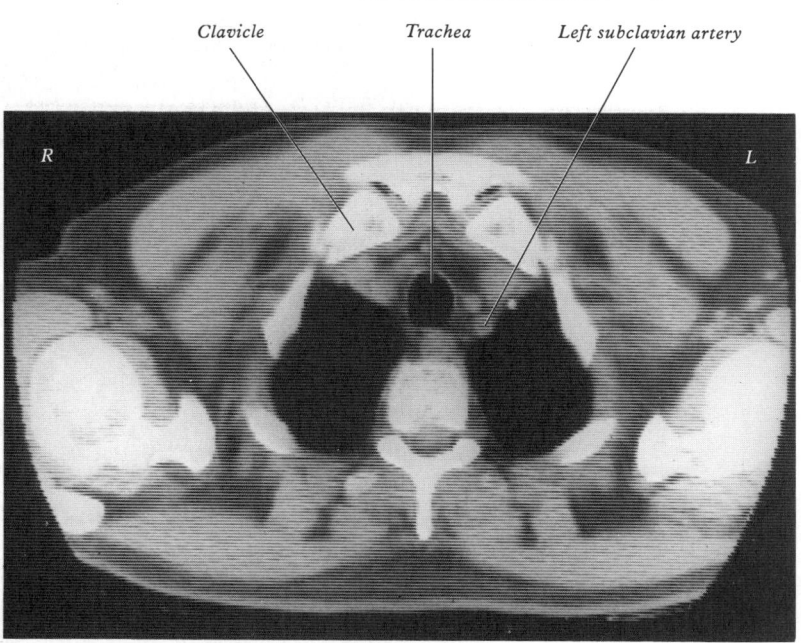

8.23A Computed tomogram in the transverse plane of the thorax at the level of the sternoclavicular joint. Provided by Shaun Gallagher, Guy's Hospital; photography by Sarah Smith, UMDS, Guy's Hospital Campus, London.

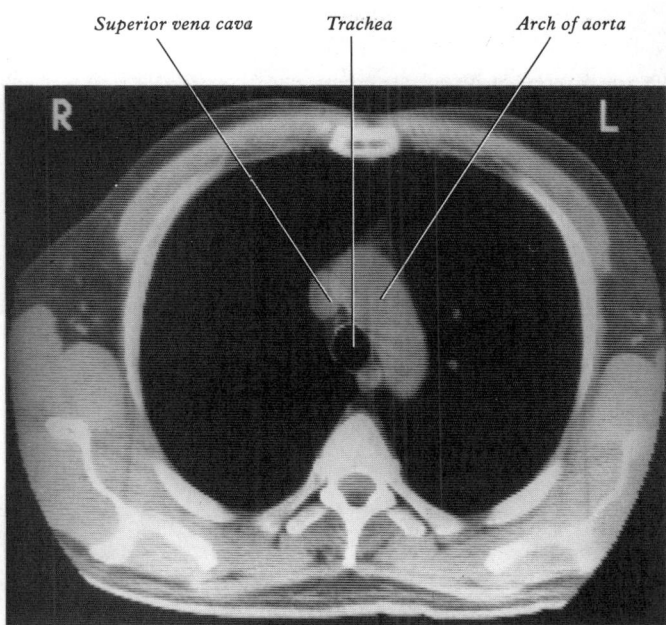

8.23B Computed tomogram in the transverse plane through the thorax at the level of the aortic arch. Supplied by Shaun Gallagher, Guy's Hospital; photography by Sarah Smith, UMDS, Guy's Hospital Campus, London.

Tracheal cartilages vary from 16 to 20, each an imperfect 'ring' surrounding approximately the anterior two-thirds of the tracheal circumference; behind, where they are deficient, the tube is flat and is completed by fibro-elastic tissue and non-striated muscle. The cartilages are horizontally stacked, separated by narrow intervals and are about 4 mm vertically and 1 mm in thickness; their external surfaces are vertically flat, their internal convex. Two or more cartilages often unite, partially or completely, and sometimes bifurcate at their ends. They are hyaline but may become calcified in the aged. In extrapulmonary bronchi, cartilages are shorter, narrower and

more bronchopulmonary segments, malignant neoplasms and tuberculosis are not so confined. A knowledge of bronchial branching is essential during bronchoscopy and for the correct interpretation of bronchograms. It is also determinative in the postural drainage of infected pulmonary regions. The superior (apical) segment of the inferior lobe is a common site of abscess following aspiration of material by supine patients. Inhaled foreign bodies may obstruct a main, lobar, segmental or smaller bronchus according to their size. The interpretation of their effects and the design of surgical treatment inevitably involve considerations of bronchial branching patterns. Resection of a single segment is practicable, while more extensive procedures may include the removal of several segments, lobectomy or a complete pneumonectomy.

STRUCTURE OF THE TRACHEA AND MAJOR BRONCHI

The trachea and extrapulmonary bronchi have a framework of incomplete rings of hyaline cartilage, united by fibrous tissue and non-striated muscle (**8.26**) and lined by mucosa (**8.27**).

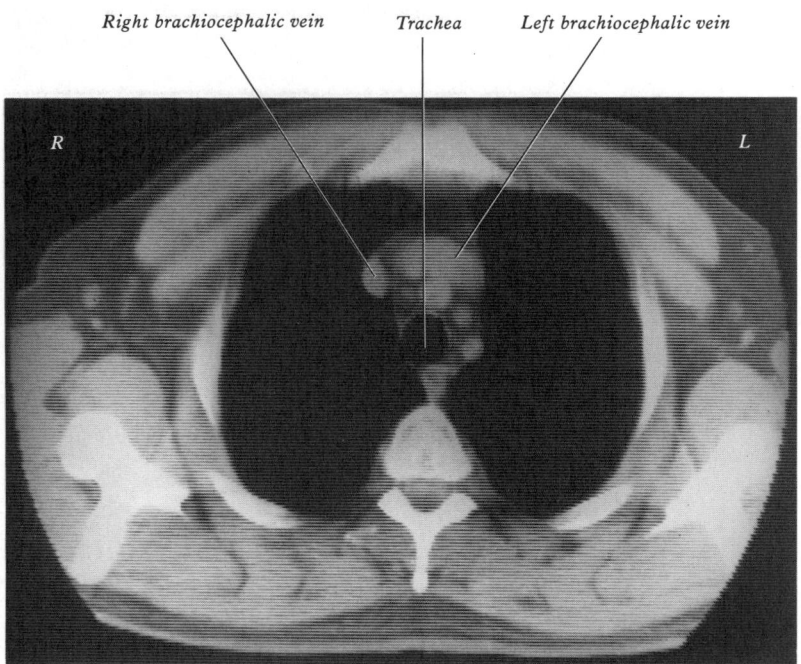

8.23C Computed tomogram in the transverse plane through the thorax at the level of the commencement of the superior vena cava. Supplied by Shaun Gallagher, Guy's Hospital; photography by Sarah Smith, UMDS, Guy's Hospital Campus, London.

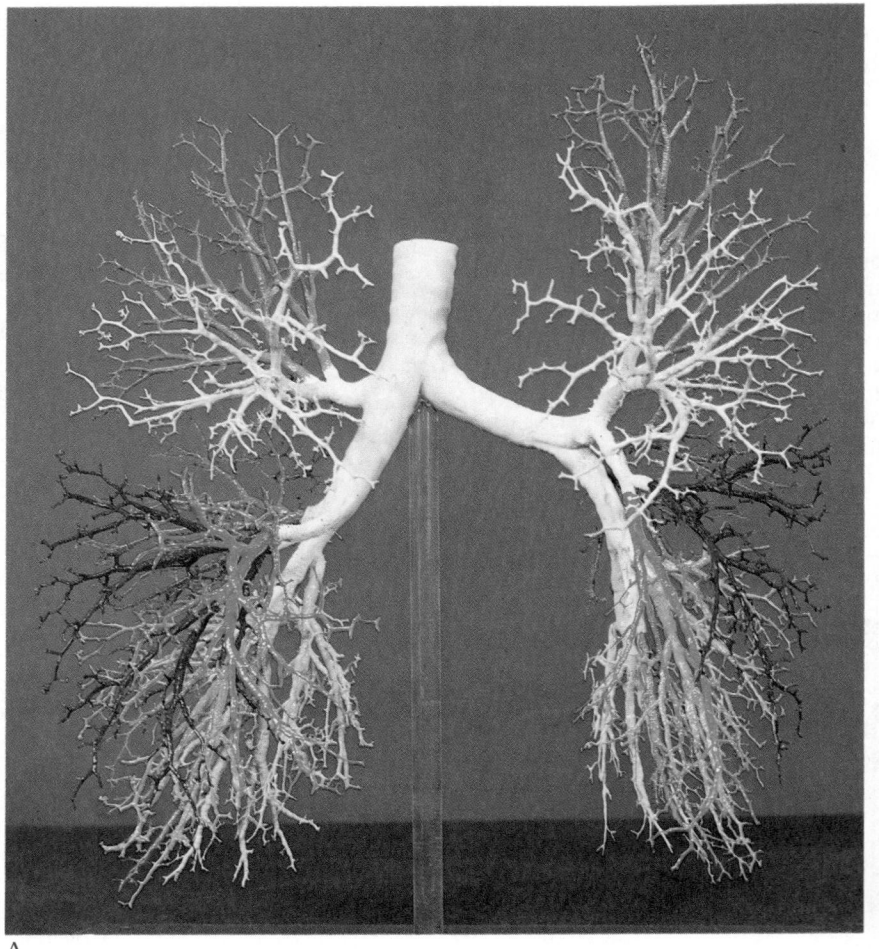

A

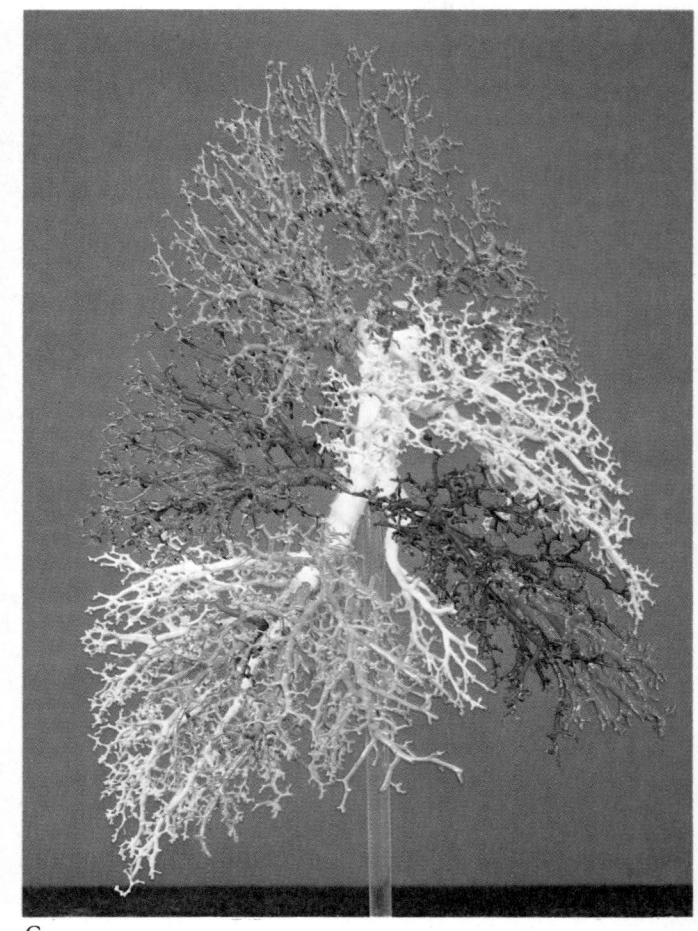

B   C

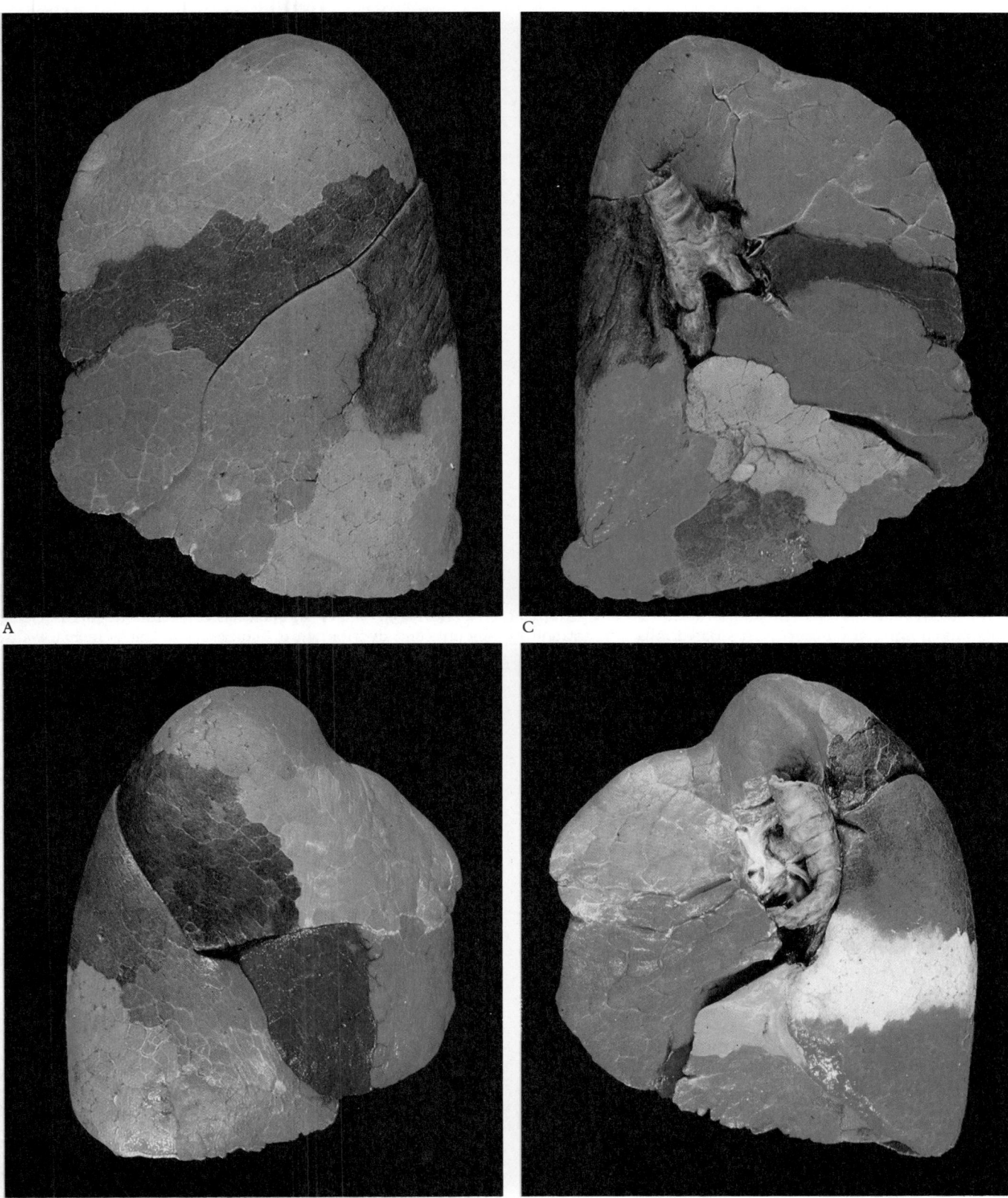

A

C

B

D

8.24 (*opposite*) A. A resin corrosion cast of the adult human lower trachea and bronchial tree photographed from the anterior aspect. The segmental bronchi and their main branches have been coloured: apical (brown), posterior (grey/blue), anterior (pink), lateral (middle lobe) and superior lingular (dark blue), medial (middle lobe) and inferior lingular (red), superior (apical) of inferior lobe (dark green), medial basal (yellow), anterior basal (orange), lateral basal (blue), posterior basal (light green).

B. Corrosion cast, left lung: lateral view (colour coding as above).

C. Corrosion cast, right lung: lateral view (colour coding as above). Note that in the final preparation of such specimens large amounts of finer detail have been carefully removed to reveal the major arrangement of the bronchial 'tree'.

8.25 (*above*) Specimens of human lungs in which the segmental bronchi have been injected with coloured gelatin. Unfixed specimens of the human lungs were obtained at post-mortem. The bronchial trees were washed out with water and individual segmental bronchi, identified by dissection, were cannulated separately and injected with gelatin of contrasting colours. The lungs were subsequently fixed in 10% formaldehyde solution. A. costal surface of left lung; B. costal surface of right lung; C. mediastinal surface of left lung; D. mediastinal surface of right lung. Approximately the same colour convention has been used as for the corrosion cast shown in 8.24A. In these specimens, however, in the right lung a subsuperior segment is present in the inferior lobe and is coloured white. All specimens were prepared by M C E Hutchinson and photographed by Kevin Fitzpatrick of the Department of Anatomy, Guy's Hospital Medical School, London.

less regular but generally similar in shape and arrangement.

The first and last tracheal cartilages differ (8.21); the *first cartilage* is the broadest, often bifurcate at one end, connected by a cricotracheal ligament to the inferior cricoid border and sometimes blended with the cricoid or second cartilage. The *last cartilage* is centrally thick and broad and its lower border, the *carina*, is a triangular unciform process curving down and backwards between the bronchi. On each side it forms an imperfect ring enclosing the start of a principal bronchus. The penultimate cartilage is centrally broader than the others.

The irregularity of the cartilaginous plates in the extrapulmonary bronchi increases distally; as major bronchi approach their lungs and lobes, plates invade their dorsal aspects but never quite encompass their bifurcations. In intrapulmonary bronchi plates of cartilage progressively form less and less of the bronchial wall, disappearing where the bronchioles begin (8.19A, 28A, Reid 1976.)

The Fibrous Membrane

Each cartilage is enclosed in perichondrium, continuous with a dense fibrous membrane between the adjacent cartilages which also fills in the back of the trachea. The perichondrium and membrane are mainly composed of collagen with some elastin fibres; fibres cross diagonally, allowing changes in luminal diameter, the elastic component providing some recoil from stretching. Non-striated muscle fibres appear posteriorly in the membrane; most are transverse, being attached to the perichondrium at the ends of the cartilages and forming a transverse sheet between them. Contraction therefore alters the cross-sectional area of the trachea and bronchi. A few external longitudinal fibres also occur. Non-striated muscle in intrapulmonary bronchi is not attached to cartilages and, where these begin to disappear, i.e. in smaller bronchi, contraction may actually obliterate the lumen (Reid 1976).

The Mucous Membrane (Tunica mucosa)

This is continuous with and closely resembles that of the larynx above and the intrapulmonary bronchi below, being a layer of pseudostratified ciliated columnar epithelium interspersed with goblet cells (Dalen 1983), both lying on a basal lamina (8.26, 27). Some pseudostratified cells possess unusually large nuclei and may be polytene in chromosomal content. Numerous lymphocytes usually occur deep in the epithelium. The cilia impel mucus towards the laryngeal inlet (aditus). Deep to the basal lamina are a lamina propria with abundant longitudinal elastic fibres and a submucosa of loose connective tissue containing larger blood vessels, nerves and most of the tubular (tracheal) seromucous glands and lymphoid nodules; external to the submucosa are the perichondrium and the fibrous membrane. Most external of all is the deep fascia, merging with the fascial planes of the surrounding muscles, oesophagus and associated structures.

Vessels and Nerves

The trachea is supplied with blood mainly by the inferior thyroid *arteries*, while its thoracic end is also supplied by the bronchial arteries, whose branches ascend to anastomose with the former; all the vessels also supply the oesophagus. The *veins* draining the trachea end in the inferior thyroid venous plexus. The *lymph vessels* pass to the pretracheal and paratracheal lymph nodes. The *nerve supply* is from the tracheal branches of the vagi, recurrent laryngeal nerves and the sympathetic trunks and is distributed to the tracheal muscle and mucosa. Sympathetic nerve endings evoke bronchodilatation by releasing catecholamines; they may also exert a direct aminergic effect on glandular acini in the bronchi (Pack & Richardson 1984). Parasympathetic activity which is cholinergic causes bronchoconstriction. Many small postsynaptic ganglia occur in the autonomic plexuses of the tracheal and bronchial walls (Feyler 1965). In cats, paraganglia composed of chromaffin cells and arteriovenous anastomoses appear in the bronchial wall

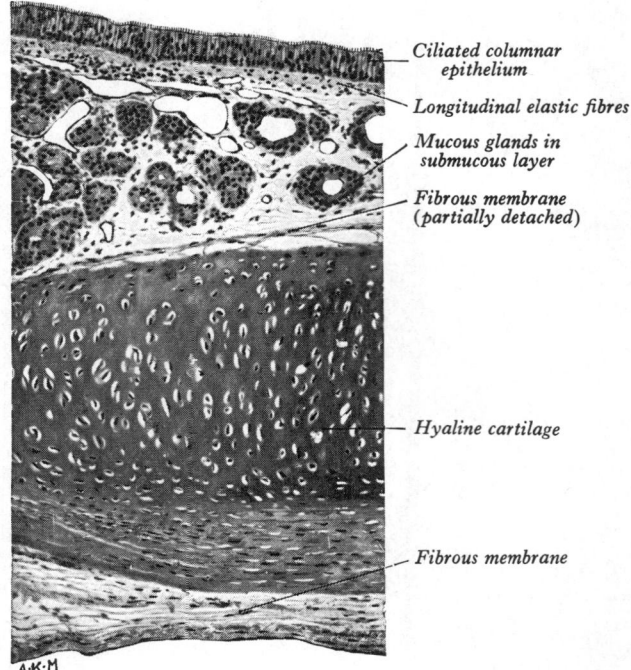

8.26 Transverse section through a part of the wall of a human trachea.

Ciliated columnar epithelium

Longitudinal elastic fibres

Mucous glands in submucous layer

Fibrous membrane (partially detached)

Hyaline cartilage

Fibrous membrane

(Muratori 1965). For a quantitative technique of assessing amounts and distribution of the tissues in the bronchial walls, normal or pathological, see Hale et al (1968).

Surface and Applied Anatomy. The *trachea* is about 2 cm wide and extends almost vertically in the midline from the cricoid cartilage to the sternal angle, inclining slightly to the right. The *right principal bronchus* runs from the trachea down to the right for 2.5 cm to the right hilum behind the sternal end of the right third costal cartilage. The *left principal bronchus* runs for 5 cm more obliquely to its left and down to the hilum behind the left third costal cartilage, 3.5 cm from midline. The trachea may be opened by median vertical incision above the thyroid isthmus

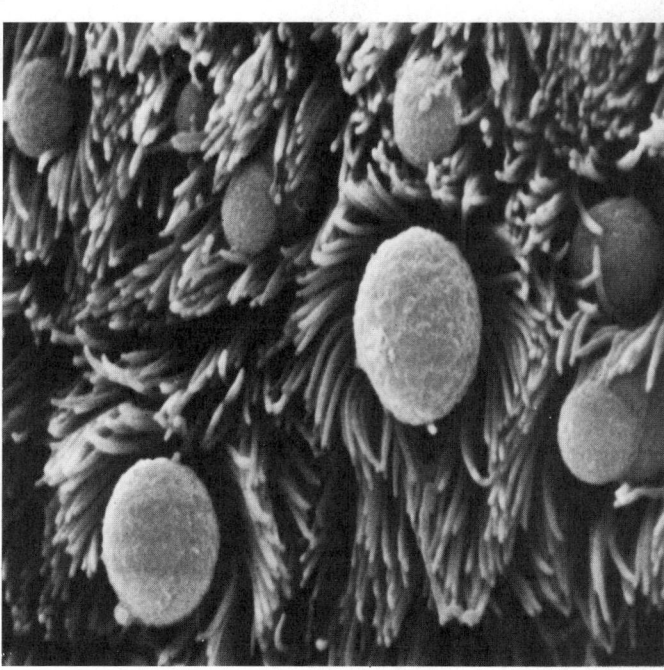

8.27 Surface view of the ciliated epithelium of the trachea (rat). Mucussecreting goblet cells occur between the ciliated cells. Prepared by Michael Crowder, Department of Anatomy, Guy's Hospital Medical School, London. Magnification × 5100.

(high tracheotomy) or below it (low tracheotomy), the latter being more difficult because the trachea recedes as it descends and has hazardous anterior relations (p. 1259). The trachea may be compressed by pathological enlargements of the thyroid gland, thymus and aortic arch. The radiological appearances of the trachea, bronchi and lungs are dealt with on p. 1286.

Bronchoscopy. By means of a bronchoscope passed through the mouth and larynx under local anaesthesia, the interior of the larynx, trachea and bronchi, with the openings of the main segmental branches (pp. 1259–1263), may all be examined and biopsies taken; foreign bodies or accumulations of fluid may also be removed.

The Pleurae

Each lung is covered by *pleura*, a serous membrane arranged as a closed invaginated sac. Part of the pleura adheres to the pulmonary surface and its interlobar fissures as *visceral* or *pulmonary pleura*. Its continuation lines the corresponding half of the thoracic wall, covers much of the diaphragm and structures occupying the middle region of the thorax; this is *parietal pleura*. The pulmonary and parietal pleurae are continuous around the hilar structures. They remain in close though sliding contact at all phases of respiration, the potential space between them being the *pleural cavity*. When air or fluid collects between the two layers, the pleural cavity expands. The right and left pleural sacs are distinct from each other and touch only behind the upper half of the sternal body (**8.28**A), although they are also close to each other behind the oesophagus at the mid-thoracic level. The region between them is the *mediastinum* (interpleural space). The left pleural cavity is the smaller of the two because the heart extends further left. Upper and lower pleural limits are about the same on the two sides, but the left sometimes descends lower in the mid-axillary line.

The pulmonary pleura is adherent to the lung over all its surfaces including those in the fissures, but absent from the hilar

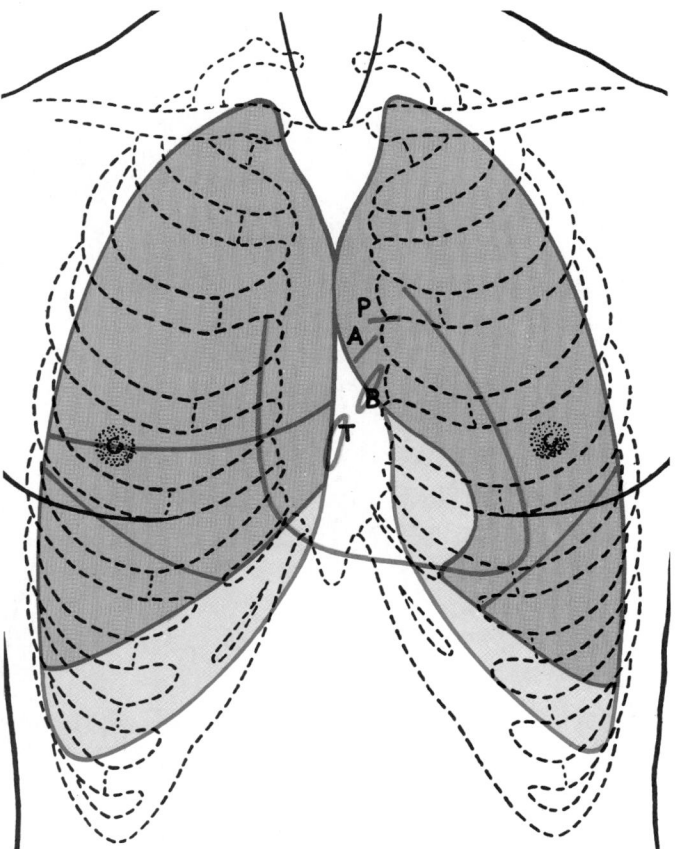

8.28A Ventral aspect of the thorax, showing surface projections: pulmonary (purple), pleural (blue) and cardiac (red outline). A = orifice of aorta, B = left atrioventricular (mitral) orifice, P = orifice of pulmonary trunk. T = right atrioventricular (tricuspid) orifice. Skeletal structures are indicated by broken lines.

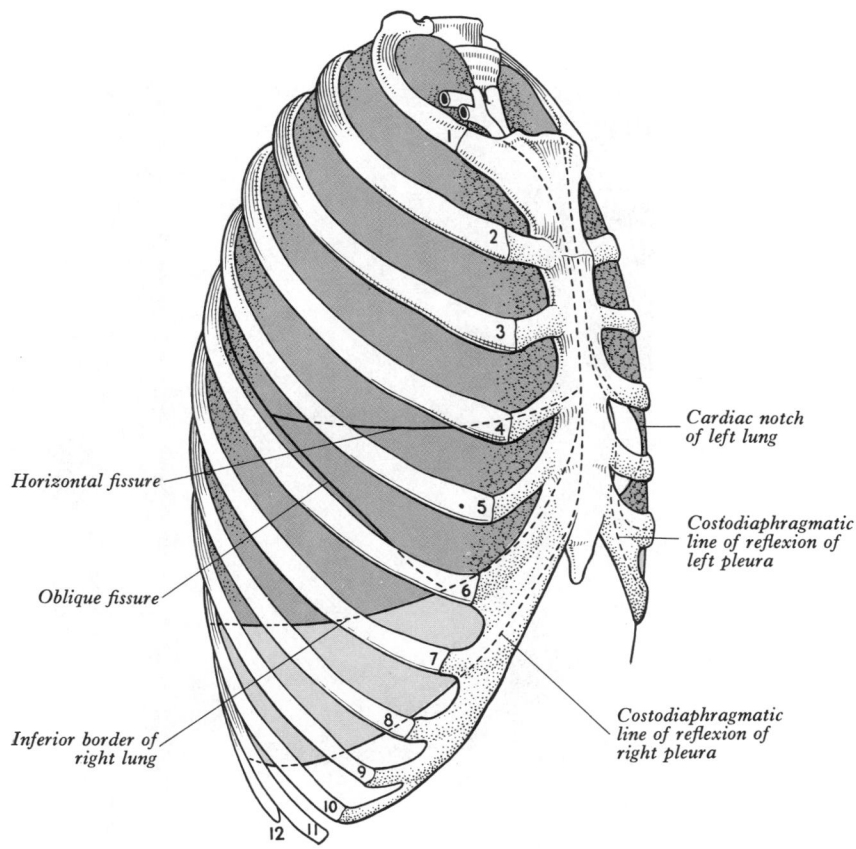

8.28B The relations of the pleurae and lungs to the chest wall: right lateral aspect. Purple = lungs, covered with the pleural sacs. Blue = pleural sac, with no underlying lung.

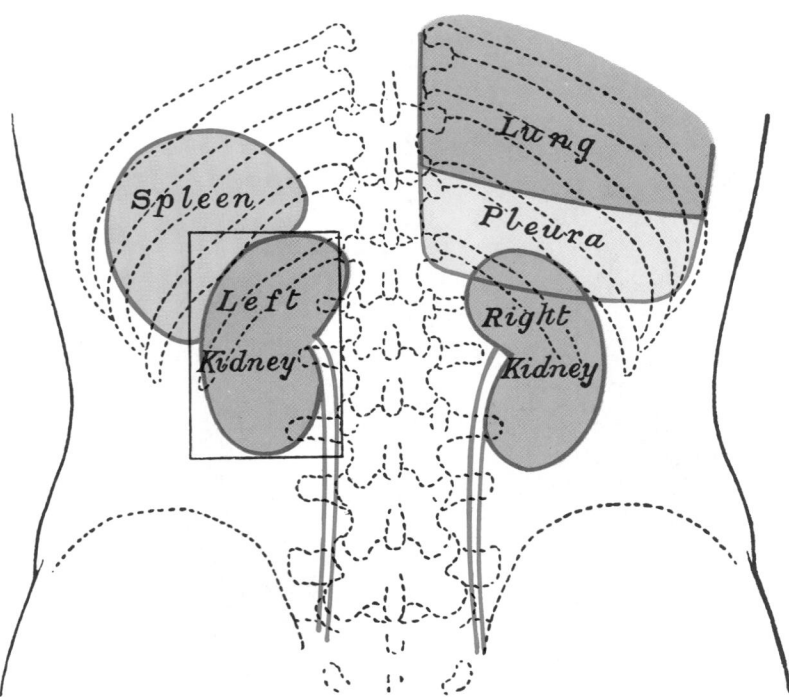

8.28C The lower limits of the lung and pleura: posterior view. The lower portions of the lung and pleura are shown on the right side.

area and along a line descending from this, marking the attachment of the pulmonary ligament (**8.34A,B**).

The parietal pleura. Different regions of parietal pleura are customarily distinguished: the part internal to the thoracic wall (p. 591) and the vertebral bodies is the *costovertebral pleura*, that on the thoracic surface of the diaphragm is the *diaphragmatic*

pleura, the part over the pulmonary apices (in the neck) is the *cervical pleura (domes of the pleura)* and that applied to the structures between the lungs is the *mediastinal pleura*.

The *costovertebral (costal) pleura* (**8.28A–C**) lines the sternum, ribs, transversus thoracis and intercostal muscles and the sides of the vertebral bodies; it is easily separated from them. Outside it a thin areolar layer of *endothoracic fascia* corresponds to the transversalis fascia of the abdominal wall. The costal pleura begins behind the sternum where it is continuous with the mediastinal pleura, at a junction extending from behind the sternoclavicular joint down and medially to the midline behind the sternal angle, where right and left junctions descend in contact to the level of the fourth costal cartilages, below which the lines of junction differ. On the right it descends to the back of the xiphisternal joint. On the left it diverges laterally and descends 2–25 mm from the sternal margin (Woodburne 1947) to the sixth costal cartilage. On each side the costal pleura sweeps laterally, lining the internal surfaces of the costal cartilages, ribs, transversus thoracis and intercostal muscles; posteriorly it passes over the sympathetic trunk and branches and to the sides of the vertebral bodies, where it is again continuous with the mediastinal pleura. The costal pleura is continuous with the cervical at the inner margin of the first rib and below it becomes continuous with the diaphragmatic along a line differing slightly on the two sides. On the right this *line of costodiaphragmatic reflexion* begins behind the xiphoid process, passes behind the seventh costal cartilage to reach the eighth rib in the mid-clavicular line, the tenth rib in the mid-axillary line and, ascending slightly, crosses the twelfth level with the upper border of the twelfth thoracic spine (**8.28C**). On the left the line follows at first the *sixth* costal cartilage but is otherwise as on the right, although it may be slightly lower.

The thin *diaphragmatic pleura* covers most of the upper surface of its own half of the diaphragm. Its circumference is largely the line described above, along which it is continuous with the costal pleura; it is continuous with the mediastinal pleura along the line of attachment of the pericardium to the diaphragm.

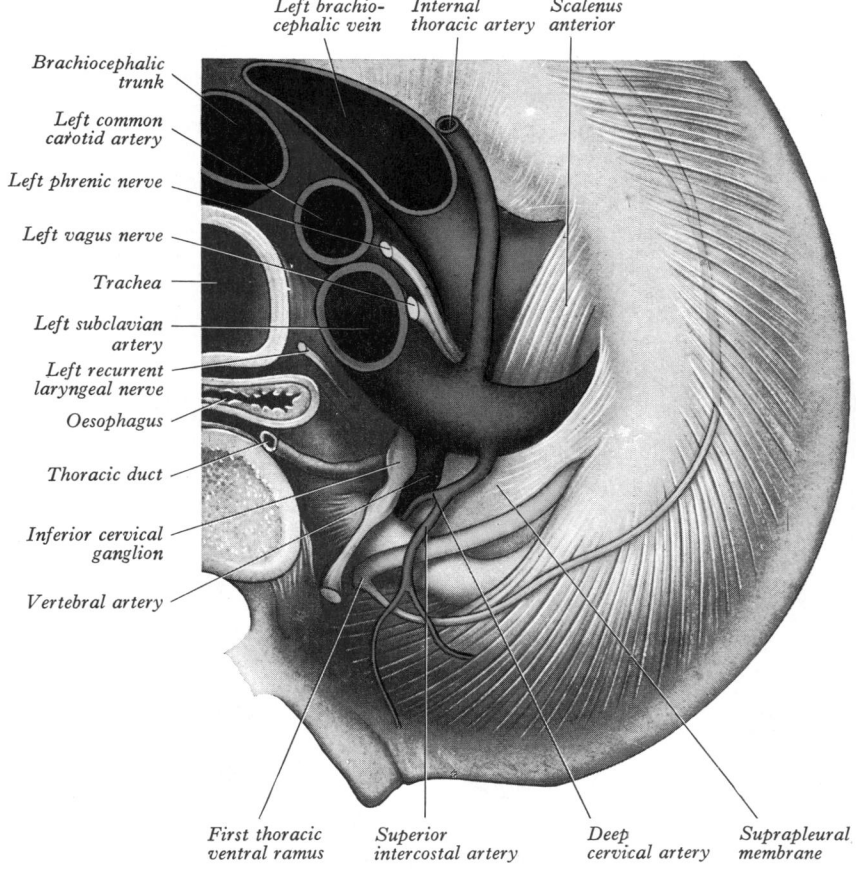

Left brachio-cephalic vein
Internal thoracic artery
Scalenus anterior
Brachiocephalic trunk
Left common carotid artery
Left phrenic nerve
Left vagus nerve
Trachea
Left subclavian artery
Left recurrent laryngeal nerve
Oesophagus
Thoracic duct
Inferior cervical ganglion
Vertebral artery
First thoracic ventral ramus
Superior intercostal artery
Deep cervical artery
Suprapleural membrane

8.29 Structures in relation to the cervical pleura of the left side: inferior aspect.

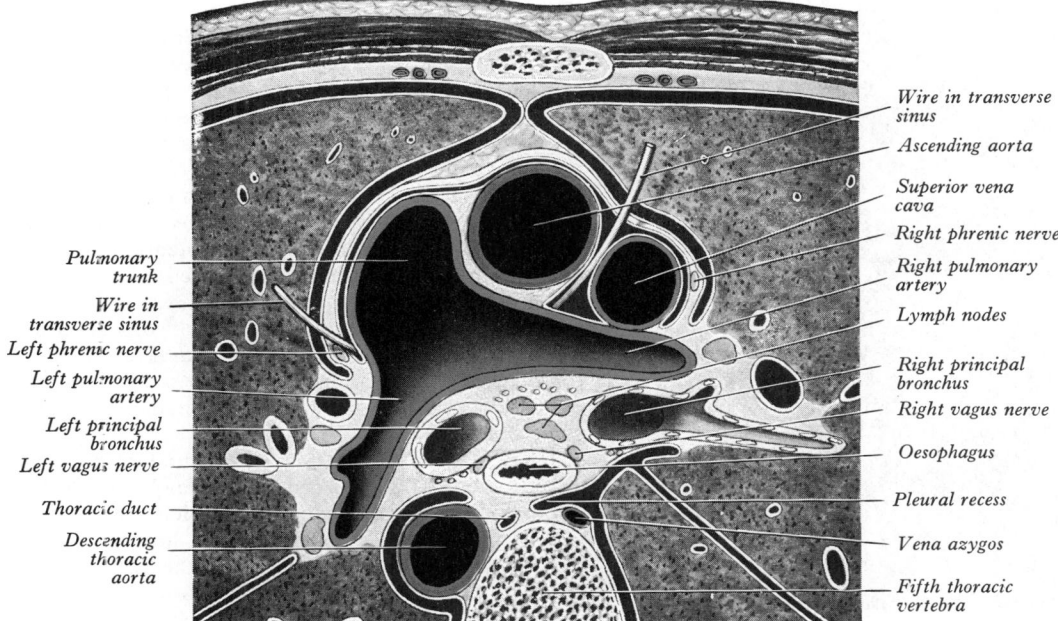

8.30A Transverse section of the mediastinum at the level of the lower part of the body of the fifth thoracic vertebra, viewed from the superior aspect.

The *cervical pleura* is the continuation of the costal pleura over the pulmonary apex (**8.**28B). It ascends medially from the first rib's internal border to the apex of the lung, as high as the lower edge of the neck of the first rib, descending lateral to the trachea to become the mediastinal pleura. Due to the rib's obliquity, the cervical pleura extends 3–4 cm above the first costal cartilage but not above the neck of the first rib. The cervical pleura is strengthened by a fascial *suprapleural membrane*, attached in front to the first rib's internal border, behind to the anterior border of the seventh cervical transverse process. It contains a few muscular fibres spreading from the scaleni (p. 586). A *scalenus minimus* muscle often extends from the anterior border of the seventh cervical transverse process to the inner border of the first rib behind its subclavian groove, spreading also into the pleural dome, which it therefore tenses. Some regard the suprapleural membrane as the tendon of this muscle. The cervical pleura (like the pulmonary apex) reaches the level of the seventh cervical spine 2.5 cm from the midline. Its projection is a curved line from the sternoclavicular joint to the junction of the medial and middle thirds of the clavicle, its summit being 2.5 cm above it. The subclavian artery ascends laterally in a furrow below the summit of the cervical pleura, the relations of which are like those of the apex of the lung (p. 1272, and **8.**29).

The *mediastinal pleura* is the lateral mediastinal boundary (p. 1271). Above the hilum of the lung it is continuous from sternum to vertebral column. On the right it covers: the right brachiocephalic vein, the upper part of the superior vena cava, the terminal part of the azygos vein, the right phrenic and vagus nerves, the trachea and oesophagus; and on the left the aortic arch, left phrenic and vagus nerves, left brachiocephalic and superior intercostal veins, left common carotid and subclavian arteries,

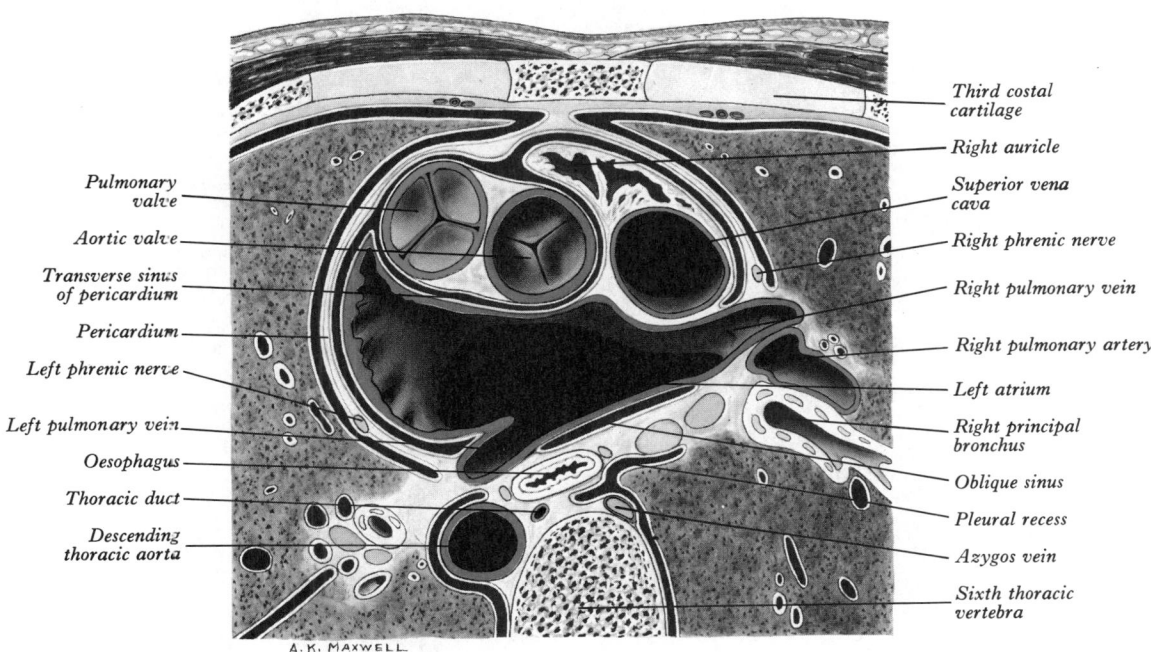

8.30B Transverse section of the mediastinum at the level of the body of the sixth thoracic vertebra, viewed from the superior aspect.

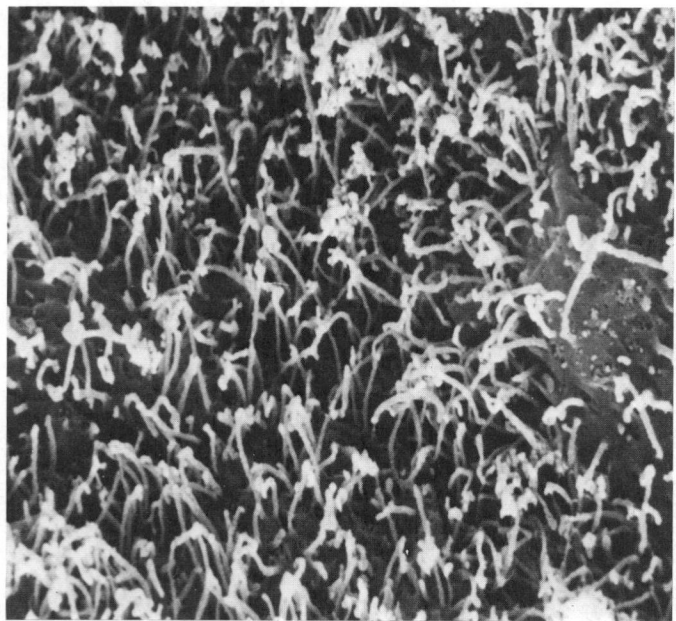

8.31 The surface of the pleura viewed by scanning electron microscopy, showing the numerous processes of the mesothelial cells lining this structure (murine). Magnification × 7000.

thoracic duct and oesophagus. At the hilum it turns laterally as a tube enclosing the hilar structures and continuous with the pulmonary pleura. Below the hilum the mediastinal pleura extends as a double layer from the lateral surface of the oesophagus to the mediastinal pulmonary surface, where it is continuous with the

pulmonary pleura; this double layer is the *pulmonary ligament* (**8.34A,B**), continuous above with the pleura around the hilar structures; below it ends in a free falciform border.

The pleura extends beyond the inferior pulmonary border (**8.23A–C**) but not to the diaphragm's attachments. Hence below the line of pleural reflexion from the thoracic wall to the diaphragm the latter is in contact with the costal cartilages and intercostal muscles. In quiet inspiration the lung's inferior margin does not reach this reflexion, the costal and diaphragmatic pleurae being separated merely by a potential *costodiaphragmatic recess*. In quiet inspiration the lung is about 5 cm above the lower pleural limit. A similar *costomediastinal recess* exists behind the sternum and the costal cartilages, where the thin anterior pulmonary margin falls short of the line of pleural reflexion. The extent of this recess, the anterior costomediastinal line of pleural reflexion and the position of the anterior pulmonary margin all vary individually.

Microscopic structure. The pleural surface is smooth, moistened by serous fluid and consists of a single layer of flat mesothelial cells on a basal lamina beneath which are networks of elastic and collagen fibres embedded in ground substance, containing fibroblasts, macrophages and other cells typical of areolar tissue (p. 58). The deeper layers of fibrous tissue are continuous with tissue around the pulmonary lobules. Blood and lymph vessels and nerves are distributed in the pleura. In ultrastructural details pleural mesothelial cells are like those of the peritoneum (p. 1346). Material from human biopsies (de Gasperis & Miani 1969) shows that their basal lamina has a lamina densa 30–40 nm thick, but their highly folded basal plasma membranes are separated from it by a lamina lucida 20–30 nm. Adjacent cell surfaces interdigitate and are joined by desmosomes. Their luminal surfaces bear microvilli and numerous cilia (see **8.31**). Micropinocytotic vesicles are common in their cytoplasm. Structural squamous variants are recognized, one probably being a stem cell.

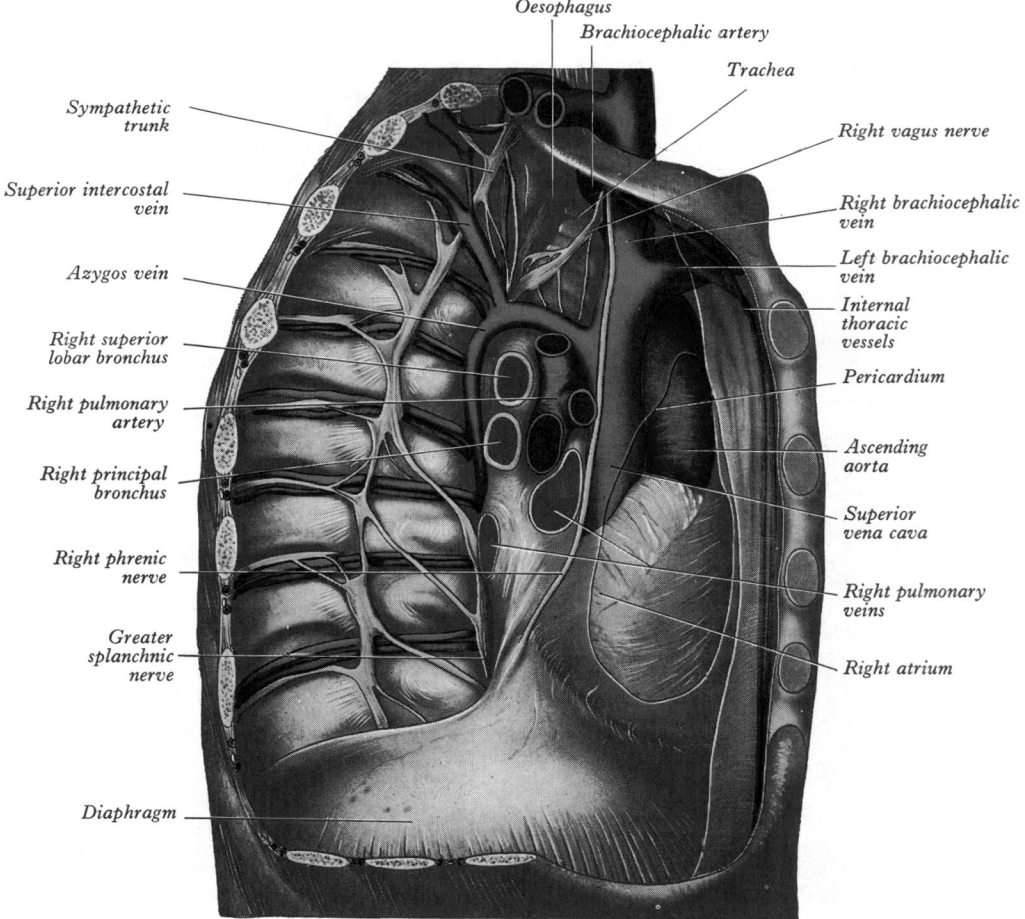

8.32A The mediastinum: right lateral aspect. A part of the pericardial sac has been removed to expose the lateral surface of the right atrium.

Labels (clockwise):
Oesophagus
Brachiocephalic artery
Trachea
Right vagus nerve
Right brachiocephalic vein
Left brachiocephalic vein
Internal thoracic vessels
Pericardium
Ascending aorta
Superior vena cava
Right pulmonary veins
Right atrium
Diaphragm
Greater splanchnic nerve
Right phrenic nerve
Right principal bronchus
Right pulmonary artery
Right superior lobar bronchus
Azygos vein
Superior intercostal vein
Sympathetic trunk

Vessels and Nerves

Parietal and visceral pleurae are developed respectively from somatopleural and splanchnopleural layers of the lateral plate mesoderm (pp. 132, 238). The parietal pleura therefore derives its blood from somatic arteries (intercostal, internal thoracic and musculophrenic); its veins join systemic veins in the thoracic wall; its lymphatics also join those in the wall, draining into intercostal, parasternal, posterior mediastinal and diaphragmatic nodes; and its nerve supply is from spinal sources (intercostal and phrenic nerves). The visceral pleura, which is integral with the lung, has a vascular supply from bronchial vessels; its lymphatics join those of the lung and its nerve supply is autonomic, reaching it along bronchial vessels. (For details of the lymphatics of visceral pleura consult Pennell 1966.) Whereas tactile or thermal stimuli to the parietal pleura elicit pain, these stimuli are inadequate when applied to the visceral pleura (cf. peritoneum, p. 1346). Costal and peripheral diaphragmatic pleurae are supplied by intercostal nerves; mediastinal and central diaphragmatic pleurae by the phrenic. Irritation of the former results in pain referred along intercostal nerves to the thoracic or abdominal wall; irritation of the latter causes pain referred to the lower neck and shoulder tip, i.e. to the area of skin supplied by the same spinal segments as the phrenic nerve (C.3,4,5).

Applied Anatomy. Normally visceral pleura glides on parietal during respiration without sound or pain; but when the pleura is inflamed friction sounds can auscultated. If fluid then distends the pleural cavity, the sounds disappear and the lung gradually collapses, the heart and mediastinum being displaced towards the opposite side. Air in the cavity (pneumothorax), whether admitted by a penetrating wound, rupture of the lung or as a therapeutic measure (e.g. to rest the lung in tuberculosis), also entails pulmonary collapse, as the elastic tissue of the lung recoils. Normally such recoil is prevented by a negative intrapleural pressure and the cohesion between the opposed parietal and visceral pleura. In renal incision the relation of the costal pleura to the twelfth rib is a hazard; usually the pleura crosses the rib at the lateral border of the erector spinae, so that the kidney's medial region is above the reflexion (8.28c). If this rib does not project beyond this muscle, the eleventh may be mistaken for the twelfth in palpation; an incision prolonged to this level will damage the pleura. Whether the lowest palpable rib is the eleventh or twelfth can be ascertained by counting from the second (identified at its junction with the sternal angle).

The Mediastinum

The mediastinum is the *partition* between the lungs and includes the mediastinal pleura; but it is commonly applied to the *interval* between the two pleural sacs. It is situated between the sternum and the thoracic vertebral column (8.23A,B,C, 30A,B), extending from the thoracic inlet to the diaphragm. For description it is divided into *superior* and *inferior* mediastina, the latter being subdivided into *anterior, middle* and *posterior* parts. The plane of division into upper and lower mediastina traverses the manubriosternal joint and the lower surface of the fourth thoracic vertebra.

The superior mediastinum (8.18, 19), lying between the manubrium sterni and the upper four thoracic vertebrae, is bounded below by the sternal plane already noted, above by the plane of the thoracic inlet and laterally by the mediastinal pleurae. It contains: the lower ends of the sternohyoid, sternothyroid and longus colli muscles, the aortic arch, the brachiocephalic artery, left common carotid and subclavian arteries, the brachiocephalic veins and upper half of the superior vena cava, the left superior intercostal vein, the vagus, cardiac, phrenic and left recurrent laryngeal nerves, the superficial part of the cardiac plexus, the

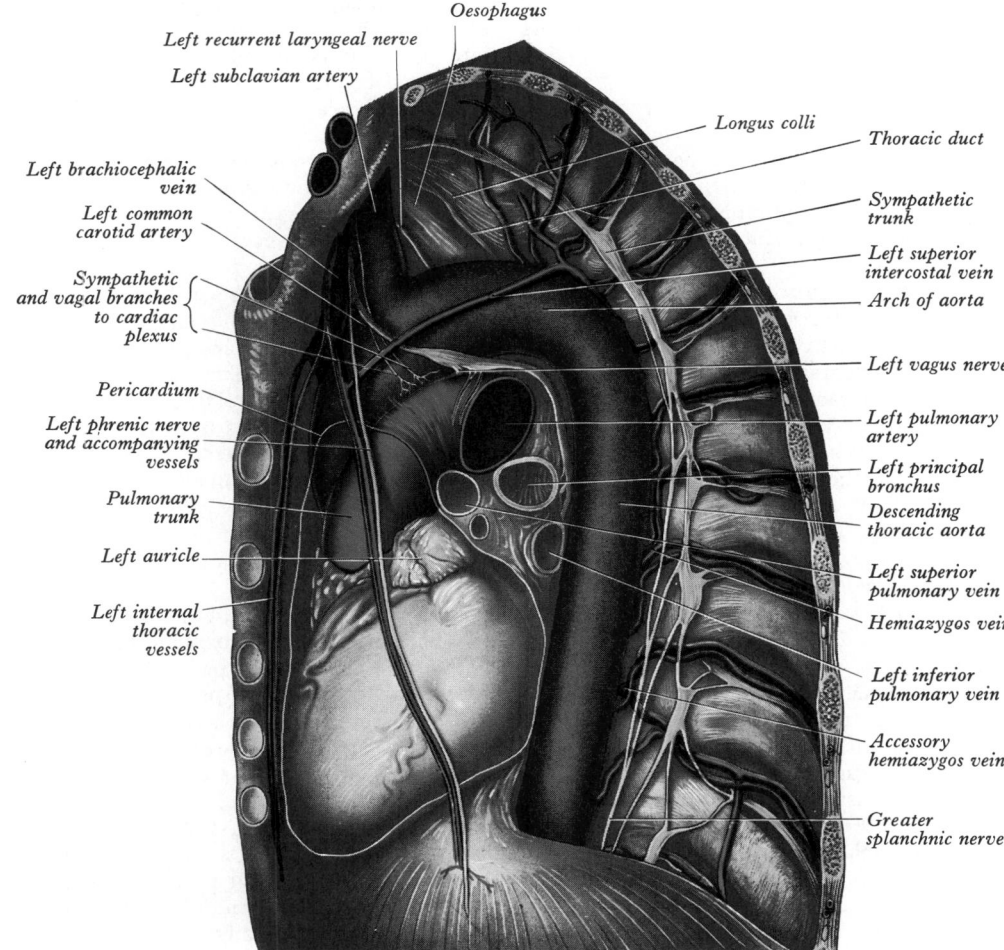

8.32B The mediastinum: left lateral aspect.

8.33 Transverse section through the mediastinum at the level of the body of the seventh thoracic vertebra, viewed from the superior aspect.

trachea, oesophagus, thoracic duct, thymic remnants and the paratracheal, brachiocephalic and tracheobronchial lymph nodes.

The anterior mediastinum, lying between the sternal body and pericardium (**8.**30A), is narrow above the fourth costal cartilages, where the pleural sacs are close. It contains loose areolar tissue, the sternopericardial ligaments, a few lymph nodes and the mediastinal branches of the internal thoracic artery and sometimes part of the thymus gland or its degenerated remains.

The middle mediastinum (**8.**30B, 33), the broadest part of the inferior mediastinum, contains: the heart, pericardium, ascending aorta, the lower half of the superior vena cava, the terminal azygos vein, tracheal bifurcation and both main bronchi, the pulmonary trunk dividing into right and left pulmonary arteries, both pulmonary veins, the phrenic nerves, the deep part of the cardiac plexus and the tracheobronchial lymph nodes.

The posterior mediastinum (**8.**30A,B, 32A,B, 33) is bounded: *in front* by the tracheal bifurcation, pulmonary vessels, pericardium and the posterior part of the upper surface of the diaphragm; *behind* by the vertebral column from the fourth (lower border) to the twelfth thoracic vertebrae and *on each side* by the mediastinal pleura. It contains: the descending thoracic aorta, the azygos and hemiazygos veins, the vagus and splanchnic nerves, the oesophagus, thoracic duct and the posterior mediastinal lymph nodes.

Mediastinal radiology (**6.**43, 44, 45). Viewed in antero-posterior radiograms (**6.**43) the heart and large blood vessels appear as the 'mediastinal shadow'. Forming its left border, from above down, are the left subclavian artery, arch of aorta ('aortic knuckle'), left auricle and ventricle. Below the arch the right ventricle's infundibulum or pulmonary trunk may be recognizable. On the right border are the right brachiocephalic vein, superior vena cava, right atrium and thoracic inferior vena cava. Enlargements or displacements of these structures accentuate the normal bulges on the shadow's borders. On both sides opacities due to pulmonary vessels form *hilar shadows*. In the upper thorax the less dense tracheal shadow is median. In lateral or oblique views the cardiac shadow is above the anterior part of the diaphragm. In front of it is the retrosternal space (anterior mediastinum); behind is the retrocardiac space (posterior mediastinum) containing the oesophagus, visualized by a barium swallow (**8.**105), and the descending thoracic aorta. Above, the less dense trachea and bronchi are recognizable; the

aortic arch and large vessels produce faint shadows in the superior mediastinum.

The Lungs

The lungs are separated by the heart and other mediastinal contents (**6.**29, **8.**33). Except for attachment to the heart and trachea by hilar structures (and the pulmonary ligament) each lung is free in its pleural cavity. A lung, being spongy, floats in water and crepitates when handled, due to air in its alveoli; it is also highly elastic and hence retracts when removed from the thorax. Its surface is smooth, shiny and marked by fine, dark lines into numerous polyhedral areas, indicating its lobules, each crossed by numerous finer lines.

At birth the lungs are pink, in adults a dark grey and patchily mottled; as age advances this maculation becomes black, due to granules of inhaled carbonaceous material deposited in loose connective tissue near the surface; it increases as age advances, often more abundantly in men and is greater in those who have dwelt in industrial areas and in smokers. The posterior pulmonary border is usually darker than the anterior. In the upper, less movable parts, surface pigmentation tends to be opposite the intercostal spaces. Lungs from fetuses or stillborn infants who have not respired differ from those of infants who have, being firm, non-crepitant and unable to float in water.

The adult right lung weighs about 625 g, the left 565 g, but they vary greatly. Weight also depends on the content of blood or serous fluid. Lungs are heavier in males, absolutely and relative to stature.

Each lung has an apex, base, three borders and two surfaces. In shape they each approximate to half a cone.

The apex is rounded, protruding above the thoracic inlet, in contact with the cervical pleura. Owing to the inlet's obliquity (p. 336), the apex is 3–4 cm above the level of the first costal cartilage but level with the neck of the rib. Its summit is about 2.5 cm above the medial third of the clavicle; the apex is therefore in the root of the neck (**8.**28B). It has been asserted that the apex is intrathoracic, close to the neck of the first rib, and that it is the anterior surface which ascends highest in inspiration (Andreassi 1967); this requires confirmation. The cervical pleura lies between the apex and suprapleural membrane (p. 1269), on which the subclavian artery arches up and laterally, grooving the anterior surface near its summit and separating it from scalenus anterior. Posterior to the apex are the cervicothoracic sympathetic ganglion, the

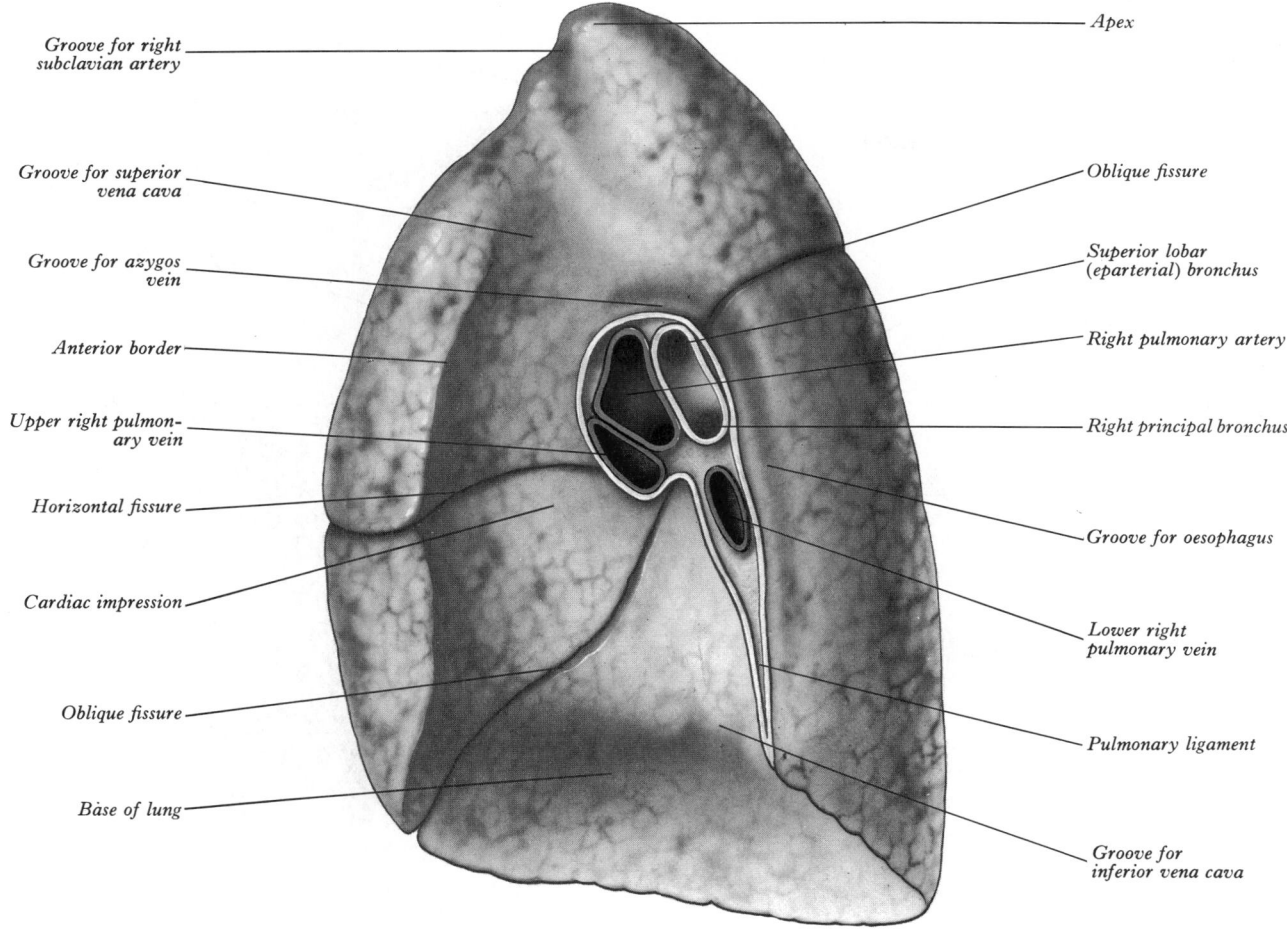

Groove for right
subclavian artery

Groove for superior
vena cava

Groove for azygos
vein

Anterior border

Upper right pulmon-
ary vein

Horizontal fissure

Cardiac impression

Oblique fissure

Base of lung

Apex

Oblique fissure

Superior lobar
(eparterial) bronchus

Right pulmonary artery

Right principal bronchus

Groove for oesophagus

Lower right
pulmonary vein

Pulmonary ligament

Groove for
inferior vena cava

8.34A The medial surface of the right lung.

ventral ramus of the first thoracic spinal nerve and the superior intercostal artery (**8.**29). Scalenus medius is lateral and the brachiocephalic artery, right brachiocephalic vein and trachea are on the right, while the left subclavian artery and left brachiocephalic vein are on the left.

The base is semilunar and concave, resting upon the superior surface of the diaphragm, which separates the right lung from the right lobe of the liver and the left lung from the left hepatic lobe, gastric fundus, and spleen. Since the diaphragm extends higher on the right than on the left, the concavity is deeper on the right lung's base. Posterolaterally the base has a sharp margin adapted to the costodiaphragmatic recess (p. 1270). **The costal surface** is convex, adapted to the thoracic wall which is deeper behind. It is in contact with the costal pleura and exhibits, in specimens preserved in situ, grooves corresponding with the overlying ribs. **The medial surface** has posterior, *vertebral* and anterior, *mediastinal* parts. The former is in contact with the sides of the thoracic vertebrae and inter-vertebral discs, the posterior intercostal vessels and the splanchnic nerves. The mediastinal area, deeply concave, is adapted to the heart as the *cardiac impression*, which is much larger and deeper on the left lung, the heart being more to the left of the median plane. Posterosuperior to this is the somewhat triangular *hilum*, where structures (p. 1275) enter and leave the viscus, collectively surrounded by pleura, which also extends below the hilum and behind the cardiac impression as the pulmonary ligament. In other features markings on the mediastinal surfaces differ when seen in preserved specimens (**8.**32A,B, 34A,B). On the *right lung* the cardiac impression is related to the anterior surface of top right auricle, the anterolateral surface of the right atrium and partially to the anterior surface of the right ventricle. The impression ascends anterior to the hilum as a wide groove for the superior vena cava and the end of the right brachiocephalic vein (**8.**34A). Posteriorly this groove is joined by

a deep sulcus arching forwards above the hilum and occupied by the azygos vein. The right side of the oesophagus makes a shallow vertical groove behind the hilum and the pulmonary ligament; nearing the diaphragm it inclines left and leaves the right lung; the groove, therefore, does not reach the lower limit of this surface. Postero-inferiorly the cardiac impression is confluent with a short wide groove adapted to the inferior vena cava. Between the apex and azygos groove the trachea and right vagus are close to the lung, but do not mark it. On the *left lung* (**8.**34B) the cardiac impression is related to the anterior and left surfaces of the left ventricle and auricle and the anterior infundibular surface and adjoining part of right ventricle; it ascends in front of the hilum to accommodate the pulmonary trunk. A large groove arches over the hilum, descending and behind it and the pulmonary ligament, adapted to the aortic arch and descending aorta; from its summit a narrower groove ascends to the apex for the left subclavian artery. Behind this, above the aortic groove, the lung contacts the thoracic duct and oesophagus. In front of the subclavian groove is a faint linear depression for the left brachiocephalic vein. Inferiorly, the oesophagus may mould the surface in front of the lower end of the pulmonary ligament.

The *inferior border* is thin where it separates the base from the costal surface and extends into the costodiaphragmatic recess; medially, where it divides the base from the mediastinal surface, it is rounded. It corresponds, in quiet respiration, to a line from the anterior border (vide infra) to the eighth rib in mid-axillary line (nearly 10 cm above the costal margin) continued medially and slightly up to a point 2 cm lateral to the tenth thoracic *spine* (**8.**28C). The *posterior border* separates the costal surface from the mediastinal, corresponding to the heads of the ribs. It has no recognizable markings and is really a rounded junction of costal and vertebral (medial) surfaces. The thin, sharp, *anterior border* overlaps the pericardium; on the *right* it corresponds closely to the

1273

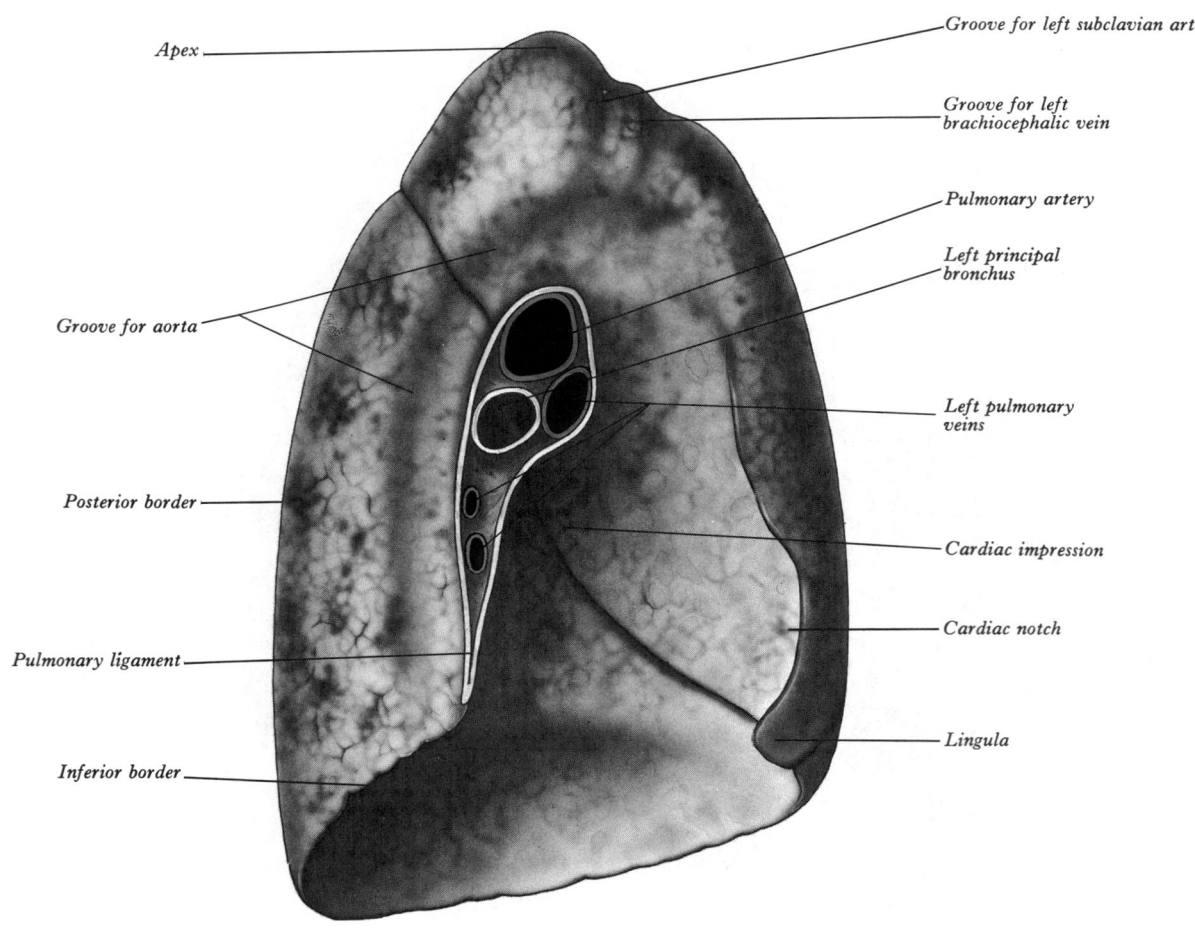

Apex

Groove for aorta

Posterior border

Pulmonary ligament

Inferior border

Groove for left subclavian artery

Groove for left brachiocephalic vein

Pulmonary artery

Left principal bronchus

Left pulmonary veins

Cardiac impression

Cardiac notch

Lingula

8.34B The medial surface of the left lung.

costomediastinal line of pleural reflexion, being almost vertical; on the *left* it approaches the same line above, but below the fourth costal cartilage it shows a variable *cardiac notch*, the edge of which passes laterally for about 3.5 cm before curving down and medially to the sixth costal cartilage about 4 cm from midline. It thus does not reach the line of pleural reflexion here (8.28A), leaving the pericardium covered by a double layer of pleura (area of superficial cardiac dullness, p. 713). However, surgical experience suggests that the line of pleural reflexion, the anterior pulmonary margin and the costomediastinal pleural recess are all most variable.

PULMONARY FISSURES AND LOBES

The left lung is divided into a superior and an inferior lobe by an *oblique fissure* (8.34B) extending from the costal to the medial surfaces of the lung both above and below the hilum. Superficially this fissure begins on the medial surface at the posterosuperior part of the hilum and ascends obliquely back to cross the posterior border of the lung about 6 cm below the apex. It then descends forwards across the costal surface (8.28A), reaching the lower border almost at its anterior end. It finally ascends on the medial surface to the lower part of the hilum. At the posterior border the fissure is indicated by a surface point 2 cm to the side of the midline between the spines of the third and fourth thoracic vertebrae, but may be above or below this level. Traced around the chest, the fissure reaches the fifth intercostal space (at or near the mid-axillary line) and follows this to intersect the inferior border close to or just below the sixth costochondral junction (7.5 cm from the midline). The left oblique fissure is usually more vertical than the right, and is indicated approximately by the medial scapular border when the arm is fully abducted above the shoulder. The *superior lobe*, anterosuperior to this fissure, includes the apex, anterior border, much of the costal and most of

the medial surfaces of the lung. At the lower end of the cardiac notch is usually a small process, the *lingula*. The larger, *inferior lobe*, postero-inferior to the fissure, comprises almost the whole of the *base*, much of the costal surface and most of the posterior border. (See pp. 1261–1265 for details of bronchopulmonary segmentation.)

The right lung is divided into superior, middle and inferior lobes by two fissures (8.34A). The upper, separating the inferior from the middle and superior lobes, corresponds closely to the left *oblique fissure* but is less vertical, crossing the inferior border of the lung about 7.5 cm behind its anterior end. On the posterior border it is level with the fourth thoracic vertebral spine or slightly lower. It descends across the fifth intercostal space and follows the sixth rib to the sixth costochondral junction. A short *horizontal fissure* separates the superior and middle lobes, passing from the oblique fissure, near the mid-axillary line, horizontally forwards to the anterior border level with the sternal end of the fourth costal cartilage; on the mediastinal surface it reaches the hilum. The small *middle lobe* is thus cuneiform and includes some of the costal surface, the lower part of the anterior border and the anterior part of the base of the lung. Sometimes the medial part of the upper lobe is partially separated by a fissure of variable depth containing the terminal part of the azygos vein enclosed in the free margin of a mesentery derived from the mediastinal pleura, forming the '*lobe of the azygos vein*'. This varies in size and sometimes includes the apex; it is always supplied by one or more branches of the apical bronchus. Radiographically, pleural effusion may be limited to the azygos fissure.

Attempts to equate the lobation in the two lungs, particularly the right middle lobe and lingula, are not supported by developmental data (p. 239). For discussion and bibliography consult Yokoh (1977).

Since the diaphragm rises higher on the right to accommodate the liver, the right lung is vertically shorter (by 2.5 cm) but due to

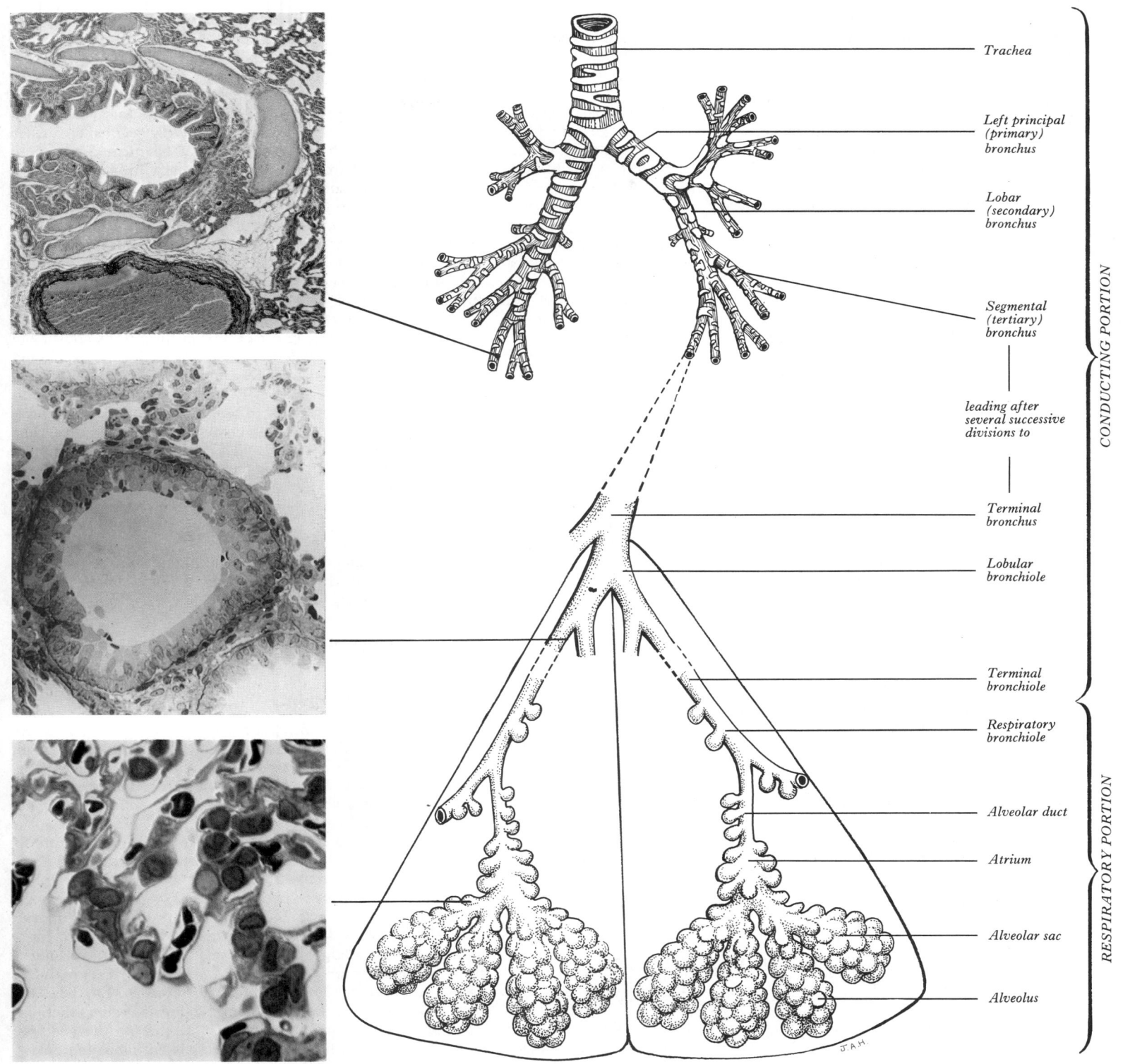

Trachea

*Left principal
(primary)
bronchus*

*Lobar
(secondary)
bronchus*

*Segmental
(tertiary)
bronchus*

*leading after
several successive
divisions to*

*Terminal
bronchus*

*Lobular
bronchiole*

*Terminal
bronchiole*

*Respiratory
bronchiole*

Alveolar duct

Atrium

Alveolar sac

Alveolus

CONDUCTING PORTION

RESPIRATORY PORTION

8.35 Bronchopulmonary structure. The diagram on the right shows the gross architecture of the conducting and respiratory parts of the trachea and lungs at segmental levels. On the left are three sections at the levels indicated. Top: A small bronchus lined by highly convoluted epithelium, surrounded by plates of hyaline cartilage. A pulmonary arteriole is visible below. Middle: A bronchiole. Note absence of cartilage and conspicuous non-striated muscle external to the epithelium. Epoxy resin section prepared by Susan Smith; photographed by Marina Morris, Guy's Hospital Medical School, London. Bottom: An interalveolar septum. Note capillaries containing erythrocytes (dark blue) separated from air spaces by thin alveolar epithelial cells. Epoxy resin section stained by toluidine blue.

cardiac asymmetry it is broader, and in capacity and weight greater, than the left.

PULMONARY HILA AND ROOTS

The pulmonary root (8.32A,B), connecting the medial surface to the heart and trachea, is formed by a group·of structures entering or leaving the hilum. These are: the principal bronchus, pulmonary artery, two pulmonary veins, bronchial arteries and veins, a pulmonary autonomic plexus, lymph vessels, bronchopulmonary lymph nodes and loose connective tissue, all enveloped by pleura. The pulmonary roots, or pedicles, are level with the fifth to seventh thoracic vertebrae. The right is behind the superior vena cava and right atrium and below the terminal part of the azygos vein. The left root is below the aortic arch and in front of the descending thoracic aorta. Common to both are: *anterior*, the phrenic nerve, pericardiacophrenic artery and vein, and anterior pulmonary plexus; *posterior*, the vagus nerve and posterior pulmonary plexus; *inferior*, the pulmonary ligament.

The major structures in both roots are similarly arranged, as follows: the upper pulmonary vein in front and the pulmonary artery and principal bronchus behind, the bronchial vessels most posterior. Their vertical arrangement differs slightly on the two sides. On the right, from above down, the sequence is: superior lobar bronchus, pulmonary artery, principal bronchus, lower pulmonary vein; on the left: pulmonary artery, principal bronchus, lower pulmonary vein. The lower left pulmonary vein is inferior to the principal bronchus which is the lowest hilar structure (**8**.34A,B).

Pulmonary regions do not all move equally in respiration. In quiet respiration the juxtahilar part of the lung scarcely moves and the central region only slightly. The superficial parts of the lung expand most, while the mediastinal surface, posterior border and apex move less, being related to less movable structures. The diaphragmatic and costomediastinal regions expand most of all (see Movements of Respiration, p. 594). Most of the lung's volumetric change during respiration occurs in the alveoli, which number more than 20 million in a neonatal lung, increasing to 300 million or more during childhood. An alveolus varies from 200–300 μm in diameter and numerous capillary segments (1800 is suggested) may contact each alveolus. The air-epithelium-blood interface is enormous even in childhood. Figures of 70–100 m² have been estimated (Peters 1969, Fisherman & Pietra 1974) and even higher values are suggested (Gehr et al 1978).

Microscopic Structure of the Lung

Lungs are epithelial outgrowths of the anterior foregut (pp. 231, 238), forming tubes which branch as they grow, and surrounded by very vascular mesenchymal tissue. This developing mass, protruding into the anterior coelomic cavity (later pleural cavities), is covered by coelomic mesothelium which eventually becomes the epithelium of the pulmonary pleurae. Within each lung the larger proximal tubes become *conducting* airways and the more distal form cavities for *respiratory exchange* between the atmosphere and adjacent capillaries. The individual components of this complex will now be considered in detail.

THE SEROUS TUNIC AND ASSOCIATED CONNECTIVE TISSUE

The *tunica serosa* forms the visceral pleura, a thin, transparent layer composed of a single-layered mesothelium and its subjacent lamina propria, inseparably connected to the underlying lung tissue which it invests except at the hilum. Subserous areolar tissue covers the entire pulmonary surface and extends along the conducting tubes and blood vessels from the hilum to finally delineate numerous small *lung lobules*. Each lobule is a small polyhedral mass receiving a lobular bronchiole and the terminal rami of arterioles, venules, lymphatics and nerves. Lobules vary in size, the superficial being large and pyramidal, their bases visible as polygonal areas about 5–15 mm across separated by thin layers of connective tissue. Internal lobules are smaller and vary in shape. In structure the lung resembles a lobulated gland, consisting of terminal lobules for respiratory exchange and extralobular ducts for ventilation (Miller 1947, von Hayek 1960, Engel 1962, Krahl 1964). The terminal lobules are also called *secondary lobules*, delineated on the surface by substantial septa enclosing areas of 1–2 cm², subdivided by delicate septa into areas of about 1 mm², the surfaces of the *primary lobules*.

RESPIRATORY PASSAGES AND SPACES

The 'lower' respiratory tract includes the larynx, trachea, extrapulmonary bronchi (vide supra) and various orders of intrapulmonary tubes which repeatedly dichotomize. *Secondary (lobar) bronchi* supply whole lobes; *tertiary* (or quaternary), *segmental bronchi* supply bronchopulmonary segments; further subdivisions branch repeatedly within segments, becoming increasingly narrower. All intrapulmonary bronchi are kept patent by cartilaginous plates which decline in size and number and disappear from tubes less than 1 mm in diameter, termed *bronchioles*. After repeated branching, one *lobular bronchiole* enters each lobule, dividing at once into about six *terminal bronchioles*; these subdivide into one to three generations of *respiratory bronchioles* (Spencer 1977). Terminal bronchioles are the most distal air passages to be lined solely by simple columnar epithelium. Respiratory bronchioles in contrast have a few small alveoli arising directly from their walls and finally end in two or three *alveolar ducts*, thin walled tubes expanding terminally into *atria*, which lead into *alveolar sacs*. The thin walls of the alveolar ducts, atria and sacs are studded with *alveoli* or *acini*, separated from adjacent alveoli by thin *interalveolar septa* of connective tissue and capillary plexuses. The above arrangement, shown diagrammatically in **8**.35 and **8**.37, 38, is normal in adults; variations occur during active growth in childhood (Engel 1947).

N.B. The terminology used here for the orders of branches is not universally agreed and confusion persists in the naming of conducting passages smaller than bronchi. For example, some authors define a terminal bronchiole as serving a lobule (often of unspecified type) and others adopt a simpler view of the passages leading to the alveoli.

MICROSCOPIC STRUCTURE OF THE BRONCHIAL TREE

Air Passage Epithelium

The epithelium of the intrapulmonary bronchi and bronchioles resembles that of the trachea and the extrapulmonary bronchi (Collet et al 1967). In larger intrapulmonary passages it is pseudostratified and largely ciliated; in terminal and respiratory bronchioles it is a single layer of lower columnar to cuboidal cells, many non-ciliated.

There has been a systematic survey of the ultrastructure of the epithelial cell types present at different air passage levels in the rat (Jeffrey & Reid 1975) and in the mouse (Pack et al 1981). The following description is of that found in the rat. Ten types, eight epithelial and two mesenchymal (probably migratory), have been identified. All the epithelial cells rest on a basal lamina but in the larger bronchi some do not reach the lumen. Epithelial cells proper were classed as: *ciliated, serous, Clara, goblet, intermediate, brush, basal* and *Kulchitsky*. Of these the serous had not been described before in air passage epithelium. Mesenchymal cells are either *'globule' leucocytes* or *lymphocytes*. The ultrastructural features of these cells are tabulated below. In rats goblet cells occur in the epithelium of all air passages but as less than 1% of the cells; in some species they are absent from smaller bronchioles. Basal cells decrease distally and are almost absent from smaller bronchioles, where the epithelium is simple. Clara cells occur throughout the system. Even in the most proximal regions over 40% of cells are non-ciliated (Jeffery & Reid 1975).

Ciliated cells, essential to the ciliary rejection current in the bronchial tree, have electronlucent cytoplasm containing numerous mitochondria, a euchromatic nucleus and cilia with apical, claw-like projections (**8**.36B,C) described by Jeffery & Reid (1975). The cilia extend into a low-viscosity liquid partly produced by the serous cells (vide infra); above this are flakes and droplets of more viscous mucus (Van As & Webster 1972) from goblet cells and the exocrine cells of bronchial mucous glands. According to Sleigh (1974) only the ciliary tips contact this mucus, which they 'claw' along towards the larynx. The mucus-secreting *goblet cells* are

A

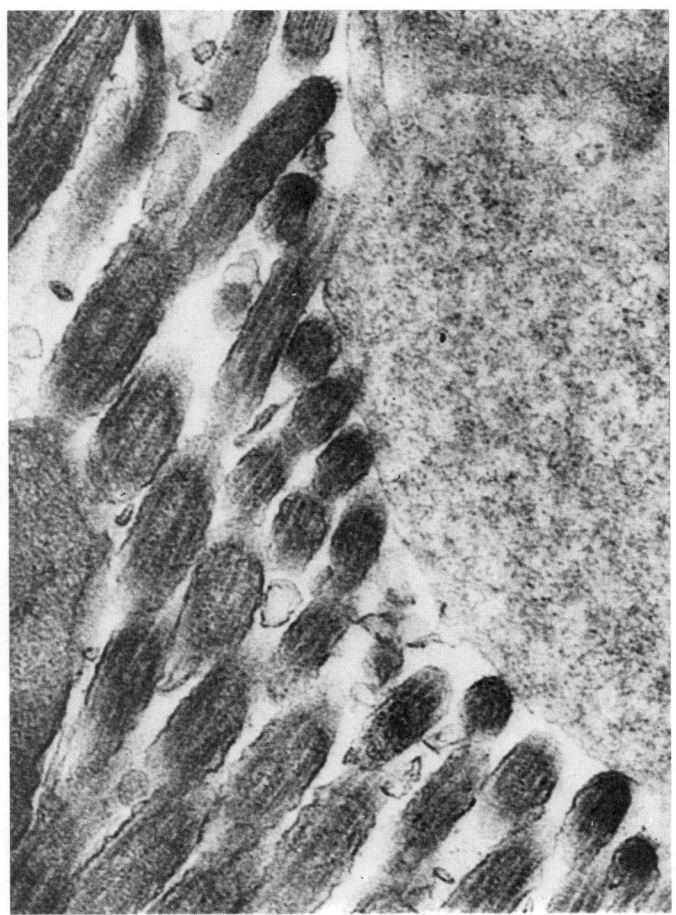

B

8.36 A. Electron micrograph showing the epithelium lining the bronchus of a rat. Clustered between the ciliated cells are a variety of non-ciliated cells, including goblet cells and Clara cells. Magnification × 1800. B. Electron micrograph showing bronchial cilia with a droplet of the mucus which they transport in contact with their tips. Apical claw-like projections can be seen on one cilium. Magnification × 35 000. C. Claw-like projections at the apex of a cilium of a bronchial epithelial cell (rat). Magnification × 90 000. Preparations by Michael Crowder, Department of Anatomy, Guy's Hospital Medical School, London.

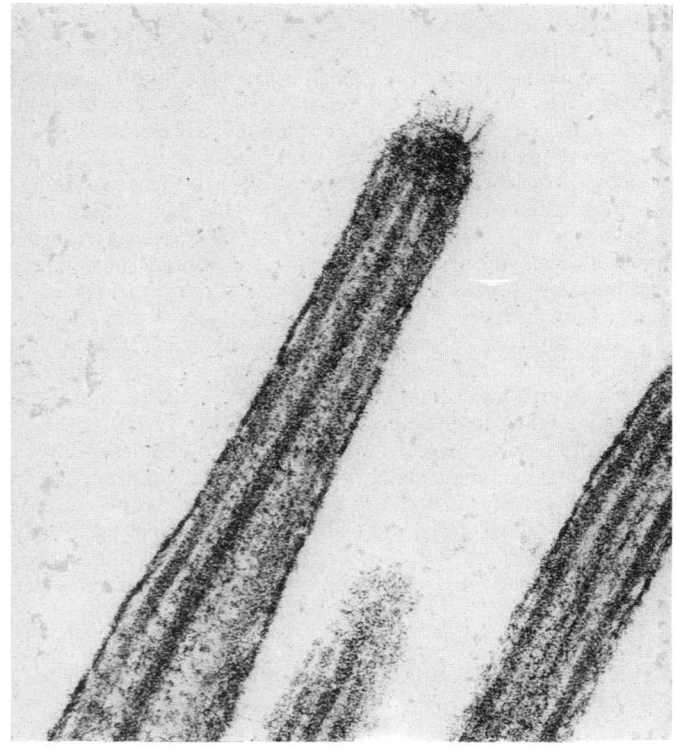

C

distended by supranuclear electron-lucent granules often with electron-dense cores.

Serous cells have an irregular nucleus, copious granular endoplasmic reticulum and homogeneously electron-dense supranuclear granules. Most numerous in the trachea and extrapulmonary bronchi, they also occur in intrapulmonary passages. Similar cells exist in human bronchial submucosal glands (Meyrick & Reid 1970), probably contributing to the low-viscosity circum-ciliary fluid.

Clara cells, originally considered typical of terminal bronchioles (Clara 1937), also occur at more proximal levels and even in nasal mucosa (Matulionis & Parks 1973). They are non-ciliated, with blunt luminal projections, irregular homogeneously electron-dense secretory granules, many lysosomes, osmiophilic lamellar bodies and much agranular endoplasmic reticulum, often apically concentrated, which may be connected to the rest by a narrow 'waist' (Spencer 1977). Niden & Yamada (1966) first suggested Clara cells as a source of surfactant material, a view supported by Etherton & Conning (1971), whose localization of radioactively labelled dipalmitoyl lecithin (a constituent of alveolar surfactant) to secretory granules of Clara cells, within five minutes of intraperitoneal injection, weakened the case for type II alveolar cells (p. 1280) as the sole source of surfactant.

'*Intermediate*' epitheliocytes have been observed in the bronchial epithelium of rats (Rhodin & Dalhamn 1956, Jeffery & Reid 1975) and in human tracheal epithelium (Rhodin 1966). They have long, luminal filopodia, electron-dense cytoplasm with no secretory granules, but sometimes contain fibrogranular accumulations, suggesting ciliogenesis (Sorokin 1968). They may be undifferentiated forms of ciliated or secretory cells (Rhodin 1966).

Brush cells are non-ciliated, with a distinct luminal border of microvilli. Numerous pinocytotic vesicles near the bases of the microvilli suggest a specialization for absorption. Most numerous in the tracheal epithelium, where they amount in rats to about 1%

of the epithelial cells; they are also dispersed throughout the other air passages (Jeffery & Reid 1975).

Bronchial Kulchitsky (argentaffin) cells (Bensch et al 1965) are the rarest type, with electron-lucent cytoplasm containing dense-cored secretory granules which accumulate between the nucleus

DISTINGUISHING ULTRASTRUCTURAL FEATURES OF CELLS WITHIN THE EPITHELIUM LINING THE AIR PASSAGES

Cell type	Granules and density	Endoplasmic reticulum	Cytoplasmic density	Surface projections
Epithelial				*Luminal*
Ciliated	−	R	L	Cilia + filiform
Serous	+D	R	D	Filiform
Clara	+D	S	L	Blunt
Goblet	+L	R	D	Filiform
Intermediate	−	R	L or D	Filiform
Brush	−	R	Intermediate	Microvillous
				Intercellular
Basal	−	R (sparse)	D	Blunt
Kulchitsky	Neurosecretory	R	L or D	Elongated
Mesenchymal				
'Globule' leucocyte	+D	R (sparse)	Intermediate	Filiform
Lymphocyte	−	R (sparse)	L	Blunt

| | + present | R rough (granular) | L electron lucent | |
| | − absent | S smooth (agranular) | D electron dense | |

(Modified from Jeffery & Reid 1975)

and cell base, discharging their contents in response to hypoxia (Moosavi et al 1973, Lauweryns & Cokelaere 1973). Termed *Feyrter* cells by Moosavi et al (1973) and P$_a$ *cells* by Capella et al (1978) in the human lung, they resemble the D$_1$ cells (p. 1376) of the gastro-entero-pancreatic system in storing and secreting amines (Ericson et al 1972), and should thus be included in the diffuse endocrine system of APUD cells (Pearse & Polack 1971, Hage 1973). Innervated groups of similar cells form neuro-epithelial bodies in human bronchial and bronchiolar mucosa (Lauweryns & Peuskens 1972). Suggested functions include involvement in the regulation of lobular growth (Rosan & Lauweryns 1971) and chemoreception (Lauweryns et al 1972).

Undifferentiated *basal cells*, in parts of the airway lined by pseudostratified epithelium, divide to replace other epithelial types (Blenkinsopp 1967). In smaller bronchi, undifferentiated cells amongst the mature epitheliocytes perform this function.

The remaining cells are either 'globule' leucocytes or lymphocytes, both being migratory cells and not epithelial cells proper. *'Globule' leucocytes* have very electron-dense granules, numerous filopodia and resemble mast cells, from which they may be derived (Murray et al 1968). The *lymphocytes*, with sparse electron-lucent cytoplasm, are most numerous in extrapulmonary bronchi. Their ultrastructural appearance in human bronchi has been described by Meyrick & Reid (1970). They are typically T cells derived from the circulation and stemming from mucosal lymphoid tissue and are probably concerned with immune surveillance of the epithelium and the destruction of virus-infected cells, etc. (see p. 1281).

The epithelium of smaller bronchi and bronchioles is folded into conspicuous longitudinal ridges which allow for changes in luminal diameter (8.37). In the respiratory bronchioles it progressively thins towards the alveolar termination, eventually being composed of cuboidal, non-ciliated cells; the lateral pouches are lined with squamous cells, providing an accessory respiratory surface.

Connective Tissue and Muscle of the Bronchial Tree (8.37,38,39)

The epithelium of the bronchi and bronchioles rests on a basal lamina attached to a lamina propria in which broad, longitudinal bands of elastin follow the course of the bronchial tree, branching at bifurcations and finally joining the elastin networks of the respiratory spaces. This elastic framework is a vital mechanical element, responsible for much elastic recoil during expiration, although in respiratory regions surface tension may be more important (vide infra and Peters 1969). Non-striated

myocytes in the submucosa (within the confines of the visceral cartilages where present) form two opposed helical tracts along the entire intrapulmonary bronchial tree, becoming thinner and finally absent at the alveolar bases (Stephens & Kroeger 1980). They are under nervous and hormonal control and narrow the airway, their relaxation permitting bronchodilatation. Normally there is some tone in the muscular bands, which relax slightly during inspiration and contract during expiration, assisting the tidal flow of air. Abnormal contraction may be caused by circulating stimulants of non-striated muscle or by local release of excitants such as serotonin, histamine and leukotrienes, producing bronchospasm and impaired respiration.

Diverticula and Glands

Diverticula of the mucosa extend outside the muscle layers in some bronchi and bronchioles, forming deep mucosal crypts which may develop into canals providing additional connections between peribronchiolar alveoli and the more proximal air spaces (Lambert 1955). Tubular glands also occur in the submucosa of bronchi and bronchioles, both *mucous* and *serous* in type, as in the extrapulmonary passages.

ALVEOLAR STRUCTURE

The alveoli are thin-walled pouches which provide collectively a large respiratory area that is a minimal barrier to gaseous exchange between the atmosphere and the blood of the capillaries lying close to their surface lining (Karrer 1956, Collet et al 1967, Klika & Petřick 1965).

The alveolar walls consist of a thin luminal epithelium resting on a basal lamina adjacent to a layer of connective tissue (8.38, 40A–D). Since the walls are thin and in contact with those of adjacent alveoli, the connective tissue and its blood vessels frequently form a thin irregular sandwich between two layers of epithelium, an *interalveolar septum*. Apertures occur in such septa, passing air between adjacent alveoli and between the alveoli and respiratory bronchioles (Macklin 1936, Lambert 1955, Suarez 1979). The alveolar epithelium is generally a single layer of attenuated squamous cells, *type I alveolar epithelial cells* or *type I pneumonocytes*, forming a continuous layer as thin as 0.05 μm. Since the basal laminae of this epithelium and of the subjacent capillaries may be fused and measure only about 0.1 μm and capillary endothelium about 0.05 μm, the total barrier to diffusion between air and blood may be as little as 0.2 μm (Collet et al 1967). However, the

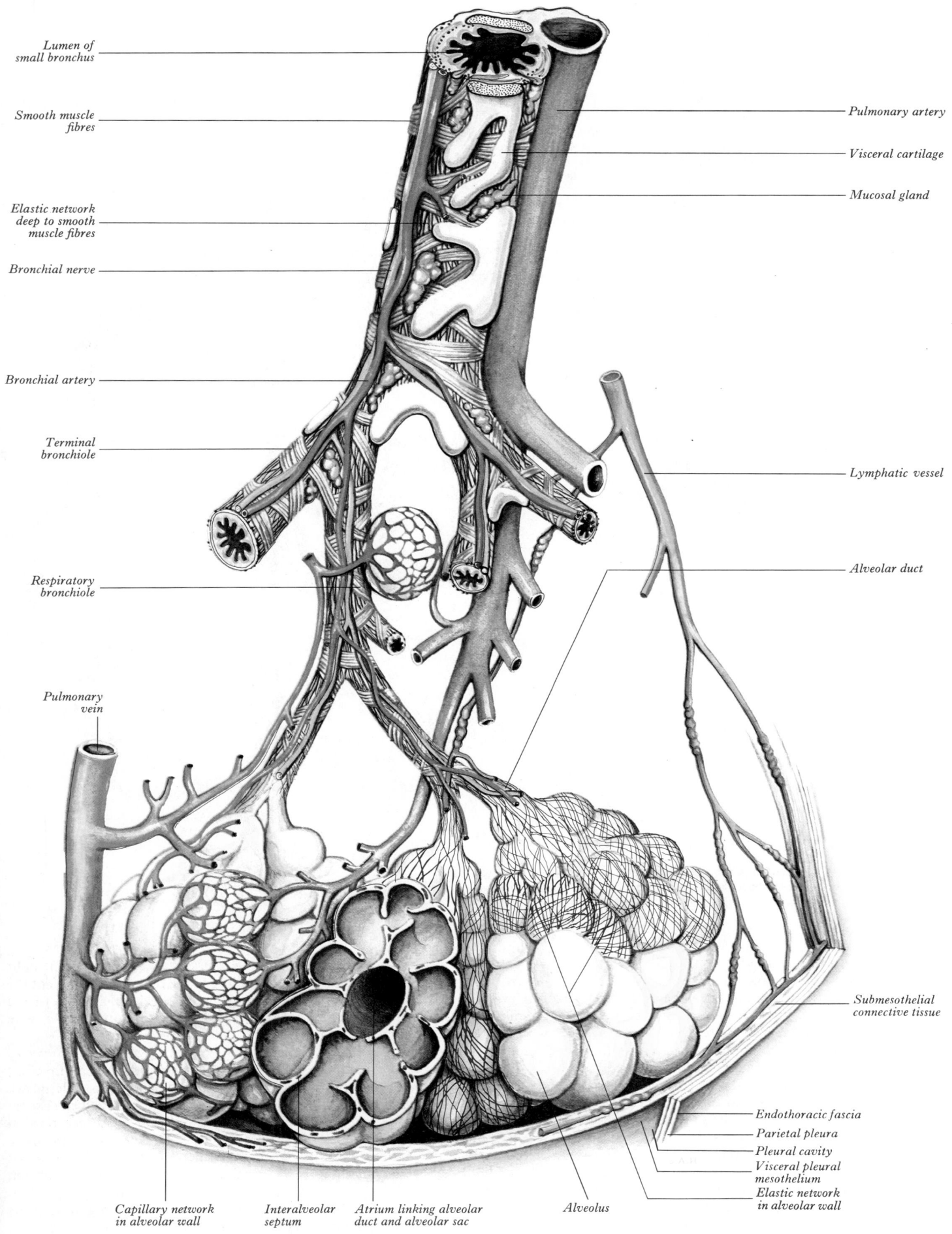

Lumen of small bronchus

Smooth muscle fibres

Elastic network deep to smooth muscle fibres

Bronchial nerve

Bronchial artery

Terminal bronchiole

Respiratory bronchiole

Pulmonary vein

Pulmonary artery

Visceral cartilage

Mucosal gland

Lymphatic vessel

Alveolar duct

Submesothelial connective tissue

Endothoracic fascia

Parietal pleura

Pleural cavity

Visceral pleural mesothelium

Elastic network in alveolar wall

Capillary network in alveolar wall

Interalveolar septum

Atrium linking alveolar duct and alveolar sac

Alveolus

8.37 Diagram of the detailed structure of the respiratory tree and its blood supply and drainage, lymphatic drainage and nerve supply. Vessels shown in blue contain de-oxygenated blood, those shown in red contain oxygenated blood.

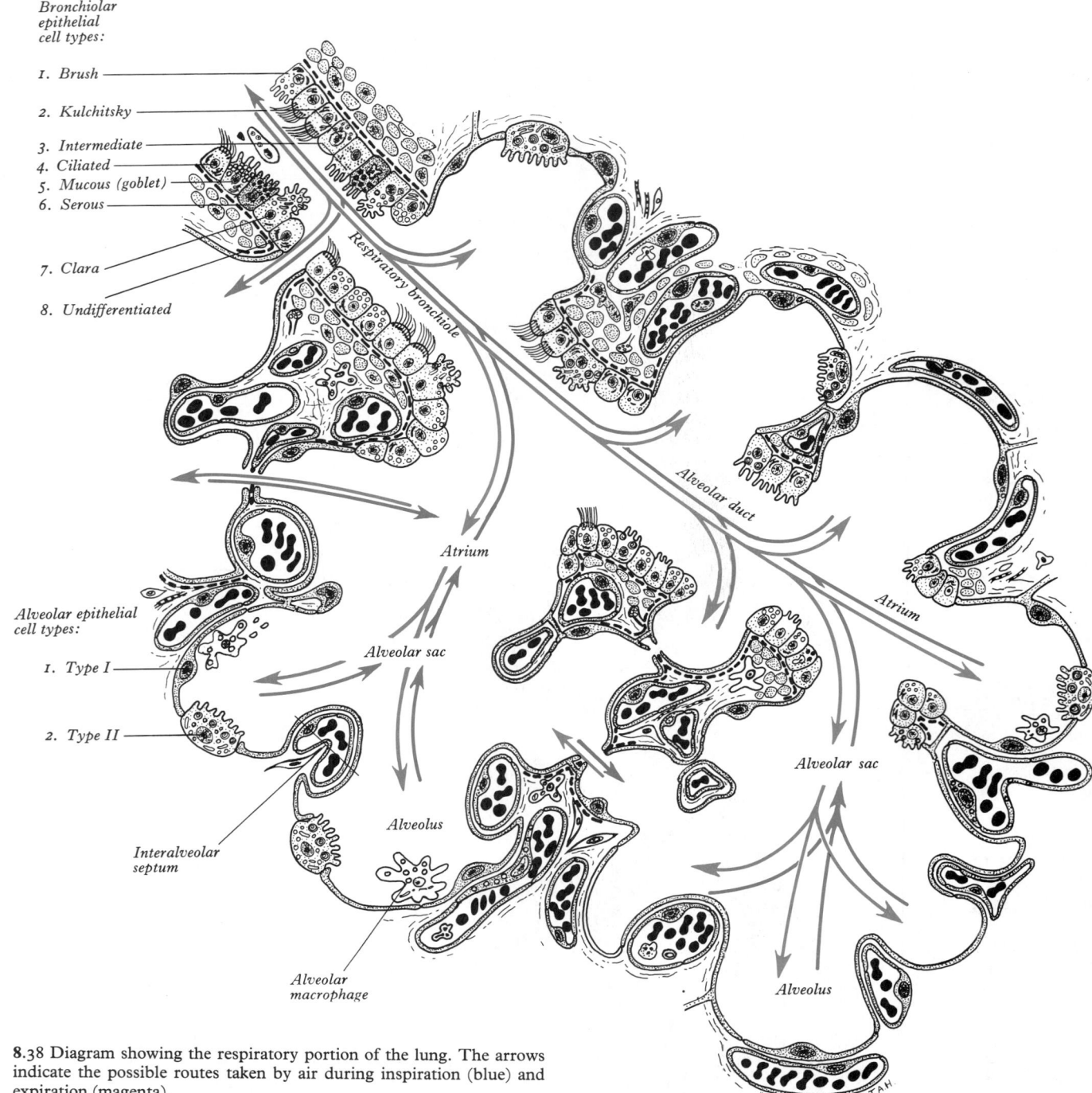

Bronchiolar epithelial cell types:

1. *Brush*
2. *Kulchitsky*
3. *Intermediate*
4. *Ciliated*
5. *Mucous (goblet)*
6. *Serous*
7. *Clara*
8. *Undifferentiated*

Respiratory bronchiole

Alveolar duct

Atrium

Atrium

Alveolar epithelial cell types:

1. *Type I*
2. *Type II*

Alveolar sac

Alveolar sac

Alveolus

Interalveolar septum

Alveolus

Alveolar macrophage

8.38 Diagram showing the respiratory portion of the lung. The arrows indicate the possible routes taken by air during inspiration (blue) and expiration (magenta).

mean barrier thickness is 2.2 μm in the normal human lung (Gehr et al 1978), much exceeding the range of 0.6–1.5 μm in other mammals (Meban 1980), due mainly to a broader interstitial space rich in connective tissue fibres found in the interalveolar septa. Over about half the total barrier area, the separation between the alveolar epithelium and capillary endothelium is reduced to two fused basal laminae, but in large areas there are also much wider separating spaces, containing free fluid, fixed and free cells and collagen and elastic fibres (**8**.38). The greater thickness, over much of its area, of the human air–blood barrier, compared with other mammals, is unexplained, but Wang & Ying (1977) have suggested that the large human lung needs more fibrous tissue to support the weight of the contained blood. This cannot be the full explanation, for the barrier is much thinner in the still larger lungs of cattle and horses (Gehr et al 1978). Perhaps bipedalism is a factor in this phenomenon.

The stereological evaluation of electron micrographs of random samples of normal human lungs has led to revised estimates of the total alveolar area (Gehr et al 1978); the mean in eight pairs of lungs was $143 \pm 12m^2$, a value 75% higher than light microscopic estimates, due mainly to the higher resolution of the electron microscope resulting in more precise definition of the alveolar surface. Capillary surface area and volume were estimated at 126 ± 12 m² and 231 ± 31 ml respectively.

The alveolar surface is normally covered by a film of *pulmonary surfactant*, composed mainly of a phospholipid (tubular 'myelin') secreted mainly by type II alveolar epithelial cells (vide infra), although Clara cells of bronchioles may also be involved (Etherton et al 1973). It is a recognized reducer of surface tension (Goerke 1974, King 1974, Tierney 1974, Kikkawa & Smith 1983). Likened to emulsifying agents which stabilize bubbles in an aqueous phase, it is essential for normal alveolar expansion (Macklin 1954), preventing alveolar collapse during expiration (Pattle 1965). The introduction of polar solvents and aqueous embedding media, cryosectioning, or the use of tannic acid-containing fixatives in electron microscopy has improved the retention of lipids and carbohydrates in ultrathin sections, leading to re-assessment of the ultrastructure of the pulmonary surfactant

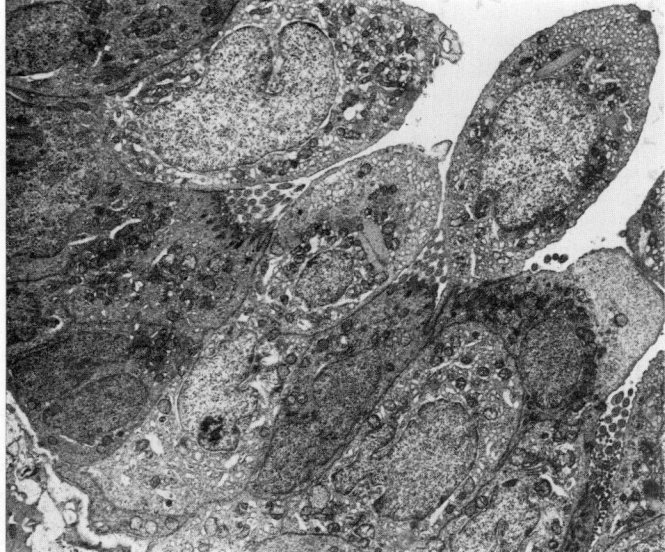

A

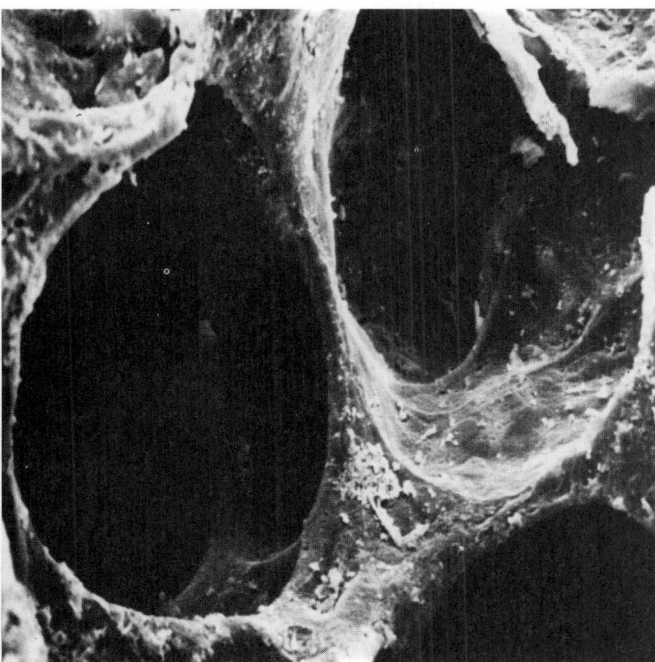

B

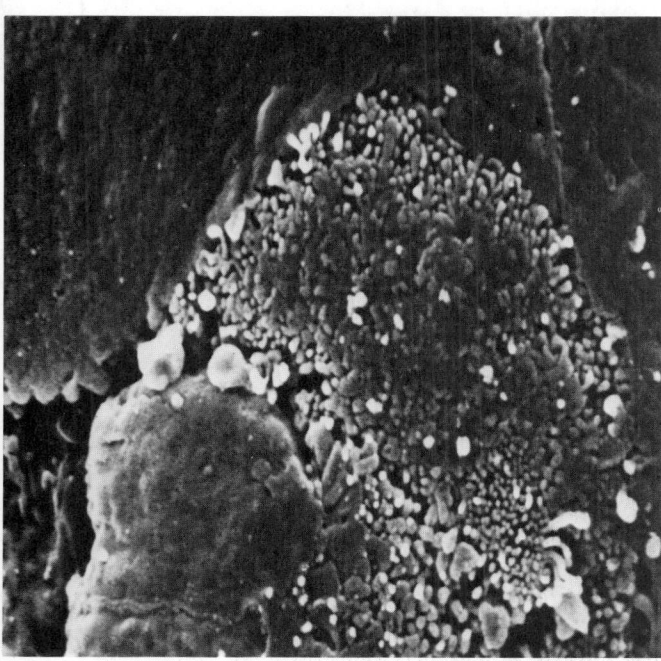

C

8.39 A. Transmission electron micrograph of obliquely sectioned bronchiolar epithelium (rat). Many of the cells are non-ciliated and the majority seen here are mucus-secreting. The basal lamina is seen at the lower left and the lumen at the upper right corner of the micrograph. Magnification × 3500. B. Scanning electron micrograph showing adjacent interalveolar septa in section and the lining epithelium of several alveoli in surface view. Magnification × 900. C. Scanning electron micrograph showing the surface of a type II alveolar epithelial cell surrounded by several type I alveolar epithelial cells. Magnification × 7000. Specimens prepared and photographed by Michael Crowder, Department of Anatomy, Guy's Hospital Medical School, London.

(Kalina & Pease 1974, Stratton 1977, Stratton 1978, Shimura et al 1986) and that of its precursors, the multilamellar bodies of type II alveolar epithelial cells (Douglas et al 1975, Stratton 1976). The pulmonary surfactant appears to be five-layered, each layer representing a stage in transformation of the multilamellar bodies. The layers typical of human surfactant are: (1) recently secreted multilamellar bodies nearest to the alveolar epithelium, (2) paired lamellae expanded and re-arranged as tubules, (3) mature tubular 'myelin' surfactant, (4) a surfactant–air interface, usually a single lipid bilayer, and (5) degraded surfactant, consisting of lipid bilaminar spheres formed at the air interface (Stratton 1978). Alveolar macrophages remove degraded surfactant, leaving some areas denuded of this covering until it is replaced by further secretion of multilamellar bodies. Localized lack of surfactant is part of the normal cycle of secretion and degradation.

The cells of the interalveolar septa can be grouped into those of three principal components: alveolar epithelium, capillary endothelium and interstitial space.

Alveolar epithelium is a mosaic of at least two cell types. Forming over 90% of the alveolar area are simple squamous *type I alveolar epithelial cells*, which resemble endothelial cells, having a highly attenuated cytoplasm about 0.05–0.2 μm wide extending from a thicker perinuclear region. The thinness of the cytoplasm facilitates gaseous diffusion between the lumen of the alveolus and its capillaries. Deep cytoplasmic extensions occasionally traverse the interstitial space between the capillaries to reach and expand on the opposite side of the septum (Weibel 1971). Since one cell extends on two surfaces, while its nucleus is only on one, cell division is impeded and type I cells do not proliferate. If damaged they are replaced by the multiplication of primitive type II cells which may later differentiate into type I pneumonocytes (Spencer 1977). The attenuated cytoplasm of type I cells contains pinocytotic vesicles, while the perinuclear cytoplasm contains a few mitochondria, a little agranular endoplasmic reticulum and occasional lysosomes. Edges of adjacent cells overlap, bound by non-leaky tight junctions. The lack of diffusion between the intercellular junctions is a vital feature which, with a similar endothelial barrier, limits the movement of fluid from blood and intercellular spaces into the alveolar lumen (the *blood–air barrier*) (see Simionescu 1985).

Type II alveolar epithelial cells, first reported by Reinhardt in 1847, are rounded secretory cells mainly engaged in the production and secretion of surfactant (Gil & Weibel 1969, King & Clements 1972, Gil & Reiss 1973). They are more numerous than type I cells, but are much less extensive and form only about 3% of the total alveolar surface. They project into the alveolar lumen and their free surfaces are covered by numerous short microvilli. In the human lung they are often associated with alveolar septal gaps (Takaro et al 1987). Their cytoplasm contains abundant mitochondria, granular endoplasmic reticulum, lysosomes and multilamellar bodies or secretory granules, composed of dipalmitoylphosphatidylcholine (50%), cholesterol (10%) and some protein, a precursor of pulmonary surfactant (Gil & Reiss 1973, Golde 1985, Bourbon & Rieutort 1987), stacked in regular lamellae, each a phospholipid bilayer. The ultrastructure of the multilamellar bodies has been described in detail by Stratton (1976), who found a lamellar width (and hence periodicity, there being no interlamellar interval) of 5.5 n or 6.6 n (depending on processing) and a tendency for the lamellae to fracture at bends, to divide and to join. Freeze-fracture preparations have shown that the lamellae are arranged in cup-shaped layers around a core (Weibel et al 1976). In mice, radioactive palmitate, a component of surfactant injected intraperitoneally, is rapidly incorporated into the multilamellar bodies of both type I and Clara cells; hence both may be involved in surfactant production (Niden 1967). Although primarily secretory, type II cells may be phagocytic (Corrin 1970, Suzuki et al 1972). Type II cells respond to injury by rapid proliferation and may replace damaged type I cells (Weibel 1974A).

A third type, the *brush cell* (Meyrick & Reid 1968), is a rare alveolar epitheliocyte occurring in some species; also called the *type III cell*, it was not observed by Gehr et al (1978) in an exhaustive study of human pulmonary ultrastructure but its occurrence was not excluded.

In addition to a respiratory role, *pulmonary endothelial cells* have other activities: they can clear up emboli and thrombi (Heinemann & Fishman 1969), assist in the metabolism of chylomicrons (Schoefl & French 1968), convert angiotensin I to angiotensin II, produce thromboplastin (Zeldis et al 1972), process hormones and prohormones arriving via the circulation (Vane 1969), synthesize prostaglandins and related substances (Ryan & Ryan 1977), inactivate serotinin and carry out a wide range of other metabolic activities. Their microanatomy and ultrastructure accord with these functions. The capillary endothelium is a simple continuous layer of squamous cells (Smith & Ryan 1973), whose particular features are abundant transcytotic vesicles (Smith & Ryan 1970, Simionescu 1985), extreme thinness (less than 0.1 μm in some areas) and an extensive array of luminal projections, thus presenting a huge surface area to blood, an area which in man must be many times greater (Ryan & Ryan 1977) than the commonly quoted 70 m² (Fishman & Pietra 1974). According to Ryan & Ryan (1977) the adult human lung may have 3×10^8 alveoli, each with a thousand or more capillary segments in its walls; the pulmonary capillary bed may measure 1500 miles or more, 1.0 ml of blood occupying 10 miles of capillaries. The capillaries are enclosed in alveolar walls or interalveolar septa and course through the interstitial space, interlacing with connective tissue fibres and bulging alternately into one or other of the alveoli on either side of each interalveolar septum. Thus, while on one side the capillary endothelium is separated by connective tissue fibres from the alveolar epithelium, on the opposite aspect their fused basal laminae are a minimal barrier, on about half the capillary surface, to gas and other solutes in blood. Where the endothelium is closely apposed to an epithelial cell, the endothelial wall is particularly thin (as little as 35 nm) and devoid of vesicles (Simionescu 1985).

The *interstitial spaces* of the interalveolar septa contain, in addition to fluid, some collagen and elastic fibres and various cells, both fixed and free (migratory). Fixed interstitial cells, fibroblasts and pericytes are the most numerous, the last demonstrated in the human lung by Weibel (1974a). All of these are capable of contraction (Kapanci et al 1974) and are structurally quite similar (p. 60), but are distinguished by their position, fibroblasts being associated with the connective tissue fibres whose components they manufacture, while pericytes lie close to the capillary basal lamina. Interstitial cells located around the openings into alveolar sacs resemble non-striated myocytes. They are stellate in form and are anchored to the basal lamina of epithelial cells; they are rich in actin and myosin and are contractile. It is thought that these cells may be able to alter the extensibility of alveoli and so modify lung compliance (see Kapanci et al 1974). Macrophages, mast cells and lymphocytes are the most numerous free cells in the septa; plasma cells also occasionally are present.

Within the alveoli, *alveolar phagocytes* (dust cells) are frequent; they are similar to connective tissue macrophages found beneath the epithelia and to circulatory monocytes and form part of the mononuclear phagocyte system (p. 669). They migrate from capillaries through the epithelium to wander about on the alveolar surfaces (Collet & Normand-Reuet 1967). Such phagocytes can clear the respiratory spaces of inhaled particles small enough to reach them, migrating back into the lymphatic channels or moving to the base of the bronchial tree to be swept up by the ciliary rejection current (vide infra) for elimination via the larynx. Whenever erythrocytes escape from pulmonary capillaries, the phagocytes become brick-red, due to engulfed cells, and are

detectable in 'rusty' sputum; such cells are typical of congestive heart failure and are often termed '*heart failure cells*'. Human alveolar phagocytes (macrophages) have been shown to produce both a collagenase and a collagenase inhibitor (Welgus et al 1985); they may thus produce a profound effect on the connective tissue matrix of the lungs and on any fibrous scar tissue produced in response to damage to the respiratory surface (vide infra). Other blood cells may invade the alveoli (and other parts of the respiratory tree) in pathological conditions, e.g. neutrophil leucocytes and lymphocytes in infections, giving sputum a characteristic yellowish appearance. The presence of these cell types in sputum is important in diagnosis.

PULMONARY DEFENSIVE MECHANISMS

The problems associated with the protection and maintenance of the respiratory tract are great. Such a huge exposed area is vulnerable to desiccation, microbial invasion and the mechanical and chemical effects of inhaled particles. Inhaled air is humidified chiefly in the upper respiratory tract where it passes, with some turbulence, over the nasal and buccopharyngeal mucosae. Secretions of the various glands of the bronchial tree also help to prevent desiccation. The composition of these secretions is hence important. Goblet cells secrete sulphated acid mucosubstances, whereas cells in mucous glands beneath the epithelial surface contain mainly carboxylated mucosubstances, particularly those associated with sialic acid, though sulphated groups also occur (de Haller 1969). In contrast, cells of serous glands contain neutral

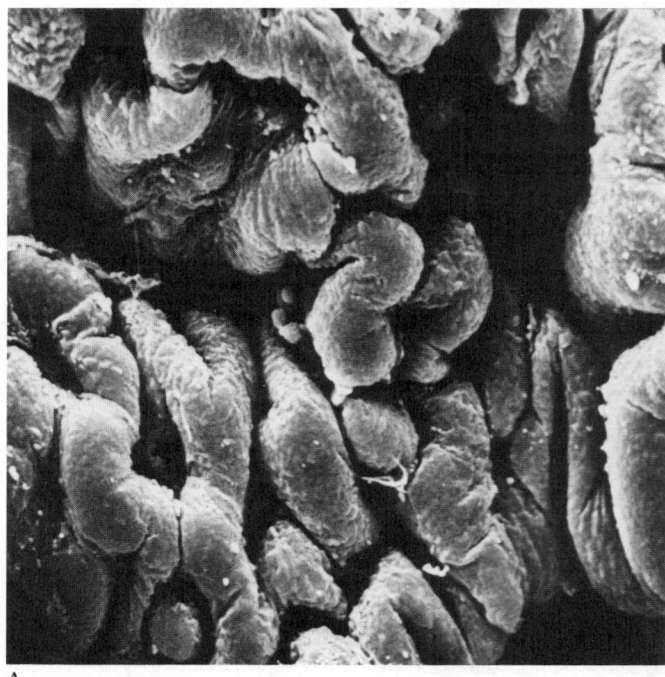

A

8.40A–D A. Scanning electron micrograph showing part of the capillary plexus of a rat alveolus. The walls of the capillaries are covered by alveolar epithelial cells. Magnification × 2500. B. Electron micrograph of a transverse section of a pulmonary capillary, situated in a murine interalveolar septum. A red blood cell (rbc) is present in the capillary lumen; the

mucosubstances. Goblet cells respond mainly to local irritation and tubular glands, both mucous and serous, to neural and hormonal control. Excessive or altered secretions may obstruct the flow of air. In addition to mucosubstances secreted by bronchial glands, antibacterial and antiviral substances also appear in the secreted fluid, e.g. lysozyme, antibodies of the IgA type and possibly interferon (Havez et al 1966).

Another defence against inhaled particles is the ciliary rejection current; cilia sweep the fluid overlying the surfaces of bronchioles, bronchi and trachea upwards at about 1 cm/minute and much inhaled matter trapped in the viscous fluid may be thus

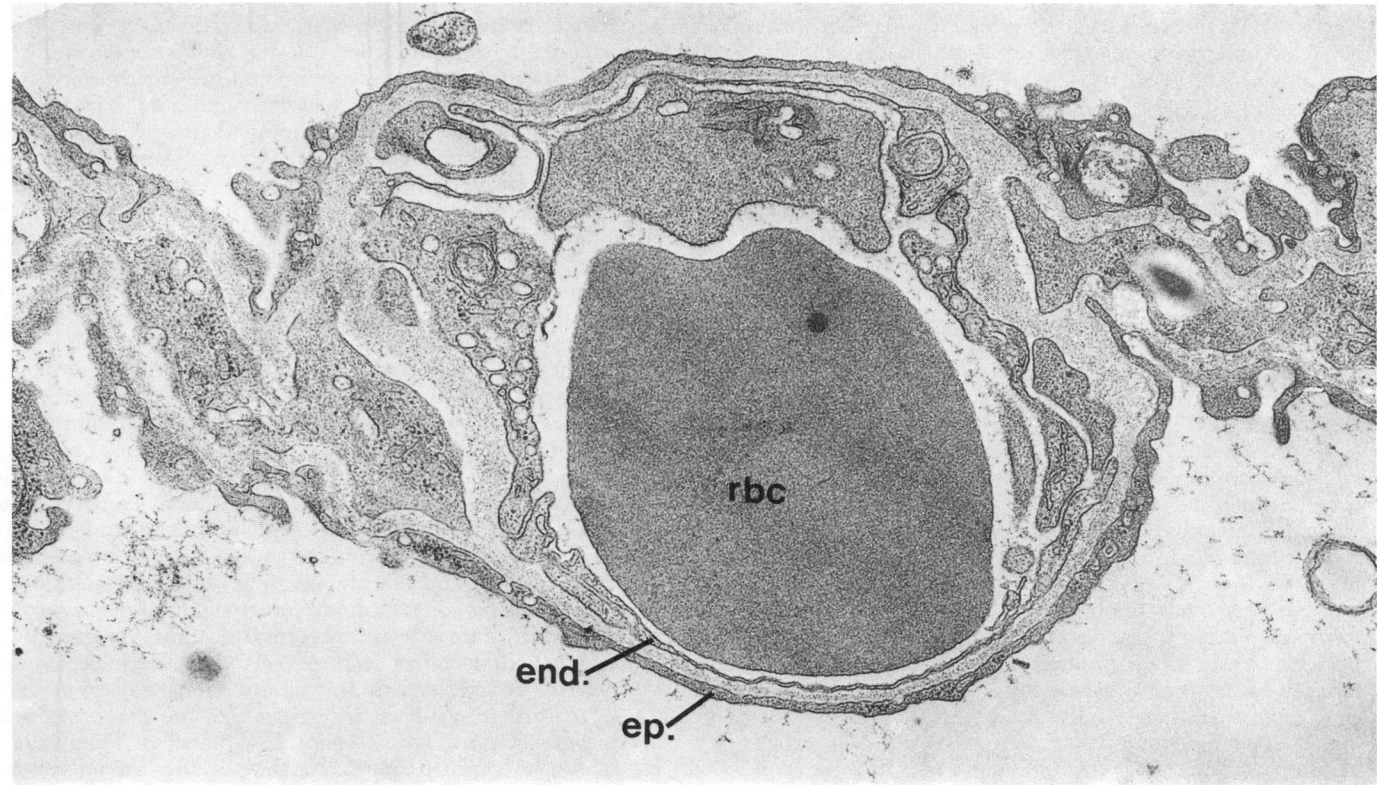

B

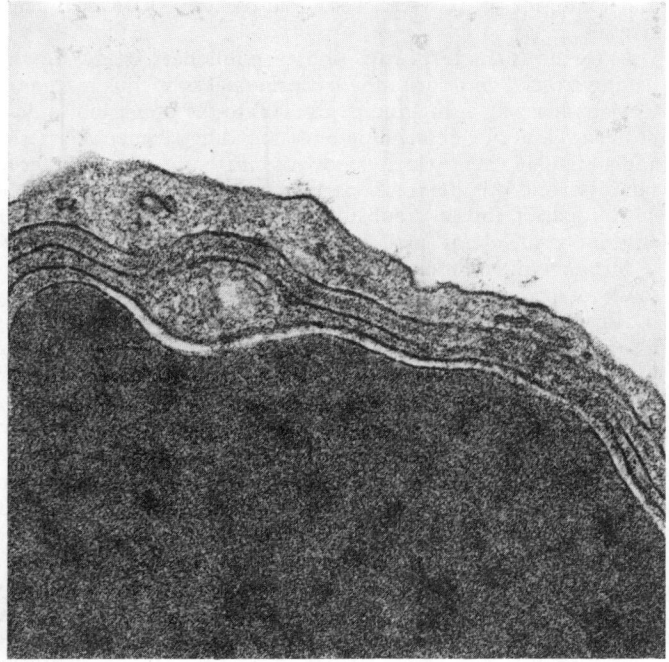

C

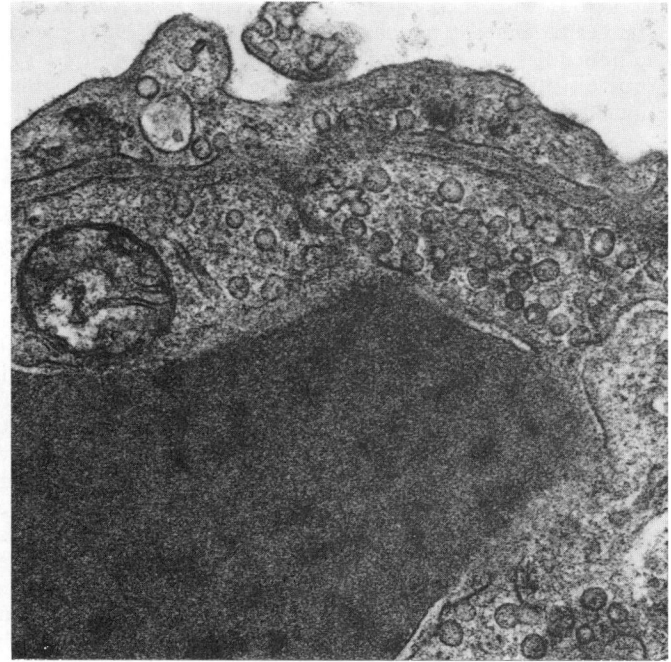

D

attenuated endothelium (end.) and the alveolar epithelium (ep.) are indicated. Magnification × 10 000. C. Transmission electron micrograph of part of an interalveolar septum showing the fused basal laminae separating a type I alveolar epithelial cell and an endothelial cell. Magnification × 40 000. D. Transmission electron micrograph of part of an interalveolar septum showing many micropinocytotic vesicles in adjacent type I alveolar epithelial and endothelial cells. Magnification × 35 000. Preparations A,C,D by Michael Crowder, Department of Anatomy, Guy's Hospital Medical School, London.

removed. However, particles small enough to reach the alveoli may be removed by alveolar phagocytes (vide supra); if particles are abrasive they may nevertheless damage the respiratory surface, with consequent fibrosis and reduction in the respiratory area, as in many industrial diseases, such as pneumoconiosis due to coal dust. Alveolar epithelium has limited powers of regeneration but normally is continually replaced. The life-span of alveolar squamous cells is about three weeks, that of alveolar phagocytes about four days (Bertalanffy 1966). Finally, numerous lymphoid nodules (bronchus-associated lymphoid tissue, BALT, see p. 832) occur in the bronchial lining, providing foci for the production of lymphocytes and giving local immunological protection against infection both by cell mediated (T cell) activities and the production of immunoglobulins (mainly IgA) from B cells to be passed on to gland cells for secretion to the epithelial surface.

In addition to these cellular mechanisms, coughing to clear pulmonary obstructions is, of course, essential. Initiation of such muscular responses to irritation involves stimulation of sensory endings. The identity of these afferent terminals is still somewhat uncertain except at the laryngeal aditus, where epithelial receptors similar to taste buds are considered responsible. However, at least some of the 'brush cells' of the respiratory epithelium appear to have neural contacts and may represent sensory cells with basal synaptic outputs.

PERINATAL PULMONARY DEVELOPMENT

Histologically the lung changes much during development, which is divisible for description into three phases: glandular, canalicular and alveolar (Engel 1947, de Reuck & Porter 1967, Emery 1969, Reid 1976). In the *glandular period*, during the first 16 weeks (approximately) of embryonic development, the tubular epithelial outgrowth ducts of the pulmonary rudiments are lined by tall columnar epithelial cells rich in glycogen, the cells almost obliterating the ductual lumina; between these tubes are closely packed mesenchymal cells and vascular tissue, the whole lung bud being a concentrated mass of cells resembling glandular tissue. At about the sixteenth week a *canalicular period* commences, marked by a rapid proliferation of the ducts, which branch profusely and become canalized as many cells become cuboidal and often ciliated. Some cells become squamous and begin to associate with capillaries, a change predominant in the *alveolar period*, which lasts from about 24 weeks onwards to birth, during which alveoli and the bronchial tree finally differentiate. Expansion of the future respiratory spaces is associated with the passage of fluid from lung tissue into their lumina, although incipient respiratory movements, causing inhalation of amniotic fluid, may also aid expansion. By full term and indeed some time before, the lungs are able to support normal respiration once fluid is expelled by expiration in air. After birth minor structural changes continue, alveolar surfaces becoming more complex and alveoli more numerous with increase in body size. Alveolar surfactant is synthesized first and premature births, particularly, may give rise to difficulties in expanding the lungs (respiratory distress syndrome); this can usually be corrected by glucocorticoid administration, which stimulates the synthesis of surfactant.

PULMONARY VASCULATURE AND INNERVATION

The *pulmonary artery* returns deoxygenated blood to the lungs; it divides into branches which accompany segmental and subsegmental bronchi and lie mostly dorsolateral to them (8.24B). They end in the dense capillary networks in the walls of alveolar sacs and alveoli (8.37). The arteries of neighbouring segments are independent. Elliot & Reid (1965) have demonstrated a fairly constant relation between the luminal and mural dimensions in the pulmonary arterial tree, this ratio (total diameter: wall thickness) being less than in most systemic arteries, as might be expected from the pressure differences. No regional differences have been detected (Simons & Reid 1969), despite evidence suggesting a large difference between blood flow in the upper and lower pulmonary regions. (See Van Meurs-Van Woezik et al 1987 for changes in the internal diameters in the pulmonary arterial tree with increasing age in infants and children.)

The *pulmonary capillaries* form plexuses immediately outside the epithelium in the walls and septa of alveoli and alveolar sacs. In interalveolar septa the network is a single layer, with meshes smaller than the capillaries, whose walls are exceedingly thin (vide supra and Suarez 1979).

Arteriovenous shunts have been demonstrated in human lungs near the terminal bronchioles. Experiments show that such shunts may pass particles of 500 μm diameter; the functional implications of this are not clear (Tobin 1966).

Pulmonary veins, two from each lung, drain the pulmonary capillaries, the radicles coalescing into larger branches which traverse the lung independently of the pulmonary arteries and bronchi (vide infra). Communicating freely they form large vessels, which ultimately accompany the arteries and bronchial tubes to the pulmonary hilum, the bronchi often separating the dorsolateral artery and the ventromedial vein. The pulmonary veins open into the left atrium, conveying oxygenated blood for systemic distribution by the left ventricle.

At the hilum the pulmonary arteries and veins accompany the main bronchial divisions but peripheral to this, in the bronchopulmonary segments, relations change (pp. 1261–1265). In

general a bronchus and its branches in a segment are central and accompanied by branching arteries but many tributaries of the pulmonary veins run *between* segments, serving adjacent segments, which drain into more than one vein. Some veins also lie beneath the visceral pleura, including that in the interlobar fissures. Thus a bronchopulmonary segment is not a complete vascular unit, with an individual bronchus, artery *and* vein. In the resection of segments it is obvious that the planes between them are not avascular but crossed by pulmonary veins and sometimes by branches of arteries. This pattern of bronchi, arteries and veins varies much, the veins being the most variable and arteries more variable than bronchi (Brock 1942, 1943, 1944, 1954, Boyden 1955, Cory & Valentine 1959, Bloomer et al 1960, Volpe et al 1969). Therefore the following general account requires amplification by reference to the above classic and to later reports.

Arteries of the right lung. Before reaching the hilum, the right pulmonary artery (p. 726) sends a large superior branch to the upper lobe; this is at first in front of the superior lobar bronchus but soon passes posterolateral to it. The main artery continues laterally between the superior lobar bronchus and the continuation of the principal bronchus to become posterolateral to the latter in the oblique fissure, dividing into middle and lower lobar arteries. This superior branch of the right pulmonary artery gives off apical, anterior descending and ascending and posterior descending and ascending segmental arteries. The middle lobar artery arises from the stem of the right pulmonary artery close to (or at) the origin of the superior branch to the lower lobe and divides into lateral and medial segmental branches. The remaining stem of the right pulmonary artery supplies the lower lobe with superior (apical), subsuperior (subapical), medial basal (cardiac), anterior basal, lateral basal and posterior basal segmental arteries.

Arteries of the left lung. The left pulmonary artery (p. 726), having crossed anterior to the left principal bronchus to its posterolateral aspect at the hilum, gives off to the upper lobe apical, posterior, anterior descending and ascending segmental arteries and a lingular artery which subdivides into superior and inferior rami. Beyond this the pulmonary artery supplies the lower lobe with superior (apical), subsuperior (subapical), medial basal, anterior basal, lateral basal and posterior basal segmental arteries.

Although the pulmonary arteries rarely depart from the above mode of primary (lobar) branching, segmental branches vary much in their sites of origin and numbers, reduplication being frequent. From observations at 521 partial pneumonectomies, Cory & Valentine (1959) recorded many variations: the right and left upper lobes showed the greatest range of variation, with 14 types of pattern in 152 right-sided operations, and 29 types in 107 on the left; other lobes displayed less variation, the patterns noted being five in the right middle, six in the right lower and four in the left lower lobe. Only the commonest patterns will be noted here.

Right upper lobe: Type 1: 3 segmental arteries, the apical and anterior arising from a common branch (91 of 152 = 60%). Type 2: 3 segmental arteries from a common branch (26 of 152 = 17%). Type 3: like Type 1, but with a reduplicated posterior segmental artery (13 of 152 = 8.5%). Type 4: like Type 2, but with a second, separate, posterior segmental branch (9 of 152 = 6%). These accounted for 91.5% of the observed patterns; the remainder (13 of 152, classified as Types 5–14) were mostly single instances. Direct branches from the right pulmonary artery varied from one to four in number and actual segmental arteries (often by common trunks) varied from the classic three (118 = 77%) to six.

Right middle lobe: Type 1: 2 segmental arteries (25 of 51 = 49%). Type 2: a common stem for 2 segmental arteries (23 of 51 = 45%). Types 3–5 (1 instance each of 51) all showed multiplicity (numbering 3,4 and 5) of middle lobe segmental branches.

Right lower lobe: Type 1: 3 segmental branches, 1 superior, 2 basal (31 of 43 = 72%). Types 3–6 (8 of the 43 lobes): a reduplicated superior branch and its atypical origin from a basal branch. Segmental arteries always numbered three or four, individual basal segments usually sharing common trunks.

Left upper lobe: Types 1–7 (78 of 107 = 70%) showed 4 or 5 segmental branches, most displaying early bifurcation, the resultant vessels being almost always distributed to two segments (with much variation in those supplied). Remaining types (8–29)

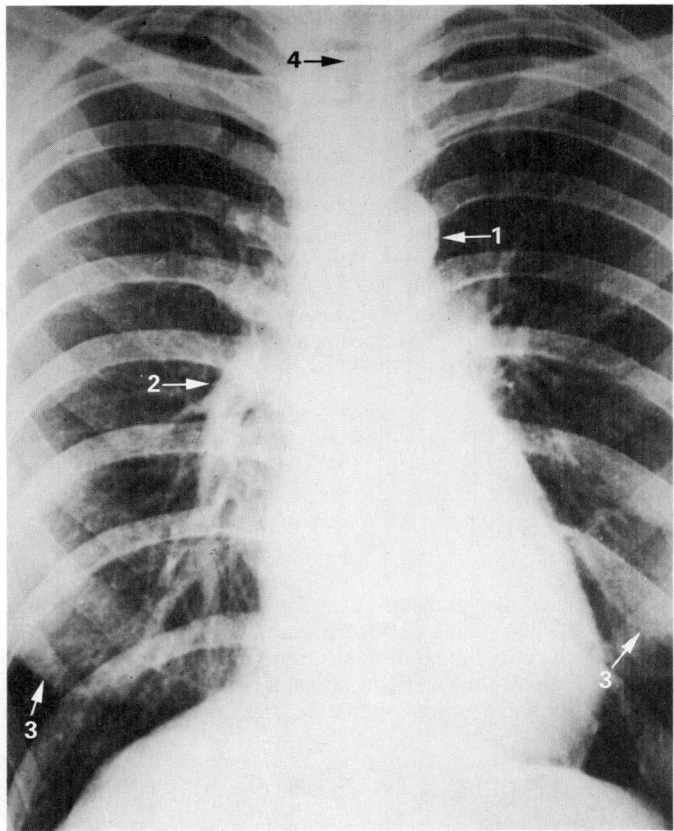

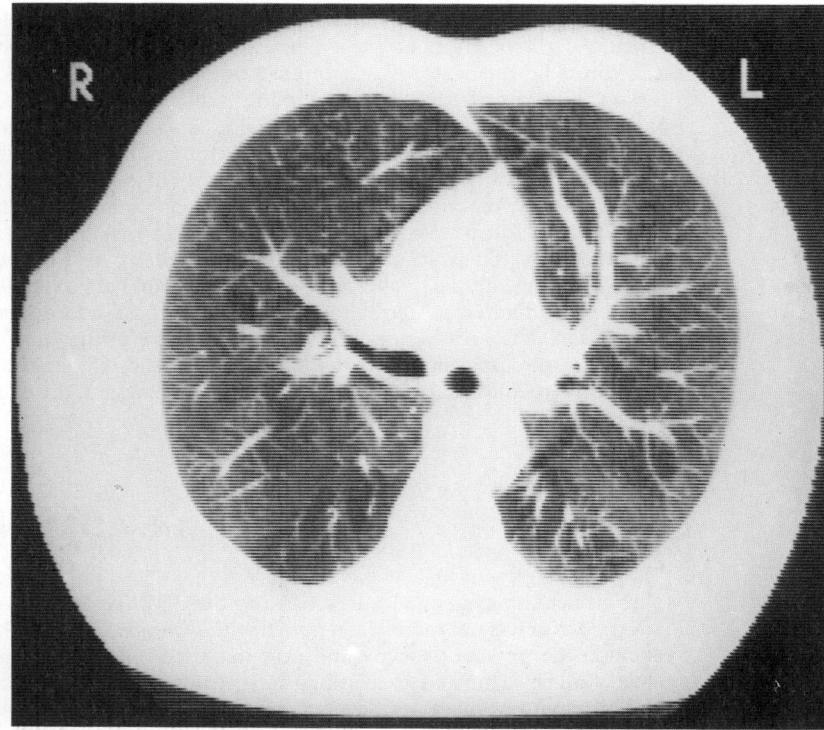

8.41A Radiograph of chest: postero-anterior view of adult female. Note the difference in level of the right and left halves of the diaphragm. 1. Aortic 'knuckle'. 2. Pulmonary vessels of right side. 3. Edge of shadow caused by breast. 4. Position of trachea.

8.41B Computed tomogram on lung settings of the thorax in the transverse plane immediately inferior (caudal) to the carina, showing the pulmonary vessels. Provided by Shaun Gallagher, Guy's Hospital; photography by Sarah Smith, UMDS, Guy's Hospital Campus, London.

were single instances, showing variation in number (2–6 segmental branches) and sites of branching from the left pulmonary artery; bifurcation was less common than in Types 1–7 and trifurcation occurred in three lobes. A common pattern for this lobe cannot be defined.

Left lower lobe: Type 1: showed 1 apical and 2 basal segmental arteries (44 of 57 = 77%). Type 2: like Type 1 but with 2 apical branches (9 of 57 = 15%). Type 3: like Type 1 but with 3 basal branches (3 of 57). Type 4: showed 2 basal and 3 apical branches.

Veins of the right lung (see also p. 791). In the hilum the superior right pulmonary vein is formed by the union of apical, anterior and posterior veins (upper lobe) with a middle lobar vein (formed by lateral and medial tributaries). The inferior right pulmonary vein is formed by the hilar union of superior (apical) and common basal veins from the lower lobe, the common basal being formed by the union of superior and inferior basal tributaries.

Veins of the left lung (p. 791). In the hilum the superior left pulmonary vein, which drains the upper lobe, is formed by the union of apicoposterior (draining the apical and posterior segments), anterior and lingular veins, the last-named consisting of an upper and a lower part. The inferior left pulmonary vein, which drains the lower lobe, is formed by the hilar union of two veins, superior (apical) and common basal, the latter formed by the union of a superior and an inferior basal vein.

All the main tributaries of the pulmonary veins receive smaller tributaries, some intrasegmental and others intersegmental (p. 1284).

Bronchial arteries supply oxygenated blood to maintain the pulmonary tissues. They are derived from the descending thoracic aorta or the upper posterior intercostal arteries. They accompany the bronchial tree and supply bronchial glands and the walls of the bronchial tubes and larger pulmonary vessels. The bronchial branches form, in the muscular tunic of the air passages, a capillary plexus supporting a second, mucosal plexus which communicates with branches of the pulmonary artery and drains into the pulmonary veins. Other arterial branches ramify in interlobular areolar tissue and end partly in deep and partly in superficial, bronchial veins. Some also ramify on the surface of the lung, forming subpleural capillary plexuses. Bronchial arteries supply the bronchial wall as far as the respiratory bronchioles. They anastomose with branches of the pulmonary arteries in the walls of the smaller bronchi and in the visceral pleura. Such bronchopulmonary anastomoses may be more numerous in the newborn, then later obliterated to a marked degree (Wagenvoort & Wagenvoort 1967). In addition to the main bronchial arteries, smaller bronchial branches arise from the descending thoracic aorta; one of these may lie in the pulmonary ligament and may cause bleeding during inferior lobectomy.

Bronchial veins form two distinct systems (Marchand et al 1950). *Deep bronchial veins* commence as intrapulmonary bronchiolar plexuses, communicating freely with the pulmonary veins and eventually joining a single trunk which ends in a main pulmonary vein or in the left atrium. *Superficial bronchial veins* drain extrapulmonary bronchi, visceral pleura and the hilar lymph nodes; they also communicate with the pulmonary veins and end on the right in the azygos vein and on the left in the left superior intercostal or the accessory hemiazygos veins. Bronchial veins do not receive all the blood conveyed by bronchial arteries; some enters the pulmonary veins. The main bronchial arteries and veins run on the dorsal aspect of the extrapulmonary bronchi.

The pulmonary lymph vessels are described on p. 857. For reconstructions of the bronchial lymphatic networks consult Bastianini (1967, 1968). Major ultrastructural features of pulmonary lymphatics include a thin basal lamina and abundant vesicles in the cytoplasm of the endothelial cells, with other evidence of pinocytosis, suggesting considerable passage of fluid across the lymphatic wall (Lauweryns & Boussaw 1967).

Pulmonary innervation. The anterior and posterior pulmonary plexuses are largely formed by sympathetic and vagal branches. Rami from these accompany the bronchial tubes, carrying efferent fibres to the bronchial muscles and glands and afferent fibres from the bronchial mucous membrane and alveoli. Small ganglia occur along these nerves. It is generally accepted that bronchoconstrictors are supplied by the vagus (pp. 1118, 1156), the sympathetic supply being inhibitory and relaxing the non-striated bronchial musculature, as does also the withdrawal or reduction of parasympathetic (vagal) stimulation. The actual stimulus for bronchodilatation is the pressure of inspired air. The innervation of the alveolar walls in the human lung has been studied at the ultrastructural level by Fox et al (1980).

Radiology. The trachea, because of its content of air, is more radio-translucent than neighbouring structures and hence visible in lateral and anteroposterior radiograms of the neck and upper thorax as a dark area in negatives (light in positive prints, 3.63). Similarly, the lungs appear as dark areas in the thorax, flanking the central mediastinal opacity; they also appear darker at the end of inspiration and in diseased conditions in which the alveoli are permanently distended (emphysema), but are more radio-opaque in pathological states which reduce the air in them (e.g. pneumonia). Pulmonary shadows have superimposed on them the light shadows of pulmonary blood vessels branching from the hilum (see 6.45 and Lodge 1946). These are sometimes mistaken for bronchi but the latter (because of contained air) obviously appear dark. Where a blood vessel is photographed end-on, it appears as a white circle (8.41A), a bronchus appearing as a dark circle surrounded by a white line (its wall). Hilar lymph nodes, if enlarged or calcified, appear as mottled areas near the mediastinum. The lumina of bronchi and bronchioles can be made radiologically visible by injecting a suitable radio-opaque oil into the trachea. Effusions of fluid into pleural cavities appear opaque in radiograms.

THE ALIMENTARY SYSTEM

The alimentary system, also described as the *digestive* or *gastrointestinal tract,* is not merely concerned with *digestion*, nor does it exclude the mouth, pharynx and oesophagus. It is a muscular canal, lined by a mucosa, extending from the lips to the anal orifice and displaying at successive levels a series of accessory structures, dental, glandular and muscular, for the ingestion, mastication, transport, digestion and absorption of foodstuffs of every kind and for the elimination of unabsorbed residues.

Throughout the *Chordata*, the alimentary and respiratory organs have shared a common history. In the primary aquatic era of

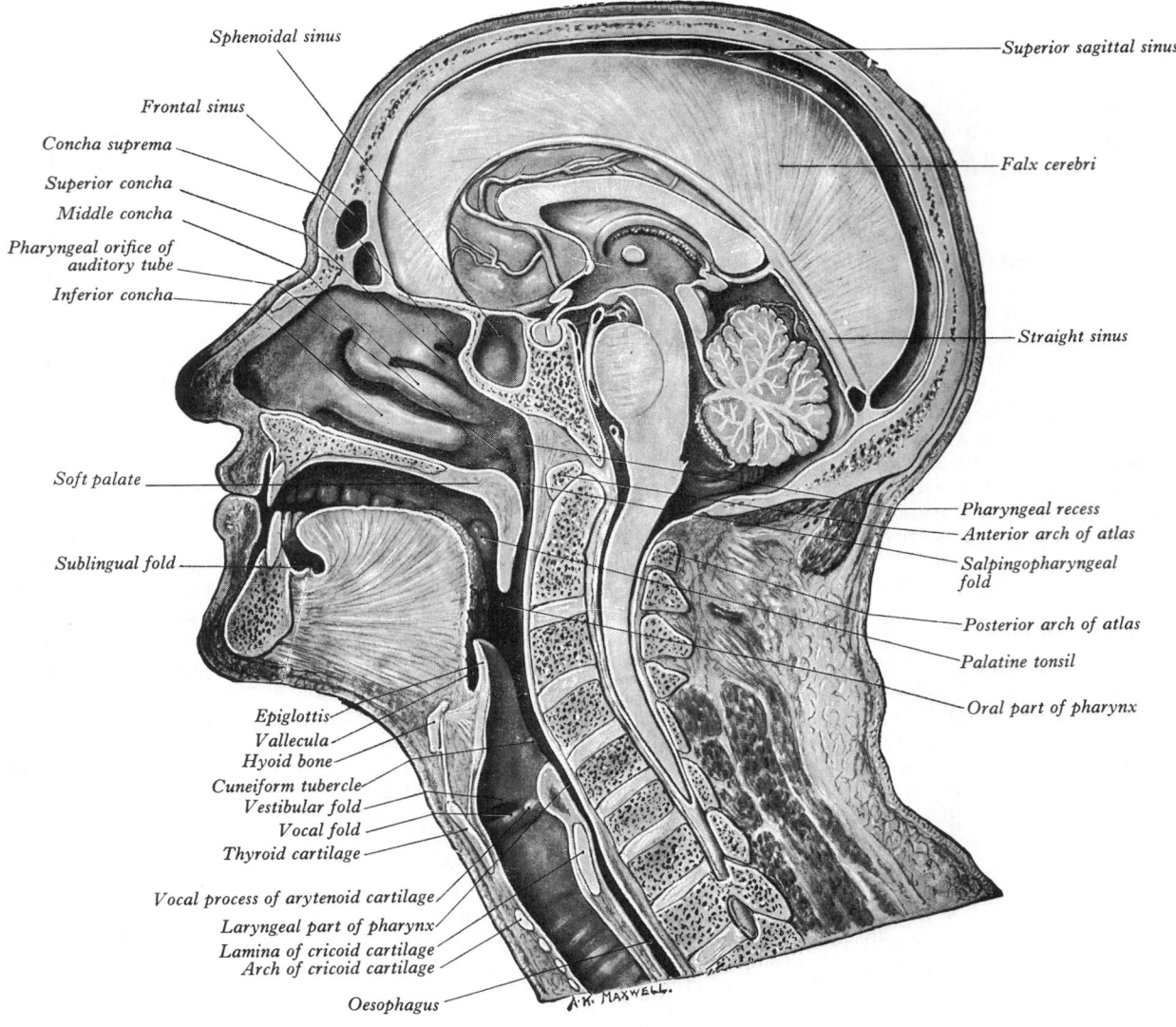

8.42A Median sagittal section through the head and neck. Where it divides the skull and the brain, the section passes slightly to the left of the median plane but, below the base of the skull, it passes slightly to the right of the median plane.

their evolution the two functions were intimately associated, as they still are in extant representatives. Dissolved oxygen and suspended particulate matter are absorbed or filtered from the same stream of water, entering through a single opening, the mouth, and escaping via a branchial apparatus of some kind, as in *Amphioxus*. The largest aquatic vertebrates of today, e.g. the basking shark *Cetorhinus* and baleen whales, the *Mysticeti*, still use filtration for food-gathering, though the filter is palatal in the latter. But many fishes, including some teleosts, have developed an ability to gulp in air to fill either a hydrostatic 'swim-bladder' or a respiratory 'air-bladder', which commonly loses continuity with the pharynx, deriving its contained gas from arterial blood (via the *dorsal* aorta). It is usually accepted that the respiratory system of terrestrial vertebrates is developed from a *ventral* 'air-bladder', supplied by the sixth branchial rami of the *ventral* aorta, as in extant *Dipnoi*, the lung fishes. *Amphibia* are similarly derived and they have the lungs and bronchial tree typical of terrestrial vertebrates, though many amphibians also depend on transcutaneous respiration. (Some urodeles, such as the newt *Necturus*, depend on this to such an extent that their lungs are said to have reverted to a flotation function!) Nevertheless, the buccal cavity and pharynx retain their initially dual function, even after division of the primitive mouth into definitive buccal and nasal chambers by palatal development in reptiles and in all subsequent vertebrates. Only with such specializations as the crocodilian larynx, which can be elevated into the nasopharynx during breathing, is complete *structural* separation of the upper respiratory and alimentary tracts achieved. In land vertebrates *functional* separation is nevertheless complete — into an apparatus conducting inspired and expired air with a regular periodicity, and another concerned in the intake and unidirectional passage of a highly varied pabulum, at intervals which vary enormously in different species and individuals. The early anatomical association persists, however; the dorsal position of the *respiratory* nasal chamber, relative to the *alimentary* buccal cavity, and the reversal of this relationship where the larynx diverges *ventrally* from the pharynx, entails a crossover between routes for air and food. Since both utilize the pharynx, except its nasal part, mechanisms have evolved to shut off the nasopharynx (by the soft palate) and respiratory tree (by the larynx) during swallowing. Because of their respiratory function, the nasal and oral parts of the pharynx, though muscular, are able to preserve their patency. (These complications are considered further on pp. 1248, 1328.)

The alimentary tract, being unidirectional, has a separate, caudal exit. This is essentially so even in the saccular urochordate *Ciona*, the sea-squirt, despite its sessile habit; its short gut with distinct mouth, pharynx and intestine opens into the atriopore, although this is, in fact, near the mouth. In the cephalochordate *Amphioxus*, differentiation of the gut has advanced, not only in the

development of a caudal anus separate from the atrial chamber (into which the pharynx strains its aqueous intake) but also in the development of a diverticulum of the midgut, sometimes regarded as 'hepatic' but more probably a source of digestive enzymes. Both these chordates use the pharynx, with its numerous branchial clefts, to strain food particles from imbibed water, the concentrated solids being passed on into the intestine. Tracts of cilia and mucous glands aid this mechanism and transport in the intestine, a method still used in mammalian respiratory systems, though lost in the alimentary canal. These are examples of the ability of the lining epithelia (and associated mesoderm) of the alimentary tract to adapt most variously to the functions of its successive parts. Even in the Agnatha (e.g. *Petromyzon*, the lamprey) the buccal cavity has primitive teeth, a rasping tongue and salivary glands; the pharynx has a branchial complex and the gut is differentiated into oesophagus (foregut) and intestine (midgut) into which a liver delivers its secretion. There is no true stomach but, while no pancreas is identifiable, some epithelial cells in the proximal intestine resemble those in the exocrine pancreas of later forms of vertebrate. (Closed follicles of intestinal cells have also been described at the same level; these may represent the pancreatic endocrine element.) The lamprey's intestine also shows a surface-increasing specialization, the *typhlosole*, a helical fold of mucosa like the spiral valve of the elasmobranch intestine. Elasmobranchs also have a well-defined stomach, pepsin-producing and divisible into cardiac (glandular) and pyloric (muscular) parts. A definite pancreas also appears, secreting into the midgut, and a large liver. A rectum is also recognized. In teleostean fishes the chief advance is intestinal elongation and coiling, to a degree varying with feeding habits. In the amphibian alimentary tract the intestine is clearly separated into intestine proper and colon, a muscular valve intervening.

Except in *Eutherian* mammals, the vertebrate intestine ends in a *cloaca*, which also receives the urinary and usually the genital ducts. Complete separation of the end of the alimentary tube at an anus, though foreshadowed in Agnatha, was not achieved until the mammals appeared, though monotremes must be excluded. A similar peculiarity exists at the *commencement* of the tract, for salivary glands, considered present in Agnatha, are also characteristic of mammals but are absent from intervening classes.

It is not appropriate to amplify this superficial account of the evolution in the alimentary canal, even in mammals; but a few examples must be added to emphasize its adaptiveness to feeding habits. Birds, of course, display peculiar specializations: the teeth

8.42B Sagittal section through the lower lip of a neonate, showing thin non-hairy skin of the vermilion border (left) grading into thicker stratified squamous epithelium of the vestibule, with mucosal labial glands (above). Small fasciculi of circumoral muscles are also visible. Haematoxylin and eosin. Provided by A Hayward; photography: Sarah Smith.

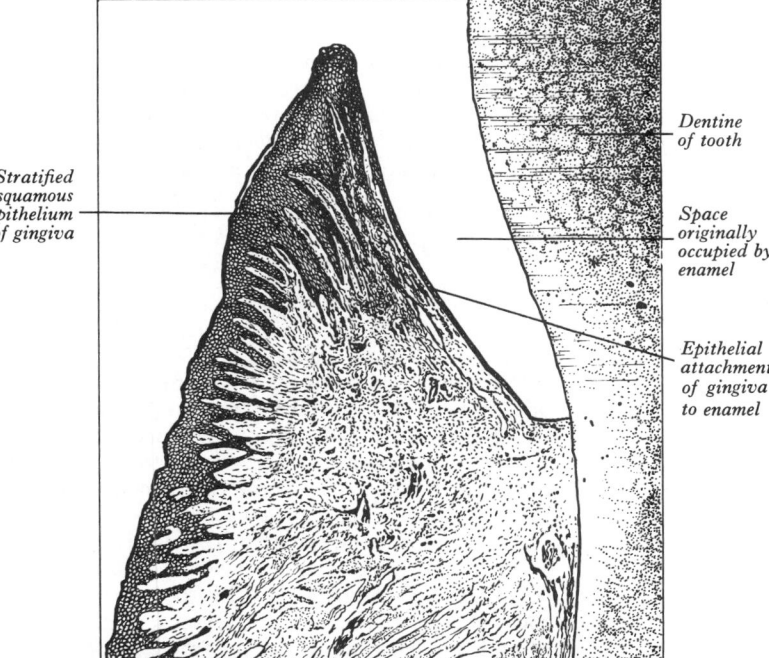

Stratified squamous epithelium of gingiva

Dentine of tooth

Space originally occupied by enamel

Epithelial attachment of gingiva to enamel

S.W.W.

8.43 A section across a tooth and the adjoining part of the gum. The enamel has been decalcified and removed, leaving a space between the dentine and the thin layer of stratified epithelium of the covering gum.

or their Jurassic ancestors (e.g. *Archaeopteryx*) have been lost; a hard, horny beak of highly variable form is the alternative feeding tool. The avian oesophagus is commonly dilated into a receptacle, the crop, and the stomach consists of a glandular proventriculus followed by a highly muscular gizzard, with a lining, often horny, and a content of swallowed stones to assist in grinding up hard foodstuffs. Among all mammals the general pattern of the alimentary tract is similar to the human; but the size and proportions of its different organs greatly vary. There is much variation in the relative size of the jaws and their musculature and even more marked differentiation in numbers, sizes and structure of teeth. All this reflects the plasticity of epithelial tissues and their associated mesodermal derivatives. Especially in the mucosa of the alimentary canal, this protean ability is manifest in production not only of varied epithelia (columnar, cuboidal, ciliated, stratified, keratinized and enamel-generating), but also a wide range of glandular structures, either as single cells or vast complexes such as the liver and pancreas, secreting a wide variety of digestive enzymes and hormones. Moreover, a range of endocrine cells is produced, including not only those of the thyroid gland and pancreatic islet tissue, long recognized, but also more recently identified systems of enterochromaffin and other endocrine cells (e.g. GEP and APUD series, p. 1376), scattered through much of the alimentary tract.

With these generalities in mind the detailed anatomy of the human alimentary tract must be considered seriatim at various levels, including its accessory organs, teeth, tongue, salivary glands, liver and pancreas. The microscopic structure of all these is included at appropriate points. It is also convenient and appropriate to describe the peritoneum with the subdiaphragmatic parts of the gut with which it is associated.

The alimentary canal is about 9 m long (see Underhill 1955 and p. 1355) and extends from mouth to anus, lined almost throughout by mucous membrane. It commences at the *mouth*, where provision is made for mechanical diminution of food (*mastication*) and admixture with the secretions of the salivary glands (*insalivation*); organs of swallowing (*deglutition*), the *palate*, *pharynx* and *oesophagus*, convey the mixture to the *stomach*, where the major digestive processes begin; the stomach empties into the *small intestine*, where digestion continues and many resulting products are absorbed into the blood and lymph. The small intestine leads to the *large intestine*, chiefly concerned in re-absorption of fluid, and opens finally at the *anus*. At many points the alimentary tube has sphincteric muscles, to be described seriatim (DiDio & Anderson 1968). For the general organization of the postpharyngeal alimentary tract see p. 1330. **The accessory digestive organs** are the *teeth*, which break up food, three pairs of main *salivary glands*—the *parotid, submandibular* and *sublingual*—whose secretion aids mastication in the mouth and the *liver* (with the *gallbladder*) and *pancreas*, large abdominal glands greatly concerned in digestion.

The Oral Cavity

The oral or buccal cavity, the 'mouth', consists of a smaller *vestibule* outside the teeth and an inner, larger *oral cavity proper*.

The *vestibule* is a slit-like space, bounded externally by the lips and cheeks and internally by the gums and teeth (**8.42**), communicating with the exterior by the *oral fissure*. Above and below, it is limited by the reflexion of the mucosa from the lips and cheeks on to the gums. When the teeth are apposed it communicates with the general oral cavity behind the third molar teeth and by clefts between adjacent teeth. On the cheek's internal surface, opposite the second upper molar crown, a small papilla bears the opening of the duct of the parotid salivary gland. The *oral cavity proper* (**8.89, 90**) is bounded anterolaterally by the alveolar arches, teeth and gums; behind, it communicates with the pharynx at the *oropharyngeal isthmus*, between the palatoglossal folds. Its roof is formed by the hard and soft palates, most of its floor by the anterior region of the tongue and the remainder by the reflexion of the mucosa from its sides and inferior surface to the gum on the internal mandibular

surface. A median crescentic *frenulum linguae* connects the inferior surface of the anterior part of the tongue to the oral floor. Inferiorly, on each side of the frenulum, a small *sublingual papilla* bears the orifice of the duct of the submandibular salivary gland. From this papilla a ridge extends posterolaterally in the floor, produced by the subjacent sublingual salivary gland and hence termed the *sublingual fold*; minute openings of the gland's ducts appear on the edge of the fold. The oral *mucosa* is continuous with the skin at the labial margins and with the pharyngeal mucosa at the oropharyngeal isthmus; it is roseate pink and very thick where it overlies bone. It has a stratified squamous epithelium which, except on the anterior tongue, is never completely cornified and is in many places non-keratinized. The *lymph vessels* of the mouth are described on p. 844.

The lips (8.42E) are two fleshy folds around the oral orifice, formed externally of skin and internally of mucous membrane enclosing the orbicularis oris, labial vessels and nerves, fibro-adipose connective tissue and numerous small labial salivary glands secreting into the oral cavity. The line of contact between the lips (*the oral fissure*) is opposite the cutting edges of the superior incisor teeth; on each side a *labial commissure* forms the *angle of the mouth*, usually near the first premolar tooth. External and central in the upper lip is a shallow vertical groove, the *philtrum*, ending below in a slight *tubercle* and limited by lateral ridges. Internally each lip is connected to the gum by a median *labial frenulum*, the upper being the larger.

The *labial glands*, situated between the mucosa and orbicularis oris, are about the size of small peas and in structure resemble mucous salivary glands (p. 1290). Their ducts open into the vestibule.

The cheeks are continuous in front with the lips, the junction indicated externally by *the nasolabial sulcus*, which descends laterally from the side of the nose to each oral angle. They contain a muscular stratum and variable but usually considerable amount of adipose tissue, together with areolar tissue, vessels, nerves and buccal glands, covered externally by skin and internally by mucous membrane. The *mucosa* of the cheek is reflected on to the external surfaces of the maxilla and mandible and thence to the gums; it is continuous behind with the palatal mucosa. The parotid duct, as already noted, opens opposite the crown of the second upper molar tooth. The principal muscle is buccinator, but others are also involved, e.g. zygomaticus major, risorius and platysma (see p. 573). The *buccal mucous glands* lie mainly between the mucosa and buccinator. Four or five larger ones (*molar glands*) lie external to the buccinator around the termination of the parotid duct; their ducts pierce the buccinator to open opposite the last molar tooth. *Lymph vessels* of the cheeks and lips are described on p. 845.

The gums (*gingivae*) (**8.43**) are composed of dense, vascular fibrous tissue covered by stratified, parakeratinized squamous epithelium. They consist of free parts which surround the neck of each tooth like a collar and attached parts which are firmly anchored to the alveolar processes of the mandible and maxillae. The fibrous tissue is continuous with the alveolar periosteum. In young people the stratified squamous epithelium is attached to the enamel of the teeth (epithelial attachment) but later the gums recede from the enamel and are attached to the cementum. Near the teeth, the mucosa on the buccal aspect of the gums develops tall papillae but becomes smooth where related to the enamel (**8.43**). *The nerves* of the upper gum come from the maxillary nerve via its anterior palatine, nasopalatine and anterior, middle and posterior superior alveolar branches. The mandibular nerve innervates the lower gum by the inferior alveolar, lingual and buccal branches. The buccal nerve supplies the external aspects of the lower gum as far forwards as the mental foramen. Vessels usually accompany the nerves. Lymphatics of the upper gum drain to the submandibular nodes; those from the anterior part of the lower gum pass to the submental nodes and from its posterior part to the submandibular nodes.

The gingival stratified squamous epithelium, like that of skin, contains melanocytes which are more obvious in pigmented races. Dendritic (Langerhans) cells are also present within the epithelium. The incidence of these two types of cells has been assessed (Barker 1967); the dendritic cells are more superficially located than the melanocytes and more frequent in the crevices between adjacent teeth. (For a detailed study of the oral cavity and tongue, consult Combelles 1972.)

THE PALATE

The palate, or oral roof, is divisible into two regions: the *hard palate* in front and *soft palate* behind.

The hard palate (8.62) is formed by the palatine processes of the maxillae and the horizontal plates of the palatine bones. It is bounded anterolaterally by the alveolar arch and gums and is continuous posteriorly with the soft palate, being covered by a dense mucoperiosteum. It has a median, linear raphe, ending anteriorly in a small papilla over the incisive fossa, on each side of which the mucosa is thick, pale and corrugated; posteriorly it is thin, smooth and redder. Its epithelium is keratinized stratified squamous. It has numerous palatine mucous glands which lie between the mucosa and periosteum in its posterior half. The upper surface of the hard palate is part of the nasal floor and is covered by ciliated epithelium.

The soft palate (8.42), a mobile flap suspended from the posterior border of the hard palate, slopes down and back between the oral and nasal parts of the pharynx. It is a fold of mucosa enclosing an aponeurosis, muscular tissue, vessels, nerves, lymphoid tissue and mucous glands. In its usual position, relaxed and pendant, its anterior (oral) surface is concave, with a median raphe. Its posterior aspect is convex and continuous with the nasal floor. Its superior border is attached to the hard palate's posterior margin, its sides blended with the pharyngeal wall and its inferior border is free, hanging between the mouth and pharynx.

A median conical process, the *uvula*, projects from its posterior border; the palatal arches, two curved folds of mucosa containing muscle, descend laterally from each side of the uvular base (8.90). The anterior, *palatoglossal arch* contains palatoglossus and descends to the side of the tongue at the junction of its oral and pharyngeal parts, forming the lateral limits of the oropharyngeal isthmus. The posterior, *palatopharyngeal arch* containing the palatopharyngeus descends on the lateral wall of the oropharynx (p. 1321). The *isthmus of the fauces* includes both arches.

Flanking the median palatine raphe, near the junction of the hard and soft parts of the palate, there is often a pair of depressions, a few millimetres apart. These *palatine foveae* may extend into pits, a few millimetres deep, with openings sagittally elongated. They are the common superficial orifices of converging ducts from the palatine mucous glands. The thin *palatine mucosa* has non-keratinized, stratified squamous epithelium, except on its upper pharyngeal surface and near the orifices of the auditory tubes, where it is columnar ciliated ('respiratory epithelium'), as in the nasal cavities. Beneath the mucosa on both surfaces are palatine mucous glands; they are most abundant around the uvula and on the oral aspect of the soft palate, where taste buds also occur.

Vessels and Nerves

The *arteries* of the palate are the greater palatine branch of the maxillary artery, the ascending palatine branch of the facial artery and the palatine branch of the ascending pharyngeal artery. The *veins* end largely in the pterygoid and tonsillar plexuses. The *lymph vessels* pass to the deep cervical lymph nodes. The *sensory nerves* issue from the greater and lesser palatine, nasopalatine and glossopharyngeal nerves; the lesser palatine contain taste fibres from the oral surface of the soft palate.

The Palatine Aoneurosis

A thin, fibrous *palatine aponeurosis* supports the muscles and strengthens the soft palate; it is attached to the posterior border and inferior surface of the hard palate behind the palatine crest. It is thick in the anterior two-thirds of the soft palate but very thin further back. It is composed of the expanded tendons of the tensores veli palatini; near the midline it encloses the musculus uvulae. All the other palatine muscles are attached to it. The

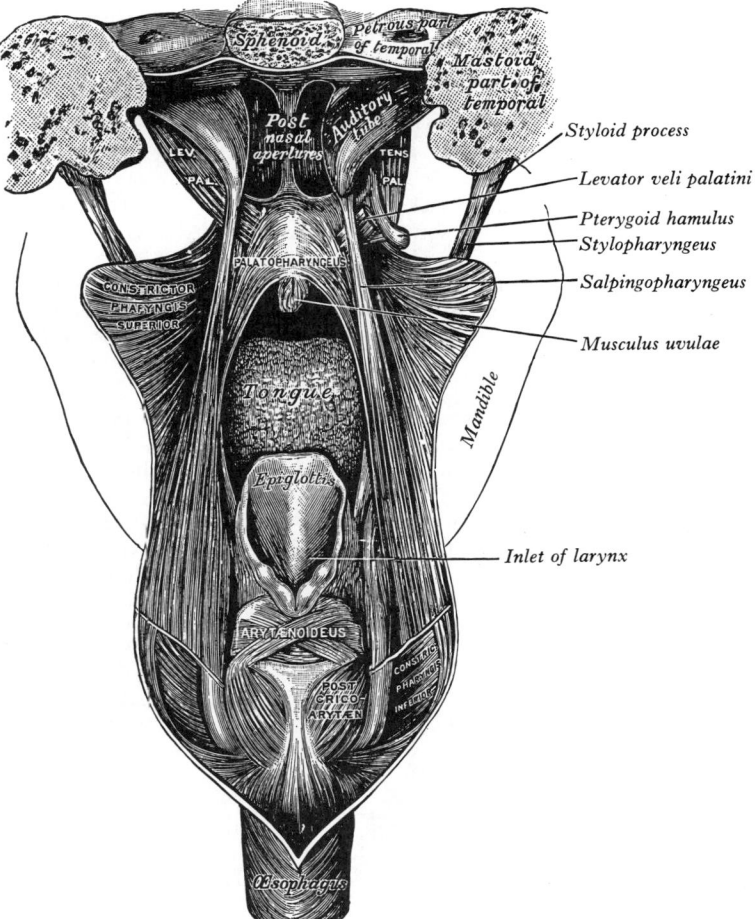

8.44 The muscles of the palate, exposed from the posterior aspect.

juxta-osseous part of the soft palate contains little muscle, consisting mainly of the palatine aponeurosis, inferior to which are many mucous glands; it is less mobile and more horizontal than the rest of the soft palate and the tensores veli palatini act mainly upon it.

THE PALATINE MUSCULATURE

The palatine muscles (8.44) include levator veli palatini, tensor veli palatini, palatoglossus and palatopharyngeus and musculus uvulae. For their activities in swallowing and speech see p. 1328.

Levator veli palatini (8.44, 99, 100) is cylindrical and lies lateral to the posterior nasal aperture. According to Rohan & Turner (1956) it is attached: (1) by a small tendon to a rough area on the inferior surface of the petrous temporal bone in front of the lower opening of the carotid canal; (2) by muscle fibres to a sheet of fascia descending from the tympanic vaginal process to form the upper part of the carotid sheath; and (3) by a few fibres to the inferior aspect of the cartilaginous part of the auditory tube. At its origin the muscle is inferior rather than medial to the auditory tube and only crosses medial to it at the level of the medial pterygoid lamina. Passing above the upper margin of the superior constrictor and in front of the salpingopharyngeus, it spreads in the soft palate between the two strands of the palatopharyngeus, its fibres being attached to the upper surface of the palatine aponeurosis as far as the midline where they blend with those of its fellow.

Action. The levator veli palatini elevates the soft palate.

The tensor veli palatini (8.44, 94, 100), a thin, triangular muscle, is lateral to the medial pterygoid plate, auditory tube and levator veli palatini. Its lateral surface contacts the upper anterior part of the medial pterygoid, the mandibular, auriculo-temporal and chorda tympani nerves, the otic ganglion and the middle meningeal artery. It is attached to the scaphoid fossa of the pterygoid process, the lateral lamina of the cartilage of the auditory tube and the medial aspect of the sphenoidal spine; from this point it descends, its fibres converging into a delicate tendon turning medially round the pterygoid hamulus to pass through the origin of the buccinator to the palatine aponeurosis and the osseous surface behind the palatine crest on the horizontal plate of the palatine bone. Between the tendon and the pterygoid hamulus is a small bursa.

Actions. Alone, the muscle pulls the soft palate to one side; with its fellow it tautens the soft palate (principally its anterior part), depressing it by flattening its arch.

The musculus uvulae, a bilateral structure, is attached to the posterior nasal spine of the palatine bone and to the palatine aponeurosis, between the two laminae of which the uvular muscles lie; it descends to be inserted into the uvular mucosa.

Action: elevation and retraction of the uvula.

The palatoglossus is a small fasciculus narrower at its middle than at its ends and forming, with the mucosa over it, the palato-glossal arch or fold. From the oral surface of the palatine aponeurosis, where it is continuous with its fellow, it extends antero-inferiorly and laterally in front of the tonsil to the side of the tongue, some of its fibres spreading over the lingual dorsum, others passing deeply into its substance to intermingle with the transversus linguae.

Actions. Palatoglossus elevates the root of the tongue and approximates the palatoglossal arch to its fellow, thus shutting off the oral cavity from the oropharynx.

The palatopharyngeus (8.44, 100) forms, with its overlying mucosa, the palatopharyngeal arch. In the palate its two fasciculi are separated by levator veli palatini. The *posterior* fasciculus is in contact with the mucosa of the pharyngeal aspect of the palate; it joins the posterior band of the opposite muscle in the midline. The thicker *anterior* fasciculus passes between the levator and tensor veli palatini and is attached to the posterior border of the hard palate and to the palatine aponeurosis, some fibres inter-digitating with their fellows across the midline. Both fasciculi are attached to the upper aspect of the palatine aponeurosis in the same plane. At the soft palate's posterolateral border the two layers unite and are joined by fibres of salpingopharyngeus (p. 1324). Passing laterally and downwards behind the tonsil, the palatopharyngeus descends posteromedial to the stylopharyn-geus, to be attached with it to the posterior border of the thyroid cartilage; some fibres end on the side of the pharynx, attached to pharyngeal fibrous tissue and others cross the midline posteriorly, decussating with those of the opposite muscle. The palatopharyn-geus thus forms an incomplete internal longitudinal muscular layer in the wall of the pharynx.

Actions. The palatopharyngei pull the pharynx up, forwards and medially, shortening it during swallowing. They also approximate the palatopharyngeal arches drawing them forwards.

Nerve supply. Except for tensor veli palatini, which is inner-vated by the mandibular nerve (p. 1103), all the palatine muscles are supplied by nerve fibres which leave the medulla in the cranial part of the accessory nerve and reach the pharyngeal plexus via the vagus nerve.

In the soft palate the muscles are arranged as follows: the palatine aponeurosis (tendon of the tensores veli palatini) is inter-mediate, enclosing the uvular muscles near the midline; the levatores veli palatini and palatopharyngi are attached to its upper surface, the two fasciculi of the latter separated by the former; the palatoglossi are attached to the inferior surface of the aponeurosis. (For a description of *the palatopharyngeal sphincter,* see p. 1327.)

Applied Anatomy. The condition of congenital cleft palate has been noted already as a developmental defect (p. 168). Paralysis of the soft palate may follow diphtheria due to the action of the toxin on the nerve cells of the medulla oblongata; in this state, the voice becomes nasal and fluids regurgitate into the nose during swal-lowing; the palate is visibly flaccid and motionless and also anaesthetic. Other pathological processes involving the glossopharyngeal, vagus and accessory nerves or their nuclei in the medulla oblongata also cause palatal paralysis.

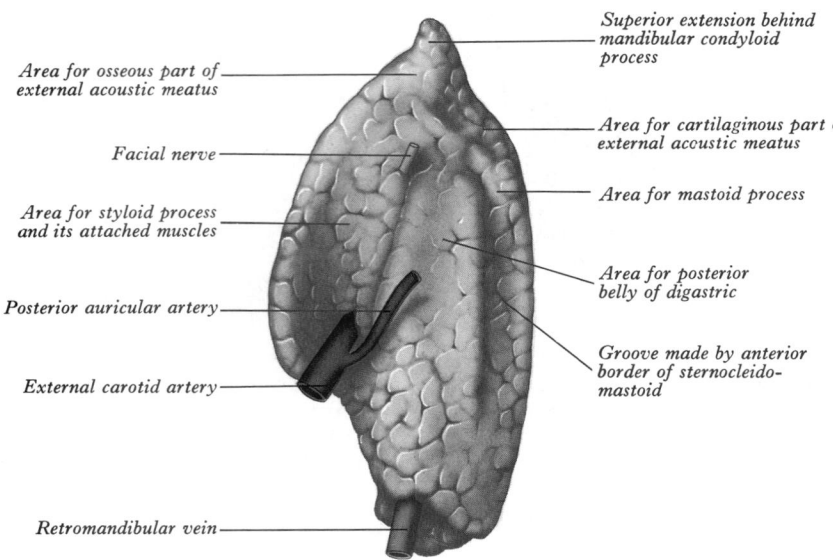

Area for osseous part of external acoustic meatus

Facial nerve

Area for styloid process and its attached muscles

Posterior auricular artery

External carotid artery

Retromandibular vein

Superior extension behind mandibular condyloid process

Area for cartilaginous part of external acoustic meatus

Area for mastoid process

Area for posterior belly of digastric

Groove made by anterior border of sternocleido-mastoid

8.45A The right parotid gland: posteromedial aspect.

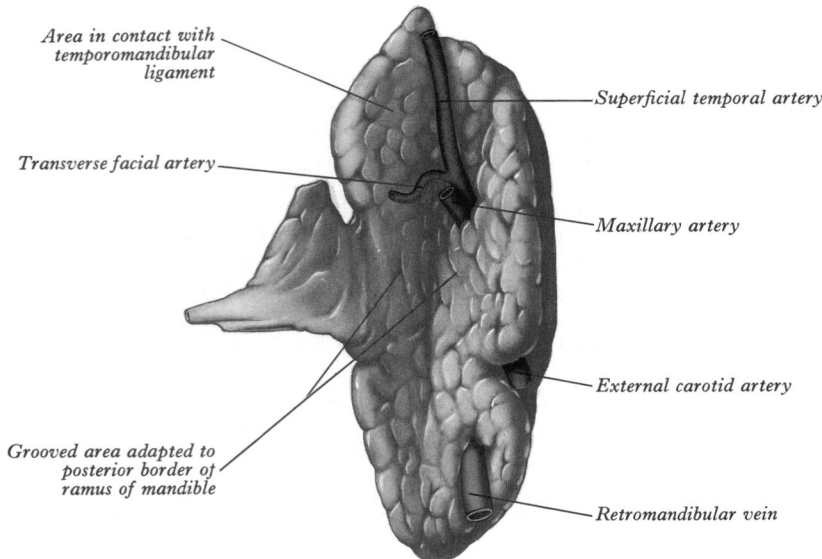

Area in contact with temporomandibular ligament

Transverse facial artery

Grooved area adapted to posterior border of ramus of mandible

Superficial temporal artery

Maxillary artery

External carotid artery

Retromandibular vein

8.45B The right parotid gland: anteromedial aspect.

The Salivary Glands

A salivary gland is any cell or organ discharging a secretion into the oral cavity; but a distinction is customarily made between the *major salivary glands,* located at some distance from the oral mucosa, with which they connect by extraglandular ducts, and the *minor salivary glands* which lie in the mucosa or submucosa, opening directly through the mucosa or indirectly via many short ducts. In humans the major salivary glands comprise the paired parotid, submandibular and sublingual glands; the minor salivary group includes the anterior glands of the tongue, numerous small lingual glands (including von Ebner's glands, p. 1322) and small labial, buccal and palatal glands (p. 1298). Their functions in-clude: lubrication of food to assist deglutition, moistening the buccal mucosa, aiding speech, provision of an aqueous solvent necessary for taste and as a fluid seal for sucking and suckling, secretion of digestive enzymes such as salivary amylase and of hormones and other compounds, such as a glucagon-like protein (Lawrence et al 1977) and possibly serotonin (Feyrter 1961) and secretion of antimicrobial agents (IgA and lysozyme).

THE PAROTID GLANDS

The paired parotid glands (8.45A, B) are the largest of the salivary glands; each has an average weight of about 25 g and is an ir-regular, lobulated, yellowish mass, lying largely below the external

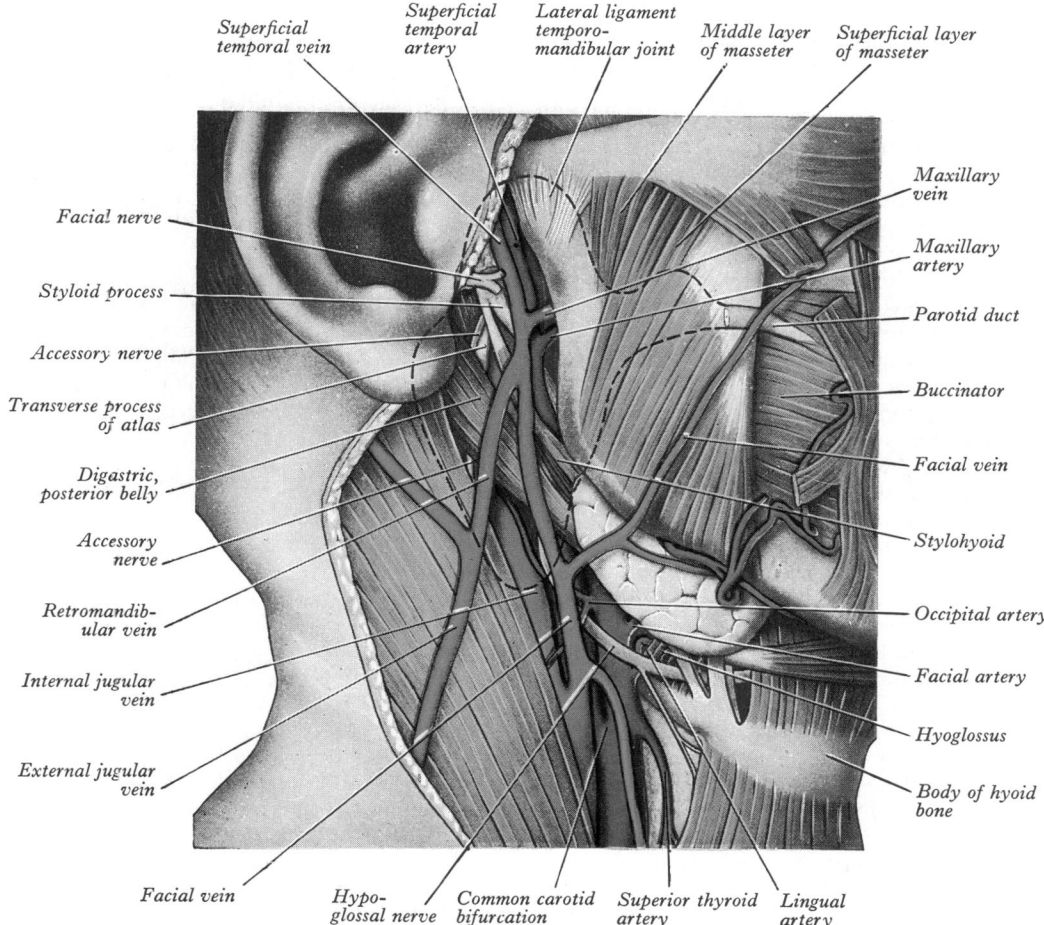

Superficial temporal vein · Superficial temporal artery · Lateral ligament temporo-mandibular joint · Middle layer of masseter · Superficial layer of masseter

Facial nerve
Styloid process
Accessory nerve
Transverse process of atlas
Digastric, posterior belly
Accessory nerve
Retromandib-ular vein
Internal jugular vein
External jugular vein

Maxillary vein
Maxillary artery
Parotid duct
Buccinator
Facial vein
Stylohyoid
Occipital artery
Facial artery
Hyoglossus
Body of hyoid bone

Facial vein · Hypo-glossal nerve · Common carotid bifurcation · Superior thyroid artery · Lingual artery

8.46 Drawing of a dissection to show the principal immediate deep relations of the parotid gland. The outline of the parotid gland is indicated by the interrupted black line.

acoustic meatus between the mandible and sternocleidomastoid; it also projects forwards on the surface of the masseter, where a small, usually detached part lies between the zygomatic arch above and the parotid duct below, the *pars accessoria* or *socia parotidis*.

The glandular capsule is derived from the deep cervical fascia; its superficial layer is dense, closely adherent to the gland and attached to the zygomatic arch; medial to the gland it is attached to the styloid process, mandible and tympanic plate, blending with the fibrous sheaths of related muscles; the fascia extending from the styloid process to the mandibular angle forms a stylomandibular ligament between the parotid and submandibular glands. The parotid gland is like an inverted, flat, three-sided pyramid, presenting a small superior and superficial, anteromedial and posteromedial surfaces; it tapers inferiorly to a blunt apex.

The concave *superior surface* is related to the cartilaginous part of the external acoustic meatus and posterior aspect of the temporomandibular joint; here the auriculotemporal nerve curves round the mandibular neck, embedded in the gland's capsule. The *apex* overlaps the posterior belly of the digastric and the carotid triangle to a variable extent.

The *superficial surface* is covered by skin and superficial fascia, which contains facial branches of the great auricular nerve, superficial parotid lymph nodes (p. 843) and the posterior border of platysma. It extends upwards to the zygomatic arch, back to overlap the sternocleidomastoid, down to its apex postero-inferior to the mandibular angle and forwards superficial to the masseter below the parotid duct (**8.46**).

The *anteromedial surface* is grooved by the posterior border of the mandibular ramus. It covers the postero-inferior part of the masseter, the lateral aspect of the temporomandibular joint and the adjoining part of the mandibular ramus, passing forwards

medial to the ramus to reach the medial pterygoid. Branches of the facial nerve emerge on the face from the anterior margin of this surface.

The *posteromedial surface* is moulded to the mastoid process, sternocleidomastoid, posterior belly of the digastric and the styloid process and its muscles. The external carotid artery grooves this surface before entering the gland. The internal carotid artery and internal jugular vein are separated from the gland by the styloid process and its muscles (**8.46**). The anteromedial and posteromedial surfaces meet at a medial margin which may project so deeply as to be in contact with the lateral wall of the pharynx.

Several structures traverse the gland partly or wholly and even branch within it. The external carotid artery enters the posteromedial surface, dividing into the maxillary artery which emerges from the anteromedial surface and the superficial temporal artery which gives off its transverse facial branch in the gland and ascends to leave its upper limit (**8.45B**). The posterior auricular artery may also branch from the external carotid within the gland, leaving by its posteromedial surface. The retromandibular vein (p. 794), formed by the union of the maxillary and superficial temporal veins (which enter near the points of exit of the corresponding arteries), is superficial to the external carotid artery and emerges behind the gland's apex to join the posterior auricular vein, forming the external jugular; it has a communicating branch which leaves anterior to the apex to join the facial vein. Most superficial is the facial nerve, entering high on the posteromedial surface (**8.45A**) and passing forwards and down behind the mandibular ramus in two main divisions, from which its terminal branches diverge (7. 118) to leave by the anteromedial surface, passing medial to its anterior margin.

The parotid gland develops as an outgrowth from the buccal cavity (p. 227), spreading back towards the ear and covering the facial nerve; prolongations penetrate medially between the branches of the nerve to form its deeper part, the largest part being between the nerve's main temporal and cervical divisions (Bailey 1947, McKenzie 1948). These processes finally engulf the nerve and its branches, which are sometimes considered to divide the gland into a superficial and a deep lobe.

The parotid duct (8.47, 54A), about 5 cm long, begins by the confluence of two main tributaries within the anterior part of the gland (p. 1298), then crosses the masseter and at its anterior border turns medially at almost a right angle, traversing the corpus adiposum (suctorial pad of infants) and the buccinator. It then runs obliquely forwards for a short distance between the buccinator and the oral mucosa to open upon a small papilla opposite the second upper molar crown. While crossing the masseter it receives the accessory parotid duct and here lies between the upper and lower buccal branches of the facial nerve; the accessory part of the gland and the transverse facial artery are above it. The buccal branch of the mandibular nerve, emerging from beneath the temporalis and masseter, is just below the duct at the masseter's anterior border.

The wall of the parotid duct is thick, with an external fibrous layer containing non-striated muscle and a mucosa lined by low columnar epithelium (vide infra). Its calibre is about 3 mm, but smaller at its oral orifice.

Surface Anatomy. The parotid duct can be felt on the face or more easily in the vestibule and rolled on the anterior border of the masseter, by pressing the finger *backwards* on it (with teeth clenched to make the muscle tense). The anterior border of the parotid gland is represented by a line descending from the mandibular condyle to about the middle of the masseter and then to a point about 2 cm inferoposterior to the mandibular angle. Its concave upper border corresponds to a curve from the mandibular condyle across the ear's lobule to the mastoid process. The posterior border is indicated by a straight line between the ends of the anterior and upper borders. The parotid duct corresponds to the middle third of a line drawn from the lower border of the tragus to a point midway between the nasal ala and upper labial margin.

Vessels and Nerves

The parotid *arterial supply* is from the external carotid and its branches in and near the gland. The *veins* drain to the external jugular, through local tributaries. The *lymph vessels* end in the superficial and deep cervical lymph nodes, interrupted by two or three nodes lying on and within the gland. The *efferent innervation* is autonomic, consisting of sympathetic fibres from the external carotid plexus and parasympathetic fibres which reach it via the tympanic branch of the glossopharyngeal nerve relaying in the otic ganglion and then travelling along the auriculotemporal nerve. Clinical observations suggest that the human gland also receives secretomotor fibres through the chorda tympani (Reichert & Poth 1933, Diamant & Wiberg 1965). Holmberg (1972) has shown that in *dogs* secretomotor fibres pass to the parotid gland from the maxillary plexus and the facial and auriculotemporal nerves, a supply unconfirmed in man. The termination of these supplies is still controversial. Studies in *cats* suggest that both parasympathetic and sympathetic fibres end in relation to glandular cells (Genvis-Galvey et al 1966, and vide infra).

THE SUBMANDIBULAR GLANDS

The paired submandibular glands (8.47) are irregular in shape and about the size of a walnut. Each consist of a large superficial and a smaller deep part, continuous with each other around the posterior border of the mylohyoid.

The *superficial part*, situated in the digastric triangle, reaches forward to the anterior belly of the digastric and back to the stylomandibular ligament, which separates it from the parotid gland. Above, it extends medial to the mandible's body; below, it usually overlaps the intermediate tendon of the digastric and the hyoidean attachment of stylohyoid. It has an inferior, a lateral and a medial surface and is partially enclosed between two layers of deep cervical fascia extending from the hyoid's greater cornu; one layer passes to the mandible's lower border, covering the gland's inferior surface, the other passes to the

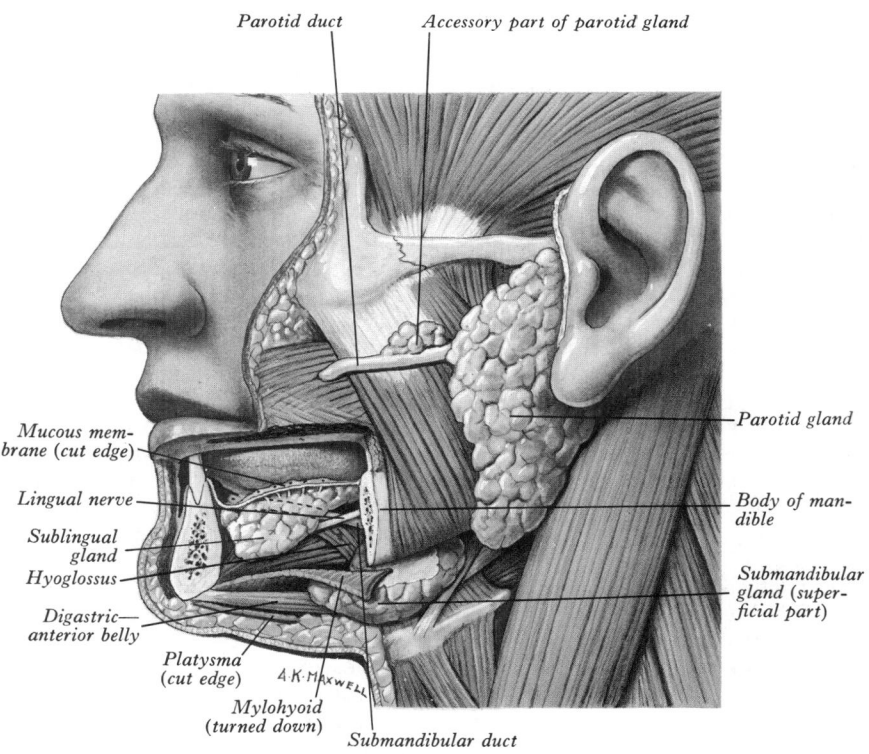

Parotid duct   *Accessory part of parotid gland*

Mucous membrane (cut edge)

Lingual nerve

Sublingual gland

Hyoglossus

Digastric anterior belly

Platysma (cut edge)

Mylohyoid (turned down)

Submandibular duct

Parotid gland

Body of mandible

Submandibular gland (superficial part)

8.47 Dissection showing the salivary glands of the left side. The cranial region of the superficial part of the submandibular gland has been excised and the cut mylohyoid has been turned down to expose a portion of the deep part of the gland.

mylohyoid line on the medial surface of the mandible and covers the gland's medial surface.

The *inferior surface*, covered by skin, platysma and deep fascia, is crossed by the facial vein and facial nerve's cervical branch; near the mandible the submandibular lymph nodes are in contact with the gland and some may be embedded in it (p. 844).

The *lateral surface* is in relation with the submandibular fossa (on the medial surface of the mandibular body) and the mandibular attachment of the medial pterygoid; the facial artery grooves its posterosuperior part, lying at first deep to the gland and then emerging between its lateral surface and the mandibular attachment of the medial pterygoid to reach the mandible's lower border.

The *medial surface* is related anteriorly to the mylohyoid, separated from it by the mylohyoid nerve and vessels and branches of the submental vessels. More posteriorly, it is related to the styloglossus, stylohyoid ligament and glossopharyngeal nerve, which separate it from the pharynx; in its intermediate part the medial surface is related to the hyoglossus, separated from it by the styloglossus, the lingual nerve, submandibular ganglion, hypoglossal nerve and deep lingual vein (sequentially from above down). Below, the medial surface is related to the stylohyoid and the posterior belly of the digastric.

The *deep part* of the gland extends forwards to the posterior end of the sublingual gland and lies between the mylohyoid inferolaterally and the hyoglossus and styloglossus medially; above it is the lingual nerve and, below it, the hypoglossal nerve and deep lingual vein.

The gland is palpable between an index finger placed on the floor of the mouth and a thumb placed anteromedial to the angle of the mandible, below the floor.

The submandibular duct (8.54B) is about 5 cm long and has a thinner wall than the parotid duct. It begins from numerous tributaries in the superficial part of the gland and emerges from the medial surface of this part of the gland behind the posterior border of the mylohyoid; it traverses the deep part, passing at first up and slightly back for 4 or 5 mm and then forwards between the mylohyoid and hyoglossus. Passing between the sublingual gland and genioglossus it opens in the oral floor on the summit of the sublingual papilla at the side of the frenulum of the tongue (8.89). On the hyoglossus it lies between the lingual and hypoglossal nerves, but at the muscle's anterior border it is crossed laterally by the lingual nerve, terminal branches of which ascend on its medial side (8.95). As it traverses the gland's deep part it receives small tributaries draining this part of the gland. Salivary calculi occasionally occur in the submandibular duct; they are radio-opaque and may be palpable.

Vessels and Nerves

The *arteries* supplying the submandibular gland are branches of the facial and lingual; the *veins* correspond. The *lymph vessels* drain into the deep cervical group of lymph nodes, particularly the jugulo-omohyoid node, interrupted by the submandibular nodes, some of which are in close relation with the anterior end and medial aspect of the superficial part of the gland and may be embedded in it. The *nerves* are derived from the submandibular ganglion, through which it receives fibres from the chorda tympani (parasympathetic), the lingual branch of the mandibular nerve and the sympathetic trunk. In dogs and cats the gland receives its nerve supply through the subsidiary ganglion of Langley (p. 1155), but in humans the parasympathetic nerve supply (chorda tympani) relays mainly in the submandibular ganglion. Small parasympathetic ganglia may occur in the gland's hilum and in the branches of the submandibular ganglion passing to it. Garrett & Kemplay (1977), using catecholamine fluorescence and electron micrography, observed that in cats all adrenergic fibres are derived from the superior cervical sympathetic ganglion.

THE SUBLINGUAL GLANDS

The paired sublingual glands (8.47), the smallest of the main salivary glands, lie beneath the oral mucosa, each in contact with the sublingual fossa on the lingual aspect of the mandible, close to

the symphysis. Each is narrow, flat, shaped like an almond, and weighs 3–4 g. *Above* it is the mucosa of the oral floor, raised as a sublingual fold; *below* the mylohyoid; *in front* the anterior end of its fellow; *behind*, the deep part of the submandibular gland; *lateral* is the mandible above the anterior part of the mylohyoid line and *medial* is the genioglossus separated from it by the lingual nerve and submandibular duct. It has 8–20 excretory ducts; of the *smaller sublingual ducts* most open separately on the summit of the sublingual fold and a few sometimes into the submandibular duct. From the anterior part of the gland small rami sometimes form a *major sublingual duct*, opening with or near to the submandibular duct.

Vessels and Nerves

The *arterial supply* is by the sublingual and submental arteries. The *veins* correspond. *Innervation* is by the lingual nerve, chorda tympani and sympathetic fibres. The parasympathetic relay is in the submandibular ganglion; neurons may occur among its distal fibres to the lingual nerve forming a distinct sublingual ganglion. The precise terminations of the nerve supply are not fully known (vide infra).

STRUCTURE OF MAJOR SALIVARY GLANDS

These compound racemose glands (8.48A,B, 49A,B, 50, 51, 52) have numerous lobes composed of lobules linked by dense connective tissue containing excretory (collecting) ducts, blood vessels, lymph vessels, nerve fibres and small ganglia. Each lobule has a single duct, whose branches end as dilated secretory 'end-pieces' or acini. Their primary secretion is modified as it traverses intercalated, striated and excretory ducts into one or more main ducts which discharge saliva into the oral cavity.

Salivary Acini or 'Endpieces'

Variation in the terms describing the form and cytology of secretory acini or endpieces has caused much confusion. Garrett (1976) and Young & van Lennep (1978) have appraised the problems involved. (The official term is *portio terminalis*— of acinar, alveolar or tubulo-alveolar types—inelegantly translated as 'endpiece'.)

In the following description a secretory 'endpiece' is termed an *acinus* if approximately spheroidal, a *tubule* if elongate; *tubulo-acini* are intermediate in shape. Secretory cells of acini are pyramidal, with narrow luminal apices and broad bases; those of tubules are more cylindrical. Such cells, the main producers of salivary protein and glycoprotein, are usually described as *serous*, *seromucous* or *mucous*, unfortunately without unanimity in use. A serous secretion of low viscosity and a highly viscous mucus were correlated initially with serous and mucous cells distinguished by simple staining. Improved techniques identified a seromucous variety. It is now established that products of these cells form an almost continuous series, from serous secretions with negligible amounts of protein-associated acidic carbohydrates to mucous secretions rich in them. Attempts to define the terms serous, seromucous and mucous histochemically (Leblond 1950, Munger 1964, Shackleford & Wilborn 1968) have proved indecisive. Though definition may be desirable, it is more convenient in practice to use observable cytological features, such as the appearance of the secretory granules. Applying the criteria of Young & van Lennep (1978) glandular cells are here dubbed: *serous* if the granules are small, discrete, homogeneous, generally eosinophilic and electron-dense; *mucous* if their granules are larger, close-packed and ill-defined, with low eosinophilia and a homogeneous, fairly electron-translucent matrix; or *seromucous* if they are intermediate in appearance, with granules either close-packed, eosinophilic and homogeneous, or more discrete, larger and heterogeneous. A gland or secretory acinus with only one type of cell is *homocrine*, while one containing more than one type is *heterocrine*.

The secretory 'endpieces' of the human *parotid gland* are mainly seromucous (or serous) acini; mucous acini are rare. In the *submandibular gland* acini are usually seromucous with some mucous. In the *sublingual gland* they are typically mucous tubules, but seromucous cells also occur, frequently as acini or as *demilunes*

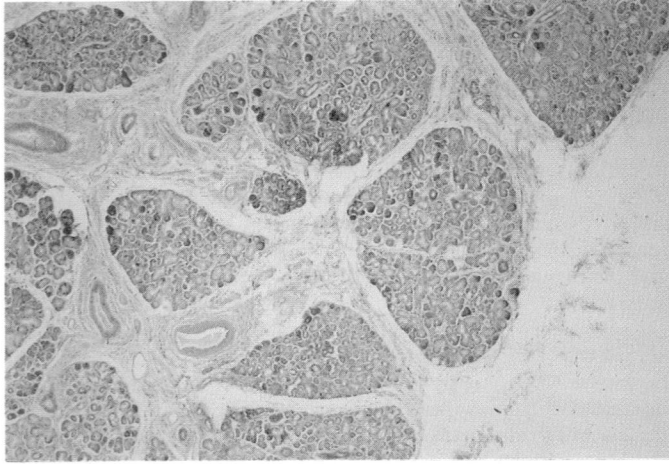

8.48A Section through the sublingual gland, stained with haematoxylin and eosin. Magnification × 80. Compare with **8.48B**.

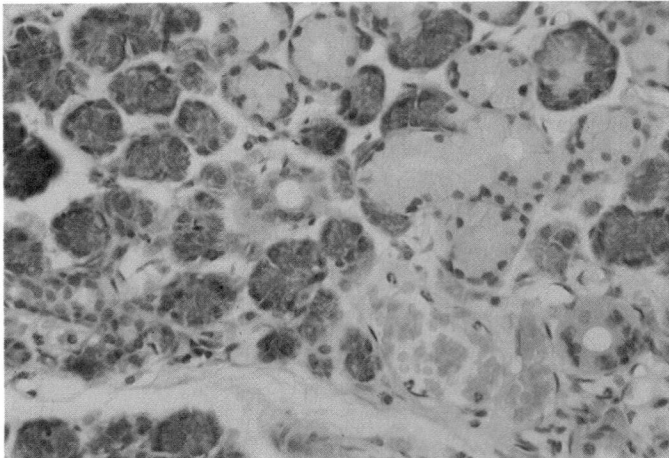

8.49A Section of the submandibular gland, stained with haematoxylin and eosin. Magnification × 350. Compare with **8.49B**.

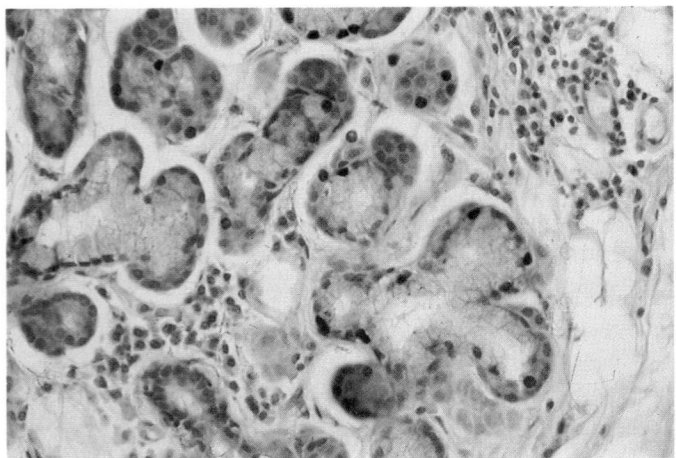

8.48B Section through the sublingual gland, stained with haematoxylin and eosin. Magnification × 350.

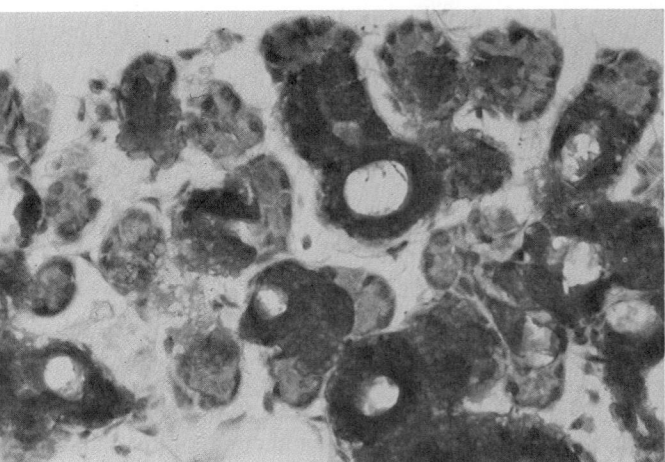

8.49B Section through the submandibular gland, stained with haematoxylin and alcian blue. Magnification × 350.

(Young & van Lennep 1978). Seromucous demilunes, which also exist in submandibular glands of many mammalian species, are crescentic groups of glandular cells found at the bases of some mucous endpieces (**8.51**); sited between the mucous cells and basal lamina, they apparently communicate with the lumen by fine canaliculi which pass between the mucous cells. Using the criteria described above all the major human salivary glands are thus heterocrine.

The ultrastructure of these glandular cells is shown diagrammatically in **8.51**. *Seromucous* (and *serous) cells* are roughly pyramidal. The basal plasmalemma is usually smooth, while the lateral is plicated and interdigitates with that of the adjacent cells. The apical (luminal) plasmalemma commonly forms microvilli, between which are often pinocytotic vesicles. Discrete, secretory canaliculi lie between the cells; limited basilaterally by junctional complexes, they open into the lumen of the 'endpiece'; their zonulae occludentes are often assumed to form a continuous seal around each cell but freeze-fracture micrographs of parotid 'end-piece' tight junctions in rats (de Camilli et al 1976) suggest that the 'seals' may be incomplete. Nuclei vary in shape and position, but are more spheroidal and less basal than in mucous cells. Apical cytoplasm is filled by secretory granules of variable form (Tandler 1972, Riva & Riva-Testa 1973); a conspicuous feature of the infranuclear cytoplasm is an abundant granular endoplasmic reticulum arranged in stacks of parallel, flat cisternae. Golgi complexes are supranuclear with adjacent small coated and smooth vesicles. Elongated mitochondria, lysosomes, microfilaments and occasional large lipid droplets also appear. In contrast *mucous cells* are cylindrical; their luminal plasma membrane is smoother and secretory canaliculi between the cells are rare. The supranuclear

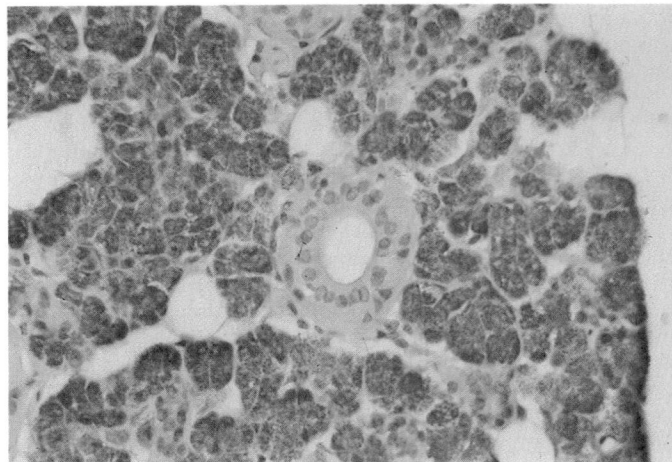

8.50 Section through the parotid gland, stained with haematoxylin and eosin. Magnification × 350.

cytoplasm is typically packed with large, electron-translucent, frequently fused, secretory droplets. Granular endoplasmic reticulum and Golgi complexes resemble those of serous and seromucous cells, but the nucleus is flatter and more basal.

Ducts of Salivary Glands

Leading consecutively from the secretory 'endpieces' are *inter-calated, striated* and *excretory ducts*.

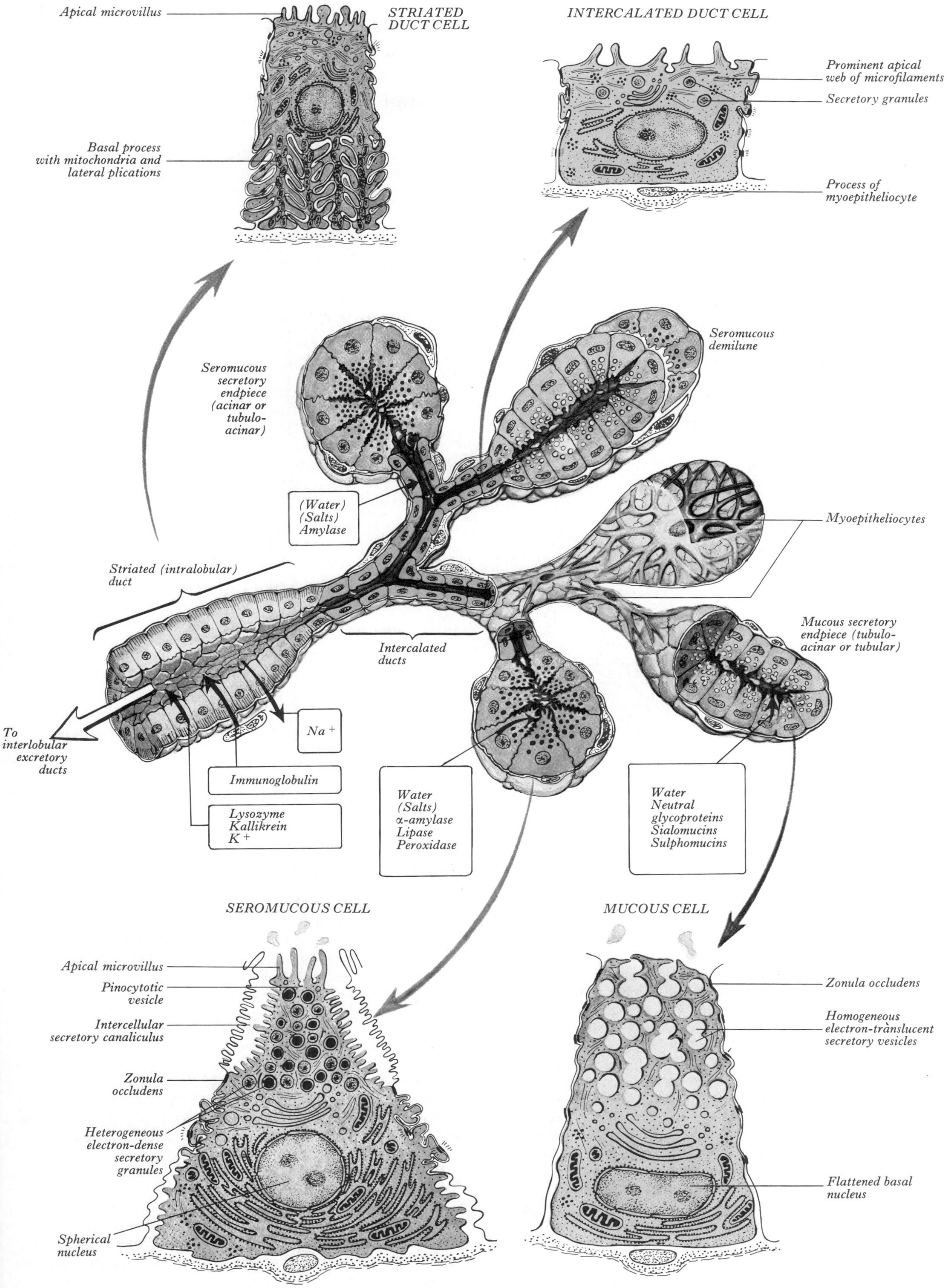

STRIATED DUCT CELL

Apical microvillus

Basal process with mitochondria and lateral plications

INTERCALATED DUCT CELL

Prominent apical web of microfilaments

Secretory granules

Process of myoepitheliocyte

Seromucous secretory endpiece (acinar or tubulo-acinar)

Seromucous demilune

(Water) (Salts) Amylase

Striated (intralobular) duct

Myoepitheliocytes

Mucous secretory endpiece (tubulo-acinar or tubular)

Intercalated ducts

To interlobular excretory ducts

Na⁺

Immunoglobulin

Lysozyme Kallikrein K⁺

Water (Salts) α-amylase Lipase Peroxidase

Water Neutral glycoproteins Sialomucins Sulphomucins

SEROMUCOUS CELL

Apical microvillus

Pinocytotic vesicle

Intercellular secretory canaliculus

Zonula occludens

Heterogeneous electron-dense secretory granules

Spherical nucleus

MUCOUS CELL

Zonula occludens

Homogeneous electron-translucent secretory vesicles

Flattened basal nucleus

8.51 Diagram of the architecture of a generalized salivary gland including ultrastructural details.

Solid and outlined black arrows indicate direction of transport.

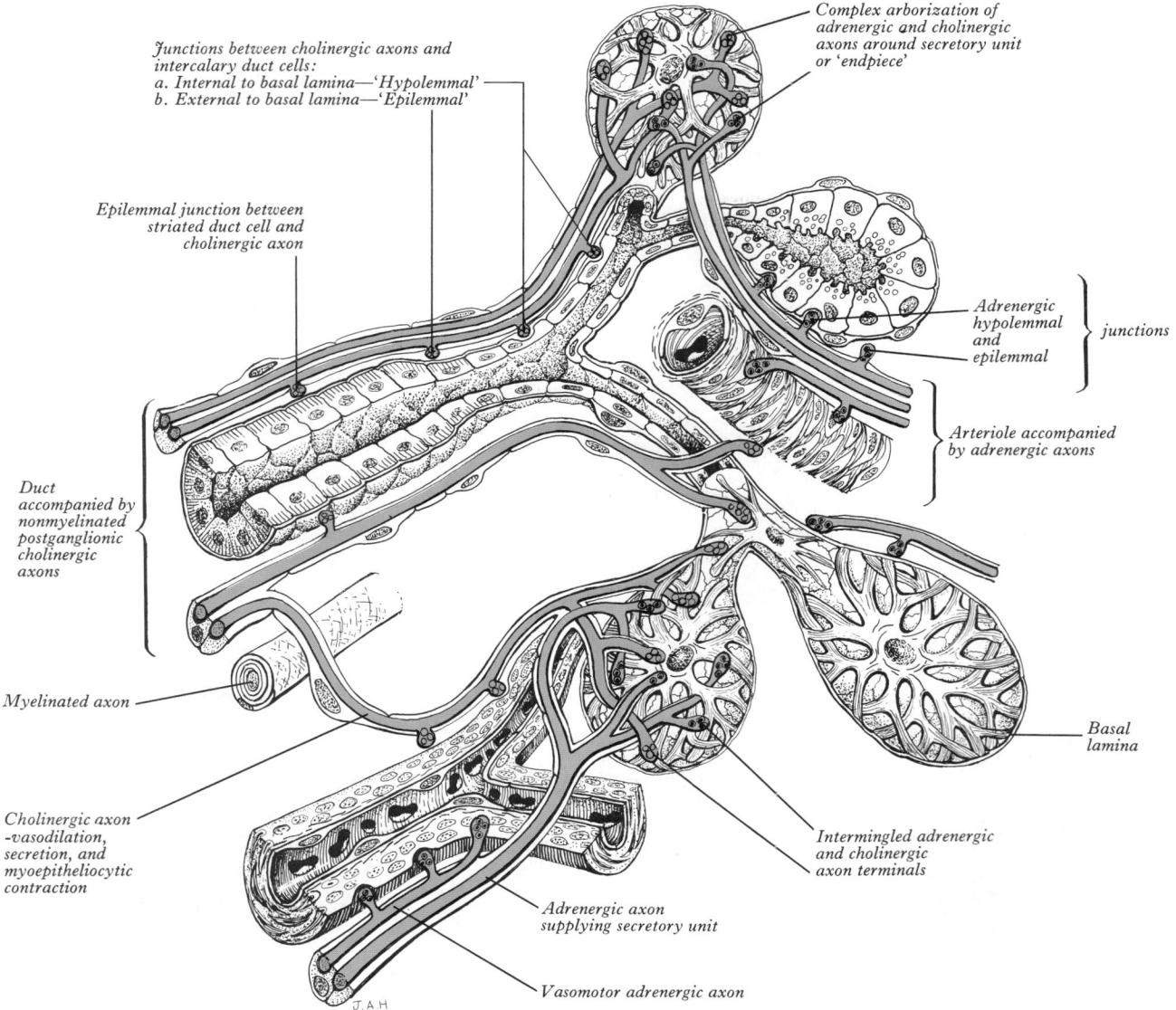

Junctions between cholinergic axons and intercalary duct cells:
a. Internal to basal lamina—'Hypolemmal'
b. External to basal lamina—'Epilemmal'

Complex arborization of adrenergic and cholinergic axons around secretory unit or 'endpiece'

Epilemmal junction between striated duct cell and cholinergic axon

Adrenergic hypolemmal and epilemmal } *junctions*

Duct accompanied by nonmyelinated postganglionic cholinergic axons

Arteriole accompanied by adrenergic axons

Myelinated axon

Basal lamina

Cholinergic axon -vasodilation, secretion, and myoepitheliocytic contraction

Intermingled adrenergic and cholinergic axon terminals

Adrenergic axon supplying secretory unit

J.A.H

Vasomotor adrenergic axon

8.52 The innervation of the ducts, secretory units and arterioles in a generalized salivary gland.

The cells lining the intercalated ducts are cuboidal or flat. Their cytoplasm contains long mitochondria, a few cisternae of granular endoplasmic reticulum, juxtanuclear Golgi complexes, lysosomes and secretory granules. While this suggests little participation in protein synthesis, it does not preclude involvement in the addition of water and electrolytes to saliva; the cells responsible for this, in either 'endpiece' or duct, are unknown but, although this function is often assigned to glandular cells, cells of the intercalated ducts may also be involved.

Cells lining striated ducts are basally striated, 'like a thick lawn' according to Pfluger (1866). In 1909 Regaud & Mawas concluded that this is due to columns of packed, elongate mitochondria, as confirmed by ultrastructural studies (Pease 1956). Between the columns the cytoplasm is divided into basal processes by infolded basal plasmalemma, the lateral aspects of which may be extensively plicated to interdigitate with adjacent processes. Desmosomes often link adjacent plications. This folding and local abundance of mitochondria is typical of many epithelial cells engaged in electrolyte transport, as these cells certainly are; they transport potassium into saliva and by reabsorbing sodium ions in excess of water they render saliva hypotonic (Thaysen et al 1954). The lateral plasmalemmae of cells lining striated ducts are linked by slightly pervious junctional complexes (Garrett & Parsons 1974). The luminal plasmalemma forms microvilli and cytoplasm often extends into the lumen as apical blebs considered to enter saliva by this apocrine process (Garrett 1976). The supranuclear

cytoplasm contains a few cisternae of granular endoplasmic reticulum, mitochondria, lysosomes, vesicles, microtubules and numerous microfilaments, the last conspicuous in the terminal web or 'separating zone' (Takano 1969). As well as modifying electrolyte composition, striated ducts secrete immunoglobulin A (Kraus & Mestecky 1971), lysozyme (Kraus & Mestecky 1971) and kallikrein (Garrett & Kidd 1975). Immunoglobulin A is produced by subepithelial plasma cells.

Few ultrastructural details of the excretory ducts have been recorded; they are lined in rats by simple columnar or pseudostratified epithelium, mainly of tall columnar cells with basal striations containing packed, elongate mitochondria (Tamarin & Sreebny 1965). These striations, and the more intimate relation to capillaries than elsewhere in the ductal system, suggest that excretory ducts are more than passive conduits; an involvement in electrolyte transport is possible (Young & van Lennep 1978).

Myoepitheliocytes of Salivary Glands

These contractile cells are associated not only with intercalated ducts and usually also with secretory endpieces, lying between the basal lamina and the epithelial cells proper, but also with intra- and extralobular ducts (Chaudhry et al 1987). Garrett & Emmelin (1979) summarized the effects of myoepithelial contraction as follows: outflow of saliva is accelerated, luminal volume of intercalated ducts and endpieces is reduced, secretory pressure is

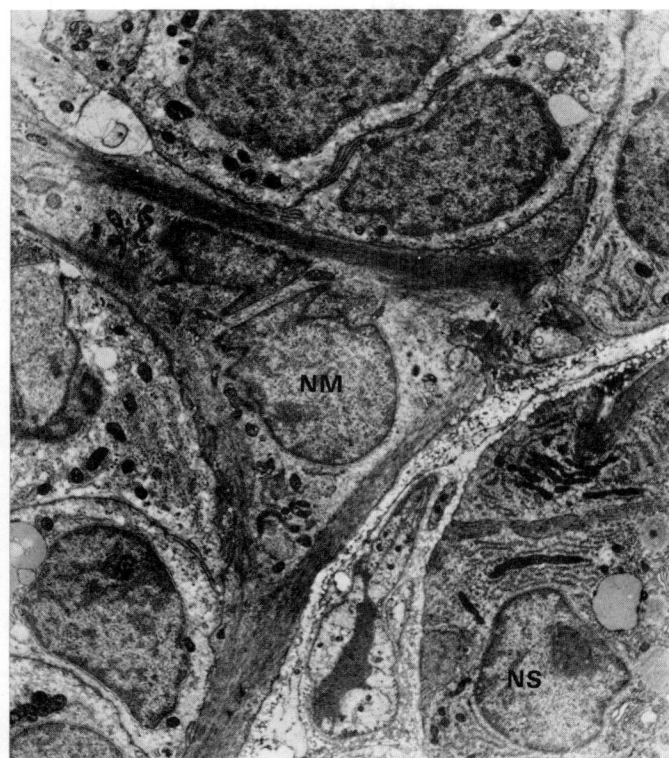

8.53 Transmission electron micrograph of a myoepitheliocyte of a salivary gland. The superficial filamentous and juxtanuclear non-filamentous compartments of the cytoplasm can be seen. NM = nucleus of myoepitheliocyte; NS = nucleus of secretory cell. Provided by R M Palmer, UMDS, Guy's Campus, London.

aided, underlying parenchyma supported, salivary flow is helped to overcome peripheral resistance and, in certain circumstances, discharge from associated secretory cells is aided.

The shape of salivary myoepitheliocytes depends on their location: in endpieces they are stellate, dendritic, with long overlapping processes which, with those of other such cells, form a reticulum around each endpiece. Those in ducts are fusiform, with fewer branches, and extend along the intercalated ducts longitudinally. 'Endpiece' myoepitheliocytes have a central perikaryon with four to eight radial processes, each with two or more successions of branches which cross but do not fuse or extend on to the ducts. In contrast, myoepitheliocytic processes in intercalated ducts seldom branch and often overlap on to the endpieces.

Myoepitheliocytic cytoplasm (8.53) may be divided into filamentous and non-filamentous compartments, the latter containing the nucleus, juxtanuclear Golgi complexes, lysosomal bodies and mitochondria. Globules of neutral fat may occur in human cells (Garrett 1963). The filaments, conspicuous in processes, and their rami resemble the myofilaments of non-striated myocytes (p. 546). Both thin (4 nm) and thick (10 nm) filaments are described, the former arranged longitudinally in processes, with the less numerous thick filaments scattered amongst them. Filaments often pass to attachment plaques on the basal plasmalemma, causing indentations. Basal lamina opposite the plaques appears thicker and may be linked to the cells at these sites. Bannerjee et al (1977) have shown that the basal lamina strengthens and supports the adjacent epithelium. When myoepitheliocytes contract, the basal lamina is probably tensed at the attachment plaques (Garrett & Emmelin 1979). Numerous caveolae are commonly associated with the stromal plasmalemma, but less frequently so with the plasmalemma adjacent to the epithelial cells. Myoepitheliocytes are linked to secretory and ductual cells by desmosomes and occasional cilia extend from them into indentations of the adjacent epithelial cells. Cilia were first observed by Tandler (1965) in myoepitheliocytes of human submandibular glands and subsequently described in other human salivary glands (Tandler et al 1970) and in glands of other species (Cutler & Chaudhry 1973, Kidd 1978). Garrett &

Emmelin (1979) suggest that there may be a cilium on each salivary myoepitheliocyte, as proposed by Stirling & Chandler (1976), for human mammary myoepitheliocytes (p. 1447). The cilia may have a chemoreceptive and/or mechanoreceptive role.

Salivary myoepitheliocytes appear to have sympathetic and parasympathetic innervation (Garrett 1972, 1976), with several axons of either or both supplying a single cell. But in the sublingual gland of the rat myoepitheliocytes may have only cholinergic innervation (Templeton & Thulin 1978). Stimulation by sympathetic or by parasympathetic nerves may result in myoepitheliocyte contraction (Garrett & Emmelin 1979).

CONTROL OF SALIVARY GLAND ACTIVITY

The observed wide and rapid variation in the composition, quantity and rate of salivary secretion in response to various stimuli suggests an elaborate control mechanism (Emmelin 1972). In some glands salivary secretion is spontaneous; in others secretion follows different types of sensory stimulation: gustatory, nociceptive, olfactory and tactile. Secretion may be continuous, but at a low resting level, partly spontaneous but mainly in response to the drying of the oral and pharyngeal mucosae. A rapid increase can be superimposed on the resting level, e.g. during mastication or by autonomic stimulation. Controlled variation in the activity of the many types of salivary effector cells (serous, as well as seromucous and mucous secretory cells, myoepitheliocytes, epithelial cells of all the ductual elements and the non-striated myocytes of local blood vessels) affects the quantity and quality of saliva. The control of these is both hormonal and neural.

Hormonal Control

The effects of circulating hormones on salivary secretion were reviewed by Blair-West et al (1967). There is no clear evidence that they evoke secretion directly at physiological levels, but they may alter the response of glandular cells to neural stimuli. Local hormones, however, have profound effects in feline submandibular glands; e.g. vasodilatation, though neurally initiated, is maintained by plasma-kinins formed locally when kallikrein is released from secretory cells stimulated by sympathetic amines (Gautvik et al 1972). Vasodilatation and secretion are functionally related (Hilton & Lewis 1956).

Neural Control

Most salivary glands, except those secreting spontaneously, depend on autonomic nerves to evoke secretion and in all salivary flow is mainly under nervous control. The nerves involved are cholinergic (parasympathetic) and adrenergic (sympathetic); there is no evidence for any other neural involvement (Garrett 1976).

The typical pattern of innervation is shown in Fig. 8.52, but this varies in different glands and species (Garrett 1972); variations may also occur with age (Yohro 1971). Only the more constant features are illustrated and described here. Cholinergic nerves often accompany ducts and arborize freely around secretory endpieces, but adrenergic nerves usually enter glands along arteries and ramify with them. The main glandular nerves contain largely non-myelinated axons, but a few myelinated axons occur, presumably either preganglionic efferent or afferent. Postganglionic efferent axons, like those elsewhere (Norberg 1967), show periodic dilatations containing mitochondria and vesicles, the latter electron-lucent in cholinergic axons and with electron-dense cores in adrenergic axons. Within the glands the nerve fibres intermingle, cholinergic and adrenergic axons often lying in the adjacent invaginations of one Schwann cell (Eneroth et al 1969, Garrett 1972, 1976).

At *neuro-effector junctions* (8.52) the synaptic regions of axons and the effector cells they supply are functionally related, the axonal surfaces closest to the effector cells being free of Schwann cell covering. Where effector cells are epithelial, the junctions may be *epilemmal* or *hypolemmal* (Garrett 1975), terms introduced by Arnstein (1889, 1895). At epilemmal sites axonal and effector

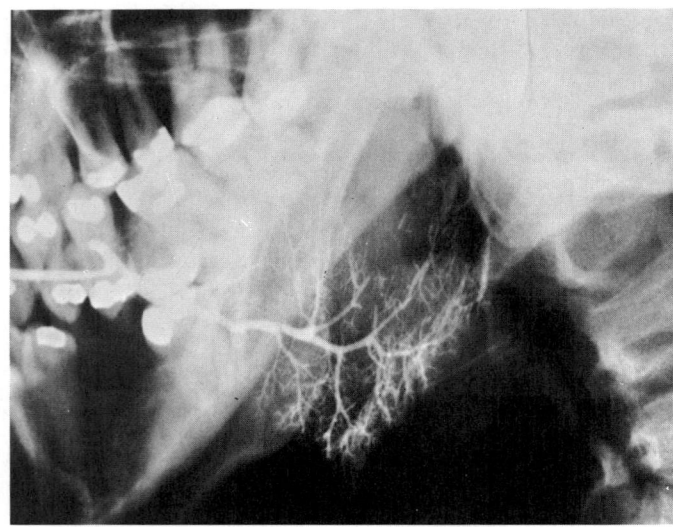

A

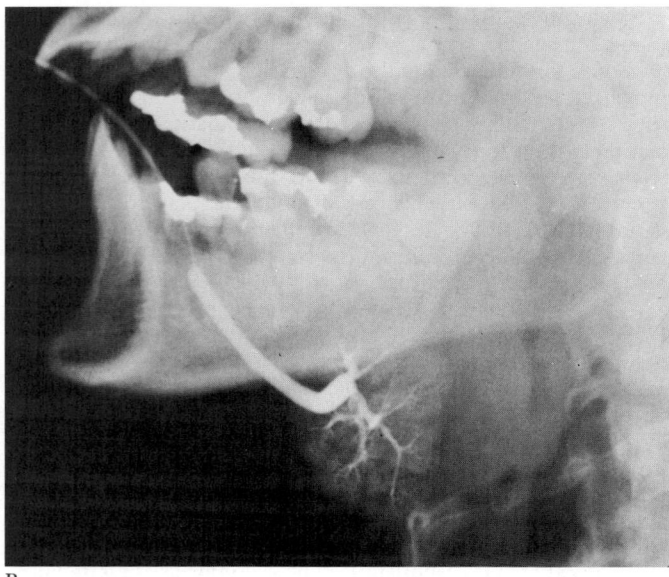

B

8.54 A. Parotid sialogram; B. Submandibular sialogram. In each case the shadow of the cannula used to introduce the radio-opaque medium into the duct of the gland is visible. See text for description.

surfaces are separated by about 100 nm, with the basal lamina intervening. At hypolemmal sites the axon penetrates the basal lamina and is separated from the effector cell by only 20 nm. One axon may supply several effector cells directly and many more indirectly through electrical coupling of adjacent cells (Lowenstein & Kanno 1964); group activity thus occurs. One effector cell may also receive several axons, both cholinergic and/or adrenergic. Single axons may act on several types of effector. Although there are separate sympathetic axons for secretion and vasoconstriction (Emmelin & Engström 1960), the former may also induce myoepitheliocyte contraction and a single parasympathetic axon may, through serial neuro-effector junctions of the *en passant* type (p. 882), induce vasodilatation, secretion and myoepitheliocytic contraction (Emmelin 1972).

Secretory endpieces usually have the most innervation, cholinergic and adrenergic, individual cells often having both. Cholinergic axons have long been accepted as the secretomotor innervation; however, in 1974 it was shown that in the parotid gland of the rat, at least, sympathetic nerves are also secretomotor (Harrop & Garrett 1974, Hodgson & Spiers 1974). Adrenergically evoked saliva differs in quantity and composition but by what mechanism is uncertain. In heterocrine glands (p. 1293) adrenergic and cholinergic nerves might activate different types of cell but in homocrine glands (p. 1293), where all cells appear similar, they presumably affect the same cells differently (Garrett 1972).

In some situations sympathetic activity may modify saliva produced in response to parasympathetic stimulation, rather than directly inducing flow.

The ductal elements of salivary glands can markedly modify saliva (p. 1296) and, though less intensely innervated than secretory endpieces, their activity is also under neural influences, in part at least. In 1958 Lundberg showed that cells, assumed to be ductual, in the feline submaxillary gland, responded electrically to both parasympathetic *and* sympathetic stimulation. Cholinergic fibres lie adjacent to the striated ducts of most species, occasionally hypolemmal in position, but more commonly so in intercalary ducts. In some species, including mankind, adrenergic nerves are also associated with striated ducts (Garrett 1967). The main excretory ducts appear to have only cholinergic nerves but Schneyer (1976) observed that sympathetic stimulation alters electrolyte transport across the epithelium of the submaxillary main duct in the rat, suggesting an adrenergic supply.

The innervation of myoepitheliocytes adjoining secretory endpieces and intercalary ducts is physiologically obscure, but electron microscopy suggests a sympathetic and parasympathetic hypolemmal supply (Kagayama & Nishiyama 1972). Myoepitheliocytes are stimulated to contract by adrenergic axons; they may respond to a single impulse, suggesting high sensitivity. The role of cholinergic axons is less certain, but they also may cause contraction (Garrett 1975), thus aiding salivation.

Structural evidence shows that salivary arterioles are innervated by both adrenergic and cholinergic axons (Young & van Lennep 1978). The former, the more numerous, maintain vasoconstrictor tone; the latter may induce vasodilatation but this is maintained by local plasma-kinins (vide supra).

Little information is available on the afferent nerves of salivary glands. Pain due to obstruction of salivary ducts and sialography suggests a nociceptive function, but this awaits anatomical investigation. Sensory endings occur in the main ducts and presumably elsewhere in the glands. Afferent axons are postulated to occur in the main parasympathetic and sympathetic nerve trunks to the glands. Increasing pressure in the submandibular ducts in dogs enhances afferent activity in the chorda typani (Garrett 1975); intraglandular baroceptors are presumed to be involved in this response. Detailed studies of salivary sensory innervation are clearly needed.

Accessory glands. Besides the main salivary glands many others exist: some in the tongue (p. 1322), others around and in the palatine tonsil between its crypts, and large numbers in the soft palate, the posterior part of the hard palate, the lips and cheeks. These are similar in structure to larger salivary glands and are mainly of the mucous type (p. 1293).

Sialography. Cannulae can be introduced into the parotid and submandibular ducts and used to inject radio-opaque substances (e.g. lipiodol) to outline the ramifications of the ductal systems of these glands. The pattern and calibres of the components of these systems are thus revealed. The *parotid duct*, as seen in lateral sialograms, is formed near the centre of the posterior border of the mandibular ramus by the union of two ducts which respectively ascend and descend at right angles to its course (8.54A). As it crosses the face, it receives from above five or six ductules from the accessory parotid gland; as it curves round the anterior border of the masseter it is often compressed, its shadow being attenuated here. The intraglandular part of the main duct receives an alternating series of descending and ascending tributaries, each formed from an arborization of fine ductules receiving acini. The acini usually do not show as dilatations in sialograms but are represented by the 'free' endings of the smallest ducts. The *submandibular duct* starts from that gland's lowest part, below the mandible in lateral views, ascends vertically above the mandible's lower border and turns sharply forwards, gradually ascending to its opening. The duct's vertical part receives anterior and posterior tributaries and, as it turns sharply forwards, it receives a large tributary from the posterior region (8.54B). Each tributary is formed from ductules (with their terminal acini) visualized as in the parotid gland. Contrast medium injected into the submandibular duct may also enter the major sublingual duct, revealing the ductules of the anterior part of the sublingual gland.

THE TEETH

Introduction

Teeth are vital to most mammals and other vertebrates, except mankind. Loss is incompatible with life; longevity is related to the endurance of the dentition under the abrasive process of mastication. In non-mammalian vertebrates, teeth are constantly replaced, a condition known as *polyphyodonty*, related to the need for successively larger teeth in animals which grow throughout life. Mammals of limited growth typically have two dentitions, *deciduous* and *permanent*, the condition of *diphyodonty*, although some, e.g. the rat, are *monophyodont*. The emergence and success of diphyodonty was probably related to the evolution of occlusion during mastication.

In non-mammalian vertebrates, the jaw joint is between the *quadrate* bone of the upper and the *articular* bone of the lower jaw, homologous respectively with the incus and malleus. The lower teeth are set on a curve so far inside the upper that, due also to restricted lateral movement, they cannot meet the upper teeth. The evolution of mammals was associated with the posterosuperior growth of the *dentary* (one of several lower jaw bones existing in all non-mammalian vertebrates) towards the *squamosal bone*, homologous with the squamous temporal in most mammals. In accord with these skeletal changes, jaw muscles were also rearranged to move the mandible (dentary) transversely. Together with these trends there was a change in dental shape; from the simple conical teeth of reptilian ancestors, mammals evolved teeth with complex shearing planes. Lateral movements now possible in the new mandible allowed the lower teeth to grind across the upper, producing a more effective trituration. Although deficiencies in the fossil record make some details uncertain, the principal events in this evolutionary sequence are known and the cusps of human molars can be homologized with those of early mammals (**8.55**).

It has been argued that in reptiles ancestral to mammals the articular and quadrate bones of the jaw joint conducted sound vibrations to the stapes, received by a tympanum partly attached to the *angular bone* of the lower jaw (Allin 1976). Backward growth of the dentary had the selective advantage of reducing the mass of these post-dentary bones, which vibrated with the tympanum, thereby enhancing their auditory function. Finally, with evolution of a new jaw joint, the angular bone with its tympanum moved to the cranium, becoming the tympanic part of the temporal bone. The quadrate and articular bones remained in contact with each other and with the stapes, forming the incus and malleus. The three auditory ossicles then became encompassed by a definitive tympanic cavity. Evolution of a single-boned mammalian mandible was thus associated with an increasingly efficient auditory apparatus.

After a few months of active use, newly erupted mammalian teeth are worn to produce precisely matching upper and lower shearing edges. Continued eruption of new teeth would constantly disrupt this relationship and therefore be disadvantageous. However, because of the need to accommodate teeth in small, young jaws, a deciduous dentition is an almost universal requirement in mammals. With reduction of replacement, dental tissues evolved to minimize the effects of wear. Thus a harder, thicker *prismatic enamel* emerged in mammals to replace the thinner *non-prismatic enamel* of reptiles.

The teeth in all reptiles, apart from Crocodilia, are rigidly bound to the jaws, but in mammals each is suspended in a socket by a *periodontal ligament*. This allows further eruption and a degree of migration, impossible for ankylosed reptilian teeth, to compensate for wear. Innervation of the periodontal tissues also provides a more comprehensive flow of proprioceptive data to nervous centres concerned with control of masticatory patterns.

The introduction of refined carbohydrates made human teeth susceptible to caries and periodontal disease. Outside human culture such conditions would probably have led to extinction. But by appropriately preparing and in a sense predigesting food,

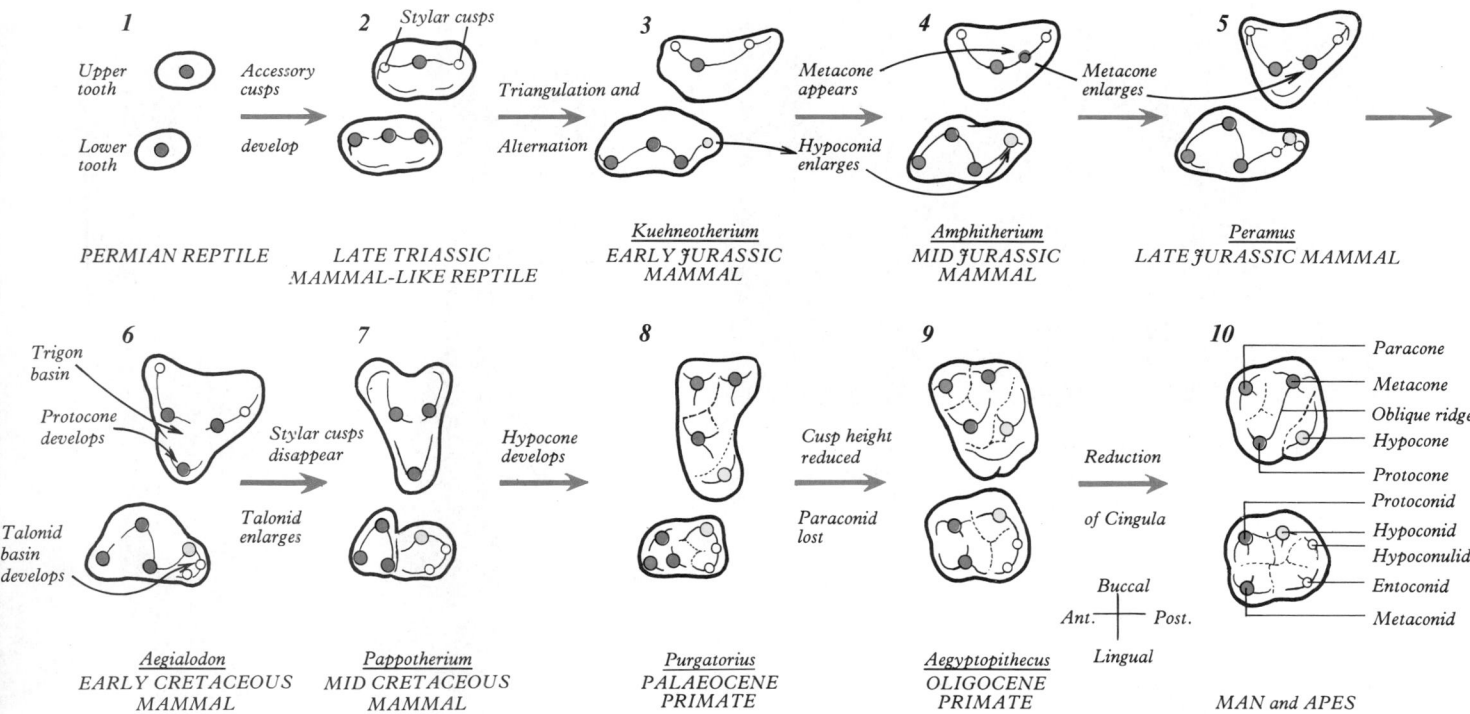

PERMIAN REPTILE   LATE TRIASSIC MAMMAL-LIKE REPTILE

Kuehneotherium
EARLY JURASSIC MAMMAL

Amphitherium
MID JURASSIC MAMMAL

Peramus
LATE JURASSIC MAMMAL

Aegialodon
EARLY CRETACEOUS MAMMAL

Pappotherium
MID CRETACEOUS MAMMAL

Purgatorius
PALAEOCENE PRIMATE

Aegyptopithecus
OLIGOCENE PRIMATE

MAN and APES

KEY • PARACONE (upper tooth) PARACONID (lower tooth)
     • METACONE (upper tooth) METACONID (lower tooth)
     • PROTOCONE (upper tooth) PROTOCONID (lower tooth)
     • HYPOCONE (upper tooth) HYPOCONID (lower tooth)

8.55 Occlusal views of the left upper and right lower teeth showing a series of steps in the evolution of complex molar occlusion. Large cusps (coloured) and smaller cusps are joined together by raised cutting edges. The cutting edges of opposing teeth pass each other during occlusion and certain cusps (e.g. the protocone and hypoconid) then crush into opposing basins (e.g. talonid and trigon basins).

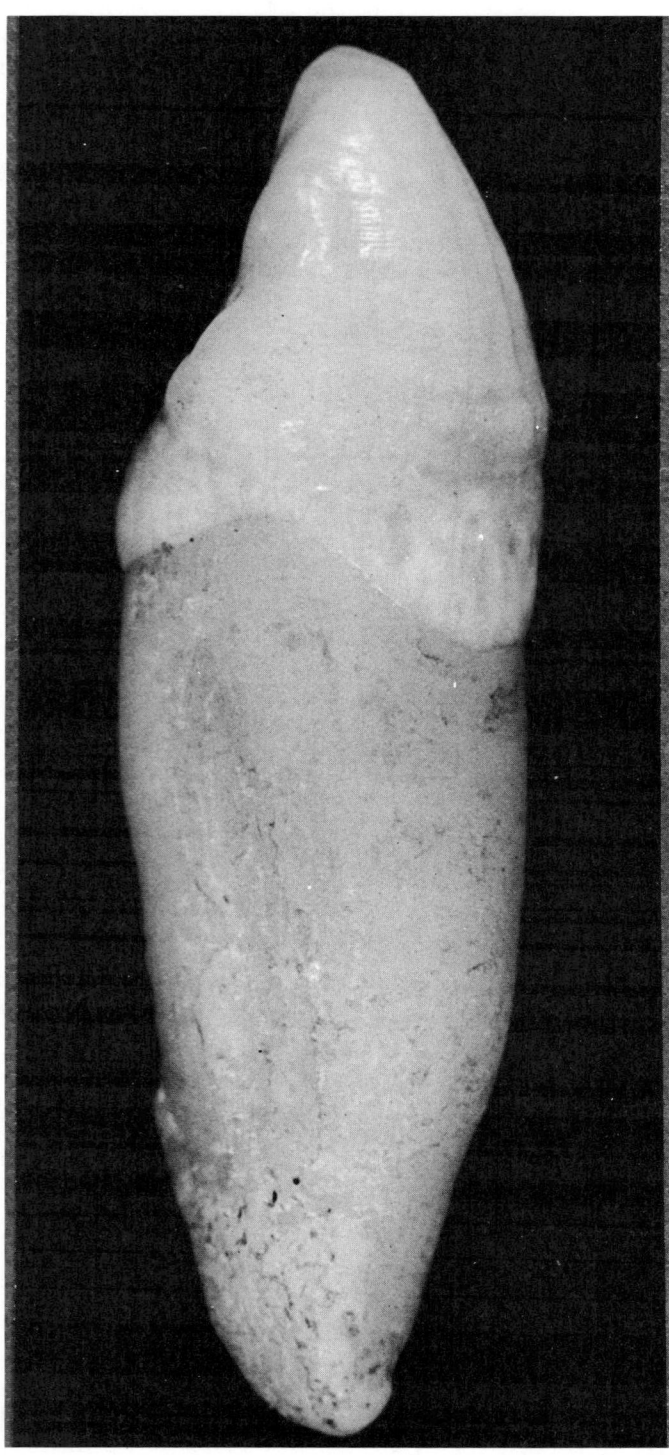

8.56 An extracted upper right canine tooth viewed from its mesial aspect. Note the root covered by cement and the shiny enamel-covered crown. The sinuous cervical margin projects towards the cusp of the tooth both mesially and distally.

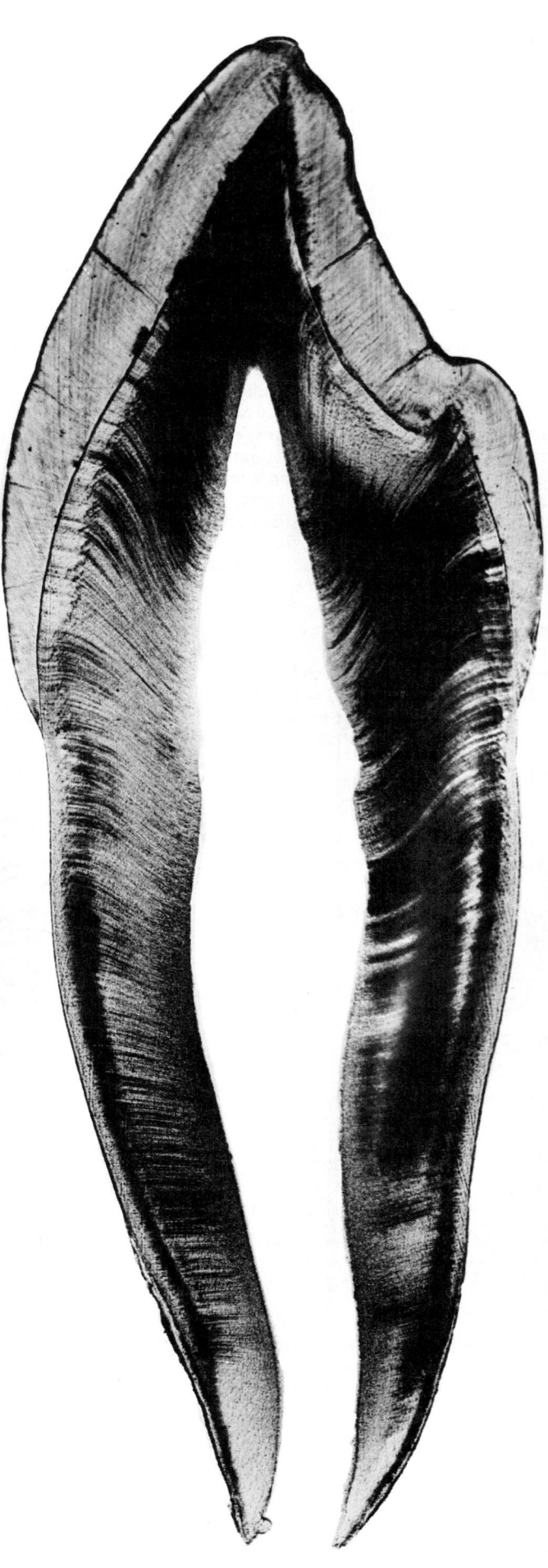

8.57 A bucco-lingual longitudinal ground section of a young lower first premolar tooth, photographed with transmitted light. Note the large buccal and small lingual cusps; the enamel tapers to a knife-edge at the cervical margin. The coarse dark lines perpendicular to the enamel surface are cracks produced during grinding; the fine lines parallel to the cracks indicate the long axes of the enamel prisms; the oblique lines which curve towards the apex of the cusp are the striae of Retzius (the incremental lines of enamel, compare with **8**.70). The primary (S-shaped) curvatures of the coronal dentinal tubules become less pronounced in the root. A thin layer of cement covers the dentine. The apical foramen is wide open and the pulp cavity is large, as in this young tooth.

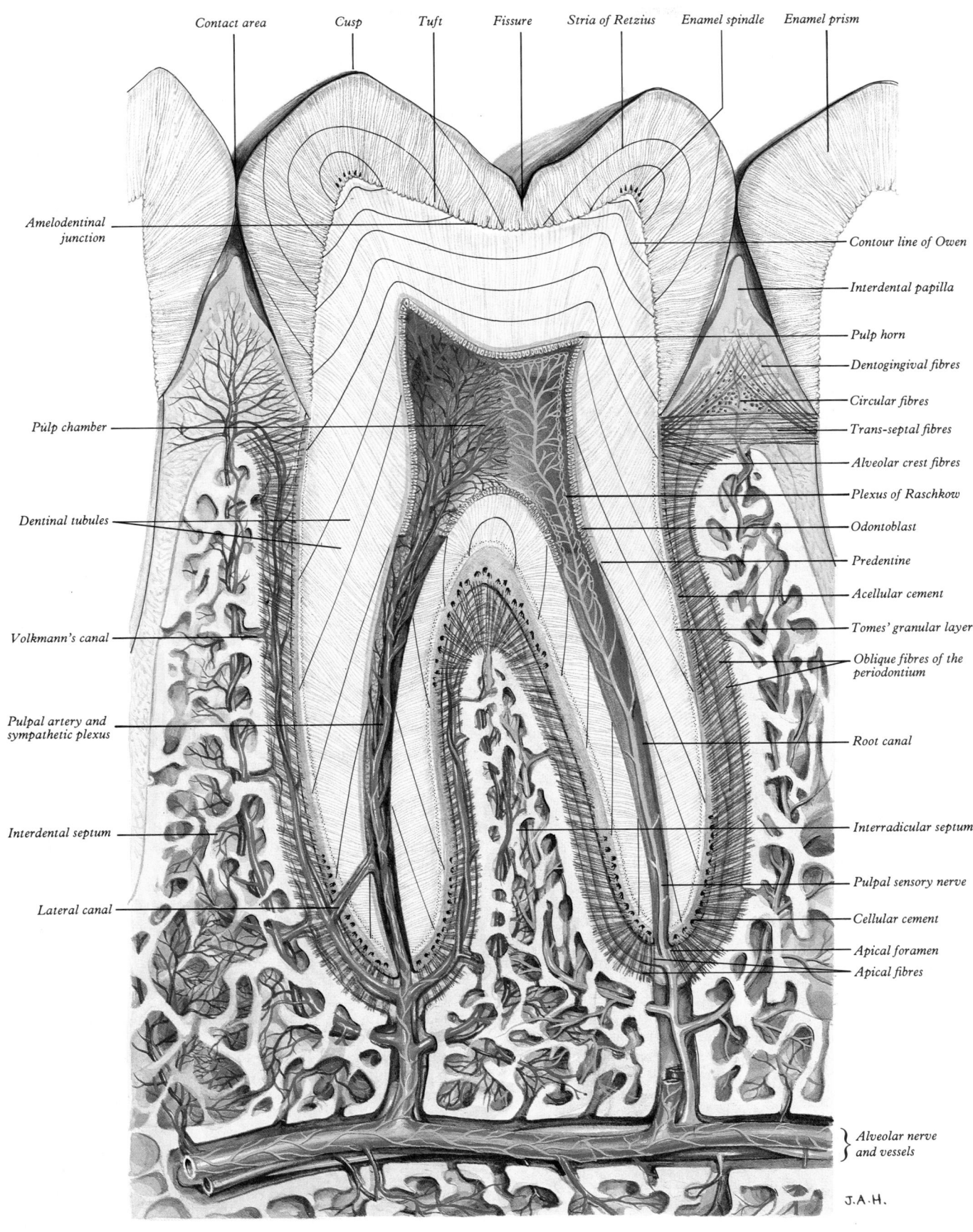

Contact area   Cusp   Tuft   Fissure   Stria of Retzius   Enamel spindle   Enamel prism

Amelodentinal junction

Contour line of Owen

Interdental papilla

Pulp horn

Dentogingival fibres

Circular fibres

Trans-septal fibres

Alveolar crest fibres

Plexus of Raschkow

Odontoblast

Predentine

Acellular cement

Tomes' granular layer

Oblique fibres of the periodontium

Pulp chamber

Dentinal tubules

Volkmann's canal

Pulpal artery and sympathetic plexus

Root canal

Interradicular septum

Interdental septum

Pulpal sensory nerve

Cellular cement

Apical foramen

Apical fibres

Lateral canal

Alveolar nerve and vessels

J.A.H.

8.58 Diagram of a longitudinal section of a tooth and its environs.

mankind has overcome the problem of passing suitably fragmented material to the action of catabolic enzymes in the alimentary canal. Nevertheless, chewing does facilitate the digestion of most foods, including cooked meat and vegetables (Farrell 1956) and the natural dentition comminutes food much more efficiently than an artificial replacement (Lucas et al 1986). However, teeth are no longer vital to survival and therefore selective pressure leading to further evolutionary change in the human detition will be limited.

Because they are the hardest and most stable of tissues, teeth are selectively preserved and fossilized, providing by far the best evolutionary record. Hence teeth are excellent models for studying relations between ontogeny and phylogeny. In modern societies, the durability of teeth to fire and bacterial decomposition makes them invaluable in identification of otherwise unrecognizable bodies, a point of great forensic importance.

General Arrangement of Dental Tissues

A tooth (8.56) consists of a crown, covered by very hard translucent *enamel* and a root covered by yellowish bone-like *cement*. These meet at the neck or *cervical margin*. A longitudinal section (8.57, 58) reveals that a tooth is mostly *dentine* (ivory) with an enamel covering about 1.5 mm thick, while the cement is usually much thinner. The dentine contains a central *pulp cavity*, expanded at its coronal end into a *pulp chamber* and narrowed in the root as a *pulp canal*, opening at or near its tip by an *apical foramen*, occasionally multiple. The root is surrounded by *alveolar bone*, its cement separated from the osseous socket (*alveolus*) by the soft *periodontal ligament*, about 0.2 mm thick. Coarse bundles of collagen fibres, embedded at one end in cement, cross the periodontal ligament to enter the osseous alveolar wall. In most non-mammalian vertebrates (vide supra) teeth are rigidly connected (ankylosed) directly to bone, a rather brittle attachment. Only in mammals (and Crocodilia) does a periodontal ligament provide an independent, tough suspension for each tooth. Near the cervical margin, the tooth, periodontal ligament and adjacent bone are covered by the *gingiva* (gum), clearly recognizable in health by its pale pink, stippled appearance (8.59). This is continuous at the *mucogingival junction* with the red, smooth oral mucosa lining much of the oral cavity and is adherent to the tooth near the cervical margin by an *epithelial attachment*. The pulp is a connective tissue, continuous with the periodontal ligament via the apical foramen. It contains vessels for the support of the dentine and sensory nerves.

Dental Morphology

The curvature of the dental arches renders the terms of descriptive anatomy, such as anterior and posterior, inappropriate. The aspect of teeth adjacent to lips or cheeks is therefore termed *labial* or *buccal*, that adjacent to the tongue being *lingual* or *palatal*. Labial and lingual surfaces of an incisor meet medially at a *mesial* surface and laterally at a *distal* surface, terms also used to describe equivalent surfaces of premolar and molar (*postcanine*) teeth (8.62, 63). Mesial surfaces of postcanine teeth are, of course, directed anteriorly and distal surfaces posteriorly. Thus the point of contact between the central incisors is the datum point for *mesial* and *distal*. The biting or *occlusal* surfaces of postcanine teeth are tuberculated by *cusps* separated by *fissures* forming a pattern characteristic of each tooth. The biting surface of an incisor is the *incisal edge*.

THE PERMANENT TEETH

The names for teeth in all mammals are based on appearance, function or position of the equivalent human teeth: they are *incisors, canines, premolars* and *molars* (8.60, 61, 62, 63).

There are two *incisors*, central and lateral, in each half jaw or *quadrant*. In labial aspect, the crowns are trapezoid, the maxillary, particularly the central incisor, being larger than the mandibular. The incisal edges originally have three small tubercles or *mam-

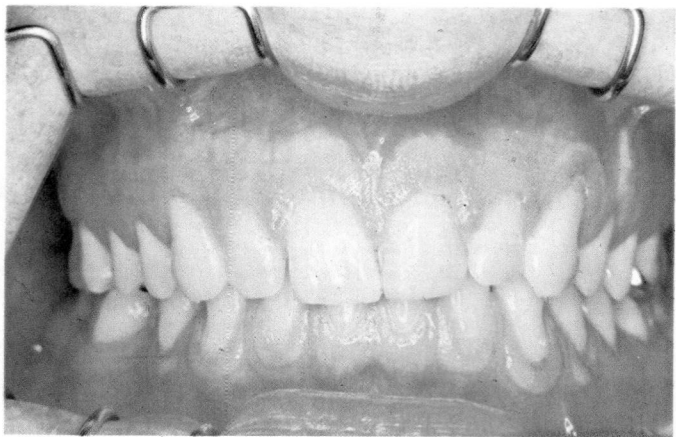

8.59 Anterior view of the dentition in centric occlusion, with the lips retracted. Note the pale pink, stippled gingivae and the red, shiny, smooth alveolar mucosa. The degree of overbite is rather pronounced and the gingiva and its epithelial attachment have receded on to the root of the upper left canine.

melons which are rapidly removed by wear. In mesial or distal view their labial profiles are convex; lingual surfaces are concavo-convex (ogival); the convexity near the cervical margin is due to a low horizontal ridge or *cingulum*, prominent only on upper incisors. The roots of incisors are single, being rounded in maxillary and flattened mesiodistally in mandibular teeth. The maxillary lateral incisor fails to develop in about 1% of people (Brothwell et al 1963) or it may be peg-shaped.

Distal to each lateral incisor is a rather larger *canine* with a single cusp (hence the term *cuspid*) instead of an incisal edge. Apart from this, canines resemble incisors, particularly when their cusps are worn flat. The canine root is the longest in the jaws and produces a bulge (*canine eminence*) on the bone externally, particularly of the maxilla. Although canines usually have single roots, triangular in cross section, that of the lower may be bifid (Kraus et al 1969).

Distal to the canines are two *premolars* each with a buccal and lingual cusp (hence the term *bicuspid*). Occlusal surfaces of upper premolars are oval (the long axis is buccopalatal) with a mesiodistal fissure separating the two cusps. In buccal view, premolars resemble canines but are smaller. The *upper first premolar* usually has two roots (one buccal, one palatal) but may have one and very rarely three roots (two buccal and one palatal). The upper second premolar usually has one root. The occlusal surfaces of lower premolars are more circular or square than those of the uppers. The buccal cusp of the *lower first premolar* towers above the lingual to which it is connected by a ridge separating mesial and distal occlusal pits. In the *lower second premolar* a mesiodistal fissure usually separates a buccal from two smaller lingual cusps. Each lower premolar has one root, but very rarely, the root of the first is bifid. Lower second premolars fail to develop in about 2% of individuals (Garn & Lewis 1962).

Posterior to the premolars are three *molars* whose size decreases distally; each has a large rhomboid (upper jaw) or rectangular (lower jaw) occlusal surface with four or five cusps. The *upper first molar* has a cusp at each corner of its occlusal surface and the mesiopalatal cusp is connected to the distobuccal by an *oblique ridge*, a primitive feature shared with many lower primates. A fifth cusp, the *cusp of Carabelli*, may appear on the mesiopalatal aspect, most commonly in Caucasoid races (Kraus 1959, Alvesalo et al 1975). The tooth has three widely separated roots, two buccal and one palatal. The smaller *upper second molar* may have a disto-palatal cusp which is reduced or occasionally absent; its roots are divergent and two may be fused. The *upper third molar*, the smallest, usually has three cusps (the distopalatal being absent) and commonly one root. The *lower first molar* has three buccal and two lingual cusps on its rectangular occlusal surface, the smallest being distobuccal. The cusps are all separated by fissures. It has two widely separated roots, one mesial and one distal. The smaller *lower second molar* is like the first but usually lacks the distobuccal

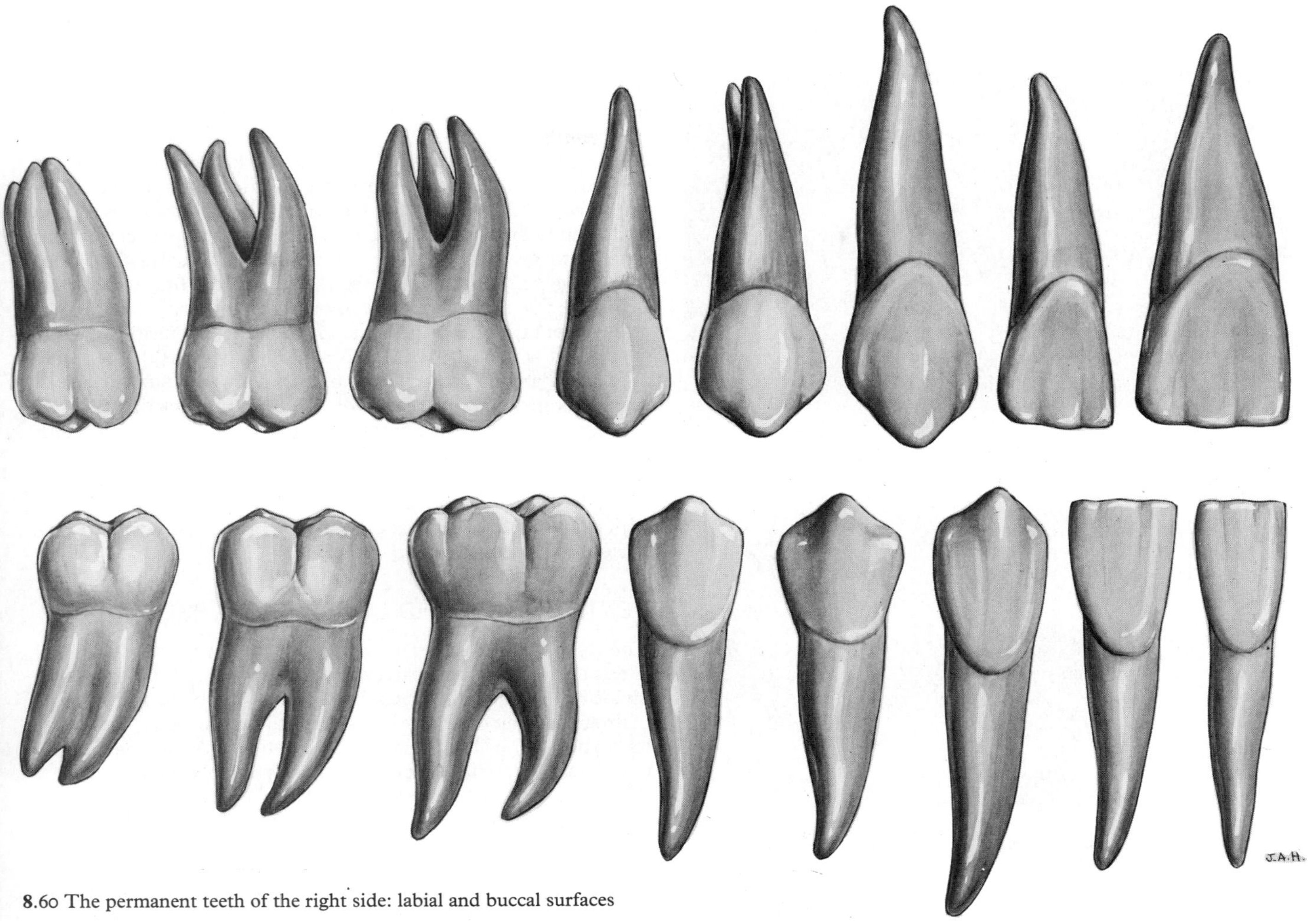

8.60 The permanent teeth of the right side: labial and buccal surfaces

cusp and its roots are closer together. The *lower third molar* is smaller still and like the upper third molar it is variable in form. Its crown may resemble that of the lower first or second molar and its roots are frequently fused. Because it erupts anterosuperiorly it is often impacted against the second molar whereas the upper third molar erupts postero-inferiorly and is rarely impacted. In various populations, one or more third molars, upper or lower, fail to develop in 0.2–25% of individuals (Brothwell et al 1963). In general, absence of third molars is commoner in Mongoloid and Caucasoid than in Negroid races.

THE DECIDUOUS TEETH

The incisors, canine and premolars of the permanent dentition (**8.60**) replace two deciduous incisors, a deciduous canine and two deciduous molars in each jaw quadrant (**8.61**). The deciduous incisors and canine are shaped like their successors but are smaller and whiter and become extremely worn in older children. The deciduous molars resemble permanent ones rather than their successors, the premolars. Each second deciduous molar has a crown almost identical to that of the posteriorly adjacent first permanent molar. The *upper first deciduous molar* has a triangular occlusal surface (its rounded 'apex' being palatal) and a fissure separates a double buccal cusp from the palatal cusp. The *lower first deciduous molar* is long and narrow; its two buccal cusps are separated from the two lingual cusps by a zigzagging mesiodistal fissure. Like permanent molars, upper deciduous molars have three roots and lower deciduous molars have two roots; these diverge more than those of permanent teeth since each developing premolar is accommodated directly under the crown of its deciduous predecessor. The roots of deciduous teeth are progressively resorbed by osteoclasts prior to being shed. An extracted deciduous tooth may thus have very short roots.

DENTAL OCCLUSION

It is possible to bring the jaws together so that the teeth meet or *occlude* in many positions (Kraus et al 1969). When opposing occlusal surfaces meet with maximal '*intercuspation*' (i.e. maximum contact), the teeth are said to be in *centric occlusion* (**8.64, 65**). In this position the lower teeth are normally opposed symmetrically and lingually with respect to the upper.

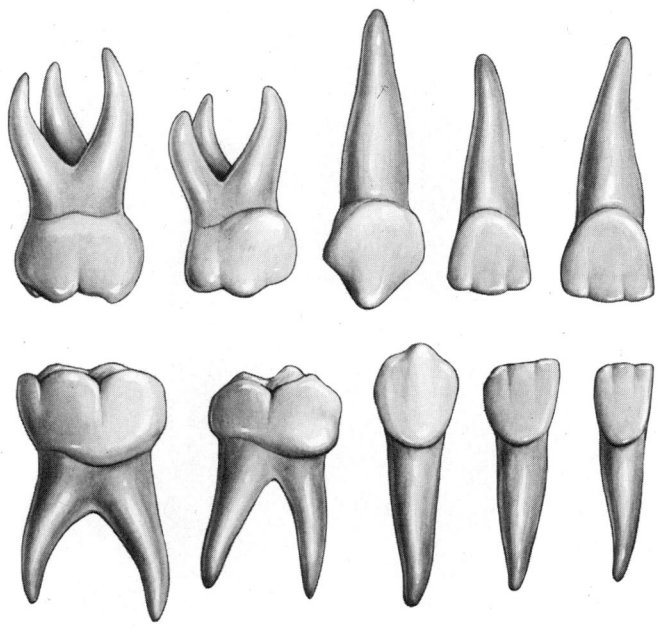

8.61 The deciduous teeth of the right side: labial and buccal surfaces.

1303

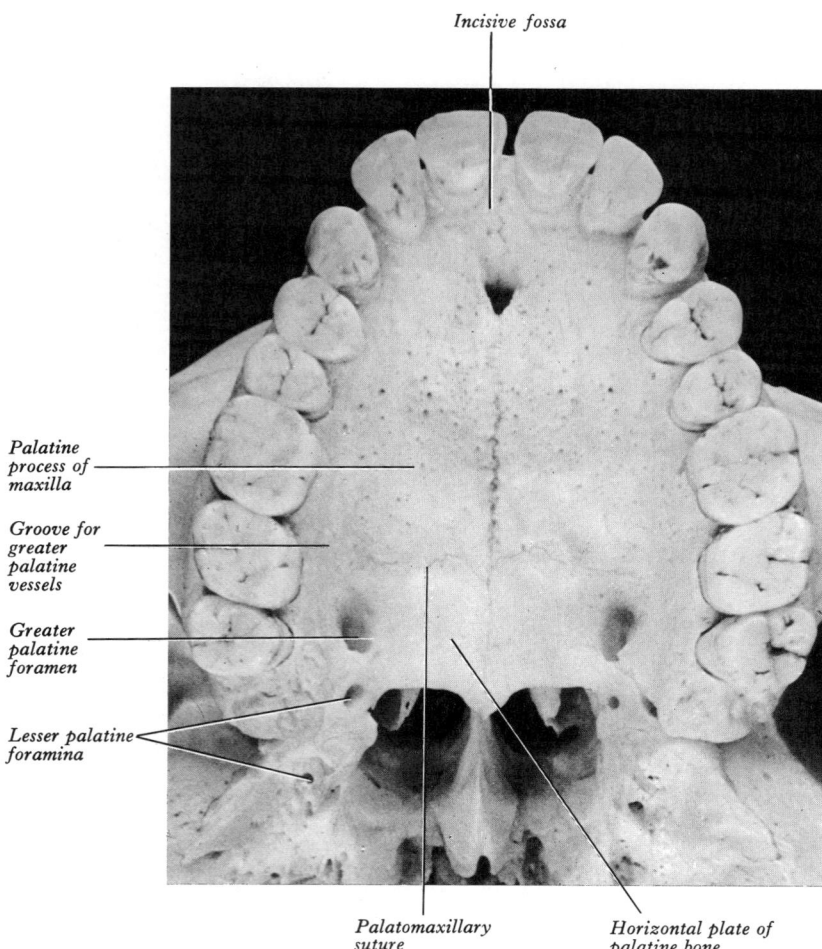

Incisive fossa

Palatine process of maxilla

Groove for greater palatine vessels

Greater palatine foramen

Lesser palatine foramina

Palatomaxillary suture

Horizontal plate of palatine bone

8.62 The permanent teeth of the upper dental arch: inferior aspect.

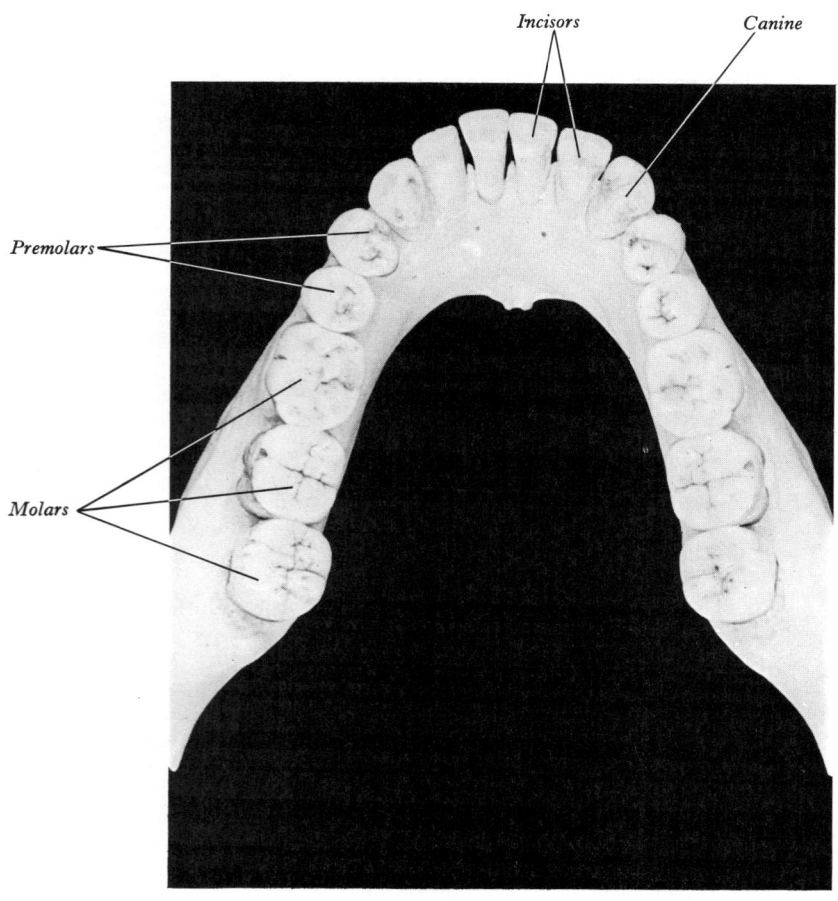

Incisors

Canine

Premolars

Molars

1304   **8.63** The permanent teeth of the lower dental arch: superior aspect.

Some important features of centric occlusion in a normal dentition must be noted. Each lower postcanine tooth is slightly in front of its upper equivalent and the lower canine is in front of the upper. Buccal cusps of the lower postcanine teeth lie between the buccal and palatal cusps of the upper teeth. Thus the lower postcanine teeth are slightly lingual and mesial to their upper equivalents. Lower incisors bite against the lingual surfaces of upper incisors, the latter normally obscuring about one-third of the crowns of the lower. This vertical overlap of incisors in centric occlusion is the *overbite*. The extent to which upper incisors are anterior to lowers is the *overjet*. In the most habitual jaw position, the *resting posture*, the teeth are slightly apart, the gap between being the *free-way space* or *interocclusal clearance*.

Each dental arch is approximately *catenary*, the form of a chain suspended at both ends (MacConaill & Scher 1949), the lower arch being slightly narrower (**8.62,63**). Viewed from the side, a line joining the buccal cusps of the upper postcanine teeth is curved (*curve of Spee*), concave upwards. The lower molar teeth are tilted slightly lingually so that a line joining the buccal and lingual cusps of the left and right lower first molars is curved (*curve of Monson*), concave upwards. These curvatures accord with movements of the mandible during mastication and are important in the construction of dentures.

DENTAL BLOOD AND LYMPHATIC VESSELS

The *inferior alveolar artery*, a branch of the maxillary artery, enters the mandibular foramen and travels forwards in its canal to divide into *incisive* and *mental* branches to supply the lower teeth, their supporting structures and the mandibular body, including its cortical bone (Saunders & Rockert 1967). About eight to 12 main rami and variable finer ones supply alveolar bone and teeth (Castelli 1963). Few anastomotic vessels cross the symphysis (Howkins 1963). Veins from alveolar bone and teeth collect either into larger vessels in the interdental septa or into plexuses around the dental apices and thence into several *inferior alveolar veins*; some of these drain through the mental foramen to the facial vein, others via the mandibular foramen to the pterygoid venous plexus (Cohen 1959). The upper jaw is supplied by *anterior* and *posterior superior alveolar arteries*. The latter, from the maxillary artery, ramify over the maxillary tuberosity, supplying alveolar bone, mucosa and teeth in the molar region and adjacent buccal mucosa, where they anastomose with the penetrating branches of the facial artery. Other rami supply the lateral wall of the maxillary sinus. The *anterior superior alveolar artery*, a branch of the infraorbital, curves through the *canalis sinuosus* (Wood Jones 1939), which swerves laterally from the infraorbital canal and inferomedially below it in the wall of the maxillary sinus, following the rim of the anterior nasal aperture, between the alveoli of canine and incisor teeth and the nasal cavity; it ends near the nasal septum where its termination emerges. The canal may be up to 55 mm long. Occasionally a small *middle superior alveolar artery* forms anastomotic arcades with the anterior and posterior vessels. On the palatal aspect of the upper teeth, the *greater palatine artery* supplies the palatal gingiva and its terminal branch ascends through the incisive canal to anastomose with septal branches of the nasopalatine artery. Veins accompanying the superior alveolar arteries drain anteriorly into the facial vein, or posteriorly into the pterygoid venous plexus.

The periodontal ligaments are supplied by *dental branches* of alveolar arteries (Melcher & Bowen 1969). One branch enters the alveolus apically and, of its small rami, two or three pass into the dental pulp through the apical foramen, others ascending in the periodontal ligament. *Interdental arteries* ascend in the interdental septa, sending branches at right angles into the ligament, and terminate by communicating with gingival vessels. Thus the ligament receives its blood from three sources: from the apical region, ascending interdental arteries and descending vessels from the gingiva; all anastomose with each other. Veins drain the periodontal ligament either into the *interdental veins* or into the *periapical plexus*. Longer vessels seen in the ligament are probably veins rather than anastomosing arteries (Folke & Stallard 1967).

Lymphatic drainage of human jaws and teeth is uncertain (Saunders & Rockert 1967). Injection techniques in monkeys

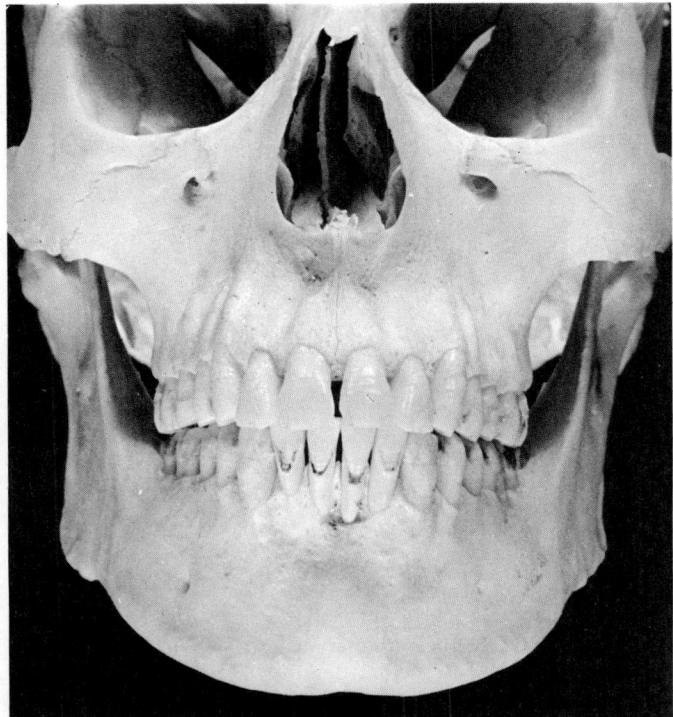

8.64 Anterior view of the dentition in centric occlusion. There has been some resorption of bone around the lower incisors.

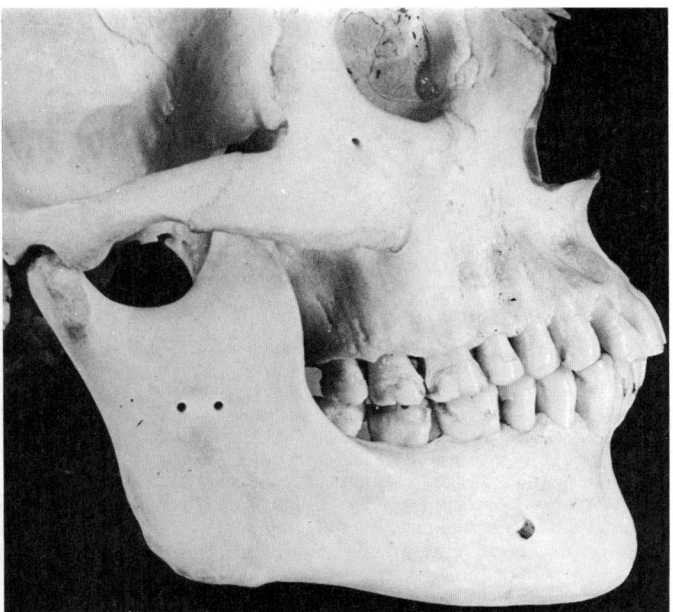

8.65 Lateral view of the dentition in centric occlusion.

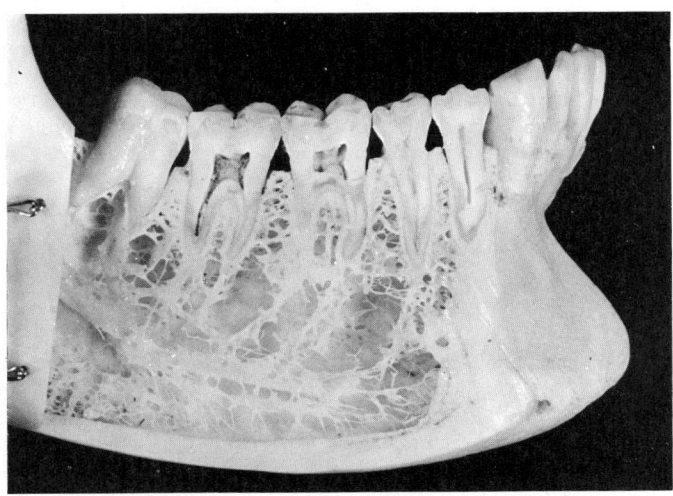

8.66 Vertical section through the right half of the body of the mandible and its dentition. Note: (1) the pulp cavities in the molar teeth; (2) the flat table of bone surmounting the interdental bony septa; (3) the cancellous nature of much of the bone; (4) the cortical plate of compact bone lining the sockets of the teeth (the lamina dura of radiographs, see **8.85,86,87**); (5) the compact bone forming the base of the mandible; (6) the inferior dental canal which, in this specimen, is widely separated from the roots of the teeth.

(MacGregor 1936) suggest that the upper jaw drains mainly to the submandibular and thence to supraclavicular lymph nodes, the lower to submental and on to the paratracheal nodes. Many dental abscesses lead to enlargement of the submandibular and upper deep cervical lymph nodes, indicating a common path for lymphatic drainage of the upper and lower teeth. Buccal lymph nodes may be affected by infection of the upper teeth. Lower incisors drain to the submental nodes and thence either to the submandibular or lower deep cervical nodes. It is presumed that alveolar bone, periodontal ligament and gingiva share the same route.

DENTAL INNERVATION

Upper teeth are supplied by the *superior alveolar nerves, anterior* and *posterior*; in 80% of individuals a *middle* nerve is also present (Fitzgerald & Scott 1958). These nerves supply a plexus lying above the apices of the teeth, partly on the posterior surface of the maxilla and partly within canals in the lateral and anterior surfaces of the bone. The *buccal nerve* provides a variable contribution to the buccal molar gingiva; the *greater palatine* and *nasopalatine* nerves pass to the palatal gingiva, overlapping in the region of the canine tooth.

The posterior superior alveolar nerves are two or three trunks from the postorbital section of the maxillary nerve. They divide into several rami within the periosteum and enter widely scattered foramina on the maxilla's posterior surface. Higher branches descend outside the antral mucosa to meet lower branches passing forwards above the teeth. The variable middle alveolar nerve, which may branch anywhere along the orbital part of the maxillary nerve, runs down the antral wall; the anterior alveolar nerve occupies the *canalis sinuosus* (p. 1304). The bony canals of the superior alveolar nerves also contain corresponding arteries forming the *superior alveolar neurovascular bundles*.

The **lower jaw** and its alveolar bone are largely supplied by the *inferior alveolar nerve*, with branches of the *buccal nerve* to the buccal gingiva of the molar and premolar teeth and branches of the *lingual nerve* to the lingual gingiva of all the lower teeth.

In its commonest form (six out of eight mandibles studied by Carter & Keen 1971), the inferior alveolar nerve is single, travelling through a well-defined osseous canal close to the dental roots, supplying individual branches to these and the interdental septa.

Between the premolar teeth the mental nerve, often multiple, leaves via the mental foramen. Intra-osseous *incisive nerves* continue to supply the first premolar, canine and incisor teeth. Branches leave the mental nerve at its origin to form an *incisor plexus* labial to the teeth, probably supplying their labial periodontium and gingiva. From this plexus and the dental branches, rami turn down and then lingually to emerge on the lingual surface of the mandible on the posterior aspect of the symphysis or opposite the premolar teeth, probably communicating with the lingual or mylohyoid nerve.

Less commonly (two out of eight mandibles), the inferior alveolar nerve was close to the lower border of the mandible, well below the roots of the teeth (**8.66**) with a variable number of large rami passing anterosuperiorly towards the roots before dividing to supply the teeth and interdental septa.

In three out of eight dissected mandibles nerves passed from the temporal muscle to enter the mandible through the retromolar fossa, communicating with branches of the inferior alveolar

A

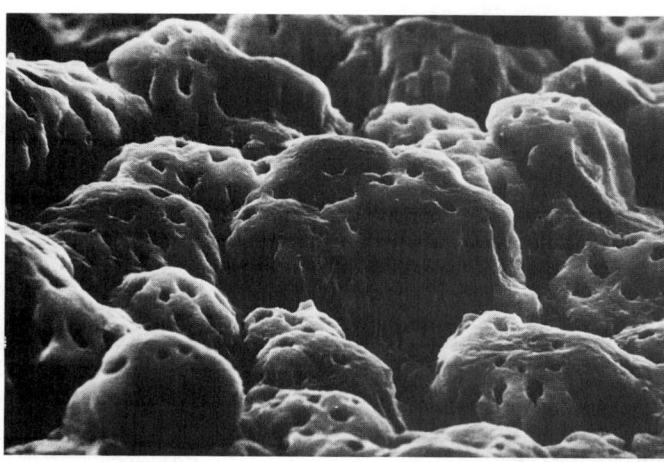

B

8.67 A. Longitudinal ground section of a tooth viewed by transmitted light showing the dentinal tubules, which appear as dark lines. One of the tubules shows lateral branches. B. A scanning-electron micrograph of the dentine/predentine junction after removal of predentine to expose the surfaces of calcospherites. The holes (about 1 μm wide) are dentinal tubules. Provided by D Whittaker, The Dental School, University of Wales, Cardiff.

nerve. Foramina occur in about 10% of retromolar fossae (Azaz & Lustmann 1973) and infiltration in this region can abolish sensation occasionally remaining after an inferior alveolar nerve block. Similarly, branches from the buccal, mylohyoid and lingual nerves which enter the mandible may provide additional routes of sensory transmission from the teeth.

The lower central incisor teeth receive a bilateral innervation, fibres probably crossing the midline within the periosteum to re-enter the bone via numerous canals in the labial cortical plate (Rood 1977).

Dental Histology

DENTINE

Dentine is yellowish avascular tissue forming the bulk of the tooth. It is a tough (work of fracture, $Wf = 270–550 \text{ J/m}^2$) and compliant (stiffness = 12 GN/m^2) composite material, about 70% by weight mineral (largely crystalline hydroxyapatite and fluorapatite but some calcium carbonate) and 20% organic matrix (type I collagen fibres, glycosaminoglycans and phosphoprotein, Weinstock & Leblond 1973). Its conspicuous feature is the regular pattern of microscopic dentinal tubules, about 1–2 μm in diameter, extending from the pulpal surface (about 50 000 tubules per 1 sq mm cross-sectional area) to the enamel–dentine junction (about 20 000 per sq mm). Tubules have a single sinuous primary curvature (8.57) oriented apically and more pronounced in the crown. A spiral secondary curvature, less regular, has a periodicity and amplitude of a few microns. Near the enamel–dentine junction tubules bifurcate, some with short extensions into enamel. Abundant lateral branches interconnect adjacent tubules (8.67A). Each tubule encloses a single cytoplasmic process of an odontoblast, containing microtubules, microfilaments and mitochondria but few ribosomes. Odontoblast cell bodies are in a pseudostratified layer lining the pulpal surface. In newly erupted teeth, processes are believed to extend the full thickness of dentine (Sigal et al 1984, Holland 1985) but in older teeth may be partly withdrawn so as to occupy only the pulpal third, the outer regions containing extracellular fluid (Thomas 1979). Lining most tubules is a heavily mineralized cylinder of peritubular dentine, devoid of collagen fibres, separated from the plasma membrane of the process by a glycosaminoglycan-rich lamina limitans (Thomas & Carella 1983). It is uncertain whether the process directly abuts the lamina limitans or whether there is a fluid-filled periodontoblastic space separating them.

Between the odontoblasts and the dentine is a layer of non-mineralized matrix, the *predentine* (8.67B,68). The predentine-dentine border is irregularly scalloped (8.67B) because dentine mineralizes as microscopic spherical aggregates of crystals (calcospherites). The enamel–dentine junction is more regularly scalloped, with convexities towards the dentine, a pattern unrelated to mineralization. Next to the enamel–dentine junction is a 30–40 μm layer (mantle dentine) which is less mineralized and has collagen fibres arranged parallel to the tubules. In the remaining circumpulpal dentine, fine collagen fibres are perpendicular to and interwoven around the tubules.

Dentine, like enamel, is deposited incrementally and is not remodelled. Both tissues carry a permanent record of changing shape, rhythmical patterns of formation and disturbances during development. Dentine has two orders of incremental line. The *contour lines of Owen* are formed by the coincidence of exaggerated secondary curvatures of adjacent tubules and may also represent a brief period of lower mineralization. A prominent Owen line is formed in those teeth whose mineralization spans birth (all deciduous teeth and the first permanent molars); this is the neonatal line, a result of the abrupt change in environment and nutrition. Fine *incremental lines of von Ebner* record a diurnal rhythm in mineralization level; they are about 5–8 μm apart in the crown and 2–5 μm apart in the root. General slight disturbances during dentinal development result in failure of fusion of calcospherites and produce interglobular areas of unmineralized matrix, so common in a layer 100–300 μm deep near the enamel–dentine junction that they must be considered normal.

The outermost 1 μm of root dentine, the hyaline layer (Owens 1972), may incorporate enamel matrix proteins secreted by the epithelial root sheath (Schonfeld & Slavkin 1977). Internal to this is the granular layer of Tomes (8.69), whose granularity may be due to minute interglobular areas or to small terminal expansions and anastomoses of adjacent odontoblast processes (Ten Cate 1972).

Primary dentine formation proceeds at a steady but declining rate as first the crown and then the root is completed. Further reduction in the size of the pulp chamber continues throughout life with the very slow and intermittent deposition of secondary dentine, sometimes distinguished from primary dentine by an Owen line and by a sudden change in direction of dentinal tubules. If dentine receives a severe stimulus (e.g. rapidly advancing caries or wear, tooth breakage) the odontoblasts of the affected region die, leaving a dead tract. This is sealed pulpally by a thin zone of sclerosed dentine and the deposition by newly differentiated pulp cells of reparative dentine, a poorly mineralized and sporadically formed tissue with few and irregular tubules. A less severe stimulus results in the odontoblasts increasing the deposition of peritubular dentine so as to fill the tubules. In ground sections this dentine appears translucent because it has assumed a near uniform refractive index. Translucent dentine also develops with age near the root apices.

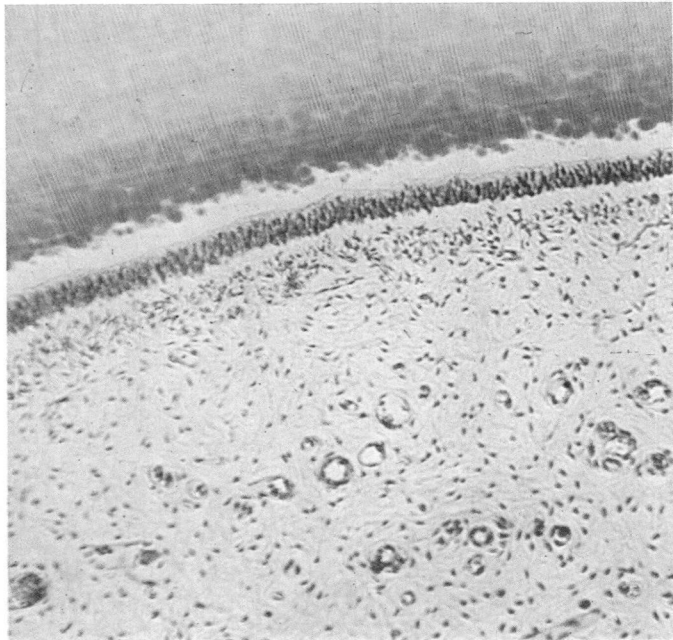

8.68 Transverse section of a demineralized tooth showing the pulp below, followed sequentially above by: (1) the layer of odontoblasts, whose stained nuclei show as a purple line; (2) the pale predentine, which adjoins (3) the darkly stained calcospherites which border the dentine at the top of the field.

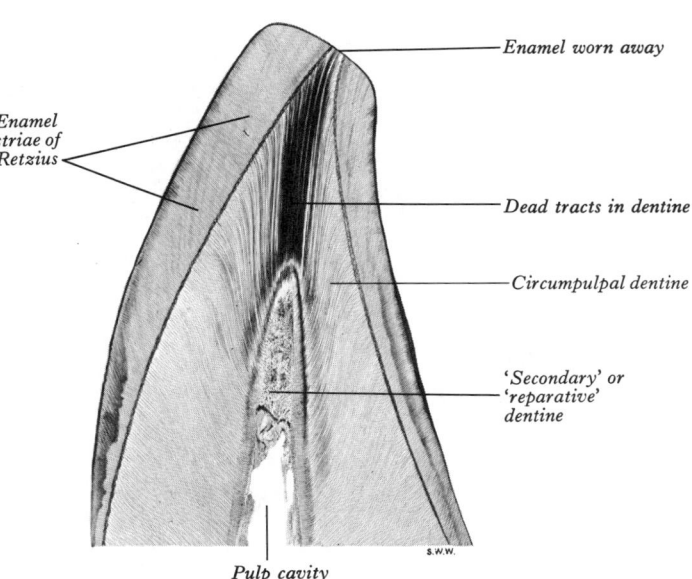

8.70 Longitudinal ground section of an incisor tooth. Compare the brown striae labelled on this section with those visible on 8.57.

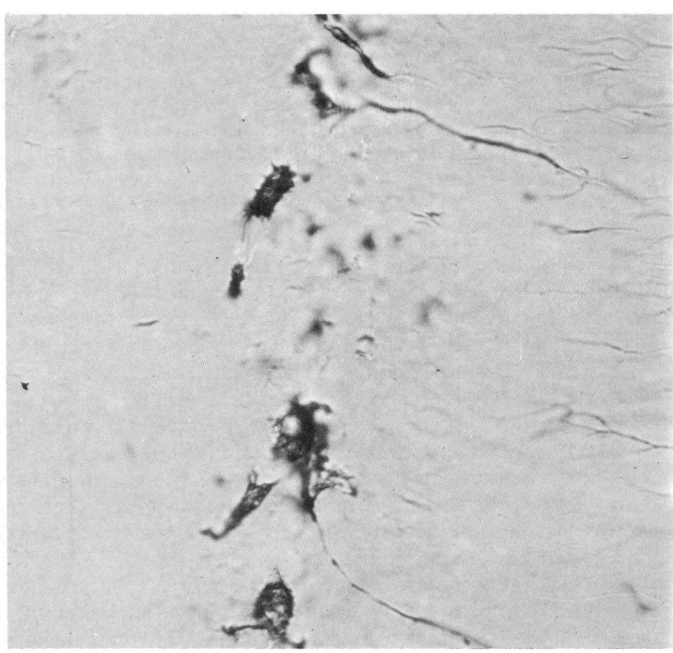

8.69 Longitudinal ground section of a tooth showing dentine (on the right), the dark spaces of the granular layer of Tomes (centre) and the structureless acellular layer of cement (on the left). Magnification c. × 2000.

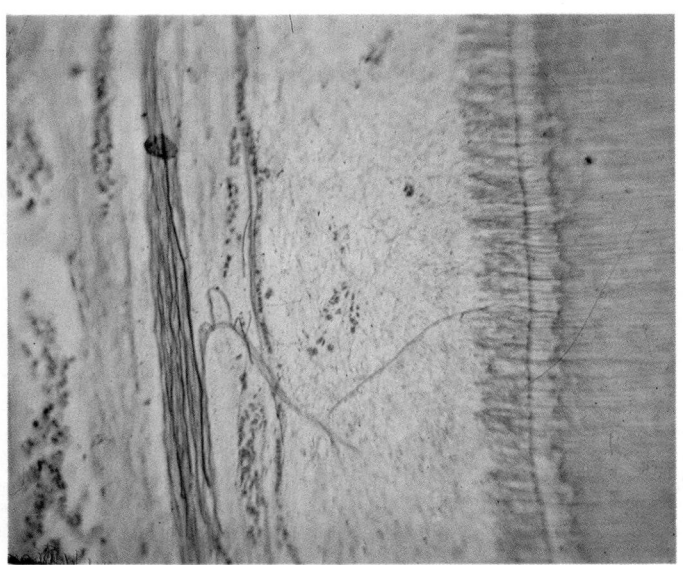

8.71 Longitudinal demineralized section of a tooth stained with a silver impregnation technique. Note the vertical nerve trunk (left of centre) within the pulp, with fine nerve fibres, one of which crosses transversely to pass between the odontoblasts lining the surface of the predentine (the pale-staining vertical layer, right of centre).

THE DENTAL PULP

The pulp is a well-vascularized loose connective tissue, enclosed by dentine. It is continuous with the periodontal ligament via the apical and accessory foramina. Several arterioles enter by the apical foramen to run longitudinally, giving branches to a subodontoblastic plexus in young teeth (Kramer 1960). Capillaries loops are seen in the odontoblast layer. The arterioles have thin walls, possibly related to their inclusion within a rigid cavity whose interstitial pressure is high (Van Hassel & McMinn 1972). Several small veins and lymphatic vessels (Bernick & Patek 1969) emerge from the pulp. Unmyelinated postganglionic sympathetic nerve fibres from the superior cervical ganglion enter the pulp with arterioles. Myelinated and unmyelinated sensory nerve fibres from the trigeminal ganglion traverse the pulp (8.71) longitudinally giving branches to ramify in the *plexus of Raschkow* (Scheinin & Light 1969) in the cell-rich parietal zone. Here fibres lose their myelin sheaths and continue into the odontoblast layer, some entering the dentinal tubules. Intratubular nerves are distinguishable from odontoblast processes because they contain numerous mitochondria (Frank 1968); they are more numerous beneath the cusps (where one in four tubules is occupied) than elsewhere (Lilja 1979). Ultrastructural studies have failed to reveal nerve fibres beyond 100 μm into human dentine but autoradiography of rats' teeth following injection of tritiated proline into the trigeminal ganglion and axonal transport of labelled proteins has revealed innervation in dentinal tubules near the enamel–dentine junction (Pimendis & Hinds 1977).

Stimulation of dentine by thermal, mechanical or osmotic stimuli evokes a diffuse pain response. The mechanism of stimulus transduction is unknown but is unlikely to involve the

1307

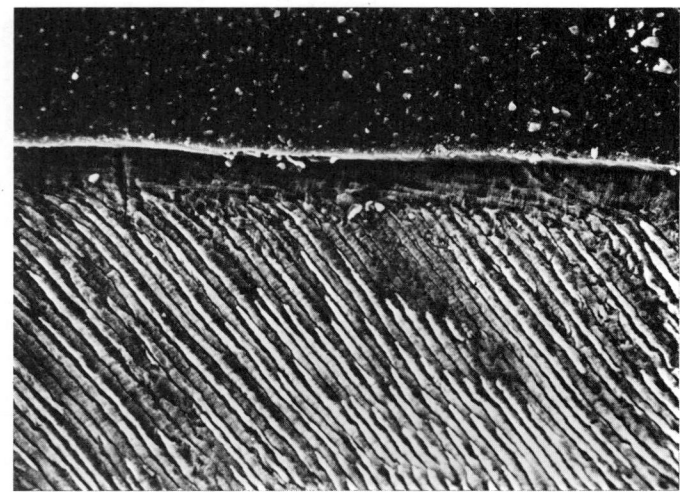

8.72 Scanning electron micrograph of enamel prisms. Each prism is about 5 μm wide and separated from adjacent prisms by interprismatic material which has been removed from this specimen by acid etching. A thick structureless surface layer is present. Provided by D Whittaker, The Dental School, University of Wales, Cardiff.

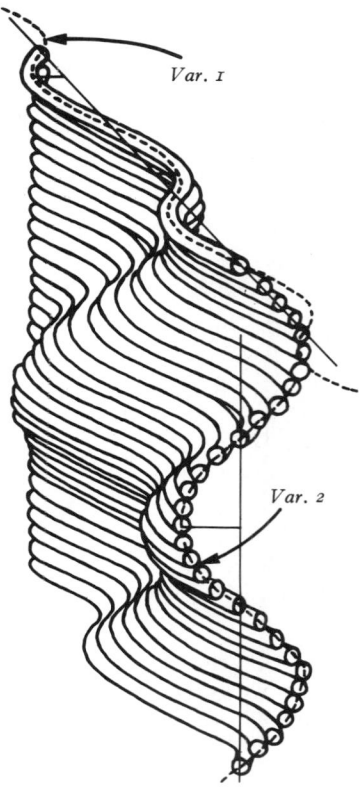

8.74A Diagram illustrating the relationships existing between a vertical stack of enamel prisms. Each prism undulates in the transverse plane of the tooth but its undulations are out of phase with those of vertically adjacent prisms. Hence, when a section is viewed by reflected light, the undulations are responsible for the characteristic alternation of dark and light bands which cross the prisms obliquely (the Hunter-Schreger bands). Var. 1 and 2 indicate the sine-wave undulations in the transverse and vertical planes, which vary in amplitude and periodicity in the enamel of different species. (From Osborn 1970 with permission of the author and publishers, Springer.)

direct stimulation of nerve endings in dentine. Newly erupted teeth are equally sensitive yet do not have a plexus of Raschkow or intratubular nerves. Pain-producing chemicals and local anaesthetics show little ability to stimulate or anaesthetize exposed dentine. One possibility is that the odontoblast process is capable of propagating some kind of impulse and of exciting nerve endings in contact with the proximal region of the process or the cell body. Gap junctions, known to give electric coupling, have been identified between adjacent odontoblasts and between odontoblasts and nerve fibres (Koling & Rask-Andersen 1983). Vesiculated terminal boutons have been noted in close (20 nm) apposition to odontoblast cell membranes (Arwill 1968), suggesting the possibility of a chemical synapse. An alternative hypothesis suggests that stimuli generate movement of intra- or extracellular fluid along the dentinal tubules, causing in turn a local distortion of the pulp, sensed by free nerve endings in the plexus (Gysi 1900, Anderson et al 1970, Brannstrom 1964). Recent evidence that odontoblasts are joined by continuous tight junctions (Bishop 1985) suggests that the odontoblast may be directly involved in relaying intratubular fluid movements to nerve endings. This 'hydrodynamic' theory would explain the ineffectiveness of neuroactive agents and why pain is produced by drying and by solutions of high osmotic pressure. Solutions equally effective at producing pain, however, create very different rates of fluid flow (Anderson & Matthews 1967, Horiuchi & Anderson 1973).

THE ENAMEL

Enamel (Osborn 1973, Boyde 1976) is an extremely hard (Knoop number = 300+) and rigid (stiffness = 40–80 GN/m²) ceramic material covering the crowns of teeth. It is a heavily mineralized cell secretion, containing 95–96% by weight crystalline apatites

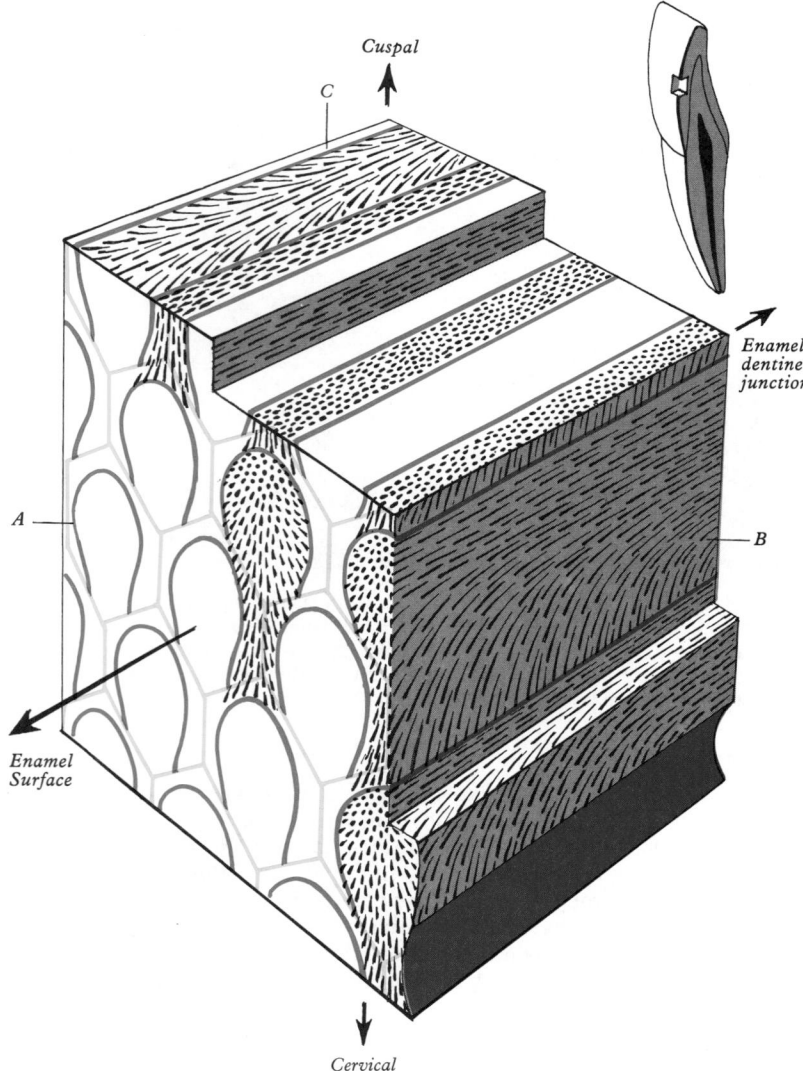

8.73 Diagram of a block of enamel showing how the orientation of the crystallites determines their appearance when the prisms are cut transversely (face A), longitudinally (face B) or at right angles to both these planes (face C). Prism boundaries or sheaths (blue) are formed wherever crystallites meet at highly discordant angles. Superimposed on the transversely cut face (A) are cross-sectional outlines of the ameloblasts (yellow).

8.74B Reconstruction of enamel prisms in the dog. Each prism is represented by a thread. Note the gradual change in direction between vertically adjacent prisms which are viewed from the cuspal aspect in this photograph. Only a part of one complete prism undulation is shown in each case. Compare with **8.74A**. Preparation by J W Osborn, Department of Anatomy, Guy's Hospital Medical School, London.

8.75 Scanning electron micrograph of the enamel surface showing perikymata. The holes (about 4 μm wide) were occupied by Tomes' processes of ameloblasts when the development of the enamel was completed. Provided by D Whittaker, The Dental School, University of Wales, Cardiff.

(88% by volume) and less than 1% organic matrix. Since its formative cells are lost from the surface (**8.75**) during eruption it is incapable of further growth or repair. It reaches a maximum thickness of 2.5 mm over cusps and thins to knife edge at the cervical margins. Enamel is composed of closely packed enamel *prisms* (or rods), U-shaped in cross section (**8.72**), extending from close to the enamel–dentine junction to within 6–12 μm of the surface. Each prism is partially delineated by a matrix-rich *prism sheath*, 70 nm thick, and is separated from neighbouring prisms by a continuous *interprismatic region*. Prisms are about 3–4 μm wide in inner enamel, increasing to about 6 μm near the surface. Prisms are packed with flattened hexagonal hydroxyapatite crystallites, 26 nm × 68 nm in cross section (Daculsi & Kerebel 1978). Hexagonal transverse profiles of ribbon-like crystallites are randomly oriented. In the cuspal region of a prism (plane C in **8.73**) these are almost parallel to the prism's long axis and may be as long as the enamel is thick (i.e. up to 2.5 mm); but in the cervical region of a prism (plane B in **8.73**) and in interprismatic regions, the crystallites have a pronounced cervical inclination and end at the cervically-adjacent prism sheath. A sudden change always exists between crystallite orientation on the two sides of a prism sheath. In surface enamel, crystallites are packed with their long axes parallel so that prism sheaths do not form.

At intervals of about 4 μm along its length, each prism is crossed by a dark *striation*, the light microscope manifestation of a rhythmic swelling and shrinking of prism diameter during one day's growth. Higher order incremental lines in enamel are *striae of Retzius* (**8.70**), passing from the enamel–dentine junction obliquely to the surface where they end in shallow furrows, *perikymata*, visible on newly erupted teeth. Each stria represents a period of 7–8 days' enamel growth (Bromage & Dean 1985). Striae are produced by a sudden double right-angle translocation of the prisms in the longitudinal plane (Weber et al 1974) and may be clear or brown in transmitted light. Tyndall scattering of short wavelengths is due to accumulations of matrix in the prism translocations. A prominent stria, the *neonatal line* (Whittaker & Richards 1978), is formed in teeth whose mineralization spans

birth. Neonatal lines in enamel and dentine are of forensic importance, indicating that an infant has survived for a few days.

Each prism is sinuous in the tooth's transverse plane with a wavelength of about 1.5 mm, undulations of one prism matched by those lateral to it but slightly (2°) out of phase with those above or below (**8.74**). Prism sheaths are comparatively weak interfaces in enamel (work of fracture, Wf = 200 J/m² perpendicular to prisms but only 13 J/m² parallel to prisms, Waters 1980). Decussation of prisms in the tooth's longitudinal plane is an adaptation which increases the toughness of enamel by enlarging the surface area of potential cracks between prisms in that plane. Similar regular undulations over cusps produce the appearance of gnarled enamel in sections.

Prism sheaths in the inner enamel are considerably thickened to form tuft-like projections from the enamel–dentine junction, extending for a considerable distance in the longitudinal plane of the tooth. Longitudinal sheets of organic material penetrating the full thickness of enamel are *enamel lamellae*. Extensions into the enamel of dentinal tubules are *enamel spindles*, prominent over cusps.

THE CEMENT

Cement is a bone-like tissue covering the dental roots, about 50% by weight hydroxyapatite and amorphous calcium phosphates (Selvig 1965, Frank & Steuer 1977). In newly erupted teeth, the cement generally overlaps the enamel slightly but may just meet the cervical margin or fall short, leaving dentine exposed at the periodontal ligament. All three situations may prevail around the

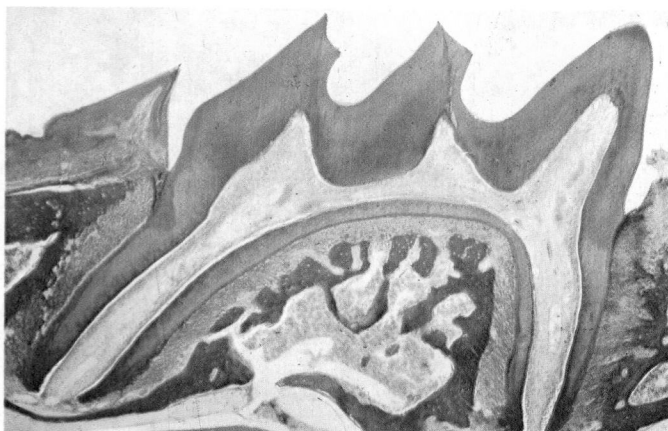

8.76 Demineralized vertical section of a young rat's molar tooth in situ. Note: (1) the alveolar bone with marrow spaces; (2) the fibres of the periodontal ligament which connect the alveolar bone to the cement of the root; (3) the large pulp cavity typical of a young tooth; (4) the absence of enamel in this demineralized section; (5) the gingiva adjacent to the enamel space at the neck of the tooth.

neck of a single tooth. In older teeth, when the root becomes exposed in the mouth through occlusal drift and gingival recession, cement is often worn away and dentine revealed.

Cement is perforated by *Sharpey's fibres*, attachment bundles of periodontal ligament collagen fibres (extrinsic fibres). New layers of cement are deposited incrementally throughout life to compensate for tooth movements, incorporating new Sharpey's fibres. Incremental lines (of Salter) are irregularly spaced.

The first formed cement is thin (up to 200 μm), acellular and contains only extrinsic fibres; but cement formed later is produced more rapidly and contains cementocytes in lacunae joined by canaliculi mainly directed towards the periodontal ligament. This cement contains both extrinsic fibres and matrix (intrinsic) fibres of cementoblastic origin. With increasing age cellular cement may reach a thickness of a millimetre or more around the apices and at the furcations of the roots, where it compensates for the loss of the periodontal attachment area through occlusal drift. Cement is not usually remodelled but will repair both small areas of resorption and fractures of the dentine. Cement deposition within the apical foramen is a cause of vascular strangulation of the pulp which progresses with age.

THE PERIODONTAL LIGAMENT (MEMBRANE)

The periodontal ligament (Melcher & Bowen 1969) is a dense connective tissue (50% dry weight is collagen types I and III)

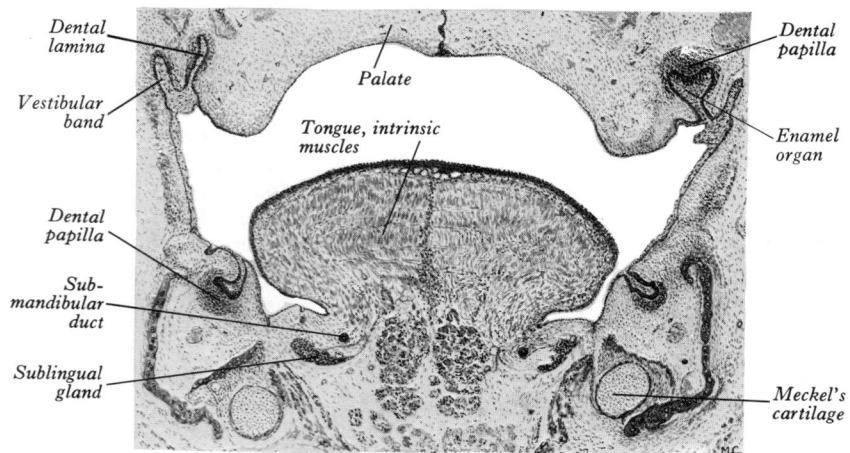

Dental lamina
Vestibular band
Dental papilla
Sub-mandibular duct
Sublingual gland
Palate
Tongue, intrinsic muscles
Dental papilla
Enamel organ
Meckel's cartilage

8.77 Coronal section of the head of a human embryo (C R length 34 mm), showing developing teeth. The pointer line to Meckel's cartilage passes through the developing mandible.

between 0.15 and 0.3 mm wide. It contains cell types typical of connective tissue with the addition of a network of epithelial cells, the *epithelial debris of Malassez*, remnants of the root sheath. These have no evident function but may produce commonly occurring *dental cysts*.

The principal functions of the periodontal ligament are to attach teeth in sockets and to provide sensory information about tooth movements (Anderson et al 1970). The majority of collagen fibres are arranged in a number of *principal groups* which connect alveolar bone and cement. *Horizontal fibres* at the alveolar crest and near the apex restrict tilt; between these groups are the *oblique fibres*, restricting intrusive movement. Radiating from the apex are apical fibres, resisting extrusive movement. Gingival fibres pass from the cervical region of the root and from the osseous alveolar crest into the gingival lamina propria, anchoring it firmly to the tooth, aided by a *circular group* arranged concentrically around the central neck. The collagen fibres compartmentalize the proteoglycan-rich hydrophilic ground substance (Sloan 1978) which provides a compressive visco-elastic support (Melcher & Walker 1976).

Each periodontal ligament has a nerve supply from several sources (vide supra); the chief role of the innervation seems to be proprioception. Various endings have been described: irregularly branched, knob-like, Meissner's corpuscle-like and spindle-shaped. Structural variations of the mechanoreceptors appear to be less important than their spatial arrangement in determining the response (Hannan 1976). Impulses from such endings probably provide an input to many brain-stem centres and the cerebellum, where masticatory cycles may, in part, be integrated.

Turnover of collagen in periodontal ligaments is remarkably rapid. Fibroblasts are involved in both fibre synthesis and degradation, processes which may occur simultaneously within the same individual cell (Ten Cate & Deporter 1975).

THE GINGIVAE (GUMS)

The gingiva is a specialized region of the oral mucosa surrounding the necks of the teeth (Squier et al 1976). In a healthy mouth it is distinguished from the oral mucosa by its pale pink, stippled appearance (**8.**59), the adjacent alveolar mucosa being red, shiny and smooth; gingival, palatal and dorsal lingual epithelia are keratinized (or parakeratinized), while alveolar epithelium is non-keratinized. Gingival lamina propria is firmly connected to the underlying alveolar bone, forming a virtually immovable mucoperiosteum.

At the gingival crest, the epithelium is reflected towards the root so that its outer surface is attached to the tooth, forming the *epithelial attachment*. Surrounding the tooth there may be a shallow *gingival sulcus* between tooth and gingiva, its floor being the epithelial attachment. The epithelium attached to the tooth is termed the *junctional epithelium*, that lining the sulcus is the *sulcular epithelium*. Junctional epithelium is non-keratinized, has wide intercellular spaces, few tonofilaments and few desmosomes (Schroeder & Listgarten 1971); it is permeable, weakly cohesive and easily ruptured. Its superficial cells (equivalent to the prickle cells of normal keratinized epithelium) adhere tightly by hemidesmosomes to an outer basal lamina covering the adjoining dental surface (Listgarten 1970) to which they are firmly bonded; this is enamel in newly erupted teeth but in older individuals the junctional epithelium extends onto the cement. Junctional epithelial cells have a very high turnover and move rapidly up the dental surface, whence they are shed into the sulcus.

Early Dental Development

DECIDUOUS TEETH

At the 9 mm embryonic stage primitive oral epithelium (**8.**78A) begins to bulge into the underlying mesenchyme where teeth will form (**8.**77–81). From these separate ingrowths and the mesenchyme associated with them, the four anterior deciduous teeth (central and lateral incisors, canine and first molar) will arise (Ooe 1957, Nery et al 1970). In amphibian embryos, odontogenic

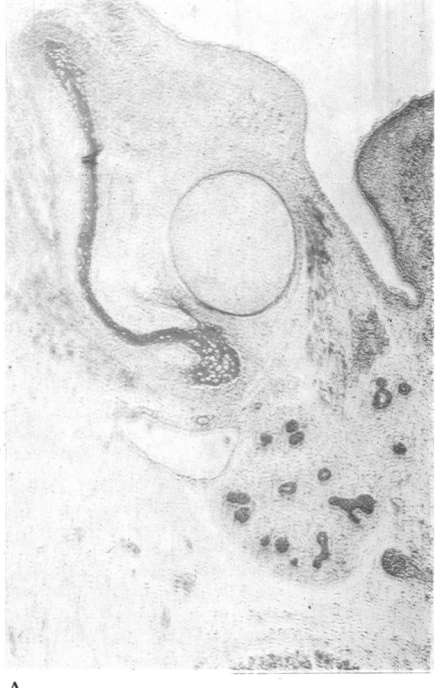

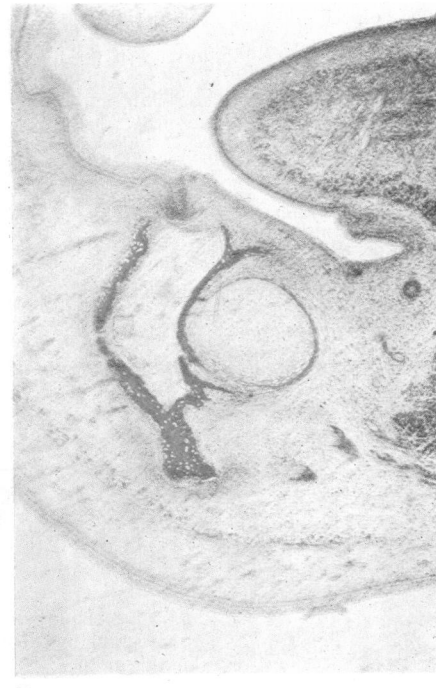

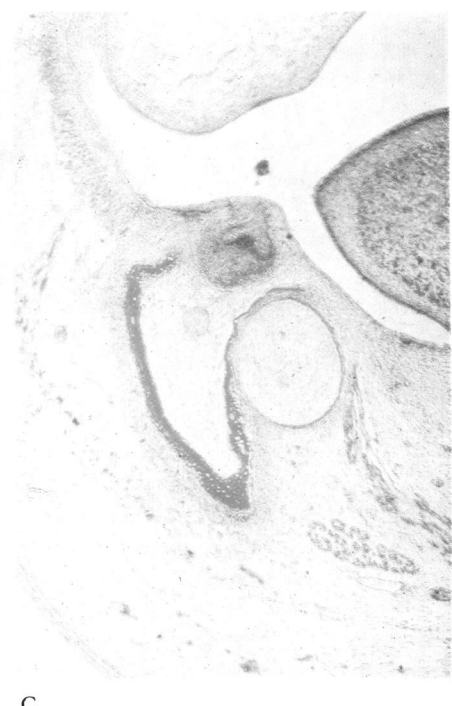

A

B

C

8.78A–C A series of stages illustrating the early development of teeth. These are all coronal sections through the right half of the body of the mandible, showing the tongue in the top right-hand corner. The mandible is mineralizing to the left of the circular profile of Meckel's cartilage.

A. The stage of development before the ingrowth of the dental lamina from the oral epithelium.

B. The dental lamina is growing between the buccal and lingual plates of the ossifying mandible.

C. The cap stage of development. The enamel organ is growing from the dental lamina around the condensation of cells which forms the dental papilla.

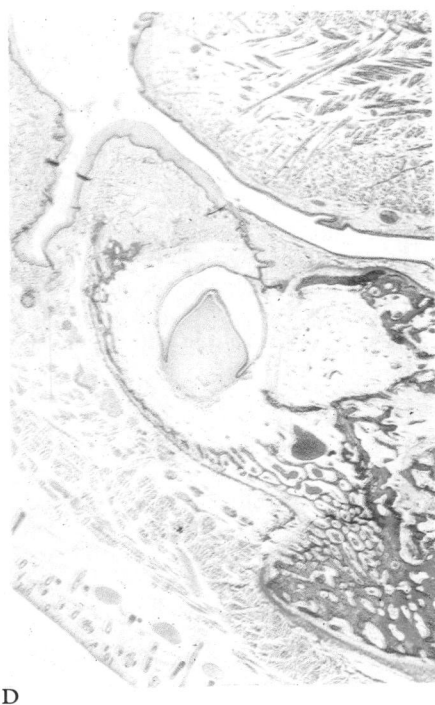

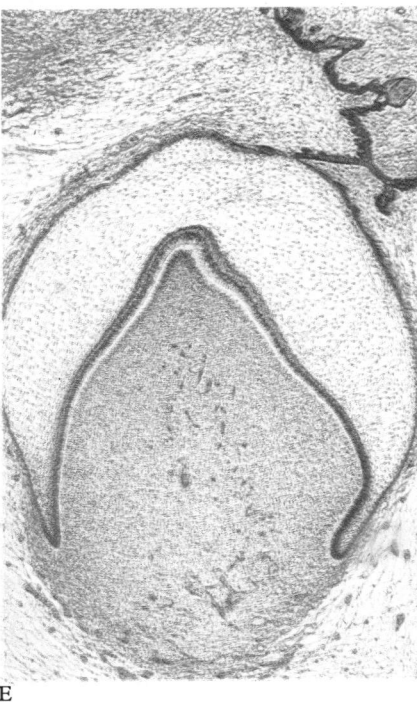

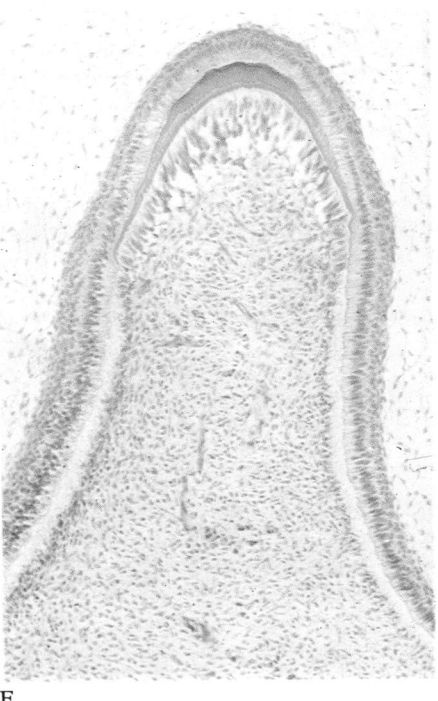

D

E

F

8.78D–F Slightly later stages in tooth development than shown in **8.**78A–C.

D. The bell stage of development. The external enamel epithelium of the enamel organ is connected to the oral mucosa by an irregularly-stranded dental lamina. Lateral to the buccal plate of the mandible the vestibular band has atrophied centrally to initiate the oral vestibule. The tooth germ is separated from the bone by the tooth follicle.

E. A photograph at higher magnification of the bell stage. Note from above downwards: (1) the degenerating dental lamina, top right; (2) the fibrous tooth follicle surrounding the developing tooth; (3) the external enamel epithelium; (4) the delicate stippled appearance produced by the nuclei of the stellate reticulum; (5) the darkly stained, somewhat flattened cells of the stratum intermedium, which is seen more clearly in F; (6) the columnar cells of the internal enamel epithelium; (7) the more closely packed cells of the dental papilla which extend outside the cervical loop; (8) the capillaries of the pulp and tooth follicle.

F. Dentine formation beginning at the cuspal tip. From above downwards note: (1) the loose stellate reticulum; (2) the stratum intermedium; (3) a layer of columnar ameloblasts; (4) a thin strip of enamel matrix (mauve); (5) mineralized dentine (pink); (6) predentine (pale blue); (7) a layer of odontoblasts.

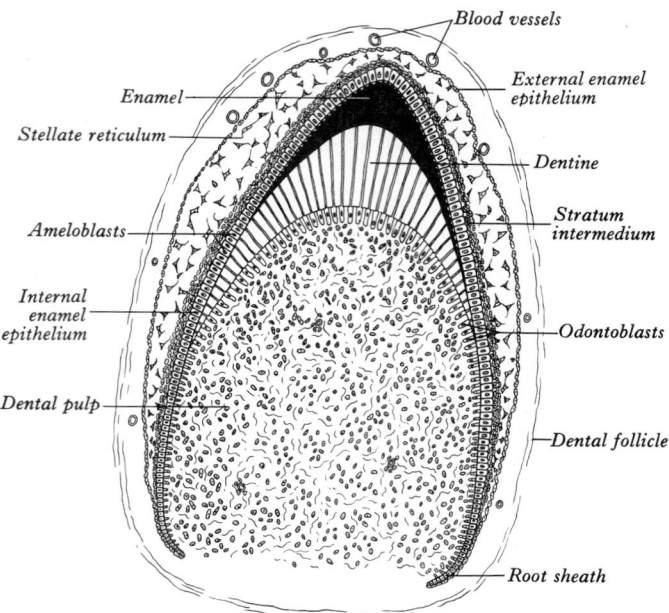

8.79 Simplified diagram of a developing tooth to show the approximate arrangement of its principal components. Compare with 8.78F.

8.81 Developing tooth with mandible and lip in situ. Drawn from a photograph by F Harrison, Dental Department, University of Sheffield, and reproduced from Pedley & Harrison with permission from Blackie.

mesenchyme originates from the mesencephalic levels of the cranial neural crests, migratory ectomesenchyme entering and expanding the branchial arches (de Beer 1947, Chibon 1967) under the inductive influence of oral epithelium (Wagner 1949, 1955, Henzen 1957). The tooth inductive potency of mandibular arch ectoderm on cranial neural crest cells has been demonstrated in the mouse embryo (Lumsden 1987).

At about the 20 mm stage the ingrown epithelial dental laminae have expanded into knob-like swellings (Ooe 1956), each surrounded by a dense aggregate of vascular mesenchyme (Gaunt 1959). The combined organ rudiment is the *tooth bud*. Ectoderm starts to grow around the mesenchymal aggregate; the ectodermal part is now an *enamel organ*, the mesenchymal part a *dental papilla*. Peripheral cuboidal cells of the enamel organ are soon distinguished from central polygonal cells. At this stage (48 mm) the bud of the second deciduous molar appears on the posteriorly growing dental lamina.

The spherical dental papilla enlarges but the encircling edge of the enamel organ (the *cervical loop*) continues to surround more of

its periphery until it sits on the papilla like a cap, the *cap stage* of development (8.78C). Meanwhile the central polygonal cells of the enamel organ have been secreting glycosaminoglycans into intercellular spaces, which attract water, swelling the enamel organ and compressing the cells. Since desmosomal connections persist, the central cells become stellar, forming a *stellate reticulum* (8.78E, 79). The originally cuboidal cells adjacent to the dental papilla lengthen to form the columnar cells of the *inner enamel epithelium* (8.78E). Cells forming the outer surface of the enamel organ are the *outer enamel epithelium* (8.78E), continuous via the dental lamina with the oral epithelium. By continued growth the cervical loop surrounds about three-quarters of the enlarging dental papilla, the *bell stage* of tooth development. Now a layer of flatter cells develops between the inner enamel epithelium and the stellate reticulum; this *stratum intermedium* (8.78E, 79) derives from the original polygonal cells of the enamel organ (the *enamel knot*). Tissue interactions between the peripheral cells of the dental papilla and the adjacent cells of the inner enamel epithelium (Thesleff & Hurmerinta 1981) result in the differentiation of odontoblasts from the former and of ameloblasts from the latter.

The development of a nerve supply to deciduous teeth has attracted little attention. Alveolar nerves enter into maxillary and mandibular processes during the fifth week, before the dental laminae form (Pearson 1977). A close association has been noted between peripheral nerve branches and the sites of prospective tooth development in the mouse (Kollar & Lumsden 1979) but initiation of tooth development does not appear to depend on innervation (Lumsden & Buchanan 1986). At cap and bell stages bundles of nerve fibres have entered the dense mesenchyme of dental papillae and follicles.

The Dental Follicle

The *dental follicle* (8.78E) is the layer of cells which surrounds the tooth germ ultimately to adjoin developing alveolar bone (8.78D, 81); the bony cavity containing the tooth germ and follicle is the *dental crypt*. The follicle cells adjacent to the outer enamel epithelium form a dense *investing layer* from which develops the cement of the root. The periodontal ligament and bone develop respectively from the loose intermediate layer and outer osteogenic layer of the follicle.

The Vestibular Band

As the dental laminae appear, a similar but continuous horseshoe-shaped ingrowth of epithelium develops external (buccal) to them. This *vestibular band* (8.77, 80) grows deeply into the

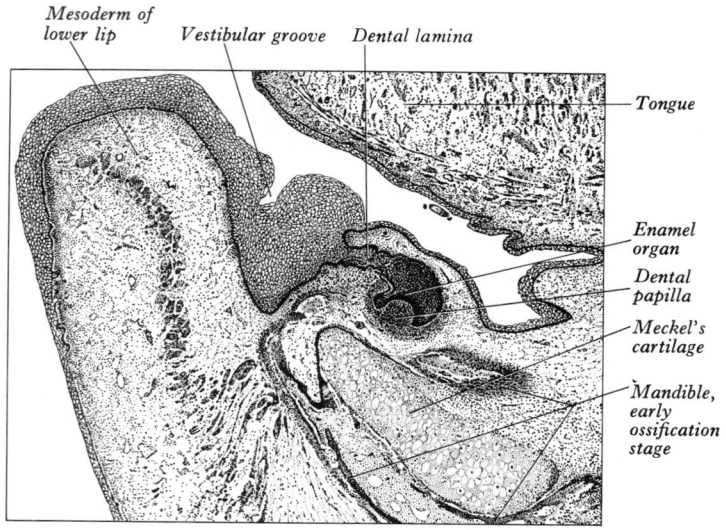

8.80 Part of a sagittal section through the head of a human embryo (CR length 60 mm), passing through the right lower central incisor tooth germ. Drawn from a photomicrograph given by C H Tonge, Anatomy Department, King's College, Newcastle-upon-Tyne. Stained with haematoxylin and eosin. Magnification × 12.

mesenchyme of the primitive jaws, separating prospective lips and cheeks from the tooth forming regions. It subsequently thickens and cleaves (**8.**80) to form the *oral vestibule*. In contrast to the dental lamina, the vestibular lamina is not associated with an aggregation of mesenchyme cells.

PERMANENT TEETH

As the jaws lengthen the dental lamina grows posteriorly from the distal aspect of the second deciduous molar germ as a solid cord of epithelium, not connected with the surface. From the deep border of this 'burrowing' lamina, buds for the three permanent molars develop in mesiodistal sequence, the first molar bud appearing in the 16-week fetus, the second at about one year and the third at five years. Each bud is initiated in the ramus of the lower jaw but, with progressive resorption of the anterior border of the coronoid process, they come to occupy the body of the mandible.

From each deciduous tooth germ at its bell stage (about 16 weeks) a lingual *successional lamina* grows from the site of continuity between outer enamel epithelium and dental lamina (**8.**81). Each grows down into mesenchyme lingual to a deciduous tooth and from its end a bud develops for a permanent successor which becomes surrounded by its own follicle and crypt. The follicle maintains fibrous continuity with the lamina propria of oral mucosa by *gubernacular cords*, whose original positions are visible in young skulls as *gubernacular canals*. Gubernacular canals are said to guide erupting permanent teeth into their correct positions (Scott 1967).

The Fate of the Dental Laminae

As dentine and enamel start to develop, the dental laminae begin to degenerate (**8.**78E), separating into clumps, many with a whorled appearance over developing deciduous teeth. These persist as epithelial rests but may sometimes proliferate to form cystic cavities, known as *eruption cysts*, recognizable as bluish swellings over erupting teeth.

Crown Pattern Morphogenesis

During the late bell stage the amelodentinal membrane, formed by the inner enamel epithelium, the peripheral cell layer of the dental papilla and the interposed basement membrane, folds in a genetically determined pattern to assume the definitive outline of the future enamel–dentine junction. Because regional variations in enamel thickness are slight, this folding determines the ultimate shape of a tooth, i.e. the number and positions of cusps. Cap stage incisor and molar tooth germs of mouse embryos separated into their epithelial and mesenchymal components and reciprocally recombined in organ culture develop the morphology expected of the mesenchyme (Kollar 1972). How the dental papilla mesenchyme acquires positional specification and how the information for shape is encoded and relayed to the apparently indifferent enamel epithelium are unknown.

DEVELOPMENT OF DENTINE

At the tip of a presumptive cusp, cells of the inner enamel epithelium lengthen and mesenchyme cells of the adjacent dental papilla extend fine processes through the reticular lamina of the basement membrane to contact the epithelial basal lamina. An extracellular matrix-mediated cell to cell interaction (Thesleff 1977) induces the mesenchyme cells to differentiate into odontoblasts, which will lay down dentine (**8.**78F, 82). Newly differentiated odontoblasts, with well developed endoplasmic reticulum and Golgi apparatus, secrete dentine matrix into the space between the basal ends of the inner enamel cells and their own secreting ends. Collagenase-containing vesicles in this early matrix (Sorgente et al 1977) may be involved in digestion of the epithelial basal lamina which permits the odontoblast processes to push up between the inner enamel epithelial cells where they may form direct cell contacts, mediating the differentiation of ameloblasts. Accumulating matrix pushes the odontoblasts back, their processes lengthening as their perikarya recede, becoming enclosed within tubules of the matrix. As soon as a few microns of matrix are formed, the matrix adjacent to the inner enamel epithelium begins to mineralize (Silva & Kailis 1972), possibly with the agency of alkaline phosphatase-rich matrix vesicles (Bernard 1972, see also p. 302).

From this region, the summit of a presumptive dentine cusp, a wave of differentiation of odontoblasts from papillary cells slowly spreads to the growing cervical loop. As soon as each differentiates, the matrix is formed, pushing the layer of odontoblasts, united by desmosomes, into the papilla. The layer of unmineralized matrix adjacent to the odontoblasts is termed predentine (**8.**82). First-formed collagen fibres lie parallel to the odontoblast processes; after this thin layer of *mantle dentine* (vide supra) is formed, fine collagen fibres of *circumpulpal dentine* are elaborated, interlacing at right angles to the processes. In mantle dentine each odontoblast has two or more processes but, as it recedes from the enamel–dentine junction, its processes unite into a single main process. This accounts for the bifurcation of the tubules near the junction. During circumpulpal dentinogenesis odontoblasts constantly extend short lateral processes at the base of the main process; these are later embedded by mineralized dentine to become fine lateral tubules. Dentinal tubules are much thinner in mineralized dentine than in predentine; this constriction starts at the level where predentine mineralization is beginning and is due to the deposition of a highly mineralized cylinder of *peritubular dentine* around the inside of the tubule, progressively reducing its lumen. It does not develop in interglobular areas (vide supra) where the tubule is walled by unmineralized intertubular dentine matrix.

DEVELOPMENT OF ENAMEL

Ameloblasts differentiate from cells of the inner enamel epithelium under the inductive influence of newly differentiated odontoblasts; the interaction between these cells is thought to involve direct cell contacts and/or the extracellular matrix. The first signs of differentiation are the manufacture of organelles required for enamel matrix production and the reversal of cell polarity; ameloblasts secrete from their original basal ends. Mitochondria, originally dispersed throughout the cytoplasm, congregate at the non-secreting pole where they cluster around the nucleus. The Golgi apparatus is located centrally; cisternae of the extensive endoplasmic reticulum are stacked in rows parallel to the cell's long axis. The mature cell is about 40 μm long and about 5 μm wide. In cross section ameloblasts are regular hexagons, accounting for the classic honeycomb appearance, and are interconnected by junctional complexes at both secreting and non-secreting poles. Their non-secreting poles are attached by desmosomes to the stratum intermedium cells, which may elaborate and transport materials to the ameloblasts (Kurahashi & Yoshiki 1972). Alkaline phosphatase, found in other hard-tissue forming cells, exists in the stratum intermedium but not in secretory ameloblasts.

Enamel matrix is secreted between mineralizing dentine and ameloblasts; a rise or potential rise in hydrostatic pressure produced by the accumulation of enamel matrix in this enclosed region probably provides the force to push ameloblasts away from the enamel–dentine junction (Osborn 1973). Since ameloblast differentiation depends on and shortly follows odontoblast differentiation, developing enamel spreads down the sides of the presumptive enamel–dentine junction in the same way as developing dentine, just behind it (**8.**82).

At the start of amelogenesis, in each region of the tooth germ, adjacent stellate reticulum seems to collapse, the enamel organ being progressively reduced in thickness until it has only three layers (outer enamel epithelium, stratum intermedium and ameloblast layer). It is widely assumed that this brings the ameloblasts, inside an avascular enamel organ, closer to the capillaries which have invaded the investing layer of the follicle adjacent to the external enamel epithelium. Meanwhile, cells of the latter, originally cuboidal, become squamous, throwing the outer surface of the enamel organ into microscopic folds to increase the area for diffusion.

When ameloblasts have moved about 10 microns from the enamel–dentine junction they develop conical extensions into the

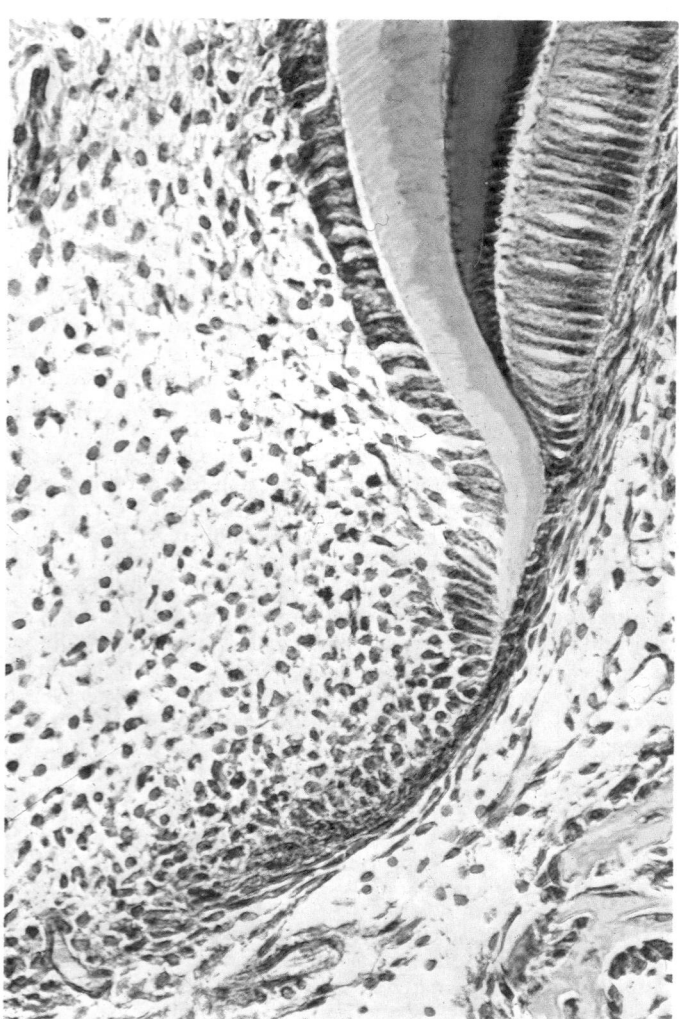

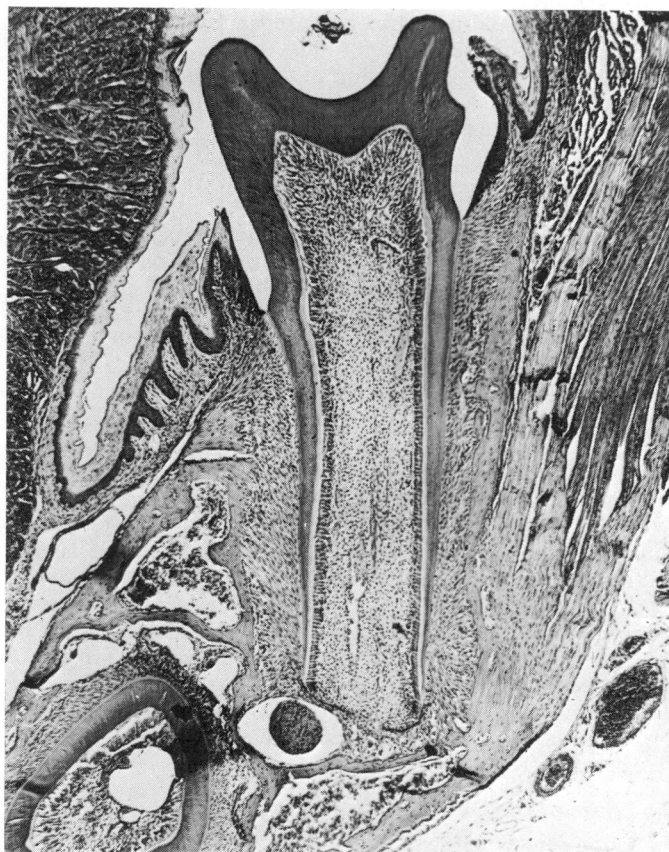

8.83 A longitudinally sectioned developing tooth showing advanced root formation. See text for further details.

8.82 Vertical section through the neck of a developing tooth with the crown above and the developing root below. Note above, from right to left: (1) the columnar ameloblasts; (2) enamel matrix (red); (3) mineralizing dentine (pale mauve); (4) predentine (pale blue); (5) odontoblasts; (6) the fibroblasts of the pulp. The columnar ameloblasts terminate at the tooth neck where the latter is continuous with the developing root. In this region note from right to left: (7) the fibrous tooth follicle; (8) the layer, two cells thick, of Hertwig's root sheath; (9) the developing dentine; (10) the odontoblasts. Beneath this the root sheath extends to the left beneath the dental papilla.

accumulating enamel. These *Tomes' processes*, whose bases are limited by the junctional complex at the secreting pole of the cell, bear a peripheral collar of microvilli (see Reith 1970). Tomes' processes give the developing front of enamel a pitted appearance (Boyde 1969); adjacent to each is an unmineralized layer about 50–100 nm thick, the *enamel matrix*. This is stippled under the electron microscope and similar material is seen in membrane-bound vesicles within Tomes' processes. On the enamel side of this stippled material are long ribbon-like crystallites. First formed enamel is non-prismatic or contains irregular prisms. At the enamel-dentine junction, enamel (recognized by long crystallites) is intermixed with dentine (recognized by collagen fibres and small crystallites).

The mineralizing enamel front shows little change until nearly the full thickness of enamel has been secreted. This is immature enamel, containing narrow crystallites about 3 nm × 29 nm in cross section, and has a composition of about 40% mineral by weight. As ameloblasts approach the final surface, deeper crystallites thicken by accretion of ions from the surrounding matrix. Diminishing calcium-rich matrix is replenished by ameloblasts, water and protein being resorbed. The matrix can travel long distances through the developing enamel. Ultimately the crystallites can widen no more, no further space being available between

them. Theoretical analysis suggests that about 12% by volume of unmineralized enamel matrix and water would thus remain (Carlstrom 1964), according well with the observation that mature enamel is 96% by weight mineral (i.e. 88% by volume). No new crystallites appear to be added to the enamel except at the mineralizing front (Ronnholm 1962); crystals therefore grow in length as the enamel is deposited and achieve their final length as maturation begins. As secretion ends the ameloblast shortens, withdraws its previously conical Tomes' process and forms a ruffled membrane with numerous microvilli which endocytose the matrix (Reith 1967a,b). The maturative ameloblast contains numerous lysosomes. Over some of the enamel surface developed at this time, crystallites are parallel and prism sheaths are not formed (vide supra). This non-prismatic surface layer is about 6–12 µm deep (Gwinnett 1967, Osborn 1973) and is about 1% by weight more mineralized than the rest, possibly because of closer packing of crystallites permitted by parallel orientation.

ROOT DEVELOPMENT

As enamel maturation proceeds towards completion of the tooth crown, the cervical loop of the enamel organ starts to recede from the cervical margin as a double-layered cylinder, Hertwig's root sheath, around the lengthening dental papilla. The outer layer of this sheath is continuous with the outer enamel epithelium (8.82). The inner layer, like the inner enamel epithelium with which it is continuous, induces the differentiation of odontoblasts from contiguous mesenchyme cells. Odontoblasts now deposit a layer of dentine against the basal surface of the epithelium (8.83). The sheath continues to grow, outlining the final shape of the roots and inducing differentiation of odontoblasts, surrounding vessels and nerves supplying the dental papilla. These vessels remain and it is because of them, particularly those located centrally near the future apex, that the foramina open through canals in the dentine into the pulp of a fully developed tooth.

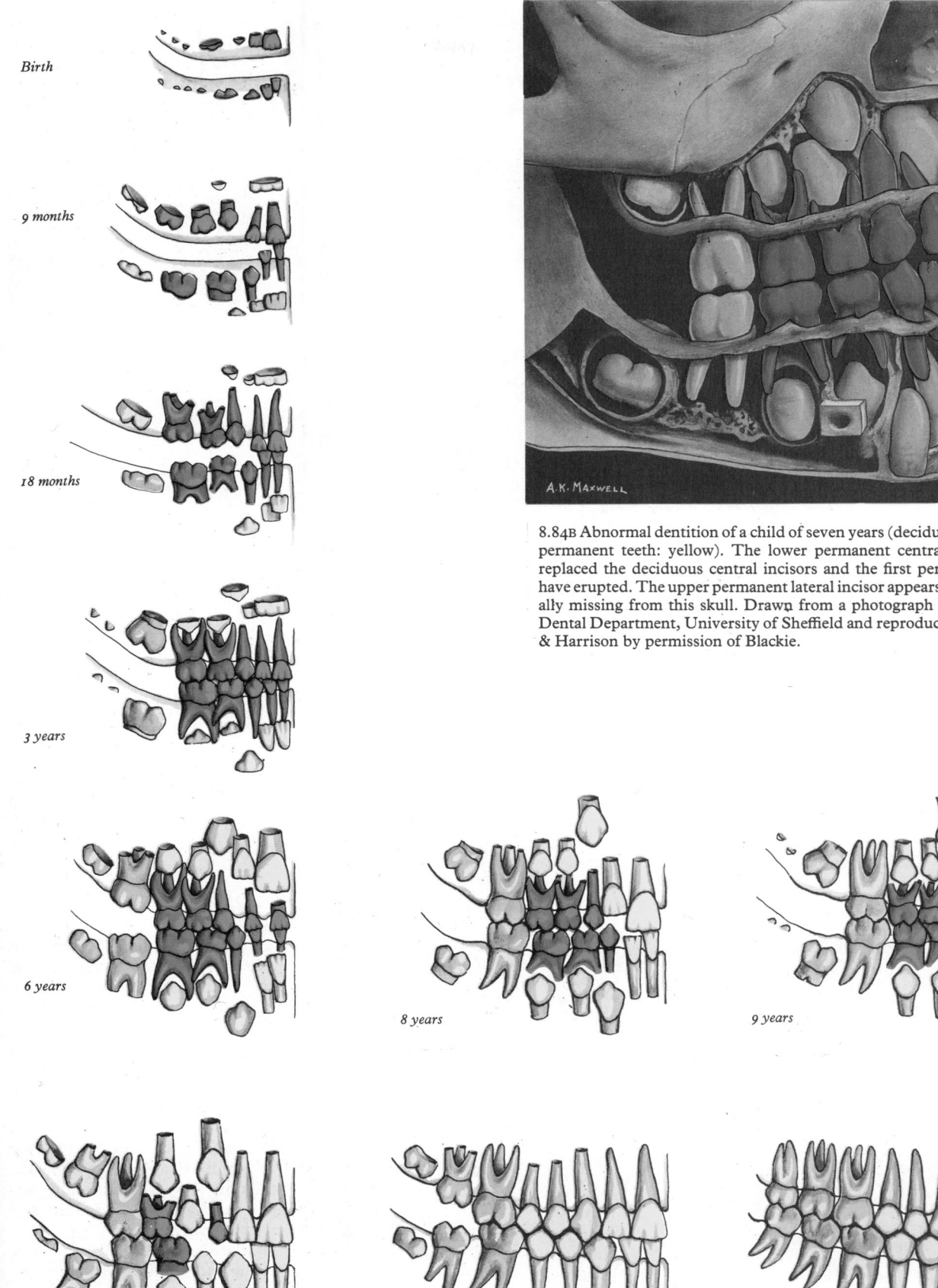

Birth

9 months

18 months

3 years

6 years

8 years

9 years

10 years

12 years

21 years

8.84B Abnormal dentition of a child of seven years (deciduous teeth: blue, permanent teeth: yellow). The lower permanent central incisors have replaced the deciduous central incisors and the first permanent molars have erupted. The upper permanent lateral incisor appears to be congenitally missing from this skull. Drawn from a photograph by F Harrison, Dental Department, University of Sheffield and reproduced from Pedley & Harrison by permission of Blackie.

8.84A Development of the deciduous (blue) and permanent (yellow) dentitions from birth to maturity. Modified from Schour & Massler (1941).

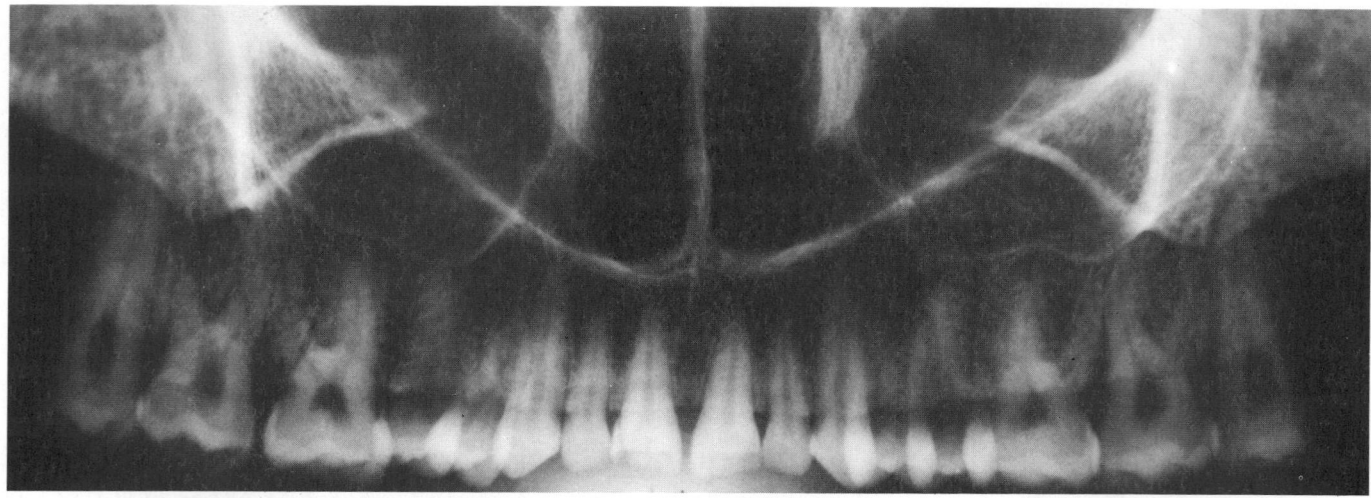

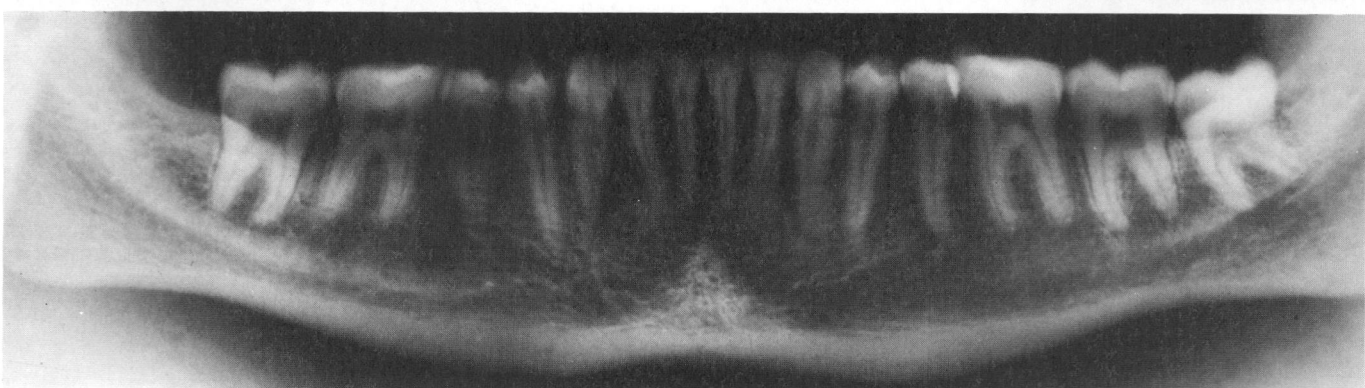

A

B

8.85 Pan-oral radiographs of the whole dentition of the upper and lower jaws of a mature human skull. Note the right lower third molar is missing. To achieve these radiographs, an X-ray source is introduced into the buccal cavity, directed towards the palate for the upper jaw and towards the floor of the cavity for the lower jaw. The X-ray beam is deflected magnetically to disperse anterolaterally and the film is wrapped around the external aspect of the jaws. The clarity of the radiographs is much greater than that possible during clinical radiography of the living head. The radiographs were prepared by D White of the X-ray Department of the Royal Dental Hospital, London.

During crown development the capillaries are most numerous beneath the cuspal growth centres. The papilla grows more rapidly near the capillaries, less rapidly elsewhere. In the latter regions a growing tip of root sheath is able to penetrate under the papilla to meet a growing tip from the other side, separating the two roots. They may, however, just fail to meet, in which case a single flattened root is formed with two root canals. Thus the principal cusps of a multicusped tooth are each supported by a root, which is either separate or incompletely separated from its neighbours.

THE DEVELOPMENT OF CEMENT

Shortly after initial dentinogenesis in the root, the adjacent root sheath epithelium becomes fenestrated (vide infra), exposing the unmineralized exterior of the dentine to the vascularized follicular mesenchyme, from whose investing layer cementoblasts are differentiated. These early cementoblasts do not synthesize collagen but secrete ground substance onto the dentine matrix and around bundles of collagen fibres developed in the follicle which have fanned out on the dentine surface. This composite matrix (which may also include epithelial cell secretions, vide supra) is mineralized under the influence of cementoblasts (Owens 1975). The collagen fibre bundles are about 6 μm wide and their cores mineralize more rapidly than their peripheries. Because it is unmineralized at the time of its formation, the junction between dentine and cement has crystalline continuity and is hence not easy to define in electron micrographs.

First-formed cement, containing only extrinsic collagen fibres, develops while the tooth is erupting and is therefore present on the cervical third of the root dentine. Cementoblasts differentiating over the remainder of the root and, in all later cement, contribute collagen fibres to the matrix (Boyde & Jones 1972). This mixed fibre cement therefore contains both *extrinsic* (Sharpey's) *fibres* of periodontal origin and *intrinsic fibres* from cementoblasts, the former being perpendicular to the developing surface, the latter parallel. Intrinsic fibres are mineralized first. Sharpey's fibres are then mineralized around their peripheries,

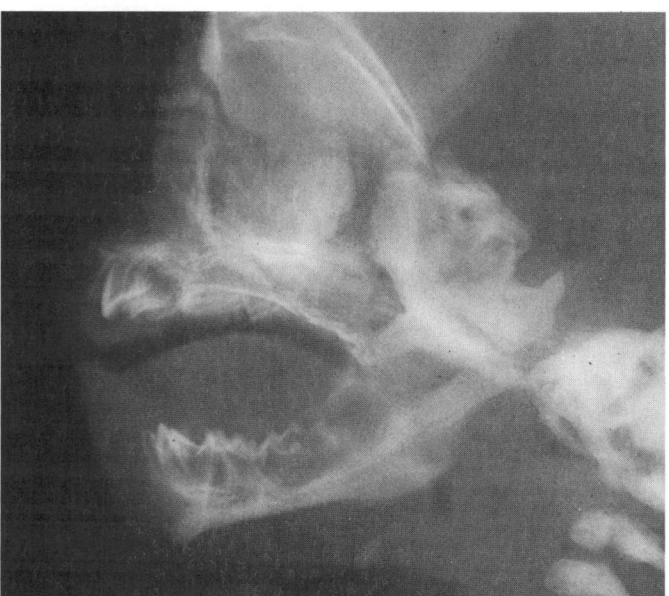

8.86 A lateral radiograph of the jaws of a newborn child. Note the state of development of the jaws and teeth.

while the cores are unmineralized. In later or more rapid cementogenesis, the cementoblasts are frequently trapped within the matrix; cellular cement is most commonly formed around the apical two-thirds of the root surface.

DEVELOPMENT OF THE PERIODONTIUM

During the cap and bell stages of tooth development some cells of the dental papilla are displaced by the growing cervical loop to lie adjoining the outer enamel epithelium. Cementoblasts and fibroblasts of the tooth-related periodontal ligament are probably derived from these cells (Ten Cate 1974, Palmer & Lumsden 1987). The remaining, bone-related, periodontal ligament and alveolar bone are developed from the outer layers of the follicle.

The principal oblique fibre group of the ligament becomes organized at the time of root development and eruption, with an orientation which is visible first as an oblique array of fibroblasts. Once formed, their oblique orientation is maintained as the tooth moves relative to the alveolar bone. Based on its appearance in sections, it was formally considered that the principal fibres were formed, broken and reformed at an intermediate plexus in the central zone of the ligament (Hindle 1967). Autoradiographic studies following tritiated proline uptake, together with ultrastructural studies of collagen phagocytosis suggest, however, that fibres are replaced more rapidly close to alveolar bone (Melcher 1976).

The epithelial debris of Malassez is formed by the remains of Hertwig's root sheath which, following its fenestration and the translocation of presumptive cementoblasts from the follicle to the dentine surface, moves away from cement into the tooth-related periodontal ligament.

DENTAL ERUPTION

When the deciduous dentition is initiated, the five tooth germs in each quadrant occupy jaws which are about 1 mm long. By the time the teeth have erupted into the oral cavity they occupy about 3 or 4 cm of jaw; during development the teeth have migrated apart. It is not known how this movement is brought about.

Very soon after root formation starts, the teeth begin to move towards the oral cavity (Darling & Levers 1976). The origin of the forces which move teeth is not certainly known. A tooth could be *pushed* by growth of its root, proliferation of cells in the pulp or tissue hydrostatic pressure. In support of the latter, van Hassel & McMinn (1972) have shown in the dog that the tissue pressure above an erupting tooth is about 10 mmHg and within the tooth it is about 23 mmHg, creating an eruptive pressure of 13 mmHg. Alternatively, a tooth could be *pulled* out of its socket by shrinkage of the obliquely placed collagen fibres (Thomas 1976) or by the movement of fibroblasts within the periodontal ligament (Beertsen et al 1974). It has also been suggested that the remodelling of alveolar bone may cause eruption (Marks & Cahill 1984). In reality, several of the above factors are likely to co-operate in the eruptive process (Moxham & Berkovitz 1983).

The permanent incisors and canines initially develop lingual to their predecessors (8.81) but erupt along labially-inclined paths. This movement is associated with the intermittent but progressive osteoclastic resorption of the roots of deciduous teeth (Furseth 1968). In periods of quiescence, the resorbed tissues are temporarily repaired by the deposition of cement.

Early in its development, each premolar moves directly deep to its predecessor to become lodged between widely-divergent roots (8.61). As the premolar erupts it induces resorption of the deciduous molar. The enamel of the underlying permanent tooth is protected from resorption by its *reduced enamel epithelium* (see p. 1313).

Usual times of eruption of deciduous teeth. (These vary. See **8.84A,B.**)

Central incisors	6–8 months
Lateral incisors	8–10 months
First molars	12–16 months
Canines	16–20 months
Second molars	20–30 months

Calcification of permanent teeth proceeds as follows: first molars at birth, incisors (except upper lateral) and canines about 4 months, upper lateral incisors about 1 year, premolars about 2 years, second molars about 3 years and third molars about 8 years.

Usual times of eruption of permanent teeth. (Upper teeth erupt a little later than lower. See **8.84A,B.**)

First molars	6–7 years
Central incisors	6–8 years
Lateral incisors	7–9 years
Canines	9–12 years
First and second premolars	10–12 years
Second molars	11–13 years
Third molars	17–21 years

ALVEOLAR DEVELOPMENT

Jaws begin to develop when the dental lamina is forming. In the mandible the growing margin of membrane bone lateral to Meckel's cartilage passes back caudal to the inferior alveolar nerve; from it lateral and medial plates grow upwards (Dixon 1958, Melcher & Bowen 1969). Developing teeth thus appear to descend between the plates (8.81). An osseous horizontal partition divides the teeth from the inferior alveolar nerve and vertical septa later isolate each tooth in its own crypt. Bone does not develop over deciduous teeth and in skeletonized neonatal jaws teeth are usually lost. A similar process is involved in the development of maxillary crypts.

When the teeth start to move towards the oral cavity, their sockets also grow, deepening the crypts and increasing the height of the jaw. The rate of eruption outstrips the rate of upward bone growth until the teeth meet their opponents. Developing roots have wide open apices but these later close around nerves and vessels to form foramina. The lengthening of osseous sockets for the teeth much increases the depth of the face up to and during puberty (Scott 1967).

CUTICLES AND EPITHELIAL ATTACHMENT

At the end of amelogenesis, the cells of the enamel organ revert to a squamous shape, becoming a thin stratified layer covering the whole surface of the enamel. This is referred to as the *reduced enamel epithelium*. During eruption it fuses with the oral epithelium to provide an epithelium-lined path for the tooth. The reduced enamel epithelium is rapidly worn away from the exposed surface of the tooth except at the neck where it forms the

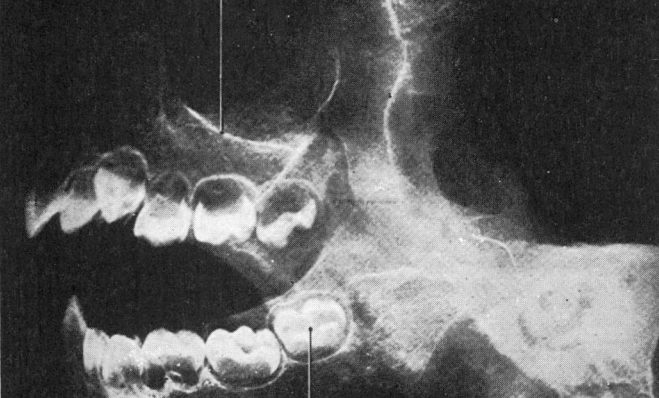

Floor of maxillary sinus

First permanent molar

8.87A Radiograph of the jaws of an infant, nine months old (from Symington & Rankin *Atlas of Skiagrams*). Only the lower central incisor has erupted; the roots of the first lower deciduous molar are just beginning to form; the crown of the first lower permanent molar faces inwards.

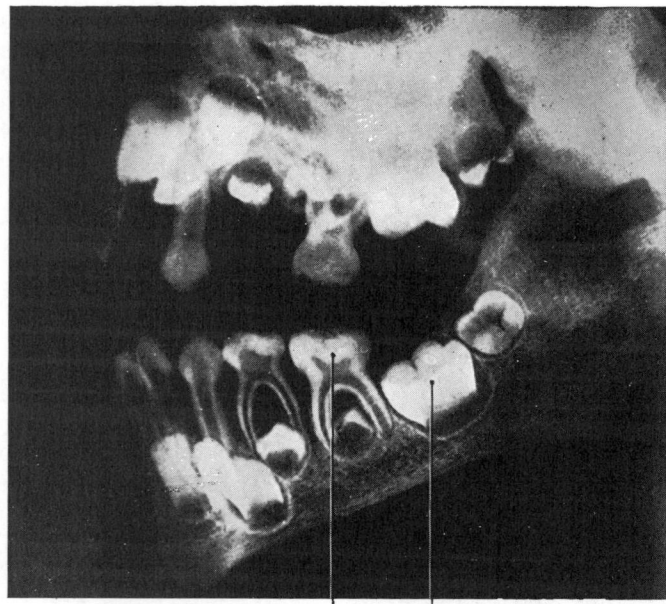

Second deciduous molar   *First permanent molar*

8.87B Radiograph of the teeth of a boy, aged five years (from Symington & Rankin *Atlas of Skiagrams*). In the maxilla the lateral deciduous incisor and the first deciduous molar have been lost, but all the deciduous teeth are present in the mandible. No absorption of the roots of the deciduous teeth has occurred. No permanent teeth have erupted.

epithelial attachment (**8.**43,83) at the dentogingival junction. The reduced enamel epithelium is separated from the enamel surface by a structureless layer, the *primary enamel cuticle*, about 1 μm thick, which may be the final, unmineralized product of ameloblasts. This cuticle together with the reduced enamel epithelium is called *Nasmyth's membrane*. A form of cement known as *afibrillar cement* has been observed over the enamel around the necks of teeth (Listgarten 1966). Afibrillar cement is only about 100 nm thick, contains no banded collagen and is probably produced by cementoblasts following premature disruption of part of the reduced enamel epithelium. Occasionally a cuticle about 4 μm thick is found between the epithelial attachment and the tooth. This is the *secondary enamel cuticle*. It may be a product of the epithelial cells or it may be the remains of blood which has leaked through the epithelial attachment following some slight trauma (Hodson 1966). Finally, salivary protein and carbohydrate form an adherent film on tooth surfaces. This is known as the *pellicle* and is the foundation of *dental plaque*.

APPLIED ANATOMY

The cortical plate of bone in each jaw is continuous over the alveolar crest with the cortical plate lining the tooth socket (the *cribriform plate* or *lamina dura* of radiographs). On the labial and buccal aspects of *upper teeth*, these two cortical plates usually fuse with very little trabecular bone between them, except where the buccal bone thickens over the molar teeth near the root of the zygomatic arch (DuBrul 1980). It is easier and more convenient to extract upper teeth by fracturing the buccal than the palatal plate. Anteriorly in the *lower jaw*, labial and lingual plates are thin but in the molar region the buccal plate is thickened as the external oblique line. Near the lower third molar, the lingual bone is much thinner than the buccal and it is mechanically easier to remove this tooth, when impacted, via the lingual plate. However, the lingual nerve is here exposed to damage.

Abscesses developing in relation to the apices of roots ultimately penetrate the surrounding bone where it is thinnest. The position of the resultant swelling in the soft tissues is largely determined by the relationship between muscle attachments and the sinus (the path taken by the infected material) in the bone.

Thus, in the lower incisor region, because the labial bone is thin, abscesses generally appear as a swelling in the labial sulcus, above the attachment of the mentalis. But the abscess may open below the mentalis and point beneath the chin. If an abscess from a lower postcanine tooth opens below the attachment of the buccinator, the swelling is in the neck; if it opens above, the swelling is in the buccal sulcus. If an abscess opens lingually above the mylohyoid, the swelling is in the lingual sulcus; if it is below, the swelling is in the neck. Because the mylohyoid ascends posteriorly, third molar abscesses tend to track into the neck rather than the mouth.

Apart from canine teeth, which have long roots, abscesses on upper teeth usually open buccally below, rather than above the attachment of buccinator. Because its root apex is occasionally curved towards the palate, abscesses of the upper lateral incisors may track into the palatal submucosa. Abscesses of upper canines often open facially just below the orbit. Here the swelling may obstruct drainage in the angular part of the facial vein (p. 793) which has no valves; it is therefore possible for infected material to travel via the angular and ophthalmic veins into the cavernous sinus. Abscesses on the palatal roots of upper molars usually open on the palate.

Upper second premolars and first and second molars are related to the maxillary sinus. When this is large, the root apices of these teeth may be separated from its cavity solely by the lining mucosa. Sinus infections may stimulate the nerves entering the teeth, simulating toothache. Upper first premolars and third molars may be closely related to the maxillary sinus.

With loss of teeth, alveolar bone is extensively resorbed. Thus in the edentulous mandible the mental nerve, originally inferior to premolar teeth, may lie near the crest of the bone. In the edentulous maxilla, its sinus may enlarge to approach the bone's oral surface. Lingual to the lower premolars or molars, the upper molars and in the midline of the palate, there are occasional bony prominences termed the *torus mandibularis*, *torus maxillaris* and *torus palatinus* (pp. 367, 354, 389). They may need surgical removal before satisfactory dentures can be fitted.

Severe systemic infections during the time the teeth are developing may lead to faults in enamel, visible as horizontal lines (cf. Harris's growth lines p. 283).

DENTAL RADIOLOGY

Due to dense mineralization, enamel and dentine are radio-opaque whereas pulp appears as a radiolucent region (**8.**85A,B, 86,87A,B). Caries, which attacks teeth through the enamel or root surface, is easily diagnosed by intra-oral radiographs because it demineralizes teeth. The root of a tooth is separated from cortical bone (*lamina dura*) by the radiolucent periodontal ligament. Chronic infections of the pulp spread into this ligament, leading to resorption of the lamina dura around the dental apex. Thus continuity of the lamina dura around the apex of a tooth usually indicates a healthy apical region, except in acute infections where resorption of bone has not yet begun.

Anteriorly, interdental septa form sharp crests between the teeth; between molariform teeth they form tables. In periodontal disease, the eventual sequel to gingival disease, bony crests or tables are resorbed and the extent of the condition can therefore be determined radiographically.

Radiographs are commonly used to estimate separation between the roots of maxillary teeth and the sinus; roots of the second premolar and first and second molar often project into the sinus. Mandibular third molars are commonly prevented from erupting by impaction against the second molars and radiographs are indispensable in assessing the degree of difficulty in extracting such teeth. Superior to the mental spines (genial tubercles), many mandibles exhibit a well-defined pit ending in a canal, presumably containing a median blood vessel or perhaps penetrating collagen fibres connected to the lingual frenulum. This pit, together with a radio-opaque ring of compact bone, can usually be seen in radiographs of incisors and is a useful median landmark in an edentulous mandible.

THE TONGUE, PHARYNX AND OESOPHAGUS

The Tongue

The tongue (**8.88–90**) is a highly muscular organ of deglutition, taste and speech; it is partly oral and partly pharyngeal in position. It is attached by its muscles to the hyoid bone, mandible, styloid processes, soft palate and pharyngeal wall. It has a root, an apex, a curved dorsum and an inferior surface. Its mucosa is normally pink and moist.

The *root of the tongue* (**8.89**) is attached to the hyoid and mandible and between them it is in contact inferiorly with the geniohyoid and mylohyoid muscles. The *dorsum* (posterosuperior surface) is generally convex at rest and is divided into an anterior part facing upwards and a posterior part facing posteriorly. These are separated by a V-shaped *sulcus terminalis*, whose limbs run anterolaterally from a median *foramen caecum* to the palatoglossal arches (**8.88**). The foramen indicates the site of the upper end of the thyroid diverticulum (p. 228) and the sulcus separates the oral (anterior two-thirds) and pharyngeal (posterior one-third) regions of the tongue, which differ in their mucosa, nerve supply and development (pp. 228, 1322). They are more suitably referred to as the *presulcal* and *postsulcal* parts.

The oral (presulcal) part (**8.88,89**), located in the oral cavity and on its floor, has an *apex* touching the incisor teeth; its *margin* is in contact with the gums and teeth; its *superior surface* (dorsum) is related to the hard and soft palates. On each side, in front of the palatoglossal arch, are four or five vertical folds, the *foliate papillae* (**8.88**). The dorsal mucosa has a median sulcus (**8.88,89**), is adherent to the subjacent muscle of the tongue and is papillated. The inferior mucosa is smooth, purplish and reflected on to the oral floor and gums, being connected to the former by the median *frenulum linguae* (**8.89**); lateral to this, the deep lingual vein is visible, and lateral to the vein is a fringed *plica fimbriata*, directed anteromedially towards the lingual apex. The oral part of the tongue develops from lingual swellings of the mandibular arch and also the tuberculum impar (p. 228). Its general sensory nerve is the lingual, the chorda tympani mediating taste. (For the origin of lingual musculature see p. 175.)

The pharyngeal (postsulcal) part (**8.88**) lies posterior to the palatoglossal arches (the *base* of the tongue) and forms the anterior wall of the oropharynx. Its mucosa is reflected laterally on to the palatine tonsils and pharyngeal wall and posteriorly on to the epiglottic folds. Devoid of papillae, it has low elevations, due to lymphoid nodules embedded in the submucosa, collectively termed the *lingual tonsil*. The postsulcal tongue develops from the hypobranchial eminence (p. 228). Its sensory nerve, including taste, is the glossopharyngeal, whose rami also extend anterior to the sulcus terminalis to supply taste organs in the vallate papillae, an arrangement explained by the anterior extension of the hypobranchial eminence (derived from the third pharyngeal arch) over part of the lingual swellings (p. 228).

The lingual papillae (**8.88**) are projections of lamina propria which elevate the epithelium above the general level. They are numerous but limited to the presulcal part of the dorsum, producing its characteristic roughness; there are *papillae vallatae, fungiformes, filiformes* and *simplices*. *Foliate* papillae (**8.88**) lie laterally. Papillae are visible modifications of mucous membrane which increase the area of contact between the tongue with the contents of the mouth. *Taste buds* (see p. 1169) are microscopic specialized cellular arrangements around the gustatory nerve endings; they are much more widespread than papillae, being scattered over the entire lingual dorsum and sides, epiglottis and lingual aspect of the soft palate; each region is hence innervated by the appropriate gustatory nerves (facial, glossopharyngeal or vagus). The papillae are more visible in the living when the tongue is dry.

Vallate papillae (**8.88**) are large, varying in number from eight to 12 on the dorsum of the tongue; they form a V-shaped row immediately in front of the sulcus terminalis. Each papilla, 1–2 mm in diameter, is encompassed by a circular depression of mucosa and surrounded by a wall (*vallum*) separated from the papilla by a circular sulcus (**8.91**). The papilla is a truncated cone,

its smaller end attached, its broader superficial aspect projecting and bearing numerous small secondary subepithelial papillae. The entire structure is covered with stratified squamous epithelium; in both sulcal walls taste buds abound.

Fungiform papillae (**8.88,92**), more numerous than vallate, occur mainly on the lingual margin but also occur irregularly on the

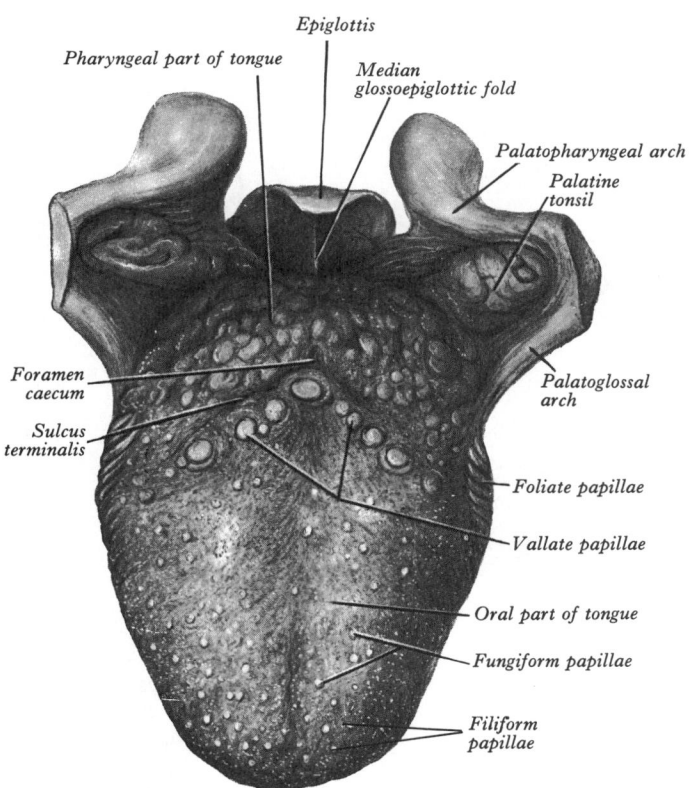

8.88 The dorsum of the tongue.

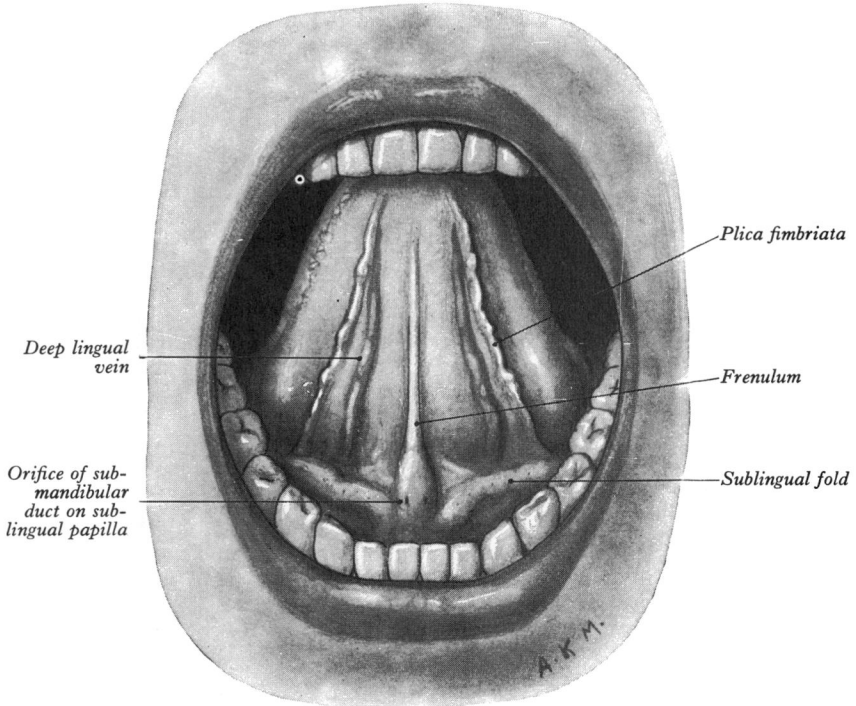

8.89 The cavity of the mouth. The tip of the tongue is turned upwards. In the person from whom the drawing was made the two sublingual papillae formed a single median elevation (see p. 1293).

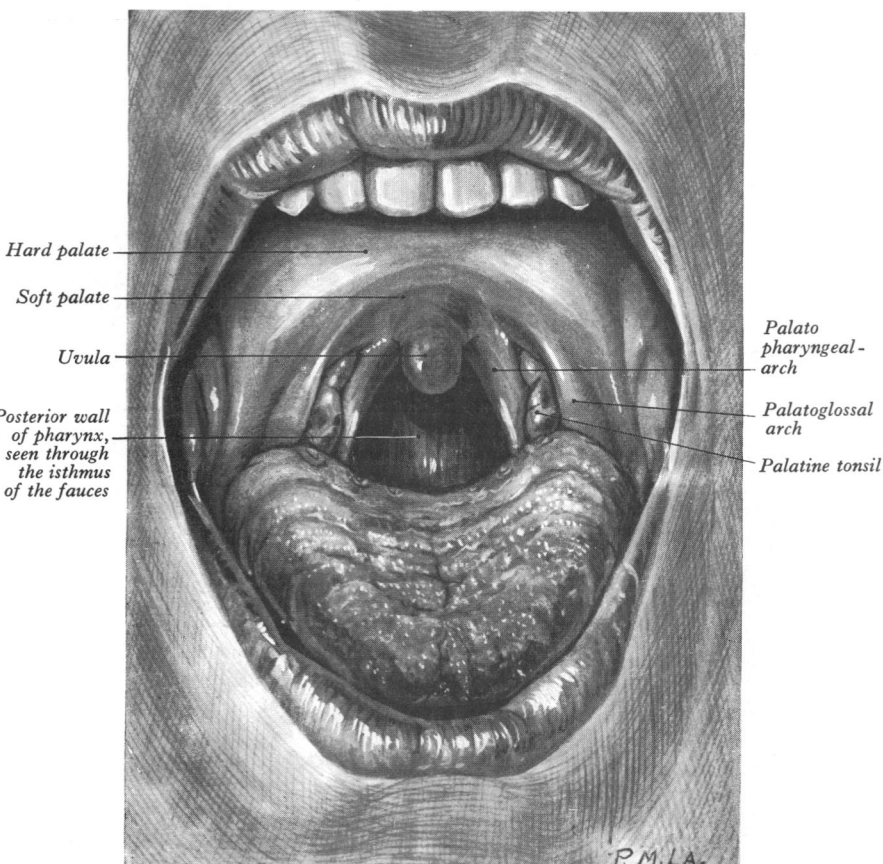

8.90 The cavity of the mouth.

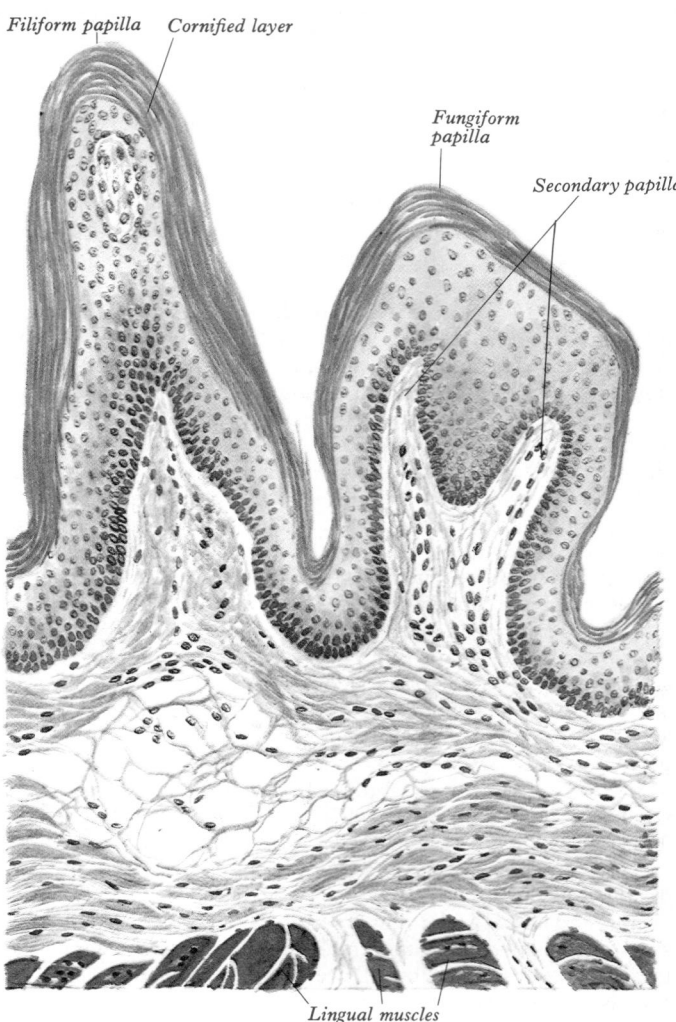

8.92 Section through a filiform papilla and an adjoining fungiform papilla. Stained with haematoxylin and eosin. Drawn from a preparation lent by E E Hewer. Magnification × 30.

dorsum, where they may occasionally be numerous. They differ from filiform papillae by their larger size, round shape and deep red colour; each has secondary subepithelial papillae and bears taste buds.

Filiform papillae (**8**.92) cover most of the presulcal dorsal area (Kullaa-Mikkonen et al 1987). They are minute, conical or cylindrical and arranged in rows parallel with the vallate papillae, except at the lingual apex where they are transverse. They bear many secondary papillae, which are more pointed than those of

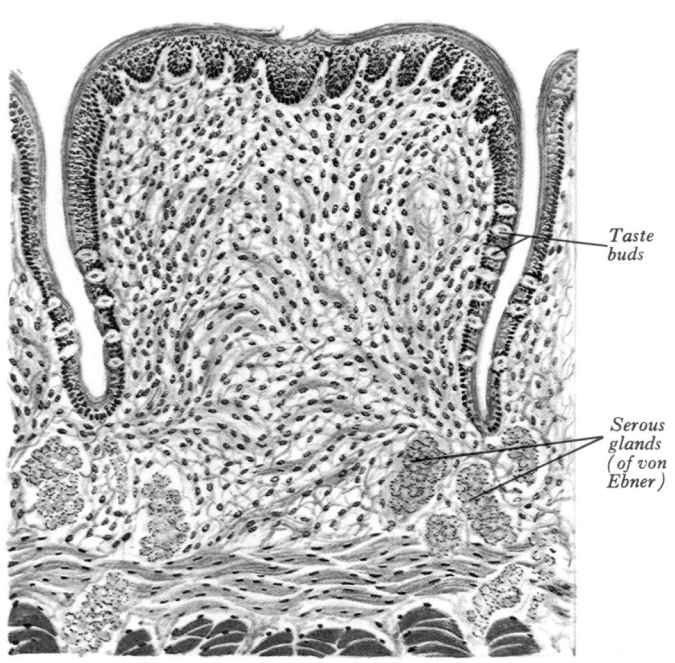

8.91 Section through a vallate papilla. (after Sobotta). Stained with haematoxylin and eosin. Magnification × 32.

vallate and fungiform papillae. Their epithelium, which is keratinized, may split into fine processes, each being the apex of a secondary papilla; these processes are whitish, owing to thickened epithelium, its elongated cells being keratinized. These papillae do not bear taste buds but increase the friction between the tongue and food, facilitating the movement of particles by the tongue within the oral cavity.

Papillae simplices are like those of the dermis (p. 79) and cover the whole mucous membrane of the tongue, including the larger papillae.

Foliate papillae (**8**.88) or *folia linguae* are red, leaf-like projections found at the sides of the tongue near the sulcus terminalis. Their epithelium contains taste buds.

THE LINGUAL MUSCULATURE

The tongue is divided by a median fibrous septum, attached to the body of the hyoid bone. In each half are both the extrinsic and intrinsic muscles (**8**.93,94), the former extending outside the tongue, the latter wholly within it.

Extrinsic Muscles

These (**8**.94) include the genioglossus, hyoglossus, styloglossus, chondroglossus and palatoglossus. The palatoglossus, closely associated with the soft palate in function and innervation, is described with the other palatal muscles (p. 1289).

The genioglossus is triangular, lying near and parallel to the midline; it arises from a short tendon attached to the superior genial tubercle behind the mandibular symphysis above the origin of the geniohyoid. From this it fans out backwards and

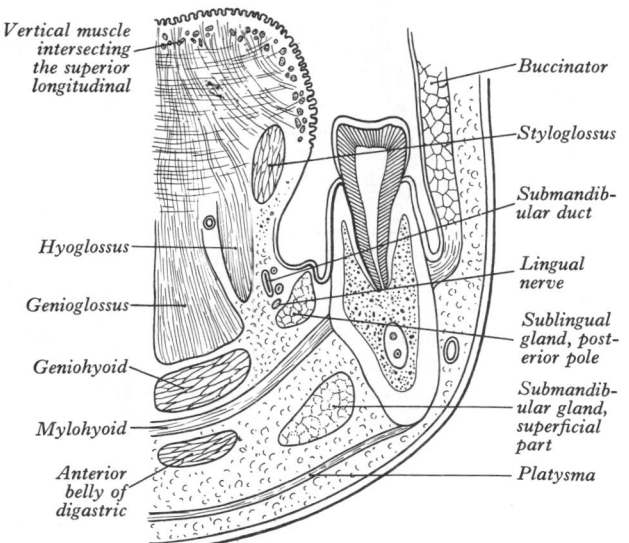

8.93 Diagram of a coronal section through the tongue, the mouth and the body of the mandible opposite the first molar tooth.

upwards. The inferior fibres of genioglossus are attached by a thin aponeurosis to the upper anterior surface of the hyoid body near the midline, a few passing between hyoglossus and chondroglossus to blend with the pharyngeal middle constrictor; intermediate fibres pass backwards, and superior fibres ascend forwards to enter the whole length of the ventral surface of the tongue, from root to apex, intermingling with the intrinsic lingual muscles. The muscles of opposite sides are separated posteriorly by the lingual septum (p. 1322); anteriorly they are variably blended by decussation of fasciculi across the midline. Doran & Baggett (1972) considered that no fibres reach the lingual apex in man or other mammals.

Actions. Genioglossus brings about the forward traction of the tongue to protrude its apex from the mouth. Acting bilaterally, the two muscles depress the central part of the tongue, making it concave from side to side.

The hyoglossus, thin and quadrilateral, is attached to the whole length of the greater cornu and the front of the body of the hyoid bone, passing almost vertically up to enter the side of the tongue between the styloglossus laterally and the inferior longitudinal muscle medially. Fibres arising from the hyoid body overlap those from the greater cornu.

Relations. Hyoglossus is related at its *superficial surface* with:

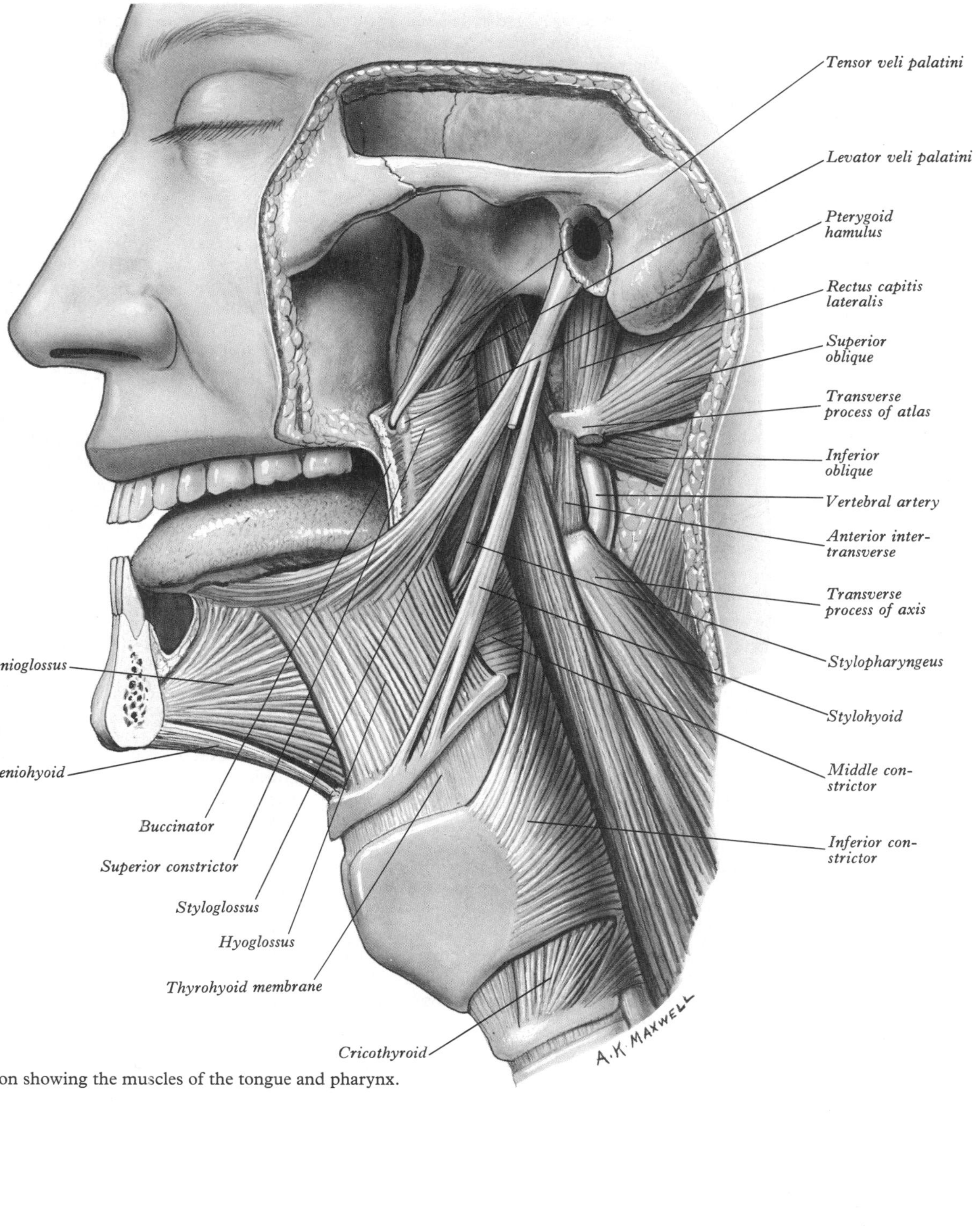

8.94 Dissection showing the muscles of the tongue and pharynx.

the digastric tendon, stylohyoid, styloglossus and mylohyoid, the lingual nerve and submandibular ganglion, the sublingual gland, the deep part of the submandibular gland and duct, the hypoglossal nerve and the deep lingual vein. By its *deep surface* it is related with: the stylohyoid ligament, genioglossus, inferior longitudinal muscle, lingual artery and glossopharyngeal nerve. Postero-inferiorly it is separated from the middle constrictor by the lingual artery; this part of the muscle is in the lateral pharyngeal wall, below the palatine tonsil. Passing deep to the muscle's posterior border are, in descending order: the glossopharyngeal nerve, stylohyoid ligament and lingual artery.

Action. The hyoglossus depresses the tongue.

The chondroglossus, sometimes described as a part of hyoglossus, is separated from it by fibres of the genioglossus which pass to the side of the pharynx. It is about 2 cm long, arising from the medial side and base of the lesser cornu and the adjoining part of the hyoid body and ascending to merge into the intrinsic musculature between the hyoglossus and genioglossus. A small slip occasionally springs from the cartilago triticea and enters the tongue with the posterior fibres of hyoglossus.

Action. Chondroglossus assists hyoglossus in depressing the tongue.

The styloglossus, shortest and smallest of the three styloid muscles, arises from the anterolateral aspect of the styloid process, near its apex, and from the styloid end of the stylomandibular ligament. Passing down and forwards, it divides at the side of the tongue into a longitudinal part, which enters the tongue dorsolaterally to blend with the inferior longitudinal muscle in front of the hyoglossus, and an oblique part, overlapping the hyoglossus and decussating with it.

Action. Styloglossus draws the tongue up and backwards.

Nerve supply. Excepting palatoglossus (p. 1290) all extrinsic lingual muscles are supplied by the hypoglossal nerve.

Intrinsic Muscles

These are the bilateral superior and inferior longitudinal, the transverse and the vertical.

The **superior longitudinal muscle** is a thin stratum of oblique and longitudinal fibres lying beneath the dorsal lingual mucosa, extending forwards from the submucous fibrous tissue near the epiglottis and from the median lingual septum to the lingual margins, some fibres being inserted into the mucous membrane. The **inferior longitudinal muscle** is a narrow band close to the inferior lingual surface between genioglossus and hyoglossus. It extends from the lingual root to the apex, some of its posterior fibres being connected to the body of the hyoid bone; anteriorly it blends with the styloglossus.

The transverse muscle passes laterally from the median fibrous septum to the submucous fibrous tissue at the lingual margin, blending with the palatopharyngeus (p. 1290).

The vertical muscle extends from the dorsal to the ventral aspects of the tongue in its anterolateral regions.

Nerve supply. All intrinsic lingual muscles are supplied by the hypoglossal nerve.

Actions. The intrinsic muscles alter the shape of the tongue; thus, the superior and inferior longitudinal muscles tend to shorten it; but the former also turns the apex and sides upwards to make the dorsum concave, while the latter pulls the apex down to make the dorsum convex. The transverse muscle narrows and elongates the tongue; the vertical muscle makes it flatter and wider. Acting alone or in pairs and in endless combination, they give the tongue precise and highly varied mobility, important not only in alimentary function but also in speech (p. 1257).

LINGUAL STRUCTURE

The tongue consists largely of skeletal muscular tissue, partly invested by mucous membrane. The *lingual mucosa* of the inferior surface is thin, smooth and like that in much of the rest of the oral cavity. The mucosa of the pharyngeal part of the dorsum contains many lymphoid follicles, each follicle forming a rounded eminence, central in which is the minute orifice of a funnel-shaped recess. Many round or oval lymphoid nodules, each encapsulated by submucous fibrous tissue, surround each recess,

which receives the ducts of mucous glands in its floor. In the oral part the dorsal mucosa is somewhat thicker than ventrally and laterally; it is adherent to muscular tissue, and covered by numerous *papillae* (p. 79). It consists of connective tissue (*lamina propria*) and stratified squamous epithelium, which also covers each papilla. The *lamina propria* is a dense fibrous connective tissue, with numerous elastic fibres, united to similar tissue between the lingual muscle fasciculi. It contains the ramifications of numerous vessels and nerves from which the papillae are supplied, large lymph plexuses and lingual glands. The epithelium varies from parakeratinized stratified squamous epithelium posteriorly, to fully keratinized epithelium overlying the filiform papillae more anteriorly; these features appear to be related to the fact that the apex of the tongue is subject to greater dehydration than the posterior and ventral parts and is more abraded during mastication.

The *lingual glands* are mucous, serous and mixed. *Mucous glands* are like labial and buccal glands in structure; they are numerous in the postsulcal region but are also present at the apex and margins. *Anterior lingual salivary glands* lie at the ventral surface of the apex (**8.**95), one on each side of the frenulum, where they are covered by mucosa and a muscular fasciculus from the styloglossus and inferior longitudinal muscles. From 12 mm to 20 mm long and about 8 mm broad, each has mucous and serous alveoli and opens by three or four ducts on the inferior apical surface. The fine structure of the human *deep posterior lingual glands* has been described in detail by Testa Riva et al (1985). *Serous glands* (of von Ebner, **8.**91) occur near the taste buds, their ducts opening mostly into vallate sulci. They are racemose, the main duct dividing into several channels ending in acini. Their secretion is watery, probably assisting in gustation by spreading and then dispersing substances. The pyramidal shape and ultrastructure of the secretory cells of the acini are similar to those of serous cells elsewhere (Testa Riva et al 1985).

The *lingual septum*, a median fibrous partition of the whole organ, does not quite reach the dorsum. It is an attachment of the transverse lingual muscles and appears prominently in coronal sections. Posteriorly it widens to form the *hyoglossal membrane*, connecting the lingual root to the hyoid bone; the inferior fibres of the genioglossi are also attached to it.

Vessels and Nerves

The main *artery* is the lingual branch of the external carotid (p. 736) but the tonsillar and ascending palatine branches of the facial and ascending pharyngeal arteries also supply the lingual root. In the vallecula (p. 1252) epiglottic rami of the superior laryngeal artery anastomose with the inferior dorsal branches of the lingual artery. Lingual muscles are supplied from this rich anastomotic network and there is a very dense submucosal plexus (Combelles 1974). The *veins* are described on p. 797 and *lymph vessels* (**6.**203) on p. 845.

The *sensory nerves* are: (1) the lingual branch of the mandibular nerve for general sensation in the presulcal region; (2) the chorda tympani branch of the facial nerve (p. 1110), running in the sheath of the lingual nerve, for gustation in the presulcal region exclusive of the vallate papillae (p. 1170); this is partly derived from the nervus intermedius (p. 1107); (3) the lingual branch of the glossopharyngeal nerve (p. 1113), distributed to the postsulcal mucosa of the lingual base and sides and to the vallate papillae and mediating general and gustatory sensation; (4) the superior laryngeal nerve (vagus) (p. 1117), which sends fine rami to the root immediately in front of the epiglottis.

The problem of proprioception in the tongue has been reviewed by Fitzgerald & Sachithanadan (1979). Muscle spindles occur in monkeys (Bowman 1968) and in mankind (e.g. Nakayama 1944, Cooper 1953, Kubota et al 1975), as also confirmed in extensive simian material by the above reviewers. The peripheral route from these undoubted receptors is not clear, though it is perhaps in the lingual or hypoglossal nerves and by cervical spinal nerves communicating with the latter. In monkeys Fitzgerald & Sachithanadan presented strong evidence indicating the hypoglossal nerve as the main proprioceptor route for the intrinsic and extrinsic musculature, many fibres leaving it to enter the second and third cervical anterior primary rami.

In the postsulcal region, isolated nerve cells have been observed, perhaps postganglionic parasympathetic neurons, *probably* innervating glandular tissue and non-striated vascular myocytes (Chu 1968).

Applied Anatomy. Congenital cysts and fistulae may develop from the persistent remains of the thyroglossal duct (p. 229). The attachment of the genioglossi to the genial tubercles behind the mandibular symphysis prevents the tongue from sinking back and obstructing respiration; therefore, anaesthetists pull forward the mandible to obtain the full benefit of this connection.

THE OROPHARYNGEAL ISTHMUS

The aperture of communication between the mouth and pharynx, the *oropharyngeal isthmus* (**8.90**) is between the soft palate and the lingual dorsum, joined at the sides by the palatoglossal arches. Each *palatoglossal arch* runs down, laterally and forwards from the soft palate to the side of the tongue; it is formed by the projecting palatoglossus (p. 1296) with its covering mucous membrane. Approximation of the arches, to shut off the mouth from the oropharynx, is essential to deglutition (p. 1328).

The Pharynx

The pharynx (**8.96–102**), situated behind the nasal cavities, mouth and larynx, is a musculomembranous tube, 12–14 cm long, extending from the cranial base to the level of the sixth cervical

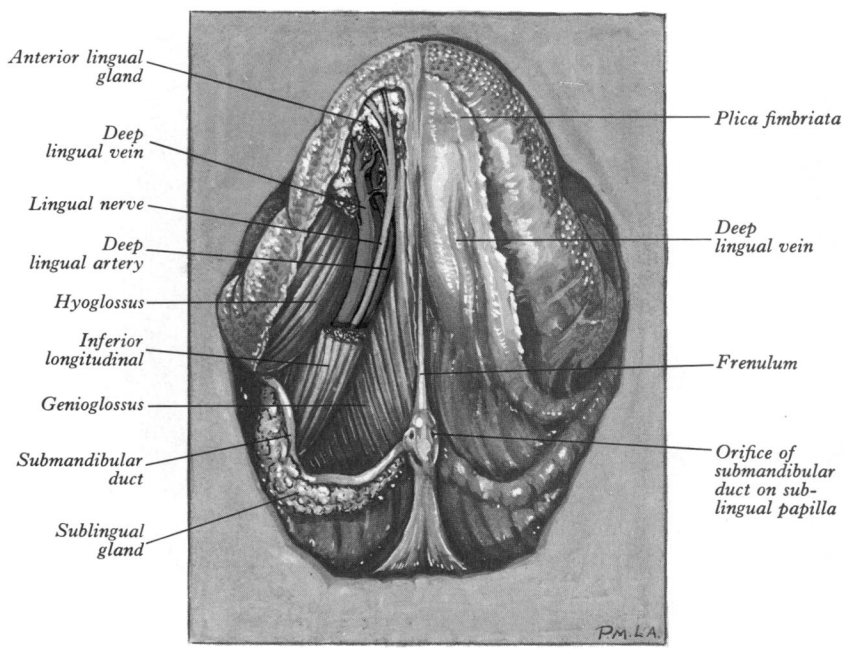

8.95 Dissection of the inferior lingual surface. On the right side (left side of figure) the mucous membrane has been removed and the inferior longitudinal muscle has been divided and partially resected.

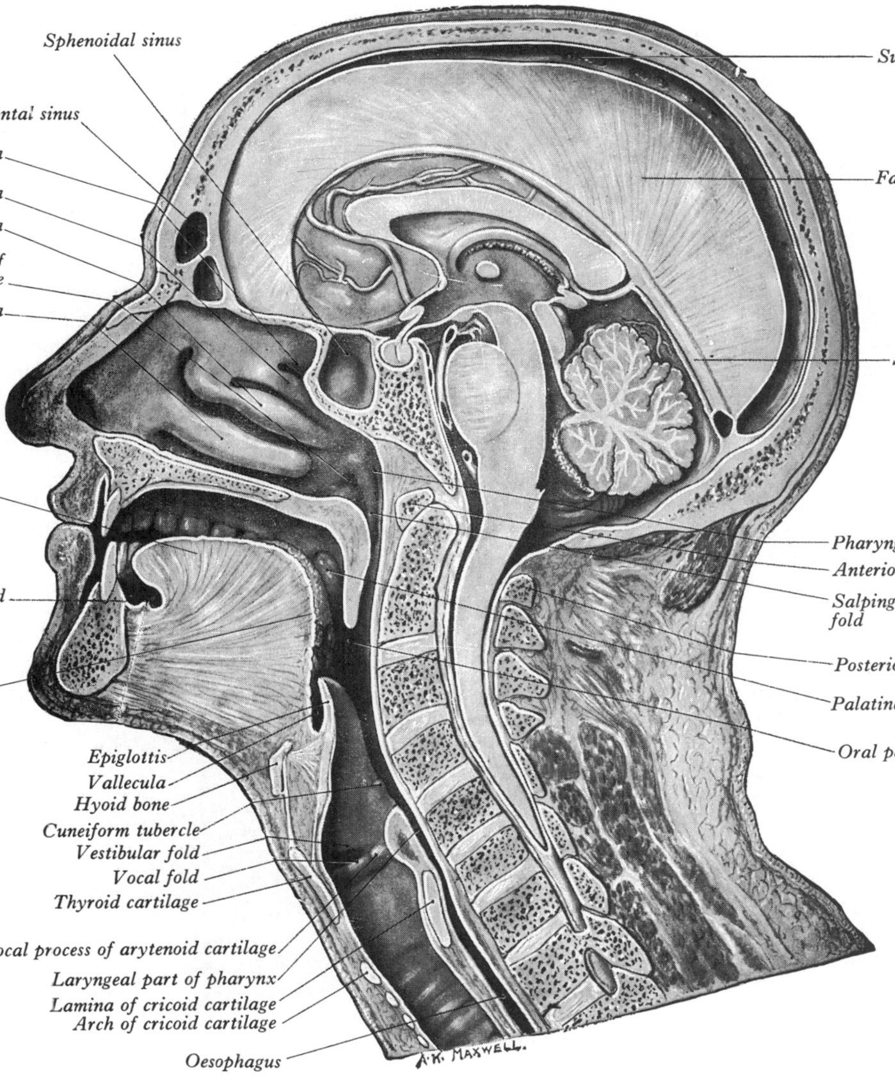

8.96 Sagittal section through the nose, mouth, pharynx and larynx. Where it divides the skull and the brain, the section passes slightly to the left of the median plane but, below the base of the skull, it passes slightly to the right of the median plane.

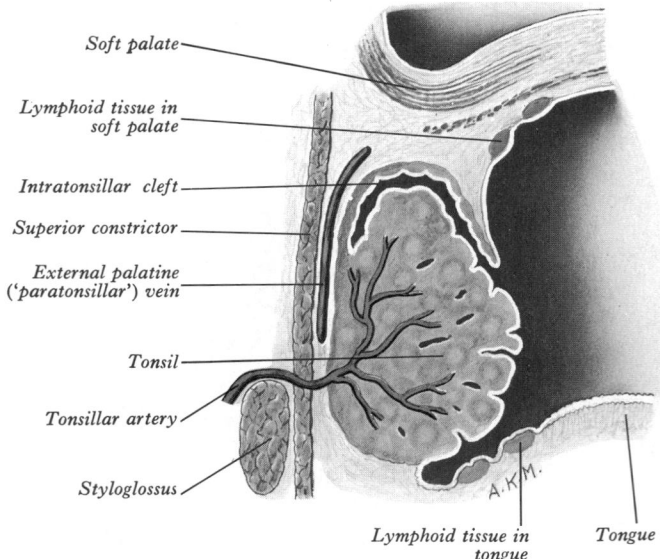

Soft palate

Lymphoid tissue in
soft palate

Intratonsillar cleft

Superior constrictor

External palatine
('paratonsillar') vein

Tonsil

Tonsillar artery

Styloglossus

Lymphoid tissue in
tongue

Tongue

A.K.M.

8.97 Coronal section through the palatine tonsil.

vertebra and the lower border of the cricoid cartilage. Its width is greatest superiorly, measuring 3.5 cm; at its junction with the oesophagus it is reduced to about 1.5 cm, this being the narrowest part of the alimentary canal (except for the vermiform appendix). It is limited *above* by the posterior part of the sphenoid body and the basilar part of the occipital bone; *below*, it is continuous with the oesophagus; *behind*, loose connective tissue separates it from the cervical part of the vertebral column and prevertebral fascia covering longus colli and capitis; *in front*, it opens into the nasal cavity, mouth and larynx, its anterior wall being therefore incomplete. It is attached, from above downwards on each side to: the medial pterygoid plate, pterygomandibular raphe, mandible, tongue, hyoid bone, thyroid and cricoid cartilages; *laterally*, it communicates with the tympanic cavities via the auditory tubes and is related to the styloid processes and their muscles, the common, internal and external carotid arteries, and some of the branches of the last named. The pharynx has three parts: nasal, oral and laryngeal (**8.**96).

NASAL PART OF THE PHARYNX

The nasal part (nasopharynx) is behind the nose and above the soft palate. Wood Jones (1940) suggested that the human nasopharynx is morphologically 'nasal', the soft palate really separating the pharynx from the nasal chambers, a view criticized by Leela et al (1974), who consider the transitional zone between nasal and pharyngeal territories to be just anterior to the orifices of the auditory tubes, although they confirm the concept of nasal and pharyngeal contributions to the region. Except for the soft palate its walls are static; its cavity is never obliterated, in which respect it differs from the oral and laryngeal parts and resembles the nasal cavities. It communicates with the nasal cavity (**8.**96) through the two posterior nasal apertures, which are each 25 mm vertically, 12.5 mm transversely, and are separated by the posterior edge of the nasal septum. Between the border of the soft palate and posterior pharyngeal wall the nasal and oral parts communicate through the *pharyngeal isthmus*, which is closed during swallowing by the elevation of the palate and contraction of the palatopharyngeal sphincter (p. 1290). Each *lateral wall* presents a *pharyngeal opening of the auditory tube*, 10–12.5 mm behind and a little below the posterior end of the inferior nasal concha. Somewhat triangular in shape, this opening is bounded above and behind by the *tubal elevation*, formed over the underlying pharyngeal end of the cartilage of the tube (p. 1226); its prominent posterior margin facilitates the introduction of catheters passed along the floor of the nasal cavity for auditory intubation. A vertical *salpingopharyngeal fold* of mucosa descends from the tubal elevation, covering the *salpingopharyngeus* in the wall of the pharynx. A smaller *salpingopalatine fold*, extends from the anterosuperior

angle of the tubal elevation to the soft palate. The levator veli palatini, entering the soft palate, produces an elevation of the mucosa immediately below the tubal opening (**8.**100). Behind the tubal elevation the mucosa lines a variable *pharyngeal recess* (extensively surveyed by Khoo et al 1969). The nasopharyngeal *roof* and *posterior wall* form a continuous postero-inferior slope, supported mainly by the basilar occipital bone posteriorly, to a lesser extent by the posterior part of the body of the sphenoid anteriorly, and the anterior arch of the atlas below. A lymphoid mass, the *pharyngeal tonsil*, lies in the submucosa of the upper part of this surface and is best developed in childhood.

The pharyngeal tonsil is visible during the later fetal months, increases in size up to six or seven years and then usually begins to atrophy. In a child of 18 months it is a forward-directed pyramidal prominence, with its apex near the nasal septum and its base at the junction of the nasopharyngeal roof and posterior wall. It consists of folds radiating anterolaterally from a median recess, the *pharyngeal bursa*, which ascends backwards into its substance. The mucosal folds are mainly diffuse lymphoid tissue, but also contain deep mucous glands. The bursa, in the base of the tonsil, is a blind recess. In the embryo, the notochord lies for a short distance inferior to the cranial base, near the developing basilar part of the occipital bone (p. 164); it is attached here to the endoderm of the primitive pharyngeal roof. Subsequently, as this region grows, the notochordal attachment draws out an angled recess of endoderm (the *pouch of Luschka*) forming the pharyngeal bursa. (For discussion see Cave 1965, Slipka 1972). The lateral prolongation of the pharyngeal tonsil behind the opening of the auditory tube produces a *tubal tonsil*, consisting of unencapsulated lymphoid tissue.

THE ORAL PART OF THE PHARYNX

The oropharynx reaches from the soft palate to the upper border of the epiglottis. Through the oropharyngeal isthmus it opens into the mouth, facing the pharyngeal aspect of the tongue. Its lateral wall consists of the palatopharyngeal arch and palatine tonsil. Posteriorly, it is level with the second cervical vertebral body and upper part of the third. The *palatopharyngeal arch* lies behind and projects medially more than the palatoglossal; it descends posterolaterally from the uvula to the side of the pharynx, as a fold of mucosa over the palatopharyngeus (p. 1290). On each side the palatopharyngeal and palatoglossal arches flank the triangular *tonsillar sinus* containing the palatine tonsil.

The palatine tonsils (**8.**90,97), two masses of lymphoid tissue in the lateral walls of the oropharynx, are each in a triangular *tonsillar sinus* between the diverging palatoglossal and palatopharyngeal arches. Its medial surface projects conspicuously into the pharynx during childhood, but the extent of this projection is not a true indication of tonsillar size. The deep, lateral aspect extends upwards, downwards and forwards beyond the medial surface and is embedded below the mucous membrane. Inferiorly, it invades the dorsum of the tongue and, superiorly, the soft palate; anteriorly, it may extend for some distance under the

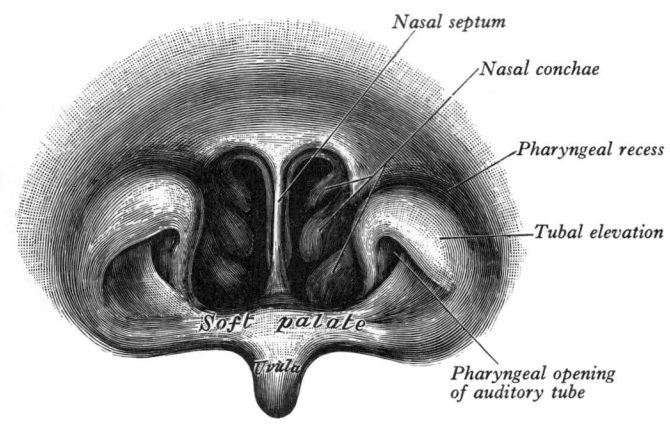

Nasal septum

Nasal conchae

Pharyngeal recess

Tubal elevation

Soft palate

Uvula

Pharyngeal opening
of auditory tube

8.98 Ventral boundary of the nasal part of the pharynx, as seen in posterior rhinoscopy.

palatoglossal arch. The tonsil is variable in size and is frequently infected, leading to inflammatory changes and hypertrophy. It is hence difficult to define a normal appearance. In late fetal life a mucous fold, the *plica triangularis*, extends back from the palato-glossal arch, covering the antero-inferior aspect of the tonsil. In childhood this fold is usually invaded by lymphoid tissue and fuses with the tonsil; it rarely persists in adults.

The upper part of the tonsil contains a deep *intratonsillar cleft*, often erroneously termed the *supratonsillar fossa*. It is not above the tonsil but in its substance; its upper wall contains lymphoid tissue (**8**.97) extending into the soft palate. The mouth of the cleft is semilunar and lies parallel to the dorsal curve of the tongue. After puberty the deep part of the tonsil diminishes considerably and the medial surface flattens.

On the *medial surface* of the tonsil are 12–15 orifices of deep, narrow *tonsillar crypts*, penetrating almost through the whole thickness of the tonsil and from which numerous follicles branch. The *lateral* or *deep aspect* is covered by a fibrous *capsule*. The tonsil and its capsule are separable with ease for most of their extent from the muscular wall of the pharynx, formed here by the superior constrictor with the styloglossus laterally (**8**.82,97,100). Antero-inferiorly the capsule is adherent to the side of the tongue and behind this point it receives the insertion of some muscular fibres of palatoglossus and palatopharyngeus. Here the tonsillar artery, a branch of the facial, pierces the superior constrictor to enter the tonsil with two venae comitantes. An important and sometimes large vein (the *external palatine* or 'paratonsillar vein') descends from the soft palate lateral to the tonsillar capsule before piercing the pharyngeal wall (**8**.97); from this vessel severe venous haemorrhage from the upper angle of the tonsillar sinus may complicate tonsillectomy (Browne 1928). The muscular wall of the tonsillar sinus separates the tonsil from the ascending palatine artery and occasionally from the facial artery itself (p. 736), which, if very tortuous, may be extremely near the pharyngeal wall at the lower tonsillar level. The internal carotid artery is 25 mm posterolateral to the tonsil.

The tonsils are part of a protective annulus of lymphoid tissue (*Waldeyer's ring*). Its antero-inferior part is formed by the lingual tonsil; lateral are the palatine and tubal tonsils, and pos-terosuperior the pharyngeal tonsil (p. 1324). Minor lymphoid masses exist between these major groups.

Surface Anatomy. The palatine tonsil lies behind the third lower molar tooth and is represented by an oval area over the lower part of the masseter, a little anterosuperior to the man-dibular angle.

Structure. The tonsillar crypts are lined by stratified squamous epithelium, continuous with that of the pharyngeal mucosa and invaded by numerous lymphocytes, some of which less enter the mouth and form salivary corpuscles. The tonsillar lymphoid tissue is arranged in nodules or follicles, in the centres of which less close-packed lymphocytes form *germinal centres* (p. 822) of prolifera-tion. The crypts may contain caseous plugs of lymphocytes, bac-teria and desquamated epithelium which are gradually eliminated or may become calcified; the bacteria may cause local or general infection. Unlike lymph nodes, there are no lymph sinuses (p. 824) and the tonsil has no *afferent* vessels; but a dense plexus of vessels surrounds each follicle and from it *efferent* vessels pass to the upper deep cervical lymph nodes, especially the jugulodigastric group (p. 843). This and the other lymphoid masses of Waldeyer's ring belong to the category of *mucosa-associated lymphoid tissue* (see p. 832) which provides humoral and cellular defence against infec-tions of the oral and nasal cavities and pharynx and related areas of the alimentary and respiratory tracts.

Vessels and nerves. The main *artery* is the tonsillar branch of the facial. It may also receive rami from dorsal lingual branches of the lingual artery, the ascending palatine branch of the facial, the ascending pharyngeal artery and the greater palatine branch of the maxillary. One or more *veins* leave the inferior deep aspect of the tonsil, traversing the superior constrictor to join the external palatine ('paratonsillar'), pharyngeal or facial veins. Its *nerves* are from the pterygopalatine ganglion via the lesser palatine and from the glossopharyngeal nerve which, by its tympanic branch, also supplies the mucosa of the tympanic cavity; hence tonsillitis may cause pain referred to the ear.

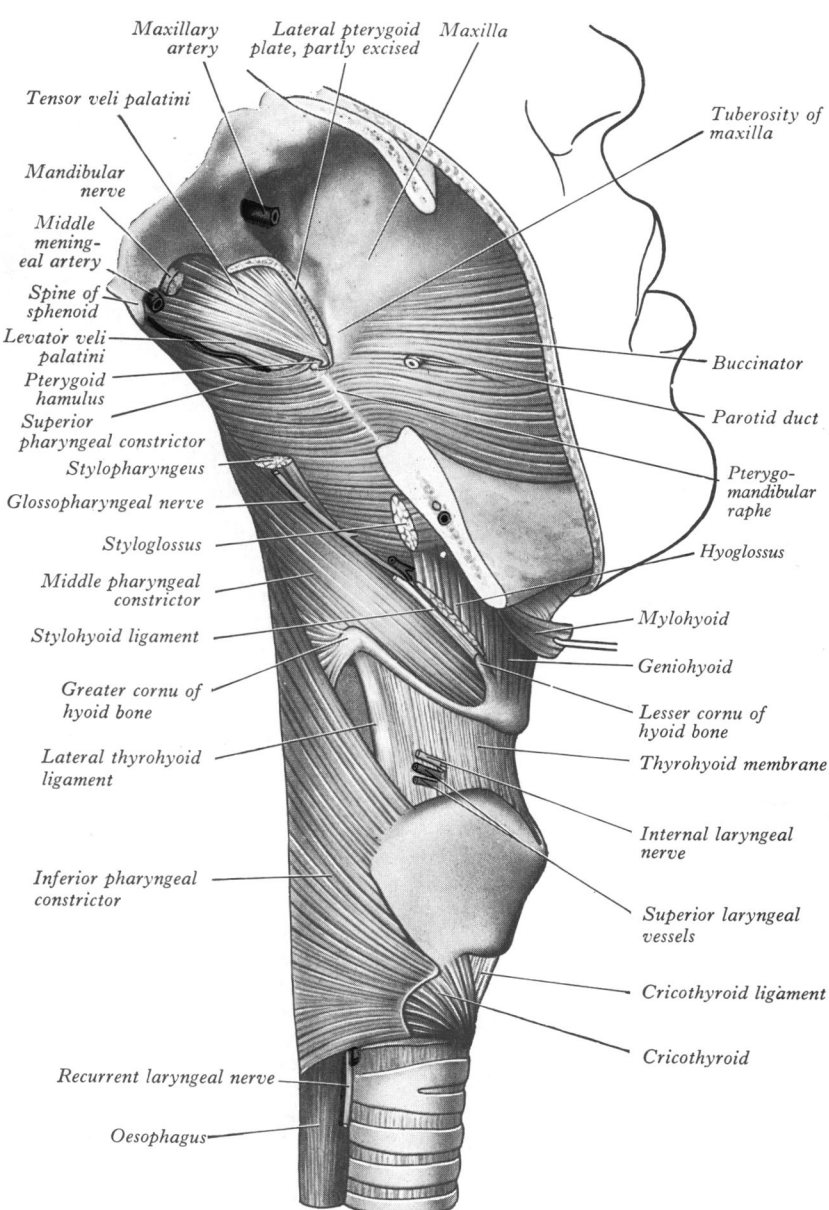

8.99 The buccinator and the muscles of the pharynx.

THE LARYNGEAL PART OF THE PHARYNX

The laryngeal part of the pharynx (laryngopharynx) extends from the cranial border of the epiglottis to the inferior border of the cricoid cartilage, where it becomes continuous with the oesophagus. In its incomplete anterior wall is the laryngeal inlet (p. 1253) and below this the posterior surfaces of the arytenoid and cricoid cartilages. A small *piriform fossa* on each side of the inlet is bounded medially by the aryepiglottic fold and laterally by the thyroid cartilage and thyrohyoid membrane. Beneath its mucous membrane are the branches of the internal laryngeal nerve which have pierced the thyrohyoid membrane. Foreign bodies may lodge in the fossa and, if the mucous membrane is pierced during their removal, the nerve may be damaged, with consequent anaesthesia of the region. Posteriorly the laryngopharynx extends from the lower part of the third cervical vertebral body to the upper part of the sixth.

STRUCTURE OF THE PHARYNX

The pharynx is described as having, from within outwards, mucous, fibrous and muscular laminae and finally a thin buccopharyngeal fascia external to the constrictor muscles and passing forwards over the pterygomandibular raphe on to the buccinator.

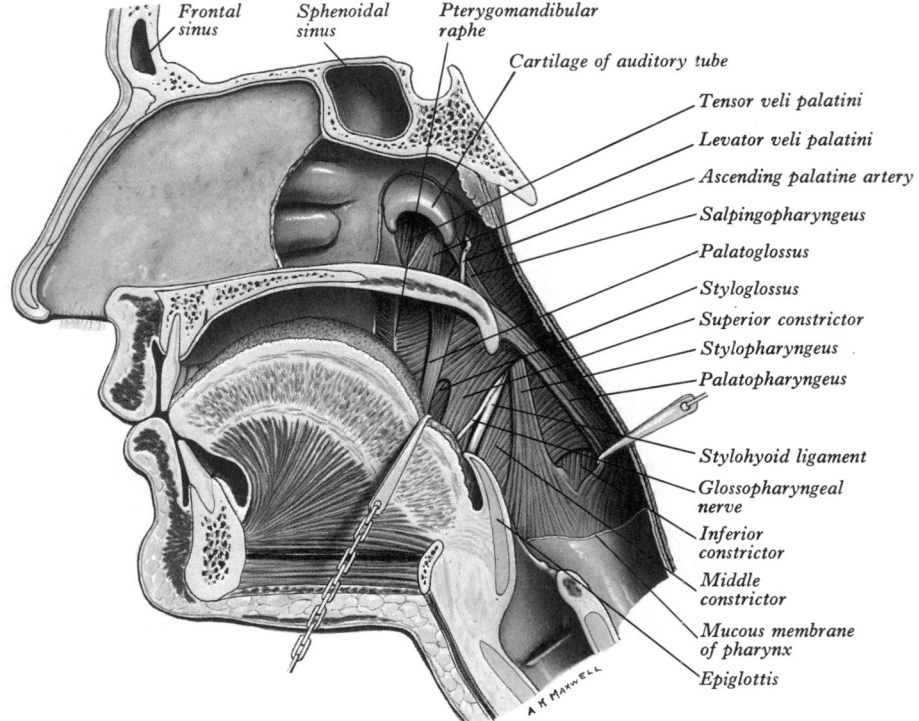

Frontal sinus
Sphenoidal sinus
Pterygomandibular raphe
Cartilage of auditory tube
Tensor veli palatini
Levator veli palatini
Ascending palatine artery
Salpingopharyngeus
Palatoglossus
Styloglossus
Superior constrictor
Stylopharyngeus
Palatopharyngeus
Stylohyoid ligament
Glossopharyngeal nerve
Inferior constrictor
Middle constrictor
Mucous membrane of pharynx
Epiglottis

8.100 Median sagittal section of the head, showing a dissection of the interior of the pharynx, after the removal of the mucous membrane. The bodies of the cervical vertebrae have been removed and the cut posterior wall of the pharynx then retracted dorsolaterally. The palatopharyngeus is drawn dorsally to show the cranial fibres of the inferior constrictor; the dorsum of the tongue is drawn ventrally to display a part of the styloglossus in the angular interval between the mandibular and the lingual fibres of origin of the superior constrictor.

The *mucous membrane* is continued into the auditory tubes, nasal cavity, mouth and larynx. Its nasopharyngeal epithelium is ciliated, pseudostratified 'respiratory' epithelium containing goblet cells and receiving submucous glands; in the oro- and laryngopharynx it is non-keratinized stratified squamous. Between the two regions is a narrow transitional zone of cuboidal epithelium, the cilia being imperfect or absent. Superiorly, this zone adjoins the nasal septum; laterally it passes over the auditory tube orifices and turns posteriorly at the union of the soft palate and the lateral wall (pp. 1289, 1327, 1328). Mucous glands are numerous around the auditory tube orifices.

The intermediate *fibrous* layer is thick above (*pharyngobasilar fascia*), where muscular fibres are absent, and is firmly connected to the basilar occipital and petrous temporal bones medial to the carotid canal, curving under the auditory tube and forwards to the posterior border of the medial pterygoid plate and pterygomandibular raphe. As it descends it diminishes in thickness but is strengthened posteriorly by a fibrous band, attached to the occipital's pharyngeal tubercle and descending as the median *pharyngeal raphe* of the constrictors. The pharyngeal muscles are described as external to this fibrous lamina, but really it is the thick, deep epimysial covering of the muscles, the thinner external epimysium being the *buccopharyngeal fascia*. The *muscular coat* is described below.

Pharyngeal Musculature

This consists of: three *constrictor* muscles, superior, middle and inferior, and a trio of muscles descending from the styloid process, the cartilaginous torus of the auditory tube, and the soft palate, respectively the *stylo-, salpingo-* and *palatopharyngei*, passing obliquely into the muscular wall. Palatopharyngeus is described on p. 1290.

The inferior pharyngeal constrictor, the thickest of the constrictors, has two parts, the cricopharyngeus and the thyropharyngeus. It is attached to (**8**.99) the side of the cricoid cartilage between the attachment of cricothyroid and, posteriorly, the articular facet for the inferior thyroid cornu (*cricopharyngeus*). It also arises: from the oblique line of the thyroid lamina, from a strip of the lamina behind this, from a fine tendinous band crossing cricothyroid from the inferior thyroid tubercle to the cricoid cartilage and, by a small slip, from the inferior cornu (*thyropharyngeus*). Both parts spread posteromedially to join the opposite muscle in the median pharyngeal raphe. The inferior fibres, which are horizontal, blend with the circular oesophageal fibres round the narrowest part of the pharynx; the rest ascend obliquely and overlap the middle constrictor. During swallowing the cricopharyngeus is 'sphincteric' (Fuller et al 1959) and the thyropharyngeus 'propulsive'; failure of relaxation of the cricopharyngeus may cause posterior mucosal herniation between the two parts of the muscle (Killian's dehiscence).

Relations. Buccopharyngeal fascia is external to the inferior constrictor. *Posterior* are the prevertebral fascia and muscles, *lateral* the thyroid gland, common carotid artery and the sternothyroid, *internal* are the middle constrictor, stylopharyngeus, palatopharyngeus and the fibrous lamina. The internal laryngeal nerve and laryngeal branch of the superior thyroid artery reach the thyrohyoid membrane between the inferior and middle constrictors. The external laryngeal nerve descends on the superficial surface of the muscle, just behind its thyroid attachment and piercing its lower part. The recurrent laryngeal nerve and the laryngeal branch of the inferior thyroid artery ascend internal to its lower border to enter the larynx.

The middle constrictor is a fan-shaped sheet attached anteriorly to the lesser hyoid cornu and the lower part of the stylohyoid ligament (the *chondropharyngeal* part of the muscle) and to the whole upper border of the greater cornu of the hyoid (the *ceratopharyngeal* part). Inferior fibres descend internal to the inferior constrictor to the lower end of the pharynx, middle fibres pass transversely and superior fibres ascend and overlap the superior constrictor. It is inserted posteriorly into the median pharyngeal raphe with its opposite fellow.

Relations. Between the middle and superior constrictors pass the glossopharyngeal nerve and the stylopharyngeus muscle and between the middle and inferior constrictors pass the internal laryngeal nerve and laryngeal branch of the superior thyroid artery. *Posterior* are the prevertebral fascia, longus colli and longus capitis; *lateral* are the carotid vessels, pharyngeal plexus of nerves and some lymph nodes. Near its hyoid attachment the constrictor is deep to the hyoglossus, the lingual artery lying between them.

Internal are the superior constrictor, stylopharyngeus, palatopharyngeus and the fibrous lamina.

The superior constrictor is a quadrilateral sheet, thinner and paler than the others. It is attached anteriorly to the pterygoid hamulus (and sometimes to the adjoining posterior margin of the medial pterygoid plate), the pterygomandibular raphe, the posterior end of the mylohyoid line of the mandible and by a few fibres to the side of the tongue (**8**.100). These attachments differentiate *pterygopharyngeal*, *buccopharyngeal*, *mylopharyngeal* and *glossopharyngeal* parts of the superior constrictor. Its fibres curve back into the median pharyngeal raphe; some are also prolonged by an aponeurosis to the pharyngeal tubercle on the basilar part of the occipital bone, the superior fibres curving under levator veli palatini and the auditory tube and leaving an interval below the cranial base for passage of the auditory tube. This interval is limited anteriorly by the medial pterygoid plate and closed by the pharyngobasilar fascia (p. 1326).

A constant band of muscle sweeps backwards from the anterolateral part of the upper surface of the palatine aponeurosis, lateral to levator veli palatini, to blend internally with the superior constrictor near its superior border (**8**.102); this band is the *palatopharyngeal sphincter* and it ridges the pharyngeal wall (*ridge of Passavant*) visibly when the soft palate is elevated (Whillis 1930). It is hypertrophied in cases of complete cleft palate. The

change from columnar, ciliated, 'respiratory' epithelium to stratified, squamous epithelium on the superior palatal aspect occurs at the attachment of the palatopharyngeal sphincter to the palate.

Relations. *External* to the superior constrictor are the prevertebral fascia and muscles, the ascending pharyngeal artery and the pharyngeal venous plexus, glossopharyngeal and lingual nerves, styloglossus, middle constrictor, medial pterygoid, stylohyoid ligament and stylopharyngeus; the internal carotid artery, sympathetic trunk, hypoglossal nerve, interal jugular vein and styloid process are more distant relations. *Internal* are the palatopharyngeus, the tonsillar capsule and pharyngobasilar fascia. *Superiorly* it is separated from the cranial base by a crescentic interval containing the levator veli palatini, tensor veli palatini and the auditory tube. *Inferiorly* its border is separated from the middle constrictor by stylopharyngeus and the glossopharyngeal nerve. *Anteriorly* it is separated from buccinator by the pterygomandibular raphe.

Nerve supply. The constrictors are supplied by the *pharyngeal plexus* (p. 1117), the inferior also by rami of the external and recurrent laryngeal nerves. The pharynx is largely a branchial derivative (p.227), hence its motor and sensory supply is through the trigeminal, glossopharyngeal and vagus nerves. Additionally, glandular tissue in pharyngeal mucosa and vascular non-striated

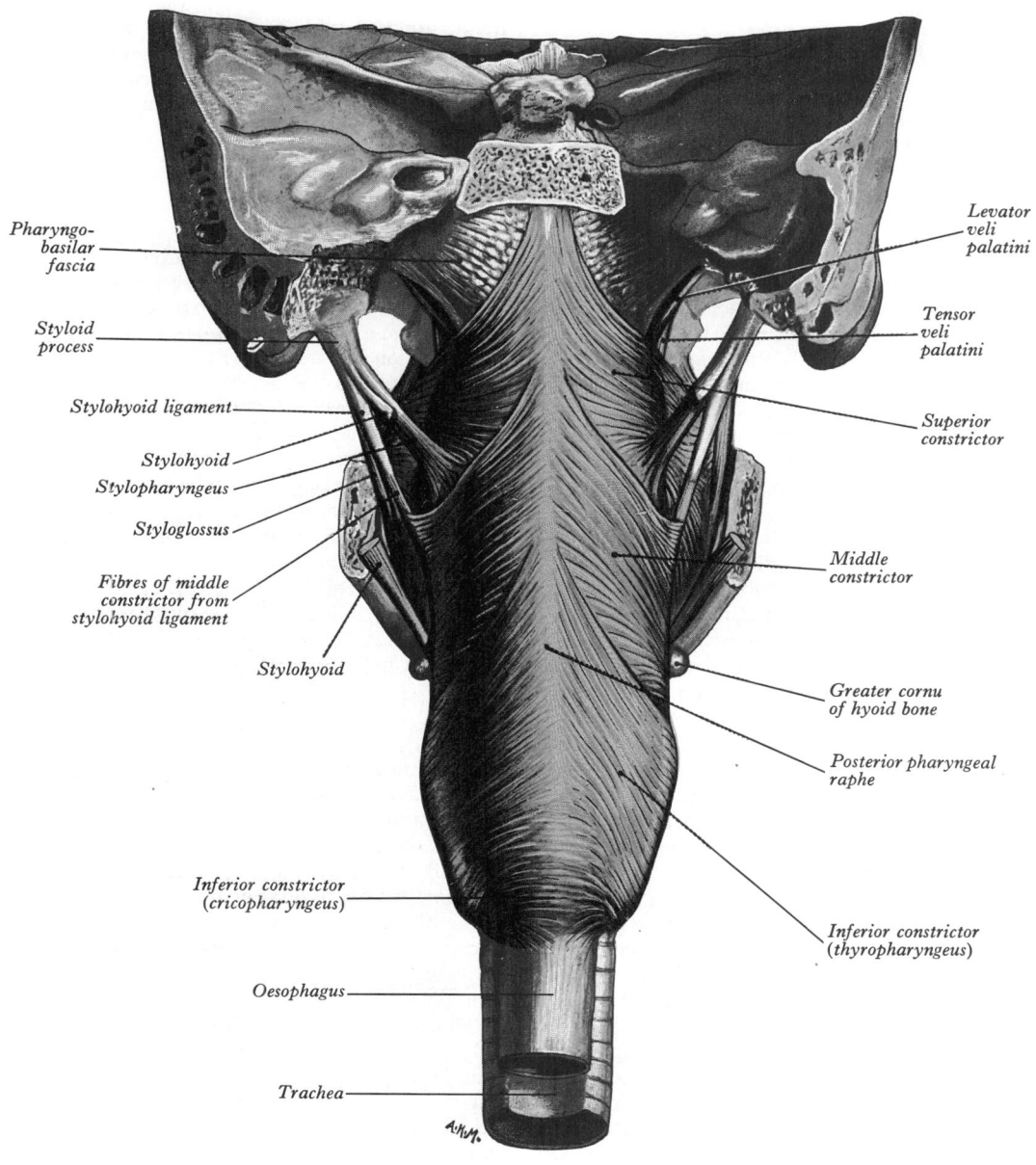

Pharyngo-basilar fascia

Styloid process

Stylohyoid ligament

Stylohyoid

Stylopharyngeus

Styloglossus

Fibres of middle constrictor from stylohyoid ligament

Stylohyoid

Inferior constrictor (cricopharyngeus)

Oesophagus

Trachea

Levator veli palatini

Tensor veli palatini

Superior constrictor

Middle constrictor

Greater cornu of hyoid bone

Posterior pharyngeal raphe

Inferior constrictor (thyropharyngeus)

8.101 The muscles of the pharynx: posterior view (from Quain's *Anatomy*, 11th edn).

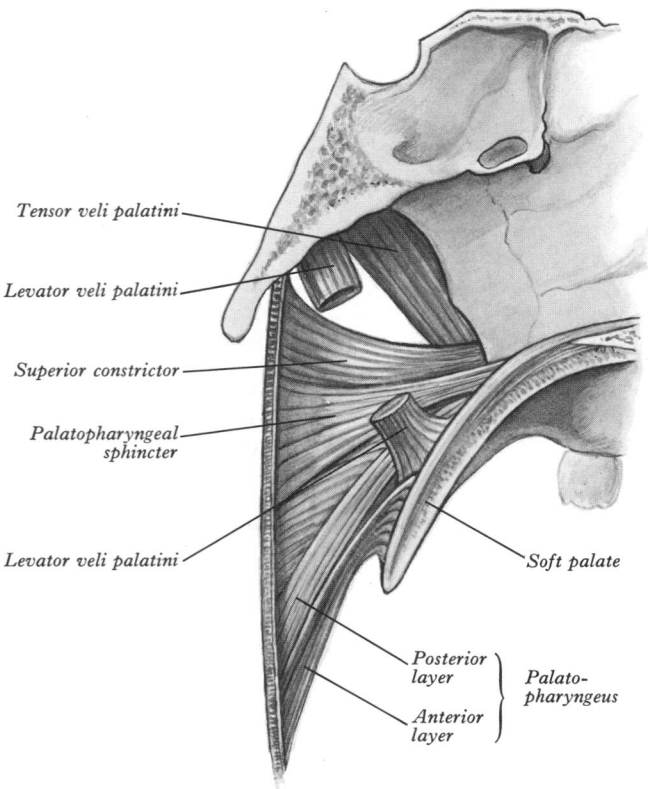

Tensor veli palatini

Levator veli palatini

Superior constrictor

Palatopharyngeal sphincter

Levator veli palatini

Soft palate

Posterior layer
Anterior layer
} Palato-pharyngeus

8.102 The muscles of the left half of the soft palate and adjoining part of the pharyngeal wall. Part of the levator veli palatini has been removed to reveal the palatopharyngeal sphincter. The soft palate is cut sagittally. Dissection by the late James Whillis, Department of Anatomy, Guy's Hospital Medical School, London.

muscle receive an autonomic supply through the pharyngeal plexus. Postganglionic sympathetic fibres reach this plexus from the superior cervical ganglion via specific rami; the preganglionic parasympathetic issue from the medulla oblongata, chiefly in the glossopharyngeal nerve, which also contains afferents from the oral and laryngeal mucosae, the nasal mucosa being trigeminal territory. The vagus nerve carries branchial efferent fibres for pharyngeal striated musculature but most of these fibres probably emerge from the brain stem in the bulbar part of the accessory nerve.

Pharyngeal rami of the glossopharyngeal and vagus nerves and of the superior cervical ganglion form a plexus in the connective tissue external to the constrictors, particularly the intermediate muscle (Hovelacque 1927). From the plexus, in which autonomic (sympathetic and parasympathetic) and branchial (efferent and afferent) fibres intermingle, mixed rami ascend and descend exterior to the superior and inferior constrictors, branching into the muscular layer and mucosa. This pattern is common to most primates (Sprague 1944), including mankind; but in some lower primates the plexus is absent or simplified and may lack glossopharyngeal or vagal components. In other mammals arrangements vary; in many there is no plexus, but precise information on the main supply rami and their exact brain-stem sources is lacking. The marked development of the plexus in man and some other primates has been ascribed to phonation; perhaps this is too facile a view, considering the lack of factual evidence.

Actions. The constrictors exercise a general sphincteric and peristaltic action in swallowing. For details, vide infra.

The stylopharyngeus (8.94,100), is a long slender muscle, cylindrical above and flat below. It arises from the medial side of the base of the styloid process, descending along the side of the pharynx and passing between the superior and middle constrictors to spread out beneath the mucous membrane. Some fibres merge into the constrictors and the lateral glosso-epiglottic fold, others are attached with palatopharyngeus to the posterior border of the thyroid cartilage. The glossopharyngeal nerve curves round the posterior border and the lateral side of stylopharyn-

geus, passing between the superior and middle constrictors to the tongue.

Nerve supply: a branch of the glossopharyngeal nerve.

Action: elevation of the pharynx in swallowing and speech (vide infra).

The salpingopharyngeus (8.100) arises from the inferior part of the cartilage of the auditory tube near the tube's pharyngeal opening to pass downwards and blend with palatopharyngeus.

Nerve supply: the pharyngeal plexus.

Action: elevation of the upper lateral pharyngeal wall, i.e. the part above the attachment of stylopharyngeus. For its role in swallowing, see below.

Vessels and Nerves

The *pharyngeal arterial supply* is from the ascending pharyngeal, ascending palatine and tonsillar rami of the facial artery, rami of the maxillary artery (greater palatine, pharyngeal and artery of the pterygoid canal) and dorsal lingual rami of the lingual artery. The *veins* form a plexus connected above with the pterygoid plexus and draining below into the internal jugular and facial veins.

The *lymph vessels* are described on pp. 844, 855.

The *nerve supply* is mainly from the pharyngeal plexus (p. 1117). The principal *motor* element is the cranial part of the accessory nerve, which, through vagal branches, supplies all pharyngeal and palatal muscles except stylopharyngeus (glossopharyngeal nerve) and tensor veli palatini (mandibular nerve). The main *sensory* nerves are the glossopharyngeal and vagal; much nasopharyngeal mucosa is supplied by the maxillary nerve (via the pterygopalatine ganglion); mucosa of the soft palate and tonsil is supplied by the lesser palatine and glossopharyngeal nerves. Proprioceptor endings have not been convincingly identified in the pharynx (Bossy & Vidić 1967).

PALATINE MOVEMENTS

Movements of the palate are essential to swallowing, blowing and speech; all require variable degrees of closure of the pharyngeal isthmus (p. 1323). Closure is maximal in oral blowing to prevent the escape of air through the nose. In deglutition closure prevents regurgitation into the nasopharynx. In speech closure is maximal in the production of explosive consonants (e.g. *b,p*). Closure of the isthmus is effected as follows: the levatores veli palatini pull the soft palate up and back towards the posterior pharyngeal wall, while simultaneously the palatopharyngeal sphincter raises the wall to meet the palatine nasopharyngeal surface over a wide area. (It is at the upper limit of this contact area that the epithelium on the upper surface of the soft palate changes from respiratory to stratified squamous.)

The palatal tensors are active in deglutition rather than speech; by producing a localized anterior depression in the soft palate (p. 1329) they squeeze the bolus against the tongue, aiding its descent in the oropharynx.

MECHANISM OF DEGLUTITION

The *first stage* of swallowing is *voluntary*: the anterior part of the tongue is raised and pressed against the hard palate, the movement commencing at the lingual apex and spreading rapidly back. A bolus, formed behind the apex, is thus pushed posteriorly. At the end of this stage the soft palate descends on to the lingual dorsum, helping to grip the bolus. Lingual movements are effected by the intrinsic muscles, especially the superior longitudinal and transverse. Simultaneously the hyoid bone is moved up and forwards by the geniohyoid, mylohyoid, digastric and stylohyoid. The postsulcal part of the tongue is drawn up and back by the styloglossi and the palatoglossal arches are approximated by the palatoglossi, pushing the bolus through the oropharyngeal isthmus into the oropharynx, where the second, *involuntary*, stage begins. In swallowing fluids, the intrinsic lingual muscles squirt liquid backwards in the mouth, after which mylohyoid contraction bulges the lingual base into the oropharynx. In swallowing solids only the mylohyoid action is needed, except in cleansing the mouth of saliva and debris after a bolus is swallowed (Whillis

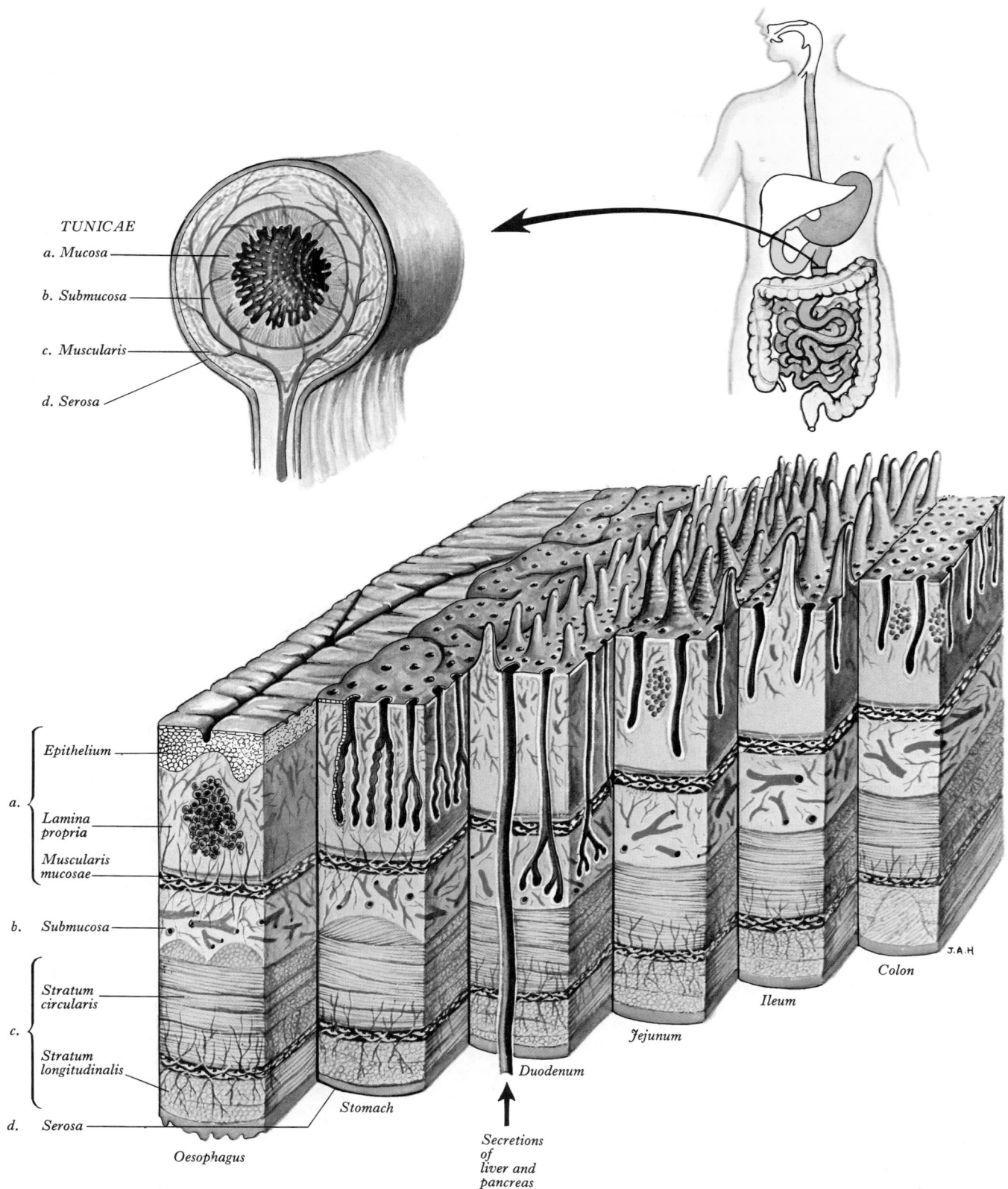

TUNICAE

a. *Mucosa*

b. *Submucosa*

c. *Muscularis*

d. *Serosa*

a. {

Epithelium

Lamina
propria

Muscularis
mucosae

b. Submucosa

c. {

Stratum
circularis

Stratum
longitudinalis

d. Serosa

Oesophagus

Stomach

Secretions
of
liver and
pancreas

Duodenum

Jejunum

Ileum

Colon

J.A.H

8.103 The general arrangement of the alimentary canal, its mural tunicae and (below) the general histology at the levels indicated (highly diagrammatic). The transverse colon (above right) has been displaced downwards to reveal the duodenum.

1946). Hiiemae et al (1978) analysed lingual and hyoid activity in swallowing, using the cat and the opossum as models. By cineradiography and electromyography they described the cyclic activities of hyoid muscles (p. 541), delineating 'envelopes' of movement. They regard lingual movements as largely transportive, the precise combination of lingual and hyoid movements depending on the nature of the ingested material. The extension of these studies to primates is awaited with interest.

In the *second stage*, the soft palate is elevated (by levator muscles), tightened (tensor muscles) and firmly approximated to the posterior pharyngeal wall by the palatopharyngeal sphincter (p. 1290) and the upper part of the superior constrictor. The pharyngeal isthmus closes tightly to prevent food from ascending into the nasopharynx. Meanwhile the larynx and the pharynx are drawn up, behind the hyoid bone, by the stylopharyngeus, salpingopharyngeus, thyrohyoid and palatopharyngeus muscles. Simultaneously the aryepiglottic folds are approximated and the arytenoid cartilages drawn up and forwards by the aryepiglottic, oblique arytenoid and thyro-arytenoid muscles, excluding the bolus from the larynx. Partly by gravity and partly by successive contractions of superior and middle constrictors, the bolus slips over the epiglottis (now bent back on to the laryngeal aditus), the closed laryngeal inlet and posterior arytenoid surfaces into the lowest part of the pharynx. Its passage is facilitated by the palatopharyngei, which shorten the pharynx by elevating it; on contraction, they make the posterior pharyngeal wall into an inclined plane directed postero-inferior and under this the bolus descends. The aryepiglottic folds provide lateral channels leading from the sides of the epiglottis through the piriform fossae into the oesophagus. They are kept tense and vertical by the backward pull of the posterior crico-arytenoids on the arytenoid cartilages and by the muscles (aryepiglottic and thyro-epiglottic) within them, assisted by the cuneiform cartilages which act as passive props. In paralysis of these muscles (which are supplied by the recurrent laryngeal nerves) the laryngeal inlet is not closed in swallowing, the folds sink medially and fluids tend to enter the larynx.

The *last stage* in swallowing is the expulsion, by the inferior constrictors, of the now compressed bolus into the oesophagus (p. 1326). The pump-like action of the pharyngeal muscles in this activity has been described by Buthitaya et al (1987).

These stages follow on each other, but it is easy to ascertain by palpation of the hyoid bone and laryngeal prominence that, during swallowing, elevation and forward movement of the hyoid precede laryngeal elevation. Note that the thyroid cartilage, and hence the whole larynx, ascends also *relative* to the hyoid, shortening the larynx and causing structures between the hyoid bone and thyroid cartilage to bulge posteriorly into the larynx above the vestibular folds. This also increases the curvature of the epiglottis, especially in its lower part, aiding stenosis of the laryngeal aditus during swallowing (see Fink & Martin 1977).

The evidence for the foregoing analysis is not mere hypothecation from muscular geometry but has been adduced from radiological studies, the effects of known paralyses, electromyography, ultrasound analysis (Shawker 1984) and cineradiography. Swallowing is, however, a highly complex neuromuscular integration and disagreement over details persists. For full critiques of the literature consult Bosma (1957), Doty (1968), Hiiemae (1978), Buthitaya et al (1987).

Applied Anatomy. In young children lymphoid hypertrophy in the nose and nasopharynx (adenoids), with or without enlargement of the palatine tonsils, may obstruct nasal respiration. The mouth has to be kept open to breathe: the hard palate and alveolar arch are then habitually out of contact with the lingual dorsum and, lacking this pressure, they develop an abnormally high arch and forward projection. The hard palate becomes transversely narrow and projecting alveolar processes afford little room for permanent teeth, which are crowded and irregular and overhang the lower teeth. The facial maxillary surfaces appear pinched together, with narrowing of the nasal cavities and maxillary air sinuses. The upper lip is drawn up, exposing the projecting upper incisors. The face is lengthened by dropping of the lower jaw. The child's expression ('adenoid facies') is characteristic, suggesting vacuity and inattention, the latter being partly due to deafness often associated with nasal obstruction and due to blockage of the pharyngeal openings of the auditory tubes.

General Alimentary Organization

Despite structural variations along the alimentary tract, there is a common basic plan which is best appreciated by reference to development (p. 100). Much of the alimentary canal develops initially as a tube of endoderm enveloped in the splanchnopleuric mesoderm, its external surface facing the intra-embryonic coelom. The endodermal lining forms the epithelium of the tract and also the secretory and ductual cells of various glands secreting into the lumen, including the pancreas and liver. The surrounding splanchnopleuric mesoderm forms the connective tissue, muscle layers, blood vessels and lymphatics of the wall, its external surface becoming visceral mesothelium; this is of course absent throughout the neck and thorax and where the hindgut traverses the pelvic floor. The disposition and development of the dorsal and ventral mesenteries of the subdiaphragmatic foregut

and of the dorsal mesentery of the midgut and hindgut have been described (p. 232). Associated neural elements invade the gut from neighbouring neural crest tissue (p. 199). Cranially, the skeletal muscle of branchial arch origin and caudally of less certain origin, contribute to the musculature of the gut's extremities.

The mature gut wall (**8**.103) is a series of layers; arranged from inside outwards as follows: (1) *epithelium*, (2) *lamina propria*, (3) *muscularis mucosae*, which with (1) and (2) constitutes the *mucosa*, (4) *submucosa*, (5) *muscularis externa*, the main muscle coat, and (6) *serosa*, or *visceral peritoneum*. In the pharynx and most of the oesophagus, the serosa is replaced by a dense connective tissue *adventitia*, blending with the fascial planes of adjacent structures (**8**.103). Mesenteries carry blood vessels, autonomic nerves and lymphatic channels to and from the gut and accompany the ducts of outlying glands.

The epithelium is the region of interaction between the body and ingested food. Its cytology varies with its regional functions: in the pharynx, oesophagus and distal anal canal it is a nonkeratinizing, stratified, squamous epithelium which protects the underlying tissue and facilitates passage of material by lubricant secretions. Elsewhere the epithelium is simple, composed largely of secretory cells, providing mucus, enzymes and various ions, and of columnar absorptive cells, with striated or brush borders. The total amount of secretion and absorption depends partly on the epithelial surface area, which is increased by the presence of large folds, villi, crypts and glands. Some glands lie in the lamina

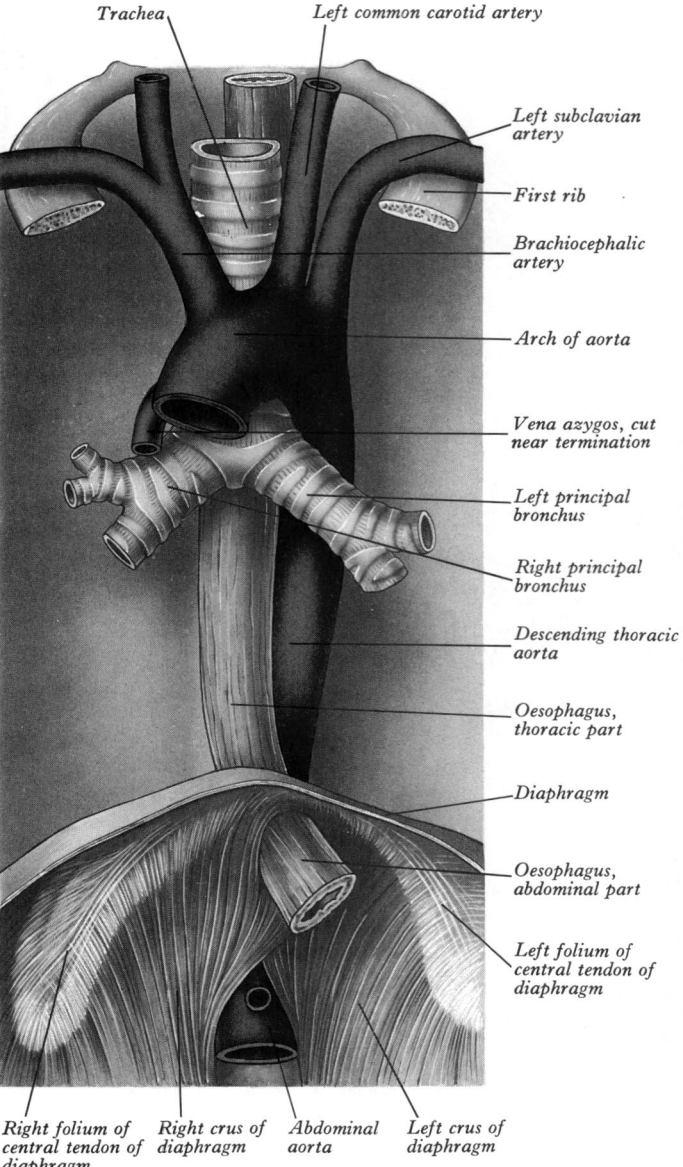

Trachea — Left common carotid artery

Left subclavian artery

First rib

Brachiocephalic artery

Arch of aorta

Vena azygos, cut near termination

Left principal bronchus

Right principal bronchus

Descending thoracic aorta

Oesophagus, thoracic part

Diaphragm

Oesophagus, abdominal part

Left folium of central tendon of diaphragm

Right folium of central tendon of diaphragm   Right crus of diaphragm   Abdominal aorta   Left crus of diaphragm

8.104A Dissection to expose the oesophagus in the posterior mediastinum and in the abdomen.

propria, some in the submucosa and some completely outside the tract, e.g. the liver and pancreas. Their secretions also maintain the correct environment for optimal activity of digestive enzymes; they are usually under neural or hormonal control.

The lamina propria is made of compact connective tissue, often rich in elastin fibres, which supports the epithelium; on its external aspect is the **muscularis mucosae**, a plexus of non-striated muscle fibres which can alter the local conformation of the mucosa by contraction of its inner circular and outer longitudinal layers.

The submucosa is rich in blood vessels, lymphatics and nerves; variations in thickness create macroscopic folds on the inner aspect of the gut, particularly in the small intestine.

The muscularis externa usually consists of distinct inner circular and outer longitudinal layers, the antagonistic activities of which create waves of peristalsis responsible for movement of ingested material through the lumen of the gut. In the stomach, where movements are more complex, there is a partial oblique layer, internal to the other two layers. The muscularis externa is chiefly non-striated muscle, except where striated muscle of the upper oesophagus and anal canal blend with it. Its layers are separated by connective tissue containing vascular and nervous plexuses. Some controversy has existed over the direction of fibres in the layers of the muscularis externa, one suggestion being that the inner circular muscle is a tight helix and the longitudinal one an open spiral (Carey 1921). Other observations (Elsen & Arey 1966) on various mammals, including man, indicated that, despite some deviations from precisely circular and longitudinal directions, fibres do not follow spiral pathways. In the human small intestine adjacent 'rings' of circular muscle are often connected by oblique slips, with some interchange between muscle groups (Schofield 1968).

The neural element consists of three plexuses: between the circular and longitudinal layers of the external muscle (*myenteric* or *Auerbach's plexus*), between the submucosa and external muscle (*submucosal* or *Meissner's plexus*) and a third associated with the muscularis mucosae (*mucosal plexus*). All three are interconnected. The first two contain scattered autonomic ganglion cells, whereas the mucosal plexus appears to be derived from the submucosal plexus. The epithelium is also innervated from the mucosal plexus; several types of sensory cells have been described (see **8.142** and pp. 1373, 1467).

The Oesophagus

The oesophagus (**8.101, 104, 105**), a muscular tube about 25 cm (10 in.) long, connects the pharynx to the stomach. It begins in the neck, level with the lower cricoid border and the sixth cervical vertebra; descending largely anterior to the vertebral column through the superior and posterior mediastina, it traverses the diaphragm, level with the tenth thoracic vertebra, and ends at the gastric cardiac orifice level with the eleventh thoracic vertebra. Generally vertical, it has two shallow curves. At its beginning it is median but inclines to the left as far as the root of the neck, gradually returns to the median plane near the fifth thoracic vertebra, and at the seventh deviates left again, finally turning anterior to traverse the diaphragm at the tenth. The tube also bends in an anteroposterior plane to follow the cervical and thoracic curvatures of the vertebral column. It is the narrowest part of the alimentary tract, except for the vermiform appendix, and is constricted (1) at its commencement, 15 cm (6 in.) from the incisor teeth, (2) where crossed by the aortic arch, 22.5 cm (9 in.) from the incisor teeth, (3) where crossed by the left principal bronchus, 27.5 cm (11 in.) from the incisors, and (4) as it traverses the diaphragm, 40 cm (16 in.) from the incisors. These data are important clinically with regard to the passage of instruments along the oesophagus.

The cervical part is *anterior* to the trachea and attached to it by areolar tissue; the recurrent laryngeal nerves ascend on each side in or near the groove between the trachea and the oesophagus; *posterior* are the vertebral column, longus colli and prevertebral layer of deep cervical fascia; *lateral* on each side are the common carotid artery and posterior part of the thyroid gland. In the lower neck, where the oesophagus deviates left, it is closer

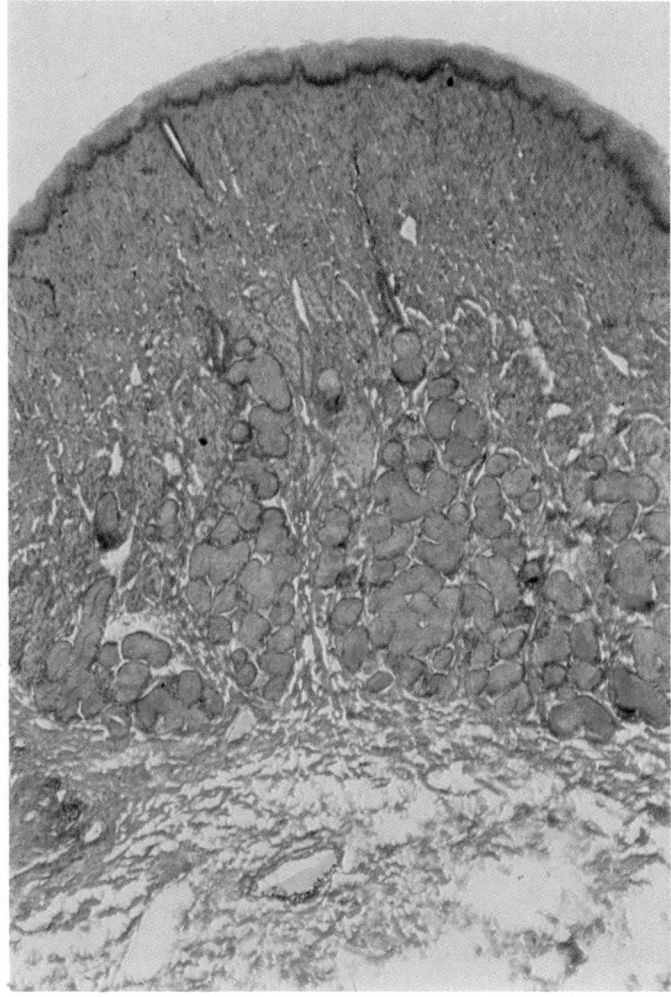

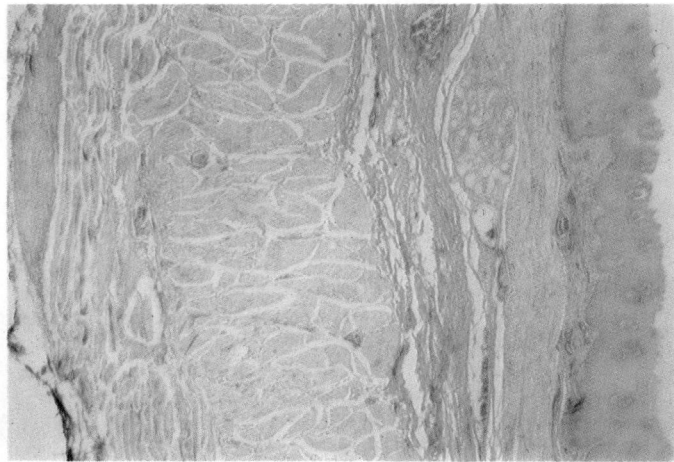

8.104B A low-power micrograph of a vertical section through the wall of a human oesophagus taken in the upper thorax. Visible in this section are: the epithelium (blue-grey on the right), the lamina propria, muscularis mucosae, submucosa with a group of mucous glands, the external muscle with the circular fibres more deeply placed and the longitudinal fibres placed externally. See text for further description. Mallory's triple stain. Magnification × 6. Prepared by W Owen, Department of Anatomy, Guy's Hospital Medical School, London.

8.104C Low magnification light micrograph of a section of the oesophagus showing the stratified squamous non-keratinized epithelium lining it and mucous glands (stained turquoise) of the submucosa. Weigert and Van Gieson, with alcian blue. Prepared by David Ristow; photographed by Marina Morris, Department of Anatomy, Guy's Hospital Medical School, London.

8.105 An oblique radiograph of the thorax during the oesophageal transit of part of a 'meal' of barium sulphate paste. At (A) the translucency of the air-containing right principal bronchus is visible. The concave ventral aspect of the oesophagus (B) is topographically related to the pericardium covering the left atrium of the heart. Longitudinal mucosal folds are visible (C) immediately proximal to the soft tissue shadow of the diaphragm (D). Some barium sulphate is already admixed with the gastric contents (E). The oesophagus in the lower thorax curves ventrally away from the vertebral column to reach the oesophageal orifice in the diaphragm.

to the left carotid sheath and thyroid gland than it is on the right. The thoracic duct ascends for a short distance along its left side.

The thoracic part (**8**.18, 19, 29, 30, 104A), at first a little to the left in the *superior mediastinum* between the trachea and the vertebral column, passes behind and to the right of the aortic arch to descend in the posterior mediastinum along the right side of the descending thoracic aorta. Below, as it inclines left, it crosses anterior to the aorta to enter the abdomen through the diaphragm at the level of the tenth thoracic vertebra. *Anterior* (from above downwards) are: the trachea, right pulmonary artery, left principal bronchus, pericardium (separating it from the left atrium) and the diaphragm; *posterior* are the vertebral column, longus colli muscles, right posterior (aortic) intercostal arteries, thoracic duct, azygos vein and the terminal parts of the hemiazygos and accessory hemiazygos veins and, near the diaphragm, the aorta. In the posterior mediastinum there is a long recess of the right pleural sac between the oesophagus in front and the vena azygos and vertebral column behind. *Left lateral*, in the superior mediastinum, are the end of the aortic arch, the left subclavian artery, thoracic duct, the left pleura and the left recurrent laryngeal nerve which ascends in or near the groove between the oesophagus and trachea. In the *posterior mediastinum* it is related to the descending thoracic aorta and left pleura. *Right lateral* are the right pleura, the azygos vein intervening as it arches forwards above the right principal bronchus to join the superior vena cava. Below the pulmonary roots the vagus nerves descend in contact with it, the right chiefly behind and the left in front, uniting to form an oesophageal plexus around it (pp.1118, 1125). Low in the *posterior mediastinum* the thoracic duct is behind and to the right; higher, it is posterior, crossing to the left at about the level of the fifth thoracic vertebra and then ascending on the left. On the right of the oesophagus, just above the diaphragm, a small serous *infracardiac bursa* may occur, representing the detached apex of the right pneumato-enteric recess (p. 233).

The abdominal part, emerging from the right diaphragmatic crus (p. 592), slightly left of the midline and level with the tenth thoracic vertebra, grooves the posterior surface of the left lobe of the liver. It forms a truncated cone, 1.25 cm long, curving sharply left (**8**.104A), its base continuous with the gastric cardiac orifice; its right side continues smoothly into the lesser curvature, while the left is separated from the gastric fundus by the cardiac notch (**8**.120). Covered by peritoneum on its front and left side, it is contained in the upper left part of the lesser omentum; the peritoneum reflected from its posterior surface to the diaphragm is part of the gastrophrenic ligament (p. 1340), through which oesophageal branches of the left gastric vessels reach it. Posterior are the left crus and left inferior phrenic artery. The vagus nerves vary in relation as the oesophagus traverses the diaphragm (Doubilet et al 1948). Sometimes one trunk (mainly of left vagal fibres) is anterior and another (mainly right vagal fibres) is posterior; but each vagus may consist of two or three trunks at this level.

Structure (**8**.104A,B). Oesophageal structure follows the general pattern of the adventitious (serosal below the level of the diaphragm), muscular, submucous and mucous layers, the last adjoining the lumen. The *fibrous adventitia* is irregular, dense connective tissue with many elastin fibres. Its fibres also penetrate and surround the fasciculi of muscle in the deeper layers.

The *muscular layer* (muscularis externa) has the usual outer longitudinal and inner circular strata, here particularly thick. The *longitudinal fibres* surround almost the whole length of the oesophagus with a continuous coat; but posterosuperiorly, 3–4 cm below the cricoid cartilage, they diverge as two fasciculi ascending obliquely to the front of the tube, where they pass deep to the lower border of the inferior constrictor to end in a tendon attached to the upper part of the ridge on the back of the cricoid lamina (**8**.14). The V-shaped interval between these fasciculi is filled by circular fibres of the oesophagus, thinly covered below by some decussating longitudinal fibres and above by the overlapping inferior constrictor. The longitudinal layer is generally thicker than the circular. *Accessory slips* of non-striated muscle sometimes pass between the oesophagus and left pleura or the root of the left principal bronchus, trachea, pericardium or aorta. These are sometimes considered to fix the oesophagus to these structures. Superiorly the *circular fibres* are continuous behind with the inferior constrictor; in front, the uppermost are attached to the lateral margins of the tendon of the two longitudinal fasciculi. Inferiorly, they are continuous with the oblique gastric fibres. Skeletal muscle is limited to the upper two-thirds of the muscularis externa in the human oesophagus; the lower third contains only non-striated. In the upper quarter both layers are skeletal; in the second non-striated muscle appears, at first internally and below this it gradually replaces the skeletal muscle.

Whitmore (1982), using primate (including human) material, has identified 'fast' and 'slow' twitch fibres in oesophageal muscle; he also observed that the striated musculature gave way to non-striated more gradually and proximally in primates than in rodents.

Radiological studies show that swallowed food stops momentarily in the gastric end of the oesophagus, before entering the stomach (**8.105**). A sphincteric mechanism, capable of contraction and relaxation, must be present at the oesophagogastric junction. Vaithilingham et al (1984) have emphatically described a sphincter muscle with a rich innervation in macaque monkeys. However, failure to identify aggregations of muscle for the control of entry into the stomach or regurgitation in man, has lead to the vague phrase, 'a physiological cardiac sphincter'. The frequency of pathological disturbances of this sphincteric control has exacerbated the controversy (see Bombeck et al 1944, Botha 1962, DiDio & Anderson 1968, Code 1968). Various factors have been invoked to explain the undoubted sphincteric effect. Circular muscle fibres in both diaphragm and oesophagus near their junction are reinforced by the splitting of the right crus round the oesophagus (p. 593), though this has been denied (Atkinson et al 1957). The phrenico-oesophageal ligament, a layer of connective tissue extending from the inferior diaphragmatic surface across the oesophageal orifice to blend with the interfascicular septa and submucosa of the terminal oesophagus, has also been invoked. The obliquity of this junction, the effects of spiral and longitudinal muscle (Stelzner & Lierse 1967) and various mucosal folds (Creamer 1955), have all been mooted as valvular mechanisms. Hence some *structures* may, alone or collectively, exert '*physiological*' sphincteric control at this site of controversy; however, the actual mechanism is still uncertain (p. 1349).

The *submucosa* loosely connects mucous and muscular layers and contains larger blood vessels, nerves and mucous glands.

The *mucosa*, thick and pink above, pale below, shows longitudinal folds which disappear in distension. It consists of: (1) lining non-keratinizing stratified squamous epithelium, (2) connective tissue, papillae from which project into the epithelium, and (3) non-striated muscularis mucosae. Where the oesophagus begins the muscularis mucosae is absent or represented merely by sparse scattered bundles; below this it is a considerable stratum. At intermediate levels its fascicles are mainly longitudinal but become more plexiform towards the gastro-oesophageal junction, where the stratified squamous epithelium is abruptly succeeded by gastric simple columnar epithelium. The junction is visible as a crenated line, the greyish-pink, smooth, oesophageal mucosa contrasting with the red, mamillated, gastric mucosa.

There are small *oesophageal glands* of compound racemose mucous type in the submucosa external to the muscularis mucosae, each with a long duct traversing it and the other layers of the mucosa. Glands of the abdominal oesophagus resemble gastric cardiac glands and lie between the muscularis mucosae and the lumen (Johns 1952).

Vessels and Nerves

Oesophageal *arteries* come from the inferior thyroid branch of the thyrocervical trunk, descending thoracic aorta, bronchial arteries, left gastric branch of the coeliac artery and left inferior phrenic. Their direction of supply is generally longitudinal. *Veins* from the cervical part drain to the inferior thyroid veins, from the thoracic part into azygos, hemiazygos and accessory hemiazygos veins, and from the abdominal part to the azygos and left gastric vein. The latter is a tributary of the portal vein, this being one of the anastomoses between portal and systemic veins (p. 820). In portal obstruction (e.g. in hepatic cirrhosis), these anastomatic veins may become varicose and burst into the lower oesophagus, leading to haematemesis and even fatal haemorrhage. *Lymph vessels* are described on pp. 821, 850.

The *nerve supply* is parasympathetic (vagal) and sympathetic. The cervical part receives rami from the recurrent laryngeal nerves and from cervical sympathetic trunks (via the plexus around the inferior thyroid artery). The thoracic part is innervated by branches from the vagal trunks and oesophageal plexus, the sympathetic trunks and greater splanchnic nerves. The abdominal part (Mitchell 1938) is supplied by the vagal trunks (anterior and posterior gastric nerves), the thoracic sympathetic trunks, greater (and sometimes lesser) splanchnic nerves and the plexuses around the left gastric and inferior phrenic arteries. These rami form a plexus containing groups of ganglion cells between the two layers of the muscular coat, and a second, submucous, plexus (Semenova 1962). In addition to these, free-ending nerve fibres have been observed in the oesophageal epithelium (Robles-Chillida 1981). Rodrigo et al (1975) have surveyed these plexuses, describing also intraganglionic nerve endings, possibly afferent.

Radiology

In oblique lateral views (**8.105**) the thoracic part of the oesophagus is shown by a barium swallow to be mainly in the 'retrocardiac space'; the entire thoracic part is seen to be near the vertebral column, its lower end inclining forwards. Its anterior wall is indented by the aortic arch, left bronchus and left atrium. A thin layer of the barium contrast medium may outline longitudinal mucosal grooves, producing a striated shadow.

THE ABDOMEN

The abdomen extends from the diaphragm to the base of the pelvis (**8.107**), comprising the *abdomen proper* and the *lesser pelvis*, continuous at the plane of the inlet into the lesser pelvis, which is bounded by the sacral promontory, arcuate lines of the innominate bones, pubic crests and the upper border of the symphysis pubis. The abdomen is largely enclosed by muscles, its shape and size varying with degrees of distension of the contained hollow organs and the phases of respiration. Muscular tone is a large factor in maintaining the positions of the abdominal and pelvic viscera (pp. 602, 606).

The *abdomen proper* is bounded: *in front* by the rectus abdominis muscles, the pyramidales and the aponeurotic parts of the obliqui externus, internus and transversus abdominis; *laterally* by the fleshy parts of these flat muscles, the iliacus muscles and iliac bones, *behind* by the lumbar vertebral column, diaphragmatic crura, paired psoas and quadratus lumborum muscles and the posterior parts of the iliac bones; *above* by the diaphragm, while *below* it is continuous with the lesser pelvis through its superior aperture (p. 429). Since the diaphragm, the domed roof of the abdominal cavity, is convex upwards, part of the cavity is within the osseous framework of the thorax (pp. 594–595). The abdomen proper contains most of the digestive tube, liver, pancreas, spleen, kidneys, ureters (in part), suprarenal glands and numerous blood and lymph vessels, lymph nodes and nerves.

The *lesser pelvis* is approximately funnel-shaped, like an inverted, truncated cone; it extends postero-inferiorly from the abdominal cavity proper (**8.110**, **3.116**) and is bounded: *anterolaterally* by the innominate (hip) bones below their pubic crests and arcuate lines, and by the obturatores interni, *posterosuperiorly* by the sacrum, coccyx, piriformes and coccygei; *inferiorly* by the levatores ani which, with their covering fasciae, form the pelvic diaphragm (pp. 604–606), and by the transversi perinei profundi and sphincter urethrae which, with *their* fascial coverings, constitute the urogenital diaphragm. The lesser pelvis contains: the urinary bladder, terminal parts of the ureters, the sigmoid colon, rectum, some ileal coils, internal genitalia, blood and lymph vessels, lymph nodes and nerves.

The abdominal and pelvic muscles are ensheathed in fascia, which receives regional names from adjacent structures, e.g. on the internal surface of transversus abdominis is the *transversalis fascia* (p. 603), inferior to the diaphragm is the *diaphragmatic fascia*, covering the psoas and iliacus is the *iliac fascia* (p. 604); anterior to the quadratus lumborum is the *anterior layer of the thoracolumbar fascia* (p. 604) and over the muscles in the pelvis is the *pelvic fascia* (p. 604). Most abdominal and pelvic organs are largely covered by a serous membrane, the *visceral peritoneum* (pp. 1336–1343).

ABDOMINAL REGIONS

For the location of viscera in clinical practice, the abdomen is divided into nine regions by imaginary planes, two horizontal and two parasagittal, their edges indicated by lines projected to the surface of the body (**8.106**). The upper, horizontal, *transpyloric plane* (of Addison) is indicated by a line encircling the body midway between the suprasternal notch and the symphysis pubis (or midway between the umbilicus and inferior end of the sternal *body* or a hand's breadth below the xiphisternal joint); it intersects the first lumbar vertebral body near its lower border and meets the costal margins at the tips of the ninth costal cartilages. The lower horizontal, *transtubercular plane* corresponds to a line round the trunk level   with the iliac tubercles (**8.106**, pp. 430–431); it cuts the front of the fifth lumbar vertebral body near its upper border. The abdomen is thus divided into three arbitrary zones; each is further subdivided into three by the *right* and *left lateral planes*, indicated on the surface by vertical lines through points midway between the anterior superior iliac spines and the symphysis pubis (these lines are also called 'midclavicular' or 'mammary' lines).

The median upper *epigastric* is flanked by *right* and *left hypochondriac* regions; the median region of the middle zone is the *umbilical* region, flanked by *right* and *left lumbar* or *lateral* regions. The lower median *hypogastric* or *pubic* region is between the *right* and *left iliac* or *inguinal* regions (**8.106**) (see Addison 1899–1901). A third horizontal plane, often used in abdominal topography, the *subcostal plane*, corresponds to a line level with the lowest limits of the tenth costal cartilages. It cuts the front of the third lumbar vertebral body near its upper border. It often replaces the transpyloric plane in descriptions of abdominal regions.

The *umbilicus* is variable in position, being usually level with the disc between the third and fourth lumbar vertebrae in young adults but, as age advances and in conditions of deficient abdominal tone, it sinks lower. It is also lower in children because of the under-developed condition of the pelvic region.

On the body's posterior surface a transverse line between the highest points on the iliac crests and level with the fourth lumbar vertebral spinous process delineates a *supracristal plane*. The fourth spine is a useful landmark in identifying other vertebral spinous processes.

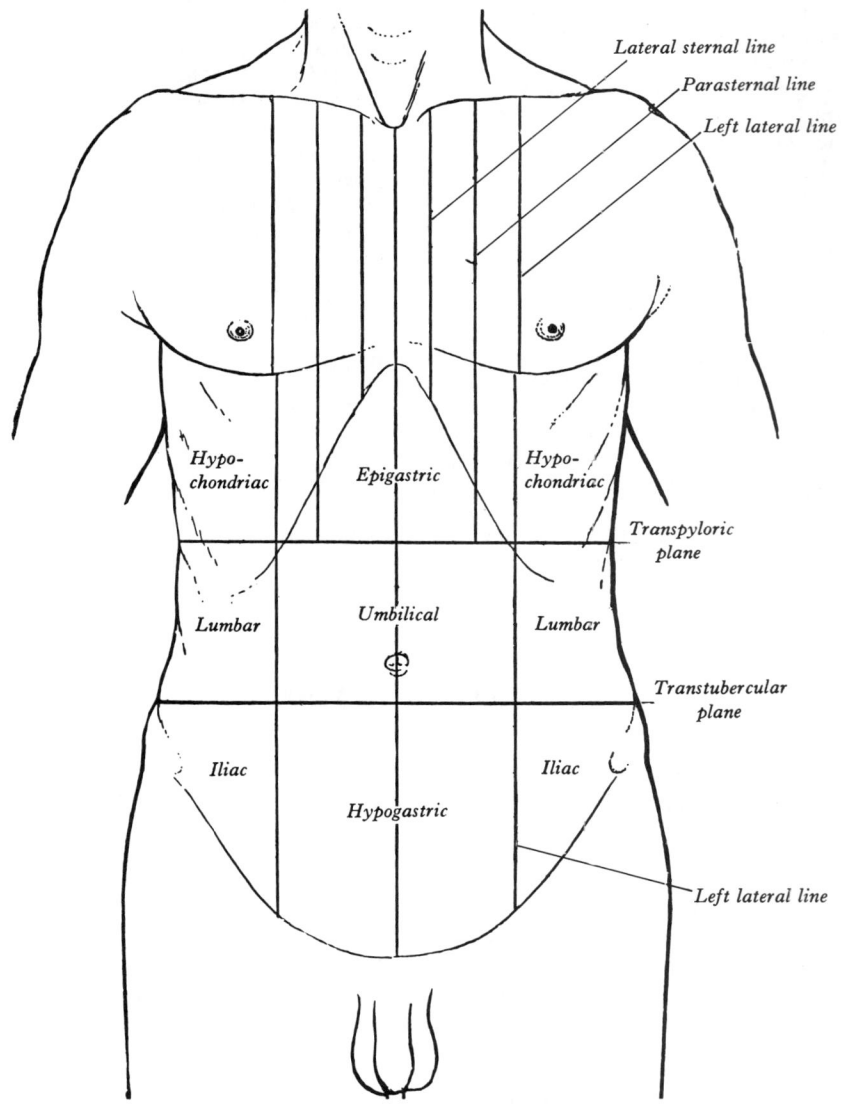

8.106 Reference lines on the anterior aspect of the thorax and abdomen for use in delineating surface projections.

LOCATION OF ABDOMINAL VISCERA

After removal of the anterior abdominal wall (8.107), the viscera are partly displayed as follows: above and right is the liver, largely in the shelter of the lower right ribs and cartilages, extending across the midline, where it descends to the transpyloric plane. The stomach is partly exposed in the angle between the left costal margin and the lower hepatic border; from its lower border an apron-like peritoneal fold, the *greater omentum*, descends for a varying distance anterior to the other viscera. Below this, however, some coils of the small intestine are usually visible. In the right iliac region the caecum and, in the left iliac region, the lower part of the descending colon are partly exposed (p. 1368). The urinary bladder, in the anterior part of the pelvis, ascends above the symphysis pubis into the hypogastric region when distended; the rectum is in the sacral concavity, usually hidden by coils of the small intestine. The sigmoid colon may appear between the rectum and the bladder.

Followed to the right, the stomach is continuous with the *duodenum*, their junction being marked by a thick, palpable *pyloric sphincter*. The duodenum reaches the liver and curves down under its cover. If, however, the greater omentum and the transverse colon behind it are turned up over the chest, the horizontal part of the duodenum can be seen crossing the vertebral column from right to left. The duodenum then ascends to the second

lumbar vertebra to become continuous with the coils of the *jejunum* and *ileum*, which are about six metres long (p. 1355) and form the rest of the small intestine. Followed inferiorly, the ileum ends in the right iliac region, opening into the colon at the junction of the *caecum* and *ascending colon*. From the caecum the colon ascends on the right, then loops to the left across the median plane below the liver and stomach and turns downwards; thus it is composed of *ascending*, *transverse* and *descending* parts. In the pelvis it forms a loop, the *sigmoid colon*, ending in the rectum. The *spleen* is posterolateral to the stomach in the left hypochondriac region and is partly exposed by displacing the stomach to the right.

The sheen on the surfaces of the abdominal wall and exposed viscera is due to their covering of *peritoneum*, a serous membrane.

The relations of the organs described here obtain in the recumbent position, but visceral relations depend on posture, respiratory movements and the degree of distension of the the hollow organs. The shapes of chest, abdomen and pelvis also vary, as do organs in the same individual and at different times, depending on physiological activity and mobility. Therefore the surface outlines of viscera, particularly the hollow organs described here, must be regarded as highly variable within wide limits.

In physique, individuals have been classified into two extremes: *hypersthenic* (pyknic) and *asthenic* (leptosomatic) with intermediate grades, *sthenic* and *hyposthenic* (Mills 1917, 1922). In the

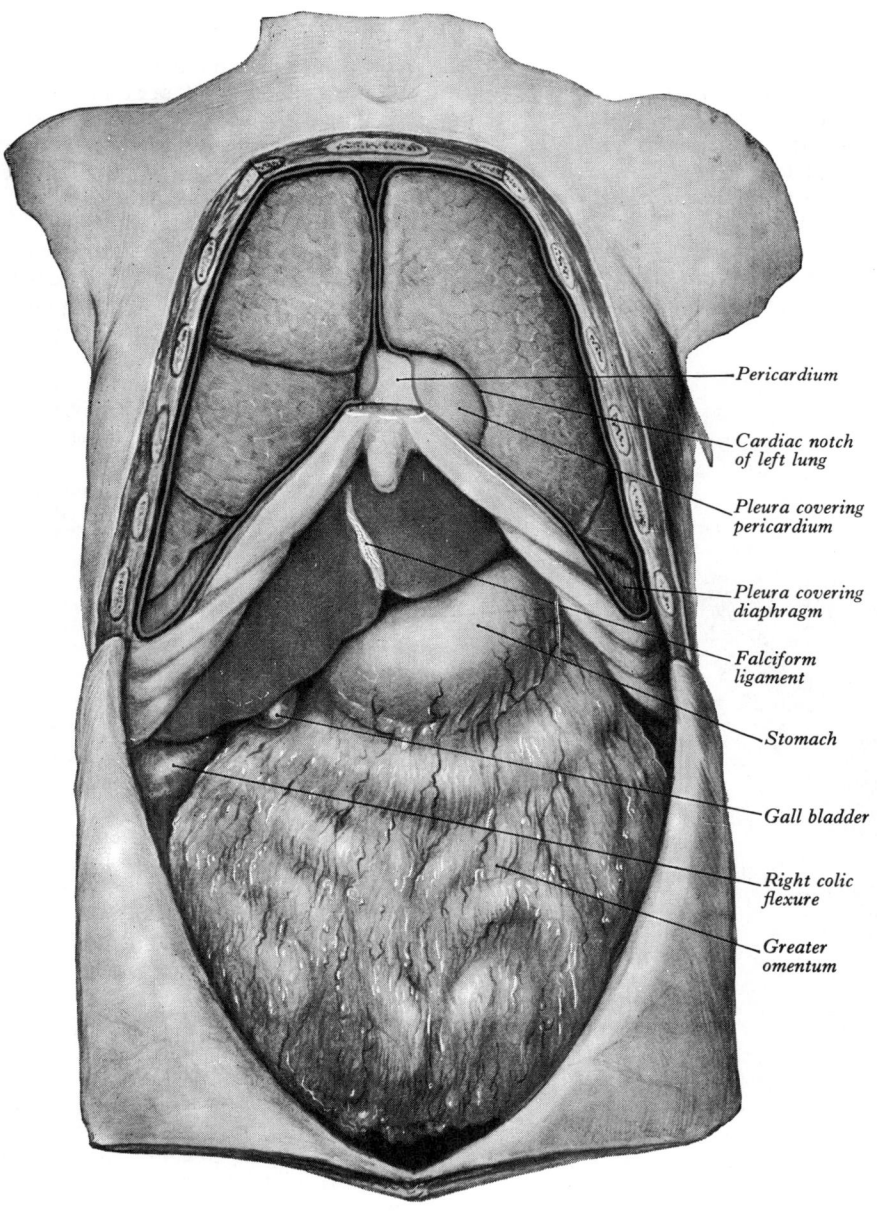

Pericardium

Cardiac notch of left lung

Pleura covering pericardium

Pleura covering diaphragm

Falciform ligament

Stomach

Gall bladder

Right colic flexure

Greater omentum

8.107 Anterior aspect of the thoracic and abdominal viscera.

hypersthenic, with massive physique, the thorax is wide and short and the subcostal angle very obtuse, so that the heart and lungs are wide transversely; the abdomen is widest superiorly and the stomach less elongated vertically, with the pylorus relatively high; while in the asthenic type, with a light and slender physique, the thorax is long and narrow and the subcostal angle acute, so that the heart and lungs are long and narrow; the abdomen is widest inferiorly, the

stomach being long with a relatively low pylorus and the colon long with a V-shaped transverse colon descending to the pelvis. Varieties of physique (somatotypes) have also been classified as endomorphic (massive), mesomorphic (intermediate) and ectomorphic (slender) with intermediate grades, each reputed to have predominant psychological characteristics (Sheldon et al 1940, Sheldon & Stevens 1942; see current physical anthropology monographs for details).

THE PERITONEUM

The peritoneum (Brizon et al 1956), the largest and most complexly arranged of the serous membranes, is an empty and intricately folded sac, lining the abdomen and reflected over the viscera. In males it is a closed sac; in females the lateral ends of the uterine tubes open into the sac's potential cavity. Where it lines the abdominal wall (parieties) it is named the *parietal* peritoneum and is reflected over the viscera as the *visceral* peritoneum. Its free surface is covered by a layer of *mesothelium*, kept moist and smooth by a film of serous fluid. Hence mobile viscera glide freely on the abdominal wall and each other within limits dictated by their attachments. Sessile organs are covered by peritoneum wherever they are in contact with mobile viscera.

The peritoneal cavity is a *coelom*—a discontinuity in the mesoderm lined by an epithelium (mesothelium) which maintains the surface. Loss of this mesothelium leads to the adherence of underlying tissues and interference with visceral function, which may be serious and even lethal (p. 1346), providing convincing evidence of an essential function of the serosa, separating of the viscera sufficiently for unimpeded activity. This basic biological arrangement is used to classify all animals as either *Coelomata* or *Acoelomata*.

Many functions are served by coelomic spaces in invertebrates and vertebrates (Jones 1913, Romer 1970). Excretory organs such

as nephridia drain fluid and excretory products from a general coelom and vertebrate nephric systems are partly derived from it. As animals evolved to greater size, a coiled gut became essential. Such coiling could not evolve and develop without the emergence of a coelom. In early chordates such as *Amphioxus*, gametes are extruded from the gonads into a coelom and in some early vertebrates this also occurs in both sexes, persisting in the females of later forms, including mankind. Special ducts have evolved to connect the kidneys and testes to the cloaca (and its derivatives) and are in part formed from coelomic epithelium, which is also involved in formation of the gonads themselves (p. 248).

GENERAL STRUCTURE

A considerable amount of *extraperitoneal connective tissue* separates the parietal peritoneum from the abdominal walls, blending with their fascial lining. Its thickness and content of fat varies in different regions. While parietal peritoneum is generally loosely attached to the abdominal and pelvic walls and so is easily stripped from them, the tissue is denser on the inferior surface of the diaphragm and behind the linea alba, the parietal peritoneum being here more firmly adherent. Its attachment is especially loose in some places to allow alteration in the size of certain organs; e.g. in the pelvis and the adjoining anterior abdominal wall it allows the urinary bladder to distend upwards behind the wall, from which it temporarily strips the peritoneum as it ascends. There is usually much perinephric extraperitoneal fat on the posterior abdominal wall. The visceral peritoneum, in contrast, is firmly united to the underlying tissues and cannot be easily detached; its connective tissue layer (*tela subserosa*) is continuous with the fibrous visceral stroma; and the visceral peritoneum must therefore be considered as part of its viscus, a concept of significance in pathology.

THE PERITONEAL CAVITY

The parietal and visceral layers of peritoneum are in contact, the potential space between them being the *peritoneal cavity*. This consists of: (1) a main region, the *greater sac*, and (2) a diverticulum, the *omental bursa* or *lesser sac* behind the stomach and adjoining structures; the two communicate via the *epiploic foramen* (*aditus to the lesser sac*).

The complex arrangement of the peritoneum can best be rationalized by the study of alimentary development (pp. 232–236) and by examination in cadavers before they are made unnaturally rigid by preservative fluids. To trace it from one viscus to another and from viscera to parieties, it is useful to follow its continuity in vertical and horizontal directions and simpler to describe the greater and lesser sacs separately.

VERTICAL DISPOSITION OF THE PERITONEUM

The ensuing descriptions will be more clearly comprehended if frequent reference is made to the development of the alimentary tract (pp. 232–236) and to illustrations, models and dissections.

It is convenient to commence tracing the arrangement of the greater sac in the vertical plane (8.112) from the anterior abdominal wall at the umbilical level. A fibrous *ligamentum teres* (*obliterated left umbilical vein*, p. 724) ascends from this to the

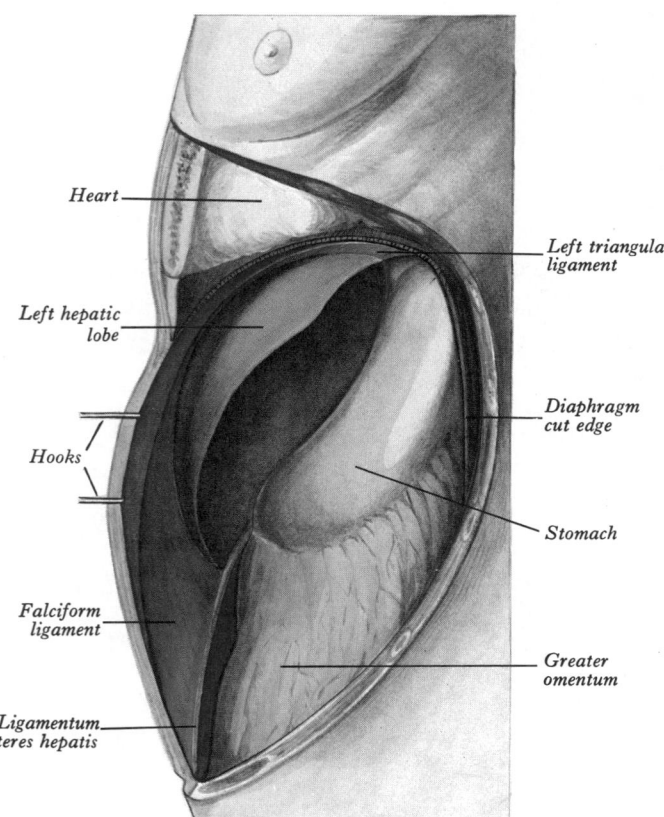

Heart

Left triangular ligament

Left hepatic lobe

Hooks

Diaphragm cut edge

Stomach

Falciform ligament

Greater omentum

Ligamentum teres hepatis

8.108 Dissection to expose the left side of the falciform fold or ligament of the liver.

inferior surface of the liver, inclining slightly right and receding from the abdominal wall as it ascends, raising a triangular *falciform hepatic ligament* (p. 1388 and **8**.108) of parietal peritoneum from the wall and inferior diaphragmatic surface. The falciform ligament has right and left peritoneal layers with intervening connective tissue (**8**.114A,B). Its juxta-umbilical region has a posterior free border from the umbilicus to the inferior hepatic surface, containing the *ligamentum teres*. Superiorly the falciform ligament extends from the diaphragm to become continuous with the visceral peritoneum on the hepatic anterosuperior surface (**8**.108). At the site of reflexion from the diaphragm to the liver, the two layers diverge (**8**.154), the right passing transversely to the right as the *superior layer of the hepatic coronary ligament* (from the diaphragm to the upper surface of the right hepatic lobe), the left layer passing left as the *anterior layer of the left hepatic triangular ligament* (from the diaphragm to the upper surface of the left hepatic lobe).

The visceral peritoneum on the anterosuperior hepatic surface continues round the sharp inferior hepatic border to the inferior (visceral) surface, where it is arranged as follows: right of the gallbladder it covers the inferior surface of the right lobe of the liver and is reflected posteriorly to the right suprarenal gland and the upper pole of the right kidney, forming the *inferior layer of the coronary ligament*; it often passes direct from the liver to the kidney as the *hepatorenal ligament*. From the right kidney it descends to the front of the first part of the duodenum and right colic flexure; it also passes medially in front of a short segment of the inferior vena cava (between the duodenum and liver), continuing on to the posterior wall of the omental bursa (**8**.113). Between the two layers of the coronary ligament is a large, triangular, posterior area on the right hepatic lobe devoid of peritoneal covering, the *bare area of the liver*, where the liver is attached to the diaphragm by areolar tissue.

Near the right hepatic margin layers of the coronary ligament converge fusing to form the *right triangular ligament* which connects the right hepatic lobe to the diaphragm (**8**.154) and forms the apex of the bare area, the base being the *groove for the inferior vena cava*.

Visceral peritoneum covers the inferior aspect and sides of the gallbladder, the inferior surfaces of the quadrate lobe of the liver as far back as the anterior margin of the porta hepatis and of the left lobe, from whose posterior surface it reaches the diaphragm as the *posterior layer of the left triangular ligament*. Along the anterior margin of the porta hepatis the peritoneum is continuous at its right end with the peritoneum of the omental bursa, the latter being reflected from the posterior portal margin (**8**.154). The visceral peritoneum plunges into the fissure for the ligamentum venosum (**8**.154), between the caudate and left hepatic lobes, in two layers, anterior and posterior. The anterior layer merges with peritoneum reflected from the anterior portal margin (**8**.154). From this L-shaped line, formed by the left margin of the fissure for the ligamentum venosum and the anterior margin of the porta hepatis, peritoneum is reflected to the gastric lesser curvature and approximately the first 2 cm of the duodenum, as the anterior layer of the *lesser omentum*.

The region of lesser omentum connecting liver to stomach is the *hepatogastric ligament*, the part passing from the liver to the duodenum being the *hepatoduodenal ligament*. The anterior layer, traced to the right, passes anterior to the hepatic artery, bile duct and portal vein, turning round their right side to continue behind them into the posterior omental layer, which here forms the anterior surface of the bursa. Thus the lesser omentum has a free right border, in which lie the hepatic artery, bile duct and portal vein; behind this border is the *epiploic foramen* (foramen of Winslow) (**8**.114A). The anterior layer of the lesser omentum is continuous below with the visceral peritoneum of the anterior gastric surface and the first 2 cm of the duodenum. This layer then descends from the greater curvature and neighbouring duodenum to become the most anterior layer of the *greater omentum*. Reaching the lower edge of this large fold, it ascends as its most *posterior* layer, running to the anterosuperior aspect of the transverse colon (at the *taenia omentalis*). It then turns back, adherent to but separable from the *upper* layer of the transverse mesocolon, to the anterior aspect of the head and the anterior border of the body of the pancreas; it leaves the latter as the upper layer of the transverse mesocolon (**8**.112), passing to the posterior surface of the transverse colon (at the *taenia mesocolica*) and covering all but its posterior aspect, returning thence to the pancreatic head and body as the *inferior* layer of the transverse mesocolon. It then descends over the pancreas to the front of the horizontal and ascending parts of the duodenum, from there turning downwards on the posterior abdominal wall. It is also carried forward on the superior mesenteric vessels to the jejenum and ileum as the *right layer of the mesentery*. It invests them and reaches the posterior

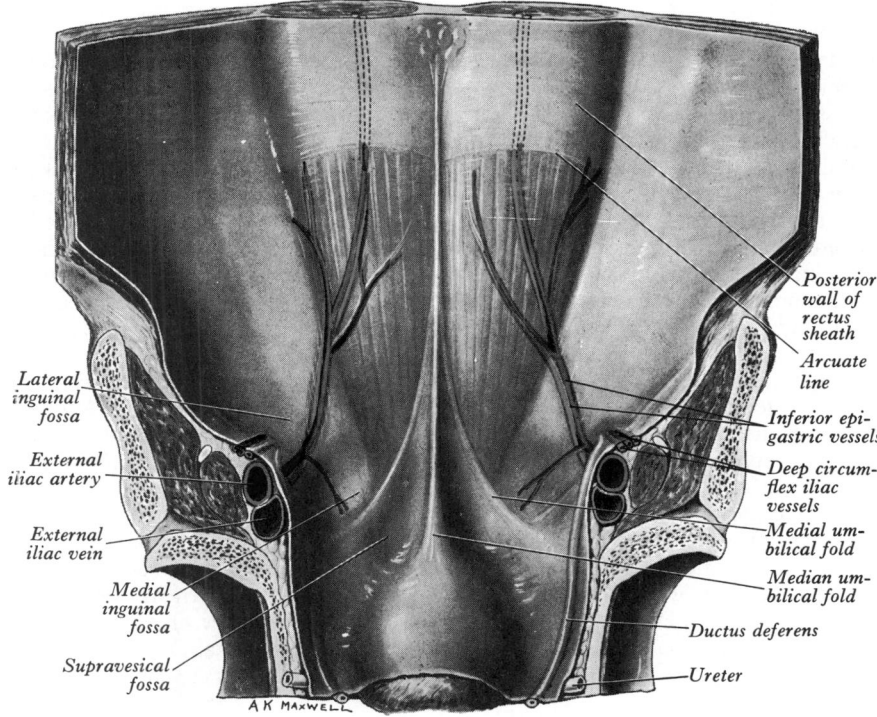

Lateral inguinal fossa

External iliac artery

External iliac vein

Medial inguinal fossa

Supravesical fossa

Posterior wall of rectus sheath

Arcuate line

Inferior epigastric vessels

Deep circumflex iliac vessels

Medial umbilical fold

Median umbilical fold

Ductus deferens

Ureter

A K MAXWELL

8.109 The infra-umbilical part of the anterior abdominal wall of a male subject: posterior surface, with the peritoneum in situ.

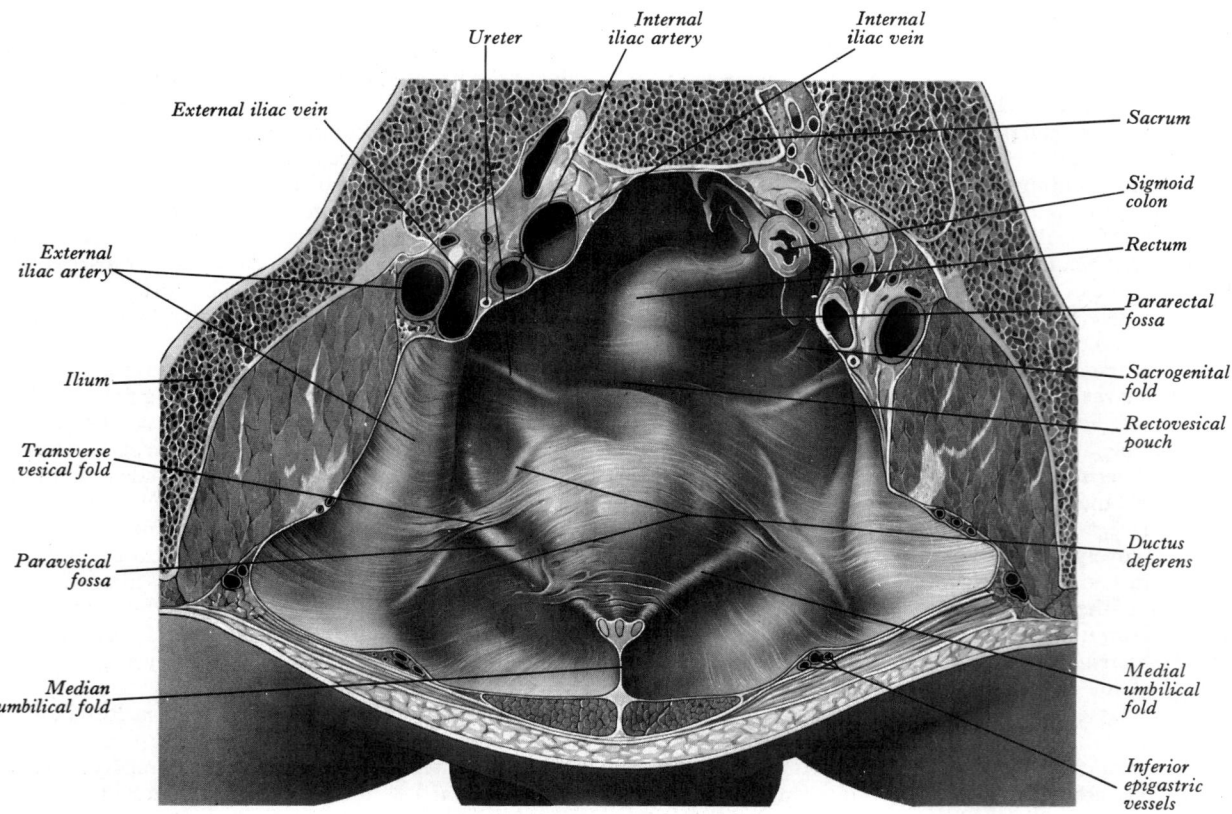

Ureter

Internal iliac artery

Internal iliac vein

External iliac vein

Sacrum

Sigmoid colon

Rectum

External iliac artery

Pararectal fossa

Ilium

Sacrogenital fold

Transverse vesical fold

Rectovesical pouch

Paravesical fossa

Ductus deferens

Median umbilical fold

Medial umbilical fold

Inferior epigastric vessels

8.110 The peritoneum of the male pelvis: anterosuperior view. The median umbilical fold contains both the unpaired median and the paired medial umbilical ligaments in the plane of section in this subject.

abdominal wall as the *left layer of the mesentery*, descending over the abdominal aorta, inferior vena cava, ureters and psoas major muscles into the lesser pelvis. Reflected from the posterior pelvic wall as the *anterior layer of the sigmoid mesocolon*, it encloses the sigmoid colon and returns to the pelvic wall as the *posterior layer* of that mesocolon, descending then to cover the front and sides of the rectum's upper third and the front of its middle third.

In males, the peritoneum leaves the junction of the middle and lower thirds of the rectum, passing forwards to the upper poles of the seminal vesicles and superior aspect of the bladder. Between the rectum and bladder it forms the *rectovesical pouch*, descending slightly below the upper seminal poles to a level about 7.5 cm from the anal orifice. From the apex of the bladder it returns along the median and medial umbilical ligaments (**8**.109) to the anterior abdominal wall and umbilicus. When the bladder distends, the peritoneum is lifted from the lower anterior abdominal wall so that part of the bladder's anterior surface is in *direct* contact with the wall (p. 1416). Instruments can then be passed through the wall into the bladder without traversing the peritoneum.

In females, the peritoneum passes from the rectum to the posterior vaginal fornix and then to the back of the uterine cervix and body, as the *recto-uterine fold*, which descends to form the *recto-uterine pouch* (of Douglas), the base of which is only 5.5 cm from the anal orifice. The peritoneum spreads over the uterine fundus to its anterior (vesical) surface as far as the junction of the body and cervix, from which it is reflected forwards to the upper surface of the bladder, forming a shallow *vesico-uterine pouch*. Peritoneum on the anterior and posterior uterine surfaces leaves the organ to reach the lateral pelvic walls as the *broad ligaments of the uterus*, each consisting of antero-inferior and posterosuperior layers continuous at the upper border of the ligament; between them at this border is the uterine tube. Peritoneum is reflected from the bladder to the anterior abdominal wall as in males.

HORIZONTAL DISPOSITION OF THE PERITONEUM

Below the transverse colon the arrangement is simple, but differs at pelvic, lower abdominal and upper abdominal levels.

(1) **In the lesser pelvis**, the peritoneum follows the surfaces of the pelvic viscera and walls, with differences in the sexes. **In males** (**8**.110) it almost encircles the sigmoid colon, passing to the posterior pelvic wall as the *sigmoid mesocolon*. It leaves the sides and finally the front of the rectum, continuing over the upper poles of the seminal vesicles to the bladder; lateral to the rectum it forms right and left *pararectal fossae*, varying with rectal distension, and anteriorly a *rectovesical pouch*, limited laterally by peritoneal folds from the sides of the bladder posteriorly to the anterior aspect of the sacrum, the *sacrogenital folds*, each lateral to its pararectal fossa. Anteriorly it covers the superior surface of the bladder, forming on each side a *paravesical fossa*, limited laterally by a ridge containing the ductus deferens. The size of these fossae depends on the state of the bladder; when it is empty, a variable *transverse vesical* fold bisects each fossa and the anterior ends of the sacrogenital folds may sometimes be joined by a ridge separating a *middle fossa* from the main rectovesical pouch (**8**.110). Between the paravesical and pararectal fossae the only elevations are due to the ureters and internal iliac vessels. **In females**, pararectal and paravesical fossae also appear, the lateral limit of the latter being the peritoneum investing the round ligament of the uterus. The rectovesical pouch is, of course, divided by the uterus and vagina into a small, anterior, vesico-uterine and a deep, posterior, recto-uterine pouch (**8**.209). Marginal recto-uterine folds of the latter (p. 1339) correspond to the sacrogenital folds in males and pass back to the sacrum from the sides of the cervix lateral to the rectum. The *broad ligaments* extend from the sides of the uterus to the lateral pelvic walls, with the uterine tubes contained in their free superior margins and the ovaries attached to their posterior layers. Below, they are continuous with the lateral pelvic parietal peritoneum. Between the elevations over the obliterated umbilical artery and the ureter on the lateral pelvic wall is a shallow *ovarian fossa* containing the ovary in nulliparous females. It lies behind the lateral attachment of the broad ligament.

(2) **In the lower abdomen**, the peritoneum of the lower anterior abdominal wall is raised into five ridges converging upwards (**8**.109). A *median umbilical fold* extends from the apex of the bladder to the umbilicus. It contains the urachus (p. 1417). On

each side of it the obliterated umbilical artery raises a *medial umbilical fold*, ascending from pelvis to umbilicus. Between these three folds are the two *supravesical fossae*. Further laterally each inferior epigastric artery raises a *lateral umbilical fold*, below its entry into the rectus sheath. *Medial inguinal fossae* exist between the lateral and medial umbilical folds. A *lateral inguinal fossa* over the deep inguinal ring is lateral to the lateral umbilical fold and indicates the site of descent of the processus vaginalis and testis into the anterior abdominal wall. A *femoral fossa*, inferomedial to the lateral inguinal and separated from it by the medial end of the inguinal ligament, overlies the femoral ring (p. 781).

From the linea alba, caudal to the level of the transverse colon, the peritoneum, followed horizontally to the right, lines the abdominal wall almost to the lateral border of quadratus lumborum; it is reflected over the sides and front of the ascending colon, enclosing the caecum and vermiform appendix and passing medially over the duodenum, psoas major and inferior vena cava towards the median plane, whence it passes along the superior mesenteric vessels to invest the small intestine and back again to the large vessels anterior to the vertebral column, forming the *mesentery* (8.112, 113). This encloses: the jejunum, the ileum, superior mesenteric blood vessels, nerves, lacteals and lymph nodes. The peritoneum then continues left across the abdominal aorta and left psoas major, covering the sides and front of the descending colon, and then returns round the abdominal wall to the midline.

(3) **In the upper abdomen** (8.105, 114A), above the transverse colon, the arrangement of the peritoneum in the greater sac is more complex. From the front of the inferior vena cava, just above the first part of the duodenum, it passes *left* behind the epiploic foramen to form the posterior wall of the omental bursa (8.114A); it passes *right* over the front of the right suprarenal gland and the upper pole of the right kidney to the anterolateral abdominal wall. From the anterior median line a double fold, the *falciform ligament*, passes back to the right around the liver. To the left the peritoneum lines the anterolateral abdominal wall, covers the lateral part of the left kidney and passes to the splenic hilum as the posterior (lateral) layer of the *splenicorenal* (lienorenal) or *phrenicolienal ligament* (8.114A). It then invests the spleen, returning to the front of its hilum and thence to the cardiac end of the greater curvature as the left layer of the *gastrosplenic ligament*. Covering the anterosuperior gastric surface and adjacent duodenum, it ascends from the lesser curvature to the liver as the anterior layer of the lesser omentum, whose right free border has been described (p.1337); this anterior layer of the lesser omentum (peritoneum of the greater sac) continues as the posterior layer of the omental bursa (peritoneum of the omental bursa).

THE OMENTAL BURSA (LESSER SAC)

The omental bursa or lesser sac is a large, irregular, potential recess behind the stomach and beyond its limits. Its name is related to the concept that it forms a bursa (p.563) facilitating movements of the posterior aspect of the stomach. However, it is not closed; its connection with the greater sac is narrowed by embryological rather than functional factors. The stomach expands or contracts just as freely in vertebrates with no special narrow-necked diverticulum. (In any case movements of the *anterior aspect* of the stomach also occur.) The extensive anterior and posterior walls of the 'bursa' are limited by variable borders (right, left, inferior and superior). It is separated from the greater sac except at its upper *right border* where they communicate by a vertical slit, the *epiploic foramen*. Its upper posterior wall is a single peritoneal layer closely applied to the posterior abdominal wall (8.112) but below the pancreas its potential cavity projects into the greater omentum, whose posterior wall is formed by two layers which, above the transverse colon, blend with the transverse mesocolon (8.112). The greater omentum is traditionally described as having *four* peritoneal layers; but it must be understood that mesothelial peritoneum is lost, except where a true surface persists. Where two separate 'folds' of peritoneum adhere and blend, opposed mesothelia disappear but recognizable layers of subepithelial connective tissue often remain.

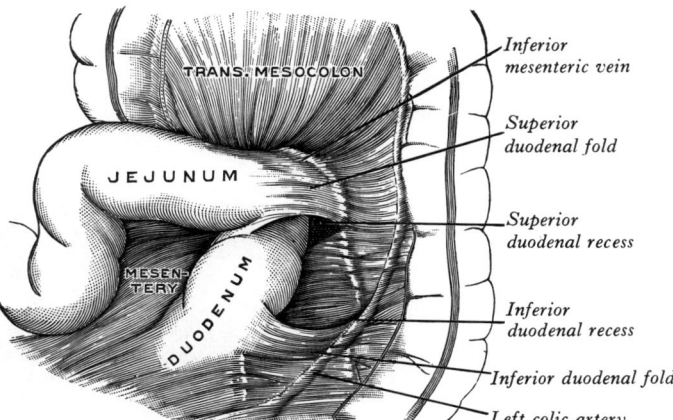

8.111 The superior and inferior duodenal recesses. The transverse colon and jejunum have been displaced (after Jonnesco, from Poirier & Charpy *Traité d'Anatomie humaine*. Masson et Cie).

The epiploic foramen (aditus to the lesser sac), a short, vertical slit of about 3 cm, leads from the upper part of the right border of the lesser into the greater sac, this border forming the foramen's *anterior margin* and containing between its layers the bile duct (on the right), portal vein (posterior) and hepatic artery (left) (8.114A). Superiorly the two layers separate, the posterior covering the caudate process of the liver in the *roof* of the foramen (8.115) and descending anterior to the inferior vena cava as the foramen's *posterior margin*. At the upper border of the first duodenal segment this layer passes forwards from the inferior vena cava, above the head of the pancreas, into the posterior layer of the lesser omentum, forming the foramen's *floor* which is medially continued down into the right border of the sac (8.105). Passing forwards below the medial end of this floor the hepatic artery passes between the two layers of the lesser omentum (8.114A). A narrow passage, the *vestibule* of the omental bursa, is left of the foramen between the caudate process and the first part of the duodenum. To the right the rim of the foramen is continuous with the peritoneum of the greater sac: the roof is continuous with the peritoneum on the inferior surface of the right hepatic lobe (8.154), the posterior wall with the peritoneum on the right suprarenal gland (8.105), its anterior wall with the anterior layer of the lesser omentum round the portal vein and bile duct (8.114A), the floor with the peritoneum on the lower part of the right suprarenal gland and on adjacent parts of the duodenum and right kidney (8.113). Anterior and posterior walls of the foramen are normally apposed.

The omental bursa has an *anterior wall* formed (1) by peritoneum over the postero-inferior aspect of the stomach and about the first 2 cm of the duodenum, which *descends* to become the posterior of the anterior two layers of (2), the greater omentum, and *ascends* to the right to leave the lesser gastric curvature and duodenum at its upper border, becoming the posterior layer of (3), the lesser omentum. The bursa is often described as ascending *behind* the caudate lobe, but this projects into the bursa from its right border and is covered by peritoneum on both anterior and posterior surfaces (8.112,154).

The *posterior wall* is formed by the anterior of two posterior layers of the greater omentum. Above, the posterior of these is fused, but not inseparably, with the upper peritoneum of the transverse colon and mesocolon. Surgical separation of the greater omentum from these provides posterior access to the stomach through the posterior wall of the greater omentum. Dissection where the omentum and transverse colon meet opens up an embryological 'bloodless plane' between vessels of the greater omentum (from the gastro-epiploic) and the middle colic vessels in the transverse mesocolon (Freder 1905, Lardennois & Okinczyc 1913, Grégoire 1922, Ogilvie 1935). There are no anastomoses across this plane. Above the anterior pancreatic border the posterior bursal peritoneum covers the posterior abdominal wall, a small part of the head and the whole neck and body of the pancreas, part of the anterior aspect of the left kidney, most of the left

1339

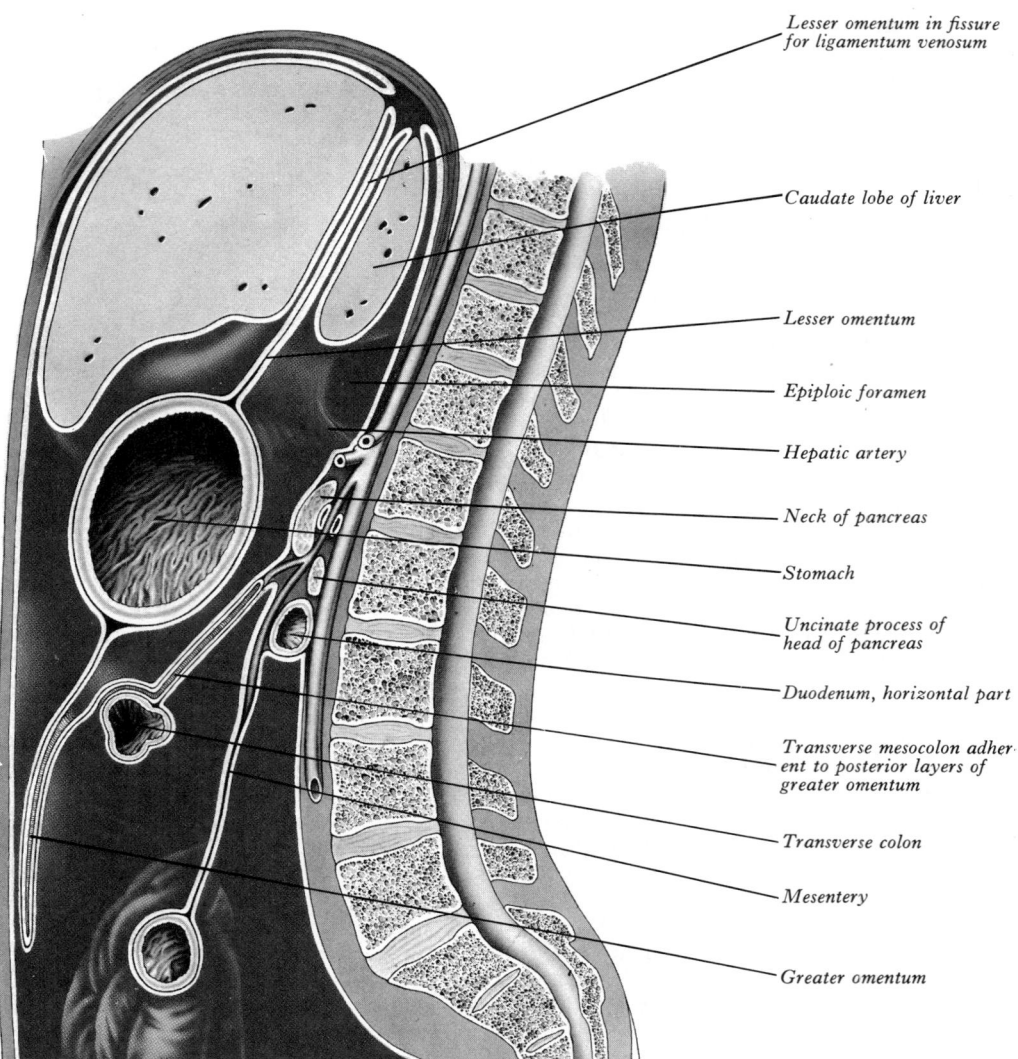

*Lesser omentum in fissure
for ligamentum venosum*

Caudate lobe of liver

Lesser omentum

Epiploic foramen

Hepatic artery

Neck of pancreas

Stomach

*Uncinate process of
head of pancreas*

Duodenum, horizontal part

*Transverse mesocolon adher-
ent to posterior layers of
greater omentum*

Transverse colon

Mesentery

Greater omentum

8.112 Sagittal section through the abdomen, approximately in the median plane. Compare with **8.113**. The section cuts the posterior abdominal wall along the line YY in **8.113**. The peritoneum is shown in blue except along its cut edges, which are left white.

suprarenal gland, the commencement of the abdominal aorta and coeliac artery and part of the diaphragm. The inferior phrenic, splenic, left gastric and hepatic arteries lie partly behind the bursa (**8.113, 114A**).

The *limits of the bursa* are the lines where its posterior peritoneal wall is reflected to be continuous with its anterior; their positions vary somewhat. The *inferior border* is, developmentally (p. 235), the lower limit of the greater omentum; but partial fusion of the latter's layers usually occurs after birth, so that the bursa's *cavity* in adults does not usually extend very far below the transverse colon. The internally apposed peritoneal surfaces lose their mesothelium. The narrow *upper border* is between the right side of the oesophagus and the upper end of the fissure for the ligamentum venosum (**8.154**). Here peritoneum of the posterior omental wall is reflected anteriorly from the diaphragm to join the posterior layer of the lesser omentum.

The *right border* corresponds, below, to that of the greater omentum; above, its upper part is formed by reflexion of the peritoneum from the pancreatic neck and head on to the inferior aspect of the beginning of the duodenum (**8.114B**); the line of this reflexion ascends to the left, along the medial side of the gastroduodenal artery. Near the upper duodenal margin the right border joins the floor of the epiploic foramen round the hepatic artery proper (**8.113**). Above this interruption the border is formed by the reflexion of peritoneum from the diaphragm to the right margin of the liver's caudate lobe and along the left side of the inferior vena cava (**8.113**).

The *left border* again corresponds, below, to that of the greater omentum. Above the root of the transverse mesocolon (**8.113**) the border broadens and is formed by the *splenico-* (lieno-) *renal* and *gastrosplenic ligaments* (**8.114A**), both formed from a part of the original dorsal mesogastrium (p. 232). The splenicorenal ligament extends from the front of the left kidney to the splenic hilum as a bilaminar fold, enclosing the splenic vessels and pancreatic tail (**8.113, 114A**). From the hilum these two layers proceed to the greater curvature of the stomach as the gastrosplenic ligament. The inner (right) layer of the splenicorenal ligament joins the corresponding layer of the gastrosplenic; but its outer (left) layer joins the peritoneum covering the spleen at the back of the hilum. The latter is reflected from the front of the hilum, as the outer (left) layer of the gastrosplenic ligament. The spleen thus projects left into the greater sac (**8.114A**). The part of the bursa projecting towards it, between the splenic ligaments, is the *splenic recess*. Superiorly the two ligaments merge into a short *gastrophrenic ligament*, passing forwards from the diaphragm to the posterior aspect of the fundus of the stomach. Its two layers diverge near the oesophagus, leaving part of the posterior gastric surface devoid of peritoneum. The upper end of the left border is continuous with the left end of the roof; the left gastric artery turns forwards here into the lesser omentum. (Many peritoneal *folds* are misleadingly termed 'ligaments'. With little in common in structure or function with skeletal ligaments, they are more often neurovascular pedicles of organs which *must* be covered by peritoneum. Some may have a supportive function, but the evidence for this usually is tenuous.)

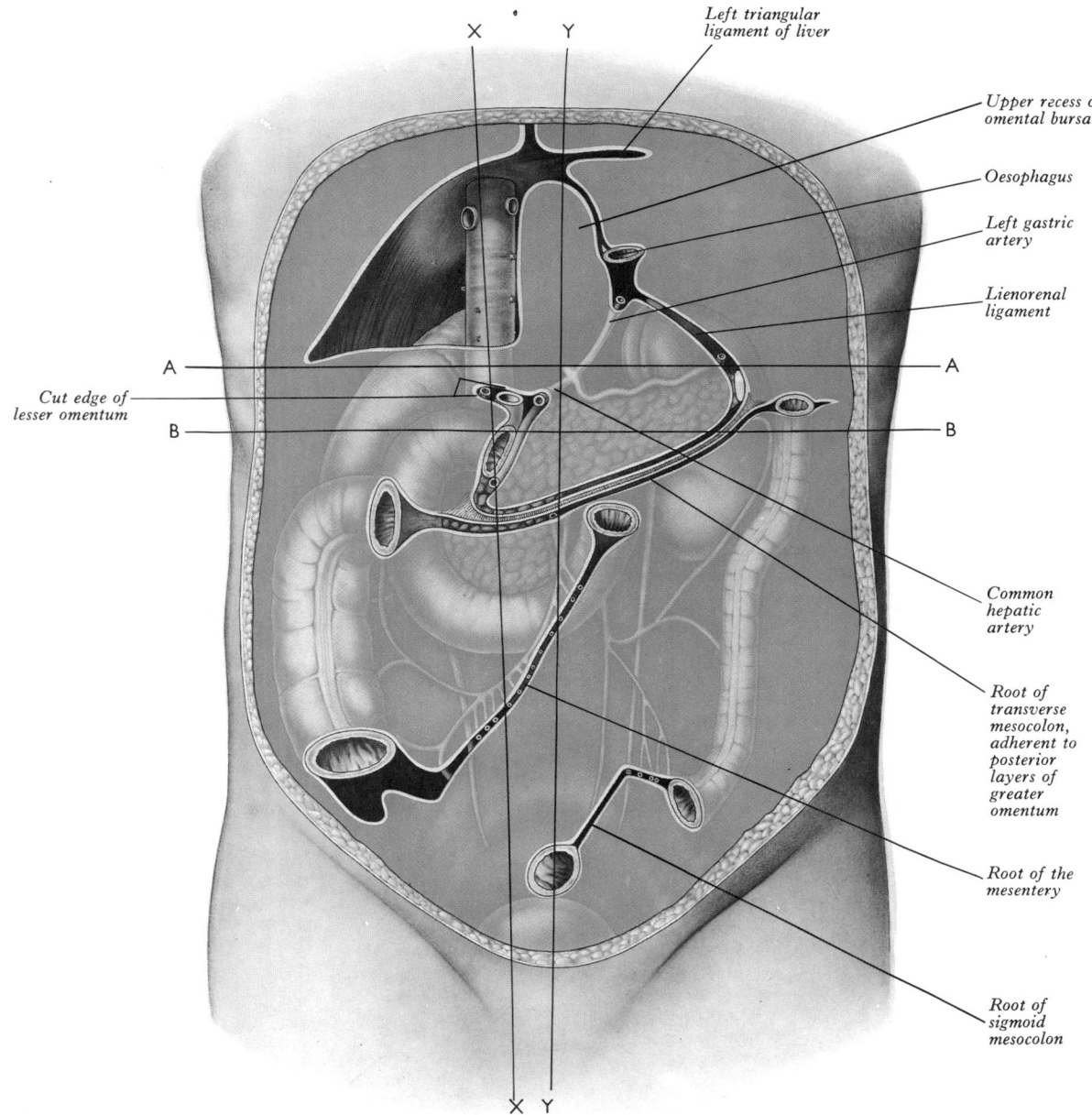

Left triangular ligament of liver

Upper recess of omental bursa

Oesophagus

Left gastric artery

Lienorenal ligament

X Y

A ————————————————————————————— A

Cut edge of lesser omentum

B ————————————————————————————— B

Common hepatic artery

Root of transverse mesocolon, adherent to posterior layers of greater omentum

Root of the mesentery

Root of sigmoid mesocolon

X Y

8.113 The posterior abdominal wall, showing the lines of peritoneal reflexion, after removal of the liver, spleen, stomach, jejunum, ileum, caecum, transverse colon and sigmoid colon. The various sessile (retroperitoneal) organs are seen shining through the posterior parietal peritoneum. Note: the ascending and descending colon, duodenum, kidneys, suprarenals, pancreas and inferior vena cava. Line YY represents the plane of **8.**112. Line AA represents the plane of **8.**114A. Line XX represents the plane of **8.**115. Line BB represents the plane of **8.**114B.

The omental bursa is narrowed by two crescentic peritoneal folds drawn into the sac by the hepatic and left gastric arteries. The *left gastropancreatic fold* transmits the left gastric artery from the posterior abdominal wall to the lesser curvature of the stomach; the *right gastropancreatic fold*, at a lower level, transmits the hepatic artery from the posterior abdominal wall to the lesser omentum (**8.**113). The folds vary, but when well marked they constrict the lesser sac to form a *foramen bursae omenti majoris*. The *superior recess* of the omental bursa is above this, joined through it to the *inferior recess* (representing the embryonic pancreatico-enteric recess, p. 235). The superior recess lies posterior to the lesser omentum and liver, the inferior behind the stomach and in the greater omentum.

During much of fetal life the transverse colon is attached to the posterior abdominal wall by its own mesentery, the posterior two layers of the greater omentum passing in front of the colon (**8.**154), a condition which may persist; usually, however, the mesentery of the transverse colon and posterior omental layer adhere; even so, these layers are separable in adults, especially in the living (pp. 1343, 1344), though their mesothelial elements disappear where the layers have fused. In its final form the oment-al bursa separates the stomach from structures which form the 'stomach bed' (p. 1349) and therefore facilitates movements of the stomach over these structures.

Peritoneal folds between various organs, or connecting them to the abdominal and pelvic walls, enclose the vessels and nerves proceeding to the viscera; though clearly not designed to sustain much weight, they may help to retain certain viscera in contact with each other. They are named ligaments, omenta and mesenteries. (The inappropriate nature of the term 'ligament' has been alluded to above.) An 'omentum' is a cover; the word may have been used to denote an apron and is thus suitable for the greater omentum, but is less suitably extended to other gastric peritoneal folds. The peritoneal 'ligaments' will be described with their respective organs.

THE OMENTA

The lesser omentum is the fold of peritoneum that extends to the liver from the lesser gastric curvature and the commencement of the duodenum. It is continuous with the two layers covering the anterosuperior and postero-inferior gastric surfaces and about

1341

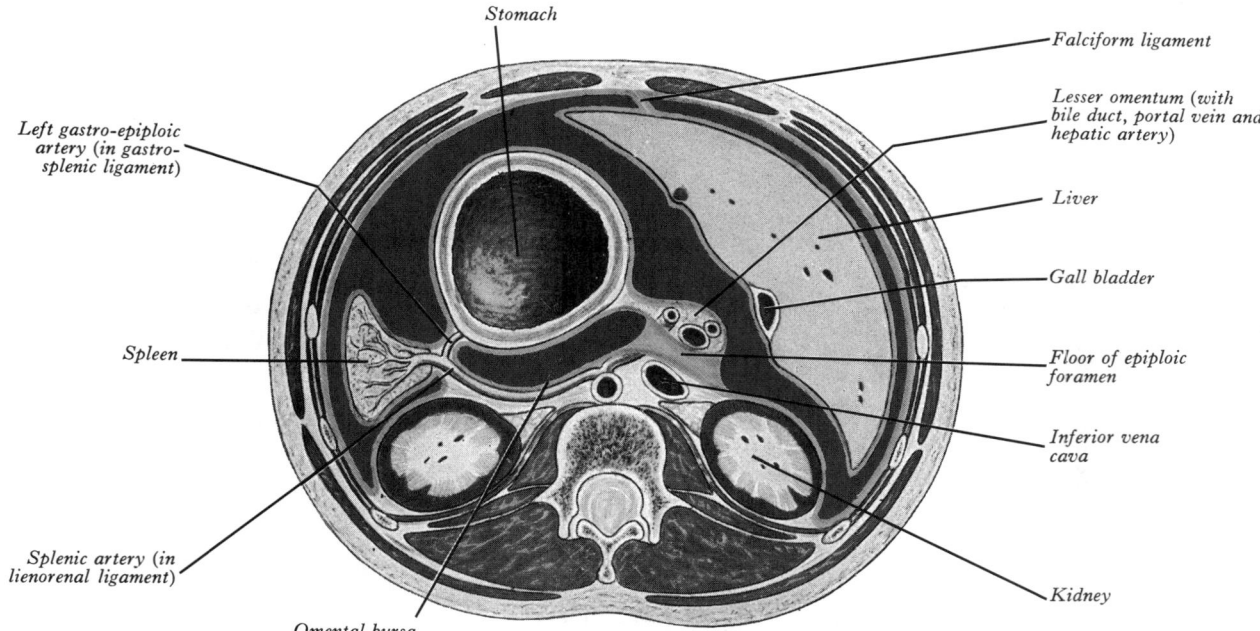

Stomach

Falciform ligament

Lesser omentum (with bile duct, portal vein and hepatic artery)

Left gastro-epiploic artery (in gastro-splenic ligament)

Liver

Gall bladder

Spleen

Floor of epiploic foramen

Inferior vena cava

Splenic artery (in lienorenal ligament)

Kidney

Omental bursa

8.114A Transverse section through the abdomen, at the level of line AA in 8.113, viewed from above. The peritoneal cavity is shown in dark blue; the peritoneum and its cut edges in lighter blue.

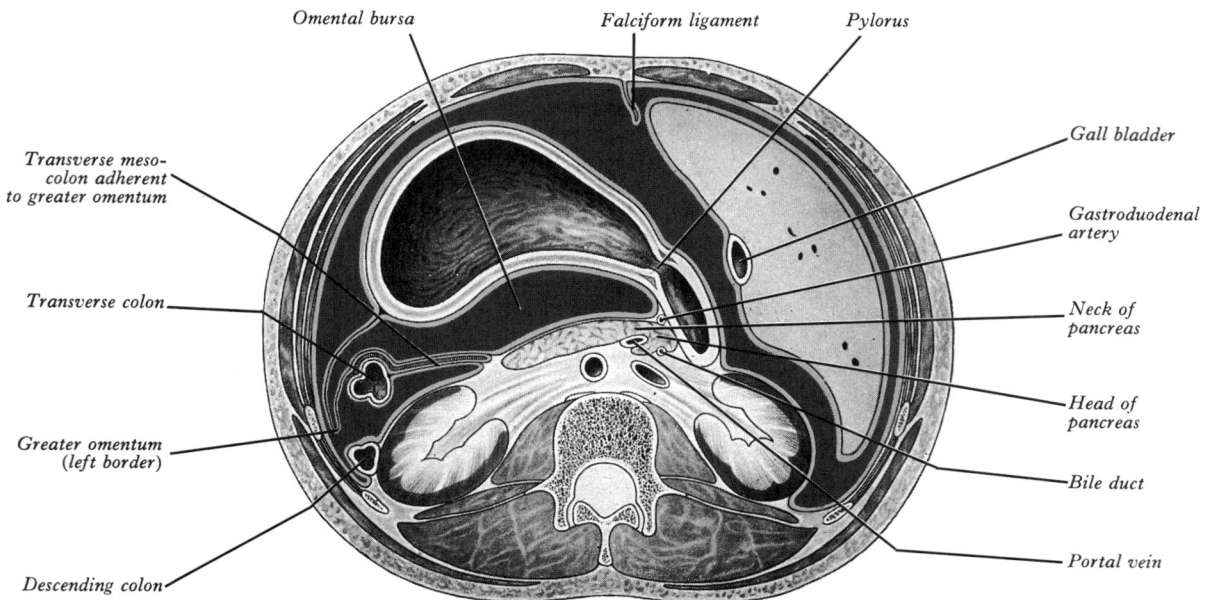

Omental bursa

Falciform ligament

Pylorus

Transverse meso-colon adherent to greater omentum

Gall bladder

Gastroduodenal artery

Transverse colon

Neck of pancreas

Head of pancreas

Greater omentum (left border)

Bile duct

Descending colon

Portal vein

8.114B Transverse section through the abdomen at the level of the line BB in 8.113, viewed from above. Colours as in 8.114A.

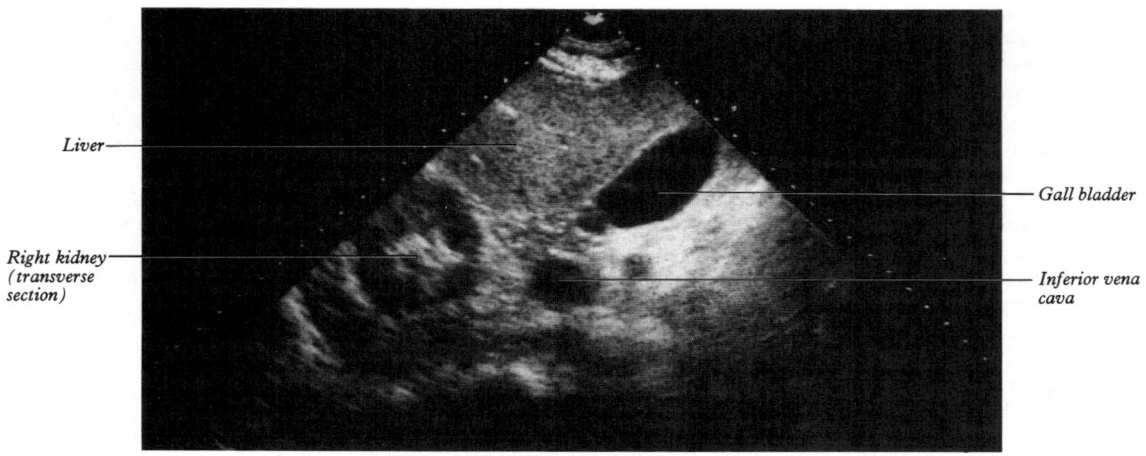

Liver

Gall bladder

Right kidney (transverse section)

Inferior vena cava

8.114C Transverse ultrasound sector scan of the abdomen to show the gallbladder and right kidney. Supplied by Shaun Gallagher, Guy's Hospital; photography by Sarah Smith, UMDS, Guy's Hospital Campus, London.

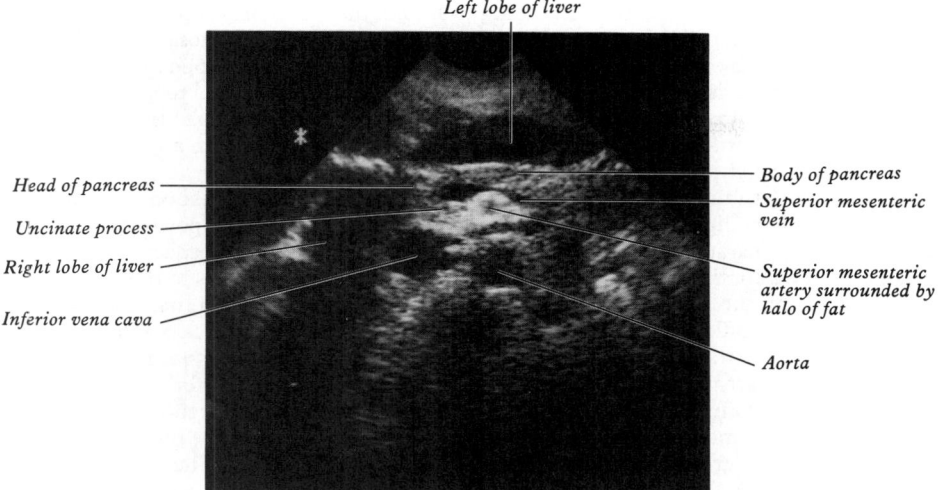

Left lobe of liver

Head of pancreas

Uncinate process

Right lobe of liver

Inferior vena cava

Body of pancreas

Superior mesenteric vein

Superior mesenteric artery surrounded by halo of fat

Aorta

8.114D Transverse ultrasound sector scan of the abdomen to show the pancreas and related structures. Supplied by Shaun Gallagher, Guy's Hospital; photography by Kevin Fitzpatrick, UMDS, Guy's Hospital Campus, London.

the first 2 cm of the duodenum. From the lower part of the lesser curvature and upper border of the duodenum these two layers ascend as a double fold to the porta hepatis; from the upper part of the lesser curvature they pass to the bed of the fissure for the ligamentum venosum. This hepatic attachment is J-shaped, with a hook-like limb corresponding to the margins of the porta hepatis and a vertical ascending along the roof of the fissure (**8.**112, 154), at the superior limit of which the lesser omentum reaches the diaphragm where its two layers separate around the abdominal part of the oesophagus. At the right omental border the two layers are continuous and this free margin is anterior to the epiploic foramen. (The omentum may be described as consisting of *hepatogastric* and *hepatoduodenal ligaments*, but these form a sing-

le entity.) Close to its right free margin the two omental layers enclose the hepatic artery, portal vein and bile duct, a few lymph nodes and lymph vessels and the hepatic plexus of nerves, all enclosed in a *perivascular fibrous capsule* (**8.**114A). The right and left gastric vessels, rami of gastric (vagus) nerves (p. 1118), and some of the left gastric lymph nodes and their lymph vessels are all contained between the two layers near their gastric attachment. The lesser omentum is thinner on the left and may be fenestrated. This variation in thickness is dependent upon the amount of connective tissue, especially fat.

The greater omentum, the largest of the peritoneal folds (**8.**107,108,112,115), is a double sheet, folded on itself to make four layers. The anterior double-layered fold descends from the

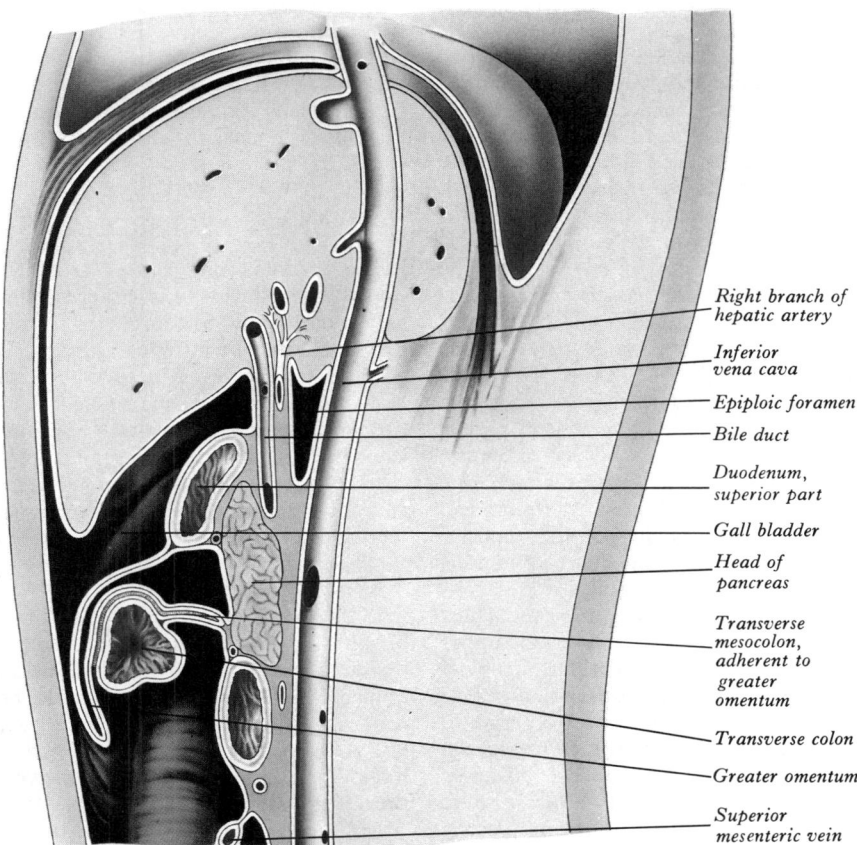

Right branch of hepatic artery

Inferior vena cava

Epiploic foramen

Bile duct

Duodenum, superior part

Gall bladder

Head of pancreas

Transverse mesocolon, adherent to greater omentum

Transverse colon

Greater omentum

Superior mesenteric vein

8.115 Section through the upper part of the abdominal cavity, along the line XX in **8.**113. The boundaries of the epiploic foramen are shown and a small recess of the omental bursa is displayed in front of the head of the pancreas. Note that the transverse colon and its mesocolon are adherent to the posterior two layers of the greater omentum.

stomach and first part of the duodenum in front of the succeeding part of the small intestine for a variable distance and ascends behind itself as far as the transverse colon (opposite the taenia omentalis). It adheres to, though it is separable from the peritoneum on the superior surface of the transverse colon and mesocolon (pp. 1341, 1344). The left border is continuous above with the gastrosplenic ligament; the right border extends to the commencement of the duodenum. (It must be emphasized that the greater omentum and gastrosplenic ligament are not merely continuous but the same structure, separated only by descriptive convenience and terminology. Their continuity has been further obscured by *changes* in terminology; the gastrosplenic *ligament* was formerly an *omentum* but is now officially the gastro-*lienal* ligament (an inadvisable use of the stem 'lien', the spleen, in a region where all else is 'splenic'!). The greater omentum is usually thin and cribriform but it always contains some adipose tissue, which in the obese may be massive in amount. Between the two layers of its anterior fold, close to the greater curvature of the stomach, the right and left gastro-epiploic vessels form a wide anastomotic arc. Variations in distribution and anastomoses of arteries in the omentum were recorded by Jiang Dian-fu (1978).

Apart from storing fat, the greater omentum may limit peritoneal infection. When the abdomen is opened without disturbance it is frequently found wrapped about organs in the upper abdomen; it is rarely evenly dependent anterior to the intestines. It is less absorptive than the general peritoneum. It may be congenitally absent and may be removed without apparent ill effect and is hence not physiologically vital. It contains numerous fixed macrophages which are easily mobilized. These may accumulate into dense, oval or round visible '*milky-spots*'. Similar spots may occur on other serous membranes (pleura p. 1267), pericardium and sometimes the leptomeninges.

THE MESENTERIES

Peritoneal folds, designated *mesenteries*, include the mesentery of the small intestine (the mesentery proper), mesoappendix, transverse mesocolon, sigmoid mesocolon, (sometimes) an ascending or descending mesocolon and occasionally a mesentery for the gallbladder.

The mesentery (of the small intestine), a broad, fan-shaped fold, connects the coils of the jejunum and ileum to the posterior abdominal wall. The attached, parietal border is the *root of the mesentery*, (8.113) about 15 cm (6 in.) long and directed obliquely down from the duodenojejunal flexure (left of the second lumbar vertebra) to the upper part of the right sacro-iliac joint. (Schmidt 1974 measured the mesenteric 'root' in 44 cadavers, finding a mean length of 13.9 cm, with extremes of 7.4 and 19.3 cm.) It passes successively in front of the horizontal part of the duodenum (where the superior mesenteric vessels enter it), the abdominal aorta, inferior vena cava, right ureter and right psoas major. The intestinal border is about 6 m (20 ft) long and compactly plicated. (For variations in length, see p. 1355.) The plication diminishes towards the posterior abdominal wall where the attachment is almost along a straight line. The central part is longest (measured from its root to the intestinal border), attaining a maximum of about 20 cm (8 in.); it shortens towards each end. The mesentery consists of two layers of peritoneum, a right and a left, enclosing the jejunal and ileal branches of the superior mesenteric vessels, with their accompanying neural plexuses, lymph vessels (here called *lacteals*), mesenteric lymph nodes, connective and adipose tissue. Fat is most abundant in its caudal part and here extends from the root to the intestinal border; the upper mesentery contains less fat, with a tendency to accumulate near the root, leaving rounded, translucent, fat-free areas adjoining the upper jejunum. At the intestinal border, the layers separate to enclose the gut, as its visceral peritoneum. At the mesenteric root the right layer is reflected in its lower part to the posterior abdominal wall and ascending colon and in its upper part to become continuous with the inferior layer of the transverse mesocolon; the left layer passes to the posterior abdominal wall and descending colon. (This arrangement helps to distinguish between proximal and distal coils of small intestine when in situ.)

The mesoappendix (8.117) is a triangular fold of peritoneum

around the vermiform appendix, attached to the back of the lower end of the mesentery close to the ileocaecal junction. It usually reaches the tip of the appendix but sometimes fails to reach the distal third, being then represented by a low peritoneal ridge containing fat. It encloses the blood vessels, nerves and lymph vessels of the vermiform appendix, together with a lymph node (p. 853).

The transverse mesocolon is a broad fold connecting the transverse colon to the posterior abdominal wall. Its two layers pass from the anterior aspect of the head and anterior border of the body of the pancreas to the posterior aspect of the transverse colon (opposite the *taenia mesocolica*), where they separate to surround it. The upper layer is adherent to, but separable from, the greater omentum (pp. 1340, 1341 and 8.112). Posteriorly, the inferior layer covers the inferior aspect of the pancreas and passes in front of the horizontal and ascending parts of the duodenum. Between its layers are the blood vessels, nerves and lymphatics of the transverse colon. The middle colic artery descends to the right, leaving a large avascular area to its left and a smaller one to its right.

The sigmoid mesocolon is a peritoneal fold attaching the sigmoid colon to the pelvic wall, the attachment being an inverted V, with an apex near the division of the left common iliac artery (8.113); the left limb descends medial to the left psoas major and the right passes into the pelvis to end in the midline at the level of the third sacral vertebra. Sigmoid and superior rectal vessels run between its layers and the left ureter descends into the pelvis behind its apex.

Peritoneum usually covers only the front and sides of the ascending and descending parts of the colon, but sometimes these are virtually surrounded by peritoneum and attached to the posterior abdominal wall by an ascending and descending mesocolon respectively (p. 1367). The *phrenicocolic ligament* is a peritoneal fold extending from the left colic flexure to the diaphragm level with the tenth and eleventh ribs; it has an anterior edge and passes inferolateral to the lateral end of the spleen, sometimes being misleadingly named the *sustentaculum lienis* or 'splenic shelf', implying a hypothetical supportive role.

Appendices epiploicae are small peritoneal appendages filled with adipose connective tissue and situated along the colon; they are most conspicuous on the transverse (8.116) and sigmoid parts, absent from the rectum and rudimentary on the caecum and appendix. Many contain a small arteriole from the wall of the gut. In the colon they are most numerous along the line of the *taenia libera* (p. 1374).

PERITONEAL RECESSES

Peritoneal folds may create fossae or recesses of the peritoneal cavity. These are of clinical interest because a segment of intestine may enter one and be constricted by the fold at the entrance to the recess and may be a site of a kind of 'internal' hernia. Since the entrance to a recess may need to be cut to relieve strangulation and to withdraw the gut, the degree of vascularization of the fold becomes important. Surgically the omental bursa belongs to this category, with its opening at the epiploic foramen. Much smaller recesses sometimes occur, sometimes related to the duodenum, caecum and sigmoid mesocolon. For a discussion of the developmental origins of these recesses, see p. 242.

1. Duodenal Recesses

(a) The *superior duodenal recess* (8.111), present in about 50% of people, may exist alone but usually occurs with an inferior duodenal recess. It is to the left of the distal end of the duodenum, opposite the second lumbar vertebra, behind a crescentic *superior duodenal fold* (*duodenojejunal fold*), which has a semilunar free lower edge, merging on the left with peritoneum anterior to the left kidney. The inferior mesenteric vein is behind the junction of the left end of this fold and the posterior parietal peritoneum. The recess is about 2 cm deep, admitting a fingertip; it opens downwards, its orifice being in the angle formed by the left renal vein as it crosses the abdominal aorta.

(b) The *inferior duodenal recess* (8.111) is present in about 75% of subjects, usually associated with a superior recess with which it may share an orifice; it lies left of the distal end of the duodenum, opposite the third lumbar vertebra, behind a non-vascular, triangular *inferior duodenal fold* (*duodenomesocolic fold*), which has a sharp upper edge. It is about 3 cm deep, admits one or two fingers and opens upwards towards the superior duodenal recess. It sometimes extends behind the ascending part of the duodenum and to the left, in front of the ascending branch of the left colic artery and inferior mesenteric vein. This large fossa is liable to become the site of an internal hernia.

(*c*) The *paraduodenal recess* (**8.116**) may occur with the superior and inferior duodenal recesses. It is more frequent in the fetus and newborn than in adults, in whom it occurs in about 2%. It is a little to the left of the ascending part of the duodenum, behind a falciform *paraduodenal fold*, the free right edge of which contains the inferior mesenteric vein and ascending branch of the left colic artery, the fold being their mesentery. Its free edge lies in front of the wide orifice of the recess, which faces right.

(*d*) The *retroduodenal recess* is rarely present. The largest of the duodenal recesses, it lies behind the horizontal and ascending parts of the duodenum in front of the abdominal aorta. It ascends nearly to the duodenojejunal junction, being about 8–10 cm deep and bounded on both sides by *duodenoparietal folds*; its orifice faces down to the left.

(*e*) The *duodenojejunal* or *mesocolic recess*, present in about 20% and rarely or never accompanied by any other duodenal recess, is about 3 cm deep and lies on the left side of the abdominal aorta, between the duodenojejunal junction and the root of the transverse mesocolon. Bounded above by the pancreas, on the left by the kidney, and below by the left renal vein, it has a circular opening between two peritoneal folds, which faces down to the right.

(*f*) The *mesentericoparietal recess* (of Waldeyer) is more frequent in the fetus and newborn, occurring in only about 1% of adults. It lies just below the horizontal part of the duodenum, invaginating the upper part of the mesentery towards the right. Its orifice is large and faces left behind a fold of mesentery raised by the superior mesenteric artery.

2. Caecal Recesses

The *superior ileocaecal recess* (**8.117**), usually present and best developed in children, is often reduced and absent in the aged, especially the obese. It is formed by the *vascular fold of the caecum*, which arches over the anterior caecal artery, supplying the anterior part of the ileocaecal junction, and its accompanying vein. It is a narrow slit bounded in front by the vascular fold, behind by the ileal mesentery, below by the terminal ileum and on the right by the ileocaecal junction. Its orifice opens downwards to the left.

The *inferior ileocaecal recess* (**8.117**) is well marked in youth but frequently obliterated by fat in later years. It is produced by the *ileocaecal fold*, extending from the antero-inferior aspect of the terminal ileum to the front of the mesoappendix (or to the appendix or caecum). It is also known as the '*bloodless fold of Treves*', although it sometimes contains blood vessels; if inflamed, especially when the appendix and its mesentery are retrocaecal, it may be mistaken for the mesoappendix. (For the source of vessels in this fold, see Cabanie & Javelle 1966.) The recess is bounded in front by the ileocaecal fold, above by the posterior ileal surface and its mesentery, to the right by the caecum, and behind by the upper mesoappendix. Its orifice opens downwards to the left.

The *retrocaecal recess* (**8.117**), behind the caecum, varies in size and extent and ascends behind the ascending colon, being large enough to admit an entire finger. It is bounded in front by the caecum (and sometimes the lower ascending colon), behind by the parietal peritoneum and on each side by *caecal folds* (parietocolic folds) passing from the caecum to the posterior abdominal wall. The vermiform appendix frequently occupies this recess (pp. 1365, 1366).

3. The Intersigmoid Recess

This recess is constant in fetal life and infancy, but may later disappear. It lies behind the apex of the V-shaped parietal attachment of the sigmoid mesocolon, is funnel-shaped and directed upwards and opens downwards. It varies in size from a dimple to a fossa admitting the fifth finger. Its posterior wall of posterior

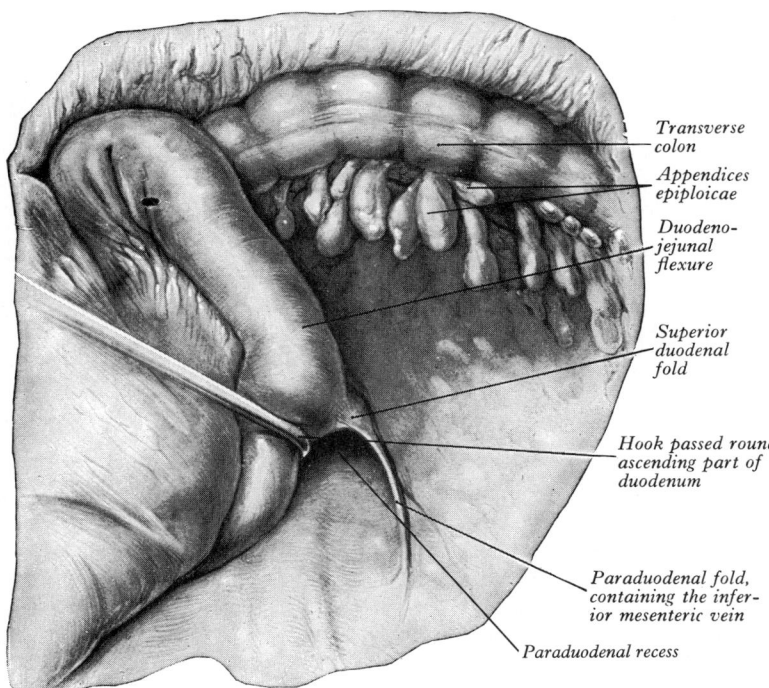

Transverse colon

Appendices epiploicae

Duodeno-jejunal flexure

Superior duodenal fold

Hook passed round ascending part of duodenum

Paraduodenal fold, containing the inferior mesenteric vein

Paraduodenal recess

8.116 The paraduodenal recess.

parietal peritoneum covers the left ureter as this crosses the bifurcation of the left common iliac artery. Occasionally the recess is within the layers of the sigmoid mesocolon nearer gut than root. Its presence is due to an imperfect blending of the mesocolon with the posterior parietal peritoneum.

ANOMALOUS PERITONEAL FOLDS

Certain other folds, bands or ligaments sometimes occur in the abdomen. Some are considered to cause obstruction by pulling on or kinking sections of intestine; others may limit the spread of peritoneal effusions. Their exact modes of origin are doubtful; they have been attributed to errors in development, previous inflammation (peritonitis), mechanical traction by the gut and even (and improbably) linked with the evolution of upright posture. These anomalous folds must be distinguished from pathological adhesions definitely due to peritonitis; also, when coils of intestine are pulled out of normal position during examination they may be artificially kinked, with resultant simulation of bands. Anomalous folds are only clinically important if proved to interfere with normal function; their presence should not halt a search for other possible causes of the symptoms. The commonest anomalous folds encountered are as follows:

(*a*) Occasionally the lesser omentum is prolonged to the right of the usual site of the epiploic foramen by a peritoneal fold extending from the gallbladder to the superior part of the duodenum (*cystoduodenal ligament*), or in front of the superior part of the duodenum to the greater omentum or right colic flexure, or from the inferior surface of the right hepatic lobe to the right colic flexure (*hepatocolic ligament*).

(*b*) The duodenojejunal junction is sometimes joined to the transverse mesocolon by a peritoneal band.

(*c*) The greater omentum may be attached to the front of the ascending colon or extend over it to the lateral abdominal wall. A thin sheet of peritoneum (*Jackson's membrane*), containing small blood vessels, may spread from the front of the ascending colon and caecum to the posterolateral abdominal wall and may merge on the left with the greater omentum. Occasionally a band passes from the right side of the ascending colon to the lateral abdominal wall near the level of the iliac crest. Sometimes termed a 'sustentaculum hepatis', it is closely related to the liver only in fetal and early postnatal life, when the liver is relatively larger. Other folds between the ascending colon and posterolateral abdominal wall may divide the right lateral paracolic gutter (between the

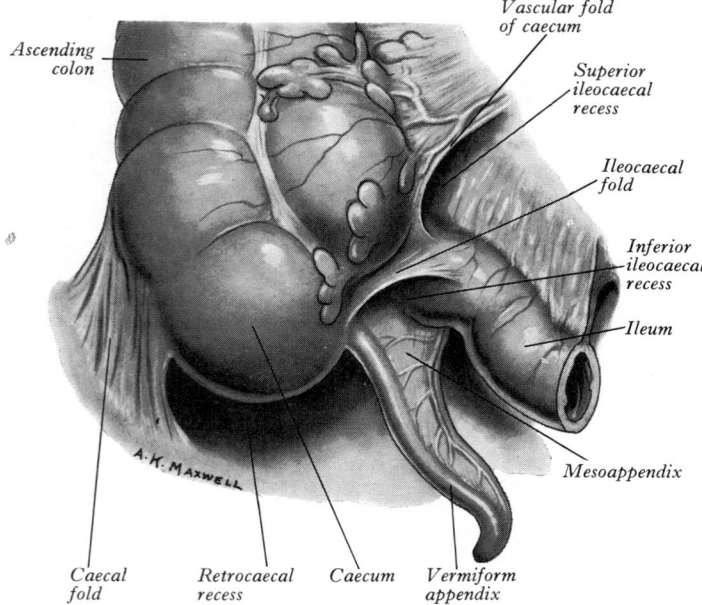

8.117 The peritoneal folds and recesses in the caecal region.

Labels on figure: Ascending colon; Vascular fold of caecum; Superior ileocaecal recess; Ileocaecal fold; Inferior ileocaecal recess; Ileum; Mesoappendix; A. K. Maxwell; Caecal fold; Retrocaecal recess; Caecum; Vermiform appendix

ascending colon and posterior abdominal wall) into several small recesses.

(d) The ascending and less frequently the descending colon may have a mesentery.

(e) Proximal and distal ends of the sigmoid colon may be tied together by a fibrous band.

(f) Frequently a fan-shaped *presplenic fold* extends from the front of the gastrosplenic ligament (near the greater gastric curvature), below the inferolateral pole of the spleen, to blend with the phrenicocolic ligament; it may adhere to the spleen or diaphragm and contains rami of the splenic or the left gastro-epiploic artery; the omental bursa may enter it. It is more obvious in the fetus, often appearing as just part of the phrenicocolic ligament in adults. It may limit peritoneal effusions in the left supracolic space (vide infra), and, if adherent to the spleen or diaphragm, it may form a vascular obstruction in splenectomy.

(g) A fibrous band, described as passing from the terminal ileum to the posterior abdominal wall, and a similar one, from the proximal sigmoid colon to the same wall, were once thought to be causes of partial obstruction by kinking of these parts of the gut, but this view is no longer popular.

SPECIAL PERITONEAL REGIONS

Regarding the spread of pathological collections of fluid, certain potential peritoneal spaces or recesses, normally in communication with each other, must be mentioned because they may be sealed off by pathological adhesions. These are as follows:

(1) The *supracolic space (subphrenic region)* between the diaphragm and the transverse colon and mesocolon is subdivided into: *(a)* the *right subphrenic space*, between the diaphragm and the anterior, superior and right lateral surfaces of the right lobe of the liver, bounded on the left side by the falciform ligament and behind by the coronary ligament's upper layer; *(b)* the *left subphrenic space* between the diaphragm, the anterior and superior surfaces of the left hepatic lobe, the anterosuperior surface of the stomach and the diaphragmatic surface of the spleen; it is limited to the right by the falciform ligament and behind by the left triangular ligament's anterior layer; *(c)* the *right subhepatic space* (*hepatorenal recess*, or *Morison's pouch*), bounded above and in front by the inferior surface of the right hepatic lobe and gallbladder, below and behind by the right suprarenal gland, the upper part of the right kidney, the descending part of the duodenum, right colic flexure, transverse mesocolon and part of the head of the pancreas; above and behind, it extends between the right kidney and the liver as far as the inferior layer of the

coronary ligament and right triangular ligament; *(d)* the *left subhepatic space*, i.e. the omental bursa.

(2) The *right infracolic space* lies postero-inferior to the transverse colon and mesocolon and to the right of the mesentery, whose obliquity narrows the lower part of the space. The vermiform appendix is often in the lower part of this space.

(3) The *left infracolic space* is below and behind the transverse colon and mesocolon and left of the mesentery; it is widest below, continuing into the pelvis.

(4) The *pelvic cavity* (p. 604).

(5) The *paracolic gutters* are alongside the ascending and descending colon (which are normally sessile), where the peritoneum turns dorsally on the medial and lateral aspects of the gut. *Medial* and *lateral paracolic gutters* flank both left and right parts of the colon. Of commonest clinical interest is the *right lateral paracolic gutter*, which skirts the superolateral aspect of the hepatic flexure of the colon and continues into the hepatorenal pouch (of Morison) and, through the epiploic foramen, into the omental bursa and its superior recess. Inferiorly, around the caecum it curves over the pelvic brim into the *rectovesical* (male) or *recto-uterine* (female) *pouch* (of Douglas). Related to this gutter, and its superior extension, are the vermiform appendix, right kidney, gallbladder, lesser gastric curvature and the first and second parts of the duodenum; all are common sites of acute abdominal disease. Seepage of infected fluid along this channel, from place to place, may cause puzzling symptoms and signs and errors in diagnosis. In supine patients infected fluid in the right lateral gutter may *ascend* to, enter and accumulate in the superior recess of the omental bursa, with grave consequences, because of its inaccessibility and nearness to the pleural and pericardial cavities. In patients nursed in a sitting posture, fluid *descends* to the relatively accessible rectovesical pouch or to the recto-uterine pouch, approachable surgically through the rectum or vagina.

(6) Two extraperitoneal subphrenic regions may become infected; *(a)* the *right extraperitoneal space*, between the two layers of the coronary ligament i.e. the 'bare area' of the liver and the diaphragm; and *(b)* the *left extraperitoneal space*, which contains extraperitoneal connective tissue around the left suprarenal gland and upper pole of left kidney.

PERITONEAL ABSORPTION

Substances in *solution* are probably absorbed into the capillaries, whereas suspended matter probably passes into the lymph vessels, aided by phagocytes. After abdominal or pelvic operations, it has been customary to prop up patients to encourage intraperitoneal effusions to gravitate into the pelvis. One reason for this was the belief that the subphrenic peritoneum was more absorptive than elsewhere and hence inflammatory products would enter the circulation more rapidly here. It was supposed that gaps or (peritoneal stomata) between the mesothelial cells of the subphrenic peritoneum and similar gaps (endothelial stigmata) between the endothelial cells of lymph vessels greatly facilitated absorption. Such gaps have been often considered to be histological artefacts, but scanning electron microscopy supports the existence of slit-like orifices. However absorption is believed to be much the same in all parts of the peritoneum. Greater absorption in the upper abdomen may be due to the larger subphrenic peritoneal area and to respiratory movements.

PERITONEAL STRUCTURE

The peritoneum is composed of a single layer of flat mesothelial cells lying on a layer of loose connective tissue. Mesothelium usually forms a continuous surface, adjacent cells being joined by junctional complexes but probably allowing the passage of macrophages, just as leucocytes pass between endothelial cell junctions. In some areas, e.g. the greater omentum, the peritoneum may be discontinuous, having fenestrations which are sometimes visible to the unaided eye; but mesothelium continues over the trabeculae of connective tissue interlacing around such fenestrae.

Submesothelial connective tissue contains cells typical of loose connective tissue (p. 58) but macrophages, lymphocytes and in some regions adipocytes are particularly numerous. Macrophages and lymphocytes may aggregate as submesothelial 'milky spots'. Mesothelial cells may also transform into fibroblasts; the fusion between layers of fibroblasts of mesothelial origin may cause macroscopic adhesions between the peritoneal surfaces, interfering with intestinal motility or even leading to complete obstruction of the gut.

The mesothelium resembles vascular endothelium in being a dialysing membrane which fluids and small molecules may traverse. Numerous endocytic vesicles occur near the cell surfaces, the remaining cytoplasm being poor in organelles, indicating low metabolic activity (Tesi & Forssmann 1970). Normally the volumes of fluid transmitted by peritoneal surfaces are small, but large volumes may be administered via the intraperitoneal route; conversely, substances such as urea can be dialysed from blood into fluid circulated through the peritoneal cavity.

PERITONEAL FLUID

The fluid covering the peritoneal surfaces contains water, electrolytes and other solutes derived from interstitial fluid in the adjacent tissues and from the plasma of local vessels. It also contains proteins and cells (Carr 1967), the latter varying in number, structure and type in different diseases; hence it is of diagnostic importance. Normally they are mesothelial desquamated elements, nomadic peritoneal macrophages, mast cells, fibroblasts, lymphocytes and some other leucocytes. Some, particularly macrophages, migrate freely between the peritoneal cavity and the surrounding connective tissue; intraperitoneally injected particles are ingested by them and transported to various tissue sites. Lymphocytes provide both cellular and humoral immunological defence mechanisms.

PERITONEAL VESSELS AND NERVES

Parietal and visceral peritoneum are developed from, respectively, somatopleural and splanchnopleural layers of lateral plate mesoderm (p. 138). Parietal peritoneum is therefore supplied by somatic blood vessels of the abdominal and pelvic walls; its lymphatics join those in the body wall and drain to parietal lymph nodes; its nerve supply is derived from nerves supplying the muscles and skin of the parietes. Visceral peritoneum, however, as an integral part of the viscera, derives its blood vessels from those supplying viscera. Its lymphatics join the visceral vessels and its nerve supply is autonomic. Differences in the sensibility of the two layers correlate with their innervations. Whereas pain is elicited by mechanical, thermal or chemical stimulation of the parietal peritoneum, the visceral peritoneum and viscera are not affected; e.g. the liver, stomach or intestine can be injured without evoking pain, insensibility extending from the mid-oesophagus to the junction of endoderm and ectoderm in the anal canal. However, tension *does* evoke pain when applied to viscera or visceral peritoneum by over-distension or traction on mesenteries, stretching various neural elements in the visceral walls or mesenteries. Also effective are spasms of visceral muscle and ischaemia. Somatic nerves of the parietal peritoneum also supply the corresponding segmental areas of skin and muscles and, when the parietal peritoneum is irritated, muscles are reflexly contracted, causing rigidity of the abdominal wall. The parietal peritoneum on the underside of the diaphragm is supplied centrally by the phrenic nerves and peripherally by the lower six intercostal and subcostal nerves. Hence peripheral irritation may result in pain, tenderness and muscular rigidity in the distribution of the lower thoracic spinal nerves, while central irritation may result in pain in the cutaneous distribution of the third to fifth cervical spinal nerves, i.e. the shoulder region.

ABDOMINAL VISCERA

The Stomach

The stomach (ventriculus or preferably *gaster*) is the most dilated part of the alimentary canal and is situated between the oesophagus and the small intestine; it lies in the epigastric, umbilical and left hypochondriac areas of the abdomen, occupying a recess bounded by the upper abdominal viscera and completed above and anterolaterally by the anterior abdominal wall and diaphragm. Its shape and position are modified by intrinsic changes and by changes in the surrounding viscera. Its mean capacity varies from about 30 ml at birth, increasing to 1000 ml at puberty and about 1500 ml in adults. It has two openings and is described as if it had two borders or curvatures and two surfaces. In reality its external surface is a continuum and not divided by perceptible 'borders'. However, since its peritoneal surface is interrupted by the attachments of the greater and lesser omenta along profiles which define the gastric radiographic shadow, these curvatures may be conveniently regarded as 'borders' separating the surfaces. (Similar arbitrary borders are assigned to the heart, p. 700.)

THE GASTRIC ORIFICES

The opening from the oesophagus into the stomach is the *cardiac orifice*, situated to the left of the midline behind the seventh costal cartilage, 2.5 cm (1 in.) from its sternal junction at the level of the eleventh thoracic vertebra. It is about 10 cm (4 in.) from the anterior abdominal wall and 40 cm (16 in.) from the incisor teeth. The short abdominal part of the oesophagus, shaped like a truncated cone, curves sharply left, the base of the cone being continuous with the cardiac orifice. The right side of the oesophagus is continued as the *lesser curvature*, while its left side joins the *greater curvature* at an acute angle, the *cardiac notch* or *incisure*. The part of the stomach above the level of the cardiac orifice is the *fundus*, an inappropriate term, but it is the *bottom* of the stomach, when entered surgically from below.

The *pyloric orifice*, the opening into the duodenum, is usually indicated (8.118) by a circular *pyloric constriction* on the surface of the organ, indicating the pyloric sphincter; it can be identified by the prepyloric vein crossing its anterior surface vertically. The pyloric orifice is about 1.2 cm (0.5 in.) right of the midline in the transpyloric plane (level of the lower border of the first lumbar vertebra), with the body supine and the stomach empty.

THE GASTRIC CURVATURES

The lesser curvature, extending between the cardiac and pyloric orifices, is the right (posterosuperior) border of the stomach. It descends from the right side of the oesophagus in front of the decussating fibres of the right crus, curving to the right below the omental tuberosity of the pancreas to end at the pylorus (8.118, 120). In the most dependent part may be the *angular incisure*, varying in position with gastric distension; it is sometimes used to define the right and left parts of the stomach. The lesser omentum is attached to the lesser curvature and contains the right and left gastric vessels near the curvature.

The greater curvature is directed antero-inferiorly and is four or five times as long as the lesser; it starts from the cardiac incisure and arches upwards posterolaterally and to the left; its highest convexity, the *fundus*, is level with the left fifth intercostal space just below the left nipple in males, though varying with respiration (pp. 594–595). From this level it sweeps down and forwards, slightly convex to the left, almost as far as the tenth costal cartilage in the supine body; it finally turns right to end at the pylorus. Opposite the angular incisure of the lesser curvature the greater curvature presents a bulge, taken as the left limit of the *pyloric part* of the stomach, its right limit being a slight groove (*sulcus intermedius*) indicating subdivision into the *pyloric antrum* and *canal*, the latter only 2–3 cm in length and terminating at the pyloric constriction. The start of the greater curvature is covered by peritoneum continuous with that anterior

1347

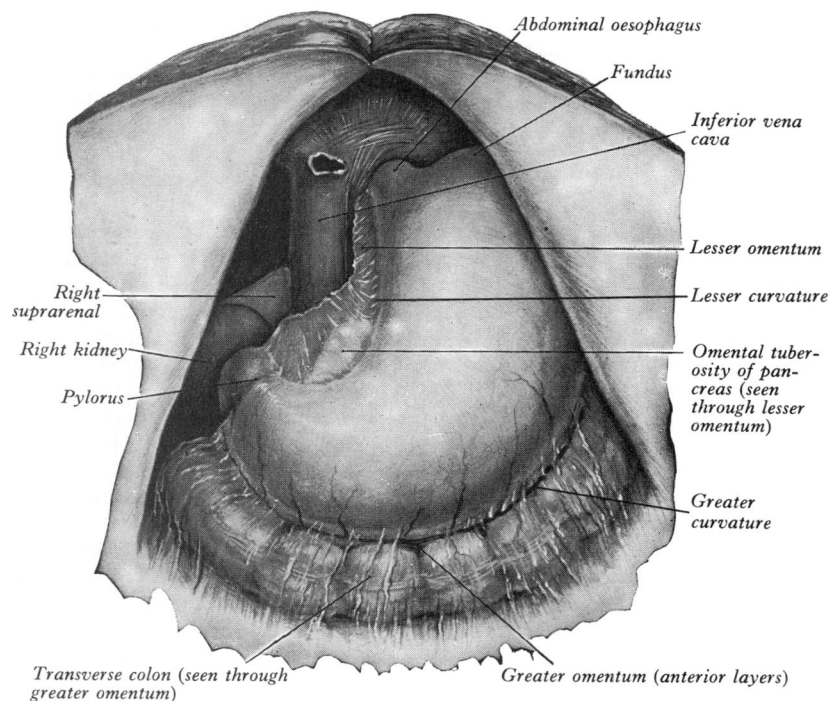

Abdominal oesophagus

Fundus

Inferior vena cava

Lesser omentum

Lesser curvature

Omental tuberosity of pancreas (seen through lesser omentum)

Greater curvature

Right suprarenal

Right kidney

Pylorus

Transverse colon (seen through greater omentum)

Greater omentum (anterior layers)

8.118 The stomach in situ, after removal of the liver.

to the stomach. Left of the fundus and the adjoining gastric body the greater curvature gives attachment to the gastrosplenic ligament and beyond this to the greater omentum, its two layers being separated by the gastro-epiploic vessels. The gastrosplenic ligament and the greater omentum (with the gastrophrenic and splenicorenal ligaments see p. 1339, **8.**114A) are continuous parts

of the original dorsal mesogastrium (p. 232). The names merely indicate regions of the same fold.

THE GASTRIC SURFACES

When the stomach is empty and contracted its surfaces are almost superior and inferior; but in distension they become anterior and

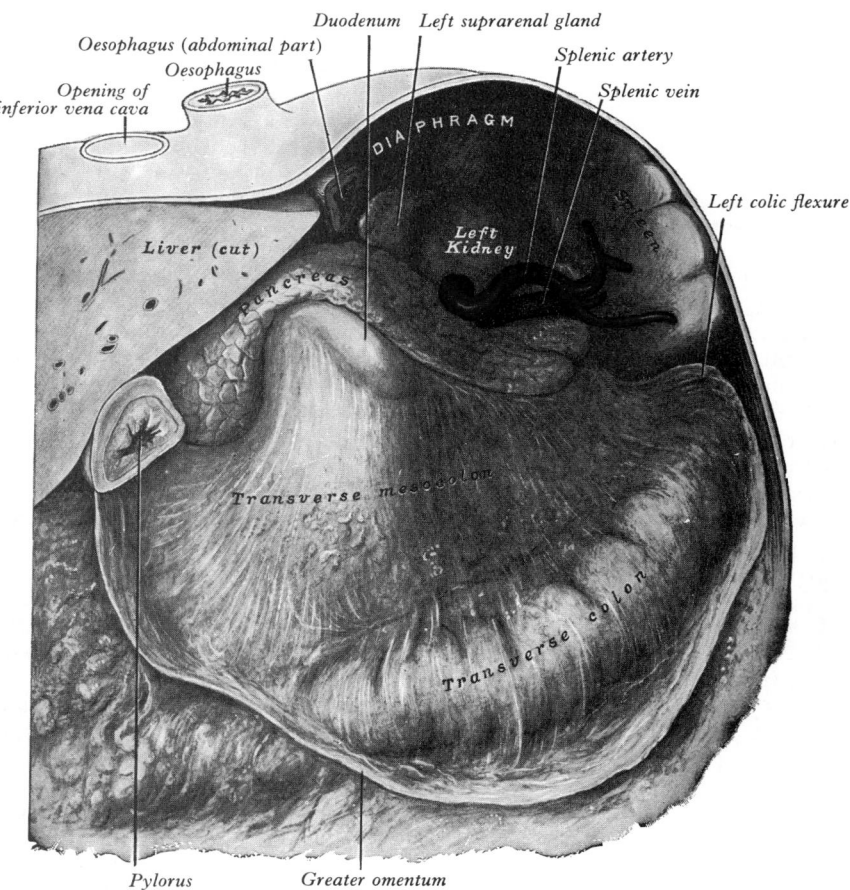

Duodenum   *Left suprarenal gland*

Oesophagus (abdominal part)

Oesophagus

Splenic artery

Splenic vein

Opening of inferior vena cava

DIAPHRAGM

Left colic flexure

Liver (cut)

Left Kidney

Spleen

Pancreas

Transverse mesocolon

Transverse colon

Pylorus

Greater omentum

8.119 The stomach bed: a dissection in which the stomach has been removed to show its posterior relations.

posterior and are therefore described here as anterosuperior and postero-inferior.

Anterosuperior surface. The left part of this surface is posterior to the left costal margin and in contact with the diaphragm which separates it from the left pleura, the base of the left lung, the pericardium and the left sixth to ninth ribs and intercostal spaces. It is related to the costal attachments of the upper fibres of the transversus abdominis, separating it from the seventh to ninth costal cartilages. The upper and left part of this surface becomes posterolateral and is in contact with the spleen's gastric surface. The right part of the anterosuperior surface is related to the left and quadrate lobes of the liver and the anterior abdominal wall. When the stomach is empty, the transverse colon may be anterior to it. The whole surface is covered by peritoneum and part of the greater sac separates it from the above structures.

Postero-inferior surface. This is related to the diaphragm, the left suprarenal gland, upper part of the front of the left kidney, the splenic artery, anterior pancreatic surface, left colic flexure and the transverse mesocolon (upper layer), which together form the shallow *stomach bed* (8.119), over which the stomach slides, due to the intervening lesser sac. The spleen's gastric surface is usually included in the stomach bed, though separated from the stomach by the greater sac. The greater omentum and the transverse mesocolon separate the stomach from the duodenojejunal flexure and small intestine. The postero-inferior surface is covered by peritoneum, except near the cardiac orifice, where a small, triangular area contacts the left diaphragmatic crus and sometimes the left suprarenal gland. The left gastric vessels reach the lesser curvature at the right extremity of this area in the left gastropancreatic fold (p. 1341); from its left side the *gastrophrenic ligament*, continuous below with the splenicorenal and gastro-splenic ligaments, passes to the diaphragm's inferior surface.

A plane through the angular incisure of the lesser curvature and the left limit of the opposed bulge on the greater curvature arbitrarily divides the stomach into a large *body* (left) and a smaller *pyloric part* (right).

RADIOLOGY

The form and position of the stomach can be studied after swallowing a suitable 'meal' containing barium sulphate (8.122). During digestion it is divided by a muscular constriction in its body into a large, dilated, left region and a narrow, contracted, tubular right one. The constriction in the body follows no anatomical landmarks but moves gradually left as digestion progresses. The position of the stomach varies with posture, contents and the state of the intestines, it is also influenced by the tone of the abdominal wall and gastric musculature and by the individual's build. Most commonly the empty organ is J-shaped and, in the erect posture, the pylorus descends to the level of the second or the third lumbar vertebra, its most dependent part being subumbilical. The fundus usually contains gas. Variation in content mainly affects the body, the pyloric part remaining contracted during digestion. As the stomach fills it expands forwards and downwards but, when the colon or intestines are distended, the fundus presses on the liver and diaphragm and may evoke discomfort. When hardened in situ the contracted stomach is crescentic, the fundus directed backwards. Surfaces are superior and inferior, the former sloping gradually to the right, the greater curvature being anterior to and at a slightly higher level than the lesser.

The position of the full stomach varies. When the intestines are empty the fundus expands vertically and forwards, the pylorus is displaced right and the whole organ becomes oblique. Its surfaces are then directed more forwards and backwards, the lowest part being the pyloric antrum which extends below the umbilicus. When intestinal distension interferes with downward expansion of the fundus, the stomach retains the horizontal position characteristic of the contracted viscus. Less commonly it may lie almost transversely, even in the erect posture, as the '*steer-horn*' type. Intermediate types of stomach, between J-shaped and 'steer-horn', also occur (Barclay 1936).

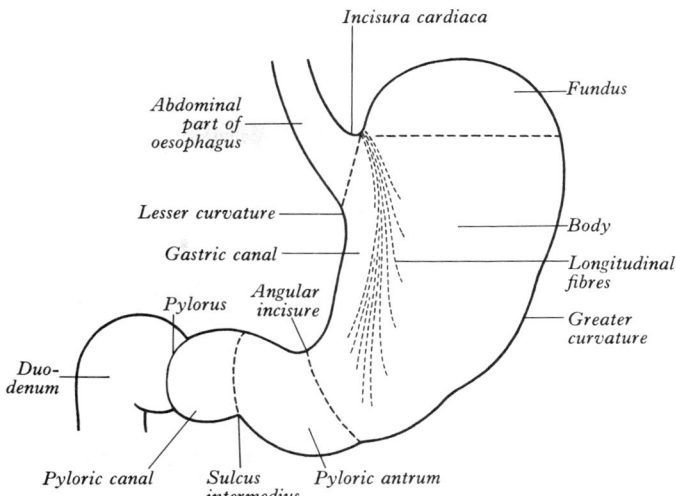

8.120 The parts of the stomach.

INTERIOR OF THE STOMACH

After death the stomach is usually fixed at some stage of the digestion, commonly as shown in 8.123. When it is bisected in the plane of its curvatures, it shows two segments: (1) a large globular left part and (2) a narrow tubular right part; the transition is gradual, so their division is arbitrary. The cardiac incisure lies to the left of the abdominal part of the oesophagus and its projection into the cavity increases as the organ distends, supposedly acting as a valve preventing oesophageal regurgitation. The elevation opposite the angular incisure is at the beginning, and the circular thickening of the pyloric sphincter at the end, of the pyloric region.

Modelling of human fetal gastric epithelium (Lewis 1912) has shown that a *gastric canal* extends along the lesser curvature from the cardiac orifice to the angular incisure (8.120). Jefferson (1915) demonstrated such a canal radiologically in adults and examination during the act of swallowing radio-opaque fluid showed that it was first confined to the region adjacent to the lesser curvature, suggesting that contraction of oblique muscle fibres causes temporary separation of a canal along the lesser curvature.

The pyloric sphincter is a muscular ring formed by a marked thickening of the circular gastric muscle, some longitudinal fibres also interlacing with it (DiDio & Anderson 1968.)

A *cardiac sphincter* is sometimes described, formed from the circular fibres of the gastric wall. Radiological observation strongly suggests some force capable of delaying the entry of oesophageal contents but the histological evidence is inconclusive. Muscle fibres of the right crus of the diaphragm, decussating obliquely round the termination of the oesophagus, may compress or kink it at this level; but this is merely an anatomical speculation. Bowden & El-Ramli (1967) have described this part of the diaphragm (p. 592); they concluded that, while the right crus invariably embraces the oesophagus, this arrangement does not suggest an effective sphincter, though in some positions of the trunk a kinking effect might occur. That the sphincter may be oesophageal has also been suggested (see pp. 594, 1333).

THE GASTRIC STRUCTURE

The gastric wall consists of the usual serous, muscular, submucous and mucous layers, together with their vessels and nerves.

The serosa or visceral peritoneum covers the entire surface except: (1) along the greater and lesser curvatures at the attachment of the greater and lesser omenta, where the peritoneal layers leave space for vessels and nerves; and (2) a small postero-inferior area, near the cardiac orifice, where the stomach contacts the diaphragm at the reflexions of the gastrophrenic and left gastropancreatic folds.

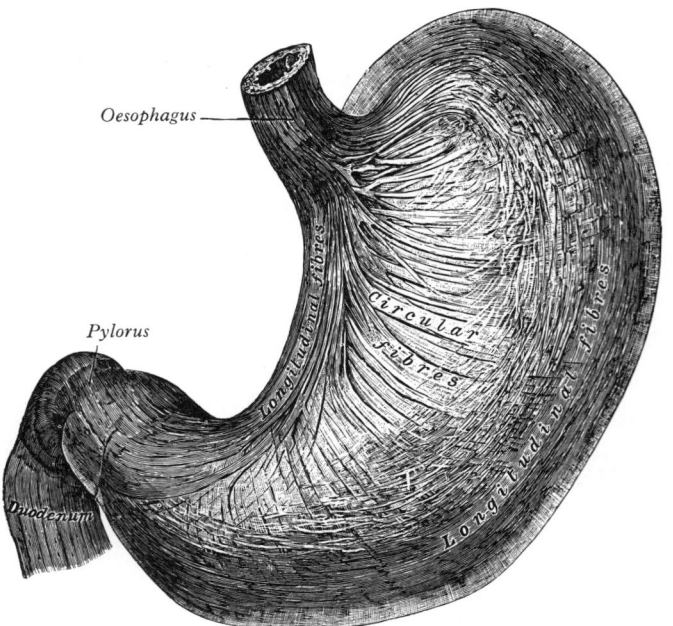

8.121A The longitudinal and circular gastric muscular fibres: anterosuperior aspect (Spalteholz).

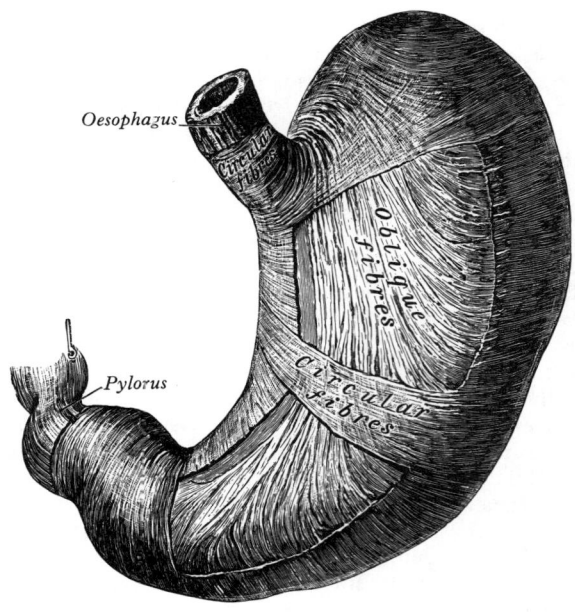

8.121B The oblique muscular fibres of the stomach, shown by partial dissection of its wall: anterosuperior aspect.

The muscularis externa (8.121A,B) is immediately under the serosa, with which it is closely connected by subserous areolar tissue; it has longitudinal, circular and oblique layers of non-striated muscle fibres.

Longitudinal fibres are the most superficial and arranged in two groups. The first set is continuous with the longitudinal oesophageal fibres; radiating from the cardiac orifice, they are best developed near the curvatures and end proximal to the pyloric region. The fibres of the other group commence in the body and pass to the right, becoming more thickly arranged as they approach the pylorus; some superficial fibres pass to the duodenum, deeper ones turning in to interlace with the fibres of the pyloric sphincter.

Circular fibres form a uniform layer over the whole stomach internal to the longitudinal fibres. At the pylorus they are most abundant, aggregated into the annular *pyloric sphincter*; they are also continuous with the circular oesophageal fibres but sharply separated from those of the duodenum by a septum of connective tissue.

Oblique fibres, internal to the circular, are limited to the gastric body and are most developed near the cardiac orifice. They sweep

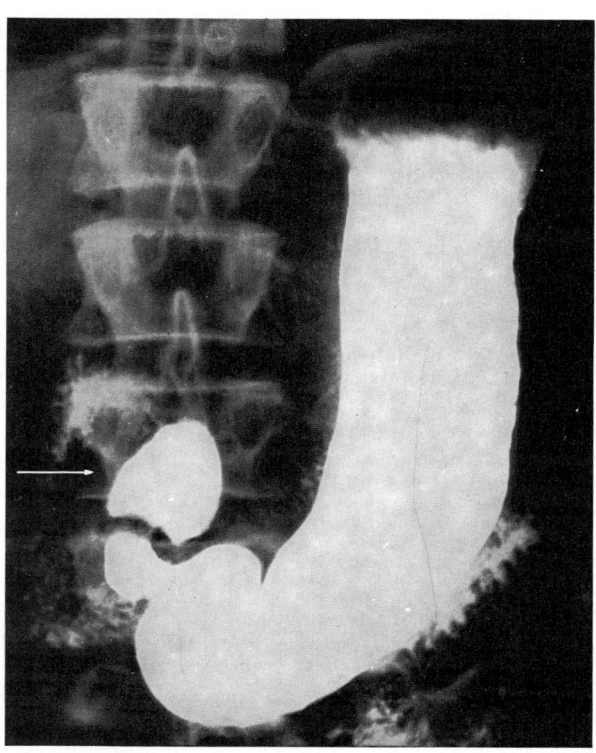

8.122A Radiograph of a normal stomach after a barium meal. The tone of the muscular wall is good and supports the weight of the column in the body of the organ. The arrow points to the duodenal cap, below which a gap in the barium indicates the position of the pylorus.

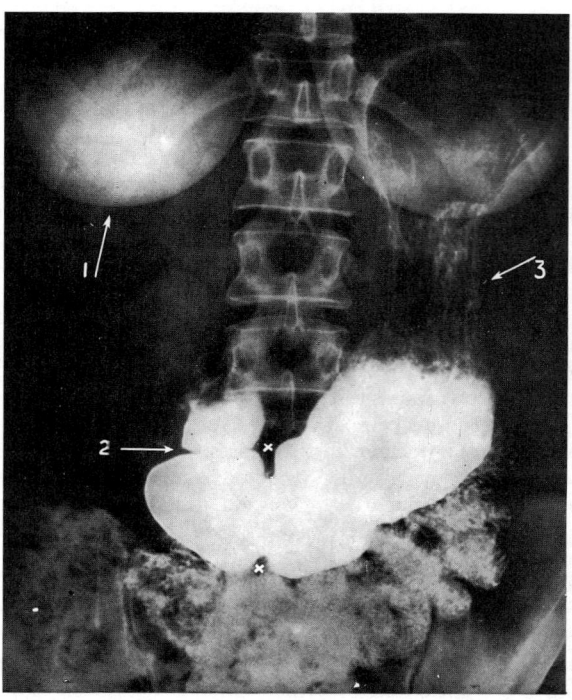

8.122B Radiograph of an atonic stomach after a barium meal. Note that this stomach contains the same amount of barium suspension as the stomach in 8.122A. Arrow 1 points to the shadow of the right breast, arrow 2 to the pylorus, arrow 3 to the upper part of the body of the stomach, where longitudinal folds can be seen in the mucous membrane. XX marks a wave of peristalsis.

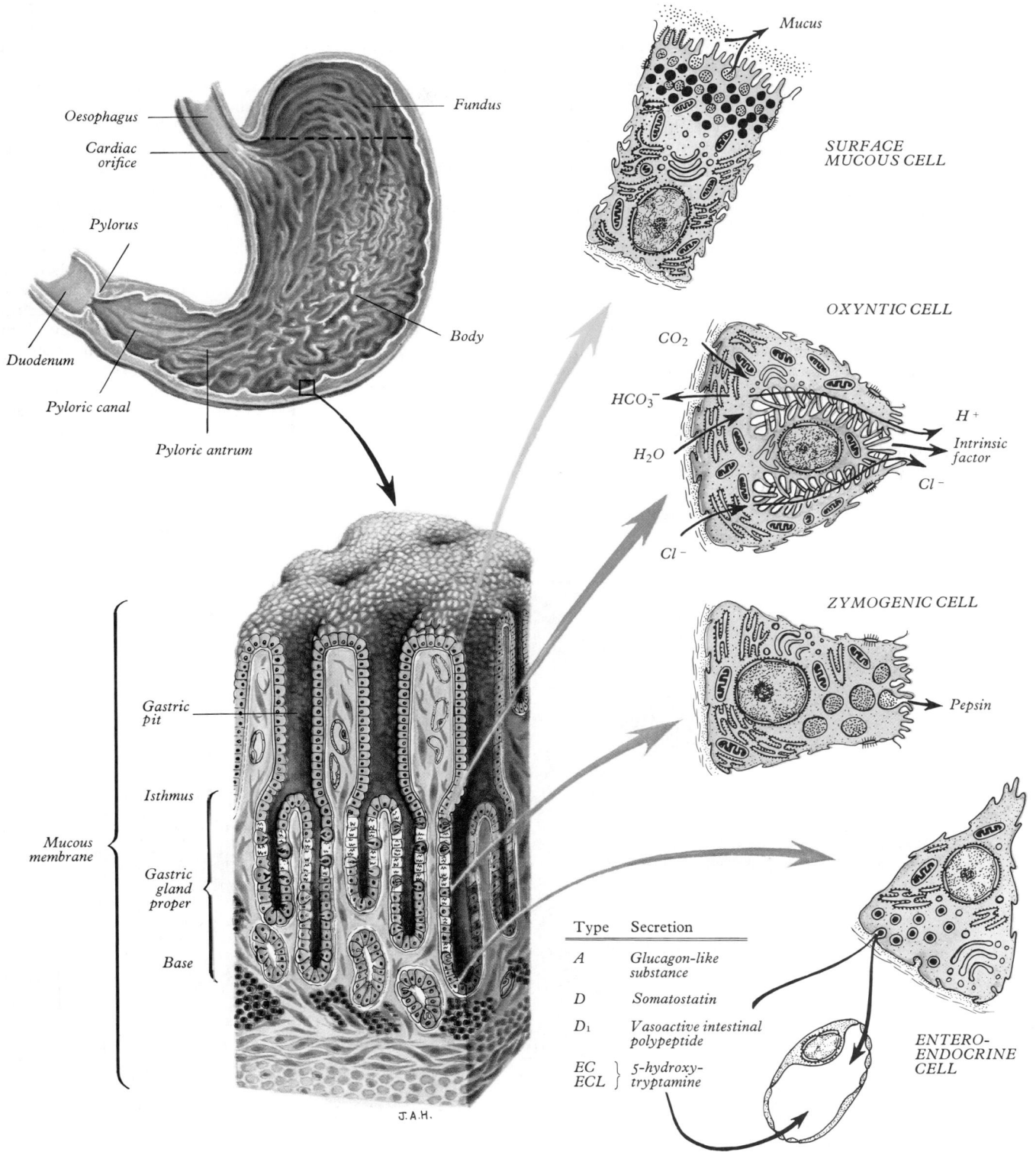

8.123 Diagram showing the principal regions of the interior of the stomach and the histology and ultrastructure of its mucous membrane. Undifferentiated, dividing cells are shown in white.

down from the cardiac incisure more or less parallel with the lesser curvature, near which they present a free and well-defined margin (**8.121**B); on the left they blend with the circular fibres nearer the greater curvature.

Peristalsis in the pyloric antrum thoroughly mixes its contents, returning some of them to the body of the stomach and propelling some into the duodenum. The pyloric sphincter contracts intermittently during antral activity; but in the resting stomach it is relaxed, leaving the pylorus open (Atkinson et al 1957, Spira 1957). The control of transfer of the stomach contents through the pylorus is still

not fully clarified. For an authoritative summary consult Hunt & Knox (1968).

The submucosa is composed of loose, areolar tissue, connecting the mucosa to the muscularis externa.

The mucosa is thick, its surface smooth, soft, velvety and, over most of its surface, reddish brown; it is pink in the pyloric region. During contraction it is folded into numerous folds or rugae, mostly longitudinal and best marked towards the pyloric end and along the greater curvature (**8.123**). The rugae are obliterated by gastric distension. The detailed structure of the mucosa is described below.

Structure of the Gastric Mucosa

At low magnification the luminal surface of the mucosa (**8.**123) appears honeycombed by small depressions or alveoli, polygonal or slit-like, about 0.2 mm in diameter. These *gastric pits* contain the orifices of gastric glands. The whole surface, including the pits, is covered by simple secretory columnar epithelium, the *surface mucous cells*, which liberate mucus from their apices as a lubricant and also to protect the gastric lining against its own secretions of acid and enzymes. This epithelium commences abruptly at the cardiac orifice, where there is a sudden transition from the oesophageal stratified epithelium. The *gastric glands* comprise three categories: (1) cardiac, (2) main (in the body and fundus) and (3) pyloric (Ito & Winchester 1963).

Cardiac glands (**8.**124A) are confined to a small area near the cardiac orifice; some are simple tubular glands, others are compound racemose. Mucus-secreting cells predominate and oxyntic and zymogenic cells are few.

The main gastric glands are in the body and fundus, three to seven opening into each gastric pit; they are cytologically the most highly differentiated (**8.**124B), with at least four distinct cell types as well as undifferentiated cells.

(1) The *chief (peptic* or *zymogenic) cells*, usually basal in position, are cuboidal and protein-synthesizing, containing prominent secretory vacuoles, also much granular endoplasmic reticulum and a prominent Golgi complex; they are strongly basophilic because of their contained RNA. They are the source of gastric digestive enzymes (**8.**123).

(2) The *oxyntic (parietal) cells*, large, round and eosinophilic, are most numerous on the side walls of the glands and near their ducts. They occur only at intervals and with light microscopy seem to be applied to the exterior of other cells or partly intercalated between them. They bulge into the adjacent lamina propria, giving it a beaded appearance, and are connected with the lumina of the glands by fine processes between adjacent cells. Oxyntic cells have tortuous invaginations forming what appear to be intracellular canaliculi the surfaces of which are covered with microvilli, and which open directly into the lumen of the gastric gland on the apices of fine processes passing centrally between adjacent cells. Each possesses a large, round, central nucleus and the surrounding cytoplasm displays sparse endoplasmic reticulum, no secretory granules but very numerous, densely packed mitochondria. Minute smooth-walled membranous tubules converge on and open into the intracellular canaliculi (**8.**123). These cells are the source of chloride and hydrogen ions in the acid secretions of the stomach. They also elaborate and secrete *intrinsic factor* necessary for the absorption of vitamin B_{12} (Hoedemseker et al 1966).

(3) *Mucous ('neck') cells* are dispersed among other the types and are numerous at the necks of the glands. They are typical mucus-secreting cells, but their secretions are distinct histochemically from those of the superficial mucous cells.

(4) *Argentaffin cells* occur in all types of gastric gland but more commonly in those of the body and fundus. They are more usual in the deeper parts of the glands, lying between the zymogenic cells and the basal lamina; they rarely reach the lumen. Their

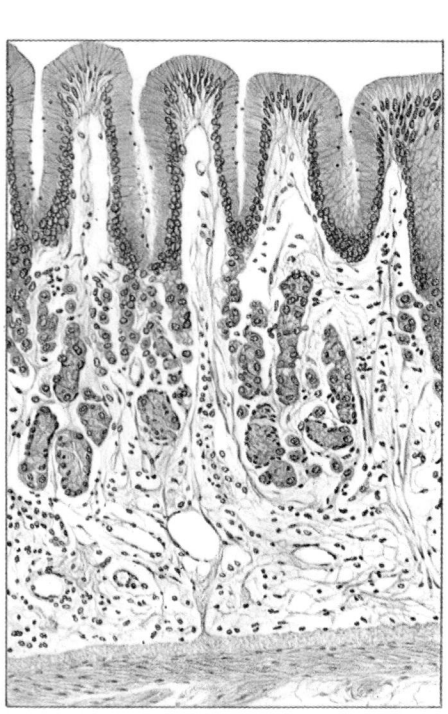

A

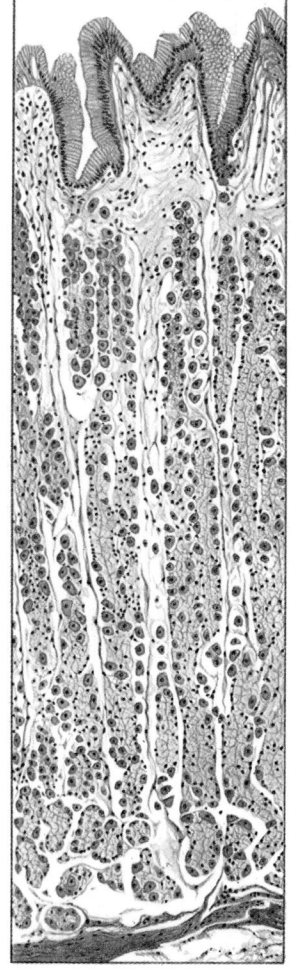

B

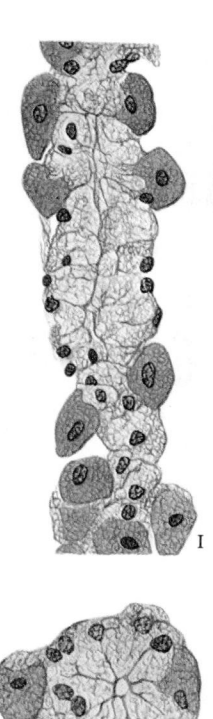

I

II

C

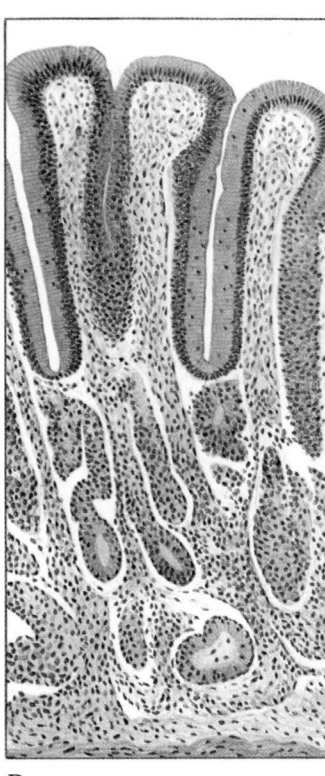

D

8.124 A. Vertical section through the mucous membrane of the cardiac part of the stomach (human). Stained with haematoxylin and eosin. Magnification *c.* × 150.

B. Vertical section through the mucous membrane of the fundus of the stomach (cat). Note the beaded appearance given by the oxyntic cells. Stained with haematoxylin and eosin. Magnification *c.* × 100.

C. (I) Gland from the fundus of the stomach (cat). (II) Lower part of the gland cut transversely. stained with haematoxylin and eosin. The peripherally placed cells staining deeply with eosin are the oxyntic cells. Magnification *c.* × 530.

D. Vertical section through the mucous membrane of the pyloric part of the stomach (cat). Stained with haematoxylin and eosin. Magnification *c.* × 75.

irregular nuclei are surrounded by granular cytoplasm stained strongly by silver salts. The cytoplasm, particularly that nearest to the basal lamina, contains many dense membrane-bound vacuoles of varying size, some large (0.3 μm diameter) and responsible for the staining reaction. They are part of the *gastro-enteropancreatic endocrine system* (p. 1376) and are subdivided into a variety of types on the basis of differences in their ultrastructure and secretions (8.123, 146).

(5) *Undifferentiated columnar cells* also exist in smaller numbers, at least some of them being stem cells from which other cells are continually replaced. Superficial mucous cells are replaced after about three days and mucous neck cells after about one week. Other cell types appear to live considerably longer.

Pyloric glands each consist of two or three short convoluted tubes opening into a conical pit, which occupies about two-thirds of the mucosal depth (8.124D). Epithelial cells are mostly mucous, oxyntic cells being few; enteroendocrine cells are, however, present. The enteric hormone *gastrin* has been isolated from human pyloric glands; it is released by mechanical stimuli and

increases gastric motility and secretion by chief and oxyntic cells (see also p. 1376). Although oxyntic cells are few in pyloric glands, they are always present in both fetal and postnatal material; in adults they may appear in the duodenal mucosa but only proximally, near the pylorus (Leela & Kanagasuntheram 1968). Between the glands the lamina propria forms a connective tissue framework and contains lymphoid tissue which, especially in early life, collects in small masses, termed *gastric lymphatic follicles*, which resemble solitary intestinal follicles. In the mucosa, deep to the glands, is a thin stratum of non-striated muscle fibres, the *muscularis mucosae*, arranged as inner circular and outer longitudinal layers, with a third external, circular layer in places. The inner layer sends strands of myocytes between the glands, contraction probably aiding their emptying.

Vessels and Nerves

The arterial supply of the stomach comes from the left gastric artery (from the coeliac artery), right gastric and right gastro-epiploic (from the common hepatic) and the left gastro-epiploic

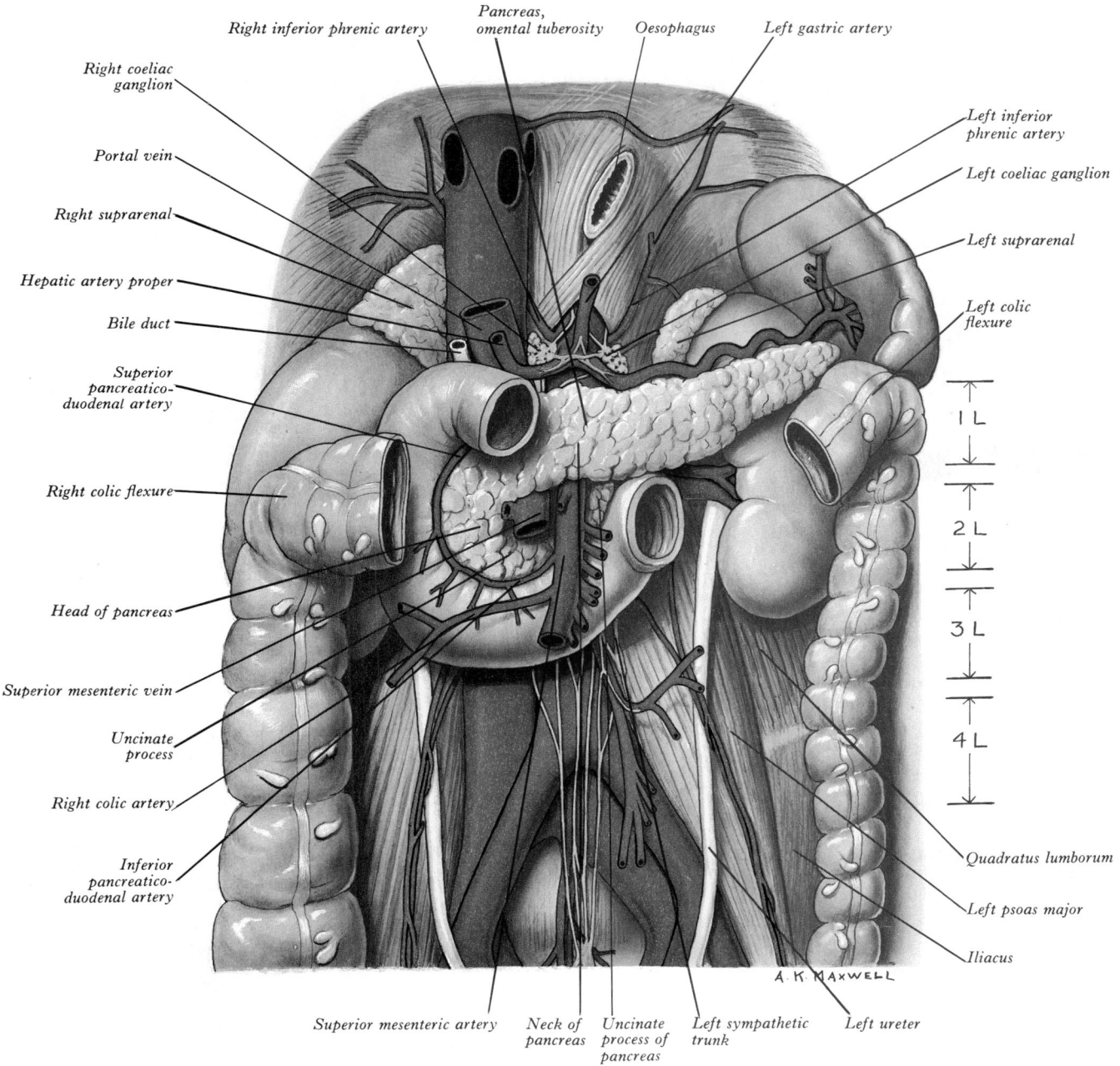

Right inferior phrenic artery
Pancreas, omental tuberosity
Oesophagus
Left gastric artery
Right coeliac ganglion
Portal vein
Right suprarenal
Hepatic artery proper
Bile duct
Superior pancreatico-duodenal artery
Right colic flexure
Head of pancreas
Superior mesenteric vein
Uncinate process
Right colic artery
Inferior pancreatico-duodenal artery
Left inferior phrenic artery
Left coeliac ganglion
Left suprarenal
Left colic flexure
1 L
2 L
3 L
4 L
Quadratus lumborum
Left psoas major
Iliacus
A. K. MAXWELL
Superior mesenteric artery
Neck of pancreas
Uncinate process of pancreas
Left sympathetic trunk
Left ureter

8.125 Dissection to show the duodenum, pancreas, stem arterial rami of the gastrointestinal tract and surrounding structures. The right and left hepatic veins have been cut away at their points of entry into the inferior vena cava. The superior hypogastric plexus is shown in front of the sacral promontory and the sympathetic nerves which form it are seen descending across the bifurcation of the aorta, the left common iliac vein and the body of the fifth lumbar vertebra. (In this specimen the left renal artery is situated anterior to the left renal vein at the hilum of the kidney.)

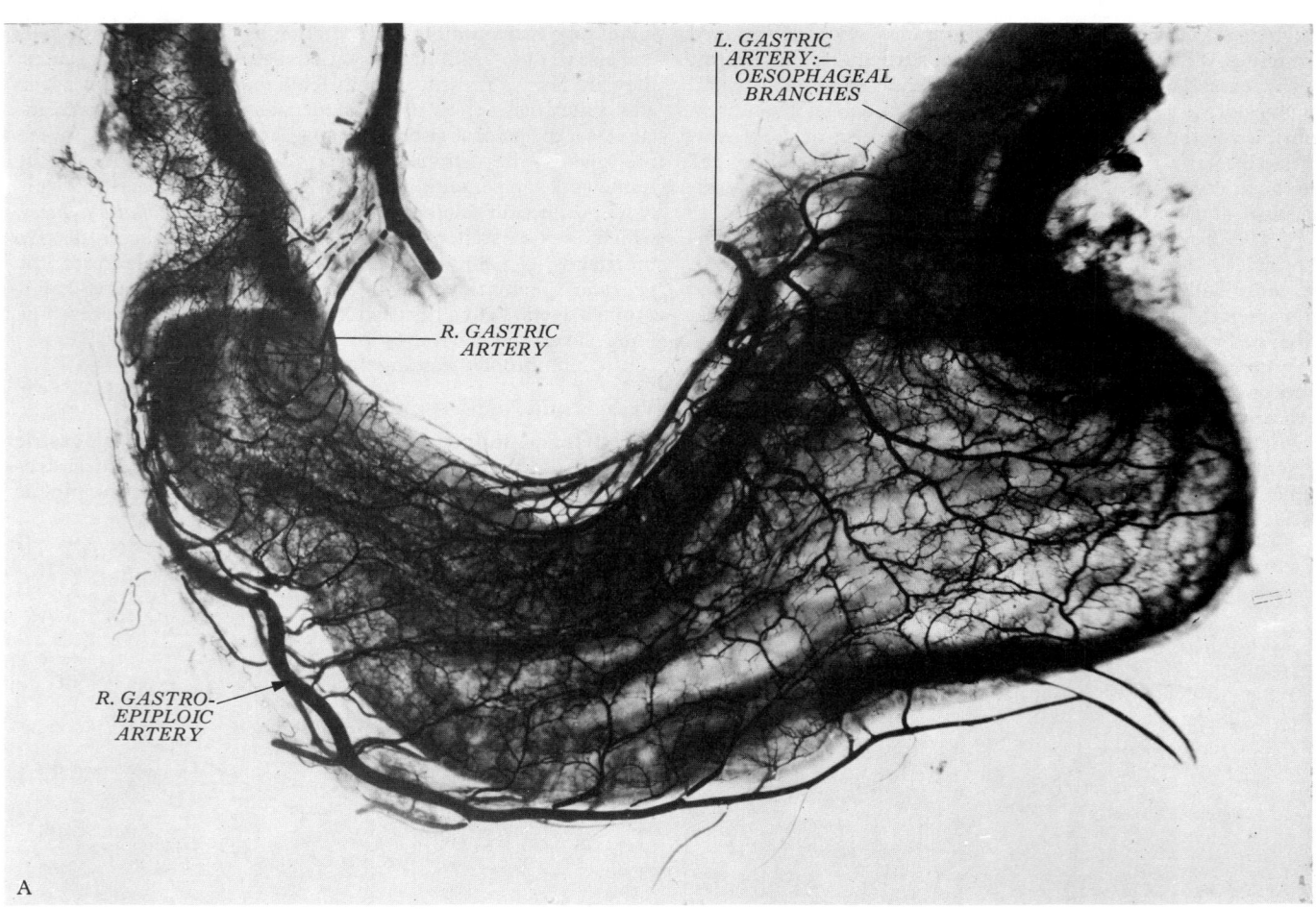

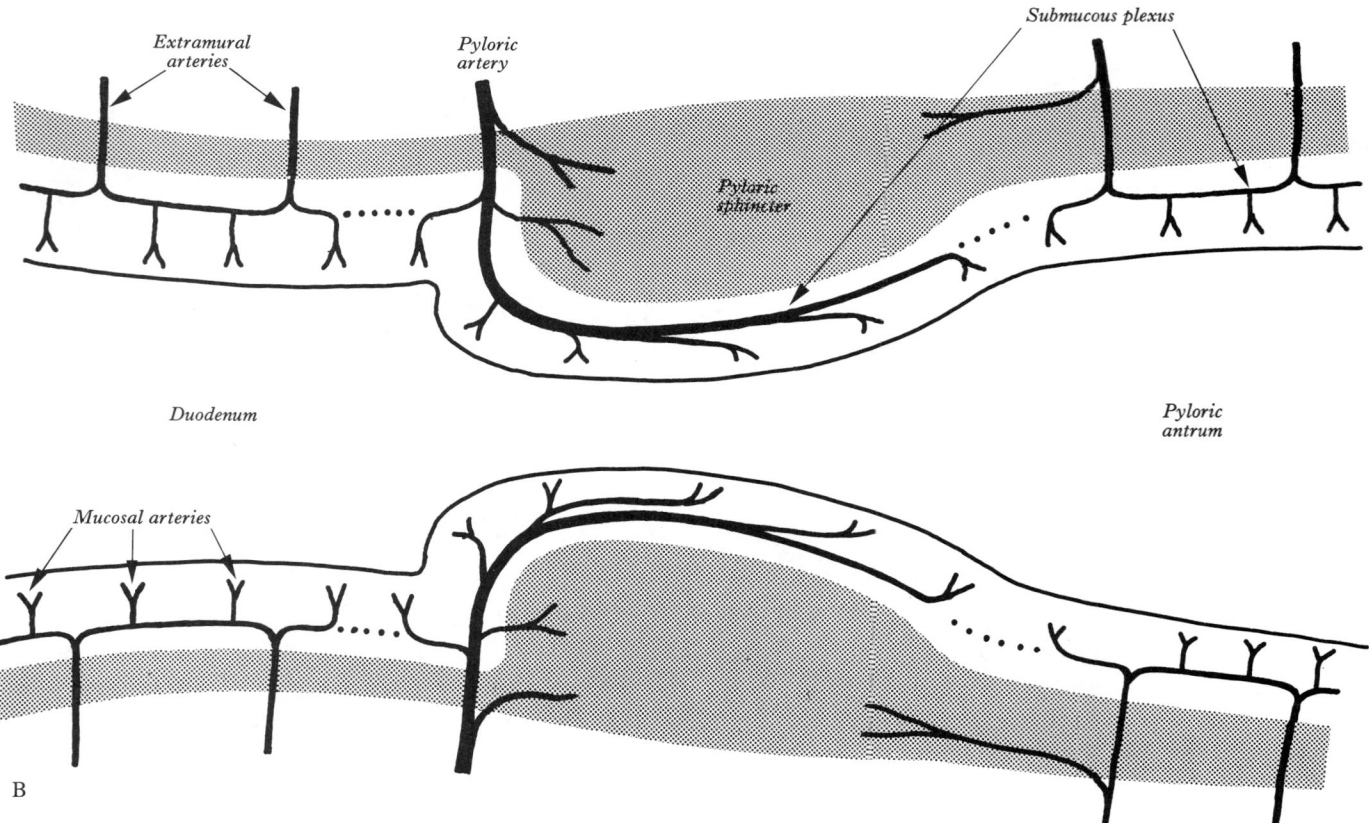

8.126 Blood supply of the stomach and the proximal duodenum. A. Arterial system in a fetal human stomach. The muscle layer has been removed. Note double arcade along the lesser curvature. The arteries have been injected with a mixture of 2% gelatin and India ink and subsequently cleared by the Spalteholz technique. Magnification × 6. B. A scheme of arterial arrangements at the gastroduodenal junction. Dotted lines indicate sites where the submucous plexus may be deficient in continuity. Shaded areas represent the muscular layer of the visceral wall. By courtesy of C Piasecki, Department of Anatomy, Royal Free Hospital School of Medicine, London and the *Journal of Anatomy*.

and short gastric (from the splenic) artery. These vessels not only anastomose extensively on the serosal aspect of the stomach (p. 768) but also form anastomotic networks within its walls at intramuscular, submucosal and mucosal levels; a true plexus of small arteries and arterioles is present in the submucosa. This *submucosal plexus*, from which the mucosa is supplied, shows considerable regional variation both in the gastric wall and also in the proximal duodenum. In view of a possible vascular factor in the genesis of peptic ulcers, the local details of angio-architecture are of interest. That of the mucosa of the body of the stomach was described recently by Raschke et al (1987). Arteriovenous anastomoses are said to occur in the gastroduodenal mucosa (Spanner 1932, De Busscher 1948, Barlow et al 1951, Boulter & Parkes 1963); dysfunction in these might produce local ischaemia and ulceration. Mucosal end-arteries have also been described and larger vessels such as the supraduodenal artery (p.769) have been similarly designated end-arterial. Such problems have been discussed anatomically by Piasecki (1974, 1977) and the results of his observations on fetal, neonatal and adult human stomachs, injected with India ink, are the basis of this account of gastric intramural arteries. From anastomotic arcades along the greater and lesser curvatures, formed by the main arteries of supply described above, many *anterior* and *posterior gastric arteries* pass to the anterior and posterior aspects, approximately transverse to the organ's long axis. Smaller rami, often paired, also pass directly to parts of the gastric wall subjacent to the omental attachments. All these vessels ramify on the external surface and penetrate the muscular layers to reach the submucosa and mucosa, forming subserosal, intramuscular and submucosal plexuses, the second being the best developed (8.126A,B). This muscular plexus is supplied by branches from the subserous and submucosal plexuses; the muscular vessels vary in direction in different muscular laminae, perhaps adapting to directions of contraction. Submucosal arteries anastomose freely, but the incidence of anastomoses varies. Counts by Piasecki (1974) showed that while anastomoses along, e.g. the lesser curvature, increased in *number* from cardia to pylorus; the mean *calibre* of anastomosing arteries showed a reverse tendency. Mucosal arteries, which fill the capillary networks supplying the epithelium and its glands, are mainly from the submucosal plexus; but along both curvatures a few mucosal arteries come directly from subserosal sources, traversing the muscular layers and submucosa, often without lateral junctions with submucosal arteries; their frequency apparently increases from the cardiac to the pyloric regions; the capillary networks supplied by them are largely independent of those fed by adjacent submucosal arteries and the patch of mucosa supplied by such a vessel is perhaps more vulnerable to vascular obstruction. Mucosal arteries in general do not form lateral anstomoses but, since their submucosal feeders do, vascular obstruction is less likely. Piasecki also showed a different pattern of supply in the pyloric canal and sphincter. 'Pyloric arteries', rami of the right gastric and gastro-epiploic arteries, pierce the duodenum distal to the sphincter around its entire circumference, passing through the muscular layer to the submucosa where each divides into two or three rami, which turn into the pyloric canal, internal to the sphincter; they traverse the submucosa to the end of the pyloric antrum (8.126B), supplying the whole mucosa of the pyloric canal. Branches of these pyloric submucosal arteries may anastomose at their commencement with the duodenal submucosal arteries and, by their terminal rami, with corresponding gastric arteries. The pyloric sphincter is supplied by the gastric and pyloric arteries, whose rami leave their parent vessels in the subserosal and submucosal levels to penetrate the sphincter. In a more recent study, Piasecki (1986) has described arrangements in a large number of animals, including rodents, swine, cats, dogs and monkeys, with little modification from the pattern he found earlier in primates.

Gastric veins commence as straight vessels between the mucosal glands and drain into the submucosal veins. Their further arrangment has not received as much attention as that of the corresponding arteries; but larger veins generally accompany main arteries to their ultimate drainage into the splenic and superior mesenteric veins, while some pass directly to the portal vein. Smaller **lymphatic vessels** are said to resemble the veins in

distribution. Regional lymph nodes and their drainage are described on p. 851.

Nerves are from several sources (Kyosola et al 1980). The *sympathetic supply* is mainly from the coeliac plexus through its extensions around the gastric and gastro-epiploic arteries. Some rami from the hepatic plexus reach the lesser curvature between the layers of the hepatogastric ligament (p. 1343). Rami of the left phrenic plexus pass to the cardiac end of the stomach, as does one from the left phrenic branch to the right crus of the diaphragm. Inconstant gastric branches come from the left thoracic splanchnic nerves and the thoracic and lumbar sympathetic trunks.

The *parasympathetic supply* is from the vagus nerves (Mackay & Andrews 1983). Usually one or two rami branch on the anterior and posterior aspects of the gastro-oesophageal junction; the anterior nerves are mostly from the left vagus and the posterior from the right vagus, emerging from the oesophageal plexus. The anterior nerves supply filaments to the cardiac orifice and divide near the oesophageal end of the lesser curvature into branches: (1) *the gastric branches* (4-10) radiate on the anterior surface of the body and fundus; one, larger than the others, lies in the lesser omentum near the lesser curvature (*greater anterior gastric nerve*); (2) *the pyloric branches*, generally two, one of which traverses the lesser omentum almost horizontally to the right, towards its free edge, then turns down on the left side of the hepatic artery to reach the pylorus, while the other, usually arising from the greater anterior gastric nerve, passes obliquely to the pyloric antrum. The posterior nerves produce two groups of branches, gastric and coeliac: (1) *gastric branches* radiate over the posterior surface of the body and fundus and extend to the pyloric antrum but do not reach the pyloric sphincter; the largest (*greater posterior gastric nerve*) passes posteriorly along the lesser curvature, giving branches to the *coeliac plexus*; (2) *coeliac branches*, larger than the gastric, pass in the lesser omentum to the coeliac plexus. No true plexuses occur on either the anterior or posterior gastric surfaces, but they do in the submucosa and between the layers of the muscularis externa. The latter corresponds to the myenteric (Auerbach's) plexus and contains many neurons. They distribute many axons containing a variety of neurotransmitters and neuromodulators (Burnstock 1986) and also sensory fibres to the muscular tissue and the mucosa.

The vagus has both secretory and motor effects on the stomach; vagal stimulation evokes a secretion rich in pepsin and increases gastric motility; after vagotomy the stomach is flaccid and empties only slowly. The sympathetic nerve supply is vasomotor to the gastric blood vessels and provides the main pathway for gastric pain fibres.

The Small Intestine

The small intestine, a coiled tube, extends from the pylorus to the ileocaecal valve, where it joins the large intestine. It is usually said to be 6–7 m long, gradually diminishing in diameter towards its termination. However, it is longer after death owing to the loss of muscle tone; its average length in living adults is perhaps about 5 m (vide infra). In 109 adult subjects shortly after death it ranged from 3.35–7.16 m in women and from 4.88–7.85 m in men, the average being 5.92 m in women and 6.37 m in men (Underhill 1955). Length was correlated with the height of the individual but was independent of age; the large intestine was much more constant in length. Jit & Grewal (1975), reviewing the topic, reported findings in 137 Indian subjects confirming these associations with height and noting lack of a correlation with weight. They observed that fixation in formalin caused contraction which sometimes reached 44%. Various observers have also passed flexible tubes through the alimentary tract, recording total lengths of 2.7–4.5 m (see Jit & Grewal 1975).

The small intestine occupies the central and lower parts of the abdominal cavity, usually within the colonic loop; it is related in front to the greater omentum and abdominal wall; a portion may reach the pelvis in front of the rectum. It consists of a short, curved sessile section, the *duodenum*, and a long, greatly coiled part attached to the posterior abdominal wall by the mesentery (p. 1344), the proximal two-fifths being the *jejunum*, the distal three-fifths, the *ileum*.

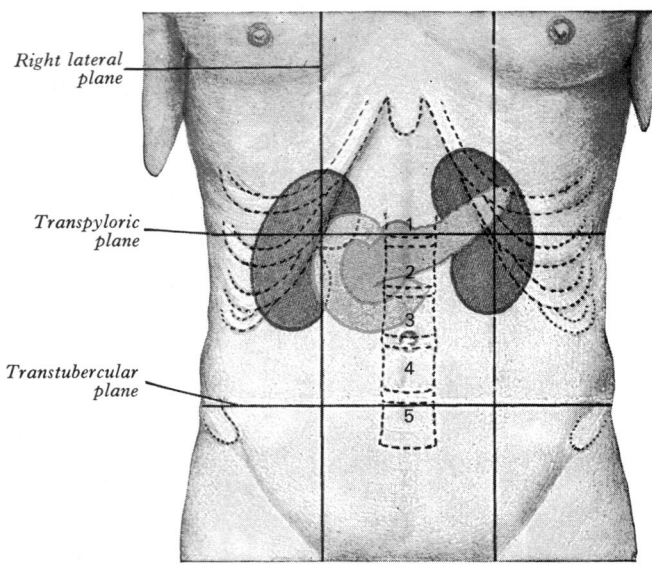

Right lateral plane

Transpyloric plane

Transtubercular plane

8.127 Surface projection of the duodenum, pancreas and kidneys. The lumbar vertebrae are numbered.

THE DUODENUM

The duodenum (8.125) is 20–25 cm long (*12* in., hence the name) and is the shortest, widest and most sessile part of the small intestine. It has no mesentery, and is thus only partially covered by peritoneum. It is constantly curved in an incomplete circle, enclosing the head of the pancreas. It is situated entirely above the level of the umbilicus. Arising from the pylorus, it passes backwards, up and to the right for about 5 cm, inferior to the posterior part of the quadrate lobe, to the neck of the gallbladder; its direction varies slightly according to the distension of the stomach. It then curves abruptly (*superior duodenal flexure*) to descend about 7.5 cm anterior to the medial part of the right kidney, usually to the level of the lower border of the third lumbar vertebral body, just medial to the lateral plane (8.127). At a second bend (*inferior duodenal flexure*), it turns horizontally left across the vertebral column for about 5–10 cm, just above the umbilical level, with a slight upward slope; it then ascends in front and to the left of the abdominal aorta for about 2.5 cm, ending opposite the second lumbar vertebra in the jejunum. At this union it turns abruptly forwards; this *duodenojejunal flexure* is about 2.5 cm left of the midline and 1 cm below the transpyloric plane. For descriptive

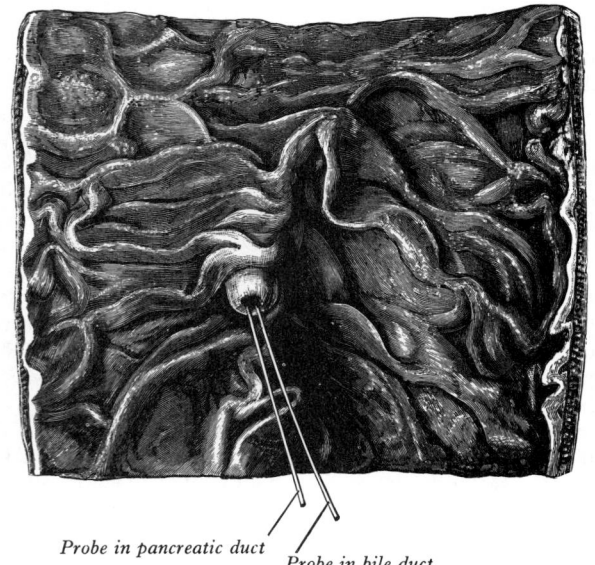

Probe in pancreatic duct

Probe in bile duct

8.128 Interior of the descending (second) part of the duodenum, showing the major duodenal papilla.

purposes it is hence divided into parts: first (superior), second (descending), third (horizontal) and fourth (ascending).

Duodenal Relations

The superior (first) part, about 5 cm long, is the most mobile section, extending from the pylorus to the neck of the gallbladder. Peritoneum covers its anterior aspect but it is bare of this posteriorly, *except near the pylorus* where it takes a small part in the formation of the anterior wall of the omental bursa; here the lesser omentum is attached to its upper border and the greater omentum to its lower (proximal half). It is related above and in front with the quadrate lobe of the liver and gallbladder and more posteriorly above with the epiploic foramen, behind with the gastroduodenal artery, bile duct and portal vein and posteroinferiorly with the head and neck of the pancreas. It is usually stained by bile after death especially on its anterior surface where it is related to the gallbladder.

The descending (second) part, from 8–10 cm long, descends from the neck of the gallbladder along the right side of the vertebral column to the lower border of the third lumbar vertebral body. Crossed by the transverse colon, it is connected to it by some areolar tissue and above and below this attachment it is covered in front with peritoneum. It is related in front, from above downwards: to the right lobe of the liver, transverse colon and the root of its mesocolon and to the jejunum; behind it is variably related to the right kidney near to its hilum, being connected to it by areolar tissue, to the right renal vessels, the edge of the inferior vena cava and psoas major. Medial to it are the head of the pancreas and bile duct, while lateral is the right colic flexure. A small part of the pancreatic head is sometimes embedded in the duodenal wall. The bile and pancreatic ducts come into contact at its medial side, entering its wall obliquely and uniting to form the *hepatopancreatic ampulla* (p. 1394). The narrow, distal end of this opens on the summit of the *major duodenal papilla*, sited posteromedially in the descending duodenum (8.128, 150), 8–10 cm distal to the pylorus. An accessory pancreatic duct may open about 2 cm proximal to the major papilla on a *minor duodenal papilla*.

The horizontal (inferior or third) part, about 10 cm long, passes from the right of the lower border of the third lumbar vertebra, sloping slightly up and to the left across the inferior vena cava, to end in the fourth part in front of the abdominal aorta. Its anterior surface is covered with peritoneum, except in the median plane where it is crossed by the superior mesenteric vessels and mesenteric root. Its posterior surface is covered by peritoneum only at its left end, where the left layer of the mesentery sometimes covers it. The posterior surface rests upon: the right ureter, right psoas major, right testicular (or ovarian) vessels, the inferior vena cava and the abdominal aorta (with the origin of the inferior mesenteric artery). Its superior aspect is related to the head of the pancreas, its inferior to coils of the jejunum.

The ascending (fourth) part, about 2.5 cm long, ascends on or immediately to the left of the aorta, to the level of the upper border of the second lumbar vertebra, where it turns forwards into the jejunum at the *duodenojejunal flexure*; it is anterior to the left sympathetic trunk, left psoas major, left renal and gonadal vessels and the inferior mesenteric vein. To the right it gives attachment to the upper part of the root of the mesentery, its left layer being continued over the duodenum's anterior surface and left side. To its left are the left kidney and ureter; above is the body of the pancreas; in front are the transverse colon and transverse mesocolon, the latter separating the duodenojejunal flexure from the omental bursa and stomach.

The superior part of the duodenum is slightly mobile, while the rest is almost fixed, being sessile upon neighbouring structures. Radiologically, after a barium meal, the superior part appears as a triangular, homogeneous shadow, the 'duodenal cap' (8.122).

The terminal part and the duodenojejunal flexure are said to be positioned by the '*suspensory muscle of the duodenum*' (suspensory muscle, or ligament of Treitz), often described as being in two parts: (1) a slip of *striated* muscle derived from the diaphragm near its oesophageal opening, ending in connective tissue near the

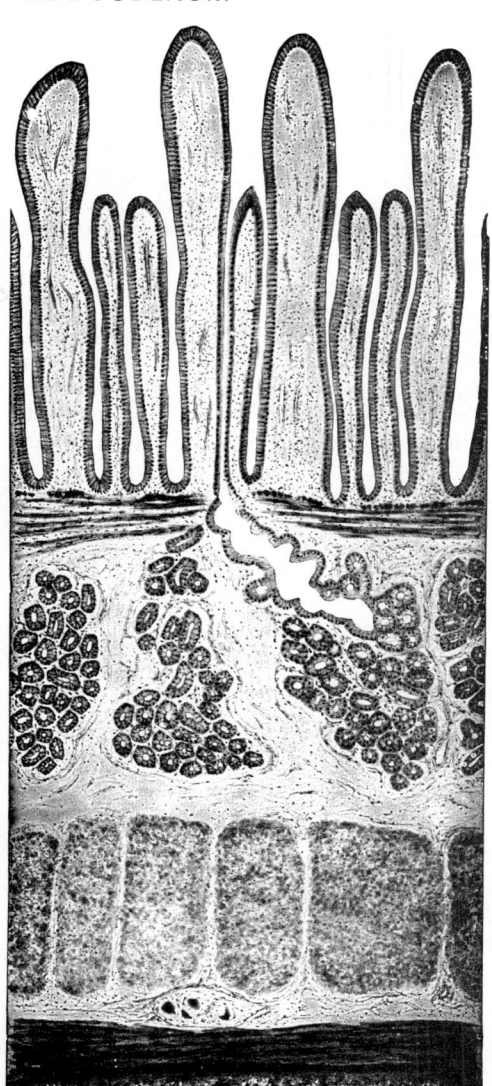

Villi

Intestinal crypts

Muscularis mucosae

Duodenal glands in submucosa

Circular muscular layer

Longitudinal muscular layer

Serous layer

A

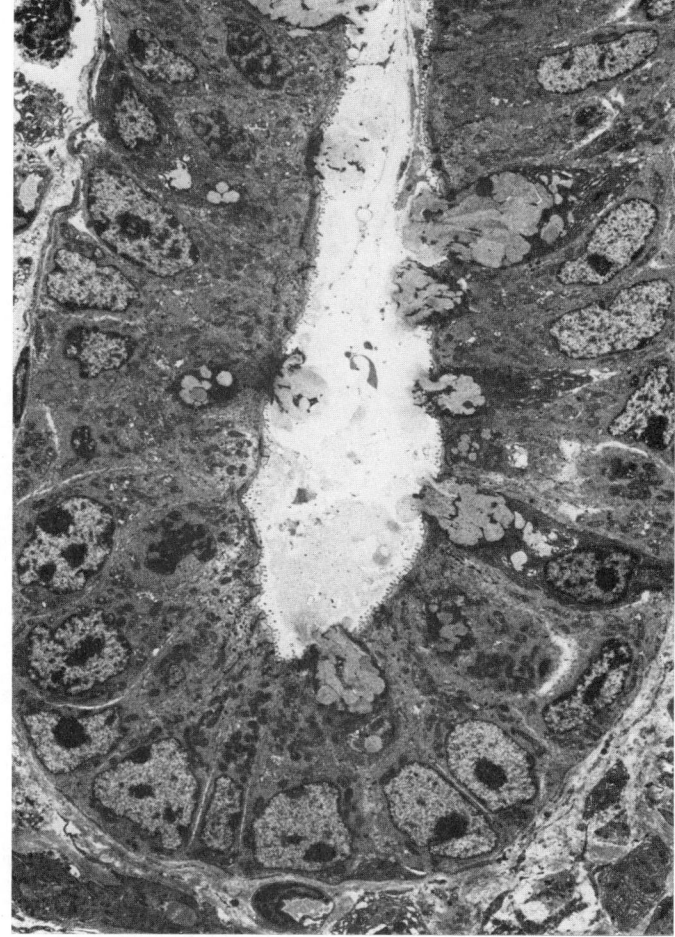

B

8.129 A. Longitudinal section of the feline duodenal wall. Magnification *c.* × 60. B. Electron micrograph of the base of a duodenal crypt showing absorptive columnar epithelial cells interspersed with mucus-secreting goblet cells (rat). Magnification × 1700. C. Electron micrograph showing absorptive columnar epithelial cells of a duodenal villus (rat). Magnification × 3700. Specimens B and C prepared and photographed by Susan Smith, Department of Anatomy, Guy's Hospital Medical School, London.

coeliac artery and (2) a fibromuscular band of *non-striated* muscle, passing from the duodenum (third and fourth parts and duodenojejunal flexure) to blend with the same pericoeliac connective tissue. Treitz (1853) described both entities, naming the former *der Hilfsmuskel* (the accessory muscle). Subsequent authorities (Lockwood 1886, Low 1907) regarded them as a digastric muscle, naming the whole the suspensory *muscle* of Treitz, a misnomer perpetuated in most textbooks. Confusion was increased by Haley & Peden (1943), who derived the 'suspensory muscle' from the right crus, and by Argeme et al (1970), who described an intermediate tendon but regarded this as part of a 'false' digastric muscle. Jit (1952, 1977) has persistently repeated the dual nature of the original description by Treitz, supporting it by embryological and histological evidence. The diaphragmatic slip (Hilfsmuskel) has no satisfactory official name. It is supplied, according to Jit, by myelinated nerve fibres probably from the phrenic nerve (pp. 594, 1128) and is sometimes considered an aberrant part of iliocostalis thoracis. The suspensory muscle proper (non-striated) is supplied by autonomic fibres from the coeliac and superior mesenteric plexuses (Jit & Grewal 1977). Descriptions of the duodenal attachments of the muscle vary; none of these accounts contain a convincing view of its function,

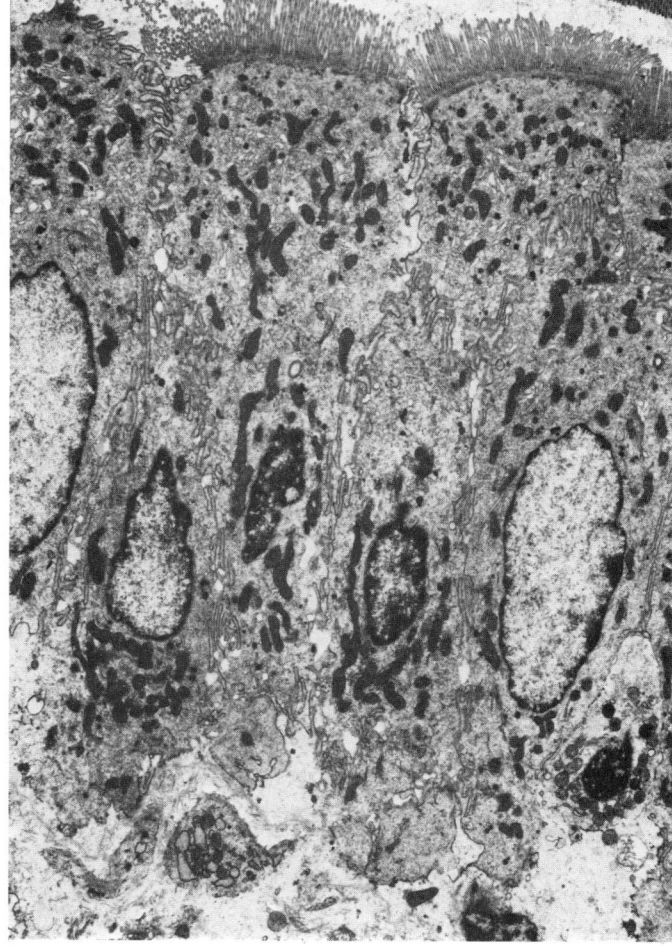

C

the usual suggestion being that it augments duodenojejunal flexure, acting like a valve.

Vessels and Nerves

Arteries supplying the duodenum are from the right gastric, supraduodenal, right gastro-epiploic and superior and inferior pancreaticoduodenal arteries (pp. 769, 772; **6**.109.) The superior part receives two leashes of small rami, one from the hepatic artery proper and one from the gastroduodenal artery. These rami also supply the adjacent pyloric canal, with some anastomosis in the muscular layer across the pyloroduodenal junction (p. 1355). *Veins* end in the splenic, superior mesenteric and portal veins. *Nerves* are from the coeliac plexus.

THE JEJUNUM AND ILEUM

The rest of the small intestine extends from the duodenojejunal flexure to the ileocaecal valve, ending at the junction of the caecum and ascending colon. It is arranged in a series of coils attached to the posterior abdominal wall by the mesentery. It is completely covered by peritoneum, except along its mesenteric border where the two mesenteric layers diverge to enclose it. Its proximal two-fifths is the *jejunum*, the rest the *ileum*; the division is arbitrary, as the character of the intestine changes only gradually, but samples from these two 'parts' show characteristic differences.

The jejunum, with a diameter of about 4 cm, is thicker walled, redder and more vascular. Its circular mucosal folds (p. 1360) are large and frequent and its villi larger. Aggregated lymphatic follicles (p. 1363) are almost absent from the proximal (upper) jejunum; distally they are still fewer and smaller than in the ileum and are often discoidal. The circular folds can be felt through its wall and, since they are absent from the distal ileum, palpation allows a crude distinction between upper and lower intestinal levels. The jejunum lies largely in the umbilical region but may extend into surrounding areas. The first coil occupies a recess between the left part of the transverse mesocolon and the left kidney.

The ileum has a diameter of 3.5 cm; its wall is thinner than in the jejunum. A few circular folds occur proximally but these are small and disappear almost entirely in its distal part. Aggregated lymphatic follicles are, however, larger and more numerous than in the jejunum. The ileum is mainly in the hypogastric (pubic) and pelvic regions. Its terminal part usually lies in the pelvis, from which it ascends over the right psoas major and right iliac vessels to end in the right iliac fossa, opening into the medial side of the junction between the caecum and colon.

The fan-like mesenteric attachment of the jejunum and ileum to the posterior abdominal wall allows free movement, each coil adapting to changes in form and position.

The mesentery (p. 1344), like a complex fan, has a root, about 15 cm long, attached to the posterior abdominal wall along a line running diagonally from the left side of the second lumbar vertebral body to the right sacro-iliac joint, crossing successively: the horizontal part of the duodenum, aorta, inferior vena cava, right ureter and right psoas major (**8**.113). Its average breadth from its root to the intestinal border is about 20 cm, but is greater at intermediate levels. Its two peritoneal layers contain: the jejunum, ileum, jejunal and ileal branches of the superior mesenteric vessels, nerves, lacteals and lymph nodes, together with a variable amount of fat.

The ileal diverticulum (of Meckel) projects from the antimesenteric border of the distal ileum in about 3% of subjects, its average position being about 1 m above the ileocaecal valve and its average length about 5 cm. Its calibre is like that of the ileum, its blind extremity being free or connected with the abdominal wall or some other part of the intestine by a fibrous band. It represents the vitelline (yolk) duct's persistent proximal part (pp. 141, 233, 235); its mucosa is ileal in type but small areas may have a gastric structure, with oxyntic cells secreting acid. Sometimes heterotopic areas of pancreatic or other tissues occur in its wall. In a study of 1816 late fetal and neonatal cadavers Miyabara et al (1974) found a diverticulum in 61 individuals (3.4%). Of these, gastric mucosa was present in 11, jejunal mucosa in two, colonic mucosa in two and pancreatic tissue in one.

STRUCTURE OF THE SMALL INTESTINE

The intestinal wall has the usual serous, muscular, submucous and mucous layers (**8**.131). The serosa is visceral peritoneum with a subserous stratum of loose connective tissue. The *muscularis externa* is thicker in the proximal intestine, consisting of a thin external longitudinal and a thick internal circular layer of non-striated myocytes. The submucosa is loose connective tissue carrying blood vessels, lymphatics and nerves. The mucosa is thick and very vascular in the proximal small intestine, but thinner and less vascular in the distal. It forms *circular folds*

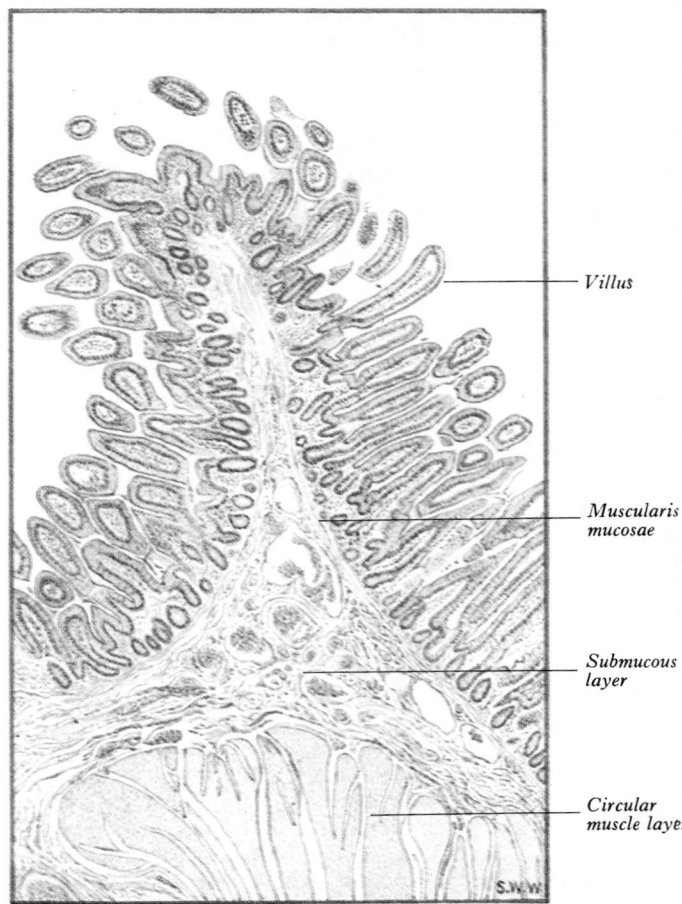

Villus

Muscularis mucosae

Submucous layer

Circular muscle layer

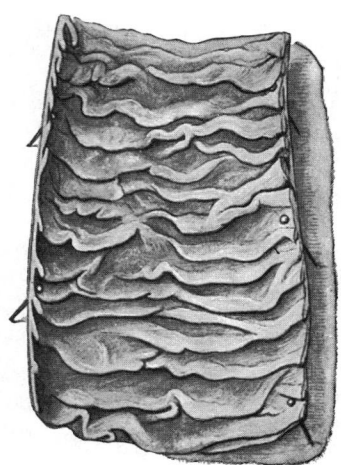

8.130A Internal aspect of a representative sample of the proximal jejunum, showing circular folds.

8.130B Section through a circular fold from the human small intestine. Stained with haematoxylin and eosin. Magnification × 19.

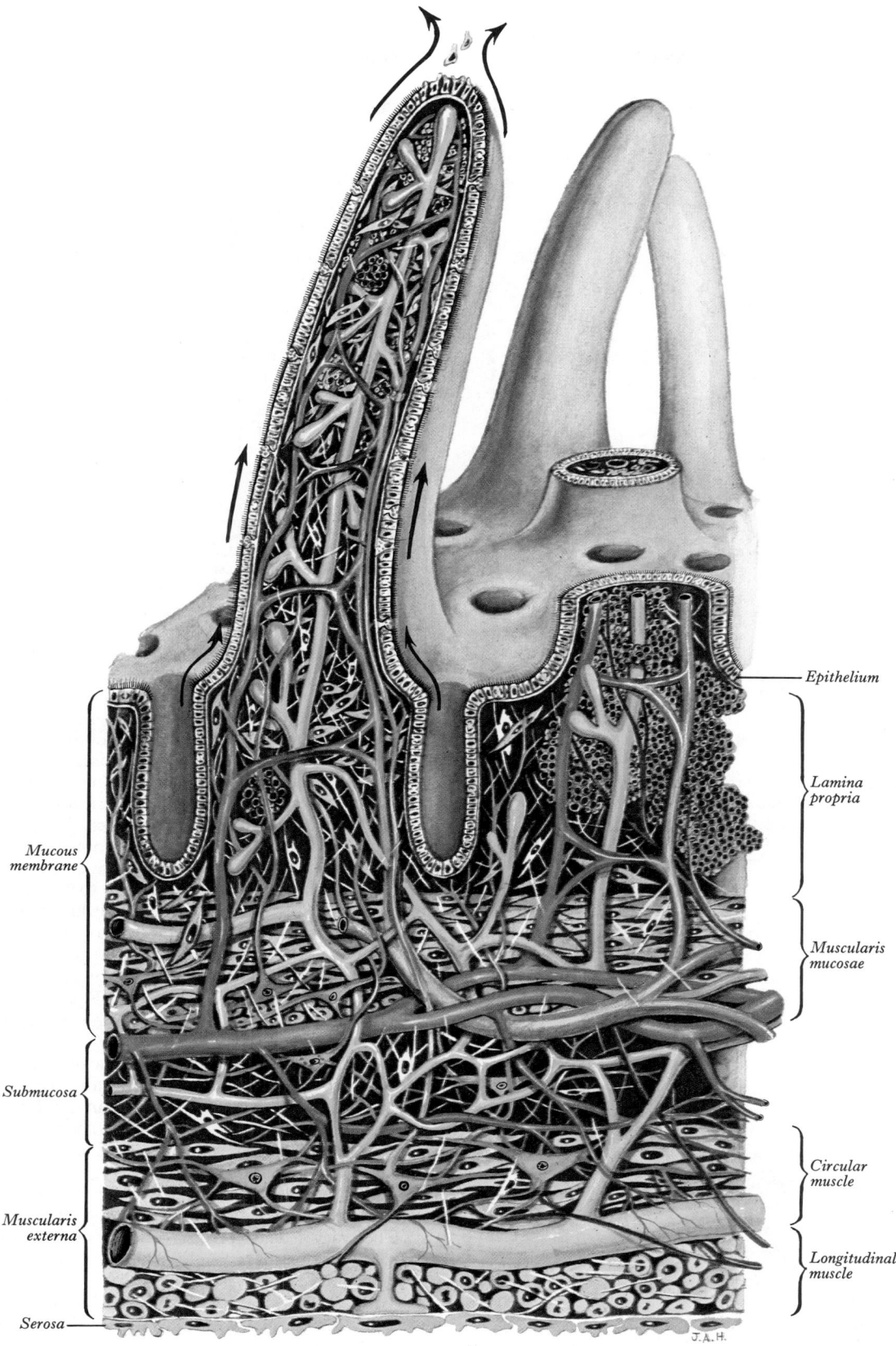

8.131 A three-dimensional reconstruction of the architecture of the intestinal villi and subjacent wall (the principal layers of the latter are indicated): arteries and arterioles (red), veins and venules (blue), central lacteals and other lymphatic channels (orange), aggregations of lymphocytes (yellow), neural elements (green), non-striated muscle fibres (magenta), fibroblasts (white). Note the orifices of the intestinal crypts (of Lieberkühn). Types of cells in the epithelium include absorptive cells, goblet cells and entero-endocrine cells. Arrows indicate the direction of cell migration. The various layers are not drawn to scale.

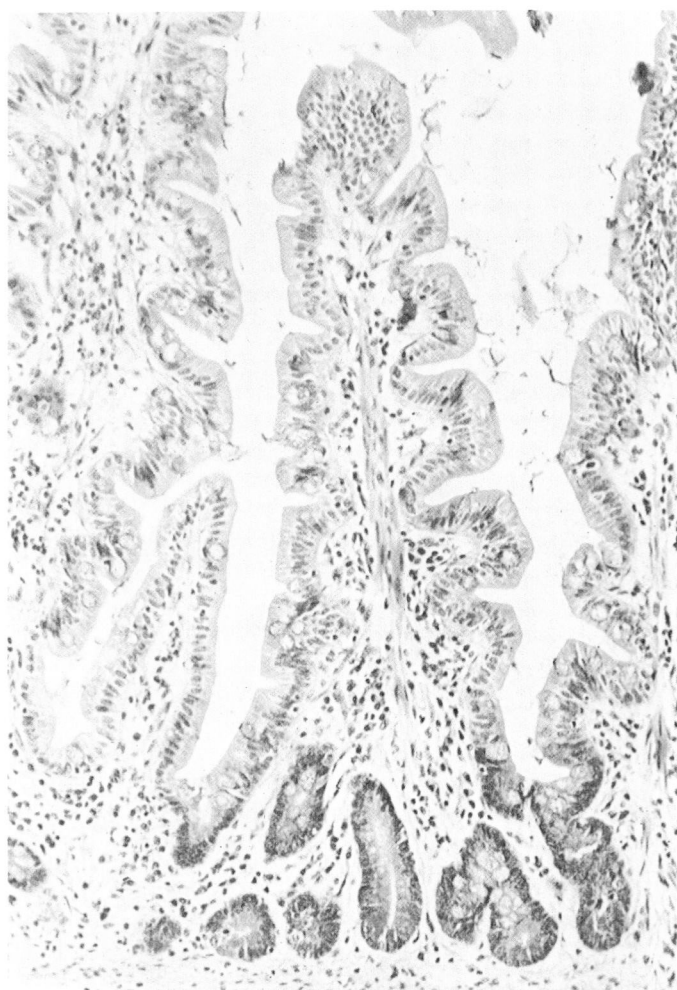

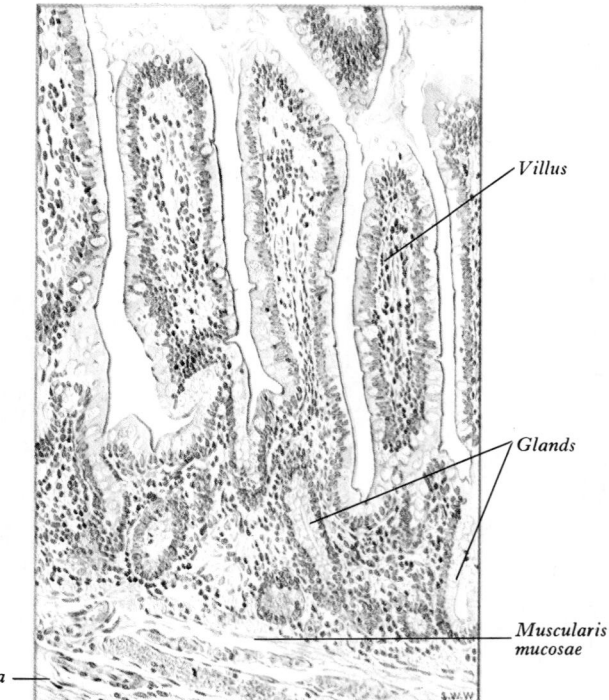

8.132C Intestinal glands and villi in the human small intestine. Stained with haematoxylin and eosin. Magnification × 120.

8.132A Light micrograph of part of the mucosa of the murine small intestine, showing a villus in longitudinal section. Non-striated myocytes (pink) can be seen in the lamina propria of the villus. Stained with haematoxylin, eosin, and periodic acid/Schiff. Prepared and photographed by Stephen Sitch, Department of Anatomy, Guy's Hospital Medical School, London. Magnification × 250.

(sometimes spiral) and the whole surface is covered by filiform or linguiform *intestinal villi*.

The circular folds (*plicae circulares* or 'valves' of *Kerkring*, 8.130, 134) are large, crescentic folds of mucosa which project into the intestinal lumen transverse to the long axis. Unlike gastric folds they are not obliterated by distension. Most extend round about one-half or two-thirds of the luminal circumference; some are complete circles, some bifurcate and join adjacent folds, some are spiral but extend little more than once round the lumen, though occasionally two or three times. Larger folds are about 8 mm deep at their broadest, but most are smaller. Larger and smaller folds often alternate. Plicae begin to appear about 2.5–5 cm beyond the pylorus. Distal to the major duodenal papilla they are large and close together, as they also are in the proximal half of the jejunum; but from here to midway along the ileum they diminish, disappearing almost wholly in distal ileum, hence the thinness of this part of the intestinal wall. The circular folds slow the passage of contents and increase the absorptive surface (8.134).

Intestinal villi are highly vascular processes just visible to the naked eye and project from the entire intestinal mucosa, giving it a velvety texture. Large and numerous in the duodenum and jejunum, they are smaller and fewer in the ileum. In the first part of the duodenum they are broad ridges, changing to tall foliate villi in the distal duodenum and proximal jejunum, beyond which they gradually shorten to a finger-like form in the distal jejunum and ileum (Verzár & McDougall 1936, McMinn & Mitchell 1954). They vary in density from 10–40 per square millimetre and from about 0.5–1.0 mm in height. They increase the surface area about eightfold.

The mucosa (8.131) has three layers. Next to the submucosa is the *muscularis mucosae*, with external longitudinal and internal circular layers of non-striated myocytes; it extends into the circular folds and follows their surface profile. Internal to it is connective tissue, the lamina propria, in which are fibroblasts and connective tissue fibres, lymphocytes, eosinophilic leucocytes, macrophages, mast cells, capillaries, lymphatic vessels and nonmyelinated nerve fibres. Plasma cells are numerous and lymphocytes in many regions are clustered in solitary and aggregated lymphatic follicles, some extending through the muscularis mucosae into the submucosa. Internal to the lamina propria a basement membrane supports an epithelium composed mainly of

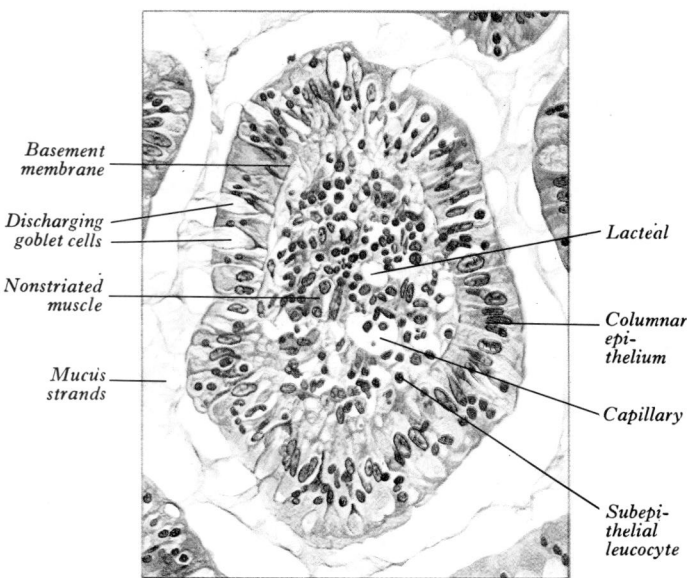

8.132B Transverse section through a villus in the human jejunum. Stained with haematoxylin and eosin. Magnification × 380.

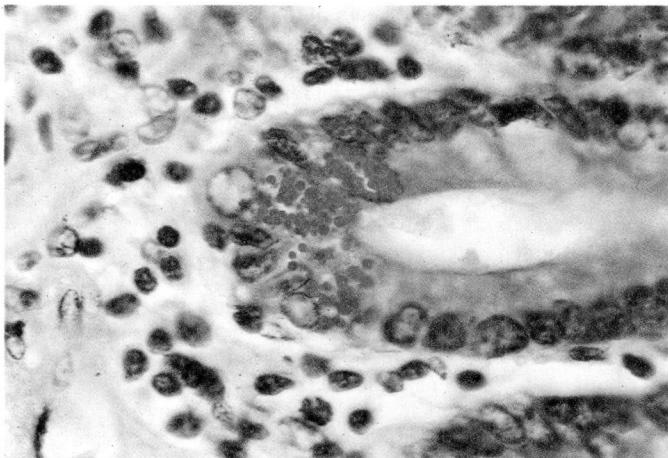

8.132D Part of a transverse section of the ileum, showing zymogenic (Paneth) cells containing orange-stained zymogen granules at the base of an intestinal gland. 'Undifferentiated' epithelial cells are also visible. Mallory's azan stain. Magnification × 400.

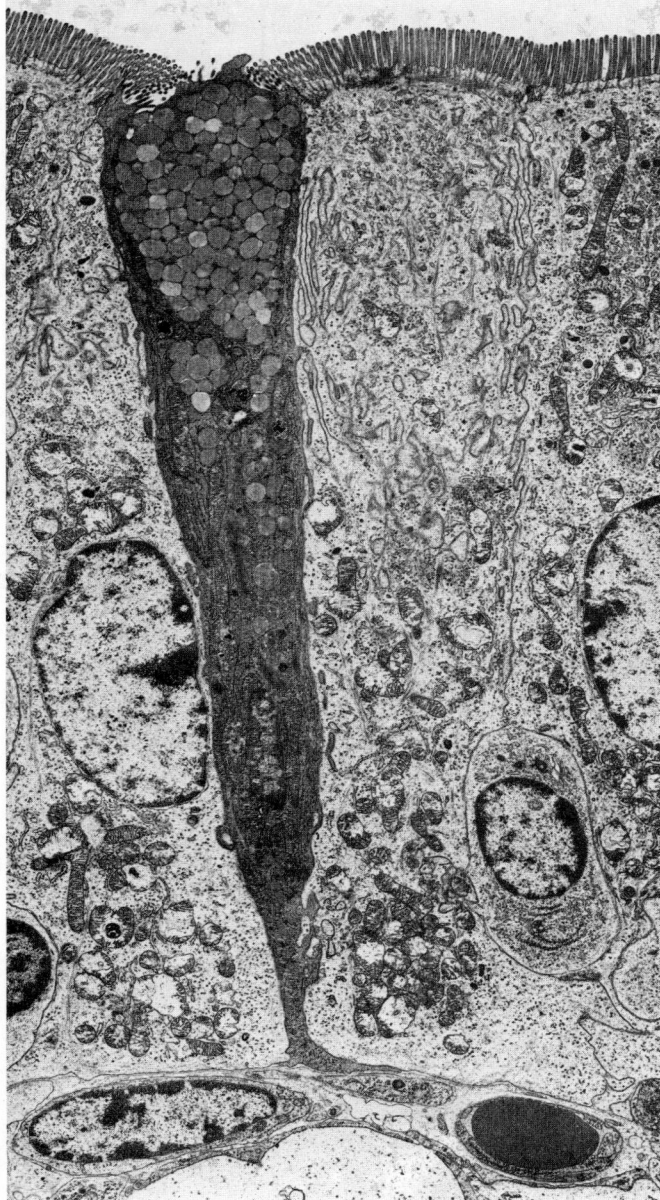

columnar cells, except over the lymphatic follicles where they are partly replaced by follicle-associated epithelial cells (*microfold* or *membrane* (M) cells, pp. 833, 1362), which may be specialized for transport of antigens (Owen & Nemamic 1978). Extending into the mucosa between the intestinal villi are simple, tubular, *intestinal glands (crypts of Lieberkühn)*, reaching almost to the muscularis mucosae. In the duodenum there are also branched tubulo-alveolar *duodenal mucous glands (of Brunner)*, their ducts extending through the muscularis mucosae to expand into the submucosa; their secretions contain mucosubstances, bicarbonate ions and a small amount of enzyme helping to activate trypsinogen from the pancreas.

Structure of Intestinal Villi

A villus has a core of reticular tissue containing a lymph vessel or lacteal, blood vessels, nerves and non-striated myocytes, covered by columnar epithelium on a basement membrane (8.131, 132A).

The lacteal, usually single but occasionally double, starts in a closed, dilated extremity near the villous summit and descends to empty into a lymphatic plexus in the lamina propria. Its wall is a single layer of endothelial cells.

Myocytes derived from the muscularis mucosae cluster around the lacteal from the base to the summit of the villus, some being attached to both the basement membrane of the epithelium and the lacteal. Contraction of these myocytes therefore 'milks' the lacteals.

Blood vessels form a capillary plexus in the lamina propria, enclosed in reticular tissue; capillaries are lined by fenestrated endothelium, probably to ensure the rapid uptake of nutrients diffusing from epithelium (Clementi & Palade 1969).

The epithelium of the villi is mainly of (1) absorptive columnar cells and (2) interspersed goblet cells.

(1) **The columnar cells, enterocytes** (1.4), are granular, with a clear oval basal nucleus. At the surface is a refractile, vertically striated zone, about 1 μm in depth, the *striated* or *brush border*, rich in alkaline phosphatase and concerned in active absorption; electron microscopy shows it to be composed of minute parallel

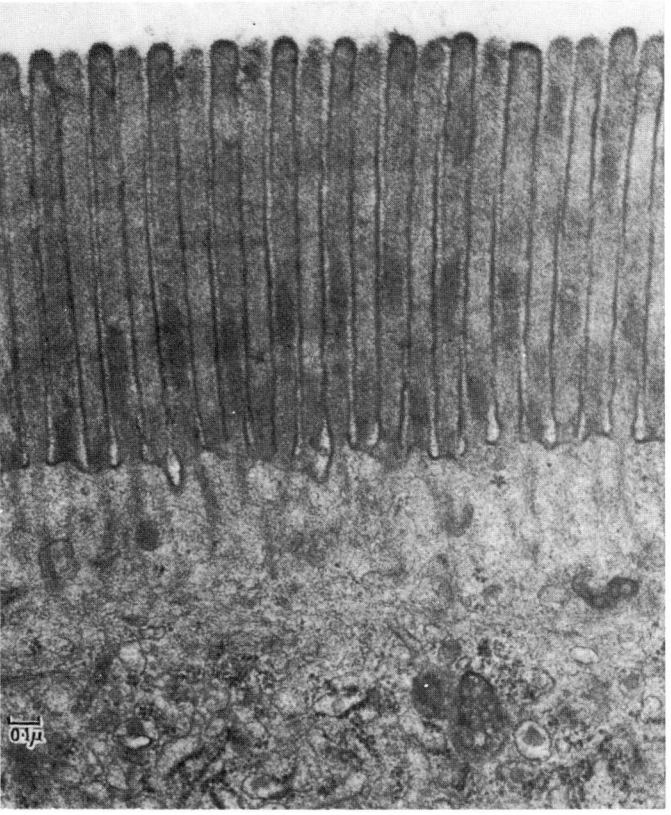

8.133A Transmission electron micrograph of the columnar epithelium lining the murine small intestine, showing a mucus-secreting goblet cell between two absorptive cells which bear microvilli. The cells rest on a delicate basal lamina deep to which is the vascular lamina propria. Prepared and photographed by Derrick J Lovell, Department of Anatomy, Guy's Hospital Medical School, London. Magnification × 4800.

8.133B Electron micrograph of the apical region in a columnar cell from the jejunum (rat) showing the regular series of microvilli which constitutes the striated border of light microscopy. Filaments can be seen passing from the microvilli to the terminal web. Magnification *c.* × 32 500.

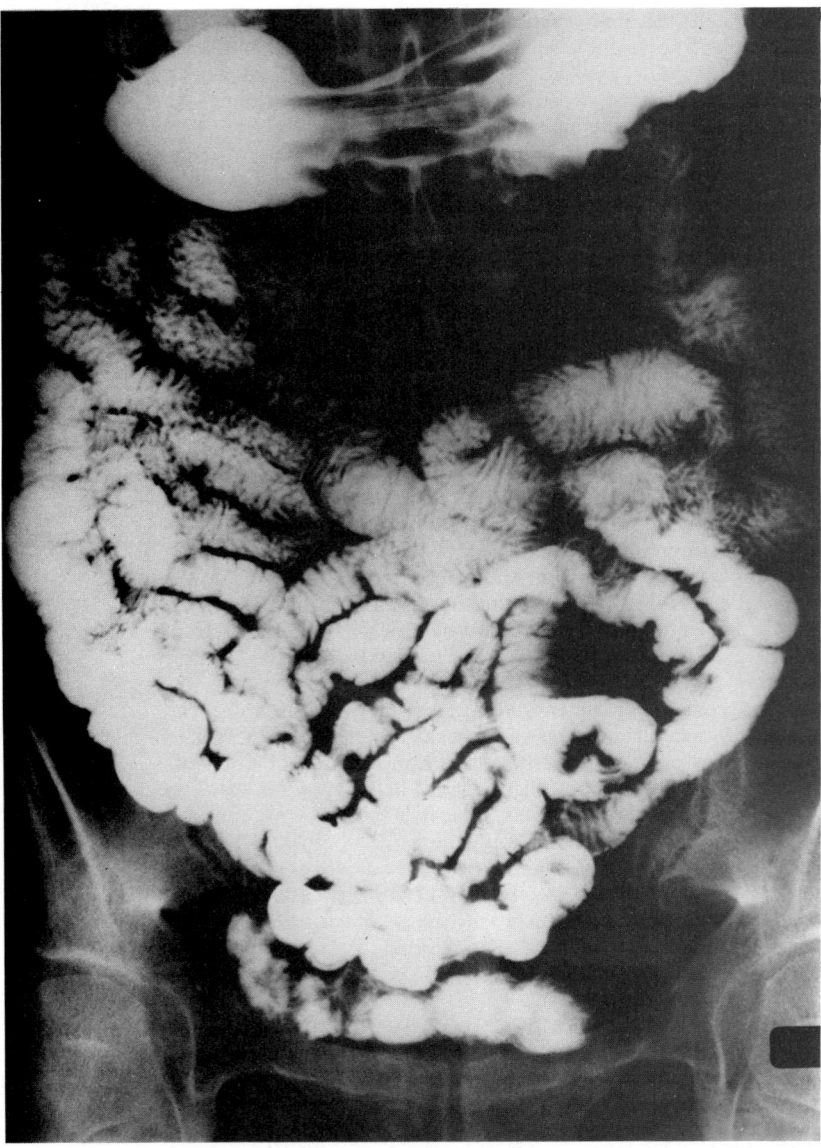

8.134A Radiograph showing the small intestine, taken during a barium follow-through. The feathery appearance of the profile of the small intestine is due to the plicae circulares; constrictions due to peristalsis can also be seen. Provided by Shaun Gallagher; photography by Sarah Smith, UMDS, Guy's Hospital Campus, London.

cylindrical microvilli, about 1 μm long and 0.1 μm broad. External to the villar plasma membrane, electron microscopy show a thick glycocalyx (p. 20) composed of fine filaments, rich in mucosubstances. In the cytoplasmic core of each microvillus are fine filaments basally continuous with a plexiform band of similar filaments in the cell's apical cytoplasm, the *terminal web*, containing occasional vesicles (**8.**133B). Lateral plasma membranes of columnar cells are often interdigitated; they have typical juxtaluminar junctional complexes (p. 24, **1.**13), and scattered desmosomes; basally they are separated from adjacent cells by intercellular canaliculi. The basal plasma membranes are usually smooth and adjoin a basement membrane about 20 nm wide. Rod-shaped mitochondria are scattered in the cytoplasm with some granular and agranular endoplasmic reticulum. The Golgi apparatus is supranuclear and the apical region, beneath the terminal web, contains numerous lysosomes. At the summits of intestinal villi these cells are often in the process of disintegrating, their cytoplasm darkly staining and their microvilli stunted and degenerate. Changes in the microvillous surface area have also been shown to occur during reproduction (in female rats) and during ageing (Penzes & Regius 1985).

The roles of these cells in digestion and absorption have attracted much attention. The absorption of amino acids and simple

carbohydrates is probably facilitated by diffusion across cell membranes, materials traversing the cells to subjacent capillary arrays in the lamina propria. Lipid absorption appears to be by diffusion of small molecules (fatty acids, etc.) through the luminal membrane, the lipid accumulating in vacuoles in the apical cytoplasm, before reaching the subjacent lymphatics. Examination of striated borders isolated by fractionation and centrifugation has shown that enzymes such as disaccharidases are bound to their surfaces, where much intestinal digestion may occur, perhaps in the cell coat in close proximity to the site of absorption.

(2) **Goblet cells** have elongated, basal nuclei and an apical region distended with membrane-bound mucin granules. The golgi apparatus is well developed and supranuclear in position; the perinuclear granular endoplasmic reticulum is abundant. Microvilli are less frequent and irregular and their terminal web is rudimentary. The mucin is said to be discharged by fusion of the granules with the plasma membrane in a merocrine manner (**8.**1A, 133A).

Intestinal glands are numerous throughout the intestinal mucosa. They are tubular, perpendicular pits, opening at small circular apertures and visible with a simple lens as minute dots between the villi. Their thin walls consist of columnar epithelium on a basement membrane, associated with which is a rich capillary plexus (**8.**131). The epithelium consists of (a) undifferentiated stem cells, (b) zymogenic (Paneth) and (c) argentaffin cells.

(a) **Undifferentiated stem cells** are the most numerous and proliferate by mitotic division, their progeny ascending out of the intestinal glands along the sides of the villi, where they differentiate into columnar or goblet cells; eventually they reach the apices of villi and are shed. Thus the villous epithelium is continually renewed. The apical surfaces of these cells, when not dividing, have fewer and more irregular microvilli than do the columnar cells, with occasional pseudopodia. Their lateral plasma membranes are smooth, but intercellular junctions are similar in both types of cell. Their nuclei are basal and the terminal webs poorly developed. Membrane-bound secretory granules are believed to be discharged by both apocrine and merocrine methods. These cells multiply at the rate of 1 cell per 100 per hour, one of the most rapid proliferation rates in the body (Lipkin et al 1963, MacDonald et al 1964).

(b) **Zymogenic cells** (of Paneth) are numerous in the deeper parts of the intestinal crypts, particularly in the duodenum. They are rich in zinc and contain acidophilic granules (**8.**132D) staining with e.g. phosphotungstic haematoxylin. Electron microscopy shows irregular apical microvilli and prominent membrane-bound vacuoles containing a granular matrix with crystalline inclusions in the supranuclear cytoplasm. These vacuoles are PAS positive and they contain some carbohydrate. Scattered mitochondria, lysosomes and much granular endoplasmic reticulum are present, especially in the basal region. The functions of zymogenic cells are not certain, but there is evidence that they secrete lysozyme, an antibacterial substance. For further details, see e.g. Rodning et al (1982).

(c) **Argentaffin or enterochromaffin cells** are present among the cells lining the intestinal glands, and less commonly over the villi. They are pyramidal or columnar cells containing cytoplasmic granules which stain black with silver salts (hence *argentaffin*) and brown with chromium salts (hence *enterochromaffin*). They have been classified as one or more types of APUD cell (Pearse 1980, see also pp. 1376, 1379). Distributions of these and other cells of the gastro-enteroendocrine system are shown in **8.**146C.

Duodenal glands (of Brunner), limited to the duodenum (**8.**129), are sited *in the submucous connective tissue*, i.e. their ducts traverse the muscularis mucosae. Largest and most numerous near the pylorus, and forming an almost complete layer in the superior part and proximal half of the descending duodenum, they gradually diminish and disappear at the duodenojejunal junction. They are small, compound and acinotubular, each having several alveoli lined by short columnar epithelium and apparently containing a single type of exocrine cell in humans, a typical mucous element. Nuclei of these cells, small and basal, vary during the secretory cycle. The golgi apparatus is extensive and mucin droplets numerous. Many argentaffin cells (vide supra) are present among the mucinogenic cells. These glands secrete a watery fluid rich in bicarbonate, which helps to neutralize the acid secretions of the stomach as the food enters the duodenum. The cells may also secrete a trypsinogen-activating factor which converts this enzyme to trypsin after secretion from the pancreas.

Solitary lymphatic follicles, scattered throughout the intestinal mucosa, are most numerous in the distal ileum, their luminal surfaces being covered by rudimentary villi, except at their summits. Each follicle, surrounded by the openings of intestinal glands, consists of dense reticular tissue packed with lymphocytes and a dense capillary network. Reticular spaces are continuous with larger lymph spaces around the follicle, communicating thus with the lacteals. They are partly submucosal, partly mucosal and have 'M' cells (p. 1361) in their surface epithelium.

Aggregated lymphoid follicles (Peyer's patches, 8.135A) are circular or oval, each containing 10–260 follicles, and varying in length from 2 to 10 cm. Like other lymphoid tissue (except lymph nodes) solitary and aggregated lymphoid follicles are most prominent around puberty, when they may number up to 300, thereafter diminishing in number and size although many persist into old age (Cornes 1965). Aggregated follicles, largest and most numerous in the ileum, are small, circular and few in the distal jejunum and only occasional in the duodenum. They are usually antimesenteric in site and arrayed axially. Each is a group of solitary follicles covered by columnar epithelial and 'M' cells (p. 1361); villi are usually absent. An abundant vascular plexus surrounds each follicle from which fine branches permeate the lymphoid tissue. Efferent lymphatic plexuses are especially abundant. These and similar masses of unencapsulated lymphoid tissue provide B- and T-lymphocytes for the defence of the alimentary tract and form part of the system known as *gut-associated lymphoid tissue* (see p. 832). In typhoid fever follicles may ulcerate, such ulcers being oval, their long axes along the gut; hence subsequent fibrosis does not constrict the intestine.

Vessels and nerves. Jejunal and ileal *arteries* (8.136A,B) stem from the superior mesenteric, branches of which, reaching the mesenteric border, extend between the serosal and muscular layers. From these, numerous rami traverse the muscle, supplying it and forming an intricate submucosal plexus from which minute vessels pass to glands and villi (see p.1361). Anastomoses between the terminal intestinal arteries are few and alternate vessels are often distributed to opposite sides of the gut. The *veins* follow the arteries. (For a detailed investigation of the distribution and variations in coeliac and superior mesenteric arteries consult Nesebar et al 1969.) The *lymph vessels* (lacteals) are arranged at two levels, mucosal and muscular. Lymph vessels of villi commence, as described on p. 1361, form an intricate plexus in mucosa and submucosa, are joined by vessels from lymph spaces at the bases of solitary follicles and drain to larger vessels at the mesenteric aspect of the gut. Lymph vessels of the muscular tunic form a close plexus running mostly between the muscle layers; they communicate everywhere with mucosal vessels and open like them into the lacteal drainage at the attached border of the gut.

Innervation is by the vagi and thoracic splanchnic nerves through the coeliac ganglia and superior mesenteric plexuses. Fibres pass to the *myenteric plexus* (p. 1331) of nerves and ganglia between the circular and longitudinal layers of the muscularis externa, which they supply. From this a secondary, *submucous plexus* is derived, formed by branches perforating the circular muscular layer; it also contains ganglionic neurons from which fibres pass to the muscularis mucosae and the rest of the mucosa. Nerve bundles in the submucous plexus are finer. Ganglion cells in both plexuses are essentially parasympathetic (vagal). An old controversy on the source of postganglionic neurons in the enteric ganglia was renewed by Andrew (1971); endodermal and mesodermal origins have been suggested, but the evidence (though largely equivocal) indicates the neural crest as their source. In general the sympathetic system inhibits peristalsis but stimulates the sphincters and muscularis mucosae. The parasympathetic generally augments peristalsis and inhibits the sphincters, the results of parasympathetic stimulation depending on the state of contraction or relaxation of the organ at the time of stimulation. The parasympathetic also augments intestinal

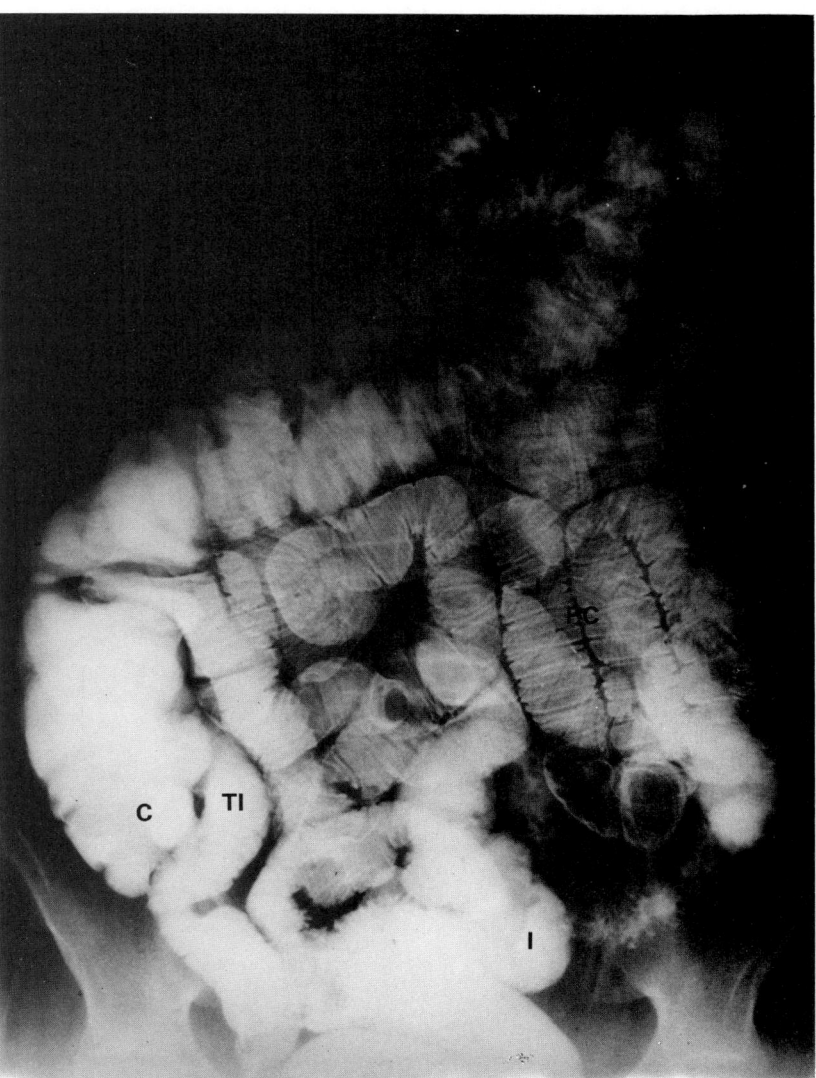

8.134B Radiograph showing part of the small and large intestine, taken after the administration of a small bowel enema which outlines the intestine, followed by methyl cellulose which distends it and produces a double contrast image. The plicae circulares are clearly demonstrated by this technique. C = caecum; I = ileum; J = jejunum; PC = plicae circulares; TI = terminal part of ileum. Supplied by Shaun Gallagher; photography by Sarah Smith, UMDS, Guy's Hospital Campus, London.

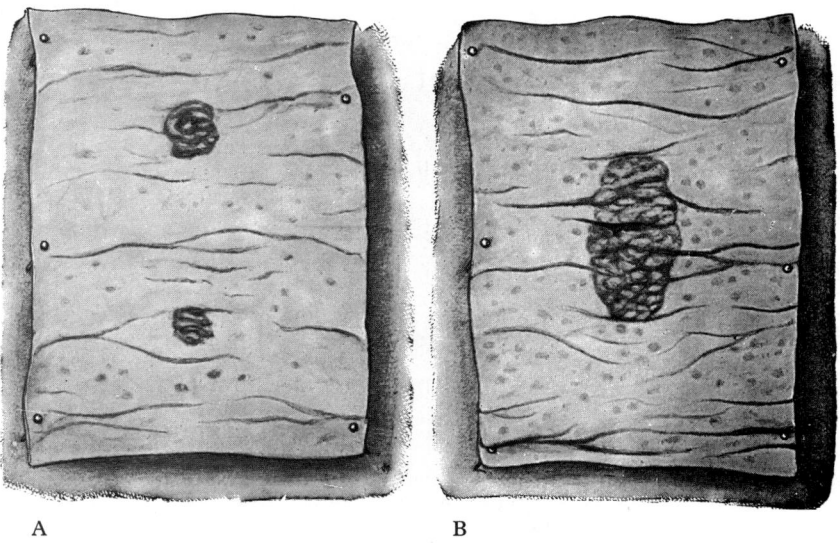

A   B

8.135 Aggregated lymphatic follicles in the proximal (A) and distal (B) parts of the ileum.

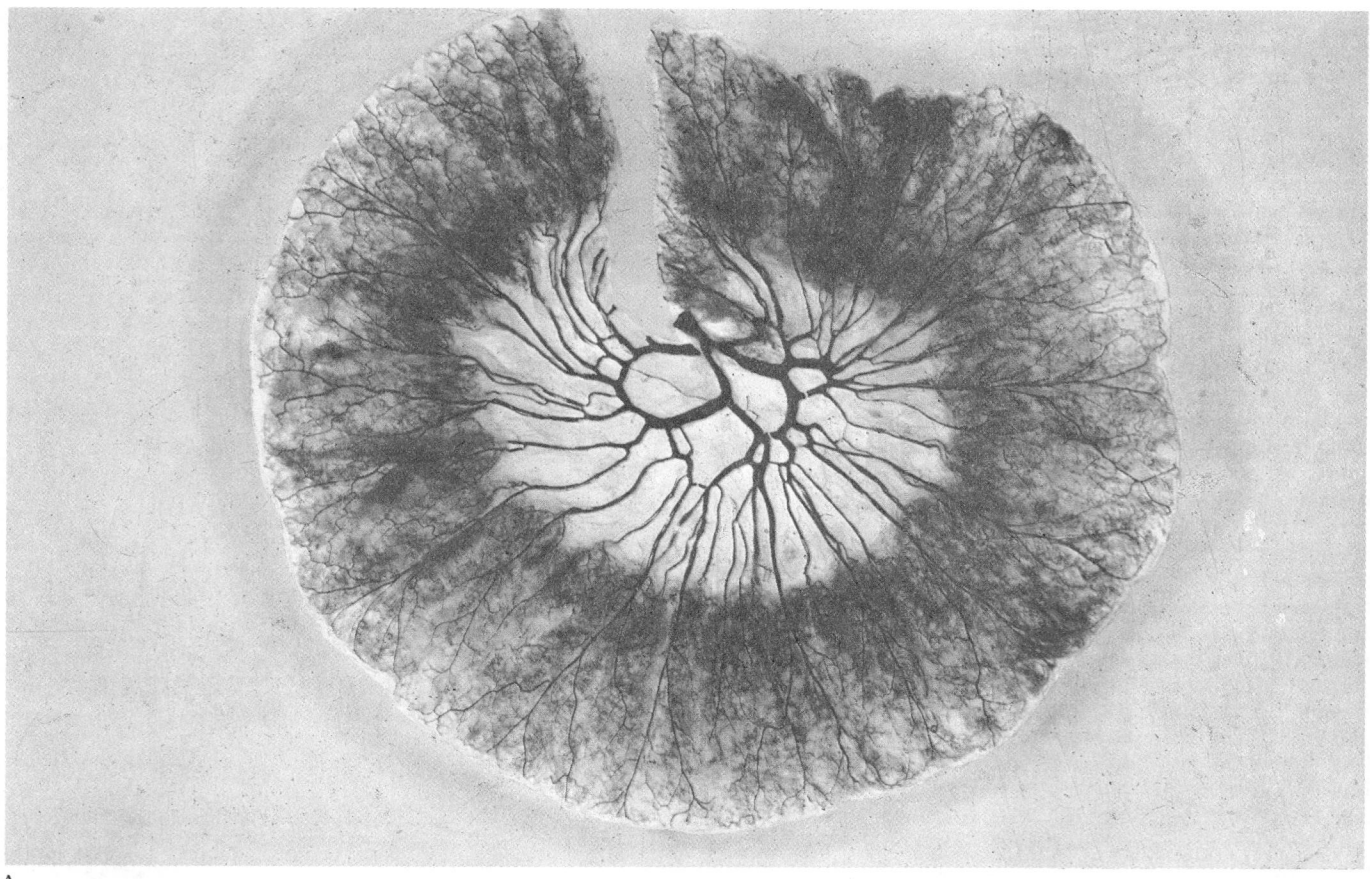

A

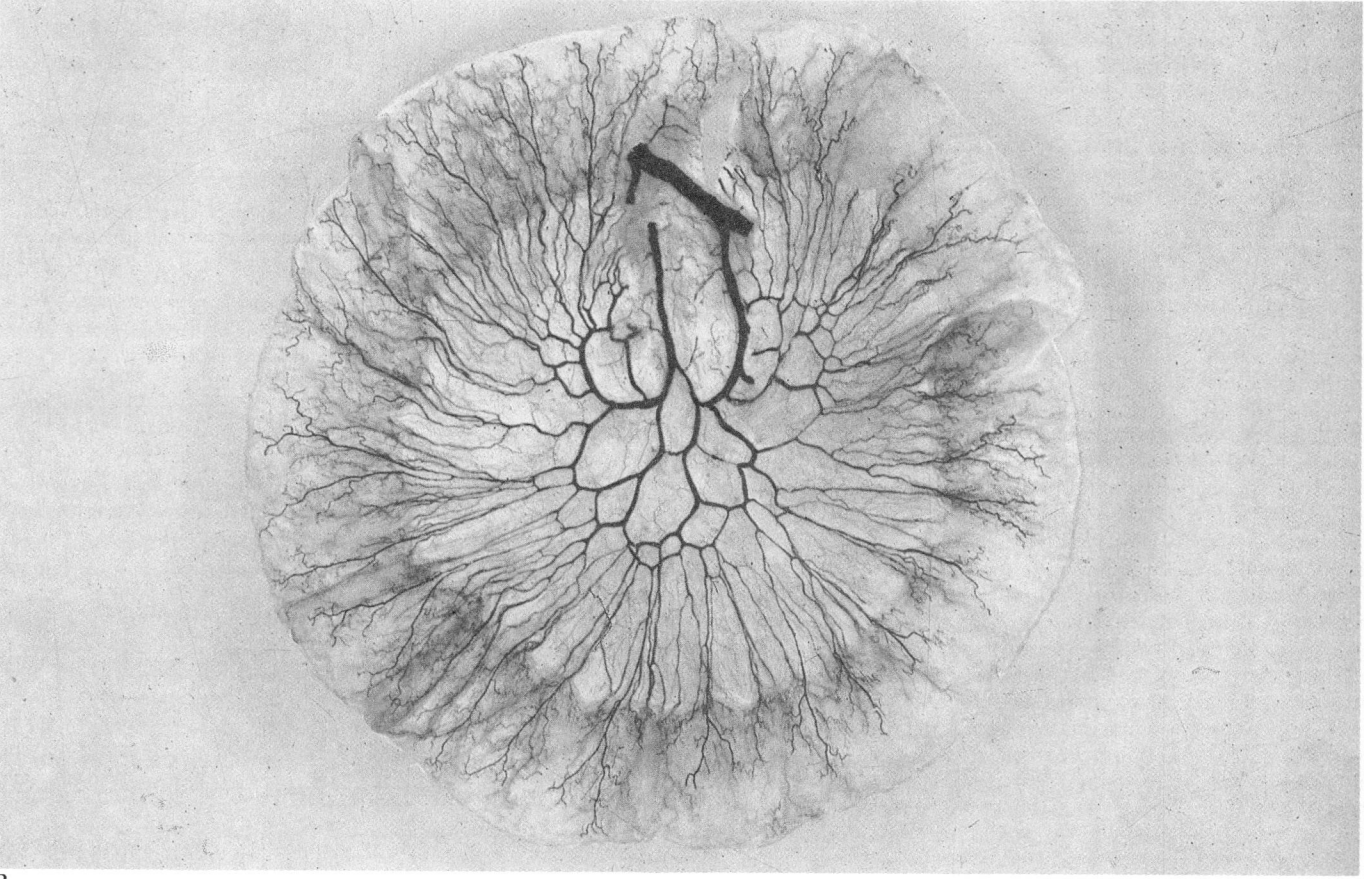

B

8.136 Specimens of the jejunum (A) and ileum (B) from a subject in whom the superior mesenteric artery was injected with a red coloured mass of gelatin before fixation. Subsequently the specimens were dehydrated and then cleared in benzene followed by methyl salicylate. The largest vessels present are the jejunal and ileal branches of the superior mesenteric artery and these are succeeded by anastomotic arterial arcades, which are rela-

tively few in number (1–3) in the jejunum, becoming more numerous (5–6) in the ileum. From the arcades, straight arteries pass towards the gut wall; frequently, successive straight arteries are distributed to opposite sides of the gut. Note the denser vascularity of the jejunal wall. Specimens prepared by Michael C E Hutchinson and photographed by K Fitzpatrick, Department of Anatomy, Guy's Hospital Medical School, London.

secretion. For evaluation of the 'interstitial' cells (of Cajal) in the intestinal wall, consult Rogers & Burnstock (1966) and for a comparative study see Faussone-Pellegrini (1987). It seems probable that these are connective tissue cells and not neurons. Evidence suggests that some neurons in the intestinal ganglia may be afferent (Féher & Vajda 1974).

The Large Intestine

The large intestine, extending from the distal end of the ileum to the anus, is about 1.5 m long; its calibre is greatest near the caecum and gradually diminishes to the rectum, where it enlarges just above the anal canal. Its function being chiefly absorption of fluid and solutes, it differs in structure, size and arrangement from the small intestine. (1) It has a greater calibre. (2) It is for the most part more fixed in position. (3) Its longitudinal muscle, though a *complete* layer, is concentrated into three longitudinal *taeniae coli*. (4) The colonic wall is puckered into *sacculations (haustrations)* by the taeniae (so it is said) but sacculation is probably not thus fully explained (p. 1374; see also Hamilton 1946, Pace 1968). (5) Small adipose projections, *appendices epiploicae*, are scattered over the free surface of the whole colon, but are absent from the caecum, vermiform appendix and rectum.

The large intestine curves around the coils of the small intestine, commencing in the right iliac region as a dilated *caecum (intestinum crassum caecum)*. (The term *caecum*, like rectum, duodenum, etc. is an adjective, used by linguistic abbreviation as a noun.) The caecum leads to the *vermiform appendix* and *colon*, the latter ascending in the right lumbar and hypochondriac regions to the inferior aspect of the liver; here it bends (*right colic flexure*, 8.125) to the left and, with an antero-inferior convexity, loops across the abdomen as the *transverse colon* to the left hypochondriac region, where it curves again (*left colic flexure*, 8.125) to descend through the left lumbar and iliac regions to the lesser pelvis. Here it forms a sinuous loop, the *sigmoid colon* (8.142), continuing along the lower posterior pelvic wall as the *rectum* and *anal canal*.

THE CAECUM

The caecum (8.137A,B) lies in the right iliac fossa; its surface projection occupies the triangular area between the right lateral and transtubercular planes and the inguinal ligament. It is a large cul-de-sac continuous with the ascending colon at the level of the ileal opening on the medial side and below this with the vermiform appendix. Its average axial length is about 6 cm and its breadth about 7.5 cm. It is superior to the lateral half of the inguinal ligament, resting posteriorly on the right iliacus (with the lateral cutaneous nerve of the thigh interposed) and psoas major, separated from both by covering fasciae and peritoneum. Posterior to it is the *retrocaecal recess* (p. 1345) frequently containing the vermiform appendix. Anteriorly, it usually contacts the anterior abdominal wall, but the greater omentum and, when it is empty, some coils of the small intestine may intervene. Usually it is entirely covered by peritoneum, but sometimes incompletely, when the upper part of the posterior surface is sessile and connected to the iliac fascia by areolar tissue. Commonly, however, the caecum is mobile, and may even herniate through the right inguinal canal. It can also usually be delivered through an appropriate incision in the anterior abdominal wall at appendicectomy.

Caecal Variations

The caecum has been classified into four types (Treves 1885). In early fetal life it is short, conical and broad at the base, with an apex turned superomedially towards the ileocaecal junction. As the fetus grows, the caecum increases more in length than breadth, to form a longer tube with a narrower base but retaining the same inclination. Distal growth later ceases, but the proximal part continues to grow in breadth, so that at birth a narrow vermiform appendix extends from the apex of a conical caecum. This *infantile form* persists throughout life in about 2%, regarded by Treves as the *first* type; the three taeniae coli (p. 1374) start from the appendix and are equidistant from each other. In the *second*

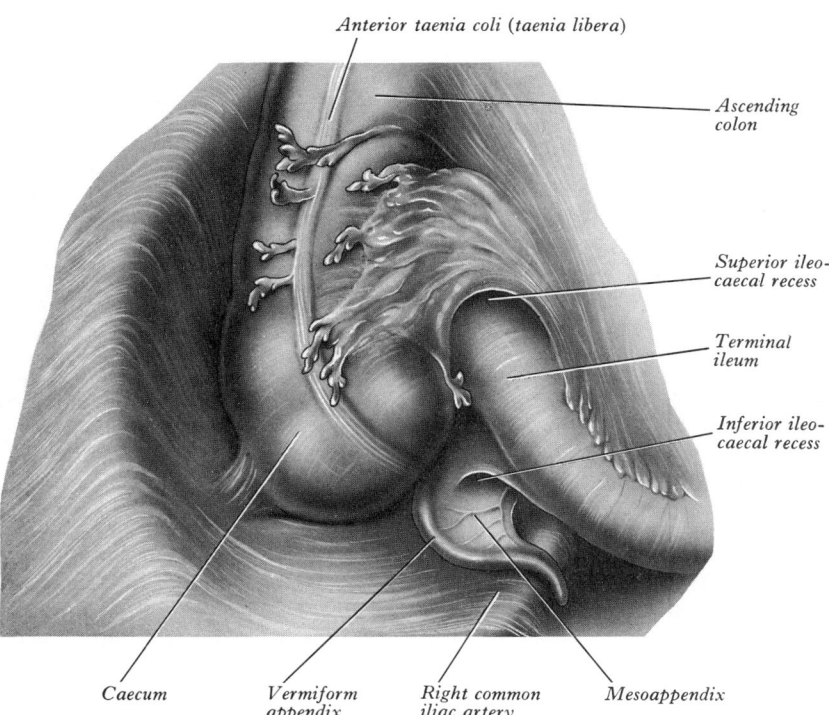

Anterior taenia coli (taenia libera)

Ascending colon

Superior ileo-caecal recess

Terminal ileum

Inferior ileo-caecal recess

Caecum *Vermiform appendix* *Right common iliac artery* *Mesoappendix*

8.137A The terminal ileum, caecum and vermiform appendix: anterior aspect.

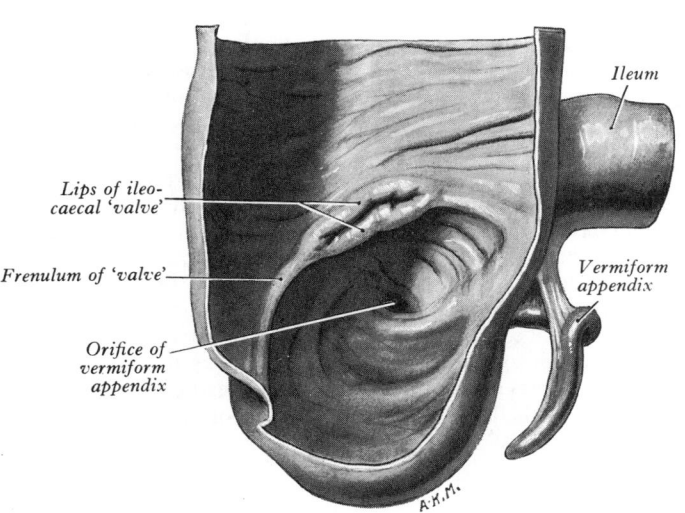

Ileum

Lips of ileo-caecal 'valve'

Vermiform appendix

Frenulum of 'valve'

Orifice of vermiform appendix

8.137B The interior of the caecum and commencement of the ascending colon, showing the ileocaecal 'valve'. See text for discussion.

type, the caecum becomes quadrate by outgrowth of a saccule on each side of the anterior taenia; these saccules are of equal size and the appendix arises from the depression between them instead of from the apex of a cone. This type occurs in about 3%. In the *third type* (normal in humans) the two saccules grow at unequal rates, the right more rapidly, forming a new 'apex'; the original apex, with the appendix attached, is pushed towards the ileocaecal junction; the taeniae still start from the appendicular base but are not equidistant, the growth of the right saccule pushing between the anterior and posterolateral taeniae. This type occurs in about 90%. The *fourth type* is merely an exaggeration of the third, the right saccule growing still further and the left atrophying so that the original caecal apex and appendix are near the ileocaecal junction, the anterior taenia also turning medially to it. This type occurs in about 4%. In a more recent study (Pavlov & Pétrov 1968) of 82 males and 44 females (adolescent and adult), the third type was designated *ampullary*, accounting for 78%. An *infundibular* type, approximating to the infantile conical category, occurred in 13%; 9% were intermediate. The caecum was mobile 20% more often in females. (For further analyses consult Balthazar & Gade 1976.)

THE ILEOCAECAL VALVE

The ileum opens into the posteromedial aspect of the large intestine, at the junction of the caecum and colon (8.137B). A surface marking of this structure is the intersection of the right lateral and transtubercular planes; about 2 cm below this the vermiform appendix opens into the caecum. The ileocaecal orifice has a so-called 'valve', consisting of two flaps projecting into the lumen of the large intestine (Decarvalho et al 1987). In the distended, fixed caecum the flaps are semilunar. The upper, approximately horizontal, is attached to the junction of the ileum and colon, the lower, longer and more concave, to the junction of the ileum and caecum. At their ends the flaps coalesce, continuing as narrow membranous ridges, the *frenula* of the valve. The anterior or left end of the aperture is rounded, the right or posterior is narrow and pointed. In the natural state the valvular lips project as thick folds into the caecal lumen, the orifice appearing like a slit or oval. Circular and longitudinal muscle layers of the terminal ileum continue into the valve to form a sphincter. However, direct observation of the living ileocaecal 'valve' does not corroborate this description (Rosenberg & DiDio 1969); in nine cases, studied by caecostomy, the ileal projection was papillary in shape. Radiological evidence also contradicts the concept of an effective ileocaecal valve at this junction.

Accumulations of circular fibres, sometimes described as sphincters, have been observed at various levels in all parts of the colon (DiDio & Anderson 1968, Rosenberg & DiDio 1969). The functional reality of most of these remains doubtful. Such sphincteric mechanisms must, of course, be balanced by antagonistic, dilatatory actions.

The margin of the ileocaecal valve is a reduplication of the intestinal mucosa and *circular* muscle; longitudinal fibres are partly reduplicated as they enter the valve (Jit & Singh 1956), but the more superficial and the peritoneum continue uninterruptedly from the small to the large intestine. The ileal valvular surfaces are covered with villi and have the structure of the mucosa of the small intestine; their caecal aspects display no villi but numerous orifices of tubular glands peculiar to the colonic mucosa. It is usually said that the valve not only prevents reflux from the caecum to the ileum but is probably also a sphincter regulating the passage of ileal contents into the caecum; the valve is kept in tonic contraction by sympathetic innervation. Entry of food into the stomach initiates contraction of the small intestine, expelling ileal contents into the large intestine (the gastro-ileal reflex).

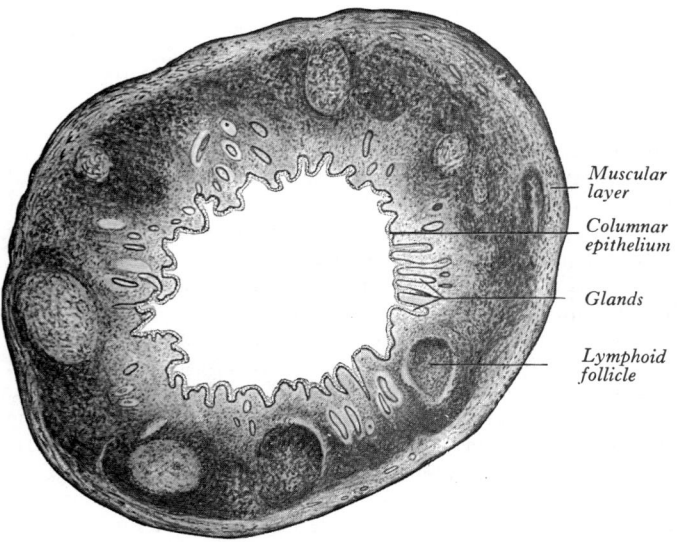

Muscular layer

Columnar epithelium

Glands

Lymphoid follicle

8.138 Transverse section of human vermiform appendix. Magnification × 20.

THE VERMIFORM APPENDIX

The vermiform appendix (8.137A,B) is a narrow, vermian (worm-shaped) tube, arising from the posteromedial caecal wall, 2 cm or less below the end of the ileum. It may occupy one of several positions: (1) behind the caecum and lower ascending colon *(retrocaecal and retrocolic)*; (2) dependent over the pelvic brim *(pelvic* or *descending)*, in females in close relation to the right uterine tube and ovary; (3) lying below the caecum *(subcaecal)*; (4) in front of the terminal ileum when it may be in contact with the anterior abdominal wall; or (5) behind the terminal ileum. In 10 000 subjects (Wakeley 1933) the vermiform appendix was retrocaecal and retrocolic (65.28%), pelvic (31.01%), subcaecal (2.26%), pre-ileal (1.0%) and post-ileal (0.4%). Subsequent literature, anatomical and surgical, shows much contradiction of this classic study. Buschard & Kjaeldgaard (1973), reporting a short series (234 autopsies), compared the results of several studies dating from 1885–1973, Wakeley's remaining by far the largest. They classified all positions as either *Anterior* (pelvic and ileocaecal) or *Posterior* (retrocaecal and subcaecal). All but three of 11 series quoted found *anterior* positions more frequent. Like Wakeley they observed *posterior* positions more commonly in their own Danish series; in German autopsies the finding was reversed. Collins (1932), in the second largest series (4680), returned percentages the reverse of Wakeley's, the ratio of anterior to posterior being 78.5% to 21.5% (Collins) and 32.4% to 67.6% (Wakeley). In view of these disagreements, such figures are of dubious value. Perhaps observers have used differing criteria or possibly there are demographic variations. For the present, however, such percentages remain unreliable.

The usual *surface marking* for the appendicular base is the junction of the lateral and middle thirds of the line joining the right anterior superior iliac spine to the umbilicus *(McBurney's point)*; but this is merely a useful surgical approximation, with considerable variation. The three taeniae coli on the ascending colon and caecum converge on the base of the appendix, merging into its longitudinal muscle. The anterior caecal taenia is usually distinct and traceable to the appendix, affording a guide to it. The appendix varies from 2–20 cm in length, the average being about 9 cm. It is longer in children and may atrophy or diminish after mid-adult life. It is connected by a short *mesoappendix* to the lower part of the ileal mesentery. This fold is usually triangular, extending almost to the appendicular tip along the whole tube.

The main *appendicular artery*, a branch from the lower division of the ileocolic (p. 773), runs behind the terminal ileum to enter the mesoappendix a short distance from the appendicular base. Here it gives off a recurrent branch which anastomoses at the base of the appendix with a branch of the posterior caecal artery, the anastomosis sometimes being large. The main appendicular artery approaches the tip of the organ, at first near to and then in the edge of the mesoappendix. The terminal part of the artery, however, lies on the wall of the appendix and may be thrombosed in appendicitis, resulting in distal gangrene or necrosis. The arterial supply of the appendix may vary considerably. Accessory arteries are common; in 80% of subjects there are two or more arteries of supply (Solanke 1968).

The canal of the appendix is small and opens into the caecum by an orifice lying below and a little behind the ileocaecal opening. The orifice is sometimes guarded by a semilunar mucosal fold forming a valve. The lumen may be partially or wholly obliterated in the later decades of live. In view of its rich vascularity and histological differentiation, the appendix is probably a specialized rather than a degenerate or vestigial structure. The caecum and appendix in man and anthropoid apes is considered to be less primitive than in monkeys. A comparative study of the primate vermiform appendix has been made by Scott (1980).

Structure. The mural layers are essentially as in the rest of the large intestine. The *serosa* is a complete investment, except along the mesenteric attachment; there is a subserous layer of connective tissue. The *longitudinal muscular fibres* form a complete, uniformly thick layer, except over a few small areas where both muscular layers are deficient, leaving the serosa and submucosa in contact. At the base the longitudinal muscle thickens to form rudimentary taeniae continuous with those of the caecum and

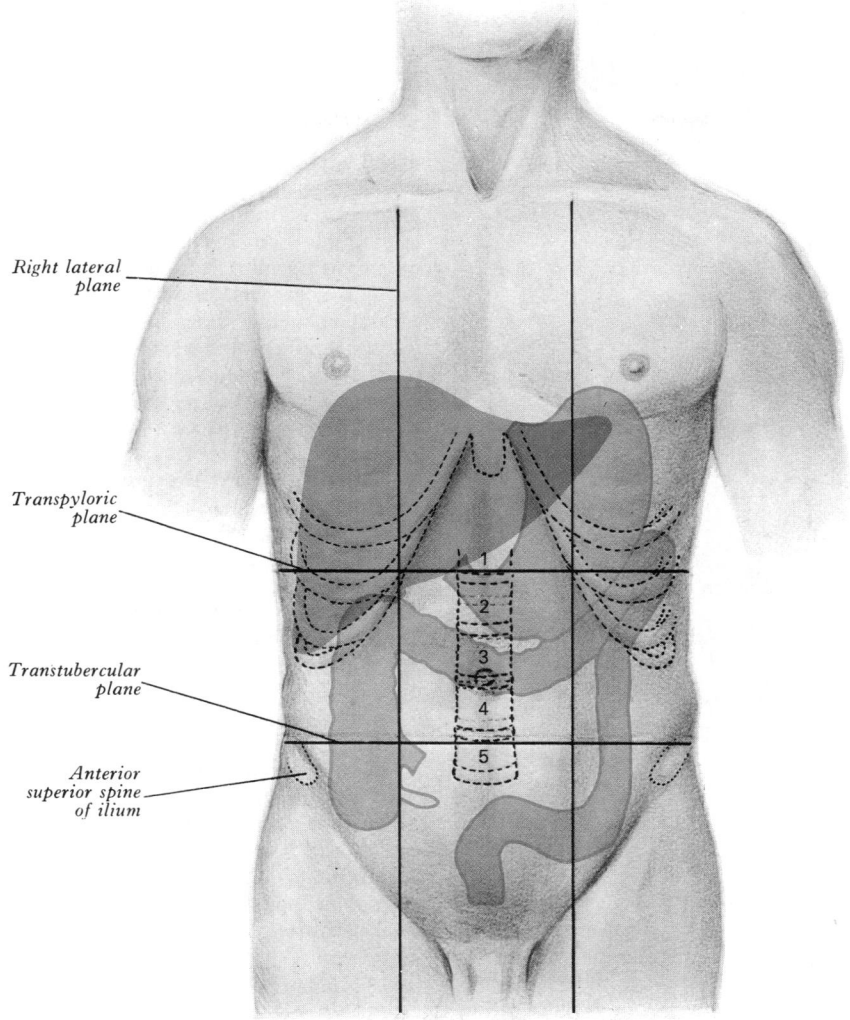

Right lateral plane

Transpyloric plane

Transtubercular plane

Anterior superior spine of ilium

8.139 Surface projection of the stomach, liver and colon. The outlines of the lumbar vertebral bodies and intervertebral discs, lower ribs, xiphoid process and parts of the iliac crests are indicated.

colon. The *circular muscular fibres* form a thicker layer separated from the longitudinal by connective tissue. The *submucosa* is well developed, containing many lymphoid masses which cause the mucosa to bulge into the lumen, narrowing it irregularly. The *mucosa* is covered by columnar epitheliocytes and attenuated antigen-transporting 'M' cells (Owen & Nemanic 1978). Glands (crypts similar to those of the colon) are few and penetrate deeply into the lymphoid tissue (**8.138**), which in the normal human appendix is situated primarily in the lamina propria and extends into the submucosa; follicular and parafollicular zones containing B- and T-lymphocytes (p. 824) can be distinguished; clustered lymphocytes also appear between the epithelial cells, where some may possibly differentiate into plasma cells (Gorgollón 1978). Lymphoid tissue in the lamina propria contains many plasma cells, with lymphocytes, eosinophil and other leucocytes, mast cells and macrophages embedded in a fibrocellular reticulum. The submucosal follicles (germinal centres) are organized like those of other examples of gut-associated lymphoid tissue (p. 824; see also Kaiserling et al 1974). The *lymphoid masses* are a local defence against infection; it has also been suggested that they may be a homologue of the avian *bursa of Fabricius* (p. 671) concerned in the acquisition of immunological competence by certain lymphocytes. However, experimental evidence argues against this function (p. 761). In many mammals, particularly herbivores, the caecum and appendix are large and constitute a highly important site of digestion of cellulose by symbiotic bacteria.

THE COLON

The colon is conveniently considered in four parts: ascending, transverse, descending and sigmoid.

The ascending colon, about 15 cm long and narrower than the caecum, ascends to the inferior surface of the right lobe of the liver, on which it makes a shallow depression; here it turns abruptly forwards and to the left, at the *right colic flexure* (**8.125**). In surface projection it ascends lateral to the right lateral plane (**8.139**) from the transtubercular to midway between the subcostal and transpyloric planes. It is covered by peritoneum except where its posterior surface is connected by areolar tissue to the iliac fascia, and to the iliolumbar ligament, quadratus lumborum, aponeurosis of transversus abdominis and the perirenal fascia on the front of the inferolateral area of the right kidney. The lateral femoral cutaneous nerve, usually the fourth lumbar artery, and sometimes the ilio-inguinal and iliohypogastric nerves, cross behind it. Sometimes it possesses a distinct but narrow mesocolon. In a series of 100 subjects, 52% had neither an ascending nor descending mesocolon, 14% had both, 12% an ascending and 22% a descending mesocolon (Treves 1885). Anteriorly it is in contact with the coils of the ileum, the greater omentum and the abdominal wall.

The right colic flexure is at the junction of the ascending and transverse colon; the latter turns down, forwards and to the left. Posterior is the inferolateral part of the anterior surface of the right kidney; above and anterolaterally is the right lobe of the liver; anteromedially is the descending part of the duodenum and fundus of the gallbladder. Its posterior aspect is not covered by peritoneum and is in direct contact with renal fascia. It is not so acute as the left colic flexure.

The transverse colon (**8.118, 140A**), about 50 cm long, extends from the right colic flexure in the right lumbar region, across into the left hypochondriac region, here curving sharply

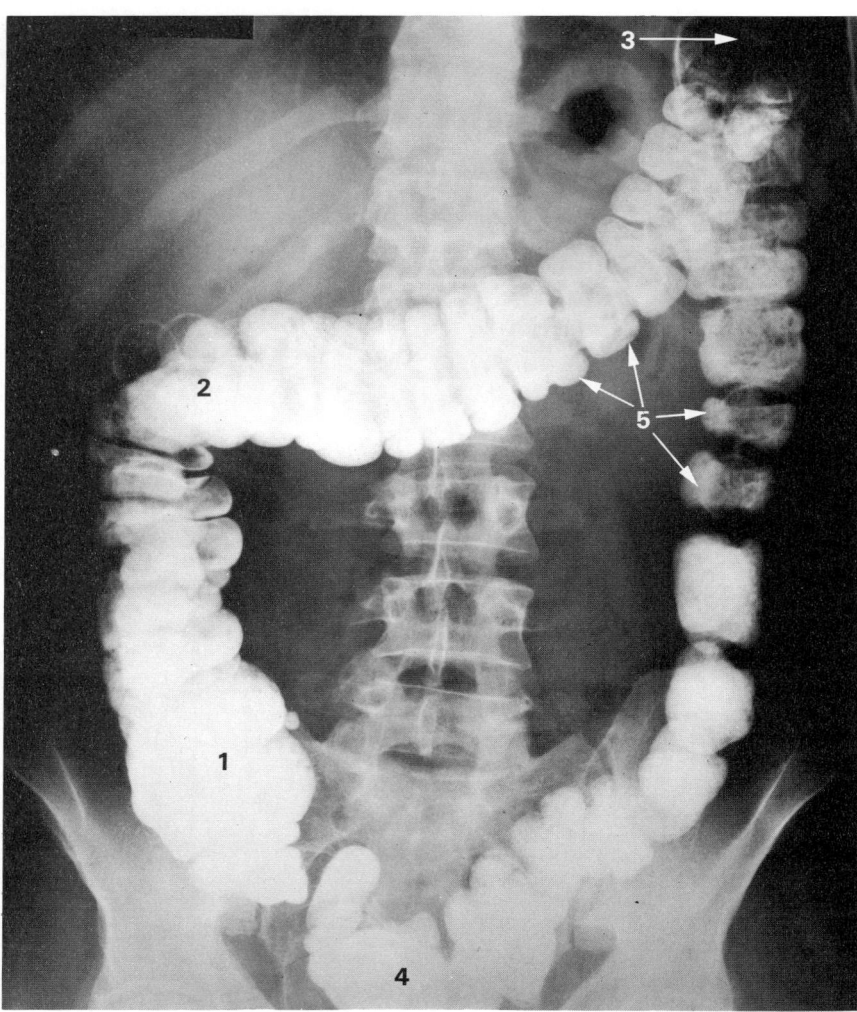

8.140A A radiograph of the abdomen after the administration of a barium enema which has filled the whole of the large intestine as far as the caecum and ileocaecal valve. (1) the caecum; (2) the right or hepatic flexure of the colon, which is much inferior to (3) the left or splenic flexure of the colon; (4) the sigmoid colon; (5) the sacculations, or haustrations, which are clearly visible throughout most of the colon.

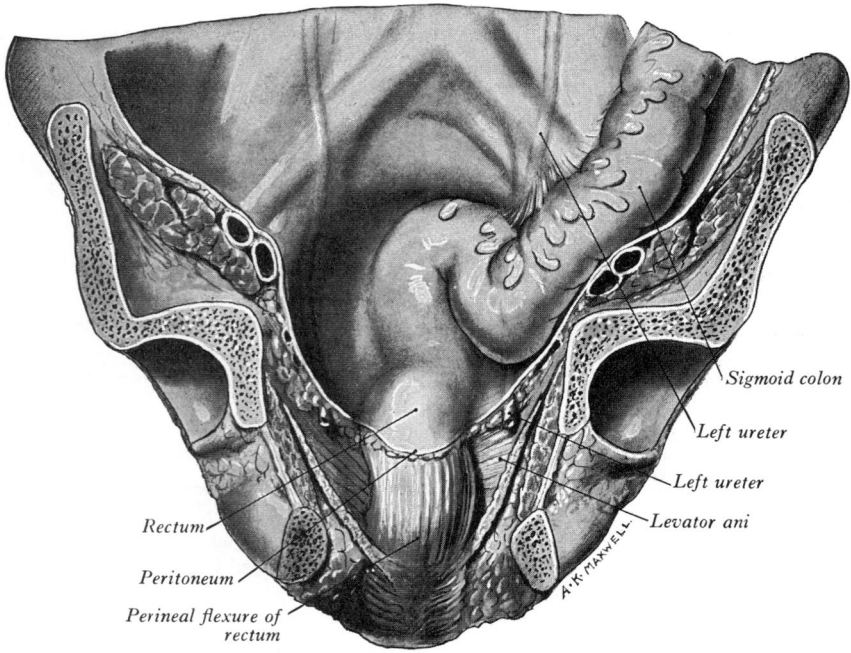

Sigmoid colon

Left ureter

Left ureter

Levator ani

Rectum

Peritoneum

Perineal flexure of rectum

8.140B Oblique coronal section through the pelvis to expose the anterior aspect of the rectum.

down and backwards below the spleen as the *left colic flexure*. The transverse colon describes an arch, its concavity usually directed back and up; near its splenic end an abrupt U-shaped curve may descend lower than the main arch. Its surface projection (8.139) extends from a point, situated just lateral to the right lateral plane and midway between the subcostal and transpyloric planes, to the umbilicus and then up and left to a point just superolateral to the intersection of the left lateral and transpyloric planes. A precise projection is difficult to define, varying much even in the same individual. Commonly it is in the lower umbilical or upper hypogastric region. It frequently descends in a V-shaped manner, the apex being well below the level of the iliac crests (p. 1334). In a radiological assessment in the upright position, its lowest level in 1000 young adults was found to vary much, even reaching the true pelvis; levels varied as much as 17 cm in the same individual between upright and recumbent positions (Moody 1927).

The posterior surface at its right end is devoid of peritoneum and is attached by areolar tissue to the front of the descending part of the duodenum and the head of the pancreas; but from the latter to the left colic flexure it is almost completely invested by peritoneum, connecting it to the anterior border of the body of the pancreas by the *transverse mesocolon*. Above the transverse colon are the liver and gallbladder, the greater gastric curvature and the lateral end of the spleen; below is the small intestine, in front are the posterior layers of the greater omentum and behind are the descending part of the duodenum, the head of the pancreas, the upper end of the the mesentery, the duodenojejunal flexure and coils of the jejunum and ileum.

The left colic flexure (8.125), the junction of the transverse colon and descending colon in the left hypochondriac region, is related to the lower part of the spleen and pancreatic tail above and medially with the front of the left kidney. It is so acute that the end of the transverse colon usually overlaps the front of the descending colon. The left flexure is above and on a more posterior plane than the right flexure and is attached to the diaphragm level with the tenth and eleventh ribs by the *phrenicocolic ligament*, which lies below the anterolateral pole of the spleen (p. 1344).

The descending colon (8.125), about 25 cm long, descends through the left hypochondriac and lumbar regions, at first following the lower part of the lateral border of the left kidney and then descending in the angle between the psoas major and quadratus lumborum to the iliac crest; it then curves downwards and medially in front of the iliacus and psoas major to end in the sigmoid colon at the inlet of the lesser pelvis. (It is sometimes described as ending at the iliac crest, the part between this and the pelvic inlet being named the *iliac colon*.) In surface projection (8.139) it descends just lateral to the left lateral plane, from a little above and left of the intersection of the transpyloric and left lateral planes as far as the inguinal ligament. Peritoneum covers all but its posterior surface, which is connected by areolar tissue to fascia over the inferolateral region of the left kidney, the aponeurosis of transversus abdominis, the quadratus lumborum, iliacus and psoas major (8.125). Crossing behind it are the following left structures: subcostal vessels and nerve, iliohypogastric and ilio-inguinal nerves, fourth lumbar artery (usually), the lateral femoral cutaneous, femoral and genitofemoral nerves, the testicular (or ovarian) vessels and the external iliac artery. The descending colon is smaller in calibre, more deeply placed, and more frequently covered behind by peritoneum than the ascending colon (p. 1367). Anteriorly are the coils of the jejunum, except for its lower part which is palpable when the abdominal muscles are relaxed.

The sigmoid colon (pelvic colon) (8.140B) begins at the pelvic inlet, continuing in the descending part; it forms a variable loop of about 40 cm and is normally in the lesser pelvis. The loop first descends in contact with the left pelvic wall, then crosses the pelvic cavity between the rectum and bladder in males, and rectum and uterus in females, and may reach the right pelvic wall; finally it turns back to the midline level with the third piece of the sacrum, where it bends downwards and ends in the rectum. It is closely surrounded by peritoneum, forming a mesentery, the *sigmoid mesocolon* (p. 1344), which diminishes in length from the centre towards its ends, where it disappears; the loop is fixed at its junctions with the descending colon and rectum but quite mobile between them. Its relations are therefore variable. *Laterally* are: the left external iliac vessels, the obturator nerve, ovary or ductus deferens and the lateral pelvic wall, *posteriorly* the left internal iliac vessels, ureter, piriformis and sacral plexus; *inferiorly* the bladder in males or uterus and bladder in females; *superiorly* and to the *right* it is in contact with terminal coils of the ileum.

The position and shape of the sigmoid colon vary much, depending on (1) its length, (2) the length and mobility of its mesocolon, (3) the degree of distension (when distended it rises into the abdominal cavity, sinking again into the lesser pelvis when empty), (4) the condition of the rectum, bladder and uterus. (When these are distended the sigmoid colon tends to rise and to fall when they are empty.) Racial variation has been noted (Lisowski 1969): in some groups, particularly Ethiopians, the incidence of a suprapelvic loop, perhaps conducive to volvulus, is particularly high.

THE RECTUM

The rectum (8.140B, 141A,B) is continuous with the sigmoid colon at the level of the third sacral vertebra, the junction being at the lower end of the sigmoid mesocolon. It descends along the sacrococcygeal concavity, with an anteroposterior curve, the *sacral flexure*. It thus curves down and back, then downwards, and finally down and forwards to join the anal canal by passing through the pelvic diaphragm (p. 604). The *anorectal junction* is 2–3 cm in front of and slightly below the coccygeal tip; from this level (in males opposite the apex of the prostate) the anal canal passes down and backwards from the lower end of the rectum, this backward bend of the gut being termed the *perineal flexure* of the rectum. The rectum also deviates in three lateral curves: the upper is convex to the right, the middle (the most prominent) bulges to the left and the lower is convex to the right. Both ends of the rectum are in the median plane (8.141A). The rectum is about 12 cm long, with the same diameter as the sigmoid colon above (about 4 cm in the empty state), but its lower part is dilated as the *rectal ampulla*. The rectum differs from the sigmoid colon in having no sacculations, appendices epiploicae or mesentery; the taeniae blend about 5 cm above the rectosigmoid junction, forming two wide muscular bands which descend, anterior and posterior, in the rectal wall. Peritoneum is related only to the upper two-thirds, covering its front and sides above, and lower down, only its front, from which it is reflected on to the bladder in males, forming the rectovesical pouch, and on to the posterior vaginal wall in females, forming the recto-uterine pouch. The level of this reflexion is higher in males, the rectovesical pouch being about 7.5 cm (about the length of the index finger) from the anus; in females the recto-uterine pouch is about 5.5 cm from the anus. In the male fetus, peritoneum extends on to the front of the rectum as far as the lower limit of the prostate (p. 1433). On the sigmoid colon, peritoneum is firmly attached to the muscle layer by fibrous connective tissue but as it descends on to the rectum it is more loosely attached by fatty areolar tissue, allowing for considerable expansion.

In the empty rectum, the mucosa in its lower part presents a number of longitudinal folds, effaced during distension. There are also permanent semilunar *transverse* or *horizontal folds*, most marked in rectal distension. Two forms of horizontal fold have been recognized (Jit 1961); one consists of the mucosa, a circular muscle layer and part of the longitudinal muscle, and an indentation on the rectal exterior; the other is devoid of longitudinal muscle and has no external marking. Their number is variable but there are commonly three folds. An upper one, near the beginning of the rectum, may be on the left or right; occasionally it encircles the gut, constricting its lumen. The middle fold is largest, most constant and lies immediately above the ampulla, projecting from the anterior and right wall just below the level of the anterior peritoneal reflexion; circular muscle is more marked in it. The lowest fold, inconstant and on the left, is about 2.5 cm below the middle fold. Sometimes a fourth occurs on the left about 2.5 cm above the middle fold.

It has been suggested (Paterson 1912) that the rectum consists of two functional parts, above and below the middle fold, the upper containing faeces and being free to distend into the peritoneal cavity, the lower more confined, enclosed in a tube of condensed extraperitoneal tissue and (except during defaecation) normally empty; in chronic constipation or after death it may contain faeces. (Note that the rectum above the middle fold is considered to develop from the hindgut and the part below with the upper anal canal, from the cloaca or post-allantoic gut.) Others (O'Beirne 1833, Hurst 1919) have considered the sigmoid colon a faecal reservoir, the rectum being normally empty and the entry of faeces into it exciting defaecation. Experimental distension of the rectum and anal canal results in the desire to defaecate and causes the relaxation of the anal sphincters (Denny-Brown & Robertson 1935).

Relations of the Rectum

Posterior to the rectum in the median plane are: the lower three sacral vertebrae, coccyx, median sacral vessels, ganglion impar and branches of the superior rectal vessels; while on each side, particularly on the left, are: the piriformis, the anterior rami of the lower three sacral and coccygeal nerves, sympathetic trunk, lower lateral sacral vessels, the coccygei and the levatores ani. The rectum is attached to the sacrum along the lines of the anterior sacral foramina by fibro-areolar tissue enclosing: the sacral nerves and the pelvic splanchnic nerves from the anterior rami of the second to fourth sacral nerves, which join the pelvic plexuses on the rectal wall; rami of the superior rectal vessels, lymphatic vessels, lymph nodes, and loose perirectal fat. *Anterior in males* above the site of the peritoneal reflexion from the rectum, are the upper parts of the base of the bladder and of the seminal vesicles, the rectovesical pouch and its contents (terminal coils of the ileum and sigmoid colon); below the reflexion are: the lower parts of the base of the bladder and of the seminal vesicles, deferent ducts, terminal parts of the ureters and the prostate. *In females*, above the reflexion are: the uterus, upper vagina, recto-uterine pouch and contents (terminal coils of the ileum and sigmoid colon), while below the reflexion is the lower part of the vagina. *Laterally*, the upper part of the rectum is related to the pararectal fossa and contents (sigmoid colon or lower ileum), while below the peritoneal reflexion laterally are the pelvic sympathetic plexuses, coccygei and levatores ani and branches of the superior rectal vessels.

THE ANAL CANAL

The anal canal (Milligan et al 1937, Gabriel 1945, Wilde 1949, Goligher et al 1955, Fowler 1957) begins where the rectal ampulla suddenly narrows, passing down and backwards to the anus (8.141B,C). It is about 4 cm long in adults, its anterior wall being slightly shorter than its posterior. When empty its lumen is a sagittal or triradiate longitudinal slit. *Posterior* is a mass of fibromuscular tissue, the *anococcygeal ligament*, separating it from the tip of the coccyx; *anteriorly* it is separated by the *perineal body* (p. 607) from the membranous urethra and penile bulb or from the lower vagina; *laterally* are the ischiorectal fossae. Over its whole length it is surrounded by sphincters which normally keep it closed.

The lining of the canal varies along its course. The mucosa of the lower part of the rectum is pale pink and semitransparent, the branching pattern of the superior rectal vessels being visible through it. The upper half (15 mm) of the anal canal is also lined by mucosa, plum-red in colour due to blood in the subjacent internal rectal venous plexus. The epithelium is variable. In the upper part it is similar to that of the rectum, consisting of simple columnar cells, some secretory and others absorptive, with numerous tubular glands or crypts. In the lower half, this gives way to non-keratinized stratified squamous epithelium, finally merging with the keratinized stratified squamous epithelium of the perianal epidermis (Walls 1958). In this part of the canal are six to 10 vertical folds, the *anal columns*, well marked in children but sometimes less defined in adults. Each column contains a terminal radicle of the superior rectal artery and vein, these radicles being largest in the left-lateral, right-posterior and right-anterior quadrants of the wall of the canal; enlargements of venous radicles in these three sites constitute primary internal haemorrhoids. The lower ends of the columns are linked by small crescentic mucous folds, the *anal valves*, above each of which is a small recess or *anal sinus*. The sinuses, deepest in the posterior wall, may retain faecal matter and become infected, leading to abscess formation in the anal canal wall; anal valves may be torn by hard faeces, producing an anal fissure (p. 1373). Anal valves are situated along the *pectinate line*, opposite the middle of the sphincter ani internus and commonly considered to be the site of the anal membrane in early fetal life, thus representing the junction of the endodermal (cloacal) and ectodermal (proctodeal) parts of the canal. Small epithelial anal papillae may occur on the edges of the anal valves, perhaps remnants of the anal membrane. However, the junction of ectodermal and endodermal parts may be at the lower border of the pecten (Johnson 1914).

The anal canal extends about 15 mm below the anal valves, as the *transitional zone* or *pecten*, whose epithelium is non-keratinized, stratified squamous and intermediate in thickness between that of the mucosa of the upper part of the canal and the epidermis in its lowest part; only the latter contains sweat glands. The transitional zone overlies part of the internal rectal venous plexus and is shiny and bluish. Its submucosa contains dense connective tissue, contrasting with the lax connective tissue in the upper half of the anal canal and suggesting the firm support and anchorage of the pectineal lining to the surrounding anal muscle. The transitional zone ends below at a narrow sinuous zone, the 'white line' (of Hilton). In the living this 'line' is bluish pink and rarely visible (Ewing 1954), its only interest being that it is at a level between the subcutaneous part of the external sphincter and the lower border of the internal sphincter; digital examination in the living reveals an *anal intersphincteric groove* at this site. Below the white line, the final 8 mm or so of the anal canal is lined by true skin, dull white or brown in colour and containing sweat and sebaceous glands. There is much variation in the epithelial zones

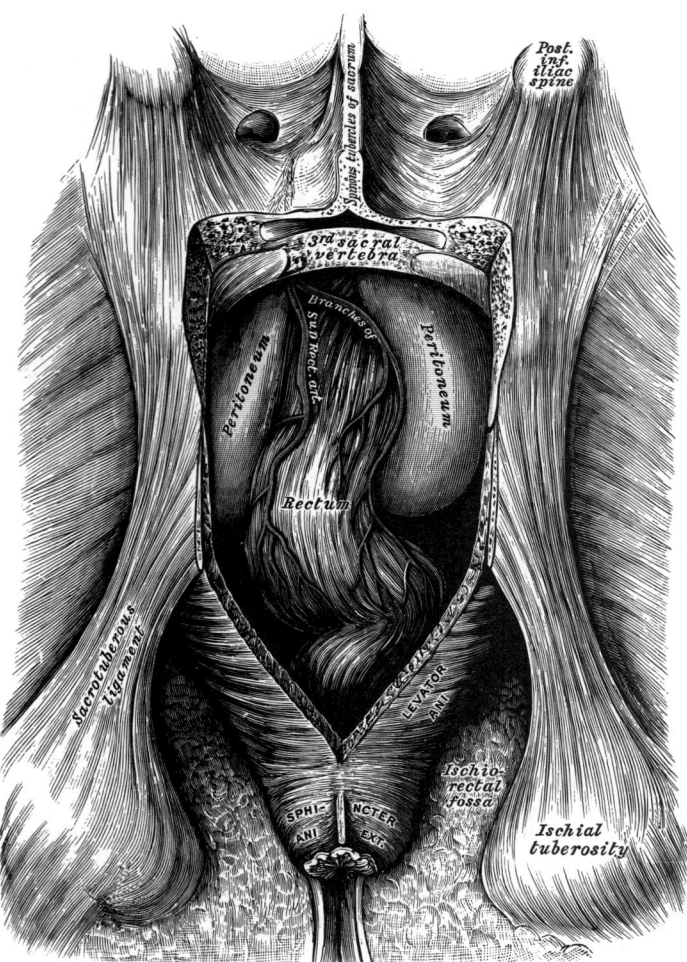

8.141A Posterior aspect of the rectum exposed by removal of the lower part of the sacrum and coccyx. Note the superior rectal artery (red) and peritoneum of the pararectal fossae (blue).

described above and frequently the various types interpenetrate, the zones being poorly defined.

Near the anal sinuses, *anal glands* (Fowler 1957, McColl 1967) extend upwards or downwards into the submucosa, occasionally penetrating deeply into the internal sphincter. Each consists of one to six spiral or straight tubules, sometimes branched, and lined by two or three layers of mucous secretory cells. The duct of each gland, lined by stratified columnar epithelium, opens into a small depression, an *anal crypt*. The glands are surrounded by lymphocytes in a form similar to lymphatic follicles and the submucosal non-striated muscle is thick in their vicinity. Occasionally the termination of a duct is not canalized and secretions may then form a cyst. The glands are sometimes infected, producing an abscess or fistula. They vary widely in number and depth of penetration, even extending into the submucosa above the anorectal junction. (For details of comparative anatomy and pathology see McColl 1967.) In this series, 50 normal anal canals were examined; half had anal glands passing right through the internal sphincter; the average number of such extensions was four but they may reach 16. McColl considered that these human glands were not homologous with the *anal scent glands* of some other mammals.

ANAL MUSCULATURE

The anal walls are surrounded by a complex of sphincters (8.141), divisible into internal and external parts. At the anorectal junction, the rectal circular muscle is thickened (5–8 mm) as a non-striated **sphincter ani internus** around the upper three-quarters (30 mm) of the anal canal and ends below at the white line (vide supra et infra).

The sphincter ani externus (8.141B) surrounds the whole anal canal; it is usually described as consisting of three parts, all composed of skeletal muscle. These are, from inferior to superior, the subcutaneous, the superficial and the deep portions. The *subcutaneous part* is a flat band, about 15 mm broad, around the lower anal canal and lies horizontally below the lower borders of the internal sphincter and superficial part of the external sphincter; it lies beneath the skin at the anal orifice and is inferior to the white line. Anteriorly a few fibres join the perineal body (or the superficial transverse perineal muscles); posteriorly some are usually attached to the anococcygeal ligament. The *superficial part* is elliptical and superior to the subcutaneous; it is the only part attached to bone, arising from the posterior surface of the terminal coccygeal segment by the median anococcygeal raphe; anteriorly it surrounds the *lower* part of the internal sphincter and is chiefly attached to the perineal body. The *deep part* is a thick annular band around the *upper* part of the internal sphincter; its deeper fibres blend inseparably with the puborectalis (p. 605); anterior to the anal canal many fibres decussate into the superficial transverse perineal muscles, especially in females. Some posterior fibres are usually attached to the anococcygeal raphe. However, a clear separation into three parts has also been denied (Goligher et al 1955). In females according to Oh & Kark (1972) the muscle is a single band, anteriorly at least. (See also Wendell-Smith & Wilson 1977.)

Muscle tone in both internal and external sphincters keeps the canal and anus closed; during defaecation (Shafik 1986) they relax and the lower anal canal opens out and flattens, the mucosa of its upper part appearing on the surface. The external sphincter can be voluntarily contracted to occlude the anus more firmly. Its nerve supply is derived from the inferior rectal branch of the pudendal nerve (S.2 and 3) and the perineal branch of the fourth sacral nerve.

At the anorectal junction the pubococcygeal fibres of levator ani fuse with the longitudinal non-striated rectal muscle to form a *conjoint longitudinal coat* for the anal canal between the internal and external sphincters (8.141B,C). Distally, this layer is increasingly fibro-elastic; at the white line it breaks up into 9–12 circumferential septa which diverge mainly through the subcutaneous part of the external sphincter to become attached to the dermis of circumanal skin. These septa are largely elastic fibres; the most peripheral passes between the subcutaneous and superficial parts of the external sphincter into the ischiorectal fat. The

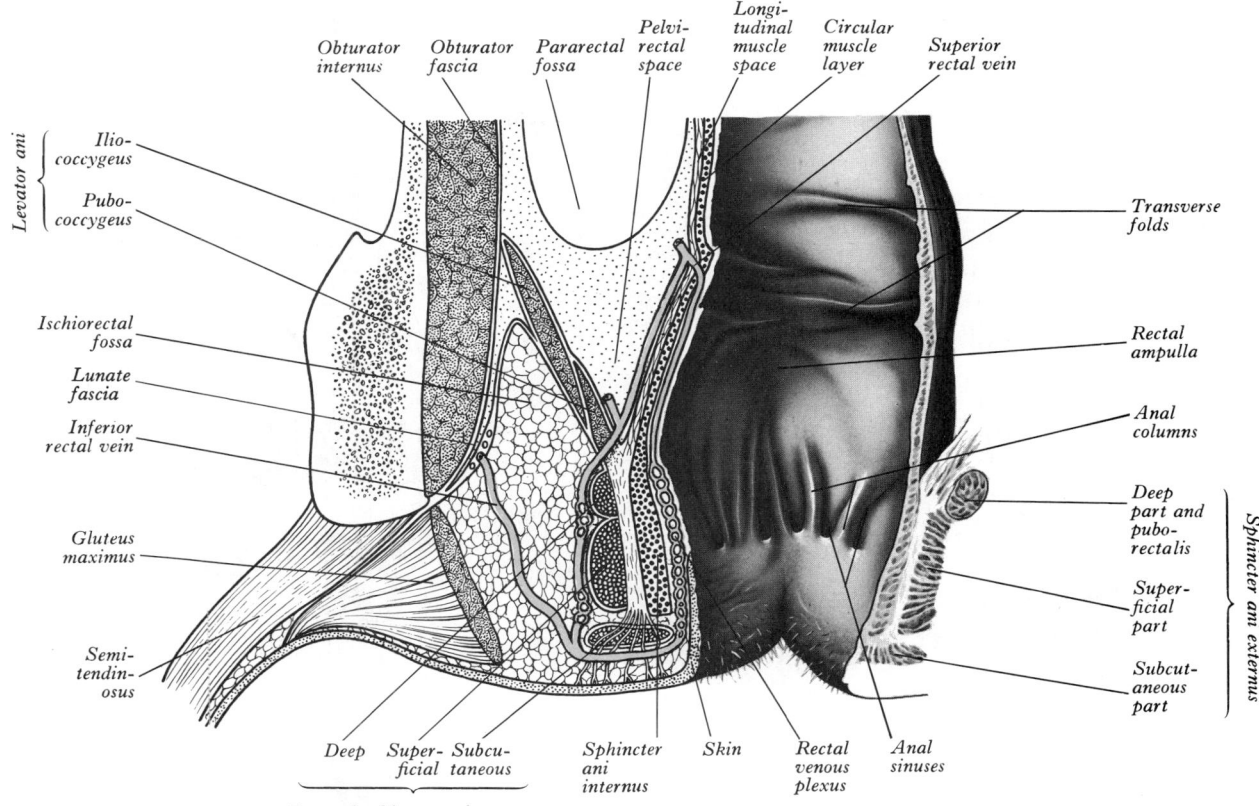

8.141B Diagram of a coronal section of the rectum and anal canal and the adjacent structures (adapted from Rauber-Kopsch, *Lehrbuch und Atlas der Anatomie des Menschen,* 1919). The internal pudendal vessels, the dorsal nerve of the penis and the perineal nerve are shown transected in the lateral wall of the ischiorectal fossa, where they are traversing the 'lunate fascia' (pudendal canal).

most central (juxta-anal) septum is said to pass between the internal sphincter and the external's subcutaneous part to reach the anal lining at the white line as the *anal intermuscular septum,* producing an *anal intersphincteric groove.* Wilde (1949), Goligher et al (1955) and Fowler (1957), however, considered that the longitudinal fibres in this position (compared with those penetrating the subcutaneous part of the external sphincter) were too scanty to warrant a name, maintaining that the groove is due to the muscle masses of the internal sphincter above and the subcutaneous part of the external sphincter below and to the contraction of the latter. In the anal submucosa, inferior to the anal sinuses, is a layer of non-striated muscle, yellow elastic fibres and fibrous connective tissue, derived mainly from strands of the conjoint longitudinal coat, which descends inwards between fascicles of the internal sphincter (8.141C); some of the strands end by turning laterally around the lower edge of the internal sphincter to

rejoin the main longitudinal layer; but most continue inferolaterally, superficial to the subcutaneous part of the external sphincter, to the dermis of the skin from the white line to well beyond the anus. These attachments corrugate the region; hence the term *corrugator cutis ani muscle.* Wilde (1949) considered these fibres to be exclusively elastic, but Goligher et al (1955) noted nonstriated muscle fibres among them. Fowler (1957), finding no muscle fibres here, ascribed puckering of the perianal skin to the combined effects of levator ani and the subcutaneous part of the external sphincter.

A *muscularis mucosae* has been described in the anal canal immediately above the pectinate line and possibly extending below it (Jit 1974).

Between the subcutaneous part of the external sphincter and the skin of the anal canal is the inferior part of the internal rectal venous plexus; veins link the external and internal rectal plexuses,

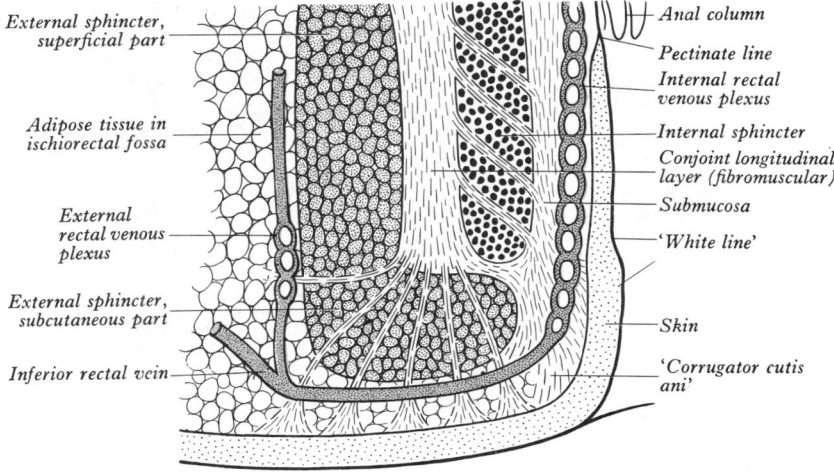

8.141C Part of 8.141B enlarged to show greater detail.

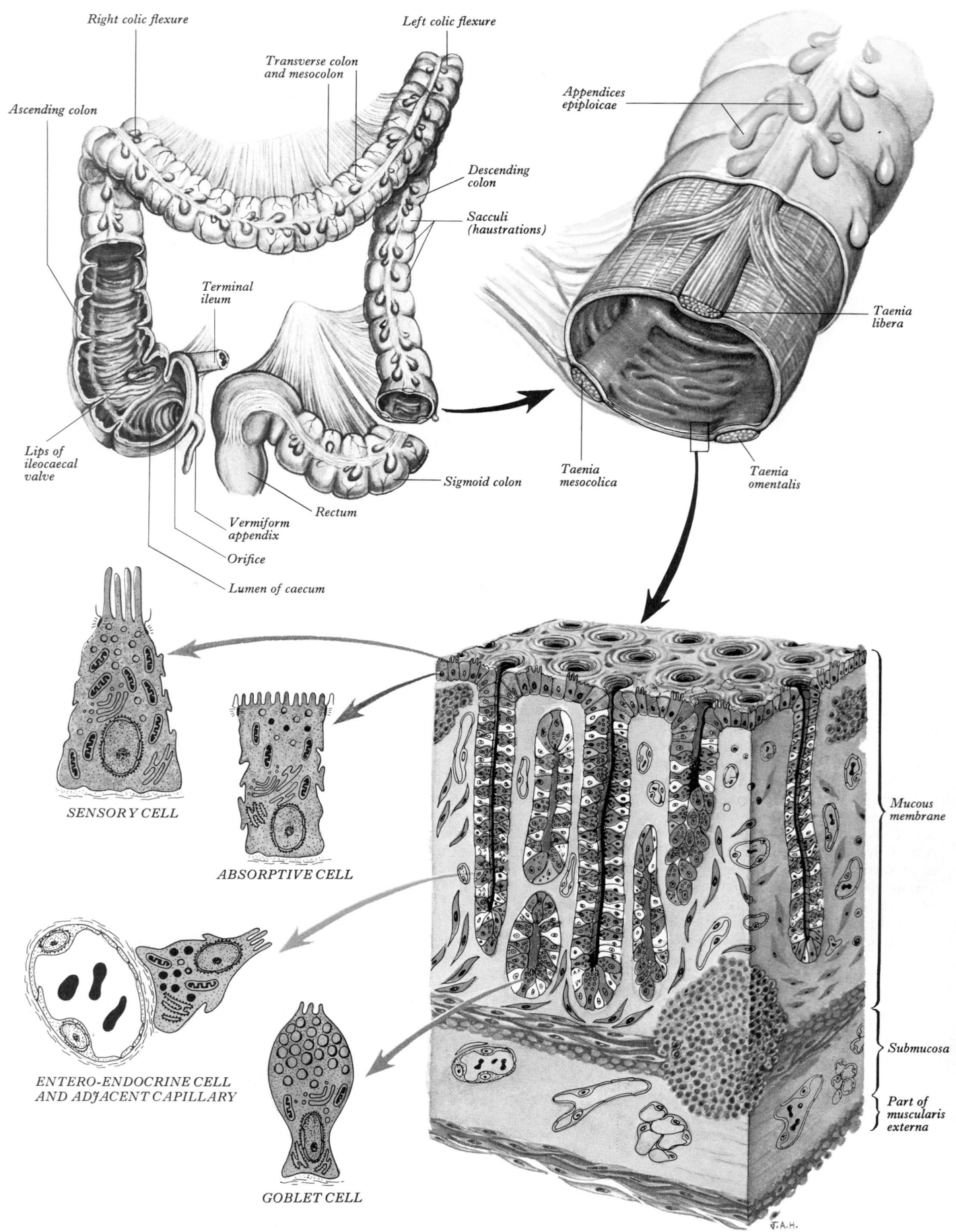

Right colic flexure

Left colic flexure

Transverse colon and mesocolon

Ascending colon

Descending colon

Appendices epiploicae

Sacculi (haustrations)

Terminal ileum

Taenia libera

Lips of ileocaecal valve

Taenia mesocolica

Taenia omentalis

Sigmoid colon

Vermiform appendix

Rectum

Orifice

Lumen of caecum

SENSORY CELL

ABSORPTIVE CELL

Mucous membrane

ENTERO-ENDOCRINE CELL AND ADJACENT CAPILLARY

Submucosa

GOBLET CELL

Part of muscularis externa

8.142 Diagrams of the disposition of the major regions of the large intestine, the micro-architecture and histology of the colonic wall and the ultrastructure of its epithelial cells. Note the aggregations of lymphocytes (shown in yellow) and undifferentiated epithelial cells (shown in white).

thus connecting the portal and systemic venous systems. The radiating elastic septa end in a network dividing the narrow cleft between the subcutaneous part of the external sphincter and the skin into a compact honeycomb, which may explain the severe pain produced by pus or blood collecting here, and the localization of a haemorrhage following the rupture of a vein from the external rectal plexus. The submucosa above the white line, containing the upper part of the internal rectal venous plexus, forms a *submucous space* and that below the white line, containing its lower part, forms a *perianal space*; they are separated by the dense submucous layer of non-striated muscle and fibrous connective tissue referred to above, which is especially well marked below the anal valves.

Fowler (1957) concluded that the anal canal proper extends between two palpable muscular landmarks, the *anorectal ring* above and the *intersphincteric ring* below. In the relaxed state, the lower border of the internal sphincter and intersphincteric groove lie at the anal orifice, the subcutaneous external sphincter being lateral to it; only when the external sphincter contracts is the orifice withdrawn around the lower part of the 'apparent' anal canal. At the anorectal junction the puborectalis, deep external and internal sphincters collectively form the *anorectal ring* of muscle, which is palpable in the canal; surgical division of this results in rectal incontinence. Its anterior part is less well marked, since relatively few fibres of the puborectalis pass in front of the anorectal junction, most forming posterolateral loops around the gut at this site, slinging the anorectal junction forwards towards the pubis.

Correlated with the *dual development* of the anal canal, the part above the anal valves arising from the endodermal cloaca and the part below this from the ectodermal proctodeum (p. 233), the following facts emerge. The lining of the *ectodermal part* is skin, supplied by spinal nerves (inferior rectal), the 'somatic' inferior rectal artery, a venous drainage via the inferior rectal passing to the internal pudendal vein (systemic) and by lymphatics draining with the perianal skin to the superficial inguinal lymph nodes. In the *endodermal part* the mucosa is supplied by autonomic nerves, the arterial supply (Griffiths 1961) coming mainly from the superior rectal artery, the venous drainage by the superior rectal vein, a tributary (via the inferior mesenteric) of the portal venous system; the lymphatics drain with those of the rectum (p. 854). In portal obstruction, the collateral circulation opened up by anastomosis between the portal and systemic veins in the anal canal may cause these veins to dilate. The differing nerve supply of the two parts accords with the response to stimuli; the lower part responds to touch, pain and heat like skin; the upper part, like the gut at large, has a high threshold for such stimuli but responds readily to tension. The effects of the difference in innervation of the two parts are apparent in haemorrhoids, which may be covered by skin inferiorly *and* mucosa superiorly; to thrombose such varicose veins by injection, a needle is inserted into the insensitive *mucosal* part. Fissure *in ano* (tearing of anal valves) is very painful because it involves the lower, sensitive, *dermal* part of the anal canal.

RECTAL EXAMINATION

On inserting the index finger through the anal orifice in rectal examination, the finger is first resisted by the subcutaneous external sphincter and then by the internal sphincter, superficial and deep parts of the external sphincter and the puborectalis; beyond this it may reach the inferior (or even middle) transverse rectal fold. Many structures related to the canal and lower rectum may be palpated.

In *males* through the anterior rectal wall, the penile bulb and (particularly with a catheter in the urethra) the membranous urethra are first identified; about 4 cm from the anus the prostate can be felt and beyond this the seminal vesicles (if enlarged) and the base of the bladder (especially if distended). Posteriorly pelvic surfaces of the lower sacrum and coccyx are palpable and laterally the ischial spines and tuberosities and (if enlarged) the internal iliac lymph nodes. Pathological thickening

of the ureters, swellings in the ischiorectal fossa and abnormal contents of the rectovesical recess may also be detected.

In *females* the uterine cervix is palpable through the anterior rectal wall; its degree of dilatation during parturition may be assessed in this manner. Pathological conditions causing tenderness or changes in the shape, size, consistency or position of the ovaries, uterine tubes, broad ligaments and recto-uterine pouch may be detected.

In both sexes, tenderness of an inflamed vermiform appendix (if pelvic) can be elicited.

RECTAL FASCIAE AND 'SPACES'

Parts of the pararectal pelvic fascia are composed of loose connective tissue, whilst others are denser, with particular orientations and attachments; the latter are often considered to be rectal 'supports' requiring surgical division to mobilize the organ. From the lower sacrum's anterior surface a strong avascular condensation proceeds to the posterior aspect of the anorectal junction (*fascia of Waldeyer*). Around the middle rectal vessels fascia extends from the posterolateral pelvic wall (level with the third sacral vertebra) to the rectum as the *lateral rectal ligaments*. Anteriorly, between the rectum and the seminal vesicles and prostate, the *rectovesical fascia* (p. 1433) is more loosely attached to the seminal vesicles and prostate than to the rectum and in rectal excision it must be separated from them.

In addition to the ischiorectal fossae (p. 606) several '*spaces*' of surgical import are related to the rectum and anal canal. The *pelvirectal space* comprises the loose extraperitoneal connective tissue above the levator ani; it is divided into anterior and posterior regions by the *lateral rectal ligaments*. The *submucous space* of the anal canal is between the mucosa (above the white line) and the internal sphincter; it contains the superior part of the internal rectal venous plexus and lymphatics; above, it is continuous with the rectal submucosa, below with the *perianal space* (p. 1370), the lateral part of which is bounded above by the most lateral elastic septum traversing the subcutaneous part of the external sphincter. The septum divides the ischiorectal fossa into a superior part containing coarsely lobulated fat and a smaller, lower, perianal space containing fine, compact fat. The perianal space contains the subcutaneous part of the external sphincter, the external rectal venous plexus and terminal rami of the inferior rectal vessels and nerves. The radiating septa traversing the subcutaneous part of the external sphincter tend to divert pus in the perianal space to the anal canal at the white line or to the surface of the perianal skin, rather than to the main ischiorectal fossa. Since the perianal space surrounds the lower anal canal, pus on one side may spread around it.

STRUCTURE OF THE LARGE INTESTINE

The layers of mural tissue in the large intestine are like those in the small (p. 1358). The *serosa* or *visceral peritoneum* is variable in extent. Along the colon the peritoneum forms small fat-filled *appendices epiploicae*, most numerous on the sigmoid and transverse colon but absent from the rectum. Subserous loose connective tissue attaches the peritoneum to the *muscularis externa*, which has outer longitudinal and inner circular layers of non-striated muscle. The *longitudinal fibres* form a continuous layer (Hamilton 1946) but are also aggregated as longitudinal bands or *taeniae coli*, between which the longitudinal layer is less than half the circular layer in thickness. In the caecum and colon three taeniae appear, varying from 6–12 mm in width. The *taenia libera* is anterior in the caecum, the ascending, descending and sigmoid colon, but inferior in the transverse. The *taenia mesocolica* is posteromedial in the caecum, the ascending, descending and sigmoid colon, but posterior in the transverse, being located at the attachment of the transverse mesocolon. The *taenia omentalis* is posterolateral in the caecum, the ascending, descending and sigmoid colon, but anterosuperior in the transverse colon, being located where the posterior (ascending) layers of the greater omentum meet

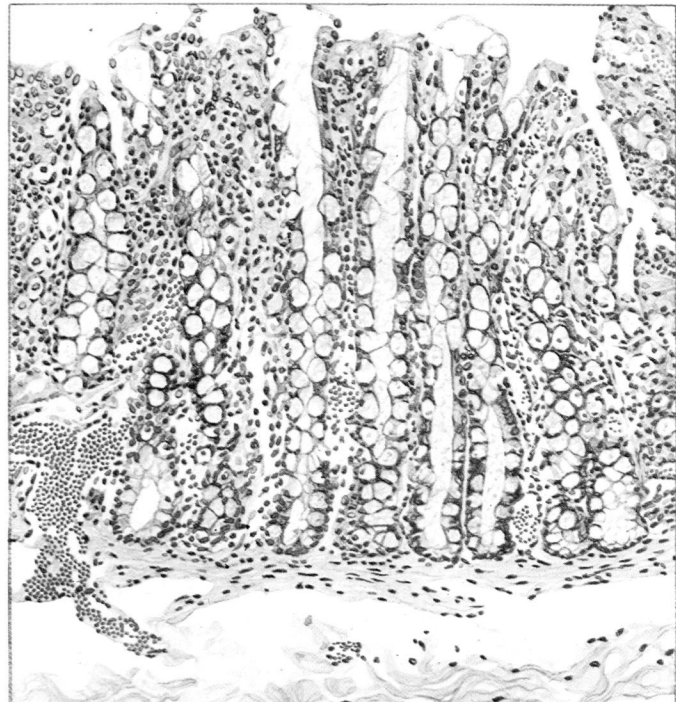

8.143A Section of the mucous membrane of the feline large intestine. Note the presence of large numbers of goblet cells and the vascularity of the mucosa. Stained with haematoxylin and eosin. Magnification *c.* × 100.

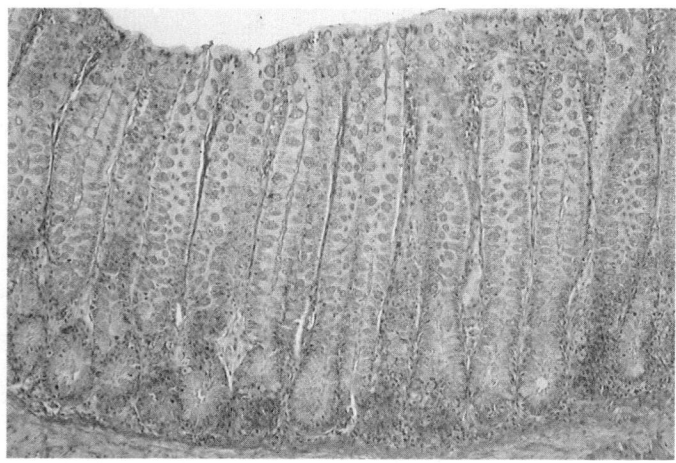

8.143B Medium-power light micrograph of the rectal mucosa showing crypts containing goblet cells. Stained with alcian blue/light green. Material provided by D Ristow, Department of Anatomy, UMDS, Guy's Campus, London.

this part of the large intestine. These bands are said to be shorter than the other intestinal layers, thus producing puckering or haustration of the caecum and colon into *sacculi*. When they are removed, the tube lengthens and loses its sacculation. In the descending colon the taeniae thicken at the expense of the rest of the longitudinal layer, while there is a real increase in its total bulk in the sigmoid colon, where the longitudinal fibres are more scattered. They form a layer which completely encircles the rectum, but are thicker on its anterior and posterior aspects, producing recognizable broad anterior and posterior bands. At the rectal ampulla a few strands of the anterior longitudinal fibres pass forwards to the perineal body (p. 607), as the *musculus recto-urethralis*. In addition, two fasciculi of non-striated muscle pass antero-inferiorly from the front of the second and third coccygeal vertebrae to blend with the longitudinal muscle fibres on the posterior wall of the anal canal, forming the *rectococcygeal muscles* (Wesson 1951). The *circular fibres* form a thin layer over the caecum and colon, aggregated particularly in the intervals

between the sacculi; in the rectum they are a thick layer; in the anal canal they form the sphincter ani internus. Older observations of an interchange of fascicles between circular and longitudinal layers have been confirmed in a study of 112 cadavers (from early fetal life to 88 years); interchanges of fibres, especially near the taenia coli, are commonplace. Deviation of longitudinal fibres from the taeniae to the circular layer may, in some instances, explain the haustration of the colon (Pace 1968).

The *submucosa* and *muscularis mucosae* are like those in the small intestine. The *mucosa* is pale, smooth, non-villous and raised into numerous crescentic folds between the sacculi; in the rectum it is thicker, darker, more vascular, and more loosely attached to the muscularis mucosae. The *epithelium* of the caecum, colon and upper rectum (8.145) consists of scattered mucus-secreting (goblet) cells, columnar absorptive cells (Pittman & Pittman 1966) and M cells (p. 1366) overlying lymphoid follicles in the lamina propria. It extends into numerous tubular glands or crypts providing a large surface for mucous secretion and absorption of water, salts and other materials, functions important in lubricating the passage of contents and in resorption of many substances secreted into the alimentary canal in its more proximal zones. The large intestine also provides an environment for a large bacterial flora, essential in many animals for digestion and for the metabolism of organic compounds to supplement vitamin intake. The columnar cells, which have luminal microvilli, are linked by juxta-luminal junctional complexes; their cytoplasm contains a subapical terminal web and the usual cytoplasmic organelles. Many also contain secretory granules in their apical cytoplasm. Their secretion appears to be mucoid, but is also rich in antibodies of the IgA type, providing some protection against pathogenic micro-organisms (Schofield & Atkins 1970). Other columnar cells possess an apical tuft of long microvilli (8.144A,B) and may be sensory (Silva 1966).

Glands of the large intestine are minute, perpendicular tubules of mucosal epithelium, longer, more numerous and in closer apposition than those of the small intestine. Their openings (8.145) produce a cribriform appearance; they are lined by short columnar epithelium, consisting mostly of goblet cells (8.143), between which are columnar absorptive cells.

Solitary lymphoid follicles, most abundant in the caecum, vermiform appendix and rectum, are scattered over the rest of the large intestine (Langman & Rowlands 1986). They are like those of the small intestine (p. 1363).

VESSELS AND NERVES OF THE LARGE INTESTINE

The *arteries* which supply the parts of the large intestine derived from the mid-gut (caecum, appendix, ascending colon and right two-thirds of the transverse colon) are derived from colic branches of the superior mesenteric artery; those supplying hind-gut derivatives (left part of the transverse, descending and sigmoid colon, rectum and upper anal canal) are derived from the inferior mesenteric (and its terminal branch, the superior rectal) and the middle rectal arteries (a branch of the internal iliac). Their large branches ramify between and supply the muscular layers, divide into small submucosal rami and enter the mucosa. Rectal and anal canal arteries are: (1) The superior rectal (the continuation of the inferior mesenteric). This is the main rectal vessel, dividing into two rami descending one each side of the rectum, their terminal branches piercing the muscular coat to enter the rectal submucosa and descending into the anal columns as far as the anal valves, where they form looped anastomoses. (2) The middle rectal arteries which traverse the 'lateral rectal ligaments' to supply the muscle of the lower rectum, anastomosing freely with each other but forming only poor anastomoses with the superior and inferior rectal arteries. (3) The inferior rectal arteries (from the internal pudendals), which supply the internal and external sphincters, the anal canal below its valves and the perianal skin. (4) The median sacral artery which supplies the posterior wall of the anorectal junction and of the anal canal.

The veins of the large intestine are the superior and inferior mesenteric, draining the regions supplied by the corresponding arteries. The veins of the rectum and anal canal are: (1) The superior rectal veins, which pass from the internal rectal plexus in the anal canal and ascend in the rectal submucosa as about six vessels of considerable size to pierce the rectal wall about 7.5 cm above the anus, uniting to form the superior rectal vein, which continues as the inferior mesenteric. (2) The middle rectal veins, from the submucosa of the rectal ampulla which drain chiefly its muscular walls. (3) The inferior rectal veins, which drain the external rectal plexus and lower anal canal. Anastomoses occur between portal and systemic veins in the wall of the anal canal (pp. 816, 820).

The *nerve supply* is (except in the lower anal canal) sympathetic and parasympathetic. The caecum, appendix, ascending colon and right two-thirds of the transverse colon (derivatives of the mid-gut) have a sympathetic supply from the coeliac and superior mesenteric ganglia, and a parasympathetic supply from the vagus; the nerves are distributed in plexuses around the rami of the superior mesenteric artery. The left third of the transverse colon, the descending and sigmoid colon, rectum and upper anal canal (derivatives of the hind-gut) take their sympathetic supply from the lumbar part of the trunk and the superior hypogastric plexus by means of periarterial plexuses on rami of the inferior mesenteric artery. The sympathetic supply of the colon is largely vasomotor. The parasympathetic supply is from the pelvic splanchnic nerves (nervi erigentes), from which rami pass to the inferior hypogastric plexuses to supply the rectum and upper half of the anal canal: some fibres ascend through the superior hypogastric plexus to accompany the inferior mesenteric artery to the transverse, descending and sigmoid colon (p. 1166). Rami of the pelvic splanchnic nerves ascend on the posterior abdominal wall behind the peritoneum, independently of the inferior mesenteric artery, to be distributed directly to the left colic flexure and descending colon (Mitchell 1953). The ultimate distribution in the wall of the large intestine is as in the small intestine (p. 1563). Adrenergic and cholinergic activity in the nerve supply of the taenia coli, and distribution of the nerve fibres, suggest that (in guinea-pigs) few non-striated myocytes are directly innervated, propagation of excitation being chiefly through gap junctions between myocytes (Bennett & Rogers 1967).

Sympathetic nerves to the rectum and upper anal canal pass mainly along the inferior mesenteric and superior rectal arteries and partly via the superior and inferior hypogastric plexuses, the latter supplying the lower part of the rectum and the internal anal sphincter. Parasympathetic rami from the pelvic splanchnic nerves pass forwards as long strands (about 3 cm long) from the sacral nerves to join the inferior hypogastric plexuses on the sides of the rectum, being motor to the rectal musculature and inhibitory to the internal anal sphincter. The external sphincter ani is supplied by the inferior rectal ramus of the pudendal nerve (S2,3) and the perineal ramus of the fourth sacral nerve (p. 1149). In rectal surgical excision, dissection must be kept close to its wall to avoid damage to these nerves with consequent bladder dysfunction and, in males, loss of penile erection. Afferent impulses mediating sensations of distension pass in afferent fibres in the parasympathetic nerves, pain impulses in the sympathetic *and* parasympathetic nerves supplying the rectum and the upper part of the anal canal. In colonic *aganglionosis (megacolon)* postganglionic autonomic neurons are reduced or absent in the colonic wall (Bodian et al 1961, Bodian 1966, Soltero-Harrington et al 1969). Garrett et al (1969) have studied the myenteric plexus and ganglionic neurons by electron

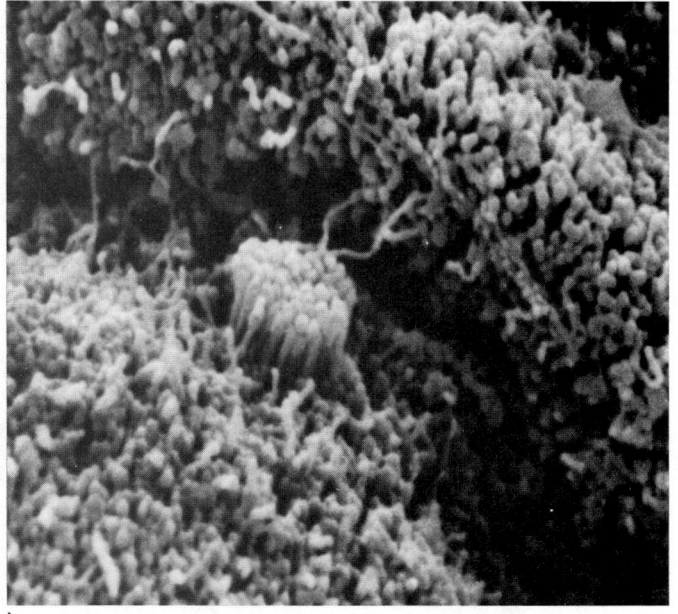

A

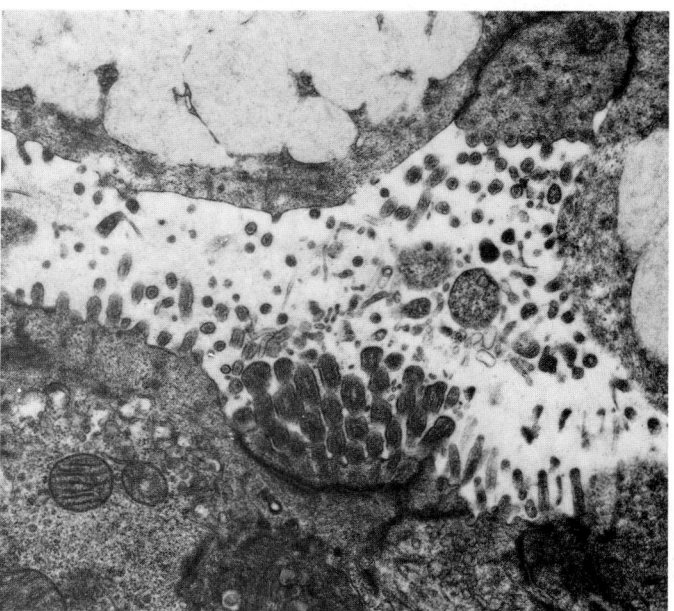

B

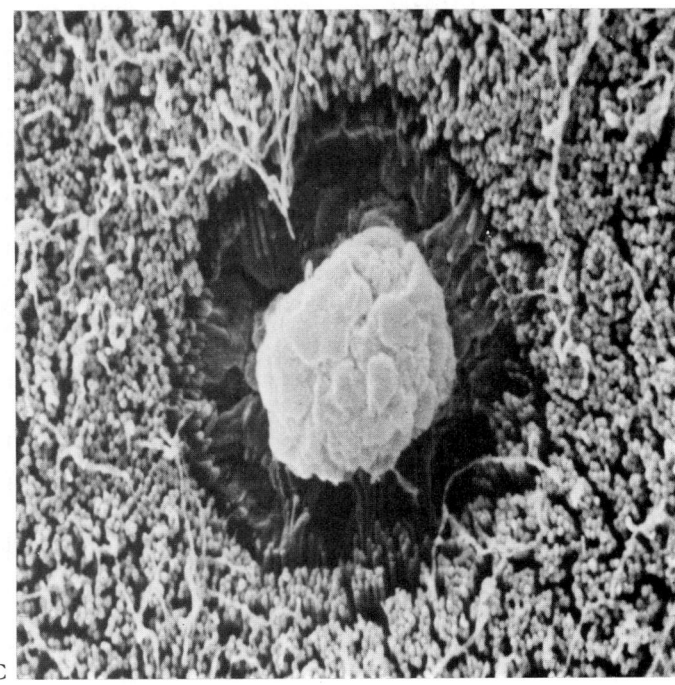

8.144 A. Scanning electron micrograph of the epithelium lining the colon (rat), showing a cell of suggested sensory function, bearing an apical tuft of particularly long microvilli. Magnification × 12 000. B. Transmission electron micrograph showing part of a colonic epithelial cell bearing sensory microvilli. The smaller absorptive microvilli of adjacent cells can also be seen. Magnification × 14 000. C. Scanning electron micrograph of the epithelium lining the colon (rat), showing a goblet cell surrounded by absorptive cells bearing microvilli. Prepared and photographed by Michael Crowder, Guy's Hospital Medical School, London. Magnification × 8000.   C

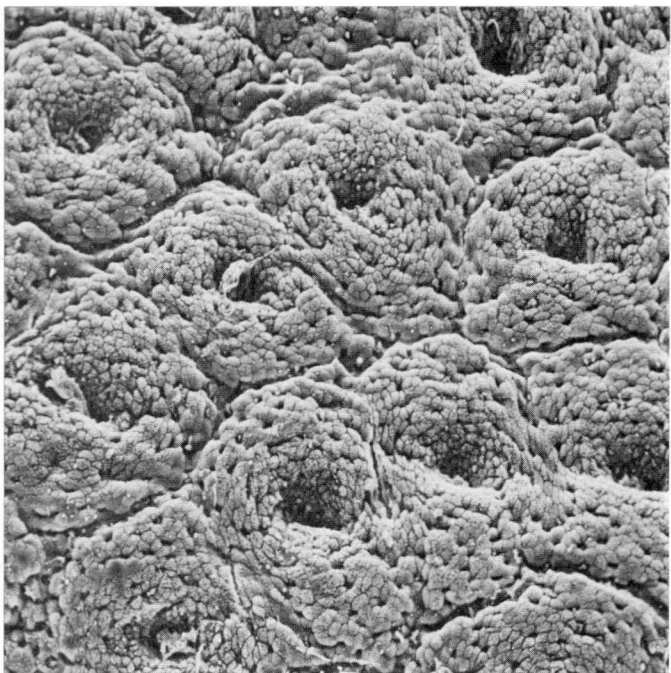

8.145 Scanning electron micrograph of the luminal surface of the human rectal mucosa. The outlines of cells bearing microvilli and the openings of rectal crypts can be seen. Material supplied by D S Rampton; prepared and photographed by Michael Crowder, Guy's Hospital Medical School, London. Magnification × 240.

microscopy and histochemical techniques for transmitter substances reporting that in megacolon a variable diminution and sometimes absence of ganglion cells occurred, but that innervation of the muscle layers was defective even when ganglionic neurons were present.

The *lymph nodes and vessels* of the large intestine are described on p. 854.

The Gastro-entero-pancreatic Endocrine System

The gastro-entero-pancreatic (GEP) endocrine system (Fujita 1973) consists of scattered, often solitary, hormone-producing cells of the gastrointestinal mucosa and pancreas. Electron microscopy of ultra-thin (0.07 μm) sections and immunocytochemistry of adjacent semi-thin (1 μm) sections of resin-embedded tissue has led to the identification of many types of GEP endocrine cell types and their secretory products (p. 1377). Of at least 15 human types so far recognized (p. 1377), one, type B, is restricted to the pancreatic islets and at least three occur in both gastrointestinal mucosa and pancreas.

The ultrastructure of human GEP endocrine cells has been detailed by Rubin (1972), Sasagawa et al (1973), Capella et al (1976) and Cavallero et al (1976) and summarized by Solcia (1981). Endocrine cells are scattered in the gastrointestinal mucosa, with their bases resting on the basal lamina. Those of the mucosa in the main part of the stomach (the 'oxyntic mucosa') do not usually reach the lumen and are referred to as 'closed', while most of those in the pyloric and intestinal mucosae do so and are therefore termed 'open'. Their osmiophilic secretory granules vary in shape, size and ultrastructure in the different cell types, being usually infranuclear, while the Golgi complexes are supranuclear; luminal aspects of 'open' cells display microvilli of variable number, length and shape. Typical entero-endocrine cells are shown in Fig. **8.**146A,B. Human *P cells* contain very small (100–140 nm) secretory granules slightly reactive to Grimelius' silver stain (Capella et al 1977); rare in normal adult tissues, they may contain a bombesin-like polypeptide (Polak et al 1976). *EC cells*, which contain osmiophilic, argentaffin, Grimelius' silver-reactive granules, are classified as EC_1, EC_2 and EC_n; in addition

to 5-hydroxytryptamine, the EC_1 cells store substance P (Heitz et al 1977), EC_2 cells store motilin (Polak et al 1975) and EC_n cells an unidentified material. The D_1 *cells* contain argyrophilic granules about 140–190 nm in diameter (Capella et al 1977) and store a vaso-active intestinal peptide-like material. *PP cells*, common in the pancreatic islets but rare in the exocrine pancreas, store pancreatic polypeptide in granules of 150-170 nm; they are equivalent to F cells identified in other mammals (Baetens et al 1976). Human D, B and A cells are described with the endocrine pancreas (p. 1383). *X cells*, identified in human oxyntic mucosa by Solcia et al (1977), are of unknown function. Human *ECL cells* (Vassallo et al 1971) store a reducing amine, possibly 5-hydroxytryptamine, in granules with intensely argyrophilic cores; histamine may also occur. Human *G cells* (Vassallo et al 1971) manufacture gastrin and possibly enkephalin (Polak et al 1978), and have slightly argyrophilic granules with floccular contents. *S cells* (Capella et al 1976) are scattered in duodenojejunal mucosa and produce secretin (Larsson et al 1977); they are similar to D_1 cells but differ in their secretory product. Human *I cells* (Capella et al 1976), commonest in the duodenum and jejunum but rare in the ileum, are sources of cholecystokinin-pancreozymin (pancreaticozymin) (Buchan et al 1977). Human *K cells* (Capella et al 1976) contain large granules (approximately 350 nm in diameter) with osmiophilic, argyrophobic cores; like I cells, they are commonest in the duodenum and jejunum. K cells produce gastric inhibitory peptide. Human *N and L cells* are difficult to distinguish cytologically; granules of N cells are, however, generally homogeneous and about 300 nm in diameter, while those of L cells sometimes have argyrophilic cores and tend to be smaller (about 260 nm). N cells produce neurotensin (Orci et al 1976), L cells enteroglucagon or glicentin.

Concentrations of endocrine cells in the gastrointestinal mucosa are minute and generally decrease progressively in an aboral direction (**8.**146C).

Certain common features of GEP endocrine cells allow other classifications. They all produce peptides and/or amines active as hormones or neurotransmitters and contain neuron-specific enolase, an isoenzyme of the glycolytic enzyme enolase (Polak et al 1980). They thus belong to the *APUD cell series* (Pearse 1968, 1977, 1980) and modulate not only autonomic activity but also each other. The APUD concept is detailed elsewhere (p. 1466). The discovery of supposedly similar neurohormones and neurotransmitter peptides in cerebral neurons and some GEP endocrine cells, with other neuronal characteristics of the latter, has led to their designation as *paraneurons* (Fujita 1976). Peptides common to brain and gastrointestinal mucosa include: substance P, somatostatin, vaso-active intestinal peptide, bombesin, neurotensin, cholecystokinin and the opiatoid enkephalin (Bloom & Polak 1978). GEP endocrine cells are presumed to have superficial receptor sites, stimulation of which by 'secretogogues' triggers stimulus-secretion coupling (Kanno 1973), as in chromaffin cells (Douglas 1968); they can thus also be termed *receptosecretory cells* (Fujita 1976).

The mode of action of endocrine cells restricted to the gastrointestinal mucosa remains in doubt. Their ultrastructure and proximity to capillaries suggest that their secretory products are endocrinal, exerting distant, diffuse effects via the blood. However, of their many products, only the following have been shown to act as circulating hormones: gastrin, secretin, cholecystokinin-pancreozymin, gastric inhibitory peptide, motilin and enteroglucagon (Bloom & Polak 1978). Basic differences exist between endocrine cells in gastric and intestinal mucosae and those in most other endocrine organs: they are not aggregated into glands but scattered among their local targets; most are close to the alimentary lumen, allowing their specialized plasmalemmae to detect and respond to luminal stimuli; there are no common hyposecretory or hypersecretory syndromes; relations between plasma hormone levels and functional response (e.g. modulation of neural control of gut motility) is not stoichiometric. Wingate (1976) has suggested that gastrointestinal hormones may have local 'paracrine' *and* distant 'endocrine' effects; direct actions on gastrointestinal myocytes, on adjacent endocrine cells, on other enterocytes and on local neurons are speculative but feasible. Although the gut might be regarded as the largest

HUMAN GASTRO-ENTERO-PANCREATIC ENDOCRINE CELLS AND THEIR PRODUCTS

	LOCATION						SECRETORY PRODUCTS
	Pancreas	Stomach		Small Intestine		Large Intestine	
		Oxyntic	Pyloric	Upper	Lower		
	P*	P	P	P	—	—	Bombesin-like?
	—	EC_n	EC_n	EC_2	EC_1	EC_1	EC_1: 5-HT, Substance P EC_2: 5-HT, Motilin EC_n: 5-HT, Undetermined
	D_1	D_1	D_1	D_1	D_1	D_1	VIP-like?
C E L L T Y P E S	PP	—	—	PP?	—	—	Pancreatic Polypeptide
	D	D	D	D	—	—	Somatostatin
	B	—	—	—	—	—	Insulin
	A	A?*	—	A?	—	—	Glucagon
	—	X	—	—	—	—	Undetermined
	—	ECL	—	—	—	—	H or 5-HT
	—	—	G	G	—	—	Gastrin, Enkephalin
	—	—	—	S	S	—	Secretin
	—	—	—	I	I	—	Cholecystokinin-pancreozymin
	—	—	—	K	K	—	GIP
	—	—	—	—	N	—	Neurotensin
	—	—	—	L	L	L	Enteroglucagon or GLI

*Found in the fetus or new-born, rare in adults.
Abbreviations used for secretory products:
5-HT = 5-hydroxytryptamine
VIP = Vaso-active intestinal peptide
H = Histamine
GIP = Gastric inhibitory peptide
GLI = Glicentin

(Based mainly on the Lausanne 1977 classification
of gastro-entero-pancreatic endocrine cells
reported by Solcia et al in 1978).

endocrine organ (Pearse 1974), it is perhaps more properly regarded as a region in which neural, paracrine and endocrine controls of activity are intimately linked.

Applied Anatomy of the Intestine: Hernia

The rarity of rupture of the jejunum and ileum by external injury is due to their elasticity and mobility; the more fixed duodenum, particularly its horizontal part across the vertebral column, is more vulnerable. In external hernia the ileum is most frequently involved; in the large intestine it is usually the caecum or sigmoid colon. The chief sites of hernia are inguinal, femoral and umbilical.

Inguinal Hernia
Here a viscus is protruded through the inguinal region of the abdominal wall. The principal varieties are oblique and direct.
(1) *Oblique (indirect) inguinal hernia.* Here the intestine is pushed through the lateral inguinal fossa (behind the deep inguinal ring), preceded by a pouch of parietal peritoneum and extraperitoneal areolar tissue. It enters the inguinal canal at its deep ring and is invested by internal spermatic fascia enclosing the spermatic cord. As it traverses the canal it pushes up the arching fibres of transversus abdominis and obliquus internus, is covered by the cremasteric fascia and muscle and lies anterior to the cord.

Emerging at the superficial inguinal ring, it is invested by the external spermatic fascia and also, as it descends into the scrotum, by superficial fascia and skin. The hernia may become constricted at the deep ring, with interference to its blood supply (strangulation). When this is relieved the deep ring should be cut superolaterally to avoid the inferior epigastric vessels. Most oblique inguinal hernias follow congenital defects in the processus vaginalis (p. 258). Obliteration of this may be complete at birth or it may begin late and be completed only after birth; closure begins at the deep inguinal ring and epididymal head, extending until all of the intervening region becomes a fibrous cord. Complete or partial failure of closure of the processus entails variations in the relation of the hernial protrusion to the testis and tunica vaginalis; e.g. if the processus is fully patent, the herniated gut descends in front of the testis into the tunica vaginalis (*complete congenital hernia*); in this case, the processus and tunica form the hernial sac. In *incomplete congenital hernia* (hernia into the funicular process), the herniating gut descends to the top of the testis, where the processus is sealed off from the tunica vaginalis (8.147). Although the above types are called congenital, actual extrusion into a pre-existing peritoneal sac may not occur until adult life and then be produced by an increased intra-abdominal pressure or sudden muscular strain. (For a critique of the anatomy of inguinal hernia consult Lytle 1979.)
(2) *Direct inguinal hernia.* Here the protrusion is through some part of the inguinal triangle, which is bounded inferiorly by the

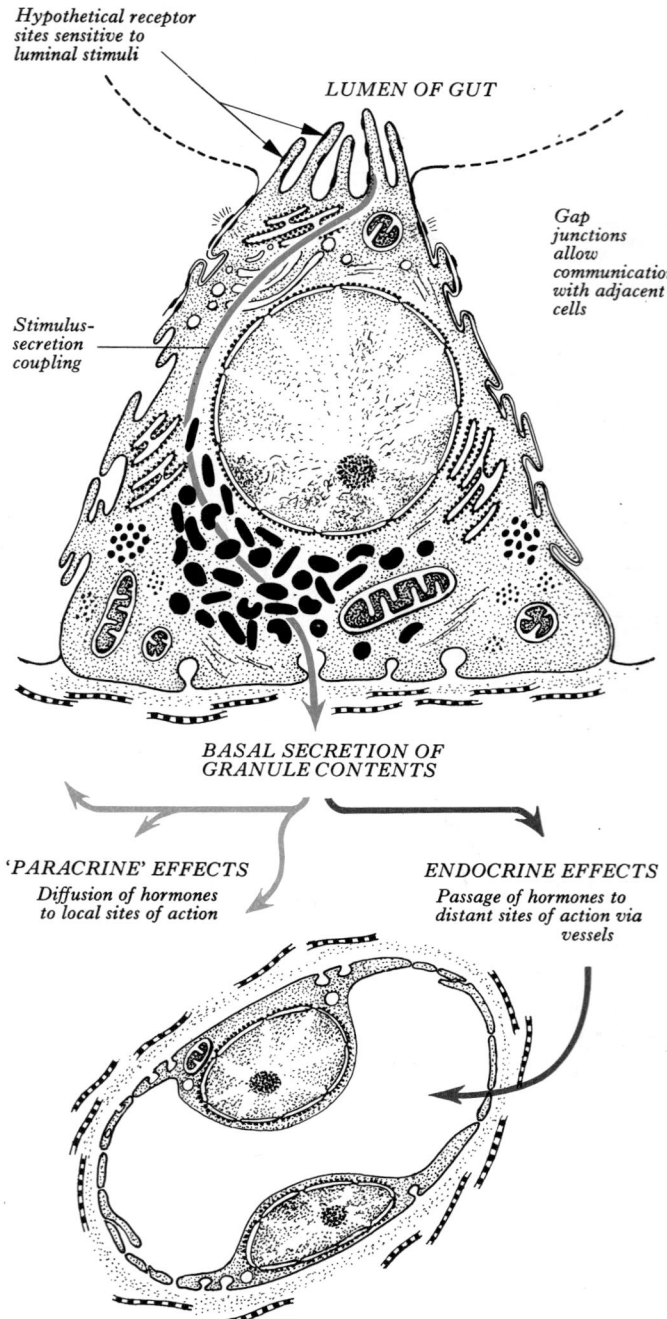

8.146A Diagram showing the ultrastructure and possible modes of action of an entero-endocrine cell.

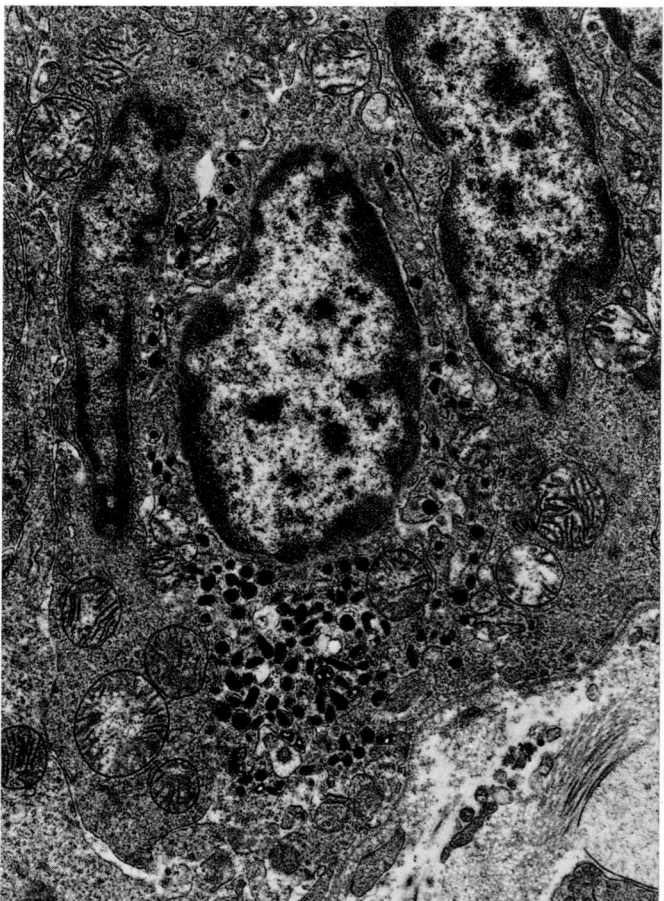

8.146B Transmission electron micrograph of an entero-endocrine (APUD) cell of the epithelium lining the colon (rat). Secretory vesicles can be seen towards the basal aspect of the cell. Absorptive columnar cells lie on either side of the APUD cell. Prepared and photographed by Michael Crowder, Guy's Hospital Medical School, London. Magnification × 11 000.

medial half of the inguinal ligament, medially by the lower lateral border of rectus abdominis and laterally by the inferior epigastric artery. It overlies the medial inguinal fossa and, partly, the supravesical fossa (p. 1338). A direct hernia is either through *(a)* the *medial inguinal fossa*, where only extraperitoneal tissue and transversalis fascia separate the peritoneum from the aponeurosis of the external oblique, or *(b)* the *supravesical fossa* and *falx inguinalis* (conjoint tendon), which lies in front of the fossa. In the first form, herniation is lateral to the conjoint tendon, propelling before it the peritoneum, extraperitoneal tissue and transversalis fascia to enter the inguinal canal, which it traverses to emerge from the superficial ring, covered by external spermatic fascia. Its coverings are like those of the oblique form, except that a part of the general layer of transversalis fascia replaces the internal spermatic fascia, the hernia being between the innermost and middle coverings of the spermatic cord. In the second, more frequent form, the hernia is either between the fibres of the falx inguinalis, or the falx is gradually distended to form a complete covering. The hernia thus enters the lower end of the canal, escapes at the superficial ring medial to the cord and is covered by external spermatic fascia, superficial fascia and skin. Its coverings differ from those of oblique hernia, the conjoint tendon replacing the cremaster and part of the transversalis fascia replacing the internal spermatic fascia. In all varieties the most superficial covering is the external spermatic fascia, the outermost covering of the cord. An oblique inguinal hernia is within the cord, sharing all its coverings; a direct hernia acquires an additional covering from the transversalis fascia.

Direct inguinal hernia is much less frequent than the oblique type, and occurs more often in males. Its main peculiarities are: *(a)* it is sited above the body of the pubic bone, *(b)* the inferior epigastric artery is *lateral* (not medial) to the neck of the sac; *(c)* the spermatic cord is posterolateral, not directly posterior as in oblique hernia. A direct hernia is always of the acquired type. The stricture in both varieties of direct hernia is usually at the neck of the sac or at the superficial ring. Where the conjoint tendon is split, constriction may occur at the edges of the fissure. In all cases of inguinal hernia, whether oblique or direct, it is correct to divide the stricture upwards, parallel to the inferior epigastric artery to avoid damaging that vessel.

Femoral Hernia

A femoral hernia protrudes through the femoral ring (p. 781), which is normally closed by a femoral septum of modified extraperitoneal tissue and is therefore a weak spot, especially in females, where the ring is larger and subject to profound changes during pregnancy. Femoral hernia is hence more common in women. When a section of intestine bulges through the ring, it pushes out a hernial sac of peritoneum. It is covered by extraperitoneal tissue (the femoral septum) and descends along the

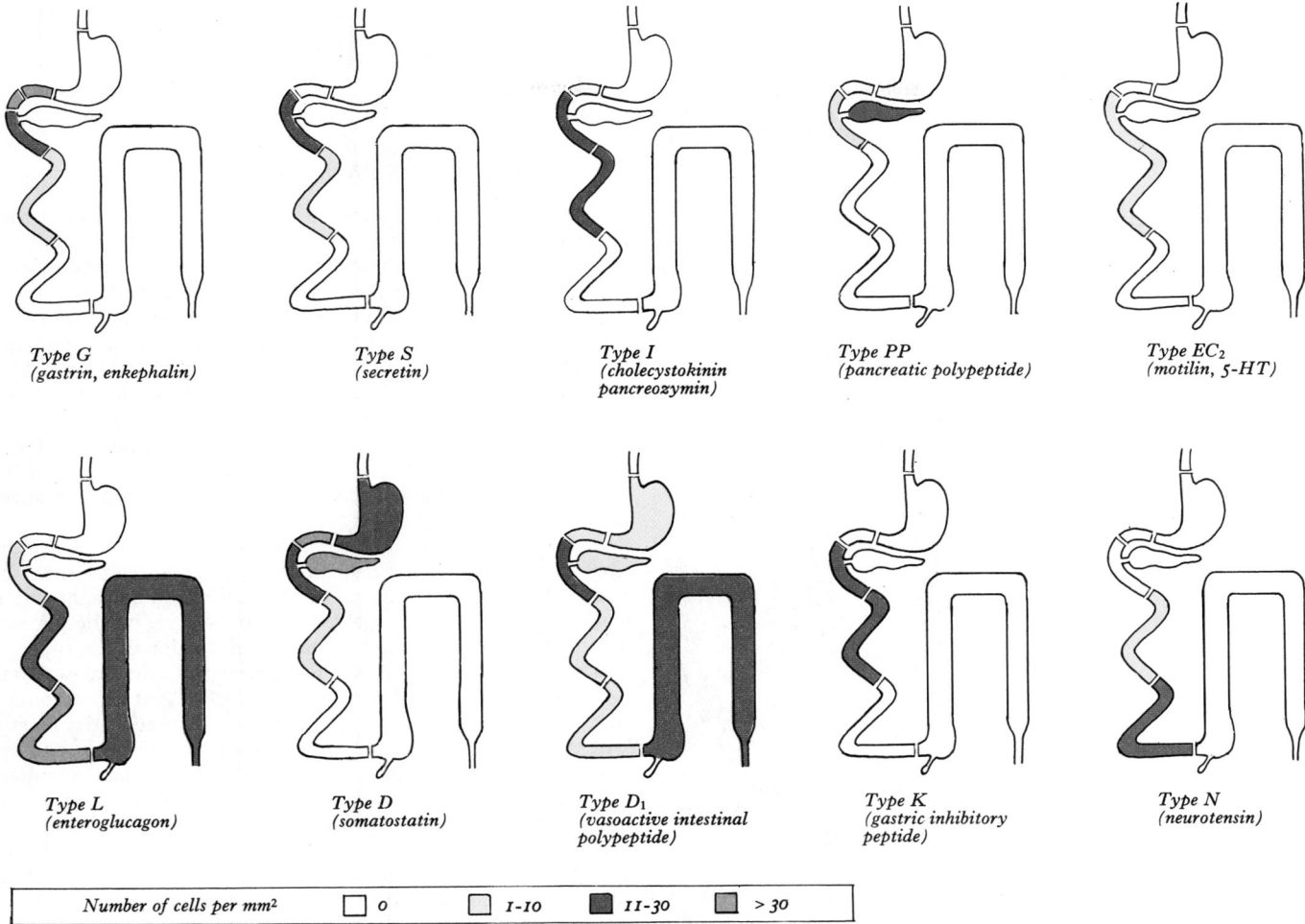

Type G
(gastrin, enkephalin)

Type S
(secretin)

Type I
(cholecystokinin
pancreozymin)

Type PP
(pancreatic polypeptide)

Type EC₂
(motilin, 5-HT)

Type L
(enteroglucagon)

Type D
(somatostatin)

Type D₁
(vasoactive intestinal
polypeptide)

Type K
(gastric inhibitory
peptide)

Type N
(neurotensin)

| Number of cells per mm² | ☐ 0 | ☐ 1-10 | ■ 11-30 | ▨ > 30 |

8.146C Approximate quantitative distribution of a selection of human gastro-entero-pancreatic (GEP) endocrine cells (highly diagrammatic, after Bloom & Polak 1978, with permission from Churchill Livingstone).

femoral canal to the saphenous opening, where it is prevented from descending along the femoral sheath by the narrowing of the latter, by the vessels and by the close attachment of the superficial fascia and sheath to the lower part of the rim of the saphenous opening (p. 638). The hernia hence turns forwards, distending the cribriform fascia and curving upwards over the inguinal ligament and the lower part of the external oblique aponeurosis. While in the canal the hernia is usually small, due to the resistance of its surrounds; but with escape into the inguinal loose connective tissue it enlarges. Thus a femoral hernia first descends, then ascends forwards; hence pressure to reduce it should be directed in the reverse order, with the thighs passively flexed for greatest relaxation.

Covering a femoral hernia are (from within outwards): the peritoneum, femoral septum, femoral sheath, cribriform fascia, superficial fascia and skin. A fibrous covering, the *fascia propria*, just outside the peritoneal sac but frequently separated from it by adipose tissue, may be easily mistaken for the sac, and its contained extraperitoneal fat for omentum; the fat may resemble a lipoma, but dissection will reveal the true hernial sac in its centre. The fascia propria is merely a femoral septum thickened to form a membranous sheet by hernial pressure. The intestine reaches only to the saphenous opening in *incomplete femoral hernia* in contradistinction to *complete hernia* where it passes through the opening. The small size of an incomplete hernia renders it difficult to detect and therefore dangerous, especially in the corpulent. The site of strangulation varies: it may be at the hernial sac's neck; more often it is at the junction of the falciform margin of the saphenous opening with the free edge of the pectineal part of the inguinal ligament; or it may be at the saphenous opening (p. 638). The stricture should be divided superomedially for a distance of 4–6 mm to avoid all normally positioned vessels and

other important structures. (However, an abnormal obturator artery may be a complication, see p. 778.)

The pubic tubercle is an important landmark in distinguishing inguinal from femoral hernias; the hernia's neck is superomedial to it in inguinal hernia but inferolateral in the femoral form.

Umbilical Hernia

There are three varieties of umbilical hernia.

(1) *Congenital umbilical hernia.* This is due to the failure of retraction of the umbilical loop of the gut (p. 232).

(2) *Infantile umbilical hernia.* This is due to stretching of umbilical scar tissue, usually within three years of birth, and is associated with increased intra-abdominal pressure.

(3) *Acquired umbilical hernia.* This is really through the linea alba, usually just above the umbilicus (para-umbilical hernia) and occurs most frequently in obese multiparous females.

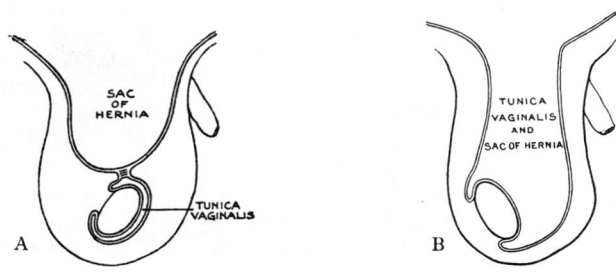

8.147 Diagrams representing varieties of oblique inguinal hernia: A. incomplete congenital; B. complete congenital.

Rarely, hernia may occur at other sites, e.g. through the *lumbar triangle* (p. 610), *obturator foramen*, *greater* or *lesser sciatic foramen* or *ischiorectal fossa*. *Incisional hernia* may occur at the sites. of abdominal scars, particularly if the wound becomes infected.

For variations of medical and surgical importance in the small intestine and also in the colon consult Goligher (1967) and Kanagasuntheram (1970).

The Pancreas

The pancreas is a soft, lobulated, greyish-pink gland, 12–15 cm long, extending nearly transversely across the posterior abdomi-

nal wall from the duodenum to the spleen, behind the stomach. Its broad, right extremity or *head* is connected to the *body* by a slightly constricted *neck*; its narrow, left extremity is the *tail*. It ascends slightly to the left in the epigastric and left hypochondriac regions.

RELATIONS OF THE PANCREAS

The structures related to the pancreas are best considered with respect to its different parts (**8.125, 148, 149, 150**).

The head, flattened anteroposteriorly, lies within the duodenal curve. Its upper border is overlapped by the superior duodenal segment, the other borders being grooved by the adjacent margin of the duodenum. which they variably overlap in front and behind. Sometimes a small part of the head is actually embedded in the wall of the descending duodenum. From the lower left part of the head the *uncinate process* projects upwards and to the left behind the superior mesenteric vessels. In or near the groove between the duodenum and the right and lower borders of the head are the anastomosing superior and inferior pancreatico-duodenal arteries (pp. 769, 772; **6.109**).

Anterior surface. From the head's anterosuperior aspect the neck juts forwards, upwards and to the left, merging with the body. The boundary between head and neck, on the right and in front, is a groove for the gastroduodenal artery; on the left and behind it is a deep incisure containing the union of the superior mesenteric and splenic veins to form the portal vein. Below and to the right of the neck the head's anterior surface is at first in contact with the transverse colon, separated only by areolar tissue; still lower the surface is covered by peritoneum continuous with the inferior layer of the transverse mesocolon (**8.93**) and in contact with the jejunum. The uncinate process is crossed anteriorly by the superior mesenteric vessels.

Posterior surface. The head is related posteriorly to the inferior vena cava which ascends behind it, covering almost all of this surface. Also it is related to the terminal parts of the renal veins and the right crus of the diaphragm. The uncinate process lies in front of the aorta. The bile duct is lodged either in a superolateral

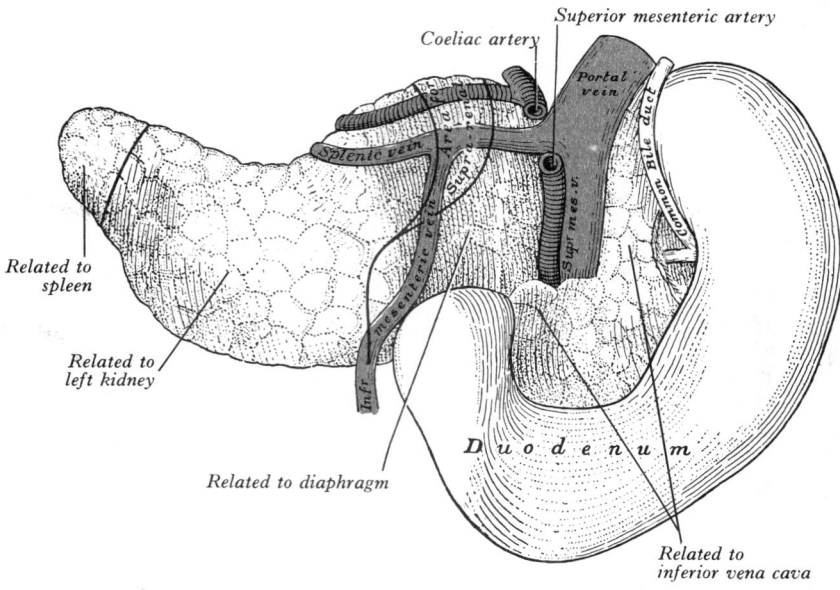

8.148 Posterior aspect of the pancreas and duodenum.

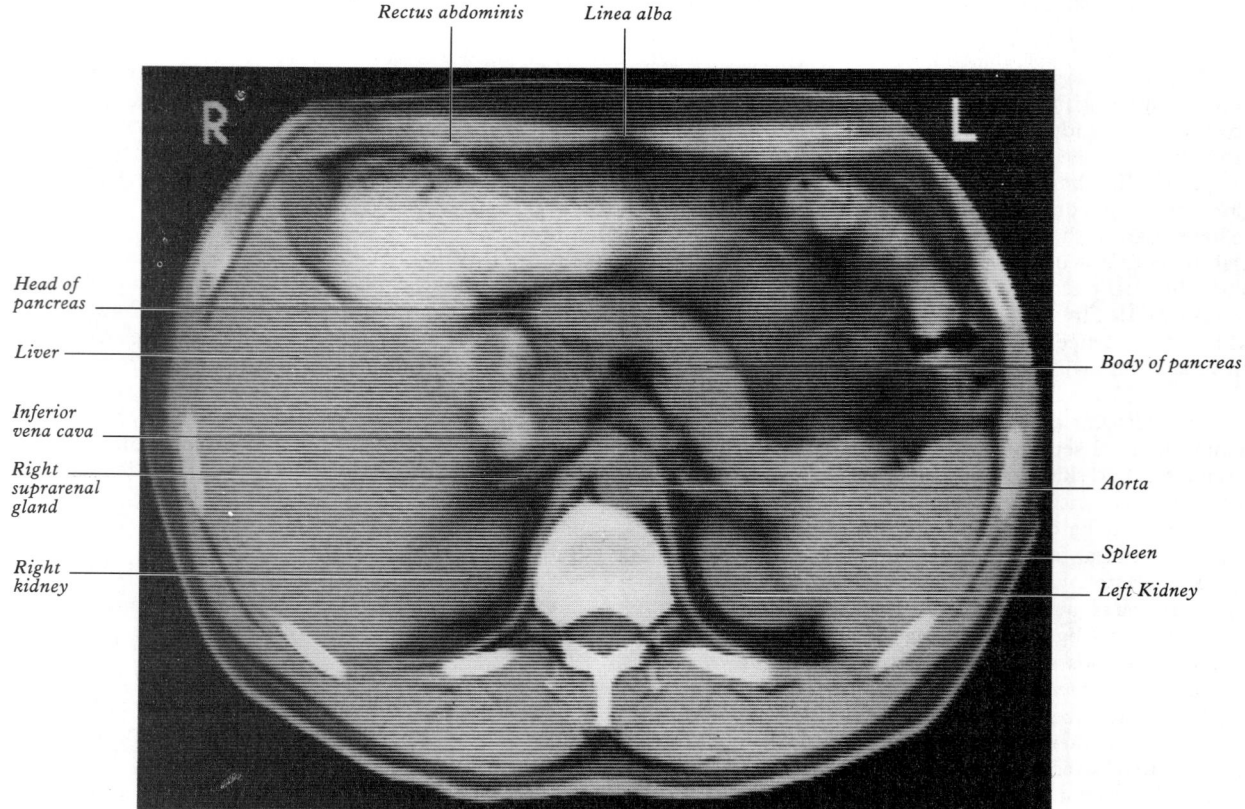

8.149 Computed tomogram of the abdomen in the transverse plane at the level of the pancreas. Supplied by Shaun Gallagher, Guy's Hospital; photography by Sarah Smith, UMDS, Guy's Hospital Campus, London.

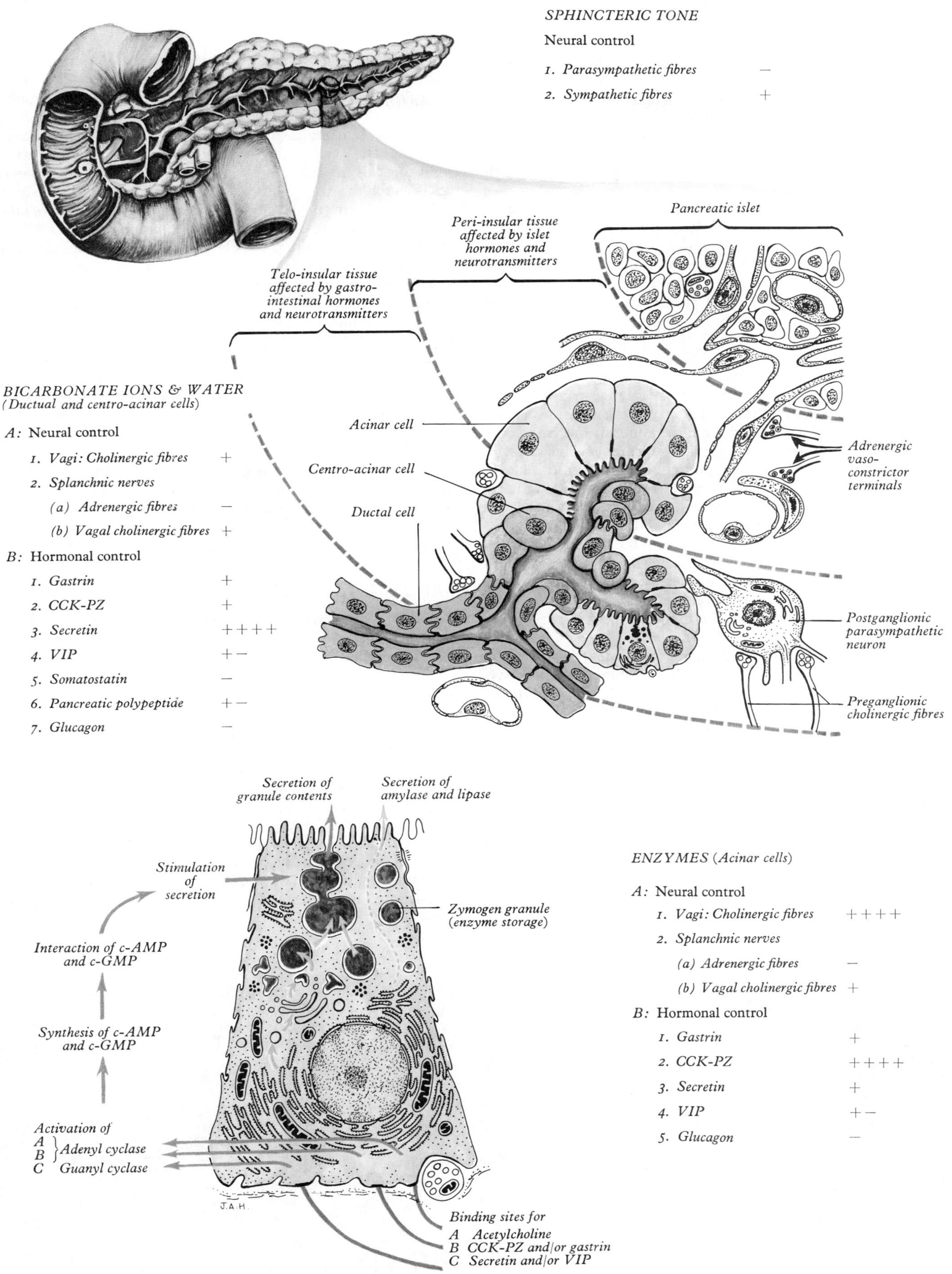

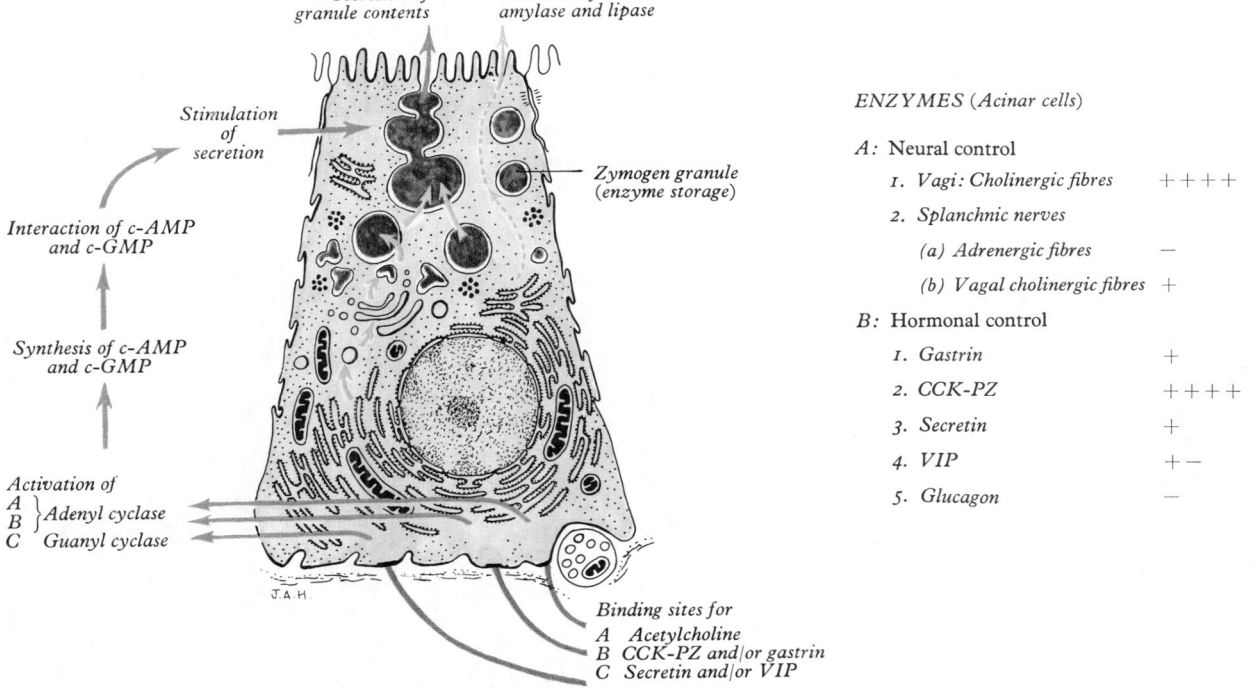

BICARBONATE IONS & WATER
(Ductual and centro-acinar cells)

A: Neural control

1. *Vagi: Cholinergic fibres*   +
2. *Splanchnic nerves*
    (a) *Adrenergic fibres*   —
    (b) *Vagal cholinergic fibres*   +

B: Hormonal control

1. *Gastrin*   +
2. *CCK-PZ*   +
3. *Secretin*   + + + +
4. *VIP*   + −
5. *Somatostatin*   —
6. *Pancreatic polypeptide*   + −
7. *Glucagon*

Labels on central diagram:
Pancreatic islet
Peri-insular tissue affected by islet hormones and neurotransmitters
Telo-insular tissue affected by gastro-intestinal hormones and neurotransmitters
Acinar cell
Centro-acinar cell
Ductal cell
Adrenergic vaso-constrictor terminals
Postganglionic parasympathetic neuron
Preganglionic cholinergic fibres

Labels on lower diagram:
Secretion of granule contents
Secretion of amylase and lipase
Stimulation of secretion
Interaction of c-AMP and c-GMP
Synthesis of c-AMP and c-GMP
Zymogen granule (enzyme storage)
Activation of
A
B } *Adenyl cyclase*
C *Guanyl cyclase*
J.A.H.
Binding sites for
A Acetylcholine
B CCK-PZ and/or gastrin
C Secretin and/or VIP

ENZYMES (Acinar cells)

A: Neural control

1. *Vagi: Cholinergic fibres*   + + + +
2. *Splanchnic nerves*
    (a) *Adrenergic fibres*   —
    (b) *Vagal cholinergic fibres*   +

B: Hormonal control

1. *Gastrin*   +
2. *CCK-PZ*   + + + +
3. *Secretin*   +
4. *VIP*   + −
5. *Glucagon*   —

8.150 Diagram of the ultrastructure of the exocrine pancreas and the mechanisms by which its secretion is controlled. The hormones referred to by acronyms are as follows: CCK-PZ = cholecystokinin-pancreatico-zymin; VIP = vaso-active intestinal polypeptide.

groove on the posterior surface or within the gland's substance (p. 1394).

The neck, about 2 cm long, projects anterosuperiorly and to the left from the head, merging into the body. Its anterior surface, covered with peritoneum, adjoins the pylorus, with part of the omental bursa intervening; the gastroduodenal and anterior superior pancreaticoduodenal arteries descend in front and to the right of the junction of the neck and head; the posterior surface is related to the superior mesenteric vein and the beginning of the portal vein.

The body, prism-like in section, has three surfaces: anterior, posterior and inferior (more precisely anterosuperior, posterior and antero-inferior, all obliquely orientated).

The concave *anterior surface* faces anterosuperiorly, is covered by peritoneum continuous antero-inferiorly with the *anterior ascending layer* of the greater omentum (8.112) and is separated from the stomach by the omental bursa. On reaching the taenia mesocolica, the greater omentum's *posterior* ascending layer fuses with the anterosuperior surface of the transverse mesocolon, while the anterior layer continues up to the mesocolon's root and is then reflected up over the anterior pancreatic surface.

The *posterior surface*, which is devoid of peritoneum, is in contact with the aorta and the origin of the superior mesenteric artery, the left crus of the diaphragm, left suprarenal gland and with the left kidney and renal vessels, particularly the vein. It is closely related to the splenic vein which courses from left to right and separates it from the structures mentioned. The left kidney is also separated from the pancreas by perirenal fascia and fat.

The narrow *inferior surface* broadens to the left and is covered by the peritoneum of the postero-inferior layer of the transverse mesocolon; inferior to it are the duodenojejunal flexure and coils of the jejunum; its left end rests on the left colic flexure.

The *superior border* is blunt and flat to the right, narrow and sharp to the left near the tail. An *omental tuberosity* usually projects from the right end of the superior border above the level of the lesser curvature of the stomach, in contact with the posterior surface of the lesser omentum. It is related above to the coeliac artery, its common hepatic branch coursing to the right just above the gland, while its sinuous splenic ramus runs to the left along this border.

The *anterior border* separates anterior and inferior surfaces; along it the two layers of the transverse mesocolon diverge, one passing up over the anterior surface, the other backwards over the inferior surface. The *inferior border* separates posterior and inferior surfaces, the superior mesenteric vessels emerging from under its right end.

The tail is narrow, usually reaching the inferior part of the gastric surface of the spleen. It is contained between the two layers of the splenorenal (lienorenal) ligament with the splenic vessels.

The main pancreatic duct traverses the gland from left to right, being nearer its posterior than its anterior surface (8.150). It begins by the junction of lobular ducts in the tail and, running to the right in the body, receives further lobular ducts which join it almost at right angles (a 'herringbone pattern'). Much enlarged, it reaches the neck of the gland, turning down, backwards and right towards the bile duct, which lies on its right side. The two ducts enter the wall of the descending part of the duodenum obliquely and unite in a short dilated *hepato-pancreatic ampulla* or ampulla of the bile duct (p. 1394); the narrow distal end of this opens on the summit of the *major duodenal papilla*, posteromedial in this part of the duodenum, and 8–10 cm distal to the pylorus. Usually they do not unite until very near the orifice on the papilla. Sometimes they open separately. Frequently an *accessory pancreatic duct* drains the lower part of the head (8.150), ascending in front of the main duct, with which it communicates, and opening on a small rounded *minor duodenal papilla*, about 2 cm antero-superior to the major. The duodenal end of the accessory duct may fail to expand; secretion is then diverted along the connecting channel into the main duct (Dawson & Langmann 1961).

Surface Anatomy

The head of the pancreas is in the duodenal curve. The neck lies in the transpyloric plane, behind the pylorus. The body passes obliquely up and left for about 10 cm, its left part lying a little above the transpyloric plane. The tail is a little above and to the left of the intersection of the transpyloric and left lateral planes.

PANCREATIC STRUCTURE

The pancreas is composed of two different types of glandular tissues in intimate association with each other. The main mass is *exocrine*, embedded in which are *pancreatic islets* of *endocrine cells*.

THE EXOCRINE PANCREAS

The exocrine pancreas is a branched acinar (acino-racemose) gland, surrounded and incompletely lobulated by delicate loose connective tissue (de Reuck & Cameron 1962, Beck & Sinclair 1971). Its pyramidal, acinar, secretory cells are arranged in flask-shaped or tubular groups. A narrow intercalated (intralobular) duct lies in each secretory mass, the initial parts of its walls being lined by cuboidal *centro-acinar cells*, later replaced by taller cuboidal and eventually columnar cells more distally. Larger, interlobular ducts are surrounded by areolar tissue containing non-striated myocytes and autonomic nerve fibres. Argentaffin

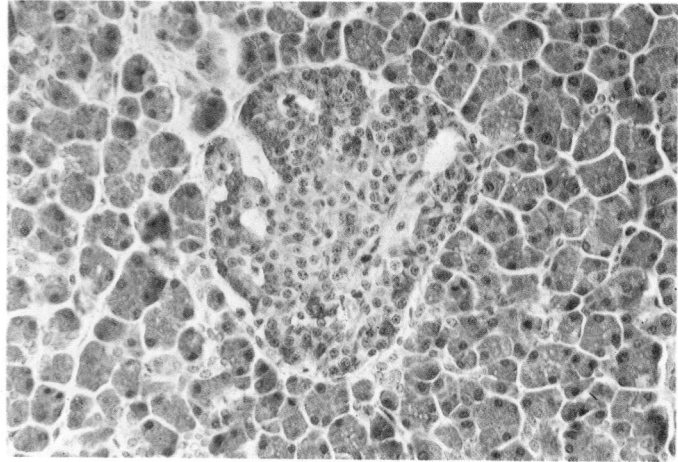

8.151 An islet of Langerhans and surrounding exocrine glandular tissue in the pancreas of a rhesus monkey, stained with orange G and aldehyde fuchsin. Within the islet, the B cells stain purple, whereas the A cells are pale yellow in colour. Provided by J Henderson, Department of Physiology, Guy's Hospital Medical School, London.

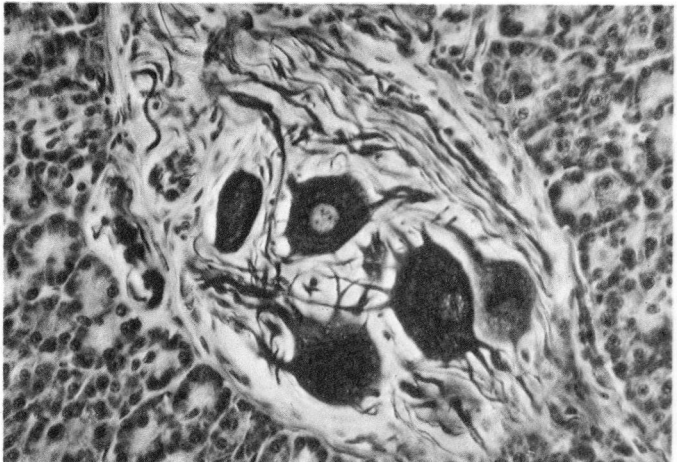

8.152 A low-power micrograph showing a cluster of autonomic ganglionic neurons with dendritic trees and axonal bundles, situated amongst pancreatic acinar cells of the goat. Palmgren silver impregnation. Provided by J Henderson, Department of Physiology, Guy's Hospital Medical School, London.

cells (p. 1376) are present amongst the undifferentiated columnar ductal cells; mast cells are numerous in the surrounding areolar tissue.

The acinar cells are typical zymogenic cells, with a basal nucleus and basophilic cytoplasm consisting of regular arrays of granular endoplasmic reticulum with mitochondria and dense secretory granules. A prominent supranuclear Golgi complex is surrounded by many larger, membranous granules containing enzymic constituents of pancreatic secretion, only active after release. The orderly contents of the acinar cells have provided a widely used model for investigation of routes of secretory synthesis and transport in protein-secreting cells at large (p. 25). After death, the action of pancreatic hydrolytic enzymes rapidly obscures cellular detail.

Ganglionic neurons (**8.152**) and cords of undifferentiated epitheliocytes also appear in the exocrine pancreas; the latter may provide stem cells for replacement of exocrine and perhaps endocrine cells. The structure of the exocrine pancreas and its control are summarized in **8.150**. For further details consult Webster et al (1977), Singh & Webster (1978), Case (1979) and Wormsley (1979).

THE ENDOCRINE PANCREAS

This consists of *pancreatic islets* or *insulae* (of Langerhans), consisting of spheroidal or ellipsoidal clusters of cells dispersed in the exocrine tissue (Laguesse 1906, Lane 1907), together with scattered, often solitary, endocrine cells (Heitz et al 1976).

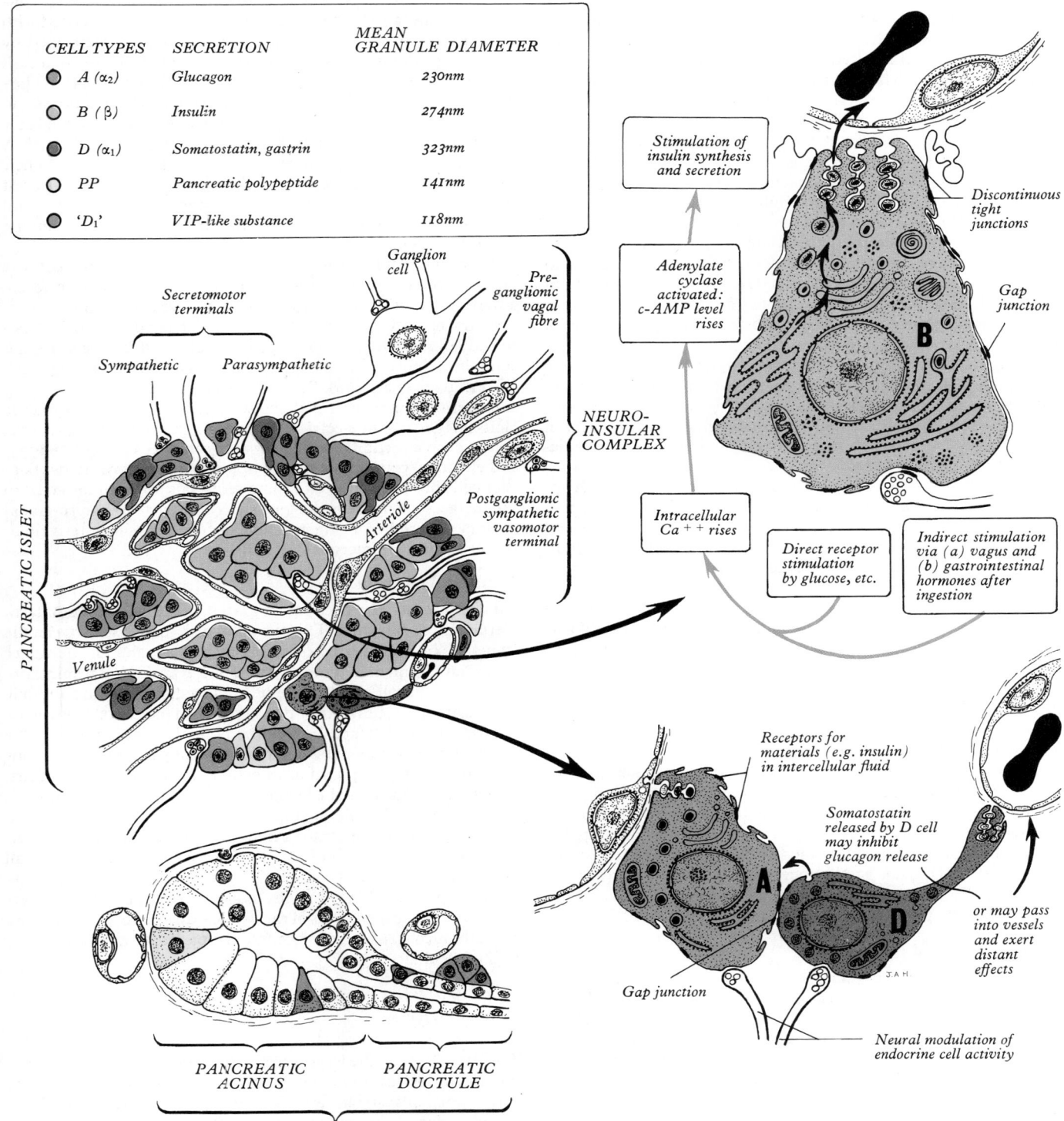

CELL TYPES	SECRETION	MEAN GRANULE DIAMETER
A (α₂)	Glucagon	230nm
B (β)	Insulin	274nm
D (α₁)	Somatostatin, gastrin	323nm
PP	Pancreatic polypeptide	141nm
'D₁'	VIP-like substance	118nm

8.153 Diagram of the histology, ultrastructure and mode of operation of the endocrine pancreas (VIP = vaso-active intestinal polypeptide).

The human pancreas may contain more than a million islets, usually most numerous in the tail (Findlay & Ashcroft 1975). Each is a mass of polyhedral cells pervaded by fenestrated capillaries (Goldstein & Davies 1968) and a rich autonomic innervation (Gerich & Lorenzi 1978). Staining procedures distinguish the major types of cell (Lane 1907, Bensley 1911, Bloom 1931, Kito & Hosoda 1977). Immunofluorescence microscopy and immuno-electron microscopy (Heitz et al 1976, Baetens et al 1977) have confirmed the identity of their secretory products and revealed other types of endocrine cell (8.153).

The most numerous cells, types A (alpha) and B (beta), respectively secrete glucagon (Baum et al 1962) and insulin (Lacy & Davies 1957). Though interspecific variation exists (Findlay & Ashcroft 1975), human A cells tend to be peripheral in islets and B cells more central (Orci 1976). Cytoplasmic storage granules of A cells are fixed by alcohol, are generally smaller than those of B cells, stain brilliant orange or red with Orange G and Mallory-Azan, and are aldehyde-fuchsin negative; in B cells they are alcohol-soluble and aldehyde-fuchsin positive. A third type, the D cell, discovered in human islets by Bloom (1931) and generally said to make gastrin, contains somatostatin or a similar peptide (Orci et al 1975). Human D cells are peripherally placed within the islets, like A cells. Orci & Unger (1975) suggested that islets may have two functional regions: a medulla mainly of B cells (where insulin is secreted at a constant rate in response, e.g., to the presence of glucose in the intercellular fluid) and a mixed cortex of A, B and D cells, rich in neurovascular elements (where secretory activity responds rapidly to various environmental changes). In the cortex somatostatin released by D cells may inhibit secretory activity in adjacent A or B cells (Orci 1976). In many mammals, including humans, D cells more often contact A cells than B, suggesting that pancreatic somatostatin may chiefly inhibit glucagon release. Organ culture studies by Barden et al (1977) corroborate this; when antisomatostatin serum was incubated with rat islets, glucagon release increased tenfold with no significant change in the release of insulin. How somatostatin inhibits the release of glucagon and possibly insulin is not clear; it may act intracellularly, passing through gap junctions from adjacent cells (Gerich & Lorenzi 1978). Another suggested hormonal modifier of islet activity is gastric inhibitory peptide, which appears to potentiate insulin secretory response to glucose (Dupre et al 1973). The autonomic 'neurohormones', acetylcholine and noradrenalin, also affect secretion, acetylcholine augmenting insulin and glucagon release, noradrenalin inhibiting glucose-induced insulin release; they may also affect somatostatin and pancreatic polypeptide secretion. The roles of the circulating noradrenalin and adrenalin or of neurogenous noradrenalin acting locally on islet cell secretion remain obscure. Control of islet activity is summarized in 8.153.

Peptide-secreting cells, with smaller granules than those in A, B and D cells, occur in human pancreas, in at least two forms: one contains pancreatic polypeptide (PP cell); another has an ultrastructure like that of D_1 cells of the gastric mucosa (p. 1377). Pancreatic 'D_1' cells differ from PP cells in their granules; those of the former do not react with antibovine pancreatic polypeptide serum, those of the latter do (Baetens et al 1977). Although the product of gastro-enteric D_1 cells is uncertain, it may be related to vaso-active intestinal polypeptide (Buffa et al 1977); the product of pancreatic 'D_1' cells is still uncertain.

'D_1' and PP cells are not restricted to islets, being also scattered throughout the predominantly exocrine tissue (8.153).

Islet Vessels and Nerves

Pancreatic *arteries* are rami of the splenic and pancreaticoduodenal arteries (pp. 769, 772 and 6.109). *Venous drainage* is into the portal, splenic and superior mesenteric veins. *Lymph drainage* is described elsewhere (p. 853). Larger blood and lymph vessels travel with the exocrine ducts and nerves in the interlobular connective tissue, supplying lobular branches. Bunnag et al (1963) have shown that in mice one to three afferent arterioles arise from arterial rami to supply each islet, before which they may supply the acini. In each islet they feed a capillary network almost as dense as in a renal glomerulus; the network is drained by one to six venules which join to enter an intralobular vein. McCuskey &

Chapman (1969) reported an intermittent flow in islet capillaries, local interruption being due to luminal bulging of the endothelial cells. The capillaries are fenestrated.

The pancreatic *nerve supply*, from the coeliac plexus, enters along with the arteries of supply. Little is known of the afferent nerves; the efferents consist of sympathetic postganglionic fibres from the coeliac ganglion and parasympathetic preganglionic from the right vagus. The fibres, mainly nonmyelinated (Benscome 1959), are vasomotor (sympathetic) and parenchymal (sympathetic and parasympathetic) in their distribution. Fine branches ramify among the cells, from peri-insular plexuses (Findlay & Ashcroft 1975). Fibres frequently synapse with acinar cells before innervating the islets, suggesting a close linkage between neural control of exocrine and endocrine components. Many fibres enter the islets with the arterioles (Coupland 1958).

Parasympathetic ganglia lie in the inter- and intralobular connective tissue, and in the latter case are frequently associated with insular cells, forming *neuro-insular complexes*, first described by van Campenhout (1925) as 'complexes sympathico-insulaires', revised to 'complexes neuro-insulaires' by Simard (1937). They were classified by Fujita (1959) in two groups: one of neurons and insular cells (8.153), one of nerve fibres and insular cells, the latter being described in detail by Kobayashi & Fujita (1969). Both A and B cells are involved in the neuro-insular complexes. Their roles remain speculative; but if they are from the neural crest (Pearse 1969, Weichert 1970), the adjacency of insular and neural cells is not surprising, nor an autonomic involvement in the control of islet secretion.

Three types of nerve terminal are noted in islets (Smith & Porte 1976): cholinergic (with 30–50 nm diameter agranular vesicles), adrenergic (with 30–50 nm dense-cored vesicles) and a third, uncharacterized, type (with 60–200 nm dense-cored vesicles). No selective link with any one type of insular cell has been found; sometimes more than one type of terminal contacts a single cell (Esterhuizen et al 1968). Some of the chemical synapses between axon terminal and islet cell show narrow areas in the synaptic clefts suggesting an electrical synapse or gap junction (Orci et al 1973); such junctions also occur between islet cells (Orci 1974) and electrical coupling of nerve supply to a functional network of islet cells has been mooted. Some terminals appear remote from the surfaces of islet cells (Kobayashi & Fujita 1969); neuro-transmitters released from them could diffuse through intercellular spaces to affect numerous islet cells.

Applied Anatomy

Pancreatic cysts may become very large, pressing on the stomach, diaphragm or bile duct. They usually push forwards between the stomach and transverse colon, becoming palpable in the upper abdomen as a median tumour, immobile even during respiration. Carcinoma usually affects the head, speedily involving the bile duct, leading to jaundice; or it may press on the portal vein, causing ascites, or obstruct the pylorus. The descending part of the duodenum is occasionally encircled by the pancreatic head (annular pancreas); neoplasm or infection may then cause duodenal obstruction. (Kasai et al 1974 have surveyed the literature on this occurrence.) If the bile duct is embedded in the pancreatic head (p. 1394), chronic pancreatitis may obstruct it and produce jaundice. Accessory nodules of pancreatic tissue may exist in the wall of the duodenum (most commonly), jejunum, ileum or ileal diverticulum (p. 1356). They may be associated with duodenal diverticula, as small protrusions of the whole wall or only of the mucosa and submucosa, usually adjacent to the pancreas and the opening of the bile duct.

The Liver

The liver (hepar), the largest gland, lies in the upper right part of the abdominal cavity, occupying most of the right hypochondrium and epigastrium and extending into the left hypochondrium as far as the left lateral line (Rouiller 1964). In males it generally weighs 1.4–1.8 kg, and in females 1.2–1.4 kg, with a range of 1.0–2.5 kg. It is somewhat cuneiform, is reddish brown in colour in the fresh state and, though firm and pliant, is easily lacerated. Wound cannot be tightly sutured and bleeding may be severe, due to the organ's great vascularity. Despite its weight, it is widely believed that, like various other viscera, its position is not maintained by peritoneal (p. 1388) or fibrous attachments, but mainly by intra-abdominal pressure due to tonus in the abdominal muscles. The continuity of hepatic veins with the inferior vena cava may provide some support. However, such dogma should be viewed with caution. Without systematic studies of intra-abdominal pressure gradients and their variations with posture, respiration, gastrointestinal dilatation and so forth, and studies of the statics and dynamics of the peritoneal folds, connective tissues, adjacent viscera and vascular pedicles, any theories as to the mechanisms maintaining the position of the liver (or any other abdominal organ) remain highly speculative.

THE BORDERS OF THE LIVER

The superior, anterior and right surfaces are continuous at rounded 'borders', but a sharp *inferior border* (8.154) separates the right and anterior surfaces from the inferior surface. This border is rounded between the right and inferior surfaces, but becomes thin and angular at the lower limit of the anterior surface and is notched along this edge by the *ligamentum teres*, to the right of the midline. Lateral to the fundus of the gallbladder, which often corresponds to a second notch 4–5 cm to the right of the midline, the inferior border largely follows the costal margin. Left of the fundus it ascends less obliquely than the costal margin, crossing the infrasternal angle to pass behind the left costal margin near the tip of the eighth costal cartilage. It then ascends sharply to merge with the thin margin of the left lobe. At the infrasternal angle the inferior border adjoins the anterior abdominal wall and is accessible to examination by percussion, but is not usually palpable; in the midline the inferior border is near the transpyloric plane, about a hand's breadth below the xiphisternal joint (8.139). In women and children the border often projects a little below the right costal margin.

THE HEPATIC LOBES

The liver is divisible into a right and a much smaller left lobe. The original basis for demarcation was a concatenation of superficial features (attachment of the falciform fold, the position of fissures for ligamenta teres et venosum, etc.) and such descriptions persist in many texts. However, classic studies by Hjortsjö (1948, 1951, 1956) on the intrahepatic segmental branching of the bile ducts, hepatic artery and portal vein, have emphasized that primary anatomical and functional lobation are better defined as the territories of the right and left hepatic ducts. Despite individual variation and some interdigitation between the smaller tributaries of the right and left ducts, the approximate dividing zone between the two lobes is a plane considerably to the right of the traditional demarcation (8.156). Inferiorly and posteriorly the division extends from the fossa for the gallbladder towards the inferior vena cava, making the quadrate and caudate lobes (vide infra) part of the *left* lobe, *not* the right, except for the processus caudatus which is served by the *right* hepatic duct. The plane of division is not quite sagittal, being tilted in the horizontal and vertical planes.

Other workers have confirmed Hjortsjö's classification and have made further subdivisions of lobes into segments, as described below (e.g. Healey & Schroy 1953, Goldsmith & Woodburne 1957, Stucke 1959), by corrosion cast techniques, dissection and radiology. In such investigations, particularly those involving examination of casts of ducts or vessels, so-called 'fissures' are visible (8.155A) between the territories of the right and left branches, and less clearly between lobar subdivisions (segments). These fissures, even in the main interlobar zone, do not produce reliable surface indications of use in lobectomy.

The right lobe, much the greater in volume, contributes to all surfaces, as described later, including the entire costal aspect. Its arbitrarily described surfaces (anterior, superior, inferior and posterior) all pass uninterruptedly on to the *left lobe*, except where shallow grooves partially demarcate the quadrate and caudate 'lobes', really parts of the left lobe, as already mentioned.

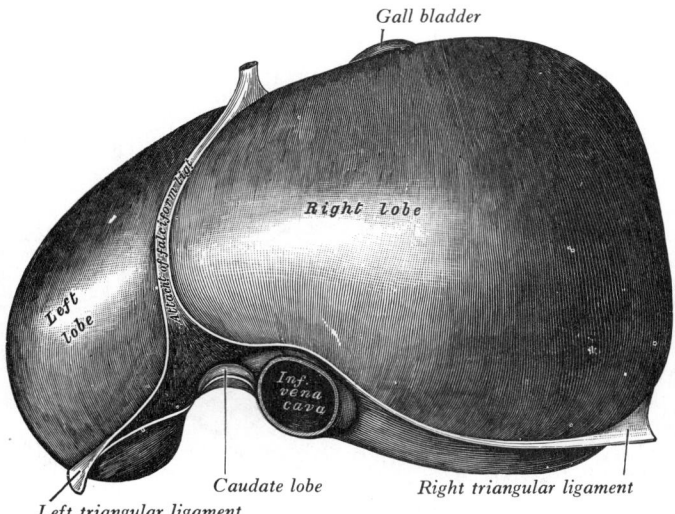

8.154A The superior, anterior and right lateral surfaces of the liver.

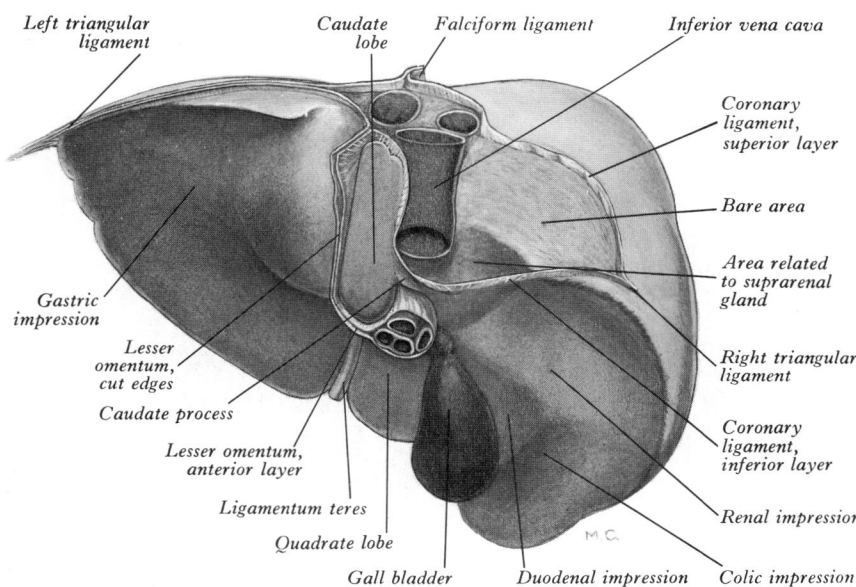

8.154B Posterior aspect of the liver, showing its peritoneal connections divided close to its surfaces.

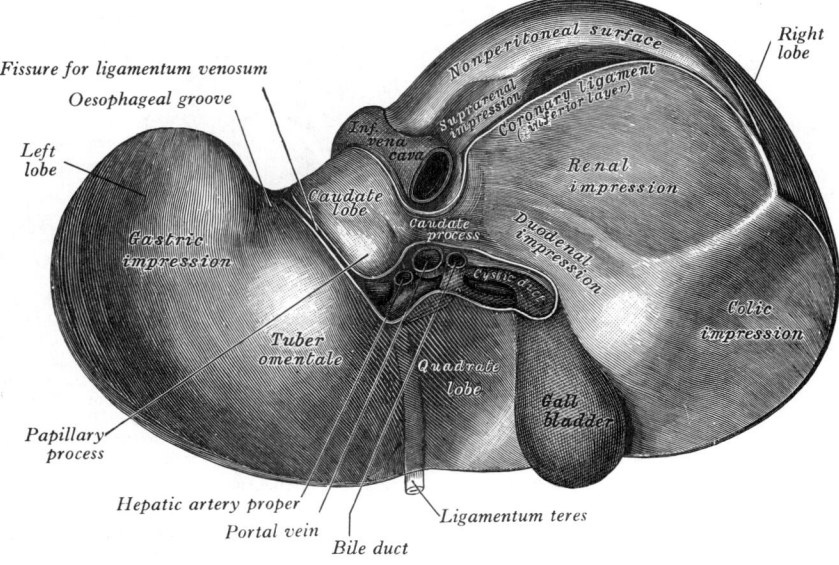

8.154C The inferior surface of the liver.

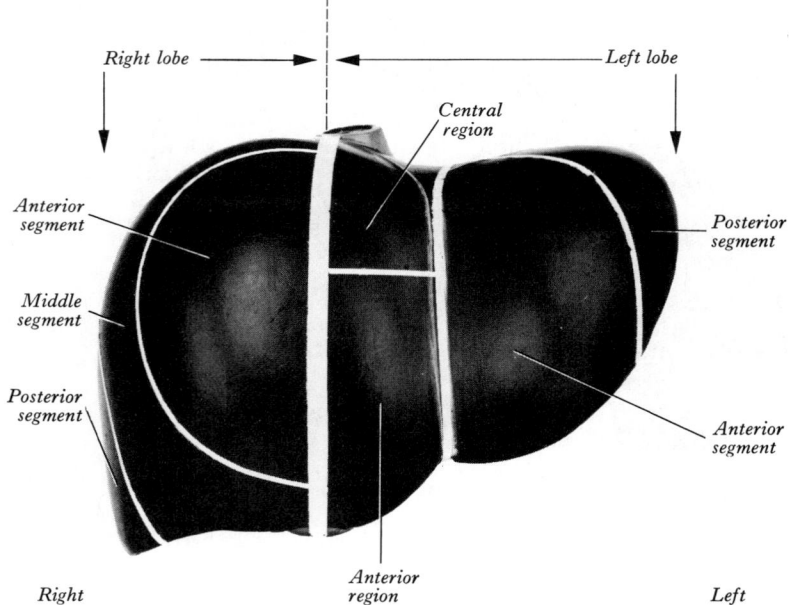

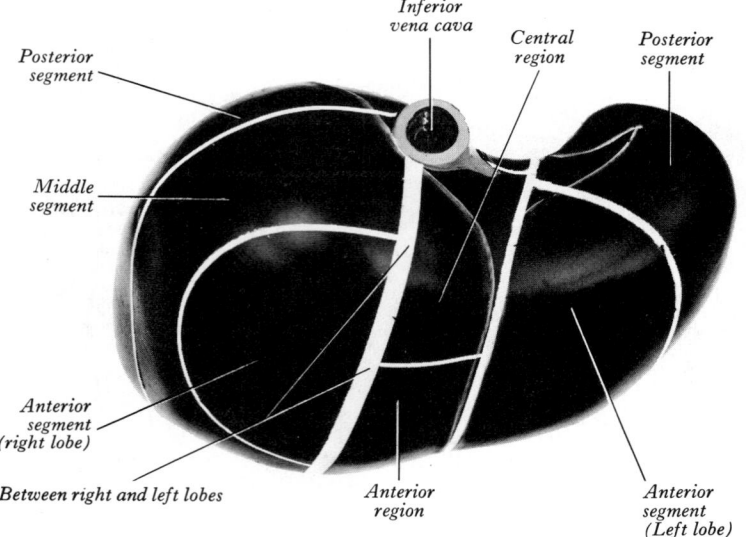

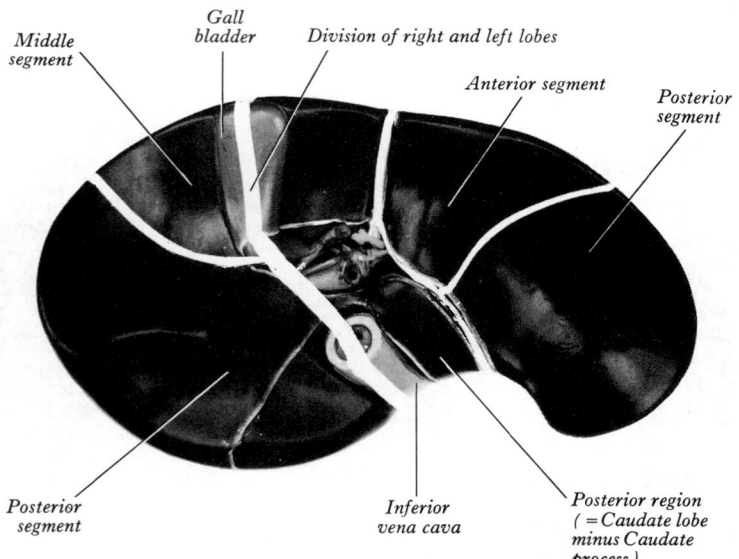

8.155A Hepatic segmentation. Surface projection of the boundaries between hepatic segments based on the researches of Professor Carl-Herman Hjortsö, University of Lund, Sweden. See text for comment and references. Top: anterior view; middle: anterosuperior view; bottom: inferior visceral view.

The *quadrate lobe*, visible on the inferior surface, appears somewhat rectangular and is bounded in front by the inferior border, on the left by the fissure for ligamentum teres, behind by the porta hepatis and on the right by the fossa for the gallbladder.

The caudate lobe is visible on the posterior surface, bounded on the left by the fissure for the ligamentum venosum, below by the porta hepatis and on the right by the groove for the inferior vena cava. Above, it continues into the superior surface on the right of the upper end of the fissure for the ligamentum venosum. Below and to the right, it is connected to the right lobe by a narrow *caudate process*, which is immediately behind the porta hepatis and above the epiploic foramen. Below and to the left, the caudate lobe has a small rounded *papillary process*. Due to the depth of the fissure for the ligamentum venosum, the caudate lobe has an anterior surface, which forms the posterior wall of the fissure and is in contact with the lesser omentum (hepatic part).

HEPATIC SEGMENTATION

As in other organs with a group of hilar structures (e.g. the lungs), investigations of the branching patterns of blood supply and biliary drainage in the liver has revealed a system of lobes and further subdivisions (sectors or segments). The mode of growth of such organs, with the branching and rebranching of their vessels and ducts, makes it inevitable that territories are associated, in the liver, with individual rami of the *portal triad* of tubes (i.e. branches of the portal vein, hepatic artery and hepatic ducts). The size and number of such territories depends on the size of the tube designated as 'segmental', i.e. upon the level in the arborization arbitrarily selected. The hepatic artery, portal vein and common bile duct divide and subdivide with a common pattern, as implied in the classic observations of Glisson (1654) and confirmed by many subsequent workers. No evidence of significant intrahepatic anastomosis in these dendriform systems has been recorded. For example, despite variation in origin (and occurrence of accessory vessels), the hepatic arterial system consists of end-arteries (Michels 1966), a feature typical in any organ developed from a single vascularized blastema of dichotomizing potential. The implications of this pattern were not followed up until the late nineteenth century, apart from the recognition of right and left hepatic lobes (see McIndoe & Counsellor 1927, Hjortsö 1948 for earlier literature). Division into two lobes, based on the primary divisions of the triadic system (hepatic arteries, portal veins and biliary tree) rather than on surface features, is generally accepted but confusing descriptions persist in many texts. Inevitably, further divisions have been suggested, if only in the interests of effective partial hepatectomy. The pioneer in this field was Hjortsö (1948, 1951, 1956, 1975), who was the first to propose a complete segmental model (8.155A), based on dissections, injections and radiography, particularly applied to the biliary ducts and portal vein. Many have subsequently modified his scheme of segmentation, subdividing some segments, redefining and renaming others, (Elias & Petty 1952, Healey & Schroy 1953, Couinaud 1954, Goldsmith & Woodburne 1957, Bilbey & Rappaport 1960.) The main extension of Hjortsö's scheme is the division of his segments into superior and inferior parts, chiefly by Healey & Schroy 1953 (8.155B) and Couinaud 1954. Apart from this and minor differences in delineating the major segments, there is general consensus on the division of the right lobe into approximately 'anterior', intermediate and 'posterior' segments; and the left into lateral and medial parts. Most reports have been based on injections and casts, principally upon corrosion casts of one or more of the components of the triad's ramifications. In such casts is an easily discernible (8.156) sagittal zone between the lobes, the *fissura principalis*, picturesquely described by Hjortsö (1956) as lying in the plane of the left tympanic membrane. All other 'fissures' described (with some variation) by different workers are intersegmental. The term '*spatium*' was suggested for 'fissure', and Hjortsö recognized several of these (8.155A). Such 'spaces' in corrosion casts are due to the absence of all but the smallest rami of the portal triad, usually too fragile to preserve. These demarcations do *not* correspond to substantial zones of connective tissue, which might provide superficial or internal indications for purposes of subtotal resection (cf.

bronchopulmonary segments). Surface projections of segmental fissures are shown in **8.155A** and B, according to Hjörtsö (1948) and Healey & Schroy (1953), the chief difference being that the latters' scheme shows superior and inferior regions in each major segment (as also recognized by Elias & Petty 1952, Couinaud 1954, and Platzer & Maurer 1966). More recently Gupta et al (1978) have put forward a similar scheme (9 segments); in 1981 further observations by Gupta et al on hepatovenous segmentation in the human liver were published in which they recognized five segments: left, middle, right, paracaval and caudate. To equate the segments described, with their differing names and delimitations, by all the investigators quoted would not improve the reliability of this information in practical application. As Gupta et al (1978) stated, segmental pattern is in itself variable; of 41 corrosion casts more than half showed marked variation in the volume of one segment or another, as noted by others. Although most regard segments as functionally independent and uncomplicated by, e.g. intrahepatic arterial anastomosis, occasionally this has been noted. Surgical opinion is divided upon the usefulness of such patterns in hepatic resections. Dawson (1974) and others consider resection of less than a lobe to be hazardous; others, such as Ryncki (1974), have recorded successful segmental resections. Individual variation requires portal venography and cholangiography to define segmental patterns before operation, wherever feasible. It should be noted that the disposition of the *hepatic veins* and their tributaries is not a reliable guide; these veins, also studied by corrosion casts (Goldsmith & Woodburne 1957), do not follow the pattern of the hepatic triads; they drain parts of adjoining segments. It is therefore difficult to plan a resection plane which is optimal in respect of both triadic structures and hepatic veins.

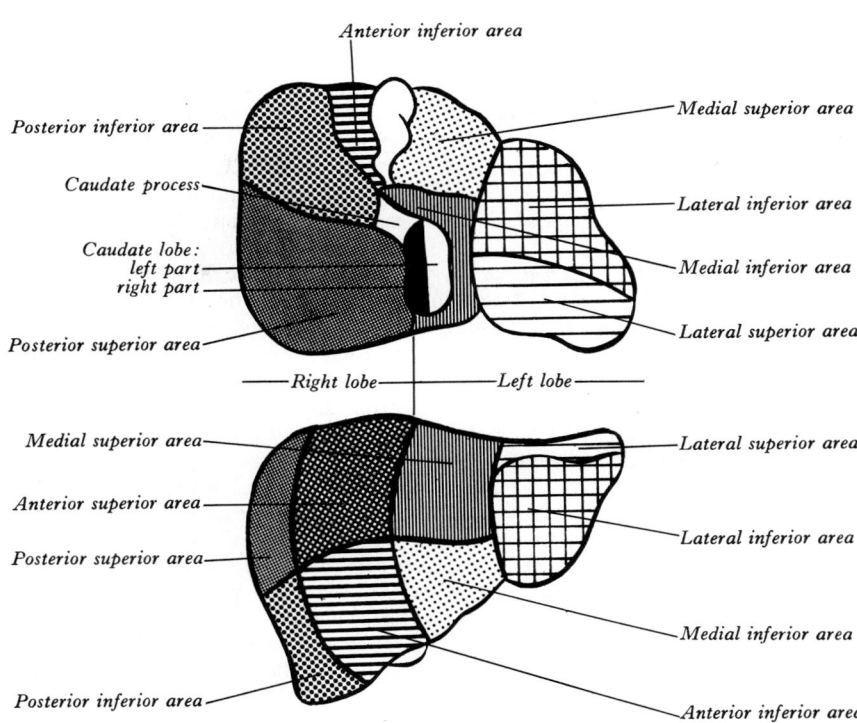

8.155B The segmentation of the liver, based upon the principal divisions of the hepatic artery and accompanying hepatic ducts. The upper drawing is of the visceral surface of the liver, the lower drawing is of the diaphragmatic surface. See text for further description.

8.156 A resin corrosion cast of the blood vessels and duct systems of the liver of a woman: bile duct, cystic duct, gallbladder and their tributaries (yellow); the hepatic artery and its branches (red); the portal vein and its tributaries (light blue); the inferior vena cava, hepatic veins and their tributaries (dark blue). The photograph is of the visceral surface of the organ and was taken before the finer blood vessels and ducts were removed by trimming. The posterior aspect is above. Prepared by D H Tompsett of the Royal College of Surgeons of England.

PERITONEAL CONNECTIONS OF THE LIVER

Except for a triangular area on its posterior surface, the liver is almost completely covered by peritoneum, which connects it to the stomach, duodenum, diaphragm and anterior abdominal wall by several folds, their lines of attachment, of course, being devoid of peritoneum. These folds include the falciform ligament, right and left triangular and coronary ligaments and the lesser omentum.

The *falciform ligament* (8.108), a crescentic fold, consists of two applied layers of peritoneum, connecting the liver to the diaphragm and the supra-umbilical part of the anterior abdominal wall. Its convex base is fixed to the inferior diaphragmatic surface and to the posterior surface of the anterior abdominal wall, down to the umbilicus; as it ascends from this it inclines slightly right. It is attached to the notch for the ligamentum teres on the inferior hepatic border and to the anterior and superior hepatic surfaces. Its concave, free edge, from the umbilicus to the notch for the ligamentum teres, contains the latter and the small para-umbilical veins and is anterior to the pyloric region of the stomach. At its diaphragmatic end its layers separate to expose a triangular area on the superior hepatic surface devoid of peritoneum. The left layer continues into the anterior layer of the left triangular ligament, the right into the upper layer of the coronary ligament.

The *coronary ligament* (8.154), a reflexion of the peritoneum from the diaphragm to the right lobe's superior and posterior surfaces, has upper and lower layers, continuous with the right triangular ligament and diverging left to enclose posteriorly a triangular 'bare' area of the right lobe devoid of peritoneum. Its upper layer merges into the right falciform layer, skirts anteriorly the upper end of the caval groove and descends from the upper to the posterior surface, continuing into the anterior layer of the right triangular ligament. The lower coronary layer becomes the posterior layer of the right triangular ligament, following almost horizontally the lower limit of the right posterior hepatic surface. Here peritoneum may be reflected to the anterosuperior aspect of the right kidney (*hepatorenal ligament*) instead of to the diaphragm. The left end of the lower layer passes in front of the caval groove's lower end into the line of peritoneal reflexion from the right caudate border, i.e. the right margin of the upper omental recess.

The *left triangular ligament* ascends back from the left lobar superior surface to the diaphragm. Its closely applied layers become fused at the left edge of the ligament. To the right the anterior layer merges with the left falciform layer, the posterior with the anterior lesser omental layer at the upper end of the fissure for the ligamentum venosum. The left triangular ligament lies in front of the abdominal part of the oesophagus, the upper end of the lesser omentum and part of the fundus of the stomach. It varies much and may contain large blood vessels (Outrequin et al 1967).

The *right triangular ligament* is a short V-shaped fold connecting the lateral and posterior aspects of the right lobe to the diaphragm. At its right margin its two layers are continuous. The ligament is really the right extremity of the coronary ligament.

The *lesser omentum* has been described (p. 1341); at the upper end of the fissure for the ligamentum venosum, its anterior layer merges with the posterior layer of the left triangular ligament and its posterior layer with the line of reflexion of the peritoneum from the upper end of the right caudate lobe and so, indirectly, with the lower coronary layer (8.154B).

HEPATIC SURFACES

The superior surface (8.154A) includes parts of the right and left lobes. It fits closely under the diaphragm, separated from it by peritoneum except for a small triangular area where the two layers of the falciform ligament diverge. Right and left it is convex, but centrally it presents a shallow *cardiac impression* corresponding with the position of the heart above the diaphragm. It is related to the right diaphragmatic pleura and right pulmonary base, to the pericardium and ventricular part of the heart and to part of the left diaphragmatic pleura and left pulmonary base. It should be noted that the superior surface curves directly into the so-called anterior

surface and the peritonealized part of the posterior surface of the right lobe. No definable border separates superior, anterior, right lateral and right posterior *aspects* of the liver and it would be more appropriate to group these as the *diaphragmatic surface*, mostly separated from the *visceral surface* by a narrow edge or border.

The anterior surface, which is triangular and convex, is covered by peritoneum except at the attachment of the falciform ligament. Much of it is in contact with the diaphragm, which separates it on the right from the pleura and sixth to tenth ribs and cartilages and on the left from the seventh and eighth costal cartilages. The thin margins of the base of the lungs are thus quite close to the upper part of this surface, more extensively so on the right. The median area of the anterior hepatic surface lies behind the xiphoid process and the anterior abdominal wall in the infracostal angle (8.107).

The hepatic profile is superficially projected as follows (8.139): its upper limit corresponds to a line through the xiphisternal joint, ascending to a point below the right nipple (fourth intercostal space) and to the left to a point inferomedial to the left nipple; its right border corresponds to a curved line, convex to the right, running from the right end of the upper border to a point 1 cm below the costal margin at the tip of the tenth costal cartilage; its lower limit is the line completing this triangle (8.139), crossing the midline at the transpyloric plane (slightly concave near the right linea semilunaris).

The right surface, covered by peritoneum, adjoins the right dome of the diaphragm which separates it from the right lung and pleura and the seventh to eleventh ribs. Above its upper third, both lung and pleura are inserted between the diaphragm and ribs; over its middle third only the costodiaphragmatic pleura is interposed; over its lower third the diaphragm and thoracic wall are in contact.

The posterior surface is convex and wide on the right but narrow on the left, with a deep median concavity corresponding to the forward convexity of the vertebral column (8.154A). Much of this surface is devoid of peritoneum, being attached to the diaphragm by loose connective tissue, forming the so-called 'bare area', triangular and limited above and below by the layers of the coronary ligament. The base of the posterior hepatic surface to the left is the caval groove; its apex, directed down and right, corresponds to the right triangular ligament. The *groove for the inferior vena cava* (caval groove), which is deep and occasionally a tunnel, lies at the posterior surface and is bare of peritoneum and adapted to the upper part of the vessel it contains; its floor is pierced by the hepatic veins (p. 818). Infero-anteriorly the caudate process separates it from the porta hepatis. Lateral to its lower end the 'bare area' adjoins the upper pole of the left suprarenal gland. Left of the groove the *caudate lobe* forms the posterior surface in the superior omental recess; the peritoneum on its posterior aspect curves round its left border to its anterior aspect, which is the posterior wall of the fissure for the ligamentum venosum (8.154). The caudate lobe projects into the superior omental recess from the right; its posterior surface is related to the diaphragmatic crura (above the aortic opening) and the right inferior phrenic artery, separated by them from the descending thoracic aorta. The *papillary process* often descends in front of the origin of the coeliac artery.

The *fissure for the ligamentum venosum* (8.154) separates the posterior aspect of the caudate from the main part of the left lobe. The fissure cuts deeply in front of the caudate lobe and contains the two layers of the lesser omentum. Below, it curves laterally in front of the papillary process to the left end of the porta hepatis. The *ligamentum venosum*, the fibrous remnant of the ductus venosus (p. 724), is attached below to the left branch of the portal vein's posterior aspect; ascending in the floor of the fissure and passing laterally at the upper end of the caudate lobe it joins the left hepatic vein near its entry into the inferior vena cava, or sometimes the vena cava itself.

The left lobe's posterior aspect has a shallow *oesophageal impression* near the upper end of the fissure for the ligamentum venosum, occupied by the abdominal part of the oesophagus. Left of this the left lobe is related to part of the fundus of the stomach.

The inferior or visceral surface (8.154), facing down, back and to the left, bears the imprint, when preserved in situ, of the adjacent viscera. It is covered by visceral peritoneum except at the porta hepatis, the fissure for the ligamentum teres and the fossa for the gallbladder. On the left lobe, continuous with the oesophageal groove, is a *gastric impression*. To the right of this the rounded *omental tuberosity*, in the concavity of the lesser curvature, is in contact with the lesser omentum. The *fissure for the ligamentum teres*, of variable depth, ascends backwards from its notch on the inferior hepatic border to the left end of the porta hepatis, meeting the lower end of the fissure for the ligamentum venosum. It is the left boundary of the quadrate lobe and may be, partially or wholly, bridged by a band of liver. In its floor is the *ligamentum teres*, the obliterated vestige of the left umbilical vein (p. 724). From the umbilicus this ligament ascends in the edge of the falciform ligament to the inferior hepatic border, where it traverses the fissure to join the left branch of the portal vein at the left end of the porta hepatis, opposite the attachment of the ligamentum venosum.

The gastric impression may invade anteriorly the *quadrate lobe*, which is moulded to the pyloric region and the beginning of the duodenum. The posterior part of the quadrate lobe adjoins the right border of the lesser omentum and its contained structures. When the stomach is empty the quadrate lobe is related to the first (superior) part of the duodenum and part of the transverse colon.

The *porta hepatis*, situated between the quadrate lobe in front, and the caudate process behind, is a deep transverse fissure between the upper ends of the fissure for the ligamentum teres and the fossa for the gallbladder. At the porta hepatis the portal vein, hepatic artery and hepatic nervous plexus enter and the right and left hepatic ducts and some lymph vessels emerge. The hepatic ducts are anterior, the portal vein and its branches posterior and the hepatic artery proper and its branches lie intermediate in position.

The *caudate process* connects the inferolateral part of the caudate lobe (left lobe) to the right lobe. It lies behind the porta hepatis, in front of the inferior vena cava, and roofs the epiploic foramen. It is often assigned to the right lobe but lies within the territory of the left hepatic duct, i.e. it forms part of the *left* lobe (vide supra).

The *fossa for the gallbladder*, forming the right limit of the quadrate lobe, extends from the inferior hepatic border to the right end of the porta hepatis. Usually shallow, it is variably bare

of peritoneum. The inferior hepatic surface to the right of the fossa adjoins the colon, kidney and duodenum. A *colic impression* fits the right colic flexure near the inferior border. A *renal impression*, usually well marked, lies behind the colic and is separated from the neck and adjoining part of the gallbladder by a duodenal impression; it is related to the right upper renal pole and superomedially to the lower pole of the right suprarenal gland. When the lower coronary layer is reflected from the liver to the right kidney, the renal and suprarenal impressions extend to the lower part of the 'bare area'. The *duodenal impression* is lateral to the neck of the gallbladder and related to the junction of the first (superior) and second (descending) parts of the duodenum.

Hepatic relations vary with posture and respiration. In the above description the body is assumed to be supine.

The branches of the portal vein and tributaries of the hepatic veins are more numerous before birth, after which they are reduced by fusion or degeneration. The fetal portal vein joins the umbilical in a smooth right-hand curve, maintained after birth, with a sharp angle between the portal trunk and its left branch; the left vascular lobe may therefore be at a circulatory disadvantage and unable to keep pace in growth with the right. At the left end of the adult left lobe a fibrous band (*fibrous appendix of the liver*) may appear as an atrophied remnant of the more extensive left lobe in children; it contains atrophied bile ducts, the *hepatic vasa aberrantia*. Similar remnants may occur in the left lobe's edges and near the inferior vena cava. Occasionally the right lobe's lower border, to the right of the gallbladder, projects down as a broad linguiform process (*Riedel's lobe*).

HEPATIC VESSELS AND NERVES

The vessels connected with the liver are the portal vein, hepatic artery proper and hepatic veins. The *portal vein* and *hepatic artery proper* ascend in the lesser omentum to the porta hepatis, where each dichotomizes; the *bile duct* and *lymphatic vessels* descend from the porta in the same omentum. All these structures are enveloped in the *perivascular fibrous capsule (hepatobiliary capsule of Glisson)*, a sheath of loose connective tissue, which also surrounds the vessels as they course through the portal canals in the liver. It is also continuous with the fibrous hepatic capsule.

The *hepatic artery* and its branches, after variable courses in the porta, divide and subdivide in the liver, their smaller rami being associated with those of the portal vein with which they are

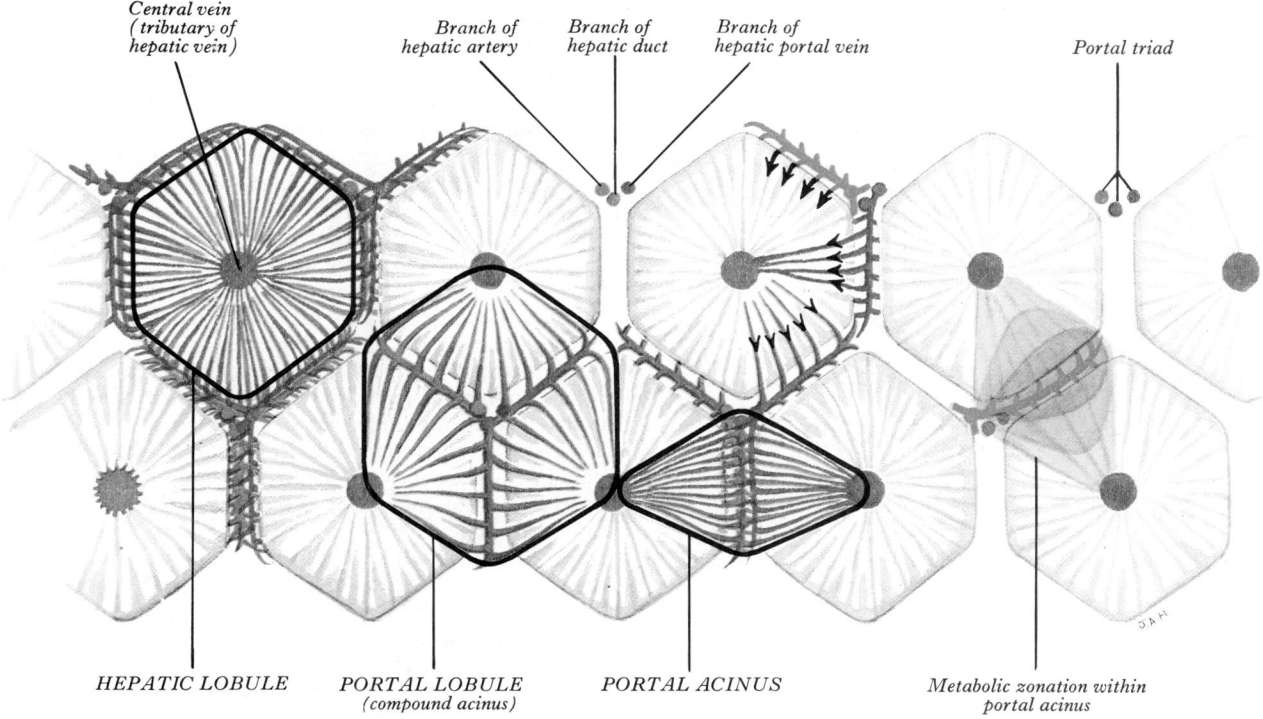

Central vein (tributary of hepatic vein)   Branch of hepatic artery   Branch of hepatic duct   Branch of hepatic portal vein   Portal triad

HEPATIC LOBULE   PORTAL LOBULE (compound acinus)   PORTAL ACINUS   Metabolic zonation within portal acinus

8.157 Diagram of the histological organization of the liver, showing the principal types of subdivisions which have been proposed. For purposes of clarity, the territories of the classic hepatic lobules are shown as regular hexagons, unlike their real appearance which is highly variable (see text).

Portal canal   *Central vein of lobule*

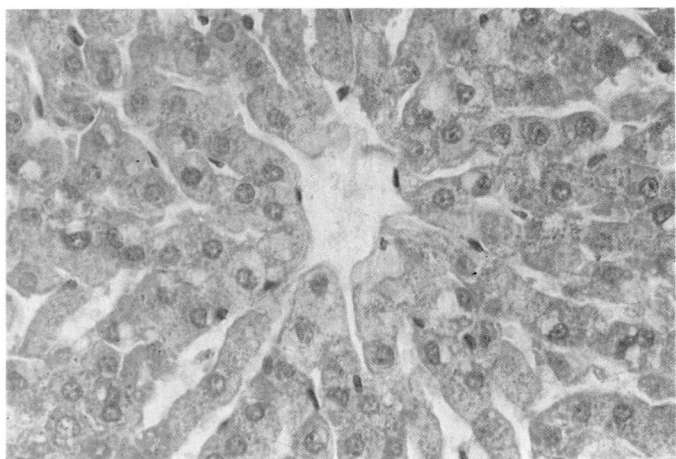

8.158A Section through a hepatic lobule (human) (after Sobotta). Stained with haematoxylin and eosin. Magnification × 70.

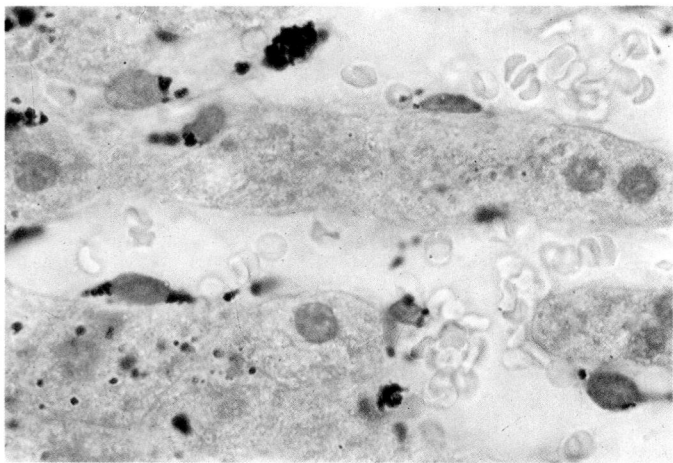

8.158B Section through a number of hepatic cords radiating from a central hepatic venule of the rabbit. The hepatocytes appear cuboidal and the Kupffer and endothelial cells, which line the sinusoids, have flattened, densely staining nuclei. Haematoxylin and eosin.

8.158C Section similar to that shown in (B), but taken from the liver of a rabbit previously injected intravenously with carbon particles. The Kupffer cell nuclei are outlined with phagocytosed particles which demonstrate the limits of the cell cytoplasm.

distributed. There are no (or very few) anastomoses between their territories; each is an end-artery (Glauser 1953).

The *hepatic veins* (8.156) convey blood from the liver to the inferior vena cava (p. 818). They have only a thin tunica adventitia, binding them to the walls of their canals; hence, in sections, they are widely open and solitary, and so easily distinguished from the branches of portal veins, which tend to collapse post mortem and are always accompanied by an artery and a biliary duct.

Hepatic *lymph vessels* are described on p. 851. Lymph from the liver has an abundant protein content. Obstruction of the hepatic venous drainage increases the flow of lymph in the thoracic duct. The importance of the transdiaphragmatic lymph drainage of the liver into internal mammary and diaphragmatic lymph nodes has been emphasized by Nidden et al (1973), who confirm that lymph reaches the right lymphatic duct, partly via the tracheobronchial lymph nodes.

Innervation is by an hepatic plexus (p. 1165) containing sympathetic and parasympathetic (vagal) fibres. These enter at the porta hepatis and largely accompany blood vessels and bile ducts; very few run amongst the liver cells and their terminations are uncertain. Both myelinated and non-myelinated fibres reach the liver from nerves in its various peritoneal folds (Sutherland 1965).

THE MICROSCOPIC STRUCTURE OF THE LIVER

The liver consists of a connective tissue network of great complexity, in the meshes of which lie the great majority of its cells. These are epithelial in origin and carry out the major metabolic activities of this organ. Permeating the whole structure are great numbers of blood vessels of different types, perfusing and draining the liver with a rich flow of hepatic portal venous blood, and arterial blood derived from the hepatic arteries.

Lobulation of the liver

Under the peritoneal lining which covers most of the liver's surface, lies a thin layer of connective tissue (‘*Glisson's capsule*’) from which extensions pass into the organ, branching and rebranching as connective tissue septa and trabeculae.

As classically described in the adult pig and some other mammals, this connective tissue component separates and encloses a multitude of polyhedral *hepatic lobules* (often hexagonal in histological sections), each about 1 mm in diameter with a small *central vein* (hepatic tributary) as its axis, surrounded by *portal triads* (8.157). Each triad contains a branch of the portal vein and hepatic artery and an interlobular biliary ductule, all ensheathed in connective tissue (the *portal canal* or *perivascular fibrous capsule*). In man (and many other mammals) hepatic lobules are not readily discernible as discrete entities and indeed three-dimensional reconstruction indicates that the concept of a lobule with definite geometrically-defined boundaries, surrounded by connective tissue and centred on an efferent vein, cannot be sustained. Instead, the more useful idea of a functional unit has been proposed, reflecting centres of biliary secretion and considerations of blood flow, oxygenation, metabolic gradients and, in pathology, hepatic degeneration. This is the *portal lobule*, consisting of the adjoining parts of (in pigs) at least three ‘classic’ lobules, bile from which drains into a biliary ductule in the portal canal between three such *hepatic* lobules; hence, in sections, a portal lobule is a polygonal territory centred on a portal triad, its boundary passing through adjacent central veins. A third unit, the *portal acinus*, is an even more useful concept in considerations of metabolic organization. It is centred on a preterminal branch of an hepatic arteriole and includes the hepatic tissue served by this, bounded by the territories of other acini and by two adjacent central veins (8.157). Because such phenomena as zones of anoxic damage, glycogen deposition and removal and toxic trauma are related to arterial flow, they tend to follow the acinar pattern. There are also real structural and physiological differences within the acinar lobules and they are customarily divided into three zones: zone 1 (periportal) i.e. nearest the portal radicles; zone 2 further away from these; and zone 3 around the central venous drainage (vide infra). Rappaport (1969) has proposed the concept of *compound portal acini*, which are more complex groups of

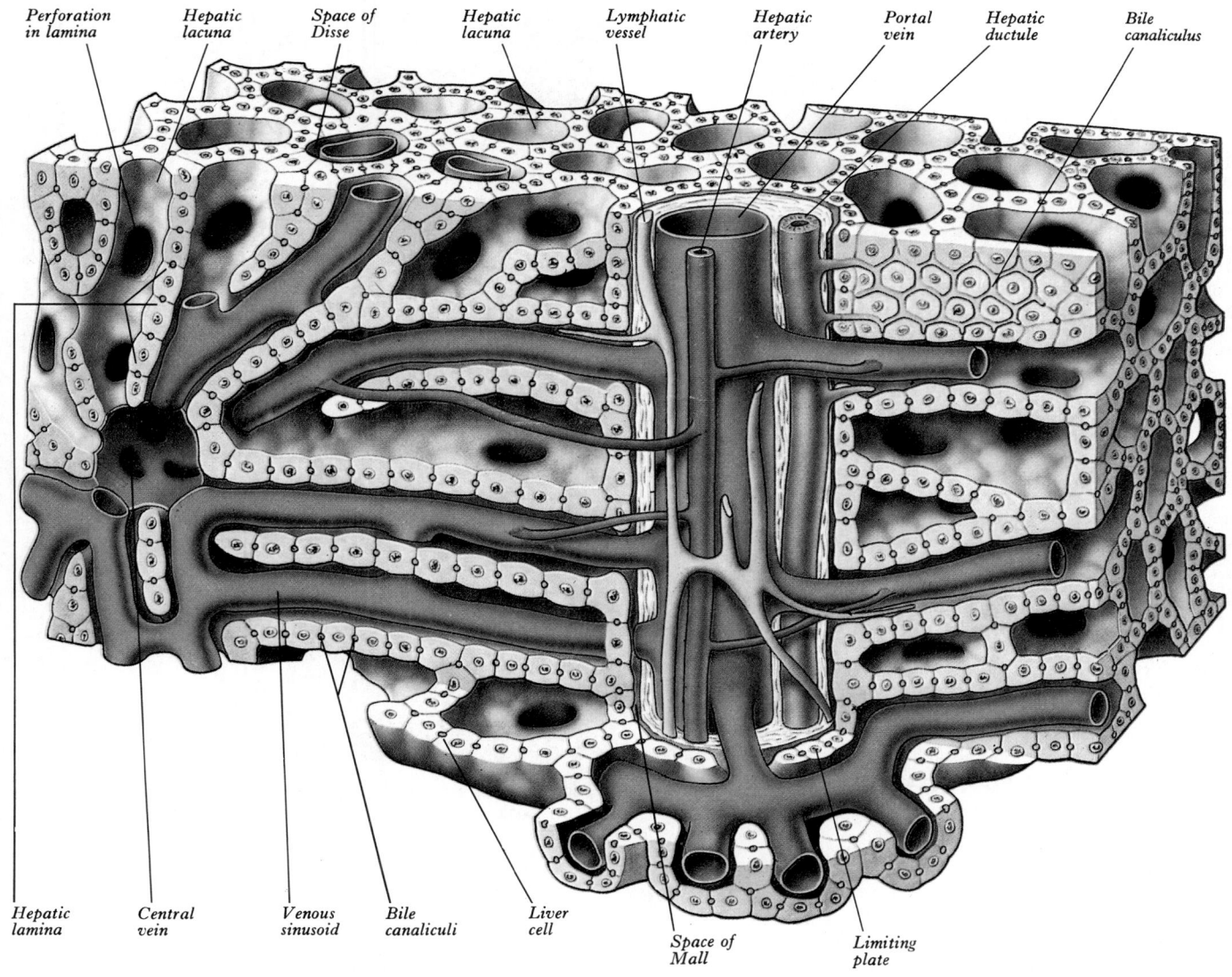

Perforation in lamina *Hepatic lacuna* *Space of Disse* *Hepatic lacuna* *Lymphatic vessel* *Hepatic artery* *Portal vein* *Hepatic ductule* *Bile canaliculus*

Hepatic lamina *Central vein* *Venous sinusoid* *Bile canaliculi* *Liver cell* *Space of Mall* *Limiting plate*

8.159 Diagram of hepatic structure (after H Elias, Department of Anatomy, Chicago Medical School). Note that in this picture, perisinusoidal endothelial cells and macrophages (Kupffer cells) are not shown.

hepatic units. Neither hepatic nor portal lobules are fixed anatomical entities; the *'hepatic lobular structure'* is usually evident but can be changed to a *'portal lobular structure'* by alteration in the relative blood pressures in the portal and hepatic veins (e.g. by raising hepatic venous or lowering portal pressure), these changes being reversible. Such alterations in venous pressures occur in pathological conditions.

The cells of the liver include hepatocytes (the main metabolic element), endotheliocytes, macrophages (Kupffer cells), lipocytes, the cells of the biliary tree (cuboidal to columnar epitheliocytes) and connective tissue cells of the capsule and portal tracts. For further details, see Phillips et al (1987), Jones & Spring-Mills (1983).

Hepatocytes. About 60% of cells in the liver (and 80% of its volume) are formed by *hepatocytes (parenchymal cells),* derived from endoderm of the caudal part of the foregut, with which they retain connection by biliary ductules and hepatic ducts. Hepatocytes are polyhedral, with five to twelve sides and are from 12–25 μm across. Their nuclei are spheroidal and euchromatic and often polyploid or multiple (2 or more) in each cell (Doljansky 1960). Their cytoplasm typically displays much granular and agranular endoplasmic reticulum, many mitochondria, lysosomes and many well-developed Golgi bodies, features indicating a high metabolic activity. Glycogen granules and lipid vacuoles are usually prominent. Numerous, particularly large peroxisomes and vacuoles containing enzymes such as urease (uricosomes) in distinctive crystalline forms indicate the complex metabolism of these cells. Their role in iron metabolism is shown by storage

vacuoles containing crystals of ferritin and haemosiderin. Where hepatocytes adjoin bile canaliculi they carry microvilli and numerous membrane-bound vesicles cluster near the lumen (**8.160A**). The borders of these canaliculi are marked by tight junctions where adjacent hepatocytes lie in contact, preventing their secretions from entering the general intercellular spaces of the liver and confining them to the canalicular system.

Their structure and metabolism varies within the portal acini, according to their distance from the portal inflow (vide supra); in zone 1, cells have more mitochondria and granular endoplasmic reticulum, whilst in zone 3 (nearest the central veins) the agranular endoplasmic reticulum is even more extensive, but mitochondria are fewer and more spheroidal. In zone 2, there is a gradient between the other two conditions.

Hepatocytes mediate many metabolic activities; they synthesize and release various plasma proteins such as albumins, clotting factors and complement components; they deaminate amino acids by the urea cycle, liberating urea for excretion; convert bilirubin to biliverdin for secretion into bile; synthesize bile salts, biliary secretions essential to the emulsification of fats in digestion; they eliminate many endogenous and exogenous toxic substances from the blood and convert tetra-iodothyronin to more active tri-iodothyronin. They also store carbohydrates as glycogen and triglycerides as lipid droplets, metabolizing these as required to release glucose and lipid into the blood. Most lipid thus secreted leaves the perisinusoidal surface of the cell in secretory vesicles, having been conjugated with protein in the endoplasmic reticulum and Golgi apparatus, to form the 'very low density

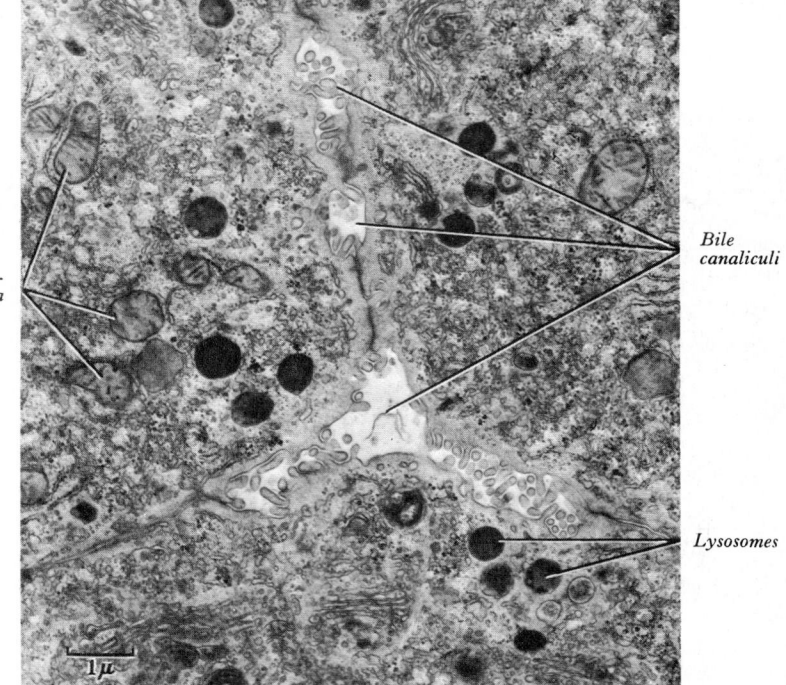

*Mito-
chondria*

*Bile
canaliculi*

Lysosomes

1μ

8.160A Electron micrograph showing portions of three adjacent hepatocytes and the intervening bile canaliculi. Magnification × 10 000.

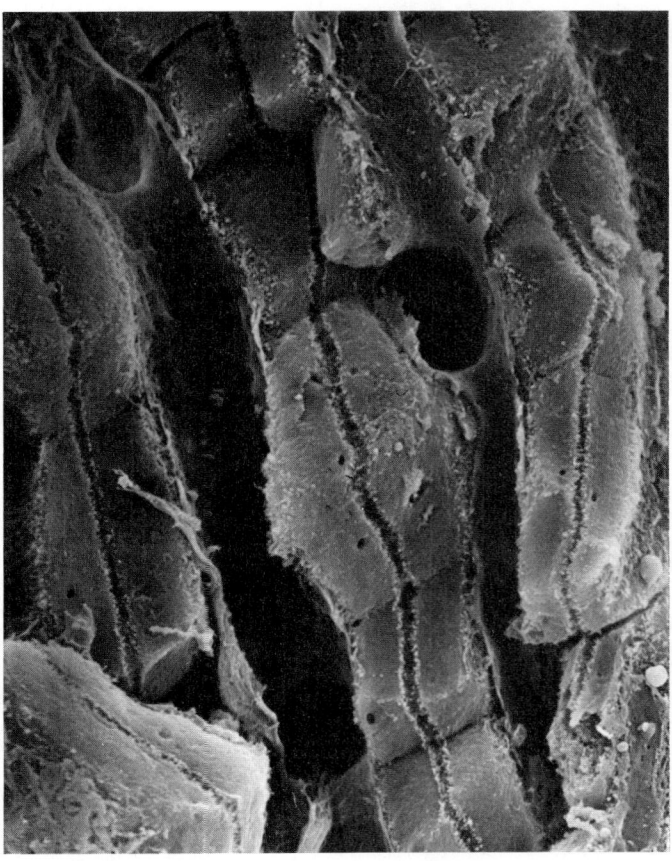

8.160C Scanning electron micrograph of the cut surface of the liver (rodent) showing sinusoids with endothelial linings and adjacent hepatocytes, grooved by bile canaliculi (see **8.159**).

lipoprotein' moiety of blood plasma (Claude 1970). Some cholesterol also enters bile, apparently as an excretion. Iron is stored in the cells as ferritin. All these processes produce heat, important in the maintenance of body temperature.

The multitude of activities in hepatocytes is reflected in their structural complexity, which varies with metabolic demand; e.g. agranular endoplasmic reticulum proliferates during barbiturate detoxification (Jones & Fawcett 1966), while in starvation glycogen and lipid reserves disappear. Because of involvement in many such processes, hepatocytes are most vulnerable to anoxia, various toxins and carcinogens, which cause characteristic patterns of degeneration in the portal acini.

Hepatic laminae. The manner in which hepatocytes are arranged in the liver has been much debated. The most recent studies show that in the mature liver, hepatocytes are arranged mainly in plates of hepatocytes, one cell thick *(hepatic laminae)*, forming a continuous system or *muralium* which may branch or anastomose, as interlaminar bridges of cells connect adjacent laminae. Between the laminae lie *hepatic lacunae* containing

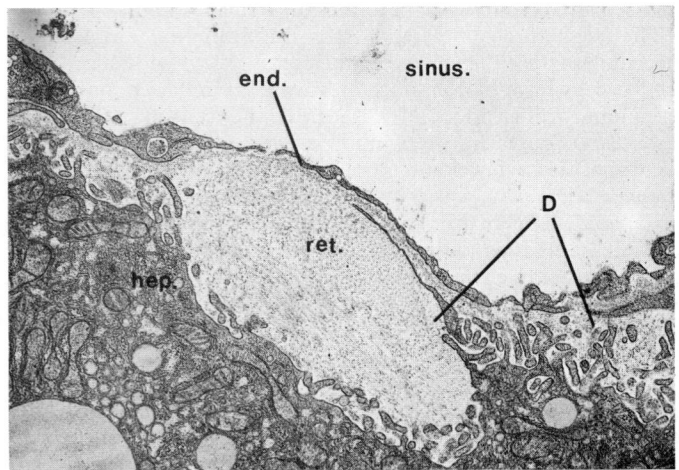

end.   **sinus.**

D

ret.

hep.

8.160B Transmission electron micrograph of the border of a hepatic sinusoid (sinus.) showing part of an hepatocyte (hep.) and the tenuous fenestrated endothelium (end.) separated by the space of Disse (D) in part of which lies a reticulin bundle (ret.). Magnification × 5000.

venous sinusoids, which anastomose via perforations in the hepatic laminae. Where hepatocytes adjoin portal canals or hepatic venous tributaries they form a *limiting plate*, surrounding the vessels and perforated by their radicles and by rami of the hepatic artery and biliary ductules; a similar limiting plate composed of a single layer of liver cells underlies the hepatic capsule.

In histological sections (**8.158**) rows of liver cells seen radiating from the central vein to the lobular periphery are really sections through hepatic laminae. These rows, with intervening sinusoids, do not pass straight to the periphery like the spokes of a wheel but run irregularly, because the laminae themselves are irregular and branched. Some rows of cells (and sinusoids) arrive at the periphery of lobules between adjacent portal canals; others pursue indirect routes between the central vein and portal canals.

Perisinusoidal cells. Intralobular venous sinusoids are wider than blood capillaries and are lined by a thin but highly fenestrated endothelium. The fenestrae are grouped in clusters ('sieve plates'), the endothelium elsewhere being continuous. Attached to the luminal side of the endothelium are the *hepatic macrophages (stellate cells of von Kupffer or Kupffer cells)*; the latter form a major part of the mononuclear phagocyte system (p. 668), responsible for removing from the circulation much debris, cellular and microbial. These activities are shared with the spleen and circulating leucocytes (Wisse 1970). In the liver they also remove aged and damaged red cells from the circulation (as in the spleen). Sinusoids are fed at one end with blood from fine branches of the hepatic portal venules (*inlet venules*) and hepatic arterioles, which pass through the limiting plate of hepatocytes to enter a hepatic lobule. Blood from these sources percolates between the walls of the sinusoids to the central veins and is exposed to the activities of the cells around the sinusoids. The endothelial linings of the sinusoids are separated from hepatocytes of the hepatic laminae by a narrow gap, the *perisinusoidal space of Disse* (**8.160B**) which is normally about 0.2–0.5 μm wide, but distends in

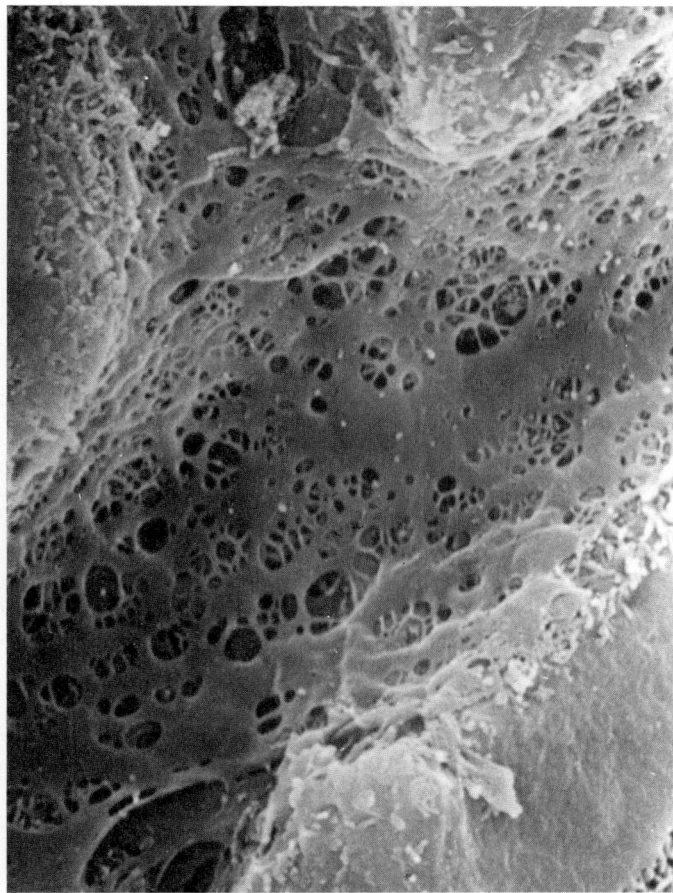

8.160D Scanning electron micrograph of the internal surface of a hepatic sinusoid, showing endothelial fenestrations. Magnification × 8000.

anoxic conditions: it contains reticulin fibres, and the irregular microvilli of adjacent hepatocytes; this space is continuous at the lobular periphery with a *space of Mall* surrounding the vessels and ductules in portal canals. In the latter space lymph vessels begin as cul-de-sac capillaries, as elsewhere; only a very few reach the periphery of the lobule. *Hepatic adipocytes* (Ito cells) are also present in small numbers in the space of Disse. *Central veins* from adjacent lobules form *interlobular veins*, which unite as *hepatic veins*, draining blood to the inferior vena cava.

Minute *bile canaliculi* (8.159, 160A,C) form nets with polygonal meshes in the hepatic lobules; each hepatocyte is surrounded by canaliculi except on its juxtasinusoidal sides. Hepatic laminae thus enclose a network of canaliculi which pass to the lobular periphery where they join to form narrow intralobular ductules *(terminal ductules* or the *canals of Hering*) lined by cuboidal epithelium; these exit through the terminal laminae to enter interlobular hepatic ductules in the portal canals. Intralobular ductules differ from the other biliary canals in their structure and reaction to injury; e.g. they proliferate when the flow of bile is obstructed outside the liver (Biava 1964, Jones et al 1975). Hepatic ductules in portal canals are lined by cuboidal or columnar cells which may contain cholesterol crystals and lipid droplets.

The permeability barriers between blood, hepatocytes and canaliculi are demonstrable with colloidal tracers, e.g. injected horseradish peroxidase diffuses freely from the sinusoids through the large gaps between hepatic endothelial cells into the intercellular spaces between hepatocytes, which are held together largely by gap (communicating) junctions; but diffusion is halted at the perimeters of the canaliculi by tight junctions between hepatocytes. It is inferred that materials have free access from the bloodstream to hepatocytes but can only enter the canaliculi through their *cytoplasm*; they therefore resemble exocrine cells secreting into a duct system.

Preterminal hepatic arterioles in the portal canals branch to convey arterial blood to the sinusoids by several routes, the chief

being via a fine capillary plexus around the interlobular ductules and ducts which drains to branches of the portal veins, inlet venules and hepatic sinusoids. Some arterial blood passes directly to the hepatic sinusoids, bypassing these capillary plexuses; but this is apparently only a small part of the total flow (Burkel 1970, Healey 1970, Jones & Spring-Mills 1977). Sinusoids thus contain mixed venous and arterial blood to sustain their cells. The composition, volume and velocity of blood through any local region may be regulated according to changing needs by the sphincters around the entry points of inlet venules and hepatic arterioles, and by the contractile walls of the sinusoids. Each portal triad supplies a distinct territory; normally there are no anastomoses between territories. Hepatic veins run quite separately with respect to the triadic system, freely crossing the boundaries of triadic territories.

The fetal liver is a major haemopoietic organ, erythrocytes, leucocytes and platelets developing from the mesenchyme covering the sinusoidal endothelium (p. 678).

Biliary Ducts and the Gallbladder

The hepatic ductual apparatus consists of: (1) the *common hepatic duct*, formed by the junction of the *right* and *left hepatic ducts*; (2) the *gallbladder*, a reservoir for bile; (3) the *cystic duct* of the gallbladder; (4) the *bile duct*, formed by the junction of the common hepatic and cystic ducts.

THE COMMON HEPATIC DUCT

The main right and left hepatic ducts issue from the liver and unite near the right end of the porta hepatis as the common hepatic duct, which descends about 3 cm before being joined on its right at an acute angle by the cystic duct to form the main bile duct (8.161). The common hepatic duct lies to the right of the hepatic artery and anterior to the portal vein.

THE GALLBLADDER

The gallbladder (8.154C, 161, 162) is a slate-blue, piriform sac partly sunk in a fossa in the right hepatic lobe's inferior surface. It

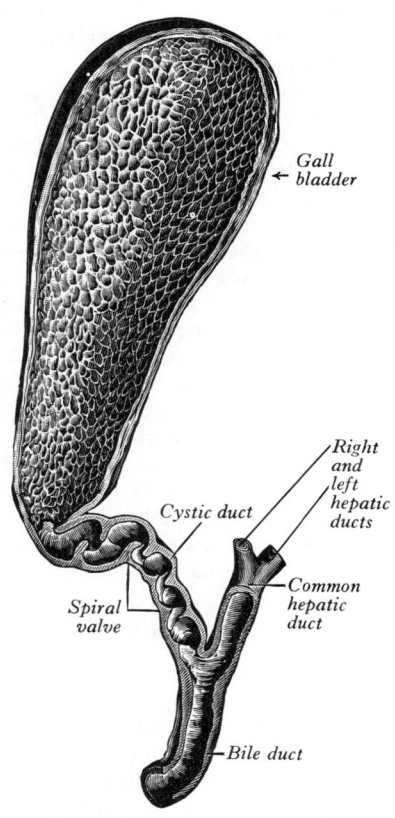

8.161 Interior of the gallbladder and bile ducts.

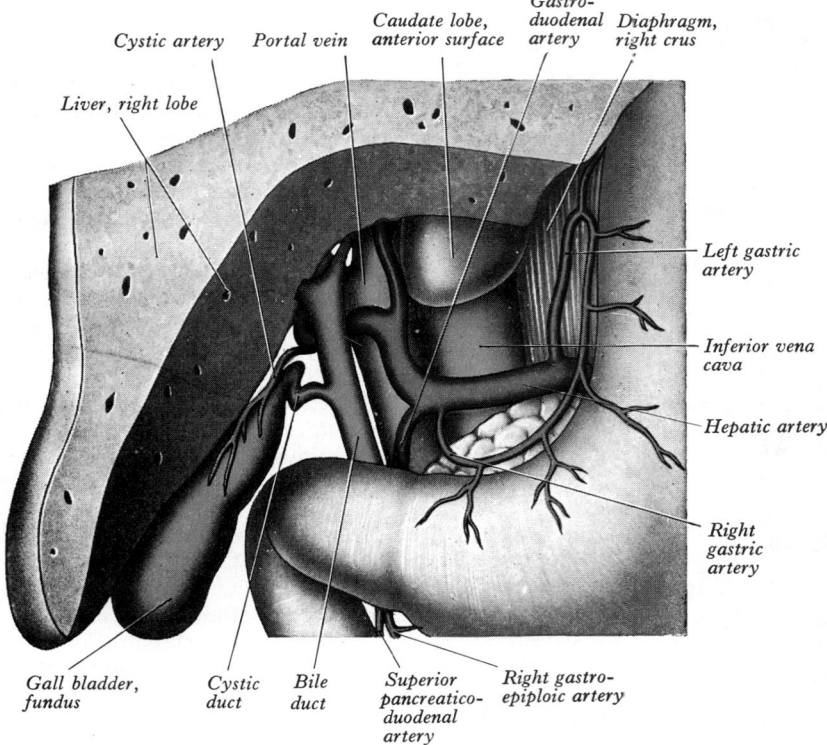

Cystic artery *Portal vein* *Caudate lobe, anterior surface* *Gastro-duodenal artery* *Diaphragm, right crus*

Liver, right lobe

Left gastric artery

Inferior vena cava

Hepatic artery

Right gastric artery

Gall bladder, fundus *Cystic duct* *Bile duct* *Superior pancreatico-duodenal artery* *Right gastro-epiploic artery*

8.162 Dissection to show the relations of the hepatic artery, bile duct and portal vein to each other in the lesser omentum: anterior aspect.

extends forwards from a point near the right end of the porta hepatis to the inferior hepatic border. Its upper surface is attached to the liver by connective tissue; elsewhere it is completely covered by peritoneum continued from the hepatic surface. Occasionally it is completely invested by peritoneum and even connected to the liver by a short mesentery. It is 7–10 cm long, 3 cm broad at its widest and 30–50 ml in capacity. It is described as having a fundus, body and neck. The *fundus*, the expanded end, projects down, forwards and to the right, extending beyond the inferior border to contact the anterior abdominal wall behind the ninth right costal cartilage, where the lateral edge of the right rectus abdominis crosses the costal margin. Posteriorly it is related to the transverse colon, near its commencement. (These relations change when the gallbladder is lower, as it often is in slender females, see Fleischner & Sayegh 1958.) the *body* is directed up, back and to the left; near the right end of the porta it is continuous with the neck. It is related above to the liver, below to the transverse colon and, further back, to the first and upper end of the second segments of the duodenum. The *neck* (cervix) is narrow, curving up and forwards and then abruptly back and downwards, to become the cystic duct, at which transition there is a constriction. The neck is attached to the liver by areolar tissue containing the cystic artery. The mucosa of the neck is obliquely ridged, forming a spiral valve; when the neck is distended, this gives its surface a spiral groove. From the right side of the neck a small recess may project down and back towards the duodenum. Often termed Hartmann's pouch (but originally described by Broca), it has been widely regarded as a constant feature, but Davies & Harding (1942) have shown that it is always a sequela of pathological states, especially dilatation; when it is large the cystic duct arises from its upper left aspect and not from what appears to be the gallbladder's apex.

THE CYSTIC DUCT

The cystic duct (**8.**161, 162) is 3–4 cm long; it passes back, down and to the left from the neck of the gallbladder, joining the common hepatic duct to form the bile duct. It is adherent to the common hepatic duct for a short distance before joining it, usually near the porta hepatis but sometimes lower, in which case the cystic duct lies along the lesser omentum's right edge. Its mucosa

bears five to 12 crescentic folds, like those in the gallbladder's neck. They project obliquely in regular succession, appearing like a *spiral valve* (**8.**161). When the duct is distended, the spaces between the folds dilate and externally it appears twisted like the neck of the gallbladder.

THE BILE DUCT

The bile duct is formed near the porta hepatis, by the junction of the cystic and common hepatic ducts; it is usually about 7.5 cm long and 6 mm in diameter (vide infra). It descends posteriorly and slightly to the left, anterior to the epiploic foramen, at the right border of the lesser omentum, in front and to the right of the portal vein and to the right of the hepatic artery proper (**8.**162). It passes behind the first (superior) part of the duodenum, with the gastro-duodenal artery on its left, and then runs in a groove on the superolateral part of the posterior surface of the head of the pancreas (**8.**149), anterior to the inferior vena cava and sometimes embedded in pancreatic tissue (pp. 1382, 1384). Lytle (1959) has shown that the duct may be close to the left aspect of the second (descending) part of the duodenum or as much as 2 cm from it and that, even when it is embedded in pancreas, a superficial groove marking its position can be palpated behind the descending part of the duodenum, stones in the duct being thus detected. Left of the descending part of the duodenum the bile duct reaches the pancreatic duct; together they enter the duodenal wall where they usually unite to form the *hepatopancreatic ampulla* (p. 1382), the distal, constricted end of which opens into the descending part of the duodenum on the summit of the major duodenal papilla (**8.**150), about 8–10 cm from the pylorus (p. 1394). The position of the bile duct is indicated on the anterior abdominal surface by a line starting 5 cm above the transpyloric plane and 2 cm right of the median plane and descending vertically for 7.5 cm.

Vessels and Nerves

The *cystic artery* is described on p. 769 and the *cystic veins* on p. 820. The lower part of the bile duct receives rami from the *posterior superior pancreaticoduodenal artery* (p. 769), while its upper part and the hepatic ducts receive rami from the *cystic artery*. The *right hepatic artery* supplies its intermediate part through very small rami, the main supply being from the *cystic* and *posterior superior pancreaticoduodenal arteries.* These supplies vary (Shapiro & Robillard 1948, Michels 1962). The posterior superior pancreatico-duodenal artery anastomoses with the posterior branch of the inferior pancreaticoduodenal near the hepatopancreatic ampulla; where this anastomosis is poor, ligation of the posterior superior pancreaticoduodenal artery may result in gangrene or stricture of the bile duct (Henley 1955). *Veins* from the upper part of the bile duct and hepatic ducts and from the gallbladder and cystic duct usually enter the liver, while those from the lower part of the bile duct enter the portal vein. *Lymph vessels* of the gallbladder and bile ducts are described on p. 852. Sympathetic and parasympathetic *innervation* is from the coeliac plexus along the hepatic artery and its branches. Autonomic plexuses exist in the muscular and submucous layers and ganglion cells, presumably parasympathetic, have been demonstrated in these plexuses in monkeys (Sutherland 1966, 1967). Fibres from the right phrenic nerve, through communications between the phrenic and coeliac plexuses, appear to reach the gallbladder via the hepatic plexus, thus explaining referred 'shoulder pain' in gallbladder pathology.

VARIATIONS IN GALLBLADDER AND BILE DUCTS

The gallbladder varies in size and shape; in rare cases it is duplicated (Mincsev 1967), with two or combined cystic ducts, or is absent, though its duct may be present. The junction of the cystic and common hepatic ducts varies in its level from the porta hepatis to behind or even below the duodenum's first (superior) part; when the junction is low, the two ducts may be connected by fibrous tissue and in cholecystectomy clamping the cystic duct without injuring the common hepatic (or main bile) duct is dif-

ficult. Occasionally the cystic duct joins the right hepatic duct; it may pass behind or in front of the common hepatic duct, joining it on its left surface. Accessory hepatic ducts may emerge, more often from the right lobe, to join the main hepatic ducts or, rarely, the gallbladder itself. Failure of canalization of bile ducts during development leads to a rare congenital atresia or stenosis with rapidly fatal results. The bile and pancreatic ducts may open separately into the duodenum or join even before entering its wall. Variations of the ductual arteries are much more common (p. 768). For further information consult Santulli & Blanc (1961), Boyden et al (1967).

MICROSCOPIC STRUCTURE OF THE GALLBLADDER AND BILIARY DUCTS

The gallbladder's wall has serous, fibromuscular and mucous layers. The *serosa* completely covers the fundus but only coats the inferior surfaces and sides of the body and neck of the gallbladder; beneath it is subserous loose connective and adipose peritoneal tissue. The *fibromuscular layer* is composed of fibrous tissue mixed with non-striated myocytes arranged loosely in longitudinal, circular and oblique bundles. Internally, the *mucosa* is loosely connected with the fibrous layer, is generally yellowish-brown and elevated into minute rugae with a honeycomb appearance

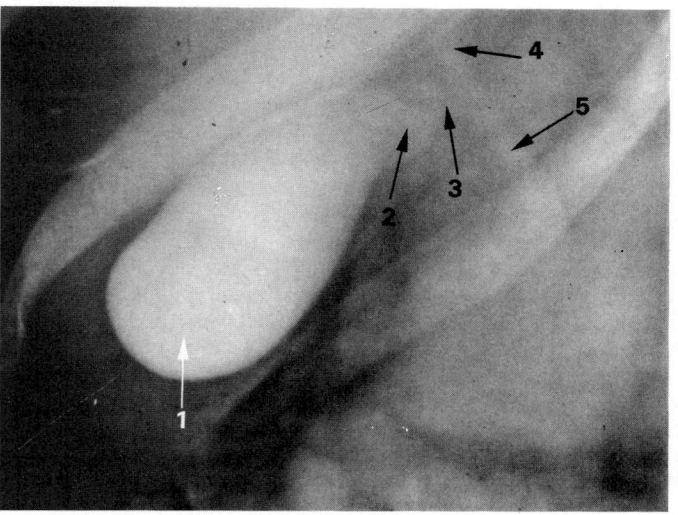

8.164A Anteroposterior radiograph of the gallbladder and biliary ducts after the oral administration of sodium tetra-iodophenolphthalein the previous day.

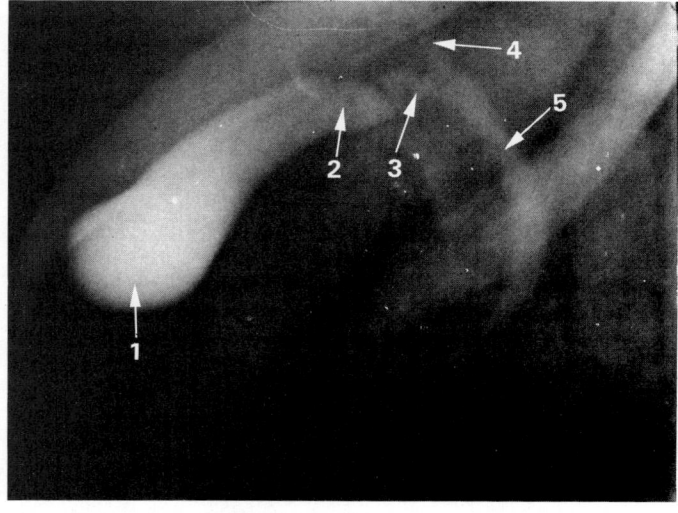

8.164B The same field as 8.164A but taken 20 minutes after a fatty meal, demonstrating the contraction and partial emptying of the gallbladder. In both A and B: (1) fundus of gallbladder; (2) neck of gallbladder; (3) cystic duct; (4) common hepatic duct; (5) the bile duct.

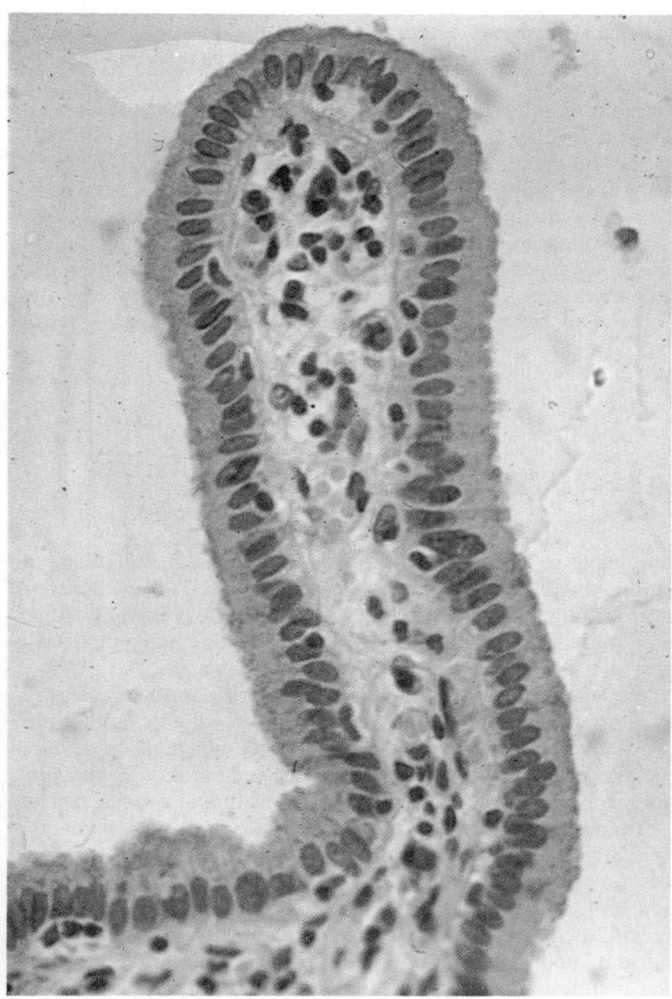

8.163 Section through a surface projection (ruga) of the gallbladder showing the columnar epithelial cells and lamina propria. Haematoxylin and eosin. Magnification × 500.

(8.161, 163). Its epithelium is a single layer of columnar cells which vary with species. Electron microscopy (Chapman et al 1966) in dogs shows microvilli on their apical surface, irregularly arranged, with endocytic or pinocytotic vesicles between their bases. Basally, spaces between epithelial cells are dilated and many capillaries adjoin the basement membrane. These features indicate active absorption of water and solutes from the bile to concentrate it. Basal spaces are large during absorption of water (Kaye et al 1966). Biliary concentration appears to involve the active transport of sodium and calcium, making an osmotic gradient from the lumen of the gallbladder to the capillaries of the lamina propria. In the apical parts of some cells, particularly those near the ducts, mucous granules appear; they are secreted into the lumen (Johnson et al 1962, Mueller et al 1972).

The large biliary ducts have external fibrous and internal mucous layers. The former is fibro-areolar tissue with a few longitudinal, oblique and circular non-striated myocytes. (Myocytes appeared in only 12 of 100 human common bile ducts examined by Mahour et al 1967.) The mucosa is continuous with that of the hepatic ducts, gallbladder and duodenum; like theirs, its epithelium is columnar; many lobulated mucous glands occur in the wall of these ducts. In the bile duct are many tubulo-alveolar glands arranged in clusters, secreting mucin, some at least of which is sulphated (McMinn & Kugler 1961). Electron microscopy of the bile duct epithelium in guinea-pigs showed microvilli on its luminal surfaces. Secretory granules, some of mucinogen, appear in the apical cytoplasm. Epithelial cells in rats, which lack a gall bladder, are termed either light or dark according to their electron density; light cells have longer, more regular microvilli but the basal intercellular spaces between adjacent dark cells are larger; the mucosa appears to modify bile, compensating for the absent gallbladder (Riches & Palfrey 1966).

Circular muscle around the lower part of the main bile duct, including the ampulla and terminal pancreatic duct, forms the *sphincter of the hepatopancreatic ampulla (sphincter of Oddi)*, comprising muscle at three levels: (1) at the end of the bile duct *(sphincter ductus choledoci)*; (2) at the end of the pancreatic duct *(sphincter ductus pancreatici)*; and (3) around the ampulla. Only (1) is constant. Expulsion of gallbladder contents appears to be under hormonal control. Fat or acid in the duodenum probably causes the liberation of cholecystokinin, stimulating the gallbladder to contract. Myocytes in its wall have surface-borne receptors for this hormone, which can therefore stimulate them to contract directly. In any case, when storage pressure exceeds 100 mm of bile, the gallbladder contracts, the sphincter of Oddi relaxes and bile enters the duodenum. Kirk (1944) denied the presence of sphincteric musculature around the openings of the bile and

pancreatic ducts, describing the sphincter of Oddi as submucosal and continuous with the circular muscle of the duodenum. However, subsequent studies suggest that in man and other primates a common sphincter surrounds both ducts and that the common bile duct has a second sphincter as described above (Boyden 1957, 1966). The termination of the united bile and pancreatic ducts is packed with villous, valvular folds of mucosa and myocytes enter their connective tissue cores. This suggests that contraction results in retraction and clumping of the folds, preventing reflux of duodenal contents and controlling the exit of bile. In cats stimulation of vagal rami to this region relaxes the biliary opening; the human *myenteric (Auerbach's) plexus* is well developed at the ends of the ducts. Inflammatory swelling of villous folds may obstruct them.

Applied Anatomy. On account of its large size, fixed position, and friability, the liver is sometimes ruptured; haemorrhage may be severe, because the hepatic veins lie in rigid canals and are unable to contract. The organ may be torn by a broken rib, perforating the diaphragm. Clinical evidence suggests that the bloodstreams of the superior mesenteric and splenic veins remain largely separate in the portal vein, passing respectively along the right and left portal branches to the right and left physiological (vascular) lobes (p. 1385); thus malignant or infective emboli may be more pronounced in the right lobe if the primary disease is in the territory drained by the superior mesenteric vein, or in the left if it is in the splenic or inferior mesenteric territory. This correlation is also supported in experiments in living animals. Vascular hepatic segmentation (p. 1385) is a vital factor in *partial hepatectomy*.

The gallbladder may be distended by calculi or by obstruction of the cystic duct and may project down and forwards towards the umbilicus. It moves with respiration. Obstruction of the bile duct, apart from lithiasis, is often due to pressure of malignant tumours, especially in the pylorus or pancreas. It also follows cicatricial contraction after ulceration in the duct. Enormous distension of the bile duct and its radicles may also occur.

Cholecystography (8.164). The gallbladder is not radio-opaque, but radio-opaque substances introduced into the bloodstream are excreted by the liver into bile, and concentration in the gallbladder renders it much more strongly radio-opaque. The form, position and emptying of the gallbladder can be demonstrated by X-rays; its position and form vary with the general build of the body (or somatotype) (Davies 1927); in broad (hypersthenic) types it is broad, high up and far laterally (at the level of the first lumbar vertebra), whereas in narrow (asthenic) types it is narrow, more medial and may reach as low as the fourth lumbar vertebral level.

THE UROGENITAL SYSTEM

Introduction

So customary has it become to link the organs of urinary excretion and reproduction as a *urogenital system* that the suitability of this concept is rarely even questioned. Yet functionally the two have nothing in common and to associate them in description may seem at first absurd (Romer 1970). In mature human organs the structural overlap is little more than a shared use of the male urethra as a urinary *and* seminal duct. Of course, in development, gonadal ducts have a *nephric* origin; but in human females even the *oviduct* ('female', paramesonephric or Müllerian duct) is no longer formed from nephric tissue. Indeed, no adequate reason remains to continue defining a 'urogenital' system in female mammals. Even in males, where the duct systems of the testes (as in other mammals) are derived from nephric tubules and the *mesonephric duct*, these are not excretory structures. The reptilian functional kidney is no longer a mesonephros, and the intromittent *penis*, developed from cloacal tissues, has bilateral parts enclosing a purely seminal groove when erected for copulation, urine entering the cloaca. Apart from development, therefore, the genital system is already

completely separated from the urinary organs in both sexes in reptiles. Birds, like amphibians, generally have no intromittent organ, though this is probably a secondary loss in the former, some more primitive birds possessing a rudimentary penis of a reptilian type.

In all vertebrates (excluding elementary chordates) nephric excretory tubules and gonads develop from coelomic epithelium, together with their collecting ducts as these evolve. Both develop in dorsal sites flanking the major vessels; and perhaps this propinquity, combined with the more medial location of gonads, has led to the adoption of urinary ducts and tubules to serve the gonads, especially in males. In primitive *Chordata*, such as *Amphioxus*, excretory tubules, or *nephridia*, have capillary *glomeruli* close to them and secrete from the adjoining bloodstream and coelom into a peribranchial atrium. These excretory organs are ectodermal and unrelated to the mesodermal nephric tubules of vertebrates. The evolutionary relation of *Amphioxus* to early vertebrates is, in any case, uncertain; it has even been suggested that this chordate is a paedomorphic form of a primitive vertebrate (Grassé 1948, Young 1962), matters beyond this discussion.

In vertebrates living in water, saline or fresh, the problems of preserving internal osmotic constants in blood and body fluids, despite variable and sometimes very large intakes of water and salts, have led to the continuous evolution of more efficient and elaborate excretory tubules, derived from mesoderm intermediate between the somites and the lateral plate in embryos, forming *nephrotomes* (p. 248). Though originally segmental, their large numbers overshadow and in most lower vertebrates they form elongated masses or cords of nephrogenic tissue projecting into the coelom from a dorsolateral position. Gonads develop *medial* to these *nephrogenic cords* (p. 248); and since *primary excretory ducts* (p. 248), carrying urine to the exterior, evolve *lateral* to the nephros, gonadal access to them is effected by adaptation of the nephric tubules. The degree of development and functional status of different parts of this elongate kidney or vertebrate *holonephros*, extending through many body segments, has varied much. The cranial part generally tends to disappear, except where it functions in embryos or in larval forms, as in some cyclostomes (e.g. *Petromyzon*, the lamprey). In anamniotic gnathostomes (fish and amphibia) an intermediate nephrogenic region becomes a functional kidney and, while still elongate, this occupies a comparatively small number of segments, with which its vessels approximately correspond. In reptiles and mammals the most caudal mass of nephrotomes differentiates into a definitive kidney, which becomes more localized and rounded, of familiar reniform shape particularly in mammals, losing almost all indications of plurisegmental origin when fully developed.

This progressive caudal shift of the functioning nephros has prompted the division of the holonephros into pro-, meso-, and meta-nephros, and has given rise to some misleading concepts; e.g. amphibian kidneys are often termed mesonephric and mammalian organs metanephric. Nevertheless, the formation of nephric tubules is substantially alike at all levels and it cannot be assumed that these regional terms are applicable over large and diverse vertebrate groups with anything more than approximate segmental uniformity. The terms are more of a descriptive convenience than vehicles of biological precision. (For a discussion of problems in the development of the urogenital organs, consult Fraser 1950.) Distinctions between the three regional levels of the nephrogenic ridge depend more upon their functional status and ducts than their intrinsic morphology. In cyclostomes, excretory tubules discharge, as do gonads, into the coelom, from which their products escape through the *abdominal pores* or through short ducts situated caudally. In gnathostomes, *nephric tubules* lose continuity with the coelom, each developing a saccular end, with which *glomeruli* are closely related. The tubules join a longitudinal *archinephric duct (primary excretory duct)*, which first serves the pronephros but is taken over by the more caudal mesonephros. Through mesonephric tubules testes find outlets via the urinary ducts, an arrangement typical of amphibians, both urine and spermatozoa reaching the cloaca by such *mesonephric (Wolffian) ducts*. Ovaries also acquire ducts, in some forms from the mesonephric duct; but this *paramesonephric, female* or *Müllerian duct* becomes a separate entity and does not transmit urine. Moreover, oviducts never establish direct continuity with the ovaries, which retain a primitive habit of shedding ova into the coelom, though very close to the open, coelomic beginnings of the oviducts. This relation is constant and occurs in human females.

Oviducts of mammals and of tetrapods in general appear to develop independently by evagination from the coelomic lining.

In reptiles, birds and mammals, the functioning kidney is a *metanephros*, developing caudal to the mesonephros. Its main distinction is its *dissociation from genital function*; though in part an outgrowth of the caudal end of the mesonephric duct, it ultimately opens separately into the bladder by its own duct, the *ureter*. The mesonephros is thus relieved of all excretory function, its tubules and ducts persisting only in modified form as testicular *vasa efferentia, epididymis* and *ductus deferens*.

To summarize: the pronephros is an embryonic or larval excretory organ with a purely excretory duct. The mesonephros, which takes over this duct and excretion, also provides a route for spermatozoa and, by the division of its duct in some vertebrates, is a separate exit for the ova. The metanephros is again purely excretory, male genital ducts continuing as mesonephric derivatives and the oviduct as an increasingly independent development.

The complete segregation of urinary and genital organs in reptiles (apart from their embryonic development and the opening of their separate ducts in a common cloaca) becomes modified again by the evolution of a male intromittent organ for internal fertilization, necessitated by full emancipation from reproduction in water. Production of eggs, or rather embryos, protected from desiccation in a terrestrial environment by shells impervious to water (though not to respiratory gases), requires fusion of gametes in the oviduct proximal to specializations of its wall to produce such coverings. Similarly, production of offspring at a sufficient stage of development for life exposed to the external environment also entails internal fertilization. Apart from monotremes, all mammals have this viviparous habit. Internal fertilization is accomplished without an intromittent organ in most birds, but an erectile *penis* is universal in marsupials and eutherian mammals. Since this is traversed by the urethra, used to void both urine and semen, male mammals, including the human, exhibit to this extent an association of the excretory and reproductive organs. Both systems display many adaptations and specializations, concerned in mammals with a much improved ability to preserve a steady internal environment in terms of total body water volume and concentrations of ions and other dissolved substances. This entails an increasing complexity in glomeruli and renal tubules, forming *nephrons* or kidney units, capable not only of removing variable and sometimes large volumes of water, but also of selective reabsorption of substances such as glucose and salts and retention of others, such as urea, in the urine.

The renal ducts open into various forms of reservoir in many vertebrates and in mammals into a cloacal derivative, the *bladder*. In mammalian reproductive organs, accessory structures have evolved in connection with intromission and the retention and nurture of developing embryos; these include seminal vesicles and the prostate, uterus and vagina. All have already been considered in embryological development (p. 248); they must now be studied in their mature primate morphology. The customary association of the two groups of organs has been partially retained, since even in description of completely differentiated urinary and genital systems reference to ontogeny and phylogeny is occasionally necessary. It is, nevertheless, worthy of repetition that their association in the postnatal human being is wholly limited to the dual function of the penile urethra.

THE URINARY ORGANS

The urinary organs are: (1) two *kidneys* (renes) producing urine, (2) *ureters*, conveying it to (3) the *urinary bladder* (vesica urinaria) for temporary storage, and (4) the *urethra* by which the bladder empties. The female urethra is purely urinary; the male is urinary and reproductive, the latter essential, the former merely the use of a tube almost entirely genital. Loss of the penile urethra affects urination little, whereas genital function is abolished. Almost the whole male urethra is therefore genital (p. 1421) but for the sake of completeness it is described in this section.

The Kidneys

The kidneys excrete the final products of metabolic activities and excess water, both of these actions being essential to the control of concentrations of various substances in the body fluids, e.g. maintaining electrolyte and water balance approximately constant in tissue fluids. They also have endocrine functions producing and releasing *erythropoietin* which affects blood formation, *renin* influencing blood pressure and *1,25-hydroxycholecalciferol*, involved in the control of calcium metabolism and a derivative of vitamin D, perhaps modifying the action of the parathyroid hormone (O'Riordan 1978), and various other soluble factors with metabolic actions. The kidneys in the fresh state are reddish-brown, are situated posteriorly behind the peritoneum on each side of the vertebral column and are surrounded by adipose connective tissue. Superiorly they are level with the upper border of the twelfth thoracic vertebra, inferiorly with the third lumbar. The right is usually slightly inferior, probably due to the volume of the liver on this side; the left is a little longer and narrower and lies nearer the median plane. The long axis of each is directed inferolaterally and the transverse axis posterolaterally. Hence the anterior and posterior aspects usually described are in fact *antero-lateral* and *posteromedial*. The transpyloric plane (p. 1334) passes through the superior part of the right renal hilum and the inferior part of the left.

Each kidney is about 11 cm in length, 6 cm in breadth and 3 cm in anteroposterior dimension. In adult males the average weight is about 150 g, in adult females 135 g. In thin individuals with a lax abdominal wall the lower pole may just be felt in full inspiration by bimanual lumbar examination; usually it is impalpable.

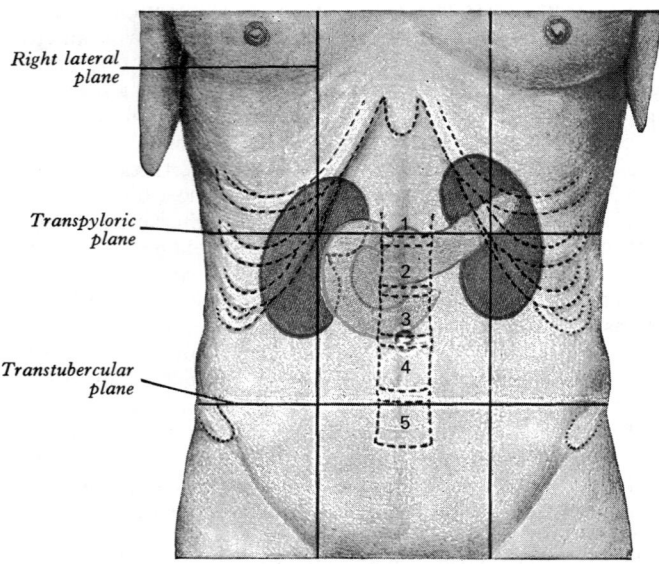

Right lateral plane

Transpyloric plane

Transtubercular plane

8.165 Surface projections of the duodenum, pancreas and kidneys. The lower ribs and the lumbar vertebrae are also indicated, the latter being numbered.

Renal Surface Projections

In a recumbent posture, each renal profile can be projected to the anterior or posterior surface of the body as follows, the right kidney being about 1.25 cm lower: (1) *Anterior surface.* The hilar centre is at the transpyloric plane, about 5 cm from the midline and slightly medial to the apex of the ninth costal cartilage. The left hilum is just above the plane, the right just below it (**8.165**). In relation to the position of the hilum, a reniform profile can be drawn 11 cm long and 4.5 cm broad, the upper pole being about 2.5 cm and the lower 7.5 cm from the midline. Since the transverse axis is oblique, the width thus shown is 1.5 cm less than the actual width. (2) *Posterior surface.* The hilar centre is opposite the first lumbar spine's lower edge, and about 5 cm from the midline. In relation to this point, a reniform profile can be similarly traced, the lower pole being usually about 2.5 cm above the summit of the iliac crest. The kidneys are about 2.5 cm lower in the standing position; they ascend and descend a little with respiration.

Renal Relations

The convex *anterior surface* (**8.**125, 166) in reality faces antero-laterally. Its relations differ on the two sides of the body.

(1) *Right anterior surface.* A small area at the superior pole contacts the right suprarenal gland, which may overlap it or the upper medial border. A large area below this (about three-quarters of the surface) adjoins the renal impression on the right lobe of the liver and a narrow medial area is related to the descending duodenum. Inferiorly the anterior surface is related laterally to the right colic flexure and medially to the small intestine. The areas related to the small intestine and almost all those in contact with the liver are covered by peritoneum (with the renal fascia subjacent); the suprarenal, duodenal and colic areas are devoid of peritoneum.

(2) *Left anterior surface.* A small medial area at the superior pole is related to the left suprarenal gland and approximately the upper two-thirds of the lateral half of the anterior surface is related to the spleen. A central quadrilateral area lies in contact with the pancreatic body and splenic vessels. Above this a small variable triangular region, between the suprarenal and splenic areas, is in contact with the stomach. Below the pancreatic and splenic areas the lateral region is related to the left colic flexure and the beginning of the descending colon, while the medial region adjoins the coils of the jejunum. The latter region is extensive but the colic area is an irregular, narrow strip adjoining the lateral border of the kidney. The gastric area is covered with the peritoneum of the

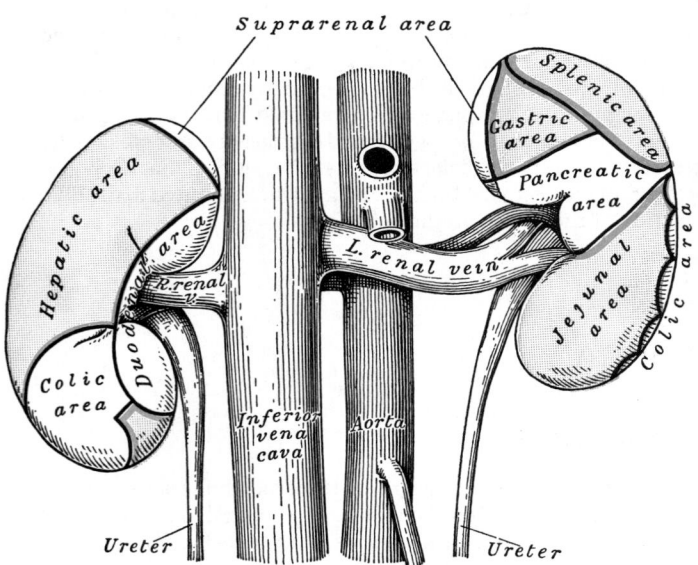

8.166 The anterior surfaces of the kidneys, showing the areas related to neighbouring viscera. Areas coloured pale blue are separated from adjacent viscera by the peritoneum.

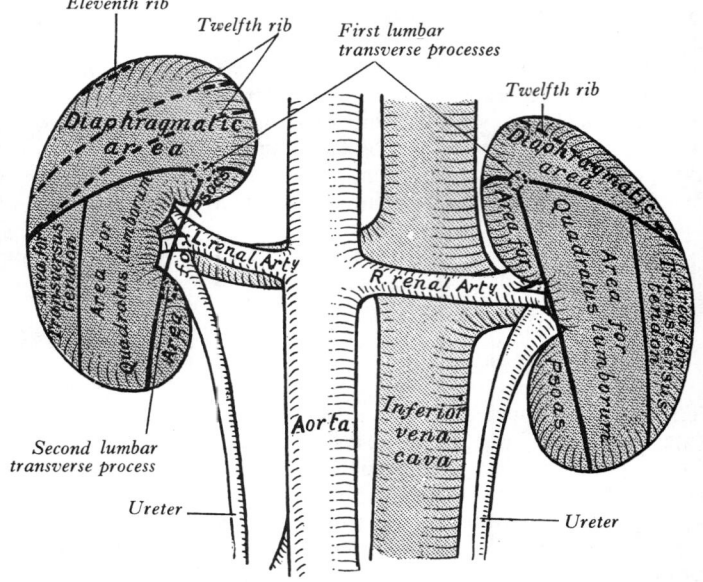

8.167 The posterior surfaces of the kidneys, showing the areas of relation to the posterior abdominal wall.

omental bursa and the splenic and jejunal areas are covered by the peritoneum of the greater sac; behind the jejunal area's peritoneum, branches of the left colic vessels are related to the kidney. Suprarenal, pancreatic and colic areas are devoid of peritoneum, there being no mutual movement of organs here.

The *posterior surface* (**8**.167, 168, 170), in reality posteromedial, is embedded in fat and devoid of peritoneum. It is anterior to the diaphragm and to the medial and lateral arcuate ligaments, psoas major, quadratus lumborum and the aponeurotic tendon of transversus abdominis, to the subcostal vessels and subcostal, iliohypogastric and ilio-inguinal nerves. The right upper pole is level with the twelfth rib, the left with the eleventh and twelfth. The diaphragm separates the kidney from the pleura descending in the costodiaphragmatic recess (**8**.168); sometimes its muscle is defective or absent in a triangle immediately above the lateral arcuate ligament, allowing perirenal adipose tissue to contact the diaphragmatic pleura.

The *superior poles* are thick, round and nearer the midline, each related to its suprarenal gland. The *inferior poles*, smaller and thinner, extend to within 2.5 cm of the iliac crests. The *lateral borders* are convex, the left covered superiorly by greater sac peritoneum, so separating it from the spleen, and below this is in contact with the descending colon; the right lateral border is separated by the peritoneum of the greater sac from the liver (right lobe). The *medial borders* are convex adjacent to the poles, concave between them and slope inferolaterally. In each a deep vertical fissure opens anteromedially as the *hilum*, bounded by anterior and posterior lips and containing the renal vessels and nerves and the *renal pelvis* of the ureter. The relative positions of the main hilar structures are: the renal vein anterior, the renal artery intermediate and the pelvis posterior. Commonly an arterial branch enters behind the renal pelvis and a renal venous tributary often leaves the hilum in the same plane. Above the hilum the medial border is related to the suprarenal gland and below to the commencement of the ureter.

The hilum leads into a central *renal sinus*, lined by the renal capsule and almost filled by the renal pelvis and vessels; numerous *renal papillae* project on the wall of the sinus. The collecting tubules open onto the summits of the renal papillae and drain into the funnel-shaped expansions of the upper urinary tract, named the minor calices (**8**.169B, 172). The minor calices terminate in

two or three major calices which in turn open into the renal pelvis. The renal calices and pelvis are described in detail on p. 1411.

The kidney and its vessels are embedded in *perirenal (perinephric) fat*, which is thickest at the renal borders and prolonged at the hilum into the renal sinus. Fibrous connective tissue surrounding this fat is condensed as *renal fascia* (**8**.170).

The Renal Fascia

At the lateral renal borders the two layers of renal fascia fuse; the anterior extends medially in front of the kidney and its vessels to merge with connective tissue enclosing the aorta and inferior vena cava, but it is thin and does not ascend above the superior mesenteric artery. The posterior layer passes medially between the kidney and the fascia on quadratus lumborum and psoas major, attaching to this fascia at the lateral and medial borders of the psoas and to the vertebrae and intervertebral discs. A deeper stratum (not shown in **8**.170) unites the anterior and posterior layers at the medial renal border and is pierced by renal vessels (Martin 1942); this may account for the failure of perirenal effusions to cross the midline (although injected air *does* diffuse along this route, Grossman 1954). Above the suprarenal gland the two layers of renal fascia fuse and blend with the diaphragmatic fascia; it is generally agreed that below the kidney they are separate, enclosing the ureter, the anterior fading into the extraperitoneal tissue of the iliac fossa, the posterior blending with the iliac fascia (although this has also been denied, Mitchell 1950). Renal fascia joins the renal capsule by numerous trabeculae traversing the perirenal fat and strongest near the lower pole. Behind the renal fascia is a mass of fat, the *pararenal (paranephric) body*. The kidney is held in position partly by renal fascia but principally by the apposition of neighbouring viscera.

The fetal kidney has about 12 lobules (**8**.171), but these are fused in adults to present a smooth surface, though traces of lobulation may remain.

General Renal Structure (**8**.172)

The kidney has a thin capsule, easily removed, composed of collagen-rich tissue with some elastic and non-striated muscle

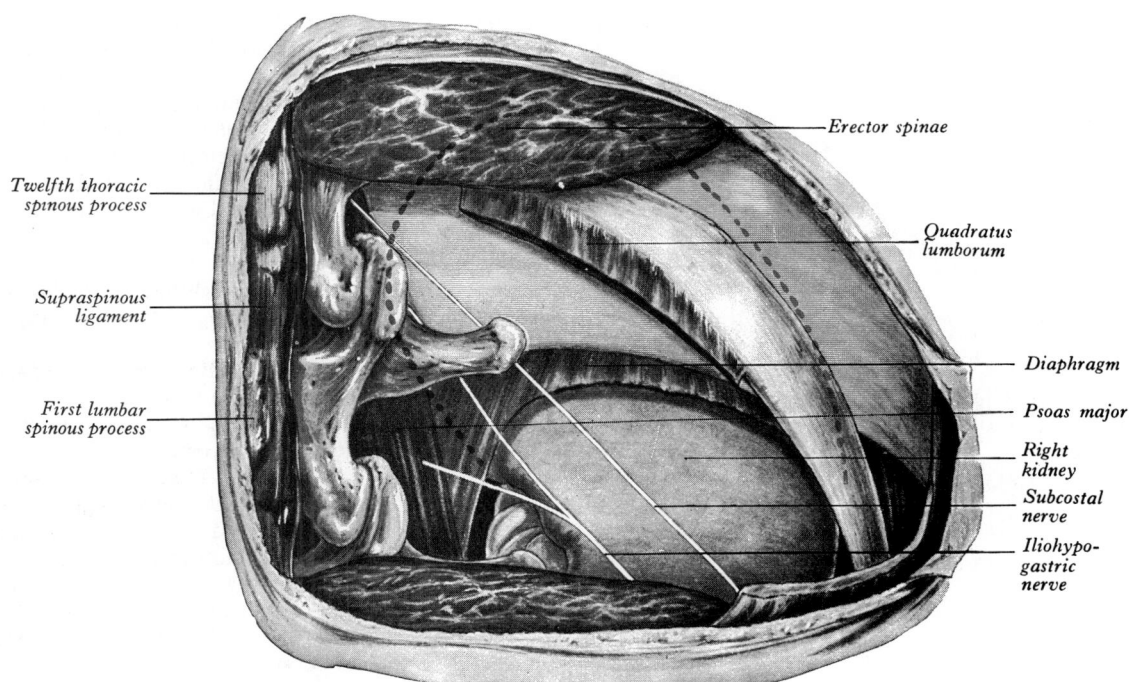

Twelfth thoracic spinous process

Supraspinous ligament

First lumbar spinous process

Erector spinae

Quadratus lumborum

Diaphragm

Psoas major

Right kidney

Subcostal nerve

Iliohypogastric nerve

8.168 The right kidney (posterior exposure). The blue area represents the pleura, the broken red line the upper part of the kidney. The subcostal nerve has been displaced downwards. Parts of the diaphragm and the quadratus lumborum have been resected.

fibres. In renal disease it may become adherent. The kidney itself has an internal *medulla* and external *cortex*.

The *renal medulla* consists of pale, striated, conical *renal pyramids*, their bases peripheral, their apices converging to the renal sinus where they project into calices as papillae, each minor calix receiving from one to three of these structures. Each pyramid is capped by cortical tissue to form a renal *lobe*. Estimates of papillae, and hence of pyramids or renal lobes, are variable. Counts in 375 human kidneys (Lake et al 1966) ranged from 5–11 in 89% with a most frequent value of 8 (in 26%). Total numbers of terminal uriniferous ducts (of Bellini) opening on papillae

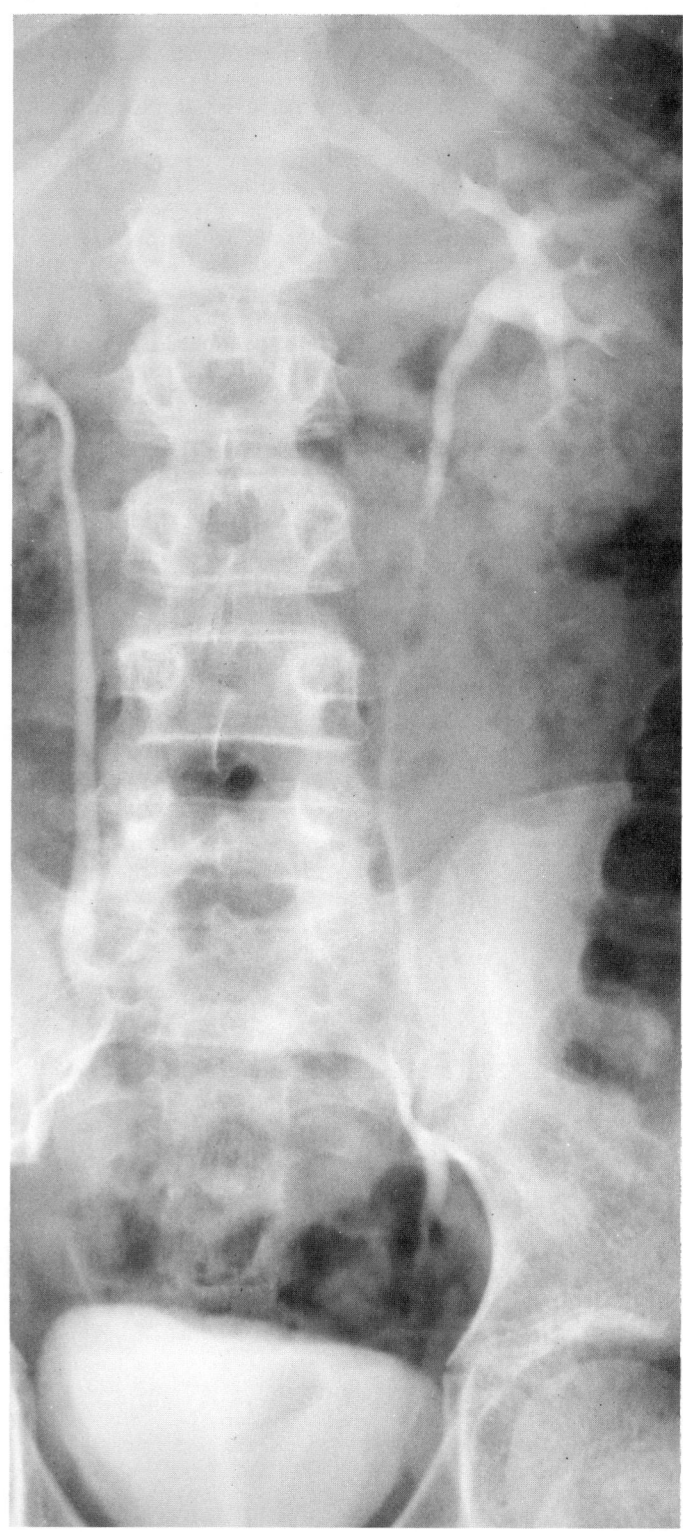

8.169A An intravenous pyelogram. Both ureters can be visualized throughout much of their length, except where peristalsis has temporarily displaced the excreted contrast medium, which also fills the urinary bladder. Compare with the retrograde pyelogram illustrated in **8.169B**.

varied (in 208 kidneys) from 116 to 776; no peak of frequency was observed, but 23% of kidneys displayed about 275 such openings. In another series of 54 kidneys (Arvis 1969) the number of papillae was 6 to 14.

The *renal cortex* (**8.172A**) is subcapsular, arching over the bases of pyramids and extending between them towards the renal sinus as *renal columns* (**8.172C**); the peripheral regions are *cortical arches* and are traversed by radial, lighter-coloured *medullary rays* (**8.172C**), separated by darker tissue, the *convoluted part*. The rays taper towards the renal capsule and are peripheral prolongations from the bases of renal pyramids. The cortex is also histologically divisible into *outer* and *inner zones*; the inner is demarcated from the medulla by tangential blood vessels (arcuate arteries and veins) which lie at the junction of the two, but a thin layer of cortical tissue ('*subcortex*') appears on the medullary side of this zone. The cortex close to the medulla is sometimes termed *juxta-medullary*.

Renal Histology

The kidney is composed of many tortuous, closely packed *uriniferous tubules*, bound by a little connective tissue in which run blood vessels, lymphatics and nerves. Each tubule consists of two embryologically distinct parts (p. 248): (1) the secreting *nephron* which elaborates urine and (2) a *collecting tubule*. The *nephron* comprises *(a) a renal corpuscle*, concerned with filtration from the plasma and *(b) a renal tubule*, concerned with selective resorption from the glomerular filtrate to form the urine. *Collecting tubules* carry fluid from several renal tubules to a terminal *papillary duct (of Bellini)*, opening into a minor calix at the apex of a renal papilla. Papillary surfaces (vide supra) show numerous minute orifices of these ducts and pressure on a fresh kidney expresses urine from them.

The Renal Corpuscle (8.172D, 173A, 175B)

Renal (Malpighian) corpuscles, small rounded masses averaging about 0.2 mm in diameter, are visible in the renal cortex and columns except from a narrow peripheral cortical zone (*cortex corticis*). There are one to two million renal corpuscles in each kidney, decreasing with age (Dunnil & Halley 1973). Each has a central *glomerulus* of vessels and a membranous *glomerular capsule*, the commencement of a renal tubule.

A glomerulus is a lobulated collection of convoluted, capillary blood vessels, united by scant connective tissue etc., and fed by an *afferent arteriole* which usually enters the capsule opposite the exit into the tubule; an *efferent arteriole* emerges from the same *vascular pole (mesangium)* of the capsule. Glomeruli are simple until late prenatal life, some remain so for about six months after birth, the majority maturing by six years and all by 12 (Macdonald & Emery 1959). In the fetal rabbit, guinea-pig and sheep, the glomerular precursor is a solid sphere of mesodermal cells in which vessels canalize until the glomerulus is a compact anastomosing plexus, not showing independent capillary loops (Lewis 1958).

The glomerular capsule (of Bowman), the expanded end of a renal tubule, deeply invaginated by the glomerulus, is lined by simple squamous epithelium on its outer (parietal) wall; its glomerular, juxtacapillary (visceral) wall is composed of specialized epithelial *podocytes* (Latta 1973, Spinelli 1974, Arakawa & Tokunaga 1974). Thus, between glomerulus and external capsular layer, is a flattened *urinary space*, varied in size by secretory activity (vide infra) and continuous with the proximal convoluted tubule. The basal lamina of the capsular cells fuses with that of glomerular endothelium.

Podocytes surrounding the capillary loops are flat, stellate cells, their major processes curved around capillaries, interdigitating tightly with each other and attached to the basal lamina by numerous pedicles; narrow gaps lie between these cellular extensions. Podocytes contain many mitochondria, microtubules, microfilaments and vesicles of various types, all of these being signs of active metabolism. The glomerular *endothelium* is finely fenestrated and the sole barriers to the passage of fluid from capillary lumen to urinary space are therefore the fused endothelial and podocytic basal laminae. This *glomerular basement membrane* is about 0.33 μm thick in man; it is finely fibrillar and shows three layers, the first and third pale-staining (*laminae rarae interna et externa*) and the middle dense and fibrous (*lamina densa*). Unlike many fenestrated endothelia, a regular dense lamina appears absent from the *fenestrae* of glomerular endothelial cells; but at the distal aspect of glomerular membrane areas between adjacent pedicles are covered by a clear dense *glomerular slit diaphragm*, through which filtrate must pass to enter the urinary space. This diaphragm is composed of fine filaments arranged in two regular rows like a zip-fastener (Rodewald & Karnovsky 1974) which may form an important part of the glomerular filter. The glomerular basement membrane certainly acts as a selective filter, allowing the passage from blood, under pressure, of water and various small molecules and ions in the circulation. Haemoglobin may pass, but larger molecules are held back. Selectivity appears related to the physicochemical nature of the filter. Tracer experiments show that some materials withheld from urine by the glomerulus traverses all but the outer surface of the filter, which is therefore more complex than a simple physical filter (Rennke et al 1975). The glomerular filter has chemical features similar to those of basal laminae elsewhere, including type IV collagen in the lamina densa and negatively charged proteoglycans elsewhere (see Farquhar 1981). The presence of fixed negative charges tends to inhibit the filtration of charged macromolecules from the blood, particularly in the laminae rarae. Irregular *mesangial cells*, with phagocytic and contractile capacities, have been demonstrated, particularly around bases of glomerular capillary tufts; these may help to clear the glomerular filter of enmeshed substances which might clog it, e.g. immune complexes, cellular debris, etc (Caulfield & Farquhar 1974; see also Brenner & Rector 1981, Bulger & Dobyan 1982).

The Renal Tubule

A renal tubule (**8.**172D) consists of: (1) a *glomerular capsule*, already described, (2) a *proximal (first) convoluted tubule*, connected to the capsule by a short *neck* (in humans a little narrower than the tubule) leading to a first sinuous or partially coiled *convoluted part*, then to a terminally straight or slightly spiral *straight part*; this approaches the medulla to become (3) the *descending limb* of the *loop of Henle* connected by (4) a 'U' turn to (5) the *ascending limb*. The limbs of Henle's loop are for some distance narrower and thin-walled, forming the *thin segment*; but the upper part of the ascending limb (*thick segment*) has the same diameter as the proximal convoluted tubule. The ascending limb continues into (6) the *distal (second) convoluted tubule*, divisible into *straight* and *convoluted sections* with an intermediate thickened *macula densa* where the tubule comes close to a glomerulus. The nephron finally straightens as (7) the *junctional (connecting) tubule*, which ends in (8) a *collecting (straight) duct*. Between the distal convoluted and junctional tubules, there is often a short angular *zigzag* (or *irregular*) *tubule*.

Collecting ducts (**8.**172) commence in the cortical medullary rays and unite at short intervals, finally opening into wider *papillary ducts* (*of Bellini*), which end on a papillary summit, their numerous orifices forming an *area cribrosa*.

STRUCTURE OF THE RENAL TUBULE (**8.**172D, 173A)

Renal tubules are lined by a simple epithelium with a basal lamina of varying thickness (Maunsbach 1973). The type of epithelial cell varies according to the functional roles of the different regions, being concerned mainly with active transport and passive diffusion of various ions and water into and out of tubules, with reabsorption of organic substances such as glucose and amino acids and with the uptake of any proteins which may leak through the glomerular filter (Bulger 1977).

Proximal tubules are lined by cuboidal epithelium with a brush border of tall microvilli on their luminal aspect. The shape of the cells depends on tubular fluid pressure, which in life distends the

lumen and stretches cells to a slightly flattened shape, becoming taller when glomerular blood pressure falls at death or at biopsy. The cytoplasm of these cells is strongly eosinophilic and their nuclei euchromatic and central. By light microscopy their bases show faint striae, which by electron microscopy appear as a complex series of pleats between which numerous mitochondria are

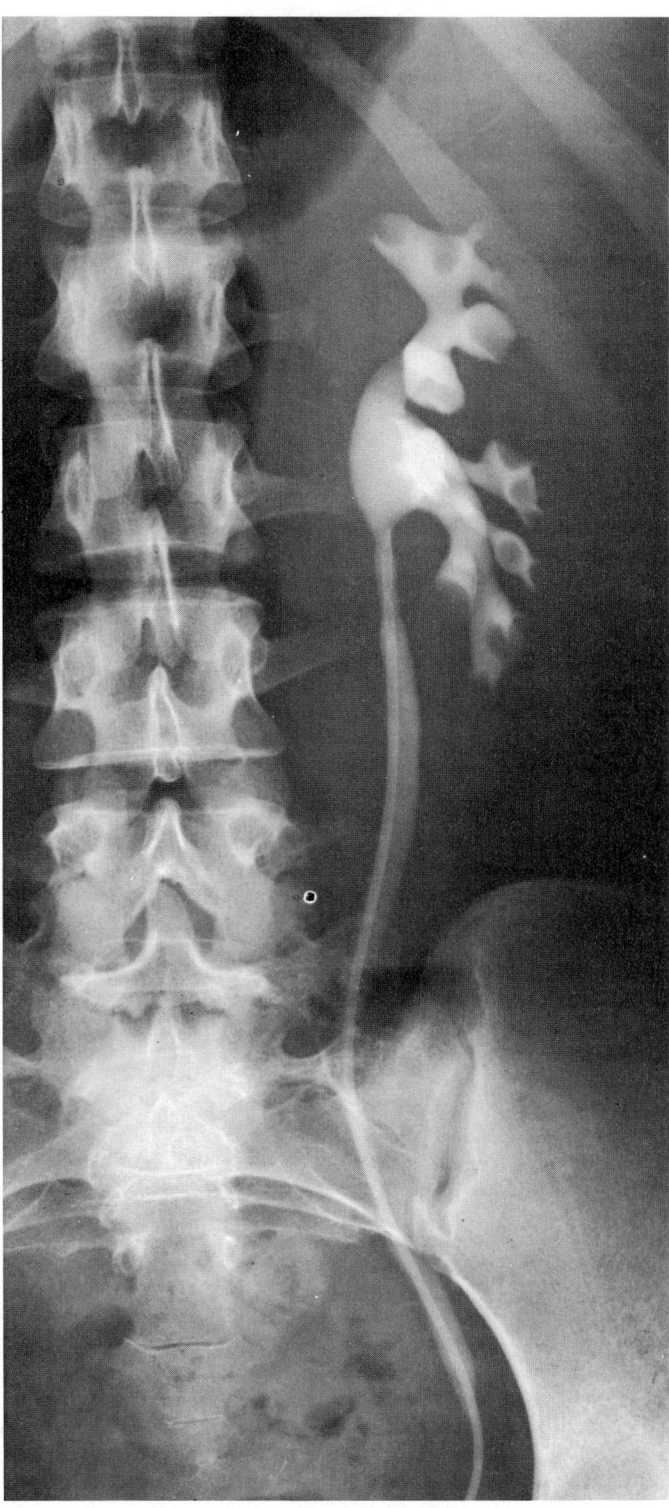

8.169B A left retrograde pyelogram. The contrast medium has been introduced into the calices, pelvis and upper ureter via a ureteric catheter, which is still in position and can be clearly identified. There is considerably greater density of the shadows of the calices than that achieved by the intravenous method. Note the relation of the ureter to the tips of the lumbar transverse processes and the characteristic 'cupping' or 'champagne glass' profiles of the tips of the lesser calices where they surround the renal pyramids. 'Calix' means a cup and such 'cupping' of the minor calices is the normal appearance. (Note that the major calices are not cusps and are hence inappropriately named.)

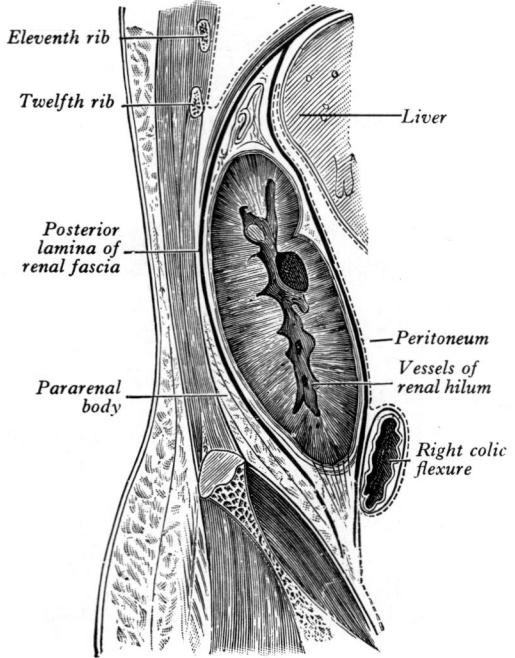

8.170A Sagittal section through the posterior abdominal wall showing the relations of the renal fascia of the right kidney.

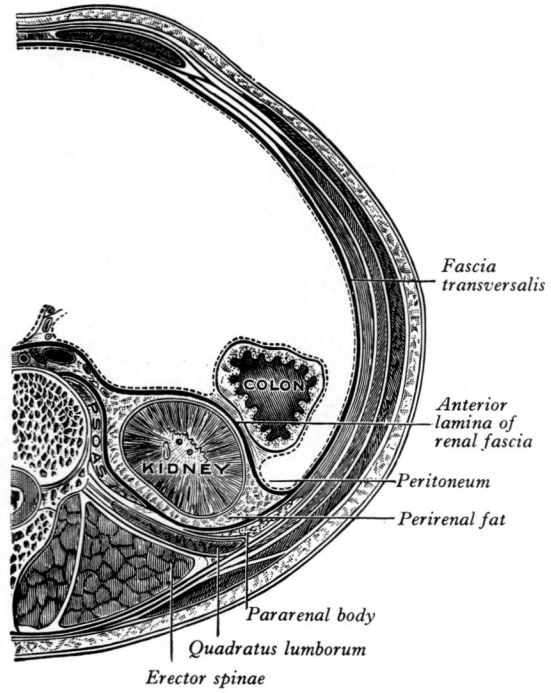

8.170B Transverse section, showing the relations of the renal fascia.

orientated perpendicular to the basal lamina. Reconstruction has shown this complex to consist of processes inter-digitating between the lateral aspects of adjacent epithelial cells, creating a labyrinth of cytoplasmic pedicles in the outer part of the tubular wall (Bulger 1965, 1977). Taking into account also the microvilli on the luminal surface, such cells provide relatively vast areas of plasma membrane in contact with tubular fluid and the extratubular space, an arrangement facilitating the transport of ions and small molecules against steep concentration gradients and energized by abundant mitochondria. Sodium/potassium-stimulated adenosine triphosphatase (Na/K ATPase) is located in apical and basal membranes (Wachstein & Bradshaw 1965), with numerous other cytoplasmic enzymes associated with ionic transport. Apart from such ionic transport through epithelial cells,

water and other ions may pass between cells passively along osmotic and electrical gradients, probably through 'leaky' tight junctions at their apices (Fromter 1979). These cells also contain extensive invaginations of their plasma membranes, channels and lysosomes engaged in the uptake and hydrolysis of proteins which escape in small amounts into the glomerular filtrate (Caulfield & Farquhar 1975). Peroxisomes and lipid droplets are also frequent and basal microfilaments assist to maintain the shape of the tubule. Among many cytoplasmic enzymes demonstrable by histochemistry are cytochrome oxidase, succinic dehydrogenase and other respiratory enzymes, acid phosphatase in lysosomes and glucose-6-phosphatase and leucine aminopeptidase, reflecting the highly energetic nature of these cells (vide infra).

Variations occur in cell structure, enzyme content and functions along the proximal tubules. The short, initial *neck* is the simple squamous epithelial extension of the outer wall of the renal corpuscle; the rest of the proximal tubule is divided into three (sometimes four) regions: the first typified by long microvilli, the second by microvilli of medium length, the third (in man) again by long microvilli (Moffat 1975). Patterns of lysosomal and transport activities appear to follow these divisions, too, but the functional significance of this distribution is not clear.

The *ansa nephroni* (renal loop of Henle) consists of the *thin segment* (about 30 μm in diameter), lined by low cuboidal to squamous cells, and the *thick segment* (about 60 μm across) composed of cuboidal cells like those in the distal tubule, with which it is sometimes considered, as its *straight part*. The *thin segment* forms most of the loop in juxtamedullary and deep cortical nephrons which reach far into medulla; here, the descending limb of the thin segment has low cuboidal epithelial cells with complex interdigitations and narrow belts of tight junctions between cells, whereas the ascending limb is lined by squamous cells with wider junctional zones. These differences may be related to the different permeabilities to water and ions (Bulger 1971). Few organelles appear in either cell type, indicating a passive rather than an active role in ionic movements. The *thick segment* is composed of cuboidal epithelium with many mitochondria, deep basolateral folds and short apical microvilli, indicative of a more active metabolic state.

Cells of the *distal tubule* resemble those in the proximal, but microvilli are few, small and irregularly spaced; the basolateral folds containing mitochondria are so deep that they almost reach the luminal aspect (**8.**172D). Enzymes concerned in active transport of sodium, potassium and other ions are richly present and are known to be important in ionic regulation. At the junction of the straight and convoluted regions the distal tubule approaches the vascular pole of the renal corpuscle where tubular cells form a sensory structure, the *macula densa*, concerned in the regulation of blood flow and ionic exchange (vide infra). In the terminal part of the distal tubule, cells have fewer basal folds and mitochondria, constituting what is sometimes termed a *connecting duct*, in part like a distal collecting duct but formed from nephrogenic rather than ductal tissue during embryogenesis.

Collecting ducts are of simple cuboidal or columnar epithelium, increasing in height from the cortex, where they receive the contents of distal tubules, to the wide papillary ducts discharging at the area cribrosa. The cells have relatively few organelles or lateral interdigitations and only occasional microvilli; scattered among them are *dark cells* (also present in smaller numbers in the distal convoluted tubule) with longer microvilli and more mitochondria. The functions of these remain uncertain.

The Structure and Function of Nephrons

Three processes co-operate in nephrons to determine renal excretory and regulatory functions (**8.**172D): *filtration*, at the glomerular level, *selective resorption* from the filtrate passing along the renal tubules and *secretion* by the cells of the tubules into this filtrate. These processes have been intensively studied and can be generally correlated with structural variations along the nephron. For further details, see Brenner & Rector (1981).

Glomerular filtration is the passage of water containing various dissolved small molecules from the blood to the urinary

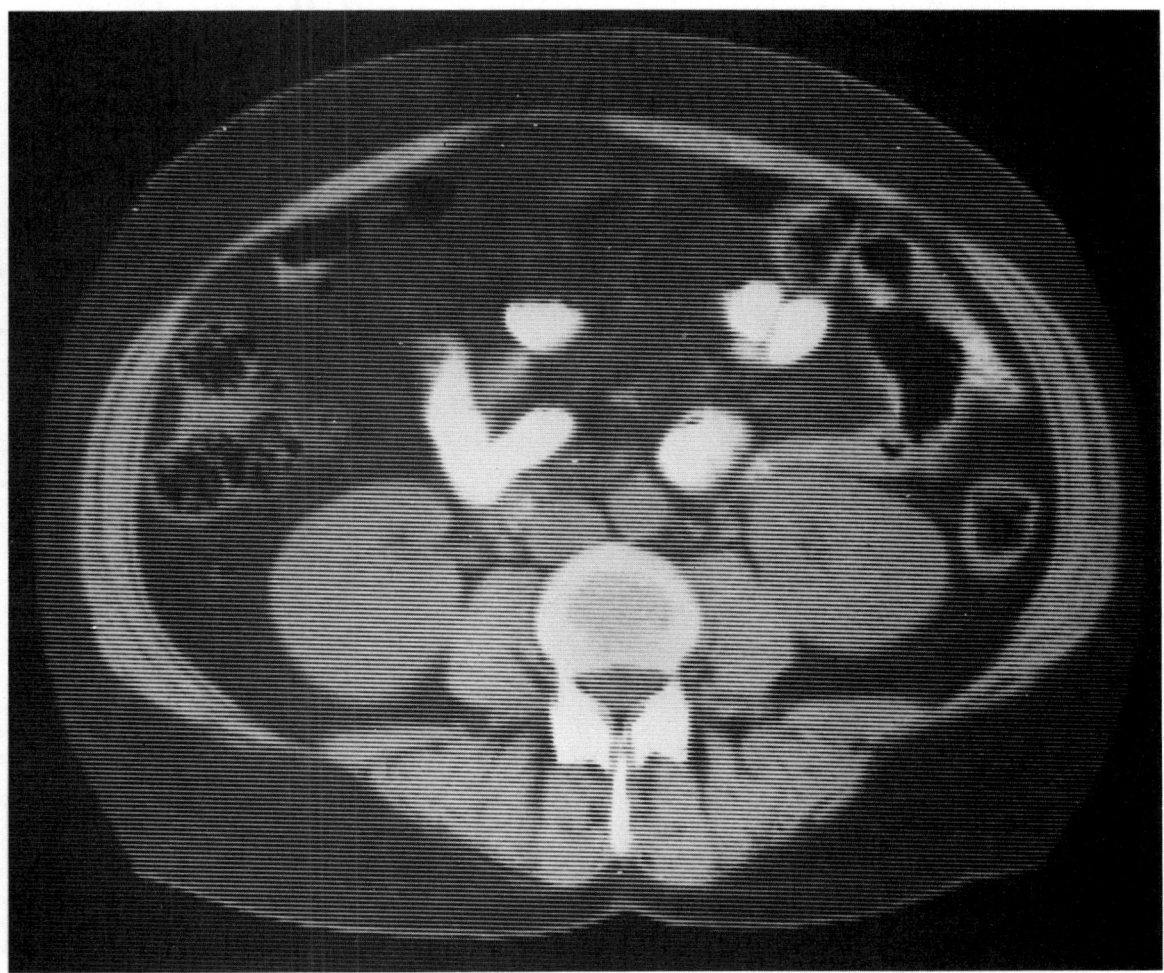

8.170C Computed tomogram through the abdomen in the transverse plane at the level of the kidneys (see **8.170D**, below). Contrast material is present in the small intestine. Supplied by Shaun Gallagher, Guy's Hospital; photography by Sarah Smith, UMDS, Guy's Campus, London.

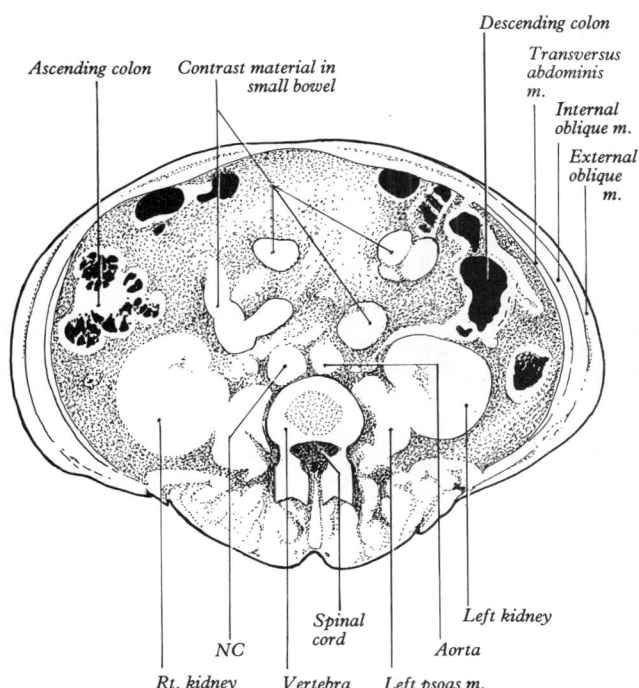

8.170D Diagram illustrating the major features demonstrated in **8.170C**.

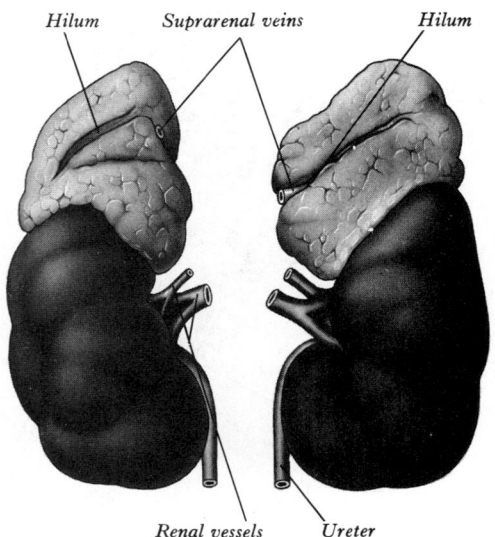

8.171 The kidneys and suprarenal glands of a newborn infant: anterior aspect. Note the lobulation of the renal surface and relative size of the organs.

1403

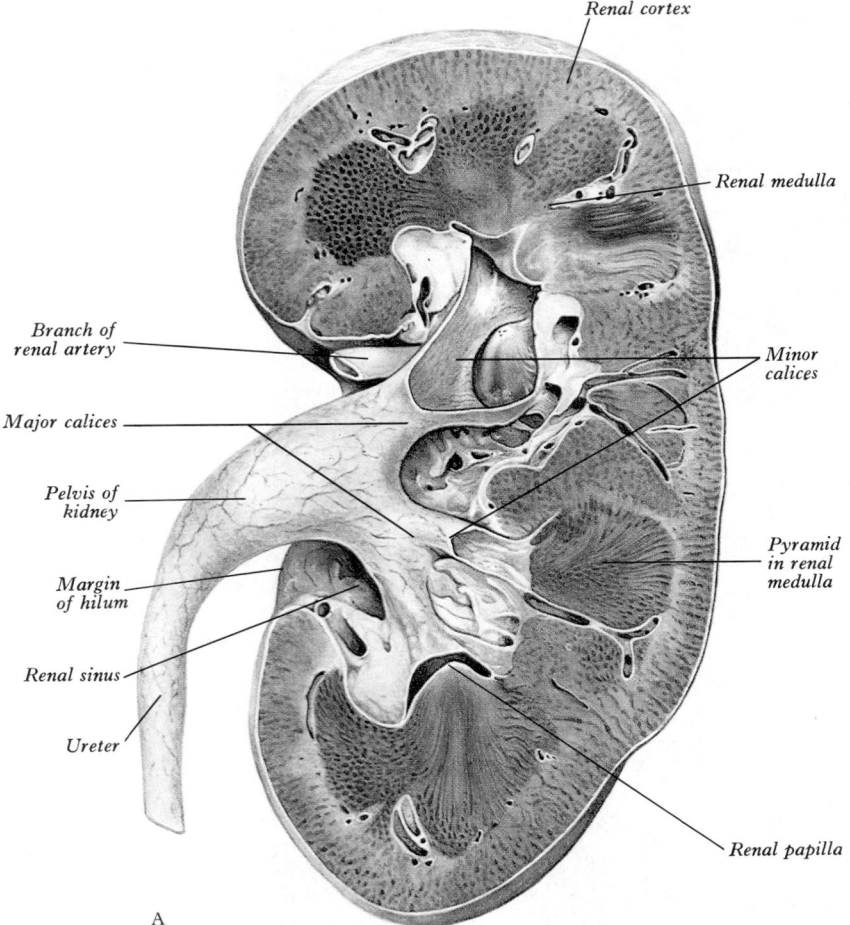

Renal cortex

Renal medulla

Branch of
renal artery

Minor
calices

Major calices

Pelvis of
kidney

Pyramid
in renal
medulla

Margin
of hilum

Renal sinus

Ureter

Renal papilla

A

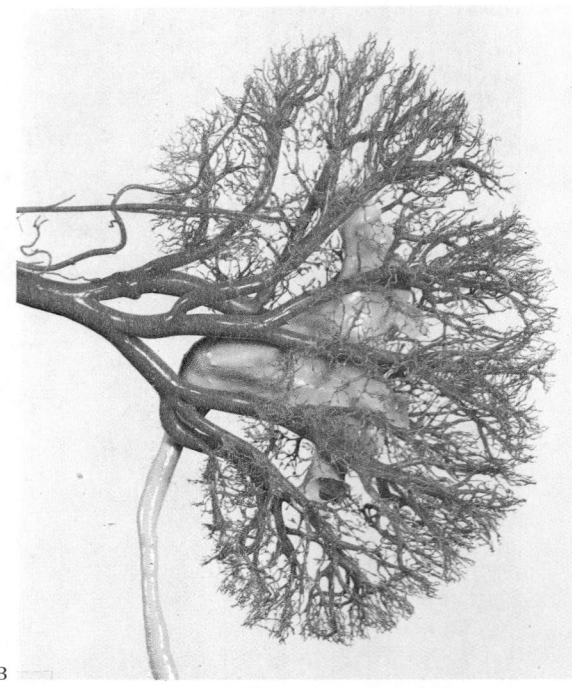

B

8.172 The structural and functional organization of the kidney. A. Long-itudinal section through a kidney; note the pelvis of the ureter and its division into calices; also the macroscopic appearance of the normal kidney. The pelvis and major calices have not been opened. B. A corrosion cast of a human kidney, showing minor and major calices, ureteric pelvis and upper ureter (all in yellow) and the renal arterial tree (in red). Note also suprarenal branches from the renal artery and direct from the aorta. Prepared by M C E Hutchinson; photographed by Kevin Fitzpatrick, Anatomy Dept, Guy's Hospital Medical School, London. C. Diagram illustrating the major structures in the kidney cortex and medulla (left), the position of cortical and juxtamedullary nephrons (middle) and the major blood vessels (right). D. Schematic diagram of the regional structure and principal activities of a kidney nephron and collecting duct. For clarity, a nephron of the long loop (juxtamedullary) type is depicted.

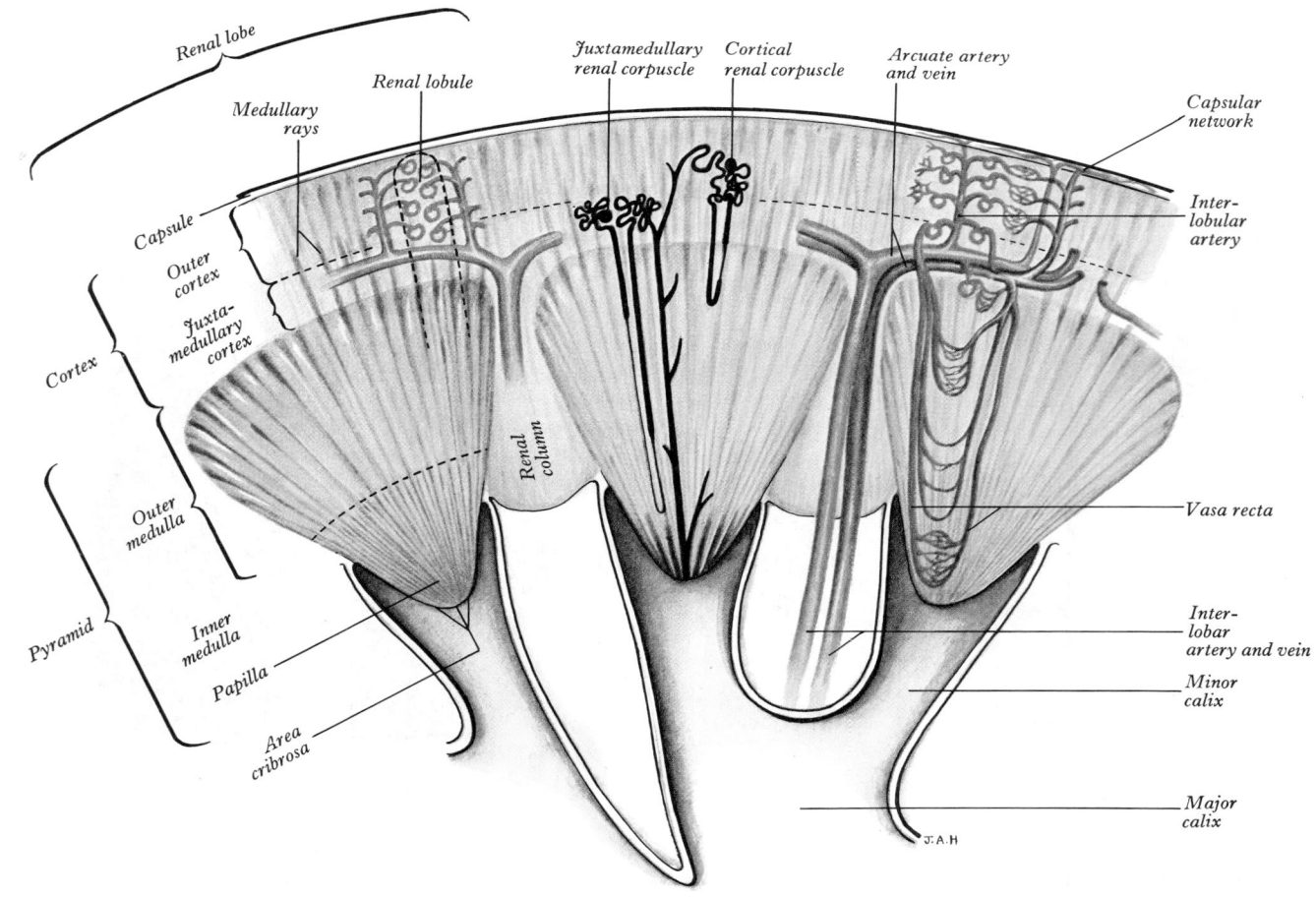

Renal lobe

Juxtamedullary
renal corpuscle

Cortical
renal corpuscle

Arcuate artery
and vein

Renal lobule

Capsular
network

Medullary
rays

Capsule

Inter-
lobular
artery

Outer
cortex

Juxta-
medullary
cortex

Cortex

Renal
column

Outer
medulla

Vasa recta

Pyramid

Inner
medulla

Inter-
lobar
artery and vein

Papilla

Minor
calix

Area
cribrosa

Major
calix

C

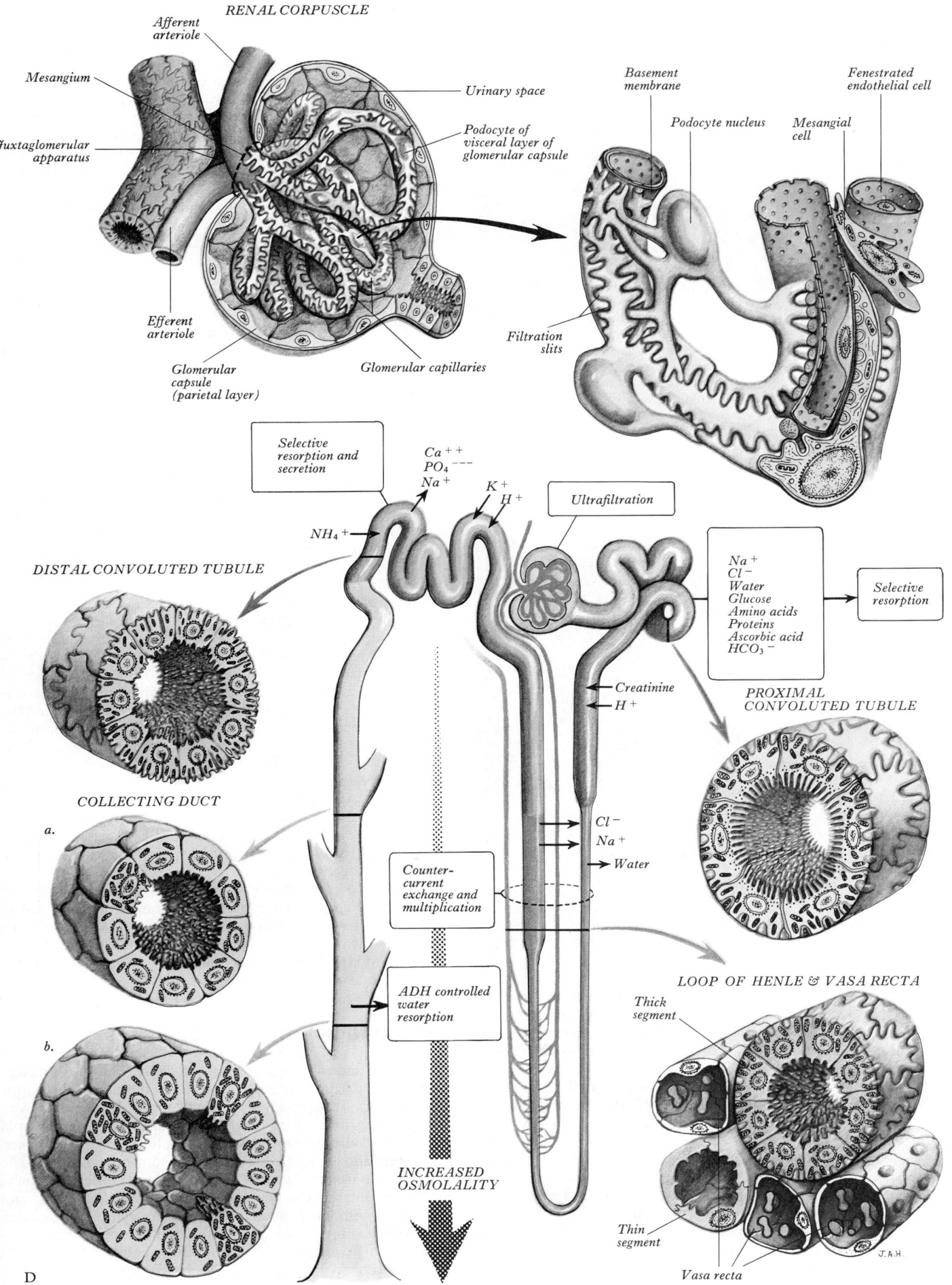

RENAL CORPUSCLE

Afferent arteriole

Mesangium

Juxtaglomerular apparatus

Efferent arteriole

Glomerular capsule (parietal layer)

Urinary space

Podocyte of visceral layer of glomerular capsule

Glomerular capillaries

Basement membrane

Podocyte nucleus

Mesangial cell

Fenestrated endothelial cell

Filtration slits

Selective resorption and secretion

Ca^{++}
PO_4^{---}
Na^+
K^+
H^+

Ultrafiltration

NH_4^+

DISTAL CONVOLUTED TUBULE

Na^+
Cl^-
Water
Glucose
Amino acids
Proteins
Ascorbic acid
HCO_3^-

Selective resorption

Creatinine
H^+

PROXIMAL CONVOLUTED TUBULE

COLLECTING DUCT

a.

Counter-current exchange and multiplication

Cl^-
Na^+
Water

ADH controlled water resorption

b.

LOOP OF HENLE & VASA RECTA

Thick segment

INCREASED OSMOLALITY

Thin segment

Vasa recta

J.A.H.

D

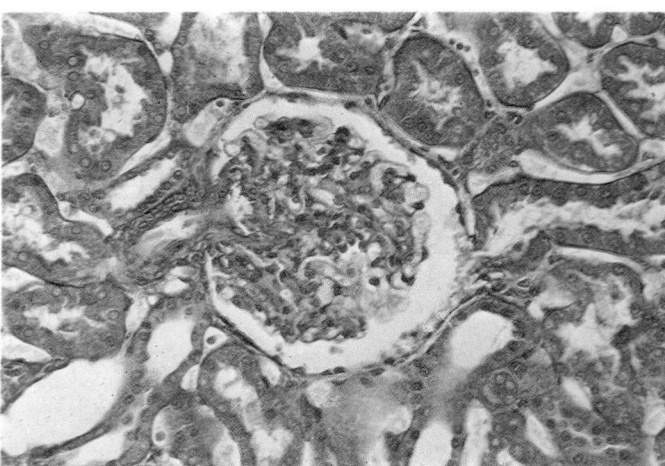

8.173A A renal corpuscle (simian kidney) in section, showing a glomerulus (centre) within its capsule, the urinary space of which is continuous with the proximal convoluted tubule (right). A glomerular arteriole is visible at the vascular pole of the capsule, where it is associated with a *macula densa* of a distal convoluted tubule (left), denoted by a group of closely spaced nuclei. Profiles of proximal convoluted tubules (with brush borders) and distal convoluted tubules (lacking brush borders) are also visible. Masson's trichrome stain. Magnification × 150.

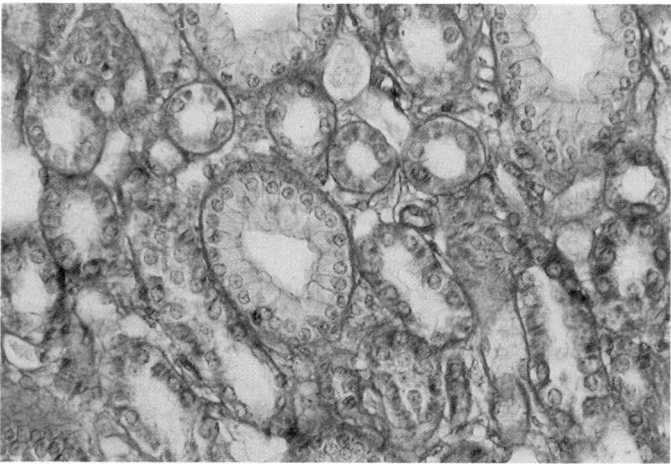

8.173B Part of a medullary ray in cross-section, showing large collecting ducts and small thin segments, interspersed with vasa recta distended with erythrocytes. Tissue and stain as in A. Magnification × 250.

space in the glomerular capsule; larger molecules, e.g. plasma proteins, polysaccharides and lipids, are largely retained in blood by the selective permeability of the glomerular basement membrane.

Filtration occurs along a steep pressure gradient existing between the large glomerular capillaries and the urinary space, the only structure separating the two being the glomerular basement membrane. This gradient far exceeds the colloid osmotic pressure of blood which opposes the outward flow of filtrate. In the peripheral renal cortex the arteriolar pressure gradient is enhanced by an inequality in the calibres of afferent and efferent glomerular arterioles, the former having larger diameters (vide infra). In all glomeruli the rate of filtration can be altered by changes in the tonus of the glomerular arterioles. When first formed, the *glomerular filtrate* is isotonic with glomerular blood and has an identical concentration of ions and small molecules.

Selective resorption of many substances from the filtrate is an active process and occurs mainly in the proximal convoluted tubules, particularly the resorption of glucose, amino acids, phosphate, chloride, sodium, calcium and bicarbonate. Cells of the proximal tubules are permeable to water, which leaves the tubule along an osmotic gradient created by resorption of these

solutes, particularly sodium and chloride ions, so that the filtrate remains locally isotonic with blood. Numerous microvilli, the folded lateral and basal surfaces and profuse mitochondria indicate that absorption by proximal tubular cells is an energy-dependent process. Further selective absorption, particularly of sodium ions, also occurs in distal convoluted tubules.

The rest of the tubule reabsorbs most of the filtrate's water (95%); when it reaches the calices, urine is thus much reduced in volume and is *hypertonic* to blood. Along the *descending* limb of the renal loop, sodium and water pass freely between the tubular lumen and adjacent extratubular spaces of the renal medulla, within which lie many loops. In part of the *ascending* limb (the thick segment), chloride ions are transported from the tubule lumen to interstitial spaces, sodium ions following passively; but the lining cells of the tubules do not allow water to follow sodium and chloride ions, some of which diffuse back into the descending limbs, adding to that already in filtrate passing along them; in ascending limbs, sodium is again extracted from this enriched solution, increasing intercellular ion concentrations and causing further diffusion of these into the descending limbs. Alternatively, sodium and chloride may not be recycled into the descending limb, water being merely withdrawn from the tubular loop because of the raised tonicity in the extratubular space, thus concentrating the filtrate. In either manner, high concentrations of sodium and chloride ions are built up in the renal medulla by this *countercurrent multipler system*. Rapid removal of ions from the renal medulla by the circulation of blood is minimized by another looped *countercurrent exchange system*, in which arterioles entering the medulla pass for long distances parallel to the venules leaving it before ending in capillary beds around tubules; this close apposition of oppositely flowing blood allows the direct diffusion of ions from outflowing to inflowing blood, so that these vessels (*vasa recta*) conserve a general high osmotic pressure in the medulla.

Because of the selective secretion of sodium and chloride ions by the cells of the ascending limbs and distal tubules (under aldosterone control), the filtrate at the distal end of the convoluted tubules is *hypotonic*; but, as it proceeds into the collecting ducts, descending again through the medulla, it re-enters a region of high osmotic pressure. Here, since ductual cells are, under the influence of the neurohypophysial antidiuretic hormone (ADH, vasopressin), variably permeable to water, the latter follows an osmotic gradient into the adjacent extratubular spaces (Gottschalk & Mylle 1959). Thus, along collecting ducts, the tonicity of the filtrate gradually rises until at the renal pyramids it is above that of blood. As much as 95% of water in the original glomerular filtrate is thus resorbed into blood. This complex system is highly flexible and the balance between the rate of filtration and absorption can be varied to accommodate to the current general physiological needs.

Secretion of various ions occurs at several sites, control of hydrogen and ammonium ion concentrations being essential to the regulation of acids and bases in the blood. Other secreted substances include various organic acids and antibiotics, these being passed into the filtrate especially in the proximal and distal tubules (8.172D).

The Juxtaglomerular Apparatus

The afferent and efferent arterioles of a glomerulus pass through the mesangium (the vascular stalk of the glomerulus) nearly opposite the exit of the proximal tubule (8.172D). In each nephron, the ascending limb of its renal loop returns from the medulla towards its glomerulus and the distal tubule commences between the afferent and efferent vessels, in close contact with them. The cells of the tunica media of the afferent arteriole differ from nonstriated myocytes elsewhere in being large, rounded and 'epitheloid' with large spherical nuclei; their cytoplasm contains dense vesicles, 10–40 nm in diameter, and also many mitochondria. These *juxtaglomerular cells* contact cells of the slightly dilated distal convoluted tubule, closely aggregated together here as the *macula densa*; the latter cells are clustered in a group of up to 40, with large, oval nuclei, each cell containing a concentration

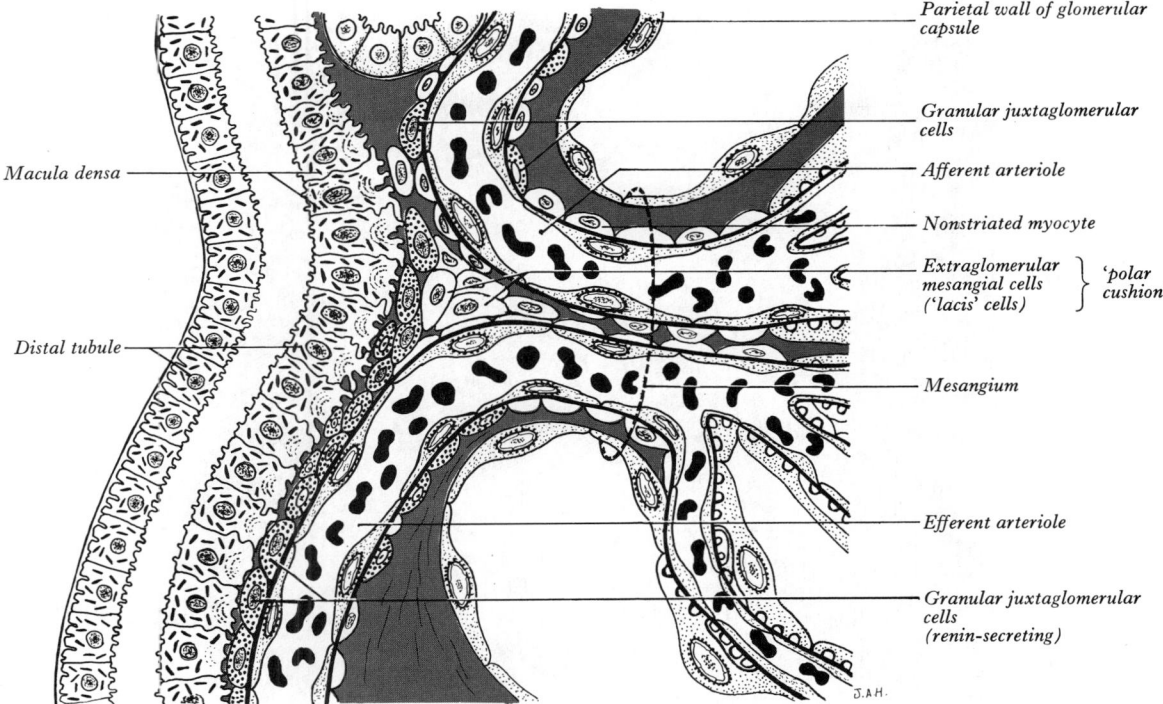

Parietal wall of glomerular capsule

Granular juxtaglomerular cells

Afferent arteriole

Nonstriated myocyte

Extraglomerular mesangial cells ('lacis' cells) } *'polar cushion'*

Macula densa

Distal tubule

Mesangium

Efferent arteriole

Granular juxtaglomerular cells (renin-secreting)

8.174 Diagram showing the organization of the juxtaglomerular complex including the macula densa (left), granular juxtaglomerular cells (middle) and the vascular pole of the glomerular capsule (right).

of mitochondria apically (see Barajas 1971, Sikri & Foster 1981) and have a few short microvilli on their luminal surface. The two groups, with various mesangial elements, constitute the renal *juxta-glomerular apparatus* (Edelman & Hartroft 1961, Barajas 1970). In many animals, interspersed between the macula densa and mesangium are *'lacis cells'* (Polkissen cells, extraglomerular mesangial cells, see Spanidis et al 1982). These cells are stellate in form and their processes create an irregular network (lacis). Each cell has a surrounding basal lamina and internally contains occasional granules of a secretory type, but few organelles otherwise. Adrenergic nerve fibres occur in small numbers among these cells. Recently, another cell type, present as a minor component of this intermediate region, has been described. This is the *granular peripolar cell* reported in sheep (Ryan et al 1979) and humans (Gardiner et al 1986). Their main feature is the presence of dense membrane-bound vacuoles 0.4–2.1 μm in diameter. These cells are situated close to the epithelial cells of the renal corpuscle wall, near the mesangial root. Their functions are not yet known but they would appear to be secretory.

In experimental diminution in renal blood flow, with consequent increase of blood pressure and sometimes in hypertension associated with renal disease, juxtaglomerular cells hypertrophy; they contain *renin* (Cook 1971) within their granules, an enzyme converting a polypeptide in blood, *angiotensinogen*, to *angiotensin I*. This is converted by other enzymes (notably in the lungs) into *angiotensin II*, a polypeptide whose actions include the elevation of blood pressure and stimulation of aldosterone release from the adrenal cortex, increasing resorption of sodium ions from the distal convoluted tubules. Details of renin secretion and consequent changes in tubular activity are not yet clear but the juxtaglomerular apparatus appears to be a feedback device regulating flow of fluid through the glomerular filter and the ionic resorption in renal tubules, thus determining the final concentration of the urine. Renin secretion may be controlled by at least three factors (Davies & Freeman 1976): (1) the activity of the macula densa cells which react to changes in fluid passing them in the distal tubules; (2) pressure in glomerular arterioles affecting the secretory activity of their granule cells; (3) stimulation by sympathetic fibres ending near juxtaglomerular cells. These and other agents (perhaps including other cell types in this region) appear to assist the juxtaglomerular apparatus in correlating

blood flow, filtration rate and osmoregulation and in mediating the appropriate actions needed for homeostasis.

Other Renal Cells

Between the renal tubules and blood vessels lie other cells essential to renal structure and function, such as connective tissue elements, inconspicuous in the cortex but prominent in the medulla, where they secrete the proteoglycans and collagen of the connective tissue matrix, particularly visible in the papillae. Medullary *interstitial cells*, some apparently modified fibroblasts, form vertical piles of tangentially orientated cells between the more distal collecting ducts, like the rungs of a ladder. These cells secrete prostaglandins (Muirhead et al 1972).

Renal Blood Vessels

The complex renal vascular patterns show regional specializations, closely adapted to the spatial organization and functions of renal corpuscles, tubules and ducts (8.172, 176, 178). The large literature concerning renal angioarchitecture, haemodynamics and controlling mechanisms cannot be analysed here, nor can variation between species, or minor human variations be considered. For these admirable reviews exist by Trueta et al (1947), Fourman & Moffat (1971) and Moffat (1975), among others.

Renal vasculature may be studied at various levels, commencing with the principal and accessory *renal arteries*. Their primary patterns of branching and areas of distribution suggest the presence of *vascular segmentation*. From the primary stems branch *lobar, interlobar, arcuate* and *interlobular* arteries, *afferent* and *efferent glomerular arterioles* and cortical *intertubular capillary plexuses*; cortical venous radicles drain them and also the *vasa recta* and associated capillary plexuses of the medulla to the renal vein 8.172, 176).

A single renal artery to each kidney (p. 737), is present in about 70% of individuals but they vary in their level of origin (the *right* often being superior) and in their calibre, obliquity and precise relations. (For a review of these features in almost 11 000 kidneys, see Merklin & Michels 1958.) In its extrarenal course (Schneider

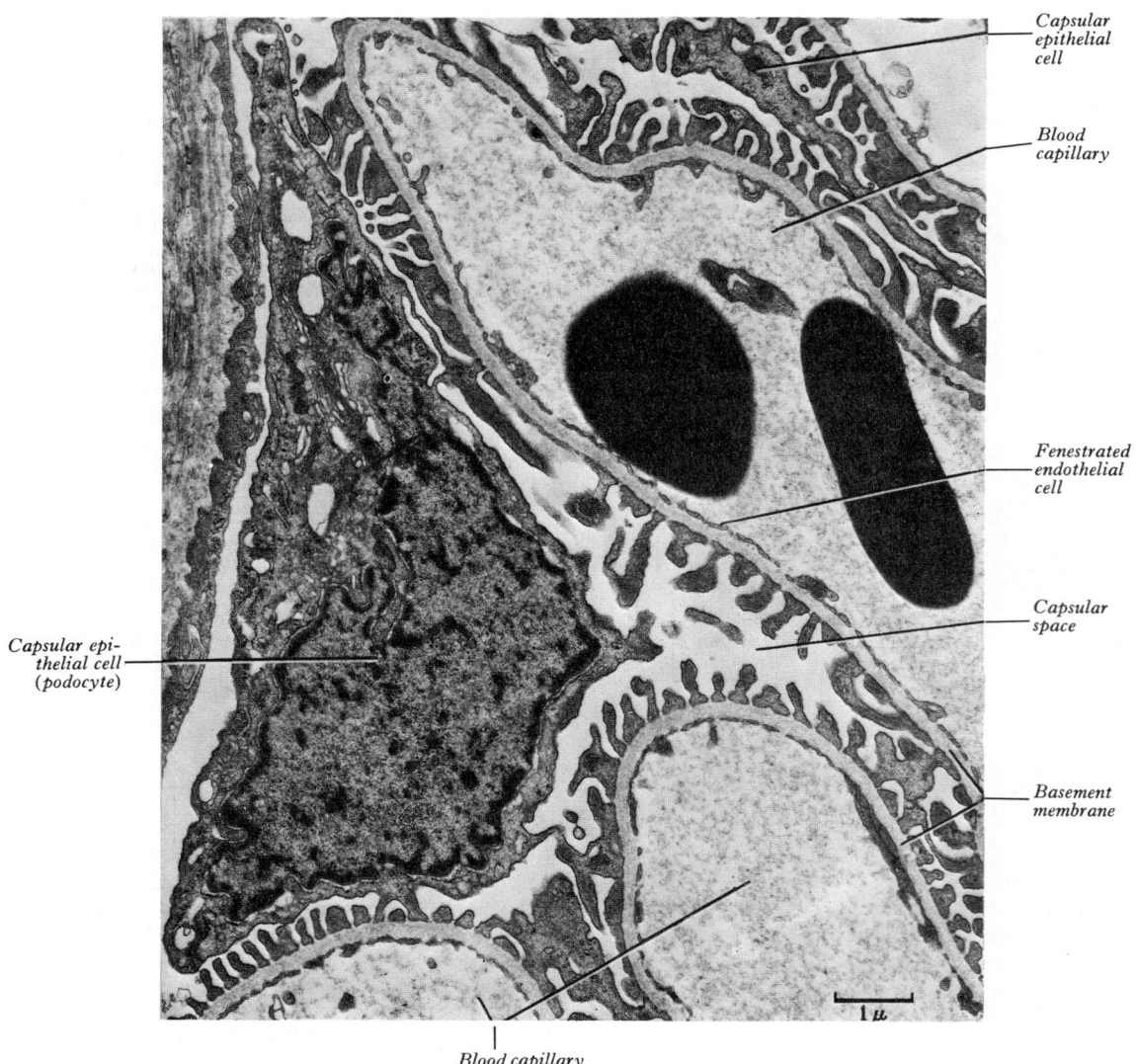

Capsular epithelial cell

Blood capillary

Fenestrated endothelial cell

Capsular space

Basement membrane

Capsular epithelial cell (podocyte)

1 μ

Blood capillary

8.175A Electron micrograph showing capillaries and epithelial cells of a renal corpuscle (rat). Note the basement membrane and fenestrated endothelial cells as well as the foot processes of the epithelial cell (podocyte). Magnification × 9200.

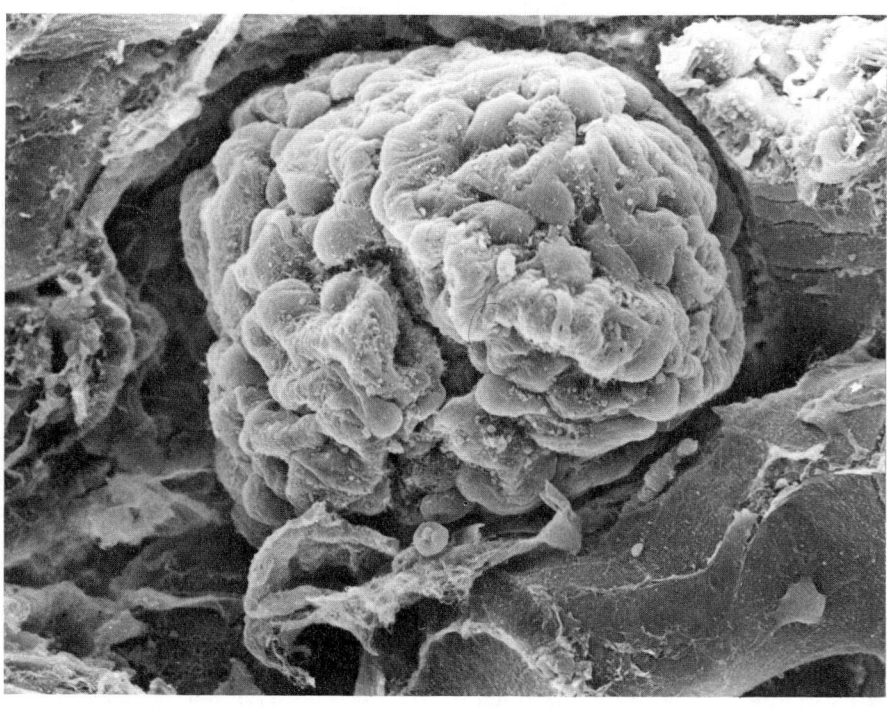

8.175B Scanning electron micrograph of a renal corpuscle (guinea pig) fractured to expose the glomerulus and podocytes and surrounding tubules. Magnification × 1000.

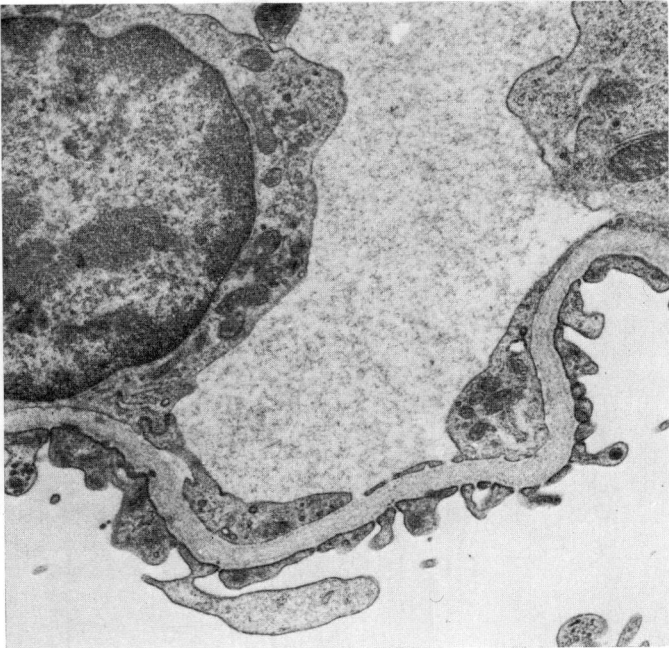

8.175C Electron micrograph of part of a glomerulus, showing the edge of a capillary (above) with fenestrated endothelial cells, one with a prominent nucleus (left). Also visible are the basal lamina with lamina densa and flanking laminae rarae (interna and externa), and below these, podocyte pedicels and urinary space. Magnification × 15 000.

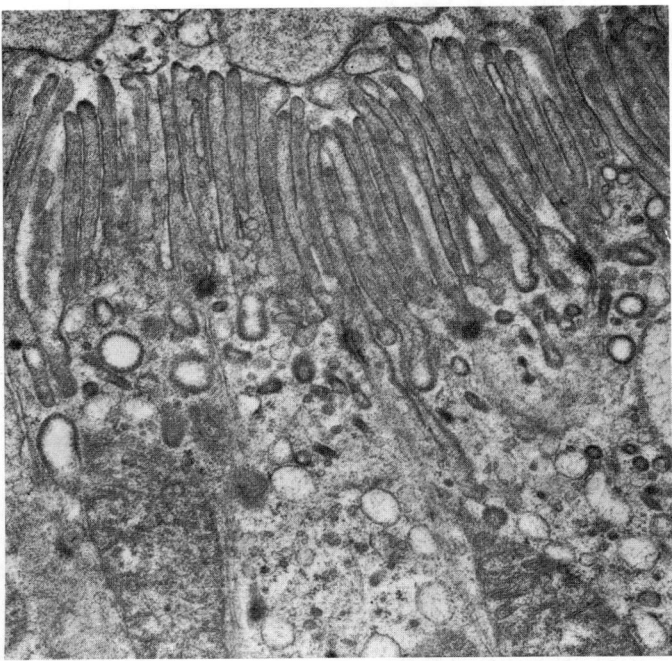

8.175D A high-power electron micrograph of the apical (luminal) parts of four proximal convoluted tubular epithelial cells. The apical array of microvilli, subapical junctional complexes between adjacent cells and the highly active cytoplasm containing lysosomes, parts of very large mitochondria and a wide variety of granules, vesicles and vacuoles are visible. Magnification × 22 000.

et al 1969) each renal artery gives off one or more inferior suprarenal arteries and branches which supply perinephric tissue, the renal capsule, pelvis and the proximal part of the ureter; near the renal hilum, each artery divides into an *anterior* and *posterior division*, the primary branches of which (*segmental arteries*) supply renal *vascular segments*. *Accessory renal arteries* are common (30% of individuals), usually arising from the aorta above or below the main renal artery and following it to the renal hilum. Higher or

lower origins are not uncommon, an accessory artery or leash of arteries passing to the superior or inferior renal pole. They are regarded as persistent embryonic *lateral splanchnic* arteries. Accessory vessels to the inferior pole cross anterior to the ureter and may, by its obstruction, cause hydronephrosis. Rarely, accessory renal arteries arise from the coeliac or superior mesenteric arteries near the aortic bifurcation or from the common iliac arteries.

Renal vascular segmentation was originally recognized by John Hunter in 1794 but the first detailed account of the primary pattern was by Graves (1954, 1956a, b) from casts and radiograms of injected kidneys. He described five segments: (1) *apical*, occupying the anteromedial region of the superior pole, (2) *superior (anterior)*, including the rest of the superior pole and the central anterosuperior region, (3) *inferior*, encompassing the whole lower pole, (4) *middle (anterior)* between the anterior and inferior segments and (5) *posterior* including the whole posterior region between the apical and inferior segments. Graves' terminology has been adopted internationally and was used by some researchers (Smith 1963, Sykes 1963, 1964), but others have proposed more complex schemes, e.g. three posterior segments (Faller & Ungvary 1962); some have emphasized the great variability of the regions supplied by segmental arteries (Fine & Keen 1966). Whatever the divergences, it must be emphasized that vascular segments are supplied by virtual end-arteries. In contrast, larger *intrarenal veins* have no segmental organization and anastomose freely.

Brodel (1901) described a relatively avascular longitudinal zone (*the 'bloodless' line of Brodel*) along the convex renal border, proposed as the most suitable site for surgical incision. However, many vessels cross this zone, which is far from 'bloodless'; planned radial or intersegmental incisions are said to be preferable.

Initial branches of segmental arteries are *lobar*, usually one to each renal pyramid, but before entry they subdivide into two or three *interlobar arteries*, extending towards the cortex around each pyramid. At the junction of the cortex and medulla, interlobar arteries dichotomize into *arcuate arteries* which diverge at right angles; as they arch between cortex and medulla, each divides further and, from its branches, *interlobular arteries* diverge radially into the cortex. The terminations of adjacent arcuate arteries do not anastomose but end in the cortex as additional interlobular arteries. Though most interlobular arteries come from arcuate branches, some arise directly from arcuate or even terminal interlobar arteries (8.176B).

Interlobular arteries ascend towards the superficial cortex or may branch a few times en route (8.172C); some are more tortuous, recurring towards the medulla once or more before proceeding towards the renal surface. Some traverse the surface as *perforating arteries* (Hammersen & Staubesand 1961) to anastomose with the *capsular plexus* (also supplied from the *inferior suprarenal, renal* and *testicular* or *ovarian* arteries).

Afferent glomerular arterioles are mainly the lateral rami of interlobular arteries but a few arise from arcuate and interlobular arteries, when they vary their direction and angle of origin: deeper ones incline obliquely back towards the medulla, the intermediate pass out horizontally, while the more superficial approach the renal surface obliquely before ending in a glomerulus (8.176A,B).

From most glomeruli (excepting the juxtamedullary and a few at intermediate cortical levels) *efferent glomerular arterioles* soon divide to form a dense *peritubular capillary plexus* around the proximal and distal convoluted tubules. In the main renal cortical circulation there are thus *two* sets of *capillaries* in series, glomerular and peritubular, linked by efferent glomerular arterioles. From the venous ends of the peritubular plexuses fine radicles converge to join *interlobular veins*, one with each interlobular artery. Many interlobular veins commence beneath the fibrous renal capsule by convergence of several *stellate veins*, draining the most superficial zone of the renal cortex and so named from their surface appearance. Proceeding to the corticomedullary junction, interlobular veins also receive some ascending vasa recta (vide infra) and end in *arcuate veins* which accompany arcuate arteries but, unlike them, they anastomose with neighbouring veins. Arcuate veins drain into *interlobar veins*, which anastomose and converge to form the renal vein.

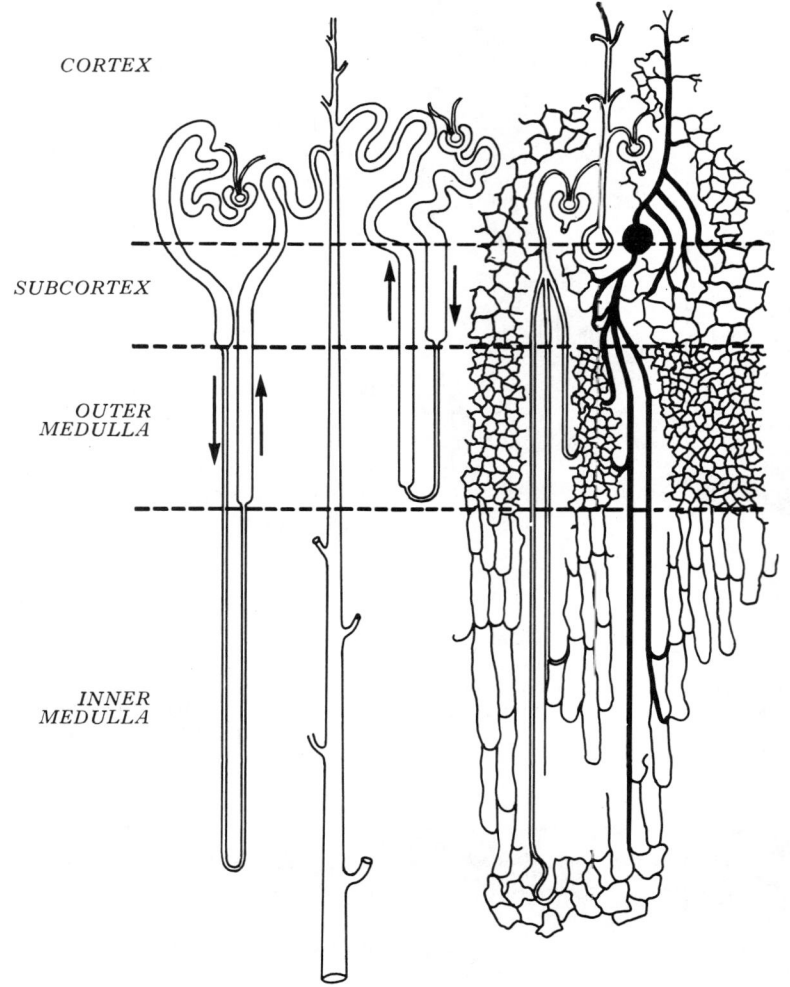

CORTEX

SUBCORTEX

OUTER MEDULLA

INNER MEDULLA

8.176A Diagram to illustrate the arrangement of the tubules (left) and blood vessels (right) in various structural zones of the kidney. Note the variations in the pattern of the tubules with either long or short medullary loops; tubules of intermediate length also occur. Compare the different structural and functional segments of the tubules which occur together in the cortical, subcortical and outer and inner medullary zones with their related vascular patterns. Arteries are black outlines; capillaries are single black lines; veins are full black. Compare with 8.176B. (From Fourman & Moffat 1971 with permission of the authors and Blackwell Scientific Publications.)

The vascular supply of the *renal medulla* is largely from efferent arterioles of juxtamedullary glomeruli, supplemented by some from more superficial glomeruli, and 'aglomerular' arterioles (probably from degenerated glomeruli). Efferent glomerular arterioles passing into the medulla are relatively long, wide vessels, contributing side branches to neighbouring capillary plexuses before entering the medulla, where each divides into 12–25 *descending vasa recta*; these, as their name suggests, run straight to varying depths in the renal medulla, contributing side branches to a radially elongated capillary plexus (8.176A) applied to the descending and ascending limbs of renal loops and to collecting ducts. The venous ends of capillaries converge to the *ascending vasa recta*, which drain into arcuate or interlobular veins. An essential feature of the vasa recta is that, particularly in outer medulla, both ascending and descending vessels are grouped into *vascular bundles*, within which the external aspects of both types are closely apposed, bringing them close to the limbs of renal loops and collecting ducts. As these bundles converge centrally into the renal medulla they contain fewer vessels, some terminating at successive levels in neighbouring capillary plexuses. This proximity of descending and ascending vessels with each other and adjacent ducts is the structural basis for the countercurrent exchange and multiplier phenomena previously mentioned (p. 1406, 8.172D, 177A,C).

In *ultrastructure*, renal vessels show the regional features described elsewhere; renal, interlobar and arcuate arteries are typical 'large muscular arteries' (p. 686); the interlobular vessels are like 'small muscular arteries' and afferent glomerular vessels have typical arteriolar structure with a muscular coat two to three cells thick; this and the connective tissue components of the wall diminish near a glomerulus until a point 30–50 μm proximal to it where arteriolar cells begin to show modifications typical of the *juxtaglomerular apparatus* (p. 1406). The efferent arterioles from most cortical glomeruli have thicker walls and a narrower calibre than corresponding afferents. The peritubular and medullary capillaries possess a well-defined basal lamina and their endothelial cells have typically fenestrated cytoplasm (p. 689), as in ascending vasa recta; in contrast, the descending vasa recta have thicker, continuous endothelium. The structure of glomerular capillaries is considered above (p. 1409). For brief comments on the functional association between the regions of renal tubules and associated blood vessels see p. 1391 and Fourman & Moffat (1971). A detailed account of renal microvasculature has also recently been given for monkeys (Horacek et al 1986, 1987).

Renal innervation. The general sources of renal nerves are described on p. 1165. Direct nerve fibres from plexuses around arcuate arteries to innervate juxtamedullary efferent arterioles and vasa recta have been described (Munkacsi & Newstead 1971);

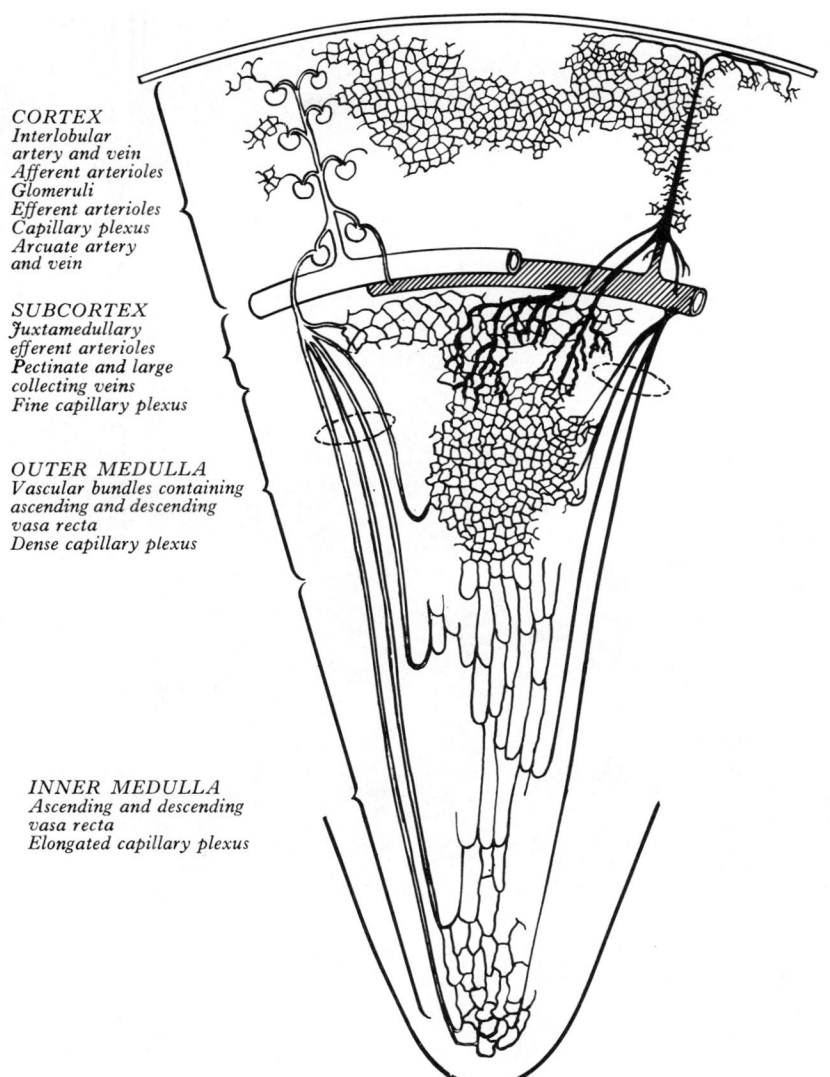

CORTEX
*Interlobular
artery and vein
Afferent arterioles
Glomeruli
Efferent arterioles
Capillary plexus
Arcuate artery
and vein*

SUBCORTEX
*Juxtamedullary
efferent arterioles
Pectinate and large
collecting veins
Fine capillary plexus*

OUTER MEDULLA
*Vascular bundles containing
ascending and descending
vasa recta
Dense capillary plexus*

INNER MEDULLA
*Ascending and descending
vasa recta
Elongated capillary plexus*

8.176B Diagram of the basic arrangements of the blood vessels in the mammalian kidney. Arteries are black outlines, capillaries are single black lines, veins are cross-hatched or full black. Note the variations in the pattern of the meshes in the capillary networks. See accompanying text for further description. (From Fourman & Moffat 1971 with permission of the authors and Blackwell Scientific Publications.)

these might control blood flow between the cortex and medulla without affecting the glomerular circulation.

Lymph vessels are described on p. 855.

Renal anomalies. The early pelvic position of the kidneys (p. 250) may persist; they are then usually supplied from the common iliac arteries and the hila are anterior. Occasionally kidneys are connected by a transverse bridge of renal tissue, forming a 'horseshoe kidney', the commissure usually lying between the inferior poles and rarely the superior. The ureters curve anterior to the connection and may be here partially obstructed. A congenital absence or imperfect development of one kidney may be compensated by an enlargement of the opposite organ. Two kidneys may also occur on the same side. Single or multiple congenital renal cysts and widespread congenital polycystic disease may also be present.

The Upper Urinary Tract

This term is generally used as a collective name for the main urinary outflow conduits from the kidney (renal calices, renal pelvis and ureter), as distinct from the bladder and urethra, which constitute the *lower urinary tract*. This separation is of course arbitrary.

RENAL CALICES AND PELVIS

Within the renal sinus the proximal parts of the urinary tract consist of the minor and major calices and the renal pelvis. The minor calices are attached to the renal parenchyma around the bases of a variable number (7–14) of conical renal papillae which form the tips of the renal pyramids (p. 1400). The renal capsule covers the external surface of the kidney and continues through the hilus to line the sinus (p. 1399) and fuse with the adventitial coverings of the minor calices. Each minor calix is a trumpet-shaped structure which surrounds either a single papilla or, more rarely, groups of two or three papillae (8.169B, 172A,B). The minor calices unite with their neighbours to form two or possibly three larger chambers, the major calices. The latter usually fuse with each other to form a single funnel-shaped renal pelvis, which tapers as it passes inferomedially, traversing the renal hilus to become continuous with the ureter (8.169A,B, 172A). Normally it is not possible to determine precisely the position where the renal pelvis ceases and the ureter begins. Consequently the precision implied by the phrase 'pelviureteric junction' to describe this area is anatomically invalid and the term 'pelviureteric region' is preferable. This region is usually extrahilar in location and normally lies adjacent to the lower part of the medial border of the kidney. In some individuals, however, the entire renal pelvis has been found to lie

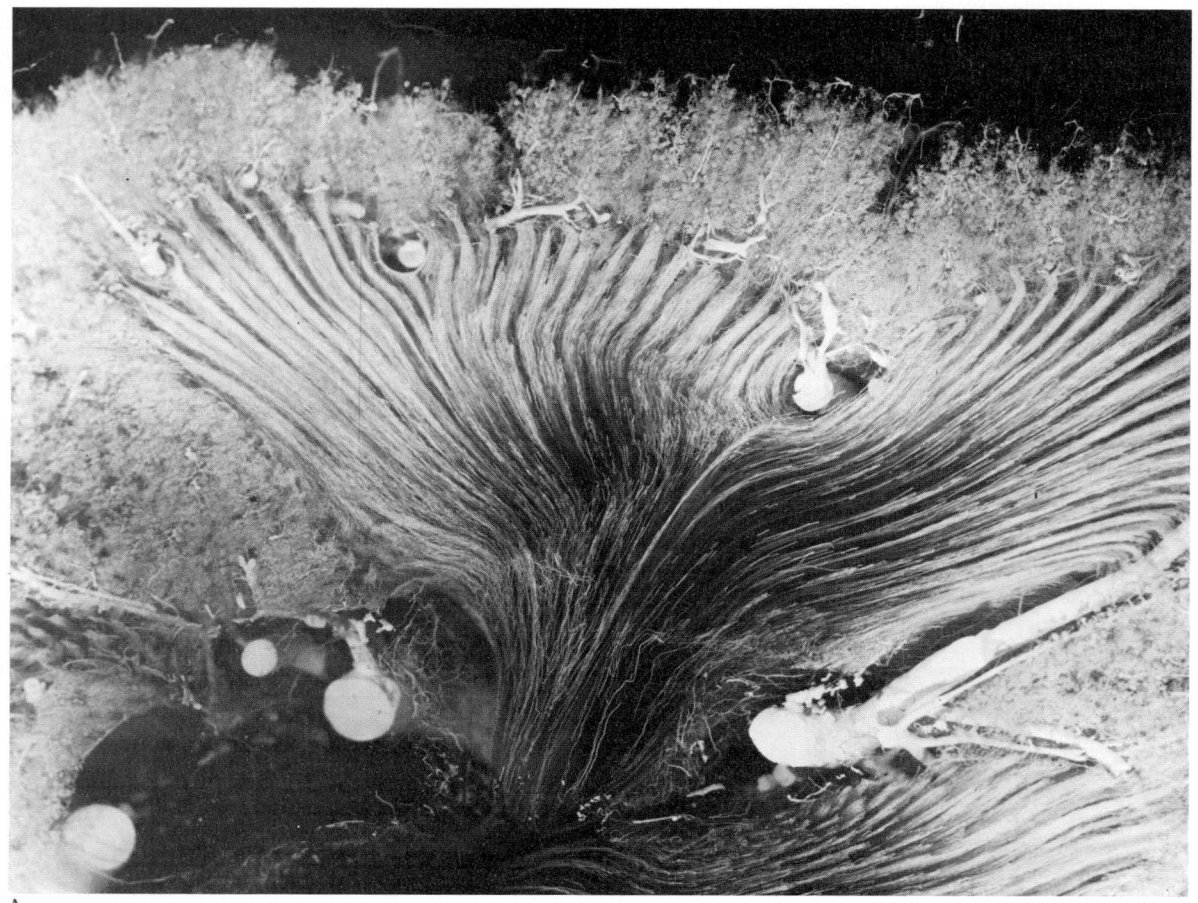

A

B

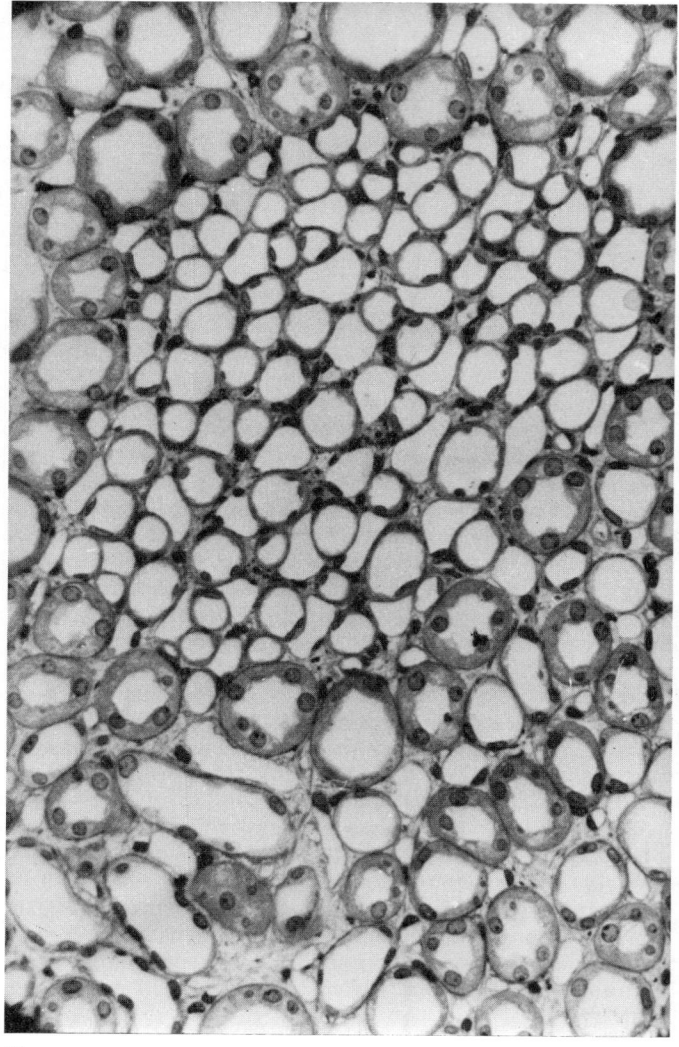

C

8.177 (*opposite*) A. Low-power, survey micrograph of a single nephric lobe following 'microfil' injection of the arterial tree of the human kidney. Note the clear distinction between the cortex peripherally and the medulla converging on the renal papilla. The lobe is flanked by large interlobar arteries. Note also the arcuate arteries at the corticomedullary junction, the interlobular arteries ascending into the cortex, where the glomeruli are also visible, and the converging vascular bundles of the medulla. Compare with **8.176B**.

B. A higher magnification of the same preparation at the corticomedullary junction. Note the juxtamedullary efferent arterioles, leaving the glomeruli to form medullary vascular bundles (descending vasa recta).

C. Transverse section through a vascular bundle (consisting of descending and ascending vasa recta). Surrounding the vascular bundle are sections of large collecting ducts and thin and thick segments of the medullary loops (of Henle). Preparation provided by D B Moffat, Department of Anatomy, University College of Wales, Cardiff.

inside the sinus of the kidney and, as a consequence, the pelviureteric region is situated either in the vicinity of the renal hilus or completely within the renal sinus.

Structure of the Renal Calices and Pelvis

The wall of the proximal part of the urinary tract is composed of three histological layers, namely a connective tissue adventitia (fibrous coat), a coat of smooth muscle and an inner mucosa. (The mucosal lining of the renal calices and pelvis is identical in structure to that of the ureter (p. 1414) and will not be considered further here.) The adventitia forms the outermost layer and consists of loose fibro-elastic connective tissue which merges with retroperitoneal areolar tissue. Proximally the coat fuses with the fibrous capsule of the kidney lining the renal sinus. The muscle coat of the renal calices and pelvis is composed of two morphologically and histochemically distinct types of smooth muscle cell (Dixon & Gosling 1982). One type of muscle cell is identical to that described for the ureter and can be traced proximally through the pelviureteric region and renal pelvis as far as the minor calices. The other type possesses a number of unusual structural features; these cells form the muscle coat of each minor calix and continue into the major calices and pelvis where they form a distinct inner layer. The cells also form a thin sheet of muscle which covers each minor caliceal fornix and extends across the renal parenchyma between the attachments of neighbouring minor calices, thereby linking each minor calix to its neighbours. Further out in the wall of each minor calix, the muscle cells are arranged longitudinally and form a discrete layer confined to the inner aspect of the muscle coat, the remainder of which is formed of bundles of typical smooth muscle. Such a configuration continues throughout the wall of the major calices and renal pelvis but ceases in the pelviureteric region so that the proximal ureter is devoid of a morphologically distinct inner layer. Individually, these muscle cells are separated from one another by extensive amounts of connective tissue and differ histochemically from typical ureteric nonstriated muscle. In addition the muscle cells possess unusual fine structural features when examined by means of the electron microscope (Dixon & Gosling 1982). The functional significance of these structural features is considered on p. 1415.

The Ureters

Ureters are muscular tubes whose peristaltic contractions convey urine from the kidneys to the urinary bladder. Each measures 25–30 cm in length and is thick-walled, narrow and continuous superiorly with the funnel-shaped renal pelvis (p. 1411); a slight constriction may mark this junction (**8.172A**). Each descends slightly medially anterior to psoas major, entering the pelvic cavity to open into the base of the urinary bladder. Its surface projection is an almost vertical line from a point on the transpyloric plane, 5 cm from the midline to the pubic tubercle. Its diameter is about 3 mm but slightly less at its junction with the renal pelvis,

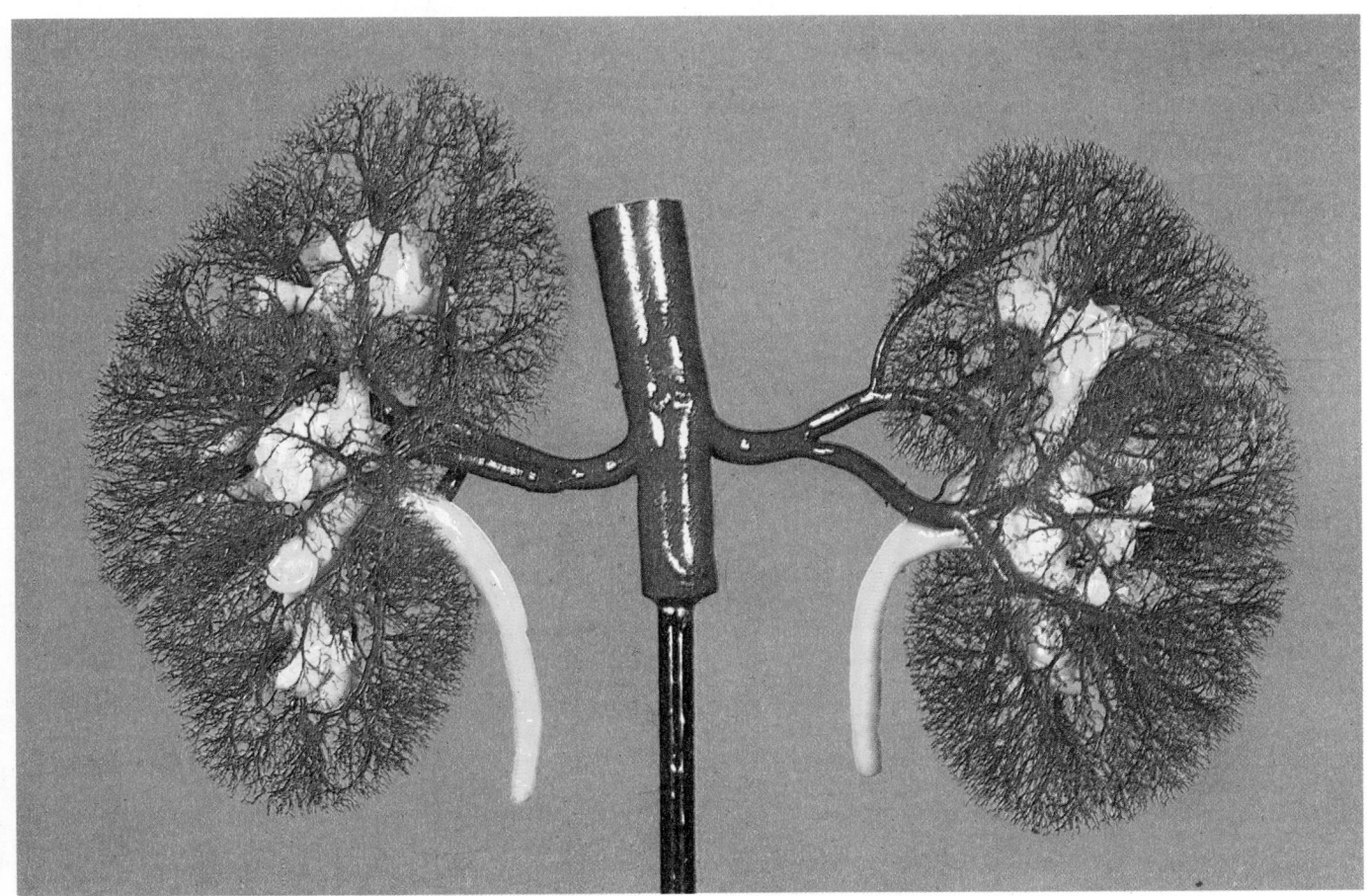

8.178 Resin corrosion cast of human kidneys prepared by D H Tompsett of the Royal College of Surgeons of England. Ureter, pelvis and calices are yellow; aorta, renal arteries and their branches are red. See text for a detailed description of the renal blood vessels.

the brim of the lesser pelvis near the medial border of psoas major, and where it traverses the vesical (bladder) wall (its narrowest part). The *renal pelvis* has already been described (p. 1411).

The ureter's *abdominal part* descends posterior to the peritoneum on the medial part of psoas major, which separates it from the tips of the lumbar transverse processes (8.169B). Anterior to the muscle it crosses in front of the genitofemoral nerve and is obliquely crossed by the gonadal vessels. It enters the lesser pelvis anterior to either the end of the common or the beginning of the external iliac vessels.

At its origin the *right* ureter is usually overlapped by the descending part of the duodenum; it descends lateral to the inferior vena cava, crossed anteriorly by the right colic and ileocolic vessels; near the superior aperture of the lesser pelvis it passes behind the lower part of the mesentery and terminal ileum. The *left ureter*, crossed by the left colic vessels, passes posterior to the sigmoid colon and its mesentery in the posterior wall of the intersigmoid recess. At operation, the abdominal part of the left ureter is hence easier to expose than the right.

The *pelvic part*, about the same length as the abdominal, lies in both sexes in extraperitoneal areolar tissue. At first it descends posterolaterally on the lateral wall of the lesser pelvis along the anterior border of the greater sciatic notch. Opposite the ischial spine it turns anteromedially into fibrous adipose tissue above the levator ani to reach the base of the bladder. On the pelvic wall it is anterior to the internal iliac artery and the beginning of its anterior trunk, posterior to which are the internal iliac vein, lumbosacral nerve and sacro-iliac joint. Laterally it lies on the fascia of obturator internus. It progressively crosses medial to the umbilical artery, the obturator nerve, artery and vein, the inferior vesical and middle rectal arteries. *In males*, in the anteromedial part of its descent, the pelvic ureter is crossed anterosuperiorly, from lateral to medial, by the ductus deferens. Then it passes in front of and slightly above the upper pole of the seminal vesicle to traverse the bladder wall obliquely before opening at the ipsilateral trigonal angle (8.184). Its terminal part is surrounded by tributaries of the vesical veins. *In females*, the pelvic part at first has the same relations as in males, but anterior to the internal iliac artery it is immediately behind the ovary, forming the posterior boundary of the ovarian fossa (p. 1435). In the anteromedial part of its course to the bladder it is related to the uterine artery, uterine cervix and vaginal fornices. It is in extraperitoneal connective tissue in the inferomedial part of the broad ligament of the uterus (parametrium, p. 1445); here the uterine artery is anterosuperior to the ureter for 2.5 cm and then crosses to its medial side to ascend alongside the uterus. The ureter turns forwards slightly above the lateral vaginal fornix and is here generally 2 cm lateral to the supravaginal uterine cervix (p. 1441). It then inclines medially to reach the bladder, with a variable relation to the front of the vagina. As the uterus is commonly deviated to one side, one ureter may be more extensively apposed to the vagina, usually the left, which may cross the midline; the reverse may occur and sometimes one ureter is not anterior to the vagina, a much longer part of the other then being in front of it.

In the distended bladder, in both sexes, the ureteric openings are about 5 cm apart, somewhat less in the empty viscus. In its oblique course through the vesical wall, the ureter is compressed and flattened as the bladder distends, perhaps preventing regurgitation, though ureteric peristalsis is also a factor.

Radiography

The ureter, renal pelvis and calices are easily demonstrated in the living by radiography (1) after intravenous injection of radioopaque substances excreted in urine (descending or *excretion pyelography*), or (2) after the introduction of similar solutions into the ureter by catheterization through an operating cystoscope (ascending or *retrograde pyelography*), the results being *pyelograms* (8.169B). Normal cupping of the minor calices by projecting renal papillae is clinically important; it may be obliterated by conditions such as hydronephrosis, associated with chronic distension of the ureter and renal pelvis due to urinary back pressure.

STRUCTURE OF THE URETERS

The wall of the human ureter is composed of three histological layers, namely an external adventitia, a non-striated muscle layer and an inner mucosal layer. The latter consists of two components: the transitional epithelium (or urothelium) and the underlying connective tissue (the lamina propria). The ureteric *adventitia* contains elongated fibrocytes and interlacing bundles of collagen and elastic fibrils, together with numerous blood vessels, lymphatics, nerves and occasional fat cells. The majority of the adventitial blood vessels and the connective tissue elements are orientated parallel to the long axis of the ureter.

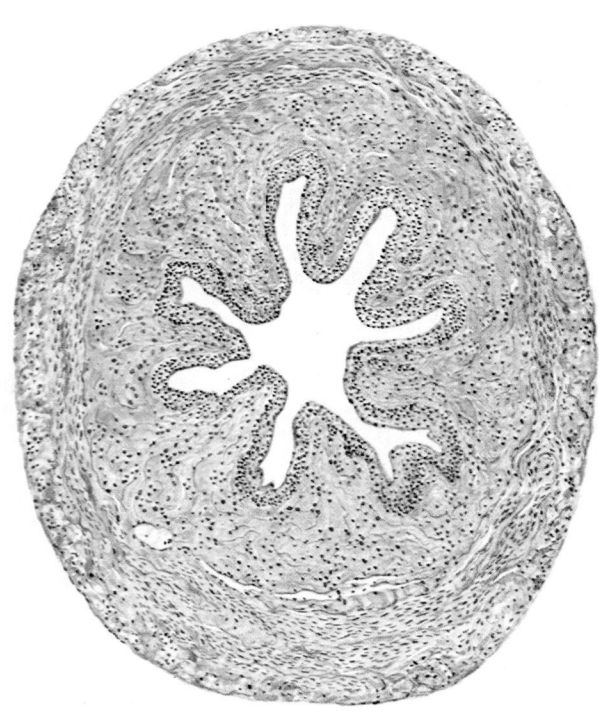

8.179 Transverse section through the lower third of a human ureter. Note the great thickness of muscular wall. Stained with haematoxylin and eosin. Magnification × 30.

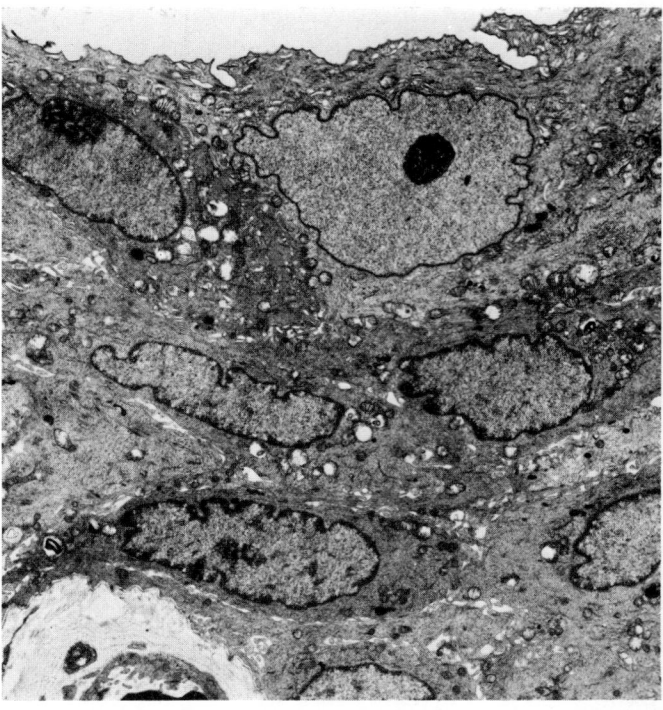

8.180 Transmission electron micrograph showing the transitional epithelium lining the ureter (rat). Prepared and photographed by Susan Smith, Department of Anatomy, Guy's Hospital Medical School, London. Magnification × 4000.

Throughout its length, the *muscle coat* of the ureter is fairly uniform in thickness and in cross-section measures about 750–800 µm in width. The muscle bundles which constitute this coat are frequently separated from one another by relatively large amounts of connective tissue. Branches which interconnect muscle bundles are common and result in frequent interchange of muscle fibres between adjacent bundles. Due to this extensive branching individual muscle bundles do not spiral around the ureter (Gosling 1970). Hence, the ureteric muscle coat consists of a complex meshwork of interweaving and interconnecting smooth muscle bundles. In addition, unlike the gut, the muscle bundles are so arranged that morphologically distinct longitudinal and circular layers cannot be distinguished. However, in the upper part of the ureter, the inner muscle bundles tend to lie longitudinally while those on the outer aspect have a circular or oblique orientation. In its middle and lower parts, additional outer longitudinally-orientated fibres can also be discerned. As the ureterovesical junction is approached, the muscle coat consists predominantly of longitudinally-orientated muscle bundles. However, there are distinct differences between ureteric and vesical muscles (Gilpin & Gosling 1983). Ureteric muscle bundles are composed of closely-packed spindle-shaped cells which are approximately 300–400 µm in length and 4–7 µm at their widest diameter. Each non-striated muscle cell is surrounded by a trilaminar plasma membrane on the outer aspect of which is a layer of amorphous electron-dense basal lamina material approximately 40 nm in thickness. The latter is separated from the underlying plasma membrane by an electron-lucent zone about 4 nm in width (the lamina lucida).

Within any one muscle bundle the plasma membranes of adjacent cells are frequently observed in mutual opposition. The most common form of junction between neighbouring ureteric muscle cells is the 'region of close approach' at which the apposed plasma membranes converge to within 10–20 nm of each other. Basal lamina material is reflected from one cell to the other at the margins of the junction and does not extend into the narrow intercellular cleft. However neither the plasma membranes nor the adjacent subsarcolemmal cytoplasm show any form of specialization. Intercellular junctions of this type often occur between the tips of narrow cytoplasmic protrusions from adjacent non-striated muscle cells. Another common type of association between ureteric non-striated muscle cells is the so-called 'peg and socket' junction. This type consists of an elongated or bulbous projection from one cell which fits snugly into a depression in an adjacent cell. At the interdigitation, basal lamina material is absent from the 10 nm gap separating the membranes of the two adjacent muscle cells. Occasionally, the apposing cell membranes are more closely related in the narrowed 'stalk' of the projection. It has been suggested that these interdigitations provide a mechanical linkage between neighbouring cells. Alternatively, they may represent a special type of 'close approach' and permit myogenic conduction.

The *mucosa* of the ureter consists of an epithelium (the urothelium) on the external aspect of which is a layer of subepithelial fibro-elastic connective tissue (the lamina propria). The latter varies in thickness from 350–700 µm and consists mainly of bundles of collagen and elastic fibres, fibrocytes and small blood vessels. Many of the latter are accompanied by bundles of non-myelinated nerve fibres. Occasional lymphocytes may be present in the lamina propria but their aggregation into definitive lymph nodules is rare. The urothelium is usually extensively folded, giving the ureteric lumen a stellate outline. Although the urothelium appears to consist of four to five separate layers of cells, it has been shown that all cells reach down to the base and as a consequence the urothelium is by definition a pseudostratified epithelium (Martin 1973).

The surface cells of the urothelium are large and polyhedral, the underlying intermediate cells are smaller and spindle-shaped, and the basal cells are mainly cuboidal in form. The luminal cells of the urothelium are characterized by the presence of a specialized apical membrane (Hicks 1965, Staehelin et al 1972) and are attached to one another near the ureteric lumen by a junctional complex composed of the three typical components (p. 25) of zonula occludens, zonula adherens and macula ad-

herens. The lateral cell membranes of the luminal cells are often extensively interdigitated, although such infoldings presumably flatten out to enable the urothelium to accommodate distension of the ureteric lumen.

Vessels and Nerves

The arteries supplying the ureter (Daniel & Shackman 1952) are derived from the renal, abdominal aortic, testicular (or ovarian), common iliac, internal iliac, vesical and uterine vessels, their branches supplying the different parts of the ureter in its course and being subject to much variation. The longitudinal anastomosis between these branches on the wall of the ureter is good. The branches from the inferior vesical artery (p. 778) are constant in their occurrence and supply the lower part of the ureter as well as a large part of the trigone of the bladder.

The lymph vessels of the ureter are described on p. 855.

The ureteric nerves (pp. 1165, 1166) are derived from the renal, aortic and superior and inferior hypogastric plexuses; through these plexuses fibres are derived from the lower three thoracic and first lumbar and the second to fourth sacral segments of the spinal cord. In the adventitia the nerves consist of relatively large axon bundles which form an irregular plexus, from which numerous smaller branches penetrate the ureteric muscle coat. Some of the adventitial nerves accompany the blood vessels and branch with them as they extend into the muscle layer. Others are unrelated to the vascular supply and lie free in the adventitial connective tissue around the circumference of the ureter. Autonomic ganglion cells occur only at the extreme lower end of the ureter; such cells are absent from all other regions of the ureter. In the muscle coat of the upper urinary tract, nerve fibres can be identified both between and within the non-striated muscle bundles. However, regional differences in the density of these nerves result in a gradual increase in innervation from the renal pelvis and upper ureter (which has a sparse distribution of autonomic nerves) to a maximum in the juxtavesical segment. The non-striated muscle of the upper urinary tract is supplied with at least three different types of autonomic nerve. Some nerves are rich in acetylcholinesterase and the presence of this enzyme has been taken to be indirect evidence in support of a cholinergic innervation to ureteric non-striated muscle. Noradrenergic nerves also occur in the ureteric muscle coat where they are evident as finely beaded fibres, some of which run parallel to the muscle bundles. Others accompany the vascular supply and their branches occasionally penetrate the muscle bundles and course amongst the smooth muscle cells. A third type of nerve, characterized by its content of substance P, has been demonstrated in the muscle coat of the ureter (Alm et al 1978). The distribution of this so-called peptidergic innervation within the ureter appears to be similar to the other types of autonomic nerve. Other neurotransmitters also exist (see Burnstock 1986), so the control mechanisms are likely to be complex (see also p. 889). Using the electron microscope, occasional axon bundles, some containing up to 50 unmyelinated axons, can be seen in the connective tissue separating the ureteric muscle bundles and, very rarely, varicose vesicle-packed terminal axons run between individual smooth muscle cells. However, neuromuscular relationships suggestive of synaptic regions are exceedingly sparse by comparison with many other autonomically innervated smooth muscle systems (e.g. the urinary bladder and ductus deferens).

The functional significance of these different types of autonomic nerve in relation to ureteric non-striated muscle activity is not fully understood, particularly since nerves are not essential for the initiation and propagation of ureteric contraction waves (see p. 1416). However, the contractile response of the ureter has been shown to be modified by the use in vivo of exogenous transmitter substances. Thus, while nerves are not essential for the normal occurrence of ureteric peristaltic activity, they may exert a modulating influence upon the contractility of ureteric non-striated muscle cells.

A branching plexus of fine nerve fibres occurs within the lamina propria and extends from the inner aspect of the muscle coat towards the base of the urothelium. These nerves are acetylcholinesterase positive and, while some form perivascular plexuses, others lie in isolation from the vascular supply. A similar distribution of noradrenergic and peptidergic nerves has been observed throughout the lamina propria. The functional significance of these nerves, which are not related to blood vessels, remains unclear. It seems unlikely that the urothelium receives a direct efferent innervation, particularly when most of the nerve fibres are more than 100 µm away from the basal urothelial cells and thus beyond the accepted range for effective synaptic transmission. It seems probable therefore that some of the nerves in the lamina propria are sensory in function.

Ureteric Peristalsis

Under normal conditions contraction waves originate in the proximal part of the upper urinary tract and propagate in an anterograde direction towards the bladder. However, the mechanism involved in the initiation of these contractile events has been the subject of much controversy in the past. One theory proposed that contraction of the non-striated muscle of the calices and pelvis was stimulated by stretching forces produced by the luminal contents. However, there is now a considerable body of experimental evidence to show that the urinary tract possesses spontaneously active regions which initiate the peristaltic activity of the ureter and exercise a controlling influence on ureteric contraction (Constantinou & Djurhuus 1981). From a morphological viewpoint the unusual non-striated muscle cells which occur in the wall of the proximal parts of the upper urinary tract (p. 1415) and which are structurally distinct from those elsewhere, may act collectively as one or more pacemaker sites. Since each minor calix possesses these cells (and is linked across the renal parenchyma to other calices by similar cells), the number of pacemaker sites within a given kidney will be related to the number of minor calices. It seems probable that the normal sequence of events begins with the initiation of a contraction wave at one (or possibly more) of the several minor caliceal pacemaker sites. Once initiated, the contraction is propagated through the wall of the adjacent major calix and activates the non-striated muscle of the renal pelvis. To what extent the inner layer of cells acts as pacemaker (or as a preferential conduction pathway) remains to be determined. The proximal pacemaker site for successive contractions may change from one minor calix to another, although it has been shown that a single minor calix can initiate several successive contraction waves. Functionally, the proximal location of these pacemaker sites ensures that, once initiated, contraction waves are propagated away from the kidney, thereby avoiding undesirable pressure rises directed against the renal parenchyma. In addition, since several potential pacemaker sites exist, the initiation of contraction waves is unimpaired by partial nephrectomy; the minor calices spared by the resection remain in situ to continue their pacemaking function.

Concerning the means by which contractions are propagated across the renal pelvis and along the ureter, considerable experimental evidence indicates that autonomic nerves do not play a major part in the propagation of upper urinary tract peristalsis. It seems more likely that autonomic nerves may have only a modulatory role on the contractile events occurring in the musculature of the upper urinary tract. The most likely mechanism to account for impulse propagation is that of myogenic conduction resulting from electrotonic coupling of one muscle cell to its immediate neighbours. Regions of close approach are extremely numerous between ureteric non-striated muscle cells and also between both types of muscle cell in the renal pelvis and calices. Thus in the upper urinary tract this type of intercellular junction may be responsible for the conduction of excitation from one myocyte to the next.

To conclude, contraction waves arising from the upper end of the urinary tract are thought to be propagated from muscle cell to muscle cell by means of intercellular junctions. This process is essentially a property of non-striated muscle and does not require the direct involvement of autonomic nerves. The direction of the propagated contractile wave is normally from the renal calices and pelvis towards the bladder, as dictated by the pacemaker mechanism in these proximal regions. However, since electrotonic excitation can spread from one muscle cell to another in either direction, retrograde peristalsis can also occur.

Applied Anatomy

As in the case of the gut (p. 1347), excessive ureteric distension or spasm of its muscle provokes severe pain (renal colic), e.g. by a stone (calculus) causing incomplete and intermittent obstruction, particularly if it is gradually forced down by the muscle spasm. The pain, spasmodic and agonizing, is referred to cutaneous areas innervated from spinal segments which supply the ureter, mainly T11–L2 (p. 1168). It shoots down and forwards from the loin to the groin and scrotum or labium majus and may extend into the proximal anterior aspect of the thigh by projection to the genitofemoral nerve (L1,2); the cremaster (which has the same innervation) may reflexly retract the testis. A ureteric stone is liable to impact at one of the ureteric constrictions, namely its cranial end, where it crosses the brim of lesser pelvis or as it passes through the vesical wall; radiologically it would appear near the tip of the second lumbar transverse process, overlying the sacroiliac joint, or slightly medial to the ischial spine. Sometimes the ureter is duplicated on one or both sides even as far as the bladder; separate vesical openings are rare in such cases.

PELVIC VISCERA

The Urinary Bladder

The urinary bladder (8.181), is solely a reservoir and varies in size, shape, position and relations, according to its content and the state of neighbouring viscera. When empty it is entirely in the lesser pelvis but as it distends it expands anterosuperiorly into the abdominal cavity. When empty, it is somewhat tetrahedral and has a fundus, neck, apex, a superior and two inferolateral surfaces. The *fundus* or *base* is triangular and postero-inferior. In females (8.183) it is closely related to the anterior vaginal wall, in males to the rectum but its upper part is separated from the rectum by the rectovesical pouch and below that by the seminal vesicles and deferent ducts (8.182). In a triangular area between the deferent ducts, the bladder and rectum are separated only by rectovesical fascia (p. 1433); its inferior part may be obliterated by approximation of the deferent ducts above the prostate. Although the vesical fundus should be, by definition, the lowest region, the *neck* is in fact lowest and also the most fixed; it is 3–4 cm behind lower part of the symphysis pubis (i.e. a little above the plane of the inferior aperture of the lesser pelvis). It is pierced by the internal urethral orifice and alters little in position with varying conditions of the bladder and rectum. There is no special vesical constriction at its neck. In males the neck rests on its wall and is in direct continuity with, the base of the prostate; in females it is related to the pelvic fascia which surrounds the upper urethra. The vesical *apex* in both sexes faces towards the upper part of the symphysis pubis; from it the median umbilical ligament (urachus, p. 1417) ascends behind the anterior abdominal wall to the umbilicus, the peritoneum over it being the median umbilical fold. The triangular *superior surface* is bounded by lateral borders from the apex to the ureteric entrances and by a posterior border joining them. In males the superior surface is completely covered by peritoneum, extending slightly on to the base and continued posteriorly into the rectovesical pouch, laterally into the paravesical fossae and anteriorly into the median umbilical fold. It is in contact with the sigmoid colon and the terminal coils of the ileum. In females the superior surface is also largely covered by peritoneum but posteriorly this is reflected to the uterus at the level of the internal os (i.e. the junction of the uterine body and cervix), forming the vesico-uterine pouch. The posterior part of the superior surface, devoid of peritoneum, is separated from the supravaginal cervix by fibro-areolar tissue. Each *inferolateral surface* in males is separated anteriorly from the pubis and puboprostatic ligaments by an adipose retropubic pad and posteriorly by fascia from levator ani and obturator internus. In females the relations are similar, but the puboprostatic are replaced by pubovesical ligaments. The inferolateral surfaces are not covered by peritoneum.

As the bladder fills it becomes ovoid. In front it displaces the parietal peritoneum from the suprapubic region of the abdominal wall, so that the surfaces become anterior and rest against it without intervening peritoneum for a distance above the symphysis pubis varying with the degree of distension and commonly about 5 cm; in excessive distension the bladder may reach the umbilicus or beyond. Thus a full bladder may be approached surgically through the anterior abdominal wall above the symphysis pubis without traversing the peritoneum. The full bladder's apex points up and forwards above the attachment of the median umbilical ligament, so that the peritoneum forms a supravesical recess of varying depth between the apex and the anterior abdominal wall; this recess often contains coils of small intestine.

At birth (8.183), the bladder is relatively higher than in the adult, the internal urethral orifice being level with the *upper* symphysial border; the bladder is abdominal rather than pelvic, extending about two-thirds towards the umbilicus. It progressively descends, reaching the adult position shortly after puberty.

THE LIGAMENTS OF THE BLADDER

In both sexes stout bands of fibromuscular tissue extend from the bladder neck to the inferior aspect of the pubic bones. These structures are the pubovesical ligaments and constitute the superior extensions of the pubo-urethral ligaments in the female or the puboprostatic ligaments in the male. The two pubovesical ligaments lie one on each side of the median plane, leaving a midline hiatus through which pass numerous small veins. In addition to the pubovesical ligaments a number of other so-called ligaments have been described in relation to the base of the urinary bladder (Zacharin 1972). These are formed by condensation of connective tissue around neurovascular structures and as such do not merit the distinction of having specific names assigned to them.

The apex of the bladder is joined to the umbilicus by the remains of the urachus, which forms the median umbilical ligament. The lumen of the lower part of the urachus may persist throughout life and communicate with the cavity of the bladder (Begg 1930). From the superior surface of the bladder the peritoneum is carried off in a series of folds which are termed the 'false' ligaments of the bladder. Anteriorly there are three folds: the median umbilical fold over the median umbilical ligament and two medial umbilical folds over the obliterated umbilical arteries (8.109). The reflexions of the peritoneum from the bladder to the side walls of the pelvis form the lateral false ligaments while the sacrogenital folds (p. 1338) constitute the posterior false ligaments.

THE VESICAL INTERIOR

The vesical mucosa (8.185, 186), attached only loosely to subjacent muscle for the most part, folds when the bladder empties, the folds being effaced as it fills. Over the *trigone* (8.184), immediately above and behind the internal urethral orifice, it is adherent to the subjacent muscle layer and always smooth. The trigone's anteroinferior angle is formed by the internal urethral orifice, its posterolateral angles by the ureteric orifices. The superior trigonal boundary is a slightly curved *interureteric crest*, connecting the two ureteric orifices and produced by the continuation into the vesical wall of the ureteric internal longitudinal muscle. (For details of trigonal musculature, consult Uhlenhuth et al 1952, Woodburne 1965, 1968.) Laterally this ridge extends beyond the ureteric openings as *ureteric folds*, produced by the terminal parts of the ureters running obliquely through the bladder wall. At

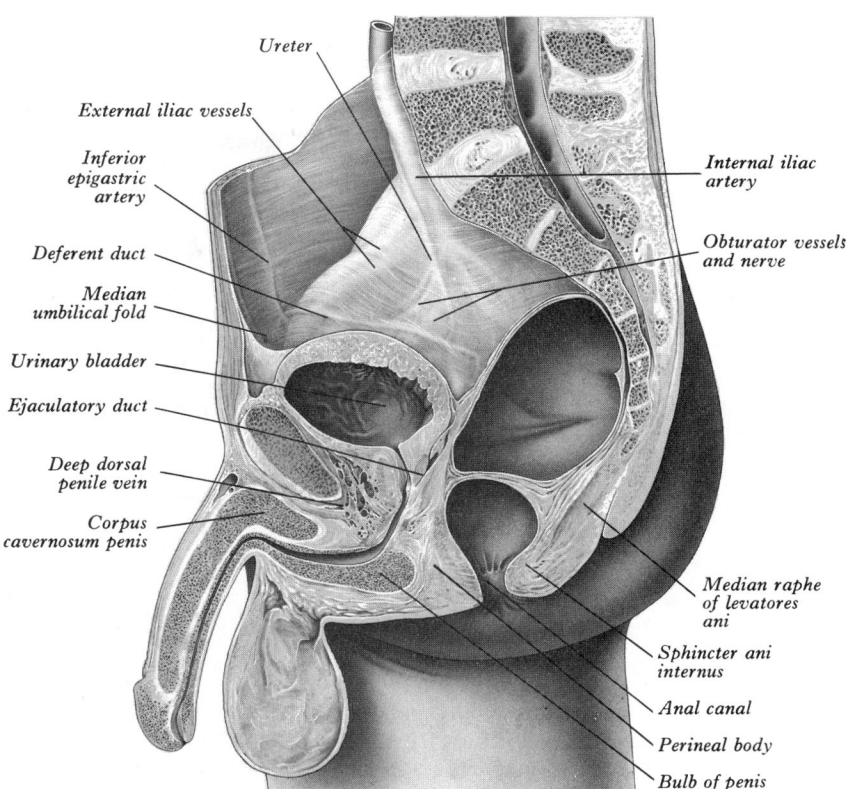

8.181 Median sagittal section to show male internal and external genitalia, bladder etc. A number of structures (e.g. obturator vessels, ureter) are only faintly visible through the overlying peritoneum.

cystoscopy the interureteric crest appears as a pale band and is a guide to the ureteric orifices in catheterization.

The ureteric orifices, placed at the posterolateral trigonal angles (8.184), are usually slit-like. In empty bladders they are about 2.5 cm apart and about the same from the internal urethral orifice; in distension these measurements may be doubled.

The internal urethral orifice, at the trigonal apex, the lowest part of the bladder, is usually somewhat crescentic in

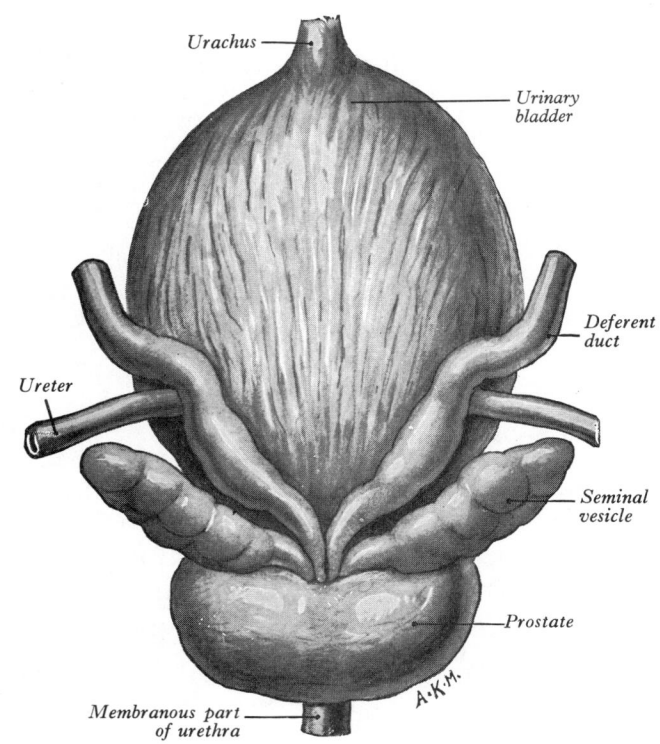

8.182 Posterosuperior aspect of the male internal urogenital organs.

1417

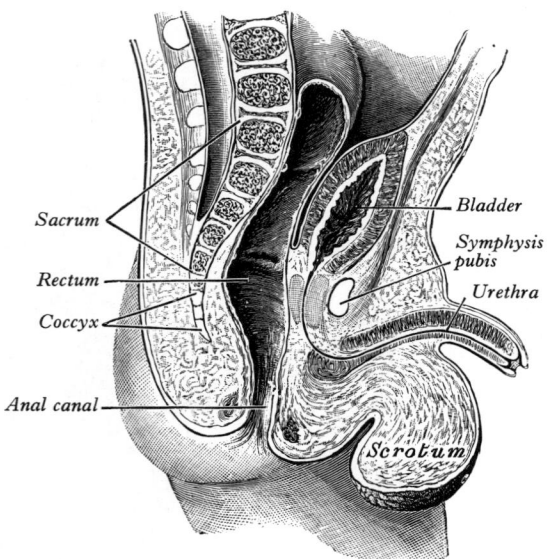

8.183 Sagittal section of the pelvis of a newborn male infant. Note the abdominal position of the urinary bladder.

section; in adult males, particularly past middle age, immediately behind it is a slight elevation caused by the median prostatic lobe, the *uvula* of the bladder.

Bladder Capacity

Mean vesical capacity in male adults varies from 120–320 ml (Thompson 1919); micturition commonly occurs at about 280 ml. Filling to about 500 ml may be tolerated but beyond this pain is caused by tension in the wall, leading to reflex contractions and the urgent desire to micturate. Pain is referred to the cutaneous areas supplied by spinal segments supplying the bladder (T11–L2, S2–4), including the lower anterior abdominal wall, perineum and penis.

BLADDER STRUCTURE

Histologically the wall of the urinary bladder consists of three layers (**8.185**): an outer adventitial layer of connective tissue which in some regions possesses a serosal covering of peritoneum, a non-striated muscle coat (the detrusor muscle) and an inner layer of mucous membrane which lines the interior of the bladder.

The Serous Layer

This is restricted to the superior and in males part of the posterior surfaces of the bladder, the rest being devoid of peritoneum (p. 1417). It consists of mesothelium and underlying connective tissue as elsewhere in the peritoneum.

The Muscular Layer

This is composed of relatively large (in diameter) interlacing bundles of non-striated muscle cells arranged as a complex meshwork. Three ill-defined layers are present and arranged in such a way that longidutinally orientated muscle bundles predominate on the inner and outer aspects of a substantial middle circular layer. Posteriorly some of the outer longitudinal bundles pass over the bladder base and fuse with the capsule of the prostate or with the anterior vaginal wall. Other bundles are carried on to the anterior aspect of the rectum and named the rectovesical muscle. Anteriorly some of the outer longitudinal bundles continue into the pubovesical ligaments and contribute to the muscular component of these structures. As in the muscle coat of the ureter, exchange of fibres between adjacent muscle bundles within the bladder wall frequently occurs so that from a functional viewpoint, the detrusor comprises a single unit of interlacing smooth muscle. An electron-dense basal lamina surrounds each bladder non-striated myocyte except at certain junctional regions. The most frequently observed type of junction between muscle cells is the region of close approach at which an intercellular separation of 10–20 nm occurs over distances occasionally in excess of 1 μm in length. Junctions of the 'peg and socket' and 'intermediate' types are observed occasionally but gap junctions (nexuses) have not been reported in the detrusor. Since electrotonic spread excitation occurs in the non-striated muscle of the bladder wall, the

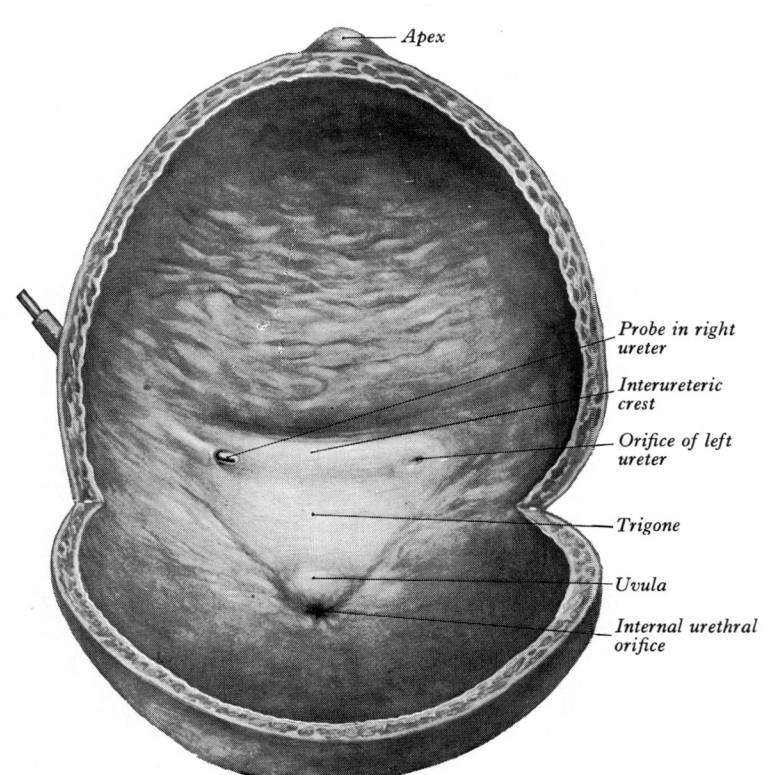

8.184 Anterior aspect of the interior of the urinary bladder.

regions of close approach may represent the morphological feature which enables this physiological event to take place, although it is more likely that gap junctions exist, albeit few of them. Within muscle bundles, the smooth muscle cells are closely packed together such that the basal lamina of one cell very often becomes confluent with that of its neighbours.

Trigone. The smooth muscle of this region consists of two distinct layers, sometimes termed the superficial and deep trigonal muscles. The latter is composed of muscle cells which are indistinguishable from the muscle cells of the detrusor. Hence this deep trigonal muscle is simply the postero-inferior portion of the detrusor muscle proper and confusion might be avoided if the term deep trigonal muscle was abandoned in favour of the more accurate definition as trigonal detrusor muscle. The superficial trigonal muscle represents a morphologically distinct component of the trigone which, unlike the detrusor, is composed of relatively small diameter muscle bundles continuous proximally with those of the intramural ureters. The muscle layer comprising the superficial trigone is relatively thin but becomes thickened along its superior border to form the interureteric crest. Similar thickenings occur along the lateral edges of the superficial trigone (Bell 1812). In both sexes the superficial trigone muscle become continuous with the smooth muscle of the proximal urethra, extending in the male along the urethral crest as far as the openings of the ejaculatory ducts.

Ureterovesical junction. The distal 1–2 cm of each ureter is surrounded by an incomplete collar of the detrusor non-striated muscle which forms a sheath (of Waldeyer) separated from the ureteric muscle coat by a connective tissue sleeve. The ureters pierce the posterior aspect of the bladder and run obliquely through its wall for a distance of 1.5–2.0 cm before terminating at the ureteric orifices. This arrangement is believed to assist in the prevention of reflux of urine into the ureter, since the intramural ureters are thought to be occluded during increases in bladder pressure. The longitudinally orientated muscle bundles of the terminal ureter continue into the bladder wall and at the ureteric orifices become continuous with the superficial trigonal muscle (Tanagho et al 1968).

Bladder neck. The non-striated muscle of this region is histologically, histochemically and pharmacologically distinct from that which comprises the detrusor proper (Nergardh & Boreus 1972, Kluck 1980). Hence the bladder neck should be considered as a separate functional unit. The arrangement of non-striated muscle in this region is quite different in males and females and consequently each sex will be described separately.

Male. In the male bladder neck, the non-striated muscle cells form a complete circular collar which extends distally to surround the preprostatic portion of the urethra. Because of the location and orientation of its constituent fibres, the terms internal, proximal or preprostatic urethral sphincter are suitable alternatives for this particular component of urinary tract smooth muscle. Distally, the bladder neck muscle merges with and becomes indistinguishable from the musculature in the stroma and capsule of the prostate gland.

Female. The female bladder neck also consists of morphologically distinct non-striated muscle, since the large diameter fasciculi characteristic of the detrusor are replaced in the region of the bladder neck by those of small diameter. However, unlike the circularly orientated preprostatic non-striated muscle, the muscle fasciculi in the female extend obliquely or longitudinally into the urethral wall. The female does not therefore possess a smooth muscle sphincter at the bladder neck and it is unlikely that active contraction of this region plays a significant part in the maintenance of female urinary continence.

The Mucosa

This has a structure similar to that of the ureters and consists of an epithelium (urothelium, transitional epithelium, p. 56) supported by a layer of loose connective tissue, the lamina propria. The latter consists of loose fibro-elastic connective tissue and forms a relatively thick layer, varying in depth from 500 µm in the fundus and inferolateral walls to about 100 µm in the trigone. Small-diameter bundles of non-striated muscle cells also occur in the subepithelial connective tissue forming an incomplete and

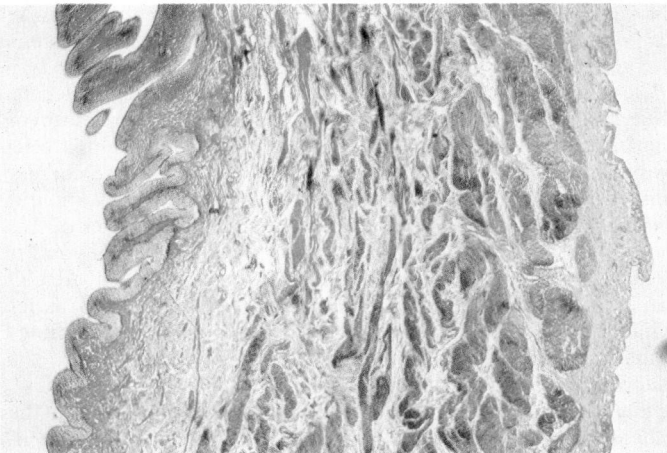

8.185 Transverse section through the urinary bladder wall of a monkey. Masson's trichrome stain. The section shows the folded transitional epithelium (grey-brown on the left), the connective tissue (blue-green), and fasciculi of the detrusor muscle (green-brown, to the right), surrounded externally by the connective tissue adventitia. Magnification × 20.

rudimentary muscularis mucosae. The connective tissue elements immediately beneath the urothelium, particularly in the region of the trigone, are densely packed. At deeper levels they are more loosely arranged, thus allowing the bladder mucosa to form numerous thick folds when the volume of fluid contained within the lumen is small. An extensive network of blood vessels is present throughout the lamina propria and supplies a plexus of thin-walled fenestrated capillaries lying in grooves at the base of the urothelium.

Non-trigonal urothelium is often up to six cells in thickness. These cells can be classified according to position and consist of highly differentiated superficial or luminal cells, one or more layers of smaller intermediate cells and a layer of undifferentiated basal cells. The large superficial cells frequently bulge into the bladder lumen and are often binucleate. In contrast the intermediate and basal cells are smaller and each contains a single darkly-staining nucleus. The flattened urothelium of the trigone

8.186 Scanning electron micrograph of the luminal surface of the bladder (murine) showing the plate-like organization of the urothelial plasma membranes of its lining cells. Magnification × 7000.

usually consists of only two or three layers of cells and a similar appearance prevails throughout the bladder when in the distended state.

In addition to the basal, intermediate and superficial cells described above, a fourth type of cell occurs in the urothelium of the human bladder neck and trigone. These flask-shaped cells extend throughout the depth of the urothelium and are characterized by the presence of numerous large membrane-bound vesicles each containing a central dense granule. Vesicles of this type are believed to be involved in the storage of amines and it seems likely that these cells belong to the so-called APUD (amine-precursor-uptake and decarboxylation, see p. 1466) series which have a wide distribution throughout the body. The functional significance of cells of this type in the urothelium of the bladder remains a mystery.

Several morphological variations have been described in the mucosa of the bladder which, because of their occurrence in otherwise normal healthy adults, are not considered to represent pathological conditions. One of the commonest epithelial variants found in bladder biopsy samples or at post-mortem is the occurrence of so-called Brunn's nests. these consist of proliferations of morphologically normal basal urothelial cells which project into the underlying connective tissue of the lamina propria and are particularly frequent in the trigone. Mucus-secreting glands with single or branched ducts are another frequently observed feature of the bladder mucosa. When present these structures are particularly numerous near the ureteric and internal urethral orifices. Non-keratinizing squamous metaplasia of the vaginal type also frequently occurs in the urinary bladder mucosa, especially over the trigone. This histological appearance, whilst occasionally observed in males and in children, is more common in adult females.

The Vascular Supply of the Bladder

The *principal arteries* of supply to the bladder are the superior and inferior vesical, derived from the anterior trunk of the internal iliac artery. The obturator and inferior gluteal arteries also send small branches to it and in the female additional branches are derived from the uterine and vaginal arteries.

The *veins* form a complicated plexus on the inferolateral surfaces and pass backwards in the posterior ligaments of the bladder to end in the internal iliac veins.

The *lymph vessels* are described on p. 855.

The Nerve Supply of the Bladder

The nerves supplying the bladder form the vesical plexus (see pp. 1167, 1168) and consist of both sympathetic and parasympathetic components, each of which contains both efferent and afferent fibres.

Efferent fibres. Parasympathetic fibres arise from the second to the fourth sacral segments of the spinal cord (*nervi erigentes*); the sympathetic fibres are derived from the lower two thoracic and upper two lumbar segments of the spinal cord. In addition to the branches from the vesical plexus, small groups of autonomic neurons occur throughout all regions of the bladder wall. These multipolar intramural neurons are rich in acetylcholinesterase and occur in ganglia consisting of five to 20 nerve cell bodies. Numerous preganglionic autonomic fibres form both axosomatic and axodendritic synapses with the ganglion cells. The majority of these preganglionic nerve terminals correspond morphologically to presumptive cholinergic fibres. Noradrenergic terminals also relay on cell bodies in the pelvic plexus although it is unknown whether similar nerves synapse on intramural bladder ganglia.

The urinary bladder (including trigonal detrusor muscle) is profusely supplied with nerves which form a dense plexus among the detrusor smooth muscle cells. The majority of these nerves contain acetylcholinesterase and occur in abundance throughout the muscle coat of the bladder. Axonal varicosities adjacent to detrusor non-striated muscle cells possess features which are considered to typify cholinergic nerve terminals and contain clusters of small (50 nm diameter) agranular vesicles together with occasional large (80–160 nm diameter) granulated vesicles and small mitochondria. Terminal regions approach to within 20 nm of the muscle cells' surface and are either partially surrounded by

or more often totally denuded of Schwann cell cytoplasm. The human detrusor possesses a sparse supply of sympathetic noradrenergic nerves (Sundin et al 1977). Nerves of this type generally accompany the vascular supply and only rarely extend among the non-striated myocytes of the urinary bladder. A further component plays a part in the autonomic innervation of the urinary bladder (Ambache & Zar 1970), which has been classified as a non-adrenergic, non-cholinergic nerve mediated effect. A number of other neurotransmitters or neuromodulators have been detected in intramural ganglia, including the peptide somatostatin (see p. 891). Superficial trigonal muscle is associated with few cholinergic (parasympathetic) nerves while those of the noradrenergic (sympathetic) variety occur relatively frequently. These differences support the view that the superficial trigonal muscle should be regarded as 'ureteric' rather than 'vesical' in origin. It should be emphasized that the superificial trigonal muscle forms a very minor part of the total muscle mass of the bladder neck and proximal urethra in either sex and is probably of little significance in the physiological mechanisms which control these regions.

In the male, bladder neck non-striated muscle is sparsely supplied with cholinergic (parasympathetic) nerves but possesses a rich noradrenergic (sympathetic) innervation (Gosling et al 1977). A similar distribution of autonomic nerves also occurs in the non-striated muscle of the prostate gland, seminal vesicles and ducti deferentes. From a functional standpoint, sympathetic nerves on stimulation cause contraction of non-striated muscle in the wall of the genital tract resulting in seminal emission. Concomitant sympathetic stimulation of the proximal urethral muscle causes sphincteric closure of the bladder neck, thereby preventing reflux of ejaculate into the bladder. Although this genital function of the male bladder neck is well established it is not known whether the non-striated muscle of this region plays an active role in maintaining urinary continence.

In contrast with this rich sympathetic innervation in the male, the non-striated muscle of the female bladder neck receives relatively few noradrenergic nerves but is richly supplied with presumptive cholinergic fibres. The sparse supply of sympathetic nerves presumably relates to the absence of a functioning 'genital' portion incorporated within the wall of the female urethra.

The lamina propria of the fundus and inferolateral walls of the bladder is virtually devoid of autonomic nerve fibres, apart from some noradrenergic and occasional presumptive cholinergic perivascular nerves. However, as the urethral orifice is approached the density of nerves unrelated to blood vessels increases. At the bladder neck and the trigone a nerve plexus extends throughout the lamina propria. The constituent nerves are cholinesterase positive and run through the connective tissue unassociated with blood vessels. Some of the larger diameter axons are myelinated and others lie adjacent to the basal urothelial cells. As in the ureter, the subepithelial nerve plexus of the bladder is assumed to subserve a sensory function in the absence of any obvious effector target sites (Gosling & Dixon 1974).

Afferent fibres. Vesical nerves are also concerned with pain and awareness of distension. Pain fibres are stimulated by distension or spasm due to a stone, inflammation or malignant disease; they are in sympathetic *and* parasympathetic nerves, predominantly the latter. Hence, simple division of the sympathetic paths (e.g. 'presacral neurectomy'), or of the superior hypogastric plexus (p. 1166), does not materially relieve vesical pain. The spinal path for pain is in the anterolateral white columns and considerable relief follows bilateral anterolateral cordotomy. Since nerve fibres mediating awareness of distension are in the posterior columns (fasciculus gracilis), after anterolateral cordotomy the patient still retains awareness of the need to micturate. The nerve endings detecting noxious stimuli are probably of more than one type; a subepithelial plexus of fibres containing dense vesicles, probably afferent endings, has been described by Dixon & Gilpin (1987).

Applied Anatomy

A distended bladder may be ruptured in lower abdominal or pelvic injuries, either extraperitoneally or, if the superior surface is involved, with tearing of the peritoneum and escape of vesical contents into the peritoneal cavity. In progressive chronic obstruction to micturition, e.g. by prostatic enlargement (p. 1435) or urethral stricture, vesical musculature hypertrophies, its fasciculi increasing in size and interlacing in all directions to produce an enlarged, 'trabeculated bladder'. Mucosa between the fascicles forms 'diverticula', which may contain phosphatic concretions. When outflow is thus obstructed, emptying is not complete; some urine remains and may become infected; infection may extend to the ureters and kidneys. Back pressure from a distended bladder may gradually dilatate the ureters, renal pelves and even the renal collecting tubules. Lesions of the fasciculus gracilis (e.g. tabes dorsalis) cause loss of desire to micturate; the distended bladder may empty merely by overflow. Severe spinal cord lesions above its sacral segments, interrupting efferent and afferent tracts involved in normal micturition, may result in 'automatic' emptying.

The vesical interior can be examined with a cystoscope, introduced via the urethra after distending the bladder with fluid. A special cystoscope is used to catheterize the ureters, to obtain a direct specimen of urine from either kidney or to inject radio-opaque fluid for retrograde pyelography (**8.169B**). The vesical outline can be similarly demonstrated.

The distended bladder may be punctured just above the symphysis pubis without traversing the peritoneum *(suprapubic cyst-ostomy)*. When the bladder contains about 300 ml its antero-inferior surface contacts the anterior abdominal wall directly for about 7.5 cm above the pubis. Surgical access to the bladder is usually by this route. In females, owing to the shorter and more dilatable urethra, small calculi, foreign bodies and growths may be removed through it.

Congenital abnormalities of the bladder are described on p. 260.

The Male Urethra

The male urethra (**8.**181, 187) from 18–20 cm long, extends from an internal orifice in the urinary bladder to an external opening, or meatus, at the end of the penis. It may be considered in four regional parts: pre-prostatic, prostatic, membranous and spongiose and presents a double curve while the penis is in its ordinary flaccid state (**8.**181). Except during the passage of fluid along it, the urethral canal is a mere slit; in the prostatic part the slit is transversely arched in transverse section, in the membranous portion it is stellate, in the spongiose portion transverse, while at the external orifice it is sagittal in orientation.

The Pre-prostatic Part

The pre-prostatic urethra possesses a stellate lumen and is approximately 1–1.5 cm in length, extending almost vertically from the bladder neck to the superior aspect of the prostate gland. The non-striated muscle bundles surrounding the bladder neck and pre-prostatic urethra are arranged as a distinct circular collar which becomes continuous distally with the capsule of the prostate gland. The bundles which form this pre-prostatic or *internal sphincter (sphincter vesicae)* are separated by connective tissue containing many elastic fibres (Gilpin & Gosling 1983). Unlike the detrusor muscle, the non-striated muscle surrounding the proximal urethra is almost totally devoid of parasympathetic cholinergic nerves but is richly supplied with sympathetic noradrenergic nerves. Similar nerves also supply the non-striated muscle of the prostate, ducti deferentes and seminal vesicles and are involved in causing muscle contraction at the time of ejaculation (Learmonth 1931). Contraction of the pre-prostatic sphincter serves to prevent the retrograde flow of ejaculate through the proximal urethra into the bladder.

The Prostatic Part

The prostatic urethra is approximately 3–4 cm in length and tunnels through the substance of the prostate closer to the anterior than the posterior surface of the gland. It is continuous above with the pre-prostatic part and emerges from the prostate slightly anterior to its apex (its most inferior point). Throughout most of its length the posterior wall possesses a midline ridge, the *urethral crest* which projects into the lumen causing it to appear crescentic

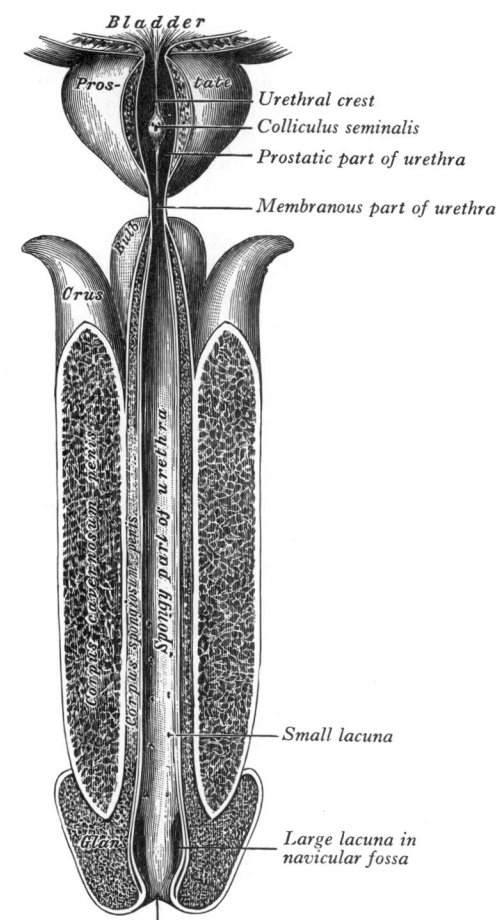

8.187 The whole length of the lumen of the male urethra exposed by an incision extending into it from its dorsal aspect. Note openings of prostatic utricle and ejaculatory ducts on the colliculus seminalis.

in transverse section. On each side of the crest there is a shallow depression, termed the *prostatic sinus*, the floor of which is perforated by the orifices of the prostatic ducts. About the middle of the length of the urethral crest the *colliculus seminalis* or *(verumontanum)* forms an elevation on which the slit-like orifice of the *prostatic utricle* is situated; on both sides of or just within this orifice there are the two small openings of the ejaculatory ducts. The prostatic utricle is a cul-de-sac about 6 mm long, which runs upwards and backwards in the substance of the prostate behind its medial lobe. Its walls are composed of fibrous tissue, muscular fibres and mucous membrane; the latter presents the openings of numerous small glands. Developed from the paramesonephric ducts or urogenital sinus, it is thought to be homologous with the vagina of the female (p. 260). The prostatic utricle is therefore called by some '*the vagina masculina*', but the more usual view is that it is a uterine homologue and hence the term 'utricle'. The ejaculatory ducts are described on p.1430. Distally the prostatic urethra possesses an outer layer of circularly disposed skeletal muscle cells which are continuous with a prominent collar of striated muscle (the external urethral sphincter) within the wall of the membranous urethra.

The Membranous Part

The membranous part is the shortest, least dilatable and, with the exception of the external orifice, the narrowest section of the urethra. It descends with a slight ventral concavity from the prostate to the bulb of the penis (**8.**181), passing through the perineal membrane about 2.5 cm postero-inferior to the pubic symphysis. The hind part of the bulb of the penis is closely apposed to the inferior aspect of the urogenital diaphragm (perineal membrane) but anteriorly it is slightly separated from the latter, so that the wall of the urethra is related anteriorly neither to the perineal membrane nor the penile bulb. This part of the anterior wall of the urethra is regarded as the 'membranous' part; anteriorly it is about 2 cm long, whilst posteriorly it is only 1.2 cm. The wall of the membranous urethra consists of a muscle coat which is separated from the epithelial lining by a narrow layer of fibro-elastic connective tissue. This muscle coat consists of a relatively thin layer of non-striated muscle bundles continuous proximally with those of the prostatic urethra and a prominent outer layer of circularly orientated skeletal muscle fibres forming the external urethral sphincter. The skeletal muscle fibres which comprise this external sphincter are unusually small in cross-section, with diameters of only 15–20 μm (Hayek 1969). The fibres are physiologically of the slow twitch type (Gosling et al 1981), unlike the pelvic floor musculature which is a heterogeneous mixture of slow and fast twitch fibres of larger diameter (Parks et al 1977). Moreover the external sphincter is devoid of muscle spindles and is supplied by the pelvic splanchnic nerves, further distinguishing it from the periurethral levator ani muscle (Donker et al 1976). The slow twitch fibres of the external sphincter are capable of sustained contraction over relatively long periods of time and actively contribute to the tone which closes the urethra and maintains urinary continence.

The Spongiose Part

The spongiose part is contained in the corpus spongiosum penis (p.1432). It is about 15 cm long and extends from the end of the membranous urethra to the external urethral orifice on the glans penis. Commencing below the perineal membrane, it continues the ventrally concave curve of the membranous urethra to a point anterior to the lowest level of the symphysis pubis. From here, when the penis is flaccid, the urethra curves downwards in the 'free' part of the penis. It is narrow, with a uniform diameter of about 6 mm in the penis; it is dilated at its commencement as the *intrabulbar fossa* and again within the glans penis, where it becomes the *navicular fossa*. The enlargement of the intrabulbar fossa affects the floor and side walls but not the roof of the urethra. The *bulbo-urethral glands* open into the spongiose section of the urethra about 2.5 cm below the perineal membrane (p. 1435).

The *external urethral orifice* is the narrowest part of the urethra: it is a sagittal slit, about 6 mm long, bounded on each side by a small labium.

The epithelium of the urethra, except in its most anterior part, presents the orifices of numerous small mucous glands and follicles situated in the submucous tissue and named the *urethral glands*. Besides these there is a number of small pit-like recesses, or *lacunae*, of varying sizes; the orifices of these are directed forwards and may intercept the point of a catheter in its passage along the canal. One, larger than the rest, the *lacuna magna*, is situated on the roof of the navicular fossa.

Mucous Membrane of the Male Urethra

The epithelium lining the preprostatic urethra and the proximal part of the prostatic urethra is of the urothelial type and is in continuity with that lining the bladder; it is also continuous with the ducts of the prostate and bulbo-urethral glands and with the linings of the seminal vesicles, deferent ducts and ejaculatory ducts, a relationship which is important in the spread of urinary tract infections. However, below the openings of the ejaculatory ducts this epithelium changes to a patchily pseudostratified or stratified columnar variety which lines the membranous urethra and the major part of the penile urethra (p. 56). Mucus-secreting cells are common throughout this epithelium and frequently occur in small clusters in the penile urethra. The mucous membrane of the penile urethra shows many recesses which continue into deeper branching tubular mucous glands (of Littré) which are especially numerous on the dorsal aspect. In older individuals many of the recesses of the urethral mucosa contain concretions similar to those found within the substances of the prostate. Towards the distal end of the penile urethra the epithelium changes once again, becoming stratified squamous in character with well-defined connective tissue papillae. This type of epithelium lines the navicular fossa and becomes keratinized at the external meatus.

Urethral Sphincters

Of the two urethral sphincters, the internal *sphincter vesicae* (p. 1418) controls the vesical neck and the prostatic urethra above the ejaculatory ducts. It is composed of non-striated muscle and supplied by sympathetic and parasympathetic fibres from the vesical plexus (vide supra and p. 1166). The external *sphincter urethrae* (p. 608) surrounds the membranous urethra; it consists of striated muscle and is supplied by the perineal branches of the pudendal nerve (S2,3 and 4); it is voluntary after early infancy.

The existence of an internal sphincter is, however, controversial. Many consider that its non-striated muscle fibres are multi-directional as in the vesical wall, no true circumferential fibres being identifiable (Woodburne 1961, Angell 1969). However, all agree that a substantial muscular aggregation, with an admixture of elastic and collagenous fibres, exists at the vesical outlet (Vincent 1966a). Its significance in micturition is noted on p. 1423.

THE FEMALE URETHRA

The female urethra is about 4 cm long and 6 mm in diameter. It begins at the internal urethral orifice of the bladder, approximately opposite the middle of the symphysis pubis, and runs antero-inferiorly behind the symphysis pubis, embedded in the anterior wall of the vagina. It traverses the perineal membrane and ends at the external urethral orifice, an anteroposterior slit with rather prominent margins, which is situated directly anterior to the opening of the vagina and about 2.5 cm behind the glans clitoridis. Except during the passage of urine the anterior and posterior wall of the urethra are in apposition and the epithelium is thrown into longitudinal folds, one of which, on the posterior wall of the canal, is termed the *urethral crest*. Many small mucous *urethral glands* and minute pit-like recesses or *lacunae* open into the urethra. On each side, near the lower end of the urethra, a number of these glands are grouped together and open into a duct, named the *para-urethral duct*; each duct runs down in the submucous tissue and ends in a small aperture on the lateral margin of the external urethral orifice.

Microscopic Structure of the Female Urethra

The wall of the female urethra comprises an outer muscle coat and an inner mucous membrane which lines the lumen and is continuous with that of the bladder. The muscle coat consists of an outer sleeve of striated muscle (*the external urethral sphincter*) together with an inner coat of smooth muscle fibres. The female external urethral sphincter is anatomically separate from the adjacent peri-urethral striated muscle of the anterior pelvic floor. The constituent fibres of this sphincter are circularly disposed and form a sleeve which is thickest in the middle one-third of the urethra. In this region skeletal muscle completely surrounds the urethra although the posterior portion lying between the urethra and vagina is relatively thin. The skeletal muscle extends into the anterior wall of both the proximal and distal thirds of the urethra but is deficient posteriorly in these regions. The myocytes forming the external urethral sphincter are all of the slow twitch variety. As in the male, muscle fibres of the external urethral sphincter are unusually small and have diameters of 15–20 μm on average. Although the thickness of the external urethral sphincter in the female is less than that of the male, its constituent fibres are able to exert tone upon the urethral lumen over prolonged periods, especially in relation to the middle third of its length. Peri-urethral skeletal muscle (pubococcygeus) aids urethral closure during events which require rapid, albeit short-lived, elevation of urethral resistance. The *smooth muscle coat* extends throughout the length of the urethra and consists of slender muscle bundles, the majority of which are orientated obliquely or longitudinally. A few circularly arranged muscle fibres occur in the outer aspect of the non-striated muscle layer and intermingle with the skeletal muscle fibres forming the inner part of the external urethral sphincter. Proximally the urethral smooth muscle extends as far as the bladder neck where it is replaced by fascicles of detrusor non-striated muscle. This region in the female is devoid of a well-defined circular non-striated muscle component comparable with the pre-prostatic sphincter of the male. When traced distally, urethral smooth muscle bundles terminate in the subcutaneous adipose tissue surrounding the external urethral meatus. The smooth muscle of the female urethra is associated with relatively few noradrenergic nerves but receives an extensive presumptive cholinergic parasympathetic nerve supply identical in appearance to that which supplies the detrusor (Ek et al 1977). From a functional viewpoint it seems unlikely that competence of the female bladder neck and proximal urethra is solely the result of non-striated muscle activity, in the absence of an anatomical sphincter. The innervation and longitudinal orientation of most of the muscle fibres suggest that urethral smooth muscle in the female is active during micturition, serving to shorten and widen the urethral lumen. For further details, see Gilpin & Gosling 1983. The *mucous membrane* lining the female urethra consists of a stratified epithelium and a supporting layer of loose fibro-elastic connective tissue (the lamina propria). The lamina propria contains an abundance of elastic fibres orientated both longitudinally and circularly around the urethra. Numerous thin-walled veins are another characteristic feature and in the past have been falsely likened to erectile tissue. A fine plexus of acetylcholinesterase-positive nerves is present throughout the lamina propria and these fibres are believed to be sensory branches of the pudendal nerves. The proximal part of the urethra is lined by urothelium, identical in appearance to that of the bladder neck. Distally the epithelium changes into a non-keratinizing stratified squamous type which lines the major portion of the female urethra. This epithelium is keratinized at the external urethral meatus and becomes continuous with the skin of the vestibule.

FUNCTIONAL ANATOMY OF THE LOWER URINARY TRACT

The urinary bladder performs a dual function, acting at times as a reservoir for fluid accumulating within its lumen and at others as a contractile organ actively expelling its contents into the urethra. In the following account the tissue components and, where appropriate, their neurological control will be considered under the headings *continence of urine* and *micturition*.

Continence of Urine

To achieve urinary continence, the bladder acts as a passive reservoir retaining fluid because the forces acting on the urethra produce an intra-urethral pressure greater than bladder pressure. Several tissue components play a part in generating this urethral resistance and make either an active or passive contribution. Since the non-striated muscle of the bladder is replaced in the bladder neck region by a different type of non-striated muscle, the detrusor does not play a part in closing the proximal urethra.

In the male, a distinct collar of circularly orientated non-striated muscle occurs in the bladder neck and pre-prostatic urethra, continuous distally with the muscular components of the genital tract. This smooth muscle sphincter is supplied by a rich plexus of sympathetic nerve fibres which, on stimulation, cause the sphincter to contact, thereby preventing retrograde flow of semen into the urinary bladder at ejaculation. During seminal emission, the sympathetic nervous system also prevents coincidental contraction of detrusor smooth muscle. This inhibitory effect on bladder contractility is mediated by noradrenergic nerves which synapse in the vesical plexus upon parasympathetic motor neurons. Despite this well defined genital role, it is not known whether the non-striated muscle of the bladder neck region and pre-prostatic urethra plays an active part in the maintenance of continence. Intramural collagen and elastic fibres within the wall of the bladder neck, proximal urethra and prostate generate passive forces which help to close the urethral lumen. However, postoperative incontinence of urine does not usually follow radical surgical excision of the bladder neck, pre-prostatic urethra and prostate, suggesting that these regions make only a minor contribution to urinary continence.

In the female, a non-striated muscle sphincter cannot be anatomically recognized in the wall of the bladder neck and proximal urethra. Consequently it is even less likely that active smooth muscle contraction can be considered as an important factor in the continence of urine. However, the bladder neck and

proximal urethra possess within their walls innumerable elastic fibres which are of particular importance in producing passive occlusion of the urethral lumen (Lapides 1958). Indeed, it has been suggested that the passive elastic resistance offered by the urethral wall is the most important single factor responsible for the closure of the bladder neck and proximal urethra in the continent woman.

In both sexes the urethra contains within its walls the external urethral sphincter, the location of which corresponds anatomically to the zone where maximal urethral closure pressures are normally recorded. This skeletal muscle sphincter is morphologically adapted to maintain tone over relatively long periods without fatigue and plays an important active role in producing urethral occlusion at rest. It remains to be determined, however, whether the force exerted by the sphincter is maximal at all times between two consecutive acts of micturition or whether additional motor units are recruited during coughing, sneezing, etc. to enhance the occlusive force on the urethra during these events. The external sphincter is innervated by nerve fibres which travel via the pelvic splanchnic nerves (Gil Vernet 1968) and *not* the pudendal nerves as often described. The clinical relevance of this arrangement is that pudendal blockade or neurectomy performed in order to reduce urethral resistance will not achieve the desired effect since the motor innervation of the striated sphincter remains unaffected by these procedures.

Concerning the role of peri-urethral muscle in the maintenance of continence, the medial parts of the levator ani muscles in both sexes are related to (but structurally separate from) the urethral wall. These peri-urethral fibres are innervated by the pudendal nerve and consist of an admixture of large diameter fast and slow twitch fibres. Therefore, unlike the external sphincter, peri-urethral muscle possesses morphological features which are similar to other 'typical' voluntary muscles. This pelvic floor musculature plays an important part (especially in the female) by providing an additional occlusive force on the urethral wall, particularly during events which are associated with an increase in intra-abdominal pressure. In addition, the muscles provide support for the pelvic viscera.

Micturition

To enable fluid to flow along the urethra it is necessary for the pressure in the urinary bladder to exceed that within the urethral lumen. Under normal circumstances, in order to initiate micturition, a fall in urethral resistance immediately precedes a rise in pressure within the lumen of the bladder. This pressure rise is usually produced by active contraction of detrusor smooth muscle at the onset of micturition. The detrusor muscle coat consists of numerous interlacing bundles forming a complex meshwork of smooth muscle which, on contraction, reduces all dimensions of the bladder. The muscle coat is collectively involved and it is unnecessary to attach special significance to the precise orientation of individual bundles within the wall of the viscus.

The preganglionic nerve supply travels in the pelvic splanchnic nerves before synapsing on neurons located within the vesical part of the pelvic plexuses and within the wall of the bladder. These peripheral neurons supply nerve fibres which ramify throughout the thickness of the detrusor smooth muscle coat. The profuse distribution of these motor nerves emphasizes the importance of the autonomic nervous system in initiating and sustaining bladder contracting during micturition. For micturition to occur the pressure differential between the bladder and urethra must overcome the elastic resistance of the bladder neck. Immediately prior to the onset of micturition, the tonus of the external sphincter is reduced by central inhibition of its motor neurons located in the second, third and fourth sacral spinal segments. This inhibition is mediated by descending spinal pathways originating in higher centres of the central nervous system. Concomitantly, other descending pathways activate (either directly or via sacral interneurons) the preganglionic parasympathetic motor outflow to the urinary bladder. This central integration of the nervous control of the bladder and urethra is essential for normal micturition.

Applied Anatomy. After urethral rupture, the extravasation of urine may complicate micturition; urine usually extends between the perineal membrane and the membranous layer of the superficial fascia. As both of these are attached firmly to the ischiopubic rami, extravasated fluid cannot pass posteriorly because the two layers are continuous around the superficial transverse perineal muscles. Laterally, the spread of urine is blocked by the pubic and ischial rami; it cannot enter the lesser pelvis through the perineal membrane and, if this remains intact, fluid can make its way only anteriorly into the scrotal and penile

loose connective tissue and thence to the anterior abdominal wall. When the lesser pelvis is crushed the urethra may be ruptured between the prostatic and membranous parts; extravasation of urine then occurs into the pelvic extraperitoneal tissue.

The lower urinary tract is subject to many congenital anomalies (p. 260), some amenable to surgical correction. For an anatomical survey, consult Paul & Kanagasuntheram (1956).

REPRODUCTIVE ORGANS OF THE MALE

The male genital organs include the *testes, epididymides, deferent* and *ejaculatory ducts* and *penis*, with the accessory glandular structures: *seminal vesicles, prostate* and *bulbo-urethral glands*.

The Testes and Epididymes

These, the *primary male reproductive organs* or *gonads*, are suspended in the scrotum by scrotal tissues including the non-striated dartos muscles and the spermatic cords, the left testis usually being about 1 cm lower than the right. Average testicular dimensions are 4–5 cm in length, 2.5 cm in breadth and 3 cm in antero-posterior diameter; their weight varies from 10.5–14 g. Each testis, ellipsoidal (8.188) and compressed laterally, is obliquely set in the scrotum, its upper pole tilted anterolaterally and the lower posteromedially. The anterior aspect is convex, the posterior nearly straight, with the spermatic cord attached to it. Anterior, medial and lateral surfaces and both poles are convex, smooth and covered by the visceral layer of the serosal tunica vaginalis, which separates them from the parietal layer and the scrotal tissues external to this. The posterior aspect is only partly covered by tunica serosa; the epididymis adjoins its lateral part.

The epididymis, a tortuous canal and the first part of the efferent route from the testis, is much folded and tightly packed to form a long, narrow mass attached posterolaterally to the testis. It has a central *body*, a superior enlarged *head* and an inferior pointed *tail*. The head is connected to the upper testicular pole by *efferent ductules* and the tail to the lower pole by areolar tissue and the reflected tunica vaginalis. Laterally the head and tail are covered by the tunica and are here 'free'; the body is also so invested, except on its posterior aspect. The tunica vaginalis is recessed between the epididymal body and the lateral surface of the testis, forming the *sinus of the epididymis*.

Testicular and epididymal appendices. At the upper pole of the testis, just inferior to the epididymal head, is a minute, oval,

sessile *appendix of the testis*, a remnant of the upper end of the paramesonephric duct. On the epididymal head is a small, stalked appendage (sometimes double), the *appendix of the epididymis*, usually considered a mesonephric vestige (p. 250, 2.116).

The testis is invested by three coats, from outside inwards: the tunica vaginalis, tunica albuginea and tunica vasculosa.

The tunica vaginalis (8.188, 197) is the lower end of the peritoneal processus vaginalis, which precedes the descent of the fetal testis from abdomen to scrotum (p. 257); after this migration, the tunica's proximal part, from the internal inguinal ring almost to the testis, contracts and obliterates, leaving a closed distal sac invaginated by the testis and reflected to the internal scrotal surface, thus forming the visceral and parietal layers of the tunica.

The *visceral layer* covers all aspects of the testis except the posterior border. Posteromedially it is reflected forwards as the parietal layer; posterolaterally it passes to the medial aspect of the epididymis, lining its sinus, and then laterally to its posterior border where it is reflected forwards into the parietal layer. The visceral and parietal layers are also continuous at both poles but at the upper the visceral layer surmounts the head of the epididymis before reflexion.

The *parietal layer*, more extensive than the visceral, reaches below the testis and ascends in front of and medial to the spermatic cord. Internally the tunica vaginalis has a smooth, moist mesothelium, the potential space between its visceral and parietal layers being its cavity.

In the embryo, the male and femal gonads project into the coelom, covered by so-called 'germinal' epithelium; even in the adult the mesothelium covering the testis (and ovary) has been considered atypical peritoneum, a remnant of the original 'germinal' epithelium. The parietal tunica vaginalis would then be considered continuous with the germinal epithelium at the posterior aspect of the testis. Structurally, however, in the testis the parietal and visceral epithelia are similar and like that of the general peritoneum, although the epithelium covering the ovary is a specialized, cuboidal layer. Doubtless the term germinal epithelium reflects the misconception that generations of gonocytes develop from the specialized mesothelium of the genital ridge.

The obliterated part of the processus vaginalis is often seen as a fibrous thread in the anterior part of the spermatic cord, extending from the internal end of inguinal canal and connected here to peritoneum, the tunica vaginalis. Sometimes it disappears in the cord or its proximal part remains patent, the peritoneal cavity then communicating with the tunica, or the proximal processus may persist, although shut off distally from the tunica (pp. 257, 1377). Occasionally its cavity may persist at an intermediate level as a cyst.

The tunica albuginea is a dense, bluish-white covering for the testis, composed mainly of interlacing bundles of collagen fibres, covered externally by the visceral layer of the tunica vaginalis, except at the epididymal head and tail and the posterior testicular aspect, where vessels and nerves enter the testis. It covers the tunica vasculosa and, at the posterior border of the testis, projects into it as a thick, vertical but incomplete septum, the **mediastinum testis** (8.189); this extends from the upper almost to the lower end of the testis and is wider above than below. From its front and sides numerous incomplete *septula testis* radiate towards the surface of the testis, attached there to the deep aspect of the tunica albuginea; these divide the organ

8.188 The right testis, exposed by incising and laying open the cremasteric fascia and parietal layer of the tunica vaginalis on the lateral aspect of the testis.

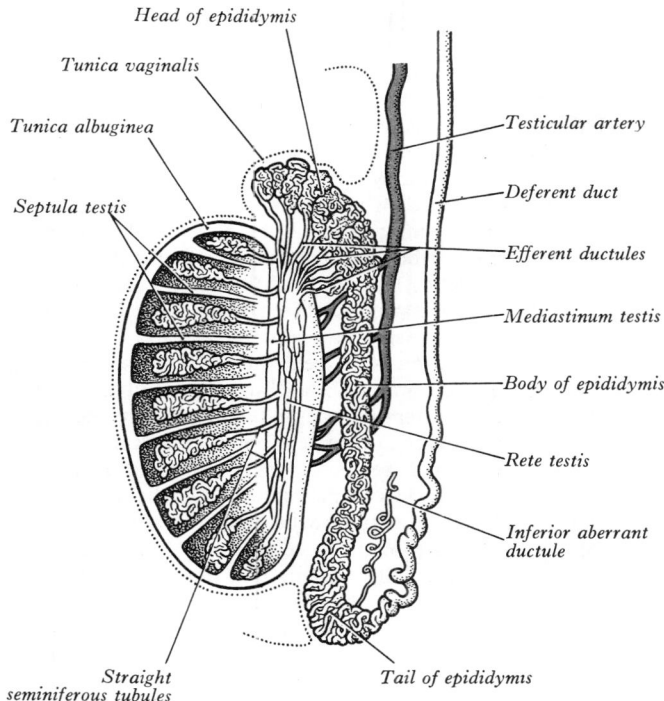

Head of epididymis

Tunica vaginalis

Tunica albuginea

Septula testis

Testicular artery

Deferent duct

Efferent ductules

Mediastinum testis

Body of epididymis

Rete testis

Inferior aberrant
ductule

Straight
seminiferous tubules

Tail of epididymis

8.189 Vertical section through the testis and epididymis, showing, diagrammatically, the arrangement of the ducts of the testis and the mode of formation of the deferent duct.

incompletely into cone-shaped lobules, their bases at the surface and their apices converging upon the mediastinum. For a detailed comparative account of the mediastinum consult Dhingra (1977). Its connective tissue is very variable; in some species, including man (Holstein 1967), non-striated myocytes occur amid the collagen fibres (p. 1429). It is traversed by many vessels and by

efferent tubules of the rete testis. Leydig cells are absent from the human mediastinum testis but are present in some ungulates.

The tunica vasculosa contains a plexus of blood vessels and delicate loose connective tissue, extending over the internal aspect of the tunica albuginea and covering the septa and therefore all lobules.

STRUCTURE OF THE TESTIS

At the surface is a layer of flat mesothelial cells resembling those of the surrounding peritoneum; some consider them to be remnants of a 'germinal' epithelium (see, however, p. 255). Internally the testicular architecture is dominated by the lobules (**8.**189), their number in a human testis being 200–300. They differ in size, those centrally placed being larger and longer. Each contains one to three or more minute *convoluted seminiferous tubules*. When unravelled these are seen to begin blindly or by anastomotic loops. Clermont & Huckins (1961) and Roosen-Runge (1961) have described them (in rats) as closed loops, opening at both ends into the tubuli recti and thus into the rete testis, and this pattern is now generally accepted. Their loose supporting connective tissue contains grouped *interstitial cells* (**8.**190) of yellow pigment granules. The tubules in each testis total 400–600 and the length of each is 70–80 cm. Their diameter varies from 0.12–0.3 mm. They are pale in early life, but in old age they contain much fat and are deep yellow. Each (**8.**190) has a basement membrane of laminated connective tissue containing numerous elastin fibres, with flat cells between the layers, and covered externally by flat epithelioid cells. Nicander (1967) has described intercellular contacts between these cells. As it ages the basement membrane becomes thicker and denser. Internal to it the seminiferous epithelium consists of *spermatogenic* and *supportive* cells. The former (**8.**191, 194), when active, include an array of types from spermatogonia through their derived forms, spermatocytes and spermatids, to mature spermatozoa. Among the spermatids may be *residual bodies* (pp. 120, 122), spherical structures containing membranous and mitochondrial residues and numerous free ribosomes, derived from spermatids from which they have separated. Their role in

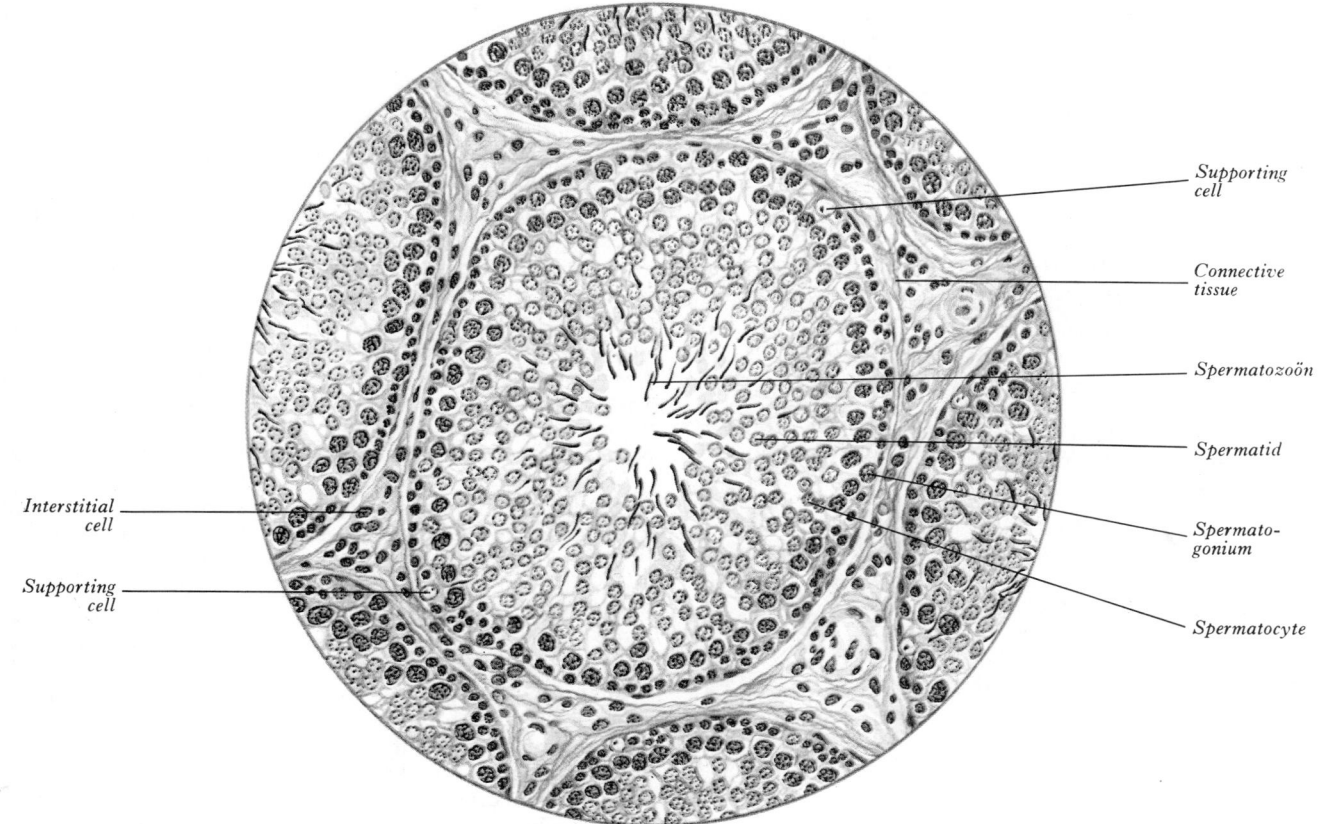

Supporting
cell

Connective
tissue

Spermatozoön

Spermatid

Spermato-
gonium

Spermatocyte

Interstitial
cell

Supporting
cell

8.190 Transverse section through a part of a human testis. Stained with iron haematoxylin and Van Gieson's stain. Magnification × 350.

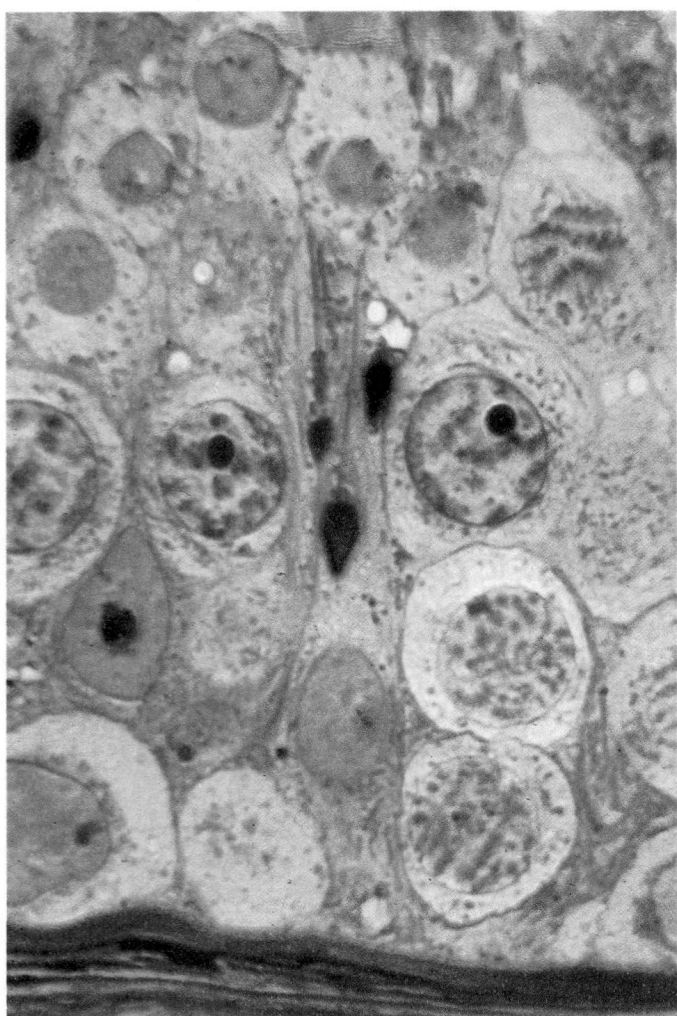

8.191 The epithelium of a seminiferous tubule in an adult man. A thin section of glutaraldehyde-fixed material, stained with toluidine blue. Spermatogonium (bottom left), primary spermatocytes (bottom right), and late spermatids (central) are visible. The spermatids are attached to a supportive cell of Sertoli (below them). For further details compare with **8.194**. (From Holstein & Wulfhekel 1971 with permission of Grosse Verlag, Berlin.)

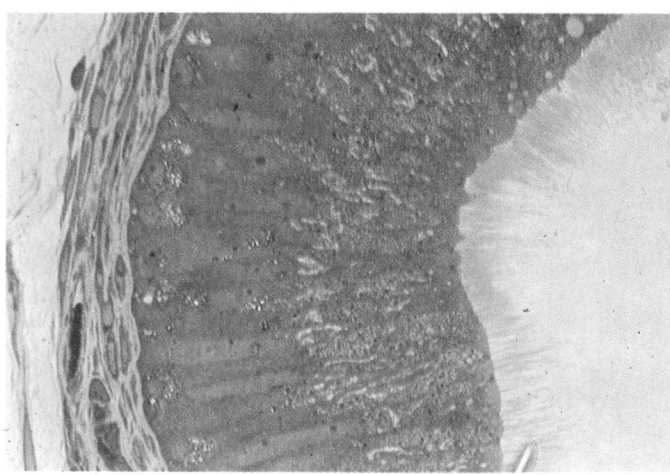

8.192 Section through the corpus of the human adult epididymis, showing deep columnar ciliated epithelium, small contractile cells, arranged circumferentially, immediately external to the epithelium, and larger non-striated muscle cells more externally. A thin glutaraldehyde-fixed section stained with toluidine blue. Supplied by A F Holstein, University of Hamburg. (For details see Baumgarten et al 1971.)

spermatogenesis is not yet clarified (Vaughn 1966) but may be regulatory; they are autolysed and perhaps phagocytosed by sustentacular cells, as the mature spermatozoa separate.

Spermatogonia, the stem cells for all spermatozoa, are descended from primordial germ cells which reach and multiply in the genital cords of developing testis, becoming *gonocytes*. In fully differentiated testis they appear along the basal laminae of the seminiferous tubules. Changes in the structure and appearance of human spermatogonia from birth to the onset of puberty have been described by Paniagua & Nistal (1984). Their cytological characteristics, like those of the generations of spermatocytes and spermatids produced from them, have been noted (p. 122). Several types of spermatogonia were early recognized (Regaud 1901) on the basis of cell and nuclear dimensions, distribution of nuclear chromatin and histochemical and ultrastructural data, as is generally accepted, though disagreements over minor details persist (Courot et al 1970). Similarly, various stages in the maturation of spermatocytes and spermatids (p. 122) have been recognized and a complex lineage of cells, from spermatogonia to spermatozoa, can now be defined and identified in the seminiferous tubules of most vertebrates. Mammals have been most studied, especially rodents and domestic cattle, the latter under the impetus of artificial insemination.

Three basic groups, *dark type A, pale type A,* and *type B* have been noted (p. 122); the first divides to maintain the basic store of spermatogonia, producing also some *pale* type A cells which divide and differentiate into type B, the immediate precursors of spermatocytes. More complex classifications are current; e.g. a *pro-spermatogonial stage* (A_0) is recognized, and a series, designated A_1–A_4, is utilized in describing the intermediate types in active seminiferous epithelium. Similarly, several grades of B type are recognized. Much of the complexity of current nomenclature is merely due to the successive divisions of spermatogonia, which can be investigated by labelling with radioactive thymidine. Extensive studies have been carried out in primates (Clermont 1963, 1969) and the findings are probably applicable to man. The dark A spermatogonia (A_1) form a reserve of resting cells, whereas the pale type (A_2) form further A_2 cells, one of which repeats this divisional step while the other divides into two type B spermatogonia. Since the latter produces, by serial division, several generations, these are designated B_1, B_2, etc. In monkeys (*Cercopithecus aethiops*) five generations of B spermatogonia occur, the final product being *primary spermatocytes*. There are four B spermatogonial generations in man and, since the cells (ignoring degeneration) are doubled at each stage, very large number of spermatozoa are formed during the active reproductive life of male vertebrates. The number of spermatogonial generations varies, being high in fish (six to 14), low in birds (one to three) and from four to seven in the mammals so far investigated (p. 122, Holstein 1976).

Primary spermatocytes with a diploid chromosome content divide into **secondary spermatocytes** with a haploid complement, the nuclear division designated as meiosis I. Secondary spermatocytes undergo the second meiotic division to form spermatids. Both types of spermatocytes and spermatids pass through a series of changes recognized chiefly by their nuclear activities; the latter also show profound modifications in cell structure to become the mature flagellated spermatozoa. The events of meiosis are described on p. 42.

Spermatids do not divide again but gradually mature into spermatozoa by a series of nuclear and cytoplasmic modifications (p. 122); several stages in this process have been described, according to their nuclear shape, including 'round' and 'elongating' spermatids. The range of cell types in the seminiferous epithelium is thus quite complex and to this must be added the supportive element.

Sustentacular, supporting cells (of Sertoli) **or sustentocytes,** the non-germinal element in the complex cell population of the seminiferous tubules, are variable in overall cell shape and nuclear configuration. Changes in the form and distribution of the human Sertoli cells from birth to the onset of puberty have been described by Nistal et al (1982). They observed a progressive diminution in their numbers from three years onwards, as seen in transverse sections of the tubules; but they ascribe this to testicular growth and consider the total number constant. They have been the subject of much controversy; in developing seminiferous tubules various supporting cells beside the 'Sertolian' category are also described; but it is generally assumed that most non-germinal cells in the adult functioning tubules are of this limited type. Sustentacular cells, sited on the tubular limiting membrane, occupy it almost to the exclusion of all but occasional spermatogonia. In tangential section they are polygonal, in transverse irregular, but approximately columnar. Their apices are complicated by recesses into which spermatids and spermatozoa fit until the latter are mature enough for release. Long cytoplasmic processes also extend among the spermatogonia and spermatocytes, suggesting that sustentacular cells maintain structural cohesion in the epithelium. Electron microscopy (Fawcett 1975) has disproved the view that they are syncytial, revealing that they are in fact held together by specialized contacts similar to 'tight' junctions (p. 24) but of at least three types (Nicander 1967, Flickinger & Fawcett 1967). Sertoli demonstrated the individuality of these cells in 1865. The nucleus is irregular, often indented, weakly positive to Feulgen staining, and contains one or two nucleoli which may be tripartite, consisting of a nucleolonema (p. 40) and two associated electron-dense bodies (p. 40). Cytoplasmic organelles are numerous, particularly mitochondria, endoplasmic reticulum, secretory granules, Golgi complexes, ribosomes, microfilaments and microtubules, the last named perhaps helping to form a cytoskeleton and being also a factor in the cohesive effect of sustentacular cells on the total epithelium. These details suggest that the cells exercise a metabolic influence in relation to the germinal elements. These cells are also considered to be phagocytic (Carr et al 1968) and perhaps exert an endocrine influence too. They change considerably during the spermatogenic cycle (vide infra) and are influenced by the hypophysial hormones, LH and FSH. They are also concerned in the 'blood–testis barrier'. For a general review consult Johnson & Gomes (1977).

THE SPERMATOGENIC CYCLE

The elements of the seminiferous epithelium noted above form an intimately related and dynamic population, details of which have been extensively investigated in several laboratory animals (particularly rats), in species of commercial or 'domestic' importance and in man. In addition to much detail on the cytological progression from spermatogonia to spermatozoa, quantitative aspects of this cycle have been assessed, including the timing of its phases, relative numbers of different germinal cells at definable temporal points and so on. It is clearly established that, at any locus in a seminiferous tubule, activities in the several generations of germinal cells and hence in the qualitative and quantitative associations of cell types pass through a *cycle*, the length and stages of which vary in different species (Clermont 1972). The cycle occupies 12–13 days in rats, only 8.6 in mice, 10.5–11.6 days in monkeys (Arsenieva et al 1961) and about 16 in mankind (Heller & Clermont 1964). Stages have been classified in two ways, one involving the development of the spermatid acrosomal

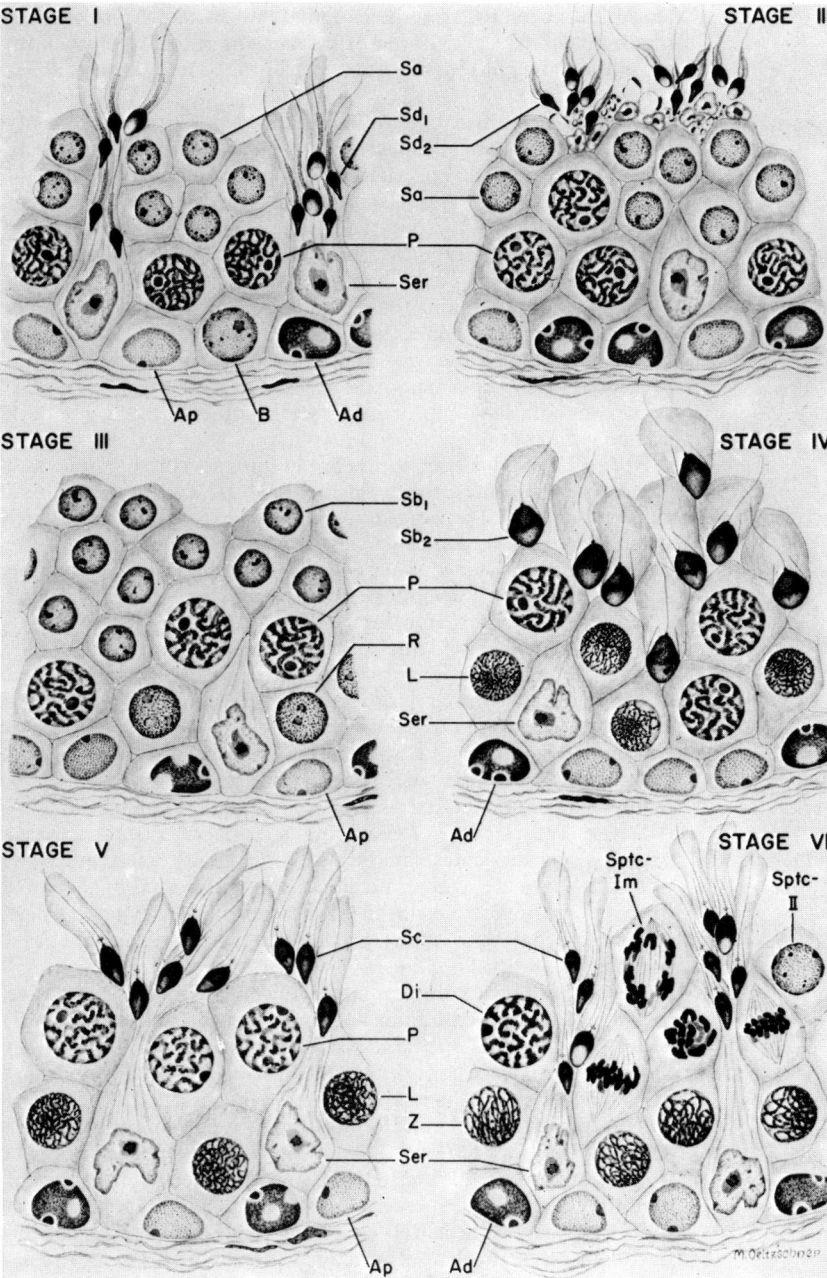

8.194 The cycle of spermatogenesis in the human seminiferous tubule. For details of the six stages see accompanying text. Ser = sustentacular cells of Sertoli. Ad, Ap and B = A type (dark and pale) and B type of spermatogonia. R = resting primary spermatocyte. L, Z, P and Di = primary spermatocytes in leptotene, zygotene, pachytene and diplotene stages. Sptc-Im = dividing primary spermatocyte. Sptc-II = secondary spermatocyte in interphase. Sa, b, etc. = generations of spermatids. (From Clermont 1963 with permission of the author and the Wistar Institute of Anatomy and Biology.)

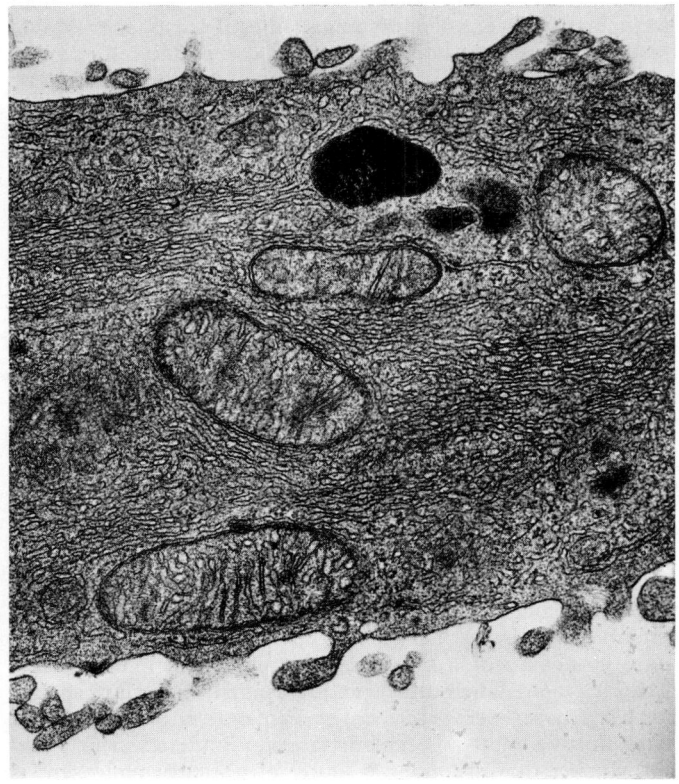

8.193 Transmission electron micrograph of part of an interstitial cell of the testis (Leydig cell) showing the profuse agranular endoplasmic reticulum and prominent mitochondria (murine). Magnification × 20 000.

system (Leblond & Clermont 1952) and the other based on changes in cell nuclei (Roosen-Runge & Giesel 1950). Both methods were elaborated in rats, but have been applied to other mammals, birds and men. The two methods respectively indicate 14 and eight stages in rats. The *acrosomalic technique* identifies only six human stages, which will be detailed here. In many mammals the whole circumference of a short length of seminiferous tubule is in the same stage of the cycle, with adjoining, cylindrical regions displaying preceding or succeeding stages. All stages can thus be recognized in numerical succession along the tubules, the last being followed by a cylinder undergoing activity of the first stage. Each region of the tubule therefore passes through all stages and begins again. Hence there exists not only a *local* succession of stages, spread over the period of the cycle characteristic of species involved, but also a procession of stages undulating along the tubule. Such a *'spermatogenic wave'* has been measured in some species (e.g. rats and bulls, see Mochereau 1963) but the 'wavelength' has been found to be irregular and variable. Not even all quadrants of a local cylinder of human seminiferous epithelium are in the same phase at any moment. These different stages are as follows (**8.194**).

Stage I is typified by two spermatid generations: one newly formed with spherical nuclei and an older one with elongating nuclei and deeply invaginated in groups into the cytoplasm of sustentacular cells. Associated with these are primary spermatocytes starting the long pachytene stage of the first meiotic prophase division; type A (dark and pale) and type B spermatogonia adjoin the basal lamina.

Stage II also contains these two phases of spermatid maturation but they are further advanced; older spermatids are near their final change to spermatozoa and hence there are residual bodies present. Primary spermatocytes are still in pachytene but type B spermatogonia now appear in greater numbers among the dark and pale A types.

Stage III. Older spermatids are ending spermatogenesis and escaping as spermatozoa (spermiation). Nearest to them is a generation of well-advanced spermatids, external to which are two generations of primary spermatocytes, the older still in the pachytene stage, the younger cells in the proleptotene or resting stage of meiosis. Type A spermatogonia predominate, type B having become spermatocytes.

Stage IV. Spermatid nuclei are elongating, but are irregular in shape. Older primary spermatocytes are in the pachytene stage, the younger ones moving into leptotene. Type A spermatogonia are gathered external to the above cells.

Stage V. Groups of spermatids display nuclei elongated centrifugally towards the tubular periphery. Primary spermatocytes are now in *late* pachytene and some in leptotene.

Stage VI. Groups of maturing spermatids occur between primary spermatocytes dividing into secondary spermatocytes, of which some are in the interphase stage. External to them are primary spermatocytes in the zygotene stage. Type A spermatogonia (pale and dark) predominate again near the basal lamina.

Details of these complex stages accord with the successive divisions and differentiations known to occur in the germinal cell series. The process appears less regular, particularly in human biopsy material, perhaps due to mechanical disturbance; an unknown number of cells retrogress and degenerate.

The epithelium of seminiferous tubules is perhaps unique in its organization (Fawcett 1975), consisting of a permanent population of non-proliferative sustentacular cells seated on its basement membrane and interspersed proliferating spermatogonia and their derivatives; the latter are related to the basal parts of the Sertolian cells, the lateral junctions of which (**8.191**) are considered to create a *'basal compartment'* for preleptotene spermatocytes. These junctions must break down to allow spermatocytes to move towards the tubular lumen into a so-called *'adluminal compartment'*. In primates the 'blood–testis barrier' may reside in these junctions, perhaps under hormonal control; spermatogonia and early spermatocytes may be accessible to all substances crossing capillary walls, the later spermatocytes and spermatids being screened from substances incapable of passing the 'barrier'. As spermatids mature they move towards the lumen

from the lateral aspects of sustentacular cells, at first embedded largely in their cytoplasm, and attached by a junctional specialization which changes (Ross 1976) until the spermatozoa are released into the lumen of the seminiferous tubule.

THE EFFERENT DUCTULES AND EPIDIDYMIS

The events described above occur in the highly coiled parts of the seminiferous tubules. As these reach the lobular apices they are less convoluted, assume an almost straight course and unite into 20–30 larger straight ducts, about 0.5 mm in diameter (**8.189**). Straight seminiferous tubules enter the fibrous tissue of the mediastinum testis, ascending backwards as a close network (*rete testis*) of anastomosing tubes lined by a flat epithelium. At the upper pole of the mediastinum, 12–20 *efferent ductules* perforate the tunica albuginea to pass from the testis to the epididymis. They are at first straight, becoming enlarged and very convoluted and forming conical *lobules of the epididymis*, which make up its head (caput). Each lobule is a convoluted duct, 15–20 cm in length. Opposite the lobular bases the ducts open into a single *duct of the epididymis*, whose coils form the epididymal body (corpus) and tail (cauda). With the coils unravelled the tube measures more than 6 metres, increasing in thickness as it approaches the epididymal tail, where it becomes the deferent duct. The coils are held together by bands of fibrous connective tissue. The epididymal body and tail are thus a single tube.

Efferent ductules are lined by ciliated columnar epithelium, with a thin circular layer of non-striated myocytes in their walls. In the epididymal duct the muscle is thicker and the epithelium composed of columnar pseudostratified cells, the superficial cells having long (15 μm) regular microvilli about 0.2 μm in diameter. Called *'stereocilia'* from a superficial resemblance to cilia, they lack the ultrastructure of true cilia. Tracer experiments have demonstrated the uptake of proteins by epithelial cells of the efferent ductules by phagocytic vesicle formation; presumably they modify the composition of seminal fluid. Electron microscopy shows inclusion bodies in the nuclei of human epididymal epithelial cells (Horstmann 1962, 1966). In man fragments of ingested spermatozoa appear in small round cells, which have cytoplasmic processes projecting into the epididymal duct (Holstein 1967). In several species the presence of lysosomal enzymes indicates an absorptive function in the epididymal epithelium (Moniem & Glover 1972). Baumgarten et al (1970) suggest that three ultrastructurally distinct types of contractile cell occur in both ductules and epididymis. These receive a sparse adrenergic innervation; few nerve fibres penetrate between the myocytes, unlike the more profuse innervation of non-striated muscle in the caput epididymis and ductus deferens. These differences accord with the slow, spontaneous, local contractions of the ductules and proximal epididymis and the rapid, reflex contractions of the caput and ductus during seminal emission.

THE TESTICULAR INTERSTITIAL TISSUE

Interstitial cells of the testis, usually equated with *cells of Leydig* (1850), also include others, some of them cells of the connective tissue lying among vessels and nerves between the seminiferous tubules. The interstitial cells of Leydig are probably mesenchymal in origin but may arise from mesonephric blastema (Witschi 1951). Such cells (**8.193**), isolated or clustered, occur in the intertubular tissue of most vertebrates including man, being large and polyhedral, with an eccentric nucleus containing: one to three nucleoli, a scanty, poorly staining cytoplasm (Hooker 1971), much agranular endoplasmic reticulum (rich in ascorbic acid), vacuoles containing fats, phospholipids and cholesterol, thus resembling the interstitial cells of the ovary (p. 1436), luteal cells of the corpus luteum (p. 1437) and secretory cells in the adrenal cortex (p. 1469). Their ultrastructural features in the human ageing testis have been described by Paniagua et al 1986. Their masculinizing effect was demonstrated by Bouin & Ancel (1903) but whether they are the sole source of testicular androgens is uncertain. Most experimental evidence demonstrates a lack of endocrine function in the tubules, rather than proving that interstitial cells are a source of androgens. However, that they do secrete androgens is accepted; no other function can be attributed to them (Christensen 1975). They are stimulated by interstitial-cell-stimulating hormone (ICSH—identical with luteinizing hormone, LH) and possibly follicle-stimulating hormone (FSH) of the anterior lobe of the hypophysis cerebri (p. 1451). In cryptorchidism, where testes are retained in the inguinal canal and hence in a warmer environment, the rate of production of androgens and spermatozoa is depressed, without visible changes in the interstitial cells. Androgens stimulate the growth and activity of male accessory reproductive glands (prostate, seminal vesicles and bulbo-urethral glands) and also secondary sexual changes at puberty (growth of facial, axillary and pubic hair, enlargement of the larynx and paranasal air sinuses and greater skeletal growth), as well as a plethora of other actions on the general metabolism of the body, nervous system, etc. Most of these changes are inhibited or altered by oestrogens (in sufficient amounts), especially when androgenic output is depressed, as in some eunuchoid conditions (p. 1452.)

The proportion of interstitial cells of Leydig to other intertubular components varies with age and in some diseases. After initial multiplication in early fetal life, they appear to atrophy at birth; but whether they disappear or are merely depressed in number in the immediate postnatal period is uncertain (Cooper 1929, Sohval 1954). It is accepted that they reappear before puberty, perhaps developing from mesenchyme (Hisakayu & Harrison 1971). They are not reduced in the elderly (55–65 years) according to Kothari & Gupta (1974). In cryptorchid testes they are absent but total intertubular tissue is increased. For techniques of quantitative assessment in human and other mammalian testes see Dykes (1969), Heller et al (1971), Christensen (1975), and Kothari et al (1978).

THE TESTICULAR CAPSULE

Testicular coverings, already noted (p. 1424), may be summarized, proceeding internally from the surface, as composed of: skin, dartos muscle, superficial perineal fascia, external and internal spermatic fasciae with cremasteric fascia interposed, parietal and visceral layers of the tunica vaginalis (with an intervening capillary interval), tunica albuginea and tunica vasculosa. The last three form the so-called 'testicular capsule', a concept perhaps justified by the demonstration of a contractile element in this otherwise inert, largely fibrous tissue structure (Davis & Langford 1969). In rabbits and rats isolated preparations of testicular capsule may show spontaneous contractions, reacting also to both cholinergic and adrenergic agents. Abundant autonomic nerve endings near blood vessels have been described (Norbert et al 1967); non-striated myocytes have been demonstrated in the tunica in rodents and man (Holstein 1967); the distribution of these myocytes varies and, while widely scattered in the human tunica albuginea, they are concentrated posteriorly near the epididymis. This contractile capsule probably massages or pumps the ducts in the testis, impelling spermatozoa onwards. For a review consult Davis et al (1971).

TEMPERATURE REGULATION IN THE TESTIS

Testes are external or *scrotal* in most marsupials and eutherians, but *abdominal* in fish, amphibians, reptiles, birds and monotremes. There are exceptions in some mammals: the testes are abdominal in sloths, elephants and hyraces. In aquatic mammals they are either abdominal (whales and dolphins) or in the inguinal region (seals). They may be scrotal during breeding seasons but inguinal or abdominal at other times (rodents, bats and insectivores). Such variations in mammals are puzzling but it is generally accepted that the predominantly scrotal suspension is associated with homeothermic specialization; this does not, however, explain the abdominal position of the avian testes. Regulation of internal temperature (Hammel 1968) appears to create too warm an environment for spermatogenesis. Experimental evidence, reviewed by Bishop & Walton (1960), confirms this. Scrotal temperature is usually several degrees below abdominal (three degrees in man); but the optimal range for spermatogenesis, though lower, must be maintained. Mechanisms for

this are numerous (Waites 1970) and testicular thermoregulation can only be summarized here. The scrotal skin is usually hairy but this heat-conserving cover varies; it is scant in man, profuse in marsupials. Sweat glands are well developed and numerous, while subcutaneous fat is generally absent. The radiant area of scrotal skin, well vascularized, is much reduced in cold conditions by contraction of the dartos muscle, probably by direct stimulation or perhaps by local reflexes. Primate scrotal skin has many nerve endings responsive to chilling or warming, the latter type being more numerous (Iggo 1969). The cremaster muscles, by approximating the testes to the perineum, may conserve heat. *Countercurrent heat exchange* is said to exist between the arteries and veins of the spermatic cord, where both are coiled and in close apposition (Harrison & Weiner 1949). In dogs blood delivered to the testis is pre-cooled by 3°C by this mechanism (Waites & Moule 1961). Respiratory gases and other substances (including, experimentally, labelled testosterone and prostaglandin-F) are also transferred between these arteries and veins in several species, including man (Free 1977). Thus a surprising array of factors are concerned in testicular thermoregulation and maintenance of fertility; while these appear jointly involved in this function, it is suggested that the scrotum may also act in a reciprocal control of *general* body temperature.

Vessels and Nerves

The *testicular artery and veins* are described on pp. 774 and 818. Intratesticular arteries have been much studied in man (Hundeiker & Keller 1963). Kormano & Suoranta (1971) noted that they are markedly coiled. Capillaries adjoining seminiferous tubules penetrate the layers of interstitial tissue and are of interest as part of the 'blood–testis barrier' (Setchell & Waites 1975, Neaves 1977). They run either parallel to the tubules or across them but do not enter their walls, being separated from germinal and supporting cells by a basement membrane and variable amounts of fibrous tissue containing interstitial cells, at which level occur selective exchange phenomena involving androgens and immune substances.

Lymph vessels of the testis end in lateral and pre-aortic lymph nodes (p. 855).

Nerves accompany the testicular vessels and are derived from the tenth and eleventh thoracic spinal segments through the renal and also the aortic autonomic plexuses (pp. 1163, 1165). Catecholaminiferous nerve fibres occur in the human testis and epididymis, forming plexuses around smaller blood vessels and among the interstitial cells (Baumgarten & Holstein 1967). Hodson (1970) has reviewed this topic.

Applied Anatomy

At an early fetal period the testes are located posteriorly in the abdominal cavity. Their descent to the scrotum (p. 256) appears to be under hormonal control (gonadotropins and androgens). In the scrotum the testes are cooler, favouring spermatogenesis (vide supra). Testicular descent may be arrested (1) in the abdomen, (2) at the deep inguinal ring, (3) in the inguinal canal, (4) between the superficial inguinal ring and the scrotum. A retained testis is probably infertile; a man with both retained (*anorchism*) is sterile but may not be impotent. Absence of one is *monorchism*. Retention in the inguinal canal is often complicated by congenital hernia, the processus vaginalis remaining patent. The testis may traverse the canal but reach an abnormal site (*ectopia testis*, p. 257). It may be inverted in the scrotum, with its normally posterior or attached border anterior and the tunica vaginalis posterior.

Serous fluid often accumulates in the scrotum, as a *hydrocele*. Usually such fluid is in the sac of the tunica vaginalis, forming a *vaginal hydrocele*. In *congenital hydrocele* the fluid is in the tunical sac but this communicates with the peritoneal cavity through a non-obliterated processus vaginalis. *Infantile hydrocele* occurs when the processus is obliterated only at or near the deep inguinal ring; it resembles vaginal hydrocele but fluid extends up the cord into the inguinal canal. If the processus is obliterated both at the deep inguinal ring *and* above the epididymis, leaving a central open part, this may distend as an *encysted hydrocele of the cord*. *Encysted hydrocele of the epididymis*, or *spermatocele*, is a cyst related to the caput epididymis; it may contain spermatozoa and it is probably a retention cyst of one of the seminiferous tubules.

The Ductus Deferens (Vas Deferens)

The deferent duct (ductus or vas deferens), the distal continuation of the epididymis (**8.189**), starts at the epididymal tail and is at first very tortuous but, becoming straighter, ascends along the posterior aspect of the testis, medial to the epididymis. From the superior pole of the testis it ascends in the posterior part of the spermatic cord, to traverse the inguinal canal. At the internal (deep) inguinal ring it leaves the cord, curves round the lateral side of the inferior epigastric artery and ascends for about 2.5 cm anterior to the external iliac artery. It turns back and slightly down obliquely across the external iliac vessels to enter the lesser pelvis, where it continues posteriorly between the peritoneum and lateral pelvic wall medial to the obliterated umbilical artery,

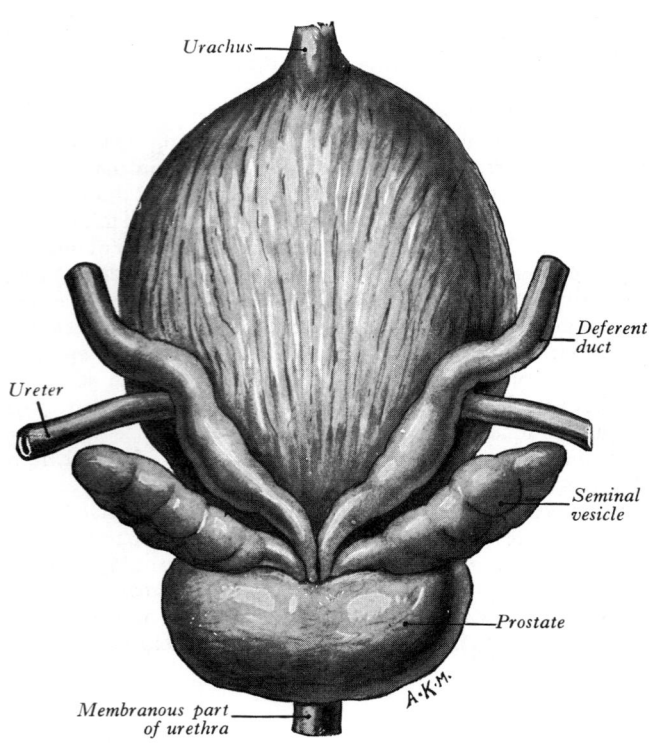

8.195 Posterosuperior aspect of the male internal urogenital organs.

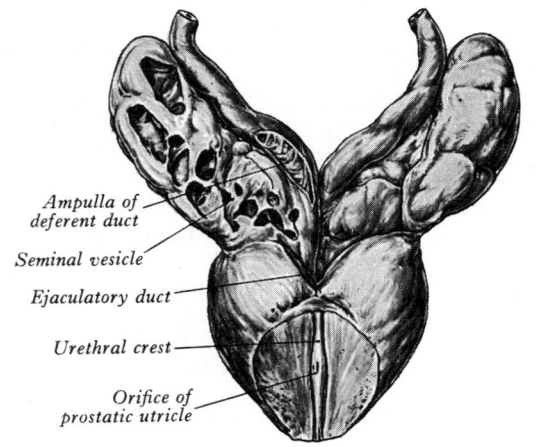

8.196 Anterior aspect of the seminal vesicles, terminal parts of the deferent ducts and the prostate. The lamina of the right seminal vesicle, the ampulla of the right deferent duct and of the prostatic part of the urethra have been exposed by appropriate removal of tissues.

the obturator nerve and vessels and the vesical vessels (**8.181**). It crosses the ureter (**8.195**) and medial to it bends acutely to pass anteromedially between the posterior surface of the bladder and the upper pole of the seminal vesicle, medial to which it descends in contact with it, gradually approaching the opposite duct. Here it lies between the vesical base and the rectum, separated by rectovesical fascia. It finally descends to the base of the prostate, joining at an acute angle the duct of the seminal vesicle to form the ejaculatory duct (**8.196**). Due to its thick wall and small lumen it feels cord-like when grasped. Its lumen is generally small, but posterior to the bladder it becomes dilated and tortuous, as the *ampulla*; beyond this, where it joins the seminal vesicular duct, it is again greatly diminished in calibre (**8.195**).

Structure. The wall of the deferent duct has external loose connective tissue, intermediate muscular and internal mucosal layers. The thick muscular layer is composed of non-striated myocytes arranged in external longitudinal and internal circular strata; an additional internal longitudinal layer exists at the duct's commencement. All muscle strata intermingle. The mucosa is longitudinally folded and its epithelium is columnar and non-ciliated through most of the duct; towards its distal end a bilaminar columnar epithelium appears, the superficial tier displaying non-motile stereocilia. Many of the columnar cells are secretory; for histochemical studies consult Wendler (1968). Elastic fibres are present in its connective tissue (Paniagua et al 1983).

Vessels

The deferent duct has its own artery (p. 777), usually derived from the superior vesical artery. It anastomoses with the testicular artery, thus also supplying the epididymis and testis. The anastomosis is of especial interest in connection with the toxic effects of cadmium on the mammalian testis (Gunn & Gould 1975, Johnson 1977). These appear to be due to interference with vascularization, but the artery to the ductus deferens (in rats) is apparently immune from this effect.

Aberrant Ductules

A narrow, blind, *caudal aberrant ductule* often occurs, connected with the caudal part of the epididymal duct or with the commencement of the deferent duct. Uncoiled, it varies in length from 5 to 35 cm; it may be dilated near its end, but is otherwise uniform. In structure it is like the deferent duct; occasionally it is not connected with the epididymis. A *rostral aberrant ductule* occurs in the epididymal head, connected with the rete testis. Aberrant ductules are derived from mesonephric tubules (p. 250).

Paradidymis

This is a small collection of convoluted tubules, found anterior in the spermatic cord above the epididymal head. They are lined by ciliated columnar epithelium and probably represent the remains of the mesonephros (p.250).

Seminal Vesicles and Ejaculatory Ducts

The two seminal vesicles (**8.195**) are sacculated, contorted tubes placed between the bladder and rectum. Each vesicle is about 5 cm long, somewhat pyramidal, the base being directed up and posterolaterally, and is a single coiled tube with irregular diverticula (**8.196**); the coils and diverticula are connected by fibrous tissue. The diameter of the tube is 3–4 mm and its uncoiled length is from 10–15 cm. The upper pole is a cul-de-sac, the lower narrowing to a straight duct, which joins the deferent duct to form the ejaculatory duct. The *anterior surface* contacts the posterior aspect of the bladder, extending from near the entry of the ureter to the prostatic base. The *posterior surface* is related to the rectum, separated from it by rectovesical fascia. The seminal vesicles diverge superiorly, are related to the deferent ducts and the terminations of the ureters and are partly covered by peritoneum; each has a dense, fibromuscular sheath. Along the medial margin of each vesicle is the ampulla of a deferent duct. Lateral are the veins of the prostatic venous plexus draining posteriorly to the internal iliac veins.

Structure

The wall of the seminal vesicle has three layers: external connective tissue, middle *muscular* (thinner than in the deferent duct and arranged in external longitudinal and internal circular layers) and an internal *mucosal* layer with a reticular structure. The columnar epithelium of the mucosa contains goblet cells in the diverticula, the secretion of which forms much of the seminal fluid; it is alkaline, and contains fructose (an energy source for spermatozoa) and a coagulating enzyme (vesiculase). The vesicles are *not* reservoirs for spermatozoa, these being stored in the epididymis (and possibly the ampulla of the deferent duct). The vesicles contract during ejaculation, their secretion forming most of the ejaculate.

Electron microscopy of human vesicular mucosa reveals a second epithelial type among the columnar, a small stellate cell, usually between the basal parts of the columnar cells, containing few cytoplasmic organelles. Columnar cells have microvilli, contain numerous mitochondria, a well-developed granular endoplasmic reticulum and a Golgi apparatus with numerous secretory vacuoles (Riva 1967).

Vessels and Nerves

The *arteries* of the seminal vesicles are derived from the inferior vesical and middle rectal arteries. *Veins* and *lymph vessels* accompany the arteries. *Nerves* are derived from the pelvic plexuses.

Applied anatomy. The seminal vesicles can be palpated per rectum. Abscesses in them may rupture into the peritoneal cavity.

The Ejaculatory Ducts (8.181, 196)

These are formed on each side by union of the duct of a seminal vesicle with a deferent duct. Each is almost 2 cm in length, starts from the base of the prostate, runs antero-inferiorly between its median and right or left lobes, and skirts the prostatic utricle to end on the colliculus seminalis at two slit-like orifices on, or just within, the utricular opening (p. 1421). They diminish and converge towards their ends.

Structure. The walls of the ejaculatory ducts are thin, containing an *outer fibrous layer*, almost absent beyond their entrance into the prostate, a layer of *non-striated myocytes* in thin outer circular and inner longitudinal strata and a *mucosa* having a columnar epithelium. For details of the ejaculatory musculature, see Schlager (1967).

The Spermatic Cord and its Coverings

As the testis traverses the abdominal wall into the scrotum, it carries its vessels, nerves and deferent duct with it. These meet at the deep inguinal ring to form the *spermatic cord*, suspending the testis in the scrotum and extending from the deep inguinal ring to the posterior aspect of the testis; the left cord is a little longer than the right. Between the superficial ring and testis the cord is anterior to the rounded tendon of adductor longus and is crossed here *anteriorly* by the *superficial* and *posteriorly* by the *deep* external pudendal arteries respectively. The cord traverses the inguinal canal (p. 603) with its walls as relations, the ilio-inguinal nerve being inferior. In the canal it acquires coverings from the layers of the abdominal wall, which extend into the scrotal wall as internal spermatic, cremasteric and external spermatic fasciae.

The *internal spermatic fascia* is a thin, loose layer around the spermatic cord, derived from the transversalis fascia (p. 603). The *cremasteric fascia* contains fasciculi of skeletal muscle united by loose connective tissue to form the cremaster and is continuous with the obliquus internus abdominis (p. 599). The *external spermatic fascia*, a thin fibrous stratum continuous above with the aponeurosis of the obliquus externus abdominis, descends from the crura of the superficial ring (p. 604).

Structure of the Vessels and Nerves of the Spermatic Cord

The cord is composed of arteries, veins, lymph vessels, nerves and the deferent duct, conjoined by loose connective tissue. The *arteries* are the testicular (p. 774), cremasteric (p. 781) and

deferential (p. 777). *Testicular veins* are described on p. 815. *Lymph vessels* of the testis are described on p. 855. The *nerves* are: the genital branch of the genitofemoral (p. 1141), the cremasteric nerve and the testicular sympathetic plexus (p. 1165), joined by filaments from the pelvic plexus accompanying the deferential artery.

The Scrotum

The scrotum, a cutaneous fibromuscular sac containing the testes and lower parts of the spermatic cords, depends below the pubic symphysis between the anteromedial aspects of the thighs. It is divided into right and left halves by a cutaneous *raphe*, continued ventrally to the inferior penile surface and dorsally along the midline of the perineum to the anus; its left side is usually lower, due to the greater length of the left spermatic cord. The raphe indicates the bilateral origin of the scrotum from the genital swellings (p. 259). The external appearance varies: thus, when warm and in the elderly and debilitated, the scrotum is smooth, elongated and flaccid; but when cold and in the young and robust, it is short, corrugated and closely applied to the testes. It consists of skin, dartos muscle and external spermatic, cremasteric and internal spermatic fasciae. The internal spermatic fascia is loosely attached to the parietal layer of the tunica vaginalis (8.197).

The *scrotal skin*, thin, pigmented and often rugose, bears thinly scattered, crisp hairs, their roots visible through the skin; it has sebaceous glands, whose secretion has a characteristic odour and also numerous sweat glands, pigment cells and nerve endings responding to mechanical stimulation of the hairs and skin and to variations in temperature. Subcutaneous adipose tissue is lacking.

The *dartos muscle* is a thin layer of non-striated myocytes, continuous beyond the scrotum with the superficial inguinal and perineal fasciae, extending into a *scrotal septum*, connecting the raphe to the inferior surface of the penile radix and dividing the scrotum into two cavities. The septum contains all the layers of scrotal wall except skin. The dartos muscle is closely united to the skin, but is connected to subjacent parts by delicate loose connective tissue, giving it marked independence. Shafik (1977) has described a fibromuscular 'scrotal ligament' extending from the dartos sheet to the inferior testicular pole, regarding this as thermoregulatory (p. 1429).

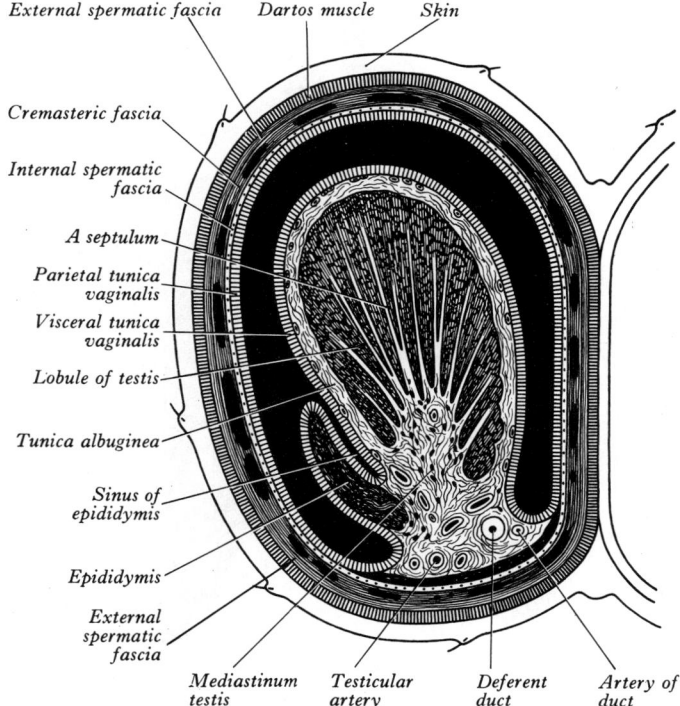

8.197 Diagrammatic transverse section through the left half of the scrotum and the left testis. The tunica vaginalis is represented as artificially distended to show its visceral and parietal layers.

Vessels and Nerves

The *arteries* supplying the scrotum are the external pudendal branches of the femoral (p. 782), the scrotal branches of the internal pudendal (p. 778) and a cremasteric branch from the inferior epigastric (p. 781). Dense subcutaneous plexuses of scrotal vessels carry a substantial blood flow effecting loss of heat (Esser 1932). Arteriovenous anastomoses of a simple but large-calibre type are also prominent (Molyneux 1965 and p. 1429). The *veins* follow the corresponding arteries. The *lymph vessels* end in the inguinal lymph nodes (p. 856). The *nerves* are: the ilio-inguinal and the genital branch of the genitofemoral (p. 1141), the two posterior scrotal branches of the perineal (p. 1149) and the perineal branch of the posterior femoral cutaneous nerve (p. 1145). The scrotum's anterior third is supplied mainly from the first lumbar spinal segment (by way of the ilio-inguinal and genitofemoral nerves), while the posterior two-thirds come mainly from the third sacral (via the perineal and posterior femoral cutaneous nerves). The ventral axial line of the lower limb (p. 1151) passes between these areas. A spinal anaesthetic therefore must be injected much higher to anaesthetize the anterior region.

The Penis

The penis, the male copulatory organ, comprises an attached *radix* or *root* in the perineum and a free, normally pendulous *corpus* or *body* completely enveloped in skin.

The radix of the penis comprises the three masses of erectile tissue in the urogenital triangle: the two crura and the bulb of the penis, firmly attached to the pubic arch and perineal membrane respectively. The crura and bulb are posterior regions of the corpora cavernosa and corpus spongiosum. Each *crus penis* (**8**.198)

commences behind as a blunt, elongate but round process, attached firmly to the everted edge of the ischiopubic ramus and covered by ischiocavernosus (p. 608). Anteriorly it converges towards its fellow and is slightly enlarged posterior to this. Near the inferior symphysial border the two crura bend sharply down and forwards to become the corpora cavernosa. The *bulb of the penis* (**8**.181, 198) is between the crura and firmly connected to the inferior aspect of the perineal membrane, from which it receives a fibrous covering. Oval in section, it narrows anteriorly into the corpus spongiosum, bending sharply down and forwards here. Its convex superficial surface is covered by bulbospongiosus; its flat deep surface is pierced by the urethra, which traverses it to reach the corpus spongiosum. This part of the urethra has an intrabulbar fossa (p. 1422).

The corpus of the penis contains three elongated erectile masses, capable of much enlargement when engorged with blood during erection. When flaccid it is cylindrical, but when erect it is triangular with rounded angles (**8**.199). The surface which is posterosuperior during erection is termed the *dorsum of the penis* and the opposite aspect the *urethral surface*. The erectile masses are termed the right and left *corpora cavernosa* and the median *corpus spongiosum penis*, continuations of the crura and bulbus penis.

The *corpora cavernosa penis* form most of its body. In close apposition throughout, they have a common fibrous envelope and are separated only by a median fibrous septum. On the urethral surface their combined mass has a wide median groove, adjoining the corpus spongiosum (**8**.199); dorsally a similar but narrower groove contains the deep dorsal vein. The corpora end in the hollow, proximal aspect of the glans penis in a rounded cone, on which each has a small terminal projection (**8**.198). They are enclosed in a strong fibrous *tunica albuginea*, consisting of superficial and deep strata. The superficial fibres are longitudinal, forming a single tube round both corpora; the deep fibres are circular and surround each corpus separately, joining as a median *septum of the penis*, which is thick and complete proximally but imperfect distally where it is a pectiniform series of bands; hence the term *pectiniform septum*.

The *corpus spongiosum penis*, traversed by the urethra, adjoins the median groove on the urethral surface of the conjoined corpora cavernosa. It is cylindrical, tapering slightly distally, and surrounded by the tunica albuginea. Near the end of the penis it expands into a somewhat conical enlargement like an acorn, whence its name, *glans penis* (**8**.198).

The *glans penis* projects dorsally over the end of the corpora cavernosa, with a shallow concave surface to which they are attached. Its base has a projecting *corona glandis*, overhanging an obliquely grooved *neck of the penis*. The navicular fossa (p. 1151) of the urethra is in the glans and opens by a sagittal slit on or near its apex.

Penile skin is remarkably thin, dark and loosely connected to the tunica albuginea. At the neck of the penis it is folded to form the *prepuce* or foreskin, variably overlapping the glans. The internal preputial layer is confluent at the neck with the thin skin covering and firmly adherent to the glans and by this with the urethral mucosa at the external urethral orifice. On the urethral aspect of the glans a median fold passes from the deep surface of the prepuce to the glans immediately proximal to the orifice; this is the *frenulum*. Cutaneous sensitivity, high on the general surface of the glans, is accentuated near the frenulum. The prepuce and

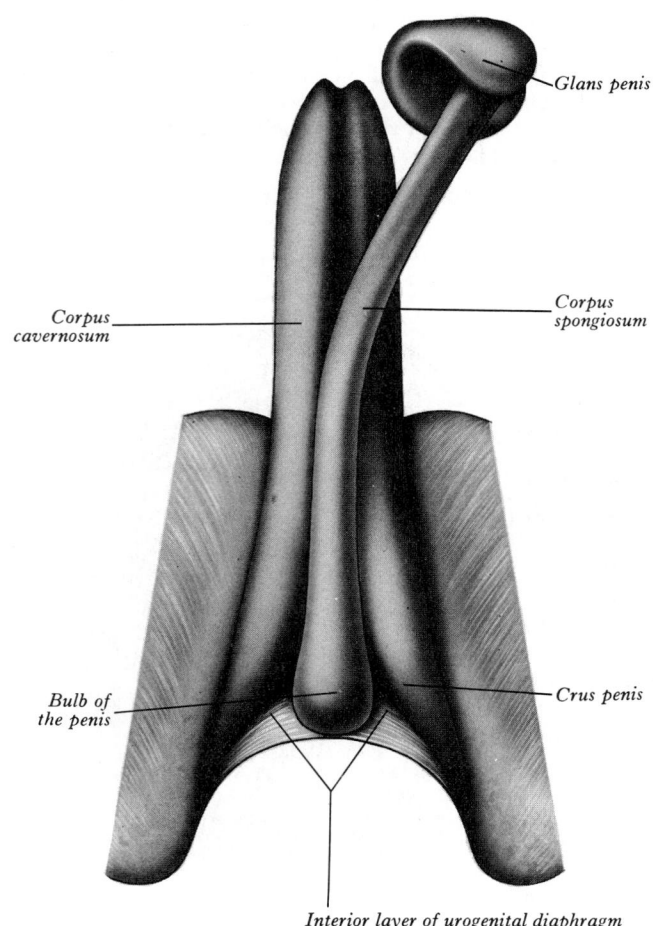

Corpus cavernosum

Glans penis

Corpus spongiosum

Crus penis

Bulb of the penis

Interior layer of urogenital diaphragm (perineal membrane)

8.198 Ventral aspect of the constituent erectile masses of the penis in erect position. The glans penis and the distal part of the corpus spongiosum are shown detached from the corpora cavernosa penis and turned to the left.

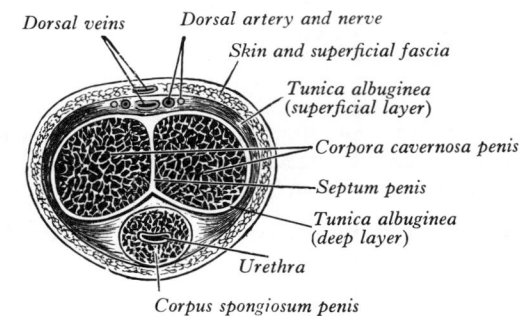

Dorsal veins   *Dorsal artery and nerve*

Skin and superficial fascia

Tunica albuginea (superficial layer)

Corpora cavernosa penis

Septum penis

Tunica albuginea (deep layer)

Urethra

Corpus spongiosum penis

8.199 Transverse section of human penis.

glans penis enclose a potential cleft, the *preputial sac*, with two shallow fossae flanking the frenulum. On the corona glandis and penile neck are numerous small *preputial glands*, secreting sebaceous *smegma*.

The *superficial penile fascia*, devoid of fat, consists of loose connective tissue, invaded by a few fibres of dartos muscle from the scrotum (p. 1431). As in the suprapubic abdominal wall, the deepest layer is condensed, here as the *fascia penis* surrounding both the corpora cavernosa and corpus spongiosum and separating the superficial and deep dorsal veins. At the penile neck it blends with the fibrous covering of all three corpora. Proximally, it is continuous with the dartos muscle and with the fascia covering the urogenital region of the perineum (p. 1431).

The corpus penis is supported by two ligaments continuous with its fascia and consisting largely of elastin fibres. The *fundiform ligament* (**5.45**), from the lowest part of the linea alba, splits into two lamellae which skirt the penis and unite below with the scrotal septum. The triangular *suspensory ligament*, deep to the fundiform ligament, is attached above to the front of the pubic symphysis, blending below, on each side, with the fascia penis.

Structure of the Penis

From the inside of the fibrous sheaths of the corpora cavernosa and the sides of their septum numerous *trabeculae* arise, crossing them in all directions and creating *cavernous spaces*, which give them a spongy form (**8.199**). These trabeculae are of collagen and elastin fibres and non-striated myocytes and contain numerous vessels and nerves. The cavernous spaces are filled with blood during erection, but most are empty in the flaccid penis. They are lined by flat endothelial cells without fenestrae (Leeson & Leeson 1965).

The fibrous tunica albuginea of the corpus spongiosum is thinner, whiter, more elastic than that of the corpora cavernosa and is formed partly of non-striated myocytes; a layer of the same tissue surrounds the urethral epithelium (p. 1422).

Vessels and Nerves

The *arteries* supplying the cavernous spaces are the deep penile arteries (p. 779) and branches of the dorsal penile arteries which perforate the tunica albuginea along the dorsum, especially near the glans. On entry the cavernous arteries divide into branches in the trabeculae, some ending in capillary networks opening into the cavernous spaces; others become convoluted and somewhat dilated *helicine arteries*, which open into the cavernous spaces and from them capillary branches supply the trabeculae. Helicine arteries are most abundant in the posterior region of the corpora cavernosa (Alvarez-Morujo 1968).

The veins drain the cavernous spaces by means of a series of vessels, some emerging from the base of the glans and converging on the dorsum penis to form the deep dorsal vein, others leaving the corpora cavernosa to join the same vein; some also emerge inferiorly from the corpora cavernosa and, receiving tributaries from the corpus spongiosum, curve round to the deep dorsal vein, but many leave at the root of the penis to join the prostatic plexus (p. 816).

Erection is purely vascular and independent of compression by the ischiocavernosi and bulbospongiosus. Rapid inflow from the helicine arteries fills the cavernous spaces and the resulting distension also contributes to erection by pressure on the veins which drain the erectile tissue.

The lymph vessels are described on p. 855.

The *nerves* are from the second, third and fourth sacral spinal segments via the pudendal nerve and pelvic plexuses (p. 1115). On the penile glans and bulb some cutaneous filaments connect with lamellated corpuscles and many end in characteristic end bulbs (p. 910). Cutaneous stimulation of the glans and frenulum contributes much to the maintaining erection and to initiation of orgasm and ejaculation.

The Prostate

The prostate (**8.187,195,196**), a firm, partly glandular, partly fibromuscular body, surrounds the beginning of the male urethra. It lies at a low level in the lesser pelvis, behind the inferior border of the symphysis pubis and pubic arch and anterior to the rectal ampulla, through which it may be palpated. Being somewhat conical, it presents: above, a base or vesical aspect; below, an apex and also a posterior, an anterior and two inferolateral surfaces.

The *base* is largely contiguous with the neck of the bladder above it; the urethra enters here, nearer its anterior border. The *apex* is inferior and in contact with the fascia on the superior aspects of the sphincter urethrae and tranversi perinei profundi (p. 607).

The *posterior surface*, transversely flat and vertically convex, is separated from the rectum by the prostatic sheath and loose connective tissue external to the sheath. Near its superior (juxtavesical) border is a depression where the two ejaculatory ducts penetrate the gland, dividing this surface into a superior and an inferior, larger part. The superior part is variable in size and usually regarded as the external aspect of the *median lobe*; the inferior part shows a shallow, median sulcus, usually considered to mark a partial separation into *right and left lateral lobes*, forming the main prostatic mass and continuous behind the urethra. A band of fibromuscular tissue, ventral to the urethra, joins these lobes together and is often referred to as the anterior lobe; it contains less glandular tissue than the rest of the gland. This simplified view of prostatic lobation, based mainly on the classic work of Lowsley (1912), is retained here despite many modifications introduced by subsequent observers with persistent confusions and inconsistencies which have not yet been completely resolved. Some deny any topographical lobation and those who favour it differ on boundaries and terminology. McNeal (1975) has analysed these disagreements. Lobation became important through the conviction among some investigators that malignant tumours occur in particular prostatic regions. This interest prompted the work of Tisell & Salander (1975) who claim to have confirmed a recognizable lobar structure after dissection of more than 100 human prostate glands. They recognize two large *lateral lobes* but consider that these do not appear on the dorsal (rectal) aspect, which they describe as occupied by paired *dorsal lobes* extending laterally to form the apex. They recognize *median lobes*, around the urethra (except at the apex), deep to the dorsal and lateral lobes; their median lobes may be equated with the *internal zone* described below. All three pairs of lobes are, they affirm, separable by dissection. It is impossible to reconcile the differences between this and other descriptions and the question of lobation must be left sub judice (Goland 1975).

The *anterior surface*, transversely narrow and convex, extends from the apex to the base, about 2 cm behind the pubic symphysis from which it is separated by a venous plexus and loose adipose tissue. Near its superior limit it is connected to the pubic bones by the puboprostatic ligaments. The urethra emerges from this surface anterosuperior to the apex.

The *inferolateral surfaces* are related to the anterior parts of the levatores ani, which are separated from them by a plexus of veins embedded in the fibrous prostatic sheath.

The prostatic base measures about 4 cm transversely, the gland being about 2 cm in anteroposterior and 3 cm in its vertical diameters. It weighs about 8 g. It has a fibrous sheath, partly vascular; on each side this consists of fibrous tissue containing the prostatic venous plexus (**6.170**); anteriorly it blends with the puboprostatic ligaments (p. 1417) and inferiorly with fascia on the deep surfaces of the sphincter urethrae, the deep transverse perineal muscles and with the perineal body (**5.54**). Posteriorly the sheath has a different origin and is avascular. In male fetuses, at the fourth month, the rectovesical peritoneal pouch descends to the pelvic floor, separating prostate from rectum; its lower part is obliterated and the fused peritoneal layers here form the posterior prostatic sheath (Smith 1908), sometimes termed the *rectovesical fascia*. Traces of its separate layers persist as a plane of cleavage. Above, it ascends over the posterior aspects of the seminal vesicles and deferent ducts and is connected to the floor of the rectovesical pouch (**5.54**); on each side, it joins with the posterior vesical ligament (p. 1417). Below, adherent to the prostate, it joins the perineal body. Some have denied this peritoneal fusion and consider the rectovesical fascia to be a condensation of areolar tissue (Silver 1956). The anterior parts of the levatores ani pass back from the pubis around the prostate as *levatores prostatae*.

The prostate is traversed by the urethra and ejaculatory ducts and contains the prostatic utricle. The urethra usually passes between its anterior and middle thirds. The ejaculatory ducts pass antero-inferiorly through its posterior region to open into the prostatic urethra (p. 1421).

Structure (8.200, 201)

The prostate is grey to reddish according to its activity and very dense. It is enveloped in a thin, dense, fibrous capsule within a sheath derived from pelvic fascia, the latter containing a venous plexus. The capsule is firmly adherent to the gland and continuous with a median septum in the urethral crest separating the lateral masses below the level of the colliculus seminalis. It is also continuous with numerous fibromuscular septa enmeshing the glandular tissue.

The *muscular tissue* is mainly non-striated (Hutch & Rambo 1970); ventral to the urethra a layer of non-striated myocytes

curves to merge with the main mass of muscle in the fibromuscular septa; superiorly it is continuous with vesical non-striated muscle. However, anterior to this a transversely crescentic mass of skeletal muscle is continuous inferiorly with the sphincter urethrae in the deep perineal pouch. Its fibres pass transversely internal to the capsule, attached to it laterally by diffuse collagen bundles; other collagen bundles pass posteromedially, merging with the prostatic fibromuscular septa and the septum of the urethral crest. This muscle, supplied by the pudendal nerve, probably compresses the urethra (Haines 1969) but it is suggested that it may pull the urethral crest back and the prostatic sinuses forwards, dilating the urethra; glandular contents may be expelled simultaneously into the urethra, thus expanded to contain seminal fluid (3–5 ml) during the period of sexual excitement prior to ejaculation.

The *glandular tissue* consists of numerous follicles with frequent internal papillae. Follicles open into elongated canals, which join to form 12–20 excretory ducts. The follicles are conjoined by loose connective tissue, supported by extensions of the fibrous capsule and muscular stroma and enclosed in a delicate capillary plexus. Canalicular and follicular epithelium is columnar; electron microscopy reveals luminal microvilli, relatively few short profiles of granular endoplasmic reticulum and some free ribosomes. Lysosomal vesicles, containing acid phosphatase, also occur. The apposed borders of these cells are mainly straight, the plasma membranes being united by desmosomes (Fisher & Jeffrey 1965). *Prostatic ducts* open mainly into the prostatic sinuses in the floor of the prostatic urethra; they have a bilaminar epithelium, the luminal layer is columnar and the basal has small cubical cells. Small colloid *amyloid bodies* often occur in the follicles. The prostatic and seminal vesicular secretions form the bulk of the seminal fluid. The former is slightly acid, containing acid phosphatase and fibrinolysin.

Histological sections (**8.200**) do not show a lobar pattern, but two concentric zones of glandular tissue, being partially circumurethral, are distinguishable (Le Duc 1939, Franks 1954, Fergusson & Gibson 1956). The larger *peripheral zone* has long, branched glands, whose ducts curve posteriorly to open mainly into the prostatic sinuses, though a few open on the lateral urethral walls. The *internal zone* consists of 'submucosal' glands, with ducts opening on the floor of the prostatic sinuses and colliculus seminalis and an *innermost* group of simple 'mucosal' glands, surrounding the upper prostatic urethra. Anteriorly, in the prostatic isthmus (**8.200A**), the peripheral zone and 'submucosal' glands are absent. Peripheral and internal zones are said to be separated by an ill-defined 'capsule'. Carcinoma affects almost exclusively the peripheral zone, the internal being prone to benign hypertrophy, such enlargements projecting into the bladder, displacing the peripheral zone postero-inferiorly and thus accentuating the 'capsule' between outer and inner zones. This provides a 'cleavage plane' which is used in surgical enucleation.

Vessels and nerves. The *arteries* (Clegg 1955, 1956) are rami of the internal pudendal, inferior vesical and middle rectal arteries. The *veins* form a plexus around the prostatic sides and base (p. 816), receiving in front the deep dorsal penile vein and draining to the internal iliac veins. The *lymph vessels* are described on p. 856. The *nerves* come from the inferior hypogastric (pelvic) plexus (p. 1116).

Age Changes in the Prostate

At birth (Swyer 1944) the prostate has a system of ducts embedded in a stroma which forms a large part of the gland. Follicles are represented by small end buds on the ducts. There is hyperplasia and squamous metaplasia of epithelium of the ducts, colliculus seminalis and prostatic utricle, possibly due to maternal oestrogens in the fetal blood. These changes subside within six or seven weeks and little structural change follows until about the ninth year, when hyperplasia of the ductal epithelium and formation of side buds causes an elaboration of the system of ducts; up to puberty the prostate grows slowly and continuously.

At puberty, over a period of six to twelve months, the gland more than doubles in size, due almost entirely to follicular development, partly from end buds on ducts, partly from modification of the ductal branches. This growth is associated with

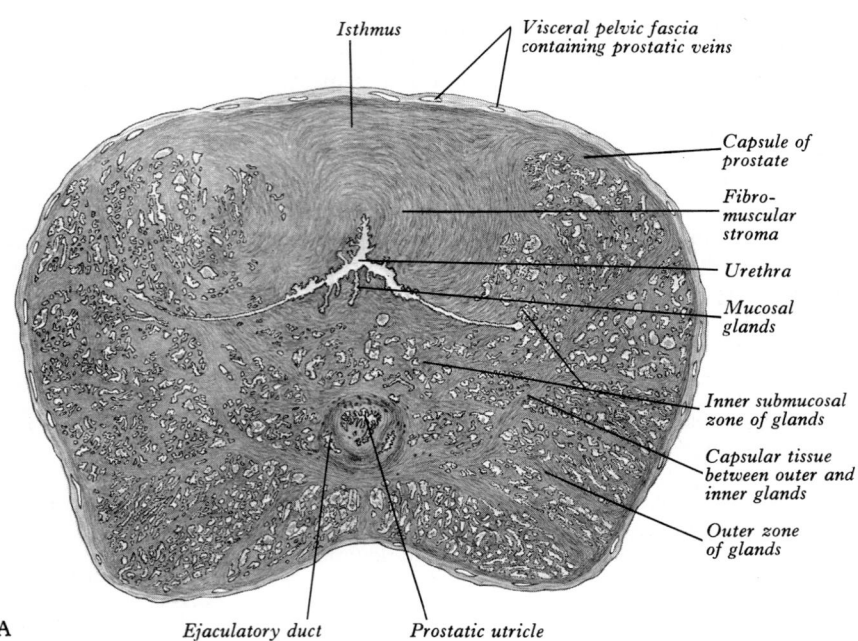

Isthmus — Visceral pelvic fascia containing prostatic veins

Capsule of prostate

Fibro-muscular stroma

Urethra

Mucosal glands

Inner submucosal zone of glands

Capsular tissue between outer and inner glands

Outer zone of glands

A

Ejaculatory duct   Prostatic utricle

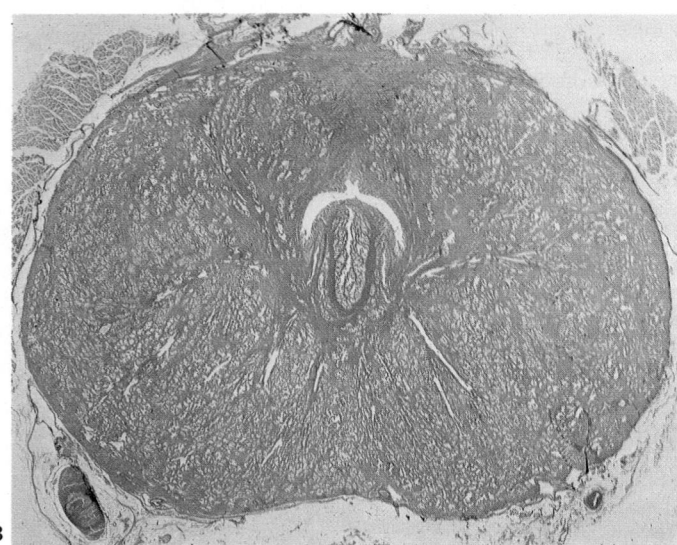

B

8.200 Human prostate. A. Transverse section at a level to show the prostatic utricle and ejaculatory ducts. B. Transverse section to show the urethral crest. Stained by haematoxylin and eosin and lent by L M Franks, Imperial Cancer Research Institute, London and photographed by K Fitzpatrick. Compare with A and see text for further details. Both A and B are slightly more than twice the natural size. Ventral is to the top in both figures.

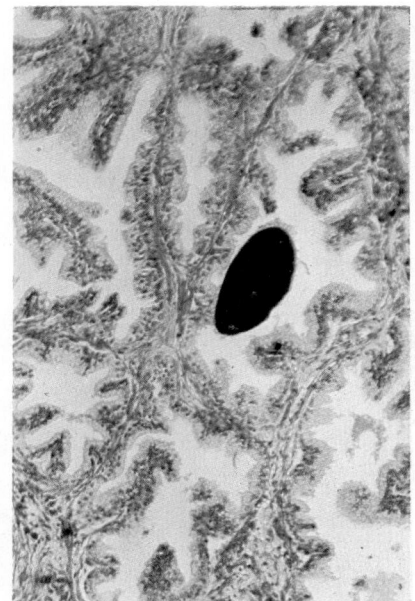

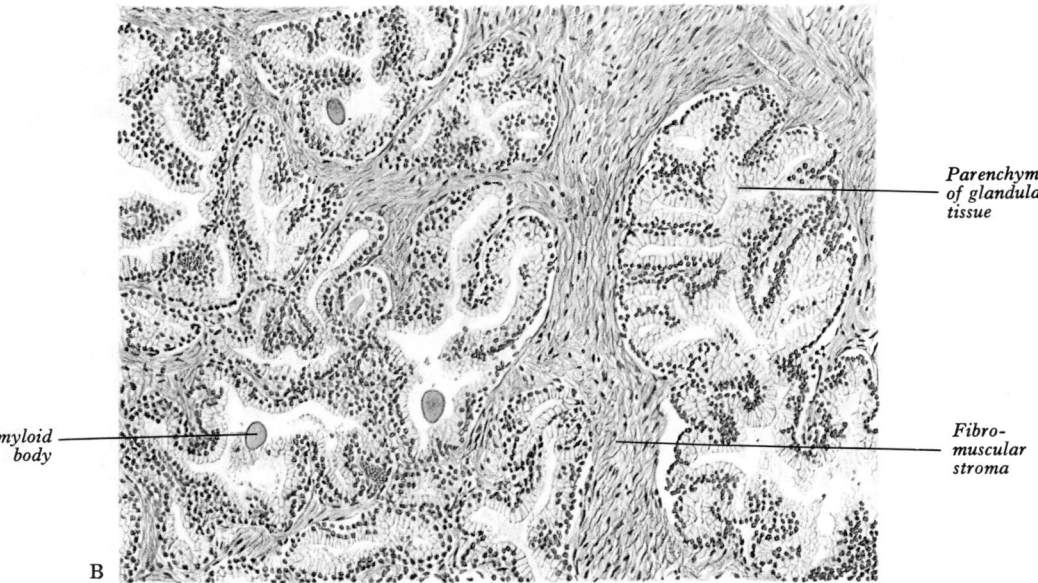

Parenchyma
of glandular
tissue

Amyloid
body

Fibro-
muscular
stroma

A   B

8.201 Characteristic fields of the human prostate. Note the highly convoluted columnar epithelium, forming papillary projections into the
lumen of a follicle, and abundant interfollicular fibromuscular tissue.
Specimen shown in A provided by Abla el Nasser (1980).

a condensation of the stroma, which diminishes relative to the
glandular tissue, probably due to the secretion of testosterone into
the bloodstream by the testis.

During the third decade the glandular epithelium grows by irregular multiplication of the epithelial infoldings into the lumen
of the follicles. *After the third decade* the size remains unaltered
until 45–50 years. Epithelial foldings tend to disappear, follicular
outlines becoming more regular, and amyloid bodies increase in
number. All of these changes are signs of prostatic involution.

After 45–50 years the prostate may undergo benign hypertrophy, its size increasing until death, or alternatively it may
undergo progressive atrophy.

Applied Anatomy

After middle age the prostate often enlarges, projecting into the
bladder to impede urination by distorting the prostatic urethra.
The median lobe may enlarge the most, with even a small enlargement obstructing the internal urethral orifice; the more the
patient strains, the more the prostatic mass, acting like a valve,
blocks the opening. The hypertrophied part may be removed
surgically (prostatectomy).

Valveless venous communications between the prostatic and
extradural venous plexuses normally occur, probably an important factor in the metastasis of prostatic neoplasms to the vertebral
bodies (Batson 1940, Franks 1953).

THE BULBO-URETHRAL GLANDS

The two bulbo-urethral glands (6.112), small, round, yellow,
somewhat lobulated and about a centimetre in diameter, lie lateral
to the membranous urethra above the perineal membrane and
penile bulb and are enclosed by fibres of the sphincter urethrae.
They gradually diminish in size in the later decades. The excretory duct of each, almost 3 cm long, passes obliquely forwards
external to the mucosa of the membranous urethra and penetrates
the inferior urogenital fascia (perineal membrane) to open by a
minute orifice on the floor of the spongiose urethra about 2.5 cm
below the perineal membrane.

Structure. The gland has several lobules held together by a
fibrous capsule. The lobules consist of acini of columnar epithelial
cells. Their secretion is added to the seminal fluid but human
bulbo-urethral glands are comparatively small and their role is
uncertain.

REPRODUCTIVE ORGANS OF THE FEMALE

The female reproductive system consists of: *internal organs*
situated within the lesser pelvis, namely the ovaries, uterine
tubes, uterus and vagina; and *external organs* lying antero-inferior
to the pubic arch, namely the mons pubis, labia majora et minora,
pudendi, clitoris, vestibular bulbs, greater vestibular glands and
vestibule.

The Ovaries

The ovaries (8.202, 209), are homologous with the testes and like
these they develop from the genital ridges (p. 255). Situated one
on each side of the uterus close to the lateral pelvic wall, they are
attached to the posterosuperior aspect of the broad uterine ligament, postero-inferior to the uterine tube (8.202). They are
greyish-pink and smooth before ovulation begins but thereafter
are distorted by the degeneration of successive corpora lutea.
Each is almond-shaped (amygdaloid), about 3 cm long, 1.5 cm

wide and 1 cm thick. Its position varies much in parous women,
because it is displaced in the first pregnancy and probably never
returns to its original location. Also variably mobile, it may
change position according to the state of surrounding organs,
such as the intestines. In nulliparous women, in the upright
position, its long axis is vertical; it has lateral and medial surfaces,
tubal and uterine extremities and mesovarian and free borders. It
occupies the *ovarian fossa*, on the lateral pelvic wall, bounded
anteriorly by the obliterated umbilical artery and posteriorly by
the ureter and internal iliac artery. To the *tubal* (superior) *extremity*, near the external iliac vein, are attached the ovarian fimbria of
the uterine tube and a peritoneal *ovarian suspensory ligament*,
which contains the ovarian vessels and nerves and passes over the
external iliac vessels (8.209) to join the peritoneum on the psoas
major, posterior to the caecum or descending colon. The *uterine*
(inferior) *extremity* faces downwards towards the pelvic floor; it is
usually narrower than the tubal extremity and is attached to the
lateral uterine angle postero-inferior to the uterine tube by a

Ovarian ligament

Uterine tube

Mesosalpinx

Epoöphoron

Mesovarium

Uterine fundus

Uterine body

Vesicular appendix

Abdominal ostium

Suspensory ligament of ovary

Ovarian fimbria

Ovary

Cut edge of broad ligament

Meso-metrium

Intravaginal cervix of uterus

Vagina

8.202 Posterosuperior aspect of the uterus and the left broad ligament. The 'ligament' has been spread out and the ovary is displaced downwards.

rounded *ovarian ligament* which lies in the the broad ligament and contains non-striated myocytes. The *lateral surface* contacts parietal peritoneum in the ovarian fossa and is separated by it from extraperitoneal tissue and the obturator vessels and nerve. The uterine tube largely covers the *medial surface*; the peritoneal recess here, between the ovary and overlapping mesosalpinx (p. 1445), is termed the *ovarian bursa*. The *mesovarian border*, straight and facing the obliterated umbilical artery, is attached to the back of the broad ligament by a short peritoneal fold, the *mesovarium*, in which blood vessels and nerves reach the ovarian hilum. The convex *free border* faces the ureter. The uterine tube arches over the ovary, ascending in relation to its mesovarian border, to curve over its tubal end and pass down on its free border and medial surface (**8**.209).

In embryonic and early fetal life the ovaries are, like the testes, juxtarenal in the lumbar region but they gradually descend into the lesser pelvis (p. 429). Accessory ovaries may occur in the mesovarium or in the adjacent broad ligament.

Structure of the Ovary

The surface is covered, in young females, by a layer of cuboidal cells which flatten in later life. This so-called *germinal epithelium* gives to the ovary a dull grey colour, contrasting with the shining, smooth peritoneum; the transition between peritoneum and ovarian epithelium is usually marked by a white line around the anterior, mesovarian border.

After puberty the ovary has a thick cortex containing ovarian follicles and corpora lutea surrounding a vascular medulla, except at the hilum. The dense *cortical stroma* contains woven reticular fibres and many fusiform cells, resembling non-striated myocytes but possibly mesenchymal. These cells contribute to growth of the theca folliculi (vide infra) and may secrete oestrogens. *Medullary* stroma consists of a looser connective tissue, with many elastin fibres, non-striated myocytes and numerous blood vessels, particularly veins. At the hilum, strands of myocytes enter the medulla from the mesovarium. The cortex is much less vascular than the medulla. Beneath the germinal epithelium, the cortical connective tissue condenses into a delicate *tunica albuginea*, increasing in density with age. It is collagenous, unlike the general ovarian stroma. The prenatal cortical stroma contains small groups of *interstitial cells*, persisting after puberty only in the thecae of atretic follicles.

Ovarian Follicles

At birth, the ovarian cortex contains many *primary ovarian follicles*. Each has a large central *oögonium* surrounded by a single layer of small cuboidal or flat *follicular cells*. Their prenatal development is described on p. 255. Many degenerate during childhood and after puberty, some developing each month as *vesicular ovarian (Graafian) follicles*, one usually maturing and rupturing (ovulation). During the child-bearing period (i.e. until menopause) the cortex contains ovarian follicles, corpora lutea and atretic follicles (**8**.203).

Primary follicles become vesicular as follows: follicular cells multiply to form many layers and a cavity (the *antrum folliculi*) containing *liquor folliculi* appears among them between an outer group, forming the *membrana granulosa* and an inner around the presumptive ovum, attaching it to one pole of the follicle and forming the *cumulus ovaricus* (**8**.204). Cortical stromal cells form a *theca folliculi* around the follicle; this consists of an inner *tunica interna*, which is cellular and permeated by capillaries, and an outer, fibrous *tunica externa*. The tunica interna is separated from the membrana granulosa by a basal lamina. As follicles mature, the tunica interna is well-defined as the '*thecal gland*'. Its cells and those of the membrana granulosa produce oestrogenic hormones (primarily oestradiol), the follicular development being stimulated by the hypophysial gonadotropic hormone (follicle stimulating hormone, FSH). Meanwhile the oögonium becomes a *primary oöcyte*, dividing meiotically into a haploid *secondary oöcyte* and first polar cell. When a mature follicle ruptures (ovulation), the secondary oöcyte is extruded; a second polar cell is only formed if fertilization occurs. During maturation a presumptive ovum grows from 30 µm in the primordial follicle to 110 µm in the mature follicle, largely due to the accumulation of *deutoplasm* or yolk within the cell.

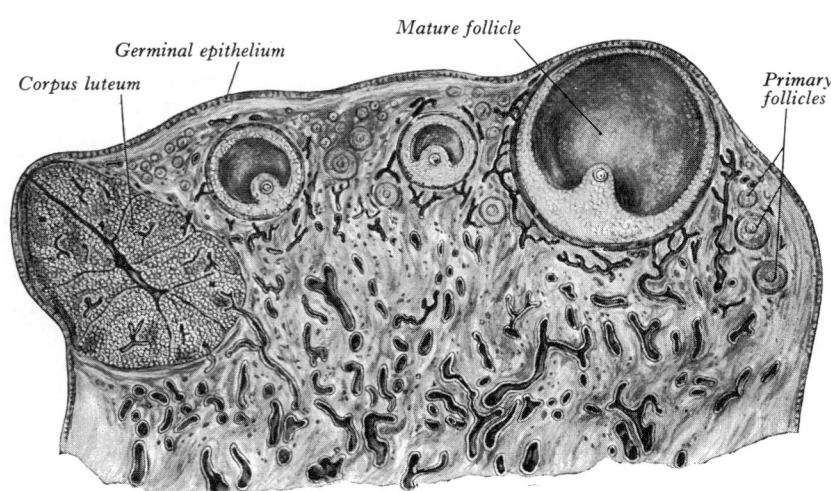

Germinal epithelium

Corpus luteum

Mature follicle

Primary follicles

8.203 Semi-diagrammatic section of an ovary.

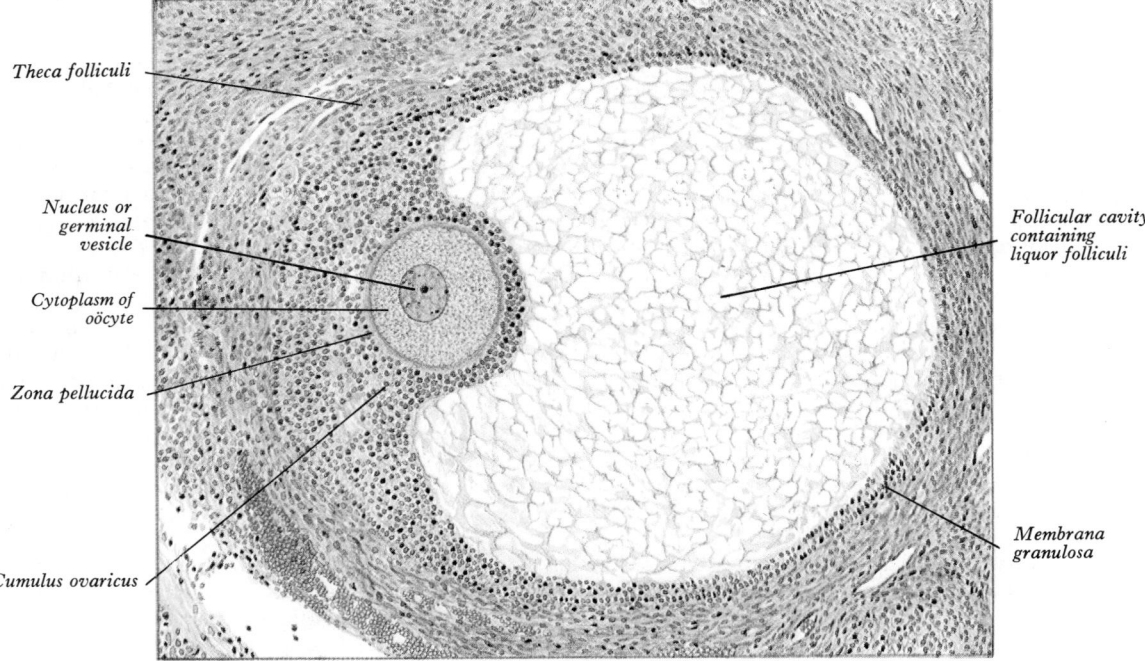

Theca folliculi

Nucleus or germinal vesicle

Cytoplasm of oöcyte

Zona pellucida

Cumulus ovaricus

Follicular cavity containing liquor folliculi

Membrana granulosa

8.204 Section through an ovarian follicle from a woman aged 28 years. Stained with haematoxylin and eosin. Magnification × 90.

A mature follicle is about 10 mm in total diameter; usually one matures and ruptures in each monthly cycle; *anovulatory* cycles may occur. However, several other small vesicular follicles may also develop but degenerate to form *atretic follicles*, cells of their tunicae internae becoming interstitial ovarian cells like those of the testis (p. 1425).

Ovulation

When a mature vesicular follicle ruptures at the ovarian surface the secondary oöcyte, surrounded by the cumulus ovaricus, is expelled and enters the uterine tube at its fimbriated end. If fertilization occurs, it usually does so in the lateral third of the uterine tube; if it does not, the oöcyte begins to degenerate after 24–48 hours. Ovulation is usually 12–16 days before the expected onset of a menstrual cycle.

Corpus Luteum

After ovulation, the walls of the empty follicle collapse and fold, cells of the membrana granulosa increasing in size and forming a cytoplasmic carotenoid pigment (lutein). Such *luteal cells* form most of the *corpus luteum* (**8.**203). Smaller *paraluteal cells*, derived from the tunica interna, appear among the superficial luteal cells. The basement membrane outside the granulosa cells vanishes, allowing capillaries to grow in between the luteal cells from vessels in the tunica interna. Blood coagulum forms in the corpus luteum; if fertilization fails, the corpus luteum functions for about 12–14 days, then degenerates into a *corpus luteum of menstruation*. Changes include fatty degeneration of the luteal cells and gradual replacement by fibrous tissue; eventually, after about two months, a small cicatricial *corpus albicans* remains. Hypophysial follicle stimulating (FSH), luteinizing (LH) and luteotropic (LTH, prolactin or lactogen) hormones activate the conversion of granulosal into luteal cells, which produce progesterone; luteal and paraluteal cells also produce some oestradiol. (For the action of these ovarian hormones on uterine endometrium see pp. 146–149.)

If fertilization occurs, implantation of the blastocyst into the endometrium usually begins on the seventh day thereafter; the trophoblast begins to produce FSH, LH, progesterone and oestradiol. These chorionic gonadotropins (FSH and LH) stimulate the corpus luteum of menstruation to grow and become a *corpus luteum of pregnancy*. Whereas the former is active for 12–14 days and is usually about 1 cm in diameter, the latter increases to about 2.5 cm in diameter at mid-gestation and remains active. By the end of pregnancy it is reduced to about 1 cm and in the next few months it degenerates like a corpus luteum of menstruation.

As females age, ovaries are increasingly fibrosed by corpora albicantia; after menopause, usually near 50 years, ovarian activity involving the formation of follicles and corpora lutea ends.

Vessels and Nerves

The *arteries* of the ovaries and uterine tubes are the ovarian branches of the abdominal aorta (p. 776). The *veins* emerge from the ovarian hila as a *pampiniform plexus*, drained by the ovarian veins (p. 818). *Lymph vessels* are described on p. 856. The *nerves*, derived from the ovarian plexuses (p. 1165), are postganglionic sympathetic, parasympathetic and autonomic afferent fibres, but little is known of their actual distribution or function, particularly in humans. Bulmer (1965) and Owman et al (1967) have demonstrated cholinergic nerve fibres in the stroma histochemically. (See Neilson et al 1970, for a review.) Balboni (1971) has described the distribution of adrenergic fibres. The structures innervated, apart from the vessels, remain uncertain.

Vestigial Structures

The *epoöphoron* (**8.**202), in the lateral part of the mesosalpinx (p. 250) between the ovary and uterine tube, consists of 10–15 short, blind-ended *transverse ductules* converging towards the ovary; their other ends open into a rudimentary *longitudinal duct*, running medially in the broad ligament, parallel with the lateral part of the uterine tube. Between the epoöphoron and the fimbriated end of the tube are often one or more small cystic *vesicular appendices*. Occasionally the longitudinal epoöphorontic duct (*duct of Gartner*) runs along the uterus nearly to the internal os, penetrating its muscular wall to descend in the wall of the cervix uteri, gradually approaching the mucosa though remaining superficial to it and descending in the lateral vaginal wall to end near or at the free margin of the hymen. The *paroöphoron* consists of a few rudimentary tubules scattered in the broad ligament between the epoöphoron and uterus, most easily seen in children. Both the epoöphoron and paroöphoron tubules are remnants of mesonephric tubules; the duct of the epoöphoron is a persistent part of the mesonephric duct (p. 250).

The Uterine Tubes

The two **uterine (Fallopian) tubes** (8.202, 208), which carry ova to the uterine cavity, lie in the upper margins (mesosalpinges) of the broad ligaments. Each is about 10 cm long and opens medially into the uterine cavity's superior angle and laterally into the peritoneal cavity near its ovary. The uterine ostium is minute; the peritoneal opening or *abdominal ostium*, when relaxed, has a diameter of about 3 mm; it is deep within a trumpet-shaped expansion of the uterine tube, the *infundibulum* and its circumference is prolonged by a varying number of *fimbriae*, hence the term *fimbriated end*. The fimbriae are lined by mucosa which, in larger fimbriae, has longitudinal folds continuous with those in the infundibulum. One *ovarian fimbria*, longer and more deeply grooved than the others, is applied to the tubal pole of the ovary. The infundibulum opens into the *ampulla*, which is thin-walled, has a tortuous lumen and is rather more than half the tube's length; this leads into the *isthmus*, rounded and firm, and forms approximately the tube's medial third. The terminal, intramural, *uterine* part is about 1 cm long. The tube extends laterally as far as the inferior (uterine) pole of the ovary, ascends along the mesovarian border to the tubal pole, arches over this and turns down to end close to the ovary's free border and medial surface. On the fimbriae, or adjacent broad ligament, one or more small, pedunculated *vesicular appendices* often occur (vide supra and **8.202**).

Structure (8.205, 206)

The uterine tube has an external *serosa*, an intermediate *muscular* stratum and an internal *mucosa*. The *serosa* is peritoneum with subjacent connective tissue. The *muscular layer* has external longitudinal and internal circular layers of non-striated muscle; additional *internal* longitudinal fibres appear in some parts. The uterine or intramural region has attracted attention (Lisa et al 1954, Sweeney 1962), because of the notion that a sphincter might exist to cut off the uterine cavity from the tube and peritoneum, guarding against infection. No such sphincter has been found but internal longitudinal fibres have been confirmed. The isthmal part has the thickest muscular layer, most reduced where the isthmus joins the ampulla, the thickening largely involving the circular muscle, and usually has the smallest lumen here, varying from 0.1 to 1.0 mm. Isthmal mucosal folds are less complex than in the infundibulum and ampulla, with usually three to six primary folds. Evidence suggests that this stretch of the tube acts as a sphincter, delaying progress of the segmenting zygote perhaps to ensure that it reaches a sufficiently advanced state of development to implant in the endometrium. (An agent able to overrule this sphincter might be a valuable postcoital contraceptive, Mastroianni 1962.) In the ampulla, internal longitudinal muscle is absent and the external longitudinal and circular one much intermingled. Infundibular musculature is similar but attempts to explain fimbrial movements are not yet very convincing. A sphincter at the abdominal ostium has been described, on the basis of experimental distension (Whitelaw 1933, Woodruff & Pauerstein 1969) but has not been histologically confirmed.

The tubal *mucosa* has an epithelium and underlying connective tissue containing blood and lymph vessels and nerves. The epithelium is classically described as columnar ciliated but at least three types of cells occur, only one ciliated, the others secretory and intercalated. They have an incomplete basement membrane. Intercalary cells and another undifferentiated type with a dark nucleus (Pauerstein & Woodruff 1967) may be secretory precursors or their exhausted remnants. The ultrastructure of the ciliated cells shows no peculiarities (pp. 31–32); the secretory cells vary during the menstrual cycle, as seen by light (Snyder 1924) and electron (Hashimoto et al 1960) microscopy, indicating increased secretory preparation during the follicular, and augmented secretory discharge during the luteal, phases of the menstrual cycle.

Tubal mucosa is invaginated to form the major plicae, each with secondary and even tertiary folds. In transverse section (**8.**205, 206) the lumen appears much invaded by plicae, with cores of connective tissue, capillaries and supporting vessels, particularly in the ampulla. The tube transports ova, spermatozoa, zygote, the pre-implantation morula and blastocyst (p. 126). Spermatozoa must be maintained viable prior to fertilization; similarly, the blastocyst, though perhaps more dependent on its own resources, must be maintained in a suitable fluid environment. The dynamics of these activities have been much studied but are beyond the scope of this text. Why so many spermatozoa are required initially remains unexplained; or why so few reach the ampulla, where they await the presumptive ovum. Tubal fluid is an intermediary for respiratory gases, spermatozoal demands for oxygen being high. Capacitation (p. 121) also occurs in the tube. The mode of entry of presumptive ova into the tube is still uncertain; conveyance by fimbriae, suction by ciliary action and so on, have been suggested, but evidence is scant. However, in general it is agreed that muscular movements are the most important factor in the transport of ova and spermatozoa; this is demonstrated by the fact that women suffering from immobile cilium syndrome are still able to bear children. The highly plicated tubal epithelium, the isthmic sphincter and other details accord with the needs of interchange of gases, nutrients and hormones and with the delay necessary to ensure blastocystic readiness for implantation; but the mechanisms involved are not clear. Contractile patterns and the effects on them of nerves, hormones (oestrogens and progesterone) and transmitters have been extensively investigated in experimental animals. The transient arrest of the zygote at the

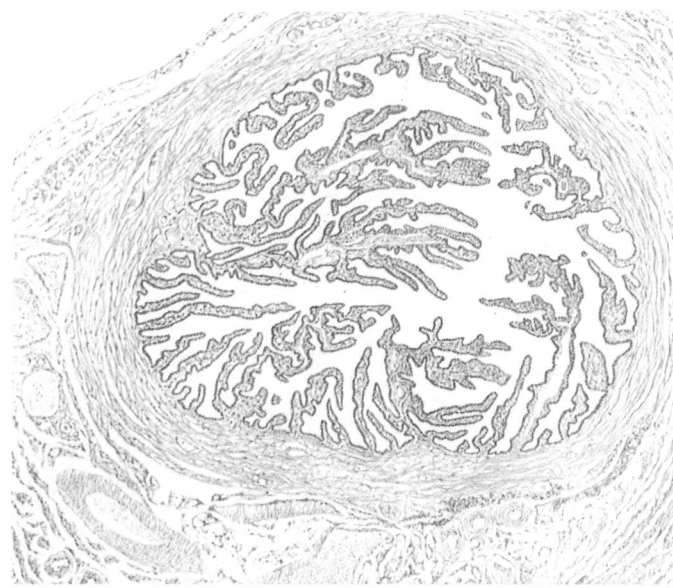

8.205 Transverse section through the ampulla of a human uterine tube. Stained with haematoxylin and eosin. Magnification × 15.

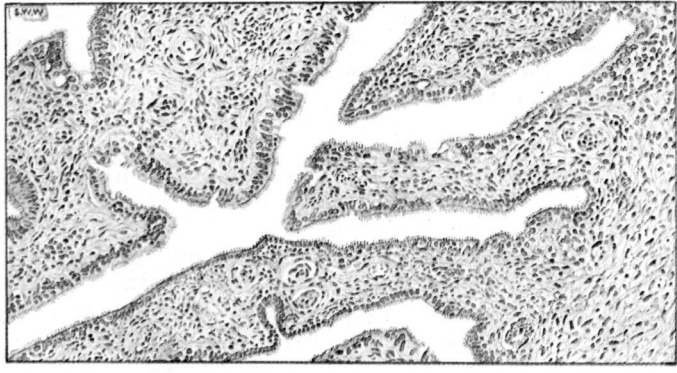

8.206 Section showing the plicated mucous membrane in the ampullary part of human uterine tube. Note the columnar epithelium. Stained with haematoxylin and eosin. Magnification × 75.

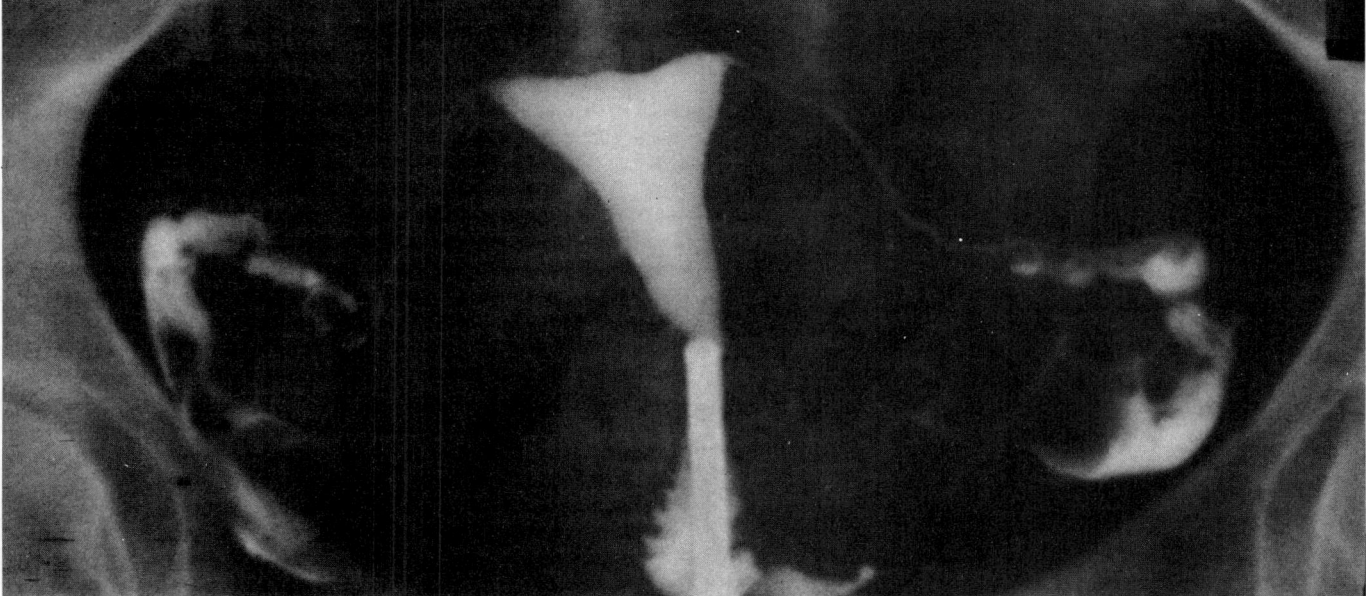

8.207 A radiograph of the uterus and uterine tubes after the introduction of a radio-opaque contrast medium through a cannula passed through the vagina into the cervix uteri. The shadow of the cannula is visible; above this is the triangular cavity of the uterine body and fundus. From the superior angles the lumina of the narrow (less than 1.0 mm diameter) intramural and isthmic parts of the uterine tubes may be traced in inferolaterally, where they expand into the wider (2–4 mm diameter) ampullary parts of the tubes. Some contrast medium has escaped into the pelvic cavity from the abdominal ostia. Radiograph provided by J Hilliger Smitham, Chelsea Hospital for Women, London.

isthmo-ampullary junction (the so-called isthmic block) has attracted special attention; the isthmic sphincter is probably adrenergic, like many others. Histochemical findings suggest an absence of cholinergic control in some parts, at least, of the human uterine tube (Nakanishi et al 1967). Prostaglandins (pp. 121, 1407), occurring in human semen, stimulate the muscle in the medial part of the tube and inhibit it in the ampullary region.

Vessels and Nerves

The vessels of the uterine tube arise from *ovarian* and *uterine* stems. Some disagreement on the regions supplied by the two arteries is apparent in the literature on this subject; it has even been claimed that the intramural tube is supplied by the ovarian artery, but arteriographic evidence (Borell & Fernstrom 1953) shows that the uterine artery usually supplies its medial two-thirds, the ovarian the remainder, partition between the two (which in any case anastomose) being variable (consult Koritke et al 1967). The veins are arranged similarly; intrinsic mucosal, muscular and subserous networks have been described (Gatsalov 1963, Koritke et al 1967). *Lymphatic drainage* follows the veins (p. 856); its intrinsic details have been extensively described (Sampson 1937, Gatsalov 1963). The *nerve supply*, largely along the ovarian and uterine arteries, shows a similar pattern of distribution. Most of the tube has sympathetic and parasympathetic supplies; vagal fibres reach the lateral half, pelvic splanchnic fibres the medial moiety. Sympathetic supply is from the tenth thoracic to the second lumbar spinal segments. Afferent fibres travel with the sympathetic motor supply, entering the cord through corresponding dorsal roots. Modified Pacinian corpuscles appear in the ampullary submucosa (Chiara 1959). Intrinsic innervation has been studied by metallic impregnations (Chiara 1959, Damiani et al 1961), and fluorescence microscopy (Owman & Capodacqua 1967, Kubo 1970). In the wall of the tube are postganglionic sympathetic fibres and visceral afferents, and preganglionic parasympathetic fibres which must form synapses, although ganglia have not been reliably reported in the wall or near it; the relay may be in paracervical ganglia. Afferent autonomic fibres may also accompany parasympathetic nerves.

Applied Anatomy

Pelvic peritonitis is said to occur more frequently in females because infection of the vagina, uterus or of the uterine tube may spread directly to the peritoneum, via the abdominal ostium. Pus in the recto-uterine pouch is palpable through the posterior vaginal fornix. Tubal inflammation (*salpingitis*) is usually secondary, having spread from the vagina or the uterus. The fimbriated end may be closed by adhesions and pus then collects in the tube (*pyosalpinx*).

Fertilization (p. 123) usually occurs in the ampulla; normally the segmenting zygote enters the uterus but it may adhere to and develop in the tube, the commonest variety of *ectopic gestation* (p. 131); an amnion and a chorion are formed, but true decidua never develop; such gestation usually ends by extrusion through the abdominal ostium, but sometimes by tubal rupture into the peritoneal cavity accompanied by severe haemorrhage (Pauerstein & Woodruff 1967, Woodruff & Pauerstein 1969).

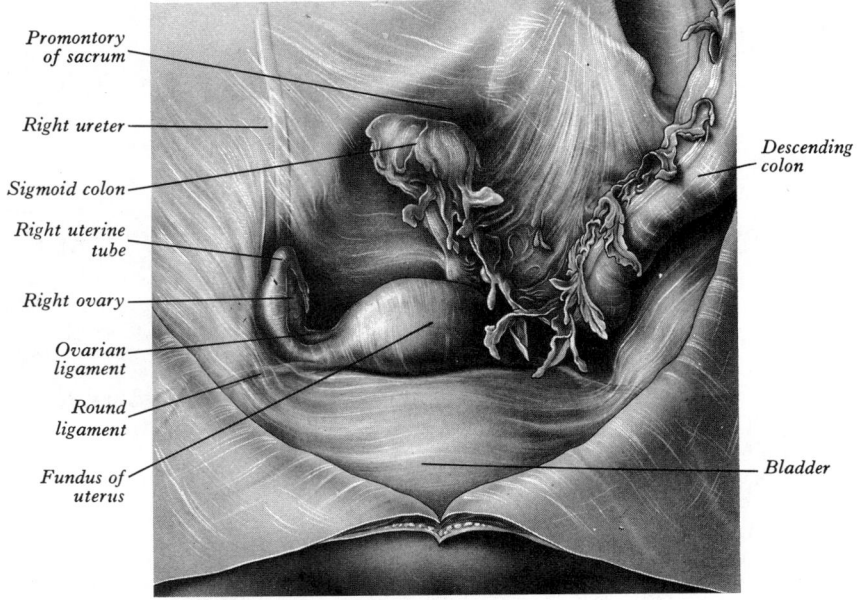

Promontory of sacrum

Right ureter

Sigmoid colon

Right uterine tube

Right ovary

Ovarian ligament

Round ligament

Fundus of uterus

Descending colon

Bladder

8.208 A female pelvis and its contents: anterosuperior view.

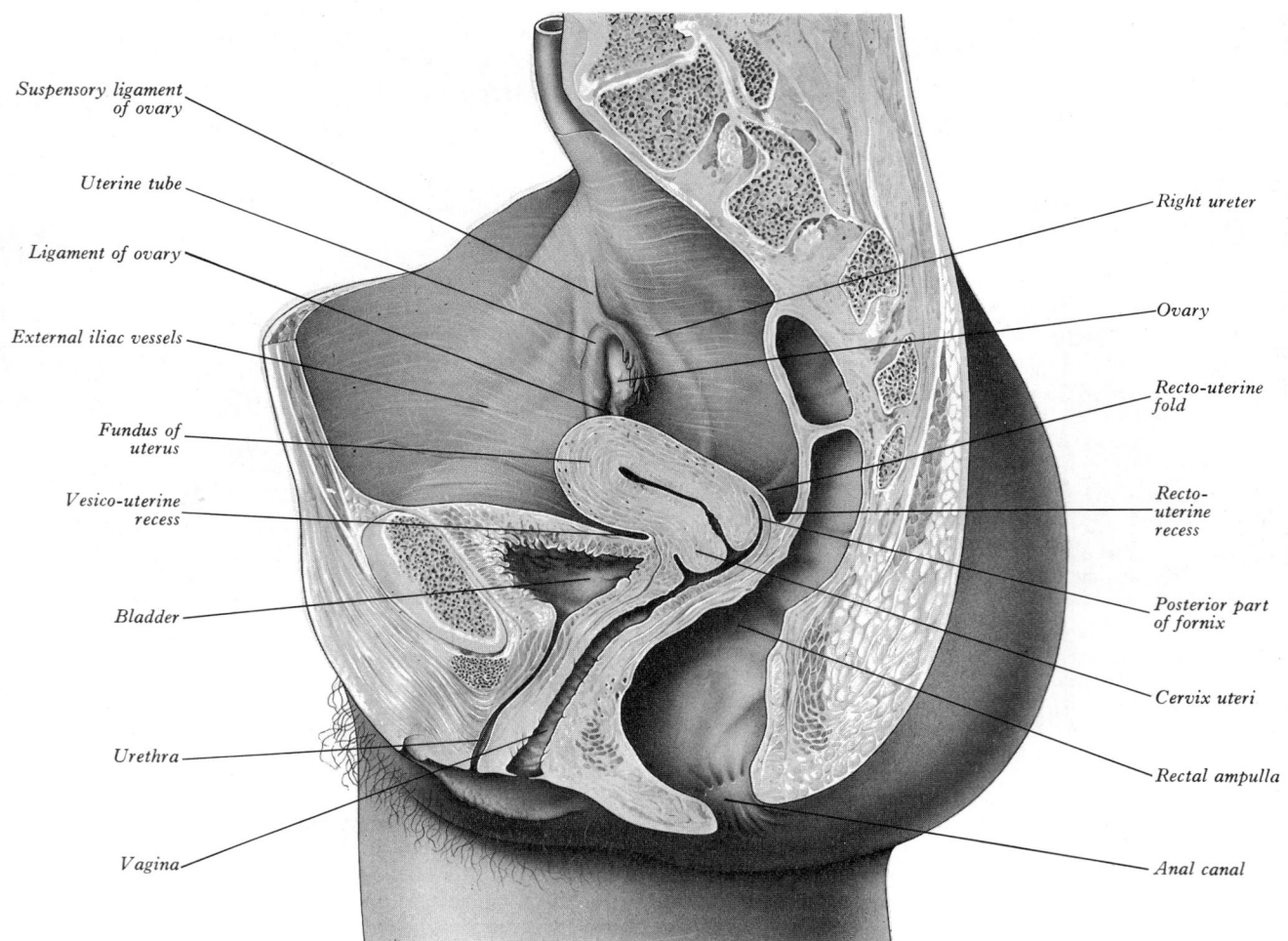

Suspensory ligament of ovary

Uterine tube

Ligament of ovary

External iliac vessels

Fundus of uterus

Vesico-uterine recess

Bladder

Urethra

Vagina

Right ureter

Ovary

Recto-uterine fold

Recto-uterine recess

Posterior part of fornix

Cervix uteri

Rectal ampulla

Anal canal

8.209 Median sagittal section through a human female pelvis. The peritoneum is shown in blue.

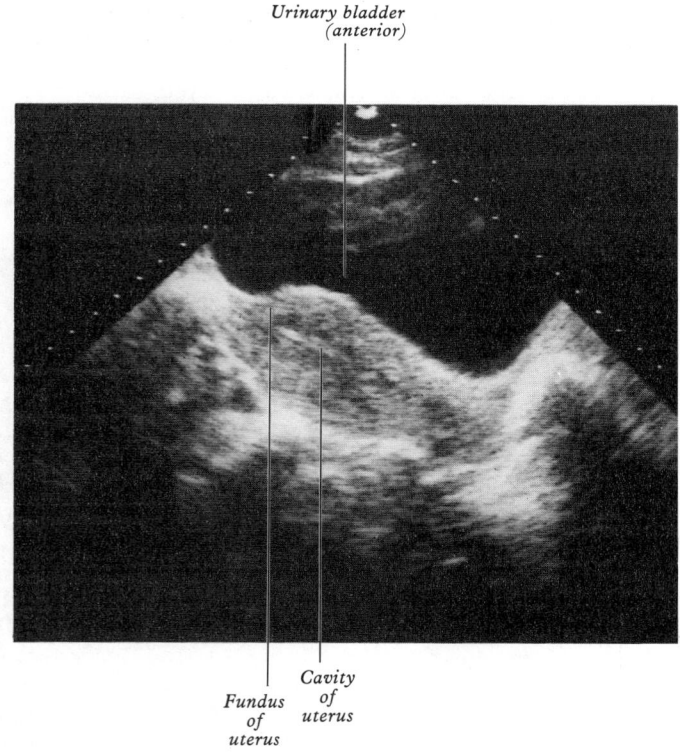

Urinary bladder (anterior)

Fundus of uterus

Cavity of uterus

8.210 Sagittal ultrasonogram of the pelvis of an adult woman. Supplied by Shaun Gallagher, Guy's Hospital; photography by Sarah Smith, UMDS, Guy's Hospital Campus, London.

The Uterus

The uterus (**8.**202, 208, 209), hollow, thick-walled and muscular, is normally situated in the lesser pelvis between the urinary bladder and rectum. Into its upper part open the uterine tubes, one on each side; below, it continues into the vagina. Presumptive ova, ova or their derivatives are carried to the uterine cavity by the tubes; if fertilized, a blastocyst forms which embeds in the uterine lining and is normally retained until development is complete, the uterus adapting in size and structure to the needs of the growing embryo and fetus. After parturition it returns almost to its former condition, though somewhat larger than in its nulliparous state. In the *adult nulliparous state* it is flat anteroposteriorly and piriform, its narrow end postero-inferior in position. It lies between the bladder antero-inferiorly and the sigmoid colon and rectum posterosuperiorly; it lies completely below the pelvic inlet. The long axis is usually approximately in the axis of the pelvic *inlet* (p. 429), but since it is movable its position varies with the distension of the bladder and rectum. Except when displaced by a much distended bladder, it forms almost a right angle with the vagina, since the latter's axis corresponds to that of the *pelvic outlet* (p. 430). It is about 7.5 cm in length, 5 cm in breadth at its widest, nearly 2.5 cm in thickness, weighs 30–40 g and is divisible into two regions. On the surface, a little below the uterine midpoint, is a slight constriction corresponding to narrowing of the cavity at the *internal os*, the part above being the *corpus* or *body* and, below, the *cervix*. The part of the corpus above the entry-points of the uterine tubes is the *fundus*. (It is, of course, the *highest* part but the *deepest* when approached via the cervix.)

The corpus uteri. The body gradually narrows from fundus to internal os. The *anterior (vesical) surface*, apposed to the

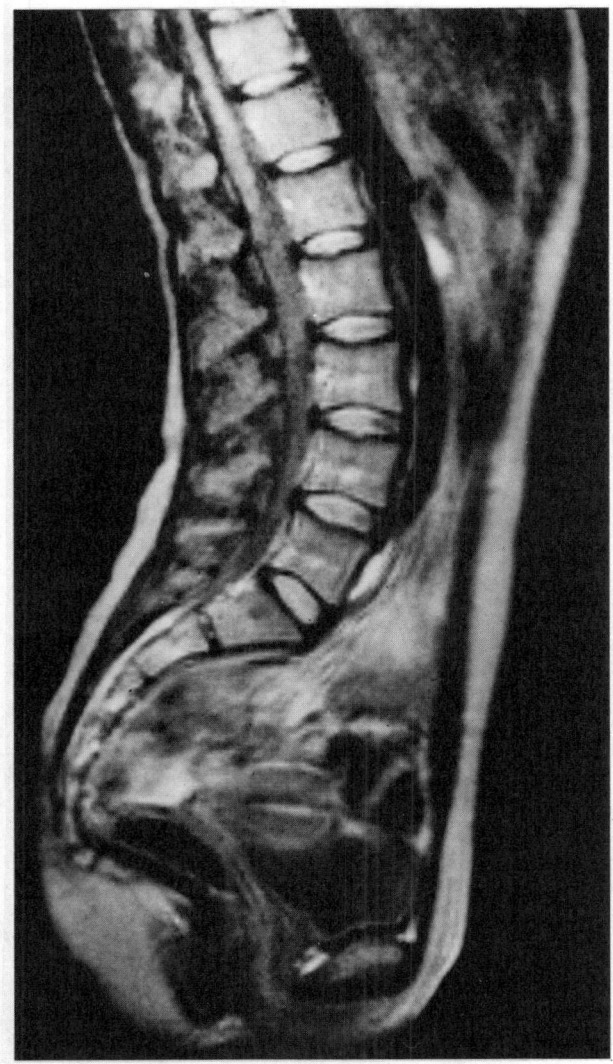

8.211A Sagittal magnetic resonance image of the lumbar spine and pelvis of an adult woman. Supplied by Philips Medical Systems; photography by Sarah Smith, UMDS, Guy's Hospital Campus, London.

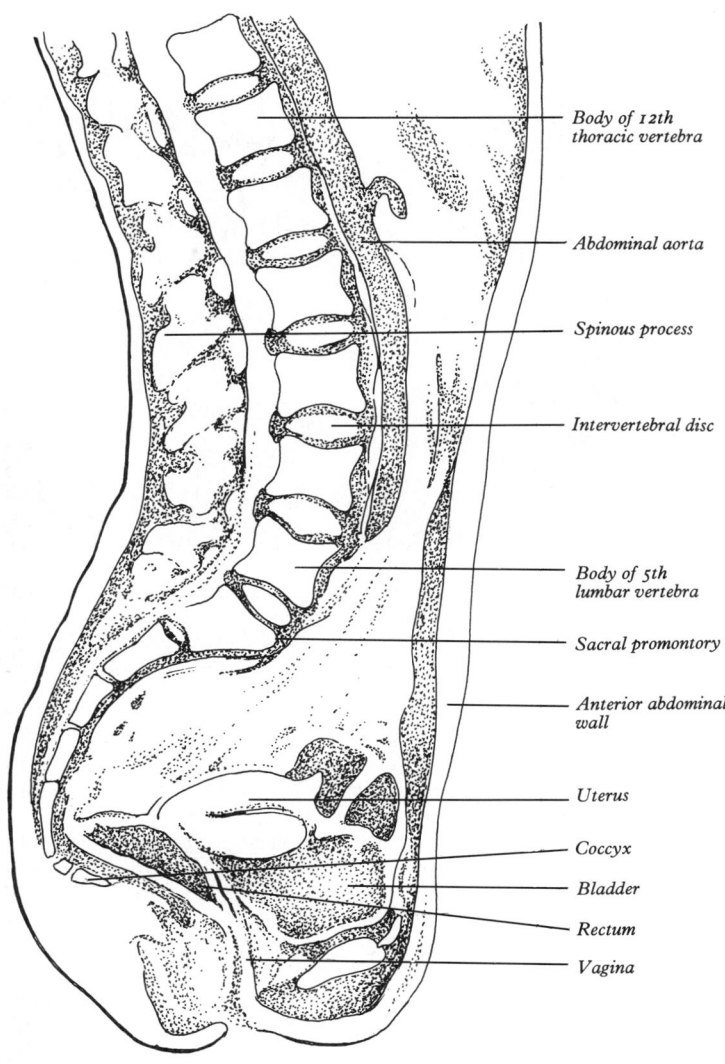

Body of 12th thoracic vertebra

Abdominal aorta

Spinous process

Intervertebral disc

Body of 5th lumbar vertebra

Sacral promontory

Anterior abdominal wall

Uterus

Coccyx

Bladder

Rectum

Vagina

8.211B Diagram illustrating the major features demonstrated in 8.211A.

urinary bladder, is flattened and covered by peritoneum, reflected on to the bladder as the uterovesical fold level with the internal os. Between the bladder and uterus is the *vesico-uterine pouch*, usually empty but sometimes occupied by part of the intestine (p. 1338).

The *posterior (intestinal) surface*, convex transversely, is covered by peritoneum, which is continued down to the cervix and upper vagina and then reflected back to the rectum (8.209). It is related to the sigmoid colon and usually separated from it by the terminal ileal coil. The dome-like *fundus* is covered by peritoneum continuous with that on neighbouring surfaces. Coils of small intestine and occasionally distended sigmoid colon contact it. The *lateral margins* or sides are convex; at their upper ends the uterine tubes traverse the uterine wall. Antero-inferior to this are attached the round ligament and postero-inferior to it the ligament of the ovary, both running in the broad ligament and stretching from the lateral uterine margin to the lateral pelvic wall (p. 1444).

The cervix uteri, about 2.5 cm in length, is narrower and more cylindrical than the corpus, and is widest at its mid-level; it is also less mobile than the body of the uterus, so that their axes are seldom in line. The uterine long axis is concave forwards, described as *anteflexed*; occasionally there is an angular bend at the level of the internal os (*acute* anteflexion). With the bladder empty the cervix meets the vagina at an angle facing antero-inferiorly, the whole uterus being turned anteriorly on it or *anteverted*. The cervix bulges into the anterior vaginal wall, which divides it into supravaginal and vaginal regions (8.209).

The *supravaginal part* of the cervix is separated *in front* from the bladder by cellular connective tissue, the *parametrium*, which passes also to the sides of the cervix and laterally between the two layers of the broad ligaments. The uterine arteries flank the cervix in this tissue and the ureters descend forwards in it about 2 cm from the cervix (p. 1414). The relation of the arteries to the ureters is not always symmetrical; one ureter may be anterior to the cervix. *Posteriorly* the supravaginal cervix is covered by peritoneum, prolonged below on to the posterior vaginal wall and then reflected on to the rectum, forming a *recto-uterine recess* (8.209); it is related to the rectum, but may be separated from it by a terminal ileal coil.

The *vaginal part* of the *cervix* projects into the anterior vaginal wall, forming grooves termed *vaginal fornices* (p. 1445). On its rounded end the small, circular *external os* connects its cavity and the vagina. In parous women the external os has anterior and posterior lips, the anterior shorter, thicker and projecting lower than the posterior. Normally both contact the posterior vaginal wall.

The uterine cavity (8.207, 214) is relatively small, partly due to its thick wall. The *cavity of the body* is a mere transverse slit in sagittal section, the anterior and posterior walls being almost in contact. In coronal section it is triangular, its base being formed by the internal fundal surface between the openings of the uterine tubes; its apex is the internal os, leading to the cervical canal (8.209).

The *cervical canal* is somewhat fusiform, flattened transversely and broadest at mid-level. It communicates by the *internal os* with the main uterine cavity, by the *external os* with the vagina.

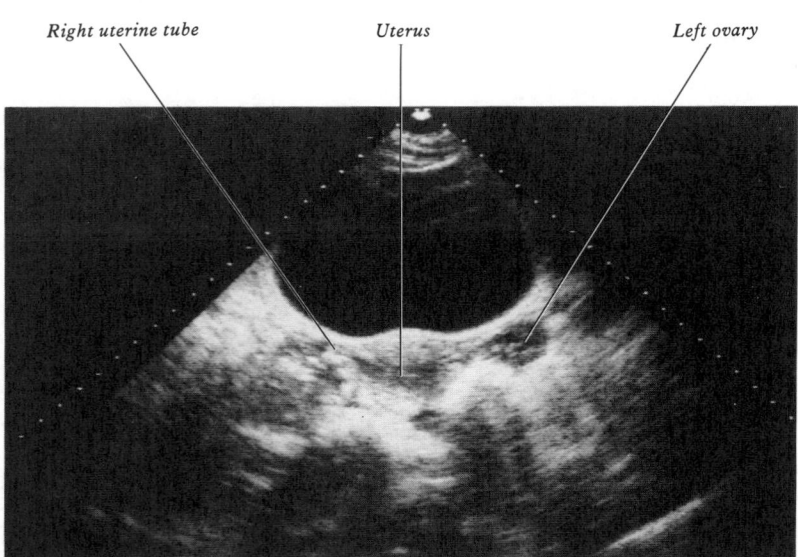

Right uterine tube   *Uterus*   *Left ovary*

8.212 Transverse ultrasonogram of the female pelvis through the level of the uterus and left ovary. Supplied by Guy's Hospital; photography by Sarah Smith, UMDS, Guy's Hospital Campus, London.

From a longitudinal ridge on its anterior and posterior walls small oblique *palmate folds* ascend laterally, like branches of a tree (*arbor vitae uteri*). Folds on opposing walls interdigitate to close the canal. The total length of the uterine cavity from external os to the fundus is about 6 cm.

The *isthmus of the cervix*, the cervical upper third, has distinctive features (Stieve 1927, Frankl 1933). Although unaffected in the first month of pregnancy, it is gradually taken up into the uterine body during the second month to form the '*lower uterine segment*'. Fetal membranes, firmly blended elsewhere with the uterine mucosa, are not attached to this lower segment. In non-pregnant women the isthmus undergoes menstrual changes, though less pronouncedly than in the uterine body. Histologically it resembles the body: its epithelium is low, cylindrical and ciliated, its mucosa thinner and its glands fewer than in the cervix.

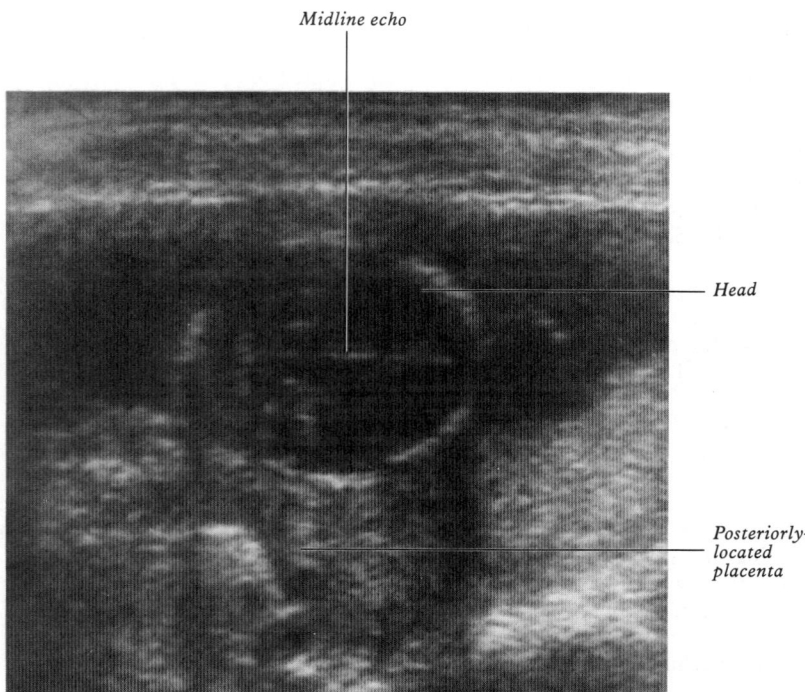

Midline echo

— *Head*

Posteriorly-located placenta

8.213A Ultrasonogram through the uterus showing the head and placenta of a 16-week fetus. Supplied by Shaun Gallagher, Guy's Hospital; photography by Sarah Smith, UMDS, Guy's Hospital Campus, London.

Age and Reproductive Changes

The functional anatomy of the human uterus has been described in details by Lopes & Barriere (1986). Uterine form, size and position vary in different periods and circumstances.

In fetal life the uterus projects above the lesser pelvis and the cervix is considerably larger than the body. *At puberty* it is piriform and weighs 14–17 g; the fundus is just below the superior pelvic aperture; palmate folds are distinct and extend to the upper part of the uterine cavity. *In adults* its position varies, depending chiefly on the contents of the bladder and rectum. With an empty bladder the entire uterus bends anteriorly, curved at the junction of the body and cervix, the body contacting the bladder. As the bladder fills, the uterus gradually becomes more erect until, with a full bladder, the fundus may turn towards the sacrum.

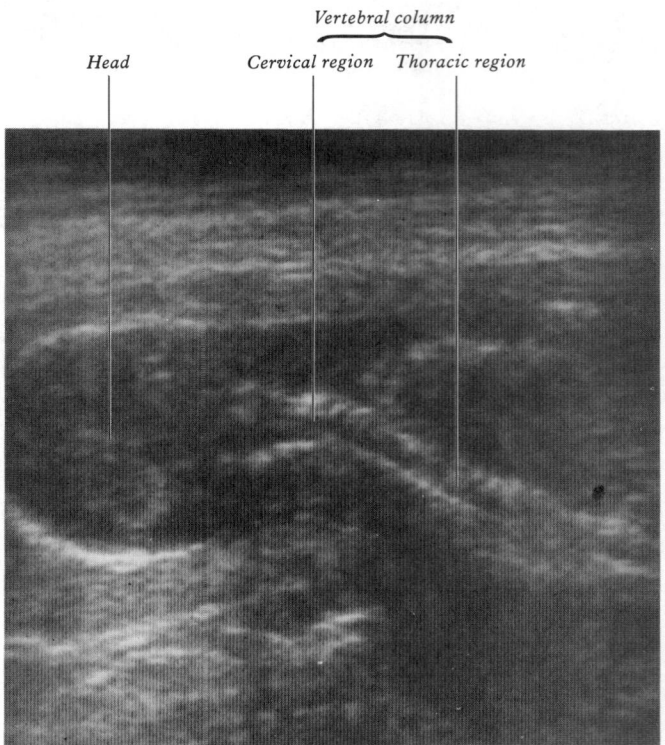

Vertebral column

Head   *Cervical region* *Thoracic region*

8.213B Ultrasonogram through the uterus showing the head and upper part of the vertebral column of a 16-week fetus, in sagittal section. Provided by Shaun Gallagher, Guy's Hospital; photography by Sarah Smith, UMDS, Guy's Hospital Campus, London.

During menstruation the organ is slightly enlarged and more vascular; its surfaces are rounder; the external os is rounded, its lips swollen; the endometrium is darker. *During pregnancy* the uterus becomes greatly enlarged and reaches the epigastric region in the eighth month. The increase in wall thickness is mainly due to hypertrophy of existing myocytes, but partly to formation of new fibres. As pregnancy proceeds, the wall progressively thins. *After parturition* the uterus nearly regains resting size, weighing about 42 g; but its cavity remains larger, its vessels tortuous and its muscular layers thicker and more defined; the external os is more prominent, its edges variably fissured.

In old age the uterus becomes atrophied, paler and denser in texture; a more distinct constriction separates the body and cervix. The internal os is frequently obliterated, the external os occasionally, when the latter's lips almost disappear.

Structure

The uterine wall has an external serosal, middle muscular and an internal mucosal layer.

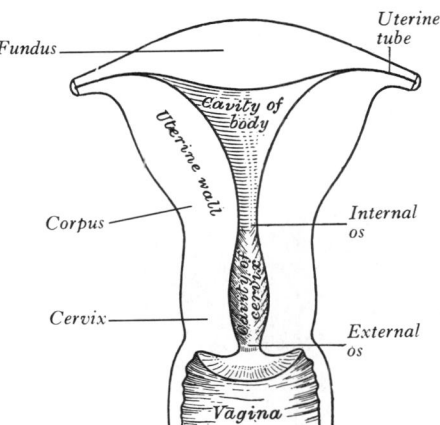

8.214A Sectional diagram showing the interior divisions of the uterus and its continuity with the vagina.

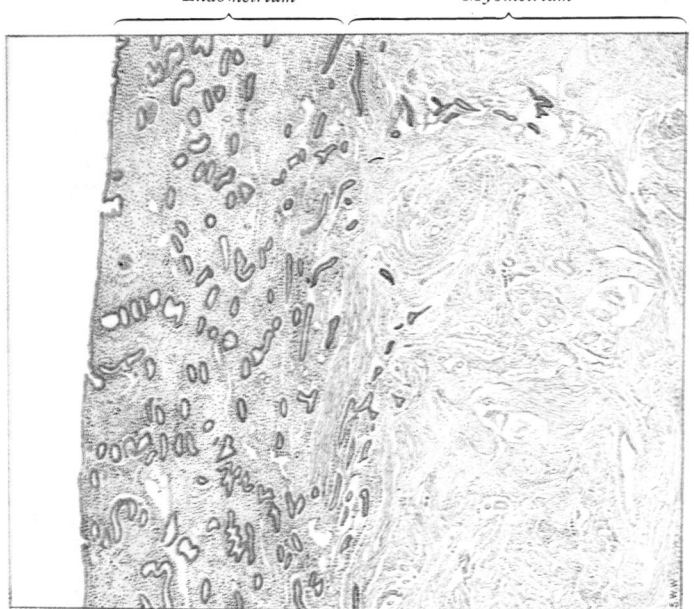

Endometrium   Myometrium

8.214B Section of human endometrium and underlying musculature in the interval phase. Note that some glands extend into the more internal layers of the myometrium. Stained with haematoxylin and eosin. Magnification × 20.

The serosa (perimetrium) is peritoneum, which posteriorly covers the uterine body and supravaginal cervix but in front only the body. In the lower posterior quarter the peritoneum is separated from the uterus by loose cellular tissue and large veins. Beneath is a subserous layer of areolar tissue.

The muscular layer (myometrium) forms most of the uterine wall. In nulliparae it is dense, firm, greyish and (in the fixed state) cuts almost like cartilage. It is about 1.25 cm thick at the mid-level and fundus and thin at the tubal orifices. Its bundles of non-striated myocytes intermix with loose connective tissue, blood vessels, lymph vessels and nerves. During pregnancy the muscle hypertrophies, the myocytes being much enlarged. Although these interlace in all directions, they form variably distinct external, middle and internal layers. The cervical muscle contains more collagen and elastin than does the body. The external layer is mostly longitudinal, fibres passing over the fundus to converge at the lateral angles and continuing into the uterine tubes and the round and ovarian ligaments; some enter the broad ligaments, others turn back into the uterosacral ligaments. The intermediate layer is thickest and contains longitudinal, oblique and transverse fibres, together with larger blood vessels. The myocytes of the internal layer are longitudinal and circular; deep parts of the uterine glands are related to it. An infundibular array of elastin and non-striated myocytes has been described in the cervix, its narow extremity adjacent to the external os (Hamper 1970); this may act as a valve in keeping the os closed. Toth (1977) has described bilateral non-striated longitudinal fascicles in the uterine wall, extending in the lateral regions from the fundal angle to the cervix and largely submucosal. Each fascicle, in its juxtafundal part, consists of dispersed subfascicles, but is more compact near the cervix; epithelial rests were observed in them, suggesting a mesonephric origin; their myocytes differ structurally from those of myometrium proper, suggesting that they may provide fast conducting pathways via gap junctions (p. 24) which might integrate uterine activity.

The mucosa (endometrium) (**8.214B**) is continuous through the uterine tubes with the peritoneum and, through the external os, with the vaginal mucosa. In the uterine body its epithelium is columnar, ciliated before puberty but usually non-ciliated over large areas in the adult uterus. The subepithelial layer consists of embryonic, highly cellular connective tissue containing blood vessels, lymphatic spaces and many tubular *uterine glands* lined by columnar cells, patchily ciliated, and opening into the uterine cavity (see p. 149).

Cyclical endometrial changes. Between puberty and the menopause, in each lunar month, the endometrium undergoes a *menstrual cycle*, with a duration varying even in the same individual. Histological details and hormonal factors are described on pp. 146–150.

In the upper two-thirds of the cervix, the mucosa has numerous, deep, glandular follicles secreting a clear, viscid, alkaline mucus; throughout the canal small cysts (*ovula Nabothi*) occur, presumably follicles occluded and distended by secretion. The mucosa of the lower half of the canal has numerous papillae. Its epithelium in the upper two-thirds is cylindrical and ciliated, but below this it loses cilia and, near the external os, becomes non-keratinized stratified squamous, as it is on the vaginal aspect of the cervix. The mucosa of the cervical lower two-thirds does not show menstrual changes (pp. 146–150). See Ferenczy (1980) for a review of the ultrastructure of the human cervix.

Vessels and Nerves

The main arterial supply to the whole uterus is by way of the uterine branch of the internal iliac artery on each side (p. 778). This anastomoses with ovarian and vaginal arteries, but the dominance of the uterine arteries is indicated by their marked hypertrophy during pregnancy, which affects them alone. They

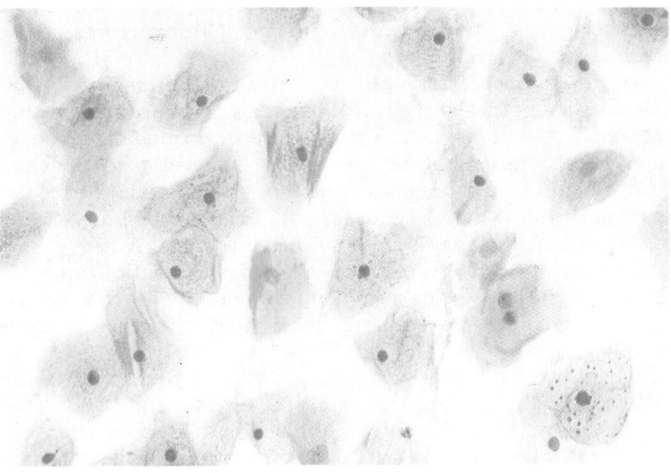

8.215 Cells of the mature vaginal epithelium in a diagnostic smear preparation stained by Papanicolaou's technique (Shorr modification). All the cells shown are superficial keratinizing squames, some of which show actual keratin granules. The pink cells are the oldest and most superficial. Preparation provided by Max Levene, St Helier Hospital, Carshalton.

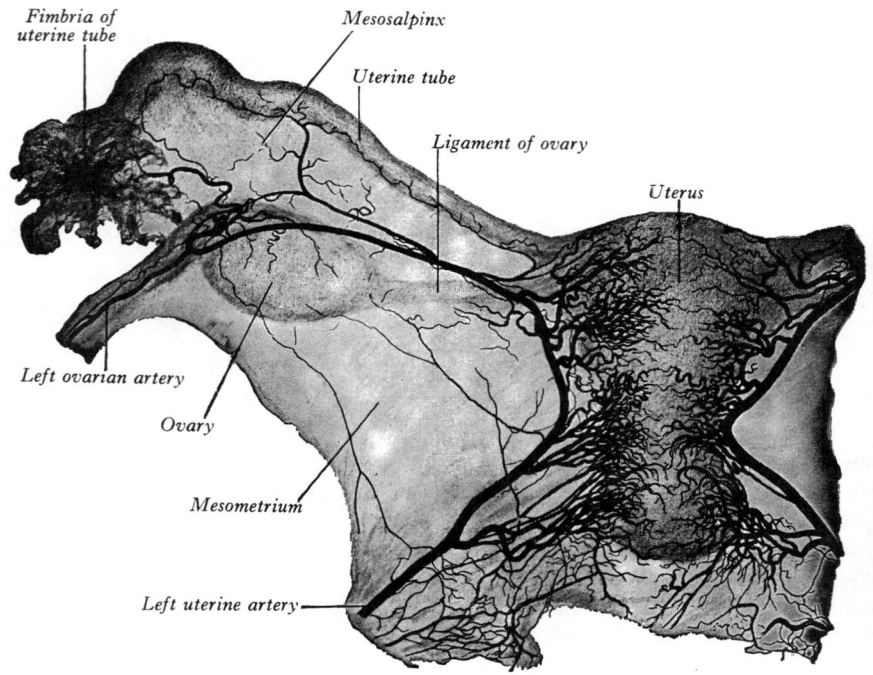

8.216 Posterior aspect of a cleared injected specimen (prepared by Hamilton Drummond) to show the distribution of the left uterine and ovarian arteries of a female aged $17\frac{1}{2}$ years.

anastomose extensively with each other; one can be ligated without serious interference (Siegel & Mengert 1961) and even more extensive ligation has succeeded. Their tortuosity as they ascend in the broad ligaments is repeated in their branches within the uterine wall, but all sinuosities disappear as the pregnant uterus expands; however, they are obvious in nulliparae. Each uterine artery gives numerous branches which immediately enter the uterine wall, where they divide into groups of *anterior* and *posterior arcuate arteries*, passing transversely in the myometrium; their terminal branches anastomose across the midline (**8.216**). Anterior and posterior regions hence have median zones devoid of larger vessels, but which are *not* avascular. The arcuate arteries supply many tortuous *radial branches* which pass centripetally through the deeper myometrial layers, supplying these en route, to reach the endometrium. As microradiography and injections show, these provide a series of dense capillary plexuses in the myometrium and endometrium (Farrer-Brown et al 1970). From the former many helical arteriolar rami of the radial branches pass to supply the endometrium, their form being affected by menstrual cycles; during the proliferative phase helical arterioles are less prominent; in the secretory phase they grow in length and calibre, becoming even more tortuous (Ramsey 1955).

Uterine veins are arranged like arteries (Farrer-Brown et al 1970). Volumetrically they are, however, greater. Minute venous endometrial sinuses, a constant feature in pregnant uteri, have also been noted in the resting organ (Schlegel 1946), but not confirmed. During pregnancy *ovarian veins*, unlike their companion arteries, may be enlarged (O'Leary & O'Leary 1966). *Uterine lymphatic vessels* are described on p. 856. Intrinsic lymphatic plexuses of the uterine wall have been little studied in human females; Wislocki & Dempsey (1939) have, however, made a detailed study of arrangements in primates (macaque monkeys).

The *nerves* supplying the uterus are described on p. 1167. The autonomic supply is directly from the ovarian and hypogastric plexuses; sympathetic preganglionic fibres proceed from the twelfth thoracic and first lumbar spinal segments while parasympathetic preganglionic axons issue in the second to fourth ventral sacral spinal roots. Despite general agreement on these pathways, details of their distribution and physiological effects are still ill-defined. Cholinergic and adrenergic fibres have been identified in cervical muscular and submucous strata, the former predominating. Some observers consider that the corpus uteri is innervated only by sympathetic nerves. In a histochemical study (Owman et al 1967) only adrenergic terminals occurred throughout female internal genitalia, including the uterus. The effects of uterine innervation are complicated by hormonal influences and by the evocation of different responses by the same agent (e.g. adrenalin) in differing conditions (Wansbrough et al 1967).

Applied Anatomy

Small degrees of *anteversion* or *retroversion* are not pathological; but when flexion at the junction of the body with the cervix is marked it must be so regarded, especially when retroversion is combined with retroflexion. *Retroversion* is defined as a posterior inclination of the whole uterus, so that the cervix faces forwards; *retroflexion* is a posterior curvature of body alone, at the junction of body and cervix. These conditions are usually combined. Prolapse is another common condition; the uterus sinks abnormally low and may protrude at the vulva, usually after imperfect repair of the pelvic floor following damage sustained during parturition (p. 605).

THE UTERINE LIGAMENTS

The uterus is connected to the bladder, rectum and pelvic walls by 'ligaments'; some are merely peritoneal folds and can provide little mechanical support, others contain non-striated muscle and fibrous tissue and may be real ties, with also some measure of dynamic control.

The **anterior ligament** or *uterovesical fold* consists of peritoneum reflected on to the bladder from the uterus at the junction of its cervix and body. The **posterior ligament** or *rectovaginal fold* is composed of peritoneum reflected from the posterior vaginal fornix on to the front of the rectum, forming the deep *recto-uterine pouch*, bounded anteriorly by the uterus, supravaginal cervix uteri and posterior vaginal fornix, posteriorly by the rectum, laterally by two peritoneal crescentic folds passing back from the cervix uteri on each side of the rectum to the posterior pelvic wall; these *recto-uterine folds* contain much fibrous tissue and non-striated muscle and are attached to the front of the sacrum to form the **uterosacral ligaments**, which can be palpated per rectum lateral to it.

The two **broad ligaments** (**8.202**) extend from the sides of the uterus to the lateral pelvic walls, together with the uterus forming a septum across the lesser pelvic cavity, dividing it into an anterior part containing the bladder and a posterior part containing the rectum, usually the terminal ileal coil and part of the sigmoid colon. With the bladder empty or almost so, the surfaces of each ligament are superior and inferior, with a free anterior and an attached posterior border. As the bladder fills, the plane of the ligaments rises, their free borders becoming superior and their layers anterior and posterior, continuous with each other at the free edge and diverging below near the superior surfaces of levatores ani. Each uterine tube lies in a free border, the part between this and the ovarian mesentery and ligament being the *mesosalpinx*. The tubal infundibulum projects from the free border near its lateral end. The ovary is attached posteriorly by a *mesovarium*. Between the infundibulum and upper (tubular) pole of the ovary and the lateral pelvic wall the broad ligament contains ovarian vessels and nerves, forming a *suspensory ligament of the ovary (infundibulopelvic ligament)*, continued laterally over the external iliac vessels as a distinct fold. Between the ovary and uterine tube the mesosalpinx contains the epoöphoron (p. 1437) and medially the paroöphoron (p. 1437), and anastomoses between the uterine and ovarian vessels. The *mesometrium* is the part of broad ligament extending from the pelvic floor to the ovarian ligament and uterine body. The uterine artery passes between the two layers of broad ligament about 1.5 cm lateral to the cervix, after crossing the ureter (p. 778) and ascending in it; it turns laterally below the uterine tube to anastomose with the ovarian artery. The broad ligament also encloses the ovarian ligament (p. 1435), the proximal part of the round ligament of the uterus, non-striated muscle and areolar tissue.

The two uterine **round ligaments** (8.202) are narrow, flat bands 10–12 cm long lying between the layers of the broad ligament antero-inferior to the uterine tubes. From the lateral uterine angle each passes laterally across the vesical, obturator and external iliac vessels, the obturator nerve and the obliterated umbilical artery. The round ligament enters the deep inguinal ring, round the start of the inferior epigastric artery, traverses the inguinal canal and finally splits into strands which merge with connective tissue in the labium majus. Near the uterus it contains much non-striated muscle which gradually diminishes until its terminal part is purely fibrous. It is accompanied by blood vessels, nerves and lymphatics; the last drain the uterine region around the entry of the uterine tube to the superficial inguinal lymph nodes (p. 856); uterine neoplasms may spread by this route. In the fetus a peritoneal *processus vaginalis* is carried with the round ligament for a short distance into the inguinal canal, but it is generally obliterated in adults, although sometimes patent even in old age. In the canal the ligament receives the same coverings as the spermatic cord (p. 1431), although they are thinner and blend with the ligament itself, which may not reach the labium majus but ends by fusing with these coverings. The round and ovarian ligaments are together homologous with the gubernaculum testis (p. 257).

The **transverse cervical ligament** (of Mackenrodt) extends from the side of the cervix to the vault and lateral fornix of the vagina and is continuous with the fibrous tissue around the pelvic blood vessels; it probably plays a part in stabilizing the uterus (hence its gynaecological importance). Other dense strands of pelvic fascia connect the cervix and upper vagina to the back of the pubis.

While the above ligaments (and vagina) may act in varying measure as mechanical uterine supports, the levatores ani and coccygei, urogenital diaphragm and perineal body appear at least as important in this respect (Wendell-Smith & Wilson 1977).

The Vagina

The vagina (8.209), the female copulatory organ, is a fibromuscular tube lined by stratified epithelium, extending from the vestibule (the cleft between the labia minora) to the uterus; the bladder and urethra are anterior, the rectum and anal canal posterior. The vagina ascends posterosuperiorly, at an angle of over 90° to the uterine axis, but which varies with the contents of the bladder and rectum. The inner surfaces of its walls are ordinarily in contact, its lumen forming an H in transverse section in its lower part, the transverse line being slightly convex forwards or backwards and the lateral limbs medially convex; at its intermediate level the lumen is a transverse slit. Its anterior wall is 7.5 cm in length, the posterior 9 cm; its width increases as it ascends. Its upper end surrounds the vaginal projection of the cervix uteri near the external os, attached higher on the posterior cervical wall than on the anterior. The annular recess between the cervix and vagina is the *fornix*; anterior, posterior and lateral parts of this recess are often named, but the recess is essentially continuous.

The *anterior* vaginal wall is related to the urethra, which is embedded in it, and to the base of the bladder. The *posterior* wall, covered by peritoneum in its upper quarter, is separated from the rectum by the recto-uterine pouch above, and by moderately loose connective tissue in its middle half; in its lower quarter it is separated from the anal canal by the musculofibrous *perineal body*. Laterally are the levatores ani (p. 605) and pelvic fascia. As the ureters pass anteromedially to reach the fundus of the bladder, they pass close to the lateral fornices and as they enter the bladder they are usually in front of the vagina (p. 1414). Each ureter is crossed transversely here by a uterine artery (p. 778).

Structure

The vagina has an inner mucosal and an external muscular stratum, the lamina propria of the former containing many thin-walled veins.

The mucosa adheres firmly to the muscular layer; on its epithelial surface are two median longitudinal ridges, one anterior and the other posterior. From these *vaginal columns* numerous transverse bilateral rugae extend, divided by sulci of variable depth, giving an appearance of conical papillae which are most numerous on the posterior wall and near the orifice. These are especially well developed before parturition. The epithelium is non-keratinized stratified squamous. After puberty it thickens and is rich in glycogen. Unlike many other mammals, human vaginal epithelium does not change markedly during the menstrual cycle, but glycogen increases after ovulation and then diminishes towards the cycle's end. The fermentative action of certain bacteria (Döderlein's bacillus) on the glycogen renders vaginal fluid acid. There are no mucous glands; mucus from the cervical glands lubricates the vagina.

Cyclical variations in the vaginal epithelium have been much studied in many animals, including the human, by taking 'smears' to examine cell types and their relative frequencies (Stockard & Papanicolaou 1917). The cells are mostly epitheliocytes and leucocytes (8.215). This technique has been useful in reproductive studies of experimental animals and is also applied in clinical gynaecology. Effects of hormonal therapy include changes in vaginal cellular debris, which also occur in uterine and especially cervical carcinoma (Papanicolaou & Traut 1943, Koss 1968).

The muscular layers are non-striated and consist of a stronger external longitudinal, and an internal circular layer. Longitudinal fibres are continuous with the superficial muscle fibres of the uterus, the strongest fasciculi being those attached to the rectovesical fascia on each side. The two layers are not distinct but connected by oblique decussating fasciculi. The lower vagina is also surrounded by the skeletal muscle fibres of *bulbospongiosus* (p. 607). Most external is a layer of loose connective tissue, containing extensive vascular plexuses.

Vessels and Nerves

The *arteries* are derived from the vaginal, uterine, internal pudendal and middle rectal branches of the internal iliac arteries (p. 777; 6.121). The *veins* form lateral plexuses, drained through the vaginal veins to the internal iliac veins. *Lymph vessels* are described on p. 856. The *nerves* are derived from the vaginal plexuses and pelvic splanchnic nerves (p. 1166). The lower vagina is supplied by the pudendal nerve (p. 1149). Many nerve fibres in the lamina propria and muscle react strongly to tests for cholinesterase and are probably cholinergic.

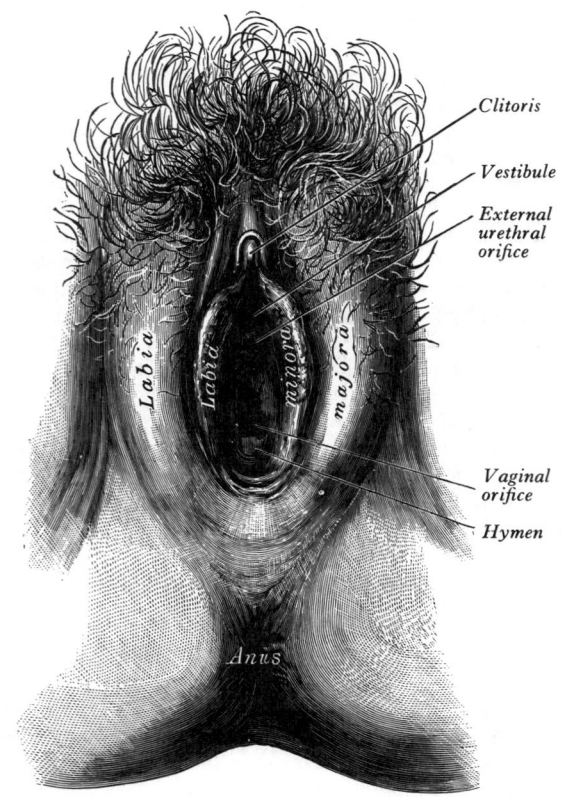

8.217 Female external genitalia, with the labia majora et minora separated.

Female External Genital Organs

The female external genitalia (8.217) include: the mons pubis, labia majora et minora pudendi, the clitoris, vestibule, vestibular bulb and the greater vestibular glands. The term *pudendum* or *vulva* includes all these parts.

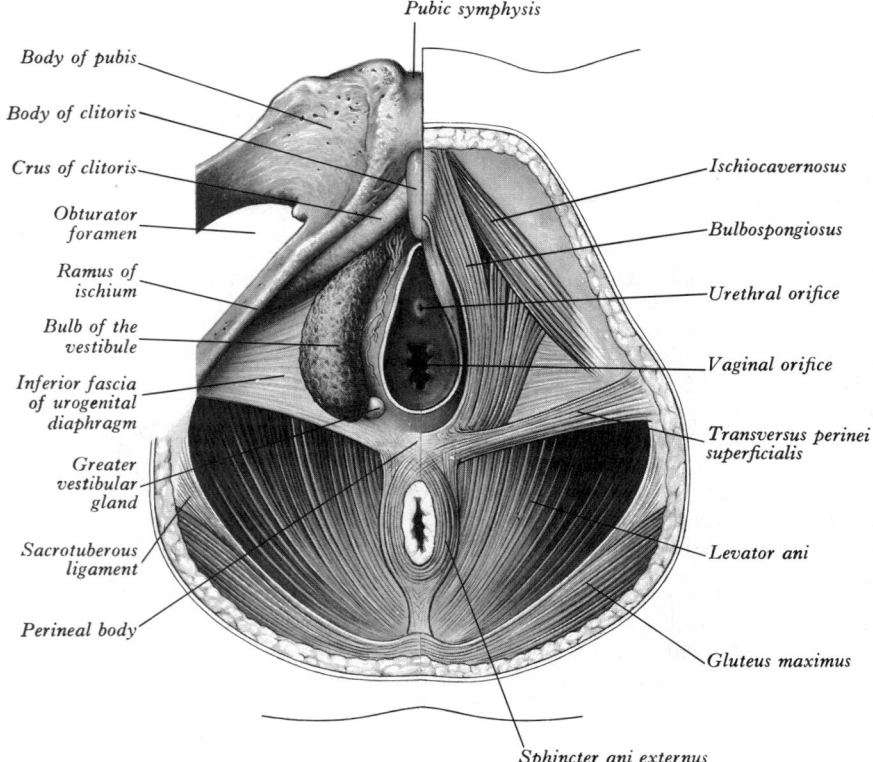

8.218 Dissection of the female perineum to show the bulb of the vestibule and greater vestibular gland on the right; on the left side of the body the muscles superficial to these structures have been left in situ.

The mons pubis, the rounded eminence anterior to the pubic symphysis, is formed by a mass of subcutaneous adipose connective tissue, covered by coarse hair at the time of puberty over an area usually limited above by an approximately horizontal boundary. (In males this upper limit is similar; its apparent continuation to the umbilicus consists of ordinary body hair.)

The labia majora (8.217), two prominent, longitudinal, cutaneous folds from mons pubis to perineum, form the lateral boundaries of the *pudendal cleft*, into which the vagina and urethra open. Each labium has an external, pigmented surface, covered with crisp hairs and a pink, smooth, internal surface with large sebaceous follicles. Between these surfaces is much loose connective and adipose tissue, intermixed with non-striated muscle resembling the scrotal dartos muscle, together with vessels, nerves and glands. The uterine round ligament ends in the adipose tissue and skin in the front part of the labium. A peristent processus vaginalis and congenital inguinal hernia may reach a labium. Labia are thicker in front, where they join to form the *anterior commissure*. Posteriorly they do not join but merge into neighbouring skin, ending near and almost parallel to each other; with the connecting skin between they form a *posterior commissure*, the posterior limit of the vulva; the interval between this and the anus, 2.5–3 cm, is the 'gynaecological' perineum.

The labia minora (8.217), two small cutaneous folds, devoid of fat, between the labia majora, extend from the clitoris obliquely down, laterally and back for about 4 cm, flanking the vaginal orifice. In virgins their posterior ends may be joined by the cutaneous *frenulum of the labia minora*. Anteriorly, each labium minus bifurcates, its upper layer passing above the clitoris to form with its fellow a fold, the *prepuce*, overhanging the glans clitoridis. The lower layer passes below the clitoris to form with its fellow the *frenulum clitoridis*. Sebaceous follicles are numerous on the apposed labial surfaces.

The vestibule (8.218), between the labia minora, contains the vaginal and external urethral orifices and those of numerous mucous *lesser vestibular glands*. Between the vaginal orifice and the frenulum of the labia minora is a shallow *vestibular fossa*.

The clitoris (8.217,218) is an erectile structure, homologous with the penis, and lies postero-inferior to the anterior commissure, partially enclosed by the anterior, bifurcated ends of the labia minora. The *corpus clitoridis* has two corpora cavernosa, composed of erectile tissue and enclosed in dense fibrous tissue separated medially by an incomplete fibrous *pectiniform septum*; each corpus cavernosum is connected to its ischiopubic ramus by a *crus*. The *glans clitoridis* is a small round tubercle of spongy erectile tissue; its epithelium has high cutaneous sensitivity, important in sexual responses. The clitoris, like the penis, has a 'suspensory' ligament and two small muscles, the ischiocavernosi (p. 608), attached to its crura. In many details it is a small version of the penis, but differs from it basically in being separate from the urethra.

The vaginal orifice (*introitus*) (8.218) is usually a sagittal slit positioned postero-inferior to the urethral meatus; its size varies inversely with that of the *hymen*; like all the vagina it is capable of great distension during parturition and to a lesser degree during coitus.

The hymen vaginae, a thin fold of mucous membrane, is just within the vaginal orifice; the internal surfaces of the fold are normally folded to contact each other and the vaginal orifice appears as a cleft between them. The hymen varies greatly in shape and area; when stretched, it is annular and widest posteriorly; sometimes it is semilunar, concave towards the pubes; occasionally it is cribriform or fringed. It may be absent or form a complete, *imperforate* hymen. When it is ruptured, small round *carunculae hymenales* are its remnants. It has no established function.

The external urethral orifice (urinary meatus) (8.218) is about 2.5 cm postero-inferior to the glans clitoridis, anterior to the vaginal orifice: it is usually a short, sagittal cleft with slightly raised margins and very distensible.

The bulbs of the vestibule (8.218), homologues of the single penile bulb and corpus spongiosum, are two elongate erectile masses, flanking the vaginal orifice and united in front of it by a narrow *commissura bulborum* (pars intermedia). Each lateral mass is about 3 cm in length; their posterior ends are expanded in contact with the greater vestibular glands, their anterior ends tapered and joined to one another by a commissure and to the glans clitoridis by two slender bands of erectile tissue. Their deep surfaces contact the inferior aspect of the urogenital diaphragm; superficially each is covered by a muscle, the bulbospongiosus. Thus the female corpus spongiosum is cleft into bilateral masses, except in its most anterior region, by the vestibule and the vaginal and urethral orifices.

The greater vestibular glands (8.218), homologues of the male bulbo-urethral glands, are two small, round or oval reddish-yellow bodies, flanking the vaginal orifice, in contact with and often overlapped by the posterior end of the vestibular bulb. Each opens, by a duct of about 2 cm, in the groove between the hymen and a labium minus.

Vessels and nerves

The arterial blood supply, venous and lymph drainage and nerve supply of the female external genitalia resemble those of homologous structures in males. The arterial supply, from two external and one internal pudendal artery on each side, is massive. Hence, haemorrhage from vulval injuries may be severe. The sensory innervation of the anterior and posterior parts of the labium majus differ, as in the scrotum (p. 1432).

The Mammae or Mammary Glands

The mammae exist in both sexes, though they are rudimentary in males thoughout life. In females they are undeveloped before puberty but grow and differentiate considerably at and after puberty. They attain their greatest development during the later months of pregnancy and lactation. A mamma consists of glandular tissue (*mammary gland* proper), capable of secreting milk, and fibro-adipose tissue between its glandular lobes and lobules, together with blood and lymph vessels and nerves and covered by skin.

THE FEMALE MAMMA (8.219)

In young adult females, each breast is a rounded eminence lying within the superficial fascia chiefly anterior to the upper thorax but spreading laterally to a variable extent. Its shape varies greatly in individuals and races and at different ages, being hemispherical, conical, variably pendulous, piriform or thin and flattened. It is largely composed of adipose tissue, except during lactation; its consistency, shape and size primarily depend on this. In the lateral plane (8.106) its base extends vertically from the second to the sixth rib; at the level of the fourth costal cartilage it extends transversely from the parasternal region almost to the mid-axillary line. The superolateral quadrant is prolonged superolaterally towards the axilla, as the *axillary tail*, along the lower border of pectoralis major, sometimes passing through the deep fascia nearly as far as the pectoral group of axillary lymph nodes. Its deep aspect is slightly concave and is anterior to pectoralis major, serratus anterior, obliquus externus abdominis and its aponeurosis as it forms the anterior wall of the sheath of rectus abdominis. It is separated from these by the deep pectoral fascia, with superficially intervening loose connective tissue, the *retromammary (submammary) 'space'*, which allows the breast some degree of movement on the deep pectoral fascia. (Advanced mammary carcinoma may, by invasion, fix the breast to pectoralis major.) Occasionally, small projections of glandular tissue may pass through the deep fascia into the muscle in normal subjects. The *mammary papilla (nipple)*, the cylindrical or conical projection below the centre of the anterior mammary aspect, is commonly level with the fourth intercostal space in nulliparous females. It is pink or light brown or darker and is traversed by 15–20 lactiferous ducts, their minute orifices opening on to its wrinkled tip. The papilla contains many non-striated myocytes, their arrangement being largely circular; their contraction (when stimulated, e.g. by suckling) erects the papilla; other muscle fibres are longitudinal and may retract it. Occasionally the papilla may not evert during prenatal development (p. 177), remaining permanently retracted

and so causing difficulty in suckling. Its base is encircled by a cutaneous discoidal area, the *areola*, rose-pink in nulliparous Caucasian females and usually strongly pigmented in more melanized races and individuals. In the second month of pregnancy the areola becomes larger and darker, dark brown in pregnancy, the depth of colour varying with complexion; the degree of pigmentation diminishes after lactation, but the areola never returns to its original hue. It contains many sebaceous *areolar glands*, much enlarged in pregnancy and lactation as subcutaneous 'tubercles', whose oily secretion is a protective lubricant during lactation. Some appear intermediate in structure between sebaceous and sweat glands; others (Montgomery's glands) are of the mammary type although small. There is no fat immediately beneath the skin of the areola and papilla.

Structure of the Breast

The mammary gland (8.219, 220A,B) contains (1) glandular tissue of the tubulo-alveolar type, (2) fibrous tissue connecting its lobes, (3) interlobar adipose tissue (Cowie 1974). Subcutaneous tissue encloses the gland (but not as a distinct capsule), sending in many septa to support its lobules. From its covering fascia fibrous processes extend to the skin and the papilla, being more developed over the upper part of the breast as *suspensory ligaments* (of Cooper). These may be contracted by fibrosis in carcinoma, causing retraction or pitting of the overlying skin. Normal mammary tissue is a pale red, firm, lobulated mass, antero-posteriorly flattened and thicker centrally than at its periphery. It has 15–20 lobes containing many lobules held together by loose connective tissue which supports blood vessels and ducts. The smallest lobules, when fully developed, consist of clusters of rounded alveoli opening into ductules which unite to form larger tributaries of the terminal *lactiferous ducts*; each of the latter drains a lobe and so are the same in number (15–20), converging to the areola and forming beneath it variable *lactiferous sinuses*, which may serve as reservoirs. (Some authorities, however, consider a lactiferous 'sinus' to be an artefact.) At the papillary base these narrow and pass straight to its summit, ending at still narrower orifices; they are surrounded by connective tissue containing longitudinal and transverse elastic fibres and lined by columnar epithelium, with an external layer of longitudinal myoepitheliocytes on a basal lamina. In larger ducts the epithelium has two or more layers, becoming, near the papillary openings, keratinizing stratified squamous. Mammary carcinoma is usually caused by neoplastic change in the ductal cells, forming tumours in the ductal tissues or infiltrating the surrounding connective tissue.

Mammary structure varies with *age*, *pregnancy* and *lactation*. (Prenatal development is described on p. 177.) At *birth* there are lactiferous ducts but no alveoli and *until puberty* little branching of the ducts occurs, slight mammary enlargement being due to growth of fibrous stroma and fat. *After puberty*, stimulated by ovarian oestrogens, the ducts develop branches whose ends form solid, spheroidal masses of granular polyhedral cells, which are potential alveoli. In the resting state the glandular epithelium is separated from the vascular stroma by a thin avascular zone of fibroblasts. This 'epithelio-stromal junction' may control the passage of materials to the secretory cells (Ozzello 1974). Secreting alveoli appear only in pregnancy, when the ducts branch markedly, their ends enlarging as milk is secreted; this growth is due to the rising output of placental oestrogen and progesterone. Adipose tissue also increases and blood flow becomes richer. Secretory activity in alveolar cells augments progressively in the latter half of pregnancy. The secretion in late pregnancy and for a few days after parturition is different from the later milk and is known as *colostrum*; it contains many cytoplasmic fat globules and *colostral corpuscles*, whose nature is uncertain; some consider them to be the lining cells of primitive alveoli, which undergo fatty degeneration and are shed; others regard them as macrophages which engulf the small amounts of fat present in the secretion at this time. True milk secretion begins a few days after parturition due to a reduction of circulating oestrogen and progesterone, which appears to stimulate production of prolactin by the anterior hypophysis (Wolstenholme & Knight 1972 and p. 1451). Milk distends the alveoli, at first lined by a single layer of granular, short columnar cells with stellate myoepitheliocytes

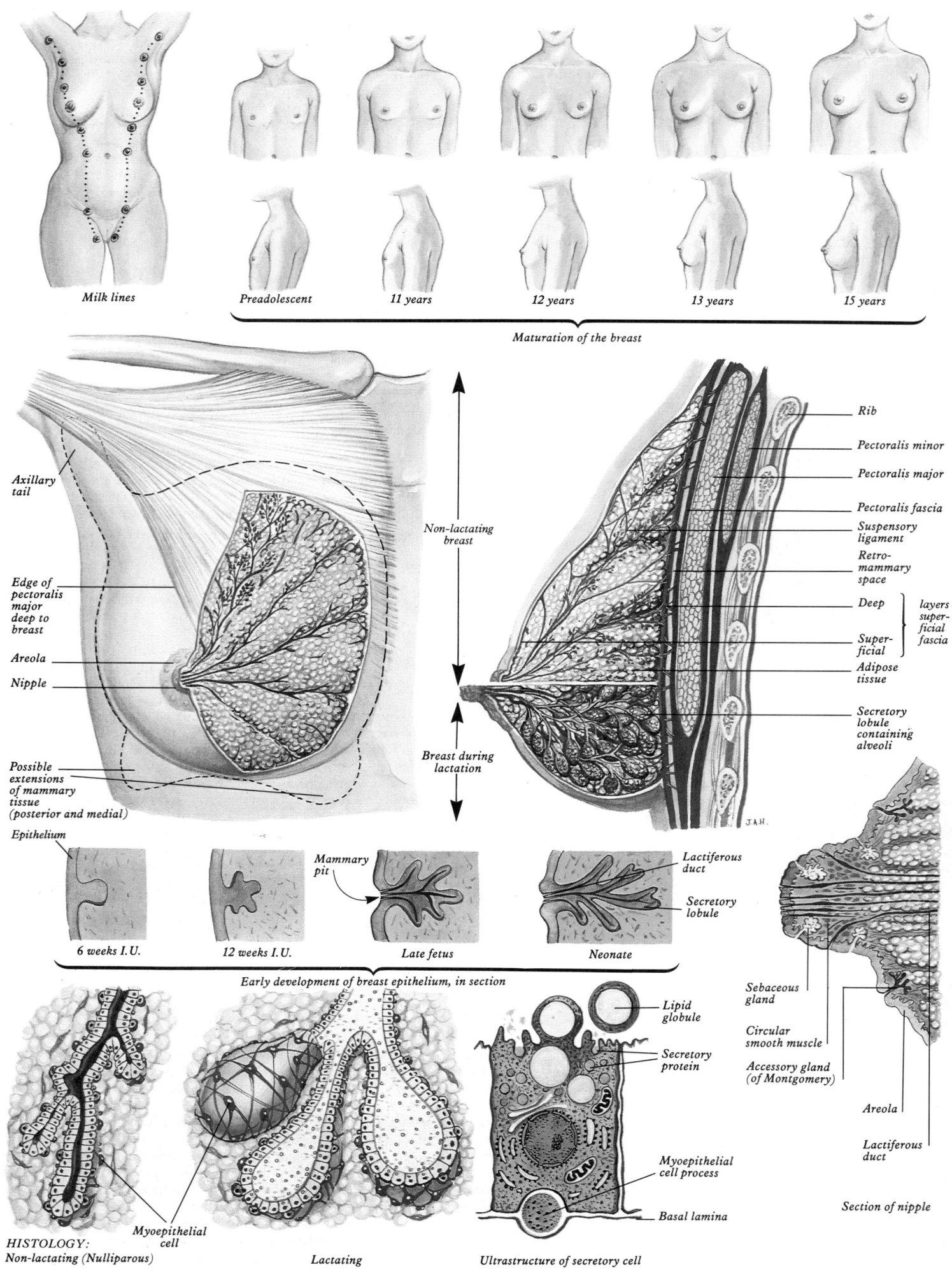

Milk lines

Maturation of the breast

Preadolescent 11 years 12 years 13 years 15 years

Axillary tail

Edge of pectoralis major deep to breast

Areola

Nipple

Non-lactating breast

Breast during lactation

Possible extensions of mammary tissue (posterior and medial)

Rib

Pectoralis minor

Pectoralis major

Pectoralis fascia

Suspensory ligament

Retro-mammary space

Deep

Superficial

layers of superficial fascia

Adipose tissue

Secretory lobule containing alveoli

Epithelium

Mammary pit

Lactiferous duct

Secretory lobule

6 weeks I.U. 12 weeks I.U. Late fetus Neonate

Early development of breast epithelium, in section

Sebaceous gland

Circular smooth muscle

Accessory gland (of Montgomery)

Areola

Lactiferous duct

Lipid globule

Secretory protein

Myoepithelial cell process

Basal lamina

HISTOLOGY: Non-lactating (Nulliparous)

Myoepithelial cell

Lactating

Ultrastructure of secretory cell

Section of nipple

near the basement membrane; the cells flatten as secretion increases. Fat droplets accumulate superficially in the cells before entering the alveolar lumen in apocrine fashion.

Alveolar cells vary in form and content according to their role in the mammary secretory cycle, being cuboidal in the resting state and columnar in lactation; as they distend with milk they may become cuboidal again, by stretching of the alveolar lining. Then, however, they are much distended by huge apical secretory vacuoles. When secreting milk their cytoplasm is basophilic and contains abundant granular endoplasmic reticulum, mitochondria, lysosomes and free ribosomes. Apical to the basal nucleus are a Golgi complex and large secretory vacuoles of two types; proteinaceous and lipid-containing. Protein vacuoles contain granules of micellar casein and other lactic proteins formed in the granular endoplasmic reticulum and passed to the Golgi apparatus to form larger vacuoles, eventually secreted in a merocrine manner. Lipid vacuoles do not traverse the Golgi apparatus, but accumulate apically, fusing into large 'milk vacuoles' up to 10 μm across, frequently protruding from the cell's surface before release. Lipid vacuoles are eventually discharged with a thin surround of cell membrane and cytoplasm (Saacke & Heald 1974, Hollmann 1974, Pitelka 1977, Tobon & Salazar 1975, Kesinger et al 1986). This may be considered *apocrine* secretion, since actual cytoplasm is lost, though not to a large extent.

Alveolar cells are joined apically by occluding tight junctions which prevent the passage of substances from the lumen into the intercellular spaces and vice versa; many desmosomes and communicating junctions also occur between cells (Pitelka et al 1973). In pregnant rats, other types of protein granule are synthesized as precursors of normal granules; these may appear in colostrum (Murad 1970).

Lactation

Passage of milk from alveoli into and along the ducts may be aided by *myoepitheliocytes* (p. 55) located between the lining cells and basement membrane. Flow is initiated by suckling, which stimulates nerves abundant in the papilla and areola; such afferent impulses evoke secretion of oxytocin from the neurohypophysis, apparently stimulating contraction of the ductual myoepitheliocytes and non-striated myocytes. Without suckling, secretion of milk soon ceases. Oestrogens stimulate growth of ducts, while progesterone (from the corpus luteum) directly or indirectly stimulates alveolar formation at their ends. Formation of true secretory alveoli in pregnancy is due to the synergic action of oestrogens, progesterone (mainly placental) and hypophysial hormones (prolactin and growth hormone, which also maintain lactation). Commonly lactation continues for five or six months after birth, then progressively diminishes, infants usually being weaned at about nine months. When lactation stops, the glandular tissue returns to the 'resting' condition, remaining milk is absorbed and the alveoli shrink, many losing their lumina. Glandular tissue may fail to produce milk throughout pregnancy or secretion may cease within a few weeks of birth. After menopause, mammary glands atrophy, the alveolar cells and ducts degenerating, but a few ducts may remain (Ozzello 1974); the stroma becomes much less cellular and collagenous fibres decrease. The amount of fat varies. In neonates of both sexes, in the first week or two after birth, cells lining ducts may be stimulated by maternal oestrogens, via the placenta, to secrete a little fat-free fluid, sometimes termed 'witch's milk'.

Milk

Milk is a complex fluid, composed in humans of about 88% water, 7% lactose, 4% fat, 1% protein and various ions, notably calcium, sodium, potassium, phosphate and chloride. Vitamins and antibodies, mainly of the IgA (secretory) class are present, the latter being largely responsible for the sterility of milk during lactation (Jennesse 1974). The proteins are chiefly caseins and a lactalbumen; these, with lactose and several triglycerides, are synthesized from circulating precursors which enter the gland and are unique to it. *Colostral milk* is markedly different, poor in

nutrients with an ionic composition like blood plasma. For details of lactation and human mammary structure see the review by Larson (1978).

The Male Mamma

The male mamma remains rudimentary. It is formed of small ducts (without alveoli) and a little supporting fibro-adipose tissue. Sometimes the 'ducts' are largely solid cellular cords. Slight temporary enlargement may occur at puberty. Generally the ducts do not extend beyond the areola, which, however, is well developed. The papilla and areola are relatively small.

Vessels and Nerves

The *arteries* supplying the mammary glands are from thoracic branches of the axillary and from the internal thoracic and intercostal arteries. The *veins* form an anastomotic *circulus venosus* at the papillary base. From this and from the glandular tissue, veins carry blood to the periphery to end in the axillary and internal thoracic veins. *Lymph vessels* are described on p. 847. The *nerves* are from the anterior and lateral cutaneous branches of the fourth to sixth thoracic spinal nerves, which also convey sympathetic fibres to the mamma; however secretory activities of the gland are

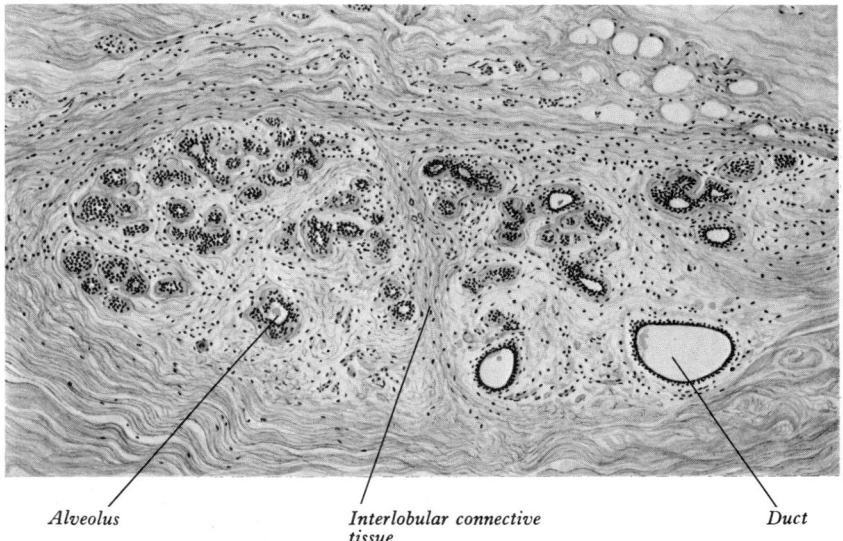

Alveolus　　　　*Interlobular connective tissue*　　　　*Duct*

8.220A A sample section of human non-lactating breast from a young adult female.

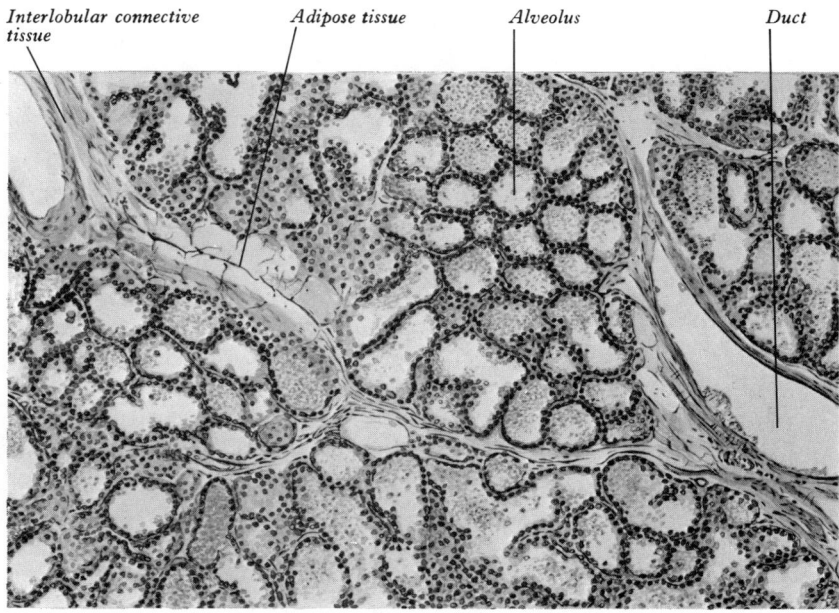

Interlobular connective tissue　　　*Adipose tissue*　　　*Alveolus*　　　*Duct*

8.220B A sample section of an actively lactating human breast. Magnification × 60.

8.219 (*opposite*) Postnatal development and structure of the female breast.

largely controlled by ovarian and hypophysial hormones. The papilla has a dense nervous plexus supplying many receptors, such as Meissner's corpuscles and Merkel's discs and also 'free' terminals; these are essential in signalling suckling to the central nervous system (vide supra).

Applied Anatomy

The papillary ducts are radially orientated and incisions should hence also be radial. An obstructed lactiferous duct may distend as a *galactocele*. Abscesses may occur between the septa in the glandular tissue, subcutaneously near the papilla or between gland and deep fascia anterior to the pectoralis major.

Supernumerary mammae (*polymastia*) or papillae (*polythelia*) occur in males and females, usually along a line extending from the axilla to the pubic region, the *milk line* (**8.219**). The male mamma may hypertrophy after puberty (*gynaecomastia*), usually due to imbalance between oestrogenic and androgenic hormones. (See pp. 1451, 1452, 1463 and 1472 for endocrine influences.)

THE ENDOCRINE SYSTEM

For a multicellular organism to survive and maintain its integrity in a varying, often adverse environment, regulatory mechanisms are vital. Metazoa have attained some freedom from the vagaries of the external environment (over which only the most advanced have much control, although this is only of a short-term nature) by developing the capacity to regulate, with varying success, the composition and properties of their immediate cellular environments which comprise intercellular or tissue fluid and, in vascularized metazoa, the fluid components of blood and lymph. The effective stabilization of these media is directly related to an organism's success and longevity.

Tissue fluid and plasma are stabilized by the co-ordinated regulatory activity of the autonomic and endocrine systems, the latter including the diffuse neuro-endocrine system (p. 1466) and the endocrine system proper. All operate by intercellular communication but differ in the mode and speed, and in the degree of localization of the effects produced. The *autonomic nervous system* utilizes conduction and neurotransmitter release to transmit information; it is swift and localized in the responses induced. The diffuse *neuro-endocrine system* uses only secretions, is slower and

the induced responses are less localized, because the secretions e.g. neurotransmitters, can act on contiguous cells, on groups of nearby cells reached by diffusion (a paracrine secretion) or on distant cells via blood like a hormone. The *endocrine system* proper, comprising isolated or clustered cells and discrete ductless glands producing hormones (organic molecules transported by blood to distant effector cells), is even slower and less localized, though its effects are specific and often prolonged. These regulatory systems overlap in form and function, with a gradation from the neural autonomic system, through intermediate diffuse neuro-endocrine mechanisms to the endocrine system proper.

The close interrelationship of autonomic and endocrine systems, both structural and functional, is exemplified by the hypothalamus (p. 1007). This integrates both systems and is the major site at which their activities combine. Apart from its nervous functions the hypothalamus is also endocrine, producing by neurosecretion a wide range of peptide hormones including releasing and inhibiting factors which control the activity of the adenohypophysis, itself a major endocrine gland. Though conveniently considered separately, the autonomic, diffuse

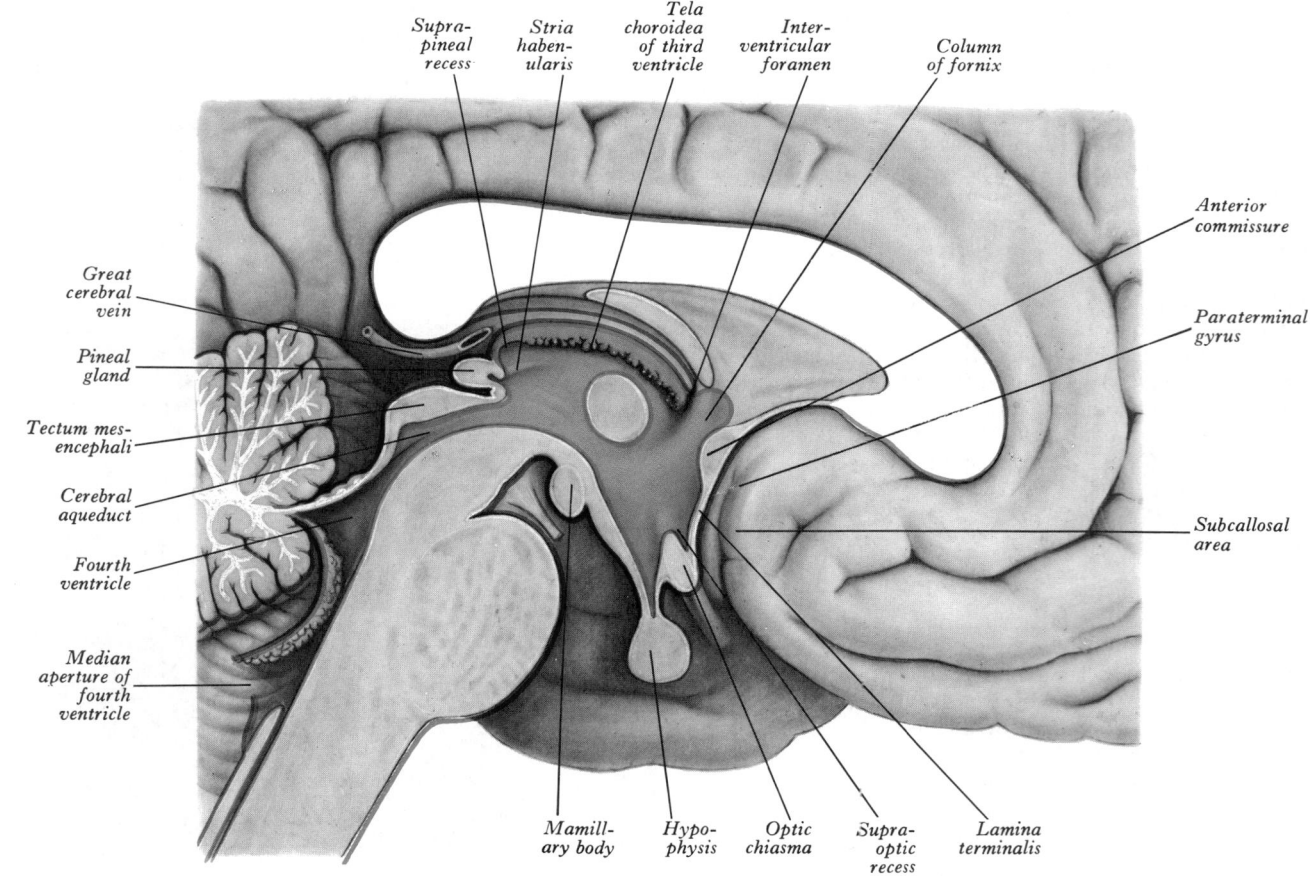

8.221 Part of a median section through the brain to demonstrate the location of the hypophysis and its immediate surroundings, particularly hypothalamic structures. The cut edge of the pia mater is shown in red, the ependyma in blue. Compare with 7.140A,B.

neuro-endocrine and endocrine systems are really a single neuro-endocrine regulator of the metabolic activities and internal environment of the organism, providing conditions in which it can function successfully.

There are, in addition to the endocrine glands and diffuse neuro-endocrine system described here, other hormone-producing cells which form minor components of other systems and are described with them. These include: pancreatic islets (p. 1383), gastro-entero-endocrine cells (p. 1466), certain thymic (p. 833) and renal (p. 1409) cells, pulmonary endocrine cells (p. 1278), interstitial testicular cells (p. 1428), interstitial, follicular and luteal ovarian cells (p. 1437) and, in pregnancy, placental cells (p. 156). Some cardiac myocytes, particularly in the walls of the atria, also have endocrine functions. In addition to these various cell types, it has become apparent over the last few years that many secretory substances which were hitherto thought to be exclusive to the endocrine glands also occur widely in the central nervous system, where they may act as neurotransmitters or neuromodulators (p. 887). The ability to synthesize and secrete such substances is therefore a very widespread activity, emphasizing the complex interdependence of the systems of chemical control throughout the body.

The Hypophysis Cerebri

The hypophysis cerebri (pituitary gland, 8.221–227) is a reddish-grey, ovoid body, about 12 mm in transverse and 8 mm in antero-posterior diameter and weighing about 500 mg. It is continuous with the infundibulum, a hollow, conical, inferior process from the tuber cinereum of the hypothalamus (p. 1007). It lies within the hypophysial fossa of the sphenoid, covered superiorly by a circular *diaphragma sellae* of dura mater; this is pierced centrally by an aperture for the infundibulum and separates the anterior superior aspect of the hypophysis from the optic chiasma (8.222). The hypophysis is flanked by the cavernous sinuses and their contained structures (p. 802). Inferiorly it is separated from the floor of its fossa by a venous sinus communicating with the circular sinus (p. 803). Meninges blend with the hypophysial capsule and are not separate layers.

The terminology of the hypophysial divisions and subdivisions are somewhat confused. The major division is into two regions, differing in origin, structure and function: one is a diencephalic down-growth connected with the hypothalamus; the other is an ectodermal derivative of the stomatodeum. These are respectively the *neurohypophysis* and *adenohypophysis*. Both include parts of the *infundibulum* (the older terms 'anterior' and 'posterior lobes' do not). The infundibulum has a central *infundibular stem*, containing neural hypophysial connections and continuous with the *median eminence* of the tuber cinereum (p. 1007). Thus, the term *neurohypophysis* includes the median eminence, infundibular stem and *neural lobe* or *pars posterior* and also an extension of the adenohypophysis, largely surrounding the neural infundibular stem, the *pars tuberalis* or *infundibularis*. The main mass of the adenohypophysis is divisible into the *pars anterior* (pars distalis) and *pars intermedia*, separated in fetal and early postnatal life by the hypophysial cleft, a vestige of Rathke's pouch (p. 231) from which it develops. Usually obliterated in childhood, it may persist in the form of cystic cavities often present near the adeno-neurohypophysial frontier and sometimes invading the neural lobe. The human pars intermedia is rudimentary; because it may also be partially displaced into the neural lobe, it has been included in the anterior *and* posterior parts by different observers. Apart from this equivocation, of little significance in view of the exiguous status of the human pars intermedia, the partes anterior et posterior (nervosa) may be equated with the anterior and posterior lobes. When the associated infundibular parts continuous with these lobes are included, the names adenohypophysis and neurohypophysis become appropriate. The terms will hence be used as follows:

(1) *Neurohypophysis*: includes the pars posterior (pars nervosa, posterior or neural lobe), infundibular stem and median eminence.

(2) *Adenohypophysis*: includes the pars anterior (pars distalis or glandularis), pars intermedia and pars tuberalis.

The hypophysial divisions are shown in 8.222, 223A, 227. They differ in the types and arrangement of cells and the details of their vascular and neural supplies. Studies of these structures have engendered a formidable literature, which will be summarized here only briefly.

The Adenohypophysis

The adenohypophysis (8.223, 224) is highly vascular and consists of epithelial cells of varying size and shape arranged in cords or irregular follicles, separated from each other by thin-walled vascular sinusoids and supported by a delicate skeleton of reticular tissue. At least seven hormones are synthesized and released by the adenohypophysis, most of them being *tropic*, such as: *somatotropin* (STH) involved in the control of body growth, *mammotropin* (*lactogenic hormone* LTH) stimulating growth and

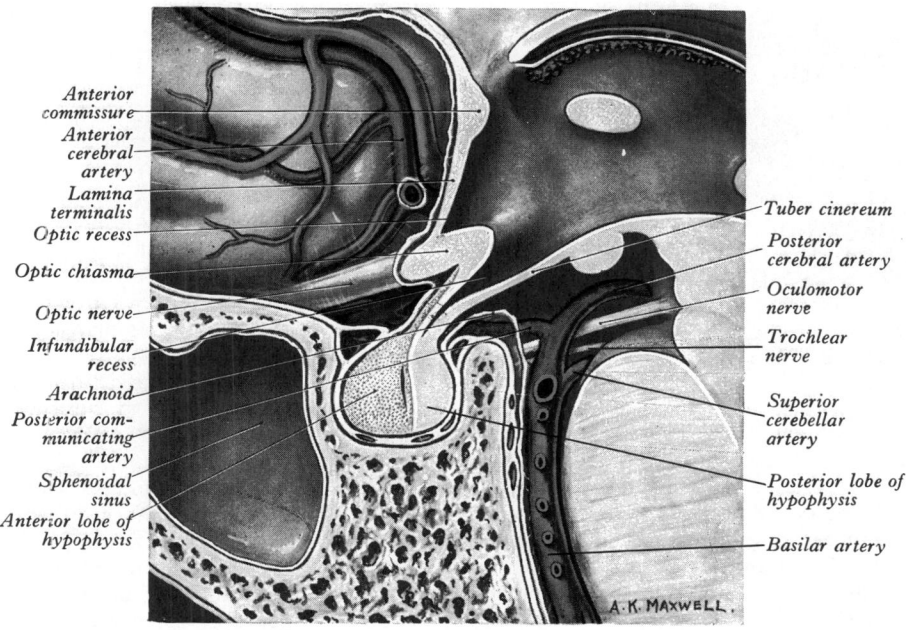

Anterior
commissure
Anterior
cerebral
artery
Lamina
terminalis
Optic recess
Optic chiasma
Optic nerve
Infundibular
recess
Arachnoid
Posterior communicating
artery
Sphenoidal
sinus
Anterior lobe of
hypophysis

Tuber cinereum
Posterior
cerebral artery
Oculomotor
nerve
Trochlear
nerve
Superior
cerebellar
artery
Posterior lobe of
hypophysis
Basilar artery

A. K. MAXWELL.

8.222 A median section through the hypophysis, in situ.

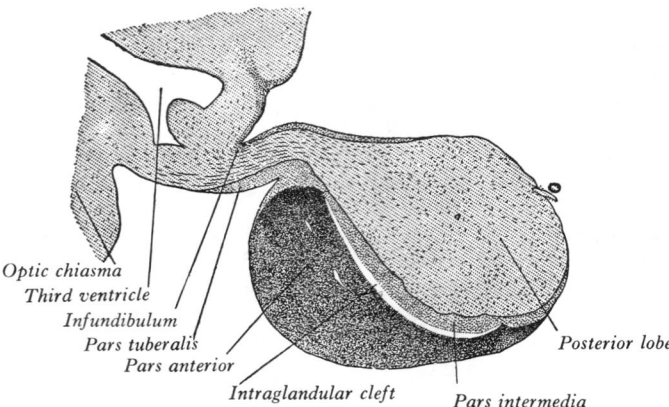

Optic chiasma
Third ventricle
Infundibulum
Pars tuberalis
Pars anterior
Posterior lobe
Intraglandular cleft
Pars intermedia

8.223A Diagram of a median section through the hypophysis cerebri of an adult monkey.

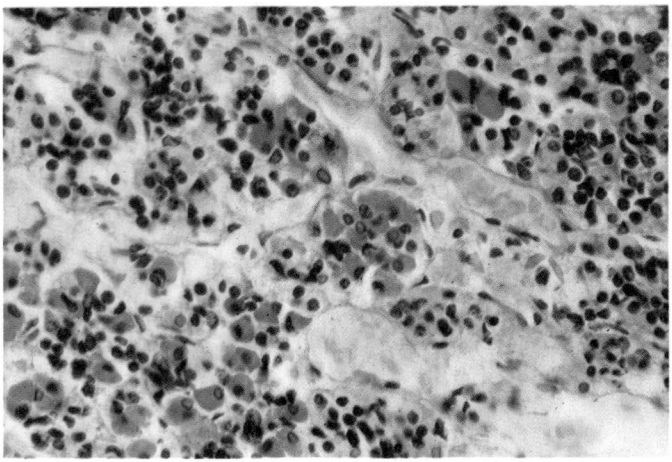

8.223B Medium-power micrograph of adenohypophysial cells stained with the PAS-orange G technique, to distinguish between α cells (yellow), β cells (red to brown) and chromophobic cells (pale grey with little cytoplasm). A number of large vascular sinusoids containing yellow-stained erythrocytes are also prominent. Magnification × 400.

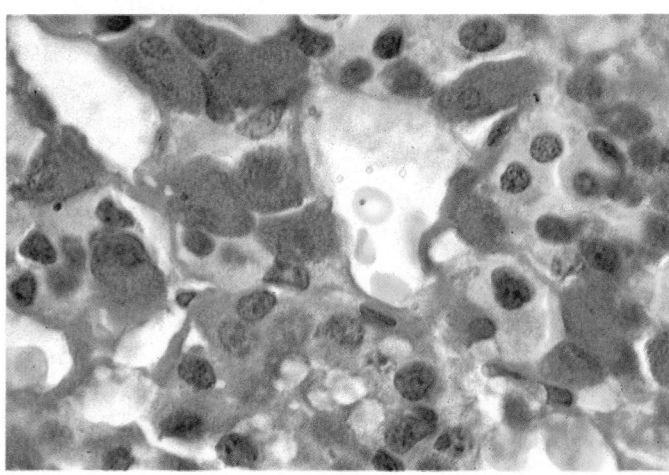

8.223C High-power micrograph of a group of cells stained as for B, grouped around a vascular sinusoid. Note α cells (yellow) and β cells (red). Magnification × 1000.

secretion by the female breast, *adrenocorticotropin* (ACTH) governing the secretion of some adrenal cortical hormones, *thyrotropin* (TSH) stimulating thyroid activity, *follicle-stimulating hormone* (FSH) stimulating growth and secretion of oestrogens in the ovarian follicles and spermatogenesis in the testis, *interstitial-cell-stimulating hormone* (ICSH) activating androgen secretion by the testis, *luteinizing hormone* (LH), inducing progesterone secretion by the corpus luteum and *melanocyte-*

stimulating hormone (MSH) increasing cutaneous pigmentation. Most of these hormones also have other complex metabolic effects.

CELL TYPES IN THE ADENOHYPOPHYSIS

The identities of endocrine cells secreting the different adenohypophysial hormones have been intensively sought by various staining techniques to distinguish between different cell types (Harris & Donovan 1966); earlier techniques used the simple criterion of the affinities of cells for acidic and basic dyes, such as orange-G and aldehyde fuchsin, respectively. Cells staining strongly were *chromophilic*, those with little affinity for dyes being *chromophobic*. Chromophilic cells were classed as *acidophils* (α-cells) or *basophils* (β-cells) for obvious reasons. Modifications using complex, multi-stage staining further distinguished many subcategories of cells, each with its own tinctorial nuance. Such techniques, applied particularly to experimental or pathological pituitary glands from individuals with known hormonal defects, were used to associate specific cell types with particular hormones. However, many uncertainties remained, due to difficulty in standardizing stains and in classifying endless minor variations in different individuals and species.

More recently, immunochemical methods have been applied (Nakane 1970) and immunofluorescence techniques have identified the sites of synthesis of individual hormones. Such methods, combined with electron microscopy and cell fractionation, have made it possible to consign all major hormones to particular cell types, although the terminology of older classifications has in part been retained. Ultrastructural studies of the normal human adenohypophysis are relatively scarce and much of our information has come from non-human primates. However, species differences do exist in the fine structural and dye-binding properties of cells of this kind. For detailed accounts of adenohypophysial structure, see Costoff (1977), Pelletier et al (1978), Bhatnagar (1983), Motta (1984) and Martin (1985).

The pars anterior contains the following cell varieties:

A. Chromophil Cells

1. *Acidophils (α cells)*
(a) *Somatotropes*. These are ovoid and usually grouped along sinusoids; they are the largest and most abundant class of adenohypophysial chromophils, secreting the protein somatotropin. They stain strongly with orange G, and ultrastructurally are seen to contain numerous electron-dense, spherical, secretory granules, 350–500 nm in diameter, and a well-developed Golgi complex but, in the secretory phase, relatively small amounts of granular endoplasmic reticulum; the nucleus is central. Cells of similar fine structure characterize human eosinophilic adenomata associated with acromegaly or giantism (Zambrano et al 1968).

(b) *Mammotropes*, secreting the polypeptide hormone prolactin, are dominant in pregnancy and hypertrophy during lactation. They are distinguished by their affinity to the dye erythrosin and also azocarmine. Their secretory granules are the largest in any hypophysial cell (over 600 nm in diameter) in pregnant and lactating females, although smaller (200 nm) and fewer in non-pregnant females and in males; the granules are evenly dense, ovoid or irregular, the latter form resulting from fusion. Excess granules fuse with lysosomes to form autophagic vacuoles which degrade unused granules. In active cells, granular endoplasmic reticulum and a Golgi complex are prominent.

2. *Basophils (β cells)*
(a) *Corticotropes*. The identification of the cells which secrete adrenocorticotropin (ACTH) was difficult to achieve until it was realized that a precursor molecule, pro-opio-melanocorticotropin, is cleaved into a number of different molecules including ACTH, beta lipotropin and beta-endorphin; the functions of the latter two substances in the pituitary are not known, although the opio-melanocorticotropin complex is also synthesized in neurons of the central nervous system and has neuromodulator functions there (p. 887). In humans, though not in some other species, this precursor is glycosylated, making the granules PAS-positive. They are also weakly basophilic. These cells are

irregular in shape and have short dendritic processes which are inserted among other neighbouring cells. Their granules are also small (about 200 nm) and difficult to detect by light microscopy.

(*b*) *Thyrotropes* (beta-basophils) secrete TSH. They are elongate, polygonal and lie in clusters towards the adeno-hypophysial centre. They usually form cellular cords and are not in direct contact with sinusoids; they stain selectively with aldehyde fuchsin. Their granules, peripheral and irregular, are less electron-dense than in other basophils, being 100–150 nm in diameter and among the smallest granules in adenohypophysial cells.

(*c*) *Gonadotropes* (*delta-basophils*) are larger than thyrotrophs, are rounded and usually situated next to sinusoids; they have secretory granules with an affinity for the periodic acid/Schiff stain. In some cells, usually peripheral in the lobe, granules stain purple while in others, more central, they stain red, suggesting that the former secrete FSH and the latter LH or ICSH (Purves 1961). Electron microscopy of the rat adenohypophysis has distinguished two cell types (Kurosumi & Oota 1968); but immunocytochemical studies of Phifer et al (1973) in man and of Moriarity (1975) in rats suggest that FSH and LH may coincide in the same cell, within the same secretory granules. Gonadotrophs have pleomorphic nuclei and spherical granules about 200 nm in diameter which tend to gather in lines under the cell surface during secretory activity, a vesicular granular endoplasmic reticulum and a well-developed Golgi complex (von Lawzewitsch et al 1972).

B. Chromophobe Cells

Chromophobe cells constitute the majority of the cells of the adenohypophysis (about half the population of epithelial cells) but, because of their small size and lack of reaction to routine stains, they are not a conspicuous feature of the pituitary. They appear to consist of a number of different types of cells, including degranulated secretory cells of the types described above, stem cells capable of giving rise to chromophils (the turnover of these cells has been described) and follicular cells containing numbers of lysosomes and forming cell clusters around cysts of various sizes; cysts are often present in the junctional area with the neurohypophysis and are filled with a PAS-positive substance of unknown significance.

The pars intermedia has many β-cells and follicles of chromophobe cells surrounding PAS-positive colloidal material, some of these being derived from pouches of the embryonic intrahypophysial cleft (of Rathke); in some species, where this pouch remains large, a ciliated lining epithelium has been noted.

Secretory cells of the pars intermedia have granules containing either α-endorphin or β-endorphin scattered uniformly (Bloom et al 1977, Guillerman 1978); these cells have been shown to contain various peptide hormones including ACTH and α-MSH (melanocyte-stimulating hormone), which are assigned to the APUD series (p. 1466), as are other adenohypophysial secretory cells. Endorphin, also occurring in the pars anterior, is localized here to peripheral granules in discrete groups of cells.

The pars tuberalis is remarkable for its large number of blood vessels, between which cords or balls of undifferentiated cells are admixed with some α- and β-cells.

The release of adenohypophysial hormones into the circulation appears to be by exocytosis of their vesicle contents into the perivascular spaces of the neighbouring sinusoids; these are lined by a fenestrated endothelium, facilitating diffusion into the blood. The signal for secretion is the liberation of releasing factors from neurons (McKelvy 1974) in the median eminence, nucleus infundibularis (arcuate nucleus) and other hypothalamic nuclei, into the upper capillary bed of the venous portal system which carries them to the adenohypophysis to act on its endocrine cells. These neurosecretory cells are *neuroendocrine transducers* (Wurtman 1970), receiving neural and hormonal signals and responding by secreting hormones, thus transducing one type of signal into another. The neurons which produce releasing factors are peptidergic, whereas neurons modulating endocrine activity are largely monoaminergic, making axosomatic or axo-axonic synapses with them.

Tanycytes (p. 894, 8.224, 225) may also control secretion, possibly transporting hormones from the cerebrospinal fluid to

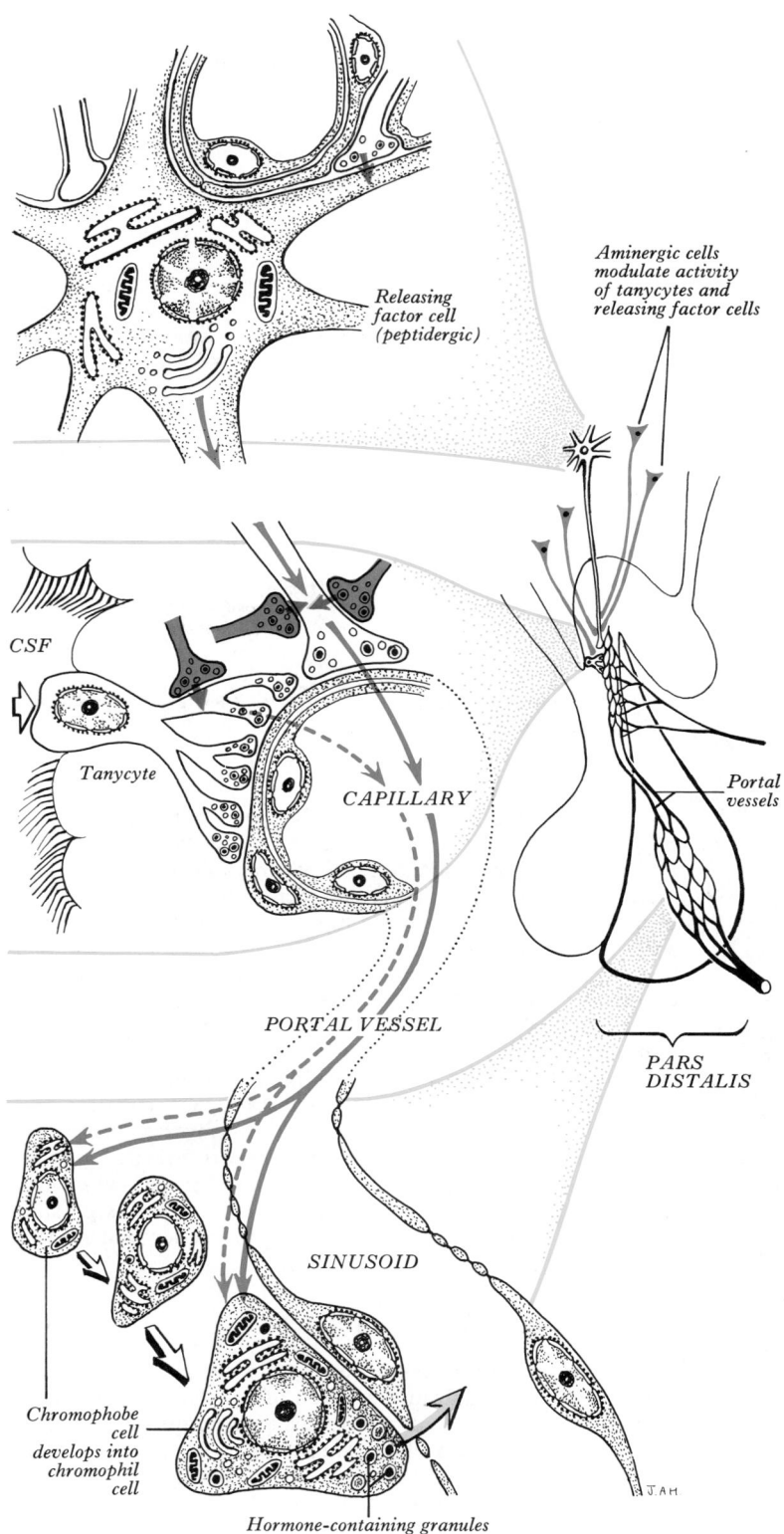

8.224 Diagram illustrating the control systems which influence the hormonal output of the adenohypophysis. The small diagram on the right shows the ventral hypothalamus, median eminence and adenohypophysis (pars distalis) and their associated neurons and vasculature. On the left three zones are shown in greater details. The concatenation of events is as follows: under the influence of blood-borne factors and neural stimuli, the hypothalamic neuron shown above liberates specific releasing factors into capillaries of the median eminence. Modulation at this point is mediated by aminergic neurons and tanycytes. Onward transport of these factors through portal vessels is indicated by blue arrows. Stimulation and hormone production and release on the part of the adenohypophysial cell are also indicated.

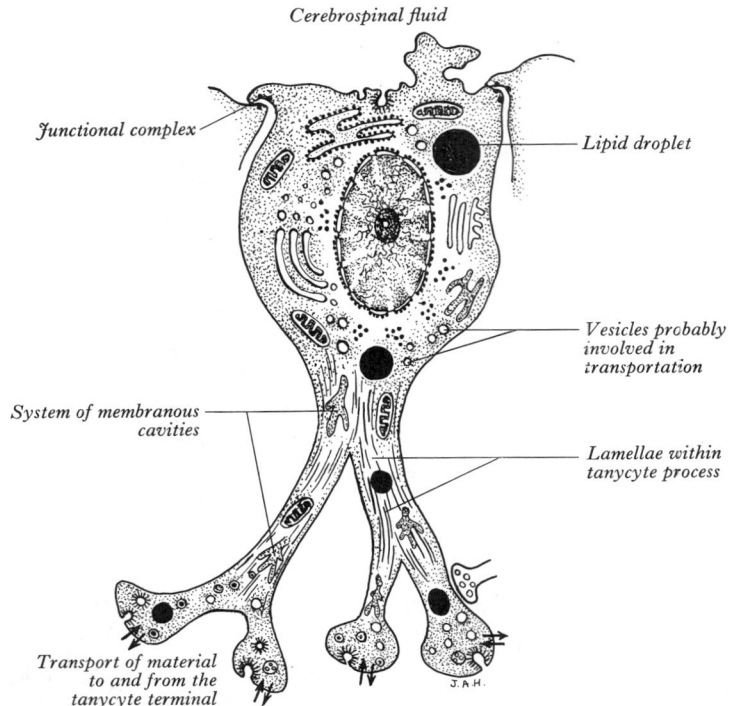

Cerebrospinal fluid

Junctional complex

Lipid droplet

Vesicles probably involved in transportation

System of membranous cavities

Lamellae within tanycyte process

Transport of material to and from the tanycyte terminal

8.225 The general ultrastructure of a tanycyte. (See text on pp. 894 and 1453 for explanatory details.)

capillaries of the portal system (Knigge 1976) and/or from hypothalamic neurons to the cerebrospinal fluid (Joseph & Knigge 1978). Tight junctions between them and ependymal cells (Brightman et al 1975), though rudimentary compared with those in the choroid plexus, impede movement of peptides and even some amino acids (Weindl & Joynt 1972), suggesting that materials such as releasing factors would need to cross the ependyma via, not between, the tanycytes. The endocytotic (presumably also exocytotic) activity of tanycytes appears to be under monoaminergic neuronal control (Kobayashi 1975, Nozaki et al 1975). Tanycytes in the walls of the ventricular recess appear suitable for the transport of releasing factors from neurosecretory cells to cerebrospinal fluid, because they traverse the arcuate nucleus, linked to its neurons by junctions permitting the passage of small molecules such as peptide hormones between cells which they unite (del Cerro & Knigge 1977). However, this possible involvement of tanycytes in the control of secretion by transport of releasing factors awaits verification.

The Neurohypophysis

The posterior hypophysial lobe is a downgrowth from the diencephalic floor, and in early fetal life contains a cavity continuous with the third ventricle, which in some animals, e.g. cats, persists. The posterior lobe, infundibular stem and median eminence are often grouped as the *neurohypophysis* (vide supra and consult Harris 1955, Heller & Clark 1962, Scharrer & Scharrer 1963, Gabe 1966, Donovan 1970, Knowles 1974).

Axons arising from groups of hypothalamic neurons (supra-optic and para-ventricular nuclei, etc.) terminate in the neurohypophysis; some are short, ending in the median eminence and infundibular stem among the superior capillary beds of the venous portal circulation, possibly providing neural control of adenohypophysial function (vide infra). Longer axons pass to the main neurohypophysis thereby forming the neurosecretory hypothalamohypophysial tract (p. 1009), terminating near the sinusoids. The hormones concerned are *vasopressin (antidiuretic hormone,* ADH) controlling reabsorption of water by renal tubules (p. 1406) and *oxytocin* which promotes the contraction of uterine and mammary non-striated muscle. Their sites of production have been described (p. 1009); from their neuronal perikarya

they traverse axons of the tract for release at nerve terminals. Ligation of the tract causes proximal damming of secretion granules, whose high glycoprotein content facilitates PAS staining. The active hormones are simple polypeptides, elaborated and transported in cells with a glycoprotein, *neurophysin*; after release into the circulation this association is broken, the hormone being carried to its cellular targets by plasma glycoproteins.

Other substances demonstrated in these endings include met-enkephalin (co-localized with vasopressin) and cck (p. 891) (with oxytocin).

The neurohypophysis contains thin, non-myelinated nerve fibres and associated cells; these are the terminal ramifications of the hypothalamohypophysial tract. Proximally, in the infundibulum, the fibres are ensheathed by typical astrocytes but near the posterior lobe *pituicytes* appear, dendritic cells of variable appearance, often with long processes running parallel to adjacent axons. Pituicytes constitute most of the non-excitable tissue in the neurohypophysis. Cytoplasmic processes of many pituicytes end on or near the walls of adjacent capillaries and sinusoids between nerve terminals. Axons end in perivascular spaces and, though they approach close to the walls of sinusoids, they remain separated from them by two basal laminae, one applied to the nerve endings and the other to the external surface of the endothelial cells. Fine collagen fibres often exist between the two. The cytoplasm of endothelial cells is generally extremely attenuated and has regular fenestrations which facilitate the passage of hormones into the bloodstream (8.226).

Three types of nerve terminal are described in the posterior hypophysial lobe: (1) terminal axonal swellings adjacent to the sinusoids, in which are large (200–300 nm), dense vesicles containing hormones bound to glycoproteins, and also small (40–60 nm), clear, spherical vesicles; (2) peri-axonal endings with small (80 nm) dense-cored vesicles, like those containing catecholamine in sympathetic nerve endings (p. 905); (3) peri-axonal endings containing small (40–60 nm), clear, spherical vesicles forming synapses with large hormone-containing endings (Bargmann 1966). Interactions between these diverse terminals are not clear. The hormones are probably released by exocytosis from endings with large dense vesicles. Excitation or inhibition of release is mediated by excitation of hormone-containing endings by action potentials carried along the same axon.

Vessels of the Hypophysis

These have been studied by injection with neoprene latex and Berlin blue; also by staining erythrocytes (Popa & Fielding 1930 a, b, Green & Harris 1949, Xuereb et al 1954 a, b, Stanfield 1960) and examination by scanning electron microscopy of corrosion casts to determine details of angio-architecture of the median eminence and the vascular connections in and between the neurohypophysis and adenohypophysis (Page & Bergland 1977).

The *arteries* of the hypophysis arise from the internal carotids via a single *inferior* and several *superior hypophysial arteries* on each side, the former coming from the cavernous part of the internal carotid, the latter from its supra-clinoid part and from the anterior and posterior cerebral arteries. The inferior hypophysial arteries divide into medial and lateral branches which anastomose across the midline to form an infundibular *arterial ring*. From this fine branches enter the neurohypophysis, supplying its capillary bed. The superior hypophysial arteries supply the median eminence, upper infundibulum and, via *arteries of the trabeculae*, its lower part. A confluent capillary net, extending through the neurohypophysis (median eminence, infundibular stem and pars nervosa), is supplied by both sets of hypophysial vessels. *Reversal* of flow can occur in cerebral capillary beds lying between the two supplies (Gillilan 1974) and has been suggested in the neurohypophysial network (Page & Bergland 1977).

The arteries of the median eminence and infundibulum end in characteristic sprays of capillaries, which are most complex in the upper infundibulum. In the median eminence these are an *external* or '*mantle' plexus* (Green 1951) and an *internal* or '*deep' plexus* (Duvernoy 1972); the external plexus, fed by the superior hypophysial arteries, is continuous with the infundibular plexus and is drained by *long portal vessels* descending to the pars anterior; the internal plexus projects within the external plexus which is supplied by it and is continuous posteriorly with the infundibular capillary bed and, like the external plexus, is drained by *long portal vessels*. *Short portal vessels* run from the lower infundibulum to the pars anterior. Both types of portal vessel open into vascular sinusoids lying between the secretory cords in the adenohypophysis, providing most of its blood; there is no direct arterial supply (Wislocki & King 1936). The *portal system* is considered to carry hormone-releasing factors, probably elaborated in parvocellular groups of hypothalamic neurons to control the secretory cycles of cells in the pars anterior, supply of which is not clear; it appears to be avascular in corrosion casts (Page & Bergland 1977).

The venous drainage of the neurohypophysis is by three possible routes: (1) to the adenohypophysis via long and short portal vessels; (2) via the large inferior hypophysial veins into the dural venous sinuses; (3) to the hypothalamus via capillaries passing to the median eminence. The venous drainage carries hypophysial hormones from the gland to their targets and also facilitates feedback control of secretion. In contrast, the venous drainage of the adenohypophysis appears restricted. Few vessels connect it directly to the systemic veins so the routes by which blood leaves remain obscure. If flow in the short portal veins running between the adeno- and neurohypophysis were reversible (Adams et al 1969), these portal vessels could constitute drainage channels; adenohypophysial hormones could enter the neuro-hypophysial capillaries before reaching the systemic veins, providing a 'short feedback' loop postulated from the evidence of endocrinology. The reversed flow in neurohypophysial capillaries (from neurohypophysis to hypothalamus, according to Torok 1954) could be a vascular route for neurohypophysial hormones to the ventricular tanycytes and hence to the cerebrospinal fluid. The demonstration of neurohypophysial hormones in the cerebrospinal fluid (Robinson & Zimmerman 1973) and of releasing factors there and in tanycytes (Knigge & Joseph 1974) strengthens this hypothesis. Cushing's original (1912) suggestion that the neurohypophysis secretes into the third ventricle may yet be confirmed.

This model of hypophysial blood supply has far-reaching implications; instead of the median eminence being the final common pathway for neural control of the adenohypophysis (Harris 1947), the entire neurohypophysis may be involved; its capillary

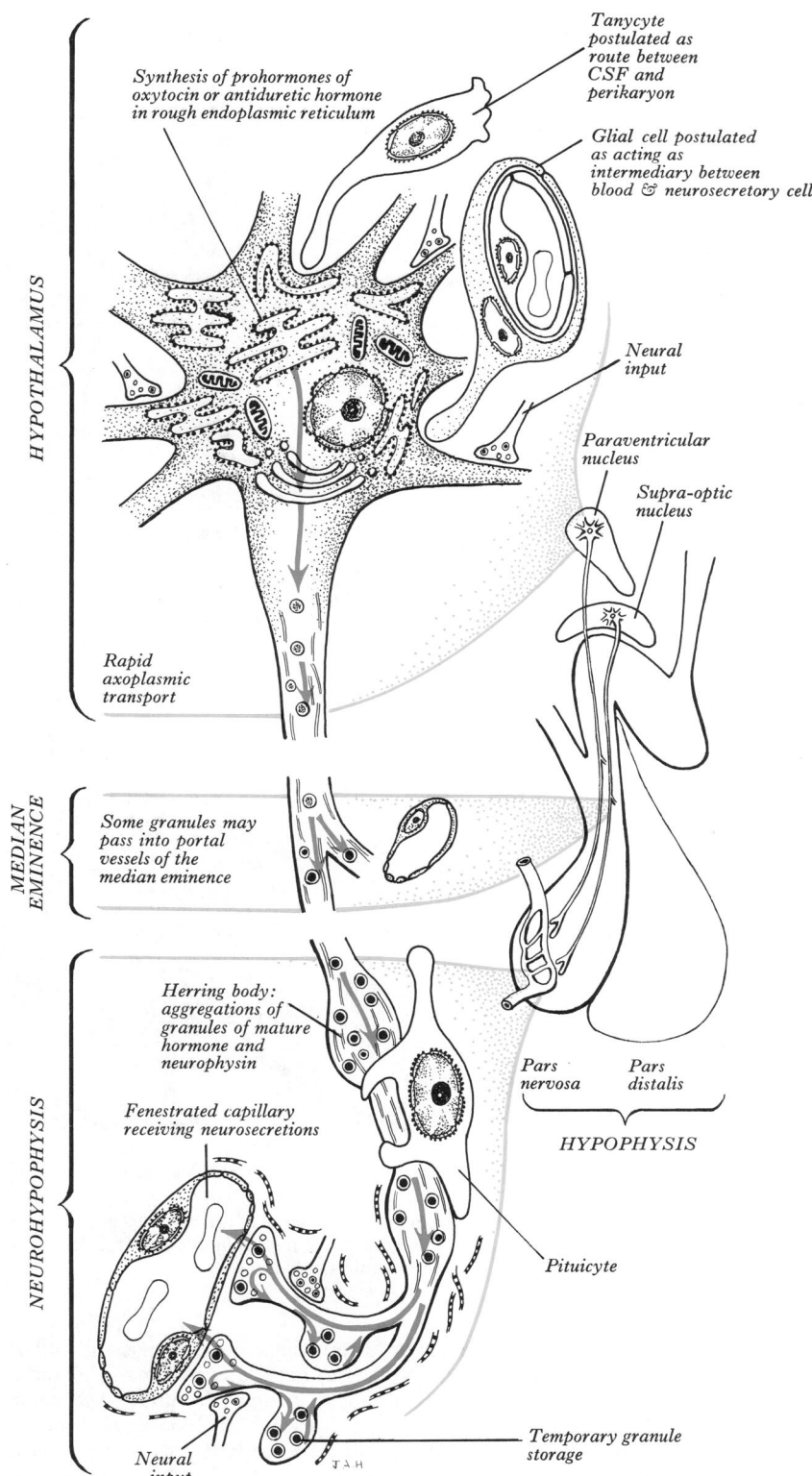

8.226 Diagram summarizing the control systems which influence the production of neurohypophysial hormones.

bed may selectively 'determine the destination of both hypothalamic and pituitary secretions, conveying some to the glandular pituitary, others to distant target organs and yet others to the brain' (Page & Bergland 1977).

THE PHARYNGEAL HYPOPHYSIS

This small collection of adenohypophysial tissue lies in the mucoperiosteum of the human nasopharyngeal roof (Boyd 1956) as in many other mammals (McGrath 1974). By 28 weeks in utero it is well vascularized and capable of secretion, receiving blood from the systemic vessels of the nasopharyngeal roof. At this stage

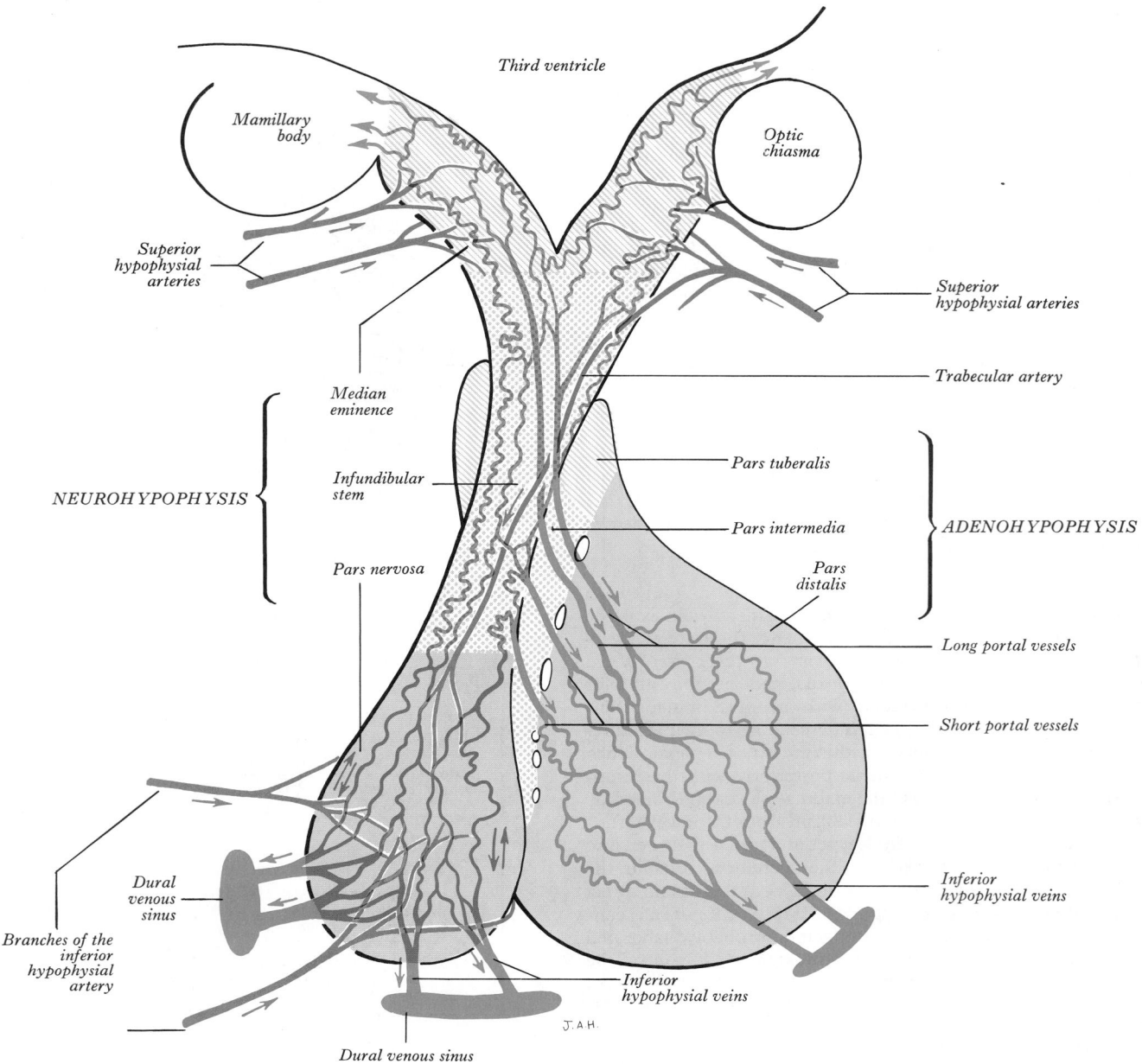

8.227 Diagram summarizing the vasculature of the hypothalamic median eminence, infundibulum and the rest of the hypophysis cerebri. (See text for a detailed account and possible significances.)

it is covered posteriorly by fibrous tissue but in the second half of fetal life this is replaced by venous sinuses and a trans-sphenoidal portal venous system develops, bringing it under the same hypothalamic control as the cranial adenohypophysial tissue (McGrath 1978). The peripheral vascularity of the pharngeal hypophysis persists until about the fifth year; the organ is then reinvested by fibrous tissue and presumed to be again controlled by factors present in systemic blood. Though it does not change in size after birth in males, in females it becomes temporarily smaller, returning to natal volume during the fifth decade (McGrath 1971), when once again it may be controlled via a trans-sphenoidal extension of the hypothalamo-hypophysial portal venous system. The human pharyngeal hypophysis may be a reserve of potential adenohypophysial tissue which may be stimulated, particularly in females, to synthesize and secrete adenohypophysial hormones in middle age, when intracranial adenohypophysial tissue is beginning to fail.

The Pineal Gland

The pineal gland or *epiphysis cerebi* (**8.**221), a small, piriform, reddish-grey organ, occupies a depression between the superior

colliculi. (For classic reviews see Gladstone & Wakeley 1940, Kappers 1960, Wolfe et al 1962, Kappers & Schade 1965, Wurtman et al 1968, Wolstenholme & Knight 1971.) It is inferior to the splenium of the corpus callosum, separated from it by the tela choroidea of the third ventricle and its cerebral veins and enveloped by the tela's lower layer, which is reflected thence to the tectum (**8.**221). The gland is about 8 mm in length; its base, directed anteriorly, is attached by a *peduncle* dividing into inferior and superior laminae, separated by the *pineal recess* of the third ventricle (**8.**221), and respectively containing the posterior (p. 1005) and habenular (p. 1006) commissures. Aberrant commissural fibres may invade the gland but do not terminate near parenchymal cells. Nerve fibres enter its dorsolateral aspects from the tentorium cerebelli as a single or paired *nervus conarii*, which is subendothelial in the wall of the straight sinus (p. 801), its fibres derived from neurons of the superior cervical ganglia. These fibres are adrenergic sympathetic elements associated with blood vessels and parenchymal cells (Kappers 1960, Wolfe et al 1962). Bjorklund et al (1972) have shown in rats that postganglionic sympathetic fibres from the *nervus conarii* reach neurons in the habenular nuclei, from which some fibres of the *habenulopineal tract* (p. 1006) may start. Møller (1978) reported that, in human fetuses, fibres of this tract reach the *ganglion conarii* (Pastori ganglion) at the pineal apex (**8.228**); the neurons of origin of these fibres need verification. Møller (1978, 1979) has also reported an unpaired nerve containing non-myelinated fibres and bipolar neurons with axosomatic and axodendritic synapses, the *nervus pinealis*, in human fetuses; it is in the subarachnoid space near the median plane just caudal to the pineal, connecting it to the posterior commissure. It is ephemeral, presumed to degenerate late in intrauterine life and its function is uncertain; but since it is presumed to be homologous with the sensory pineal nerve of fish and amphibians (Møller 1978), whose pineal complexes contain photoreceptors, it might transmit light-generated impulses. Though in some mammals (e.g. neonatal rats) pinealocytes transiently resemble photoreceptors (Zimmerman & Tsi 1975), those in human fetuses do not. The human fetal nervus pinealis may be involved in pineal differentiation. At present it can only be regarded as a phylogenetic vestige; its loss may be associated with evolution of the pineal as a secretory rather than photoreceptive organ. Møller (1978) has also described a ganglion, presumed to be parasympathetic, rostral to the pineal gland and near the choroid plexus of the third ventricle. An anterior *intrapineal ganglion* has also been found (**8.228**).

STRUCTURE OF THE PINEAL GLAND

The pineal gland contains cords and follicles of pinealocytes and neuroglial cells among which ramify many blood vessels and nerves. Septa extend into it from the surrounding pia mater.

Pinealocytes form the pineal parenchyma. Extending from each cell body, with a spherical, oval or lobulated nucleus, are one or more tortuous basophilic processes, containing parallel microtubules (Knight et al 1973) which end in expanded terminal buds near capillaries or, less frequently, ependymal cells of the pineal recess. The terminal buds contain granular endoplasmic reticulum, mitochondria and electron-dense cored vesicles, which store monoamines and polypeptide hormones (Sheridan & Sladek 1975), release of which appears to require sympathetic innervation. Lukaszyk & Reiter (1975) have proposed that the polypeptide hormones (which they consider could be produced by pineal neuroglia and neurons) combine with specific protein carriers, termed *neuro-epiphysins* to distinguish them from hypophysial neurophysins. They are released by exocytosis, together with fragments of vesicular membrane, the latter forming exocytotic debris. When released the complex is believed to dissociate, hormones being exchanged for calcium ions. The calcium-carrier complex so formed is, in the pineal, deposited concentrically around exocytotic debris as *corpora arenacea* or 'brain sand' (**8.229, 230**). It is often supposed that the pineal atrophies with age, corpora arenacea being a sign of atrophy; on the contrary, these corpuscles may indicate continued secretion. Wildi & Frauchiger (1965) found no evidence of pineal degeneration in the elderly.

Pinealocytes contain both granular and agranular endoplasmic reticulum, with extensive Golgi complexes, lipid droplets and numerous mitochondria, according with secretory function. An unusual organelle (groups of microtubules and perforated lamellae) sometimes occurs near granular endoplasmic reticulum and lipid droplets. Termed '*canaliculate lamellar bodies*' by Lin (1967) and McNeil (1977), '*annulate lamellae*' by Friere & Cardinali (1975) and '*mikrotubuli*' by Gusek (1976), they may be involved in secretion (McNeil 1977) but their function remains uncertain.

Pinealocytes of some mammals contain synaptic ribbons, perhaps involved in transmission; vesicles near them contain neurotransmitters such as gamma-amino-butyric acid (Krstic 1976). These ribbons may arise from microtubular sheaths and in turn from centrioles (Karasek 1976). Similar organelles occur in mammalian retinal photoreceptors and simpler submammalian photoreceptors, supporting the theory that mammalian pinealocytes are derived from photoreceptors (Kappers 1976, Relkin 1976). Transient similarities between pinealocytes and retinal photoreceptors in neonatal rats lend further support to this concept (Zimmerman & Tso 1975).

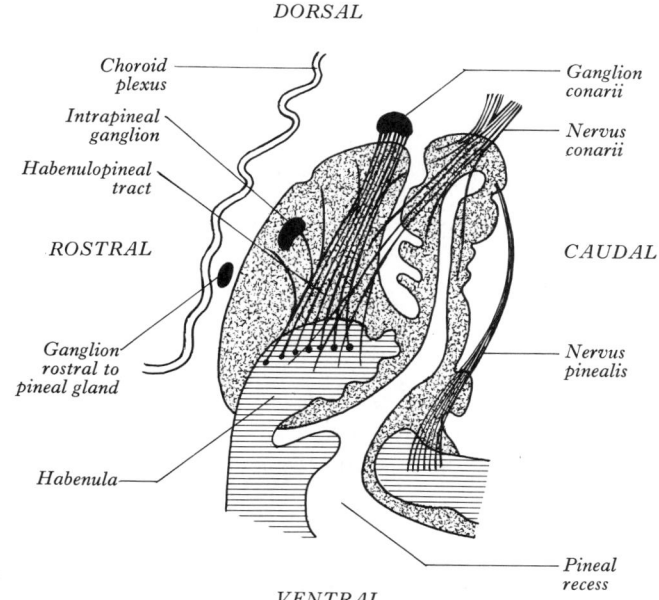

8.228 Diagram showing the principal neural pathways which have been described in connection with the human fetal pineal gland (epiphysis cerebri).

The ultrastructure of human fetal pinealocytes indicates their secretory function in early intrauterine life (Møller 1974). As in adults, they contain all the appropriate organelles, together with abundant microfilaments, microtubules and a few cilia with a $9 + 0$ microtubular pattern, the latter also associated with secretory cells in other endocrine glands, such as the hypophysis (Barnes 1961, Andersen et al 1970). Fetal pinealocytes have gap junctions, desmosomes and 'intermediate-like junctions' (Møller 1976); the first are for electrotonic coupling between adjacent cells enabling group activity, the others of an adhesive nature. Synaptic ribbons do not appear in human fetal pinealocytes.

Neuroglial cells, partially separating the pinealocytes, are like astrocytes and in the pineal stalk many have extensive longitudinal processes, composing most of the stalk. Ultrastructurally they display numerous filaments extending throughout their processes.

Vessels and Nerves

Pineal *capillaries* are lined by thin, sometimes fenestrated, endothelial cells, with a tenuous external basal lamina, sometimes incomplete. The non-myelinated autonomic *nerve fibres* are mostly noradrenergic, with dense-cored vesicles in terminal and preterminal expansions. Arising in the superior cervical ganglion they penetrate the pineal either via perivascular spaces or through the nervi conarii (vide supra) or else by the leptomeningeal pineal surface. They either terminate in perivascular spaces between the pinealocytes or synapse with them. Other nerve fibres in the stalk are of uncertain origin but habenular nuclei are strongly mooted as their source.

FUNCTIONS OF THE PINEAL GLAND

Once considered to be a phylogenetic relic, a vestige of a dorsal third eye and of little functional significance, the mammalian pineal is now accepted as an endocrine gland of major regulatory importance, modifying the activity of: the adenohypophysis, neurohypophysis, endocrine pancreas, parathyroids, adrenal cortex, adrenal medulla and gonads (De Vries & Kappers 1971, Klein 1978, Haulica & Coculescu 1981, Reiter 1983, 1984, 1985, 1987, Malendowicz 1985). Its effects are largely inhibitory: indoleamine and polypeptide hormones secreted by pinealocytes are believed to reduce synthesis and release of hormones of the pars anterior, e.g. by direct action on its secretory cells and indirectly by inhibiting production of hypothalamic releasing factors.

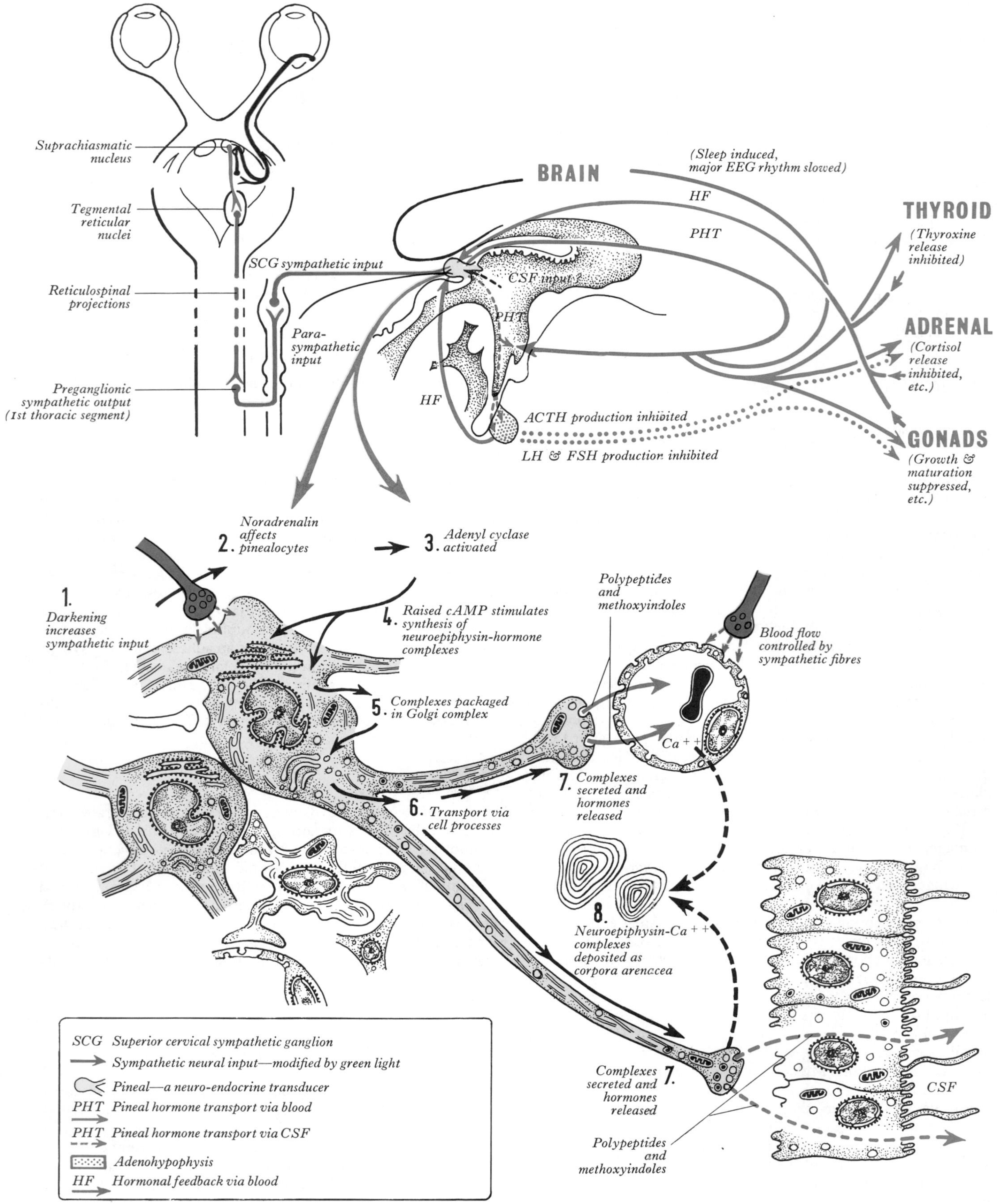

Suprachiasmatic nucleus

Tegmental reticular nuclei

Reticulospinal projections

Preganglionic sympathetic output (1st thoracic segment)

SCG sympathetic input

Para-sympathetic input

BRAIN

(Sleep induced, major EEG rhythm slowed)

HF

PHT

CSF input?

PHT

HF

ACTH production inhibited

LH & FSH production inhibited

THYROID
(Thyroxine release inhibited)

ADRENAL
(Cortisol release inhibited, etc.)

GONADS
(Growth & maturation suppressed, etc.)

Noradrenalin affects pinealocytes

2.

3. *Adenyl cyclase activated*

1. *Darkening increases sympathetic input*

4. *Raised cAMP stimulates synthesis of neuroepiphysin-hormone complexes*

5. *Complexes packaged in Golgi complex*

6. *Transport via cell processes*

Polypeptides and methoxyindoles

Blood flow controlled by sympathetic fibres

Ca++

7. *Complexes secreted and hormones released*

8. *Neuroepiphysin-Ca++ complexes deposited as corpora arenacea*

7. *Complexes secreted and hormones released*

Polypeptides and methoxyindoles

CSF

SCG   Superior cervical sympathetic ganglion

  Sympathetic neural input—modified by green light

  Pineal—a neuro-endocrine transducer

PHT   Pineal hormone transport via blood

PHT   Pineal hormone transport via CSF

  Adenohypophysis

HF   Hormonal feedback via blood

8.229 Diagrammatic review of current hypotheses regarding the control and effects of pineal function. (See text for details.)

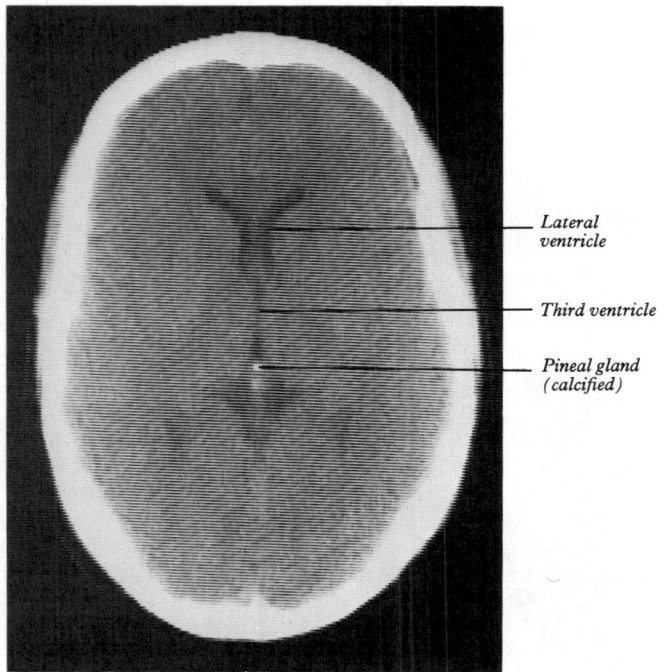

8.230 Computed tomogram of the head in the horizontal plane at the level of the pineal gland. Supplied by Shaun Gallagher, Guy's Hospital; photography by Sarah Smith, UMDS, Guy's Hospital Campus, London.

Lateral ventricle

Third ventricle

Pineal gland (calcified)

Pineal secretions may reach their target cells via cerebrospinal fluid (Sheridan et al 1969, Knight et al 1973) or the bloodstream.

Some pineal indole-amines, including melatonin and enzymes for their biosynthesis (e.g. serotonin N-acetyltransferase) show circadian rhythms in concentration and activity in many mammalian pineal glands. In rats there may be an *endogenous circadian oscillator* in the suprachiasmatic nucleus (hypothalamus), whose intrinsic rhythmicity governs cyclical pineal behaviour (Klein 1978). Photic stimuli, in particular intensity changes of yellow-green light acting on rhodopsinoid retinal pigment, seem to be involved in this rhythm (Cardinali et al 1971, Minneman et al 1974), the pineal being most *active* in *darkness*. With the visual pathway and pineal sympathetic supply intact, exposure to *light* after several hours of darkness *depresses* pineal activity (Klein & Weller 1972). Suprachiasmatic neurons are supplied directly by axons of retinal ganglion cells, allowing response to changes in illumination (Nishino et al 1976). In rats, the path from the suprachiasmatic nucleus to the pineal gland is claimed to include the tegmental nuclei and the upper thoracic spinal intermediolateral column (Saper et al 1976): preganglionic sympathetic fibres from the latter are said to reach the superior cervical ganglia, from which post-ganglionic fibres can be traced to the pinealocytes. Release of catecholamines from these fibres causes a receptor-mediated increase in the production of cyclic AMP in pinealocytes, evoking a 70- to 100-fold increase in serotonin N-acetyltransferase activity, which regulates daily changes in melatonin production and hence its plasma levels. Interest has centred on sympathetic innervation; the role of its parasympathetic supply is as yet unknown. These hypothecations are summarized in 8.229. Evidence has been found of a circadian rhythm in *human* pineal activity, in changes of plasma melatonin levels (Vaughan et al 1976). As in other mammals, the level rises during darkness, falling during the day. Whether the control mechanisms demonstrated in other animals also operate in man is unresolved.

The Thyroid Gland

The thyroid gland (8.231), brownish-red and highly vascular, is placed anteriorly in the lower neck, level with the fifth cervical to the first thoracic vertebrae. Ensheathed by the pretracheal layer of deep cervical fascia, it has right and left lobes connected by a narrow, median *isthmus*. Its weight is usually about 25 g but varies, being slightly heavier in females, and enlarging during menstruation and pregnancy.

The lobes are approximately conical; their ascending apices diverge laterally to the level of the oblique lines on the laminae of the thyroid cartilages; their bases are level with the fourth or fifth tracheal cartilages. Each lobe is about 5 cm long, its greatest transverse and anteroposterior extents being about 3 cm and 2 cm respectively. Its *posteromedial aspect* is attached to the side of the cricoid cartilage by a *lateral thyroid ligament* (p. 1118). The *lateral (superficial) surface* is convex and covered by the sternothyroid, whose attachment to the oblique thyroid line prevents the upper pole of the gland from extending onto the thyrohyoid muscle. More anteriorly are the sternohyoid and superior belly of omohyoid, overlapped inferiorly by the anterior border of sternocleidomastoid. The *medial surface* is adapted to the larynx and trachea, contacting at its superior pole the inferior pharyngeal constrictor and the posterior part of the cricothyroid, which separate it from the posterior part of the thyroid lamina and the side of the cricoid cartilage. The external laryngeal nerve is medial to this part of the gland, on its way to the cricothyroid. Below the trachea, and posterior to this, the recurrent laryngeal nerve (p. 1118) and oesophagus (closer on the left) are medial relations. The *posterolateral surface* is next to the carotid sheath, overlapping the common carotid artery. The thin *anterior border*, near the anterior branch of the superior thyroid artery, slants down medially. The rounded *posterior border* is related below to the inferior thyroid artery and its anastomosis with the posterior branch of the superior thyroid artery. The parathyroid glands are usually related to this border (p. 1463), whose lower end on the left is near the thoracic duct.

The isthmus connects the lobe's lower parts; it measures about 1.25 cm transversely and vertically and is usually anterior to the second and third tracheal cartilages, though often higher or sometimes lower; its site and size vary greatly. Pretracheal fascia separates it from the sternothyroid muscles; more superficially are the sternohyoids, anterior jugular veins, fascia and skin. The superior thyroid arteries anastomose along its upper border; at the lower border the inferior thyroid veins leave the gland. Occasionally the isthmus is absent.

A conical *pyramidal lobe* often ascends towards the hyoid bone from the isthmus or the adjacent part of either lobe (more often the left). It is occasionally detached or in two or more parts. A fibrous or fibromuscular band, the *levator of the thyroid gland*, sometimes descends from the hyoid body to the isthmus or pyramidal lobe. Small detached masses of thyroid tissue may occur above the lobes or isthmus as *accessory thyroid glands*. Vestiges of the thyroglossal duct (p. 229) may persist between the isthmus and the lingual foramen caecum, sometimes as accessory nodules or cysts of thyroid tissue near the midline or even in the tongue.

STRUCTURE AND FUNCTIONS

The gland has a thin capsule of connective tissue, whose extensions divide it into masses of irregular form and size. The parenchyma derives mainly from the endoderm of the thyroglossal duct (p. 229), usually a transient embryonic structure extending between the gland and the tongue. From its distal end solid epithelial cords and sheets branch and lumina filled with yellow, viscid colloid appear in them. This endodermal epithelium is generally considered to develop into *separate* follicles, approximately spherical and about 0.02–0.9 mm in diameter, each with a central colloid core surrounded by a simple epithelium and basal lamina. But three-dimensional reconstructions of mature thyroid glands show that follicles are typically aggregated, with a common epithelium bordering several colloid masses and linking follicles together (Isler et al 1968). Solid cell nests have also been described (Harach 1985). The colloid, staining pink with eosin, contains an iodinated glycoprotein, iodothyroglobulin, a precursor of the thyroid hormones, tri-iodothyronine (T_3) and tetra-iodothyronine or thyroxine (T_4), and a product of the follicular epithelial cells. Aggregated follicles nestle in a delicate connective

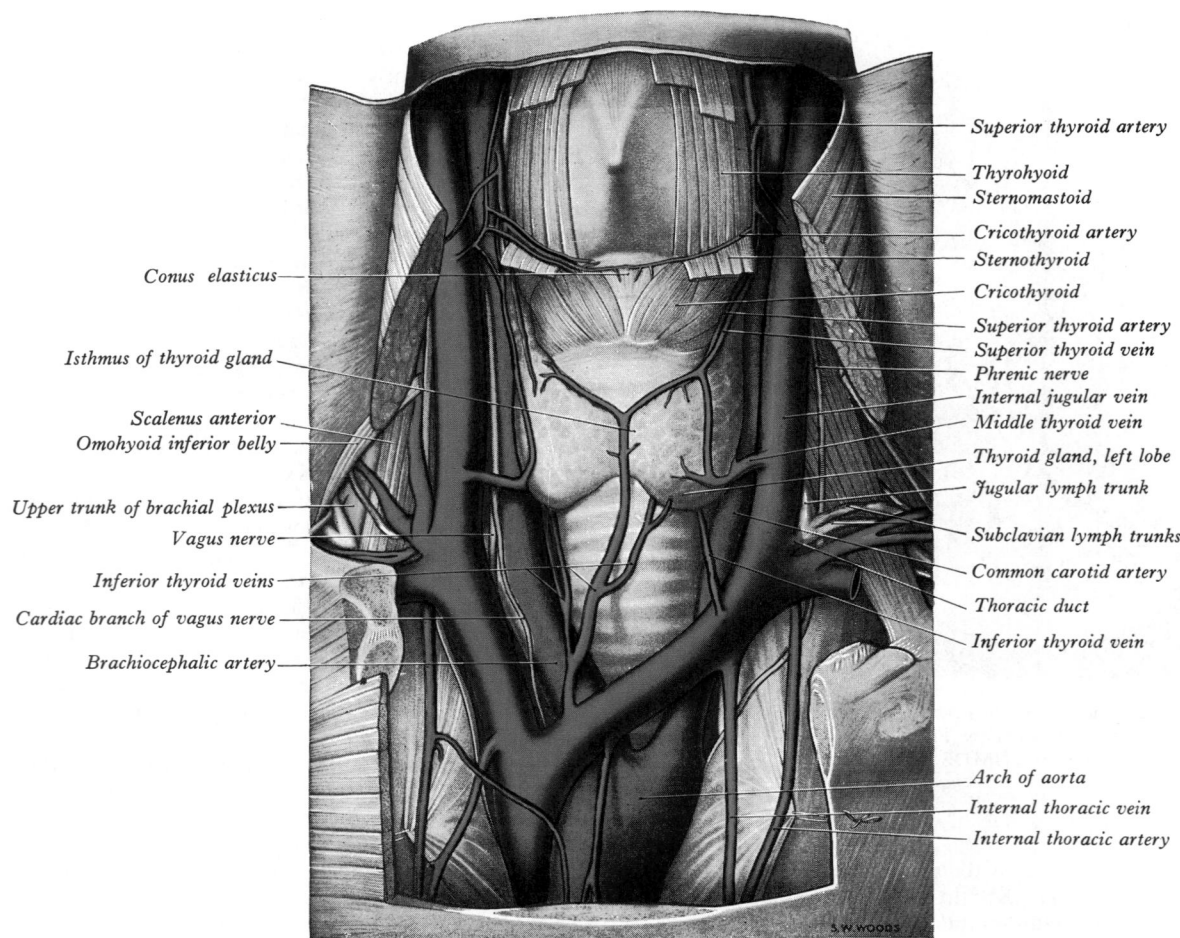

Superior thyroid artery

Thyrohyoid
Sternomastoid

Cricothyroid artery
Sternothyroid

Cricothyroid

Superior thyroid artery
Superior thyroid vein
Phrenic nerve
Internal jugular vein
Middle thyroid vein

Thyroid gland, left lobe

Jugular lymph trunk

Subclavian lymph trunks

Common carotid artery

Thoracic duct

Inferior thyroid vein

Arch of aorta
Internal thoracic vein
Internal thoracic artery

Conus elasticus

Isthmus of thyroid gland

Scalenus anterior
Omohyoid inferior belly

Upper trunk of brachial plexus
Vagus nerve

Inferior thyroid veins
Cardiac branch of vagus nerve
Brachiocephalic artery

8.231 The thyroid gland and its environs. The manubrium sterni and the sternal ends of the clavicles and first costal cartilages have been removed and the pleural sac and lung have been retracted on each side.

tissue stroma, surrounded by dense plexuses of fenestrated capillaries, extensive lymphatic networks and sympathetic nerve fibres which supply the arterioles and capillaries, some nerve fibres ending close to the follicular epithelial cells (Melander et al 1975, Melander 1977).

Thyroid parenchyma contains a second, parafollicular type: 'light', 'clear' or C cells, producing the peptide hormone thyrocalcitonin. C cells, an APUD element (p. 1446), are derived from the ultimobranchial bodies (p. 229). They are a minor component but improved techniques (Solcia et al 1968, 1969) suggest that in some species they have been underestimated; e.g. in dogs Kameda (1971, 1976) found a ratio of 30–90 parafollicular to 100 follicular cells. Branched tubules containing desquamated cells may occur; like C cells, these are ultimobranchial in origin (Halmi 1978). Other cells, some ciliated, have been described in rodent thyroid glands (Wollman & Neve 1971).

Follicular cells proper vary from squamous to columnar, depending on their activity, mainly controlled by circulating hypophysial thyrotropin (TSH) (**8.232**), in the absence of which the follicular cells are squamous and 'resting', luminal colloid being abundant and reflecting increased storage of iodinated thyroglobulin. The secretion of TSH leads to endocytosis of colloidal droplets at the luminal aspects of follicular cells, cavities appearing in the luminal colloid near the epithelium. Prolonged high levels of circulating TSH induce follicular hypertrophy and even hyperplasia, with progressive resorption of colloid and increased stromal vascularity.

Follicular cells have a striking ultrastructural and functional polarity: when activated by TSH, or possibly adrenergic nerve terminals (vide infra), they engage in apical synthesis and luminal exocytosis of thyroglobulin, with basally directed thyroglobulin endocytosis, degradation and liberation of thyroid hormones (T_3

and T_4) into the blood capillaries. This polarized dual function is indicated by the arrangement of their organelles. To form a continuous wall around each mass of colloidal thyroglobulin, the follicular cells are linked by apical junctional complexes and have their bases resting on a basal lamina (**8.232, 233**). Each cell has a basal nucleus, prominent granular endoplasmic reticulum and supranuclear Golgi complex, the last named being particularly prominent in TSH-activated cells. Apical to the Golgi complex and derived from it, secretory vesicles transport glycoprotein, assembled by serial activity of the granular endoplasmic reticulum and the Golgi complex, to the apical plasmalemma for luminal release by exocytosis. Iodine required to complete thyroglobulin formation enters the follicular cells basally by active transport from the blood capillaries across the basal plasmalemma; labelled iodide rapidly appears in the lumina (Wolff 1964) and is oxidized to iodine, mainly by thyroid peroxidase in the apical plasmalemma (Taurog 1970). Iodine is then attached to the tyrosyl groups of the secreted glycoprotein manufactured by the follicular cells. to form mono- and di-iodotyrosyls. These are then coupled into iodothyronyl groups (thyroid hormones in peptide linkage) to complete the formation of iodinated thyroglobulin, the precursor of thyroid hormones (Bjorkman & Ekholm 1973, Robbins et al 1974, Haeberli et al 1975). In follicular cells endogenous peroxidase occurs in perinuclear cisternae, granular endoplasmic reticulum, Golgi complex, apical vesicles, apical plasmalemma and particularly on microvilli (Strum & Karnovsky 1970); but it is likely that only in the apical plasmalemma is it involved in iodide oxidation, its distribution indicating its route to this site. Apical microvilli, numerous but short in resting cells (Wetzel et al 1965), elongate and often branch on stimulation by TSH, which also provokes extension of cytoplasmic processes into the luminal colloid; these fuse around portions of the colloid

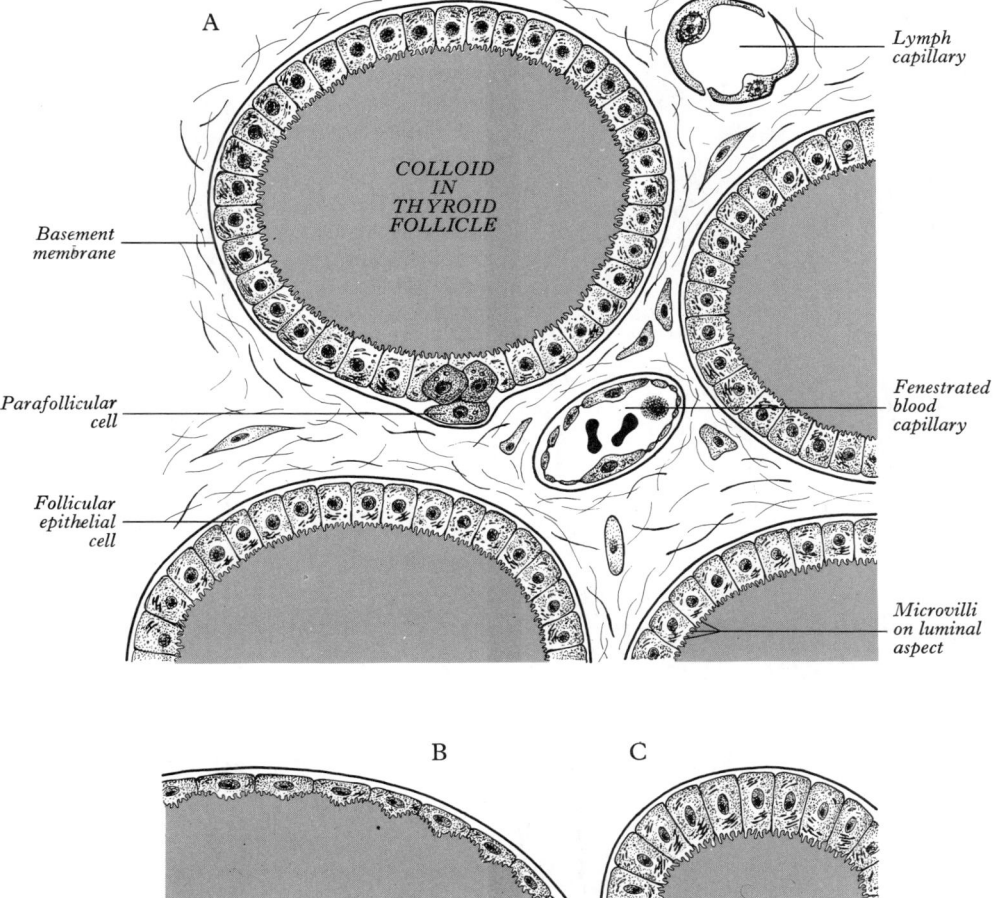

8.232 Diagrams of thyroid follicular structure. A. Normal structure under average physiological conditions. B. 'Resting' state. C. Highly active state. (See text for details.)

and take it into the cell. After colloid endocytosis, lysosomes, located basally in the resting state, migrate towards the lumen to fuse with the intracellular droplets of colloid, forming secondary lysosomes or phagolysosomes, which return to the cell's base. During this period the colloid gradually disappears as acid proteases and peptidases in the phagolysosomes degrade the iodinated thyroglobulin, releasing the thyroid hormones T_3 and T_4, which pass basally for release, leaving the gland mainly via the blood capillaries and lymphatics. More numerous precursor molecules, such as 3-mono–iodotyrosine and 3,5-di-iodotyrosine, are de-iodinated by a dehalogenase; the released iodine migrates apically to be recycled in the iodination of newly synthesized thyroglobulin molecules. Microtubules and microfilaments are probably involved in thyroid hormone secretion (Wolff & Williams 1973); treatment with colchicine and cytochalasin B (interfering respectively with microtubule and microfilament activity) inhibits TSH and dibutyryl cAMP-induced secretion. Some iodinated thyroglobulin escapes intact from the follicles and can be radio-immunoassayed in blood, probably reaching it via the lymphatic vessels. Follicular activity in thyroglobulin synthesis and thyroid hormone release is illustrated in **8.233**. Greer & Haibach (1974), DeGroot & Niepomniszcse (1977) and Taurog (1978) provide detailed accounts of these processes.

Of the two thyroid hormones tri-iodothyronine is probably the main stimulator of cellular metabolic rate, its action being very powerful and immediate, whereas tetra-iodothyronine (*thyroxine*) is powerful but less rapid. Over-production of both causes *thyrotoxicosis* (exophthalmic goitre); hyposecretion in adults produces *myxoedema* and in infants *cretinism*. Their synthesis and release are mainly controlled by adenohypophysial thyrotropic hormone (TSH). However, in *Graves' disease*, antibodies to human thyroid-stimulating immunoglobulin (HTSI) bind to TSH-receptor sites in the follicular

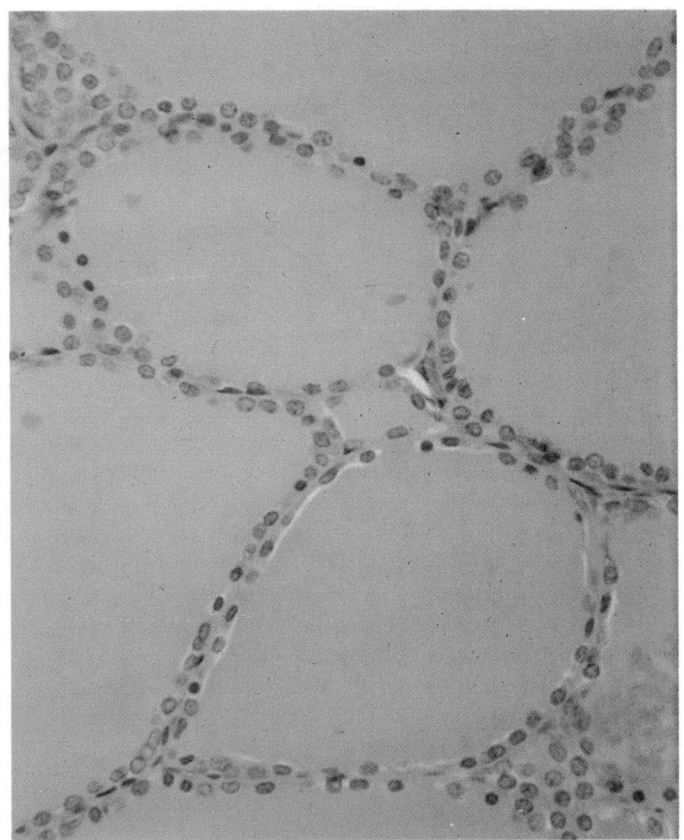

8.232D Section through a thyroid gland showing thyroid follicles surrounding central masses of thyroid colloid. Provided by P Shepherd and S Liebowitz, Guy's Campus, UMDS, London. Haematoxylin and eosin. Magnification × 500.

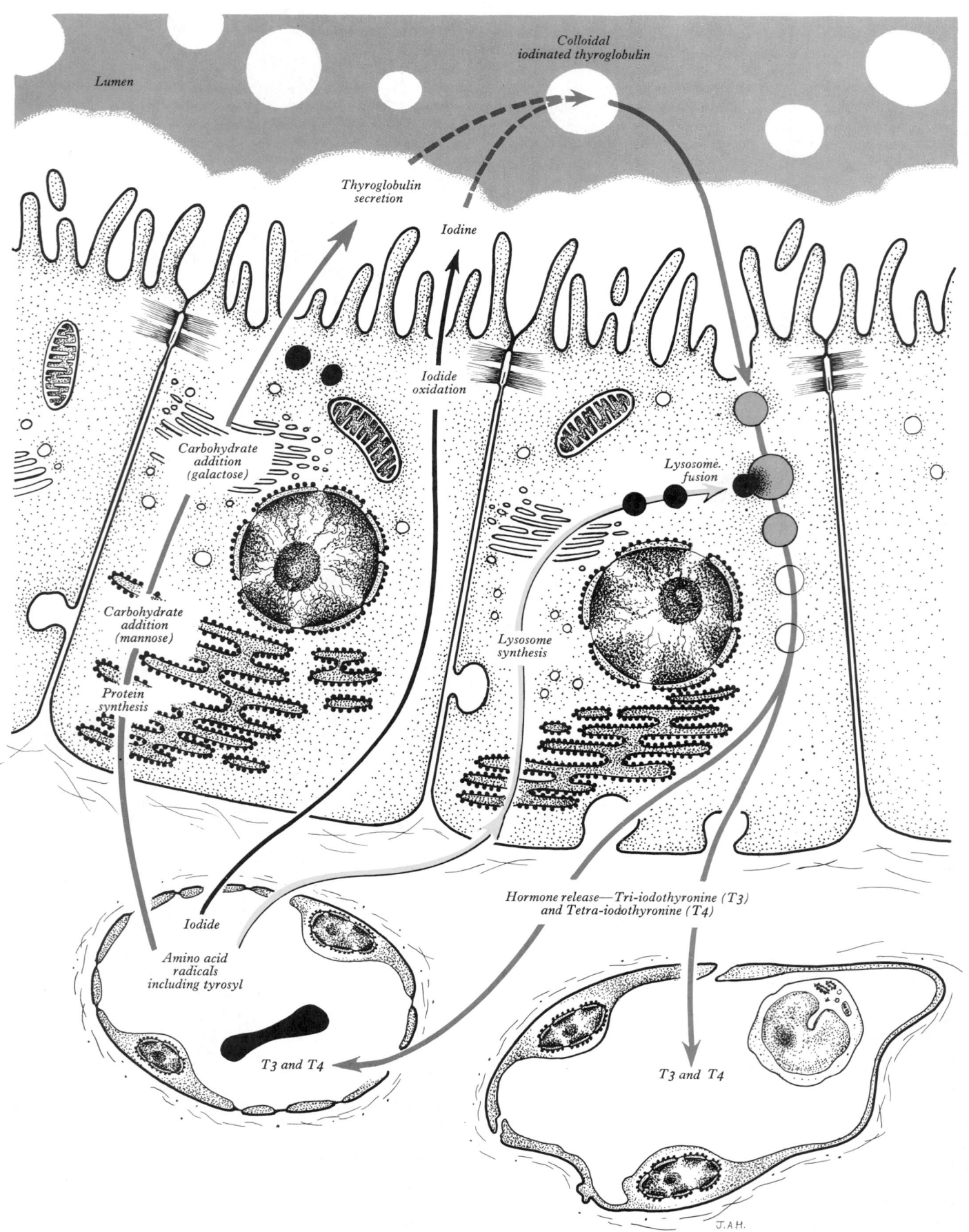

8.233 The functional architecture of the thyroid follicular cells, showing on one aspect the colloid-containing cavity and on the other blood and lymphatic capillaries. Arrows indicate metabolic flow pathways.

cells, interfering with this control and causing excessive hormone production (Werner 1978). Thyroid hormones increase the sensitivity of tissues to adrenalin and noradrenalin. Recent advances in understanding the action of the thyroid hormones at the cellular level have been reviewed by Oppenheimer et al (1987).

Although follicular activity is mainly controlled by circulating TSH, evidence in several mammals, including mankind, suggests a direct sympathetic influence on follicular cells, separately from sympathetic control of the gland's circulation (Melander 1978). Fluorescence histochemistry and electron autoradiography have demonstrated adrenergic terminals near blood vessels and follicular cells. In mice, with TSH secretion eliminated to avoid indirect effects on thyroid secretion, unilateral sympathetic stimulation evokes secretion of the thyroid hormones only in those regions supplied by the stimulated nerve (Melander et al 1972). Studies of human thyroid tissue in vitro also suggest that noradrenalin directly induces follicular changes associated with hormone secretion, including colloid droplet formation and lysosome migration (Melander 1978). In vitro investigations on isolated calf's thyroid cells show that catecholamines can enhance the incorporation of iodine and synthesis of hormones (Melander et al 1973), effects abolished in vivo and in vitro by drugs which block adrenergic receptors (Melander 1970, Maayan & Ingbar 1978) but which do not affect the ability of cells to respond to TSH. Therefore catecholamines and TSH interact with different receptors in follicular cells and can act independently, however, with similar effects. Both activate adenyl cyclase, increasing the formation of cyclic-AMP, which augments hormonal release (Melander 1978). Although anatomicophysiological evidence leaves little doubt that follicular cells and their vessels have direct sympathetic innervation, its effect on thyroid secretion is uncertain. TSH may be a greater influence in sustained regulation; the link between the nervous system and follicular cells may mediate rapid, transient responses to external influences.

Parafollicular ('C') cells, an APUD type (p. 1466), occur singly or in small groups, close to the outer follicular borders but within the follicular basement membrane. Often partly insinuated between adjacent follicular cells, they do not reach the follicular lumen; they are oval or polyhedral and larger than follicular cells. Grouping is sometimes more marked, e.g. in canine thyroid glands, and when perfused with fixative for electron microscopy, where folliculoid groups of such cells appear with expanded central extracellular spaces, perhaps storing hormone. Parafollicular, unlike follicular, cells have no apparent nerve supply; their cytoplasm has numerous membrane-bound secretory granules, probably containing a stored form of thyrocalcitonin, as supported by immunocytochemical studies (Pearse 1966, Wolfe et al 1974). As befits cells engaged in making exportable protein, they contain granular endoplasmic reticulum varying with the level of activity, obvious Golgi complexes, many mitochondria and free ribosomes. The oval nucleus is generally eccentric and has a smooth or slightly irregular membrane.

The main factor controlling release of thyrocalcitonin is usually said to be the concentration of serum calcium; a rise of this in blood perfusing the thyroid gland stimulates thyrocalcitonin secretion, while hypocalcaemia suppresses it, there being a reciprocal relation between secretion of thyrocalcitonin and parathyroid hormone (**8.234A**). Thyrocalcitonin regulates calcium metabolism in many species, largely by suppressing bone resorption, but its human role is uncertain; difficulties arise in detecting thyrocalcitonin in human plasma even by sensitive radio-immunoassay; no pathological condition has been associated with thyrocalcitonin deficiency (O'Riordan 1978).

Vessels and Nerves

The *arteries* supplying the gland are the superior (p. 736) and inferior thyroid (p. 755) and sometimes an *arteria thyroidea ima* from the brachiocephalic trunk or aortic arch. The arteries are remarkably large with frequent anastomoses on and in the gland. The *veins* form a plexus on its surface and in front of the trachea; from this plexus superior, middle and inferior thyroid veins arise: the first two end in the internal jugular, the inferior in the left brachio-cephalic vein. A dense blood capillary plexus surrounds the follicles, between their epithelium and the endothelium of

lymph capillaries, which also surround much of the follicular peripheries. *Lymph vessels* run in the interlobular connective tissue, often around arteries; they communicate with a capsular network, may contain colloid material and end in the thoracic and right lymphatic ducts. The *nerves* are derived from the superior, middle and inferior cervical sympathetic ganglia.

Applied Anatomy

Apart from variable enlargement during menstruation and pregnancy, any thyroid swelling is a *goitre*, which may press on related structures. Symptoms are most commonly due to pressure on the trachea or on the recurrent laryngeal nerves. These nerves and therefore the laryngeal muscles supplied by them, may be affected by pressure or damaged in thyroidectomy; if the external laryngeal nerve or the cricothyroid muscle (which is supplied by it and tenses the vocal folds) are damaged, the voice becomes incapable of varying pitch and is slightly tremulous (Harries 1955).

Partial thyroidectomy is often necessary in hyperthryoidism and thyroid enlargement. Enough is removed to relieve symptoms; except in malignant disease it is not entirely removed, since this leads to myxoedema. During tying of the inferior thyroid artery, the proximity of the recurrent laryngeal nerve (pp. 755, 1118) is a hazard. Temporary aphonia sometimes follows mere bruising of the nerve; complete division reduces the voice to a whisper. In partial thyroidectomy the posterior parts of both lobes are left intact to preserve the parathyroid glands.

The Parathyroid Glands

The parathyroid glands (**8.234**) are small, yellowish-brown, ovoid or lentiform structures, usually lying between the posterior lobar borders of the thyroid gland and its capsule. They are commonly about 6 mm long, 3–4 mm across, and 1–2 mm from back to front, each weighing about 50 mg. Usually there are two on each side, superior and inferior. The anastomotic connection between the superior and inferior thyroid arteries along the posterior thyroid border (p. 755) usually passes very close to the parathyroids.

The superior parathyroid glands are more constant in site than the inferior and are usually midway along the posterior thyroid borders but sometimes higher. The *inferior pair* (Walton 1931, Gilmour 1938, Murley & Peters 1961) may be: (1) within the fascial thyroid sheath, below the inferior thyroid arteries and near the inferior lobar poles; (2) outside the sheath, immediately above an inferior thyroid artery; or (3) in the thyroid gland near its inferior pole. These variations are surgically important; a tumour of the inferior parathyroid, if in position (1), may descend along the inferior thyroid veins anterior to the trachea into the superior mediastinum, whereas if in position (2) it may extend posteroinferiorly behind the oesophagus into the posterior mediastinum. The superior parathyroids are usually dorsal, the inferior ventral, to the recurrent laryngeal nerves (Pyrtek & Painter 1964).

The glands develop from endoderm in the pharyngeal pouches (p. 229), the inferior from the third and therefore referred to as *parathyroids III*, the superior from the fourth and termed *parathyroids IV*. An inferior parathyroid is connected in early development with the diverticulum of the third pouch, which forms the thymus and is carried with it in its caudal migration. Normally the inferior parathyroids migrate only to the inferior thyroid poles but may descend with the thymus into the thorax or not descend at all, remaining above their normal level near the carotid bifurcation. Parathyroid glands vary in number; there may be only three or many minute parathyroid islands scattered in connective tissue near the usual sites (Hintzsche 1937, Vail & Coller 1967).

Structure (**8.234B**). Each parathyroid has a thin connective tissue capsule with intraglandular septa but lacking lobules. In childhood, the gland consists of wide, irregular, interconnecting columns of *chief cells* or *principal cells*, responsible throughout life for the synthesis and secretion of parathyroid hormone (parathormone); three types of chief cell, *light*, *dark* and *clear*, can be distinguished by the depth of cytoplasmic staining; by light microscopy their cytoplasm appears homogeneous. Between the columns of cells is a dense plexus of sinusoidal capillaries, via which the

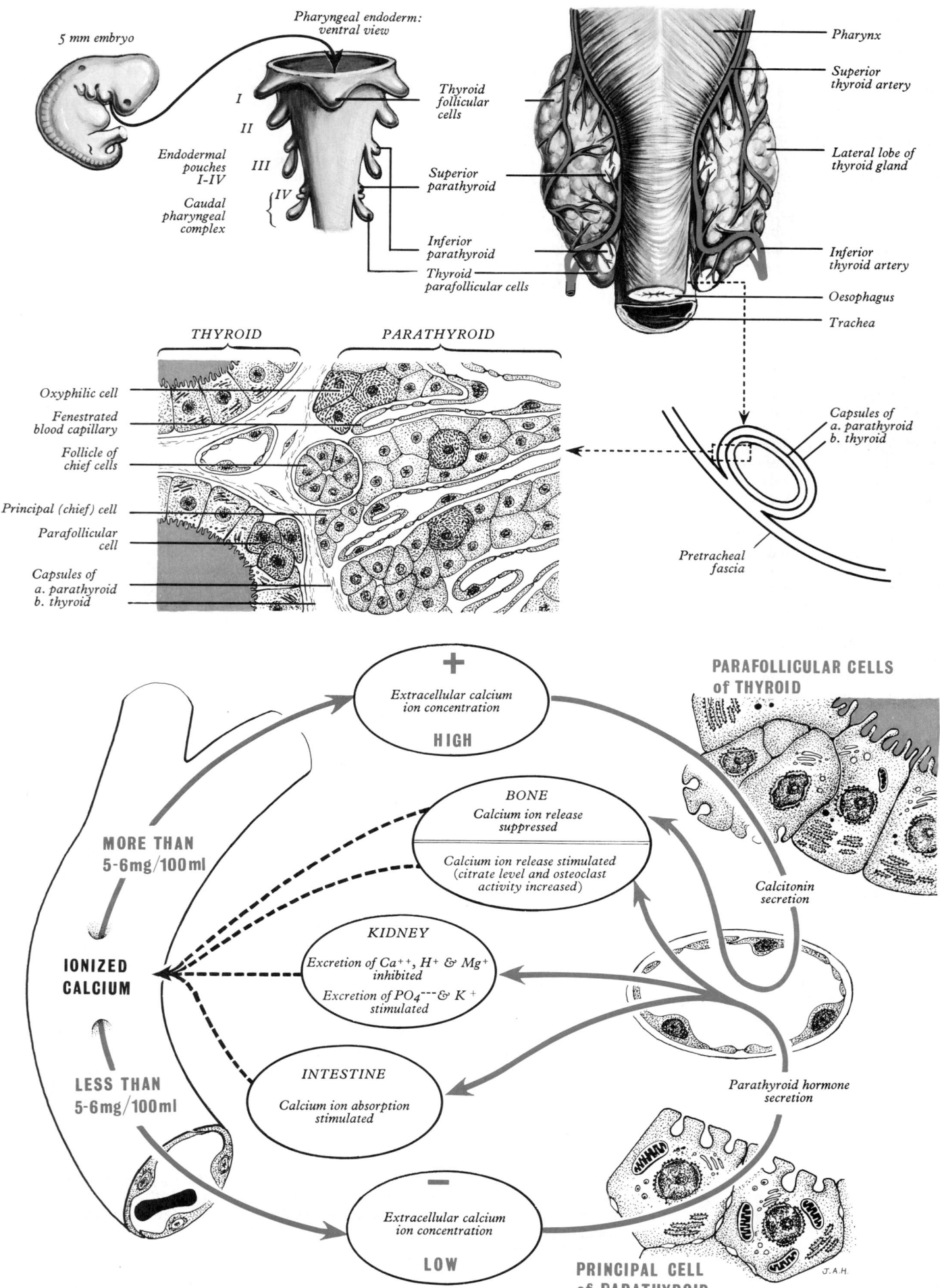

8.234A Diagrammatic representation of the roles of the parathyroid and thyroid glands in the control of calcium metabolism.

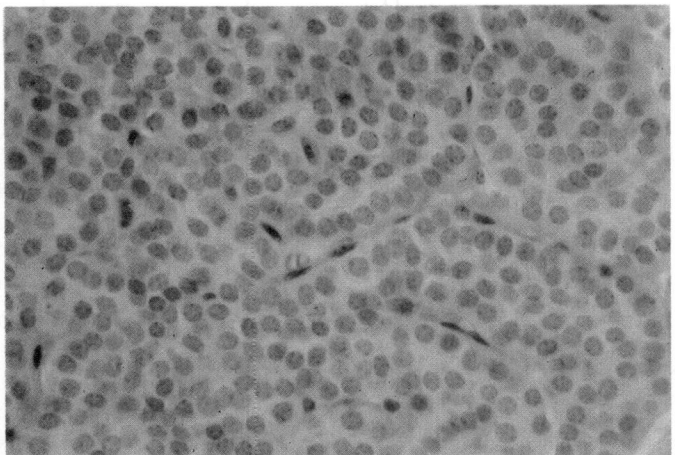

8.234B Section through a parathyroid gland showing densely-packed chief cells and capillaries (rabbit). Haematoxylin and eosin. Magnification × 800.

hormone leaves the gland. Their ultrastructure was reviewed by Capen in 1975. Human chief cells differ according to the level of their activity (Munger & Roth 1963). *Active* chief cells have large Golgi complexes with numerous vesicles and small membrane-bound granules, the latter probably being prosecretory; secretory granules are rare, cytoplasmic glycogen sparse, much of the cytoplasm being occupied by flat sacs of granular endoplasmic reticulum in parallel arrays. In contrast *inactive* chief cells contain small Golgi complexes with only a few grouped vesicles and membrane-bound secretory granules; glycogen and many lipofuscin granules abound but sacs of granular endoplasmic reticulum are rare and dispersed. In normal human parathyroid gland inactive chief cells outnumber active in ratio of 3–5:1.

Active chief cells synthesize, assemble and secrete parathyroid hormone. It is assumed that the dense-cored, membrane-bound granules in the chief cells of all mammals contain parathyroid hormone, though this awaits proof (Capen & Roth 1973). During secretion the granules first become peripheral and then, under appropriate stimulus (vide infra), their membranes fuse with the plasmalemma to release their contents, presumably parathyroid hormone. Involution continues, with increased lysosomal activity and reduction of Golgi complexes and granular endoplasmic reticulum; glycogen accumulates again; lipofuschin granules form and the cells enter a phase of temporary inaction. In contrast to the thyroid, where activity of adjacent cells is synchronized, each parathyroid chief cell appears to pass through its secretory cycle independently (Roth & Capen 1974).

A second cell type, the *oxyphil (eosinophil)* cell, appears just before puberty and multiplies with age (Roth 1962). Only in man, macaque monkey and cattle have such cells been noted. They are larger than chief cells and contain more cytoplasm which by light microscopy appears granular and which stains deeply with eosin; the nucleus is smaller and more darkly staining than in chief cells. Ultrastructural observations (Munger & Roth 1963, Gaillard et al 1965, Roth & Capen 1974, Capen 1975) show that 'granules' seen by light microscopy are actually mitochondria, extremely numerous, tightly packed and often bizarre in form. The cytoplasm also contains a few sacs of granular endoplasmic reticulum, some glycogen and, rarely, small Golgi complexes. No secretory granules have been reported. These features suggest that oxyphil cells are not involved in hormone synthesis or secretion, though abundant mitochondria suggest a high metabolic activity. Their role remains to be determined. The arrangement of chief and oxyphil cells in parathyroid glands is shown in **8.234A**.

Parathyroid hormone (PTH), a single-chain polypeptide of 84 amino acid residues (Potts et al 1971), is concerned with control of the level and distribution of calcium and phosphorus (**8.234**). Two other hormones, namely calcitonin (p. 302) and 1,25-hydroxycholecalciferol, are also involved, the latter produced by the sequential action of hepatic and renal cells on vitamin D

(O'Riordan 1978). Hormonal control of calcium metabolism has been described in detail by Copp & Talmage (1978). Secretion of PTH is dependent on the level of calcium ions in the blood traversing the parathyroid glands. PTH acts upon osteocytes and osteoclasts, its rapid initial effect being to increase the rate of release of calcium from bone into blood, apparently by stimulation of osteocytic osteolysis (Belanger 1969, see also p. 303). It also has a delayed effect if a high level of secretion is maintained, stimulating internal bone remodelling by promoting osteoclast activity; changes in the membrane potential of osteoclasts may be involved in the latter (Mears 1971). PTH also affects renal ion transport (Puschett 1978), increasing excretion of phosphate, sodium and potassium and decreasing that of calcium. It may also affect intestinal transport of calcium. 1,25-hydroxy-cholecalciferol, production of which is regulated by PTH, shares many of these effects and may 'modulate' PTH action (O'Riordan 1978). How PTH affects target cells is not clear; adenyl cyclase activation and consequent rise in intracellular cAMP appear to be involved (Chase & Aurbach 1967, Chabardes et al 1975).

Vessels and Nerves

The glands have a rich blood supply from the *inferior thyroid arteries* or from anastomoses between the superior and inferior vessels. *Lymph* vessels are numerous and associated with those of the thyroid and thymus glands. The *nerve supply* is sympathetic, either direct from the superior or middle cervical ganglia or via a plexus in the fascia on posterior lobar aspects. The nerves are not secretomotor but probably vasomotor; parathyroid activity is controlled by variation in blood calcium level: inhibited by a rise, stimulated by a fall.

Applied Anatomy

If all parathyroid tissue is removed, body muscles show convulsive spasms (*tetany*); since respiratory and laryngeal muscles are involved, death ensues. Tetany is due to a fall in blood calcium. Excess of parathyroid secretion, due to tumours, causes the removal of calcium ions from bones, which soften, a condition of *generalized osteitis fibrosa*. Calcium ions leak from the bones into the blood (hypercalcaemia), are excreted in urine and may calcify in the renal tubules, with resultant fatal renal disease.

The Chromaffin System

Chromaffin cells (phaeochromocytes) are classically defined as derived from neuro-ectoderm, innervated by preganglionic sympathetic nerve fibres and capable of synthesizing and secreting catecholamines (dopamine, noradrenalin or adrenalin), storing enough to give an intense yellow-brown coloration, the *positive chromaffin reaction*, when treated with aqueous solutions of chromium salts, particularly potassium dichromate (Coupland 1965). Groups of such cells, associated topographically and functionally with the sympathetic nervous system, comprise the *chromaffin system* which includes: (1) medullae of the suprarenal glands, (2) para-aortic bodies, (3) paraganglia proper, (4) certain cells of the carotid bodies and (5) small groups of cells irregularly dispersed among the paravertebral sympathetic ganglia, splanchnic nerves and prevertebral autonomic plexuses; they may, therefore, be related to: the heart, liver, kidney, ureter, prostate, epididymis, ovary, etc. Distribution of the main components in newborn infants is shown in a diagram in **8.235**.

Three main groups of cells give positive chromaffin reactions (Coupland 1976): (1) *true chromaffin cells*, as defined above, (2) *enterochromaffin cells* in epithelial tissue lining the gastrointestinal and respiratory tracts (p. 1466) and (3) *amine-storing mast cells* in the connective tissues of the gut, pancreas and liver. Ultrastructural similarity of (1) and (2) and the ability of all three groups to decarboxylate amino acids (p. 1466), observations that other non-chromaffin cells, in the gastrointestinal and respiratory tracts, pancreas and other endocrine glands are alike in both ultrastructure and amino-acid uptake and the discovery of paraneurons with many features of chromaffin cells in sympathetic ganglia (p. 1157), all undermine the rationale of continuing to restrict the term 'chromaffin system' to 'true' chromaffin cells. It may be

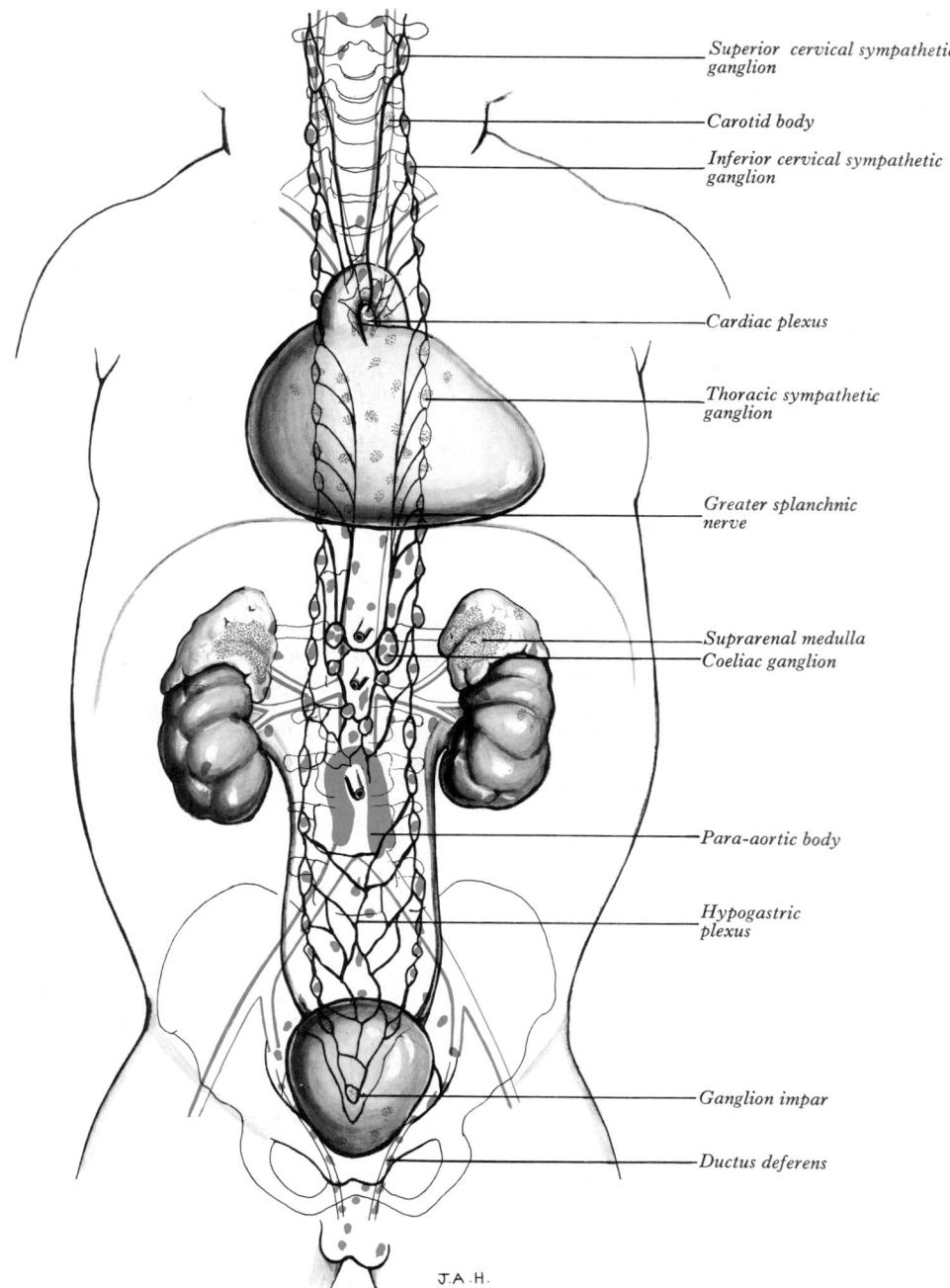

Superior cervical sympathetic ganglion

Carotid body

Inferior cervical sympathetic ganglion

Cardiac plexus

Thoracic sympathetic ganglion

Greater splanchnic nerve

Suprarenal medulla
Coeliac ganglion

Para-aortic body

Hypogastric plexus

Ganglion impar

Ductus deferens

J.A.H.

8.235 The principal aggregations of 'classic' chromaffin tissue in the human neonatal child. The aggregates in stippled blue lie deep to overlying structures.

more appropriate now to consider it as part of a diffuse neuroendocrine system (vide infra).

The Diffuse Neuro-endocrine System

Feyrter (1938) drew attention to isolated groups of hormone-secreting cells not restricted to accepted endocrine glands and widely scattered in many tissues. Terming such elements *clear cells*, he noted their prominence in the gut and pancreas. These are now classified as types of 'APUD' cells (Pearse 1966, 1980), the acronym being derived from their amine-handling properties (*amine precursor uptake and decarboxylation*). Most APUD cells synthesize related peptides which act as *hormones* or *neurotransmitters*, though in others a main secretion is an *amine* of similar action. Collectively APUD cells are a 'system' more extensive than envisaged by Feyrter, including chromaffin cells (vide supra), SIF cells (p. 1157), peptide-producing cells of the

hypothalamus (p. 1009), hypophysis (p. 1453), pineal (p. 1457), parathyroids (supra) and placenta, the Kulchitsky cells in the lungs (p. 1282), the myoendocrine cells of the cardiac atria (Forssmann & Girardier 1979, Forssmann et al 1983) and ventricles (Cantin et al 1987). Over 40 different cell types (see accompanying table) have been categorized as APUD cells and included in what has been described as the *diffuse neuro-endocrine system* (Pearse & Polak 1978). The myoendocrine cells of the heart differ from the others in being modified myocytes. These cells are listed in the accompanying table.

In 1966 Pearse described common cytochemical features in cells making peptide hormones and most notably in those which produce biogenic amines (adrenalin, noradrenalin, dopamine, 5-hydroxy-tryptamine, etc.) He suggested that uptake of 5-hydroxy-tryptophan (5-HTP) and its decarboxylation to 5-hydroxytryptamine (5-HT) might be linked to peptide hormone production in general. From this concept the designation 'APUD' cell arose (Pearse 1968). Pearse (1977) also suggested

THE APUD CELLS OF THE DIFFUSE NEURO-ENDOCRINE SYSTEM

I. APUD cells of neural crest origin

Location	Type	Main secretion Peptide	Amine
Thyroid	Parafollicular (C)	Calcitonin	5-HT, Da
Ultimobranchial body	C	Calcitonin	5-HT, DA
Carotid body	Type I Glomus	—	Da, NA
Sympathetic ganglia	SIF	—	NA
Adrenal medulla	Chromaffin	—	Ad
Adrenal medulla	Chromaffin	—	NA
Skin	Melanoblast	—	Promelanin
Urogenital tract	EC	—	5-HT
Urogenital tract	E	—	—

II. APUD cells of placodal or specialized ectodermal origin

Location	Type	Main secretion Peptide	Amine
Hypothalamus	N pv	Oxytocin, CRF	—
	N so	Vasopressin	—
	N sch	—	—
	N dm/vm	TRF	—
	N arc	LHRF	Da
	N ant/post	SRF, CRF	—
	N periv	Somatostatin	—
Pineal gland	P	LHRF	5-HT, MT
Parathyroid	Chief	PTH	—
Pituitary	Somatotroph	Somatotropin	Da
	Mammotroph	Prolactin	Da
	Gonadotroph	Follitropin	Da
	Gonadotroph	Lutropin	Da
	Corticotroph	Corticotropin	—
	M	Melanotropin	T
	Thyrotroph	Thyrotropin	Da

II. (continued)

Location	Type	Main secretion Peptide	Amine
Placenta	Endocrine	Gonadotropin	—
	Endocrine	Somato-mammotropin	—
	Endocrine	Corticotropin	—

III. APUD cells of disputed origin

Location	Type	Main secretion Peptide	Amine
Pancreas	A	Glucagon	5-HT
	B	Insulin	5-HT
	D	Somatostatin	Da
	D₁	VIP-like	Da
	P	Bombesin-like	—
	PP	Pancreatic polypeptide	Da
Stomach	A	Glucagon	—
	D	Somatostatin	—
	ECL	—	H?
	EC₁	Substance P	5-HT
	G	Gastrin, Enkephalin	—
	X	—	—
Intestine	D	Somatostatin	—
	D₁(H)	VIP	—
	EC₁	Substance P	5-HT
	EC₂	Motilin	5-HT
	ECₙ	—	5-HT
	I	Cholecystokinin	—
	K	GIP	—
	L	Enteroglucagon	—
	N	Neutrotensin	—
	S	Secretin	—
Lung	Kulchitsky (Pₐ)	—	—
Heart	Myoendocrine	Cardiodilation Atrial natriuretic factor	

ABBREVIATIONS

Ad	Adrenalin
CRF	Corticotropin releasing factor
Da	Dopamine
GIP	Gastric inhibitory peptide
H	Histamine
5-HT	5-hydroxytryptamine
LHRF	Luteotropin releasing factor (Luteinizing hormone releasing factor)
MT	Melatonin
NA	Noradrenalin
N ant/post	Anterior and posterior nuclear 'zones' of hypothalamus
N arc	Nucleus arcuatus (Nucleus infundibularis)
N dm/vm	Nucleus dorsomedialis/ventromedialis
N periv	Nuclei periventriculares
N pv	Nucleus paraventricularis
N sch	Nucleus suprachiasmaticus
N so	Nucleus supraopticus
PTH	Parathyroid hormone
SIF	Small intensely fluorescent
SRF	Somatotropin releasing factor
T	Tryptamine
TRF	Thyrotropin releasing factor
VIP	Vasoactive intestinal peptide
—	Unidentified

that all APUD cells were derived from 'neuroendocrine-programmed cells of the ectoblast' but there is now experimental evidence that many APUD cells (e.g. GEP endocrine cells, p. 1466) develop from other sources (Le Douarin 1978) and may be endodermal in origin (Andrew et al 1982). Other cells, e.g. the myoendocrine cells of the heart (Forssmann et al 1983), may be of mesodermal origin but can be grouped with the other cells of the neuro-endocrine system because of their similarities as producers of peptide hormones. Pearse considered APUD cells to be a third division of the nervous system, third-line effectors which support, modify or amplify the actions of neurons in the autonomic and somatic divisions. Their effects are slower in onset and longer in duration than those of the autonomic cells, which in turn have a similar functional relation to the faster somatic neurons. Secretions of APUD cells (diffuse neuro-endocrine system) may act upon contiguous cells, on groups of adjacent cells or on distant cells by transport in blood; they may thus be considered intermediate between the locally-acting transmitters produced by neurons and remote-acting endocrine secretions. Such a diffuse neuro-endocrine system complements and co-ordinates the nervous and endocrine systems, all three interacting to provide a precise mechanism for homeostatic control. For a critical review of the APUD concept, see Andrew (1982). It must be added that, as already stated (p. 809), the central nervous system also shares in many of the transmitters elaborated by APUD cells, further blurring the distinctions between neural and endocrine systems.

Deviations in the relative levels of secretions of different cells in this diffuse neuroendocrine system may cause disorders currently described as psychosomatic (Pearse & Polak 1978) or frankly psychotic (Webster 1978). If this proves true, the growing understanding of this system may improve their treatment.

The Suprarenal (Adrenal) Glands

The suprarenal glands (**8.**113, 237, 238), two small yellowish bodies, flat anteroposteriorly, are situated one immediately anterosuperior to each superior renal pole. Surrounded by connective tissue containing much perinephric fat (p. 1399), they are enclosed in renal fascia but separated from the kidneys by fibrous tissue. Each has a cortical zone rich in lipids but with *no* chromaffin tissue and an internal medulla staining deeply with chromic salts. Small masses of identical cortical suprarenal tissue occur often near the main gland and elsewhere as '*cortical bodies*'. Ontogenetically, phylogenetically, structurally and functionally, cortex and medulla are distinct, despite their forming a single organ.

The right gland is an irregular tetrahedron, the left semilunar and usually larger and more superior in level. Each in adults measures about 50 mm vertically, 30 mm transversely, and 10 mm in the anteroposterior dimension, weighing about 5 g. (the medulla being about one-tenth of the total weight). At birth the gland is about one-third the size of a kidney (**8.**171, 235) but in adults only a thirtieth. This change in ratio is not only due to renal growth but also to postnatal suprarenal diminution due to in-

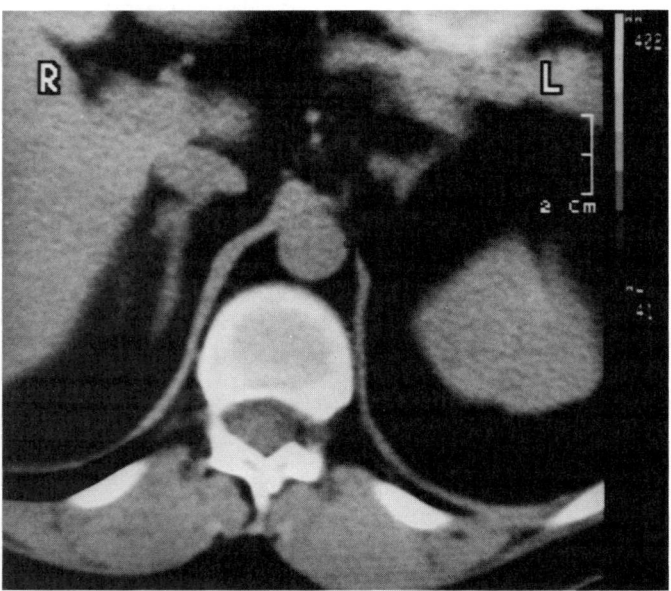

8.236A High resolution computed tomogram of the abdomen in the transverse plane at the level of the suprarenal glands. Provided by Shaun Gallagher, Guy's Hospital; photography by Sarah Smith, UMDS, Guy's Hospital Campus, London.

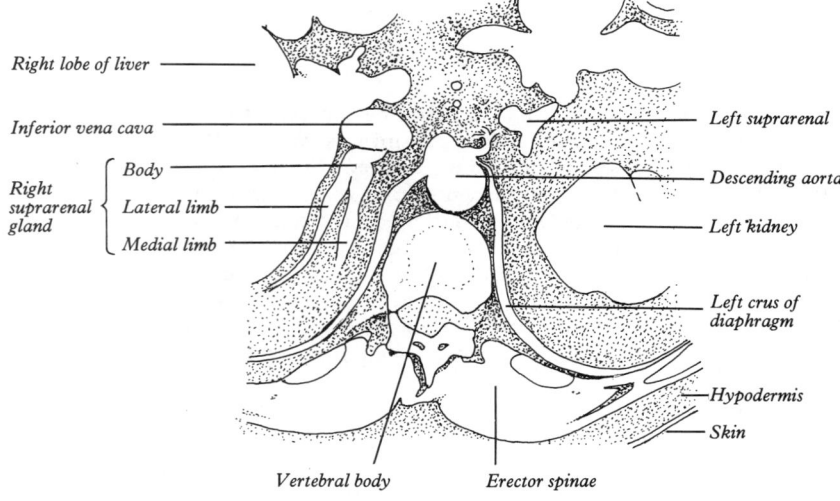

Right lobe of liver

Inferior vena cava

Right suprarenal gland { *Body* / *Lateral limb* / *Medial limb* }

Left suprarenal

Descending aorta

Left kidney

Left crus of diaphragm

Hypodermis

Skin

Vertebral body   *Erector spinae*

8.236B Diagram illustrating the major features demonstrated in **8.**236A.

volution of the fetal cortex (p. 201). By the end of the second month the weight of the suprarenal is reduced to one-half. In the latter half of the second year it begins to increase, gradually regaining its natal weight around puberty, after which its weight increases little in adult life.

RELATIONS OF THE SUPRARENAL GLAND

The right suprarenal gland (**8.**125) is posterior to the inferior vena cava and right hepatic lobe and anterior to the diaphragm and superior pole of the right kidney. It is shaped like an irregular tetrahedron. Its *base*, inferior, adjoins the anteromedial aspect of the right superior renal pole, often overlapping the upper part of the right kidney's medial border. Its *anterolateral surface* has a medial, narrow, vertical area, uncovered by peritoneum and posterior to the inferior vena cava, and a lateral triangular area in contact with the liver, devoid of peritoneum; above this it is in contact with the inferomedial angle of the bare area of the liver. Its inferior part may be covered by peritoneum, reflected on to it from the inferior layer of the coronary ligament; the duodenum may overlap this area. Below the apex, near the anterior border of the gland, is a short sulcal hilum where the right suprarenal vein emerges to join the inferior vena cava. The *posterior surface* is divided into upper and lower areas by a curved transverse ridge: its upper area, slightly convex, rests on the diaphragm; the lower, concave, contacts the superior pole and adjacent anterior surface of the right kidney. The thin *medial border* of the gland is related to the right coeliac ganglion, which is medial to it below, and to the right inferior phrenic artery, coursing superolaterally on the right crus of the diaphragm.

The left suprarenal gland (**8.**125) is crescentic, its concavity being adapted to the medial side of the superior pole of the left kidney. It is medially convex, laterally concave; its superior border is sharp, the inferior rounded. Its *anterior surface* has a superior area covered by peritoneum of the omental bursa, which separates it from the cardiac end of the stomach and sometimes from the posterior pole of the spleen; also an inferior area, not covered by peritoneum, in contact with the pancreas and splenic artery. The hilum faces ventrocaudally on the lower anterior surface; from it a left suprarenal vein emerges to join the left renal vein. Its *posterior surface* is divided by a ridge into a lateral area adjoining the kidney and a smaller medial one in contact with the diaphragm's left crus. The convex *medial border* is related to the left coeliac ganglion, which is inferomedial, and to the left inferior phrenic and left gastric arteries, which ascend on the left crus.

Small **accessory suprarenal glands**, sometimes composed only of cortical tissue, often occur in the areolar tissue near the main organ and sometimes in the spermatic cord, epididymis and broad ligament of the uterus.

STRUCTURE OF THE SUPRARENAL GLANDS

A sectioned suprarenal gland (**8.**239) reveals an outer *cortex*, yellow in colour and forming the main mass, and a thin *medulla*, forming about one-tenth of the gland, dark red or pearly grey depending on its content of blood. The medulla is completely enclosed by cortex, except at its hilum. The gland has a thick, collagenous capsule, which sends variably deep trabeculae into the cortex; the capsule contains a rich arterial plexus supplying branches to the gland.

The suprarenal cortex (**8.**240, 241) shows three cellular zones: the *zonae glomerulosa, fasciculata* and *reticularis*. The outer, subcapsular, *zona glomerulosa* consists of small polyhedral cells in rounded groups or curved columns with deeply staining nuclei, scanty basophilic cytoplasm and a few lipid droplets. Ultrastructurally (Lever 1955, Long & Jones 1967, Bloodworth & Powers 1968, Shelton & Jones 1971) the cytoplasm displays many microtubules, long mitochondria and abundant agranular endoplasmic reticulum, the latter being typical of cells elsewhere which synthesize steroids (vide infra). In humans the zona glomerulosa is poorly developed; deep to it the broader *zona fasciculata* consists of large polyhedral basophilic cells arranged in straight columns, two cells wide, with parallel fenestrated venous sinusoids between them. The cells contain many lipid droplets

and large amounts of phospholipids, fats, fatty acids and cholesterol embedded in complex agranular endoplasmic reticulum.
Mitochondria are typically spherical with tubular cristae; the
Golgi complex is extensive. The innermost part of the cortex, the
zona reticularis, consists of branching, interconnected columns of
round cells whose cytoplasm contains much agranular endoplasmic reticulum, many lysosomes and pigment bodies which may
indicate degeneration. Some consider that glomerulosar cells,
particularly those more deeply set, proliferate continuously, some
of the new cells migrating through the zona fasciculata to the
reticularis in which they may degenerate and disappear.
Autoradiography indicates that most proliferation occurs in the
zona glomerulosa and outer reticularis but mitoses also appear in
other cortical regions (Reiter & Hoffman 1967). There is, however, little ultrastructural evidence of cell death in the zona
reticularis.

The deeper part of the zona fasciculata widens in pregnancy
(Whiteley & Stoner 1957) and in women of childbearing age in
summer (MacKinnon & MacKinnon 1958). Cortical atrophy in
old males is greatest in the same region of this zone and least at its
periphery (MacKinnon & MacKinnon 1960). Cortical cells
produce several hormones and the cells of the zonae fasciculata
and reticularis are also rich in *ascorbic acid (vitamin C)*. Cells in
the zona glomerulosa produce *aldesterone*, which affects
electrolyte and water balance; cells in the zona fasciculata produce
hormones maintaining carbohydrate balance (glucocorticoids),
e.g. *cortisol (hydrocortisone)*; cells in the zona reticularis may
produce sex hormones (*progesterone, oestrogens* and *androgens*).
The cortex is essential to life; complete removal is lethal, without
replacement therapy. It also exerts considerable control over lymphocytes and lymphoid tissue (p. 838); increase in secretion of
corticosteroids can result in a marked reduction in lymphocytes
numbers. In some mammals the cortex shows cycles of
hypertrophy and regression during the oestrous cycle. Between
the cortical cells are sinusoids, into which branches from the
capsular arterial plexus and cortical arteries largely open. Some
arteries traverse the cortex to supply the medullary sinusoids.
Cortical sinusoids discharge into medullary sinusoids. Sinusoidal
endotheliocytes are phagocytic and belong to the macrophage
(mononuclear phagocyte) system (p. 669).

Suprarenal development is described on p. 201. The relatively
large size of the neonatal suprarenal gland is due to its very thick
fetal cortex (Johannisson 1968), the *definitive cortex* being a thin
peripheral zone. Around birth the fetal cortex begins to regress
and largely disappears in a few weeks thereafter. Too rapid a
cortical involution may be complicated by fatal haemorrhage.
This transient fetal cortex occurs only in anthropoids. There is no
proof that it produces androgenic hormones. It is poorly
developed in anencephalic fetuses. It does not represent the X
zone (androgenic), occurring e.g. in young mice as a zone around
the medulla (Jones 1957).

The suprarenal medulla is composed of groups and columns
of *chromaffin cells (phaeochromocytes)* separated by wide venous
sinusoids. Single or small groups of neurons occur in the medulla.

Chromaffin cells synthesize and expel noradrenalin and
adrenalin into the venous sinusoids; release is under preganglionic sympathetic control (see Coupland 1965a for review). In some
mammals these hormones have been identified in two distinct
types of cell (Yates et al 1962), cells storing noradrenalin being
more peripheral than those storing adrenalin. All cells, which are
large and columnar, form single rows along the venous sinusoids.
Cell bases and nuclei are distal from the sinusoids, adjoining
expanded extracellular spaces, in which the nerve terminals
synapse with the chromaffin cells. These may form follicles but
not like those in thyroid tissue (Al-Lami 1970). The cytoplasm is
basophilic, showing well-developed granular endoplasmic
reticulum, mitochondria, Golgi complex and many vesicles, indicating a high metabolic activity (Al-Lami 1970, Coupland
1965b). In noradrenalin-storing cells the vesicles are typically
round or ellipsoidal and, after treatment with aldehyde and osmium, highly electron-dense. In adrenalin-storing cells, treated
alike, the vesicles are paler (Coupland et al 1964), often with a
clear zone between the granular contents and membrane. In the
human suprarenal, cells with both types of vesicle have been

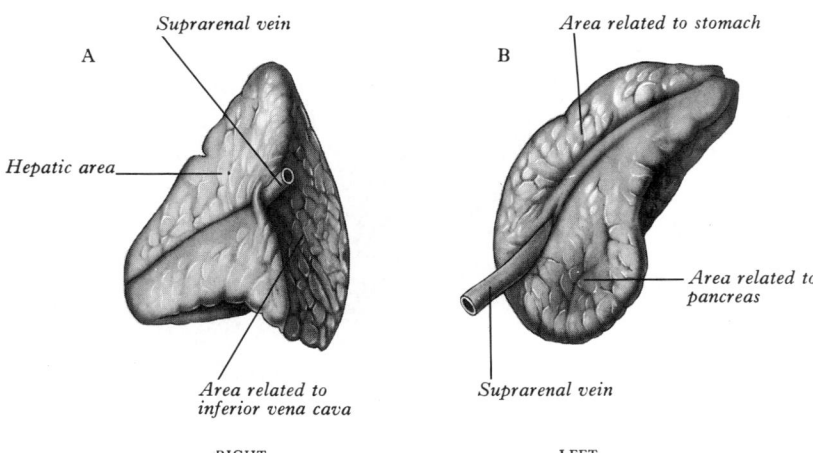

8.237 The suprarenal glands: anterior aspect.

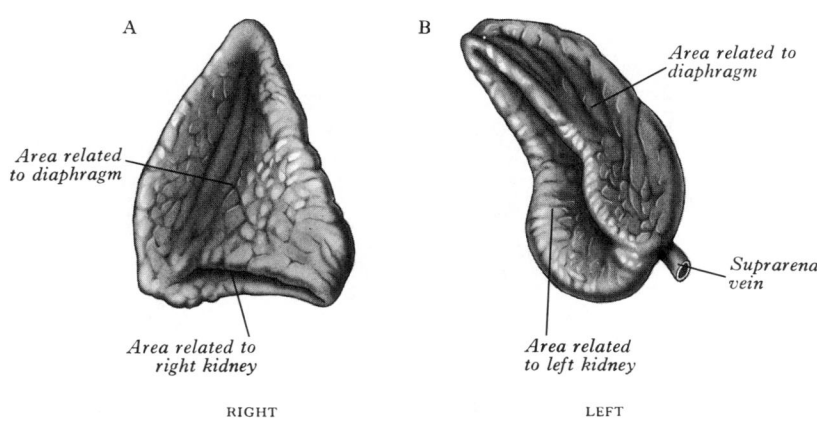

8.238 The suprarenal glands: posterior aspect.

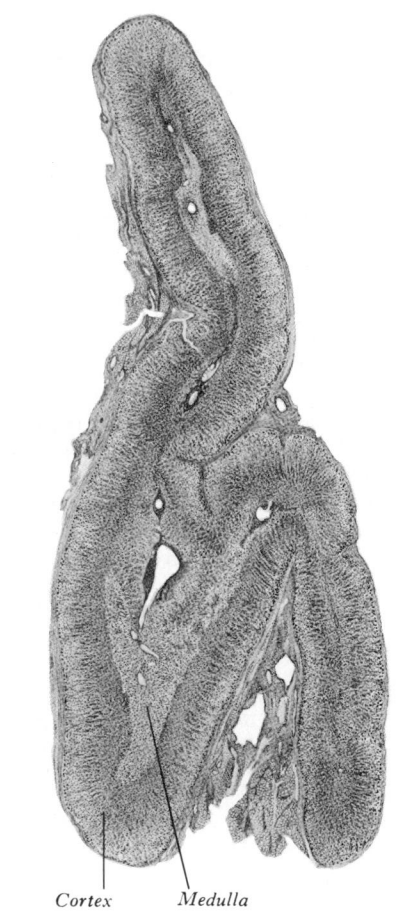

8.239 Vertical section through a whole adult human suprarenal gland.

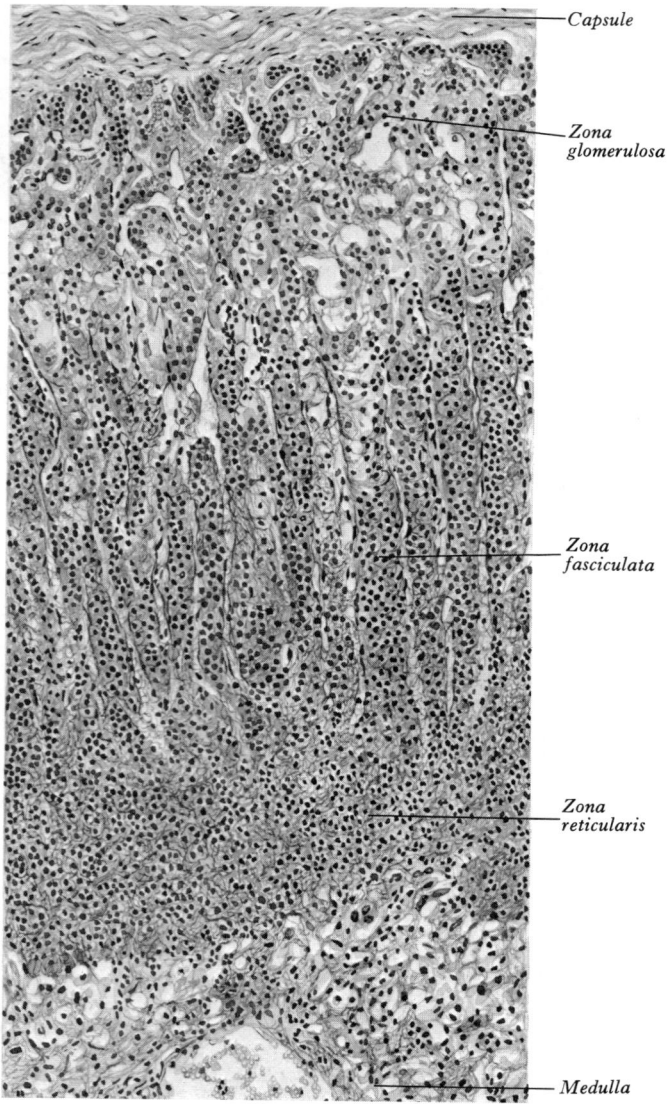

Capsule

Zona
glomerulosa

Zona
fasciculata

Zona
reticularis

Medulla

8.240A Section of adult human suprarenal gland. Magnification × 200.

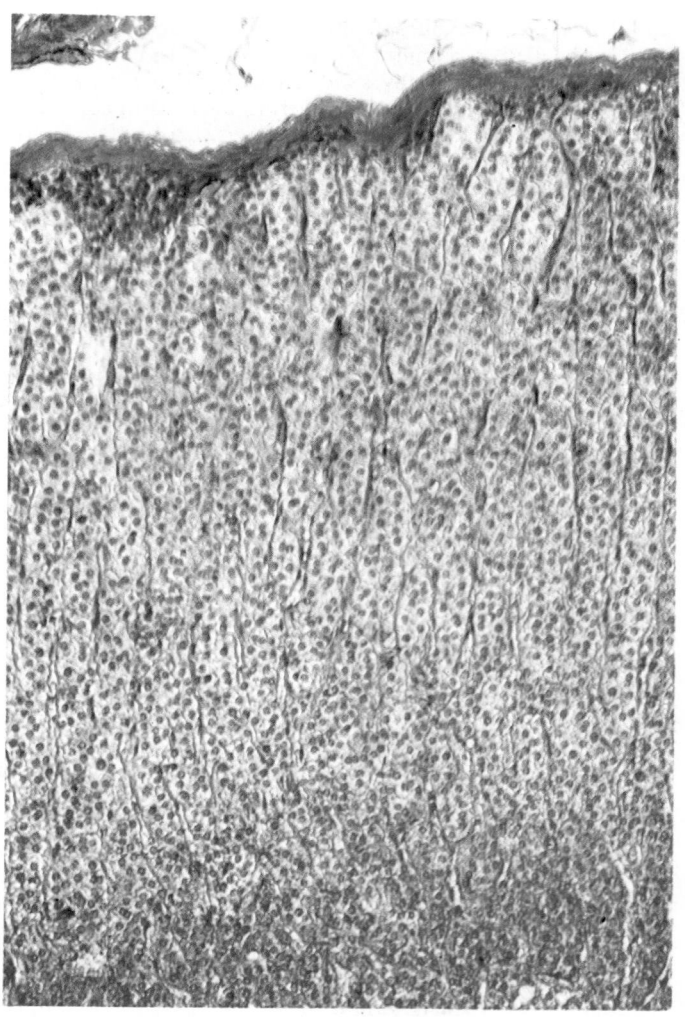

8.240B Medium magnification light micrograph of a section of the cortex of a human suprarenal gland. Beneath the capsule (blue) lie the zonae glomerulosa, fasciculata and reticularis. M.S.B. triple stain. Prepared and photographed by Stephen Sitch, Department of Anatomy, Guy's Hospital Medical School, London.

reported (Brown et al 1970) indicating that adrenalin and noradrenalin may both appear in the same cell. Vesicle contents are released at the cell apices into the perivascular spaces, entering the circulation through the fenestrated endothelium of venous sinusoids (Elfvin 1965, Al-Lami 1969).

The sinusoids drain to the hilar suprarenal vein. Normally, little adrenalin or noradrenalin is released but in fear, anger and stress secretion is augmented. Noradrenalin produces cardiac acceleration, vasoconstriction, raised blood pressure, etc., while adrenalin has a marked effect on carbohydrate metabolism. Unlike its cortex, the suprarenal medulla is not essential to life; removal has no clear effect. Medullary chromaffin cells develop in and migrate from the neural crests (sympatho-chromaffin tissue, p. 201). The chromaffin reaction (brown staining of granules by potassium bichromate, due to oxidation of adrenalin and noradrenalin) is positive from the fifth fetal month but adrenalin appears as early as the third (Keene & Hewer 1927).

Interaction between Cortex and Medulla

Only in mammals is the suprarenal chromaffin tissue almost enclosed by cortex. In elasmobranchs (e.g. dogfish) cortical tissue forms paired *inter-renal bodies*, chromaffin tissue being separate in segmental masses close to the sympathetic ganglia. In amphibia and birds, cords of chromaffin and cortical cells are intimately associated but a true medulla is not formed (Coupland 1965a). The proximity of cortical to chromaffin tissue may be associated with the formation of adrenalin by methylation of noradrenalin (vide infra).

Vessels and Nerves

The suprarenal gland is very vascular and is supplied by three groups of *arteries* (the superior, middle and inferior suprarenal) from the inferior phrenic, abdominal aorta and renal artery respectively (Harrison & Hoey 1960). Most suprarenal branches ramify over the capsule before entering the gland to form a subcapsular plexus, from which fenestrated sinusoids pass around clustered glomerulosal cells and between columns in the zona fasciculata to a deep plexus in the zona reticularis. From this venules pass between medullary chromaffin cells to medullary veins, which they enter between prominent longitudinal bundles of muscle fibres; these appear to regulate flow through a 'dam' at the corticomedullary junction, i.e. the internal aspect of the zona reticularis (Dobbie & Symington 1966). Since this would also control flow through the other zones, it could also control, in part, the availability of ACTH to their secretory cells (Griffiths & Cameron 1975).

Some relatively large arteries bypass the above route by going direct to the medulla, giving it a dual supply (8.241). Blood reaching it indirectly via the cortical sinusoids probably contains enough glucocorticoid hormone to maintain the synthesis of phenylethanolamine-N-methyl-transferase for synthesis of adrenalin from noradrenalin, whereas blood arriving by the direct non-cortical route does not. Whether medullary chromaffin cells can make adrenalin or noradrenalin depends on this enzyme and may be determined by the blood supply (Wurtman & Pohorecky 1971). Changes in relative blood flow via the two routes could thus have profound consequences.

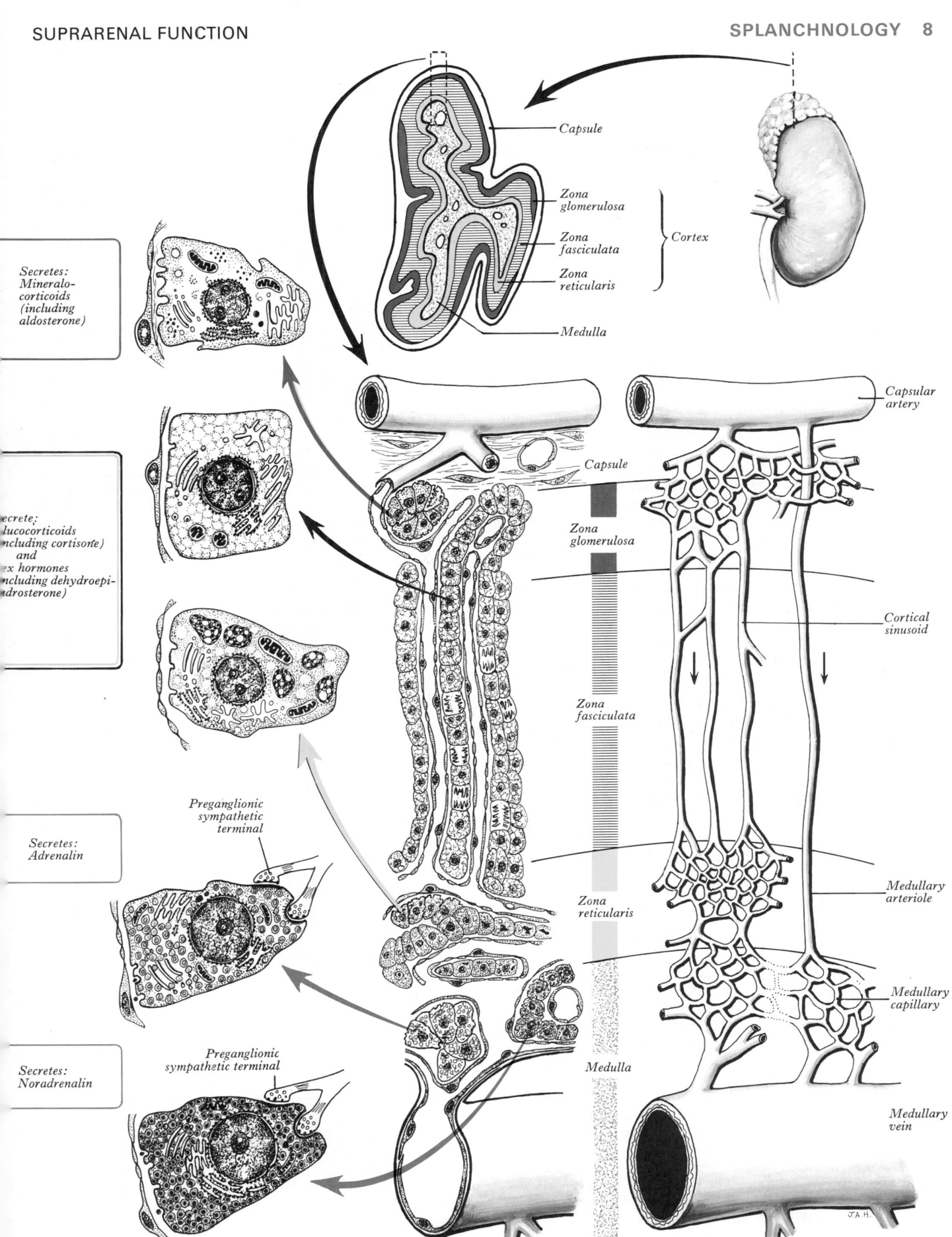

Secretes:
Mineralo-
corticoids
(including
aldosterone)

Capsule

Zona
glomerulosa

Zona
fasciculata

Zona
reticularis

Cortex

Medulla

ecrete;
lucocorticoids
ncluding cortisone)
and
ex hormones
ncluding dehydroepi-
ndrosterone)

Capsular
artery

Capsule

Zona
glomerulosa

Cortical
sinusoid

Zona
fasciculata

Preganglionic
sympathetic
terminal

Secretes:
Adrenalin

Zona
reticularis

Medullary
arteriole

Medullary
capillary

Preganglionic
sympathetic terminal

Secretes:
Noradrenalin

Medulla

Medullary
vein

J.A.H.

8.241 The suprarenal gland, displaying its gross sectional appearance,
histology, vasculature and ultrastructure. Brief functional summaries are
appended.

Medullary veins emerge from the hilum to form a *suprarenal vein*, draining to the inferior vena cava (right side) and left renal vein.

The *lymph* vessels end in lateral aortic nodes (p. 854).

The *nerves*, which are exceedingly numerous, are mainly myelinated preganglionic sympathetic fibres (p. 1165) distributed to the medullary chromaffin cells. Suprarenal cortical activities are largely controlled by hypophysial adrenocorticotropic hormone (ACTH).

Applied Anatomy

Various clinical conditions due to lesions of the suprarenal cortex or medulla are attributable to the effects of excess or deficiency of secretions in the parts of the gland affected.

Atrophy (usually tuberculous) of the suprarenal cortex, with consequent insufficiency of cortical secretion, results in Addison's disease, typified by muscular weakness, low blood pressure, anaemia, cutaneous pigmentation, changes in electrolytic and fluid balance, and terminal circulatory and renal failure. Excessive cortical secretion, due to tumours or hyperplasia, produces various effects: (1) In adults, Cushing's syndrome may result, typified by obesity, excessive hairiness of the face and trunk, diabetes mellitus, impotence and hypogonadism in males and amenorrhoea in females. (2) In women, masculinization (virilism) may occur due to excess of androgenic hormones or (3) in men, feminization, particularly mammary development. (4) Children may show precocious body growth and development of the external genital organs, with early menstruation in girls. (5) In the female fetus, cortical hyperplasia between the third and fourth months leads to female pseudohermaphroditism, the excess androgen interfering with differentiation of the urogenital sinus; the urethra and vagina open into a persistent urogenital sinus, the clitoris enlarges and the external genital organs resemble the male's. In the male fetus cortical hyperplasia causes excessive external genital development.

Bilateral removal of the suprarenal glands is a treatment for some inoperable disseminated mammary or prostatic carcinomata, when the malignant changes are considered to be dependent on androgens or oestrogens. Tumours of the suprarenal gland's medulla and the para-aortic bodies (phaeochromocytomata) may cause excessive secretion of adrenalin and noradrenalin, producing attacks of palpitations, excessive sweating, pallor, hypertension, headaches and, if of long duration, retinitis and renal vascular changes.

The suprarenal glands can be demonstrated radiologically by air injected into perirenal fat.

The Paraganglia

Paraganglia (Zuckerkandl 1901, Kohn 1903) are extra-suprarenal aggregations of chromaffin tissue (p. 1465, **8.235**), distributed near or in the autonomic nervous system (Coupland 1965a, Mascorro & Yates 1971, Hervonen et al 1978a). Cells like those in paraganglia proper (which adjoin various autonomic ganglia) also occur in the sympathetic ganglia as *small, intensely fluorescent (SIF) cells* (Williams et al 1975), in the walls of various viscera and in a variety of retroperitoneal and mediastinal sites (Ramsdale et al 1972, Hervonen et al 1976, Hervonen et al 1978b). All are neuro-ectodermal and can synthesize and store catecholamines but their functions differ with location: intraneural cells act as interneurons (p. 865), the remainder as sources of endocrine secretions including a tryptophanic protein and catecholamines (Hervonen et al 1978a). This dispersed array of extra-adrenal chromaffin tissue, often dubbed the *paraganglion system* (Mascorro & Yates 1975), is prominent in fetuses as the main source of catecholamines while the adrenal medulla is still developing (West et al 1955, Kovrishko 1964). Though many paraganglia degenerate soon after birth (Coupland 1965a), the specific fluorescent histochemical technique for the detection of catecholamines (Eranko 1967) has located many persistent, often minute, paraganglia in adult human and other mammals, contradicting the earlier assumption of general postnatal involution of paraganglia.

Paraganglia contain two typical varieties of cell: *type I* or granule-containing and *type II* or satellite cells (Mascorro & Yates 1975). Type I have a large nucleus, long mitochondria, some granular endoplasmic reticulum, well-developed Golgi complexes, glycogen deposits, numerous membrane-bound, electron-dense granules containing catecholamines and possibly the tryptophanic protein mentioned above, features which together with their neuro-ectodermal origin place them in the APUD category (Pearse 1969, Pearse & Polak 1974). By their cytoplasmic density, type I cells can be classified as 'light' or 'dark', though 'dark' cells are widely considered to be a fixation artefact (Mugnaini 1965, Benedeczky & Smith 1972) and both types may be stages in a secretory cycle. Type II cells lack cytoplasmic granules and their processes envelop type I cells partially or completely.

Type I cells of paraganglia receive a 'preganglionic' sympathetic innervation (Mascorro & Yates 1974) like the adrenal medullary chromaffin cells (Cummings 1969). The non-myelinated nerve fibres are largely separated from type I cells by Schwann cell cytoplasm and cytoplasmic projections of the type II cells, approximating only at synapses. Presynaptic endings contain mitochondria, glycogen granules and many synaptic vesicles, mostly electron-lucent, though some have electron-dense cores. In vagal paraganglia such endings may be cholinergic and efferent (Chen & Yates 1970). Nerve fibres containing vasoactive intestinal polypeptide are present in human fetal abdominal paraganglia (Hervonen et al 1985).

Paraganglia are well-vascularized and their secretory type I cells are usually next to one or more fenestrated capillaries, often with only basal lamina intervening, although sometimes fine collagen fibres and cytoplasmic processes of type II cells are present. Thus little obstructs the passage of hormones from type I cells to blood (Mascorro & Yates 1975). Evidence suggests that paraganglia are endocrine organs producing catecholamines and storing them as cytoplasmic granules until stimulated to release them. In addition to having a remote endocrine effect, these secretions may exert local paracrine action on nearby cells. Paraganglia comprise a dispersed system which throughout life may by the source of catecholamines additional to the suprarenal medulla and thus collectively having considerable metabolic importance (vide infra).

The Para-Aortic Bodies

Developing progressively during fetal life these attain maximum size in the first three postnatal years, when the largest are two brownish bodies about 1 cm long, flanking the abdominal aorta and usually united anterior to it by a horizontal mass immediately above the inferior mesenteric artery (**8.235**). They thus form an inverted crescentic or H-shaped arrangement, intimately related to the intermesenteric and superior hypogastric plexuses. Their constituent cells disperse and atrophy and by 14 years they may have completely disintegrated (Coupland 1965a). When well-developed, they are masses of polygonal chromaffin cells embedded in wide-meshed capillary plexuses and secreting noradrenalin. In fetuses chromaffin bodies are also widespread in the abdominal and pelvic prevertebral sympathetic plexuses, reaching a maximum size between the fifth and eighth fetal months and surviving in adults mainly near the coeliac and superior mesenteric arteries and as microscopic collections of cells persisting in the lower parts of the intermesenteric plexus.

Although chromaffin cells in sympathetic ganglia, as noted, may act as interneurons, those elsewhere are endocrine and probably support the suprarenal medulla as sources of catecholamines (Chen et al 1976), particularly in pre- and early postnatal life, when the suprarenal medulla and autonomic nervous system are immature (West et al 1953). This is supported by Coupland & Weakley (1970), who observed that extra-suprarenal chromaffin cells resemble those in the suprarenal medulla. Ultrastructural evidence shows that in rats chromaffin cells in nodes of the solar plexus have processes extending beyond their glial cells towards the capillaries, into which their catecholamines pass (Levkova & Kakabadze 1977).

The Carotid Bodies (Glomera Carotica)

The two carotid bodies are reddish-brown, ellipsoid and lie near the carotid sinuses in the neck (p. 734). Each is 5–7 mm in height and 2.5–4 mm in width and either posterior to the carotid bifurcation or between the start of its branches, being attached to or sometimes partly embedded in their adventitia. Occasionally it is a group of separate nodules. Aberrant 'miniglomera', microstructurally similar but with diameters of 600 μm or less, may appear in the adventitia and adipose tissue near the carotid sinus in human cadavers (Garfia 1980). Each glomus is innervated by glossopharyngeal carotid branches, including the *carotid sinus nerve*, and by a *plexus* of glossopharyngeal, vagal and sympathetic components (pp. 1113, 1158). Its abundant blood supply is derived from the adjacent external carotid rami.

Structure and Function

The carotid glomus is an *arterial chemoreceptor*. When stimulated by hypoxia, hypercapnia or increased hydrogen ion concentration, it elicits reflex increase in the rate and volume of ventilation via connections with brain-stem respiratory centres. Although this main role is certain, which of its components are chemoreceptors is obscure (Chen et al 1976). It may also be *endocrine*; its *glomus cells* (vide infra) are in the APUD series (p. 1466), which has been extended to include virtually all cells specialized for peptide hormone production (Pearse 1976); but no specific peptide hormone has yet been assigned to them. A suggested controlling role in erythropoiesis (Tramezzani et al 1971) has been disproved (Paulo et al 1973).

The glomus has a fibrous capsule whose septa lobulate it, each lobule being a collection of epithelioid *'glomus'* (type I) cells enveloped by *sustentacular (type II) cells* (**8.242**) and separating the former from extensive anastomoses of fenestrated sinusoids. Between sustentacular cells and the sinusoidal endothelium are non-myelinated nerve fibres, neurolemmocytes, attenuated processes of fibrocytes and collagen fibres (Chen et al 1976). Many of the nerve fibres are afferent, passing between the sustentacular to synapse with the 'glomus' (glomeral) cells. Preganglionic efferent fibres reach a sparse population of parasympathetic and sympathetic *ganglion cells*; these fibres are derived from the carotid sinus and sympathetic nerves (McDonald & Mitchell 1975), while the ganglion cells are either separate or in small groups near the body's surface. Axons proceed from the ganglion cells to local blood vessels, the parasympathetic efferent fibres probably being vasodilatory (Biscoe et al 1969) and the sympathetic ones vasoconstrictor (Purves 1970).

Glomeral cells (**8.243**), more numerous than sustentacular cells, are moderately large, with much cytoplasm and a few dendritic processes extending into intercellular spaces. Membrane-bound, electron-dense granules adjoin the Golgi apparatus and plasma membrane. Fusions of the limiting membranes of the granules and plasma membranes have been seen (Hansen 1977). Glomeral cells store dopamine (Kobayashi 1969), some other neurotransmitters (Fidone et al 1988) and the protein 'glomin' (Pearse 1969), presumably in their granules. They are accepted as chromaffin (Böck & Gorgas 1976) but a positive chromaffin response can be detected only ultrastructurally. They have also been regarded as paraneurons (Fujita 1976). Granular endoplasmic reticulum is not abundant but arrangements resembling neuronal Nissl substance have been described in human glomeral cells (Bock et al 1970). In rats, two types of glomeral cell are recognized (MacDonald & Mitchell 1975): type A have larger, more numerous electron-dense granules than type B and usually a smooth, globular contour with a few short dendrites; type B cells are more irregular with several long thin processes. Nerve endings seldom synapse with type B cells but at least two kinds of fibre synapse with type A, over 95% being *chemo-afferent axons* leaving in the carotid sinus nerve, their cell bodies being in the sensory glossopharyngeal ganglia. Less than 5% are *preganglionic efferent axons* from the cervical sympathetic trunk, entering the glomus with post-ganglionic axons from the superior cervical sympathetic ganglion. No efferent glossopharyngeal axons appear to synapse with glomeral cells but some are preganglionic to parasympathetic ganglion cells. In rabbits ultrastructurally

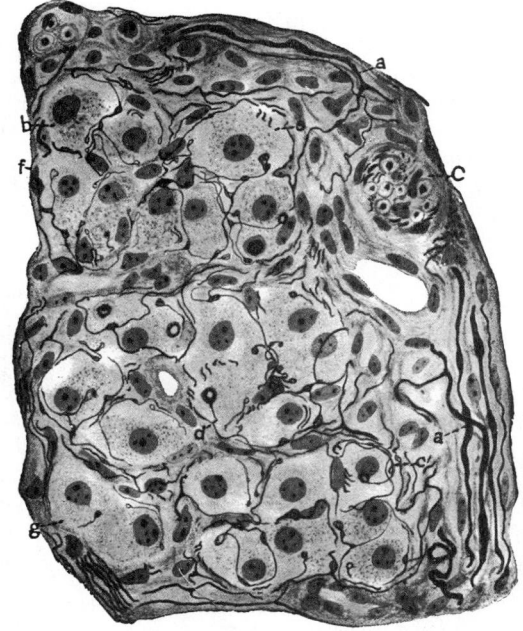

8.242 A section of the carotid body, showing nerve fibres distributed to the cells (de Castro): a, myelinated fibre dividing into two fine branches; b, cell closely surrounded by nerve fibrils; c, section of a small nerve, composed of several myelinated fibres; f, a nerve fibril apparently ending within the cytoplasm of a cell; g, a nerve fibril ending between the cells.

efferent fibres of uncertain origin, presumed inhibitory, synapse with the chemo-afferent axons (Verna 1975). Among 'afferent' nerve endings some are presynaptic to type A glomeral cells, others postsynaptic and some form *reciprocal synapses* with them. Such synapses have also been found between glomeral cells (McDonald & Mitchell 1975); they are dendrodendritic (Reese & Shepherd 1972). Similar connections occur in the central nervous system (pp. 888, 885) but not hitherto in the *peripheral*. Various neural circuits described in the glomus of rats and rabbits (McDonald & Mitchell 1975, Verna 1975, Fidone et al 1988) are shown in **8.243**. Human details are awaited with interest.

Ultrastructural and neurophysiological evidence ascribes at least three functions to glomeral cells. Firstly, they release neurotransmitters in response to hypoxia (Eyzaguire et al 1972) and may be *sensory*. Secondly, in ultrastructure they are like APUD cells (Pearse 1969) and may be *effector cells*, modifying the sensitivity of chemoreceptive nerve endings by variable release of dopamine (Biscoe et al 1970). Thirdly, having a synaptic input and output, a gliaform sheath and dendritic processes, they may act as *interneurons* (McDonald & Mitchell 1975). Involvement in several functions, as postulated above, is reminiscent of the presumed phylogenetic progenitor neuronal cell.

As stated above there is no consensus on which carotid glomeral cell is a chemoreceptor; *glomeral cells* (Lever et al 1959, Eyzaguire et al 1972), *afferent nerve terminals* (McDonald & Mitchell 1975) and *sustentacular cells* (Mills & Jöbsis 1972) have all been favoured. McDonald & Mitchell (1975) hypothecate co-operation between afferent axons and glomeral cells, proposing that: (1) afferent nerve endings, connected with glomeral cells by reciprocal synapses, are the true chemoreceptors; (2) glomeral cells are dopaminergic interneurons which modify the sensitivity of the chemoreceptive endings; (3) the reciprocal synapses between the glomeral cells and the afferent axons may form an inhibitory feedback, the glomerular cells inhibiting the activities of the axons with dopamine, the axons releasing an excitatory transmitter when stimulated, e.g. by hypoxia; (4) preganglionic sympathetic neurons may decrease chemoreception by evoking dopamine release from glomeral cells; and (5) synaptic interconnections between adjacent glomeral cells could mediate reciprocation. Another hypothesis (Mills & Jöbsis 1972), implicating sustentacular cells, suggests that hypoxia may affect these cells, initiating 'firing' of adjacent 'free' afferent nerve endings; glomeral cells, with their efferent sympathetic innervation, could be part of a feedback modifying the activity of the afferent axons. Half the cytochrome (a_3) of the glomus has a low affinity for oxygen and appears to be located in sustentacular cell mitochondria, which would make such cells sensitive to hypoxia, cytochrome remaining reduced in oxygen shortage and depressing oxidative metabolism. This might make them release e.g. potassium, activating adjacent afferent axons whose many

mitochondria, containing a cytochrome a_3 with a high affinity for oxygen, might sustain normal function even in hypoxia. It is interesting to speculate on the role of local haemodynamic control in such a monitoring system.

The carotid glomus develops from mesenchyme in the third pharyngeal arch (Boyd 1937), as a condensation around its artery; its nerve supply is mainly glossopharyngeal, the nerve being of that arch.

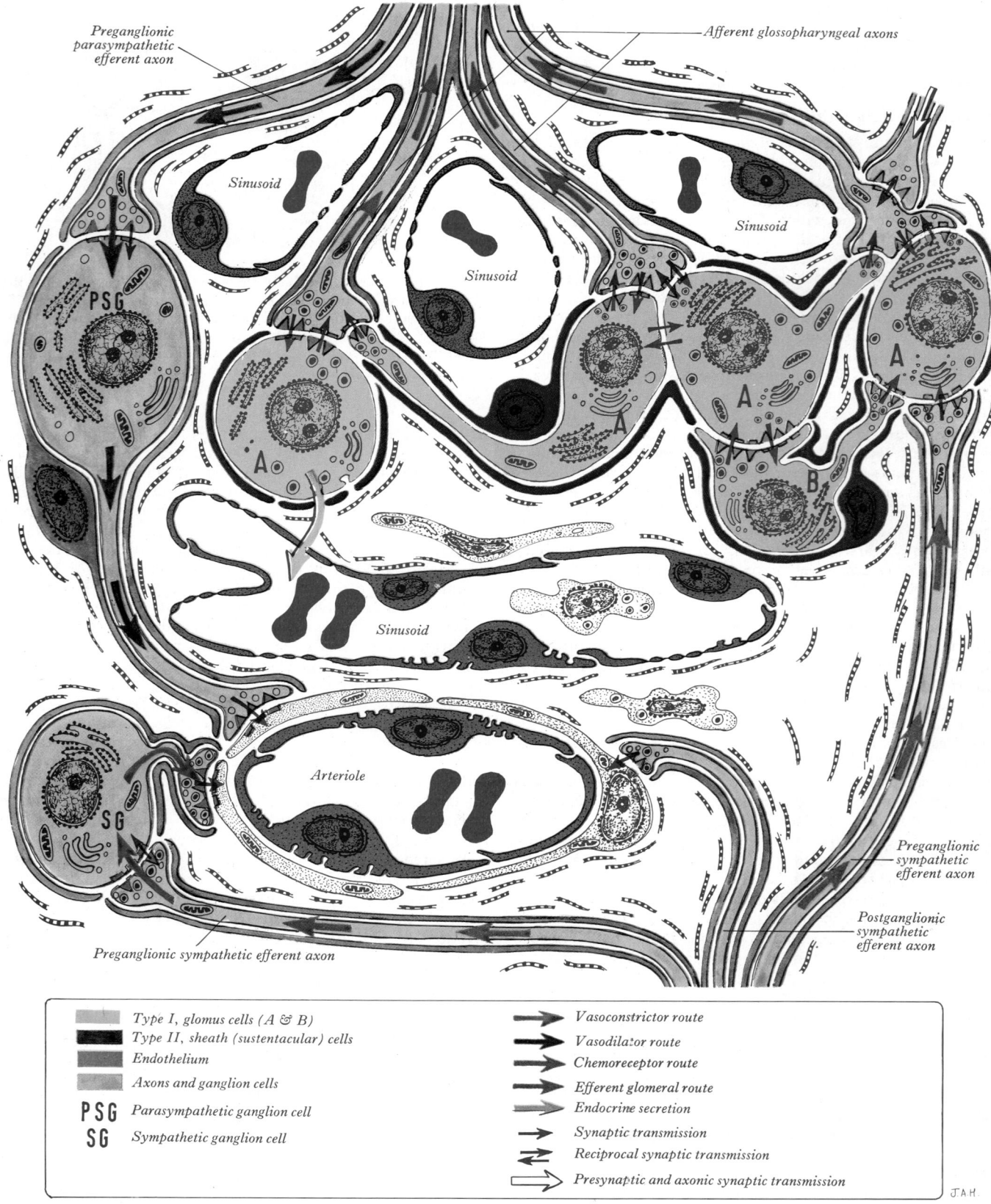

8.243 The cellular, neural and vascular architecture of the carotid body. Functional pathways are indicated. (Consult text for detailed discussion.)

Other small bodies, resembling carotid glomera, occur near the arteries of the fourth and sixth pharyngeal arches and hence near the aortic arch, ductus arteriosus and right subclavian artery and are supplied by the vagus nerve. They are also considered to be chemoreceptors.

The Tympanic Body

The tympanic body (*jugular glomus*) is ovoid, about 0.5 mm long and 0.25 mm broad, and occurs in the adventitia of the upper part of the superior bulb of the internal jugular vein. It is similar in structure to the carotid body, with presumably a similar function (Guild 1941, Kjaegaard 1944,1973). It may be present in two or more parts near the glossopharyngeal tympanic branch or vagal auricular branch, within their canals in the petrous temporal bone. Tumours of these bodies may involve the adjacent cranial nerves and the middle ear.

The Coccygeal Body

The coccygeal body (*glomus coccygeum*) is anterior or inferior to the apex of the coccyx, at the end of the median sacral vessels which supply it. It is close to the sympathetic ganglion impar, is about 2.5 mm in diameter and irregularly oval; smaller nodules of similar structure occur near the main mass of the body. The coccygeal glomus consists of irregular masses of spherical or polyhedral 'epithelioid' cells, grouped around sinusoidal capillaries. Each cell has a large, round or oval nucleus; its cytoplasm is clear and not stained by chromic salts, indicating no clear affiliation to the chromaffin system. The sinusoidal vessels are part of a complex of arteriovenous anastomoses; the epithelioid cells appear to be modified non-striated myocytes of the vessels' walls. Similar structures have long been identified in many mammals, ventral to the caudal vertebrae. Detailed analyses of innervation, ultrastructure and function are awaited with anticipation.

BIBLIOGRAPHY

A

Abbé E 1878 Die optischen Hülfsmittel der Mikroscopie. F Vieweg: Braunschweig

Abbie A A 1933 The blood supply of the lateral geniculate body, with a note on the morphology of the choroidal arteries. J Anat 67: 491–521

Abbie A A 1934 The morphology of the fore-brain arteries, with especial reference to the evolution of the basal ganglia. J Anat 68: 433–470

Abbie A A 1938 Blood supply of visual pathways. Med J Aust 2: 199–202

Abbie A A 1950 Closure of cranial articulations in the skull of the Australian aborigine. J Anat 84: 1–12

Abbie A A 1952 A new approach to the problem of human evolution. Trans R Soc S Aust 75: 70–88

Abbot M E 1936 Atlas of Congenital Heart Disease. American Heart Association: New York

Abdel-Maguid T E, Bowsher D 1985 The grey matter of the dorsal horn of the adult human spinal cord, including comparisons with general somatic and visceral afferent cranial nerve nuclei. J Anat 142: 33–58

Abe J, Atsuji K, Tsunawaki A, Nishida T, Inoue T, Yamasaki F, Tasaki K, Iwanaga S, Sato K 1981 Scanning electron microscopic observations of the myoepithelial cells in the prelacting and lactating mammary glands of the rat. Kurume Med J 28: 233–239

Abercrombie M 1961 The bases of the locomotory behaviour of fibroblasts. Exp Cell Res (Suppl 8): 188–198

Abercrombie M, Heaysman J E M 1953 Observations on the social behaviour of cells in tissue culture. I. Speed of movement of chick heart fibroblasts in relation to their mutual contacts. Exp Cell Res 5: 111–131

Abercrombie M, Heaysman J E M 1954 Observations on the social behaviour of cells in tissue culture. II. 'Monolayering' of fibroblasts. Exp Cell Res 6: 293–306

Abercrombie M, Heaysman J E M, Karthauser H M 1957 Social behaviour of cells in tissue culture. III. Mutual influence of sarcoma cells and fibroblasts. Exp Cell Res 13: 276–291

Acheson R M 1960 Effects of nutrition and disease on human growth. In: Tanner J M (ed) Human growth (Symposia of the Society for the Study of Human Biology Vol 3). Pergamon: Oxford: pp 73–92

Adachi N, Hasche K 1928 Das Arteriensystem der Japaner. Vol 1 Maruzen: Kyoto

Adams J, Daniel P M, Prichard M M 1969 The blood supply of the pituitary gland of the ferret with special reference to infarction after stalk section. J Anat 104: 209–225

Adams C W M ed 1965 Neurohistochemistry. Elsevier: Amsterdam

Adams W E 1957 On the possible homologies of the occipital artery in mammals, with some remarks on the phylogeny and certain anomalies of the subclavian and carotid systems. Acta Anat 29: 90–113

Adams W E 1958 The comparative morphology of the carotid body and carotid sinus. Thomas: Springfield Illinois

Ades H W 1944 Midbrain auditory mechanisms in cats. J Neurophysiol 7: 415–424

Adinolfi A M, Pappas G D 1968 The fine structure of the caudate nucleus of the cat. J Comp Neurol 133: 167–184

Adrian E D 1941 Afferent discharges to the cerebral cortex from peripheral sense organs. J Physiol 100: 159–191

Afifi A, Kaelber W W 1965 Efferent connections of the substantia nigra in the cat. Exp Neurol II: 474–482

Afzelius B 1979 The immotile-cilia syndrome and other ciliary diseases. Int Rev Exp Pathol 19: 1–43

Aguayo A, Peyronnard J M, Bray G M 1973 A quantitative ultrastructural study of regeneration from isolated proximal stumps of transected unmyelinated nerves. J Neuropathol Exp Neurol 32: 256–269

Aguayo A, Perkins S, Duncan I, Bray G 1978 Human and animal neuropathies studied in experimental nerve transplants. In: Canal N, Pozza G (eds) Peripheral neuropathies. Elsevier/North-Holland Biomedical Press: Amsterdam: pp 37–48

Aho A 1950 On the venous network of the human heart and its arterio-venous anastomoses. Ann Med Exp Biol Fenn 29: (Suppl 1) 1–90

Aitken J T, Bridger J E 1961 Neuron size and population density in lumbosacral region of the cat's spinal cord. J Anat 95: 38–53

Ajmani M L, Mittal R K, Jain S P 1983 Incidence of the metopic suture in adult Nigerian skulls. J Anat 137: 177–183

Ajmone Marsan C 1965 The thalamus. Data on its functional anatomy and on some aspects of thalamocortical integration. Archs Ital Biol 103: 847–882

Akelaitis A J 1942 Studies on corpus callosum; orientation (temporalspatial gnosis) following section of corpus callosum. Arch Neurol Psychiat 48: 914–937

Akerblom B 1948 Standing and sitting posture (Synge A trans). Nordiska Bokhandeln: Stockholm

Akert K, Pfenninger K, Sandri C, Moor H 1972 Freeze-etching and cytochemistry of vesicles and membrane complexes in synapses of the central nervous system. In: Pappas G D, Purpura D P (eds) Structure and function of synapses. Raven Press: New York: pp 67–86

Akisada T, Oda M 1978 Taste buds in the vallate papillae of the rat studied with freeze-fracture preparation. Arch Histol Jpn 41: 87–98

Albe-Fessard D, Delacour J 1968 Notions anatomo-physiologiques sur les voies et les centres d'intégration des messages douloureux. J Psychol Norm Pathol 65: 1–44

Albe-Fessard D, Liebeskind J 1966 Origine des messages somato-sensitifs activant les cellules du cortex moteur chez le singe. Exp Brain Res 1: 127–246

Albert M L, Obler L K 1978 The bilingual brain. Academic Press: New York

Alberts B, Bray D, Lewis J, Raff M, Roberts K, Watson J D 1983 Molecular Biology of the Cell. Garland: New York

Alcolado R, Weller R O, Parrish E P, Garrod D 1988 The cranial arachnoid and pia mater in man: anatomical and structural observations. Neuropath Appl Neurobiol 14: 1–17

Aleksandrowicz R, Gosek M, Prorok M 1974 Normal and pathologic dimensions of the abdominal aorta. Folia Morphol 33: 309–315

Alexander R McN 1977 Terrestrial locomotion. In: Alexander R McN, Goldspink G (eds) Mechanics and energetics of animal locomotion. Chapman and Hall: London: pp 168–203

Alexander R McN 1983 Animal mechanics, 2nd edition. Blackwell: Oxford

Alexander R McN 1984a Elastic energy stores in running vertebrates. Am Zool 24: 85–94

Alexander R McN 1984b The gaits of bipedal and quadruped animals. Int J Robot Res 3(2): 49–59

Alexander R McN 1984c Human walking and running. J Biol Educ 18: 135–140

Alexander R McN 1984d Optimum strengths for bones liable to fatigue and accidental fracture. J Theor Biol 109: 621–636

Alexander R McN 1984e Walking and running. Am Sci 72: 348–354

Alexander R McN, Bennet-Clark H C 1977 Storage of elastic strain energy in muscle and other tissues. Nature 265: 114–117

Alexander R McN, Goldspink G (eds) 1977 Mechanics and energetics of animal locomotion. Chapman & Hall: London

Alexander R McN, Jayes A S 1980 Fourier analysis of forces exerted in walking and running. J Biomech 13: 383–390

Alexander R McN, Vernon A 1975 The mechanics of hopping by kangaroos. J Zool 177: 265–303

Alexandre J H, Hamonet O, Lacert P, Moulin A M 1968 Arch Anat Path 16

Al-Lami F 1969 Light and electron microscopy of the adrenal medulla of Macaca mulatta monkey. Anat Rec 164: 317–332

Al-Lami F 1970 Follicular arrangements in hamster adrenomedullary cells: light and electron microscopic studies. Anat Rec 168: 161–178

BIBLIOGRAPHY

Allanson J T, Whitfield I C 1955 Third London symposium on information theory. Butterworths: London

Allen C, Sievers J, Berry M, Jenner S 1981 Experimental studies on cerebellar foliation. II. A morphometric analysis of cerebellar fissuration defects and growth retardation after neonatal treatment with 6-OHDA in the rat. J Comp Neurol 203: 771–783

Allen E, Pratt J P, Newell Q U, Bland L J 1930 Human tubal ova; related early corpora lutea and uterine tubes. Contrib Embryol Carnegie Inst Washington 22: 45–75

Allen L 1967 Lymphatics and lymphoid tissues. Annu Rev Physiol 29: 197–224

Allen M, Wright P, Reid L 1972 The human lacrimal gland. A histochemical and organ culture study of the secreting cells. Arch Ophthalmol 88: 493–497

Allen R M 1969 The mental age-visual perception issue assessed. Exceptional Child 35: 748–749

Allin E F 1975 Evolution of the mammalian middle ear. J Morphol 147: 403–437

Allison A C 1954 The secondary olfactory areas in the human brain. J Anat 88: 481–488

Allison P R 1951 Reflux esophagitis, sliding hiatal hernia and anatomy of repair. Surg Gynecol Obstet 92: 419–431

Alm P, Alumets J, Brodin E, Håakanson R, Nilsson G, Sjöberg N O, Sundler F 1978 Peptidergic (substance P) nerves in the genito-urinary tract. Neuroscience 3: 419–421

Alpern M 1969 Movements of the eyes. In: Davson H, (ed) The eye Vol 3, 2nd edition. Academic Press: New York: pp 1–214

Alpern M, Wolter J R 1956 The relation of horizontal saccadic and vergence movements. Arch Ophthal 56: 685–690

Alphen G W H M van, see van Alphen G W H M

Altman J 1972a Postnatal development of the cerebellar cortex in the rat. I. The external germinal layer and the transitional molecular layer. J Comp Neurol 145: 353–398

Altman J 1972b Postnatal development of the cerebellar cortex in the rat. II. Phases in the maturation of Purkinje cells and of the molecular layer. J Comp Neurol 145: 399–464

Altman J 1972c Postnatal development of the cerebellar cortex in the rat. III. Maturation of the components of the granular layer. J Comp Neurol 145: 465–514

Altman J, Bayer S 1975 Postnatal development of the hippocampal dentate gyrus under normal and experimental conditions. In: Isaacson R L, Pribram K H (eds) The hippocampus Vol 1. Plenum Press: New York: pp 95–122

Altman J, Bayer S 1978 Prenatal development of the cerebellar system in the rat. 1. Cytogenesis and histogenesis of the deep nuclei and the cortex of the cerebellum. J Comp Neurol 179: 23–48

Al-Turk M, Metcalf W K 1984 A study of the superficial palmar arteries using the Doppler Ultrasonic Flowmeter. J Anat 138: 27–32

Alvarez-Moruja Suarez A J 1980 Los Conductos Biliares Intrahepaticos. University of Salamanca: Salamanca

Alvarez-Morujo A 1968 Terminal arteries of the penis. Acta Anat 67: 387–398

Alvesalo L, Osborne R H, Kari M 1975 The 47, XYY male, Y chromosome, and tooth size. Am J Hum Genet 27: 53–61

Ambache N, Zar M A 1970 Non-cholinergic transmission by post-ganglion motor neurones in the mammalian bladder. J Physiol 210: 761–783

Amici G B 1818 De Microscopi Catadiottrici Memoria. Modena

Amonoo-Kuofi H S 1982 Maximum and minimum lumbar interpedicular distances in normal adult Nigerians. J Anat 135: 225–233

Amonoo-Kuofi H S 1985 The sagittal diameter of the lumbar vertebral canal in normal adult Nigerians. J Anat 140: 69–78

Amoore J E 1971 Olfactory genetics and anosmia. In: Beidler L M (ed) Handbook of sensory physiology. Vol 4 Chemical Senses Part 1. Springer: Berlin: pp 245–256

Amoroso E C 1952 Placentation. In: Parkes A S (ed) Marshall's physiology of reproduction 3rd edition Vol 2. Longmans Green: London: pp 127–311

Amos L A, Klug A 1974 Arrangement of subunits in flagellar microtubules. J Cell Sci 14: 523–537

Amprino R 1948 Recherches et considérations sur la structure du cartilage hyalin. Acta Anat 5: 123–146

Amprino R 1968 Bone histophysiology. Guy's Hosp Rep 116: 51–69

Amprino R, Bairati A 1933 Studi sulle trasformazioni delle cartilagini dell'uomo nell'accrescimento e nella senescenza: cartilagini jaline. Z Zellforsch 20: 143–205

Amprino R, Camosso M 1955 Richerche sperimentali sulla morfogenesi degli arti nel pollo. J Exp Zool 129: 453–494

Amprino R, Engström A 1952 Studies on X-ray absorption and diffraction of bone tissue. Acta Anat 15: 1–22

Anand B K 1961 Nervous regulation of food intake. Physiol Rev 41: 677–708

Anand B K 1970 Regulation of visceral activities by the central nervous system. In: Wolstenholme G E W, Knight J (eds) Control processes in multicellular organisms (Ciba Foundation Symposium). Churchill: London: pp 356–380

Anand B K, Brobeck J R 1951 Hypothalamic control of food intake in rats and cats. Yale J Biol Med 24: 123–140

Anberg A 1957 The ultrastructure of the human spermatozoon. Acta Obstet Gynecol Scand 36 (Suppl 2): 1–33

Andersen H, Bulow F A von, Møllgärd K, 1970 The histochemical and ultrastructural basis of the cellular function of the human foetal adenohypophysis. Prog Histochem Cytochem 1: 153–184

Andersen H T ed 1969 The Biology of marine mammals. Academic Press: New York

Andersen P, Eccles J C, Løyning Y 1964a Location of postsynaptic inhibitory synapses on hippocampal pyramids. J Neurophysiol 27: 592–607

Andersen P, Eccles J C, Løyning Y 1964b Pathway of postsynaptic inhibition in the hippocampus. J Neurophysiol 27: 608–619

Andersen P, Eccles J C, Schmidt R F, Yokota T 1964 Identification of relay cells and interneurons in the cuneate nucleus. J Neurophysiol 27: 1080–1095

Andersen P, Eccles J C, Sears T A 1964 The ventro-basal complex of the thalamus: Types of cells, their responses and their functional organisation. J Physiol 174: 370–399

Andersen P, Blackstad T W, Lømo T 1966 Location and identification of excitatory synapses on hippocampal pyramidal cells. Exp Brain Res 1: 236–248

Anderson C E 1962 The structure and function of cartilage. J Bone Jt Surg 44A: 777–786

Anderson C E, Parker J 1966 Invasion and resorption in endochondral ossification. An electron microscopic study. J Bone Jt Surg 48A: 899–914

Anderson D J, Matthews B 1967 Osmotic stimulation of human dentine and the distribution of dental pain thresholds. Arch Oral Biol 12: 417–426

Anderson D J, Hannam A G, Matthews B 1970 Sensory mechanisms in mammalian teeth and their supporting structures. Physiol Rev 50: 171–195

Anderson D R, Hoyt W 1969 Ultrastructure of intraorbital portion of human and monkey optic nerve. Arch Ophthal 82: 506–530

Anderson E, Beams H W 1960 Cytological observations on the fine structure of the guinea pig ovary with special reference to the oogonium, primary oocyte and associated follicle cells. J Ultrastruct Res 3: 432–446

Anderson H C 1967 Electron microscopic studies of induced cartilage development and calcification. J Cell Biol 35: 81–101

Anderson H C 1969 Vesicles associated with clacification in the matrix of epiphyseal cartilage. J Cell Biol 41: 59–72

Anderson H C 1980 Calcification processes. In: Sommers S C, Rosen P P (eds) Pathology annual Vol 15. Appleton-Century-Crofts: New York: pp 47–75

Anderson H C 1985 Matrix vesicle calcification: review and update. In: Peck W A (ed) Bone and mineral research annual 3. Elsevier: Amsterdam: pp 109–149

Anderson H C, Reynolds J J 1973 Pyrophosphate stimulation of calcium uptake into cultured embryonic bones. Fine structure of matrix vesicles and their role in calcification. Dev Biol 34: 211–227

Anderson R H, Becker A E 1980 Cardiac Anatomy. An integrated text and colour atlas. Gower: London

Anderson R H, Wilkinson J L, Arnold R, Lubkiewicz K 1974 Morphogenesis of bulboventricular anomalies. (1) Consideration of embryogenesis in the normal heart. Br Heart J 36: 242–256

Anderson R H, Becker A E, Brechenmacher C, Davies M J, Rossi L 1975 The human atrioventricular junctional area: a morphological study of the A-V node and bundle. Eur J Cardiol 3: 11–25

Anderson R H, Ho S Y, Becker A E 1984 The clinical anatomy of the cardiac conducting system. In: Rowlands D J (ed) Recent Advances in Cardiology 9. Churchill Livingstone: Edinburgh: pp 1–26

Andreassi G 1967 Sur la topographie de l'apex pulmonaire chez l'homme. C R Assoc Anat 137: 141–147

Andres K H 1966 Der feinbau der Regio olfactoria von Makrosmatikern. Z Zellforsch 69: 140–154

Andrew A 1971 The origin of intramural ganglia. 4. The origin or enteric ganglia; a critical review and discussion of the present state of the problem. J Anat 103: 169–184

Andrew A 1982 The APUD concept: where has it led us? Br Med Bull 38: 221–225

Andrew A, Kramer B, Rawden B B 1982 The embryonic origin of endocrine cells of the gastrointestinal tract. Gen Comp Endocrinol 47: 249–265

Andrews J 1971 Streak gonads and the Y chromosome. J Obstet Gynecol 78: 448–457

Andrews P M 1975 Microplicae. J Cell Biol 67: IIa

Andy O J, Stephan H 1968 The septum in the human brain. J Comp Neurol 133: 383–410

Angaut P, Brodal A 1967 The projection of the 'vestibulocerebellum' into the vestibular nuclei in the cat. Arch Ital Biol 105: 441–479

Angaut P, Repérant J 1976 Fine structure of the optic fibre termination layers in the pigeon optic tectum. Neuroscience I: 93–105

Angell J C 1969 Treatment of benign prostatic hyperplasia by phenol injection. Br J Urol 41: 735–738

Angevine J B Jr 1975 Development of the hippocampal region. In: Isaacson R L, Pribram K H (eds) The hippocampus vol 1. Structure and Development. Plenum Press: New York: pp 61–94

Anglo-American Symposium 1958 on the brain and its functions. Blackwell: Oxford

Annett M 1964 A model of the inheritance of handedness and cerebral dominance. Nature 204: 59–60

Annett M 1967 The binomial distribution of right, mixed and left handedness. J Exp Psychol 19: 327–333

Annis D 1962 A study of the regenerative ability of the epithelial lining of the urinary bladder. Ann R Coll Surg 31: 23–45

Anson B H 1963 The aortic arch and its branches. In: Luisada A (ed) Cardiology Vol I. McGraw-Hill: New York: p 119

Anson B J, McVay C B 1971 Surgical anatomy 5th edition. Saunders: Philadelphia

Anson B J, Beaton L E, McVay C B 1938 Pyramidalis muscle. Anat Rec 72: 405–411

Anson B J, Jamieson R W, O'Connor V J, Beaton L E 1953 The pectoral muscles. Q Bull North West Univ Med Sch 27: 211–218

Anson B J, Morgan E H, McVay C B 1960 Surgical anatomy of the inguinal region based upon a study of 500 body-halves. Surg Gynecol Obstet III: 707–725

Appenzeller O 1964 Electron microscope study of the innervation of the auricular artery in the rat. J Anat 98: 87–91

Applebaum A E, Beall J E, Foreman R D, Willis W D 1975 Organization and receptive fields of primate spinothalamic tract neurons. J Neurophysiol 38: 572–586

Appleton A B 1922 On the hypotrochanteric fossa and accessory adductor groove of the primate femur, J Anat 56: 295–306

Appleton A B 1934 Postural deformities and bone growth; experimental study. Lancet 1: 451–454

Arakawa M, Tokunaga J 1974 Further scanning electron microscope studies of the human glomerulus. Lab Invest 31: 436–440

Arbib M 1964 Brains, machines and mathematics. McGraw-Hill: New York

Arbuthnott E R, Ballard K J, Boyd I A, Gladden M H, Sutherland F I 1982 The ultrastructure of cat fusimotor endings and their relationship to foci of sarcomere convergence in intrafusal fibres. J Physiol 331: 285–309

Arem A J, Kischer C W 1980 Analysis of striae. Plast Reconstruct Surg 65: 22–29

Arendt J 1985 Mammalian pineal rhythms. In: Reiter R J (ed) Pineal research reviews. Vol 3. Liss: New York: pp 161–213

Arey L B 1932 Retina choroid and sclera. In: Penfield W (ed) Cytology and cellular pathology of the nervous system. Hoeber: New York: pp 743–836

Arey L B 1949 The craniopharyngeal canal re-interpreted on the basis of its development. Anat Rec 103: 420

Argème M, Mambrini A, Lebreuil G et al 1970 Dissection du muscle de Treitz. C R Assoc Anat 147: 76–86

Ariens Kappers C U see Kappers C U A

Armstrong J, Richardson K C, Young J Z 1956 Staining neural end feet and mitochondria after postchroming and carbowax embedding. Stain Technol 31: 263–270

Arnold F 1851 Handbuch der Anatomie des Menschen, Vol 2. Emmerling und Herder: Freiburg

Arnold G 1974 Festigkeit und Kraft-Längenänderungs-Verhalten der Strecksehnen des menschlichen Fusses. Res Exp Med 164: 123–136

Arnstein K 1889 Über die Nerven der Schweissdrüsen. Anat Anz 4: 378–383

Arnstein K 1895 Zur Morphologie der sekretorischen Neverendapparate. Anat Anz 10: 410–419

Arsenieva N A, Dubinin N P, Orlova N N, Bakulina E D 1961 A radiation analysis of the duration of meiosis phases in the spermatogenesis of Macaca mulatta. Dokl Akad Nauk S S R 141: 1486–1490

Arslan M 1960 The innervation of the middle ear. Proc R Soc Med 53: 1068–1074

Arvis G 1969 Considérations anatomiques sur le hile et le sinus du rein. Ann Radiol 12: 75–106

Arwill T 1968 The ultrastructure of the pulpodentinal broder zone. In: Symons N B B (ed) Dentine and pulp: their structure and relations. A Symposium. University of Dundee: Dundee: pp 147–167

Asanuma H, Sakata H 1967 Functional organisation of a cortical efferent system examined with focal depth stimulation in cats. J Neurophysiol 30: 35–54

Ascenzi A, Bell G H 1972 Bone as a mechanical and engineering problem. In: Bourne G H (ed) The biochemistry and physiology of bone. Academic Press: London, New York: 2nd edition

Ascenzi A, Bonucci E 1968 The compressive properties of single osteons. Anat Rec 161: 377–392

Ascenzi A, Bonucci E, Ripamonti A, Roveri N 1978 X-ray diffraction and electron microscope study of osteons during calcification. Calcif Tissue Res 25: 133–143

Asch R H, Ellsworth L R, Balmaceda J P, Wong P C 1984 Pregnancy after translaparoscopic gamete intrafallopian transfer. Lancet 2: 1034–1035

Asch R H, Balmaceda J P, Ellsworth L R, Wong P C 1986 Gamete intrafallopian transfer (GIFT): a new treatment for infertility. Int J Fertil 30: 41–45

Aschoff K A L 1924 Das reticulo-edotheliale System. Ergebn Inn Med 26: 1–118

Ash P, Loutit J F, Townsend K M 1980 Osteoclasts derived from haematopoietic stem cells. Nature 283: 669–670

Ashby W R 1960 Design for a Brain. Wiley: New York

Ashley G T 1952 The manner of insertion of the pectoralis major muscle in man. Anat Rec 113: 301–308

Ashley G T 1954 Morphological and pathological significance of synostoses at the manubrio-sternal joint. Thorax 9: 159–166

Ashley G T 1956 The relationship between the pattern of ossification and the definitive shape of the mesosternum in man. J Anat 90: 87–105

Ashton E H, Moore W J 1980 Cranial shape in the hominoidea – exploratory considerations. J Anat 131: 744–745

Ashton E H, Moore W J, Spence T F 1976 Growth changes in endocranial capacity in the cercopithecoidea and hominoidea. J Zool 180: 355–365

Ashworth J M 1973 Cell Differentiation. Chapman and Hall: London

Aspden R M, Hukins D W L 1979 The lamina splendens of articular cartilage is an artefact of phase contrast microscopy. Proc R Soc Lond [Biol] 206: 109–113

Aspden R M, Yarker Y E, Hukins D W L 1985 Collagen orientations in the meniscus of the knee joint. J Anat 140: 371–380

Assali N S, Rauramo L, Peltonen T 1960 Measurement of uterine blood flow and uterine metabolism. VIII. Uterine and fetal blood flow and oxygen consumption in early human pregnancy. Am J Obstet Gynecol 79: 86–98

Asscher A W, Moffat D B, Sanders E, eds 1982 Nephrology illustrated. An integrated text and colour atlas. Pergamon Press: Oxford

Athias M 1897 Recherches sur l'histogénèse de l'écorce du cervelet. J Anat Physiol 33: 372–404

Atkinson M, Edwards D A, Honour A J, Rowlands E N 1957 Comparison of cardiac and pyloric sphincters: a manometric study. Lancet 2: 918–922

Atkinson P J, Hallsworth A S 1982 The spatial structure of bone. In: Harrison R J, Navaratum V (eds) Progress in anatomy. Vol 2. Cambridge University Press: Cambridge: pp 179–199

Attenborough D 1979 Life on Earth. Collins: London

Auerbach R 1960 Morphogenetic interactions in the development of the mouse thymus gland. Dev Biol 2: 271–284

Auerbach R 1961 Experimental analysis of the origin of cell types in the development of the mouse thymus. Dev Biol 3: 336–354

Augustine J R 1985 The insular lobe in primates including humans. Neurol Res 7: 2–10

Austin C R 1951 Observations on the penetration of the sperm into the mammalian egg. Aust J Sci Res Ser B 4: 581–596

Austin C R 1961 The mammalian egg. Blackwell Scientific: Oxford

Austin C R 1963 Fertilization and transport of the ovum. In: Hartman C G (ed) Mechanisms concerned with conception. Proceedings of a symposium under the auspices of the Population Council and the Planned Parenthood Federation of America. Pergamon: Oxford: pp 285–320

Austin C R 1965 Fertilization. Prentice-Hall: New Jersey

Austin C R 1968 Ultrastructure of fertilisation. Holt, Rinehart and Winston: New York

Austin C R, Bishop M W H 1958 Some features of the acrosome and perforatorium in mammalian spermatozoa. Proc R Soc Lond [Biol] 149: 234–240

Austin C R, Braden A W H 1956 Early reactions of the rodent egg to spermatozoon penetration. J Exp Biol 33: 358–365

Austin C R, Short R V 1984 Reproduction in mammals. 2nd edition. Books 1–3 Cambridge University Press: Cambridge

Austin C R, Walton A 1960 Fertilisation, In: Parkes A S (ed) Marshall's physiology of reproduction. 3rd edition Vol 1 Part 2. Longmans: London: pp 310–416

Avendano S, Croxatto H D, Pereda J, Croxatto H B 1975 A seven cell human egg recovered from the oviduct. Fertil Steril 26: 1167–1172

Axelsson S, Björklund A, Falck B et al 1973 Glyoxylic acid condensation: a new fluorescence method for the histochemical demonstration of biogenic monoamines. Acta Physiol Scand 87: 57–62

Azaz B, Lustmann J 1973 Anatomical configurations in dry mandibles. Br J Oral Surg 11: 1–9

Azizkhan R G, Azizkhan J C, Zetter B R, Folkman J 1980 Mast cell heparin stimulates migration of capillary endothelial cells in vitro. J Exp Med 152: 931–944

B

Babel J, Bischoff A, Spoendlin H 1970 Ultrastructure of the peripheral nervous system and sense organs. Churchill: London

Baca M, Zamboni L 1967 The fine structure of human follicular oocytes. J Ultrastruct Res 19: 354–381

Bachmann G 1916 The inter-auricular time interval. Am J Physiol 41: 309

Bachmann K 1960 On the blood flow in arterial and venous circulation. Arch Kreislaufforsch 33: 225–259

Bach-y-Rita P, Collins C C 1971 The control of eye movements. Academic Press: New York, London

Backhouse K M 1964 The gubernaculum testis Hunteri: testicular descent and maldescent. Ann R Coll Surg Engl 35: 15–33

BIBLIOGRAPHY

Backhouse K M, Butler H 1958 The development of the coverings of the testis and cord. J Anat 92: 645P

Backhouse K M, Butler H 1960 The gubernaculum testis of the pig. (Sus scorpha) J Anat 94: 107–120

Backhouse K M, Catton W T 1952 Observations on the function of the lumbrical muscles of the hand. J Anat 86: 472–498 Proc Anat Soc of Great Britain and Ireland

Bacon G E, Bacon P J, Griffiths R K 1984 A neutron diffraction study of the bones of the foot. J Anat 139: 265–273

Bacon G E, Griffiths R K 1985 Texture stress and age in the human femur. J Anat 143: 97–101

Baetens D, De Mey J, Gepts W 1977 Immunohistochemical and ultra-structural identification of the pancreatic polypeptide-producing (PP-cell) in the human pancreas. Cell Tissue Res 185: 239–246

Bagnall K M, Harris P F, Jones P R M 1977 A radiographic study of the human fetal spine. 2. The sequence of development of ossification centres in the vertebral column. J Anat 124: 791–802

Bailey H 1947 Parotidectomy; indications and results. Br Med J 1: 404–407

Bailey P, Bonin G von 1951 The isocortex of man. University of Illinois Press: Urbana, Illinois

Baillarger J G F 1840 Recherches sur la structure de la couche corticale des circonvolutions du cerveaux. Mém Acad R Méd Paris 8: 149–183

Bainton D 1981 The discovery of lysosomes. J Cell Biol 91: 66s–76s

Bainton D, Ullyot J L, Farquhar M G 1971 The development of neutro-philic polymorphonuclear leukocytes in human bone marrow. Origin and content of azurophil and specific granules. J Exp Med 134: 907–934

Baker J R 1960 Cytological Techniques. 4th edition. Methuen: London

Baker T G 1963 A quantitative and cytological study of germ cells in human ovaries. Proc R Soc Lond [Biol] 158: 417–433

Baker T G 1966 A quantitative and cytological study of oogenesis in the rhesus monkey. J Anat 100: 761–776

Baker T G, Neal P 1974 Oogenesis in human fetal ovaries maintained in organ culture. J Anat 117: 591–604

Balboni Q C 1972 Distribuzione del reticolo terminale adrenergico nell'ovaio. Boll Soc Ital Biol Sper 48: 84–86

Balinsky B I 1981 An introduction to embryology. 5th edition. Saunders: Philadelphia

Ballantyne F M 1952 The continuity of the vertebrate nervous system. Studies on lepidosiren paradoxa. Trans R Soc Edinb 53: 663–670

Baló J 1950 The dural venous sinuses. Anat Rec 106: 319–326

Balthasar K 1952 Morphologie der spinalen Tibialis-und Peronaeus-Kerne bei der Katze: Topographie, Architekonik, Axon- und Dendritenverlauf der Motoneurrone und Zwischenneurone in den segmenten L₆-S₂. Arch Psychiat Nerven Krankh 188: 345–378

Balthazar E J, Gade M 1976 The normal and abnormal development of the appendix. Radiology 121: 599–604

Banks R W, Barker D, Harker D W, Stacey M J 1975 Correlation between ultrastructure and histochemistry of mammalian intrafusal muscle fibres. J Physiol 252: 16P–17P

Bannerjee S D, Cohn R H, Bernfield M R 1977 Basal lamina of embryonic salivary epithelia. Production by the epithelium and role in maintaining lobular morphology. J Cell Biol 73: 445–463

Bannister L H 1976 Sensory terminals of peripheral nerves. In: Landon D N (ed) The peripheral nerve. Chapman & Hall: London: pp 396–463

Bannister L H, Dodson H C 1975 Proliferative units in the olfactory epithelium of the mouse. J Anat 119: 407–408

Bannister L H, Dodson H C, Astbury A R, Douek E E 1988 The cortical lattice: a highly ordered system of subsurface filaments in guinea pig cochlear outer hair cells. Prog Brain Res 74: 213–219

Baragar F A, Osborn J W 1984 A model relating patterns of human jaw movement to biomechanical constraints. J Biomech 17: 757–767

Barajas L 1970 The ultrastructure of the juxtaglomerular apparatus as disclosed by three-dimensional reconstructions from serial sections. The anatomical relationship between the tubular and vascular components. J Ultrastruct Res 33: 116–147

Barbenel J C 1972 The biomechanics of the temporo-mandibular joint. J Biomech 5: 251–256

Barbenel J C 1974 The mechanics of the temporomandibular joint–a theoretical and electromyographical study. J Oral Rehabil 1: 19–27

Barbizet J 1963 Defect of memorizing of hippocampal-mamillary origin: a review. J Neurol Neurosurg Psychiat 26: 127–135

Barclay A E 1936 The digestive tract. Cambridge University Press: London

Barclay A E, Franklin K J 1938 Time of functional closure of foramen ovale in lamb. J Physiol 94: 256–258

Barclay A E, Barcroft J, Barron D H, Franklin K J 1939 A radiographic demonstration of the circulation through the heart in the adult and in the fetus and the identification of the ductus arteriosus. Br J Radiol 12: 505–517

Barclay A E, Franklin K J, Prichard M M L 1942 Further data about circulation and about the cardiovascular system before and just after birth. Br J Radiol 15: 249–256

Barclay A E, Franklin K J, Prichard M M L 1944 The foetal circulation. Blackwell: Oxford

Barclay-Smith E 1896 The astragalo-calcaneo-navicular joint. J Anat 30: 390–412

Barcroft J 1941 Four phases of birth. Lancet 2: 91–95

Bard P, Rioch D McK 1937 A study of four cats deprived of neocortex and additional portions of the forebrain. Bull Johns Hopkins Hosp 60: 73–147

Barden H 1969 The histochemical relationship of neuromelanin and lipofuscin. J Neuropathol Exp Neurol 28: 419–441

Barden N, Lavole M, Dupont A et al 1977 Stimulation of glucogen release by addition of anti-stmatostatin serum to islets of Langerhans in vitro. Endocrinology 101: 635–638

Bargmann W 1966 Neurosecretion. Int Rec Cytol 19: 183–201

Bargmann W, Schadé J P eds 1963 The rhinencephalon and related structures. Prog Brain Res 3:

Bargmann W, Scharrer B eds 1970 Aspects of neuroendocrinology. Springer-Verlag: Berlin

Barker A T, Lunt M J 1983 The effects of pulsed magnetic fields of the type used in the stimulation of bone fracture healing. Clin Phys Physiol Meas 4: 1–27

Barker D 1974a Morphology of muscle receptors. In: Hunt C C (ed) Handbook of Sensory Physiology Vol 3 pt 2. Springer: Berlin: pp 191–234

Barker D 1974b The motor innervation of muscle spindles. In: Bellaris R, Gray E G Essays on the nervous-system. Clarendon Press: Oxford: pp 131–154

Barker D S, 1967 The dendritic cell system in human gingival epithelium. Archs Oral Biol 12: 203–208

Barland P, Novikoff A B, Hamerman D 1962 Electron microscopy of the human synovial membrane. J Cell Biol 14: 207–220

Barlow H B 1952 Eye movements during fixation. J Physiol 116: 290–306

Barlow T E, Bentley F H, Walder D N 1951 Arteries veins and arterio-venous anastromoses in the human stomach. Surg Gynecol Obstet 93: 657–671

Barlow T E, Haigh A L, Walder D N 1961 Evidence for two vascular pathways in skeletal muscle. Clin Sci 20: 367–385

Barnard J W 1940 The hypoglossal complex of vertebrates. J Comp Neurol 72: 489–524

Barnard J W, Woolsey C N 1956 A study of localisation in the cortico-spinal tracts of monkey and rat. J Comp Neurol 105: 25–50

Barnes B G 1961 Ciliated secretory cells in the pars distalis of the mouse hypophysis. J Ultrastruct Res 5: 453–467

Barnes R D 1976 In-vitro fertilisation. Lancet 1: 1016–1017

Barnett C H 1952 Locking at the knee joint. J Anat 86: 485P

Barnett C H 1954 Comparison of human knee and avian ankle. J Anat 88: 59–70

Barnett C H 1956 Phases of human gait. Lancet 2: 617–621

Barnett C H 1962 Valgus deviation of the distal phalanx of the great toe. J Anat 96: 171–177

Barnett C H, Lewis O J 1958 The evolution of some traction epiphyses in birds and mammals. J Anat 92: 593–607

Barnett C H, Napier J R 1952 The axis of rotation at the ankle joint in man. Its influence upon the form of the talus and the mobility of the fibula. J Anat 86: 1–9

Barnett C H, Davies D V, MacConaill M A 1961 Synovial joints. Their structure and mechanics. Longman: London

Baroldi G, Scomazzoni G 1967 Coronary circulation in the normal and the pathologic heart. Office of the Surgeon General: Washington DC

Baroldi G, Mantero O, Scomazzoni G 1956 The collaterals of the coronary arteries in normal and pathologic hearts. Circ Res 4: 223–229

Barr M L, Bertram E G 1949 A morphological distinction between neurones of the male and female, and the behaviour of the nucleolar satellite during accelerated nucleoprotein synthesis. Nature 163: 676–678

Barrett W C 1951 A note of the internal cremaster muscle. Anat Rec 109: 392(a)

Barrow M V 1971 A brief history of teratology to the early 20th century. Teratology 4: 119–130

Barry A 1951 The aortic arch derivatives in the human adult. Anat Rec 111: 221–238

Barson A J, Sands J 1977 Regional and segmental characteristics of the human adult spinal cord. J Anat 123: 797–803

Bartelmez G W, Dekaban A S 1962 The early development of the human brain. Contrib Embryol Carnegie Inst Washington 37: 13–32

Barth M, Tongio J, Warter P 1976 Interprétation embryologique de l'anatomie des artères digestives. Ann Radiol (Paris) 19: 305–313

Basbaum A I, Fields H L 1979 The origin of descending pathways in the dorsolateral funiculus of the spinal cord of the cat and rat: pain modulation. J Comp Neurol 187: 513–532

Basbaum A I, Fields H L 1984 Endogenous pain control systems: brainstem spinal pathways and endorphin circuitry. Annu Rev Neurosci 7: 309–338

Basbaum A I, Marley N, O'Keefe J 1976 Spinal cord pathways involved in the production of analgesia by brain stimulation. In: Bonica J J, Albe-Fessard D (eds) Advances in pain research in therapy Vol 1. Raven Press: New York: pp 511–515

Basmajian J V 1959 'Spurt' and 'shunt' muscles: An electromyographic confirmation. J Anat 93: 551–553

Basmajian J V 1967 Muscles alive. 2nd edition, Williams and Wilkins: Baltimore

Basmajian J V, Spring W B 1955 Electromyography of the male (voluntary) sphincter urethrae. Anat Rec 121: 388(a)

Basmajian J V, Travill A A 1961 Electromyography of the pronator muscles in the forearm. Anat Rec 139: 45–49

Bassett C A L 1962 Current concepts of bone formation. J Bone Jt Surg 44A: 1217–1244

Bassett C A L, Pawluk R J, Pilla A A 1974 Augmentation of bone repair by inductively coupled electromagnetic fields. Science 184: 575–577

Bast T H, Anson B J 1949 The temporal bone and the ear. Thomas: Springfield, Illinois

Bastianini A 1967a Aspetti microscopici della trama linfatica polmonare in condizioni sperimentali. I. Anossia intrauterina. Atti Acad Fisiocr 16: 1304–1308

Bastianini A 1967b Observazioni sulla morfologia microscopica e l'istotopografia dervasi linfatici del polmone umano. Boll Soc Ital Biol Sper 43: 1567–1570

Bastianini A, Comparini L 1968 Contribution a l'étude de l'histotopographie des vaisseaux lymphatiques du poumon. Essais de reconstruction graphique dans des conditions normales, pathologiques et experimentales. C R Assoc Anat 141: 520–525

Bates J A V 1969 The significance of tremor phasic units in the human thalamus. In: Gillingham F J, Donaldson I M L (eds) Third symposium on Parkinson's disease, held at the Royal College of Surgeons of Edinburgh 1968. Livingstone: Edinburgh: pp 118–124

Bates M W 1948 The early development of the hypoglossal musculature in the cat. Am J Anat 83: 329–356

Batson O V 1940 Function of vertebral veins and their role in the spread of metastases. Ann Surg 112: 138–149

Batson O V 1957 The vertebral vein system. Am J Roentgenol 78: 195–212

Bauer W, Ropes M W, Waine H 1940 Physiology of articular structures. Physiol Rev 20: 272–312

Baum J, Simons B E, Unger R H, Madison L L 1962 Localisation of glucagon in the alpha cells in the pancreatic islet by immunofluorescent technics. Diabetes 11: 371–374

Baumann J A, Gajisin S 1975 Sur la multiplicité et la dispersion des ganglions parasympathiques de la tète. Bull Assoc Anat 59: 329–332

Baumann K I, Hamann W and Leung M S 1988 Responsiveness of slowly adapting cutaneous mechanoreceptors after close arterial infusion of neomycin in cats. Prog Brain Res 74: 43–49

Baumel J J 1974 Trigeminal-facial nerve communications. Their function in facial muscle innervation and reinnervation. Arch Otolaryngol 99: 34–44

Baumgarten H G, Holstein A F 1967 Catecholaminhaltige Nervenfasern im Hoden des Menschen. Z Zellforsch 79: 389–395

Baumgarten H G, Holstein A F, Owman C 1970 Auerbach's plexus of mammals and man: electron microscopic identification of three different types of neuronal processes in myenteric ganglia from rhesus monkeys, guinea pigs and man. Z Zellforsch 106: 376–397

Baumgarten H G, Holstein A F, Rosengren E 1971 Arrangement, ultrastructure, and adrenergic innervation of smooth musculature of the ductuli efferentes, ductus epididymidis and ductus deferens of man. Z Zellforsch 120: 37–39

Baumgarten R von, Aranda Coddou L 1959 Acta Neurol Lat Am 5

Baxter J S 1953 Frazer's manual of embryology. The development of the human body 3rd edition, Baillière: London

Beard J 1896 The history of a transient nervous apparatus in certain Ichthyopsida. An account of the development and degeneration of ganglion cells and nerve fibres. Part 1 Raja Batis Zoll Jb 9: 319

Bearn J G 1961 An electromyographic study of the trapezius, deltoid, pectoralis major, biceps and triceps muscles, during static loading of the upper limb. Anat Rec 140: 103–108

Bearn J G 1967 Direct observation on the function of the capsule of the sterno-clavicular joint in clavicular support. J Anat 101: 159–170

Beatty R A 1957 Parthenogenesis and polyploidy in mammalian development. Cambridge University Press: Cambridge

Beatty R A 1967 Parthenogenesis in vertebrates. In: Metz C B, Monroy A (eds) Fertilization Vol 1. Academic Press: New York: pp 413–440

Beaudet A, Descarries L 1979 Radioautographic characterisation of a serotonin-accumulating nerve cell group in adult at hypothalamus. Brain Res 160: 231–243

Bebin J 1956 The central tegmental bundle. An anatomical and experimental study in the monkey. J Comp Neurol 105: 287–332

Beck I T, Sinclair D G eds 1971 The exocrine pancreas. Churchill: London

Becker N, Fuchs A F 1969 Further properties of the human saccadic system: eye movements and correction saccades with and without visual fixation points. Vision Res 9: 1247–1258

Beckett E B, Bourne G H 1972 Histochemistry of developing skeletal and cardiac muscle. In: Bourne G H (ed) The structure and function of muscle, 2nd edition, Vol 1 Structure Part 1. Academic Press: New York: pp 150–178

Becks H, Asling C W, Collins D A, Simpson M E, Evans H M 1948 Changes with increasing age in the ossification of the third metacarpal of the female rat. Anat Rec 100: 577–592

Bedford J M 1977 Sperm/egg interaction: the specificity of human spermatozoa. Anat Rec 188: 477–488

Bedford J M, Calvin H I 1974 The occurrence and possible functional significance of -S-S crosslinks in sperm heads with particular reference to eutherian mammals. J Exp Zool 188: 137–155

Bedford J M, Calvin H, Cooper G W 1973 The maturation of spermatozoa in the human epididymis. J Reprod Fertil (Suppl 18): 199–213

Beertsen W, Everts V, van den Hooff 1974 Fine structure of fibroblasts in the periodontal ligament of the rat incisor and their possible role in tooth eruption. Arch Oral Biol 19: 1087–1098

Begg R C 1930 The urachus: its anatomy, histology and development. J Anat 64: 170–183

Behrens J C, Walker P S, Shoji H 1974 Variations in strength and structure of cancellous bone at the knee. J Biomech 7: 201–207

Beidler L M 1970 Physiological properties of mammalian taste receptors. In: Wolstenholme G E W, Knight J (eds) Taste and smell in vertebrates (Ciba Foundation Symposium). Churchill: London: pp 51–67

Beidler L M, Smallman R L 1965 Renewal of cells within taste buds. J Cell Biol 27: 263–272

Békésy G V 1960 In: Wever E G (ed) Experiments in hearing. McGraw-Hill: New York

Bélanger L F 1969 Osteocyte osteolysis. Calcif Tissue Res 4: 1–12

Bélanger L F, Robichon J, Migicovsky B B, Copp D H, Vincent J 1963 Resorption without osteoclasts (osteolysis). In: Sognnaes R F (ed) Mechanisms of hard tissue destruction. American Association for Advancement of Science: Washington, DC: pp 531–556

Bell G H 1956 Bone as a mechanical engineering problem, In: Bourne G H (ed) The biochemistry and physiology of bone. Academic Press: New York: pp 27–52

Bell G H, Cuthbertson D P, Orr J 1941 Strength and size of bone in relation to calcium intake. J Physiol 100: 299–317

Bell J 1812 The Anatomy of the human body 4 vol 4th edition, Collins: New York

Bellairs R 1959 The development of the nervous system in chick embryos, studied by electron microscopy. J Embryol Exp Morphol 7: 94–115

Bellairs R 1971 Developmental processes in higher vertebrates. Logos Press: London

Bellairs R 1974 Early differentiation of the nervous system. In: Bellairs R, Gray E G (eds) Essays on the nervous system. A Festschrift for Professor J Z Young. Clarendon Press: Oxford

Bellairs R, Gray E G (eds) 1974 Essays on the nervous system. A Festschrift for Professor J Z Young, Clarendon Press: Oxford

Bellhouse, B J, Bellhouse F H 1968 Mechanism of closure of the aortic valve. Nature 217: 86–87

Belmonte N 1968 Estudios anatomicos sobre la vascularizacion del nervio optico. Arch Soc Oftalmol Hisp Am 28: 801–810

Belt E 1965 Leonardo the anatomist, University of Kansas Press: Laurence, Kansas

Beltrani F 1946 Considérations biologiques sur le mandibule chez l'homme. Rev Stomat 47: 1–9

Bender M B ed 1964 The oculomotor system. Hoeber: New York

Bender M B, Weinstein E A 1943 Functional representation in oculomotor and trochlear nuclei. Arch Neurol Psychiatry 49: 98–106

Benedeczky I, Smith A D 1972 Ultrastructural studies on the adrenal medulla of the golden hamster: origin and fate of secretory granules. Z Zellforsch 124: 367–386

Benjamin M, Evans E J, Copp L 1986 The histology of tendon attachments to bone in man. J Anat 149: 89–100

Bennett A G, Francis J L 1962 Refraction at plane and spherical surfaces. In: Davson H (ed) The Eye. Vol 4. Academic Press: New York: pp 19–34

Bennett H S, Luft J H, Hampton J C 1957 Morphological classification of vertebrate blood capillaries. Am J Physiol 196: 381–390

Bennett M R, Rogers D C 1967 A study of the innervation of the taenia coli. J Cell Biol 37: 573–596

Bennett M V L 1973 Function of electronic junctions in embryonic and adult tissues. Fed Proc 32: 65–72

Bensch K G, Gordon G B, Miller L R 1965 Studies on the bronchial counterpart of the Kultschitzky (Argentaffin) cell and innervation of bronchial glands. J Ultrastruct Res 12: 668–686

Benscome S 1959 Studies on the terminal autonomic nervous system with special reference to the pancreatic islets. Lab Invest 8: 629–646

Bensley R R 1911 Studies on the pancreas of the guinea pig. Am J Anat 12: 297–388

Bentfield M E, Bainton D F 1975 Cytochemical localization of lysosomal enzymes in rat megataryocytes and platelets. J Clin Invest 56: 1635–1649

Bergel D H 1961 The static elastic properties of the arterial wall. J Physiol 156: 445–457. The dynamic elastic properties of the arterial wall. J Physiol 156: 458–469

Bergland R, Ray B S 1969 The arterial supply of the human optic chiasm. J Neurosurg 31: 327–334

Bergmanson J P G 1978 Ophthalmic terminals in the iris and the ciliary body of monkeys. An electron microscopical study. Albr von Graefes Arch Klin Exp Ophthal 206: 39–47

Bergstresser P R, Chapman S L 1980 Maturation of normal epidermis without an ordered structure. Br J Dermatol 102: 641–648

Bergstresser P R, Taylor J R 1977 Epidermal 'turnover time': a new examination. Br J Dermatol 96: 503–509

Berlin N I, Waldmann T A, Weissman S M 1959 Life span of red blood cell. Physiol Rev 39: 577–616

Berman N, Jones E G 1977 A retino-pulvinar projection in the cat. Brain Res 134: 237–248

BIBLIOGRAPHY

Bernard C 1855–1856 Leçons de physiologie, experimentale appliquée à la médecine. 2 vol Baillière: Paris

Bernard G W 1972 Ultrastructural observations of initial calcification in dentine and enamel. J Ultrastruct Res 41: 1–17

Bernard G W, Pease D C 1969 An electron microscopic study of initial intramembranous osteogenesis. Am J Anat 125: 271–290

Bernhard C G, Bohm E, Petersen I 1953 Investigation on the organisation of the cortico-spinal system in monkeys. Acta Physiol Scand 29: 79–103

Bernick S, Patek R 1969 Lymphatic vessels of the dental pulp in dogs. J Dent Res 48: 959–964

Berquist H 1932 Zur Morphologie des Zwischenhirns bei neideren Wirbeltieren. Acta Zool, Stockholm 13: 57–304

Berquist H 1952 Studies on the cerebral tube in vertebrates. The neuromeres. Acta Zool, Stockholm 33: 117–191

Berquist H 1968 Uber die Differenzierung der Neuraloknes besounders des Stratum Zonale. Z Zellforsch 86: 401–421

Berridge M J 1985 The molecular basis of communication with the cell. Sci Am 253: (4) 124–134

Berry A C 1975 Factors affecting the incidence of non-metrical skeletal variants. J Anat 120: 519–535

Berry A C, Berry R J 1967 Epigenetic variation in the human cranium. J Anat 101: 361–380

Berry M 1974 Development of the cerebral neocortex of the rat. In: Gottlieb G (ed) Aspects of neurogenesis: studies on the development of behaviour and the nervous system. Vol 2. Academic Press: New York, London: pp 7–67

Berry M 1982 Cellular differentiation: development of dendritic arborizations under normal and experimentally altered conditions. Neurosci Res Program Bull 20: 451–461

Berry M 1986 Neurogenesis and gliogenesis in the human brain. Food Chem Toxicol 24: 79–89

Berry M, Eayrs J T 1966 The effects of X-irradiation on the development of the cerebral cortex. J Anat 100: 707–722

Berry M, Rogers A W 1965 The migration of neuroblasts in the developing cerebral cortex. J Anat 99: 691–709

Berry M, Bradley P, Borges S 1978 Environmental and genetic determinants of connectivity in the central nervous system – an approach through dendritic field analysis. In: Corner M A, Baker R E, van der Poll N E, Swaab D F, Uylings H B M (eds) Maturation of the nervous system. Elsevier: Amsterdam. Prog Brain Res 48: 133–146

Berry M, McConnell P, Sievers J 1980a Dendritic growth and the control of neuronal form. Curr Top Dev Biol 15: 67–101

Berry M, Sievers J, Baumgarten H G 1980b The influence of afferent fibres on the development of the cerebellum. In: Di Bendetta C, Balázs R, Gombos G, Porcellati G (eds) Multidisciplinary approach to brain development. Elsevier/North-Holland Biomedical Press: Amsterdam: pp 91–106

Berry M, Sievers J, Baumgarten H G 1980c Adaption of the cerebellum to deafferentation. In: McConnell P S, Boer G J, Romijn H J, van der Poll N E, Corner M A (eds) Adaptive capabilities of the nervous system. Elsevier: Amsterdan. Prog Brain Res 53: 65–92

Berry M, McConnell P, Sievers J, Price S, Annar A 1980d Factors influencing the growth of cerebellar neural networks. Bibl Anat 19: 1–51

Berry M, Sadler, M 1988 Factors influencing the development of dendritic form. In: Parnavalas, J G, Stern C D, Stirling R V (eds) The Making of the Nervous System. Oxford University Press: Oxford: pp 473–501

Bertalanffy F D 1964 Respiratory tissue: structure and histophysiological cytodynamics. I. Review and basic cytomorphology. Int Rev Cytol 16: 233–328

Berthold C H 1968 Ultrastructure of the node-paranode region of mature feline ventral lumbar spinal-root fibres. Acta Soc Med Upsal 73: (Suppl 9)

Berthold C H, Carlstedt T 1977 Observations on the morphology at the transition between the peripheral and the central nervous system in the cat. V. A light microscopical and histochemical study of S dorsal rootlets in developing kittens. Acta Physiol Scand 100: (Suppl 446), 73–85

Bharihoke V, Gupta M 1986 Muscular attachments along the medial border of the scapula. Surg Radiol Anat 8: 71–73

Biava C 1964a Studies on cholestasis. The fine structure and morphogenesis of hepatocellular and canalicular bile pigment. Lab Invest 13: 1099–1123

Biava C 1964b Studies on cholestasis. A re-evaluation of the fine structure of normal human bile canaliculi. Lab Invest 13: 840–864

Bickers W 1960 Sperm migration and uterine contractions. Fertil Steril 11: 286–290

Biedenbach M A 1972 Cell density and regional distribution of cell types in the cuneate nucleus of the rhesus monkey. Brain Res 45: 1–14

Bilbey D L J, Rappaport A M 1960 The segmental anatomy of human liver. Anat Rec 136: 330

Bill A 1970 In: Moses R A (ed) Adler's physiology of the eye. 5th edition. Mosby: St Louis

Billingham R E, Silvers W K 1960 The melanocytes of mammals. Q Rev Biol 35: 1–40

Billings-Gagliardi S, Chan-Palay V, Palay S L 1974 A review of lamination in Area 17 of the visual cortex of Macaca mulatta. J Neurocytol 3: 619–629

Billingsley P R, Ransom S W 1918 On the number of nerve cells in the ganglion cervicale superioris. J Comp Neurol 29: 359–366

Birdwell C R, Gospodarowicz D, Nicholson G L 1978 Identification, localization, and role of fibronectin in cultured bovine endothelial cells. Proc Natl Acad Sci USA 75: 3273–3277

Bischoff R 1975 Regeneration of single skeletal muscle fibres in vitro. Anat Rec 182: 215–235

Biscoe T J, Bradley G W, Purves M J 1969 The relation between carotid body chemoreceptor activity and carotid sinus pressure in the cat. J Physiol 203: 40P

Biscoe T J, Lall A, Samson J R 1970 Electron microscopical and electrophysiological studies on the carotid body following intracranial section of the glossopharyngeal nerve. J Physiol 208: 133–152

Bishop A 1964 The use of the hand in lower primates. In: Buettner-Janusch J (ed) Evolutionary and genetic biology of primates. Vol 2. Academic Press: New York

Bishop D W 1962 Sperm motility. Physiol Rev 42: 1–59

Bishop D W, Tyler A 1956 Fertilizins of mammalian eggs. J Exp Zool 132: 575–601

Bishop M A 1985 Vascular permeability to lanthanum in the rat incisor pulp. Comparison with endoneurial vessels in the inferior alveolar nerve. Cell Tissue Res 239: 131–136

Bishop M W H, Walton A 1960 Spermatogenesis and the structure of mammalian spermatozoa. In: Parkes A S (ed) Marshall's physiology of reproduction 3rd edition, Vol 1, Part 2. Longmans: London: pp 1–129

Bizarro A H 1921 On sesamoid and supernumerary bones of the limbs. J Anat 55: 256–268

Björklund A, Hökfelt T 1984 Handbook of chemical neuroanatomy. Vol 2 Classical Transmitters in the CNS. Part 1. Elsevier: Amsterdam

Björklund A, Nobin A 1973 Fluorescence histochemical and microspectrofluorometric mapping of dopamine, and noradrenaline cell groups in the rat diencephalon. Brain Res 51: 193–205

Björklund A, Owman Ch, West K A 1972 Peripheral sympathetic innervation of serotonin cells in the habenular region of the rat brain. Z Zellforsch 127: 570–579

Björkman U, Ekholm R 1973 Thyroglobulin synthesis and intracellular transport studied in bovine thyroid slices. J Ultrastruct Res 45: 231–253

Black D 1917 The motor nuclei of the cerebral nerves in phylogeny; a study of the phenomena of neurobiotaxis. I. Cyclostomi and pisces. J Comp Neurol 27: 117–126

Blackstad T W 1956 Commissural connections of the hippocampal region in the rat, with special reference to their mode of termination. J Comp Neural 105: 417–538

Blackstad T W 1958 On the termination of some afferents to the hippocampus and fascia dentata. Acta Anat 35: 202–214

Blackstad T W 1967 Cortical grey matter. A correlation of light and electron microscopic data. In: Hydén H (ed) The neuron. Elsevier: Amsterdam: pp 49–118

Blackstad T W, Flood P R 1963 Ultrastructure of hippocampal axo-somatic synapses. Nature 198: 542–543

Blackstad T W, Kjaerheim A 1967 Special axo-dendritic synapses in the hippocampal cortex: electron and light microscopic studies on the layer of mossy fibres. J Comp Neurol 117: 133–160

Blackwood H J J 1959 The development, growth and pathology of the mandibular condyle. M D Thesis, Queen's University: Belfast

Blackwood H J J 1965 Vascularisation of the condylar cartilage of the human mandible. J Anat 99: 551–563

Blair-West J R, Coghlan J P, Denton D A, Wright R D 1967 Effect of endocrines on salivary glands. In: Code C F, Heidel W (eds) Handbook of physiology, Section 6, Alimentary Canal, Vol 2. American Physiological Society: Washington, DC: pp 633–664

Blakemore C 1974 Development of functional connexions in the mammalian visual system. Br Med Bull 30: 152–157

Blakemore C B, Falconer M A 1967 Long-term effects of anterior temporal lobectomy on certain cognitive functions. J Neurol Neurosurg Psychiat 30: 364–367

Blalock A 1947 Use of shunt or by-pass operations in the treatment of certain circulatory disorders including portal hypertension and pulmonic stenosis. Ann Surg 125: 129–141

Blandau R J, Rumery R E 1964 The relationship of swimming movements of epididymal spermatozoa to their fertilising capacity. Fertil Steril 15: 571–579

Blanes T 1898 Sobre algunos puntos dudosos de la estructura del bulbo olfatoria. Rev Trim Microgr 3: 99–127

Blaschko A 1901 Die Nervenverteilung und der Haut in ihrer Beziehung zu den Erkrankungen der Haut. Braunmuller: Vienna

Bleier R 1977 Ultrastructure of supraependymal cells and ependyma of hypothalamic third ventricle of mouse. J Comp Neurol 174: 359–376

Blenkinsopp W K 1967 Proliferation of respiratory tract epithelium in the rat. Exp Cell Res 46: 144–154

Blevins C E 1967 Innervation patterns of the human stapedius muscle. Arch Otolaryugol 86: 136–142

Blinkov S M, Glezer I I 1968 The human brain in figures and tables. Basic Books: New York

Blix M 1882–1883 Experimentala bidrag till lösning af frågan om hudnervenas specifika energi. Uppsala läkareförhandlingar 18: 87–102

Block E 1953 Quantitative morphological investigation of follicular system in newborn female infants. Acta Anat 17: 201–206

Bloebaum R D, Wilson A S 1980 The morphology of the surface of articular cartilage in adult rats. J Anat 131: 333–346

Bloodworth J M B, Powers K L 1968 The ultrastructure of the normal dog adrenal. J Anat 102: 457–476

Bloom G, Nicander L 1961 On the ultrastructure of the protoplasmic droplet of spermatozoa. Z Zellforsch 55: 833–844

Bloom S R, Polak J M 1978 Gut hormone overview. In: Bloom S R (ed) Gut hormones. Churchill Livingstone: Edinburgh, London: pp 3–18

Bloom S R, Polak J M eds 1981 Gut hormones. 2nd edition, Churchill Livingstone: Edinburgh

Bloom W 1931 New types of granular cell in islets of Langerhans of man. Anat Rec 49: 363–371

Bloom W, Bartelmez G W 1940 Hematopoiesis in young human embryos. Am J Anat 67: 21–54

Bloom W, Fawcett D W 1975 A textbook of histology 10th edition. Saunders: Philadelphia

Bloom W, Fawcett D W 1986 A textbook of histology 11th edition. Saunders: Philadelphia

Bloomer W E, Liebow A A, Hales M R 1960 Surgical anatomy of the bronchovascular segments. Thomas: Springfield, Illinois

Bloor C M, Lowman R M 1963 Myocardinal bridges in coronary atherosclerosis. Am Heart J 65: 195–199

Blotman F, Poubelle P, Chaintreuil J, Damon M, Flandre O, Crastes de Paulet A, Simon L 1982 Mononuclear phagocytes, prostanoids and rheumatoid arthritis. Int J Immunopharmacol 4: 119–125

Blum F 1893 Der Formaldehyd als Hörtungsmittel. Vorläufige Mittheilung. Z Wiss Mikr 10: 314–315

Blum J S, Chow K C, Pribram K H 1950 Behavioral analysis of the organisation of the parieto-temporo-preoccipital cortex. J Comp Neurol 93: 53–100

Blunt M J 1951 Posterior wall of inguinal canal. Br J Surg 39: 230–233

Blunt M J 1954 The blood supply of the facial nerve. J Anat 88: 520–526

Blunt M J 1959 The vascular anatomy of the median nerve in the forearm and hand. J Anat 93: 15–22

Böck P, Gorgas K 1976 Catecholamine and granule content of carotid body type I-cells. In: Coupland R E, Fujita T (eds) Chromaffin, enterochromaffin and related cells. Elsevier: New York: pp 355–374

Böck P, Stockinger L, Vyslonzil E 1970 Die Feinstrucktur des Glomus caroticum beim Menschen. Z Zellforsch 105: 543–568

Bockman D E, Coopor M D 1973 Pinocytosis by epithelium associated with lymphoid follicles in the bursa of fabricius, appendix and Peyer's patches. An electron microscopic study. Am J Anat 136: 455–477

Bodian D 1936 A new method for staining nerve fibers and nerve endings in mounted paraffin sections. Anat Rec 65: 89–97

Bodian D 1962 Discussion in: Mountcastle V B (ed) Interhemispheric relations and cerebral dominance. Johns Hopkins University Press: Baltimore, Md

Bodian D 1970a An electron microscopic characterisation of classes of synaptic vesicles by means of controlled aldehyde fixation. J Cell Biol 44: 115–124

Bodian D 1970b A model of synaptic and behavioral ontogeny. In: Schmitt F O, Quarton G C, Melnechuk T, Adelman G (eds) The neurosciences, a second study program. Rockefeller University Press: New York: pp 129–140

Bodian D, Mellors R C 1945 Regenerative cycle of motoneurons, with special reference to phosphatase activity. J Exp Med 81: 469–488

Bodian M 1960 In: Dyke S C (ed) Recent advances in clinical pathology. 3rd series. Churchill: London

Bogduk N, Macintosh J E 1984 Applied anatomy of the thoracolumbar fascia. Spine 9: 164–170

Bogduk N, Wilson A S, Tynan W 1982 The human lumbar dorsal rami. J Anat 134: 383–397

Bogduk N, Major G A C, Carter J 1984 Lateral subluxation of the atlas in rheumatoid arthritis. Ann Rheum Dis 43: 341–346

Bogen J E, Gazzaniga M S 1965 Cerebral commissurotomy in man. J Neurosurg 23: 394–399

Boggon R P, Palfrey A J 1973 The microscopic anatomy of human lymphatic trunks. J Anat 114: 389–405

Böhme C C 1962 The fine structure of Clarke's nucleus of the spinal cord. Thesis, University of Pennsylvania

Boime I, Boguslawski S 1974 The synthesis of human placental lactogen by ribosomes derived from human placenta. Proc Natl Acad Sci USA 71: 1322–1325

Bois E, Feingold J, Benmaiz H, Briard M L 1975 Congenital urinary tract malformations: epidemiologic and genetic aspects. Clin Genet, 8: 37–47

Bojsen-Møller F 1975 Demonstration of terminalis, olfactory, trigeminal perivascular nerves in the rat nasal septum. J Comp Neurol 159: 245–256

Bojsen-Møller F, Flagstad K E 1976 Plantar aponeurosis and internal architecture of the ball of the foot. J Anat 121: 599–611

Bojsen-Møller F, Schmidt L 1974 The palmar aponeurosis and the central spaces of the hand. J Anat 117: 55–68

Bojsen-Møller F, Tranum-Jensen J 1971 Whole-mount demonstration of colinesterase containing nerves in the right atrial wall, nodal tissue and atrioventricular bundle of the pig heart. J Anat 108: 375–386

Bok H E 1966 De foetale transformatie van het middenoorgebied. Drukkerij Holland: Amsterdam

Bolger P M, Eisner G M, Shea P T, Ramwell P W, Slotkoff L M 1977 Effects of PGD on canine renal function. Nature 267: 628–630

Bolk L, Goppert G, Kallius E, Lubosch W 1931–1939 Handbuch der vergleichenden Anatomie der Wirbeltiere. Urban: Berlin

Bollobas B, Hajdu I 1975 The development of the tympanic sinus. ORL J Otorhinolaryngal Relat Spec 37: 97–102

Bombeck C T, Dillard D H, Nyhus L M 1966 Muscular anatomy of the gastroesphageal junction and role of phrenoesophageal ligament. Autopsy study of sphincter mechanism. Ann Surg 164: 643–654

Bonucci E 1967 Fine structure of early cartilage calcification. J Ultrastruct Res 20: 33–50

Bonucci E 1970 Fine structure and histochemistry of calcifying globules in epiphyseal cartilage. Z Zellforsch 103: 192–217

Borell U, Fernström I 1953 Adnexal branches of uterine artery; arteriographic study in human subjects. Acta Radiol 40: 561–582

Borell U, Fernström I 1960 Radiologic pelvimetry. Acta Radiol Suppl 191: 3–97

Borelli G A 1681 De Motu Animalium 2 parts. A Bernabo: Rome

Borg E 1973 A neuroanatomical study of the brainstem auditory system of the rabbit. Acta Morph Neerl Scand 11: 49–62

Borges A F, Alexander J E 1962 Relaxed skin tension lines, Z-plasties on scars, and fusiform excision of lesions. Br J Plast Surg 15: 242–254

Borison R, Borison H 1973 Postnatal development of the area postrema with relation to digitalis-induced vomiting in the cat. Exp Neurol 40: 138–152

Born G V R, Dawes G S, Mott J C, Widdicombe J G 1954 Changes in the heart and lungs at birth. Cold Spring Harbor Symp Quant Biol 19: 102–108

Bors J 1926 Uber das Zahlenverhaltnis zwischen Nerven und Muskelfasern. Anat Anz 60: 415–420

Bortolami R, Veggetti A, Callegari E, Lucchi M L, Palmieri G 1977 Afferent fibres and sensory ganglion cells within the oculomotor nerve in some mammals and man. 1. Anatomical investigations. Arch Ital Biol 115: 355–385

Borun T W, Stein G S 1972 The synthesis of acidic chromosome proteins during the cell cycle of HeLa S3 cells. II. The kinetics of residual protein synthesis and transport. J Cell Biol 52: 308–315

Borun T W, Gabrielli F, Ajiro K 1975 Further evidence of transcriptional and translational control of histone messenger RNA during the HeLa S3 cycle. Cell 4: 59–67

Bosma J F 1957 Deglutition: pharyngeal stage. Physiol Rec 37: 275–300

Bosma J F ed 1976 Symposium on development of the basicranium. U S Department of Health, Education, and Welfare (DHEW publication (NIH) 76–989) Bethesda, Maryland

Bossy J 1968 Sur la présence d'un noyau sensitif du annexe au tractus spinal du V. Acta Anat 70: 332–340

Bossy J, Vidić B 1967 Existe-il une innervation proprioceptive des muscles du pharynx chez l'homme? Arch Anat Histol Embryol 50: 273–284

Botha G S M 1962 The gastro-oesophageal junction. Churchill: London

Boucher B J 1957 Sex differences in the foetal pelvis. Am J Phys Anthropol N S 15: 581–600

Bouillard J P 1825 Recherches cliniques propres à démontrer que la perte de la parole correspond à la lésion des lobules antérieurs du cerveau. Arch Gén Méd 8: 24–45

Bouin P, Ancel P 1903 Le gland intestitielle du testicule chez le cheval. Arch Zool Exp Gén 3: 391

Boulder Committee 1970 Embryonic vertebrate central nervous system. Revised terminology Anat Rec 166: 257–261

Boulter P S, Parkes A G 1963 Submucosal vascular patterns of the alimentary tract and their significance. Br J Surg 47: 546–550

Bounoure L 1939 L'origine des cellules réproductrices et le problème de la lignée germinale. Gauthiers-Villars: Paris

Bourbon J R, Rieutort M 1987 Pulmonary surfactant: biochemistry, physiology, and pathology. News Physiol Sci 2: 129–132

Böving B G 1959 The biology of trophoblast. Ann NY Acad Sci 80: 21–43

Böving B G 1963 Implantation mechanisms. In: Hartman C G (ed) Mechanisms Concerned with Conception. Proceedings of a symposium under the auspices of the population council and the planned parenthood federation of America. Pergamon Press: Oxford: pp 321–396

Bowden D M, German D C, Poynter W D 1978 An autoradiographic, simistereotaxic mapping of major projections from the locus coeruleus and adjacent nuclei. Brain Res 145: 257–276

Bowden R E M 1955 Surgical anatomy of the recurrent laryngeal nerve. Br J Surg 43: 153–163

Bowden R E M 1966 The functional anatomy of striated muscle. Ann R Coll Surg Engl 38: 41–59

Bowden R E M, el-Ramli H A 1967 The anatomy of the oesophageal hiatus. Brit J Surg 54: 983–989

Bowersox J C, Sorgente N 1982 Chemotaxis of aortic endothelial cells in response to fibronectin. Cancer Res 42: 2547–2551

Bowin P 1895 J Anat Physiol. Paris 31

Bowman J P 1968 Muscle spindles in the intrinsic and extrinsic muscles of the rhesus monkey's tongue. Anat Rec 161: 483–488

Boycott B B 1974 Aspects of the comparative anatomy and physiology of the vertebrate retina. In: Bellairs R, Gray E G (eds) Essays on the nervous system. Clarendon Press: Oxford: pp 223–257

BIBLIOGRAPHY

Boycott B B, Dowling J E 1969 Origin of the primate retina: light microscopy. Phil Trans R Soc Ser B 255: 109–184

Boycott B B, Kolb H 1973 The connections between bipolar cells and photoreceptors in the retina of the domestic cat. J Comp Neurol 148: 91–114

Boyd I A 1954 The histological structure of the receptors in the knee joint of the cat correlated with their physiological response. J Physiol 124: 476–488

Boyd I A 1962 The structure and innervation of the nuclear bag muscle fibre system and nuclear chain muscle fibre system in mammalian muscle spindles. Philos Trans R Soc Lond [Biol] 245: 81–136

Boyd I A 1985 Muscle spindles and stretch reflexes. In: Swash M, Kennard C (eds) Scientific basis of clinical neurology. Churchill Livingstone: Edinburgh: pp 74–97

Boyd I A, Roberts T D M 1953 Proprioceptive discharges from stretch-receptors in the knee joint of the cat. J Physiol 122: 38–58

Boyd J D 1933 The classification of the upper lip in mammals. J Anat 67: 409–416

Boyd J D 1937 The development of the human carotid body. Contrib Embryol Carnegie Inst Washington 26: 1–31

Boyd J D 1956 Observations on the human pharyngeal hypophysis. J Endocrinol 14: 66–77

Boyd J D 1960 The embryology and comparative anatomy of the melanocyte. In: Rook A (ed) Progress in the biological sciences in relation to dermatology. Cambridge University Press: pp 3–14

Boyd J D 1961 The inferior aortico-pulmonary glomus. Br Med Bull 17: 127–131

Boyd J D 1964 Development of the human thyroid gland. In: Pitt-Rivers R, Trotter W R (eds) The thyroid gland. Vol 1. Butterworths: London: pp 9–31

Boyd J D, Hamilton W J 1966 Electron microscopic observations on the cytotrophoblastic contribution to the syncytium in the human placenta. J Anat 100: 535–548

Boyd J D, Hamilton W J 1970 The human placenta. Heffer: Cambridge

Boyd J D, Monro P A G 1949 Partial retention of autonomic function after paravertebral (intermediate lumbar sympathetic ganglia as the probable explanation). Lancet 2: 892–895

Boyd J D, Trevor J C 1953 Race, sex, age and stature from skeletal material. In: Simpson K (ed) Modern trends in forensic medicine. Part 1, Chapter 7 Butterworths: London

Boyd J D, Hamilton W J, Boyd C A 1968 The surface of the syncytium of the human chorionic villus. J Anat 102: 553–563

Boyd W, Blincoe H, Hayner J C 1965 Sequences of action of the diaphragm and quadratus lumborum during quiet breathing. Anat Rec 151: 579–582

Boyde A 1969 Electron microscopic observations relating to the nature and development of prism decussation in mammalian dental enamel. Bull Grp Int Rech Sci Stomat 12: 151–207

Boyde A 1975 Scanning electron microscopy of enamel surfaces. Br Med Bull 31: 120–124

Boyde A, Hobdell M H 1969 Scanning electron microscopy of lamellar bone. Z Zellforsch 93: 213–231

Boyde A, Jones S J 1983 Scanning electron microscopy of cartilage. In: Hall B K (ed) Cartilage. Vol. 1. Structure, Function and Biochemistry. Academic Press: New York: pp 105–148

Boyden E A 1955 Segmental anatomy of the lungs. McGraw-Hill: New York

Boyden E A 1957 The anatomy of the choledochoduodenal junction in man. Surg, Gynecol Obstet 104: 641–652

Boyden E A 1966 The pancreatic sphincters of the baboon as revealed by serial sections of the choledochoduodenal junction. Surgery 60: 1187–1194

Boyden E A 1972 Development of the human lung. In: Kelley V O (ed) Brenneman's practice of pediatrics. 2nd edition, Vol 4, Chapter 64, Harper and Row: New York

Boyden E A, Cope J G, Bill A H 1967 A new look at the etiology of duodenal atresia. Anat Rec 157: 218P

Bozler E 1948 Conduction, automaticity and tonus of visceral muscles. Experientia 4: 213–218

Braak H 1975 On the fine structure of the external glial layer in the isocortex of man. Cell Tissue Res 157: 367–390

Braak H 1977 The pigment architecture of the human occipital lobe, Anat Embryol 150: 229–250

Braak H 1978 On magnopyramidal temporal fields in the human brain. Anat Embryol 152: 141–169

Brachet J 1960 The biochemistry of development. Pergamon Press: Oxford

Brachet J 1965 The biochemistry of animal development, Academic Press: New York

Brachet J 1967 Biochemical changes during fertilization and early development. In: De Reuck A V S, Knight J (eds) Cell differentiation (Ciba Foundation Symposium). Churchill: London: pp 39–64

Brachet J 1969 Acides nucléiques et différenciation embryonnaire. Ann Embryol Morphogen Suppl 1: 21–37

Brachet J 1985 Molecular cytology, Vol 1. The cell cycle. Academic Press: New York

Brachet J, Mirsky A E eds 1960 The cell, Vol 1 Academic Press: New York

Brachet J, Quertier J 1963 Cytochemical detection of cytoplasmic deoxyribonucleic acid (DNA) in amphibian oocytes. Exp Cell Res 32: 410–413

Bradamante Z, Svajger A 1977 Pre-elastic (oxytalan) fibres in the developing elastic cartilage of the external ear of the rat. J Anat 123: 735–743

Braden A W H 1955 The reactions of isolated mucopolysaccharides to several histochemical tests. Stain Technol 30: 19–26

Bradley R L 1973 Surgical anatomy of the gastro-duodenal artery. Int Surg 58: 393–396

Brain R 1961 The neurology of language. Brain 84: 145–166

Braitenberg V 1967 Is the cerebellar cortex a biological clock in the millisecond range? Progr Brain Res 25: 334–346

Braitenberg V 1973 Remarks on the texture of brains. Int J Neurosci 6: 5–6

Braitenberg V 1977 Cortical architectonics, general and areal. In: Brazier M A B, Petsche H (eds) Architectonics of the cerebral cortex. Raven Press: New York

Braitenberg V, Atwood R P 1958 Morphological observations on the cerebellar cortex. J Comp Neurol 109: 1–33

Braithwaite F, Channel G D, Moore F T, Whillis J 1948 The applied anatomy of the lumbrical and interosseous muscles of the hand. Guy's Hosp Rep 97: 185–195

Braithwaite J L 1951 The arterial supply of the urinary bladder. J Anat 85: 413P

Braithwaite JL 1952 Variations in origin of the parietal branches of the internal iliac artery. J Anat 86: 423–430

Braithwaite J L, Adams D J 1957 The venous drainage of the rat spleen. J Anat 91: 352–357

Brånemark P I 1972 Rehabilitering med käkbensförankrad bettersättning. Lakartidningen 69: 4813–4848

Brännström M 1963 A hydrodynamic mechanism in the transmission of pain-producing stimuli through the dentine. In: Anderson D J (ed) Sensory mechanisms in dentine. Proceedings of a symposium held London 1962. Pergamon: Oxford: pp 73–79

Brantigan O C, Voshell A F 1943 Tibial collateral ligament; its functions, its bursae and its relation to medial meniscus. J Bone Jt Surg 25: 121–131

Branton D 1971 Freeze-etching studies of membrane structure. Philos Trans R Soc Lond [Biol] 261: 133–138

Brash J C 1924 The growth of the jaws and the palate. In: The growth of the jaws, normal and abnormal, in health and disease. Dental Board of the United Kingdom: London

Brash J C 1934 Some problems in the growth and developmental mechanics of bone. Edinb Med J 41: 305–387

Brash J C 1955 Neuro-vascular hila of limb muscles: an Atlas. Livingstone: Edinburgh

Brauer K, Winkelmann E 1976 Zur Rolle der Interneuronen im corpus geniculatum laterale der säugetiere. Z Mikrosk 90: 334–351

Braverman I M, Fonferko C 1982 Studies in cutaneous aging. I. The elastic fiber network. J Invest Dermatol 78: 434–443

Braverman I M, Fonferko E 1982b Studies in cutaneous aging. II. The microvasculature. J Invest Dermatol 78: 444–448

Brazier M A B 1959 The historical development of neurophysiology. In: Field E J, Magoun H W (eds) Handbook of physiology. Section 1. Neurophysiology Vol 1, pp 1–58, American Physiological Society: Washington, DC: pp 1–58

Brazier M A B ed 1969 The interneuron. University of California Press: Berkeley

Breathnach A S 1969 Normal and abnormal melanin pigmentation of the skin. In: Wolman M (ed) pigments in pathology. Academic Press: New York: pp 353–394

Breathnach A S 1971 An atlas of the ultrastucture of human skin. Churchill: London

Breathnach A S 1971 An atlas of the ultrastructure of human skin: development, differentiation, and post-natal features. Churchill: London

Breathnach A S, Wylie L M A 1966 Ultrastructure of retinal pigment epithelium of the human fetus. J Ultrastruct Res 16: 584–597

Brekelmans W A M, Poort H W, Slooff T J 1972 A new method to analyse the mechanical behaviour of skeletal parts. Acta Orthop Scand 43: 301–317

Bretscher M S 1985 The molecules of the cell membrane. Sci Am 253: (Oct) 86–90

Bretscher M S, Raff M C 1975 Mammalian plasma membranes. Nature 258: 43–49

Briggaman R A 1983 The epidermal-dermal junction and genetic disorders of this area. In: Goldsmith L A (ed) Biochemistry and physiology of the skin. Vol 2. Oxford University Press: New York: pp 1001–1024

Brightman M W, Palay S L 1963 The fine structure of ependyma in the brain of the rat. J Cell Biol 19: 415–439

Brightman M W, Reese T S 1969 Junctions between intimately apposed cell membranes in the vertebrate brain. J Cell Biol 40: 648–677

Brightman M W, Klatzo I, Olsson Y, Reese T 1970 The blood brain barrier to proteins under normal and pathological conditions. J Neurol Sci 10: 215–239

Brightman M W, Prescott L, Reese T S 1975 Intercellular junctions of special ependyma. In: Knigge K M, Scott D E, Kobayashi H, Ishii S (eds) Brain-endocrine interaction. II. The ventricular system in neuroendocrine mechanisms. 2nd International Symposium. Karger: Basel: pp 146–165

Brighton C T 1981 The treatment of non-unions with electricity. J Bone Jt Surg 63A: 847–851

Brini A, Porte A, Stoeckel M E 1968 Biology and surgery of the vitreous body. Masson: Paris

Brizon J, Castaing J, Hourtaille F C 1956 Le péritoine. Maloine: Paris

Brizzee K R, Neal L M 1954 Re-evaluation of cellular morphology of area postrema in view of recent evidence for chemorceptor function. J Comp Neurol 100: 41–61

Brizzi E, Serantoni C, Ciani P A, Orlandini A, Pernice L 1973 The distribution of the vagus nerves in the stomach. Chir Gastroentol (Engl Ed) 7: 17–34

Broca P 1878 Anatomie comparée des circonvolutions cérébrales. Le grand lobe limbique et la scissure dans la série des mammifères. Rev Anthropol 1: 385–498

Broca P P 1861 Perte de la parole, ramollissement chronique et destruction partielle du lobe antérieur gauche du cerveau. Bull Soc Anthropol 2: 235–238; 301–321

Broca P P 1875 Instructions craniologiques et craniométriques de la société d'anthropologie de Paris. Bull Soc Anthropol 6: 534–536

Brock R C 1942 The use of silver nitrate in the production of aseptic obliterative pleuritis. Guy's Hosp Rep 91: 111–130

Brock R C 1943 Observations on the anatomy of the bronchial tree, with special reference to the surgery of lung abscess. Guy's Hosp Rep 92: 26–37

Brock R C 1944 A note on secondary diphtheritic infection of empyema and thoracotomy wounds. Guy's Hosp Rep 93: 62–66

Brock R C 1949 Surgery of pulmonary stenosis. Br Med J 2: 399–406

Brock R C 1952 Congenital pulmonary stenosis. Am J Med 12: 706–719

Brock R C 1954 The anatomy of the bronchial tree. Oxford University Press: London

Brock R C 1955 Control mechanisms in the outflow tract of the right ventricle in health and disease. Guy's Hosp Rep 104: 356–377

Brodal A 1940 Modification of the Gudden method for the study of cerebral localisation. Arch Neurol Psychiat 43: 46–58

Brodal A 1947 Central course of afferent fibers or pain in facial, glossopharyngeal, and vagus nerves. Clinical observations. Arch Neurol Psychiatry 57: 292–306

Brodal A 1949 Spinal afferents to the lateral reticular nucleus of the medulla oblongata in the cat. An experimental study. J Comp Neurol 91: 259–295

Brodal A 1957 The reticular formation of the brain stem. Anatomical aspects and functional correlations. Oliver and Boyd: Edinburgh

Brodal A 1969 Neurological anatomy. Oxford University Press: London

Brodal A, Angaut P 1967 The termination of spino vestibular fibres in the cat. Brain Res 5: 494–500

Brodal A, Pompeiano O 1957 The vestibular nuclei inthe cat. J Anat 91: 438–454

Brodal A, Walberg F 1952 Ascending fibres in the pyramidal tract of the cat. A Arch Neurol Psychiatry 68: 755–775

Brodal A. Pompeiano O, Walberg F 1962 The vestibular nuclei and their connexions, anatomical and functional correlations. Oliver and Boyd: Edinburgh

Brodal P 1968a The coricopontine projection in the cat. I. Demonstration of a somatotopically organized projection from the primary sensorimotor cortex. Exp Brain Res 5: 212–237

Brodal P 1968b The corticopontine projection in the cat. 2. Demonstration of a somatotopically organized projection from the second somatosensory cortex. Arch Ital Biol 106: 310–332

Brödel M 1901 The intrinsic blood-vessels of the kidney and their significance in nephrotomy. Johns Hopkins Hosp Bull 12: 10–13

Brodie A G 1941 On the growth pattern of the human head. From the third month to the eighth year of life. Am J Anat 68: 209–262

Brodmann K 1909 Vergliechende Lokaltastionslehre der Grosshirnrinde in ihren Prinzipien dargestellt auf Grund des Zellenbaues. Barth: Leipzig

Brokaw C J 1975 Molecular mechanism for oscillation in flagella and muscle. Proc Natl Acad Sci USA 72: 102–106

Bromage T G, Dean M C 1985 Re-evalution of the age at death of immature fossil hominids. Nature 317: 525–527

Broman I 1904 Die Entwickelungsgeschichte der Bursa Omentalis und ähnlicher Rezessbildungen bei den Wirbeltieren. Bergmann: Wiesbaden

Broman I 1938 Warum wird die Entwicklung der Bursa omentalis in Lehrbüchern fortwährend unrichtig beschrieben? Anat Anz 86: 195–202

Bronk D W, Ferguson L K 1935 Nervous control of intercostal respiration. Am J Physiol 110: 700–707

Brooke M H, Engel W K 1969 The histographic analysis of human muscle biopsies with regard to fiber types. 1. Adult male and female. Neurology 19: 221–233

Brooke M H, Kaiser K K 1970 Three human myosin ATPase systems and their importance in muscle pathology. Neurology 20: 404–405

Brooker B E 1984 An ultrastructural study of the sinus epithelium in the mammary gland of the lactating ewe. J Anat 138: 287–296

Brookes M 1958 The vascularisation of long bones in the human fetus. J Anat 92: 261–267

Brookes M 1963 Cortical vascularisation and growth in foetal tubular bones. J Anat 97: 597–609

Brookes M 1964 The blood supply of bones. In: Clark J M P (ed) Modern trends in orthopaedics Vol 4, Science of Fractures. Butterworths: London

Brookes M 1967 The osseous circulation. Biomed Eng 2: 294–299

Brookes M 1971 The blood supply of bone. An approach to bone biology. Butterworths: London

Brookes M, Harrison R G 1957 The vascularisation of the rabbit femur and tibio-fibula. J Anat 91: 61–72

Brookes M, Landon D N 1964 The juxta-epiphysial vessels in the long bones of foetal rats. J Bone Jt Surg 46B: 336–345

Brooks S T 1955 Skeletal age at death: reliability of cranial and pubic age indicators. Am J Phys Anthropol N S 13: 567–597

Broom N D 1984 Further insights into the structural principles governing the function of articular cartilage. J Anat 139: 275–294

Broom N D, Marra D L 1986 Ultrastructural evidence of fibril-to-fibril associations in articular cartilage and their functional implications. J Anat. 146: 185–200

Broom R 1901 On the structure and affinities of Udenodon. Proc Zool Soc Lond 2: 162–190

Broom R 1930. The origin of the human skeleton: An introduction to human osteology witherby: London

Brothwell D R 1967 Some problems and objectives related to the study of dental variation in human populations. J Dent Res 46: 938–941

Brothwell D R ed 1968 The skeletal biology of earlier human populations. (Symposia of the Society for the Study of Human Biology, Vol 8), Pergamon: Oxford

Brothwell D R, Carbonell V M, Goose D H 1963 In: Brothwell D R (ed) Dental anthropology. Pergamon: Oxford

Brouwer B 1918 Klinisch-anatomische untersuchung über den oculometoriuskern. Zentbl ges neurol psychiat 40: 152–189

Brouwer B, Zeeman W P C 1926 The projection of the retina in the primary optic neuron in monkeys. Brain 49: 1–35

Browder J, Kaplan H A 1976 Cerebral dural sinuses and their tributaries. Thomas: Springfield, Illinois

Brown A G 1973 Ascending and long spinal pathways: dorsal columns, spinocervical tract and spinothalamic tract. In: Iggo A (ed) Handbook of sensory physiology. Vol 2. Springer: Berlin: pp 315–338

Brown A G 1986 Sensory function. In: Thody A J, Friedmann P S (eds) Scientific basis of dermatology. A physiological approach. Churchill Livingstone: Edinburgh: pp 74–88

Brown A G, Gordon G 1977 Subcortical mechanisms concerned in somatic sensation. Br Med Bull 33: 121–128

Brown D D 1966 The nucleolus and synthesis of ribosomal RNA during oogenesis and embryogenesis of Xenopus laevis. Natl Cancer Inst Monogr 23: 297–309

Brown G M, Grota L J, Pulido O, Burns T G, Niles L P, Snieckus U 1983 Application of immunologic techniques to the study of pineal indolealkylamines. In: Reiter R J (ed) Pineal Research Reviews. Vol 1. Liss: New York: pp 207–246

Brown J R 1949 Localising cerebellar syndromes. J Am Med Assoc 141: 518–521

Brown J W 1950 Congenital heart disease. 2nd edition, Staples Press: London

Brown L T 1974 Coticorubral projections in the rat. J Comp Neurol 154: 149–168

Brown R L 1944 Rate of transport of spermia in human uterus and tubes. Am J Obstet Gynec 47: 407–411

Brown W J, Barajas L, Latta H 1970 The ultrastructure of the human adrenal medulla: with comparative studies of the white rat. Anat Rec 169: 173–183

Browne D 1928 The surgical anatomy of the tonsil. J Anat 63: 82–86

Browne D 1938 Diagnosis of undescended testicle. Br med J 2: 168–171

Brownell W E, Bader C R, Bertrand D, de Ribaupierre Y 1985 Evoked mechanical responses of isolated cochlear outer hair cells. Science 227: 194–196

Bruch C W 1967 Microbes in the upper atmosphere and beyond. In: Gregory PH, Monteith J L (eds) Airborne microbes. (Symp Soc Gen Microbiol 17)

Bruesch S R, Arey L B 1942 The number of myelinated and unmyelinated fibres in the optic nerve of vertebrates. J Comp Neurol 77: 631–665

Brun A 1965 The subpial granular layer of the foetal cerebral cortex in man. Acta Pathol Microbiol Immunol Scand [Suppl] 179: 1–98

Bruner J M 1951 Incisions for plastic and reconstructive (non-septic) surgery of the hand. Br J Plast Surg 4: 48–55

Bruns R R, Palade G E 1968 Studies on blood capillaries. I. General organization of blood capillaries in muscles. J Cell Biol 37: 244–276

Bryce T H, Teacher J H 1908 Contributions to the study of the early development and imbedding of the human ovum. An early ovum imbedded in the Decidua. J Maclehose: Glasgow

Buchan A M J, Polak J M, Facer P, Bloom S R, Szelke M, Hudson D, Pearse A G E 1977 Abstract: Use of synthetic fragments for specific immunostaining of CCK cells. Presented at the British Society of Gastroenterology, April 1977

Buckland-Wright J C 1977 Microradiographic and histological examination of the split-line formation in bone. J Anat 124: 193–203

Buckland-Wright J C 1978 Bone structure and the patterns of force transmission in the cat skull (Felis catus) J Morphol 155: 35–62

Bucy P C ed 1944 The precentral motor cortex. University of Illinois Press: Urbana, Ill

Buffa R, Capella C, Solcia E, Frigerio B, Saio S I 1977 Vasoactive intestinal peptide (VIP) cells in the pancreas and gastrointestinal mucosa. An immunohistochemical and ultrastructural study. Histochemistry 50: 217–227

Bülbring E, Shuba M F (eds) 1976 Physiology of smooth muscle. Raven Press: New York

Bulger R E 1965 The shape of rat kidney tubule cells. Am J Anat 116: 237–255

Bulger R E 1971 Ultrastructure of the junctional complexes from the descending thin limbs of the loop of Henle from rats. Anat Rec 171: 471–475

Bulger R E, Croning R F, Dobyan D C 1979 Survey of the morphology of the dog kidney. Anat Rec 194: 41–48

Bulger R E, Dobyan D C 1982 Recent advances in renal morphology. Annu Rev Physiol 44: 147–179

Buller A J 1970 The neural control of the contractile mechanism in skeletal muscle. Endeavour 29: 107–111

Buller A J 1974 The physiology of the motor unit. In: Walton J N (ed) disorders of voluntary muscle, 3rd edition. Churchill Livingstone: Edinburgh: pp 20–30

Bullough P, Goodfellow J 1968 The significance of the fine structure of articular cartilage. J Bone Jt Surg 50B: 852–857

Bullough W S 1967 The evolution of differentiation. Academic Press: New York

Bullough W S, Laurence E B 1960 The control of epidermal mitotic activity in the mouse. Proc R Soc Lond [Biol] 151: 517–536

Bullough W S, Laurence E B 1961 The study of mammalian epidermal mitosis in vitro. A critical analysis of technique. Exp Cell Res 24: 287–297

Bullough W S, Laurence E B, Iverson O H, Elgjo K 1967 The vertebrate epidermal chalone. Nature 214: 578–580

Bulmer D 1957 Observations on the development of the vaginal wall. J Anat 91: 599P

Bulmer D 1959 Histochemical observations on the foetal vaginal epithelium. J Anat 93: 36–42

Bulmer D 1965 A histochemical study of ovarian cholinesterases. Acta Anat 62: 254–265

Bulmer M G 1970 The biology of twinning in man. Clarendon Press: Oxford

Bumke O, Foerster O eds 1936 Handbuch der Neurologie. Springer-Verlag: Berlin

Bunge R P 1968 Glial cells and the central myelin sheath. Physiol Rev 48: 197–251

Bunge R P 1976 The expression of neuronal specificity in tissue culture. In: Barondes S H (ed) Neuronal recognition. Plenum Press: New York: pp 109–128

Bunnag S C, Bunnag S, Warner N E 1963 Microcirculation in the islets of langerhans of the mouse, Anat Rec 146: 117–123

Bunnell S 1970 Bunnell's Surgery of the Hand, 5th edition. Revised by J H Boyes. Lippincott: Philadelphia

Bunning P S C, Barnett C H 1963 Variations in the talocalcaneal articulation. J Anat 97: 643P

Bunning P S C, Barnett C H 1965 A comparison of adult and foetal talocalcaneal articulations. J Anat 99: 71–76

Buntine J A 1970 The omohyoid muscle and fascia: morphology and anomalies. Aust N Z J Surg 40: 86–88

Buranarugsa M, Houghton P 1981 Polynesian head form: an interpretation of a factor analysis of Cartesian co-ordinate data. J Anat 133: 333–350

Burdi A R, Faist K 1967 Morphogenesis of the palate in normal human embryos with special emphasis on the mechanisms involved. Am J Anat 120: 149–160

Burgen A S V, Mitchell J F 1978 Central nervous system transmitters. In: Burgen A S V, Mitchell J F M (eds) Gaddum's pharmacology, 8th edition. Oxford University Press: London: pp 24–32, 63–64

Burgess A M C 1981 Orientation of the cells of the paraxial mesoderm and somite formation. J Anat 132: 449P

Burgess A M C 1983 On the role of the notochord in somite formation and the possible evolutionary significance of the concomitant cell reorientation. J Anat 136: 829–835

Burgess P R, Perl E R 1973 Cutaneous mechanoreceptors and nociceptors. In: Iggo A (ed) Handbook of sensory physiology Vol 2 Somatosensory system. Springer-Verlag: Berlin: pp 29–78

Burke R E, Tsairis P 1973 Anatomy and innervation ratios in motor units of cat gastrocnemius. J Physiol 234: 749–765

Burkel W E 1970 The fine structure of the terminal branches of the hepatic arterial system of the rat. Anat Rec 167: 329–349

Burkitt A N, Lightoller G H S 1926–1927 The facial musculature of the Australian aboriginal. J Anat 61: 14–39, 62: 33–35

Burn J H 1968 The mechanism of the release of noradrenaline. In: Wolstenholme G E W, O'Conner M (eds) Adrenergic neurotransmission (CIBA Foundation Study Group 33). Churchill: London: pp 16–25

Burnard E D 1959 The cardiac murmur in relation to symptoms in the newborn. Br Med J 1: 134–138

Burnet F M 1969 Rok 1960 nagroda dia F M Burneta i P B Medawara za odkrycie nabytej tolerancji immunologiczne. Wlad Lek, 22: 505–506

Burns R K 1955 Urogenital system. In: Willier B H, Weiss P A, Hamburger V (eds) Analysis of development. Saunders: Philadelphia: pp 462–491

Burnside M B 1971 Microtubules and microfilaments in newt neurulation. Dev Biol 26: 416–441

Burnstock G 1970 Structure of smooth muscle and its innervation. In: Bülbring E, Brading A F, Jones A W, Tomita T (eds) Smooth Muscle. Arnold: London: pp 1–69

Burnstock G 1976 Control of smooth muscle activity in vessels by adrenergic nerves and circulating catecholamines. In: Noval M, Vassori G (eds) Smooth muscle pharmacology and physiology. INSERM Paris 50: pp 251–264

Burnstock G ed 1981a Current approaches to development of the autonomic nervous system: clues to clinical problems, In: Elliott K, Lawrenson G (eds) Development of the autonomic nervous system (Ciba Foundation Symposium 83). Pitman Medical: London: pp 1–14

Burnstock G 1981b Development of smooth muscle and its innervation. In: Bülbring E, Brading A F, Jones W, Tomita T (eds) Smooth muscle: an assessment of current knowledge. Arnold: London: pp 431–457

Burnstock G 1986 Autonomic neuromuscular junctions: current developments and future directions. J Anat 146: 1–30

Burt A M 1968 Acetylcholinesterase and choline acetyltransferase activity in the developing chick spinal chord. J Exp Zool 169: 107–112

Burt A M 1970 A histochemical procedure for the localization of choline acetyltransferase activity. J Histochem Cytochem 18: 408–415

Burton H, Loewy A D 1977 Projections to the spinal cord from medullary somatosensory relay nuclei. J Comp Neurol 173: 773–792

Buschard K, Kjaeldgaard A 1973 Investigations and analysis of the position, fixation, length and embryology of the vermiform appendix. Acta Chir Scand 139: 293–298

Bussolati G, Pearse A G 1967 Immunofluorescent localisation of calcitonin in the 'C' cells of pig and dog thyroid. J Endocrinol 37: 205–209

Buthpitiya A G, Stroud D, Russell C O H 1987 Pharyngeal pump and esophageal transit. Dig Dis Sci 32: 1244–1248

Butler C R 1966 Cortical lesions and interhemisphere communication in monkeys (Macaca mulatta). Nature 209: 59–61

Butler J 1957 The development of certain human dural venous sinuses. J Anat 91: 510–526

Butler H 1967 The development of mammalian dural venous sinuses with special reference to the post-glenoid vein. J Anat 102: 33–36

Büttner-Ennever J A 1977 Pathways from the pontine reticular formation to structurees controlling horizontal and vertical eye movements in the monkey. In: Baker R, Berthoz A Control of gaze by brain stem neurons proceedings of the symposium held. Paris 1977. (Developments in Neurosciences, Vol 1) Elsevier: Amsterdam

C

Cabanie M H, Javelle J 1966 La frange péritonéale sous-iléo-terminale. C R Hebd Séanc Acad Sci Paris 131: 248–252

Cairns J 1963 The bacterial chromosome and its manner of replication as seen by autoradiography. J Molec Biol 6: 208–213

Cajal S R see Ramón y Cajal S

Calasans O M 1953 Arquitetura do músulo ciliar no homen. An Fac Med Univ S Paulo 27: 3–98

Caldwell W E, Molcy H C 1933 Anatomical variations in the female pelvis and their effect in labor with suggested classification. Am J Obstet Gynecol 26: 479–505

Caldwell W E, Moloy H C, D'Esopo D A 1940 The more recent conceptions of the pelvic architecture. Am J Obstet Gynecol 40: 558–565

Calne D B 1970 Parkinsonism. Arnold: London

Cameron D A 1961 Erosion of the epiphysis of the rat tibia by capillaries. J Bone Jt Surg 43B: 590–594

Campian R, Minckler J 1976 A note on the gross configurations of the human auditory cortex. Brain and language 3: 318–323

Campbell A H, Liddelow A G 1967 Significant variations in the shape of the trachea and large bronchi. Med J Aust 54: 1017–1020

Campbell A W 1905 Histological studies on the localisation of cerebral function. Cambridge University Press: Cambridge

Campbell E J M 1955 The role of the scalene and sternomastoid muscles in breathing in normal subjects. An electromyographic study. J Anat 89: 378–386

Campbell E J M 1958 The respiratory muscles and the mechanics of breathing. Lloyd-Luke: London

Campbell M 1965 Causes of malformation of the heart. Br Med J 2: 895–904

Campos-Ortega J A, Glees P, Neuhoff V 1968 Ultrastructural analysis of individual layers in the lateral geniculate body of the monkey. Z Zellforsch 87: 82–100

Campos-Ortega J A, Hayhow W R, Clüver P F 1972 A note on the problem of retinal projections to the inferior pulvinar of primates. Brain Res 22: 126–130

Candiollo L 1965 Richerche anatomo-comparative sul musculo tensore del timpano, con riferimento alla innergazione propriocettiva. Z Zellforsch 67: 34–56

Candiollo L, and Levi A C 1969 Studies on the morphogenesis of the middle ear muscles in man. Arch Ohr-Nas Kehlkopf Heilk 195: 55–67

Candy J M, Perry R H, Thompson J E, Johnson M, Oakley A E 1985 Neuropeptide localisation in the substantia innominata and adjacent regions of the human brain. J Anat 140: 309–327

Cann R L, Stoneking M, Wilson A C 1986 Mitochondrial DNA and human evolution. Nature 325: 31–36

Cannieu A 1886 Recherche sur l'innervation de l'éminence thénar par le cubital. J Méd Bordeaux: 377–379

Cannon B, Nedergaard J 1985 Brown adipose tissue: molecular mechanisms controlling activity and thermogenesis. In: Cryer A, Van R L R (eds) New perspectives in adipose tissues: structure, function and development. Butterworths: London

Cannon W B, Britton S W 1925 Studies on the conditions of activity in endocrine glands. XV. Pseudaffective medulliadrenal secretion. Am J Physiol 72: 283–294

Cantin M, Tibault G, Ding J, Gutkousta J, Garcia R, Hamet P, Genest J 1987. Int J Nucl Med Biol 14: 313–322

Capella C, Solcia E, Frigerio B, Buffa R 1976 Endocrine cells of the human intestine. An ultrastructural study. In: Fujita T (ed) Endocrine gut and pancreas. Elsevier: New York: pp 43–60

Capella C, Solcia E, Frigerio B, Buffa R, Usellini L, Fontana P 1977 The endocrine cells of the pancreas. Ultrastructural study and classification. Virchows Arch Abt A 373: 327–352

Capen C C 1975 Functional and fine structural relationships of parathyroid glands. Adv Vet Sci Comp Med 19: 249–286

Capen C C, Roth S I 1973 Ultrastructural and functional relationships of normal and pathologic parathyroid cells. In: Ioachim H L (ed) Pathobiology annual 1973 Vol 3. Appleton-Century-Crofts: New York: pp 129–175

Cardinali D P, Larin F, Wurtman R J 1971 Action spectra for effects of light on hydroxyindole-o-methyltransferases in rat pineal, retina and harderian glands. Endocrinology 91: 877–886

Carey E J 1921 Studies on the structure and function of the small intestine. Anat Rec 21: 189–216

Carey E J, Zeit W, McGrath B F 1927 Studies in the dynamics of histogenesis. XII. The regeneration of the patellae of dogs. Am J Anat 40: 127–158

Carlon N, Stahl A 1973 Les premiers stades du développement des gonades chez l'homme et les vertébrés supérieurs. Pathol Biol (Paris) 21: 903–914

Carlsen F, Behse F 1980 Three dimensional analysis of Schwann cells associated with unmyelinated nerve fibres in human aural nerve. J Anat 130: 545–547

Carlson B M 1973 The regeneration of skeletal muscles. A review. Am J Anat 137: 119–149

Carlstedt T 1977 Observations on the morphology at the transition between the peripheral and the central nervous system in the cat. I–V. Acta Physiol Scand 100: Suppl 446, 5–85

Carlström D 1964 Polarization microscopy of dental enamel with reference to incipient carious lesions. Adv Oral Biol 1: 255–296

Carlström D, Engström H, Hjorth S 1953 Electron microscopic and X-ray diffraction studies of statoconia. Laryngoscope 63: 1052–1057

Carman J B, Cowan W M, Powell T P S 1963 The organisation of the cortico-striate connexions in the rabbit. Brain 86: 525–562

Carman J B, Cowan W M, Powell T P S 1964 Cortical connexions of the thalamic reticular nucleus. J Anat 98: 587–598

Carman J B, Cowan W M, Powell T P S, Webster K E 1965 A bilateral cortico-striate projection. J Neurol Neurosurg Psychiat 28: 71–77

Carmel P, Starr A 1963 Acoustic and non-acoustic factors modifying middle-ear muscle activity in waking cats. J Neurophysiol 26: 598–616

Carpenter D, Lundberg A, Norrsell U 1963 Primary afferent depolarisation evoked from the sensorimotor cortex. Acta Physiol Scand 59: 126–142

Carpenter M B 1950 Athetosis and basal ganglia; review of the literature and a study of forty-two cases. Arch Neurol Psychiatry 63: 875–901

Carpenter M B, Britten G M 1958 Subthalamic hyperkinesia in the rhesus monkey: effects of secondary lesions in the red nucleus and brachium conjunctivum. J Neurophysiol 21: 400–413

Carpenter M B, Carpenter C S 1951 Analysis of somatopic relations of the corpus Luysi in man and monkey. Relation between the site of dyskinesia and distribution of lesions within the subthalamic nucleus. J Comp Neurol 95: 349–370

Carpenter M B, Peter P 1971 Accessory oculomotor nuclei in the monkey. J Hirnforsch 12: 405–418

Carpenter M B, Whittier J R 1952 Study of methods for producing experimental lesions of the central nervous system with special reference to stereotaxic technique. J Comp Neurol 97: 73–132

Carpenter M B, Whittier J R, Mettler F A 1950 Analysis of choreoid hyperkinesia in the rhesus monkey. Surgical and pharmacological analysis of hyperkinesia resulting from lesions in the subthalamic nucleus of Luys. J Comp Neurol 92: 293–332

Carpenter M B, Correll J W, Hinman A 1960 Spinal tracts mediating subthalmic hyperkinesia. Physiological effects of selective partial cordotomies upon dyskinesia in the rhesus monkey. J Neurophysiol 23: 288–304

Carpenter M B, Stein B M, Shriver J E 1968 Central projections of spinal dorsal roots in the monkey. II. Lower thoracic, lumbosacral and coccygeal dorsal roots. Am J Anat 123: 75–118

Carpenter M B, Nakano K, Kim R 1976 Nigrothalamic projections in the monkey demonstrated by autoradiographic technics. J Comp Neurol 165: 401–415

Carr I 1967 Nuclear membranous whorls. Z Zellforsch 80: 140–144

Carr I 1970 The fine structure of the mammalian lymphoreticular system. Int Rev Cytol 27: 283–348

Carr J, Clegg E J, Meek G A 1968 Sertoli cells as phagocytes: an electron microscopic study. J Anat 102: 501–510

Carter R B, Keen E N 1971 The intramandibular course of the inferior alveolar nerve. J Anat 108: 433–440

Case R M 1979 Pancreatic secretion: cellular aspects. In: Duthi H L, Wormsley K G (eds) Scientific basis of gastroenterology. Churchill Livingstone: Edinburgh: pp 163–198

Caspersson T 1936 Ueber den chemischen Aufbau der Strukturen des Zellkernes. Skand Arch Physiol Suppl 8 73:

Caspersson T, Farber S, Foley G E, Kudynowski J, Modest E J, Simonsson E, Wagh V, Zech L 1968 Chemical differentiation along metaphase chromosomes. Exp Cell Res 49: 219–222

Castelli W A 1963 Vascular architecture of the human adult mandible. J Dent Res 42: 786–792

Castleman B 1966 The pathology of the thymus gland in myasthenia gravis. Ann N Y Acad Sci 135: 496–505

Castor C W 1962 The microscopic structure of normal human synovial tissue. Arthritis Rheum 3: 140–151

Catton W T, Gray J E 1951 Electromyographic study of the action of the serratus anterior muscle in respiration. J Anat 85: 412P

Cauldwell E W, Siekert R G, Lininger R E, Anson B J 1948 Bronchial arteries. Anatomic study of 150 human cadavers. Surg Gynecol Obstet 86: 395–412

Caulfield J P, Farquhar M G 1974 The permeability of glomerular capillaries to graded dextrans. Identification of the basement membrane as the primary filtration barrier. J Cell Biol 63: 883–903

Cauna N 1966 Fine structure of the receptor organs and its probable functional significance. In: de Reuck A V S, Knight J (eds) Touch, heat and pain (Ciba Foundation Symposium). Churchill: London: pp 117–127

Cauna N, Hinderer K H 1969 Fine structure of blood vessels of the human nasal respiratory epithelium. Ann Otol Rhinol Laryngol 78: 865–879

Cauna N, Mannan G 1959 Developmental and post-natal changes of digital Pacinian corpuscles (Corpuscula lamellosa) in the human hand. J Anat 93: 271–286

Cauna N, Manzetti G W, Hinderer K H, Swanson E W 1972 Fine structure of nasal polyps. Ann Otol Rhinol Laryngol 81: 41–58

Cavallero C, Spagnoli L G, Villaschi S 1976 An electron microscopic study of human pancreatic islets. In: Fujita T (ed) Endocrine gut and pancreas. Elsevier: New York: pp 61–72

Cavatori P 1908 Il tipo normale e le variazioni delle arterie della base dell'encefalo nell'uomo. Monitore Zool Ital 19: 248–258

Cave A J 1965 The bursa pharyngea in the giant panda (Ailuropoda melanoleuca). Nature 208: 865–867

Cave A J E 1929 The distribution of the first intercostal nerve and its relation to the first rib. J Anat 63: 367–379

Cave A J E 1934 On the occipito-atlanto-axial articulations. J Anat 68: 416–423

Cave A J E 1937 The innervation and morphology of the cervical intertransverse muscles. J Anat 71: 497–515

Cave A J E 1958 Ancient Egypt and the origin of anatomical science. Proc R Soc Med 43: 568–571

Cave A J E 1961 The nature and morphology of the costoclavicular ligament. J Anat 95: 170–179

Cave A J E 1975 The morphology of the mammalian cervical pleurapophysis. J Zool Lond 177: 377–393

Cave A J E, Brown R W 1952 On the tendon of the subclavius muscle. J Bone Jt Surg 34B 466–469

Cave A J E, Porteous C J 1958 The attachments of m semimembranosus. J Anat 92: 638P

Cave A J E, Griffiths J D, Whiteley M M 1955 Osteo-arthritis deformans of Luschka joints. Lancet 1: 176–179

Cawley J C, Hayhoe F G J 1973 Ultrastructure of haemic cells: a cytological atlas of normal and leukaemic blood and bone marrow. Saunders: London

Celtis A, Porter A J 1952 Lymphatics of the thorax. Acta Radiol 38: 461–470

Chabardes D, Imbert M, Clique A, Montegut M, Morel F 1975 PTH sensitive adenyl cyclase activity in different segments of the rabbit nephron. Pflügers Arch 354: 229–239

Chacko L W 1955 The lateral geniculate body of the chimpanzee. J Anat Soc Ind 4: 10–13

Challice C E, Viragh S 1973 Ultrastructure of the mammalian heart. Academic Press: New York

Chamberlain D W, Nopajaroonsri C, Simon G T 1973 Ultrastructure of the pulmonary lymphoid tissue. Am Rev Respir Dis 108: 621–631

Chamberlain H D 1928 The inheritance of left-handedness. J Hered 19: 557–559

BIBLIOGRAPHY

Chambers M R, Andres K H, Duering M von, Iggo A 1972 The structure and function of the slowly adapting type II receptor in hairy skin. Q J Exp Physiol 57: 417–445

Chambers T J, Dunn C J 1984 Pharmacological control of osteoclastic motility. Calcif Tissue Int 35: 566–570

Chambers W W, Liu C-N 1957 Cortico-spinal tract of the cat. An attempt to correlate the pattern of degeneration with deficits in reflex activity following neocortical lesions. J Comp Neurol 108: 23–56

Chambers W W, Sprague J M 1955a Functional localization in cerebellum; organization in longitudinal cortico-nuclear zones and their contribution to control of posture, both extrapyramidal and pyramidal. J Comp Neurol 103: 105–129

Chambers W W, Sprague J M 1955b Functional localization in cerebellum; somatotopic organization in cortex and nuclei. Arch Neurol Psychiatry 74: 653–680

Champetier J, Descours C 1968 The branches of the posterior tibial nerve in the tibiotarsal joint. C R Assoc Anat 141: 677–685

Chang H T 1951 Caudal extension of Clarke's column in the spider monkey. J Comp Neurol 95: 43–51

Chang L, Blair W F 1985 The origin and innervation of the adductor pollicis muscle. J Anat 140: 381–388

Chang M C 1951 Fertilizing capacity of spermatozoa deposited into fallopian tubes. Nature 168: 697–698

Chang M C, Hunter R H F 1975 Capacitation of mammalian sperm: biological and experimental aspects. In: Hamilton D W, Greep R O (eds) Handbook of physiology Section 7, Endocrinology Vol V. American Physiological Society: Washington, DC: pp 339–352

Chan-Palay V 1983 Gamma-aminobutyric acid pathways in the cerebellum studied by retrograde and anterograde transport of glutamic acid decarboxylase (GAD) antibody after in vivo injections. Progr Brain Res 55: 51–76

Chan-Palay V, Palay S L, Billings-Gagliardi S M 1974 Maynert cells in the primate visual cortex. J Neurocytol 3: 631–658

Chang H T 1959 The evoked potentials. In: Field J, Magoun H W (eds) Handbook of physiology, Section 1 Neurophysiology, Vol 1. American Physiological Society: Washington DC: pp 299–313

Chao L P, Wolfgram F 1973 Purification and some properties of choline acetyltransferase (EC 2.3.1.6) from bovine brain. J Neurochem 20: 1075–1081

Chaplin D M, Greenlee T K 1975 The development of human digital tendons. J Anat 120: 253–274

Chapman G B, Chiardo A J, Coffey R J, Weineke K 1966 The fine structure of mucosal epithelial cells of a pathological human gall bladder. Anat Rec 154: 579–616

Charcot J M 1883 Lecture on localization of cerebral and spinal disease. Edited and translated by M A Hodden. New Sydenham Society: London

Charnley J 1959 The lubrication of animal joints. In: Proceedings of the Symposium on Biomechanics, held London 1959. Institution of Mechanical Engineers: London: pp 12–19

Chase L R, Aurbach G D 1967 Parathyroid function and the renal excretion of 3' 5'-adenylic acid. Proc Natl Acad Sci USA 58: 518–525

Chaudhry A P, Cutler L S, Yamane G M, Labay G R, Sunderraj M, Manak J R Jr 1987 Ultrastructure of normal human parotid gland with special emphasis on myoepithelial distribution. J Anat 152: 1–11

Chau-Pham T T 1978 The oplate receptors and the discovery of opioid-like peptides. Drug Metab Rev 7: 255–294

Chayen D, Nathan H 1974 Anatomical observations of the subgaleotic fascia. Acta Anat 87: 427–432

Chen I-Li, Yates R D 1970 Ultrastructural studies of vagal paraganglia in Syrian hamsters. Z Zellforsch 108: 309–323

Chen I-Li, Mascorro J A, Yates R D 1976 Morphology and functional considerations of the carotid body and paraganglia. In: Coupland R E, Fujita T (eds) Chromaffin, enterchromaffin and related cells. Elsevier: New York: pp 333–353

Chen J M 1952 Studies on the morphogenesis of the mouse sternum. I. Normal embryonic development. J Anat 86: 373–386

Chen L, Weiss L 1972 Electron microscopy of the red pulp of the human spleen. Am J Anat 134: 425–458

Chen L, Weiss L 1973 The role of the sinus wall in the passage of erythrocytes through the spleen. Blood 41: 529–537

Chiara F 1959 Study of the fine innervation of the female genitalia. I. Uterus. Annali Ostet Ginecol 81: 553–576

Chiba T, Yamauchi A 1970 On the fine structure of the nerve terminals in the human myocardium. Z Zellforsch 108: 324–338

Chibon P 1967 Etude expérimentale par ablations, greffes et autoradiographie, de l'origine des dents chez l'amphibien urodele pleurodeles waltlii. Arch Oral Biol 12: 745–753

Chiechi M A, Lees W M, Thompson R 1956 Functional anatomy of normal mitral valve. J Thorac Surg 32: 378–398

Child C M 1927 Modification of polarity and symmetry in Corymorpha palma by means of inhibiting conditions and differential exposure; forms resulting from modification. J Exp Zool 47: 343–383

Child C M 1941 Patterns and problems of development. Chicago University Press: Chicago

Childress D S, Jones R W 1967 Mechanics of horizontal movement of the human eye. J Physiol 188: 273–284

Chiquoine A D 1959 Electron microscopic observations on the developmental cytology of the mammalian ovum. Anat Rec 133: 258–259

Chisholm G D, Williams D I eds 1982 Scientific foundations of urology. 2nd edition. Heinemann Medical: London

Chomsky N 1965 Aspects of the Theory of Syntax. M I T Press: Cambridge, Massachusetts

Chow K-L, Blum J S, Blum R A 1950 Cell ratios in the thalamo-corti-calvisual system of Macaca mulatta. J Comp Neurol 92: 227–239

Christie G A 1963 The development of the limbus fossae ovalis in the human heart – a new septum. J Anat 97: 45–54

Chrzanowska G, Beben A 1973 Weight of the brain and body weight in man between the ages of 20 and 89 years. Folia Morphol 32: 391–406

Chrzanowska G, Krechowiecki A 1975 Hängt das Gehirngewicht von der Körperlänge ab? Gegenbaurs Morphol Jahrb 121: 192–208

Chu C H U 1968 Solitary neurons in human tongue. Anat Rec 1962: 505–510

Chung S-H, Gaze R M, Keating M J 1972 The technique and results of mapping the visual projection in Xenopus tadpoles of various stages. J Physiol (Lond) 222: 137P

Church R B, Schultz G A 1974 Differential gene activity in the pre- and postimplantation mammalian embryo. Curr Top Dev Biol 8: 179–202

Chu-Wu T, Wen-Kuei W 1965 Further observations on the development and connections of the mesencephalic nucleus of the trigeminal nerve in human brain. Acta Anat Sin 8: 352–372

Čihák R 1970 Variations of limbosacral joints and their morphogenesis. Acta Univ Carol [Med] (Praha) 16: 145–165

Čihák R 1972 Ontogeny of the skeleton and intrinsic muscles of the human hand and foot. Ergeb Anat Entwicklungsgesch 46: 5–194

Čihák R, Popelka S 1961 Částečné defekty velkého svalu prsního. Morfologická a kliniká studie. Acta Chir Orthop Traum Cech 28: 185–194

Ciochon R L, Corruccini R S 1977 The coraco-acromial ligament and projection index in man and other anthropoid primates. J Anat 124: 627–632

Clara M 1937 Zur Histobiologie des Bronchalepithels. Z Mikrosk 41: 321–347

Clark E R 1938 Arterio-venous anastomoses. Physiol Rev 18: 229–247

Clark G L ed 1961 The encyclopedia of microscopy. Reinhold: New York

Clark R A F 1985 Cutaneous tissue repair: basic biologic considerations. I. J Am Acad Dermatol 13: 701–725

Clark R A F, Lanigan J M, DellaPelle P, Manseau E, Dvorak H F, Colvin R B 1982 Fibronectin and fibrin provide a provisional matrix for epidermal cell migration during wound reepithelialization. J Invest Dermatol 79: 264–269

Clark S L 1934 Innervation of the choroid plexuses and the blood vessels within the central nervous system. J Comp Neurol 60: 21–36

Clark W E le G 1920 On the Pacchionian bodies. J Anat 55: 40–48

Clark W E le G 1926 The mammalian oculomotor nucleus. J Anat 60: 426–448

Clark W E le G 1932 A morphological study of the lateral geniculate body. Br J Ophthalmol 16: 264–284

Clark W E le G 1933 The medial geniculate body and the nucleus isthmi. J Anat 67: 536–548

Clark W E le G 1936a The thalamic connections of the temporal lobe of the brain in the monkey. J Anat 70: 447–464

Clark W E le G 1936b The topography and homologies of the hypothalamic nuclei in man. J Anat 70: 203–214

Clark W E le G 1937 The termination of ascending tracts in the thalamus of the macaque monkey. J Anat 71: 7–40

Clark W E le G 1940 Vascular mechanism related to great vein of Galen. Br Med J 1: 476

Clark W E le G 1941 The laminar organisation and cell content of the lateral geniculate body in the monkey. J Anat 75: 419–433

Clark W E le G 1945 Deformation patterns in the cerebral cortex. In: Clark W E le G, Medawar P B (eds) Essays on growth and form. Clarendon Press: Oxford: pp 1–22

Clark W E le G 1949 Laminar pattern of lateral geniculate nucleus considered in relation to colour vision. Documenta Ophthalmol 3: 57–64

Clark W E le G 1951 The projection of the olfactory epithelium on the olfactory bulb in the rabbit. J Neurol Neurosurg Psychiatry 14: 1–10

Clark W E le G 1957 Inquiries into the anatomical basis of olfactory discrimination. Proc R Soc Lond [Biol] 146: 299–319

Clark W E le G, Boggon R H 1933 On the connections of the medial cell groups of the thalamus. Brain 56: 83–98

Clark W E le G, Meyer M 1947 The terminal connexions of the olfactory tract in the rabbit. Brain 70: 304–328

Clark W E le G, Penman G C 1934 The projection of the retina in the lateral geniculate body. Proc R Soc Lond [Biol] 114: 291–313

Clark W E le G, Powell T P S 1953 On the thalamocortical connexions of general sensory cortex of Macaca. Proc R Soc Lond [Biol] 141: 467–487

Clark W E le G, Russell W R 1939 Observations on the efferent connexions of the centre median nucleus. J Anat 73: 255–262

Clark W E le G, Beattie J, Riddoch G, Dott N M 1938 The hypothalamus, morphological, functional, clinical and surgical aspects. Oliver and Boyd: Edinburgh

Clarke C A 1975 Rhesus haemolytic disease. Selected papers and extracts. Medical and Technical Publishing: Lancaster

Clarke E, O'Malley C D 1968 The human brain and spinal cord. University of California Press: Los Angeles, Calif

Clarke E C, Bearn J G 1968 The brain 'glands' of Malpighi elucidated by practical history. J Hist Med Allied Sci 23: 309–330

Clarke I C 1973a Correlation of SEM replication and light microscopy studies of the bearing-surfaces in human joints. In: Scanning Elecron Mocroscopy/1973 (Part III). Proceedings of the workshop on scanning electron microscopy in pathology (Johari O, Corvin I eds) pp 659–666, IIT Research Institute: Chicago, Illinois

Clarke I C 1973b Quantitative measurement of human articular surface topography 'in vitro' by profile recorder and stereomicroscopy techniques. J Microsc 97: 309–314

Claude A 1970 Growth and differentiation of cytoplasmic membranes in the course of lipoprotein granule synthesis in the hepatic cell. J Cell Biol 47: 745–766

Clausen H, Vedtofte P, Moe D, Dadelsteen E, Sun T-T, Dale B 1986 Differentiation-dependent expression of keratins in human oral epithelia. J Invest Dermatol 86: 249–254

Clawson R C, Domm L V 1969 Origin and early migration of primordial germ cells in the chick embryo: A study of the stages definitive primitive streak through eight somites. Am J Anat 125: 87–112

Clay R S, Court T H 1932 The history of the microscope. Griffin: London

Clegg E J 1955 The arterial supply of the human prostate and seminal vesicles. J Anat 89: 209–216

Clegg E J 1956 The vascular arrangements within the human prostate gland. Br J Urol 28: 428–435

Clementi F, Palade G E 1969 Intestinal capillaries. I. Permeability to peroxidase and ferritin. J Cell Biol 41: 33–58

Clermont Y 1963 The cycle of the seminiferous epithelium in man. Am J Anat 112: 35–52

Clermont Y 1966 Renewal of spermatogonia in man. Am J Anat 118: 509–524

Clermont Y 1969 Two classes of spermatogonial stem cells in the monkey (Cercopithecus aethiops) Am J Anat 126: 57–72

Clermont Y 1972 Kinetics of spermatogenesis in mammals: seminiferous epithelium cycle and spermatogonial renewal. Physiol Rev 52: 198–236

Clermont Y, Huckins C 1961 Microscopic anatomy of the sex cords and seminiferous tubules in growing and adult male albino rats. J Anat 108: 79–97

Clermont Y, Leblond C P 1955 Spermiogenesis of man, monkey, ram and other mammals as shown by the 'periodic acid-Schiff' technique. Am J Anat 96: 229–254

Cliff W J 1976 Blood vessels. Cambridge University Press: Cambridge

Close R I 1972 Dynamic properties of mammalian skeletal muscles. Physiol Rev 52: 129–197

Cnóckaert J C, Pertuzon E 1974 Sur la géometrie musculosquelettique du triceps brachii. Application à la détermination dynamique de sa compliance. Europ J Appl Physiol 32: 149–158

Coakley J B, King T S 1959 Cardiac muscle relation of the coronary sinus, the oblique vein of the left atrium and the left precaval vein in mammals. J Anat 93: 30–35

Cobb J L S, Bennett T 1969 A study of nexuses in visceral smooth muscle. J Cell Biol 41: 287–297

Cockett F B 1956 Diagnosis and surgery of high-pressure venous leaks in the leg. Br Med J 2: 1399–1403

Code C F ed 1968 Handbook of physiology, Section 6, alimentary canal, Vol 4 Motility, American Physiological Society: Washington DC

Coërs C, Woolf A L 1959 The innervation of muscle. A biopsy study. Blackwell Scientific: Oxford

Cogan D G 1956 Neurology of the ocular muscles. 2nd edition. Thomas: Springfield, Illinois

Cogan D G, Kuwabara T 1967 The sphingolipidoses and the eye. Arch Ophthal 79: 437–452

Coggeshall R E, Coulter J D, Willis W D 1973 Unmyelinated fibres in the ventral root. Brain Res 57: 229–233

Coghill G E 1929 Anatomy and the problem of behaviour. Cambridge University Press: Cambridge

Cohen A I 1970 Further studies on the question of the patency of saccules in outer segments of vertebrate photoreceptors. Vision Res 10: 445–453

Cohen A I 1972 Rods and cones. In: Fuortes M G F (ed) Handbook of sensory physiology. Vol 7. Springer-Verlag: Berlin: pp 63–110

Cohen A S, McNeill M, Calkins E, Sharp J T, Schubart A 1967 The 'normal' sacroiliac joint: analysis of 88 sacroiliac roentgenograms. Am J Roentgenol 100: 559–563

Cohen D, Chambers W W, Sprague J M 1958 Experimental study of the efferent projections from the cerebellar nuclei to the brain stem of the cat. J Comp Neurol 109: 233–259

Cohen J, Harris W H 1958 The three dimensional anatomy of Haversian systems. J Bone Jt Surg 40A: 419–434

Cohen L 1959 Venous drainage of the mandible. Oral Surg 12: 1447–1449

Cohen S 1958 A nerve growth-promoting protein. In: McElroy W D, Glass B (eds) A symposium on the chemical basis of development. Johns Hopkins Press: Baltimore: pp 665–679

Cohen S 1965 The stimulation of epidermal proliferation by a specific protein (EGF) 1965. Develop Biol 12: 394–407

Cohen S, Levi-Montalcini R 1956 A nerve growth-stimulating factor isolated from snake venom. Proc Natl Acad Sci USA 42: 571–574

Cokkinis A J 1930 Observations on the mesenteric circulation. J Anat 64: 200–205

Code F J 1944 A history of comparative anatomy. From Aristotle to the Eighteenth Century. Macmillan: London

Cole P 1954 Recordings of respiratory air temperature. J Laryngol Otol 68: 295–307

Coleman S S, Anson B J 1961 Arterial patterns in the hand based upon a study of 650 specimens. Surg Gynecol Obstet 113: 409–424

Collet A, Normand-Reuet C 1967 Aspects infrastructureaux de la traversée de la paroi alvéolaire du poumon par des cellules migratrices. Sém Hôp Paris 43: 1928–1937

Collet A, Basset F, Normand-Reuet C 1967 Etude au microscope électronique du poumon humain normal et pathologique. Poumon Coeur 23: 747–785

Collier J, Buzzard E F 1903 The degenerations resulting from lesions of posterior nerve roots and from transverse lesions of the spinal cord in man. A study of 20 cases. Brain 26: 559–591

Collins D C 1932 Adenomatous polyps of vermiform appendix. S Clin N Am 12: 1063–1067

Collins P, Woollam D H M 1981 The circumventricular organs. In: Hamson R J, Holmes R L (eds) Progress in anatomy. Vol 1. Cambridge University Press: Cambridge: pp 123–139

Collis J L, Satchwell L M, Abrams L D 1954 Nerve supply to the diaphragm. Thorax 9: 22–25

Colonnier M 1964 The tangential organization of the visual cortex. J Anat 98: 327–344

Colonnier M 1968 Synaptic patterns on different cell types in the different laminae of the cat visual cortex. Brain Res 9: 268–287

Colonnier M 1974 Spatial inter-relationships as physiological mechanisms in the central nervous system. In: Bellairs R, Gray E G (eds) Essays on the nervous system. Clarendon Press: Oxford: pp 344–366

Colonnier M, Guillery R W 1964 Synaptic organisation in the lateral geniculate nucleus of the monkey. Z Zellforsch 62: 333–335

Colonnier M L 1966 The structural design of the neocortex. In: Eccles J C (ed) Brain and conscious experience. Spring-Verlag: Berlin: pp 1–18

Combelles R 1972 Vascularization de la cavité buccale. Arch Anat Histol Embyrol 55: 179–208

Comings D E, Okada T 1972 Architecture of meiotic cells and mechanisms of chromosome pairing. In: Dupraw E J (ed) Advances in cell and molecular biology. Vol 2. Academic Press: New York: pp 309–384

Condé F, Condé H 1973 Etude de la morphologie des cellules du noyau rouge du chat par la méthode de Golgi-Cox. Brain Res 53: 249–271

Conel J L 1939–1959 The post-natal development of the humal cerebral cortex. Vol I–VI. Harvard University Press: Cambridge, Massachusetts

Conel J L 1942 The origin of the neural crest. J Comp Neurol 76: 191–216

Congdon E D 1922 Transformation of the aortic-arch system during the development of the human embryo. Contrib Embryol Carnegie Inst Washington 14: 47–110

Conklin J L 1962 Cytogenesis of the human fetal pancreas. Am J Anat 111: 181–193

Conklin J L 1968 The development of the human fetal adenohypophysis. Anat Rec 160: 79–92

Conn H J 1948 The history of staining. Biotechnical Publications: Geneva

Connell C J, Christensen A K 1975 The ultrastructure of the canine testicular interstitial tissue. Biol Reprod 12: 368–382

Connor J D 1968 Caudate unit responses to nigral stimuli: evidence for a possible nigro-neostriatal pathway. Science 160: 899–900

Connors T A 1975 Cytotoxic agents in teratogenic research. In: Teratology trends and applications (Berry C L, Poswillo D E eds) pp 49–79 Springer-Verlag: Berlin

Constantinesco A 1974 Etude biomécanique des mouvements du sternum chez l'homme adulte. Arch Anat 57: 153–199

Constantinou C E, Djurhuus J C 1981 Pyeloureteral dynamics in the intact and chronically obstructed multicalyceal kidney. Am J Physiol 241: R398–411

Conway J H 1938 Notes on cutaneous healing in wounds. Surg Gynec Obstet 66: 140–144

Cooke J 1975 Experimental analysis and a theory of the control of somite number during amphibian morphogenesis. Squaw valley winter conference on developmental biology, Walter Benjamin: Menlo Park, Calif

Cooke J, Zeeman E C 1976 A clock and wavefront model for control of the number of repeated structures during animal morphogenesis. J Theor Biol 58: 455–476

Cooke P 1976 A filamentous cytoskeleton in vertebrate smooth muscle fibers. J Cell Biol 68: 539–556

Coombs J S, Curtis D R, Landgren S 1956 Spinal cord potentials generated by impulses in muscle and cutaneous afferent fibres. J Neurophysiol 19: 452–467

Cooper E R A 1929 The histology of the retained tests in the human subject at different ages, and its comparison with the scrotal testis. J Anat 64: 5–27

Cooper E R A 1945 The development of the human lateral geniculate body. Brain 68: 222–242

Cooper E R A 1946 Development of human red nucleus and corpus striatum. Brain 69: 34–43

Cooper E R A 1950 The development of the thalamus. Acta Anat 9: 201–226

BIBLIOGRAPHY

Cooper E R A 1958 Nerves of the meninges and choroid plexus. In: Wolstenholme E E W, O'Connor C M (eds) The cerebrospinal fluid (Ciba Foundation Symposium). Churchill: London: pp 80–91

Cooper R R 1968 Nerves in cortical bone. Science 160: 327–328

Cooper S 1953 Muscle spindles in intrinsic muscles of the human tongue. J Physiol 122: 193–202

Cooper S, Daniel P D, Whitteridge D 1955 Muscle spindles and other sensory endings in the extrinsic eye muscles; the physiology and anatomy of these receptors and of their connexions with the brain stem. Brain 78: 564–583

Cope V Z 1917 The internal structure of the sphenoidal sinus. J Anat 51: 127–136

Copp D H, Talmage R V eds 1978 Endocrinology of calcium metabolism (International Congress Series No 421) Elsevier: New York

Copp D H, Cockcroft D W, Kueh Y 1967 Calcitonin from ultimobranchial glands of dogfish and chickens. Science 158: 924–925

Corazza R, Fadiga E, Parmeggiani P L 1963 Patterns of pyramidal activation of cat's motoneurons. Arch Ital Biol 101: 337–364

Corbin K B, Harrison F 1940 Function of mesencephalic root of the fifth cranial nerve. J Neurophysiol 3: 423–435

Cordier A C, Haumont S M 1980 Development of thymus, parathyroids, and ultimo-branchial bodies in NMRI and nude mice. Am J Anat 157: 227–263

Corner G W 1938 Quantitative studies of experimental menstruation-like bleeding due to hormone deprivation. Am J Physiol 124: 1–12

Cornes J S 1965 Number, size and distribtuion of Peyer's patches in the human small intestine. II. The development of Peyer's patches. Gut 6: 225–229

Corrin B 1970 Phagocytic potential of pulmonary alveolar epithelium with particular reference to surfactant metabolism. Thorax 25: 110–115

Corsellis J A, Alston R L, Miller A K 1975 Cell counting in the human brain: traditional and electronic methods. Postgrad Med J 51: 722–726

Cory R A, Valentine E J 1959 Varying patterns of the lobar branches of the pulmonary artery. A study of 524 lungs and lobes seen at operation on 426 patients. Thorax 14: 267–280

Costanzo R M, Graziadei P P C 1983 A quantitative analysis of changes in the olfactory epithelium following bulbectomy in hamster. J Comp Neurol 215: 370–381

Costoff A 1977 Ultrastructure of the pituitary gland. In: Allen M B Jr, Mahesh V B (eds) The pituitary, a current review. Academic Press: New York: pp 59–76

Cottle M H 1960 Concepts of nasal physiology as related to corrective nasal surgery. Arch Otolaryngol 72: 11–20

Couinaud C 1954 Les enveloppes vasculo-biliaires du foie ou capsule de Glisson. Lyon Chir 49: 489–607

Coulombre A J 1964 Problems in corneal morphogenesis. In: Abercrombie M, Brachet J (eds) Advances in morphogenesis. Vol 4. Academic Press: New York: pp 81–109

Coulton L A, Henderson B, Bitensky L, Chayen J 1980 DNA synthesis in human rheumatoid and nonrheumatoid synovial lining. Ann Rheum Dis 39: 241–247

Coupland R E 1958 The innervation of the pancreas of the rat, cat and rabbit as revealed by the cholinesterase technique. J Anat 92: 143–149

Coupland R E 1965a Electron microscope observations on the structure of the rat adrenal medulla. I. The ultrastructure and organisation of chromaffin cells in the normal adrenal medulla. II. Normal innervation. J Anat 99: 231–254, 255–272

Coupland R E 1965b The natural history of the chromaffin cell. Longman: London

Coupland R E, Fujita T ed 1976 Chromaffin enterochromaffin and related cells. Elsevier Scientific: Amsterdam

Coupland R E, Weakley B S 1970 Electron microscopic observation on the adrenal medulla and extra-adrenal chromaffin tissue of the post-natal rabbit. J Anat 106: 213–231

Coupland R E, Pyper A S, Hopwood D 1964 A method for differentiating between noradrenaline- and adrenaline-storing cells in the light and electron microscope. Nature 201: 1240–1242

Courot M, Hochereau de Reviers M-T, Ortevant R 1970 Spermatogenesis. In: Johnson A D, Gomes W R, Vandemark N L (eds) The testis. Vol 1. Academic Press: New York: pp 339–432

Courpron P, Meunier P, Edouard C et al 1973 Données histologiques quantitatives sur le vieillissement osseux humain. Rev Rhum Mal Osteoarctic 40: 469–483

Courtney H 1949 Posterior subsphincteric space; its relation to posterior horseshoe fistula. Surg Gynecol Obstet 89: 222–226

Courville J 1966a The nucleus of the facial nerve; the relation between cellular groups and peripheral branches of the nerve. Brain Res 1: 338–354

Courville J 1966b Rubrobulbar fibres to the facial nucleus and the lateral reticular nucleus (nucleus of the lateral funiculus). An experimental study in the cat with silver impregnation methods. Brain Res 1: 317–337

Courville J, Brodal A 1966 Rubrocerebellar connections in the cat. An experimental study with silver impregnation methods. J Comp Neurol 126: 471–485

Cowan W M 1970 Anterograde and retrograde transneuronal degeneration in the central and peripheral nervous system. In: Nauta W J H, Ebbeson S O E (eds) Contemporary research methods in neuroanatomy. Springer: Berlin: pp 217–251

Cowan W M, Clarke P G H 1976 The development of the isthmo-optic nucleus. Brain Behav Evol 13: 354–375

Cowan W M, Gottlieb D I, Hendrickson A E, Price J L, Woolsey T A 1972 The autoradiographic demonstration of axonal connections in the central nervous system. Brain Res 37: 21–51

Cowie A T 1974 Overview of the mammary gland. J Invest Dermatol 63: 2–9

Cox R W, Peacock M A 1977 The fine structure of developing elastic cartilage. J Anat 123: 283–296

Cragg B G 1967 The density of synapses and neurones in the motor and visual areas of the cerebral cortex. J Anat 101: 639–654

Cragg B G 1969 The topography of the afferent projections in the circumstriate visual cortex of the monkey studied by the Nauta method. Vision Res 9: 733–747

Cragg B G 1970 What is the signal for chromatolysis? Brain Res 23: 1–21

Cragg B G 1976 Ultrastructural features of human cerebral cortex. J Anat 121: 331–362

Cralley J, Fitch K, McGonagle W 1975 Lumbrical muscles and contracted toes. Anat Anz 138: 348–353

Crandall W R 1938 A quantitative study of the influence of the ovarian hormones on hyperplasia by mitosis in the rabbit uterus in early pregnancy. Anat Rec 72: 195–210

Creamer B 1955 Oesophageal reflux. Lancet 1: 279–281

Crescitelli F 1972 The visual cells and visual pigments of the vertebrate eye. In: Dartnall H J A (ed) Handbook of sensory physiology. Vol 8. Part I. Springer: Berlin: pp 245–263

Crick F 1982 DNA today. Perspect Biol Med 25: 512–517

Crick F H C, Orgel L E 1973 Directed panspermia. Icarus 19: 341–348

Critchley M 1953 The parietal lobes. Arnold: London

Crock H V 1965 A revision of the anatomy of the arteries supplying the upper end of the human femur. J Anat 99: 77–88

Crock H V 1967 The blood supply of the lower limb bones in man. Livingstone: London

Croll J 1976 Is catastrophe theory dangerous? New Scientist 17: 630–632

Crompton A W, Hiiemäe K 1969 How mammalian molar teeth work. Discovery 5: 23–34

Crosby E C 1953 Relations of brain centers to normal and abnormal eye movements in the horizontal plane. J Comp Neurol 99: 437–480

Crosby E C 1960 Anatomie du lobe occipital et anatomie comparée des voies visuelles. In: Alajouanine Th (ed) Les grandes activités du lobe occipital. Masson: Paris

Crosby E C, Dejonge B R 1963 Experimental and clinical studies of the central connections and central relations of the facial nerve. Ann Atol Rhinol Larngol 72: 735–755

Crosby E C, Henderson J W 1948 The mammalian midbrain and isthmus regions. Part II. Fiber connections of the superior colliculus. B. Pathways concerned in automatic eye movements. J Comp Neurol 88: 53–92

Crosby E C, Humphrey T 1941 Studies of vertebrate telencephalon; nuclear pattern of anterior olfactory nucleus, tuberculum olfactorium and amygdaloid complex in adult man. J Comp Neurol 74: 309–352

Crosby E C, Lauer E W 1959 Anatomy of the midbrain. In: Schaltenbrand G, Bailey P (eds) Introduction to stereotaxis: with an atlas of the human brain. 3 vol. Grune and Stratton: New York: pp 88–118

Crosby E C, Woodburne R T 1940 Comparative anatomy of the pre-optic area and hypothalamus. Res Publ Assoc Res Nerv Ment Dis Proc 20: 52–169

Crosby E C, Woodburne R T 1943 Nuclear pattern of non-tectal portions of midbrain and isthmus in primates. J Comp Neurol 78: 441–482

Crosby E C, Woodburne R T 1951 Mammalian midbrain and isthmus regions: fiber connections; hypothalamo-tegmental pathways, J Comp Neurol 94: 1–32

Crosby E C, Yoss R E, Henderson J W 1952 The mammalian midbrain and isthmus regions. II. The fiber connections. D. The pattern for eye movements on the frontal eyefield and the discharge of specific portions of this field to and through midbrain levels. J Comp Neurol 97: 357–384

Crosby E C, Humphrey T, Showers M J 1959 In: Bauer K F (ed) Einige anordnungen und funktionen der supplementären motorischen rinden. Thieme: Stuttgart

Crosby E C, Humphrey R, Lauer E W 1962 Correlative anatomy of the nervous system. Macmillan: New York

Crowder R E 1957 The development of the adrenal gland in man, with special reference to origin and ultimate location of cell types and evidence in favour of the 'cell migration' theory. Contrib Embryol Carnegie Inst Washington 36: 193–210

Crowell R M, Morawetz R B 1977 The anterior communicating artery has significant branches. Stroke 8: 272–273

Csapo A 1962 Smooth muscle as a contractile unit. Physiol Rev 42: Suppl 5, 7–33

Cserr H F 1971 Physiology of the choroid plexus. Physiol Rev 51: 273–311

Cuajunco F 1940 Development of the neuromuscular spindle in human fetuses. Contrib Embryol Carnegie Inst Washington 28: 95–128

Cuajunco F 1942 Development of the human motor end-plate. Contrib Embryol Carnegie Inst Washington 30: 127–152

Cullis W, Tribe E 1913 Distribution of nerves in the heart. J Physiol 46: 141–150

Cummings J F 1969 Thoracolumbar preganglionic neurons and adrenal innervation in the dog. Acta Anat 73: 27–37

Cummins H 1926 Epidermal-ridge configurations in developmental defects, with particular reference to the ontogenetic factors which condition ridge direction. Am J Anat 38: 39–151

Cummins H, Midlo C 1961 Finger prints, palms and soles. An introduction to dermatoglyphics. Dover: New York

Cunha G R, Donjacour A A, Cooke P S, Mee S, Bigsby R M, Higgins S J, Sugimura Y 1987 The endocrinology and developmental biology of the prostate. Endocr Rev 8: 338–362

Currey J D 1984a Can strains give adequate information for adaptive bone remodeling. Calcif Tissue Int 36: (Suppl 1) 18–22

Currey J D 1984b Effects of differences in mineralization on the mechanical properties of bone. Philos Trans R Soc Lond [Biol] 304: 509–518

Currey J D 1984c What should bones be designed to do? Calcif Tissue Int 36: (Suppl 1) 7–10

Curtis A S G 1964 The mechanism of adhesion of cells to glass. A study by interference reflection microscopy. J Cell Biol 20: 199–215

Curtis A S G 1967 The cell surface: its molecular role in morphogenesis. Logos Press: London

Curtis A S G 1973 Cell adhesion. Prog Biophys Mol Biol 27: 316–386

Curtis A S G, Pitts J D eds 1980 Cell adhesion and motility. Cambridge University Press: Cambridge

Cuschieri A, Bannister L H 1975a The embryonic development of the olfactory organ of the mouse. I. Light microscopy. J Anat 119: 277–286

Cuschieri A, Bannister L H 1975b The embryonic development of the olfactory organ of the mouse. 2. Electron microscopy. J Anat 119: 471–498

Cushing H 1912 The pituitary body and its disorders. Lippincott: Philadelphia

Cutler L S, Chaudhry A P 1973 Differentiation of the myoepithelial cells of the rat submandibular gland in vivo and in vitro: an ultrastructural study. J Morphol 140: 343–354

Czarnetzki A 1971 Epigenetische Skeletlmerkmale im Populationsvergleich. I. Rechts-links-Unterschiede bilateral angelegter Merkmale. Z Morphol Anthropol 63: 238–254

Czernielewski J M, Demarchez M 1987 Further evidence for the self-reproducing capacity of Langerhans cells in human skin. J Invest Dermatol 88: 17–20

D

Daculsi G, Kerebel B 1978 High-resolution electron microscope study of human enamel crystallites: size, shape and growth. J Ultrastruct Res 65: 163–172

Dahlström A, Fuxe K 1964 Evidence for the existence of monamine-containing neurons in the central nervous system. Acta Physiol Scand Suppl 232: 1–55

Dahlström A, Fuxe K 1965 Evidence for the existence of monoamine neurons in the central nervous system. II. Experimentally induced changes in the intraneuronal amine levels of bulbospinal neuron systems. Acta Physiol Scand Suppl 247: 1–36

Dail W G, Evans A P 1974 Neural and vascular development in the human phallus. Invest Urol 11: 427–438

Dalcq A M 1954 Nouvelles données structurales et cytochimiques sur l'oeuf des mammifères. Rec Gén Sci Pur Appl 61: 19–41

Dalen H 1983 An ultrastructural study of the tracheal epithelium of the guinea-pig with special reference to the ciliary structure. J Anat 136: 47–67

Dallenbach-Hellweg G, Nette G 1964 Morphological and histochemical observations on trophoblast and decidua of the basal plate of the human placenta at term. Am J Anat 115: 309–326

Dal Pont G 1960 Contribution à l'étude de la structure fonctionelle du maxillaire. Ann Stomatol 9: 921–932

Damiani N, Capodacqua A 1961 On the intrinsic innervation of the fallopian tube. Annali Ostet Ginecol 83: 436–446

Dancis J 1959 The placenta. J Pediat 55: 85–101

Daniel J C Jr, Olson J D 1966 Cell movement, proliferation and death in the formation of the embryonic axis of the rabbit. Anat Rec 156: 123–127

Daniel O, Shackman R 1952 Blood supply of human ureter in relation to utero-colic anastomoses. Br J Urol 24: 334–343

Danielli J F, Davson H 1935 A contribution to the theory of the permeability of thin films. J Comp Cell Physiol 5: 495–508

Daoust R, Clermont Y 1955 Distribution of nucleic acids in germ cells during the cycle of the seminiferous epithelium in the rat. Am J Anat 96: 255–284

Darazs B, Taylor H R, Van Gelder A L 1988 The relevant of left ventricular bands. S Afr Med J 74: 68–71

D'Arcy Thompson W see Thompson W D'A

Darling A I, Levers B G H 1976 The pattern of eruption. In: Poole D F G, Stack M V (eds) The eruption and occlusion of teeth. Proceedings of the 27th Symposium of the Colston Research Society 1975. Butterworths: London: pp 80–96

Darnell J E Jr 1985 R N A Sci Am 253: (Oct) 54–64

Dart A M 1971 Cells of the dorsal column nuclei projecting down into the spinal cord. Physiol 219: 29–30

Daseler E H, Anson B J 1943 The planataris muscle. An anatomical study of 750 specimens. J Bone Jt Surg 25: 822–827

Davenport H A, Ranson S W 1930 The red nucleus and adjacent cell groups. A topographical study in the cat and rabbit. Arch Neurol Psychiat 24: 257–266

Davidowitz J, Philips G, Breinin G M 1977 Organization of the orbital surface layer in rabbit superior rectus. Invest Ophthalmol 16: 711–729

Davies D V 1950 Structure and function of synovial membrane. Br Med J 1: 92–95

Davies D V 1951 Blood supply of the tendon of extensor pollicis longus. Br Med J 2: 56

Davies D V, Edwards D A W 1948 Blood supply of synovial membrane and intra-articular structures. Ann R Coll Surg 2: 142–156

Davies D V, Barnett C H, Cochrane W, Palfrey A J 1962 Electron microscopy of articular cartilage in the young adult rabbit. Ann Rheum Dis 21: 11–22

Davies F 1927 Normal cholecystography Br Med J 1: 1138–1140

Davies F 1935 A note on the first lumbar nerve (anterior ramus) J Anat 70: 177–178

Davies F 1944 A previllous human ovum, aged nine to ten days (the Davies-Harding ovum). Trans R Soc Edinb 61: 315–328

Davies F, Harding H E 1942 Pouch of Hartmann, Lancet 1: 193–195

Davies F, Francis E T B, King T S 1952 Neurological studies of the cardiac ventricles of mammals. J Anat 86: 130–143

Davies F, Gladstone R J, Stibbe E P 1932 The anatomy of the intercostal nerves. J Anat 66: 323–333

Davies J 1960 Survey of research in gestation and the developmental sciences. Williams and Wilkins: Baltimore Maryland

Davies J, Routh J I 1957 Comparison of the foetal fluids of the rabbit. J Embryol Exp Morphol 5: 32–39

Davis C L 1923 Description of a human embryo having twenty paired somites. Contrib Embryol Carnegi Inst Washington 15: 1–51

Davis C L 1927 Development of the human heart from its first appearance to the stage found in embryos of twenty paired somites. Contrib Embryol Carnegie Inst Washington 19: 245–284

Davis J O, Freeman R H, Watkins B E, Stephens G A, Williams G M 1976 Angiotension II blockade and the functions of the renin-angiotensin system. In: Stokes G S, Edwards K D G (eds) Drugs affecting the renin-angiotensin-aldosterone system. Use of angiotensin inhibitors. Progress in Biochemical Pharmacology 12: 1–15 Karger: Basel

Davis J R, Langford G A 1969 Response of the testicular capsule to acetylcholine and noradrenaline. Nature 222: 386–387

Davis J R, Langford G A 1971 Comparative responses of the isolated testicular capsule and parenchyma to autonomic drugs. J Reprod Fertil 26: 241–245

Davis J R, Langford G A, Kirby P J 1970 The testicular capsule Vol 1. In: Johnson A D, Gomes W R, Vandemark N L (eds) The Testis. Academic Press: New York: pp 282–338

Davis P R 1955 The thoracolumbar mortice joint. J Anat 89: 370–377

Davis P R 1959 The medial inclination of the human thoracic intervertebral articular facets. J Anat 93: 68–74

Davis P R 1961 Human lower lumbar vertebrae: some mechanical and osteological considerations. J Anat 95: 337–344

Davis P R 1963 Some effects of lifting, pulling and pushing on the human trunk. Ergonomics 6: 303–304

Davis P R, Troup J D G, Burnard J H 1965 Movements of the thoracic and lumbar spine when lifting: a chrono-cyclophotographic study. J Anat 99: 13–26

Davson H 1970 Physiology of the cerebro-spinal fluid. Churchill: London

Davson H 1980 Physiology of the eye. 4th edition, Churchill Livingstone: Edinburgh

Davson H, Graham L T Jr eds 1974 The eye, Vol 5 Comparative physiology. Academic Press: New York

Dawes G S 1961 Changes in the circulation at birth. Br Med Bull 17: 148–153

Dawes G S 1969 Foetal and neonatal physiology. A comparative study of the changes at birth, year book: Chicago

Dawson J L 1974 Tumours of the liver. In: Smith R (ed) Surgical forum—the liver. Butterworths: London

Dawson W, Langman J 1961 An anatomical-radiological study on the pancreatic duct pattern in man. Anat Rec 139: 59–68

Dax M 1865 Lésions de la moitié gauche de l'encéphale coincident avec l'oubli désignés de la pensée. Lu au congrès méridional tenu à Montpellier en 1836. Gaz Hebdom Med 2nd ser 2: 259–260

Day M A 1964 Postural reflex patterns. Nurs Res 13: 139–147

Day M H, Napier J R 1961 The two heads of flexor pollicis brevis. J Anat 95: 123–130

Dearden L C, Bonucci E, Cuicchio M 1974 An investigation of ageing in human costal cartilage. Cell Tissue Res 152: 305–337

BIBLIOGRAPHY

de Beer G R 1937 The development of the vertebrate skull. Oxford University Press: London

de Beer G R 1947 The differentiation of neural crest cells into visceral cartilages and odontoblasts in amblystoma, and a re-examination of the germ-layer theory. Proc R Soc B 134: 377–398

de Busscher G 1948 Les anastomoses artériveneuses. Acta Neerl Morphol 6: 87–105

de Camilli P, Peluchetti D, Meldolesi J 1976 Dynamic changes in the luminal plasmalemma in stimulated parotid acinar cells. A freeze-fracture study. J Cell Biol 70: 59–74

de Castro F 1932 Sensory ganglia of the cranial and spinal nerves. In: Penfield W G (ed) Cytology and cellular pathology of the nervous system. Hoeber: New York

de Castro F, Herreros M L 1945 Actividad del ganglio cervical superior. Trab Inst Cajal Invest Biol 37: 287–342

de Duve C 1963 The lysosome. Sci Am 208: May 64–72

de Duve C 1973 Biochemical studies on the occurrence, biogenesis and life history of mammalian peroxisomes. J Histochem Cytochem 21: 941–948

de Duve C 1983 Microbodies in the living cell. Sci Am 248: May 52–62

Defendini R, Zimmerman E A 1978 The magnocellular and neurosecretory system of the mammalian hypothalamus. In: Reichlin S, Baldessarini R J, Martin J B (eds) The hypothalamus. (Research Publications Vol 56) Association for research in nervous and mental disease. Raven Press: New York: pp 137–152

De Feudis F V 1974 Central cholinergic systems and behaviour. Academic Press: London, New York

Deiters O F C 1865 Untersuchungen über Gehirn und Rückenmark des Menschen und der Säugethiere. F Vieweg und Sohn: Braunschweig

de Gasperis C, Miani A 1969 Observations sur l'ultrastruture du mesothelium pleural de l'homme. Bull Assoc Anat Paris 145: 188–202

DeGroot L J, Niepomniszcze H 1977 Biosynthesis of thyroid hormone, basic and clinical aspects. Metab Clin Exp 26: 665–718

Dejean C, Hervouët F, Leplat G 1958 L'embryologie de l'oeil et sa tératologie. Masson: Paris

Déjerine J, Déjerine-Klumpke H 1901 Anatomie des centres nerveux Vol 2, Rueff: Paris

Dekaban A 1953 Human thalamus: An anatomical, developmental and pathological study. J Comp Neurol 99: 639–683

De Lara Galindo S, Cuspinera E De G, Cardenas Ramirez L 1977 Anatomical and functional account on the lateral nasal cartilages. Acta Anat 97: 393–399

Del Cerro M, Knigge K M 1977 A system of gap junctions and some unusual glial-neuronal contacts in the rat arcuate nucleus. Cell Tissue Res

Delmas A, Senecail B 1977 Aspects biométriques de la mécanique hyöidienne chez l'homme. Bull Assoc Anat (Nancy), 61: (173), 189–198

del Rio Hortega P 1924 Le névroglie et le troisième élément des centres nerveux. Bull Soc Sci Méd Biol, Montpellier 5:

de Marsh Q B, Windle W F, Alt H L 1942 Blood volume of newborn infants in relation to early and late clamping of the umbilical cord. Am J Dis Child 63: 1123–1129

De Meyer W 1959 Number of axons and myelin sheaths in adult human medullary pyramids. Study with silver impregnation and iron hematoxylin staining methods. Neurology 9: 42–47

Dennison M 1971 Electron steroscopy as a means of classifying synaptic vesicles. J Cell Sci 8: 525–540

Denny-Brown D, Robertson E G 1935 Investigation of nervous control of defaecation. Brain 58: 256–310

de Palma A F 1957 Degenerative changes in the sternoclavicular and acromioclavicular joints in various decades. Thomas: Springfield Illinois

De Reuck A V S, Cameron M P eds 1962 The exocrine pancreas: normal and abnormal functions (Ciba Foundation Symposium) Churchill: London

De Reuck A V S, Knight J eds 1967 Cell Differentiation. (Ciba Foundation Symposium) Churchill: London

De Reuck A V S, Knight J eds 1968 Hearing mechanisms in vertebrates. (Ciba Foundation Symposium) Churchill: London

De Reuck A V S, Porter R eds 1967 Development of the lungs (Ciba Foundation Symposium) Churchill: London

de Reuck J 1972 The cortico-subcortical arterial anglo-architecture in the human brain. Acta Neurol Belg 72: 232–239

De Robertis E 1960 Some observations on the ultrastructure and morphogenesis of photoreceptors. J Gen Physiol 43: 1–13

De Robertis E D P, Bennett H S 1955 Some features of the sub-microscopic morphology of synapses in frog and earthworm. J Biophys Biochem Cytol 1: 47–58

Derry D E 1923 On the sexual and racial characters of the human ilium. J Anat 58: 71–83

de Sousa E 1955 Alcoolização do espaço extradural: tratamento em certas algias rebeldes. Rev Brasil Cir 30: 353–360

de Sousa O M 1964 Estudo electromiogràfico do m. platysma. Folia Clin Biol 33: 42–52

de Sousa O M, Vitti M 1966 Estudio electromiogràfico de los músculos adductores largo y mayor. Arch Mex Anat 7: 52–53 (abstract)

de Sousa O M, Furlani J, Vitti M 1973 Étude électromyographique du m. sternocleidomastoideus. Electromyogr Clin Neurophysiol 13: 93–106

Detwiler S M 1936 Neuroembryology, an experimental study. Macmillan: New York

Detwiler S R 1955 Experiments on the origin of the ventrolateral trunk musculature in the urodele (Amblystoma). J Exp Zool 129: 45–76

Deuchar E M 1958 Regional differences in cathepetic activity in xenopus iaevis embryos. J Embryol Exp Morphol 6: 223–237

Deuchar E M, Burgess A M C 1967 Somite segmentation in amphibian embryos: is there a transmitted control mechanism? J Embryol Exp Morphol 17: 349–358

De Vries P A, Saunders J B de C M 1962 Development of the ventricles and spiral outflow tract in the human heart. A contribution to the development of the human heart from age group IX to age group XV. Contrib Embryol Carnegie Inst Washington 37: 87–114

de Vries R A C, Kappers J A 1971 Influence of the pineal gland on the neurosecretory activity of the supraoptic hypothalmic nucleus in the male rat. Neuroendocrinology 8: 359–366

Dhingra L D 1977 Mediastinum testis. In: Johnson A D, Gomes W R (eds) The testis Vol 4. Advances in physiology biochemistry and function. Academic Press: New York: pp 451–460

Diamant H, Wiberg A 1965 Does the chorda tympani in man contain secretory fibres for the parotid gland? Acta Oto-Laryngol 60: 255–264

Diamond I T, Jones E G, Powell T P S 1969 The projection of the auditory cortex upon the diencephalon and brain stem in the cat. Brain Res 15: 305–340

Diamond J, Mills L R, Mearow K M 1988 Evidence that the Merkel cell is not the transducer in the mechanosensory Merkel cell – neurite complex. Prog Brain Res 74: 51–56

Dible J H, West C M 1940 A human ovum at the previllous stage. J Anat 75: 269–281

Di Chiro G 1962 Angiographic patterns of cerebral convexity veins. Am J Roentgenol 87: 308–321

Dickie J K M 1920 Note on the anatomy of the membranous labyrinth. J Laryngol 35: 76–81

Dickson A D 1957 The development of the ductus venosus in man and the goat. J Anat 91: 358–368

DiDio L J, Anderson M C 1968 The 'sphincters' of the digestive system. Williams and Wilkins: Baltimore

DiDio L J, Zappalá A, Carney W P 1967 Anatomico-functional aspects of the musculus articularis genus in man. Acta Anat 67: 1–23

Dieterich C E, Rohen J W 1970 Uber die Receptoren der menslichen Netzhaut. Albrecht v graefes arch klin ophthalmol 179: 235–258

Dijkstra C 1969 Structure of the autonomic terminal network of the thoracic organs after visualisation with the osmium zinc iodide method. Mikroskopie 24: 161–171

Dilly P N, Wall P D, Webster K E 1968 Cells of origin of the spinothalamic tract in the cat and the rat. Exp Neurol 21: 550–562

Dimond S J, Beaumont J G eds 1973 Hemisphere function in the human brain. Wiley: New York

Dingle J T 1962 Aetiological factors in the collagen diseases. Lysosomal enzymes and the degradation of cartilage matrix. Procs R Soc Med 55: 109–111

Dintenfass L 1963 Lubrication in synovial joints: a theoretical analysis. J Bone Jt Surg 45A: 1241–1256

Dixon A D 1958 Development of the jaws. Dent Pract 9: 10–12

Dixon A F 1920 Note on the vertebral epiphyseal discs. J Anat 55: 38–39

Dixon J S, Gosling J A 1982 The musculature of the human renal calices, pelvis and upper ureter. J Anat 135: 129–137

Dobbie J W, Symington T 1966 The human adrenal gland with special reference to the vasculature. J Endocrinol 34: 479–489

Dobrin P B 1978 Mechanical properties of arteries. Physiol Rev 58: 397–460

Dodd H 1959 The varicose tributaries of the superficial femoral vein passing into Hunter's canal. Postgrad Med J 35: 18–23

Dodd H, Cockett F B 1976 The pathology and surgery of the veins of the lower limb. 2nd edition. Churchill Livingstone: Edinburgh

Dodo Y 1986 Observations on the bony bridging of the jugular foramen in man. J Anat 144: 153–165

Dofferhof A S M, Vink P 1985 The stabilizing function of the mm iliocostales and mm multifidi during walking. J Anat 140: 329–336

Dogson M C H 1962 The growing brain. Wright: Bristol

Dohlman C H 1971 The function of the corneal epithelium in health and disease. Invest Ophthalmol 10: 303–407

Dohr G, Ebner I, Gallasch E 1986 Morphological and biomechanical studies of the ligamentum arteriosum. Acta Anat 126: 97–102

Doljanski F 1960 The growth of the liver with special reference to mammals. Int Rev Cytol 10: 217–241

Domnić-Stošić T, Jeličić N 1974 Morphological differences between meningeal arteries and the arteries of the scalp in the fetus and neonate. Srpski Arkhiv 102: 175–180

Donaldson P J, Mahan J T 1983 Fibrinogen and fibronectin as substrates for epidermal cell migration during wound closure. J Cell Sci 62: 117–127

Donker A J, Arisz L, Brentjens J R et al 1976 The effect of indomethacin on kidney function and plasma renin activity in man. Nephron 17: 288–296

Donoff R B, McLennan J E, Grillo H C 1971 Preparation and properties of collagenases from epithelium and mesenchyme of healing mammalian wounds. Biochim Biophys Acta 227: 639–653

Donovan B T 1970 Mammalian Neuroendocrinology. McGraw-Hill: London

Doran G A, Baggett H 1972 The genioglossus muscle: a reassessment of its anatomy in some mammals, including man. Acta Anat 83: 403–410

Doty R 1968 Neural organization of deglutition. In: Code C F (ed) Handbook of physiology, section 6, alimentary canal, vol 4. American Physiological Society: Washington DC: pp 1861–1902

Doubilet H, Shafiroff, B G P, Mulholland J H 1948 Anatomy of periesophageal vagi. Ann Surg 127: 128–135

Douek E D, Bannister L H, Dodson H C 1975 Olfaction and its disorders. Proc R Soc Med 68: 467–470

Douglas W H J, Redding R A, Stein M 1975 The lamellar substructure of osmiophilic inclusion bodies in rat type II alveolar pneumocytes. Tissue & cell 7: 137–142

Douglas W W 1968 Stimulus-secretion coupling: the concept and clues from chromaffin and other cells. Br J Pharmacol 34: 451–474

Døving K B 1967 Problems in the physiology of olfaction. In: Schultze H W, Day E A, Libbey L M (eds) Chemistry and physiology of flavors. AVI: Westport, Connecticutt: pp 52–94

Dow R S 1942 The evolution and anatomy of the cerebellum. Biol Rev 17: 179–220

Dow R S 1969 Cerebellar syndromes including vermis and hemispheric syndromes. In: Vinken P J, Bruyn G W (eds) Handbook of clinical neurology, vol 2. North Holland Publishing: Amsterdam

Dow R S, Moruzzi G 1958 The physiology and pathology of the cerebellum. University of Minnesota Press: Minneapolis

Dowling J E 1965 Foveal receptors of the monkey retina: fine structure. Science 147: 57–59

Dowling J E 1987 The Retina: an Approachable Part of the Brain. Harvard University Press: Cambridge Massachusetts

Dowling J E, Boycott B B 1966 Organisation of the primate retina: electron microscopy. Proc R Soc B 166: 80–111

Dowling J E, Ehinger B 1975 Synaptic organization of the amine containing interplexiform cells of the goldfish and cebus monkey retinas. Science 188: 270–273

Dowson D, Wright V, Longfield M D 1969 Human joint lubrication. Bio-Med Eng 4: 160–165

Doyle J F 1970 The perforating veins of the gluteus maximus. Ir J Med Sci 3: 285–288

Doyle J L, Watkins H O, Halbert D S 1967 Undescended laryngeal nerve. Tex Med J 63: 53–56

Doyle J R, Blythe W F 1975 The finger flexor tendon sheath and pulleys: anatomy and reconstruction. In: American Academy of Orthopaedic Surgeons Symposium on tendon surgery in the hand, Philadelphia 1974. Mosby: St Louis, Minnesota: pp 81–87

Draper M H, Ladefoged P, Whitteridge D 1960 Expiratory pressures and air flow during speech. Br Med J 1: 1837–1843

Dreyer B, Budtz-Olson O E 1952 Splenic venography; demonstration of portal circulation with diodone. Lancet 1: 530–531

Droz B 1963 Dynamic conditions of proteins in the visual cells of rats and mice as shown by radioautography with labeled amino acids. Anat Rec 145: 157–168

Druckman R 1952 A critique of 'suppression', with additional observations in the cat. Brain 75: 226–243

Drumheller G 1969 Anatomical observations of the lower lateral nasal cartilages. Arch Otolaryngol 89: 599–601

Drummond H 1914 The arterial supply of the rectum and pelvic colon. Br J Surg 1: 677–685

Duarte L R 1983 The stimulation of bone growth by ultrasound. Arch Orthop Trauma Surg 101: 153–159

Dubois P 1967 Etude au microscope électronique de la pars distalis de l'hypophyse de l'embryon humain. C R Assoc Anat Nancy 138: 429–433

DuBrul E L 1980 Sicher's oral anatomy. 7th edition. Mosby: St Louis

Duce I R, Keen P 1977 An ultrastructural classification of the neuronal cell bodies of the rat dorsal root ganglion using zinc iodideosmium impregnation. Cell Tissue Res 185: 263–277

Duchen L W, Gale A N 1985 The motor end plate. In: Swash M, Kennard C (eds) Scientific basis of clinical neurology. Churchill Livingstone: Edinburgh: pp 400–409

Duckworth W L H 1947 Some complexities of human structure. Oxford University Press: London

Duenhoelter J H, Pritchard J A 1973 Human fetal respiration. Obstet Gynecol 42: 746–750

Duke-Elder J S 1969 The practice of refraction, 8th edition, Churchill: London

Duke-Elder J S, Wybar K C 1961 A system of ophthalmology, vol 2. Kimpton: London

Duke-Elder W S 1932 Textbook of ophthalmology, vol 1. Kimpton: London

Duke-Elder Sir W S 1963 System of ophthalmology. volume 3 normal and abnormal development part 1, Embryology. Kimpton: London

Dunnill M S, Halley W 1973 Some observations on the quantitative anatomy of the kidney. J Pathol 110: 113–121

DuPraw E J 1968 Cell and molecular biology. Academic Press: New York

Dupré J, Ross S A, Watson D, Brown J C 1973 Stimulation of insulin secretion by gastic inhibitory polypeptide in man. J Clin Endocr Metab 37: 826–828

Durcan D J, Shea J J, Sleeckx J P 1967 Bifurcation of the facial nerve. Arch Octolaryngol 86: 619–631

Dusser de Barenne J G 1933 'Corticalization' of function and functional localization in cerebral cortex. Arch Neurol Psychiatry 30: 884–901

Dustin P 1980 Microtubules Sci Am 243: (Aug) 59–68

Duval M M 1879 Technique de l'emploi du collodior humide pour la pratique des coupes microscopiques. J Anat Physiol (Paris) 15: 185–188

Duvall A J, Quick C A 1969 Tracers and endogenous debris in delineating cochlear barriers and pathways. An experimental study. Ann Otol Rhinol Laryngol 78: 1041–1057

Duvernoy J 1972 The vascular architecture of the median eminence. In: Knigge K M, Scott D E, Weindl A (eds) Brain-endocrine interaction. Median eminence: structure and function. Karger: Basel: pp 79–108

Duvernoy J, Koritké J G, Monnier G 1969 On the vascularisation of the lamina terminalis in the human. Z Zellforsch 102: 49–77

Duvernoy H M 1975 The superficial veins of the human brain. Springer-Verlag: Berlin

Duvernoy H M 1978 Human brainstem vessels. Springer-Verlag: Berlin

Duvernoy H M, Delon S, Vannson J L 1981 Cortical blood vessels of the human brain. Brain Res Bull 7: 519–579

Duvernoy H M, Delon S, Vannson J L 1983 The vascularization of the human cerebellar cortex. Brain Res Bull 11: 419–480

Dykes J R W 1969 Histometric assessment of human testicular biopsies. J Pathol 97: 429–440

Dylevský I 1968 Tendons of the m flexor digitorum superficialis et profundus in the ontogenesis of the human hand. Folia Morphol (Praha) 16: 124–130

Dyson M 1987 Mechanisms involved in therapeutic ultrasound. Physiotherapy, 73: 116–120

Dyson M, Brookes M 1983 Stimulation of bone repair by ultrasound. In: Lerski R A, Morley P (eds) Ultrasound '82. Pergamon Press: Oxford: pp 61–66

Dyson M, Young S 1986 Effect of laser therapy on wound contraction and cellularity in mice. Lasers Med Sci 1: 125–130

Dyson M, Young S, Pendle C L, Webster D F, Lang S M 1988 Comparison of the effects of moist and dry conditions on dermal repair. J Invest Dermatol 91: 434–439

E

Eager R P 1966 Patterns and mode of termination of cerebellar corticonuclear pathways in the monkey (Macaca mulatta) J Comp Neurol 126: 551–565

Earle K M 1952 The tract of Lissauer and its possible relation to the pain pathway. J Comp Neurol 96: 93–111

Eayrs J T 1955 Cerebral cortex of normal and hypothyroid rats. Acta Anat 25: 160–183

Ebashi S 1983 Regulation of muscle contraction. Cell Muscle Motil 3: 79–87

Ebbesson S O E 1968 Quantitative studies of superior cervical sympathetic ganglia in a variety of primates including man. II. Neuronal packing density. J Morphol 124: 181–186

Eccles J C 1957 The physiology of the nerve cell. Johns Hopkins Press: Baltimore

Eccles J C 1964a The excitatory responses of spinal neurons. Progr Brain Res 12: 65–91

Eccles J C 1964b The physiology of synapses. Springer-Verlag: Berlin

Eccles J C 1970 Neurogenesis and morphogenesis in the cerebellar cortex. Proc Natl Acad Sci USA 66: 294–301

Eccles J C 1973 The understanding of the brain. McGraw-Hill: New York

Eccles J C, Schadé J P eds 1964a Organisation of the spinal cord. Progr Brain Res 11:

Eccles J C, Schadé J P eds 1964b Physiology of spinal neurons. Progr Brain Res 12:

Eccles J C, Eccles R M, Iggo I, Lundberg A 1960 Electrophysiological studies on gamma motoneurons. Acta Physiol Scand 50: 32–40

Eccles J C, Fatt P, Landgren S, Winsbury G J 1954 Spinal cord potentials generated by volleys in large muscle afferents. J Physiol 125: 590–606

Eccles J C, Ito M, Szentágothai J 1967 The cerebellum as a neuronal machine. Springer-Verlag: Berlin

Eccles J C, Sasaki K, Strata P 1967a A comparison of the inhibitory action of Golgi cells and of basket cells. Exp Brain Res 3: 81–94

Eccles J C, Sasaki K, Strata P 1967b Interpretation of the potential fields generated in the cerebellar cortex by a mossy fibre volley. Exp Brain Res 3: 58–80

Eckert-Mobius A 1924 Über die Rolle der gefässhaltigen Knorpelkanäle bei der enchondralen Verknöcherung. Dtsch Med Wochenschr 1: 1798

Ede D A 1976 Cell interactions in vertebrate limb development. In: Poste G, Nicolson G L (eds) The cell surface: in animal embryogenesis and development. North-Holland: Amsterdam: pp 495–543

Edelman G M 1974 Antibody structure and cellular specificity in the immune response. Harvey Lectures 68: 149–184

Edelman G M 1985 Molecular recognition of neural morphogenesis. In: Edelman G M, Gall W E, Cowan W M (eds) Molecular bases of neural development. John Wiley: New York: pp 35–49

Edelman G M, Rutishauser U 1981 Molecules involved in cell-cell adhesion during development. J Supramolec Struct Cell Biochem 16: 259–268

Edelman R, Hartcroft P M 1961 Localisation of renin in juxtaglomerular cells of rabbit and dog through the use of the fluorescent-antibody technique. Circ Res 9: 1069–1077

Edvinsson L, Londvall M, Nielsen K C, Owman Ch 1973 Are brain vessels innervated also by central (non-sympathetic) adrenergic neurones? Brain Res 63: 496–499

Edwards D A W 1946 The blood supply and lymphatic drainage of tendons. J Anat 80: 147–152

Edwards J C W 1982 The origin of the type A synovial lining cells. Immunobiology 161: 227–231

Edwards J C W, Willoughby D A 1982 Demonstration of bone marrow derived cells in cynovial lining by means of giant intracellular granules as genetic markers. Ann Rheum Dis 41: 177–182

Edwards R G, Bavister B D, Steptoe P C 1969 Early stages of fertilisation in vitro of human oocytes matured in vitro. Nature 221: 632–635

Edwards R G, Donahue R P, Baramki T A, Jones H W Jr 1966 Preliminary attempts to fertilize human oocytes matured in vitro. Am H Obstet Gynec 96: 192–200

Edwards S B, Henkel C K 1978 Superior colliculus connections with the extra ocular motor nuclei in the cat. J Comp Neurol 179: 451–467

Egeberg J, Jensen O A 1969 The ultrastructure of the acini of the human lacrimal gland. Acta Ophthal 47: 400–410

Eggli P, Schaffner T, Gerber H A, Hess M W, Cottier H 1986 Accessibility of thymic cortical lymphocytes to particles translated from the peritoneal cavity to parathymic lymph nodes. Thymus 8: 129–139

Eggli P S, Herrman W, Hunziker E B, Schenk R K 1985 Matrix compartments in the growth plate of the proximal tibia of rats. Anat Rec 211: 246–257

Ehinger B, Falck B 1966 Concomitant adrenergic and parasympathetic fibres in the rat iris. Acta Physiol Scand 67: 201–207

Ehrenberg C G 1833 Notwendigkeit einer feineren mechanischen Zerlegung des Gehirns und der Nerven. Poggendorff's Ann Phys Chem 22: 449–465

Ehrlich P 1886 Über die Methylenblaureaction der lebenden Nervensubstanz. Dtsch Med Wochenschr 12: 49–52

Eichner D 1957 Über histologie und Topochemie der sehschicht in der Netzhaut des menschen. Z Mikrosk Anat Forsch 63: 82–93

Einarson L 1932 Method for progressive selective staining of Nissl and nuclear substances in nerve cells. Am J Path 8: 295–308

Eisenberg B R, Kuda A M 1976 Discrimination between fiber populations in mammalian skeletal muscle by using ultrastructural parameters. J Ultrastruct. Res 54: 76–88

Ek A, Alm P, Andersson K E et al 1977a Adrenergic and cholinergic nerves of the human urethra and urinary bladder. A histochemical study. Acta Physiol Scand 99: 345–352

Ek A, Alm P, Andersson K E et al 1977b Adrenoreceptor and cholinoceptor mediated responses of the isolated human urethra. Scand J Urol Nephrol 111: 97–102

Eleftheriou B E, Elias M F, Norman R L 1972 Effects of amygdaloid lesions on reversal learning in the deermouse. Physiol Behav 9: 69–73

Elftman H 1966 Biomechanics of muscle with particular application to studies of gait. J Bone Jt Surg 48A: 363–377

Elfvin L G 1965 The fine structure of the cell surface of chromaffin cells in the rat adrenal medulla. J Ultrastruct Res 12: 263–286

Elias H 1955 Liver morphology. Biol Rev 30: 263–310

Elias H 1967 Recruitment in human bile duct formation. Acta Hepatosplen (Stuttgart) 14: 253–260

Elias H, Petty D 1952 Gross anatomy of the blood vessels and ducts within the human liver. Am J Anat 90: 59–111

Elišková M 1973 Blood vessels of the ciliary ganglion in man. Br J Ophthal 57: 766–772

Elliot F M, Reid L 1965 Some new facts about the pulmonary artery and its branching pattern. Clin Radiol 16: 193–199

Elliott H C 1942 Studies on the motor cells of the spinal cord. II. Distribution in the normal human cord. Am J Anat 70: 95–117

Elliott H C 1943 Studies on the motor cells of the spinal cord. II. Distribution in the normal human fetal cord. Am J Anat 72: 29–38

Ellis LeG C, Hargrove J L 1977 Prostaglandins. In: Johnson A D, Gomes W R (eds) The testis. Vol 4. Advances in physiology, biochemistry and function. Academic Press: New York: pp 289–313

Ellis R A 1965 Fine structure of the myoepithelium of the eccrine sweat glands of man. J Cell Biol 27: 551–564

Ellison J P, Williams T H 1969 Sympathetic nerve pathways to the human heart and their variations. Am J Anat 124: 149–162

El-Najjar M Y, Dawson G L 1977 The effect of artificial cranial deformation on the incidence of Wormian bones in the lambdoidal suture. Am J Phys Anthropol 46: 155–160

Elsdale T, Pearson M 1979 Somitogenesis in amphibia. II. Origins in early embryogenesis of two factors involved in somite specification. J Embryol Exp Morphol 53: 245–267

Elsen J, Arey L B 1966 On spirality in the intestinal wall. Am J Anat 118: 11–20

Emery J ed 1969 The anatomy of the developing lung. Spastics International Medical Publications: New York

Emmelin N 1972 Control of salivary glands. In: Emmelin N, Zotterman Y (eds) Oral physiology. Pergamon: Oxford: pp 1–16

Emmelin N, Engström J 1960 On the existence of specific secretory sympathetic fibres for the cat's submaxillary gland. J Physiol 153: 1–8

Emson P C 1982 CNS neuropeptides: implications for neuropharmacology. Bull Mem Acad R Med Belg 137: Suppl 78–92

Enders A C, Schlatke S J 1965 The fine structure of the blastocyst: some comparative studies. In: Wolstenholme G E W, O'Connor M (eds) Preimplantation stages of pregnancy. (Ciba Foundation Symposium). Churchill: London: pp 29–59

Eneroth C-M, Hökfelt T, Norberg K-A 1969 The role of the parasympathetic and sympathetic innervation for the secretion of human parotid and submandibular glands. Acta Oto-laryngol 68: 369–375

Eng L F, Uyeda C T, Chao L P, Wolfgram F 1974 Antibody to bovine choline acetyltransferase and immunofluorescent localisation of the enzyme in neurones. Nature 250: 243–245

Engel R, Bogduk N 1982 The menisci of the lumbar zygapophysial joints. J Anat 135: 795–809

Engel S 1947 The child's lung. Arnold: London

Engel S 1962 Lung structure. Thomas: Springfield, Illinois

English D T, Blevins C E 1969 Motor units of laryngeal muscles. Arch Otolaryngol 89: 778–784

English K B 1977 The ultrastructure of cutaneous type I mechanoreceptors (Haarscheiben) in cats following denervation. J Comp Neurol 172: 137–164

Engström H, Wersäll J 1958 Myelin sheath structure in nerve fibre demyelinization and branching regions. Exp Cell Res 14: 414–425

Engström H, Ades H W, Hawkins J E J 1965 In: Neff W D (ed) Contributions to sensory physiology. Academic Press: New York

Enlow D H, Harris D B 1964 Study of the post-natal growth of the human mandible. Am J Orthop 50: 25–50

Ennabli E, Niveiro M 1967 Etude embryonnaire des artères intercostales. Reconstruction par la méthode de Born de deux embryons humains (14 et 17 mm) Pathol Biol (Paris) 15: 92–98

Enoch D M, Kerr F W L 1967 Hypothalamic vasopressor and vesicopressor pathways. II. Anatomic study of their course and connections. Arch Neurol Psychiat 16: 307–320

Eränkö O 1967 The practical histochemical demonstration of catecholamines by formalin-induced fluorescence. J R Microsc Soc 87: 259–276

Eränkö O 1978 Small intensely fluorescent (SIF) cells and neurotransmission in sympathetic ganglia. Ann Rev Pharmacol Toxicol 18: 417–430

Eränkö O, Härkönen M 1965 Moncamine-containing cells in the superior cervical ganglion of the rat. Acts Physiol Scand 63: 511–512

Ericksen M F 1976 Some aspects of aging in the lumbar spine. Am J Phys Anthropol 45: 575–580

Ericson E, Hakanson R, Larson B, Owman Ch, Sundler F 1972 Fluorescence and electron microscopy of amine-staining enterochromaffin-like cells in tracheal epithelium of mouse. Z Zellforsch 124: 532–545

Erlanger J, Gasser H S 1937 Electrical signs of nervous activity. University of Pennsylvania Press: Philadelphia

Ernster L, Schatz G 1981 Mitochondria: a historical review. J Cell Biol 91: 227s–255s

Esser P H 1932 Über die Funktion und den Bau des Scrotums. Z Zellforsch 31: 108–174

Estable C, Acosta-Ferreira W, Sotelo J R 1957 An electron microscope study of the regenerating nerve fibres. Z Zellforsch 46: 387–399

Esterhuizen A, Spriggs T, Lever J 1968 Nature of islet cell innervation in the cat pancreas. Diabetes 17: 33–36

Etherton J E, Conning D M 1971 Early incorporation of labelled palmitate into mouse lung. Experientia 27: 554–555

Etherton J E, Conning D M, Corrin B 1973 Autoradiographical and morphological evidence for apocrine secretion of dipalmitoyl lecithin in the terminal bronchiole of mouse lung. Am J Anat 138: 11–35

Evans D H L, Murray J G 1954 Regeneration of non-medullated nerve fibres. J Anat 88: 465–480

Evans E 1949 Congenital heart disease. J Florida Med Assoc 35: 487–491

Evans E F 1968 Cortical representation. In: De Renck W S, Knight J (eds) Heavy mechanisms in vertebrates. (Ciba Foundation Symposium). Churchill: London: pp 272–287

Evans E J, Benjamin M 1984 Fibrocartilage at the insertion of the human supraspinatus tendon. J Anat 139: 727 (Proc Anat Soc GB & I)

Evans F G 1973 Mechanical properties of bone. Thomas: Springfield

Ewing M R 1954 White line of Hilton. Proc R Soc Med 47: 525–530

Eyzaguirre C, Nishi K, Fidone S 1972 Chemoreceptor synapses in the carotid body. Fed Proc Fed Am Soc Exp Biol 31: 1385–1393

F

Falck B 1962 Observations on the possibilities of the cellular localization of monoamines by a fluorescence method. Acta Physiol Scand 56: suppl 197

Falck B, Hillarp, N-Å 1959 A note on the chromaffin reaction. J Histochem Cytochem 7: 149

Falck B, Owman C 1965 A detailed methodological description of the fluorescence method for the cellular demonstration of biogenic monoamines. Acta Univ Lund Sect II 7: 1–23

Falck B, Hillarp N åA, Thieme G, Torp A 1962 Fluorescence of catecholamines and related compounds condensed with formaldehyde. J Histochem Cytochem 10: 348–354

Falconer D, Spinner M 1985 Anatomic variations in the motor and sensory supply of the thumb. Clin Orthop 195: 83–96

Falconer M 1967 Brain mechanisms suggested by neurophysiologic studies. In: Millikan C H, Darley F L (eds) Brain mechanisms underlying speech and language. Grune and Stratton: New York: pp 185–190

Falconer M A 1949 Intramedullary trigeminal tractotomy and its place in the treatment of facial pain. J Neurol Neurosurg Psychiat 12: 297–311

Falconer M A, Wilson J L 1958 Visual field changes following anterior temporal lobectomy, their significance in relation to 'Meyer's loop' of the optic radiation. Brain, 81: 1–14

Faller J, Ungváry G 1962 Die arterielle Segmentation der Niere. Zbl Chir 87: 972–984

Farbman A I 1980 renewal of tast cells in rat circumvallate papillae. Cell Tiss Kinet 13: 349—357

Farbman A I, Gesteland R C 1975 Developmental and electrophysiological studies of olfactory mucosa in organ culture. In: Denton D, Coghlan J P (eds) Olfaction and taste V. Academic Press: New York: pp 107–110

Farnarier G, Planche D, Rohner J J 1977 Blocage des afférences nociceptives par stimulation périphérique percuntanee chez le chat. C R Soc Biol 171: 1054–1058

Farquhar M G, Palade G E 1963 Junctional complexes in various epithelia. J Cell Biol 17: 375–412

Farquhar M, Palade G 1981 The Golgi apparatus (Complex)–1954–1981–from artifact to center stage. J Cell Biol 91: 77s–103s

Farquhar M G 1981 Membrane recycling in secretory cells: implications for traffic of products and specialized membranes within the Golgi complex. Methods Cell Biol 23: 399–427

Farquhar M G 1985 Progress in unravelling pathways of Golgi traffic. Annu Rev Cell Biol 1: 447–488

Farrell J H 1956 The effect of mastication on the digestion of food. Br Dent J 100: 149–155

Farrer-Brown G, Beilby J O W, Tarbit M M 1970a The blood supply of the uterus. I. Arterial vasculature. J Obstet Gynaec Br Commonw 77: 673–681

Farrer-Brown G, Beilby J O W, Tarbit M M 1970b The blood supply of the uterus. II. Venous pattern. J Obstet Gynaec Br Commonw 77: 682–689

Fatani J A, Qayyum M A, Mehta L, Singh U 1986 Parasympathetic innervation of the thymus: a histochemical and immunocytochemical study. J Anat 147: 115–119

Fawcett D W 1961a Cilia and flagella. In: Brachet J, Mirsky A E (eds) The cell. Biochemistry, physiology, morphology, volume II, cells and their component parts. Academic Press: New York: pp 217–297

Fawcett D W 1961b Intercellular bridges. Exp Cell Res Suppl 8: 174–187

Fawcett D W 1965 The anatomy of the mammalian spermatozoon with particular reference to the guinea pig. Z Zellforsch 67: 279–296

Fawcett D W 1968 The topographical relationship between the plane of the central pair of flagellar fibrils and the transverse axis of the head in guinea-pig spermatozoa. J Cell Sci 3: 187–198

Fawcett D W 1975 Ultrastructure and function of the Sertoli cell. In: Hamilton D W, Greep R O (eds) Handbook of Physiology, section 7. Endocrinology. Vol 5. American Physiological Society: Washington DC: pp 21–55

Fawcett D W 1981 The cell 2nd edition. Saunders: Philadelphia

Fawcett D W, Burgos M H 1956 A comparison of the structural organisation of mammalian and amphibian sperm tails. Anat Rec 124: 289P

Fawcett D W, Ito S 1965 The fine structure of bat spermatozoa. Am J Anat 116: 567–609

Fawcett D W, McNutt N S 1969 The ultrastructure of the cat myocardium. I. Ventricular papillary muscle. J Cell Biol 42: 1–45

Fawcett D W, Ito S, Slautterback D 1959 The occurrence of intercellular bridges in groups of cells exhibiting synchronous differentiation. J Biophys Biochem Cytol 5: 453–460

Fawcett E 1895 The structure of the inferior maxilla with special reference to the position of the inferior dental canal. J Anat 29: 355–366

Fawcett E 1907 On the completion of ossification of the human sacrum. Anat Anz 30: 414–421

Fawcett E 1911a The development of the human maxilla, vomer and paraseptal cartilage. J Anat 45: 378–406

Fawcett E 1911b Some notes on the epiphyses of the ribs. J Anat 45: 172–178

Fawcett E 1917 The primordial cranium of microtus amphibius (water rat) as determined by sections and a model of the 25 mm stage, with comparative remarks. J Anat 51: 309–359

Fawcett E 1918 The primordial cranium of erinaceus europaeus. J Anat 52: 211–250

Fawcett E, Blachford J V 1906 The circle of Willis: an examination of 700 specimens. J Anat 40: 63–70

Fawcett E, Brasch J C, Northcroft G, Keith A 1924 The growth of the jaws. Dental Board of the United Kingdom: London

Feeney L, Grieshaber J, Hogan M J 1965 In: Rohen J (ed) The structure of the eye. Shattaner: Stuttgart

Fehér E, Vajda J 1974 Degeneration analysis of the extrinsic nerve elements of the small intestine. Acta Anat (Basel) 87: 97–109

Feldman M, Globerson A 1974 Reception of immunogenic signals by lymphocytes. Curr Top Dev Biol 8: 1–40

Fell H B 1939 The origin and developmental mechanics of the avian sternum. Philos Trans R Soc Lond [Biol] 229: 407–463

Fell H B 1978 Synoviocytes. J Clin Pathol 31: (Suppl R Coll Pathol 12) 14–24

Fell H B, Canti R G 1934 Observations on the early development of the knee joint in vivo and in vitro. Proc R Soc Lond [Biol] 116: 316–351

Felten D L, Laties A M, Carpenter M B 1974 Monoamine-containing cell bodies in the squirrel monkey brain. Am J Anat 139: 153–165

Feltz P, Mackenzie J S 1969 Properties of caudate unitary responses to repetitive nigral stimuli. Brain Res 13: 612–616

Ferenczy A 1980 The ultrastructure of the human cervix. In: Naftolin F, Stubblefield P G (eds) Dilatation of the uterine cervix. Connective tissue biology and clinical management. Raven Press: New York: pp 27–44

Ferguson M W 1977 The mechanism of palatal shelf elevation and the pathogenesis of cleft palate. Virchows Arch [Pathol Anat] 375: 97–113

Fergusson J D, Gibson E C 1956 Prostatic smear diagnosis. Br Med J 1: 822–825

Ferraro A, Barrera S E 1936 Lamination of the medial lemniscus in macacus rhesus. J Comp Neurol 64: 313–324

Ferraro J A, Minckler J 1977a The brachium of the inferior colliculus. The human auditory pathways: a quantitative study. Brain and Language 4: 156–164

Ferraro J A, Minckler J 1977b The human lateral lemniscus and its nuclei. Brain and Language 4: 277–294

Ferraz de Carvalho C A, Rodrigues de Souza R, Henrique A, Henrique A, Nogueira de Lima M A 1987 Functional anatomy of the tela submucosa of the valva ileocecalis in the adult man. Anat Anz 164: 63–76

Ferrier D 1874 The localisation of function in the brain. Proc R Soc Lond [Biol] 22: 229–232

Feyler K P 1965 Quantative Untersuchungen über die vegetativen Ganglien im Paries membranaceus tracheae des Menschen. Anat Anz 117: 371–379

Feyrter F 1938 Über diffuse endokrine epitheliale organe. Zentbl Innere Med 545: 31–41

Feyrter F 1961 Zur Frage der Endokrinie des sogenannten Speicheldrüsenmischtumors. Dtsch Med Wochenschr 86: 335–339

Fidone S J, Gonzalez C, Dinger B G and Hanson G R 1986 Mechanisms of chemotransmission in the mammalian carotid body, Prog Brain Res 74: 169–179

Field E J 1951 The development of the conducting system in the heart of sheep. Br Heart J 13: 129–147

Fields W S, Bruetman M E, Weibel J 1965 Collateral circulation of the brain. Williams and Wilkins: Baltimore

Findlay J A, Ashcroft J J H 1975 Cells of the islets of Langerhans. In: Beck F, Lloyd J B (eds) The cell in medical science. Vol 3. Academic Press: New York: pp 243–306

Fine B S, Tousimis A J 1961 The structure of the vitreous body and the suspensory ligaments of the lens. Arch Ophthal 65: 95–110

Fine B S, Zimmerman L E 1962 Miller's cells and the 'middle limiting membrane' of the human retina. An electron microscopic study. Invest Ophthal 1: 304–326

Fine H, Keen E N 1966 The arteries of the human kidney. J Anat 100: 881–894

Fink B, Demarest R J 1978 Laryngeal Biomechanics. Harvard University Press: Cambridge, Mass

Fink B R 1974 Folding mechanism of the human larynx. Acta Oto-laryng (Stockh), 78: 124–128

Fink B R 1978 Fourth Daniel C Baker Jr, memorial lecture. Energy and the larynx. Ann Otol Rhinol Laryngol 87: (5 Pt 1), 595–605

Fink B R, Martin R, LaVigne A B 1975 Spring mechanics of the human larynx. In: Bosma J F, Showacre J (eds) Symposium on development of upper respiratory anatomy and function. Implications for sudden death syndrome. (Fogarty International Center Proceedings No 29) US Government Printing Office: Washington DC: pp 63–75

Fink R P, Heimer L 1967 Two methods for selective silver impregnation of degenerative axons and their synaptic endings in the central nervous system. Brain Res 4: 369–374

Finnegan M 1972 Population definition on the North West Coast by analysis of discrete character variation. Ph D Dissertion, University of Colorado, Boulder

Finnegan M 1978 Non-metric variation of the infracranial skeleton. J Anat 125: 23–37

Finnegan M, Faust M A 1974 Bibliography of human and non-human non-metric variation. (Research Report No 14) Dept Anthropology, University of Massachusetts: Cambridge, Mass

Finney K J, Ince P, Appleton D R, Sunter J P, Watson A J 1986 A comparison of crypt-cell proliferation in rat colonic mucosa in vivo and in vitro. J Anat 149: 177–188

Fischer A 1977 Autonomy for a specific gene product in oocytes: experimental evidence in the polychaetous annelid Platynereis dumerilii. Dev Biol 55: 46–58

1495

Fischer J 1956 The labyrinth: Physiology and functional tests. Grune and Stratton: New York

Fischer L, Machenaud A, Morin A 1974 Contribution à l'étude de la vascularisation du cubitus. Arch Anat Path 22: 261–265

Fischman D A 1970 The synthesis and assembly of myofibrils in embryonic muscle. Curr Top Dev Biol 5: 235–280

Fisher E R, Jeffrey W 1965 Ultrastructure of human normal and neoplastic prostate; with comments related to prostatic effects of hormonal stimulation in the rabbit. Am J Clin Pathol 44: 119–134

Fisher S K, Linberg K A 1975 Intercellular junctions in the early human embryonic retina. J Ultrastruct Res 5: 69–78

Fishman A P 1982a Endothelium: a distributed organ of diverse capabilities. Ann N Y Acad Sci 401: 1–8

Fishman A P (ed) 1982b Endothelium. Ann N Y Acad Sci 401

Fishman A P, Hecht H H 1969 The pulmonary circulation and interstitial space eds, University of Chicago Press: Chicago

Fishman A P, Pietra G G 1974 Handling of bioactive materials by the lung. N Engl J Med 291: 884–889

Fitzgerald M J T 1956 The occurrence of a middle superior alveolar nerve in man. J Anat 90: 520–522

Fitzgerald M J T, Sachithanadan S R 1979 The structure and source of lingual proprioceptors in the monkey. J Anat 128: 523–552

Fitzgerald M J T, Scott J H 1958 Observations on the anatomy of the superior dental nerves. Br Dent J 104: 205–208

Fitzgerald M J T, Comerford P T, Tuffery A R 1982 Sources of innervation of the neuromuscular spindles in sternomastoid and trapezius. J Anat 134: 471–490

Fitzke F W 1986 Spatial properites of scotopic sensitivity. PhD Thesis: University of London

Flechsig P 1876 Die Leitungsbahnen in Gehirn und Rückenmark des Menschen auf Grund entwicklungsgeschichtlicher Untersuchungen. Engelmann: Leipzig

Flecker H 1929 Röntgenographic study of movements of abduction at normal shoulder joint. Med J Aust 2: 123–124

Fleischner F G, Sayegh V 1958 Assessment of the size of the liver: roentgenologic considerations. New Engl J Med 259: 271–274

Flickinger C, Fawcett D W 1967 The junctional specialisations of Sertoli cells in the seminiferous epithelium. Anat Rec 158: 207–222

Flock A 1988 Do sensory cells in the ear have a motile function? Prog Brain Res 74: 297–304

Flood S, Jansen J 1966 The efferent fibres of the cerebellar nuclei and their distribution on the cerebellar peduncles in the cat. Acta Anat 63: 137–166

Florey L 1966 The endothelial cell. Br Med J 2: 487–490

Florian J 1930 The formation of the connecting stalk and the extension of the amniotic cavity towards the tissue of the connecting stalk in young human embryos. J Anat 64: 454–476

Floyd W F, Silver P H S 1950 Electromyographic study of patterns of activity of the anterior abdominal wall muscles in man. J Anat 84: 132–145

Floyd W F, Silver P H S 1955 The function of the erectores spinae muscles in certain movements and postures in man. J Physiol Lond 129: 184–203

Flyger G, Hjelmquist U 1957 Normal variations in the caliber of the human cerebral aqueduct. Anat Rec 127: 151–162

Foerster O 1933 The dermatomes in man. Brain 56: 1–39

Foerster O 1936 Motorische Felder und Bahnen. In: Bumke O, Foerster O (eds) Handbuch der Neurologie. Vol 6. Springer-Verlag: Berlin: pp 1–357

Foh E, Haug H, König M et al 1973 Quantitative Bestimmung zum feineren Aufbau der Sehrinde der Katze, zugleich ein methodischer Beitrag zur Messung des Neuropils. Microsc Acta 75: 148–168

Foley J M, Baxter D 1958 On the nature of pigment granules in the cells of the locus coerulus and substantia nigra. J Neuropathol Exp Neurol 17: 586–598

Foley J O, DuBois F S 1937 Quantitative studies of the vagus nerve in the cat. J Comp Neurol 67: 49–67

Folke L E A, Stallard R E 1967 Periodontal microcirculation as revealed by plastic microspheres. J Periodont Res 2: 53–63

Folkman J, Klagsbrun M 1987 Angiogenic factors. Science 235: 442–447

Foltz E L, White L E Jr 1962 Pain 'relief' by frontal cingulotomy. J Neurosurg 19: 89–100

Fontana F 1781 Traité sur le vénin de la vipère, sur les poisons Americains, sur le laurier-cerise, etc. Florence

Forbes G 1938 Lymphatic drainage of the skin, with a note on lymphatic watershed areas. J Anat 72: 399–410

Ford E H R 1956 The growth of the foetal skull. J Anat 90: 63–72

Forrest W J 1967 Motor innervation of human thenar and hypothenar muscles in 25 hands: a study combining E M G and percutaneous nerve stimulation. Canad J Surg 10: 196–199

Forrester J C 1973 Mechanical biochemical and architectural features of surgical repair. Adv Biol Med Phys 14: 1–34

Forsberg J G 1963 The derivation and differentiation of the vaginal epithelium Dissertation, Lund

Forslind B 1970 Biophysical studies of the normal nail. Acta Dermatol Venereol (Stockh) 50: 161–168

Forssmann W G, Girardier L 1970 A study of the T system in rat heart. J Cell Biol 44: 1–19

Forssmann W G, Hock D, Lottspeich F, Henschen A, Kreye V, Christmann M, Reinecke M, Metz J, Carlquist M, Mutt V 1983 The right auricle of the heart is an endocrine organ. Cardiolatin as a peptide hormone candidate. Anat Embryol 168: 307–313

Forssner H 1928 Uber den Deszensus der Geschlechtsdrüsen beim Menschen. Acta Obstet Gynecol Scand 7: 379–406

Foster C A, Holbrook K A, Farr A G 1986 Ontogeny of Langerhans cells in human embryonic and fetal skin: expresion of HLA-DR and OKT 6 determinants. J Invest Dermatol 86: 240–243

Fountain F P, Minear W L, Allison R D 1966 Function of longus colli and longissimus cervicis muscles in man. Archs Phys Med 47: 665–669

Fourman J, Moffat D B 1971 The blood vessels of the kidney. Blackwell: Oxford

Fowler R Jr 1957 Primary peritonitis. Aust N Z J Surg 26: 204–213

Fox B, Bull, T B, Guz A 1980 Innervation of alveolar walls in the human lung: an electron microscopic study. J Anat 131: 683–692

Fox C A, Barnard J W 1957 A quantitative study of the Purkinje cell dendritic branchlets and their relationship to afferent fibres. J Anat 91: 299–313

Fox C A, Snider R S eds 1967 The cerebellum. Progr Brain Res 25:

Fox C A, Hillman D E, Siegesmund K A, Sether L A 1966 The primate globus pallidus and its feline and avian homologues: a Golgi and electron microscopic study. In: Hasler R, Stephan H (eds) Evolution of the forebrain. Thieme: Stuttgart: pp 237–248

Fox C A, Hillman D E, Siegesmund K A, Dutta C R 1967 The primate cerebellar cortex: a Golgi and electron microscopic study. Progr Brain Res 25: 174–225

Fox C A, Ubeda-Purkiss M, Ihrig H K, Biagioli D 1951 Zinc-chromate modification of Golgi technic. Stain Technol 26: 109–114

Foxon G E H 1955 Problems of the double circulation in the vertebrates. Biol Rev 30: 196–226

Foxon G E H, Bannister L H 1974 Circulation and circulatory systems. In: Encyclopaedia Britannica 15th edition. Encyclopaedia Britannica: Chicago: pp 618–634

Frable M A 1961 Computation of motion at the crico-arytenoid joint. Arch Otolaryngol 73: 551–556

Fraenkel L, Papanicolaou G N 1938 Growth, desquamation and involution of the vaginal epithelium of fetuses and children with a consideration of the related hormonal factors. Am J Anat 62: 427–452

Fraley E E, Weiss L 1961 An electron microscopic study of the lymphatic vessels in the penile skin of the rat. Am J Anat 109: 85–102

François J, Neetens A 1969 Physioanatomy of the axial vascularisation of the optic nerve. Doc Ophthalmol 26: 38–49

François R J, Dhem A 1974 Microradiographic study of the normal human vertebral body. Acta Anat 89: 251–265

Frank J S, Langer G A 1974 The myocardial interstitium: its structure and its role in ionic exchange. J Cell Biol 60: 586–601

Frank R M 1968 Ultrastructural relationship between the odontoblast, its process and the nerve fibres. In: Symons N B B (ed) Dentine and pulp; their structure and reactions. A symposium. University of Dundee: Dundee: pp 115–145

Frank R M, Steuer P 1977 Etude ultrastructurale du cément cellulaire chez le rat. J Biol Buccale 5: 121–135

Franke W W, Scheer U, Krohne G, Jarasch E-D 1981 The nuclear envelope and the architecture of the nuclear periphery. J Cell Biol 91: 39s–50s

Franke W W, Schmid E, Freudenstein C, Appelhans B, Osborn M, Weber K 1980 Intermediate-sized filaments of the prekeratin type in myoepithelial cells. J Cell Biol 84: 633–654

Franklin K J 1937 A monograph on veins. Baillière, Tindall and Cox: London

Frankfurter A, Weber J T, Royce G J, Strominger N L, Harting J K 1976 An autoradiographic analysis of the tecto-olivary projection in primates. Brain Res 118: 245–257

Frankl O 1933 On the physiology and pathology of the isthmus uteri. J Obstet Gynaecol Br Commonw 40: 397–422

Franklin K J 1939 Radiographic demonstration of circulation through the heart in the adult and in the foetus, and identity of the ductus arteriosus. Br J Radiol 12: 505–517

Franks L M 1953 Spread of prostatic cancer to bone. J Pathol Bact 66: 91–93

Franks L M 1954 Benign nodular hyperplasia of prostate; review. Ann R Coll Surg 14: 92–106

Franzini-Armstrong C 1970 Natural variability in the length of thin and thick filaments in single fibres from a crab Portunus depurator. J Cell Sci 6: 559–582

Franzini-Armstrong C 1975 Membrane particles and transmission at the triad. Fed Proc 34: 1382–1389

Frasca P, Harper R A, Katz J L 1978 A new technique for studying collagen fibers and ground substance in bone with scanning electron microscopy. Microsc Acta 80: 211–214

Frasca P, Harper R A, Katz J L 1981 Scanning electron microscopy studies of collagen, mineral and ground substance in human cortical bone. Scan Electron Microsc 3: 339–346

Fraser E A 1950 The development of the vertebrate excretory system. Biol Rev 25: 159–187

Fraser G R, Mayo O 1975 Textbook of Human Genetics. Blackwell Scientific: Oxford

Fraser H M, Gunn A, Jeffcoate S L 1975 Effect of active immunisation to luteinizing hormone releasing hormone on serum and pituitary gondotrophins, testes and accessory sex organs in the male rat. J Endocrinol 63: 399–406

Fraser J S, Dickie J K M 1914 A reconstruction model of the right middle and inner ear. J Anat 49: 119–135

Fraser L R, Zanellotti H M, Paton G R, Drury L M 1976 Increased incidence of triploidy in embryos derived from mouse eggs fertilised in vitro. Nature 260: 39–40

Fraser S E 1985 Gap junctions and cell interactions during development. Trends Neurosci 8: 3–4

Frazer J E 1914 The second visceral arch and groove in the tubo-tympanic region. J Anat 48: 391–408

Frazer J E 1926 The disappearance of the pre-cervical sinus. J Anat 61: 132–143

Frazer J E 1928 Development of the region of the isthmus rhombencephali. J Anat 63: 7–18

Frazer J E, Robbins R H 1915 On the factors concerned in causing rotation of the intestine in man. J Anat 50: 75–110

Freder T 1905 A propos de la communication de M M Quénu et Heitz-Boyer sur l'anatomie du caecum et de l'appendice. Bull Mém Soc Anat Par 80: 188–190

Free M J 1977 Blood supply to the testis and its role in local exchange and transport of hormones. In: Johnson A D, Gomes W R (eds) The testis vol 4. Advances in physiology, biochemistry and function. Academic Press: New York: pp 39–90

Freedman R, Foote S L, Bloom F E 1975 Histochemical characterization of a neocortical projection of the nucleus locus coeruleus in the squirrel monkey. J Comp Neurol 164: 209–231

Freeman M A R ed 1979 Adult Articular Cartilage. 2nd edition, Pitman Medical: London

Freeman M A R, Wyke B 1967 The innervation of the knee joint. J Anat 101: 505–532

Friede R L, Brzoska J, Hartmann U 1985 Changes in myelin sheath thickness and internode geometry in the rabbit phrenic nerve during growth. J Anat 143: 103–113

Friedenberg Z B, Roberts P G Jr, Didizian N H, Brighton C T 1971 Stimulation of fracture healing by direct current in the rabbit fibula. J Bone Jt Surg 53A: 1400–1408

Friedmann P S 1986 Immune functions of skin. In: Thody A J, Friedmann P S (eds) Scientific basis of dermatology. A physiological approach. Churchill Livingstone: Edinburgh: pp 58–73

Friendenstein A J 1976 Precursor cells of mechanocytes. Int Rev Cytol 47: 327–359

Friere F, Cardinali D F 1975 Effects of melatonin treatment and environmental lighting on the ultrastructure, appearance, melatonin synthesis, norepinephrine turnover and microtubule protein content of the rat pineal gland. J Neurol Transmission 37: 237–257

Frisch D 1967 Ultrastructure of mouse olfactory mucosa. Am J Anat 121: 87–120

Fritsch G, Hitzig E 1870 Ueber die elektrische Erregbarkeit des Grosshirns. Arch Anat Physiol 37: 300–332

Frömter E 1979 The Feldberg lecture 1976. Solute transport across epithelia: what can we learn from micropuncture studies in kidney tubules? J Physiol 288: 1–31

Frutiger P 1969 Zur frühentwicklung der ductus paramesonephrici und des müllerrerschen hugels beim menschen. Acta Anat 72: 233–245

Fry D B 1979 The physics of speech, Cambridge University Press: Cambridge

Fry G N, Devine C E, Burnstock G 1977 Freeze-fracture studies of nexuses between smooth muscle cells. Close relationship to sarcoplasmic reticulum. J Cell Biol 72: 26–34

Fujita H, Fujita S 1963 Electron microscopic studies on neuroblast differentiation in the central nervous system of domestic fowl. Z Zellforsch 60: 463–478

Fujita S 1963 The matrix cell and cytogenesis in the developing central nervous system. J Comp Neurol 120: 37–42

Fujita S 1967 Quantitative analysis of cell proliferation and differentiation in the cortex of the postnatal mouse cerebellum. J Cell Biol 32: 277–287

Fujita S, Shimada M, Nakamura T 1966 ^{3}H-thymidine autoradiographic studies on the cell proliferation and differentiation in the external and internal granular layers of the mouse cerebellum. J Comp Neurol 128: 191–208

Fujita T 1959 Histological studies on the neuro-insular complex in the pancreas of some mammals. Z Zellforsch 50: 94–109

Fujita T ed 1973 Gastro-entero-pancreatic system. A cell-biological approach. Igaku Shoiu: Tokyo

Fujita T 1976 The gastro-enteric endocrine cell and its paraneuronic nature. In: Coupland R E, Fujita T (eds) Chromaffin, enterochromaffin and related cells. A NAITO Foundation Symposium. Elsevier: New York, Amsterdam: pp 191–208

Fujita T, Miyoshi M, Tokunaga J 1970 Scanning and transmission electron microscopy of human ejaculate spermatozoa with special reference to their abnormal forms. Z Zellforsch 105: 483–497

Fukuda T 1976 Ultrastructrue of primordial germ cells in human embryo. Virchows Arch B Cell Pathol 20: 85–89

Fukuda T, Hedinger C 1975 Ultrastructure of developing germ cells in the fetal human testis. Cell Tissue Res 161: 55–70

Fuld H, Irwin D T 1954 Clincial application of portal venography. Br Med J 1: 312–313

Fuller A P, Fozzard J A, Wright G H 1959 Spincteric action of cricopharyngeus: radiographic demonstration. Br J Radiol 32: 32–35

Fulton J F, Sheenan D 1935 The uncrossed lateral pyramidal tract in higher primates. J Anat 69: 181–187

Fürbinger M 1873 Zur vergleichenden Anatomie der Schulter-muskeln. Jena Z Naturw 7: 237–320

Furcht L T, Wendelschafer-Crabb G, Mosher D F, Foidart J M 1980 An axial periodic fibrillar arrangement of antigenic deterinmants for fibronectin and procollagen on ascorbate-treated human fibroblasts. J Supramol Struct 13: 15–33

Furseth R 1968 The resorption process of human deciduous teeth studied by light microscopy, microradiography and electron microscopy. Arch Oral Biol 13: 417–431

Fusijawa H, Morioka H, Watanabe K, Nakamura H 1976 A decay of gap junctions in association with cell differentiation of neural retina in chick embryonic development. J Cell Sci 22: 585–596

Fuxe K, Jonsson G 1974 Further mapping of central 5-hydroxytryptamine neurons: studies with the neurotoxic dihydroxytryptamines. Adv Biochem Psychopharmacol 10: 1–12

Fuxe K, Hökfelt T, Jonsson G, Ungerstedt U 1970 Fluorescence microscopy in neuroanatomy. In: Nauta W J H, Ebbesson S O E (eds) Contemporary research methods in neuroanatomy. Springer-Verlag: Berlin: pp 275–314

G

Gabbiani G, Ryan G B, Lamelin J P et al 1973 Human smooth muscle autoantibody. Its identification as antiactin antibody and a study of its binding to 'nonmuscular' cells. Am J Pathol 72: 473–488

Gabbiani G, Chaponnier C, Hüttner I 1978 Cytoplasmic filaments and gap junctions in epithelial cells and myofibroblasts during wound healing. J Cell Biol 76: 561–568

Gabe M 1966 Neurosecretion. Pergamon: Oxford

Gabella G 1973 Cellular structures and electrophysiological behaviour. Fine structure of smooth muscle. Philos Trans R Soc Lond 265B: 7–16

Gabella G 1974 Special muscle cells and their innervation in the mammalian small intestine. Cell Tissue Res 153: 63–77

Gabella G 1976 Structure of the autonomic nervous system. Chapman & Hall: London

Gabella G 1981 Structure of smooth muscles. In: Bülbring E, Brading A F, Jones A W, Tomita T (eds) Smooth muscle: an assessment of current knowledge. Edward Arnold: London: pp 1–46

Gabella G 1984 Hypertrophic smooth muscle. V. Colagen and other extracellular materials. Vascularization. Cell Tissue Res 235: 275–283

Gabriel A C 1958 Some anatomical features of the mandible. J Anat 92: 580–586

Gabriel W B 1945 The principles and practice of rectal surgery. Lewis: London

Gacek R R 1974 Transection of the post-ampullary nerve for the relief of benign paroxysmal postural vertigo. Ann Otol Rhinol Laryngol 83: 596–605

Gaddum P 1968 Sperm maturation in the male reproductive tract: Development of motility. Anat Rec 161: 471–482

Gadow H F 1933 The evolution of the vertebral column. In: Gaskell J F, Green H L H H (eds) A contribution to the study of vertebrate phylogeny. Cambridge University Press: Cambridge

Gaillard P J, Talmage R V, Budy A M eds 1965 The parathyroid glands: ultrastructure, secretion and function. University of Chicago Press: Chicago

Gajisin S, Zbrodowski A, Grodecki J 1983 Vascularization of the extensor apparatus of the fingers. J Anat 137: 315–322

Galilei, Galileo 1638 Discorsi e dimostrazioni matematiche intorno à due nuove scienze. Attenenti alla mecanica e i movimenti locali . . . con una appendice del centro di gravità d'alcuni solidi. Elzevirs: Leiden

Galindo see De Lara Galindo

Gall F J, Spurzheim J C 1810–1819 Anatomie et Physiologie du Système Nerveux en Générale, et du Cerveau en Particulier. 4 Vol and atlas F Schoell: Paris

Galli G, Ottaviani G, Galli S 1976 Studio Biometrico dell'apofisi mastoidea nell'uomo. I. Indagini ad indirizzo antropologico: l'atezza della mastoide correlata a parametri cranici e statura. II. Catt Ist Anat Um Norm Univ Modena Ita–Quadranat Prat 32: 1–4 123–132

Gamble H J, Eames R A J 1964 An electron microscope study of the connective tissue of human peripheral nerve. J Anat 98: 655–663

Gamble H J, Fenton J, Allsopp G 1978 Electron microscope observations on the changing relationships between unmyelinated axons and Schwann cells in human fetal nerves. J Anat 127: 363–378

Ganguly D N, Roy K K 1964 A study on the cranio-vertebral joint in the man. Anat Anz 114: 433–452

BIBLIOGRAPHY

Ganguly D N, Singh-Roy K K 1965 A study on the cranio-vertebral joint in the vertebrates. I. In the mammals, as illustrated by its structure in the guinea pig and development in the guinea pig and Talpa. Anat Anz 117: 421–429

Gardiner D S, More I A R, Lindop G B M 1986 The granular peripolar cell of the human glomerulus: an ultrastructural study. J Anat 146: 31–43

Gardner D L, Woodward D 1969 Scanning electron microscopy and replica studies of articular surfaces of guinea pig synovial joints. Ann Rheum Dis 28: 379–391

Gardner E 1967 Spinal cord and brain stem pathways for afferents from joints. In: Reuck A V S, Knight J (eds) Myotatic, kinesthetic and vestibular mechanisms. (Ciba Foundation Symposium). Churchill: London: pp 56–76

Gardner E, Gray D J 1950 Prenatal development of the human hip joint. Am J Anat 87: 162–212

Gardner E, Gray D J 1953 Prenatal development of the human shoulder and acromioclavicular joints. Am J Anat 92: 219–276

Gardner E, Lenn N J 1977 Fibres in monkey posterior articular nerves. Anat Rec 187: 99–106

Gardner E, O'Rahilly R 1968 The early development of the knee joint in staged human embryos. J Anat 102: 289–299

Gardner E, O'Rahilly R 1976 The nerve supply and conducting system of the human heart at the end of the embryonic period proper. J Anat 121: 571–587

Gardner E D 1948a The innervation of the knee joint. Anat Rec 101: 109–130

Gardner E D 1948b The innervation of the shoulder joint. Anat Rec 102: 1–18

Gardner E D 1950 Physiology of movable joints. Physiol Rev 30: 127–176

Gardner F 1949 Angiocardiography. Postgrad Med J 25: 553–564

Garey L J, Fisken R A, Powell T P S 1976 Cellular changes in the lateral geniculate nucleus of the cat and monkey after section of the optic tract. J Anat 121: 15–27

Garey L J, Powell T P S 1967 The projection of the lateral geniculate nucleus upon the cortex in the cat. Proc R Soc Lond B 169: 107–126

Garey L J, Powell T P S 1968 The projection of the retina in the cat. J Anat 102: 189–222

Garfia A 1980 Glomus tissue in the vicinity of the human carotid sinus. J Anat 130: 1–12

Garn S M, Lewis A B 1962 The relationship between third molar agenesis and reduction in tooth number. Angle Orthod 33: 14–18

Garn S M, Rohmann C G 1960 Variability in the order of ossification of the bony centers of the hand and wrist. Am J Phys Anthropol N S 18: 219–230

Garrett F D 1948 Development of the cervical vesicles in man. Anat Rec 100: 101–114

Garrett J R 1963 The ultrastructure of intracellular fat in the parenchyma of human submandibular salivary glands. Arch Oral Biol 8: 729–734

Garrett J R 1967 The innervation of normal human submandibular and parotid salivary glands demonstrated by cholinesterase histochemistry, catecholamine fluorescence and electron microscopy. Arch Oral Biol 12: 1417–1436

Garrett J R 1972 Neuro-effector sites in salivary glands. In: Emmelin N, Zotterman Y (eds) Oral physiology. Pergamon: Oxford: pp 83–97

Garrett J R 1975 Recent advances in physiology of salivary glands. Br Med Bull 31: 152–155

Garrett J R 1976 Structure and innervation of salivary glands. In: Cohen B, Kramer J R H (eds) Scientific foundations of denistry. Heinemann: London: pp 499–516

Garrett J R, Emmelin N 1979 Activities of salivary myoepithelial cells: a review. Med Biol 57: 1–28

Garrett J R, Kemplay S K 1977 The adrenergic innervation of the submandibular gland of the cat and the effects of various surgical denervations on these nerves. A histochemical and ultrastructural study including the use of 5-hydroxydopamine. J Anat 124: 99–115

Garrett J R, Kidd A 1975 Effects of nerve stimulation and denervation on secretory material in submandibular striated duct cells of cats, and the possible role of these cells in the secretion of salivery kallikrein. Cell Tissue Res 161: 71–84

Garrett J R, Parsons P A 1974 Movement of horseradish peroxidase in submandibular glands of rabbits after arterial injection. J Physiol 237: 3–4P

Garrett J R, Howard E R, Nixon H H 1969 Autonomic nerves in rectum and colon in Hirschsprung's disease. A cholinesterase and catecholamine histochemical study Arch Dis Childh 44: 406–417

Garrod A E 1963 Inborn errors of metabolism. Oxford University Press: Oxford

Gasser H S 1956 Olfactory nerve fibres. J Gen Physiol 39: 473–498

Gastaut H, Lammers H J 1960 Anatomie du rhinencéphale. Les grandes activités du rhinencéphale. Masson: Paris

Gatsalov M D 1963 Limfaticheskaia sistema slizistoĭ obolochki fallopievoĭ truby cheloveka. Akush Ginek 39: 85–90

Gatter K C, Powell T P 1977 The projection of the locus coeruleus upon the neocortex in the macaque monkey. Neuroscience 2: 441–445

Gauer O H 1968 Osmocontrol versus volume control. Fed Proc Fed Am Socs Exp Biol 27: 1132–1136

Gaughran G R L 1963 Mylohyoid boutonnière and sublingual bouton. J Anat 97: 565–568

Gaunt P N, Gaunt W A 1978 Three dimensional reconstruction in biology. (First published as Microreconstruction by Gaunt W A in 1971), Pitman Medical: Tunbridge Wells

Gaunt W A 1959 The vascular supply to the dental lamina during early developement. Acta Anat 37: 232–252

Gaunt W A 1971 Microreconstruction. Pitman Medical: London (cf Gaunt P N, Gaunt W A 1978)

Gautvik K M, Kriz M, Lund-Larsen K 1972 Adrenergic vasodilation in the cat submandibular salivary gland. In: Emmelin N, Zotterman Y (eds) Oral physiology. Pergamon: Oxford: pp 161–162

Gaze R M 1970 The formation of nerve connections. Academic Press: London

Gazzaniga M S 1970 The bisected brain. Appleton-Century-Crofts: New York

Gebhardt W 1901 Über funktionell wichtige Anordnungsweisen der grösseren und feineren Bauelemente des Wirbeltierknochens. Arch EntwMech Org 11: 383–498; 20: 187–334

Gehr P, Bachofen M, Weibel E R 1978 The normal human lung: ultrastructure and morphometric estimation of diffusion capacity. Resp Physiol 32: 121–140

Geiger R S 1963a In: Pfeiffer C C, Smythies J R (eds) Neurobiology. Vol 5. Academic Press: New York

Geiger R S 1963b The behaviour of adult mammalian brain cells in culture. Int Rev Biol 5: 1–52

Gelfan S, Rapisarda A F 1964 Synaptic density on spinal neurons of normal dogs and dogs with experimental hind-limb rigidity. J Comp Neurol 123: 73–96

Gelfan S, Kao G, Ruchkin D S 1970 The dendritic tree of spinal neurons. J Comp Neurol 139: 355–412

Geller M, Barbato D 1970 Nervus peronaeus profundus. A study of the terminal branches and their variations. Hospital Rio de J 77: 679–698

Gemzell C A, Roos P 1966 Pregnancies following treatment with human gonadotrophin, with special reference to the problem of multiple births. Am J Obstet Gynecol 94: 490–496

Geniec P, Morest D K 1971 The neuronal architecture of the human posterior colliculus. A study with the Golgi method. Acta Otolaryngol suppl 295

Génis-Gálvez J M, Santos Gutierrez L, Martin Lopez M 1966 On the double innervation of the parotid gland. An experimental study. Acta Anat 63: 398–403

Genovese S T 1959 Diferencias sexuales en el huesco coxal. Universidad Nacional Autonoma de Mexico No 49: Mexico

Genovese S T, Messmacher M 1959 Valor de los patrones tradicionales para la determinacion de la edad por medio de las suturas en craneos Mexicanos (Indigenas y Mestizos) Cuad Inst Hist Méx No 7

George W C 1942 A presomite human embryo with chorda canal and prochordal plate. Contrib Embryol Carnegie Inst Washington 30: 1–7

Gerdy P N 1823 Recherches, Discussions et propositions, etc. Thèse, Paris

Gerfen C R 1984 The neostriatal mosaic: compartmentalization of corticostriatal input and striatonigral output systems. Nature 311: 461–464

Gerich J E, Lorenzi M 1978 The role of the autonomic nervous system and somatostatin in the control of insulin and glucagon secretion. In: Ganong W F, Martini L (eds) Frontiers in neuroendocrinology. Vol 5. Raven Press: New York

Gerlach, Joseph von 1858 Mikroskopische Studien aus dem Gebiete der menschlichen morphologie. F Enke: Erlangen

Gerlis L M, Wright H M, Wilson N, Erzengin F, Dickinson D F 1984 Left ventricular bands. A normal anatomical feature. Br Heart J 52: 641–647

German J 1964 The pattern of DNA synthesis in the chromosomes of human blood cells. J Cell Biol 20: 37–55

Gern W A, Karn C M 1983 Evolution of melatonin's functions and effects. In: Reiter R J (ed) Pineal research reviews. Vol 1. Liss: New York: pp 49–90

Gernandt B E, Iranyi M, Livingston R B 1959 Vestibular influences on spinal mechanisms. Exp Neurol 1: 248–273

Geschwandtner W R 1973 Striae cutis atrophicae nach Lokalbehandlung mit Corticosteroiden. Hautarzt 24: 70–73

Geschwind N, Levitsky W 1968 Human brain: left-right asymmetries in temporal speech region. Science 161: 186–187

Gesteland R C, Lettvin J Y, Pitts W, Rojas A 1963 Odor specificies of the frog's olfactory receptors. In: Zotterman Y (ed) Olfaction and taste. Pergamon: Oxford: pp 19–34

Getz B, Sirnes T 1949 The localisation within the dorsal motor vagal nucleus. An experimental investigation. J Comp Neurol 90: 95–110

Ghabriel M N, Alt G 1982 The node of Ranvier. In: Harrison R J, Navaratnam V (eds) Progress in Anatomy. Vol 2. Cambridge University Press: Cambridge: pp 138–160

Ghadially F N 1983 Fine structure of synovial joints. A text and atlas of the ultrastructure of normal and pathological articular tissues. Butterworths: London

Ghadially J A, Ghadially F N 1975 Evidence of cartilage flow in deep defects in articular cartilage. Virchows Arch [B] 18: 193–204

Ghadially F N, Roy S 1969 Ultrastructure of synovial joints in health and disease. Butterworths: London

Ghadially F N, Ailsby R L, Oryschak A F 1974 Scanning electron microscopy of superficial defects in articular cartilage. Ann Rheum Dis 44: 327–332

Ghadially F N, Ghadially J A, Oryschak A F, Yong N K 1976 Experimental production of ridges on rabbit articular cartilage. J Anat 121: 119–132

Ghadially F N, Ghadially J A, Oryschak A F, Yong N K 1977a The surface of dog articular cartilage: a scanning electron microscopic study. J Anat 123: 527–536

Ghadially F N, Lalonde J-M A, Wedge J H 1983 Ultrastructure of normal and torn menisci of the human knee joint. J Anat 136: 773–791

Ghadially F N, Moshurchak E M, Thomas I 1977b Humps on young human and rabbit articular cartilage. J Anat 124: 425–435

Ghadially F N, Yong N K, Lalonde J M 1982 A transmission electron microscopic comparison of the articular surface of cartilage processed attached to bone and detached from bone. J Anat 135: 685–706

Ghebrehiwet B, Silverberg M, Kaplan A P 1981 Activation of the classical pathway of complement by Hageman factor fragment. J Exp Med 153: 665–676

Gianelli F 1970 Human chromosomes and DNA synthesis (monographs in human genetics vol 5) Karger: Basel

Gibbons I R, Grimstone A V 1960 On flagellar structures in certain flagellates. J Biophys Biochem Cytol 7: 697–716

Gil J, Reiss O K 1973 Isolation and characterisation of lamellar bodies and tubular myelin from rat lung homogenates. J Cell Biol 58: 152–171

Gil J, Weibel E R 1969 Improvements in demonstration of lining layer of lung alveoli by electron microscopy. Resp Physiol 8: 13–36

Gil Vernet S 1969 Plan structural y dinámico de la musculatura del uréter. An Acad Nac Med (Madrid) 86: 347–364

Gilad I, Nissan M 1985 Sagittal evaluation of elemental geometrical dimensions of human vertebrae. J Anat 143: 115–120

Gilbert P W 1952 The origin and development of the head cavities in the human embryo. J Morphol 90: 149–187

Gilbert P W 1957 The origin and development of the human extrinsic ocular muscles. Contrib Embryol Carnegie Inst Washington 36: 59–78

Giles E, Elliot O 1960, Negro-white identity from the skull. Proceedings of the 6th International Anthropological Congress, Paris. Quoted in: Krogman W M (ed) The human skeleton in forensic medicine 1962. Thomas: Springfield, Illinois: pp 195–196

Gillilan L 1958 The arterial blood supply of the human spinal cord. J Comp Neurol 110: 75–104

Gillilan L A 1941 The connexions of the basal optic root (posterior accessory optic tract) and its nucleus in various mammals. J Comp Neurol 74: 367–408

Gillilan L A 1972 Anatomy and embryology of the arterial system of the brain stem and cerebellum. In: Vinken I J, Bruyn G W (eds) Handbook of clinical neurology. Vol 2. North Holland: Amsterdam: pp 24–44

Gillilan L A 1974 Potential collateral circulation to the human cerebral cortex. Neurology 24: 941–948

Gillman J 1948 The development of the gonads in man, with a consideration of the role of fetal endocrines and the histogenesis of ovarian tumors. Contrib Embryol Carnegie Inst Washington, 32: 81–131

Gilman S (1985) The cerebellum: its role in posture and movement. In: Swash M, Kennard C (eds) Scientific Basis of Clinical Neurology. Churchill Livingstone: Edinburgh: pp 36–55

Gilmour J R 1938 Gross anatomy of the parathyroid glands. J Pathol Bact 46: 133–149

Gilpin S A, Gosling J A 1983 Smooth muscle in the wall of developing human urinary bladder and urethra. J Anat 137: 503–512

Gilula N B, Satir P 1972 The ciliary necklace: a ciliary membrane specialization. J Cell Biol 53: 494–509

Gingerich P D 1971 Functional significance of mandibular translation in vertebrate jaw mechanics. Postilla 152: 1–10

Giok S P 1956 Localisation of fibre systems within the white matter of the medulla oblongata and the cervical cord in man. Ijido: Leiden

Girgis F G, Marshall J L, Al Monajem A R S 1975 The cruciate ligaments of the knee joint. Clin Orthop 106: 216–231

Gladstone R J 1929 Development of the inferior vena cava in the light of recent research, with especial reference to certain abnormalities, and current descriptions of the ascending lumbar and azygos veins. J Anat 64: 70–93

Gladstone R J Wakeley C P G 1940 The pineal organ. Baillière, Tindall and Cox: London

Glaister J, Brash J C 1937 Medicolegal aspects of the Buck Ruxton case. Livingstone: Edinburgh

Glauser F 1953 Studies on intrahepatic arterial circulation. Surgery 33: 333–341

Glees P 1941 The termination of optic fibres in the lateral geniculate body of the cat. J Anat 75: 434–440

Glees P 1944 The anatomical basis of cortico-striate connexions. J Anat 78: 47–51

Glees P 1946 Terminal degeneration within the central nervous system as studied by a new silver method. J Neuropath Exp Neurol 5: 54–59

Glees P 1961 Experimental neurology. Clarendon Press: Oxford

Glees P 1963 Neuroglia, morphology and function, Blackwell: Oxford

Glees P, Cole J, Whitty C W M, Cairns H 1950 The effects of lesions in the cingular gyrus and adjacent areas in monkeys. J Neurol Neurosurg Psychiat 13: 178–190

Glenister T W 1954 The origin and fate of the urethral plate in man. J Anat 88: 413–425

Glenister T W 1962 The development of the utricle and of the so-called 'middle' or 'median' lobe of the human prostate. J Anat 96: 443–455

Glenister T W 1976 An embryological view of cartilage. J Anat 122: 323–330

Glickstein M, King R A, Miller J, Berkley M 1967 Cortical projections from the dorsal lateral geniculate nucleus of cats. J Comp Neurol 130: 55–76

Glisson F 1654 Anatomia hepatis subjiciuntur nonnulla de lymphaeductibus nuper repertis. Du-Gardianis: London

Gloor P 1960 Amygdala. In: Field J, Magoun H W, Hall V E (eds) Handbook of physiology section 1, neurophysiology. Vol 2. American Physiological Society: Washington DC: pp 1395–1420

Gloster J, Perkins E S, Pommier M 1957 Extensibility of strips of sclera and cornea. Br J Ophthalmol 41: 103–110

Glücksmann N A 1951 Cell deaths in normal vertebrate ontogeny. Biol Rev 26: 59–86

Glynn L E 1977 Hypothesis. Primary lesion in osteoarthritis. Lancet 1: 574–575

Godlewski G, Hedon B, Castel C 1975 Contribution à l'étude de l'origine de l'artère hépatique chez l'embryon et le foetus. Bull Assoc Anat Nancy 59: 411–418

Goel V K, Svensson N L 1977 Forces on the pelvis. J Biomech 10: 195–200

Goerke J 1974 Lung surfactant. Biochim Biophys Acta 344: 241–261

Goland M ed 1975 Normal and abnormal growth of the prostate. Thomas: Springfield, Illinois

Goldberg J M, Moore R Y 1967 Ascending projections of the lateral lemniscus in the cat and monkey. J Comp Neurol 129: 143–156

Goldberg S 1970 The origin of the lumbrical muscles in the hand of the South African native. The hand 2: 168–171

Golde D W, Takaku F eds 1985 Haemopoietic Stem Cells. Dekker: New York

Goldman B D 1983 The physiology of melatonin in mammals. In: Reiter R J (ed) Pineal research reviews. Vol 1. Liss: New York: pp 145–182

Goldman P, Nauta W J H 1977 Columnar distribution of cortico-cortical fibers in the frontal association, limbic, and motor cortex of the developing Rhesus monkey. Brain Res 122: 393–413

Goldman-Rakic P S 1984 Modular organization of prefrontal cortex. Trends Neurosci 7: 419–424

Goldman-Rakic P S, Schwartz M L 1982 Interdigitation of contralateral and ipsilateral columnar projections to frontal association cortex in primates. Science 216: 755–757

Goldsmith N A, Woodburne R T 1957 The surgical anatomy pertaining to liver resection. Surgery Gynecol Obstet 105: 310–318

Goldstein M B, Davies E A Jr 1968 The three-dimensional architecture of the islets of Langerhans. Acta Anat 71: 161–171

Golgi C 1878 . . . R C Ist Lombardo Sci 2nd series 12

Goligher J C 1967 Surgery of the anus, rectum and colon. Baillière, Tindall and Cox: London

Goligher J C, Leacock A G, Brossy J-J 1955 Surgical anatomy of anal canal. Br J Surg 43: 51–61

Goltz F 1891-2 Der hund ohne Grosshirn; siebente Abhandlung über die Verrichtungen des Grosshirns. Pflügers Arch Ges Physiol 51: 570–614

Gomes F A 1969 . . . O Medico

Gomori G 1943 Calcification and phosphatase. Am J Pathol 19: 197–209

Goodenough D A, Revel J P 1970 A fine structural analysis of intercellular junctions in the mouse liver. J Cell Biol 45: 272–290

Goodman D C, Hallett R E, Welch R B 1963 Patterns of localisation in the cerebellar corticonuclear projections of the albino rat. J Comp Neurol 121: 51–67

Goodrich E S 1958 Studies on the structure and development of vertebrates. Macmillan: London (republication of 1930 edition)

Gorbsky G, Steinberg M S 1981 Isolation of the intracellular glycoproteins of desmosomes. J Cell Biol 90: 243–248

Gordon B 1973 Receptive fields in deep layers of cat superior colliculus. J Neurophysiol 36: 157–178

Gordon K C D 1967 A comparative anatomical study of the distribution of the cystic artery in man and other species. J Anat 101: 351–359

Gorgollón P 1978 The normal human appendix: a light and electron microscopic study. J Anat 126: 87–101

Gorp P E V, Kennedy W R 1974 Localization of muscle spindles in the human extensor indicis muscle for biopsy purposes. Anat Rec 179: 447–452

Gorter E, Grendel F J 1925 On bimolecular layers of lipoids on the chromocytes of the blood. J Exp Med 41: 439–443

Gosling J A 1970 The musculature of the upper urinary tract. Acta Anat 75: 408–422

Gosling J A, Dixon J S 1974 Species variation in the location of upper urinary tract pacemaker cells. Invest Urol 11: 418–423

Gosling J A, Dixon J S, Critchley H O, Thompson S A 1981 A comparative study of the human external sphincter and periurethral levator ani muscles. Br J Urol 53: 35–40

Gosling J A, Dixon J S, Lendon R G 1977 The autonomic innervation of the human male and female bladder neck and proximal urethra. J Urol 118: 302–305

Gosling R G, Newman D L, Bowden N L R, Twinn K W 1971 The area ration of normal aortic junctions. Aortic configuration and pulse-wave reflection. Br J Radiol 44: 850–853

Goss C M 1942 The physiology of the embryonic mammalian heart before circulation. Am J Physiol 137: 146–152

Göthlin G, Ericsson J L E 1976 The osteoclast: review of ultrastructure, origin and structure-function relationship. Clin Orthop 120: 201–231

Goto F 1959 Histological and histochemical studies on human fetal membranes. Acta Med Okayama 13: 276–300

Gotsev T 1939 Blood volume in lambs. J Physiol 94: 539–549

Gottlieb G ed 1973–1974 Behavioral embryology. Studies on the development of Behavior and the nervous system. Vol 1 [and 2] Academic Press: New York

Gottschaldt K M, Iggo A, Young D W 1973 Functional characteristics of mechanoreceptors in sinus hair follicles of the cat. J Physiol 235: 287–315

Gottschalk C W, Mylle M 1959 Micropuncture studies of the mammalian urinary concentrating mechanism: evidence for the counter current hypothesis. Am J Physiol 196: 927–936

Gowans J L 1966 Lifespan recirculation and transformation of lymphocytes. Int Rev Exp Pathol 5: 1–24

Gowans J L, Knight E J 1964 The route of recirculation of lymphocytes in the rat. Proc R Soc Lond [Biol] 159: 257–282

Gower D B, Nixon A, Mallet A I, Jackman P J H 1987 The significance of odorous steroids in axillary odour. In: Dodd G H, van Toller S (eds) Perfumery: the Psychology and Biology of Fragrance. Chapman and Hall: London: pp 96–105

Graabaek P M 1984 Characteristics of the two types of synoviocytes in rat synovial membrane. An ultrastructural study. Lab Invest 50: 690–702

Grainger F, James D W 1970 Association of glial cells with the terminal parts of neurite bundles extending from the chick spinal cord in vitro. Z Zellforsch 108: 93–104

Grainger F, James D W, Tresman R L 1968 An electron microscopic study of the early outgrowth from chick spinal cord in vitro. Z Zellforsch 90: 53–67

Granger D N, Perry M A, Kvietys P R 1983 The microcirculation and fluid transport in digestive organs. Fed Proc 42: 1667–1672

Granger D N, Perry M A, Kvietys P R, Taylor D E 1983 A new method for estimating intestinal capillary pressure. Am J Physiol 244: G 341–344

Granit R 1962 Receptors and sensory perception. Yale University Press: New Haven, Connecticut

Granit R 1970 The basis of motor control. Academic Press: New York

Granit R, Henatsch H D, Steg G 1956 Tonic and phasic ventral horn cells differentiated by post-tetanic potentiation in cat extensors. Acta Physiol Scand 37: 114–126

Grant G, Oscarsson O 1966a Functional organisation of the spinoreticulocerebellar path with identification of its spinal component. Exp Brain Res 1: 306–319

Grant G, Oscarsson O 1966b Mass discharges evoked in the olivocerebellar tract on stimulation of muscle and skin nerves. Exp Brain Res 1: 329–337

Grant P G 1973 Lateral pterygoid: two muscles? Am J Anat 138: 1–10

Grant R T 1930 Observations on direct connections between arteries and veins in the rabbit's ear. Heart 15: 281–303

Grant R T, Bland F F 1931 Observations on arterio-venous anastomoses in human skin and in the bird's foot with special reference to reaction to cold. Heart 15: 385–407

Grant R T, Regnier M 1926 The comparative anatomy of the cardiac coronary vessels. Heart 13: 285–317

Grant R T, Wright H P 1968 Further observations on the blood vessels of skeletal muscle (rat cremaster). J Anat 103: 553–565

Grant R T, Wright H P 1970 Anatomical basis for non-nutritive circulation in skeletal muscle exemplified by blood vessels of rat biceps femoris tendon. J Anat 106: 125–134

Grant R T, Wright H P 1971 The peculiar vasculature of the external spermatic fascia in the rat: possibilities subserving thermoregulation. J Anat 109: 293–305

Grassé P P ed 1948 Traité de zoologie Vol 11. Masson: Paris

Grassé P P ed 1954 Traité de zoologie Vol 12. Masson: Paris

Graves F T 1954 Anatomy of intrarenal arteries and its application to segmental resection of the kidney. Br J Surg 42: 132–139

Graves F T 1956a The aberrant renal artery. J Anat 90: 553–558

Graves F T 1956b Ganglion in muscle belly of peroneus longus. Br J Surg 43: 438–439

Gray D J, Gardner E D 1943 The human sternochondral joints. Anat Rec 87: 235–254

Gray D J, Gardner E 1951 Prenatal development of human elbow joint. Am J Anat 88: 429–469

Gray D J, Gardner E, O'Rahilly R 1957 The prenatal development of the skeleton and joints of the human hand. Am J Anat 101: 169–224

Gray E G 1959 Axo-sematic and axo-dendritic synapses of the cerebral cortex: an electron microscope study. J Anat 93: 420–433

Gray E G 1961 The granule cells, mossy synapses and Purkinje spine synapses of the cerebellum: light and electron microscope observations. J Anat 95: 345–356

Gray E G 1969 Electron microscopy of excitatory and inhibitory synapses: a brief review. Progr Brain Res 31: 141–155

Gray E G 1974 Synaptic morphology with special references to microneurons. In: Bellairs R, Gray E G (eds) Essays on the nervous system. A festschrift for Professor J Z Young. Clarendon Press: Oxford

Gray E G 1975 Presynaptic microtubules and their association with synaptic vesicles. Proc R Soc Lond [Biol] 190: (1100), 367–372

Gray E G 1976 Problems of understanding the substructure of synapses. Progr Brain Res 45: 207–234

Gray E G 1978 Synaptic vesicles and microtubules in frog motor end plates. Proc R Soc Lond [Biol] 203: 219–227

Gray E G, Basmajian J V 1968 Electromyography and cinematography of leg and foot ('normal' and flat) during walking. Anat Rec 161: 1–16

Gray E G, Guillery R W 1966 Synaptic morphology in the normal and degenerating nervous system. Int Rev Cytol 19: 111–182

Gray J 1958 The movement of the spermatozoa of the bull. J Exp Biol 35: 96–108

Gray J A B, Sato M 1953 Properties of receptor potentials in Pacinian corpuscles. J Physiol 122: 610–636

Graybiel A M 1975 Anatomical organization of retinotectal afferents in the cat: an autoradiographic study. Brain Res 96: 1–23

Graybiel A M, Hartwieg E A 1974 Some afferent connections of the oculomotor complex in the cat: an experimental study with tracer techniques. Brain Res 81: 543–551

Graziadei P P C 1971 The olfactory mucosa of vertebrates. In: Beidler L M (ed) Handbook of sensory physiology. Vol 4. Chemical senses. Springer-Verlag: Berlin

Graziadei P P C 1974 The olfactory organ of vertebrates: a survey. In: Bellairs R, Gray E G (eds) Essays on the nervous system. Clarendon Press: Oxford: pp 191–222

Graziadei P P C, Gagne H T 1973 Extrinsic innervation of olfactory epithelium. Z Zellforsch 138: 315–326

Graziadei P P C, Monti-Graziadei G A 1978 Continuous nerve cell renewal in the olfactory system. In: Jacobson M (ed) Handbook of sensory physiology. Vol 9. Springer: Berlin: pp 55–83

Green J D 1951 The comparative anatomy of the hypophysis with special references to its local blood supply and innervation. Am J Anat 88: 225–311

Green J D, Harris G W 1949 Observations of hypophysioportal vessels of living rat. J Physiol 108: 359–361

Green N A, Griffiths J D, Lavy G A D 1958 Venous drainage of anterior tibio-fibular compartment of leg, with reference to varicose veins. Br Med J 1: 1209–1210

Greengard P, Kebabian J W 1974 Role of cyclic AMP in synaptic transmission in the mammalian peripheral nervous system. Fed Proc 33: 1059–1067

Greenwald A S, Haynes D W 1972 Weight-bearing areas in the human hip joint. J Bone Jt Surg 54B: 157–163

Greer M, Haibach H 1974 Thyroid secretion. In: Greep R O, Astwood E B (eds) Handbook of physiology. Section 7. Endocrinology vol 3. Thyroid. American Physiological Society: Washington DC: pp 135–146

Grégoire R 1922 Anatomo médico-chirurgicale de l'abdomen. Région sous-thoracique. Baillière: Paris

Gregson N A 1975 The chemistry of myelin. In: Landon D N (ed) The peripheral nerve. Chapman & Hall: London: pp 512–604

Greulich W W 1951 The growth and development of Guamarian school-children. Am J Phys Anthropol N S 9: 55–70

Greulich W W, Pyle S I 1959 Radiographic atlas of skeletal development of the hand and wrist, 2nd edition, Stanford University Press: Stanford, Calif

Greulich W W, Thomas H 1938 The dimensions of the pelvic inlet of 789 white females. Anat Rec 72: 45–52

Greulich W W, Thomas H 1939 Study of pelvic type and its relationship to body build in white women. JAMA 112: 485–493

Grieve D W 1979 The postural stability diagram (PSD); personal constraints on the static exertion of force. Ergonomics 10: 1165–1175

Greive D W, Cavanagh P R 1974 The validity of quantitative statements about surface electromyograms recorded during locomotion. Scand J Rehab Med suppl 3: 19–25

Grieve D W, Leggett D, Wetherstone B 1978 The analysis of normal stepping movements as a possible basis for locomotor assessment of the lower limbs. J Anat 127: 515–532

Griffiths H E 1943 Treatment of injured workmen (Hunterian lecture, abridged) Lancet 1: 729–733

Griffiths J D 1961 Extramural and intramural blood supply of the colon. Br Med J 1: 323–326

Griffiths K, Cameron E H D 1975 The adrenal cortex. In: Beck F, Lloyd J B (eds) The cell in medical science. Vol 3. Academic Press: London: pp 155–192

Grillo M A, Jacobs L, Comroe J H Jr 1974 A combined fluorescence histochemical and electron microscopic method for studying special monoamine-containing cells (SIF cells). J Comp Neurol 153: 1–14

Grobstein C 1967 The problem of the chemical nature of embryonic inducers. In: De Reuck A V S, Knight J (eds) Cell differentiation (Ciba Foundation Symposium). Churchill: London: pp 131–138

Gross L 1921 The blood supply to the heart. Oxford University Press: London

Gross M S, Gelbermann M D 1985 The anatomy of the distal ulnar tunnel. Clin Orthop 196: 238–247

Grosser O 1936 Ueber hergleichende Anatomie und Phylogenese der Placenta. Stufenfolge der Placenten, Amnionbildung, misch-placenta (Katze), alterung der placenta. Verh Anat Ges Jena 81: 15–33

Grossman J 1954 A note on the radiological demonstration of perirenal space. J Anat 88: 407–409

Grotte G 1956 Passage of dextran molecules across blood-lymph barrier. Acta Chir Scand suppl 211: 1–84

Grottel K 1968 The innervation of the suprarenal gland of man, dog, cat and rabbit. Pr Tow Przyjac Nauk Poznañ 37: 41–79

Grüneberg H 1973 A ganglion probably belonging to the N terminalis system in the nasal mucosa of the mouse. Z Anat EntwGesch 140: 39–52

Grzybiak M, Szostakiewics-Sawicka H, Treder A 1975 Remarks on pathways of venous drainage from the left upper intercostal spaces in man. Folia Morphol 34: 301–313

Guasp F T, See Torrent Gausap F

Gudden B A von 1870 Experimentaluntersuchungen über das peripherische und centrale Nervensystem. Arch Psychiat NervKrankh 2: 693–723

Guerrier Y 1975 Le nerf facial. Quelques points d'anatomie topographique. Ann Oto-Laryngol 92: 161–171

Guffarth A, Graumann W 1975 Über die Lagebezielung der Arteria carotis externa zur Glandula parotis. Archs Oto-Rhino-Laryngol 211: 17–23

Guild S R 1941 A hitherto unrecognised structure, the glomus jugularis, in man. Anat Rec 79: 28P

Guillemin R 1978 Peptides in the brain: the new endocrinology of the neuron. Science, 202: 390–402

Guillery R W 1966 A study of Golgi preparations from the dorsal lateral geniculate nucleus of the adult cat. J Comp Neurol 128: 21–41

Guillery R W 1969 The organization of synaptic interconnections in the laminae of the dorsal lateral geniculate nucleus of the cat. Z Zellforsch 96: 1–38

Guillery R W 1971 Survival of large cells in the dorsal lateral geniculate laminae after interruption of retinogeniculate afferents. Brain Res 28: 541–544

Guillery R W 1973 Quantitative studies of transneuronal atrophy in the lateral geniculate nucleus of cats and kittens. J Comp Neurol 149: 423–438

Guillery R W 1974 Visual pathways in albinos. Sci Am 230: May 44–54

Guillery R W, Colonnier M 1970 Synaptic patterns in the dorsal lateral geniculate nucleus of the monkey. Z Zellforsch 103: 90–108

Guiot G, Albe-Fessard D, Arfel G, Derome P 1964 Dérivations d'activités limitaires en cours d'interventions stéréotaxiques. Neurochirurgia 10: 427–435

Guiot G, Hardy J, Albe-Fessard D 1962 Precise delimitation of the subcortical structures and identification of thalamic nuclei in man by stereotatic electrophysiology. Neurochirurgia 5: 1–18

Gunn S A, Gould T C 1975 Vasculature of the testis and adnexa. In: Male reproductive system. Handbook of physiology. Section 7. Endocrinology. Vol 5. (Hamilton D W, Greep R O eds) pp 117–142, American Physiological Society: Washington D C

Gupta K K, Knoell A C 1973 Mathematical modelling and structural analysis of the mandible. Biomat Med Dev Art Org 1: 469–479

Gupta S C, Gupta C O, Arora A K 1977 Subsegmentation of the human liver. J Anat 124: 413–423

Gupta C D, Gupta S C, Arora A K, Singh J P 1976 Vascular segments in the human spleen. J Anat 121: 613–616

Gupta S C, Gupta C D, Gupta S B 1981 Hepatovenous segments in the human liver. J Anat 133: 1–6

Guraya S S 1963 Histochemistry of the cytoplasmic droplet in the mammalian spermatozoon. Experientia 19: 94–95

Gurdjian F S, Lissner H R 1945 Deformations of the skull in head injury studied by the 'stresscoat' technique. Surg Gynecol Obstet 83: 219–233

Gurdon J B 1967 Nuclear transplantation and cell differentiation. In: De Reuck A V S, Knight J (eds) Cell differentiation. (Ciba Foundation Symposium). Churchill: London: pp 65–74

Gurdon J B 1968 Nucleic acid synthesis in embryos and its bearing on cell differentiation. Essays Biochem 4: 25–68

Gurr E 1965 The rational use of dyes in biology. Hill: London

Gusek W 1976 Die feinstruktur der Rattenzirbel und ihr Verhalten unter Einfluss von Antiadrogen und nach Kastration. Endokrinologie 67: 129–154

Gustafson T 1965 Morphogenetic significance of biochemical patterns in sea urchin embryos. In: Weber R (ed) The biochemistry of animal development. Vol 1. Academic Press: New York: pp 140–202

Gustafson T, Wolpert L 1963 The cellular basis of morphogenesis and sea urchin development. Int Rev Cytol 15: 139–214

Gwinnett A S 1967 The ultrastructure of the 'prismless' enamel of permanent human teeth. Arch. Oral Biol 12: 381–387

Gwyn D G, Waldron H A 1968 A nucleus in the dorsolateral funiculus of the spinal cord of the rat. Brain Res 10: 342–351

Gwyn D G, Nicholson G P, Flumerfelt B A 1977 The inferior olivary nucleus of the rat: a light and electron microscopic study. J Comp Neurol 174: 489–520

Gysi A 1900 An attempt to explain the sensitiveness of dentine. Br J Dent Sci 43: 865–868

H

Ha H, Liu C N 1966 Organisation of the spino-cervico-thalamic system. J Comp Neurol 127: 445–469

Ha H, Liu C N 1968 Cell origin of the ventral spinocerebellar tract. J Comp Neurol 133: 185–206

Haas L L 1952 Roentgenological skull measurements and their diagnostic application. Am J Roentgenol 67: 197–209

Haberich F J 1968 Osmoreception in the portal circulation. Fed Proc Fed Am Socs Exp Biol 27: 1137–1141

Hadden J W 1987 Neuroendocrine modulation of the thymus-dependent immune system. Agonists and mechanisms. Ann N Y Acad Sci 446: 39–48

Haeberli A, Studer H, Kohler H et al 1975 Autoradiographic localization of slow turnover iodocompounds within the follicular cells of the rat thyroid gland. Endocrinology 97: 978–984

Hage E 1973 Electron microscopic identification of several types of endocrine cells in branchial epithelium of human foetuses. Z Zellforsch 141: 401–412

Häggqvist G 1936 Analyse der Faserverteilung in einem Rückenmarkquerschnitt (Th 3) Z Zellforsch 39: 1–34

Haines R W 1933–1934 Cartilage canals. J Anat 68: 45–64

Haines R W 1935 A consideration of the constancy of muscular nerve supply. J Anat 70: 33–55

Haines R W 1937 The primitive form of epiphysis in the long bones of tetrapods. J Anat 72: 323–343

Haines R W 1942a Eudiarthrodial joints in fishes. J Anat 77: 12–19

Haines R W 1942b The tetrapod knee joint. J Anat 76: 270–301

Haines R W 1944 The mechanism of rotation at the first carpometacarpal joint. J Anat 78: 44–46

Haines R W 1947 The development of joints. J Anat 81: 33–55

Haines R W 1951 The extensor apparatus of the finger. J Anat 85: 251–259

Haines R W 1969 The striped compressor of the prostatic urethra. Br J Urol 41: 481–493

Haines R W 1974 The pseudoepiphysis of the first metacarpal of man. J Anat 117: 145–158

Haines R W 1975 The histology of epiphyseal union in mammals. J Anat 120: 1–25

Haines R W, Mohuiddin A 1968 Metaplastic bone. J Anat 103: 527–538

Halasz B, Rethelyi M, Szentágothai J 1968 Electron microscopic examination of the isolated median eminence (neurally deafferented). Arch Anat (Strasbourg) 51: 287–298

Hale A R 1952 Morphogenesis of volar skin in the human fetus. Am J Anat 91: 147–181

Hale F C, Olsen C R, Mickey M R 1968 The measurement of bronchial wall components. Am Rev Resp Dis 98: 978–987

Haley J C, Peden J K 1943 Suspensory muscle of the duodenum. Am J Surg 59: 546–550

Hall B K ed 1983 Cartilage. Vols 1–3 Academic Press: New York

Hall D A ed 1968 International review of connective tissue research. Vol 4. Academic Press: New York

Hall M C 1965 The locomotor system: Functional anatomy. Thomas: Springfield, Illinois

Hall M C 1966 The architecture of bone. Thomas: Springfield, Illinois

Hall S M, Williams P L 1970 Studies on the 'incisures' of Schmidt and Lauterman. J Cell Sci 6: 767–791

Hall-Craggs E C B 1974 The regeneration of skeletal muscle fibres per continuum. J Anat 117: 171–178

Haller R de 1969 Development of mucus-secreting elements. In: Emery J (ed) The anatomy of the developing lung. Spastics International Medical Publications: London: pp 94–115

Hallermann H 1934 Die Beziehungen der Werkstoffmechanik und Werstofforschung zur allgemeinen Knocken-Mechanik. Verh Deutsch Orthop Gesh 62: 347–360

Hallett P E, Lightstone A D 1976 Saccadic eye movements towards stimuli triggered by prior saccades. Vision Res 16: 99–106

Hallpike, C S 1931 The precise anatomy of the Hensen's cells in the cochlea of the guinea pig and its physiological significance. J Physiol 73: 8P

Hallpike C S 1935 Function of the tympanic membrane. Proc R Soc Med 28: 226–231

Halmi N S 1986 The normal thyroid. Anatomy and histochemistry. In: Ingbar S H, Braverman L E (eds) Werner's The thyroid. A fundamental and clinical text. 5th edition. Lippincott: Philadelphia: pp 24–36

Halprin K M 1972 Epidermal 'turnover time'–a re-examination. Br J Dermatol 86: 14–19

Ham A W 1974 Histology, 7th edition, Pitman: London

Hamburger V 1952 Development of the nervous system. Ann N Y Acad Sci 55: 117–132

Hamburger V 1960 A manual of experimental embryology. 2nd impression. Chicago University Press: Chicago, Illinois

Hamburger V 1968 Origins of integrated behavior. In: Locke M (ed) The emergence of order in developing systems. IV. Emergence of nervous coordination. (Dev Biol Suppl 2) pp 251–271

Hamburger V 1975 Cell death in the development of the lateral motor column of the chick embryo. J Comp Neurol 160: 535–546

Hamilton G F 1946 The longitudinal muscle coat of the human colon. J Anat 80: 230P

Hamilton W J 1944 Phases of maturation and fertilisation in the human ova. J Anat 78: 1–4

Hamilton W J 1949 Early stages of human development. Ann R Coll Surg 4: 281–294

Hamilton W J, Boyd J D 1951 Observations on the human placenta. Proc R Soc Med 44: 489–496

Hamilton W J, Boyd J D 1960 Development of the human placenta in the first three months of gestation. J Anat 94: 297–328

Hamilton W J, Mossman H W 1972 Hamilton Boyd and Mossman's human embryology. Prenatal development of form and Function. 4th edition Heffer: Cambridge

Hamilton W J, Boyd J D, Mossman H W 1962 Human embryology. Prenatal development of form and function. 3rd edition Heffer: Cambridge

Hammar J A 1935 Konstitutionsanatomische Studien über die Neurotisierung des Menschenembryos; über die Innervationsverhältnisse der Inkretorgane und des Thymus bis in den 4, Fötalmonat. Z Mikrosp-Anat Forsch 38: 253–293

Hammel H T 1968 Regulation of internal body temperature. Annu Rev Physiol 30: 641–710

Hammelbo T 1972 On the development of the cerebral fissures in cetacea. Acta Anat (Basel) 82: 606–618

Hammer G, Råadberg C 1961 The sphenoidal sinus. An anatomical and roentgenologic study with reference to trans-sphenoidal hypophysectomy. Acta Radiol 56: 401–422

Hammersen F, Staubesand J 1961 Arteries and capillaries of the human renal pelvis, with special reference to the so-called spiral arteries. I. Angio-architectural studies on the kidneys. Z Anat EntwGesch 122: 314–347

Hammond B T, Charnley J 1967 The sphericity of the femoral head. Med Biol Eng 5: 445–453

Hammond J 1952 Fertility. In: Parkes A S (ed) Marshall's physiology of reproduction. 3rd edn. Vol 2. Longmans, Green: London: pp 648–740

Hammond W S 1941 The development of the aortic arch bodies in the cat. Am J Anat 69: 265–294

Hámori J 1972 Developmental morphology of dendritic postsynaptic specialisations. In: Lissák K (ed) Recent developments of neurobiology in Hungary. Akadémiai Kiadó: Budapest

Hámori J, Szentágothai J 1965 Purkinje cell baskets: ultrastructure of an inhibitory synapse. Act Biol Hung 15: 465–479

Hámori J, Szentágothai J 1966a Identification under the electron microscope of climbing fibers and their synaptic contacts. Exp Brain Res 1: 65–81

Hámori J, Szentágothai J 1966b Participation of Golgi neuron processes in the cerebellar golmeruli: an electron microscope study. Exp Brain Res 2: 35–48

Hamperl H 1970 Bau und mögliche Funktioneiner elastisch-muskulären Struktur in der Portio uteri. Geburtshilfe Frauenheilkd 30: 953–959

Hampson J L, Harrison C R, Woolsey C N 1952 Cerebro-cerebellar projections and the somatotopic localization of motor function in the cerebellum. Res Pub Assoc Res Nerv Ment Dis 30: 299–316

Han S S, Kim S K, Cho M I 1976 Cytochemical characterization of the myoepithelial cells in palatine glands. J Anat 122: 559–570

Hanak H, Böck P 1971 Die Feinstruktur der Muskelsehnenverbindung von Skelett-und Herzmuskel. J Ultrastruct Res 36: 68–75

Hanaway J, Young R R 1977 Localization of the pyramidal tract in the internal capsule of man. J Neurol Sci 34: 63–70

Hancox N M 1972 The osteoclast. In: Bourne G H (ed) The biochemistry and physiology of bone. Vol 1. Academic Press: New York: pp 45–67

Hancox N M, Boothroyd B 1963 Structure-function relationships in the osteoclast. In: Sognnaes R F (ed) Mechanisms of hard tissue destruction. Am Assoc Adv Sci: Washington DC: pp 497–514

Hand P J and Liu C-N 1966 Efferent projections of the nucleus gracilis. Anat Rec 154: 353–354

Hanna R E, Washburn S L 1953 Determination of sex of skeletons, as illustrated by a study of the Eskimo plevis. Hum Biol 25: 21–27

Hannam A G 1976 Periodontal mechanoreceptors. In: Anderson D J, Matthews B (eds) Mastication. Wright: Bristol: pp 42–49

Hannover A 1844 Recherches microscopiques sur le système nerveux. P G Philipsen: Copenhagen

Hansen J T 1977 Freeze-fracture study of the carotid body. Am J Anat 148: 295–300

Hanson J R, Anson B J, Strickland E M 1962 Branchial sources of the auditory ossicles in man. Part II. Observations of embryonic stages from 7mm to 28mm. (CR length) Arch Otolaryngol 76: 200–215

Harach H R 1985 Solid cell nests of the thyroid. An anatomical survey and immunohistochemical study for the presence of thyroglobulin. Acta Anat 122: 249–253

Hardisty M W 1967 The numbers of vertebrate primordial germ cells. Biol Rev 42: 265–287

Harken D E, Ellis L B, Dexter L, Farrand R E, Dickson J F 1952 The responsibility of the physician in the selection of patients with mitral stenosis for surgical treatment. Circulation 5: 349–362

Harker D W 1972 The structure and innervation of sheep superior rectus and levator palpebrae muscles. I. Extrafusal muscle fibres. Invest Ophthal 11: 956–969

Harness D, Sekeles E 1971 The double anastomotic innervation of thenar muscles. J Anat 109: 461–466

Harness D, Sekeles E, Chaco J 1974 The double motor innervation of the opponens pollicis muscles: an electromyographic study. J Anat 117: 329–331

Harper W F 1947 Observations on the blood vasculature of the turbinate mucosa in man and other mammals. J Anat 81: 392P

Harpman J A, Woollard H H 1938 The tendon of the lateral pterygoid muscle. J Anat 73: 112–115

Harries D J 1955 Thyroid enlargements and cricothyroid muscle. Br Med J 1: 1012–1013

Harris A E 1965 Differentiation and degeneration in the motor horn of the foetal mouse. PhD Thesis. University of Cambridge

Harris G W 1947 Blood vessels of rabbit's pituitary gland, and significance of pars and zona tuberalis. J Anat 81: 343–351

Harris G W 1955 Neural control of the pituitary gland. Arnold: London

Harris G W, Donovan B T eds 1966 The pituitary gland. Butterworths: London

Harris H 1970a Cell fusion. The Dunham lectures. Clarendon Press: Oxford

Harris H 1970b Nucleus and Cytoplasm. Clarendon Press: Oxford

Harris H A 1933 Bone growth in health and disease. Oxford University Press: London

Harris P F, Jones P R M 1976 A radiological study of morphology and growth in the human fetal colon. Br J Radiol 49: 316–320

Harris W 1939 The morphology of the brachial plexus. Oxford University Press: London

Harris W 1952 Fifth and seventh cranial nerves in relation to nervous mechanism of taste sensation; new approach. Br Med J 1: 831–836

Harrison R G 1906 Further experiments on the development of peripheral nerves. Am J Anat 5: 121–132

Harrison R G 1907a Experiments in transplanting limbs and their bearing upon the problems of the development of nerves. J Exp Zool 4: 239–282

Harrison R G 1907b Observations on the living developing nerve fiber. Anat Rec 1: 116–118

Harrison R G 1910 The outgrowth of the nerve fiber as a mode of protoplasmic movement. J Exp Zool 9: 787–848

Harrison R G 1914 The reaction of embryonic cells to solid structures. J Exp Zool 17: 521–544

Harrison R G 1924 Neuroblast versus sheath cell in the development of peripheral nerves. J Comp Neurol 37: 123–206

Harrison R G, Barclay A E 1948 Distribution of testicular artery (internal spermatic artery) to human testis. Br J Urol 20: 57–66

Harrison R G, Hoey M J 1960 The adrenal circulation. Oxford University Press: Oxford

Harrison R G, Weiner J S 1949 Vascular patterns of the mammalian testis and their functional significance. J Exp Biol 26: 304–316

Harrison R G, Connolly R C, Abdalla A 1969 Kinship of Smenkhkare and Tutankamen demonstrated serologically. Nature 224: 325–326

Harrison T J 1957 Pelvic growth. PhD Thesis, Queen's University, Belfast

Harrop T J, Garrett J R 1974 Effects of preganglionic sympathectomy on secretory changes in parotid acinar cells of rats on eating. Cell Tissue Res 154: 135–150

Hartman B K 1973 Immunofluorescence of dopamine-hydroxylase. Application of improved methodology to the localization of the peripheral and central noradrenergic nervous system. J Histochem Cytochem 21: 312–332

Hartroft P M 1968 The juxtaglomerular complex as an endocrine gland. In: Bloodworth J M B (ed) Endocrine pathology. Williams and Wilkins: Baltimore, Maryland: pp 641–677

Harvey S C, Burr H S 1926 The development of the meninges. Arch Neurol Psychiatr Chicago 15: 545–567

Hasan M, Das A C 1969 A note on the falx cerebelli. Acta Anat (Basel) 74: 624–628

Hascall V C, Hascall G K 1981 Proteoglycans. In: Hay E D (ed) Cell biology of extracellular matrix. Plenum: New York: pp 39–64

Hashimoto K 1971 Ultrastructure of the human toenail. II. Keratinisation and foramtion of marginal band. J Ultrastruct Res 36: 391–410

Hashimoto K 1974 New methods for surface ultrastructure: comparative studies of scanning electron microscopy, transmission electron microscopy and replica methods. Int J Dermatol 13: 357–381

Hashimoto K 1978 The apocrine gland. In: Jarrett A (ed) The physiology and pathophysiology of the skin. Vol 5. Academic Press: New York: pp 1575–1596

Hashimoto M, Komori A, Kosaka M, Mori Y, Shimoyama R, Akashi H 1960 Electron microscopic studies on the smooth muscle of the human uterus. J Jap Obstet Gynecol Soc 7: 115–121

Hassler R 1950 Projections of the cerebellum to the midbrain and the thalamus. Dtsch Z Nerveneilk 163: 629–671

Hassler R 1959 Anatomy of the thalamus. In: Schaltenbrand G, Bailey P (eds) Introduction to stereotaxis with an atlas of the human brain. Vol I. Rune and Stratton: New York: pp 230–290

Hast M H, Perkins R E 1984 Secondary supinator muscles of the human elbow (musculi revivi). J Anat 139: 745–746 (Proceedings of the Anatomical Society of Great Britain and Ireland. 19 July)

Hast M H, Fischer J M, Wetzel A B, Thompson V E 1974 Cortical motor representation of the laryngeal muscles in Macaca mulatta. Brain Res 73: 229–240

Haug H 1956 Remarks on the determination and significance of the grey cell coefficient. J Comp Neurol 104: 473–492

Hăulică I, Coculescu M 1981 Is angiotensin a new pineal hormone? Rev Roum Méd Endocrinol 19: 3–21

Havers C 1691 Osteologia nova, or some new observations of the bones. S Smith: London

Havez R, Decaud P, Roussel P, Voisin C, Biserte C, Gernez-Rieux Ch 1966 Identification des gamma-globulines, de la kallikreine, de la transferrine dans la muquesque bronchique humaine. C R Hebd Séanc Acad Sci Paris 262d: 1777

Haxton H A 1954 Sympathetic nerve supply of the upper limb in relation to sympathectomy. Ann R Coll Surg 14: 247–266

Hay E D 1981 Extracellular matrix. J Cell Biol 91: 205s–223s

Hayashi H, Harrison R G 1971 The development of the interstitial tissue of the human testis. Fertil Steril 22: 351–355

Hayes G M, Woodroofe M N, Cuzner M L 1988 Characterization of microglia isolated from adult human brain. J Neuroimmunol 19: 177–189

Haymaker W, Anderson E, Nauta W J H eds 1969 The Hypothalamus. Thomas: Springfield, Illinois

Hayrch S S 1969 Blood supply of the optic nerve head and its role in optic atrophy, glaucoma and oedema of the optic disc. Br J Ophthal 53: 721–748

Hayward A F 1979 Membrane coating granules. Int Rev Cytol 59: 97–127

Hayward J 1961 The lower end of the oesophagus. Thorax 16: 36–41

Headon M P, Powell T P S 1973 Cellular changes in the lateral geniculate nucleus of infant monkeys after suture of the eyelids. J Anat 166: 135–145

Headon M P, Powell T P S 1978 The effect of bilateral eye closure upon the lateral geniculate nucleus in infantile monkeys. Brain Res 143: 147–154

Healey J E, Schroy P C 1953 Anatomy of biliary ducts within the human liver; analysis of prevailing pattern of branchings and major variations of biliary ducts. Arch Surg 66: 599–616

Heathcote J G, Grant M E 1981 The molecular organization of basement membranes. In: Hall D A, Jackson D S (eds) International review of connective tissue research. Vol 9. Academic Press: New York: pp 191–264

Hécaen H, Ajuriaguerra J de 1964 Left-handedness: manual superiority and cerebral dominance. Grune & Stratton: New York

Heimer L 1968 Synaptic distribution of centripetal and centrifugal nerve fibres in the olfactory system of the rat. An experimental anatomical study. J Anat 103: 413–32

Heimer L, Wall P D 1968 The dorsal root distribution to the substantia gelatinosa of the rat with a note on the distribution in the cat. Exp Brain Res 6: 89–99

Heinemann H O, Fishman A P 1969 Nonrespiratory functions of mammalian lung. Physiol Rev 49: 1–47

Heintzberger C F M 1974 The development of the sinu-atrial node in the mouse. Acta Morphol. Neerl-Scand. 12: 317–330

Heiple K G, Lovejoy C O 1971 The distal femoral anatomy of Australopithecus. Am J Phys Anthropol 35: 75–84

Heitz Ph, Polak J M, Bloom S K, Pearse A G E 1976 Identification of the D$_1$-cell as the source of human pancreatic polypeptide (HPP). Gut 17: 755–758

Held H 1863 Die centrale Gehörleitung Arch Mikrosk anat EntwMech, 201–248

Hell E A, Cruickshank C N 1963 The effect of injury upon the uptake of 3-H-thymidine by guinea pig epidermis. Exp Cell Res 31: 128–139

Heller C G, Lalli M F, Pearson J E, Leach D R 1971 A method for the quantification of Leydig cells in man. J Reprod Fertil 25: 177–184

Heller C H, Clermont Y 1964 Kinetics of the germinal epithelium in man. Recent Prog Horm Res 20: 545–575

Heller H, Clark R B 1962 Neurosecretion. Academic Press: New York

Hellmer H 1935 Röntgenologische Beobachtungen über die Ossifikation der Patella. Acta Radiol 27: Suppl 1–82

Henderson B, Pettipher E R 1985 The synovial lining cell: biology and pathobiology. Semin Arthritis Rheum 15: 1–32

Henderson R G 1978 The position of the nutrient foramen in the growing tibia and femur of the rat. J Anat 125: 593–599

Hendrickson A 1969 Electron microscopic radioautography: identification of origin of synaptic terminals in normal nervous tissue. Science 165: 194–196

Hendrickson A 1975 Technical modifications to facilitate tracing synapses by electron microscopic autoradiography. Brain Res 85: 241–247

Hendrickson A, Kupfer C 1976 The histogenesis of the fovea in the macaque monkey. Invest Ophthal 15: 746–756

Hendry N G C 1958 The hydration of th enucleus pulposus and its relation to intervertebral disc derangement. J Bone Jt Surg 40B: 132–44

Henkart P A 1985 Mechanism of lymphocyte-mediated cytotoxicity. Annu Rev Immunol 3: 31–58

Henkind P, Levitsky M 1969 Angioarchitecture of the optic nerve. I. The papilla. Am J Opthal 68: 979–986

Henle F G J 1856-7 Handbuch der Systematischen Anatomie des Menschen. 3 vol. Vieweg: Braunschweig

Henle J 1876 Handbuch der Gefasslebre des Menschen. Vieweg: Braunschweig

Henle W, Henle G, Chambers L A 1938 Studies on ontogenic structure of some mammalian spermatozoa. J Exp Med 68: 335–352

Henley F A 1955 Blood supply of common bile duct and its relationship to the duodenum. Br J Surg 43: 75–80

Henzen W 1957 Transplantationen zur entwicklungsphysiologischen Analyse der larvalen Mundorgane bei Bombinator und Triton. Arch EntwMech Org 149: 387–442

Heppenstall R B 1980 Fracture healing. In: Heppenstall R B (ed) Fracture treatment and healing. Saunders: Philadelphia: pp 35–64

Heppenstall R B, Grislis G, Hunt T K 1975 Tissue gas tensions and oxygen consumption in healing bone defects. Clin Orthop 106: 357–365

Herberman R B, Reynolds C W, Ortaldo J R 1986 Mechanism of cytotoxicity by natural killer (NK) cells. Ann Rev Immunol 4: 651–680

Herbert M C, Graham C F 1974 Cell determination and biochemical differentiation of the early mammalian embryo. Curr Top Dev Biol 8: 151–178

Herman C J, Lapham L W 1968 DNA content of neurons in the cat hippocampus. Science 160: 537

Herman P G, Yamamoto I, Mellins H Z 1972 Blood microcirculation in the lymph node during the primary immune response. J Exp Med 136: 697–714

Herman P G, Yamamoto I, Mellins H Z 1973 Microcirculation of the aortic wall in experimental atheromatosis. Radiology 107: 265–271

Herndon R M 1963 The fine structure of the Purkinje cell. J Cell Biol 18: 167–180

Herrick C J 1922 Introduction to Neurology. Saunders: Philadelphia, Pa

Hertig A T 1935 Angiogenesis in the early human chorion and in the primary placenta of the macaque monkey. Contrib Embryol Carnegie Inst. Washington 25: 37–81

Hertig A T, Rock J 1941 Two human ova of the pre-villous stage, having an ovulation age of about eleven and twelve days respectively. Contrib Embryol Carnegie Inst. Washington 29: 127–156

Hertig A T, Rock J 1945 Two human ova of the pre-villous stage, having a developmental age of about seven and nine days respectively. Contrib Embryol Carnegie Inst. Washington 31: 65–84

Hertig A T, Rock J, Adams E C 1956 A description of 34 human ova within the first 17 days of development. Am J Anat 98: 435–493

Hertig A T, Rock J, Adams E C, Mulligan W J 1954 On the preimplantation stages of the human ovum: a description of four normal and four abnormal specimens ranging from the second to the fifth day of development. Contrib Embryol Carnegie Inst. Washington 35: 199–220

Hertwig W A O 1881 Die Coelomtheorie: versuch einer Erklärung des mittleren Keimblatts. G Fischer: Jena

Hertzberg E L, Lawrence T S, Gilula N B 1981 Gap junctional communication. Annu Rev Physiol 43: 479–491

Hervonen A, Linnoila I, Tainio H, Vaalasti A, Mascorro J A 1985 Immunohistochemical evidence for the occurrence of vasoactive intestinal polypeptide (VIP)–containing nerve fibres in human fetal abdominal paraganglia. J Anat 143: 121–128

Hervonen A, Vaalasti A, Vaalasti T, Partenen M, Kanerva L 1976 Paraganglia in the urogenital tract of man. Histochemistry 48: 307–13

Hervonen A, Vaalasti A, Partanen M, Kanerva L 1978a The endocrine nature of the paraganglia of man. Experientia 34: 111–112

Hervonen A, Vaalasti A, Partanen M 1978b Paraganglia of the bladder. J Urol 119: 335–337

Hess A, Young J Z 1952 The nodes of Ranvier. Proc R Soc Lond [Biol] 140: 301–320

Hess W R, Brügger M, Bucher V 1946 Zur Physiologie von Hypothalamus, Area praeoptica und Septum, sowie angrenzender Balken- und Stirnhirnberache. Monatsschr Psychiat Neurol 111: 17–59

Heuser C H, Corner G W 1957 Developmental horizons in human embryos. Description of age group X, 4 to 12 somites. Contrib Embryol Carnegie Inst. Washington 36: 29–39

Heuser C H, Streeter G L 1941 Development of the macaque embryo. Contrib Embryol Carnegie Inst. Washington 29: 15–56

Heuser J E, Reese T S, Landis D M D 1974 Functional changes in frog neuromuscular junctions studies with freeze-fracture. J Neurocytol 3: 109–131

Heuser J E, Reese T S, Landis D M 1975 Preservation of synaptic structure by rapid freezing, Cold Spring Harbor Symp Quant Biol 40: 17–24

Hewitt A B 1977 An investigation using holographic interferometry, of surface strain in bone induced by orthodontic forces: a preliminary report. Br J Orthodont 4: 39–41

Hewitt A T, Kleinman H K, Pennypacker T P, Martin G R 1980 Identification of an adhesion factor for chondrocytes. Proc Natl Acad Sci USA 77: 385–388

Hewitt W 1958 The development of the human caudate and amygdaloid nuclei. J Anat 92: 377–382

Hewitt W 1961 The development of the human internal capsule and lentiform nucleus. J Anat 95: 191–199

Hewitt W 1962 The development of the human corpus callosum. J Anat 96: 355–358

Heylings D J A 1978 Supraspinous and interspinous ligaments of the human lumbar spine. J Anat 125: 127–131

Hickey D S, Hukins D W L 1981 Collagen fibril diameters and elastic fibres in the annulus fibrosus of human fetal intervetebral disc. J Anat 133: 351–357

Hicks J H 1953a The mechanics of the foot. I. The joints. J Anat 87: 345–357

Hicks J H 1953b The mechanics of the foot. II. The plantar aponeurosis and the arch. J Anat 88: 25–30

Hicks J H 1955 The foot as a support. Acta Anat 25: 34–45

Hicks R M 1965 Permeability barriers in the rat ureter transitional epithelium. J Anat 99: 932P

Hicks S P, Amoto C J D', Lowe M J 1959 The development of the mammalian nervous system. I. Malformations of the brain, especially the cerebral cortex, induced in rats by radiation. J Comp Neurol 113: 435–469

Highstein S M 1977 Abducens to medial rectus pathway in the MLF. In: Brooks B A, Bajandas F J (eds) Eye movements. Plenum Press: New York

Hiiemae K M 1978 Mammalian mastication: a review of the activity of the jaw muscles and the movements they produce in chewing. In: Butler P M, Joysey K A (eds) Development, function and evolution of teeth. Academic Press: New York: pp 359–398

Hiiemae K, Thexton A J, Crompton A W 1978 Intra-oral food transport: the fundamental mechanism of feeding. In: Carlson D S, McNamara J A (eds) Muscle adaptation in the craniofacial region. (Monograph 8, Craniofacial Growth Series) University of Michigan Center for Human Growth and Development: Michigan

Hill A V 1970 First and last experiments in muscle mechanics. Cambridge University Press: Cambridge

Hill J P 1932 The developmental history of the primates. Philos Trans R Soc Lond [Biol] 221: 45–178

Hill J T, Hirsh A V, Pryor J P, Kellett M J 1982 Changes in the appearance of venography after ligation of a varicocele. J Anat 135: 47–52

Hillman P, Wall P D 1969 Inhibitory and excitatory factors influencing the receptive fields of lamina 5 spinal cord cells. Exp Brain Res 9: 284–306

Hilton S M, Lewis G P 1956 The relationship between glandular activity, bradykinin formation and functional vasodilatation in the submandibular salivary gland. J Physiol 134: 471–483

Himstedt H W, Schumacher G H, Menning A et al 1974 Zur Topographie der muskulüren Nervenausbreitungen. 5. Untere Extremität Ischiokrurale Muskeln. M. biceps femoris. Anat Anz 135: 235–244

Hinchliffe J R 1982 Cell death in vertebrate limb morphogenesis. Prog Anat 2: 1–17

Hinchliffe J R, Johnson D R 1980 The development of the vertebrate limb. An approach through experiment, genetics, and evolution. Clarendon Press: Oxford

Hinchcliffe R, Harrison D 1976 Scientific foundations of otolaryngology. Heinemann Medical: London

Hindle M O 1967 The intermediate plexus of the periodontal membrane. In: The mechanisms of tooth support. A symposium. Oxford 1965. Wright: Bristol: pp 66–71

Hinds J W 1971 Early neuroblast differentiation in the mouse olfactory bulb. Anat Rec 169: 340–341

Hinds J W, Hinds P L 1972 Reconstruction of dendritic growth cones in neonatal mouse olfactory bulb. J Neurocytol 1: 169–187

Hinrichsen C F L, Larramendi L M H 1969 Features of trigeminal mesencephalic nucleus structure and organization. Am J Anat 126: 497–506

Hinssen H, D'Haese J, Small J V, Sobieszek A 1978 Mode of filament assembly of myosins from muscle and nonmuscle cells. J Ultrastruct Res 64: 282–302

Hintzsche E 1937 Uber den Einfluss der Schilddrüsengrosse auf die Lage der Epithelkörperchen. Anat Anz 84: 18–25

Hirsch S 1960 Morphologie et Physiologie des Anastomses Arterioveneuses, Acta Tertii Europaei de Cordis Scientia Conventus, Romae. Excerpta Medica i: 61–64

Hirsch-Hoffman H U 1976 Licht- und elektronenmikroskopische Untersuchungen an der Tränendrüse des Affen (Macaca mulatta) Z Mkrosk 90: 369–384

Hirsch-Hoffman H U 1978 Die Ultrastruktur von Drüsenendstücken der menschlichen Tränendrüse, Klin Mbl Augenheilk 172: 80–87

His W 1887a On the development of the roots of the nerves and on their propagation to the central organs and to the periphery. Rep Br Assoc Adv Sci 57: 773–775

His W 1887b Zur Geschichte des menschlichen Rückenmarkes und der Nervenwurzeln. Abh Math-phys Cl K Sächs Ges Wiss Leipzig 13ch: 477–514

His W 1879 Ueber die Anfänge des peripherischen Nervensystemes. Arch Anat Physiol Leipzig Anat Abt: 455–482

His W 1883 Ueber das Auftreten der weissen Substanz und der Wurzelfasern am Rückenmark menschlicher Embryonen. Arch Anat Physiol Leipzig Anat Abt: 163–170

His W 1890 Histogenese und Zusammenhang der Nervenelemente. Arch Anat Physiol Leipzig Anat Abt: Suppl 95–117

His W Jr 1893 Die Thätigkeit des embryonalen Herzens und deren Bedeutung für die Lehre von der Herzbewegung beim Erwachsenen. Arb Med Klin Leipzig: 14–50

His Wulhelm Sr 1874 Unsere Körperform und das physiologische Problem ihrer Entstetung. Vogel: Leipzig

Hjarnø J, Jørgensen J B, Vesely M 1974 Archeological and anthropological investigations of late heathen graves in Upernarvik district. Reitzels Forlag: København

Hjortsjö C-H 1948 Die Anatomie der Intrahepatischen Gallengänge beim Menschen; Mittels Röntgen- und Injections- Technik stadiert. Lunds Univ Arssk N F Avd 2, 44: 1–112

Hjortsjö C-H 1950 Studies on the earliest pulmonary development in mammals. Lunds Univ Arsskrift NF Avd 2, 46: nr 4

Hjortsjö C-H 1951 The topography of the intrahepatic duct systems. Acta Anat 11: 599–615

Hjortsjö C-H 1956 The intrahepatic ramification of the portal vein. Lunds Univ Årssk N F Avd 2, 52: 1–30

Hjortsjö C-H 1975 Den Segmentella Indelingen av Levern. Comm Dept Anat Univ Lund 4

Hockereau M-T 1963 Etude comparée de la vague spermatogénétique chez le taureau et chez le rat. Ann Biol Anim Biochem Biophys 3: 5–20

Hodgkin A L 1964 The conduction of the nervous impulse. Thomas: Springfield, Illinois

Hodgkin A L, Huxley A F 1952 Currents carried by sodium and potassium ions through membrane of giant axon of Loligo. J Physiol 116: 449–472

Hodgson C, Spiers R L 1974 The effect of preganglionic cervical sympathectomy on the amylase content of parotid glands in fasted and fed rats. J Physiol 237: 56–57P

Hodson J J 1966 The distribution, structure, origin and nature of the dental cuticle of Gottlieb. I and II. Periodontics 5: 237–250, 296–302

Hodson N 1970 The nerves of the testis, epididymus and scrotum. In: Johnson A D, Gomes W R, Vandemark N L (eds) The testis. Vol 1. Academic Press: New York: pp 47–100

Hoefsmit E C 1982 Macrophages, Langerhans cells, interdigitating and dendritic accessory cells: a summary. Adv Exp Med Biol 149: 463–468

Hoerr N L, Pyle S J, Francis C C 1962 Radiographic Atlas of Skeletal Development of Foot and Ankle. Thomas: Springfield, Illinois

Hoff E C, Hoff H E 1934 Spinal termination of the projection fibers from the motor cortex of primates. Brain 57: 454–474

Hoffman B F, Cranefield F P 1960 Electrophysiology of the heart. McGraw-Hill: New York

Hoffman E, Theil W 1956 Untersuchungen vermeintlicher und wirkincher Abflusswege aus dem Subdural- und Subarachnoide-atraum. Z Anat EntwGesch 119: 283–310

Hoffman H H, Kuntz A 1957 Vagus nerve components. Anat Rec 127: 551–568

Hogan M J 1972 Role of the retinal pigment epithelium in macular disease. Trans Am Acad Ophthal Otol 76: 64–80

Hogan M J, Alvarado J A, Weddell J E 1971 Histology of the Human Eye. Saunders: Philadelphia

Hökfelt T 1968 In vitro studies on central and peripheral monoamine neurons at the ultrastructural level. Z Zellforsch 91: 1–74

Hökfelt T, Elde R, Johannson O, Luft R, Nilsson G, Arimura A 1976 Immunohistochemical evidence for separate populations of somatostatin-containing and substance P-containing primary afferent neurons in rat. Neuroscience 1: 131–136

Hökfelt T, Elfvin L G, Elde R, Schultzberg M, Goldstein M, Luft R 1977 Occurrence of somatostatin-like immunoreactivity in some peripheral sympathetic noradrenergic neurons. Proc Natl Acad Sci USA 74: 3587–3591

Hökfelt T, Johansson O, Fuxe K, Goldstein M, Park D 1977 Immunohistochemical studies on the localization and distribution of monoamine neuron systems in the rat brain. II Tyrosine hydroxylase in the telencephalon. Med Biol 55: 21–40

Hökfelt T, Ljungdahl Å, Terenius L, Elde R, Nilsson G 1977 Immunohistochemical analysis of peptide pathways possibly related to pain and analgesia; enkephalin and substance P. Proc Natl Acad Sci USA 74: 3081–3085

Holgate S T 1983 Mast cells and their mediators. In: Holborrow E J, Reeves W G (eds) Immunology in medicine. 2nd edition. Academic Press: London: pp 79–94

Holland G R 1985 The odontoblast process; form and function. J Dent Res 64: 499–514

Hollenberg M J, Spira A W 1973 Human retinal development. Am J Anat 137: 357–386

Holley M, Ashmore J F 1988 On the mechanism of a high frequency force generator in outer hair cells isolated from the guinea pig cochlea. Proc Roy Soc Lond B 232: 413–429

Holliday R 1964 A mechanism for gene conversion in fungi. Genet Res 5: 282–304

Hollingworth T, Berry M 1975 Network analysis of dendritic fields of pyramidal cells in neocortex and Purkinje cells in the cerebellum of the rat. Philos Trans R Soc Lond [Biol] 270: 227–64

Hollinshead W H 1971 Anatomy for Surgeons. Harper & Row: New York

Hollmann K H 1974 Cytology and fine structure of the mammary gland. In: Larson B L, Smith V R (eds) Lactation. A comprehensive treatise. Vol 1. Academic Press: New York: pp 3–95

Holloway R L 1968 The evolution of the primate brain: some aspects of quantitative relations. Brain Res 7: 121–172

Holmberg J 1972 On the nerves of the parotid gland. In: Emmelin N, Zotterman Y (eds) Oral physiology. Pergamon: Oxford: pp 17–19

Holmdahl D E, Ingelmark B E 1950 The contact between the articular cartilage and the medullary cavities of the bone. Acta Orthopaed Scand 20: 156–165

Holmes G 1939 The cerebellum of man. Brain 6: 1–30

Holmes G, Lister W T 1916 Disturbances of vision, from cerebral lesions, with special reference to the cortical representation of the macula. Brain 39: 34–73

Holstein A F 1967a Zur Nervenzellverteilung in glattmuskeligen Sphinkteren. Verh Anat Ges Jena 61: 269–275

Holstein A F 1967b Spermiophages in the human epididymis (Spermiophagen im Nebenhoden des Menschen). Naturwissenschaften 54: 98–99

Holstein A F 1976 Ultrastructural observations on the differentiation of spermatids in man. Andrologia 8: 57–65

Holtzer H, Abbott J 1968 Oscillations of the chondrogenic phenotype in vitro. In: Ursprung H (ed) The stability of the differentiated state. Springer-Verlag: Berlin: pp 1–16

Holtzer H, Strahs K, Biehl J 1975 Thick and thin filaments in postmitotic mononucleated myoblasts. Science 188: 943–945

Holtzer H, Weintraub H, Mayne R et al 1972 The cell cycle, cell lineages, and cell differentiation. Curr Top Dev Biol 7: 229–256

Holtzman E 1976 Lysosomes: A survey. Cell Biology Monographs Vol 3. Springer-Verlag: Vienna

Hooke Robert 1665 Micrographia, or some physiological descriptions of minute bodies made by magnifying glasses; with Observations and Inquiries thereupon. J Martyn, J Allestry: London

Hooker C W 1971 The intertubular tissue of the testis. In: Johnson A D, Gomes W R, Vandemark N L (eds) The testis. Vol 1. Academic Press: New York: pp 483–550

Hoopen M ten, Reuver H A 1970 Probabilistic analysis of dendritic branching patterns of cortical neurones. Kybernetik 6: 176–188

Hope J 1965 The fine structure of the developing follicle of the rhesus ovary. J Ultrastruct Res 12: 592–610

Horacek M J, Earle A M, Gilmore J P 1986 The renal microvasculature of the monkey: an anatomical investigation. J Anat 148: 205–231

Horacek M J, Earle A M, Gilmore S P 1987 The renal vascular system of the monkey: a gross anatomical description. J Anat 153: 123–137

Horiguchi M 1980 The cutaneous branch of some human suprascapular nerves. J Anat 130: 191–195

Horiguchi M 1981 The recurrent branch of the lateral cutaneous nerve of the forearm. J Anat 132: 243–247

Horiuchi H, Matthews B 1973 In-vitro observations on fluid flow through human dentine caused by pain-producing stimuli. Arch Oral Biol 18: 275–294

Horridge G A 1968 Interneurons. Freeman: London

Horsley V A H, Clarke R H 1908 The structure and functions of the cerebellum examined by a new method. Brain 31: 45–124

Hörstadius S 1950 The neural crest. Its properties and derivatives in the light of experimental research. Oxford University Press: London

Horstmann E 1962 Electron microscopy of the human epididymis. Z Zellforsch 57: 692–718

Horstmann E 1966 Uber das Endothel der Zottenkapillaren im Dünndarm des Meerschwanchens und des Menschen. Z Zellforsch 72: 364–369

Horton R C 1952 The gastro-epiploic arteries. Guy's Hosp Rep 101: 108–110

Hoshiko M 1962 Electromyographic investigation of the intercostal muscles during speech. Arch Phys Med 43: 115–119

Hovelacque A 1927 Anatomie des Nerfs Graniens et Rachidiens et du Système Grand Sympathétique chez l'Homme. Doin: Paris

Howarth F, Cooper E R A 1949 Departure of substances from spinal theca. Lancet 2: 937–940

Howkins C H 1935 Blood supply of the lower jaw. Proc R Soc Med 29: 506–507

Hoyle F 1983 The intelligent universe: a new view of creation and evolution. Michael Joseph: London

Hoyle F, Wickramasinghe N C 1981 Evolution from space. Dent: London

Hoyte D A N 1960 Alizarin as an indicator of bone growth. J Anat 94: 432–442

Hoyte D A N 1966 Experimental investigations of skull morphology and growth. Int Rev Gen Exp Zool 2: 345–408

Hoyte D A N 1975 A critical analysis of the growth in length of the cranial base. Birth Defects 11: 255–282

Hrdlička A 1939 Practical Anthropometry. Wistar Institute: Philadelphia

Hsu T C, Arrighi F E 1971 Distribution of constitutive heterochromatin in mammalian chromosomes. Chromosoma 34: 243–253

Hubbard J E, Di Carlo V 1973 Fluorescence histochemistry of monoamine-containing cell bodies in the brain stem of the squirrel monkey (Saimiri sciureus). I. The locus caeruleus. J Comp Neurol 147: 553–566

Hubel D H, Wiesel T N 1961 Integrative action in the cat's lateral geniculate body. J Physiol 155: 385–398

Hubel D H, Wiesel T N 1962 Receptive fields, binocular interaction and functional architecture in the cat's visual cortex. J Physiol 160: 106–154

Hubel D H, Wiesel T N 1963 Receptive fields of cells in striate cortex of very young, visually inexperienced kittens. J Neurophysiol 26: 994–1002

Hubel D H, Wiesel T N 1965 Binocular interaction in striate cortex of kittens reared with artificial squint. J Neurophysiol 28: 1041–59

Hubel D H, Wiesel T N 1968 Receptive fields and functional architecture of the monkey striate cortex. J Physiol 195: 215–243

Hubel D G, Wiesel T N 1969 Anatomical demonstration of columns in the monkey striate cortex. Nature 221: 747–750

Hubel D H, Wiesel T N 1971 Aberrant visual projections in the Siamese cat. J Physiol 218: 33–62

Hubel D H, Wiesel T N 1977 Functional architecture of the macaque monkey visual cortex. (Ferrier Lecture). Proc R Soc Lond [Biol] 198: 1–59

Hubel D H, Wiesel T, LeVay S 1976 Functional architecture of area 17 in normal and monocularly deprived macaque monkeys. Cold Spring Harbor Symp Quant Biol 40: 581–589

Hubel D H, Wiesel T N, LeVay S 1977 Plasticity of ocular dominance columns in monkey striate cortex. Philos Trans R Soc Lond B 278: 377–409

Huber A C, Crosby E C 1933 The reptilian optic tectum. J Comp Neurol 57: 57–164

Huddart H, Hunt S 1975 Visceral muscle: its structure and function. Blackie: Glasgow

Hudgson P, Field E J 1973 Regeneration of cardiac muscle. In: Bourne G H (ed) The structure and function of muscle, 2nd edition, vol 2 Structure, Part 2. Academic Press: New York: pp 355–357

Hudson C L, Moritz A R, Wearn J T 1932 The extracardiac anastomoses of the coronary arteries. J Exp Med 56: 919–925

Hudson R E B 1965 Cardiovascular pathology. Arnold: London

Hudson R E B, Wendell-Smith C P 1976 Congenital anomalies of the heart and great vessels. In: Symmers W St C (ed) Systemic pathology. 2nd edition. Churchill Livingstone: Edinburgh: pp 73–117

Hughes A 1974 Endocrines, neural development and behavior. In: Gottlieb G (ed) Aspects of neurogenesis. Studies on the development of behavior and the nervous system. Vol 2. Academic Press: New York: pp 223–243

Hughes A 1976 The development of the dorsal funiculus in the human spinal cord. J Anat 122: 169–175

Hughes A F W 1968 Aspects of Neural Ontogeny. Academic Press: New York

Hughes H 1952 The factors determining the direction of the canal for nutrient artery in the long bones of mammals and birds. Acta Anat 15: 261–280

Hughes J, Kosterlitz H W 1977a The enkephalins: endogenous peptides with opiate receptor agonist activity. In: Usdin E, Hamburg D A, Barchas J D (eds) Neuroregulators and psychiatric disorders. Oxford University Press: New York: pp 329–336

Hughes J, Kosterlitz H W 1977b Opioid peptides. Br Med Bull 33: 157–161

Hughlings Jackson J, see Jackson J H

Hugli T E, Müller-Eberhard H J 1978 Anaphylatoxins, C3a and C5a. Adv Immunol 26: 1–53

Hultkrantz W 1898 Über die Spaltrichtungen der Gelenkknorpel. Verhandl anat Gesellsch Jena 12: 248–256

Humphrey T 1944 Primitive neurons in the embryonic central nervous system. J Comp Neurol 81: 1–45

Humphrey T 1947 Sensory ganglion cells within the central canal of the embryonic human spinal cord. J comp Neurol 86: 1–36

Humphrey T 1960 The development of the pyramidal tracts in human fetuses, correlated with cortical differentiation. In: Tower D B, Schadé J P (eds) Structure and Function of the cerebral cortex. Proceedings of the Second International Meeting of Neurobiologists, Amsterdam 1959. Elsevier: Amsterdam: pp 93–103

Humphrey T 1964 Some observations on the development of the human hippocampal formation. Trans Am Neurol Assoc 89: 207–9

Humphrey T 1967 The development of the human hippocampal fissure. J Anat 101: 655–676

Humphrey T 1969 The central relations of the trigeminal nerve. In: Kahn E A (ed) The surgery of pain. 2nd edition, Thomas: Springfield, Illinois

Hundeiker M, Keller L 1963 Die Gefassarchitektur des menschlichen Hodens. Morphol Jb 105: 26–73

Hunt C C 1974 The physiology of muscle receptors. In: Hunt C C (ed) Handbook of sensory physiology. Vol 3, pt 2. Springer: Berlin: pp 191–234

Hunt J N, Knox M T 1968 Regulation of gastric emptying. In: Code C F (ed) Handbook of physiology. Section 6, alimentary canal. Vol 4. American Physiological Society: Washington, DC: 1917–1936

Hunt R K, Jacobson M 1974 Neuronal specificity revisited. In: Moscona A A, Monroy A (eds) Current topics in developmental biology. Vol 8. Academic Press: New York: pp 203–259

BIBLIOGRAPHY

Hunter A J, Fleming D, Dixson A F 1984 The structure of the vomeronasal organ and nasopalatine ducts in Aotus trivirgatus and some other primate species. J Anat 138: 217–225

Hunter J 1762 Observations on the state of the testis in the foetus, and on the hernia congenita. In: Hunter W Medical Commentaries. Part I, London

Hunter J 1794 A treatise on the blood, inflammation, and gun-shot wounds; with his life by Sir Everard Home, Nicol: London

Huntingford P J 1959 Pudendal nerve block; the results of its routine use, with special reference to the trans-vaginal technique. J Obstet Gynaecol Br Commonw 66: 26–31

Huntington G S 1908 The genetic interpretation of the development of the mammalian lymphatic system. Am J Anat 2: 19–45

Huntington G S 1920 The morphology of the pulmonary artery in the Mammalia. Anat Rec 17: 165–202

Hurrell D J 1934 The vascularisation of cartilage. J Anat 69: 47–61

Hurst A F 1919 Chronic Constipation. Oxford University Press: London

Hutch J A, Rambo O S Jr 1970 A study of the anatomy of the prostate, prostatic urethra and urinary sphincter system. J Urol 104: 443–452

Hutchings M, Weller R O 1986 Anatomical relations of pia mater to cerebral blood vessels in man. J Neurosurg 65: 316–325

Hutchinson M C E 1978 A study of the atrial arteries in man. J Anat 125: 39–54

Huxley A F, Niedergerke R 1954 Structural changes in muscle during contraction. Interference microscopy of living muscle fibres. Nature 173: 971–973

Huxley H E 1969 The mechanism of muscular contraction. Science 164: 1356–1366

Huxley H E 1972 Molecular basis of contraction in cross-striated muscles. In: Bourne G H (ed) The structure and function of muscle. Vol 1. Academic Press: New York: pp 301–387

Huxley H E 1983 Molecular basis of contraction in cross-striated muscles and relevance to motile mechanisms in other cells. In: Stracher A (ed) Muscle and non-muscle motility. Vol 1. Academic Press: New York: pp 1–104

Huxley H E, Brown W 1967 The low angle X-ray diagram of vertebrate striated muscle and its behaviour during contraction and rigor. J Molec Biol 30: 383–434

Huxley H E, Hanson J 1954 Changes in the cross striations of muscle during contraction and stretch and their interpretation. Nature 173: 973–976

Huxley J S 1942 Evolution. The Modern Synthesis. Harper: New York

Huxley J S, De Beer G R 1934 The elements of experimental embryology. Cambridge University Press: Cambridge

Hyde J B, Akeson E J, Berinstein E 1973 Asymmetrical growth of superior temporal gyri in man. Experientia 29: 1131

Hydén H 1960 The neuron. In: Brachet J, Mirsky A E (eds) The cell. Vol 4. Academic Press: New York: pp 215–324

Hydén H ed 1967 The neuron. Elsevier: Amsterdam

Hyndman O R, Wolkin J 1943 Anterior cordotomy; further observations on physiologic results and optimum manner of performance. Arch Neurol Psychiatry 50: 129–148

I

Ibuka N, Kawamura H 1975 Loss of circadian rhythm in sleep-wakefulness cycle of the rat by suprachiasmatic nucleus lesions. Brain Res 96: 76–81

Ichikawa A, Nakajima Y 1962 Electron microscope study on the lacrimal gland of the rat. Tohuku J Exp Med 77: 136–149

Iggo A 1969 Cutaneus thermoreceptors in primates and sub-primates. J Physiol 200: 403–430

Iggo A 1974 Activation of cutaneous receptors and their actions on dorsal horn neurons. Adv Neurol 4: 1–9

Iggo A 1978 The physiological interpretation of electrical stimulation of the nervous system. Electroencephalog Clin Neurophysiol. (Suppl 34): 335–341

Iggo A, Muir A R 1969 The structure and function of a slowly adapting touch corpuscle in hairy skin. J Physiol 200: 763–796

Iklé F A 1961 Trophoblast cells in the circulating blood. Schweiz Med Wochenschr 91: 943–945

Iklé F A 1964 Dissemination von Syncytiotrophoblastzellen im mutterlichen Blut während der Gravidität. Bull Scheiz Akad Med Wiss 20: 62–72

Illis L 1964 Spinal Cord Synapses in the cat: the normal appearances by the light microscope. Brain 87: 543–554

Imamoto K, Leblond C P 1977 Presence of labelled monocytes, macrophages and microglia in a stab wound of the brain following an injection of bone marrow cells labelled with ³H-urdine into rats. J Comp Neurol 174: 255–280

Imamoto K, Leblond C P 1978 Radioautographic investigation of gliogenesis in the corpus callosum of young rats. II. Origin of microglial cells. J Comp Neurol 180: 139–164

Imre G 1964 Studies on the mechanism of retinal neovascularization. Role of lactic acid. Br J Ophthalmol 48: 75–82

Ince H, Young M 1940 The bony pelvis and its influence on labour: a radiological and clinical study of 500 women. J Obstet Gynaecol Br Commonw 47: 130–190

Ingle D, Schneider G E eds 1969 Subcortical Visual Systems. Karger: Basel

Ingram W R 1940 Nuclear organisation and chief connections of the primate hypothalamus. Res Pub Assoc Res Nerv Ment Dis 20: 195–244

Ingram W R 1976 A Review of Anatomical Neurology. University Park Press: Baltimore Md

Ingram W R, Ranson S W 1935 The nucleus of Darkschewitsch and the nucleus interstitialis in the brain of man. J Nerv Ment Dis 81: 125–137

Inman V T 1969 The influence of the foot-ankle complex on the proximal skeletal structures. Artif Limbs 13: 59–65

Inman V T, Saunders J B de C M 1937 The ossification of the human frontal bone, with special reference to its presumed pre- and post-frontal elements. J Anat 71: 383–394

Inman V T, Saunders J B de C M, Abbot L C 1944 Observations on the function of the shoulder joint. J Bone Jt Surg 26: 1–30

Inoue H 1973 Three-dimensional observation of collagen framework of intervertebral discs in rats, dogs and humans. Arch Histol Jpn 36: 39–56

Inoué S 1981 Cell division and the mitotic spindle J Cell Biol 91–132S–147S

Inoué S, Sato H 1967 Cell motility by labile association of molecules: the nature of mitotic spindle fibres and their role in chromosome movement. J Gen Physiol 50: 259–292

Intaglietta M, Sweifach B W 1971 Measurement of blood plasma Bolloid osmotic pressure. I. Technical aspects. Microvasc Res 3: 72–82

International Anatomical Nomenclature Committee 1977 Nominà Anatomica. 4th edition. Excerpta Medica: Amsterdam

International Anatomical Nomenclature Committee 1983 Nomina Anatomica 5th edition. Williams and Wilkins: Baltimore, Maryland

Irving M H 1964 The blood supply of the growth cartilage in young rats. J Anat 98: 931–939

Irving R, Harrison K M 1967 The superior olivary complex and audition: A comparative study. J Comp Neurol 130: 77–86

Irwin M H, Mayne R 1986 Use of monoclonal antibodies to locate the chondroitin sulfate chain(s) in type IX collagen. J Biol Chem 261: 1681–1683

Isaacson R L 1974 The limbic system, Plenum Press: New York

Isaacson R L, Pribram K H eds 1975 The hippocampus. Plenum Press: New York

Ishikawa H J 1983 Fine structure of skeletal muscle. In: Dowben R M, Shay J W (eds) Cell and muscle motility. Vol 4. Plenum Press: New York: pp 1–84

Isler H, Sarkar S K, Thompson B, Tonkin R 1968 The architecture of the thyroid gland: a 3-dimensional investigation. Anat Rec 161: 325–335

Isotupa K 1972 Alizarin trajectories in experimental studies of skull growth. Proc Finn Dent Soc 68: Suppl 2, 1–49

Isseroff R P, Fusenig N E, Rifkin D B 1982 Plasminogen activator in differentiating mouse keratinocytes. J Invest Dermatol 80: 217–222

Ito M 1984 The cerebellum and Neural Control. Raven Press: New York

Ito M, Yoshida M 1966 The origin of cerebellar-induced inhibition of Deiter's neurones. I. Monosynaptic initiation of the inhibitory postsynaptic potentials. Exp Brain Res 2: 330–349

Ito M, Yeshida M, Obata K 1964 Monosynaptic inhibition of the intracerebellar nuclei induced from the cerebellar cortex. Experienta 20: 575–576

Ito S 1967 Anatomic structure of the gastric mucosa. In: Code C F (ed) Handbook of physiology, section 6, alimentary canal. Vol 2. American Physiological Society: Washington, DC: pp 705–742

Ito S, Winchester R J 1963 The fine structure of the gastric mucosa in the bat. J Cell Biol 16: 541–577

Ito T, Shibasaki S 1964 Lichtmikroskopische Untersuchungen über die Glandula lacrimalis des Menschen. Arch Hist Jpn 25: 117–143

Iurato S 1967 Submicroscopic structure of the inner ear. Pergamon: Oxford

Iversen L L, Kelly J S 1975 Uptake and metabolism of gamma-aminobutyric acid by neurones and glial cells. Biochem. Pharmacol 24: 933–938

Iwayama T 1970 Ultrastructural changes in the nerves innervating the cerebral artery after sympathectomy. Z Zellforsch 109: 465–480

J

Jabbur S J, Towe A L 1961 Cortical excitation of neurons in dorsal column nuclei of cat, including an analysis of pathways. J Neurophysiol 24: 499–509

Jackson J H 1868 Observations on the physiology and pathology hemichorea. Oliver and Boyd: Edinburgh

Jackson K M, Joseph J, Wyard S J 1977 Sequential muscular contraction. J Biomech 10: 97–106

Jackson S A, Cartwright A G, Lewis D 1978 The morphology of bone mineral crystals. Calcif Tissue Res 25: 217–222

Jacob F, Monod J 1961 Genetic regulatory mechanisms in the synthesis of proteins. J Molec Biol 3: 318–356

Jacob F, Monod J 1963 Genetic repression, allosteric inhibition and cellular differentiation. In: Locke M (ed) Cytodifferentiation and macromolecular synthesis. Academic Press: New York: pp 30–64

Jacobowitz D M 1970 Catecholamine fluorescence studies of adrenergic neurons and chromaffin cells in sympathetic ganglia. Fed Proc 29: 1929–1944

Jacobowitz D M, Palkovits M 1974 Topographic atlas of catecholamine and acetylcholinesterase-containing neurons in the rat brain. I. Forebrain (telecephalon, diencephalon). J Comp Neurol 157: 13–28

Jacobs L, Comroe H J 1971 Reflex apnoea, bradycardia and hypotension produced by serotonin on the nodose ganglia of the cat. Circ Res 29: 145–155

Jacobsen K 1974 Area intercondylaris tibiae: osseous surface structure and its relation to soft tissue structures and applications to radiography. J Anat 117: 605–618

Jacobson M 1969 Development of specific neuronal connections. Science 163: 543–547

Jacobson M 1970 Development, specification and diversification of neuronal connections. In: Schmitt F O, Quarton G C, Meinechuk T, Adelman G (eds) The neurosciences. Second study program. Rockefeller University Press: New York: pp 116–129

Jacobson M 1974 A plentitude of neurons. In: Gottlieb G (ed) Aspects of neurogenesis. Studies on the development of behavior and the nervous system. Vol 2. Academic Press: New York: pp 151–166

Jacobson M 1978 Developmental neurobiology. 2nd edition, Holt, Rinehart and Winston: New York:

Jakubovic H R, Ackerman A B 1985 Structure and function of skin. Development, morphology and physiology. In: Moschella S L, Hurley H J (eds) Dermatology. 2nd edition. Vol 1. Saunders: Philadelphia: pp 1–74

Jakus M A 1964 Ocular fine structure. Churchill: London

James B 1977 see Walker A, James B 1977

James D W 1974 Growth cones and synaptic connections in tissue culture. In: Bellairs R, Gray E G (eds) Essays on the nervous system. A festschrift for professor J Z Young. Clarendon Press: Oxford

James D W, Tresman R L 1969 An electron microscopic study of the de nova formation of neuromuscular junctions in tissue culture. Z Zellforsch 100: 126–140

James T N 1961 Anatomy of the human sinus node, Anat Rec 141: 109–116

James T N 1978 Anatomy of the coronary arteries and veins. In: Hurst J W (ed) The heart, arteries and veins. 4th edition. McGraw-Hill: New York: pp 32–47

James T N, Burch G E 1958 The atrial coronary arteries in man. Circulation 17: 90–98

James T N, Sherf L, Schlant R C, Silverman M E 1982 Anatomy of the heart. In: Hurst J W (ed) The heart, arteries and veins, 5th edition. McGraw-Hill: New York: pp 22–74

James T N, Sherf L, Urthaler F 1974 Fine structure of the bundle-branches. Br Heart J 36: 1–18

Jamieson J D, Palade G E 1967 Intracellular transport of secretory proteins in the pancreatic exocrine cell. I. role of the peripheral elements of the Golgi complex. J Cell Biol 34: 577–596

Jamieson J K, Dobson J F 1907 Lectures on the lymphatic system of the caecum and appendix. Lancet I: 1061–1066

Jamieson J K, Dobson J F 1908 The lymphatics of the colon. Proc R Soc Med 2: 149–174

Jamieson J K, Dobson J F 1910a The lymphatics of the testicle. Lancet I: 493–495

Jamieson J K, Dobson J F 1910b On the injection of lymphatics by Prussian blue. J Anat 45: 7–10

Jamieson J K, Dobson J F 1920 The lymphatics of the tongue: with particular reference to the removal of lymphatic glands in cancer of the tongue. Br J Surg 8: 80–87

Jancsó G, Kiraly E, Jancsó-Gábor A 1977 Pharmacologically-induced selective degeneration of chemosensitive primary sensory neurones. Nature 270: 741–743

Janda V, Stará V 1965 Therole of thigh adductors in movement patterns of the hip and knee joint. Courier 15: 1–3

Janes R D, Brandys J C, Hopkins D A, Johnstone D E, Murphy D A, Armour J A 1986 Anatomy of human extrinsic cardiac nerves and ganglia. Am J Cardiol 57: 299–309

Janossy G, Prentice H G, Grob J P, Ivory K, Tidman N, Grundy J, Favrot M, Brenner M K, Campana D, Blacklock H A et al 1986 T lymphocyte regeneration after transplantation of R cell depleted allogeneic bone marrow. Clin Exp Immunol 63: 577–586

Jansen J, Brodal A ed 1954 Aspects of cerebellar anatomy. Grundt Tanum: Oslo

Jansen J, Jansen J Jr 1955 On the efferent fibers of the cerebellar nuclei in the cat. J Comp Neurol 102: 607–632

Jansen J, Korneluissen H 1977 Morphogenesis and morphology of the brainstem nuclei of cetacea. The hypoglossal nucleus. J Hirnforsch 18: 253–269

Jasin H E, Dingle J T 1981 Human mononuclear cell factors mediate cartilege matrix degradation through chondrocyte activation, J Clin Invest 68: 571–581

Jasper H H 1966 Recording from microelectrodes in stereotatic surgery for Parkinson's disease. J Neurosurg 24: Suppl 11, 219–221

Jasper H, Proctor L D, Knighton R S, Costello R T eds 1958 Reticular formation of the brain, Henry Ford Hospital International Symposium, Little, Brown Boston, Massachusetts

Jaworek T E 1973 The intrinsic vascular supply to the first metatarsal. Surgical considerations. J Am Podiatry Assoc 63: 189–197

Jayaraman A, Batton R R, Carpenter M B 1977 Nigrotectal projections in the monkey: an autoradiographic study. Brain Res 135: 147–152

Jayatilika A D P 1965 An electron microscopic study of sheep arachnoid granulations. J Anat 99: 635–649

Jdanov D A 1959 Anatomie du canal thoracique et des principaux collecteurs lymphatiques du tronc chez l'homme. Acta Anat 37: 20–47

Jefferson G 1915 The human stomach and the canalis gastricus (Lewis). J Anat 49: 165–181

Jefferson N C, Ogawa T, Syleos C, Zambetoglou A, Necheles H 1960 Restoration of respiration by nerve anastomosis. Am J Physiol 198: 931–933

Jeffery P K, Reid L 1975 New features of rat airway epithelium: a quantitative and electron microscopic study. J Anat 120: 295–320

Jeffrey P K, Reid L 1975 New features of rat airway epithelium: a quantitative and electron microscopic study. J Anat 120: 295–300

Jenkins F A 1969 The evolution and development of the dens of the mammalian axis. Anat Rec 164: 173–184

Jenness R 1974 Biosynthesis and composition of milk. J Invest Dermatol 63: 109–118

Jerison H J 1970 Gross brain indices and the analysis of fossil endocasts. In: Noback C R, Montagna W (eds) The primate brain. Advances in primatology. Vol 1. Appleton-Century-Crofts: New York: pp 225–244

Jit I 1952 The development and the structure of the suspensory muscle of the duodenum. Anat Rec 113: 395–407

Jit I 1961 The structure and development of the valves of Houston. Ind J Med Res 49: 635–647

Jit I 1974 Muscularis submucosae ani and its development. J Anat 118: 11–17

Jit I, Bakshi V 1984 Incidence of sterna foramina in North India. J Anat Soc Ind 33: 77–84

Jit I, Charnalia J 1959 The vertebral level of the termination of the spinal cord. J Anat Soc Ind 8: 93–101

Jit I, Gandhi O P 1966 The value of pre-auricular sulcus in sexing bony pelves. J Anat Soc Ind 15: 104–107

Jit I, Grewal S S 1975 Lengths of the small and large intestines in north Indian subjects. J Anat Ind 24: 89–100

Jit I, Grewal S S 1977 The suspensory muscle of the duodenum and its nerve supply. J Anat 123: 397–405

Jit I, Harjeet 1982 Sternal angle. J Anat Soc Ind 31: 115–117

Jit I, Kulkaria M 1976 Times of appearance and fusion of epiphysis at the medial end of the clavicle. Ind J Med Res 64: 773–782

Jit I, Mukerjee R N 1960 Observations on the anatomy of the human thoracic sympathetic chain and its branches, with an anatomical assessment of operations for hypertension. J Anat Soc Ind 9: 55–82

Jit I, Sahni D 1983 Sexing the North Indian clavicles. J Anat Soc Ind 32: 61–72

Jit I, Singh S 1956 Estimation of stature from clavicles. Ind J Med Res 44: 137–155

Jit I, Singh S 1966 The sexing of the adult clavicles. Ind J Med Res 54: 551–571

Jit I, Jhingan V, Kulkarni M 1980 Sexing the human sternum. Am J Phys Anthropol 53: 217–224

Johannisson E 1968 The foetal adrenal cortex in the human. Its ultrastructure at different stages of development and in different functional states. Acta Endocr Copnh 58: (Suppl 130)

Johns B A 1952 Developmental changes in the oesophageal epithelium in man. J Anat 86: 431–442

Johnson A D 1977 The influence of cadmium on the testis. In: Johnson A D, Gomes L R (eds) The testis. Vol 4. Advances in physiology, biochemistry and function. Academic Press: New York: pp 565–576

Johnson A D, Gomes W R eds 1977 The testis. Vol 4. Advances in physiology, biochemistry and function, Academic Press: New York

Johnson D A, Roth G M, Craig W M 1952 Autonomic pathways in the spinal cord. J Neurosurg 9: 599–605

Johnson E F, Berryman H, Mitchell R, Wood W B 1985 Elastic fibres in anulus fibrosus of the adult human intervertebral disc. A preliminary report. J Anat 143: 57–63

Johnson F P 1914 The development of the rectum in the human embryo. Am J Anat 16: 1–58

Johnson F R, McMinn R M H, Artfield G N 1968 Ultrastructural and biochemical observations on the tympanic membrane. J Anat 103: 297–310

Johnson F R, McMinn R M H, Birchenough R F 1962 The ultrastructure of the gall bladder epithelium of the dog. J Anat 96: 477–487

Johnson K E 1970 The role of changes in cell contact behavior in amphibian gastrulation. J Exp Zool 175: 391–427

Johnson M A, Polgar J, Weightman D, Appleton D 1973 Data on the

BIBLIOGRAPHY

distribution of fibre types in thirty-six human muscles: an autopsy study. J Neurol Sci 18: 111–129

Johnston J, Parkinson D 1974 Intracranial sympathetic pathways associated with the sixth nerve. J Neurosurg 40: 236–243

Johnston J B 1909 The morphology of the forebrain vesicle in vertebrates. J Comp Neurol 19: 458–539

Johnston M C, Pratt K M 1975 A developmental approach to teratology. In: Berry C L, Poswillo D E (eds) Teratology trends and applications. Springer-Verlag: Berlin: pp 2–16

Jollie M 1962 Chordate Morphology. Reinhold: Pitsburg

Jones A L, Fawcett D W 1966 Hypertrophy of the agranular endoplasmic reticulum in hamster liver induced by plenobarbital (with a review on the functions of this organelle in liver). J Histochem Cytochem 14: 215–232

Jones A W 1981 Vascular smooth muscle and alterations during hypertension. In: Bülbring E, Brading A F, Jones A W, Tomito T (eds) Smooth muscle: an assessment of current knowledge. Edward Arnold: London: pp 397–429

Jones D G 1978 Some current concepts of synaptic organization. Adv Anat Embryol Cell Biol 55: 3–69

Jones D S, Beargie R J, Pauly J E 1953 An electromyographic study of some muscles of costal respiration in man. Anat Rec 117: 17–24

Jones E G 1970 On the mode of entry of blood vessels into the cerebral cortex. J Anat 106: 507–520

Jones E G, Hartman, B K 1978 Recent advances in neuroanatomical methodology. Annu Rev Neurosc I: 215–296

Jones E G, Powell T P S 1968 The ipsilateral cortical connexions of the somatic sensory areas in the cat. Brain Res 9: 71–94

Jones E G, Powell T P S 1969 Connexions of the somatic sensory cortex of the rhesus monkey. I. Ipsilateral cortical connexions. Brain 92: 447–502

Jones E G, Powell T P S 1970 An anatomical study of converging sensory pathways within the cerebral cortex of the monkey. Brain 93: 793–820

Jones E G, Burton H, Porter R 1975 Commissural and cortico-cortical 'columns' in the somatic sensory cortex of primates. Science 190: 572–574

Jones E G, Coulter J D, Burton H, Porter R 1977 Cells of origin and terminal distribution of corticostriatal fibres arising in the sensori-motor cortex of monkeys. J Comp Neurol 173: 53–80

Jones F W 1911 On the grooves upon the ossa parietalia commonly said to be caused by the arteria meningea media. J Anat 46: 228–238

Jones F W 1912 Some nerve markings on lumbar vertebrae. J Anat 47: 118–120

Jones F W 1913 The function of the coelom and the diaphragm. J Anat 47: 282–318

Jones F E 1931 The non-metrical morphological characters of the skull as criteria for racial diagnosis. I. General discussion of the morphological characters employed in racial diagnosis. II. The non-metrical morphological characters of the Hawaiian skull. III. the non-metrical morphological characters of the prehistoric inhabitants of Guam. J Anat 65: 179–195, 368–378, 438–445

Jones F W 1939a The anterior superior alveolar nerve and vessels. J Anat 73: 583–591

Jones F W 1939b The so-called maxillary antrum of the gorilla. J Anat 74: 116–119

Jones F W 1940 The nature of the soft palate. J Anat 74: 147–170

Jones F W 1941 The principles of anatomy as seen in the hand. Baillière, Tindall and Cox: London

Jones F W 1949 Structure and function as seen in the foot. Baillière, Tindall and Cox: London

Jones I C 1957 The adrenal cortex. Cambridge University Press: Cambridge

Jones J P, Fox H 1977 Syncytial knots and intervillous bridges in the human placenta: an ultrastructural study. J Anat 124: 275–286

Jones M M, Amis A A 1988 The fibrous flexor sheaths of the fingers. J Anat 156: 185–196

Jones R L 1937 Cell fibre ratios in the vagus nerve. J Comp Neurol 67: 469–482

Jones R L 1941 The human foot. An experimental study of its mechanics, and the role of its muscles and ligaments in the support of the arch. Am J Anat 68: 1–40

Jones S J, Boyde A 1972 A study of human root cementum surfaces as prepared for and examined in the scanning electron microscope. Z Zellforsch. Mikrosk Anat 130: 318–337

Jones-Seaton A 1950 Étude de l'organisation cytoplasmique de l'oeuf des rongeurs, principalement quant à la basophilie ribonucléique. Arch Biol Paris 61: 291–444

Jonsson B 1974 Function of the erector spinae muscle on different working levels. Acta Morph Neerl Scand 12: 211–214

Jonsson B, Hagberg M 1974 The effect of different working heights on the deltoid muscle. Scand J Rehab Med Suppl 3: 26–32

Jonsson B, Steen B 1962 Function of the hip and thigh muscles in Romberg's test and 'standing at ease'. Acta Morph Neerl Scand 5: 267–276

Jordan R K, McFarlane B, Scothorne R J 1973 An electron microscopic study of the histogenesis of the ultimobranchial body and of the C-cell system in the sheep. J Anat 114: 115–136

Joseph J 1951 Further studies of the metacarpophalangeal and interphalangeal joints of the thumb. J Anat 85: 221–229

Joseph J 1960 Man's posture: electromyographic studies. Thomas: Springfield, Illinois

Joseph J 1973 Sequential contraction of muscles producing the same movement at a joint. In: Desmedt J E (ed) New developments in EMG and clinical neurophysiclogy. Karger: Basel: pp 665–674

Joseph J 1975 Movements at the hip joint. Am R Coll Surg 56: 192–201

Joseph J, Williams P L 1957 Electromyography of certain hip muscles. J Anat 91: 286–294

Joseph J, Nightingale A, Williams P L 1955 Detailed study of electric potentials recorded over some postural muscles while relaxed and standing. J Physiol 127: 617–625

Joseph S A, Knigge K M 1978 The endocrine hypothalamus: recent anatomical studies. In: Reichlin S, Baldessarine R J, Martin J B (eds) The hypothalamus. Research publications. Vol 56. Association for research in nervous and mental disease. Raven Press: New York: pp 15–47

Jost A 1961 The role of fetal hormones in pre-natal development. Harvey Lect 55: 201–226

Joubert D M 1955 Growth of muscle fibre inthe foetal sheep. Nature 175: 936–937

Jouvet M 1962 Recherches sur les structures nerveuses et les mécanismes responsables des différentes phases du sommeil physiologique. Arch Ital Biol 100: 125–206

Jouvet M 1964 Etude neurophysiologique unique des troubles de la conscience. Acta Neurochir 12: 258–269

Jouvet M 1965 Paradcxical sleep–a study of its nature and mechanisms. Progr Brain Res 18: 20–62

Jouvet M 1969 Biogenic amines and the states of sleep. Science 163: 32–41

Jouvet M 1972 The role of monoamines and acetylocholine-containing neurons in the regulation of the sleep-waking cycle. Ergeb Physiol 64: 166–307

Jovanović S, Živanović S 1965 The establishment of the sex by the great sciatic notch. Acta Anat 61: 101–107

Junqueira L C, Carneiro J, Long J A 1986 Basic histology. 5th edition, Lange: Los Altos, California

K

Kaada B R 1960 Cingulate, posterior orbital, anterior insular and temporal pole cortex. In: Field J, Magoun H, Hall V E (eds) Handbook of physiology. Vol 2. Section 1. Neurophysiology. American Physiologicial Society: Washington, DC: pp 1345–72

Kaar G F, Fraher J P 1986 The sheaths surrounding the attachments of rat lumbar ventral roots to the spinal cord: a light and electron microscopical study. J Anat 148: 137–146

Kagayama M, Nishiyama A 1972 Comparative aspect on the innervation of submandibular glands in cat and rabbit; an electron microscopic study. Tohuku J Exp Med 108: 179–193

Kahn A J Fallon M D, Teitelbaum S L 1984 Structive-function relationships in bone: an examination of events at the cellular level. In: Peck W A (ed) Bone and mineral research, annual 2. Elsevier: Amsterdam: pp 125–174

Kaiser O 1891 Die Funktionen der Ganglionzellen des Halsmarkes. Nijhoff: The Hague

Kaiserling E, Stein H, Müller-Hermelink H K 1974 Interdigitating reticulum cells in the human thymus. Cell Tissue Res 155: 47–55

Kalebic T, Garbisa S, Glaser B, Liotta L A 1983 Basement membrane collagen: degradation by migrating endothelial cells. Science, 221: 281–283

Kamath S 1981 Observations on the length and diameter of vessels forming the circle of Willis. J Anat 133: 419–423

Kameda Y 1971 The occurrence and distribution of the parafollicular cells in the thyroid, parathyroid IV and thymus IV in some mammals. Arch Histol Jpn 33: 283–299

Kameda Y 1976 Fine structural and endocrinological aspects of thyroid parafollicular cells. In: Coupland R E, Fujita T (eds) Chromaffin, enterochromaffin and related cells. Elsevier: New York: pp 155–170

Kampmeier O F 1969 Evolution and comparative morphology of the lymphatic system. Thomas: Springfield, Illinois

Kanagasuntheram R 1957 Development of the human lesser sac. J Anat 91: 188–206

Kanagasuntheram R 1960 Some observations on the development of the human duodenum. J Anat 94: 231–240

Kanagasuntheram R 1967 A note on the development of the tubotympanic recess in the human embryo. J Anat 101: 731–742

Kanagasuntheram R, Kin L S 1970 Observations on some anomalies of the colon. Singapore Med J 11: 110–117

Kanaseki T, Kadota K 1969 The 'vesicle in a basket'. J Cell Biol 42: 202–220

Kane E C 1974 Synaptic organization in the dorsal cochlear nucleus of the cat: a light and electron microscopic study. J Comp Neurol 155: 301–329

Kanno T 1973 Unidirectional cellular processes in stimulus-secretion coupling in cells of the GEP system. In: Fujita T (ed) Gastro-entero-pancreatic endocrine system. A cell-biological approach. Igaku Shoiu: Tokyo: pp 64–70

Kapanci Y, Assimacopoulos A, Irle C, Zwahlen A, Gabbiani G 1974 'Contractile interstitial cells' in pulmonary alveolar septa: a possible regulator of venitlation/perfusion ratio? Ultrastructure, immunofluorescence and 'in vitro' studies. J Cell Biol 60: 375–392

Kapandji I A 1963 Physiologie articulaire. Fascicule I. Membre supérieur. Maloine: Paris

Kapandji I A 1970–1974 The physiology of the joints. Annotated diagrams of the mechanics of the human joints. 2nd edition, 3 vol translated by L H Honoré. (Vol 1 upper limb. Vol 2 lower limb. Vol 3 the trunk and the vertebral column.) Churchill Livingstone: Edinburgh

Kapeller K, Mayor D 1967 The accumulation of noradrenaline in constricted sympathetic nerves as studies by fluorescence and electron microscopy. Proc R Soc Lond [Biol] 167: 282–292

Kaplan E B 1958 The iliotibial tract. Clinical and morphological significance. J Bone Jt Surg 40A: 817–831

Kaplan E B 1965 Functional and surgical anatomy of the hand., 2nd edition. Pitman Medical: London

Kaplan H A 1956 Arteries of the brain; anatomic study. Acta Radiol 46: 364–470

Kaplan H A, Ford D H 1966 The brain vascular system. Elsevier: Amsterdam

Kaplan H A, Browder A, Browder J 1973 Nasal venous drainage and the foramen caecum. Laryngoscope 83: 327–9

Kaplan H A, Browder J, Krieger A J 1976 Intercavernous connections of the cavernous sinuses. The superior and inferior circular sinuses. J Neurosurg 45: 166–168

Kaplan M S, Hinds J W 1980 Gliogenesis of astrocytes and oligodendrocytes in the neocortical grey and white matter of the adult rat: electron microscopic analysis of light radioautographs. J Comp Neurol 193: 711–727

Kappers C U A 1920–1921 Die vergleichende Anatomie des Nervensystems der Wirbeltiere und des menschen. Bohn: Haarlem

Kappers C U A 1921 On structural laws in the nervous system; the principles of neurobiotaxis. Brain 44: 125–149

Kappers C U A 1934 Differences in the effect of various impulses on the structure of the central nervous system. Ir J Med Sci 105: 495–519

Kappers C U A 1947 Anatomie comparée du systeme nerveux. Masson: Paris

Kappers C U A, Huber G C, Crosby E C 1936 The comparative anatomy of the nervous system of vertebrates, including man. Macmillan: New York

Kappers J A 1960 The development, topographical relations and innervation of the epiphysis cerebri in the albino rat. Z Zellforsch 52: 163–215

Kappers J A 1976 The mammalian pineal gland. A survey. Acta Neurochir Genesskd 120: 109–149

Kappers J A 1979 Short history of pineal discovery and research. In: Kappers J A, Péret P (eds) The pineal gland of vertebrates including man. Proceedings of the 1st colloquim of the european study group (EPSG), Amsterdam, 1978. Elsevier/North-Holland: pp 3–22. Biomedical Press: Amsterdam (Progress in Brain Research Vol 52)

Karasek M 1976 Quantitative changes in number of 'synaptic' ribbons in rat pinealocytes after orchidectomy and in organ culture. J Neurol Transm 38: 149–157

Karasek M 1983 Ultrastructure of the mammalian pineal gland: its comparative and functional aspects. In: Reiter R J (ed) Pineal research reviews vol 1. Liss: New York: pp 1–48

Karfunkel P 1971 The role of microtubules and microfilaments in neurulation in Xenopus. Dev Biol 25: 30–56

Kariniemi A L, Lehto V P, Vartio T, Virtanen I 1982 Cytoskeleton and pericellular matrix organization of pure adult human keratinocytes cultured from suction-blister roof epidermis. J Cell Sci 58: 49–61

Karnovsky M L 1967 Energetics of transport. Protoplasma 63: 76–85

Karnovsky M L 1968 The metabolism of leukocytes. Seminars Hemat 5: 156–165

Karrer H E 1956 The ultrastructure of mouse lung. General architecture of capillary and alveolar walls. J Biophys Biochem Cytol 2: 241–252

Kása P, Mann S P, Hebb C 1970 Localization of choline acetyltransferase. Nature 226: 812–814

Kasai T, Chiba S 1977 True nature of the muscular arch of the axilla and its nerve supply. Kaibogaku Zasshi 25: 657–669

Kasai T, Takahashi G, Aiyama S 1974 A case report of annular pancreas. Acta Anat Nippon 49: 103–119

Kashef R 1966 The node of Ranvier. PhD thesis. University of London

Kashibara M, Ueda M, Horiguchi Y, Fukumi F, Hanaoka M, Imamura S 1986 A monoclonal antibody specifically reactive to human Langerhans cells. J Invest Dermatol 87: 6902–607

Katchburian E 1973 Membrane-bound bodies as initiators of mineralization of dentine. J Anat 116: 285–302

Kate B R 1968 The torsion of the humerus in central India. J Ind Anthropol Soc 3: 17–30

Kate B R, Robert S L 1965 Some observations on the upper end of the tibia in squatters. J Anat 99: 137–142

Katz B 1966 Nerve, muscle and synapse. McGraw-Hill: New York

Katz B, Miledi R 1965 The effect of calcium on acetylcholine release from motor nerve terminals. Proc R Soc Lond [Biol] 161: 496–503

Katz S I, Tamaki K, Sachs D H 1979 Epidermal Langerhans' cells are derived from cells which orginate in bone marrow. Nature 282: 324–326

Kauer J M G 1974 The interdependence of carpal articulation chains. Acta Anat 88: 481–501

Kawabata I, Paparella M M 1969 Ultrastructe of normal human middle ear mucosa. Preliminary report. Ann Otol Rhinol Laryngol 78: 125–138

Kawamura K, Brodal A, Hoddevik G 1974 The projection of the superior colliculus onto the reticular formation of the brain stem. An experimental anatomical study in the cat. Exp Brain Res 19: 1–19

Kaye G I, Wheeeler H O, Whitlock R T, Lane N 1966 Fluid transport in the rabbit gall bladder. A combined physiological and electron microscopic study. J Cell Biol 30: 237–268

Keagy R D, Brumlik J, Bergan J L 1966 Direct electromyography of the psoas major muscle in man. J Bone Jt Surg 48A: 1377–1382

Keech M K 1960 Electron microscope study of the normal rat aorta. J Biophys Biochem Cytol 7: 533–538

Keegan J J, Garrett F D 1948 The segmental distribution of the cutaneous nerves in the limbs of man. Anat Rec 102: 409–437

Keen E N 1981 Origin of renal arteries from the aorta. Acta Anat 110: 285–286

Keen J A 1950 Study of differences between male and female skulls. Am J Phys Anthropol 8: 65–79

Keene M F L 1961 Muscle spindles in human laryngeal muscles. J Anat 95: 25–29

Keene M F L, Hewer E E 1927 Observations on the development of the human suprarenal gland. J Anat 61: 302–324

Keene M F L, Hewer E E 1935 The sub-commissural organ and the mesocoelic recess in the human brain, with a note on Reissner's fibre. J Anat 69: 501–507

Kefalides N A 1973 Structure and biosynthesis of basement membranes. In: Hall D A, Jackson D S (eds) International review of connective tissue research. Vol 6. Academic Press: New York: pp 63–104

Keith Sir A 1948 Human embryology and morphology. 6th edition. Arnold: London

Keith Sir A, Flack M W 1907 The form and nature of the muscular connnections between the primary divisions of the vertebrate heart. J Anat 41: 172–189

Kelley A E, Domesick V B, Nauta W J H 1982 The amygdalostriatal projection in the rat–an anatomical study by anterograde and retrograde tracing methods. Neuroscience 7: 615–630

Kelley R O 1973 Fine structure of the apical rim-mesenchyme complex during limb morphogenesis in man. J Embryol Exp Morphol 29: 117–131

Kellgren J H, Samuel E P 1950 Sensitivity and innervation of the articular capsule. J Bone Jt Surg 32B: 84–92

Kelly A M, Zacks S I 1969 The fine structure of motor end plate morphogenesis. J Cell Biol 42: 154–169

Kelly D E, Cahill M A 1972 Filamentous and matrix components of skeletal muscle Z-disks. Anat Rec 172: 623–642

Kelly P J 1968 Anatomy, physiology and pathology of the blood supply of bones. J Bone Jt Surg 50A: 766–783

Kemali M, Casale E, Guglielmotti V 1977 On the impregnation by the Golgi method of an entire human brain. Brain Res 126: 345–349

Kemp J M 1968a An electron microscopic study of the termination of afferent fibres in the caudate nucleus. Brain Res 11: 484–487

Kemp J M 1968b Observations on the caudate nucleus of the cat impregnated with the Golgi method. Brain Res 11: 467–470

Kemp J M 1970 The termination of strio-pallidal and strio-nigral fibres. Brain Res 17: 125–128

Kemp J M, Powell T P S 1970 The cortico-striate projections in the monkey. Brain 93: 525–546

Kempson G E, Freeman M A, Swanson S A 1968 Tensile properties of articular cartilage. Nature 220: 1127–1128

Kendall M D (ed) 1981 The Thymus Gland (Anatomical Society of Great Britain and Ireland, Symposium 1). Academic Press: London

Kendall M D 1988 Anatomical and physiological functions influence the thymic microenvironment. Thymic Update 1: 27–65

Kendall M D, Johnson H R M, Singh J 1980 The weight of the human thymus gland at necropsy. J Anat 131: 485–499

Kennedy W P 1967 Epidemiological aspects of the problem of congenital malformations. Birth Defects 3: 1–18

Kenny M 1944 The clinically suspect pelvis and its radiographical investigation in 1,000 cases. J Obstet Gynaecol Br Commonw 51: 277–292

Kensinger R S, Collier R J, Bazer F W 1986 Ultrastructural changes in porcine mammary tissue during lactogenesis. J Anat 145: 49–59

Kerckring T T 1970 Spicilegium anatomicum. A Frisius: Amsterdam

Kerjaschki D, Hörandner H 1976 The development of mouse olfactory vesicles and their cell contacts. A freeze-etching study. J Ultrastruct Res 54: 420–444

Kerr F W L 1962 Facial, vagal and glossopharyngeal nerves in the cat. Afferent connexions. Arch Neurol Psychiatry 6: 264–281

Kerr F W L 1969 Preserved vagal visceromotor function following destruction of the dorsal motor nucleus. J Physiol 202: 755–769

Kessel R G 1968 Annulate lamellae. J Ultrastruct Res Suppl 10: 1–82

BIBLIOGRAPHY

Keswani N H, Hollinshead W H 1956 Localisation of the phrenic nucleus in the spinal cord of man. Anat Rec 125: 683–700

Kettlekamp D B, Jacobs A W 1972 Tibiofemoral contact area: determination and implications. J Bone Jt Surg 54A: 349–356

Key J A 1932 The synovial membrane of joints and bursae. In: Cowdry E V (ed) Special cytology. The form and functions of the cell in health and disease, 2nd editiion. Paul B Hoeber: New York: pp 1055–1086

Keynes G 1954 The physiology of the thymus gland. Br Med J 2: 659–663

Khaledpour C 1984 Eine anatomische Variation des Nervus alveolaris inferior beim Menschen. Anat Anz 156: 403–456

Khoo F Y, Kanagasuntheram R, Chia K B 1969 Variations of the lateral recesses of the naso-pharynx. Archs Otolaryngol 88: 456–462

Kielbasinski G 1976 Arteries of the inferior part of the vermis cerebelli in man. Folia Morphol 25: 149–157

Kier E L 1966 Embryology of the normal optic canal and its anomalies. Invest Radiol 1: 346–362

Kier E L 1977 The cerebral ventricles: a phylogenetic and ontogenetic study. In: Newton T H, Potts D G (eds) Radiology of the skull and brain. Anatomy and pathology. Vol 3. Mosby: St Louis: pp 2787–2914

Kievit J, Kuypers H G J M 1975 Subcortical afferents to the frontal lobe in the rhesus monkey studied by means of retrograde horseradish peroxidase transport. Brain Res 85: 261–266

Kikkawa Y, Smith F 1983 Cellular and biochemical aspects of pulmonary surfactant in health and disease. Lab Invest 49: 122–139

Kimmel D L 1961a Innervation of spinal dura mater and dura mater of the posterior cranial fossa. Neurology II: 800–809

Kimmel D L 1961b The nerves of the cranial dura mater and their significance in dural headache and referred pain. Chicago Med Sch Q 22: 16–26

Kimura K 1977 Foramina and noches on the supraorbital margin in some racial groups. Acta Anat 52: 203–209

Kimura R S 1969 Distribution, structure and function of dark cells in the vestibular labyrinth. Ann Otol Rhinol Laryngol 78: 542–561

King R H M, Thomas P K 1971 Aberrant regeneration of unmyelinated axons in the vagus nerve of the rabbit. J Anat 108: 596P

King R J 1974 The surfactant system of the lung. Fed Proc Fed Am Socs Exp Biol 33: 2238–2247

King R J, Clements J A 1972 Surface active materials from dog lung. I. Composition and physiological correlations. Am J Physiol 223: 715–726

King T J, Briggs R 1965 Serial transplantation of embryonic nuclei. In: Bell E (ed) Molecular and cellular aspects of development. Harper and Row: New York: pp 171–193

King T S 1954 The anatomy of hare-lip in man. J anat 88: 1–12

King T S, Coakley J B 1958 The intrinsic nerve cells of the cardiac atria of mammals and man. J Anat 92: 353–376

King T S, Steinlacher S 1985 Pineal indolalkylamine synthesis and metabolism: kinetic considerations. Vol 3. In: Reiter R J (ed) Pineal research reviews. Liss: New York: pp 69–113

Kinman J 1977 Surgical aspects of the anatomy of the sphenoidal sinuses and the sella turcica. J Anat 124: 541–553

Kinmonth J B 1964 Some general aspects of the investigation and surgery of the lymphatic system. J Cardiovasc Surg 5: 680–682

Kinmonth J B, Taylor G W 1964 Chylous reflux. Br Med J I: 529–532

Kinnaert P 1973 Anatomical variations of the cervical part of the thoracic duct in man. J Anat 115: 45–52

Kinnamon J C, Taylor B J, Delay R J, Roper S D 1985 Ultrastructure of mouse vallate taste buds. I. Taste cells and their associated synapses. J Comp Neurol, 235: 48–60

Kinsbourne M, Hicks R F 1978 Mapping cerebral functional space: competition and collaboration in human performance. In: Kinsbourne M (ed) Asymmetrical function of the brain. Cambridge University Press: pp 267–273

Kinsella J D, Baum J, Ziff M 1970 Studies of isolated synovial cells of rheumatoid and nonrheumatoid synovial membranes. Arthritis Rheum 13: 734–753

Kirchner J A, Wyke B D 1965 Articular reflex mechanisms in the larynx. Ann Otol Rhinol Laryngol 74: 749–768

Kirk J 1944 Observations on the histology of the choledochduodenal junction and papilla duodeni, with particular reference to the ampulla of vater and sphincter of Oddi. J Anat 78: 118–120

Kisch B 1958 New investigations on cardiac nerves. An electron microscopic study. Exp Med Surg 16: 81–95

Kiss F 1932 Sympathetic elements in the cranial and spinal ganglia. J Anat 66: 488–498

Kisvárday Z F, Martin K A C, Whitteridge D, Somogyi P 1985 Synaptic connections of intracellularly filled clutch cells; a type of small basket cell in the visual cortex of the cat. J Comp Neurol 241: 111–37

Kitchin I C 1949 The effects of notochordectomy in amblystoma mexicanum. J Exp Zool 112: 393–415

Kito H, Hosoda S 1977 Triple staining for simultaneous visualization of cell types in islets of Langerhans of pancreas. Successive application of argyrophil, aldehyde-fuchsin and lead-hematoxylin stains in a single tissue section. J Histochem Cytochem 25: 1019–1020

Kjaegaard J 1974 An electron microscopic study of the tympanojugular glomus. Acta Otolaryngol (Stockh) 78: 84–89

Klareskog L, Forsum U, Kabelitz D, Plöen L, Sundström C, Nilsson K,

Wigren A, Wigzell H 1982 Immune functions of human synovial cells. Phenotypic and T cell regulatory properties of macrophage-like cells that express HLA-DR. Arthritis Rheum 25: 488–501

Klebs T A E 1869 Die Einschmelzungs-Methode, ein Beitrag sur mikroskopischen Technik. Arch Mikr Anat 5: 164–6

Kleiger B, Mankin H J 1961 A roentgenographic study of the development of the calcaneus by means of the posterior tangential view. J Bone Joint Surg 43A: 961–969

Klein D C 1978 The pineal gland: a model of neuroendocrine regulation. Res Publ Assoc Res Nerv Ment Dis 56: 303–327

Klein D C, Weller J L 1972 A rapid light-induced decrease in pineal serotonin N-acetyltransferase activity. Science 177: 532–533

Klekamp J, Riedela A, Harper C, Kretschmann H J 1987 A quantitative study of Australian aboriginal and caucasian brains. J Anat 150: 191–210

Klika E, Petrík P 1965 A study of the structure of the lung alveolar and bronchiolar epithelium (a histological and histochemical study using the method of membranous preparations). Acta Histochem 20: 331–342

Klimer W L, McCulloch W S, Blum J, Craighill E, Peterson D 1968 In: Corning W C, Balaban M (eds) The mind, biological approaches to its functions. Interscience: New York

Klineberg I J, Wyke B D 1975 Articular reflex control of mastication. In: Kay L W (ed) Oral surgery IV. Munksgaard: Copenhagen

Klintworth G K 1967 The ontogeny and growth of the human tentorium cerebelli. Anat Rec 158: 433–442

Klintworth G K 1968 The comparative anatomy and phylogeny of the tentorium cerebelli. Anat Rec 160: 635–642

Klosovskii B N 1963 The development of the brain and its disturbance by harmful factors (Haigh B ed and trans) Pergamon Press: Oxford

Klück P 1980 The autonomic innervation of the human urinary bladder, bladder neck and urethra: a histochemical study. Anat Rec 198: 439–447

Klug W S, Cummings M R 1986 Concepts of genetics. 2nd edition. Merrill: Columbus

Klüver H, Barrera E 1953 Method for combined staining of cells and fibers in the nervous system. J Neuropathol Exp Neurol 12: 400–403

Klüver H, Bucy P C 1937 Psychic blindness and other symptoms following temporal lobectomy in rhesus monkeys. Am J Physiol 119: 352–353

Knese K-H 1979 Stützgewebe und Skeletsystem. Handbuch der mikroscopischen Anatomie des Menschen. Vol 2. Die Gewebe, part 5, pp 225–428, Springer-Verlag: Berlin

Knigge K M, Joseph S A 1974 Thyrotropin releasing factor (TRF) in cerebrospinal fluid of the third ventricle of rat. Act Endocrinol 76: 209–213

Knigge K M, Scott D E, Kobayashi H, Ishii S eds 1975 Brain-endocrine interaction. II. The ventricular system in neuroendocrine mechanisms. 2nd International symposium . . . shizuoka . . . 1974 Karger: Basel

Knight B K, Hayes M M M, Symington R B 1973 The pineal gland–a synopsis of present knowledge with particular emphasis on its possible role in control of gondadotrophin function. S Afr J Anim Sci 3: 143–146

Knisely M H 1936 Spleen studies. I. Microscopic observations of the circulatory system of living unstimulated mammalian spleens. Anat Rec 65: 23–50

Knowles F 1974 Ependyma of the third ventricle in relation to pituitary function. Prog Brain Res 38: 255–270

Knussman R, Finke E 1977 Studies on the sex-specificity of the human spinal profile. Acta Med Auxol 9: 16

Ko J S, Bernard G W 1981 Osteoclast formation in vitro from bone marrow mononuclear cells in osteoclast-free bone. Am J Anat 161: 415–425

Kobayashi H 1975 Absorption of cerebrospinal fluid by ependymal cells of the median eminence. In: Brain-endocrine interaction. II. 2nd international symposium (Knigge K M, Scott D E, Kobayashi H, Ishi S eds) pp 109–122 Karger: Basel

Kobayashi S 1969 On the fine structure of the carotid body of the bird, Uroloncha domestica. Arch Histol Jpn 31: 9–19

Kobayashi S, Fujita T 1969 Fine structure of mammalian and avian pancreatic islets with special reference to D cells and nervous elements. Z Zellforsch 100: 340–363

Koch C, Poggio T 1983. A theoretical analysis of electrical properties of spines. Proc R Soc Lond [Biol] 298: 227–263

Koch J C 1917 The laws of bone architecture. Am J Anat 21: 177–298

Kocher E T 1892 Zur Radicalcorder Hernien. Korres Bl Schweiz Aerzte 22: 561–76

Koelle G B, Friedenwald J S 1949 Histochemical method for localizing cholinesterase activity. Proc Soc Exp Biol Med 70: 617–622

Köhn A 1903 Die paraganglien. Ark Mikrosk Anat 62: 263

Kohnstamm O 1898 Zur Anatomie und Physiologie des Phrenicuskernes. Fortschr Med 16: 643–653

Kokott W 1934 Das Spaltlinienbild der Sklera (Ein Beitrag zum funktionellen Bau der Sklera) Klin Monatsbl Augenheilkol 92: 177–185

Kolb H, Nelson R 1984 Neural architecture of the cat retina. Prog Retinal Res 3. Pergamon: New York

Kolb H, West R W 1977 Synaptic connection of the interplexiform cell in the retina of the cat. J Neurocytol 6: 155–170

Köling A, Rask-Andersen H 1983 Membrane junctions in the subodontoblastic region. A freeze-fracture study of the human dental pulp. Acta Odontol Scand 41: 99–109

Kollar E J 1972 Histogenetic aspects of dermal-epidermal interactions. In: Slavkin H C, Bavetta L A (eds) Developmental aspects of oral biology. Academic Press: New York: pp 125–149

Kollar E J, Lumsden A G S 1979 Tooth morphogenesis: the role of the innervation during induction and pattern formation. J Biol Buccale 7: 49–60

Kölliker R A von 1852 Handbuch der Gewebelehre des Menschen. Engelmann: Leipzig

Kölliker R A von 1896 Handbuch der Gewebelehre des Menschen, 6th edition. Engelmann: Leipzig

Kolmodin G M 1957 Integrative processes in single spinal interneurones with proprioceptive connections. Acta Physiol Scand 40: Suppl 139, 1–89

Kolmodin G M, Skoglund C R 1960 Analysis of spinal interneurons activated by tactile and nociceptive stimuli. Acta Physiol Scand 50: 337–355

Kolnberger I, Altner H 1971 Cilary-structure precursor bodies as stable constituents in the sensory cells of the vomero-nasal organ of reptiles and mammals. Z Zellforsch 118: 254–262

Komai T, Fukoka G 1934 A study of the frequency of left-handedness and left-footedness among Japanese school children. Hum Biol 6: 33–42

Konigsmark B W 1970 Methods for the counting of neurons. In: Nauta W J H, Ebbesson S O E (eds) Comtemporary research methods in neuroanatomy. Springer-Verlag: Berlin: pp 315–339

Konigsmark B W, Kalyanarama U P, Corey P, Murphy E A 1969 An evaluation of techniques in neuronal population estimates: the sixth nerve nucleus. Johns Hopkins Hosp Bull 125: 146–158

Konishi A, Sato M, Mizuno N, Hon K, Nomura S, Sugimoto T 1978 An electron microscope study of the areas of the Onuf's nucleus in the cat. Brain Res 156: 333–338

Koornneef L 1977 New insights in the human orbital connective tissue. Arch Ophthal 95: 1269–1273

Kopac M J 1959 Micrurgical studies on living cells. Biochemistry, physiology, morphology. In: Brachet J, Mirsky A E (eds) The cell. Vol 1. Academic Press: New York

Koritké J G, Gillet J Y, Pietri J 1967 Les artères de la trompe uterine chez la femme. Arch Anat Histol Embryol 50: 47–70

Kormano M, Suoranta H 1971 Microvascular organisation of the adult human testis. Anat Rec 170: 31–40

Kornberg R D, Klug A 1981 The nucleosome. Sci Am 244: (Feb), 52–64

Korneliussen A K 1968 On the ontogenetic development of the cerebellum (nuclei, fissures and cortex) of the rat, with special reference to regional variations in corticogenesis. J Hernforsch 10: 379–412

Kornguth S E, Anderson J W, Scott G 1966 Observations on the ultrastructure of the developing cerebellum of the Macaca mulatta. J Comp Neurol 130: 1–23

Kornguth S E, Anderson J W, Scott G 1968 the development of synaptic contacts in the cerebellum of Macaca mulatta. J. Comp Neurol 132: 531–546

Kos J 1970 L'ultrastructure des franges et des plis synoviaux, C R Assoc Anat 149: 802–813

Kosinski C 1926 Observations on the superficial venous system of the lower extremity. J Anat 60: 131–142

Kothari L K, Gupta A S 1974 Effect of ageing on the volume, structure and total Leydig cell content of the human testis. Int J Fertil 19: 140–146

Kovrishko N M 1964 Postnatal development and structural characteristic of the principal paraganglia in man. Fed Proc Trans Suppl) 22: 740

Kozielec T, Józwa H 1977 Variation in the course of the facial artery in the prenatal period in man. Folia Morphol Warsz 36: 55–61

Koskinen L, Isotupa K, Koski K 1976 A note on craniofacial sutural growth. Am J Phys Anthropol 45: 511–516

Koss L G 1968 Diagnostic Cytology 2nd edition. Pitman: London

Kostick E L 1963 Facets and imprints on the upper and lower extremities of femora from a Western Nigerian population. J Anat 97: 393–402

Kostović I, Molliver M E 1974 A new interpretation of the laminar development of cerebral cortex: synaptogenesis in different layers of neopallium in the human fetus. Anat Rec 178: 395

Kothari L K, Patni M K, Jain M L 1978 The total Leydig cell volume of the testis in some common mammals. Andrologia 10: 218–222

Kozielec T, Józwa H 1976 The supoticial temporal artery in human fetuses. Folia Morphol 35: 79–84

Krahl V E 1944 An apparatus for measuring the torsion angle in long bones. Science 99: 498

Krahl V E 1964 Anatomy of the mammalian lung. In: Fenn W O, Rahn H (eds) Handbook of physiology Section 3. Respiration vol 1. American Physiological Society: Washington, DC: pp 213–284

Krahl V E 1976 The phylogeny and ontogeny of humeral torsion. Am J Phys Antropol 45: 595–599

Kraissl C J 1951 The selection of appropriate lines for elective surgical incisions. Plast Reconstruct Surg 8: 1–28

Kramer I R H 1960 The vascular architecture of the human dental pulp. Arch Oral Biol 2: 177–189

Kraus B 1959 Occurrence of the Carabelli trait in southwest ethnic groups. Am J Phys Anthropol 17: 117–124

Kraus B S 1961 The Western Apache: some anthropometrics observations. Am J Phys Anthropol 19: 227–236

Kraus F W, Mestecky J 1971 Immunohistochemical localization of amylase, lysozyme and immunoglobins in the human parotid gland. Arch Oral Biol 16: 781–789

Kraus K S, Jordan R E, Abrams L A 1969 Dental anatomy and occlusion. Williams and Wilkins: Baltimore, Md

Krayenbühl H A 1967 Cerebral venous and sinus thrombosis. Clin Neurosurg 14: 1–24

Krehl L 1891 Beitrage zur Kentnisse der Fullung und entleerung des Herzens, abhandlungen der Mathematische-physischen Klassen der kåoliglichen Sachsischen Fesellschalt der Wissenschaften 17: 341–362

Kretschmann H-J 1988 Localization of the corticospinal fibres in the internal capsule in man. J Anat (Lond) 160: 219–225

Krieckhaus E E 1967 The mamillary bodies: their function and anatomical connections. Acta Biol Exp Warsz 27: 319–337

Krieckhaus E E, Randall D 1968 Lesions of the mamillothalmic tract in the rat produce no decrements in recent memory. Brain 91: 369–378

Kristensson K, Olsson Y 1971a Age-dependant variations in the spread of exogenous proteins in the peripheral nerve. Acta Pathol Microbiol Scand [A] 79: 310–11

Kristensson K, Olsson Y 1971b The perineurium as a diffusion barrier to protein tracers. Differences between mature and immature animals. Acta Neuropathol (Berl) 17: 127–138

Krmpotić-Nemanić J, Keros P 1973 Funktionale Bedeutung der Adaptation des Dens axis beim Menschen Verh Anat Ges 67: 393–397

Krmpotić-Nemanić J, Kostović I, Nemanić D, Kelović Z 1979 The laminar organization of the prospective auditory cortex in the human fetus (11–13.5 weeks of gestation), Acta Otolaryngol (Stockh) 87: 1241–126

Krnjević K 1975 Coupling of neuronal metabolism and electrical activity. In: Ingvar D H, Lassen N A (eds) Brain work: the coupling of function metabolism and blood flow in the brain. Munksgaard: Copenhagen: pp 65–78

Krogh A 1929 The anatomy and physiology of capillaries. Yale University Press: New Haven, Conn

Krogman W M 1941 Bibliography of human morphology 1914–1939, Chicago University Press: Chicago

Krogman W M 1962 The human skeleton in forensic medicine. Thomas: Springfield, Illinois

Krompecher S 1967 Local tissue metabolism and the quality of callus. Symp Biol Hung 7: 275–281

Krstić R 1976 Ultracytochemistry of the synaptic ribbons in the rat pineal organ. Cell Tissue Res 166: 135–143

Krstić R V 1979 Ultrastructure of the mammalian cell. An atlas. Translated by A R von Hochstetter. Springer Verlag: Berlin

Kruger L 1987 Morphological correlates of 'free' nerve endings: a reappraisal of thin sensory axon classification. In: Schmidt R F, Schaible H-G, Vahle-Hinz C (eds) Fine Afferent Nerve Fibres and Pain. VCH Verlag Chemie: Weinheim: pp 1–13

Kubik S 1967 The efferent lymph vessels and the regional lymph nodes of the female genital organs. In: Rüttimann A (ed) Progress in lymphology. Proceedings of the international symposium, Zurich, 1966. Thieme: Stuttgart: pp 196–197

Kubik S 1970 Lung lymphatics. Prog Lymphol 11: 29–31

Kubik S 1974 Anatomische Voraussetzungen zur endolymphatischen Radionuklidtherapie. Med Welt 25: 1011–1016

Kubik S, Müntener M 1969 Zur Topographie der spinalen Nervenwurzeln. II. Der Einfuss des Wachstums des Duralsackes, sowie der Krümmagen und der Bewegungen der spinalen nervenwurzeln. Acta Anat 74: 149–168

Kubo J 1970 Some observations on the autonomic innervation of the human oviduct. Int J Fertil 15: 30–35

Kubo M, Norris D A, Howell S E, Clark R A F 1984 Humam keratinocytes synthesize secrete and deposit fibronectin in the pericellular matrix. J Invest Dermatol 82: 580–586

Kubota K, Negishi T, Nasegi T 1975 Topological distribution of muscle spindles in the human tongue. Bull Tokyo Med Dent Univ 22: 235–242

Kuczynski K 1974 Carpometacarpal joint of the human thumb. J Anat 118: 119–126

Kudo H, Nori S 1974 Topography of the facial nerve in the human temporal bone. Acta Anat 90: 467–480

Kuettner K E, Pauli B U 1983 Inhibition of neurovascularization by a cartilage factor. In: (Ciba Foundation Symposium 100) Symposium on the development of the vascular system. Pitman: London: pp 163–173

Kuffler S W, Nicholls J G 1976 From neuron to brain. A cellular approach to the function of the nervous system. Sinauer Associates: Sunderland, Mass

Kugel M A 1927 Anatomical studies on the coronary arteries and their branches. I. Arteria anastomotica auricularis magna. Am Heart J 3: 260–270

Kuhar M J, Yamamura H I 1975 Light autoradiographic localization of cholinergic muscarinic receptors in rat brain by specific binding of a potent antagonist. Nature 253: 560–561

Kuhar M J, de Haven R T, Yamamura H I, Rommelspacher H, Simon J R 1975 Further evidence for cholinergic habenulo-inter-peduncular neurons: pharmacologic and functional characteristics. Brain Res 97: 265–275

Kuhlenbeck H 1954 Human diencephalon; summary of development, structure, function and pathology. Confinia Neurol Suppl 14: 1–230

BIBLIOGRAPHY

Kuhlenbeck H, Miller R N 1949 The pretectal region of the human brain. J Comp Neurol 91: 369–408

Kühnel W 1968 Vergleichende histologische histochemische und elektronenmikroskopische Untersuchungen an Tränendrüsen. VI. Menschliche Tränendrüsen. Z Zellforsch 89: 550–572

Kullaa-Mikkonen A, Hynynen M, Hyvönen P 1987 Filiform papillae of human, rat and swine tongue. Acta Anat 130: 280–284

Kummer B K F 1966 Photoelastic studies on the functional structure of bone. Folia Biotheoret 6: 31–40

Kummer B K F 1972 Biomechanics of bone. In: Fung Y-C B, Perrone N, Anliker M (eds) Biomechanics. Chapter 10, Prentice-Hall: New Jersey

Kuntscher G 1934 Die Darstellung des Kraftflusses im Knocken. Z Beit Chir 61: 2130–2136

Kuntz A 1953 The autonomic nervous system 4th edition, Lea and Febiger: Philadelphia

Kupfer C, Chumbley L, Downer J de C 1067 Quantitative histology of optic nerve, optic tract and lateral geniculate nucleus of man. J Anat 101: 393–402

Kurahashi Y, Yoshiki S 1972 Electron microscopic localisation of alkaline phosphatase in the enamel organ of the young rat. Arch Oral Biol 17: 155–163

Kuré K, Murakami S, Okinaka S 1934 Die Spinalpara-sympathetischen ganglionzellen im den spinalganglien und der Spinalparasympathetiens des Halssegmentes. Z Zellforsch M Anat 22: 54–79

Kuré K, Saégusa G, Kawaguchi K, Shiraishi K 1930 On the parasympathetic (spinal parasympathetic) fibres in the dorsal or posterior roots of the lumbar region of the spinal cord. Q J Exp Physiol 20: 333–44

Kurosumi K, Oota Y 1968 Electron microscopy of two types of gonadotrophs in the anterior pituitary glands of persistent estrus and diestrus rats. Z Zellforsch 85: 34–46

Kurrat H J, Oberländer W 1978 The thickness of the cartilage in the hip joint. J Anat 126: 145–55

Kuru Y 1967 Meningeal branches of the ophthalmic artery. Acta Radiol 6: 241–251

Kuwabara T 1975 The maturation of the lens cell. Exp Eye Res 20: 427–443

Kuypers H G J M 1958 Cortico-bulbar connexions to the pons and lower brain stem in man. An anatomical study. Brain 81: 364–388

Kuypers H G J M 1960 Central cortical projections to motor and somatosensory cell groups. Brain 83: 161–184

Kuypers H G J M 1985 The anatomical and functional organization of the motor system. In: Swash M, Kennard C (eds) Scientific Basis of Clinical Neurology. Churchill Livingstone: Edinburgh: pp 3–18

Kuypers H G J M, Lawrence D G 1967 Cortical projections to the red nucleus and the brain stem in the rhesus monkey. Brain Res 4: 151–188

Kuypers H G J M, Maisky V A 1975 Retrograde axonal transport of horseradish peroxidase from spinal cord to brainstem cell groups in the cat. Neurosci Lett 1: 9–14

Kuypers H G J M, Tuerk J D 1964 The distribution of the cortical fibres within the nuclei cuneatus and gracilis in the cat. J Anat 98: 143–162

Kvinnsland S, Kvinnsland S 1975 Growth in craniofacial cartilages studied by ³H-thymidine incorporation. Growth 39: 305–14

Kyber E 1870 Uber die Milz des Menschen und einiger Saugeticre. Arch Mikrosk Anat EntwMech 6: 540–570

Kyösola K, Rechardt L, Veijola L, Waris T, Penttilä O 1980 Innervation of the human gastric wall. J Anat 131: 453–470

L

Labriola A R, Laemle L K 1977 Cellular morphology in the visual layer of the developing rat superior colliculus. Exp Neurol 55: 247–268

Laey P E, Davies J 1957 Preliminary studies on the demonstration of insulin in the islet by the fluorescent antibody technic. Diabetes 6: 354–357

Lackie J M 1986 Cell movement and cell behaviour. Allen and Unwin: London

Lacroix P 1951 The organisation of bones. Churchill: London

Lacy D 1960 Light and electron microscopy and its use in the study of factors influencing spermatogenesis in the rat. J R Microscop Soc 79: 209–225

Laguesse E 1906 La glande nouvelle ou endocrine (Ilots de Langerhans) Revue Gén Hist 2: 1–286

Laitio M, Lev R, Orlic D 1974 The developing human fetal pancreas: an ultrastructural and histochemical study with special reference to exocrine cells. J Anat 117: 619–634

Lake G, Schneider W, Schneider U 1966 Anat Anz 118:

Lake J A 1981 The ribosome. Sci Am 245: (Aug) 84–97

Lallemand R C, Newman D L 1973 Role of the bifurcation in athermatosis of the abdominal aorta. Surg Gynecol Obstet 137: 987–990

Lam J H, Ranganathan N, Wigle E D 1970 Morphology of the human mitral valve. I. Chordae tendineae: a new classification. circulation 41: 449–458

Lamb T D 1986 Transduction in vertebrate photoreceptors: the roles of cyclic GMP and calcium. Trans Neurosci 9: 224–228

Lambert M W 1955 Accessory bronchio-alveolar communications. J Pathol Bact 70: 311–314

Lamberty B G H, Živanovic S 1973 The retro-articular vertebral artery ring of the atlas and its significance. Acta Anat 85: 113–122

Landgren S, Silfvenius H 1971 Nucleus Z, the medullary relay in the projection path to the cerebral cortex of group I muscle afferents from the cat's hind limb. J Physiol 218: 551–571

Landgren S, Phillips C G, Porter R 1962 Cortical fields of origin of the monosynaptic pyramidal pathways to some alpha motoneurons of the baboon's hand and forearm. J Physiol 161: 112–125

Landis E M, Pappenheimer J R 1963 Exchange of substance through the capillary walls. In: Hamilton W F, Dow P (eds) Handbook of physiology. Vol 2. American Physiological Society: Washington, DC: pp 961–1034

Landis D M D, Reese T S 1974 Differences in membrane structure between excitatory and inhibitory synapses in the cerebellar cortex. J Comp Neurol 155: 93–126

Landon D N 1985 Structure and function of nerve fibres. In: Swash M, Kennard C (eds) Scientific Basis of Clinical Neurology. Churchill Livingstone: Edinburgh: pp 375–389

Landon D N 1966 Electron microscopy of muscle spindles. In: Andrew B L (ed) Control and innervation of skeletal muscle. Thompson: Dundee: pp 96–111

Landon D N 1970 The influence of fixation upon the fine structure of the Z disk of rat striated muscle. J Cell Sci 6: 257–276

Landon D N 1972 The fine structure of the equatorial regions of developing muscle spindles in the rat. J Neurocytol I: 189–210

Landon D N ed 1976 The peripheral nerve. Chapman and Hall: London

Landon D N, Langley O K 1971 The local chemical environment of nodes of Ranvier: a study of cation binding. J Anat 108: 419–432

Landon D N, Williams P L 1963 Ultrastructure of the node of Ranvier. Nature 199: 575–577

Landsmeer J M F 1949 The anatomy of the dorsal aponeurosis of the human finger and its functional significance. Anat Rec 104: 31–44

Landsmeer J M F 1955 Anatomical and functional investigations on the articulation of the fingers. Acta Anat Suppl 24:

Landsmeer J M F 1976 Atlas of anatomy of the hand. Churchill Livingstone: Edinburgh

Lane M A 1907 The cytological characters of the areas of Langerhans. Am J Anat 7: 709–722

Lang J 1977 Structure and postnatal organization of heretofore uninvestigated and infrequent ossifications of the sella turcica region. Acta Anat 99: 121–130

Lang J, Brunner F X 1978 Über Rie rami centrales der Aa. cerebri anterior und media. Gegenb Morphol Jahrb Leipzig 124: 364–374

Lang J, Schäfer K 1977 Über Form, Grösse und Variabilität des Plexus choroideus ventriculi III. Gegenb Morphol Jahrb Leipzig 123: 727–741

Langer K 1861 Zur Anatomie und Physiologie der Haut. Uber die Spaltbarkeit der Cutis, S B Akad Wiss Wien 44: 19–46 Translated into English in Br J Plast Surg 1978 31: 3–8

Langham M E ed 1969 The cornea. Macromolecular organization of a connective tissue. papers from a symposium held in Kyoto, Japan, 1967 under the auspices of the department of ophthalmology, Osaka University, etc. Johns Hopkins Press: Baltimore, Maryland

Langman J M, Rowland R 1986 The number and distribtuion of lymphoid follicles in the human large intestine. J Anat 149: 189–194

Lanyon L E 1973 Analysis of surface bone strain in the calcaneus of sheep during normal locomotion. J Biomech 6: 41–49

Lapides J 1958 A simplified modification of the Johanson urethroplasty for structures of the deep bulbous urethra. Proc North Centr Sect Am Urol Assoc 166–169

Lardennois G, Okinczyc J 1913 La typhlosigmoidöstomie en Y dans la traitement des coutes rebelles et de la stase du gros intestin. Bull Mem Soc Anat Paris 39: 858–872

Larsell O 1937 The cerebellum. A review and interpretation. Arch Neurol Psychiatry 38: 580–607

Larsell O 1947 The development of the cerebellum in man in relation to its comparative anatomy. J Comp Neurol 87: 85–130

Larsell O 1953 The cerebellum of the cat and monkey. J Comp Neurol 99: 135–200

Larson B L ed 1978 Lactation. A comprehensive treatise. Vol 4. The mammary gland. Human lactation. Milk syndrome. Academic Press: New York

Larsson L, Sjödin B, Karlsson J 1978 Histochemical and biochemical changes in human skeletal muscle with age in sedentary males, age 22–65 years. Acta Physiol Scand 103: 31–39

Larsson L-I, Sundler F, Alumets G, Håkanson R, Schaffalitzky de Muckadell O, Fahrenkrug J 1977 Distribution, ontogeny and ultrastructure of the mammalian secretin cell. Cell Tissue Res 181: 361–368

Laruelle L, Reumont M 1933 Etude de l'anatomie microscopique de la moelle épinère par la méthode des coupes longitudinales plurisegmentalés. Rev Neurol 44: 1130–141

Lasek R J 1982 Translation of the neuronal cytoskeleton and axonal locomotion. Phil Trans Roy Soc Lond B 229: 313–327

Lasek R J, Joseph B S, Whitlock D G 1968 Evaluation of a radioautographic neuroanatomical tracing method. Brain Res 8: 319–336

Lasek R J, Hoffman P N 1976 The neuronal cytoskeleton, axonal transport and axonal growth. In: Goldman R, Pollard R, Rosenbaum J (eds) Cell motility. Cold Spring Harbor Laboratories: New York

Lashley K S, Clark G 1946 The cytoarchitecture of the cerebral cortex of Ateles: A critical examination of architectonic studies. J Comp Neurol 85: 223–306

Lasi G N 1959 Pre- and post-natal changes in the thymus gland. PhD Thesis. University of London

Lassek A M 1940 The human pyramidal tract. II. A numerical investigation of the Beta cells of the motor area. Arch Neurol Psychiatry 44: 718–724

Lassek A M 1942a The human pyramidal tract. IV. A study of the mature, myelinated fibers of the pyramid. J Comp Neurol 76: 217–225

Lassek A M 1942b The pyramidal tract. The effects of pre- and post-central cortical lesions on the fiber components of the pyramids in the monkey. J Nerv Ment Dis 95: 721–729

Lassek A M 1954 The pyramidal tract. Thomas: Springfield, Illinois

Lassek A M, Rasmussen G L 1939 The human pyramidal tract. A fiber and numerical analysis. Arch Neurol Psychiatry 42: 872–876

Lassen N A, Ingvar D H, Skinhaj E 1978 Brain function and blood flow. Eur Neurol 17: Suppl 1, 4–8

Last R J 1948 Some anatomical details of knee joint. J Bone Jt Surg 30B: 683–688

Last R J 1950 The popliteus muscle and the lateral meniscus. J Bone Jt Surg 32B: 93–99

Last R J 1951 Specimens from Hunterian collection: synovial cavity of knee joint (specimen S 110A); ligaments of knee (specimen S 95A). J Bone Jt Surg 33B: 442–445

Latarjet M, Neidhart J H, Morrin A, Autissier J-M 1967 L'entrée du nerf musculo-cutané dans le muscle coraco-brachial. C R Assoc Anat 138: 755–765

Latham R A 1966 Observations on the growth of the cranial base in the human skull. J Anat 100: 435P

Latham R A 1973 Development and structure of the premaxillary deformity in bilateral cleft lip and palate. Br J Plast Surg 26: 1–11

Latham R A, Deaton T G 1976 The structural basis of the philtrum and the contour of the vermilion border: a study of the musculature of the upper lip. J Anat 121: 151–160

Laties A M, Liebman P A 1970 Cones of living amphibian eyes: selective staining. Science 165: 1475–1477

Latif A 1957 An electromyographic study of the temporalis muscle in normal persons during selected positions and movements of the mandible. Am J Orthodont 43: 577–591

Latta H 1973 Ultrastructure of the glomerulus and juxtaglomerular apparatus. In: handbook of physiology. Section 8. Renal physiology (Orloff J, Berliner R W ed) pp 1–29, American Physiological Society: Washington, DC

Laurie W, Woods J D 1958 Anastomoses of the coronary circulation. Lancet 2: 812–816

Lauweryns J-M, Boussaw L 1967 L'ultrastructure des vaisseaux lymphatiques pulmonaires. C R Assoc Anat 138: 766–775

Lauweryns J M, Cokelaere M 1973 Hypoxia-sensitive neuro-epithelial bodies, intrapulmonary secretory neuro-receptors modulated by the CNS. Z Zellforsch Mikrosk Anat 145: 521–546

Lauweryns J M, Peuskens J C 1972 Neuro-epithelial bodies (neuroreceptor or secretory organs?) in human infant bronchial and bronchiolar epithelium. Anat Rec 172: 471–482

Lauweryns J M, Cokelaere M, Theunynck P 1972 Neuro-epithelial bodies in the respiratory mucosa of various mammals. Z Zellforsch 135: 569–592

LaVail J H, LaVail M M 1972 Retrograde axonal trnsport in the central nervous system. Science 176: 1416–1417

Lavelle C L 1974 The effect of age on human third molar and rat molar teeth. Acta Anat 87: 110–118

Lavker R M, Sun T T 1982 Heterogeneity of epidermal basal keratinocytes: morphological and functional correlations. Science 215: 1239–1241

Lawn A M 1966 The localisation, in the nucleus ambiguus of the rabbit, of the cells of origin of motor nerve fibers in the glossopharyngeal nerve and various branches of the vagus nerve by means of retrograde degeneration. J Comp Neurol 127: 293–305

Lawrence A M, Tan S, Hojvat S et al 1977 Salivary gland hyperglycemic factor: an extrapancreatic source of glucagon-like material. Science. 195: 70–72

Lazarides E 1976 Actin, alpha-actinin and tropomyosin interaction in the structural organisation of actin filaments in non-muscle cells. J Cell Biol 68: 202–219

Lazorthes G 1949 Le système neurovasculaire. Masson: Paris

Lazorthes G, Gouaze A, Zadeh J O, Santini J J, Lazorthes Y, Burdin P 1971 Arterial vascularization of the spinal cord. J Neurosurg 35: 253–262

Leak L V, Burke J F 1968 Ultrastructural studies on the lymphatic anchoring filaments. J Cell Biol 36: 129–149

Leao A A P 1944 Spreading depression of activity in cerebral cortex. J Neurophysiol 7: 359–390

Learmonth J R 1931 Contribution to neurophysiology of urinary bladder in man. Brain 54: 147–176

Leblond C P 1950 Distribution of periodic acid-reactive carbohydrates in the adult rat. Am J Anat 86: 1–50

Leblond C P, Cheng H 1976 Identification of stem cells in the small intestine of the mouse. In: Cairnie A B et al (eds) Stem cells of renewing cell populations. Academic Press: New York: pp 7–31

Leblond C P, Clermont Y 1952a Definition of the stages of the cycle of the seminiferous epithelium in the rat. Ann NY Acad Sci 55: 548–573

Leblond C P, Clermont Y 1952b Spermiogenesis of rat, mouse, hamster and guinea pig as revealed by the 'periodic acid-fuchsin sulfurous acid' technique. Am J Anat 90: 167–216

Leblond P F, Wright G M 1981 Steps in the elaboration of collagen by odontoblasts and osteoblasts. Methods Cell Biol 23: 167–189

Leborgne J, Letenneur J, Pannier M, Visset J, Bainvel J-V, Barbin J-Y 1973 Considérations sur la vascularisation artérielle du muscle carré crural. Arch Anat Path 21: 359–363

Lebourg L, Champagne G 1951 A propos du développement mandibulaire post-natal. Précisions sur la chronologie de la suture symphysaire. Rev Stomatol Chir Maxillofac 52: 891–897

Lecco V, Balli R 1968 Alcune considerazioni etiopatogenetiche e clinche su due casi di diverticulosi multipla dell'esofago. Arch Ital Otol 79: 497–506

Le Douarin N M 1978 The embryological origin of the endocrine cells associated with the digestive tract. Experimental analysis based on the use of a stable cell marking technique. In: Bloom S R (ed) Gut hormones. Churchill Livingstone: Edinburgh: pp 49–56

Le Douarin N M 1980 The ontogeny of the neural crest in avian embryo chimaeras. Nature 186: 663–669

Le Douarin N M 1981 Plasticity in the development of the peripheral nervous system. In: Elliott K, Lawrenson G (eds) Development of the autonomic nervous system. (Ciba Foundation Symposium 83:). Pitman Medical: London: pp 19–46

Le Douarin N M, Teillet M A 1974 Experimental analysis of the migration and differentiation of neuroblasts of the autonomic nervous system and of neuroectodermal mesenchymal derivatives, using a biological marker technique. Dev Biol 41: 162–184

Le Double, A-F 1903 Les variations des os du crâne human. Rev Sci 4s 20: 641–649

LeDuc I E 1939 Anatomy of the prostate and pathology of early benign hyperplasia. J Urol 42: 1217–1241

Lee C-S, Tsai T-L 1974 The relation of the sciatic nerve to the piriformis muscle. J Formosan Med Assoc 73: 75–80

Lee W R 1964 Appositional bone formation in canine bone: a quantitative microscopic study using tetracycline markers. J Anat 98: 655–677

Leela K, Kanagasuntheram R 1968 A microscopic study of the human pyloro-duodenal junction and proximal duodenum. Acta Anat 71: 1–12

Leela K, Kanagasuntheram R, Khoo F Y 1974 Morphology of the primate nasopharynx. J Anat 117: 333–340

Leeson T S 1957 The fine structure of the mesonephros of the 17-day rabbit embryo. Exp Cell Res 12: 670–672

Leeson T S, Leeson C R 1964 A light and electron microscope study of developing respiratory tissues in the rat. J Anat 98: 183–193

Leeson T S, Leeson C R 1965 The fine structure of cavernous tissue in the adult rat penis. Invest Urol 3: 144–154

Leeuwenhoek A 1674 Microscopical observations concerning blood, milk, bones, the brain, spittle and cuticula, etc. Philos Trans R Soc Lond [Biol] 9: 121–128

Legait H, Contet-Audonneau J L, Burlet C, Floquet J 1973 Etude des rapports des volumes du cerveau de l'hypothalamus et des lobes hypophysaires chez divers mammifères. Bull Assoc Anat 57: No 158

Le Gros Clark W E see Clark W E le G

Lehninger A L 1965 Bioenergetics. Benjamin: New York, Amsterdam

Lehrer S 1985 Puberty and menopause in the human: possible relation to gonadotropin-releasing hormone pulse frequency and the pineal gland. In: Reiter R J (ed) Pineal Research Reviews. Vol 3. Liss: New York: pp 237–257

Leibovich S J, Ross R 1975 The role of the macrophage in wound repair. A study with hydrocortisone and antimacrophage serum. Am J Pathol 78: 71–100

Leichnetz G R, Spencer R F, Hardy S G and Astruc J 1981 The prefrontal corticotectal projection in the monkey; an anterograde and retrograde horseradish peroxidase study. Neuroscience, 6: 1023–1041

Leikola A 1976 The neural crest: migrating cells in embryonic development. Folia Morphol, 24: 155–172

Leithner C, Sinzinger H, Hohenecker J, Wicke L, Olbert F, Feigl W 1975 Radiologic anatomy of the abdominal aorta and their large branches. Okajimas Folia Anat Jpn, 52: 119–150

Le May M, Culebras A 1972 Human brain-morphologic differences in the heispheres demonstrable by carotid arteriography. New Engl J Med, 287: 168–170

Le Minor J M 1987 Comparative anatomy and significance of the sesamoid bone of the peroneus longus muscle (os peroneum). J Anat 151: 85–99

Lenneberg E H 1964 New directions in the study of language. Massachusetts Institute of Technology Press: Cambridge, Massachusetts

Leontovich T A, Zhukova G P 1963 The specificity of the neuronal structure and topography of the reticular formation in the brain and spinal cord of carnivora. J Comp Neurol 121: 347–379

Lepor H, Gregerman M, Crosby R, Mostofi F K, Walsh P C 1985 Precise localization of the autonomic nerves from the pelvic plexus to the corpora cavernosa: a detailed anatomical study of the adult male pelvis. J Urol 133: 207–212

BIBLIOGRAPHY

Léranth C, Hámori J 1970 'Dark' Purkinje cells of the cerebellar cortex. Acta Biol Acad Sci Hung 21: 405–419

Lerche W 1965 Elektronenmikroskopische Beobachtungen über die Histogenese der bruchschen Membran des Menschen. Z Zellforsch 65: 163–175

Leslie D R 1954 The tendons on the dorsum of the hand. Aust N Z J Surg 23: 253–256

Leuchtenberger C, Schrader F 1950 The chemical nature of the acrosome in the male germ cells. Proc Natl Acad Sci USA 36: 677–683

Lev R, Orlic D 1974 Histochemical and radioautographic studies of normal human fetal colon. Histochemistry 39: 301–311

LeVay S 1971 On the neurons and synapses of the lateral geniculate nucleus of the monkey. Z Zellforsch 113: 396–419

LeVay S, Stryker M P, Shatz C J 1978 Ocular dominance columns and their development in layer IV of the cat's visual cortex. J Comp Neurol 179: 233–244

Levene C 1964 The patterns of cartilage canals. J Anat 98: 515–538

Leveque J I, de Rigal J, Agache P G, Monneur C 1980 Influence of aging on the in vivo extensibility of human skin at a low stress. Arch Dermatol Res 269: 127–135

Lever J D 1955 Electron microscopic observations on the adrenal cortex. Am J Anat 97: 409–430

Lever J D, Lewis P R, Boyd J D 1959 Observations on the fine structure and histochemistry of the carotid body in the cat and rabbit. J Anat 93: 478–490

Lever J D, Irvine G, Chick W J 1965 The vesiculated axons in relation to arteriolar smooth muscle in the pancreas. A fine structural and quantitative study. J Anat 99: 299–313

Lever J D, Spriggs T L B, Graham J D P 1968 A formol-fluorescence, fine-structural and autoradiographic study of the adrenergic innervation of the vascular tree in the intact and sympathectomised pancreas of the cat. J Anat 103: 15–34

Levi-Montalcini R 1950 The origin and development of the visceral system in the spinal cord of the chick embryo. J Morphol 86: 253–283

Levi-Montalcini R 1952 Effects of mouse tumor transplants on the nervous system. Ann N Y Acad Sci 55: 330–343

Levi-Montalcini R 1960 Destruction of the sympathetic ganglia in mammals by an antiserum to the nerve growth-promoting factor. Proc Natl Acad Sci USA 46: 384–391

Levi-Montalcini R 1967 Differentiation and growth control mechanisms in the nervous system. In: Morphological and Biochemical Aspects of Cytodifferentiation (Exsp Biol Med I: 170–182) Karger: Basel

Levi-Montalcini R, Angeletti P U 1968 Nerve growth factor. Physiol Rev 48: 534–569

Levi-Montalcini R, Chen J S 1971 Selective outgrowth of nerve fibers in vitro from embryonic ganglion of Periplaneta americana. Arch Ital Biol 109: 307–337

Levin P M, Bradford F K 1938 The exact origin of the corticospinal tract in the monkey. J Comp Neurol 68: 411–422

Levinger I M, Kedem J 1974 A method for the evaluation of the surface area of cerebral ventricles in animals. J Anat 117: 481–485

Levitt P, Rakic P 1980 Immunoperoxidase localization of glial fibrillary acid protein in radial glial cells and astrocytes of the developing rhesus monkey brain. J Comp Neurol 193: 815–840

Levitt P, Cooper M L, Rakic P 1981 Coexistence of neuronal and glial precursor cells in the dorsal ventricular zone of the fetal monkey: an ultrastructural immunoperoxidase analysis. J Nerosci 1: 27–39

Levkova N A, Kakabadze S A 1977 Ultrastructural organisation of chromaffin paraganglia in the nodes of the solar plexus. Arkh Anat Gistol Embriol 72: 53–58

Levy J, Nagylaki T 1972 A model for the genetics of handedness. Genetics 72: 117–128

Lewin B 1980 Gene expression. Vol 2, Eucaryotic chromosomes. 2nd edition. Wiley: New York

Lewin B M 1970 The molecular basis of gene expression. Wiley: London

Lewin W 1961 Observations on selective leucotomy. J Neurol Neurosurg Psychiat 24: 37–44

Lewis D J, Prentice D E 1980 The ultrastructure of rat laryngeal epithelia. J Anat 130: 617–632

Lewis F T 1912 The form of the stomach in human embryos with notes upon the nomenclature of the stomach. Am J Anat 13: 477–503

Lewis O J 1956 The development of the circulation in the spleen of the foetal rabbit. J Anat 90: 282–289

Lewis O J 1957a The blood vessels of the adult mammalian spleen. J Anat 91: 245–250

Lewis O J 1957b The formation and development of the blood vessels of the mammalian cerebral cortex. J Anat 91: 40–46

Lewis O J 1958a The development of the blood vessels of the metanephros. J Anat 92: 84–97

Lewis O J 1958b The vascular arrangement of the mammalian renal glomerulus as revealed by a study of its development. J Anat 92: 433–440

Lewis O J 1958c The tubercle of the tibia. J Anat 92: 587–592

Lewis O J 1959 The coraco-clavicular joint. J Anat 93: 296–303

Lewis O J 1962 The comparative morphology of M flexor accessorius and the associated flexor tendons. J Anat 96: 321–333

Lewis O J 1964a The homologies of the mammalian tarsal bones. J Anat 98: 195–208

Lewis O J 1964b The tibialis posterior tendon in the primate foot. J Anat 98: 209–218

Lewis O J 1965 The evolution of the Mm Interossei in the primate hand. Anat Rec 153: 275–288

Lewis O J 1977 Joint remodelling and the evolution of the human hand. J Anat 123: 157–201

Lewis O J 1983 The evolutionary emergence and refinement of the mammalian pattern of foot architecture. J Anat 137: 21–45

Lewis O J 1985a Derived morphology of the wrist articulations. Part I. The lorisine joints and theories of hominoid evolution. J Anat 140: 447–460

Lewis O J 1985b Derived morphology of the wrist articulations and theories of hominoid evolution. Part II. The midcarpal joints of higher primates. J Anat 142: 151–172

Lewis O J, Hamshere R J, Bucknill T M 1970 The anatomy of the wrist joint. J Anat 106: 539–552

Lewis P R, Shute C C D 1959 Selective staining of visceral efferents in the rat brain stem by a modified Koelle technique. Nature 183: 1743–1744

Lewis P R, Scott J A, Navaratnam V 1970 Localization in the dorsal motor nucleus of the vagus in the rat. J Anat 107: 197–208

Lewis W B 1878 On the comparative structure of the cortex cerebri. Brain I: 79–96

Lewis W H 1945 Axon growth and regneration. Anat Rec 91: 287 (abstract)

Lewis W H, Hartman C G 1933 Early cleavage stages of the egg in the monkey (Macaracus rhesus). Contrib Embryol Carnegie Inst Wash 24: 287–201

Lewy F H, Kobrak H 1936 The neural projection of the cochlear spirals on the primary acoustic centers. Arch Neurol. Psychiatry 35: 839–852

Leydig F 1850 Zur Anatomie der mannlichen Geschelechtsorgane und der Saugethiere. Z Wiss Zool 2: 1–10

Leyton A S F, Sherrington C S 1917 Observations on the excitable cortex of the chimpanzee, orang utan and gorilla. O J Exp Physiol 11: 135–222

Libet B, Owman C 1974 Concomitant changes in formaldehyde-induced fluorescence of dopamine interneurons and in slow inhibitory post-synaptic potentials. J Physiol 237: 636–662

Liddell E G T, Phillips C G 1950 Thresholds of cortical representation. Brain 73: 125–140

Lieb F J, Perry J 1968 Quadriceps function. An anatomical and mechanical study using amputated limbs. J Bone Jt Surg 50A: 1535–1548

Lieberman A R 1968 An investigation by light and electron microscopy of chromatoloytic and other phenomena induced in mammalian nerve cells by experimental lesions. PhD Thesis. University of London

Lieberman A R 1969 Absence of ultrastructural changes in ganglionic neurons after supranodose vagotomy. J Anat 104: 49–54

Lieberman A R 1973 Neurons with presynaptic perikarya and presynaptic dendrites in the rat lateral geniculate nuclus. Brain Res 59: 35–59

Lieberman A R 1974a Comments on the fine structural organization of the dorsal lateral geniculate nucleus of the mouse. Z Anat Entwicklungsgesch 145: 261–267

Lieberman A R 1974b Some factors affecting retrograde neuronal responses to axonal lesions. In: Bellairs R, Gray E G (eds) Essays on the Nervous System. Clarendon Press: Oxford: pp 71–104

Lieberman A R 1976 Sensory ganglia. In: Landon D N (ed) The Peripheral Nerve. Chapman & Hall: London: pp 188–278

Lieberman A R, Webster K E 1974 Aspects of the synaptic organization of the intrinsic neurons in the dorsal lateral geniculate nucleus. An ultrastructural study of the normal and of the experimentally deafferented nucleus in the rat. J Neurocytol 3: 677–710

Lightoller G H S 1925 Facial muscles. The modiolus and muscles surrounding the rima oris with some remarks about the panniculus adiposus. J Anat 60: 1–85

Lilja J 1979 Innervation of different parts of the predentin and dentin in young human premolars. Acta Odontol Scand 37: 339–346

Lillie F R 1913 The mechanism of fertilisation. Science 38: 524–528

Lillie F R 1971 The free-martin; a study of the action of sex hormones in the foetal life of cattle. J Exp Zool 23: 371–452

Lim C H, Ruskell G L 1978 Corneal nerve access in monkeys. Albr von Graefes Arch Klin Exp Ophthalmol 208: 15–23

Lim-Howe D, Studd J, Dooley M 1987 Gamete intrafallopian transfer (GIFT). Br J Hosp Med 37: 241–244

Lin H S 1967 A peculiar configuration of agranular reticulum (canaliculate lamellar body) in the rat pinealocyte. J Cell Biol 33: 15–25

Linc R, Fleischmann J 1968 K současnému stavu anatomického názvosloví. Cas Lek Cesk 107: 107–121

Linck G, Porte A 1981 Cytophysiology of the synovial membrane: distinction of two cell types of the intima revealed by their reaction with horseradish peroxidase and iron saccharate in the mouse. Biol Cell 42: 147–152

Lind J, Wergelius C 1954 Human fetal circulation: Change in the cardiovascular system at birth and disturbances in the post-natal closure of the foramen ovale and ductus arteriosus. Cold Spring Harbor Symp Quant biol 19: 109–125

Lindahl P E, Drevius L O 1964 Observations on bull spermatozoa in a hypotonic medium related to sperm mobility mechanisms. Exp Cell Res 36: 632–646

1514

Lindsay R D, Scheibel A B 1974 Quantitative analysis of the dendrite branching pattern of small pyramidal cells from adult rat somesthetic and visual cortex. Exp Neurol 45: 424–435

Lindvall O, Björklund A 1974 The organizations of the ascending catecholamine neuron systems in the rat brain as revealed by the glyoxylic acid fluorescence method. Acta Physiol Scand 412: 1–48

Lindvall M, Edvinsson L, Owman C 1977 Histochemical study on regional differences in the cholinergic nerve supply of the choroid plexus from various laboratory animals. Exp Neurol 55: 152–159

Ling E A 1978 Evidence for a haematogenous origin of some of the mascrophages appearing in the spinal cord of the rat after dorsal rhizotomy. J Anat 128: 43–154

Ling E A, Paterson J A, Privat A, Mori S, Leblond C P 1973 Investigation of glial cells in semithin sections. I. Identification of glial cells in the brain of young rats. J Comp Neurol 149: 43–72

Linkevich V R, 1969 Embryogenesis of female internal genitalia. Akush Ginekol (Mosk) 7: 43–47

Lipkin M, Sherlock P, Bell B 1963 Cell proliferation kinetics in the gastrointestinal tract of man. II. Cell renewal in stomach, ileum, colon and rectum. Gastroenterology 45: 721–729

Lippert H, Kafer H 1974 Biomechanik des Schädeldachs 2. Dicken der Knochenschichten. Mschr Unfallheik 77: 329–339

Lisa J R, Gioia J D, Rubin I C 1954 Observations on interstitial portion of the fallopian tube. Surgery Gynecol Obstet 99: 159–169

Lisowski F F P 1969 Ethiop Med J 7:

Lissäk K 1967a Recent developments of neurobiology in Hungary (editor), Akadémiai Kiadö: Budapest

Lissäk K ed 1967b Results in neuroanatomy, neurochemistry, neuropharmacology and neurophysiology. Akadémiai Kiado: Budapest

Lissauer H 1885 Beitrag zur pathologischen Anatomie der Tabes dorsalis und zum Faserverlauf in menschlichen Rückenmark. Neurol Zbl 4: 245–246

Lissmann H W, Machin K E 1958 The mechanism of object location in Gymnarchus niloticus and similar fish. J Exp Biol 35: 451–486

Lister G 1985 Indications and techniques for repair of the flexor tendon sheath. Hand Clin 1: 85–95

Lister J J 1830 On some properties in achromatic object-glasses applicable to the improvement of the microscope. Philos Trans R Soc Lond [Biol], 187–200

Lister U M 1968 Ultrastructure of the human amnion, chorion and fetal skin. J Obstet Gynaecol Br Commonw 75: 327–331

Listgarten M A 1966 Phase contrast and electron microscopic study of the junction between reduced enamel epithelium and enamel in unerupted human teeth. Arch Oral Biol II: 99–1016

Listgarten M A 1970 Changing concepts about the dentoepithelial junction. J Can Dent Assoc 36: 70–75

Liu C-N, Chambers W W 1964 An experimental study of the cortico-spinal system in the monkey (Macaca mulatta). The spinal pathways and preterminal distribution of degenerating fibers following discrete lesions of the pre- and postcentral gyri and bulbar pyramid. J Comp Neurol 123: 257–284

Livingston K E, Hornykeiwiecz O, eds 1978 Limbic Mechanisms. Plenum Press: New York

Livingston R B 1970 Some general integrative aspects of brain function. In: Wolstenholme G E W, Knight J (eds) Control Processes in Multicellular Organisms. Ciba Foundation Symposium. Churchill: London: pp 384–400

Ljunggren A E 1976 The tuberositas tibiae and extension in the knee joint. Acta Morphol Neerl Scand 14: 215–239

Lloyd D, Poole P K, Edwards S W 1982 The cell division cycle. Temporal Organization and Control of Cellular Growth and Reproduction Academic Press: London

Lloyd D P C 1943 Reflex action in relation to pattern and peripheral source of afferent stimulation. J Neurophysiol 6: 111–119

Locke S 1967 Thalamic connections to insular and opercular cortex of monkey. J Comp Neurol 129: 219–240

Lockwood C B 1886 The anatomy of the muscles, ligaments and fasciae of the orbit, including an account of the capsule of tenon, the check ligaments of the recti and of the suspensory ligaments of the eye. J Anat 20: 1–25

Lodge T (1946) Anatomy of blood vessels of the human lung as applied to chest radiology. Br J Radiol 19: 1–7

Loewenfeld I E 1958 Mechanisms of reflex dilation of the pupil. Historical review and experimental analysis. Documenta Ophthal 12: 185–448

Loewnstein O, Loewenfeld I E 1969 The pupil. In: Davson H (ed) The eye, 2nd edition. Vol 3. Muscular Mechanisms. Academic Press: London: pp 255–337

Loewenstein W R 1971 Mechano-electric transduction in the Pacinian corpuscle. Initiation of sensory impulse in mechanoreceptors. In: Loewenstein W R (ed) Handbook of Sensory Physiology. Vol 1. Springer: Berlin: pp 269–290

Lohmander L S, Hascall V C, Yanagishita M, Kuettner K E, Kimura J H 1986 Post-translational events in proteoglycan synthesis: kinetics of synthesis of chondroitin sulfate and oligosaccharides on the core protein. Arch Biochem. Biophys 250: 211–227

Løken A C, Brodal A 1970 A somatotopical pattern in the human lateral vestibular nuclei. Arch Neurol Psychiat 23: 350–357

Lombard R E, Bolt J R 1979 Evolution of the tetrapod ear: an analysis and reinterpretation. Biol J Linn Soc II: 19–76

Lømo T, Westguard R H 1974 Contractile properties of muscle: control by pattern of muscle activity in the rat. Proc. R Soc Lond 187B: 99–103

Long C 1968 Intrinsic-extrinsic muscle control of the fingers. Electomyographic studies. J Bone Jt Surg. 50A: 973–984

Long C, Brown M E 1964 Electromyographic kinesiology of the hand: muscles moving the long finger. J Bone Jt Surg 46A 1683–1706

Long C 2d, Brown M E, Weiss G 1961 Electromyographic kinesiology of the hand. Part II. Third dorsal interosseus and extensor digitorum of the long finger. Arch. Phys. Med. 42: 559–565

Long J A, Jones A L 1967 Observations on the fine structure of the adrenal cortex of man. Lab Invest 17: 355–370

Longfield M D, Dowson D, Walker P S, Wright V 1969 'Boosted lubrication' of human joints by fluid enrichment and entrapment. Bio-Med Eng 4: 517–522

Longmore R B 1976 Reflected light interference microscopy (RLIM) of load-bearing human articular cartilage. Proc R Microsc Soc Lond II: 60–61

Longmore R D, Gardner D L 1978 The surface structure of ageing human articular cartilage: a study by reflected light interference microscopy (RLIM). J Anat 126: 353–365

Longo D 1972 Placental transfer mechanisms–an overview. Obstet Gynocol Annu 1: 103–138

Lopashov G V, Stroeva O G 1961 Morphogenesis of the vertebrate eye. In: Abercrombie M, Brachet J (eds) Advances in Morphogenesis. Vol 1. Academic Press: New York: pp 331–377

Lorente de Nó R 1933 Anatomy of the eighth nerve. I. The central projections of the nerve endings of the internal ear. Laryngoscope, 43: 1–38

Lorente de Nó R 1934 Studies on the striation of the cerebral cortex. II. Continuation of the study of the ammonic system. F Psychol Neurol Lpz 46: 113–177

Lorente de Nó R 1949 Cerebral cortex: architecture, intracortical connections, motor projections. In: Fulton's Physiology of the Nervous System 3rd edition. Oxford University Press: London: pp 288–312

Lovell C R, Smolenski K A, Duance V C, Light N D, Young S, Dyson M 1987 Type I and III collagen content and fibre distribution in normal human skin during ageing. Br J Dermatol 117: 419–428

Low A 1907 A note on the crura of the diaphragm and the muscle of Treitz J Anat 42: 93–96

Lowenstein W R, Kanno Y 1964 Studies on an epithelial (gland) cell junction. I. Modification of surface membrane permeability. J Cell Biol 22: 565–586

Lower R 1669 Tractatus de Corde. J Allestry: London

Lowsley O S 1912 The development of the human prostate gland with reference to the development of other structures at the neck of the urinary bladder. Am J Anat 13: 299–346

Lowy J, Small J V 1970 The organization of myosin and actin in vertebrate smooth muscle. Nature 227: 46–51

Lozanoff S, Sciulli P W, Schneider K N 1985 Third trochanter incidence and metric trait covariation in the human femur. J Anat 143: 149–159

Lubińska L 1964 Axoplasmic streaming in regenerating and in normal nerve fibres. In: Singer M, Schadé J P (eds) Mechanisms of Neural Regeneration. Elsevier: Amsterdam (Prog Brain Res 13: 1–71)

Lucas P W, Luke D A, Voon F C T, Chew C L, Ow R 1986 Food breakdown patterns produced by human subjects possessing artificial and natural teeth. J Oral Rehabil 13: 205–214

Luckett W P 1975 The development of primordial and definitive amniotic cavities in early rhesus monkey and human embryos. Am J Anat 144: 149–167

Luckett W P 1978 Origin and differentiation of the yolk sac and extraembryonic mesoderm in presomite human and rhesus monkey embryos. Am J Anat 152: 59–97

Luk S C, Nopajaroonsri C, Simon G T 1973 The architecture of the normal lymph node and hemolymph node. A scanning and transmission electron microscopic study. Lab Invest 29: 258–265

Lukaszyk A, Reiter R J 1975 Histophysiological evidence for the secretion of polypetptides by the pineal gland. Am J Anat 143: 451–464

Lumsden A G S 1987 Neural crest contribution to tooth development in the mammalian embryo. In: Maderson P F A (ed) Developmental and Evolutionary Aspects of the Neural Crest. Wiley: New York: pp 261–300

Lumsden A G S, Buchanan J A G 1986 An experimental study of the timing and topography of early tooth development in the mouse embryo with an analysis of the role of innervation. Arch Oral Biol 31: 310–311

Lundberg A 1958 Electrophysiology of salivary glands. Physiol Rev 38: 21–40

Lundborg G, Myrhage R 1977 The vascularization and structure of the human digital tendon sheath as related to flexor tendon function. Scand J Plast Reconstruct Surg 11: 195–203

Lutfi A M 1970 Mode of growth, fate and functions of cartilate canals. J Anat 106: 135–146

Lutfi A M 1974 The role of cartilage in long bone growth: a reappraisal. J Anat 117: 413–417

Lyon M F 1962 Sex chromatin and gene action in the mammalian X-chromosome. Am J Hum Genet 14: 135–148

Lytle W J 1957 Femoral hernia. Ann R Coll Surg Engl 21: 244–262

Lytle W J 1959 The common bile-duct groove in the pancreas. Br J Surg 47: 209–212

Lytle W J 1970 The deep inguinal ring, development, function, and repair. Br J Surg 57: 531–536

Lytle W J 1974 The inguinal and lacunar ligaments. J Anat 118: 241–251

Lytle W J 1979 Inguinal anatomy. J Anat 128: 581–594

M

Maayan M L, Ingbar S H 1970 Effects of epinephrine on iodine and intermediary metabolism on isolated thyroid cells. Endocrinology 87: 588–595

Mabuchi M, Kusama T 1966 The corticorubral projection in the cat. Brain Res 2: 254–273

MacCabe J A, Errick I, Saunders J W Jr 1974 Ectodermal control of the dorsoventral axis in the leg bud of the chick embryo. Dev Biol 39: 69–82

McCall J G 1968 Scanning electron microscopy of articular surfaces. Lancet 2: 1194

MacCallum J B 1897 On the histology and histogenesis of the heart muscle cell. Anat Anz 13: 609–620

MacCallum J B 1900 On the musculature architecture and growth of the ventricles of the heart. In: Contributions to the Science of Medicine. Dedicated to W H Welch. Baltimore: pp 307–335

McCloskey D I, Torda T A 1975 Corollary motor discharges and kinaesthesia. Brain Res 100: 467–470

McClure C F W, Butler E G 1925 The development of the vena cava inferior in man. Am J Anat 35: 331–384

McColl I 1967 The comparative anatomy and pathology of anal glands. Ann R Coll Surg 40: 36–67

MacConaill M A 1932 The function of intra-articular ficrocartilages, with special reference to the knee and inferior radio-ulnar joints, J Anat 66: 210–227

MacConaill M A 1941 The mechanical anatomy of the carpus and its bearings on some surgical problems, J Anat 775: 166–175

MacConaill M A 1945 The postural mechanism of the human foot. Proc R Ir Acad L Sect B, 14: 265–278

MacConaill M A 1946 Some anatomical factors affecting the stabilising functions of muscles. Ir J Med Sci 6th series, 245: 160–164

MacConaill M A 1949 Movements of bones and joints; function of musculature, J Bone Jt Surg 31B: 100–104

MacConaill M A 1950 Rotary movements and functional decalage, with some references to rehabilitation, Br J Phys Med Ind Hyg 13: 50–56

MacConaill M A 1951 The movements of bones and joints. 4. The mechanical structure of articulating cartilage, J Bone Jt Surg 33B: 251–257

MacConaill M A 1953 The movements of bones and joints. V. The significance of shape, J Bone Jt Surg 35B: 290–297

MacConaill M A 1964 Joint movements, Physiotherapy, November, 359–367

MacConaill M A 1966a The compound polarizer, Lab Pract, 15: 659–663

MacConaill M A 1966b The geometry and alegbra of articular kinematics, Bio-Med Eng 1: 205–212

MacConaill M A 1975 The muscular slings of the mandible J Ir Dent Ass 21: 22–24

MacConaill M A 1978A Anatomical note. Spurt and shunt muscles. J Anat 126: 619–621

MacConaill M A 1978b A generalized mechanics of articular swings. I. From Earth to outer space. J Anat 127: 577–587

MacConaill M A 1978c The strange physics of moving bones. Ir J Med Sci 147: 140–144

MacConaill M A, Basmajian J V 1977 Muscles and movements. A basis for human kinesiology. 2nd edition. Krieger: New York

McCotter R E 1915 A note on the course and distribution of the nervus terminalis in man. Anat Rec 9: 243–246

McCuskey R S, Chapman T M 1969 Microscopy of the living pancreas in situ. Am J Anat 126: 395–407

McCutchen C W 1959 Sponge-hydrostatic and weeping bearings, Nature, 184: 1284–1285

McCutchen C W 1983 Lubrication of and by articular cartilage. In: Hall B K (ed) Cartilage. Vol 3. Biomedical Aspects. Academic Press: New York: pp 87–107

McDonald D M, Mitchell R A 1975 The innervation of glomus cells, ganglion cells and blood vessels in the rat carotid body: a quantitative ultrastructural analysis. J Neurocytol 4: 177–230

MacDonald M S, Emery J L 1959 The late intrauterine and postnatal development of human renal glomeruli. J Anat 93: 331–340

MacDonald W C, Trier J S, Everett N B 1964 Cell proliferation and migration in the stomach, duodenum and rectum of man, radioautographic studies. Gastroenterology 46: 405–417

MacDougall J D B 1955 The attachments of the masseter muscle. Br dent J 98: 193–199

McEwen W K, Goodner E T 1962 Secretion of tears and blinking. In: Davson H (ed) The Eye. Vol 3. Academic Press: New York: pp 341–378

McGeer P L, McGeer E G, Fibiger H C 1975 Choline acetylase and glutamine acid decarboxylase in Huntington's chorea. Neurology, 23: 912–917

McGrath P 1971 The volume of the human pharyngeal hypophysis in relation to age and sex. J Anat 110: 275–282

McGrath P 1974 The pharyngeal hypophysis in some laboratory animals. J Anat 117: 95–115

McGrath P 1977 The cavernous sinus: an anatomical survey, Aust NZ J Surg 47: 601–613

McGrath P 1978 Aspects of the human pharyngeal hypophysis in normal and anencephalic fetuses and neonates and their possible significance in the mechanism of its control. J Anat 127: 65–81

MacGregor A 1936 An experimental investigation of the lymphatic system of the teeth and jaws. Proc R Soc Med 29: 1237–1272

Machado A B, DiDio L J 1967 Frequency of the musculus palmaris longus studied in vivo in some Amazon indians, Am J phys Anthropol 27: 11–20

Maciewicz R J, Eagen K, Kaneko C R S, Highstein S M 1977 Vestibular and medullary brain stem afferents to the abducent nucleus in the cat. Brain Res 123: 229–240

Maciewicz R J, Kaneko C R, Highstein S M 1975 Morphophysiological identification of interneurons in the oculomotor nuclei that project to the abducent nucleus in the cat. Brain Res 96: 60–65

McIndoe A H, Counsellor V S 1927 The bilaterality of the liver. Arch Surg 13: 589–612

McIntosh J R 1979 Cell division. In: Roberts K, Hyams J S (eds) Microtubules. Academic Press: London: pp 381–441

MacIntosh S R 1974 The innervation of the conjuctiva in monkeys, Albrechit v Graefes Arch Ophthalmol 192: 105–116

McIntyre A K, Holman M E, Veale J L 1967 Cortical responses to impulses from single Pacinian corpuscles in the cat's hind limb, Exp Brain Res 4: 243–255

McKay D G, Hertig A T, Bardawil W A, Velardo J T 1956 Histochemical observations on the endometrium. I. Normal endometrium. Obstet Gynecol 8: 22–39

Mackay T W, Andrews P L R 1983 A comparative study of the vagal innervation of the stomach in man and the ferret. J Anat 136: 449–481

McKelvy J F 1974 Biochemical neuroendocrinology. I. Biosynthesis of thyrotropin releasing hormone (TRH) by organ culture of mammalian hypothalamus. Brain Res 65: 489–502

MacKenzie D W Jr, Whipple A O, Winterstiener M P 1941 Studies on the microscopic anatomy and physiology of living transilluminated mammalian spleens, Am J Anat 68: 397–456

McKenzie J 1948 The parotid gland in relation to the facial nerve. J Anat 82: 183–186

McKenzie J 1955 A Bronze Age burial near Stonehaven, Kincardineshire. J Anat 89: 579P

McKern T W, Stewart T D 1957 Skeletal age changes in young American males. (Tech Rep EP 45 Environmental Protection Research Div Natick. Mass)

McKibbin B 1968 The action of the iliopsoas muscle in the newborn, J Bone Jt Surg 50B: 161–165

MacKinnon I L, MacKinnon P C B 1958 Seasonal rhythm in the morphology of the suprarenal cortex in women of child-bearing age. J Endocrinol 17: 456–467

MacKinnon P C B, MacKinnon I L 1960 Morphologic features of the human suprarenal cortex in men aged 20–86 years. J Anat 94: 183–191

Macklin C C 1936 Alveolar pores and their significance in the human lung. Arch Pathol 21: 202–216

Macklin C C 1954 The pulmonary alveolar mucoid film and the pneumonocyte. Lancet i: 1099–1104

McKusick V A 1986 Mendelian inheritance in man. 7th edition. Johns Hopkins Press: Baltimore

McLachlan E M 1974 The formation of synapses in mammalian sympathetic ganglia reinnervated with preganglionic or somatic nerves. J Physiol 237: 217–242

McLean F C, Urist M R 1968 Bone: Fundamentals of the Physiology of Skeletal Tissue. 3rd edition. University of Chicago Press: Chicago, Illinois

McLean F C, Urist M R 1969 Bone: an introduction to the physiology of skeletal tissue. 2nd edition, revised and enlarged. Chicago University Press: Chicago

MacLean P D 1958 The limbic system with respect to self-preservation and the preservation of the species. J Nerv Ment Dis 127: 1–11

MacLean P D 1969 The hypothalamus and emotional behavior. In: Haymaker W, Anderson E, Nauta W J H (eds) The Hypothalamus. Thomas: Springfield, Illinois: pp 659–678

McMahon T A 1977 Scaling quadrupedal galloping: frequencies, stresses and joint angles. In: Pedley T J (ed) Scale effects in animal locomotion. Based on the Proceedings of an International Symposium held at Cambridge University. 1975. Academic Press: London: pp 143–151

McMasters R E, Weiss A H, Carpenter M B 1966 Vestibular projections to the nuclei of the extraocular muscles. Degeneration resulting from discrete partial lesions of the vestibular nuclei in the monkey. Am J Anat 118: 163–194

McMinn R M H 1969 Tissue repair. Academic Press: New York

McMinn R M H, Kugler J H 1961 The glands of the bile and pancreatic ducts: autoradiographic and histochemical studies. J Anat 95: 1–11

McMinn R M H, Mitchell J E 1954 The formation of villi following artificial lesions of the mucosa in the small intestine of the cat. J Anat 88: 99–107

McNamara J A J 1972 Dual functions of the lateral pterygoid muscle: a study of Macaca mulatta, Anat Rec 172: 360 [Abstract]

McNeal J E 1975 Structure and pathology of the prostate. In: Goland M (ed) Normal and Abnormal Growth of the Prostate. Thomas: Springfield Illinois: pp 55–65

McNeill M E 1977 An unusual organelle in the pineal gland of the rat. Cell Tissue Res 184: 133–137

Macrides F, Davis B J 1983 The olfactory bulb. In: Emson P C (ed) Chemical Neuroanatomy. Raven Press: New York: pp 391–426

McVay C B, Anson B J 1940a Aponeurotic and fascial continuities in the abdomen, pelvis and thigh, Anat Rec 76: 213–232

McVay C B, Anson B J 1940b Composition of the rectus sheath, Anat Rec 77: 213–225

Madri J A, Stenn K S 1982 Aortic endothelial cell migration. I. Matrix requirements and composition. Am J Pathol 106: 180–186

Maeda T, Pin C, Salvert D et al 1973 Les neurones contenant des catécholamines du tegmentum pontique et leurs voies de projection chez le chat. Brain Res 57: 119–152

Mage R W 1982 Venae comitantes van het onderbeen, Academisch Proefschrift. Rodopsi: Amsterdam

Maggio E 1965 Microhemocirculation: observable variables and their biologic control. Thomas: Springfield, Illinois

Maguire A, Dayal V S 1974 Supraglottic anatomy: the pre- or the periepiglottic space? Can J Otolaryngol 3: 432–445

Mahour G H, Wakim K G, Soule F H, Ferris D O 1967 The common bile duct after cholecystectomy: comparison of common bile ducts in patients who have intact biliary systems with those in patients who have undergone cholecystectomy. Ann Surg 166: 964–967

Maillot C, Koritke J G, Laude M 1976 La vascularisation de la toile choroidienne inférieure chez l'homme. Arch Anat Histol Embryol 59: 33–70

Maisey M N 1978 Nuclear medicine. Practitioner 220: 445–451

Majno G 1979 The story of the myofibroblasts. Am J Surg Pathol 3: 535–542

Malacinski G M (ed), Bryant S V (consulting ed) 1984 Pattern formation. A Primer in developmental biology, Macmillan: New York and Collier Macmillan: London

Malendowicz L K 1985 Stereological studies on the effects of pinealectomy, melatonin and oestradiol on the adrenal cortex of oviarectomised rats. J Anat 141: 115–120

Malgaigne J F 1838 Traité d'Anatomie Chirurgicale et de Chirurgie Expérimentale. J B Baillière: Paris p 1

Malis L I, Loevinger R, Kruger L, Rose J E 1957 Production of laminar lesions in the cerebral cortex by heavy ionising particles. Science 126: 302–303

Mall F P 1911 On the muscular architecture of the ventricles of the human heart, Am J Anat 11: 211–278

Mall F P 1912 On the development of the human heart, Am J Anat 13: 249–298

Malpighi M 1666 De Viscerum Structura Exercitatio Anatomica. J Montiji: Bononia

Mangoushi M A 1975 Branches of the inferior mesenteric artery: 'a rare anomaly'. Ethiop Med J 13: 23–26

Mankin H J, Lippiello L 1969 The turnover of adult rabbit articular cartilage, J Bone Jt Surg 51A: 1591–1600

Mann I C 1924 Notes on the anatomy of the living eye, as revealed by the Gullstrand slitlamp J Anat 59: 155–165

Mann I C 1927 The relations of the hyaloid canal in the foetus and in the adult. J Anat 62: 290–296

Mann I C 1964 The development of the human eye. 3rd edition, Grune & Stratton: New York

Mann T 1949 Metabolism of semen. Adv Enzymol 9: 329–390

Mann T 1967 Sperm metabolism. In: Metz C B, Monroy A (eds) Fertilization. Vol I. Academic Press: New York: pp 99–116

Mannen H 1968 Neuronal stereophotogrammetry. A new approach to the problem of nerve cell shape and dendritic domain. Med Biol Illus 18: 96–102

Mannen T, Iwata M, Toyokura Y, Nagashima K 1977 Preservation of a certain motoneurone group of the sacral cord in amyotrophic lateral sclerosis: its clinical significance. J Neurol Neurosurg Psychiat 40: 464–469

Manni E, Bortolami R, Deriu P L 1970 Presence of cell bodies of the afferents from the eye muscles in the semilunar ganglion. Arch Ital Biol 108: 106–120

Manni E, Bortolami R, Pettorossi V E, Lucchi M L, Callegari E 1978 Afferent fibers and sensory ganglion cells within the oculomotor nerve in some mammals and man. II. Electrophysiological investigations. Arch Ital Biol 116: 16–24

Manotaya T, Potter E L 1963 Oocytes in prophase of meiosis from squash preparations of human fetal ovaries. Fertil Steril 14: 378–392

Marchand P, Gilroy J C, Wilson V H 1950 Anatomical study of bronchial vascular system and its variations in disease. Thorax 5: 207–221

Marchesi V T 1961 The site of leucocyte emigration during inflammation. Quart J Exp Physiol 46: 115–118

Marchesi V T 1962 The passage of colloidal carbon through inflamed endothelium, Proc R Soc Lond [Biol] 156: 550–552

Marchesi V T, Gowans J L 1964 The migration of lymphocytes through the endothelium of venules in lymph nodes, Proc R Soc Lond [Biol] 159: 283–290

Marchi V, Algeri G 1815–1816 Sulle degenerazioni discendenti consecutive a lesioni sperimentale in diverse zone delle corteccia cerebrale. Riv Sper Freniatria Med Legal II: 492–494; 12: 208–252

Margulis L 1981 Symbiosis in cell evolution. Life and its environment on the early earth. Freeman: Oxford

Marikovsky Y, Danon D 1969 Electron microscope analysis of young and old red blood cells stained with colloidal iron for surface charge evaluation, J Cell Biol 43: 1–7

Marinesco G 1909 La Cellule Nerveuse. Dom: Paris

Mark R 1974 Memory and Nerve Cell Conections. Clarendon Press: Oxford

Markee J E, Logue J T Jr, Williams M, Stanton W B, Wrenn R N, Walker L B 1955 Two joint muscles of the thigh, J Bone Jt Surg 37A: 125–142

Markowski J 1911 Über die Entwicklung der Sinus durae matris und der Hirnvenen bei menschlichen Embryonen von 15 5–49 mm Scheitel-Steisslange. Bull Int Acad Sci LeH Cracov B. 590–611

Marks S C Jr, Cahill D R 1984 Experimental study in the dog of the nonactive role of the tooth in the eruptive process. Arch Oral Biol 29: 311–322

Marneffe R de 1951 Recherches morphologiques et expérimentales sur la vascularisation osseuse. Acta Chir Belge 50: 469–488, 568–599, 681–704

Maroudas A, Stockwell R, Nachemson A, Urban J 1975 Factors involved in the nutrition of the human lumbar intervertebral disc: cellularity and diffusion of glucose in vitro, J Anat 120: 113–130

Marrack P, Kappler J 1986 The antigen-specific, major histocompatibility complex-restricted receptor or T cells. Adv Immunol 38: 1–30

Marsden C D 1985 The basal ganglia. In: Swash M, Kennard C (eds) Scientific Basis of Clinical Neurology. Churchill Livingstone: Edinburgh: pp 56–73

Marsden C D 1961 Pigmentation in the nucleus substantiae nigrae of mammals. J Anat 95: 256–261

Marshall J, Ansell P L 1971 Membranous inclusions in the retinal pigmented epithelium: phagosomes and myeloid bodies. J Anat 110: 91–104

Martin B F 1958 The annular ligament of the superior radio-ulnar joint. J Anat 92: 473–482

Martin B M, Gimbrone M A Jr, Unanue E R, Cotran R S 1981 Stimulation of nonlymphoid mesenchymal cell proliferation by a macrophage-derived growth factor. J Immunol 126: 1510–1515

Martin C B Jr 1965 Uterine blood flow and placental circulation. Anaesthesiology 26: 447–459

Martin C P 1932 The cause of torsion of the humerus and of the notch on the anterior edge of the glenoid cavity of the scapula. J Anat 67: 573–582

Martin C P 1942 A note on the renal fascia. J Anat 77: 101–103

Martin G M, Sprague C A, Epstein C J 1970 Replicative lifespan of cultured human cells: effects of donor's age, tissue and genotype. Lab Invest 23: 86–92

Martin R 1928 Lehrbuch der Anthropologie. 3 vol. 2nd edition. Fischer: Jena

Martin R, Saller K 1961 Lehrbuch der Anthropologie. 3rd edition. Fischer: Stuttgart

Martin X, Dolivo M 1983 Neuronal and transneuronal tracing in the trigeminal system of the rat using the herpes virus suis. Brain Res 273: 253–276

Martins R 1961 Lateral cervical and pre-auricular sinuses: their transmission as dominant characters. Br Med J 1: 255–256

Mascorro J A, Yates R D 1971 Ultrastructural studies of the effects of reserpine on mouse abdominal sympathetic paraganglia. Anat Rec 170: 269–280

Mascorro J A, Yates R D 1974 Intervention of abdominal paraganglia. An ultrastructural study. J Morphol 142: 153–164

Mascorro J A, Yates R D 1975 A review of abdominal paraganglia. Ultrastructure, mitotic cells, catecholamine release, innervation, light and dark cells, vascularity. In: Hess M (ed) Electron microscopic concepts of secretion. Ultrastructure of endocrinal and reproductive organs. Wiley: New York: pp 435–452

Masland, R II, Tauchi M 1986 The cholinergic amacrine cell. Trends Neurosci 9: 218–223

Mason C A, Lincoln D W 1976 Visualization of the retinohypothalamic projection in the rat by cobalt precipitation, Cell Tissue Res 168: 117–131

Massopust L C Jr, Daigle H J 1960 Cortical projection of the medial and spinal vestibular nuclei in the cat. Exp Neurol 2: 179–185

Masterton R B, Jane J A, Diamond I T 1967 Role of brainstem auditory structures in sound localisation. I. Trapezoid body superior olive, and lateral lemniscus, J Neurophysiol 30: 341–359

1517

BIBLIOGRAPHY

Mastroianni L 1962 The structure and function of the fallopian tube. A correlative review. Clin Obstet Gynecol 5: 781–790

Matano S 1970 Experimental studies on the medial longitudinal fasciculus in the rabbit. V. Ascending fibers from the reticular formation and the oculomotor system. J Hirnforsch 12: 241–253

Mathiasen M S 1973 Determination of bone age and recording of minor skeletal hand anomalies in normal children, Dan med Bull 20: 80–85

Matsuda H 1968 Electron microscopic study on the corneal nerve with special reference to its endings, Acta-Soc Ophthal Jpn 12: 163–173

Matsumoto S, 1913 Uber eine eigentumliche Pigment verteilung an der voigtechen Linien. Arch. Dermatol. Syph. 118: 157–161

Matsumoto T 1920 The granules, vacuoles and mitochnodria in sympathetic nerve-fibers cultivated in vitro. Johns Hopkins Hosp Bull 31: 91–93

Matthews M A 1968 An electron microscopic study of the relationship between axon diameter and the initiation of myelin production in the peripheral nervous system, Anat Rec 161: 337–352

Matthews M R, Nelson V H 1975 Detachment of structurally intact nerve endings from chromatolytic neurones of rat superior cervical ganglion during the depression of synaptic transmission induced by post-ganglionic axotomy. J Physiol 245: 91–135

Matthews M R, Raisman G 1969 The ultrastructure and somatic efferent synapses of small granule-containing cells in the superior cervical ganglion, J Anat 105: 255–282

Matthews P B C 1971 Recent advances in the understanding of the muscle spindle. In: Gilliland I, Francis J (eds) British postgraduate medical federation, scientific basis of medicine. Annual Reviews. Athlone Press: London: pp 99–128

Matthews P B C 1972 Mammalian muscle receptors and their central actions, Arnold: London

Matthews P B C 1977 Muscle afferents and kinaesthesia. Br Med Bull 33: 137–142

Matulionis D H, Parks H F 1973 Ultrastructural morphology of the normal nasal respiratory epithelium of the mouse, Anat Rec 175: 68–84

Maul G G 1971 Structure and formation of pores in fenestrated capillaries, J Ultrastruct Res 36: 768–782

Maxwell D S, Pease D C 1956 The electron microscopy of the coroid plexus. J Biophys Biochem Cytol 2: 467–474

Maxwell L C, Faulkner J A, Lieberman D A 1973 Histochemical manifestations of age and endurance training in skeletal muscle fibers. Am J Physiol 224: 356–361

Mayfield J K, Johnson R P, Kilcoyne R F 1976 The ligaments of the human wrist and their functional significance, Anat Rec 1986: 417–428

Mayhall J T, Dahlberg A A, Owen D G 1970 Torus mandibularis in an Alaskan Eskimo population. Am J Phys Anthropol 33: 57–60

Mayne R, von der Mark K 1983 Collagens of cartilage. In: Hall B K (ed) Cartilage Vol 1. Structure, function and biochemistry. Academic Press: New York: pp 181–214

Mays E T 2d, Mays E T 1983 Are hepatic arteries end-arteries? J Anat 137: 637–644

Meachim G 1982 Age-related degeneration of patellar articular cartilage. J Anat 134: 365–371

Meachim G, Allibone R 1984 Topographical variation in calcified zone of upper femoral articular cartilage, J Anat 139: 341–352

Meachim G, Stockwell R A 1978 The matrix. In: Freeman M A R (ed) Adult articular cartilage. 2nd edition. Pitman Medical: London: pp 1–67

Meachim G, Denham D, Emergy I H, Wilkinson P H 1974 Collagen alignments and artificial splits at the surface of human articular cartilage, J Anat 188: 101–118

Mears D C 1971 Effects of parathyroid hormone and thyrocalcitonin on the membrane potential of osteoclasts. Endocrinology, 88: 1021–1028

Meban C 1980 Thickness of the air-blood barriers in vertebrate lungs. J Anat 131: 299–307

Mechanik N 1934 Das Venesystem der Herzwände Z Anat EntwGesch 103: 813–843

Meckel J F 1832 Manual of Anatomy. Vol 3, Carey & Lea: Philadelphia

Medical Research Council 1941 Aids to the investigation of peripheral nerve injuries. (War Memorandum No 7). HMSO: London

Medical Research Council 1976 Aids to the examination of the peripheral nervous system. (Memorandum No 45, superseding War Memorandum No 7). HMSO: London

Meessen H, Olszewski J 1949 A cytoarchitectonic atlas of the rhombencephalon of the rabbit. Karger: Basel

Mehler W R 1962 The anatomy of the so-called 'pain tract' in man: An analysis of the course and distribution of the ascending fibers of the fasciculus anterolateralis. In: French J D, Porter R W (eds) Basic research in paraplegia. Thomas: Springfield, Illinois: pp 26–55

Mehler W R 1966 Further notes on the centre median nucleus of Luys. In: Purpura D P, Yahr M D (eds) The thalamus. Columbia University Press: New York: pp 109–127

Mei N, Dussardier M 1966 Etudes des Lésions pulmonaires produites par la section des fibres sensitives vagales, J Physiol, Paris, 58: 427–431

Mehler W R, Feferman M E, Nauta W J H 1960 Ascending axon degeneration following anterolateral cordotomy. An experimental study in the monkey. Brain 83: 718–750

Metha H J, Gardner W U 1961 A study of lumbrical muscles in the human hand, Am J Anat 109: 227–238

Meikle T H, Sprague J M 1964 The neural organisation of the visual pathways in the cat. Int Rev Neurobiol 6: 150–191

Meininger V, Baudrimont M 1977 The cytoarchitecture of the inferior colliculus in the cat. J Neurol Sci 34: 25–36

Melander A 1970 Amines and mouse thyroid activity: release of thyroid hormone by catecholamines and indoleamines and its inhibition by adrenergic blocking drugs. Acta Endocrinol 65: 371–384

Melander A 1977 Aminergic regulation of thyroid activity: importance of the sympathetic innervation and of the mast cells of the thyroid gland. Acta Med Scand 201: 257–262

Melander A 1978 Sympathetic nervous-adrenal medullary system. In: Werner S C, Ingbar S H (eds) The thyroid–a fundamental and clinical text. Harper & Row: New York: pp 216–221

Melander A, Nilsson E, Sundler F 1972 Sympathetic activation of thyroid hormone secretion in mice. Endocrinology 90: 194–199

Melander A, Sundler F, Westgren U 1943 Intrathyroidalamines and the synthesis of thyroid hormone. Endocrinology 93: 193–200

Melander A, Ericson L E, Sundler F, Westgren U 1975 Intrathyroidal amines in the regulation of thyroid activity. Rev Physio 73: 39–71

Melcher A H 1976 On the repair potential of periodontal tissues. J Periodontol 47: 256–260

Melcher A H, Bowen W H eds 1969 The biology of the periodontium. Academic Press: New York, London

Melcher A H, Walker T W 1976 The periodontal ligament in attachment and as a shock absorber. In: Poole D F G, Stack M V (eds) The eruption and occlusion of teeth. Butterworths: London: pp 183–192

Meller K, Breipohl W, Glees P 1969 Ontogony of the mouse motor cortex. The polymorph layer or layer VI. A Golgi and electronmicroscopical study. Z Zellforsch 99; 443–458

Melzack R 1973 The puzzle of pain. Basic Books: New York

Melzack R, Southmayd S E 1974 Dorsal column contributions to anticipatory motor behavior. Exp Neurol 42: 274–281

Melzack R, Wall P D 1965 Pain mechanisms: a new theory. Science 150: 971–979

Menco B P 1984 Ciliated and microvillous structures of rat and olfactory and nasal respiratory epitelia. A study using ultra-rapid cryo-fixation followed by freeze-substitution or freeze-etching. Cell Tissue Res 235: 225–241

Menco B P M, Dodd G H, Davey M, Bannister L H 1976 Presence of membrane particles in freeze-etched bovine olfactory cilia. Nature 263: 597–599

Mendell L M 1966 Physiological properties of unmyelinated fiber projections to the spinal cord. Exp Neurol 16: 316–332

Mendell L M, Wall P D 1964 Presynaptic hyperpolarization: a role for fine afferent fibres. J Physiol 172: 274–294

Menning A von, Schumacher G H, Lau H, Schultz M, Himstedt H W 1974 Zur Topographie der muskularen Nervenausbreitungen 6. Untere Extremität, Glutealmuskeln, Anat Anz 135: 302–314

Menton D N, Simmons D J, Orr B Y, Plurad S B 1982 A cellular investment of bone marrow. Anat Rec 203: 157–164

Merchan M A, Collia F P, Merchan J A, Saldana E 1985 Distribution of primary afferent fibres in the coclear nuclei. A silver and horseradish peroxidase (HRP) study. J Anat 141: 121–130

Mercier R, Vanneuville G, Bresson P et al 1970a Etude de la structure osseuse de la branche horizontale du maxillaire in inférieur apport des techniques radiographiques. Lab Anat Clermont Ferrand. C R Ass Anat 149, 891–901

Mercier R, Vanneuville G, Bresson P et al 1970b Etude des lignes de force des corticales du maxillaire inférieur par la méthode des lignes de fissuration colorées. Lab Anat Clermont Ferrand. C R Ass Anat 149: 902–913

Merendino K A, Johnson R J, Skinner H H, Maguire R X 1956 Intradiaphragmatic distribution of phrenic nerve with particular reference to placement of diaphragmatic incisions and controlled segmental paralysis, Surg Gynec Obstet 39: 189–198

Merklin R J, Michels N A 1958 The variant renal and suprarenal blood supply with data on the inferior phrenic, ureteral and gonadal arteries: a statistical analysis based on 185 dissections and a review of the literature. J Int Coll Surg 29: 41–76

Metchnikoff M E 1900 Récherches sur l'influence de l'organisme sur les toxines. Sur la spermotoxine et l'antispermatoxine; quatrième mémoire. Ann Inst Pasteur 14: 1

Mettler F A 1967 Cortical subcortical relations in abnormal motor functions. In: Yahr M D, Purpura D P (eds) Neurophysiological basis of normal and abnormal motor functions. Raven Press: New York: pp 445–497

Mettler F A 1968 In: Vinken P J, Bruyn G W (eds) Handbook of clinical neurology. North Holland: Amsterdam

Metz C B, Monroy A eds 1967 and 1969 Fertilization 2 vol. Academic Press: New York

Meyer A 1971 Historical aspects of cerebral anatomy. Oxford University Press: London

Meyer A W 1917 Studies on hemal nodes. VII. The development and function of hemal nodes. Am J Anat 21: 375–406

Meyer D B, O'Rahilly R 1976 The onset of ossification in the human calcaneus, Anat Embryol (Berl) 150; 19–33

Meyer H 1867 Die Architektur der Spongiosa. Arch Anat Physiol 47:

Meyer R 1938 Zur Frage der Entwicklung der menschlichen Vagina: Vagina infima septa und andere Besonderheiten. Arch Gynecol 167: 306–338

Meynert T 1867–1868 Der Bau der Gross-Hirnrinde und seine ortlichen Verschiedenheiten nebst einem pathologisch-anatomischen Corollarium. Vjschr Psychiat Vienna 1: 77–93, 198–217; 2: 88–113

Meyrick B, Reid L 1968 The alveolar brush cell in rat lung–a third pneumonocyte, J Ultrastruct Res 23: 71–80

Meyrick B, Reid L 1970 Ultrastructure of cells in the human bronchial submucosal glands. J Anat 107: 281–299

Miale I L, Sidman R L 1961 An autoradiographic analysis of the histogenesis of the mouse cerebellum. Exp Neurol 4: 227–296

Michels N A 1962 The anatomic variations of the arterial pancreaticoduodenal arcades: their import in regional resection involving the gall bladder, bile ducts, liver, pancreas and parts of the small and large intestines. J Int Coll Surg 37: 13–40

Midgley A R, Pierce G B Jr, Deneau G A, Gosling J R 1963 Morphogenesis of syncitiotrophoblast in vivo: an autoradiographic demonstration. Science 141: 349–350

Mierzwa J, Kozielec T 1975 Variation of the anterior cardiac veins. Folia Morphol 34: 125–133

Miki H, Bellhorn M B, Henkind P 1975 Specializations of the retinochoroid juncture, Invest Ophthal 14: 701:7

Mikić Z Dj 1978 Age changes in the triangular fibrocartilage of the wrist joint. J Anat 126: 367–384

Milaire J 1957 Contribution à la connaissance morphologique et cytochimique des bourgeons de membres chez quelques reptiles. Arch Biol, Liége 68: 429–512

Milaire J 1965 Aspects of limb morphogenesis in mammals. In: De Haan R L, Ursprung H (eds) Organogenesis. Holt: New York: pp 283–300

Milburn A 1973 The early development of muscle spindles in the rat. J Cell Sci 12: 175–195

Miledi R, Molinoff P, Potter L T 1971 Isolation of the cholinergic receptor protein of Torpedo electric tissue. Nature 229: 554–557

Millar J, Basbaum A I 1975 Topography of the projection of the body surface of the cat to cuneate and gracile nuclei. Exp Neurol 49: 281–290

Millen J W, Woollam D H M 1953 Vascular patterns in the choroid plexus. J Anat 87: 114–123

Millen J W, Woollam D H M 1961 On the nature of the pia mater. Brain 84: 514–520

Millen J W, Woollam D H M 1962 The anatomy of the cerebrospinal fluid, Oxford University Press: London

Miller E J, Gay S 1987 The collagens: an overview and update. Methods Enzymol 144: 3–41

Miller L H, Carter R 1976 Innate resistance in malaria. Exp Parasitol 40: 132–146

Miller R H, Raff M C 1984 Fibrous and protoplasmic astrocytes are biochemically and developmentally distinct. J Neurosci 4: 585–592

Miller R L, Chaudhry A P 1976 Comparative ultrastructure of vallate, foliate and fungiform taste buds of golden Syrian hamster, Acta Anat 95: 75–92

Miller W S 1947 The lung. 2nd edition, Thomas: Springfield, Illinois

Milligan E T C, Morgan C N, Jones, L E, Officer R 1937 Surgical anatomy of anal canal and operative treatment of haemorrhoids. Lancet 2: 1119–1124

Millikan C H, Darley F L eds 1967 Brain mechanisms underlying speech and language, Grune and Stratton: New York

Millington P F, Wilkinson R 1983 Skin. Biological Structure and Function 9). (Harrison R J, McMinn R M H eds), Cambridge University Press: Cambridge

Mills E, Jöbsis F F 1972 Mitochondrial respiratory chain of carotid body and chemoreceptor response to changes in oxygen tension. J Neurophysiol 35: 405–428

Mills R W 1917 The relation of bodily habits to visceral form, position, tonus and motility. Am J Roentg 4: 155–169

Mills R W 1922 X-ray evidence of abdominal small intestinal states embodying an hypothesis of the transmission of gastro-intestinal tension. Am J Roentg 9: 199–225

Milner B 1958 Psychological defects products by temporal lobe excision. Res Publ Assoc Res Nerv Ment Dis 36: 244–257

Milner B 1967 Brain mechanisms suggested by studies of temporal lobes. In: Millikan C H, Darley F L (eds) Brain mechanisms underlying speech and language. Grune and Stratton: New York: pp 122–131

Milner B 1974 Hemispheric specialization: scope and limits. In: Schmitt F O, Worden F G (eds) The neurosciences, third study program. MIT Press: Cambridge, Massachusetts

Milner B, Teubér H L 1968 Alteration of perception and memory in man. In: Weiskrantz G L (ed) Analysis of behavioural change. Harper & Row: New York

Milner B, Branch C, Rasmussen T 1964 Observations on cerebral dominance. In: Wolstenholme D W, O'Connor M (eds) Disorders of Language. Churchill: London

Mincsev M 1967 Bilocular gall bladder. Orvoskepzes 42: 286–298

Minkoff E C 1974 The Fürbinger hypothesis of nerve-muscle specificity re-examined, Can J Zool 52: 525–532

Minkowski M 1913 Experimentelle Untersuchungen über die Beziehungen der Grosshirnrinde und der Netzhaut zu den primären optischen Zentren, besonders zum Corpus geniculatum externum. Arb Hirnanat Inst Zürich 7: 259–362

Minneman R P, Lynch H, Wurtman R J 1974 Relationship between environmental light intensity and retina-mediated suppression of rat pineal serotonin N-acetyltransferase. Life Sci 15: 1791–1796

Minns R J, Stevens F S 1977 The collagen fibre organisation in human articular cartilage J Anat 123: 437–457

Minsky B D, Chlapowski F J 1978 Morphometric analysis of the translocation of lumenal membrane between cytoplasm and cell surface of transitional epithelial cells during the expansion-contraction cycles of mammalian urinary bladder. J Cell Biol 77: 685–697

Minsky M 1965 Matter, mind and models. Spartan Books: Washington, DC

Mintz B 1960 Embryological phases of mammalian gameto-genesis. J Cell Comp Physiol 56: (Suppl 1): 31–47

Mintz B 1962 Experimental study of the developing mammalian egg: removal of the zona pellucida. Science 138: 594–595

Mintz B 1964 Synthetic processes and early development in the mammalian-egg. J Exp Zool 157: 267–272

Mintz B 1965 Experimental genetic mosaicism in the mouse. In: Wolstenholme G E W, O'Connor M (eds) Preimplanation stages of pregnancy. Ciba Foundation Symposium. Churchill: London: pp 194–216

Mintz B 1970 Do cells fuse in vivo? In Vitro 5: 40–47

Mishima S, Maurice D M 1961 The effect of normal evaporation on the eye, Exp Eye Res 1: 46–52

Misotten L, Van den Dooren E 1966 L'ultrastructure de la rétine humaine. Les contacts latéraux des pédoncules de cônes de la fovea. Bull Soc Belg Ophthalmol 144: 800–805

Mitchell G A G 1935 Innervation of distal colon, Edinb Med J 42: 11–20

Mitchell G A G 1938 Nerve supply of the gastro-oesophageal junction. Br J Surg 26: 33–45

Mithcell G A G 1950 Renal fascia. Br J Surg 37: 257–266

Mitchell G A G 1952 Rostral extremities of sympathetic trunks, Nature 129: 533–534

Mitchell G A G 1953 Anatomy of the autonomic nervous system. Livingstone: Edinburgh

Mitchell G A G 1956 Cardiovascular Innervation. Livingstone: Edinburgh.

Mitchell G A G, Warwick R 1955 The dorsal vagal nucleus, Acta Anat 25: 371–395

Mitchell P 161 Coupling of phosphorylation to electron and hydrogen transfer by a chemi-osmotic type of mechanism. Nature 191: 144–148

Mitchell R E 1967 Chronic solar dermatosis: a light and electron microscopic study of the dermis, J Invest Dermatol 48: 203–220

Mitchison J M 1971 The biology of the cell cycle. Cambridge University Press: Cambridge

Miyabara S, Okamoto N, Akimoto N, Satow Y, Hidaka N 1974 Meckel's diverticulum found at autopsy. Hiroshima J Med Sci 23: 179–190

Mizeres N J 1963 The cardiac plexus in man. Am J Anat 112 141–151

Mizuno N 1966 An experimental study of the spino-olivary fibers in the rabbit and the cat. J Comp Neurol 127: 267–291

Moar J J, Tobias P V 1985 Multiple renal arteries. S Afr Med J 67: 399

Mochereau M T 1963 Ann Biol Anim Biochem Biophys 3:

Modi J P 1957 Jurisprudence and toxicology, Tripathi Private: Bombay

Moe G K, Preston J B, Burlington H 1956 Physiologic evidence for a dual A V transmission system Circ Res 4: 357

Moens P 1974 Quantitative electron microscopy of chromosome organization at meiotic prophase, Cold Spring Harb Symp Quant Biol 28: 99–107

Moffat D B 1959 Developmental changes in the aortic arch system of the rat. Am J Anat 105: 1–36

Moffat D B 1961a The development of the anterior cerebral artery and its related vessels in the rat. Am J Anat 108: 17–29

Moffat D B 1961b The development of the ophthalmic artery in the rat. Anat Rec 140: 217–222

Moffat D B 1975 The mammalian kidney. Cambridge University Press

Moffat D B 1982 Developmental abnormalities of the urogenital system. In: Chisholm G D, Williams D I (eds) Scientific foundations of urology. 2nd edition. Heinemann Medical: London: pp 357–372

Moffett B C Jr 1957 The prenatal development of the human temporomandibular joint. Contrib Embryol Carnegie Inst Washington 36: 19–28

Moffett B C, Johnson L C, McCabe J B, Askew H C 1964 Articular remodelling in the adult human temporomandibular joint. Am J Anat, 115: 119–142

Mohuiddin A 1953 Vagal preganglionic fibres to the ailmentary canal. J Comp Neurol 99: 289–318

Møller M 1974 The ultrastructure of the human fetal pineal gland. I. Cell types and blood vessels. Cell Tissue Res 152: 13–30

Møller M 1976 The ultrastructure of the human fetal pineal gland. II. Innervation and cell junctions. Cell Tissue Res 169: 7–21

Møller M 1978 Presence of a pineal nerve (nervus pinealis) in the human fetus; a light and electron microscopical study of the innervation of the pineal gland. Brain Res 154: 1–12

Møller M 1979 Presence of a pineal nerve (nervus pinealis) in fetal mammals. Prog Brain Res 52: 103–106

Møllgard K, Møller M, Kimble J 1973 Histochemical investigations on the human fetal sub-commissural organ. Histochemie 37: 61–74

BIBLIOGRAPHY

Molliver M E, Kostović I, Van der Loos H 1973 The development of synapses in cerebral cortex of the human fetus. Brain Res 50: 403–407

Molliver M E, Van der Loos H 1970 The ontogenesis of cortical circuitry: the spatial distribution of synapses in somesthetic cortex of newborn dog. Ergeb Anat Entwicklungsgesch 42: 5–53

Molyneux G S 1965 Observations on the structure, distribution and significance of arterio-venous anastomoses in sheepskin. In: Lyne A G, Short B F (eds) Biology of the skin and hair growth. Proceedings of a symposium held at Canberra 1964. Angus and Robertson: Sydney

Monakow C von 1882 Weitere Mitteilungen über durch Exstirpation circumscripter Hirnrindenreigionem bedingte Entwickelungshemmungen des Kaninchegehirns. Arch Psychiat. NervKrankh 12: 141–156, 535–549

Monakow C von 1905 Gehirnpathologie, 2nd edition, Hölder: Vienna

Moniem K A, Glover T D 1972 Alkaline phosphatase in the cytoplasmic droplet of mammalian spermatozoa. J Reprod Fertil 29: 65–69

Monkhouse W S, Khalique A 1986 The adrenal and renal veins of man and their connections with azygos and lumbar veins. J Anat 146: 105–115

Monro A 1746 The anatomy of the human bones and nerves. Hamilton & Balfour: Edinburgh

Monroy A, Tyler A 1967 The activation of the egg. In: Metz C B, Monroy A (eds) Fertilization. Vol I. Academic Press: New York: pp 369–412

Montagna W 1968 The structure and function of skin 2nd edition, Academic Press: New York, London

Montagna W, Carlisle K, 1979 Structural changes in aging human skin, J Invest Dermatol 73: 47–53

Montagna W, Ellis R A (eds) 1961 Advances in biology of skin. Vol 2. Blood vessels and circulation. Pergamon: Oxford

Montagna W, Lobitz W C Jr eds 1964 The epidermis. Academic Press: London

Montagna W, Parakkal P F 1974 The structure and function of skin. 3rd edition. Academic Press: New York

Montagu M F A 1951 Wallbrook frontal bone, Am J phys Anthropol 9: 5–14

Montagu M F A 1960 An introduction to physical anthropology, 3rd edition. Thomas: Springfield, Illinois

Moody R O 1927 The position of the abdominal viscera in healthy young British and American adults, J Anat 61: 223–231

Moore C R 1947 Embryonic sex hormones and sexual differentiation. Thomas: Springfield, Illinois

Moore M A, Owen J J 1967 Experimental studies on the development of the thymus J Exp Med 126: 715–726

Moore R Y 1973 Retinohypothalamic projection in mammals: a comparative study, Brain Res 49: 403–409

Moore R Y 1975 Monoamine neurons innervating the hippocampal formation and septum. In: Isaacson R L, Pribram K H (eds) The hippocampus. 2 vol. Plenum: New York

Moore R Y, Eichler V B 1972 Loss of a circadian adrenal corticosterone rhythm following suprachiasmatic lesions in the rat, Brain Res 42: 201–206

Moore R Y, Goldberg J M 1963 Ascending projections of the inferior coliculus in the cat. J Comp Neurol 121: 109–136

Moore R Y, Goldberg J M 1966 Projections of the inferior colliculus in the monkey, Exp Neurol, 14: 429–438

Moore R Y, Klein D C 1974 Visual pathways and the central neural control of a circadian rhythm in pineal serotonin n-acetyltransferase activity, Brain Res 71: 17–34

Moore R Y, Lenn N J 1972 A retinohypothalamic projection in the rat. J Comp Neurol, 146: 1–14

Moores G R, Partridge T A 1974 The cell surface. In: Beck F, Lloyd J B (eds) The cell in medical science. Vol 1. The cell and its organelles. Academic Press: London: pp 75–104

Moosavi H, Smith P, Heath D 1973 The Feyrter cell in hypoxia, Thorax 28: 729–741

Mooseker M S 1985 Organization, chemistry and assemby of the cytoskeletal apparatus of the intestinal brush border. Annu Rev Cell Biol 1: 209–241

Mooseker M S, Tilney L G 1975 Organization of an actin filament-membrane, complex. Filament polarity and membrane attachment in the microvilli of intestinal epithelial cells, J Cell Biol 67: 725–743

Moran D T, Rowley J C III, Jafek B W, Lovell M A 1982 The fine structure of the olfactory mucosa in man. J Neurocytol, 11: 721–746

Morant G M 1936 A biometric study of the human mandible. Biometrics 28: 84–122

Morest D K 1964 The neuronal architecture of the medial geniculate body of the cat. J Anat 98: 611–630

Morest D K 1965 The lateral tegmental system of the mid brain and the medial geniculate body: study with Golgi and Nauta methods in cat. J Anat 99: 611–634

Morest D K 1967 Experimental study of the projections of the nucleus of the tractus solitarius and the area postrema in the cat. J Comp Neurol 130: 277–300

Morest D K 1971 Dendrodenritic synapses of cells that have axons: the fine structure of the Golgi type II cell in the medial geniculate body of the cat, Z Anat EntwGesch 133: 216–246

Morgan D L, Proske U 1984 Vertebrate slow muscle: its structure, pattern of innervation, and mechanical properties. Physiol Rev 64: 103–169

Morgan J D 1959 Blood supply of the growing rabbit's tibia, J Bone Jt Surg 41B: 185–203

Morgan M W 1944 Accommodation and its relation to convergence Am J Optom 21: 183–195

Morgan-Hughes J A 1986 Mitochondrial diseases. Trends Neurosci 9: 15–19

Mori S, Leblond C P 1970 Electron microscopic identification of three classes of oligodendrocytes and a preliminary study of their proliferative activity in the corpus callosum of young rats. J Comp Neurol 139: 1–60

Moriarity G C 1975 Electron microscopic-immunocytohistochemical studies of rat pituitary gonadotrophs: a sex difference in morphology and cytochemistry of LH cells, Endocrinology, 97: 1215–1225

Morin F, Catalano J V 1955 Central connections of a cervical nucleus (nucleus cervicalis lateralis of the cat). J Comp Neurol 103: 17–32

Morin F, Schwartz H G, O'Leary J L 1951 Experimental study of the spino-thalamic and related tracts. Acta Psychiat Neurol Scand 26: 371–396

Morris E W T 1976 Observations on the source of embryonic myoblasts, J Anat 121: 47–64

Morrison A B 1954 The levatores costarum and their nerve supply, J Anat 88: 19–24

Moruzzi G, Magoun H W 1949 Brain stem reticular formation and activation of the EEG. Electroenceph Clin Neurophysiol 1: 455–473

Moscatelli D, Gross J L, Rifkin D B 1981 Angiogenic factors stimulate plasminogen activator and collagenase production by capillary endothelial cells. J Cell Biol 91: 201a

Moscona A A 1976 Cell recognition in embryonic morphogenesis and the problem of neuronal specificities. In: Barondes S H (ed) Neuronal Recognition. Plenum Press: New York: pp 205–226

Mossman H W 1937 Comparative morphogenesis of the fetal membranes and accessory uterine structures. Contrib Embryol Carnegie Inst Wagh 26: 128–247

Moss-Saletnijn L 1975 Cartilage canals in the human spheno-occipital synchondrosis during fetal life. Acta Anat 92: 595–606

Moss-Salentijn A G M 1976 The epiphyseal vascularization of growth plates. A developmental study in the rabbit. Doctoral Thesis. Rijks Universiteit: Utrecht

Moss-Salentijn L, Moss M L, Shinozuka M, Skalak R 1987 Morphological analysis and computer-aided, three dimensional reconstruction of chondrocytic columns in rabbit growth plates. J Anat 151: 157–167

Mott J C 1982 Control of the foetal circulation. J Exp Biol 100: 129–146

Moulton D G, Beidler L M 1967 Structure and function of the peripheral olfactory system, Physiol Rev 47: 1–52

Moulton D G, Celebi G, Fink R P 1970 Olfaction in mammals–two aspects: proliferation of cells in the olfactory epithelium and sensitivity to odours. In: Wolstenholme G E W, Knight J (eds) Taste and smell in vertebrates. (Ciba Foundation Symposium). Churchill: London: pp 227–245

Mountcastle V B 1957 Modality and topographic properties of single neurons of cat's omatic sensory cortex. J Neurophysiol 20: 408–434

Mountcastle V B 1978 An organizing principle for cerebral function: the unit module and the distributed system. In: Edelman G M, Mountcastle V B (eds) The mindful brain: cortical organization and the group-selective theory of higher brain function. MIT Press: Cambridge, Massachusetts: pp 7–50

Mountcastle V B 1980 Medical physiology 14th edition. 2 vol. Mosby: St Louis

Mountcastle V B, Powell T P S 1959a Central nervous mechanisms subserving position sense and kinesthesis. Bull Johns Hopkins Hosp 105: 173–200

Mountcastle V B, Powell T P S 1959b Neural mechanisms subserving cutaneous sensibility with special reference to the role of afferent inhibition in sensory perception and discrimination. Bull Johns Hopkins Hosp 105: 201–232

Mountcastle V B, Lynch C J, Georgopoulos A, Sakata H, Acuna A 1975 Posterior parietal association cortex of the monkey, J Neurophysiol 38: 871–908

Mourant A E, Kopeć A C, Domaniewska-Sobczak K, 1976 The distribution of the human blood groups and other polymorphisms. 2nd edition. Oxford University Press: London

Mow V C, Lai W M, Redler I 1974 Some surface characteristics of articular cartilage. I. A scanning electron microscopy study and a theoretical model for the dynamic interaction of synovial fluid and articular cartilage, J Biomech 7: 449–456

Moxham B J, Berkovitz B K B 1983 Interactions between thyroxine, hydrocortisone and cyclophosphamide in their effects on the eruption of the rat mandibular incisor. Arch Oral Biol 28: 1083–1087

Moyers R E 1950 Electromyographic analysis of certain muscles involved in temporomandibular movement, Am J Orthodont 36: 481–515

Muchmore W B 1951 Differentiation of the trunk mesoderm in amblystoma maculatum. J Exp Zool 118: 137–185

Mueller J C, Jones A L, Long J A 1972 topographic and subcellular anatomy of the guinea pig gallbladder. Gastroenterology 63: 856–868

Mueller K H, Trias A, Ray R D 1966 Bone density and composition. Age-related and pathological changes in water and mineral content, J Bone Jt Surg 48A: 140–148

Mugnaini E 1965 'Dark cells' in electron micrographs from the central nervous system of vertebrates. J Ultrastruct Res 12: 235–236

Mugnaini E 1970 The relationship between cytogenesis and the formation of different types of synaptic contact. Brain Res 17: 169–179

Mugnaini E, Forstrønen P F, 1967 Ultrastructural studies on cerebellar histogenesis. I. Differentiation of granule cells and development of glomeruli in the chick embryo. Z Zellforsch. Mikrosk Anat 77: 115–143

Muir A R 1954 The development of the ventricular part of the conducting tissue in the heart of the sheep. J Anat 88: 381–391

Muirhead E E, Germain G, Leach B E, Pitcock J A, Stephenson P, Brooks B, Brosius W L, Daniels E G, Hinman J W 1972 Production of renomedullary prostaglandins by renomedullary interstitial cells grown in tissue culture. Circ Res 31: Suppl 2, 161–172

Mukai N 1970 Axonal reaction of the optic nerve following heat coagulation. Histochemical evidence for antidromic conduction. Can J Ophthal 5: 78–90

Müller F 1977 The development of the anterior falcate and lacrimal arteries in the human. Anat Embryol (Berl) 150: 207–227

Muller T 1959 Variations in the abductor pollicis longus and extensor pollicis brevis in the South African Bantu. S Afr J Lab Clin Med 5: 56–62

Mulnard J G 1964 Obtention in vitro du developpement continu de l'oeuf de souris du stade II au stade du blastocyste. C R Acad Sci, Paris 258: 6228–6229

Mulnard J G 1965 Studies of regulation of mouse ova in vitro. In: Wolstenholme G E W, O'Connor M (eds) Preinplantation stages of pregnancy. Ciba Foundation Symposium. Churchill: London: pp 123–144

Munger B L 1964 Histochemical studies on seromucous and mucous-secreting cells of human salivary glands. Am J Anat 115: 411–429

Munger B L, Halata Z 1983 The sensory innervation of the primate facial skin. I. Hairy skin. Brain Res Rev 5: 48–80

Munger B L, Roth S I 1963 The cytology of the normal parathyroid glands of man and Virginia deer; a light and electron microscopic study with morphologic evidence of secretory activity. J Cell Biol 16: 379–400

Munkacsi K, Newstead J D 1971 Direct autonomic nerve fibers to the renal medulla in man. Experientia 27: 175–177

Münzer E, Wiener M 1902 Das Zwischen- und Mittel- hirn des Kaninchens. Mschr Psychiat. Neurol 12: 241–279

Murad T M 1970 Ultrastructural study of rat mammary gland during pregnancy. Anat Rec 167: 17–36

Muratori G 1965 Struttura microscopica del seno carotideo nel gatto, cane e coniglio, Boll Soc Ital Biol Sper 42: 301–303

Murley R S, Peters P M 1961 Inadvertent parathyroidectomy. Proc R Soc Med 54: 487–489

Murray J R, Bock R D, Roche A F 1971 The measurement of skeletal maturity. Am J Phys Anthropol 35: 327–330

Murray M, Miller H R P, Jarrett W F H 1968 The globule leucocyte and its derivation from the subepithelial mast cell, Lab Invest 19: 222–234

Murray M R 1965 Nervous tissues in vitro. In: Willmer E B (ed) Cells and tissues in culture. Vol 2. Academic Press: New York: pp 373–455

Murray P D F 1936 Bones. A study of the development and structure of the vertebrate skeleton. Cambridge University Press: Cambridge

Murray P D F, Huxley J S 1924 Self differentiation of the grafted limb-bud of the chick. J Anat 59: 379–384

Muskens L J J 1914 An anatomico-physiological study of the posterior longitudinal bundle in its relation to forced movement. Brain 36: 352–426

Mustafa G Y, Gamble H J 1979 Changes in axonal numbers in developing human trochlear nerve. J Anat 128: 323–330

Myers R D 1969 Temperature regulation: neurochemical systems in the hypothalamus. In: Haymaker W, Anderson E, Nauta W J H (eds) The hypothalamus. Thomas: Springfield, Illinois: pp 506–523

Myers R E 1959 Localisation of functions in the corpus callosum. Visual gnostic transfer. Arch Neurol Psychiat I: 74–77

Myers R E, Henson C O 1960 Role of corpus callosum in transfer of tactuokinesthetic learning in chimpanzee. Arch Neurol Psychiat 3: 404–409

Mygind N 1975 Scanning electron microscopy of the human nasal mucosa, Rhinology 13: 57–75

Mygind N 1978 Immunohistopathology of allergic rhinitis and conditions allied. Clin Otolaryngol 3: 325–342

Mygind S H 1948 Further labyrinthine studies; on labyrinthine transformation of acoustic vibrations to pitch-differentiated nervous impulses, Acta Otolaryngol, Suppl 68: 53–80

Mysorekar V R 1967 Diaphysial nutrient foramina in human long bones. J Anat 101: 813–822

N

Nabeshima S, Reese T S, Landis D M D, Brightman M W 1975 Junctions in the meninges and marginal glia. J Comp Neurol 164: 127–169

Nachemson A 1960 Lumbar intradiscal pressure. Experimental studies on post-mortem material. Act Orthop Scand Suppl 43: 1–104

Naessen R 1971 The 'receptor surface' of the olfactory organ (epithelium) of man and guinea pig. A descriptive and experimental study. Acta Otolaryngol 71: 335–348

Nagato T, Yoshida H, Yoshida A, Uehara Y 1980 A scanning electron microscope study of myoepithelial cells in exocrine glands. Cell Tissue Res 209: 1–10

Nagylaki T, Levy J 1973 'The sound of one paw clapping' is not sound, Behav Genet 3: 279–292

Nakai J 1960 Studies on the mechanism determining the course of nerve fibers in tissue culture. II. The mechanism of fasciculation. Z Zellforsch Mikrosk Anat 52: 427–449

Nakai J, Kawasaki Y 1959 Studies on the mechanism determining the course of nerve fibres in tissue culture. I. The reaction of the growth cone to various obstructions. Z Zellforsch, 51: 108–122

Nakaizumi Y 1964 The ultrastructure of Bruch's membrane. I. The human, monkey, rabbit, guinea pig and rat eyes. Arch Ophthal 72: 380–387

Nakane P K 1970 Classifications of anterior pituitary cell types with immuonenzyme histochemistry. J Histochem Cytochem 18: 9–20

Nakane P K, Kawaoi A 1974 Peroxidase-labelled antibody. A new method of conjugation. J Histochem Cytochem 22: 1084–1091

Nakanishi H, Wansbrough H, Wood C 1967 Postganglionic sympathetic nerve innervating the human fallopian tube. Am J Physiol 213: 613–619

Nakanishi T 1967 Studies on the pudendal nerve. I. Macroscopic observations on the pudendal nerve in humans. Acta Anat Nippon 42: 223–239

Nakayama M 1944 Nerve terminations in the muscle spindle of the human lingual muscles. Tohuku Med J 34: 367–377

Napier J R 1955 The form and function of the carpo-metacarpal joint of the thumb. J Anat 89: 362–369

Napier J R 1956 The prehensile movements of the human hand. J Bone Jt Sug 38B: 902–913

Napier J R 1966 Functional aspects of the anatomy of the hand. In: Pulvertaft R G (ed) The hand. Butterworths: London: pp 1–31

Napolitano L M, Scallen T J 1969 Observations on the fine structure of peripheral nerve myelin. Anat Rec 163: 1–6

Napper R M A, Harvey R J 1988 Number of parallel fiber synapses on an individual Purkinje cell in the cerebellum of the rat. J Comp Neurol 274: 168–177

Nathan H, Gloobe H 1974 Flexor digitorum brevis–anatomical variations, Anat Anz 135: 295–301

Nathan H, Levy J 1982 The course and relations of the hypoglossal nerve and the occipital artery. Am J Otolaryngol 3: 128–132

Nathan P W 1963 Results of antero-lateral cordotomy for pain in cancer. J Neurol Neurosurg Psychiatry 26: 353–362

Nathan P W 1976 The gate-control theory of pain. A critical review, Brain 99: 123–158

Nathan P W, Smith M C 1955a The Babinski response. A review and new observations. J Neurol Neurosurg Psychiat 18: 250–259

Nathan P W, Smith M C 1955b Long descending tracts in man. I. Review of present knowledge. Brain 78: 248–303

Nathan P W, Smith M C 1959 Fasciculi proprii of the spinal cord in man. Review of present knowledge. Brain 82: 610–668

Nauta W J H 1950 Über die sogenante terminale Degeneration im Zentralnervensystem und ihre Darstellung durch Silberimprägnation. Schweiz Arch Neurol Psychiat 66: 353–376

Nauta W J H, Ebesson O E eds 1970 Contemporary research methods in Neuroanatomy. Springer-Verlag: Berlin

Nauta W J H, Gygax P A 1951 Silver impregnation of degenerating axon terminals in central nervous system: technic, chemical notes. Stain Technol 26: 5–11

Nauta W J H, Haymaker W 1969 Hypothalamic nuclei and fiber connections. In: Haymaker W, Anderson E, Nauta W J H (eds) The hypothalamus. Thomas: Springfield, Illinois: pp 136–209

Nauta W J H, Mehler W R 1966 Projections of the lentiform nucleus in the monkey, Brain Res I: 3–42

Nauta W J H, Mehler W R 1969 Fiber connections of the basal ganglia. In: Crane G E, Gardner R (eds) Psychotropic drugs and dysfunctions of the basal ganglia. US Govt Printing Office: Washington DC: pp 68–72 (US Public Service Health Publication No 1938)

Navaratnam V 1963 Observations on the right pulmonary arch artery and its nerve supply in human embryos. J Anat 97: 569–573

Navaratnam V 1965 Development of the nerve supply to the human heart. Br Heart J 27: 640–650

Neal H V 1918 The history of the eye muscles, J Morphol 30: 433–453

Neaves W B 1977 The blood-testis barrier. In: Johnson A D, Gomes W R (eds) The testis. Vol 4. Advances in physiology, biochemistry, and function. Academic Press: New York: pp 126–162

Needham J 1942 Biochemistry and morphogenesis. Cambridge University Press: Cambridge

Needham J 1959 A history of embryology. 2nd edition, Cambridge University Press: Cambridge

Negami S 1964 Dynamic mechanical properties of synovial fluid, MSc Thesis, Lehigh University: Bethlehem, Penn

Negus V E 1928 The mechanism of the larynx. Heinemann: London

Negus V E 1947 Intrinsic carcinoma of larynx; a review of a series of cases. Proc R Soc Med 40: 515–524

BIBLIOGRAPHY

Negus V E 1949 The comparative anatomy of physiology of the larynx. Heinemann: London

Negus V E 1958 The comparative anatomy and physiology of the nose and paranasal sinuses. Livingstone: Edinburgh

Nelson D S 1968 Macrophages and immunity. North Holland: Amsterdam

Nelson E, Blinziger K, Hager H 1961 Electron microscopic observations on subarachnoid and perivascular spaces of the Syrian hamster brain, Neurology, 11: 285–295

Nelson E, Rennels M 1970 Innervation of intracranial arteries. Brain 93: 475–490

Nelson L 1967 Sperm motility. In: Metz C B, Monroy A (eds) Fertilization. Vol 1. Academic Press: New York: pp 27–97

Nergårdh A, Boréus L 1972 Autonomic receptor function in the lower urinary tract of man and cat. Scand J Urol Nephrol 6: 32–36

Nery E B, Kraus B S, Croup M 1970 Timing and topography of early human tooth development. Arch Oral Biol 15: 1315–1326

Nesebar R A, Kornblith P L, Pollard J J, Michels N A 1969 Celiac and superior mesenteric arteries: a correlation of angiograms and dissections. Little, Brown, Boston, Massachusetts

Neuman W F, Neuman M W 1953 The nature of the mineral phase of bone. Chem Rev 53: 1–45

Newman D L, Gosling R G, Bowden R 1971 Changes in aortic distensibility and area ratio with the development of atherosclerosis, Atheroscleroisis 14: 231–240

Nicander L 1967 An electron microscopical study of cell contacts in the seminiferous tubules of some mammals. Z Zellforsch 83: 375–397

Nicholls J G, Kuffler S W 1964 Extracellular space as a pathway for exchange between blood and neurons in the central nervous system of the leech: ionic composition of glial cells and neurons. J Neurophysiol 27: 645–671

Nicholson G W 1937 Studies on tumour formation; sacro-coccygeal teratoma with 3 metacarpal bones and digits, Guy's Hosp Rep 87: 46–106

Nicholson G L 1976 Transmembrane control of the receptors on normal and tumour cells. I. Cytoplasmic influence over cell surface components, Biochem, Biophys Acta 457: 57–108

Nicolson G L, Poste G 1976 The cancer cell: dynamic aspects and modifications in cell-surface organisation, New Engl J Med 295: 197–203

Niden A H 1967 Bronchiolar and large alveolar cell in pulmonary phospholipid metabolism. Science 158: 1323–1324

Niden A H, Yamada E 1966 Some observations on the fine structure and function of the non-ciliated bronchiolar cells. In: VIth International Congress for Electron Microscopy. Maruzen: Tokyo: p 599

Nielsen K C, Owman C 1968 Difference in cardiac adrenergic innervation between hibernators and non-hibernating mammals. Acta Physiol Scand Suppl 316: 1–30

Niemineva K 1950 Observations on the development of the hypophysial-portal system. Acta Paediat 39: 366–377

Nieuwenhuys R 1967 Comparative anatomy of the cerebellum. Prog Brain Res 25: 1–93

Nieuwenhuys R 1985 Chemoarchitecture of the brain. Springer-Verlag: Berlin

Nieuwenhuys R, Pouwels E, Venning J C 1978 Structure and composition of the medial forebrain bundle. Neurosci Lett 8: (supply 1), S 190

Nieuwenhuys R, Voogd J, van Huijzen, Chr 1978 The human central nervous system. A synopsis and atlas. Springer: Heidelberg

Nieuwkoop P D 1967 Problems of embryonic induction and pattern formation in amphibians and birds. In: Hagen E, Wechsler W, Zilliken P (eds) Morphological and biochemical aspects of cytodifferentiation. (Exp Biol Med 1) Karger: Basel: pp 22–36

Niimi K, Naito F 1974 Cortical projections of the medial geniculate body in the cat. Exp Brain Res 19: 326–342

Niimi K, Fujiwara N, Takimoto T, Matsugi S 1962 The course and termination of the ascending fibers of the brachium conjunctivum in the cat as studied by the Nauta method. Tokushima J Exp Med 8: 269–284

Niimi K, Kanaseki T, Takimoto T 1963 The comparative anatomy of the ventral nucleus of the lateral geniculate body in mammals. J Comp Neurol 121: 313–324

Nilsson O 1962 Electron microscopy of the glandular epithelium in the human uterus. I. Follicular phase. J Ultrastruct Res 6: 413–421

Nilsson O 1967 Attachment of rat and mouse blastocysts onto uterine epithelium. Int J Fertil 12: 5–13

Nishino N, Koizumi K, Brooks C Mc, 1976 The role of the suprachiasmatic nuclei of the hypothalamus in the production of circadian rhythm. Brain Res 112: 45–59

Nissl F 1892 Ueber experimentell erzeugte Veränderungen an den Vorderhornzellen des Rückenmarks bei Kaninchen mit Demonstrationen mikroskopischer Präparate. Allg Zt Psychiat 48: 675–682

Nistal M, Abaurrea M A, Paniagua R 1982 Morphological and histometric study on the human Sertoli cell from birth to the onset of puberty. J Anat 134: 351–363

Nithiuthai S, Allen J R 1984 Significant changes in epidermal Langerhans' cells infested with ticks (Dermacentor andersoni). Immunology 51: 133–142

Noback C R, Moss M L 1953 The topology of the human pre-maxillary bone. Am J phys Anthropol II: 181–187

Nomina Anatomica see International Anatomical Nomenclature Committee

Nomura M 1984 The control of ribosome synthesis. Sci Am 250: (Jan) 102–114

Nopajaroonsri C, Luk S C, Simon G T 1971 Ultrastructure of the normal lymph node. Am J Pathol 65: 1–24

Nopajaroonsri C, Luk S C, Simon G T 1974 The passage of intravenously injected colloidal carbon into lymph node parenchyma. Lab Invest. 30: 533–538

Norberg K-A 1967 Transmitter histochemistry of the sympathetic adrenergic nervous sytem. Brain Res 5: 125–170

Norberg K-A, Risley P L, Ungerstedt U 1967 Adrenergic innervation of the male reproductive ducts in some mammals. I. The distribution of adrenergic nerves. Z Zellforsch. 76: 278–286

Nordling J J, Ackles A A, Traynor F F 1981 The proliferative and toxic effects of ultraviolet light and inflammation on epidermal pigment cells. J Invest Dermatol 77: 361–368

Norris E H 1916 The morphogenesis of the follicles in the human thyroid gland. Am J Anat 20: 411–448

Norris E H 1938 The morphogenesis and histogenesis of the thymus gland in man: in which the origin of the Hassall's corpuscles of the human thymus is discovered. Contrib Embryol Carnegie Inst Wash 27: 191–207

Norton A C 1968 UCLA brain information service, updated review project-cutaneous sensory pathways: Dorsal Column-Medial Lemniscus System. University of California Press: Berkeley

Norvell J E 1968 The aorticorenal ganglion and its role in renal innervation, J Comp Neurol 133: 101–112

Nozaki M, Kobayashi H, Yanagisawa M et al 1975 Monoamine fluroescence in the median eminence of the Japanese quail, Coturnix coturnix japonica, following medial basal hypothalamic deafferentiation. Cell Tissue Res 164: 425–434

Nussbaum A 1912 Über das Gefassystem des Herzens, Arch Mikrosk Anat Entw Mech 80: 450–477

Nutt A B 1955 Significance and surgical treatment of congenital ocular palsies, Ann R Coll Surg 16: 30–59

Nyberg-Hansen R 1964a The location and termination of tectospinal fibers in the cat. Exp Neurol 9: 212–227

Nyberg-Hansen R 1964b Origin and termination of fibers from the vestibular nuclei descending in the medial longitudinal fasciculus. An experimental study with silver impregnation methods in the cat. J Comp Neurol 122: 355–367

Nyberg-Hansen R 1965a Anatomical demonstration of gamma moto-neurons in the cat's spinal cord. Exp Neurol 13: 71–81

Nyberg-Hansen R 1965b Sites and mode of termination of reticulo-spinal fibers in the cat. An experimental study with silver impregnation methods. J Comp Neurol 124: 71–100

Nyberg-Hansen R 1966a Functional organisation of descending supraspinal fibre systems to the spinal cord. Anatomical observations and physiological correlations. Ergebn Anat EntwGesch 39: Heft 2, 1–48

Nyberg-Hansen R 1966b Sites of termination of interstitiospinal fibers in the cat. An experimental study with silver impregnation methods. Archs Ital Biol 104: 98–111

Nyberg-Hansen R 1969 Cortico-spinal fibres from the medial aspect of the cerebral hemisphere in the cat. An experimental study with the Nauta method. Exp Brain Res 7: 120–132

Nyberg-Hansen R, Brodal A 1963 Sites of termination of cortico-spinal fibers in the cat. An experimental study with silver impregnation methods. J Comp Neurol 120: 369–391

Nyberg-Hansen R, Brodal A 1964 Sites and mode of termination of rubrospinal fibres in the cat. An experimental study with silver impregnation methods. J Anat 98: 235–253

Nyberg-Hansen R, Mascitti T 1964 Sites and mode of termination of fibers of the vestibulospinal tract in the cat. An experimental study with silver impregnation methods. J Comp Neurol 122: 369–387

Nyberg-Hansen R, Rinvik E 1963 Some comments on the pyramidal tract, with special reference to its individual variations in man. Acta Neurol Scand 39: 1–30

Nyby O, Jansen J 1951 An experimental investigation of the cortico-pontine projection in Macaca mulatta Skr Norske Vidensk-Akad 1: Mat-nat Kl No 3, 1–47

O

Oakes B W, Bialkower B 1977 Biomechanical and ultrastructural studies on the elastic wing tendon from the domestic fowl. J Anat 123: 369–387

Obayashi T 1959 Electron microscope study on lacrimal gland of normal rabbit. Acta Soc ophthalol Jpn, 63: 2631–2645

O'Beirne J 1833 New views of the process of defecation and their application to the pathology and treatment of diseases of the stomach, bowels and other organs; with an analytical correction of Sir Charles Bell's views respecting the nerves of the face. Dublin

Obersteiner H 1883 Der feinere Bau der Kleinhirnrinde beim Menschen und bei Tieren. biol Zentralbl 3: 145–155

Obletz B E, Halbstein B M 1938 Non-union of fractures of carpal navicular. J Bone Jt Surg 20: 424–428

Ochoa J 1971 The sural nerve of the human foetus: electron microscope observations and counts of axons. J Anat 108: 213–245

Ochoa J, Mair W E P 1969a The normal sural nerve in man. I. Ultrastructure and numbers of fibres and cells, Acta Neuropathol 13: 197–216

Ochoa J, Mair W E P 1969b The normal sural nerve in man. II. Changes in the axons and Schwann cells due to ageing. Acta Neuropathol 13: 217–239

Ockleford C D, Wakely J 1982 The skeleton of the placenta. Prog Anat 2: 19–47

Odensten M, Gillquist J 1985. Functional anatomy of the anterior cruciate ligament and a rationale for reconstruction. J Bone Jt Surg 67A: 257–262

Odgers P N B 1930 Some observations on the development of the ventral pancreas in man. J Anat 65: 1–7

Odgers P N B 1934 The formation of the venous valves, the foramen secundum and the septum secundum in the human heart. J Anat 69: 412–422

Odgers P N B 1937 An early human ovum (Thomson) in situ. J Anat 71: 161–168

Odgers P N B 1938 The development of the pars membranacea septi in the human heart. J Anat 72: 247–259

Odland G, Ross R 1968 Human wound repair. 1. Epidermal regeneration. J Cell Biol 39: 135–151

Odor D L 1960 Electron microscope studies on ovarian oocytes and unfertilised tubal ova in the rat. J Biophys Biochem Cytol 7: 567–574

Offer G, Moos C, Starr R 1973 A new protein of the thick filaments of vertebrate skeletal myofibrils. Extraction, purification and characterisation. J Mol Biol 74: 653–676

Ogawa T, Jefferson N C, Toman J E, Chiles T, Zambetoglou A, Necheles H 1960 Action potentials of accessory respiratory muscles in dogs. Am J Physiol 199: 569–572

Ogden J A 1974 Changing patterns of proximal femoral vascularity. J Bone Joint Surg 56A: 941–950

Ogilvie W H 1935 Some points in the operation of gastrectomy. Br Med J 1: 457–462

Ogura J H, Bellow J A 1952 Laryngectomy and radical neck dissection for carcinoma of the larynx, Laryngoscope, 62: 1–52

Oh C, Kark A E 1972 Anatomy of the external anal sphincter. Br J Surg 59: 717–723

Ohno S, Smith J B 1964 Role of fetal follicular cells in meiosis of mammalian oöcytes. Cytogenetics 3: 324–333

Ohta M, Offord K, Dyck P J 1974 Morphometric evaluation of first sacral ganglia of man, J Neurol Sci 22: 73–82

Ojemann G A, Fedio P, van Buren J M 1968 Anomia from pulvinar and subcortical parietal stimulation, Brain 91: 99–116

Okamoto E, Ueda T 1967 Embryogenesis of intramural ganglia of the gut and its relation to Hirschsprung's disease. J Pediatr Surg 2: 437–443

O'Keefe E J, Woodley D, Castillo G et al 1984 Production of soluble and cell-associated fibronectin by cultured keratinocytes. J Invest Dermatol 82: 150–155

Olds J 1976 Brain stimulation and the motivation of behavior. Prog Brain Res 45: 401–426

O'Leary J L, O'Leary J A 1966 Uterine artery ligature in the control of intractable post-partum hemorrhage. Am J Obstet Gynecol 94: 920–924

Olins A L, Carlson R D, Wright E B et al 1976 Chromatin nu bodies: isolation, subfractionation and physical characterization. Nucleic Acids Res 3: 3271–3291

Olivier G 1951 Anthropologie de la clavicule du francais. Bull Soc Anthropol Paris, 2: 4–6

Olivier G 1975 Biometry of the human occipital bone. J Anat 120: 507–518

Olsen B R 1981 Collagen biosynthesis. In: Hay E D (ed) Cell biology of extracellular matrix. Plenum Press: New York: pp 139–177

Olson L, Boreus L O, Seiger A 1973a Histochemical and mapping of 5-hydroxytryptamine and catecholamine containing neuron systems in the human fetal brain. Zanat Entwickl-Gesch 139: 259–282

Olson L, Nystrom B, Seiger A 1973b Monoamine fluorescence histochemistry of human post mortem brain. Brain Res 63: 231–247

Olsson Y 1969 Studies on vascular permeability in peripheral nerves. Acta Neuropath 17: 114–146

Olsson Y, Kristensson K 1971 Permeability of blood vessels and connective tissue sheaths in the peripheral nervous system to exogenous proteins. Acta Neuropathol (Berl) 5: Suppl, 5 61–69

Olszewski J 1954 The cytoarchitecture of the human reticular formation. In: Adrian E D, Bremer F, Jasper H H (eds) Brain mechanisms and consciousness. Blackwell: Oxford

Olszewski J, Baxter D 1954 Cytoarchitecture of the human brain stem. Karger: Basel

O'Malley B W, Schrader W T 1976 The receptors of steroid hormones. Sci Am 242: Feb 32–43

Oman C, Rosenbren, E, Sjöberg N O 1967 Adrenergic innervation of the human female reproductive organs: a histochemical and chemical investigation. Obstet Gynecol 30: 763–773

Onufrowicz B 1899 Notes on the arrangement and function of the cell groups in the sacral region of the spinal cord. J Nerv Ment Dis 26: 498–504

Ooé T 1956 On the development of position of the tooth germs in the human deciduous front teeth. Okajimas Folia Anat Jpn 28: 317–340

Ooé T 1957 On the early development of human dental lamina. Okajimas Folia Anat Jpn 30; 198–210

Oppel O 1963 Microscopic investigations of the number and caliber of the medullated nerve fibers of the optic fasciculus in man, Albrecht v Graefes Arch Ophthalmol 166: 19–27

Oppenheimer J H, Schwartz H L, Mariash C N, Kinlaw W B, Wong N C W, Freake H C 1987 Advances in our understanding of thyroid hormone action at the cellular level. Endocr Rev 8: 288–308

O'Rahilly R 1953 Survey of carpal and tarsal anomalies. J Bone Joint Surg 35-A: 626–642

O'Rahilly R 1956 Developmental deviations in the carpus and tarsus, Clin Orthop 10: 9–18

O'Rahilly R 1966 The early development of the eye in staged human embryos, Contrib Embryol Carnegie Inst Washington 38: 1–42

O'Rahilly R 1973 Developmental stages in human embryos. Part A: Embryos of the first three weeks (Stages 1–9). Carnegie Institution: Washington DC

O'Rahilly R 1975 The prenatal development of the human eye. Exp Eye Res 21: 92–112

O'Rahilly R, Boyden E A 1973 The timing and sequence of events in the development of the human respiratory system during the embryonic period proper. Z Anat Entwicklungsgesch 141: 237–250

O'Rahilly R, Gardner E 1971 The timing and sequence of events in the development of the human nervous system during the embryonic period proper. Z Anat Entwicklungsgesch 134: 1–12

O'Rahilly R, Gardner E 1972 The initial appearance of ossification in staged human embryos. Am J Anat 134: 291–301

O'Rahilly R, Gardner E 1975 The timing and sequence of events in the development of the limbs in the human embryo. Anat Embryol 148: 1–23

O'Rahilly R, Meyer D B 1959 The early development of the eye in the chick Gallus domesticus (stages 8 to 25) Acta Anat 36: 20–58

O'Rahilly R, Muecke E C 1972 The timing and sequence of events in the development of the human urinary system during the embryonic period proper. Z Anat Entwicklungsgesch 138: 99–109

O'Rahilly R, Tucker J A 1973 The early development of the larynx in staged human embryos. I Embryos of the first five weeks (to stage 15). Ann Otol Rhinol Laryngol 82: (suppl 7) 1–27

O'Rahilly R, Gardner E, Gray D J 1956 The ectodermal thickening and ridge in the limbs of staged human embryos. J Embryol Exp Morphol 4: 254–264

Orci L 1974 A portrait of a pancreatic B-cell. Diabetologia 10: 163–187

Orci L 1976 Morphofunctional aspects of the islets of Langerhans. The microanatomy of the islets of Langerhans. Metabolism 25: (Suppl 1) 1303–1313

Orci L, Unger R H 1975 Functional subdivision of islets of Langerhans and possible role of D-cells. Lancet 2: 1243–1244

Orci L, Like A A, Amherdt M, Blondell B, Kanazawa Y, Marliss E B, Lambert A E, Wollheim C B, Renold A E 1973 Monolayer cell culture of neonatal rat pancreas: an ultrastructural and biochemical study of functioning endocrine cells. J Ultrastruct Res 43: 270–297

Orci L, Baetens D, Dubois M P, Rufener C 1975 Evidence for the D-cell of the pancreas secreting somatostatin. Horm Metab Res 7: 400–402

Orci L, Baetens D, Ravazzola M, Malaisse-Lagae F, Amherdt M, Rufener C 1976 Somatostatin in the pancreas and gastrointestinal tract. In: Fujita T (ed) Endocrine gut and pancreas. Elsevier: New York: pp 73–78

O'Riordan J L H 1978 Hormonal control of mineral metabolism. In: O'Riordan J L H (ed) Recent advances in endocrinology and metabolism. Churchill Livingstone: Edinburgh: pp 217–217

Ormerod F C 1960 The physiology of the endolymph. J Laryngol Otol 74: 659–667

Orofino C, Sherman M S, Schechter D 1960 Luschka's joint–a degenerative phenomenon, J Bone Jt Surg 42A: 853–858

Ortmann R 1975 Use of polarized light for quantitative determination of the adjustment of the tangential fibres in articular cartilage, Anat Embryol (Berl) 148: 109–120

Ortner D J 1975 Aging effects on osteon remodelling, Calcif Tissue Res 17: 169–172

Orts-Llorca F, Puerta Fonolla J, Sobrado J 1982 The formation, septation and fate of the truncus arteriosus in man, J Anat: 134: 41–56

Orzalesi N, Riva A, Testa F 1971 Fine structure of human lacrimal gland. I. The normal gland, J Submicroscop Cytol 3: 283–298

Osborn J W 1973 Variations in structure and development of enamel. Oral Sci Rev 3: 3–83

Osborn J W 1985 The disc of the human temporomandibular joint: design, function and failure. J Oral Rehabil 12: 279–293

Oscarsson O 1965 Functional organisation of the spino- and cuneo-cerebellar tracts. Physiol Rev 45: 495–522

Oscarsson O 1967 Termination and functional organisation of a dorsal spino-olivocerebellar path. Brain Res 5: 531–534

Oscarsson O, Uddenberg N 1964 Identification of a spinocerebellar tract activated from forelimb afferents in the cat. Acta Physiol Scand 62: 125–136

BIBLIOGRAPHY

Osen K K 1969 The intrinsic organisation of the cochlear nuclei, Acta Otolaryngol 67: 352–359

Osen K K 1970 Coruse and termination of the primary afferents in the cochlear nuclei of the cat. An experimental anatomical study, Archstal Biol 108: 21–51

Osterberg G A 1935 Topography of the layers of the rods and cones in the human retina, Acta Ophthal, Suppl 6

Otsuka R, Hassler R 1962 On the striation and segmentation of the cortical center of vision in the cat, Arch Psychiat NervKrankh, 203: 212–234

Outrequin G, Caix M, Casanova G 1967 Variations du ligament triangulaire gauche du foie en fonction du type morphologique. CR Assoc Anat 136: 756–762

Owen R 1868 On the anatomy of vertebrates. Longmans, Green: London

Owen R L, Jones A L 1974 Epithelial cell specialization within human Peyer's patches. An ultrastructural study of intestinal lymphoid follicles, Gastroenterology 66: 189–203

Owen R L, Nemanic P 1978 Antigen processing structures of the mammalian intestinal tract: an SEM study of lymphoepithelial organs. Scanning Electron Microscopy. Vol 2. SEM Inc AMF O'Hare, Illinois 60666 USA

Owens P D 1972 Light microscopic observations on the formation of the layer of Hopewell-Smith in human teeth. Arch Oral Biol 17: 1785–1788

Owens P D 1975 patterns of mineralization in the roots of premolar teeth in dogs. Arch Oral Biol 20: 709–712

Owman Ch, Rosengren E, Sjöberg N-O, 1967 Adrenergic innervation of the human female reproductive organs: a histochemical and chemical investigation. Obstet Gynecol 30: 763–773

Ozzello L 1974 Electron microscopic study of functional and dysfunctional human mammary glands. J Invest Dermatol 63: 19–26

P

Pace J L 1968 Stereoscopic micro-anatomy of human colonic mucosa and its blood vessels. J Anat 103: 602P

Pacini A, Gremigni D 1975 Alcune modalità nella distribuzione del nervo mascellare. Arch Ital Anat Embriol 80: 29–35

Pack R J, Richardson P S 1984 The aminergic innervation of human bronchus: a light and electron microscopic study. J Anat 138: 493–502

Pack R J, Al-Ugaily L H, Morris G 1981 The cells of the trachiobronchial epithelium of the mouse: a quantitative light and electron microscope study. J Anat 132: 71–84

Padget D H 1948 The development of the cranial arteries in the human embryo. Contrib Embryol Carnegie Inst Washington 32: 205–261

Padget D H 1957 The development of the cranial venous system in man, from the viewpoint of comparative anatomy. Contrib Embryol Carnegie Inst Washington 36: 79–140

Padykula H A, Gauthier G F 1970 The ultrastructure of the neuromuscular junctions of mammalian red, white and intermediate skeletal muscle fibers. J Cell Biol 46: 27–41

Page R B, Bergland R M 1977 Pituitary vasculature. In: Allen M B, Makesh V B (eds) The pituitary. A current review. Academic Press: New York: pp 9–17

Page R B, Rosenstein J M, Dovey B J, Leure–duPree A E 1979 Ependymal changes in experimental hydrocephalus. Anat Rec 194: 83–103

Pakkenberg H 1966 The number of nerve cells in the cerebral cortex of man. J Comp Neurol 128: 17–20

Pal G P, Routal R V 1986 A study of weight transmission through the cervical and upper thoracic regions of the vertebral column in man. J Anat 148: 245–261

Pal, G P, Tamankar B P 1983 Preliminary study of age changes in Gujarati (Indian) pubic bones. Ind J Med Res 78: 694–701

Pal G P, Tamankar B P, Routal R V, Bhagwat SS 1984 The ossification of the membraneous part of the squamous occipital bone in man. J Anat 138: 259–266

Palade G 1975 Intracellular aspects of the process of protein synthesis. Science 189: 347–358

Palay S L, Chan-Palay V 1977 General morphology of neurons and neuroglia. In: Kandel E R (ed) Handbook of physiology. Volume 1 part 2. Physiological Society: Bethesda: pp 803–853

Palay S L, Palade G E 1955 The fine structure of neurons. J Biophys Biochem Cytol 1: 69–88

Palfrey A J, Davies D V 1966 The fine structure of chondrocytes. J Anat 100: 213–226

Palkovits M, Jacobowitz D M (1974) Topographic atlas of catecholamine and acetylcholinesterase-containing neurons in the rat brain. II. Hindbrain (mesencephalon rhombencephalon). J Comp Neurol 157: 29–42

Palkovits M, Magyar P, Szentágothai J 1971a Quantitative histological analysis of the cerebellar cortex in the cat. I. Number and arrangement in space of the Purkinje cells. Brain Res 32: 1–13

Palkovits M, Magyar P, Szentágothai J 1971b Quantitative histological anaylsis of the cerebellar cortex in the cat. II. Cell numbers and densities in the granular layers. Brain Res 31: 15–30

Palkovits M, Magyar P, Szentágothair J 1971c Quantitative histological analysis of the cerebellar cortex in the cat. III. Structural organization of the molecular layer. Brain Res 34: 1–18

Palkovits M, Magyar P, Szentágothai J 1972 Quantitative histological analysis of the cerebellar cortex in the cat. IV. Mossy fiber–Purkinje cell numerical transfer. Brain Res 45: 15–29

Pallie W, Manuel J K 1968 Intersegmental anastomoses between dorsal spinal rootlets in some vertebrates, Acta Anat 70: 341–351

Palmer R M, Lumsden A G S 1987 Development of periodontal ligament and alveolar bone in homografted recombinations of enamel organs and papillary, pulpal and follicular mesenchyme in the mouse. Arch Oral Biol 32: 281–289

Pandya P N, Kuypers H G J M 1969 Cortico-cortical connections in the rhesus monkey. Brain Res 13: 13–36

Pandya P N, Vignolo L A 1969 Interhemisperic projection of the parietal lobe in the rhesus monkey. Brain Res 106: 365–370

Pang S F 1985 Melatonin concentrations in blood and pineal gland. Vol 3. In: Reiter R J (ed) Pineal research reviews. Liss: New York: pp 115–159

Paniagua R, Amat P, Nistal M, Martin A 1986 Ultrastructure of Leydig cells in human ageing testes. J Anat 146: 173–183

Paniagua R, Nistal M 1984 Morphological and histometric study of human spermatogonia from birth to the onset of puberty. J Anat 139: 535–552

Paniagua R, Regadera J, Nistal M, Santamaría L 1983 Elastic fibres of the human ductus deferens. J Anat 137: 467–476

Papanicolau G N, Traut H F 1943 Diagnosis of uterine cancer by the vaginal smear. Commonwealth Fund: London

Papathanassion B T 1968 A variant of the motor branch of the median nerve in the hand. J Bone Jt Surg 50B: 156–157

Papez J W 1927 Subdivisions of the facial nucleus. J Comp Neurol 43: 159–191

Papez J W 1937 A proposed mechanism of emotion. Arch Neurol Psychiat 38: 725–743

Pappas G D, Purpura D P eds 1972 Structure and function of synapses. Raven Press: New York

Pappas G D, Tennyson V M 1962 An electron microscopic study of the passage of colloidal particles from the blood vessels of the ciliary processes and choroid plexus of the rabbit. J Cell Biol 15: 227–239

Parkes A S ed 1952–66 Marshall's physiology of reproduction. 3rd edition. Vol 1 part 1 1956. Vol 1 part 2 1960. Vol 2 1952. Vol 3 1966. Longmans Green: London

Parkin I G, Harrison G R 1985 The topographical anatomy of the lumbar epidural space. J Anat 141: 211–217

Parkinson D 1973 Carotid cavernous fistula: direct repair with preservation of the carotid artery. Technical note. J Neurosurg 38: 99–106

Parkinson D, Johnston J, Chaudhuri A 1978 Sympathetic connections to the fifth and sixth cranial nerves. Anat Rec 191: 221–226

Parry E W 1970 Some electron microscope observations on the mesenchymal structures of full-term umbilical cord. J Anat 107: 505–518

Parsons F G 1903 On the meaning of some of the epiphyses, J Anat 37: 315–323

Parsons F G 1904 Observations on traction epiphyses. J Anat 38: 248–258

Parsons F G 1905 On pressure epiphyses. J Anat 39: 402–412

Partlow G D, Colonnier M, Szabo J 1977 Thalamic projections of the superior colliculus in the rhesus monkey (Macaca mulatta). A light and electron microscope study. J Comp Neurol 171: 285–318

Patake S M, Mysorekar V R 1977 Diaphyseal nutrient foramina in human metacarpals and metatarsals. J Anat 124: 299–304

Paterson A M 1904 The human sternum, Williams and Norgate: London

Paterson A M 1912 The form of the human stomach. J Anat 47: 356–359

Patten B M 1956 The development of the sinuventricular conduction system. Univ Michigan Med Bull 22: 1–21

Patterson J F Jr 1946 Cervical ribs and scalenus anticus syndrome: review of literature and report of case. N Carolina Med J 7: 13–20

Pattle R E 1965 Surface lining of lung alveoli. Physiol Rev 45: 48–79

Pauerstein C J, Woodruff J D 1967 The role of the 'indifferent' cell of the tubal epithelium. Am J Obstet Gynecol 98: 121–125

Paul M, Kanagasuntheram R 1956 Congenital anomalies of lower urinary tract. Br J Urol 28: 64–74

Paula–Barbosa M M, Sousa–Pinto A 1973 Auditory cortical projections to the superior colliculus in the cat, Brain Res 50: 47–61

Paulo L G, Fink G D, Roh B L, Fischer J W 1973 Influence of carotid body ablation on erythropoietin production in rabbits. Am J Physiol 224: 442–444

Pauwels F 1965 Gesammelte abhandlungen zur funktionellen Anatomie des Bewegungsapparates. Springer-Verlag: Berlin

Pavlov S, Petrov V 1968 Sur l'anse sous-clavière de l'artère sous-clavière droite rétro–oesophagienne. Folia Med Plovdiv 10: 73–78

Payton C G 1934 The position of the nutrient foramen and direction of the nutrient canal in the long bones of the madder–fed pig, J Anat 68: 500–510

Peacock A 1951 Observations on the pre-natal development of the intervertebral disc in man. J Anat 85: 260–274

Peacock A 1952 Observations on the postnatal structure of the intervertebral disc in man. J Anat 86: 162–179

Pearce G W 1960 Some cortical projections to the midbrain reticular formation. In: Tower D B, Schadé J P (eds) Structure and function of the cerebral cortex. Elsevier: Amsterdam: pp 131–137

Pearse A G E 1966a The cytochemistry of the thyroid C cells and their relationship to calcitonin. Proc R Soc Lond [Biol] 164: 478–487

Pearse A G E 1966b 5-Hydroxtryptophan uptake by dog thyroid 'C' cells and its possible significance in polypeptide hormone production. Nature 211: 598–600

Pearse A G E 1968 Histochemistry. 3rd edition, Churchill: London

Pearse A G E 1969 The cytochemistry and ultrastructure of polypeptide hormone-producing cells (the APUD series) and the embryologic, physiologic and pathologic implications of the concept. J Histochem Cytochem 17: 303–313

Pearse A G E 1974 The gut as an endocrine organ. Br J Hosp Med 11: 697–704

Pearse A G E 1976 Neurotransmission and the APUD concept. In: Coupland R E, Fujita T (eds) Chromaffin, enterochromaffin and related cells. Elsevier: New York: pp 147–154

Pearse A G E 1977a The apudomas; with particular reference to those of gastroenteropancreatic origin. In: Yardley J H, Morson B C, Abell M R (eds) The gastrointestinal tract. Williams & Wilkins: Baltimore: pp 206–218

Pearse A G E 1977b The diffuse neuroendocrine system and the APUD concept: related 'endocrine' peptides in brain, intestine, pituitary, placenta and anuran cutaneous glands. Med Biol 55: 115–125

Pearse A G E 1980 The APUD concept and hormone production. Clin Endocrinol Metab 9: 211–222

Pearse A G E, Polak J M 1971 Cytochemical evidence for the neural crest origin of mammalian C cells. Histochemie 37: 96–102

Pearse A G E, Polak J M 1974 Endocrine tumours of neural crest origin: neurolophomas, apudomas and the APUD concept. Med Biol 52: 3–18

Pearse A G E, Polak J M 1978 The diffuse neuroendocrine system and the APUD concept. In: Bloom S R (ed) Gut hormones. Churchill Livingstone: Edinburgh

Pearse A G E, Polak J M, Bloom S R et al 1974 Enterochromaffin cells of the mammalian small intestine as a source of motilin. Virchows Arch Abt B Zell Path 16: 111–120

Pearse B M 1982 Structure of coated pits and vesicles. (Ciba Foundation Symposium 92. Pitman: London: pp 246–265

Pearse B M F, Bretscher M S 1981 Membrane recycling by coated vesicles. Annu Rev Biochem 50: 85–101

Pearson A A 1941 The development of the nervus terminalis in man. J Comp Neurol 75: 39–66

Pearson A A 1944 The oculomotor nucleus in the human fetus. J Comp Neurol 80: 47–68

Pearson AA 1949 The development and connections of the mesencephalic root of the trigeminal nerve in man. J Comp Neurol 90: 1–46

Pearson A A 1977 The early innervation of the developing deciduous teeth. J Anat 123: 563–577

Pearson A A, Sauter R W 1971a The internal thoracic (mammary) nerve. Thorax 26: 354–356

Pearson A A, Sauter R W 1971b Observations on the caudal end of the spinal cord. Am J Anat 131: 463–470

Pearson J, Goldstein M, Markey K, Brandeis L 1983 Human brainstem catecholamine neuronal anatomy as indicated by immunocytochemistry with antibodies to tyrosine hydroxylase. Neuroscience 8: 3–32

Pearson R C A, Powell T P S 1978 The cortico-cortical connections to area 5 of the parietal lobe from the primary somatic sensory cortex of the monkey. Proc R Soc Lond [Biol] 200: 103–108

Pease D C 1956 An electron microscopic study of red bone marrow. J Hematol 11: 501–526

Pease D C, Quilliam T A 1957 Electron microscopy of the Pacinian corpuscle. J Biophys Biochem Cytol 3: 331–342

Pech P, Haughton V M 1985 Lumbar intervertebral disk: correlative MR and anatomic study. Radiology 156: 699–701

Peck H M, Hoerr N L 1951 The effect of environmental temperature changes on the circulation of the mouse spleen. Anat Rec 109: 479–494

Pedersen H 1969 Ultrastructure of the ejaculated human sperm. Z Zellforsch 94: 542–554

Pedley R D, Harrison F pre-1936 Our teeth. Blackie: Glasgow

Pelletier G, Robert F, Hardy H 1978 Identification of human anterior pituitary cells by immunoelectron microscopy. J Clin Endocrinol Metab 46: 534–542

Pembrey M E, Winter R M, Davies K E 1986 Fragile X mental retardation: current controversies. Trends Neurosci 9: 58–62

Penfield W 1957 Vestibular sensation and the cerebral cortex. Ann Otol Rhinol Laryngol 66: 691–698

Penfield W 1966 Speech, perception and the uncommitted cortex. In: Eccles J C (ed) Brain and conscious experience. Springer-Verlag: Berlin: pp 217–237

Penfield W, Jasper H 1954 Epilepsy and the functional anatomy of the human brain. Little, Brown: Boston Massachusetts

Penfield W, McNaughton F 1940 Dural headache and innervation of the dura mater. Arch Neurol Psychiat 44: 43–75

Penfield W, Rasmussen T 1950 The cerebral cortex of man, Macmillan: New York

Penfield W, Roberts L 1959 Speach and Brain Mechanisms, Princeton University Press: Princeton NJ

Penfield W, Welch K 1951 The supplementary motor area of the cerebral cortex. A clinical and experimental study. Arch Neurol Psychiat 66: 289–317

Pennell T C 1966 Anatomical study of the peripheral pulmonary lymphatics. J Thorac Cardiovasc Surg 52: 629–634

Penrose L S, Loesch D 1969 Dermatoglyphic sole patterns: a new attempt at classification. Hum Biol 41: 427–448

Pénzes L, Regius O 1985 Changes in the intestinal microvillous surface area during reproduction and ageing in the female rat. J Anat 140: 389–396

Percheron G 1977 Les artères du thalamus humain. Les artères choroïdiennes. I–V, Rev Neurol 133: 533–558

Perera H, Edwards F R 1957 Intradiaphragmatic course of the left phrenic nerve in relation to diaphragmatic incisions. Lancet 2: 75–77

Pernkopf E 1963 Atlas of topographical and applied human anatomy, Saunders: Philadelphia

Perren S M, Huggler A, Russenberger M et al 1969 The reaction of cortical bone to compression. Acta Orthop Scand Suppl 125 Suppl 19–29

Perry J S, Crombie P R 1982 Ultrastructure of the uterine glands of the pig. J Anat 134: 339–350

Persaud T V N 1977 Problems of birth defects. From Hippocrates to thalidomide and after. MTP Press: Lancaster

Person R S, Roshtchina N A 1958 Elektromiograficheskow issledovanie koordinatsil deiatel'nostimyshtsantagonistov pri dvizhenii pal'tsev ruti cheloveka. [Electromyographic investigation of coordinated activity of antagonistic muscles of fingers of the human hand.] Fizol Zh SSSR 44: 455–462

Perutz M F, Rossmann M G, Cullis A F, Muirhead H, Will G, North A C T 1960 Structure of haemoglobin. A three-dimensional Fourier synthesis at 5.5 Å resolution, obtained by X-ray analysis, Nature, 185: 416–422

Peters A, Palay S L 1966 The morphology of the laminae A and A_1 of the dorsal nucleus of the lateral geniculate body of the cat, J Anat 100: 451–486

Peters A, Palay S L, Webster H deF 1976 The fine structure of the nervous system. The neurons and supporting cells, 2nd edition. Saunders: Philadelphia

Peters H 1899 Ueber die Einbettung des menschlichen Eies und das fruheste bisher bekannte menschliche Placentationsstadium Deutcke: Leipzig

Peters R M 1969 The mechanical basis of respiration. Churchill: London

Peterson M, Leblond C P 1964 Synthesis of complex carbohydrates in the Golgi region, as shown by radioautography after injection of labelled glucose. J Cell Biol 21: 143–148

Peterson M R, Leblond C P 1964 Uptake by the Golgi region of glucose labelled with tritium in the 1 or 6 position, as an indicator of synthesis of complex carbohydrates. Exp Cell Res 34: 420–423

Petkov P 1968 Anatomopotographic investigations on the anterior hepatic nerve plexus. Scripta Scient Med Varna 7: 95–97

Petras J M 1977 Spinocerebellar neurons in the rhesus monkey. Brain Res 130: 146–151

Petras J M, Cummings J F 1977 The origin of spinocerebellar pathways: II. The nucleus centrobasalis of the cervical enlargement and the nucleus dorsalis of the thoracolumbar spinal cord. J Comp Neurol 173: 693–716

Petrovic A G 1972 Mechanisms and regulation of mandibular condylar growth, Acta Morphol Neerl Scand 10: 25–34

Pettigrew J 1864 On the arrangement of the muscular fibres in the ventricles of the vertebrate heart: with Physiological remarks, London

Pfeifer R A 1940 Die angioarchtecktonische areale Gliederung der Grosshirnrinde. Thieme: Leipzig

Pfenninger K H, Rees R P 1976 From the growth cone to the synapse: properties of membranes involved in synapse formation. In: Barondes S H (ed) Neuronal Recognition. Plenum: New York: pp 131–178

Pfaffman C 1970 Physiological and behavioural processes of the sense of taste. In: Wolstenholme G E W, Knights J (eds) Taste and smell in vertebrates. Ciba Foundation Symposium. Churchill: London: pp 31–44

Pfenninger K H, Rees R P 1976 From the growth cone to the synapse: properties of membranes involved in synapse formation. In: Neuronal recognition (Barondes S H ed) pp 131–78, Plenum: New York

Pflug H D, Jaeschke-Boyer H, Sattler E L 1979 Microscop Acta, 82: 255

Pflüger E 1866 Ueber die Epithelien der glandula submaxillaris. Zbl Med Wiss 4: 193–195

Phifer R F, Midgley A R, Spicer S S 1973 Immunologic and histologic evidence that follicle stimulating and luteinizing hormones are present in the same cell types in the human pars distalis. J Clin Endocr Metab 36: 125–141

Phillips C G 1981 Microarchitecture of the motor cortex of primates. Prog Anat 1: 61–94

Phillips C G, Porter R 1977 Corticospinal neurones. Their role in movement. (Monographs of the Physiological Society No 34) Academic Press: London

Phillips D M 1975 Mammalian sperm structure. In: Hamilton D W, Greep R O (eds) Handbook of physiology, section 7. Endocrinology. Vol V. American Physiological Society: Washington DC: pp 405–420

Phillips D M, Olson G 1975 Mammalian sperm motility-structure in relation to function. In: Afzelius B A (ed) The functional anatomy of the spermatozoon. Proceedings of the Second International Symposium.

Wenner–Cren Center, Stockholm 1973. Pergamon: Oxford: pp 117–126

Philipson T, Hanninen L, Balazs E A 1975 Cell contacts in human and bovine lenses. Exp Eye Res 21: 205–219

Piasecka–Kacperska K, Gladyskowska-Rzeczycka J 1972 Splot krzyżowy u naczelnych. Folia Morphol 31: 21–33

Piasecki C 1974 Blood supply to the human gastro–duodenal mucosa. J Anat 118: 295–335

Piasecki C 1977 Role of ischaemia in the initiation of peptic ulcer. Ann R Coll Surg 59: 476–478

Piasecki C, Wyatt C 1986 Patterns of blood supply to the gastric mucosa. A comparative study revealing an end-artery model. J Anat 149: 21–39

Pick J 1970 The autonomic nervous system, Lippincott: Philadelphia

Pick J, Sheehan D 1946 Sympathetic rami in man. J Anat 80: 12–20

Pick J W, Anson B J, Ashley F L 1942 The origin of the obturator artery. A study of 640 body halves. Am J Anat 70: 317–344

Pickel V M, Joh T H, Reis D J, Leeman S E, Miller R J 1979 Electron microscope localization of substance P and enkephalin in axon terminals related to dendrites of catecholaminergic neurons. Brain Res 160: 387–400

Pickel V M, Segal M, Bloom F E 1974 A radioautographic study of the efferent pathways of the nucleus locus coeruleus. J Comp Neurol 155: 15–42

Pickett-Heaps J D 1975 Aspects of spindle evolution. Ann N Y Acad Sci 253: 352–361

Pieron A P 1973 The mechanism of the first carpo-metacarpal joint. Acta Orthop Scand Suppl 148:

Pierson R, Carpenter M B 1974 Anatomical analysis of pupillary reflex pathways in the rhesus monkey. J Comp Neurol 158: 121–144

Pikó L 1969 Gamete structure and sperm entry in mammals. In: Metz C B, Monroy A (eds) Fertilization. Comparative morphology, biochemistry and immunology. Vol 2. Academic Press: New York: pp 325–403

Pikó L, Tyler A 1964 Fine structural studies of sperm penetration in the rat. Proc Vth Congr Internazionale per la Riproduzione Animale e la Fecundacione Artificale, Trento, section 1, volume II, 372–377

Pimendis M Z, Hinds J W 1977 An autoradiographic study of the sensory innervation of teeth. I. Dentin, II. Dental pulp and periodontium. J Dent Res 56: 827–834; 835–840

Pin C, Jones B, Jouvet M 1968a Les neurones contenant des monoamines du tronc cerebral du chat. I. Etude topographique par histofluorescence et histochimie. J Physiol (Paris), 60: suppl 2, 519

Pin C, Jones B, Jouvet M 1968b Topographie des neurones monoaminergiques du tronc cérébral du chat: étude par histoflourescence. C R Soc Biol (Paris). 162: 2136–2141

Pinching A J, Powell T P S 1971 The neuron types of the glomerular layer of the olfactory bulb. J Cell Sci 9: 305–346

Pinkerton J H M, McKay D G, Adams E C, Hertig A T 1961 Development of the human ovary–a study using histochemical technics. Obstet Gynec 18: 152–181

Pinkus H 1958 Embryology of hair. In: Montagna W, Ellis R A (eds) The biology of hair growth. Academic Press: New York: pp 1–32

Pinto da Silva P, Kachar B 1982 On tight-junction structure. Cell 28: 441–450

Pirenne M 1967 Vision and the eye, 2nd edition, Chapman and Hall: London

Pismenov I A, Zapetski E V 1977 Regularities and distinctions in the structure of the circulation bed of the sternum. Arkh Anat Gistol Embriol 72: 61–67

Pitelka D R, Hamamoto S T 1977 Form and function in mammary epithelium: the interpretation of ultrastructure. J Dairy Sci 60: 643–654

Pitelka D R, Hamamoto S T, Duofala J G, Nemanic M K 1973 Cell contacts in the mouse mammary gland. 1. Normal gland in postnatal development and the secretory cycle. J Cell Biol 56: 797–818

Pittman F E, Pittman J C 1966 An electron microscopic study of the epithelium of normal human sigmoid colonic mucosa. Gut 7: 644–661

Platz F, Adelmann G 1976 Zur Anatomie der 'Vena arcuata cruris posterior', und ihrer Tiefen–anastomosen (Vv communicantes sive perforantes). Verh Anat Ges Jena 70: 709–714

Platzer W, Maurer H 1966 Zur Segmenteinteilung der Leber. Acta Anat 63: 8–31

Plentl A A 1958 The origin of amniotic fluid. In: Villee C A (ed) Gestation transactions of conference. 5th and final. Josiah Macy Jr Foundation: New York: pp 71–114

Plets C 1969 The arterial blood supply and angioarchitecture of the posterior wall of the third ventricle. Acta Neurochir 21: 309–317

Plets C, de Reuck J, Vander Eecken H, Van Den Bergh R 1970 The vascularization of the human thalamus. Acta Neurol Belg 70: 687–770

Poggio G F, Mountcastle V B 1960 A study of the functional contributions of the lemniscal and spinothalamic systems to somatic sensibility. Bull Johns Hopkins Hosp 106: 266–316

Poidevin L O S 1959 Striae gravidarum: their relation to adrenal cortical hyperfunction. Lancet 2: 436–439

Poirier L J, Bertrand C 1955 Experimental and anatomical investigations of lateral spinothalamic and spinotectal tracts. J Comp Neurol 102: 745–757

Poiriraer N P, Chernikov Y F 1965 Mat Teoret Klin Med 5:

Poláček P 1961 Relation of myocardial bridges and loops on the coronary arteries to coronary occlusion. Am Heart J 61: 44–62

Poláček P, Halata Z 1970 Development of simple encapsulated corpuscles in the nasolabial region of the cat. Ultrastructural study. Folia Morphol (Praha), 18: 359–368

Polak J M, Bloom S R, Hobbs S, Solcia E, Pearse A G E 1976 Distribution of a bombesin-like peptide in human gastrointestinal tract. Lancet 1: 1109–1110

Polak J M, Buchan A M J, Czykowska W, Solcia E, Bloom S R, Pearse A G E 1978 Bombesin in the gut. In: Bloom S R (ed) Gut hormones. Churchill Livingstone: Edinburgh: pp 541–543

Polak J M, Buchan A M, Probert L, Tapia F, de Mey J, Bloom S R 1980 Regulatory peptides in endocrine cells and autonomic nerves: electron immunocytochemistry. Scand J Gastroenterol (Suppl) 1981 70: 11–23

Polani P 1973 The incidence of developmental and other genetic abnormalities. Guy's Hosp Rep 122: 53–63

Polge C 1957 Low–temperature storage of mammalian spermatozoa. Proc R Soc B 147: 498–508

Policard A 1950 Sur quelques caractères histophysiologiques des formations lymphoides brochiques. Bull Hist Appl 27: 118

Politzer G 1952 Zur normalen und abnormen Entwicklung des menschlichen Gesichtes. Z Anat Entwicklungsgesch 116: 332–346

Polóyni J, Kapeller K, Mráz P 1977 SGG (SIF) cells in autonomic ganglion: evidence for a possible secretion of their contents into the blood vessels. Z Mikrosk Anat Forsch 91: 581–589

Polyak S 1941 The retina. University of Chicago Press: Chicago

Polyak S 1957 The vertebrate visual system, University of Chicago Press: Chicago

Pomeranz B, Wall P D, Weber W V 1968 Cord cells responding to fine myelinated afferents from viscera, muscle and skin. J Physiol 199: 511–532

Pomerat C M, Hendelman W J, Raiborn C W Jr, Massey J F 1967 Dynamic activities of nervous tissue in vitro. In: Hydén H (ed) The neuron. Elsevier: Amsterdam: pp 119–178

Pompeiano O, Brodal A 1957b The origin of vestibulospinal fibres in the cat. An experimental-anatomical study, with comments on the descending medial longitudinal fasciculus. Arch Ital Biol 95: 166–195

Pontes C, Reis F F, Sousa-Pinto A 1975 The auditory cortical projections on to the medial geniculate body in the cat. Brain Res 91: 43–63

Popa G T, Fielding U 1930a A portal circulation from the pituitary to the hypothalamic region. J Anat 65: 88–91

Popa G T, Fielding U 1930b Vascular link between pituitary and hypothalamus. Lancet 2: 238–240

Popescu L M, Diculescu I 1975 Calcium in smooth muscle sarcoplasmic reticulum in situ. Conventional and X-ray analytical electron microscopy. J Cell Biol 67: 911–918

Popoff N W 1934 Digital vascular system, with reference to the state of the glomus in inflammation, arteriosclerotic gangrene, diabetic gangrene, thrombo-angitis obliterans and supernumerary digits in man. Archs Path 18: 295–330

Popper K R, Eccles J C 1977 The self and its brain, Springer International: Berlin

Porter R 1981 Internal organization of the motor cortex for input–output arrangements. Section 1. The nervous system. In: Brooks V B (ed) Handbook of physiology. Vol 2. Motor control. Part 2. American Physiological Society: Bethesda Maryland: pp 1063–1081

Porter R 1985 The cerebral cortex and control of movement performance. In: Swash M, Kennard C (eds) Scientific Basis of Clinical Neurology. Churchill Livingstone: Edinburgh: pp 19–35

Porteous C J 1960 The olecranon epiphyses. J Anat 94: 286P

Portmann A 1950 In: Grassé P P (ed) Traité de Zoologie. Masson: Paris

Posselt U 1952 Studies in mobility of human mandible. Acta Odont Scand 10: Suppl 10: 3–160

Poste G, Nicolson G L eds 1977 Dynamic aspects of cell surface organization. Cell surface reviews. Vol 3. Elsevier: Amsterdam

Poston T, Stewart I 1978 Catastrophe theory and its applications. Pitman: London

Potenza A D 1963 Critical evaluation of flexor-tendon healing and adhesion formation within artificial digital sheaths. An experimental study. J Bone Jt Surg 45A: 1217–1233

Potten C S 1974 The epidermal proliferative unit: the possible role of the central basal cell. Cell Tissue Kinet 7: 77–88

Potten C S 1983 Stem Cells: Their Identification and Characterization. Churchill Livingstone: Edinburgh

Potten C S, Schofield R, Lajtha L G 1979 A comparison of cell replacement in bone marrow, testis and three regions of surface epithelium. Biochim Biophys Acta 560: 281–299

Potts T K 1925 The main peripheral connections of the human sympathetic nervous system. J Anat 59: 129–135

Potts T R Jr, Murray T, Peacock M, Niall H D, Tregear G W, Keutmann H T, Powell D, Deftos L J 1971 Parathyroid hormone: sequence, synthesis, immunosassay studies. Am J Med 50: 639–649

Powell T P S 1981 Certain aspects of the intrinsic organisation of the cerebral cortex. In: Pompeiano O, Marsan C A (eds) Brain mechanisms of perceptual awareness and purposeful behaviour. Raven Press: New York: pp 1–19

Powell T P S, Cowan W M 1956 A study of thalamostriate relations in the monkey. Brain 79: 364–390

Powell T P S, Cowan W M 1963 Centrifugal fibres in the lateral olfactory tract. Nature 199: 1296–1297

Powell T P S, Mountcastle V 1959a The cytoarchitecture of the postcentral gyrus of the monkey Macaca mulatta. Bull Johns Hopkins Hosp 105: 108–131

Powell T P S, Mountcastle V B 1959b Some aspects of the functional organisation of the cortex of the postcentral gyrus of the monkey. A correlation of findings obtained in a single unit analysis with cytoarchitectonics. Bull Johns Hopkins Hosp 105: 173–200

Powell T P S, Hendrickson A E 1981 Similarity in number of neurons through the depth of the cortex in the binocular and monocular parts of area 17 of the monkey. Brain Res 216: 409–413

Powell T P S, Cowan W M, Raisman G 1965 The central olfactory connexions. J Anat 99: 791–813

Präder A 1947 Die frühembryonale Entwicklung der menschlichen Zwischenwirbelscheibe. Acta Anat 3: 68–83

Prakash S, Chopra S R K, Jit I 1979 Ossification of the human patella. J Anat. Ind 28: 78–83

Prescott D M 1976 Reproduction of eukaryotic cells. Academic Press: New York

Pressman J J 1942 Physiology of vocal cords in phonation and respiration. Archs Otolaryngol 35: 355–398

Prestige M C 1965 Cell turnover in the spinal ganglia of Xenopus laevis tadpoles. J Embryol Exp Morphol 13: 63–72

Prestige M C 1967 The control of cell number in the lumbar spinal ganglia during the development of Xenopus laevis tadpoles. J Embryol Exp Morphol 17: 453–471

Prestige M C 1970 Differentiation, degeneration, and the role of the periphery: quantitative considerations. In: Schmitt F O, Quarton G C, Melnechuk T, Adelman G (eds) The neurosciences. Second study program. Rockefeller University Press: New York: pp 73–82

Preston J B, Whitlock D G 1961 Intracellular potentials recorded from motoneurons following precentral gyrus stimulation in primates. J Neurophysiol 24: 91–100

Price J L 1968 The termination of centrifugal fibres in the olfactory bulb. Brain Res 7: 483–486

Price J L, Powell T P S 1970a The morphology of the granule cells of the olfactory bulb. J Cell Sci 7: 91–123

Price J L, Powell T P S 1970b The synaptology of the granule cells of the olfactory bulb. J Cell Sci 7: 125–156

Printz R H, Hall J L 1974 Evidence for a retinohypothalamic pathway in the golden hamster. Anat Rec 179: 57–66

Prinzmetal M, Simkin B, Bergman H C, Kruger H E 1947 Studies on the coronary circulation. Am Heart J 33: 420–442

Pritchard H, Micklem H S 1973 Haimopoietic stem cells and progenitors of functional T–lymphocytes in the bone marrow of 'nude' mice. Clin Exp Immunol 14: 597–607

Pritchard J J, Scott J H, Girgis F G 1956 The structure and development of cranial and facial sutures. J Anat 90: 73–87

Privat A 1975 Postnatal gliogenesis in the mammalian brain. Int Rev Cytol 40: 281–323

Prockop D J, Kivirikko K I, Tuderman L, Guzman N A 1979 The biosynthesis of collagen and its disorders (first of two parts). N Engl J Med 301: 13–23

Proctor D, Fletcher R D, Del Negro A A 1978 Temporary cardiac pacing: causes, recognition, and management of failure to pace. Nurs Clin North Am 13: 409–422

Pšenicka P 1966 Beitrag zur Kenntnis der Innervation der Kehlkopfgelenke. Anat Anz 118: 1–6

Puchades–Orts A, Nombela-Gomez M, Ortuño–Pacheco G 1976 Variation in form of the circle of Willis. Some anatomical and embryological considerations. Anat Rec. 185: 119–123

Pulec J L, Kamio T, Graham M D 1975 Eustachian tube lymphatics. Ann Otol Rhinol Otolaryngol 84: 483–492

Purkinje J E 1837 Ueber die gangliösen körperchen in verschiedenen Theilen des Gehirns. Ber Versamml Dtsch Naturf Aerzte Prague 1837, 1838, 15: 179–180

Purpura D P, Schadé J P 1964 Growth and maturation of the brain, Elsevier: Amsterdam

Purpura D P, Yahr M D 1966 The Thalamus Columbia University Press: New York

Purpura D P, Frigyesi, T L, Malliani A 1967 Intrinsic synaptic organization and relations of the corpus striatum. In: Yahr M D, Purpura D P (eds) Neurophysiological basis of normal and abnormal motor activities. Raven Press: New York: pp 177–214

Purves D, Hadley R D, Voyvodic J T 1986 Dynamic changes in the dendritic geometry of individual neurons visualized over periods of up to three months in the superior cervical ganglion of living mice. J Neurosci 6: 1051–1060

Purves H D 1961 Morphology of the hypophysis related to its function. In: Young W C (ed) Sex and internal secretion. William & Wilkins: Baltimore: pp 161–238

Purves M J 1970 The role of the cervical sympathetic nerve in the regulation of oxygen consumption of the carotid body of the cat. J Physiol 209: 417–431

Purves M J 1972 The physiology of the cerebral circulation, Cambridge University Press: Cambridge

Püschell J 1930 Wassergehalt normaler und degenerierter Zwischenwirbelscheiben, Beitr Pathol Anat 84: 123–130

Puschett J B 1978 Renal tubular effects of parathyroid hormone. An update. Clin Orthop 135: 249–259

Putschar W 1931 Entwicklung, Wachstum und Pathologie der Beckenverbindungen des Menschen mit besonderer Berücksichtigung von Schwangerschaft, Geburt und ihren Folgen. Aus dem Pathologischen Institute der Universität Göttingen, Fischer Verlag: Stuttgart

Putz R 1976 Zur morphologie und Rotationsmechanik der kleinen Gelenke der Ledenwirbel. Z Orthop 114: 902–912

Putz R 1985 The functional morphology of the superior articular processes of the lumbar vertebrae. J Anat 143: 181–187

Pyle S I, Hoerr N L 1955 Radiographic atlas of skeletal development of the knee, Thomas: Springfield Illinois

Pyrtek L J, Painter R L 1964 An anatomic study of the relationship of the parathyroid glands to the recurrent laryngeal nerve. Surgery Gynecol Obstet 119: 509–512

Q

Quain J 1908–29 Elements of descriptive and practical anatomy, 11th edition. Longmans, Green: London

Quain R 1844 Anatomy of the arteries of the human body, with its applications to pathology and operative surgery. London, 1844

Quay W B 1984 Pineal function in mammals: a reconsideration. Vol 2. In: Reiter R J (ed) Pineal research reviews. Liss: New York: pp 87–112

Quesenberry P, Levitt L 1979 Hemapoietic stem cells. N Engl J Med 301: 755–761

Quilliam T A 1966 Unit design and array patterns in receptor organs. In: de Reuck A V S, Knight J (eds) Touch, heat and pain. (Ciba Foundation Symposium). Churchill: London: pp 86–112

Quintarelli G, Dellovo M C 1966 Age changes in the localization and distribution of glycosaminoglycans in human hyaline cartilage. Histochemie, 7: 141–167

Qvist G 1977 The course and relations of the left phrenic nerve in the neck. J Anat 124: 803–805

R

Rabey G P 1968 Morphanalysis. Lewis: London

Rabey G P 1971 Craniofacial morphanalysis. Proc R Soc Med 64: 103–111

Rabey G P 1977 Morphanalysis of craniofacial dysharmoney. Br J Oral Surg. 15: 110–120

Rabinowicz T 1964 The cerebral cortex of the premature infant of the 8th month. Prog Brain Res 4: 39–92

Rabinowicz T 1967 Quantitative appraisal of the cerebral cortex of the premature infant of 8 months. In: Minkowski A (ed) Regional Development of the brain in early life. A symposium organized by the Council for International Organizations of Medical Sciences, [etc]. Blackwell Scientific: Oxford: pp 91–124

Race R R, Sanger R 1975 Blood groups in man. 6th edition. Blackwell Scientific: Oxford

Radnót M, Lovas B 1968 Data on the ultrastructure of Miller's cells of the retina. Archs Soc Am Opthalmol Optom 6: 393–404

Raff M C, Mirsky R, Fields K L, Lisak R P, Dorfman S H, Silverberg D H, Gregson N A, Leibowitz S, Kennedy M C 1978 Galactocerebroside is a specific cell-surface antigenic marker for oligodendrocytes in culture. Nature, 274: 813–816

Rafferty N S, Esson E A 1974 An electron microscope study of adult mouse lens. J Ultrastruct Res 46: 239–253

Rafferty N S, Goosens W 1978 Cytoplasmic filaments in the crystalline lens of various species. Exp Eye Res 26: 177–190

Raisman G 1966 Neural connections of the hypothalamus. Br Med Bull 22: 197–201

Raisman G, Cowan W M, Powell T P S 1965 The extrinsic afferent, commissural and association fibers of the hippocampus. Brain 88: 963–996

Raisman G, Cowan W M, Powell T P S 1966 An experimental anaalysis of the efferent projection of the hippocampus. Brain 89: 83–108

Rajendran K 1985 Mechanism of locking at the knee joint. J Anat 143: 189–194

Rakic P 1971a Guidance of neurons migrating to the foetal monkey neocortex. Brain Res 33: 471–476

Rakic P 1971b Neuronglia relationship during granule cell migration in developing cerebellar cortex. A Golgi and electron microscopic study in Macacus rhesus. J Comp Neurol 141: 283–312

Rakic P 1975 Local circuit neurons. Neurosci Res Program Bull 13: 295–416

Rakic P 1976 Prenatal genesis of connexions subserving ocular dominance in the rhesus monkey. Nature, 261: 467–471

Rakic P 1977 Genesis of the dorsal lateral geniculate nucleus in the rhesus monkey. J Comp Neurol 176: 23–52

Rakic P 1981 Neuronal-glial interaction during brain development. TINS 4: 184–187

Rakic P 1982 Early developmental events: cell lineages, acquisition of neuronal positions, and areal and laminar development. Neurosci Res Program Bull 20: 439–451

Rakic P 1984a Organizing principles for development of primate cerebral cortex. In: Sharma S C (ed) Organizing principles of neural development. Plenum Press: New York: pp 21–48

Rakic P 1984b Emergence of neuronal and glial cell lineages in primate brain. In: Black I B (ed) Cellular and molecular biology of neuronal development. Plenum Press: New York: pp 29–50

Rakic P, Goldman-Rakic P S 1982 Development and modifiability of the cerebral cortex. Based on an NRP Work Session . . . 1980, etc. Overview. Neurosci Res Program Bull 20: 433–438

Rakic P, Yakovlev P I 1968 Development of the corpus callosum and cavum septi in man. J Comp Neurol 132: 45–72

Rall W 1977 Core conductor theory and cable properties of neurons. In: Kandel E R (ed) Handbook of physiology. Section 1. Vol 1 pt 1. American Physiological Society: Bethesda Maryland: pp 39–97

Rall W, Shepherd G M, Reese T, Brightman M W 1966 Dendrodendritic synaptic pathway for inhibition in the olfactory bulb. Exp Neurol 14: 44–56

Ralston H J 1965 The organization of the substantia gelatinosa Rolandi in the cat lumbrosacral cord. Z Zellforsch 67: 1–23

Ralston H J 1974 On the neuronal organization of the spinal cord. In: Bellairs R, Gray E G (eds) Essays on the nervous system. Clarendon Press: Oxford

Ramón-Moliner E, Dansereau J A 1974 The peribrachial region of the cat. Cell Tissue Res 149: 173–190; 191–204

Ramón-Moliner E, Nauta W J H 1966 The iso-dendritic core of the brain stem. J Comp Neurol 126: 311–335

Ramón y Cajal S 1890 Origen y terminación de las fibras nerviosas olfatorias. Gac San Barcelona, pp 1–21

Ramón y Cajal S 1900 Textura del sistema nervioso del hombre y de los vertebrados. vol 1. Moya: Madrid

Ramón y Cajal S 1908 Structure et connexion des neurons. In: Les prixs nobel en 1906. Norstedt and Söner: Stockholm: pp 1–25

Ramón y Cajal S 1909 Histologie du système nerveux de l'homme et des vertébrés. Maloine: Paris

Ramón y Cajal S 1911 Histologie du système nerveux de l'homme et des vertébrés. Maloine: Paris

Ramón y Cajal S 1919 Acción neurotrópica de los epitelios. Algunos detalles sobre el mecanismo genético de las ramificaciones nerviosas intraepiteliales sensitivas y sensoriales. Trab Lab Invest Biol 17: 65–86

Ramón y Cajal S 1928 Degeneration and regeneration of the nervous system. 2 vol. (May R M ed and trans) Oxford University Press: London

Ramón y Cajal S 1955 Studies on the cerebral cortex (Kraft L M trans). Lloyd–Luke: London

Ramón y Cajal S 1960 Studies on vertebrate neurogenesis (Guth I trans) 35th edition, Thomas: Springfield Illinois

Ramsdale D R, Dixon J S, Gosling J A 1972 Chromaffin cells in the mammalian urethra. J Anat 113: 290–291

Ramsey E M 1955 Vascular patterns in endometrium and placenta. Angiology, 6 321–339

Ramsey G M, Corner G W Jr, Donner M W 1963 Serial and cineradioangiographic visualisation of maternal circulation in the primate (hemochorial) placenta. Am J Obstet Gynecol 86: 213–225

Rang M 1969 The growth plate and its disorders. Livingstone: Edinburgh.

Ranganathan N, Lam J H, Wigle E D 1970 Morphology of the human mitral valve. II. The valve leaflets. Circulation, 41: 459–467

Ranson S W, Magoun W W 1933 The central path of the pupilloconstriction reflex in response to light. Arch Neurol Psychiat 30: 1193–1202

Ranvier L A 1871 Contributions à l'histologie et à la physiologie des nerfs périphériques. C R Acad Sci (Paris) 73: 1168–1171

Rao G S, Sahu S 1974 The localization within the dorsal motor nucleus of the vagus in the buffalo (Bubalus bubalis). Acta Anat 90: 388–393

Rappaport, A M 1972 The normal microcirculation of the mammalian liver. J Anat 113: 302

Raschke, M, Lierse W, van Ackeren H 1987 Microvascular architecture of the mucosa of the gastric corpus in man. Acta Anat 130: 185–190

Rascol M M, Izard J Y 1974 The subdural neurothelium of the cranial meninges in man. Anat Rec 186: 429–436

Rasmussen A T 1932 Secondary vestibular tracts in the cat. J Comp Neurol 54: 143–172

Rasmussen A T, Peyton W T 1946 Origin of ventral external arcuate fibres and their continuity with the striae medullares of the fourth ventricle in man. J Comp Neurol 84: 325–337

Rasmussen G L 1942 An efferent chochlear bundle, Anat Rec 82: 441P

Rasmussen G L 1946 The olivary peduncle and other fiber projections of the superior olivary complex. J Comp Neurol 84: 141–220

Rasmussen G L 1957 Selective silver impregnation of synaptic endings. In: Windle W F (ed) New research techniques of neuroanatomy. Thomas: Springfield Illinois: pp 27–39

Rasmussen G L 1960 Efferent fibers of the cochlear nerve and cochlear nucleus. In: Rasmussen G L, Windle W F (eds) Neural mechanisms of the auditory and vestibular systems. Thomas: Springfield Illinois: pp 105–115

Rasmussen G L 1967 Efferent connections of the cochlear nerve. In: Graham A B (ed) Sensori-neural hearing processes and disorders. (Henry Ford Hospital International Symposium). Little, Brown: Boston Massachusetts: pp 61–75

Rasmussen G L, Windle W F 1960 Neural mechanisms of the auditory and vestibular systems. Thomas: Springfield, Illinois

Ratcliffe F F 1981 Arterial anatomy of the developing human dorsal and lumbar vertebral body. A microarteriographic study. J Anat 133: 625–638

Rattner, J B, Phillips S G 1973 Independence of centriole formation and DNA synthesis. J Cell Biol 57: 359–372

Raveau, F 1968 . . . Arch Anat Path 16:

Raven C P 1958 Morphogenesis: The analysis of molluscan development. Pergamon, New York

Raven C P 1961 Oogenesis. Pergamon, Oxford

Raven C P 1963 Differentiation in mollusc eggs. In: Cell differentiation. (Society for Experimental Biology Symposia). Cambridge University Press: Cambridge: 17: 274–284

Raven C P 1966a Analysis of molluscan development. Pergamon Press: Oxford

Raven C P 1966b An outline of developmental physiology. Pergamon Press, Oxford

Raviola E, Gilula N B 1973 Gap junctions between photoreceptor cells in the vertebrate retina. Proc Nat Acad Sci, USA 70: 1677–1681

Raviola E, Karnovsky M J 1972 Evidence for a blood–thymus barrier using electron opaque tracers. J Exp Med 136: 466–498

Rawles M E 1948 Origin of melanophores and their role in development of color patterns in vertebrates. Physiol Rev 28: 383–408

Ray R D, Johnson R J, Jameson R M 1951 Rotation of the forearm. J Bone Jt Surg 33A: 993–996

Raybuck H E 1952 The innervation of the parathyroid glands. Anat Rec 112: 117–124

Redler I 1974 A scanning electron microscopic study of human normal and osteoarthritic articular cartilage. Clin Orthop Rel Res 103: 262–268

Redler I, Zimny M L 1970 Scanning electron microscopy of normal and abnormal articular cartilage and synovium. J Bone Jt Surg 52A: 1395–1404

Reed C I, Reed B P 1948 Comparative study of human and bovine sperm by electron microscopy. Anat Rec 100: 1–8

Rees L A 1954 The structure and function of the mandibular joint. Br Dent J 96: 125–133

Rees L 1976 A quantitative electron microscopic study of the aging human cerebral cortex. Acta Neuropathol 36: 347–362

Reese T 1965 Olfactory cilia in the frog. J Cell Biol 25: 209–230

Reese T, Brightman M W 1970 Olfactory surface and central olfactory connexions in some vertebrates. In: Wolstenholme G E W, Knight J (eds) Taste and smell in vertebrates. (Ciba Foundation Symposium). Churchill: London: pp 115–143

Reese T S, Shepherd G M 1972 Dendrodendritic synapses in the central nervous system. In: Pappas G D, Purpura D P (eds) Structure and function of synapses. Raven Press: New York: pp 121–136

Regaud C 1897 Les vaisseaux lymphatiques du testicle, et des faux endothéliums de la surface des tubes séminifères, [No publisher]: Lyon

Regaud C, Mawas J 1909 Sur les mitochondries des glandes salivaires chez les mammiferes. C R Séanc Soc Biol 66: 97–100

Régnier M, Vaigot P, Darmon M, Prunieras M 1986 Onset of epidermal differentiation in rapidly proliferating basal keratinocytes. J Invest Dermatol 87: 472–476

Reichert F L, Poth E J 1933 Pathways for the secretory fibres of the salivary glands in man. Proc Soc Exp Biol Med 30: 973–977

Reichlin S, Baldessarini R J, Martin J B eds 1978 The hypothalamus. (Research publications vol 56 Association for research in nervous and mental disease) Raven Press: New York

Reid L 1970 Evaluation of model systems for study of airway epithelium, cilia and mucus. Arch Intern Med (Chicago) 126: 428–434

Reid L 1976 Visceral cartilage. J Anat 122: 349–355

Reil J C 1807 Fragmenten über die bildung des kleinen gehirns in menschen. Arch Physiol Halle, 8: 1–58

Reilly D T, Burstein A H 1974 The mechanical properties of cortical bone. J Bone Jt Surg 56: 1001–1022

Reimann A F, Daseler E H, Anson B J, Beaton L E 1944 The palmaris longus muscle and tendon. Anat Rec 89: 495–505

Reinherz E L 1987 T-cell receptors. Who needs more? Nature 325: 660–663

Reinholt F P, Engfeldt B, Hjerpe A, Jansson K 1982 Stereological studies on the epiphyseal growth plate with special reference to the distribution of matrix vesicles. J Ultrastruct Res 80: 270–279

Reinius S 1967 Ultrastructure of blastocyst attachment in the mouse. Z Zellforsch 57: 255–266

Reiter R J 1983 The pineal gland: an intermediary between the environment and the endocrine system. Psychoneuroendocrinology 8: 31–40

Reiter R J 1985a Action spectra, dose–response relationships, and temporal aspects of light's effects on the pineal gland. Ann N Y Acad Sci 453: 215–230

Reiter R J 1985b Impact of photoperiodic information on pineal metabolism and physiology. Int J Biometeorol 29: (suppl 2), 178–187

Reiter R J 1987 The melatonin message: duration versus coincidence hypotheses. Life Sci 40: 2119–2131

Reiter R J, Hoffman R A 1967 Adrenocortical cytogenesis in the adult male golden hamster. A radioautographic study using tritiated thymidine. J Anat 101: 723–730

Reith E J 1967a The absorptive activity of ameloblasts during the maturation of enamel. Anat Rec 157: 577–588

Reith E J 1967b The early stages of amelogenesis as observed in molar teeth of young rats. J Ultrastruct Res 17: 503–526

Reith E J 1970 The stages of amelogenesis as observed in the molar teeth of young rats. J Ultrastruct Res 30: 111–151

Relkin R 1976 The pineal. Annual research review. Eden Press: Montreal

Remak R 1836 Vorläufige Mitteilung mikroscopischen Beobachtungen über den innern Bau der Cerebrospinalnerven und über die Entwicklung ihrer Formelemente. Arch Anat Physiol 145–161

Remensnyder I P, Majno G 1968 Oxygen gradients in healing wounds. Am J Pathol 52: 301–319

Renaud L P 1978 Neurophysiological organization of the endocrine hypothalamus. In: Reichlin S, Baldessarini R J, Martin J B (eds) The hypothalamus. (Research Publications Vol 56 Association for research in nervous and mental disease) Raven Press: New York: pp 269–301

Renfrew S, Melville I D 1960 The somatic sense of space (choraesthesia) and its threshold. Brain 83: 93–112

Rennke H G, Cotran R S, Venkatachalam, M A 1975 Role of molecular charge in glomerular permeability. Tracer studies with cationized ferritins. J Cell Biol 67: 638–646

Renshaw B 1941 Influence of discharge of motoneurons upon excitation of neighbouring motoneurons. J Neurophysiol 4: 167–183

Renshaw B 1946 Central effects of centripetal impulses in axons of spinal ventral roots. J Neurophysiol 9: 191–204

Repesh L A, Fitzgerald T J, Furcht L T 1982 Fibronectin involvement in granulation tissue and wound healing in rabbits. J Histochem Cytochem 30: 351–358

Réthelyi M, Szentágothai J 1969 The large synaptic complexes of the substantia gelatinosa. Exp Brain Res 7: 258–274

Rexed B 1952 The cytoarchitectonic organization of the spinal cord in the cat. J Comp Neurol 96: 415–495

Rexed B 1954 A cytoarchitectonic atlas of the spinal cord in the cat. J Comp Neurol 100: 297–379

Rexed B 1964 Some aspects of the cytoarchitectonics and synaptology of the spinal cord. Progr Brain Res 11: 58–92

Rexed B, Brodal A 1951 Nucleus cervicalis lateralis. A spino-cerebellar relay nucleus. J Neurophysiol 14: 399–407

Reyners H, Gianfelici de Reyners E, Regniers L, Maisin J R 1986 A glial progenitor cell in the cerebral cortex of the adult rat. J Neurocytol, 15: 53–61

Reynolds E L 1945 Bony pelvic girdle in early infancy; roentgenometric study. Am J Phys Anthropol 3: 321–354

Reynolds E L 1947 The bony pelvis in prepuberal childhood. Am J Phys Anthropol N S 5: 165–200

Reynolds J J 1974 The role of 1,25-Dihydroxy-cholecalciferol in bone metabolism. (Biochemical Society Special Publication 3) pp 91–192, Biochemical Society: London

Reynolds J J 1976 Calcification of bone. J Anat 122: 92

Reynolds S R M, Zweifach B W eds 1959 The microcirculation. Symposium on factors influencing exchange of substances across capillary Wall. Proceedings of the 5th Conference on Microcirculatory Physiology and Pathology, Buffalo . . . 1958. University of Illinois Press: Urbana Illinois

Rhinelander F W 1968 The normal microcirculation of diaphysial cortex and its response to fracture. J Bone Jt Surg 50A: 784–800

Rhodin J A 1971 The ultrastructure of the arterial cortex of the rat under normal and experimental conditions. J Ultrastruct Res 34: 23–71

Rhodin J A G 1962a Fine structure of vascular walls in mammals, with special reference to the smooth muscle component. Physiol Rev 42: (suppl 5) 48–81

Rhodin J A G 1962b The diaphragm of capillary endothelial fenestrations, J Ultrastruct Res 6: 171–185

Rhodin J A G 1966 Ultrastructure and function of the human tracheal mucosa Annu Rev Respir Dis 93: 1–15

Rhodin J A G 1968 Ultrastructure of mammalian venous capillaries, venules and small collecting veins. J Ultrastruct Res 25: 452–500

Rhodin J, Dalhamn T 1956 Electron microscopy of the tracheal ciliated mucusa in rat. Z Zellforsch 44: 345–412

Rhodin J A G, Terzakis J 1962 The ultrastructure of the human full-term placenta. J Ultrastruct Res 6: 88–106

Rhoton A L Jr, Kobayashi S, Hollinshead W H 1968 Nervus intermedius. J Neurosurg 29: 609–618

Richardson A P, Hinsey J C 1933 Functional study of the nodose ganglion of the vagus with degeneration methods. Proc Soc Exp Biol Med 30: 1141–1143

Richardson K C 1962 The fine structure of autonomic nerve endings in smooth muscle of the rat vas deferens. J Anat 96: 427–442

Riches D J, Palfrey A J 1966 The ultrastructure of the bile duct epithelium of the rat. J Anat 100: 429–430P

Riggs H E, Rupp C 1963 Variation in form of circle of Willis. Arch Neurol Psychiat 8: 8–14

Riggs L A, Niehl E W 1960 Involuntary motions of the eye during monocular fixation. J Exp Psychol 40: 687–701

Riggs L A, Ratliff F, Cornsweet J C, Cornsweet T N 1953 The disappearance of steadily fixated visual test objects. J Opt Soc Am 43: 495–501

Riley D A, Allin E F 1973 The effects of inactivity, programmed stimulation and denervation on the histochemistry of skeletal muscle fibre types, Exp Neurol 40: 391–413

Riley H A 1930 Lobules of mammalian cerebellum and cerebellar nomenclature. Arch Neurol Psychiatry 24: 227–256

Ringvist M 1974 Fibre types in human masticatory muscles. Relation to function. Scand J Dent Res 82: 333–355

Ringvold A 1975 Distribution of ascorbic acid in the ciliary body of albino rabbit, guinea pig and rat. Acta Ophthalmol 53: 751–759

Rinvik E 1968a The corticothalamic projection from the pericruciate and coronal gyri in the cat. An experimental study with silver-impregnation methods. Brain Res 10: 79–119

Rinvik E 1968b The corticothalamic projection from the second somatosensory cortical area in the cat. An experimental study with silver-impregnation methods. Exp Brain Res 5: 153–172

Ripps H, Weale R A 1976 The visual photoreceptors. In: Davson H (ed) The eye. Vol 2A, 2nd edition. Academic Press: New York

Ritchie J M, Rogart R B 1977 The density of sodium channels in mammalian myelinated nerve fibers and the nature of the axonal membrane under the myelin sheath. Proc Natl Acad Sci USA 74: 211–215

Ritter M A, Schuurman H J, Mackenzie W A, de Maagd R A, Price K M, Broekhuizen R, Kater L 1985 Heterogeneity of human thymus epithelial cells revealed by monoclonal anti-epithelial cell antibodies. Adv Exp Med Biol 186: 283–288

Riva A 1967 Fine structure of human seminal vesicular epithelium. J Anat 102: 71–86

Riva A, Testa Riva F 1973 Fine structure of acinar cells of human parotid gland. Anat Rec 176: 149–165

Rizk N N 1980 A new description of the anterior abdominal wall in man and mammals. J Anat 131: 373–385

Robb G P, Steinberg I 1939 Visualisation of the chambers of the heart, pulmonary circulation, and great blood vessels in man; practical method. Am J Roentgenol 41: 1–17

Robb H J 1965 Microembolism in the pathophysiology of shock. Angiology 16: 405–411

Robbins J 1973 Radioassay and the thyroid gland. Metabolism 22: 1021–1026

Roberts J A F, Pembrey M E 1985 An introduction to medical genetics, 8th edition. Oxford University Press: London

Roberts K, Hyams J S eds 1979 Microtubules. Academic Press: London

Roberts W, Taylor W H 1973 Inferior rectal nerve variation as it relates to prudendal block. Anat Rec 177: 461–463

Robertson J D 1981 Membrane structure. J Cell Biol 91: 189s–204s

Robinson A G, Zimmerman E A 1973 Cerebrospinal fluid and ependymal neurophysin. Clin Invest 52: 1260–1267

Robinson C, Lowe N R, Weatherall J A 1975 Amino acid composition, distribution and origin of 'tuft' protein in human and bovine dental enamel. Arch Oral Biol 20: 29–42

Robinson D A 1964 The mechanics of human saccadic eye movement. J Physiol 174: 245–264

Robinson R A 1964 Observations regarding compartments for tracer calcium in the body. In: Frost H M (ed) Bone biodynamics. Henry Ford Hospital International Symposium. Churchill: London: pp 423–439

Robles-Chillada E M, Rodrigo J, Mayo I, Anedo A, Gómez A 1981 Ultrastructure of free-ending nerve fibres in the oesophageal epithelium. J Anat 133: 227–233

Robson J A, Hall W C 1977 The organization of the pulvinar in the grey squirrel (Sciurus carolinensis). II. Synaptic organization and comparisons with the dorsal lateral geniculate nucleus. J Comp Neurol 173: 389–416

Roche A F, Lewis A B 1974 Sex differences in the elongation of the cranial base during pubescence. Angle Orthodont 44: 279–294

Rock J, Hertig A T 1942 Some aspects of early human development. Am J Obstet Gynecol 44: 973–983

Rock J, Hertig A T 1944 Information regarding time of human ovulation derived from study of 3 unfertilized and 11 fertilized ova. Am J Obstet Gynecol 47: 343–356

Rockel A J, Hiorns R W, Powell T P S 1980 The basic uniformity in structure of the neocortex. Brain, 103: 221–244

Rockell A J, Jones E G 1973 The neuronal organization of the inferior colliculus of the adult cat. J Comp Neurol 147: 11–60

Rodan G A, Rodan S B 1984 Expression of the osteoblastic phenotype. In: Peck W A (ed) Bone and mineral research annual 2. Elsevier: Amsterdam: pp 244–285

Rodewald R, Karnovsky M J 1974 Porous substructure of the glomerular slit diaphram in the rat and mouse. J Cell Biol 60: 423–433

Rodrigo J, Hernández C J, Vidal M A, Pedrosa J A 1975 Vegetative innervation of the oesophagus. Acta Anat 92: 79–100

Rogers A W 1967 Techniques of autoradiography. Elsevier: Amsterdam

Rogers A W 1973 Techniques of autoradiography. 2nd edition, Elsevier: Amsterdam

BIBLIOGRAPHY

Rogers D C 1965 The development of the rat carotid body. J Anat 99: 89–101

Rogers D C, Burnstock G 1966 Multiaxonal autonomic junctions in intestinal smooth muscle of the toad (Bufo marinus). J Comp Neurol 126: 626–652

Rohan R F, Turner L 1956 The levator palati muscle. J Anat 90: 153–154

Rohen J 1964 Ciliarkörper. In: von Möllendoff W (ed) Das Auge und seine Hilfsorgane Ergänzung zu Band 111/2. Haut und Sinnesorgane. 4 Teil. Handbuch der mikroskopischen Anatomie des Menschen. Springer: Berlin: pp 189–238

Rohon J V 1884 Zur histiogenese des rückenmarkes der forelle. Sitz ber mathemat -phys Klasse Königl Bayr Akad Wiss 14: 301–356

Roitt I M 1984 Essential immunology, 5th edition, Blackwell Scientific: Oxford

Rokx J T M, Jüch P J W, Van Willigen J D 1985 On the bilateral innervation of masticatory muscles: a study with retrograde tracers. J Anat 140: 237–243

Rollet F 1899 De la mensuration de os longs des membres, Thèse pur le doc en méd, 1st ser, 43: 1–128

Rollin H 1977 Course of the peripheral gustatory nerves. Ann Otol Rhinol Laryngol 86: 251–258

Romanes G J, 1941a Cell columns in the spinal cord of a human foetus of fourteen weeks. J Anat 75: 145–152

Romanes G J 1941b The development and significance of the cell columns in the ventral horn of the cervical and upper thoracic spinal cord of the rabbit. J Anat 76: 112–130

Romanes G J 1942 The spinal cord in a case of congenital absence of the right limb below the knee. J Anat 77: 1–5

Romanes G J 1946 Motor localisation and the effects of nerve injury on the ventral horn cells of the spinal cord. J Anat 80: 117–131

Romanes G J 1951 The motor cell columns of the lumbosacral cord of the cat. J Comp Neurol 94: 313–363

Romanes G J 1953 The motor cell groupings of the spinal cord. In: Wolstenholme G E W (ed) The spinal cord. (Ciba Foundation Symposium) Churchill: London: pp 24–38

Romanes G J 1964 The motor pools of the spinal cord. Progr Brain Res 11: 93–119

Romani N, Schuler G, Frisch P 1986 Ontogeny of Ia–positive and thy–1 positive leukocytes of murine epidermis. J Invest Dermatol 86: 129–133

Romer A S 1942 Cartilage–an embryonic adaptation. Am Nat 76: 394–404

Romer A S 1970 The vertebrate body. 4th edition, Saunders: Philadelphia

Rood J P 1977 The nerve supply of the mandibular incisor region. Br Dent J 143: 227–230

Roosen-Runge E C 1952 The third maturation division in mammalian spermatogenesis. Anat Rec 112: 453 (Demonstration 50)

Roosen–Runge E C 1961 The rete testis in the albino rat: its structure, development and morphological significance. Acta Anat (Basel) 45: 1–30

Roosen–Runge E C, Giesel L O 1950 Quantitative studies on spermatogenesis in the albino rat. Am J Anat 87: 1–30

Rosa M, Borzone M 1973 The venous system of the corpus striatum. Angiographic study. Neuroradiology 6: 219–230

Rosan R C, Lauweryns J 1971 Secretory cells in the premature human lung lobule. Nature 232: 60–61

Rose G C ed 1963 Cinematography in biology. Academic Press: New York

Rose J E, Malis L I 1965a Geniculo-striate connections in the rabbit. I. Retrograde changes in the dorsal lateral geniculate body after destruction of cells in various layers of the striate region. J Comp Neurol 125: 95–120

Rose J E, Malis L I 1965b Geniculo-striate connections in the rabbit. II. Cytoarchitectronic structure of the striate region and of the dorsal lateral geniculate body; organisation of the geniculo–striate projections. J Comp Neurol 125: 121–140

Rose J E, Woolsey C N 1958 Cortical connexions and functional organisation of the thalamic and Auditory system of the cat. University of Wisconsin Press: Madison Wisconsin

Rose M 1927 Die sogenannte Riechrinde beim Menschen und Affen. J Psychol Neurol Lpz 35: 261–401

Rose M D 1975 Functional proportions of primate lumbar vertebral bodies. J Hum Evol 4: 21–38

Rose V, Izukawa T, Moës C A F 1975 Syndromes of asplenia and polysplenia. A review of cardiac and non–cardiac malformations in 60 cases with special reference to diagnosis and prognosis. Br Heart J 37: 840–852

Rosenberg J C, DiDio L J A 1969 In vivo appearance and function of the termination of the ileum as observed directly through a cecostomy. Am J Gastroenterol 52: 411–419

Rosenthal J 1977 Trophic interactions of neurons. In: Kandel E R (ed) Handbook of physiology. Section 1. The nervous system. Vol 1. Pt 2. American Physiological Society: Bethesda Maryland: pp 775–801

Ross K F A 1967 Phase Contrast and interference microscopy for cell biologists. Arnold: London

Ross M A 1971 Functional anatomy of the tensor paliti. Its relevance in cleft palate surgery. Arch Otolaryngol 93: 1–3

Ross M D 1969 The general visceral efferent component of the eighth cranial nerve. J Comp Neurol 135: 453–478

Ross M H, Dobler J 1975 The Sertoli cell junctional specializations and their relationship to the germinal epithelium as observed after efferent ductule ligation. Anat Rec 183: 267–291

Ross R 1968 The fibroblast and wound repair. Biol Rev 43: 57–96

Ross R 1975 Connective tissue cells, cell proliferation and synthesis of extracellular matrix. A review. Philos Trans R Soc Lond [Biol] 271: 247–259

Ross R, Klebanoff S J 1967 Fine structural changes in uterine smooth muscle and fibroblasts in response to estrogen. J Cell Biol 32: 155–167

Ross R, Klebanoff S J 1971 The smooth muscle cell. I. In vivo synthesis of connective tissue proteins. J Cell Biol 50: 159–171

Rossi G F 1964 A hypothesis on the neural basis of consciousness. Considerations based upon some experimental work. Acta Neurochir, 12: 187–197

Rossignol S, Colonnier M 1971 Vision Res (Suppl 3)

Roth S I 1962 Pathology of the parathyroids in hyperparathyroidism. Discussion of recent advances in the anatomy and pathology of the parathyroid glands. Arch Pathol 73: 495–510

Roth S I, Capen C C 1974 Ultrastructural and functional correlations of the parathyroid gland. Int Rev Exp Path 13: 161–221

Rothblat L A, Schwarz M L 1979 The effect of monocular deprivation on dendritic spines in visual cortex of young and adult albino rats: evidence for a sensitive period. Brain Res 161: 156–161

Rother P, Hunger H, Leopold D et al 1977 Zur bestimmung des lebensalters und des geschlechts aus humerusmassen. Anat Anz 142: 243–254

Rother P, Hunger H, Liebert U et al 1975 Die Geschlechtsunterschiede des menschlichen Sternums Gengenbaurs. Morphol Jahrb 121: 29–37

Rothman S 1981 The Golgi apparatus: two organelles in tandem. Science 213: 1212–1219

Rothschild Lord 1957 The fertilising spermatozoon. Discovery 18: 64–65

Rouiller C ed 1964 The liver: morphology, biochemistry, physiology. 2 vol. Academic Press: New York

Routal R R, Pal G P, Bhagwat S S, Tamankar B P 1984 Metrical studies with sexual dimorphism in foramen magnum of human crania. J Anat Soc Ind 33: 85–89

Routtenberg A, Santos-Anderson R 1977 The central role of prefrontal cortex in intracranial self-stimulation: a case history of anatomical localization of motivational substrates. In: Iversen L L, Iversen S, Synder S H (eds) Handbook of psychopharmacology. Plenum Press: New York

Rowbotham G F, Little E 1962 The circulation and reservoir of the brain. Br J Surg 50: 244–250

Rowden G 1981 The Langerhans' cell. CRC Crit Rev Immunol 3: 95–180

Rowe R W D, Goldspink G 1969 Muscle fibre growth in five different muscles of both sexes of mice. I. Normal mice. J Anat 104: 519–530

Rowntree T 1949 Anomalous innervation of the hand muscles. J Bone Jt Surg 31B: 505–510

Roy S, Ghadially F N 1967 Ultrastructure of normal rat synovial membrane. Ann Rheum Dis 26: 26–38

Royle G 1973 A groove in the lateral wall of the orbit. J Anat 115: 461–465

Royle J P, Eisner R 1981 The saphenofemoral junction. Surg Gynecol Obstet 152; 282–284

Rozendal R H, Molen N H 1972 The relevancy of the concept of 'shunt' and 'spurt' muscles in functional anatomy. Acta Morphol Nearl Scand 10: 347–350

Rubin W 1972A Endocrine cells in the human stomach. A fine structural study. Gastro-enterol 63: 784–800

Rubin W 1972b An unusual intimate relationship between endocrine cells and other types of epithelial cells in the human stomach. J Cell Biol 52: 219–230

Rudolph A M, Drorbaugh J E, Auld P H, Rudolph A J, Nades A S, Smith C A, Aubbell J P 1961 Studies on the circulation in the neo-natal period. The circulation in respiratory distress syndrome. Pediatrics 27: 551–566

Ruoslahti E, Engvall E, Hayman E G 1981 Fibronectin: current concepts of its structure and functions. Coll Rel Res Clin Exp 1: 95–128

Rushmer R F 1976 Structure and function of the cardiovascular system. 2nd edition. Saunders: London, Philadelphia

Rushton W A H 1962 Visual pigments in man. Liverpool University Press: Liverpool

Rushton W A H, Henry G H 1968 Bleaching and regeneration of cone pigments in man. Vision Res 8: 617–631

Ruskell G L 1968 The fine structure of nerve terminations in the lacrimal glands of monkeys. J Anat 103: 65–76

Ruskell G L 1969 Changes in nerve terminals and acini of the lacrimal gland and changes in secretion induced by autonomic denervation. Z Zellforsch 94: 261–281

Ruskell G L 1970 An ocular parasympathetic nerve pathway of facial nerve origin and its influence on intraocular pressure. Exp Eye Res 10: 319–330

Ruskell G L 1971a The distribution of autonomic postganglionic nerve fibres in the lacrimal gland in the rat. J Anat 109: 229–242

Ruskell G L 1971b Ocular autonomic nerves of facial nerve origin. J Anat 107: 374P

Ruskell G L 1974 Form of the choroidocapillaris. Exp Eye Res 18: 411–412

Ruskell G L 1975 Nerve terminals and epithelial cell variety in the human lacrimal gland. Cell Tissue Res 158: 121–136

Ruskell G L 1983 Fibre analysis of the nerve to the interior oblique muscle in monkeys. J Anat 137: 445–455

Ruskell G L 1984 Spiral nerve endings in human extraocular muscles terminate in motor end plates. J Anat 139: 33–43

Russell G V 1955 Schematic presentation of thalamic morphology and connections. Tex Rep Biol Med 13: 989–992

Russell J G 1969 Radiology in the diagnosis of fetal abnormalities. Br J Obstet Gynaecol 76: 345–350

Rusted I E, Scheifley C H, Edwards J E 1952 Studies of mitral valve; anatomic features of normal mitral value and associated structures. Circulation 6: 825–831

Rustioni A 1973 Non-primary afferents to the nucleus gracilis from the lumbar cord of the cat. Brain Res 51: 81–95

Rustioni A, Dekker J J 1974 Non-primary afferents to the dorsal column nuclei of cat: distribution pattern and cells of origin. Anat Rec 178: 454–455

Rustioni A, Sotelo C 1974 Synaptic organization of the nucleus gracilis of the cat. Experimental identification of dorsal root fibers and cortical afferents. J Comp Neurol 155: 441–468

Rusznyák I, Földi M, Szabo G 1960 Lymphatics and lymph circulation physiology and pathology. Pergamon Press: Oxford

Rutherford W 1873 A new freezing microtome. Mon Microsc J 10: 186–189

Ryan G B, Coghlan J P, Scoggins B A 1979 The granulated peripolar epithelial cell: a potential secretory component of the renal juxtaglomerular complex. Nature 277: 655–656

Ryan J W, Ryan U S 1977 Pulmonary endothelial cells. Fed Proc Fed Am Soc Exp Biol 36: 2683–2691

Ryan R B, Cliff W J, Babbiani G et al 1974 Myofibroblasts in human granulation tissue. Hum Pathol 5: 55–67

Rybicki E F, Smonen F A, Weis E B 1972 On the mathematical analysis of stress in the human femur. J Biomech. 5: 203–215

Ryncki P V 1974 Anatomie chirugicale du foie. Helv Chir Acta 41: 543–574

S

Saacke R G, Heald C W 1974 Cytological aspects of milk formation and secretion. In: Larson B L, Smith V R (eds) Lactation. A comprehensive treatise. Vol 2. Academic Press: New York: pp 147–189

Saavedra J P, Mascitti T A, Vaccarezza O L 1969 Lamination of the Cebus lateral geniculate nucleus. Z Zellforsch 94: 346–351

Sabin F R 1912 On the origin of the abdominal lymphatics in mammals from the vena cava and the renal glands. Anat Rec 6: 335–342

Sabin F R 1920 Studies on the origin of blood vessels and of red blood corpuscles in the living blastoderm of chicks during the second day of incubation. Contrib Embryol Carnegie Inst Washington 9: 213

Sager R 1965 Genes outside the chromosomes. Sci Am 212: 70–79

Saha A K 1961 Theory of shoulder mechanism: descriptive and applied: Thomas: Springfield Illinois

St Helen R, McEwen W K 1691 Rheology of the human sclera. 1. Anelastic behaviour. Am J Ophthal 52: 539–548

Salmon J A, Higgs G A, Vane J R, Bitensky L, Chayen J, Henderson B, Cashman B 1983 Synthesis of arachidonate cyclo-oxygenase products by rheumatoid and nonrheumatoid synovial lining in nonproliferative organ culture. Ann Rheum Dis 42: 36–39

Salsbury C R 1937 The interosseous muscles of the hand. J Anat 71: 395–403

Salter N 1955 Methods of measurement of muscle and joint function. J Bone Jt Surg 37B: 474–491

Salzmann J A ed 1961 Roentgenographic cephalometrics. Proceedings of the 2nd research workshop conducted by the special committee of the American Association of Orthodontists. Lippincott: Philadelphia.

Samarasinghe D D 1965 The innervation of the cerebral arteries in the rat: an electron microscope study. J Anat 99: 815–828

Sammarco J 1977 Biomechanics of the ankle. I. Surface velocity and instant center of rotation in the sagittal plane. Am J Sports Med 5: 231–234

Sampson J A 1937 Lymphatics of mucosa of fimbriae of the fallopian tube. Am J Obstet Gynecol 33: 911–930

Samuel E P 1953 Chromidial studies on the superior cervical ganglion of the rabbit; (a) caudally projected postganglionic axons; (b) intercalary 'commissural' neurons. J Comp Neurol 98: 93–112

Sandberg L B, Soskel N T, Leslie J G 1981 Elastin structure, biosynthesis and relation to disease states. N Engl J Med 304: 566–577

Sanides F 1964 The cyto-myeloarchitecture of the human frontal lobe and its relation to phylogenetic differentiation of the cerebral cortex. J Hirnforsch 6: 269–282

Santer R M, Lu K S, Lever J D, Presley R 1975 A study of the distribution of chromaffin-positive (CH+) and small intensely flourescent (SIF) cells in sympathetic ganglia of the rat at various ages. J Anat 119: 589–599

Santo Neto H, de Carvalho V C, Penteado C V 1984 Calculation of force and work developed by human abductor digiti minimi. Rev Bras Cien Morf 1: 9–11

Santo Neto H, Penteado C V, de Carvalh. V C 1984 Presence of a groove in the lateral wall of the human orbit. J Anat 138: 631–633

Santulli T V, Blanc W A 1961 Congenital atresia of the intestine: pathogenesis and treatment. Ann Surg 154: 939–948

Saper C B, Loewi A D, Swanson L W, Cowan W M 1976 Direct hypothalamoautonomic connections. Brain Res 117: 305–312

Sapin M R, Borziak E I 1974 Anatomie des ganglions lymphatiques du médiastin. Acta Anat 90: 200–225

Sarnat B G ed 1951 The temporomandibular joint. 2nd edition, Thomas: Springfield, Illinois

Sarton G 1954 Galen of Pergamon, University of Kansas Press: Laurence, Kansas

Sasagawa T, Kobayashi S, Fujita T 1973 Electron microscope studies on the endocrine cells of the human gut and pancreas. In: Fujita T (ed) Gastro-entero-pancreatic endocrine system. A cell-biological approach. Igaku Shoiu: Tokyo: pp 17–38

Sathian K, Devanandan M S 1983 Receptors of the metacarpophalangeal joints: a histological study in the bonnet monkey and man. J Anat 137: 601–613

Satiukova G S, Rassokhina-Volkova L J 1972 Compensatory-adaptive changes in the lymphatic system of organs in experiment and disease. Bibl Anat 11: 481–487

Sato T 1973a A new classification of the transverso-spinalis system. Proc Jap Acad 49: 51–56

Sato T 1973b Innervation and morphology of the musculi levatores costarum longi. Proc Jap Acad 49: 555–558

Sato J 1974 The Mm subcostales in man and monkeys, Okajimas Fol Anat Jpn 50: 345–358

Sauer F C 1935a The cellular structure of the neural tube. J Comp Neurol 63: 13–23

Sauer F C 1935b Mitosis in the neural tube. J Comp Neurol 62: 337–405

Sauer F C 1936 The interkinetic migration of embryonic epithelial nuclei. J Morphol 60: 1–11

Sauer M E, Chittenden A C 1959 Deoxyribonucleic acid content of cell nuclei in the neural tube of the chick embryo: evidence for intermitotic migration of nuclei. Exp Cell Res 16: 1–6

Saunders J W Jr 1948 The proximo-distal sequence of origin of the parts of the chick wing and the role of the ectoderm. J Exp Zool 108: 363–404

Saunders J W Jr, Fallon J F 1966 Cell death in morphogenesis. In: Locke M (ed) Major problems in developmental biology. Academic Press: New York: pp 289–314

Saunders J W Jr, Gasseling M T 1968 Ectodermal-mesenchymal interactions in the origin of limb symmetry. In: Fleischmajer R, Billingham R R (eds) Epithelial-mesenchymal interactions. Williams and Wilkins: Baltimore: pp 78–97

Saunders R L de C H, Röckert H O E 1967 Vascular supply of dental tissues including lymphatics. In: Miles A E W (ed) Structural and chemical organisation of teeth. Academic Press: New York: pp 199–245

Sawaki Y 1977 Retinohypothalamic projection: electrophysiological evidence for the existence in female rats. Brain Res 120: 336–341

Saxén L 1970 Defective regulatory mechanisms in teratogenesis. Int J Gynaec Obstet 8: 798–804

Sayfi Y 1967 Note sur l'innervation du dos de la main. Arch Anat Pathol 15: 139–140

Scapinelli R 1968 Studies on the vasculature of the human knee joint. Acta Anat 70: 305–331

Schachner M 1982 Cell type-specific surface antigens in the mammalian nervous system. J Neurochem 39: 1–8

Schachner M, Hedley-White E T, Hsu D W, Schoonmaker G, Bignami A 1977 Ultrastructural localization of glial fibrillary acidic protein in mouse cerebellum by immunoperoxidase labeling. J Cell Biol 75: 67–73

Schadé J P 1964 On the volume and surface area of spinal neurons. Progr Brain Res 11: 261–277

Schaefer K P, Schneider H 1968 Reizversuche im Tectum opticum des Kaninchens. Ein experimenteller Beitrag zur seno-mororischen Koordination des Hirnstammes. Arch Psychiat NervKrankh 211: 118–137

Schaltenbrand G, Bailey P 1959 Introduction to stereotaxis: with an atlas of the human brain. 3 vol. Grune and Stratton: New York

Schaper A 1897 Die frühesten Differenzirungsvorgänge im central nervensystem; kritische Studie und Versuch einer Geschichte der Entwicklung nervöser Substanz. Arch Entwcklngsmechn Organ Leipzig 5: 81–132

Scharf J H 1958 Snesible ganglien. In: Handbuch der Mikroskopischen Anatomie des Menschen. Bd 4/3, Springer: Berlin

Scharrer E, Scharrer B 1963 Neuroendocrinology, Columbia University Press: New York

Schaumann B, Alter M A 1976 Dermatoglyphics in medical disorders. Springer: New York

Scheibel M E, Scheibel A B 1966a The organisation of the nucleus reticularis thalami: A Golgi study. Brain Res 1: 43–62

Scheibel M E, Scheibel A B 1966b Patterns of organisation in specific and non-specific thalamic fields. In: Purpura D P, Yahr M D (eds) The thalamus. Columbia University Press: New York: pp 12–46

Scheibel M E, Scheibel A B 1966c Spinal motoneurons, interneurons and Renshaw cells: A Golgi study. Archs Ital Biol 104: 328–353

Scheibel M E, Scheibel A B 1966d Terminal axonal patterns in cat spinal cord. I. The lateral corticospinal tract. Brain Res 2: 333–350

BIBLIOGRAPHY

Scheibel M E, Scheibel A B 1967 Structural organisation of nonspecific thalamic nuclei and their projection toward cortex. Brain Res 6: 60–94

Scheibel M E, Scheibel A B 1968 Terminal exonal patterns in cat spinal cord. II. The dorsal horn. Brain Res 9: 32–58

Scheibel M E, Scheibel A B 1970 Elementary processes in selected thalamic and cortical subsystems–the structural substrates. In: Schmitt F O, Quarton G C, Melnechuk T, Adelman G (eds) The neurosciences, a second study program. Rockefeller University Press: New York: pp 443–457

Scheiner C 1619 Oculus, hoc est: fundamentum opticum. D Agricola: Oenipontus

Scheinin A, Light E I 1969 Innervation of the dental pulp. Acta Odont Scand 27: 313–319

Scheuer J L 1964 Fibre size frequency distribution in normal human laryngeal nerves. J Anat 98: 99–104

Schiaffino S, Pierobon Bormiolo S 1976 Morphogenesis of rat muscle spindles after nerve lesion during early postnatal development. J Neurocytol, 5: 319–336

Schiefferdecker E F P 1882 Uber die Verwendung des Celloidins in der anatomischen Technik. Arch Anat Physiol, 199–203

Schippel K, Reissig D 1968 Zur Feinstruktur des Muskel-Sehnenüberganges. Z Mikrost Anat Forsch 78: 235–255

Schlager F 1967 Uber die muskalatur der Ductus ejaculatorii beim Menschen. Z Zellforsch 76: 268–274

Schlegel J V 1946 Arterio-venous anastomoses in the endometrium in man. Acta Anat 1: 284–325

Schleiden M J 1838 Beiträge zur phytogenesis. Arch Anat Physiol Wiss Med 137–176

Schliwa M 1986 The cytoskeleton. An introductory survey. (Cell Biol Monographs 13), Springer: Vienna

Schlossman A, Priestley B S eds 1966 Strabismus. International ophthalmology clinics. 6. No 3. Little, Brown: Boston: pp 397–749

Schmalbruch H, Hellhammer U 1976 The number of satellite cells in normal human muscle. Anat Rec 185: 279–287

Schmidt C F, Hendrix J P 1937 Action of chemical substances on cerebral blood vessels. A. Research. Nerve Ment Dis Proc 18: 229–276

Schmidt H M 1974 Über den Verlauf der Radix mesenterii beim Menschen. Z Anat EntwGesch 144: 187–193

Schmidt R F 1973 Control of the access of afferent activity to somatoscensory pathways. In: Iggo A (ed) Handbook of sensory physiology. Vol 2. Springer: Berlin: pp 151–206

Schmidt W J 1938 Polarisationsoptische analyse eines eiweiss-lipoid systems. Kolloidzschr 85; 137–148

Schmitt F O, Worden F G 1974 The neurosciences. Third study program (eds) MIT Press: Cambridge Massachusetts

Schmitt F O, Worden F G eds 1979 The neurosciences. Fourth study program. MIT Press: Cambridge Massachusetts

Schmitt F O, Parvati D, Smith B H 1976 Electronic processing of information by brain cells. Science 193: 114–120

Schmitt F O, Quarton G C, Melnechuk T, Adelman G eds 1967 The neurosciences, a study program, Rockefeller University Press: New York

Schmitt F O, Quarton G C, Melnechuk T, Adelman G (eds) 1970 The Neurosciences. Second study program, Rockefeller University Press: New York

Schnapp B J, Vale R D, Sheetz M P, Reese T S 1986 Microtubules and the mechanism of directed organelle movement. Ann N Y Acad Sci 466: 909–918

Schneider L H, Hunter J M 1982 Flexor tendons–late reconstruction. In: Green D P (ed) Operative hand surgery. Vol 2. Churchill Livingstone: New York: pp 1375–1440

Schneider U, Inke G, Schneider I G 1969 Zhale, Abstand der Verzweigungsstellen vom Rand des Sinus renalis und Kaliber der extrarenalen Nierengefässe des menschen. Anat Anz 124: 278–291

Schneyer L H 1976 Sympathetic control of Na$^+$, K$^+$ transport in perfused submaxilliary main duct of rat. Am J Physiol 230: 341–345

Schnitzlein H N, Rowe L C, Hoffman H H 1958 The myelinated component of the vagus nerves in man. Anat Rec 131: 649–667

Schoefl G T 1972 The migration of lymphocytes across the vascular endothelium in lymphoid tissue. A re-examination. J Exp Med 136: 568–588

Schoefl G T, French J E 1968 Vascular permeability to particulate fat: morphological observations on vessels of lactating mammary glands and of lung. Proc R Soc Lond 169B: 153–165

Schoenen J 1973 Organisation cytoarchitectonique de la moelle épinière de différents mammifères et de l'homme. Acta Neurol Belg 73: 348–358

Schofield G C 1968 Anatomy of muscular and neural tissue in the alimentary canal. In: Code C F (ed) Handbook of physiology. Section 6, Alimentary canal, vol 4. American Physiological Society: Washington: pp 1579–1628

Schofield G C, Atkins A M 1970 Secretory immunoglobulin in columnar epithelial cells of the large intestine. J Anat 107: 491–504

Schon F, Kelly J S 1975 Selective uptake of (3H) B-alanine by glia: association with glial uptake system for GABA. Brain Res 86: 243–257

Schonfeld S E, Slavkin H C 1977 Demonstration of enamel matrix proteins on root-analogue surfaces of rabbit permanent incisor teeth. Calcif Tissue Res 24: 223–229

Schoultze T W, Swett J E 1972 The fine structure of the Golgi tendon organ. J Neurocytol 1: 1–26

Schoultze T W, Swett J E 1974 Ultrastructural organization of the sensory fibres innervating the Golgi tendon organ, Anat Rec 179: 147–162

Schour I, Massler M 1941 The development of the human dentition. J Am Dent Assoc 28: 1153–1160

Schramm L P, Stribling J M, Adair J R 1976 Developmental reorientation of sympathetic preganglionic neurons in the rat. Brain Res 106: 166–171

Schrödinger E 1967 What is life?: mind and matter. Cambridge University Press: Cambridge

Schroeder H E, Listgarten M A 1971 Fine structure of the developing epithelial attachment of human teeth. (Monographs in Developmental Biology. vol 2), Karger: Basel

Schroeder T E 1973 Actin in dividing cells: contractile ring filaments bind heavy meromyosin. Proc Natl Acad Sci USA, 70: 1688–1692

Schuknecht H F 1960 Neuroanatomical correlates of auditory sensitivity and pitch discrimination in the cat. In: Rasmussen G L, Windle W F (eds) Neural mechanisms of the auditory and vestibular systems. Thomas: Springfield Illinois: pp 76–90

Schulman C C 1975 Development of the innervation of the ureter. Eur Urol 1: 46–48

Schultze A H 1956 Postembryonic age changes. Primatologia 1: 887–964

Schultze A H 1969 The life of primates, Weidenfeld and Nicolson: London

Schulter F P 1976 Studies of the basicranial axis: a brief review. Am J Phys Anthrop 45: 545–552

Schumacher G H 1961 Funkionelle Morphologie der Kaumuskulatur. Fischer: Jena

Schumacher G H, Lau H, Freund E, Schultz M, Himstedt H W, Menning A 1976 Zur Topographie der muskulären Nervenausbreitungen. 9. Kaumuskeln M pterygoideus medialis und lateralis. Anat Anz 139: 71–87

Schunke G B 1938 The anatomy and development of the sacro-iliac joint in man. Anat Rec 72: 313–331

Schwalbe G 1876. Z Anat EntwGesch 1: 307–352

Schwann T, 1839 Mikroskopische Untersuchungen über die Uebereinstimmung in der Struktur und dem Wachsthum der Thiere und Pflanzen, Sander: Berlin

Schwartz H, Weddell G 1938 Observations on pathways transmitting the sensation of taste. Brain 61: 99–115

Schwartz H G, Roulhac G E, Lam R L, O'Leary J 1951 Organisation of the fasciculus solitarius in man. J Comp Neurol 94: 221–237

Schwarz W 1961 Electron microscopic observations on the human vitreous body. In: Smelser G K (ed) The structure of the eye. Academic Press: New York: pp 283–291

Scollay R, Jacobs S, Jerabek L, Butcher E, Weissman I 1980 T cell maturation: thymocyte and thymus migrant subpopulations defined with monoclonal antibodies to MHC region antigens. J Immunol 124: 2845–2853

Scollo-Lavizzari G, Akert K 1963 Cortical area 8 and its thalamic projection in Macaca mulatta. J Comp Neurol 121: 259–270

Scott B L, Pease D C 1959 Electron microscopy of the salivary and lacrimal glands of the rat. Am J Anat 104: 115–161

Scott D E, Dudley G K, Knigge K M 1974 The ventricular system in neuroendocrine mechanisms. 2. In vivo monoamine transport by ependyma of the median eminence. Cell Tissue Res 154: 1–16

Scott D E, Van Dyke D H, Paull W K et al 1974a Ultrastructural analysis of the human cerebral ventricular system. 3. The choroid plexus. Cell Tissue Res 150: 389–397

Scott D M, Harwood R, Grant M E, Jackson D S 1977 Characterization of the major collagen species present in porcine aortae and the synthesis of their precursors by smooth muscle cells in culture. Connect Tissue Res 5: 7–13

Scott G B D 1980 The primate caecum and appendix vermiformis: a comparative study. J Anat 131: 549–563

Scott J E 1988 Proteoglycan-fibrillar collagen interactions. Biochem J 252: 313–323

Scott J H 1955 A contribution to the study of the mandibular joint function. Br Dent J 98: 345–348

Scott J H 1967 Dento-facial development and growth. Pergamon Press: London

Scott J H, Symons N B B 1977 Introduction to dental anatomy, 7th edition, Churchill Livingstone: Edinburgh

Sebuwufu P H 1968 Ultrastructure of human fetal thymic cilia. J Ultrastruct Res 24: 171–180

Sedzmir C B 1959 An angiographic test of collateral circulation through the anterior segment of the circle of Willis. J Neurol Neurosurg Psychiat 22: 64–68

Seedhom B B 1979 Transmission of the load in the knee joint with special reference to the role of the menisci. Part I. Anatomy, analysis and apparatus. Eng Med 8: 207–219

Seedhom B B, Tsubuku M 1977 A technique for the study of contact between visco-elastic bodies with special reference to the patello-femoral joint. J Biomech 10: 253–260

Seedhom B B, Dowson D, Wright V 1974 Functions of the menisci. A preliminary study. Heberden Society Proceedings. Ann Rheum Dis 33: 111

Seiju M 1967 Subcellular particles and melanu formation in melanocytes. In: Montagna W, Hu F (eds) Advances in biology of the skin. Vol 8. The pigmentary system. Pergamon Press: Oxford: pp 189–222

Seipel M 1948 Studies on the structure of the mandible. Acta Odontol Scand 8: 81–191

Sellars I E, Keen E N 1978 The anatomy and movements of the crico-arytenoid joint. Laryngoscope 88: 667–674

Selye H 1965 The mast cells. Butterworths: Washington DC

Semenova G S 1962 Innervation of the pharynx and the cervical portion of the oesophagus in man and certain animals. Trudy Smolensk Med Inst 15: 18–27

Semeraro D, Davies J D 1986 The arterial blood supply of human inguinal and mesenteric lymph nodes. J Anat 144: 221–233

Sem-Jacobsen C W, Petersen M C, Dodge H W, Jacks Q D, Lazarte J A, Holman C B 1956 Electric activity of the olfactory bulb in man. Am J Med Sci 232: 243–251

Senior H D 1919 The development of the arteries of the human lower extremity. Am J Anat 25: 55–96

Senior H D 1920 The development of the human femoral artery, a correction. Anat Rec 17: 271–280

Senior P V, Pritchett C J, Sunter J P, Appleton D R, Watson A J 1982 Crypt regeneration in adult human colonic mucosa during prolonged organ culture. J Anat 134: 459–469

Sensenig E C 1949 The early development of the human vertebral column. Contrib Embryol Carnegie Inst Washington 33: 21–41

Sensenig E C 1951 The early development of the meninges of the spinal cord in human embryos. Contrib Embryol Carnegie Inst Washington 34: 145–157

Sensenig E C 1957 The development of the occipital and cervical segments and their associated structures in human embryos. Contrib Embryol Carnegie Inst Washington 36: 141–152

Serafini-Fracassini A, Smith J W 1974 The structure and biochemistry of cartilage, Churchill Livingstone: Edinburgh

Sertoli E 1865 Dell 'esistenza di particolari cellule ramificate nei canalicoli seminiferi del testicolo unamo. Morgagni 7: 31–39

Setchell B P, Waites G M H 1975 The blood-testis barrier. In: Hamilton D W, Greep R O (eds) Handbook of physiology. Section 7. Endocrinology Vol 5. American Physiological Society: Washington DC: pp 143–172

Severn C B, Holyoke E A 1973 Human acardiac anomalies. Am J Obstet Gynec 116: 358–365

Shackleford J M, Wilborn W H 1968 Structural and histochemical diversity in mammalian salivary glands. Ala J Med Sci 5: 180–203

Shafik A 1977 The cremasteric muscle. In: Johnson A D, Gomes W R (eds) The testis. Academic Press: New York

Shafik A 1986 A new concept of the anatomy of the anal sphincter. Mechanism and the physiology of defecation. Reversion to normal defecation after combined excision operation and end colostomy for rectal cancer. Am J Surg 151: 278–284

Shah P M, Scarton H A, Tsapogas M J 1978 Geometric anatomy of the aortic-common iliac bifurcation. J Anat 126: 451–458

Shakir A, Zaini S 1974 Skeletal maturation of the hand and wrist of young children in Baghdad. Ann Hum Biol 1: 189–199

Shaner R F 1921 The development of the pharynx and aortic arches of the turtle, with a note on the fifth and pulmonary arches of mammals. Am J Anat 29: 407–430

Shaner R F 1929 The development of the atrioventricular node, bundle of His and sinoatrial node in the calf; with a description of a third embryonic node-like structure. Anat Rec 44: 85–100

Shanks M F, Pearson R C A, Powell T P S 1978 The intrinsic connexions of the primary somatic sensory cortex of the monkey. Proc R Soc Lond [Biol] 200: 95–101

Shanks M F, Rockel A J, Powell T P S 1975 The commissural fibre connexions of the primary somatic sensory cortex. Brain Res 98: 166–171

Shapiro A L, Robillard G L 1948 Arterial blood supply of common and hepatic bile ducts with references to problems to common duct injury and repair: based on a series of 23 dissections. Surgery 23: 1–11

Shariff G A 1953 Cell counts in the primate cerebral cortex. J Comp Neurol 98: 381–400

Sharpey R, Perry V H 1986 Cat and monkey retinal ganglion cells and their visual functional roles. Trends Neurosci 9: 229–235

Sharrard W J W 1955 The distribution of the permanent paralysis in the lower limb in poliomyelitis. J Bone Jt Surg 37B: 540–558

Sharrard W J W 1956 Poliomyelitis: The distribution of the paralysis. In: British Surgical Progress, p 83

Shaw N E, Martin B F 1962 Histological and histochemical studies on mammalian knee-joint tissues. J Anat 96: 359–373

Shawker T H, Sonies B, Hall T E, Baum B F 1984 Ultrasound analysis of tongue, hyoid, and larynx activity during swallowing. Invest Radiol 19: 82–86

Sheehan D 1933 On unmyelinated fibres in the spinal nerves. Anat Rec 55: 111–116

Sheehan D 1936 Discovery of the autonomic nervous system. Arch Neurol Psychiat 35: 1081–1115

Sheehan D, Mulholland J H, Shariroff B 1941 Surgical anatomy of the carotid sinus nerve. Anat Rec 80: 431–442

Shehata R 1966 The crura of the diaphragm and their nerve supply. Acta Anat 63: 49–54

Sheldon H 1983a Transmission electron microscopy of cartilage. In: Hall B K (ed) Cartilage. Vol 1. Structure, function and biochemistry. Academic Press: New York: pp 87–104

Sheldon H 1983b Transmission electron microscopy. In: Hall B K (ed) Cartilage. Vol 1. Structure, function and biochemistry. Academic Press: New York: pp 87–104

Sheldon H, Robinson R A 1958 Studies on cartilage: electron microscopic observations on normal rabbit ear cartilage. J Biosphys Biochem Cytol 4: 401–406

Sheldon W H, Stevens S S 1942 The variety of temperament. Harper: New York

Sheldon W H, Stevens S S, Tucker W B 1940 The varieties of human physique. Harper: London

Shelton J H, Jones A L 1971 The fine structure of the mouse adrenal cortex and the ultrastructural changes in the zona glomerulosa with low and high sodium diets. Anat Rec 170: 147–182

Shepard E 1951 Tarsal movements. J Bone Jt Surg 33B: 258–263

Shepherd G M 1974 Synaptic organization of the brain. Oxford University Press: London

Shepherd G M 1978 Microcircuits in the nervous system. Sci Am 238: (Feb), 92–103

Sheps J G 1945 Nuclear configuration and cortical connections of human thalamus. J Comp Neurol 83: 1–56

Sherf L, James T N 1979 Fine structure of cells and their histologic organization within internodal pathways of the heart: clinical and electrocardiagraphic implications. Am J Cardiol 44: 345–369

Sheridan M N, Sladek J R Jr 1975 Histofluorescence and ultrastructural analysis of hamster and monkey pineal. Cell Tissue Res 164: 145–152

Sheridan M N, Reiter R J, Jacobs J J 1969 An interesting anatomical relationship between the hamster pineal gland and the ventricular system of the brain. J Endocr 45: 131–132

Sherrington C S 1905 On recripocal innervation of antagonistic muscles. Proc R Soc Lond B 76: 160–163

Sherrington C S 1906 The integrative action of the nervous system. Scribner: New York, reprinted by Yale University Press in 1947

Sherrington Sir C S 1947 The integrative action of the nervous system. 2nd edition, Yale University Press: New Haven Connecticut

Shettles L B 1953 Observations on human follicular and tubal ova. Am J Obstet Gynec 66: 235–247

Shettles L B 1955 A morula stage of human ovum developed in vitro. Fertil Steril 6: 287–289

Shettles L B 1958 The living human ovum. Am J Obstet Gynec 76: 398–406

Shettles L B 1960 Ovum Humanum: growth, muturation, nourishment. Fertilisation and early development. Hafner: New York

Shimaguchi S 1974 Tenth rib is floating in Japanese. Anat Anz 135: 72–82

Sholl D 1953 Dendritid organization in the neurons of the visual and motor cortices of the cat. J Anat 87: 387–406

Sholl D A 1955 The organization of the visual cortex in the cat. J Anat. 89: 33–46

Sholl D A 1956 The organisation of the cerebral cortex. Methuen: London

Showers M J C 1959 The cingulate gyrus: additional motor area and cortical autonomic regulator. J Comp Neurol 112: 231–310

Showers M J C, Crosby E C 1958 Somatic and visceral responses from the cingulate gyrus. Neurology 8: 561–565

Showers M J C, Lauer E W 1961 Somatovisceral motor pathways in the insula. J Comp Neurol 117: 107–116

Shrewsbury M M, Kuzynski K 1974 Flexor digitorum superficialis tendon in the fingers of the human hand. Hand 6: 121–133

Shriver J E, Stein B M, Carpenter M B 1968 Central projections of spinal dorsal roots in the monkey. 1. Cervical and upper thoracic dorsal roots. Am J Anat 123: 27–74

Shuster S, Black M M, McVitie E 1975 The influence of age and sex on skin thickness, skin collagen and density. Br J Dermatol 93: 639–643

Shute C C D 1956 The evolution of the mammalian ear drum and the tympanic cavity. J Anat 90: 261–281

Shute C C D, Lewis P R 1965 Cholinesterase-containing pathways of the hindbrain: afferent cerebellar and centrifugal cochlear fibres. Nature 205: 242–246

Shute C C D, Lewis P R 1967a The ascending cholinergic reticular system: neocortical olfactory and subcortical projections. Brain 90: 497–520

Shute C C D, Lewis P R 1967b The cholinergic limbic system : projections to hippochampal formation, medial cortex, nuclei of the ascending cholingergic reticular system, and the subfornical organ and supra-optic crest. Brain 90: 521–540

Sicher H 1962 Temporomandibular articulation: concepts and misconceptions. J Oral Surg 20: 281–284

Sicher H 1980 Sicher's oral anatomy. 7th edition by E L DuBrul. Mosby: St Louis

Sick H, Koritke J G 1976 La 7e articulation sterno-costale. Arch Anat Histol Embryol (Strasb) 59: 151–164

Sick H, Ring P 1976 La vascularisation de l'articulation sterno-costo-claviculaire. Arch Anat Histol Embryol (Strasb) 59: 71–78

Sidman R L 1970 Cell proliferation, migration and interaction in the developing mammalian central nervous system. In: Schmitt F O, Quarton G C, Melnechuk T, Adelman G (eds) The neurosciences. Second study program. Rockefeller University Press: New York: pp 100–107

BIBLIOGRAPHY

Sidman R L 1974a Cell-cell recognition in the central nervous system. In: Schmitt F O and Worden F G (eds) The neurosciences. Third study program. MIT Press: Cambridge Massachusetts: pp 743–768

Sidman R L 1974b Contact interaction among developing mammalian brain cells. In: Moscona A H (ed) The cell surface in development. Wiley: New York: pp 221–253

Sidman R L, Miale I L, Feder N 1959 Cell proliferation and migration in the primitive ependymal zone. Exp Neurol 1: 322–333

Siegel P, Mengert W F 1961 Internal iliac artery ligation in obstetrics and gynecology. J Am Med Assoc 178: 1059–1062

Siegel R C 1979 Lysyl oxidase. Int Rev Conn Tissue Res 8: 73–118

Siekevitz P, Palade G E 1960 A cytochemical study on the pancreas of the guinea pig. 6. Release of enzymes and ribonucleic acid from ribonucleoprotein particles. J Biophys Biochem Cytol 7: 631–644

Sierociński W 1975 Arteries supplying the left colic flexure in man. Folia Morphol 34: 117–124

Sierociński W 1976 Anastomoses of the arteries supplying the descending and sigmoid colon in man. Folia Morphol 35: 467–479

Sievers J, Mangold U, Berry M, Allen C, Schlossberger H G 1981 Experimental studies on cerebellar foliation. I. A qualitative morphological analysis of cerebellar fissuration defects after neonatal treatment with 6-OHDA in the rat. J Comp Neurol 203: 751–769

Siffert R S 1956 The effect of staples and longitudinal wires on epiphysial growth: an experimental study. J Bone Jt Surg 38A: 1077–1088

Sigal M J, Pitaru S, Aubin J E, Ten Cate A R 1984 A combined scanning electron microscopy and immunofluorescence study demonstrating that the odontoblast process extends to the dentinoenamel junction in human teeth. Anat Rec 210: 453–462

Sikri K L, Foster C L 1981 Light and electron microscopical observation on the macula dense of the Syrian hamster kidney. J Anat 132: 57–69

Silva D G 1966 The fine structure of multivesicular cells and large microvilli in the epithelium of the mouse colon. J Ultrastruct Res 16: 693–705

Silva D G, Hart J A L 1967 Ultrastructural observations on the mandibular condyle of the guinea pig. J Ultrastruct Res 20: 227–243

Silva D G, Kailis D G 1972 Ultrastructural studies on the cervical loop and the development of the amelo-dentinal junction in the cat. Arch Oral Biol 17: 279–290

Silver I A 1973 Some factors affecting wound healing. Equine Vet J 5: 47–51

Silver J A 1984 The physiology of wound healing. Schweiz Rundschau Med (praxis) 73: 30

Silver M D, Lam J H C, Ranganathan N, Wigle E D 1971 Morphology of the human tricuspid valve. Circulation 43: 333–348

Silver P H 1956 The role of the peritoneum in the formation of the septum recto-vesicale. J Anat 90: 538–546

Silverman A J, Zimmerman E A 1983 Magnocellular neurosecretory system. Ann Rev Neurosci 6: 357–380

Silverstein H, Davies D G, Griffin W L Jr 1969 Cochlear aqueduct obstruction: changes in perilymph biochemistry. Ann Otol Rhinol Laryngol 78: 532–541

Simantov R, Kuhar M J, Uhl G R, Snyder S H 1977 Opioid peptide enkephalin: immunohistochemical mapping in the rat nervous system. Proc Natl Acad Sci USA 74: 2167–2171

Simard L C 1937 Les complexes neuro-insulaires du pancréas humain. Arch Anat Microsc 33: 49–64

Simionescu M 1980 Ultrastructural organization of the alveolar-capillary unit. In: Metabolic activities of the lung. (Ciba Foundation Symposium 78 (NS)). Excerpta Medica: Amsterdam

Simionescu M, Simionescu N, Palade G E 1982 Differentiated microdomains on the luminal surface of capillary endothelium. Distribution of lectin receptors. J Cell Biol 94: 406–413

Simionescu M, Simionescu N, Palade G E 1984 Partial chemical characterization of the anionic sites in the basal lamina of fenestrated capillaries. Microvasc Res 28: 352–367

Simionescu N 1983 Cellular aspects of transcapillary exchange. Physiol Rev 63: 1536–1568

Simionescu N, Simionescu M, Palade G E 1972 Permeability of intestinal capillaries. Pathway followed by dextrans and glycogens. J Cell Biol. 53: 365–392

Simionescu N, Simionescu M, Palade G E 1973 Permeability of muscle capillaries to exongenous myoglobin. J Cell Biol 57: 424–452

Simionescu N, Simionescu M, Palade G E 1975 Permeability of muscle capillaries to small hemepeptides. Evidence for the existence of patent transendothelial channels. J Cell Biol 64: 586–607

Simons K, Fuller S D 1985 Cell surface polarity in epithelia. Annu Rev Cell Biol 1: 243–288

Simons P, Reid L 1969 Muscularity of pulmonary arterial branches in the upper and lower lobes of the normal young and aged lung. Br J Dis Chest 63: 38–44

Sinclair D 1967 Cutaneous sensation, Oxford University Press: London

Sinclair D 1969 Human growth after birth, Oxford University Press: London

Sinclair D 1973 Normal anatomy of sensory nerves and receptors. In: Jarrett A (ed) The physiology and pathophysiology of the skin. Academic Press: New York: pp 348–402

Sindou M, Quoex O, Baleydier C 1974 Fibre organization at the posterior spinal cord-rootlet junction in man. J Comp Neurol 153: 15–26

Singer C 1931 A short history of biology, Clarendon Press: Oxford

Singer C 1956 Galen on anatomical proceedings (translation of the surviving books and introduction and notes), Oxford University Press: London

Singer C, Underwood E A 1962 A short history of medicine. 2nd edition. Clarendon Press: Oxford

Singer I I 1979 The fibronexus; a transmembrane association of fibronectin-containing fibers and bundles of 5 nm microfilaments in hamster and human fibroblasts. Cell 16: 675–685

Singer I I, Kawka D W, Kazarzis D M, Clark R A F 1984 In vivo co-distribution of fibronection and actin fibers in granulation tissue: immunofluorescence and electron microscope studies of the fibronexus at the myo-fibroblast surface. J Cell Biol 98: 2091–2106

Singer K H, Haynes B 1987 Epithelial-thymocyte interactions in human thymus. Human Immunol 20: 127–144

Singer R 1953 Estimation of age from cranial suture closure: report on its unreliability. J For Med 1: 52–59

Singer S J, Nicolson G L 1972 The fluid mosaic model of the structure of cell membranes. Science 175: 720–731

Singh I 1959 Variations in the metacarpal bones. J Anat 93: 262–267

Singh I 1960 Variations in the metatarsal bones. J Anat 94: 345–350

Singh M, Webster P D 1978 Neurohormonal control of pancreatic secretion. A review. Gastroenterology 74: 294–309

Singh S 1965 Variations of the superior articular facets of atlas vertebrae. J Anat 99: 565–571

Singh S, Dass R 1960 The central artery of the retina. I. Origin and course. II. A study of its distribution and anastomoses. Br J Ophthalmol 44: 193–212, 280–299

Singh S, Potturi B R 1978 Greater sciatic notch in sex determination. J Anat 125: 619–624

Singh S, Singh S P 1974 Weight of the femur–a useful measurement for identification of sex. Acta Anat 87: 141–145

Singh S, Singh S P 1975 Identification of sex from tarsal bones. Acta Anat 93: 568–573

Sinha D N 1985 Cancellous structure of tarsal bones. J Anat 140: 111–117

Sirang H 1973 Ein canalis alae ossis ilii und seine bedeutrung. Anat Anz 133: 225–238

Širca A, Kostevc V 1985 The fibre type composition of thoracic and lumbar paravertebral muscles in man. J Anat 141: 131–137

Sisca R F, Provenza D V 1972 Initial dentin formation in human deciduous teeth. An electron microscopic study. Calcif Tiss Res 9: 1–16

Sissons H A 1969 Anatomy of the motor unit. In: Walton J N (ed) Disorders of voluntary muscle. Churchill: London: pp 1–16

Siwe S A 1931 The cervical part of the ganglionated cord, with special reference to its connections with the spinal nerves and certain cerebral nerves. Am J Anat 48: 479–497

Sjöstrand F S 1953a Ultrastructure of inner segments of retinal rods of the guinea pig eye as revealed by the electron microscope. J Cell Comp Physiol 42: 45–70

Sjöstrand F S 1953b Ultrastructure of outer segments of rods and cones of the eye as revealed by the electron microscope. J Cell Comp Physiol 42: 15–44

Sjöstrand F S 1961 Electron microscopy of the retina. In: Smelser G K (ed) The structure of the eye. Academic Press: New York: pp 1–20

Sjöstrand F S 1967 Electron miscroscopy of cells and tissues. Vol 1. Academic Press: New York

Skoglund S 1956 Anatomical and physiological studies of knee joint innervation in the cat. Acta Physiol Scand 36: Suppl 124, 1–101

Skoglund S 1973 Joint receptors and kinaesthesis. In: Iggo A (ed) Handbook of sensory physiology. Vol 2. Springer: Berlin: pp 110–136

Skórnicki R, Zieniánski A, Orebbwski A 1968 Galáz zewnetrzna nerwu krtaniowego górnego u czlowieka i psa. Folia Morphol 27: 79–87

Slavin B G 1985 The morphology of adipose tissue. In: Cryer A, Van R L R (eds) New Perspectives in adipose tissue: structure, function and development. Butterworths: London: pp 23–43

Sledzinski Z, Tyszkiewicz T 1975 Hepatic veins of the right part of the liver in man. Folia Morphol 34: 315–322

Sleigh M A 1974 Cilia and flagella. editor. Academic Press: London

Sleigh M A 1977 The nature and action of respiratory tract cilia. In: Brain J D, Proctor D F, Reid L M (eds) Respiratory defense mechanisms. Part 1. Dekker: New York: pp 247–288

Slipka J 1972 Early development of the bursar pharyngea. Folia Morphol 20: 138–140

Small J V 1974 Contractile units in vertebrate smooth msucle cells. Nature 249: 324–327

Smart I 1971 Location and orientation of mitotic figures in the developing mouse olfactory epithelium. J Anat 109: 243–251

Smart I H M 1982 Radial unit analysis of hippocampal hitogenesis in the mouse. J Anat 135: 763–793

Smart I H M 1983 Three dimensional growth of the mouse isocortex. J Anat 137: 683–694

Smart I H M, McSherry G M 1982 Growth patterns in the lateral wall of the mouse telencephalon. II. Histological changes during and subsequent to the period of isocortical neuron production. J Anat 134: 415–442

Smelser G K 1966 Electron microscopy of a typical epithelial cell and of the normal human ciliary process. Trans Am Acad Ophthal Otolaryngol 70: 738–754

Smith C A 1959 The physiology of the newborn infant. 3rd editon. Blackwell Scientific: Oxford

Smith C A, Dempsey E W 1957 Electron microscopy of the organ of Corti. Am J Anat 100: 337–368

Smith C G, Richardson F G 1966 The course and distribution of the arteries supplying the visual (striate) cortex. Am J Ophthal 61: 1391–1396

Smith D S 1972 Muscle. Academic Press: New York

Smith G E 1903 Notes on the morphology of the cerebellum. J Anat 37: 329–332

Smith G E 1907 New studies on the folding of the visual cortex and the significance of the occipital sulci in the human brain. J Anat 41: 198–207

Smith G E 1908a The cerebral cortex in Lepidosiren with comparative notes on the interpretation of certain features of the forebrain with other vertebrates. Anat Anz 33: 513–550

Smith G E 1908b Studies in the anatomy of the pelvis, with special reference to the fasciae and visceral supports. I and II. J Anat 42: 191–218, 251–270

Smith G T 1963 The renal vascular patterns in man. J Urol 89: 274–288

Smith J M, Savage R J G 1959 The mechanics of mammalian jaws. School Sci Rev 141: 289–301

Smith J W 1953 The act of standing. Acta Orthopaed Scand 23: 159–168

Smith J W 1954 Muscular control of the arches of the foot in standing: an electromyographic assessment. J Anat 88: 152–163

Smith J W 1956 Observations on the postural mechanism of the human knee joint. J Anat 90: 236–260

Smith J W 1958 The ligamentous structures in the canalis and sinus tarsi. J Anat 92: 616–620

Smith J W 1962a The relationship of epiphysial plates to stress in some bones of the lower limb. J Anat 96: 58–78

Smith J W 1962b The structure and stress relation of fibrous epiphysial plates. J Anat 96: 209–225

Smith J W, Walmsley R 1951 Experimental incision of the intervertebral disc. J Bone Jt Surg 33B: 612–625

Smith J W, Walmsley R 1959 Factors affecting the elasticity of bone. J Anat 93: 505–523

Smith M C 1951 Use of Marchi staining in late stages of human tract degeneration. J Neurol Neurosurg Psychiat 14: 222–225

Smith M C 1957 The anatomy of the spino-cerebellar fibers in man. 1. The course of the fibers in the spinal cord and brain stem. J Comp Neurol 108: 285–352

Smith M C 1967 Stereotatic operations for Parkinson's disease–anatomical considerations. In: Williams D (ed) Modern trends in neurology. Butterworths: London: pp 21–52

Smith M W, Peacock M A 1980 'M' cell distribution in follicle-associated epithelium of mouse Peyer's patch. Am J Anat 159: 157–166

Smith P, Porte D 1976 Neuropharmacology of the pancreatic islets. Annu Rev Pharmacol Toxicol 16: 269–285

Smith R, Sanders W J, Stewart K C 1974 Blood supply to the levator scapulae muscle relative to carotid artery protection. Trans Am Acad Ophthalmol Otolaryngol 78: 128–134

Smith R B 1970 The development of the intrinsic innervation of the human heart between the 10 and 70 mm stages. J Anat 107; 271–280

Smith R B 1971 Intrinsic innervation of the human heart in foetuses between 70 mm and 420 mm crown-rump length. Acta Anat 78: 200–209

Smith U, Ryan J W 1970 An electron microscopic study of the vascular endothelium as a site for bradykinin and adenosine-5'-triphospate inactivation in the rat lung. Adv Exp Med Biol 8: 249–261

Smith U, Ryan J W 1973 Electron microscopy of endothelial cells collected on cellulose acetate paper. Tissue Cell 5: 333–336

Smout C F V, Jacoby F, Lillie E W 1969 Gynaecological and obstetrical anatomy and functional histology, Arnold: London

Smyth G E 1939 Systemisation and central connections of spinal tract and nucleus of trigeminal; clinical and pathological study. Brain 62: 41–87

Sneath R S 1955 The insertion of the biceps femoris, J Anat 89: 550–553

Snider R S 1936 Alterations which occur in mossy terminals of the cerebellum, following transection of the brachium points. J Comp Neurol 64: 417–431

Snider R S 1940 Morphology of the cerebellar nuclei in the rabbit and the cat. J Comp Neurol 72: 399–415

Snider R S 1945 Electro-anatomical studies on a tecto-cerebellar pathway. Anat Rec 91: 299

Snider R S 1952 Inter-relations of cerebellum and brain stem. Res Publ Assoc Res Nerv Ment Dis 30: 267–281

Snider R S 1967 Functional alterations of cerebral sensory areas by the cerebellum. Progr Brain Res 25: 322–333

Snider R S, Eldred E 1951 Electro-anatomical studies on cerebrocerebellar connections in the cat. J Comp Neurol 95: 1–16

Snider R S, Eldred E 1952 Cerebro-cerebellar relationships in the monkey. J Neurophysiol 15: 27–40

Snook T 1950 A comparative study of the vascular arrangements in mammalian spleens. Am J Anat 87: 31–78

Snyder F F 1924 Changes in the human oviduct during the menstrual cycle and pregnancy. Bull Johns Hopkins Hosp 35: 141–146

Snyder S H 1978 Peptide neurotransmitter candidates in the brain: focus on enkephalin, angiotensin II, and neurotensin. In: Reichlin S, Baldessarini R J, Martin J B (eds) The hypothalamus. Research Publications.

Vol 56, Association for Research in Nervous and Mental Disease. Raven Press: New York: pp 233–253

Snyder S H, Bennett J P Jr 1976 Neurotransmitter receptors in the brain: biochemical identification. Annu Rev Physiol 38: 153–175

Sohval A R 1954 Histopathology of cryptorchidism; studies based upon comparative histology of retained and scrotal testes from birth to maturity. Am J Med 16: 346–362

Soifer D ed 1986 Dynamic aspects of microtubule biology. Ann N Y Acad Sci 466:

Solanke T F 1968 The blood supply of the vermiform appendix in Nigerians. J Anat 102: 353–362

Solari A J, Tres L L 1967 The ultrastructure of the human sex vesicle. Chromosoma 22: 16–31

Solicia E, Vassalo G, Capella C 1968 Selective staining of endocrine cells by basic dyes after acid hydrolyisis. Stain Technol 43: 257–263

Socia E, Sampietro R, Capella C 1969 Different staining of catecholamines, 5 hydroxytryptamine and related compounds in aldehyde fixed tissues. Histochemie 17: 273–283

Solcia E, Capella C, Buffa R, Usellini L, Fontana P, Frigerio B 1978 Endocrine cells of the gastrointestinal tract: general aspects, ultrastructure and tumor pathology. Adv Exp Med Biol 106: 11–22

Solcia E, Capella C, Buffa R, Usellini L, Fiocca R, Frigerio B, Tenti P, Sessa F 1981 The diffuse endocrine-paracrine system of the gut in health and disease: ultrastructural features. Scand J Gastroenterol (Suppl) 70: 25–36

Solter M, Paljan D 1973 Variations in shape and dimensions of sigmoid groove, venous portion of jugular foramen, jugular fossa, condylar and mastoid foramina classified by age, sex and body size. Z Anat EntwGesch 140: 319–335

Soltero-Harrington L R, Garcia-Rinaldi R, Albe L W 1969 Total aganglionosis of the colon: recognition and management. J Pediat Surg 4: 330–338

Somlyo A V 1979 Bridging structures spanning the junctional gap at the triad of skeletal muscle. J Cell Biol 80: 743–750

Somlyo A V, Somlyo A P 1968 Electromechanical and pharmacomechanical coupling in vascular smooth muscle. J Pharmacol Exp Ther 159: 129–145

Somlyo A P, Devine C E, Somlyo A V, Rice R V 1973 Filament organization in vertebrate smooth muscle. Philos Trans R Soc London [Biol], 265: 223–229

Sommer J R, Johnson E A 1968 Cardiac muscle. A comparative study of Purkinje fibers and ventricular fibres. J Cell Biol 36: 497–526

Somogyi B, Undi F, Kausz M 1973 Blood supply of the spinal ganglia. Morphol Igazsagugyi Orv Sz 13: 191–195

Somogyi P, Soltész I 1986 Immunogold demonstration of GABA in synaptic terminals of intracellularly recorded, horseradish peroxidase-filled basket cells and clutch cells in the cat's visual cortex. Neuroscience 19: 1051–1065

Somogyi P, Kisvárday Z F, Martin K A C, Whitteridge D 1983 Synaptic connections of morphologically identified and physiologically characterized large basket cells in the striate cortex of the cat. Neuroscience 10: 261–294

Sonesson B 1959 The functional anatomy of the cricoarytenoid joint. Z Anat Entwicklungsgesch 121: 292–303

Sordat B, Hess M W, Cottier H 1971 IgG immunoglobin in the wall of post-capillary venules: possible relationship to lymphocyte recirculation. Immunology 20: 115–118

Sorgente N, Brownell A, Slavkin H C 1977 Basal lamina degradation: the identification of mammalian-like collagenase activity in mesenchymal-derived matrix vesicles. Biochem Biophys Res Commun 74: 448–454

Sorokin S P 1968 Reconstructions of centriole formation and ciliogenesis in mammalian lungs. J Cell Sci 3: 207–230

Sorsby A, Sheridan M 1960 The eye at birth: measurements of the principal diameters in forty-eight cadavers. J Anat 94: 192–197

Sotelo C, Llinas R, Baker R 1974 Structural study of inferior olivary nucleus of the cat: morphological correlates of electrotonic coupling. J Neurophysiol 37: 541–559

Sotelo J R, Porter K R 1959 An electron microscope study of the rat ovum. J Biophys Biochem Cytol 5: 327–342

Soupart P, Clewe T H 1965 Sperm penetration of rabbit zona pellucida inhibited by treatment of ova with neuraminidase. Fertil Steril 16: 677–689

Soupart P, Noyes R W 1964 Sialic acid as a component of the zona pellucida of the mammalian ovum. J Reprod Fertil 8: 251–253

Soupart P, Strong P A 1975 Ultrastructural observations on polyspermic penetration of zona pellucida-free human oöcytes inseminated in vitro. Fertil Steril 26: 523–537

Sousa-Pinto A 1973 Cortical projections of the medial geniculate body in the cat. Adv Anat Embryol Cell Biol 48: 1–42

Sousa-Pinto A, Brodal A 1969 Demonstration of a somatotopical pattern in the cortico-olivary projection in the cat. An experimental-anatomical study. Exp Brain Res 8: 364–386

Southam J A 1959 The inferior mesenteric ganglion. J Anat 93: 304–308

Sow M L, Dintimille H, Padonov N, Sylla S, Argenson C 1975 La vascularisation veineuse du pancréas. Bull Assoc Anat (Nancy) 59: 255–264

Spaček J, Lieberman A R 1974 Ultrastructure and three dimensional organization of synaptic glomeruli in rat somatosensory thalamus. J Anat 117: 487–516

BIBLIOGRAPHY

Spalteholtz W 1924 Die Arterien der Herzwand. Anatomische Untersuchungen an Menschen and Tieren. Hirzel: Leipzig

Spanidis A, Wunsch H, Kaissling B, Kris W 1982 Three-dimensional shape of a Goormaghtigh cell and its contact with a granular cell in the rabbit kidney. Anat Embryol (Berl) 165: 239–252

Spanner R 1932 Neue Befunde über die Blutwege der Darmwand und ihre funktionelle Bedeutung. Morphol Jahrb 69: 394–454

Speidel C C 1932 Studies of living nerves. I. The movements of individual sheath cells and nerve sprouts correlated with the process of myelinsheath formation in amphibian larvae. J Exp Zool 61: 279–331

Speidel C C 1933 Studies of living nerves. II. Activities of ameboid growth cones, sheath cells, and myelin segments, as revealed by prolonged observations of individual nerve fibers in frog tadpoles. Am J Anat 52: 1–80

Speidel C C 1935 Studies of living nerves. III. Phenomena of nerve irritation and recovery, degeneration and repair. J Comp Neurol 6:1; 1–82

Speller A M, Moffat D B 1977 Tubulo-vascular relationships in the developing kidney. J Anat 123: 487–500

Spemann H 1938 Embryonic development and induction. Yale University Press: New Haven

Spencer H 1977 Pathology of the lung, vol 1, 3rd edition, Pergamon Press: Oxford

Spencer L M, Foos R W, Straatsma B R 1969 Meridional folds and meridional complexes in the peripheral retina. Trans Am Acad Ophthal Otolaryngol 73: 204–221

Sperber G H 1976 Craniofacial embryology. 2nd edition. Year Book: Chicago

Sperry R W 1951a Mechanisms of neural maturation. In: Stevens S S (ed) Handbook of experimental psychology. Wiley: New York: pp 236–280

Sperry R W 1951b Regulative factors in the orderly growth of neural circuits. Growth Symp 10: 63–87

Sperry R W 1958 Developmental basis of behavior. In: Roe A, Simpson G T (eds) Behavior and evolution. Yale University Press: New Haven: pp 128–139

Sperry R W 1961a Cerebral organization and behavior. Science. 133: 1749–1757

Sperry R W 1961b Some developments in brain lesion studies and learning. Fed Proc 20: 609–616

Sperry R W 1963 Chemoaffinity in the orderly growth of nerve fiber patterns and connections. Proc Natl Acad Sci U S A 50: 703–710

Sperry R W 1965 Embryogenesis of behavioral nerve nets. In: De Haan R L, Ursprung H (eds) Organogenesis. Holt: New York: pp 161–186

Sperry R W 1970 Perception in the absence of the neocortical commissures. Percept Disord (A R N M D) 48: 123–138

Sperry R W 1971 How a developing brain gets itself properly wired for adaptive function. In: Tobach E, Aronson L R, Shaw E (eds) The biopsychology of development. Academic Press: New York: pp 27–44

Sperry R W 1974 Lateral specialization in the surgically separated hemispheres. In: Schmitt F O, Worden F G (eds) The neurosciences. Third study program. MIT Press: Cambridge Massachusetts

Sperry R W 1977 Problems outstanding in the evolution of brain function. In: Duncan R, Weston-Smith M (eds) The encyclopaedia of ignorance. Pergamon Press: Oxford

Sperry R W, Gazzaniga M S, Bogen J E 1969. Interhemispheric relationships. In: Vinken P J, Bruyn G W (eds) Clinical neurology. Volume 4. North Holland: Amsterdam

Spinelli F 1974 Structure and development of the renal glomerulus as revealed by scanning electron microscopy. Int Rev Cytol 39: 345–378

Spira A 1962 Die lymphknotengruppen (lymphocentra) bei der Säugernein ein homologisierungsversuch. Anat Anz 111: 294–364

Spira A W, Hollenberg M J 1973 Human retinal development ultrastructure of the inner retinal layers. Dev Biol 31: 1–21

Spira J J 1957 Comparison of cardiac and pyloric sphincters. Lancet 2: 1008

Spitznas M 1970 Zur feinstruktur der sog. Membrana limitans externa der meschlichen retina. Albrecht v Graefes Arch Ophthal 180: 44–56

Spitznas, M, Hogan M J 1970 Outer segments of photo-receptors and the pigmented epithelium interrelationships in the human eye. Arch Ophthal 84: 810–819

Spoendlin H 1968 In: De Reuck A V S, Knight J S (eds) Hearing Mechanisms in vertebrates. Ciba Foundation Symposium. Churchill: London

Spoerri P E, Glees P 1977 Subsurface cisterns in the cynomolgus retina. Cell Tissue Res 181: 33–38

Sprague J M 1948 A study of motor cell localization in the spinal cord of the rhesus monkey. Am J Anat 82: 1–26

Sprague J M 1958 The distribution of dorsal root fibres on motor cells in the lumbosacral spinal cord of the cat, and the site of excitatory and inhibitory terminals in monosynaptic pathways. Proc R Soc Lond [Biol] 149: 534–556

Sprague J M 1944 The innervation of the pharynx in the rhesus monkey, and the formation of the pharyngeal plexus in primates. Anat Rec 90: 197–208

Sprague J M, Hongchien H A 1964 The terminal fields of dorsal root fibers in the lumbosacral spinal cord of the cat, and the dendritic organization of the motor nuclei. Progr Brain Res 11: 120–54

Spudich J A, Lin S 1972 Cytochalasin B, its interaction with actin and actomyosin from muscle. Proc Natl Acad Sci U S A 69: 442–446

Squier C A, Johnson N W, Hopps R M 1976 Human oral mucosa: development, structure and function, Blackwell Scientific: Oxford

Srivastava H C 1977 Development of ossification centres in the squamous portion of the occipital bone in man. J Anat 124: 643–649

Srivastava P N, Adams C E, Hartree E F 1965 Enzymic action of acrosomal preparations on the rabbit ovum in vitro. J Reprod Fertil 10: 61–617

Stack H G 1962 Muscle function in the fingers. J Bone Jt Surg 44B: 899–1022

Stack H G 1973 The Palmar fascia, Churchill Livingstone: Edinburgh

Staehelin L A 1974 Structure and functions of intercellular junctions. Int Rev Cytol 39: 191–283

Staehelin L A 1975 A new occludens-like junction linking endothelial cells of small capillaries (probably venules) of rat jejunum. J Cell Sci 18: 545–551

Stalcup S A, Turino G M, Mellins R B 1982 Critical issues in the use of vasoactive substances to assess lurd microvascular injury. Ann N Y Acad Sci 384: 433–457

Stalsberg H, DeHaan R L 1968 Endodermal movements during foregut formation in the chick embryo. Dev Biol 18: 198–215

Stamm T T 1931 The constitution of the ligamentum cruciatum cruris. J Anat 66: 80–83

Stämpfli R 1954 Saltatory conduction in nerve. Physiol Rev 34: 101–112

Stanfield J P 1960 The blood supply of the human pituitary gland. J Anat 94: 257–273

Stanier M W 1977 The function of muscles around a simple joint. J Anat 123: 827–830

Stanley J R, Alvarez O M, Berke E W Jr et al 1981 Detection of basement membrane zone antigens during epidermal wound healing in pigs. J Invest Dermatol 77: 240–243

Stanley J R, Woodley D T, Katz S I, Martin G R 1982 Structure and function of basement membrane. J Invest Dermatol 79: 69s–72s

Staprans I, Dirksen E R 1974 Microtubule protein during ciliogenesis in the mouse oviduct. J Cell Biol 62: 164–174

Stark D 1965 Embryologie, 2nd edition, Thieme: Stuttgart

Starkie C, Stewart D 1931 Intra-mandibular course of inferior dental nerve. J Anat 65: 319–323

Steel F L D, Tomlinson J D W 1958 The 'carrying angle' in man. J Anat 92: 315–317

Steele D G 1970 Estimation of stature from fragments of long limb bones. In: Stewart T D (ed) Personal identification in mass disasters. Smithsonian Institute: Washington DC

Steele D G 1976 The estimation of sex on the basis of the talus and calcaneous. Am J Phys Anthropol 45: 581–588

Steele E J, Blunt M J 1956 The blood supply of the optic nerve and chiasma in man. J Anat 90: 486–493

Stegner H E, Wartenburg H 1961 Electron microscopic and histopochemical studies on the structure and formation of the zone pellucida of human ova. Z Zellforsch 53: 702–713

Steiger H-J, Büttner-Ennever J 1978 Relationship between motor neurons and internuclear neurons in the abducens nucleus. Brain Res 148: 181–188

Stein B M, Carpenter M B 1967 Central projections of portions of the vestibular ganglia innervating specific parts of the labyrinth in the rhesus monkey. Am J Anat 120: 281–318

Stein G S, Borun T W 1972 The synthesis of acidic chromosomal proteins during the cell cycle of HeLa S3 cells. I. The accelerated accumulation of acidic residual nuclear protein before initiation of DNA replication. J Cell Biol 52: 292–307

Steinberger E 1974 Darstellung der somatotropen Zellen der Hypophyse an Semidünnschnitten mit Luxol-Fast-Blue. Mikroskopie 30: 34–37

Steindler A 1955 Kinesiology of the human body under normal pathological conditions. 1. Movement. 2 Muscles. Thomas: Springfield Illinois

Steinhardt G 1958 Anatomy and physiology of the temporomandibular joint: effect of function. Int Dent J 8: 155–156

Steinmann G G, Müller-Hermelink H K 1984a Immunohistological demonstration of terminal transferase (TaT) in the age-involuted human thymus. Immunobiology 166: 45–52

Steinmann G G, Müller-Hermelink H K 1984b Lymphocyte differentiation and its microenvironment in the human thymus. Monogr Dev Biol 17: 142–155

Steinman R M, Lustig D S, Cohn Z A 1974 Identification of a novel cell type in peripheral lymphoid organs of mice. II. Functional properties in vitro. J Exp Med 139: 380–397

Steinmann G G, Klaus B, Müller-Hermelink H K 1986 The involution of the ageing human thymic epithelium is independent of puberty. A morphometric study. Scand J Immunol 22; 563–575

Stelzner F, Lierse W 1967 Über das Verschluss-system der terminalen Speiseröhre. Thoraxchirurgie 15: 676–679

Stenström S 1946 Untersuchungen über die Variation and Kovariation der optischen Elemente des menschlichen Auges. Acta Ophthal Suppl 26: 1–103

Stephan F K, Zucker I 1972 Circadian rhythms in drinking behavior and locomotor activity of rats are eliminated by hypothalamic lesions. Proc Natl Acad Sci U S A 69: 1583–1586

Stephan H, Andy O J 1962 The septum (a comparative study in its size in insectivores and primates). J Hirnforsch 5: 229–244

Stephan H, Bauchet R, Andy O J 1970 Data on size of the brain and of various brain parts in insectivores and primates. In: Noback C R, Montagna W (eds) The primate brain. Advances in primatology. Vol 1. Appleton-Century-Crofts: New York: pp 289–310

Stephens N L, Kroeger E A 1980 Ultrastructure, biophysics, and biochemistry of airway smooth muscle. In: Nadel J A (ed) Physiology and pharmacology of the airways. Dekker: New York (Lung Biology in Health and Disease 15): pp 31–121

Steptoe P C, Edwards R G 1970 Laparoscopic recovery of preovulatory human oocytes after priming of ovaries with gonadotrophins. Lancet 1: 683–689

Steptoe P C, Edwards R G 1976 Reimplantation of a human embryo with subsequent tubal pregnancy. Lancet 1: 880–882

Sterling P, Kuypers H G J M 1967a Anatomical organisation of the brachial spinal cord of the cat. I. The distribution of dorsal root fibers. Brain Res 4: 1–15

Sterling P, Kuypers H G J M 1967b Anatomical organisation of the brachial spinal cord of the cat. II. The motoneuron plexus. Brain Res 4: 16–32

Sterling P, Freed M, Smith R G 1986 Microcircuitry and functional architecture of the cat retina. Trends Neurosci 9: 186–192

Stern J T Jr 1971 Investigations concerning the theory of 'spurt' and 'shunt' muscles. J Biomech 4: 437–453

Sternberger L 1986 Immunocytochemistry, Second edition. John Wiley: New York

Stet R J, Wagenaar-Hilbers J P, Niewenhuis P 1987 Thymus localization of monoclonal antibodies circumventing the blood-thymus barriers. Scand J Immunol 25: 441–446

Stevenson P H 1924 Age order of epiphysial union in man. Am J Phys Anthrop 7: 53–93

Stewart M E, Downing D T, Strauss J S 1983 Sebum secretion and sebaceous lipids. Dermatol Clinics 3: 335–345

Stewart T D 1954 Evaluation of evidence from the skeleton. In: Gradwohl R E H (ed) Legal medicine. Mosby: St Louis: pp 407–450

Stewart W B, Shepherd G M 1985 The chemical senses: taste and smell. In: Swash M, Kennard C (eds) Scientific Basis of Clinical Neurology. Churchill Livingstone: Edinburgh: pp 214–224

Stickland N C 1981 Muscle development in the human fetus as exemplified by m. sartorius: a quantitative study. J Anat 132: 557–579

Stieve H 1926 Ein 13½ tage altes in der Gebärmutter ergaltenes und durch Eingriff gewonnenes menschliches ei. Arch Mikrosk Anat EntwMech 7: 295–402

Stieve H 1927 Der Halsteil der Menschlichen Gebärmutter sein Bau und seine Aufgaben während der Schwangerschaft, der Geburt und des Wochenbeltes. Akadem. Verlag-sgesellschaft: Leipzig

Stieve H 1936 Ein ganz junges, in der Gebärmutter erhaltenes menschliches Ei (Keimling Werner). Z Zellforsch 40: 281–322

Stieve H 1948 Der Bau der Primatenplacenta. Anat Anz 96: 299–328

Stilling B 1846 Disquisitiones de structura et functionibus Cerebri: de structura pontis, Varolii. F Maukius: Jena

Stilling B, Wallach J 1842-43 Untersuchungen über den Bau des Nervensystems. 2 vol 1. Untersuchungen über die Textur des Rückenmarks. 2. Uber die Medulla Oblongata. O Wigand: Leipzig

Stillwell D L Jr 1957 The innervation of tendons and aponeuroses. Am J Anat 100: 289–318

Stimler N P, Bach M K, Bloor C M, Hugli T E 1982 Release of leukotrienes from guinea pig lung stimulated by C5a$_{desArg}$ anaphylatoxin. J Immunol 128: 2247–2257

Stingl G, Tamaki K, Katz S I 1980 Origin and function of epidermal Langerhans' cells. Immunol Revs 53: 149–174

Stirling J W, Chandler J A 1976 Ultrastructural studies of the female breast. I 9+0 cilia in myoepithelial cells. Anat Rec 186: 413–416

Stockard C R, Papanicolau G N 1917 The existence of a typical oestrus cycle in guinea pigs and its histology. Anat Rec 11: 411P

Stockwell R A 1967 Lipid content of human costal and articular cartilages. J Anat 101: 607P

Stockwell R A 1978 Chondrocytes. J Clin Pathol [Suppl] 12: 7–13

Stolinski C, Breathnach A S 1975 Freeze-fracture replication of biological tissues. Techniques, interpretation and applications. Academic Press: London

Stopford J S B 1916 The arteries of the pons and medulla oblongata. Part II. J Anat 50: 255–280

Stopford J S B 1921–22 The nerve supply of the interphalangeal and metacarpo-phalangeal joints, J Anat, 56: 1–11

Stopford J S B 1922–23 The anatomy of so-called deep sensibility. J Anat 57: 199–202

Stratton C J 1976 The high resolution ultrastructure of the periodicity and architecture of lipid-retained and extracted lung multilamellar body laminations. Tissue and Cell 8: 713–728

Stratton C J 1977 The three dimensional aspect of the mammalian lung surfactant myelin figure. Tissue and Cell 9: 285–300

Stratton C J 1978 The ultrastructure of multilamellar bodies and surfactant in the human lung. Cell Tissue Res 193: 219–229

Straus W L 1927 Human ilium: sex and stock. Am J Phys Anthrop 11: 1–28

Straus W L Jr, Rawles M E 1953 Effects of fluorine on calcium metabolism and bone growth in pigs. Am J Anat 92: 361–390

Strauss J S, Fochi P E, Downing D T 1976 The sebaceous glands: twenty-five years of progress. J Invest Dermatol 67: 90–97

Streeter G L 1917 The development of the scala tympani, scala vestibuli and perioticular cistern in the human embryo. Am J Anat 21: 299–320

Streeter G L 1918 The developmental alterations in the vascular system of the brain of the human embryo. Contrib Embryol Carnegi. Inst Washington 8: 5–38

Streeter G L 1919 Factors involved in the formation of the filum terminale. Am J Anat 25: 1–12

Streeter G L 1922 Development of the auricle in the human embryo. Contrib Embryol Carnegie Inst Washington 14: 111–138

Streeter G L 1942 Developmental horizons in human embryos. Descriptions of age group XI, 13 to 20 somites, and age group XII, 21 to 29 somites. Contrib Embryol Carnegie Inst Washington 30: 211–245

Streeter G L 1945 Developmental horizons in human embryos. Description of age group XIII, embryos of about 4 or 5 millimeters long, and age group XIV, period of indentation of the lens vesicle. Contrib Embryol Carnegie Inst Washington 31: 27–63

Streeter G L 1948 Developmental horizons in human embryos. Description of age groups XV, XVI, XVII, and XVIII, being the third issue of a survey of the Carnegie collection. Contrib Embryol Carnegie Inst Washington 32: 133–203

Streeter G L 1949 Developmental horizons in human embryos (fourth issue): A review of the histiogenesis of cartilage and bone. Contrib Embryol Carnegie Inst Washington 33: 149–167

Strum J M, Karnovsky M J 1970 Cytochemical localization of endogenous peroxidase in thyroid follicular cells. J Cell Biol 44: 655–666

Stucke K 1959 Current problems of liver surgery. Munch Med Wochenschr 102: 975–978

Sturrock R R 1974 A light microscope study of glial necrosis with age in the anterior limb of the anterior commissure of pre and postnatal mouse. J Anat 117: 469–474

Sturrock R R 1975 A quantitative electron microscopic study of myelination in the anterior limb of the anterior commissure of the mouse brain. J Anat 119: 67–75

Sturrock R E 1976a Changes in the total number of neuroglia, mitotic cells and necrotic cells in the anterior limb of the mouse anterior commissure following hypoxic stress. J Anat 122: 447–453

Sturrock R R 1976b Quantitative changes in neuroglia in the white matter of the mouse brain following hypoxic stress. J Anat 121: 7–13

Sturrock R R 1982 A scanning and transmission electron microscopic study of the embryonic mouse telencephalon. J Anat 134: 25–40

Sturrock R R 1984 Microglia in the human embryonic optic nerve. J Anat 139: 81–91

Stuurman F J 1916 Die lokalisation der zungenmuskeln im nucleus hypoglossi. Anat Anz 48: 593–610

Suchey J M, Wiseley D V, Green R F, Noguchi T T 1979 Analysis of dorsal pitting in the os-pubis in an extensive sample of modern American females. Am J Phys Anthropol 51: 517–540

Sudeck P 1907 Uber die Gefässversorgung des Mastdarmes in Hinsicht auf die operative Gangrän. Münch Med Wochenschr 54: 1314–1317

Sugimoto T, Itoh K, Mizuno N 1978 Localization of neurons giving rise to the oculomotor parasympathetic outflow: a HRP study in cat. Neurosci Lett 7: 301–305

Sugimoto T, Itoh K, Mizuno N, Nomura S, Konishi A 1979 The site of origin of cardiac preganglionic fibers of the vagus nerve: an HRP study in the cat. Neurosci Lett 12: 53–58

Sugiura S 1957 Physico-chemical studies of the vitreous. Jap J Ophthal 1: 7–13

Sullivan F M 1975 Effects of drugs on fetal development. In: Beard R W, Nathanielsz P W (eds) Fetal physiology and medicine. Saunders: London

Sunderland S 1938 The production of cortical lesions by devascularisation of cortical areas. J Anat 73: 120–129

Sunderland S 1945a Arterial relations of the internal auditory meatus. Brain 68: 56–72

Sunderland S 1945b The actions of the extensor digitorum communis, interosseous and lumbrical muscles. Am J Anat 77: 189–209

Sunderland S 1946 The innervation of the first dorsal interosseous muscle of the hand. Anat Rec 95: 7–10

Sunderland S 1974 Meningeal-neural relations in the intervertebral foramen. J Neurosurg 40: 756–763

Sunderland S, Bedbrook G M 1949 Relative sympathetic contribution to individual roots of the brachial plexus in man. Brain 72: 297–310

Sunderland S, Hughes E S R 1946 The pupilloconstrictor pathway and the nerves to the ocular muscles in man. Brain 69: 301–309

Sundin T, Dahlström A, Norlén L et al 1977 The sympathetic innervation and adrenoreceptor function of the human lower urinary tract in the normal state and after parasympathetic denervation. J Invest Urol 14: 322–328

Sutherland G 1984 The fragile X chromosome. Int Rev Cytol 81: 107–143

Sutherland S D 1965 The intrinsic innervation of the liver. Rev Int Hepat 15: 569–578

Sutherland S D 1966 The intrinsic innervation of the gall bladder in Macaca rhesus and Cavia porcellus. J Anat 100: 261–268

Sutherland S D 1967 The neurons of the gall bladder and gut. J Anat 101: 701–710

Sutton R N 1974 The practical significance of mandibular accessory foramina. Aust Dent J 19: 167–173

BIBLIOGRAPHY

Suzuki N 1972 An electromyographic study of the role of the muscles in arch support of the normal and flat foot. Nagoya Med J 17: 57–79

Suzuki Y, Churg J, Ono T 1972 Phagocytic activity of the alveolar epithelial cells in pulmonary asbestosis. Am J Path 69: 373–388

Swammerdam J 1675 Ephermi vita of afbeeldingh van's menschen leven, vertoont in de wonderbaarelijcke en nooyt gehoorde historie van het vliegent ende een-dagh-levent haft of oever-aas. Een dierken, ten aansien van sijn naam, overal in Neerlandt bekent: maarhet welck binnen de tijt, van vijfouren groegt, geboren wordt, jongh is, twee mal verwelt, teelt, eyeren leght, zaat schiet, out wordt, end sterft. Wolfgang: Amsterdam.

Swann D A 1978 Macromolecules of synovial fluid. In: Sokoloff L (ed) The joints and synovial fluid. Vol 1. Academic Press: New York: pp 407–435

Swann D A 1982 Structure and function of lubricin, the glycoprotein responsible for the boundary lubrication of articular cartilage. In: Franchimont P (ed) Articular synovium. Anatomy, physiology, pathology, pharmacology and therapy. Karger: Basel: pp 45–58

Swarz J R, del Cerro M 1975 Lack of evidence for glial cells originating from the external granular layer in the mouse cerebellum. Neurosci Abs 1: 760

Swarz J R 1976 The presence of Bergmann fibres in prenatal mouse cerebellum and its implications in cerebellar histogenesis. Anat Rec 184: 543

Swarz J R, del Cerro M 1977 Lack of evidence for glial cells originating from the external granular layer in mouse cerebellum. J Neurocytol 6: 241–250

Sweeney W J 1962 The interstitial portion of the uterine tube – its gross anatomy, course and length. Obstet Gynecol 19: 1–8

Swyer G I M 1944 Post natal growth changes in the human prostate. J Anat 78: 130–145

Sykes D 1963 The arterial supply of the human kidney with special reference to accessory renal arteries. Br J Surg 50: 368–374

Sykes D 1964 The correlation between renal vascularisation and lobulation of the kidney. Br J Urol 36: 549–555

Sylvén B 1951 On the biology of the nucleus pulposus. Acta Orthop Scand 20: 275–279

Szabo T, Dussardier M 1964 Les noyaux d'origine du nerf vague chez le mouton, Z Zellforsch Mikrosk Anat 63: 247–276

Székely G 1974 Problems of neuronal specificity in the development of some behavioral patterns in amphibia. In: Gottlieb G (ed) Aspects of neurogenesis. Studies on the development of behavior and the nervous system. Vol 2. Academic Press: New York: pp 115–150

Szentágothai J 1942 Die innere Gliederung des Oculomotoriuskernus. Arch Psychiat NervKrankh 115: 127–135

Szentágothai J 1943 Die Lokalisation der Kehlkopfmuskulatur in den Vaguskernen. Z Anat EntwGesch 112: 704–710

Szentágothai J 1948 Anatomical considerations of monosynaptic reflex arcs. J Neurophysiol 11: 445–454

Szentágothai J 1949 Functional representation in the motor trigeminal nucleus. J Comp Neurol 90: 111–120

Szentágothai J 1950 Recherches expérimentales sur les voies oculogyres Sem Hop Paris 26: 2989–2995

Szentágothai J 1952 The general visceral efferent column of the brain stem. Acta Morphol Acad Sci Hung 2: 313–327

Szentágothai J 1963 The structure of the synapse in the lateral geniculate body. Acta Anat 55: 166–185

Szentágothai J 1964 Neuronal and synaptic arrangements in the substantia gelantinosa Rolandi. J Comp Neurol 122: 219–239

Szentágothai J 1965 The use of degeneration methods in the investigation of short neuronal connexions. Progr Brain Res 14: 1–32

Szentágothai J 1967 The anatomy of complex integrative units in the nervous system. In: Lissák K (ed) Results in neuroanatomy, neurochemistry, neuropharmacology and neurophysiology. Recent developments of neurobiology in Hungary. Vol 1. Akadémiai Kiadó: Budapest: pp 9–45

Szentágothai J 1967b Models of specific neuron arrays in the thalamic relay nuclei. Acta Morph Acad Sci Hung 15: 113–124

Szentágothai J 1967c Synaptic architecture of the spinal motoneuron pool. Electroencephalogr Clin Neurophysiol (suppl 25), 4–19

Szentágothai J 1969a The anatomical substrates of nervous inhibitory functions. Acta Morph Acad Sci Hung 17: 325–327

Szentágothai J 1969b Architecture of the cerebral cortex. In: Jasper H H, Ward A A Jr, Pope A (eds) Basic mechanisms of the epilepsies. Churchill: London: pp 13–28

Szentágothai J 1969c The synaptic architecture of the hypothalamo-hypophyseal neuron systems. Acta Neurol Belg 69: 453–468

Szentágothai J 1970 Glomerular synapses, complex synaptic arrangements and their operational significance. In: Schmitt F O, Quarton G C, Melnechuck T, Adelman G (eds) The neurosciences, a second study program. Rockefeller University Press: New York: pp 427–443

Szentágothai J 1974 The 'module-concept' in cerebral cortex architecture. Brain Res 95: 475–496

Szentágothai J 1978 The neuron network of the cerebral cortex: a functional interpretation. (Ferrier Lecture, 1977). Proc R Soc Lond [B] 201: 219–248

Szentágothai J 1985 The neuronal architectonic principle of the neocortex. An Acad Bras Cienc 57: 249–258

Szentágothai J 1987 The architecture of neural centres and understanding neural organization. In: McLennan H, Ledsome J R, McIntosh C H S, Jones D R (eds) Advances in physiological research. Plenum Press: New York

Szentágothai J, Albert A 1955 The synaptology of Clarke's column. Acta Morphol Acad Sci Hung 5: 43–51

Szentágothai J, Flerkó B, Mess B 1968 Halász B, Hypothalamic control of the anterior pituitary. 2nd edition. Akadémiai Kiadó: Budapest

Szentágothai-Schimert J 1941 Die Bedeutung des Faserkalibers und der Markscheidendicke im Zentralnervensystem. Z Anat Entw Gesch 111: 201–223

Szollosi D 1967 Modification of the endoplasmic reticulum in some mammalian oöcytes. Anat Rec 158: 59–73

Szollosi D G, Ris H 1961 Observations on sperm penetration in the rat. J Biophys Biochem Cytol 10: 275–283

T

Taber E 1961 The cytoarchitecture of the brain stem of the cat. I. Brain stem nuclei of the cat. J Comp Neurol 116: 27–70

Taboada R P 1927 Note sur la structure du corps genouille externe. Trab Lab Invest Biol Univ Madr. 25: 319–329

Tachi S, Tachi C, Lindner H R 1970 Ultrastructural features of blastocyst attachment and trophoblastic invasion in the rat. J Reprod Fertil 21: 37–56

Takahashi D 1913 Zur vergleichenden Anatomoie des Seitenharns im Rückenmark der Vertebraten. Arb Neurol Inst Univ Wien. 20: 62–83

Takahashi Y 1975 Anthropological studies on the humerus of the recent Japanese. J Anthropol Soc Nippon 83: 219–232

Takahashi Y 1976 Anthropological studies on the humerus of the recent Japanese. Acta Anat Nippon 51: 79–88

Takano K 1969 Electron microscopic study of the so-called 'separating zone' in the striated duct cell of the parotid gland. Okajimas Folia Anat Jpn. 46: 201–229

Takaro T, Parra S C, Peduzzi P N 1985 Anatomical relationships between type II pneumonocytes and alveolar septal gaps in the human lung. Anat Rec 213: 540–550

Takashima A, Grinnell F 1984 Human keratinocyte adhesion and phagocytosis promoted by fibronectin. J Invest Dermatol 83: 352–358

Takashima A, Grinnell F 1985 Fibronectin-mediated keratinocyte migration and initiation of fibronectin receptor function in vitro. J Invest Dermatol 85: 304–308

Takei Y, Ozanics V 1975 Origin and development of Bruch's membrane in monkey fetuses: an electron microscopic study. Invest Ophthal 14: 903–916

Talland G A 1965 Deranged memory: a psychonomic study of the amnesic syndrome. Academic Press: New York

Tamarin A, Sreebny L M 1965 The rat submaxillary salivary gland. A correlative study by light and electron microscopy. J Morphol 117: 296–352

Tanagho E A, Miller F R 1970 Initiation of voiding. Br J Urol 42: 175–183

Tanaka O 1976 Time of the appearance of cartilage centers in human embryos with special reference to its individual difference. Okajimas Folia Anat Jpn. 53: 173–198

Tandler B 1965 Ultrastructure of the human submaxillary gland. III. Myoepithelium. Z Zellforsch 65: 852–863

Tandler B 1972 Microstructure of salivary glands. In: Rowe N H (ed) Proceedings of symposium of salivary glands and their secretions. University of Michigan: Ann Arbor, Michigan: pp 8–21

Tandler B, Denning C R, Mandel J D, Kutscher A H 1970 Ultrastructure of human labial salivary glands. III. Myoepithelium and ducts. J Morphol 130: 227–246

Tandler J 1912 In: Keibel and Mall's manual of embryology. 2 vol 1910–1912. Lippincott: Philadelphia, Pa.

Tandler J 1913 Anatomie des Herzens. Gustav Fischer: Jena

Tani M, Yamamoto K, Mishima Y 1980 Apocrine acrosyringeal complex in the human skin. J Invest Dermatol 75: 431–435

Tanner J M 1962 Growth at adolescence. 2nd edition. Blackwell Scientific: Oxford

Tanner J M 1978 Fetus into man: physical growth from conception to maturity. Harvard University Press: Cambridge, Mass.

Tappen N C 1954 A comparative functional analysis of primate skulls by the split-line technique. Human Biol 26: 220–238

Tarkowski A K 1961 Mouse chimaeras developed from fused eggs. Nature 190: 857–860

Tarkowski A K 1963 Studies on mouse chimeras developed from eggs fused in vitro. Natl Cancer Inst Monogr 11: 51–71

Tarkowski A K 1965 Embryonic and postnatal development of mouse chimeras. In: Wolstenholme G E W, O'Connor M (eds) Pre-implantation stages of pregnancy. (Ciba Foundation Symposium). Churchill: London: pp 183–193

Tarlov E 1972 Anatomy of the two vestibulo-oculomotor projection systems. Progr Brain Res 37: 489–491

Tarlov E 1975 Synopsis of current knowledge about association projections from the vestibular nuclei. In: Naunton R F (ed) The vestibular system. Academic Press: New York: pp 55–69

Taton R 1966 A general history of the sciences. Thames & Hudson: London

Taurog A 1970 Thyroid peroxidase and thyroxine biosynthesis. Recent Progr Horm Res 26: 189–247

Taurog A 1978 Thyroid hormone synthesis and release. In: Werner S C, Ingbar S H (eds) The thyroid. A fundamental and clinical text. Harper & Row: New York: pp 31–61

Taussig H B 1961 Congenital malformations of the heart. 2nd edition. Vols 1 and 2. Harvard University Press: Cambridge, Mass.

Tautz C, Rohen H W 1967 Ueber den konstruktiven bau des M. vocalis beim menschen. Anat Anz 120: 409–429

Tavasolli M, Yoffey J M 1983 Bone marrow: structure and function. Liss: New York

Tawara S 1906 Das reizleitungssystem des säugethierherzens. G Fischer: Jena

Tay S S W, Wong W C, Ling E A 1984 An ultrastructural study of the effects of right cervical sympathectomy on the sinuatrial and atrioventricular nodes in the heart of the monkey (Macaca fascicularis). J Anat 139: 449–461

Taylor A 1960 The contribution of the intercostal muscles to the effort of respiration in man. J Physiol 151: 390–402

Taylor A 1972 Muscle receptors in the control of voluntary movement. Paraplegia 9: 167–172

Taylor E W 1965 Control of DNA synthesis in mammalian cells in culture. Exp Cell Res 40: 316–332

Taylor J F, Warrell E, Evans R A 1987 The response of the rat tibial growth plates to distal periosteal division. J Anat 151: 221–231

Taylor J H 1958 The mode of chromosome duplication of crepis capillaris. Exp Cell Res 15: 350–357

Taylor J H, Hozier S C 1976 Evidence for a four micron replication unit in CHO cells. Chromosoma 57: 341–350

Taylor J R 1975 Growth of human intervertebral discs and vertebral bodies. J Anat 120: 49–268

Taylor J R, Twomey L T 1984 Sexual dimorphism in human vertebral shape. J Anat 138: 281–286

Taylor K J W 1978 Atlas of grey scale ultrasonography. Churchill Livingstone: Edinburgh

Taylor S (ed) 1968 Calcitonin. Proceedings of the symposium on thyrocalcitonin and the C cells. London . . . 1967. Heinemann Medical: London

Taylor W K 1964 Cortico-thalamic organization and memory. Proc R Soc Lond. [Biol] 159: 466–478

Telford D, Stopford J S B 1934 Autonomic nerve supply of distal colon; anatomical and clinical study. Br Med J 1: 572–574

Templeton D, Thulin A 1978 Secretory, motor and vascular effects in the sublingual gland of the rat caused by autonomic nerve stimulation. Q Jl Exp Physiol 63: 59–66

Ten Cate A R 1971 Physiological resorption of connective tissue associated with tooth eruption. An electron microscopic study. J Periodont Res 6: 168–181

Ten Cate A R 1972 An analysis of Tomes' granular layer. Anat Rec 172: 137–147

Ten Cate A R 1975 Formation of supporting bone in association with periodontal ligament organization in the mouse. Arch Oral Biol 20: 137–138

Ten Cate A R, Deporter D A 1975 The degradative role of the fibroblast in the remodelling and turnover of collagen in soft connective tissue. Anat Rec 182: 1–14

Ten Cate A R, Freeman E 1974 Collagen remodelling by fibroblasts in wound repair. Preliminary observations. Anat Rec 179: 543–546

Tench E N 1936 Development of the anus in the human embryo. Am J Anat 59: 333–346

Tennyson V M 1969 The fine structure of the developing nervous system. In: Himwich H (ed) Developmental Neurobiology. Part 2, Chapter 3. Thomas: Springfield, Illinois

Tennyson V M 1970 The fine structure of the axon and growth cone of the dorsal root neuroblast of the rabbit embryo. J Cell Biol 44: 62–79

Termine J D, Kleinman H K, Whitson S W, Conn K M, McGarvey M L, Martin G R 1981 Osteonectin, a bone-specific protein linking mineral to collagen. Cell 26: 99–105

Terni T 1922 Ricerche sulla struttura e sull'evoluzione del simpatico dell'uomo. Monitore Zool Ital. 33: 63–72

Terracol J, Calvet J, Granel F, Ardouin P, Fabre L 1965 L'anatomie fonctionelle du larynx. Biol Med 54: 180–255

Terzakis J A 1963 The ultrastructure of normal human first trimester placenta. J Ultrastruct Res 9: 268–284

Terzian H, Ore G D 1955 Syndrome of Klüver and Bucy. Reproduced in man by bilateral removal of the temporal lobes. Neurology, Minneapolis 5: 373–380

Tesi D, Forssmann W G 1970 Untersuchungen am mesenterium der ratte. Anat Anz 126: 365–373

Testa Riva F, Cossu M, Lantini M S, Riva A 1985 Fine structure of human deep posterior lingual glands. J Anat 142: 103–115

Teubér H L 1974 Why two brains? In: Schmitt F O, Worden F G (eds) The neurosciences, third study program. MIT Press: Cambridge, Massachusetts

Teubér H L, Battersby W S, Bender M B 1960 Visual field defects after penetrating missile wounds of the brain. Harvard University Press: Cambridge, Mass.

Thaemert J C 1969 Fine structure of neuromuscular relations in mouse heart. Anat Rec 163: 575–586

Thaysen J H, Thorn N A, Schwartz I L 1954 Excretion of sodium, potassium, chloride and carbon dioxide in human parotid saliva. Am J Physiol 178: 155–159

Thebesius A C 1708 Disputatio medica inauguralis de circulo sanguinis in corde. A Elzevier: Lugduni Batavorum

Theiler K 1957 Uber die differenezierung der rumpfmyotome beim menschen und die herkunft der bauchwandmuskeln. Acta Anat 30: 842–864

Theofilopoulos A N, Carson D A, Tavassoli M, Slovin S F, Speers W C, Jensen F B, Vaughan J H 1980 Evidence for the presence of receptors for C3 and IgG Fe on human synovial cells. Arthritis Rheum 23: 1–9

Thesleff I 1977 Tissue interactions in tooth development in vitro. In: Karkinen-Jääskeläinen M, Saxén L, Weiss L (eds) Cell interactions in differentiation. Sixth Sigrid Jusélius Foundation Symposium in Helsinki, 1976. Academic Press: London: pp 191–207

Thesleff I, Hurmerinta K 1981 Tissue interactions in tooth development. Differentiation 18: 75–88

Thiel G A, Downey H 1921 The development of the mammalian spleen, with special reference to its hematopoietic activity. Am J Anat 28: 279–333

Thody A J, Friedmann P S (eds) 1986 Scientific basis of dermatology. A physiological approach. Churchill Livingstone: Edinburgh

Thom R 1969 A mathematical approach to morphogenesis: archetypal morphologies. In: Defendi V (ed) Heterospecific genome interaction. A Symposium held at the Wistar Institute of Anatomy and Biology . . . 1968. (Wistar Institute Symposium Monograph No 9): pp 165–174

Thom R 1975 Structural stability and morphogenesis: an outline of a general theory of models. Translated from the French original text (1972) by D H Fowler. Benjamin: Reading, Mass.

Thom R, Zeeman E C 1975 Catastrophe theory: its present state and future perspectives. In: Manning A (ed) Dynamical systems, Warwick, 1974. Springer Lecture Notes in Mathematics 468. Springer-Verlag: Berlin: pp 366–401

Thomas C E 1957 The muscular architecture of the ventricles of hog and dog hearts. Am J Anat 101: 17–57

Thomas C E 1965 The ultrastructure of human amnion epithelium. J Ultrastruct Res 13: 65–83

Thomas H F 1979 The effect of various fixatives on the extent of the odontoblast process in human dentine. Arch Oral Biol 28: 465–469

Thomas H F, Carella P 1983 A scanning electron microscope study of dentinal tubules from un-erupted human teeth. Arch Oral Biol 28: 1125–1130

Thomas N R 1976 Collagen as the generator of tooth eruption. In: Poole D F G, Stack M V (eds) The eruption and occlusion of teeth. Proceedings of the 27th Symposium of the Colston Research Society . . . 1975. Butterworths: London: pp 290–301

Thomas P K 1963 The connective tissue of peripheral nerve: an electron microscope study. J Anat 97: 35–44

Thompson W D'A 1942 On growth and form. 2nd edition. Cambridge University Press: Cambridge

Thompson J S, Thompson M W 1980 Genetics in medicine. 3rd edition. Saunders: Philadelphia

Thompson M W, Bandler E 1973 Finger pattern combinations in normal individuals and in Down's Syndrome. Hum Biol 45: 563–570

Thompson R 1919 The capacity of, and the pressure of fluid in, the urinary bladder. J Anat 53: 241–253

Thompson R H S, Wootton I D M eds 1970 Biochemical disorders in human disease. 3rd edition. Churchill: London

Thoms H 1940 Roentgen pelvimetry as a routine prenatal procedure. Am J Obstet Gynecol 40: 891–905

Thorel C 1909 Vorläufige Mitteilung über eine besondere Muskeln Verbindung zwischen der Cava superior und dem Hisschen Bündel. Munch Med Wochenschr 56: 2159

Thorel C 1910 Über den Aufbaum des Sinusknotens und seine Verbindung mit der Cava superior und den Wenckebachschen Bündeln. Munch Med Wochenschr 57: 183

Thornton M W, Schweisthal M R 1969 The phrenic nerve: its terminal divisions and supply to the crura of the diaphragm. Anat Rec 164: 283–290

Thorstensson A 1977 Observations on strength training and detraining. Acta Physiol Scand 100: 491–493

Thuma B D 1928 Studies on the diencephalon of the cat. I. The cytoarchitecture of the corpus geniculatum laterale. J Comp Neurol 46: 173–200

Thyberg J, Friberg U 1970 Ultrastructure and acid phosphatase activity of matrix vesicles and cytoplasmic dense bodies in the epiphyseal plate. J Ultrastruct Res 33: 554–573

Tierney D F 1974 Lung metabolism and biochemistry. Ann Rev Physiol 36: 209–231

BIBLIOGRAPHY

Tiedemann F 1860 Anatomie und Bildungsgeschichte des Gehirns im Foetus des Menschen. Steinishen: Nuremburg

Tillmann B, Pietzsch-Rohrschneider I, Hoenges H L 1977 The human vocal cord surface. Cell Tissue Res 185: 279–283

Tills D, Kopeć A C, Ellis R E 1983 The distribution of the human blood groups and other polymorphisms (suppl 1). Oxford University Press: Oxford

Tilney F 1933 Behavior in its relation to development of the brain. Part II. Correlation between the development of the brain and behavior in the albino rat from embryonic states to maturity. Bull Neurol Inst New York 3: 252–358

Tilney F, Riley H A 1928 The brain from ape to man. Hoeber: New York

Tisell L E, Salander H 1975 The lobes of the human prostate. Scand J Urol Nephrol 9: 185–191

Tobias P V 1970 Brain size, grey matter and race–fact or fiction? Am F Phys Anthropol 32: 3–25

Tobias P V 1971 The brain in hominid evolution. Columbia University Press: New York

Tobin C E 1966 Arteriovenous shunts in the peripheral pulmonary circulation in the human lung. Thorax 21: 197–204

Tobon H, Salazar H 1974 Ultrastructure of the human mammary gland. I. Development of the fetal gland throughout gestation. J Clin Endocrinol Metab 39: 443–456

Todd T W 1920a Age changes in the pubic bone. I. The male white pubis. Am J Phys Anthropol 3: 285–334

Todd T W 1920b Age changes in the pubic bone. II. The pubis of the male negro-white hybrid. III. The pubis of the white female. IV. The pubis of the female negro-white hybrid. Am J Phys Anthropol 4: 1–70

Todd T W 1921a Age changes in the pubic bone. V. Mammalian pubic metamorphosis. Am J Phys Anthropol 4: 333–406

Todd T W 1921b Age changes in the pubic bone. VI. The interpretation of variations in the symphyseal area. Am J Phys Anthropol 4: 407–424

Todd T W 1931 Differential skeletal maturation in relation to sex, race, variability and disease. Child Dev 2: 49–56

Todd T W 1937 Atlas of skeletal maturation of the wrist. Mosby: St Louis

Todd T W, D'Erico J Jr 1928 The clavicular epiphyses. Am J Anat 41: 25–50

Todd T W, Lindàla A 1928 Dimensions of the body; whites and American negros of both sexes. Am J Phys Anthropol 12: 35–119

Todd T W, Lyon D W Jr 1924 Endocranial suture closure, its progress and age relationship. I. Adult males of white stock. Am J Phys Anthropol 7: 325–384

Todd T W, Lyon D W Jr 1925a Endocranial suture closure, its progress and age relationship. II. Ectocranial closure in adult males of white stock. Am J Phys Anthropol 8: 23–45

Todd T W, Lyon D W Jr 1925b Endocranial suture closure, its progress and age relationship. III. Endocranial closure in adult males of negro stock. Am J Phys Anthropol 8: 47–71

Todd T W, Lyon D W Jr 1925c Endocranial suture closure, its progress and age relationship. IV. Ectocranial closure in adult males of negro stock. Am J Phys Anthropol 8: 149–168

Todd T W, Tracy B 1930 Racial features in American negro cranium. Am J Phys Anthropol 15: 53–110

Toivonen S 1967 Mechanism of primary embryonic induction. In: Hagen E, Wechsler W, Zilliken P (eds) Morphological and biochemical aspects of cytodifferentiation. (Exp Biol Med 1) Karger: Basel: pp 1–7

Tolbert N E, Essner E 1981 Microbodies: peroxisomes and glyoxysomes. J Cell Biol 91: 271s–283s

Tolles R B 1875 On the measurement of air-angle. [no publisher]: Boston, Massachusetts.

Tömböl T 1967 Short neurons and their synaptic relations in the specific thalamic nuclei. Brain Res 3: 307–326

Tominaga Y L, Ikui H 1964 The fine structure of the arteriovenous crossing parts in the human retina. Acta Soc Ophthal Jpn. 68: 148–150

Toncray J E, Kreig N J S 1946 Nuclei of the human thalamus: comparative approach. J Comp Neurol 85: 421–459

Töndury G 1943 Zur anatomie der Halswirbelsäule. Gibt es Uncovertebralgelenke? Z Anat EntwGesch 112: 448–459

Töndury G 1958 Entwicklungsgeschichte und Fehlbildungen der Wirbesaule. Thieme: Stuttgart

Tongerson J 1951 Developmental, genetic and evolutionary meaning of metopic suture. Am J Phys Anthropol 9: 193–210

Tonna E A, Pentel L 1972 Chondrogenic cell formation via osteogenic cell progeny transformation. Lab Invest 27: 418–426

Török B 1954 Lebeudbeobachtung des hypophysenkreislaufes an hunden. Acta Morphol Acad Sci Hung. 4: 83–89

Torr J B D 1957 The blood supply of the human cord. M.D. Thesis University of Manchester

Torrent Guasp F 1957 Anatomía functional del corazón. Paz Montalvo: Madrid

Torrent Guasp F 1970 The electrical circulation. Torrent Guasp: Denia

Torrent Guasp F (ed) 1987 Estructura y mecánica del corazón. Ediciones Grass: Barcelona

Torrey T W 1954 The early development of the human nephros. Contrib Embryol Carnegie Inst Washington 35: 175–197

Torvik A 1957 The spinal projections from the nucleus of the solitary tract. An experimental study in the cat. J Anat 91: 314–322

Torvik A, Brodal A 1954 The cerebellar projection of the perihypoglossal nuclei (nucleus intercalatus, nucleus praepositus hypoglossi and nucleus of Roller) in the cat. J Neuropathol Exp Neurol 13: 515–527

Torvik A, Brodal A 1957 The origin of reticulospinal fibers in the cat. An experimental study. Anat Rec 128: 113–137

Toth A 1977 Studies on the muscular structure of the human uterus. II. Fasciculi cervicoangulares: vestigial or functional remnant of the mesonephric duct? Obstet Gynecol 49: 190–196

Tournade A, Maillot C, Koritke J G 1972 Les veines superficielles du tronc cérébral chez l'homme. Arch Anat Histol Embryol 55: 233–281

Tournay A, Paillard J 1953 Electromyographie des muscles radiaux a l'état normal. Revue Neurol 89: 277–279

Tousimis A J, Fine B S 1959 Ultrastructure of the iris: intercellular stromal components. Arch Ophthal 62: 974–976

Tow P M, Whitty C W M 1953 Personality changes after operations on the cingulate gyrus in man. J Neurol Neurosurg Psychiatry 16: 186–193

Tower S S, Richter C P 1931 Injury and repair within the sympathetic nervous system; preganglionic neurons. Archs Neurol Psychiatry 26: 485–495

Townes P L, Holtfreter J 1965 Directed movements and selective adhesion of embryonic amphibian cells. In: Bell E (ed) Molecular and cellular aspects of development. Harper and Row: New York: pp 3–39

Townes-Anderson E, Raviola G 1978 Degeneration and regeneration of autonomic nerve endings in the anterior part of rhesus monkey ciliary muscle. J Neurocytol 7: 583–600

Toyoda Y, Chang M C 1974 Fertilisation of rat eggs in vitro by epididymal spermatozoa and the development of eggs following transfer. J Reprod Fertil 36: 9–22

Tozer F M 1911 On the presence of ganglion cells in the roots of III, IV and VI cranial nerves. Proc Physiol Soc Lond. 1910–1911 p xv

Tramezzani J H, Morita E, Chiocchio S R 1971 The carotid body as a neuroendocrine organ involved in the control of erythropoiesis. Proc Natl Acad Sci USA 68: 52–55

Traquair H M 1948 An introduction to clinical perimetry. 5th edition. Kimpton: London

Travill A A 1964 Transmission of pressures across the elbow joint. Anat Rec 150: 243–247

Travis A M 1955 Neurological deficiencies following supplementary motor area lesions in Macaca mulatta. Brain 78: 174–198

Treitz W 1853 Über einen neuen Muskel am Duodenum. Vjschr Pract Heikunde (Prague) 37: 113–144

Trenouth M J 1984 Shape changes during human fetal craniofacial growth. J Anat 139: 639–651

Treves Sir F 1885a The anatomy of the intestinal canal and peritoneum in man. Lewis: London

Treves Sir F 1885b Lectures on the anatomy of the intestinal canal and peritoneum in man. Br Med J 1: 415, 470, 527, 580

Trier J S 1968 Morphology of the epithelium of the small intestine. In: Code C F (ed) Handbook of physiology. Section 6. Alimentary canal. Vol 3. Intestinal absorption. American Physiological Society: Washington DC: pp 1125–1175

Trolle D 1947 Accessory bones of the human foot. Munskgaard: Copenhagen

Trotter M 1937 Accessory sacro-iliac articulations. Am J Phys Anthropol 22: 247–261

Trotter M 1971 The density of bones in the young skeleton. Growth 35: 221–231

Trotter M, Gleser G C 1958 A re-evaluation of estimation of stature based on measurements of stature taken during life and of long bones after death. Am J Phys Anthropol NS 16: 79–123

Trueta J 1957 The normal vascular anatomy of the femoral head during growth. J Bone Jt Surg 39B: 353–358

Trueta J, Morgan J D 1960 The vascular contribution to osteogenesis. I. Studies by the injection method. J Bone Jt Surg 42B: 97–109

Trueta J, Barclay A E, Daniel P M, Franklin K J, Prichard M M L 1947 Studies of the renal circulation. Blackwell: Oxford

Truex R C, Bishof J K 1958 Conducting system in human hearts with interventricular septal defect. J Thorac Surg 35: 421–439

Truex R C, Taylor M J, Smythe M Q, Gildenberg P 1970 The lateral cervical nucleus of cat, dog and man. J Comp Neurol 139: 93–104

Tsanev R 1975 Cell differentiation and the structure of chromatin. In: Hidvegi E J (ed) Biochemistry of the cell nucleus: Mechanism and regulation of gene expression. North Holland Press: Amsterdam: pp 409–417

Tsanev R, Sendov B 1971 Possible molecular mechanism for cell differentiation in multicellular organisms. J Theor Biol 30: 337–393

Tschumi P 1957 The growth of the hindlimb bud of Xenopus laevis and its dependence upon the epidermis. J Anat 91: 149–173

Ts'o M O, Friedman E 1967 The retinal pigment epithelium. I. Comparative histology. Arch Ophthalmol 78: 641–649

Ts'o M O, Friedman E 1968 The retinal pigment epithelium. III. Growth and development. Arch Ophthalmol 80: 214–216

Tsuchida U 1906 Ueber die Ursprungskerne der Augenbewegungsnerven und über die mit diesen in Beziehung stehenden Bahnen im Mittel- und Zwishenhirn; normal-anatomische, embryologische, pathologisch-anatomische und vergleichend-anatomische Untersuchungen. Arb Hirnanat Inst Zürich 2: 1–205

Turnbull I M, Brieg A, Hassler O 1966 Blood supply of cervical spinal cord in man. A microangiographic cadaver study. J Neurosurg 24: 951–965

Turnbull W D 1970 Mammalian masticatory apparatus. Fieldiana Geology 18: 149–356

Turner D R 1969a The vascular tree of the haemal node in the rat. J Anat 104: 481–494

Turner D R 1969b The vascular tree of the haemal node in the rat. J Anat 104: 481–494

Turner R S 1943 Chromatolysis and recovery of efferent neurons. J Comp Neurol 79: 73–78

Turner W 1886 The index of the pelvic brim as a basis of classification. J Anat 20: 125–143

Turner-Warwick R T 1959 The lymphatics of the breast. Br J Surg 46: 574–582

Twomey L, Taylor J, Furniss B 1983 Age changes in the bone density and structure of the lumbar vertebral column. J Anat 136: 15–25

Tyler A 1967a Masked messenger RNA and cytoplasmic DNA in relation to protein synthesis and processes of fertilisation and determination in embryonic development. Dev Biol (suppl 1): 170–226

Tyler A 1967b Problems and procedures of comparative gametology and syngamy. In: Metz C B, Monroy A (eds) Fertilization. Vol 1. Academic Press: New York: pp 1–26

Tyler A, Bishop D W 1963 Immunological phenomena. In: Hartman C G (ed) Mechanisms concerned with conception. Proceedings of a Symposium under the auspices of the Population Council and the Planned Parenthood Federation of America. Pergamon Press: Oxford: pp 397–482

Tytell M, Black M M, Garner J, Lasek R J 1981 Axonal transport: each rate component reflects the movement of distinct macromolecular complexes. Science 214: 170–181

Tzagoloff A 1982 Mitochondria. Plenum Press: New York

U

Uchizono K 1965 Characterisation of excitatory and inhibitory synapses in the central nervous system of the cat. Nature 207: 642–643

Uchizono K 1967 Synaptic organization of the Purkinje cells in the cerebellum of the cat. Exp Brain Res 4: 97–113

Uhlenhuth E, Hunter D W T, Loechel W E 1952 Problems in the anatomy of the pelvis. Lippincott: Philadelphia

Ullah M 1978 Localization of the phrenic nucleus in the spinal cord of the rabbit. J Anat 125: 377–386

Underhill B M L 1955 Intestinal length in man. Br Med J 2: 1243–1246

Undi F, Somogyi B, Kausz M 1973 Data on blood supply of spinal nerve roots. Acta Morphol Acad Sci Hung. 21: 311–318

Ungerstedt U 1971 Stereotaxic mapping of the monoamine pathways in the rat brain. Acta Physiol Scand. 367: 1–48

Ungváry Gy, Faller J 1962 The arterial segmentation of the kidneys. Morph Igazs Orv Szemle 2: 252–259

Unwin P N T, Ennis P D 1984 Two configurations of a channel-forming membrane protein. Nature 307: 609–613

Unwin P N T, Zampighi G 1980 Structure of the junction between communicating cells. Nature 283: 545–549

Uitto J, Lichtenstein J R 1976 Defects in the biochemistry of collagen in diseases of connective tissue. J Invest Dermatol 66: 59–79

Urist M R 1966 Origins of current ideas about calcification. Clin Orthop Rel Res 44: 13–39

V

Vail A D, Coller F C 1967 The parathyroid glands: clinicopathologic correlation of parathyroid disease as found in 200 unselected autopsies. Missouri Med 64: 234–238

Vaithlingam U D, Wong W C, Ling E A 1984 Light and electron microscopic features of the structure and innervation of the gastro-oesophageal junction of the monkey (Macaca fascicularis). J Anat 138: 471–484

Vallbo Å B 1974a Afferent discharge from human muscle spindles in noncontracting muscles. Steady state impulse frequency as a function of joint angle. Acta Physiol Scand. 90: 303–318

Vallbo Å B 1974b Human muscle spindle discharge during isometric voluntary contractions. Amplitude relations between spindle frequency and torque. Acta Physiol Scand. 90: 319–336

Vallbo Å B, Hagbarth K E, Torebjörk H E, Wallin B G 1979 Somatosensory, proprioceptive, and sympathetic activity in human peripheral nerves. Physiol Rev 59: 919–957

Valverde F 1961a A new type of cell in the lateral reticular formation of the brain stem. J Comp Neurol 117: 189–195

Valverde F 1961b Reticular formation of the pons and medulla oblongata: A Golgi study. J Comp Neurol 116: 71–100

Valverde F 1962 Reticular formation of the albino rat's brain stem; cytoarchitecture and corticofugal connections. J Comp Neurol 119: 25–53

Valverde F 1965 Studies on the piriform lobe. Harvard University Press: Cambridge, Mass.

van Alphen G W H M 1963 The structural changes in miosis and mydriasis of the monkey eye. Arch Ophthal Chicago 69: 802–814

van As A, Webster I 1972 The organisation of ciliary activity and mucus transport in pulmonary airways. S Afr. Med J 46: 347–350

van Beusekom G T 1955 Fibre analysis of the anterior and lateral funiculi of the cord in the cat. Thesis. University of Leiden

Van Buren J M, Borke R C 1972 The variations and connections of the human thalamus. 2 vol. Springer: New York

van Buskirk C 1945 The seventh nerve complex. J Comp Neurol 82: 303–333

van Campenhout E 1925 Etude sur le développement et la signification morphologique des ilots endocrine du pancréas chez l'embryon de mouton. Arch Biol Liège 35: 45–88

Van Campenhout E 1956 Le développement embryonnaire comparé des nerfs olfactif et auditif. Acta Otorhinolaryngol Belg. 11: 279–287

Van den Berg J 1968 Sound production in isolated human larynges. Ann NY Acad Sci 155: 18–27

Vane J R 1969 The release and fate of vaso-active hormones in the circulation. Br J Pharmacol 35: 209–242

Van Gehuchten 1892 Contribution à l'étude des ganglions cérébrospinaux. Cellule 8: 209–231; 233–254

Van Hassel H J, McMinn R G 1972 Pressure differential favouring tooth eruption in the dog. Arch Oral Biol 17: 183–190

van Meurs-Van Woezik H, Krediet P 1982 Measurements of the descending aorta in infants and children: comparison with other aortic dimensions. J Anat 135: 273–279

van Meurs-van Woezik H, Debets T, Klein H W 1987 Growth of the internal diameters in the pulmonary arterial tree in infants and children. J Anat 151: 107–115

van Rhede van der Kloot E, Drukker J, Lemmens H A J, Greep J M 1986 The high thoracic sympathetic nerve system–its anatomical variability. J Surg Res 40: 112–119

Van Valen L 1974 Brain size and intelligence in man. Am J Phys Anthropol 40: 417–424

Vassallo G, Capella C, Solcia E 1971 Endocrine cells of the human gastric mucosa. Z Zellforsch 118: 49–67

Vastesaeger M M, van der Straeten P P, Friart J, Candaele G, Ghys A, Bernard M 1957 Les anastomoses intercoronariennes telles qu'elles apparaissent à la coronarographie postmortem. Acta Cardiol 12: 365–401

Vaughan G M 1984 Melatonin in humans. In: Reiter R J (ed) Pineal research reviews. Vol 2. Liss: New York: pp 141–201

Vaughan G M, Pelham R W, Pang S F, Loughlin L L, Wilson K M, Sandock K L, Vaughan M K, Koslow S H, Reiter R J 1976 Nocturnal elevation of plasma melatonin and urinary 5-hydroxyindoleacetic acid in young men: attempts at modification by brief changes in environmental lighting and sleep and by autonomic drugs. J Clin Endocr Metab 42: 752–764

Vaughn J C 1966 The relationship of the 'sphere chromatophile' to the fate of displaced histones following histone transition in rat spermiogenesis. J Cell Biol 31: 257–278

Veggetti A, Mascarello F, Carpenè E 1982 A comparative histochemical study of fibre types in middle ear muscles. J Anat 135: 333–352

Veleanu C, Grün U, Diaconescu M et al 1972 Structural peculiarities of the thoracic spine. Their functional significance. Acta Anat (Basel) 82: 97–107

Venning P 1956a Radiological studies of variations in the segmentation and ossification of the digits of the human foot. I. Variations in the number of phalanges and centers of ossification of the toes. Am J Phys Anthropol 14: 1–34

Venning P 1956b Radiological studies of variations in the segmentation and ossification of the digits of the human foot. II. Variations in length of the digit segments correlated with differences of segmentation and ossification of the toes. Am J Phys Anthropol 14: 129–151

Verbout A J 1976 A critical review of the 'Neugliederung' concept in relation to the development of the vertebral column. Acta Biotheor (Leiden) 25: 219–258

Verbout A J 1985 The development of the vertebral column. Adv Anat Embryol Cell Biol 90: 1–118

Verhaart W J C 1953 The fibre structure of the cord in the cat. Acta Anat 18: 88–100

Verhaart W J C 1954 Fiber tracts and fiber patterns in anterior and lateral funiculus of cord in Macaca ira. Acta Anat 20: 330–373

Verhaart W J C, Mechelse K 1954 The pedunculus cerebri and the capsula interna. Monatsschr Psychiatry Neurol 127: 65–80

Verhaart W J C, van Beusekom G T 1958 Fibre tracts in the cord in the cat. Acta Physiol Neurol Scand. 33: 359–376

Verna A 1975 Observations on the innervation of the carotid body of the rabbit. In: Purves M J (ed) Peripheral arterial chemoreceptors. Cambridge University Press: Cambridge: pp 75–99

Verney E B 1947 The antidiuretic hormone and the factors which determine its release. Proc R Soc Lond. [Biol] 135: 25–106

BIBLIOGRAPHY

Vernon-Roberts B, Doré J L, Jessop J D, Henderson W J 1976 Selective concentration and localization of gold in macrophages of synovial and other tissues during and after crysotherapy in rheumatoid patients. Ann Rheum Dis 35: 477–486

Verzár F, McDougall E J 1936 Absorption from the small intestine. Longmans, Green: London

Vesalius A 1543 De humani corporis fabrica libri septem. ex off J Oporini: Basel

Vesely T M, Cahill D R 1986 Cross-sectional anatomy of the pericardial sinuses, recesses, and adjacent structures. Surg Radiol Anat 8: 221–227

Vidal F 1940 Pallidohypothalamic tract, or x bundle of Meynert, in the rhesus monkey. Arch Neurol Psychiatry 44: 1219–1223

Vidal G 1984 The oldest eukaryotic cells. Sci Am 249: (Feb) 32–41

Vidić B, Young P A 1967 Gross and microscopic observations on the communicating branch of the facial nerve to the lesser petrosal nerve. Anat Rec 158: 257–261

Vieussens R 1705 Novum vasorum corporis humani systema. P Marret: Amsterdam

Vilas E 1932 Über die Entwicklung der menschlichen Scheide. Z Anat Entwicklungsgesch 98: 263–292

Vilas E 1933 Über die Entwicklung des Utriculus prostaticus beim Menschen. Z Anat Entwicklungsgesch 99: 399–421

Villegas G M 1964 Ultrastructure of the human retina. J Anat 98: 501–513

Villegas J 1975 Effects of cholinergic compounds on the axon-Schwann cell relationship in the squid nerve fiber. Fed Proc 34: 1370–1373

Villiger E 1946 Die Periphere Innervation. 10th edition. Schwabe: Basel

Vincent S A 1966 Postural control of urinary incontinence. The curtsey sign. Lancet 2: 631–632

Virchow R L K 1846 Uber das granulierte Ansehen der Wandungen der Gehirnventrikel. Allg Z Psychiatry 3: 424–450

Vitti M, Basmajian J V 1977 Integrated actions of masticatory muscles. Anat Rec 187: 173–190

Vivien-Roels B, Péret P 1983 The pineal gland and the synchronization of reproductive cycles with variations of the environmental climatic conditions, with special reference to temperature. In: Reiter R J (ed) Pineal Research Reviews. Vol 1. Liss: New York: pp 91–143

Vizoso A D, Young J Z 1948 Internode length and fibre diameter in developing and regenerating nerves. J Anat 82: 110–134

Vlahovitch B, Fuentes J M, Verger A C 1973 Angioarchitecture insulaire chez l'homme et chez les primates. Arch Anat Path 21: 395–399

Voetmann E 1949 On striations and surface area of human choroid plexuses, quantitative anatomical study. Acta Anat 8: 1–116

Vogel H G 1983 Effects of age on biomechanical and biochemical properties of rat and human skin. J Soc Cosmet Chem 34: 453–463

Vogelberg K 1957 Die Lichtungsweite der Koronarostein an normalen und hypertrophen herzen. Z Kreislaufforsch 46: 101–115

Vogt A 1942 Lehrbuch und Atlas der Spaltlampenmikroskopie des lebenden Auges. Teil 3. Iris, Glaskörper, Bindehaut. Enke: Stuttgart

Vogt C, Vogt O 1926 Die vergleichendarchitektonische und die vergleichendreizphysiologische Felderung der Grosshirnrinde unter besonderer Berücksichtigung der menschlichen. Naturwissenschaften 14: 1190–1194

Volpe E, Bellissimo U, Lamberti A 1969. Arch Oral Biol 17:

von Bonin G 1950 Essay on the cerebral cortex. Thomas: Springfield, Ill.

von Bonin G 1962 Anatomical asymmetries of the cerebral hemispheres. In: Mountcastle V B (ed) Interhemispheric relations and cerebral dominance. Johns Hopkins University Press: Baltimore

von der Mark K, Mollenhauer J, Müller P K, Pfäffle M 1985 Anchorin CII, a type II collagen-binding glycoprotein from chondrocyte membranes. Ann NY Acad Sci 460: 214–223

von Düring M, Andres K H 1988 Structure and functional anatomy of visceroreceptors in the mammalian respiratory system. Prog Brain Res 74: 139–154

von Economo C, Koskinas G N 1925 Die Cytoarchitecktonik der hirnrinde. Springer: Berlin

Voneida T J 1965 Visual loss following midline section through the mesencephalic tegmentum in cats. Anat Rec 151: 429 (abstract)

von Euler U S 1936 On specific vasodilating and plain muscle stimulating substances from accessory genital glands in man and certain animals (prostaglandin and vesiglandin). J Physiol 88: 213–234

Von Gudden J B A 1875 Über ein neues Mikrotom. Arch Psychiat Nervenkrank 5: 229–244

Von Haller A 1764 Elementa physiologiae corporis humani. 8 vol Lausanne 1757–1766. (Soc Typograph: Bern, 1st vol 1757)

von Hayek H 1960 The human lung. Hafner: New York

von Herrath E 1958 Bau und Funktion der Normalen Milz. De Gruyter: Berlin

von Holt C 1985 Histones in perspective. Bioessays 3: 120–124

Von Inke G, Schneider W, Schneider U 1966 Anzahl der Papillen und der Pori uriniferi der menschlichen Niere. Anat Anz 118, 241–246

von Knief J-J 1967 Quantitative Untersuchung der Verteilung der Hartsubstanzen im knochen in ihrer Beziehung zur lokalen mechanischen Beanspruchung. Methodik und biomechanische Problematik, dargestellt am Beispiel des coxalen Femurendes. Z Anat Entwickl-Gesch 126: 55–80

von Lawzewitsch I, Dickmann G H, Amezua L, Pardal C 1972 Cytobiological and ultrastructural characterisation of the human pituitary. Acta Anat 81: 286–316

von Luschka H 1850 Die Nerven in der harten Hirnhaut. Laupp: Tubingen

von Luschka H 1860 Icones nervorum capitis. Mohr: Heidelberg

von Monakow C see Monakow C von

Von Nooren G K 1973 Histological studies of the visual system in monkeys with experimental amblyopia. Invest Ophthal 12: 727–738

von Wettstein D, Rasmussen S W, Holm P B 1984 The synaptonemal complex in genetic segregation. Annu Rev Genet 18: 331–413

Voogd J 1964 The cerebellum of the cat. Structure and fibre connexions. Proefschr, Van Gorcum: Assen.

Voss H 1966 Untersuchungen über Vorkommen, Zahne und individuelle Variation der Muskelspindeln in den Muskeln des menschlichen Kehlkopfes. Anat Anz 118: 306–309

Vraa-Jensen G F 1942 The motor nucleus of the facial nerve. Munksgaard: Copenhagen

Vrabec F 1952 Sur la question de l'endothélium de la surface antérieure de l'iris humain. Ophthalmologica 123: 20–30

Vriend J 1983 Pineal-thyroid interactions. In: Reiter R J (ed) Pineal research reviews. Vol 1. Liss: New York: pp 183–206

W

Wachstein M, Bradshaw M 1965 Histochemical localisation of enzymes acting in the kidneys of three mammalian species during their postnatal development. J Histochem Cytochem 13: 44–56

Wada J 1949 A new method for the determination of the side of cerebral speech dominance. Igakuto Seibutsugaku 14: 221–222

Waddington C H 1940 Organisers and genes. Cambridge University Press: Cambridge

Waddington C H 1956 Principles of embryology. Allen & Unwin: London

Waddington C H 1962 New patterns in genetics and development. Columbia University Press: New York

Waddington C H 1966a Fields and gradients. In: Locke M (ed) Major problems in developmental biology. Academic Press: New York: pp 105–124

Waddington C H 1966b The nucleolus-retrospect and prospect. Natl Cancer Inst Monogr 23: 563–572

Wagenvoort C A, Wagenvoort N 1967 Arterial anastomoses, bronchopulmonary arteries and pulmobronchial arteries in perinatal lungs. Lab Invest 16: 13–24

Waggener J D, Beggs J 1967 The membranous coverings of neural tissues: an electron microscopy study. J Neuropathol Exp Neurol 26: 412–426

Wagner D D, Marder V J, Urban-Pickering M 1984 Von Willebrand protein binds to extracellular matrices independently of collagen. Proc Natl Acad Sci USA 81: 471–475

Wagner G 1949 Die Bedeutung der Neuralleiste für die Kopfgestaltung der Amphibienlarven. Untersuchungen and Chimaeren von Triton und Bombinator. Rev Suisse Zool 56: 519–620

Wagner G 1955 Chimaerische Zahnanlagen aus Triton-Schmelzorgan und Bombinator- Papille. Mit Beobachtungen über die Entwicklung von Kiemenzähnchen und Mundsinnesknospen in den Triton-Larven. J Embryol Exp Morphol 3: 160–188

Waite P M E 1977 Normal nerve fibres in the barrel region of developing and adult mouse cortex. J Comp Neurol 173: 165–174

Waites G M H 1970 Temperature regulation and the testis. In: Johnson A D, Gomes W R, Vandemark N L (eds) The testis. Vol I. Academic Press: New York: pp 241–281

Waites G M H, Moule G R 1961 Relation of vascular heat exchange to temperature regulation in the testis of the ram. J Reprod Fertil 2: 213–220

Wakeley C P C 1929 A note on the architecture of the ilium. J Anat 64: 109–110P

Wakeley C P C 1933 The position of the vermiform appendix as ascertained by an analysis of 10,000 cases. J Anat 67: 277–283

Walberg F 1960 Further studies on the descending connections to the inferior olive. Reticulo-olivary fibers: an experimental study in the cat. J Comp Neurol 114: 79–87

Walberg F, Bowsher D, Brodal A 1958 The termination of primary vestibular fibers in the vestibular nuclei in the cat. An experimental study with silver methods. J Comp Neurol 110: 391–419

Walberg F, Brodal A 1953a Pyramidal tract fibers from temporal and occipital lobes. An experimental study in the cat. Brain 76: 491–508

Walberg F, Brodal A 1953b Spino-pontine fibers in the cat. An experimental study. J Comp Neurol 99: 251–288

Waldeyer H 1888 Das Gorilla-Rückenmark. Akademie der Wissenschaften: Berlin

Waldeyer W 1891 Ueber einige neuere forschungen im gebiete des anatomie des centralnervensystems. Dtsch Med Wochenschr 17: 1213–1218, 1244–1246, 1267–1289, 1287–1289, 1331–1332, 1352–1356

Walker A, James B 1977 Elementary cerebral cartography. Chiltern Med Soc Gaz, Jan 42–49

Walker A E 1934 The thalamic projection to the central gyri in Macacus rhesus. J Comp Neurol 60: 161–184

Walker A E 1937 Experimental anatomical studies of the topical localisation within the thalamus of the chimpanzee. Proc K Med Akad Wet 40: 198–206

Walker A E 1938 The primate thalamus. Chicago University Press: Chicago

Walker A E 1942 Somatotopic localisation of spinothalamic and secondary trigeminal tracts in mesencephalon. Arch Neurol Psychiatry Chicago 48: 884–889

Walker A E 1943 Central representation of pain. In: Wolff H G, Gasser H S, Hinsey J S (eds) Pain. (Res Publ Assoc Res Nerv Ment Dis 23: 63–85)

Walker A E, Fulton J F 1938 The thalamus of the chimpanzee. III. Metathalamus. Normal structure and cortical connections. Brain 61: 250–268

Walker P S, Sikorski J, Dowson D, Longfield M D, Wright V, Buckley T 1969 Behaviour of synovial fluid on surfaces of articular cartilage. A scanning electron microscope study. Ann Rheum Dis 28: 1–14

Wall P D 1964 Presynaptic control of impulses at the first central synapse in the cutaneous pathway. Progr Brain Res 12: 92–118

Wall P D 1967 The laminar organization of dorsal horn and effects of descending impulses. J Physiol 188: 403–423

Wall P D 1970 The sensory and motor role of impulses travelling in the dorsal columns towards cerebral cortex. Brain 93: 505–524

Wall P D 1973 Dorsal horn electrophysiology. In: Iggo A (ed) Handbook of sensory physiology. Vol 2. Springer: Berlin: pp 253–270

Wall P D 1976 Modulation of pain by non-painful events. In: Bonica J J, Albe-essard D G eds) Advances in pain research and therapy. Raven Press: New York: pp 1–16

Wall P D 1978 The gate control theory of pain mechanisms. A re-examination and re-statement. Brain 101: 1–18

Wall P D 1985 Pain. In Swash M, Kennard C (eds) Scientific Basis of Clinical Neurology. Churchill Livingstone: Edinburgh: pp 163–171

Wall P D, Gutnick M 1974 Ongoing activity in peripheral nerves: properties of afferent nerve impulses originating from a neuroma. Nature 248: 740–743

Wall R T, Harker L A, Quadracci L J, Striker G E 1978 Factors influencing endothelial cell proliferation in vitro. J Cell Physiol 96: 203–214

Wall R T, Harker L A, Striker G E 1978 Human endothelial cell migration: stimulation by a released platelet factor. Lab Invest 39: 523–529

Waller A V 1850 Experiments on the section of the glossopharyngeal and hypoglossal nerves of the frog, and observations of the alterations produced thereby in the structure of their primitive fibres. Philos Trans R Soc Lond. [Biol] 140: 423–429

Walls E W 1947 Development of specialized conducting tissue of human heart. J Anat 81: 93–110

Walls E W 1958 Observations on the microscopic anatomy of the human anal canal. Br J Surg 45: 504–512

Walls G L 1963 The vertebrate eye. Hafner: New York

Walmsley R 1937 The sheath of the rectus abdominis. J Anat 71: 404–414

Walmsley R 1953 The development and growth of intervertebral disc. Edinb. Med J 60: 341–364

Walmsley R, Watson H 1978 Clinical anatomy of the heart. Churchill Livingstone: Edinburgh

Walmsley T 1915 The costal musculature. J Anat 50: 165–171

Walmsley T 1928 Articular mechanism of diarthroses. J Bone J Surg 10: 40–45

Walmsley T 1929 The heart. In: Quain J Elements of descriptive and practical anatomy. Vol 4, Pt 3. Longmans, Green: London

Walter W G 1953 The living brain. Duckworth: London

Waltner J G 1948 Barrier membrane of cochlear aqueduct: histologic studies on the patency of the cochlear aqueduct. Arch Otolaryngol 47: 656–669

Walton A J 1931 Surgical treatment of parathyroid tumours. Br J Surg 19: 285–291

Walton J, Yoshigama J M, Vanderlaan M 1982 Ultrastructure of the rat urothelium in en face section. J Submicrosc Cytol 1: 1–15

Wang K 1985 Sarcomere-associated cytoskeletal lattices in striated muscle. In: Shay J W (ed) Cell and muscle motility. Vol 6. Plenum Press: New York: pp 315–369

Wang K P, Tai H P 1965 An analysis of variations of the segmental vessels of the right lower lobe in 50 Chinese lungs. Acta Anat Sin 8: 408–423

Wang N S, Ying W L 1977 A scanning electron microscopic study of alkali-digested human and rabbit alveoli. Am Rev Resp Dis 115: 449–460

Wanko T, Gavin M A 1961 Cell surfaces in the crystalline lens. In: Smelser G K (ed) The structure of the eye. Academic Press: New York: pp 221–234

Wanquier A, Rolls E T (ed) 1976 Brain-stimulation reward. North-Holland: Amsterdam

Wansbrough H, Nakanishi H, Wood C 1967 Effect of epinephrine on human uterine activity in vitro and in vivo. Obstet Gynecol 30: 779–789

Warbrick J C 1960 The early development of the nasal cavity and upper lip in the human embryo. J Anat 94: 351–362

Ward F O 1838 Outlines of human osteology. Renshaw: London

Warner F D, Mitchell D R 1980 Dynein the mechanochemical coupling adenosine triphosphatase of microtubule-based sliding filament mechanisms. Int Rev Cytol 66: 1–43

Warner F D, Satir P 1974 The structural basis of ciliary bend formation. Radial spoke positional changes accompanying microtubule sliding. J Cell Biol 63: 35–63

Warr W B 1966 Fiber degeneration following lesions in the anterior ventral cochlear nucleus of the cat. Exp Neurol 14: 453–474

Warren J M, Akert K (eds) 1964 The frontal granular cortex and behaviour. McGraw-Hill: New York

Wartenberg H, Holstein A F 1975 Morphology of the spindle shaped body in the developing tail of human spermatids. Cell Tissue Res 159: 435–443

Wartenburg H, Stegner H E 1960 On the electron microscopic fine structure of the human ovarian egg. Z Zellforsch 52: 450–474

Warwick R 1950a The relation of the direction of the mental foramen to the growth of the human mandible. J Anat 84: 116–120

Warwick R 1950b Study of retrograde degeneration in oculomotor nucleus of rhesus monkey, with a note on method of recording its distribution. Brain 73: 532–543

Warwick R 1951 A juvenile skull exhibiting duplication of the optic canals. J Anat 85: 289–291

Warwick R 1953a Observations upon certain reputed accessory nuclei of the oculomotor complex. J Anat 87: 46–53

Warwick R 1953b Representation of the extra-ocular muscles in the oculomotor nuclei of the monkey. J Comp Neurol 98: 449–503

Warwick R 1954a The ocular parasympathetic nerve supply and its mesencephalic sources. J Anat 88: 71–93

Warwick R 1954b The peculiarities of ciliary ganglion neurons. J Anat 88: 555P

Warwick R 1955 The so-called nucleus of convergence. Brain 78: 92–114

Warwick R 1964 Oculomotor organization. In: Bender M (ed) The oculomotor system. Harper & Row: New York: pp 173–202

Warwick R 1968 The skeletal remains. In: Wenham L P (ed) The Romano-British Cemetery at Trentholme Drive, York. (Ministry of Public Building and Works, Archaeological Report 5). HMSO: London: pp 113–178

Warwick R, Mitchell G A G 1956 Localization of the phrenic nucleus in the spinal cord of man. J Comp Neurol 105: 683–700

Warwick R, Pond J B 1968 Trackless lesions in nervous tissues produced by high intensity focused ultrasound (high frequency mechanical waves). J Anat 102: 387–406

Warwick R, Marshall J, Bron A J (eds) (In press) Eugene Wolff's anatomy of the eye and orbit including the central connexions and comparative anatomy of the visual apparatus (8th edition in press). H K Lewis: London. [see also Wolff E (1976) 7th edition]

Washburn S L 1947 The relation of the temporal muscle to the form of the skull. Anat Rec 99: 239–248

Washburn S L 1949 Sex differences in the pubic bone of Bantu and Bushman. Am J Phys Anthropol 7: 425–432

Wasserman G D 1973 Molecular genetics and developmental biology. Nature, New Biol 245: 163–165

Watanabe H, Yamamoto T Y 1974 Freeze-etch study of smooth muscle cells from vas deferens and taeniae coli. J Anat 117: 553–564

Watanabe Y 1960 An experimental study on the coronary luminal communicating channels in coronary circulation. Jpn. Circ J 24: 11–26

Waters N E 1980 Some mechanical and physical properties of teeth. Symp Soc Exp Biol 34: 99–135

Watson D M S 1917 The evolution of tetrapod shoulder girdle and forelimb. J Anat 52: 1–63

Watson J D, Crick F H C 1953a Genetical implications of the structure of deoxyribonucleic acid. Nature 171: 964–967

Watson J D, Crick F H C 1953b Molecular structure of nucleic acids. A structure for deoxyribose nucleic acid. Nature 171: 737–738

Watterson R L 1965 Structure and mitotic behavior of the early neural tube. In: DeHaan R L, Ursprung H (eds) Organogenesis. Holt: New York: pp 129–159

Watterson R L, Veneziano P, Bartha A 1956 Absence of a true germinal zone in neural tubes of young chick embryos as demonstrated by the colchicine technique. Anat Rec 124: 379

Watzka M 1955 Die Leydigschen Zwischenzellen im Funiculus spermaticus des Menschen. Z Zellforsch 43: 206–213

Weale R A 1970 Optical properties of photoreceptors. Br Med Bull 26: 134–137

Wearn J T 1941 Morphological and functional alterations of coronary circulation. Harvey Lect 17: 754–777

Wearn J T, Mettier S R, Klumpp T G, Zschiesche L J 1933 The nature of the vascular communications between the coronary arteries and the chambers of the heart. Am Heart J 9: 143–164

Webb S M, Lewiński A K, Reiter R J 1985 Somatostatin: its possible relevance to pineal function. In: Reiter R J (ed) Pineal Research Reviews. Vol 3. Liss: New York: pp 215–236

Webber C E, Garnett E S 1976 Density of os calcis and limb dominance. J Anat 121: 203–205

Weber D F 1974 Human dentine sclerosis: a microradiographic survey. Arch Oral Biol 19: 163–169

Weber E H 1834 De Pulsu, resorptione, auditu et tactu. Koehler: Leipzig

Webster H de F 1964 Some ultrastructural features of segmental demyelination and myelin regeneration in peripheral nerve. Prog Brain Res 13: 151–172

Webster H de F 1971 The geometry of peripheral myelin sheaths during their formation and growth in rat sciatic nerves. J Cell Biol 48: 348–367

BIBLIOGRAPHY

Webster K E 1965 The cortico-striate projection in the cat. J Anat 99: 329–337

Webster K E 1974 Changing concepts of the organization of the central visual pathways in birds. In: Bellairs R, Gray E G (eds) Essays on the nervous system. Clarendon Press: Oxford: pp 258–298

Webster K E 1975 Structure and function of the basal ganglia. A non-clinical view. Proc R Soc Med 68: 203–210

Webster K E 1977 Somaesthetic pathways. Br Med Bull 33: 113–120

Webster K E 1978 The brainstem reticular formation. In: Hennings G, Hemmings W A (eds) The biological basis of Schizophrenia. M T P: Lancaster

Webster P D, Black U, Mainz D L, Singh M 1977 Pancreatic acinar cell metabolism and function. Gastroenterology 73: 1434–1449

Weddell G 1941 The pattern of cutaneous innervation in relation to cutaneous sensibility. J Anat 75: 346–367

Weddell G, Palmer E, Pallie W 1955 Nerve endings in mammalian skin. Biol Rev 30: 159–195

Wedgwood M 1966 The peripheral course of the inferior dental nerve. J Anat 100: 639–650

Wee E L, Wolfson L G, Zimmerman E F 1976 Palate shelf movement in mouse embryo culture: evidence for skeletal and smooth muscle contractility. Dev Biol 48: 91–103

Weed L H 1920 The experimental production of an internal hydrocephalus. Contrib Embryol Carnegie Inst Washington 9: 425–446

Weed L H 1938 Meninges and cerebrospinal fluid. J Anat 72: 181–215

Weibel E R 1971 The mystery of 'non-nucleated plates' in the alveolar epithelium of the lung explained. Acta Anat 78: 425–443

Weibel E R 1974a A note on differentiation and divisibility of alveolar epithelial cells. Chest 65 (suppl 4): 19S–21S

Weibel E R 1974b On pericytes, particularly their existence on lung capillaries. Microvasc Res 8: 218–235

Weibel E R, Elias H 1969 Quantitative methods in morphology. Springer-Verlag: Berlin

Weibel E R, Gehr P, Haies D, Gil J, Bachofen M 1976 The cell population of the normal lung. In: Bouhuys A (ed) Lung cells in disease. North-Holland: Amsterdam: pp 3–16

Weigert C 1882 Über eine neue Untersuchungs-methode des Zentralnervensystems. In: Gesammelte Abhandlungen (1906), Vol 2, pp. 533–538

Weil A J 1965 The spermatozoa-coating antigen (SCA) of the seminal vesicle. Ann NY Acad Sci 124: 267–269

Weindl A, Joynt R J 1972 Ultrastructure of the ventricular walls. Three-dimensional study of regional specialization. Arch Neurol 26: 420–427

Weiner N, Schadé J P (eds) 1963 Nerve, brain and memory models. Elsevier: Amsterdam

Weinmann J P, Sicher H 1955 Bone and bones. Fundamentals of bone biology. 2nd edition. Kimpton: London

Weinstock M, Leblond C P 1973 Radioautographic visualization of the deposition of a phosphoprotein at the mineralization front in the dentin of the rat incisor. J Cell Biol 56: 838–845

Weintraub W 1953 Thèse de l'Université de Genève, Paris.

Weisberg J A, Rustioni A 1977 Cortical cells projecting to the dorsal column nuclei of Rhesus monkey. Exp Brain Res 28: 521–528

Weisengreen H H 1975 Observation of the articular disc. Oral Surg 40: 113–121

Weisl H 1953 The relation of movement to structure in the sacroiliac joint. PhD Thesis, University of Manchester

Weisl H 1954a The articular surfaces of the sacro-iliac joint and their relation to the movements of the sacrum. Acta Anat 22: 1–14

Weisl H 1954b The ligaments of the sacro-iliac joint examined with particular reference to their function. Acta Anat 20: 201–213

Weisl H 1955 Movements of the sacro-iliac joint. Acta Anat 23: 80–91

Weiss C, Rosenberg L, Helfert A J 1968 An ultrastructural study of normal young adult human articular cartilage. J Bone Jt Surg 50A: 663–674

Weiss L 1957 A study of the structure of splenic sinuses in man and the albino rat with the light microscope and the electron microscope. J Biophys Biochem Cytol 3: 599–610

Weiss L (ed) 1983a Histology: Cell and tissue biology. 5th edition. Elsevier Science: New York

Weiss L 1983b The spleen. In: Weiss L (ed) Histology. Cell and tissue biology. 5th edition. Macmillan: New York: pp 544–568

Weiss P 1941 Nerve patterns: The mechanics of nerve growth. Growth (suppl 5) 163–203

Weiss P 1950 An introduction to genetic neurology. In: Weiss P (ed) Genetic neurology. Problems of the development, growth, and regeneration of the nervous system and of its functions. Chicago University Press: Chicago: pp 1–39

Weiss P 1961 Guiding principles in cell locomotion and cell aggregation. Exp Cell Res (suppl 8) 260–281

Weiss P 1970 Neural development in biological perspective. In: Schmitt F O, Quarton G C, Melnechuk T, Adelman G (eds) The neuroscience. Second study program. Rockefeller University Press: New York: pp 53–61

Weiss P, Hiscoe H B 1948 Experiments on the mechanism of nerve growth. J Exp Zool 107: 315–396

Welcker H 1856 Ueber Aufbewahrung mikroskopischer Objecte, nebst Mittheilunge über das Mikroskop und dessen Zubehör. Ricker: Giessen

Welgus H G, Campbell E J, Bar-Shavit Z, Senior R M, Teitelbaum S L

1985 Human alveolar macrophages produce a fibroblast-like collagenase and collagenase inhibitor. J Clin Invest 76: 219–224

Weller L G Jr 1933 Development of the thyroid parathyroid and thymus glands in man. Contrib Embryol Carnegie Inst Washington 24: 93–139

Wells L J 1954 Development of the human diaphragm and pleural sacs. Contrib Embryol Carnegie Inst Washington 35: 107–134

Welsh M G 1985 Pineal calcification: Structural and functional aspects. In: Reiter R J (ed) Pineal Research Reviews. Vol 3. Liss: New York: pp 41–68

Wen C Y, Tan C K, Wong W C 1977 Presynaptic dendrites in the cuneate nucleus of the monkey (Macaca fascicularis). Neurosci Lett 5: 129–132

Wen C Y, Wong W C, Tan C K 1978 The fine structural organization of the cuneate nucleus in the monkey (Macaca fascicularis). J Anat 127: 169–180

Wen C Y, Wong W C, Tan C K 1980 Experimental degeneration of motor and sensory cortical terminals in the cuneate nucleus of the monkey (Macaca fascicularis). J Anat 130: 13–23

Wenckebach K F 1908 Beitrage zur Kenntnis der menschlichen Herztätigkeit. Arch Anat Physiol 3: 53

Wende S, Nakayama N, Schwerdtfeger P 1975 The internal auditory artery (embryology, anatomy, angiography, pathology). J Neurol 210: 21–31

Wendell-Smith C P 1967 Studies on the morphology of the pelvic floor. PhD Thesis. University of London

Wendell-Smith C P, Wilson P M 1977 Musculature of pelvic floor. In: Philipp E E, Barnes J, Newton M (eds) Scientific foundations of obstetrics and gynaecology. 2nd edition. Heinemann Medical: London: pp 78–84

Wendell-Smith C P, Williams P L, Treadgold S 1984 Basic human embryology. 3rd edition. Urban and Schwarzenberg: Baltimore

Wendler D 1968 Histologisch-histochemische Befunde an der Schleimhaut des Ductus deferens (pars funicularis) beim geschlectsreifen Mann. Acta Histochem 31: 48–69

Wendt R, Albe-Fessard D 1962 Sensory responses of the amygdala with special reference to somatic afferent pathways. In: Physiologie de l'Hippocampe. Centre National de la Recherche Scientifique: Paris: pp 171–200

Wenink A C G 1971 Some details on the final stages of heart septation in the human embryo. Thesis. Leiden.

Wenink A C G 1976a Development of the human cardiac conducting system. J Anat 121: 617–631

Wenisch H J 1976 Retino-hypothalamic projections in the mouse: electron microscopic and iontophoretic investigations of hypothalamic and optic centres. Cell Tissue Res 167: 547–561

Wennberg E, Weiss L 1969 The structure of the spleen and hemolysis. Ann Rev Med 20: 29–40

Wentink G H 1976 The action of the hind limb musculature of the dog in walking. Acta Anat 96: 70–80

Werneck H J L 1957 Contribuição paro o estudo de alguns aspectos morfologicos de M fibularis tertius. An Fac Med Univ Minas Gerais 17: 417–520

Werner S 1978 Immune system. III. Role in thyroid disease. In: Werner S, Ingbar S H (eds) The thyroid. Harper and Row: New York: pp 615–623

Werner S, Ingbar S H (eds) 1978 The thyroid. 5th edition. Harper and Row: New York

Wertheim M G 1847 Mémoire sur l'élasticité et la cohésion des principaux tissus du corps humain. Ann Chim Phys 21: 385

Wessels N K, Spooner B S, Ash J F, Bradley M O, Luduena M A, Taylor L E, Wrenn J T, Yamada K M 1971 Microfilaments in cellular and developmental processes. Science 171: 135–143

Wesson M B 1951 Rationale of prostatectomy. Am J Surg 82: 714–719

West G B, Shepherd D M, Hunter R B, MacGregor A R 1953 The function of the organs of Zuckerkandl. Clin Sci 12: 317–326

Westheimer G 1954 Mechanism of saccadic eye movements. Arch Ophthal 52: 710–724

Weston J A 1970 The migration and differentiation of neural crest cells. In: Abercrombie M, Brachet J (eds) Advances in morphogenesis. Vol 8. Academic Press: New York: pp 41–114

Wetzel B K, Spicer S S, Wollman S H 1965 Changes in fine structure and acid phosphatase localization in rat thyroid cells following thyrotropin administration. J Cell Biol 25: 593–618

Wheatley V R (ed) 1986 The sebaceous glands. The physiology and pathophysiology of the skin. Vol 9. (Jarrett A ed). Academic Press: London

Whillis J 1930 A note on the muscles of the palate and the superior constrictor. J Anat 65: 92–95

Whillis J 1931 Lower end of the oesophagus. J Anat 66: 132–133P

Whillis J 1946 Movements of the tongue in swallowing. J Anat 80: 115–116

White J C, Smithwick R H, Simeone F A 1952 The autonomic nervous system. 3rd edition. Kimpton: London

White J W 1943 Torsion of achilles tendon; its surgical significance. Archs Surg 46: 784–787

White L E 1959 Ipsilateral afferents to the hippocampal formation in the albino rat. I. Cingulum projections. J Comp Neurol 113: 1–42

White L E Jr 1965 A morphological concept of the limbic lobe. Int Rev Neurobiol 8: 1–34

White E G 1935 Die struktur des glomus caroticum. Beitr Path Anat 96: 177–227

Whitehouse H L K 1973 Towards an understanding of the mechanism of heredity. 3rd edition. Arnold: London

Whitehouse H L K, Hastings P J 1965 The analysis of genetic recombination on the polaron hybrid DNA model. Genet Res 6: 27–92

Whitehouse W J 1975 Scanning electron micrographs of cancellous bone from the human sternum. J Pathol 116: 213–224

Whitehouse W J 1977 Cancellous bone in the anterior part of the iliac crest. Calcif Tissue Res 23: 67–76

Whitehouse W J, Dyson E D 1974 Scanning electron microscope studies of trabecular bone in the proximal end of the human femur. J Anat 118: 417–444

Whitelaw M J 1933 Tubal contractions in relation to estrus cycle as determined by uterotubal insufflation. Am J Obstet Gynecol 25: 475–484

Whiteley H J, Stoner H B 1957 The effect of pregnancy on the human adrenal cortex. J Endocrinol 14: 325–334

Whitfield I C 1967 The auditory pathway. Arnold: London

Whitfield I C, Evans E F 1965 Responses of auditory cortical neurons to changing frequency. J Neurophysiol 28: 655–672

Whitlock D G, Nauta W J H 1956 Subcortical projections from the temporal neocortex in Macaca mulatta. J Comp Neurol 106: 183–212

Whitmore I 1982 Oesophageal striated muscle arrangement and histochemical fibre types in guinea-pig, marmoset, macaque and man. J Anat 134: 685–695

Whitmore I, Notman J A A 1987 A quantitative investigation into some ultrastructural characteristics of guinea-pig oesophageal striated muscle. J Anat 153: 233–239

Whitnall S E 1911 The relation of the lacrimal fossa to the ethmoidal cells. Ophthal Rev 30: 321–325

Whitnall S E 1932 Anatomy of the human orbit. 2nd edition. Oxford University Press: London

Whittaker D K, Adams D 1971 The surface layer of human foetal skin and oral mucosa: a study by scanning and transmission electron microscopy. J Anat 108: 453–464

Whittaker D K, Richards D 1978 Scanning electron microscopy of the neonatal line in human enamel. Arch Oral Biol 23: 45–50

Whittier J R, Mettler F A 1949 Studies on the subthalamus of the rhesus monkey. I. Anatomy and fibre connections of the subthalamic nucleus of Luys. J Comp Neurol 90: 281–318

WHO see World Health Organization

Wiesel T N, Hubel D H 1963a Effects of visual deprivation on morphology and physiology of cells in the cat's lateral geniculate body. J Neurophysiol 26: 978–993

Wiesel T N, Hubel D H 1963b Single cell responses in striate cortex of kittens deprived of vision in one eye. J Neurophysiol 26: 1003–1017

Wiesenhaan P F 1972 Fetography. Am J Obstet Gynecol 113: 819–822

Wijngaert F P van de, Kendall M D, Schuurman H-J, Rademakers L H P M, Kater L 1984 Heterogeneity of epithelial cells in the human thymus. An ultrastructural study. Cell Tissue Res 237: 227–237

Wilde F R 1949 Anal intermuscular spasm. Br J Surg 36: 279–285

Wilde F R 1951 Perivascular neural pattern of femoral region. Br J Surg 39: 97–105

Wildi E, Frauchiger E 1965 Modifications histologiques de l'epiphyse humaine pendant l'enfance, l'age adulte et le viellissement. Progr Brain Res 10: 218–233

Wiles P 1935 Movements of lumbar vertebrae during flexion and extension. Proc R Soc Med 28: 647–651

Wilkin P, Bursztein M 1958 Etude quantitative de l'évolution au cours de la grossesse, de la superficie de la membrane d'échange du placenta humain. In: Snoek J (ed) Le placenta humain. Masson: Paris: pp 211–248

Wilkinson H J 1927 Argyll-Robertson pupil: contribution toward its understanding. Med J Aust. I: 267–272

Wilkinson J L 1953 The insertions of the flexores pollicis longus et digitorum profundus. J Anat 87: 75–88

Wilkinson J L 1954 The terminal phalanx of the great toe. J Anat 88: 537–541

Williams A F 1951 Nerve supply of laryngeal muscles. J Laryngol Otol 65: 343–348

Williams A F 1954 Recurrent laryngeal nerve and the thyroid gland. J Laryngol Otol 68: 719–725

Williams D J 1936 Origin of posterior cerebral artery. Brain 59: 175–180

Williams E E 1959 Gadow's arcualia and the development of tetrapod vertebrae. Q Rev Biol 34: 1–32

Williams J F, Svensson N L 1968 A force analysis of the hip joint. Bio-Med Eng 3: 365–370

Williams P L, Hall S M 1970 In vivo observations on mature myelinated nerve fibres of the mouse. J Anat 107: 31–38

Williams P L, Hall S M 1971a Prolonged in vivo observations of normal peripheral nerve fibres and their acute reactions to crush and deliberate trauma. J Anat 108: 397–408

Williams P L, Hall S M 1971b Chronic Wallerian degeneration–an in vivo and ultrastructural study. J Anat 109: 487–503

Williams P L, Kashef R 1968 Asymmetry of the node of Ranvier. J Cell Sci 3: 341–356

Williams P L, Wendell-Smith C P 1971 Some parametric variations between peripheral nerve fibre populations. J Anat 109: 505–526

Williams P L, Wendell-Smith C P, Treadgold S 1969 Basic human embryology. 2nd edition. Lippincott: Philadelphia

Williams T H 1967 Electron microscopic evidence for an autonomic interneuron. Nature 214: 309–310

Williams T H 1971 Morphological interactions of sheath cells and neurites, using the Murray and Thompson model. J Anat 110, 158P

Williams T H, Palay S L 1969 Ultrastructure of the small neurones in the superior cervical ganglion. Brain Res 15: 17–34

Williams T H, Black A C Jr, Chiba T, Bhalla R C 1975 Morphology and biochemistry of small, intensely fluorescent cells of sympathetic ganglia. Nature 256: 315–317

Williams T J, Jose P J 1981 Mediation of increased vascular permeability after complement activation: histamine independent action of C5a. J Exp Med 153: 136–153

Willier B H, Weiss P A, Hamburger, V (eds) 1955 Analysis of Development. Saunders: Philadelphia

Willis A G, Tange J D 1959 Studies of the innervation of the carotid sinus of man. Am J Anat 104: 87–114

Willis R A 1936 Growth of embryo bones transplanted whole in rat's brain. Proc R Soc Lond. [Biol] 120: 496–498

Willis T 1664 Cerebri anatome: cui accessit nervorum descriptio et usus. J Flesher: London

Willis T A 1949 Nutrient arteries of the vertebral bodies. J Bone Jt Surg 31A: 538–540

Willis W D, Willis J C 1966 Properties of interneurons in the ventral spinal cord. Archs Ital Biol 104: 354–386

Wilsman N J, Van Sickle D C 1972 Cartilage canals, their morphology and distribution. Anat Rec 173: 79–93

Wilson G H 1920 A manual of dental prosthetics. 4th edition. Kimpton: London

Wilson H G 1928 Postnatal development of the lung. Am J Anat 41: 97–122

Wilson M E, Cragg B G 1967 Projections from the lateral geniculate nucleus in the cat and monkey. J Anat 101: 677–692

Winckler G 1960 Remarks on the histological structure of the leptomeninges in man. Arch Anat Histol Embryol 43: 259–277

Winckler G 1961 Arch Anat Histol Embryol 44:

Winckler G 1972 Remarques sur la structure de l'artère vertébrale. Quad Anat Pract 28: 105–115

Winckler G, Cochet B 1968 La systematisation du nerf tympanique (nerf de Jacobson). Inst d'Anat Norm Lausanne-CR Assoc Anat 139: 1215–1221

Windle B C A 1888 On the arteries forming the Circle of Willis. J Anat Physiol 22: 289–293

Winer J W 1977 A review of the status of the horseradish peroxidase method in neuroanatomy. Behavioral Rev 1: 45–54

Winer J A, Diamond I T, Raczkowski D 1977 Subdivisions of the auditory cortex of the cat: the retrograde transport of horseradish peroxidase to the medial geniculate body and posterior thalamic nuclei. J Comp Neurol 176: 387–418

Winfield D A, Neal J W, Powell T P S 1983 The basal dendrites of Meynert cells in the striate cortex of the monkey. Proc R Soc Lond. [Biol] 217: 129–139

Winfield D A, Rivera-Dominguez M, Powell T P S 1981 The number and distribution of Meynert cells in area 17 of the macaque monkey. Proc R Soc Lond. [Biol] 213: 27–40

Wingate D 1976 The eupeptide system: a general theory of gastrointestinal hormones. Lancet 1: 529–532

Wingerd J, Peritz E, Sproul A 1974 Race and stature differences in the skeletal maturation of the hand and wrist. Ann Hum Biol 1: 201–209

Winter G D 1962 Formation of the scab and the rate of epithelialization of superficial wounds in the skin of the young domestic pig. Nature 193: 293–294

Winter G D 1964 Movement of epidermal cells over the wound surface. Adv Biol Skin 5: 113–127

Wintrobe M M et al 1981 Clinical Hematology. 8th edition. Lea & Febiger: Philadelphia

Wischnitzer S 1973 The submicroscopic morphology of the interphase nucleus. Int Rev Cytol 34: 1–48

Wislocki G B 1929 On the placentation of primates, with a consideration of the phylogeny of the placenta. Contrib Embryol Carnegie Inst Washington 20: 51–80

Wislocki G B 1937 The meningeal relations of the hypophysis cerebri. II. An embryological study of the meninges and blood vessels of the human hypophysis. Am J Anat 61: 95–130

Wislocki G B, Dempsey E W 1939 Remarks on the lymphatics of the reproductive tract of female rhesus monkey (Macaca mulatta). Anat Rec 75: 341–363

Wislocki G B, King L S 1936 The permeability of the hypophysis and hypothalamus to vital dyes, with a study of the hypophyseal vascular supply. Am J Anat 58: 421–472

Wislocki G B, Ladman A J 1958 The fine structure of the mammalian choroid plexus. In: The cerebrospinal fluid (Wolstenholme G E W, O'Connor C M eds). pp 55–74. Ciba Foundation Symposium, Churchill: London

Wislocki G B, Leduc E 1953 The cytology and histochemistry of the subcommissural organ and Reissner's fiber in rodents. J Comp Neurol 97: 515–544

Wislocki G B, Streeter G L 1938 On the placentation of the macaque (Macaca mulatta), from the time of implantation until the formation of the definitive placenta. Contrib Embryol Carnegie Inst Washington 27: 1–66

Wisse E 1970 An electron microscopic study of the fenestrated endothelial lining of rat liver sinusoids. J Ultrastruct Res 31: 125–150

Witkovsky P, Shakib M, Ripps H 1974 Inter-receptoral junctions in the teleost retina. Invest Ophthalmol 13: 996–1009

Witschi E 1948 Migration of the germ cells of human embryos from the yolk sac to the primitive gonadal folds. Contrib Embryol Carnegie Inst Washington 32: 67–80

Witschi E 1951 Gonad development and function. Embryogenesis of the adrenal and reproductive glands. In: Pincus G (ed) Recent progress in hormone research. Vol 6. Academic Press: New York: pp 1–27

Wladmirow B 1968 Arterial sources of blood supply of the knee joint in man. Acta Med 47: 1–10

Wolfe D E, Potter L T, Richardson K C, Axelrod J 1962 Localising tritiated norepinephrine in sympathetic axons by electron microscope autoradiography. Science 138: 440–442

Wolfe H J, Voelkel E F, Tashjian A H Jr 1974 Distribution of calcitonin-containing cells in the normal adult human thyroid gland: a correlation of morphology with peptide content. J Clin Endocrinol Metab 38: 688–694

Wolff C F 1781 Acta Acad Sci Petropol. Part 1, 211

Wolff E 1976 Eugene Wolff's anatomy of the eye and orbit, including the central connexions, development, and comparative anatomy of the visual apparatus. 7th edition. Revised by R Warwick, Lewis: London.

Wolff J 1964 Transport of iodide and other anions in the thyroid gland. Physiol Rev 44: 45–90

Wolff J, Williams J A 1973 The role of microtubules and microfilaments in thyroid secretion. Rec Progr Horm Res 29: 229–285

Wolff J R 1976 An ontogenetically defined angioarchitecture of the neocortex. (Proceedings). Arzneim Forsch 26: 1239

Wolffson D M 1950 Scapula shape and muscle function, with special reference to vertebral border. Am J Phys Anthropol 8: 331–342

Wolinsky H, Glagov S 1964 Structural basis for the static mechanical properties of the aortic media. Circ Res 14: 400

Wolinsky H, Glagov S 1967a A lamellar unit of aortic medial structure and function in mammals. Circ Res 20: 99–111

Wollman S H, Nève P 1971 Ultimobranchial follicles in the thyroid glands of rats and mice. Rec Progr Horm Res 27: 213–234

Wolpert L 1969 Positional information and the spatial pattern of cellular differentiation. J Theor Biol 25: 1–47

Wolpert L 1971a Cell movement and cell contact. Sci Basis Med Annu Rev, 81–98

Wolpert L 1971b Positional information and pattern formation. Curr Top Dev Biol 6: 183–224

Wolpert L 1978 Pattern formation in biological development. Sci Am 239: (Oct) 124–137

Wolpert L 1981 Positional information and pattern formation. Philos Trans R Soc Lond. [Biol] 295: 441–450

Wolstenholme G E W, Knight J (eds) 1970a Control processes in multicellular organisms. (Ciba Foundation Symposium). Churchill: London

Wolstenholme G E W, Knight J (eds) 1970b Sensorineural hearing loss. (Ciba Foundation Symposium). Churchill: London

Wolstenholme G E W, Knight J (eds) 1971 The pineal gland. (Ciba Foundation Symposium). Churchill Livingstone: Edinburgh

Wolstenholme G E W, Knight J (eds) 1972 Lactogenic hormones. A Ciba Foundation Symposium in memory of Professor S J Folley. Churchill Livingstone: Edinburgh

Wolter J R 1959 Glia of the human retina. Am J Ophthalmol 48: 370–393

Wolter J R, Liss L 1956 Zentrifugale (antidrome) nervenfasern im menschlichen sehnerven. Albrect v Graefes Arch Ophthalmol 158: 1–7

Wong W C, Ling E A, Yick T Y, Tay S S W 1987 Effects of bilateral vagotomy on the ultrastructue of the cardiac ganglia in the monkey (Macaca fascicularis). J Anat 150: 75–88

Wong W C, Tan C K 1980 The fine structure of the intermediolateral nucleus of the spinal cord of the monkey (Macaca fascicularis). J Anat 130: 263–277

Woo J-K 1949 Ossification and growth of the human maxilla, pre-maxilla and palate bone. Anat Rec 105: 737–762

Wood N K, Wragg L E, Stuteville O H 1967 The premaxilla. Embryological evidence that it does not exist in man. Anat Rec 158: 485–490

Wood N K, Wragg L E, Stuteville O H, Oglesby R J 1969 Osteogenesis of the human upper jaw. Proof of the non-existence of a separate premaxillary centre. Arch Oral Biol 14: 1331–1339

Wood P, Bunge R 1975 Evidence that sensory axons are mitogenic for Schwann cells. Nature 256: 662–664

Woodburne R T 1947 Costomediastinal border of left pleura in precordial area. Anat Rec 97: 197–210

Woodburne R T 1956 The sacral parasympathetic innervation of the colon. Anat Rec 124: 67–76

Woodburne R T 1960 The accessory obturator nerve and the innervation of the pectineus muscle. Anat Rec 136: 367–369

Woodburne R T 1961 The sphincter mechanism of the urinary bladder and the urethra. Anat Rec 141: 11–20

Woodburne R T 1962 Segmental anatomy of the liver: blood supply and collateral circulation. Univ Mich. Med Bull 28: 189–199

Woodburne R T 1965 The ureter, ureterovesical junction, and vesical trigone. Anat Rec 151: 243–249

Woodburne R T 1968 Anatomy of the bladder and bladder outlet. J Urol 100: 474–487

Woodburne R T, Crosby E C, McCotter R E 1946 The mammalian hindbrain and isthmus region. II. The fiber connections. A. The relations of the tegmentum of the midbrain with the basal ganglia in Macaca mulatta. J Comp Neurol 85: 67–92

Wood Jones F see Jones F W

Woodley D I, O'Keefe E J, Reese M J, Mechanic G H, Briggaman R A, Gammon W R 1986 Epidermolysis bullosa acquisita antigen, a new major component of the cutaneous basement membrane, is a glycoprotein with collagenous domains. J Invest Dermatol 86: 668–672

Woodruff J D, Pauerstein C J 1969 The fallopian tube. Williams and Wilkins: Baltimore

Woollard H H 1926 The innervation of the heart. J Anat 60: 345–373

Woollard H H, Weddell G, Harpman J A 1940 Observations on the neurohistological basis of cutaneous pain. J Anat 74: 413–440

Woolsey T A, van der Loos H 1970 The structural organisation of layer IV in the somatosensory region (SI) of mouse cerebral cortex. The description of a cortical field composed of discrete cytoarchitecture units. Brain Res 17: 205–242

Woolsey C N 1964 Cortical localization as defined by evoked potential and electrical stimulation studies. In: Schaltenbrand G, Woolsey C N (eds) Cerebral localization and organization. University of Wisconsin Press: Madison, Wisconsin: pp 17–32

Woolsey C N, Walzl E M 1942 Topical projection of nerve fibers from local regions of the cochlea to the cerebral cortex of the cat. Bull Johns Hopkins Hosp 71: 315–344

Woolsey C N, Settlage B H, Meyer D R, Spencer W, Hamuy T P, Travis A M 1952 Patterns of localisation in precentral and 'supplementary' motor areas and their relation to the concept of a premotor area. Res Publ Assoc Res Nerv Ment Dis 30: 238–264

World Health Organization 1972 Genetic disorders: Prevention, treatment, and rehabilitation. (WHO Technical Report Series 497). World Health Organization: Geneva

Wormsley K G 1979 Pancreatic secretion: physiological control. In: Duthie H L, Wormsley K G (eds) Scientific basis of gastroenterology. Churchill Livingstone: Edinburgh: pp 199–248

Woźniak W 1966 Odcinki krzyzowe pni wzpolczulnych (pies, kot, czlowiek). Folia Morphol 25: 433–440

Wray J B 1963 Vascular regeneration in the healing fracture. An experimental study. Angiology 14: 134–138

Wright D M, Moffett B C Jr 1974 The postnatal development of the human temporomandibular joint. Am J Anat 141: 235–250

Wright N L 1969 Dissection study and mensuration of the human aortic arch. J Anat 104: 377–385

Wulle K G, Lerche W 1967 Zur feinstruktur der embryonalen menschlichen linsenblase. Graefas Arch Clin Exp Ophthalmol 173: 141–152

Wurtman R J 1970 The pineal gland: endocrine interrelationships. Adv Intern Med 16: 155–169

Wurtman R J, Pohorecky L A 1971 Adrenocortical control of epinephrine synthesis in health and disease. Adv Metab Disord 5: 53–76

Wurtman R J, Axelrod J, Kelly D E 1968 The pineal. Academic Press: New York

Wyburn G M 1937 The development of the infra-umbilical portion of the abdominal wall, and remarks on the aetiology of ectopia vesicae. J Anat 71: 201–231

Wyburn G M 1956 Uncertainties of anatomy of vascular innervation of lower limb. Scot. Med J 1: 201–205

Wyckoff R W G, Young J Z 1956 The motoneurone surface. Proc R Soc Lond. [Biol] 144: 440–450

Wyke B 1980 The neurology of low back pain. In: The lumbar spine and back pain. 2nd edition. (Jayson M I V ed). pp 265–339. Pitman Medical: Tunbridge Wells

Wyke B 1981 The neurology of joints: a review of general principles. Clin Rheum Dis 7: 223–239

Wyke B D 1947 Clinical physiology of the cerebellum. Med J Aust. 2: 533–540

Wyke B D 1974 Proceedings: laryngeal myotatic reflexes and phonation. Folia Phoniatr (Basel) 26: 249–264

Wynn R M, French G L 1968 Comparative ultrastructure of the mammalian amnion. Obstet Gynecol 31: 759–774

Wynne-Roberts C R, Anderson C 1978 Light- and electron-microscopic studies of normal juvenile human synovium. Semin Arthritis Rheum 7: 279–302

X

Xuereb G P, Prichard M M L, Daniel P M 1954a The arterial supply and venous drainage of the human hypophysis cerebri. Q Jl Exp Physiol 39: 199–218

Xuereb G P, Prichard M M L, Daniel P M 1954b The hypophyseal portal system of vessels in man. Q Jl Exp Physiol 39: 219–27

Y

Yagita K 1910 Experimentelle Untersuchungen über den Ursprung des Nervus facialis. Anat Anz 37: 195–218

Yamada K M 1981 Fibronectin and other structural proteins. In: Hay E D (ed) Cell biology of extracellular matrix. Plenum: New York: pp 95–114

Yamamoto M, Shimoyama I, Highstein S M 1978 Vestibular nucleus neurons relaying excitation from the anterior canal to the oculomotor nucleus. Brain Res 148: 31–42

Yamashiro S, Harris W H, Stopps T P 1984 Ultrastructural study of developing rabbit diaphragm. J Anat 139: 67–79

Yamauchi A 1973 Ultrastructure of the innervation of the mammalian heart. In: Challice C E, Viragh S (eds) Ultrastructure of the mammalian heart. Academic Press: New York: pp 127–166

Yanishevsky R M, Stein G H 1981 Regulation of the cell cycle in eukaryotic cells. Int Rev Cytol 69: 223–259

Yasuda Y 1973 Differentiation of human limb buds in vitro. Anat Rec 175: 561–578

Yates R D, Wood J G, Duncan D 1962 Phase and electron microscope observations on two cell types in the adrenal medulla of the Syrian hamster. Tex Rep Biol Med 20: 494–502

Yeterian E H, Hoesen G W van 1978 Cortico-striate projections in the rhesus monkey: the organization of certain cortico-caudate connections. Brain Res 139: 43–63

Yoffey J M 1962 The present status of the lymphocyte problem. Lancet 1: 206–211

Yohro T 1971 Nerve terminals and cellular junctions in young and adult mouse submandibular glands. J Anat 108: 409–417

Yokoh Y 1977 Early development of human lung. Acta Anat 97: 317–320

Yoshikawa T, Suzuki T 1969 Comparative anatomical study of the masseter of the mammal. Anat Anz 125: 363–387

Yoss R E 1952 Studies of the spinal cord. I. Topographic localization within the dorsal spinocerebellar tract in Macaca mulatta. J Comp Neurol 97: 5–20

Yoss R E 1953 Studies of the spinal cord. II. Topographic localization within the ventral spinocerebellar tract in the macaque. J Comp Neurol 99: 613–638

Young J 1940 Relaxation of pelvic joints in pregnancy: pelvic arthropathy of pregnancy. J Obstet Gynaec Br Commonw 47: 493–524

Young J A, van Lennep E E 1978 The morphology of salivary glands. Academic Press: London, New York

Young J Z 1958 Anatomical considerations. Electroencephalogr Clin Neurophysiol (suppl 10), 9–11

Young J Z 1962 The life of vertebrates. Oxford University Press: London

Young J Z 1964 A model of the brain. Clarendon Press: Oxford

Young J Z 1971 An introduction to the study of man. Clarendon Press: Oxford

Young J Z 1978 Programs of the brain. Oxford University Press: London

Young J Z, Zuckerman S 1936 The course of fibres in the dorsal roots of Macaca mulatta, the rhesus monkey. J Anat 71: 447–457

Young M, Turnbull H M 1931 Analysis of data collected by status lymphaticus investigating committee. J Path Bact 34: 213–258

Young M W 1952 The termination of the perilymphatic duct. Anat Rec 112: 404–405

Young M W 1953 The perilymphatic sac. Anat Rec 115: 419–420

Young R W 1971 Shedding of discs from rod outer segments in the rhesus monkey. J Ultrastruct Res 34: 190–203

Young R W, Bok D 1969 Participation of the retinal pigment epithelium in the rod outer segment renewal process. J Cell Biol 42: 392–403

Yunis J J, Sawyer J R, Ball D W 1978 The characterisation of high resolution G banded chromosomes of man. Chromosoma 67: 293–307

Z

Zaias N 1963 Embryology of the human nail. Arch Dermatol 87: 37–53

Zaki W 1960 The trochlear nerve in man. Study relative to its origin, its intracerebral traject and its structure. Arch Anat Histol Embryol 43: 105–120

Zaki W 1973 Aspect morphologique et fonctionnel de l'annulus fibrosus du disque intervertébral de la colonne dorsale. Arch Anat Path 21: 401–403

Zamboni L, De Martino C 1968 Embryogenesis of the human renal glomerulus. I. A histologic study. Arch Path (Chicago) 86: 279–291

Zamboni L, Thompson R S, Moore-Smith D 1972 Fine morphology of human oocyte maturation in vitro. Biol Reprod 7: 425–457

Zambrano D, Amezuo L, Dickmann G, Franke E 1968 Ultrastructure of human pituitary adenomata. Acta Neurochir 18: 78–94

Zangwill O L 1967 Speech and the minor hemisphere. Acta Neurol Psychiatry Belg. 67: 1013–1020

Zauberman H, Michaelson I C, Bergmann F, Maurice D M 1969 Stimulation of neovascularization of the cornea by biogenic amines. Exp Eye Res 8: 77–83

Zbrodowski A, Gajisin S, Grodecki J 1982 Vascularization of the tendons of extensor pollicis longus, extensor carpi radialis longus and extensor carpi radialis brevis muscles. J Anat 135: 235–244

Zeeman E C 1974 Primary and secondary waves in developmental biology. Lectures on mathematics in the life sciences. Am Math Soc 7: 69–101

Zeeman E C 1975 Differentiation and pattern formation. Ann Rev Biophys Bioeng 4: 210. (This is appendix to Cooke J 1975.) Some current theories of the emergence and regulation of spatial organisation in early animal development. Ann Rev Biophys Bioeng

Zeeman E C 1976a Catastrophe theory. Sci Am 234: (Apr) 65–83

Zeeman E C 1976b Gastrulation and formation of somites in amphibia and birds. In: Structural stability, the theory of catastrophes and applications in the sciences, Springer lecture notes in mathematics, 525: 396–401

Zeki S M 1969 Representation of central visual fields in prestriate cortex of monkey. Brain Res 14: 271–291

Zeki S M 1970 Interhemispheric connections of prestriate cortex in monkey. Brain Res 19: 63–75

Zeki S M 1974 The mosaic organization of the visual cortex in the monkey. In: Bellairs R, Gray E G (eds) Essays on the nervous system. Clarendon Press: Oxford: pp 327–343

Zeki S M 1977 Simultaneous anatomical demonstration of the representation of the vertical and horizontal meridians in areas V_2 and V_3 of rhesus monkey visual cortex. Proc R Soc Lond. [Biol] 195: 517–523

Zeldis S M, Nemerson Y, Pitlick F A, Lentz T L 1972 Tissue factor (thromboplastin): localization to plasma membranes by peroxidase-conjugated antibodies. Science 175: 766–768

Zelená J, Szentágothai J 1957 Verlagerung der Lokalisation spezifischer Cholinesterase während der Entwicklung der Muskelinnervation. Acta Histochem 3: 284–296

Zelickson A S 1971 Ultrastructure of the human epidermis. In: Borrie P F (ed) Modern trends in dermatology. Vol 4. Butterworths: London: pp 31–52

Zimmerman A M, Forer A (eds) 1981 Mitosis/Cytokinesis. Academic Press: New York

Zimmerman B L, Tsi M O M 1975 Morphological evidence of photoreceptor differentiation of pinealocytes in the neonatal rat. J Cell Biol 66: 60–75

Zimmerman J 1966 The functional and surgical anatomy of the heart. Ann R Coll Surg Eng 39: 348–366

Živanović S 1973 The menisco-fibular ligament of the knee joint. Acta Vet Beograd 23: 89–94

Živanović S 1974 Menisco-meniscal ligaments of the human knee joint. Anat Anz 135: 35–42

Zrenner C 1985 Theories of pineal function from classical antiquity to 1900. A history. In: Reiter R J (ed) Pineal Research Reviews. Vol 3. Liss: New York: pp 1–40

Zubay G L (ed) 1968 Papers in biochemical genetics. Holt, Rinehart and Winston: New York

Zucker-Franklin D 1980 Eosinophil structure and maturation. In: Mahmoud A A F, Austen K F (eds) The eosinophil in health and disease. Grune and Stratton: New York: pp 43–59

Zucker-Franklin D 1985 Eosinophils. In: Golde D W, Takaku F (eds) Hemopoietic stem cells. Dekker: New York: pp 45–63

Zucker-Franklin D, Lavie G, Franklin E C 1981 Demonstration of membrane-bound proteolytic activity on the surface of mononuclear leukocytes. J Histochem Cytochem 29: (suppl 3A), 451–464

Zucker-Franklin D, Greaves M F, Grossi C E, Marmont A M 1981 Atlas of Blood Cells. Function and Pathology. Lea and Febiger: Philadelphia

Zuckerkandl E 1901 Ueber Nebenorgane des Sympatheticus in Retroperitoneaolraum des Menschen. Anat Anz 15: 97

Zuckerman S 1940 The histogenesis of tissues sensitive to oestrogens. Biol Rev 15: 231–272

Zuckerman S 1951 Gonad development and function. The number of oöcytes in the mature ovary. In: Pineus G (ed) Recent progress in hormone research. Vol 6. Academic Press: New York: pp 63–109

Zuckerman S 1956 The regenerative capacity of ovarian tissue. In: Wolstenholme G E W, Millar E C P (eds) Colloquia on ageing, Vol 2, Ageing in transient tissues. Ciba Foundation Symposium. Churchill: London: pp 31–52

Zuckerman S, Ashton E H, Flinn R M, Oxnard C E, Spence T E 1973 Some locomotor features of the pelvic girdle in primates. Symp Zool Soc Lond. 33: 71–166

Zweifach B W 1959 The microcirculation of the blood. Scient Am 200: 54–60

BIBLIOGRAPHY

Zweifach B W 1961 The structural basis of the microcirculation. In: Luisada A A (ed) Development and structure of the cardiovascular system. McGraw-Hill: New York: pp 198–205

Zweifach B W 1973 Microcirculation. Annu Rev Physiol 35: 117–150

Zweifach B W, Kossmann C E 1937 Micromanipulation of small blood vessels in mouse. Am J Physiol 120: 23–35

Zwilling E 1961 Limb morphogenesis. In: Abercrombie M, Brachet J (eds) Advances in Morphogenesis. Vol 1. Academic Press: New York: pp 301–330

INDEX

Main references are given in bold type

INDEX